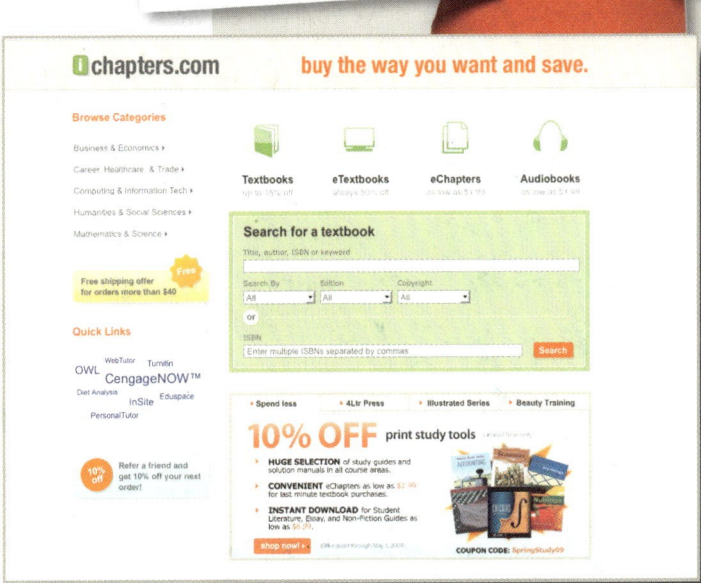

RULES OF DIFFERENTIATION

Basic Formulas

1. $\dfrac{d}{dx}(c) = 0$

2. $\dfrac{d}{dx}(cu) = c\dfrac{du}{dx}$

3. $\dfrac{d}{dx}(u \pm v) = \dfrac{du}{dx} \pm \dfrac{dv}{dx}$

4. $\dfrac{d}{dx}(uv) = u\dfrac{dv}{dx} + v\dfrac{du}{dx}$

5. $\dfrac{d}{dx}\left(\dfrac{u}{v}\right) = \dfrac{v\dfrac{du}{dx} - u\dfrac{dv}{du}}{v^2}$

6. $\dfrac{d}{dx}f(g(x)) = f'(g(x))g'(x)$

7. $\dfrac{d}{dx}(u^n) = nu^{n-1}\dfrac{du}{dx}$

Exponential and Logarithmic Functions

8. $\dfrac{d}{dx}(e^u) = e^u\dfrac{du}{dx}$

9. $\dfrac{d}{dx}(a^u) = (\ln a)a^u\dfrac{du}{dx}$

10. $\dfrac{d}{dx}\ln|u| = \dfrac{1}{u}\dfrac{du}{dx}$

11. $\dfrac{d}{dx}(\log_a u) = \dfrac{1}{u\ln a}\dfrac{du}{dx}$

Trigonometric Functions

12. $\dfrac{d}{dx}(\sin u) = \cos u\dfrac{du}{dx}$

13. $\dfrac{d}{dx}(\cos u) = -\sin u\dfrac{du}{dx}$

14. $\dfrac{d}{dx}(\tan u) = \sec^2 u\dfrac{du}{dx}$

15. $\dfrac{d}{dx}(\csc u) = -\csc u\cot u\dfrac{du}{dx}$

16. $\dfrac{d}{dx}(\sec u) = \sec u\tan u\dfrac{du}{dx}$

17. $\dfrac{d}{dx}(\cot u) = -\csc^2 u\dfrac{du}{dx}$

Inverse Trigonometric Functions

18. $\dfrac{d}{dx}(\sin^{-1} u) = \dfrac{1}{\sqrt{1-u^2}}\dfrac{du}{dx}$

19. $\dfrac{d}{dx}(\cos^{-1} u) = -\dfrac{1}{\sqrt{1-u^2}}\dfrac{du}{dx}$

20. $\dfrac{d}{dx}(\tan^{-1} u) = \dfrac{1}{1+u^2}\dfrac{du}{dx}$

21. $\dfrac{d}{dx}(\csc^{-1} u) = -\dfrac{1}{|u|\sqrt{u^2-1}}\dfrac{du}{dx}$

22. $\dfrac{d}{dx}(\sec^{-1} u) = \dfrac{1}{|u|\sqrt{u^2-1}}\dfrac{du}{dx}$

23. $\dfrac{d}{dx}(\cot^{-1} u) = -\dfrac{1}{1+u^2}\dfrac{du}{dx}$

Hyperbolic Functions

24. $\dfrac{d}{dx}(\sinh u) = \cosh u\dfrac{du}{dx}$

25. $\dfrac{d}{dx}(\cosh u) = \sinh u\dfrac{du}{dx}$

26. $\dfrac{d}{dx}(\tanh u) = \text{sech}^2 u\dfrac{du}{dx}$

27. $\dfrac{d}{dx}(\text{csch}\, u) = -\text{csch}\, u\coth u\dfrac{du}{dx}$

28. $\dfrac{d}{dx}(\text{sech}\, u) = -\text{sech}\, u\tanh u\dfrac{du}{dx}$

29. $\dfrac{d}{dx}(\coth u) = -\text{csch}^2 u\dfrac{du}{dx}$

Inverse Hyperbolic Functions

30. $\dfrac{d}{dx}(\sinh^{-1} u) = \dfrac{1}{\sqrt{1+u^2}}\dfrac{du}{dx}$

31. $\dfrac{d}{dx}(\cosh^{-1} u) = \dfrac{1}{\sqrt{u^2-1}}\dfrac{du}{dx}$

32. $\dfrac{d}{dx}(\tanh^{-1} u) = \dfrac{1}{1-u^2}\dfrac{du}{dx}$

33. $\dfrac{d}{dx}(\text{csch}^{-1} u) = -\dfrac{1}{|u|\sqrt{u^2+1}}\dfrac{du}{dx}$

34. $\dfrac{d}{dx}(\text{sech}^{-1} u) = -\dfrac{1}{u\sqrt{1-u^2}}\dfrac{du}{dx}$

35. $\dfrac{d}{dx}(\coth^{-1} u) = \dfrac{1}{1-u^2}\dfrac{du}{dx}$

TABLE OF INTEGRALS

Basic Forms

1. $\displaystyle\int u^n\,du = \frac{u^{n+1}}{n+1} + C, \quad n \neq -1$

2. $\displaystyle\int \frac{du}{u} = \ln|u| + C$

3. $\displaystyle\int \sin u\,du = -\cos u + C$

4. $\displaystyle\int \cos u\,du = \sin u + C$

5. $\displaystyle\int \tan u\,du = \ln|\sec u| + C$

6. $\displaystyle\int e^u\,du = e^u + C$

7. $\displaystyle\int a^u\,du = \frac{a^u}{\ln a} + C$

8. $\displaystyle\int \sec u\,du = \ln|\sec u + \tan u| + C$

9. $\displaystyle\int \csc u\,du = \ln|\csc u - \cot u| + C$

10. $\displaystyle\int \cot u\,du = \ln|\sin u| + C$

11. $\displaystyle\int \sec^2 u\,du = \tan u + C$

12. $\displaystyle\int \csc^2 u\,du = -\cot u + C$

13. $\displaystyle\int \sec u \tan u\,du = \sec u + C$

14. $\displaystyle\int \csc u \cot u\,du = -\csc u + C$

15. $\displaystyle\int \frac{du}{\sqrt{a^2 - u^2}} = \sin^{-1}\frac{u}{a} + C$

16. $\displaystyle\int \frac{du}{u\sqrt{u^2 - a^2}} = \frac{1}{a}\sec^{-1}\frac{u}{a} + C$

17. $\displaystyle\int \frac{du}{a^2 + u^2} = \frac{1}{a}\tan^{-1}\frac{u}{a} + C$

18. $\displaystyle\int \frac{du}{a^2 - u^2} = \frac{1}{2a}\ln\left|\frac{u+a}{u-a}\right| + C$

Forms Involving $a + bu$

19. $\displaystyle\int \frac{u\,du}{a+bu} = \frac{1}{b^2}\left(a + bu - a\ln|a+bu|\right) + C$

20. $\displaystyle\int \frac{u^2\,du}{a+bu}$
$$= \frac{1}{2b^3}\left[(a+bu)^2 - 4a(a+bu) + 2a^2\ln|a+bu|\right] + C$$

21. $\displaystyle\int \frac{u\,du}{(a+bu)^2} = \frac{a}{b^2(a+bu)} + \frac{1}{b^2}\ln|a+bu| + C$

22. $\displaystyle\int \frac{u^2\,du}{(a+bu)^2} = \frac{1}{b^3}\left(a + bu - \frac{a^2}{a+bu} - 2a\ln|a+bu|\right) + C$

23. $\displaystyle\int \frac{du}{u(a+bu)} = \frac{1}{a}\ln\left|\frac{u}{a+bu}\right| + C$

24. $\displaystyle\int \frac{du}{u^2(a+bu)} = -\frac{1}{au} + \frac{b}{a^2}\ln\left|\frac{a+bu}{u}\right| + C$

25. $\displaystyle\int \frac{du}{u(a+bu)^2} = \frac{1}{a(a+bu)} - \frac{1}{a^2}\ln\left|\frac{a+bu}{u}\right| + C$

26. $\displaystyle\int \frac{du}{u^2(a+bu)^2} = -\frac{1}{a^2}\left[\frac{a+2bu}{u(a+bu)} + \frac{2b}{a}\ln\left|\frac{u}{a+bu}\right|\right] + C$

Forms Involving $\sqrt{a + bu}$

27. $\displaystyle\int u\sqrt{a+bu}\,du = \frac{2}{15b^2}(3bu - 2a)(a+bu)^{3/2} + C$

28. $\displaystyle\int \frac{u\,du}{\sqrt{a+bu}} = \frac{2}{3b^2}(bu - 2a)\sqrt{a+bu} + C$

29. $\displaystyle\int \frac{u^2\,du}{\sqrt{a+bu}} = \frac{2}{15b^3}(8a^2 + 3b^2u^2 - 4abu)\sqrt{a+bu} + C$

30. $\displaystyle\int \frac{du}{u\sqrt{a+bu}} = \begin{cases} \dfrac{1}{\sqrt{a}}\ln\left|\dfrac{\sqrt{a+bu} - \sqrt{a}}{\sqrt{a+bu} + \sqrt{a}}\right| + C & \text{if } a > 0 \\[3mm] \dfrac{2}{\sqrt{-a}}\tan^{-1}\sqrt{\dfrac{a+bu}{-a}} + C & \text{if } a < 0 \end{cases}$

31. $\displaystyle\int \frac{\sqrt{a+bu}}{u}\,du = 2\sqrt{a+bu} + a\int \frac{du}{u\sqrt{a+bu}}$

32. $\displaystyle\int \frac{\sqrt{a+bu}}{u^2}\,du = -\frac{\sqrt{a+bu}}{u} + \frac{b}{2}\int \frac{du}{u\sqrt{a+bu}}$

33. $\displaystyle\int u^n\sqrt{a+bu}\,du$
$$= \frac{2}{b(2n+3)}\left[u^n(a+bu)^{3/2} - na\int u^{n-1}\sqrt{a+bu}\,du\right]$$

34. $\displaystyle\int \frac{u^n\,du}{\sqrt{a+bu}} = \frac{2u^n\sqrt{a+bu}}{b(2n+1)} - \frac{2na}{b(2n+1)}\int \frac{u^{n-1}\,du}{\sqrt{a+bu}}$

35. $\displaystyle \int \frac{du}{u^n \sqrt{a + bu}} = -\frac{\sqrt{a + bu}}{a(n - 1)u^{n-1}} - \frac{b(2n - 3)}{2a(n - 1)} \int \frac{du}{u^{n-1}\sqrt{a + bu}}$

36. $\displaystyle \int \frac{\sqrt{a + bu}}{u^n}\, du = \frac{-1}{a(n - 1)}\left[\frac{(a + bu)^{3/2}}{u^{n-1}} + \frac{(2n - 5)b}{2} \int \frac{\sqrt{a + bu}}{u^{n-1}}\, du \right], \quad n \neq 1$

Forms Involving $\sqrt{a^2 + u^2},\ a > 0$

37. $\displaystyle \int \sqrt{a^2 + u^2}\, du = \frac{u}{2}\sqrt{a^2 + u^2} + \frac{a^2}{2}\ln\!\left(u + \sqrt{a^2 + u^2}\right) + C$

38. $\displaystyle \int u^2 \sqrt{a^2 + u^2}\, du = \frac{u}{8}(a^2 + 2u^2)\sqrt{a^2 + u^2}$
$$- \frac{a^4}{8}\ln\!\left(u + \sqrt{a^2 + u^2}\right) + C$$

39. $\displaystyle \int \frac{\sqrt{a^2 + u^2}}{u}\, du = \sqrt{a^2 + u^2} - a\ln\left|\frac{a + \sqrt{a^2 + u^2}}{u}\right| + C$

40. $\displaystyle \int \frac{\sqrt{a^2 + u^2}}{u^2}\, du = -\frac{\sqrt{a^2 + u^2}}{u} + \ln\!\left(u + \sqrt{a^2 + u^2}\right) + C$

41. $\displaystyle \int \frac{du}{\sqrt{a^2 + u^2}} = \ln\!\left(u + \sqrt{a^2 + u^2}\right) + C$

42. $\displaystyle \int \frac{u^2\, du}{\sqrt{a^2 + u^2}} = \frac{u}{2}\sqrt{a^2 + u^2} - \frac{a^2}{2}\ln\!\left(u + \sqrt{a^2 + u^2}\right) + C$

43. $\displaystyle \int \frac{du}{u\sqrt{a^2 + u^2}} = -\frac{1}{a}\ln\left|\frac{\sqrt{a^2 + u^2} + a}{u}\right| + C$

44. $\displaystyle \int \frac{du}{u^2 \sqrt{a^2 + u^2}} = -\frac{\sqrt{a^2 + u^2}}{a^2 u} + C$

45. $\displaystyle \int \frac{du}{(a^2 + u^2)^{3/2}} = \frac{u}{a^2\sqrt{a^2 + u^2}} + C$

Forms Involving $\sqrt{a^2 - u^2},\ a > 0$

46. $\displaystyle \int \sqrt{a^2 - u^2}\, du = \frac{u}{2}\sqrt{a^2 - u^2} + \frac{a^2}{2}\sin^{-1}\frac{u}{a} + C$

47. $\displaystyle \int u^2 \sqrt{a^2 - u^2}\, du = \frac{u}{8}(2u^2 - a^2)\sqrt{a^2 - u^2} + \frac{a^4}{8}\sin^{-1}\frac{u}{a} + C$

48. $\displaystyle \int \frac{\sqrt{a^2 - u^2}}{u}\, du = \sqrt{a^2 - u^2} - a\ln\left|\frac{a + \sqrt{a^2 - u^2}}{u}\right| + C$

49. $\displaystyle \int \frac{\sqrt{a^2 - u^2}}{u^2}\, du = -\frac{1}{u}\sqrt{a^2 - u^2} - \sin^{-1}\frac{u}{a} + C$

50. $\displaystyle \int \frac{u^2\, du}{\sqrt{a^2 - u^2}} = -\frac{u}{2}\sqrt{a^2 - u^2} + \frac{a^2}{2}\sin^{-1}\frac{u}{a} + C$

51. $\displaystyle \int \frac{du}{u\sqrt{a^2 - u^2}} = -\frac{1}{a}\ln\left|\frac{a + \sqrt{a^2 - u^2}}{u}\right| + C$

52. $\displaystyle \int \frac{du}{u^2 \sqrt{a^2 - u^2}} = -\frac{1}{a^2 u}\sqrt{a^2 - u^2} + C$

53. $\displaystyle \int (a^2 - u^2)^{3/2}\, du$
$$= -\frac{u}{8}(2u^2 - 5a^2)\sqrt{a^2 - u^2} + \frac{3a^4}{8}\sin^{-1}\frac{u}{a} + C$$

54. $\displaystyle \int \frac{du}{(a^2 - u^2)^{3/2}} = \frac{u}{a^2\sqrt{a^2 - u^2}} + C$

Forms Involving $\sqrt{u^2 - a^2},\ a > 0$

55. $\displaystyle \int \sqrt{u^2 - a^2}\, du = \frac{u}{2}\sqrt{u^2 - a^2} - \frac{a^2}{2}\ln\left|u + \sqrt{u^2 - a^2}\right| + C$

56. $\displaystyle \int u^2 \sqrt{u^2 - a^2}\, du$
$$= \frac{u}{8}(2u^2 - a^2)\sqrt{u^2 - a^2} - \frac{a^4}{8}\ln\left|u + \sqrt{u^2 - a^2}\right| + C$$

57. $\displaystyle \int \frac{\sqrt{u^2 - a^2}}{u}\, du = \sqrt{u^2 - a^2} - a\cos^{-1}\frac{a}{|u|} + C$

58. $\displaystyle \int \frac{\sqrt{u^2 - a^2}}{u^2}\, du = -\frac{\sqrt{u^2 - a^2}}{u} + \ln\left|u + \sqrt{u^2 - a^2}\right| + C$

59. $\displaystyle \int \frac{du}{\sqrt{u^2 - a^2}} = \ln\left|u + \sqrt{u^2 - a^2}\right| + C$

60. $\displaystyle \int \frac{u^2\, du}{\sqrt{u^2 - a^2}} = \frac{u}{2}\sqrt{u^2 - a^2} + \frac{a^2}{2}\ln\left|u + \sqrt{u^2 - a^2}\right| + C$

61. $\displaystyle \int \frac{du}{u^2 \sqrt{u^2 - a^2}} = \frac{\sqrt{u^2 - a^2}}{a^2 u} + C$

62. $\displaystyle \int \frac{du}{(u^2 - a^2)^{3/2}} = -\frac{u}{a^2\sqrt{u^2 - a^2}} + C$

Forms Involving sin u, cos u, tan u

63. $\displaystyle\int \sin^2 u \, du = \frac{1}{2} u - \frac{1}{4} \sin 2u + C$

64. $\displaystyle\int \cos^2 u \, du = \frac{1}{2} u + \frac{1}{4} \sin 2u + C$

65. $\displaystyle\int \tan^2 u \, du = \tan u - u + C$

66. $\displaystyle\int \sin^3 u \, du = -\frac{1}{3}(2 + \sin^2 u) \cos u + C$

67. $\displaystyle\int \cos^3 u \, du = \frac{1}{3}(2 + \cos^2 u) \sin u + C$

68. $\displaystyle\int \tan^3 u \, du = \frac{1}{2} \tan^2 u + \ln|\cos u| + C$

69. $\displaystyle\int \sin^n u \, du = -\frac{1}{n} \sin^{n-1} u \cos u + \frac{n-1}{n} \int \sin^{n-2} u \, du$

70. $\displaystyle\int \cos^n u \, du = \frac{1}{n} \cos^{n-1} u \sin u + \frac{n-1}{n} \int \cos^{n-2} u \, du$

71. $\displaystyle\int \tan^n u \, du = \frac{1}{n-1} \tan^{n-1} u - \int \tan^{n-2} u \, du$

72. $\displaystyle\int \sin au \sin bu \, du = \frac{\sin(a-b)u}{2(a-b)} - \frac{\sin(a+b)u}{2(a+b)} + C$

73. $\displaystyle\int \cos au \cos bu \, du = \frac{\sin(a-b)u}{2(a-b)} + \frac{\sin(a+b)u}{2(a+b)} + C$

74. $\displaystyle\int \sin au \cos bu \, du = -\frac{\cos(a-b)u}{2(a-b)} - \frac{\cos(a+b)u}{2(a+b)} + C$

75. $\displaystyle\int u \sin u \, du = \sin u - u \cos u + C$

76. $\displaystyle\int u \cos u \, du = \cos u + u \sin u + C$

77. $\displaystyle\int u^n \sin u \, du = -u^n \cos u + n \int u^{n-1} \cos u \, du$

78. $\displaystyle\int u^n \cos u \, du = u^n \sin u - n \int u^{n-1} \sin u \, du$

79. $\displaystyle\int \sin^n u \cos^m u \, du$

$$= -\frac{\sin^{n-1} u \cos^{m+1} u}{n+m} + \frac{n-1}{n+m} \int \sin^{n-2} u \cos^m u \, du$$

$$= \frac{\sin^{n+1} u \cos^{m-1} u}{n+m} + \frac{m-1}{n+m} \int \sin^n u \cos^{m-2} u \, du$$

Forms Involving cot u, sec u, csc u

80. $\displaystyle\int \cot^2 u \, du = -\cot u - u + C$

81. $\displaystyle\int \cot^3 u \, du = -\frac{1}{2} \cot^2 u - \ln|\sin u| + C$

82. $\displaystyle\int \sec^3 u \, du = \frac{1}{2} \sec u \tan u + \frac{1}{2} \ln|\sec u + \tan u| + C$

83. $\displaystyle\int \csc^3 u \, du = -\frac{1}{2} \csc u \cot u + \frac{1}{2} \ln|\csc u - \cot u| + C$

84. $\displaystyle\int \cot^n u \, du = \frac{-1}{n-1} \cot^{n-1} u - \int \cot^{n-2} u \, du$

85. $\displaystyle\int \sec^n u \, du = \frac{1}{n-1} \tan u \sec^{n-2} u + \frac{n-2}{n-1} \int \sec^{n-2} u \, du$

86. $\displaystyle\int \csc^n u \, du = \frac{-1}{n-1} \cot u \csc^{n-2} u + \frac{n-2}{n-1} \int \csc^{n-2} u \, du$

Forms Involving Inverse Trigonometric Functions

87. $\displaystyle\int \sin^{-1} u \, du = u \sin^{-1} u + \sqrt{1 - u^2} + C$

88. $\displaystyle\int \cos^{-1} u \, du = u \cos^{-1} u - \sqrt{1 - u^2} + C$

89. $\displaystyle\int \tan^{-1} u \, du = u \tan^{-1} u - \frac{1}{2} \ln(1 + u^2) + C$

90. $\displaystyle\int u \sin^{-1} u \, du = \frac{2u^2 - 1}{4} \sin^{-1} u + \frac{u\sqrt{1 - u^2}}{4} + C$

91. $\displaystyle\int u \cos^{-1} u \, du = \frac{2u^2 - 1}{4} \cos^{-1} u - \frac{u\sqrt{1 - u^2}}{4} + C$

92. $\displaystyle\int u \tan^{-1} u \, du = \frac{u^2 + 1}{2} \tan^{-1} u - \frac{u}{2} + C$

93. $\displaystyle\int u^n \sin^{-1} u \, du = \frac{1}{n+1}\left[u^{n+1} \sin^{-1} u - \int \frac{u^{n+1} \, du}{\sqrt{1 - u^2}} \right]$, $n \neq -1$

94. $\displaystyle\int u^n \cos^{-1} u \, du = \frac{1}{n+1}\left[u^{n+1} \cos^{-1} u + \int \frac{u^{n+1} \, du}{\sqrt{1 - u^2}} \right]$, $n \neq -1$

95. $\displaystyle\int u^n \tan^{-1} u \, du = \frac{1}{n+1}\left[u^{n+1} \tan^{-1} u - \int \frac{u^{n+1} \, du}{\sqrt{1 + u^2}} \right]$, $n \neq -1$

RULES OF DIFFERENTIATION

Basic Formulas

1. $\dfrac{d}{dx}(c) = 0$

2. $\dfrac{d}{dx}(cu) = cu\dfrac{du}{dx}$

3. $\dfrac{d}{dx}(u \pm v) = \dfrac{du}{dx} \pm \dfrac{dv}{dx}$

4. $\dfrac{d}{dx}(uv) = u\dfrac{dv}{dx} + v\dfrac{du}{dx}$

5. $\dfrac{d}{dx}\left(\dfrac{u}{v}\right) = \dfrac{v\dfrac{du}{dx} - u\dfrac{dv}{du}}{v^2}$

6. $\dfrac{d}{dx}f(g(x)) = f'(g(x))g'(x)$

7. $\dfrac{d}{dx}(u^n) = nu^{n-1}\dfrac{du}{dx}$

Exponential and Logarithmic Functions

8. $\dfrac{d}{dx}(e^u) = e^u\dfrac{du}{dx}$

9. $\dfrac{d}{dx}(a^u) = (\ln a)a^u\dfrac{du}{dx}$

10. $\dfrac{d}{dx}\ln|u| = \dfrac{1}{u}\dfrac{du}{dx}$

11. $\dfrac{d}{dx}(\log_a u) = \dfrac{1}{u\ln a}\dfrac{du}{dx}$

Trigonometric Functions

12. $\dfrac{d}{dx}(\sin u) = \cos u\dfrac{du}{dx}$

13. $\dfrac{d}{dx}(\cos u) = -\sin u\dfrac{du}{dx}$

14. $\dfrac{d}{dx}(\tan u) = \sec^2 u\dfrac{du}{dx}$

15. $\dfrac{d}{dx}(\csc u) = -\csc u\cot u\dfrac{du}{dx}$

16. $\dfrac{d}{dx}(\sec u) = \sec u\tan u\dfrac{du}{dx}$

17. $\dfrac{d}{dx}(\cot u) = -\csc^2 u\dfrac{du}{dx}$

Inverse Trigonometric Functions

18. $\dfrac{d}{dx}(\sin^{-1} u) = \dfrac{1}{\sqrt{1 - u^2}}\dfrac{du}{dx}$

19. $\dfrac{d}{dx}(\cos^{-1} u) = -\dfrac{1}{\sqrt{1 - u^2}}\dfrac{du}{dx}$

20. $\dfrac{d}{dx}(\tan^{-1} u) = \dfrac{1}{1 + u^2}\dfrac{du}{dx}$

21. $\dfrac{d}{dx}(\csc^{-1} u) = -\dfrac{1}{|u|\sqrt{u^2 - 1}}\dfrac{du}{dx}$

22. $\dfrac{d}{dx}(\sec^{-1} u) = \dfrac{1}{|u|\sqrt{u^2 - 1}}\dfrac{du}{dx}$

23. $\dfrac{d}{dx}(\cot^{-1} u) = -\dfrac{1}{1 + u^2}\dfrac{du}{dx}$

Hyperbolic Functions

24. $\dfrac{d}{dx}(\sinh u) = \cosh u\dfrac{du}{dx}$

25. $\dfrac{d}{dx}(\cosh u) = \sinh u\dfrac{du}{dx}$

26. $\dfrac{d}{dx}(\tanh u) = \operatorname{sech}^2 u\dfrac{du}{dx}$

27. $\dfrac{d}{dx}(\operatorname{csch} u) = -\operatorname{csch} u\coth u\dfrac{du}{dx}$

28. $\dfrac{d}{dx}(\operatorname{sech} u) = -\operatorname{sech} u\tanh u\dfrac{du}{dx}$

29. $\dfrac{d}{dx}(\coth u) = -\operatorname{csch}^2 u\dfrac{du}{dx}$

Inverse Hyperbolic Functions

30. $\dfrac{d}{dx}(\sinh^{-1} u) = \dfrac{1}{\sqrt{1 + u^2}}\dfrac{du}{dx}$

31. $\dfrac{d}{dx}(\cosh^{-1} u) = \dfrac{1}{\sqrt{u^2 - 1}}\dfrac{du}{dx}$

32. $\dfrac{d}{dx}(\tanh^{-1} u) = \dfrac{1}{1 - u^2}\dfrac{du}{dx}$

33. $\dfrac{d}{dx}(\operatorname{csch}^{-1} u) = -\dfrac{1}{|u|\sqrt{u^2 + 1}}\dfrac{du}{dx}$

34. $\dfrac{d}{dx}(\operatorname{sech}^{-1} u) = -\dfrac{1}{u\sqrt{1 - u^2}}\dfrac{du}{dx}$

35. $\dfrac{d}{dx}(\coth^{-1} u) = \dfrac{1}{1 - u^2}\dfrac{du}{dx}$

TABLE OF INTEGRALS

Basic Forms

1. $\int u^n \, du = \dfrac{u^{n+1}}{n+1} + C, \quad n \neq -1$

2. $\int \dfrac{du}{u} = \ln|u| + C$

3. $\int \sin u \, du = -\cos u + C$

4. $\int \cos u \, du = \sin u + C$

5. $\int \tan u \, du = \ln|\sec u| + C$

6. $\int e^u \, du = e^u + C$

7. $\int a^u \, du = \dfrac{a^u}{\ln a} + C$

8. $\int \sec u \, du = \ln|\sec u + \tan u| + C$

9. $\int \csc u \, du = \ln|\csc u - \cot u| + C$

10. $\int \cot u \, du = \ln|\sin u| + C$

11. $\int \sec^2 u \, du = \tan u + C$

12. $\int \csc^2 u \, du = -\cot u + C$

13. $\int \sec u \tan u \, du = \sec u + C$

14. $\int \csc u \cot u \, du = -\csc u + C$

15. $\int \dfrac{du}{\sqrt{a^2 - u^2}} = \sin^{-1}\dfrac{u}{a} + C$

16. $\int \dfrac{du}{u\sqrt{u^2 - a^2}} = \dfrac{1}{a}\sec^{-1}\dfrac{u}{a} + C$

17. $\int \dfrac{du}{a^2 + u^2} = \dfrac{1}{a}\tan^{-1}\dfrac{u}{a} + C$

18. $\int \dfrac{du}{a^2 - u^2} = \dfrac{1}{2a}\ln\left|\dfrac{u+a}{u-a}\right| + C$

Forms Involving $a + bu$

19. $\int \dfrac{u \, du}{a + bu} = \dfrac{1}{b^2}\big(a + bu - a\ln|a + bu|\big) + C$

20. $\int \dfrac{u^2 \, du}{a + bu}$

$\qquad = \dfrac{1}{2b^3}\big[(a + bu)^2 - 4a(a + bu) + 2a^2\ln|a + bu|\big] + C$

21. $\int \dfrac{u \, du}{(a + bu)^2} = \dfrac{a}{b^2(a + bu)} + \dfrac{1}{b^2}\ln|a + bu| + C$

22. $\int \dfrac{u^2 \, du}{(a + bu)^2} = \dfrac{1}{b^3}\left(a + bu - \dfrac{a^2}{a + bu} - 2a\ln|a + bu|\right) + C$

23. $\int \dfrac{du}{u(a + bu)} = \dfrac{1}{a}\ln\left|\dfrac{u}{a + bu}\right| + C$

24. $\int \dfrac{du}{u^2(a + bu)} = -\dfrac{1}{au} + \dfrac{b}{a^2}\ln\left|\dfrac{a + bu}{u}\right| + C$

25. $\int \dfrac{du}{u(a + bu)^2} = \dfrac{1}{a(a + bu)} - \dfrac{1}{a^2}\ln\left|\dfrac{a + bu}{u}\right| + C$

26. $\int \dfrac{du}{u^2(a + bu)^2} = -\dfrac{1}{a^2}\left[\dfrac{a + 2bu}{u(a + bu)} + \dfrac{2b}{a}\ln\left|\dfrac{u}{a + bu}\right|\right] + C$

Forms Involving $\sqrt{a + bu}$

27. $\int u\sqrt{a + bu} \, du = \dfrac{2}{15b^2}(3bu - 2a)(a + bu)^{3/2} + C$

28. $\int \dfrac{u \, du}{\sqrt{a + bu}} = \dfrac{2}{3b^2}(bu - 2a)\sqrt{a + bu} + C$

29. $\int \dfrac{u^2 \, du}{\sqrt{a + bu}} = \dfrac{2}{15b^3}(8a^2 + 3b^2u^2 - 4abu)\sqrt{a + bu} + C$

30. $\int \dfrac{du}{u\sqrt{a + bu}} = \begin{cases} \dfrac{1}{\sqrt{a}}\ln\left|\dfrac{\sqrt{a + bu} - \sqrt{a}}{\sqrt{a + bu} + \sqrt{a}}\right| + C & \text{if } a > 0 \\[2mm] \dfrac{2}{\sqrt{-a}}\tan^{-1}\sqrt{\dfrac{a + bu}{-a}} + C & \text{if } a < 0 \end{cases}$

31. $\int \dfrac{\sqrt{a + bu}}{u} \, du = 2\sqrt{a + bu} + a\int \dfrac{du}{u\sqrt{a + bu}}$

32. $\int \dfrac{\sqrt{a + bu}}{u^2} \, du = -\dfrac{\sqrt{a + bu}}{u} + \dfrac{b}{2}\int \dfrac{du}{u\sqrt{a + bu}}$

33. $\int u^n\sqrt{a + bu} \, du$

$\qquad = \dfrac{2}{b(2n + 3)}\left[u^n(a + bu)^{3/2} - na\int u^{n-1}\sqrt{a + bu} \, du\right]$

34. $\int \dfrac{u^n \, du}{\sqrt{a + bu}} = \dfrac{2u^n\sqrt{a + bu}}{b(2n + 1)} - \dfrac{2na}{b(2n + 1)}\int \dfrac{u^{n-1} \, du}{\sqrt{a + bu}}$

35. $\displaystyle \int \frac{du}{u^n \sqrt{a + bu}} = -\frac{\sqrt{a + bu}}{a(n - 1)u^{n-1}} - \frac{b(2n - 3)}{2a(n - 1)} \int \frac{du}{u^{n-1}\sqrt{a + bu}}$

36. $\displaystyle \int \frac{\sqrt{a + bu}}{u^n} \, du = \frac{-1}{a(n - 1)} \left[\frac{(a + bu)^{3/2}}{u^{n-1}} + \frac{(2n - 5)b}{2} \int \frac{\sqrt{a + bu}}{u^{n-1}} \, du \right], \quad n \neq 1$

Forms Involving $\sqrt{a^2 + u^2}, \, a > 0$

37. $\displaystyle \int \sqrt{a^2 + u^2} \, du = \frac{u}{2} \sqrt{a^2 + u^2} + \frac{a^2}{2} \ln\left(u + \sqrt{a^2 + u^2} \right) + C$

38. $\displaystyle \int u^2 \sqrt{a^2 + u^2} \, du = \frac{u}{8}(a^2 + 2u^2) \sqrt{a^2 + u^2}$
$\displaystyle \qquad\qquad\qquad - \frac{a^4}{8} \ln\left(u + \sqrt{a^2 + u^2} \right) + C$

39. $\displaystyle \int \frac{\sqrt{a^2 + u^2}}{u} \, du = \sqrt{a^2 + u^2} - a \ln\left| \frac{a + \sqrt{a^2 + u^2}}{u} \right| + C$

40. $\displaystyle \int \frac{\sqrt{a^2 + u^2}}{u^2} \, du = -\frac{\sqrt{a^2 + u^2}}{u} + \ln\left(u + \sqrt{a^2 + u^2} \right) + C$

41. $\displaystyle \int \frac{du}{\sqrt{a^2 + u^2}} = \ln\left(u + \sqrt{a^2 + u^2} \right) + C$

42. $\displaystyle \int \frac{u^2 \, du}{\sqrt{a^2 + u^2}} = \frac{u}{2} \sqrt{a^2 + u^2} - \frac{a^2}{2} \ln\left(u + \sqrt{a^2 + u^2} \right) + C$

43. $\displaystyle \int \frac{du}{u\sqrt{a^2 + u^2}} = -\frac{1}{a} \ln\left| \frac{\sqrt{a^2 + u^2} + a}{u} \right| + C$

44. $\displaystyle \int \frac{du}{u^2 \sqrt{a^2 + u^2}} = -\frac{\sqrt{a^2 + u^2}}{a^2 u} + C$

45. $\displaystyle \int \frac{du}{(a^2 + u^2)^{3/2}} = \frac{u}{a^2 \sqrt{a^2 + u^2}} + C$

Forms Involving $\sqrt{a^2 - u^2}, \, a > 0$

46. $\displaystyle \int \sqrt{a^2 - u^2} \, du = \frac{u}{2} \sqrt{a^2 - u^2} + \frac{a^2}{2} \sin^{-1} \frac{u}{a} + C$

47. $\displaystyle \int u^2 \sqrt{a^2 - u^2} \, du = \frac{u}{8}(2u^2 - a^2)\sqrt{a^2 - u^2} + \frac{a^4}{8} \sin^{-1} \frac{u}{a} + C$

48. $\displaystyle \int \frac{\sqrt{a^2 - u^2}}{u} \, du = \sqrt{a^2 - u^2} - a \ln\left| \frac{a + \sqrt{a^2 - u^2}}{u} \right| + C$

49. $\displaystyle \int \frac{\sqrt{a^2 - u^2}}{u^2} \, du = -\frac{1}{u} \sqrt{a^2 - u^2} - \sin^{-1} \frac{u}{a} + C$

50. $\displaystyle \int \frac{u^2 \, du}{\sqrt{a^2 - u^2}} = -\frac{u}{2} \sqrt{a^2 - u^2} + \frac{a^2}{2} \sin^{-1} \frac{u}{a} + C$

51. $\displaystyle \int \frac{du}{u\sqrt{a^2 - u^2}} = -\frac{1}{a} \ln\left| \frac{a + \sqrt{a^2 - u^2}}{u} \right| + C$

52. $\displaystyle \int \frac{du}{u^2 \sqrt{a^2 - u^2}} = -\frac{1}{a^2 u} \sqrt{a^2 - u^2} + C$

53. $\displaystyle \int (a^2 - u^2)^{3/2} \, du$
$\displaystyle \qquad = -\frac{u}{8}(2u^2 - 5a^2)\sqrt{a^2 - u^2} + \frac{3a^4}{8} \sin^{-1} \frac{u}{a} + C$

54. $\displaystyle \int \frac{du}{(a^2 - u^2)^{3/2}} = \frac{u}{a^2 \sqrt{a^2 - u^2}} + C$

Forms Involving $\sqrt{u^2 - a^2}, \, a > 0$

55. $\displaystyle \int \sqrt{u^2 - a^2} \, du = \frac{u}{2} \sqrt{u^2 - a^2} - \frac{a^2}{2} \ln\left| u + \sqrt{u^2 - a^2} \right| + C$

56. $\displaystyle \int u^2 \sqrt{u^2 - a^2} \, du$
$\displaystyle \qquad = \frac{u}{8}(2u^2 - a^2)\sqrt{u^2 - a^2} - \frac{a^4}{8} \ln\left| u + \sqrt{u^2 - a^2} \right| + C$

57. $\displaystyle \int \frac{\sqrt{u^2 - a^2}}{u} \, du = \sqrt{u^2 - a^2} - a \cos^{-1} \frac{a}{|u|} + C$

58. $\displaystyle \int \frac{\sqrt{u^2 - a^2}}{u^2} \, du = -\frac{\sqrt{u^2 - a^2}}{u} + \ln\left| u + \sqrt{u^2 - a^2} \right| + C$

59. $\displaystyle \int \frac{du}{\sqrt{u^2 - a^2}} = \ln\left| u + \sqrt{u^2 - a^2} \right| + C$

60. $\displaystyle \int \frac{u^2 \, du}{\sqrt{u^2 - a^2}} = \frac{u}{2} \sqrt{u^2 - a^2} + \frac{a^2}{2} \ln\left| u + \sqrt{u^2 - a^2} \right| + C$

61. $\displaystyle \int \frac{du}{u^2 \sqrt{u^2 - a^2}} = \frac{\sqrt{u^2 - a^2}}{a^2 u} + C$

62. $\displaystyle \int \frac{du}{(u^2 - a^2)^{3/2}} = -\frac{u}{a^2 \sqrt{u^2 - a^2}} + C$

Forms Involving sin u, cos u, tan u

63. $\displaystyle\int \sin^2 u \, du = \frac{1}{2} u - \frac{1}{4} \sin 2u + C$

64. $\displaystyle\int \cos^2 u \, du = \frac{1}{2} u + \frac{1}{4} \sin 2u + C$

65. $\displaystyle\int \tan^2 u \, du = \tan u - u + C$

66. $\displaystyle\int \sin^3 u \, du = -\frac{1}{3}(2 + \sin^2 u) \cos u + C$

67. $\displaystyle\int \cos^3 u \, du = \frac{1}{3}(2 + \cos^2 u) \sin u + C$

68. $\displaystyle\int \tan^3 u \, du = \frac{1}{2} \tan^2 u + \ln|\cos u| + C$

69. $\displaystyle\int \sin^n u \, du = -\frac{1}{n} \sin^{n-1} u \cos u + \frac{n-1}{n} \int \sin^{n-2} u \, du$

70. $\displaystyle\int \cos^n u \, du = \frac{1}{n} \cos^{n-1} u \sin u + \frac{n-1}{n} \int \cos^{n-2} u \, du$

71. $\displaystyle\int \tan^n u \, du = \frac{1}{n-1} \tan^{n-1} u - \int \tan^{n-2} u \, du$

72. $\displaystyle\int \sin au \sin bu \, du = \frac{\sin(a-b)u}{2(a-b)} - \frac{\sin(a+b)u}{2(a+b)} + C$

73. $\displaystyle\int \cos au \cos bu \, du = \frac{\sin(a-b)u}{2(a-b)} + \frac{\sin(a+b)u}{2(a+b)} + C$

74. $\displaystyle\int \sin au \cos bu \, du = -\frac{\cos(a-b)u}{2(a-b)} - \frac{\cos(a+b)u}{2(a+b)} + C$

75. $\displaystyle\int u \sin u \, du = \sin u - u \cos u + C$

76. $\displaystyle\int u \cos u \, du = \cos u + u \sin u + C$

77. $\displaystyle\int u^n \sin u \, du = -u^n \cos u + n \int u^{n-1} \cos u \, du$

78. $\displaystyle\int u^n \cos u \, du = u^n \sin u - n \int u^{n-1} \sin u \, du$

79. $\displaystyle\int \sin^n u \cos^m u \, du$

$\displaystyle = -\frac{\sin^{n-1} u \cos^{m+1} u}{n+m} + \frac{n-1}{n+m} \int \sin^{n-2} u \cos^m u \, du$

$\displaystyle = \frac{\sin^{n+1} u \cos^{m-1} u}{n+m} + \frac{m-1}{n+m} \int \sin^n u \cos^{m-2} u \, du$

Forms Involving cot u, sec u, csc u

80. $\displaystyle\int \cot^2 u \, du = -\cot u - u + C$

81. $\displaystyle\int \cot^3 u \, du = -\frac{1}{2} \cot^2 u - \ln|\sin u| + C$

82. $\displaystyle\int \sec^3 u \, du = \frac{1}{2} \sec u \tan u + \frac{1}{2} \ln|\sec u + \tan u| + C$

83. $\displaystyle\int \csc^3 u \, du = -\frac{1}{2} \csc u \cot u + \frac{1}{2} \ln|\csc u - \cot u| + C$

84. $\displaystyle\int \cot^n u \, du = \frac{-1}{n-1} \cot^{n-1} u - \int \cot^{n-2} u \, du$

85. $\displaystyle\int \sec^n u \, du = \frac{1}{n-1} \tan u \sec^{n-2} u + \frac{n-2}{n-1} \int \sec^{n-2} u \, du$

86. $\displaystyle\int \csc^n u \, du = \frac{-1}{n-1} \cot u \csc^{n-2} u + \frac{n-2}{n-1} \int \csc^{n-2} u \, du$

Forms Involving Inverse Trigonometric Functions

87. $\displaystyle\int \sin^{-1} u \, du = u \sin^{-1} u + \sqrt{1 - u^2} + C$

88. $\displaystyle\int \cos^{-1} u \, du = u \cos^{-1} u - \sqrt{1 - u^2} + C$

89. $\displaystyle\int \tan^{-1} u \, du = u \tan^{-1} u - \frac{1}{2} \ln(1 + u^2) + C$

90. $\displaystyle\int u \sin^{-1} u \, du = \frac{2u^2 - 1}{4} \sin^{-1} u + \frac{u\sqrt{1 - u^2}}{4} + C$

91. $\displaystyle\int u \cos^{-1} u \, du = \frac{2u^2 - 1}{4} \cos^{-1} u - \frac{u\sqrt{1 - u^2}}{4} + C$

92. $\displaystyle\int u \tan^{-1} u \, du = \frac{u^2 + 1}{2} \tan^{-1} u - \frac{u}{2} + C$

93. $\displaystyle\int u^n \sin^{-1} u \, du = \frac{1}{n+1} \left[u^{n+1} \sin^{-1} u - \int \frac{u^{n+1} \, du}{\sqrt{1 - u^2}} \right]$, $n \neq -1$

94. $\displaystyle\int u^n \cos^{-1} u \, du = \frac{1}{n+1} \left[u^{n+1} \cos^{-1} u + \int \frac{u^{n+1} \, du}{\sqrt{1 - u^2}} \right]$, $n \neq -1$

95. $\displaystyle\int u^n \tan^{-1} u \, du = \frac{1}{n+1} \left[u^{n+1} \tan^{-1} u - \int \frac{u^{n+1} \, du}{\sqrt{1 + u^2}} \right]$, $n \neq -1$

Forms Involving Exponential and Logarithmic Functions

96. $\displaystyle\int ue^{au}\,du = \frac{1}{a^2}\,(au-1)e^{au} + C$

97. $\displaystyle\int u^n\,e^{au}\,du = \frac{1}{a}\,u^n\,e^{au} - \frac{n}{a}\int u^{n-1}e^{au}\,du$

98. $\displaystyle\int e^{au}\sin bu\,du = \frac{e^{au}}{a^2+b^2}\,(a\sin bu - b\cos bu) + C$

99. $\displaystyle\int e^{au}\cos bu\,du = \frac{e^{au}}{a^2+b^2}\,(a\cos bu + b\sin bu) + C$

100. $\displaystyle\int \frac{du}{1+be^{au}} = u - \frac{1}{a}\ln(1+be^{au}) + C$

101. $\displaystyle\int \ln u\,du = u\ln u - u + C$

102. $\displaystyle\int u^n \ln u\,du = \frac{u^{n+1}}{(n+1)^2}\,[(n+1)\ln u - 1] + C$

103. $\displaystyle\int \frac{1}{u\ln u}\,du = \ln|\ln u| + C$

Forms Involving Hyperbolic Functions

104. $\displaystyle\int \sinh u\,du = \cosh u + C$

105. $\displaystyle\int \cosh u\,du = \sinh u + C$

106. $\displaystyle\int \tanh u\,du = \ln\cosh u + C$

107. $\displaystyle\int \coth u\,du = \ln|\sinh u| + C$

108. $\displaystyle\int \text{sech}\,u\,du = \tan^{-1}|\sinh u| + C$

109. $\displaystyle\int \text{csch}\,u\,du = \ln\left|\tanh\tfrac{1}{2}u\right| + C$

110. $\displaystyle\int \text{sech}^2\,u\,du = \tanh u + C$

111. $\displaystyle\int \text{csch}^2\,u\,du = -\coth u + C$

112. $\displaystyle\int \text{sech}\,u\,\tanh u\,du = -\text{sech}\,u + C$

113. $\displaystyle\int \text{csch}\,u\,\coth u\,du = -\text{csch}\,u + C$

Forms Involving $\sqrt{2au^2 - u^2},\ a > 0$

114. $\displaystyle\int \sqrt{2au-u^2}\,du = \frac{u-a}{2}\sqrt{2au-u^2}$
$$+ \frac{a^2}{2}\cos^{-1}\left(\frac{a-u}{a}\right) + C$$

115. $\displaystyle\int u\sqrt{2au-u^2}\,du = \frac{2u^2-au-3a^2}{6}\sqrt{2au-u^2}$
$$+ \frac{a^3}{2}\cos^{-1}\left(\frac{a-u}{a}\right) + C$$

116. $\displaystyle\int \frac{\sqrt{2au-u^2}}{u}\,du = \sqrt{2au-u^2} + a\cos^{-1}\left(\frac{a-u}{a}\right) + C$

117. $\displaystyle\int \frac{\sqrt{2au-u^2}}{u^2}\,du = -\frac{2\sqrt{2au-u^2}}{u} - \cos^{-1}\left(\frac{a-u}{a}\right) + C$

118. $\displaystyle\int \frac{du}{\sqrt{2au-u^2}} = \cos^{-1}\left(\frac{a-u}{a}\right) + C$

119. $\displaystyle\int \frac{u\,du}{\sqrt{2au-u^2}} = -\sqrt{2au-u^2} + a\cos^{-1}\left(\frac{a-u}{a}\right) + C$

120. $\displaystyle\int \frac{u^2\,du}{\sqrt{2au-u^2}} = -\frac{(u+3a)}{2}\sqrt{2au-u^2}$
$$+ \frac{3a^2}{2}\cos^{-1}\left(\frac{a-u}{a}\right) + C$$

121. $\displaystyle\int \frac{du}{u\sqrt{2au-u^2}} = -\frac{\sqrt{2au-u^2}}{au} + C$

SINGLE VARIABLE

CALCULUS

EARLY TRANSCENDENTALS

SINGLE VARIABLE
CALCULUS

EARLY TRANSCENDENTALS

SOO T. TAN

STONEHILL COLLEGE

BROOKS/COLE
CENGAGE Learning™

Australia • Brazil • Japan • Korea • Mexico • Singapore • Spain • United Kingdom • United States

Single Variable Calculus: Early Transcendentals
Soo T. Tan

Senior Acquisitions Editor: Liz Covello

Publisher: Richard Stratton

Senior Developmental Editor: Danielle Derbenti

Developmental Editor: Ed Dodd

Associate Editor: Jeannine Lawless

Editorial Assistant: Lauren Hamel

Media Editor: Maureen Ross

Marketing Manager: Jennifer Jones

Marketing Assistant: Erica O'Connell

Marketing Communications Manager: Mary Anne Payumo

Content Project Manager: Cheryll Linthicum

Creative Director: Rob Hugel

Art Director: Vernon Boes

Print Buyer: Becky Cross

Rights Acquisitions Account Manager, Text: Roberta Broyer

Rights Acquisitions Account Manager, Image: Amanda Groszko

Production Service: Martha Emry

Text Designer: Diane Beasley

Art Editors: Leslie Lahr, Lisa Torri, Martha Emry

Photo Researcher: Kathleen Olsen

Copy Editor: Barbara Willette

Illustrators: Precision Graphics, Matrix Art Services, Network Graphics

Cover Designer: Terri Wright

Cover Image: Nathan Fariss for *Popular Mechanics*

Compositor: Graphic World

For product information and technology assistance, contact us at
Cengage Learning Customer & Sales Support, 1-800-354-9706.
For permission to use material from this text or product,
submit all requests online at **www.cengage.com/permissions**.
Further permissions questions can be e-mailed to
permissionrequest@cengage.com.

Library of Congress Control Number: 2009941226

Student Edition:
ISBN-13: 978-0-534-46570-4

ISBN-10: 0-534-46570-6

Brooks/Cole
20 Davis Drive
Belmont, CA 94002-3098
USA

Cengage Learning is a leading provider of customized learning solutions with office locations around the globe, including Singapore, the United Kingdom, Australia, Mexico, Brazil, and Japan. Locate your local office at **www.cengage.com/global**.

Cengage Learning products are represented in Canada by Nelson Education, Ltd.

To learn more about Brooks/Cole, visit **www.cengage.com/brookscole**
Purchase any of our products at your local college store or at our preferred online store **www.ichapters.com**.

Printed in Canada
1 2 3 4 5 6 7 13 12 11 10

To Olivia, Maxwell, Sasha, Isabella, and Ashley

About the Author

SOO T. TAN received his S.B. degree from the Massachusetts Institute of Technology, his M.S. degree from the University of Wisconsin–Madison, and his Ph.D. from the University of California at Los Angeles. He has published numerous papers on optimal control theory, numerical analysis, and the mathematics of finance. He is also the author of a series of textbooks on applied calculus and applied finite mathematics.

One of the most important lessons I have learned from my many years of teaching undergraduate mathematics courses is that most students, mathematics and non-mathematics majors alike, respond well when introduced to mathematical concepts and results using real-life illustrations.

This awareness led to the intuitive approach that I have adopted in all of my texts. As you will see, I try to introduce each abstract mathematical concept through an example drawn from a common, real-life experience. Once the idea has been conveyed, I then proceed to make it precise, thereby assuring that no mathematical rigor is lost in this intuitive treatment of the subject. Another lesson I learned from my students is that they have a much greater appreciation of the material if the applications are drawn from their fields of interest and from situations that occur in the real world. This is one reason you will see so many examples and exercises in my texts that are drawn from various and diverse fields such as physics, chemistry, engineering, biology, business, and economics. There are also many exercises of general and current interest that are modeled from data gathered from newspapers, magazines, journals, and other media. Whether it be global warming, brain growth and IQ, projected U.S. gasoline usage, or finding the surface area of the Jacqueline Kennedy Onassis Reservoir, I weave topics of current interest into my examples and exercises to keep the book relevant to all of my readers.

Contents

Author's Commitment to Accuracy

As with all of my projects, accuracy is of paramount importance. For this reason, I solved every problem myself and wrote the solutions for the solutions manual. In this accuracy checking process, I worked very closely with several professors who contributed in different ways and at different stages throughout the development of the text and manual: Jason Aubrey (*University of Missouri*), Kevin Charlwood (*Washburn University*), Jerrold Grossman (*Oakland University*), Tao Guo (*Rock Valley College*), Selwyn Hollis (*Armstrong Atlantic State University*), Diane Koenig (*Rock Valley College*), Michael Montano (*Riverside Community College*), John Samons (*Florida Community College*), Doug Shaw (*University of Northern Iowa*), and Richard West (*Francis Marion University*).

Accuracy Process

First Round
- The first draft of the manuscript was reviewed by numerous calculus instructors, all of whom either submitted written reviews, participated in a focus group discussion, or class-tested the manuscript.

Second Round
- The author provided revised manuscript to be reviewed by additional calculus instructors who went through the same steps as the first group and submitted their responses.
- Simultaneously, author Soo Tan was writing the solutions manual, which served as an additional check of his work on the text manuscript.

Third Round
- Two calculus instructors checked the revised manuscript for accuracy while simultaneously checking the solutions manual, sending their corrections back to the author for inclusion.
- Additional groups of calculus instructors participated in focus groups and class testing of the revised manuscript.
- First drafts of the art were produced and checked for accuracy.
- The manuscript was edited by a professional copyeditor.
- Biographies were written by a calculus instructor and submitted for copyedit.

Fourth Round
- Once the manuscript was declared final, a compositor created galley pages, whose accuracy was checked by several calculus instructors.
- Revisions were made to the art, and revised art proofs were checked for accuracy.
- Further class testing and live reviews were completed.
- Galley proofs were checked for consistency by the production team and carefully reviewed by the author.
- Biographies were checked and revised for accuracy by another calculus instructor.

Fifth Round
- First round page proofs were distributed, proofread, and checked for accuracy again. As with galley proofs, these pages were carefully reviewed by the author with art seen in place with the exposition for the first time.
- The revised art was again checked for accuracy by the author and the production service.

Sixth Round
- Revised page proofs were checked by a second proofreader and the author.

Seventh Round
- Final page proofs were checked for consistency by the production team and the author performed his final review of the pages.

Throughout my teaching career I have always enjoyed teaching calculus and helping students to see the elegance and beauty of calculus. So when I was approached by my editor to write this series, I welcomed the opportunity. Upon reflecting, I see that I started this project from a strong vantage point. I have written an *Applied Mathematics* series, and over the years I have gotten a lot of feedback from many professors and students using the books in the series. The wealth of suggestions that I gained from them coupled with my experience in the classroom served me well when I embarked upon this project.

In writing the *Calculus* series, I have constantly borne in mind two primary objectives: first, to provide the instructor with a book that is easy to teach from and yet has all the content and rigor of a traditional calculus text, and second, to provide students with a book that motivates their interest and at the same time is easy for them to read. In my experience, students coming to calculus for the first time respond best to an intuitive approach, and I try to use this approach by introducing abstract ideas with concrete, real-life examples that students can relate to, wherever appropriate. Often a simple real-life illustration can serve as motivation for a more complex mathematical concept or theorem. Also, I have tried to use a clear, precise, and concise writing style throughout the book and have taken special care to ensure that my intuitive approach does not compromise the mathematical rigor that is expected of an engineering calculus text.

In addition to the applications in mathematics, engineering, physics, and the other natural and social sciences, I have included many other examples and exercises drawn from diverse fields of current interest. The solutions to all the exercises in the book are provided in a separate manual. In keeping with the emphasis on conceptual understanding, I have included concept questions at the beginning of each exercise set. In each end-of-chapter review section I have also included fill-in-the-blank questions for a review of the concepts. I have found these questions to be an effective learning tool to help students master the definitions and theorems in each chapter. Furthermore, I have included many questions that ask for the interpretation of graphical, numerical, and algebraic results in both the examples and the exercise sets.

Unique Approach to the Presentation of Limits

Finally, I have employed a unique approach to the introduction of the limit concept. Many calculus textbooks introduce this concept via the slope of a tangent line to a curve and then follow by relating the slope to the notion of the rate of change of one quantity with respect to another. In my text I do precisely the opposite: I introduce the limit concept by looking at the rate of change of the maglev (magnetic levitation train). This approach is more intuitive and captures the interest of the student from the very beginning—it shows immediately the relevance of calculus to the real world. I might add that this approach has worked very well for me not only in the classroom; it has also been received very well by the users of my applied calculus series. This intuitive approach (using the maglev as a vehicle) is carried into the introduction and explanation of some of the fundamental theorems in calculus, such as the Intermediate Value Theorem and the Mean Value Theorem. Consistently woven throughout the text, this idea permeates much of the text—from concepts in limits, to continuity, to integration, and even to inverse functions.

Soo T. Tan

■ Tan *Calculus* Series

The Tan *Calculus* series includes the following textbooks:

- *Calculus: Early Transcendentals* © 2011 (ISBN 0-534-46554-4)
- *Single Variable Calculus: Early Transcendentals* © 2011 (ISBN 0-534-46570-6)
- *Calculus* © 2010 (ISBN 0-534-46579-X)
- *Single Variable Calculus* © 2010 (ISBN 0-534-46566-8)
- *Multivariable Calculus* © 2010 (ISBN 0-534-46575-7)

■ Features

An Intuitive Approach . . . Without Loss of Rigor

Beginning with each chapter opening vignette and carrying through each chapter, Soo Tan's intuitive approach links the abstract ideas of calculus with concrete, real-life examples. This intuitive approach is used to advantage to introduce and explain many important concepts and theorems in calculus, such as tangent lines, Rolles's Theorem, absolute extrema, increasing and decreasing functions, limits at infinity, and parametric equations. In this example from Chapter 5 the discussion of the area between two curves is motivated with a real-life illustration that is followed by the precise discussion of the mathematical concepts involved.

■ A Real-Life Interpretation

Two cars are traveling in adjacent lanes along a straight stretch of a highway. The velocity functions for Car A and Car B are $v = f(t)$ and $v = g(t)$, respectively. The graphs of these functions are shown in Figure 1.

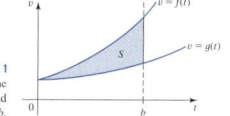

FIGURE 1
The shaded area S gives the distance that Car A is ahead of Car B at time $t = b$.

The area of the region under the graph of f from $t = 0$ to $t = b$ gives the total distance covered by Car A in b seconds over the time interval $[0, b]$. The distance covered by Car B over the same period of time is given by the area under the graph of g on the interval $[0, b]$. Intuitively, we see that the area of the (shaded) region S between the graphs of f and g on the interval $[0, b]$ gives the distance that Car A will be ahead of Car B at time $t = b$.

■ The Area Between Two Curves

Suppose f and g are continuous functions with $f(x) \geq g(x)$ for all x in $[a, b]$, so that the graph of f lies on or above that of g on $[a, b]$. Let's consider the region S bounded by the graphs of f and g between the vertical lines $x = a$ and $x = b$ as shown in Figure 2. To define the *area* of S, we take a regular partition of $[a, b]$,

$$a = x_0 < x_1 < x_2 < x_3 < \cdots < x_n = b$$

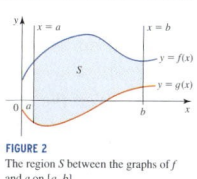

FIGURE 2
The region S between the graphs of f and g on $[a, b]$

and form the Riemann sum of the function $f - g$ over $[a, b]$ with respect to this partition:

$$\sum_{k=1}^{n} [f(c_k) - g(c_k)]\Delta x$$

where c_k is an evaluation point in the subinterval $[x_{k-1}, x_k]$ and $\Delta x = (b - a)/n$. The kth term of this sum gives the area of a rectangle with height $[f(c_k) - g(c_k)]$ and width Δx. As you can see in Figure 3, this area is an approximation of the area of the subregion of S that lies between the graphs of f and g on $[x_{k-1}, x_k]$.

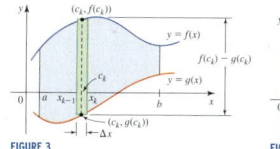
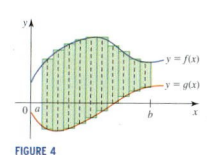

FIGURE 3
The kth term of the Riemann sum of $f - g$ gives the area of the kth rectangle of width Δx.

FIGURE 4
The Riemann sum of $f - g$ approximates the area of S.

Unique Applications in the Examples and Exercises

Our relevant, unique applications are designed to illustrate mathematical concepts and at the same time capture students' interest.

69. Constructing a New Road The following figures depict three possible roads connecting the point $A(-1000, 0)$ to the point $B(1000, 1000)$ via the origin. The functions describing the dashed center lines of the roads follow:

$$f(x) = \begin{cases} 0 & \text{if } -1000 \leq x \leq 0 \\ x & \text{if } 0 < x \leq 1000 \end{cases}$$

$$g(x) = \begin{cases} 0 & \text{if } -1000 \leq x \leq 0 \\ 0.001x^2 & \text{if } 0 < x \leq 1000 \end{cases}$$

$$h(x) = \begin{cases} 0 & \text{if } -1000 \leq x \leq 0 \\ 0.000001x^3 & \text{if } 0 < x \leq 1000 \end{cases}$$

Show that f is not differentiable on the interval $(-1000, 1000)$, g is differentiable but not twice differentiable on $(-1000, 1000)$, and h is twice differentiable on $(-1000, 1000)$. Taking into consideration the dynamics of a moving vehicle, which proposal do you think is most suitable?

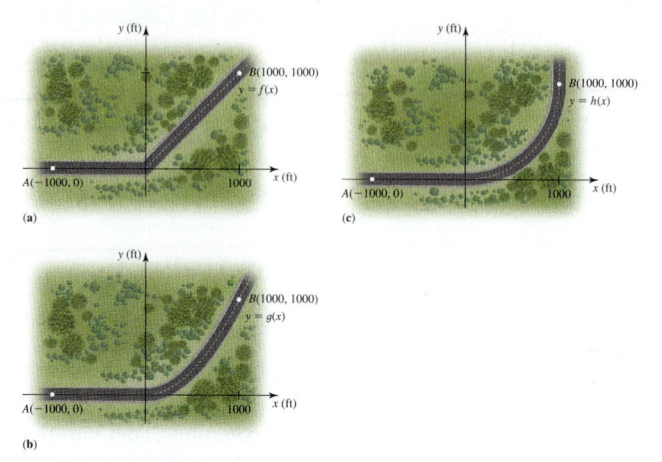

Connections

One particular example—the maglev (magnetic levitation) train—is used as a common thread throughout the development of calculus from limits through integration. The goal here is to show students the connection between the important theorems and concepts presented. Topics that are introduced through this example include the Intermediate Value Theorem, the Mean Value Theorem, the Mean Value Theorem for Definite Integrals, limits, continuity, derivatives, antiderivatives, initial value problems, inverse functions, and indeterminate forms.

■ A Real-Life Example

A prototype of a maglev (magnetic levitation train) moves along a straight monorail. To describe the motion of the maglev, we can think of the track as a coordinate line. From data obtained in a test run, engineers have determined that the maglev's displacement (directed distance) measured in feet from the origin at time t (in seconds) is given by

$$s = f(t) = 4t^2 \qquad 0 \leq t \leq 30 \qquad (1)$$

where f is called the position function of the maglev. The position of the maglev at time $t = 0, 1, 2, 3, \ldots, 30$, measured in feet from its initial position, is

$$f(0) = 0, \qquad f(1) = 4, \qquad f(2) = 16, \qquad f(3) = 36, \qquad \ldots, \qquad f(30) = 3600$$

(See Figure 1.)

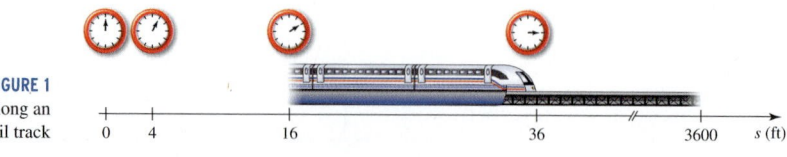

FIGURE 1
A maglev moving along an elevated monorail track

Precise Figures That Help Students Visualize the Concepts

Carefully constructed art helps the student to visualize the mathematical ideas under discussion.

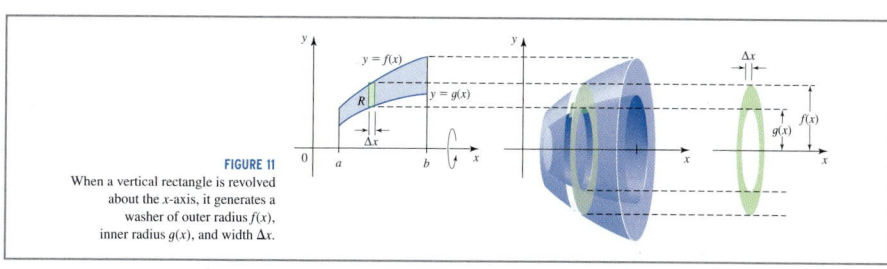

FIGURE 11
When a vertical rectangle is revolved about the x-axis, it generates a washer of outer radius $f(x)$, inner radius $g(x)$, and width Δx.

172   **Chapter 2**   The Derivative

2.1 CONCEPT QUESTIONS

1. a. Give a geometric and a physical interpretation of the expression

$$\frac{f(x + h) - f(x)}{h}$$

b. Give a geometric and a physical interpretation of the expression

$$\lim_{h \to 0} \frac{f(x + h) - f(x)}{h}$$

2. Under what conditions does a function fail to have a derivative at a number? Illustrate your answer with sketches.

2.1 EXERCISES

In Exercises 1–14, use the definition of the derivative to find the derivative of the function. What is its domain?

1. $f(x) = 5$ **2.** $f(x) = 2x + 1$

3. $f(x) = 3x - 4$ **4.** $f(x) = 2x^2 + x$

5. $f(x) = 3x^2 - x + 1$ **6.** $f(x) = x^3 - x$

7. $f(x) = 2x^3 + x - 1$ **8.** $f(x) = 2\sqrt{x}$

9. $f(x) = \sqrt{x + 1}$ **10.** $f(x) = \dfrac{1}{x}$

11. $f(x) = \dfrac{1}{x + 2}$ **12.** $f(x) = -\dfrac{2}{\sqrt{x}}$

13. $f(x) = \dfrac{3}{2x + 1}$ **14.** $f(x) = x + \sqrt{x}$

In Exercises 15–20, find an equation of the tangent line to the graph of the function at the indicated point.

Function	Point
15. $f(x) = x^2 + 1$	$(2, 5)$
16. $f(x) = 3x^2 - 4x + 2$	$(2, 6)$
17. $f(x) = 2x^3$	$(1, 2)$
18. $f(x) = 3x^3 - x$	$(-1, -2)$
19. $f(x) = \sqrt{x - 1}$	$(4, \sqrt{3})$
20. $f(x) = \dfrac{2}{x}$	$(2, 1)$

21. a. Find an equation of the tangent line to the graph of $f(x) = 2x - x^3$ at the point $(1, 1)$.

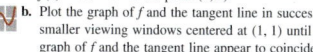 **b.** Plot the graph of f and the tangent line in successively smaller viewing windows centered at $(1, 1)$ until the graph of f and the tangent line appear to coincide.

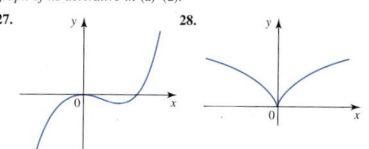 **22. a.** In Example 6 we showed that $f(x) = |x|$ is not differentiable at $x = 0$. Plot the graph of f using the viewing window $[-1, 1] \times [-1, 1]$. Then **ZOOM IN** using successively smaller viewing windows centered at $(0, 0)$. What can you say about the existence of a tangent line at $(0, 0)$?

b. Plot the graph of

$$f(x) = \begin{cases} x + 1 & \text{if } x \le 1 \\ \dfrac{2}{x} & \text{if } x > 1 \end{cases}$$

using the viewing window $[-2, 4] \times [-2, 3]$. Then **ZOOM IN** using successively smaller viewing windows centered at $(1, 2)$. Is f differentiable at $x = 1$?

In Exercises 23–26, find the rate of change of y with respect to x at the given value of x.

23. $y = -2x^2 + x + 1$; $x = 1$

24. $y = 2x^3 + 2$; $x = 2$

25. $y = \sqrt{2x}$; $x = 2$

26. $y = x^2 - \dfrac{1}{x}$; $x = -1$

In Exercises 27–30, match the graph of each function with the graph of its derivative in (a)–(d).

27.

28.

V Videos for selected exercises are available online at **www.academic.cengage.com/login.**

Concept Questions

Designed to test student understanding of the basic concepts discussed in the section, these questions encourage students to explain learned concepts in their own words.

Exercises

Each exercise section contains an ample set of problems of a routine computational nature, followed by a set of application-oriented problems (many of them sourced) and true/false questions that ask students to explain their answer.

Graphing Utility and CAS Exercises

Indicated by and **cas** icons next to the corresponding exercises, these exercises offer practice in using technology to solve problems that might be difficult to solve by hand. Sourced problems using real-life data are often included.

CHAPTER 2 REVIEW

CONCEPT REVIEW

In Exercises 1–14, fill in the blanks.

1. a. The derivative of a function with respect to x is the function f' defined by the rule _____.

b. The domain of f' consists of all values of x for which the _____ exists.

c. The number $f'(a)$ gives the slope of the _____ to the graph of f at _____.

d. The number $f'(a)$ also measures the rate of change of _____ with respect to _____ at _____.

e. If f is differentiable at a, then an equation of the tangent line to the graph of f at $(a, f(a))$ is _____.

2. a. A function might not be differentiable at a _____. For example, the function _____ fails to be differentiable at _____.

b. If a function f is differentiable at a, then f is _____ at a. The converse is false. For example, the function _____ is continuous at _____ but is not differentiable at _____.

3. a. If c is a constant, then $\dfrac{d}{dx}(c) =$ _____.

b. If n is any real number, then $\dfrac{d}{dx}(x^n) =$ _____.

4. If f and g are differentiable functions and c is a constant, then the Constant Multiple Rule states that _____, the Sum Rule states that _____, the Product Rule states that _____, and the Quotient Rule states that _____.

5. If $y = f(x)$, where f is a differentiable function, and α denotes the angle that the tangent line to the graph of f at $(x, f(x))$ makes with the positive x-axis, then $\tan \alpha =$ _____.

6. Suppose that $f(t)$ gives the position of an object moving on a coordinate line.

a. The velocity of the object is given by _____, its acceleration is given by _____, and its jerk is given by _____. The speed of the object is given by _____.

b. The object is moving in the positive direction if $v(t)$ _____ and in the negative direction if $v(t)$ _____. It is stationary if $v(t) =$ _____.

7. If C, R, P, and $\overline{C}$ denote the total cost function, the total revenue function, the profit function, and the average cost function, respectively, then the marginal total cost function is given by _____, the marginal total revenue function by _____, the marginal profit function by _____, and the marginal average cost function by _____.

8. If f is differentiable at x and g is differentiable at $f(x)$, then the function $h = g \circ f$ is differentiable at _____, and $h'(x) =$ _____.

9. a. The General Power Rule states that $\dfrac{d}{dx}[f(x)]^n =$ _____.

b. If f is differentiable, then $\dfrac{d}{dx}[\sin f(x)] =$ _____,

$\dfrac{d}{dx}[\cos f(x)] =$ _____, $\dfrac{d}{dx}[\tan f(x)] =$ _____,

$\dfrac{d}{dx}[\sec f(x)] =$ _____, $\dfrac{d}{dx}[\csc f(x)] =$ _____,

and $\dfrac{d}{dx}[\cot f(x)] =$ _____.

10. a. If u is a differentiable function of x, then $\dfrac{d}{dx}e^u =$ _____.

b. If $a > 0$ and $a \neq 1$, then $a^x =$ _____; if u is a differentiable function of x, then $\dfrac{d}{dx}a^u =$ _____.

11. If u is a differentiable function of x, then $\dfrac{d}{dx}\log_a|u| =$ _____.

REVIEW EXERCISES

In Exercises 1 and 2, use the definition of the derivative to find the derivative of the function.

1. $f(x) = x^2 - 2x - 4$

2. $f(x) = 2x^3 - 3x + 2$

In Exercises 3 and 4, sketch the graph of f' for the function f whose graph is given.

3.

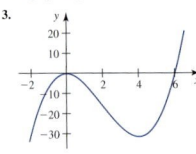

4.

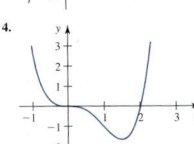

5. The amount of money on fixed deposit at the end of 5 years in a bank paying interest at the rate of r per year is given by $A = f(r)$ (dollars).

a. What does $f'(r)$ measure? Give units.

b. What is the sign of $f'(r)$? Explain.

c. If you know that $f'(6) = 60,775.31$ estimate the change _____ terest rate changes

6. Use the graph of the function f to find the value(s) of x at which f is not differentiable.

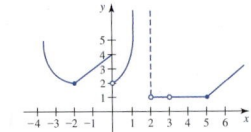

7. Find a function f and a number a such that

$$\lim_{h \to 0} \frac{3(4 + h)^{3/2} - 24}{h} = f'(a)$$

8. Evaluate $\displaystyle\lim_{h \to 0} \frac{2(1 + h)^3 + (1 + h)^2 - 3}{h}$.

In Exercises 9–64, find the derivative of the function.

9. $f(x) = \dfrac{1}{3}x^6 - 2x^4 + x^2 - 5$

10. $g(x) = 2x^4 + 3x^{1/2} - x^{-1/3} + x^{-4}$

11. $s = 2t^2 - \dfrac{4}{t} + \dfrac{2}{\sqrt{t}}$

12. $f(x) = \dfrac{x + 1}{x - 1}$

13. $g(t) = \dfrac{t - 1}{2t + 1}$

14. $h(x) = \dfrac{x}{2x^2 + 3}$

15. $h(u) = \dfrac{\sqrt{u}}{u^2 + 1}$

16. $u = \dfrac{t^2}{1 - \sqrt{t}}$

17. $g(\theta) = \cos \theta - 2 \sin \theta$

18. $f(x) = x \tan x + \sec x$

PROBLEM-SOLVING TECHNIQUES

The following example shows that rewriting a function in an alternative form sometimes pays dividends.

EXAMPLE Find $f^{(n)}(x)$ if $f(x) = \dfrac{x}{x^2 - 1}$.

Solution Our first instinct is to use the Quotient Rule to compute $f'(x)$, $f''(x)$, and so on. The expectation here is either that the rule for $f^{(n)}$ will become apparent or that at least a pattern will emerge that will enable us to guess at the form for $f^{(n)}(x)$. But the futility of this approach will be evident when you compute the first two derivatives of f.

Let's see whether we can transform the expression for $f(x)$ before we differentiate. You can verify that $f(x)$ can be written as

$$f(x) = \frac{x}{x^2 - 1} = \frac{\frac{1}{2}(x - 1) + \frac{1}{2}(x + 1)}{(x + 1)(x - 1)} = \frac{1}{2}\left[\frac{1}{x + 1} + \frac{1}{x - 1}\right]$$

Concept Review Questions

Beginning each end of chapter review, these questions give students a chance to check their knowledge of the basic definitions and concepts from the chapter.

Review Exercises

Offering a solid review of the chapter material, these exercises contain routine computational exercises as well as applied problems.

Problem-Solving Techniques

At the end of selected chapters the author discusses problem-solving techniques that provide students with the tools they need to make seemingly complex problems easier to solve.

CHALLENGE PROBLEMS

1. Find $\lim\limits_{x \to 2} \dfrac{x^{10} - 2^{10}}{x^5 - 2^5}$.

2. Find the derivative of $y = \sqrt{x + \sqrt{x + \sqrt{x}}}$.

3. **a.** Verify that $\dfrac{2x + 1}{x^2 + x - 2} = \dfrac{1}{x + 2} + \dfrac{1}{x - 1}$.

   b. Find $f^{(n)}(x)$ if $f(x) = \dfrac{2x + 1}{x^2 + x - 2}$.

4. Find the values of x for which f is differentiable.
   a. $f(x) = \sin |x|$ **b.** $f(x) = |\sin x|$

5. Find $f^{(10)}(x)$ if $f(x) = \dfrac{1 + x}{\sqrt{1 - x}}$.

   Hint: Show that $f(x) = \dfrac{2}{\sqrt{1 - x}} - \sqrt{1 - x}$.

6. Find $f^{(n)}(x)$ if $f(x) = \dfrac{ax + b}{cx + d}$.

7. Suppose that f is differentiable and $f(a + b) = f(a)f(b)$ for all real numbers a and b. Show that $f'(x) = f'(0)f(x)$ for all x.

8. Suppose that $f^{(n)}(x) = 0$ for every x in an interval (a, b) and $f(c) = f'(c) = \cdots = f^{(n-1)}(c) = 0$ for some c in (a, b). Show that $f(x) = 0$ for all x in (a, b).

9. Let $F(x) = f(\sqrt{1 + x^2})$, where f is a differentiable function. Find $F'(x)$.

10. Determine the values of b and c such that the parabola $y = x^2 + bx + c$ is tangent to the graph of $y = \sin x$ at the point $\left(\frac{\pi}{6}, \frac{1}{2}\right)$. Plot the graphs of both functions on the same set of axes.

11. Suppose f is defined on $(-\infty, \infty)$ and satisfies $|f(x) - f(y)| \le (x - y)^2$ for all x and y. Show that f is a constant function.
    Hint: Look at $f'(x)$.

12. Use the definition of the derivative to find the derivative of $f(x) = \tan ax$.

13. Find y'' at the point $(1, -2)$ if
$$2x^2 + 2xy + xy^2 - 3x + 3y + 7 = 0$$

Challenge Problems

Providing students with an opportunity to stretch themselves, the Challenge Problems develop their skills beyond the basics. These can be solved by using the techniques developed in the chapter but require more effort than the problems in the regular exercise sets do.

Guidance When Students Need It

The caution icon advises students how to avoid common mistakes and misunderstandings. This feature addresses both student misconceptions and situations in which students often follow unproductive paths.

Historical Biography

BLAISE PASCAL
(1623-1662)

A great mathematician who was not acknowledged in his lifetime, Blaise Pascal came extremely close to discovering calculus before Leibniz (page 157) and Newton (page 179), the two people who are most commonly credited with the discovery. Pascal was something of a prodigy and published his first important mathematical discovery at the age of sixteen. The work consisted of only a single printed page, but it contained a vital step in the development of projective geometry and a proposition called *Pascal's mystic hexagram* that discussed a property of a hexagon inscribed in a conic section. Pascal's interests varied widely, and from 1642 to 1644 he worked on the first manufactured calculator, which he designed to help his father with his tax work. Pascal manufactured about 50 of the machines, but they proved too costly to continue production. The basic principle of Pascal's calculating machine was still used until the electronic age. Pascal and Pierre de Fermat (page 307) also worked on the mathematics in games of chance and laid the foundation for the modern theory of probability. Pascal's later work, *Treatise on the Arithmetical Triangle*, gave important results on the construction that would later bear his name, *Pascal's Triangle*.

⚠ Theorem 1 states that a relative extremum of f can occur only at a critical number of f. It is important to realize, however, that the converse of Theorem 1 is false. In other words, you may *not* conclude that if c is a critical number of f, then f must have a relative extremum at c. (See Example 3.)

Biographies to Provide Historical Context

Historical biographies provide brief looks at the people who contributed to the development of calculus, focusing not only on their discoveries and achievements, but on their human side as well.

Videos to Help Students Draw Complex Multivariable Calculus Artwork

Unique to this book, Tan's *Calculus* provides video lessons for the multivariable sections of the text that help students learn, step-by-step, how to draw the complex sketches required in multivariable calculus. Videos of these lessons will be available at the text's companion website.

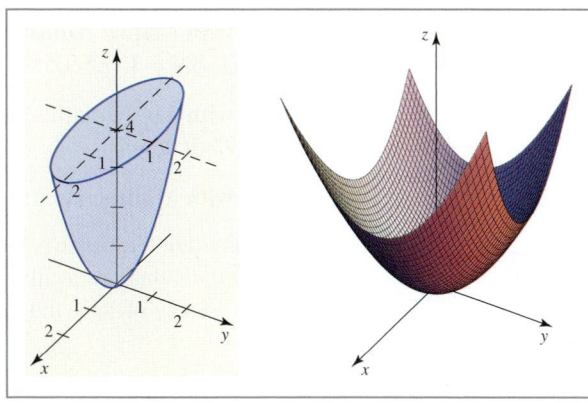

■ Instructor Resources

Instructor's Solutions Manual for Single Variable Calculus:
Early Transcendentals (ISBN 0-534-46572-2)

Instructor's Solutions Manual for Multivariable Calculus (ISBN 0-534-46578-1)
Prepared by Soo T. Tan

These manuals provide worked-out solutions to all problems in the text.

PowerLecture CD (ISBN 0-495-11482-0)
This comprehensive CD-ROM includes the *Instructor's Solutions Manual;* PowerPoint slides with art, tables, and key definitions from the text; and ExamView computerized testing, featuring algorithmically generated questions to create, deliver, and customize tests. A static version of the test bank will also be available online.

Solution Builder (ISBN 0-534-41831-7)
The online Solution Builder lets instructors easily build and save personal solution sets either for printing or for posting on password-protected class websites. Contact your local sales representative for more information on obtaining an account for this instructor-only resource.

Enhanced WebAssign (ISBN 0-495-39345-2)
Instant feedback and ease of use are just two reasons why WebAssign is the most widely used homework system in higher education. WebAssign allows instructors to assign, collect, grade, and record homework assignments via the Web. Now this proven homework system has been enhanced to include links to textbook sections, video examples, and problem-specific tutorials. Enhanced WebAssign is more than a homework system—it is a complete learning system for math students.

■ Student Resources

Student Solutions Manual for Single Variable Calculus:
Early Transcendentals (ISBN 0-534-46573-0)

Student Solutions Manual for Multivariable Calculus (ISBN 0-534-46577-3)
Prepared by Soo T. Tan

Providing more in-depth explanations, this insightful resource includes fully worked-out solutions for the answers to select exercises included at the back of the textbook, as well as problem-solving strategies, additional algebra steps, and review for selected problems.

CalcLabs with Maple: Single Variable Calculus, 4e by Phil Yasskin and Art Belmonte (ISBN 0-495-56062-6)

CalcLabs with Maple: Multivariable Calculus, 4e by Phil Yasskin and Art Belmonte (ISBN 0-495-56058-8)

CalcLabs with Mathematica: Single Variable Calculus, 4e by Selwyn Hollis (ISBN 0-495-56063-4)

CalcLabs with Mathematica: Multivariable Calculus, 4e (ISBN 0-495-82722-3)

Each of these comprehensive lab manuals helps students learn to effectively use the technology tools that are available to them. Each lab contains clearly explained exercises and a variety of labs and projects to accompany the text.

■ Acknowledgments

I want to express my heartfelt thanks to the reviewers for their many helpful comments and suggestions at various stages during the development of this text. I also want to thank Kevin Charlwood, Jerrold Grossman, Tao Guo, Selwyn Hollis, Diane Koenig, and John Samons, who checked the manuscript and text for accuracy; Richard West, Richard Montano, and again Kevin Charlwood for class testing the manuscript; and Andrew Bulman-Fleming for his help with the production of the solutions manuals. Additionally, I would like to thank Diane Koenig and Jason Aubrey for writing the biographies and also Doug Shaw and Richard West for their work on the projects. A special thanks to Tao Guo for his contribution to the content and accuracy of the solutions manuals.

I feel fortunate to have worked with a wonderful team during the development and production of this text. I wish to thank the editorial, production, and marketing staffs of Cengage Learning: Richard Stratton, Liz Covello, Cheryll Linthicum, Danielle Derbenti, Ed Dodd, Terri Mynatt, Leslie Lahr, Jeannine Lawless, Peter Galuardi, Lauren Hamel, Jennifer Jones, Angela Kim, and Mary Ann Payumo. My editor, Liz Covello, who joined the team this year, has done a great job working with me to finalize the product before publication. My development editor, Danielle Derbenti, as in the many other projects I have worked with her on, brought her enthusiasm and expertise to help me produce a better book. My production manager, Cheryll Linthicum, coordinated the entire project with equal enthusiasm and ensured that the production process ran smoothly from beginning to end. I also wish to thank Martha Emry, Barbara Willette, and Marian Selig for the excellent work they did in the production of this text. Martha spent countless hours working with me to ensure the accuracy and readability of the text and art. Without the help, encouragement, and support of all those mentioned above, I wouldn't have been able to complete this mammoth task.

I wish to express my personal appreciation to each of the following colleagues whose many suggestions have helped to make this a much improved book.

Arun Agarwal
Grambling State University

Mazenia Agustin
Southern Illinois University–Edwardsville

Mike Albanese
Central Piedmont Community College

Robert Andersen
University of Wisconsin–Eau Claire

Daniel Anderson
George Mason University

Joan Bell
Northeastern State University

David Bradley
University of Maine–Orono

Bob Bradshaw
Ohlone College

Paul Britt
Louisiana State University

Bob Buchanon
Millersville University

Christine Bush
Palm Beach Community College

Nick Bykov
San Joaquin Delta College

Janette Campbell
Palm Beach Community College

Kevin Charlwood
Washburn University

S.C. Cheng
Creighton University

Vladimir Cherkassky
Diablo Valley College

Charles Cooper
University of Central Oklahoma

Kyle Costello
Salt Lake Community College

Katrina Cunningham
Southern University–Baton Rouge

Eugene Curtin
Texas State University

Wendy Davidson
Georgia Perimeter College–Newton

Steven J. Davidson
San Jacinto College

John Davis
Baylor University

Ann S. DeBoever
Catawba Valley Community College

John Diamantopoulos
Northwestern State University

John Drost
University of Wisconsin–Eau Claire

Joe Fadyn
Southern Polytechnic State University

Tom Fitzkee
Francis Marion University

James Galloway
Collin County Community College

Jason Andrew Geary
Harper College

Don Goral
Northern Virginia Community College

Alan Graves
Collin County Community College

Elton Graves
Rose-Hulman Institute

Ralph Grimaldi
Rose-Hulman Institute

Ron Hammond
Blinn College–Bryan

James Handley
Montana Tech of the University of Montana

Patricia Henry
Drexel University

Irvin Hentzel
Iowa State University

Alfa Heryudono
University of Massachusetts–Dartmouth

Guy Hinman
Brevard Community College–Melbourne

Gloria Hitchcock
Georgia Perimeter College–Newton

Joshua Holden
Rose-Hulman Institute

Martin Isaacs
University of Wisconsin–Madison

Mic Jackson
Earlham College

Hengli Jiao
Ferris State College

Clarence Johnson
Cuyahoga Community College–Metropolitan

Cindy Johnson
Heartland Community College

Phil Johnson
Appalachian State University

Jack Keating
Massasoit Community College

John Khoury
Brevard Community College–Melbourne

Raja Khoury
Collin County Community College

Rethinasamy Kittappa
Millersville University

Carole Krueger
University of Texas–Arlington

Don Krug
Northern Kentucky University

Kouok Law
Georgia Perimeter College–Lawrenceville

Richard Leedy
Polk Community College

Suzanne Lindborg
San Joaquin Delta College

Tristan Londre
Blue River Community College

Ann M. Loving
J. Sargeant Reynolds Community College

Cyrus Malek
Collin County Community College

Robert Maynard
Tidewater Community College

Phillip McCartney
Northern Kentucky University

Robert McCullough
Ferris State University

Shelly McGee
University of Findlay

Rhonda McKee
Central Missouri State University

George McNulty
University of South Carolina–Columbia

Martin Melandro
Sam Houston State University

Mike Montano
Riverside Community College

Humberto Munoz
Southern University–Baton Rouge

Robert Nehs
Texas Southern University

Charlotte Newsom
Tidewater Community College

Jason Pallett
Longview Community College

Joe Perez
Texas A&M University–Kingsville

Paul Plummer
University of Central Missouri

Tom Polaski
Winthrop University

Tammy Potter
Gadsden State Community College

Linda Powers
Virginia Polytechnic Institute

David Price
Tarrant County College–Southeast

Janice Rech
University of Nebraska–Omaha

Ellena Reda
Dutchess Community College

Michael Reed
Tennessee State University

Lynn Foshee Reed
Maggie L. Walker Governor's School

James Reynolds
Clarion University of Pennsylvania

Joe Rody
Arizona State University

Jorge Sarmiento
County College of Morris

Rosa Seyfried
Harrisburg Area Community College

Kyle Siegrist
University of Alabama–Huntsville

Nathan Smale
University of Utah

Teresa Smith
Blinn College

Shing So
University of Central Missouri

Sonya Stephens
Florida A&M University

Andrew Swift
University of Nebraska–Omaha

Arnavaz P. Taraporevala
New York City College of Technology

W.E. Taylor
Texas Southern University

Beimet Teclezghi
New Jersey City University

Tim Teitloff
Clemson University

Jim Thomas
Colorado State University

Fred Thulin
University of Illinois–Chicago

Virginia Toivonen
Montana Tech of the University of Montana

Deborah Upton
Malloy College

Jim Vallade
Monroe Community College

Kathy Vranicar
University of Nebraska–Omaha

Susan Walker
Montana Tech of the University of Montana

Lianwen Wang
University of Central Missouri

Pam Warton
University of Findlay

Rich West
Francis Marion University

Jen Whitfield
Texas A&M University

Board of Advisors

Shawna Haider
Salt Lake Community College

James Handley
Montana Tech of the University of Montana

Carol Krueger
University of Texas at Arlington

Alice Eiko Pierce
Georgia Perimeter College

Joe Rody
Arizona State University

Scott Wilde
Baylor University

■ Note to the Student

The invention of calculus is one of the crowning intellectual achievements of mankind. Its roots can be traced back to the ancient Egyptians, Greeks, and Chinese. The invention of modern calculus is usually credited to both Gottfried Wilhelm Leibniz and Isaac Newton in the seventeenth century. It has widespread applications in many fields, including engineering, the physical and biological sciences, economics, business, and the social sciences. I am constantly amazed not only by the wonderful mathematical content in calculus but also by the enormous reach it has into every practical field of human endeavor. From studying the growth of a population of bacteria, to building a bridge, to exploring the vast expanses of the heavenly bodies, calculus has always played and continues to play an important role in these endeavors.

In writing this book, I have constantly kept you, the student, in mind. I have tried to make the book as interesting and readable as possible. Many mathematical concepts are introduced by using real-life illustrations. On the basis of my many years of teaching the subject, I am convinced that this approach makes it easier for you to understand the definitions and theorems in this book. I have also taken great pains to include as many steps in the examples as are needed for you to read through them smoothly. Finally, I have taken particular care with the graphical illustrations to ensure that they help you to both understand a concept and solve a problem.

The exercises in the book are carefully constructed to help you understand and appreciate the power of calculus. The problems at the beginning of each exercise set are relatively straightforward to solve and are designed to help you become familiar with the material. These problems are followed by others that require a little more effort on your part. Finally, at the end of each exercise set are problems that put the material you have just learned to good use. Here you will find applications of calculus that are drawn from many fields of study. I think you will also enjoy solving real-life problems of general interest that are drawn from many current sources, including magazines and newspapers. The answers often reveal interesting facts.

However interesting and exciting as it may be, reading a calculus book is not an easy task. You might have to go over the definitions and theorems more than once in order to fully understand them. Here you should pay careful attention to the conditions stated in the theorems. Also, it's a good idea to try to understand the definitions, theorems, and procedures as thoroughly as possible before attempting the exercises. Sometimes writing down a formula is a good way to help you remember it. Finally, if you study with a friend, a good test of your mastery of the material is to take turns explaining the topic you are studying to each other.

One more important suggestion: When you write out the solutions to the problems, make sure that you do so neatly, and try to write down each step to explain how you arrive at the solution. Being neat helps you to avoid mistakes that might occur through misreading your own handwriting (a common cause of errors in solving problems), and writing down each step helps you to work through the solution in a logical manner and to find where you went wrong if your answer turns out to be incorrect. Besides, good habits formed here will be of great help when you write reports or present papers in your career later on in life.

Finally, let me say that writing this book has been a labor of love, and I hope that I can convince you to share my love and enthusiasm for the subject.

Soo T. Tan

Clark County in Nevada—dominated by greater Las Vegas—was the fastest-growing metropolitan area in the United States from 1990 through the early 2000s. In this chapter, we will construct a mathematical model that can be used to describe how the population of Clark County grew over that period.

Jose Fuste/Raga/Corbis

0

Preliminaries

LINES PLAY AN important role in calculus, albeit indirectly. So we begin our study of calculus by looking at the properties of lines in the plane. Next, we turn our attention to the discussion of functions. More specifically, we will see how functions can be combined to yield other functions; we will see how functions can be represented graphically; and finally, we will see how functions afford us a way to describe real-world phenomena in mathematical terms.

In this chapter we also look at some of the ways in which graphing calculators and computer algebra systems can help us in our study of calculus. Finally, we consider two families of very important functions: the exponential and logarithmic functions. They play an important role in both mathematics and its applications.

V This symbol indicates that one of the following video types is available for enhanced student learning at **www.academic.cengage.com/login**:
- Chapter lecture videos
- Solutions to selected exercises

0.1   Lines

0.1   SELF-CHECK DIAGNOSTIC TEST

1. Find an equation of the line passing through the points $(-1, 3)$ and $(2, -4)$.
2. Find an equation of the line that passes through the point $(3, -2)$ and is perpendicular to the line with equation $2x + 3y = 6$.
3. Determine whether the points $A(-1, 4)$, $B(1, 1)$, and $C(3, -4)$ lie on a straight line.
4. Find an equation of the line that has an x-intercept of 4 and a y-intercept of 6.
5. Find an equation of the line that is parallel to the line $3x + 4y = 6$ and passes through the point of intersection of the lines $4x - 5y = 1$ and $2x + 3y = -5$.

Answers to Self-Check Diagnostic Test 0.1 can be found on page ANS 1.

Figure 1a depicts a ladder leaning against a vertical wall, and Figure 1b depicts the trajectory of an aircraft flying along a straight line shortly after takeoff. How do we measure the steepness of the ladder (with respect to the ground) and the steepness of the flight path of the plane (with respect to the horizontal)? To answer these questions, we need to define the steepness or the slope of a straight line. (We will solve the problems posed here in Examples 2 and 3, respectively.)

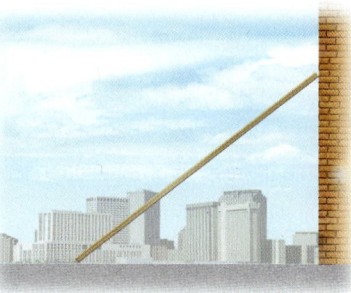

FIGURE 1   (a) How steep is the ladder?   (b) How steep is the path of the plane?

■ Slopes of Lines

DEFINITION   **Slope**

Let L be a nonvertical line in a coordinate plane. If $P_1(x_1, y_1)$ and $P_2(x_2, y_2)$ are any two distinct points on L, then the **slope** of L is

$$m = \frac{\Delta y}{\Delta x} = \frac{y_2 - y_1}{x_2 - x_1} \qquad (1)$$

(See Figure 2.) The slope of a vertical line is undefined.

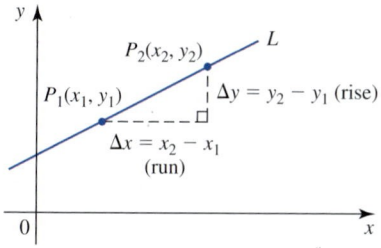

FIGURE 2

The slope of the line L is

$$m = \frac{\Delta y}{\Delta x} = \frac{y_2 - y_1}{x_2 - x_1} = \frac{\text{rise}}{\text{run}}.$$

The quantity $\Delta y = y_2 - y_1$ (Δy is read "delta y") measures the change in y from P_1 to P_2 and is called the **rise**; the quantity $\Delta x = x_2 - x_1$ measures the change in x from P_1 to P_2 and is called the **run**. Thus, the slope of a line is the ratio of its rise to its run.

Since the ratios of corresponding sides of similar triangles are equal, we see from Figure 3 that the slope of a line is independent of the two distinct points that are used to compute it; that is,

$$m = \frac{y_2 - y_1}{x_2 - x_1} = \frac{y_2' - y_1'}{x_2' - x_1'}$$

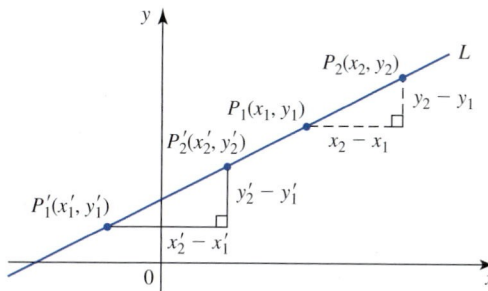

FIGURE 3

The slope of a nonvertical straight line is independent of the two distinct points used to compute it.

The slope of a straight line is a numerical measure of its steepness with respect to the positive x-axis. In fact, if we take $\Delta x = x_2 - x_1$ to be equal to 1 in Equation (1), then we see that

$$m = \frac{\Delta y}{\Delta x} = \frac{\Delta y}{1} = \Delta y = y_2 - y_1$$

gives the *change in y per unit change in x*.

Figure 4 shows four lines with different slopes. By taking a run of 1 unit to compute each slope, you can see that the larger the absolute value of the slope is, the larger the change in y per unit change in x is and, therefore, the steeper the line is. We also see that if $m > 0$, the line slants upward; if $m < 0$, the line slants downward; and finally, if $m = 0$, the line is horizontal.

FIGURE 4

The slope of a line is a numerical measure of its steepness.

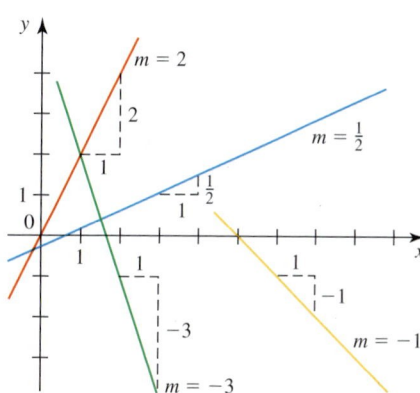

EXAMPLE 1 Find the slope of the line passing through (a) the points $P_1(1, 1)$ and $P_2(3, 5)$ and (b) the points $P_1(1, 3)$ and $P_2(3, 2)$.

Solution

a. Using Equation (1), we obtain the required slope as

$$m = \frac{5 - 1}{3 - 1} = 2$$

This tells us that y increases by 2 units for each unit increase in x (see Figure 5a).

b. Equation (1) gives the required slope as

$$m = \frac{2 - 3}{3 - 1} = -\frac{1}{2}$$

This tells us that y decreases by $\frac{1}{2}$ unit for each unit increase in x or, equivalently, y decreases by 1 unit for each increase of 2 units in x (see Figure 5b).

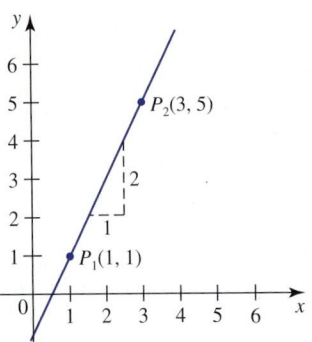

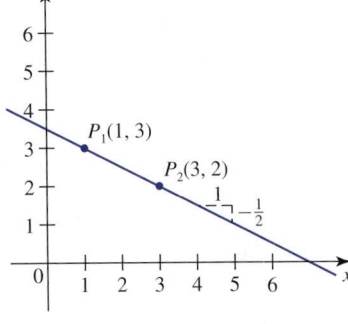

FIGURE 5   (a) The slope of the line is 2.   (b) The slope of the line is $-\frac{1}{2}$.

Note   In Example 1 we arbitrarily labeled the point $P_1(1, 1)$ and the point $P_2(3, 5)$. Suppose we had labeled the points $P_1(3, 5)$ and $P_2(1, 1)$ instead. Then Equation (1) would give

$$m = \frac{1 - 5}{1 - 3} = 2$$

as before. In general, relabeling the points P_1 and P_2 simply changes the sign of both the numerator and denominator of the ratio in Equation (1) and therefore does not change the value of m. Therefore, when we compute the slope of a line using Equation (1), it does not matter which point we label as P_1 and which point we label as P_2.

EXAMPLE 2 A 20-ft ladder leans against a wall with its top located 12 ft above the ground. What is the slope of the ladder?

Solution   The situation is depicted in Figure 6, where x denotes the distance of the base of the ladder from the wall. By the Pythagorean Theorem we have

$$x^2 + 12^2 = 20^2$$

$$x^2 = 256$$

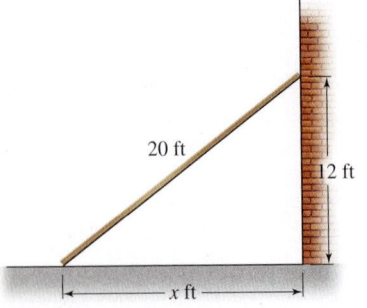

FIGURE 6
A ladder leaning against a wall

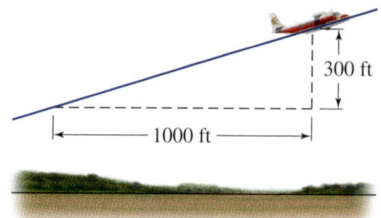

FIGURE 7
The flight path of the plane along a straight line

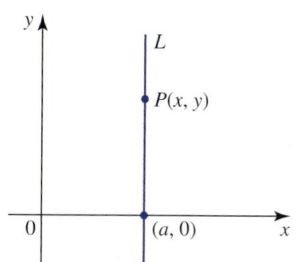

FIGURE 8
Every point on the vertical line L has an x-coordinate that is equal to a.

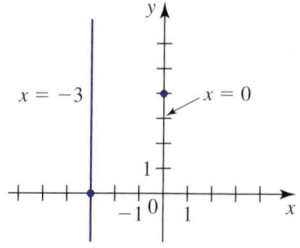

FIGURE 9
The graphs of the equations $x = -3$ and $x = 0$

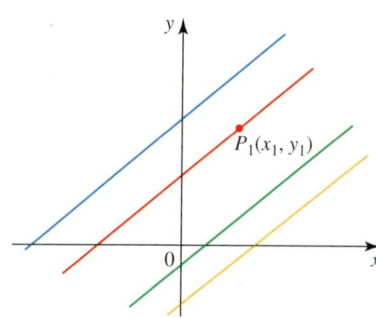

FIGURE 10
There are infinitely many lines with slope m but only one that passes through the point $P_1(x_1, y_1)$ with slope m.

or $x = 16$. The slope of the ladder is

$$\frac{12}{16} = \frac{3}{4} \qquad \frac{\text{rise}}{\text{run}}$$

EXAMPLE 3 Shortly after takeoff a plane climbs along a straight path. The plane gains altitude at the rate of 300 ft for each 1000 ft it travels horizontally, that is, parallel to the ground. What is the slope of the trajectory of the plane? What is the altitude gained by the plane after traveling 5000 ft horizontally?

Solution The flight path is depicted in Figure 7. We see that the slope of the flight path of the plane is

$$\frac{300}{1000} = \frac{3}{10} \qquad \frac{\text{rise}}{\text{run}}$$

This tells us that the plane gains an altitude of $\frac{3}{10}$ ft for each foot traveled by the plane horizontally. Therefore, the altitude gained after traveling 5000 ft horizontally is

$$\frac{3}{10} \cdot 5000 = 1500$$

or 1500 ft.

Equations of Vertical Lines

Let L be a vertical line in the xy-plane. Then L must intersect the axis at some point $(a, 0)$ as shown in Figure 8. If $P(x, y)$ is any point on L, then x must be equal to a, whereas y may take on any value, depending on the position of P. In other words, the only conditions on the coordinates of the point (a, y) on L are $x = a$ and $-\infty < y < \infty$. Conversely, we see that the set of all points (x, y) where $x = a$ and y is arbitrary is precisely the vertical line L. We have found an algebraic representation of a vertical line in a coordinate plane.

> **DEFINITION   Equation of a Vertical Line**
>
> An equation of the vertical line passing through the point (a, b) is
>
> $$x = a \qquad (2)$$

EXAMPLE 4 The graph of $x = -3$ is the vertical line passing through $(-3, 0)$. An equation of the vertical line passing through $(0, 4)$ is $x = 0$. This is an equation of the y-axis (see Figure 9).

Equations of Nonvertical Lines

If a line L is nonvertical, then it has a well-defined slope m. But specifying the slope of a line alone is not enough to pin down a particular line, because there are infinitely many lines with a given slope (Figure 10). However, if we specify a point $P_1(x_1, y_1)$ through which a line L passes in addition to its slope m, then L is uniquely determined.

To derive an equation of the line passing through a given point $P_1(x_1, y_1)$ and having slope m, let $P(x, y)$ be *any* point distinct from P_1 lying on L. Using Equation (1)

and the points $P_1(x_1, y_1)$ and $P(x, y)$, we can write the slope of L as

$$\frac{y - y_1}{x - x_1}$$

But the slope of L is m. So

$$\frac{y - y_1}{x - x_1} = m$$

or, upon multiplying both sides of the equation by $x - x_1$,

$$y - y_1 = m(x - x_1) \tag{3}$$

Observe that $x = x_1$ and $y = y_1$ also satisfy Equation (3), so all points on L satisfy this equation. We leave it as an exercise to show that only the points that satisfy Equation (3) can lie on L.

Equation (3) is called the point-slope form of an equation of a line because it utilizes a point on the line and its slope.

DEFINITION **Point-Slope Form of an Equation of a Line**

An equation of the line passing through the point $P_1(x_1, y_1)$ and having slope m is

$$y - y_1 = m(x - x_1)$$

EXAMPLE 5 Find an equation of the line passing through the point $(2, 1)$ and having slope $m = -\frac{1}{2}$.

Solution Using Equation (3) with $x_1 = 2$, $y_1 = 1$, and $m = -\frac{1}{2}$, we find

$$y - 1 = -\frac{1}{2}(x - 2)$$

or

$$y = -\frac{1}{2}x + 2$$

EXAMPLE 6 Find an equation of the line passing through the points $(-1, -2)$ and $(2, 3)$.

Solution We first calculate the slope of the line, obtaining

$$m = \frac{3 - (-2)}{2 - (-1)} = \frac{5}{3}$$

Then using Equation (3) with $P_1(-1, -2)$ (the other point will also do, as you can verify) and $m = \frac{5}{3}$, we obtain

$$y - (-2) = \frac{5}{3}[x - (-1)]$$

$$y = \frac{5}{3}x + \frac{5}{3} - 2$$

or

$$y = \frac{5}{3}x - \frac{1}{3}$$

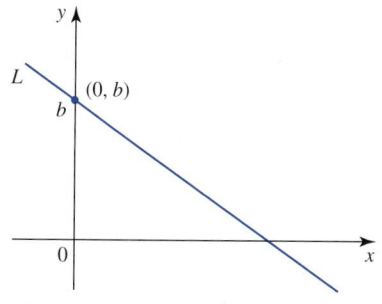

FIGURE 11
The line L with y-intercept b and slope m has equation $y = mx + b$.

A nonvertical line L crosses the y-axis at some point $(0, b)$. The number b is called the y-intercept of the line. (See Figure 11.) If we use the point $P_1(0, b)$ in Equation (3), we obtain

$$y - b = m(x - 0)$$

or

$$y = mx + b$$

which is called the **slope-intercept** form of an equation of a line.

DEFINITION **Slope-Intercept Form of an Equation of a Line**

An equation of the line with slope m and y-intercept b is

$$y = mx + b \tag{4}$$

EXAMPLE 7 Find an equation of the line with slope $\frac{3}{4}$ and y-intercept 4.

Solution We use Equation (4) with $m = \frac{3}{4}$ and $b = 4$, obtaining the equation

$$y = \frac{3}{4}x + 4$$

■ The General Equation of a Line

An equation of the form

$$Ax + By + C = 0 \tag{5}$$

where A, B, and C are constants and A and B are not both zero, is called a **first-degree equation** in x and y. You can verify the following result.

THEOREM 1 **General Equation of a Line**

Every first-degree equation in x and y has a straight line for its graph in the xy-plane; conversely, every straight line in the xy-plane is the graph of a first-degree equation in x and y.

Because of this theorem, Equation (5) is often referred to as a **general equation of a line** or a **linear equation** in x and y.

EXAMPLE 8 Find the slope of the line with equation $2x + 3y + 5 = 0$.

Solution Rewriting the equation in the slope-intercept form by solving it for y in terms of x, we obtain

$$3y = -2x - 5$$

or

$$y = -\frac{2}{3}x - \frac{5}{3}$$

Comparing this equation with Equation (4), we see immediately that the slope of the line is $m = -\frac{2}{3}$. ■

Note Example 8 illustrates one advantage of writing an equation of a line in the slope-intercept form: The slope of the line is given by the coefficient of x. ■

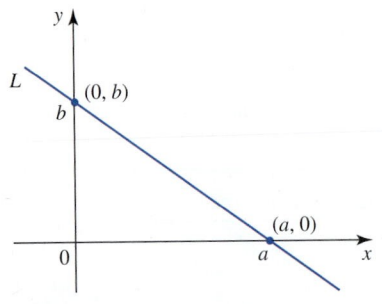

FIGURE 12
The x-intercept of L is a, and the y-intercept of L is b.

■ Drawing the Graphs of Lines

We have already mentioned that the y-intercept of a straight line is the y-coordinate of the point $(0, b)$ at which the line crosses the y-axis. Similarly, the x-intercept of a straight line is the x-coordinate of the point $(a, 0)$ at which the line crosses the x-axis (see Figure 12). To find the x-intercept of a line L, we set $y = 0$ in the equation for L because every point on the x-axis must have its y-coordinate equal to zero. Similarly, to find the y-intercept of L, we set $x = 0$. The easiest way to sketch a straight line is to find its x- and y-intercepts, when possible, as the following example shows.

EXAMPLE 9 Sketch the graphs of

a. $2x + 3y - 6 = 0$ **b.** $x - 3y = 0$

Solution
a. Setting $y = 0$ gives the x-intercept as 3. Next, setting $x = 0$ gives the y-intercept as 2. Plotting the points $(3, 0)$ and $(0, 2)$ and drawing the line passing through them, we obtain the desired graph (see Figure 13a).
b. Setting $y = 0$ gives $x = 0$ as the x-intercept. Next, setting $x = 0$ gives $y = 0$ as the y-intercept. Thus, the line passes through the origin. In this situation we need to find another point through which the line passes. If we pick, say, $x = 3$ and substitute this value of x into the equation $x - 3y = 0$ and solve the resulting equation for y, we obtain $y = 1$ as the y-coordinate. Plotting the points $(0, 0)$ and $(3, 1)$ and drawing the line through them, we obtain the desired graph (Figure 13b).

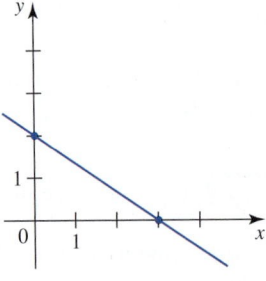

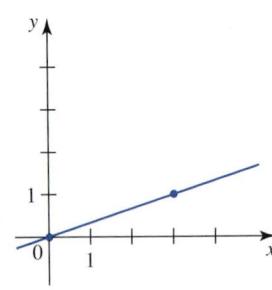

FIGURE 13

(a) The graph of $2x + 3y - 6 = 0$

(b) The graph of $x - 3y = 0$

■

■ Angles of Inclination

> **DEFINITION** **Angle of Inclination**
> The **angle of inclination** of a line L is the smaller angle ϕ (the Greek letter *phi*) measured in a counterclockwise direction from the direction of the positive x-axis to L (see Figure 14).

FIGURE 14
The angle of inclination is measured in a counterclockwise direction from the direction of the positive x-axis.

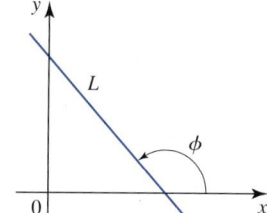

Note The angle of inclination ϕ satisfies $0° \le \phi < 180°$ or, in radian measure, $0 \le \phi < \pi$.

The relationship between the slope of a line and the angle of inclination of the line can be seen from examining Figure 15. Letting m denote the slope of L and ϕ its angle of inclination, we have

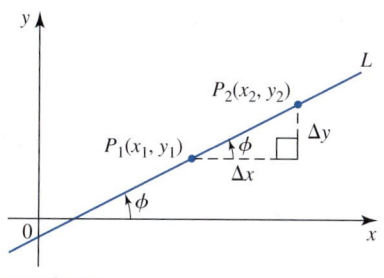

FIGURE 15
The slope of L is $m = \dfrac{\Delta y}{\Delta x} = \tan \phi$.

$$m = \tan \phi \qquad (6)$$

Notes
1. Although Figure 15 illustrates Equation (6) for the case in which $0° \le \phi < 90°$, it can be shown that the equation also holds when $90° < \phi < 180°$. We leave it as an exercise.
2. Observe that the angle of inclination of a vertical line is 90°. Since $\tan 90°$ is undefined, we see that the slope of a vertical line is undefined, as was noted earlier.

EXAMPLE 10 Refer to Example 3. Find the angle of the flight path of the plane. (Note: This angle is referred to as the *angle of climb*.)

Solution From the result of Example 3 we see that $m = \frac{3}{10} = 0.3$. Therefore, the angle of climb, ϕ, satisfies

$$\tan \phi = 0.3$$

from which we deduce that the angle of climb is

$$\phi = \tan^{-1} 0.3 \approx 0.29 \text{ rad}$$

or approximately 17°.

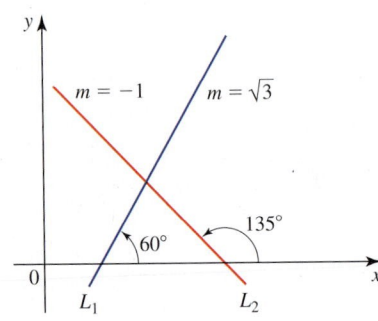

FIGURE 16
L_1 has slope $m = \sqrt{3}$, and L_2 has angle of inclination 135°.

EXAMPLE 11

a. Find the slope of a line whose angle of inclination is 60° ($\pi/3$ radians).
b. Find the angle of inclination of a line with slope $m = -1$.

Solution
a. Equation (6) immediately yields

$$m = \tan 60° = \sqrt{3}$$

as the slope of the line (see Figure 16).
b. Equation (6) gives

$$-1 = \tan \phi$$

and we see that $\phi = 3\pi/4$ radians, or 135° (see Figure 16). ■

▪ Parallel Lines and Perpendicular Lines

Two lines are parallel if and only if they have the same angle of inclination (see Figure 17).

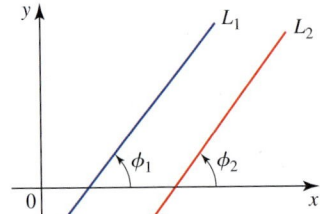

Slope of $L_1 = m_1 = \tan \phi_1$
Slope of $L_2 = m_2 = \tan \phi_2$

FIGURE 17
L_1 and L_2 are parallel if and only if their slopes are equal or both lines are vertical.

Therefore, using Equation (6), we have the following result.

THEOREM 2

Two nonvertical lines are parallel if and only if they have the same slope.

Note   If two lines are vertical, then they are parallel. ▪

Suppose that L_1 and L_2 are two nonvertical perpendicular lines with slopes m_1 and m_2 and angles of inclination ϕ_1 and ϕ_2, respectively. The case in which ϕ_1 is acute and ϕ_2 is obtuse is shown in Figure 18.

Since $90° < \phi_2 < 180°$, m_2 is negative, so the length of the side BC is $-m_2$. The two right triangles $\triangle ABC$ and $\triangle DAC$ are similar, and since the ratios of corresponding sides of similar triangles are equal, we have

$$\frac{m_1}{1} = \frac{1}{-m_2}$$

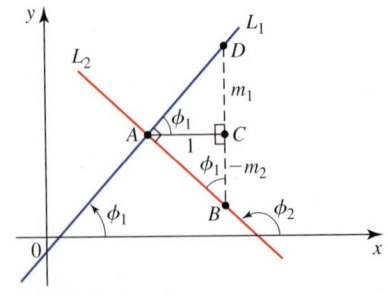

FIGURE 18
$\triangle ABC$ and $\triangle DAC$ are similar.

which may be rewritten as

$$m_1 = -\frac{1}{m_2} \qquad \text{or} \qquad m_1 m_2 = -1$$

This argument can be reversed to prove the converse: The lines are perpendicular if $m_1 m_2 = -1$.

THEOREM 3 **Slopes of Perpendicular Lines**

Two nonvertical lines L_1 and L_2 with slopes m_1 and m_2, respectively, are perpendicular if and only if $m_1 m_2 = -1$ or, equivalently, if and only if

$$m_1 = -\frac{1}{m_2} \qquad \text{or} \qquad m_2 = -\frac{1}{m_1} \tag{7}$$

Thus, the slope of each is the negative reciprocal of the slope of the other.

Note If a line L_1 is vertical (and hence has no slope), then another line L_2 is perpendicular to it if and only if L_2 is horizontal (has zero slope), and vice versa. ■

EXAMPLE 12 Find an equation of the line that passes through the point $(6, 7)$ and is perpendicular to the line with equation $2x + 3y = 12$.

Solution First we find the slope of the given line by rewriting the equation in the slope-intercept form:

$$y = -\frac{2}{3}x + 4$$

From this we see that its slope is $-\frac{2}{3}$. Since the required line is perpendicular to the given line, its slope is

$$-\frac{1}{-\frac{2}{3}} = \frac{3}{2}$$

Therefore, using the point-slope form of an equation of a line with $m = \frac{3}{2}$ and $P_1(6, 7)$, we obtain the required equation as

$$y - 7 = \frac{3}{2}(x - 6)$$

or

$$y = \frac{3}{2}x - 2$$ ■

■ The Distance Formula

Another benefit that arises from using the Cartesian coordinate system is that the distance between any two points in the plane may be expressed solely in terms of their coordinates. Suppose, for example, that (x_1, y_1) and (x_2, y_2) are any two points in the plane (see Figure 19). Then the distance between these two points can be computed using the following formula.

Distance Formula

The distance d between two points $P_1(x_1, y_1)$ and $P_2(x_2, y_2)$ in the plane is given by

$$d = \sqrt{(x_2 - x_1)^2 + (y_2 - y_1)^2} \tag{8}$$

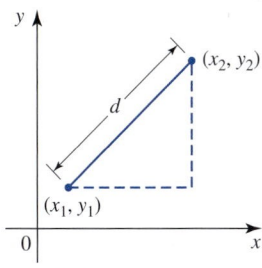

FIGURE 19
The distance d between the points (x_1, y_1) and (x_2, y_2)

In what follows, we give several applications of the distance formula.

EXAMPLE 13 Find the distance between the points $(-4, 3)$ and $(2, 6)$.

Solution Let $P_1(-4, 3)$ and $P_2(2, 6)$ be points in the plane. Then, we have

$$x_1 = -4, \qquad y_1 = 3, \qquad x_2 = 2, \qquad y_2 = 6$$

Using Formula (8), we have

$$d = \sqrt{[2 - (-4)]^2 + (6 - 3)^2}$$
$$= \sqrt{6^2 + 3^2}$$
$$= \sqrt{45} = 3\sqrt{5} \qquad \blacksquare$$

EXAMPLE 14 Let $P(x, y)$ denote a point lying on the circle with radius r and center $C(h, k)$. (See Figure 20.) Find a relationship between x and y.

Solution By the definition of a circle, the distance between $C(h, k)$ and $P(x, y)$ is r. Using Formula (8), we have

$$\sqrt{(x - h)^2 + (y - k)^2} = r$$

which, upon squaring both sides, gives the equation

$$(x - h)^2 + (y - k)^2 = r^2$$

that must be satisfied by the variables x and y. $\qquad \blacksquare$

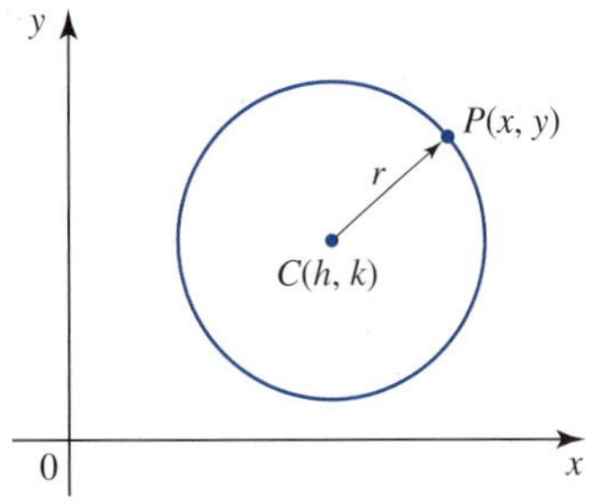

FIGURE 20
A circle with radius r and center $C(h, k)$

A summary of the result obtained in Example 14 follows.

Equation of a Circle
An equation of the circle with center $C(h, k)$ and radius r is given by

$$(x - h)^2 + (y - k)^2 = r^2 \qquad \textbf{(9)}$$

EXAMPLE 15 Find an equation of the circle with

a. Radius 2 and center $(-1, 3)$.
b. Radius 3 and center located at the origin.

Solution
a. We use Formula (9) with $r = 2$, $h = -1$, and $k = 3$, obtaining

$$[x - (-1)]^2 + (y - 3)^2 = 2^2 \qquad \text{or} \qquad (x + 1)^2 + (y - 3)^2 = 4$$

(See Figure 21a.)
b. Using Formula (9) with $r = 3$ and $h = k = 0$, we obtain

$$x^2 + y^2 = 3^2 \qquad \text{or} \qquad x^2 + y^2 = 9$$

(See Figure 21b.)

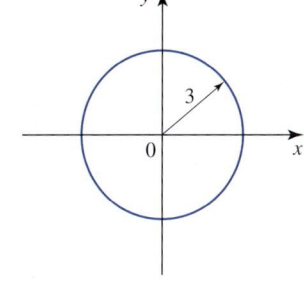

FIGURE 21

(a) The circle with radius 2 and center $(-1, 3)$

(b) The circle with radius 3 and center $(0, 0)$

0.1 EXERCISES

In Exercises 1–4, find the slope of the line passing through the pair of points.

1. $(1, -2)$ and $(2, 4)$
2. $(-4, -2)$ and $(-1, 3)$
3. $(1.2, 3.6)$ and $(3.2, 1.4)$
4. $(-3, -3)$ and $(-3, \sqrt{3})$

5. Refer to the figure below.

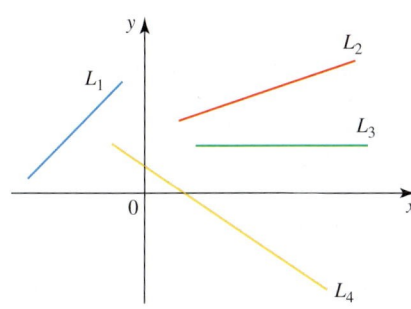

   a. Give the sign of the slope of each of the lines.
   b. List the lines in order of increasing slope.

6. Find the slope of each of the lines shown in the accompanying figure.

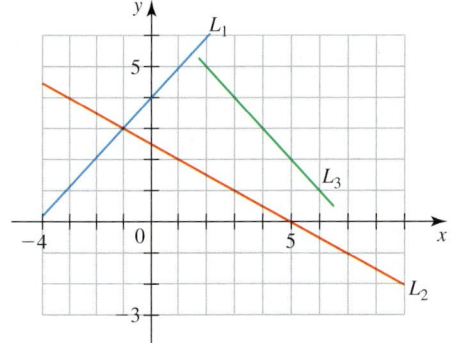

7. Find a if the line passing through $(1, 3)$ and $(-4, a)$ has slope 5.

8. Find a if the line passing through $(2, a)$ and $(-9, 3)$ has slope -3.

In Exercises 9–14, find the slope of the line that has the angle of inclination.

9. $45°$ 10. $135°$ 11. $30°$

12. $\dfrac{\pi}{4}$ 13. $\dfrac{\pi}{3}$ 14. $\dfrac{2\pi}{3}$

In Exercises 15–20, find the angle of inclination of a line with the given slope. You may use a calculator.

15. -1 16. $\dfrac{1}{2}$ 17. $\sqrt{3}$

18. 10 19. $-\dfrac{1}{\sqrt{3}}$ 20. 20

In Exercises 21–24, sketch the line through the given point with the indicated slope.

21. $(1, 2)$;   3   22. $(2, 3)$; -2
23. $(-1, -2)$; -1 24. $(-2, 3)$;   4

In Exercises 25–28, determine whether the lines through the given pairs of points are parallel or perpendicular to each other.

25. $(1, -2)$, $(-3, -10)$ and $(1, 5)$, $(-1, 1)$
26. $(4, 6)$, $(4, -2)$ and $(-1, 5)$, $(-1, 8)$
27. $(-2, 5)$, $(4, 2)$ and $(-1, -2)$, $(3, 6)$
28. $(-1, -2)$, $(3, 4)$ and $(9, -6)$, $(3, -2)$
29. If the line passing through the points $(-1, a)$ and $(3, -1)$ is parallel to the line passing through the points $(3, 6)$ and $(-5, a + 2)$, what must the value of a be?

30. If the line passing through the points $(-2, 4)$ and $(1, a)$ is perpendicular to the line passing through the points $(a + 4, 8)$ and $(3, -4)$, what must the value of a be?

31. The point $(-5, k)$ lies on the line passing through the point $(1, 3)$ and perpendicular to a line with slope 3. Find k.

32. Show that the triangle with vertices $A(-2, -8)$, $B(2, 2)$, and $C(-3, 4)$ is a right triangle.

33. A line passes through $(3, 4)$ and the midpoint of the line segment joining $(-1, 1)$ and $(3, 9)$. Show that this line is perpendicular to the line segment.

In Exercises 34 and 35, determine whether the given points lie on a straight line.

34. $A(-2, 1)$, $B(1, 7)$, and $C(4, 13)$

35. $A(-3, 6)$, $B(3, 3)$, and $C(6, 0)$

In Exercises 36–41, write the equation in the slope-intercept form, and then find the slope and y-intercept of the corresponding lines.

36. $2x - 3y - 12 = 0$ **37.** $-3x + 4y - 8 = 0$

38. $y + 4 = 0$ **39.** $Ax + By = C$, $B \neq 0$

40. $\dfrac{x}{3} + \dfrac{y}{4} = 1$ **41.** $\sqrt{2}x - \sqrt{3}y = 4$

In Exercises 42–45, find the angle of inclination of the line represented by the equation.

42. $4x - 7y - 8 = 0$ **43.** $\sqrt{3}x - y + 4 = 0$

44. $x + \sqrt{3}y - 5 = 0$ **45.** $x + y - 8 = 0$

In Exercises 46–59, find an equation of the line satisfying the conditions. Write your answer in the slope-intercept form.

46. Is perpendicular to the x-axis and passes through the point (π, π^2)

47. Passes through $(4, -3)$ with slope 2

48. Passes through $(-3, 3)$ and has slope 0

49. Passes through $(2, 4)$ and $(3, 8)$

50. Passes through $(-1, -2)$ and $(3, -4)$

51. Passes through $(2, 5)$ and $(2, 28)$

52. Has slope -2 and y-intercept 3

53. Has slope 3 and y-intercept -5

54. Has x-intercept 3 and y-intercept -5

55. Passes through $(3, -5)$ and is parallel to the line with equation $2x + 3y = 12$

56. Is perpendicular to the line with equation $y = -3x - 5$ and has y-intercept 7

57. Passes through $(-2, -4)$ and is perpendicular to the line with equation $3x - y - 4 = 0$

58. Passes through $(3, -4)$ and is perpendicular to the line through $(-1, 2)$ and $(3, 6)$

59. Passes through $(2, 3)$ and has an angle of inclination of $\pi/6$ radians

In Exercises 60–63, determine whether the pair of lines represented by the equations are parallel, perpendicular, or neither.

60. $3x - 4y = 8$ and $6x - 8y = 10$

61. $x - 3 = 0$ and $y - 5 = 0$

62. $2x - 3y - 12 = 0$ and $3x + 2y - 6 = 0$

63. $\dfrac{x}{a} + \dfrac{y}{b} = 1$ and $\dfrac{x}{b} - \dfrac{y}{a} = 1$

In Exercises 64–65, find the point of intersection of the lines with the given equations.

64. $2x - y = 1$ and $3x + 2y = 12$

65. $x - 3y = -1$ and $4x + 3y = 11$

66. Find the distance between the points.
   a. $(1, 3)$ and $(4, 7)$
   b. $(1, 0)$ and $(4, 4)$
   c. $(-1, 3)$ and $(4, 9)$
   d. $(-2, 1)$ and $(10, 6)$

67. Find an equation of the circle that satisfies the conditions.
   a. Radius 5 and center $(2, -3)$
   b. Center at the origin and passes through $(2, 3)$
   c. Center $(2, -3)$ and passes through $(5, 2)$
   d. Center $(-a, a)$ and radius $2a$

68. Show that the two lines with equations $a_1x + b_1y + c_1 = 0$ and $a_2x + b_2y + c_2 = 0$, respectively, are parallel if and only if $a_1b_2 - b_1a_2 = 0$.

69. Show that an equation of the line L that passes through the points $(a, 0)$ and $(0, b)$ with $a \neq 0$ and $b \neq 0$ can be written in the form

$$\frac{x}{a} + \frac{y}{b} = 1$$

This is called the **intercept form** of the equation of L.

70. Use the result of Exercise 69 to find an equation of the line with x-intercept 2 and y-intercept 5.

71. Use the result of Exercise 69 to find an equation of the line passing through the points $(-4, 0)$ and $(0, -1)$.

72. Find an equation of the line passing through $(5, 2)$ and the midpoint of the line segment joining $(-1, 1)$ and $(3, 9)$.

73. Find the distance from the point $(5, 3)$ to the line with equation $2x - y + 3 = 0$.
   Hint: Find the point of intersection of the given line and the line perpendicular to it that passes through $(5, 3)$.

74. The top of a ladder leaning against a wall is 9 ft above the ground. The slope of the ladder with respect to the ground is $3\sqrt{7}/7$. What is the length of the ladder?

75. A plane flying along a straight path loses altitude at the rate of 1000 ft for each 6000 ft covered horizontally. What is the angle of descent of the plane?

76. A plane flies along a straight line that has a slope of 0.22. If the plane gains altitude of 1000 ft over a certain period of time, what will be the horizontal distance covered by the plane over that period?

77. Truss Bridges Simple trusses are common in bridges. The following figure depicts such a truss superimposed on a coordinate system. Find an equation of the line containing the line segments (a) *OD*, (b) *AD*, (c) *AC*, and (d) *BC*.

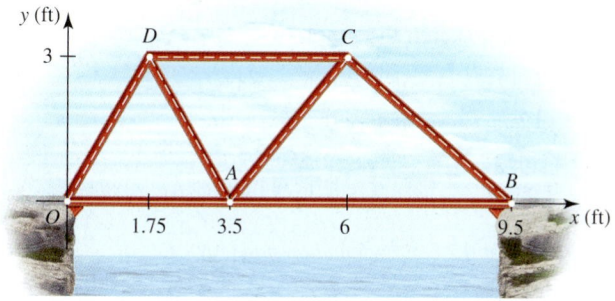

78. Temperature Conversion The relationship between the temperature in degrees Fahrenheit (°F) and the temperature in degrees Celsius (°C) is

$$F = \frac{9}{5}C + 32$$

a. Sketch the line with the given equation.
b. What is the slope of the line? What does it represent?
c. What is the *F*-intercept of the line? What does it represent?

79. Nuclear Plant Utilization The United States is not building many nuclear plants, but the ones that it has are running full tilt. The output (as a percent of total capacity) of nuclear plants is described by the equation

$$y = 1.9467t + 70.082$$

where *t* is measured in years, with *t* = 0 corresponding to the beginning of 1990.
a. Sketch the line with the given equation.
b. What are the slope and the *y*-intercept of the line found in part (a)?
c. Give an interpretation of the slope and the *y*-intercept of the line found in part (a).
d. If the utilization of nuclear power continued to grow at the same rate and the total capacity of nuclear plants in the United States remained constant, by what year were the plants generating at maximum capacity?
Source: Nuclear Energy Institute.

80. Social Security Contributions For wages less than the maximum taxable wage base, Social Security contributions by employees are 7.65% of the employee's wages.
a. Find an equation that expresses the relationship between the wages earned (*x*) and the Social Security taxes paid (*y*) by an employee who earns less than the maximum taxable wage base.
b. For each additional dollar that an employee earns, by how much is his or her Social Security contribution increased? (Assume that the employee's wages are less than the maximum taxable wage base.)
c. What Social Security contributions will an employee who earns $75,000 (which is less than the maximum taxable wage base) be required to make?
Source: Social Security Administration.

81. Weight of Whales The equation $W = 3.51L - 192$, $70 \le L \le 100$, which expresses the relationship between the length *L* (in feet) and the expected weight *W* (in British tons) of adult blue whales, was adopted in the late 1960s by the International Whaling Commission.
a. What is the expected weight of an 80-ft whale?
b. Sketch the straight line that represents the equation.

82. The Narrowing Gender Gap Since the founding of the Equal Employment Opportunity Commission and the passage of equal-pay laws, the gap between men's and women's earnings has continued to close gradually. At the beginning of 1990 (*t* = 0), women's wages were 68% of men's wages; and by the beginning of 2000 (*t* = 10), women's wages were projected to be 80% of men's wages. If this gap between women's and men's wages continued to narrow *linearly,* what percent of men's wages were women's wages at the beginning of 2004?
Source: Journal of Economic Perspectives.

83. Show that only those points satisfying Equation (3) can lie on the line *L* passing through $P_1(x_1, y_1)$ with slope *m*.

84. Show that Equation (6) also holds when $90° < \phi < 180°$.

In Exercises 85–88, determine whether the statement is true or false. If it is true, explain why. If it is false, explain why or give an example that shows it is false.

85. Suppose the slope of a line *L* is $-\frac{1}{2}$ and *P* is a given point on *L*. If *Q* is the point on *L* lying 4 units to the left of *P*, then *Q* lies 2 units above *P*.

86. The line with equation $Ax + By + C = 0$, where $B \ne 0$, and the line with equation $ax + by + c = 0$, where $b \ne 0$, are parallel if $Ab - aB = 0$.

87. If the slope of the line L_1 is positive, then the slope of a line L_2 perpendicular to L_1 must be negative.

88. The lines with equations $ax + by + c_1 = 0$ and $bx - ay + c_2 = 0$, where $a \ne 0$ and $b \ne 0$, are perpendicular to each other.

0.2 | Functions and Their Graphs

0.2 SELF-CHECK DIAGNOSTIC TEST

1. If

$$f(x) = \begin{cases} \sqrt{-x} & \text{if } x < 0 \\ \sqrt{x} + 1 & \text{if } x \geq 0 \end{cases}$$

find $f(-4)$, $f(0)$, and $f(9)$.

2. If $f(x) = x^2 + 2x$, find and simplify $\dfrac{f(x + h) - f(x)}{h}$.

3. Find the domain of $f(x) = \dfrac{\sqrt{2x + 1}}{x^2 + x - 2}$.

4. Find the domain and range, and sketch the graph of

$$f(x) = \begin{cases} -2x + 1 & \text{if } x < 0 \\ 2x - 1 & \text{if } x \geq 0 \end{cases}$$

5. Determine whether $f(x) = \dfrac{2x^3 - x}{x^2 + 1}$ is odd, even, or neither.

Answers to Self-Check Diagnostic Test 0.2 can be found on page ANS 1.

■ Definition of a Function

In many situations, one quantity depends on another. For example:

- The area of a circle depends on its radius.
- The distance fallen by an object dropped from a building depends on the length of time it has fallen.
- The initial speed of a chemical reaction depends on the amount of substrate used.
- The size of the population of a certain culture of bacteria after the introduction of a bactericide depends on the time elapsed.
- The profit of a manufacturer depends on the company's level of production.

To describe these situations, we use the concept of a function.

DEFINITION Function

A **function** f from a set A to a set B is a rule that assigns to each element x in A one and only one element y in B.

Let's consider an example that illustrates why there can be only one element y in B for each x in A. Suppose that A is the set of items on sale in a department store and f is a "pricing" function that assigns to each item x in A its selling price y in B. Then for each x there should be exactly one y. Note that the definition does not preclude the possibility of more than one element in A being associated with an element in B. In the context of our present example, this could mean that two or more items would have the same selling price.

The set A is called the **domain** of the function. The element y in B, called the value of f at x, is written $f(x)$ and read "f of x." The set of all values $y = f(x)$ as x varies over the domain of f is called the **range of f.** If A and B are subsets of the set of real numbers, then both x and $f(x)$ are also real numbers. In this case we refer to the function f as a *real-valued* function of a *real variable*.

We can think of a function f as a machine or processor. In this analogy the domain of f consists of the set of "inputs," the rule describes how the "inputs" are to be processed by the machine, and the range is made up of the set of "outputs" (see Figure 1).

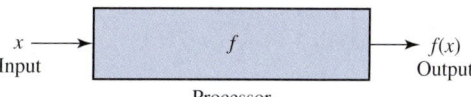

FIGURE 1

A function machine

As an example, consider the function that associates with each nonnegative number x its square root, $\sqrt{x}$. We can view this function as a square root extracting machine. Its domain is the set of all nonnegative numbers, and so is its range. Given the input 4, for example, the function extracts its square root $\sqrt{4}$ and yields the output 2.

Another way of viewing a function is to think of the function f from a set A to a set B as a *mapping* or *transformation* that maps an element x in A onto its image $f(x)$ in B (Figure 2). For example, the "square root" function is a function from the set of nonnegative real numbers to the set of real numbers. This function maps the number 4 onto the number 2, the number 7 onto the number $\sqrt{7}$, and so on.

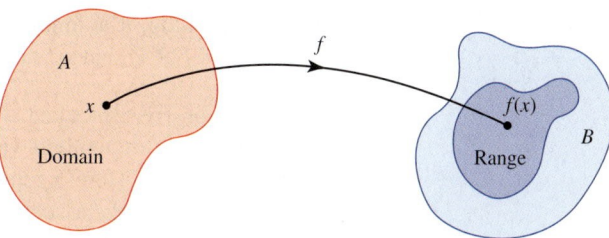

FIGURE 2

f maps a point x in its domain onto its image $f(x)$ in its range.

Note The range of f is contained in the set B but need not be equal to B. For example, consider the function f that associates with each real number x its square, x^2, from the set of real numbers R to the set of real numbers R (so $A = B = R$). Then the range of f is the set of nonnegative numbers, a proper subset of B. ■

Describing Functions

Functions can be described in many ways. Earlier, we defined the square root function by giving a verbal description of the rule. Functions can also be described by giving a table of values describing the relationship between x and $f(x)$. This method of describing a function is particularly effective when both the domain and the range of f contain a small number of elements. For example, the function f giving the Manhattan hotel occupancy rate in each of the years 1999 ($x = 0$) through 2006 can be defined by the data given in Table 1.

Here, the domain of f is $A = \{0, 1, 2, 3, 4, 5, 6, 7\}$ and the range of f is $B = \{74.5, 75.0, \ldots, 85.1\}$. Observe that we can also describe the rule for f by writing $f(0) = 81.1, f(1) = 83.7, \ldots, f(7) = 85.1$.

TABLE 1   The function f giving the Manhattan hotel occupancy rate in year x

x (year)	$y = f(x)$ (percent)
0	81.1
1	83.7
2	74.5
3	75.0
4	75.9
5	83.2
6	84.9
7	85.1

Source: PricewaterhouseCoopers LLP.

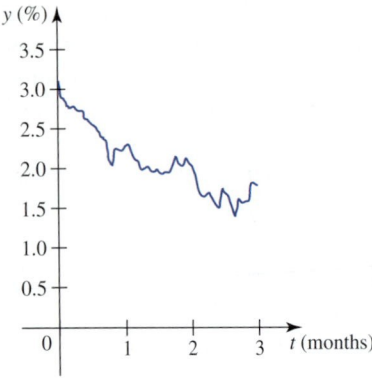

FIGURE 3

The function f gives the annual yield for two-year Treasury notes in the first three months of 2008.

Source: Financial Times.

A function can also be described graphically, as shown in Figure 3. Here, the function f gives the annual yield in percent for two-year Treasury notes, $f(t)$, for the first three months of 2008.

EXAMPLE 1   The function f defined by the formula $y = \sqrt{x}$, or $f(x) = \sqrt{x}$, is just the square root function mentioned earlier. The domain of this function is the set of all values of x in the interval $[0, \infty)$. For example, if $x = 16$, then $f(16) = \sqrt{16} = 4$ is the square root of 16. The range of f consists of all the square roots of nonnegative numbers and is therefore the set of all numbers in $[0, \infty)$. (See Figure 4.)

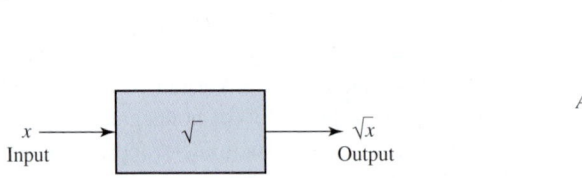

FIGURE 4   (**a**) The square root machine   (**b**) The function f maps x onto $\sqrt{x}$.

Notes

1. We often use letters other than f to denote a function. For example, we might speak of the area function A, the population function P, the function F, and so on.

2. Strictly speaking, it is improper to refer to a function f as $f(x)$ (recall that $f(x)$ is the value of f at x), but it is conventional to do so.

If a function f is described by an equation $y = f(x)$, we call x the **independent variable** and y the **dependent variable** because y (the value of f at x) is dependent upon the choice of x. Here, x represents a number in the *domain* of f and y the unique number in the *range* of f associated with x.

■ Evaluating Functions

Let's look again at the square root function f defined by the rule $f(x) = \sqrt{x}$. We could very well have defined this function by giving the rule as $f(t) = \sqrt{t}$ or $f(u) = \sqrt{u}$. In other words, it doesn't matter what letter we choose to represent the independent variable when describing the rule for a function. Indeed, we can describe the rule for f using the expression

$$f(\ ) = \sqrt{(\ )} = (\ )^{1/2}$$

To find the value of f at x, we simply insert x into the blank spaces inside the parentheses! As another example, consider the function g defined by the rule $g(x) = 2x^2 + x$. We can describe the rule for g by

$$g(\ ) = 2(\ )^2 + (\ )$$

obtained by replacing each x in the expression for $g(x)$ by a pair of parentheses. To find the value of g at $x = 2$, insert the number 2 in the blank spaces inside each pair of parentheses to obtain

$$g(2) = 2(2)^2 + 2 = 10$$

EXAMPLE 2 Let $f(x) = x^2 + 2x - 1$. Find

a. $f(-1)$
b. $f(\pi)$
c. $f(t)$, where t is a real number
d. $f(x + h)$, where h is a real number
e. $f(2x)$

Solution We think of $f(x)$ as

$$f(\ ) = (\ )^2 + 2(\ ) - 1$$

Then

a. $f(-1) = (-1)^2 + 2(-1) - 1 = -2$
b. $f(\pi) = (\pi)^2 + 2(\pi) - 1 = \pi^2 + 2\pi - 1$
c. $f(t) = (t)^2 + 2(t) - 1 = t^2 + 2t - 1$
d. $f(x + h) = (x + h)^2 + 2(x + h) - 1 = x^2 + 2xh + h^2 + 2x + 2h - 1$
e. $f(2x) = (2x)^2 + 2(2x) - 1 = 4x^2 + 4x - 1$

■ Finding the Domain of a Function

Sometimes the domain of a function is determined by the nature of a problem. For example, the domain of the function $A(r) = \pi r^2$ that gives the area of a circle in terms of its radius is the interval $(0, \infty)$, since r must be positive.

EXAMPLE 3 A man wants to enclose a vegetable garden in his backyard with a rectangular fence. If he has 100 ft of fencing with which to enclose his garden, find a function that gives the area of the garden in terms of its length x (see Figure 5). (Assume that he uses all of the fencing.) What is the domain of this function?

Solution From Figure 5, we see that the perimeter of the rectangle, $(2x + 2y)$ ft, must be equal to 100 ft. Thus, we have the equation

$$2x + 2y = 100 \tag{1}$$

The area of the rectangle is given by

$$A = xy \tag{2}$$

Solving Equation (1) for y in terms of x, we obtain $y = 50 - x$. Substituting this value of y into Equation (2) yields

$$A = x(50 - x)$$
$$= -x^2 + 50x$$

Since the sides of the rectangle must be positive, we have $x > 0$ and $50 - x > 0$, which is equivalent to $0 < x < 50$. Therefore, the required function is

$$A(x) = -x^2 + 50x$$

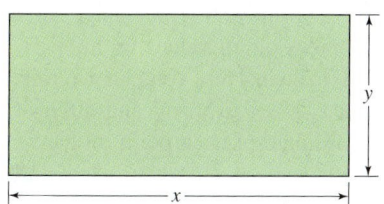

FIGURE 5
A rectangular garden with dimensions x ft by y ft

with domain $(0, 50)$.

   Unless we specifically mention the domain of a function f, we will adopt the convention that the domain of f is the set of all numbers for which $f(x)$ is a real number.

EXAMPLE 4   Find the domain of each function:

a. $f(x) = \dfrac{2x + 1}{x^2 - x - 2}$ **b.** $f(x) = \dfrac{x + \sqrt{x + 1}}{2x - 1}$

Solution

a. Since division by zero is prohibited and the denominator of $f(x)$ is equal to zero if $x^2 - x - 2 = (x - 2)(x + 1) = 0$, or $x = -1$ or $x = 2$, we conclude that the domain of f is the set of all numbers except -1 and 2. Equivalently, the domain of f is the set $(-\infty, -1) \cup (-1, 2) \cup (2, \infty)$.

b. We begin by looking at the numerator of $f(x)$. Because the expression under the radical sign must be nonnegative, we see that $x + 1 \geq 0$, or $x \geq -1$. Next, since division by zero is not allowed, we see that $2x - 1 \neq 0$. But $2x - 1 = 0$ if $x = \frac{1}{2}$, so $x \neq \frac{1}{2}$. Therefore, the domain of f is the set $\left[-1, \frac{1}{2}\right) \cup \left(\frac{1}{2}, \infty\right)$.

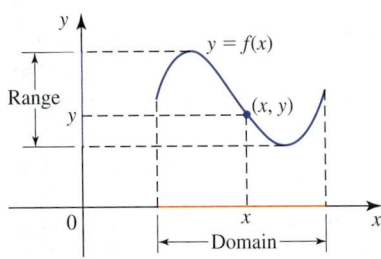

FIGURE 6
The graph of a function f

DEFINITION   Graph of a Function

The graph of a function f is the set of all points (x, y) such that $y = f(x)$, where x lies in the domain of f.

The graph of f provides us with a way of visualizing a function (see Figure 6).

Note   If the function f is defined by the equation $y = f(x)$, then the domain of f is the set of all x-values, and the range of f is the set of all y-values.

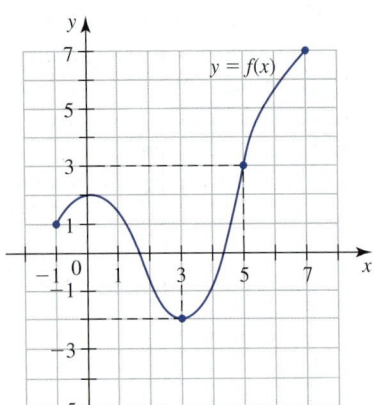

FIGURE 7
The graph of a function f

EXAMPLE 5   The graph of a function f is shown in Figure 7.

a. What is $f(3)$? $f(5)$?
b. What is the distance of the point $(3, f(3))$ from the x-axis? The point $(5, f(5))$ from the x-axis?
c. What is the domain of f? The range of f?

Solution

a. From the graph of f, we see that $y = -2$ when $x = 3$, and we conclude that $f(3) = -2$. Similarly, we see that $f(5) = 3$.

b. Since the point $(3, -2)$ lies below the x-axis, we see that the distance of the point $(3, f(3))$ from the x-axis is $-f(3) = -(-2) = 2$ units. The point $(5, f(5))$ lies above the x-axis, and its distance is $f(5)$, or 3 units.

c. Observe that x may take on all values between $x = -1$ and $x = 7$, inclusive, so the domain of f is $[-1, 7]$. Next, observe that as x takes on all values in the domain of f, y takes on all values between -2 and 7, inclusive. (You can see this by running your index finger along the x-axis from $x = -1$ to $x = 7$ and observing the corresponding values assumed by the y-coordinate of each point on the graph of f.) Therefore, the range of f is $[-2, 7]$.

EXAMPLE 6   Sketch the graph of the function $f(x) = \dfrac{1}{x}$. What is the range of f?

Solution   The domain of f is $(-\infty, 0) \cup (0, \infty)$. From the following table of values for $y = f(x)$ corresponding to some selected values of x, we obtain the graph of f shown in Figure 8.

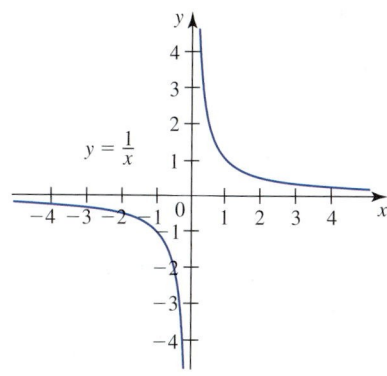

FIGURE 8
The graph of $f(x) = \dfrac{1}{x}$

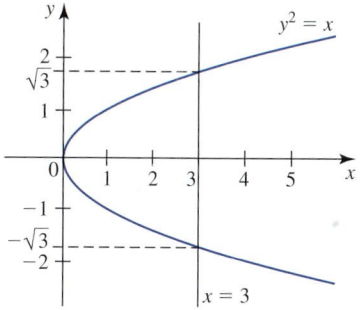

FIGURE 9
The number 3 has two images, $-\sqrt{3}$ and $\sqrt{3}$.

x	$\frac{1}{3}$	$\frac{1}{2}$	1	2	3	-3	-2	-1	$-\frac{1}{2}$	$-\frac{1}{3}$
y	3	2	1	$\frac{1}{2}$	$\frac{1}{3}$	$-\frac{1}{3}$	$-\frac{1}{2}$	-1	-2	-3

Setting $f(x) = y$ gives $1/x = y$, or $x = 1/y$, where $y \neq 0$. This shows that corresponding to any nonzero value of y there is an x in the domain of f that is mapped onto y. So the range of f is $(-\infty, 0) \cup (0, \infty)$. ■

■ The Vertical Line Test

Consider the equation $y^2 = x$. Solving for y in terms of x, we obtain

$$y = \pm\sqrt{x} \qquad (3)$$

Since each positive value of x is associated with *two* values of y—for example, the number 3 is mapped onto the two images $-\sqrt{3}$ and $\sqrt{3}$—we see that the equation $y^2 = x$ does not define y as a function of x. The graph of $y^2 = x$ is shown in Figure 9.

Note that the vertical line $x = 3$ intersects the graph of $y^2 = x$ at the *two* points $(3, -\sqrt{3})$ and $(3, \sqrt{3})$, verifying geometrically our earlier observation that the number $x = 3$ is associated with the two values $y = -\sqrt{3}$ and $y = \sqrt{3}$. These observations lead to the following criterion for determining when the graph of an equation is a function.

> **The Vertical Line Test**
>
> A curve in the xy-plane is the graph of a function f defined by the equation $y = f(x)$ if and only if no vertical line intersects the curve at more than one point.

■ Piecewise Defined Functions

In certain situations, a function is defined by several equations, each valid over a certain portion of the domain of the function.

EXAMPLE 7   Sketch the graph of the absolute value function $f(x) = |x|$.

Solution   We can plot a few points lying on the graph of f and draw a suitable curve passing through them. Alternatively, we can proceed as follows. Recall that

$$|x| = \begin{cases} x & \text{if } x \geq 0 \\ -x & \text{if } x < 0 \end{cases}$$

This shows that the function $f(x) = |x|$ is defined piecewise over its domain $(-\infty, \infty)$. In the subdomain $[0, \infty)$ the rule for f is $f(x) = x$. So the graph of f coincides with that of $y = x$ for $x \geq 0$. But the latter is the right half of the line with equation $y = x$. In the subdomain $(-\infty, 0)$ the rule for f is $f(x) = -x$, and we see that the graph of f over this portion of its domain coincides with the left half of the line with equation $y = -x$. The graph of f is sketched in Figure 10. ■

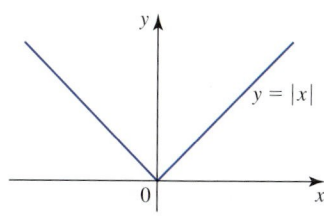

FIGURE 10
The graph of $f(x) = |x|$ consists of the left half of the line $y = -x$ and the right half of the line $y = x$.

EXAMPLE 8 Sketch the graph of the function

$$f(x) = \begin{cases} x & \text{if } x < 1 \\ \frac{1}{4}x^2 - 1 & \text{if } x \geq 1 \end{cases}$$

Solution   The function f is defined piecewise and has domain $(-\infty, \infty)$. In the sub-domain $(-\infty, 1)$ the rule for f is $f(x) = x$, so the graph of f over this portion of its domain is the half-line with equation $y = x$. In the subdomain $[1, \infty)$ the rule for f is $f(x) = \frac{1}{4}x^2 - 1$. To sketch the graph of f over this subdomain, we use the following table.

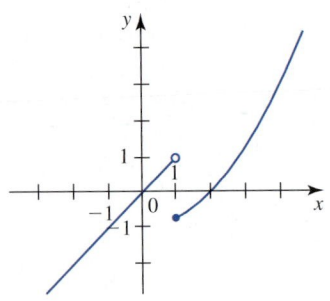

FIGURE 11

x	1	2	3	4
$f(x) = \frac{1}{4}x^2 - 1$	$-\frac{3}{4}$	0	$\frac{5}{4}$	3

The graph of f is shown in Figure 11.

Note   Be sure that you use the correct equation when you evaluate a function that is defined piecewise. For instance, to find $f\left(\frac{1}{2}\right)$ in the preceding example, we note that $x = \frac{1}{2}$ lies in the subdomain $(-\infty, 1)$. So the correct rule here is $f(x) = x$ giving $f\left(\frac{1}{2}\right) = \frac{1}{2}$. To compute $f(5)$, we use the rule $f(x) = \frac{1}{4}x^2 - 1$, which gives $f(5) = \frac{21}{4}$.

Even and Odd Functions

A function f that satisfies $f(-x) = f(x)$ for every x in its domain is called an **even function**. The graph of an even function is symmetric with respect to the y-axis (see Figure 12a). An example of an even function is $f(x) = x^2$, since $f(-x) = (-x)^2 = x^2 = f(x)$.

A function f that satisfies $f(-x) = -f(x)$ for every x in its domain is called an **odd function**. The graph of an odd function is symmetric with respect to the origin (see Figure 12b). An example of an odd function is $f(x) = x^3$, since $f(-x) = (-x)^3 = -x^3 = -f(x)$.

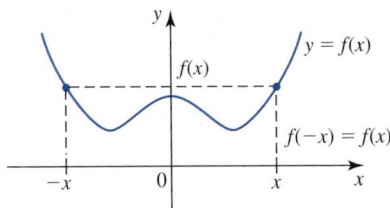

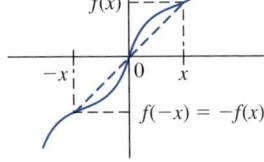

FIGURE 12   **(a)** f is even.   **(b)** f is odd.

EXAMPLE 9 Determine whether the function is even, odd, or neither even nor odd:

a. $f(x) = x^3 - x$ **b.** $g(x) = x^4 - x^2 + 1$ **c.** $h(x) = x - 2x^2$

Solution

a. $f(-x) = (-x)^3 - (-x) = -x^3 + x = -(x^3 - x) = -f(x)$. Therefore, f is an odd function.

b. $g(-x) = (-x)^4 - (-x)^2 + 1 = x^4 - x^2 + 1 = g(x)$, and we see that g is even.
c. $h(-x) = (-x) - 2(-x)^2 = -x - 2x^2$, which is neither equal to $h(x)$ nor $-h(x)$, and we conclude that h is neither even nor odd.

The graphs of the functions f, g, and h are shown in Figure 13.

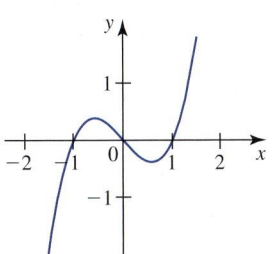

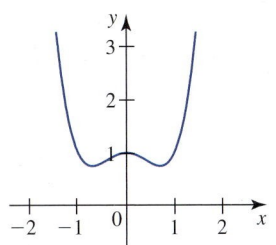

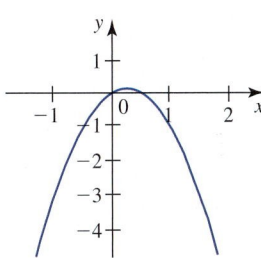

FIGURE 13   **(a)** $f(x) = x^3 - x$ **(b)** $g(x) = x^4 - x^2 + 1$ **(c)** $h(x) = x - 2x^2$

0.2 EXERCISES

1. If $f(x) = 3x + 4$, find $f(0), f(-4), f(a), f(-a), f(a + 1), f(2a), f(\sqrt{a})$, and $f(x + 1)$.

2. If $f(x) = 2x - 1$, find $f(-\sqrt{2}), f(t + 1), f(2t - 1), f\left(\dfrac{x}{2}\right)$, and $f(a + h)$.

3. If $g(x) = -x^2 + 2x$, find $g(-2), g(\sqrt{3}), g(a^2), g(a + h)$, and $\dfrac{1}{g(3)}$.

4. If $f(t) = \dfrac{2t^2}{\sqrt{t - 1}}$, find $f(2), f(x + 1)$, and $f(2x - 1)$.

5. If $f(x) = 2x^3 - x$, find $f(-1), f(0), f(x^2), f(\sqrt{x})$, and $f\left(\dfrac{1}{x}\right)$.

6. If $f(x) = \dfrac{\sqrt{x}}{x^2 + 1}$, find $f(4), f(x + h), f(x - h)$, and $f(x + 2h)$.

7. If
$$f(x) = \begin{cases} x^2 + 1 & \text{if } x \leq 0 \\ \sqrt{x} & \text{if } x > 0 \end{cases}$$
find $f(-2), f(0)$, and $f(1)$.

8. If
$$f(x) = \begin{cases} \dfrac{x}{x + 1} & \text{if } x < -1 \\ 1 + \sqrt{x + 1} & \text{if } x \geq -1 \end{cases}$$
find $f(-2), f(-1)$, and $f(0)$.

9. If $f(x) = x^2$, find and simplify $\dfrac{f(x) - f(1)}{x - 1}$, where $x \neq 1$.

10. If $f(x) = 2x^2 + 1$, find and simplify $\dfrac{f(1 + h) - f(1)}{h}$, where $h \neq 0$.

11. If $f(x) = x - x^2$, find and simplify $\dfrac{f(x + h) - f(x)}{h}$, where $h \neq 0$.

12. If $f(x) = \sqrt{x}$, find and simplify $\dfrac{f(a + h) - f(a)}{h}$, where $h \neq 0$.

13. a. If $f(x) = x^2 - 2x + k$ and $f(1) = 3$, find k.
   b. If $g(t) = |t - 1| + k$ and $g(-1) = 0$, find k.

14. If $f(x) = ax^3 + b$, find a and b if it is known that $f(1) = 1$ and $f(2) = 15$.

In Exercises 15–26, find the domain of the function.

15. $f(x) = \dfrac{3x + 1}{x^2}$ **16.** $f(x) = \dfrac{2x + 1}{x - 1}$

17. $g(t) = \dfrac{t + 1}{2t^2 - t - 1}$ **18.** $h(x) = \sqrt{2x + 3}$

19. $f(x) = \sqrt{9 - x^2}$ **20.** $F(x) = \sqrt{x^2 - 2x - 3}$

21. $f(x) = \sqrt{x - 2} + \sqrt{4 - x}$

22. $f(x) = \dfrac{\sqrt{x - 1}}{x^2 - x - 6}$

23. $f(x) = \dfrac{\sqrt{x + 2} + \sqrt{2 - x}}{x^3 - x}$

24. $f(x) = \dfrac{\sqrt[3]{x^2 - x + 1}}{x^2 + 1}$

25. $f(x) = \dfrac{x^3 + 1}{x\sqrt{x^2 - 1}}$ **26.** $f(x) = \dfrac{1}{\sqrt{|x| - x}}$

27. Refer to the graph of the function f in the following figure.

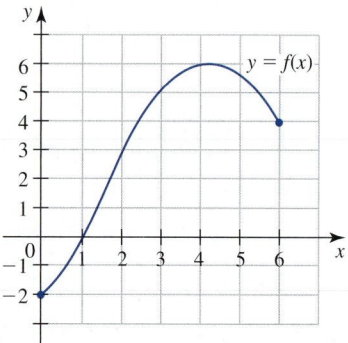

a. Find $f(0)$.
b. Find the value of x for which (i) $f(x) = 3$ and (ii) $f(x) = 0$.
c. Find the domain of f.
d. Find the range of f.

28. Refer to the graph of the function f in the following figure.

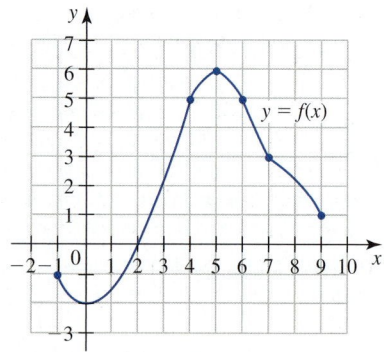

a. Find $f(7)$.
b. Find the values of x corresponding to the point(s) on the graph of f located at a height of 5 units above the x-axis.
c. Find the point on the x-axis at which the graph of f crosses it. What is $f(x)$ at this point?
d. Find the domain and range of f.

In Exercises 29–30, determine whether the point lies on the graph of the function.

29. $P(3, 3)$; $f(x) = \dfrac{x + 1}{\sqrt{x^2 + 7}} + 2$

30. $P\left(-3, -\tfrac{1}{13}\right)$; $f(t) = \dfrac{|t + 1|}{t^3 + 2}$

In Exercises 31–38, find the domain and sketch the graph of the function. What is its range?

31. $f(x) = -2x + 1$ **32.** $f(x) = \dfrac{1}{2}x^2 + 1$

33. $g(x) = \sqrt{x - 1}$ **34.** $f(x) = |x| - 1$

35. $h(x) = \sqrt{x^2 - 1}$ **36.** $f(t) = \dfrac{|t - 1|}{t - 1}$

37. $f(x) = \begin{cases} -x + 1 & \text{if } x \le 1 \\ x^2 - 1 & \text{if } x > 1 \end{cases}$

38. $f(x) = \begin{cases} -x - 1 & \text{if } x < -1 \\ 0 & \text{if } -1 \le x \le 1 \\ x + 1 & \text{if } x > 1 \end{cases}$

In Exercises 39–42, use the vertical line test to determine whether the curve is the graph of a function of x.

39.
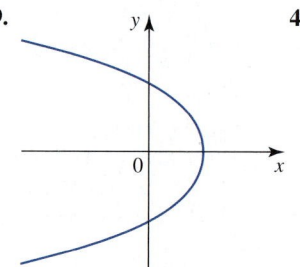

40.
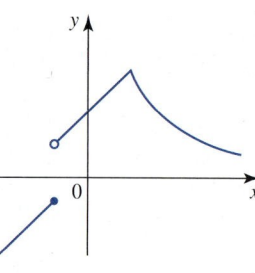

41.
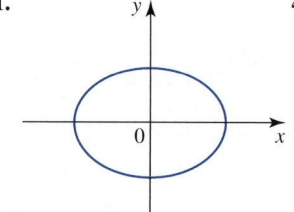

42.
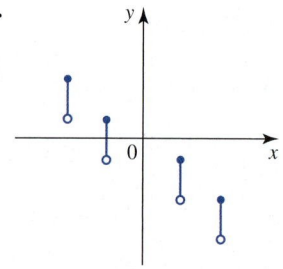

43. Refer to the curve for Exercise 39. Is it the graph of a function of y? Explain.

44. Refer to the curve for Exercise 40. Is it the graph of a function of y? Explain.

In Exercises 45–48, determine whether the function whose graph is given, is even, odd, or neither.

45.
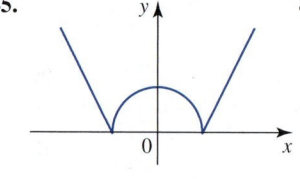

46.
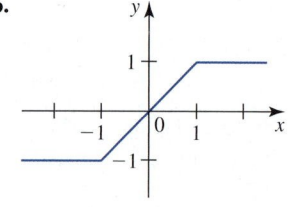

47.
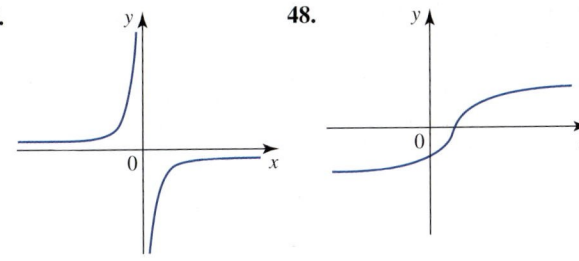
48.

In Exercises 49–54, determine whether the function is even, odd, or neither.

49. $f(x) = 1 - 2x^2$

50. $f(x) = \dfrac{x}{x^2 + 1}$

51. $f(x) = 2x^3 - 3x + 1$

52. $f(x) = 2x^{1/3} - 3x^2$

53. $f(x) = \dfrac{|x| + 1}{x^4 - 2x^2 + 3}$

54. $f(x) = \sqrt{x^2 + x + 1} - \sqrt{x^2 - x + 1}$

55. The following figure shows a portion of the graph of a function f defined on the interval $[-2, 2]$. Sketch the complete graph of f if it is known that (a) f is even, (b) f is odd.

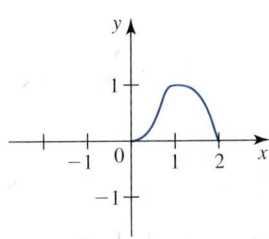

56. The following figure shows a portion of the graph of a function f defined on the interval $[-2, 2]$.
   a. Can f be odd? Explain. If so, complete the graph of f.
   b. Can f be even? Explain. If so, complete the graph of f.

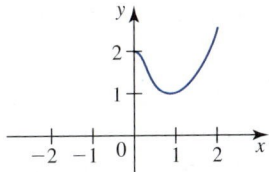

57. The function $y = f(t)$, whose graph is shown in the following figure, gives the distance the Jacksons were from their home on a recent trip they took from Boston to Niagara Falls as a function of time t ($t = 0$ corresponds to 7 A.M.).

The 500-mi trip took a total of 8 hr. What does the graph tell us about the trip?

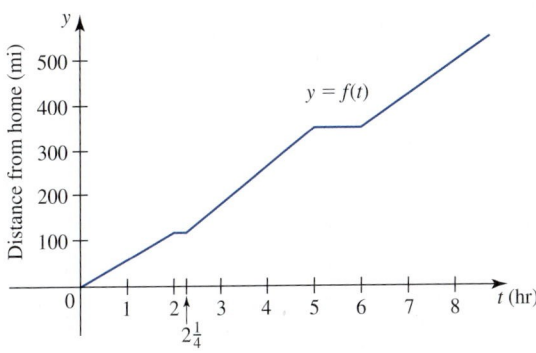

58. A plane departs from Logan Airport in Boston bound for Heathrow Airport in London, a 6-hr, 3267-mi flight. After takeoff, the plane climbs to a cruising altitude of 35,000 ft, which it maintains until its descent to the airport. While at its cruising altitude, the plane maintains a ground speed of 550 mph. Let $D = f(t)$ denote the distance (in miles) flown by the plane as a function of time (in hours), and let $A = g(t)$ denote the altitude (in feet) of the plane.
   a. Sketch a graph of f that could describe the situation.
   b. Sketch a graph of g that could describe the situation.

59. **Oxygen Content of a Pond** When organic waste is dumped into a pond, the oxidation process that takes place reduces the pond's oxygen content. However, given time, nature will restore the oxygen content to its natural level. Let $P = f(t)$ denote the oxygen content (as a percentage of its normal level) t days after organic waste has been dumped into the pond. Sketch a graph of f that could depict the process.

60. **The Gender Gap** The following graph shows the ratio of women's earnings to men's from 1960 through 2000.

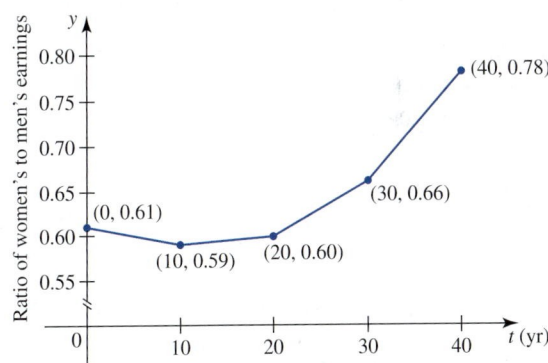

   a. Write the rule for the function f giving the ratio of women's earnings to men's in year t, with $t = 0$ corresponding to 1960.
   Hint: The function f is defined piecewise and is linear over each of four subintervals.

b. In what decade(s) was the gender gap expanding? Shrinking?

c. Refer to part (b). How fast was the gender gap (the ratio per year) expanding or shrinking in each of these decades?

Source: U.S. Bureau of Labor Statistics.

61. Prevalence of Alzheimer's Patients On the basis of a study conducted in 1997, the percentage of the U.S. population by age afflicted with Alzheimer's disease is given by the function

$$P(x) = 0.0726x^2 + 0.7902x + 4.9623 \qquad 0 \le x \le 25$$

where x is measured in years, with $x = 0$ corresponding to age 65. What percentage of the U.S. population at age 65 is expected to have Alzheimer's disease? At age 90?

Source: Alzheimer's Association.

62. U.S. Health Care Information Technology Spending As health care costs increase, payers are turning to technology and outsourced services to keep a lid on expenses. The amount of health care information technology spending by payer is approximated by

$$S(t) = -0.03t^3 + 0.2t^2 + 0.23t + 5.6 \qquad 0 \le t \le 4$$

where $S(t)$ is measured in billions of dollars and t is measured in years with $t = 0$ corresponding to 2004. What was the amount spent by payers on health care IT in 2004? What amount was spent by payers in 2008?

Source: U.S. Department of Commerce.

63. Hotel Rates The average daily rate of U.S. hotels from 2001 through 2006 is approximated by the function

$$f(t) = \begin{cases} 82.95 & \text{if } 1 \le t \le 3 \\ 0.95t^2 - 3.95t + 86.25 & \text{if } 3 < t \le 6 \end{cases}$$

where $f(t)$ is measured in dollars and $t = 1$ corresponds to 2001.

a. What was the average daily rate of U.S. hotels from 2001 through 2003?

b. What was the average daily rate of U.S. hotels in 2004? In 2005? In 2006?

c. Sketch the graph of f.

Source: Smith Travel Research.

64. Postal Regulations In 2007 the postage for packages sent by first-class mail was raised to $1.13 for the first ounce or fraction thereof and 17¢ for each additional ounce or fraction thereof. Any parcel not exceeding 13 oz may be sent by first-class mail. Letting x denote the weight of a parcel in ounces and letting $f(x)$ denote the postage in dollars, complete the following description of the "postage function" f:

$$f(x) = \begin{cases} 1.13 & \text{if } 0 < x \le 1 \\ 1.30 & \text{if } 1 < x \le 2 \\ \vdots & \\ ? & \text{if } 12 < x \le 13 \end{cases}$$

a. What is the domain of f?

b. Sketch the graph of f.

65. Harbor Cleanup The amount of solids discharged from the Massachusetts Water Resources Authority sewage treatment plant on Deer Island (near Boston Harbor) is given by the function

$$f(t) = \begin{cases} 130 & \text{if } 0 \le t \le 1 \\ -30t + 160 & \text{if } 1 < t \le 2 \\ 100 & \text{if } 2 < t \le 4 \\ -5t^2 + 25t + 80 & \text{if } 4 < t \le 6 \\ 1.25t^2 - 26.25t + 162.5 & \text{if } 6 < t \le 10 \end{cases}$$

where $f(t)$ is measured in tons/day and t is measured in years, with $t = 0$ corresponding to 1989.

a. What amount of solids were discharged per day in 1989? In 1992? In 1996?

b. Sketch the graph of f.

Source: Metropolitan District Commission.

66. Rising Median Age Increased longevity and the aging of the baby boom generation—those born between 1946 and 1965—are the primary reasons for a rising median age. The median age (in years) of the U.S. population from 1900 through 2000 is approximated by the function

$$f(t) = \begin{cases} 1.3t + 22.9 & \text{if } 0 \le t \le 3 \\ -0.7t^2 + 7.2t + 11.5 & \text{if } 3 < t \le 7 \\ 2.6t + 9.4 & \text{if } 7 < t \le 10 \end{cases}$$

where t is measured in decades, with $t = 0$ corresponding to the beginning of 1900.

a. What was the median age of the U.S. population at the beginning of 1900? At the beginning of 1950? At the beginning of 1990?

b. Sketch the graph of f.

Source: U.S. Census Bureau.

67. Suppose a function has the property that whenever x is in the domain of f, then so is $-x$. Show that f can be written as the sum of an even function and an odd function.

68. Prove that a nonzero *polynomial function*

$$f(x) = a_n x^n + a_{n-1}x^{n-1} + \cdots + a_2 x^2 + a_1 x + a_0$$

where n is a nonnegative integer and $a_0, a_1, \ldots, a_n$ are real numbers with $a_n \ne 0$, can be expressed as the sum of an even function and an odd function.

In Exercises 69–72, determine whether the statement is true or false. If it is true, explain why. If it is false, explain why or give an example that shows that it is false.

69. If $a = b$, then $f(a) = f(b)$.

70. If $f(a) = f(b)$, then $a = b$.

71. If f is a function, then $f(a + b) = f(a) + f(b)$.

72. A curve in the xy-plane can be simultaneously the graph of a function of x and the graph of a function of y.

0.3   The Trigonometric Functions

0.3   SELF-CHECK DIAGNOSTIC TEST

1. Given that $\sec\theta = \frac{5}{3}$ and $0 \le \theta < \frac{\pi}{2}$, find $\tan\theta$.
2. Determine whether the function $f(x) = \dfrac{\sin 2x}{\sqrt{1 + \cos^2 x} + 1}$ is even, odd, or neither.
3. Verify the identity $\dfrac{\cot x - 1}{1 - \tan x} = \cot x$.
4. Using the substitution $x = a \sin\theta$, where $a > 0$ and $-\frac{\pi}{2} \le \theta \le \frac{\pi}{2}$, express $a^2 - x^2$ in terms of θ.
5. Solve the equation $\cos\theta - 2\sin^2\theta + 1 = 0$, where $0 \le \theta < 2\pi$.

Answers to Self-Check Diagnostic Test 0.3 can be found on page ANS 2.

In this section we review the basic properties of the trigonometric functions and their graphs. The emphasis is placed on those topics that we will use later in calculus.

■ Angles

An angle in the plane is generated by rotating a ray about its endpoint. The starting position of the ray is called the initial side of the angle, the final position of the ray is called the terminal side, and the point of intersection of the two sides is called the **vertex** of the angle (see Figure 1a).

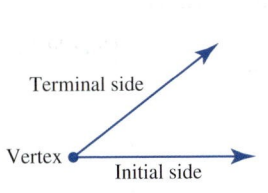

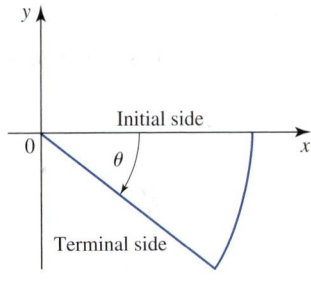

(**a**) An angle

(**b**) A positive angle in standard position

(**c**) A negative angle in standard position

FIGURE 1

In a rectangular coordinate system an angle θ (the Greek *theta*) is in **standard position** if its vertex is centered at the origin and its initial side coincides with the positive x-axis. An angle is **positive** if it is generated by a counterclockwise rotation and **negative** if it is generated by a clockwise rotation (Figure 1b–c).

■ Radian Measure of Angles

We can express the magnitude of an angle in either degrees or radians. In calculus, however, we prefer to use the radian measure of an angle because it simplifies our work.

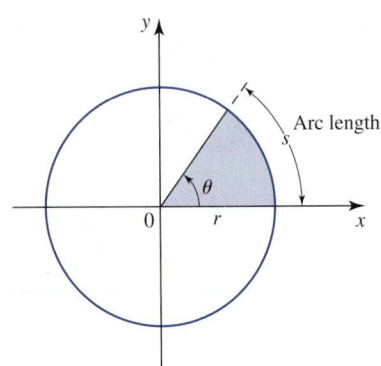

FIGURE 2

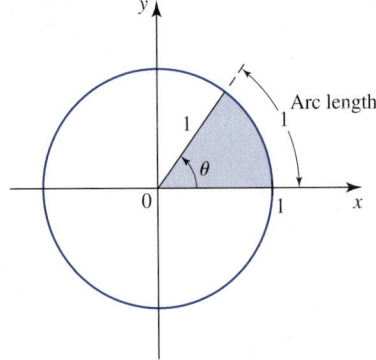

FIGURE 3
The unit circle $x^2 + y^2 = 1$

DEFINITION **Radian Measure of an Angle**

If s is the length of the arc subtended by a central angle θ in a circle of radius r, then

$$\theta = \frac{s}{r} \tag{1}$$

is the **radian measure** of θ (see Figure 2).

For convenience we often work with the **unit circle,** that is, the circle of radius 1 centered at the origin. On the unit circle, an angle of 1 radian is subtended by an arc of length 1 (see Figure 3). To specify the units of measure for the angle θ in Figure 3, we write $\theta = 1$ radian or $\theta = 1$. By convention, if the unit of measure is not specifically stated, we assume that it is radians.

Since the circumference of the unit circle is 2π and the central angle subtended by one complete revolution is $360°$, we see that

$$2\pi \text{ radians (rad)} = 360°$$

or

$$1 \text{ rad} = \left(\frac{180}{\pi}\right)^\circ \tag{2}$$

and

$$1° = \frac{\pi}{180} \text{ rad} \tag{3}$$

These relationships suggest the following useful conversion rules.

Converting Degrees and Radians

To convert degrees to radians, multiply by $\dfrac{\pi}{180}$.

To convert radians to degrees, multiply by $\dfrac{180}{\pi}$.

EXAMPLE 1 Convert each of the following to radian measure:

a. $60°$ **b.** $300°$ **c.** $-225°$

Solution

a. $60 \cdot \dfrac{\pi}{180} = \dfrac{\pi}{3}$, or $\dfrac{\pi}{3}$ rad

b. $300 \cdot \dfrac{\pi}{180} = \dfrac{5\pi}{3}$, or $\dfrac{5\pi}{3}$ rad

c. $-225 \cdot \dfrac{\pi}{180} = -\dfrac{5\pi}{4}$, or $-\dfrac{5\pi}{4}$ rad

EXAMPLE 2 Convert each of the following to degree measure:

a. $\dfrac{\pi}{3}$ rad   **b.** $\dfrac{3\pi}{4}$ rad   **c.** $-\dfrac{7\pi}{4}$ rad

Solution

a. $\dfrac{\pi}{3} \cdot \dfrac{180}{\pi} = 60$, or $60°$

b. $\dfrac{3\pi}{4} \cdot \dfrac{180}{\pi} = 135$, or $135°$

c. $-\dfrac{7\pi}{4} \cdot \dfrac{180}{\pi} = -315$, or $-315°$ ■

More than one angle may have the same initial and terminal sides. We call such angles **coterminal**. For example the angle $4\pi/3$ has the same initial and terminal sides as the angle $\theta = -2\pi/3$ (see Figure 4).

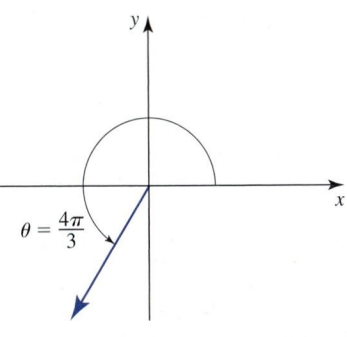

FIGURE 4
Coterminal angles

     (a) $\theta = \dfrac{4\pi}{3}$ **(b)** $\theta = -\dfrac{2\pi}{3}$

An angle may be greater than 2π rad. For example, an angle of 3π rad is generated by rotating a ray in a counterclockwise direction through one and a half revolutions (Figure 5a). Similarly, an angle of $-5\pi/2$ radians is generated by rotating a ray in a clockwise direction through one and a quarter revolutions (Figure 5b).

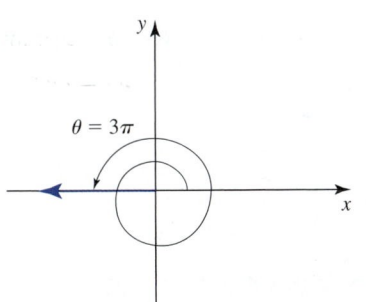
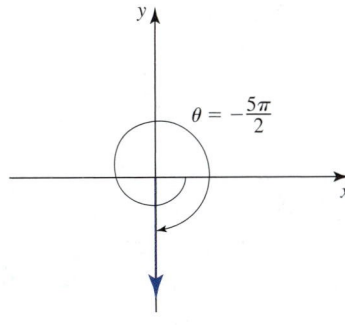

FIGURE 5
Angles generated by more than one revolution

     (a) $\theta = 3\pi$ **(b)** $\theta = -\dfrac{5\pi}{2}$

The radian and degree measures of several common angles are given in Table 1. Be sure that you familiarize yourself with these values.

TABLE 1

Degrees	0°	30°	45°	60°	90°	120°	135°	150°	180°	270°	360°
Radians	0	$\dfrac{\pi}{6}$	$\dfrac{\pi}{4}$	$\dfrac{\pi}{3}$	$\dfrac{\pi}{2}$	$\dfrac{2\pi}{3}$	$\dfrac{3\pi}{4}$	$\dfrac{5\pi}{6}$	π	$\dfrac{3\pi}{2}$	2π

By rewriting Equation (1), $\theta = s/r$, we obtain the following formula, which gives the length of a circular arc.

Length of a Circular Arc

$$s = r\theta \tag{4}$$

Another related formula that we will use later in calculus gives the area of a circular sector.

Area of a Circular Sector

$$A = \frac{1}{2}r^2\theta \tag{5}$$

Note   In Equations (4) and (5) θ must be expressed in radians.   ■

EXAMPLE 3   What is the length of the arc subtended by $\theta = 7\pi/6$ radians in a circle of radius 3? What is the area of the circular sector determined by θ?

Solution   To find the length of the arc, we use Equation (4) to obtain

$$s = 3\left(\frac{7\pi}{6}\right) = \frac{7\pi}{2}$$

The area of the sector is obtained by using Equation (5). Thus,

$$A = \frac{1}{2}r^2\theta = \frac{1}{2}(3)^2\left(\frac{7\pi}{6}\right)$$

$$= \frac{21\pi}{4}$$ ■

■ The Trigonometric Functions

Two approaches are generally used to define the six trigonometric functions. We summarize each approach here.

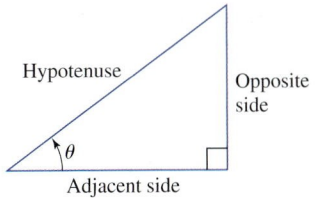

Hypotenuse

Opposite side

θ

Adjacent side

FIGURE 6

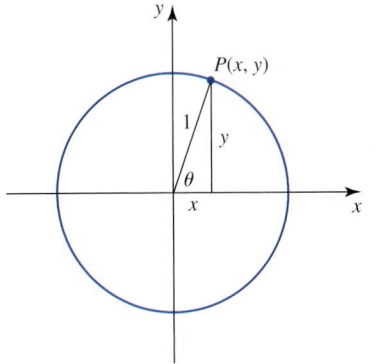

$P(x, y)$

1

y

θ

x

x

y

FIGURE 7
The unit circle

THE TRIGONOMETRIC FUNCTIONS

The Right Triangle Definition

For an acute angle θ (see Figure 6),

$$\sin \theta = \frac{\text{opp}}{\text{hyp}} \qquad \cos \theta = \frac{\text{adj}}{\text{hyp}} \qquad \csc \theta = \frac{\text{hyp}}{\text{opp}}$$

$$\sec \theta = \frac{\text{hyp}}{\text{adj}} \qquad \tan \theta = \frac{\text{opp}}{\text{adj}} \qquad \cot \theta = \frac{\text{adj}}{\text{opp}}$$

The Unit Circle Definition

Let θ denote an angle in standard position, and let $P(x, y)$ denote the point where the terminal side of θ meets the unit circle. (See Figure 7.) Then

$$\sin \theta = y \qquad\qquad \cos \theta = x$$

$$\csc \theta = \frac{1}{y}, \quad y \neq 0 \qquad \sec \theta = \frac{1}{x}, \quad x \neq 0$$

$$\tan \theta = \frac{y}{x}, \quad x \neq 0 \qquad \cot \theta = \frac{x}{y}, \quad y \neq 0$$

Referring to the point $P(x, y)$ on the unit circle (Figure 7), we see that the coordinates of P can also be written in the form

$$x = \cos \theta \quad \text{and} \quad y = \sin \theta \tag{6}$$

Note $\tan \theta$ and $\sec \theta$ are not defined when $x = 0$. Also, $\csc \theta$ and $\cot \theta$ are not defined when $y = 0$. ■

Table 2 lists the values of the trigonometric functions of certain angles. Since these values occur very frequently in problems involving trigonometry, you will find it helpful to memorize them. The right triangles shown in Figure 8 can be used to help jog your memory.

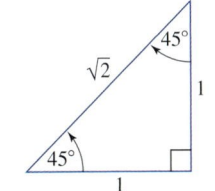

$45°$

$\sqrt{2}$

1

$45°$

1

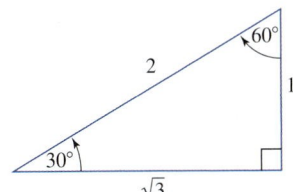

$60°$

2

1

$30°$

$\sqrt{3}$

FIGURE 8

TABLE 2

θ (radians)	θ (degrees)	$\sin \theta$	$\cos \theta$	$\tan \theta$
$\dfrac{\pi}{6}$	$30°$	$\dfrac{1}{2}$	$\dfrac{\sqrt{3}}{2}$	$\dfrac{\sqrt{3}}{3}$
$\dfrac{\pi}{4}$	$45°$	$\dfrac{\sqrt{2}}{2}$	$\dfrac{\sqrt{2}}{2}$	1
$\dfrac{\pi}{3}$	$60°$	$\dfrac{\sqrt{3}}{2}$	$\dfrac{1}{2}$	$\sqrt{3}$

The sign of a trigonometric function of an angle θ is determined by the quadrant in which the terminal side of θ lies. Figure 9 shows a helpful way of remembering the functions that are positive in each quadrant. The signs of the other functions are easy to remember, since they are all negative.

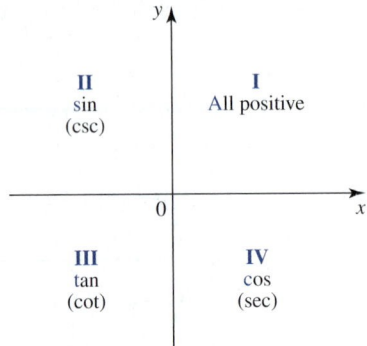

FIGURE 9
The trigonometric functions that are positive in each quadrant can be remembered with the mnemonic device ASTC: **A**ll **S**tudents **T**ake **C**alculus. The functions that are not listed in each quadrant are negative.

To evaluate the trigonometric functions in quadrants other than the first quadrant, we use a reference angle. A **reference angle** for an angle θ is the acute angle formed by the x-axis and the terminal side of θ. Reference angles for each quadrant are depicted in Figure 10.

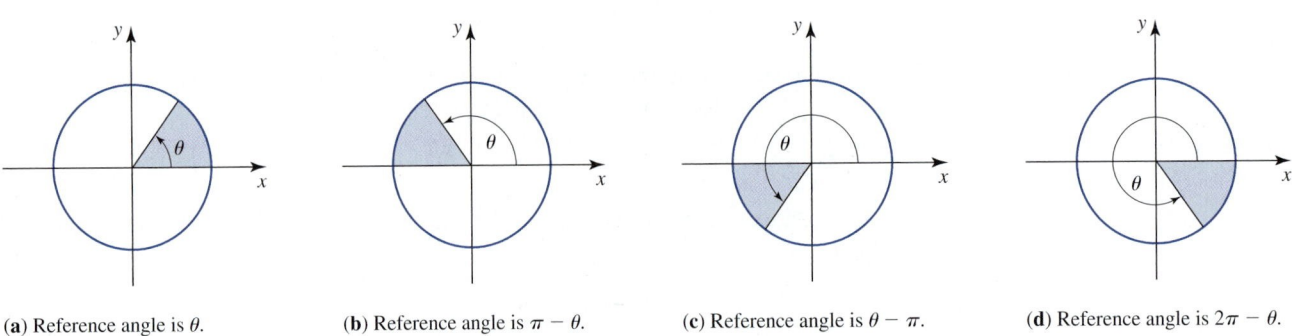

(**a**) Reference angle is θ.   (**b**) Reference angle is $\pi - \theta$.   (**c**) Reference angle is $\theta - \pi$.   (**d**) Reference angle is $2\pi - \theta$.

FIGURE 10

The next example illustrates how we find the trigonometric functions of an angle.

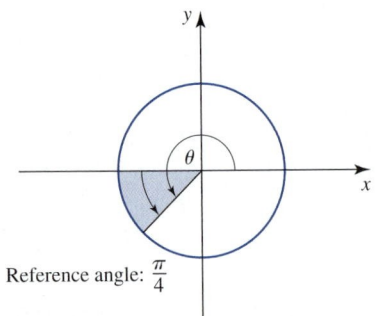

Reference angle: $\frac{\pi}{4}$

FIGURE 11
The reference angle for $\theta = 5\pi/4$ is $\pi/4$, or $45°$.

EXAMPLE 4   Find the sine, cosine, and tangent of $5\pi/4$.

Solution   We first determine the reference angle for the given angle. As is indicated in Figure 11, the reference angle is $(5\pi/4) - \pi = \pi/4$, or $45°$. Since $\sin 45° = \sqrt{2}/2$ and the sine is negative in Quadrant III, we conclude that $\sin(5\pi/4) = -\sqrt{2}/2$. Similarly, since $\cos 45° = \sqrt{2}/2$ and the cosine is negative in Quadrant III, we conclude that $\cos(5\pi/4) = -\sqrt{2}/2$. Finally, since $\tan 45° = 1$ and the tangent is positive in Quadrant III, we conclude that $\tan(5\pi/4) = 1$.   ■

The values of the trigonometric functions that we found in Example 4 are *exact*. The *approximate* value of any trigonometric function can be found by using a calculator. If you use a calculator, be sure to set the mode correctly. For example, to find

$\sin(5\pi/4)$, first set the calculator in radian mode and then enter $\sin(5\pi/4)$. The result will be

$$\sin \frac{5\pi}{4} \approx -0.7071068$$

The number of digits in your answer will depend on the calculator that you use. As we saw in Example 4, the *exact* value of $\sin(5\pi/4)$ is $-\sqrt{2}/2$. Notice that we do not need to use reference angles when we use a calculator.

■ Graphs of the Trigonometric Functions

Referring once again to the unit circle, which is reproduced in Figure 12, we see that an angle of 2π rad corresponds to one complete revolution on the unit circle. Since $P(x, y) = (\cos \theta, \sin \theta)$ is the point where the terminal side of θ intersects the unit circle, we see that the values of $\sin \theta$ and $\cos \theta$ repeat themselves in subsequent revolutions.

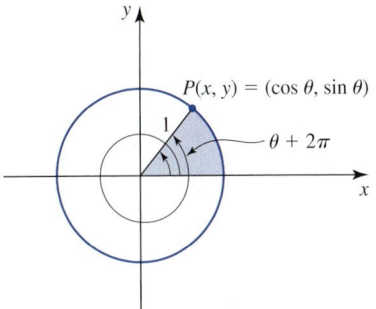

FIGURE 12
The x and y coordinates of the point P are the same for θ and $\theta + 2\pi$.

Therefore,

$$\sin(\theta + 2\pi) = \sin \theta \quad \text{and} \quad \cos(\theta + 2\pi) = \cos \theta \qquad \text{(7a)}$$

and

$$\sin(\theta + 2n\pi) = \sin \theta \quad \text{and} \quad \cos(\theta + 2n\pi) = \cos \theta \qquad \text{(7b)}$$

for every real number θ and every integer n, and we say that the sine and cosine functions are periodic with period 2π.

More generally, we have the following definition of a periodic function.

DEFINITION Periodic Function
A function f is **periodic** if there is a number $p > 0$ such that

$$f(x + p) = f(x)$$

for all x in the domain of f. The smallest such number p is called the **period** of f.

The graphs of the six trigonometric functions are shown in Figure 13. Note that we have denoted the independent variable by x instead of θ. Here, the real number x denotes the radian measure of an angle. As their graphs indicate, the six trigonometric functions are all periodic. The sine and cosine functions, as well as their reciprocals, the

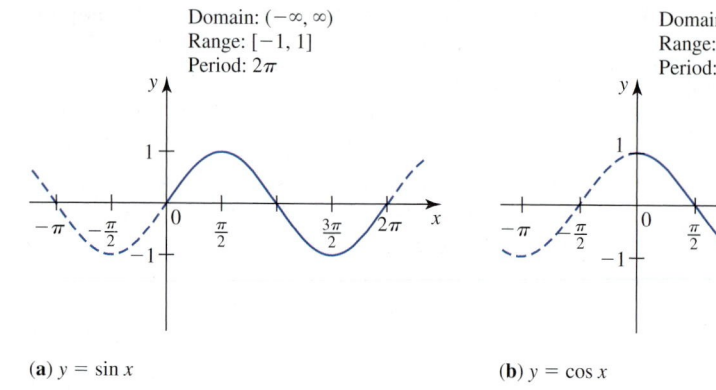

Domain: $(-\infty, \infty)$
Range: $[-1, 1]$
Period: 2π

(a) $y = \sin x$

Domain: $(-\infty, \infty)$
Range: $[-1, 1]$
Period: 2π

(b) $y = \cos x$

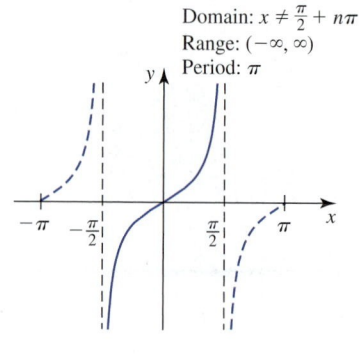

Domain: $x \neq \frac{\pi}{2} + n\pi$
Range: $(-\infty, \infty)$
Period: π

(c) $y = \tan x$

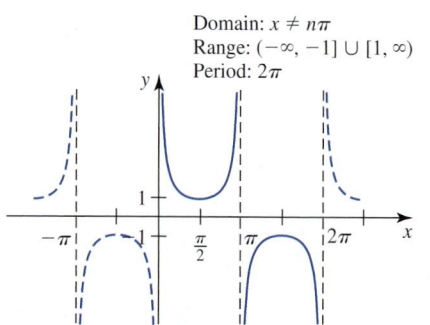

Domain: $x \neq n\pi$
Range: $(-\infty, -1] \cup [1, \infty)$
Period: 2π

(d) $y = \csc x$

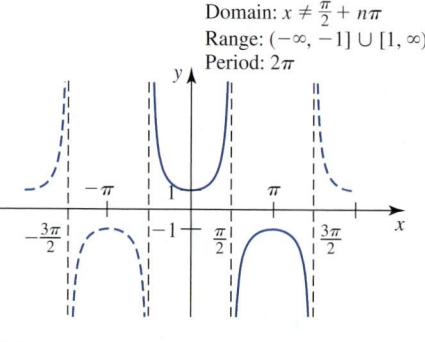

Domain: $x \neq \frac{\pi}{2} + n\pi$
Range: $(-\infty, -1] \cup [1, \infty)$
Period: 2π

(e) $y = \sec x$

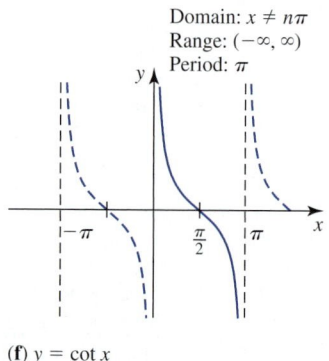

Domain: $x \neq n\pi$
Range: $(-\infty, \infty)$
Period: π

(f) $y = \cot x$

FIGURE 13
Graphs of the six trigonometric functions

cosecant and secant functions, have period 2π. The period of the tangent and cotangent functions, however, is π.

Let's look more closely at the graphs shown in Figure 13a–b. Notice that the graphs of $y = \sin x$ and $y = \cos x$ oscillate between $y = -1$ and $y = 1$. In general, the graphs of the functions $y = A \sin x$ and $y = A \cos x$ oscillate between $y = -A$ and $y = A$, and we say that their amplitude is $|A|$. The graphs of $y = 4 \sin x$ and $y = \frac{1}{4} \sin x$ are shown in Figure 14a–b. Observe that the factor 4 in $y = 4 \sin x$ has the effect of "stretching" the graph of $y = \sin x$ between the values of -4 and 4, whereas the factor $\frac{1}{4}$ in $y = \frac{1}{4} \sin x$ has the effect of "compressing" the graph between $-\frac{1}{4}$ and $\frac{1}{4}$.

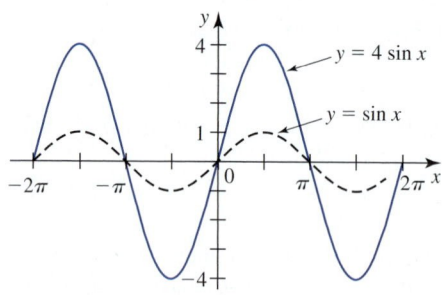

(a) The graph of $y = 4 \sin x$ superimposed upon the graph of $y = \sin x$

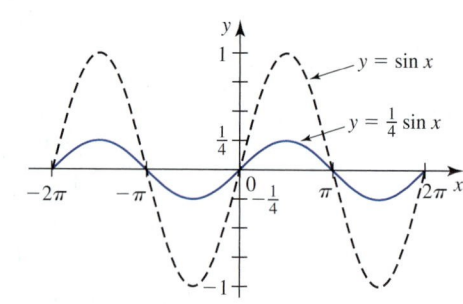

(b) The graph of $y = \frac{1}{4} \sin x$ superimposed upon the graph of $y = \sin x$

FIGURE 14

Next, let's compare the graphs of $y = \cos 2x$ and $y = \cos(x/2)$ with the graph of $y = \cos x$ (see Figure 15a–b). Notice here that the factor of 2 has the effect of "speeding up" the graph of the cosine: The period is decreased from 2π to π. In contrast, the factor of $\frac{1}{2}$ has the effect of "slowing down" the graph of the cosine: The period is increased from 2π to 4π. In general, the period of both $y = \sin Bx$ and $y = \cos Bx$ is $2\pi/|B|$ if $B \neq 0$.

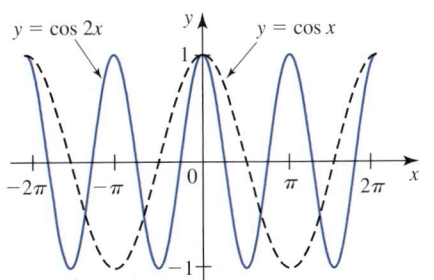

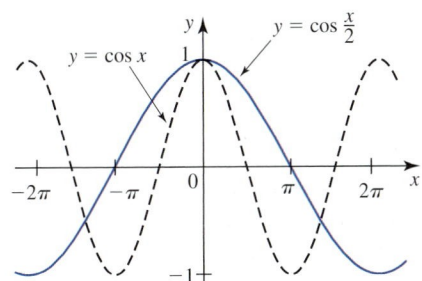

(a) The graph of $y = \cos 2x$ superimposed upon the graph of $y = \cos x$

(b) The graph of $y = \cos \frac{x}{2}$ superimposed upon the graph of $y = \cos x$

FIGURE 15

We now summarize these definitions.

DEFINITION   Period and Amplitude of $A \sin Bx$ and $A \cos Bx$

The graphs of

$$f(x) = A \sin Bx \qquad \text{and} \qquad f(x) = A \cos Bx$$

where $A \neq 0$ and $B \neq 0$, have period $2\pi/|B|$ and **amplitude** $|A|$.

EXAMPLE 5   Sketch the graph of $y = 3 \sin \frac{1}{2} x$.

Solution   The function $y = 3 \sin \frac{1}{2} x$ has the form $y = A \sin Bx$, where $A = 3$ and $B = \frac{1}{2}$. This tells us that the amplitude of the graph is 3 and the period is $2\pi/|\frac{1}{2}| = 4\pi$. Using the graph of the sine curve, we sketch the graph of $y = 3 \sin \frac{1}{2} x$ over one period $[0, 4\pi]$. (See Figure 16.) Next, the periodic properties of the sine function allow us to extend the graph in either direction by completing another cycle as shown.

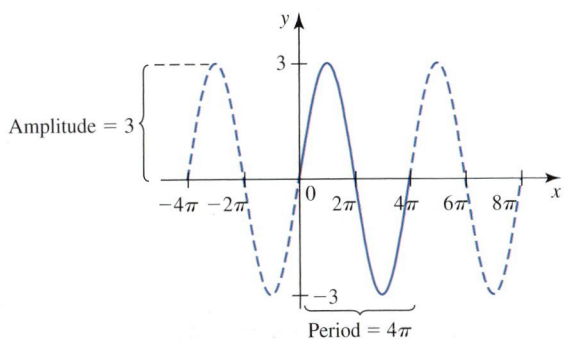

FIGURE 16
The graph of $y = 3 \sin \frac{1}{2} x$ has amplitude 3 and period 4π.

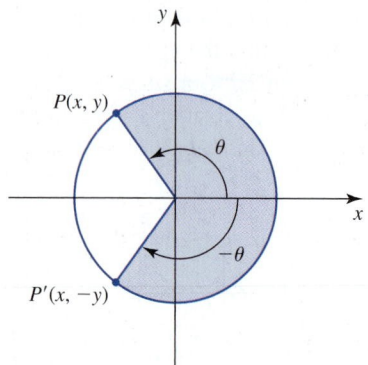

FIGURE 17
The angles θ and $-\theta$ have the same magnitude but opposite signs.

▪ The Trigonometric Identities

By comparing the angles θ and $-\theta$ in Figure 17, we see that the points P and P' have the same x-coordinates and that their y-coordinates differ only in sign. Thus,

$$\cos \theta = x = \cos(-\theta) \tag{8}$$

and

$$\sin \theta = y = -\sin(-\theta) \tag{9}$$

We conclude that the cosine function is even and the sine function is odd. Similarly, we can show that the cosecant, tangent, and cotangent functions are odd, while the secant function is even. These results are also confirmed by the symmetry of the graph of each function (see Figure 13).

Equations such as Equations (8) and (9) that express a relationship between trigonometric functions are called **trigonometric identities.** Each identity holds true for every value of θ in the domain of the specified trigonometric functions.

Referring once again to the point $P(x, y)$ on the unit circle (see Figure 7), we see that the equation $x^2 + y^2 = 1$ can also be written in the form

$$\cos^2 \theta + \sin^2 \theta = 1 \tag{10}$$

Note   Recall that $\sin^2 \theta = (\sin \theta)^2$. In general, $(\sin \theta)^n$ is usually written $\sin^n \theta$. The same convention applies to the other trigonometric functions.   ▪

The **addition** and **subtraction formulas** for the sine and cosine are

$$\sin(A \pm B) = \sin A \cos B \pm \cos A \sin B \tag{11}$$

and

$$\cos(A \pm B) = \cos A \cos B \mp \sin A \sin B \tag{12}$$

If we let $A = B$ in Formulas (11) and (12), we obtain the **double-angle** formulas

$$\sin 2A = 2 \sin A \cos A \tag{13}$$

and

$$\cos 2A = \cos^2 A - \sin^2 A \tag{14a}$$

$$= 2 \cos^2 A - 1 \tag{14b}$$

$$= 1 - 2 \sin^2 A \tag{14c}$$

Solving (14b) and (14c) for $\cos^2 A$ and $\sin^2 A$, respectively, we obtain the **half-angle formulas**

$$\cos^2 A = \frac{1}{2}(1 + \cos 2A) \tag{15}$$

and

$$\sin^2 A = \frac{1}{2}(1 - \cos 2A) \tag{16}$$

These and several other trigonometric identities are summarized in Table 3.

TABLE 3 Trigonometric Identities

Pythagorean identities	Half-angle formulas	Addition and subtraction formulas
$\cos^2 \theta + \sin^2 \theta = 1$	$\cos^2 A = \frac{1}{2}(1 + \cos 2A)$	$\sin(A \pm B) = \sin A \cos B \pm \cos A \sin B$
$\tan^2 \theta + 1 = \sec^2 \theta$	$\sin^2 A = \frac{1}{2}(1 - \cos 2A)$	$\cos(A \pm B) = \cos A \cos B \mp \sin A \sin B$
$\cot^2 \theta + 1 = \csc^2 \theta$		

Double-angle formulas	Cofunctions of complementary angles
$\sin 2A = 2 \sin A \cos A$	$\sin \theta = \cos\left(\frac{\pi}{2} - \theta\right)$
$\cos 2A = \cos^2 A - \sin^2 A = 2 \cos^2 A - 1$	$\cos \theta = \sin\left(\frac{\pi}{2} - \theta\right)$
$\qquad\qquad = 1 - 2 \sin^2 A$	

A more complete list of trigometric identities can be found in the reference pages at the back of the book.

EXAMPLE 6 Find the solutions of the equation $\cos 2x - \cos x = 0$ that lie in the interval $[0, 2\pi]$.

Solution Using the identity (14b), we make the substitution $\cos 2x = 2 \cos^2 x - 1$, obtaining

$$\cos 2x - \cos x = 0$$
$$(2 \cos^2 x - 1) - \cos x = 0$$
$$2 \cos^2 x - \cos x - 1 = 0$$
$$(2 \cos x + 1)(\cos x - 1) = 0$$
$$2 \cos x + 1 = 0 \quad \text{or} \quad \cos x - 1 = 0$$

Thus,

$$\cos x = -\frac{1}{2} \quad \text{or} \quad \cos x = 1$$

and $x = 2\pi/3, 4\pi/3, 0$, and 2π are the solutions in the interval $[0, 2\pi]$.

0.3 EXERCISES

In Exercises 1–8, convert each angle to radian measure.

1. $150°$ **2.** $210°$ **3.** $330°$ **4.** $405°$

5. $-120°$ **6.** $-225°$ **7.** $-75°$ **8.** $-495°$

In Exercises 9–16, convert each angle to degree measure.

9. $\dfrac{\pi}{3}$ **10.** $\dfrac{3\pi}{4}$ **11.** $\dfrac{5\pi}{6}$ **12.** $\dfrac{9\pi}{4}$

13. $-\dfrac{\pi}{2}$ **14.** $-\dfrac{11\pi}{6}$ **15.** $-\dfrac{13\pi}{4}$ **16.** $-\dfrac{11\pi}{3}$

In Exercises 17–24, find the exact value of the trigonometric functions at the indicated angle.

17. $\sin \theta$, $\cos \theta$, and $\tan \theta$ for $\theta = \pi/3$

18. $\sin \theta$, $\cos \theta$, and $\csc \theta$ for $\theta = -\pi/4$

19. $\cos x$, $\tan x$, and $\sec x$ for $x = 2\pi/3$

20. $\sin x$, $\cot x$, and $\csc x$ for $x = 5\pi/6$

21. $\sin \alpha$, $\tan \alpha$, and $\csc \alpha$ for $\alpha = \pi$

22. $\cos \alpha$, $\cot \alpha$, and $\csc \alpha$ for $\alpha = -3\pi/2$

23. $\csc t$, $\sec t$, and $\cot t$ for $t = 17\pi/6$

24. $\sin t$, $\tan t$, and $\cot t$ for $t = -11\pi/3$

25. Given that $\sin \theta = \frac{3}{5}$ and $\frac{\pi}{2} \le \theta \le \pi$, find the five other trigonometric functions of θ.

26. Given that $\cot \theta = -\frac{5}{3}$ and $\frac{\pi}{2} \le \theta \le \pi$, find the five other trigonometric functions of θ.

27. If $f(x) = \sin x$, find $f(0), f(\frac{\pi}{4}), f(-\frac{\pi}{3}), f(3\pi)$, and $f(a + \frac{\pi}{2})$.

28. If

$$f(x) = \begin{cases} 2 + \sqrt{1 - x} & \text{if } x \le 1 \\ 2 \cos 2\pi x & \text{if } x > 1 \end{cases}$$

find $f(0), f(1)$, and $f(2)$.

In Exercises 29 and 30, find the domain of the function.

29. $f(t) = \sqrt{\sin t - 1}$ **30.** $f(x) = \dfrac{x}{2 + \sin x}$

In Exercises 31–32, determine whether the functions are even, odd, or neither.

31. a. $y = 2 \sin x$

   b. $y = -\dfrac{\cos^2 x}{x}$

   c. $y = -\csc x$

32. a. $y = \cot x$

   b. $y = 2 \sin \dfrac{x}{2}$

   c. $y = 2 \sec x$

In Exercises 33–42, verify the identity.

33. $\sec t - \cos t = \tan t \sin t$

34. $2 \csc 2u = \sec u \csc u$

35. $\dfrac{\sin y}{\csc y} + \dfrac{\cos y}{\sec y} = 1$

36. $(\sin x)(\csc x - \sin x) = \cos^2 x$

37. $\tan A + \tan B = \dfrac{\sin(A + B)}{\cos A \cos B}$

38. $\dfrac{\cos \theta \tan \theta + \sin \theta}{\tan \theta} = 2 \cos \theta$

39. $\csc t - \sin t = \cos t \cot t$

40. $\sin 3t = 3 \sin t - 4 \sin^3 t$

41. $\sin 2\theta = 2 \sin^3 \theta \cos \theta + 2 \sin \theta \cos^3 \theta$

42. $\tan \dfrac{t}{2} = \dfrac{1 - \cos t}{\sin t}$

In Exercises 43 and 44, find the domain and sketch the graph of the function. What is its range?

43. $h(\theta) = 2 \sin \pi\theta$ **44.** $f(t) = |\cos t|$

In Exercises 45–58, determine the amplitude and the period for the function. Sketch the graph of the function over one period.

45. $y = \sin(x - \pi)$ **46.** $y = \cos(x + \pi)$

47. $y = \sin(x + \frac{\pi}{2})$ **48.** $y = \cos(x - \frac{\pi}{4})$

49. $y = \cos x + 2$ **50.** $y = 2 - \sin x$

51. $y = 2 \sin(2x + \frac{\pi}{2})$ **52.** $y = \cos(2x + \frac{\pi}{4})$

53. $y = 2 \sin x \cos x$ **54.** $y = \cos^2 x - \sin^2 x$

55. $y = -2 \cos 3x$ **56.** $y = -3 \sin(-4x)$

57. $y = 3 \cos 2x$ **58.** $y = -3 \sin(\pi x + \pi)$

In Exercises 59–66, find the solutions of the equation in $[0, 2\pi)$.

59. $\sin 2x = 1$ **60.** $\tan 2\theta = -1$

61. $\cos t + 2 \sec t = -3$ **62.** $\tan^2 x - \sec x - 1 = 0$

63. $\cos^2 x - \sin x \cos x = 0$ **64.** $\csc^2 x - \cot x - 1 = 0$

65. $2 \cos^2 x - 3 \cos x + 1 = 0$

66. $(\sin 2x)(\sin x) = 0$

67. After takeoff, an airplane climbs at an angle of $20°$ at a speed of 200 ft/sec. How long does it take for the airplane to reach an altitude of 10,000 ft?

68. A man located at a point A on one bank of a river that is 1000 ft wide observed a woman jogging on the opposite bank. When the jogger was first spotted, the angle between the river bank and the man's line of sight was $30°$. One minute later, the angle was $40°$. How fast was the woman running if she maintained a constant speed?

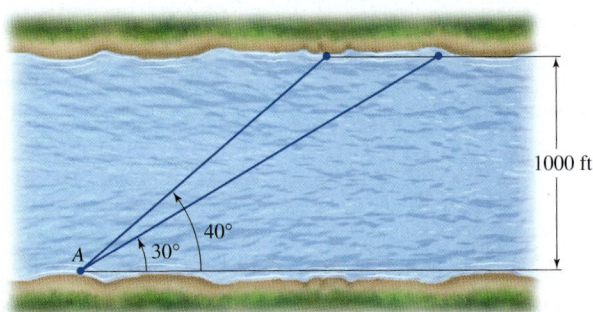

In Exercises 69–74, determine whether the statement is true or false. If it is true, explain why. If it is false, explain why or give an example that shows it is false.

69. The graph of $y = \cos x$ is the same as the graph of $y = \cos(-x)$.

70. The product $y = (\sin x)(\cos x)$ is an odd function of x.

71. The graph of $y = \cos(x + \pi)$ is the same as the graph of $y = -\cos x$.

72. The graph of $y = \cos(x + \frac{\pi}{4})$ is the same as the graph of $y = -\sin(x - \frac{\pi}{4})$.

73. The graph of $y = \csc x$ is symmetric with respect to the y-axis.

74. The function $y = \sin^2 x$ is an odd function.

0.4 Combining Functions

■ Arithmetic Operations on Functions

Many functions are built up from other, and generally simpler, functions. Consider, for example, the function h defined by $h(x) = x + (1/x)$. Note that the value of h at x is the sum of two terms. The first term, x, may be viewed as the value of the function f defined by $f(x) = x$ at x, and the second term, $1/x$, may be viewed as the value of the function g defined by $g(x) = 1/x$ at x. These observations suggest that h can be viewed as the *sum* of the functions f and g, $f + g$, defined by

$$(f+g)(x) = f(x) + g(x) = x + \frac{1}{x}$$

The domain of $f + g$ is $(-\infty, 0) \cup (0, \infty)$, the intersection of the domains of f and g. Note that the plus sign on the left side of this equation denotes an operation (addition in this case) on two *functions*.

Since the value of $h = f + g$ at x is the sum of the values of f and g at x, we see that the graph of h can be obtained from the graphs of f and g by adding the y-coordinates of f and g at x to obtain the corresponding y-coordinate of h at x. This technique is used to sketch the graph of h, the sum of $f(x) = x$ and $g(x) = 1/x$, discussed above (see Figure 1). We show the graph of h only in the first quadrant.

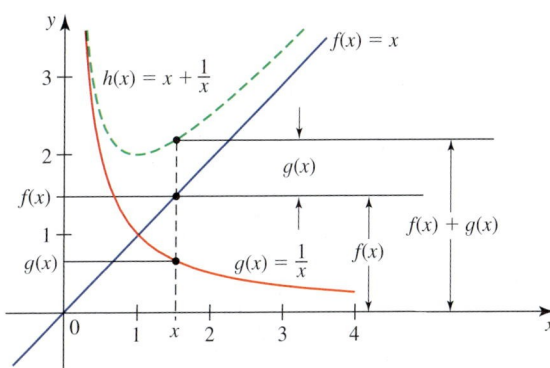

FIGURE 1
The graphs of f, g, and h

The difference, product, and quotient of two functions are defined in a similar manner.

DEFINITION **Operations on Functions**

Let f and g be functions with domains A and B, respectively. Then their sum $f + g$, difference $f - g$, product fg, and quotient f/g are defined as follows:

$$(f + g)(x) = f(x) + g(x) \quad \text{with domain } A \cap B \tag{1a}$$

$$(f - g)(x) = f(x) - g(x) \quad \text{with domain } A \cap B \tag{1b}$$

$$(fg)(x) = f(x)g(x) \quad \text{with domain } A \cap B \tag{1c}$$

$$\left(\frac{f}{g}\right)(x) = \frac{f(x)}{g(x)} \quad \text{with domain } \{x \mid x \in A \cap B \text{ and } g(x) \neq 0\} \tag{1d}$$

EXAMPLE 1 Let f and g be functions defined by $f(x) = \sqrt{x}$ and $g(x) = \sqrt{3 - x}$. Find the domain and the rule for each of the functions $f + g$, $f - g$, fg, and f/g.

Solution The domain of f is $[0, \infty)$, and the domain of g is $(-\infty, 3]$. Therefore, the domain of $f + g$, $f - g$, and fg is

$$[0, \infty) \cap (-\infty, 3] = [0, 3]$$

The rules for these functions are

$$(f + g)(x) = f(x) + g(x) = \sqrt{x} + \sqrt{3 - x} \qquad \text{By Equation (1a)}$$

$$(f - g)(x) = f(x) - g(x) = \sqrt{x} - \sqrt{3 - x} \qquad \text{By Equation (1b)}$$

and

$$(fg)(x) = f(x)g(x) = \sqrt{x}\sqrt{3 - x} = \sqrt{3x - x^2} \qquad \text{By Equation (1c)}$$

For the domain of f/g we must exclude the value of x for which $g(x) = \sqrt{3 - x} = 0$ or $x = 3$. Therefore, f/g is defined by

$$\left(\frac{f}{g}\right)(x) = \frac{f(x)}{g(x)} = \frac{\sqrt{x}}{\sqrt{3 - x}} = \sqrt{\frac{x}{3 - x}} \qquad \text{By Equation (1d)}$$

with domain $[0, 3)$.

Notes

1. To determine the domain of the product or quotient of two functions, begin by examining the domains of the functions to be combined. One common mistake is to try to deduce the domain of the combined function by studying its rule. For example, suppose $f(x) = \sqrt{x}$ and $g(x) = 2\sqrt{x}$. Then, if $h = fg$, we have $h(x) = f(x)g(x) = (\sqrt{x})(2\sqrt{x}) = 2x$. On the basis of the rule for h alone, we might be tempted to conclude that its domain is $(-\infty, \infty)$. But bearing in mind that h is a product of the functions f with domain $[0, \infty)$ and g with domain $[0, \infty)$, we see that the domain of h is $[0, \infty)$.

2. Equations (1a–d) can be extended to the case involving more than two functions. For example, $fg - h$ is just the function with rule

$$(fg - h)(x) = f(x)g(x) - h(x)$$

■ Composition of Functions

There is another way in which certain functions are built up from simpler functions. For example, consider the function $h(x) = \sqrt{2x + 1}$. Let f be the function defined by $f(x) = 2x + 1$, and let g be the function defined by $g(x) = \sqrt{x}$. Then

$$h(x) = \sqrt{2x + 1} = \sqrt{f(x)} = g(f(x))$$

In other words, the value of h at x can be obtained by *evaluating* the function g at $f(x)$. This method of combining two functions is called **composition.** More specifically, we say that the function h is the **composition** of g and f, and we denote it by $g \circ f$ (read "g circle f").

DEFINITION **Composition of Two Functions**

Given two functions g and f, the composition of g and f, denoted by $g \circ f$, is the function defined by

$$(g \circ f)(x) = g(f(x)) \tag{2}$$

The domain of $g \circ f$ is the set of all x in the domain of f for which $f(x)$ is in the domain of g.

Figure 2 shows an interpretation of the composition $g \circ f$, in which the functions f and g are viewed as machines. Notice that the output of f, $f(x)$, must lie in the domain of g for $f(x)$ to be an input for g.

FIGURE 2

The output of f is the input for g (in this order).

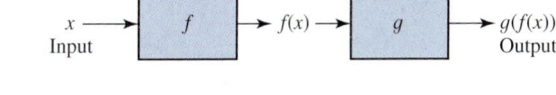

Figure 3 shows how the composition $g \circ f$ can be viewed in terms of transformations or mappings. The point x in the domain of $g \circ f$ is mapped onto the image $f(x)$ that lies in the domain of g. The function g then maps $f(x)$ onto its image $g(f(x))$. Thus, we may view the function $g \circ f$ as a transformation that maps a point x in its domain onto its image $g(f(x))$ in two steps: from x to $f(x)$ via the function f, then from $f(x)$ to $g(f(x))$ via the function g.

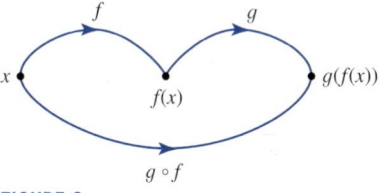

FIGURE 3

$g \circ f$ maps x onto $g(f(x))$ in two steps: via f, then via g.

EXAMPLE 2 Let f and g be functions defined by $f(x) = x + 1$ and $g(x) = \sqrt{x}$. Find the functions $g \circ f$ and $f \circ g$. What is the domain of $g \circ f$?

Solution The rule for $g \circ f$ is found by evaluating g at $f(x)$. Thus,

$$(g \circ f)(x) = g(f(x)) = \sqrt{f(x)} = \sqrt{x + 1}$$

To find the domain of $g \circ f$, recall that $f(x)$ must lie in the domain of g. Since the domain of g consists of all nonnegative numbers and the range of f is the set of all numbers $f(x) = x + 1$, we require that $x + 1 \geq 0$ or $x \geq -1$. Therefore, the domain of $g \circ f$ is $[-1, \infty)$. Note that all x are in the domain of f.

The rule for $f \circ g$ is found by evaluating f at $g(x)$. Thus,

$$(f \circ g)(x) = f(g(x)) = g(x) + 1 = \sqrt{x} + 1$$

We leave it to you to show that the domain of $f \circ g$ is $[0, \infty)$.

Note In general, $g \circ f \neq f \circ g$, as was demonstrated in Example 2. Thus, the order in which functions are composed is important. For example, in the composition $g \circ f$, remember that f is applied first, followed by g. ■

EXAMPLE 3 Let $f(x) = \sin x$ and $g(x) = 1 - 2x$. Find the functions $g \circ f$ and $f \circ g$. What are their domains?

Solution $(g \circ f)(x) = g(f(x)) = 1 - 2f(x) = 1 - 2\sin x$. Since the range of f is $[-1, 1]$ and this interval lies in $(-\infty, \infty)$, the domain of g, we see that the domain of $g \circ f$ is given by the domain of f, namely, $(-\infty, \infty)$. Next,

$$(f \circ g)(x) = f(g(x)) = f(1 - 2x) = \sin(1 - 2x)$$

The range of g is $(-\infty, \infty)$, and this is also the domain of f. So the domain of $f \circ g$ is given by the domain of g, namely, $(-\infty, \infty)$. ■

EXAMPLE 4 Find two functions f and g such that $F = g \circ f$ if $F(x) = (x + 2)^4$.

Solution The expression $(x + 2)^4$ can be evaluated in two steps. First, given any value of x, add 2 to it. Second, raise this result to the fourth power. This suggests that we take

$$f(x) = x + 2 \qquad \text{Remember that } f \text{ is applied first in } g \circ f.$$

and

$$g(x) = x^4$$

Then

$$(g \circ f)(x) = g(f(x)) = [f(x)]^4 = (x + 2)^4 = F(x)$$

so $F = g \circ f$, as required. ■

Note There is always more than one way to write a function as a composition of functions. In Example 4 we could have taken $f(x) = (x + 2)^2$ and $g(x) = x^2$. However, there is usually a "natural" way of decomposing a complicated function. ■

Composite functions play an important role in describing practical situations in which one variable quantity depends on another, which in turn depends on a third, as the following example shows.

EXAMPLE 5 **Oil Spills** In calm waters, the oil spilling from the ruptured hull of a grounded tanker spreads in all directions. Assuming that the area polluted is a circle and that its radius is increasing at the rate of 2 ft/sec, find the area as a function of time.

Solution The circular polluted area is described by the function $g(r) = \pi r^2$, where r is the radius of the circle, measured in feet. Next, the radius of the circle is described by the function $f(t) = 2t$, where t is the time elapsed, measured in seconds. Therefore, the required function A describing the polluted area as a function of time is $A = g \circ f$ defined by

$$A(t) = (g \circ f)(t) = g(f(t)) = \pi[f(t)]^2 = \pi(2t)^2 = 4\pi t^2$$ ■

The composition of functions can be extended to include the composition of three or more functions. For example, the composite function $h \circ g \circ f$ is found by applying f, g, and h in that order. Thus,

$$(h \circ g \circ f)(x) = h(g(f(x)))$$

EXAMPLE 6 Let $f(x) = x - (\pi/2)$, $g(x) = 1 + \cos^2 x$, and $h(x) = \sqrt{x}$. Find $h \circ g \circ f$.

Solution $(h \circ g \circ f)(x) = h(g(f(x))) = \sqrt{g(f(x))}$. But

$$g(f(x)) = 1 + \cos^2[f(x)] = 1 + \cos^2\left(x - \tfrac{\pi}{2}\right)$$

So

$$(h \circ g \circ f)(x) = \sqrt{1 + \cos^2\left(x - \tfrac{\pi}{2}\right)}$$

EXAMPLE 7 Suppose $F(x) = \dfrac{1}{\sqrt{2x + 3} + 1}$. Find functions f, g, and h such that $F = h \circ g \circ f$.

Solution The rule for F says that as a first step, we multiply x by 2 and add 3 to it. This suggests that we take $f(x) = 2x + 3$. Next, we take the square root of this result and add 1 to it. This suggests that we take $g(x) = \sqrt{x} + 1$. Finally, we take the reciprocal of the last result, so let $h(x) = 1/x$. Then

$$F(x) = (h \circ g \circ f)(x) = h(g(f(x)))$$
$$= h(g(2x + 3)) = h(\sqrt{2x + 3} + 1) = \frac{1}{\sqrt{2x + 3} + 1}$$

■ Graphs of Transformed Functions

Sometimes it is possible to obtain the graph of a relatively complicated function by transforming the graph of a simpler but related function. We will describe some of these transformations here.

1. Vertical Translations

The graph of the function g defined by $g(x) = f(x) + c$, where c is a positive constant, is obtained from the graph of f by shifting the latter vertically upward by c units (see Figure 4). This follows by observing that for each x in the domain of g (which is the same as the domain of f) the point $(x, f(x) + c)$ on the graph of g lies precisely c units above the point $(x, f(x))$ on the graph of f. Similarly, the graph of the function g defined by $g(x) = f(x) - c$, where c is a positive constant, is obtained from the graph of f by shifting the latter vertically downward by c units (see Figure 4). These results are also evident if you think of g as the sum of the function f and the constant function $h(x) = c$ and use the graphical interpretation of the sum of two functions described earlier.

2. Horizontal Translations

The graph of the function g defined by $g(x) = f(x + c)$, where c is a positive constant, is obtained from the graph of f by shifting the latter horizontally to the left by c units

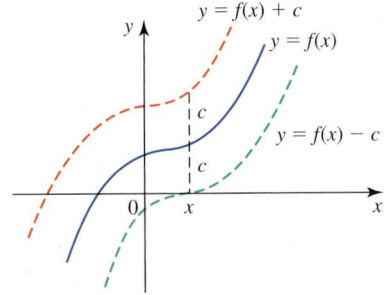

FIGURE 4
The graphs of $y = f(x) + c$ and $y = f(x) - c$, where $c > 0$, are obtained by translating the graph of $y = f(x)$ vertically upward and downward, respectively.

(see Figure 5a). To see this, observe that the number $x + c$ lies c units to the right of x. Therefore, for each x in the domain of g, $(x, f(x + c))$ on the graph of g has precisely the same y-coordinate as the point on the graph of f located c units to the *right* of x (measured horizontally). Similarly, the graph of the function $g(x) = f(x - c)$, where c is a positive constant, is obtained from the graph of $y = f(x)$ by shifting the latter horizontally to the right by c units (see Figure 5b). We summarize these results in Table 1.

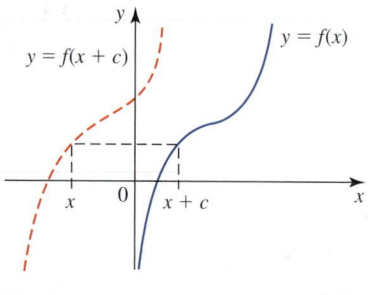
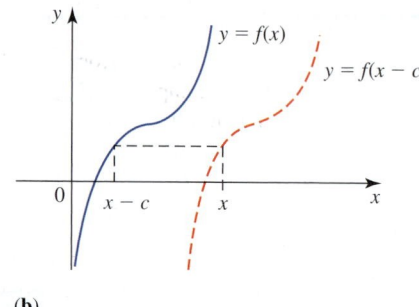

(a)   **(b)**

FIGURE 5

The graphs of $y = f(x + c)$ and $y = f(x - c)$, where $c > 0$, are obtained by shifting the graph of $y = f(x)$ horizontally to the left and right, respectively.

TABLE 1   Vertical and Horizontal Translations

If $c > 0$, then we have the following:	
Function g	**The graph of g is obtained by shifting the graph of f**
$g(x) = f(x) + c$	Upward by a distance of c units
$g(x) = f(x) - c$	Downward by a distance of c units
$g(x) = f(x + c)$	To the left by a distance of c units
$g(x) = f(x - c)$	To the right by a distance of c units

3. Vertical Stretching and Compressing

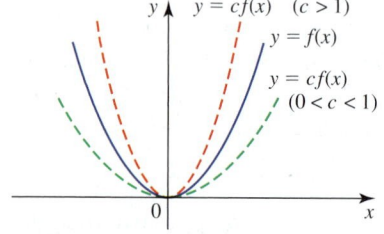

FIGURE 6

The graph of $y = cf(x)$ is obtained from the graph of $y = f(x)$ by stretching it (if $c > 1$) or compressing it (if $0 < c < 1$).

The graph of the function g defined by $g(x) = cf(x)$, where c is a constant with $c > 1$, is obtained from the graph of f by stretching the latter vertically by a factor of c. This can be seen by observing that for each x in the domain of g (and therefore in the domain of f), the point $(x, cf(x))$ on the graph of g has a y-coordinate that is c times as large as the y-coordinate of the point $(x, f(x))$ on the graph of f (see Figure 6). Similarly, if $0 < c < 1$ then the graph of g is obtained from that of f by compressing the latter vertically by a factor of $1/c$ (see Figure 6).

4. Horizontal Stretching and Compressing

The graph of the function g defined by $g = f(cx)$, where c is a constant with $0 < c < 1$, is obtained from the graph of f by stretching the graph of the latter horizontally by a factor of $1/c$ (see Figure 7). To see this, observe that if $x > 0$, the number cx lies to the left of x. Therefore, for each x in the domain of g, the point $(x, g(x)) = (x, f(cx))$ on the graph of g has precisely the same y-coordinate as the point on the graph of f located at the point with x-coordinate cx. (We leave it to you to analyze the case in which $x < 0$.) Similarly if $c > 1$, then the graph of g is obtained from that of f by compressing the latter horizontally by a factor of c. We summarize these results in Table 2.

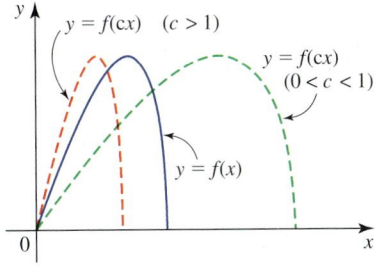

FIGURE 7
The graph of $y = f(cx)$ is obtained from the graph of $y = f(x)$ by compressing it if $c > 1$ and stretching it if $0 < c < 1$.

TABLE 2 Vertical and Horizontal Stretching and Compressing

a. If $c > 1$ then we have the following:

Function g	The graph of g is obtained by
$g(x) = cf(x)$	Stretching the graph of f vertically by a factor of c
$g(x) = f(cx)$	Compressing the graph of f horizontally by a factor of c

b. If $0 < c < 1$, then we have the following:

Function g	The graph of g is obtained by
$g(x) = cf(x)$	Compressing the graph of f vertically by a factor of $1/c$
$g(x) = f(cx)$	Stretching the graph of f horizontally by a factor of $1/c$

5. Reflecting

The graph of the function defined by $g(x) = -f(x)$ is obtained from the graph of f by reflecting the latter with respect to the x-axis (see Figure 8a). This follows from the observation that for each x in the domain of g, the point $(x, -f(x))$ on the graph of g is the mirror reflection of the point $(x, f(x))$ with respect to the x-axis. Similarly, the graph of $g(x) = f(-x)$ is obtained from the graph of f by reflecting the latter with respect to the y-axis (see Figure 8b). These results are summarized in Table 3.

FIGURE 8
The graphs of $y = -f(x)$ and $y = f(-x)$ are obtained from the graph of $y = f(x)$ by reflecting it with respect to the x-axis and with respect to the y-axis, respectively.

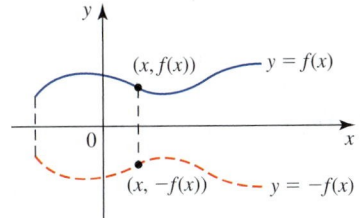

(a) $g(x) = -f(x)$

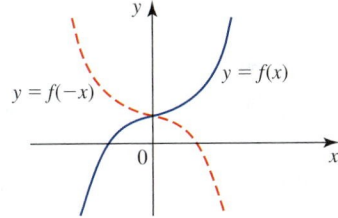

(b) $g(x) = f(-x)$

TABLE 3 Reflecting

Function g	The graph of g is obtained by reflecting the graph of f
$g(x) = -f(x)$	With respect to the x-axis
$g(x) = f(-x)$	With respect to the y-axis

EXAMPLE 8 By translating the graph of $y = x^2$, sketch the graphs of $y = x^2 + 2$, $y = x^2 - 2$, $y = (x + 2)^2$, and $y = (x - 2)^2$.

Solution The graph of $y = x^2$ is shown in Figure 9a. The graph of $y = x^2 + 2$ is obtained from the graph of $y = x^2$ by translating the latter vertically upward by 2 units (see Figure 9b). The graph of $y = x^2 - 2$ is obtained by translating the graph of $y = x^2$ vertically downward by 2 units (see Figure 9c). The graph of $y = (x + 2)^2$ is obtained by translating the graph of $y = x^2$ horizontally to the left by 2 units (see Figure 9d). Finally, the graph of $y = (x - 2)^2$ is obtained by translating the graph of $y = x^2$ to the right by 2 units (see Figure 9e).

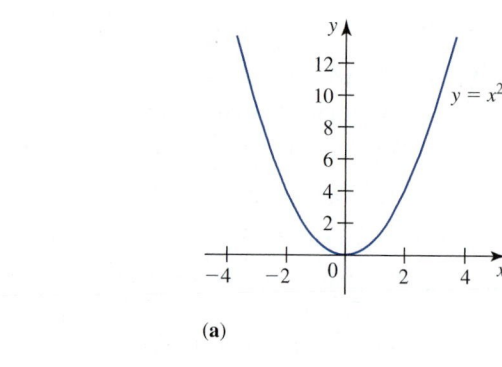

(a)

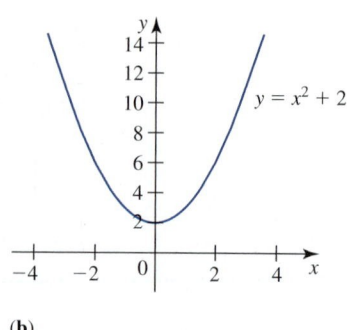

(b)

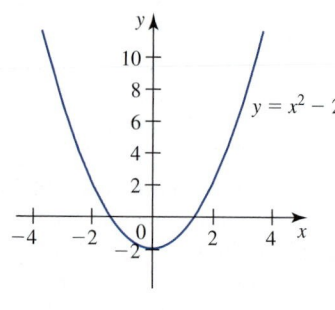

(c)

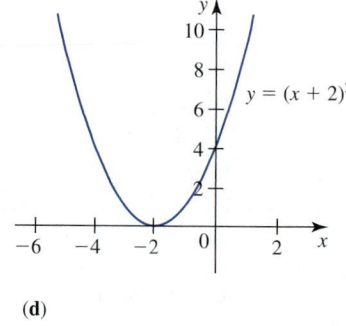

(d)

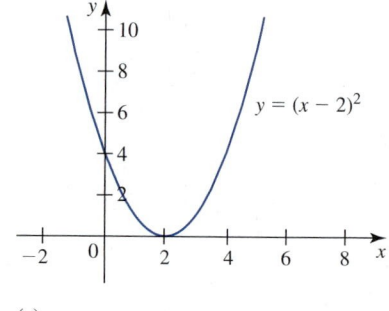

(e)

FIGURE 9

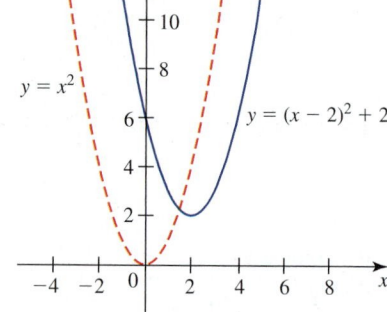

FIGURE 10
The graph of $y = (x - 2)^2 + 2$ can be obtained by shifting the graph of $y = x^2$.

EXAMPLE 9   Sketch the graph of the function f defined by $f(x) = x^2 - 4x + 6$.

Solution   By completing the square, we can rewrite the given equation in the form

$$y = [x^2 - 4x + (-2)^2] + 6 - (-2)^2$$

$$= (x - 2)^2 + 2$$

We see that the required graph can be obtained from the graph of $y = x^2$ by shifting it 2 units to the right and 2 units upward (see Figure 10). Compare this with Example 8.

EXAMPLE 10   By stretching or compressing the graph of $y = \sin x$, sketch the graphs of $y = 2 \sin x$, $y = \frac{1}{2} \sin x$, $y = \sin 2x$, and $y = \sin(x/2)$.

Solution   The graph of $y = \sin x$ is shown in Figure 11a. The graph of $y = 2 \sin x$ is obtained from the graph of $y = \sin x$ by stretching the latter vertically by a factor of 2 (see Figure 11b). The graph of $y = \frac{1}{2} \sin x$ is obtained by compressing the graph of $y = \sin x$ vertically by a factor of 2 (see Figure 11c). The graph of $y = \sin 2x$ is obtained from the graph of $y = \sin x$ by compressing the graph of the latter horizontally by a factor of 2. In fact, the period of $\sin x$ is 2π, whereas the period of $\sin 2x$ is π (see Figure 11d). Finally, the graph of $y = \sin(x/2)$ is obtained from the graph of $y = \sin x$ by stretching the latter horizontally by a factor of 2 (see Figure 11e).

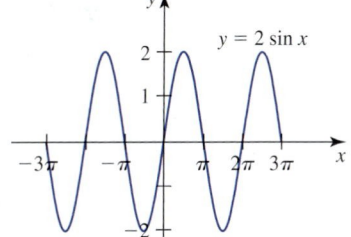

(**a**)

(**b**) Vertical stretching

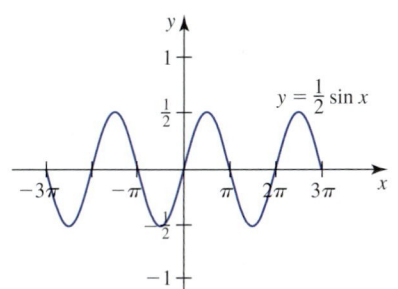

(**c**) Vertical compression

FIGURE 11

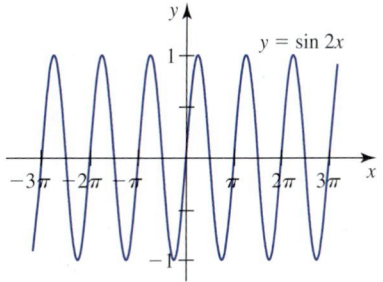

(**d**) Horizontal compression

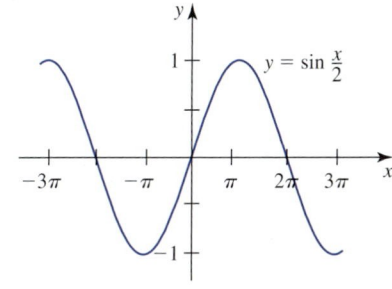

(**e**) Horizontal stretching

EXAMPLE 11 By reflecting the graph of $y = \sqrt{x}$, sketch the graphs of $y = -\sqrt{x}$ and $y = \sqrt{-x}$.

Solution The graph of $y = \sqrt{x}$ is shown in Figure 12a. To obtain the graph of $y = -\sqrt{x}$, we reflect the graph of $y = \sqrt{x}$ with respect to the x-axis (see Figure 12b). To obtain the graph of $y = \sqrt{-x}$, we reflect the graph of $y = \sqrt{x}$ with respect to the y-axis (see Figure 12c).

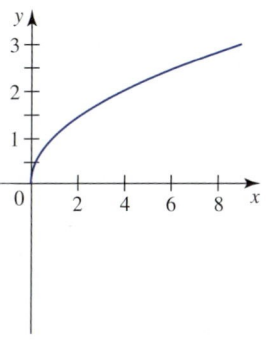

(**a**) The graph of $y = \sqrt{x}$

FIGURE 12

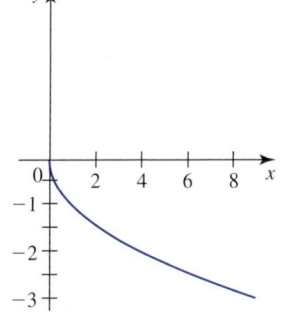

(**b**) The graph of $y = -\sqrt{x}$

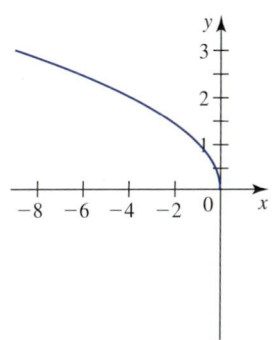

(**c**) The graph of $y = \sqrt{-x}$

The next example involves the use of another transformation of interest.

EXAMPLE 12

a. Explain how you can obtain the graph of $y = |f(x)|$ given the graph of $y = f(x)$.
b. Use the method you devised in part (a) to sketch the graph of $y = ||x| - 1|$.

Solution

a. By the definition of the absolute value, we have

$$|f(x)| = \begin{cases} f(x) & \text{if } f(x) \geq 0 \\ -f(x) & \text{if } f(x) < 0 \end{cases}$$

So to obtain the graph of $y = |f(x)|$ from that of $y = f(x)$ (Figure 13a), we retain the portion of the graph of $y = f(x)$ that lies above the axis and reflect the portion of the graph of $y = f(x)$ that lies below the x-axis with respect to the x-axis (see Figure 13b).

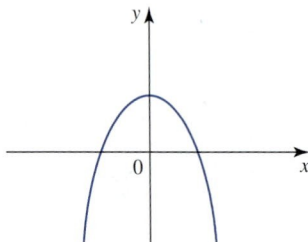

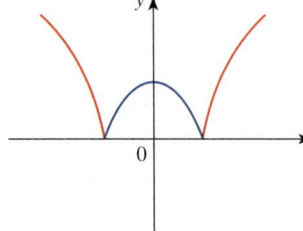

FIGURE 13   **(a)** $y = f(x)$ **(b)** $y = |f(x)|$

b. We begin by sketching the graph of $y = |x|$ as shown in Figure 14a. Next, we sketch the graph of $y = |x| - 1$ by translating the graph of $y = |x|$ vertically downward by 1 unit (see Figure 14b). Finally, using the method of part (a), we obtain the desired graph (see Figure 14c).

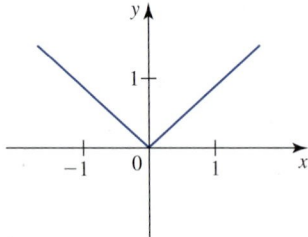

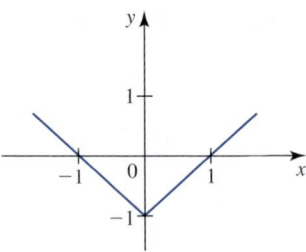

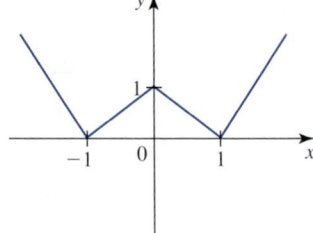

(a) $y = |x|$ **(b)** $y = |x| - 1$ **(c)** $y = ||x| - 1|$

FIGURE 14

0.4 EXERCISES

In Exercises 1–4, find (a) $f + g$, (b) $f - g$, (c) fg, *and* (d) f/g.
What is the domain of the function?

1. $f(x) = 3x$, $g(x) = x^2 - 1$

2. $f(x) = x^2 + 1$, $g(x) = 1 + \sqrt{x}$

3. $f(x) = \sqrt{x + 1}$, $g(x) = \sqrt{x - 1}$

4. $f(x) = \dfrac{1}{x + 1}$, $g(x) = \dfrac{x}{x - 1}$

In Exercises 5–8, find f ∘ g and g ∘ f, and give their domains.

5. $f(x) = x^2, \quad g(x) = 2x + 3$

6. $f(x) = \sqrt{x}, \quad g(x) = 1 - x^2$

7. $f(x) = \dfrac{1}{x}, \quad g(x) = \dfrac{x + 1}{x - 1}$

8. $f(x) = \sqrt{x + 1}, \quad g(x) = \dfrac{1}{x - 1}$

In Exercises 9–10, evaluate h(2), where h = g ∘ f.

9. $f(x) = \sqrt[3]{x^2 - 1}, \quad g(x) = 3x^3 + 1$

10. $f(x) = \dfrac{\pi x}{4}, \quad g(x) = 2 \sin x + 3 \cos x$

11. Let

$$f(x) = \begin{cases} x + 1 & \text{if } x < 0 \\ x - 1 & \text{if } x \geq 0 \end{cases}$$

and let $g(x) = x^2$. Find
a. $g \circ f$, and sketch its graph.
b. $f \circ g$, and sketch its graph.

12. Suppose the function f is defined on the interval $[0, 1]$. Find the domain of h if (a) $h(x) = f(2x + 3)$ and (b) $h(x) = f(2x^2)$.

13. Let $f(x) = x + 2$ and $g(x) = 2x^2 + \sqrt{x}$. Find
a. $(g \circ f)(0)$ **b.** $(g \circ f)(2)$
c. $(f \circ g)(4)$ **d.** $(g \circ g)(1)$

14. Let $f(x) = \dfrac{\pi}{2} - x$ and $g(x) = \dfrac{2 \sin x}{1 + \cos x}$. Find
a. $g(f(0))$ **b.** $(g \circ f)\left(\dfrac{\pi}{2}\right)$
c. $f\left(g\left(\dfrac{\pi}{2}\right)\right)$ **d.** $(f \circ f)\left(\dfrac{\pi}{2}\right)$

In Exercises 15–16, find f ∘ g ∘ h.

15. $f(x) = \sqrt{x}, \quad g(x) = 2x + 1, \quad h(x) = x^2 - 1$

16. $f(x) = \dfrac{1}{x}, \quad g(x) = a - bx, \quad h(x) = \cos x$

In Exercises 17–22, find functions f and g such that h = g ∘ f. (Note: The answer is not unique.)

17. $h(x) = (3x^2 + 4)^{3/2}$

18. $h(x) = |x^2 - 2x + 3|$

19. $h(x) = \dfrac{1}{\sqrt{x^2 - 4}}$

20. $h(x) = \sqrt{2x + 1} + \dfrac{1}{\sqrt{2x + 1}}$

21. $h(t) = \sin(t^2)$

22. $h(t) = \dfrac{\tan t}{1 + \cot t}$

In Exercises 23–24, find functions f, g, and h such that F = f ∘ g ∘ h. (Note: The answer is not unique.)

23. a. $F(x) = \sqrt{1 - \sqrt{x}}$ **b.** $F(x) = \sin^3(2x + 3)$

24. a. $F(x) = \dfrac{1}{(2x^2 + x + 3)^3}$
     b. $F(x) = \dfrac{\sqrt{x + 1} - 1}{\sqrt{x + 1} + 1}$

25. Use the following table to evaluate each composite function.
a. $(f \circ g)(1)$ **b.** $(g \circ f)(2)$
c. $f(g(2))$ **d.** $g(f(0))$
e. $f(f(2))$ **f.** $g(g(1))$

x	0	1	2	3	4	5
$f(x)$	1	$\sqrt{2}$	2	4	3	1
$g(x)$	2	3	5	6	7	9

26. Use the graphs of f and g to estimate the values of $(g \circ f)(x)$ for $x = -2, -1, 0, 1, 2,$ and 3. Then use these values to make a rough sketch of the graph of $g \circ f$.

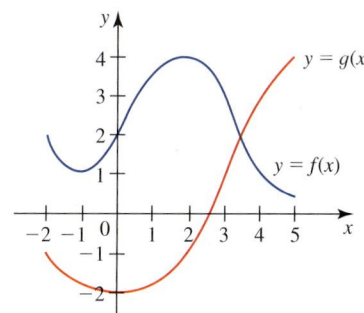

In Exercises 27–30 the graph of f is given. Match the other graphs with the given function(s).

27. $y = f(x) + 1, \quad y = f(x) - 1$

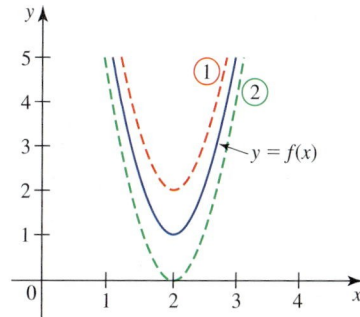

28. $y = f(x + 2), \quad y = f(x - 2)$

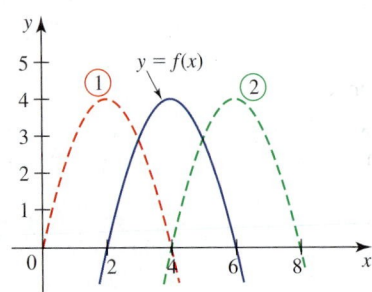

29. $y = f(2x), \quad y = f\left(\dfrac{x}{2}\right)$

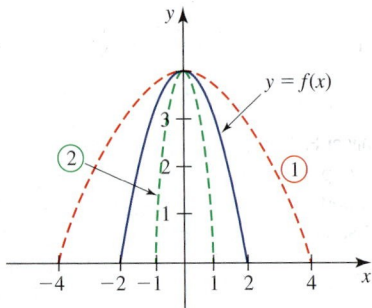

30. $y = f(-x), \quad y = -f(x), \quad y = 2f(x), \quad y = \dfrac{1}{2}f(-x)$

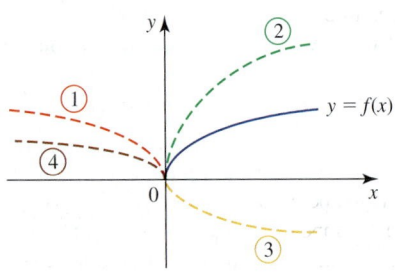

In Exercises 31–40, the graph of the function f is to be transformed as described. Find the function for the transformed graph.

31. $f(x) = x^3 + x - 1$; shifted vertically upward by 3 units

32. $f(x) = x + \sqrt{x + 1}$; shifted vertically downward by 2 units

33. $f(x) = x + \dfrac{1}{\sqrt{x}}$; shifted horizontally to the left by 3 units

34. $f(x) = \dfrac{\sin x}{1 + \cos x}$; shifted horizontally to the right by 4 units

35. $f(x) = \dfrac{\sqrt{x}}{x^2 + 1}$; stretched vertically by a factor of 3

36. $f(x) = \sqrt{x^2 + 4}$; compressed vertically by a factor of 2

37. $f(x) = x \sin x$; stretched horizontally by a factor of 2

38. $f(x) = 5 \sin 4x$; compressed horizontally by a factor of 3

39. $f(x) = \sqrt{4 - x^2}$; shifted horizontally to the right by 2 units, compressed horizontally by a factor of 2, and shifted vertically upward by 1 unit

40. $f(x) = \sqrt{x} + 1$; shifted horizontally to the left by 1 unit, compressed horizontally by a factor of 3, stretched vertically by a factor of 3, and shifted vertically downward by 2 units

41. The graph of the function f follows.

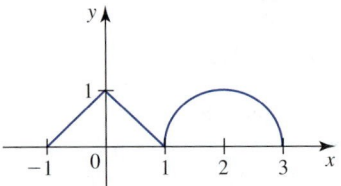

Use it to sketch the following graphs.

a. $y = f(x) + 1$ **b.** $y = f(x + 2)$
c. $y = 2f(x)$ **d.** $y = f(2x)$
e. $y = -f(x)$ **f.** $y = f(-x)$
g. $y = 2f(x - 1) + 2$ **h.** $y = -2f(x + 1) + 3$

42. The graph of the function f follows.

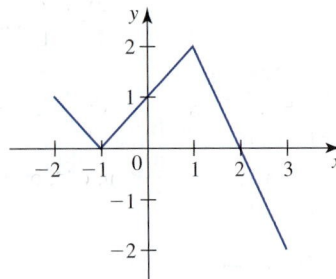

Use it to sketch the following graphs.

a. $y = f(x - 1)$ **b.** $y = f\left(\dfrac{x}{2}\right)$

c. $y = |f(x)|$ **d.** $y = \dfrac{|f(x)|}{f(x)}$

e. $y = f(-x)$ **f.** $y = \dfrac{(|f(x)| + f(x))}{2}$

g. $y = -2f(-x) + 1$

In Exercises 43–54, sketch the graph of the first function by plotting points if necessary. Then use transformation(s) to obtain the graph of the second function.

43. $y = x^2, \quad y = x^2 - 2$

44. $y = x^2, \quad y = (x - 2)^2$

45. $y = \dfrac{1}{x}$, $y = \dfrac{1}{x-1}$

46. $y = \sqrt{x}$, $y = 2\sqrt{x-1} + 1$

47. $y = |x|$, $y = 2|x+1| - 1$

48. $y = |x|$, $y = |2x-1| + 1$

49. $y = x^2$, $y = 2x^2 - 4x + 1$

50. $y = x^2$, $y = |x^2 - 1|$

51. $y = \sin x$, $y = 2\sin\dfrac{x}{2}$

52. $y = \cos x$, $y = \dfrac{1}{2}\cos\left(x - \dfrac{\pi}{4}\right)$

53. $y = x^2$, $y = |x^2 - 2x - 1|$

54. $y = \tan x$, $y = \tan\left(x + \dfrac{\pi}{3}\right)$

55. a. Describe how you would construct the graph of $f(|x|)$ from the graph of $y = f(x)$.
   b. Use the result of part (a) to sketch the graph of $y = \sin|x|$.

56. Find $f(x)$ if $f(x + 1) = 2x^2 + 7x + 4$.

57. a. If $f(x) = x - 1$ and $h(x) = 2x + 3$, find a function g such that $h = g \circ f$.
   b. If $g(x) = 3x + 4$ and $h(x) = 4x - 8$, find a function f such that $h = g \circ f$.

58. Let $g(x) = \dfrac{x+1}{2x-1}$, and let $h(x) = \dfrac{2x+2}{4x+1}$. Find a function f such that $h = g \circ f$.

59. Let $f(x) = 2x^2 + x$, and let $h(x) = 6x^2 + 3x - 1$. Find a function g such that $h = g \circ f$.

60. Determine whether $h = g \circ f$ is even, odd, or neither, given that
   a. both g and f are even.
   b. g is even and f is odd.
   c. g is odd and f is even.
   d. both g and f are odd.

61. Let f be a function defined by $f(x) = \sqrt{x} + \sin x$ on the interval $[0, 2\pi]$.
   a. Find an even function g defined on the interval $[-2\pi, 2\pi]$ such that $g(x) = f(x)$ for all x in $[0, 2\pi]$.
   b. Find an odd function h defined on the interval $[-2\pi, 2\pi]$ such that $h(x) = f(x)$ for all x in $[0, 2\pi]$.

62. a. Show that if a function f is defined at $-x$ whenever it is defined at x, then the function g defined by $g(x) = f(x) + f(-x)$ is an even function and the function h defined by $h(x) = f(x) - f(-x)$ is an odd function.
   b. Use the result of part (a) to show that any function f defined on an interval $(-a, a)$ can be written as a sum of an even function and an odd function.

c. Rewrite the function

$$f(x) = \frac{x+1}{x-1} \qquad -1 < x < 1$$

as a sum of an even function and an odd function.

63. Spam Messages The total number of email messages per day (in billions) between 2003 and 2007 is approximated by

$$f(t) = 1.54t^2 + 7.1t + 31.4 \qquad 0 \le t \le 4$$

where t is measured in years, with $t = 0$ corresponding to 2003. Over the same period the total number of spam messages per day (in billions) is approximated by

$$g(t) = 1.21t^2 + 6t + 14.5 \qquad 0 \le t \le 4$$

   a. Find the rule for the function $D = f - g$. Compute $D(4)$, and explain what it measures.
   b. Find the rule for the function $P = g/f$. Compute $P(4)$, and explain what it means.

Source: Technology Review.

64. Global Supply of Plutonium The global stockpile of plutonium for military applications between 1990 ($t = 0$) and 2003 ($t = 13$) stood at a constant 267 tons. On the other hand, the global stockpile of plutonium for civilian use was $2t^2 + 46t + 733$ tons in year t over the same period.
   a. Find the function f giving the global stockpile of plutonium for military use from 1990 through 2003 and the function g giving the global stockpile of plutonium for civilian use over the same period.
   b. Find the function h giving the total global stockpile of plutonium between 1990 and 2003.
   c. What was the total global stockpile of plutonium in 2003?

Source: Institute for Science and International Security.

65. Motorcycle Deaths Suppose that the fatality rate (deaths per 100 million miles traveled) of motorcyclists is given by $g(x)$, where x is the percentage of motorcyclists who wear helmets. Next, suppose that the percentage of motorcyclists who wear helmets at time t (t measured in years) is $f(t)$, where $t = 0$ corresponds to the year 2000.
   a. If $f(0) = 0.64$ and $g(0.64) = 26$, find $(g \circ f)(0)$, and interpret your result.
   b. If $f(6) = 0.51$ and $g(0.51) = 42$, find $(g \circ f)(6)$, and interpret your result.
   c. Comment on the results of parts (a) and (b).

Source: NHTSA.

66. Fighting Crime Suppose that the reported serious crimes (crimes that include homicide, rape, robbery, aggravated assault, burglary, and car theft) that end in arrests or in the identification of suspects is $g(x)$ percent, where x denotes the total number of detectives. Next, suppose that the total

number of detectives in year t is $f(t)$, where $t = 0$ corresponds to 2001.
a. If $f(1) = 406$ and $g(406) = 23$, find $(g \circ f)(1)$, and interpret your result.
b. If $f(6) = 326$ and $g(326) = 18$, find $(g \circ f)(6)$, and interpret your result.
c. Comment on the results of parts (a) and (b).
Source: Boston Police Department.

67. **Overcrowding of Prisons** The 1980s saw a trend toward old-fashioned punitive deterrence of crime in contrast to the more liberal penal policies and community-based corrections that were popular in the 1960s and early 1970s. As a result, prisons became more crowded, and the gap between the number of people in prison and the prison capacity widened. The number of prisoners (in thousands) in federal and state prisons is approximated by the function

$$N(t) = 3.5t^2 + 26.7t + 436.2 \qquad 0 \le t \le 10$$

where t is measured in years, with $t = 0$ corresponding to 1983. The number of inmates for which prisons were designed is given by

$$C(t) = 24.3t + 365 \qquad 0 \le t \le 10$$

where $C(t)$ is measured in thousands and t has the same meaning as before.
a. Find an expression that shows the gap between the number of prisoners and the number of inmates for which the prisons were designed at any time t.
b. Find the gap at the beginning of 1983 and at the beginning of 1986.
Source: U.S. Department of Justice.

68. **Hotel Occupancy Rate** The occupancy rate of the all-suite Wonderland Hotel, located near an amusement park, is given by the function

$$r(t) = \frac{10}{81}t^3 - \frac{10}{3}t^2 + \frac{200}{9}t + 55 \qquad 0 \le t \le 11$$

where t is measured in months and $t = 0$ corresponds to the beginning of January. Management has estimated that the monthly revenue (in thousands of dollars) is approximated by the function

$$R(r) = -\frac{3}{5000}r^3 + \frac{9}{50}r^2 \qquad 0 \le r \le 100$$

where r (percent) is the occupancy rate.
a. What is the hotel's occupancy rate at the beginning of January? At the beginning of July?
b. What is the hotel's monthly revenue at the beginning of January? At the beginning of July?

In Exercises 69–74, determine whether the statement is true or false. If it is true, explain why. If it is false, explain why or give an example that shows it is false.

69. If f and g are both linear functions of x, then so are $f \circ g$ and $g \circ f$.

70. If f is a polynomial function of x and g is a rational function, then $g \circ f$ and $f \circ g$ are rational functions.

71. If f and g are both even (odd), then $f + g$ is even (odd).

72. If f is even and g is odd, then $f + g$ is neither even nor odd.

73. If f and g are both even, then fg is even.

74. If f and g are both odd, then fg is odd.

0.5 Graphing Calculators and Computers

The graphing calculator and the computer are indispensable tools in helping us to solve complex mathematical problems. In this book we will use them to help us explore ideas and concepts in calculus both graphically and numerically. But the amount and accuracy of the information obtained by using a graphing utility depend on the experience and sophistication of the user. As you progress through this text, you will see that the more knowledge of calculus you gain, the more effective the graphing utility will prove to be as a tool for problem solving. But there are pitfalls in using the graphing utility, and we will point them out when the opportunity arises.

In this section we will look at some basic capabilities of the graphing calculator and the computer that we will use later.

■ Finding a Suitable Viewing Window

The first step in plotting the graph of a function with a graphing utility is to select a suitable viewing window $[a, b] \times [c, d]$ that displays the portion of the graph of the function in the rectangular set $\{(x, y) \mid a \le x \le b, c \le y \le d\}$. For example, you might first plot the graph using the *standard viewing window* $[-10, 10] \times [-10, 10]$. If necessary, you then might adjust the viewing window by enlarging it, reducing it, or even changing it altogether to obtain a sufficiently complete view of the graph or at least the portion of the graph that is of interest.

EXAMPLE 1 Plot the graph of $f(x) = 2x^2 - 4x - 5$ in the standard viewing window $[-10, 10] \times [-10, 10]$.

Solution   The graph of f, shown in Figure 1a, is a parabola. Figure 1b shows a typical window screen, and Figure 1c shows a typical equation screen.

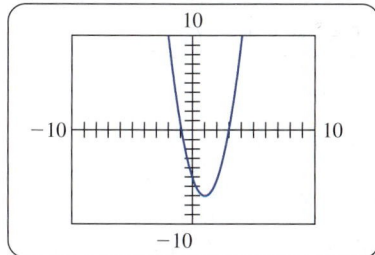

(**a**) The graph of $f(x) = 2x^2 - 4x - 5$ in $[-10, 10] \times [-10, 10]$

WINDOW
$X_{min} = -10$
$X_{max} = 10$
$X_{scl} = 1$
$Y_{min} = -10$
$Y_{max} = 10$
$Y_{scl} = 1$
$X_{res} = 1$ ■

(**b**) A window screen on a graphing calculator

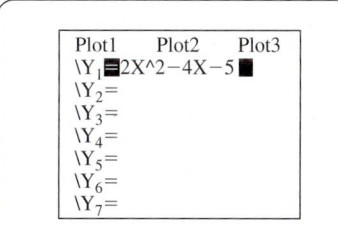

(**c**) An equation screen on a graphing calculator

FIGURE 1

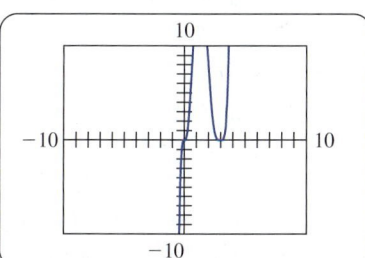

FIGURE 2

An incomplete sketch of $f(x) = x^3(x - 3)^4$ on $[-10, 10] \times [-10, 10]$

EXAMPLE 2 Let $f(x) = x^3(x - 3)^4$.

a. Plot the graph of f in the standard viewing window.
b. Plot the graph of f in the window $[-1, 5] \times [-40, 40]$.

Solution

a. The graph of f in the standard viewing window is shown in Figure 2. Since the graph does not appear to be complete, we need to adjust the viewing window.

b. The graph of f in the window $[-1, 5] \times [-40, 40]$, shown in Figure 3, is an improvement over the previous graph. (Later, we will be able to show that the figure does in fact give a rather complete view of the graph of f.)

Evaluating a Function

A graphing utility can be used to find the value of a function with minimal effort, as the following example shows.

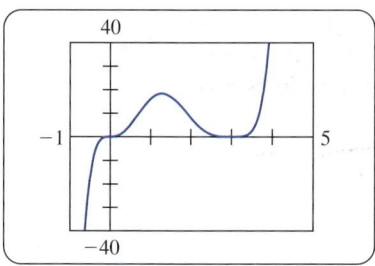

FIGURE 3

A more complete sketch of $f(x) = x^3(x - 3)^4$ is shown by using the window $[-1, 5] \times [-40, 40]$.

EXAMPLE 3 Let $f(x) = x^3 - 4x^2 + 4x + 2$.

a. Plot the graph of f in the standard viewing window.
b. Find $f(3)$ using a calculator, and verify your result by direct computation.
c. Find $f(4.215)$.

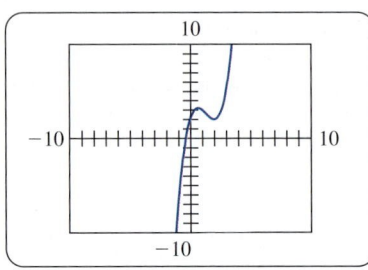

FIGURE 4
The graph of
$f(x) = x^3 - 4x^2 + 4x + 2$ in
the standard viewing window

Solution

a. The graph of f is shown in Figure 4.

b. Using the evaluation function of the graphing utility and the value 3 for x, we find $y = 5$. This result is verified by computing

$$f(3) = 3^3 - 4(3)^2 + 4(3) + 2 = 27 - 36 + 12 + 2 = 5$$

c. Using the evaluation function of the graphing utility and the value 4.215 for x, we find $y = 22.679738375$. Thus, $f(4.215) = 22.679738375$. The efficacy of the graphing utility is clearly demonstrated here! ∎

EXAMPLE 4   **Number of Alzheimer's Patients**   The number of patients with Alzheimer's disease in the United States is approximated by

$$f(t) = -0.0277t^4 + 0.3346t^3 - 1.1261t^2 + 1.7575t + 3.7745 \qquad 0 \le t \le 6$$

where $f(t)$ is measured in millions and t is measured in decades, with $t = 0$ corresponding to the beginning of 1990.

a. Use a graphing utility to plot the graph of f in the viewing window $[0, 6] \times [0, 12]$.

b. What is the anticipated number of Alzheimer's patients in the United States at the beginning of 2010 ($t = 2$)? At the beginning of 2030 ($t = 4$)?

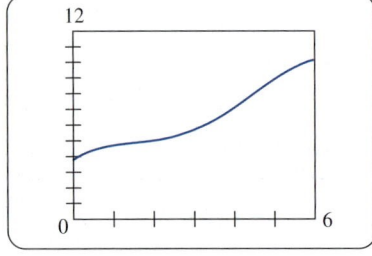

FIGURE 5
The graph of f in the viewing window $[0, 6] \times [0, 12]$

Solution

a. The graph of f is shown in Figure 5.

b. Using the evaluation function of the graphing utility and the value 2 for x, we see that the anticipated number of Alzheimer's patients at the beginning of 2010 is given by $f(2) = 5.0187$, or approximately 5 million. The anticipated number of Alzheimer's patients at the beginning of 2030 is given by $f(4) = 7.1101$, or approximately 7.1 million. ∎

■ Finding the Zeros of a Function

There will be many occasions when we need to find the zeros of a function. This task is greatly simplified if we use a graphing calculator or a computer algebra system (CAS).

EXAMPLE 5   Let $f(x) = x^3 + x + 1$. Find the zero of f using (a) a graphing calculator and (b) a CAS.

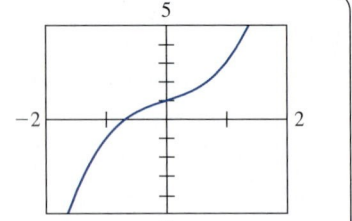

FIGURE 6
The graph of f intersects the x-axis at $x \approx -0.6823278$.

Solution

a. The graph of f in the window $[-2, 2] \times [-5, 5]$ is shown in Figure 6. Using TRACE and ZOOM or the function for finding the zero of a function, we find the zero to be approximately -0.6823278.

b. In Maple we use the command

$$\text{solve(x\^3+x+1=0,x);}$$

and in Mathematica we use the command

$$\text{Solve[x\^3+x+1==0,x]}$$

to obtain the solution $x \approx -0.682328$. ∎

■ Finding the Point(s) of Intersection of Two Graphs

A graphing calculator or a CAS can be used to find the point(s) of intersection of the graphs of two functions. Although the points of intersection of the graphs of the functions f and g can be found by finding the zeros of the function $f - g$, it is often more illuminating to proceed as in Example 6.

EXAMPLE 6 Find the points of intersection of the graphs of $f(x) = 0.3x^2 - 1.4x - 3$ and $g(x) = -0.4x^2 + 0.8x + 6.4$.

Solution The graphs of both f and g in the standard viewing window are shown in Figure 7a. Using **TRACE** and **ZOOM** or the function for finding the points of intersection of two graphs on your graphing utility, we find the point(s) of intersection, accurate to four decimal places, to be $(-2.4158, 2.1329)$ (Figure 7b) and $(5.5587, -1.5125)$ (Figure 7c).

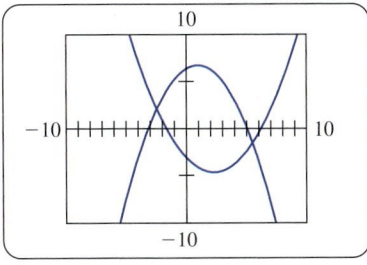

(**a**) The graphs of f and g in the standard viewing window

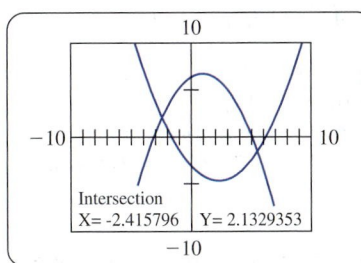

(**b**) An intersection screen

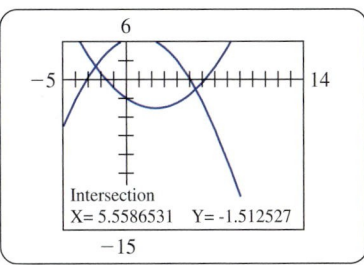

(**c**) An intersection screen

FIGURE 7

■ Constructing Functions from a Set of Data

A graphing calculator or a CAS can often be used to find the function that fits a given set of data points "best" in some sense. For example, if the points corresponding to the given data are scattered about a straight line, then we use linear regression to obtain a function that approximates the data at hand. If the points seem to be scattered about a parabola (the graph of a quadratic function), then we use second-degree polynomial regression, and so on.

We will exploit these capabilities of graphing calculators and computer algebra systems in Section 0.6, where we will see how "mathematical models" are constructed from raw data. The solution to the following example is obtained by using linear regression. (Consult the manual that accompanies your calculator for instructions for using linear regression. If you are using a CAS, consult your HELP menu for instructions.)

EXAMPLE 7

a. Use a graphing calculator or computer algebra system to find a linear function whose graph fits the following data "best" in the sense of least squares:

x	1	2	3	4	5
y	3	5	5	7	8

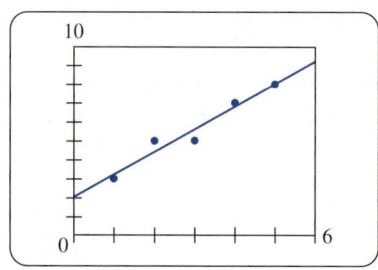

FIGURE 8
The scatter diagram and least-squares
line for the data set.

b. Plot the data points (x, y) for the values of x and y given in the table (the graph is called a *scatter diagram*) and the graph of the least-squares line (called the *regression line*) on the same set of axes.

Solution

a. We first enter the data and then use the linear regression function on the calculator or computer to obtain the graph shown in Figure 8. We also find that the equation of the least-squares regression line is $y = 1.2x + 2$.

b. See Figure 8.

0.5 EXERCISES

 In Exercises 1–4, *plot the graph of the function f in (a) the standard viewing window and (b) the indicated window.*

1. $f(x) = x^3 - 20x^2 + 8x - 10;$ $[-20, 20] \times [-1200, 100]$

2. $f(x) = x^4 - 2x^2 + 8;$ $[-2, 2] \times [6, 10]$

3. $f(x) = x\sqrt{4 - x^2};$ $[-3, 3] \times [-2, 2]$

4. $f(x) = \dfrac{4}{x^2 - 8};$ $[-5, 5] \times [-5, 5]$

In Exercises 5–16, *plot the graph of the function f in an appropriate viewing window. (Note: The answer is not unique.)*

5. $f(x) = 2x^4 - 3x^3 + 5x^2 - 20x + 40$

6. $f(x) = -2x^4 + 5x^2 - 4$ **7.** $f(x) = \dfrac{x^3}{x^3 + 1}$

8. $f(x) = \dfrac{2x^4 - 3x}{x^2 - 1}$ **9.** $f(x) = \sqrt[3]{x} - \sqrt[3]{x + 1}$

10. $f(x) = \dfrac{5x}{x - 1} + 5x$ **11.** $f(x) = x^2 \sin \dfrac{1}{x}$

   Hint: Stay close to the origin.

12. $f(x) = \dfrac{1}{2 + \cos x}$ **13.** $f(x) = \dfrac{\sin \sqrt{x}}{\sqrt{x}}$

14. $f(x) = \dfrac{1}{2} \sin 2x + \cos x$ **15.** $f(x) = x + 0.01 \sin 50x$

16. $f(x) = x^2 - 0.1x$

In Exercises 17–22, *find the zero(s) of the function f to five decimal places.*

17. $f(x) = 2x^3 - 3x + 2$ **18.** $f(x) = x^3 - 9x + 4$

19. $f(x) = x^4 - 2x^3 + 3x - 1$ **20.** $f(x) = 2x^4 - 4x^2 + 1$

21. $f(x) = \sin 2x - x^2 + 1$ **22.** $f(x) = x^2 - 2x - 2 \sin x + 1$

In Exercises 23–28, *find the point(s) of intersection of the graphs of the functions. Express your answers accurate to five decimal places.*

23. $f(x) = 0.3x^2 - 1.7x - 3.2;$ $g(x) = -0.4x^2 + 0.9x + 6.7$

24. $f(x) = -0.3x^2 + 0.6x + 3.2;$ $g(x) = 0.2x^2 - 1.2x - 4.8$

25. $f(x) = 0.3x^3 - 1.8x^2 + 2.1x - 2;$ $g(x) = 2.1x - 4.2$

26. $f(x) = -0.2x^3 + 1.2x^2 - 1.2x + 2;$
   $g(x) = -0.2x^2 + 0.8x + 2.1$

27. $f(x) = 2 \sin x;$ $g(x) = 2 - \dfrac{1}{2} x^2$

28. $f(x) = \sin^2 x;$ $g(x) = \sqrt{x^2 - x^4}$

29. Let $f(x) = x + \dfrac{1}{100} \sin 100x.$

   a. Plot the graph of f using the viewing window $[-10, 10] \times [-10, 10]$.

   b. Plot the graph of f using the viewing window $[-0.1, 0.1] \times [-0.1, 0.1]$.

   c. Explain why the two displays obtained in parts (a) and (b) taken together give a complete description of the graph of f.

30. a. Plot the graph of $f(x) = \cos(\sin x)$. Is f odd or even?

   b. Verify your answer to part (a) analytically.

31. a. Plot the graph of $f(x) = x/x$ and $g(x) = 1$.

   b. Are the functions f and g identical? Why or why not?

32. a. Plot the graph of $f(x) = \sqrt{x}\sqrt{x - 1}$ using the viewing window $[-5, 5] \times [-5, 5]$.

   b. Plot the graph of $g(x) = \sqrt{x(x - 1)}$ using the viewing window $[-5, 5] \times [-5, 5]$.

   c. In what interval are the functions f and g identical?

   d. Verify your observation in part (c) analytically.

33. Let $f(x) = 2x^3 - 5x^2 + x - 2$ and $g(x) = 2x^3$.

   a. Plot the graph of f and g using the same viewing window: $[-5, 5] \times [-5, 5]$.

   b. Plot the graph of f and g using the same viewing window: $[-50, 50] \times [-100,000, 100,000]$.

   c. Explain why the graphs of f and g that you obtained in part (b) seem to coalesce as x increases or decreases without bound.

   Hint: Write $f(x) = 2x^3\left(1 - \dfrac{5}{2x} + \dfrac{1}{2x^2} - \dfrac{1}{x^3}\right)$ and study its

   behavior for large values of x.

34. Let $f(x) = \left(1 + \dfrac{1}{x}\right)^x$, where $x > 0$.

 a. Plot the graph of f using the window $[0, 10] \times [0, 3]$, and then using the window $[0, 100] \times [0, 3]$. Does $f(x)$ appear to approach a unique number as x gets larger and larger?

 b. Use the evaluation function of your graphing utility to fill in the accompanying table. Use the table of values to estimate, accurate to five decimal places, the number that $f(x)$ seems to approach as x increases without bound.

 Note: We will see in Section 2.8 that this number, written e, is given by 2.71828 . . .

x	$f(x)$
10	
100	
1000	
10,000	
100,000	
1,000,000	
10,000,000	
100,000,000	
1,000,000,000	

0.6 | Mathematical Models

0.6   SELF-CHECK DIAGNOSTIC TEST

1. Give an example of each of the following.

   a. a linear function

   b. a polynomial function of degree 4

   c. a rational function

   d. a power function

   e. an algebraic function

   f. a trigonometric function

2. The book value of an asset at time t (measured in years) being depreciated linearly over a period of n years is given by

$$V(t) = C - \frac{C - S}{n}t$$

where C and S (in dollars) give the initial and scrap value of the asset, respectively.

   a. What is the V-intercept? Interpret your result.

   b. By how much is the asset being depreciated annually?

3. By cutting away identical squares from each corner of a square piece of cardboard with sides 12 in. long and then folding up the resulting flaps, an open box can be made. If the square cutaways have dimensions x in. by x in., find a function giving the volume of the resulting box.

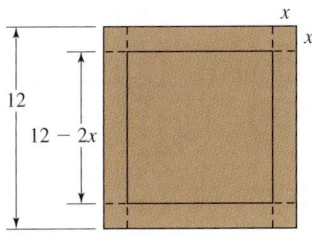

Answers to Self-Check Diagnostic Test 0.6 can be found on page ANS 6.

Mathematical modeling is a process that enables us to use mathematics as a tool to analyze and understand real-world phenomena. The four steps in this process are illustrated in Figure 1.

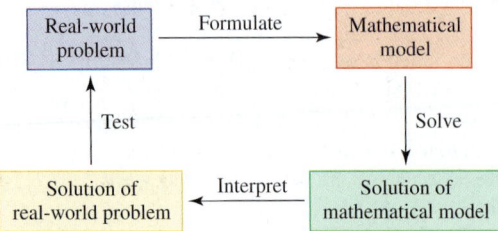

FIGURE 1

1. **Formulate.** Given a real-world problem, our first task is to formulate the problem using the language of mathematics. This mathematical description of the real-world phenomenon is called a **mathematical model.** The many techniques that are used in constructing mathematical models range from theoretical consideration of the problem on the one extreme to an interpretation of data associated with the problem on the other. For example, the mathematical model that gives the accumulated amount at any time after a certain sum of money has been deposited in the bank can be derived theoretically (see Section 0.8, pp. 87–89). On the other hand, the mathematical models in Examples 2 and 3 of this section are constructed by requiring that they fit the data associated with the problem "best" according to some specified criterion. In calculus we are primarily concerned with how one (dependent) variable depends on one or more (independent) variables. Consequently, most of our mathematical models will involve functions of one or more variables or equations defining these functions (implicitly).

2. **Solve.** Once a mathematical model has been constructed, we can use the appropriate mathematical techniques, which we will develop throughout this text, to solve the problem.

3. **Interpret.** Bearing in mind that the solution obtained in Step 2 is just the solution of the mathematical model, we need to interpret these results in the context of the original real-world problem.

4. **Test.** Some mathematical models of real-world applications describe the situations with complete accuracy. For example, the model describing a deposit in a bank account gives the exact accumulated amount in the account at any time. But other mathematical models give, at best, an approximate description of the real-world problem. In such cases we need to test the accuracy of the model by observing how well it describes the original real-world problem and how well it predicts past and/or future behavior. If the results are unsatisfactory, then we might have to reconsider the assumptions that were made in the construction of the model or, in the worst case, return to Step 1.

■ Modeling with Functions

Many real-world phenomena, such as the speed at which a screwdriver falls after being accidentally dropped from a building under construction, the speed of a chemical reaction, the population of a certain strain of bacteria, the life expectancy of a female infant at birth in a certain country, and the demand for a product, can be modeled by an appropriate function.

In what follows, we will recall some familiar functions and give examples of real-world phenomena that are modeled by using these functions.

■ Polynomial Functions

A **polynomial function of degree n** is a function of the form

$$f(x) = a_n x^n + a_{n-1} x^{n-1} + \cdots + a_2 x^2 + a_1 x + a_0 \qquad a_n \neq 0$$

where n is a nonnegative integer and the numbers $a_0, a_1, \ldots, a_n$ are constants called the **coefficients** of the polynomial function. For example, the functions

$$f(x) = 2x^5 - 3x^4 + \frac{1}{2}x^3 + \sqrt{2}x^2 - 6$$

$$g(x) = 0.001x^3 - 0.2x^2 + 10x + 200$$

are polynomial functions of degree 5 and 3, respectively. Observe that a polynomial function is defined for every value of x, so its domain is $(-\infty, \infty)$.

A polynomial function of degree 1 ($n = 1$) has the form

$$y = f(x) = a_1 x + a_0 \qquad a_1 \neq 0$$

and is an equation of a straight line in the slope-intercept form with slope $m = a_1$ and y-intercept $b = a_0$ (see Section 0.1). For this reason a polynomial function of degree 1 is called a **linear function.**

Linear functions are used extensively in mathematical modeling for two important reasons. First, some models are linear by nature. For example, the formula for converting temperature from Celsius (°C) to Fahrenheit (°F) is $F = \frac{9}{5}C + 32$, and F is a linear function of C for C in any feasible prescribed domain (see Figure 2a). Second, some natural phenomena exhibit linear characteristics over a small range of values and can therefore be modeled by a linear function that is restricted to a small interval. For example, according to Hooke's Law, the magnitude of a force F required to stretch a spring by an elongation x beyond its unstretched length is given by $F = kx$, provided that the elongation x is not too great. If stretched beyond a certain point, called the elastic limit, the spring will become permanently deformed and will not return to its natural length when the force is removed. The constant k is called the *spring constant* or the *stiffness* of the spring. In this instance we have to restrict our interest to the portion of the graph that is linear (see Figure 2b).

FIGURE 2
The graph of a linear function and the graph of a function that is linear over a small interval

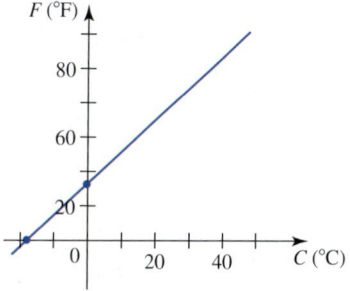

(a) F is linear in C.

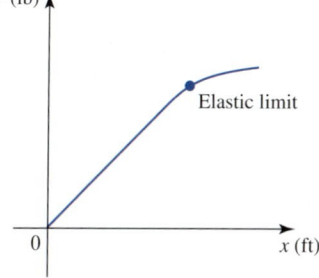

(b) F is linear for small values of x.

In the following example we assume that Hooke's Law applies.

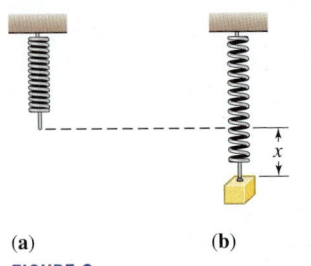

(a)   **(b)**

FIGURE 3
The spring in part (a) is stretched by an elongation of x feet beyond its natural length by a weight in part (b).

EXAMPLE 1 **Force Required to Stretch a Spring** A force of 3.18 lb is required to stretch a spring by 2.4 in. beyond its unstretched length (see Figure 3).

a. Use Hooke's Law to find a mathematical model that describes the force F required to stretch the spring by x feet beyond its unstretched length.
b. What is the spring constant?
c. Find the force required to stretch the spring by 1.8 in. beyond its unstretched length.

Solution

a. By Hooke's Law, $F = kx$, where k is the spring constant. Next, using the given data, we find

$$3.18 = 0.2k \qquad \text{2.4 in. is equal to 0.2 ft}$$

from which we deduce that $k = 15.90$. Therefore, the required mathematical model is $F = 15.9x$.
b. From the result of part (a) we see that the spring constant is 15.9 lb/ft.
c. We first note that 1.8 in. is equal to 0.15 ft. Then, using the model obtained in part (a), we see that the required force is

$$F = (15.9)(0.15) = 2.385$$

or approximately 2.39 lb. ◼

    In Example 1 the model was constructed by using the data obtained from one measurement. In practice, one normally takes a set of measurements and then uses these data to construct a mathematical model. This practice generally results in a more accurate model.

EXAMPLE 2 **Force Required to Stretch a Spring** Table 1 gives the force F required to stretch the spring (Example 1) by an elongation x ft beyond its unstretched length. As Hooke's Law predicts, the data points in the *scatter plot* associated with these data appear to lie close to a straight line passing through the origin (see Figure 4).

TABLE 1

x (ft)	0	0.1	0.2	0.3	0.4	0.5
F (lb)	0	1.68	3.18	4.84	6.36	8.02

FIGURE 4
The data points are scattered about a line through the origin.

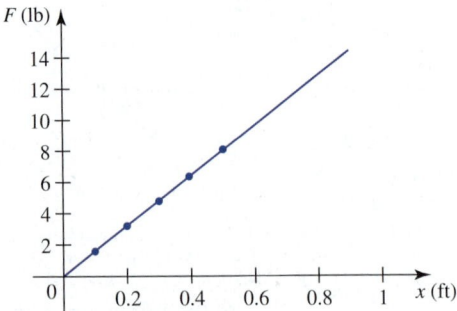

To find a mathematical model based on these data, we use the *method of least squares* to find a function of the form $f(x) = kx$ (as suggested by Hooke's Law) that fits the data "best" in the sense of least squares. (See Exercises 3.7, Problems 71 and 72.) We obtain the function

$$f(x) = 16.02x$$

as the required model. Incidentally, this model also tells us that the spring constant is approximately 16.02 lb/ft.

Notes

1. If you use the linear least-squares regression program that is built into most graphing calculators and computers to find a mathematical model using the data in Example 2, you will obtain a different model, namely, $g(x) = 15.94x + 0.028$. This occurs because the program finds the "best" fit for the data (in the sense of least squares) using the most general linear function, that is, one having the form $f(x) = ax + b$.

2. Since F must be equal to zero if x is equal to zero, we see that the class of functions chosen to fit the data should have the form $f(x) = ax$, that is, with $b = 0$. Therefore, the model $F = 16.02x$ that we found in Example 2 should be regarded as being a more accurate mathematical model than the model suggested by $g(x) = 15.94x + 0.028$, in which $g(0) = 0.028 \neq 0$. As a consequence, we should accept the spring constant to be 16.02 lb/ft found in Example 2 rather than the figure of 15.94 that is found by using the function g as the model.

A polynomial function of degree 2 has the form

$$y = f(x) = a_2x^2 + a_1x + a_0 \qquad a_2 \neq 0$$

or, more simply, $y = ax^2 + bx + c$ and is called a **quadratic function.** The graph of a quadratic function is a parabola (see Figure 5). The parabola opens upward if $a > 0$ and downward if $a < 0$. To see this, we rewrite

$$f(x) = ax^2 + bx + c = x^2\left(a + \frac{b}{x} + \frac{c}{x^2}\right) \qquad x \neq 0$$

FIGURE 5
The graph of a quadratic function is a parabola.

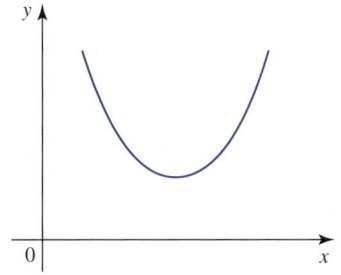

(a) If $a > 0$, the parabola opens upward.

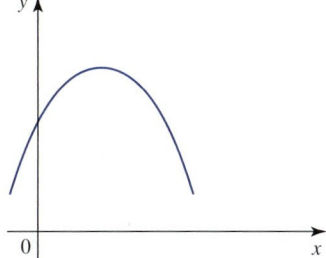

(b) If $a < 0$, the parabola opens downward.

Observe that if x is large in absolute value, then the expression inside the parentheses is close to a, so $f(x)$ behaves like ax^2 for large values of x. Therefore, for large values of x, $y = f(x)$ is large and positive if $a > 0$ (the parabola opens upward) and is large in magnitude and negative if $a < 0$ (the parabola opens downward). The highest point on a parabola that opens downward or the lowest point on a parabola that opens upward is called the *vertex* of the parabola. The vertex of the parabola with equation $y = ax^2 + bx + c$, where $a \neq 0$, is $(-b/(2a), f(-b/(2a)))$ since $y = f(x)$. You can verify this fact by using the method of completing the square (see Exercise 30).

Quadratic functions serve as mathematical models for many phenomena. For example, Newton's Second Law of Motion can be used to show that the distance covered by a falling object dropped near the surface of the earth is given by $D = \frac{1}{2} gt^2$ where g, the gravitational constant at sea level at the equator, is approximately 32.088 ft/sec^2. In fact, a model for this motion can be found, experimentally, as the following example shows.

EXAMPLE 3 A steel ball is dropped from a height of 10 ft. The distance covered by the ball at intervals of one tenth of a second is measured and recorded in Table 2. A scatter plot of the data is shown in Figure 6. You can see from the figure that the points associated with the data do lie close to a parabola with equation $y = at^2$ for some constant a, as was suggested earlier.

TABLE 2

Time (sec)	0.0	0.1	0.2	0.3	0.4	0.5	0.6	0.7
Distance (ft)	0	0.1608	0.6416	1.4444	2.5672	4.0108	5.7760	7.8614

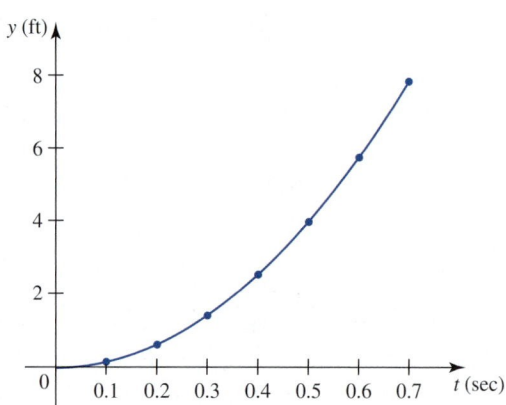

FIGURE 6

To find a mathematical model to describe this motion, we use the method of least squares to find a function of the form $y = at^2$ that fits the data "best." We obtain the function

$$y = 16.044t^2$$

(See Exercises 3.7, Problems 73 and 74.) On the basis of this model, the ball will hit the ground when $y = 10$. Solving the equation $16.044t^2 = 10$ gives $t \approx \pm 0.7895$. Rejecting the negative root, we conclude that the ball will hit the ground approximately 0.79 sec after it is dropped. Thus, a complete description of the mathematical model for this motion is

$$D = 16.044t^2 \qquad 0 \le t \le 0.79$$

where D is the distance covered by the ball after t sec.

Notes

1. Observe that even though the function $f(t) = 16.044t^2$ is defined on $(-\infty, \infty)$, we need to restrict its domain to the interval $[0, 0.79]$ to obtain a mathematical model for the motion of the ball. Once the ball reaches the ground, the function f no longer describes its motion.

2. If you use the quadratic regression program that is found in most graphing calculators and in computers, you will find the quadratic model

$$D = 16.0425t^2 + 0.00075t + 0.000075$$

which is not very satisfactory, since we know that $D = 0$ when $t = 0$. Besides, as you will be able to confirm later, this model implies that the ball started out with an initial velocity of 0.00075 ft/sec. But we know that the steel ball had an initial velocity of 0 ft/sec. ▬

A polynomial of degree three is called a **cubic polynomial,** one of degree four is called a **quartic polynomial,** and one of degree five is called a **quintic** polynomial. In general, the higher the degree of the polynomial function, the more its graph wiggles. Figure 7a–c shows the graph of a cubic, a quartic, and a quintic, respectively.

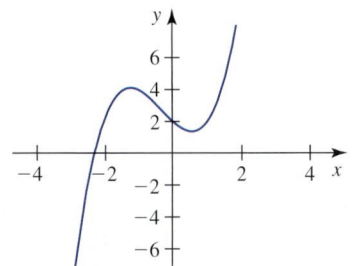

(a) $y = x^3 + x^2 - 2x + 2$
(a cubic)

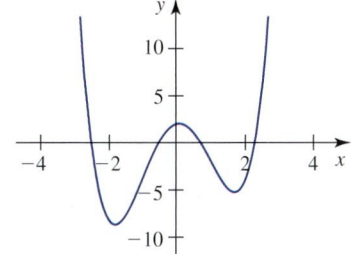

(b) $y = x^4 - 6x^2 + x + 2$
(a quartic)

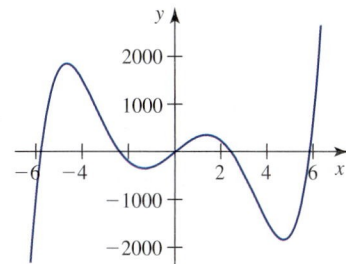

(c) $y = 2x^5 - 80x^3 + 400x$
(a quintic)

FIGURE 7

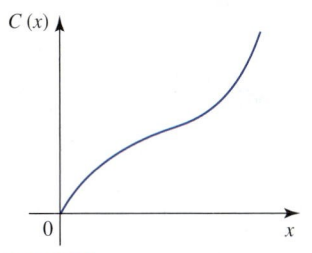

FIGURE 8
A total cost function is often modeled by using a cubic function.

Cubic polynomials lend themselves to modeling some phenomena in business and economics. For example, let $C(x)$ denote the total cost incurred when x units of a certain commodity are produced. A typical graph of the function C is shown in Figure 8. As the level of production x increases, the cost per unit drops, so C increases but at a slower pace. However, a level of production is soon reached at which the cost per unit begins to increase dramatically (because of overtime, a shortage of raw materials, and breakdown of machinery due to excessive stress and strain), so C continues to increase at a faster pace. The graph of a cubic polynomial can exhibit precisely the characteristics just described.

The following example shows how we can use a quartic function to describe the assets of the Social Security system.

EXAMPLE 4 **Social Security Trust Fund Assets** The projected assets of the Social Security trust fund (in trillions of dollars) from 2008 through 2040 are given in Table 3. The scatter plot associated with these data is shown in Figure 9a, where $t = 0$ corresponds to 2008. A mathematical model giving the approximate value of the assets in the trust fund $A(t)$ (in trillions of dollars) in year t is

$$A(t) = -0.00000268t^4 - 0.000356t^3 + 0.00393t^2 + 0.2514t + 2.4094$$

The graph of A is shown in Figure 9b.

TABLE 3

Year	2008	2011	2014	2017	2020	2023	2026	2029	2032	2035	2038	2040
Assets	$2.4	$3.2	$4.0	$4.7	$5.3	$5.7	$5.9	$5.6	$4.9	$3.6	$1.7	0

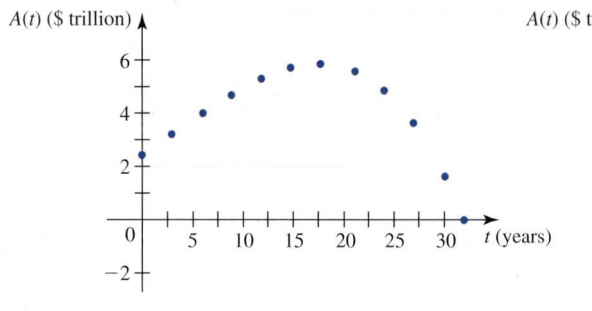

(a) Scatter plot   (b) Graph of A

FIGURE 9

Source: Social Security Administration.

a. The first baby boomers will turn 65 in 2011. What will the assets of the Social Security system trust fund be at that time? The last of the baby boomers will turn 65 in 2029. What will the assets of the trust fund be at that time?

b. Unless payroll taxes are increased significantly and/or benefits are scaled back dramatically, it is only a matter of time before the assets of the current system are depleted. Use the graph of the function $A(t)$ to estimate the year in which the current Social Security system is projected to go broke.

Solution

a. The assets of the Social Security trust fund in 2011 ($t = 3$) will be

$$A(3) = -0.00000268(3)^4 - 0.000356(3)^3$$
$$+ 0.00393(3)^2 + 0.2514(3) + 2.4094 \approx 3.19$$

or approximately $3.19 trillion. The assets of the trust fund in 2029 ($t = 21$) will be

$$A(21) = -0.00000268(21)^4 - 0.000356(21)^3$$
$$+ 0.00393(21)^2 + 0.2514(21) + 2.4094 \approx 5.60$$

or approximately $5.60 trillion.

b. From Figure 9b we see that the graph of A crosses the t-axis at approximately $t = 32$. So unless the current system is changed, it is projected to go broke in 2040. (At this time the first of the baby boomers will be 94, and the last of the baby boomers will be 76.) ◼

Note Observe that the model in Example 4 utilizes only a small portion of the graph of f, as is often the case in practice. A more complete picture of the graph of f is shown in Figure 10. ◼

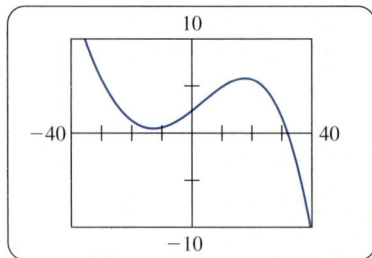

FIGURE 10

The graph of f in the viewing window $[-40, 40] \times [-10, 10]$

■ Power Functions

A **power function** is a function of the form $f(x) = x^a$, where a is a real number. If a is a nonnegative integer, then f is just a polynomial function of degree a with one term (a monomial). Examples of other power functions are

$$f(x) = x^{-2} = \frac{1}{x^2}, \quad f(x) = x^{-1} = \frac{1}{x}, \quad f(x) = x^{1/2} = \sqrt{x}, \quad \text{and} \quad f(x) = x^{1/3} = \sqrt[3]{x}$$

whose graphs are shown in Figure 11.

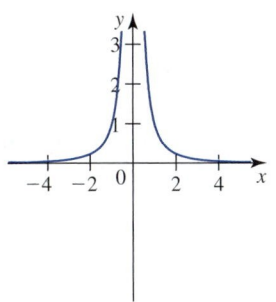

(a) $f(x) = x^{-2}$

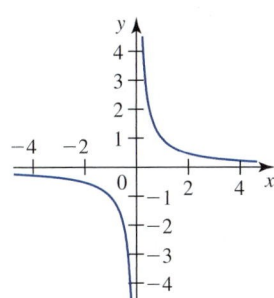

(b) $f(x) = x^{-1}$

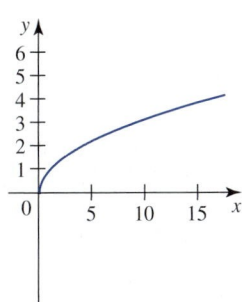

(c) $f(x) = x^{1/2}$

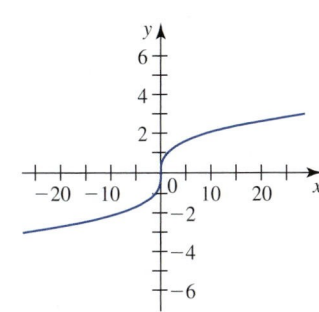

(d) $f(x) = x^{1/3}$

FIGURE 11
The graphs of some power functions

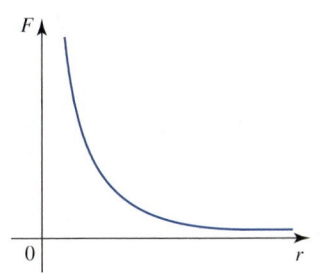

FIGURE 12
The magnitude of a gravitational force F

Power functions serve as mathematical models in many fields of study. For example, according to Newton's Law of Gravitation, the force exerted by a particle of mass m_1 on another particle of mass m_2 a distance r away is directed toward m_1 and has magnitude

$$F = \frac{Gm_1m_2}{r^2}$$

where G is the universal gravitational constant. The graph of F is similar to that of $f(x) = x^{-2}$ for $x > 0$ (see Figure 12).

■ Rational Functions

A **rational function** is a quotient of two polynomials. Examples of rational functions are

$$f(x) = \frac{3x^3 + x^2 - x + 1}{x - 2} \quad \text{and} \quad g(x) = \frac{x^2 + 1}{x^2 - 1}$$

In general, a rational function has the form

$$f(x) = \frac{P(x)}{Q(x)}$$

where P and Q are polynomial functions. The domain of a rational function is the set of all real numbers except the zeros of Q, that is, the roots of the equation $Q(x) = 0$. Thus, the domain of f is $\{x \mid x \neq 2\}$, and the domain of g is $\{x \mid x \neq \pm 1\}$. A mathematical model involving a rational function is suggested by the experiments conducted by A.J. Clark on the response $R(x)$ of a frog's heart muscle to the injection of x units of

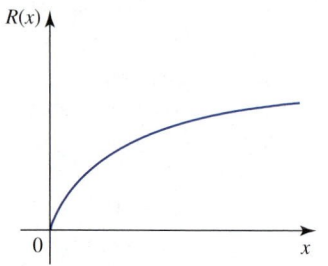

FIGURE 13

The graph of $R(x) = \dfrac{100x}{b + x}$

acetylcholine (as a percentage of the maximum possible effect of the drug). His results show that R has the form

$$R(x) = \frac{100x}{b + x} \qquad x \geq 0$$

where b is a positive constant that depends on the particular frog (see Figure 13).

■ Algebraic Functions

Algebraic functions are functions that can be expressed as sums, differences, products, quotients, or roots of polynomial functions. By definition, rational functions are algebraic functions. The function

$$f(x) = 2x^3 - 3\sqrt{x} + \frac{x\sqrt[3]{x^2 + 1}}{x(x + \sqrt{x})}$$

is another example of an algebraic function. The following example from the special theory of relativity involves an algebraic function.

EXAMPLE 5   **Special Theory of Relativity**   According to the special theory of relativity, the relativistic mass of a particle moving with a speed v is

$$m = f(v) = \frac{m_0}{\sqrt{1 - \dfrac{v^2}{c^2}}}$$

where m_0 is the rest mass (the mass at zero speed) and $c = 2.9979 \times 10^8$ m/sec is the speed of light in a vacuum. What is the speed of a particle whose relativistic mass is twice that of its rest mass?

Solution   We solve the equation

$$2m_0 = \frac{m_0}{\sqrt{1 - \dfrac{v^2}{c^2}}}$$

for v, obtaining

$$2 = \frac{1}{\sqrt{1 - \dfrac{v^2}{c^2}}}$$

$$\sqrt{1 - \frac{v^2}{c^2}} = \frac{1}{2}$$

$$1 - \frac{v^2}{c^2} = \frac{1}{4}$$

$$\frac{v^2}{c^2} = \frac{3}{4}$$

$$v = \frac{\sqrt{3}}{2}c$$

or approximately 0.866 times the speed of light (approximately 2.596×10^8 m/sec). ■

■ Trigonometric Functions

Trigonometric functions were reviewed in Section 0.3. The characteristics of the trigonometric functions make them suitable for modeling phenomena that exhibit cyclical, or almost cyclical, behavior such as the motion of sound waves, the vibration of strings, and the motion of a simple pendulum.

EXAMPLE 6 **Average Temperature**   Table 4 gives the average monthly temperature in degrees Fahrenheit recorded in Boston.

TABLE 4

Month	Jan.	Feb.	March	April	May	June	July	Aug.	Sept.	Oct.	Nov.	Dec.
Temp (°F)	28.6	30.3	38.6	48.1	58.2	67.7	73.5	71.9	64.8	54.8	45.3	33.6

Source: The Boston Globe.

To find a model describing the average temperature T in month t, we assume that T is a sine function with period 12 and amplitude given by $\frac{1}{2}(73.5 - 28.6) = 22.45$. A possible model is

$$T = 51.05 + 22.45 \sin\left[\frac{\pi}{6}(t - 4.3)\right]$$

where $t = 1$ corresponds to January. The graph of T is shown in Figure 14.

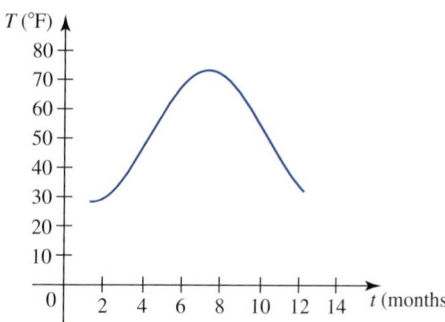

FIGURE 14
A model of the average temperature in Boston is $T = 51.05 + 22.45 \sin\left[\frac{\pi}{6}(t - 4.3)\right]$.

Other functions, such as *exponential* and *logarithmic* functions, also play an important role in modeling and will be studied in later chapters.

■ Constructing Mathematical Models

We close this section by showing how some mathematical models can be constructed by using elementary geometric and algebraic arguments.

The following guidelines can be used to construct mathematical models.

> **Guidelines for Constructing Mathematical Models**
> 1. Assign a letter to each variable mentioned in the problem. If appropriate, draw and label a figure.
> 2. Find an expression for the quantity that is being sought.
> 3. Use the conditions given in the problem to write the quantity being sought as a function f of one variable. Note any restrictions to be placed on the domain of f from physical considerations of the problem.

EXAMPLE 7 **Enclosing an Area** The owner of Rancho Los Feliz has 3000 yd of fencing with which to enclose a rectangular piece of grazing land along the straight portion of a river. Fencing is not required along the river. Letting x denote the width of the rectangle, find a function f in the variable x giving the area of the grazing land if she uses all of the fencing (see Figure 15).

FIGURE 15
The rectangular grazing land has width x and length y.

Solution
1. This information is given in the statement of the problem.
2. The area of the rectangular grazing land is $A = xy$. Next, observe that the amount of fencing is $2x + y$ and that this must be equal to 3000, since all the fencing is to be used; that is,

$$2x + y = 3000$$

3. From the equation we see that $y = 3000 - 2x$. Substituting this value of y into the expression for A gives

$$A = xy = x(3000 - 2x) = 3000x - 2x^2$$

Finally, observe that both x and y must be positive, since they represent the width and length of a rectangle, respectively. Thus, $x > 0$ and $y > 0$, but the latter is equivalent to $3000 - 2x > 0$, or $x < 1500$. So the required function is $f(x) = 3000x - 2x^2$ with domain $0 < x < 1500$. ∎

EXAMPLE 8 **Charter Flight Revenue** If exactly 200 people sign up for a charter flight, Leisure World Travel Agency charges $300 per person. However, if more than 200 people sign up for the flight (assume that this is the case), then each fare is reduced by $1 for each additional person. Letting x denote the number of passengers above 200, find a function giving the revenue realized by the company.

Solution
1. This information is given.
2. If there are x passengers above 200, then the number of passengers signing up for the flight is $200 + x$. Furthermore, the fare will be $(300 - x)$ dollars per passenger.
3. The revenue will be

$$R = (200 + x)(300 - x) \qquad \text{number of passengers} \cdot \text{fare per passenger}$$

$$= -x^2 + 100x + 60{,}000$$

Clearly, x must be positive, and $300 - x > 0$, or $x < 300$. So the required function is $f(x) = -x^2 + 100x + 60{,}000$ with domain $(0, 300)$. ∎

0.6 EXERCISES

In Exercises 1 and 2, classify each function as a polynomial function (state its degree), a power function, a rational function, an algebraic function, a trigonometric function, or other.

1. a. $f(x) = 2x^3 - 3x^2 + x - 4$

   b. $f(x) = \sqrt[3]{x^2}$

   c. $g(x) = \dfrac{x}{x^2 - 4}$

   d. $f(t) = 3t^{-2} - 2t^{-1} + 4$

   e. $h(x) = \dfrac{\sqrt{x} + 1}{\sqrt{x} - 1}$

   f. $f(x) = \sin x + \cos x$

2. a. $f(t) = 2t^4 - 3t^2 - 2\sqrt{t}$

   b. $g(x) = 2\sqrt{1 - x^2}$

   c. $f(x) = \dfrac{\sqrt{x}}{x^2 + 1}$

   d. $h(x) = \dfrac{\sin x}{1 + \tan x}$

   e. $f(x) = \tan 2x$

   f. $h(x) = (3x - 1)^2 + 4$

3. Instant Messaging Accounts The number of enterprise instant messaging (IM) accounts is approximated by the function

$$N(t) = 2.96t^2 + 11.37t + 59.7 \qquad 0 \le t \le 5$$

where $N(t)$ is measured in millions and t is measured in years with $t = 0$ corresponding to 2006.

   a. How many enterprise IM accounts were there in 2006?

   b. How many enterprise IM accounts are there projected to be in 2010?

Source: The Radical Group.

4. Average Single-Family Property Tax On the basis of data from 298 of Massachusetts' 351 cities and towns, the average single-family tax bill from 1997 through 2007 in that state is approximated by the function

$$T(t) = 7.26t^2 + 91.7t + 2360 \qquad 0 \le t \le 10$$

where $T(t)$ is measured in dollars and t is measured in years with $t = 0$ corresponding to 1997.

   a. What was the average property tax on a single-family home in Massachusetts in 1997?

   b. If the trend continued, what would the average property tax be in 2010?

Source: Massachusetts Department of Revenue.

5. Testosterone Use Fueled by the promotion of testosterone as an antiaging elixir, use of the hormone by middle-aged and older men grew dramatically. The total number of prescriptions for testosterone from 1999 through 2002 is given by

$$N(t) = -35.8t^3 + 202t^2 + 87.8t + 648 \qquad 0 \le t \le 3$$

where $N(t)$ is measured in thousands and t is measured in years with $t = 0$ corresponding to 1999. Find the total number of prescriptions for testosterone in 1999, 2000, 2001, and 2002.

Source: IMS Health.

6. Aging Drivers The number of driver fatalities due to car crashes, based on the number of miles driven, begins to climb after the driver is past age 65 years. Aside from declining ability as one ages, the older driver is more fragile. The number of driver fatalities per 100 million vehicle miles driven is approximately

$$N(x) = 0.0336x^3 - 0.118x^2 + 0.215x + 0.7 \qquad 0 \le x \le 7$$

where x denotes the age group of drivers, with $x = 0$ corresponding to those aged 50–54 years, $x = 1$ corresponding to those aged 55–59, $x = 2$ corresponding to those aged 60–64, . . . , and $x = 7$ corresponding to those aged 85–89. What is the driver fatality rate per 100 million vehicle miles driven for an average driver in the 50–54 age group? In the 85–89 age group?

Source: U.S. Department of Transportation.

7. Obese Children in the United States The percentage of obese children aged 12–19 years in the United States is approximately

$$P(t) = \begin{cases} 0.04t + 4.6 & \text{if } 0 \le t < 10 \\ -0.01005t^2 + 0.945t - 3.4 & \text{if } 10 \le t \le 30 \end{cases}$$

where t is measured in years, with $t = 0$ corresponding to the beginning of 1970. What was the percentage of obese children aged 12–19 years at the beginning of 1970? At the beginning of 1985? At the beginning of 2000?

Source: Centers for Disease Control and Prevention.

8. Rwandan Genocide The population of Rwanda in millions from 1990 through 2002 is approximated by the function

$$P(t) = \begin{cases} 0.17t + 6.99 & \text{if } 0 \le t < 3 \\ -0.9t + 10.2 & \text{if } 3 \le t < 5 \\ 0.7t + 2.2 & \text{if } 5 \le t < 7 \\ 0.12t + 6.26 & \text{if } 7 \le t \le 12 \end{cases}$$

where t is measured in years, with $t = 0$ corresponding to 1990. The genocide that the majority Hutus committed against the Tutsis and moderate Hutus resulted in almost a million deaths and mass migration of the population out of the country. Eventually, most of the refugees returned to the country.

   a. Sketch the graph of the population function P.

   b. What was the population in 1993? In 1995? In 2002?

   c. In what year was the population of Rwanda at the lowest level?

   d. Did the population eventually recover to at least its previous level?

Source: CIA World Factbook.

V Videos for selected exercises are available online at **www.academic.cengage.com/login.**

9. **Linear Depreciation** In computing income tax, businesses are allowed by law to depreciate certain assets, such as buildings, machines, furniture, and automobiles, over a period of time. The linear depreciation method, or straight-line method, is often used for this purpose. Suppose an asset has an initial value of $C and is to be depreciated linearly over n years with a scrap value of $S. Show that the book value of the asset at any time t, where $0 \leq t \leq n$, is given by the linear function

$$V(t) = C - \frac{C - S}{n} t$$

Hint: Find an equation of the straight line that passes through the points $(0, C)$ and (n, S). Then rewrite the equation in the slope-intercept form.

10. **Cricket Chirping and Temperature** Entomologists have discovered that a linear relationship exists between the number of chirps of crickets of a certain species and the air temperature. When the temperature is 70°F, the crickets chirp at the rate of 120 times/min; when the temperature is 80°F, they chirp at the rate of 160 times/min.
   a. Find an equation giving the relationship between the air temperature t and the number of chirps per minute, N, of the crickets.
   b. Find N as a function of t, and use this formula to determine the rate at which the crickets chirp when the temperature is 102°F.

11. **Reaction of a Frog to a Drug** Experiments conducted by A.J. Clark suggest that the response $R(x)$ of a frog's heart muscle to the injection of x units of acetylcholine (as a percent of the maximum possible effect of the drug) can be approximated by the rational function

$$R(x) = \frac{100x}{b + x} \qquad x \geq 0$$

where b is a positive constant that depends on the particular frog.
   a. If a concentration of 40 units of acetylcholine produces a response of 50% for a certain frog, find the response function for this frog.
   b. Using the model found in part (a), find the response of the frog's heart muscle when 60 units of acetylcholine are administered.

12. **Outsourcing of Jobs** According to a study conducted in 2003, the total number of U.S. jobs (in millions) that were projected to leave the country by year t, where $t = 0$ corresponds to 2000, is

$$N(t) = 0.0018425(t + 5)^{2.5} \qquad 0 \leq t \leq 15$$

How many jobs were projected to be outsourced in 2005? In 2010?

Source: Forrester Research.

13. **Online Video Viewers** As broadband Internet grows more popular, video services such as YouTube will continue to expand. The number of online video viewers (in millions) is projected to grow according to the rule

$$N(t) = 52t^{0.531} \qquad 1 \leq t \leq 10$$

where $t = 1$ corresponds to 2003.
   a. Sketch the graph of N.
   b. How many online video viewers will there be in 2010?

Source: eMarketer.com.

14. **Cost, Revenue, and Profit Functions** A manufacturer of indoor-outdoor thermometers has fixed costs (executive salaries, rent, etc.) of $F/month, where F is a positive constant. The cost for manufacturing its product is $c/unit, and the product sells for $s/unit.
   a. Write a function $C(x)$ that gives the total cost incurred by the manufacturer in producing x thermometers/month.
   b. Write a function $R(x)$ that gives the total revenue realized by the manufacturer in selling x thermometers.
   c. Write a function $P(x)$ that gives the total monthly profit realized by the manufacturer in selling x thermometers/month.
   d. Refer to your answer in part (c). Find $P(0)$, and interpret your result.
   e. How many thermometers should the manufacturer produce per month to have a break-even operation?
      Hint: Solve $P(x) = 0$.

15. **Global Warming** The increase in carbon dioxide in the atmosphere is a major cause of global warming. The Keeling Curve, named after Dr. Charles David Keeling, a professor at Scripps Institution of Oceanography, gives the average amount of carbon dioxide (CO_2) measured in parts per million volume (ppmv), in the atmosphere from 1958 ($t = 1$) through 2007 ($t = 50$). (Even though data were available for every year in this time interval, we will construct the curve only on the basis of the following randomly selected data points.)

Year	1958	1970	1974	1978	1985	1991	1998	2003	2007
Amount	315	325	330	335	345	355	365	375	380

   a. Use a graphing utility to find a second-degree polynomial regression model for the data.
   b. Plot the graph of the function f that you found in part (a), using the viewing window $[1, 50] \times [310, 400]$.
   c. Use the model to estimate the average amount of atmospheric carbon dioxide in 1980 ($t = 23$).
   d. Assume that the trend continues, and use the model to predict the average amount of atmospheric carbon dioxide in 2010.

Source: Scripps Institution of Oceanography.

16. Population Growth in Clark County Clark County in Nevada, dominated by greater Las Vegas, is one of the fastest-growing metropolitan areas in the United States. The population of the county from 1970 through 2000 is given in the following table.

Year	1970	1980	1990	2000
Population	273,288	463,087	741,459	1,375,765

  a. Use a graphing utility to find a third-degree polynomial regression model for the data. Let t be measured in years, with $t = 0$ corresponding to the beginning of 1970.

  b. Plot the graph of the function f that you found in part (a), using the viewing window $[0, 30] \times [0, 1,500,000]$.

  c. Compare the values of f at $t = 0, 10, 20$, and 30 with the given data.

Source: U.S. Census Bureau.

17. Hiring Lobbyists Many public entities such as cities, counties, states, utilities, and Indian tribes are hiring firms to lobby Congress. One goal of such lobbying is to place earmarks—money directed at a specific project—into appropriation bills. The amount (in millions of dollars) spent by public entities on lobbying from 1998 through 2004 is shown in the following table.

Year	1998	1999	2000	2001	2002	2003	2004
Amount	43.4	51.7	62.5	76.3	92.3	101.5	107.7

  a. Use a graphing utility to find a third-degree polynomial regression model for the data, letting $t = 0$ correspond to 1998.

  b. Plot the scatter diagram and the graph of the function f that you found in part (a).

  c. Compare the values of f at $t = 0, 3$, and 6 with the given data.

Source: Center for Public Integrity.

18. Measles Deaths Measles is still a leading cause of vaccine-preventable death among children, but because of improvements in immunizations, measles deaths have dropped globally. The following table gives the number of measles deaths (in thousands) in sub-Saharan Africa from 1999 through 2005.

Year	1999	2001	2003	2005
Number	506	338	250	126

  a. Use a graphing utility to find a third-degree polynomial regression model for the data, letting $t = 0$ correspond to 1999.

  b. Plot the scatter diagram and the graph of the function f that you found in part (a).

  c. Compute the values of f for $t = 0, 2$, and 6.

  d. How many measles deaths were there in 2004?

Source: Centers for Disease Control and Prevention, World Health Organization.

19. Nicotine Content of Cigarettes Even as measures to discourage smoking have been growing more stringent in recent years, the nicotine content of cigarettes has been rising, making it more difficult for smokers to quit. The following table gives the average amount of nicotine in cigarette smoke from 1999 through 2004.

Year	1999	2000	2001	2002	2003	2004
Yield per cigarette (mg)	1.71	1.81	1.85	1.84	1.83	1.89

  a. Use a graphing utility to find a fourth-degree polynomial regression model for the data. Let $t = 0$ correspond to 1999.

  b. Plot the graph of the function f that you found in part (a), using the viewing window $[0, 5] \times [1, 3]$.

  c. Compute the values of $f(t)$ for $t = 0, 1, 2, 3, 4$, and 5.

Source: Massachusetts Tobacco Control Program.

20. Periods of Planets The following table gives the mean distance D between a planet and the sun measured in astronomical units (an AU is the mean distance between the earth and the sun), and its period T, measured in years, of some planets of the solar system.

Planet	D	T
Mercury	0.39	0.24
Venus	0.72	0.62
Earth	1.00	1.00
Mars	1.52	1.88
Jupiter	5.20	11.9
Saturn	9.54	29.5

  a. Use a graphing utility to find a power regression model, $T(D)$, for the data.

  b. Does the model that you obtained in part (a) confirm Kepler's Third Law of Planetary Motion? (The squares of the periods of the planets are proportional to the cubes of their mean distances from the sun.)

21. Enclosing an Area Patricia wishes to have a rectangular-shaped garden in her backyard. She has 80 ft of fencing with which to enclose her garden. Letting x denote the width of the garden, find a function f in the variable x that gives the area of the garden. What is its domain?

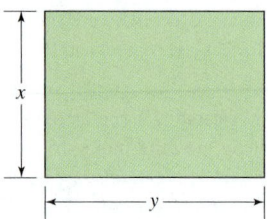

22. Enclosing an Area Ramon wishes to have a rectangular-shaped garden in his backyard. But Ramon wants his garden to have an area of 250 ft^2. Letting x denote the width of the garden, find a function f in the variable x that gives the length of the fencing required to construct the garden. What is the domain of the function?

23. Packaging By cutting away identical squares from each corner of a rectangular piece of cardboard and folding up the resulting flaps, an open box can be made. If the cardboard is 15 in. long and 8 in. wide and the square cutaways have dimensions of x in. by x in., find a function that gives the volume of the resulting box.

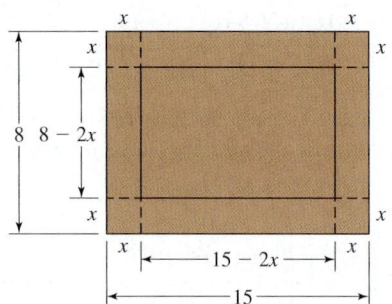

24. Construction Costs A rectangular box is to have a square base and a volume of 20 ft^3. The material for the base costs 30¢/ft^2, the material for the sides costs 10¢/ft^2, and the material for the top costs 20¢/ft^2. Letting x denote the length of one side of the base, find a function in the variable x that gives the cost of materials for constructing the box.

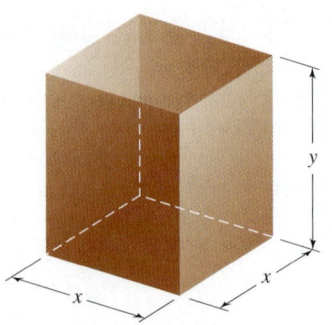

25. Area of a Norman Window A Norman window has the shape of a rectangle surmounted by a semicircle. Suppose a Norman window is to have a perimeter of 28 ft. Find a function in the variable x that gives the area of the window.

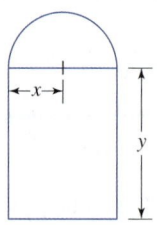

26. Yield of an Apple Orchard An apple orchard has an average yield of 36 bushels of apples per tree if tree density is 22 trees per acre. For each unit increase in tree density, the yield decreases by 2 bushels per tree. Letting x denote the number of trees beyond 22 per acre, find a function of x that gives the yield of apples.

27. Book Design A book designer decided that the pages of a book should have 1-in. margins at the top and bottom and $\frac{1}{2}$-in. margins on the sides. She further stipulated that each page should have an area of 50 in.^{2}. Find a function in the variable x, giving the area of the printed page (see the figure). What is the domain of the function?

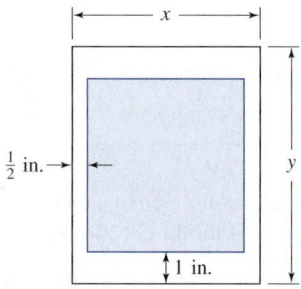

28. Profit of a Vineyard Phillip, the proprietor of a vineyard, estimates that if 10,000 bottles of wine are produced this season, then the profit will be $5 per bottle. But if more than 10,000 bottles are produced, then the profit per bottle for the entire lot will drop by $0.0002 for each bottle sold. Assume that at least 10,000 bottles of wine are produced and sold, and let x denote the number of bottles produced and sold above 10,000.
   a. Find a function P giving the profit in terms of x.
   b. What is the profit that Phillip can expect from the sale of 16,000 bottles of wine from his vineyard?

29. Charter Revenue The owner of a luxury motor yacht that sails among the 4000 Greek islands charges $600 per person per day if exactly 20 people sign up for the cruise. However, if more than 20 people (up to the maximum capacity of 90) sign up for the cruise, then each fare is reduced by $4 per day for each additional passenger. Assume at least 20 people sign up for the cruise, and let x denote the number of passengers above 20.

a. Find a function R giving the revenue per day realized from the charter.

b. What is the revenue per day if 60 people sign up for the cruise?

c. What is the revenue per day if 80 people sign up for the cruise?

30. Show that the vertex of the parabola $f(x) = ax^2 + bx + c$, where $a \neq 0$, is $(-b/(2a), f(-b/(2a)))$.
Hint: Complete the square.

0.7 Inverse Functions

0.7 SELF-CHECK DIAGNOSTIC TEST

1. Determine whether $f(x) = |x + 1| - |x|$ is one-to-one.

2. Find $f^{-1}(1)$ if $f(x) = \dfrac{1}{2} + \sqrt{x - \dfrac{3}{4}}, \quad x \geq \dfrac{3}{4}$.

3. Find the inverse of $f(x) = 3x - 2$. Then sketch the graph of f and f^{-1} on the same set of axes.

4. Find the exact value of each of the following:
   a. $\tan^{-1}(-1)$
   b. $\cot^{-1}(-\sqrt{3})$

5. Write $\tan(\sin^{-1} x)$ in algebraic form.

Answers to Self-Check Diagnostic Test 0.7 can be found on page ANS 6.

■ The Inverse of a Function

A prototype of a maglev (magnetic levitation train) moves along a straight monorail. To describe the motion of the maglev, we can think of the track as a coordinate line. From data obtained in a test run, engineers have determined that the displacement (directed distance) of the maglev measured in feet from the origin at time t (in seconds) is given by

$$s = f(t) = 4t^2 \qquad 0 \leq t \leq 30 \tag{1}$$

where f is called the position function of the maglev (see Figure 1).

FIGURE 1
A maglev moving along an elevated monorail track

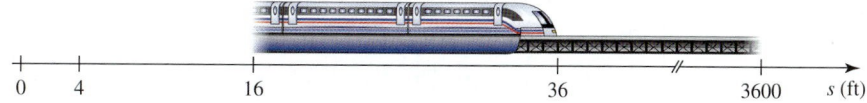

The domain of this position function is $[0, 30]$, and the graph of f is shown in Figure 2. Formula (1) enables us to compute algebraically the position of the maglev at any given time t. Geometrically, we can find the position of the maglev at any given time t by following the path indicated in Figure 2, which associates the given time t with the desired position $f(t)$.

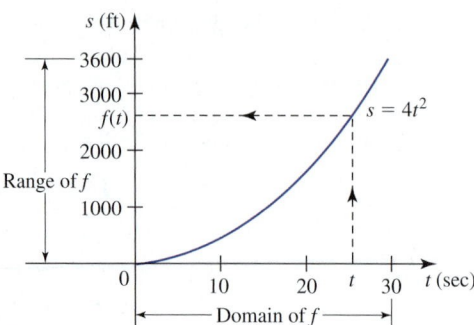

FIGURE 2
Each t in the domain of f is associated with the (unique) position $s = f(t)$ of the maglev.

Now consider the reverse problem: Knowing the position function of the maglev, can we find some way of obtaining the time it takes for the maglev to reach a given position? Geometrically, this problem is easily solved: Locate the point on the s-axis corresponding to the given position. Follow the path considered earlier but traced in the *opposite* direction. This path associates the given position s with the desired time t.

Algebraically, we can obtain a formula for the time t it takes for the maglev to get to the position s by solving Equation (1) for t in terms of s. Thus,

$$t = \frac{1}{2}\sqrt{s}$$

(we reject the negative root, since t lies in $[0, 30]$). Observe that the function g defined by

$$t = g(s) = \frac{1}{2}\sqrt{s}$$

has domain $[0, 3600]$ (the range of f) and range $[0, 30]$ (the domain of f). The graph of g is shown in Figure 3.

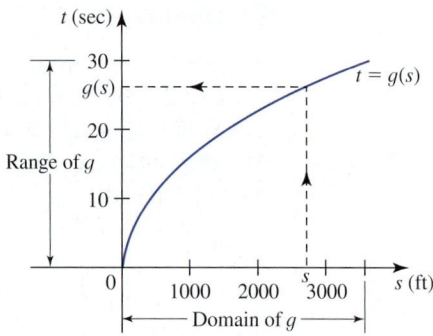

FIGURE 3
Each s in the domain of g is associated with the (unique) time $t = g(s)$.

The functions f and g have the following properties:

1. The domain of g is the range of f and vice versa.
2. They satisfy the relationships

$$(g \circ f)(t) = g[f(t)] = \frac{1}{2}\sqrt{f(t)} = \frac{1}{2}\sqrt{4t^2} = t$$

and

$$(f \circ g)(t) = f[g(t)] = 4[g(t)]^2 = 4\left(\frac{1}{2}\sqrt{t}\right)^2 = t$$

In other words, one undoes what the other does. This is to be expected because f maps t onto $s = f(t)$ and g maps $s = f(t)$ back onto t.

The functions f and g are said to be *inverses* of each other. More generally, we have the following definition.

DEFINITION   Inverse Functions

A function g is the inverse of the function f if

$$f[g(x)] = x \text{ for every } x \text{ in the domain of } g$$

and

$$g[f(x)] = x \text{ for every } x \text{ in the domain of } f$$

Equivalently, g is the inverse of f if the following condition is satisfied:

$$y = f(x) \qquad \text{if and only if} \qquad x = g(y)$$

for every x in the domain of f and for every y in its range.

Note   The inverse of f is normally denoted by f^{-1} (read "f inverse"), and we will use this notation throughout the text. ■

 Do not confuse $f^{-1}(x)$ with $[f(x)]^{-1} = \dfrac{1}{f(x)}$.

EXAMPLE 1   Show that the functions $f(x) = x^{1/3}$ and $g(x) = x^3$ are inverses of each other.

Solution   First, observe that the domain and range of both f and g are $(-\infty, \infty)$. Therefore, both the composite functions $f \circ g$ and $g \circ f$ are defined. Next, we compute

$$(f \circ g)(x) = f[g(x)] = [g(x)]^{1/3} = (x^3)^{1/3} = x$$

and

$$(g \circ f)(x) = g[f(x)] = [f(x)]^3 = (x^{1/3})^3 = x$$

Since $f[g(x)] = x$ for all x in $(-\infty, \infty)$, and $g[f(x)] = x$ for all x in $(-\infty, \infty)$, we conclude that f and g are inverses of each other. In short, $f^{-1}(x) = x^3$. ■

Interpreting Our Results

We can view f as a cube root extracting machine and g as a "cubing" machine. In this light, it is easy to see that one function does undo what the other does. So f and g are indeed inverses of each other.

■ The Graphs of Inverse Functions

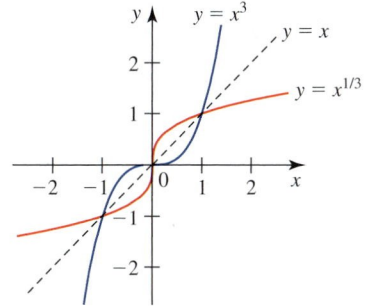

FIGURE 4
The functions $y = x^{1/3}$ and $y = x^3$ are inverses of each other.

The graphs of $f(x) = x^{1/3}$ and $f^{-1}(x) = x^3$ are shown in Figure 4. They seem to suggest that the graphs of inverse functions are mirror images of each other with respect to the line $y = x$. This is true in general, as we will now show.

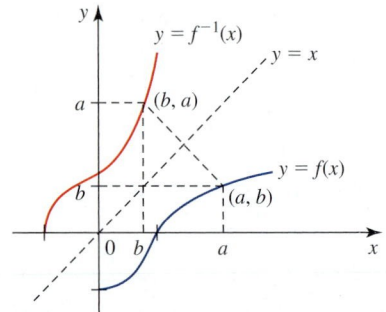

FIGURE 5
The graph of f^{-1}

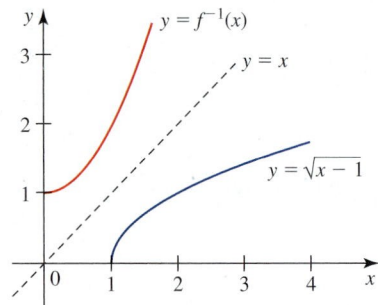

FIGURE 6
The graph of f^{-1} is obtained by reflecting the graph of f with respect to the line $y = x$.

Suppose that (a, b) is any point on the graph of a function f. (See Figure 5.) Then $b = f(a)$, and we have

$$f^{-1}(b) = f^{-1}[f(a)] = a$$

This shows that (b, a) is on the graph of f^{-1} (Figure 5). Similarly, we can show that if (b, a) lies on the graph of f^{-1}, then (a, b) must be on the graph of f. But the point (b, a), as you can see in Figure 5, is the reflection of the point (a, b) with respect to the line $y = x$. We have proved the following.

The Graphs of Inverse Functions

The graph of f^{-1} is the reflection of the graph of f with respect to the line $y = x$ and vice versa.

EXAMPLE 2 Sketch the graph of $f(x) = \sqrt{x - 1}$. Then reflect the graph of f with respect to the line $y = x$ to obtain the graph of f^{-1}.

Solution The graphs of both f and f^{-1} are sketched in Figure 6. ∎

Which Functions Have Inverses

Does every function have an inverse? Consider, for example, the function f defined by $y = x^2$ with domain $(-\infty, \infty)$ and range $[0, \infty)$. From the graph of f shown in Figure 7, you can see that each value of y in the range of $[0, \infty)$ of f is associated with exactly *two* numbers $x = \pm\sqrt{y}$ in the domain $(-\infty, \infty)$ of f (except for $y = 0$). This implies that f does not have an inverse, since the uniqueness requirement of a function cannot be satisfied in this case. Observe that any horizontal line $y = c$, where $c > 0$, intersects the graph of f at more than one point.

Next, consider the function g defined by the same rule as that of f, namely, $y = x^2$, but with domain restricted to $[0, \infty)$. From the graph of g shown in Figure 8, you can see that each value of y in the range $[0, \infty)$ of g is mapped onto exactly *one* number $x = \sqrt{y}$ in the domain $[0, \infty)$ of g. Thus, in this case we can define the inverse function of g, from the range $[0, \infty)$ of g, onto the domain $[0, \infty)$ of g. To find the rule for g^{-1}, we solve the equation $y = x^2$ for x in terms of y. Thus, $x = \sqrt{y}$, since $x \geq 0$, so $g^{-1}(y) = \sqrt{y}$, or, since y is a dummy variable, we can write $g^{-1}(x) = \sqrt{x}$. Also, observe that every horizontal line intersects the graph of g at no more than one point.

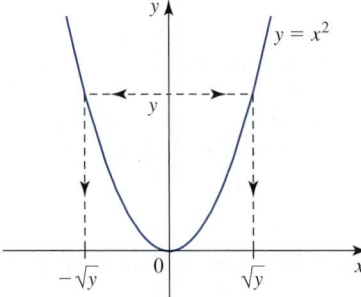

FIGURE 7
Each value of y is associated with two values of x.

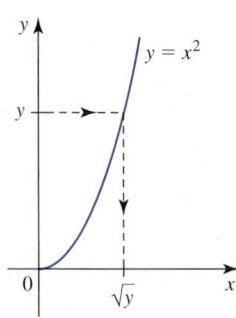

FIGURE 8
Each value of y is associated with exactly one value of x.

Our analysis of the functions f and g reveals the following important difference between the two functions that enables g to have an inverse but not f. Observe that f takes on the same value twice; that is, there are two values of x that are mapped onto each value of y (except $y = 0$). On the other hand, g never takes on the same value more than once; that is, any two values of x have different images. The function g is said to be *one-to-one*.

DEFINITION   **One-to-One Function**

A function f with domain D is **one-to-one** if no two numbers in D have the same image; that is,

$$f(x_1) \neq f(x_2) \qquad \text{whenever} \qquad x_1 \neq x_2$$

Geometrically, a function is one-to-one if every horizontal line intersects its graph at no more than one point. This is called the **horizontal line test.**

We have the following important theorem concerning the existence of an inverse function.

THEOREM 1   **The Existence of an Inverse Function**

A function has an inverse if and only if it is one-to-one.

You will be asked to prove this theorem in Exercise 60.

EXAMPLE 3   Determine whether the function has an inverse.

a. $f(x) = x^{1/3}$ **b.** $f(x) = x^3 - 3x + 1$

Solution
a. Refer to Figure 4, page 75. Using the horizontal line test, we see that f is one-to-one on $(-\infty, \infty)$. Therefore, f has an inverse on $(-\infty, \infty)$.
b. The graph of f is shown in Figure 9. Observe that the horizontal line $y = 1$ intersects the graph of f at three points, so f does not pass the horizontal line test. Therefore, f is not one-to-one. In fact, the three points $x = -\sqrt{3}$, 0, and $\sqrt{3}$ are mapped onto the point 1. Therefore, by Theorem 1, f does not have an inverse.

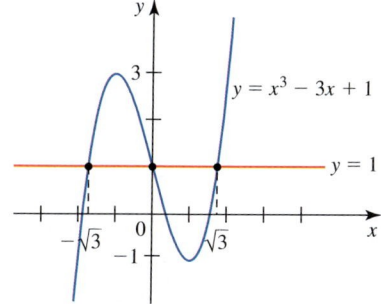

FIGURE 9
f is not one-to-one because it fails the horizontal line test.

Finding the Inverse of a Function

Before looking at the next example, let's summarize the steps for finding the inverse of a function, assuming that it exists.

Guidelines for Finding the Inverse of a Function

1. Write $y = f(x)$.
2. Solve this equation for x in terms of y (if possible).
3. Interchange x and y to obtain $y = f^{-1}(x)$.

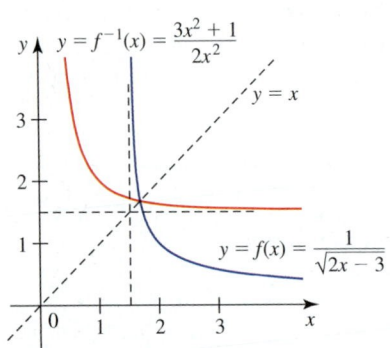

FIGURE 10
The graphs of f and f^{-1}. Notice that
they are reflections of each other with
respect to the line $y = x$.

EXAMPLE 4 Find the inverse of the function defined by $f(x) = \dfrac{1}{\sqrt{2x - 3}}$.

Solution The graph of f shown in Figure 10 shows that f is one-to-one and so f^{-1}
exists. To find the rule for this inverse, write

$$y = \frac{1}{\sqrt{2x - 3}}$$

and then solve the equation for x:

$$y^2 = \frac{1}{2x - 3} \qquad \text{Square both sides.}$$

$$2x - 3 = \frac{1}{y^2} \qquad \text{Take reciprocals.}$$

$$2x = \frac{1}{y^2} + 3 = \frac{3y^2 + 1}{y^2}$$

and

$$x = \frac{3y^2 + 1}{2y^2}$$

Finally, interchanging x and y, we obtain

$$y = \frac{3x^2 + 1}{2x^2}$$

giving the rule for f^{-1} as

$$f^{-1}(x) = \frac{3x^2 + 1}{2x^2}$$

The graphs of both f and f^{-1} are shown in Figure 10. ■

■ Inverse Trigonometric Functions

Generally speaking, the trigonometric functions, being periodic, are not one-to-one
and, therefore, do not have inverse functions. For example, you can see by examining the graph of $y = \sin x$ shown in Figure 11 that this function is not one-to-one,
since it fails the horizontal line test. But observe that by restricting the domain of
the function $f(x) = \sin x$ to the interval $\left[-\frac{\pi}{2}, \frac{\pi}{2}\right]$, it is one-to-one and its range is $[-1, 1]$
(Figure 12a). So, by Theorem 1, f has an inverse function with domain $[-1, 1]$ and
range $\left[-\frac{\pi}{2}, \frac{\pi}{2}\right]$. This function is called the **inverse sine function** or **arcsine function**

FIGURE 11
The horizontal line cuts the graph of
$y = \sin x$ at infinitely many points,
so the sine function is not one-to-one.

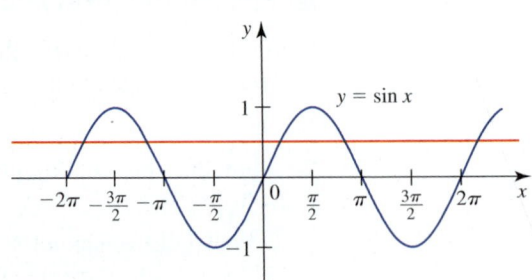

and is denoted by arcsin or $\sin^{-1}$. Thus,

$$y = \sin^{-1} x \qquad \text{if and only if} \qquad \sin y = x$$

where $-1 \le x \le 1$ and $-\frac{\pi}{2} \le y \le \frac{\pi}{2}$. (The graph of $y = \sin^{-1} x$ is shown in Figure 13a.)

Similarly, by suitably restricting the domains of the other five trigonometric functions, each function can also be made one-to-one, and therefore, each function also has an inverse. Figure 12 shows the graphs of the six trigonometric functions and their restricted domains.

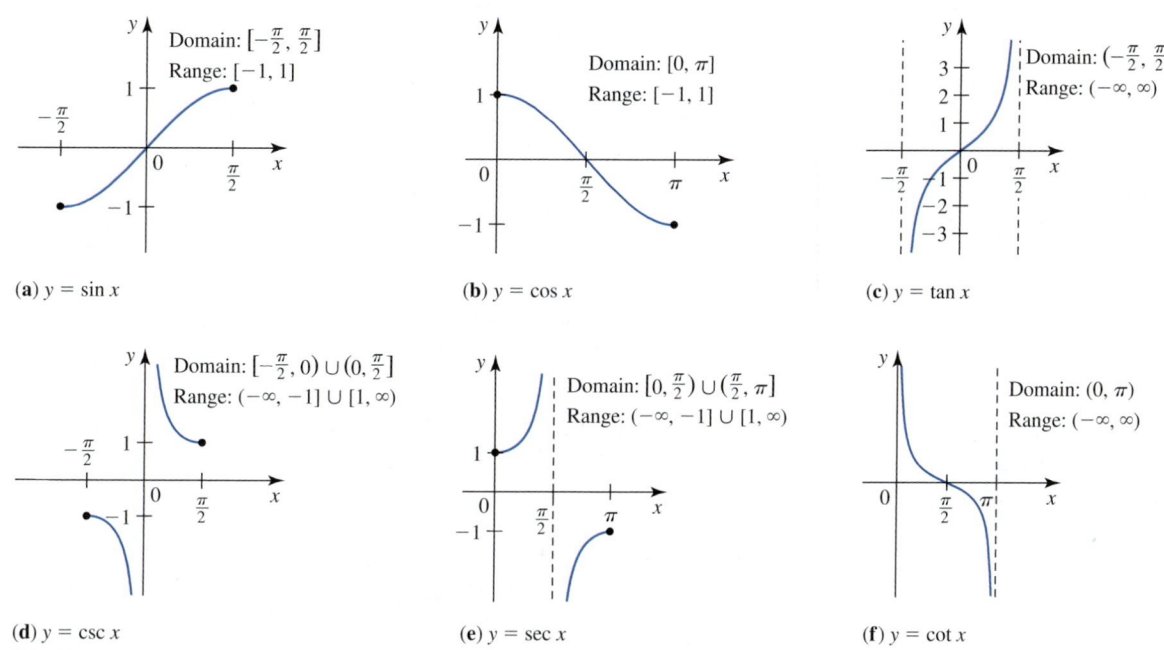

(a) $y = \sin x$ **(b)** $y = \cos x$ **(c)** $y = \tan x$

(d) $y = \csc x$ **(e)** $y = \sec x$ **(f)** $y = \cot x$

FIGURE 12

When restricted to the indicated domains, each of the six trigonometric functions is one-to-one.

With these restrictions the corresponding trigonometric inverse functions are defined as follows.

DEFINITION	Inverse Trigonometric Functions		
		Domain	
$y = \sin^{-1} x$	if and only if $x = \sin y$	$[-1, 1]$	**(2a)**
$y = \cos^{-1} x$	if and only if $x = \cos y$	$[-1, 1]$	**(2b)**
$y = \tan^{-1} x$	if and only if $x = \tan y$	$(-\infty, \infty)$	**(2c)**
$y = \csc^{-1} x$	if and only if $x = \csc y$	$(-\infty, -1] \cup [1, \infty)$	**(2d)**
$y = \sec^{-1} x$	if and only if $x = \sec y$	$(-\infty, -1] \cup [1, \infty)$	**(2e)**
$y = \cot^{-1} x$	if and only if $x = \cot y$	$(-\infty, \infty)$	**(2f)**

The graphs of the six inverse trigonometric functions are shown in Figures 13a–13f.

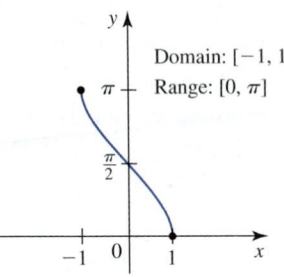

(a) $y = \sin^{-1} x$

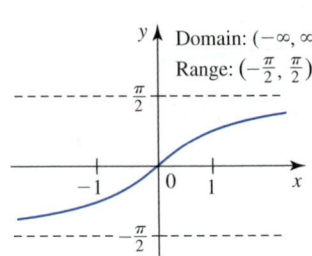

(b) $y = \cos^{-1} x$

Domain: $[-1, 1]$
Range: $[0, \pi]$

Domain: $(-\infty, \infty)$
Range: $\left(-\frac{\pi}{2}, \frac{\pi}{2}\right)$

(c) $y = \tan^{-1} x$

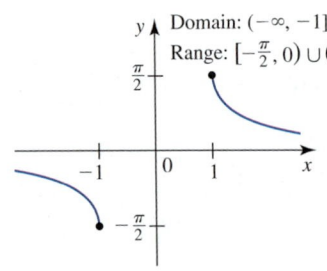

(d) $y = \csc^{-1} x$

Domain: $(-\infty, -1] \cup [1, \infty)$
Range: $\left[-\frac{\pi}{2}, 0\right) \cup \left(0, \frac{\pi}{2}\right]$

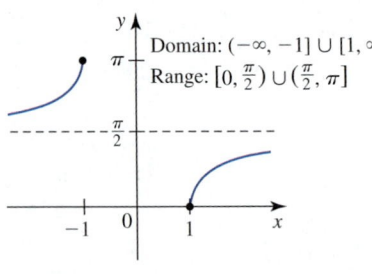

(e) $y = \sec^{-1} x$

Domain: $(-\infty, -1] \cup [1, \infty)$
Range: $\left[0, \frac{\pi}{2}\right) \cup \left(\frac{\pi}{2}, \pi\right]$

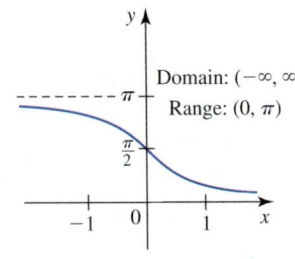

(f) $y = \cot^{-1} x$

Domain: $(-\infty, \infty)$
Range: $(0, \pi)$

FIGURE 13

EXAMPLE 5 Evaluate

a. $\sin^{-1} \dfrac{1}{2}$ **b.** $\cos^{-1}\left(-\dfrac{\sqrt{3}}{2}\right)$ **c.** $\tan^{-1} \sqrt{3}$ **d.** $\cos^{-1} 0.6$

Solution

a. Let $y = \sin^{-1} \frac{1}{2}$. Then by Formula (2a), $\sin y = \frac{1}{2}$. Since y must lie in the interval $\left[-\frac{\pi}{2}, \frac{\pi}{2}\right]$, we see that $y = \pi/6$. Therefore,

$$\sin^{-1} \frac{1}{2} = \frac{\pi}{6}$$

b. Let $y = \cos^{-1}(-\sqrt{3}/2)$ so that, by Formula (2b), $\cos y = -\sqrt{3}/2$. Since y must be in the interval $[0, \pi]$, we see that $y = 5\pi/6$. Therefore,

$$\cos^{-1}\left(-\frac{\sqrt{3}}{2}\right) = \frac{5\pi}{6}$$

c. Let $y = \tan^{-1} \sqrt{3}$ so that $\tan y = \sqrt{3}$. Since y must lie in the interval $\left(-\frac{\pi}{2}, \frac{\pi}{2}\right)$, we see that $y = \pi/3$. Therefore,

$$\tan^{-1} \sqrt{3} = \frac{\pi}{3}$$

d. Here, we use a calculator to find

$$\cos^{-1} 0.6 \approx 0.9273$$

Remember to set the calculator in the *radian mode*.

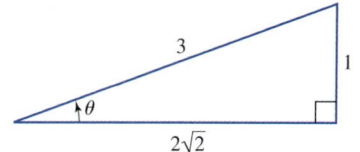

FIGURE 14
The right triangle associated with the equation $\theta = \sin^{-1}\frac{1}{3}$

EXAMPLE 6 Evaluate $\cot\left(\sin^{-1}\frac{1}{3}\right)$.

Solution Let $\theta = \sin^{-1}\frac{1}{3}$. Then θ is the angle in the right triangle with opposite side of length 1 and hypotenuse of length 3. (See Figure 14.) Therefore, by the Pythagorean Theorem the length of the adjacent side of the right triangle is

$$\sqrt{9-1} = 2\sqrt{2}$$

and

$$\cot\left(\sin^{-1}\frac{1}{3}\right) = \cot\theta = \frac{2\sqrt{2}}{1} = 2\sqrt{2} \qquad \blacksquare$$

Recall that if f and f^{-1} are inverses of each other, then

$$f(f^{-1}(x)) = x \qquad \text{and} \qquad f^{-1}(f(x)) = x$$

For the trigonometric functions sine, cosine, and tangent (and similarly for the other three trigonometric functions) these relationships translate into the following properties.

Inverse Properties of Trigonometric Functions

$\sin(\sin^{-1}x) = x$	for	$-1 \leq x \leq 1$	**(3a)**
$\sin^{-1}(\sin x) = x$	for	$-\frac{\pi}{2} \leq x \leq \frac{\pi}{2}$	**(3b)**
$\cos(\cos^{-1}x) = x$	for	$-1 \leq x \leq 1$	**(3c)**
$\cos^{-1}(\cos x) = x$	for	$0 \leq x \leq \pi$	**(3d)**
$\tan(\tan^{-1}x) = x$	for	$-\infty < x < \infty$	**(3e)**
$\tan^{-1}(\tan x) = x$	for	$-\frac{\pi}{2} < x < \frac{\pi}{2}$	**(3f)**

 Remember that these properties hold only for the specified values of x. For example, $\sin^{-1}(\sin \pi) = \sin^{-1}(0) = 0$, but a careless application of the property $\sin^{-1}(\sin x) = x$ with $x = \pi$—which does not lie in the interval $\left[-\frac{\pi}{2}, \frac{\pi}{2}\right]$—leads to the erroneous result $\sin^{-1}(\sin \pi) = \pi$.

EXAMPLE 7 Evaluate

a. $\sin(\sin^{-1} 0.7)$ **b.** $\cos^{-1}(\cos(3\pi/2))$

Solution
a. Since 0.7 lies in the interval $[-1, 1]$, we conclude, by Formula (3a), that

$$\sin(\sin^{-1} 0.7) = 0.7$$

b. Notice that $3\pi/2$ does not lie in the interval $[0, \pi]$, so we may not use Formula (3d). But observe that $\cos(3\pi/2) = 0$, and since 0 lies in the interval $[-1, 1]$, we have

$$\cos^{-1}\left(\cos\frac{3\pi}{2}\right) = \cos^{-1} 0 = \frac{\pi}{2} \qquad \blacksquare$$

0.7   CONCEPT QUESTIONS

1. **a.** What is a one-to-one function? Give an example.
    b. Explain how the horizontal line test is used to determine whether a curve in the plane is the graph of a one-to-one function. Illustrate with a figure.
2. Suppose that f is a one-to-one function with domain $[a, b]$ and range $[c, d]$.
    a. How is f^{-1} defined?
    b. What are the domain and range of f^{-1}? Illustrate with a figure.
3. Suppose that f is a one-to-one function defined by $y = f(x)$.
    a. Describe how to find the rule for f^{-1}. Give an example.

    b. Describe the relationship between the graph of f and that of f^{-1}.
4. For each of the following inverse trigonometric functions, (a) give its definition, (b) give its domain and range, and (c) sketch its graph:
    (i) $f(x) = \sin^{-1} x$ **(ii)** $f(x) = \cos^{-1} x$ **(iii)** $f(x) = \tan^{-1} x$
5. For each of the following inverse trigonometric functions, (a) give its definition, (b) give its domain and range, and (c) sketch its graph:
    (i) $f(x) = \csc^{-1} x$ **(ii)** $f(x) = \sec^{-1} x$ **(iii)** $f(x) = \cot^{-1} x$

0.7   EXERCISES

In Exercises 1–6, show that f and g are inverses of each other by verifying that f[g(x)] = x and g[f(x)] = x.

1. $f(x) = \dfrac{1}{3}x^3$; $g(x) = \sqrt[3]{3x}$

2. $f(x) = \dfrac{1}{x}$; $g(x) = \dfrac{1}{x}$

3. $f(x) = 2x + 3$; $g(x) = \dfrac{x - 3}{2}$

4. $f(x) = x^2 + 1 \ (x \le 0)$; $g(x) = -\sqrt{x - 1}$

5. $f(x) = 4(x + 1)^{2/3}$, where $x \ge -1$;

    $g(x) = \dfrac{1}{8}(x^{3/2} - 8)$, where $x \ge 0$

6. $f(x) = \dfrac{1 + x}{1 - x}$; $g(x) = \dfrac{x - 1}{x + 1}$

In Exercises 7–10, you are given the graph of a function f. Determine whether f is one-to-one.

7.
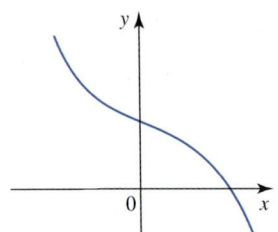

8.
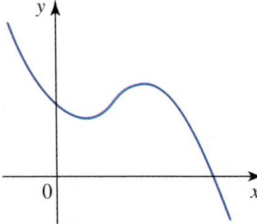

9.
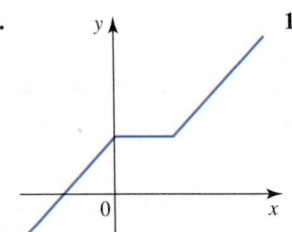

10.
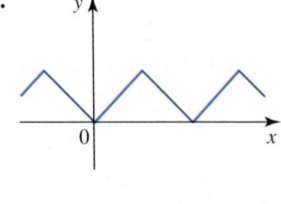

In Exercises 11–14, determine whether the function is one-to-one.

11. $f(x) = 4x - 3$

12. $f(x) = -x^2 + 2x - 3$

13. $f(x) = \sqrt{1 - x}$

14. $f(x) = -x^4 + 16$

15. Suppose that f is a one-to-one function such that $f(2) = 5$. Find $f^{-1}(5)$.

16. Suppose that f is a one-to-one function such that $f(3) = 7$. Find $f[f^{-1}(7)]$.

In Exercises 17–20, find $f^{-1}(a)$ for the function f and the real number a.

17. $f(x) = x^3 + x - 1$; $a = -1$

18. $f(x) = 2x^5 + 3x^3 + 2$; $a = 2$

19. $f(x) = \dfrac{3}{\pi}x + \sin x$; $-\dfrac{\pi}{2} < x < \dfrac{\pi}{2}$; $a = 1$

20. $f(x) = 2 + \tan\left(\dfrac{\pi x}{2}\right)$, $-1 < x < 1$; $a = 2$

21. The graph of f is given. Sketch the graph of f^{-1} on the same set of axes.

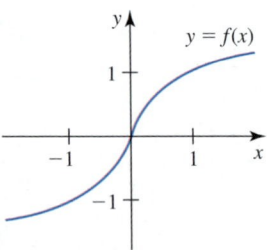

22. The graph of the inverse of a function f, f^{-1}, is given. Sketch the graph of f on the same set of axes.

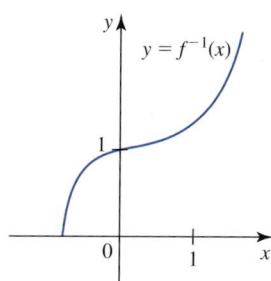

In Exercises 23–28, find the inverse of f. Then sketch the graphs of f and f^{-1} on the same set of axes.

23. $f(x) = x^3 + 1$

24. $f(x) = 2\sqrt{x + 3}$

25. $f(x) = \sqrt{9 - x^2}$, $x \geq 0$

26. $f(x) = x^{3/5} + 1$

27. $f(x) = \sin(2x - 1)$, $\frac{1}{2}\left(1 - \frac{\pi}{2}\right) \leq x \leq \frac{1}{2}\left(1 + \frac{\pi}{2}\right)$

28. $f(x) = \cot^{-1}\left(\frac{x}{3}\right)$, $0 < x < 3\pi$

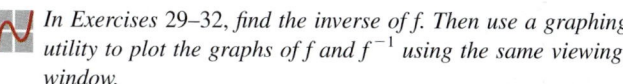

In Exercises 29–32, find the inverse of f. Then use a graphing utility to plot the graphs of f and f^{-1} using the same viewing window.

29. $f(x) = \sqrt[3]{x - 1}$

30. $f(x) = 1 - \frac{1}{x}$

31. $f(x) = \frac{x}{x^2 + 1}$, $-\frac{1}{2} \leq x \leq \frac{1}{2}$

32. $f(x) = \frac{x}{\sqrt{x^2 + 1}}$, $-1 \leq x \leq 1$

33. Let

$$f(x) = \begin{cases} 2x - 1 & \text{if } x < 1 \\ \sqrt{x} & \text{if } 1 \leq x < 4 \\ \frac{1}{2}x^2 - 6 & \text{if } x \geq 4 \end{cases}$$

Find $f^{-1}(x)$, and state its domain.

34. a. Show that $f(x) = -x^2 + x + 1$ on $\left[\frac{1}{2}, \infty\right)$ and $g(x) = \frac{1}{2} + \sqrt{\frac{5}{4} - x}$ on $\left(-\infty, \frac{5}{4}\right)$ are inverses of each other.

b. Solve the equation $-x^2 + x + 1 = \frac{1}{2} + \sqrt{\frac{5}{4} - x}$.
Hint: Use the result of part (a).

In Exercises 35–48, find the exact value of the given expression.

35. $\sin^{-1} 0$

36. $\cos^{-1} 0$

37. $\sin^{-1} \dfrac{1}{2}$

38. $\cos^{-1} \dfrac{1}{2}$

39. $\tan^{-1} \sqrt{3}$

40. $\cot^{-1}(-1)$

41. $\sin^{-1}\left(\dfrac{\sqrt{3}}{2}\right)$

42. $\cos^{-1}\left(-\dfrac{1}{\sqrt{2}}\right)$

43. $\sec^{-1} 2$

44. $\csc^{-1} \sqrt{2}$

45. $\sin^{-1}\left(-\dfrac{1}{2}\right)$

46. $\tan^{-1}\left(-\dfrac{1}{\sqrt{3}}\right)$

47. $\sin\left(\sin^{-1}\dfrac{1}{\sqrt{2}}\right)$

48. $\cos\left(\sin^{-1}\dfrac{1}{2}\right)$

In Exercises 49–54, write the expression in algebraic form.

49. $\cos(\sin^{-1} x)$

50. $\sin(\cos^{-1} x)$

51. $\tan(\tan^{-1} x)$

52. $\sec(\sin^{-1} x)$

53. $\cot(\sec^{-1} x)$

54. $\csc(\cot^{-1} x)$

55. Temperature Conversion The formula $F = f(C) = \frac{9}{5}C + 32$, where $C \geq -273.15$, gives the temperature F (in degrees) on the Fahrenheit scale as a function of the temperature C (in degrees) on the Celsius scale.
a. Find a formula for f^{-1}, and interpret your result.
b. What is the domain of f^{-1}?

56. Motion of a Hot Air Balloon A hot air balloon rises vertically from the ground so that its height after t sec is $h = \frac{1}{2}t^2 + \frac{1}{2}t$, where h is measured in feet and $0 \leq t \leq 60$.
a. Find the inverse of the function $f(t) = \frac{1}{2}t^2 + \frac{1}{2}t$ and explain what it represents.
b. Use the result of part (a) to find the time when the balloon is between an altitude of 120 ft and 210 ft.

57. Aging Population The population of Americans age 55 and over as a percent of the total population is approximated by the function

$$f(t) = 10.72(0.9t + 10)^{0.3} \qquad 0 \leq t \leq 25$$

where t is measured in years and $t = 0$ corresponds to the year 2000.
a. Find the rule for f^{-1}.
b. Evaluate $f^{-1}(25)$, and interpret your result.
Source: U.S. Census Bureau.

58. Special Theory of Relativity According to the special theory of relativity, the relativistic mass of a particle moving with speed v is

$$m = f(v) = \frac{m_0}{\sqrt{1 - \dfrac{v_2}{c^2}}}$$

where m_0 is the rest mass (the mass at zero speed) and $c = 2.9979 \times 10^8$ m/sec is the speed of light in a vacuum.
a. Find f^{-1}, and interpret your result.
b. What is the speed of a particle when its relativistic mass is four times its rest mass?

59. Prove that if f has an inverse, then $(f^{-1})^{-1} = f$.

60. Prove that a function has an inverse if and only if it is one-to-one.

In Exercises 61–66, *determine whether the statement is true or false. If it is true, explain why it is true. If it is false, explain why or give an example to show why it is false.*

61. If f is one-to-one on $(-\infty, \infty)$, then $f^{-1}(f(a)) = a$ if a is a real number.

62. The function $f(x) = 1/x^2$ has an inverse on any interval (a, b), where $a < b$, not containing the origin.

63. If $f(x) = a_{2n+1}x^{2n+1} + a_{2n-1}x^{2n-1} + \cdots + a_1x$, where $a_1, a_3, \ldots, a_{2n+1}$ are nonnegative numbers $(a_{2n+1} \neq 0)$, then f^{-1} exists.

64. $\sin^{-1} x = \dfrac{1}{\sin x}$

65. $\cot^{-1} x = \dfrac{\cos^{-1} x}{\sin^{-1} x}$

66. $(\sin^{-1} x)^2 + (\cos^{-1} x)^2 = 1$

0.8 | Exponential and Logarithmic Functions

0.8 SELF-CHECK DIAGNOSTIC TEST

1. Simplify the expression $e^{\ln x} + \ln e^{2x}$.

2. Solve the equation $\dfrac{200}{1 + 3e^{-0.3t}} = 100$.

3. Find the domain of $f(x) = \dfrac{e^{1/x}}{1 + \ln x}$.

4. Find the inverse of $f(x) = 2\ln(x + 1)$. What is its domain?

5. Express $\ln x + \dfrac{1}{2}\ln(x + 1) - \ln \cos x$ as a single logarithm.

Answers to Self-Check Diagnostic Test 0.8 can be found on page ANS 7.

■ Exponential Functions and Their Graphs

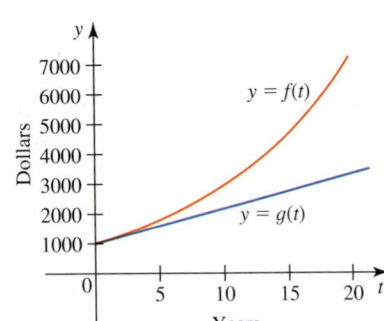

FIGURE 1
Under continuous compounding, a sum of money grows exponentially.

Suppose you deposit a sum of $1000 in an account earning interest at the rate of 10% per year *compounded continuously* (the way most financial institutions compute interest). Then, the accumulated amount at the end of t years ($0 \le t \le 20$) is described by the function f, whose graph appears in Figure 1.* This function is called an *exponential function.* Observe that the graph of f rises rather slowly at first but very rapidly as time goes by. For purposes of comparison, we have also shown the graph of the function $y = g(t) = 1000(1 + 0.10t)$, giving the accumulated amount for the same principal ($1000) but earning *simple* interest at the rate of 10% per year. The moral of the story: It is never too early to save.

Recall that if a is a real number and n is a positive integer, then

$$a^n = \underbrace{a \cdot a \cdot \cdots \cdot a}_{n \text{ factors}}$$

In the expression a^n, a is called the base and n is the exponent or power to which the base is raised. Also, by definition, $a^0 = 1$, and if n is a positive integer, then

$$a^{-n} = \frac{1}{a^n}$$

---

*We will discuss simple and compound interest later in this section. Continuous compound interest will be discussed in Section 3.5.

If p/q is a rational number, where p and q are integers with $q > 0$, then we define the expression $a^{p/q}$ with rational exponent by

$$a^{p/q} = \sqrt[q]{a^p} = (\sqrt[q]{a})^p$$

To define expressions with irrational exponents such as $2^{\sqrt{2}}$, we proceed as follows. Observe that $\sqrt{2} = 1.414213\ldots$. So $\sqrt{2}$ can be approximated successively and with increasing accuracy by the rational numbers

$$1.4, \quad 1.41, \quad 1.414, \quad 1.4142, \quad 1.41421, \quad 1.414213, \quad \ldots$$

Thus, we can expect that $2^{\sqrt{2}}$ may be approximated by the numbers

$$2^{1.4}, \quad 2^{1.41}, \quad 2^{1.414}, \quad 2^{1.4142}, \quad 2^{1.41421}, \quad 2^{1.414213}, \quad \ldots$$

In fact, from Table 1 we see that as x approaches $\sqrt{2}$, the corresponding values of 2^x approach the number $2.665143\ldots$. It can be shown that this number is unique, and we define it to be $2^{\sqrt{2}}$. Furthermore, Table 1 suggests that correct to five decimal places,

$$2^{\sqrt{2}} \approx 2.66514$$

TABLE 1

x	1.4	1.41	1.414	1.4142	1.41421	1.414213
2^x	2.639015 ...	2.657371 ...	2.664749 ...	2.665119 ...	2.665137 ...	2.665143 ...

Similarly, we can define 2^x, where x is an irrational number. In fact, this procedure can be used to define a^x, where a is any positive number and x is an irrational number. In this manner, we see that the number a^x can be defined for *all* real numbers x.

Computations involving exponentials are facilitated by the following laws of exponents.

LAWS OF EXPONENTS

If a and b are positive numbers and x and y are real numbers, then

a. $a^x a^y = a^{x+y}$ **b.** $\dfrac{a^x}{a^y} = a^{x-y}$ **c.** $(a^x)^y = a^{xy}$

d. $(ab)^x = a^x b^x$ **e.** $\left(\dfrac{a}{b}\right)^x = \dfrac{a^x}{b^x}$

EXAMPLE 1

a. $(2^{1/3})(2^{3/5}) = 2^{(1/3)+(2/5)} = 2^{11/15}$ **b.** $\dfrac{3^{1/2}}{3^{1/3}} = 3^{(1/2)-(1/3)} = 3^{1/6}$

c. $(2^x)^3 = 2^{3x}$ **d.** $(4x^3)^{-1/2} = (4^{-1/2})(x^{-3/2}) = \dfrac{1}{2x^{3/2}}$ ■

Since the number a^x $(a > 0)$ is defined for all real numbers x, we can define a function f with the rule given by

$$f(x) = a^x$$

where a is a positive constant and $a \neq 1$. The domain of f is $(-\infty, \infty)$. This function is called an **exponential function with base a.** Examples of exponential functions are

$$f(x) = 2^x, \quad f(x) = \left(\frac{1}{2}\right)^x, \quad \text{and} \quad f(x) = \pi^x$$

An alternative and more rigorous definition of exponential functions is given in Appendix C.

⚠ Do not confuse an exponential function with a power function such as $f(x) = x^2$, encountered in Section 0.6. In the case of the power function the base is a variable, and its exponent is a constant.

EXAMPLE 2 Sketch the graphs of $f(x) = 2^x$, $g(x) = 3^x$, $h(x) = \left(\frac{1}{2}\right)^x$, and $F(x) = \left(\frac{1}{3}\right)^x$ on the same set of axes.

Solution We first construct a table of values for each of the functions (see Table 2). With the help of Table 2 we obtain the graphs of f, g, h, and F shown in Figure 2.

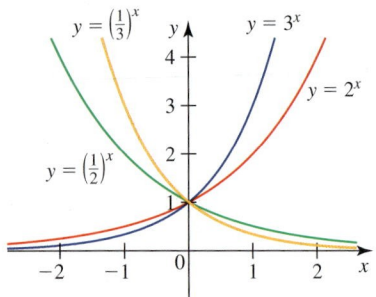

FIGURE 2
The graphs of $f(x) = 2^x$, $g(x) = 3^x$, $h(x) = \left(\frac{1}{2}\right)^x$, and $F(x) = \left(\frac{1}{3}\right)^x$

TABLE 2

x	-4	-3	-2	-1	0	1	2	3	4
2^x	$\frac{1}{16}$	$\frac{1}{8}$	$\frac{1}{4}$	$\frac{1}{2}$	1	2	4	8	16
3^x	$\frac{1}{81}$	$\frac{1}{27}$	$\frac{1}{9}$	$\frac{1}{3}$	0	3	9	27	81
$\left(\frac{1}{2}\right)^x$	16	8	4	2	1	$\frac{1}{2}$	$\frac{1}{4}$	$\frac{1}{8}$	$\frac{1}{16}$
$\left(\frac{1}{3}\right)^x$	81	27	9	3	0	$\frac{1}{3}$	$\frac{1}{9}$	$\frac{1}{27}$	$\frac{1}{81}$

The graphs of $f(x) = 2^x$, $g(x) = 3^x$, $h(x) = \left(\frac{1}{2}\right)^x$, and $F(x) = \left(\frac{1}{3}\right)^x$ obtained in Example 2 are special cases of the graphs of $f(x) = a^x$, obtained by setting $a = 2, 3, \frac{1}{2}$, and $\frac{1}{3}$, respectively. In general, the exponential function $y = a^x$ with $a > 1$ has a graph similar to that of $y = 2^x$ or $y = 3^x$, whereas the graph of $y = a^x$ for $0 < a < 1$ is similar to that of $y = \left(\frac{1}{2}\right)^x$. If $a = 1$, then the function $y = a^x$ reduces to the constant function $y = 1$. The graphs of $y = a^x$ for each of these three cases are shown in Figure 3. Observe that all the graphs pass through the point $(0, 1)$ because $a^0 = 1$. Also, as suggested by Figure 2, the larger a is $(a > 1)$, the faster the graph of $f(x) = a^x$ rises for $x > 0$.

The properties of exponential functions are summarized below.

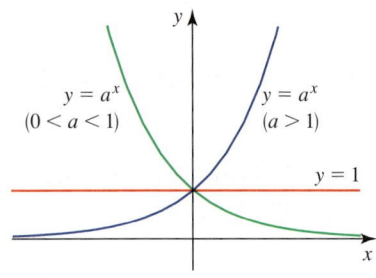

FIGURE 3
The graph of $y = a^x$ rises from left to right if $a > 1$, is constant if $a = 1$, and falls from left to right if $0 < a < 1$.

Properties of Exponential Functions

The exponential function $f(x) = a^x$ ($a > 0$, $a \neq 1$) has the following properties.

1. Its domain is $(-\infty, \infty)$.
2. Its range is $(0, \infty)$.
3. Its graph passes through the point $(0, 1)$.
4. Its graph rises from left to right on $(-\infty, \infty)$ if $a > 1$ and falls from left to right on $(-\infty, \infty)$ if $a < 1$.

EXAMPLE 3 Sketch the graph of the function $f(x) = 1 - 2^x$, and find its domain and range.

Solution The required graph is obtained by first reflecting the graph of $y = 2^x$ (see Figure 4a) to obtain the graph of $y = -2^x$ (see Figure 4b) and then translating this graph upward by 1 unit. The resulting graph is shown in Figure 4c. The domain of f is $(-\infty, \infty)$ and its range is $(-\infty, 1)$.

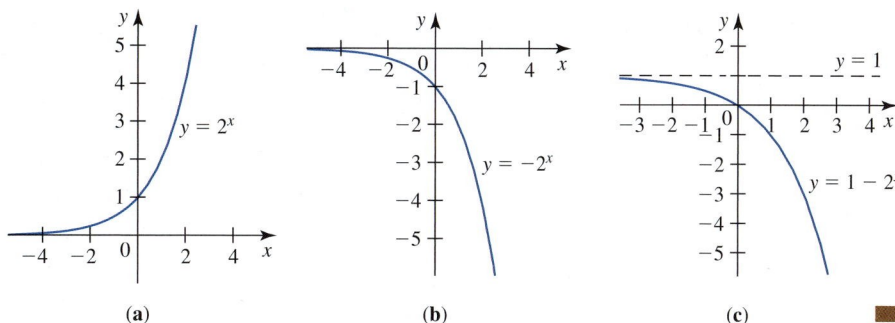

FIGURE 4   (a)   (b)   (c)

■ The Natural Exponential Function

Of all the possible choices for the base of an exponential function, there is one that plays an important role in calculus. This base, denoted by the letter e, is the irrational number whose value, correct to five decimal places, is given by

$$e \approx 2.71828$$

As you will see later on, the use of e for the base of an exponential function enables us to express some of the formulas of calculus in the simplest form possible. The rationale for this choice of the base will be given in Section 2.2.

The function $f(x) = e^x$ is called the **natural exponential function.** Since the number e lies between 2 and 3, we expect the graph of $y = e^x$ to lie between the graphs of $y = 2^x$ and $y = 3^x$, as we will see in Example 4. In the definition of the exponential function $f(x) = a^x$, the base a can be any positive constant. But as mentioned earlier, the choice of e as the base of the exponential function will lead to much simpler calculations in our work ahead. We will give a precise definition of e in Section 2.8.

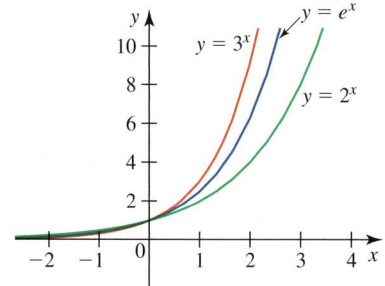

FIGURE 5
The graph of $y = e^x$ lies between the graphs of $y = 2^x$ and $y = 3^x$.

EXAMPLE 4 Sketch the graphs of $f(x) = 2^x$, $g(x) = e^x$, and $h(x) = 3^x$ on the same set of axes.

Solution   The values of $f(x)$ and $h(x)$ for selected values of x were found in Example 2. With the aid of a calculator we obtain the following table. The graphs of f, g, and h are shown in Figure 5.

x	-3	-2	-1	0	1	2	3
e	0.05	0.14	0.37	1	2.72	7.39	20.09

■ Compound Interest

An important application of exponential functions is found in computations involving interest—charges on borrowed money.

Simple interest is interest that is computed on the original principal only. Thus, if I denotes the interest on a principal P (in dollars) at an interest rate of r per year for t years, then

$$I = Prt$$

The **accumulated amount** A, the sum of the principal and interest after t years, is given by

$$A = P + I = P + Prt$$

$$= P(1 + rt) \tag{1}$$

and is a linear function of t.

In contrast to simple interest, earned interest that is periodically added to the principal and thereafter itself earns interest at the same rate is called **compound interest.** To find a formula for the accumulated amount, suppose that P dollars (the principal) is deposited in a bank for a term of t years, earning interest at the rate of r per year (called the **nominal** or **stated rate**) compounded annually. Then, using Equation (1), we see that the accumulated amount at the end of the first year is

$$A_1 = P(1 + rt)$$

To find the accumulated amount A_2 at the end of the second year, we use Equation (1) again, this time with $P = A_1$, since the principal *and* interest now earn interest over the second year. We obtain

$$A_2 = A_1(1 + rt) = P(1 + rt)(1 + rt) = P(1 + rt)^2$$

Continuing, we see that the accumulated amount A after t years is

$$A = P(1 + r)^t \tag{2}$$

Equation (2) was derived under the assumption that interest was compounded *annually*. In practice, however, interest is usually compounded more than once a year. The interval of time between successive interest calculations is called the **conversion period.**

If interest at a nominal rate of r per year is compounded m times a year on a principal of P dollars, then the simple interest rate per conversion period is

$$i = \frac{r}{m} \qquad \frac{\text{annual interest rate}}{\text{number of periods per year}}$$

For example, if the nominal rate is 8% per year ($r = 0.08$) and interest is compounded quarterly ($m = 4$), then

$$i = \frac{r}{m} = \frac{0.08}{4} = 0.02$$

or 2% per period.

To find a general formula for the accumulated amount when a principal of P dollars is deposited in a bank for a term of t years and earns interest at the (nominal) rate of r per year compounded m times per year, we proceed as before, using Equation (2) repeatedly with the interest rate $i = r/m$. We see that the accumulated amount at the end of each period is as follows:

First period: $A_1 = P(1 + i)$

Second period: $A_2 = A_1(1 + i) = [P(1 + i)](1 + i) = P(1 + i)^2$

$\vdots$ $\qquad\qquad\qquad\qquad\qquad\qquad\vdots$

nth period: $A_n = A_{n-1}(1 + i) = [P(1 + i)^{n-1}](1 + i) = P(1 + i)^n$

But there are $n = mt$ periods in t years (number of conversion periods times the term). Therefore, the accumulated amount at the end of t years is given by

$$A = P\left(1 + \frac{r}{m}\right)^{mt} \tag{3}$$

TABLE 3   The accumulated amount A after 3 years when interest is converted m times/year

m	A (dollars)
1	1259.71
2	1265.32
4	1268.24
12	1270.24
365	1271.22

EXAMPLE 5   Find the accumulated amount after 3 years if $1000 is invested at 8% per year compounded annually, semiannually, quarterly, monthly, and daily. (Assume that there are 365 days in a year.)

Solution   We use Equation (3) with $P = 1000$, $r = 0.08$, and $m = 1, 2, 4, 12$, and 365 in succession to obtain the results summarized in Table 3. ∎

■ Logarithmic Functions

If you examine the graph of the exponential function $f(x) = a^x$ where $a > 0$ and $a \neq 1$ (see Figure 3), you will see that it passes the horizontal line test, and so the function f is one-to-one and therefore possesses an inverse function f^{-1}. This function is called the **logarithmic function with base a**. The graph of $f^{-1}(x) = \log_a x$ is obtained by reflecting the graph of $f(x) = a^x$ about the line $y = x$. The graph of $y = \log_a x$ for the case $a > 1$ is given in Figure 6.

The function f^{-1} is called the **logarithmic function with base a** and is denoted by $\log_a$. Using the definition of an inverse function given in Section 0.7,

$$f^{-1}(x) = y \qquad \text{if and only if} \qquad f(y) = x$$

we are led to the following:

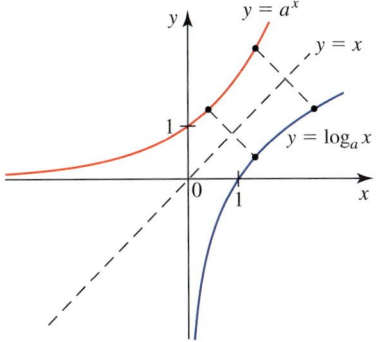

$$\log_a x = y \qquad \text{if and only if} \qquad a^y = x$$

FIGURE 6
The graphs of $f^{-1}(x) = \log_a x$ and $f(x) = a^x$ are mirror reflections about the line $y = x$.

Thus, if $x > 0$, then $\log_a x$ is the exponent to which a must be raised to obtain x. Also because $f(x) = a^x$ and $g(x) = \log_a x$ are inverses of each other, we have

$$a^{\log_a x} = x \qquad \text{for all } x \text{ in } (0, \infty)$$

and

$$\log_a(a^x) = x \qquad \text{for all } x \text{ in } (-\infty, \infty)$$

A summary of the properties of logarithmic functions follows.

Properties of Logarithmic Functions

The logarithmic function $f(x) = \log_a x$ ($a > 0, a \neq 1$) has the following properties.

1. Its domain is $(0, \infty)$.
2. Its range is $(-\infty, \infty)$.
3. Its graph passes through the point $(1, 0)$.
4. Its graph rises from left to right on $(0, \infty)$ if $a > 1$ and falls from left to right if $a < 1$.

The graphs of $y = \log_a x$ for different bases a are shown in Figure 7.

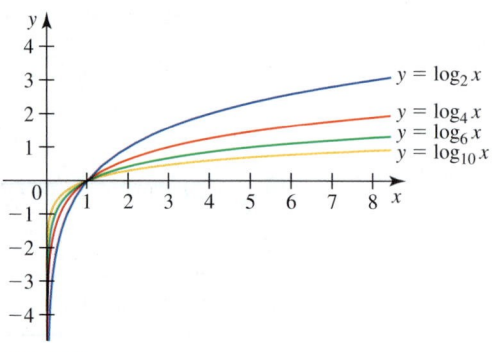

FIGURE 7
The graphs of $y = \log_a x$ for $a = 2, 4, 6,$ and 10

As in the case of exponentials, computations involving logarithms are facilitated by the following laws of logarithms. (These laws are proved in Appendix C.)

LAWS OF LOGARITHMS

If x and y are positive numbers, then

a. $\log_a xy = \log_a x + \log_a y$

b. $\log_a \dfrac{x}{y} = \log_a x - \log_a y$

c. $\log_a x^r = r \log_a x$ where r is any real number

d. $\log_a 1 = 0$

e. $\log_a a = 1$

EXAMPLE 6   Use the laws of logarithms to evaluate $\log_2 40 - \log_2 5$.

Solution   We have

$$\log_2 40 - \log_2 5 = \log_2 \frac{40}{5} \qquad \text{Use Law b.}$$

$$= \log_2 8 = \log_2 2^3$$

$$= 3 \log_2 2 \qquad \text{Use Law c.}$$

$$= 3(1) = 3 \qquad \text{Use Law e.} \qquad \blacksquare$$

Before turning to another example, we mention that the two widely used systems of logarithms are the system of **common logarithms,** which uses the number 10 as the base, and the system of **natural logarithms,** which uses the number e as the base. It is standard practice to write **log** for $\log_{10}$ and **ln** for $\log_e$. As in the case of exponentials, the use of natural logarithms rather than logarithms with other bases leads to simpler expressions.

EXAMPLE 7   Expand and simplify the following expressions:

a. $\log_2 \dfrac{x^2 + 1}{2^x}$ **b.** $\ln \dfrac{x^2\sqrt{x^2 - 1}}{e^x}$

Solution

a.
$$
\begin{aligned}
\log_2 \frac{x^2 + 1}{2^x} &= \log_2 (x^2 + 1) - \log_2 2^x && \text{Use Law b.}\\[1mm]
&= \log_2 (x^2 + 1) - x \log_2 2 && \text{Use Law c.}\\[1mm]
&= \log_2 (x^2 + 1) - x && \text{Use Law e.}
\end{aligned}
$$

b.
$$
\begin{aligned}
\ln \frac{x^2\sqrt{x^2 - 1}}{e^x} &= \ln \frac{x^2(x^2 - 1)^{1/2}}{e^x} && \text{Rewrite.}\\[2mm]
&= \ln x^2 + \frac{1}{2}\ln(x^2 - 1) - x \ln e && \text{Use Laws a, b, and c.}\\[2mm]
&= 2 \ln x + \frac{1}{2}\ln(x^2 - 1) - x \ln e && \text{Use Law c.}\\[2mm]
&= 2 \ln x + \frac{1}{2}\ln(x^2 - 1) - x && \text{Use Law e.}
\end{aligned}
$$

■ Properties Relating the Natural Exponential and the Natural Logarithmic Functions

The following properties follow as an immediate consequence of the definition of the natural logarithm of a number.

Properties Relating e^x and $\ln x$

$$e^{\ln x} = x \qquad x > 0 \tag{4}$$

$$\ln e^x = x \qquad \text{for any real number} \tag{5}$$

The relationships expressed in Equations (4) and (5) are useful in solving equations that involve exponentials and logarithms.

EXAMPLE 8   Solve the equation $2e^{x+2} = 5$.

Solution   We first divide both sides of the equation by 2 to obtain

$$e^{x+2} = \frac{5}{2} = 2.5$$

Next, taking the natural logarithm of each side of the equation and using Equation (5), we have

$$\ln e^{x+2} = \ln 2.5$$

$$x + 2 = \ln 2.5$$

$$x = -2 + \ln 2.5 \approx -1.08$$

EXAMPLE 9 Solve the equation $2 \ln(3x - 5) = 15$.

Solution We have

$$2 \ln(3x - 5) = 15$$

$$\ln(3x - 5) = 7.5$$

$$3x - 5 = e^{7.5}$$

$$x = \frac{1}{3}(e^{7.5} + 5) \approx 604.347$$

Change of Base Formula

As we mentioned earlier, it is sometimes preferable to use one base rather than another when solving a problem. More specifically, we mentioned that we often use natural logarithms to simplify formulas in calculus. The following formula enables us to write the logarithms with any base in terms of natural logarithms.

Change of Base Formula

If a is a positive number and $a \neq 1$, then

$$\log_a x = \frac{\ln x}{\ln a}$$

PROOF Let $y = \log_a x$. Then $x = a^y$. Taking the natural logarithm of both sides of this equation gives $\ln x = \ln a^y = y \ln a$, and so, solving for y, we obtain

$$y = \frac{\ln x}{\ln a}$$

and this proves the result.

EXAMPLE 10 Evaluate $\log_9 7$ correct to five decimal places.

Solution We have

$$\log_9 7 = \frac{\ln 7}{\ln 9} \approx 0.88562$$

0.8 CONCEPT QUESTIONS

1. Define the number e. What is its approximate value?
2. Define the natural exponential function $f(x) = e^x$. What are its domain and range?
3. State the laws of exponents.
4. What is the relationship between the graph of $f(x) = e^x$ and that of $g(x) = \ln x$? Sketch the graphs on the same set of axes.

5. Define the natural logarithmic function $f(x) = \ln x$. What are its domain and range?
6. State the laws of logarithms.

0.8   EXERCISES

In Exercises 1–4, given that $\ln 2 \approx 0.6931$, $\ln 3 \approx 1.0986$, *and* $\ln 5 \approx 1.6094$, *use the laws of logarithms to approximate each expression.*

1. a. $\ln 6$
   b. $\ln \dfrac{3}{2}$

2. a. $\ln \dfrac{20}{\sqrt{3}}$
   b. $\ln \left(\dfrac{15}{2} \right)^{1/3}$

3. a. $\ln 30$
   b. $\ln 7.5$

4. a. $\ln \dfrac{1}{125}$
   b. $\ln \dfrac{5}{9}$

In Exercises 5–10, use the laws of logarithms to expand the expression.

5. $\ln \dfrac{2\sqrt{3}}{5}$
6. $\ln \dfrac{xy}{z}$

7. $\ln \dfrac{x^{1/3} y^{2/3}}{z^{1/2}}$
8. $\ln\left(x^2 \sqrt{x^2 + 1} \right)$

9. $\ln \left(\dfrac{x + 1}{x - 1} \right)^{1/3}$
10. $\ln\left[\sqrt{x} \, |\cos x| \, (x + 1)^{-1/3} \right]$

In Exercises 11–14, use the laws of logarithms to write the expression as the logarithm of a single quantity.

11. $\ln 4 + \ln 6 - \ln 12$
12. $\ln(x^2 - 1) - 2\ln(x + 1)$

13. $3 \ln 2 - \dfrac{1}{2} \ln(x + 1)$

14. $\dfrac{1}{2}[2 \ln(x + 1) + \ln x - \ln(x - 1)]$

In Exercises 15–18, simplify the expression.

15. a. $\ln e^3$
   b. $\ln e^{x^2}$

16. a. $\ln \sqrt{e}$
   b. $\ln e^{\sqrt{e}}$

17. a. $e^{2 \ln 3}$
   b. $e^{\ln \sqrt{x}}$

18. a. $\ln e^{x^2 + 1}$
   b. $e^{2 \ln x + \cos x}$

In Exercises 19–26, find the domain of the function.

19. $f(x) = xe^{-x}$
20. $g(x) = \dfrac{x}{1 - e^x}$

21. $h(t) = \sqrt{2^t - 1}$
22. $f(x) = \sin^{-1}(|x| - 3)$

23. $f(x) = \ln(2x + 1)$
24. $g(x) = \ln(-x)$

25. $g(x) = \ln(\cos x)$
26. $h(x) = \ln\left(\dfrac{x + 1}{x - 1} \right)$

In Exercises 27–32, solve the equation.

27. a. $e^{\ln x} = 2$
   b. $\ln e^{-2x} = 3$

28. a. $\ln(2x + 1) = 3$
   b. $\ln x^2 = 5$

29. a. $2e^{x+2} = 5$
   b. $\ln \sqrt{x + 1} = 1$

30. a. $\ln x + \ln(x - 1) = \ln 2$
   b. $2e^{-0.2x} - 2 = 8$

31. a. $\dfrac{50}{1 + 4e^{0.2x}} = 20$
   b. $e^{2x} - 5e^x + 6 = 0$

32. a. $\ln(x + \sqrt{x^2 + 1}) = 2$
   b. $x^{1/\ln x} - x^2 + 1 = 0$

In Exercises 33–36, determine whether f is even, odd, or neither even nor odd.

33. $f(x) = \dfrac{1 + e^{kx}}{1 - e^{kx}}$
34. $f(x) = \dfrac{2^x}{(1 + 2^x)^2}$

35. $f(x) = \ln \dfrac{1 - x}{1 + x}$
36. $f(x) = \ln(x + \sqrt{1 + x^2})$

In Exercises 37–42, use the graph of $y = \ln x$ *as an aid to sketch the graph of the function.*

37. $f(x) = 2 \ln x$
38. $g(x) = -\ln x$

39. $y = 1 + \ln x$
40. $f(x) = \ln 2x$

41. $g(x) = \ln(x + 1)$
42. $h(x) = \ln|x|$

 43. a. Plot the graphs of $f(x) = \ln x + \ln(x - 1)$ and $g(x) = \ln x(x - 1)$ using the same viewing window.
   b. For what values of x is $f = g$? Prove your assertion.

 44. a. For what values of x is $f = g$ if $f(x) = \ln \sqrt{x}/(x - 1)$ and $g(x) = \frac{1}{2}[\ln x - \ln(x - 1)]$?
   b. Verify the result of part (a) graphically by plotting the graphs of f and g.

In Exercises 45–48, show that the functions are inverses of each other. Sketch the graphs of each pair of functions on the same set of axes.

45. $f(x) = e^{2x}$ and $g(x) = \ln \sqrt{x}$

46. $f(x) = e^{-x}$ and $g(x) = -\ln x$

47. $f(x) = e^{x/2}$ and $g(x) = 2 \ln x$

48. $f(x) = e^{x-1}$ and $g(x) = 1 + \ln x$

 *In Exercises 49–52, find the inverse of f. Then use a graphing utility to plot the graphs of f and* f^{-1} *on the same set of axes.*

49. $f(x) = e^x + 1$
50. $f(x) = \ln(2x + 3)$

51. $f(x) = \dfrac{e^x + 1}{e^x - 1}$
52. $f(x) = 2^{\ln x}$

 53. a. Plot the graph of $f(x) = \tan^{-1}(\tan x)$ using the viewing window $[-10, 10] \times [-2, 2]$.
   b. Is f periodic? Prove your assertion.

54. Sketch the graph of $f(x) = x^{1/\log x}$.

55. Are the functions $f(x) = x$ and $g(x) = e^{\ln x}$ identical? Explain.

56. Over-100 Population On the basis of data obtained from the Census Bureau, the number of Americans over age 100 years is expected to be

$$P(t) = 0.07e^{0.54t} \qquad 0 \le t \le 4$$

where $P(t)$ is measured in millions and t is measured in decades, with $t = 0$ corresponding to the beginning of 2000. What was the population of Americans over age 100 years at the beginning of 2000? What will it be at the beginning of 2030?
Source: U.S. Census Bureau.

57. World Population Growth After its fastest rate of growth ever during the 1980s and 1990s, the rate of growth of world population is expected to slow dramatically, in the twenty-first century. The function

$$G(t) = 1.58e^{-0.213t}$$

gives the projected average percent population growth/decade in the tth decade, with $t = 1$ corresponding to the beginning of 2000. What will the projected average percent population growth rate be at the beginning of 2020 ($t = 3$)?
Source: U.S. Census Bureau.

58. Epidemic Growth During a flu epidemic the number of children in the Woodhaven Community School System who contracted influenza by the tth day is given by

$$N(t) = \frac{5000}{1 + 99e^{-0.8t}}$$

($t = 0$ corresponds to the date when data were first collected.) How many students were stricken by the flu on the first day?

59. Blood Alcohol Level The percentage of alcohol in a person's bloodstream t hr after drinking 8 fluid oz of whiskey is given by

$$A(t) = 0.23te^{-0.4t} \qquad 0 \le t \le 12$$

What is the percentage of alcohol in a person's bloodstream after $\frac{1}{2}$ hr? After 8 hr?
Source: Encyclopedia Britannica.

60. Von Bertalanffy Functions The mass $W(t)$ (in kilograms) of the average female African elephant at age t (in years) can be approximated by a *von Bertalanffy function*

$$W(t) = 2600(1 - 0.51e^{-0.075t})^3$$

a. What is the mass of a newborn female elephant?
b. If a female elephant has a mass of 1600 kg, what is her approximate age?

61. Death Due to Strokes Before 1950, little was known about strokes. By 1960, however, risk factors such as hypertension had been identified. In recent years, CAT scans used as a diagnostic tool have helped to prevent strokes. As a result, deaths due to strokes have fallen dramatically. The function

$$N(t) = 130.7e^{-0.1155t^2} + 50 \qquad 0 \le t \le 6$$

gives the number of deaths per 100,000 people from the beginning of 1950 to the beginning of 2010, where t is measured in decades, with $t = 0$ corresponding to the beginning of 1950.
a. How many deaths due to strokes per 100,000 people were there at the beginning of 1950?
b. If the trend continues, how many deaths due to strokes per 100,000 people will there be at the beginning of 2010?
Source: American Heart Association, Centers for Disease Control and Prevention, and National Institutes of Health.

62. Length of Fish The length (in centimeters) of a typical Pacific halibut t years old is approximately

$$f(t) = 200(1 - 0.956e^{-0.18t})$$

 a. Plot the graph of f using the viewing window $[0, 20] \times [0, 200]$. What is the maximum length that a typical Pacific halibut can attain?
b. What is the approximate length of a typical 10-year-old Pacific halibut?

63. Annuities At the time of retirement, Christine expects to have a sum of $500,000 in her retirement account. Her accountant pointed out to her that if she made withdrawals in monthly installments amounting to x dollars per year ($x > 25,000$), assuming that the account earns interest at the rate of 5% per year compounded continuously, then the time required to deplete her savings would be T years, where

$$T = f(x) = 20 \ln\left(\frac{x}{x - 25,000}\right) \qquad x > 25,000$$

a. Plot the graph of f, using the viewing window $[25,000, 50,000] \times [0, 100]$.
b. How much should Christine plan to withdraw from her retirement account each year if she wants it to last for 25 years?

64. Growth of a Tumor The rate at which a tumor grows with respect to time is given by

$$R = Ax \ln\frac{B}{x}$$

for $0 < x < B$, where A and B are positive constants and x is the radius of the tumor. Plot the graph of R for the case $A = B = 10$.

65. Atmospheric Pressure In the troposphere (lower part of the atmosphere), the atmospheric pressure p is related to the height y from the earth's surface by the equation

$$\ln\left(\frac{p}{p_0}\right) = \frac{Mg}{R\alpha} \ln\left(\frac{T_0 - \alpha y}{T_0}\right)$$

where p_0 is the pressure at the earth's surface, T_0 is the temperature at the earth's surface, M is the molecular mass for air, g is the constant of acceleration due to gravity, R is the ideal gas constant, and α is called the lapse rate of temperature. Find p for $y = 6194$ m (the altitude at the summit of Mount McKinley), taking $M = 28.8 \times 10^{-3}$ kg/mol, $T_0 = 300$ K, $g = 9.8$ m/sec^2, $R = 8.314$ J/mol $\cdot$ K, and $\alpha = 0.006$ K/m. Explain why mountaineers experience difficulty in breathing at very high altitudes.

66. **A Sliding Chain** A chain of length 6 m is held on a table with 1 m of the chain hanging down from the table. Upon release, the chain slides off the table. Assuming that there is no friction, the end of the chain that initially was 1 m from the edge of the table is given by the function

$$s(t) = \frac{1}{2}\left(e^{\sqrt{g/6}\,t} + e^{-\sqrt{g/6}\,t}\right)$$

where $g = 9.8$ m/sec^2 and t is measured in seconds. Find the time it takes for the end of the chain to move 1 m.

67. **Increase in Juvenile Offenders** The number of youths aged 15 to 19 years increased by 21% between 1994 and 2005, pushing up the crime rate. According to the National Council on Crime and Delinquency, the number of violent crime arrests of juveniles under age 18 in year t is given by

$$f(t) = -0.438t^2 + 9.002t + 107 \qquad 0 \le t \le 13$$

where $f(t)$ is measured in thousands and t in years, with $t = 0$ corresponding to the beginning of 1989. According to the same source, if trends such as inner-city drug use and wider availability of guns continues, then the number of violent crime arrests of juveniles under age 18 in year t will be given by

$$g(t) = \begin{cases} -0.438t^2 + 9.002t + 107 & \text{if } 0 \le t < 4 \\ 99.456e^{0.07824t} & \text{if } 4 \le t \le 13 \end{cases}$$

where $g(t)$ is measured in thousands and $t = 0$ corresponds to the beginning of 1989. Compute $f(11)$ and $g(11)$, and interpret your results.
Source: National Council on Crime and Delinquency.

68. **Percent of Females in the Labor Force** Based on data from the U.S. Census Bureau, the following model giving the percent of the total female population in the civilian labor force, $P(t)$, at the beginning of the tth decade ($t = 0$ corresponds to the year 1900) was constructed.

$$P(t) = \frac{74}{1 + 2.6e^{-0.166t + 0.04536t^2 - 0.0066t^3}} \qquad 0 \le t \le 12$$

What was the percent of the total female population in the civilian labor force at the beginning of 2010?
Source: U.S. Census Bureau.

69. **An Extinction Situation** The number of saltwater crocodiles in a certain area of northern Australia t years from now is given by

$$P(t) = \frac{300e^{-0.024t}}{5e^{-0.024t} + 1}$$

a. How many crocodiles were in the population initially?

 b. Plot the graph of P in the viewing window $[0, 200] \times [0, 70]$.
Note: This phenomenon is referred to as an extinction situation.

 70. **Income of American Families** On the basis of data from the Census Bureau, it is estimated that the number of American families y (in millions) who earned x thousand dollars in 1990 is given by the equation

$$y = 0.1584xe^{-0.0000016x^3 + 0.00011x^2 - 0.04491x} \qquad x > 0$$

Plot the graph of the equation in the viewing window $[0, 150] \times [0, 2]$.
Source: House Budget Committee, House Ways and Means Committee, and U.S. Census Bureau.

71. Find the accumulated amount after 5 years on an investment of \$5000 earning interest at the rate of 10% per year compounded (a) annually, (b) semiannually, (c) quarterly, (d) monthly, and (e) daily.

72. Find the accumulated amount after 10 years on an investment of \$10,000 earning interest at the rate of 12% per year compounded (a) annually, (b) semiannually, (c) quarterly, (d) monthly, and (e) daily.

73. **Pension Funds** The managers of a pension fund have invested \$1.5 million in U.S. government certificates of deposit that pay interest at the rate of 5.5% per year compounded semiannually over a period of 10 years. At the end of this period, how much will the investment be worth?

74. **Retirement Funds** Five and a half years ago, Chris invested \$10,000 in a retirement fund that grew at the rate of 10.82% per year compounded quarterly. What is his account worth today?

In Exercises 75–80, determine whether the statement is true or false. If it is true, explain why it is true. If it is false, give an example to show why it is false.

75. The inverse of $f(x) = e^{x/2}$ is $f^{-1}(x) = 2 \ln x$.

76. $f(x) = \dfrac{\cos x}{e^x}$ is not defined at $x = 0$.

77. $e^{3 \ln x} = x^3$ on $(0, \infty)$.

78. $\ln a - \ln b = \ln(a - b)$ for all positive numbers $a > b > 0$.

79. $(\ln x)^3 = 3 \ln x$ for all x in $(0, \infty)$.

80. The domain of $f(x) = \ln|x|$ is $(-\infty, 0) \cup (0, \infty)$.

CHAPTER 0 REVIEW

REVIEW EXERCISES

In Exercises 1–4, find the slope of the line satisfying the given condition.

1. Passes through the points $(-1, 3)$ and $(2, -4)$

2. Has the same slope as the line $2x + 3y = 8$

3. Has the same slope as the line perpendicular to the line $-2x + 4y = -6$

4. Has an angle of inclination of $120°$

In Exercises 5–10, find an equation of the line satisfying the conditions.

5. Passes through $(-2, -4)$ and is parallel to the x-axis

6. Passes through $(1, 3)$ and has slope -4

7. Passes through $(-2, 3)$ and $(4, -5)$

8. Passes through $(2, 3)$ and is parallel to the line $3x + 4y - 8 = 0$

9. Passes through $(-1, 3)$ and is parallel to the line passing through the points $(-3, 4)$ and $(2, 1)$

10. Passes through $(-2, -4)$ and is perpendicular to the line $2x - 3y - 24 = 0$

11. Find an equation of the line passing through the point $(2, -1)$ and the point of intersection of the lines $x + 2y = 3$ and $2x - 3y = 13$.

12. **Dial-up Internet Households** The number of U.S. dial-up Internet households stood at 42.5 million at the beginning of 2004 and was projected to decline at the rate of 3.9 million households per year for the next 6 years.
    a. Find a linear function f giving the projected U.S. dial-up Internet households (in millions) in year t, where $t = 0$ corresponds to the beginning of 2004.
    b. What is the projected number of U.S. dial-up Internet households at the beginning of 2010?
    Source: Strategy Analytics, Inc.

13. **Satellite TV Subscribers** The following table gives the number of satellite TV subscribers in the United States (in millions) from 1998 through 2005 ($x = 0$ corresponds to 1998).

Year, x	0	1	2	3	4	5	6	7
Number, y	8.5	11.1	15.0	17.0	18.9	21.5	24.8	27.4

   a. Plot the number of satellite TV subscribers in the United States (y) versus the year (x).

   b. Draw the line L through the points $(0, 8.5)$ and $(7, 27.4)$.
   c. Find an equation of the line L.
   d. Assuming that this trend continues, estimate the number of satellite TV subscribers in the United States in 2006.
   Sources: National Cable & Telecommunications Association, Federal Communications Commission.

14. If $f(x) = x^2 + x + 1$, find and simplify $\dfrac{f(x + h) - f(x)}{h}$.

15. If $f(x) = \tan x$, find $f(0), f\left(\frac{\pi}{6}\right), f\left(\frac{\pi}{4}\right), f\left(\frac{\pi}{3}\right)$, and $f(\pi)$.

16. Let $f(x) = \begin{cases} \sqrt{-x} & \text{if } x \le 0 \\ x^2 + x & \text{if } x > 0 \end{cases}$

    Find
    a. $f(-4)$
    b. $f(1)$
    c. $\dfrac{f(-1 - h) - f(-1)}{h}, \quad h > 0$
    d. $\dfrac{f(2 + h) - f(2)}{h}, \quad h > 0$

In Exercises 17–23, find the domain of the function.

17. $f(x) = \dfrac{x}{x^2 - 4}$

18. $g(x) = \sqrt{x^2 - 4}$

19. $h(x) = \dfrac{\sqrt{x - 1}}{x(x - 2)}$

20. $f(x) = \sec \pi x$

21. $f(x) = \dfrac{\sin x}{2 - \cos x}$

22. $f(x) = \tan^{-1}(e^{-x})$

23. $h(t) = \ln(e^t - 1)$

In Exercises 24–25, find the domain and sketch the graph of the function. What is its range?

24. $f(x) = \sqrt{1 - x}$

25. $g(t) = |\sin t| + 1$

In Exercises 26–29, determine whether the function is even, odd, or neither.

26. $f(x) = -3x^7 + 4x^3 - 2x$

27. $g(x) = \dfrac{\sin x}{x}, \quad x \ne 0$

28. $f(x) = x\dfrac{e^x + 1}{e^x - 1}$

29. $f(x) = x^3 \ln\left(\dfrac{1 + x}{1 - x}\right), \quad -1 < x < 1$

30. Convert the angle to radian measure.
    a. $120°$ **b.** $450°$ **c.** $-225°$

31. Convert the angle to degree measure.
    a. $\dfrac{11\pi}{6}$ radians   **b.** $-\dfrac{5\pi}{2}$ radians   **c.** $-\dfrac{7\pi}{4}$ radians

32. If $f(x) = \cos x$, find $f(0), f\left(\frac{\pi}{4}\right), f\left(-\frac{\pi}{4}\right), f(3\pi)$ and $f\left(a + \frac{\pi}{2}\right)$.

33. Find all values of θ that satisfy the equation over the interval $[0, 2\pi)$.
    a. $\cos\theta = \dfrac{1}{2}$ **b.** $\cot\theta = -\sqrt{3}$

34. Verify the identity.
    a. $(\sec\theta + \tan\theta)(1 - \sin\theta) = \cos\theta$
    b. $\dfrac{\sec\theta - \cos\theta}{\tan\theta} = \sin\theta$

35. Find the solutions of the equation in $[0, 2\pi)$.
    a. $\cot^2 x - \cot x = 0$ **b.** $\sin x + \sin 2x = 0$

36. If $f(x) = 2x + 3$ and $g(x) = \dfrac{x}{2x^2 - 1}$, find the functions $f + g, g - f, fg, f/g,$ and g/f.

37. Find $g \circ f$ if $f(x) = x^2 - 1$ and $g(x) = \sqrt{x + 1}$. What is its domain?

38. Find functions f and g such that $h = g \circ f$, where $h(x) = \cos^2(\pi x)$.

39. Find functions $f, g,$ and h such that $F = f \circ g \circ h$ if $F(x) = \cos^2(1 + \sqrt{x + 2})$.

40. If $f(x) = 2x$ and $h(x) = 4x^2 - 1$, find g such that $h = g \circ f$.

In Exercises 41–54, solve the equation for x.

41. $\ln x = \dfrac{2}{5}$ **42.** $e^x = 3$

43. $\log_3 x = 2$ **44.** $\log_8(x - 3) = \dfrac{2}{3}$

45. $e^{\sqrt{x}} = 4$ **46.** $e^{x^2} = 15$

47. $2 + 3e^{-x} = 6$ **48.** $\ln x = -1 + \ln(x + 2)$

49. $\ln x + \ln(x - 2) = 0$ **50.** $\dfrac{50}{1 + 4e^{0.2x}} = 20$

51. $3^{2x} - 12 \cdot 3^x + 27 = 0$ **52.** $\ln x^e = 2$

53. $\tan^{-1} x = 1$ **54.** $\cos^{-1}(\sin x) = 0$

In Exercises 55 and 56, solve the equation for x in terms of y.

55. $y = e^{2x} + 2$ **56.** $y = \dfrac{e^x - e^{-x}}{2}$

In Exercises 57 and 58, expand the expression. Assume all variables are positive.

57. $\ln x^3 \sqrt{y/z^2}$ **58.** $\ln \dfrac{\sqrt{x}}{y\sqrt[3]{x^2 + y^2}}$

In Exercises 59 and 60, write the expression as a single logarithm.

59. $2 \ln x + \ln \dfrac{x^3}{y^2} - 4 \ln\sqrt{x + y}$

60. $3 \ln x - \dfrac{1}{3}\ln(yz) + 6 \ln\sqrt{xy}$

In Exercises 61–66, use a transformation to sketch the graph of the function.

61. $y = x^3 - 2$ **62.** $y = 3(x + 2)^2$

63. $y = 2 - \sqrt{x}$ **64.** $y = \dfrac{1}{x + 1}$

65. $y = 3\cos\dfrac{x}{2}$ **66.** $y = |\sin x|$

 67. Plot the graph of $f(x) = x^5 - 3x^2 + x - 1$.

 68. Plot the graph of $f(x) = x^3 - 0.01x^2$.

69. Use a calculator or computer to find the zeros of $f(x) = x^5 - 4x^3 + x^2 - x + 1$ accurate to five decimal places.

 70. Find the point(s) of intersection of the graphs of $f(x) = \cos^2 x$ and $g(x) = 0.1x^2$ accurate to five decimal places.

 71. Find the zero(s) of $f(x) = 2x^5 - 3x^3 + x^2 - 2$ accurate to five decimal places.

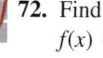 **72.** Find the point(s) of intersection of the graphs of $f(x) = \sin 2x$ and $g(x) = 3x^2 - 2$ accurate to four decimal places.

73. **Clark's Rule** Clark's Rule is a method for calculating pediatric drug dosages on the basis of a child's weight. If a denotes the adult dosage (in milligrams) and w is the weight of the child (in pounds), then the child's dosage is given by

$$D(w) = \dfrac{aw}{150}$$

If the adult dose of a substance is 500 mg, how much should a child who weighs 35 lb receive?

74. **Population Growth** A study prepared for a Sunbelt town's chamber of commerce projected that the population of the town in the next 3 years will grow according to the rule

$$P(t) = 50{,}000 + 30t^{3/2} + 20t$$

where $P(t)$ denotes the population t months from now. By how much will the population increase during the next 9 months? The next 16 months?

75. **Thurstone Learning Curve** Psychologist L.L. Thurstone discovered the following model for the relationship between the learning time T and the length of a list n:

$$T = f(n) = An\sqrt{n - b}$$

where A and b are constants that depend on the person and the task. Suppose that for a certain person and a certain task, $A = 4$ and $b = 4$. Compute $f(4), f(5), \ldots, f(12)$, and use this information to sketch the graph of the function f. Interpret your results.

76. Forecasting Sales The annual sales of Crimson Drug Store are expected to be given by

$$S_1(t) = 2.3 + 0.4t$$

million dollars t years from now, whereas the annual sales of Cambridge Drug Store are expected to be given by

$$S_2(t) = 1.2 + 0.6t$$

million dollars t years from now. When will the annual sales of Cambridge first surpass the annual sales of Crimson?

77. Oil Spills The oil spilling from the ruptured hull of a grounded tanker spreads in all directions in calm waters. Suppose that the area polluted after t sec is a circle of radius r and the radius is increasing at the rate of 2 ft/sec.
   a. Find a function f giving the area polluted in terms of r.
   b. Find a function g giving the radius of the polluted area in terms of t.
   c. Find a function h giving the area polluted in terms of t.
   d. What is the size of the polluted area 30 sec after the hull was ruptured?

78. Film Conversion Prices PhotoMart transfers movie films to DVDs. The fees charged for this service are shown in the following table. Find a function C relating the cost $C(x)$ to the number of feet x of film transferred. Sketch the graph of the function C.

Length of film, x (ft)	Cost for conversion ($)
$1 < x \le 100$	5.00
$100 < x \le 200$	9.00
$200 < x \le 300$	12.50
$300 < x \le 400$	15.00
$x > 400$	$7.00 + 0.02x$

79. Packaging An open box is made from a square piece of cardboard by cutting away identical squares from each corner of

the cardboard and folding up the resulting flaps. The length of one side of the cardboard is 10 in. Let the square cutaways have dimensions x in. by x in.
   a. Draw and label an appropriate figure.
   b. Find a function of x giving the volume of the resulting box.
   c. What is the volume of the box if the cutaway is 1 in. by 1 in.?

80. A closed cylindrical can has a volume of 54 in.^{3}. Find a function S giving the total area of the cylindrical can in terms of r, the radius of the base. What is the total surface area of a closed cylindrical can of radius 4 in.?

81. A man wishes to construct a cylindrical barrel with a capacity of 32π ft^3. The cost of the material for the side of the barrel is $4/ft^2, and the cost of the material for the top and bottom is $8/ft^2.
   a. Draw and label an appropriate figure.
   b. Find a function in terms of the radius of the barrel giving the total cost for constructing the barrel.
   c. What is the total cost for constructing a barrel of radius 2 ft?

82. Linear Depreciation A farmer purchases a new machine for $10,000. The machine is to have a salvage value of $2000 after 5 years. Assuming linear depreciation, find a function giving the book value V of the machine after t years, where $0 \le t \le 5$.

83. Cost of Housing The Brennans are planning to buy a house 4 years from now. Housing experts in their area have estimated that the cost of a home will increase at a rate of 3% per year during that 4-year period. If their predictions are correct, how much can the Brennans expect to pay for a house that currently costs $300,000?

84. Yahoo! in Europe Yahoo! is putting more emphasis on Western Europe, where the number of online households is expected to grow steadily. In a study conducted in 2004, the number of online households (in millions) in Western Europe was projected to be

$$N(t) = 34.68 + 23.88 \ln(1.05t + 5.3) \qquad 0 \le t \le 2$$

where $t = 0$ corresponds to the beginning of 2004. What was the projected number of online households in Western Europe at the beginning of 2005?

Source: Jupiter Research.

A maglev is a train that uses electromagnetic force to levitate, guide, and propel it. Compared to the more conventional steel-wheel and track trains, the maglev has the potential to reach very high speeds, perhaps 600 mph. In Section 1.1 we use the maglev as a vehicle to help us introduce the concept of the *limit* of a function. Specifically, we will see how the limit concept enables us to find the velocity of the maglev knowing only its position as a function of time. Then, generalizing, we use the limit to define the *derivative* of a function, the fundamental tool in differential calculus, which we will use to solve many practical problems in the ensuing chapters.

China Images/Alamy

1   Limits

THE NOTION OF a limit permeates much of our work in calculus. We begin with an intuitive introduction to limits. We then develop techniques that will allow us to find limits much more easily than would be the case if we had to use the definition. The limit of a function allows us to define a very important property of functions: that of continuity. Finally, the limit plays a central role in the study of the rate of change of one quantity with respect to another—the central theme of calculus.

V This symbol indicates that one of the following video types is available for enhanced student learning at **www.academic.cengage.com/login**:
- Chapter lecture videos
- Solutions to selected exercises

1.1   An Intuitive Introduction to Limits

■ A Real-Life Example

A prototype of a maglev (magnetic levitation train) moves along a straight monorail. To describe the motion of the maglev, we can think of the track as a coordinate line. From data obtained in a test run, engineers have determined that the maglev's displacement (directed distance) measured in feet from the origin at time t (in seconds) is given by

$$s = f(t) = 4t^2 \qquad 0 \le t \le 30 \tag{1}$$

where f is called the position function of the maglev. The position of the maglev at time $t = 0, 1, 2, 3, \ldots, 30$, measured in feet from its initial position, is

$$f(0) = 0, \qquad f(1) = 4, \qquad f(2) = 16, \qquad f(3) = 36, \qquad \ldots, \qquad f(30) = 3600$$

(See Figure 1.)

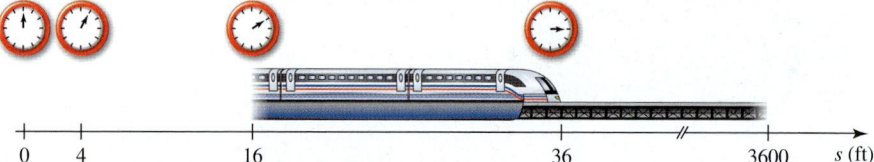

FIGURE 1
A maglev moving along an elevated monorail track

It appears that the maglev is accelerating over the time interval [0, 30] and, therefore, that its velocity varies over time. This raises the following question: Can we find the velocity of the maglev at *any* time in the interval (0, 30) using only Equation (1)? To be more specific, can we find the velocity of the maglev when, say, $t = 2$?

For a start, let's see what quantities we can compute. We can certainly compute the position of the maglev for some selected values of t by using Equation (1), as we did earlier. Using these values of f, we can then compute the *average velocity* of the maglev over any interval of time. For example, to compute the average velocity of the train over the time interval [2, 4], we first compute the **displacement** of the train over that interval, $f(4) - f(2)$, and then divide this quantity by the time elapsed. Thus,

$$\frac{\text{displacement}}{\text{time elapsed}} = \frac{f(4) - f(2)}{4 - 2} = \frac{4(4)^2 - 4(2)^2}{2} = \frac{64 - 16}{2} = 24$$

or 24 ft/sec. Although this is not quite the velocity of the maglev at $t = 2$, it does provide us with an approximation of its velocity at that time.

Can we do better? Intuitively, the smaller the time interval we pick (with $t = 2$ as the left endpoint), the more closely the average velocity over that time interval will approximate the actual velocity of the maglev at $t = 2$.*

Now let's describe this process in general terms. Let $t > 2$. Then the average velocity of the maglev over the time interval [2, t] is given by

$$v_{\text{av}} = \frac{f(t) - f(2)}{t - 2} = \frac{4t^2 - 4(2)^2}{t - 2} = \frac{4(t^2 - 4)}{t - 2} \tag{2}$$

---

*Actually, any interval containing $t = 2$ will do.

By choosing the values of t closer and closer to 2, we obtain a sequence of numbers that gives the average velocities of the maglev over smaller and smaller time intervals. As we observed earlier, this sequence of numbers should approach the *instantaneous velocity* of the train at $t = 2$.

Let's try some sample calculations. Using Equation (2) and taking the sequence $t = 2.5, 2.1, 2.01, 2.001,$ and $2.0001,$ which approaches 2, we find

$$\text{The average velocity over } [2, 2.5] \text{ is } \frac{4(2.5^2 - 4)}{2.5 - 2} = 18 \text{ ft/sec}$$

$$\text{The average velocity over } [2, 2.1] \text{ is } \frac{4(2.1^2 - 4)}{2.1 - 2} = 16.4 \text{ ft/sec}$$

and so forth. These results are summarized in Table 1. From the table we see that the average velocity of the maglev seems to approach the number 16 as it is computed over smaller and smaller time intervals. These computations suggest that the instantaneous velocity of the train at $t = 2$ is 16 ft/sec.

TABLE 1   The average velocity of the maglev

t	2.5	2.1	2.01	2.001	2.0001
v_{av} **over** $[2, t]$	18	16.4	16.04	16.004	16.0004

Note   We cannot obtain the instantaneous velocity for the maglev at $t = 2$ by substituting $t = 2$ into Equation (2) because this value of t is not in the domain of the average velocity function.   ■

■ Intuitive Definition of a Limit

Consider the function g defined by

$$g(t) = \frac{4(t^2 - 4)}{t - 2}$$

which gives the average velocity of the maglev (see Equation (2)). Suppose that we are required to determine the value that $g(t)$ approaches as t approaches the (fixed) number 2. If we take a sequence of values of t approaching 2 from the right-hand side, as we did earlier, we see that $g(t)$ approaches the number 16. Similarly, if we take a sequence of values of t approaching 2 from the left, such as $t = 1.5, 1.9, 1.99, 1.999,$ and $1.9999,$ we obtain the results in Table 2.

TABLE 2   The values of g as t approaches 2 from the left

t	1.5	1.9	1.99	1.999	1.9999
$g(t)$	14	15.6	15.96	15.996	15.9996

Observe that $g(t)$ approaches the number 16 as t approaches 2—this time from the left-hand side. In other words, as t approaches 2 from *either* side of 2, $g(t)$ approaches 16. In this situation we say that the *limit* of $g(t)$ as t approaches 2 is 16, written

$$\lim_{t \to 2} g(t) = \lim_{t \to 2} \frac{4(t^2 - 4)}{t - 2} = 16$$

The graph of the function g, shown in Figure 2, confirms this observation.

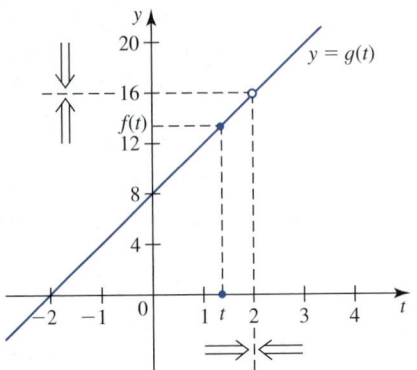

FIGURE 2
As t approaches 2, $g(t)$ approaches 16.

Note   Observe that the number 2 does not lie in the domain of g. (For this reason the point (2, 16) is not on the graph of g, and we indicate this by an open circle on the graph.) Notice, too, that the existence or nonexistence of $g(t)$ at $t = 2$ plays no role in our computation of the limit.

DEFINITION   Limit of a Function at a Number

Let f be a function defined on an open interval containing a, with the possible exception of a itself. Then the limit of $f(x)$ as x approaches a is the number L, written

$$\lim_{x \to a} f(x) = L \tag{3}$$

if $f(x)$ can be made as close to L as we please by taking x to be sufficiently close to a.

EXAMPLE 1   Use the graph of the function f shown in Figure 3 to find the given limit, if it exists.

a. $\lim_{x \to 1} f(x)$ **b.** $\lim_{x \to 3} f(x)$ **c.** $\lim_{x \to 5} f(x)$ **d.** $\lim_{x \to 7} f(x)$ **e.** $\lim_{x \to 10} f(x)$

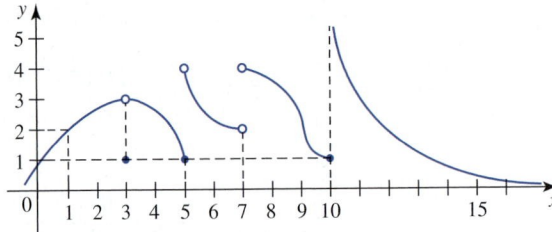

FIGURE 3
The graph of the function f

Solution
a. The values of f can be made as close to 2 as we please by taking x to be sufficiently close to 1. So $\lim_{x \to 1} f(x) = 2$.
b. The values of f can be made as close to 3 as we please by taking x to be sufficiently close to 3. So $\lim_{x \to 3} f(x) = 3$. Observe that $f(3) = 1$, but this has no bearing on the answer.

c. No matter how close x is to 5, there are values of f, corresponding to values of x smaller than 5, that are close to 1; and there are values of f, corresponding to values of x greater than 5, that are close to 4. In other words, there is no *unique* number that $f(x)$ approaches as x approaches 5. Therefore, $\lim_{x \to 5} f(x)$ does not exist. Observe that $f(5) = 1$, but, again, this has no bearing on the existence or nonexistence of the limit.

d. No matter how close x is to 7, there are values of f that are close to 2 (corresponding to values of x less than 7) and values of f that are close to 4 (corresponding to values of x greater than 7). So $\lim_{x \to 7} f(x)$ does not exist. Observe that $x = 7$ is not in the domain of f, but this does not affect our answer.

e. As x approaches 10 from the right, $f(x)$ increases without bound. Therefore, $f(x)$ cannot approach a unique number as x approaches 10, and $\lim_{x \to 10} f(x)$ does not exist. Here, $f(10) = 1$, but this fact plays no role in our determination of the limit. ■

Note   Example 1 shows that when we evaluate the limit of a function f as x approaches a, it is immaterial whether f is defined at a. Furthermore, even if f is defined at a, the value of f at a, $f(a)$, has no bearing on the existence or the value of the limit in question. ■

EXAMPLE 2   Find $\lim_{x \to 2} f(x)$ if it exists, where f is the piecewise-defined function

$$f(x) = \begin{cases} 4x + 8 & \text{if } x \neq 2 \\ 4 & \text{if } x = 2 \end{cases}$$

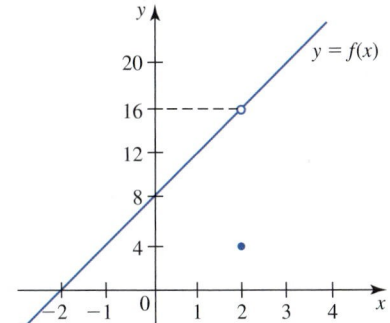

FIGURE 4
The graph of f coincides with the graph of the function g shown in Figure 2, except at $x = 2$.

Solution   From the graph of f shown in Figure 4, we see that $\lim_{x \to 2} f(x) = 16$. If you compare the function f with the function g discussed earlier (page 102), you will see that the values of f are identical to the values of g except at $x = 2$ (Figures 2 and 4). Thus, the limits of $f(x)$ and $g(x)$ as x approaches 2 are equal, as expected. We can see why the graphs of the two functions coincide everywhere except at $x = 2$ by writing

$$g(x) = \frac{4(x^2 - 4)}{x - 2} \qquad \text{\textcolor{orange}{Use x instead of t.}}$$

$$= \frac{4(x + 2)(x - 2)}{x - 2}$$

$$= 4(x + 2) \qquad \text{\textcolor{orange}{Assume that $x \neq 2$.}}$$

which is equivalent to the rule defining f when $x \neq 2$. ■

EXAMPLE 3   **The Heaviside Function**   The Heaviside function H (the unit step function) is defined by

$$H(t) = \begin{cases} 0 & \text{if } t < 0 \\ 1 & \text{if } t \geq 0 \end{cases}$$

This function, named after Oliver Heaviside (1850–1925), can be used to describe the flow of current in a DC electrical circuit that is switched on at time $t = 0$. Show that $\lim_{t \to 0} H(t)$ does not exist.

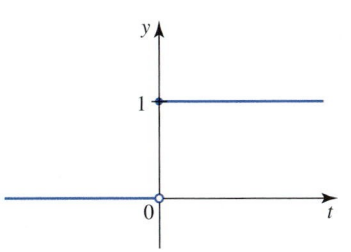

FIGURE 5
$\lim_{t \to 0} H(t)$ does not exist.

Solution   The graph of H is shown in Figure 5. You can see from the graph that no matter how close t is to 0, $H(t)$ takes on the value 1 or 0, depending on whether t is to the right or to the left of 0. Therefore, $H(t)$ cannot approach a unique number L as t approaches 0, and we conclude that $\lim_{t \to 0} H(t)$ does not exist. ■

■ One-Sided Limits

Let's reexamine the Heaviside function. We have shown that $\lim_{t \to 0} H(t)$ does not exist, but what can we say about the behavior of $H(t)$ at values of t that are close to but greater than 0? If you look at Figure 5 again, it is evident that as t approaches 0 through positive values (from the right of 0), $H(t)$ approaches 1. In this situation we say that the right-hand limit of H as t approaches 0 is 1, written

$$\lim_{t \to 0^+} H(t) = 1$$

More generally, we have the following:

DEFINITION **Right-Hand Limit of a Function**

Let f be a function defined for all values of x close to but greater than a. Then the right-hand limit of $f(x)$ as x approaches a is equal to L, written

$$\lim_{x \to a^+} f(x) = L \tag{4}$$

if $f(x)$ can be made as close to L as we please by taking x to be sufficiently close to but greater than a.

Note Equation (4) is just Equation (3) with the further restriction $x > a$. ■

The left-hand limit of a function is defined in a similar manner.

DEFINITION **Left-Hand Limit of a Function**

Let f be a function defined for all values of x close to but less than a. Then the left-hand limit of $f(x)$ as x approaches a is equal to L, written

$$\lim_{x \to a^-} f(x) = L \tag{5}$$

if $f(x)$ can be made as close to L as we please by taking x to be sufficiently close to but less than a.

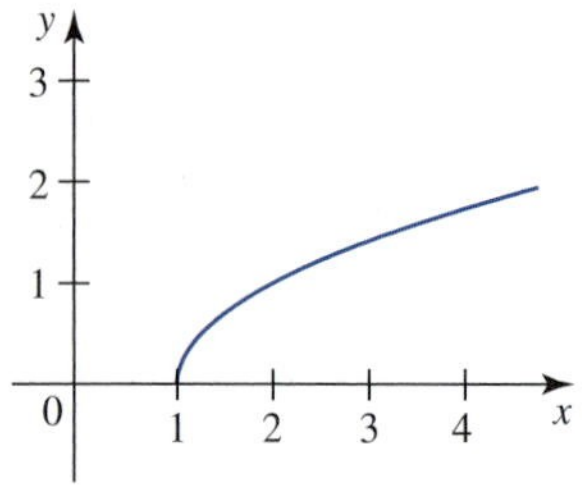

For the function H of Example 3 we have $\lim_{t \to 0^-} H(t) = 0$.

The right-hand and left-hand limits of a function, $\lim_{x \to a^+} f(x)$ and $\lim_{x \to a^-} f(x)$, are often referred to as **one-sided limits,** whereas $\lim_{x \to a} f(x)$ is called a **two-sided limit.**

For some functions it makes sense to look only at one-sided limits. Consider, for example, the function f defined by $f(x) = \sqrt{x - 1}$, whose domain is $[1, \infty)$. Here it makes sense to talk only about the right-hand limit of $f(x)$ as x approaches 1. Also, from Figure 6, we see that $\lim_{x \to 1^+} f(x) = 0$.

FIGURE 6
The right-hand limit of $f(x) = \sqrt{x - 1}$ as x approaches 1 is 0.

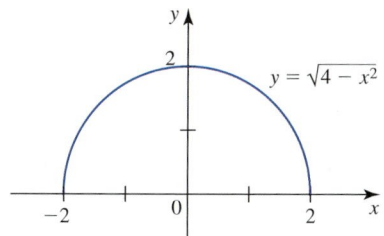

FIGURE 7
We can approach −2 only from the right and 2 only from the left.

EXAMPLE 4 Let $f(x) = \sqrt{4 - x^2}$. Find $\lim_{x \to -2^+} f(x)$ and $\lim_{x \to 2^-} f(x)$.

Solution The graph of f is the upper semicircle shown in Figure 7. From this graph we see that $\lim_{x \to -2^+} f(x) = 0$ and $\lim_{x \to 2^-} f(x) = 0$. ■

Theorem 1 gives the connection between one-sided limits and two-sided limits.

THEOREM 1 **Relationship Between One-Sided and Two-Sided Limits**
Let f be a function defined on an open interval containing a, with the possible exception of a itself. Then

$$\lim_{x \to a} f(x) = L \quad \text{if and only if} \quad \lim_{x \to a^-} f(x) = \lim_{x \to a^+} f(x) = L \qquad \textbf{(6)}$$

Thus, the (two-sided) limit exists if and only if the one-sided limits exist and are equal.

EXAMPLE 5 Sketch the graph of the function f defined by

$$f(x) = \begin{cases} 3 - x & \text{if } x < 1 \\ 1 & \text{if } x = 1 \\ 2 + \sqrt{x - 1} & \text{if } x > 1 \end{cases}$$

Use your graph to find $\lim_{x \to 1^-} f(x)$, $\lim_{x \to 1^+} f(x)$, and $\lim_{x \to 1} f(x)$.

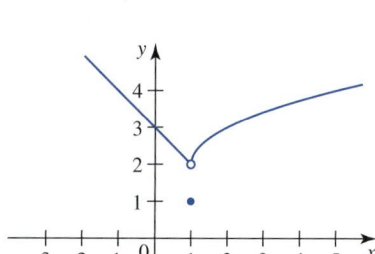

FIGURE 8
$\lim_{x \to 1^-} f(x) = \lim_{x \to 1^+} f(x) = \lim_{x \to 1} f(x) = 2$

Solution From the graph of f, shown in Figure 8, we see that

$$\lim_{x \to 1^-} f(x) = 2 \quad \text{and} \quad \lim_{x \to 1^+} f(x) = 2$$

Since the one-sided limits are equal, we conclude that $\lim_{x \to 1} f(x) = 2$. Notice that $f(1) = 1$, but this has no effect on the value of the limit. ■

EXAMPLE 6 Let $f(x) = \dfrac{\sin x}{x}$. Use your calculator to complete the following table.

x	± 1	± 0.5	± 0.1	± 0.05	± 0.01	± 0.005	± 0.001
$\dfrac{\sin x}{x}$							

Then sketch the graph of f, and use your graph to guess at the value of $\lim_{x \to 0^-} f(x)$, $\lim_{x \to 0^+} f(x)$, and $\lim_{x \to 0} f(x)$.

Solution Using a calculator, we obtain Table 3. (Remember to use radian mode!) The graph of f is shown in Figure 9. We find

$$\lim_{x \to 0^-} f(x) = 1, \qquad \lim_{x \to 0^+} f(x) = 1, \qquad \text{and so} \qquad \lim_{x \to 0} f(x) = 1$$

We will prove in Section 1.2 that our guesses here are correct.

TABLE 3

x	$\dfrac{\sin x}{x}$
± 1	0.841470985
± 0.5	0.958851077
± 0.1	0.998334166
± 0.05	0.999583385
± 0.01	0.999983333
± 0.005	0.999995833
± 0.001	0.999999833

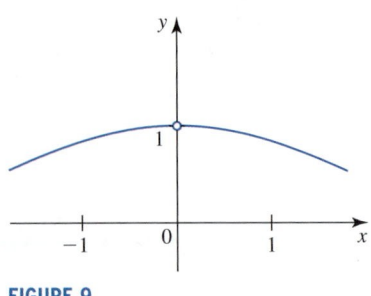

FIGURE 9

The graph of $f(x) = \dfrac{\sin x}{x}$

EXAMPLE 7 Let $f(x) = \dfrac{1}{x^2}$. Evaluate the limit, if it exists.

a. $\lim\limits_{x \to 0^-} f(x)$ **b.** $\lim\limits_{x \to 0^+} f(x)$ **c.** $\lim\limits_{x \to 0} f(x)$

Solution Some values of the function are listed in Table 4, and the graph of f is shown in Figure 10.

TABLE 4

x	$\dfrac{1}{x^2}$
± 1	1
± 0.5	4
± 0.1	100
± 0.05	400
± 0.01	10,000
± 0.001	1,000,000

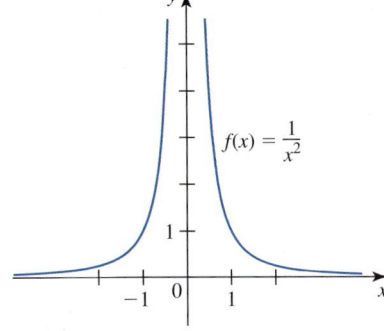

FIGURE 10

As $x \to 0$ from the left (or from the right), $f(x)$ increases without bound.

a. As x approaches 0 from the left, $f(x)$ increases without bound and does not approach a unique number. Therefore, $\lim_{x \to 0^-} f(x)$ does not exist.
b. As x approaches 0 from the right, $f(x)$ increases without bound and does not approach a unique number. Therefore, $\lim_{x \to 0^+} f(x)$ does not exist.
c. From the results of parts (a) and (b) we conclude that $\lim_{x \to 0} f(x)$ does not exist.

Note Even though the limit $\lim_{x \to 0} f(x)$ does not exist, we write $\lim_{x \to 0} (1/x^2) = \infty$ to indicate that $f(x)$ increases without bound as x approaches 0. We will study "infinite limits" in Section 3.5.

■ Using Graphing Utilities to Evaluate Limits

In Example 6 we employed both a numerical and a graphical approach to help us conjecture that

$$\lim_{x \to 0} \frac{\sin x}{x} = 1$$

Either or both of these approaches can often be used to estimate the limit of a function as x approaches a specified value. But there are pitfalls in using graphing utilities, as the following examples show.

EXAMPLE 8   Use a graphing utility to find

$$\lim_{x \to 0} \frac{\sqrt{x + 4} - 2}{x}$$

Solution   We first investigate the problem numerically by constructing a table of values of $f(x) = (\sqrt{x + 4} - 2)/x$ corresponding to values of x that approach 0 from either side of 0. Table 5a shows the values of f for x close to but to the left of 0, and Table 5b shows the values of f for x close to but to the right of 0.

If you look at f evaluated at the first nine values of x shown in each column, we are tempted to conclude that the required limit is $\frac{1}{4}$. But how do we reconcile this result with the last two values of f in each column? Upon reflection we see that this discrepancy can be attributed to a phenomenon known as *loss of significance.*

TABLE 5   Values of f for x close to 0

x	$\dfrac{\sqrt{x + 4} - 2}{x}$	x	$\dfrac{\sqrt{x + 4} - 2}{x}$
-0.001	0.250015627	0.001	0.249984377
-0.0001	0.250001562	0.0001	0.249998438
-10^{-5}	0.25000016	10^{-5}	0.24999984
-10^{-6}	0.25	10^{-6}	0.25
-10^{-7}	0.25	10^{-7}	0.25
-10^{-8}	0.25	10^{-8}	0.25
-10^{-9}	0.25	10^{-9}	0.25
-10^{-10}	0.25	10^{-10}	0.25
-10^{-11}	0.25	10^{-11}	0.25
-10^{-12}	0.3	10^{-12}	0.2
-10^{-13}	0	10^{-13}	0

(**a**) x approaches 0 from the left.   (**b**) x approaches 0 from the right.

When x is very small, the computed values of $\sqrt{x + 4}$ are very close to 2. For $x = -10^{-13}$ or $x = 10^{-13}$ (and values that are smaller in absolute value) the calculator rounds off the value of $\sqrt{x + 4}$ to 2 and gives the value of $f(x)$ as 0. Figures 11a–b show the graphs of f using the viewing windows $[-2, 2] \times [0.2, 0.3]$ and $[-10^{-3}, 10^{-3}] \times [0.2, 0.3]$, respectively. Both these graphs reinforce the earlier observation that the required limit is $\frac{1}{4}$. The graph of f using the viewing window $[-10^{-11}, 10^{-11}] \times [0.24995, 0.25005]$, shown in Figure 11c, proves to be of no help because of the problem with loss of significance stated earlier.

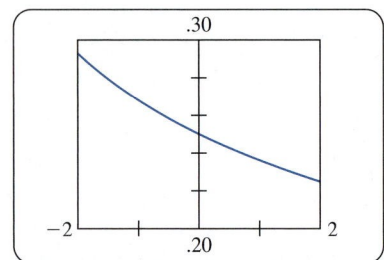

(a) $[-2, 2] \times [0.2, 0.3]$

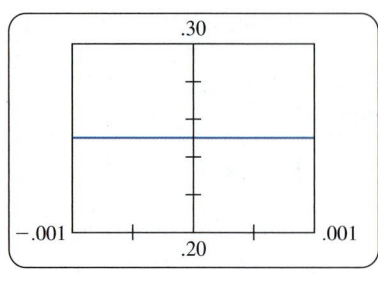

(b) $[-10^{-3}, 10^{-3}] \times [0.2, 0.3]$

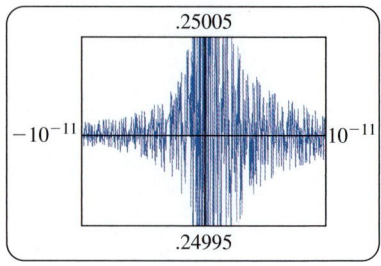

(c) $[-10^{-11}, 10^{-11}] \times [0.24995, 0.25005]$

FIGURE 11

The graphs of $f(x) = \dfrac{\sqrt{x + 4} - 2}{x}$ in different viewing windows

Having recognized the source of the difficulty, how can we remedy the situation? Let's find another expression for $f(x)$ that does not involve subtracting numbers that are so close to each other that it results in a loss of significance. Rationalizing the numerator, we obtain

$$f(x) = \frac{\sqrt{x + 4} - 2}{x} = \frac{\sqrt{x + 4} - 2}{x} \cdot \frac{\sqrt{x + 4} + 2}{\sqrt{x + 4} + 2}$$

$$= \frac{(x + 4) - 4}{x(\sqrt{x + 4} + 2)} \qquad (a + b)(a - b) = a^2 - b^2$$

$$= \frac{1}{\sqrt{x + 4} + 2} \qquad x \neq 0$$

Observe that the use of the last expression avoids the pitfalls that we encountered with the original expression. We leave it as an exercise to show that both a numerical analysis and a graphical analysis of

$$\lim_{x \to 0} \frac{1}{\sqrt{x + 4} + 2}$$

suggest that a good guess for

$$\lim_{x \to 0} \frac{\sqrt{x + 4} - 2}{x} = \lim_{x \to 0} \frac{1}{\sqrt{x + 4} + 2}$$

is $\frac{1}{4}$, a result that can be proved analytically using the techniques to be developed in the next section.

EXAMPLE 9 Find $\lim_{x \to 0} \sin \dfrac{1}{x}$.

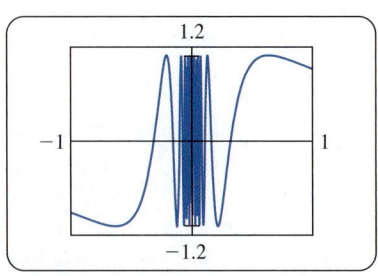

FIGURE 12

The graph of $f(x) = \sin(1/x)$ in the viewing window $[-1, 1] \times [-1.2, 1.2]$

Solution Let $f(x) = \sin(1/x)$. The graph of f using the viewing window $[-1, 1] \times [-1.2, 1.2]$ does not seem to be of any help to us in finding the required limit (see Figure 12). To obtain a more accurate graph of $f(x) = \sin(1/x)$, note that the sine function is bounded by -1 and 1. Thus, the graph of f lies between the horizontal lines $y = -1$ and $y = 1$. Next, observe that the sine function has period 2π. Since $1/x$ increases without bound (decreases without bound) as x approaches 0 from the right (from the left), we see that $\sin(1/x)$ undergoes more and more cycles as x approaches 0. Thus, the graph of $f(x) = \sin(1/x)$ oscillates between -1 and 1, as shown in Figure 13. Therefore, it seems reasonable to conjecture that the limit does not exist. Indeed, we can demonstrate this conclusion by constructing Table 6.

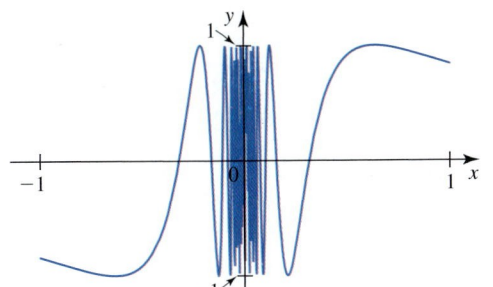

FIGURE 13
The graph of $f(x) = \sin(1/x)$

TABLE 6

x	$\dfrac{2}{\pi}$	$\dfrac{2}{3\pi}$	$\dfrac{2}{5\pi}$	$\dfrac{2}{7\pi}$	$\dfrac{2}{9\pi}$	$\dfrac{2}{11\pi}$	$\cdots$
$\sin\dfrac{1}{x}$	1	-1	1	-1	1	-1	$\cdots$

Note that the values of x approach 0 from the right. From the table we see that no matter how close x is to 0 (from the right), there are values of f corresponding to these values of x that are equal to 1 or -1. Therefore, $f(x)$ cannot approach any fixed number as x approaches 0. A similar result is true if the values of x approach 0 from the left. This shows that

$$\lim_{x \to 0} \sin\frac{1}{x}$$

does not exist.

1.1 CONCEPT QUESTIONS

1. Explain what is meant by the statement $\lim_{x \to 2} f(x) = 3$.
2. **a.** If $\lim_{x \to 3} f(x) = 5$, what can you say about $f(3)$? Explain.
   b. If $f(2) = 6$, what can you say about $\lim_{x \to 2} f(x)$? Explain.

3. Explain what is meant by the statement $\lim_{x \to 3^-} f(x) = 2$.
4. Suppose $\lim_{x \to 1^-} f(x) = 3$ and $\lim_{x \to 1^+} f(x) = 4$.
   a. What can you say about $\lim_{x \to 1} f(x)$? Explain.
   b. What can you say about $f(1)$? Explain.

1.1 EXERCISES

In Exercises 1–6, use the graph of the function f to find each limit.

1.

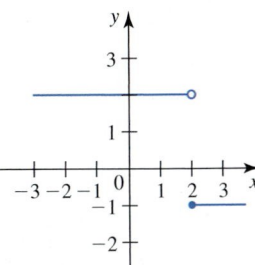

a. $\lim_{x \to 2^-} f(x)$

b. $\lim_{x \to 2^+} f(x)$

c. $\lim_{x \to 2} f(x)$

2.

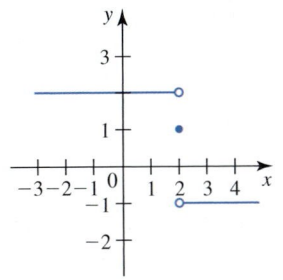

a. $\lim_{x \to 2^-} f(x)$

b. $\lim_{x \to 2^+} f(x)$

c. $\lim_{x \to 2} f(x)$

3.

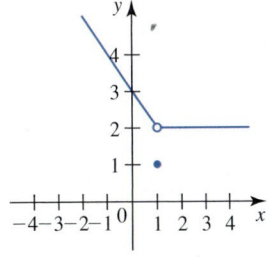

a. $\lim_{x \to 1^-} f(x)$

b. $\lim_{x \to 1^+} f(x)$

c. $\lim_{x \to 1} f(x)$

4.

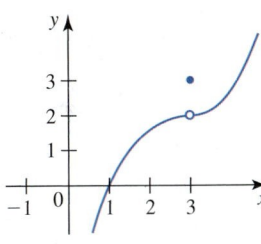

a. $\lim_{x \to 3^-} f(x)$

b. $\lim_{x \to 3^+} f(x)$

c. $\lim_{x \to 3} f(x)$

5.

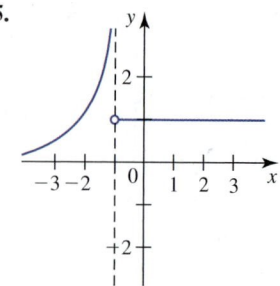

a. $\lim\limits_{x \to -1^-} f(x)$　**b.** $\lim\limits_{x \to -1^+} f(x)$　**c.** $\lim\limits_{x \to -1} f(x)$

6.

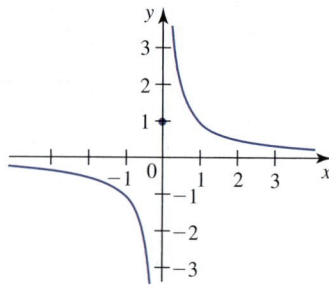

a. $\lim\limits_{x \to 0^-} f(x)$　**b.** $\lim\limits_{x \to 0^+} f(x)$　**c.** $\lim\limits_{x \to 0} f(x)$

7. Use the graph of the function f to determine whether each statement is true or false. Explain.

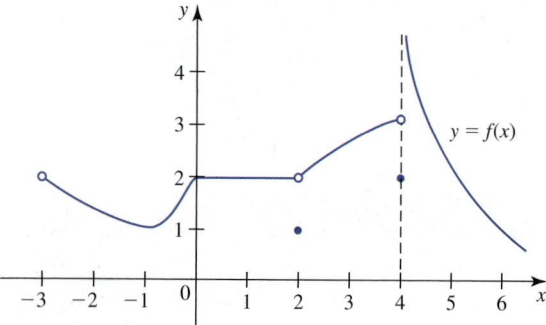

a. $\lim\limits_{x \to -3^+} f(x) = 2$　　　**b.** $\lim\limits_{x \to 0} f(x) = 2$

c. $\lim\limits_{x \to 2} f(x) = 1$　　　　**d.** $\lim\limits_{x \to 4^-} f(x) = 3$

e. $\lim\limits_{x \to 4^+} f(x)$ does not exist　**f.** $\lim\limits_{x \to 4} f(x) = 2$

8. Use the graph of the function f to determine whether each statement is true or false. Explain.

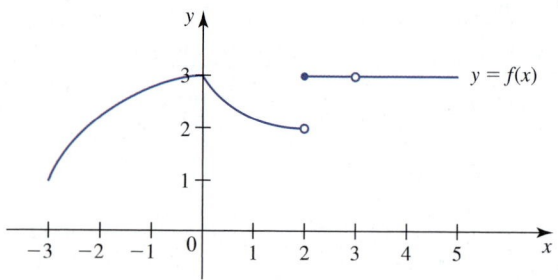

a. $\lim\limits_{x \to -3^+} f(x) = 1$　　　**b.** $\lim\limits_{x \to 0} f(x) = f(0)$

c. $\lim\limits_{x \to 2^-} f(x) = 2$　　　　**d.** $\lim\limits_{x \to 2^+} f(x) = 3$

e. $\lim\limits_{x \to 3} f(x)$ does not exist　**f.** $\lim\limits_{x \to 5^-} f(x) = 3$

In Exercises 9–16, complete the table by computing $f(x)$ at the given values of x, accurate to five decimal places. Use the results to guess at the indicated limit, if it exists.

9. $\lim\limits_{x \to 1} \dfrac{x - 1}{x^2 - 3x + 2}$

x	0.9	0.99	0.999	1.001	1.01	1.1
$f(x)$						

10. $\lim\limits_{x \to 1} \dfrac{x - 1}{x^2 + x - 2}$

x	0.9	0.99	0.999	1.001	1.01	1.1
$f(x)$						

11. $\lim\limits_{x \to 2} \dfrac{\sqrt{x + 2} - 2}{x - 2}$

x	1.9	1.99	1.999	2.001	2.01	2.1
$f(x)$						

12. $\lim\limits_{x \to 0} \dfrac{\sqrt{3 + x} - \sqrt{3 - x}}{x}$

x	-0.1	-0.01	-0.001	0.001	0.01	0.1
$f(x)$						

13. $\lim\limits_{x \to 2} \dfrac{\dfrac{1}{\sqrt{2 + x}} - \dfrac{1}{2}}{x - 2}$

x	1.9	1.99	1.999	2.001	2.01	2.1
$f(x)$						

14. $\lim\limits_{x \to 3} \dfrac{3\sqrt{x + 1} - 2x}{x(x - 3)}$

x	2.9	2.99	2.999	3.001	3.01	3.1
$f(x)$						

15. $\lim\limits_{x \to 0} \dfrac{e^x - 1}{x}$

x	−0.1	−0.01	−0.001	0.001	0.01	0.1
f(x)						

16. $\lim\limits_{x \to 0} \dfrac{\tan^{-1} 2x}{\ln(1 + 2x)}$

x	−0.1	−0.01	−0.001	0.001	0.01	0.1
f(x)						

In Exercises 17–22, sketch the graph of the function f and evaluate (a) $\lim_{x \to a^-} f(x)$, *(b)* $\lim_{x \to a^+} f(x)$, *and (c)* $\lim_{x \to a} f(x)$ *for the given value of a.*

17. $f(x) = \begin{cases} x - 1 & \text{if } x \le 3 \\ -2x + 8 & \text{if } x > 3 \end{cases}$; $\quad a = 3$

18. $f(x) = \begin{cases} 2x - 4 & \text{if } x < 4 \\ x - 2 & \text{if } x \ge 4 \end{cases}$; $\quad a = 4$

19. $f(x) = \begin{cases} -e^{-x} & \text{if } x \ne 0 \\ 1 & \text{if } x = 0 \end{cases}$; $\quad a = 0$

20. $f(x) = \begin{cases} x^2 - 1 & \text{if } x \ne 0 \\ 1 & \text{if } x = 0 \end{cases}$; $\quad a = 0$

21. $f(x) = \begin{cases} x & \text{if } x < 1 \\ 2 & \text{if } x = 1 \\ -x + 2 & \text{if } x > 1 \end{cases}$; $\quad a = 1$

22. $f(x) = \begin{cases} x^2 - 1 & \text{if } x < 1 \\ 2 & \text{if } x = 1 \\ \ln x & \text{if } x > 1 \end{cases}$; $\quad a = 1$

The symbol ⟦ ⟧ *denotes the greatest integer function defined by* ⟦x⟧ = *the greatest integer n such that n ≤ x. For example,* ⟦2.8⟧ = 2, *and* ⟦−2.7⟧ = −3. *In Exercises 23–28, use the graph of the function to find the indicated limit, if it exists.*

23. $\lim\limits_{x \to 3^-} \llbracket x \rrbracket$

24. $\lim\limits_{x \to 3^+} \llbracket x \rrbracket$

25. $\lim\limits_{x \to -1^+} \llbracket x \rrbracket$

26. $\lim\limits_{x \to -1} \llbracket x \rrbracket$

27. $\lim\limits_{x \to 3.1} \llbracket x \rrbracket$

28. $\lim\limits_{x \to 2.4} \llbracket 2x \rrbracket$

29. Let

$$f(x) = \begin{cases} 0 & \text{if } x \le 0 \\ \sin \dfrac{1}{x} & \text{if } x > 0 \end{cases}$$

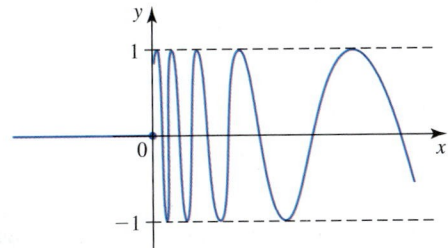

(As *x* approaches 0 from the right, *y* oscillates more and more.) Use the figure and construct a table of values to guess at $\lim_{x \to 0^+} f(x)$, $\lim_{x \to 0^-} f(x)$, and $\lim_{x \to 0} f(x)$. Justify your answer.

30. Let

$$f(x) = \begin{cases} 0 & \text{if } x = 0 \\ x \sin \dfrac{1}{x} & \text{if } x \ne 0 \end{cases}$$

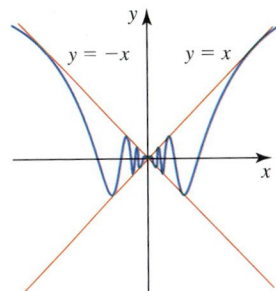

Use the figure, and construct a table of values to guess at $\lim_{x \to 0^+} f(x)$, $\lim_{x \to 0^-} f(x)$, and $\lim_{x \to 0} f(x)$. Justify your answer.

31. Let

$$f(x) = \begin{cases} \dfrac{1}{x} & \text{if } x < 0 \\ \sin x & \text{if } 0 \le x < \pi \\ 0 & \text{if } x \ge \pi \end{cases}$$

a. Sketch the graph of *f*.
b. Find all values of *x* in the domain of *f* at which the limit of *f* exists.
c. Find all values of *x* in the domain of *f* at which the left-hand limit of *f* exists.
d. Find all values of *x* in the domain of *f* at which the right-hand limit of *f* exists.

32. Let

$$f(x) = \begin{cases} -x^2 & \text{if } x < 0 \\ \tan x & \text{if } 0 \le x < \frac{\pi}{2} \\ \ln\left(x - \frac{\pi}{2} + 1\right) & \text{if } x \ge \frac{\pi}{2} \end{cases}$$

a. Sketch the graph of f.

b. Find all values of x in the domain of f at which the limit of f exists.

c. Find all values of x in the domain of f at which the left-hand limit of f exists.

d. Find all values of x in the domain of f at which the right-hand limit of f exists.

33. The Heaviside Function A generalization of the unit step function or Heaviside function H of Example 3 is the function H_c defined by

$$H_c(t - t_0) = \begin{cases} 0 & \text{if } t < t_0 \\ c & \text{if } t \ge t_0 \end{cases}$$

where c is a constant and $t_0 \ge 0$. Show that if $c \ne 0$, then $\lim_{t \to t_0} H_c(t - t_0)$ does not exist.

34. The Square-Wave Function The square-wave function f can be expressed in terms of the Heaviside function (Exercise 33) as follows:

$$f(t) = H_k(t) - H_k(t - k) + H_k(t - 2k)$$
$$- H_k(t - 3k) + H_k(t - 4k) - \cdots$$

Referring to the following figure, show that $\lim_{t \to nk} f(t)$ does not exist for $n = 1, 2, 3, \ldots$.

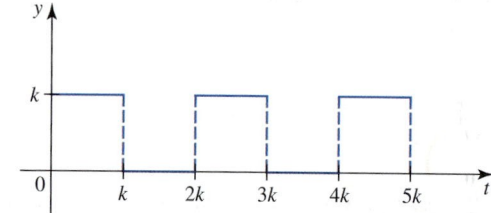

35. Let $f(h) = (1 + h)^{1/h}$, and assume that $\lim_{h \to 0}(1 + h)^{1/h}$ exists. (We will establish this in Section 2.8.) Find its value to four decimal places of accuracy by computing $f(h)$ for $h = 0.1, 0.01, 0.001, 0.0001, 0.00001, 0.000001,$ and 0.0000001.

36. Let $f(\theta) = (\tan \theta - \theta)/\theta^3$. By computing $f(\theta)$ for $\theta = \pm 0.1$, ± 0.01, and ± 0.001, accurate to five decimal places, guess at $\lim_{\theta \to 0}(\tan \theta - \theta)/\theta^3$.

In Exercises 37–42, plot the graph of f. Then zoom-in to guess at the specified limit (if it exists).

37. $f(x) = \dfrac{2x^2 - x - 6}{x - 2};\quad \lim_{x \to 2} f(x)$

38. $f(x) = \dfrac{x - 3}{\sqrt{x + 1} - 2};\quad \lim_{x \to 3} f(x)$

39. $f(x) = \dfrac{x^3 + x^2 - 3x + 1}{|x - 1|};\quad \lim_{x \to 1} f(x)$

40. $f(x) = \dfrac{\tan x}{x};\quad \lim_{x \to 0} f(x)$

41. $f(x) = \dfrac{\sin^{-1}\sqrt{x}}{1 - \cos\sqrt[4]{x}};\quad \lim_{x \to 0^+} f(x)$

42. $f(x) = \dfrac{e^{\tan 3x} - 1}{\ln(1 + \sin 2x)};\quad \lim_{x \to 0} f(x)$

In Exercises 43–46, determine whether the statement is true or false. If it is true, explain why. If it is false, explain why or give an example that shows it is false.

43. If $\lim_{x \to a} f(x) = c$, then $f(a) = c$.

44. If f is defined at a, then $\lim_{x \to a} f(x)$ exists.

45. If $\lim_{x \to a} f(x) = \lim_{x \to a} g(x)$, then $f(a) = g(a)$.

46. If both $\lim_{x \to a^+} f(x)$ and $\lim_{x \to a^-} f(x)$ exist, then $\lim_{x \to a} f(x)$ exists.

1.2 Techniques for Finding Limits

■ Computing Limits Using the Laws of Limits

In Section 1.1 we used tables of functional values and graphs of functions to help us guess at the limit of a function, if it exists. This approach, however, is useful only in suggesting whether the limit exists and what its value might be for simple functions. In practice, the limit of a function is evaluated by using the laws of limits that we now introduce.

LAW 1   Limit of a Constant Function $f(x) = c$

If c is a real number, then

$$\lim_{x \to a} c = c$$

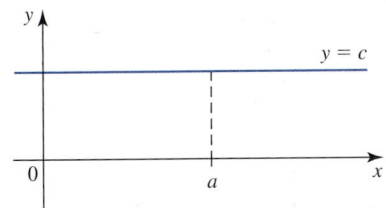

FIGURE 1
For the constant function $f(x) = c$, $\lim_{x \to a} f(x) = c$.

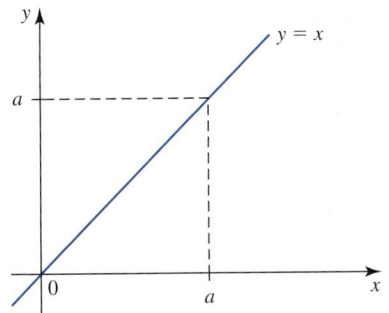

FIGURE 2
If f is the identity function $f(x) = x$, then $\lim_{x \to a} f(x) = a$.

You can see this intuitively by studying the graph of the constant function $f(x) = c$ shown in Figure 1. You will be asked to prove this law in Exercise 15, Section 1.3.

EXAMPLE 1 $\lim_{x \to 2} 5 = 5$, $\lim_{x \to -1} 3 = 3$, and $\lim_{x \to 0} 2\pi = 2\pi$. ■

LAW 2   Limit of the Identity Function $f(x) = x$

$$\lim_{x \to a} x = a$$

Again, you can see this intuitively by examining the graph of the identity function $f(x) = x$. (See Figure 2.) You will also be asked to prove this law in Exercise 16, Section 1.3.

EXAMPLE 2 $\lim_{x \to 4} x = 4$, $\lim_{x \to 0} x = 0$, and $\lim_{x \to -\pi} x = -\pi$. ■

The following limit laws allow us to find the limits of functions algebraically.

LIMIT LAWS
If $\lim_{x \to a} f(x) = L$ and $\lim_{x \to a} g(x) = M$, then

LAW 3   Sum Law

$$\lim_{x \to a} [f(x) \pm g(x)] = L \pm M$$

LAW 4   Product Law

$$\lim_{x \to a} [f(x)g(x)] = LM$$

LAW 5   Constant Multiple Law

$$\lim_{x \to a} [cf(x)] = cL, \quad \text{for every } c$$

LAW 6   Quotient Law

$$\lim_{x \to a} \frac{f(x)}{g(x)} = \frac{L}{M}, \quad \text{provided that } M \neq 0$$

LAW 7   Root Law

$$\lim_{x \to a} \sqrt[n]{f(x)} = \sqrt[n]{L}, \quad \text{provided that } n \text{ is a positive integer,}$$
$$\text{and } L > 0 \text{ if } n \text{ is even}$$

In words, these laws say the following:

3. The limit of the sum (difference) of two functions is the sum (difference) of their limits.

4. The limit of the product of two functions is the product of their limits.

5. The limit of a constant times a function is the constant times the limit of the function.

6. The limit of a quotient of two functions is the quotient of their limits, provided that the limit of the denominator is not zero.

7. The limit of the nth root of a function is the nth root of the limit of the function, provided that n is a positive integer and $L > 0$ if n is even.

(We will prove the Sum Law in Section 1.3. The other laws are proved in Appendix B.)

Although the Sum Law and the Product Law are stated for two functions, they are also valid for any finite number of functions. For example, if

$$\lim_{x \to a} f_1(x) = L_1, \qquad \lim_{x \to a} f_2(x) = L_2, \qquad \dots, \qquad \lim_{x \to a} f_n(x) = L_n$$

then

$$\lim_{x \to a}[f_1(x) + f_2(x) + \cdots + f_n(x)] = L_1 + L_2 + \cdots + L_n$$

and

$$\lim_{x \to a}[f_1(x)f_2(x) \cdots f_n(x)] = L_1 L_2 \cdots L_n \tag{1}$$

If we take $f_1(x) = f_2(x) = \cdots = f_n(x) = f(x)$, then Equation (1) gives the following result for powers of f.

LAW 8 If n is a positive integer and $\lim_{x \to a} f(x) = L$, then $\lim_{x \to a}[f(x)]^n = L^n$.

Next, if we take $f(x) = x$, then Equation (1) and Law 8 give the following result.

LAW 9 $\lim_{x \to a} x^n = a^n$, where n is a positive integer.

EXAMPLE 3 Find $\lim_{x \to 2}(2x^3 - 4x^2 + 3)$.

Solution

$$\lim_{x \to 2}(2x^3 - 4x^2 + 3) = \lim_{x \to 2} 2x^3 - \lim_{x \to 2} 4x^2 + \lim_{x \to 2} 3 \qquad \text{Law 3}$$

$$= 2 \lim_{x \to 2} x^3 - 4 \lim_{x \to 2} x^2 + \lim_{x \to 2} 3 \qquad \text{Law 5}$$

$$= 2(2)^3 - 4(2)^2 + 3 \qquad \text{Law 9}$$

$$= 3$$

■ Limits of Polynomial and Rational Functions

The method of solution that we used in Example 3 can be used to prove the following.

LAW 10 **Limits of Polynomial Functions**

If $p(x) = a_n x^n + a_{n-1} x^{n-1} + \cdots + a_0$ is a polynomial function, then

$$\lim_{x \to a} p(x) = p(a)$$

Thus, the limit of a polynomial function as x approaches a is equal to the value of the function at a.

PROOF Applying the (generalized) sum law and the constant multiple law repeatedly, we find

$$\lim_{x \to a} p(x) = \lim_{x \to a}(a_n x^n + a_{n-1} x^{n-1} + \cdots + a_0)$$

$$= a_n(\lim_{x \to a} x^n) + a_{n-1}(\lim_{x \to a} x^{n-1}) + \cdots + \lim_{x \to a} a_0$$

Next, using Laws 1, 2, and 9, we obtain

$$\lim_{x \to a} p(x) = a_n a^n + a_{n-1} a^{n-1} + \cdots + a_0 = p(a)$$

In light of this, we could have solved the problem posed in Example 3 as follows:

$$\lim_{x \to 2}(2x^3 - 4x^2 + 3) = 2(2)^3 - 4(2)^2 + 3 = 3$$

EXAMPLE 4 Find $\lim_{x \to -1}(3x^2 + 2x + 1)^5$.

Solution

$$\lim_{x \to -1}(3x^2 + 2x + 1)^5 = [\lim_{x \to -1}(3x^2 + 2x + 1)]^5 \qquad \text{Law 8}$$

$$= [3(-1)^2 + 2(-1) + 1]^5 \qquad \text{Law 10}$$

$$= 2^5 = 32$$

The following result follows from the Quotient Law for limits and Law 10.

LAW 11   Limits of Rational Functions

If f is a rational function defined by $f(x) = P(x)/Q(x)$, where $P(x)$ and $Q(x)$ are polynomial functions and $Q(a) \neq 0$, then

$$\lim_{x \to a} f(x) = f(a) = \frac{P(a)}{Q(a)}$$

Thus, the limit of a rational function as x approaches a is equal to the value of the function at a provided the denominator is not zero at a.

PROOF Since P and Q are polynomial functions, we know from Law 10 that

$$\lim_{x \to a} P(x) = P(a) \qquad \text{and} \qquad \lim_{x \to a} Q(x) = Q(a)$$

Since $Q(a) \neq 0$, we can apply the Quotient Law to conclude that

$$\lim_{x \to a} f(x) = \lim_{x \to a} \frac{P(x)}{Q(x)} = \frac{\lim_{x \to a} P(x)}{\lim_{x \to a} Q(x)} = \frac{P(a)}{Q(a)} = f(a)$$

EXAMPLE 5 Find $\displaystyle\lim_{x \to 3} \frac{4x^2 - 3x + 1}{2x - 4}$.

Solution Using Law 11, we obtain

$$\lim_{x \to 3} \frac{4x^2 - 3x + 1}{2x - 4} = \frac{4(3)^2 - 3(3) + 1}{2(3) - 4} = \frac{28}{2} = 14 \qquad \blacksquare$$

EXAMPLE 6 Find $\displaystyle\lim_{x \to 1} \sqrt[3]{\frac{2x + 14}{x^2 + 1}}$.

Solution

$$\lim_{x \to 1} \sqrt[3]{\frac{2x + 14}{x^2 + 1}} = \sqrt[3]{\lim_{x \to 1} \frac{2x + 14}{x^2 + 1}} \qquad \text{Law 7}$$

$$= \sqrt[3]{\frac{2(1) + 14}{1^2 + 1}} \qquad \text{Law 11}$$

$$= \sqrt[3]{8} = 2 \qquad \blacksquare$$

Lest you think that we can *always* find the limit of a function by substitution, consider the following example.

EXAMPLE 7 Find $\displaystyle\lim_{x \to 2} \frac{x^2 - 4}{x - 2}$.

Solution Because the denominator of the rational expression is 0 at $x = 2$, we cannot find the limit by direct substitution. However, by factoring the numerator, we obtain

$$\frac{x^2 - 4}{x - 2} = \frac{(x + 2)(x - 2)}{x - 2}$$

so if $x \neq 2$, we can cancel the common factors. Thus,

$$\frac{x^2 - 4}{x - 2} = x + 2 \qquad x \neq 2$$

In other words, the values of the function f defined by $f(x) = (x^2 - 4)/(x - 2)$ coincide with the values of the function g defined by $g(x) = x + 2$ for all values of x except $x = 2$. Since the limit of $f(x)$ as x approaches 2 depends only on the values of x other than 2, we can find the required limit by evaluating the limit of $g(x)$ as x approaches 2 instead. Thus,

$$\lim_{x \to 2} \frac{x^2 - 4}{x - 2} = \lim_{x \to 2} (x + 2) = 2 + 2 = 4 \qquad \blacksquare$$

In certain instances the technique that we used in Example 7 can be applied to find the limit of a quotient in which both the numerator and denominator of the quotient approach 0 as x approaches a. The trick here is to use the appropriate algebraic manipulations that will enable us to replace the original function by one that is identical to that function except perhaps at a. The limit is then found by evaluating this function at a.

Notes
1. If the numerator does not approach 0 but the denominator does, then the limit of the quotient does not exist. (See Example 7 in Section 1.1.)
2. A function whose limit at a can be found by evaluating it at a is said to be continuous at a. (We will study continuous functions in Section 1.4.)

EXAMPLE 8 Find $\displaystyle\lim_{x \to -3} \frac{x^2 + 2x - 3}{x^2 + 4x + 3}$.

Solution Notice that both the numerator and the denominator of the quotient approach 0 as x approaches -3, so Law 6 is not applicable. Instead, we proceed as follows:

$$\lim_{x \to -3} \frac{x^2 + 2x - 3}{x^2 + 4x + 3} = \lim_{x \to -3} \frac{(x + 3)(x - 1)}{(x + 3)(x + 1)}$$

$$= \lim_{x \to -3} \frac{x - 1}{x + 1} \qquad x \neq -3$$

$$= \frac{-3 - 1}{-3 + 1} = 2$$

EXAMPLE 9 Find $\displaystyle\lim_{x \to 0} \frac{\sqrt{1 + x} - 1}{x}$.

Solution Both the numerator and the denominator of the quotient approach 0 as x approaches 0, so we cannot evaluate the limit using Law 6. Let's rationalize the numerator of the quotient by multiplying both the numerator and the denominator by $\sqrt{1 + x} + 1$. Thus,

$$\lim_{x \to 0} \frac{\sqrt{1 + x} - 1}{x} = \lim_{x \to 0} \frac{\sqrt{1 + x} - 1}{x} \cdot \frac{\sqrt{1 + x} + 1}{\sqrt{1 + x} + 1}$$

$$= \lim_{x \to 0} \frac{(\sqrt{1 + x} - 1)(\sqrt{1 + x} + 1)}{x(\sqrt{1 + x} + 1)}$$

$$= \lim_{x \to 0} \frac{1 + x - 1}{x(\sqrt{1 + x} + 1)} \qquad \text{\textcolor{orange}{Difference of two squares}}$$

$$= \lim_{x \to 0} \frac{1}{\sqrt{1 + x} + 1} = \frac{1}{2} \qquad x \neq 0$$

All of the limit laws stated for two-sided limits in this section also hold true for one-sided limits.

EXAMPLE 10 Let

$$f(x) = \begin{cases} -x + 3 & \text{if } x < 2 \\ \sqrt{x - 2} + 1 & \text{if } x \geq 2 \end{cases}$$

Find $\lim_{x \to 2} f(x)$ if it exists.

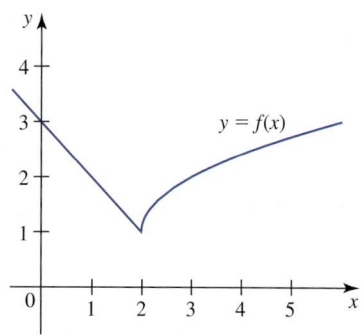

FIGURE 3
$\lim_{x \to 2^-} f(x) = \lim_{x \to 2^+} f(x) = 1$, so $\lim_{x \to 2} f(x) = 1$.

Solution   The function f is defined piecewise. For $x \geq 2$ the rule for f is $f(x) = \sqrt{x - 2} + 1$. Letting x approach 2 from the right, we obtain

$$\lim_{x \to 2^+} (\sqrt{x - 2} + 1) = \lim_{x \to 2^+} \sqrt{x - 2} + \lim_{x \to 2^+} 1 \qquad \text{Sum Law}$$

$$= 0 + 1 = 1$$

For $x < 2$, $f(x) = -x + 3$, and

$$\lim_{x \to 2^-} (-x + 3) = \lim_{x \to 2^-} (-x) + \lim_{x \to 2^-} 3 \qquad \text{Sum Law}$$

$$= -2 + 3 = 1$$

The right-hand and left-hand limits are equal. Therefore, the limit exists and

$$\lim_{x \to 2} f(x) = 1$$

The graph of f is shown in Figure 3.

The next example involves the **greatest integer** function defined by $f(x) = [\![x]\!]$, where $[\![x]\!]$ is the greatest integer n such that $n \leq x$. For example, $[\![3]\!] = 3$, $[\![2.4]\!] = 2$, $[\![\pi]\!] = 3$, $[\![-4.6]\!] = -5$, $[\![-\sqrt{2}]\!] = -2$, and so on. As an aid to finding the value of the greatest integer function, think of "rounding down."

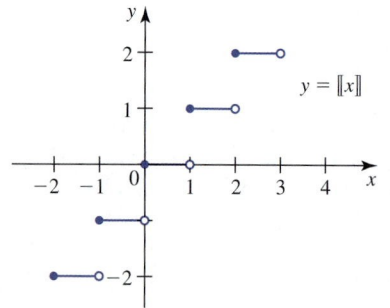

FIGURE 4
The graph of $y = [\![x]\!]$

EXAMPLE 11   Show that $\lim_{x \to 2} [\![x]\!]$ does not exist.

Solution   The graph of the greatest integer function is shown in Figure 4. Observe that if $2 \leq x < 3$, then $[\![x]\!] = 2$, and therefore,

$$\lim_{x \to 2^+} [\![x]\!] = \lim_{x \to 2^+} 2 = 2$$

Next, observe that if $1 \leq x < 2$, then $[\![x]\!] = 1$, so

$$\lim_{x \to 2^-} [\![x]\!] = \lim_{x \to 2^-} 1 = 1$$

Since these one-sided limits are not equal, we conclude by Theorem 1, Section 1.1, that $\lim_{x \to 2} [\![x]\!]$ does not exist.

■ Limits of Trigonometric Functions

So far, we have dealt with limits involving algebraic functions. The following theorem tells us that if a is a number in the domain of a trigonometric function, then the limit of that function as x approaches a can be found by substitution.

THEOREM 1   Limits of Trigonometric Functions

Let a be a number in the domain of the given trigonometric function. Then

a. $\lim_{x \to a} \sin x = \sin a$ **b.** $\lim_{x \to a} \cos x = \cos a$

c. $\lim_{x \to a} \tan x = \tan a$ **d.** $\lim_{x \to a} \cot x = \cot a$

e. $\lim_{x \to a} \sec x = \sec a$ **f.** $\lim_{x \to a} \csc x = \csc a$

The proofs of Theorem 1a and Theorem 1b are sketched in Exercises 97 and 98. The proofs of the other parts follow from Theorems 1a and 1b and the limit laws.

EXAMPLE 12   Find

a. $\lim\limits_{x\to\pi/2} x \sin x$ **b.** $\lim\limits_{x\to\pi/4} (2x^2 + \cot x)$

Solution

a. $\lim\limits_{x\to\pi/2} x \sin x = \left(\lim\limits_{x\to\pi/2} x\right)\left(\lim\limits_{x\to\pi/2} \sin x\right) = \dfrac{\pi}{2} \sin \dfrac{\pi}{2} = \dfrac{\pi}{2}$

b. $\lim\limits_{x\to\pi/4} (2x^2 + \cot x) = \lim\limits_{x\to\pi/4} 2x^2 + \lim\limits_{x\to\pi/4} \cot x$

$$= 2\left(\dfrac{\pi}{4}\right)^2 + \cot \dfrac{\pi}{4}$$

$$= \dfrac{\pi^2}{8} + 1 = \dfrac{\pi^2 + 8}{8}$$ ■

■ The Squeeze Theorem

The techniques that we have developed so far do not work in all situations. For example, they cannot be used to find

$$\lim_{x\to 0} x^2 \sin \dfrac{1}{x}$$

For limits such as this we use the Squeeze Theorem.

> **THEOREM 2**   **The Squeeze Theorem**
>
> Suppose that $f(x) \leq g(x) \leq h(x)$ for all x in an open interval containing a, except possibly at a, and
>
> $$\lim_{x\to a} f(x) = L = \lim_{x\to a} h(x)$$
>
> Then
>
> $$\lim_{x\to a} g(x) = L$$

The Squeeze Theorem says that if $g(x)$ is squeezed between $f(x)$ and $h(x)$ near a and both $f(x)$ and $h(x)$ approach L as x approaches a, then $g(x)$ must approach L as well (see Figure 5). A proof of this theorem is given in Appendix B.

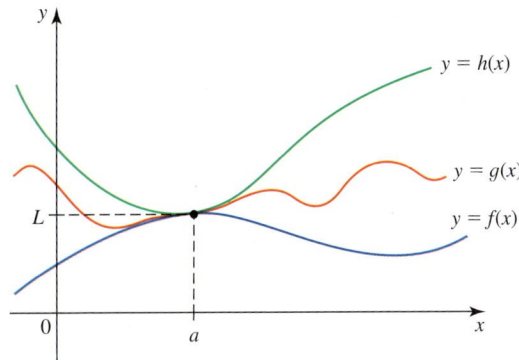

FIGURE 5
An illustration of the Squeeze Theorem

EXAMPLE 13 Find $\lim\limits_{x \to 0} x^2 \sin \dfrac{1}{x}$.

Solution   Since $-1 \le \sin t \le 1$ for every real number t, we have

$$-1 \le \sin \frac{1}{x} \le 1$$

for every $x \ne 0$. Therefore,

$$-x^2 \le x^2 \sin \frac{1}{x} \le x^2 \qquad x \ne 0$$

Let $f(x) = -x^2$, $g(x) = x^2 \sin(1/x)$, and $h(x) = x^2$. Then $f(x) \le g(x) \le h(x)$. Since

$$\lim_{x \to 0} f(x) = \lim_{x \to 0} (-x^2) = 0 \qquad \text{and} \qquad \lim_{x \to 0} h(x) = \lim_{x \to 0} x^2 = 0$$

the Squeeze Theorem implies that

$$\lim_{x \to 0} g(x) = \lim_{x \to 0} x^2 \sin \frac{1}{x} = 0$$

(See Figure 6.)

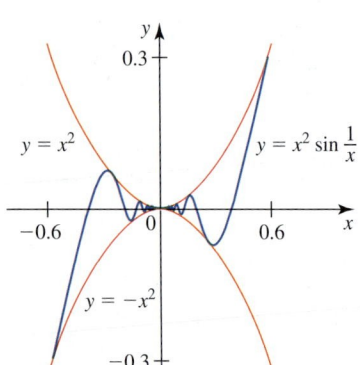

$y = x^2$

$y = x^2 \sin \dfrac{1}{x}$

$y = -x^2$

FIGURE 6

$\lim\limits_{x \to 0} g(x) = \lim\limits_{x \to 0} x^2 \sin \dfrac{1}{x} = 0$

The property of limits given in Theorem 3 will be used later. (Its proof is given in Appendix B.)

THEOREM 3

Suppose that $f(x) \le g(x)$ for all x in an open interval containing a, except possibly at a, and

$$\lim_{x \to a} f(x) = L \qquad \text{and} \qquad \lim_{x \to a} g(x) = M$$

Then

$$L \le M$$

The Squeeze Theorem can be used to prove the following important result, which will be needed in our work later on.

THEOREM 4

$$\lim_{\theta \to 0} \frac{\sin \theta}{\theta} = 1$$

PROOF   First, suppose that $0 < \theta < \frac{\pi}{2}$. Figure 7 shows a sector of a circle of radius 1. From the figure we see that

$$\text{Area of } \triangle OAB = \frac{1}{2}(1)(\sin \theta) = \frac{1}{2} \sin \theta \qquad \tfrac{1}{2}\text{base} \cdot \text{height}$$

$$\text{Area of sector } OAB = \frac{1}{2}(1)^2 \theta = \frac{1}{2} \theta \qquad \tfrac{1}{2}r^2\theta$$

$$\text{Area of } \triangle OAC = \frac{1}{2}(1)(\tan \theta) = \frac{1}{2} \tan \theta \qquad \tfrac{1}{2}\text{base} \cdot \text{height}$$

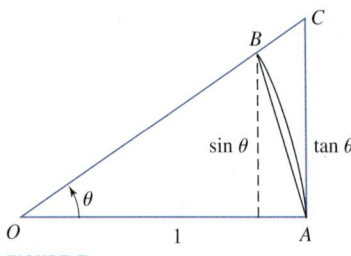

FIGURE 7

Since $0 < \text{area of } \triangle OAB < \text{area of sector } OAB < \text{area of } \triangle OAC$, we have

$$0 < \frac{1}{2}\sin\theta < \frac{1}{2}\theta < \frac{1}{2}\tan\theta$$

Multiplying through by $2/(\sin\theta)$ and keeping in mind that $\sin\theta > 0$ and $\cos\theta > 0$ for $0 < \theta < \frac{\pi}{2}$, we obtain

$$1 < \frac{\theta}{\sin\theta} < \frac{1}{\cos\theta}$$

or, upon taking reciprocals,

$$\cos\theta < \frac{\sin\theta}{\theta} < 1 \qquad\qquad \textbf{(2)}$$

If $-\frac{\pi}{2} < \theta < 0$, then $0 < -\theta < \frac{\pi}{2}$, and Inequality (2) gives

$$\cos(-\theta) < \frac{\sin(-\theta)}{-\theta} < 1$$

or, since $\cos(-\theta) = \cos\theta$ and $\sin(-\theta) = -\sin\theta$, we have

$$\cos\theta < \frac{\sin\theta}{\theta} < 1$$

which is just Inequality (2). Therefore, Inequality (2) holds whenever θ lies in the intervals $\left(-\frac{\pi}{2}, 0\right)$ and $\left(0, \frac{\pi}{2}\right)$.

Finally, let $f(\theta) = \cos\theta$, $g(\theta) = (\sin\theta)/\theta$, and $h(\theta) = 1$, and observe that

$$\lim_{\theta\to 0} f(\theta) = \lim_{\theta\to 0}\cos\theta = 1$$

and

$$\lim_{\theta\to 0} h(\theta) = \lim_{\theta\to 0} 1 = 1$$

Then the Squeeze Theorem implies that

$$\lim_{\theta\to 0} g(\theta) = \lim_{\theta\to 0}\frac{\sin\theta}{\theta} = 1$$

EXAMPLE 14   Find $\displaystyle\lim_{x\to 0}\frac{\sin 2x}{3x}$.

Solution   We first rewrite

$$\frac{\sin 2x}{3x} \qquad \text{as} \qquad \left(\frac{2}{3}\right)\frac{\sin 2x}{2x}$$

Then, making the substitution $\theta = 2x$ and observing that $\theta \to 0$ as $x \to 0$, we find

$$\lim_{x\to 0}\frac{\sin 2x}{3x} = \lim_{\theta\to 0}\left(\frac{2}{3}\right)\frac{\sin\theta}{\theta}$$

$$= \frac{2}{3}\lim_{\theta\to 0}\frac{\sin\theta}{\theta}$$

$$= \frac{2}{3} \qquad \text{Use Theorem 4.}$$

EXAMPLE 15 Find $\lim\limits_{x \to 0} \dfrac{\tan x}{x}$.

Solution

$$
\begin{aligned}
\lim_{x \to 0} \frac{\tan x}{x} &= \lim_{x \to 0}\left(\frac{\sin x}{x} \cdot \frac{1}{\cos x}\right) \\[2mm]
&= \left(\lim_{x \to 0}\frac{\sin x}{x}\right)\left(\lim_{x \to 0}\frac{1}{\cos x}\right) \\[2mm]
&= (1)(1) \\[2mm]
&= 1
\end{aligned}
$$

Theorem 5 is a consequence of Theorem 4.

THEOREM 5

$$
\lim_{\theta \to 0} \frac{\cos \theta - 1}{\theta} = 0
$$

PROOF We use the identity $\sin^2 x = \frac{1}{2}(1 - \cos 2x)$ to write

$$
1 - \cos \theta = 2 \sin^2\left(\frac{\theta}{2}\right) \qquad \text{Let } x = \frac{\theta}{2}.
$$

Then

$$
\begin{aligned}
\lim_{\theta \to 0} \frac{\cos \theta - 1}{\theta} &= \lim_{\theta \to 0}\left(\frac{-2 \sin^2\left(\frac{\theta}{2}\right)}{\theta}\right) \\[2mm]
&= \lim_{\theta \to 0}\left(-\sin \frac{\theta}{2}\right)\left(\frac{\sin \frac{\theta}{2}}{\frac{\theta}{2}}\right) \\[2mm]
&= -\left(\lim_{\theta \to 0} \sin \frac{\theta}{2}\right)\left(\lim_{\theta \to 0} \frac{\sin \frac{\theta}{2}}{\frac{\theta}{2}}\right) \qquad \text{Note: } \frac{\theta}{2} \to 0 \text{ as } \theta \to 0. \\[2mm]
&= 0 \cdot 1 = 0
\end{aligned}
$$

1.2   CONCEPT QUESTIONS

1. State the Sum, Product, Constant Multiple, Quotient, and Root Laws for limits at a number.

2. Find the limit and state the limit law that you use at each step.

   a. $\lim\limits_{x \to 2}(3x^2 - 2x + 1)$ **b.** $\lim\limits_{x \to 3}\dfrac{x^2 + 4}{2x + 3}$

3. Find the limit and state the limit law that you use at each step.

   a. $\lim\limits_{x \to 4} \sqrt{x}(2x^2 + 1)$ **b.** $\lim\limits_{x \to 1}\left(\dfrac{2x^2 + x + 5}{x^4 + 1}\right)^{3/2}$

4. State the Squeeze Theorem in your own words, and give a graphical interpretation.

1.2 EXERCISES

In Exercises 1–22, find the indicated limit.

1. $\lim_{t \to 2}(3t + 4)$

2. $\lim_{x \to 2}(3x^2 + 2x - 8)$

3. $\lim_{h \to -1}(h^4 - 2h^3 + 2h - 1)$

4. $\lim_{x \to 2}(x^2 + 1)(2x^2 - 4)$

5. $\lim_{x \to 1}(3x^2 - 4x + 2)^4$

6. $\lim_{t \to 3}(2t - 1)^2(t^2 - 2t)^3$

7. $\lim_{x \to 1} \dfrac{x - 2}{x^2 + x + 1}$

8. $\lim_{t \to -1} \dfrac{t^3 - 1}{t^3 - 2t + 4}$

9. $\lim_{x \to 2}\left(\sqrt{2x^3} - \sqrt{2}x\right)$

10. $\lim_{x \to 3} \sqrt{2x^3 - 3x + 7}$

11. $\lim_{x \to -1^+}(x^3 - 2x^2 - 5)^{2/3}$

12. $\lim_{x \to -2}(x + 3)^2\sqrt{4x^2 - 8}$

13. $\lim_{x \to 0^+} \dfrac{1 + \sqrt{x}}{\sqrt{x + 4}}$

14. $\lim_{t \to 4} t^{-1/2}(t^2 - 3t + 4)^{3/2}$

15. $\lim_{u \to -2} \sqrt[3]{\dfrac{3u^2 + 2u}{3u^3 - 3}}$

16. $\lim_{w \to 0} \dfrac{\sqrt{w + 1} - \sqrt{w^2 + 4}}{(w + 2)^2 - (w + 1)^2}$

17. $\lim_{x \to 1} \sin \dfrac{\pi x}{2}$

18. $\lim_{x \to \pi/4}(x \tan x)$

19. $\lim_{x \to \pi/4} \dfrac{\sin x}{x}$

20. $\lim_{x \to 0} \dfrac{\sec 2x}{\sqrt{x + 4}}$

21. $\lim_{x \to \pi} \sqrt{2 + \cos x}$

22. $\lim_{x \to \pi/4} \dfrac{\tan^2 x}{1 + \cos x}$

In Exercises 23–28, you are given that $\lim_{x \to a} f(x) = 2$, $\lim_{x \to a} g(x) = 4$, and $\lim_{x \to a} h(x) = -1$. Find the indicated limit.

23. $\lim_{x \to a}[2f(x) + 3g(x)]$

24. $\lim_{x \to a} \dfrac{f(x) + g(x)}{2h(x)}$

25. $\lim_{x \to a} \dfrac{f(x)}{\sqrt{g(x)}}$

26. $\lim_{x \to a} \dfrac{f(x)g(x)}{\sqrt{g(x) + 5}}$

27. $\lim_{x \to a}\{[h(x)]^2 - f(x)g(x)\}$

28. $\lim_{x \to a} \dfrac{\sqrt[3]{f(x)g(x)}}{\sqrt{f(x)g(x) + 1}}$

In Exercises 29 and 30, suppose that $\lim_{x \to -2} f(x) = 2$ and $\lim_{x \to -2} g(x) = 3$. Find the indicated limit.

29. $\lim_{x \to -2}[xf(x) + (x^2 + 1)g(x)]$

30. $\lim_{x \to -2} \dfrac{xf(x)}{1 + x^2}$

In Exercises 31–36, use the graphs of f and g that follow to find the indicated limit, if it exists. If the limit does not exist, explain why.

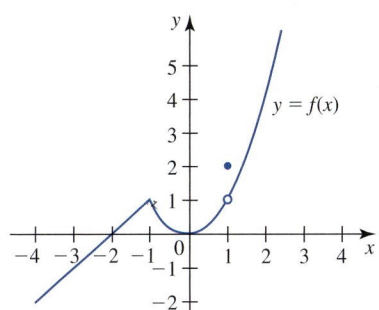

The graph of f

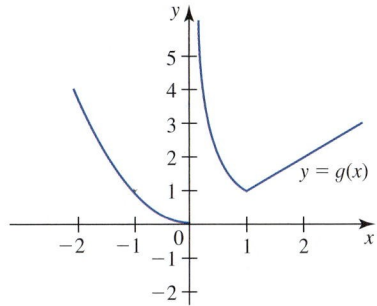

The graph of g

31. $\lim_{x \to -1}[f(x) + g(x)]$

32. $\lim_{x \to 0}[f(x) - g(x)]$

33. $\lim_{x \to 1}[f(x)g(x)]$

34. $\lim_{x \to 2} \dfrac{f(x)}{g(x)}$

35. $\lim_{x \to 0^-}[2f(x) + 3g(x)]$

36. $\lim_{x \to 0^+} \dfrac{f(x)}{g(x)}$

37. Is the following argument correct?

$$f(x) = \frac{x^2 - 9}{x + 3} = \frac{(x + 3)(x - 3)}{x + 3} = x - 3$$

Therefore, $\lim_{x \to -3} f(x) = f(-3) = -6$. Explain your answer.

38. Is the following argument correct?

$$\lim_{x \to -3} \frac{x^2 - 9}{x + 3} = \lim_{x \to -3} \frac{(x + 3)(x - 3)}{x + 3} = \lim_{x \to -3}(x - 3) = -6$$

Explain your answer. Compare it with Exercise 37.

39. Give an example to illustrate the following: If $\lim_{x \to a} f(x) = L \neq 0$ and $\lim_{x \to a} g(x) = 0$, then $\lim_{x \to a}[f(x)/g(x)]$ does not exist.

40. Give examples to illustrate the following: If $\lim_{x \to a} f(x) = 0$ and $\lim_{x \to a} g(x) = 0$, then $\lim_{x \to a}[f(x)/g(x)]$ might or might not exist.

In Exercises 41–76, find the limit, if it exists.

41. $\lim_{x \to 2} \dfrac{x^2 - 4}{x - 2}$

42. $\lim_{x \to 5} \dfrac{5 - x}{x^2 - 25}$

43. $\lim_{t \to 1} \dfrac{t + 1}{(t - 1)^2}$

44. $\lim_{x \to 2^+} \dfrac{x + 1}{x - 2}$

45. $\lim_{x \to 1} \dfrac{x^2 + 2x - 3}{x^2 - 1}$

46. $\lim_{x \to 2} \dfrac{x^2 - x - 2}{x - 2}$

47. $\lim_{x \to -1} \sqrt{\dfrac{2 + x - x^2}{x^2 + 4x + 3}}$

48. $\lim_{x \to -5} \sqrt{\dfrac{x^2 - 25}{2x^2 + 6x - 20}}$

49. $\lim_{t \to 0} \dfrac{2t^3 + 3t^2}{3t^4 - 2t^2}$

50. $\lim_{t \to 1} \dfrac{3t^3 + 4t + 1}{(t - 1)(2t^2 + 1)}$

51. $\lim_{x \to 1} \dfrac{x^3 - 1}{x - 1}$

52. $\lim_{v \to 2} \dfrac{v^4 - 16}{v^2 - 4}$

53. $\lim_{t \to 1} \dfrac{\sqrt{t} - 1}{t - 1}$

54. $\lim_{x \to 4} \dfrac{x - 4}{\sqrt{x} - 2}$

55. $\lim_{t \to 1} \dfrac{\sqrt{t} + 1}{t - 1}$

56. $\lim_{t \to 0} \dfrac{t}{\sqrt{2t + 1} - 1}$

57. $\lim_{x \to 0} \dfrac{\sqrt{x + 3} - \sqrt{3}}{x}$

58. $\lim_{h \to 0} \dfrac{\sqrt{a + h} - \sqrt{a}}{h}$

59. $\lim_{x \to 1} \dfrac{\sqrt{5 - x} - 2}{\sqrt{2 - x} - 1}$

60. $\lim_{h \to 0} \dfrac{(2 + h)^{-1} - 2^{-1}}{h}$

61. $\lim_{x \to 7^-} [\![x]\!]$

62. $\lim_{x \to -5^+} [\![x]\!]$

63. $\lim_{x \to 2^-} (x - [\![x]\!])$

64. $\lim_{x \to 3^+} [\![x + 1]\!]$

65. $\lim_{x \to 0} \dfrac{\sin x}{3x}$

66. $\lim_{x \to 0} \dfrac{\sin 2x}{x}$

67. $\lim_{h \to 0^+} \dfrac{\cos^{-1}(1 - h)}{\sqrt{h}}$

68. $\lim_{x \to 0} \dfrac{\tan 2x}{3x}$

Hint: Let $x = \cos^{-1}(1 - h)$.

69. $\lim_{x \to 0} \dfrac{\tan^2 x}{x}$

70. $\lim_{x \to 0} \dfrac{\cos x - 1}{\sin x}$

71. $\lim_{\theta \to 0} \dfrac{\cos \theta - 1}{\theta^2}$

72. $\lim_{x \to 0} \dfrac{x}{1 - \cos^2 x}$

73. $\lim_{x \to \pi/4} \dfrac{\sin x - \cos x}{1 - \tan x}$

74. $\lim_{\theta \to 0} \dfrac{\theta}{\cos\left(\theta - \frac{\pi}{2}\right)}$

75. $\lim_{x \to 0} \dfrac{\tan^{-1} x}{x}$

76. $\lim_{x \to 0^+} \sqrt{\dfrac{\tan x - \sin x}{x^2}}$

Hint: Let $u = \tan^{-1} x$.

77. Find $\lim_{x \to \pi/2} \dfrac{\cos x}{x - \frac{\pi}{2}}$.

Hint: Let $t = x - (\pi/2)$.

78. Find $\lim_{x \to \pi/2} \dfrac{\sin\left(x - \frac{\pi}{2}\right)}{2x - \pi}$.

Hint: Let $t = 2x - \pi$.

 79. Let $f(x) = \dfrac{x - 1}{\sqrt[3]{x + 7} - 2}$.

   a. Plot the graph of f, and use it to estimate the value of $\lim_{x \to 1} f(x)$.

   b. Construct a table of values of $f(x)$ accurate to three decimal places, and use it to estimate $\lim_{x \to 1} f(x)$.

   c. Find the exact value of $\lim_{x \to 1} f(x)$ analytically.

   Hint: Make the substitution $x + 7 = t^3$, and observe that $t \to 2$ as $x \to 1$.

 80. Let $f(x) = \dfrac{x + 2}{\sqrt[4]{x + 18} - 2}$.

   a. Plot the graph of f, and use it to estimate the value of $\lim_{x \to -2} f(x)$.

   b. Construct a table of values of $f(x)$ accurate to three decimal places, and use it to estimate $\lim_{x \to -2} f(x)$.

   c. Find the exact value of $\lim_{x \to -2} f(x)$ analytically.

   Hint: Make the substitution $x + 18 = t^4$, and observe that $t \to 2$ as $x \to -2$.

81. Special Theory of Relativity According to the special theory of relativity, when force and velocity are both along a straight line, resulting in straight-line motion, the magnitude of the acceleration of a particle acted upon by the force is

$$a = f(v) = \frac{F}{m}\left(1 - \frac{v^2}{c^2}\right)^{3/2}$$

where v is its speed, F is the magnitude of the force, m is the mass of the particle at rest, and c is the speed of light.

   a. Find the domain of f, and use this result to explain why we may consider only $\lim_{v \to c^-} f(v)$.

   b. Find $\lim_{v \to c^-} f(v)$, and interpret your result.

82. Special Theory of Relativity According to the special theory of relativity, the speed of a particle is

$$v = c\sqrt{1 - \left(\frac{E_0}{E}\right)^2}$$

where $E_0 = m_0 c^2$ is the rest energy and E is the total energy.

   a. Find the domain of v, use this result to explain why we may consider only $\lim_{E \to E_0^+} v$, and interpret your result.

   b. Find $\lim_{E \to E_0^+} v$, and interpret your result.

 83. Use the Squeeze Theorem to find $\lim_{x \to 0} x \sin(1/x)$. Verify your result visually by plotting the graphs of $f(x) = -x$, $g(x) = x \sin(1/x)$, and $h(x) = x$ in the same window.

84. Use the Squeeze Theorem to find $\lim_{x \to 0^+} \sqrt{x} \cos(1/x^2)$. Verify your result visually.
Hint: See Exercise 83.

85. Let
$$f(x) = \begin{cases} x + 2 & \text{if } x < -1 \\ x^2 + 2x + 3 & \text{if } x > -1 \end{cases}$$

a. Find $\lim_{x \to -1^-} f(x)$ and $\lim_{x \to -1^+} f(x)$.
b. Does $\lim_{x \to -1} f(x)$ exist? Why?

86. Let
$$f(x) = \begin{cases} \dfrac{x^3 - 16}{x} & \text{if } x < -2 \\ -x^2 - 4x + 8 & \text{if } x > -2 \end{cases}$$

Does $\lim_{x \to -2} f(x)$ exist? If so, what is its value?

87. Let
$$f(x) = \begin{cases} -x^5 + x^3 + x + 1 & \text{if } x < 0 \\ 2 & \text{if } x = 0 \\ x^2 + \sqrt{x + 1} & \text{if } x > 0 \end{cases}$$

Find $\lim_{x \to 0^-} f(x)$ and $\lim_{x \to 0^+} f(x)$. Does $\lim_{x \to 0} f(x)$ exist? Justify your answer.

88. Let
$$f(x) = \begin{cases} \sqrt{1 - x} + 2 & \text{if } x < 1 \\ 1 & \text{if } x = 1 \\ 1 + x^{3/2} & \text{if } x > 1 \end{cases}$$

Find $\lim_{x \to 1^-} f(x)$ and $\lim_{x \to 1^+} f(x)$. Does $\lim_{x \to 1} f(x)$ exist? Justify your answer.

89. Let
$$f(x) = \begin{cases} [\![x]\!] & \text{if } x < 2 \\ \sqrt{x - 2} + 1 & \text{if } x \ge 2 \end{cases}$$

Does $\lim_{x \to 2} f(x)$ exist? If so, what is its value?

90. Let
$$f(x) = \begin{cases} |x| & \text{if } x \le 1 \\ [\![x]\!] & \text{if } x > 1 \end{cases}$$

Does $\lim_{x \to 1} f(x)$ exist? If so, what is its value?

91. Let
$$f(x) = \begin{cases} x^2 & \text{if } x \text{ is rational} \\ -x^2 & \text{if } x \text{ is irrational} \end{cases}$$

Show that $\lim_{x \to 0} f(x) = 0$.

92. The Dirichlet Function The function
$$f(x) = \begin{cases} 1 & \text{if } x \text{ is rational} \\ 0 & \text{if } x \text{ is irrational} \end{cases}$$

is called the *Dirichlet function*. For example, $f\left(\frac{1}{2}\right) = 1$, $f\left(\frac{20}{21}\right) = 1$, $f(\sqrt{2}) = 0$, and $f(-\pi) = 0$. Show that for every a, $\lim_{x \to a} f(x)$ does not exist.

93. Show by means of an example that $\lim_{x \to a}[f(x) + g(x)]$ may exist even though neither $\lim_{x \to a} f(x)$ nor $\lim_{x \to a} g(x)$ exists. Does this example contradict the Sum Law of limits?

94. Show by means of an example that $\lim_{x \to a}[f(x)g(x)]$ may exist even though neither $\lim_{x \to a} f(x)$ nor $\lim_{x \to a} g(x)$ exists. Does this example contradict the Product Law of limits?

95. Suppose that $f(x) < g(x)$ for all x in an open interval containing a, except possibly at a, and that both $\lim_{x \to a} f(x)$ and $\lim_{x \to a} g(x)$ exist. Does it follow that $\lim_{x \to a} f(x) < \lim_{x \to a} g(x)$? Explain.

96. The following figure shows a sector of radius 1 and angle θ satisfying $0 < \theta < \frac{\pi}{2}$.

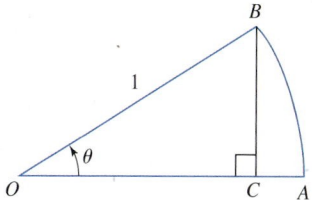

a. From the inequality $|BC| < \text{arc } AB$, deduce that $0 < \sin \theta < \theta$.
b. Use the Squeeze Theorem to prove that $\lim_{\theta \to 0^+} \sin \theta = 0$.
c. Use the result of part (a) to show that if $-\frac{\pi}{2} < \theta < 0$, then $\lim_{\theta \to 0^-} \sin \theta = 0$. Conclude that $\lim_{\theta \to 0} \sin \theta = 0$.
d. Use the result of part (c) and the trigonometric identity $\sin^2 \theta + \cos^2 \theta = 1$ to show that $\lim_{\theta \to 0} \cos \theta = 1$.

97. Use the result of Exercise 96 to prove that $\lim_{x \to a} \sin x = \sin a$.
Hint: It suffices to show that $\lim_{h \to 0} \sin(a + h) = \sin a$. Use the addition formula for the sine function.

98. Show that $\lim_{x \to a} \cos x = \cos a$. (See the hint for Exercise 97.)

In Exercises 99–102, determine whether the statement is true or false. If it is true, explain why. If it is false, explain why or give an example that shows it is false.

99. $\lim_{x \to 2}\left(\dfrac{3x}{x - 2} - \dfrac{2}{x - 2}\right) = \lim_{x \to 2} \dfrac{3x}{x - 2} - \lim_{x \to 2} \dfrac{2}{x - 2}$.

100. $\lim_{x \to 1} \dfrac{x^2 + 3x - 4}{x^2 - 2x - 3} = \dfrac{\lim_{x \to 1} x^2 + 3x - 4}{\lim_{x \to 1} x^2 - 2x - 3}$

101. If $\lim_{x \to a}[f(x) - g(x)]$ exists, then $\lim_{x \to a} f(x)$ and $\lim_{x \to a} g(x)$ also exist.

102. If $f(x) \le g(x) \le h(x)$ for all x in an open interval containing a, except possibly at a, and both $\lim_{x \to a} f(x)$ and $\lim_{x \to a} h(x)$ exist, then $\lim_{x \to a} g(x)$ exists.

1.3   A Precise Definition of a Limit

■ Precise Definition of a Limit

The definition of the limit of a function given in Section 1.1 is intuitive. In this section we give precise meaning to phrases such as "$f(x)$ can be made as close to L as we please" and "by taking x to be sufficiently close to a." We will focus our attention on the (two-sided) limit

$$\lim_{x \to a} f(x) = L \tag{1}$$

where a and L are real numbers. (The precise definition of one-sided limits is given in Exercise 28.)

Let's begin by investigating how we might establish the result

$$\lim_{x \to 2}(2x - 1) = 3 \tag{2}$$

with some degree of mathematical rigor. Here, $f(x) = 2x - 1$, $a = 2$, and $L = 3$. We need to show that "$f(x)$ can be made as close to 3 as we please by taking x to be sufficiently close to 2."

Our first step is to establish what we mean by "$f(x)$ is close to 3." For a start, suppose that we invite a challenger to specify some sort of "tolerance." For example, our challenger might declare that $f(x)$ is close to 3 provided that $f(x)$ differs from 3 by no more than 0.1 unit. Recalling that $|f(x) - 3|$ measures the distance from $f(x)$ to 3, we can rephrase this statement by saying that $f(x)$ is close to 3 provided that

$$|f(x) - 3| < 0.1 \qquad \text{Equivalently, } 2.9 < f(x) < 3.1. \tag{3}$$

(See Figure 1.)

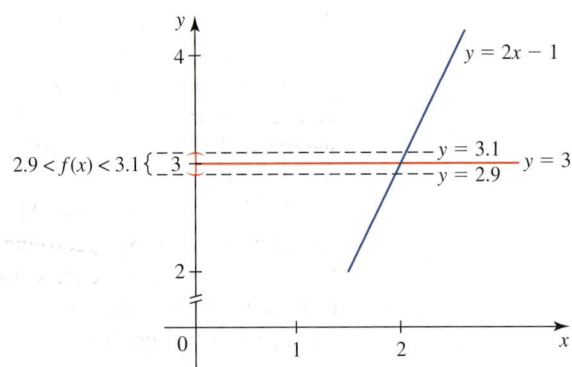

FIGURE 1
All the values of f satisfying
$2.9 < f(x) < 3.1$ are "close" to 3.

Now let's show that Inequality (3) is satisfied by all x that are "sufficiently close to 2." Because $|x - 2|$ measures the distance from x to 2, what we need to do is to show that there exists some positive number, call it δ (delta), such that

$$0 < |x - 2| < \delta \qquad \text{implies that} \qquad |f(x) - 3| < 0.1$$

(The first half of the first inequality precludes the possibility of x taking on the value 2. Remember that when we evaluate the limit of a function at a number a, we are not concerned with whether f is defined at a or its value there if it is defined.) To find δ, consider

$$|f(x) - 3| = |(2x - 1) - 3| = |2x - 4| = |2(x - 2)|$$
$$= 2|x - 2|$$

Now, $2|x - 2| < 0.1$ holds whenever

$$|x - 2| < \frac{0.1}{2} = 0.05 \qquad\qquad (4)$$

Therefore, if we pick $\delta = 0.05$, then $0 < |x - 2| < \delta$ implies that Inequality (4) holds. This in turn implies that

$$|f(x) - 3| = 2|x - 2| < 2(0.05) = 0.1$$

as we set out to show. (See Figure 2.)

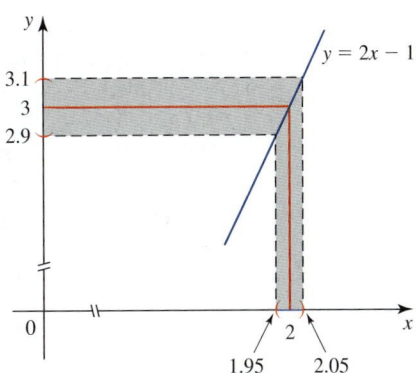

FIGURE 2
Whenever x satisfies $|x - 2| < 0.05$, $f(x)$ satisfies $|f(x) - 3| < 0.1$.

Have we established Equation (2)? The answer is a resounding no! What we have demonstrated is that by restricting x to be sufficiently close to 2, $f(x)$ can be made "close to 3" as measured by the norm, or tolerance, specified by one particular challenger. Another challenger might specify that "$f(x)$ is close to 3" if the tolerance is 10^{-20}! If you retrace these last steps, you can show that corresponding to a tolerance of 10^{-20}, we can make $|f(x) - 3| < 10^{-20}$ by requiring that $0 < |x - 2| < 5 \times 10^{-21}$. (Choose $\delta = 5 \times 10^{-21}$.)

To handle *all* such possible notions of closeness that could arise, suppose that a tolerance is given by specifying a number ε (epsilon) that may be *any* positive number whatsoever. Can we show that $f(x)$ is close to 3 (with tolerance ε) by restricting x to be sufficiently close to 2? In other words, given any number $\varepsilon > 0$, can we find a number $\delta > 0$ such that

$$|f(x) - 3| < \varepsilon \qquad \text{whenever} \qquad 0 < |x - 2| < \delta$$

All we have to do to answer these questions is to repeat the earlier computations with ε in place of 0.1. Consider

$$|f(x) - 3| = |(2x - 1) - 3| = |2x - 4| = 2|x - 2|$$

Now,

$$2|x - 2| < \varepsilon \qquad \text{provided that} \qquad |x - 2| < \frac{\varepsilon}{2}$$

Therefore, if we pick $\delta = \varepsilon/2$, then $0 < |x - 2| < \delta$ implies that $|x - 2| < \varepsilon/2$, which implies that

$$|f(x) - 3| = 2|x - 2| < 2\left(\frac{\varepsilon}{2}\right) = \varepsilon$$

Now, because ε is arbitrary, we have indeed shown that "$f(x)$ can be made as close to 3 as we please" by restricting x to be sufficiently close to 2.

This analysis suggests the following precise definition of a limit.

DEFINITION (Precise)   Limit of a Function at a Number

Let f be a function defined on an open interval containing a with the possible exception of a itself. Then the limit of $f(x)$ as x approaches a is the number L, written

$$\lim_{x \to a} f(x) = L$$

if for every number $\varepsilon > 0$, we can find a number $\delta > 0$ such that

$$0 < |x - a| < \delta \qquad \text{implies that} \qquad |f(x) - L| < \varepsilon$$

■ A Geometric Interpretation

Here is a geometric interpretation of the definition. Let $\varepsilon > 0$ be given. Draw the lines $y = L + \varepsilon$ and $y = L - \varepsilon$. Since $|f(x) - L| < \varepsilon$ is equivalent to $L - \varepsilon < f(x) < L + \varepsilon$, $\lim_{x \to a} f(x) = L$ exists provided that we can find a number δ such that if we restrict x to lie in the interval $(a - \delta, a + \delta)$ with $x \neq a$, then the graph of $y = f(x)$ lies inside the band of width 2ε determined by the lines $y = L + \varepsilon$ and $y = L - \varepsilon$. (See Figure 3.) You can see from Figure 3 that once a number $\delta > 0$ has been found, then any number smaller than δ will also satisfy the requirement.

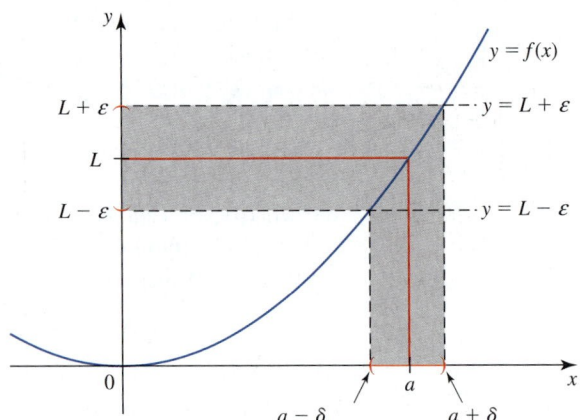

FIGURE 3
If $x \in (a - \delta, a)$ or $(a, a + \delta)$,
then $f(x)$ lies in the band defined
by $y = L + \varepsilon$ and $y = L - \varepsilon$.

■ Some Illustrative Examples

EXAMPLE 1 Prove that $\lim_{x \to 2} \dfrac{4(x^2 - 4)}{x - 2} = 16$. (Recall that this limit gives the instantaneous velocity of the maglev at $x = 2$ as described in Section 1.1.)

Solution   Let $\varepsilon > 0$ be given. We must show that there exists a $\delta > 0$ such that

$$\left| \frac{4(x^2 - 4)}{x - 2} - 16 \right| < \varepsilon$$

whenever $0 < |x - 2| < \delta$. To find δ, consider

$$\left| \frac{4(x^2 - 4)}{x - 2} - 16 \right| = \left| \frac{4(x - 2)(x + 2)}{x - 2} - 16 \right|$$

$$= |4(x + 2) - 16| = |4x - 8| \qquad x \neq 2$$

$$= 4|x - 2|$$

Therefore,

$$\left| \frac{4(x^2 - 4)}{x - 2} - 16 \right| = 4|x - 2| < \varepsilon$$

whenever

$$|x - 2| < \frac{1}{4}\varepsilon$$

So we may take $\delta = \varepsilon/4$. (See Figure 4.)

By reversing the steps, we see that if $0 < |x - 2| < \delta$, then

$$\left| \frac{4(x^2 - 4)}{x - 2} - 16 \right| = 4|x - 2| < 4\left(\frac{1}{4}\varepsilon \right) = \varepsilon$$

Thus,

$$\lim_{x \to 2} \frac{4(x^2 - 4)}{x - 2} = 16$$

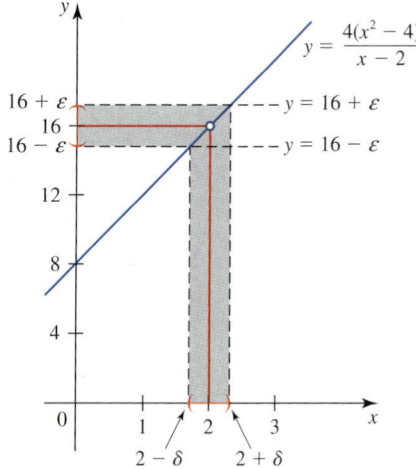

FIGURE 4
If we pick $\delta = \varepsilon/4$, then
$$0 < |x - 2| < \delta \Rightarrow$$
$$\left| \frac{4(x^2 - 4)}{x - 2} - 16 \right| < \varepsilon.$$

EXAMPLE 2   Prove that $\lim_{x \to 2} x^2 = 4$.

Solution   Let $\varepsilon > 0$ be given. We must show that there exists a $\delta > 0$ such that

$$|x^2 - 4| < \varepsilon$$

whenever $|x - 2| < \delta$. To find δ, consider

$$|x^2 - 4| = |(x + 2)(x - 2)|$$
$$= |x + 2||x - 2| \tag{5}$$

At this stage, one might be tempted to set

$$|x + 2||x - 2| < \varepsilon$$

and then divide both sides of this inequality by $|x + 2|$ to obtain

$$|x - 2| < \frac{\varepsilon}{|x + 2|}$$

and conclude that we may take

$$\delta = \frac{\varepsilon}{|x + 2|}$$

But this approach will not work because δ *cannot depend on* x. Let us begin afresh with Equation (5). On the basis of the experience just gained, we should obtain an upper bound for the quantity $|x + 2|$; that is, we want to find a positive number k such that $|x + 2| < k$ for all x "close to 2." As we observed earlier, once a δ has been found that satisfies our requirement, then any number smaller than δ will also do. This allows us to agree beforehand to take $\delta \leq 1$ (or any other positive constant); that is, we will consider only those values of x that satisfy $|x - 2| < 1$; that is $-1 < x - 2 < 1$, or $1 < x < 3$. Adding 2 to each side of this last inequality, we have $1 + 2 < x + 2 < 3 + 2$; $3 < x + 2 < 5$; thus, $|x + 2| < 5$. So $k = 5$, and Equation (5) gives

$$|x^2 - 4| = |x + 2||x - 2| < 5|x - 2|$$

Now

$$5|x - 2| < \varepsilon$$

whenever $|x - 2| < \varepsilon/5$. Therefore, if we take δ to be the smaller of the numbers 1 and $\varepsilon/5$, we are guaranteed that $|x - 2| < \delta$ implies that

$$|x^2 - 4| < 5|x - 2| < 5\left(\frac{\varepsilon}{5}\right) = \varepsilon$$

This proves the assertion (see Figure 5).

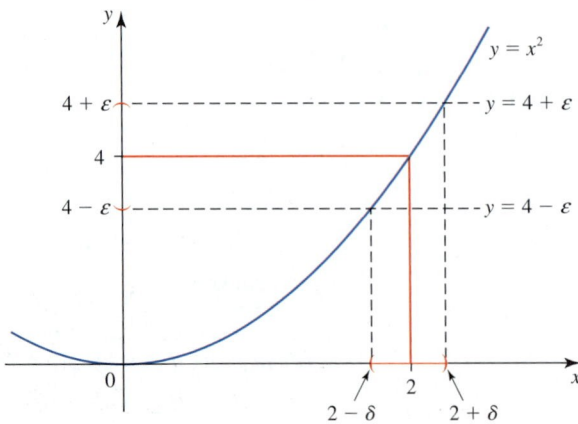

FIGURE 5
If we pick δ to be the
smaller of 1 and $\varepsilon/5$, then
$|x - 2| < \delta \Rightarrow |x^2 - 4| < \varepsilon$.

EXAMPLE 3 Let

$$f(x) = \begin{cases} 1 & \text{if } x \geq 0 \\ -1 & \text{if } x < 0 \end{cases}$$

Prove that $\lim_{x \to 0} f(x)$ does not exist.

Solution Suppose that the limit exists. We will show that this assumption leads to a contradiction. It will follow, therefore, that the opposite is true, namely, the limit does not exist.

So suppose that there exists a number L such that

$$\lim_{x \to 0} f(x) = L$$

Then, for every $\varepsilon > 0$ there exists a $\delta > 0$ such that

$$|f(x) - L| < \varepsilon \qquad \text{whenever} \qquad 0 < |x - 0| < \delta$$

In particular, if we take $\varepsilon = 1$, there exists a $\delta > 0$ such that

$$|f(x) - L| < 1 \qquad \text{whenever} \qquad 0 < |x - 0| < \delta$$

If we take $x = -\delta/2$, which lies in the interval defined by $0 < |x - 0| < \delta$, we have

$$\left| f\left(-\frac{\delta}{2} \right) - L \right| = |-1 - L| < 1$$

This inequality is equivalent to

$$-1 < -1 - L < 1$$
$$0 < -L < 2$$

or

$$-2 < L < 0$$

Next, if we take $x = \delta/2$, which also lies in the interval defined by $0 < |x - 0| < \delta$, we have

$$\left| f\left(\frac{\delta}{2} \right) - L \right| = |1 - L| < 1$$

This inequality is equivalent to

$$-1 < 1 - L < 1$$
$$-2 < -L < 0$$

or

$$0 < L < 2$$

But the number L cannot satisfy both the inequalities

$$-2 < L < 0 \qquad \text{and} \qquad 0 < L < 2$$

simultaneously. This contradiction proves that $\lim_{x \to 0} f(x)$ does not exist.

We end this section by proving the Sum Law for limits.

EXAMPLE 4 Prove the Sum Law for limits: If $\lim_{x \to a} f(x) = L$ and $\lim_{x \to a} g(x) = M$, then $\lim_{x \to a}[f(x) + g(x)] = L + M$.

Solution Let $\varepsilon > 0$ be given. We must show that there exists a $\delta > 0$ such that

$$\big|[f(x) + g(x)] - (L + M)\big| < \varepsilon$$

whenever $0 < |x - a| < \delta$. But by the Triangle Inequality,*

$$\big|[f(x) + g(x)] - (L + M)\big| = \big|(f(x) - L) + (g(x) - M)\big|$$

$$\leq |f(x) - L| + |g(x) - M| \qquad \textbf{(6)}$$

and this suggests that we consider the bounds for $|f(x) - L|$ and $|g(x) - M|$ separately.

Since $\lim_{x \to a} f(x) = L$, we can take $\varepsilon/2$, which is a positive number, and be guaranteed that there exists a $\delta_1 > 0$ such that

$$|f(x) - L| < \frac{\varepsilon}{2} \qquad \text{whenever} \qquad 0 < |x - a| < \delta_1 \qquad \textbf{(7)}$$

Similarly, since $\lim_{x \to a} g(x) = M$, we can find a $\delta_2 > 0$ such that

$$|g(x) - M| < \frac{\varepsilon}{2} \qquad \text{whenever} \qquad 0 < |x - a| < \delta_2 \qquad \textbf{(8)}$$

If we take δ to be the smaller of the two numbers δ_1 and δ_2 so that δ is itself positive, then both Inequalities (7) and (8) hold simultaneously if $0 < |x - a| < \delta$. Therefore, by Inequality (6)

$$\big|[f(x) + g(x)] - (L + M)\big| \leq |f(x) - L| + |g(x) - M|$$

$$< \frac{\varepsilon}{2} + \frac{\varepsilon}{2} = \varepsilon$$

whenever $0 < |x - a| < \delta$, and this proves the Sum Law.   ■

---

*The Triangle Inequality $|a + b| \leq |a| + |b|$ is proved in Appendix A.

1.3 CONCEPT QUESTIONS

1. State the precise definition of $\lim_{x \to 2}(x^3 + 5) = 13$.
2. Write the precise definition of $\lim_{x \to a} f(x) = L$ without using absolute values.
3. Use the figure to find a number δ such that $|x^2 - 1| < \frac{1}{2}$ whenever $|x - 1| < \delta$.

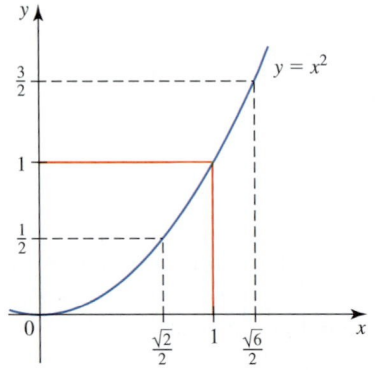

4. Use the figure to find a number δ such that $\left|\frac{1}{x} - 1\right| < \frac{1}{4}$ whenever $|x - 1| < \delta$.

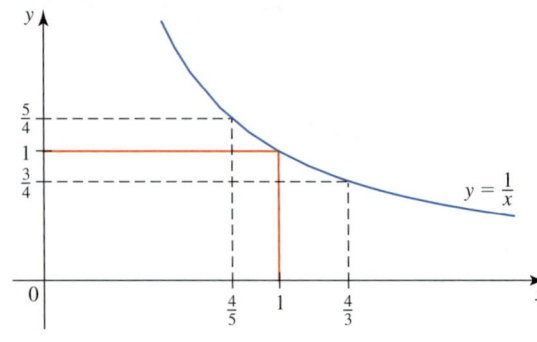

1.3   EXERCISES

In Exercises 1–10 you are given $\lim_{x \to a} f(x) = L$ *and a tolerance* ε. *Find a number* δ *such that* $|f(x) - L| < \varepsilon$ *whenever* $0 < |x - a| < \delta$.

1. $\lim_{x \to 2} 3x = 6$; $\varepsilon = 0.01$

2. $\lim_{x \to -1} 2x = -2$; $\varepsilon = 0.001$

3. $\lim_{x \to 1} (2x + 3) = 5$; $\varepsilon = 0.01$

4. $\lim_{x \to -2} (3x - 2) = -8$; $\varepsilon = 0.05$

5. $\lim_{x \to 3} \dfrac{x^2 - 9}{x - 3} = 6$; $\varepsilon = 0.02$

6. $\lim_{x \to -2} \dfrac{x^2 - 4}{x + 2} = -4$; $\varepsilon = 0.005$

7. $\lim_{x \to 3} 2x^2 = 18$; $\varepsilon = 0.01$

8. $\lim_{x \to 4} \sqrt{x} = 2$; $\varepsilon = 0.01$

9. $\lim_{x \to 2} \dfrac{x^2 + 4}{x + 2} = 2$; $\varepsilon = 0.01$

10. $\lim_{x \to 2} \dfrac{1}{x} = \dfrac{1}{2}$; $\varepsilon = 0.05$

In Exercises 11–22, use the precise definition of a limit to prove that the statement is true.

11. $\lim_{x \to 2} 3 = 3$

12. $\lim_{x \to -2} \pi = \pi$

13. $\lim_{x \to 3} 2x = 6$

14. $\lim_{x \to -2} (2x - 3) = -7$

15. $\lim_{x \to a} c = c$

16. $\lim_{x \to a} x = a$

17. $\lim_{x \to 1} 3x^2 = 3$

18. $\lim_{x \to 2} (x^2 - 2) = 2$

19. $\lim_{x \to 2} \dfrac{x^2 - 4}{x - 2} = 4$

20. $\lim_{x \to 0} \dfrac{x^2 + 2x}{x} = 2$

21. $\lim_{x \to 9} \sqrt{x} = 3$

22. $\lim_{x \to 0} (x^3 + 1) = 1$

23. Let

$$f(x) = \begin{cases} -1 & \text{if } x < 0 \\ 1 & \text{if } x \geq 0 \end{cases}$$

Prove that $\lim_{x \to 0} f(x)$ does not exist.

24. Let

$$g(x) = \begin{cases} -1 + x & \text{if } x < 0 \\ 1 + x & \text{if } x \geq 0 \end{cases}$$

Prove that $\lim_{x \to 0} g(x)$ does not exist.

25. Prove that $\lim_{x \to 0} H(x)$ does not exist, where H is the Heaviside function

$$H(x) = \begin{cases} 0 & \text{if } x < 0 \\ 1 & \text{if } x \geq 0 \end{cases}$$

26. Let

$$f(x) = \begin{cases} 0 & \text{if } x \text{ is rational} \\ 1 & \text{if } x \text{ is irrational} \end{cases}$$

Prove that $\lim_{x \to 0} f(x)$ does not exist.

27. Prove the Constant Multiple Law for limits: If $\lim_{x \to a} f(x) = L$ and c is a constant, then $\lim_{x \to a} [cf(x)] = cL$.

28. The precise definition of the left-hand limit, $\lim_{x \to a^-} f(x) = L$, may be stated as follows: For every number $\varepsilon > 0$ there exists a number $\delta > 0$ such that $|f(x) - L| < \varepsilon$ whenever $a - \delta < x < a$. Similarly, for the right-hand limit, $\lim_{x \to a^+} f(x) = L$ if for every number $\varepsilon > 0$ there exists a number $\delta > 0$ such that $|f(x) - L| < \varepsilon$ whenever $a < x < a + \delta$. Explain, with the aid of figures, why these definitions are appropriate.

29. Use the definition in Exercise 28 to prove that $\lim_{x \to 2^-} \sqrt[4]{4 - x^2} = 0$.

30. Use the definition in Exercise 28 to prove that $\lim_{x \to 2^+} \sqrt{x - 2} = 0$.

In Exercises 31–34, determine whether the statement is true or false. If it is true, explain why. If it is false, explain why or give an example that shows it is false.

31. The limit of $f(x)$ as x approaches a is L if there exists a number $\varepsilon > 0$ such that for all $\delta > 0$, $|f(x) - L| < \varepsilon$ whenever $0 < |x - a| < \delta$.

32. If $\lim_{x \to a} f(x) = L$, then given the number 0.01, there exists a $\delta > 0$ such that $0 < |x - a| < \delta$ implies that $|f(x) - L| < 0.01$.

33. The limit of $f(x)$ as x approaches a is L if for all $\varepsilon > 0$, there exists a $\delta > 0$ such that $|f(x) - L| < \varepsilon$ whenever $0 < |x - a| < \delta$.

34. The limit of $f(x)$ as x approaches a is L if for all $\delta > 0$, there exists an $\varepsilon > 0$, such that $|f(x) - L| < \varepsilon$ whenever $0 < |x - a| < \delta$.

1.4   Continuous Functions

■ Continuous Functions

The graph of the function

$$s = f(t) = 4t^2 \qquad 0 \le t \le 30$$

giving the position of the maglev at any time t (discussed in Section 1.1) is shown in Figure 1. Observe that the curve has no holes or jumps. This tells us that the displacement of the maglev must vary continuously with respect to time—it cannot vanish at any instant of time, and it cannot skip a stretch of the track to reappear and resume its motion somewhere else. The function s is an example of a *continuous function*. Observe that you can draw the graph of this function without lifting your pencil from the paper.

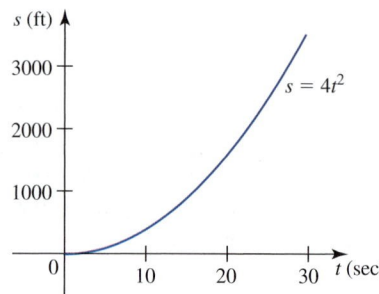

FIGURE 1
$s = f(t) = 4t^2$ gives the position of the maglev at any time t.

Functions that are *discontinuous* also occur in practical applications. Consider, for example, the Heaviside function H defined by

$$H(t) = \begin{cases} 0 & \text{if } t < 0 \\ 1 & \text{if } t \ge 0 \end{cases}$$

and first introduced in Example 3 in Section 1.1. You can see from the graph of H that it has a jump at $t = 0$ (Figure 2). If we think of H as describing the flow of current in an electrical circuit, then $t = 0$ corresponds to the time at which the switch is turned on. The function H is discontinuous at 0.

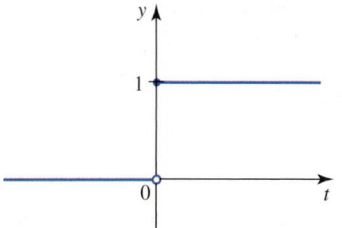

FIGURE 2
The Heaviside function is discontinuous at $t = 0$.

■ Continuity at a Number

We now give a formal definition of continuity.

DEFINITION   Continuity at a Number

Let f be a function defined on an open interval containing all values of x close to a. Then f **is continuous at a** if

$$\lim_{x \to a} f(x) = f(a) \tag{1}$$

If we write $x = a + h$ and note that x approaches a as h approaches 0, we see that the condition for f to be continuous at a is equivalent to

$$\lim_{h \to 0} f(a + h) = f(a) \tag{2}$$

Briefly, f is continuous at a if $f(x)$ gets closer and closer to $f(a)$ as x approaches a. Equivalently, f is continuous at a if proximity of x to a implies proximity of $f(x)$ to $f(a)$. (See Figure 3.)

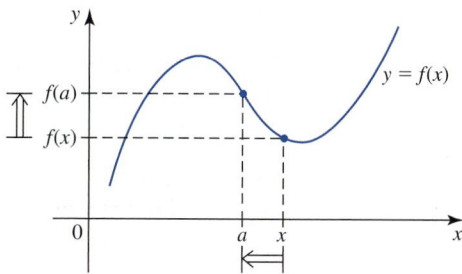

FIGURE 3
As x approaches a, $f(x)$ approaches $f(a)$.

If f is defined for all values of x close to a but Equation (1) is not satisfied, then f is **discontinuous at a** or f has a **discontinuity at a.**

Note It is implicit in Equation (1) that $f(a)$ is defined and the $\lim_{x \to a} f(x)$ exists. However, for emphasis we sometimes define continuity at a by requiring that the following three conditions hold: (1) $f(a)$ is defined, (2) $\lim_{x \to a} f(x)$ exists, and (3) $\lim_{x \to a} f(x) = f(a)$. ■

EXAMPLE 1 Use the graph of the function shown in Figure 4 to determine whether f is continuous at 0, 1, 2, 3, 4, and 5.

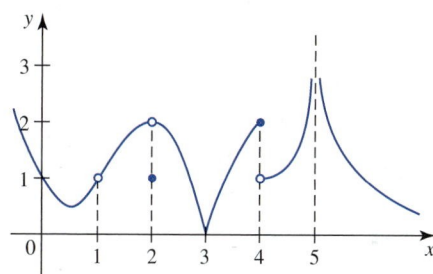

FIGURE 4
The graph of f

Solution The function f is continuous at 0 because

$$\lim_{x \to 0} f(x) = 1 = f(0)$$

It is discontinuous at 1 because $f(1)$ is not defined. It is discontinuous at 2 because

$$\lim_{x \to 2} f(x) = 2 \neq 1 = f(2)$$

Since

$$\lim_{x \to 3} f(x) = 0 = f(3)$$

we see that f is continuous at 3. Next, we see that $\lim_{x \to 4} f(x)$ does not exist, so f is not continuous at 4. Finally, because $\lim_{x \to 5} f(x)$ does not exist, we see that f is discontinuous at 5. ■

Refer to the function f in Example 1. The discontinuity at 1 and at 2, where the limit exists, is called a **removable discontinuity** because f can be made continuous at each of these numbers by defining or redefining it there. For example, if we define $f(1) = 1$, then f is made continuous at 1; if we redefine $f(2)$ by specifying that $f(2) = 2$, then f is also made continuous at 2.

The discontinuity at 4 is called a **jump discontinuity,** whereas the discontinuity at 5 is called an **infinite discontinuity.** Because the limit does not exist at a jump or at an infinite discontinuity, the discontinuity cannot be removed by defining or redefining the function at the number in question.

EXAMPLE 2   Let

$$f(x) = \begin{cases} \dfrac{x^2 - x - 2}{x - 2} & \text{if } x \neq 2 \\ 1 & \text{if } x = 2 \end{cases}$$

Show that f has a removable discontinuity at 2. Redefine f at 2 so that it is continuous everywhere.

Solution   First, let's find the limit of $f(x)$ as x approaches 2:

$$\lim_{x \to 2} \frac{x^2 - x - 2}{x - 2} = \lim_{x \to 2} \frac{(x - 2)(x + 1)}{x - 2}$$

$$= \lim_{x \to 2}(x + 1) = 3$$

Because $\lim_{x \to 2} f(x) = 3 \neq 1 = f(2)$, we see that f is discontinuous at 2. We can remove this discontinuity and thus render f continuous everywhere by redefining the value of f at 2 to be equal to 3. (See Figure 5.)

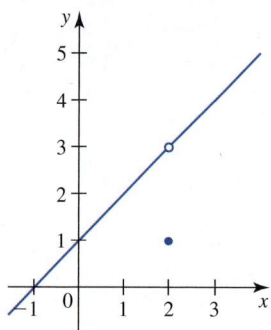

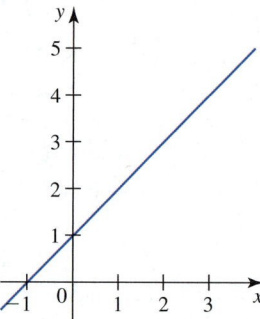

FIGURE 5
The discontinuity at 2 is removed
by redefining f at $x = 2$.

(a) f has a removable
discontinuity at 2.

(b) f is continuous at 2.

■ Continuity at an Endpoint

When we defined continuity, we assumed that $f(x)$ was defined for all values of x close to a. Sometimes $f(x)$ is defined only for those values of x that are greater than or equal to a or for values of x that are less than or equal to a. For example, $f(x) = \sqrt{x}$ is defined for $x \geq 0$, and $g(x) = \sqrt{3 - x}$ is defined for $x \leq 3$. The following definition covers these situations.

DEFINITION **Continuity from the Right and from the Left**

A function f **is continuous from the right at** a if

$$\lim_{x \to a^+} f(x) = f(a) \tag{3a}$$

A function f **is continuous from the left at** a if

$$\lim_{x \to a^-} f(x) = f(a) \tag{3b}$$

(See Figure 6.)

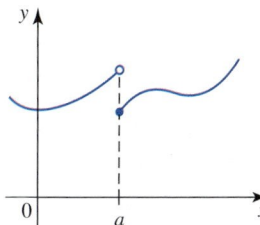 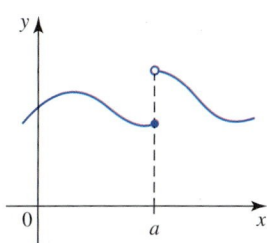

FIGURE 6

(**a**) f is continuous from the right at a.

(**b**) f is continuous from the left at a.

EXAMPLE 3 **The Heaviside Function** Consider the Heaviside function H defined by

$$H(t) = \begin{cases} 0 & \text{if } t < 0 \\ 1 & \text{if } t \geq 0 \end{cases}$$

Determine whether H is continuous from the right at 0 and/or from the left at 0.

Solution Because

$$\lim_{t \to 0^+} H(t) = \lim_{t \to 0^+} 1 = 1$$

and this is equal to $H(0) = 1$, H is continuous from the right at 0. Next, because

$$\lim_{t \to 0^-} H(t) = \lim_{t \to 0^-} (0) = 0$$

and this is not equal to $H(0) = 1$, H is not continuous from the left at 0. (See Figure 7.)

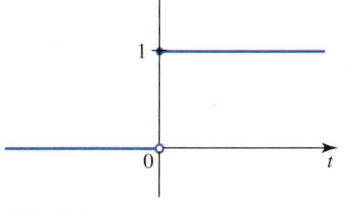

FIGURE 7
The Heaviside function H is continuous from the right at the number 0.

Note It follows from the definition of continuity that a function f is continuous at a if and only if f is simultaneously continuous from the right and from the left at a.

■ Continuity on an Interval

You might have noticed that continuity is a "local" concept; that is, we say that f is continuous at a number. The following definition tells us what it means to say that a function is continuous on an interval.

> **DEFINITION** Continuity on Open and Closed Intervals
>
> A function f **is continuous on an open interval** (a, b) if it is continuous at every number in the interval. A function f **is continuous on a closed interval** $[a, b]$ if it is continuous on (a, b) and is also continuous from the right at a and from the left at b. A function f **is continuous on a half-open interval** $[a, b)$ or $(a, b]$ if f is continuous on (a, b) and f is continuous from the right at a or f is continuous from the left at b, respectively.

EXAMPLE 4 Show that the function f defined by $f(x) = \sqrt{4 - x^2}$ is continuous on the closed interval $[-2, 2]$.

Solution We first show that f is continuous on $(-2, 2)$. Let a be any number in $(-2, 2)$. Then, using the laws of limits, we have

$$\lim_{x \to a} f(x) = \lim_{x \to a} \sqrt{4 - x^2} = \sqrt{\lim_{x \to a}(4 - x^2)} = \sqrt{4 - a^2} = f(a)$$

and this proves the assertion.

Next, let us show that f is continuous from the right at -2 and from the left at 2. Again, by invoking the limit properties, we see that

$$\lim_{x \to -2^+} f(x) = \lim_{x \to -2^+} \sqrt{4 - x^2} = \sqrt{\lim_{x \to -2^+}(4 - x^2)} = 0 = f(-2)$$

and

$$\lim_{x \to 2^-} f(x) = \lim_{x \to 2^-} \sqrt{4 - x^2} = \sqrt{\lim_{x \to 2^-}(4 - x^2)} = 0 = f(2)$$

and this proves the assertion. Therefore, f is continuous on $[-2, 2]$. The graph of f is shown in Figure 8. ▪

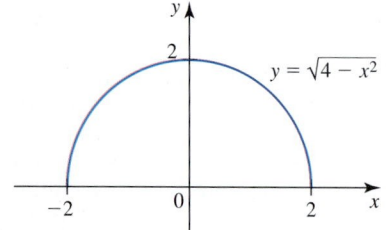

FIGURE 8

The function $f(x) = \sqrt{4 - x^2}$ is continuous on $[-2, 2]$.

> **THEOREM 1** Continuity of a Sum, Product, and Quotient
>
> If the functions f and g are continuous at a, then the following functions are also continuous at a.
>
> **a.** $f \pm g$
> **b.** fg
> **c.** cf,   where c is any constant
> **d.** $\dfrac{f}{g}$,   if $g(a) \neq 0$

We will prove Theorem 1b and leave some of the other parts as exercises. (See Exercises 94–95.)

PROOF OF THEOREM 1b

Since f and g are continuous at a, we have

$$\lim_{x \to a} f(x) = f(a) \qquad \text{and} \qquad \lim_{x \to a} g(x) = g(a)$$

By the Product Law for limits,

$$\lim_{x \to a}[f(x)g(x)] = \lim_{x \to a} f(x) \cdot \lim_{x \to a} g(x) = f(a)g(a)$$

so fg is continuous at a.

Note As in the case of the Sum Law and the Product Law, Theorems 1a and 1b can be extended to the case involving finitely many functions.

The following theorem is an immediate consequence of Laws 10 and 11 for limits from Section 1.2.

THEOREM 2 Continuity of Polynomial and Rational Functions

a. A polynomial function is continuous on $(-\infty, \infty)$.
b. A rational function is continuous on its domain.

EXAMPLE 5 Find the values of x for which the function

$$f(x) = x^8 - 3x^4 + x + 4 + \frac{x + 1}{(x + 1)(x - 2)}$$

is continuous.

Solution We can think of the function f as the sum of the polynomial function $g(x) = x^8 - 3x^4 + x + 4$ and the rational function $h(x) = (x + 1)/[(x + 1)(x - 2)]$. By Theorem 2 we see that g is continuous on $(-\infty, \infty)$, whereas h is continuous everywhere except at -1 and 2. Therefore, f is continuous on $(-\infty, -1)$, $(-1, 2)$, and $(2, \infty)$.

If you examine the graphs of the sine and cosine functions, you can see that they are continuous on $(-\infty, \infty)$. You will be asked to provide a rigorous demonstration of this in Exercises 92 and 93. Since the other trigonometric functions are defined in terms of these two functions, the continuity of the other trigonometric functions can be determined from them.

THEOREM 3 Continuity of Trigonometric Functions

The functions $\sin x$, $\cos x$, $\tan x$, $\sec x$, $\csc x$, and $\cot x$ are continuous at every number in their respective domain.

For example, since $\tan x = (\sin x)/(\cos x)$, we see that $\tan x$ is continuous everywhere except at the values of x where $\cos x = 0$; that is, except at $\pi/2 + n\pi$, where n is an integer. In other words, $f(x) = \tan x$ is continuous on

$$\ldots, \quad \left(-\frac{3\pi}{2}, -\frac{\pi}{2}\right), \quad \left(-\frac{\pi}{2}, \frac{\pi}{2}\right), \quad \left(\frac{\pi}{2}, \frac{3\pi}{2}\right), \quad \ldots$$

EXAMPLE 6 Find the values at which the following functions are continuous.

a. $f(x) = x \cos x$ **b.** $g(x) = \dfrac{\sqrt{x}}{\sin x}$

Solution
a. Since the functions x and $\cos x$ are continuous everywhere, we conclude that f is continuous on $(-\infty, \infty)$.
b. The function $\sqrt{x}$ is continuous on $[0, \infty)$. The function $\sin x$ is continuous every-where and has zeros at $n\pi$, where n is an integer. It follows from Theorem 1d, that g is continuous at all positive values of x that are not integral multiples of π; that is, g is continuous on $(0, \pi), (\pi, 2\pi), (2\pi, 3\pi), \ldots$. ∎

Because of the *reflective property* of inverse functions, we might expect that f and f^{-1} have similar properties. Thus, if f is continuous on its domain, then we might expect f^{-1} to be continuous on its domain. We give a proof of this in Appendix B. As a consequence of this result, Theorem 3, and the continuity of the exponential function (by the very way it is defined), we have the following.

THEOREM 4   **Continuity of Inverse Functions, Inverse Trigonometric Functions, Exponential Functions, and Logarithmic Functions**

If f is continuous on its domain, then f^{-1} is continuous on its domain. Also, the functions

$$\sin^{-1} x, \quad \cos^{-1} x, \quad \tan^{-1} x, \quad \sec^{-1} x, \quad \csc^{-1} x, \quad \cot^{-1} x, \quad a^x, \quad \text{and} \quad \log_a x$$

are continuous on their respective domains.

EXAMPLE 7 Find the values at which the function

$$f(x) = \frac{\tan^{-1} x + e^x}{(\log x)\sqrt{2 - x}}$$

is continuous.

Solution Since both the functions $\tan^{-1} x$ and e^x are continuous on $(-\infty, \infty)$, we see that $\tan^{-1} x + e^x$ is continuous on $(-\infty, \infty)$. Next, $\log x$ is continuous on $(0, \infty)$, and since we require that $2 - x > 0$, or $x < 2$, we see that f is continuous on $(0, 2)$. ∎

■ Continuity of Composite Functions

The following theorem shows us how to compute the limit of a composite function $f \circ g$ where f is continuous.

THEOREM 5   **Limit of a Composite Function**

If the function f is continuous at L and $\lim_{x \to a} g(x) = L$, then

$$\lim_{x \to a} f(g(x)) = f(L)$$

Intuitively, Theorem 5 is plausible because as x approaches a, $g(x)$ approaches L. Since f is continuous at L, proximity of $g(x)$ to L implies proximity of $f(g(x))$ to $f(L)$, which is what the theorem asserts. Theorem 5 is proved in Appendix B.

Note Theorem 5 states that the limit symbol can be moved through a continuous function. Thus,

$$\lim_{x \to a} f(g(x)) = f(\lim_{x \to a} g(x)) = f(L)$$

It follows from Theorem 5 that compositions of continuous functions are also continuous.

THEOREM 6   Continuity of Composite Functions

If the function g is continuous at a and the function f is continuous at $g(a)$, then the composition $f \circ g$ is continuous at a.

PROOF We compute

$$\lim_{x \to a}(f \circ g)(x) = \lim_{x \to a} f(g(x))$$

$$= f(\lim_{x \to a} g(x)) \qquad \text{Theorem 5}$$

$$= f(g(a)) \qquad \text{Since } g \text{ is continuous at } a$$

$$= (f \circ g)(a)$$

which is precisely the condition for $f \circ g$ to be continuous at a.

EXAMPLE 8

a. Show that $h(x) = |x|$ is continuous everywhere.
b. Use the result of part (a) to evaluate

$$\lim_{x \to 1} \left| \frac{-x^2 - x + 2}{x - 1} \right|$$

Solution

a. Since $|x| = \sqrt{x^2}$ for all x, we can view h as $h = f \circ g$, where $g(x) = x^2$ and $f(x) = \sqrt{x}$. Now g is continuous on $(-\infty, \infty)$, and $g(x) \geq 0$ for all x in $(-\infty, \infty)$. Also, f is continuous on $[0, \infty)$. Therefore, Theorem 6 says that $h = f \circ g$ is continuous on $(-\infty, \infty)$.

b. By the continuity of the absolute value function established in part (a) and Theorem 5, we find

$$\lim_{x \to 1} \left| \frac{-x^2 - x + 2}{x - 1} \right| = \left| \lim_{x \to 1} \frac{-x^2 - x + 2}{x - 1} \right|$$

$$= \left| \lim_{x \to 1} \frac{-(x - 1)(x + 2)}{x - 1} \right|$$

$$= \left| \lim_{x \to 1}(-1)(x + 2) \right| = |-3| = 3.$$

EXAMPLE 9 Find the intervals where the following functions are continuous.

a. $f(x) = \cos(\sqrt{3}x + 4)$ **b.** $g(x) = x^2 \sin \dfrac{1}{x}$

Solution

a. We can view f as a composition, $g \circ h$, of the functions $g(x) = \cos x$ and $h(x) = \sqrt{3}x + 4$. Since each of these functions is continuous everywhere, we conclude that f is continuous on $(-\infty, \infty)$.

b. The function $f(x) = \sin(1/x)$ is the composition of the functions $h(x) = \sin x$ and $k(x) = 1/x$. Since h is continuous everywhere and k is continuous everywhere except at 0, Theorem 6 says that the function $f = h \circ k$ is continuous on $(-\infty, 0)$ and $(0, \infty)$. Also, the function $F(x) = x^2$ is continuous everywhere. Therefore, we conclude by Theorem 1b that g, which is the product of F and f, is continuous on $(-\infty, 0)$ and $(0, \infty)$. The graph of g is shown in Figure 9.

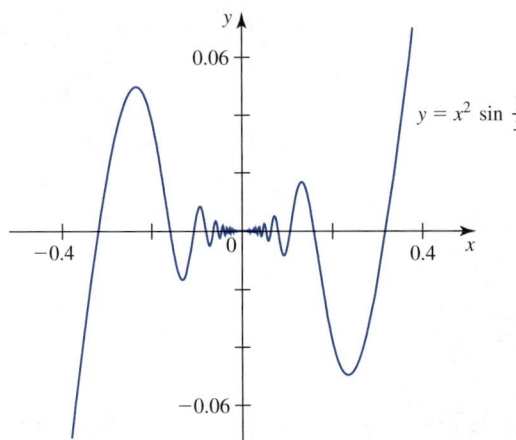

FIGURE 9
g is continuous everywhere except at 0.

■ Intermediate Value Theorem

Let's look again at our model of the motion of the maglev on a straight stretch of track. We know that the train cannot vanish at any instant of time, and it cannot skip portions of the track and reappear someplace else. To put it another way, the train cannot occupy the positions s_1 and s_2 without at least, at some time, occupying every intermediate position (Figure 10). To state this fact mathematically, recall that the position of the maglev as a function of time is described by

$$s = f(t) = 4t^2 \qquad 0 \le t \le 30$$

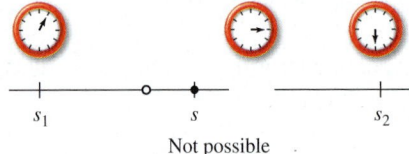
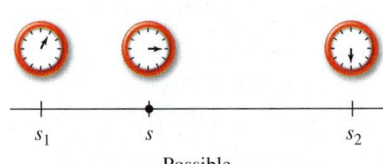

FIGURE 10
Position of the maglev

Suppose that the position of the maglev is s_1 at some time t_1 and that its position is s_2 at some time t_2. (See Figure 11.) Then if $\bar{s}$ is any number between s_1 and s_2 giving an intermediate position of the maglev, there must be at least one $\bar{t}$ between t_1 and t_2 giving the time at which the train is at $\bar{s}$; that is, $f(\bar{t}) = \bar{s}$.

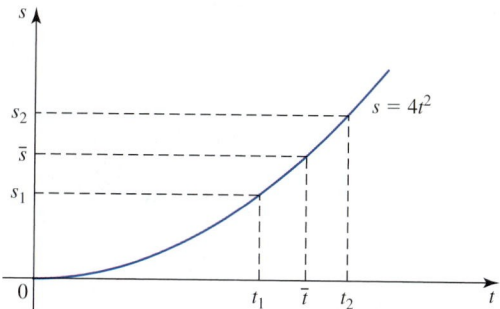

This discussion carries the gist of the Intermediate Value Theorem.

THEOREM 7   The Intermediate Value Theorem

If f is a continuous function on a closed interval $[a, b]$ and M is any number between $f(a)$ and $f(b)$, inclusive, then there is at least one number c in $[a, b]$ such that $f(c) = M$. (See Figure 12.)

FIGURE 12
If f is continuous on $[a, b]$
and $f(a) \leq M \leq f(b)$, then
there is at least one c, where
$a \leq c \leq b$ such that $f(c) = M$.

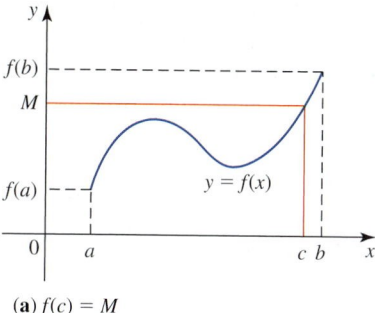

(a) $f(c) = M$

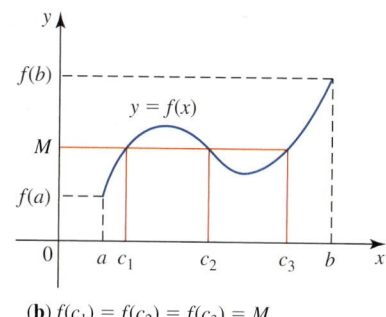

(b) $f(c_1) = f(c_2) = f(c_3) = M$

To illustrate the Intermediate Value Theorem, let's look at the example involving the motion of the maglev again (see Figure 1 in Section 1.1). Notice that the initial position of the train is $f(0) = 0$ and that the position at the end of its test run is $f(30) = 3600$. Furthermore, the function f is continuous on $[0, 30]$. So the Intermediate Value Theorem guarantees that if we arbitrarily pick a number between 0 and 3600, say, 400, giving the position of the maglev, there must be a $\bar{t}$ between 0 and 30 at which time the train is at the position $\bar{s} = 400$. To find the value of $\bar{t}$, we solve the equation $f(\bar{t}) = \bar{s}$, or

$$4\bar{t}^2 = 400$$

giving $\bar{t} = 10$. (Note that $\bar{t}$ must lie between 0 and 30.)

 Remember that when you use Theorem 7, the function f must be continuous. The conclusion of the Intermediate Value Theorem might not hold if f is not continuous (see Exercise 70).

The next theorem is an immediate consequence of the Intermediate Value Theorem. It not only tells us when a *zero of a function f* (root of the equation $f(x) = 0$) exists but also provides the basis for a method of approximating it.

> **THEOREM 8   Existence of Zeros of a Continuous Function**
>
> If f is a continuous function on a closed interval $[a, b]$ and $f(a)$ and $f(b)$ have opposite signs, then the equation $f(x) = 0$ has at least one solution in the interval (a, b) or, equivalently, the function f has at least one zero in the interval (a, b). (See Figure 13.)

FIGURE 13

If $f(a)$ and $f(b)$ have opposite signs, there must be at least one number c, where $a < c < b$, such that $f(c) = 0$.

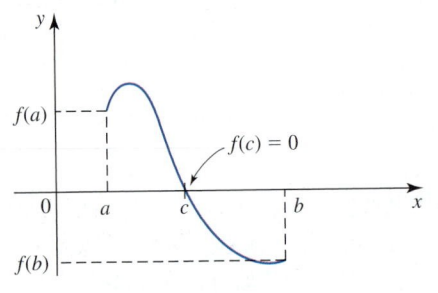

(a)

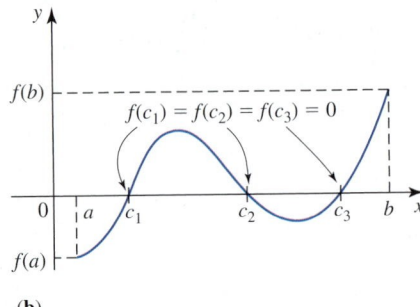

(b)

EXAMPLE 10   Let $f(x) = x^3 + x - 1$. Since f is a polynomial, it is continuous everywhere. Observe that $f(0) = -1$ and $f(1) = 1$, so Theorem 8 guarantees the existence of at least one root of the equation $f(x) = 0$ in $(0, 1)$.* We can locate the root more precisely by using Theorem 8 once again as follows: Evaluate $f(x)$ at the midpoint of $[0, 1]$. Thus,

$$f(0.5) = -0.375$$

Because $f(0.5) < 0$ and $f(1) > 0$, Theorem 8 now tells us that a root must lie in $(0.5, 1)$. Repeat the process: Evaluate $f(x)$ at the midpoint of $[0.5, 1]$, which is

$$\frac{0.5 + 1}{2} = 0.75$$

Thus,

$$f(0.75) = 0.171875$$

Because $f(0.5) < 0$ and $f(0.75) > 0$, Theorem 8 tells us that a root is in $(0.5, 0.75)$. This process can be continued. Table 1 summarizes the results of our computations through nine steps. From Table 1 we see that the root is approximately 0.68, accurate to two decimal places. By continuing the process through a sufficient number of steps, we can obtain as accurate an approximation to the root as we please.

TABLE 1

Step	Root of $f(x) = 0$ lies in
1	$(0, 1)$
2	$(0.5, 1)$
3	$(0.5, 0.75)$
4	$(0.625, 0.75)$
5	$(0.625, 0.6875)$
6	$(0.65625, 0.6875)$
7	$(0.671875, 0.6875)$
8	$(0.6796875, 0.6875)$
9	$(0.6796875, 0.68359375)$

Note   The process of finding the root of $f(x) = 0$ used in Example 10 is called the method of bisection. It is crude but effective. Later, we will look at a more efficient method, called the Newton-Raphson method, for finding the roots of $f(x) = 0$.

---

*It can be shown that f has exactly one zero in $(0, 1)$ (see Exercise 90).

1.4 CONCEPT QUESTIONS

1. Explain what it means for a function f to be continuous (a) at a number a, (b) from the right at a, and (c) from the left at a. Give examples.
2. Explain what it means for a function f to be continuous (a) on an open interval (a, b) and (b) on a closed interval $[a, b]$. Give examples.
3. Determine whether each function f is continuous or discontinuous. Explain your answer.
   a. $f(t)$ gives the altitude of an airplane at time t.
   b. $f(t)$ measures the total amount of rainfall at time t over the past 24 hr at the municipal airport.
   c. $f(t)$ is the price of admission for an adult at a movie theater as a function of time on a weekday.
   d. $f(t)$ is the speed of a pebble at time t when it is dropped from a height of 6 ft into a swimming pool.
4. **a.** Suppose that $\lim_{x \to a^-} f(x)$ and $\lim_{x \to a^+} f(x)$ both exist and f is discontinuous at a. Under what conditions does f have a removable discontinuity at a?
   b. Suppose that f is continuous from the left at a and continuous from the right at a. What can you say about the continuity of f at a? Explain.

1.4 EXERCISES

In Exercises 1–6, use the graph to determine where the function is discontinuous.

1.

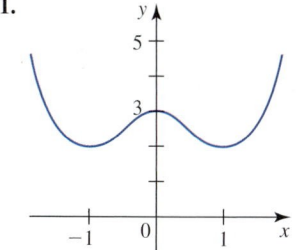

2.

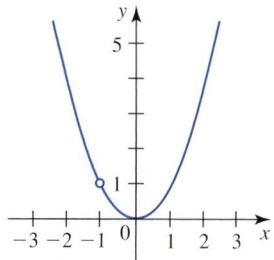

3.

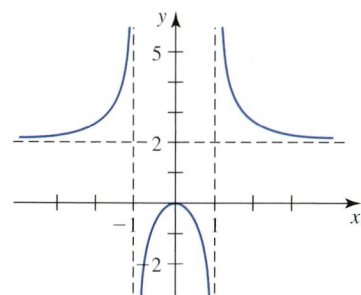

4.

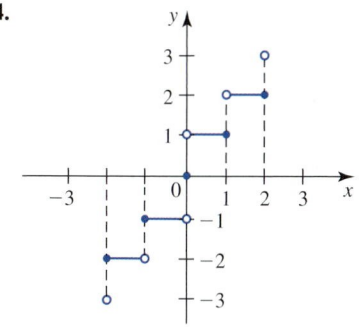

5.

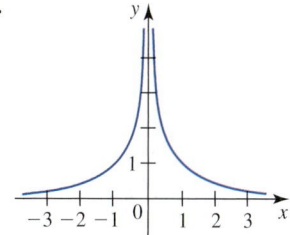

6.

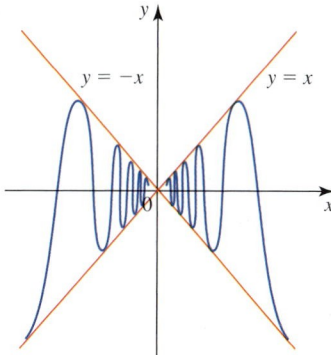

In Exercises 7–26, find the numbers, if any, where the function is discontinuous.

7. $f(x) = 2x^3 - 3x^2 + 4$

8. $f(x) = \dfrac{3}{x^2 + 1}$

9. $f(x) = \dfrac{e^x}{x - 2}$

10. $f(x) = \dfrac{\cos x}{x^2 - 1}$

11. $f(x) = \dfrac{x - 2}{x^2 - 4}$

12. $f(x) = \dfrac{x + 1}{x^2 - 2x - 3}$

13. $f(x) = \dfrac{x^2 - 3x + 2}{x^2 - 2x}$

14. $f(x) = |x^3 - 2x + 1|$

15. $f(x) = \left| \dfrac{x + 2}{x^2 + 2x} \right|$

16. $f(x) = \tan^{-1} x + \dfrac{x + 1}{|x + 1|}$

17. $f(x) = x - [\![x]\!]$

18. $f(x) = [\![x - 2]\!]$

19. $f(x) = \begin{cases} \tan^{-1}\left|\dfrac{1}{x - 5}\right| & \text{if } x \neq 5 \\ \dfrac{\pi}{2} & \text{if } x = 5 \end{cases}$

20. $f(x) = \begin{cases} x + 2 & \text{if } x < 3 \\ \ln(x - 2) + 5 & \text{if } x \geq 3 \end{cases}$

21. $f(x) = \begin{cases} \dfrac{x^2 - 1}{x + 1} & \text{if } x \neq -1 \\ 1 & \text{if } x = -1 \end{cases}$

22. $f(x) = \begin{cases} \dfrac{x^2 + x - 6}{x - 2} & \text{if } x \neq 2 \\ 5 & \text{if } x = 2 \end{cases}$

23. $f(x) = \begin{cases} e^{1/x} & \text{if } x \neq 0 \\ 1 & \text{if } x = 0 \end{cases}$

24. $f(x) = \begin{cases} -|x| + 1 & \text{if } x \neq 0 \\ 0 & \text{if } x = 0 \end{cases}$

25. $f(x) = \sec 2x$

26. $f(x) = \cot \pi x$

27. Let
$$f(x) = \begin{cases} x + 2 & \text{if } x \leq 1 \\ kx^2 & \text{if } x > 1 \end{cases}$$
Find the value of k that will make f continuous on $(-\infty, \infty)$.

28. Let
$$f(x) = \begin{cases} \dfrac{x^2 - 4}{x + 2} & \text{if } x \neq -2 \\ k & \text{if } x = -2 \end{cases}$$
Find the value of k that will make f continuous on $(-\infty, \infty)$.

29. Let
$$f(x) = \begin{cases} ax + b & \text{if } x < 1 \\ 4 & \text{if } x = 1 \\ 2ax - b & \text{if } x > 1 \end{cases}$$
Find the values of a and b that will make f continuous on $(-\infty, \infty)$.

30. Let
$$f(x) = \begin{cases} kx + \ln(x - 3)^2 & \text{if } x \leq 2 \\ 4ke^{x-2} - 3 & \text{if } x > 2 \end{cases}$$
Find the value of k that will make f continuous on $(-\infty, \infty)$.

31. Let
$$f(x) = \begin{cases} \dfrac{\sin 2x}{x} & \text{if } x \neq 0 \\ c & \text{if } x = 0 \end{cases}$$
Find the value of c that will make f continuous on $(-\infty, \infty)$.

32. Let
$$f(x) = \begin{cases} x \cot kx & \text{if } x < 0 \\ x^2 + c & \text{if } x \geq 0 \end{cases}$$
Find the value of c that will make f continuous at $x = 0$.

In Exercises 33–36, determine whether the function is continuous on the closed interval.

33. $f(x) = \sqrt{16 - x^2}, \quad [-4, 4]$

34. $g(x) = \ln(x + 3) + \sqrt{4 - x^2}, \quad [-2, 1]$

35. $f(x) = \begin{cases} x + 1 & \text{if } x < 0 \\ 2 - x & \text{if } x \geq 0 \end{cases}, \quad [-2, 4]$

36. $h(t) = \dfrac{1}{t^2 - 9}, \quad [-2, 2]$

In Exercises 37–48, find the interval(s) where f is continuous.

37. $f(x) = (3x^3 + 2x^2 + 1)^4$

38. $f(x) = \sqrt{x}(x - 5)^4$

39. $f(x) = \sqrt{x^2 + x + 1}$

40. $h(x) = \sqrt{x} + \dfrac{1}{\sqrt{x}}$

41. $f(x) = e^{\sqrt{9 - x^2}}$

42. $f(x) = \ln|x^2 - 4|$

43. $f(x) = \dfrac{1}{x\sqrt{9 - x^2}}$

44. $f(x) = \dfrac{1}{x} + \dfrac{3\sqrt{x}}{(x - 2)^2}$

45. $f(x) = \sin e^{\sqrt{x}}$

46. $f(x) = \dfrac{\ln(x + 1)}{e^x - 2}$

47. $f(x) = \tan^{-1}\dfrac{1}{x - 2}$

48. $f(x) = \dfrac{2 \cos x}{5 + 2 \sin x}$

49. Find $\displaystyle\lim_{x \to 2} \left| \dfrac{x^2 + x - 6}{x - 2} \right|$

50. Find $\displaystyle\lim_{x \to 1} \sin^{-1}\left[\dfrac{x^2 + x - 2}{6(x - 1)} \right]$

In Exercises 51–56, define the function at a so as to make it continuous at a.

51. $f(x) = \dfrac{3x^3 - 2x}{5x}, \quad a = 0$

52. $f(x) = \dfrac{2x^3 + x - 3}{x - 1}, \quad a = 1$

53. $f(x) = \dfrac{\sqrt{x + 1} - 1}{x}, \quad a = 0$

54. $f(x) = \dfrac{4 - x}{2 - \sqrt{x}}, \quad a = 4$

55. $f(x) = \dfrac{\tan x}{x}, \quad a = 0$

56. $f(x) = \dfrac{e^{-x} \sin^2 x}{1 - \cos x}, \quad a = 0$

In Exercises 57 and 58, let $f(x) = x(1 - x^2)$, and let g be the **signum** *(or* **sign***) function defined by*

$$g(x) = \begin{cases} -1 & \text{if } x < 0 \\ 0 & \text{if } x = 0 \\ 1 & \text{if } x > 0 \end{cases}$$

57. Show that $f \circ g$ is continuous on $(-\infty, \infty)$. Does this contradict Theorem 6?

58. Sketch the graph of the function $g \circ f$, and determine where $g \circ f$ is continuous.

In Exercises 59–62, use the Intermediate Value Theorem to find the value of c such that $f(c) = M$.

59. $f(x) = x^2 - x + 1$ on $[-1, 4]$; $M = 7$

60. $f(x) = x^2 - 4x + 6$ on $[0, 3]$; $M = 3$

61. $f(x) = x^3 - 2x^2 + x - 2$ on $[0, 4]$; $M = 10$

62. $f(x) = \dfrac{x - 1}{x + 1}$ on $[-4, -2]$; $M = 2$

In Exercises 63–66, use Theorem 8 to show that there is at least one root of the equation in the given interval.

63. $x^3 - 2x - 1 = 0$; $(0, 2)$

64. $x^4 - 2x^3 - 3x^2 + 7 = 0$; $(1, 2)$

65. $e^{-x} - \ln x = 0$; $(1, 2)$

66. $x^4 - 2x^3 = \sqrt{x - 1}$; $(2, 3)$

67. Let $f(x) = x^2$. Use the Intermediate Value Theorem to prove that there is a number c in the interval $[0, 2]$ such that $f(c) = 2$. (This proves the existence of the number $\sqrt{2}$.)

68. Let $f(x) = x^5 - 3x^2 + 2x + 5$.
   a. Show that there is at least one number c in the interval $[0, 2]$ such that $f(c) = 12$.
   b. Use a graphing utility to find all values of c accurate to five decimal places.
   Hint: Find the point(s) of intersection of the graphs of f and $g(x) = 12$.

69. Let $f(x) = \frac{1}{2}x^2 - \cos \pi x + 1$.
   a. Show that there is at least one number c in the interval $[0, 1]$ such that $f(c) = \sqrt{2}$.
   b. Use a graphing utility to find all values of c accurate to five decimal places.
   Hint: Find the point(s) of intersection of the graphs of f and $g(x) = \sqrt{2}$.

70. Let
$$f(x) = \begin{cases} -x^2 & \text{if } x \leq 0 \\ x + 1 & \text{if } x > 0 \end{cases}$$

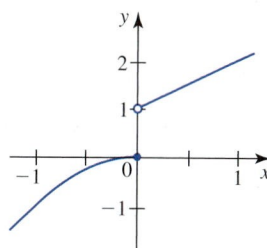

   a. Show that f is not continuous on $[-1, 1]$.
   b. Show that f does not take on all values between $f(-1)$ and $f(1)$.

71. Let
$$f(x) = \begin{cases} -x + 2 & \text{if } -2 \leq x < 0 \\ -(x^2 + 2) & \text{if } 0 \leq x \leq 2 \end{cases}$$

Does f have a zero in the interval $[-2, 2]$? Explain your answer.

72. Use the method of bisection to approximate the root of the equation $x^3 - x + 1 = 0$ accurate to two decimal places. (Refer to Example 10.)

73. Use the method of bisection to approximate the root of the equation $x^5 + 2x - 7 = 0$ accurate to two decimal places.

74. Acquisition of Failing S&L's The Tri-State Savings and Loan Company acquired two ailing financial institutions in 2009. One of them was acquired at time $t = T_1$, and the other was acquired at time $t = T_2$. ($t = 0$ corresponds to the beginning of 2009.) The following graph shows the total amount of money on deposit with Tri-State. Explain the significance of the discontinuities of the function at T_1 and T_2.

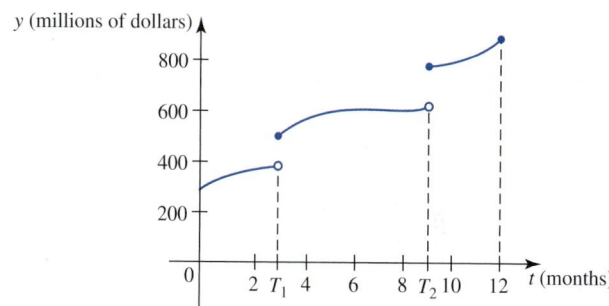

75. Colliding Billiard Balls While moving at a constant speed of v m/sec, billiard ball A collides with another stationary ball B at time t_1, hitting it "dead center." Suppose that at the moment of impact, ball A comes to rest. Draw graphs depicting the speeds of ball A and ball B (neglect friction).

76. Action of an Impulse on an Object An object of mass m is at rest at the origin on the x-axis. At $t = t_0$ it is acted upon by an impulse P_0 for a very short duration of time. The position of the object is given by
$$x = f(t) = \begin{cases} 0 & \text{if } 0 \leq t < t_0 \\ \dfrac{P_0(t - t_0)}{m} & \text{if } t \geq t_0 \end{cases}$$

Sketch the graph of f, and interpret your results.

77. Joan is looking straight out of a window of an apartment building at a height of 32 ft from the ground. A boy throws a tennis ball straight up by the side of the building where the window is located. Suppose the height of the ball (measured in feet) from the ground at time t (in sec) is $h(t) = 4 + 64t - 16t^2$.
   a. Show that $h(0) = 4$ and $h(2) = 68$.
   b. Use the Intermediate Value Theorem to conclude that the ball must cross Joan's line of sight at least once.
   c. At what time(s) does the ball cross Joan's line of sight? Interpret your results.

78. A Mixture Problem A tank initially contains 10 gal of brine with 2 lb of salt. Brine with 1.5 lb of salt per gallon enters the tank at the rate of 3 gal/min, and the well-stirred mixture leaves the tank at the rate of 4 gal/min. It can be shown that the amount of salt in the tank after t min is x lb, where

$$x = f(t) = 1.5(10 - t) - 0.0013(10 - t)^4 \qquad 0 \le t \le 3$$

Show that there is at least one instant of time between $t = 0$ and $t = 3$ when the amount of salt in the tank is 5 lb.
Note: We will find the times(s) when the amount of salt in the tank is 5 lb in Example 4 of Section 3.9.

79. Elastic Curve of a Beam The following figure shows the *elastic curve* (the dashed curve in the figure) of a beam of length L ft carrying a concentrated load of W_0 lb at its center. An equation of the curve is

$$y = f(x)$$

$$= \begin{cases} \dfrac{W_0}{48EI}(3L^2x - 4x^3) & \text{if } 0 \le x < \frac{L}{2} \\[2mm] \dfrac{W_0}{48EI}(4x^3 - 12Lx^2 + 9L^2x - L^3) & \text{if } \frac{L}{2} \le x \le L \end{cases}$$

where the product EI is a constant called the *flexural rigidity* of the beam. Show that the function $y = f(x)$ describing the elastic curve is continuous on $[0, L]$.

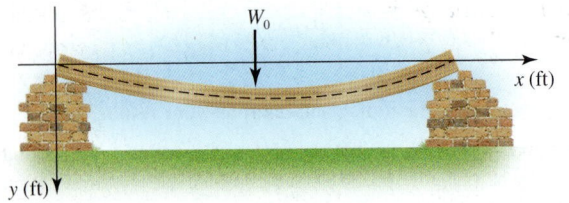

80. Newton's Law of Attraction The magnitude of the force exerted on a particle of mass m by a thin homogeneous spherical shell of radius R is

$$F(r) = \begin{cases} 0 & \text{if } r < R \\[2mm] \dfrac{GMm}{r^2} & \text{if } r \ge R \end{cases}$$

where M is the mass of the shell, r is the distance from the center of the shell to the particle, and G is the gravitational constant.

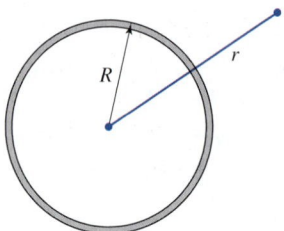

a. What is the force exerted on a particle just inside the shell? Just outside the shell?
b. Sketch the graph of F. Is F a continuous function of r?

81. A couple leaves their house at 6 P.M. on Friday for a weekend escape to their mountain cabin, where they arrive at 8 P.M. On the return trip, the couple leaves the cabin at 6 P.M. on Sunday and reverses the route they took on Friday, arriving home at 8 P.M. Use the Intermediate Value Theorem to show that there is a location on the route that the couple will pass at the same time of day on both days.

82. a. Suppose that f is continuous at a and g is discontinuous at a. Prove that the sum $f + g$ is discontinuous at a.
b. Suppose that f and g are both discontinuous at a. Is the sum $f + g$ necessarily discontinuous at a? Explain.

83. a. Suppose that f is continuous at a and g is discontinuous at a. Is the product fg necessarily discontinuous at a? Explain.
b. Suppose that f and g are both discontinuous at a. Is the product fg necessarily discontinuous at a? Explain.

84. The Dirichlet Function The Dirichlet function is defined by

$$f(x) = \begin{cases} 0 & \text{if } x \text{ is rational} \\ 1 & \text{if } x \text{ is irrational} \end{cases}$$

Show that f is discontinuous at every real number.

85. Show that every polynomial equation of the form

$$a_{2n+1}x^{2n+1} + a_{2n}x^{2n} + \cdots + a_2x^2 + a_1x + a_0 = 0$$

with real coefficients and $a_{2n+1} \ne 0$ has at least one real root.

86. Suppose that f is continuous on $[a, b]$ and has a finite number of zeros $x_1, x_2, \ldots, x_n$ in (a, b), satisfying $a < x_1 < x_2 < \cdots < x_n < b$. Show that $f(x)$ has the same sign within each of the intervals $(a, x_1), (x_1, x_2), \ldots, (x_n, b)$.

87. Let g be a continuous function on an interval $[a, b]$ and suppose $a \le g(x) \le b$ whenever $a \le x \le b$. Show that the equation $x = g(x)$ has at least one solution c in the interval $[a, b]$. Give a geometric interpretation.
Hint: Apply the Intermediate Value Theorem to the function $f(x) = x - g(x)$.

In Exercises 88 and 89, plot the graph of f. Then use the graph to determine where the function is continuous. Verify your answer analytically.

88. $f(x) = \begin{cases} \dfrac{x+1}{x\sqrt{1-x}} & \text{if } x < 1 \\ 2 & \text{if } x = 1 \\ \dfrac{x^4+1}{x^2} & \text{if } x > 1 \end{cases}$

89. $f(x) = \dfrac{|\sin x|}{\sin x}$

90. Show that $f(x) = x^3 + x - 1$ has exactly one zero in $(0, 1)$.

91. Show that there is at least one root of the equation $\sin x - x + 2 = 0$ in the interval $\left(0, \frac{3\pi}{2}\right)$.

92. Prove that $f(x) = \sin x$ is continuous everywhere.
Hint: Use the result of Exercise 97 in Section 1.2.

93. Prove that $f(x) = \cos x$ is continuous everywhere.

94. Prove that if f and g are continuous at a, then $f - g$ is continuous at a.

95. Prove that if f and g are continuous at a with $g(a) \neq 0$, then f/g is continuous at a.

In Exercises 96–100, determine whether the statement is true or false. If it is true, explain why. If it is false, explain why or give an example that shows it is false.

96. If $|f|$ is continuous at a, then f is continuous at a.

97. If f is discontinuous at a, then f^2 is continuous at a.

98. If f is defined on the interval $[a, b]$ with $f(a)$ and $f(b)$ having opposite signs, then f must have at least one zero in (a, b).

99. If f is continuous and $f + g$ is continuous, then g is continuous.

100. If f is continuous on the interval $(1, 5)$, then f is continuous on the interval $(2, 4)$.

1.5 Tangent Lines and Rates of Change

■ An Intuitive Look

One of the two problems that played a fundamental role in the development of calculus is the tangent line problem: How do we find the tangent line at a given point on a curve? (See Figure 1a.) To gain an intuitive feeling for the notion of the tangent line to a curve, think of the curve as representing a stretch of roller coaster track, and imagine that you are sitting in a car at the point P and looking straight ahead. Then the tangent line T to the curve at P is just the line parallel to your line of sight (Figure 1b).

Observe that the slope of the tangent line T at the point P appears to reflect the "steepness" of the curve at P. In other words, the slope of the tangent line at the point $P(x, f(x))$ on the graph of $y = f(x)$ provides us with a natural yardstick for measuring the rate of change of one quantity (y) with respect to another quantity (x).

Let's see how this intuitive observation bears out in a specific example. The function $s = f(t) = 4t^2$ gives the position of a maglev moving along a straight track at time t. We have drawn the tangent line T to the graph of s at the point $(2, 16)$ in Figure 2. Observe that the slope of T is $32/2 = 16$. This suggests that the quantity s is changing at the rate of 16 units per unit change in t; that is, the velocity of the maglev at $t = 2$ is 16 ft/sec. You might recall that this was the figure we arrived at in our calculations in Section 1.1!

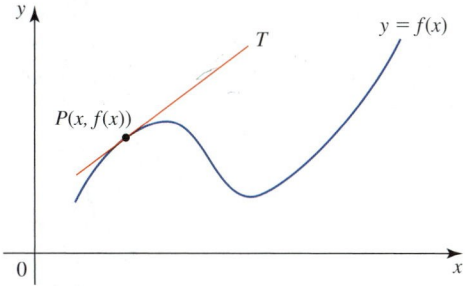

FIGURE 1   **(a)** T is the tangent line to the curve at P.

(b) The line of sight is parallel to T.

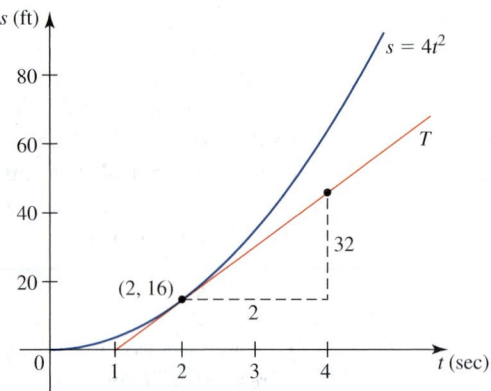

FIGURE 2
The position of the maglev at time t

■ Estimating the Rate of Change of a Function from Its Graph

EXAMPLE 1 **Automobile Fuel Economy**   According to a study by the U.S. Department of Energy and the Shell Development Company, a typical car's fuel economy as a function of its speed is described by the graph of the function f shown in Figure 3. Assuming that the rate of change of the function f at any value of x is given by the slope of the tangent line at the point $P(x, f(x))$, use the graph of f to estimate the rate of change of a typical car's fuel economy, measured in miles per gallon (mpg), when a car is driven at 20 mph and when it is driven at 60 mph.

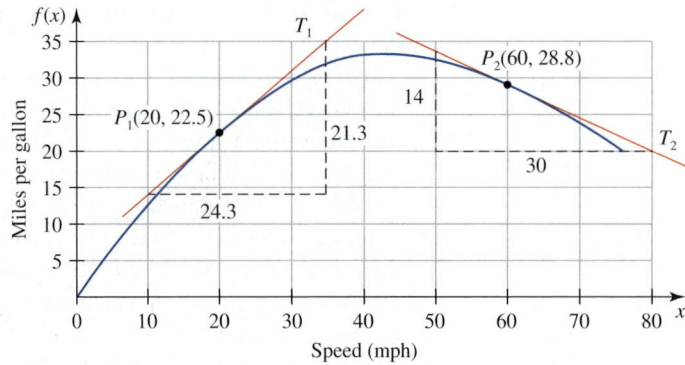

FIGURE 3
The fuel economy of a typical car
Source: U.S. Department of Energy
and Shell Development Company.

Solution   The slope of the tangent line T_1 to the graph of f at $P_1(20, 22.5)$ is approximately

$$\frac{21.3}{24.3} \approx 0.88 \qquad \frac{\text{rise}}{\text{run}}$$

This tells us that the quantity $f(x)$ is increasing at the rate of approximately 0.9 unit per unit change in x when $x = 20$. In other words, when a car is driven at a speed of 20 mph, its fuel economy typically increases at the rate of approximately 0.9 mpg per 1 mph increase in the speed of the car. The slope of the tangent line T_2 to the graph of f at $P_2(60, 28.8)$ is

$$-\frac{14}{30} \approx -0.47$$

This says that the quantity y is decreasing at the rate of approximately 0.5 unit per unit change in x when $x = 60$. In other words, when a car is driven at a speed of 60 mph, its fuel economy typically decreases at the rate of 0.5 mpg per 1 mph increase in the speed of the car.

More Examples Involving Rates of Change

The discovery of the relationship between the problem of finding the slope of the tangent line and the problem of finding the rate of change of one quantity with respect to another spurred the development in the seventeenth century of the branch of calculus called **differential calculus** and made it an indispensable tool for solving practical problems. A small sample of the types of problems that we can solve using differential calculus follows:

- Finding the velocity (rate of change of position with respect to time) of a sports car moving along a straight road
- Finding the rate of change of the harmonic distortion of a stereo amplifier with respect to its power output
- Finding the rate of growth of a bacteria population with respect to time
- Finding the rate of change of the Consumer Price Index with respect to time
- Finding the rate of change of a company's profit (loss) with respect to its level of sales

Defining a Tangent Line

The main purpose of Example 1 was to illustrate the relationship between tangent lines and rates of change. Ideally, the solution to a problem should be analytic and not rely, as in Example 1, on how accurately we can draw a curve and estimate the position of its tangent lines. So our first task will be to give a more precise definition of a tangent line to a curve. After that, we will devise an analytical method for finding an equation of such a line.

Let P and Q be two distinct points on a curve, and consider the secant line passing through P and Q. (See Figure 4.) If we let Q move along the curve toward P, then the secant line rotates about P and approaches the fixed line T. We define T to be the tangent line at P on the curve.

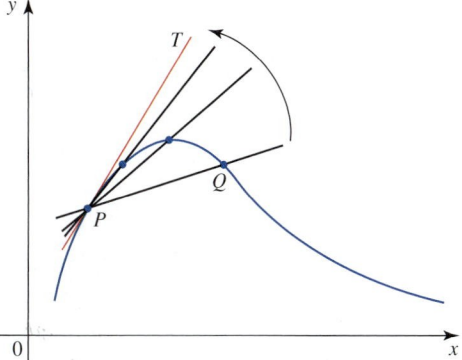

FIGURE 4
As Q approaches P along the curve, the secant lines approach the tangent line T.

Let's make this notion more precise: Suppose that the curve is the graph of a function f defined by $y = f(x)$. (See Figure 5.) Let $P(a, f(a))$ be a point on the graph of f, and let Q be a point on the graph of f distinct from P. Then the x-coordinate of Q has the form $x = a + h$, where h is some appropriate nonzero number. If $h > 0$, then Q lies to the right of P; and if $h < 0$, then Q lies to the left of P. The corresponding y-coordinate of Q is $y = f(a + h)$. In other words, we can specify Q in the usual manner by writing $Q(a + h, f(a + h))$. Observe that we can make Q approach P along the graph of f by letting h approach 0. This situation is illustrated in Figure 5b. (You are encouraged to sketch your own figures for the case $h < 0$.)

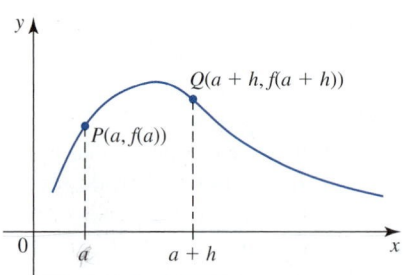
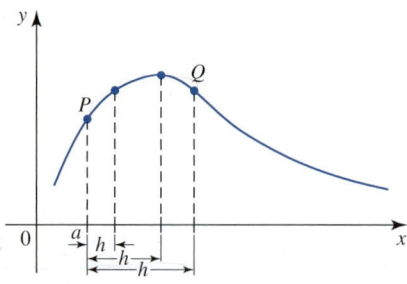

FIGURE 5   (**a**) The points $P(a, f(a))$ and $Q(a + h, f(a + h))$ (**b**) As h approaches 0, Q approaches P.

Next, using the formula for the slope of a line, we can write the slope of the secant line passing through $P(a, f(a))$ and $Q(a + h, f(a + h))$ as

$$m_{\text{sec}} = \frac{f(a + h) - f(a)}{(a + h) - a} = \frac{f(a + h) - f(a)}{h} = \frac{\text{rise}}{\text{run}} \tag{1}$$

The expression on the right-hand side of Equation (1) is called a **difference quotient.**

As we observed earlier, if we let h approach 0, then Q approaches P and the secant line passing through P and Q approaches the tangent line T. This suggests that if the tangent line does exist at P, then its slope m_{tan} should be the limit of m_{sec} obtained by letting h approach zero. This leads to the following definition.

DEFINITION   Tangent Line

Let $P(a, f(a))$ be a point on the graph of a function f. Then the **tangent line** at P (if it exists) on the graph of f is the line passing through P and having slope

$$m_{\text{tan}} = \lim_{h \to 0} \frac{f(a + h) - f(a)}{h} \tag{2}$$

Notes

1. If the limit in Equation (2) does not exist, then m_{tan} is undefined.
2. If the limit in Equation (2) exists, then we can find an equation of the tangent line at P by using the point-slope form of an equation of a line. Thus,

$$y - f(a) = m_{\text{tan}}(x - a).$$

EXAMPLE 2   Find the slope and an equation of the tangent line to the graph of $f(x) = x^2$ at the point $P(1, 1)$.

Solution   To find the slope of the tangent line at the point $P(1, 1)$, we use Equation (2) with $a = 1$, obtaining

$$m_{\text{tan}} = \lim_{h \to 0} \frac{f(1 + h) - f(1)}{h} = \lim_{h \to 0} \frac{(1 + h)^2 - 1^2}{h}$$

$$= \lim_{h \to 0} \frac{(1 + 2h + h^2) - 1}{h} = \lim_{h \to 0} \frac{2h + h^2}{h}$$

$$= \lim_{h \to 0} (2 + h) = 2$$

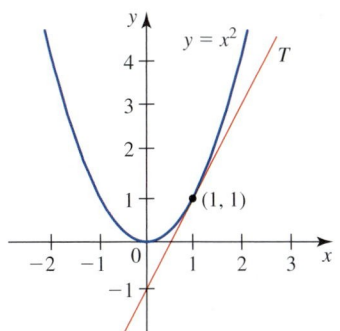

FIGURE 6
T is the tangent line at the point $P(1, 1)$ on the graph of $y = x^2$.

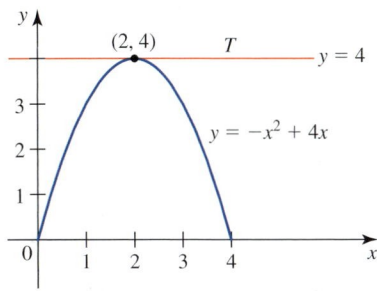

FIGURE 7
The tangent line at the point $(2, 4)$ is horizontal.

To find an equation of the tangent line, we use the point-slope form of an equation of a line to obtain

$$y - 1 = 2(x - 1)$$

or

$$y = 2x - 1$$

The graphs of f and the tangent line at $(1, 1)$ are sketched in Figure 6. ∎

EXAMPLE 3   Find the slope and an equation of the tangent line to the graph of the equation $y = -x^2 + 4x$ at the point $P(2, 4)$.

Solution   The slope of the tangent line at the point $P(2, 4)$ is found by using Equation (2) with $a = 2$ and $f(x) = -x^2 + 4x$. We have

$$m_{\text{tan}} = \lim_{h \to 0} \frac{f(2 + h) - f(2)}{h} = \lim_{h \to 0} \frac{[-(2 + h)^2 + 4(2 + h)] - [-(2)^2 + 4(2)]}{h}$$

$$= \lim_{h \to 0} \frac{-4 - 4h - h^2 + 8 + 4h + 4 - 8}{h} = \lim_{h \to 0} -\frac{h^2}{h}$$

$$= \lim_{h \to 0} (-h) = 0$$

An equation of the tangent line at $P(2, 4)$ is

$$y - 4 = 0(x - 2) \quad \text{or} \quad y = 4$$

The graphs of f and the tangent line at $(2, 4)$ are sketched in Figure 7. ∎

The solution in Example 3 is fully expected if we recall that the graph of the equation $y = -x^2 + 4x$ is a parabola with vertex at $(2, 4)$. At the vertex the tangent line is horizontal, and therefore its slope is zero.

■ Tangent Lines, Secant Lines, and Rates of Change

As we observed earlier, there seems to be a connection between the slope of the tangent line at a given point $P(a, f(a))$ on the graph of a function f and the rate of change of f when $x = a$. Let's show that this is true.

Consider the function f whose graph is shown in Figure 8a. You can see from Figure 8a that as x changes from a to $a + h$, $f(x)$ changes from $f(a)$ to $f(a + h)$. (We call h the *increment* in x.) The ratio of the change in $f(x)$ to the change in x measures the average rate of change of f over the interval $[a, a + h]$.

DEFINITION   **Average Rate of Change of a Function**
The **average rate of change of a function** f over the interval $[a, a + h]$ is

$$\frac{f(a + h) - f(a)}{h} \tag{3}$$

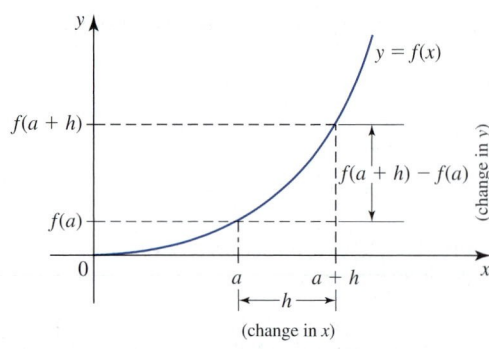

FIGURE 8   (a) The average rate of change of f over $[a, a + h]$ is given by

$$\frac{f(a + h) - f(a)}{h}$$

(b) $m_{\text{sec}} = \dfrac{f(a + h) - f(a)}{h}$

Figure 8b depicts the graph of the same function f. The slope of the secant line passing through the points $P(a, f(a))$ and $Q(a + h, f(a + h))$ is

$$m_{\text{sec}} = \frac{f(a + h) - f(a)}{(a + h) - a} = \frac{f(a + h) - f(a)}{h}$$

But this is just Equation (1). Comparing the expression in (3) and that on the right-hand side of Equation (1), we conclude that the *average rate of change of f with respect to x over the interval $[a, a + h]$ has the same value as the slope of the secant line passing through the points $(a, f(a))$ and $(a + h, f(a + h))$.*

Next, by letting h approach zero in the expression in (3), we obtain the (instantaneous) rate of change of f at a.

DEFINITION   Instantaneous Rate of Change of a Function

The **(instantaneous) rate of change of a function f with respect to x at a** is

$$\lim_{h \to 0} \frac{f(a + h) - f(a)}{h} \qquad (4)$$

if the limit exists.

But this expression also gives the slope of the tangent line to the graph of f at $P(a, f(a))$. Thus, we conclude that *the instantaneous rate of change of f with respect to x at a has the same value as the slope of the tangent line at the point $(a, f(a))$.*

Our earlier calculations suggested that the instantaneous velocity of the maglev at $t = 2$ is 16 ft/sec. We now verify this assertion.

EXAMPLE 4   The position function of the maglev at time t is $s = f(t) = 4t^2$, where $0 \le t \le 30$. Then the average velocity of the maglev over the time interval $[2, 2 + h]$ is given by the average rate of change of the position function s over $[2, 2 + h]$, where $h > 0$ and $2 + h$ lies in the interval $(2, 30)$. Using the expression in (3) with $a = 2$, we see that the average velocity is given by

$$\frac{f(2 + h) - f(2)}{h} = \frac{4(2 + h)^2 - 4(2)^2}{h} = \frac{16 + 16h + 4h^2 - 16}{h} = 16 + 4h$$

Next, using the expression in (4), we see that the instantaneous velocity of the maglev at $t = 2$ is given by

$$v = \lim_{h \to 0} \frac{f(2 + h) - f(2)}{h} = \lim_{h \to 0}(16 + 4h) = 16$$

or 16 ft/sec, as observed earlier.

1.5   CONCEPT QUESTIONS

For Questions 1 and 2, refer to the following figure.

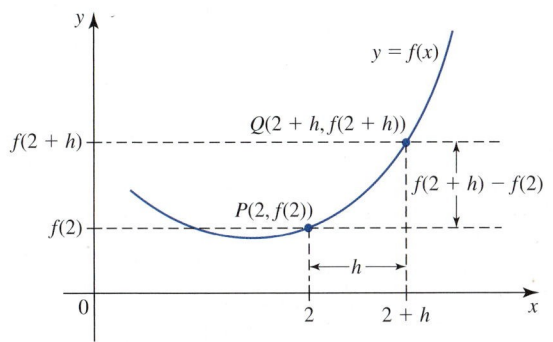

1. Let $P(2, f(2))$ and $Q(2 + h, f(2 + h))$ be points on the graph of a function f.
   a. Find an expression for the slope of the secant line passing through P and Q.
   b. Find an expression for the slope of the tangent line passing through P.
2. Refer to Question 1.
   a. Find an expression for the average rate of change of f over the interval $[2, 2 + h]$.
   b. Find an expression for the instantaneous rate of change of f at 2.
   c. Compare your answers for parts (a) and (b) with those of Question 1.

1.5   EXERCISES

1. **Traffic Flow** Opened in the late 1950s, the Central Artery in downtown Boston was designed to move 75,000 vehicles per day. The following graph shows the average speed of traffic flow in miles per hour versus the number of vehicles moved per day. Estimate the rate of change of the average speed of traffic flow when the number of vehicles moved per day is 100,000 and when it is 200,000. (According to our model, there will be permanent gridlock when we reach 300,000 cars per day!)

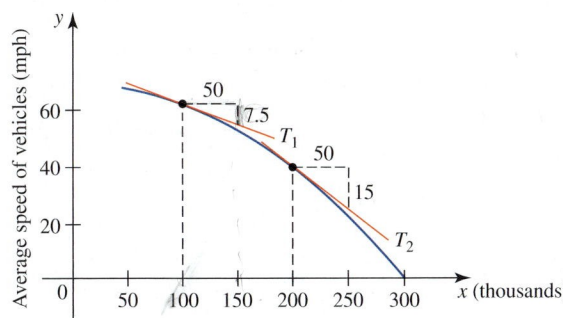

Source: The Boston Globe.

Note: Since 2003 the city of Boston has ameliorated the situation with the "Big Dig."

2. **Forestry** The following graph shows the volume of wood produced in a single-species forest. Here, $f(t)$ is measured in cubic meters per hectare, and t is measured in years. By computing the slopes of the respective tangent lines, estimate the rate at which the wood grown is changing at the beginning of year 10 and at the beginning of year 30.

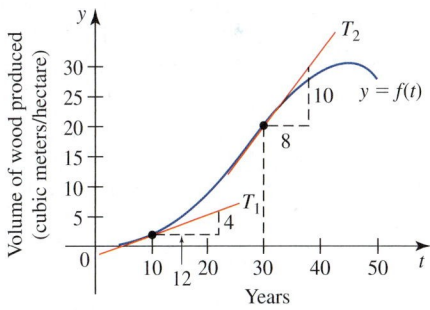

Source: The Random House Encyclopedia.

3. **TV-Viewing Patterns** The graph on the following page shows the percentage of U.S. households watching television during a 24-hr period on a weekday ($t = 0$ corresponds to 6 A.M.). By computing the slopes of the respective tangent lines,

estimate the rate of change of the percentage of households watching television at 4 P.M. and 11 P.M.

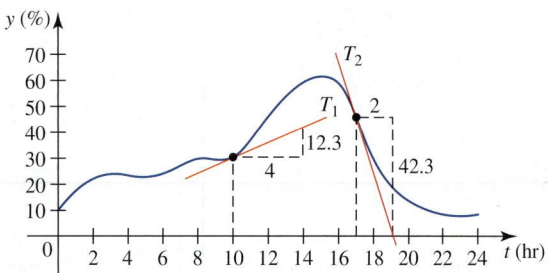

Source: A.C. Nielsen Company.

4. Crop Yield Productivity and yield of cultivated crops are often reduced by insect pests. The following graph shows the relationship between the yield of a certain crop, $f(x)$, as a function of the density of aphids x. (Aphids are small insects that suck plant juices.) Here, $f(x)$ is measured in kilograms per 4000 square meters, and x is measured in hundreds of aphids per bean stem. By computing the slopes of the respective tangent lines, estimate the rate of change of the crop yield with respect to the density of aphids if the density is 200 aphids per bean stem and if it is 800 aphids per bean stem.

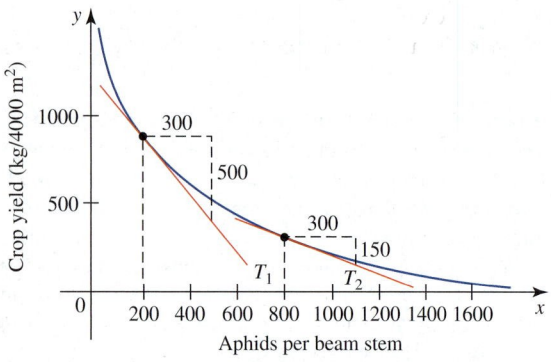

Source: The *Random House Encyclopedia.*

5. The velocities of car A and car B, starting out side by side and traveling along a straight road, are given by $v_A = f(t)$ and $v_B = g(t)$, respectively, where v is measured in feet per second and t is measured in seconds.

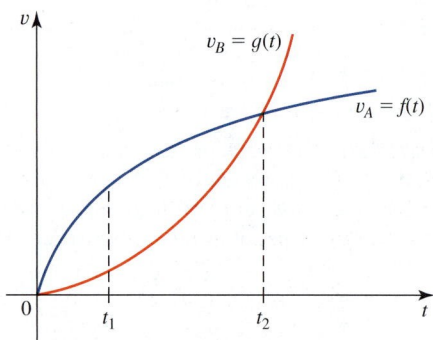

a. What can you say about the velocity and acceleration of the two cars at t_1? (Acceleration is the rate of change of velocity.)

b. What can you say about the velocity and acceleration of the two cars at t_2?

6. Effect of a Bactericide on Bacteria In the figure below, $f(t)$ gives the population P_1 of a certain bacteria culture at time t after a portion of bactericide A was introduced into the population at $t = 0$. The graph of $g(t)$ gives the population P_2 of a similar bacteria culture at time t after a portion of bactericide B was introduced into the population at $t = 0$.

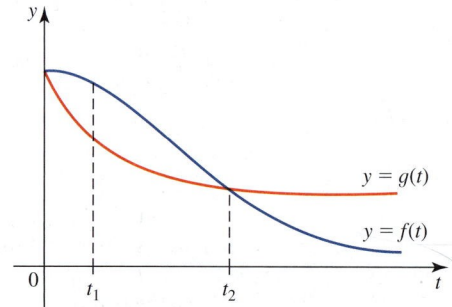

a. Which population is decreasing faster at t_1?

b. Which population is decreasing faster at t_2?

c. Which bactericide is more effective in reducing the population of bacteria in the short run? In the long run?

In Exercises 7–14, (a) use Equation (1) to find the slope of the secant line passing through the points $(a, f(a))$ and $(a + h, f(a + h))$; (b) use the results of part (a) and Equation (2) to find the slope of the tangent line at the point $(a, f(a))$; and (c) find an equation of the tangent line to the graph of f at the point $(a, f(a))$.

Function	$(a, f(a))$	Function	$(a, f(a))$
7. $f(x) = 5$	$(1, 5)$	**8.** $f(x) = 2x + 3$	$(1, 5)$
9. $f(x) = 2x^2 - 1$	$(2, 7)$	**10.** $f(x) = x^2 - x$	$(2, 2)$
11. $f(x) = x^3$	$(2, 8)$	**12.** $f(x) = x^3 + x$	$(2, 10)$
13. $f(x) = \dfrac{1}{x}$	$(1, 1)$	**14.** $f(x) = \dfrac{1}{x + 1}$	$\left(1, \frac{1}{2}\right)$

In Exercises 15–20, find the instantaneous rate of change of the given function when $x = a$.

15. $f(x) = 2x^2 + 1$; $a = 1$

16. $g(x) = x^2 - x + 2$; $a = -1$

17. $H(x) = x^3 + x$; $a = 2$

18. $f(x) = \sqrt{x}$; $a = 4$

19. $f(x) = \dfrac{2}{x} + x$; $a = 1$

20. $f(x) = \dfrac{1}{x - 2}$; $a = 1$

*In Exercises 21–24, the position function of an object moving along a straight line is given by s = f(t). The **average velocity** of the object over the time interval [a, b] is the average rate of change of f over [a, b]; its (instantaneous) **velocity at t = a** is the rate of change of f at a.*

21. The position of a car at any time t is given by $s = f(t) = \frac{1}{4}t^2$, $0 \le t \le 10$, where s is given in feet and t in seconds.

  a. Find the average velocity of the car over the time intervals [2, 3], [2, 2.5], [2, 2.1], [2, 2.01], and [2, 2.001].

  b. Find the velocity of the car at $t = 2$.

22. Velocity of a Car Suppose the distance s (in feet) covered by a car moving along a straight road after t sec is given by the function $s = f(t) = 2t^2 + 48t$.

  a. Calculate the average velocity of the car over the time intervals [20, 21], [20, 20.1], and [20, 20.01].

  b. Calculate the (instantaneous) velocity of the car when $t = 20$.

  c. Compare the results of part (a) with those of part (b).

23. Velocity of a Ball Thrown into the Air A ball is thrown straight up with an initial velocity of 128 ft/sec, so its height (in feet) after t sec is given by $s = f(t) = 128t - 16t^2$.

  a. What is the average velocity of the ball over the time intervals [2, 3], [2, 2.5], and [2, 2.1]?

  b. What is the instantaneous velocity at time $t = 2$?

  c. What is the instantaneous velocity at time $t = 5$? Is the ball rising or falling at this time?

  d. When will the ball hit the ground?

24. During the construction of a high-rise building, a worker accidentally dropped his portable electric screwdriver from a height of 400 ft. After t sec the screwdriver had fallen a distance of $s = f(t) = 16t^2$ ft.

  a. How long did it take the screwdriver to reach the ground?

  b. What was the average velocity of the screwdriver during the time it was falling?

  c. What was the velocity of the screwdriver at the time it hit the ground?

25. A hot air balloon rises vertically from the ground so that its height after t seconds is $h(t) = \frac{1}{2}t^2 + \frac{1}{2}t$ feet, where $0 \le t \le 60$.

  a. What is the height of the balloon after 40 sec?

  b. What is the average velocity of the balloon during the first 40 sec of its flight?

  c. What is the velocity of the balloon after 40 sec?

26. Average Velocity of a Helicopter A helicopter lifts vertically from its pad and reaches a height of $h(t) = 0.2t^3$ feet after t sec, where $0 \le t \le 12$.

  a. How long does it take for the helicopter to reach an altitude of 200 ft?

  b. What is the average velocity of the helicopter during the time it takes to attain this height?

  c. What is the velocity of the helicopter when it reaches this height?

27. a. Find the average rate of change of the area of a circle with respect to its radius r as r increases from $r = 1$ to $r = 2$.

  b. Find the rate of change of the area of a circle with respect to r when $r = 2$.

28. a. Find the average rate of change of the volume of a sphere with respect to its radius r as r increases from $r = 1$ to $r = 2$.

  b. Find the rate of change of the volume of a sphere with respect to r when $r = 2$.

29. Demand for Tents The quantity demanded of the Sportsman 5×7 tents, x, is related to the unit price, p, by the function

$$p = f(x) = -0.1x^2 - x + 40$$

where p is measured in dollars and x is measured in units of a thousand.

  a. Find the average rate of change in the unit price of a tent if the quantity demanded is between 5000 and 5050 tents; between 5000 and 5010 tents.

  b. What is the rate of change of the unit price if the quantity demanded is 5000?

30. At a temperature of 20°C, the volume V (in liters) of 1.33 g of O_2 is related to its pressure p (in atmospheres) by the formula $V(p) = 1/p$.

  a. What is the average rate of change of V with respect to p as p increases from $p = 2$ to $p = 3$?

  b. What is the rate of change of V with respect to p when $p = 2$?

31. Average Velocity of a Motorcycle The distance s (in feet) covered by a motorcycle traveling in a straight line at any time t (in seconds) is given by the function

$$s(t) = -0.1t^3 + 2t^2 + 24t$$

Calculate the motorcycle's average velocity over the time interval $[2, 2 + h]$ for $h = 1, 0.1, 0.01, 0.001, 0.0001$, and 0.00001, and use your results to guess at the motorcycle's instantaneous velocity at $t = 2$.

32. Rate of Change of a Cost Function The daily total cost $C(x)$ incurred by Trappee and Sons for producing x cases of TexaPep hot sauce is given by

$$C(x) = 0.000002x^3 + 5x + 400$$

Calculate

$$\frac{C(100 + h) - C(100)}{h}$$

for $h = 1, 0.1, 0.01, 0.001$, and 0.0001, and use your results to estimate the rate of change of the total cost function when the level of production is 100 cases per day.

33. a. Plot the graph of

$$g(h) = \frac{(2 + h)^3 - 8}{h}$$

using the viewing window $[-1, 1] \times [0, 20]$.

b. Zoom-in to find $\lim_{h \to 0} g(h)$.

c. Verify analytically that the limit found in part (b) is $\lim_{h \to 0} \dfrac{f(2 + h) - f(2)}{h}$ where $f(x) = x^3$.

34. Use the technique of Exercise 33a–b to find

$$\lim_{h \to 0} \frac{f(8 + h) - f(8)}{h} \text{ if } f(x) = \sqrt[3]{x}, \text{ using the viewing}$$

window $[-1, 1] \times [0, 0.1]$.

In Exercises 35–40 the expression gives the (instantaneous) rate of change of a function f at some number a. Identify f and a.

35. $\displaystyle\lim_{h \to 0} \frac{(1 + h)^5 - 1}{h}$

36. $\displaystyle\lim_{h \to 0} \frac{2\sqrt[4]{16 + h} - 4}{h}$

37. $\displaystyle\lim_{h \to 0} \left[\frac{(4 + h)^2 - 16}{h} + \frac{\sqrt{4 + h} - 2}{h} \right]$

38. $\displaystyle\lim_{h \to 0} \frac{2^{3 + h} - 8}{h}$

39. $\displaystyle\lim_{x \to 1} \frac{x^4 - 1}{x - 1}$

40. $\displaystyle\lim_{x \to \pi/2} \frac{\sin x - 1}{x - \frac{\pi}{2}}$

In Exercises 41–44, determine whether the statement is true or false. If it is true, explain why. If it is false, explain why or give an example that shows it is false.

41. The slope of the secant line passing through the points $(a, f(a))$ and $(b, f(b))$ measures the average rate of change of f over the interval $[a, b]$.

42. A tangent line to the graph of a function may intersect the graph at infinitely many points.

43. There may be more than one tangent line at a given point on the graph of a function.

44. The slope of the tangent line to the graph of f at the point $(a, f(a))$ is given by

$$\lim_{x \to a} \frac{f(x) - f(a)}{x - a}$$

CHAPTER 1   REVIEW

CONCEPT REVIEW

In Exercises 1–8, fill in the blanks.

1. a. The statement $\lim_{x \to a} f(x) = L$ means that there exists a number _____ such that the values of _____ can be made as close to _____ as we please by taking x to be sufficiently close to _____.

b. The statement $\lim_{x \to a^+} f(x) = L$ is similar to $\lim_{x \to a} f(x) = L$, but here we require that x lie to the _____ of a.

c. $\lim_{x \to a} f(x) = L$ if and only if $\lim_{x \to a^-} f(x)$ and $\lim_{x \to a^+} f(x)$ both _____ and are equal to _____.

d. The precise meaning of $\lim_{x \to a} f(x) = L$ is that given any number _____, there exists a number _____ such that $0 < |x - a| < \delta$ implies $|f(x) - L| < \varepsilon$.

2. a. If $\lim_{x \to a} f(x) = L$ and $\lim_{x \to a} g(x) = M$, then the Sum Law states _____, the Product Law states _____, the Constant Multiple Law states _____, the Quotient Law states _____ $M \neq 0$, and the Root Law states _____ provided $L > 0$, if n is even.

b. If $p(x)$ is a polynomial function, then $\lim_{x \to a} p(x) = $ _____ for every real number a.

c. If $r(x)$ is a rational function, then $\lim_{x \to a} r(x) = r(a)$, provided that a is in the domain of _____.

3. Suppose that $f(x) \leq g(x) \leq h(x)$ for all x in an interval containing a, except possibly at a, and that $\lim_{x \to a} f(x) = \lim_{x \to a} h(x) = L$. Then the Squeeze Theorem says that _____.

4. a. If $\lim_{x \to a} f(x) = f(a)$, then f is said to be _____ at a.

b. If f is discontinuous at a but it can be made continuous at a by defining or redefining f at a, then f has a _____ discontinuity at a.

c. If $\lim_{x \to a^-} f(x) = L$ and $\lim_{x \to a^+} f(x) = M$ and $L \neq M$, then f has a _____ discontinuity at a.

d. If $\lim_{x \to a^-} f(x) = f(a)$, then f is continuous from the _____ at a.

5. a. A polynomial function is continuous on _____.

b. A rational function is continuous on _____ _____.

c. The composition of two continuous functions is a _____ function.

6. a. Suppose that f is continuous on $[a, b]$ and $f(a) \leq M \leq f(b)$. Then the Intermediate Value

Theorem guarantees the existence of at least one number c in _____ such that _____.

  b. If f is continuous on $[a, b]$ and $f(a)f(b) < 0$, then there must be at least one solution of the equation _____ in the interval _____.

7. a. The tangent line at $P(a, f(a))$ to the graph of f is the line passing through P and having slope _____.

  b. If the slope of the tangent line at $P(a, f(a))$ is $m_{\tan}$, then an equation of the tangent line at P is _____.

8. a. The slope of the secant line passing through $P(a, f(a))$ and $Q(a + h, f(a + h))$ and the average rate of change of f over the interval $[a, a + h]$ are both given by _____.

  b. The slope of the tangent line at $P(a, f(a))$ and the instantaneous rate of change of f at a are both given by _____.

REVIEW EXERCISES

In Exercises 1 and 2, use the graph of the function f to find (a) $\lim_{x \to a^-} f(x)$, (b) $\lim_{x \to a^+} f(x)$, *and* (c) $\lim_{x \to a} f(x)$ *for the given value of a.*

1. $a = 4$

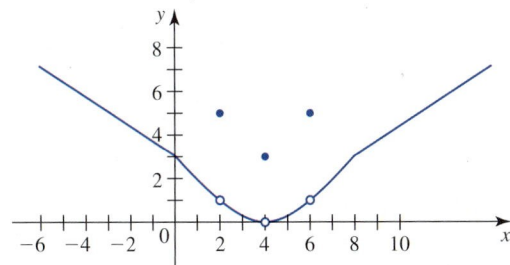

2. $a = 0$

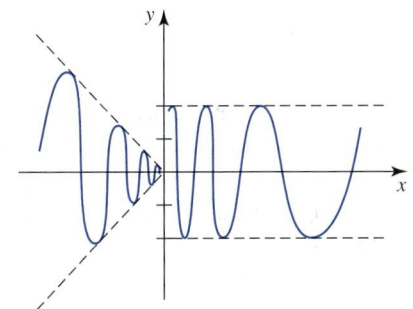

In Exercises 3–6, sketch the graph of f, and evaluate (a) $\lim_{x \to a^-} f(x)$, (b) $\lim_{x \to a^+} f(x)$, *and* (c) $\lim_{x \to a} f(x)$ *for the given value of a.*

3. $f(x) = \begin{cases} -x + 5 & \text{if } x \le 3 \\ 2x - 4 & \text{if } x > 3 \end{cases}; \quad a = 3$

4. $f(x) = \begin{cases} \dfrac{|x - 2|}{x - 2} & \text{if } x \ne 2 \\ 2 & \text{if } x = 2 \end{cases}; \quad a = 2$

5. $f(x) = \begin{cases} -x + 2 & \text{if } x < 2 \\ \sqrt{x - 2} & \text{if } x \ge 2 \end{cases}; \quad a = 2$

6. $f(x) = \begin{cases} -\dfrac{1}{2}x + 1 & \text{if } x < 0 \\ 0 & \text{if } x = 0; \quad a = 0 \\ -\dfrac{1}{x^2} & \text{if } x > 0 \end{cases}$

In Exercises 7–28, find the indicated limit if it exists.

7. $\lim\limits_{h \to 3}(4h^2 - 2h + 4)$

8. $\lim\limits_{x \to 2}(x^3 + 1)(x^2 - 1)$

9. $\lim\limits_{x \to 3}\sqrt{x^2 + 2x - 3}$

10. $\lim\limits_{t \to 1} \dfrac{t^2 - 1}{1 - t}$

11. $\lim\limits_{x \to 5}(x^2 + 2)^{2/3}$

12. $\lim\limits_{x \to 3} \dfrac{27 - x^3}{x - 3}$

13. $\lim\limits_{y \to 0} \dfrac{2y^2 + 1}{y^3 - 2y^2 + y + 2}$

14. $\lim\limits_{x \to 3^+} \dfrac{x + 1}{x - 3}$

15. $\lim\limits_{x \to 3} \dfrac{2x^2 - 5x - 3}{3x^2 - 10x + 3}$

16. $\lim\limits_{x \to 4} \dfrac{x - 4}{\sqrt{x} - 2}$

17. $\lim\limits_{h \to 0} \dfrac{(4 + h)^{-1} - 4^{-1}}{h}$

18. $\lim\limits_{x \to 2^+} \dfrac{x - 2}{|x - 2|}$

19. $\lim\limits_{x \to 3^-} \sqrt{9 - x^2}$

20. $\lim\limits_{x \to 3^+} \dfrac{1 + \sqrt{2x - 6}}{x - 2}$

21. $\lim\limits_{x \to 0} \dfrac{2 \sin 3x}{x}$

22. $\lim\limits_{x \to 0} x \cot 2x$

23. $\lim\limits_{x \to 0^+} \dfrac{\cos x}{\sqrt{x}}$

24. $\lim\limits_{x \to 0^+} \sqrt{x} \sin \dfrac{1}{x}$

25. $\lim\limits_{x \to 0} \ln(x^2 + 1)$

26. $\lim\limits_{x \to \pi/2} e^{\sin x}$

27. $\lim\limits_{x \to 0} e^{\sqrt{x}/(x+1)}$

28. $\lim\limits_{x \to 0} \ln(\sec x + x)$

29. Prove that $\lim_{x \to 0^+} x^2 \cos(1/\sqrt{x}) = 0$.

30. Suppose that $1 - x^2 \le f(x) \le 1 + x^2$ for all x. Find $\lim_{x \to 0} f(x)$.

In Exercises 31 and 32, use the graph of the function f to determine where the function is discontinuous.

31.

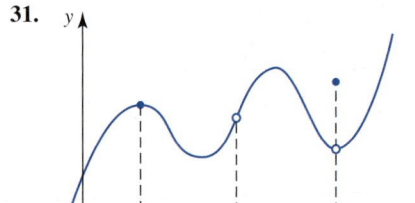

32.

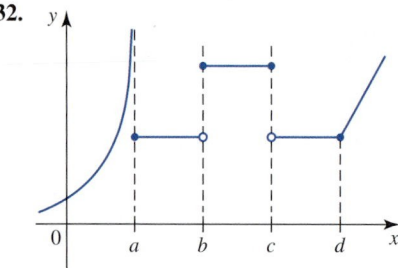

In Exercises 33 and 34, use the graph of the function f to determine whether f is continuous on the given interval(s). Justify your answer.

33.

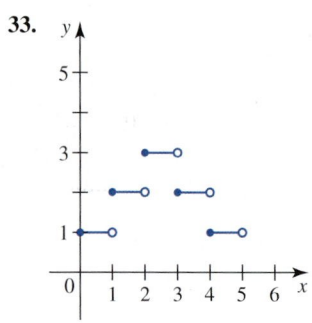

a. $[1, 2)$
b. $(0, 1)$
c. $(3, 5)$

34.

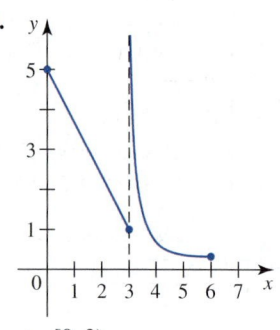

a. $[0, 3)$
b. $[0, 3]$
c. $[2, 6]$
d. $(3, 6]$

In Exercises 35–42, find the numbers, if any, where the function is discontinuous.

35. $f(x) = x^2 + 3x + \sqrt{-x}$

36. $g(x) = \dfrac{3|x - 1|}{x^2 + x - 6}$

37. $f(t) = \dfrac{(t + 2)^{1/2}}{(t + 1)^{1/2}}$

38. $h(x) = \dfrac{1}{\cos x}$

39. $f(x) = \dfrac{1}{\sin x}$

40. $f(x) = \begin{cases} x^2 + 1 & \text{if } x < 0 \\ -x + 1 & \text{if } x \geq 0 \end{cases}$

41. $f(x) = \dfrac{\ln x}{1 - \sqrt{x}}$

42. $f(x) = \dfrac{e^x}{\sqrt{1 - x}}$

43. Let

$$f(x) = \begin{cases} \dfrac{x^2 - 2x}{x^2 - 4} & \text{if } x \neq 2 \\ c & \text{if } x = 2 \end{cases}$$

Find the value of c such that f will be continuous at 2.

44. True or false? The square of a discontinuous function is also a discontinuous function. Justify your answer.

In Exercises 45–48, show that the equation has at least one zero in the given interval.

45. $x^4 + x - 5 = 0$; $(1, 2)$

46. $\sin x - x + 1 = 0$; $\left(0, \frac{3\pi}{2}\right)$

47. $x \ln x = 1$; $(1, e)$

48. $e^{-x} - x = 0$; $(0, 1)$

49. Let

$$f(x) = \begin{cases} -(x^2 + 1) & \text{if } -2 \leq x < 0 \\ x^2 + 1 & \text{if } 0 \leq x \leq 2 \end{cases}$$

Is there a number c in $[-2, 2]$ such that $f(c) = 0$? Why?

50. Find where the function

$$f(x) = \begin{cases} x \sin \dfrac{1}{x} & \text{if } x \neq 0 \\ 0 & \text{if } x = 0 \end{cases}$$

is continuous.

In Exercises 51 and 52, use the precise definition of the limit to prove the statement.

51. $\lim_{x \to -1}(2x + 3) = 1$

52. $\lim_{x \to 0} \sqrt[3]{x} = 0$

53. According to the special theory of relativity, the Lorentz contraction formula $L = L_0 \sqrt{1 - (v^2/c^2)}$ gives the relationship between the length L of an object moving with a speed v relative to an observer and its length L_0 at rest. Here, c is the speed of light.
a. Find the domain of L, and use the result to explain why one may consider only $\lim_{v \to c^-} L$.
b. Evaluate $\lim_{v \to c^-} L$, and interpret your result.

54. **Temperature Changes** The following graph shows the air temperature over a 24-hr period on a certain day in November in Chicago, with $t = 0$ corresponding to 12 midnight. Using the given data, compute the slopes of the respective tangent lines, and estimate the rate of change of the temperature at 8 A.M. and at 6 P.M.

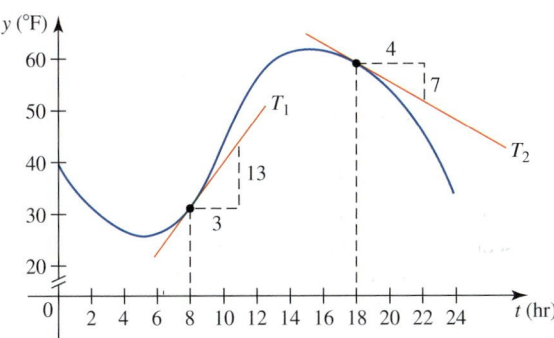

55. The position of an object moving along a straight line is $s(t) = 2t^2 + t + 1$, where $s(t)$ is measured in feet and t is measured in seconds.
  a. Find the average velocity of the object over the time intervals $[1, 2]$, $[1, 1.5]$, $[1, 1.1]$, and $[1, 1.01]$.
  b. Find the instantaneous velocity of the object when $t = 1$.

56. Gravitational Force The magnitude of the gravitational force exerted by the earth on a particle of mass m at a distance r from the center of the earth is

$$F(r) = \begin{cases} \dfrac{GMmr}{R^3} & \text{if } r < R \\[2mm] \dfrac{GMm}{r^2} & \text{if } r \geq R \end{cases}$$

where M is the mass of the earth, R is its radius, and G is the gravitational constant.
  a. Is F a continuous function of r?
  b. Sketch the graph of F.

PROBLEM-SOLVING TECHNIQUES

In this very first example in Problem-Solving Techniques, we illustrate the efficacy of the *method of substitution*. When the right substitution is used, a problem which at first glance seems impossible to solve, or as in this case, difficult to solve, is often reduced to one that is familiar or is much easier to solve. In the Problem-Solving Techniques sections throughout this book, we will showcase other problem-solving techniques.

EXAMPLE Evaluate $\displaystyle\lim_{x \to 1} \frac{3x - 3}{\sqrt[3]{x + 7} - 2}$.

Solution The obvious approach is to use the Quotient Law for limits. But since the numerator and the denominator approach zero as x approaches 1, the law is not applicable.

Drawing from experience in solving such problems, we might attempt to rationalize the denominator. Although this can be done directly, it is better to transform the expression into a simpler one. A reasonable *substitution* is to put

$$t = \sqrt[3]{x + 7}$$

so $t^3 = x + 7$ or $x = t^3 - 7$. Observe that as x approaches 1, t approaches 2. Therefore,

$$\lim_{x \to 1} \frac{3x - 3}{\sqrt[3]{x + 7} - 2} = \lim_{t \to 2} \frac{3(t^3 - 7) - 3}{t - 2} = \lim_{t \to 2} \frac{3t^3 - 24}{t - 2} = \lim_{t \to 2} \frac{3(t^3 - 2^3)}{t - 2}$$

$$= \lim_{t \to 2} \frac{3(t - 2)(t^2 + 2t + 4)}{t - 2}$$

$$= \lim_{t \to 2} 3(t^2 + 2t + 4) = 36$$

CHALLENGE PROBLEMS

1. Find $\lim\limits_{x \to 0} \dfrac{\sqrt[3]{x+1} - 1}{x}$.

2. **a.** Find $\lim\limits_{x \to 1^-} \dfrac{x^2 - 1}{|x - 1|}$ and $\lim\limits_{x \to 1^+} \dfrac{x^2 - 1}{|x - 1|}$.

   b. Find $\lim\limits_{x \to 0^+} \dfrac{|\sin x|}{\sin x}$ and $\lim\limits_{x \to 0^-} \dfrac{|\sin x|}{\sin x}$.

3. Find $\lim\limits_{x \to \pi/2} \dfrac{\cos x}{1 - \dfrac{4x^2}{\pi^2}}$.

4. Let $P\!\left(c, \sqrt{a^2 - c^2}\right)$ be a point on the upper half of the circle $x^2 + y^2 = a^2$ and located in the first quadrant, and let $Q\!\left(c + h, \sqrt{a^2 - (c + h)^2}\right)$ be another point on the circle in the same quadrant.

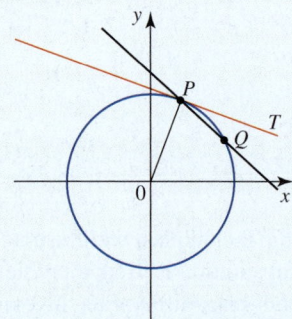

   a. Find an expression for the slope m_{sec} of the secant line passing through P and Q.

   b. Evaluate $\lim_{h \to 0} m_{\text{sec}}$, and show that this limit is the slope of the tangent line T to the circle at P.

   c. How would you establish a similar result for the case in which P and Q both lie in the third quadrant?

5. An n-sided regular polygon is inscribed in a circle of radius R, and another is circumscribed in the same circle. The figure below illustrates the case in which $n = 6$.

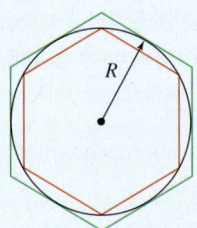

   a. Show that the perimeter of the circumscribing polygon is $2Rn \tan(\pi/n)$ and the perimeter of the inscribing polygon is $2Rn \sin(\pi/n)$

   b. Use the Squeeze Theorem and the results of part (a) to show that the circumference of a circle of radius R is $2\pi R$.

6. Find the values of x at which the function is discontinuous.
   a. $f(x) = \left[\!\left[\sqrt{x} \right]\!\right]$
   b. $g(x) = [\![x]\!] + [\![-x]\!]$

7. A function f is defined by

$$f(x) = \begin{cases} \dfrac{\tan^2 x}{1 - \cos x} & \text{if } x \neq 0 \\ c & \text{if } x = 0 \end{cases}$$

   Determine the value of c such that f is continuous at 0.

8. Show that

$$f(x) = \begin{cases} \tan x \cos \dfrac{1}{x} & \text{if } x \neq 0 \\ 0 & \text{if } x = 0 \end{cases}$$

   is continuous at 0.

9. Let f be a continuous function with domain $[1, 3]$ and range $[0, 4]$ satisfying $f(1) = 0$ and $f(3) = 4$. Show that there is at least one point c in $(1, 3)$ such that $f(c) = c$. The point c is called a *fixed point* of f.

10. Let $f(x) = \dfrac{1}{1 - x}$. Determine where the composite function $g = f \circ f \circ f$ defined by $g(x) = f\{f[f(x)]\}$ is discontinuous.

11. Determine where the composite function $h = f \circ g$ defined by $f(x) = \dfrac{1}{x^2 - x - 2}$ and $g(x) = \dfrac{1}{x - 1}$ is discontinuous.

12. Let f be a polynomial function of even degree, and suppose that there is a number c such that $f(c)$ and the leading coefficient of f have opposite signs. Show that f must have at least two real zeros.

13. Suppose that a, b, and c are positive and that $A < B < C$. Show that the equation

$$\frac{a}{x - A} + \frac{b}{x - B} + \frac{c}{x - C} = 0$$

   has a root between A and B, and a root between B and C.

14. Suppose that f is continuous on an interval (a, b) and that $x_1, x_2, \ldots, x_n$ are any n numbers in (a, b). Show that there exists a number c in (a, b) such that

$$f(c) = \frac{1}{n}\left[f(x_1) + f(x_2) + \cdots + f(x_n)\right]$$

The photograph shows a space shuttle being launched from Cape Kennedy. Suppose a spectator watches the launch from an observation deck located at a known distance from the launch pad. If the speed of the shuttle at a certain instant of time is known, can we find the speed at which the distance between the shuttle and the spectator is changing? The derivative allows us to answer questions such as this.

Matt Stroshane/Getty Images

2

The Derivative

IN THIS CHAPTER we introduce the notion of the derivative of a function. The derivative is the principal tool that we use to solve problems in differential calculus. We also develop rules of differentiation that will enable us to calculate, with relative ease, the derivatives of complicated functions. The rest of the chapter will be devoted to applications of the derivative.

V This symbol indicates that one of the following video types is available for enhanced student learning at **www.academic.cengage.com/login:**
- Chapter lecture videos
- Solutions to selected exercises

2.1 | **The Derivative**

■ The Derivative

In Section 1.5 we saw that the slope of the tangent line to the graph of a function $y = f(x)$ at the point $(a, f(a))$ has the same value as the rate of change of the quantity y with respect to x at the number a. Both values are given by

$$\lim_{h \to 0} \frac{f(a + h) - f(a)}{h}$$

provided that the limit exists. Recall that in deriving this expression, the number a was fixed but otherwise arbitrary. Therefore, if we simply replace the constant a by the variable x, we obtain a formula that gives us the slope of the tangent line at *any* point $(x, f(x))$ on the graph of f as well as the rate of change of the quantity y with respect to x for any value of x. The resulting function is called the *derivative of f*, since it is derived from the function f.

DEFINITION   The Derivative

The derivative of a function f with respect to x is the function f' defined by the rule

$$f'(x) = \lim_{h \to 0} \frac{f(x + h) - f(x)}{h} \tag{1}$$

The domain of f' consists of all values of x for which the limit exists.

Two interpretations of the derivative follow.

1. **Geometric Interpretation of the Derivative:** The derivative f' of a function f is a measure of the slope of the tangent line to the graph of f at any point $(x, f(x))$, provided that the derivative exists.
2. **Physical Interpretation of the Derivative:** The derivative f' of a function f measures the instantaneous rate of change of f at x.

(See Figure 1.)

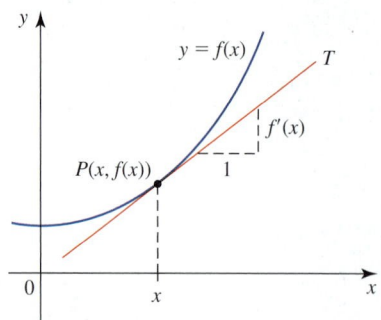

FIGURE 1
$f'(x)$ is the slope of T at P; $f(x)$ is changing at the rate of $f'(x)$ units per unit change in x at x.

■ Using the Derivative to Describe the Motion of the Maglev

Let's look at these two interpretations of the derivative via an example involving the motion of the maglev. Once again, recall that the position s of the maglev at any time t is

$$s = f(t) = 4t^2 \qquad 0 \le t \le 30$$

The derivative of the function f is

$$f'(t) = \lim_{h \to 0} \frac{f(t + h) - f(t)}{h}$$

$$= \lim_{h \to 0} \frac{4(t + h)^2 - 4t^2}{h}$$

$$= \lim_{h \to 0} \frac{4t^2 + 8th + 4h^2 - 4t^2}{h}$$

$$= \lim_{h \to 0} \frac{h(8t + 4h)}{h} = \lim_{h \to 0} (8t + 4h)$$

$$= 8t$$

Thus, the rate of change of the position of the maglev with respect to time, at time t, as well as the slope of the tangent line at the point $(t, f(t))$ on the graph of f, is given by

$$f'(t) = 8t \qquad 0 < t < 30$$

So in this setting, f' is just the velocity function giving the velocity of the maglev at any time t. In particular, the velocity of the maglev when $t = 2$ is

$$f'(2) = 8(2) = 16$$

or 16 ft/sec. Equivalently, the slope of the tangent line to the graph of f at the point $P(2, 16)$ is 16. The graph of f' is sketched in Figure 2.

From the velocity curve we see that the velocity of the maglev is steadily increasing with respect to time. We can even say more. Because the equation $v = 8t$ is a linear equation in the slope-intercept form with slope 8, we see that v is increasing at the rate of 8 units per unit change in t. Put another way, the maglev is accelerating at the constant rate of 8 ft/sec/sec, usually abbreviated 8 ft/sec^2. (Acceleration is the rate of change of velocity.)

Starting from just a formula giving the position of the maglev, we have now been able to give a complete description of the motion of the maglev, albeit just for this particular situation.

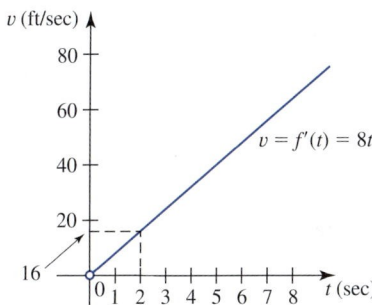

FIGURE 2
The graph of $v = f'(t) = 8t$ gives the velocity of the maglev at any time t and is called a velocity curve.

■ Differentiation

The process of finding the derivative of a function is called **differentiation.** We can view this process as an operation on a function f to produce another function f'. For example, if we let D_x denote the **differential operator,** then the process of differentiation can be written

$$D_x f = f' \qquad \text{or} \qquad D_x f(x) = f'(x)$$

Differentiation is always performed with respect to the independent variable. (Remember that we are concerned with the rate of change of the dependent variable with respect to the independent variable.) Therefore, if the independent variable is t, we write D_t instead of D_x. Another notation, and one that we will adopt, is

$$\frac{d}{dx}$$

which is read "dee dee x of." For example

$$\frac{d}{dx}f = D_x f = f' \qquad \text{or} \qquad \frac{d}{dx}f(x) = D_x f(x) = f'(x)$$

$f'(x)$ is read "f prime of x."

If we denote the dependent variable by y so that $y = f(x)$, then the derivative is written

$$\frac{dy}{dx}$$

(read "dee y, dee x") or, in an even more abbreviated form, as y' (read "y prime").

 dy/dx is not a fraction.

The value of the derivative of f at a is denoted by $f'(a)$. If the dependent variable is denoted by a letter such as y, then the value of the derivative at a is denoted by

$$\frac{dy}{dx}\bigg|_{x=a}$$

(read "dy/dx evaluated at x = a"). For example, since the position of the maglev is denoted by the letter s, where $s = f(t) = 4t^2$, the velocity of the maglev when $t = 2$ may be written as $f'(2) = 16$ or

$$\frac{ds}{dt}\bigg|_{t=2} = 8t\bigg|_{t=2} = 16$$

■ Finding the Derivative of a Function

EXAMPLE 1 Let $y = \sqrt{x}$.

a. Find dy/dx, and determine its domain.
b. How fast is y changing at $x = 4$?
c. Find the slope and an equation of the tangent line to the graph of the equation $y = \sqrt{x}$ at the point where $x = 4$.

Solution   Here, $f(x) = \sqrt{x}$.

a. $\dfrac{dy}{dx} = \lim\limits_{h \to 0} \dfrac{f(x+h) - f(x)}{h} = \lim\limits_{h \to 0} \dfrac{\sqrt{x+h} - \sqrt{x}}{h}$

$\qquad = \lim\limits_{h \to 0} \dfrac{(\sqrt{x+h} - \sqrt{x})(\sqrt{x+h} + \sqrt{x})}{h(\sqrt{x+h} + \sqrt{x})}$ Rationalize the numerator.

$\qquad = \lim\limits_{h \to 0} \dfrac{(x+h) - x}{h(\sqrt{x+h} + \sqrt{x})} = \lim\limits_{h \to 0} \dfrac{h}{h(\sqrt{x+h} + \sqrt{x})}$

$\qquad = \lim\limits_{h \to 0} \dfrac{1}{\sqrt{x+h} + \sqrt{x}} = \dfrac{1}{2\sqrt{x}}$

The domain of dy/dx is $(0, \infty)$.

b. The rate of change of y with respect to x at $x = 4$ is

$$\left.\frac{dy}{dx}\right|_{x=4} = \left.\frac{1}{2\sqrt{x}}\right|_{x=4} = \frac{1}{2\sqrt{4}} = \frac{1}{4}$$

or $\frac{1}{4}$ unit per unit change in x.

c. The slope m of the tangent line to the graph of $y = \sqrt{x}$ at the point where $x = 4$ has the same value as the rate of change of y with respect to x at $x = 4$. From the result of part (b), we find $m = \frac{1}{4}$. Next, when $x = 4$, $y = \sqrt{4} = 2$, giving $(4, 2)$ as the point of tangency. Finally, using the point-slope form of an equation of a line, we find

$$y - 2 = \frac{1}{4}(x - 4)$$

or $y = \frac{1}{4}x + 1$ as an equation of the tangent line.

The graph of $y = \sqrt{x}$ and the tangent line at $(4, 2)$ are sketched in Figure 3.

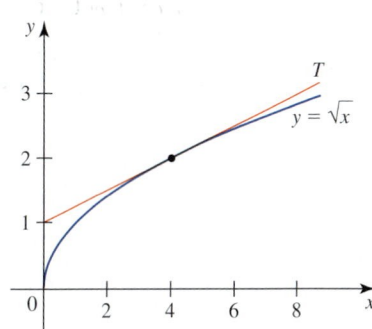

FIGURE 3
T is the tangent line to the graph of $y = \sqrt{x}$ at $(4, 2)$.

EXAMPLE 2   Let $f(x) = 2x^3 + x$.

a. Find $f'(x)$.
b. What is the slope of the tangent line to the graph of f at $(2, 18)$?
c. How fast is f changing when $x = 2$?

Solution

a. $f'(x) = \lim\limits_{h \to 0} \dfrac{f(x + h) - f(x)}{h} = \lim\limits_{h \to 0} \dfrac{[2(x + h)^3 + (x + h)] - (2x^3 + x)}{h}$

$= \lim\limits_{h \to 0} \dfrac{(2x^3 + 6x^2h + 6xh^2 + 2h^3 + x + h) - (2x^3 + x)}{h}$

$= \lim\limits_{h \to 0} \dfrac{h(6x^2 + 6xh + 2h^2 + 1)}{h} = \lim\limits_{h \to 0} (6x^2 + 6xh + 2h^2 + 1)$

$= 6x^2 + 1$

b. The required slope is given by

$$f'(2) = 6(2)^2 + 1 = 25$$

c. From the result of part (b), we see that f is changing at the rate of 25 units per unit change in x when $x = 2$.

EXAMPLE 3 Find $\dfrac{dy}{dx}$ if $y = \dfrac{1}{x + 1}$.

Solution If we write $y = f(x)$, then

$$\frac{dy}{dx} = f'(x) = \lim_{h \to 0} \frac{f(x + h) - f(x)}{h}$$

$$= \lim_{h \to 0} \frac{\dfrac{1}{(x + h) + 1} - \dfrac{1}{x + 1}}{h}$$

$$= \lim_{h \to 0} \frac{\dfrac{x + 1 - (x + h + 1)}{(x + h + 1)(x + 1)}}{h} \qquad \text{Simplify the numerator.}$$

$$= \lim_{h \to 0} -\frac{1}{(x + h + 1)(x + 1)} = -\frac{1}{(x + 1)^2}$$

Using the Graph of f to Sketch the Graph of f'

It was a simple matter to sketch the graph of the derivative function f' in the example describing the motion of a maglev, because we were able to obtain the formula $f'(t) = 8t$ from the position function f for the maglev. The next example shows how we can make a rough sketch of the graph of f' using only the graph of f. The method that is used is based on the geometric interpretation of f'.

EXAMPLE 4 **The Trajectory of a Projectile** The graph of the function f shown in Figure 4 gives the ballistic trajectory of a projectile that starts from the origin and is confined to move in the xy-plane. Use this graph to draw the graph of f'. Then use it to estimate the rate at which the altitude of the projectile (y) is changing with respect to x (the distance traveled horizontally by the projectile) when $x = 5000$ and when $x = 16{,}000$.

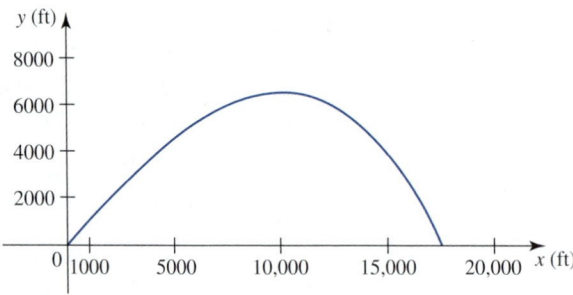

FIGURE 4
The trajectory of a projectile

Solution First we estimate the slopes $f'(x)$ of the tangent lines (drawn by sight) to some points on the graph of f using the techniques of Example 1 in Section 1.5. The results are shown in Figure 5a. Next, we plot the points $(x, f'(x))$ on the xy'-coordinate system placed directly below the xy-coordinate system. Finally, we draw a smooth curve through these points, obtaining the graph of f' shown in Figure 5b. From the graph of f' we see that the altitude of the projectile is increasing at the rate of approximately

0.7 ft/ft when $x = 5000$, and it is decreasing at the rate of approximately 1.3 ft/ft when $x = 16,000$.

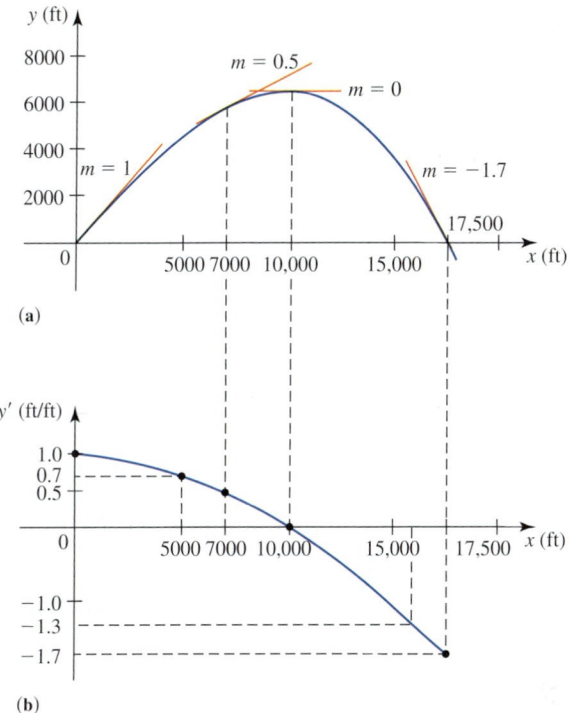

FIGURE 5
The graphs of f and f'

(b)

■ Differentiability

A function is said to be **differentiable** at a number if it has a derivative at that number. As we will soon see, a function may fail to be differentiable at one or more numbers in its domain. This should not surprise us because the derivative is the limit of a function, and we have already seen that the limit of a function does not always exist as we approach a number.

Loosely speaking, a function f does not have a derivative at a if the graph of f does not have a tangent line at a, or if the tangent line does exist, then it is vertical.

In this text we will deal only with functions whose derivatives fail to exist at a finite number of values of x. Typically, these values correspond to points where the graph of f has a discontinuity, a corner, or a vertical tangent. These situations are illustrated in the following examples.

EXAMPLE 5 Show that the Heaviside function

$$H(t) = \begin{cases} 0 & \text{if } t < 0 \\ 1 & \text{if } t \geq 0 \end{cases}$$

which is discontinuous at 0, is not differentiable at 0 (Figure 6).

Solution Let's show that the (left-hand) limit

$$\lim_{h \to 0^-} \frac{H(0 + h) - H(0)}{h} \qquad h < 0$$

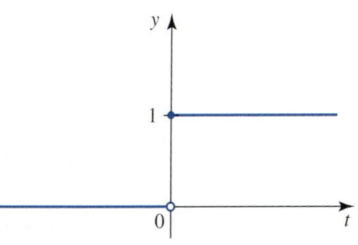

FIGURE 6
The Heaviside function is not differentiable at 0.

does not exist. This, in turn, will imply that

$$H'(0) = \lim_{h \to 0} \frac{H(0 + h) - H(0)}{h}$$

does not exist; that is, H does not have a derivative at 0. Now

$$\lim_{h \to 0^-} \frac{H(h) - H(0)}{h} = \lim_{h \to 0^-} \frac{0 - 1}{h} = \infty \qquad \textcolor{red}{\text{Since } h < 0}$$

so $H'(0)$ does not exist, as asserted.   ■

The next example shows that if f has a sharp corner at a, then f is not differentiable at a.

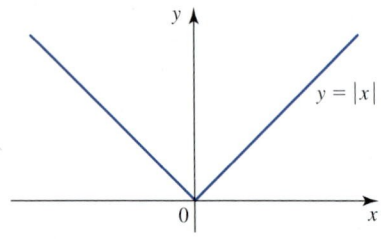

FIGURE 7
The function $f(x) = |x|$ is continuous everywhere and has a corner at 0.

EXAMPLE 6 Show that the function $f(x) = |x|$ is differentiable everywhere except at 0.

Solution The graph of f is shown in Figure 7. To prove that f is not differentiable at 0, we will show that $f'(0)$ does not exist by demonstrating that the one-sided limits of the quotient

$$\frac{f(0 + h) - f(0)}{h} = \frac{f(h) - f(0)}{h} = \frac{|h| - 0}{h} = \frac{|h|}{h}$$

as h approaches 0 are not equal. First, suppose $h > 0$. Then $|h| = h$, so

$$\lim_{h \to 0^+} \frac{|h|}{h} = \lim_{h \to 0^+} \frac{h}{h} = \lim_{h \to 0^+} 1 = 1$$

Next, if $h < 0$, then $|h| = -h$, and therefore,

$$\lim_{h \to 0^-} \frac{|h|}{h} = \lim_{h \to 0^-} \frac{-h}{h} = \lim_{h \to 0^-} (-1) = -1$$

Therefore,

$$f'(0) = \lim_{h \to 0} \frac{f(0 + h) - f(0)}{h} = \lim_{h \to 0} \frac{|h|}{h}$$

does not exist, and f is not differentiable at 0.

To show that f is differentiable at all other numbers, we rewrite $f(x)$ in the form

$$f(x) = |x| = \begin{cases} -x & \text{if } x < 0 \\ x & \text{if } x \geq 0 \end{cases}$$

and then differentiate $f(x)$ to obtain

$$f'(x) = \begin{cases} -1 & \text{if } x < 0 \\ 1 & \text{if } x > 0 \end{cases}$$

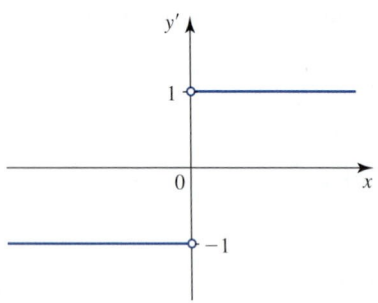

FIGURE 8
$f'(0)$ is not defined; therefore, f is not differentiable at 0.

Geometrically, this result is evident if you consider the graph of f, which consists of two rays (Figure 7). The slope of the half-line to the left of the origin is -1, and the slope of the half-line to the right of the origin is 1. The graph of f' is shown in Figure 8.   ■

The graph of a function f has a **vertical tangent line** $x = a$ at a, if f is continuous at a and

$$\lim_{x \to a} f'(x) = -\infty \qquad \text{or} \qquad \lim_{x \to a} f'(x) = \infty$$

The next example shows that the function f is not differentiable at a because the graph of f has a vertical tangent line at a.

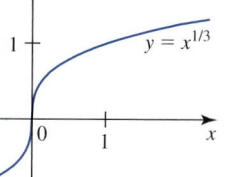

FIGURE 9
The graph of f has a vertical tangent line at $(0, 0)$.

EXAMPLE 7 Show that the function $f(x) = x^{1/3}$ is not differentiable at 0.

Solution We compute

$$\lim_{h \to 0} \frac{f(0 + h) - f(0)}{h} = \lim_{h \to 0} \frac{f(h) - f(0)}{h}$$

$$= \lim_{h \to 0} \frac{h^{1/3} - 0}{h} = \lim_{h \to 0} \frac{1}{h^{2/3}} = \infty$$

This shows that f is not differentiable at 0. (See Figure 9.) ▬

■ Differentiability and Continuity

Examples 6 and 7 show that a function can be continuous at a number yet not be differentiable there. The next theorem shows that the requirement that a function is differentiable at a number is stronger than the requirement that it be continuous there.

> **THEOREM 1**
>
> If f is differentiable at a, then f is continuous at a.

PROOF If x is in the domain of f and $x \neq a$, then we can write

$$f(x) - f(a) = \frac{f(x) - f(a)}{x - a} (x - a)$$

We have

$$\lim_{x \to a} [f(x) - f(a)] = \lim_{x \to a} \frac{f(x) - f(a)}{x - a} \cdot (x - a)$$

$$= \lim_{x \to a} \frac{f(x) - f(a)}{x - a} \cdot \lim_{x \to a} (x - a)$$

$$= f'(a) \cdot 0 = 0$$

So

$$\lim_{x \to a} f(x) = \lim_{x \to a} [f(a) + (f(x) - f(a))]$$

$$= \lim_{x \to a} f(a) + \lim_{x \to a} [f(x) - f(a)] = f(a) + 0 = f(a)$$

and this shows that f is continuous at a, as asserted. ▬

2.1   CONCEPT QUESTIONS

1. **a.** Give a geometric and a physical interpretation of the expression

$$\frac{f(x + h) - f(x)}{h}$$

   b. Give a geometric and a physical interpretation of the expression

$$\lim_{h \to 0} \frac{f(x + h) - f(x)}{h}$$

2. Under what conditions does a function fail to have a derivative at a number? Illustrate your answer with sketches.

2.1   EXERCISES

In Exercises 1–14, use the definition of the derivative to find the derivative of the function. What is its domain?

1. $f(x) = 5$

2. $f(x) = 2x + 1$

3. $f(x) = 3x - 4$

4. $f(x) = 2x^2 + x$

5. $f(x) = 3x^2 - x + 1$

6. $f(x) = x^3 - x$

7. $f(x) = 2x^3 + x - 1$

8. $f(x) = 2\sqrt{x}$

9. $f(x) = \sqrt{x + 1}$

10. $f(x) = \frac{1}{x}$

11. $f(x) = \frac{1}{x + 2}$

12. $f(x) = -\frac{2}{\sqrt{x}}$

13. $f(x) = \frac{3}{2x + 1}$

14. $f(x) = x + \sqrt{x}$

In Exercises 15–20, find an equation of the tangent line to the graph of the function at the indicated point.

	Function	Point
15.	$f(x) = x^2 + 1$	$(2, 5)$
16.	$f(x) = 3x^2 - 4x + 2$	$(2, 6)$
17.	$f(x) = 2x^3$	$(1, 2)$
18.	$f(x) = 3x^3 - x$	$(-1, -2)$
19.	$f(x) = \sqrt{x - 1}$	$(4, \sqrt{3})$
20.	$f(x) = \frac{2}{x}$	$(2, 1)$

21. **a.** Find an equation of the tangent line to the graph of $f(x) = 2x - x^3$ at the point $(1, 1)$.

    b. Plot the graph of f and the tangent line in successively smaller viewing windows centered at $(1, 1)$ until the graph of f and the tangent line appear to coincide.

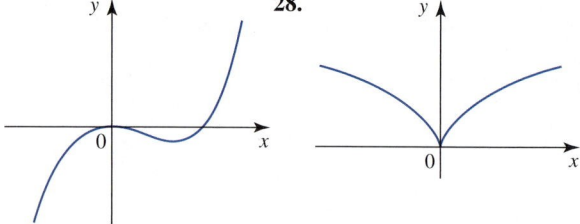 **22. a.** In Example 6 we showed that $f(x) = |x|$ is not differentiable at $x = 0$. Plot the graph of f using the viewing window $[-1, 1] \times [-1, 1]$. Then **ZOOM IN** using successively smaller viewing windows centered at $(0, 0)$. What can you say about the existence of a tangent line at $(0, 0)$?

   b. Plot the graph of

$$f(x) = \begin{cases} x + 1 & \text{if } x \leq 1 \\ \dfrac{2}{x} & \text{if } x > 1 \end{cases}$$

   using the viewing window $[-2, 4] \times [-2, 3]$. Then **ZOOM IN** using successively smaller viewing windows centered at $(1, 2)$. Is f differentiable at $x = 1$?

In Exercises 23–26, find the rate of change of y with respect to x at the given value of x.

23. $y = -2x^2 + x + 1;\quad x = 1$

24. $y = 2x^3 + 2;\quad x = 2$

25. $y = \sqrt{2x};\quad x = 2$

26. $y = x^2 - \dfrac{1}{x};\quad x = -1$

In Exercises 27–30, match the graph of each function with the graph of its derivative in (a)–(d).

27.

28.

29.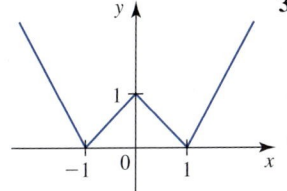

30.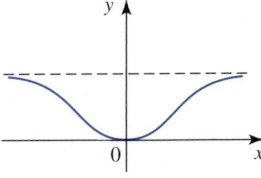

(a)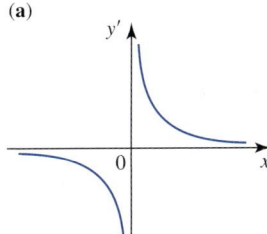

(b)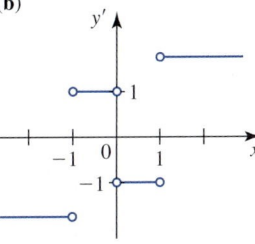

(c)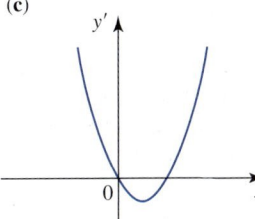

(d)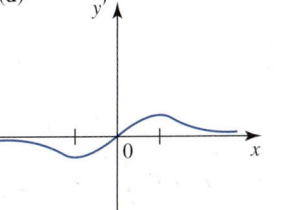

In Exercises 31–36, sketch the graph of the derivative f' of the function f whose graph is given.

31.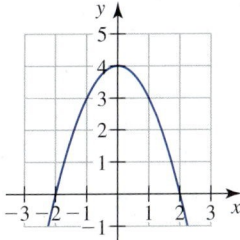

32.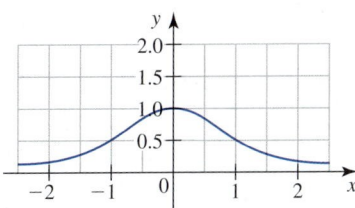

33.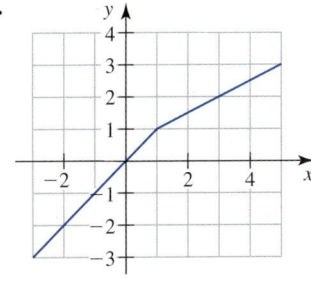

34.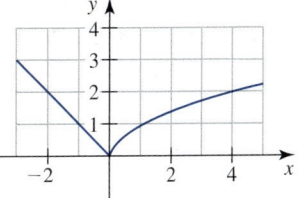

35.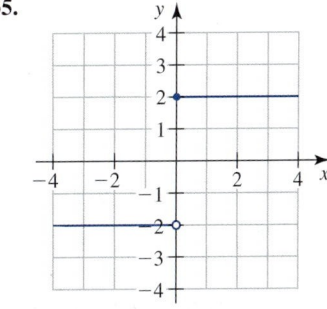

36.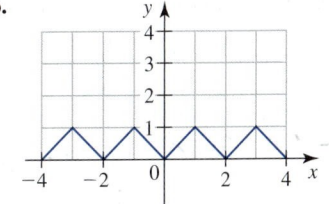

37. Air Temperature and Altitude The air temperature at a height of h feet from the surface of the earth is $T = f(h)$ degrees Fahrenheit.
   a. Give a physical interpretation of $f'(h)$. Give units.
   b. Generally speaking, what do you expect the sign of $f'(h)$ to be?
   c. If you know that $f'(1000) = -0.05$, estimate the change in the air temperature if the altitude changes from 1000 ft to 1001 ft.

38. Advertising and Revenue Suppose that the total revenue realized by the Odyssey Travel Agency is $R = f(x)$ thousand dollars if x thousand dollars are spent on advertising.
   a. What does

$$\frac{f(b) - f(a)}{b - a} \qquad 0 < a < b$$

   measure? What are the units?
   b. What does $f'(x)$ measure? Give units.
   c. Given that $f'(20) = 3$, what is the approximate change in the revenue if Odyssey increases its advertising budget from $20,000 to $21,000?

39. Production Costs Suppose that the total cost in manufacturing x units of a certain product is $C(x)$ dollars.
   a. What does $C'(x)$ measure? Give units.
   b. What can you say about the sign of C'?
   c. Given that $C'(1000) = 20$, estimate the additional cost to be incurred by the company in producing the 1001st unit of the product.

40. Range of a Projectile A projectile is fired from a cannon that makes an angle of θ degrees with the horizontal. If the muzzle velocity is constant, then the range in feet of the projectile is a function of θ, that is, $R = f(\theta)$.
  a. What is the physical meaning of $f'(\theta)$? Give units.
  b. What can you say about the sign of $f'(\theta)$, where $0° < \theta < 90°$?
  c. Given that $f(40) = 10,000$ and $f'(40) = 20$, estimate the range of a projectile if it is fired at an angle of elevation of $41°$.

41. Let $f(x) = x^2 - 2x + 1$.
  a. Find the derivative f' of f.
  b. Find the point on the graph of f where the tangent line to the curve is horizontal.
  c. Sketch the graph of f and the tangent line to the curve at the point found in part (b).
  d. What is the rate of change of f at this point?

42. Let $f(x) = \dfrac{1}{x - 1}$.
  a. Find the derivative f' of f.
  b. Find an equation of the tangent line to the curve at the point $\left(-1, -\frac{1}{2}\right)$.
  c. Sketch the graph of f and the tangent line to the curve at the point $\left(-1, -\frac{1}{2}\right)$.

In Exercises 43–48, use the graph of the function f to find the value(s) of x at which f is not differentiable.

43.
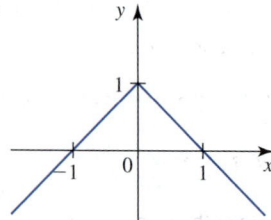

44.
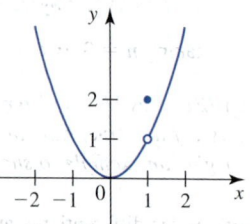

45.
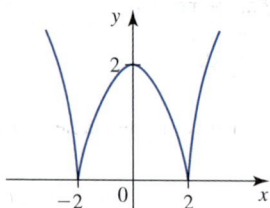

46.
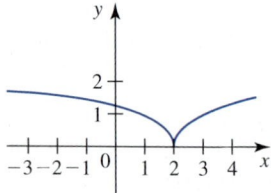

47.
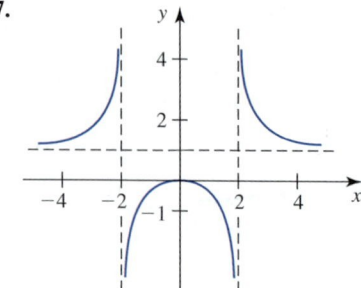

48.
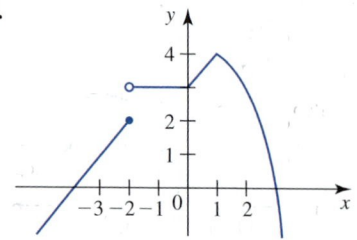

In Exercises 49–52, show that the function is continuous but not differentiable at the given value of x.

49. $f(x) = \begin{cases} x + 2 & \text{if } x \le 0 \\ 2 - 3x & \text{if } x > 0 \end{cases}; \quad x = 0$

50. $f(x) = \begin{cases} x + 1 & \text{if } x \le 0 \\ x^2 + 1 & \text{if } x > 0 \end{cases}; \quad x = 0$

51. $f(x) = |2x - 1|; \quad x = \dfrac{1}{2}$

52. $f(x) = \begin{cases} x \sin \dfrac{1}{x} & \text{if } x \ne 0 \\ 0 & \text{if } x = 0 \end{cases}; \quad x = 0$

53. R & D Expenditure The graph of the function f shown in the figure gives the Department of Energy budget for research and development for solar, wind, and other renewable energy sources over a 12-year period. Use the slopes of f at the indicated values of t and the technique of Example 4 to sketch the graph of f'. Then use the graph of f' to estimate the rate of change of the budget when $t = 1$ and when $t = 5$.

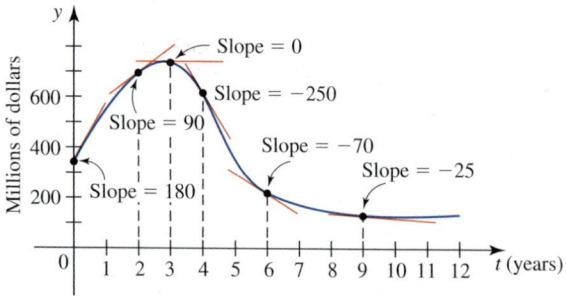

Source: U.S. Department of Energy.

54. Velocity of a Model Car The graph of the function f shown in the figure gives the position $s = f(t)$ of a model car moving along a straight line as a function of time. Use the technique of Example 4 to sketch the velocity curve for the car. (Recall that the velocity of an object is given by the rate of change (derivative) of its position.) Then use the graph of f' to estimate the velocity of the car at $t = 5$ and $t = 12$.

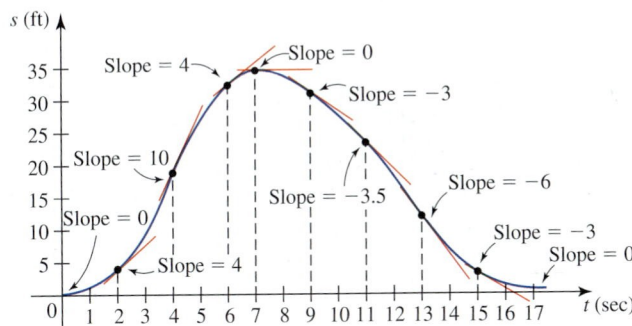

55. Let $f(x) = x^3$. For each real number $h \neq 0$, define

$$g(x) = \frac{(x + h)^3 - x^3}{h}$$

  a. For each fixed value of h, what does $g(x)$ measure?

  b. What function do you expect $g(x)$ to approach as h approaches zero?

   c. Verify your answer to part (b) visually by plotting the graph of the function you guessed at in part (b) and the graph of the function $g(x)$ for $h = 1, 0.5$, and 0.1 in a common viewing window.

56. Let $f(x) = x^3 - x$.

  a. Find $f'(x)$.

   b. Plot the graphs of f' and g, where

$$g(x) = \frac{[(x + 0.01)^3 - (x + 0.01)] - (x^3 - x)}{0.01}$$

    using a common viewing window. Is the result expected? Explain.

57. Let $f(x) = |x^3|$.

  a. Sketch the graph of f.

  b. For what values of x is f differentiable?

  c. Find a formula for $f'(x)$.

58. Let $f(x) = x|x|$.

  a. Sketch the graph of f.

  b. For what values of x is f differentiable?

  c. Find a formula for $f'(x)$.

59. Suppose that $g(x) = |x - a| f(x)$, where f is a continuous function and $f(a) \neq 0$. Show that g is continuous at a but not differentiable at a.

60. Let

$$f(x) = \begin{cases} x^{1/3} \sin \dfrac{1}{x} & \text{if } x \neq 0 \\ 0 & \text{if } x = 0 \end{cases}$$

  a. Show that f is continuous at 0, but not differentiable at 0.

   b. Plot the graph of f using the viewing window $[-0.5, 0.5] \times [-0.1, 0.1]$.

61. Let

$$f(x) = \begin{cases} x^2 \sin \dfrac{1}{x} & \text{if } x \neq 0 \\ 0 & \text{if } x = 0 \end{cases}$$

  a. Show that f is differentiable at 0. What is $f'(0)$?

  b. Plot the graph of f using the viewing window $[-0.5, 0.5] \times [-0.1, 0.1]$.

62. A function f is called *periodic* if there exists a number $T > 0$ such that $f(x + T) = f(x)$ for all x in the domain of f. Prove that the derivative of a differentiable periodic function with period T is also a periodic function with period T.

63. Show that if $f'(x)$ exists, then

$$\lim_{h \to 0} \frac{f(x + nh) - f[x + (n - 1)h]}{h} = f'(x) \qquad n \neq 0, 1$$

64. Use the result of Exercise 63 to find the derivative of (a) $f(x) = \sqrt{x}$ by taking $n = 2$ and (b) $f(x) = \dfrac{1}{x + 1}$ by taking $n = 3$. (Compare with Examples 1 and 3.)

In Exercises 65–70, determine whether the statement is true or false. If it is true, explain why it is true. If it is false, explain why or give an example to show why it is false.

65. If f is differentiable at $x = 3$, then the slope of the tangent line to the graph of f at the point $(3, f(3))$ is

$$\lim_{h \to 0} \frac{f(3 + h) - f(3)}{h}$$

66. If f is differentiable at a, and g is not differentiable at a, then the product fg is not differentiable at a.

67. If both f and g are not differentiable at a, then the product fg is not differentiable at a.

  Hint: Consider $f(x) = |x|$ and $g(x) = |x|$.

68. If both f and g are not differentiable at a, then the sum $f + g$ is not differentiable at a.

69. The domain of f' is the same as that of f.

70. If n is a positive integer, then there exists a function f such that f is differentiable everywhere except at n numbers.

2.2 Basic Rules of Differentiation

■ Some Basic Rules

Up to now we have computed the derivative of a function using its definition. But as you have seen, this process is tedious even for relatively simple functions. In this section we will develop some rules of differentiation that will simplify the process of finding the derivative of a function.

THEOREM 1   Derivative of a Constant Function

If c is a constant, then

$$\frac{d}{dx}(c) = 0$$

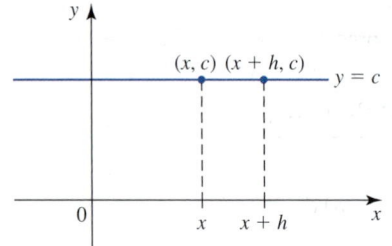

PROOF   Let $f(x) = c$. Then

$$f'(x) = \lim_{h \to 0} \frac{f(x + h) - f(x)}{h} = \lim_{h \to 0} \frac{c - c}{h} = \lim_{h \to 0} 0 = 0 \qquad ■$$

This result is also evident geometrically (see Figure 1). The tangent line to a straight line at any point on the line must coincide with the line itself. Since the constant function f defined by $f(x) = c$ is a horizontal line with slope 0, any tangent line to f must also have slope 0. Hence, $f'(x) = 0$ for every x.

FIGURE 1

The slope of the graph of $f(x) = c$ is zero at every point. Hence, $f'(x) = 0$.

EXAMPLE 1

a. If $f(x) = 19$, then $f'(x) = \dfrac{d}{dx}(19) = 0$.

b. If $f(x) = -\pi^2$, then $f'(x) = \dfrac{d}{dx}(-\pi^2) = 0$. ■

Next, we turn our attention to the rule for differentiating power functions $f(x) = x^n$ with positive integral exponents n. For the special case in which $n = 1$, we have $f(x) = x$. Its derivative is

$$f'(x) = \lim_{h \to 0} \frac{f(x + h) - f(x)}{h} = \lim_{h \to 0} \frac{(x + h) - x}{h} = \lim_{h \to 0} 1 = 1$$

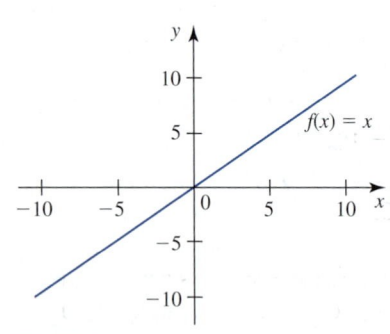

FIGURE 2

The graph of $f(x) = x$ is the line with slope 1. Hence, $f'(x) = 1$.

This result is also evident geometrically because the graph of $y = x$ is the line with slope 1 (see Figure 2) and hence $f'(x) = 1$ for every x. That is,

$$\frac{d}{dx}(x) = 1 \qquad\qquad (1)$$

We now state the general rule for finding the derivative of $f(x) = x^n$, where n is a positive integer.

THEOREM 2 **The Power Rule**

If n is a positive integer and $f(x) = x^n$, then

$$f'(x) = \frac{d}{dx}(x^n) = nx^{n-1}$$

PROOF Let $f(x) = x^n$. Then

$$f'(x) = \lim_{h \to 0} \frac{f(x+h) - f(x)}{h} = \lim_{h \to 0} \frac{(x+h)^n - x^n}{h}$$

Now observe that

$$a^n - b^n = (a - b)(a^{n-1} + a^{n-2}b + \cdots + ab^{n-2} + b^{n-1})$$

which can be verified by simply expanding the expression on the right-hand side. If we use this equation with a replaced by $x + h$ and b replaced by x, then we can write

$$f'(x) = \lim_{h \to 0} \frac{[(x+h) - x][(x+h)^{n-1} + (x+h)^{n-2}x + \cdots + (x+h)x^{n-2} + x^{n-1}]}{h}$$

$$= \lim_{h \to 0} [(x+h)^{n-1} + (x+h)^{n-2}x + \cdots + (x+h)x^{n-2} + x^{n-1}]$$

$$= \underbrace{x^{n-1} + x^{n-1} + \cdots + x^{n-1} + x^{n-1}}_{n \text{ terms}}$$

$$= nx^{n-1}$$

Theorem 2 can also be proved by using the Binomial Theorem (see Exercise 73).

EXAMPLE 2

a. If $f(x) = x^{10}$, then $f'(x) = \frac{d}{dx}(x^{10}) = 10x^{10-1} = 10x^9$.

b. If $g(u) = u^3$, then $g'(u) = \frac{d}{du}(u^3) = 3u^{3-1} = 3u^2$.

Although Theorem 2 was stated for the case in which the power n is a positive integer, the Power Rule is true for all real numbers n. For example, if we apply the more general rule *formally* to finding the derivative of $f(x) = \sqrt{x} = x^{1/2}$, we find

$$f'(x) = \frac{d}{dx}(x^{1/2}) = \frac{1}{2}x^{-1/2} = \frac{1}{2\sqrt{x}}$$

a result that we obtained in Example 1, Section 2.1, using the definition of the derivative.

We will demonstrate the validity of the Power Rule for negative integers n in Section 2.3. The rule will be extended to include rational powers n in Section 2.7. Finally, we will prove the general version of the Power Rule, where n may be any real number, in Section 2.8. But for now, we will assume that the Power Rule is *valid for all real numbers* and use it in our work.

> **THEOREM 3** **The Power Rule (General Version)**
>
> If n is any real number, then
>
> $$\frac{d}{dx}(x^n) = nx^{n-1}$$

EXAMPLE 3

a. If $f(x) = \dfrac{1}{x^3}$, then $f'(x) = \dfrac{d}{dx}\left(\dfrac{1}{x^3}\right) = \dfrac{d}{dx}(x^{-3}) = -3x^{-3-1} = -3x^{-4} = -\dfrac{3}{x^4}$.

b. If $y = x^{3/2}$, then $\dfrac{dy}{dx} = \dfrac{d}{dx}(x^{3/2}) = \dfrac{3}{2}x^{(3/2)-1} = \dfrac{3}{2}x^{1/2} = \dfrac{3\sqrt{x}}{2}$.

c. If $g(x) = x^{0.12}$, then $g'(x) = \dfrac{d}{dx}(x^{0.12}) = 0.12x^{0.12-1} = 0.12x^{-0.88} = \dfrac{0.12}{x^{0.88}}$. ∎

The next theorem tells us that *the derivative of a constant times a function is equal to the constant times the derivative of the function.*

> **THEOREM 4** **The Constant Multiple Rule**
>
> If f is a differentiable function and c is a constant, then
>
> $$\frac{d}{dx}[cf(x)] = cf'(x)$$

PROOF Let $F(x) = cf(x)$. Then

$$F'(x) = \lim_{h\to 0}\frac{F(x+h) - F(x)}{h} = \lim_{h\to 0}\frac{cf(x+h) - cf(x)}{h}$$

$$= \lim_{h\to 0} c\left[\frac{f(x+h) - f(x)}{h}\right]$$

$$= c\lim_{h\to 0}\frac{f(x+h) - f(x)}{h} \qquad \text{Constant Multiple Law for limits}$$

$$= cf'(x) \qquad\qquad\qquad\qquad ∎$$

EXAMPLE 4

a. If $f(x) = 3x^5$, then $f'(x) = \dfrac{d}{dx}(3x^5) = 3\dfrac{d}{dx}(x^5) = 3(5x^4) = 15x^4$.

b. If $y = -2u^3$, then $\dfrac{dy}{du} = \dfrac{d}{du}(-2u^3) = -2\dfrac{d}{du}(u^3) = -2(3u^2) = -6u^2$. ∎

The next theorem says that *the derivative of the sum of two functions is the sum of their derivatives.*

THEOREM 5 **The Sum Rule**

If f and g are differentiable functions, then

$$\frac{d}{dx}[f(x) + g(x)] = f'(x) + g'(x)$$

PROOF Let $F(x) = f(x) + g(x)$. Then

$$F'(x) = \lim_{h \to 0} \frac{F(x + h) - F(x)}{h}$$

$$= \lim_{h \to 0} \frac{[f(x + h) + g(x + h)] - [f(x) + g(x)]}{h}$$

$$= \lim_{h \to 0} \left[\frac{f(x + h) - f(x)}{h} + \frac{g(x + h) - g(x)}{h} \right]$$

$$= \lim_{h \to 0} \frac{f(x + h) - f(x)}{h} + \lim_{h \to 0} \frac{g(x + h) - g(x)}{h} \qquad \text{Sum Law for limits}$$

$$= f'(x) + g'(x)$$

Notes

1. Since $f(x) - g(x)$ can be written as $f(x) + [-g(x)]$, Theorem 5 implies that

$$\frac{d}{dx}[f(x) - g(x)] = \frac{d}{dx}[f(x)] + \frac{d}{dx}[-g(x)]$$

$$= \frac{d}{dx}[f(x)] - \frac{d}{dx}[g(x)] \qquad \text{By Theorem 4 with } c = -1$$

$$= f'(x) - g'(x)$$

and we see that Theorem 5 also applies to the difference of two functions.

2. The Sum (Difference) Rule is valid for any finite number of functions. For example, if f, g, and h are differentiable at x, then so is $f + g - h$, and

$$\frac{d}{dx}[f(x) + g(x) - h(x)] = f'(x) + g'(x) - h'(x)$$

EXAMPLE 5 Find the derivative of $f(x) = 4x^5 + 2x^4 - 3x^2 + 6x + 1$.

Solution Using the generalized Sum Rule, we find that

$$f'(x) = \frac{d}{dx}(4x^5 + 2x^4 - 3x^2 + 6x + 1)$$

$$= \frac{d}{dx}(4x^5) + \frac{d}{dx}(2x^4) - \frac{d}{dx}(3x^2) + \frac{d}{dx}(6x) + \frac{d}{dx}(1)$$

$$= 4\frac{d}{dx}(x^5) + 2\frac{d}{dx}(x^4) - 3\frac{d}{dx}(x^2) + 6\frac{d}{dx}(x) + \frac{d}{dx}(1)$$

$$= 4(5x^4) + 2(4x^3) - 3(2x) + 6(1) + 0$$

$$= 20x^4 + 8x^3 - 6x + 6$$

EXAMPLE 6 Find the derivative of $y = \dfrac{x^3 - 2x^2 + x - 4}{2\sqrt{x}}$.

Solution Using the generalized Sum Rule, we find

$$\frac{dy}{dx} = \frac{d}{dx}\left(\frac{x^3 - 2x^2 + x - 4}{2x^{1/2}}\right)$$

$$= \frac{d}{dx}\left(\frac{1}{2}x^{5/2} - x^{3/2} + \frac{1}{2}x^{1/2} - 2x^{-1/2}\right)$$

$$= \frac{1}{2}\left(\frac{5}{2}x^{3/2}\right) - \frac{3}{2}x^{1/2} + \frac{1}{2}\left(\frac{1}{2}x^{-1/2}\right) - 2\left(-\frac{1}{2}x^{-3/2}\right)$$

$$= \frac{5}{4}x^{3/2} - \frac{3}{2}x^{1/2} + \frac{1}{4}x^{-1/2} + x^{-3/2}$$

EXAMPLE 7 Find the points on the graph of $f(x) = x^4 - 2x^2 + 2$ where the tangent line is horizontal.

Solution At a point on the graph of f where its tangent line is horizontal, the derivative of f is zero. So we begin by finding

$$f'(x) = \frac{d}{dx}(x^4 - 2x^2 + 2) = 4x^3 - 4x = 4x(x^2 - 1)$$

Setting $f'(x) = 0$ leads to $4x(x^2 - 1) = 0$, giving $x = -1$, 0, or 1. Substituting each of the numbers into $f(x)$ gives the points $(-1, 1)$, $(0, 2)$, and $(1, 1)$ as the required points. (See Figure 3.)

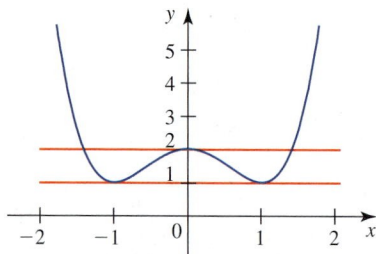

FIGURE 3
The graph of $f(x) = x^4 - 2x^2 + 2$ has horizontal tangent lines at $(-1, 1)$, $(0, 2)$, and $(1, 1)$.

EXAMPLE 8 **Carbon Monoxide in the Atmosphere** The projected average global atmospheric concentration of carbon monoxide is approximated by

$$f(t) = 0.88t^4 - 1.46t^3 + 0.7t^2 + 2.88t + 293 \qquad 0 \le t \le 4$$

where t is measured in 40-year intervals with $t = 0$ corresponding to the beginning of 1860 and $f(t)$ is measured in parts per million by volume. How fast was the projected average global atmospheric concentration of carbon monoxide changing at the beginning of the year 1900 ($t = 1$) and at the beginning of 2000 ($t = 3.5$)?
Source: Meadows et al., "Beyond the Limits."

Solution The rate at which the concentration of carbon monoxide is changing at time t is given by

$$f'(t) = \frac{d}{dt}(0.88t^4 - 1.46t^3 + 0.7t^2 + 2.88t + 293)$$

$$= 3.52t^3 - 4.38t^2 + 1.4t + 2.88$$

parts/million/(40 years). Therefore, the rate at which the concentration of carbon monoxide was changing at the beginning of 1900 was

$$f'(1) = 3.52(1) - 4.38(1) + 1.4(1) + 2.88 = 3.42$$

or approximately 3.4 parts/million/(40 years). At the beginning of the year 2000, it was

$$f'(3.5) = 3.52(3.5)^3 - 4.38(3.5)^2 + 1.4(3.5) + 2.88 = 105.045$$

or approximately 105 parts/million/(40 years).

■ The Derivative of the Natural Exponential Function

We now turn our attention to exponential functions. To find the derivative of $f(x) = a^x$, where $a > 0$ and $a \neq 1$, we use the definition of the derivative to write

$$f'(x) = \lim_{h \to 0} \frac{f(x + h) - f(x)}{h} = \lim_{h \to 0} \frac{a^{x+h} - a^x}{h}$$

$$= \lim_{h \to 0} \frac{a^x a^h - a^x}{h} = \lim_{h \to 0} \frac{a^x(a^h - 1)}{h}$$

Since a^x does not depend on h, it can be treated as a constant with respect to the limiting process, and we have

$$f'(x) = a^x \lim_{h \to 0} \frac{a^h - 1}{h} \qquad (2)$$

Thus, the derivative of $f(x) = a^x$ exists for all values of x provided that

$$\lim_{h \to 0} \frac{a^h - 1}{h}$$

exists. If we put $x = 0$ in Equation (2), we obtain

$$f'(0) = \lim_{h \to 0} \frac{a^h - 1}{h} \qquad (3)$$

so

$$f'(x) = f'(0)a^x \qquad (4)$$

Equation (4) tells us that the derivative of $f(x) = a^x$ is a constant multiple of itself, the constant being the slope of the tangent line to the graph of f at 0. Now, it can be shown that the limit in Equation (3) exists for all $a > 0$. Before proceeding further, let us consider the cases in which $a = 2$ and $a = 3$.

TABLE 1   The values of $\dfrac{a^h - 1}{h}$ for $a = 2$ and $a = 3$ correct to four decimal places

h	-0.1	-0.01	-0.001	-0.0001	0.0001	0.001	0.01	0.1
$\dfrac{2^h - 1}{h}$	0.6697	0.6908	0.6929	0.6931	0.6932	0.6934	0.6956	0.7177
$\dfrac{3^h - 1}{h}$	1.0404	1.0926	1.0980	1.0986	1.0987	1.0992	1.1047	1.1612

From Table 1 we see that if $a = 2$, then

$$f'(x) = \frac{d}{dx}(2^x) = f'(0)2^x = \lim_{h \to 0} \frac{2^h - 1}{h} \cdot 2^x \approx (0.69)2^x$$

and if $a = 3$, then

$$f'(x) = \frac{d}{dx}(3^x) = f'(0)3^x = \lim_{h \to 0} \frac{3^h - 1}{h} \cdot 3^x \approx (1.10)3^x$$

These calculations suggest that it might be possible to pick a number between 2 and 3 for which $f'(0) = 1$. The choice of this number for the base of the exponential function will lead to the simplest formula for the derivative of the function. This number is denoted by the letter e and is the same number mentioned in Section 0.8 when we introduced the natural exponential function.

> **DEFINITION**   **The Number e**
>
> The number e is the number such that
>
> $$\lim_{h \to 0} \frac{e^h - 1}{h} = 1$$

Putting $a = e$ in Equation (2) leads to the following formula:

$$f'(x) = \frac{d}{dx}(e^x) = e^x \lim_{h \to 0} \frac{e^h - 1}{h} = e^x \cdot (1) = e^x$$

In other words, the derivative of the natural exponential function is equal to itself.

> **THEOREM 6**   **Derivative of e^x**
>
> $$\frac{d}{dx} e^x = e^x$$

EXAMPLE 9   Let $f(x) = e^x + x$.

a. Find the derivative of f.

b. Find an equation of the tangent line to the graph of f at the point where $x = 0$.

Solution

a. $f'(x) = \dfrac{d}{dx}(e^x + x) = \dfrac{d}{dx}(e^x) + \dfrac{d}{dx}(x) = e^x + 1$

b. The slope of the tangent line at the point where $x = 0$ is

$$f'(0) = e^x + 1 \Big|_{x=0} = 2$$

The y-coordinate of the point of tangency is $f(0) = e^0 + 0 = 1$. So a required equation is

$$y - 1 = 2(x - 0) \qquad \text{or} \qquad y = 2x + 1$$

The graph of f and the tangent line are shown in Figure 4. ■

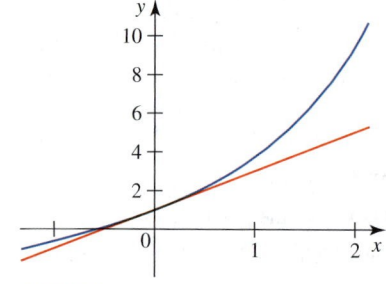

FIGURE 4
The graph of f and the tangent line at $(0, 1)$

2.2   CONCEPT QUESTIONS

1. State the rule of differentiation and explain it in your own words.
  a. The Power Rule
  b. The Constant Multiple Rule
  c. The Sum Rule

2. State the derivative of the function.
  a. $f(x) = c$, c a constant
  b. $f(x) = x^n$, n a real number
  c. $f(x) = e^x$

3. a. Give a definition of the number e.
  b. Explain why it is more desirable to use the natural exponential function $f(x) = e^x$ than the more general exponential function $f(x) = a^x$ in calculus.

4. If $f'(2) = 3$ and $g'(2) = -2$, find
  a. $h'(2)$ if $h(x) = 2f(x)$
  b. $F'(2)$ if $F(x) = 2f(x) - 4g(x)$

2.2 EXERCISES

In Exercises 1–32, find the derivative of the function.

1. $f(x) = 2.718$

2. $f(x) = 3x + 4$

3. $f(x) = 3x^2$

4. $f(x) = -2x^3 - 3e^2$

5. $f(x) = x^{2.1}$

6. $f(x) = 9x^{1/3}$

7. $f(x) = 3\sqrt{x} + 2e^x$

8. $f(u) = \dfrac{2}{\sqrt{u}}$

9. $f(x) = 7x^{-12}$

10. $f(x) = 0.3x^{-1.2}$

11. $f(x) = x^2 - 2x + 8$

12. $g(x) = -\dfrac{1}{3}x^2 + \sqrt{2}x$

13. $f(r) = \pi r^2 + 2\pi r$

14. $y = -\dfrac{1}{3}(x^3 + 2x^2 + x - 1)$

15. $f(x) = 0.03x^2 - 0.4x + 10$

16. $f(x) = 0.002x^3 - 0.05x^2 + 0.1x + 0.1e^x - 20$

17. $g(x) = x^2(2x^3 - 3x^2 + x + 4)$

18. $H(u) = (2u)^3 - 3u + 7$

19. $f(x) = \dfrac{x^3 - 4x^2 + 3}{x}$

20. $h(t) = \dfrac{t^5 - 3t^3 + 2t^2 e^t}{2t^2}$

21. $f(x) = 4x^4 - 3x^{5/2} + 2$

22. $f(x) = 5x^{4/3} - \dfrac{2}{3}x^{3/2} + x^2 - 3x + 1$

23. $f(x) = 3x^{-1} + 4x^{-2}$

24. $f(x) = -\dfrac{1}{3}(x^{-3} - x^6)$

25. $f(t) = \dfrac{4}{t^4} - \dfrac{3}{t^3} + \dfrac{2}{t}$

26. $f(x) = \dfrac{5}{x^3} - \dfrac{2}{x^2} - \dfrac{1}{x} + 200$

27. $A = 0.001x^2 - 0.4x + 5 + \dfrac{200}{x}$

28. $y = 0.0002t^3 - 0.4t^2 + 4 + \dfrac{100}{t^2}$

29. $f(x) = 2x - 5\sqrt{x} + e^{x+1}$

30. $f(t) = 2t^2 + \sqrt{t^3}$

31. $y = \sqrt[3]{x} + \dfrac{1}{\sqrt{x}}$

32. $f(u) = \dfrac{1}{\sqrt{u}} - \dfrac{3}{\sqrt[3]{u}}$

33. Let $f(x) = 2x^3 - 4x$. Find
  a. $f'(-2)$ **b.** $f'(0)$ **c.** $f'(2)$

34. Let $f(x) = 4x^{5/4} + 2x^{3/2} + x$. Find
  a. $f'(0)$ **b.** $f'(16)$

 In Exercises 35–38, (a) find an equation of the tangent line to the graph of the function at the indicated point, and (b) use a graphing utility to plot the graph of the function and the tangent line on the same screen.

35. $f(x) = 2x^2 - 3x + 4;\quad (2, 6)$

36. $f(x) = -\dfrac{5}{3}x^2 + 2x + 2;\quad \left(-1, -\dfrac{5}{3}\right)$

37. $f(x) = x^4 - 3x^3 + 2x^2 - x + e^{x-1};\quad (1, 0)$

38. $f(x) = \sqrt{x} + \dfrac{1}{\sqrt{x}};\quad \left(4, \dfrac{5}{2}\right)$

In Exercises 39–42, find the point(s) on the graph of the function at which the tangent line has the indicated slope.

39. $f(x) = 2x^3 + 3x^2 - 12x - 10;\quad m_{\tan} = 0$

40. $g(x) = \dfrac{1}{3}x^3 - \dfrac{1}{2}x^2 - x + 1;\quad m_{\tan} = -1$

41. $h(t) = 2t + \dfrac{1}{t};\quad m_{\tan} = -2$

42. $F(s) = \dfrac{2s + 1}{s};\quad m_{\tan} = -\dfrac{1}{9}$

 *A straight line perpendicular to and passing through a point of tangency of the tangent line is called a **normal line** to the curve. In Exercises 43 and 44, (a) find the equations of the tangent line and the normal line to the curve at the given point, and (b) use a graphing utility to plot the graph of the function, the tangent line, and the normal line on the same screen.*

43. The curve $y = x^3 - 3x + 1$ at the point $(2, 3)$.

44. The curve $y = 2x + (1/\sqrt{x})$ at the point $(1, 3)$.

45. Find the value(s) of x at which $y = 2x - (9/x)$ is increasing at the rate of 3 units per unit change in x.

46. Let $f(x) = \frac{2}{3}x^3 + x^2 - 12x + 6$. Find the values of x for which
  a. $f'(x) = -12$ **b.** $f'(x) = 0$ **c.** $f'(x) = 12$

47. Let $f(x) = \frac{1}{4}x^4 - \frac{1}{3}x^3 - x^2$. Find the point(s) on the graph of f where the slope of the tangent line is equal to
  a. $-2x$ **b.** 0 **c.** $10x$

48. Find the points on the graph of $y = \frac{1}{3}x^3 - 2x + 5$ at which the tangent line is parallel to the line $y = 2x + 3$.

49. Find the points on the graph of $y = \frac{1}{3}x^3 - 2x + 5$ at which the tangent line is perpendicular to the line $y = x + 2$.

50. Given that the line $y = 2x$ is tangent to the graph of $y = x^2 + c$, find c.

51. Find equations of the lines passing through the point $(3, 2)$ that are tangent to the parabola $y = x^2 - 2x$.
  Hint: Find two expressions for the slope of a tangent line.

52. Find an equation of the normal line to the parabola $y = x^2 - 6x + 11$ that is perpendicular to the line passing through the point $(1, 0)$ and the vertex of the parabola. (Refer to the directions given for Exercise 43.)

In Exercises 53–56, find the limit by evaluating the derivative of a suitable function at an appropriate value of x. (Hint: Use the definition of the derivative.)

53. $\lim\limits_{h \to 0} \dfrac{(1 + h)^3 - 1}{h}$

54. $\lim\limits_{x \to 1} \dfrac{x^5 - 1}{x - 1}$

Hint: Let $h = x - 1$.

55. $\lim\limits_{h \to 0} \dfrac{3(2 + h)^2 - (2 + h) - 10}{h}$

56. $\lim\limits_{t \to 0} \dfrac{(8 + t)^{1/3} - 2}{t}$

In Exercises 57 and 58, write the expression as a derivative of a function of x.

57. $\lim\limits_{h \to 0} \dfrac{2(x + h)^7 - (x + h)^2 - 2x^7 + x^2}{h}$

58. $\lim\limits_{h \to 0} \dfrac{\dfrac{1}{x + h} + \sqrt{x + h} - \dfrac{1}{x} - \sqrt{x}}{h}$

59. Temperature Changes The temperature (in degrees Fahrenheit) on a certain day in December in Minneapolis is given by

$$T = -0.05t^3 + 0.4t^2 + 3.8t + 19.6 \qquad 0 \le t \le 12$$

where t is measured in hours and $t = 0$ corresponds to 6 A.M. Determine the time of day when the temperature is increasing at the rate of 2.05°F/hr.

60. Traffic Flow Opened in the late 1950s, the Central Artery in downtown Boston was designed to move 75,000 vehicles per day. Suppose that the average speed of traffic flow S in miles per hour is related to the number of vehicles x (in thousands) moved per day by the equation

$$S = -0.00075x^2 + 67.5 \qquad 50 < x < 300$$

Find the rate of change of the average speed of traffic flow when the number of vehicles moved per day is 100,000; 200,000. (Compare with Exercise 1 in Section 1.5.)

Source: The Boston Globe.

61. Spending on Medicare On the basis of the current eligibility requirement, a study conducted in 2004 showed that federal spending on entitlement programs, particularly Medicare, would grow enormously in the future. The study predicted that spending on Medicare, as a percentage of the gross domestic product (GDP), will be

$$P(t) = 0.27t^2 + 1.4t + 2.2 \qquad 0 \le t \le 5$$

percent in year t, where t is measured in decades with $t = 0$ corresponding to the year 2000.

a. How fast will the spending on Medicare, as a percentage of the GDP, be growing in 2010? In 2020?

b. What will the predicted spending on Medicare, as a percentage of the GDP, be in 2010? In 2020?

Source: Congressional Budget Office.

62. Effect of Stopping on Average Speed According to data from a study by General Motors, the average speed of a trip, A (in miles per hour), is related to the number of stops per mile made on that trip, x, by the equation

$$A = \frac{26.5}{x^{0.45}}$$

Compute dA/dx for $x = 0.25$ and $x = 2$, and interpret your results.

Source: General Motors.

63. Health-Care Spending Health-care spending per person by the private sector comprising payments by individuals, corporations, and their insurance companies is approximated by the function

$$f(t) = 2.48t^2 + 18.47t + 509 \qquad 0 \le t \le 6$$

where $f(t)$ is measured in dollars and t is measured in years with $t = 0$ corresponding to the beginning of 1994. The corresponding government spending—including expenditures for Medicaid, Medicare, and other federal, state, and local government public health care—is

$$g(t) = -1.12t^2 + 29.09t + 429 \qquad 0 \le t \le 6$$

where $g(t)$ is measured in dollars and t in years.

a. Find a function that gives the difference between private and government health-care spending per person at any time t.

b. How fast was the difference between private and government expenditures per person changing at the beginning of 1995? At the beginning of 2000?

Source: Health Care Financing Administration.

64. Fuel Economy of Cars According to data obtained from the U.S. Department of Energy and the Shell Development Company, a typical car's fuel economy depends on the speed it is driven and is approximated by the function

$$f(x) = 0.00000310315x^4 - 0.000455174x^3$$
$$+ 0.00287869x^2 + 1.25986x \qquad 0 \le x \le 75$$

where x is measured in miles per hour and $f(x)$ is measured in miles per gallon (mpg).

a. Use a graphing utility to graph the function f on the interval $[0, 75]$.

b. Use a calculator or computer to find the rate of change of f when $x = 20$ and when $x = 50$.

c. Interpret your results.

Source: U.S. Department of Energy and the Shell Development Company.

65. Prevalence of Alzheimer's Patients The projected number of
Alzheimer's patients in the United States is given by

$$f(t) = -0.02765t^4 + 0.3346t^3 - 1.1261t^2$$
$$+ 1.7575t + 3.7745 \quad 0 \le t \le 6$$

where $f(t)$ is measured in millions and t in decades, with
$t = 0$ corresponding to the beginning of 1990.
a. Use a graphing utility to graph the function f on the
interval $[0, 6]$.
b. Use a calculator or computer to find the rate at which the
number of Alzheimer's patients in the United States is
anticipated to be changing at the beginning of 2010? At
the beginning of 2020? At the beginning of 2030?
c. Interpret your results.
Source: Alzheimer's Association.

 66. Hedge Fund Assets A hedge fund is a lightly regulated pool of
professionally managed money. The assets (in billions of
dollars) of hedge funds from the beginning of 1999 ($t = 0$)
through the beginning of 2004 are given in the following
table.

Year	1999	2000	2001	2002	2003	2004
Assets (billions of dollars)	472	517	594	650	817	950

a. Use the regression capability of a calculator or computer
to find a third-degree polynomial function for the data,
letting $t = 0$ correspond to the beginning of 1999.
b. Plot the graph of the function found in part (a).
c. Use a calculator or computer to find the rate at which the
assets of hedge funds were increasing at the beginning of
2000. At the beginning of 2003.
Sources: Hennessee Group, Institutional Investor.

 67. Population Decline Political and social upheaval stemming
from Russia's difficult transition from communism to capital-
ism is expected to contribute to the decline of the country's
population well into the next century. The following table
shows the total population at the beginning of each year.

Year	1985	1990	1995	2000	2005
Population (millions)	143.3	147.9	147.8	145.5	143.8

Year	2010	2015	2020	2025	2030
Population (millions)	141.2	137.5	133.2	128.7	123.3

a. Use the regression capability of a calculator or computer
to find a fourth-degree polynomial function for the data,
letting $t = 0$ correspond to the beginning of 1985, where
t is measured in 5-year intervals.

b. Plot the graph of the function found in part (a).
c. Use a calculator or computer to find the rate at which
the population was changing at the beginning of 1985?
At the beginning of 1995? At the beginning of 2030?
Sources: Population Reference Bureau, United Nations.

68. Newton's Law of Gravitation According to Newton's Law of
Gravitation, the magnitude F (in newtons) of the force
of attraction between two bodies of masses M and m
kilograms is

$$F = \frac{GmM}{r^2}$$

where G is a constant and r is the distance between the two
bodies in meters. What is the rate of change of F with
respect to r?

69. Period of a Satellite The period of a satellite in a circular orbit
of radius r is given by

$$T = \frac{2\pi r}{R}\sqrt{\frac{r}{g}}$$

where R is the earth's radius and g is the constant of acceleration. Find the rate of change of the period with respect to
the radius of the orbit.

70. Coast Guard Launch In the figure the x-axis represents a
straight shoreline. A spectator located at the point $P(2.5, 0)$
observes a Coast Guard launch equipped with a search light
execute a turn. The path of the launch is described by the
parabola $y = -2.5x^2 + 10$ (x and y are measured in hundreds of feet). Find the distance between the launch and the
spectator at the instant of time the bow of the launch is
pointed directly at the spectator.

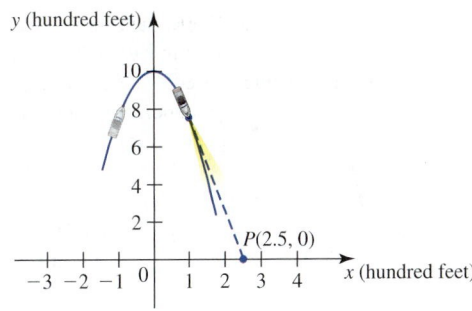

71. Determine the constants A, B, and C such that the parabola
$y = Ax^2 + Bx + C$ passes through the point $(-1, 0)$ and is
tangent to the line $y = x$ at the point where $x = 1$.

72. Let

$$f(x) = \begin{cases} x^2 & \text{if } x \le a \\ Ax + B & \text{if } x > a \end{cases}$$

Find the values of A and B such that f is continuous and
differentiable at a.

73. Prove the Power Rule $f'(x) = nx^{n-1}$ (n, a positive integer) using the Binomial Theorem

$$(a + b)^n = a^n + na^{n-1}b$$
$$+ \frac{n(n - 1)}{2} a^{n-2}b^2 + \cdots + nab^{n-1} + b^n$$

Compute

$$f'(x) = \lim_{h \to 0} \frac{f(x + h) - f(x)}{h} = \lim_{h \to 0} \frac{(x + h)^n - x^n}{h}$$

using the substitution $a = x$ and $b = h$.

74. Show that

$$f(x) = \begin{cases} \dfrac{e^x - 1}{x} & \text{if } x \neq 0 \\ 1 & \text{if } x = 0 \end{cases}$$

is continuous on $(-\infty, \infty)$.

Hint: Use the definition of the derivative.

In Exercises 75–78, determine whether the statement is true or false. If it is true, explain why it is true. If it is false, explain why or give an example to show why it is false.

75. If $f(x) = x^{2n}$, where n is an integer, then $f'(x) = 2nx^{2(n-1)}$.

76. If $f(x) = 2^x$, then $f'(x) = x \cdot 2^{x-1}$ by the Power Rule.

77. If f and g are differentiable, then

$$\frac{d}{dx}[2f(x) - 5g(x)] = 2f'(x) - 5g'(x)$$

78. If $g(x) = f(x^2)$, where f is differentiable, then $g'(x) = f'(x^2)$.

2.3   The Product and Quotient Rules

In this section we study two more rules of differentiation: the **Product Rule** and the **Quotient Rule.** We also consider higher-order derivatives.

■ The Product and Quotient Rules

In general, the derivative of the product of two functions is *not* equal to the product of their derivatives. The following rule tells us how to differentiate a product of two functions.

> **THEOREM 1   The Product Rule**
>
> If f and g are differentiable functions, then
>
> $$\frac{d}{dx}[f(x)g(x)] = f(x)g'(x) + g(x)f'(x)$$

PROOF   Let $F(x) = f(x)g(x)$. Then

$$F'(x) = \lim_{h \to 0} \frac{F(x + h) - F(x)}{h} = \lim_{h \to 0} \frac{f(x + h)g(x + h) - f(x)g(x)}{h}$$

If we add the quantity $[-f(x + h)g(x) + f(x + h)g(x)]$, which is equal to zero, to the numerator, we obtain

$$F'(x) = \lim_{h \to 0} \frac{f(x + h)g(x + h) - \mathbf{f(x + h)g(x)} + \mathbf{f(x + h)g(x)} - f(x)g(x)}{h}$$

$$= \lim_{h \to 0} \left\{ f(x + h)\left[\frac{g(x + h) - g(x)}{h}\right] + g(x)\left[\frac{f(x + h) - f(x)}{h}\right] \right\}$$

$$= \lim_{h \to 0} f(x + h) \cdot \lim_{h \to 0} \frac{g(x + h) - g(x)}{h} + \lim_{h \to 0} g(x) \cdot \lim_{h \to 0} \frac{f(x + h) - f(x)}{h} \tag{1}$$

Since f is assumed to be differentiable at x, Theorem 1 of Section 2.1 tells us that it is continuous there, so

$$\lim_{h \to 0} f(x + h) = f(x)$$

Also, because $g(x)$ does not involve h, it is constant with respect to the limiting process and

$$\lim_{h \to 0} g(x) = g(x)$$

Therefore, Equation (1) reduces to

$$F'(x) = f(x)g'(x) + g(x)f'(x)$$

In words, the Product Rule states that *the derivative of the product of two functions is the first function times the derivative of the second, plus the second function times the derivative of the first.*

EXAMPLE 1

a. Find the derivative of $f(x) = xe^x$.
b. How fast is f changing when $x = 1$?

Solution
a. Using the Product Rule, we find

$$f'(x) = \frac{d}{dx}(xe^x) = x\frac{d}{dx}(e^x) + e^x\frac{d}{dx}(x)$$

$$= xe^x + e^x \cdot 1 = (x + 1)e^x$$

b. When $x = 1$, f is changing at the rate of $f'(1) = 2e$ units per unit change in x.

EXAMPLE 2
Suppose that $g(x) = (x^2 + 1)f(x)$ and it is known that $f(2) = 3$ and $f'(2) = -1$. Evaluate $g'(2)$.

Solution   Using the Product Rule, we find

$$g'(x) = \frac{d}{dx}[(x^2 + 1)f(x)] = (x^2 + 1)\frac{d}{dx}[f(x)] + f(x)\frac{d}{dx}(x^2 + 1)$$

$$= (x^2 + 1)f'(x) + 2xf(x)$$

Therefore,

$$g'(2) = (2^2 + 1)f'(2) + 2(2)f(2)$$

$$= (5)(-1) + 4(3) = 7$$

Just as the derivative of a product of two functions is not the product of their derivatives, the derivative of a quotient of two functions is not the quotient of their derivatives! Rather, we have the following rule.

THEOREM 2 **The Quotient Rule**

If f and g are differentiable functions and $g(x) \neq 0$, then

$$\frac{d}{dx}\left[\frac{f(x)}{g(x)}\right] = \frac{g(x)f'(x) - f(x)g'(x)}{[g(x)]^2}$$

PROOF Let $F(x) = \dfrac{f(x)}{g(x)}$. Then

$$F'(x) = \lim_{h \to 0} \frac{F(x+h) - F(x)}{h}$$

$$= \lim_{h \to 0} \frac{\dfrac{f(x+h)}{g(x+h)} - \dfrac{f(x)}{g(x)}}{h}$$

$$= \lim_{h \to 0} \frac{f(x+h)g(x) - f(x)g(x+h)}{hg(x+h)g(x)}$$

Subtracting and adding $f(x)g(x)$ in the numerator yield

$$F'(x) = \lim_{h \to 0} \frac{f(x+h)g(x) - \boldsymbol{f(x)g(x)} + \boldsymbol{f(x)g(x)} - f(x)g(x+h)}{hg(x+h)g(x)}$$

$$= \lim_{h \to 0} \frac{g(x)\left[\dfrac{f(x+h) - f(x)}{h}\right] - f(x)\left[\dfrac{g(x+h) - g(x)}{h}\right]}{g(x+h)g(x)}$$

$$= \frac{\displaystyle\lim_{h \to 0} g(x) \cdot \lim_{h \to 0} \frac{f(x+h) - f(x)}{h} - \lim_{h \to 0} f(x) \cdot \lim_{h \to 0} \frac{g(x+h) - g(x)}{h}}{\displaystyle\lim_{h \to 0} g(x+h) \cdot \lim_{h \to 0} g(x)} \qquad (2)$$

As in the proof of the Product Rule, we see that

$$\lim_{h \to 0} g(x) = g(x) \qquad \text{and} \qquad \lim_{h \to 0} f(x) = f(x)$$

and, because g is continuous at x,

$$\lim_{h \to 0} g(x+h) = g(x)$$

Therefore, Equation (2) is

$$F'(x) = \frac{g(x)f'(x) - f(x)g'(x)}{[g(x)]^2} \qquad \blacksquare$$

As an aid to remembering the Quotient Rule, observe that it has the following form:

$$\frac{d}{dx}\left[\frac{f(x)}{g(x)}\right] = \frac{(\text{denominator})(\text{derivative of numerator}) - (\text{numerator})(\text{derivative of denominator})}{(\text{square of denominator})}$$

⚠ Because of the presence of the minus sign in the numerator, the order of the terms is important!

EXAMPLE 3   Find the derivative of $f(x) = \dfrac{2x^2 + x}{x^3 - 1}$.

Solution   Using the Quotient Rule, we have

$$f'(x) = \frac{(x^3 - 1)\dfrac{d}{dx}(2x^2 + x) - (2x^2 + x)\dfrac{d}{dx}(x^3 - 1)}{(x^3 - 1)^2}$$

$$= \frac{(x^3 - 1)(4x + 1) - (2x^2 + x)(3x^2)}{(x^3 - 1)^2}$$

$$= \frac{(4x^4 + x^3 - 4x - 1) - (6x^4 + 3x^3)}{(x^3 - 1)^2}$$

$$= -\frac{2x^4 + 2x^3 + 4x + 1}{(x^3 - 1)^2}$$

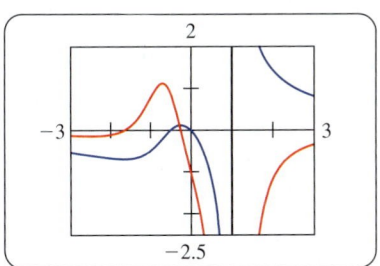

FIGURE 1
The graph of f is shown in blue, and the graph of f' is shown in red.

Note   Figure 1 shows the graph of f and f' in the same viewing window. Observe that the graph of f has horizontal tangent lines at the points where $x \approx -1.63$ and $x \approx -0.24$, the approximate roots of $f'(x) = 0$.

EXAMPLE 4   Find an equation of the tangent line to the graph of

$$f(x) = \frac{e^x}{x + 1}$$

at the point where $x = 1$.

Solution   The slope of the tangent line at any point on the graph of f is given by

$$f'(x) = \frac{(x + 1)\dfrac{d}{dx}(e^x) - e^x\dfrac{d}{dx}(x + 1)}{(x + 1)^2}$$

$$= \frac{(x + 1)e^x - e^x(1)}{(x + 1)^2} = \frac{xe^x}{(x + 1)^2}$$

In particular, the slope of the required tangent line is

$$f'(1) = \frac{(1)e^1}{(1 + 1)^2} = \frac{e}{4}$$

Also, when $x = 1$,

$$y = f(1) = \frac{e^1}{1 + 1} = \frac{e}{2}$$

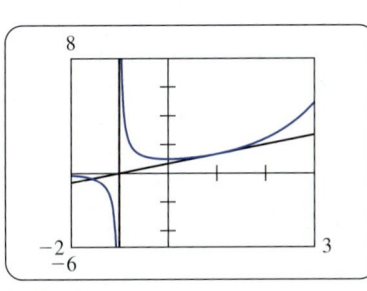

FIGURE 2
The graph of f and the tangent line to the graph of f at $\left(1, \frac{1}{2}e\right)$

Therefore, the point of tangency is $\left(1, \frac{1}{2}e\right)$, and an equation of the tangent line is

$$y - \frac{1}{2}e = \frac{1}{4}e(x - 1) \qquad \text{or} \qquad y = \frac{1}{4}e(x + 1)$$

The graph of f and the tangent line to the graph at $\left(1, \frac{1}{2}e\right)$ are shown in Figure 2.

EXAMPLE 5   Rate of Change of DVD Sales   The sales (in millions of dollars) of a DVD recording of a hit movie t years from the date of release are given by

$$S(t) = \frac{5t}{t^2 + 1} \qquad t \geq 0$$

a. Find the rate at which the sales are changing at time t.
b. How fast are the sales changing at the time the DVDs are released ($t = 0$)? Two years from the date of release?

Solution

a. The rate at which the sales are changing at time t is given by $S'(t)$. Using the Quotient Rule, we obtain

$$S'(t) = \frac{d}{dt}\left[\frac{5t}{t^2 + 1}\right] = 5\frac{d}{dt}\left[\frac{t}{t^2 + 1}\right]$$

$$= 5\left[\frac{(t^2 + 1)(1) - t(2t)}{(t^2 + 1)^2}\right]$$

$$= 5\left[\frac{t^2 + 1 - 2t^2}{(t^2 + 1)^2}\right] = \frac{5(1 - t^2)}{(t^2 + 1)^2}$$

b. The rate at which the sales are changing at the time the DVDs are released is given by

$$S'(0) = \frac{5(1 - 0)}{(0 + 1)^2} = 5$$

That is, they are increasing at the rate of $5 million per year.
      Two years from the date of release, the sales are changing at the rate of

$$S'(2) = \frac{5(1 - 4)}{(4 + 1)^2} = -\frac{3}{5} = -0.6$$

That is, they are decreasing at the rate of $600,000 per year. The graph of the function S is shown in Figure 3.

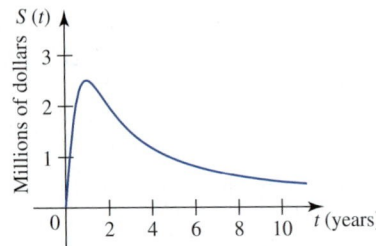

FIGURE 3
After a spectacular rise, the sales begin to taper off.

You may have observed that the domain of the function S in Example 5 is restricted, for practical reasons, to the interval $[0, \infty)$. Since the definition of the derivative of a function f at a number a requires that f be defined in an open interval containing a, the derivative of S is not, strictly speaking, defined at 0. But notice that the function S can, in fact, be defined for all values of t, and hence it makes sense to calculate $S'(0)$. You will encounter situations such as this throughout the book, especially in exercises pertaining to real-world applications. The nature of the functions appearing in these applications obviates the necessity to consider "one-sided" derivatives.

■ Extending the Power Rule

The Quotient Rule can be used to extend the Power Rule to include the case in which n is a negative integer.

THEOREM 3   The Power Rule for Integral Powers

If $f(x) = x^n$, where n is any integer, then

$$\frac{d}{dx}(x^n) = nx^{n-1}$$

PROOF   If n is a positive integer, then the formula holds by Theorem 2 of Section 2.2. If $n = 0$, the formula gives

$$\frac{d}{dx}(x^0) = \frac{d}{dx}(1) = 0$$

which is true by Theorem 1 of Section 2.2. Next, suppose $n < 0$. Then $-n > 0$, and therefore, there is a positive integer m such that $n = -m$. Write

$$f(x) = x^n = x^{-m} = \frac{1}{x^m}$$

Since $m > 0$, x^m can be differentiated using Theorem 2 of Section 2.2. Applying the Quotient Rule, we have

$$f(x) = \frac{d}{dx}(x^n) = \frac{d}{dx}\left(\frac{1}{x^m}\right)$$

$$= \frac{x^m \dfrac{d}{dx}(1) - 1 \cdot \dfrac{d}{dx}(x^m)}{x^{2m}} \qquad \text{Use the Quotient Rule.}$$

$$= \frac{0 - mx^{m-1}}{x^{2m}}$$

$$= -mx^{-m-1}$$

$$= nx^{n-1} \qquad \text{Substitute } n = -m.$$

Higher-Order Derivatives

The derivative f' of a function f is itself a function. As such, we may consider differentiating the function f'. The derivative of f', if it exists, is denoted by f'' and is called the **second derivative** of f. Continuing in this fashion, we are led to the third, fourth, fifth, and higher-order derivatives of f, whenever they exist. Notations for the first, second, third, and in general, the nth derivative of f are

$$f', \quad f'', \quad f''', \quad \ldots, \quad f^{(n)}$$

or

$$\frac{d}{dx}[f(x)], \quad \frac{d^2}{dx^2}[f(x)], \quad \frac{d^3}{dx^3}[f(x)], \quad \ldots, \quad \frac{d^n}{dx^n}[f(x)]$$

or

$$D_x f(x), \quad D_x^2 f(x), \quad D_x^3 f(x), \quad \ldots, \quad D_x^n f(x)$$

respectively.

If we denote the dependent variable by y, so that $y = f(x)$, then its first n derivatives are also written

$$y', \quad y'', \quad y''', \quad \ldots, \quad y^{(n)}$$

or

$$\frac{dy}{dx}, \quad \frac{d^2y}{dx^2}, \quad \frac{d^3y}{dx^3}, \quad \ldots, \quad \frac{d^ny}{dx^n}$$

or

$$D_x y, \quad D_x^2 y, \quad D_x^3 y, \quad \ldots, \quad D_x^n y$$

respectively.

EXAMPLE 6 Find the derivatives of all orders of $f(x) = x^4 - 3x^3 + x^2 - 2x + 8$.

Solution We have

$$f'(x) = 4x^3 - 9x^2 + 2x - 2$$

$$f''(x) = \frac{d}{dx} f'(x) = 12x^2 - 18x + 2$$

$$f'''(x) = \frac{d}{dx} f''(x) = 24x - 18$$

$$f^{(4)}(x) = \frac{d}{dx} f'''(x) = 24$$

$$f^{(5)}(x) = \frac{d}{dx} f^{(4)}(x) = 0$$

and

$$f^{(6)}(x) = f^{(7)}(x) = \cdots = 0$$

EXAMPLE 7 Find the third derivative of $y = \dfrac{1}{x}$.

Solution Rewriting the given equation in the form $y = x^{-1}$, we find

$$y' = \frac{d}{dx}(x^{-1}) = -x^{-2}$$

$$y'' = \frac{d}{dx}(-x^{-2}) = (-1)(-2x^{-3}) = 2x^{-3}$$

and hence

$$y''' = \frac{d}{dx}(2x^{-3}) = 2(-3x^{-4}) = -6x^{-4} = -\frac{6}{x^4}$$

Just as the first derivative $f'(x)$ of a function f at any point x gives the rate of change of $f(x)$ at that point, the second derivative $f''(x)$ of f, which is the derivative of f' at x, gives the rate of change of $f'(x)$ at x. The third derivative $f'''(x)$ of f gives the rate of change of $f''(x)$ at x, and so on. For example, if $P = f(t)$ gives the population of a certain city at time t, then P' gives the rate of change of the population of the city at time t and P'' gives the rate of change of the rate of change of the population at time t.

A geometric interpretation of the second derivative of a function will be given in Chapter 3, and applications of higher-order derivatives will be given in Chapter 8.

2.3 CONCEPT QUESTIONS

1. State the rule of differentiation and explain it in your own words.
   a. The Product Rule
   b. The Quotient Rule

2. If $f(1) = 3$, $g(1) = 2$, $f'(1) = -1$, and $g'(1) = 4$, find
   a. $h'(1)$ if $h(x) = f(x)g(x)$
   b. $F'(1)$ if $F(x) = \dfrac{f(x)}{g(x)}$

2.3 EXERCISES

In Exercises 1–6, use the Product Rule to find the derivative of each function.

1. $f(x) = x^2 e^x$

2. $f(x) = \sqrt{x} e^x$

3. $f(t) = \sqrt{t}(t + 2)e^t$

4. $f(x) = \dfrac{e^x}{x}$

5. $f(x) = e^{2x} + 2e^x + 4$

6. $f(w) = \dfrac{\sqrt{w} e^w + w^2}{2w}$

In Exercises 7–12, use the Quotient Rule to find the derivative of each function.

7. $f(x) = \dfrac{x}{x - 1}$

8. $g(x) = \dfrac{2x}{x^2 + 1}$

9. $h(x) = \dfrac{2x + 1}{3x - 2}$

10. $P(t) = \dfrac{e^t}{1 + t}$

11. $F(x) = \dfrac{x}{1 + xe^x}$

12. $f(s) = \dfrac{s^2 - 4}{s + 1}$

In Exercises 13 and 14, find the derivative of each function in two ways.

13. $F(x) = (x + 2)(x^2 - x + 1)$

14. $h(t) = \dfrac{t^5 - 3t^3 + 2t^2}{2t^2}$

In Exercises 15–30, find the derivative of each function.

15. $f(x) = (x + 2e^x)(x - e^x)$

16. $f(t) = (1 + \sqrt{t})(e^t - 3)$

17. $f(x) = \dfrac{2\sqrt{x}}{x^2 + 1}$

18. $f(x) = \dfrac{e^x + 1}{e^x - 1}$

19. $y = \dfrac{2x^2}{x^2 + x + 1}$

20. $y = \dfrac{2t - 1}{t^2 - 3t + 2}$

21. $f(x) = \dfrac{e^x(x + 1)}{x - 2}$

22. $f(r) = \dfrac{1 - re^r}{1 + e^r}$

23. $f(x) = \dfrac{1 + \dfrac{1}{x}}{x + 2}$

24. $y = \dfrac{1 + \dfrac{1}{x}}{1 - \dfrac{1}{x}}$

25. $f(x) = \dfrac{x + \sqrt{3x}}{3x - 1}$

26. $f(x) = \dfrac{x}{x^2 - 4} - \dfrac{x - 1}{x^2 + 4}$

27. $F(x) = \dfrac{ax + b}{cx + d}$, $\quad a, b, c, d$ constants

28. $g(t) = \dfrac{at^2}{t^2 + b}$, $\quad a, b$ constants

29. $f(x) = \dfrac{x + e^x}{1 - xe^x}$

30. $g(t) = (2t + 1)\left(t - 1 + \dfrac{2}{t - 1}\right)$

In Exercises 31–34, find the derivative of the function and evaluate $f'(x)$ at the given value of x.

31. $f(x) = (2x - 1)(x^2 + 3)$; $\quad x = 1$

32. $f(x) = \dfrac{2x + 1}{2x - 1}$; $\quad x = 2$

33. $f(x) = (\sqrt{x} + 2x)(x^{3/2} - x)$; $\quad x = 4$

34. $f(x) = \dfrac{x}{x^4 - 2x^2 - 1}$; $\quad x = -1$

In Exercises 35 and 36, find the point(s) on the graph of f where the tangent line is horizontal.

35. $f(x) = xe^{-x}$

36. $f(x) = \dfrac{x}{x^2 + 1}$

In Exercises 37 and 38, find the point(s) on the graph of the function at which the tangent line has the indicated slope.

37. $f(x) = (e^x + 1)(2 - x);\quad m_{\tan} = -1$

38. $F(s) = \dfrac{2s + 1}{s - 2};\quad m_{\tan} = -\dfrac{1}{5}$

 In Exercises 39–42, (a) find an equation of the tangent line to the graph of the function at the indicated point, and (b) use a graphing utility to plot the graph of the function and the tangent line on the same screen.

39. $y = \dfrac{e^x}{1 + x};\quad \left(1, \tfrac{1}{2}e\right)$

40. $y = \dfrac{2x}{x^2 + 1};\quad (-1, -1)$

41. $y = x^2 + 1 + \dfrac{3}{x - 1};\quad (-2, 4)$

42. $f(x) = \dfrac{\sqrt{x} - 1}{\sqrt{x} + 1};\quad \left(4, \tfrac{1}{3}\right)$

 The straight line perpendicular to and passing through the point of tangency of the tangent line is called the normal line to the curve. In Exercises 43 and 44, (a) find the equations of the tangent line and the normal line to the curve at the given point, and (b) use a graphing utility to plot the graph of the function, the tangent line, and the normal line on the same screen.

43. The curve $f(x) = (x^3 + 1)(3x^2 - 4x + 2)$ at the point $(1, 2)$.

44. The curve $y = 1/(1 + x^2)$ at the point $\left(1, \tfrac{1}{2}\right)$.

In Exercises 45–48, suppose that f and g are functions that are differentiable at $x = 1$ and that $f(1) = 2$, $f'(1) = -1$, $g(1) = -2$, and $g'(1) = 3$. Find $h'(1)$.

45. $h(x) = f(x)g(x)$

46. $h(x) = (x^2 + 1)g(x)$

47. $h(x) = \dfrac{xf(x)}{x + g(x)}$

48. $h(x) = \dfrac{f(x)g(x)}{f(x) - g(x)}$

In Exercises 49 and 50, find the limit by evaluating the derivative of a suitable function at an appropriate value of x. (Hint: Use the definition of the derivative.)

49. $\displaystyle\lim_{t \to 0} \dfrac{1 - (1 + t)^2}{t(1 + t)^2}$

50. $\displaystyle\lim_{x \to 1} \dfrac{(x + 1)^2 - 4}{x - 1}$

In Exercises 51–54, find $f''(x)$.

51. $f(x) = x^8 - x^4 + 2x^2 + 1$

52. $f(x) = x^3 e^x$

53. $f(x) = \dfrac{e^x}{x}$

54. $f(x) = \dfrac{x + 1}{x - 1}$

In Exercises 55–58, find y''.

55. $y = x^3 - 2x^2 + 1$

56. $y = e^x\left(x + \dfrac{1}{x}\right)$

57. $y = x^{5/2}e^{-x}$

58. $y = \dfrac{x}{x^2 + 1}$

59. Find
   a. $f''(2)$ if $f(x) = 4x^3 - 2x^2 + 3$
   b. $y''\Big|_{x=1}$ if $y = 2x^3 - \dfrac{1}{x}$

60. Find
   a. $f'''(0)$ if $f(x) = 8x^7 - 6x^5 + 4x^3 - x$
   b. $y'''\Big|_{x=1}$ if $y = xe^x$

61. Find the derivatives of all order of $f(x) = 2x^4 - 4x^2 + 1$.

62. Newton's Second Law of Motion Consider a particle moving along a straight line. Newton's Second Law of Motion states that the external force F acting on the particle is equal to the rate of change of its momentum. Thus,

$$F = \dfrac{d}{dt}(mv)$$

where m, the mass of the particle, and v, the velocity of the particle, are both functions of time.
   a. Use the Product Rule to show that

$$F = m\dfrac{dv}{dt} + v\dfrac{dm}{dt}$$

   b. Use the results of part (a) to show that if the mass of a particle is constant, then $F = ma$, where a is the acceleration of the particle.

63. Formaldehyde Levels A study on formaldehyde levels in 900 homes indicates that emissions of various chemicals can decrease over time. The formaldehyde level (parts per million) in an average home in the study is given by

$$f(t) = \dfrac{0.055t + 0.26}{t + 2} \qquad 0 \le t \le 12$$

where t is the age of the house in years. How fast is the formaldehyde level of the average house dropping when the house is new? At the beginning of its fourth year?
Source: Bonneville Power Administration.

64. Oxygen Content of a Pond When organic waste is dumped into a pond, the oxidization process that takes place reduces the pond's oxygen content. However, given time, nature will restore the oxygen content to its natural level. Suppose that the oxygen content t days after organic waste has been dumped into a pond is given by

$$f(t) = 100\left(\dfrac{t^2 + 10t + 100}{t^2 + 20t + 100}\right)$$

where $f(t)$ is the percentage of the oxygen content of the pond prior to dumping.

a. Derive a general expression that gives the rate of change of the pond's oxygen level at any time t.

b. How fast is the oxygen content of the pond changing one day after organic waste has been dumped into the pond? Ten days after? Twenty days after?

c. Interpret your results.

 65. Importance of Time in Treating Heart Attacks According to the American Heart Association, the treatment benefit for heart attacks depends on the time until treatment and is described by the function

$$f(t) = \frac{-16.94t + 203.28}{t + 2.0328} \qquad 0 \le t \le 12$$

where t is measured in hours and $f(t)$ is expressed as a percent.

a. Use a graphing utility to graph the function f using the viewing window $[0, 13] \times [0, 120]$.

b. Use the numerical derivative capability of a graphing utility to find the derivative of f when $t = 0$ and $t = 2$.

c. Interpret the results obtained in part (b).

Source: American Heart Association.

66. Cylinder Pressure The pressure P, volume V, and temperature T of a gas in a cylinder are related by the van der Waals equation

$$P = \frac{kT}{V - b} + \frac{ab}{V^2(V - b)} - \frac{a}{V(V - b)}$$

where a, b, and k are constants. If the temperature of the gas is kept constant, find dP/dV.

67. Constructing a New Road The following figures depict three possible roads connecting the point $A(-1000, 0)$ to the point $B(1000, 1000)$ via the origin. The functions describing the dashed center lines of the roads follow:

$$f(x) = \begin{cases} 0 & \text{if } -1000 \le x \le 0 \\ x & \text{if } 0 < x \le 1000 \end{cases}$$

$$g(x) = \begin{cases} 0 & \text{if } -1000 \le x \le 0 \\ 0.001x^2 & \text{if } 0 < x \le 1000 \end{cases}$$

$$h(x) = \begin{cases} 0 & \text{if } -1000 \le x \le 0 \\ 0.000001x^3 & \text{if } 0 < x \le 1000 \end{cases}$$

Show that f is not differentiable on the interval $(-1000, 1000)$, g is differentiable but not twice differentiable on $(-1000, 1000)$, and h is twice differentiable on $(-1000, 1000)$. Taking into consideration the dynamics of a moving vehicle, which proposal do you think is most suitable?

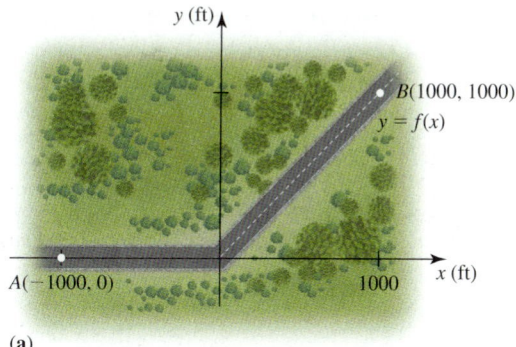

(a)

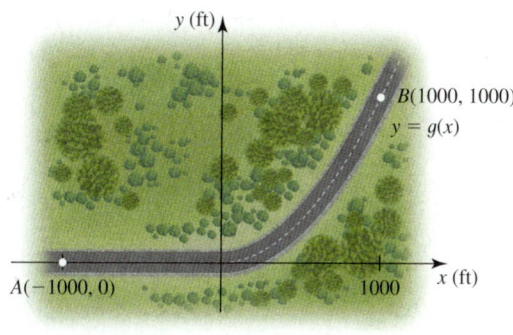

(b)

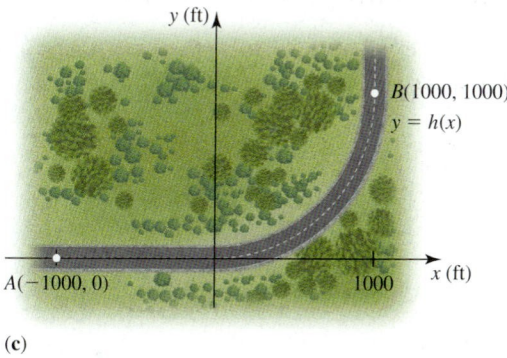

(c)

68. Obesity in America The body mass index (BMI) measures body weight in relation to height. A BMI of 25 to 29.9 is considered overweight, a BMI of 30 or more is considered obese, and a BMI of 40 or more is morbidly obese. The percent of the U.S. population that is obese is approximated by the function

$$P(t) = 0.0004t^3 + 0.0036t^2 + 0.8t + 12 \qquad 0 \le t \le 13$$

where t is measured in years, with $t = 0$ corresponding to the beginning of 1991. Show that the rate of change of the rate of change of the percent of the U.S. population that is deemed obese was positive from 1991 to 2004. What does this mean?

Source: Centers for Disease Control and Prevention.

69. Find $f''(x)$ if $f(x) = |x^3|$. Does $f''(0)$ exist?

70. a. Use the Product Rule twice to prove that if $h = uvw$, where u, v, and w are differentiable functions, then

$$h' = u'vw + uv'w + uvw'$$

b. Use the result of part (a) to find the derivative of

$$h(x) = (2x + 5)(x + 3)(x^2 + 4)$$

In Exercises 71–76, determine whether the statement is true or false. If it is true, explain why it is true. If it is false, explain why or give an example to show why it is false.

71. If f and g are differentiable, then

$$\frac{d}{dx}[f(x)g(x)] = f'(x)g'(x)$$

72. If f is differentiable, then $\dfrac{d}{dx}[xf(x)] = f(x) + xf'(x)$.

73. If f and g have second derivatives, then

$$\frac{d}{dx}[f(x)g'(x) - f'(x)g(x)] = f(x)g''(x) - f''(x)g(x)$$

74. If P is a polynomial function of degree n, then $P^{(n+1)}(x) = 0$.

75. If $g(x) = [f(x)]^2$, where f is differentiable, then $g'(x) = 2f(x)f'(x)$.

76. If $g(x) = [f(x)]^{-2}$, where f is differentiable, then

$$g'(x) = -\frac{2f'(x)}{[f(x)]^3}$$

2.4 The Role of the Derivative in the Real World

In this section we will see how the derivative can be used to solve real-world problems. Our first example calls for interpreting the derivative as a measure of the slope of a tangent line to the graph of a function. Before we look at the example, however, we make the following observation: If α denotes the angle that the tangent line to the graph of f at $P(x, f(x))$ makes with the positive x-axis, then $\tan \alpha = dy/dx$ or, equivalently, $\alpha = \tan^{-1}(dy/dx)$. (See Figure 1.)

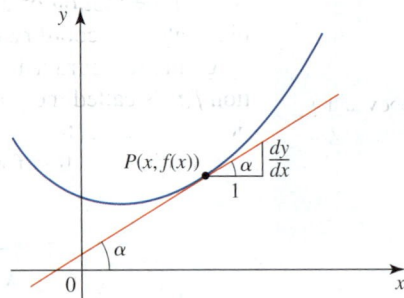

FIGURE 1
$$\tan \alpha = \frac{dy}{dx}$$

EXAMPLE 1 **Flight Path of a Plane** After taking off from a runway, an airplane continues climbing for 10 sec before turning to the right. Its flight path during that time period can be described by the curve in the xy-plane with equation

$$y = -1.06x^3 + 1.61x^2 \qquad 0 \le x \le 0.6$$

where x is the distance along the ground in miles, y is the height above the ground in miles, and the point at which the plane leaves the runway is located at the origin. Find the angle of climb of the airplane when it is at the point on the flight path where $x = 0.5$. (See Figure 2.)

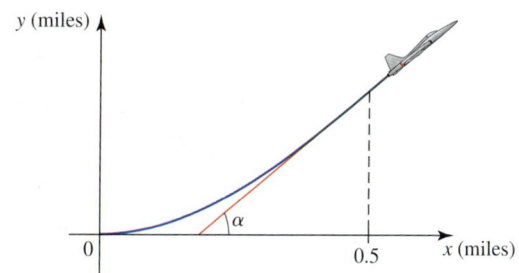

FIGURE 2
The flight path of the airplane

Solution   The required angle of climb, α, is given by

$$\tan \alpha = \frac{dy}{dx}\bigg|_{x=0.5}$$

But

$$\frac{dy}{dx} = \frac{d}{dx}(-1.06x^3 + 1.61x^2) = -3.18x^2 + 3.22x$$

so

$$\frac{dy}{dx}\bigg|_{x=0.5} = (-3.18x^2 + 3.22x)\bigg|_{x=0.5} = 0.815$$

Therefore, $\tan \alpha = 0.815$ and $\alpha = \tan^{-1} 0.815 \approx 39.18°$, giving the required angle of climb of the airplane as approximately $39°$. ∎

We now turn our attention to real-world problems that require the interpretation of the derivative as a measure of the rate of change of one quantity with respect to another.

Motion Along a Line

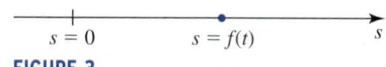

FIGURE 3
The position of a moving body at any time t is at the point $s = f(t)$ on the coordinate line.

An example of motion along a straight line was encountered in Section 1.1, where we studied the motion of a maglev. In considering such motion, we assume that it takes place along a coordinate line. Then the position of a moving body may be specified by giving its coordinate s. Since s varies with time t, we write $s = f(t)$, where the function $f(t)$ is called the **position function** of the body (see Figure 3). As we saw in Section 1.1, the (instantaneous) velocity of a body at any time t is the rate of change of the position function f with respect to t.

DEFINITION   Velocity

If $s = f(t)$, where f is the position function of a body moving on a coordinate line, then the velocity of the body at time t is given by

$$v(t) = \frac{ds}{dt} = f'(t)$$

The function $v(t)$ is called the **velocity function** of the body.

Observe that if $v(t) > 0$ at a given time t, then s is increasing, and the body is moving in the positive direction along the coordinate line at that instant of time (Figure 4a). Similarly, if $v(t) < 0$, then the body is moving in the negative direction at that instant of time (Figure 4b).

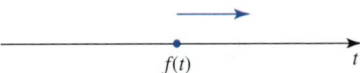

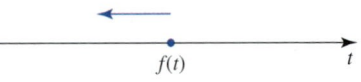

$f(t)$ t

$f(t)$ t

(**a**) If $v(t) > 0$, then the body is moving in the positive direction.

(**b**) If $v(t) < 0$, then the body is moving in the negative direction.

FIGURE 4

Sometimes we merely need to know how fast a body is moving and are not concerned with its direction of motion. In this instance we are asking for the magnitude of the velocity, or the *speed,* of the body.

DEFINITION   Speed

If $v(t)$ is the velocity of a body at any time t, then the speed of the body at time t is given by

$$|v(t)| = |f'(t)| = \left|\frac{ds}{dt}\right|$$

EXAMPLE 2   The position of a particle moving along a straight line is given by

$$s = f(t) = 2t^3 - 15t^2 + 24t \qquad t \geq 0$$

where t is measured in seconds and s in feet.

a. Find an expression giving the velocity of the particle at any time t. What are the velocity and speed of the particle when $t = 2$?

b. Determine the position of the particle when it is stationary.

c. When is the particle moving in the positive direction? In the negative direction?

Solution

a. The required velocity of the particle is given by

$$v(t) = f'(t) = \frac{d}{dt}(2t^3 - 15t^2 + 24t)$$

$$= 6t^2 - 30t + 24 = 6(t^2 - 5t + 4)$$

$$= 6(t - 1)(t - 4)$$

The velocity of the particle when $t = 2$ is

$$v(2) = 6(2 - 1)(2 - 4) = -12$$

or -12 ft/sec. The speed of the particle when $t = 2$ is $|v(2)| = 12$ ft/sec. In short, the particle is moving in the negative direction at a speed of 12 ft/sec.

b. The particle is stationary when its velocity is equal to zero. Setting $v(t) = 0$ gives

$$v(t) = 6(t - 1)(t - 4) = 0$$

and we see that the particle is stationary at $t = 1$ and $t = 4$. Its position at $t = 1$ is given by

$$f(1) = 2(1)^3 - 15(1)^2 + 24(1) = 11 \quad \text{or} \quad 11 \text{ ft}$$

Its position at $t = 4$ is given by

$$f(4) = 2(4)^3 - 15(4)^2 + 24(4) = -16 \quad \text{or} \quad -16 \text{ ft}$$

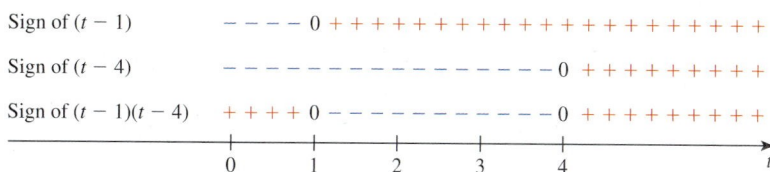

FIGURE 5
The sign diagram for
determining the sign of $v(t)$

c. The particle is moving in the positive direction when $v(t) > 0$ and is moving in the negative direction when $v(t) < 0$. From the sign diagram shown in Figure 5, we see that $v(t) = 6(t - 1)(t - 4)$ is positive in the intervals $(0, 1)$ and $(4, \infty)$ and negative in $(1, 4)$. We conclude that the particle is moving to the right in the time intervals $(0, 1)$ and $(4, \infty)$ and to the left in the time interval $(1, 4)$.

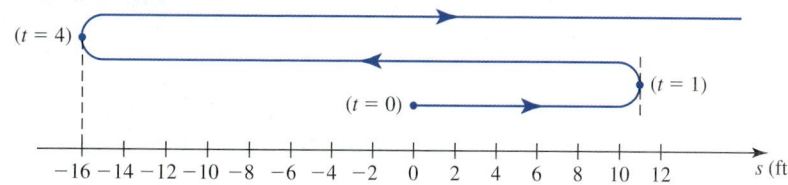

FIGURE 6
A schematic showing the
position of the particle

A schematic of the motion of the particle is shown in Figure 6. The graph of the position function $s = f(t) = 2t^3 - 15t^2 + 24t$ is shown in Figure 7. Try to explain the motion of the particle in terms of this graph. ▪

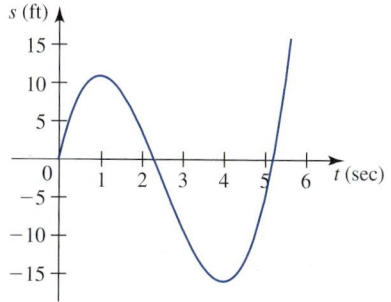

FIGURE 7
The graph of $s = 2t^3 - 15t^2 + 24t$
gives the position of the particle versus
time t. (Do not confuse this with the
path of the particle.)

If a body moves along a coordinate line, the acceleration of the body is the rate of change of its velocity, and the jerk of the body is the rate of change of its acceleration.

DEFINITIONS   Acceleration, Jerk

If $f(t)$ and $v(t)$ are the position and velocity functions, respectively, of a body moving on a coordinate line, then the **acceleration** of the body at time t is

$$a(t) = v'(t) = f''(t)$$

and the **jerk** of the body at time t is

$$j(t) = a'(t) = v''(t) = f'''(t)$$

The jerk function $j(t)$ is of particular interest to safety engineers of automobile companies who are constantly performing jerk tests on various components of motor vehicles. Large jerk conditions in automobiles not only lead to discomfort but may also cause harm to the occupants, including whiplash.

EXAMPLE 3   Consider the motion of the particle of Example 2 with position function

$$s = f(t) = 2t^3 - 15t^2 + 24t \qquad t \geq 0$$

where t is measured in seconds and s in feet.

a. Find the acceleration function of the particle. What is the acceleration of the particle when $t = 2$?
b. When is the acceleration zero? Positive? Negative?
c. Find the jerk function of the particle.

Solution

a. From the solution to Example 2 we have $v(t) = 6t^2 - 30t + 24$. Therefore,

$$a(t) = v'(t) = \frac{d}{dt}(6t^2 - 30t + 24)$$

$$= 12t - 30 = 6(2t - 5)$$

In particular, the acceleration of the particle when $t = 2$ is

$$a(2) = 6[2(2) - 5] \quad \text{or} \quad -6 \text{ ft/sec}^2$$

In other words, the particle is decelerating at 6 ft/sec^2 when $t = 2$.

b. The acceleration of the particle is zero when $a(t) = 0$, or

$$6(2t - 5) = 0$$

giving $t = \frac{5}{2}$. Since $2t - 5 < 0$ when $t < \frac{5}{2}$ and $2t - 5 > 0$ when $t > \frac{5}{2}$, we also conclude that the acceleration is negative for $0 < t < \frac{5}{2}$ and positive for $t > \frac{5}{2}$.

c. Using the result of part (b), we find

$$j(t) = \frac{d}{dt}[a(t)] = \frac{d}{dt}(12t - 30) = 12$$

or 12 ft/sec^3.

EXAMPLE 4 **The Velocity of Exploding Fireworks** In a fireworks display, a shell is launched vertically upward from the ground, reaching a height (in feet) of

$$s = -16t^2 + 256t$$

after t sec. The shell is designed to burst when it reaches its maximum altitude, simultaneously igniting a cluster of explosives.

a. At what time after the launch will the shell burst?
b. What will the altitude of the shell be at the instant it explodes?

Solution

a. At its maximum altitude the velocity of the shell is zero. But the velocity of the shell at any time t is

$$v(t) = \frac{ds}{dt} = \frac{d}{dt}(-16t^2 + 256t)$$

$$= -32t + 256 = -32(t - 8)$$

which is equal to zero when $t = 8$. Therefore, the shell will burst 8 sec after it has been launched.

b. The altitude of the shell at the instant it explodes will be

$$s = -16(8)^2 + 256(8) = 1024$$

or 1024 ft. A schematic of the motion of the shell and the graph of the function $s = -16t^2 + 256t$ are shown in Figure 8.

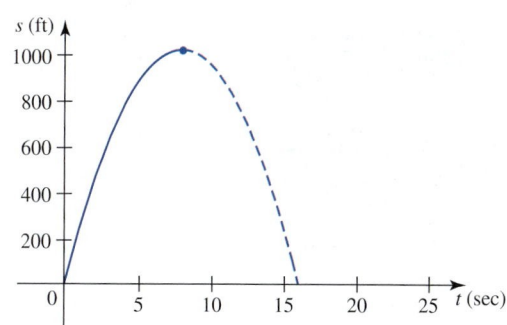

FIGURE 8

(**a**) Schematic of the position
of the shell

(**b**) Graph of the function $s = -16t^2 + 256t$
(The portion of interest is drawn with a solid line.)

◼ Marginal Functions in Economics

The derivative is an indispensable tool in the study of the rate of change of one economic quantity with respect to another. Economists refer to this field of study as **marginal analysis.** The following example will help to explain the use of the adjective **marginal.**

EXAMPLE 5 **Cost Functions** Suppose that the total cost in dollars incurred per week by the Polaraire Corporation in manufacturing x refrigerators is given by the total cost function

$$C(x) = -0.2x^2 + 200x + 9000 \qquad 0 \le x \le 400$$

a. What is the cost incurred in manufacturing the 201st refrigerator?
b. Find the rate of change of C with respect to x when $x = 200$.

Solution
a. The cost incurred in manufacturing the 201st refrigerator is the difference between the total cost incurred in manufacturing the first 201 units and the total cost incurred in manufacturing the first 200 units. Thus, the cost is

$$C(201) - C(200) = [-0.2(201)^2 + 200(201) + 9000]$$
$$- [-0.2(200)^2 + 200(200) + 9000]$$
$$= 41119.8 - 41000 = 119.8$$

or $119.80.
b. The rate of change of C with respect to x is

$$C'(x) = \frac{d}{dx}(-0.2x^2 + 200x + 9000)$$

$$= -0.4x + 200$$

In particular, when $x = 200$, we find

$$C'(200) = -0.4(200) + 200 = 120$$

In other words, when the level of production is 200 units, the total cost function is increasing at the rate of $120 per refrigerator. ◼

If you compare the results of parts (a) and (b) of Example 5, you will notice that $C'(200)$ is a pretty good approximation to $C(201) - C(200)$, the cost incurred in manufacturing an additional refrigerator when the level of production is already 200 units. To see why, let's recall the definition of the derivative of a function and write

$$C'(200) = \lim_{h \to 0} \frac{C(200 + h) - C(200)}{h}$$

Next, the definition of the limit tells us that if h is small, then

$$C'(200) \approx \frac{C(200 + h) - C(200)}{h}$$

In particular, by taking $h = 1$, we see that

$$C'(200) \approx \frac{C(200 + 1) - C(200)}{1} = C(201) - C(200)$$

as we wished to show.

Economists call the cost incurred in producing an additional unit of a commodity, given that the plant is already operating at a certain level $x = a$, the marginal cost. But as we have just seen, this quantity may be suitably approximated by $C'(a)$, where C is the total cost function associated with the process. Furthermore, as you can see from the computations in Example 5, it is often much easier to calculate $C'(a)$ than to calculate $C(a + 1) - C(a)$. For this reason, economists prefer to work with C' rather than C in marginal analysis. The derivative C' of the total cost function is called the **marginal cost function.**

The other marginal functions in economics are defined in a similar manner and have similar meanings. For example, the marginal revenue function R' is the derivative of the total revenue function R, and R' gives an approximation of the change in revenue that results when sales are increased by one unit from $x = a$ to $x = a + 1$.

A summary of these definitions follows.

Function	Marginal Function
C, cost function	C', marginal cost function
R, revenue function	R', marginal revenue function
P, profit function	P', marginal profit function
$\overline{C}$, average cost function	$\overline{C}'$, marginal average cost function

Note $\overline{C}(x) = C(x)/x$, the total cost incurred in producing x units of a commodity divided by the number of units produced.

EXAMPLE 6 **Marginal Revenue** Suppose the weekly revenue realized through the sale of x Pulsar cell phones is

$$R(x) = -0.000078x^3 - 0.0016x^2 + 80x \qquad 0 \le x \le 800$$

dollars.

a. Find the marginal revenue function.
b. If the company currently sells 200 phones per week, by how much will the revenue increase if sales increase by one phone per week?

Solution

a. The marginal revenue function is

$$R'(x) = \frac{d}{dx}(-0.000078x^3 - 0.0016x^2 + 80x)$$

$$= -0.000234x^2 - 0.0032x + 80$$

b. The company's revenue will increase by approximately

$$R'(200) = -0.000234(200)^2 - 0.0032(200) + 80$$

$$= 70$$

or $70.

■ Other Applications

We close this section by looking at a few more examples involving applications of the derivative in fields as diverse as engineering and the social sciences.

Engineering   If the shape of an electric power line strung between two transmission towers is described by the graph of $y = f(x)$, then the (acute) angle α that the cable makes with the horizontal at any point $P(x, f(x))$ on the cable is given by

$$\alpha = \left| \tan^{-1}\left(\frac{dy}{dx}\right) \right|$$

(See Figure 9.)

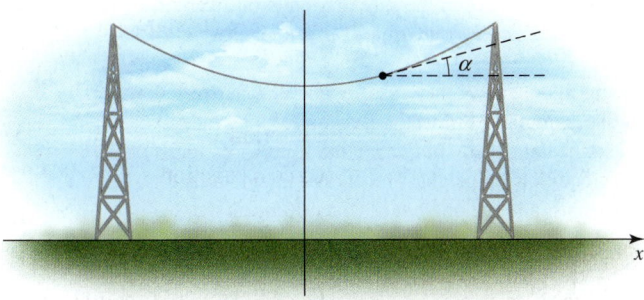

FIGURE 9
The shape of the cable is
described by $y = f(x)$.

Meteorology   If $P(h)$ is the atmospheric pressure at an altitude h, then $P'(h)$ gives the rate of change of the atmospheric pressure with respect to altitude at an altitude h.

Chemistry   Certain proteins, known as enzymes, serve as catalysts for chemical reactions in living things. If $V(x)$ gives the initial speed (in moles per liter per second) at which a chemical reaction begins as a function of x, the amount of substrate (the substance being acted upon, measured in moles per liter), then dV/dx measures the rate of change of the initial speed at which the reaction begins with respect to the amount of substrate, when the amount of substrate is x moles/liter.

Biology   If $R(I)$ denotes the rate of production in photosynthesis, where I is the light intensity, then dR/dI measures the rate of change of the rate of production with respect to light intensity, when the light intensity is I.

Epidemiology	If $p(t)$ stands for the percentage of infected students in a university in week t, then dp/dt gives the rate of change of the percentage of infected students with respect to time at time t.
Life Sciences	If $A(t)$ gives the amount of radioactive substance remaining after t years, then $A'(t)$ gives the rate of decay of that substance with respect to time at time t.
Medicine	If $C(t)$ gives the concentration of a drug in a patient's bloodstream t hours after injection, then $C'(t)$ measures the rate at which the concentration of the drug is changing with respect to time at time t.
Business	If $S(x)$ is the total sales of a company when the amount spent on advertising its products and services is x, then $S'(x)$ measures the rate of change of the sales level with respect to the amount spent on advertising when the expenditure is x.
Demographics	If $P(t)$ gives the population of the United States in year t, then $P'(t)$ gives the rate of change of the population with respect to time at time t.

2.4 CONCEPT QUESTIONS

1. Let $f(t)$ denote the position of an object moving along a coordinate line, where $f(t)$ is measured in feet and t in seconds. Explain each of the following in terms of f:
   a. average velocity
   b. velocity
   c. speed
   d. acceleration
   e. jerk

2. Suppose that P is a profit function giving the total profit $P(x)$ in dollars resulting from the sale of x units of a certain commodity. What does $P'(a)$ measure if a is a given level of sales?

3. The following figure shows the cross section of a narrow tube of radius a immersed in water. Because of a surface-tension phenomenon called *capillarity,* the water rises until it reaches an equilibrium height. The curved liquid surface is called a *meniscus,* and the angle θ at which it meets the inner wall of the tube is called the contact angle. If the meniscus is described by the function $y = f(x)$, what is the contact angle θ?

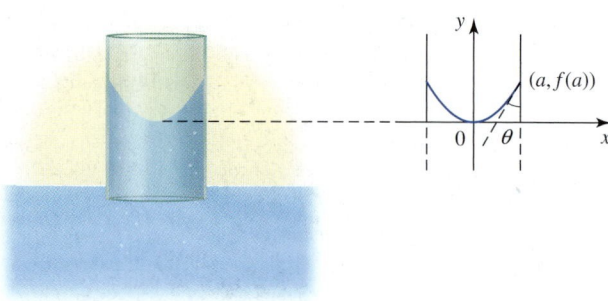

(a) Cross section of the tube and the meniscus

(b) The meniscus is described by $y = f(x)$.

2.4 EXERCISES

In Exercises 1–8, $s(t)$ is the position function of a body moving along a coordinate line; $s(t)$ is measured in feet and t in seconds, where $t \geq 0$. Find the position, velocity, and speed of the body at the indicated time.

1. $s(t) = 1.86t^2$; $t = 2$ (free fall on Mars)

2. $s(t) = 2t^3 - 3t^2 + 4t + 1$; $t = 1$

3. $s(t) = 2t^4 - 8t^2 + 4$; $t = 1$

4. $s(t) = \dfrac{t}{t + 1}$; $t = 0$ 5. $s(t) = \dfrac{2t}{t^2 + 1}$; $t = 2$

6. $s(t) = te^{-t}$; $t = 2$ 7. $s(t) = (t^2 - 1)^2$; $t = 1$

8. $s(t) = \dfrac{t^3}{t^3 + 1}$; $t = 1$

In Exercises 9–16, $s(t)$ is the position function of a body moving along a coordinate line, where $t \geq 0$, and $s(t)$ is measured in feet and t in seconds. (a) Determine the times(s) and the position(s) when the body is stationary. (b) When is the body moving in the positive direction? In the negative direction? (c) Sketch a schematic showing the position of the body at any time t.

9. $s(t) = 2t + 3$ 10. $s(t) = 4 - t^2$

11. $s(t) = 8 + 2t - t^2$ 12. $s(t) = \dfrac{1}{3}t^3 - \dfrac{3}{2}t^2 + 1$

13. $s(t) = 2t^4 - 8t^3 + 8t^2 + 1$

14. $s(t) = (t^2 - 1)^2$

15. $s(t) = \dfrac{2t}{t^2 + 1}$ **16.** $s(t) = \dfrac{t^3}{t^3 + 1}$

In Exercises 17–20, s(t) is the position function of a body moving along a coordinate line, where t ≥ 0, and s(t) is measured in feet and t in seconds. (a) Find the acceleration of the body. (b) When is the acceleration zero? Positive? Negative?

17. $s(t) = 2t^3 - 9t^2 + 12t - 2$

18. $s(t) = t^4 - 2t^2 + 2$

19. $s(t) = \dfrac{2t}{t^2 + 1}$ **20.** $s(t) = \dfrac{t^3}{t^3 + 1}$

In Exercises 21 and 22, s(t) is the position function of a body moving along a coordinate line, where t ≥ 0. If the mass of the body is 20 kg and s(t) and t are measured in meters and seconds, respectively, find (a) the momentum (mv) of the body and (b) the kinetic energy $\left(\frac{1}{2} mv^2\right)$ of the body at the indicated times.

21. $s(t) = 2t^2 - 3t + 1; \quad t = 2$

22. $s(t) = t^3 - 3t^2 + 1; \quad t = 1$

23. Tiltrotor Plane The tiltrotor plane takes off and lands vertically, but its rotors tilt forward for conventional cruising. The figure depicts the graph of the position function of a tiltrotor plane during a test flight in the vertical takeoff and landing mode. Answer the following questions pertaining to the motion of the plane at each of the times t_0, t_1, and t_2:
  a. Is the plane ascending, stationary, or descending?
  b. Is the acceleration positive, zero, or negative?

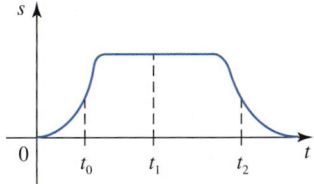 

(**a**) The graph of the position function of a tiltrotor plane   (**b**) A tiltrotor plane

24. Explosion of a Gas Main An explosion caused by the ignition of a leaking underground gas main blew a manhole cover vertically into the air. The height of the manhole cover t seconds after the explosion was $s = 24t - 16t^2$ ft.
  a. How high did the manhole cover go?
  b. What was the velocity of the manhole cover when it struck the ground?

25. Diving The position of a diver executing a high dive from a 10-m platform is described by the position function

$$s(t) = -4.9t^2 + 2t + 10 \qquad t \geq 0$$

where t is measured in seconds and s in meters.

a. When will the diver hit the water?
b. How fast will the diver be traveling at that time? (Ignore the height of the diver and his outstretched arms.)

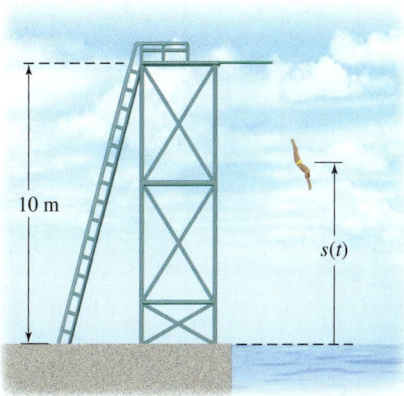

26. Stopping Distance of a Sports Car A test of the stopping distance (in feet) of a sports car was conducted by the editors of an auto magazine. For a particular test, the position function of the car was

$$s(t) = 88t - 12t^2 - \frac{1}{6}t^3$$

where t is measured in seconds and $t = 0$ corresponds to the time when the brakes were first applied.
  a. What was the car's velocity when the brakes were first applied?
  b. What was the car's stopping distance for that particular test?
  c. What was the jerk at time t? At the time when the brakes were first applied?

27. Flight of a VTOL Aircraft In a test flight of McCord Aviation's experimental VTOL (vertical takeoff and landing) aircraft, the altitude of the aircraft operating in the vertical takeoff mode was given by the position function

$$h(t) = \frac{1}{64}t^4 - \frac{1}{2}t^3 + 4t^2 \qquad 0 \leq t \leq 16$$

where $h(t)$ is measured in feet and t is measured in seconds.
  a. Find the velocity function.
  b. What was the velocity of the VTOL at $t = 0$, $t = 8$, and $t = 16$? Interpret your results.
  c. What was the maximum altitude attained by the VTOL during the test flight?

28. Rotating Fluid If a right circular cylinder of radius a is filled with water and rotated about its vertical axis with a constant angular velocity ω, then the water surface assumes a shape whose cross section in a plane containing the vertical axis is a parabola. If we choose the xy-system so that the y-axis is the axis of rotation and the vertex of the parabola passes through the origin of the coordinate system, then the equation of the parabola is

$$y = \frac{\omega^2 x^2}{2g}$$

where g is the acceleration due to gravity. Find the angle α that the tangent line to the water level makes with the x-axis at any point on the water level. What happens to α as ω increases? Interpret your result.

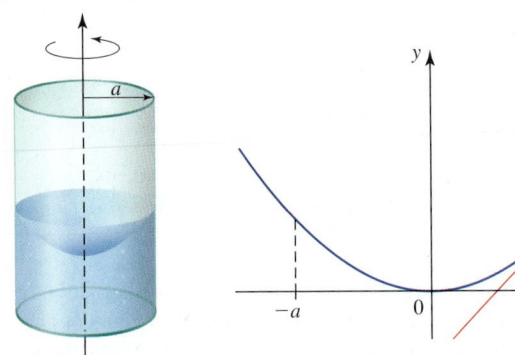

29. Motion of a Projectile A projectile is fired from a cannon located on a horizontal plane. If we think of the cannon as being located at the origin O of an xy-coordinate system, then the path of the projectile is

$$y = \sqrt{3}x - \frac{x^2}{400}$$

where x and y are measured in feet.

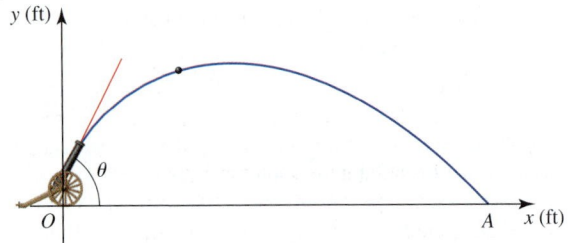

a. Find the value of θ (the angle of elevation of the gun).
b. At what point on the trajectory is the projectile traveling parallel to the ground?
c. What is the maximum height attained by the projectile?
d. What is the range of the projectile (the distance OA along the x-axis)?
e. At what angle with respect to the x-axis does the projectile hit the ground?

30. Deflection of a Beam A horizontal uniform beam of length L is supported at both ends and bends under its own weight w per unit length. Because of its elasticity, the beam is distorted in shape, and the resulting distorted axis of symmetry (shown dashed in the figure) is called the elastic curve. It

can be shown that an equation for the elastic curve is

$$y = \frac{w}{24EI}(x^4 - 2Lx^3 + L^3x)$$

where the product EI is a constant called the *flexural rigidity*.

(a) The distorted beam

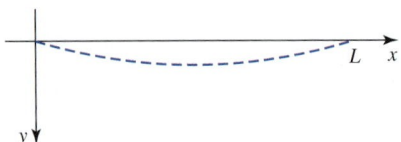

(b) The elastic curve in the xy-plane (The positive direction of the y-axis is directed downward.)

a. Find the angle that the elastic curve makes with the positive x-axis at each end of the beam in terms of w, E, and I.
b. Show that the angle that the elastic curve makes with the horizontal at $x = L/2$ is zero.
c. Find the deflection of the beam at $x = L/2$. (We will show that the deflection is maximal in Section 3.1, Exercise 74.)

31. **Flight Path of an Airplane** The path of an airplane on its final approach to landing is described by the equation $y = f(x)$ with

$$f(x) = 4.3404 \times 10^{-10}x^3 - 1.5625 \times 10^{-5}x^2 + 3000$$
$$0 \le x \le 24{,}000$$

where x and y are both measured in feet.
a. Plot the graph of f using the viewing window $[0, 24000] \times [0, 3000]$.
b. Find the maximum angle of descent during the landing approach.
   Hint: When is dy/dx smallest?

32. Middle-Distance Race As they round the corner into the final (straight) stretch of the bell lap of a middle-distance race, the positions of the two leaders of the pack, A and B, are given by

$$s_A(t) = 0.063t^2 + 23t + 15 \qquad t \ge 0$$

and

$$s_B(t) = 0.298t^2 + 24t \qquad t \ge 0$$

respectively, where the reference point (origin) is taken to be the point located 300 feet from the finish line and s is measured in feet and t in seconds. It is known that one of the two runners, A and B, was the winner of the race and the other was the runner-up.

a. Show that B won the race.

b. At what point from the finish line did B overtake A?

c. By what distance did B beat A?

d. What was the speed of each runner as he crossed the finish line?

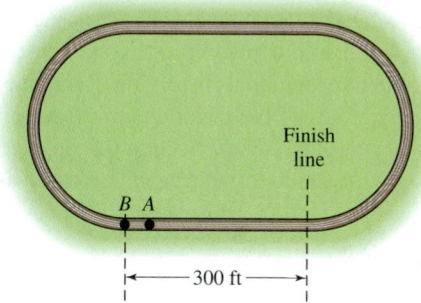

33. Acceleration of a Car A car starting from rest and traveling in a straight line attains a velocity of

$$v(t) = \frac{110t}{2t + 5}$$

feet per second after t sec. Find the initial acceleration of the car and its acceleration 10 sec after starting from rest.

34. Marginal Cost of Producing Compact Discs The weekly total cost in dollars incurred by the BMC Recording Company in manufacturing x compact discs is

$$C(x) = 4000 + 3x - 0.0001x^2 \qquad 0 \leq x \leq 10{,}000$$

a. What is the actual cost incurred by the company in producing the 2001st disc? The 3001st disc?

b. What is the marginal cost when $x = 2000$? When $x = 3000$?

35. Marginal Cost of Producing Microwave Ovens A division of Ditton Industries manufactures the "Spacemaker" model microwave oven. Suppose that the daily total cost (in dollars) of manufacturing x microwave ovens is

$$C(x) = 0.0002x^3 - 0.06x^2 + 120x + 6000$$

What is the marginal cost when $x = 200$? Compare the result with the actual cost incurred by the company in manufacturing the 201st oven.

36. Marginal Average Cost of Producing Television Sets The Advance Visual Systems Corporation manufactures a 19-inch LCD HDTV. The weekly total cost incurred by the company in manufacturing x sets is

$$C(x) = 0.000002x^3 - 0.02x^2 + 120x + 70{,}000$$

dollars.

a. Find the average cost function $\overline{C}(x)$ and the marginal average cost function $\overline{C}'(x)$.

b. Compute $\overline{C}'(5000)$ and $\overline{C}'(10{,}000)$, and interpret your results.

37. Marginal Revenue of an Airline The Commuter Air Service realizes a revenue of

$$R(x) = 10{,}000x - 100x^2$$

dollars per month when the price charged per passenger is x dollars.

a. Find the marginal revenue function R'.

b. Compute $R'(49)$, $R'(50)$, and $R'(51)$. What do your results seem to imply?

38. Marginal Profit in Producing Television Sets The Advance Visual Systems Corporation realizes a total profit of

$$P(x) = -0.000002x^3 + 0.016x^2 + 80x - 70{,}000$$

dollars per week from the manufacture and sale of x units of their 26-in. LCD HDTVs.

a. Find the marginal profit function P'.

b. Compute $P'(2000)$ and interpret your result.

39. Optics The equation

$$\frac{1}{f} = \frac{1}{p} + \frac{1}{q}$$

sometimes called a **lens-maker's equation,** gives the relationship between the focal length f of a thin lens, the distance p of the object from the lens, and the distance q of its image from the lens. We can think of the eye as an optical system in which the ciliary muscle constantly adjusts the curvature of the cornea-lens system to focus the image on the retina. Assume that the distance from the cornea to the retina is 2.5 cm.

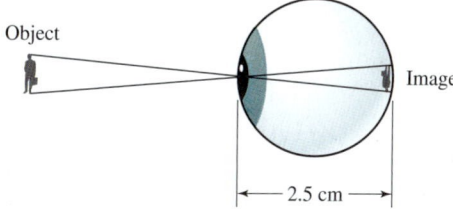

a. Find the focal length of the cornea-lens system if an object located 50 cm away is to be focused on the retina.

b. What is the rate of change of the focal length with respect to the distance of the object when the object is 50 cm away?

40. Gravitational Force The magnitude of the gravitational force exerted by the earth on a particle of mass m at a distance r from the center of the earth is

$$F(r) = \begin{cases} \dfrac{GMmr}{R^2} & \text{if } r < R \\[2ex] \dfrac{GMm}{r^2} & \text{if } r \geq R \end{cases}$$

where M is the mass of the earth, R is its radius, and G is the gravitational constant.

a. Compute $F'(r)$ for $r < R$, and interpret your result.

b. Compute $F'(r)$ for $r > R$, and interpret your result.

2.5 Derivatives of Trigonometric Functions

Many real-world problems are modeled using trigonometric functions. The motion of a pendulum, for example, is periodic (or almost periodic) and can be described by using a combination of sine and cosine functions. The motion of a shock absorber in a car can also be described by using a combination of trigonometric functions and exponential functions. You will see many other applications involving trigonometric functions throughout the book. To analyze the mathematical models involving trigonometric functions, we need to be able to find the derivatives of the trigonometric functions.

Before starting this section, you might wish to review Section 0.3 on trigonometric functions. Keep in mind that all angles are measured in radians, unless otherwise stated.

■ Derivatives of Sines and Cosines

Our first result tells us how to find the derivative of $\sin x$.

THEOREM 1 **Derivative of $\sin x$**

$$\frac{d}{dx}(\sin x) = \cos x$$

PROOF Let $f(x) = \sin x$. Then

$$f'(x) = \lim_{h \to 0} \frac{f(x + h) - f(x)}{h} \qquad \text{Definition of the derivative}$$

$$= \lim_{h \to 0} \frac{\sin(x + h) - \sin x}{h}$$

$$= \lim_{h \to 0} \frac{\sin x \cos h + \cos x \sin h - \sin x}{h} \qquad \begin{array}{l}\text{Expand } \sin(x + h) \text{ using the}\\ \text{Angle Addition Formula.}\end{array}$$

$$= \lim_{h \to 0} \left[\frac{\sin x \cos h - \sin x}{h} + \frac{\cos x \sin h}{h} \right]$$

$$= \lim_{h \to 0} \left[(\sin x)\left(\frac{\cos h - 1}{h} \right) + (\cos x)\left(\frac{\sin h}{h} \right) \right]$$

$$= \left(\lim_{h \to 0} \sin x \right)\left(\lim_{h \to 0} \frac{\cos h - 1}{h} \right) + \left(\lim_{h \to 0} \cos x \right)\left(\lim_{h \to 0} \frac{\sin h}{h} \right)$$

$$\text{Use the Sum and Product Laws for limits.}$$

But $\lim_{h \to 0} \sin x = \sin x$ and $\lim_{h \to 0} \cos x = \cos x$ because these expressions do not involve h and thus remain constant with respect to the limiting process. From Section 1.2 we have

$$\lim_{h \to 0} \frac{\cos h - 1}{h} = 0$$

and

$$\lim_{h \to 0} \frac{\sin h}{h} = 1$$

Using these results, we see that

$$f'(x) = (\sin x)(0) + (\cos x)(1) = \cos x$$

The relationship between the function $f(x) = \sin x$ and its derivative $f'(x) = \cos x$ can be seen by sketching the graphs of both functions (see Figure 1). Here, we interpret $f'(x)$ as the slope of the tangent line to the graph of f at the point $(x, f(x))$.

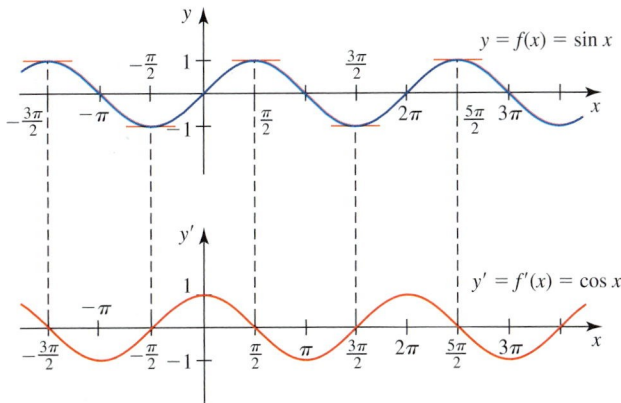

FIGURE 1
The graphs of $f(x) = \sin x$ and its derivative $f'(x) = \cos x$

EXAMPLE 1 Find $f'(x)$ if $f(x) = x^2 \sin x$.

Solution Using the Product Rule and Theorem 1, we obtain

$$f'(x) = \frac{d}{dx}(x^2 \sin x) = x^2 \frac{d}{dx}(\sin x) + (\sin x)\frac{d}{dx}(x^2)$$

$$= x^2 \cos x + 2x \sin x$$

THEOREM 2 **Derivative of $\cos x$**

$$\frac{d}{dx}(\cos x) = -\sin x$$

The proof of this rule is similar to the proof of Theorem 1 and is left as an exercise (Exercise 47).

Derivatives of Other Trigonometric Functions

The remaining trigonometric functions are defined in terms of the sine and cosine functions. Thus,

$$\tan x = \frac{\sin x}{\cos x}, \qquad \csc x = \frac{1}{\sin x}, \qquad \sec x = \frac{1}{\cos x}, \qquad \text{and} \qquad \cot x = \frac{\cos x}{\sin x}$$

Therefore, their derivatives can be found by using Theorems 1 and 2 and the Quotient Rule. For example,

$$\frac{d}{dx}(\tan x) = \frac{d}{dx}\left(\frac{\sin x}{\cos x}\right)$$

$$= \frac{(\cos x)\dfrac{d}{dx}(\sin x) - (\sin x)\dfrac{d}{dx}(\cos x)}{\cos^2 x} \qquad \text{\textcolor{orange}{Quotient Rule}}$$

$$= \frac{(\cos x)(\cos x) - (\sin x)(-\sin x)}{\cos^2 x}$$

$$= \frac{\cos^2 x + \sin^2 x}{\cos^2 x}$$

$$= \frac{1}{\cos^2 x} = \sec^2 x$$

that is,

$$\frac{d}{dx}(\tan x) = \sec^2 x$$

A complete list of the rules for differentiating trigonometric functions follows. The proofs of the remaining three rules are left as exercises. (See Exercises 48–50.)

THEOREM 3   Rules for Differentiating Trigonometric Functions

$$\frac{d}{dx}(\sin x) = \cos x \qquad \frac{d}{dx}(\cos x) = -\sin x \qquad \frac{d}{dx}(\tan x) = \sec^2 x$$

$$\frac{d}{dx}(\csc x) = -\csc x \cot x \qquad \frac{d}{dx}(\sec x) = \sec x \tan x \qquad \frac{d}{dx}(\cot x) = -\csc^2 x$$

Note   As an aid to remembering the signs of the derivatives of the trigonometric functions, observe that those functions beginning with a "c" (cos x, csc x, and cot x) have a minus sign attached to their derivatives.

EXAMPLE 2   Differentiate $y = (\sec x)(x + \tan x)$.

Solution   Using the Product Rule and Theorem 3, we have

$$\frac{dy}{dx} = \frac{d}{dx}[(\sec x)(x + \tan x)]$$

$$= (\sec x)\frac{d}{dx}(x + \tan x) + (x + \tan x)\frac{d}{dx}(\sec x)$$

$$= (\sec x)(1 + \sec^2 x) + (x + \tan x)(\sec x \tan x)$$

$$= (\sec x)(1 + \sec^2 x + x \tan x + \tan^2 x)$$

$$= (\sec x)(2 + x \tan x + 2 \tan^2 x) \qquad \text{\textcolor{orange}{$\sec^2 x = 1 + \tan^2 x$}}$$

EXAMPLE 3 Find the derivative of $y = \dfrac{\sin x}{1 - \cos x}$.

Solution Using the Quotient Rule and Theorems 1 and 2, we obtain

$$\frac{dy}{dx} = \frac{d}{dx}\left(\frac{\sin x}{1 - \cos x}\right)$$

$$= \frac{(1 - \cos x)\dfrac{d}{dx}(\sin x) - (\sin x)\dfrac{d}{dx}(1 - \cos x)}{(1 - \cos x)^2}$$

$$= \frac{(1 - \cos x)(\cos x) - (\sin x)(\sin x)}{(1 - \cos x)^2}$$

$$= \frac{\cos x - \cos^2 x - \sin^2 x}{(1 - \cos x)^2} = \frac{\cos x - 1}{(1 - \cos x)^2}$$

$$= \frac{1}{\cos x - 1}$$

EXAMPLE 4 Find an equation of the tangent line to the graph of $y = x \sin x$ at the point where $x = \pi/2$.

Solution The slope of the tangent line at any point (x, y) on the graph of $y = x \sin x$ is given by

$$\frac{dy}{dx} = \frac{d}{dx}(x \sin x) = x\frac{d}{dx}(\sin x) + (\sin x)\frac{d}{dx}(x)$$

$$= x \cos x + \sin x$$

In particular, the slope of the tangent line at the point where $x = \pi/2$ is

$$\frac{dy}{dx}\bigg|_{x=\pi/2} = (x \cos x + \sin x)\bigg|_{x=\pi/2}$$

$$= \frac{\pi}{2}\cos\frac{\pi}{2} + \sin\frac{\pi}{2}$$

$$= \frac{\pi}{2}(0) + 1 = 1$$

The y-coordinate of the point of tangency is

$$y\bigg|_{x=\pi/2} = x \sin x\bigg|_{x=\pi/2}$$

$$= \frac{\pi}{2}\sin\frac{\pi}{2} = \frac{\pi}{2}$$

Using the point-slope form of an equation of a line, we find that

$$y - \frac{\pi}{2} = x - \frac{\pi}{2} \qquad \text{or} \qquad y = x$$

The graph of $y = x \sin x$ and the tangent line are shown in Figure 2.

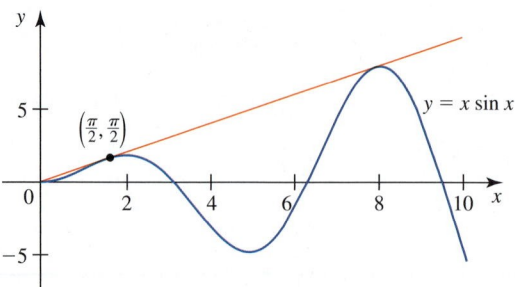

FIGURE 2
An equation of the tangent line to the graph of $y = x \sin x$ at $\left(\frac{\pi}{2}, \frac{\pi}{2}\right)$ is $y = x$.

EXAMPLE 5   **Simple Harmonic Motion**   Suppose that a flexible spring is attached vertically to a rigid support (Figure 3a). If a weight is attached to the free end of the spring, it will settle in a certain equilibrium position (Figure 3b). Suppose that the weight is pulled downward (a positive direction) and released from rest from a position that is 3 units below the equilibrium position at time $t = 0$ (Figure 3c). Then, in the absence of opposing forces such as air resistance, the weight will oscillate back and forth about the equilibrium position. This motion is referred to as *simple harmonic motion.*

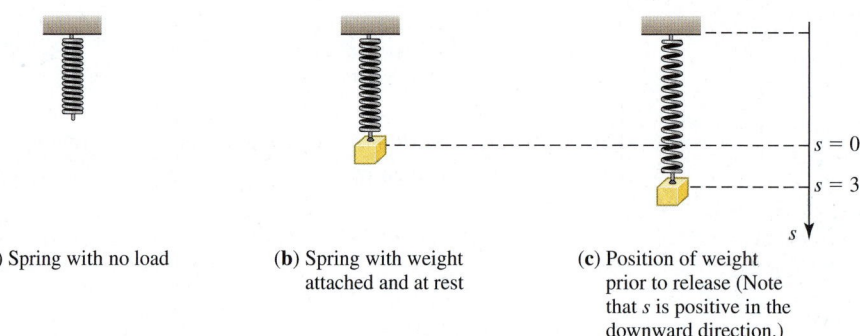

(**a**) Spring with no load

(**b**) Spring with weight attached and at rest

(**c**) Position of weight prior to release (Note that s is positive in the downward direction.)

FIGURE 3

Suppose that for a particular spring and weight, the motion is described by the equation

$$s = 3 \cos t \qquad t \geq 0$$

(See Figure 4.)

a. Find the velocity and acceleration functions describing the motion.
b. Find the values of t when the weight passes the equilibrium position.
c. What are the velocity and acceleration of the weight at these values of t?

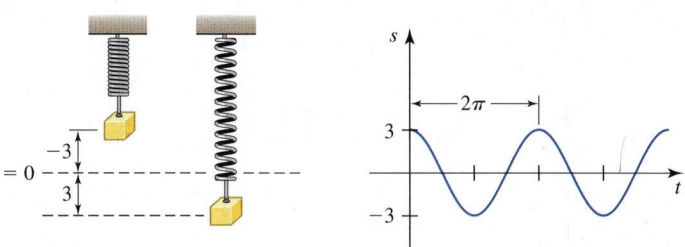

(**a**) Extreme positions of the weight

(**b**) The graph of the function $s = 3 \cos t$ describing the simple harmonic motion of the weight

FIGURE 4

Solution

a. The velocity of the weight at any time $t > 0$ is

$$v(t) = \frac{ds}{dt} = \frac{d}{dt}(3 \cos t) = -3 \sin t$$

and its acceleration at any time $t > 0$ is

$$a(t) = \frac{dv}{dt} = \frac{d}{dt}(-3 \sin t) = -3 \cos t$$

b. When $s = 0$, the weight is at the equilibrium position. Solving the equation

$$s = 3 \cos t = 0$$

we see that the required values of t are $t = \pi/2 + n\pi$, where $n = 0, 1, 2, \ldots$.

c. Using the results of parts (a) and (b), we then calculate the velocity and acceleration of the weight as it passes the equilibrium position:

t	$\dfrac{\pi}{2}$	$\dfrac{3\pi}{2}$	$\dfrac{5\pi}{2}$	$\dfrac{7\pi}{2}$	$\cdots$
$v(t)$	-3	3	-3	3	$\cdots$
$a(t)$	0	0	0	0	$\cdots$

2.5   CONCEPT QUESTIONS

1. State the rules for differentiating $\sin x$, $\cos x$, $\tan x$, $\csc x$, $\sec x$, and $\cot x$.

2. Find

a. $\displaystyle\lim_{h \to 0} \frac{\cos(a + h) - \cos a}{h}$ **b.** $\displaystyle\lim_{h \to 0} \frac{\sec\left(\frac{\pi}{4} + h\right) - \sqrt{2}}{h}$

2.5   EXERCISES

In Exercises 1–22, find the derivative of the function.

1. $f(x) = 4 \cos x - 2x + 1$

2. $g(x) = x + \tan x$

3. $h(t) = 3 \tan t - 4 \sec t$

4. $y = \sqrt{x} \sin x$

5. $f(u) = e^u \cot u$

6. $g(v) = e^v \sin v - 2v \csc v$

7. $s = \sin x \cos x$

8. $f(t) = \sec t \tan t$

9. $f(\theta) = \cos \theta(1 + \sec \theta)$

10. $g(x) = \dfrac{\cos x}{1 + x}$

11. $g(x) = e^{-x} \sin x$

12. $y = \dfrac{\cos \theta}{1 - \sin \theta}$

13. $y = \dfrac{x}{1 + \sec x}$

14. $f(x) = \dfrac{\cot x}{1 + \csc x}$

15. $f(x) = e^x \sec x + e^{-x} \cot x$

16. $y = \cos 2x$

17. $f(x) = \dfrac{1 + \sin x}{1 - \cos x}$

18. $y = \dfrac{\sin x \cos x}{1 + \csc x}$

19. $h(\theta) = \dfrac{\sin \theta + \cos \theta}{\sin \theta - \cos \theta}$

20. $s = \dfrac{1 - \tan t}{1 + \cot t}$

21. $f(x) = e^x \sin^2 x$

22. $y = \dfrac{a \sin t}{1 + b \cos t}$

In Exercises 23–28, find the second derivative of the function.

23. $f(x) = e^x \sin x$

24. $g(x) = \sec x$

25. $y = 3 \cos x - x \sin x$

26. $h(t) = (t^2 + 1) \sin t$

27. $y = \sqrt{x} \cos x$

28. $w = \dfrac{\cos \theta}{\theta}$

In Exercises 29–32, (a) find an equation of the tangent line to the graph of the function at the indicated point, and (b) use a graphing utility to plot the graph of the function and the tangent line on the same screen.

29. $f(x) = \sin x$; $\left(\frac{\pi}{6}, \frac{1}{2}\right)$

30. $f(x) = \tan x$; $\left(\frac{\pi}{4}, 1\right)$

31. $f(x) = \sec x$; $\left(\frac{\pi}{3}, 2\right)$

32. $f(x) = \dfrac{\sin x}{x}$; $\left(\frac{\pi}{2}, \frac{2}{\pi}\right)$

In Exercises 33–36, find the rate of change of y with respect to x at the indicated value of x.

33. $y = x^2 \sec x; \quad x = \dfrac{\pi}{4}$

34. $y = \csc x - 2 \cos x; \quad x = \dfrac{\pi}{6}$

35. $y = \dfrac{\sin x}{1 - \cos x}; \quad x = \dfrac{\pi}{2}$ **36.** $y = \dfrac{e^x \tan x}{\sec x}; \quad x = 0$

In Exercises 37–40, find the x-coordinate(s) of the point(s) on the graph of the function at which the tangent line has the indicated slope.

37. $f(x) = \sin x; \quad m_{\tan} = 1$ **38.** $g(x) = x + \sin x; \quad m_{\tan} = 1$

39. $h(x) = \csc x; \quad m_{\tan} = 0$ **40.** $f(x) = \cot x; \quad m_{\tan} = -2$

41. Let $f(x) = \sin x$. Compute $f^{(n)}(x)$ for $n = 1, 2, 3, \ldots$. Then use your results to show that $|f^{(n)}(x)| \leq 1$ for all x. In other words, the values of the sine function as well as all of its derivatives lie between -1 and 1.

42. Repeat Exercise 41 with the function $f(x) = \cos x$.

43. Simple Harmonic Motion The position function of a body moving along a coordinate line is

$$s(t) = 2 \sin t + 3 \cos t \qquad t \geq 0$$

where t is measured in seconds and $s(t)$ in feet. Find the position, velocity, speed, and acceleration of the body when $t = \pi/2$.

44. Pure Resonance Refer to Example 5. Suppose that the system shown in the figure is initially at rest in the equilibrium position. Further, suppose that starting at $t = 0$, the system is subjected to an external driving force that has the same frequency as the natural frequency of the system. Then the resulting motion of the body is described by the position function

$$s(t) = \sin t - t \cos t \qquad t \geq 0$$

(The frequency is just the reciprocal of the period of the position function, in this case, $1/(2\pi)$.)

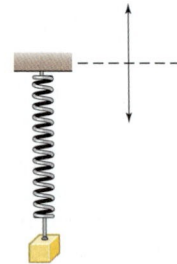

(a) The support is subject to an up-and-down motion whose frequency is the same as the natural frequency of the system.

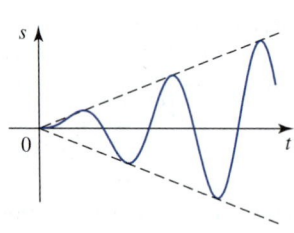

(b) The resulting motion is one in which the amplitude of the wave gets larger and larger.

a. By computing $s(t)$ for $t = n\pi$, where $n = 1, 2, 3, \ldots$, show that $|s(t)|$ gets larger and larger as t increases. This implies that the "amplitude" of the motion becomes unbounded.

b. What is the velocity of the body when $t = \pi/2 + n\pi$, $n = 0, 1, 2, 3, \ldots$?

c. What is the acceleration of the body when $t = n\pi$, $n = 1, 2, 3, \ldots$?

This phenomenon is called **pure resonance**. A mechanical system subjected to resonance will necessarily fail. For example, a singer hitting the "right note" can induce acoustic vibrations that will lead to the shattering of a wine glass.

45. Evaluate $\displaystyle\lim_{h \to 0} \dfrac{\dfrac{1}{\sin(x + h)} - \dfrac{1}{\sin x}}{h}$.

46. Evaluate $\displaystyle\lim_{h \to 0} \dfrac{\tan\left(\dfrac{\pi}{4} + h\right) - 1}{h}$.

47. Prove $\dfrac{d}{dx}(\cos x) = -\sin x$.

48. Prove $\dfrac{d}{dx}(\csc x) = -\csc x \cot x$.

49. Prove $\dfrac{d}{dx}(\cot x) = -\csc^2 x$.

50. Prove $\dfrac{d}{dx}(\sec x) = \sec x \tan x$.

In Exercises 51–52, determine whether the statement is true or false. If it is true, explain why it is true. If it is false, give an example to show why it is false.

51. If $f(x) = \dfrac{1 - \sin^2 2x}{\cos^2 2x}$, $\left(x \neq \dfrac{\pi}{4} + \dfrac{n\pi}{2},\ n \text{ is an integer}\right)$ then $f'(x) = 0$.

52. If $f(x) = \cos(x + h)$, where h is a constant, then $f'(x) = -\sin(x + h)$.

2.6 | The Chain Rule

■ Composite Functions

Suppose that we wish to differentiate the function F defined by

$$F(x) = (x^2 + 1)^{120}$$

If we use only the rules of differentiation developed so far, then a possible approach might be to expand $F(x)$ using the binomial theorem and differentiate the resulting expression term by term. But the amount of work involved would be prodigious! How about the function G defined by $G(x) = \sqrt{2x^2 - 1}$?

You can convince yourself that the same differentiation rules cannot be applied directly to compute $G'(x)$. Observe that both F and G are composite functions. For example, F is the composition of $g(x) = x^{120}$ and $f(x) = x^2 + 1$. Thus,

$$F(x) = (g \circ f)(x) = g[f(x)]$$
$$= [f(x)]^{120} = (x^2 + 1)^{120}$$

and G is the composition of $g(x) = \sqrt{x}$ and $f(x) = 2x^2 - 1$. Thus,

$$G(x) = (g \circ f)(x) = g[f(x)]$$
$$= \sqrt{f(x)} = \sqrt{2x^2 - 1}$$

Notice that each of the component functions f and g is easily differentiated by using the rules of differentiation already available to us. The question, then, is whether we can take advantage of this fact to compute the derivatives of the more complicated composite functions F and G. We will return to these examples later. But for now, let's turn our attention to the general problem of finding the derivative h' of a composite function h.

■ The Chain Rule

For each x in the domain of $h = g \circ f$, let $u = f(x)$ and $y = g(u) = g[f(x)]$. Then, as illustrated in Figure 1, we see that the composite function h maps the number x onto the number y in one step. Alternatively, we see that x is also mapped onto y in two steps—via f (x onto u) then via g (u onto y). Since it might be too difficult to compute $h' = (g \circ f)'$ directly, the following question arises: Can we find h' by somehow combining g' and f'?

FIGURE 1
The function h is composed of the functions g and f: $h(x) = g[f(x)]$.

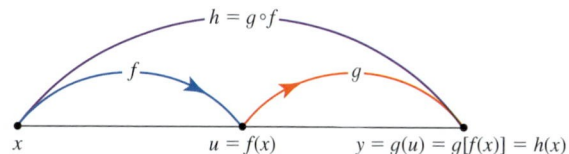

Since u is a function of x, we can compute the derivative of u with respect to x, $du/dx = f'(x)$. Next, y is a function of u, and we can compute the derivative of y with respect to u, $dy/du = g'(u)$. Because h is composed of g and f, it seems reasonable to expect that h', or dy/dx, must be a combination of f' and g' (du/dx and dy/du). But how should we combine them?

Consider the following argument: Interpreting the derivative of a function as the rate of change of that function, suppose that $u = f(x)$ changes twice as fast as x [$f'(x) = du/dx = 2$], and $y = g(u)$ changes three times as fast as u [$g'(u) = dy/du = 3$]. Then we would expect $y = h(x)$ to change six times as fast as x; that is,

$$h'(x) = g'(u)f'(x) = (3)(2) = 6$$

or, equivalently,

$$\frac{dy}{dx} = \frac{dy}{du} \cdot \frac{du}{dx} = (3)(2) = 6$$

Although it is far from being a proof, this argument does suggest how $f'(x)$ and $g'(u) = g'[f(x)]$ should be combined to obtain $h'(x)$ (that is, how du/dx and dy/du should be combined to obtain dy/dx). We simply multiply them together.

The rule for calculating the derivative of a composite function follows.

THEOREM 1 **The Chain Rule**

If f is differentiable at x and g is differentiable at $f(x)$, then the composition $h = g \circ f$ defined by $h(x) = g[f(x)]$ is differentiable at x, and

$$h'(x) = g'[f(x)]f'(x) \tag{a}$$

Also, if we write $u = f(x)$ and $y = g(u) = g[f(x)]$, then

$$\frac{dy}{dx} = \frac{dy}{du} \cdot \frac{du}{dx} \tag{b}$$

The proof of the Chain Rule is given in Appendix B.

Notes
1. The "Inside-Outside" Rule: If we label the composite function $h(x) = g[f(x)]$ in the following way

"inside function"
↓
$$h(x) = g[f(x)]$$
↑
"outside function"

then $h'(x)$ is just the derivative of the "outside function" evaluated at the "inside function" times the derivative of the "inside function."

2. When written in the form of Theorem 1b, the Chain Rule can be remembered by observing that if we "cancel" the du's on the right of the equation, we do obtain dy/dx. ∎

Applying the Chain Rule

EXAMPLE 1 Find $F'(x)$ if $F(x) = (x^2 + 1)^{120}$.

Solution As we observed earlier, F can be viewed as the composite function defined by $F(x) = g[f(x)]$, where $f(x) = x^2 + 1$ and $g(x) = x^{120}$, or $g(u) = u^{120}$ (remember that x and u are dummy variables). The derivative of the "outside function" is

$$g'(u) = \frac{d}{du}[u^{120}] = 120u^{119}$$

which, when evaluated at $f(x) = x^2 + 1$, yields

$$g'[f(x)] = g'(x^2 + 1) = 120(x^2 + 1)^{119}$$

The derivative of the "inside function" is

$$f'(x) = \frac{d}{dx}(x^2 + 1) = 2x$$

Using Theorem 1a, we obtain

$$F'(x) = g'[f(x)]f'(x) = 120(x^2 + 1)^{119} \cdot (2x)$$
$$= 240x(x^2 + 1)^{119}$$

Alternative Solution Let $u = f(x) = x^2 + 1$ and $y = g(u) = u^{120}$. Then, using Theorem 1b, we find

$$F'(x) = \frac{dy}{dx} = \frac{dy}{du} \cdot \frac{du}{dx}$$
$$= 120u^{119} \cdot (2x) = 240xu^{119}$$
$$= 240x(x^2 + 1)^{119}$$

EXAMPLE 2 Find $G'(x)$ if $G(x) = \sqrt{2x^2 - 1}$.

Solution We view $G(x)$ as $G(x) = g[f(x)]$, where $f(x) = 2x^2 - 1$ and $g(x) = \sqrt{x}$ (so $g(u) = \sqrt{u}$). Now

$$g'(u) = \frac{d}{du}[\sqrt{u}] = \frac{1}{2\sqrt{u}}$$

$$g'[f(x)] = \frac{1}{2\sqrt{f(x)}} = \frac{1}{2\sqrt{2x^2 - 1}}$$

and

$$f'(x) = \frac{d}{dx}(2x^2 - 1) = 4x$$

Therefore, if we use Theorem 1a, we obtain

$$G'(x) = g'[f(x)]f'(x) = \frac{1}{2\sqrt{2x^2 - 1}} \cdot (4x)$$

$$= \frac{2x}{\sqrt{2x^2 - 1}}$$

Alternative Solution Let $u = f(x) = 2x^2 - 1$ and $y = g(u) = \sqrt{u}$. Then, using Theorem 1b, we find

$$G'(x) = \frac{dy}{dx} = \frac{dy}{du} \cdot \frac{du}{dx}$$

$$= \frac{1}{2\sqrt{u}} \cdot (4x) = \frac{1}{2\sqrt{2x^2 - 1}} \cdot (4x)$$

$$= \frac{2x}{\sqrt{2x^2 - 1}}$$ ∎

EXAMPLE 3 Find $\dfrac{dy}{dx}$ if $y = u^3 - u^2 + u + 1$ and $u = x^3 + 1$.

Solution In this situation it is more convenient to use Theorem 1b. Thus,

$$\frac{dy}{dx} = \frac{dy}{du} \cdot \frac{du}{dx} = \frac{d}{du}(u^3 - u^2 + u + 1) \cdot \frac{d}{dx}(x^3 + 1)$$

$$= (3u^2 - 2u + 1)(3x^2)$$

If we wish, we could write dy/dx in terms of x as follows:

$$\frac{dy}{dx} = [3(x^3 + 1)^2 - 2(x^3 + 1) + 1](3x^2)$$

$$= 3x^2(3x^6 + 4x^3 + 2)$$ ∎

Note Of course, we could have worked Example 3 using Theorem 1a. In this event, simply observe that $y = g(u) = u^3 - u^2 + u + 1$ and $u = f(x) = x^3 + 1$. ∎

The General Power Rule

Although we have used the Chain Rule in its most general form to help us find the derivatives of the functions in the previous examples, in many situations we need only use a special version of the rule. For example, some functions, such as those in Examples 1 and 2, have the form $y = [f(x)]^n$. These functions are called **generalized power functions.**

To find a formula for computing the derivative of the generalized power function $y = [f(x)]^n$, where n is an integer, let $u = f(x)$ so that $y = u^n$. Using the Chain Rule, we find

$$\frac{dy}{dx} = \frac{dy}{du} \cdot \frac{du}{dx}$$

$$= nu^{n-1} \cdot f'(x)$$

$$= n[f(x)]^{n-1}f'(x)$$

> **THEOREM 2** **General Power Rule**
>
> Let $y = u^n$, where $u = f(x)$ is a differentiable function and n is a real number. Then
>
> $$\frac{dy}{dx} = nu^{n-1}\frac{du}{dx}$$
>
> Equivalently,
>
> $$\frac{dy}{dx} = n[f(x)]^{n-1} \cdot f'(x)$$

Before looking at another example, let's rework Example 1 using Theorem 2. We have $F(x) = (x^2 + 1)^{120}$, which is a generalized power function with $f(x) = x^2 + 1$. Therefore, using the General Power Rule, we obtain

$$F'(x) = \frac{d}{dx}(x^2 + 1)^{120}$$

$$= \underbrace{120(x^2 + 1)^{119}}_{nu^{n-1}} \cdot \underbrace{\frac{d}{dx}(x^2 + 1)}_{\frac{du}{dx}}$$

$$= 120(x^2 + 1)^{119}(2x) = 240x(x^2 + 1)^{119}$$

as before.

EXAMPLE 4 Find $\dfrac{dy}{dx}$ if $y = \dfrac{1}{(2x^4 - x^2 + 1)^3}$.

Solution If we rewrite the given equation as $y = (2x^4 - x^2 + 1)^{-3}$, then an application of the General Power Rule gives

$$\frac{dy}{dx} = -3(2x^4 - x^2 + 1)^{-4}\frac{d}{dx}(2x^4 - x^2 + 1)$$

$$= -3(2x^4 - x^2 + 1)^{-4}(8x^3 - 2x)$$

$$= \frac{6x(1 - 4x^2)}{(2x^4 - x^2 + 1)^4}$$

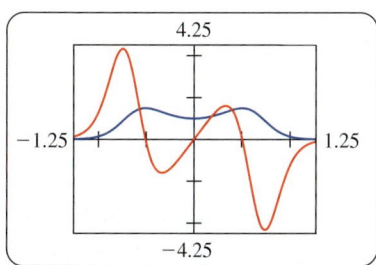

FIGURE 2
The graph of y is shown in blue, and the graph of y' is shown in red.

Observe that the graph of y has horizontal tangents at the points where $x = -\frac{1}{2}$, 0, and $\frac{1}{2}$ and that these are the numbers on the x-axis where the graph of y' crosses the axis. (See Figure 2.) ◼

EXAMPLE 5 How fast is $y = \left(\dfrac{2t - 1}{t^2 + 1}\right)^5$ changing when $t = 1$?

Solution The rate of change of y at any value of t is given by dy/dt. To find dy/dt, we use the General Power Rule, obtaining

$$\frac{dy}{dt} = \frac{d}{dt}\left(\frac{2t - 1}{t^2 + 1}\right)^5$$

$$= \underbrace{5\left(\frac{2t - 1}{t^2 + 1}\right)^4}_{nu^{n-1}} \cdot \underbrace{\frac{d}{dt}\left(\frac{2t - 1}{t^2 + 1}\right)}_{\frac{du}{dt}}$$

$$= 5\left(\frac{2t - 1}{t^2 + 1}\right)^4\left[\frac{(t^2 + 1)\dfrac{d}{dt}(2t - 1) - (2t - 1)\dfrac{d}{dt}(t^2 + 1)}{(t^2 + 1)^2}\right]$$

Use the Quotient Rule.

$$= 5\left(\frac{2t - 1}{t^2 + 1}\right)^4\left[\frac{(t^2 + 1)(2) - (2t - 1)(2t)}{(t^2 + 1)^2}\right]$$

$$= -\frac{10(t^2 - t - 1)(2t - 1)^4}{(t^2 + 1)^6}$$

In particular, when $t = 1$

$$\left.\frac{dy}{dt}\right|_{t=1} = -\frac{10(-1)(1)}{2^6} = \frac{5}{32}$$

Therefore, y is increasing at the rate of $\frac{5}{32}$ units per unit change in t when $t = 1$. ∎

■ The Chain Rule and Trigonometric Functions

In Section 2.5 we learned how to find the derivative of trigonometric functions such as $f(x) = \sin x$. How do we differentiate the function $F(x) = \sin(x^2 - \pi)$? Observe that F is the composition $g \circ f$ of the functions g and f defined by $g(x) = \sin x$ and $f(x) = x^2 - \pi$. Therefore, an application of the Chain Rule yields

$$F'(x) = \frac{d}{dx}[\sin(x^2 - \pi)]$$

$$= \underbrace{\cos(x^2 - \pi)}_{g'[f(x)]} \cdot \underbrace{\frac{d}{dx}(x^2 - \pi)}_{f'(x)} \qquad f(x) = x^2 - \pi, \quad g(x) = \sin x$$

$$= \cos(x^2 - \pi) \cdot (2x) = 2x\cos(x^2 - \pi)$$

Another approach to differentiating generalized trigonometric functions is to derive the appropriate formulas using the Chain Rule. For example, we can find the formula for differentiating the generalized sine function $y = \sin[f(x)]$ by letting $u = f(x)$ so that $y = \sin u$ and then applying the Chain Rule to obtain

$$\frac{dy}{dx} = \frac{dy}{du} \cdot \frac{du}{dx}$$

$$= \frac{d}{du}(\sin u) \cdot \frac{du}{dx}$$

$$= (\cos u)f'(x)$$

$$= \cos[f(x)] \cdot f'(x)$$

In a similar manner we obtain the following rules.

THEOREM 3 **Derivatives of Generalized Trigonometric Functions**

$$\frac{d}{dx}(\sin u) = \cos u \cdot \frac{du}{dx} \qquad \qquad \frac{d}{dx}(\cos u) = -\sin u \cdot \frac{du}{dx}$$

$$\frac{d}{dx}(\tan u) = \sec^2 u \cdot \frac{du}{dx} \qquad \qquad \frac{d}{dx}(\csc u) = -\csc u \cot u \cdot \frac{du}{dx}$$

$$\frac{d}{dx}(\cot u) = -\csc^2 u \cdot \frac{du}{dx} \qquad \qquad \frac{d}{dx}(\sec u) = \sec u \tan u \cdot \frac{du}{dx}$$

EXAMPLE 6   Find the slope of the tangent line to the graph of $y = 3 \cos x^2$ at the point where $x = \sqrt{\pi/2}$.

Solution   The slope of the tangent line at any point on the graph is given by dy/dx. To find dy/dx, we use Theorem 3, obtaining

$$\frac{dy}{dx} = \frac{d}{dx}(3 \cos x^2)$$

$$= 3\frac{d}{dx}(\cos x^2) \qquad \text{\textcolor{orange}{Constant Multiple Rule}}$$

$$= 3\underbrace{(-\sin x^2)}_{-\sin f(x)}\underbrace{(2x)}_{f'(x)}$$

$$= -6x \sin x^2$$

In particular, the slope of the tangent line to the graph of the given equation at the point where $x = \sqrt{\pi/2}$ is

$$\left.\frac{dy}{dx}\right|_{x=\sqrt{\pi/2}} = -6\left(\sqrt{\frac{\pi}{2}}\right)\sin\frac{\pi}{2} = -6\sqrt{\frac{\pi}{2}} \qquad \blacksquare$$

⚠ Do not confuse $\cos x^2$ with $(\cos x)^2$, usually written as $\cos^2 x$.

EXAMPLE 7   Find an equation of the tangent line at the point on the graph of $y = x^2 \sin 3x$, where $x = \pi/2$.

Solution   The slope of the tangent line at any point (x, y) on the graph of $y = x^2 \sin 3x$ is given by dy/dx. Using the Product Rule and Theorem 3, we obtain

$$\frac{dy}{dx} = \frac{d}{dx}(x^2 \sin 3x)$$

$$= x^2\frac{d}{dx}(\sin 3x) + (\sin 3x)\frac{d}{dx}(x^2)$$

$$= x^2(\cos 3x)\cdot\frac{d}{dx}(3x) + 2x \sin 3x$$

$$= 3x^2 \cos 3x + 2x \sin 3x$$

In particular, the slope of the tangent line at the point where $x = \pi/2$ is

$$\left.\frac{dy}{dx}\right|_{x=\pi/2} = 3\left(\frac{\pi}{2}\right)^2\cos\frac{3\pi}{2} + 2\left(\frac{\pi}{2}\right)\sin\frac{3\pi}{2} = 0 + \pi(-1) = -\pi$$

The point of tangency has y-coordinate given by

$$\left.y\right|_{x=\pi/2} = \left.x^2 \sin 3x\right|_{x=\pi/2} = \left(\frac{\pi}{2}\right)^2\sin\frac{3\pi}{2} = -\frac{\pi^2}{4}$$

Therefore, an equation of the required tangent line is

$$y - \left(-\frac{\pi^2}{4}\right) = -\pi\left(x - \frac{\pi}{2}\right) \qquad \text{or} \qquad y = -\pi x + \frac{\pi^2}{4}$$

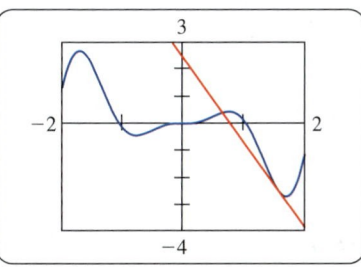

FIGURE 3

The graph of f and its tangent line at $\left(\frac{\pi}{2}, -\frac{\pi^2}{4}\right)$ are shown in Figure 3. $\blacksquare$

Next we consider an example in which the Chain Rule is applied more than once to differentiate a function.

EXAMPLE 8 Find $\dfrac{dy}{dx}$ if $y = \tan^3(3x^2 + 1)$.

Solution

$$\frac{dy}{dx} = \frac{d}{dx}[\tan^3(3x^2 + 1)] = \frac{d}{dx}[\tan(3x^2 + 1)]^3$$

$$= 3[\tan(3x^2 + 1)]^2 \cdot \frac{d}{dx}[\tan(3x^2 + 1)] \qquad \text{Use the General Power Rule.}$$

$$= 3\tan^2(3x^2 + 1) \cdot \sec^2(3x^2 + 1) \cdot \frac{d}{dx}(3x^2 + 1) \qquad \text{Use Theorem 3.}$$

$$= 3\tan^2(3x^2 + 1) \cdot \sec^2(3x^2 + 1) \cdot 6x$$

$$= 18x\tan^2(3x^2 + 1)\sec^2(3x^2 + 1) \qquad\qquad ∎$$

Notes

1. The function in Example 8 can be viewed as a composition of three functions. For example, letting $f(x) = 3x^2 + 1$, $g(u) = \tan u$, and $h(w) = w^3$, you can see that $g \circ f$ is defined by

$$(g \circ f)(x) = g[f(x)] = \tan[f(x)] = \tan(3x^2 + 1)$$

Now, if we compose h with $g \circ f$, we obtain

$$[h \circ (g \circ f)](x) = h[(g \circ f)(x)] = h[\tan(3x^2 + 1)] = \tan^3(3x^2 + 1)$$

and this is just the expression for y.

2. Suppose that $y = [h \circ (g \circ f)](x) = h\{g[f(x)]\}$. In this case the Chain Rule is

$$\frac{dy}{dx} = h'\{g[f(x)]\}g'[f(x)]f'(x)$$

Equivalently, if we let $u = f(x)$ and $v = g(u)$, then

$$\frac{dy}{dx} = \frac{dy}{dv} \cdot \frac{dv}{du} \cdot \frac{du}{dx}$$

Incidentally, the reason for calling Theorem 1 the Chain Rule becomes clear if we look at the larger "chain" that results when we have the composition of many functions. ∎

■ The Derivative of e^u

If we apply the Chain Rule to the function $y = e^{f(x)}$ by letting $u = f(x)$ (the "inside function"), we have

$$\frac{dy}{dx} = \frac{dy}{du}\frac{du}{dx}$$

$$= \frac{d}{du}(e^u)\frac{du}{dx}$$

$$= e^u\frac{du}{dx}$$

So we have the following.

THEOREM 4   **Derivative of the Generalized Natural Exponential Function**

If u is a differentiable function of x, then

$$\frac{d}{dx}(e^u) = e^u \frac{du}{dx}$$

EXAMPLE 9   Differentiate $y = e^{-x \cos x}$.

Solution   Using Theorem 4, we have

$$\frac{dy}{dx} = \frac{d}{dx}(e^{-x \cos x}) = e^{-x \cos x} \frac{d}{dx}(-x \cos x)$$

$$= e^{-x \cos x}(-\cos x + x \sin x) = (x \sin x - \cos x)e^{-x \cos x} \qquad \blacksquare$$

■ The Derivative of a^u

Thanks to the Chain Rule, we are also able to find the formula for differentiating more general exponential functions of the form a^u, where $a > 0$. We begin by recalling from Section 0.8, that if $a > 0$, then $a = e^{\ln a}$. So

$$f(x) = a^x = (e^{\ln a})^x = e^{(\ln a)x}$$

and using the Chain Rule, we obtain

$$f'(x) = \frac{d}{dx}(a^x) = \frac{d}{dx}\left(e^{(\ln a)x}\right) = e^{(\ln a)x} \cdot \frac{d}{dx}(\ln a)x$$

$$= e^{(\ln a)x} \cdot \ln a = a^x \ln a$$

To find a formula for computing the derivative of the generalized exponential function $y = a^{f(x)}$, where f is a differentiable function, let $u = f(x)$ so that $y = a^u$. Then, using the Chain Rule, we find

$$\frac{dy}{dx} = \frac{dy}{du} \frac{du}{dx} = a^u \ln a \frac{du}{dx}$$

THEOREM 5   **The Derivatives of a^x and a^u**

Let u be a differentiable function of x and $a > 0$, $a \neq 1$. then

a. $\dfrac{d}{dx} a^x = a^x \ln a$ **b.** $\dfrac{d}{dx} a^u = a^u \ln a \dfrac{du}{dx}$

EXAMPLE 10   Find the derivative of

a. $f(x) = 2^x$ **b.** $g(x) = 3^{\sqrt{x}}$ **c.** $y = 10^{\cos 2x}$

Solution

a. $f'(x) = \dfrac{d}{dx} 2^x = (\ln 2)2^x$

b. $g'(x) = \dfrac{d}{dx} 3^{\sqrt{x}} = (\ln 3)3^{\sqrt{x}} \dfrac{d}{dx} x^{1/2} = (\ln 3)3^{\sqrt{x}}\left(\dfrac{1}{2}x^{-1/2}\right) = \dfrac{(\ln 3)3^{\sqrt{x}}}{2\sqrt{x}}$

c. $\dfrac{dy}{dx} = \dfrac{d}{dx} 10^{\cos 2x} = (\ln 10)10^{\cos 2x} \dfrac{d}{dx} \cos 2x = (\ln 10)10^{\cos 2x}(-\sin 2x)\dfrac{d}{dx}(2x)$

$= -2(\ln 10)(\sin 2x)10^{\cos 2x}$ ◼

In Section 2.2 we saw that if $f(x) = a^x$ ($a > 0$), then $f'(x) = f'(0)a^x$, where $f'(0)$, the slope of the tangent line to the graph of f at $(0, 1)$, is $\lim_{h \to 0}(a^h - 1)/h$. In view of Theorem 5 we now know that

$$\lim_{h \to 0} \frac{a^h - 1}{h} = \ln a$$

For example, if $a = 2$, then $\lim_{h \to 0}(2^h - 1)/h = \ln 2 \approx 0.69$, and this agrees with our estimate,

$$\frac{d}{dx}(2^x) \approx (0.69)2^x$$

obtained in Section 2.2. Finally, observe that if we put $a = e$, then

$$\lim_{h \to 0} \frac{e^h - 1}{h} = \ln e = 1$$

as expected.

EXAMPLE 11 **A Spring System** The equation of motion of a weight attached to a spring and a dashpot damping device is

$$x(t) = e^{-t}\left(-2 \cos 3t - \frac{2}{3} \sin 3t\right)$$

where $x(t)$, measured in feet, is the displacement from the equilibrium position of the spring system and t is measured in seconds. (See Figure 4.) Find the initial position and the initial velocity of the weight.

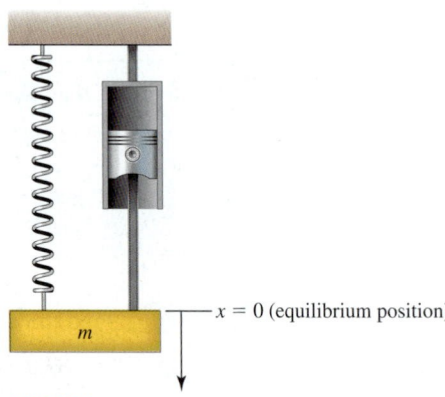

FIGURE 4
The system in equilibrium (The positive direction is downward.)

Solution   The initial position of the spring system is given by

$$x(0) = e^0\left(-2\cos 0 - \frac{2}{3}\sin 0\right) = -2$$

This tells us that the spring system is 2 ft above the equilibrium position.

The velocity of the spring system at any time t is given by

$$v(t) = \frac{d}{dt}\left[e^{-t}\left(-2\cos 3t - \frac{2}{3}\sin 3t\right)\right]$$

$$= -e^{-t}\left(-2\cos 3t - \frac{2}{3}\sin 3t\right) + e^{-t}(6\sin 3t - 2\cos 3t)$$

Use the Product Rule.

$$= \frac{20}{3}e^{-t}\sin 3t$$

In particular, its initial velocity is

$$v(0) = \frac{20}{3}e^0\sin 0 = 0$$

that is, it is released from rest.   ◼

EXAMPLE 12   **Path of a Boat**   A boat leaves the point O (the origin) located on one bank of a river traveling with a constant speed of 20 mph and always heading toward a dock at the point $A(1000, 0)$, which is directly due east of the origin (see Figure 5). The river flows north at a constant speed of 5 mph. It can be shown that the path of the boat is

$$y = 500\left[\left(\frac{1000 - x}{1000}\right)^{3/4} - \left(\frac{1000 - x}{1000}\right)^{5/4}\right] \qquad 0 \le x \le 1000$$

Find dy/dx when $x = 100$ and when $x = 900$. Interpret your results.

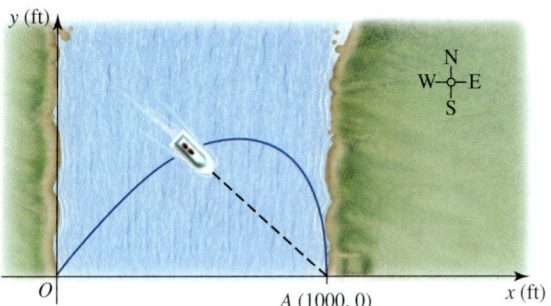

FIGURE 5
The path of the boat

Solution   We find

$$\frac{dy}{dx} = 500\left[\frac{3}{4}\left(\frac{1000 - x}{1000}\right)^{-1/4}\left(-\frac{1}{1000}\right) - \frac{5}{4}\left(\frac{1000 - x}{1000}\right)^{1/4}\left(-\frac{1}{1000}\right)\right]$$

$$= \frac{1}{2}\left[\frac{5}{4}\left(\frac{1000 - x}{1000}\right)^{1/4} - \frac{3}{4}\left(\frac{1000 - x}{1000}\right)^{-1/4}\right]$$

So

$$\left.\frac{dy}{dx}\right|_{x=100} = \frac{1}{2}\left[\frac{5}{4}\left(\frac{9}{10}\right)^{1/4} - \frac{3}{4}\left(\frac{10}{9}\right)^{1/4}\right] \approx 0.22$$

This tells us that at the point on the path where $x = 100$, the boat is drifting north at the rate of 0.22 ft/ft in the x-direction. Next,

$$\left.\frac{dy}{dx}\right|_{x=900} = \frac{1}{2}\left[\frac{5}{4}\left(\frac{1}{10}\right)^{1/4} - \frac{3}{4}(10)^{1/4}\right] \approx -0.32$$

This tells us that at the point on the path where $x = 900$, the boat is drifting south at the rate of 0.32 ft/ft in the x-direction.

2.6 CONCEPT QUESTIONS

1. State the Chain Rule for differentiating the composite function $h = g \circ f$. Explain it in your own words.
2. **a.** State the rule for differentiating the generalized power function $g(x) = [f(x)]^n$, where n is any real number.
   b. State the rule for differentiating the generalized trigonometric function
   $$h(x) = \sec[f(x)]$$

3. Suppose the population P of a certain bacteria culture is given by $P = f(T)$, where T is the temperature of the medium. Further, suppose that the temperature T is a function of time t in seconds—that is, $T = g(t)$. Give an interpretation of each of the following quantities:
   a. $\dfrac{dP}{dT}$ **b.** $\dfrac{dT}{dt}$ **c.** $\dfrac{dP}{dt}$ **d.** $(f \circ g)(t)$ **e.** $f'(g(t))g'(t)$

2.6 EXERCISES

In Exercises 1–6, identify the "inside function" $u = f(x)$ and the "outside function" $y = g(u)$. Then find dy/dx using the Chain Rule.

1. $y = (2x + 4)^3$

2. $y = \sqrt{x^2 - 4}$

3. $y = \dfrac{1}{\sqrt[3]{x^2 + 1}}$

4. $y = 2 \sin \pi x$

5. $y = \sqrt{e^x + \cos x}$

6. $y = \sec \sqrt{x}$

In Exercises 7–64, find the derivative of the function.

7. $f(x) = (2x + 1)^5$

8. $g(x) = (3x^2 + x - 1)^{4/3}$

9. $y = e^{x^2 - x}$

10. $f(t) = e^{\sqrt{t}}$

11. $y = \left(t + \dfrac{2}{t}\right)^6$

12. $f(x) = \left(\dfrac{x^2 + 3}{x}\right)^{-2}$

13. $h(u) = u^3(2u^2 - 1)^4$

14. $h(x) = (2x - 1)^2(x^2 + 1)^3$

15. $f(x) = x^2 e^{-2x}$

16. $g(t) = \sqrt{t - 2} + \sqrt{4 - t}$

17. $g(u) = u\sqrt{1 - u^2}$

18. $f(x) = \dfrac{x}{\sqrt{x^2 - x - 2}}$

19. $f(t) = \dfrac{2t + 3}{(t + 2t^2)^3}$

20. $f(x) = (x^2 + \sqrt{x})^6$

21. $y(s) = \left(1 + \sqrt{1 + s^2}\right)^5$

22. $f(x) = \left(\dfrac{x + 2}{x - 3}\right)^{3/2}$

23. $g(x) = \dfrac{e^{2x}}{1 + e^{-x}}$

24. $h(t) = \dfrac{e^t - e^{-t}}{e^t + e^{-t}}$

25. $f(x) = \sin 3x$

26. $g(x) = e^{-2x} \cos 3x$

27. $g(t) = \tan(\pi t - 1)$

28. $y = \cot(2x + 1)$

29. $f(x) = \sin^3 x$

30. $y = \cos(x^3)$

31. $f(x) = \sin 2x + \tan \sqrt{x}$

32. $y = \cos(x^2 - 3x + 1) + \tan\left(\dfrac{2}{x}\right)$

33. $f(x) = \sin^3 x + \cos^3 x$

34. $f(x) = \tan^2 x + \cot x^2$

35. $f(x) = (1 + \sin^2 3x)^{2/3}$

36. $z = (1 + \csc^2 x)^4$

37. $h(x) = (x^2 - \sec \pi x)^{-3}$

38. $g(x) = \tan^2(x^2 + x)$

39. $y = \sqrt{1 + 2 \cos x}$

40. $f(x) = \sqrt{2 + 3 \tan 2x}$

41. $f(x) = \dfrac{1 + \cos 3x}{1 - \cos 3x}$

42. $y = \dfrac{x + \sin 2x}{2 + \cos 3x}$

43. $y = e^{\cos x}$

44. $f(x) = x \sin \dfrac{1}{x}$

45. $f(x) = \sqrt{\sin 2x - \cos 2x}$

46. $g(t) = \sqrt{t + \tan 3t}$

47. $y = \sin^2\left(\dfrac{1 + x}{1 - x}\right)$

48. $y = \sec^3\left(\dfrac{\sqrt{x}}{1 + x}\right)$

49. $f(x) = \dfrac{\cos 2x}{\sqrt{1 + x^2}}$

50. $f(t) = \dfrac{\cot 2t}{1 + t^2}$

51. $y = \sec^2 x \tan 3x$

52. $y = x \tan^2(2x + 3)$

53. $f(x) = \sin(\sin x)$

54. $g(t) = \tan(\cos 2t)$

55. $f(x) = \cos^3(\sin \pi x)$

56. $y = e^{\cos x^2} \tan(e^{2x} + x)$

57. $y = x(5^{3x})$

58. $f(u) = 2^{u^2}$

59. $h(x) = (2^x + 3^{-x})^6$

60. $f(x) = x^e + e^x$

61. $g(x) = x^e e^x$

62. $f(x) = \dfrac{2^{3x}}{x}$

63. $y = 2^{\cot x}$

64. $g(x) = \dfrac{2^x}{\sqrt{3^x + 1}}$

In Exercises 65–68, find the second derivative of the function.

65. $f(x) = x(2x^2 - 1)^4$

66. $g(x) = \dfrac{1}{(2x + 1)^2}$

67. $f(t) = \sin^2 t - \sin t^2$

68. $f(x) = e^{-x} \tan e^x$

 In Exercises 69–72, (a) find an equation of the tangent line to the graph of the function at the indicated point, and (b) use a graphing utility to plot both the graph of the function and the tangent line on the same screen.

69. $f(x) = x\sqrt{x^2 + 1}$; $(1, \sqrt{2})$

70. $g(x) = \left(\dfrac{x - 1}{x + 1}\right)^3$; $\left(2, \frac{1}{27}\right)$

71. $f(x) = xe^{-x}$; $(1, e^{-1})$

72. $h(t) = 2 \cos^2 \pi t$; $\left(\frac{1}{4}, 1\right)$

73. Suppose that $F = g \circ f$ and $f(2) = 5, f'(2) = 4$, and $g'(5) = 75$. Find $F'(2)$.

74. Suppose that $F(x) = g[f(x)]$ and $f(3) = 16, f'(3) = 6$, and $g'(16) = \frac{1}{8}$. Find $F'(3)$.

75. Let $F(x) = f[f(x)]$. Does it follow that $F'(x) = [f'(x)]^2$?

76. Suppose that $h = g \circ f$. Does it follow that $h' = g' \circ f'$?

77. Find an equation of the line tangent to the graph of $y = 2^x + 1$ at the point $(0, 2)$.

78. Find an equation of the line tangent to the graph of $y = \dfrac{e^{-2x}}{x^2 + 1}$ at the point where $x = 0$.

In Exercises 79–80, refer to the following graph.

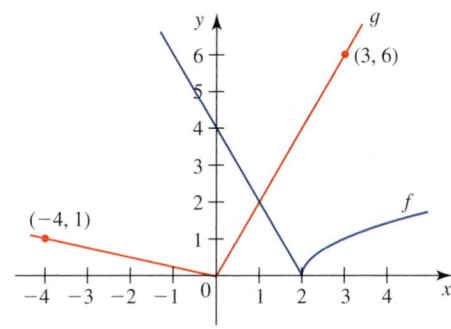

79. The graphs of f and g are shown in the figure. Let $F(x) = g[f(x)]$ and $G(x) = f[g(x)]$. Find $F'(1), G'(-1)$, and $F'(2)$. If a derivative does not exist, explain why.

80. The graphs of f and g are shown in the figure. Find
   a. $h'(1)$ if $h = g \circ g$
   b. $H'(1)$ if $H = f \circ f$
   c. $G'(1)$ if $G(x) = f(x^2 - 1)$
   d. $F'\left(\frac{\pi}{6}\right)$ if $F(x) = f(2 \sin x)$

In Exercises 81–84, find $F'(x)$. Assume that all functions are differentiable.

81. $F(x) = a[f(\sin x)] + b[g(\cos x)]$, where a and b are real numbers

82. $F(x) = a \sin[f(x)] + b \cos[g(x)]$

83. $F(x) = f(x^2 + 1) + g(x^2 - 1)$

84. $F(x) = f(x^a) + [f(x)]^a$, where a is a real number

85. Find $F''(2)$ if $F(x) = x^2 f(2x)$ and it is known that $f(4) = -2$, $f'(4) = 1$, and $f''(4) = -1$.

86. Suppose that f has second-order derivatives and $g(x) = xf(x^2 + 1)$. Find $g''(x)$ in terms of $f(x), f'(x)$, and $f''(x)$.

87. The graph of the function

$$f(x) = \dfrac{|x|}{\sqrt{2 - x^2}}$$

is called a bullet-nose curve.
   a. What is the derivative of f for $x \neq 0$? Find the equations of the tangent lines to the graph of f at $(-1, 1)$ and $(1, 1)$.
    b. Plot the graph of f and the tangent lines found in part (a) using the same viewing window.

88. Refer to Exercise 87. Explain why f is not differentiable at 0.

89. **Aging Population** The population of Americans age 55 years and over as a percent of the total population is approximated by the function

$$f(t) = 10.72(0.9t + 10)^{0.3} \qquad 0 \leq t \leq 20$$

where t is measured in years, with $t = 0$ corresponding to the year 2000. At what rate was the percent of Americans age 55 years and over changing at the beginning of 2000? At what rate will the percent of Americans age 55 years and over be changing at the beginning of 2010? What will be the percent of the population of Americans age 55 years and over at the beginning of 2010?
Source: U.S. Census Bureau.

90. **Accumulation Years** People from their mid-40s to their mid-50s are in the prime investing years. Demographic studies of this type are of particular importance to financial institutions. The function

$$N(t) = 34.4(1 + 0.32125t)^{0.15} \qquad 0 \leq t \leq 12$$

gives the projected number of people in this age group in the United States (in millions) in year t, where $t = 0$ corresponds to the beginning of 1996.

a How large was this segment of the population projected to be at the beginning of 2005?

b. How fast was this segment of the population growing at the beginning of 2005?

Source: U.S. Census Bureau.

91. World Population Growth After its fastest rate of growth ever during the 1980s and 1990s, the rate of growth of world population is expected to slow dramatically, in the twenty-first century. The function

$$G(t) = 1.58e^{-0.213t}$$

gives the projected average percent population growth/decade in the tth decade, with $t = 1$ corresponding to the beginning of 2000.

a. What will the projected average population growth rate be at the beginning of 2020 ($t = 3$)?

b. How fast will the projected average population growth rate be changing at the beginning of 2020?

Source: U.S. Census Bureau.

92. Blood Alcohol Level The percentage of alcohol in a person's bloodstream t hr after drinking 8 fluid oz of whiskey is given by

$$A(t) = 0.23te^{-0.4t} \qquad 0 \le t \le 12$$

How fast is the percentage of alcohol in a person's bloodstream changing after $\frac{1}{2}$ hr? After 8 hr?

Source: Encyclopedia Britannica.

93. Radioactivity The radioactive element polonium decays according to the law

$$Q(t) = Q_0 \cdot 2^{-(t/140)}$$

where Q_0 is the initial amount and t is measured in days.

a. If the amount of polonium left after 280 days is 20 mg, what was the initial amount present?

b. How fast is the amount of polonium changing at any time t?

94. Forensic Science Forensic scientists use the following formula to determine the time of death of accident or murder victims. If T denotes the temperature of a body t hr after death, then

$$T = T_0 + (T_1 - T_0)(0.97)^t$$

where T_0 is the air temperature and T_1 is the body temperature (in degrees Fahrenheit) at the time of death. John Doe was found murdered at midnight in his house, when the room temperature was 70°F. Assume that his body temperature at the time of death was 98.6°F.

a. Plot the graph of T using the viewing window [0, 40] × [70, 100].

b. How fast was the temperature of John Doe's body dropping 2 hr after his death?

c. If the temperature of John Doe's body was 80°F when it was found, when was he killed? Solve the problem analytically, and then verify it using a graphing calculator.

95. Simple Harmonic Motion The position function of a body moving along a coordinate line is

$$s(t) = \frac{1}{2}\cos 2t + \frac{3}{4}\sin 2t \qquad t \ge 0$$

where $s(t)$ is measured in feet and t in sec. Find the position, velocity, speed, and acceleration of the body when $t = \pi/4$.

96. Predator-Prey Population Model The wolf population in a certain northern region is estimated to be

$$P_W(t) = 9000 + 1000 \sin \frac{\pi t}{24}$$

in month t, and the caribou population in the same region is given by

$$P_C(t) = 36{,}000 + 12{,}000 \cos \frac{\pi t}{24}$$

Find the rate of change of each population when $t = 12$.

97. Stock Prices The closing price (in dollars) per share of the stock of Tempco Electronics on the tth day it was traded is approximated by

$$P(t) = 20 + 12 \sin\left(\frac{\pi t}{30}\right) - 6 \sin\left(\frac{\pi t}{15}\right)$$
$$+ 4 \sin\left(\frac{\pi t}{10}\right) - 3 \sin\left(\frac{2\pi t}{15}\right) \qquad 0 \le t \le 20$$

where $t = 0$ corresponds to the time the stock was first listed on a major stock exchange. What was the rate of change of the stock's price at the close of the fifteenth day of trading? What was the closing price on that day?

98. Shortage of Nurses The projected number of nurses (in millions) from the year 2000 through 2015 is given by

$$N(t) = \begin{cases} 1.9 & \text{if } 0 \le t < 5 \\ \sqrt{0.123t + 2.995} & \text{if } 5 \le t \le 15 \end{cases}$$

where $t = 0$ corresponds to the year 2000, while the projected number of nursing jobs (in millions) over the same period is

$$J(t) = \begin{cases} \sqrt{0.129t + 4} & \text{if } 0 \le t < 10 \\ \sqrt{0.4t + 1.29} & \text{if } 10 \le t \le 15 \end{cases}$$

a. Let $G = J - N$ be the function giving the gap between the demand and the supply of nurses from the year 2000 through 2015. Find G'.

b. How fast was the gap between the demand and the supply of nurses changing in 2008? In 2012?

Source: Department of Health and Human Services.

99. Potential Energy A commonly used potential-energy function for the interaction of two molecules is the Lennard-Jones 6-12 potential, given by

$$u(r) = u_0\left[\left(\frac{\sigma}{r}\right)^{12} - \left(\frac{\sigma}{r}\right)^{6}\right]$$

where u_0 and σ are constants. The force corresponding to this potential is $F(r) = -u'(r)$. Find $F(r)$.

100. Mass of a Body Moving Near the Speed of Light According to the special theory of relativity, the mass m of a body moving at a speed v is given by

$$m = \frac{m_0}{\sqrt{1 - \dfrac{v^2}{c^2}}}$$

where m_0 is the mass of the body at rest and $c \approx 3 \times 10^8$ m/sec is the speed of light. How fast is the mass of an electron changing with respect to its speed when its speed is $0.999c$? The rest mass of an electron is 9.11×10^{-31} kg.

101. Motion Along a Line A body moves along a coordinate line in such a way that its position function at any time t is given by

$$s(t) = t\sqrt{1 - t^2} \qquad 0 \le t \le 1$$

where $s(t)$ is measured in feet and t in seconds. Find the velocity and acceleration of the body when $t = \frac{1}{2}$.

102. Surface Area of a Cone The lateral surface area of a right circular cone is

$$S = \pi r\sqrt{r^2 + h^2}$$

where r is the radius of the base and h is the height.
a. What is the rate of change of the lateral surface area with respect to the height if the radius is constant?
b. What is the rate of change of the lateral surface area with respect to the radius if the height is constant?

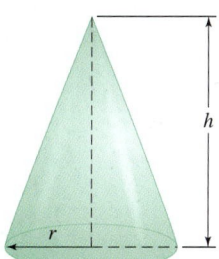

103. A Sliding Chain A chain of length 6 m is held on a table with 1 m of the chain hanging down from the table. Upon release, the chain slides off the table. Assuming that there is no friction, the end of the chain that initially was 1 m from the edge of the table is given by the function

$$s(t) = \frac{1}{2}\left(e^{\sqrt{g/6}\,t} + e^{-\sqrt{g/6}\,t}\right)$$

where $g = 9.8$ m/sec^2 and t is measured in seconds.
a. Find the time it takes for the chain to slide off the table.
b. What is the speed of the chain at the instant of time when it slides off the table?

104. Percent of Females in the Labor Force Based on data from the U.S. Census Bureau, the following model giving the percent of the total female population in the civilian labor force, $P(t)$, at the beginning of the tth decade ($t = 0$ corresponds to the year 1900) was constructed.

$$P(t) = \frac{74}{1 + 2.6e^{-0.166t + 0.04536t^2 - 0.0066t^3}} \qquad 0 \le t \le 11$$

a. What was the percent of the total female population in the civilian labor force at the beginning of 2000?
b. What was the growth rate of the percentage of the total female population in the civilian labor force at the beginning of 2000?
Source: U.S. Census Bureau.

105. Electric Current in a Circuit The following figure shows an R-C series circuit comprising a variable resistor, a capacitor, and an electromotive force. If the resistance at any time t is given by $R = k_1 + k_2t$ ohms, where $k_1 > 0$ and $k_2 > 0$, the capacitance is C farads, and the electromotive force is a constant E volts, then the charge at any time t is given by

$$q(t) = EC + (q_0 - EC)\left(\frac{k_1}{k_1 + k_2t}\right)^{1/(Ck_2)}$$

coulombs where the constant q_0 is the charge at $t = 0$. What is the current $i(t)$ at any time t?
Hint: $i(t) = dq/dt$.

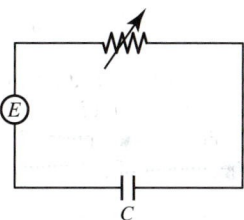

106. Simple Harmonic Motion The equation of motion of a body executing simple harmonic motion is given by

$$x(t) = A\sin(\omega t + \phi)$$

where x (in feet) is the displacement of the body, A is the amplitude, $\omega = \sqrt{k/m}$, k is a constant, and m (in slugs) is the mass of the body. Find expressions for the velocity and acceleration of the body at time t.

107. Potential of a Charged Disk The potential on the axis of a uniformly charged disk is

$$V(r) = \frac{\sigma}{2\varepsilon_0}\left(\sqrt{r^2 + R^2} - r\right)$$

where ε_0 and σ are constants. The force corresponding to this potential is $F(r) = -V'(r)$. Find $F(r)$.

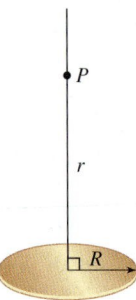

108. Electric Potential Suppose that a ring-shaped conductor of radius a carries a total charge Q. Then the electrical potential at the point P, a distance x from the center and along the line perpendicular to the plane of the ring through its center, is given by

$$V(x) = \frac{1}{4\pi\varepsilon_0} \frac{Q}{\sqrt{x^2 + a^2}}$$

where ε_0 is a constant called the permittivity of free space. The magnitude of the electric field induced by the charge at the point P is $E = -dV/dx$, and the direction of the field is along the x-axis. Find E.

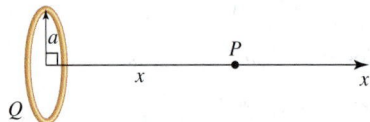

109. Motion of a Conical Pendulum A metal ball is attached to a string of length L ft and is whirled in a horizontal circle as shown in the figure. The speed of the ball is $v = \sqrt{Lg \sec \theta \sin^2 \theta}$ ft/sec, where θ is the angle the string makes with the vertical.

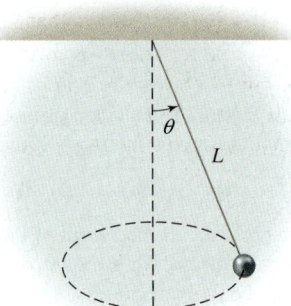

a. Show that

$$\frac{dv}{d\theta} = \frac{\sqrt{Lg}(\tan^2 \theta + 2)}{2\sqrt{\sec \theta}}$$

and interpret your result.

b. Find v and $dv/d\theta$ if $L = 4$ and $\theta = \pi/6$ rad. (Take $g = 32$ ft/sec^2.)

c. Evaluate $\lim_{\theta \to \pi/2^-} v$, and interpret your result.

d. Plot the graph of v for $0 \leq \theta < \frac{\pi}{2}$ to verify the result of part (c) visually.

110. Orbit of a Satellite An artificial satellite moves around the earth in an elliptic orbit. Its distance r from the center of the earth is approximated by

$$r = a\left[1 - e \cos M - \frac{e^2}{2}(\cos 2M - 1)\right]$$

where $M = (2\pi/P)(t - t_n)$. Here, t is time and a, e, P, and t_n are constants measuring the semimajor axis of the orbit, the eccentricity of the orbit, the period of orbiting, and the time taken by the satellite to pass the perigee, respectively. Find dr/dt, the radial velocity of the satellite.

111. Find $f''(x)$ if

$$f(x) = \begin{cases} x^2 \sin \dfrac{1}{x} & \text{if } x \neq 0 \\ 0 & \text{if } x = 0 \end{cases}$$

Does $f''(0)$ exist?

112. Suppose that u is a differentiable function of x and $f(x) = |u|$. Show that

$$f'(x) = \frac{u'u}{|u|} \qquad u \neq 0$$

Hint: $|u| = \sqrt{u^2}$.

In Exercises 113–116, use the result of Exercise 112 to find the derivative of the function.

113. $f(x) = |x + 1|$

114. $g(x) = x|x^2 + x|$

115. $h(x) = |\sin x|$

116. $f(x) = \dfrac{|x|}{x^2}$

117. Let $f(x) = \sqrt{|(x - 1)(x - 2)|}$.
a. Find $f'(x)$.
Hint: See Exercise 112.
b. Sketch the graph of f and f'.

118. A function is called *even* if $f(-x) = f(x)$ for all x in the domain of f; it is called *odd* if $f(-x) = -f(x)$ for all x in the domain of f. Prove that the derivative of a differentiable even function is an odd function and that the derivative of a differentiable odd function is an even function.

In Exercises 119–122, *determine whether the statement is true or false. If it is true, explain why it is true. If it is false, give an example to show why it is false.*

119. If f has a second-order derivative at x, g has a second-order derivative at $f(x)$, and $h(x) = g[f(x)]$, then $h''(x) = g''[f(x)]f''(x)$.

120. If f is differentiable and $h(t) = f(a + bt) + f(a - bt)$, then $h'(t) = bf'(a + bt) - bf'(a - bt)$.

121. If f is differentiable, then

$$\frac{d}{dx}f\left(\frac{1}{x}\right) = -\frac{f'\left(\dfrac{1}{x}\right)}{x^2} \qquad x \neq 0$$

122. If f is differentiable and $h = (f \circ f)^2$, then $h' = 2(f \circ f)(f' \circ f)f'$.

2.7 Implicit Differentiation

■ Implicit Functions

Up to now, the functions we have dealt with are represented by equations of the form $y = f(x)$, in which the dependent variable y has been expressed explicitly in terms of the independent variable x. Sometimes, however, a function f is defined implicitly by an equation $F(x, y) = 0$. For example, the equation

$$x^2y + y - \cos x + 1 = 0 \tag{1}$$

defines y as a function of x. (Here, $F(x, y) = x^2y + y - \cos x + 1$.) In fact, if we solve the equation for y in terms of x, we obtain the explicit representation

$$y = f(x) = \frac{\cos x - 1}{x^2 + 1} \tag{2}$$

You can verify that Equation (2) satisfies Equation (1); that is,

$$x^2f(x) + f(x) - \cos x + 1 = 0$$

Suppose we are given Equation (1) and we wish to find dy/dx. An obvious approach would be to first find an explicit representation for the function f, such as Equation (2), and then differentiate this expression in the usual manner to obtain $dy/dx = f'(x)$.

How about the equation

$$4x^4 + 8x^2y^2 - 25x^2y + 4y^4 = 0 \tag{3}$$

whose graph is shown in Figure 1? The Vertical Line Test shows that Equation (3) does not define y as a function of x. But with suitable restrictions on x and y, Equation (3) does define y as a function of x implicitly. Figure 2 shows the graphs (the solid curves) of two such functions, f and g. In this instance we would be hard pressed to find explicit representations for the functions f and g. So how do we go about computing dy/dx in this case?

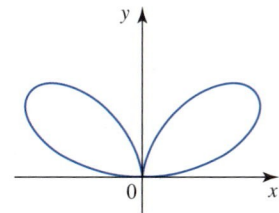

FIGURE 1
The graph of
$$4x^4 + 8x^2y^2 - 25x^2y + 4y^4 = 0$$
is a bifolium.

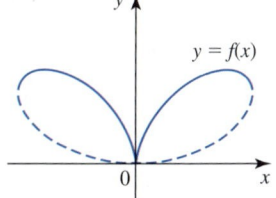

(a) The graph of f **(b)** The graph of g

FIGURE 2
f and g are defined implicitly by
$4x^4 + 8x^2y^2 - 25x^2y + 4y^4 = 0$.

Thanks to the Chain Rule, there exists a method for finding the derivative of a function directly from the equation defining it implicitly. This method is called **implicit differentiation** and will be demonstrated in the next several examples.

■ Implicit Differentiation

EXAMPLE 1

a. Find dy/dx if $x^2 + y^2 = 4$.
b. Find an equation of the tangent line to the graph of $x^2 + y^2 = 4$ at the point $(1, \sqrt{3})$.
c. Solve part (b) again, this time using an explicit representation of a function.

Solution
a. Differentiating both sides of the equation with respect to x, we obtain

$$\frac{d}{dx}(x^2 + y^2) = \frac{d}{dx}(4)$$

$$\frac{d}{dx}(x^2) + \frac{d}{dx}(y^2) = 0 \qquad \text{Use the Sum Rule for derivatives.}$$

To carry out the differentiation of the term y^2, we note that y is a function of x. Writing $y = f(x)$ to remind us of this, we see that

$$\frac{d}{dx}(y^2) = \frac{d}{dx}[f(x)]^2 \qquad \text{Write } y = f(x).$$

$$= 2f(x)f'(x) \qquad \text{Use the Chain Rule.}$$

$$= 2y\frac{dy}{dx} \qquad \text{Return to using } y \text{ instead of } f(x).$$

Therefore, the equation

$$\frac{d}{dx}(x^2 + y^2) = \frac{d}{dx}(4)$$

is equivalent to

$$2x + 2y\frac{dy}{dx} = 0$$

Solving for dy/dx yields

$$\frac{dy}{dx} = -\frac{x}{y}$$

b. Using the result of part (a), we see that the slope of the required tangent line is

$$\left.\frac{dy}{dx}\right|_{(1, \sqrt{3})} = \left.-\frac{x}{y}\right|_{(1, \sqrt{3})} = -\frac{1}{\sqrt{3}} \qquad \left.\frac{dy}{dx}\right|_{(a, b)} \text{ means } \frac{dy}{dx} \text{ evaluated at } x = a \text{ and } y = b.$$

Using the slope-intercept form of an equation of a line, we see that an equation of the tangent line is

$$y - \sqrt{3} = -\frac{1}{\sqrt{3}}(x - 1)$$

$$\sqrt{3}y - 3 = -(x - 1) \qquad \text{or} \qquad x + \sqrt{3}y - 4 = 0$$

c. Solving the equation $x^2 + y^2 = 4$ for y in terms of x gives the functions

$$y = f(x) = \sqrt{4 - x^2} \quad \text{and} \quad y = g(x) = -\sqrt{4 - x^2}$$

among others. The graph of f is the upper semicircle centered at the origin with radius 2 (here, $y \geq 0$), whereas the graph of g is the lower semicircle (here, $y \leq 0$). (See Figure 3.) Since the point $(1, \sqrt{3})$ lies on the upper semicircle, we will work with the function

$$f(x) = \sqrt{4 - x^2} = (4 - x^2)^{1/2}$$

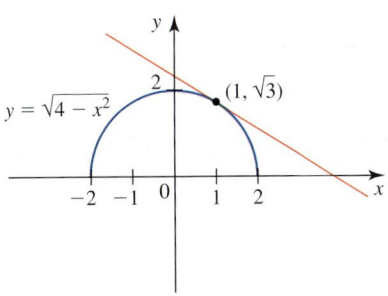
The graph of f

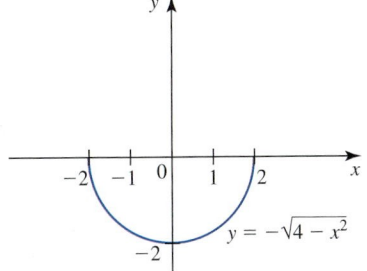
The graph of g

FIGURE 3
The graphs of $f(x) = \sqrt{4 - x^2}$
and $g(x) = -\sqrt{4 - x^2}$

Differentiating $f(x)$ with the help of the Chain Rule gives

$$f'(x) = \frac{1}{2}(4 - x^2)^{-1/2} \frac{d}{dx}(4 - x^2)$$

$$= \frac{1}{2}(4 - x^2)^{-1/2}(-2x)$$

$$= -\frac{x}{\sqrt{4 - x^2}}$$

$$= -\frac{x}{y}$$

and this gives the slope of the tangent line at any point (x, y) on the graph of f. In particular, the slope of the tangent line at $(1, \sqrt{3})$ is

$$f'(1) = -\frac{1}{\sqrt{3}}$$

as before. Continuing, we find that an equation of the tangent line is $x + \sqrt{3}y - 4 = 0$, as obtained earlier. ∎

Notes

1. You can verify that

$$g'(x) = \frac{x}{\sqrt{4 - x^2}} = -\frac{x}{y} = \frac{dy}{dx}$$

so $-x/y$ is the derivative of both f and g.

2. Even when it is possible to find an explicit representation for f, it still can be easier to find $f'(x)$ by implicit differentiation. (See Example 1.)

3. In general, if dy/dx is found by implicit differentiation, the expression for dy/dx will usually involve both x and y. ∎

Guidelines for differentiating a function implicitly follow.

> **Finding dy/dx by Implicit Differentiation**
>
> Suppose that a function $y = f(x)$ is defined implicitly via an equation in x and y. To compute dy/dx:
>
> 1. Differentiate both sides of the equation with respect to x. Make sure that the derivative of any term involving y includes the factor dy/dx.
> 2. Solve the resulting equation for dy/dx in terms of x and y.

EXAMPLE 2 Find $\dfrac{dy}{dx}$ if $y^4 - 2y^3 + x^3y^2 - \cos x = 8$.

Solution Differentiating both sides of the given equation with respect to x, we obtain

$$\frac{d}{dx}(y^4 - 2y^3 + x^3y^2 - \cos x) = \frac{d}{dx}(8)$$

$$\frac{d}{dx}(y^4) - \frac{d}{dx}(2y^3) + \frac{d}{dx}(x^3y^2) - \frac{d}{dx}(\cos x) = 0$$

or

$$\frac{d}{dx}(y^4) - 2\frac{d}{dx}(y^3) + x^3\frac{d}{dx}(y^2) + y^2\frac{d}{dx}(x^3) - \frac{d}{dx}(\cos x) = 0,$$

where we have used the Product Rule to differentiate the term x^3y^2. Next, recalling that y is a function of x, we apply the Chain Rule to the first three terms on the left, obtaining

$$4y^3\frac{dy}{dx} - 6y^2\frac{dy}{dx} + 2x^3y\frac{dy}{dx} + 3x^2y^2 + \sin x = 0$$

$$(4y^3 - 6y^2 + 2x^3y)\frac{dy}{dx} = -3x^2y^2 - \sin x$$

$$\frac{dy}{dx} = -\frac{3x^2y^2 + \sin x}{2y(2y^2 - 3y + x^3)}$$

EXAMPLE 3 Find y' if $e^x \cos y - e^{-y} \sin x = \pi$.

Solution Differentiating both sides of the given equation with respect to x, we obtain

$$\frac{d}{dx}(e^x \cos y - e^{-y} \sin x) = \frac{d}{dx}(\pi)$$

$$\frac{d}{dx}(e^x \cos y) - \frac{d}{dx}(e^{-y} \sin x) = 0$$

$$e^x\frac{d}{dx}(\cos y) + (\cos y)\frac{d}{dx}(e^x) - e^{-y}\frac{d}{dx}(\sin x) - (\sin x)\frac{d}{dx}(e^{-y}) = 0$$

$$e^x(-\sin y)y' + (\cos y)e^x - e^{-y}(\cos x) - (\sin x)e^{-y}(-y') = 0$$

$$(e^{-y}\sin x - e^x\sin y)y' = e^{-y}\cos x - e^x\cos y$$

and so

$$y' = \frac{e^{-y} \cos x - e^x \cos y}{e^{-y} \sin x - e^x \sin y}$$

If we wish to find dy/dx at a specific point (a, b) on the graph of a function defined implicitly by an equation, we need not find a general expression for dy/dx, as illustrated in Example 4.

EXAMPLE 4 Find $\dfrac{dy}{dx}$ at the point $\left(\frac{\pi}{2}, \pi\right)$ if $x \sin y - y \cos 2x = 2x$.

Solution Differentiating both sides of the equation with respect to x, we obtain

$$\frac{d}{dx}(x \sin y - y \cos 2x) = \frac{d}{dx}(2x)$$

$$\frac{d}{dx}(x \sin y) - \frac{d}{dx}(y \cos 2x) = 2$$

Using the Product Rule on each term on the left, we have

$$x \frac{d}{dx}(\sin y) + (\sin y) \frac{d}{dx}(x) - y \frac{d}{dx}(\cos 2x) - (\cos 2x) \frac{d}{dx}(y) = 2$$

Next, using the Chain Rule on the first, third, and fourth terms on the left, we obtain

$$(x \cos y) \frac{dy}{dx} + \sin y - y(-\sin 2x) \frac{d}{dx}(2x) - (\cos 2x) \frac{dy}{dx} = 2$$

or

$$(x \cos y) \frac{dy}{dx} + \sin y + 2y \sin 2x - (\cos 2x) \frac{dy}{dx} = 2$$

Replacing x by $\pi/2$ and y by π in the last equation gives

$$\left(\frac{\pi}{2} \cos \pi\right) \frac{dy}{dx} + \sin \pi + 2\pi \sin \pi - (\cos \pi) \frac{dy}{dx} = 2$$

$$-\frac{\pi}{2} \cdot \frac{dy}{dx} + \frac{dy}{dx} = 2$$

or

$$\frac{dy}{dx} = \frac{2}{1 - \dfrac{\pi}{2}} = \frac{4}{2 - \pi}$$

EXAMPLE 5 Find an equation of the tangent line to the bifolium

$$4x^4 + 8x^2y^2 - 25x^2y + 4y^4 = 0$$

at the point $(2, 1)$.

Solution   The slope of the tangent line to the bifolium at any point (x, y) is given by dy/dx. To compute dy/dx, we differentiate both sides of the equation with respect to x to obtain

$$\frac{d}{dx}(4x^4 + 8x^2y^2 - 25x^2y + 4y^4) = \frac{d}{dx}(0)$$

$$\frac{d}{dx}(4x^4) + \frac{d}{dx}(8x^2y^2) - \frac{d}{dx}(25x^2y) + \frac{d}{dx}(4y^4) = 0$$

Using the Product Rule on the second and third terms on the left, we find

$$16x^3 + 8x^2\frac{d}{dx}(y^2) + y^2\frac{d}{dx}(8x^2) - 25x^2\frac{d}{dx}(y) - y\frac{d}{dx}(25x^2) + \frac{d}{dx}(4y^4) = 0$$

With the aid of the Chain Rule, we obtain

$$16x^3 + 16x^2y\frac{dy}{dx} + 16xy^2 - 25x^2\frac{dy}{dx} - 50xy + 16y^3\frac{dy}{dx} = 0$$

By substituting $x = 2$ and $y = 1$ into the last equation, we obtain

$$16(8) + 16(4)\frac{dy}{dx} + 32 - 25(4)\frac{dy}{dx} - 100 + 16\frac{dy}{dx} = 0$$

or

$$\frac{dy}{dx} = 3$$

Using the slope-intercept form for an equation of a line, we see that an equation of the tangent line is

$$y - 1 = 3(x - 2) \qquad \text{or} \qquad y = 3x - 5$$

(See Figure 4.)

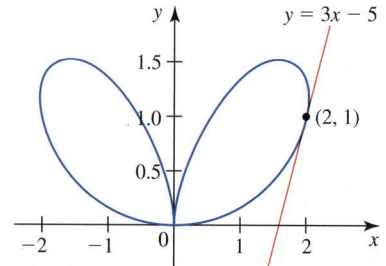

FIGURE 4
The graph of
$4x^4 + 8x^2y^2 - 25x^2y + 4y^4 = 0$

The slope of the curve at the point $(2, 1)$ is

$$\left.\frac{dy}{dx}\right|_{(2, 1)} = 3$$

Derivatives of Inverse Functions

Because of the *reflective property* of inverse functions, we might expect that f and f^{-1} have similar properties. More specifically, we might expect that if f is differentiable, then so is f^{-1}. The next theorem shows us how to compute the derivative of an inverse function, assuming that it exists.

THEOREM 1   The Derivative of an Inverse Function

Let f be differentiable on its domain and have an inverse function $g = f^{-1}$. Then the derivative of g is given by

$$g'(x) = \frac{1}{f'[g(x)]} \qquad (4)$$

provided that $f'[g(x)] \neq 0$.

The proof of Theorem 1 is given in Appendix B.

Note If we write $y = f^{-1}(x) = g(x)$, then $x = f(y)$, and we can write Equation (4) in the form

$$\frac{dy}{dx} = \frac{1}{\frac{dx}{dy}} \tag{5}$$

EXAMPLE 6 Let $f(x) = x^2$ for x in $[0, \infty)$.

a. Show that the point $(2, 4)$ lies on the graph of f.
b. Find $g'(4)$, where g is the inverse of f.

Solution
a. Since $f(2) = 4$, we conclude that the point $(2, 4)$ does lie on the graph of f.
b. Since $f'(x) = 2x$, Equation (4) gives

$$g'(4) = \frac{1}{f'[g(4)]} = \frac{1}{f'(2)} = \frac{1}{2x}\Big|_{x=2} = \frac{1}{2(2)} = \frac{1}{4}$$

Derivatives of Inverse Trigonometric Functions

The rules for differentiating the inverse trigonometric functions follow. Here, $u = g(x)$ is a differentiable function of x.

Derivatives of Inverse Trigonometric Functions

$$\frac{d}{dx}(\sin^{-1} u) = \frac{1}{\sqrt{1 - u^2}}\frac{du}{dx} \qquad \frac{d}{dx}(\csc^{-1} u) = -\frac{1}{|u|\sqrt{u^2 - 1}}\frac{du}{dx}$$

$$\frac{d}{dx}(\cos^{-1} u) = -\frac{1}{\sqrt{1 - u^2}}\frac{du}{dx} \qquad \frac{d}{dx}(\sec^{-1} u) = \frac{1}{|u|\sqrt{u^2 - 1}}\frac{du}{dx}$$

$$\frac{d}{dx}(\tan^{-1} u) = \frac{1}{1 + u^2}\frac{du}{dx} \qquad \frac{d}{dx}(\cot^{-1} u) = -\frac{1}{1 + u^2}\frac{du}{dx}$$

PROOF We will prove the first of these formulas and leave the proofs of the others as an exercise. Let $y = \sin^{-1} x$ so that $\sin y = x$ for $-\frac{\pi}{2} \le y \le \frac{\pi}{2}$. Differentiating the latter equation implicitly with respect to x, we obtain

$$(\cos y)\frac{dy}{dx} = 1$$

or

$$\frac{dy}{dx} = \frac{1}{\cos y}$$

Now $\cos y \ge 0$, since $-\frac{\pi}{2} \le y \le \frac{\pi}{2}$, so we can write

$$\cos y = \sqrt{1 - \sin^2 y} = \sqrt{1 - x^2} \qquad \text{Recall that } x = \sin y.$$

Therefore,

$$\frac{dy}{dx} = \frac{1}{\cos y} = \frac{1}{\sqrt{1 - x^2}} \qquad -1 < x < 1$$

Finally, if u is a differentiable function of x, then the Chain Rule gives

$$\frac{d}{dx} \sin^{-1} u = \frac{1}{\sqrt{1 - u^2}} \frac{du}{dx}$$

EXAMPLE 7 Find the derivative of

a. $f(x) = \cos^{-1} 3x$

b. $g(x) = \tan^{-1} \sqrt{2x + 3}$

c. $y = \sec^{-1} e^{-2x}$

Solution

a. $f'(x) = \dfrac{d}{dx} \cos^{-1} 3x$ $u = 3x$

$$= -\frac{1}{\sqrt{1 - (3x)^2}} \cdot \frac{d}{dx}(3x) = -\frac{3}{\sqrt{1 - 9x^2}}$$

b. $g'(x) = \dfrac{d}{dx} \tan^{-1}(2x + 3)^{1/2}$ $u = (2x + 3)^{1/2}$

$$= \frac{1}{1 + [(2x + 3)^{1/2}]^2} \cdot \frac{d}{dx}(2x + 3)^{1/2}$$

$$= \frac{1}{1 + 2x + 3} \cdot \frac{1}{2}(2x + 3)^{-1/2} \frac{d}{dx}(2x)$$

$$= \frac{1}{2(x + 2)\sqrt{2x + 3}}$$

c. $\dfrac{dy}{dx} = \dfrac{d}{dx} \sec^{-1} e^{-2x}$ $u = e^{-2x}$

$$= \frac{1}{e^{-2x}\sqrt{(e^{-2x})^2 - 1}} \frac{d}{dx} e^{-2x}$$

$$= \frac{-2e^{-2x}}{e^{-2x}\sqrt{e^{-4x} - 1}} = -\frac{2}{\sqrt{e^{-4x} - 1}}$$

EXAMPLE 8 **Videographing a Moving Boat** A boat is cruising at a constant speed of 20 ft/sec along a course that is parallel to a straight shoreline and 100 ft from it. A spectator standing on the shore begins to videograph the boat as soon as it passes him. Let $\theta(t)$ denote the angle of the spectator's camera at time t, where t is measured in seconds and $t = 0$ corresponds to the time that the boat passes him (see Figure 5).

a. Find an expression for θ as a function of t.

b. Find the rate at which the videographer must rotate his camera in order to keep the boat in frame.

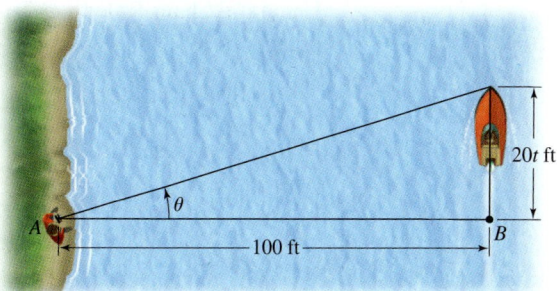

FIGURE 5
A spectator located at point A
starts videographing a boat
as it passes point B.

Solution

a. Referring to Figure 5, we find

$$\theta(t) = \tan^{-1}\left(\frac{20t}{100}\right)$$

$$= \tan^{-1}\left(\frac{t}{5}\right)$$

b. In order to keep the boat in frame, the spectator must rotate his camera at the rate given by

$$\frac{d\theta}{dt} = \frac{d}{dt}\tan^{-1}\left(\frac{t}{5}\right)$$

$$= \frac{\dfrac{d}{dt}\left(\dfrac{t}{5}\right)}{1 + \left(\dfrac{t}{5}\right)^2} = \frac{\dfrac{1}{5}}{1 + \dfrac{t^2}{25}}$$

$$= \frac{5}{25 + t^2}$$

radians/sec.   ◼

▪ Derivatives of Rational Powers of x

In Section 2.3 we proved that

$$\frac{d}{dx}(x^n) = nx^{n-1}$$

for integral values of n. Using implicit differentiation, we can now prove that this formula holds for rational powers of x. Thus, if r is a rational number, then

$$\frac{d}{dx}(x^r) = rx^{r-1}$$

PROOF   Let $y = x^r$. Since r is a rational number, it can be written in the form $r = m/n$, where m and n are integers with $n \neq 0$. Thus,

$$y = x^{m/n}$$

or

$$y^n = x^m$$

Using the Chain Rule to differentiate both sides of this equation with respect to x, we obtain

$$\frac{d}{dx}(y^n) = \frac{d}{dx}(x^m)$$

$$ny^{n-1}\frac{dy}{dx} = mx^{m-1}$$

$$\frac{dy}{dx} = \frac{m}{n}x^{m-1}y^{-n+1}$$

$$= \frac{m}{n}x^{m-1}(x^{m/n})^{-n+1} \qquad \text{Replace } y \text{ by } x^{m/n}.$$

$$= \frac{m}{n}x^{m-1}x^{-m+(m/n)}$$

$$= \frac{m}{n}x^{m-1-m+(m/n)}$$

$$= \frac{m}{n}x^{(m/n)-1} = rx^{r-1} \qquad \text{Replace } \frac{m}{n} \text{ by } r. \qquad \blacksquare$$

2.7   CONCEPT QUESTIONS

1. a. Suppose that the equation $F(x, y) = 0$ defines y as a function of x. Explain how implicit differentiation can be used to find dy/dx.
 b. What is the role of the Chain Rule in implicit differentiation?

2. Suppose that the equation $x\,g(y) + y\,f(x) = 0$, where f and g are differentiable functions, defines y as a function of x. Find an expression for dy/dx.

3. Write the derivative with respect to x of (a) $\sin^{-1} u$, (b) $\cos^{-1} u$, and (c) $\tan^{-1} u$.

2.7   EXERCISES

In Exercises 1–22, find dy/dx by implicit differentiation.

1. $2x^2 + y^2 = 4$

2. $y^2 - 3y = 2x$

3. $xy^2 + yx^2 - 2 = 0$

4. $x^2y + 2xy^2 - x + 3 = 0$

5. $x^3 - 2y^3 - y = x + 2$

6. $x^3y^2 - 2x^2y + 2x = 3$

7. $\dfrac{1}{x} + \dfrac{1}{y} = 1$

8. $\dfrac{x^3}{y} + \dfrac{y^2}{x^2} = 3$

9. $\dfrac{xy}{x^2 + y^2} = x + 1$

10. $\dfrac{x + y}{x - y} = y^2 + 1$

11. $(x + 1)^2 + (y - 2)^2 = 9$

12. $(2x^2 + 3y^2)^{5/2} = x$

13. $\sqrt{x} + \sqrt{y} = 1$

14. $\sqrt{xy} = x^2 + 2y^2$

15. $y^2 = \sin(x + y)$

16. $x + y^2 = \cos xy$

17. $\tan^2(x^3 + y^3) = xy$

18. $x = \sec 2y$

19. $\sqrt{1 + \cos^2 y} = xy$

20. $x + y^2 = \cot xy$

21. $xe^{2y} - x^3 + 2y = 5$

22. $e^{xy} - x^2 + y^2 = 5$

In Exercises 23–26, use implicit differentiation to find an equation of the tangent line to the curve at the indicated point.

23. $x^2 + 4y^2 = 4$; $\left(1, \frac{-\sqrt{3}}{2}\right)$ **24.** $x^2y + y^3 = 2$; $(-1, 1)$

25. $x^{2/3} + y^{2/3} = 2$; $(1, -1)$ **26.** $y = \sin xy$; $\left(\frac{\pi}{2}, 1\right)$

In Exercises 27–30, find the rate of change of y with respect to x at the given values of x and y.

27. $xy^2 - x^2y - 2 = 0$; $x = 1, y = -1$

28. $x^{2/3} + y^{2/3} = 5$; $x = 1, y = 8$

29. $x \csc y = 2$; $x = 1, y = \dfrac{\pi}{6}$

30. $\tan(x + 2y) - \sin x = 1$; $x = 0, y = \dfrac{\pi}{8}$

31. Find an equation of the tangent line to the curve $e^y + xy = e$ at $(0, 1)$.

32. Find an equation of the tangent line to the curve $xe^y + 2x + y = 3$ at $(1, 0)$.

In Exercises 33–36, find d^2y/dx^2 in terms of x and y.

33. $xy + x^3 = 4$

34. $x^3 - y^3 = 8$

35. $\sin x + \cos y = 1$

36. $\tan y - xy = 0$

37. Suppose that $f(x) = x^2$ for x in $[0, \infty)$, and let g be the inverse of f.
  a. Compute $g'(x)$ using Equation (4).
  b. Find $g'(x)$ by first computing $g(x)$.

38. Let $f(x) = x^{1/3}$, and let g be the inverse of f.
  a. Find $g'(x)$ using Equation (4).
  b. Find $g'(x)$ by first computing $g(x)$.

In Exercises 39–48, let g denote the inverse of the function f.
(a) Show that the point (a, b) lies on the graph of f. (b) Find $g'(b)$.

39. $f(x) = 2x + 1$; $(2, 5)$

40. $f(x) = x^3 + x + 2$; $(1, 4)$

41. $f(x) = x^5 + 2x^3 + x - 1$; $(0, -1)$

42. $f(x) = \dfrac{x + 1}{2x - 1}$; $(1, 2)$

43. $f(x) = (x^3 + 1)^3$; $(1, 8)$

44. $f(x) = 2 - \sqrt[3]{x + 1}$; $(7, 0)$

45. $f(x) = \dfrac{1}{1 + x^2}$, where $x \ge 0$; $\left(2, \frac{1}{5}\right)$

46. $f(x) = \dfrac{1}{\sqrt{x^2 + 1}}$, where $x \ge 0$; $\left(1, \frac{\sqrt{2}}{2}\right)$

47. Suppose that g is the inverse of a function f. If $f(2) = 4$ and $f'(2) = 3$, find $g'(4)$.

48. Suppose that g is the inverse of a differentiable function f and $H = g \circ g$. If $f(4) = 3$, $g(4) = 5$, $f'(4) = \frac{1}{2}$, and $f'(5) = 2$, find $H'(3)$.

In Exercises 49–72, find the derivative of the function.

49. $f(x) = \sin^{-1} 3x$

50. $g(x) = \cos^{-1}(2x - 1)$

51. $f(x) = \tan^{-1} x^2$

52. $f(t) = \sin^{-1} \sqrt{2t + 1}$

53. $g(t) = t \tan^{-1} 3t$

54. $y = \sin^{-1}\left(\dfrac{1}{x}\right)$

55. $f(u) = \sec^{-1} 2u$

56. $g(\theta) = \dfrac{\sec^{-1} \theta}{\theta}$

57. $h(x) = \sin^{-1} x + 2 \cos^{-1} x$

58. $f(x) = \sin^{-1} 2x + \cos^{-1} 3x$

59. $g(x) = \tan^{-1} x + x \cot^{-1} x$

60. $y = \sec^{-1} x + \csc^{-1} x$

61. $y = (x^2 + 1) \tan^{-1} x$

62. $f(x) = \tan^{-1} \sqrt{3x + 1}$

63. $g(t) = \tan^{-1}\left(\dfrac{t - 1}{t + 1}\right)$

64. $f(x) = \cos^{-1}(\sin 2x)$

65. $y = \tan^{-1}(\sin 2x)$

66. $h(\theta) = \tan^{-1}\left(\dfrac{\cos \theta}{2}\right)$

67. $f(x) = \sin^{-1}(e^{2x})$

68. $y = e^{\tan^{-1} 2t}$

69. $h(x) = \cot(\cos^{-1} x^2)$

70. $y = \sin^{-1}\left(\dfrac{\sin x}{1 + \cos x}\right)$

71. $f(\theta) = (\sec^{-1} \theta)^{-1}$

72. $f(x) = x \tan x \sec^{-1} x$

In Exercises 73–78, find an equation of the tangent line to the given curve at the indicated point.

73. $\dfrac{x^2}{4} + \dfrac{y^2}{9} = 1$; $\left(-1, \frac{3\sqrt{3}}{2}\right)$

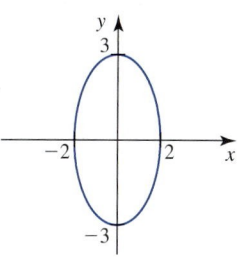

Ellipse

74. $\dfrac{x^2}{9} - \dfrac{y^2}{4} = 1$; $\left(5, \frac{8}{3}\right)$

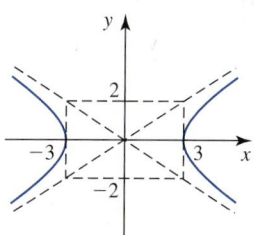

Hyperbola

75. $y^2 - xy^2 - x^3 = 0$; $\left(\frac{1}{2}, \frac{1}{2}\right)$

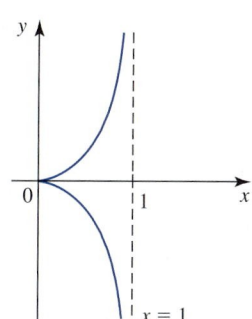

Cissoid of Diocles

76. $2y^2 - x^3 - x^2 = 0;$ $(1, 1)$

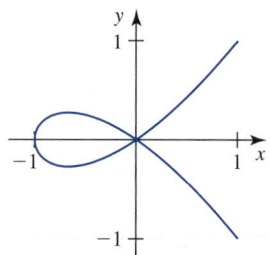

Tschirnhausen's cubic

77. $2(x^2 + y^2)^2 = 25(x^2 - y^2);$ $(3, 1)$

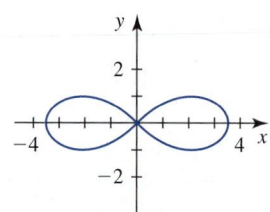

Lemniscate

78. $x^2y^2 = (y + 1)^2(4 - y^2);$ $(-2\sqrt{3}, 1)$

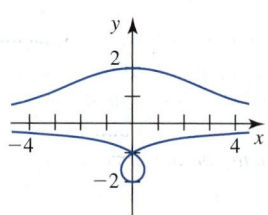

The Conchoid of Nicomedes

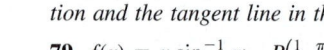 *In Exercises 79 and 80, find an equation of the tangent line to the graph of the function at the indicated point. Graph the function and the tangent line in the same viewing window.*

79. $f(x) = x \sin^{-1} x;$ $P\left(\frac{1}{2}, \frac{\pi}{12}\right)$

80. $f(x) = \sec^{-1} 2x;$ $P\left(\frac{\sqrt{2}}{2}, \frac{\pi}{4}\right)$

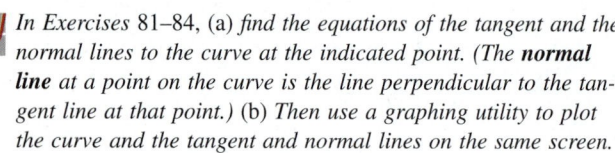 *In Exercises 81–84, (a) find the equations of the tangent and the normal lines to the curve at the indicated point. (The **normal line** at a point on the curve is the line perpendicular to the tangent line at that point.) (b) Then use a graphing utility to plot the curve and the tangent and normal lines on the same screen.*

81. $4xy - 9 = 0;$ $\left(3, \frac{3}{4}\right)$

82. $x^2 + y^2 = 9;$ $(-1, 2\sqrt{2})$

83. $4x^3 - 3xy^2 - 5xy - 8y^2 + 9x = -38;$ $(-2, 3)$

84. $x^5 - 2xy + y^5 = 0;$ $(1, 1)$

85. The graph of the equation $x^3 + y^3 = 3xy$ is called the **folium of Descartes.**

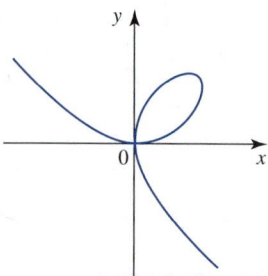

a. Find y'.
b. Find an equation of the tangent line to the folium at the point in the first quadrant where it intersects the line $y = x$.
c. Find the points on the folium where the tangent line is horizontal.

86. The curve with equation $x^{2/3} + y^{2/3} = 4$ is called an astroid. Find an equation of the tangent line to the curve at the point $(3\sqrt{3}, 1)$.

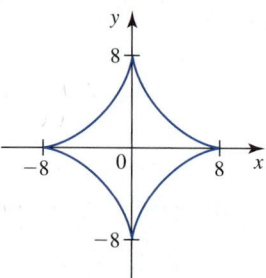

87. Water flows from a tank of constant cross-sectional area 50 ft² through an orifice of constant cross-sectional area $\frac{1}{4}$ ft² located at the bottom of the tank. Initially, the height of the water in the tank was 20 ft, and t sec later it was given by the equation

$$2\sqrt{h} + \frac{1}{25}t - 2\sqrt{20} = 0 \qquad 0 \leq t \leq 50\sqrt{20}$$

How fast was the height of the water decreasing when its height was 9 ft?

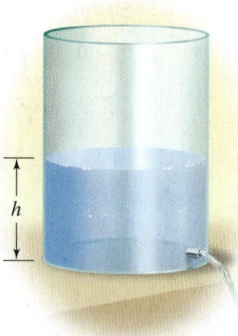

88. Watching a Rocket Launch At a distance of 2000 ft from the launch site, a spectator is observing a rocket 120-ft long being launched vertically. Let θ be her viewing angle of the rocket, and let y denote the altitude (measured in feet) of the rocket. (Neglect the height of the spectator.)

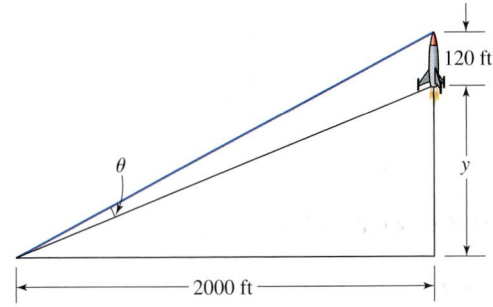

a. Show that

$$\tan \theta = \frac{240{,}000}{y^2 + 120y + 4{,}000{,}000}$$

b. What is the viewing angle when the rocket is on the launching pad? When it is at an altitude of 10,000 feet?

c. Find the rate of change of the viewing angle when the rocket is at an altitude of 10,000 feet.

d. What happens to the viewing angle when the rocket is at a very great altitude?

*Two curves are said to be **orthogonal** if their tangent lines are perpendicular at each point of intersection of the curves. In Exercises 89–92, show that the curves with the given equations are orthogonal.*

89. $x^2 + 2y^2 = 6$, $x^2 = 4y$

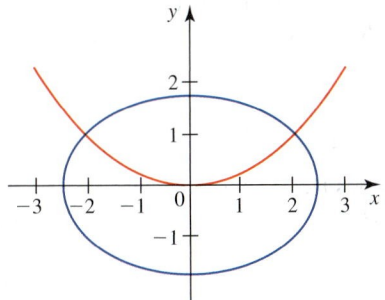

90. $x^2 - y^2 = 3$, $xy = 2$

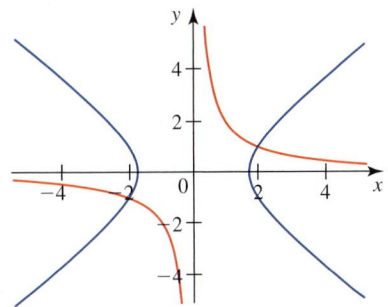

91. $x^2 + 3y^2 = 4$, $y = x^3$

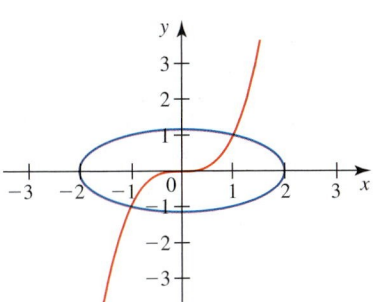

92. $y - x = \dfrac{\pi}{2}$, $x = \cos y$

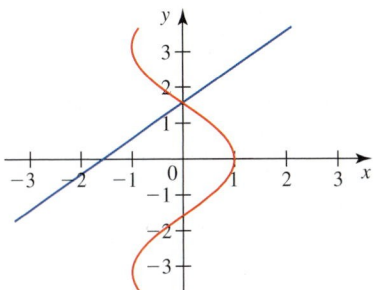

*Two families of curves are **orthogonal trajectories** of each other if every curve of one family is orthogonal to every curve in the other family. In Exercises 93–96, (a) show that the given families of curves are orthogonal to each other, and (b) sketch a few members of each family on the same set of axes.*

93. $x^2 + y^2 = c^2$, $y = kx$, c, k constants

94. $x^2 + y^2 = cx$, $x^2 + y^2 = ky$, c, k constants

95. $2x^2 + y^2 = c$, $y^2 = kx$, c, k constants

96. $9x^2 + 4y^2 = c^2$, $y^9 = kx^4$, c, k constants

97. **The Path of Steepest Descent** The contour lines of a **topographic** or **contour map** are curves that connect the contiguous points of the same altitude. The figure gives the contour map of a hill. Suppose that you start at the point A and you want to get to the point B by taking the shortest path.

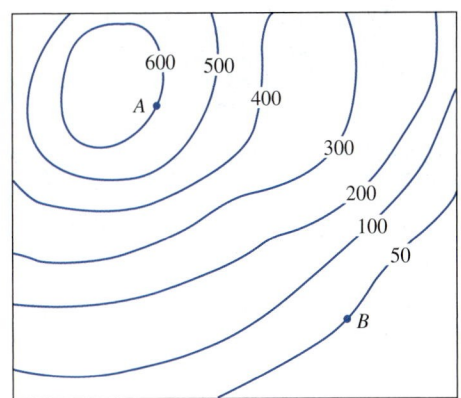

a. Explain why the direction that you start out with at A should be perpendicular to the tangent line to the contour line passing through A.

b. Using the observation made in part (a), explain why the desired path should be the curve that is orthogonal to the contour lines. Sketch this path from A to B. This path is called the *path of steepest descent*.

98. Isobars are curves on a weather map that connect points having the same air pressure. The figure shows a family of isobars.

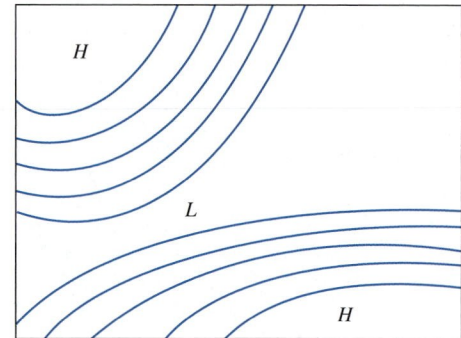

a. Sketch several members of the family of orthogonal trajectories of the family of isobars.

b. Use the fact that air flows from regions of high air pressure to those of lower air pressure to give an interpretation of the role of the orthogonal family.

99. A 20-ft ladder leaning against a wall begins to slide. How fast is the angle between the ladder and the wall changing at the instant of time when the bottom of the ladder is 12 ft from the wall and sliding away from the wall at the rate of 5 ft/sec?

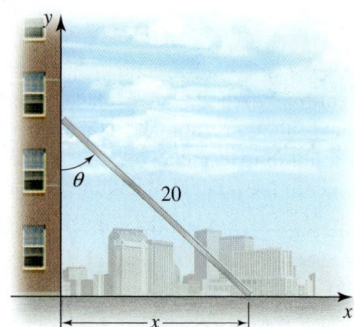

100. A trough of length L feet has a cross section in the shape of a semicircle with radius r feet. When the trough is filled with water to a level that is h feet as measured from the top of the trough, the volume of the water is

$$V = L\left[\frac{1}{2}\pi r^2 - r^2 \sin^{-1}\left(\frac{h}{r}\right) - h\sqrt{r^2 - h^2}\right]$$

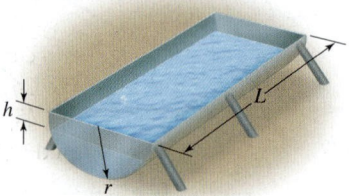

Suppose that a trough with $L = 10$ and $r = 1$ springs a leak at the bottom and that at a certain instant of time, $h = 0.4$ ft and $dV/dt = -0.2$ ft³/sec. Find the rate at which h is changing at that instant of time.

101. Verify each differentiation formula.

a. $\dfrac{d}{dx}\cos^{-1} u = -\dfrac{1}{\sqrt{1 - u^2}}\dfrac{du}{dx}$

b. $\dfrac{d}{dx}\tan^{-1} u = \dfrac{1}{1 + u^2}\dfrac{du}{dx}$

c. $\dfrac{d}{dx}\csc^{-1} u = -\dfrac{1}{|u|\sqrt{u^2 - 1}}\dfrac{du}{dx}$

d. $\dfrac{d}{dx}\sec^{-1} u = \dfrac{1}{|u|\sqrt{u^2 - 1}}\dfrac{du}{dx}$

e. $\dfrac{d}{dx}\cot^{-1} u = -\dfrac{1}{1 + u^2}\dfrac{du}{dx}$

In Exercises 102–104, determine whether the statement is true or false. If it is true, explain why it is true. If it is false, explain why or give an example to show why it is false.

102. If f and g are differentiable and $f(x)g(y) = 0$, then

$$\frac{dy}{dx} = -\frac{f'(x)g(y)}{f(x)g'(y)} \qquad f(x) \neq 0 \quad \text{and} \quad g'(y) \neq 0$$

103. If f and g are differentiable and $f(x) + g(y) = 0$, then

$$\frac{dy}{dx} = -\frac{f'(x)}{g'(y)}$$

104. $\dfrac{d}{dx}[\cos^{-1}(\cos x)] = 1$ for all x in $(0, \pi)$.

2.8 Derivatives of Logarithmic Functions

■ The Derivatives of Logarithmic Functions

The method of differentiation developed in Section 2.7 can be used to help us find the derivatives of logarithmic functions. The fact that logarithmic functions are differentiable can be demonstrated with mathematical rigor, but we will not do so here. However, if we recall that the graph of $y = \log_a x$ is the reflection of the graph of $y = a^x$ with respect to the line $y = x$, then it seems plausible that the differentiability of a^x would imply the differentiability of $\log_a x$.

To find the derivative of $y = \log_a x$, where $a > 0$, $a \neq 1$, we first recall that the equation is equivalent to

$$a^y = x$$

Differentiating the last equation implicitly with respect to x yields

$$a^y (\ln a) \frac{dy}{dx} = 1$$

so

$$\frac{dy}{dx} = \frac{1}{a^y \ln a} = \frac{1}{x \ln a}$$

provided that the derivative exists. If we put $a = e$, we obtain

$$\frac{dy}{dx} = \frac{1}{e^y \ln e} = \frac{1}{x}$$

Also, using the Chain Rule, we find that if u is a differentiable function of x, then

$$\frac{d}{dx} \ln u = \frac{1}{u} \frac{du}{dx} \qquad \text{and} \qquad \frac{d}{dx} \log_a u = \frac{1}{u \ln a} \frac{du}{dx}$$

Let's summarize these results.

THEOREM 1   The Derivatives of Logarithmic Functions

Let u be a differentiable function of x, and let $a > 0$, where $a \neq 1$. Then

a. $\dfrac{d}{dx} \ln x = \dfrac{1}{x}$

b. $\dfrac{d}{dx} \ln u = \dfrac{1}{u} \dfrac{du}{dx}$

c. $\dfrac{d}{dx} \log_a x = \dfrac{1}{x \ln a}$

d. $\dfrac{d}{dx} \log_a u = \dfrac{1}{u \ln a} \dfrac{du}{dx}$

EXAMPLE 1   Find the derivative of

a. $f(x) = \ln(2x^2 + 1)$ **b.** $g(x) = x^2 \log 2x$ **c.** $y = \ln \cos x$

Solution

a. $f'(x) = \dfrac{d}{dx} \ln(2x^2 + 1) = \dfrac{1}{2x^2 + 1} \dfrac{d}{dx} (2x^2 + 1) = \dfrac{4x}{2x^2 + 1}$

Historical Biography

JOHN NAPIER
(1550–1617)

John Napier is famous for his invention of the logarithm, which was described in two of his publications: *Mirifici logarithmorum canonis descriptio* ("A Description of the Wonderful Canon of Logarithms"), published in 1614, and *Mirifici logarithmorum canonis constructio* ("The Construction of the Wonderful Canon of Logarithms"), published in 1619. Born in 1550 at Merchiston Castle near Edinburgh, Scotland, Napier came from a line of influential noblemen. At 13 years of age he entered the University of St. Andrews in Scotland, but he left after a short time to study in Europe. It was during this time that he developed a passion for astronomy and mathematics, but he considered these pursuits a hobby, as theology was his main interest. However, astronomy so intrigued him that over the course of two decades he developed logarithms to work with the calculation of the extremely large numbers that he needed to do research in that area. Later, with Napier's consent, Henry Briggs made improvements to Napier's logarithms, such as using base 10. Napier and Briggs's important work was essential to Johannes Kepler's (page 885) study of planetary motion and therefore ultimately to the work of Isaac Newton (page 202). The work done by Napier and Briggs also led to the standard form of the logarithmic tables that remained in common use until the electronic age of calculators and computers.

b. $\displaystyle g'(x) = \frac{d}{dx}(x^2 \log 2x) = x^2 \frac{d}{dx}(\log 2x) + (\log 2x)\frac{d}{dx}(x^2)$ Use the Product Rule.

$$= x^2\left[\frac{1}{(2x)\ln 10}\right]\frac{d}{dx}(2x) + (\log 2x)(2x) = x\left(\frac{1}{\ln 10} + 2\log 2x\right)$$

c. $\displaystyle \frac{dy}{dx} = \frac{d}{dx}\ln \cos x = \frac{1}{\cos x}\frac{d}{dx}(\cos x) = -\frac{\sin x}{\cos x} = -\tan x$ ■

EXAMPLE 2 Find the derivative of $y = \ln(e^{2x} + e^{-2x})$

Solution Using the rule for differentiating a logarithmic function gives

$$\frac{dy}{dx} = \frac{d}{dx}\ln(e^{2x} + e^{-2x})$$

$$= \frac{1}{e^{2x} + e^{-2x}}\frac{d}{dx}(e^{2x} + e^{-2x})$$

$$= \frac{1}{e^{2x} + e^{-2x}}(2e^{2x} - 2e^{-2x})$$

$$= \frac{2(e^{2x} - e^{-2x})}{e^{2x} + e^{-2x}}$$ ■

If an expression contains a logarithm, it may be helpful to use the laws of logarithms to simplify the expression *before* differentiating, as illustrated in Examples 3 and 4.

EXAMPLE 3 Find the derivative of $f(x) = \ln \sqrt{x^2 + 1}$.

Solution We first rewrite the given expression as

$$f(x) = \ln(x^2 + 1)^{1/2} = \frac{1}{2}\ln(x^2 + 1)$$

Differentiating this function, we obtain

$$f'(x) = \frac{d}{dx}\left[\frac{1}{2}\ln(x^2 + 1)\right] = \frac{1}{2}\frac{d}{dx}[\ln(x^2 + 1)]$$

$$= \frac{1}{2}\cdot\frac{1}{x^2 + 1}\frac{d}{dx}(x^2 + 1) = \frac{1}{2}\cdot\frac{1}{x^2 + 1}(2x) = \frac{x}{x^2 + 1}$$ ■

EXAMPLE 4 Find the rate of change of

$$f(x) = \ln\left[\frac{x^2(2x^2 + 1)^3}{\sqrt{5 - x^2}}\right]$$

when $x = 1$.

Solution The rate of change of $f(x)$ for any value of x is given by $f'(x)$. To find $f'(x)$, we first rewrite

$$f(x) = \ln\left[\frac{x^2(2x^2 + 1)^3}{(5 - x^2)^{1/2}}\right] = 2\ln x + 3\ln(2x^2 + 1) - \frac{1}{2}\ln(5 - x^2)$$

Then we have

$$f'(x) = \frac{2}{x} + \frac{3}{2x^2 + 1} \frac{d}{dx}(2x^2 + 1) - \frac{1}{2(5 - x^2)} \frac{d}{dx}(5 - x^2)$$

$$= \frac{2}{x} + \frac{12x}{2x^2 + 1} + \frac{x}{5 - x^2}$$

from which we see that the rate of change of $f(x)$ at $x = 1$ is

$$f'(1) = 2 + \frac{12}{3} + \frac{1}{4}$$

or $\frac{25}{4}$ units per unit change in x.   ■

■ Logarithmic Differentiation

Having seen how the laws of logarithms can help simplify the work involved in differentiating logarithmic expressions, we now look at a procedure that takes advantage of these same laws to help us differentiate functions that at first blush do not necessarily involve logarithms. This method, called **logarithmic differentiation,** is especially useful for differentiating functions involving products, quotients, and/or powers that can be simplified by using logarithms.

EXAMPLE 5 Find the derivative of $y = \dfrac{(2x - 1)^3}{\sqrt{3x + 1}}$.

Solution   We begin by taking the logarithm on both sides of the equation, getting

$$\ln y = \ln \frac{(2x - 1)^3}{(3x + 1)^{1/2}}$$

or

$$\ln y = 3 \ln(2x - 1) - \frac{1}{2} \ln(3x + 1) \qquad \text{Use the laws of logarithms.}$$

Next, we differentiate implicitly with respect to x, obtaining

$$\frac{1}{y}(y') = \frac{3}{2x - 1}(2) - \frac{1}{2(3x + 1)}(3)$$

$$= \frac{6}{2x - 1} - \frac{3}{2(3x + 1)} = \frac{6(2)(3x + 1) - 3(2x - 1)}{2(2x - 1)(3x + 1)}$$

$$= \frac{15(2x + 1)}{2(2x - 1)(3x + 1)}$$

Multiplying both sides of this equation by y gives

$$y' = \frac{15(2x + 1)}{2(2x - 1)(3x + 1)} \cdot y$$

$$= \frac{15(2x + 1)}{2(2x - 1)(3x + 1)} \cdot \frac{(2x - 1)^3}{(3x + 1)^{1/2}} \qquad \text{Substitute for } y.$$

$$= \frac{15(2x + 1)(2x - 1)^2}{2(3x + 1)^{3/2}}$$ ■

Here is a summary of this procedure.

> **Finding dy/dx by Logarithmic Differentiation**
>
> Suppose we are given the equation $y = f(x)$. To compute dy/dx:
>
> **1.** Take the logarithm of both sides of the equation, and use the laws of logarithms to simplify the resulting equation.
> **2.** Differentiate implicitly with respect to x.
> **3.** Solve the equation found in Step 2 for dy/dx.
> **4.** Substitute for y.

The General Version of the Power Rule

As was promised in Section 2.2, we will now prove that the Power Rule holds for all exponents (Theorem 3). But before we prove this, we need the following result.

If $f(x) = \ln|x|$, where $x \neq 0$, then

$$f'(x) = \frac{d}{dx} \ln|x| = \frac{1}{x} \tag{1}$$

PROOF   We have

$$f(x) = \ln|x| = \begin{cases} \ln x & \text{if } x > 0 \\ \ln(-x) & \text{if } x < 0 \end{cases}$$

So

$$f'(x) = \begin{cases} \dfrac{1}{x} & \text{if } x > 0 \\ \dfrac{-1}{-x} = \dfrac{1}{x} & \text{if } x < 0 \end{cases}$$

We now prove the Power Rule for all real exponents. If n is any real number, then

$$\frac{d}{dx}(x^n) = nx^{n-1}$$

PROOF   Let $y = x^n$. Then

$$\ln|y| = \ln|x|^n = n \ln|x|$$

Using the Chain Rule and Equation (1), we obtain

$$\frac{y'}{y} = \frac{n}{x}$$

or

$$y' = \frac{ny}{x} = \frac{nx^n}{x} = nx^{n-1}$$

■ The Number e as a Limit

In Section 0.8 we mentioned that $e \approx 2.71828$, correct to five decimal places. We are now in the position to give the exact value of e, albeit in the form of a limit. If we use the definition of the derivative as a limit to compute $f'(1)$, where $f(x) = \ln x$, we obtain

$$f'(1) = \lim_{h \to 0} \frac{f(1 + h) - f(1)}{h}$$

$$= \lim_{h \to 0} \frac{\ln(1 + h) - \ln 1}{h} = \lim_{h \to 0} \frac{\ln(1 + h)}{h} \qquad \color{red}{\ln 1 = 0}$$

$$= \lim_{h \to 0} \ln(1 + h)^{1/h}$$

$$= \ln\left[\lim_{h \to 0} (1 + h)^{1/h}\right] \qquad \color{red}{\text{Use the continuity of ln.}}$$

But

$$f'(1) = \left[\frac{d}{dx} \ln x\right]_{x=1} = \left[\frac{1}{x}\right]_{x=1} = 1$$

so

$$\ln\left[\lim_{h \to 0} (1 + h)^{1/h}\right] = 1$$

or

$$\lim_{h \to 0} (1 + h)^{1/h} = e \qquad\qquad (2)$$

Table 1 shows that $e \approx 2.71828$, correct to five decimal places, as was mentioned earlier.

TABLE 1 Table of values of $(1 + x)^{1/x}$

x	$(1 + x)^{1/x}$	x	$(1 + x)^{1/x}$
-0.1	2.867972	0.1	2.593742
-0.01	2.732000	0.01	2.704814
-0.001	2.719642	0.001	2.716924
-0.0001	2.718418	0.0001	2.718146
-0.00001	2.718295	0.00001	2.718268
-0.000001	2.718283	0.000001	2.718280

2.8 CONCEPT QUESTIONS

1. State the rule for differentiating (a) $f(x) = \ln x$ and (b) $f(x) = \log_a x$.
2. Let u be a differentiable function of x. State the rule for differentiating (a) $f(x) = \ln u$ and (b) $f(x) = \log_a u$.
3. **a.** If $f(x) = \ln|x|$, what is $f'(x)$?
   b. If $f(x) = \log_a|x|$, where $a > 0$, $a \neq 1$, what is $f'(x)$?
4. Give the steps used in logarithmic differentiation.
5. Give a definition of the number e as a limit.

2.8 EXERCISES

In Exercises 1–26, differentiate the function.

1. $f(x) = \ln(2x + 3)$

2. $g(x) = \ln(x^2 + 4)^2$

3. $h(x) = \ln\sqrt{x}$

4. $y = \sqrt{\ln x}$

5. $g(u) = \ln\dfrac{u}{u + 1}$

6. $g(t) = t \ln 2t$

7. $y = x(\ln x)^2$

8. $f(x) = \ln\left(x + \sqrt{x^2 - 1}\right)$

9. $g(x) = \dfrac{\ln x}{x + 1}$

10. $y = \ln\left(\dfrac{x - 1}{x + 1}\right)^{2/3}$

11. $f(x) = \ln(\ln x)$

12. $h(t) = \dfrac{\ln t}{\ln 2t}$

13. $f(x) = \ln(x \ln x)$

14. $f(x) = \ln[x \ln(x + 2)]$

15. $g(x) = \sin(\ln x)$

16. $h(t) = t \sin(\ln 2t)$

17. $f(x) = x^2 \ln \cos x$

18. $g(\theta) = \ln|\tan 3\theta|$

19. $h(u) = \ln|\sec u|$

20. $f(x) = \sec[\ln(2x + 3)]$

21. $g(t) = \ln\left|\dfrac{\sin t + 1}{\cos t + 2}\right|$

22. $g(x) = \ln\sqrt{\dfrac{x \cos x}{(2x + 1)^3}}$

23. $f(x) = \log_2(x^2 + x + 1)$

24. $h(x) = \log_3|2x - 1|$

25. $f(t) = \log\sqrt{t^2 + 1}$

26. $y = x^2 \log_2\sqrt{x^2 - 1}$

27. Find an equation of the tangent line to the graph of $y = x \ln x$ at $(1, 0)$.

28. Find an equation of the tangent line to the curve $y - \ln(x^2 + y^2) = 0$ at $(1, 0)$.

In Exercises 29–40, use logarithmic differentiation to find the derivative of the function.

29. $y = (2x + 1)^2(3x^2 - 4)^3$

30. $y = \dfrac{x^2\sqrt{2x - 4}}{(x + 1)^2}$

31. $y = \sqrt[3]{\dfrac{x - 1}{x^2 + 1}}$

32. $y = \dfrac{\sin^2 x}{x^2\sqrt{1 + \tan x}}$

33. Find y'' if $y = x^x$.

34. Find y' if $y = x^{x^x}$.

35. $y = 3^x$

36. $y = x^{x^2}$

37. $y = (x + 2)^{1/x}$

38. $y = (x^2 + x)^{\sqrt{x}}$

39. $y = (\sqrt{\cos x})^x$

40. $y = \sin x^{\tan x}$

In Exercises 41–44, use implicit differentiation to find dy/dx.

41. $\ln y - x \ln x = -1$

42. $\ln xy - y^2 = 5$

43. $\tan^{-1}\left(\dfrac{y}{x}\right) - \ln\sqrt{x^2 + y^2} = 0$

44. $\ln(x + y) - \cos y - x^2 = 0$

45. Flight of a Rocket A rocket having mass M kg and carrying fuel of mass m kg takes off vertically from the earth's surface. The fuel is burned at the constant rate of a kg/sec, and

the gas is expelled at a constant velocity of b m/sec relative to the rocket, where $a > 0$ and $b > 0$. If the external force acting on the rocket is a constant gravitational field, then the height of the rocket t seconds after liftoff is

$$x = bt + \dfrac{b}{a}(M + m - at)\ln\left(\dfrac{M + m - at}{M + m}\right) - \dfrac{1}{2}gt^2$$

$$0 \leq t \leq \tfrac{m}{a}$$

a. Find expressions for the velocity and acceleration of the rocket at any time t after liftoff.

b. What are the velocity and acceleration of the rocket at burnout (that is, when $t = m/a$).

46. Distance Traveled by a Motorboat The distance x (in feet) traveled by a motorboat moving in a straight line t sec after the engine of the moving boat has been cut off is given by

$$x = \dfrac{1}{k}\ln(v_0 kt + 1)$$

where k is a constant and v_0 is the speed of the boat at $t = 0$.

a. Find expressions for the velocity and acceleration of the boat at any time t after the engine has been cut off.

b. Show that the acceleration of the boat is in the direction opposite to that of its velocity and is directly proportional to the square of its velocity.

c. Use the results of part (a) to show that the velocity of the boat after traveling a distance of x ft is given by

$$v = v_0 e^{-kx}$$

47. Strain on Vertebrae The strain (percentage of compression) on the lumbar vertebral disks in an adult human as a function of the load x (in kilograms) is given by

$$f(x) = 7.2956 \ln(0.0645012x^{0.95} + 1)$$

What is the rate of change of the strain with respect to the load when the load is 100 kg? When the load is 500 kg?
Source: Benedek and Villars, *Physics with Illustrative Examples from Medicine and Biology.*

48. Predator-Prey Model The relationship between the number of rabbits $y(t)$ and the number of foxes $x(t)$ at any time t is given by

$$-C \ln y + Dy = A \ln x - Bx + E$$

where A, B, C, D, and E are constants. This relationship is based on a model by Lotka (1880–1949) and Volterra (1860–1940) for analyzing the ecological balance between two species of animals, one of which is a prey species and the other of which is a predator species. Use implicit differentiation to find the relationship between the rate of change of the rabbit population in terms of the rate of change of the fox population.

49. Force Exerted by an Electric Charge An electric charge Q is distributed uniformly along a line of length $2a$, lying along the y-axis, as shown in the figure. A point charge q lies on the x-axis, at a distance x from the origin. It can be shown that the magnitude of the total force F that Q exerts on q (in the direction of the x-axis) is $F = -q\,dV/dx$, where

$$V(x) = \frac{1}{4\pi\varepsilon_0}\frac{Q}{2a}\ln\frac{\sqrt{a^2+x^2}+a}{\sqrt{a^2+x^2}-a} \qquad \varepsilon_0, \text{ a constant}$$

Show that

$$F = \frac{qQ}{4\pi\varepsilon_0}\frac{1}{x\sqrt{x^2+a^2}}$$

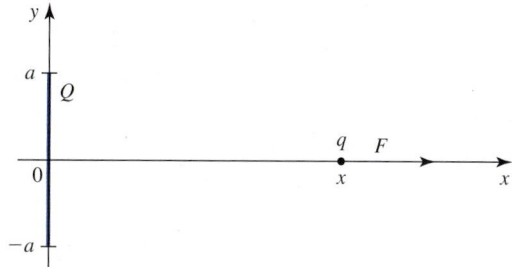

A line of charge with length $2a$ and total charge Q exerts an electrostatic force on the point charge q.

50. Rate of a Catalytic Chemical Reaction A catalyst is a substance that either accelerates a chemical reaction or is necessary for the reaction to occur. Suppose that an enzyme E (a catalyst) combines with a substrate S (a reacting chemical) to form an intermediate product X, which then produces a product P and releases the enzyme. If initially there are x_0 moles per liter of S and there is no P, then on the basis of the theory of Michaelis and Menten, the concentration of P, $p(t)$, after t hours is given by the equation

$$Vt = p - k\ln\left(1 - \frac{p}{x_0}\right)$$

where the constant V is the maximum possible speed of the reaction and the constant k is called the **Michaelis constant** for the reaction. Find the rate of change of the formation of the product P in this reaction.

51. Atmospheric Pressure In the troposphere (lower part of the atmosphere), the atmospheric pressure p is related to the height y from the earth's surface by the equation

$$\ln\left(\frac{p}{p_0}\right) = \frac{Mg}{R\alpha}\ln\left(\frac{T_0 - \alpha y}{T_0}\right)$$

where p_0 is the pressure at the earth's surface, T_0 is the temperature at the earth's surface, M is the molecular mass for air, g is the constant of acceleration due to gravity, R is the ideal gas constant, and α is called the lapse rate of temperature.

a. Find p for $y = 8882$ m (the altitude at the summit of Mount Everest), taking $M = 28.8 \times 10^{-3}$ kg/mol, $T_0 = 300$ K, $g = 9.8$ m/sec^2, $R = 8.314$ J/mol $\cdot$ K, and $\alpha = 0.006$ K/m. Explain why mountaineers experience difficulty in breathing at very high altitudes.

b. Find the rate of change of the atmospheric pressure with respect to altitude when $y = 8882$ m.
Hint: Use logarithmic differentiation.

In Exercises 52–56, determine whether the statement is true or false. If it is true, explain why it is true. If it is false, give an example to show why it is false.

52. The function $f(x) = 1/(\ln x)$ is continuous on $(1, \infty)$.

53. If $f(x) = \ln 5$, then $f'(x) = \frac{1}{5}$.

54. $\displaystyle\lim_{x\to 0}\frac{a^x - 1}{x} = \ln a$, where $a > 0$

55. $\displaystyle\frac{d}{dx}\log_a \sqrt{x} = \frac{1}{(\ln a)\sqrt{x}}$

56. $\displaystyle\lim_{x\to 0}\frac{\log(3 + x) - \log 3}{x} = \frac{1}{3\ln 10}$

2.9 Related Rates

■ Related Rates Problems

The following is a typical related rates problem: Suppose that x and y are two quantities that depend on a third quantity t and that we know the relationship between x and y in the form of an equation. Can we find a relationship between dx/dt and dy/dt? In particular, if we know one of the rates of change at a specific value of t, say, dx/dt, can we find the other rate, dy/dt, at that value of t?

As an example, consider this problem from the field of aviation: Suppose that $x(t)$ and $y(t)$ describe the x- and y-coordinates at time t of a plane pulling out of a shallow dive (Figure 1). The flight path of the plane is described by the equation

$$y^2 - x^2 = 160{,}000 \tag{1}$$

where x and y are both measured in feet.

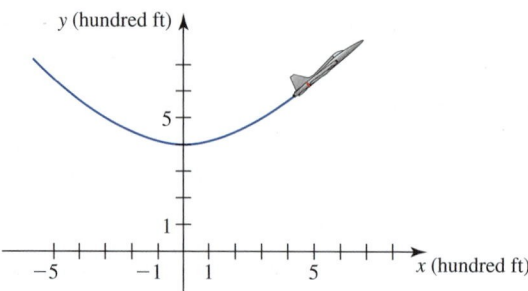

FIGURE 1
The flight path of a plane
pulling out of a shallow dive

Suppose that x and y are both differentiable functions of t, where t is measured in seconds. Then differentiating both sides of Equation (1) implicitly with respect to t, we obtain

$$2y\frac{dy}{dt} - 2x\frac{dx}{dt} = 0$$

giving a relationship between the variables x and y and their rates of change dx/dt and dy/dt. Now suppose that $dx/dt = 500$ at the point where $x = 300$ and $y = 500$. At that instant of time,

$$2(500)\frac{dy}{dt} - 2(300)(500) = 0$$

or $dy/dt = 300$. This says that the plane's altitude is increasing at the rate of 300 ft/sec.

■ Solving Related Rates Problems

In the last example we were given the relationship between x and y in the form of an equation. In certain related rates problems we must first identify the variables and then find a relationship between them before solving the problem. The following guidelines can be used to solve these problems.

> **Guidelines for Solving a Related Rates Problem**
> 1. Draw a diagram, and label the variable quantities.
> 2. Write down the *given* values of the variables and their rates of change with respect to t.
> 3. Find an equation that relates the variables.
> 4. Differentiate both sides of this equation implicitly with respect to t.
> 5. Replace the variables and derivative in the resulting equation by the values found in Step 2, and solve this equation for the required rate of change.

EXAMPLE 1 **The Speed of a Rocket During Liftoff** At a distance of 12,000 feet from the launch site, a spectator is observing a rocket being launched vertically. What is the speed of the rocket at the instant when the distance of the rocket from the spectator is 13,000 ft and is increasing at the rate of 480 ft/sec?

Solution

Step 1 Let $y =$ the altitude of the rocket and $z =$ the distance of the rocket from the spectator at any time t. (See Figure 2.)

Historical Biography

BLAISE PASCAL
(1623–1662)

A great mathematician who was not acknowledged in his lifetime, Blaise Pascal came extremely close to discovering calculus before Leibniz (page 179) and Newton (page 202), the two people who are most commonly credited with the discovery. Pascal was something of a prodigy and published his first important mathematical discovery at the age of sixteen. The work consisted of only a single printed page, but it contained a vital step in the development of projective geometry and a proposition called *Pascal's mystic hexagram* that discussed a property of a hexagon inscribed in a conic section. Pascal's interests varied widely, and from 1642 to 1644 he worked on the first manufactured calculator, which he designed to help his father with his tax work. Pascal manufactured about 50 of the machines, but they proved too costly to continue production. The basic principle of Pascal's calculating machine was still used until the electronic age. Pascal and Pierre de Fermat (page 348) also worked on the mathematics in games of chance and laid the foundation for the modern theory of probability. Pascal's later work, *Treatise on the Arithmetical Triangle*, gave important results on the construction that would later bear his name, *Pascal's Triangle*.

Step 2 We are given that at a certain instant of time

$$z = 13,000 \qquad \text{and} \qquad \frac{dz}{dt} = 480$$

and are asked to find dy/dt at that time.

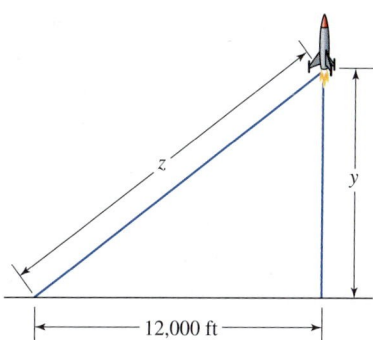

FIGURE 2
We want to find the speed of
the rocket when $z = 13,000$ ft
and $dz/dt = 480$ ft/sec.

12,000 ft

Step 3 Applying the Pythagorean Theorem to the right triangle in Figure 2, we find that

$$z^2 = y^2 + 12,000^2 \qquad (2)$$

Step 4 Differentiating Equation (2) implicitly with respect to t, we obtain

$$2z \frac{dz}{dt} = 2y \frac{dy}{dt} \qquad (3)$$

Step 5 Using Equation (2) we see that if $z = 13,000$, then

$$y = \sqrt{13,000^2 - 12,000^2} = 5000$$

Finally, substituting $z = 13,000$, $y = 5000$, and $dz/dt = 480$ in Equation (3), we find

$$2(13,000)(480) = 2(5000) \frac{dy}{dt} \qquad \text{and} \qquad \frac{dy}{dt} = 1248$$

Therefore, the rocket is rising at the rate of 1248 ft/sec. ■

⚠ Don't replace the variables in Equation (2) found in Step 3 by their values before differentiating this equation. Look at Steps 3–5 in Example 1 once again, and make sure you understand that this substitution takes place *after* the differentiation.

EXAMPLE 2 **Televising a Rocket Launch** A major network is televising the launching of the rocket described in Example 1. A camera tracking the liftoff of the rocket is located at point A, as shown in Figure 3, where ϕ denotes the angle of elevation of the camera at A. When the rocket is 13,000 ft from the camera and this distance is increasing at the rate of 480 ft/sec, how fast is ϕ changing?

Solution We are given that at a certain instant of time,

$$z = 13,000 \qquad \text{and} \qquad \frac{dz}{dt} = 480$$

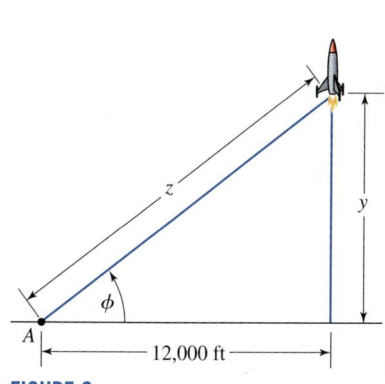

FIGURE 3
A television camera tracking a rocket
launch

and are asked to find $d\phi/dt$ at that time. From Figure 3 we see that

$$\cos \phi = \frac{12{,}000}{z}$$

Differentiating this equation implicitly with respect to t, we obtain

$$(-\sin \phi)\frac{d\phi}{dt} = -\frac{12{,}000}{z^2} \cdot \frac{dz}{dt} \qquad (4)$$

Now when $z = 13{,}000$, we find that $y = 5000$ (the same value that was obtained in Example 1). Therefore, at this instant of time,

$$\sin \phi = \frac{5{,}000}{13{,}000} = \frac{5}{13}$$

Finally, substituting $z = 13{,}000$, $\sin \phi = 5/13$, and $dz/dt = 480$ into Equation (4), we obtain

$$-\frac{5}{13}\frac{d\phi}{dt} = -\frac{12{,}000}{13{,}000^2}(480)$$

from which we deduce that

$$\frac{d\phi}{dt} \approx 0.0886$$

Therefore, the angle of elevation of the camera is increasing at the rate of approximately 0.09 rad/sec, or about 5°/sec.   ▪

EXAMPLE 3   Water is poured into a conical funnel at the constant rate of 1 in.^{3}/sec and flows out at the rate of $\frac{1}{2}$ in.^{3}/sec (Figure 4a). The funnel is a right circular cone with a height of 4 in. and a radius of 2 in. at the base (Figure 4b). How fast is the water level changing when the water is 2 in. high?

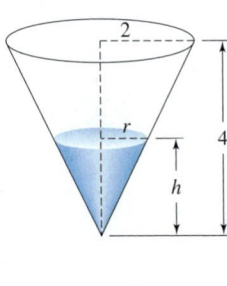

(a) Water is poured into a conical funnel.

(b) We want to find the rate at which the water level is rising when $h = 2$.

FIGURE 4

Solution

Step 1   Let

$$V = \text{the volume of the water in the funnel}$$

$$h = \text{the height of the water in the funnel}$$

and

$$r = \text{the radius of the surface of the water in the funnel}$$

at any time t (in seconds).

Step 2   We are given that

$$\frac{dV}{dt} = 1 - \frac{1}{2} = \frac{1}{2} \qquad \text{Rate of flow in minus rate of flow out}$$

and are asked to find dh/dt when $h = 2$.

Step 3   The volume of water in the funnel is equal to the volume of the shaded cone in Figure 4b. Thus,

$$V = \frac{1}{3}\pi r^2 h$$

but we need to express V in terms of h alone. To do this, we use similar triangles and deduce that

$$\frac{r}{h} = \frac{2}{4} \qquad \text{or} \qquad r = \frac{h}{2} \qquad \text{Ratio of corresponding sides}$$

Substituting this value of r into the expression for V, we obtain

$$V = \frac{1}{3}\pi\left(\frac{h}{2}\right)^2 h = \frac{1}{12}\pi h^3$$

Step 4   Differentiating this last equation implicitly with respect to t, we obtain

$$\frac{dV}{dt} = \frac{1}{4}\pi h^2 \frac{dh}{dt}$$

Step 5   Finally, substituting $dV/dt = \frac{1}{2}$ and $h = 2$ into this equation gives

$$\frac{1}{2} = \frac{1}{4}\pi(2^2)\frac{dh}{dt}$$

or

$$\frac{dh}{dt} = \frac{1}{2\pi} \approx 0.159$$

and we see that the water level is rising at the rate of 0.159 in./sec.   ■

EXAMPLE 4   A passenger ship and an oil tanker left port sometime in the morning; the former headed north, and the latter headed east. At noon the passenger ship was 40 mi from port and moving at 30 mph, while the oil tanker was 30 mi from port and moving at 20 mph. How fast was the distance between the two ships changing at that time?

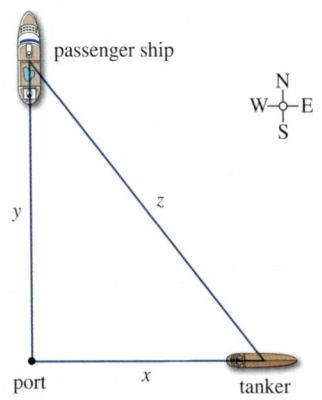

passenger ship

N
W–◇–E
S

y

z

port   x   tanker

FIGURE 5
We want to find dz/dt, the rate at which the distance between the two ships is changing at a certain instant of time.

Solution

Step 1   Let

$$x = \text{the distance of the oil tanker from port}$$

$$y = \text{the distance of the passenger ship from port}$$

and

$$z = \text{the distance between the two ships}$$

(See Figure 5.)

Step 2   We are given that at noon,

$$x = 30, \qquad y = 40, \qquad \frac{dx}{dt} = 20, \qquad \text{and} \qquad \frac{dy}{dt} = 30$$

and we are required to find dz/dt at that time.

Step 3   Applying the Pythagorean Theorem to the right triangle in Figure 5, we find that

$$z^2 = x^2 + y^2 \qquad\qquad (5)$$

Step 4   Differentiating Equation (5) implicitly with respect to t, we obtain

$$2z\frac{dz}{dt} = 2x\frac{dx}{dt} + 2y\frac{dy}{dt}$$

or

$$z\frac{dz}{dt} = x\frac{dx}{dt} + y\frac{dy}{dt}$$

Step 5   Using Equation (5) with $x = 30$ and $y = 40$, we have

$$z^2 = 30^2 + 40^2 = 2500$$

or $z = 50$. Finally substituting $x = 30$, $y = 40$, $z = 50$, $dx/dt = 20$, and $dy/dt = 30$ into the last equation of Step 4, we find

$$50\frac{dz}{dt} = (30)(20) + (40)(30)$$

and

$$\frac{dz}{dt} = 36$$

Therefore, at noon on the day in question, the ships are moving apart at the rate of 36 mph.   ■

2.9   CONCEPT QUESTIONS

1. What is a related rates problem?

2. Give the steps involved in solving a related rates problem.

2.9 EXERCISES

In Exercises 1–6, an equation relating the variables x and y, the values of x and y, and the value of either dx/dt or dy/dt at a particular instant of time are given. Find the value of the rate of change that is not specified.

1. $x^2 + y^2 = 25; \quad x = 3, y = -4, \dfrac{dx}{dt} = 2; \quad \dfrac{dy}{dt} = ?$

2. $y^3 - 2x^3 = -10; \quad x = 1, y = -2, \dfrac{dy}{dt} = -1; \quad \dfrac{dx}{dt} = ?$

3. $x^2 y = 8; \quad x = 2, y = 2, \dfrac{dx}{dt} = 3; \quad \dfrac{dy}{dt} = ?$

4. $y^2 + xy + x^2 - 1 = 0; \quad x = 1, y = -1, \dfrac{dy}{dt} = -2; \quad \dfrac{dx}{dt} = ?$

5. $\sin^2 x + \cos y = 1; \quad x = \dfrac{\pi}{4}, y = \dfrac{\pi}{3}, \dfrac{dx}{dt} = \dfrac{\sqrt{3}}{2}; \quad \dfrac{dy}{dt} = ?$

6. $4x \cos y - \pi \tan x = 0; \quad x = \dfrac{\pi}{6}, y = \dfrac{\pi}{6}, \dfrac{dx}{dt} = 1; \quad \dfrac{dy}{dt} = ?$

7. The volume V of a cube with sides of length x inches is changing with respect to time t (in seconds).
   a. Find a relationship between dV/dt and dx/dt.
   b. When the sides of the cube are 10 in. long and increasing at the rate of 0.5 in./sec, how fast is the volume of the cube increasing?

8. The volume of a right circular cylinder of radius r and height h is $V = \pi r^2 h$. Suppose that the radius and height of the cylinder are changing with respect to time t.
   a. Find a relationship between dV/dt, dr/dt, and dh/dt.
   b. At a certain instant of time, the radius and height of the cylinder are 2 in. and 6 in. and are increasing at the rate of 0.1 in./sec and 0.3 in./sec, respectively. How fast is the volume of the cylinder increasing?

9. A point moves along the curve $2x^2 - y^2 = 2$. When the point is at $(3, -4)$, its x-coordinate is increasing at the rate of 2 units per second. How fast is its y-coordinate changing at that instant of time?

10. A point moves along the curve $3y + 4y^2 + 3x = 4$. When the point is at $(1, -1)$, its x-coordinate is increasing at the rate of 3 units per second. How fast is its y-coordinate changing at that instant of time?

11. **Motion of a Particle** A particle moves along the curve defined by $y = \frac{1}{6}x^3 - x$. Determine the values of x at which the rate of change of its y-coordinate is (a) less than, (b) equal to, and (c) greater than that of its x-coordinate.

12. **Rectilinear Motion** The velocity of a particle moving along the x-axis is proportional to the square root of the distance, x, covered by the particle. Show that the force acting on the particle is constant.
    Hint: Use Newton's Second Law of Motion, which states that the force is proportional to the rate of change of momentum.

13. **Oil Spill** In calm waters, the oil spilling from the ruptured hull of a grounded tanker spreads in all directions. Assuming that the polluted area is circular, determine how fast the area is increasing when the radius of the circle is 60 ft and is increasing at the rate of $\frac{1}{2}$ ft/sec?

14. **Blowing a Soap Bubble** Carlos is blowing air into a spherical soap bubble at the rate of 8 cm³/sec. How fast is the radius of the bubble changing when the radius is 10 cm? How fast is the surface area of the bubble changing at that time?

15. If a spherical snowball melts at a rate that is proportional to its surface area, show that its radius decreases at a constant rate.

16. **Speed of a Race Car** A race car is moving along a track described by the equation

$$x^4 - 4x^2 + 2x^2 y^2 + 4y^2 + y^4 = 0$$

where both x and y are measured in miles. How fast is the car moving in the y-direction (dy/dt), when $dx/dt = -20$ (mph) and the car is at the point in the first quadrant with coordinate $x = 1$?

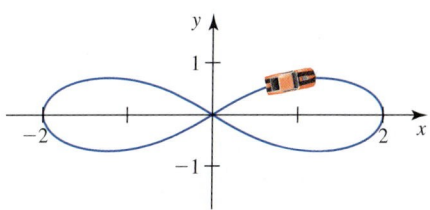

17. The base of a 13-ft ladder that is leaning against a wall begins to slide away from the wall. When the base is 12 ft from the wall and moving at the rate of 8 ft/sec, how fast is the top of the ladder sliding down the wall?

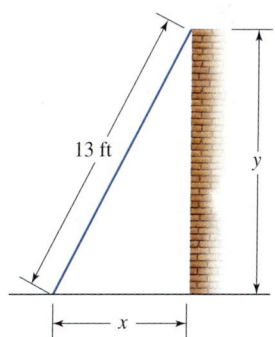

13 ft

18. A 20-ft ladder leaning against a wall begins to slide. How fast is the top of the ladder sliding down the wall at the instant of time when the bottom of the ladder is 12 ft from the wall and sliding away from the wall at the rate of 5 ft/sec?

19. Demand for Compact Discs The demand equation for the Olympus recordable compact disc is

$$100x^2 + 9p^2 = 3600$$

where x represents the number (in thousands) of 50-packs demanded per week when the unit price is p dollars. How fast is the quantity demanded increasing when the unit price per 50-pack is $14 and the selling price is dropping at the rate of 10¢ per 50-pack per week?

20. Let V denote the volume of a rectangular box of length x inches, width y inches, and height z inches. Suppose that the sides of the box are changing with respect to time t.
a. Find a relationship between dV/dt, dx/dt, dy/dt, and dz/dt.
   Hint: Write $V = x(yz)$, and use the Product Rule.
b. At a certain instant of time, the length, width, and height of the box are 3, 5, and 10 in., respectively. If the length, width, and height of the box are increasing at the rate of 0.2, 0.3, and 0.1 in./sec, respectively, how fast is the volume of the box increasing?

21. Baseball Diamond The sides of a square baseball diamond are 90 ft long. When a player who is between the second and third base is 60 ft from second base and heading toward third base at a speed of 22 ft/sec, how fast is the distance between the player and home plate changing?

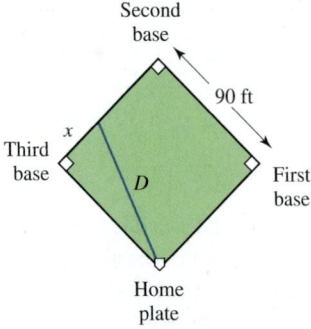

22. Docking a Boat A boat is pulled into a dock by means of a rope attached to the bow of the boat and passing through a pulley on the dock. The pulley is located at a point on the dock that is 2 m higher than the bow of the boat. If the rope is being pulled in at the rate of 1 m/sec, how fast is the boat approaching the dock when it is 12 m from the dock?

23. Tracking the Path of a Submarine The position $P(x, y)$ of a submarine moving in an xy-plane is described by the equation

$$y = 10^{-10}x^3(x - 2000) \qquad 0 \le x \le 1500$$

where both x and y are measured in feet (see the figure). How fast is the depth of the submarine changing when it is at the position $(1000, -100)$ and its speed in the x-direction is 50 ft/sec?

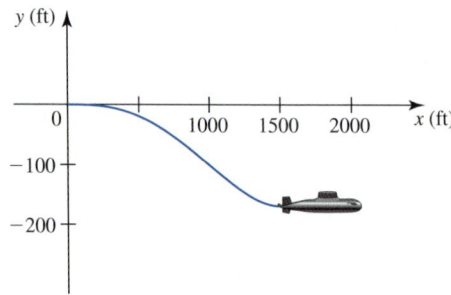

24. Length of a Shadow A man who is 6 ft tall walks away from a streetlight that is 15 ft from the ground at a speed of 4 ft/sec. How fast is the tip of his shadow moving along the ground when he is 30 ft from the base of the light pole?

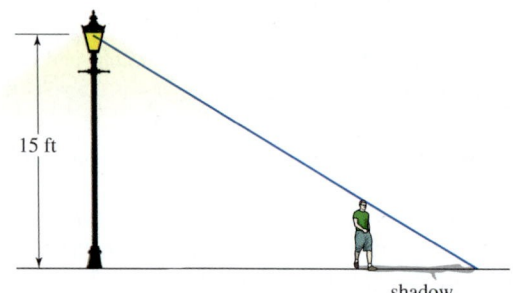

25. A coffee pot that has the shape of a circular cylinder of radius 4 in. is being filled with water flowing at a constant rate. At what rate is the water flowing into the coffee pot when the water level is rising at the rate of 0.4 in./sec?

26. A car leaves an intersection traveling west. Its position 4 sec later is 20 ft from the intersection. At the same time, another car leaves the same intersection heading north so that its position 4 sec later is 28 ft from the intersection. If the speeds of the cars at that instant of time are 9 ft/sec and 11 ft/sec, respectively, find the rate at which the distance between the two cars is changing.

27. A car leaves an intersection traveling east. Its position t sec later is given by $x = t^2 + t$ ft. At the same time, another car leaves the same intersection heading north, traveling $y = t^2 + 3t$ ft in t sec. Find the rate at which the distance between the two cars will be changing 5 sec later.

28. A police cruiser hunting for a suspect pulls over and stops at a point 20 ft from a straight wall. The flasher on top of the cruiser revolves at a constant rate of 90 deg/sec, and the light beam casts a spot of light as it strikes the wall. How fast is the spot of light moving along the wall at a point 30 ft from the point on the wall closest to the cruiser?

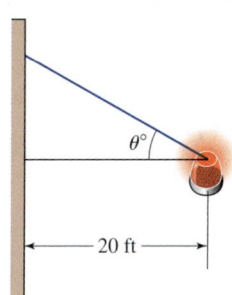

29. At 8:00 A.M. ship A is 120 km due east of ship B. Ship A is moving north at 20 km/hr, and ship B is moving east at 25 km/hr. How fast is the distance between the two ships changing at 8:30 A.M.?

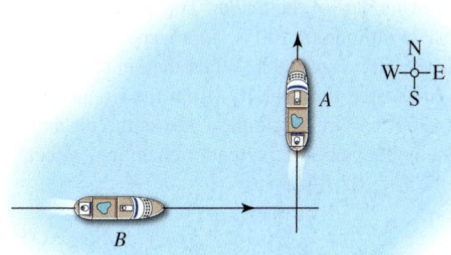

30. Two ships leave the same port at noon. Ship A moves north at 18 km/hr, and ship B moves northeast at 20 km/hr. How fast is the distance between them changing at 1 P.M.?

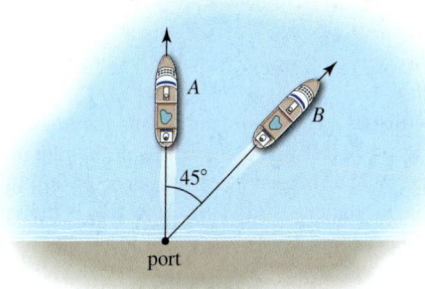

31. **Adiabatic Process** In an adiabatic process (one in which no heat transfer takes place), the pressure P and volume V of an ideal gas such as oxygen satisfy the equation $P^5V^7 = C$, where C is a constant. Suppose that at a certain instant of time, the volume of the gas is 4L, the pressure is 100 kPa, and the pressure is decreasing at the rate of 5 kPa/sec. Find the rate at which the volume is changing.

32. **Electric Circuit** The voltage V in volts (V) in an electric circuit is related to the current I in amperes (A) and the resistance R in ohms (Ω) by the equation $V = IR$. When $V = 12$, $I = 2$, V is increasing at the rate of 2 V/sec, and I is increasing at the rate of $\frac{1}{2}$ A/sec, how fast is the resistance changing?

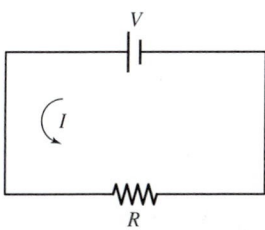

33. **Mass of a Moving Particle** The mass m of a particle moving at a velocity v is related to its rest mass m_0 by the equation

$$m = \frac{m_0}{\sqrt{1 - \dfrac{v^2}{c^2}}}$$

where c (2.98×10^8 m/sec) is the speed of light. Suppose that an electron of mass 9.11×10^{-31} kg is being accelerated in a particle accelerator. When its velocity is 2.92×10^8 m/sec and its acceleration is 2.42×10^5 m/sec^2, how fast is the mass of the electron changing?

34. Variable Resistors Two rheostats (variable resistors) are connected in parallel as shown in the figure. If the resistances of the rheostats are R_1 and R_2 ohms (Ω), then the single resistor that could replace this combination has resistance R, called the equivalent resistance, and is given by

$$\frac{1}{R} = \frac{1}{R_1} + \frac{1}{R_2}$$

Suppose that at a certain instant of time the first rheostat has a resistance of 60 Ω that is increasing at the rate of 2 Ω/sec, while the second rheostat has a resistance of 90 Ω that is decreasing at the rate of 3 Ω/sec. How fast is the resistance of the equivalent resistor changing at that time?

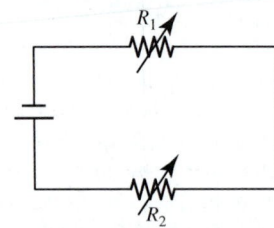

35. Coast Guard Patrol Search Mission The pilot of a Coast Guard patrol aircraft on a search mission had just spotted a disabled fishing trawler and decided to go in for a closer look. Flying in a straight line at a constant altitude of 1000 ft and at a constant speed of 264 ft/sec, the aircraft passed directly over the trawler. How fast was the aircraft receding from the trawler when the aircraft was 1500 ft from the trawler?

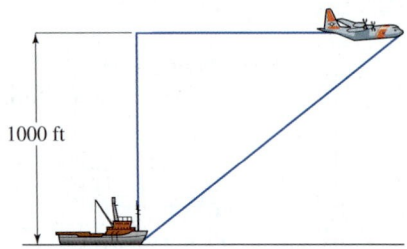

1000 ft

36. Tracking a Plane with Radar Shortly after taking off, a plane is climbing at an angle of 30° and traveling at a constant speed of 600 ft/sec as it passes over a ground radar tracking station. At that instant of time, the altitude of the plane is 1000 ft. How fast is the distance between the plane and the radar station increasing at that instant of time?

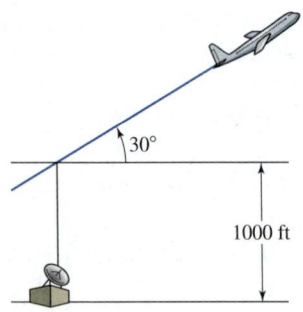

30°

1000 ft

37. A piston is attached to a crankshaft of radius 3 in. by means of a 7-in. connecting rod (see Figure a).
   a. Let x denote the position of the piston (Figure b). Use the law of cosines to find an equation relating x to θ.
   b. If the crankshaft rotates counterclockwise at a constant rate of 60 rev/sec, what is the velocity of the piston when $\theta = \pi/3$?

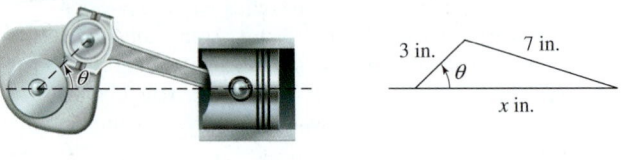

3 in. 7 in. θ x in.

(a)   **(b)**

38. An aircraft carrier is sailing due east at a constant speed of 30 ft/sec. When the aircraft carrier is at the origin ($t = 0$), a plane is launched from its deck with a flight path that is described by the graph of $y = 0.001x^2$ where y is the altitude of the plane (in feet). Ten seconds later, when the plane is at the point (1000, 1000) and $dx/dt = 500$ ft/sec, how fast is the distance between the plane and the aircraft carrier changing?

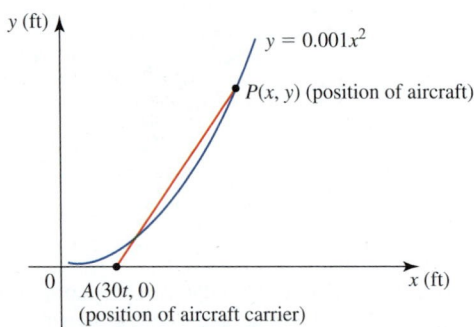

y (ft)

$y = 0.001x^2$

$P(x, y)$ (position of aircraft)

0
$A(30t, 0)$
(position of aircraft carrier)

x (ft)

39. As a tender leaves an offshore oil rig, traveling in a straight line and at a constant velocity of 20 mph, a helicopter approaches the oil rig in a direction perpendicular to the direction of motion of the tender. The helicopter, flying at a constant altitude of 100 ft, approaches the rig at a constant velocity of 60 mph. When the helicopter is 1000 ft (measured horizontally) from the rig and the tender is 200 ft from the rig, how fast is the distance between the helicopter and the tender changing? (Recall that 60 mi/hr = 88 ft/sec.)

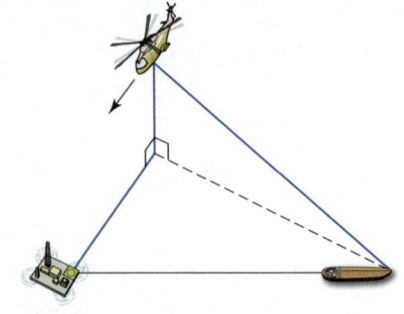

40. The following figure shows the cross section of a swimming pool that is 30 ft wide. When the pool is being filled with water at the rate of 600 gal/min and the depth at the deep end is 4 ft, how fast is the water level rising? (1 gal = 0.1337 ft^3.)

41. A hole is to be drilled into a block of Plexiglas. The 1-in. drill bit is shown in Figure (a), and the cross section of the Plexiglas block is shown in Figure (b). The drill press operator drives the drill bit into the Plexiglas at a constant speed of 0.05 in./sec. At what rate is the Plexiglas being removed 10 sec after the drill bit first makes contact with the block of Plexiglas?

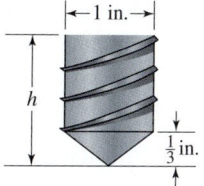
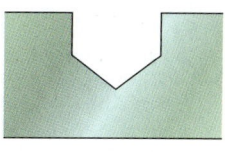

(a) Cross section of drill bit (b) Cross section of Plexiglas block

Hint: First show that the amount of material removed when the drill bit is h in. from the top surface of the Plexiglas block is $V = [\pi(9h - 2)]/36$.

42. Home Mortgage Payments The Garcias are planning to buy their first home within the next several months and estimate that they will need a home mortgage loan of $250,000 to be amortized over 30 years. At an interest rate of r per year, compounded monthly, the Garcias' monthly repayment P (in dollars) can be computed by using the formula

$$P = \frac{250{,}000r}{12\left[1 - \left(1 + \dfrac{r}{12}\right)^{-360}\right]}$$

a. If the interest rate is currently 7% per year and they secure the rate right now, what will the Garcias' monthly repayment on the mortgage be?

b. If the interest rate is currently increasing at the rate of $\frac{1}{4}$% per month, how fast is the monthly repayment on a mortgage loan of $250,000 increasing? Interpret your result.

2.10 Differentials and Linear Approximations

The Jacksons are planning to buy a house in the near future and estimate that they will need a 30-year fixed-rate mortgage of $240,000. If the interest rate increases from the present rate of 7% per year compounded monthly to 7.3% per year compounded monthly between now and the time the Jacksons decide to secure the loan, approximately how much more per month will their mortgage be? (You will be asked to answer this question in Exercise 42.)

Questions like this, in which we wish to *estimate* the change in the dependent variable (monthly mortgage payment) corresponding to a small change in the independent variable (interest rate per year), occur in many real-life applications. Here are a few more examples:

- An engineer would like to know the changes in the gaps between the rails in a railroad track due to expansions caused by small fluctuations in temperature.
- A chemist would like to know how a small increase in the amount of a catalyst will affect the initial speed at which a chemical reaction begins.
- An economist would like to know how a small increase in a country's capital expenditure will affect the country's gross domestic product.
- A bacteriologist would like to know how a small increase in the amount of a bactericide will affect a population of bacteria.

- A businesswoman would like to know how raising the unit price of a product by a small amount will affect her profits.
- A sociologist would like to know how a small increase in the amount of capital investment in a housing project will affect the crime rate.

To calculate these changes and their approximate effect, we need the concept of the *differential* of a function.

■ Increments

Let x denote a variable quantity and suppose that x changes from x_1 to x_2. Then the change in x, called the **increment in x,** is denoted by the symbol Δx (delta x). Thus,

$$\Delta x = x_2 - x_1 \qquad \text{Final value minus initial value} \qquad (1)$$

For example, if x changes from 2 to 2.1, then $\Delta x = 2.1 - 2 = 0.1$; and if x changes from 2 to 1.9, then $\Delta x = 1.9 - 2 = -0.1$.

Sometimes it is more convenient to express the change in x in a slightly different manner. For example, if we solve Equation (1) for x_2, we find $x_2 = x_1 + \Delta x$, where Δx is an increment in x. Observe that Δx plays precisely the role that h played in our earlier discussions.

Now, suppose that two quantities, x and y, are related by an equation $y = f(x)$, where f is some function. If x changes from x to $x + \Delta x$, then the corresponding change in y, or the **increment in y,** is denoted by Δy. It is the value of $f(x)$ at $x + \Delta x$ minus the value of $f(x)$ at x; that is,

$$\Delta y = f(x + \Delta x) - f(x) \qquad (2)$$

(See Figure 1.)

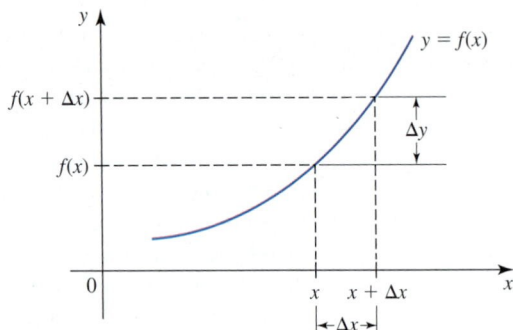

FIGURE 1
An increment of Δx in x
induces an increment of
$\Delta y = f(x + \Delta x) - f(x)$ in y.

EXAMPLE 1 Suppose that $y = 2x^3 - x + 1$. Find Δx and Δy when (a) x changes from 3 to 3.01 and (b) x changes from 3 to 2.98.

Solution

a. Here, $\Delta x = 3.01 - 3 = 0.01$. Next, letting $f(x) = 2x^3 - x + 1$, we see that

$$\Delta y = f(x + \Delta x) - f(x) = f(3.01) - f(3)$$

$$= [2(3.01)^3 - 3.01 + 1] - [2(3)^3 - 3 + 1]$$

$$= 0.531802$$

b. Here, $\Delta x = 2.98 - 3 = -0.02$. Also,

$$\begin{aligned}
\Delta y = f(x + \Delta x) - f(x) &= f(2.98) - f(3) \\
&= [2(2.98)^3 - 2.98 + 1] - [2(3)^3 - 3 + 1] \\
&= -1.052816
\end{aligned}$$

◼ Differentials

To find a quick and simple way of estimating the change in y, Δy, due to a small change in x, Δx, let's look at the graph in Figure 2.

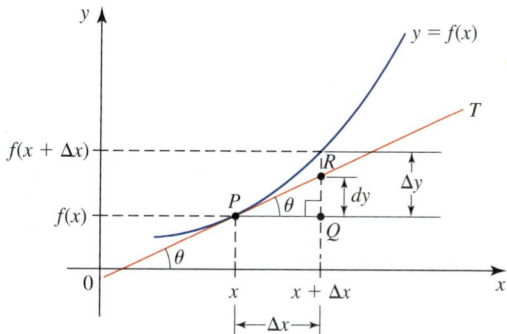

FIGURE 2
If Δx is small, dy is a good approximation of Δy.

We can see that the tangent line T lies close to the graph of f near the point of tangency at P. Therefore, if Δx is small, the y-coordinate of the point R on T is a good approximation of $f(x + \Delta x)$. Equivalently, the quantity dy is a good approximation of Δy.

Now consider the right triangle $\triangle PQR$. We have

$$\frac{dy}{\Delta x} = \tan \theta$$

or $dy = (\tan \theta)\Delta x$. But the derivative of f gives the slope of the tangent line T, so we have $\tan \theta = f'(x)$. Therefore,

$$dy = f'(x)\Delta x$$

The quantity dy is called the *differential* of y.

DEFINITION   Differential

Let $y = f(x)$ where f is a differentiable function. Then

1. The **differential dx** of the independent variable x is $dx = \Delta x$, where Δx is an increment in x.
2. The **differential dy** of the dependent variable y is

$$dy = f'(x)\Delta x = f'(x)\,dx \tag{3}$$

Notes

1. For the independent variable x, there is no difference between the differential dx and the increment Δx; both measure the change in x from x to $x + \Delta x$.

2. For the dependent variable y, the differential dy is an *approximation* of the change in y, Δy, corresponding to a small change in x from x to $x + \Delta x$.

3. The differential dy depends on both x and dx. However, if x is fixed, then dy is a *linear* function of dx.

Later, we will show that the approximation of Δy by dy is very good when dx, or Δx, is small. First, let's look at some examples.

EXAMPLE 2   Consider the equation $y = 2x^3 - x + 1$ of Example 1. Use the differential dy to approximate Δy when (a) x changes from 3 to 3.01 and (b) x changes from 3 to 2.98. Compare your results with those of Example 1.

Solution   Let $f(x) = 2x^3 - x + 1$. Then

$$dy = f'(x)\, dx = (6x^2 - 1)\, dx$$

a. Here, $x = 3$ and $dx = 3.01 - 3 = 0.01$. Therefore,

$$dy = [6(3^2) - 1](0.01) = 0.53$$

and we obtain the approximation

$$\Delta y \approx 0.53 \qquad \text{From Example 1 we know that the actual value of } \Delta y \text{ is } 0.531802.$$

b. Here, $x = 3$ and $dx = 2.98 - 3 = -0.02$. Therefore,

$$dy = [6(3)^2 - 1](-0.02) = -1.06$$

and we obtain the approximation

$$\Delta y \approx -1.06 \qquad \text{From Example 1 we know that the actual value of } \Delta y \text{ is } -1.052816.$$

EXAMPLE 3   **Estimating Fuel Costs of Operating an Oil Tanker**   The total cost incurred in operating an oil tanker on an 800-mi run, traveling at an average speed of v mph, is estimated to be

$$C(v) = \frac{1{,}000{,}000}{v} + 200v^2$$

dollars. Find the approximate change in the total operating cost if the average speed is increased from 10 mph to 10.5 mph.

Solution   Letting $v = 10$ and $dv = 0.5$, we find

$$\Delta C \approx dC = C'(10)\, dv$$

$$= -\frac{1{,}000{,}000}{v^2} + 400v \Big|_{v=10} \cdot (0.5)$$

$$= (-10{,}000 + 4000)(0.5) \approx -3000$$

So the total operating costs decrease by approximately \$3000.

EXAMPLE 4 **The Rings of Neptune**

a. A planetary ring has an inner radius of r units and an outer radius of R units, where $(R - r)$ is small in comparison to r (see Figure 3a). Use differentials to estimate the area of the ring.

b. Observations including those of Voyager I and II showed that Neptune's ring system is considerably more complex than had been believed. For one thing, it is made up of a large number of distinguishable rings rather than one continuous great ring, as had previously been thought (see Figure 3b). The outermost ring, 1989N1R, has an inner radius of approximately 62,900 km (measured from the center of the planet) and a radial width of approximately 50 km. Using these data, estimate the area of the ring.

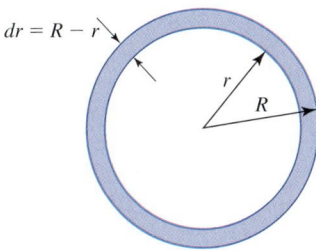

NASA

(a) The area of the ring can be approximated by the circumference of the inner circle times the thickness.

(b) Neptune and its rings

FIGURE 3

Solution

a. Since the area of a circle of radius x is $A = f(x) = \pi x^2$, we have

$$\pi R^2 - \pi r^2 = f(R) - f(r)$$

$$= \Delta A$$

Remember that $\Delta A =$ change in f when x changes from $x = r$ to $x = R$.

$$\approx dA$$

$$= f'(r)\, dr$$

where $dr = R - r$. So we see that the area of the ring is approximately $f'(r)\, dr = 2\pi r(R - r)$ square units. In words, the area of the ring is approximately equal to

circumference of the inner circle $\times$ thickness of the ring

b. Applying the results of part (a) with $r = 62,900$ and $dr = 50$, we find that the area of the ring is approximately $2\pi(62,900)(50)$, or 19,760,618 sq km, which is approximately 4% of the earth's surface. ◼

■ Error Estimates

An important application of differentials lies in the calculation of error propagation. For example, suppose that the quantities x and y are related by the equation $y = f(x)$, where f is some function; then a small error Δx or dx incurred in measuring the quantity x results in an error Δy in the calculated value of y.

EXAMPLE 5 **Estimating the Surface Area of the Moon** Assume that the moon is a perfect sphere, and suppose that we have measured its radius and found it to be 1080 mi with a possible error of 0.05 mi. Estimate the maximum error in the computed surface area of the moon.

Solution The surface area of a sphere of radius r is

$$S = 4\pi r^2$$

We are given that the error in r is $\Delta r = 0.05$ mi and are required to find the error ΔS in S. But if Δr (equivalently, dr) is small, then

$$\Delta S \approx dS = f'(r)\Delta r = 8\pi r \, dr \qquad \text{Let } f(r) = 4\pi r^2. \tag{4}$$

Substituting $r = 1080$ and $dr = \Delta r = 0.05$ in Equation (4), we obtain

$$\Delta S \approx 8\pi(1080)(0.05) \approx 1357.17$$

Therefore, the maximum error in the calculated area is approximately 1357 mi^2. ■

In Example 5 we calculated the **error Δq of a quantity q.** There are two other common error measurements. They are

$$\frac{\Delta q}{q}, \quad \text{the \textbf{relative error} in the measurement}$$

and

$$\frac{\Delta q}{q}(100), \quad \text{the \textbf{percentage error} in the measurement}$$

The error, relative error, and percentage error are often approximated by

$$dq, \qquad \frac{dq}{q}, \qquad \text{and} \qquad \frac{dq}{q}(100)$$

respectively.

The relative errors made when the surface area of the moon was calculated in Example 5 are given by

$$\text{relative error in } r \approx \frac{dr}{r} = \frac{0.05}{1080} \approx 0.0000463$$

and

$$\text{relative error in } S \approx \frac{dS}{S} = \frac{8\pi r}{4\pi r^2} \, dr = \frac{2}{r} \, dr \approx 0.0000926$$

A summary of these results and the approximate percentage errors follows.

Variable	Error	Approximate relative error	Approximate percentage error
r	0.05	0.0000463	0.00463%
S	1357.17	0.0000926	0.00926%

Note   Example 5 illustrates why the relative error is so important. The (absolute) error in S is 1357.17 mi^2. By itself, the error appears to be rather large (a little larger than the area of the state of Rhode Island). But when the error is compared to the area of the moon (approximately 14,657,415 mi^2), it is a relatively small number.   ■

EXAMPLE 6   The edge of a cube was measured and found to be 3 in. with a maximum possible error of 0.02 in. Find the approximate maximum percentage error that would be incurred in computing the volume of the cube using this measurement.

Solution   Let x denote the length of an edge of the cube. Then the volume of the cube is $V = x^3$. The error in the measurement of its volume is approximated by the differential

$$dV = 3x^2 dx \qquad \text{Let } f(x) = x^3, \text{ so } f'(x)\, dx = 3x^2 dx.$$

But we are given that

$$|dx| \le 0.02 \qquad \text{and} \qquad x = 3$$

so

$$|dV| = 3x^2 |dx| \le 3(3)^2 (0.02) = 0.54$$

Therefore, the approximate maximum percentage error that would be incurred in computing the volume of the cube is

$$\frac{|dV|}{V}(100) = \frac{0.54}{3^3}(100) = \frac{54}{27} = 2$$

or 2%.   ■

■ Linear Approximations

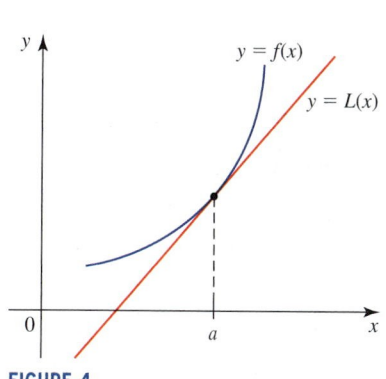

FIGURE 4

As you can see in Figure 4, the graph of f lies very close to its tangent line near the point of tangency. This suggests that the values of $f(x)$ for x near a can be approximated by the corresponding values of $L(x)$, where L is the linear function describing the tangent line.

The function L can be found by using the point-slope form of the equation of a line. Indeed, the slope of the tangent line at $(a, f(a))$ is $f'(a)$, and an equation of the tangent line is

$$y - f(a) = f'(a)(x - a)$$

or

$$y = L(x) = f(a) + f'(a)(x - a)$$

Next, if we replace x by a in Equation (2) and let $\Delta x = x - a$, then

$$\Delta y = f(x) - f(a)$$

so

$$f(x) - f(a) \approx dy = f'(a)\Delta x = f'(a)(x - a) \qquad \text{By Equation (3)}$$

or

$$f(x) \approx f(a) + f'(a)(x - a) \tag{5}$$

provided that Δx is small or, equivalently, x is close to a. But the expression on the right of Equation (5) is $L(x)$. So $f(x) \approx L(x)$ for x near a. The approximation in Equation (5) is called the **linear approximation** of f at a. The linear function L defined by

$$L(x) = f(a) + f'(a)(x - a) \qquad (6)$$

whose graph is the tangent line to the graph of f at $(a, f(a))$, is called the **linearization** of f at a. Observe that the linearization of f gives an approximation of f over a *small interval containing a*.

EXAMPLE 7

a. Find the linearization of $f(x) = \sqrt{x}$ at $a = 4$.
b. Use the result of part (a) to approximate the numbers $\sqrt{3.9}$, $\sqrt{3.98}$, $\sqrt{4}$, $\sqrt{4.04}$, $\sqrt{4.8}$, $\sqrt{6}$, and $\sqrt{8}$. Compare the results with the actual values obtained with a calculator.

Solution
a. Here, $a = 4$. Since

$$f'(x) = \frac{1}{2} x^{-1/2} = \frac{1}{2\sqrt{x}}$$

we find $f'(4) = \frac{1}{4}$. Also, $f(4) = 2$. Using Equation (6), we see that the required linearization of f is

$$L(x) = f(4) + f'(4)(x - 4)$$

or

$$L(x) = 2 + \frac{1}{4}(x - 4) = \frac{1}{4} x + 1$$

(See Figure 5.)

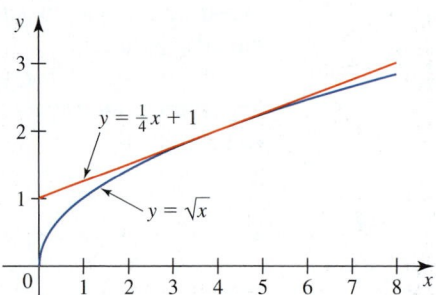

FIGURE 5
The linear approximation of
$f(x) = \sqrt{x}$ by $L(x) = \frac{1}{4} x + 1$

b. Using the result of part (a), we see that

$$\sqrt{3.9} = f(3.9) \approx L(3.9) = \frac{1}{4}(3.9) + 1 = 1.975$$

We obtain the other approximations in a similar manner. The results are summarized in the following table. You can see from the table that the approximations of $f(x)$ by $L(x)$ are good if x is close to 4 but are less accurate if x is farther away from 4.

Number	x	$L(x)$	$f(x)$ (actual value)
$\sqrt{3.9}$	3.9	1.975	1.97484177 ...
$\sqrt{3.98}$	3.98	1.995	1.99499373 ...
$\sqrt{4}$	4	2	2.00000000 ...
$\sqrt{4.04}$	4.04	2.01	2.00997512 ...
$\sqrt{4.8}$	4.8	2.2	2.19089023 ...
$\sqrt{6}$	6	2.5	2.44948974 ...
$\sqrt{8}$	8	3	2.82842712 ...

■ Error in Approximating Δy by dy

Through several numerical examples we have seen how closely the (true) increment $\Delta y = f(x + \Delta x) - f(x)$, where $y = f(x)$, is approximated by the differential dy. Let's demonstrate that this is no accident. We start by computing the error in the approximation

$$\Delta y - dy = [f(x + \Delta x) - f(x)] - f'(x)\Delta x$$

$$= \left[\frac{f(x + \Delta x) - f(x)}{\Delta x}\right]\Delta x - f'(x)\Delta x$$

$$= \left[\frac{f(x + \Delta x) - f(x)}{\Delta x} - f'(x)\right]\Delta x$$

For fixed x, the quantity in brackets depends only on Δx. Furthermore, because

$$\frac{f(x + \Delta x) - f(x)}{\Delta x}$$

approaches $f'(x)$ as Δx approaches 0, the bracketed quantity approaches 0 as Δx approaches 0. Let's denote this quantity, which is a function of Δx, by $\varepsilon(\Delta x)$.* Then we have

$$\Delta y - dy = \varepsilon(\Delta x)\Delta x$$

Therefore, if Δx is small, then

$$\Delta y - dy = (\text{small number})(\text{small number})$$

and is a *very* small number, which accounts for the closeness of the approximation.

---

*We could have called this function of Δx, g or h, say, but in mathematical literature the Greek letter ε is often used to denote a small quantity. Since the functional value $\varepsilon(\Delta x)$ is small when Δx is small, for emphasis we chose the letter ε to denote that function.

2.10 CONCEPT REVIEW

1. If $y = f(x)$, what is the differential of x? Write an expression for the differential dy.

2. Let $y = f(x)$. What is the relationship between the actual change in y, Δy, when x changes from x to $x + \Delta x$ and the differential dy of f at x? Illustrate this relationship graphically.

2.10   EXERCISES

1. Let $y = x^2 + 1$.
   a. Find Δx and Δy if x changes from 2 to 2.02.
   b. Find the differential dy, and use it to approximate Δy if x changes from 2 to 2.02.
   c. Compute $\Delta y - dy$, the error in approximating Δy by dy.

2. Let $y = 2x^3 - x$.
   a. Find Δx and Δy if x changes from 2 to 1.97.
   b. Find the differential dy, and use it to approximate Δy if x changes from 2 to 1.97.
   c. Compute $\Delta y - dy$, the error in approximating Δy by dy.

3. Let $w = \sqrt{2u + 3}$.
   a. Find Δu and Δw if u changes from 3 to 3.1.
   b. Find the differential dw, and use it to approximate Δw if u changes from 3 to 3.1.
   c. Compute $\Delta w - dw$, the error in approximating Δw by dw.

4. Let $y = 1/x$.
   a. Find Δx and Δy if x changes from 1 to 1.02.
   b. Find the differential dy, and use it to approximate Δy if x changes from 1 to 1.02.
   c. Compute $\Delta y - dy$, the error in approximating Δy by dy.

In Exercises 5–18, find the differential of the function at the indicated number.

5. $f(x) = 2x^2 - 3x + 1$; $x = 1$

6. $f(x) = x^4 - 2x^3 + 3$; $x = 0$

7. $f(x) = 2x^{1/4} + 3x^{-1/2}$; $x = 1$

8. $f(x) = \sqrt{2x^2 + 1}$; $x = 2$

9. $f(x) = x^2(3x - 1)^{1/3}$; $x = 3$

10. $f(x) = \dfrac{x^2}{x^3 - 1}$; $x = -1$

11. $f(x) = 2\sin x + 3\cos x$; $x = \dfrac{\pi}{4}$

12. $f(x) = x \tan x$; $x = \dfrac{\pi}{4}$

13. $f(x) = (1 + 2\cos x)^{1/2}$; $x = \dfrac{\pi}{2}$

14. $f(x) = \sin^2 x$; $x = \dfrac{\pi}{6}$

15. $f(x) = e^x + \ln(1 + x)$; $x = 0$

16. $f(x) = \ln(2\cos x + x)$; $x = 0$

17. $f(x) = x^3 e^{1-x}$; $x = 1$

18. $f(x) = 2^{-x^2}$; $x = 1$

In Exercises 19–22 find the linearization $L(x)$ of the function at a.

19. $f(x) = x^3 + 2x^2$; $a = 1$

20. $f(x) = \sqrt{2x + 3}$; $a = 3$

21. $f(x) = \ln x$; $a = 1$

22. $f(x) = \sin x$; $a = \dfrac{\pi}{4}$

 23. Find the linearization of $f(x) = \sqrt{x + 3}$ at $a = 1$, and use it to approximate the numbers $\sqrt{3.9}$ and $\sqrt{4.1}$. Plot the graphs of f and L on the same set of axes.

 24. Find the linearization $L(x)$ of $f(x) = \sqrt[3]{1 - x}$ at $a = 0$, and use it to approximate the numbers $\sqrt[3]{0.95}$ and $\sqrt[3]{1.05}$. Plot the graphs of f and L on the same set of axes.

In Exercises 25–28, find the linearization of a suitable function, and then use it to approximate the number.

25. 1.002^3 **26.** $\sqrt{63.8}$

27. $\sqrt[5]{31.08}$ **28.** $\sin 0.1$

29. The side of a cube is measured with a maximum possible error of 2%. Use differentials to estimate the maximum percentage error in its computed volume.

30. **Estimating the Area of a Ring of Neptune** The ring 1989N2R of the planet Neptune has an inner radius of approximately 53,200 km (measured from the center of the planet) and a radial width of 15 km. Use differentials to estimate the area of the ring.

31. **Effect of Advertising on Profits** The relationship between the quarterly profits of the Lyons Realty Company, $P(x)$, and the amount of money x spent on advertising per quarter is described by the function

$$P(x) = -\frac{1}{8}x^2 + 7x + 32 \qquad 0 \le x \le 50$$

where both $P(x)$ and x are measured in thousands of dollars. Use differentials to estimate the increase in profits when the amount spent on advertising each quarter is increased from $24,000 to $26,000.

32. Construction of a Storage Tank A storage tank for propane gas has the shape of a right circular cylinder with hemispherical ends. The length of the cylinder is 6 ft, and the radius of each hemisphere is r ft.

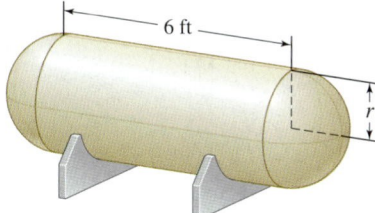

6 ft

r

a. Show that the volume of the tank is $\frac{2}{3}\pi r^2(2r + 9)$ ft³.
b. If the tank were constructed with a radius of 4.1 ft instead of a specified radius of 4 ft, what would be the approximate percentage error in its volume?

33. Unclogging Arteries Research done in the 1930s by the French physiologist Jean Poiseuille showed that the resistance R of a blood vessel of length l and radius r is $R = kl/r^4$, where k is a constant. Suppose that a dose of the drug TPA increases r by 10%. How will this affect the resistance R? (Assume that l is constant.)

34. Period of a Pendulum The period of a simple pendulum is given by

$$T = 2\pi\sqrt{\frac{L}{g}}$$

where L is the length of the pendulum in feet, g is the constant of acceleration due to gravity, and T is measured in seconds. Suppose that the length of a pendulum was measured with a maximum error of $\frac{1}{2}\%$. What will be the maximum percentage error in measuring its period?

35. Period of a Satellite The period of a satellite in a circular orbit of radius r is given by

$$T = \frac{2\pi r}{R}\sqrt{\frac{r}{g}}$$

where R is the earth's mean radius and g is the constant of acceleration. Estimate the percentage change in the period if the radius of the orbit increases by 2%.

36. Surface Area of a Horse Animal physiologists use the formula

$$S = kW^{2/3}$$

to calculate the surface area of an animal (in square meters) from its mass W (in kilograms), where k is a constant that depends on the animal under consideration. Suppose that a physiologist calculates the surface area of a horse ($k = 0.1$). If the estimated mass of the horse is 280 kg with a maximum error in measurement of 0.5 kg, determine the maximum percentage error in the calculation of the horse's surface area.

37. Child-Langmuir Law In a vacuum diode a steady current I flows between the cathode with potential 0 and anode which is held at a positive potential V_0. The Child-Langmuir Law states that $I = kV_0^{3/2}$, where k is a constant. Use differentials to estimate the percentage change in the current corresponding to a 10% increase in the positive potential.

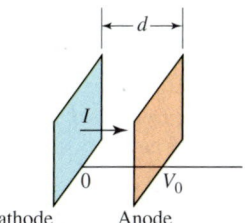

d

I

0   V_0

Cathode   Anode

38. Effect of Price Increase on Quantity Demanded The quantity x demanded per week of the Alpha Sports Watch (in thousands) is related to its unit price of p dollars by the equation

$$x = f(p) = 10\sqrt{\frac{50 - p}{p}} \qquad 0 < p \le 50$$

Use differentials to find the decrease in the quantity of watches demanded per week if the unit price is increased from $40 to $42.

39. Range of an Artillery Shell The range of an artillery shell fired at an angle of $\theta°$ with the horizontal is

$$R = \frac{1}{32}v_0^2\sin 2\theta$$

in feet, where v_0 is the muzzle speed of the shell. Suppose that the muzzle speed of a shell is 80 ft/sec and the shell is fired at an angle of 29.5° instead of the intended 30°. Estimate how far short of the target the shell will land.

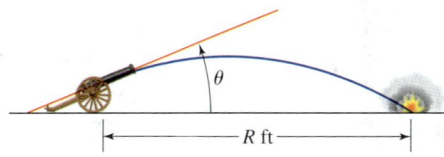

θ

R ft

40. Range of an Artillery Shell The range of an artillery shell fired at an angle of $\theta°$ with the horizontal is

$$R = \frac{v_0^2}{g}\sin 2\theta$$

in feet, where v_0 is the muzzle speed of the shell and $g = 32$ ft/sec² is the constant of acceleration due to gravity. Suppose the angle of elevation of the cannon is set at 45°. Because of variations in the amount of charge in a shell, the muzzle speed of a shell is subject to a maximum error of 0.1%. Calculate the effect this will have on the range of the shell.

41. Forecasting Commodity Crops Government economists in a certain country have determined that the demand equation for soybeans is given by

$$p = f(x) = \frac{55}{2x^2 + 1}$$

where the unit price p is expressed in dollars per bushel and x, the quantity demanded per year, is measured in billions of bushels. The economists are forecasting a harvest of 2.2 billion bushels for the year, with a possible error of 10% in their forecast. Determine the corresponding error in the predicted price per bushel of soybeans.

42. Financing a Home The Jacksons are considering the purchase of a house in the near future and estimate that they will need a loan of $240,000. Their monthly repayment for a 30-year conventional mortgage with an interest rate of r per year compounded monthly will be

$$P = \frac{20,000r}{1 - \left(1 + \dfrac{r}{12}\right)^{-360}}$$

dollars.

a. Find the differential of P.
b. If the interest rate increases from the present rate of 7% per year to 7.2% per year between now and the time the Jacksons decide to secure the loan, approximately how much more per month will their mortgage payment be? How much more will it be if the interest rate increases to 7.3% per year?

43. Period of a Communications Satellite According to Kepler's Third Law, the period T (in days) of a satellite moving in a circular orbit x mi above the surface of the earth is given by

$$T = 0.0588\left(1 + \frac{x}{3959}\right)^{3/2}$$

Suppose that a communications satellite is moving in a circular orbit 22,000 mi above the earth's surface. Because of friction, the satellite drops down to a new orbit 21,500 mi above the earth's surface. Estimate the decrease in the period of the satellite to the nearest one-hundredth hour.

44. Effect of an Earthquake on a Structure To study the effect an earthquake has on a structure, engineers look at the way a beam bends when subjected to an earth tremor. The equation

$$D = a - a\cos\left(\frac{\pi h}{2L}\right) \qquad 0 \le h \le L$$

where L is the length of a beam and a is the maximum deflection from the vertical, has been used by engineers to calculate the deflection D at a point on the beam h ft from the ground. Suppose that a 10-ft vertical beam has a maximum deflection of $\frac{1}{2}$ ft when subjected to an external force. Using

differentials, estimate the difference in the deflection between the point midway on the beam and the point $\frac{1}{10}$ ft above it.

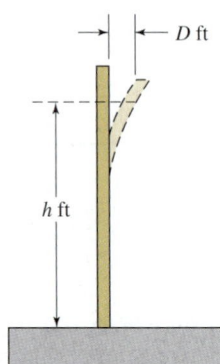

45. Relative Error in Measuring Electric Current When measuring an electric current with a tangent galvanometer, we use the formula

$$I = k\tan\phi$$

where I is the current, k is a constant that depends on the instrument, and ϕ is the angle of deflection of the pointer. Find the relative error in measuring the current I due to an error in reading the angle ϕ. At what position of the pointer can one obtain the most reliable results?

46. Percentage Error in Measuring Height From a point on level ground 150 ft from the base of a derrick, Jose measures the angle of elevation to the top of the derrick as 60°. If Jose's measurements are subject to a maximum error of 1%, find the percentage error in the measured height of the derrick.

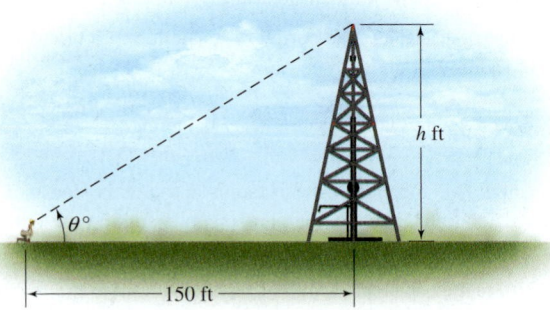

47. Heights of Children For children between the ages of 5 and 13 years, the Ehrenberg equation

$$\ln W = \ln 2.4 + 1.84h$$

gives the relationship between the weight W (in kilograms) and the height h (in meters) of a child. Use differentials to estimate the change in the weight of a child who grows from 1 m to 1.1 m.

In Exercises 48–51, *determine whether the statement is true or false. If it is true, explain why it is true. If it is false, explain why or give an example to show why it is false.*

48. If $y = ax + b$, where a and b are constants, then $\Delta y = dy$.

49. If f is differentiable at a and x is close to a, then
$$f(x) \approx f(a) + f'(a)(x - a).$$

50. If $h = g \circ f$, where g and f are differentiable everywhere, then $h(x + \Delta x) \approx g(f(x)) + g'(f(x))f'(x)\Delta x$.

51. If $y = f(x)$ and $f'(x) > 0$, then $\Delta y \geq dy$.

CHAPTER 2 REVIEW

CONCEPT REVIEW

In Exercises 1–14, *fill in the blanks.*

1. a. The derivative of a function with respect to x is the function f' defined by the rule _____.

b. The domain of f' consists of all values of x for which the _____ exists.

c. The number $f'(a)$ gives the slope of the _____ _____ to the graph of f at _____.

d. The number $f'(a)$ also measures the rate of change of _____ with respect to _____ at _____.

e. If f is differentiable at a, then an equation of the tangent line to the graph of f at $(a, f(a))$ is _____.

2. a. A function might not be differentiable at a _____. For example, the function _____ fails to be differentiable at _____.

b. If a function f is differentiable at a, then f is _____ at a. The converse is false. For example, the function _____ is continuous at _____ but is not differentiable at _____.

3. a. If c is a constant, then $\dfrac{d}{dx}(c) = $ _____.

b. If n is any real number, then $\dfrac{d}{dx}(x^n) = $ _____.

4. If f and g are differentiable functions and c is a constant, then the Constant Multiple Rule states that _____, the Sum Rule states that _____, the Product Rule states that _____, and the Quotient Rule states that _____.

5. If $y = f(x)$, where f is a differentiable function, and α denotes the angle that the tangent line to the graph of f at $(x, f(x))$ makes with the positive x-axis, then $\tan \alpha = $ _____.

6. Suppose that $f(t)$ gives the position of an object moving on a coordinate line.

a. The velocity of the object is given by _____, its acceleration is given by _____, and its jerk is given

by _____. The speed of the object is given by _____.

b. The object is moving in the positive direction if $v(t)$ _____ and in the negative direction if $v(t)$ _____. It is stationary if $v(t) = $ _____.

7. If C, R, P, and $\bar{C}$ denote the total cost function, the total revenue function, the profit function, and the average cost function, respectively, then the marginal total cost function is given by _____, the marginal total revenue function by _____, the marginal profit function by _____, and the marginal average cost function by _____.

8. If f is differentiable at x and g is differentiable at $f(x)$, then the function $h = g \circ f$ is differentiable at _____, and $h'(x) = $ _____.

9. a. The General Power Rule states that $\dfrac{d}{dx}[f(x)]^n = $ _____.

b. If f is differentiable, then $\dfrac{d}{dx}[\sin f(x)] = $ _____, $\dfrac{d}{dx}[\cos f(x)] = $ _____, $\dfrac{d}{dx}[\tan f(x)] = $ _____, $\dfrac{d}{dx}[\sec f(x)] = $ _____, $\dfrac{d}{dx}[\csc f(x)] = $ _____, and $\dfrac{d}{dx}[\cot f(x)] = $ _____.

10. a. If u is a differentiable function of x, then $\dfrac{d}{dx}e^u = $ _____.

b. If $a > 0$ and $a \neq 1$, then $a^x = $ _____; if u is a differentiable function of x, then $\dfrac{d}{dx}a^u = $ _____.

11. If u is a differentiable function of x, then $\dfrac{d}{dx}\log_a|u| = $ _____.

12. Suppose that a function $y = f(x)$ is defined implicitly by an equation in x and y. To find dy/dx, we differentiate _____ _____ of the equation with respect to x and then solve the resulting equation for dy/dx. The derivative of a term involving y includes _____ as a factor.

13. In a related rates problem we are given a relationship between a variable x and a variable _____ that depend on a third variable t. Knowing the values of x, y, and dx/dt at a, we want to find _____ at _____.

14. Let $y = f(t)$ and $x = g(t)$. If $x^2 + y^2 = 4$, then $dx/dt =$ _____. If $xy = 1$, then $dy/dt =$ _____.

15. a. If a variable quantity x changes from x_1 to x_2, then the increment in x is $\Delta x =$ _____.
b. If $y = f(x)$ and x changes from x to $x + \Delta x$, then the increment in y is $\Delta y =$ _____.

16. If $y = f(x)$, where f is a differentiable function, then the differential dx of x is $dx =$ _____, where _____ is an increment in _____, and the differential dy of y is $dy =$ _____.

REVIEW EXERCISES

In Exercises 1 and 2, use the definition of the derivative to find the derivative of the function.

1. $f(x) = x^2 - 2x - 4$

2. $f(x) = 2x^3 - 3x + 2$

In Exercises 3 and 4, sketch the graph of f' for the function f whose graph is given.

3.

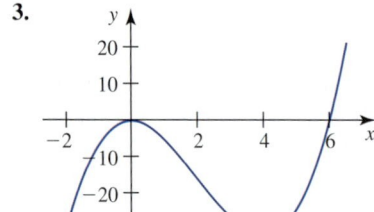

4.

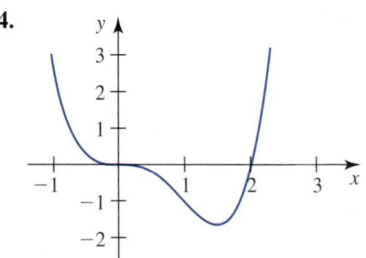

5. The amount of money on fixed deposit at the end of 5 years in a bank paying interest at the rate of r per year is given by $A = f(r)$ (dollars).
a. What does $f'(r)$ measure? Give units.
b. What is the sign of $f'(r)$? Explain.
c. If you know that $f'(6) = 60,775.31$, estimate the change in the amount after 5 years if the interest rate changes from 6% to 7% per year.

6. Use the graph of the function f to find the value(s) of x at which f is not differentiable.

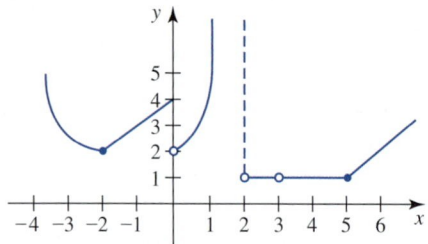

7. Find a function f and a number a such that
$$\lim_{h \to 0} \frac{3(4 + h)^{3/2} - 24}{h} = f'(a)$$

8. Evaluate $\lim_{h \to 0} \dfrac{2(1 + h)^3 + (1 + h)^2 - 3}{h}$.

In Exercises 9–64, find the derivative of the function.

9. $f(x) = \dfrac{1}{3}x^6 - 2x^4 + x^2 - 5$

10. $g(x) = 2x^4 + 3x^{1/2} - x^{-1/3} + x^{-4}$

11. $s = 2t^2 - \dfrac{4}{t} + \dfrac{2}{\sqrt{t}}$

12. $f(x) = \dfrac{x + 1}{x - 1}$

13. $g(t) = \dfrac{t - 1}{2t + 1}$

14. $h(x) = \dfrac{x}{2x^2 + 3}$

15. $h(u) = \dfrac{\sqrt{u}}{u^2 + 1}$

16. $u = \dfrac{t^2}{1 - \sqrt{t}}$

17. $g(\theta) = \cos \theta - 2 \sin \theta$

18. $f(x) = x \tan x + \sec x$

19. $f(x) = x \sin x + x^2 \cos x$

20. $y = \dfrac{1 - \sin x}{1 + \sin x}$

21. $h(t) = \dfrac{t \cos t}{1 + \tan t}$

22. $f(x) = (1 + 2x)^7$

23. $y = (t^3 + 2t + 1)^{-3/2}$

24. $g(t) = \left(t^2 + \dfrac{1}{t^2}\right)^3$

25. $f(s) = s(s^3 + s + 1)^{3/2}$

26. $y = \left(\dfrac{1 + t^2}{1 - t^2}\right)^{3/2}$

27. $y = \dfrac{2t}{\sqrt{t + 1}}$

28. $h(x) = \dfrac{1 + x}{(2x^2 + 1)^2}$

29. $f(x) = \cos(2x + 1)$

30. $g(t) = t^2 \sin(\pi t + 1)$

31. $y = x^2 + \dfrac{\sin 2x}{x}$

32. $h(x) = \sec\left(\dfrac{x + 1}{x - 1}\right)$

33. $u = \tan\dfrac{2}{x}$

34. $v = \sec 2x + \tan 3x$

35. $w = \cot^3 x$

36. $f(x) = \tan(x^2 + 1)^{-1/2}$

37. $f(\theta) = \dfrac{\cos \theta}{\theta^2}$

38. $y = \dfrac{\sin(2x + 1)}{2x + 1}$

39. $y = \ln \sqrt{x + 1}$

40. $y = \ln\dfrac{x(x - 1)}{x + 2}$

41. $y = \sqrt{x} \ln x$

42. $y = x^2 e^{\sqrt{x}}$

43. $y = e^{-x}(\cos 2x + 3 \sin 2x)$

44. $y = \dfrac{\sqrt{x}(x + 2)^3}{\sqrt{x + 3}}$

45. $x \ln y + y \ln x = 3$

46. $\ln(x - y) + \sin y - x^2 = 0$

47. $y = \ln(x^2 e^{-2x})$

48. $y = \ln(\tan x)$

49. $y = \dfrac{e^x}{\sqrt{1 + e^{-x}}}$

50. $y = \ln|\sec 2x + \tan 2x|$

51. $y = e^{\csc x}$

52. $y = x \cdot 3^{x^2 + 1}$

53. $y = e^{e^x}$

54. $y = (2e)^{x/2}$

55. $ye^{-x} + xe^{y^2} = 8$

56. $y = 2e^{\sec^{-1} x}$

57. $y = 3^{x \cot x}$

58. $y = x^2 \ln(x + \sin^{-1} x)$

59. $y = x \sec^{-1} x$

60. $y = \tan(\cos^{-1} 2x)$

61. $y = \tan^{-1} \sqrt{x^2 + 1}$

62. $y = \sin^{-1}\left(\dfrac{x + 1}{x + 2}\right)$

63. $y = \tan^{-1}(\cos^{-1}\sqrt{x})$

64. $y = (\sin x)^{\cos x}$

In Exercises 65 and 66, find $f'(a)$.

65. $f(x) = \dfrac{\sqrt{x}}{x^2 + 1}$; $a = 4$

66. $f(x) = \sin(\cos x)$; $a = \dfrac{\pi}{4}$

In Exercises 67–76, find the second derivative of the function.

67. $y = x^3 + x^2 - \dfrac{1}{x}$

68. $g(x) = \dfrac{1}{3x + 1}$

69. $y = x\sqrt{2x - 1}$

70. $f(x) = \dfrac{x - 1}{x^2 + 1}$

71. $f(x) = \cos^2 x$

72. $f(x) = \sin\dfrac{1}{x}$

73. $y = \cot\dfrac{\theta}{2}$

74. $u = \cos(\pi - 2t) + \sin(\pi + 2t)$

75. $f(t) = t \cot t$

76. $h(x) = x^2 \cos\dfrac{1}{x}$

In Exercises 77 and 78, find $f''(a)$.

77. $f(x) = \sqrt{2x + 1}$; $a = 4$

78. $f(x) = x \tan x$; $a = \dfrac{\pi}{4}$

In Exercises 79 and 80, suppose that f and g are functions that are differentiable at $x = 2$ and that $f(2) = 3$, $f'(2) = -1$, $g(2) = 2$, and $g'(2) = 4$. Find $h'(2)$.

79. $h(x) = f(x)g(x)$

80. $h(x) = \dfrac{f(x)}{g(x)}$

In Exercises 81 and 82, find $h'(x)$ in terms of f, g, f', and g'.

81. $h(x) = \sqrt[3]{\dfrac{f(x)}{g(x)}}$

82. $h(x) = g[\sin f(x)]$

In Exercises 83–92, find dy/dx by implicit differentiation.

83. $3x^2 - 2y^2 = 6$

84. $x^3 + 3xy^2 + y^3 = 1$

85. $\dfrac{1}{x^2} + \dfrac{1}{y^2} = 1$

86. $x\sqrt{y} + y\sqrt{x} - 1 = 0$

87. $(x + y)^3 + x^3 + y^3 = 0$

88. $x \sin x + y \cos y = 3$

89. $\cos(x + y) + x \sin y = 1$

90. $\csc x + x \cot y = 1$

91. $\sec xy = 8$

92. $\cos^2 x + \sin^2 y = 1$

In Exercises 93 and 94, write the expression as a function of x.

93. $\displaystyle\lim_{h \to 0} \dfrac{2(x + h)^5 - (x + h)^3 - 2x^5 + x^3}{h}$

94. $\displaystyle\lim_{h \to 0} \dfrac{\sqrt{x + h} + \dfrac{1}{x + h} - \sqrt{x} - \dfrac{1}{x}}{h}$

In Exercises 95–100, find the differential of the function at the indicated number.

95. $f(x) = \sqrt{x} + \dfrac{1}{\sqrt{x}}$; $x = 4$

96. $f(x) = \sqrt{2x^2 + x + 1}$; $x = 1$

97. $f(x) = x(2x^2 - 1)^{1/3}$; $x = 1$

98. $f(x) = \sec^2 x$; $x = \dfrac{\pi}{4}$ **99.** $f(x) = x \sin \dfrac{1}{x}$; $x = \dfrac{6}{\pi}$

100. $f(x) = \dfrac{\tan x}{1 + \cot x}$; $x = \dfrac{\pi}{4}$

101. Find an equation of the tangent line to the graph of $y = (2x)/(\ln x)$ at the point $(e, 2e)$.

102. Find an equation of the tangent line to the graph of $y = xe^{-x}$ that is parallel to the line $x - y + 3 = 0$.

In Exercises 103 and 104, find equations of the tangent line and normal line to the curve at the indicated point.

103. $x^2 + 5xy + y^2 - 7 = 0$; $(1, 1)$

104. $x + \sqrt{xy} + y = 6$; $(2, 2)$

105. Find d^2y/dx^2 by implicit differentiation, given $x^3 + y^3 = 1$.

106. Find d^2y/dx^2 by implicit differentiation, given $\sin 2x + \cos 2y = 1$.

107. Find the linearization $L(x)$ of $f(x) = \cos^2 x$ at $\pi/6$.

108. Find the linearization of a suitable function, and then use it to approximate $\sqrt[3]{0.00096}$.

109. Let $f(x) = x^2 + 1$.
 a. Find the point on the graph of f at which the slope of the tangent line is equal to 2.
 b. Find an equation of the tangent line of part (a).

110. Let $f(x) = 2x^3 - 3x^2 - 16x + 3$.
 a. Find the points on the graph of f at which the slope of the tangent line is equal to -4.
 b. Find the equation(s) of the tangent line(s) of part (a).

111. Let $y = \dfrac{\sec x}{1 + \tan x}$. How fast is y changing when $x = \pi/4$?

112. The position of a particle moving along a coordinate line is
$$s(t) = t^3 - 12t + 1 \qquad t \ge 0$$
where $s(t)$ is measured in feet and t in seconds.
 a. Find the velocity and acceleration functions of the particle.
 b. Determine the times(s) when the particle is stationary.
 c. When is the particle moving in the positive direction and when is it moving in the negative direction?
 d. Construct a schematic showing the position of the body at any time t.
 e. What is the total distance traveled by the particle in the time interval $[0, 3]$?

113. The position function of a body moving along a coordinate line is
$$s(t) = 10t^{2/3} - t^{5/3} \qquad t \ge 0$$
where $s(t)$ is measured in feet and t in seconds. Find the velocity and acceleration functions for the body.

114. The position function of a particle moving along a coordinate line is
$$s(t) = 5 \cos\left(t + \dfrac{\pi}{4}\right) \qquad t \ge 0$$
where $s(t)$ is measured in feet and t in seconds.
 a. Find the velocity and acceleration functions for the particle.
 b. At what time does the particle first reach the origin?
 c. What are the velocity and acceleration of the particle when it first reaches the origin?

115. Velocity of Blood The velocity (in centimeters per second) of blood r cm from the central axis of an artery is given by
$$v(r) = k(R^2 - r^2)$$
where k is a constant and R is the radius of the artery. Suppose that $k = 1000$ and $R = 0.2$. Find $v(0.1)$, and $v'(0.1)$ and interpret your results.

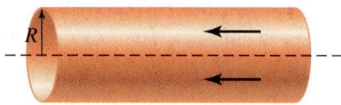

116. Traffic Flow The average speed of traffic flow on a stretch of Route 106 between 6 A.M. and 10 A.M. on a typical weekday is approximated by the function
$$f(t) = 20t - 45t^{0.45} + 50 \qquad 0 \le t \le 4$$
where $f(t)$ is measured in miles per hour and t is measured in hours with $t = 0$ corresponding to 6 A.M. How fast is the average speed of traffic flow changing at 7 A.M.? At 8 A.M.?

117. Surface Area of a Human Body An empirical formula by E.F. Dubois relates the surface area S of a human body (in square meters) to its mass W in kilograms and its height H in centimeters. The formula given by
$$S = 0.007184W^{0.425}H^{0.725}$$
is used by physiologists in metabolism studies. Suppose that a man is 1.83 m tall. How fast does his surface area change with respect to his mass when his mass is 80 kg?

118. Refer to Exercise 117. If the measurement of the mass of the man is subject to a maximum error of 0.5 kg, what is the percentage error in the calculation of the man's surface area?

119. Number of Hours of Daylight The number of hours of daylight on a particular day of the year in Boston is approximated by the function

$$f(t) = 3 \sin\left[\frac{2\pi}{365}(t - 79)\right] + 12$$

where $t = 0$ corresponds to January 1. Compute $f'(79)$, and interpret your result.

120. Projected Profit The management of the company that makes Long Horn Barbeque Sauce estimates that the daily profit from the production and sale of x cases of sauce is

$$P(x) = -0.000002x^3 + 6x - 350$$

dollars. Management forecasts that they will sell, on average, 900 cases of the sauce per day in the next several months. If the forecast is subject to a maximum error of 10%, find the corresponding error in the company's projected average daily profit.

121. The volume of a circular cone is $V = \pi r^2 h/3$, where r is the radius of the base and h is the height.

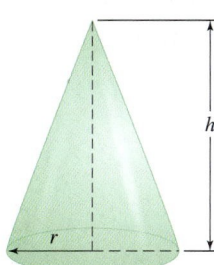

a. What is the rate of change of the volume with respect to the height if the radius is constant?
b. What is the rate of change of the volume with respect to the radius if the height is constant?

122. Depreciation of Equipment For assets such as machines, whose market values drop rapidly in the early years of usage, businesses often use the double declining balance method. In practice, a business firm normally employs the double declining balance method for depreciating such assets for a certain number of years and then switches over to the linear method. The double declining balance formula is

$$V(n) = C\left(1 - \frac{2}{N}\right)^n$$

where $V(n)$ denotes the book value of the assets at the end of n years and N is the number of years over which the asset is depreciated.
a. Find $V'(n)$.
b. What is the relative rate of change of $V(n)$?

123. Show that if the equation of motion of an object is $x(t) = ae^t + be^{-t}$, where a and b are constants, then its acceleration is numerically equal to the distance covered by the object.

124. The equation of motion of a mass attached to a spring and a dashpot damping device is

$$x(t) = e^{-2t}(2\cos 4t - 3\sin 4t)$$

where $x(t)$, measured in feet, is the displacement from the equilibrium position of the spring system and t is measured in seconds. Find expressions for the velocity and acceleration of the mass.

125. Given the equation $x^2 - y^2 = 9$, where x and y are both functions of t, find dy/dt if $x = 5$, $y = 4$, and $dx/dt = 3$.

126. Given the equation $\sin 2x + \cos 2y = 1$, where x and y are both functions of t, find dx/dt if $x = \pi/2$, $y = 0$, and $dy/dt = -1$.

127. Watching a Boat Race A spectator is watching a rowing race from the edge of a riverbank. The lead boat is moving in a straight line that is 120 ft from the river bank. If the boat is moving at a constant speed of 20 ft/sec, how fast will the boat be moving away from the spectator when it is 50 ft past her?

128. Watching a Space Shuttle Launch At a distance of 6000 ft from the launch site, a spectator is observing a space shuttle being launched. If the space shuttle lifts off vertically, at what rate is the distance between the spectator and the space shuttle changing with respect to the angle of elevation θ at the instant when the angle is 30° and the shuttle is traveling at 600 mph (880 ft/sec)?

PROBLEM-SOLVING TECHNIQUES

The following example shows that rewriting a function in an alternative form sometimes pays dividends.

EXAMPLE Find $f^{(n)}(x)$ if $f(x) = \dfrac{x}{x^2 - 1}$.

Solution Our first instinct is to use the Quotient Rule to compute $f'(x)$, $f''(x)$, and so on. The expectation here is either that the rule for $f^{(n)}$ will become apparent or that at least a pattern will emerge that will enable us to guess at the form for $f^{(n)}(x)$. But the futility of this approach will be evident when you compute the first two derivatives of f.

Let's see whether we can transform the expression for $f(x)$ before we differentiate. You can verify that $f(x)$ can be written as

$$f(x) = \frac{x}{x^2 - 1} = \frac{\frac{1}{2}(x - 1) + \frac{1}{2}(x + 1)}{(x + 1)(x - 1)} = \frac{1}{2}\left[\frac{1}{x + 1} + \frac{1}{x - 1}\right]$$

There is actually a systematic method for obtaining the last expression for $f(x)$. It is called *partial fraction decomposition* and will be taken up in Section 6.4. Differentiating, we obtain

$$f'(x) = \frac{1}{2}\frac{d}{dx}\left[\frac{1}{x + 1} + \frac{1}{x - 1}\right]$$

$$= \frac{1}{2}\left[\frac{d}{dx}(x + 1)^{-1} + \frac{d}{dx}(x - 1)^{-1}\right]$$

$$= \frac{1}{2}[(-1)(x + 1)^{-2} + (-1)(x - 1)^{-2}]$$

$$f''(x) = \frac{1}{2}[(-1)(-2)(x + 1)^{-3} + (-1)(-2)(x - 1)^{-3}]$$

$$f'''(x) = \frac{1}{2}[(-1)(-2)(-3)(x + 1)^{-4} + (-1)(-2)(-3)(x - 1)^{-4}]$$

$$= \frac{1}{2}[(-1)^3 3!(x + 1)^{-4} + (-1)^3 3!(x - 1)^{-4}]$$

$$\vdots$$

$$f^{(n)}(x) = \frac{(-1)^n n!}{2}\left[\frac{1}{(x + 1)^{n+1}} + \frac{1}{(x - 1)^{n+1}}\right]$$

where $n! = n(n - 1)(n - 2)\cdots(1)$ and $0! = 1$. ∎

CHALLENGE PROBLEMS

1. Find $\lim\limits_{x\to 2} \dfrac{x^{10} - 2^{10}}{x^5 - 2^5}$.

2. Find the derivative of $y = \sqrt{x + \sqrt{x + \sqrt{x}}}$.

3. a. Verify that $\dfrac{2x + 1}{x^2 + x - 2} = \dfrac{1}{x + 2} + \dfrac{1}{x - 1}$.

b. Find $f^{(n)}(x)$ if $f(x) = \dfrac{2x + 1}{x^2 + x - 2}$.

4. Find the values of x for which f is differentiable.
a. $f(x) = \sin|x|$ **b.** $f(x) = |\sin x|$

5. Find $f^{(10)}(x)$ if $f(x) = \dfrac{1 + x}{\sqrt{1 - x}}$.

Hint: Show that $f(x) = \dfrac{2}{\sqrt{1 - x}} - \sqrt{1 - x}$.

6. Find $f^{(n)}(x)$ if $f(x) = \dfrac{ax + b}{cx + d}$.

7. Suppose that f is differentiable and $f(a + b) = f(a)f(b)$ for all real numbers a and b. Show that $f'(x) = f'(0)f(x)$ for all x.

8. Suppose that $f^{(n)}(x) = 0$ for every x in an interval (a, b) and $f(c) = f'(c) = \cdots = f^{(n-1)}(c) = 0$ for some c in (a, b). Show that $f(x) = 0$ for all x in (a, b).

9. Let $F(x) = f(\sqrt{1 + x^2})$, where f is a differentiable function. Find $F'(x)$.

 10. Determine the values of b and c such that the parabola $y = x^2 + bx + c$ is tangent to the graph of $y = \sin x$ at the point $\left(\frac{\pi}{6}, \frac{1}{2}\right)$. Plot the graphs of both functions on the same set of axes.

11. Suppose f is defined on $(-\infty, \infty)$ and satisfies $|f(x) - f(y)| \leq (x - y)^2$ for all x and y. Show that f is a constant function.
Hint: Look at $f'(x)$.

12. Use the definition of the derivative to find the derivative of $f(x) = \tan ax$.

13. Find y'' at the point $(1, -2)$ if
$$2x^2 + 2xy + xy^2 - 3x + 3y + 7 = 0$$

14. Prove that the function $f(x) = |\ln x|$ is not differentiable at $x = 1$.

15. Let $g = f^{-1}$ be the inverse function of f. Show that if f has derivatives of order 3, then
$$g''' = \frac{3(f'')^2 - f'f'''}{(f')^5} \qquad f' \neq 0$$

16. Let f be positive and differentiable. Prove that the graphs of $y = f(x)$ and $y = f(x)\sin ax$ are tangent to each other at their points of intersection.

17. Let f be defined by
$$f(x) = \begin{cases} \dfrac{2x}{3 + e^{1/x}} & \text{if } x \neq 0 \\ 0 & \text{if } x = 0 \end{cases}$$

Is f differentiable at $x = 0$? Explain.
Hint: Use the definition of the derivative.

18. Find y' if $y = \log_{f(x)} g(x)$, where f and g are differentiable functions with $f(x) > 0$ and $g(x) > 0$ for all values of x.

Marco Simoni/Getty Images

Antarctic glaciers are calving into the ocean with greater frequency as a result of global warming. A major cause of global warming is the increase of carbon dioxide in the atmosphere. We can use the derivative to help us study the rate of change of the average amount of atmospheric CO_2.

3

Applications of the Derivative

IN THIS CHAPTER we continue to explore the power of the derivative of a function as a tool for solving problems. We will see how the first and second derivatives of a function can be used to help us sketch the graph of the function. We will also see how the derivative of a function can help us find the maximum and minimum values of the function. Determining these values is important because many practical problems call for finding one or both of these extreme values. For example, an engineer might be interested in finding the maximum horsepower a prototype engine can deliver, and a businesswoman might be interested in the level of production of a certain commodity that will minimize the unit cost of producing that commodity.

V This symbol indicates that one of the following video types is available for enhanced student learning at **www.academic.cengage.com/login**:
- Chapter lecture videos
- Solutions to selected exercises

Absolute Extrema of Functions

The graph of the function f in Figure 1 gives the altitude of a hot-air balloon over the time interval $I = [a, d]$. The point $(c, f(c))$, the lowest point on the graph of f, tells us that the hot-air balloon attains its minimum altitude, $f(c)$, at time $t = c$. The smallest value attained by f for all values of t in the domain I of f, $f(c)$, is called the *absolute minimum value* of f on I. Similarly, the point $(d, f(d))$, the highest point on the graph of f, tells us that the balloon attains its maximum altitude, $f(d)$, at time $t = d$. The largest value attained by f for all values of t in I is called the *absolute maximum value* of f on I.

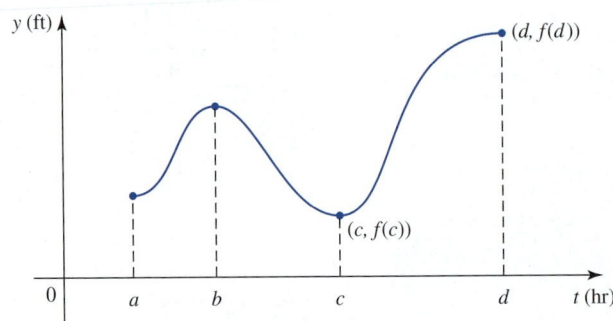

FIGURE 1
The altitude $f(t)$ of a hot-air balloon for $a \leq t \leq d$

More generally, we have the following definitions.

> **DEFINITIONS   Extrema of a Function f**
>
> A function f has an **absolute maximum** at c if $f(x) \leq f(c)$ for all x in the domain D of f. The number $f(c)$ is called the **maximum value** of f on D. Similarly, f has an **absolute minimum** at c if $f(x) \geq f(c)$ for all x in D. The number $f(c)$ is called the **minimum value** of f on D. The absolute maximum and absolute minimum values of f on D are called the **extreme values**, or **extrema**, of f on D.

EXAMPLE 1   Find the extrema of the function, if any, by examining its graph.

a. $f(x) = x^2$ **b.** $g(x) = -x^2$ **c.** $h(x) = \dfrac{1}{x}$ **d.** $k(x) = \dfrac{x}{\sqrt{x^2 + 7}}$

Solution   The graphs of the functions f, g, h, and k are shown in Figure 2.

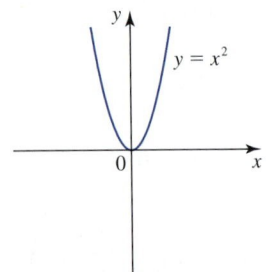

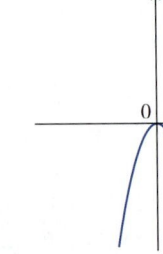

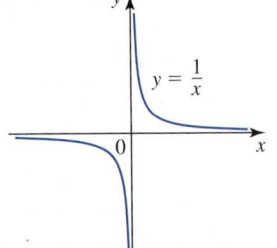

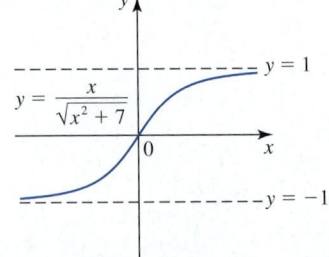

(a) f has a minimum at 0.   **(b)** g has a maximum at 0.   **(c)** h has no extrema.   **(d)** k has no extrema.

FIGURE 2

a. f has a minimum value of 0 at 0. Next, since the values of f are not bounded above, f has no maximum value.

b. g has a maximum value of 0 at 0. Also, because the values of g are not bounded below, g has no minimum value.

c. The values of h are neither bounded above nor bounded below, so h has no absolute extrema.

d. As x gets larger and larger, $k(x)$ gets closer and closer to 1. But this value is never attained; that is, a real number c does not exist such that $k(c) = 1$. Therefore, k has no maximum value. Similarly, you can show that k has no minimum value. ■

EXAMPLE 2 Find the extrema of the function:

a. $f(x) = x^2 \qquad -1 < x < 2$

b. $g(x) = x^2 \qquad -1 \le x \le 2$

Solution

a. The graph of f is shown in Figure 3a. We see that f has a minimum value of 0 at 0. Next, observe that as x approaches 2 through values less than 2, $f(x)$ increases and approaches 4. But f never attains the value 4. Therefore, f does not have a maximum.

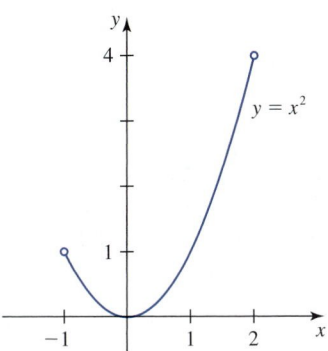

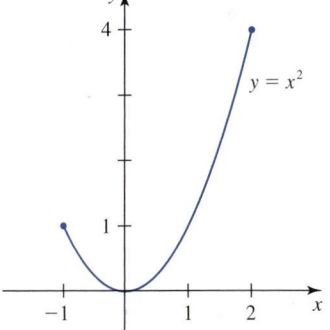

FIGURE 3 (a) f has a minimum at 0. (b) g has a minimum at 0 and a maximum at 2.

b. The graph of g is shown in Figure 3b. As before, we see that g has a minimum value of 0 at 0. Next, because 2 lies in the domain of g, we see that g does attain a largest value, namely, $g(2) = 4$. ■

■ Relative Extrema of Functions

If you refer once again to the graph of the function f giving the altitude of a hot-air balloon over the interval $[a, d]$ shown in Figure 4, you will see that the point $(b, f(b))$ is the highest point on the graph of f when compared to *neighboring points*. (For example, it is the highest point when compared to the points $(t, f(t))$, where $a < t < c$.) This tells us that $f(b)$ is the highest altitude attained by the balloon when considered over a small time interval containing $t = b$. The value $f(b)$ is called a *relative* (or *local*) *maximum value* of f.

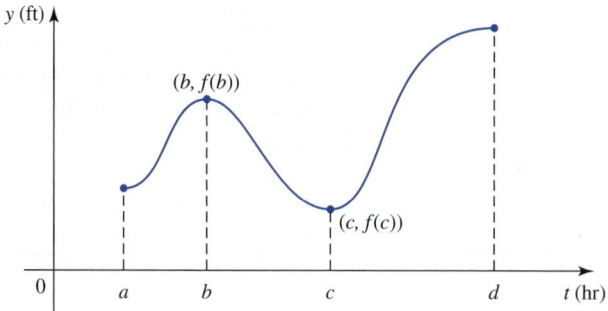

FIGURE 4
The altitude of a hot-air
balloon for $a \leq t \leq d$

Similarly, the point $(c, f(c))$ is the lowest point on the graph of f when compared to points nearby. (For example, it is the lowest point when compared to the points $(t, f(t))$, where $b < t < d$.) This tells us that the balloon attains the lowest altitude at $t = c$ when considered over a small time interval containing $t = c$. The value $f(c)$ is called a *relative* (or *local*) *minimum value* of f. Recall that $f(c)$ also happens to be the (absolute) minimum value of f, as we observed earlier.

More generally, we have the following definition.

DEFINITIONS **Relative Extrema of a Function**

A function f has a **relative** (or **local**) **maximum** at c if $f(c) \geq f(x)$ for all values of x in some open interval containing c. Similarly, f has a **relative** (or **local**) **minimum** at c if $f(c) \leq f(x)$ for all values of x in some open interval containing c.

The function f whose graph is shown in Figure 5 has a relative maximum at a and at c and a relative minimum at b and at d. The graph of f suggests that at a point corresponding to a relative extremum of f, either the tangent line is horizontal or it does not exist. Put another way, the values of x that correspond to these points are precisely the numbers in the domain of f at which f' is zero or f' does not exist.

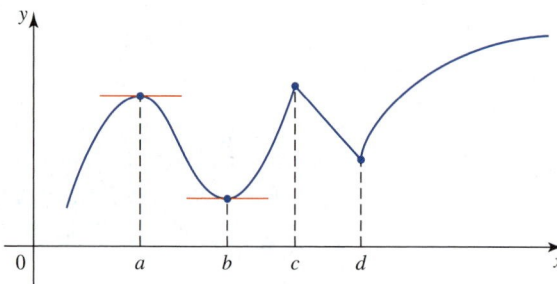

FIGURE 5
The function f has relative extrema
at a, b, c, and d. The tangent lines
at a and b are horizontal. There
are no tangent lines at c and d.

These observations suggest the following theorem, which tells us where the relative extrema of a function may occur.

THEOREM 1 **Fermat's Theorem**

If f has a relative extremum at c, then either $f'(c) = 0$ or $f'(c)$ does not exist.

PROOF   First, suppose that f has a relative maximum at c. If f is not differentiable at c, then there is nothing to prove. So let's suppose that $f'(c)$ exists. Since f has a relative maximum at c, there exists an open interval, I, such that $f(x) \leq f(c)$ for all x in I. This implies that if we pick h to be positive and sufficiently small (so that $c + h$ lies in I), then

$$f(c + h) \leq f(c) \qquad \text{or} \qquad f(c + h) - f(c) \leq 0$$

Multiplying both sides of the latter inequality by $1/h$, where $h > 0$, we obtain

$$\frac{f(c + h) - f(c)}{h} \leq 0$$

Taking the right-hand limit of both sides of this inequality gives

$$\lim_{h \to 0^+} \frac{f(c + h) - f(c)}{h} \leq \lim_{h \to 0^+} 0 = 0 \qquad \text{\textcolor{orange}{By Theorem 3 of Section 1.2}}$$

Since $f'(c)$ exists, we have

$$f'(c) = \lim_{h \to 0} \frac{f(c + h) - f(c)}{h} = \lim_{h \to 0^+} \frac{f(c + h) - f(c)}{h}$$

and we have shown that $f'(c) \leq 0$.

Next, we pick h to be negative and sufficiently small (so that $c + h$ lies in I). Then

$$f(c + h) \leq f(c) \qquad \text{or} \qquad f(c + h) - f(c) \leq 0$$

Upon multiplying this last inequality by $1/h$ and reversing the direction of the inequality (because $1/h < 0$), we have

$$f'(c) = \lim_{h \to 0} \frac{f(c + h) - f(c)}{h} = \lim_{h \to 0^-} \frac{f(c + h) - f(c)}{h} \geq 0$$

Thus, we have shown that $f'(c) \leq 0$ and $f'(c) \geq 0$, simultaneously. Therefore, $f'(c) = 0$. This proves the theorem for the case in which f has a relative maximum at c. The case in which f has a relative minimum at c can be proved in a similar manner (see Exercise 90). ∎

The values of x at which f' is zero or f' does not exist are given a special name.

DEFINITION   Critical Number of f

A **critical number** of a function f is any number c in the domain of f at which $f'(c) = 0$ or $f'(c)$ does not exist.

⚠️ Theorem 1 states that a relative extremum of f can occur only at a critical number of f. It is important to realize, however, that the converse of Theorem 1 is false. In other words, you may *not* conclude that if c is a critical number of f, then f must have a relative extremum at c. (See Example 3.)

EXAMPLE 3   Show that zero is a critical number of each of the functions $f(x) = x^3$ and $g(x) = x^{1/3}$ but that neither function has a relative extremum at 0.

Solution The graphs of f and g are shown in Figure 6. Since $f'(x) = 3x^2 = 0$ if $x = 0$, we see that 0 is a critical number of f. But observe that $f(x) < 0$ if $x < 0$ and $f(x) > 0$ if $x > 0$, and this tells us that f cannot have a relative extremum at 0.

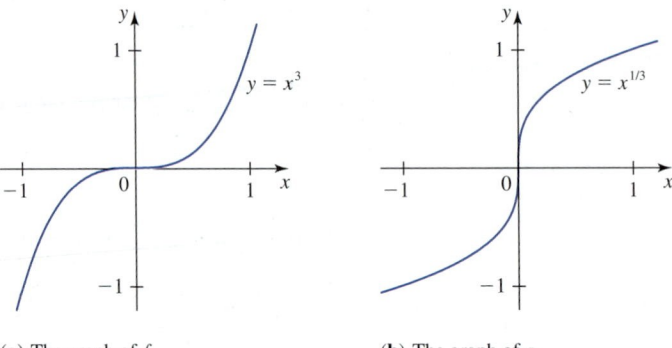

FIGURE 6
Both f and g have 0 as a critical number, but neither function has a relative extremum at 0.

(a) The graph of f (b) The graph of g

Next, we compute

$$g'(x) = \frac{1}{3}x^{-2/3} = \frac{1}{3x^{2/3}}$$

Note that g' is not defined at 0, but g is; so 0 is a critical number of g. Observe that $g(x) < 0$ if $x < 0$ and $g(x) > 0$ if $x > 0$, so g cannot have a relative extremum at 0.

EXAMPLE 4 Find the critical numbers of $f(x) = x - 3x^{1/3}$.

Solution The derivative of f is

$$f'(x) = 1 - x^{-2/3} = \frac{x^{2/3} - 1}{x^{2/3}}$$

Observe that f' is not defined at 0 and also $f'(x) = 0$ if $x = \pm 1$. Therefore, the critical numbers of f are -1, 0, and 1.

We will develop a systematic method for finding the relative extrema of a function in Section 3.3. For the rest of this section we will develop techniques for finding the extrema of continuous functions defined on closed intervals.

■ Finding the Extreme Values of a Continuous Function on a Closed Interval

As you saw in the preceding examples, an arbitrary function might or might not have a maximum value or a minimum value. But there is an important case in which the extrema always exist for a function. The conditions are spelled out in Theorem 2.

THEOREM 2 **The Extreme Value Theorem**

If f is continuous on a closed interval $[a, b]$, then f attains an absolute maximum value $f(c)$ for some number c in $[a, b]$ and an absolute minimum value $f(d)$ for some number d in $[a, b]$.

In certain applications, not only is a function continuous on a closed interval $[a, b]$, but it is also differentiable, with the possible exception of a finite set of numbers, on the open interval (a, b). In such cases, the following procedure can be used to find the extrema of the function.

> **Guidelines for Finding the Extrema of a Continuous Function f on $[a, b]$**
>
> **1.** Find the critical numbers of f that lie in (a, b).
> **2.** Compute the value of f at each of these critical numbers, and also compute $f(a)$ and $f(b)$.
> **3.** The absolute maximum value of f and the absolute minimum value of f are precisely the largest and the smallest numbers found in Step 2.

This procedure can be justified as follows: If an extremum of f occurs at a number in the open interval (a, b), then it must also be a relative extremum of f; hence it must occur at a critical number of f. Otherwise, the extremum of f must occur at one or both of the endpoints of the interval $[a, b]$. (See Figure 7.)

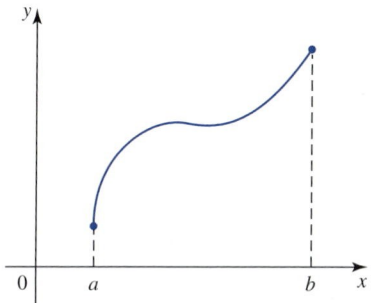

(a) The extreme values of f occur at the endpoints.

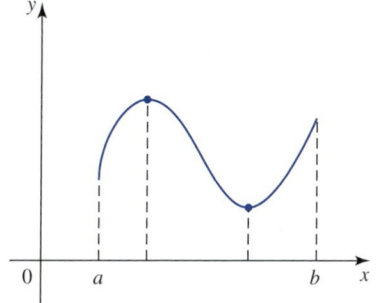

(b) The extreme values of f occur at critical numbers.

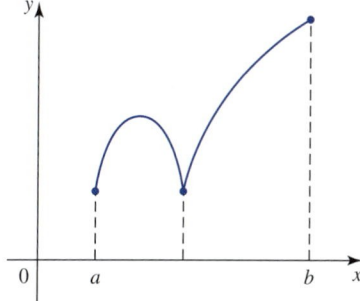

(c) The absolute minimum value of f occurs at both an endpoint and a critical number of f, whereas the absolute maximum value of f occurs at an endpoint.

FIGURE 7
f is continuous on $[a, b]$.

EXAMPLE 5 Find the extreme values of the function $f(x) = 3x^4 - 4x^3 - 8$ on $[-1, 2]$.

Solution Since f is a polynomial function, it is continuous everywhere; in particular, it is continuous on the closed interval $[-1, 2]$. Therefore, we can use the Extreme Value Theorem.

First, we find the critical numbers of f in $(-1, 2)$:

$$f'(x) = 12x^3 - 12x^2$$
$$= 12x^2(x - 1)$$

Observe that f' is continuous on $(-1, 2)$. Next, setting $f'(x) = 0$ gives $x = 0$ or $x = 1$. Therefore, 0 and 1 are the only critical numbers of f in $(-1, 2)$.

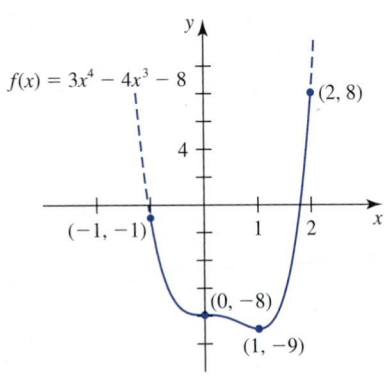

$f(x) = 3x^4 - 4x^3 - 8$

(2, 8)

(−1, −1)

(0, −8)

(1, −9)

FIGURE 8
The maximum value of f is 8, and the minimum value is −9.

Next, we compute $f(x)$ at these critical numbers as well as at the endpoints −1 and 2. These values are shown in the following table.

x	−1	0	1	2
$f(x)$	−1	−8	−9	8

From the table we see that f attains the absolute maximum value of 8 at 2 and the absolute minimum value of −9 at 1. The graph of f shown in Figure 8 confirms our results. (You don't need to draw the graph to solve the problem.)   ■

EXAMPLE 6 Find the extreme values of the function $f(x) = 2 \cos x - x$ on $[0, 2\pi]$.

Solution The function f is continuous everywhere; in particular, it is continuous on the closed interval $[0, 2\pi]$. Therefore, the Extreme Value Theorem is applicable.
  First, we find the critical numbers of f in $(0, 2\pi)$. We have

$$f'(x) = -2 \sin x - 1$$

Observe that f' is continuous on $(0, 2\pi)$. Setting $f'(x) = 0$ gives

$$-2 \sin x - 1 = 0$$

$$\sin x = -\frac{1}{2}$$

Thus, $x = 7\pi/6$ or $11\pi/6$. (Remember x lies in $(0, 2\pi)$.) So $7\pi/6$ and $11\pi/6$ are the only critical numbers of f in $(0, 2\pi)$.
  Next, we compute the values of f at these critical numbers as well as at the endpoints 0 and 2π. These values are shown in the following table.

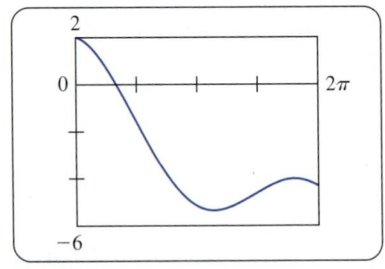

FIGURE 9
The graph of $f(x) = 2 \cos x - x$ on $[0, 2\pi]$

x	0	$\dfrac{7\pi}{6}$	$\dfrac{11\pi}{6}$	2π
$f(x)$	2	−5.40	−4.03	−4.28

From the table we see that f attains the absolute maximum value of 2 at 0 and the absolute minimum value of approximately −5.4 at $7\pi/6$. The graph of f shown in Figure 9 confirms our results.   ■

■ An Optimization Problem

The solution to many practical problems involves finding the absolute maximum or the absolute minimum of a function. If we know that the function to be optimized is continuous on a closed interval, then the techniques of this section can be used to solve the problem, as illustrated in the following example.

EXAMPLE 7 **Maximum Deflection of a Beam**   Figure 10 depicts a beam of length L and uniform weight w per unit length that is rigidly fixed at one end and simply supported at the other. An equation of the elastic curve (the dashed curve in the figure) is

$$y = \frac{w}{48EI}(2x^4 - 5Lx^3 + 3L^2x^2)$$

where the product EI is a constant called the *flexural rigidity* of the beam. Show that the maximum deflection (the displacement of the elastic curve from the x-axis) occurs at $x = (15 - \sqrt{33})L/16 \approx 0.578L$ and has a magnitude of approximately $0.0054wL^4/(EI)$.

FIGURE 10
The beam is rigidly fixed at $x = 0$ and simply supported at $x = L$. Note the orientation of the y-axis.

Solution   We wish to find the value of x on the closed interval $[0, L]$ at which the function f defined by

$$f(x) = \frac{w}{48EI}(2x^4 - 5Lx^3 + 3L^2x^2)$$

attains its absolute maximum value. Since f is continuous on $[0, L]$, this value must be attained at a critical number of f in $(0, L)$ or at an endpoint of the interval. To find the critical numbers of f, we compute

$$f'(x) = \frac{w}{48EI}(8x^3 - 15Lx^2 + 6L^2x)$$

$$= \frac{w}{48EI}x(8x^2 - 15Lx + 6L^2)$$

Setting $f'(x) = 0$ gives $x = 0$ or

$$x = \frac{15L \pm \sqrt{225L^2 - 192L^2}}{16}$$

$$= \frac{15L \pm \sqrt{33}L}{16}$$

Because $(15 + \sqrt{33})L/16 > L$, we see that the sole critical number of f in $(0, L)$ is $x = (15 - \sqrt{33})L/16 \approx 0.578L$. Evaluating f at 0, 0.578L, and L, we obtain the following table of values.

$f(0)$	$f(0.578L)$	$f(L)$
0	$\dfrac{0.0054wL^4}{EI}$	0

We conclude that the maximum deflection occurs at $x = (15 - \sqrt{33})L/16 \approx 0.578L$ and has a magnitude of approximately $0.0054wL^4/(EI)$. ■

Our final example shows how a graphing utility can be used to approximate the maximum and minimum values of a continuous function defined on a closed interval. But to obtain the *exact* values, we must solve the problem analytically.

EXAMPLE 8 Let $f(x) = 2 \sin x + \sin 2x$.

a. Use a graphing utility to plot the graph of f using the viewing window $\left[0, \frac{3\pi}{2}\right] \times [-3, 3]$. Find the approximate absolute maximum and absolute minimum values of f on the interval $\left[0, \frac{3\pi}{2}\right]$.

b. Obtain the exact absolute maximum and absolute minimum values of f analytically.

Solution

a. The required graph is shown in Figure 11. From the graph we see that the absolute maximum value of f is approximately 2.6 obtained when $x \approx 1$. The absolute minimum value of f is -2 obtained when $x = 3\pi/2$.

b. The function f is continuous everywhere and, in particular, on the interval $\left[0, \frac{3\pi}{2}\right]$. We find

$$f'(x) = 2 \cos x + 2 \cos 2x$$
$$= 2 \cos x + 2(\cos^2 x - \sin^2 x) \qquad \cos 2x = \cos^2 x - \sin^2 x$$
$$= 2 \cos x + 2(\cos^2 x - 1 + \cos^2 x) \qquad \sin^2 x = 1 - \cos^2 x$$
$$= 2(2 \cos^2 x + \cos x - 1)$$

Since

$$2 \cos^2 x + \cos x - 1 = (2 \cos x - 1)(\cos x + 1) = 0$$

if $\cos x = -1$ or $\frac{1}{2}$, we see that $x = \pi/3$ or π. From the following table we see that the absolute maximum value of f is $3\sqrt{3}/2$ and the absolute minimum value of f is -2.

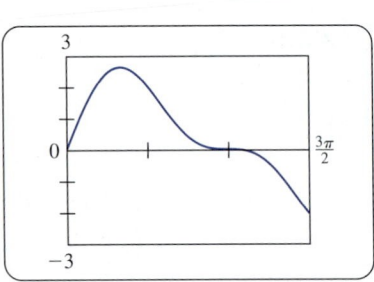

FIGURE 11

x	0	$\dfrac{\pi}{3}$	π	$\dfrac{3\pi}{2}$
$f(x)$	0	$\dfrac{3\sqrt{3}}{2}$	0	-2

3.1 CONCEPT QUESTIONS

1. Explain each of the following terms: (a) absolute maximum value of a function f; (b) relative maximum value of a function f. Illustrate each with an example.
2. **a.** What is a critical number of a function f?
   b. Explain the role of a critical number in determining the relative extrema of a function.

3. **a.** Explain the Extreme Value Theorem in your own words.
   b. Describe a procedure for finding the extrema of a continuous function f on a closed interval $[a, b]$.

3.1 EXERCISES

In Exercises 1–6, you are given the graph of a function f defined on the indicated domain. Find the absolute maximum and absolute minimum values of f (if they exist) and where they are attained.

1. *f* defined on (0, 2]

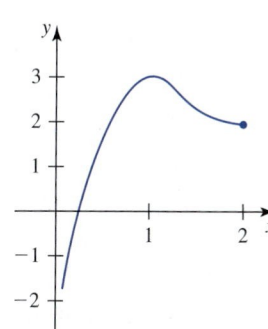

2. *f* defined on (−∞, ∞)

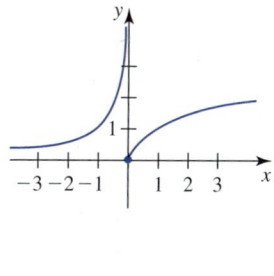

3. *f* defined on (−∞, ∞)

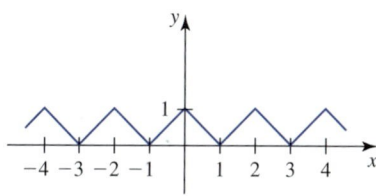

4. *f* defined on (−2, ∞)

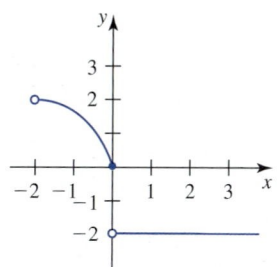

5. *f* defined on [0, 5]

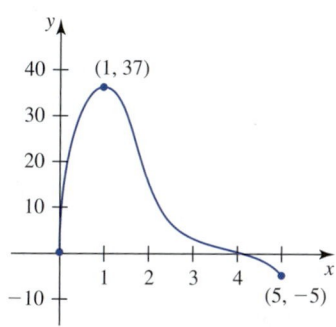

6. *f* defined on (−1, ∞)

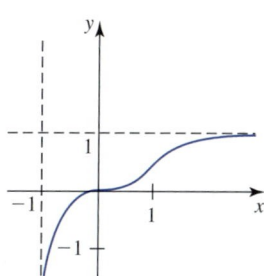

In Exercises 7–24, sketch the graph of the function and find its absolute maximum and absolute minimum values, if any.

7. $f(x) = 2x + 3$ on $[-1, \infty)$
8. $g(x) = -3x + 2$ on $(-1, 2]$
9. $h(t) = t^2 - 1$ on $(-1, 0)$
10. $f(t) = t^2 - 1$ on $[-1, 0)$
11. $g(x) = x^2 + 1$ on $(0, \infty)$
12. $h(x) = x^2 + 1$ on $(-2, 1]$
13. $f(x) = x^2 - 4x + 3$ on $(-\infty, \infty)$
14. $g(x) = 2x^2 - 3x + 1$ on $[0, 1)$
15. $f(x) = \dfrac{1}{x}$ on $(0, 1]$
16. $g(x) = \dfrac{1}{x}$ on $(-1, 1)$
17. $f(x) = |x|$ on $[-2, 1)$
18. $g(x) = |2x - 1|$ on $(0, 2]$
19. $f(t) = 2 \sin t$ on $\left(0, \frac{3\pi}{2}\right)$
20. $h(t) = \cos \pi t$ on $\left[\frac{1}{4}, 1\right)$
21. $f(x) = e^x$ on $(-\infty, 1]$
22. $g(x) = \ln x$ on $(0, e)$
23. $f(x) = \begin{cases} x & \text{if } -1 \le x \le 0 \\ 2 - x & \text{if } 0 < x \le 2 \end{cases}$
24. $f(x) = \begin{cases} \sqrt{4 - x^2} & \text{if } -2 \le x < 0 \\ -\sqrt{4 - x^2} & \text{if } 0 \le x \le 2 \end{cases}$

In Exercises 25–40, find the critical number(s), if any, of the function.

25. $f(x) = 2x + 3$
26. $g(x) = 4 - 3x$
27. $f(x) = 2x^2 + 4x$
28. $h(t) = 6t^2 - t - 2$
29. $f(x) = x^3 - 6x + 2$
30. $g(t) = 2t^3 + 3t^2 - 12t + 4$
31. $h(x) = x^4 - 4x^3 + 12$
32. $g(t) = 3t^4 + 4t^3 - 12t^2 + 8$
33. $f(x) = x^{2/3}$
34. $g(t) = 4t^{1/3} + 3t^{4/3}$
35. $h(u) = \dfrac{u}{u^2 + 1}$
36. $g(x) = \dfrac{x^2}{x^2 + 3}$
37. $f(t) = \cos^2(2t)$
38. $g(\theta) = 2 \sin \theta - \cos 2\theta$
39. $f(x) = e^{x^2} - \pi$
40. $g(t) = t^2 \ln t$

In Exercises 41–60, find the absolute maximum and absolute minimum values, if any, of the function.

41. $f(x) = x^2 - x - 2$ on $[0, 2]$

42. $f(x) = -x^2 + 4x + 3$ on $[-1, 3]$

43. $h(x) = x^3 + 3x^2 + 1$ on $[-3, 2]$

44. $f(t) = -2t^3 + 3t^2 + 12t + 3$ on $[-2, 3]$

45. $g(x) = 3x^4 + 4x^3 + 1$ on $[-2, 1]$

46. $f(x) = 2x^4 - \dfrac{8}{3}x^3 - 8x^2 + 12$ on $[-2, 3]$

47. $f(x) = \dfrac{x}{x^2 + 1}$ on $[-1, 2]$ **48.** $g(u) = \dfrac{\sqrt{u}}{u^2 + 1}$ on $[0, 2]$

49. $g(v) = \dfrac{v}{v - 1}$ on $[2, 4]$ **50.** $f(x) = 2x + \dfrac{1}{x}$ on $[-1, 3]$

51. $f(x) = x - 2\sqrt{x}$ on $[0, 9]$ **52.** $f(t) = \dfrac{1}{8}t^2 - 4\sqrt{t}$ on $[0, 9]$

53. $f(x) = x^{2/3}(x^2 - 4)$ on $[-1, 2]$

54. $g(x) = x\sqrt{4 - x^2}$ on $[0, 2]$

55. $f(x) = 2 + 3\sin 2x$ on $\left[0, \dfrac{\pi}{2}\right]$

56. $g(x) = \cos x - \sin x$ on $[0, 2\pi]$

57. $f(x) = xe^{-x}$; $[-1, 2]$ **58.** $f(x) = e^{2x} - e^x$; $[-2, 0]$

59. $f(x) = x \ln x - x$; $\left[\dfrac{1}{2}, 2\right]$ **60.** $f(x) = \dfrac{\ln x + 1}{x}$; $\left[\dfrac{1}{2}, 3\right]$

61. Maximizing Profit   The total daily profit in dollars realized by the TKK Corporation in the manufacture and sale of x dozen recordable DVDs is given by the total profit function

$$P(x) = -0.000001x^3 + 0.001x^2 + 5x - 500$$
$$0 \le x \le 2000$$

Find the level of production that will yield a maximum daily profit.

62. Reaction to a Drug   The strength of a human body's reaction to a dosage D of a certain drug is given by

$$R = D^2\left(\dfrac{k}{2} - \dfrac{D}{3}\right)$$

where k is a positive constant. Show that the maximum reaction is achieved if the dosage is k units.

63. Traffic Flow   The average speed of traffic flow on a stretch of Route 124 between 6 A.M. and 10 A.M. on a typical weekday is approximated by the function

$$f(t) = 20t - 40\sqrt{t} + 50 \qquad 0 \le t \le 4$$

where $f(t)$ is measured in miles per hour and t is measured in hours, with $t = 0$ corresponding to 6 A.M. At what time in the morning is the average speed of traffic flow highest? At what time in the morning is it lowest?

64. Foreign-Born Medical Residents   The percentage of foreign-born medical residents in the United States from the beginning of 1910 to the beginning of 2000 is approximated by the function

$$P(t) = 0.04363t^3 - 0.267t^2 - 1.59t + 14.7 \qquad 0 \le t \le 9$$

where t is measured in decades with $t = 0$ corresponding to the beginning of 1910. Show that the percentage of foreign-born medical residents was lowest in early 1970.
Source: Journal of the American Medical Association.

65. Brain Growth and IQs   In a study conducted at the National Institute of Mental Health, researchers followed the development of the cortex, the thinking part of the brain, in 307 children. Using repeated magnetic resonance imaging scans from childhood to the late teens, they measured the thickness (in millimeters) of the cortex of children of age t years with the highest IQs: 121 to 149. These data lead to the model

$$S(t) = 0.000989t^3 - 0.0486t^2 + 0.7116t + 1.46$$
$$5 \le t \le 19$$

Show that the cortex of children with superior intelligence reaches maximum thickness around age 11.
Source: Nature.

66. Brain Growth and IQs   Refer to Exercise 65. The researchers at the institute also measured the thickness (also in millimeters) of the cortex of children of age t years who were of average intelligence. These data lead to the model

$$A(t) = -0.00005t^3 - 0.000826t^2 + 0.0153t + 4.55$$
$$5 \le t \le 19$$

Show that the cortex of children with average intelligence reaches maximum thickness at age 6.
Source: Nature.

67. Maximizing Revenue   The quantity demanded per month of the Peget wristwatch is related to the unit price by the demand equation

$$p = \dfrac{50}{0.01x^2 + 1} \qquad 0 \le x \le 20$$

where p is measured in dollars and x is measured in units of a thousand. How many watches must be sold by the manufacturer to maximize its revenue?
Hint: Recall that the revenue $R = px$.

68. Poiseuille's Law   According to Poiseuille's Law, the velocity (in centimeters per second) of blood r cm from the central axis of an artery is given by

$$v(r) = k(R^2 - r^2) \qquad 0 \le r \le R$$

where k is a constant and R is the radius of the artery. Show that the flow of blood is fastest along the central axis. Where is the flow of blood slowest?

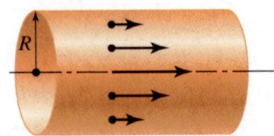

69. Chemical Reaction In an autocatalytic chemical reaction the product formed acts as a catalyst for the reaction. If Q is the amount of the original substrate that is present initially and x is the amount of catalyst formed, then the rate of change of the chemical reaction with respect to the amount of catalyst present in the reaction is

$$R(x) = kx(Q - x) \qquad 0 \le x \le Q$$

where k is a constant. Show that the rate of the chemical reaction is greatest at the point at which exactly half of the original substrate has been transformed.

70. Velocity of Airflow During a Cough When a person coughs, the trachea (windpipe) contracts, allowing air to be expelled at a maximum velocity. It can be shown that the velocity v of airflow during a cough is given by

$$v = f(r) = kr^2(R - r) \qquad 0 \le r \le R$$

where r is the radius of the trachea in centimeters during a cough, R is the normal radius of the trachea in centimeters, and k is a constant that depends on the length of the trachea. Find the radius for which the velocity of airflow is greatest.

71. A Mixture Problem A tank initially contains 10 gal of brine with 2 lb of salt. Brine with 1.5 lb of salt per gallon enters the tank at the rate of 3 gal/min, and the well-stirred mixture leaves the tank at the rate of 4 gal/min. It can be shown that the amount of salt in the tank after t min is x lb, where

$$x = f(t) = 1.5(10 - t) - 0.0013(10 - t)^4 \qquad 0 \le t \le 10$$

What is the maximum amount of salt present in the tank at any time?

72. Air Pollution According to the South Coast Air Quality Management district, the level of nitrogen dioxide, a brown gas that impairs breathing, that is present in the atmosphere between 7 A.M. and 2 P.M. on a certain May day in downtown Los Angeles is approximated by

$$I(t) = 0.03t^3(t - 7)^4 + 60.2 \qquad 0 \le t \le 7$$

where $I(t)$ is measured in pollutant standard index (PSI) and t is measured in hours, with $t = 0$ corresponding to 7 A.M. Determine the time of day when the PSI is the lowest and when it is the highest.

Source: The Los Angeles Times.

73. Office Rents After the economy softened, the sky-high office space rents of the late 1990s started to come down to earth. The function R gives the approximate price per square foot in dollars, $R(t)$, of prime space in Boston's Back Bay and Financial District from the beginning of 1997 ($t = 0$) to the beginning of 2002 ($t = 5$), where

$$R(t) = -0.711t^3 + 3.76t^2 + 0.2t + 36.5 \qquad 0 \le t \le 5$$

Show that the office space rents peaked at about the middle of the year 2000. What was the highest office space rent during the period in question?

Source: Meredith & Grew Inc./Oncor.

74. Maximum Deflection of a Beam A uniform beam of length L ft and negligible weight rests on supports at both ends. When subjected to a uniform load of w_0 lb/ft, it bends and has the *elastic curve* (the dashed curve in the figure below) described by the equation

$$y = \frac{w_0}{24EI}(x^4 - 2Lx^3 + L^3x) \qquad 0 \le x \le L$$

where the product EI is a constant called the *flexural rigidity* of the beam. Show that the maximum deflection of the beam occurs at the midpoint of the beam and that its value is $5w_0L^4/(384EI)$.

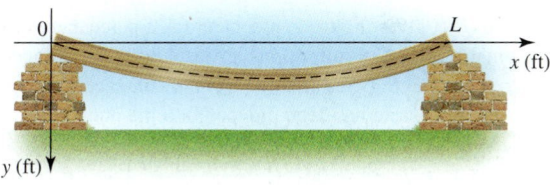

75. Use of Diesel Engines Diesel engines are popular in cars in Europe, where fuel prices are high. The percentage of new vehicles in Western Europe equipped with diesel engines is approximated by the function

$$f(t) = 0.3t^4 - 2.58t^3 + 8.11t^2 - 7.71t + 23.75$$
$$0 \le t \le 4$$

where t is measured in years, with $t = 0$ corresponding to the beginning of 1996.

 a. Plot the graph of f using the viewing window $[0, 4] \times [0, 40]$.

b. What was the lowest percentage of new vehicles equipped with diesel engines for the period in question?

Source: German Automobile Industry Association.

76. Federal Debt According to data obtained from the Congressional Budget Office, the national debt (in trillions of dollars) is given by the function

$$f(t) = 0.0022t^3 - 0.0465t^2 + 0.506t + 3.27 \qquad 0 \le t \le 20$$

where t is measured in years, with $t = 0$ corresponding to the beginning of 1990.

a. Plot the graph of f using the viewing window $[0, 20] \times [0, 14]$.

b. When was the federal debt at the highest level over the period under consideration? What was that level?

Source: Congressional Budget Office.

77. A cylindrical tank of height h is filled with water. Suppose a jet of water flows through an orifice on the tank. According to Torricelli's law, the velocity of flow of the jet of water is given by $V = \sqrt{2gx}$ where g is the gravitational constant. It can be shown that the range R (in feet) of the jet of water is given by $R = 2\sqrt{x(h - x)}$. Where should the orifice be located so that the jet of water will have the maximum range?

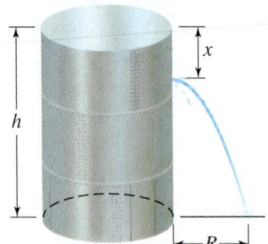

78. **Water Pollution** When organic waste is dumped into a pond, the oxidation process that takes place reduces the pond's oxygen content. However, given time, nature will restore the oxygen content to its natural level. In the accompanying graph, $P(t)$ gives the oxygen content (as a percentage of its normal level) t days after organic waste has been dumped into the pond. Suppose that the oxygen content t days after the organic waste has been dumped into the pond is given by

$$P(t) = 100\left(\frac{t^2 + 10t + 100}{t^2 + 20t + 100}\right)$$

percent of its normal level. Find the coordinates of the point P, and explain its significance.

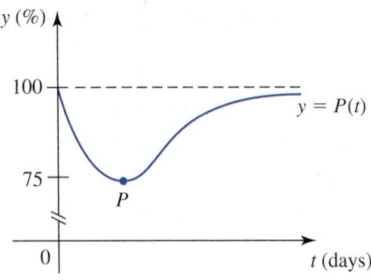

79. **Path of a Boat** A boat leaves the point O (the origin) located on one bank of a river, traveling with a constant speed of 20 mph and always heading toward a dock located at the point A $(1000, 0)$, which is due east of the origin (see the figure). The river flows north at a constant speed of 5 mph. It can be shown that the path of the boat is

$$y = 500\left[\left(\frac{1000 - x}{1000}\right)^{3/4} - \left(\frac{1000 - x}{1000}\right)^{5/4}\right]$$

$$0 \le x \le 1000$$

Find the maximum distance the boat has drifted north during its trip.

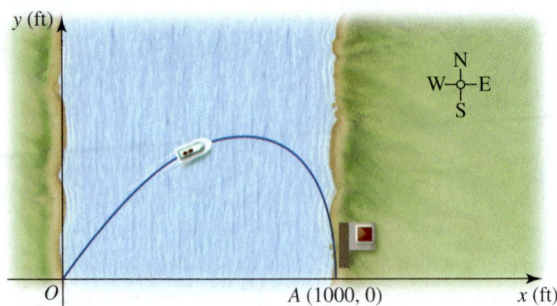

80. **A Motorcyclist's Turn** A motorcyclist weighing 180 lb traveling at a constant speed of 30 mph executes a turn on a road described by the graph of $y = 100e^{0.01x}$, where $-200 \le x \le 50$. It can be shown that the magnitude of the normal force acting on the motorcyclist is approximately

$$F = \frac{10,890e^{0.1x}}{(1 + 100e^{0.2x})^{3/2}}$$

pounds. Find the maximum force acting on the motorcyclist as he makes the turn.

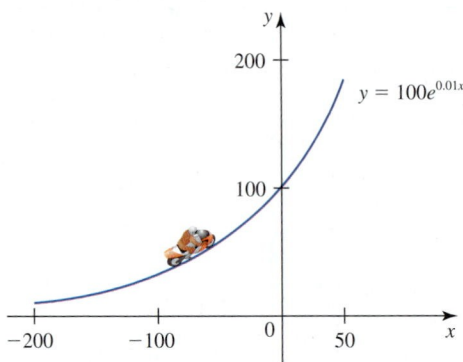

81. **Construction of an AC Transformer** In constructing an AC transformer, a cross-shaped iron core is inserted into a coil (see the figure). If the radius of the coil is a, find the values of x and y such that the iron core has the largest surface area.
Hint: Let $x = a \cos \theta$ and $y = a \sin \theta$. Then maximize the function

$$S = 4xy + 4y(x - y) = 8xy - 4y^2$$
$$= 4a^2(\sin 2\theta - \sin^2 \theta)$$

on the interval $0 \le \theta \le \frac{\pi}{4}$.

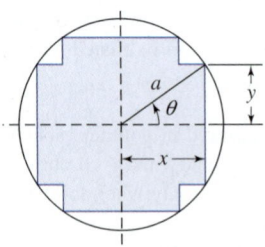

82. A body of mass m moves in an elliptical path with a constant angular speed ω (see the figure). It can be shown that the force acting on the body is always directed toward the origin and has magnitude given by

$$F = m\omega^2\sqrt{a^2 \cos^2 \omega t + b^2 \sin^2 \omega t} \qquad t \geq 0$$

where a and b are constants with $a > b$. Find the points on the path where the force is greatest and where it is smallest. Does your result agree with your intuition?

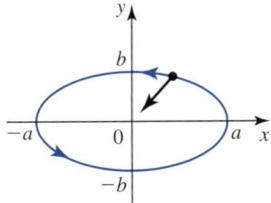

83. The object shown in the figure is a crate full of office equipment that weighs W lb. Suppose you try to move the crate by tying a rope around it and pulling on the rope at an angle θ to the horizontal. Then the magnitude F of the force that is required to set the crate in motion is

$$F = \frac{\mu W}{\mu \sin \theta + \cos \theta} \qquad 0 \leq \theta \leq \frac{\pi}{2}$$

where μ is a constant called the *coefficient of static friction*.
a. Find the angle θ at which F is minimized.
b. What is the magnitude of the force found in part (a)?
 c. Suppose $W = 60$ and $\mu = 0.4$. Plot the graph of F as a function of θ on the interval $\left[0, \frac{\pi}{2}\right]$. Then verify the result obtained in parts (a) and (b) for this special case.

84. A uniform beam of length 3 ft and negligible weight is supported at both ends. When subjected to a concentrated load W at a distance 1 ft from one end, it bends and has the elastic curve (the dashed curve in the figure) described by the equation

$$y = \begin{cases} \dfrac{W}{9EI}(5x - x^3) & \text{if } 0 \leq x \leq 1 \\[2mm] \dfrac{W}{18EI}(x^3 - 9x^2 + 19x - 3) & \text{if } 1 < x \leq 3 \end{cases}$$

where the product EI is a constant called the *flexural rigidity* of the beam. Find the maximum deflection of the beam.
Hint: Maximize $y = f(x)$ over each interval $[0, 1]$ and $[1, 3]$ separately. Then combine your results.

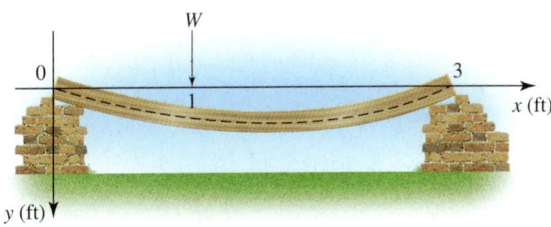

85. Let

$$f(x) = \begin{cases} -x & \text{if } -1 \leq x < 0 \\ x - 1 & \text{if } 0 \leq x \leq 1 \end{cases}$$

Show that f is discontinuous at $x = 0$ but attains an absolute maximum value and an absolute minimum value on $[-1, 1]$. Does this contradict the Extreme Value Theorem?

86. Let

$$f(x) = \begin{cases} x^2 + 1 & \text{if } -1 < x \leq 2 \\ 2 & \text{if } 2 < x < 4 \end{cases}$$

Show that f attains an absolute maximum value and an absolute minimum value on the *open* interval $(-1, 4)$. Does this contradict the Extreme Value Theorem?

87. Show that the function $f(x) = x^3 + x + 1$ has no relative extrema on $(-\infty, \infty)$.

88. Find the critical numbers of the greatest integer function $f(x) = [\![x]\!]$.

89. Find the absolute maximum value and the absolute minimum value (if any) of the function $g(x) = x - [\![x]\!]$, where $f(x) = [\![x]\!]$ is the greatest integer function.

90. a. Suppose f has a relative minimum at c. Show that the function g defined by $g(x) = -f(x)$ has a relative maximum at c.
b. Use the result of (a) to prove Theorem 1 for the case in which f has a relative minimum at c.

 In Exercises 91–94, plot the graph of f and use the graph to estimate the absolute maximum and absolute minimum values of f in the given interval.

91. $f(x) = -0.02x^5 - 0.3x^4 + 2x^3 - 6x + 4$ on $[-2, 2]$

92. $f(x) = 0.3x^6 - 2x^4 + 3x^2 - 3$ on $[0, 2]$

93. $f(x) = \dfrac{0.2x^2}{3x^4 + 2x^2 + 1}$ on $[0, 4]$

94. $f(x) = \dfrac{x + \cos x}{1 + 0.5 \sin x}$ on $[0, 2]$

In Exercises 95–98, (a) plot the graph of f in the given viewing window and find the approximate absolute maximum and absolute minimum values accurate to three decimal places, *and (b)* obtain the exact absolute maximum and absolute minimum values of f analytically.

95. $f(x) = \dfrac{1}{2}x^4 - \dfrac{3}{2}x + 2$ on $[-1, 2] \times [0, 8]$

96. $f(x) = x - \sqrt{1 - x^2}$ on $[-1, 1] \times [-2, 2]$

97. $f(x) = \dfrac{x + 1}{\sqrt{x} + 1}$ on $[0, 1] \times [0.8, 1]$

98. $f(x) = 2 \sin x - x$ on $\left[0, \frac{\pi}{2}\right] \times [0, 1]$

In Exercises 99–102, determine whether the statement is true or false. If it is true, explain why it is true. If it is false, explain why or give an example to show why it is false.

99. If $f'(c) = 0$, then f has a relative maximum or a relative minimum at c.

100. If f has a relative minimum at c, then $f'(c) = 0$.

101. If f is defined on the closed interval $[a, b]$, then f has an absolute minimum value in $[a, b]$.

102. If f is continuous on the interval (a, b), then f attains an absolute minimum value at some number c in (a, b).

3.2 The Mean Value Theorem

■ Rolle's Theorem

The graph of the function f shown in Figure 1 gives the *depth* of a radical new twin-piloted submarine during a test dive. The submarine is on the surface at $t = a$ $[f(a) = 0]$ when it commences its dive. It resurfaces at $t = b$ $[f(b) = 0]$, the end of the test run. As you can see from the graph of f, there is at least one point on the graph of f at which the tangent line to the curve is horizontal.

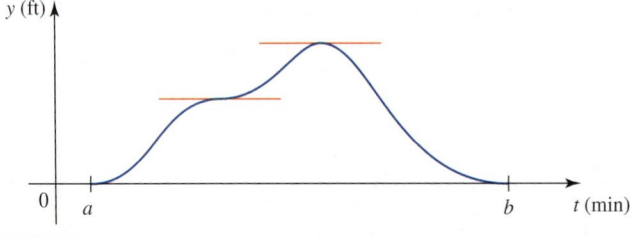

FIGURE 1
$f(t)$ gives the depth of the submarine at time t.

Historical Biography

MICHEL ROLLE
(1652-1719)

It is interesting to note that the theorem that bears Michel Rolle's name–which was originally included in a 1691 book on geometry and algebra–is the basis for so many concepts in calculus, given that Rolle himself was skeptical of the topic's validity. Rolle attacked as a set of untruths what were then the newly developing infinitesimal methods, now known as calculus. Eventually convinced of the validity of calculus by Pierre Varignon (1654-1722), Rolle later voiced his support for the subject. Shortly thereafter, the general opposition to calculus collapsed, followed by many new advances in the content area. Rolle's Theorem is now found in the development of many of the introductory topics of differential calculus.

We can convince ourselves that there must exist at least one such point on the graph of f through the following intuitive argument: Since we know that the submarine returned to the surface, there must be at least one point on the graph of f that corresponds to the time when the submarine stops diving and begins to resurface. The tangent line to the graph of f at this point must be horizontal.

A mathematical description of this phenomenon is contained in Rolle's Theorem, named in honor of the French mathematician Michel Rolle (1652–1719).

THEOREM 1   Rolle's Theorem

Let f be continuous on $[a, b]$ and differentiable on (a, b). If $f(a) = f(b)$, then there exists at least one number c in (a, b) such that $f'(c) = 0$.

PROOF   Let $f(a) = f(b) = d$. There are two cases to consider (see Figure 2).

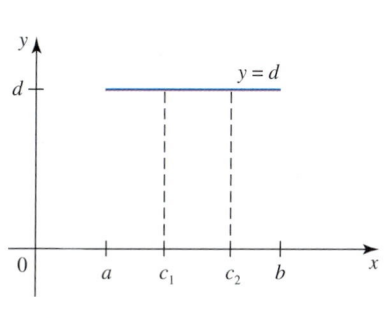   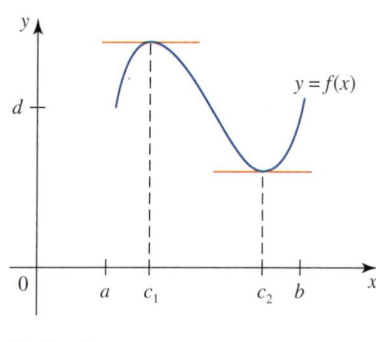

FIGURE 2
Geometric interpretations
of Rolle's Theorem

(a) Case 1   (b) Case 2

Case 1 $f(x) = d$ for all x in $[a, b]$ (see Figure 2a).
In this case, $f'(x) = 0$ for all x in (a, b), so $f'(c) = 0$ for any number c in (a, b).

Case 2 $f(x) \neq d$ for at least one x in $[a, b]$ (see Figure 2b).
In this case there must be a number x in (a, b) where $f(x) > d$ or $f(x) < d$. First, suppose that $f(x) > d$. Since f is continuous on $[a, b]$, the Extreme Value Theorem implies
that f attains an absolute maximum value at some number c in $[a, b]$. The number c
cannot be an endpoint because $f(a) = f(b) = d$, and we have assumed that $f(x) > d$
for some number x in (a, b). Therefore, c must be in (a, b). Since f is differentiable on
(a, b), $f'(c)$ exists, and by Fermat's Theorem $f'(c) = 0$.
    The proof for the case in which $f(x) < d$ is similar and is left as an exercise (Exer-
cise 40). ∎

EXAMPLE 1   Let $f(x) = x^3 - x$ for x in $[-1, 1]$.

a. Show that f satisfies the hypotheses of Rolle's Theorem on $[-1, 1]$.
b. Find the number(s) c in $(-1, 1)$ such that $f'(c) = 0$ as guaranteed by Rolle's
Theorem.

Solution
a. The polynomial function f is continuous and differentiable on $(-\infty, \infty)$. In particular, it is continuous on $[-1, 1]$ and differentiable on $(-1, 1)$. Furthermore,

$$f(-1) = (-1)^3 - (-1) = 0 \qquad \text{and} \qquad f(1) = 1^3 - 1 = 0$$

and the hypotheses of Rolle's theorem are satisfied.
b. Rolle's Theorem guarantees that there exists at least one number c in $(-1, 1)$
such that $f'(c) = 0$. But $f'(x) = 3x^2 - 1$, so to find c, we solve

$$3c^2 - 1 = 0$$

obtaining $c = \pm\sqrt{3}/3$. In other words, there are two numbers, $c_1 = -\sqrt{3}/3$ and
$c_2 = \sqrt{3}/3$, in $(-1, 1)$ for which $f'(c) = 0$ (Figure 3). ∎

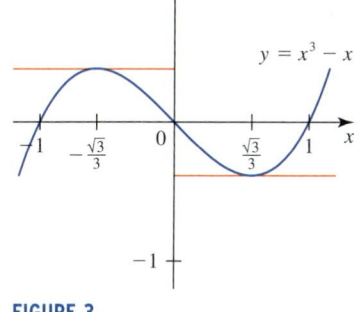

FIGURE 3
The numbers $c_1 = -\sqrt{3}/3$ and
$c_2 = \sqrt{3}/3$ satisfy $f'(c) = 0$ as
guaranteed by Rolle's Theorem.

EXAMPLE 2   **A Real-Life Illustration of Rolle's Theorem**   During a test dive of a prototype of a twin-piloted submarine, the depth in feet of the submarine at time t in minutes is given by $h(t) = t^3(t - 7)^4$, where $0 \leq t \leq 7$.

a. Use Rolle's Theorem to show that there is some instant of time $t = c$ between 0
and 7 when $h'(c) = 0$.
b. Find the number c and interpret your results.

Solution

a. The polynomial function h is continuous on $[0, 7]$ and differentiable on $(0, 7)$. Furthermore, $h(0) = 0$ and $h(7) = 0$, so the hypotheses of Rolle's Theorem are satisfied. Therefore, there exists at least one number c in $(0, 7)$ such that $h'(c) = 0$.

b. To find the value of c, we first compute

$$h'(t) = 3t^2(t - 7)^4 + t^3(4)(t - 7)^3$$
$$= t^2(t - 7)^3[3(t - 7) + 4t]$$
$$= 7t^2(t - 7)^3(t - 3)$$

Setting $h'(t) = 0$ gives $t = 0$, 3, or 7. Since 3 is the only number in the interval $(0, 7)$ such that $h'(3) = 0$, we see that $c = 3$. Interpreting our results, we see that the submarine is on the surface initially (since $h(0) = 0$) and returns to the surface again after 7 minutes (since $h(7) = 0$). The vertical component of the velocity of the submarine is zero at $t = 3$, at which time the submarine attains the greatest depth of $h(3) = 3^3(3 - 7)^4 = 6912$ ft. The graph of h is shown in Figure 4. ∎

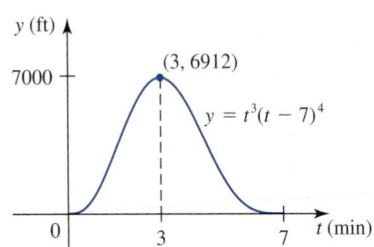

FIGURE 4
The submarine is at a depth of $h(t)$ feet at time t minutes.

Rolle's Theorem is a special case of a more general result known as the Mean Value Theorem.

> **THEOREM 2** **The Mean Value Theorem**
>
> Let f be continuous on $[a, b]$ and differentiable on (a, b). Then there exists at least one number c in (a, b) such that
>
> $$f'(c) = \frac{f(b) - f(a)}{b - a} \qquad (1)$$

To interpret this theorem geometrically, notice that the quotient in Equation (1) is just the slope of the secant line passing through the points $P(a, f(a))$ and $Q(b, f(b))$ lying on the graph of f (Figure 5). The quantity $f'(c)$ on the left, however, gives the slope of the tangent line to the graph of f at $x = c$. The Mean Value Theorem tells us that under suitable conditions on f, there is always at least one point $(c, f(c))$ on the graph of f for $a < c < b$ such that the tangent line to the graph of f at this point is parallel to the secant line passing through P and Q. Observe that if $f(a) = f(b)$, then Theorem 2 reduces to Rolle's Theorem.

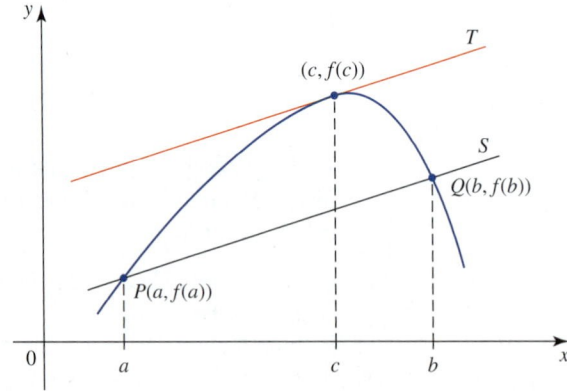

FIGURE 5
The tangent line T at $(c, f(c))$ is parallel to the secant line S through P and Q.

PROOF If you examine Figure 5, you will see that the vertical distance between the graph of f and the secant line S passing through P and Q is maximal at $x = c$. This observation gives a clue to the proof of the Mean Value Theorem: Find a function whose absolute value gives the vertical distances between the graph of f and the secant line. Then optimize this function.

Now an equation of the secant line can be found by using the point-slope form of the equation of a line with slope $[f(b) - f(a)]/(b - a)$ and the point $(b, f(b))$. Thus,

$$y - f(b) = \frac{f(b) - f(a)}{b - a} \cdot (x - b)$$

or

$$y = f(b) + \frac{f(b) - f(a)}{b - a} \cdot (x - b)$$

Define the function D by

$$D(x) = f(x) - \left[f(b) + \frac{f(b) - f(a)}{b - a} \cdot (x - b) \right] \tag{2}$$

Notice that $|D(x)|$ gives the vertical distance between the graph of f and the secant line through P and Q (Figure 6). The function D is continuous on $[a, b]$ and differentiable on (a, b), so we can use Rolle's Theorem on D. First, we note that $D(a) = D(b) = 0$. Therefore, there exists at least one number c in (a, b) such that $D'(c) = 0$. But

$$D'(x) = f'(x) - \frac{f(b) - f(a)}{b - a}$$

so $D'(c) = 0$ implies that

$$0 = f'(c) - \frac{f(b) - f(a)}{b - a}$$

or

$$f'(c) = \frac{f(b) - f(a)}{b - a}$$

as was to be shown.

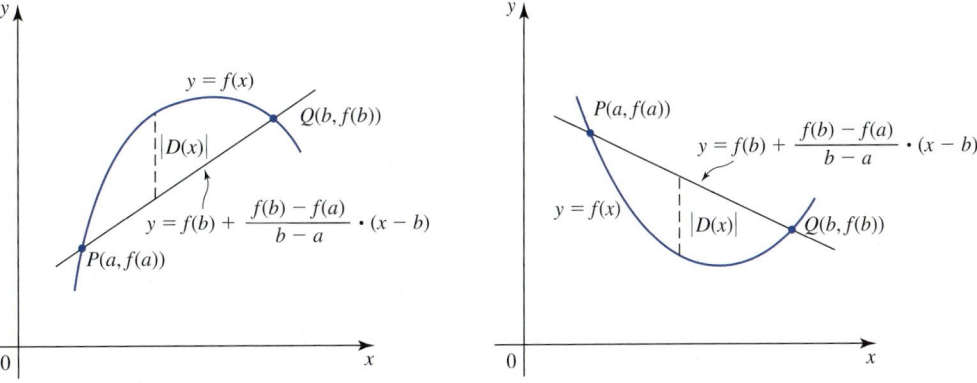

FIGURE 6
$|D(x)|$ gives the vertical distance between the graph of f and the secant line passing through P and Q.

EXAMPLE 3 Let $f(x) = x^3$.

a. Show that f satisfies the hypotheses of the Mean Value Theorem on $[-1, 1]$.
b. Find the number(s) c in $(-1, 1)$ that satisfy Equation (1) as guaranteed by the Mean Value Theorem.

Solution

a. f is a polynomial function, so it is continuous and differentiable on $(-\infty, \infty)$. In particular, f is continuous on $[-1, 1]$ and differentiable on $(-1, 1)$. So the hypotheses of the Mean Value Theorem are satisfied.

b. $f'(x) = 3x^2$, so $f'(c) = 3c^2$. With $a = -1$ and $b = 1$, Equation (1) gives

$$\frac{f(1) - f(-1)}{1 - (-1)} = f'(c)$$

or

$$\frac{1 - (-1)}{1 - (-1)} = 3c^2$$

$$1 = 3c^2$$

and $c = \pm\sqrt{3}/3$. So there are two numbers, $c_1 = -\sqrt{3}/3$ and $c_2 = \sqrt{3}/3$, in $(-1, 1)$ that satisfy Equation (1). (See Figure 7.) ◼

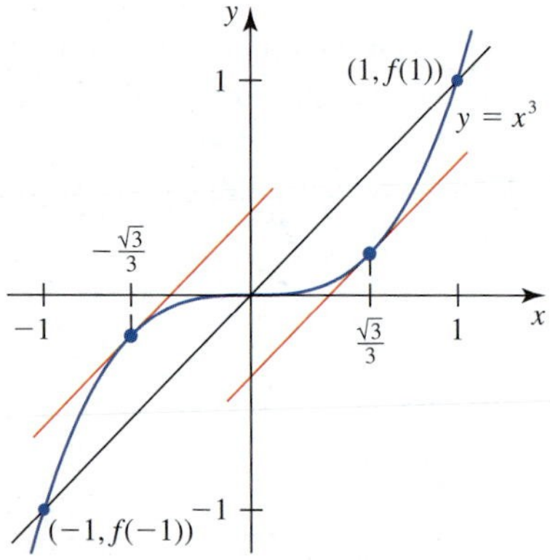

FIGURE 7
The numbers $c_1 = -\sqrt{3}/3$ and $c_2 = \sqrt{3}/3$ satisfy Equation (1), as guaranteed by the Mean Value Theorem.

The next example gives an interpretation of the Mean Value Theorem in a real-life setting.

EXAMPLE 4 **The Mean Value Theorem and the Maglev** The position of a maglev moving along a straight, elevated monorail track is given by $s = f(t) = 4t^2$, $0 \leq t \leq 30$, where s is measured in feet and t is measured in seconds. Then the average velocity of the maglev during the first 4 sec of the run is

$$\frac{f(4) - f(0)}{4 - 0} = \frac{64 - 0}{4} = 16 \tag{3}$$

or 16 ft/sec. Next, since f is continuous on $[0, 4]$ and differentiable on $(0, 4)$, the Mean Value Theorem guarantees that there is a number c in $(0, 4)$ such that

$$\frac{f(4) - f(0)}{4 - 0} = f'(c) \tag{4}$$

But $f'(t) = 8t$, so using Equation (3), we see that Equation (4) is equivalent to

$$16 = 8c$$

or $c = 2$. Since $f'(t)$ measures the instantaneous velocity of the maglev at any time t, the Mean Value Theorem tells us that at some time t between $t = 0$ and $t = 4$ (in this case, $t = 2$) the maglev must attain an instantaneous velocity equal to the average velocity of the maglev over the time interval $[0, 4]$. ◼

◼ Some Consequences of the Mean Value Theorem

An important application of the Mean Value Theorem is to establish other mathematical results. For example, we know that the derivative of a constant function is zero. Now we can show that the converse is also true.

THEOREM 3

If $f'(x) = 0$ for all x in an interval (a, b), then f is constant on (a, b).

PROOF   Suppose that $f'(x) = 0$ for all x in (a, b). To prove that f is constant on (a, b), it suffices to show that f has the same value at every pair of numbers in (a, b). So let x_1 and x_2 be arbitrary numbers in (a, b) with $x_1 < x_2$. Since f is differentiable on (a, b), it is also differentiable on (x_1, x_2) and continuous on $[x_1, x_2]$. Therefore, the hypotheses of the Mean Value Theorem are satisfied on the interval $[x_1, x_2]$. Applying the theorem, we see that there exists a number c in (x_1, x_2) such that

$$f'(c) = \frac{f(x_2) - f(x_1)}{x_2 - x_1} \tag{5}$$

But by hypothesis, $f'(x) = 0$ for all x in (a, b), so $f'(c) = 0$. Therefore, Equation (5) implies that $f(x_2) - f(x_1) = 0$, or $f(x_1) = f(x_2)$; that is, f has the same value at any two numbers in (a, b). This completes the proof. ■

COROLLARY TO THEOREM 3

If $f'(x) = g'(x)$ for all x in an interval (a, b), then f and g differ by a constant on (a, b); that is, there exists a constant c such that $f(x) = g(x) + c$ for all x in (a, b).

PROOF   Let $h(x) = f(x) - g(x)$. Then

$$h'(x) = f'(x) - g'(x) = 0$$

for every x in (a, b). By Theorem 3, h is constant; that is, $f - g$ is constant on (a, b). Thus, $f(x) - g(x) = c$ for some constant c and $f(x) = g(x) + c$ for all x in (a, b). ■

EXAMPLE 5   Prove the identity $\sin^{-1} x + \cos^{-1} x = \pi/2$.

Solution   Let $f(x) = \sin^{-1} x + \cos^{-1} x$ for all x in $[-1, 1]$. Then $f(-1) = f(1) = \pi/2$ by direct computation. For $-1 < x < 1$ we have

$$f'(x) = \frac{1}{\sqrt{1 - x^2}} - \frac{1}{\sqrt{1 - x^2}} = 0$$

Therefore, by Theorem 3, $f(x)$ is constant on $(-1, 1)$; that is, there exists a constant C such that

$$\sin^{-1} x + \cos^{-1} x = C$$

To determine the value of C, we put $x = 0$, giving

$$\sin^{-1} 0 + \cos^{-1} 0 = C \quad \text{or} \quad C = 0 + \frac{\pi}{2}$$

Thus, $\sin^{-1} x + \cos^{-1} x = \pi/2$. ■

■ Determining the Number of Zeros of a Function

Our final example brings together two important theorems—the Intermediate Value Theorem and Rolle's Theorem—to help us determine the number of zeros of a function f in a given interval $[a, b]$.

EXAMPLE 6 Show that the function $f(x) = x^3 + x + 1$ has exactly one zero in the interval $[-2, 0]$.

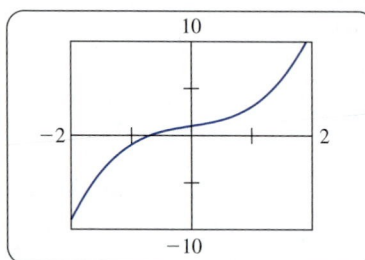

FIGURE 8
The graph shows the zero of f.

Solution First, observe that f is continuous on $[-2, 0]$ and that $f(-2) = -9$ and $f(0) = 1$. Therefore, by the Intermediate Value Theorem, there must exist at least one number c that satisfies $-2 < c < 0$ such that $f(c) = 0$. In other words, f has at least one zero in $(-2, 0)$.

To show that f has exactly one zero, suppose, on the contrary, that f has at least two distinct zeros, x_1 and x_2. Without loss of generality, suppose that $x_1 < x_2$. Then $f(x_1) = f(x_2) = 0$. Because f is differentiable on (x_1, x_2), an application of Rolle's Theorem tells us that there exists a number c between x_1 and x_2 such that $f'(c) = 0$. But $f'(x) = 3x^2 + 1 \geq 1$ can never be zero in (x_1, x_2). This contradiction establishes the result.

The graph of f is shown in Figure 8. ■

3.2 CONCEPT QUESTIONS

1. State Rolle's Theorem and give a geometric interpretation of it.
2. State the Mean Value Theorem, and give a geometric interpretation of it.
3. Refer to the graph of f.
   a. Sketch the secant line through the points $(0, 3)$ and $(9, 8)$. Then draw all lines parallel to this secant line that are tangent to the graph of f.
   b. Use the result of part (a) to estimate the values of c that satisfy the Mean Value Theorem on the interval $[0, 9]$.

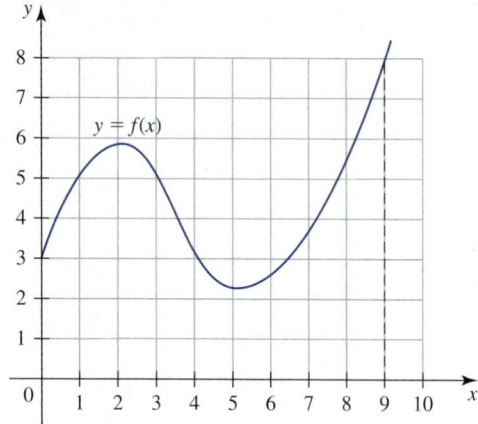

3.2 EXERCISES

In Exercises 1–8, verify that the function satisfies the hypotheses of Rolle's Theorem on the given interval, and find all values of c that satisfy the conclusion of the theorem.

1. $f(x) = x^2 - 4x + 3$; $[1, 3]$

2. $g(x) = x^3 - 9x$; $[-3, 3]$

3. $f(x) = x^3 + x^2 - 2x$; $[-2, 0]$

4. $h(x) = x^3(x - 7)^4$; $[0, 7]$

5. $f(x) = x\sqrt{1 - x^2}$; $[-1, 1]$

6. $f(t) = t^{2/3}(6 - t)^{1/3}$; $[0, 6]$

7. $h(t) = \sin^2 t$; $[0, \pi]$

8. $f(x) = \cos 2x - 1$; $[0, \pi]$

In Exercises 9–16, verify that the function satisfies the hypotheses of the Mean Value Theorem on the given interval, and find all values of c that satisfy the conclusion of the theorem.

9. $f(x) = x^2 + 1$; $[0, 2]$ **10.** $f(x) = x^3 - 2x^2$; $[-1, 2]$

11. $h(x) = \dfrac{1}{x}$; $[1, 3]$ **12.** $g(t) = \dfrac{t}{t - 1}$; $[-2, 0]$

13. $h(x) = x\sqrt{2x + 1}$; $[0, 4]$

14. $f(x) = \sin x$; $\left[0, \frac{\pi}{2}\right]$

15. $f(x) = e^{-x/2}$; $[0, 4]$ **16.** $g(x) = \ln x$; $[1, 3]$

17. Flight of an Aircraft A commuter plane takes off from the Los Angeles International Airport and touches down 30 min later at the Ontario International Airport. Let $A(t)$ (in feet) be the altitude of the plane at time t (in minutes), where $0 \le t \le 30$. Use Rolle's Theorem to explain why there must be at least one number c with $0 < c < 30$ such that $A'(c) = 0$. Interpret your result.

18. Breaking the Speed Limit A trucker drove from Bismarck to Fargo, a distance of 193 mi, in 2 hr and 55 min. Use the Mean Value Theorem to show that the trucker must have exceeded the posted speed limit of 65 mph at least once during the trip.

19. Test Flights In a test flight of the McCord Terrier, an experimental VTOL (vertical takeoff and landing) aircraft, it was determined that t sec after takeoff, when the aircraft was operated in the vertical takeoff mode, its altitude was

$$h(t) = \frac{1}{16}t^4 - t^3 + 4t^2 \qquad 0 \le t \le 8$$

Use Rolle's Theorem to show that there exists a number c satisfying $0 < c < 8$ such that $h'(c) = 0$. Find the value of c, and explain its significance.

20. Hotel Occupancy The occupancy rate of the all-suite Wonderland Hotel, located near a theme park, is given by the function

$$r(t) = \frac{10}{81}t^3 - \frac{10}{3}t^2 + \frac{200}{9}t + 56 \qquad 0 \le t \le 12$$

where t is measured in months with $t = 0$ corresponding to the beginning of January. Show that there exists a number c that satisfies $0 < c < 12$ such that $r'(c) = 0$. Find the value of c, and explain its significance.

21. Let $f(x) = |x| - 1$. Show that there is no number c in $(-1, 1)$ such that $f'(c) = 0$ even though $f(-1) = f(1) = 0$. Why doesn't this contradict Rolle's Theorem?

22. Let $f(x) = 1 - x^{2/3}$, $a = -1$, and $b = 8$. Show that there is no number c in (a, b) such that

$$f'(c) = \frac{f(b) - f(a)}{b - a}$$

Doesn't this contradict the Mean Value Theorem? Explain.

23. Let

$$f(x) = \begin{cases} x^2 & \text{if } x < 1 \\ 2 - x & \text{if } x \ge 1 \end{cases}$$

Does f satisfy the hypotheses of the Mean Value Theorem on $[0, 2]$? Explain.

24. Prove that $f(x) = 4x^3 - 4x + 1$ has at least one zero in the interval $(0, 1)$.
Hint: Apply Rolle's Theorem to the function $g(x) = x^4 - 2x^2 + x$ on $[0, 1]$.

25. Prove that $f(x) = x^5 + 6x + 4$ has exactly one zero in $(-\infty, \infty)$.

26. Prove that the equation $x^7 + 6x^5 + 2x - 6 = 0$ has exactly one real root.

27. Prove that the function $f(x) = x^5 - 12x + c$, where c is any real number, has at most one zero in $[0, 1]$.

28. Use the Mean Value Theorem to prove that $|\sin a - \sin b| \le |a - b|$ for all real numbers a and b.

29. Suppose that the equation

$$a_n x^n + a_{n-1}x^{n-1} + \cdots + a_1 x = 0$$

has a positive root r. Show that the equation

$$na_n x^{n-1} + (n - 1)a_{n-1}x^{n-2} + \cdots + a_1 = 0$$

has a positive root smaller than r.
Hint: Use Rolle's Theorem.

30. Suppose $f'(x) = c$, where c is a constant, for all values of x. Show that f must be a linear function of the form $f(x) = cx + d$ for some constant d.
Hint: Use the corollary to Theorem 3.

31. Let $f(x) = x^4 - 4x - 1$.
  a. Use Rolle's Theorem to show that f has exactly two distinct zeros.
  b. Plot the graph of f using the viewing window $[-3, 3] \times [-5, 5]$.

32. Let

$$f(x) = \begin{cases} x \sin \dfrac{\pi}{x} & \text{if } x > 0 \\ 0 & \text{if } x = 0 \end{cases}$$

Use Rolle's Theorem to prove that f has infinitely many critical numbers in the interval $(0, 1)$. Plot the graph of f using the viewing window $[0, 1] \times [-1, 1]$.

33. Prove the formula

$$\cos^2 x = \frac{1 + \cos 2x}{2}$$

34. Prove the formula

$$\cos^{-1} \frac{1 - x^2}{1 + x^2} = 2 \tan^{-1} x$$

for $0 \le x < \infty$.

35. Suppose that f and g are continuous on an interval $[a, b]$ and differentiable on the interval (a, b). Furthermore, suppose that $f(a) = g(a)$ and $f'(x) < g'(x)$ for $a < x < b$. Prove that $f(x) < g(x)$ for $a < x < b$.
Hint: Apply the Mean Value Theorem to the function $h = f - g$.

36. Let $f(x) = Ax^2 + Bx + C$, and let $[a, b]$ be an arbitrary interval. Show that the number c in the Mean Value Theorem applied to the function f lies at the midpoint of the interval $[a, b]$.

37. Let $f(x) = 2(x - 1)(x - 2)(x - 3)(x - 4)$. Prove that f' has exactly three real zeros.

38. A real number c such that $f(c) = c$ is called a *fixed point* of the function f. Geometrically, a fixed point of f is a point that is mapped by f onto itself. Prove that if f is differentiable and $f'(x) \neq 1$ for all x in an interval I, then f has at most one fixed point in I.

39. Use the result of Exercise 38 to show that $f(x) = \sqrt{x + 6}$ has exactly one fixed point in the interval $(0, \infty)$. What is the fixed point?

40. Complete the proof of Rolle's Theorem by considering the case in which $f(x) < d$ for some number x in (a, b).

41. Let f be continuous on $[a, b]$ and differentiable on (a, b). Put $h = b - a$.
  a. Use the Mean Value Theorem to show that there exists at least one number θ that satisfies $0 < \theta < 1$ such that
  $$\frac{f(a + h) - f(a)}{h} = f'(a + \theta h)$$
  b. Find θ in the formula in part (a) for the function $f(x) = x^2$.

42. Let $f(x) = x^4 - 2x^3 + x - 2$.
  a. Show that f satisfies the hypotheses of Rolle's Theorem on the interval $[-1, 2]$.
   b. Use a calculator or a computer to estimate all values of c accurate to five decimal places that satisfy the conclusion of Rolle's Theorem.
  c. Plot the graph of f and the (horizontal) tangent lines to the graph of f at the point(s) $(c, f(c))$ for the values of c found in part (b).

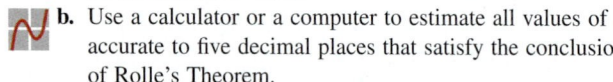 **43.** Let $f(x) = x^4 - 2x^2 + 2$.
  a. Use a calculator or a computer to estimate all values of c accurate to three decimal places that satisfy the conclusion of the Mean Value Theorem for f on the interval $[0, 2]$.
  b. Plot the graph of f, the secant line passing through the points $(0, 2)$ and $(2, 10)$, and the tangent line to the graph of f at the point(s) $(c, f(c))$ for the value(s) of c found in part (a).

44. Let $f(x) = x^2 \sin x$.
  a. Show that f satisfies the hypotheses of Rolle's Theorem on the interval $[0, \pi]$.
   b. Use a calculator or a computer to estimate all value(s) of c accurate to five decimal places that satisfy the conclusion of Rolle's Theorem.
  c. Plot the graph of f and the (horizontal) tangent lines to the graph of f at the point(s) $(c, f(c))$ for the value(s) of c found in part (b).

 45. Let $f(x) = \sin \sqrt{x}$.
  a. Use a calculator or a computer to estimate all values of c accurate to three decimal places that satisfy the conclusion of the Mean Value Theorem for f on the interval $[0, \frac{\pi^2}{4}]$.
  b. Plot the graph of f, the secant line passing through the points $(0, 0)$ and $(\frac{\pi^2}{4}, 1)$, and the tangent line to the graph of f at the point(s) $(c, f(c))$ for the value(s) of c found in part (b).

46. Prove the inequality
$$\frac{x}{x + 1} < \ln(1 + x) < x$$
for $x > 0$.
Hint: Use the Mean Value Theorem.

In Exercises 47–52, determine whether the statement is true or false. If it is true, explain why it is true. If it is false, explain why or give an example to show why it is false.

47. Suppose that f is continuous on $[a, b]$ and differentiable on (a, b). If $f'(c) = 0$ for at least one c in (a, b), then $f(a) = f(b)$.

48. Suppose that f is continuous on $[a, b]$ but is not differentiable on (a, b). Then there does not exist a number c in (a, b) such that
$$f'(c) = \frac{f(b) - f(a)}{b - a}$$

49. If $f'(x) = 0$ for all x, then f is a constant function.

50. If $|f'(x)| \leq 1$ for all x, then
$$|f(x_1) - f(x_2)| \leq |x_1 - x_2|$$
for all numbers x_1 and x_2.

51. There does not exist a continuous function defined on the interval $[2, 5]$ and differentiable on $(2, 5)$ satisfying $|f(5) - f(2)| \leq 6$ on $[2, 5]$ and $|f'(x)| > 2$ for all x in $(2, 5)$.

52. If f is continuous on $[1, 3]$, differentiable on $(1, 3)$, and satisfies $f(1) = 2$, $f(3) = 5$, then there exists a number c satisfying $1 < c < 3$, such that $f'(c) = \frac{3}{2}$.

◼ Increasing and Decreasing Functions

Among the important factors in determining the structural integrity of an aircraft is its age. Advancing age makes the parts of a plane more likely to crack. The graph of the function f in Figure 1 is referred to as a "bathtub curve" in the airline industry. It gives the fleet damage rate (damage due to corrosion, accident, and metal fatigue) of a typical fleet of commercial aircraft as a function of the number of years of service.

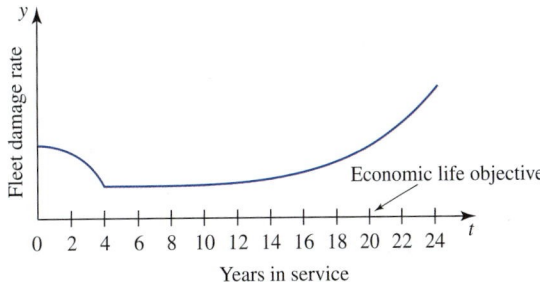

FIGURE 1

The "bathtub curve" gives the number of planes in a fleet that are damaged as a function of the age of the fleet.

The function is *decreasing* on the interval $(0, 4)$, showing that the fleet damage rate is dropping as problems are found and corrected during the initial shakedown period. The function is *constant* on the interval $(4, 10)$, reflecting that planes have few structural problems after the initial shakedown period. Beyond this, the function is *increasing,* reflecting an increase in structural defects due mainly to metal fatigue.

These intuitive notions involving increasing and decreasing functions can be described mathematically as follows.

DEFINITIONS   Increasing and Decreasing Functions

A function f is **increasing** on an interval I, if for every pair of numbers x_1 and x_2 in I,

$$x_1 < x_2 \quad \text{implies that} \quad f(x_1) < f(x_2) \qquad \text{See Figure 2a.}$$

f is **decreasing** on I if, for every pair of numbers x_1 and x_2 in I,

$$x_1 < x_2 \quad \text{implies that} \quad f(x_1) > f(x_2) \qquad \text{See Figure 2b.}$$

f is **monotonic** on I if it is either increasing or decreasing on I.

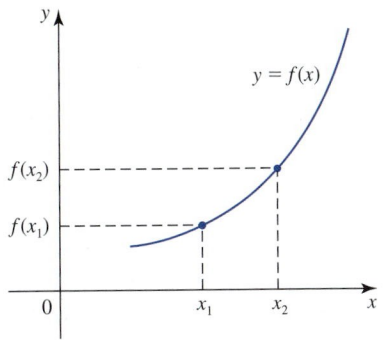

FIGURE 2   **(a)** f is increasing on I.   **(b)** f is decreasing on I.

Since the derivative of a function measures the rate of change of that function, it lends itself naturally as a tool for determining the intervals where a differentiable function is increasing or decreasing. As you can see in Figure 3, if the graph of f has tangent lines with positive slopes over an interval, then the function is increasing on that interval. Similarly, if the graph of f has tangent lines with negative slopes over an interval, then the function is decreasing on that interval. Also, we know that the slope of the tangent line at $(x, f(x))$ and the rate of change of f at x are given by $f'(x)$. Therefore, f is increasing on an interval where $f'(x) > 0$ and decreasing on an interval where $f'(x) < 0$.

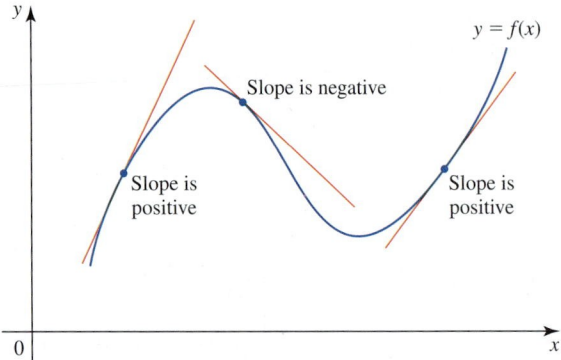

FIGURE 3
f is increasing on an interval where $f'(x) > 0$ and decreasing on an interval where $f'(x) < 0$.

These intuitive observations lead to the following theorem.

THEOREM 1 Suppose f is differentiable on an open interval (a, b).

a. If $f'(x) > 0$ for all x in (a, b), then f is increasing on (a, b).
b. If $f'(x) < 0$ for all x in (a, b), then f is decreasing on (a, b).
c. If $f'(x) = 0$ for all x in (a, b), then f is constant on (a, b).

PROOF

a. Let x_1 and x_2 be any two numbers in (a, b) with $x_1 < x_2$. Since f is differentiable on (a, b), it is continuous on $[x_1, x_2]$ and differentiable on (x_1, x_2). By the Mean Value Theorem, there exists a number c in (x_1, x_2) such that

$$f'(c) = \frac{f(x_2) - f(x_1)}{x_2 - x_1}$$

or, equivalently,

$$f(x_2) - f(x_1) = f'(c)(x_2 - x_1) \tag{1}$$

Now, $f'(c) > 0$ by assumption, and $x_2 - x_1 > 0$ because $x_1 < x_2$. Therefore, $f(x_2) - f(x_1) > 0$, or $f(x_1) < f(x_2)$. This shows that f is increasing on (a, b).
b. The proof of (b) is similar and is left as an exercise (see Exercise 66).
c. This was proved in Theorem 3 in Section 3.2.

Theorem 1 enables us to develop a procedure for finding the intervals where a function is increasing, decreasing, or constant. In this connection, recall that a function can only change sign as we move across a zero or a number at which the function is discontinuous.

Determining the Intervals Where a Function Is Increasing or Decreasing

1. Find all the values of x for which $f'(x) = 0$ or $f'(x)$ does not exist. Use these values of x to partition the domain of f into open intervals.
2. Select a test number c in each interval found in Step 1, and determine the sign of $f'(c)$ in that interval.
   a. If $f'(c) > 0$, then f is increasing on that interval.
   b. If $f'(c) < 0$, then f is decreasing on that interval.
   c. If $f'(c) = 0$, then f is constant on that interval.

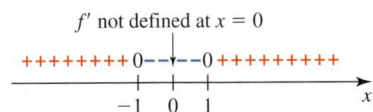

FIGURE 4
The sign diagram for f'

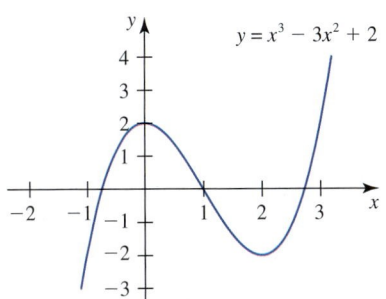

FIGURE 5
f is increasing on $(-\infty, 0)$, decreasing on $(0, 2)$, and increasing on $(2, \infty)$.

EXAMPLE 1 Determine the intervals where the function $f(x) = x^3 - 3x^2 + 2$ is increasing and where it is decreasing.

Solution We first compute

$$f'(x) = 3x^2 - 6x = 3x(x - 2)$$

from which we see that f' is continuous everywhere and has zeros at 0 and 2. These zeros of f' partition the domain of f into the intervals $(-\infty, 0)$, $(0, 2)$, and $(2, \infty)$. To determine the sign of $f'(x)$ on each of these intervals, we evaluate $f'(x)$ at a convenient test number in each interval. These results are summarized in the following table.

Interval	Test number c	$f'(c)$	Sign of $f'(c)$
$(-\infty, 0)$	-1	9	$+$
$(0, 2)$	1	-3	$-$
$(2, \infty)$	3	9	$+$

Using these results, we obtain the sign diagram for $f'(x)$ shown in Figure 4. We conclude that f is increasing on $(-\infty, 0)$ and $(2, \infty)$ and decreasing on $(0, 2)$. The graph of f is shown in Figure 5.

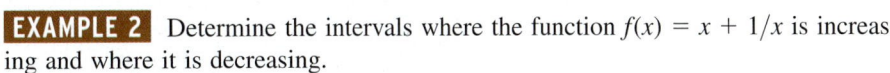

FIGURE 6
The sign diagram for f'

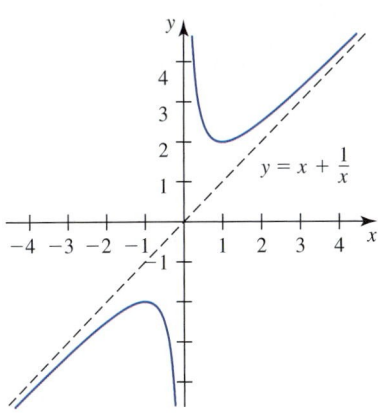

FIGURE 7
The graph of f

EXAMPLE 2 Determine the intervals where the function $f(x) = x + 1/x$ is increasing and where it is decreasing.

Solution The derivative of f is

$$f'(x) = 1 - \frac{1}{x^2} = \frac{x^2 - 1}{x^2} = \frac{(x + 1)(x - 1)}{x^2}$$

from which we see that $f'(x)$ is continuous everywhere except at $x = 0$ and has zeros at $x = -1$ and $x = 1$. These values of x partition the domain of f into the intervals $(-\infty, -1)$, $(-1, 0)$, $(0, 1)$, and $(1, \infty)$. By evaluating $f'(x)$ at each of the test numbers $x = -2, -\frac{1}{2}, \frac{1}{2}$, and 2, we find

$$f'(-2) = \frac{3}{4}, \qquad f'\left(-\frac{1}{2}\right) = -3, \qquad f'\left(\frac{1}{2}\right) = -3, \qquad \text{and} \qquad f'(2) = \frac{3}{4}$$

giving us the sign diagram of $f'(x)$ shown in Figure 6. We conclude that f is increasing on $(-\infty, -1)$ and $(1, \infty)$ and decreasing on $(-1, 0)$ and $(0, 1)$. The graph of f is shown in Figure 7.* Note that $f'(x)$ does not change sign as we move across the point of discontinuity.

---

*The graph of f approaches the dashed line as $x \to \pm\infty$. The dashed line is called a *slant asymptote* and will be discussed in Section 3.6.

■ Finding the Relative Extrema of a Function

We will now see how the derivative of a function f can be used to help us find the relative extrema of f. If you examine Figure 8, you can see that the graph of f is *rising* to the left of the relative maximum that occurs at b and *falling* to the right of it. Likewise, at the relative minima of f at a and d, you can see that the graph of f is *falling* to the left of these critical numbers and *rising* to the right of them. Finally, look at the behavior of the graph of f at the critical numbers c and e. These numbers do not give rise to relative extrema. Notice that f is either increasing or decreasing on *both* sides of these critical numbers.

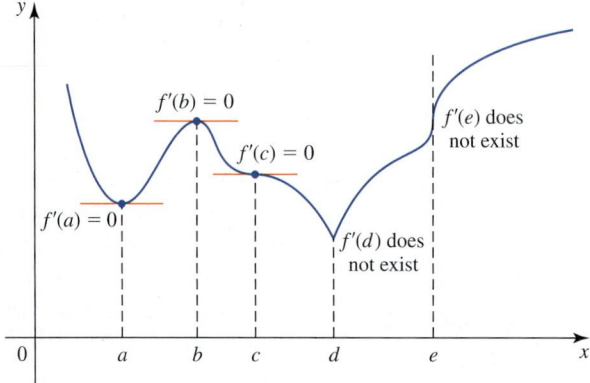

FIGURE 8
a, b, c, d, and e are critical numbers of f, but only the critical numbers a, b, and d give rise to relative extrema.

This discussion leads to the following theorem.

THEOREM 2   The First Derivative Test

Let c be a critical number of a continuous function f in the interval (a, b) and suppose that f is differentiable at every number in (a, b) with the possible exception of c itself.

a. If $f'(x) > 0$ on (a, c) and $f'(x) < 0$ on (c, b), then f has a *relative maximum* at c (Figure 9a).

b. If $f'(x) < 0$ on (a, c) and $f'(x) > 0$ on (c, b), then f has a *relative minimum* at c (Figure 9b).

c. If $f'(x)$ has the same sign on (a, c) and (c, b), then f does not have a relative extremum at c (Figure 9c).

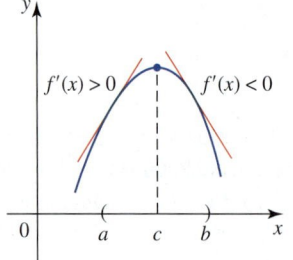

(**a**) Relative maximum at c

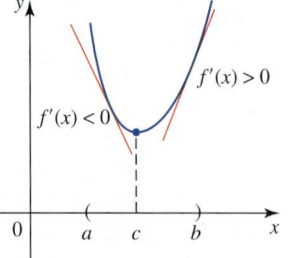

(**b**) Relative minimum at c

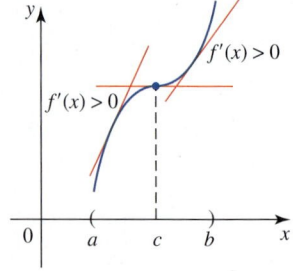

(**c**) No relative extrema at c

FIGURE 9

PROOF We will prove part (a) and leave the other two parts for you to prove (see Exercise 67). Suppose f' changes sign from positive to negative as we pass through c. Then there are numbers a and b such that $f'(x) > 0$ for all x in (a, c) and $f'(x) < 0$ for all x in (c, b). By Theorem 1 we see that f is increasing on (a, c) and decreasing on (c, b). Therefore, $f(x) \leq f(c)$ for all x in (a, b). We conclude that f has a relative maximum at c. ■

The following procedure for finding the relative extrema of a continuous function is based on Theorem 2.

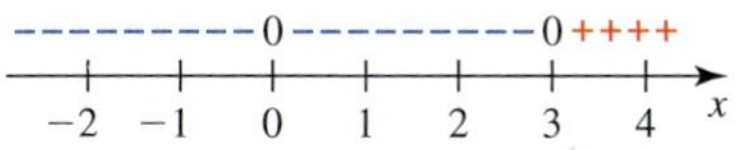

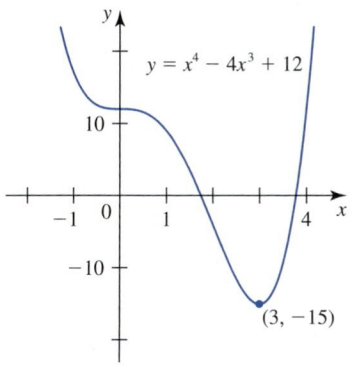

FIGURE 10
The sign diagram of f'

> **Finding the Relative Extrema of a Function**
> 1. Find the critical numbers of f.
> 2. Determine the sign of $f'(x)$ to the left and to the right of each critical number.
>   a. If $f'(x)$ changes sign from *positive* to *negative* as we move across a critical number c, then $f(c)$ is a relative maximum value.
>   b. If $f'(x)$ changes sign from *negative* to *positive* as we move across a critical number c, then $f(c)$ is a relative minimum value.
>   c. If $f'(x)$ does not change sign as we move across a critical number c, then $f(c)$ is not a relative extremum.

FIGURE 11
The graph of f

EXAMPLE 3 Find the relative extrema of $f(x) = x^4 - 4x^3 + 12$.

Solution The derivative of f,

$$f'(x) = 4x^3 - 12x^2 = 4x^2(x - 3)$$

is continuous everywhere. Therefore, the zeros of f', which are 0 and 3, are the only critical numbers of f. The sign diagram of f' is shown in Figure 10. Since f' has the same sign on $(-\infty, 0)$ and $(0, 3)$, the First Derivative Test tells us that f does not have a relative extremum at 0. Next, we note that f' changes sign from negative to positive as we move across 3, so 3 does give rise to a relative minimum of f. The relative minimum value of f is $f(3) = -15$. The graph of f is shown in Figure 11 and confirms these results. ■

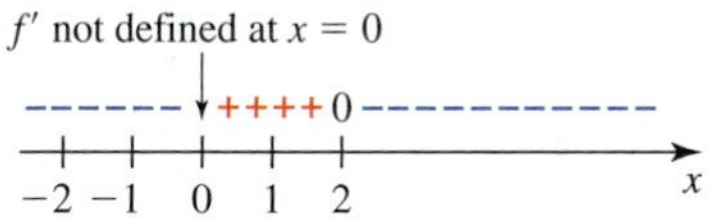

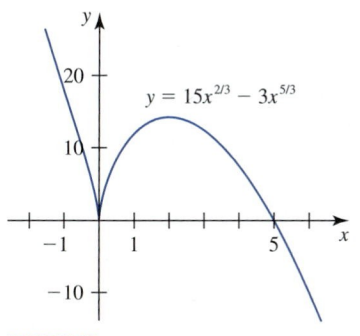

f' not defined at $x = 0$

FIGURE 12
The sign diagram of f'

EXAMPLE 4 Find the relative extrema of $f(x) = 15x^{2/3} - 3x^{5/3}$.

Solution The derivative of f is

$$f'(x) = 10x^{-1/3} - 5x^{2/3} = 5x^{-1/3}(2 - x) = \frac{5(2 - x)}{x^{1/3}}$$

Note that f' is discontinuous at 0 and has a zero at 2, so 0 and 2 are critical numbers of f. Referring to the sign diagram of f' (Figure 12) and using the First Derivative Test, we conclude that f has a relative minimum at 0 and a relative maximum at 2. The relative minimum value is $f(0) = 0$, and the relative maximum value is

$$f(2) = 15(2)^{2/3} - 3(2)^{5/3} \approx 14.29$$

The graph of f is shown in Figure 13. ■

FIGURE 13
The graph of f

EXAMPLE 5 **Motion of a Projectile**   A projectile starts from the origin of the *xy*-coordinate system, and its motion is confined to the *xy*-plane. Suppose the trajectory of the projectile is

$$y = f(x) = 1.732x - 0.000008x^2 - 0.000000002x^3 \qquad 0 \le x \le 27{,}496$$

where *y* measures the height in feet and *x* measures the horizontal distance in feet covered by the projectile.

a. Find the interval where *y* is increasing and the interval where *y* is decreasing.
b. Find the relative extrema of *f*.
c. Interpret the results obtained in part (a) and part (b).

Solution

a. Observe that

$$\frac{dy}{dx} = 1.732 - 0.000016x - 0.000000006x^2$$

is continuous everywhere. Setting $dy/dx = 0$ gives

$$0.000000006x^2 + 0.000016x - 1.732 = 0$$

Using the quadratic formula to solve this equation, we obtain

$$x = \frac{-0.000016 \pm \sqrt{(0.000016)^2 - 4(0.000000006)(-1.732)}}{2(0.000000006)}$$

$$\approx -18{,}376 \text{ or } 15{,}709$$

We reject the negative root, since *x* must be nonnegative. So the critical number of *y* is approximately 15,709. From the sign diagram for *f'* shown in Figure 14, we see that *y* is increasing on (0, 15,709) and decreasing on (15,709, 27,496).

b. From part (a) we see that *y* has a relative maximum at $x \approx 15{,}709$ with value

$$y \approx 1.732x - 0.000008x^2 - 0.000000002x^3 \big|_{x=15{,}709} \approx 17{,}481$$

c. After leaving the origin, the projectile gains altitude as it travels downrange. It reaches a maximum altitude of approximately 17,481 ft after it has traveled approximately 15,709 ft downrange. From this point on, the missile descends until it strikes the ground (after traveling approximately 27,496 ft horizontally). The trajectory of the projectile is shown in Figure 15.

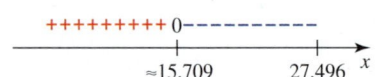

FIGURE 14
The sign diagram of *f'*

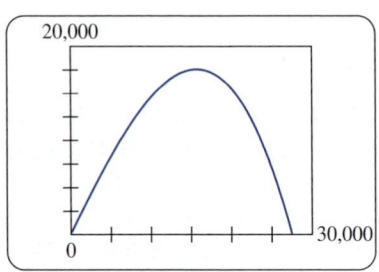

FIGURE 15
The trajectory of the projectile

3.3   CONCEPT QUESTIONS

1. Explain each of the following statements: (a) *f* is increasing on an interval *I*, (b) *f* is decreasing on an interval *I*, and (c) *f* is monotonic on an interval *I*.

2. Describe a procedure for determining where a function is increasing and where it is decreasing.

3. Describe a procedure for finding the relative extrema of a function.

3.3 EXERCISES

In Exercises 1–6 you are given the graph of a function f.
(a) Determine the intervals on which f is increasing, constant,
or decreasing. (b) Find the relative maxima and relative
minima, if any, of f.

1.

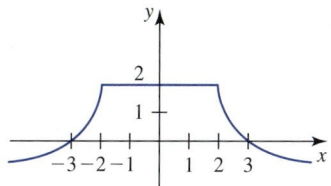

2.

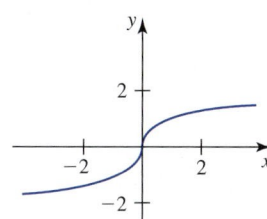

3.

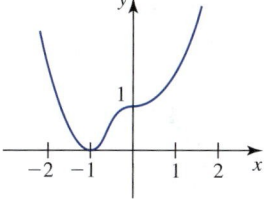

4.

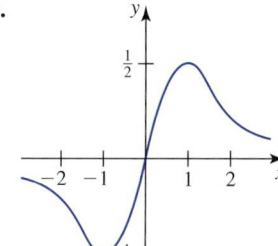

5.

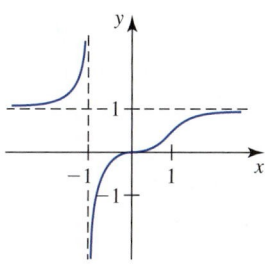

6.

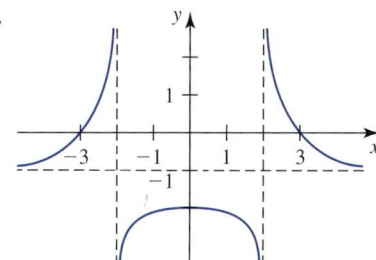

In Exercises 7 and 8 you are given the graph of the derivative f′
of a function f. (a) Determine the intervals on which f is increas-
ing, constant, or decreasing. (b) Find the x-coordinates of the
relative maxima and relative minima of f.

7.

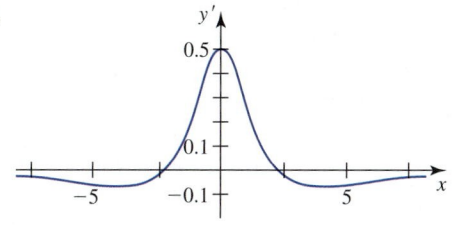

8.

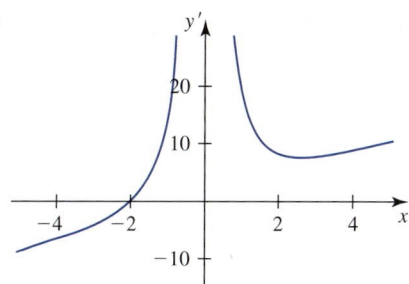

In Exercises 9–38, (a) find the intervals on which f is increasing
or decreasing, and (b) find the relative maxima and relative min-
ima of f.

9. $f(x) = x^2 - 2x$ **10.** $f(x) = -x^2 + 4x + 2$

11. $f(x) = x^3 - 6x + 1$ **12.** $f(x) = -x^3 + 3x^2 + 1$

13. $f(x) = 2x^3 + 3x^2 - 12x + 5$

14. $f(x) = x^3 - 3x^2 - 9x + 6$

15. $f(x) = x^4 - 4x^3 + 6$ **16.** $f(x) = -x^4 + 2x^2 + 1$

17. $f(x) = x^{1/3} - 1$ **18.** $f(x) = x^{1/3} - x^{2/3}$

19. $f(x) = x^2(x - 2)^3$ **20.** $f(x) = x^3(x - 6)^4$

21. $f(x) = x + \dfrac{1}{x}$ **22.** $f(x) = \dfrac{x}{x - 1}$

23. $f(x) = \dfrac{x^2}{x - 1}$ **24.** $f(x) = \dfrac{x}{x^2 + 1}$

25. $f(x) = \dfrac{2x - 3}{x^2 - 4}$ **26.** $f(x) = \dfrac{x^2 - 3x + 2}{x^2 + 2x + 1}$

27. $f(x) = x^{2/3}(x - 3)$ **28.** $f(x) = x\sqrt{4 - x}$

29. $f(x) = x\sqrt{x - x^2}$ **30.** $f(x) = \dfrac{x}{\sqrt{x^2 - 1}}$

31. $f(x) = x - 2 \sin x, \quad 0 < x < 2\pi$

32. $f(x) = x - \cos x, \quad 0 < x < 2\pi$

33. $f(x) = \cos^2 x, \quad 0 < x < 2\pi$

34. $f(x) = \sin^2 2x, \quad 0 < x < \pi$

35. $f(x) = x^2 e^{-x}$

36. $f(x) = x^2 - \ln x$

37. $f(x) = \dfrac{2x}{\ln x}$

38. $f(x) = \ln(e^x + e^{-x} - 2)$

In Exercises 39 and 40, find the relative extrema of the function.

39. $f(x) = \sin^{-1} x - 2x$ **40.** $f(x) = 3 \tan^{-1} x - 2x$

41. The Boston Marathon The graph of the function f shown in the accompanying figure gives the elevation of that part of the Boston Marathon course that includes the notorious Heartbreak Hill. Determine the intervals (stretches of the course) where the function f is increasing (the runner is laboring), where it is constant (the runner is taking a breather), and where it is decreasing (the runner is coasting).

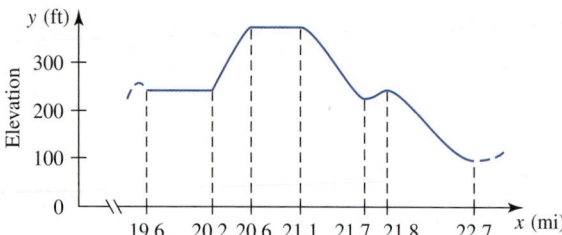

Source: The Boston Globe.

42. The Flight of a Model Rocket The altitude (in feet) attained by a model rocket t sec into flight is given by the function

$$h(t) = 0.1t^2(t - 7)^4 \qquad 0 \le t \le 7$$

When is the rocket ascending, and when is it descending? What is the maximum altitude attained by the rocket?

43. Morning Traffic Rush The speed of traffic flow on a certain stretch of Route 123 between 6 A.M. and 10 A.M. on a typical weekday is approximated by the function

$$f(t) = 20t - 40\sqrt{t} + 52 \qquad 0 \le t \le 4$$

where $f(t)$ is measured in miles per hour and t is measured in hours, with $t = 0$ corresponding to 6 A.M. Find the interval where f is increasing, the interval where f is decreasing, and the relative extrema of f. Interpret your results.

44. Air Pollution The amount of nitrogen dioxide, a brown gas that impairs breathing, that is present in the atmosphere on a certain day in May in the city of Long Beach is approximated by

$$A(t) = \frac{136}{1 + 0.25(t - 4.5)^2} + 28 \qquad 0 \le t \le 11$$

where $A(t)$ is measured in pollutant standard index (PSI) and t is measured in hours with $t = 0$ corresponding to 7 A.M. When is the PSI increasing, and when is it decreasing? At what time is the PSI highest, and what is its value at that time?

Source: The Los Angeles Times.

45. Finding the Lowest Average Cost A subsidiary of the Electra Electronics Company manufactures an MP3 player. Management has determined that the daily total cost of producing these players (in dollars) is given by

$$C(x) = 0.0001x^3 - 0.08x^2 + 40x + 5000$$

When is the average cost function $\overline{C}$, defined by $\overline{C}(x) = C(x)/x$, decreasing, and when is it increasing?

At what level of production is the average cost lowest? What is the average cost corresponding to this level of production?

Hint: $x = 500$ is a root of the equation $\overline{C}'(x) = 0$.

46. Cantilever Beam The figure below depicts a cantilever beam clamped at the left end ($x = 0$) and free at its right end ($x = L$). If a constant load w is uniformly distributed along its length, then the deflection y is given by

$$y = \frac{w}{24EI}(x^4 - 4Lx^3 + 6L^2x^2)$$

where the product EI is a constant called the *flexural rigidity* of the beam. Show that y is increasing on the interval $(0, L)$ and, therefore, that the maximum deflection of the beam occurs at $x = L$. What is the maximum deflection?

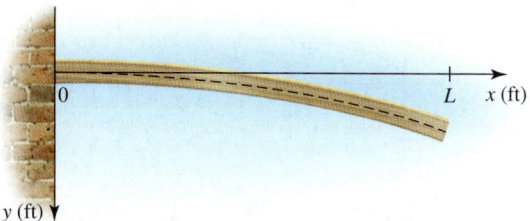

The beam is fixed at $x = 0$ and free at $x = L$.
(Note that the positive direction of y is downward.)

47. Water Level in a Harbor The water level in feet in Boston Harbor during a certain 24-hr period is approximated by the formula

$$H = 4.8 \sin\left(\frac{\pi}{6}(t - 10)\right) + 7.6 \qquad 0 \le t \le 24$$

where $t = 0$ corresponds to 12 A.M. When is the water level rising and when is it falling? Find the relative extrema of H and interpret your results.

Source: SMG Marketing Group.

 48. Spending on Fiber-Optic Links U.S. telephone company spending on fiber-optic links to homes and businesses from the beginning of 2001 to the beginning of 2006 is approximated by

$$S(t) = -2.315t^3 + 34.325t^2 + 1.32t + 23 \qquad 0 \le t \le 5$$

billion dollars in year t, where t is measured in years with $t = 0$ corresponding to the beginning of 2001.

a. Plot the graph of S in the viewing window $[0, 5] \times [0, 600]$.

b. Plot the graph of S' in the viewing window $[0, 5] \times [0, 175]$. What conclusion can you draw from your result?

c. Verify your result analytically.

Source: RHK, Inc.

49. Surgeries in Physicians' Offices Driven by technological advances and financial pressures, the number of surgeries performed in physicians' offices nationwide has been increasing over the years. The function

$$f(t) = -0.00447t^3 + 0.09864t^2 + 0.05192t + 0.8$$
$$0 \le t \le 15$$

gives the number of surgeries (in millions) performed in physicians' offices in year t, with $t = 0$ corresponding to the beginning of 1986.

 a. Plot the graph of f in the viewing window $[0, 15] \times [0, 10]$.

b. Prove that f is increasing on the interval $[0, 15]$.

Source: SMG Marketing Group.

50. Age of Drivers in Crash Fatalities The number of crash fatalities per 100,000 vehicle miles of travel (based on 1994 data) is approximated by the model

$$f(x) = \frac{15}{0.08333x^2 + 1.91667x + 1} \qquad 0 \le x \le 11$$

where x is the age of the driver in years, with $x = 0$ corresponding to age 16. Show that f is decreasing on $(0, 11)$ and interpret your result.

Source: National Highway Traffic Safety Administration.

51. Sales of Functional Food Products The sales of functional food products—those that promise benefits beyond basic nutrition—have risen sharply in recent years. The sales (in billions of dollars) of foods and beverages with herbal and other additives is approximated by the function

$$S(t) = 0.46t^3 - 2.22t^2 + 6.21t + 17.25 \qquad 0 \le t \le 4$$

where t is measured in years, with $t = 0$ corresponding to the beginning of 1997.

 a. Plot the graph of S in the viewing window $[0, 4] \times [15, 40]$.

b. Show that sales were increasing over the 4-year period beginning in 1997.

Source: Frost & Sullivan.

52. Halley's Law Halley's Law states that the barometric pressure (in inches of mercury) at an altitude of x miles above sea level is approximated by

$$p(x) = 29.92e^{-0.2x} \qquad x \ge 0$$

a. If a hot-air balloonist measures the barometric pressure as 20 in. of mercury, what is the balloonist's altitude?

b. If the barometric pressure is decreasing at the rate of 1 in./hr at that altitude, how fast is the balloon rising?

53. Polio Immunization Polio, a once-feared killer, declined markedly in the United States in the 1950s after Jonas Salk developed the inactivated polio vaccine and mass immunization of children took place. The number of polio cases in the United States from the beginning of 1959 to the beginning of 1963 is approximated by the function

$$N(t) = 5.3e^{0.095t^2 - 0.85t} \qquad 0 \le t \le 4$$

where $N(t)$ gives the number of polio cases (in thousands) and t is measured in years with $t = 0$ corresponding to the beginning of 1959.

a. Show that the function N is decreasing over the time interval under consideration.

b. How fast was the number of polio cases decreasing at the beginning of 1959? At the beginning of 1962?

Note: Since the introduction of the oral vaccine developed by Dr. Albert B. Sabin in 1963, polio in the United States has, for all practical purposes, been eliminated.

54. Find the intervals where $f(x) = e^{-x^2/2}$ is increasing and where it is decreasing.

55. Find the intervals where $f(x) = (\log x)/x$ is increasing and where it is decreasing.

56. Prove that the function $f(x) = 2x^5 + x^3 + 2x$ is increasing everywhere.

 57. a. Plot the graphs of $f(x) = x^3 - ax$ for $a = -2, -1, 0, 1,$ and 2, using the viewing window $[-2, 2] \times [-2, 2]$.

b. Use the results of part (a) to guess at the values of a such that f is increasing on $(-\infty, \infty)$.

c. Prove your conjecture analytically.

58. Find the values of a such that $f(x) = \cos x - ax + b$ is decreasing everywhere.

59. Show that the equation $x + \sin x = b$ has no positive root if $b < 0$ and has one positive root if $b > 0$.

Hint: Show that $f(x) = x + \sin x - b$ is increasing and that $f(0) > 0$ if $b < 0$ and $f(0) < 0$ if $b > 0$.

60. Prove that $x < \tan x$ if $0 < x < \frac{\pi}{2}$.

Hint: Let $f(x) = \tan x - x$ and show that f is increasing on $\left(0, \frac{\pi}{2}\right)$.

61. Prove that $2x/\pi < \sin x < x$ if $0 < x < \frac{\pi}{2}$.

Hint: Show that $f(x) = (\sin x)/x$ is decreasing on $\left(0, \frac{\pi}{2}\right)$.

62. Let $f(x) = -2x^2 + ax + b$. Determine the constants a and b such that f has a relative maximum at $x = 2$ and the relative maximum value is 4.

63. Let $f(x) = ax^3 + 6x^2 + bx + 4$. Determine the constants a and b such that f has a relative minimum at $x = -1$ and a relative maximum at $x = 2$.

64. Let $f(x) = (ax + b)/(cx + d)$, where $a, b, c,$ and d are constants. Show that f has no relative extrema if $ad - bc \ne 0$.

65. Let

$$f(x) = \begin{cases} \dfrac{1}{x^2} & \text{if } x < 0 \\ x^2 & \text{if } x \ge 0 \end{cases}$$

Show that f has a relative minimum at 0, although its first derivative does not change sign as we move across $x = 0$. Does this contradict the First Derivative Test?

66. Prove part (b) of Theorem 1.

67. Prove parts (b) and (c) of Theorem 2.

68. Prove that $x - x^3/6 < \sin x < x$ if $x > 0$.

Hint: To prove the left inequality, let $f(x) = \sin x - x + x^3/6$, and show that f is increasing on the interval $(0, \infty)$.

 69. Let $f(x) = 3x^5 - 8x^3 + x$.

a. Plot the graph of f using the viewing window $[-2, 2] \times [-6, 6]$. Can you determine from the graph of f the intervals where f is increasing or decreasing?

b. Plot the graph of f using the viewing window $[-0.5, 0.5] \times [-0.5, 0.5]$. Using this graph and the result of part (a), determine the intervals where f is increasing and where f is decreasing.

70. Let

$$f(x) = \begin{cases} \dfrac{1}{2}x + x^2 \sin\dfrac{1}{x} & \text{if } x \neq 0 \\ 0 & \text{if } x = 0 \end{cases}$$

 a. Plot the graph of f. Use **ZOOM** to obtain successive magnifications of the graph in the neighborhood of the origin. Can you see that f is not monotonic on any interval containing the origin?

b. Prove the observation made in part (a).

71. Let

$$f(x) = \begin{cases} \left(2 - \sin\dfrac{1}{x}\right)|x| & \text{if } x \neq 0 \\ 0 & \text{if } x = 0 \end{cases}$$

 a. Plot the graph of f. Use **ZOOM** to obtain successive magnifications of the graph in the neighborhood of

the origin. Can you see that f has a relative minimum at 0 but is not monotonic to the left or to the right of $x = 0$?

b. Prove the observation made in part (a).

Hint: For $x > 0$, show that $f'(x) > 0$ if $x = 1/(2n\pi)$ and $f'(x) < 0$ if $x = 1/((2n + 1)\pi)$.

72. a. Show that $e^x \geq 1 + x$ if $x \geq 0$.

b. Show that $e^x \geq 1 + x + x^2/2$ if $x \geq 0$.

Hint: Show that $f(x) = e^x - 1 - x - x^2/2$ is increasing for $x \geq 0$.

In Exercises 73–78, determine whether the statement is true or false. If it is true, explain why it is true. If it is false, explain why or give an example to show why it is false.

73. If f and g are increasing on an interval I, then $f + g$ is also increasing on I.

74. If f is increasing on an interval I and g is decreasing on the same interval I, then $f - g$ is increasing on I.

75. If f and g are increasing functions on an interval I, then their product fg is also increasing on I.

76. If f and g are positive on an interval I, f is increasing on I, and g is decreasing on I, then the quotient f/g is increasing on I.

77. If f is increasing on an interval (a, b), then $f'(x) \geq 0$ for every x in (a, b).

78. $f(x) = \cos^{-1} x$ is a decreasing function.

3.4 Concavity and Inflection Points

▪ Concavity

The graphs of the position functions s_1 and s_2 of two cars A and B traveling along a straight road are shown in Figure 1. Both graphs are rising, reflecting the fact that both cars are moving forward, that is, moving with positive velocities.

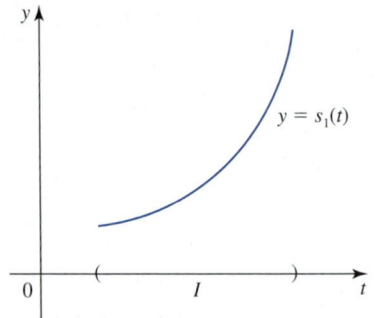

FIGURE 1   **(a)** s_1 is increasing on I.   **(b)** s_2 is increasing on I.

Observe, however, that the graph shown in Figure 1a opens upward, whereas the graph shown in Figure 1b opens downward. How do we interpret the way the curves bend in terms of the motion of the cars? To answer this question, let's look at the slopes of the tangent lines at various points on each graph (Figure 2).

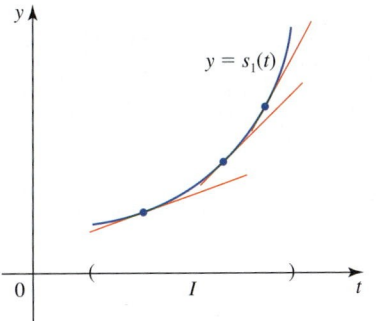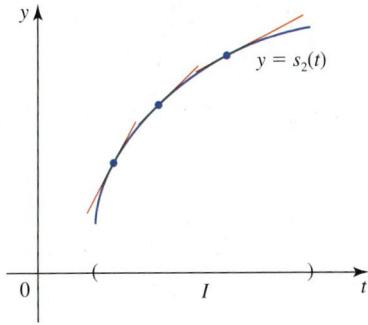

FIGURE 2

The slopes of the tangent lines to the graph of s_1 are increasing, whereas those to the graph of s_2 are decreasing.

(a) The graph of s_1 is concave upward.

(b) The graph of s_2 is concave downward.

In Figure 2a you can see that the slopes of the tangent lines to the graph increase as t increases. Since the slope of the tangent line at the point $(t, s_1(t))$ measures the velocity of car A at time t, we see not only that the car is moving forward, but also that its velocity is increasing on the time interval I. In other words, car A is accelerating over the interval I. A similar analysis of the graph in Figure 2b shows that car B is moving forward as well but decelerating over the time interval I.

We can describe the way a curve bends using the notion of concavity.

DEFINITIONS **Concavity of the Graph of a Function**

Suppose f is differentiable on an open interval I. Then

a. the graph of f is **concave upward** on I if f' is increasing on I.
b. the graph of f is **concave downward** on I if f' is decreasing on I.

Note It can be shown that if the graph of f is concave upward on an open interval I, then it lies above all of its tangent lines (Figure 2a), and if the graph is concave downward on I, then it lies below all of its tangent lines (Figure 2b). A proof of this is given in Appendix B.

Figure 3 shows the graph of a function that is concave upward on the intervals (a, b), (c, d), and (d, e) and concave downward on (b, c) and (e, g).

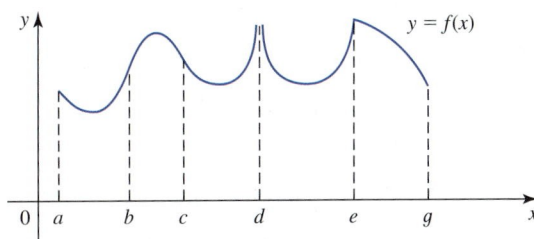

FIGURE 3

The interval $[a, g]$ is divided into subintervals showing where the graph of f is concave upward and where it is concave downward.

FIGURE 4
The sign diagram of f''

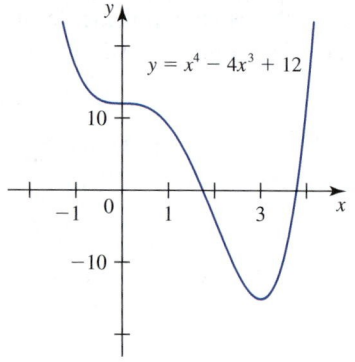

FIGURE 5
The graph of f is concave upward on $(-\infty, 0)$ and on $(2, \infty)$ and concave downward on $(0, 2)$.

If a function f has a second derivative f'', we can use it to determine the intervals of concavity of the graph of f. Indeed, since the second derivative of f measures the rate of change of the first derivative of f, we see that f' is increasing on an open interval (a, b) if $f''(x) > 0$ for all x in (a, b) and that f' is decreasing on (a, b) if $f''(x) < 0$ for all x in (a, b). Thus, we have the following result.

THEOREM 1

Suppose f has a second derivative on an open interval I.

a. If $f''(x) > 0$ for all x in I, then the graph of f is concave upward on I.
b. If $f''(x) < 0$ for all x in I, then the graph of f is concave downward on I.

The following procedure, based on the conclusions of Theorem 1, can be used to determine the intervals of concavity of a function.

Determining the Intervals of Concavity of a Function

1. Find all values of x for which $f''(x) = 0$ or $f''(x)$ does not exist. Use these values of x to partition the domain of f into open intervals.
2. Select a test number c in each interval found in Step 1 and determine the sign of $f''(c)$ in that interval.
   a. If $f''(c) > 0$, the graph of f is concave upward on that interval.
   b. If $f''(c) < 0$, the graph of f is concave downward on that interval.

Note In developing this procedure, we have once again used the fact that a function (in this case, the function f'') can change sign only as we move across a zero or a number at which the function is discontinuous. ∎

EXAMPLE 1 Determine the intervals where the graph of $f(x) = x^4 - 4x^3 + 12$ is concave upward and the intervals where it is concave downward.

Solution We first calculate the second derivative of f:

$$f'(x) = 4x^3 - 12x^2$$

$$f''(x) = 12x^2 - 24x = 12x(x - 2)$$

Next, we observe that f'' is continuous everywhere and has zeros at 0 and 2. Using this information, we draw the sign diagram of f'' (Figure 4). We conclude that the graph of f is concave upward on $(-\infty, 0)$ and on $(2, \infty)$ and concave downward on $(0, 2)$. The graph of f is shown in Figure 5. Observe that the concavity of the graph of f changes from upward to downward at the point $(0, 12)$ and from downward to upward at the point $(2, -4)$. ∎

EXAMPLE 2 Determine the intervals where the graph of $f(x) = x^{2/3}$ is concave upward and where it is concave downward.

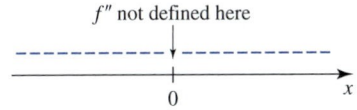

FIGURE 6
The sign diagram of f''

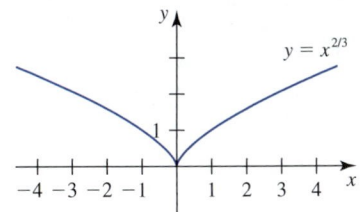

FIGURE 7
The graph of f is concave downward on $(-\infty, 0)$ and on $(0, \infty)$.

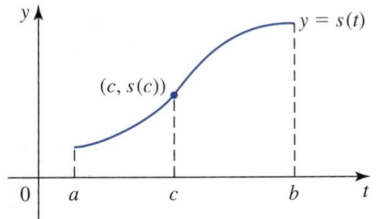

FIGURE 8
The point $(c, s(c))$ at which the concavity of the graph of s changes is called an inflection point of s.

Solution   We find

$$f'(x) = \frac{2}{3} x^{-1/3}$$

and

$$f''(x) = -\frac{2}{9} x^{-4/3} = -\frac{2}{9x^{4/3}}$$

Observe that f'' is continuous everywhere except at 0. From the sign diagram of f'' shown in Figure 6, we conclude that the graph of f is concave downward on $(-\infty, 0)$ and on $(0, \infty)$ (Figure 7). ◼

◼ Inflection Points

The graph of the position function s of a car traveling along a straight road is shown in Figure 8. Observe that the graph of s is concave upward on (a, c) and concave downward on (c, b). Interpreting the graph, we see that the car is accelerating for $a < t < c$ ($s''(t) > 0$ for t in (a, c)) and decelerating for $c < t < b$ ($s''(t) < 0$ for t in (c, b)). Its acceleration is zero when $t = c$, at which time the car also attains the maximum velocity in the time interval (a, b). The point $(c, s(c))$ on the graph of s at which the concavity changes is called an *inflection point* or *point of inflection* of s.

More generally, we have the following definition.

DEFINITION   **Inflection Point**

Let the function f be continuous on an open interval containing the point c, and suppose the graph of f has a tangent line at $P(c, f(c))$. If the graph of f changes from concave upward to concave downward (or vice versa) at P, then the point P is called an **inflection point** of the graph of f.

Observe that the graph of a function crosses its tangent line at a point of inflection (Figure 9).

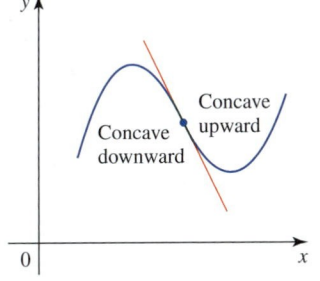

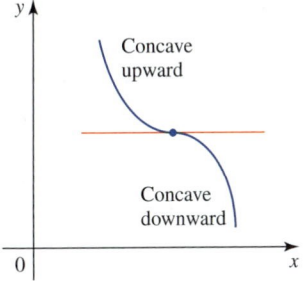

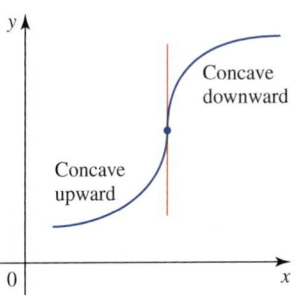

FIGURE 9
At a point of inflection the graph of a function crosses its tangent line.

The following procedure can be used to find the inflection points of a function that has a second derivative, except perhaps at isolated numbers.

> **Finding Inflection Points**
> 1. Find all numbers c in the domain of f for which $f''(c) = 0$ or $f''(c)$ does not exist. These numbers give rise to candidates for inflection points.
> 2. Determine the sign of $f''(x)$ to the left and to the right of each number c found in Step 1. If the sign of $f''(x)$ changes, then the point $P(c, f(c))$ is an inflection point of f, provided that the graph of f has a tangent line at P.

FIGURE 10
The sign diagram of f''

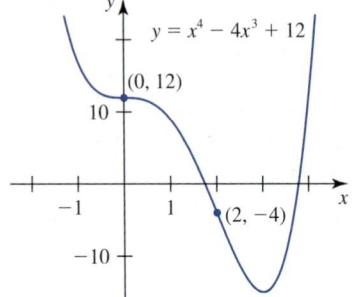

FIGURE 11
$(0, 12)$ and $(2, -4)$ are inflection points.

EXAMPLE 3 Find the points of inflection of $f(x) = x^4 - 4x^3 + 12$.

Solution We compute

$$f'(x) = 4x^3 - 12x^2 \quad \text{and} \quad f''(x) = 12x^2 - 24x = 12x(x - 2)$$

We see that f'' is continuous everywhere and has zeros at 0 and 2. These numbers give rise to candidates for the inflection points of f. From the sign diagram of f'' shown in Figure 10, we see that $f''(x)$ changes sign from positive to negative as we move across 0. Therefore, the point $(0, 12)$ is an inflection point of f. Also, $f''(x)$ changes sign from negative to positive as we move across 2, so $(2, -4)$ is also an inflection point of f. These inflection points are shown in Figure 11, where the graph of f is sketched.

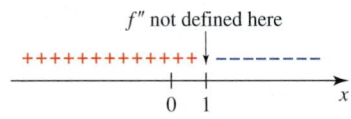

FIGURE 12
The sign diagram of f''

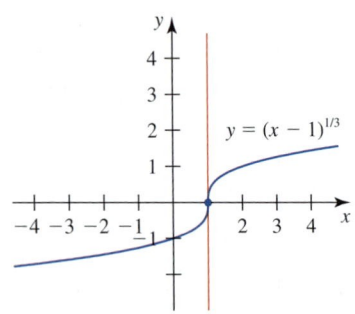

FIGURE 13
f has an inflection point at $(1, 0)$.

EXAMPLE 4 Find the points of inflection of $f(x) = (x - 1)^{1/3}$.

Solution We find

$$f'(x) = \frac{1}{3}(x - 1)^{-2/3}$$

and

$$f''(x) = -\frac{2}{9}(x - 1)^{-5/3} = -\frac{2}{9(x - 1)^{5/3}}$$

We see that f'' is continuous everywhere except at 1, where it is not defined. Furthermore, f'' has no zeros, so 1 gives rise to the only candidate for an inflection point of f. From the sign diagram of f'' shown in Figure 12, we see that $f''(x)$ does change sign from positive to negative as we move across 1. Therefore, $(1, 0)$ is indeed an inflection point of f. Observe that the graph of f has a vertical tangent line at that point. (See Figure 13.)

⚠ Remember that the numbers where $f''(x) = 0$ or where f'' is discontinuous give rise only to *candidates* for inflection points of f. For example, you can show that if $f(x) = x^4$, then $f''(0) = 0$, but the point $(0, 0)$ is not an inflection point of f (Figure 14). Also, if $g(x) = x^{2/3}$, then g'' is discontinuous at 0, as we saw in Example 2, but the point $(0, 0)$ is not an inflection point of g (Figure 15).

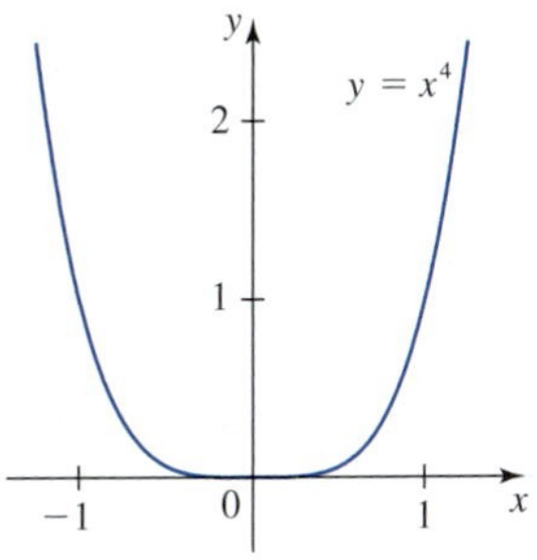

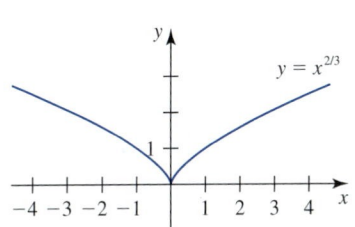

FIGURE 14
$f''(0) = 0$, but $(0, 0)$ is not an inflection point of f.

FIGURE 15
g'' is discontinuous at 0, but $(0, 0)$ is not an inflection point of g.

Examples 5 and 6 provide us with two practical interpretations of the inflection point of a function.

EXAMPLE 5 **Test Dive of a Submarine** Refer to Example 2 in Section 3.2. Recall that the depth (in feet) at time t (measured in minutes) of the prototype of a twin-piloted submarine is given by

$$h(t) = t^3(t - 7)^4 \qquad 0 \le t \le 7$$

Find the inflection points of h, and explain their significance.

Solution We have

$$h'(t) = 3t^2(t - 7)^4 + t^3(4)(t - 7)^3 = t^2(t - 7)^3(3t - 21 + 4t)$$

$$= 7t^2(t - 3)(t - 7)^3$$

$$h''(t) = \frac{d}{dt}[7(t^3 - 3t^2)(t - 7)^3]$$

$$= 7[(3t^2 - 6t)(t - 7)^3 + (t^3 - 3t^2)(3)(t - 7)^2]$$

$$= 7[3t(t - 2)(t - 7)^3 + 3t^2(t - 3)(t - 7)^2]$$

$$= 21t(t - 7)^2[(t - 2)(t - 7) + t(t - 3)]$$

$$= 42t(t - 7)^2(t^2 - 6t + 7)$$

Observe that h'' is continuous everywhere and, therefore, on $[0, 7]$. Setting $h''(t) = 0$ gives $t = 0$, $t = 7$ or $t^2 - 6t + 7 = 0$. Using the quadratic formula to solve the last equation, we obtain

$$t = \frac{6 \pm \sqrt{36 - 28}}{2} = 3 \pm \sqrt{2}$$

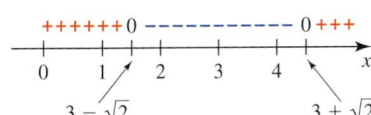

FIGURE 16
The sign diagram for h''

Since both of these roots lie inside the interval $(0, 7)$, they give rise to candidates for the inflection points of h. From the sign diagram of h'' we see that $t = 3 - \sqrt{2} \approx 1.59$ and $t = 3 + \sqrt{2} \approx 4.41$ do indeed give rise to inflection points of h (Figure 16). The graph of h is reproduced in Figure 17.

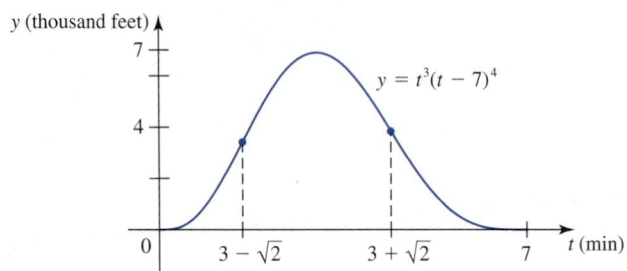

FIGURE 17
The graph of h has inflection points at $(3 - \sqrt{2}, h(3 - \sqrt{2}))$ and $(3 + \sqrt{2}, h(3 + \sqrt{2}))$.

To interpret our results, observe that the graph of h is concave upward on $(0, 3 - \sqrt{2})$. This says that the submarine is accelerating downward to a depth of $h(3 - \sqrt{2}) \approx 3427$ ft over the time interval $(0, 1.6)$. (Verify!) The graph of f is concave downward on $(3 - \sqrt{2}, 3 + \sqrt{2})$, and this says that the submarine is decelerating downward from $t \approx 1.6$ to its lowest point. Then it is accelerating upward until $t \approx 4.4$. From $t \approx 4.4$ until $t = 7$, the submarine decelerates upward until it reaches the surface, 7 min after the start of the test dive. The rate of descent of the submarine is greatest at $t = 3 - \sqrt{2} \approx 1.6$ and is approximately $h'(3 - \sqrt{2})$, or 3951 ft/min. Also the rate of ascent of the submarine is greatest at $t = 3 + \sqrt{2} \approx 4.4$ and is approximately $-h'(3 + \sqrt{2})$, or 3335 ft/min. ■

EXAMPLE 6 **Effect of Advertising on Revenue**   The total annual revenue R of the Odyssey Travel Agency, in thousands of dollars, is related to the amount of money x that the agency spends on advertising its services by the formula

$$R = -0.01x^3 + 1.5x^2 + 200 \qquad 0 \le x \le 100$$

where x is measured in thousands of dollars. Find the inflection point of R and interpret your results.

Solution

$$R' = -0.03x^2 + 3x$$

and

$$R'' = -0.06x + 3$$

which is continuous everywhere. Setting $R'' = 0$ gives $x = 50$, and this number gives rise to a candidate for an inflection point of R. Moreover, because $R'' > 0$ for $0 < x < 50$ and $R'' < 0$ for $50 < x < 100$, we see that the point $(50, 2700)$ is an inflection point of the function R. The graph of R appears in Figure 18.

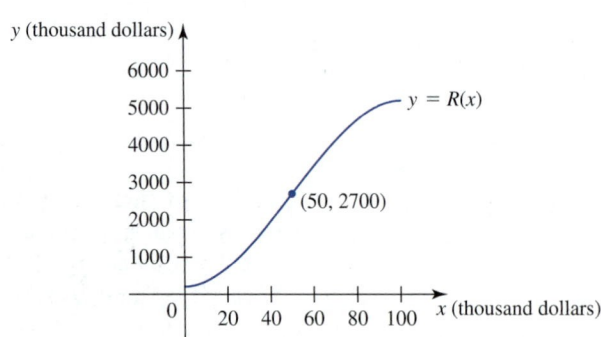

FIGURE 18
The graph of R has an inflection point at $x = 50$.

To interpret these results, observe that the revenue of the agency increases rather slowly at first. As the amount spent on advertising increases, the revenue increases rapidly, reflecting the effectiveness of the company's ads. But a point is soon reached beyond which any additional advertising expenditure results in increased revenue but at a slower rate of increase. This level of expenditure is commonly referred to as the *point of diminishing returns* and corresponds to the x-coordinate of the inflection point of R. ■

■ The Second Derivative Test

The second derivative of a function can often be used to help us determine whether a critical number gives rise to a relative extremum. Suppose that c is a critical number of f and suppose that $f''(c) < 0$. Then the graph of f is concave downward on some interval (a, b) containing c. Intuitively, we see that $f(c)$ must be the largest value of $f(x)$ for all x in (a, b). In other words, f has a relative maximum at c (Figure 19a). Similarly, if $f''(c) > 0$ at a critical number c, then f has a relative minimum at c (Figure 19b).

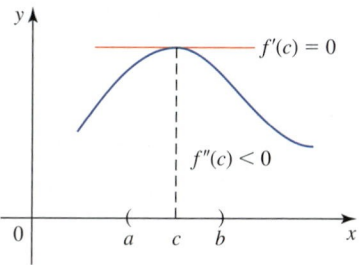

(a) f has a relative maximum at c.

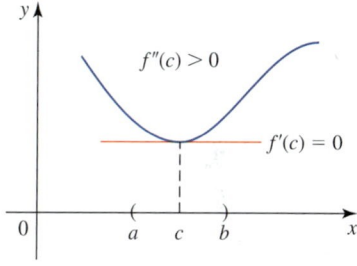

(b) f has a relative minimum at c.

FIGURE 19

These observations suggest the following theorem.

> **THEOREM 2   The Second Derivative Test**
>
> Suppose that f has a continuous second derivative on an interval (a, b) containing a critical number c of f.
>
> **a.** If $f''(c) < 0$, then f has a relative maximum at c.
> **b.** If $f''(c) > 0$, then f has a relative minimum at c.
> **c.** If $f''(c) = 0$, then the test is inconclusive.

PROOF   We will give an outline of the proof for (a). The proof for (b) is similar and will be omitted. So suppose that $f''(c) < 0$. Then the continuity of f'' implies that $f''(x) < 0$ on some open interval I containing c. This means that the graph of f is concave downward on I. Therefore, the graph of f lies below its tangent line at the point $(c, f(c))$. (See the note on page 315.) But this tangent line is horizontal because $f'(c) = 0$, and this shows that $f(x) \leq f(c)$ for all x in I (Figure 19a). So f has a relative maximum at c as asserted. ■

EXAMPLE 7   Find the relative extrema of $f(x) = x^3 - 3x^2 - 24x + 32$ using the Second Derivative Test.

Solution

$$f'(x) = 3x^2 - 6x - 24 = 3(x - 4)(x + 2)$$

Setting $f'(x) = 0$, we see that -2 and 4 are critical numbers of f. Next, we compute

$$f''(x) = 6x - 6 = 6(x - 1)$$

Evaluating $f''(x)$ at the critical number -2, we find

$$f''(-2) = 6(-2 - 1) = -18 < 0$$

and the Second Derivative Test implies that -2 gives rise to a relative maximum of f. Also

$$f''(4) = 6(4 - 1) = 18 > 0$$

so 4 gives rise to a relative minimum of f. The graph of f is shown in Figure 20.   ∎

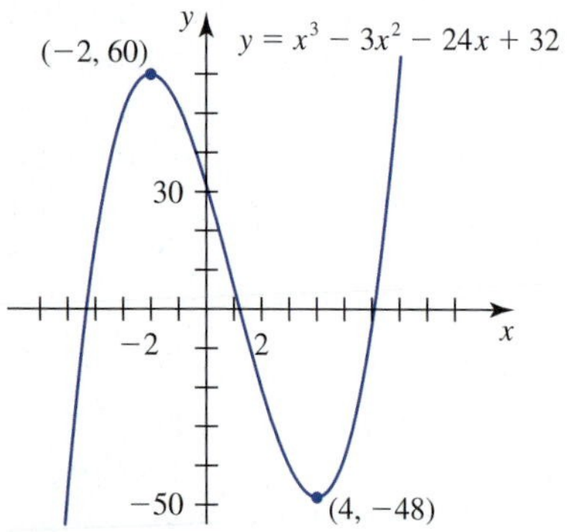

FIGURE 20
f has a relative maximum at $(-2, 60)$ and a relative minimum at $(4, -48)$.

EXAMPLE 8   Watching a Helicopter Take Off   A spectator standing 200 ft from a helicopter pad watches a helicopter take off. The helicopter rises vertically with a constant acceleration of 8 ft/sec² and reaches a height (in feet) of $h(t) = 4t^2$ after t sec, where $0 \leq t \leq 10$. (See Figure 21.) As the helicopter rises, $d\theta/dt$ increases, slowly at first, then faster, and finally it slows down again. The spectator perceives the helicopter to be rising at the greatest speed when $d\theta/dt$ is maximal. Determine the height of the helicopter at the moment the spectator perceives it to be rising at the greatest speed.

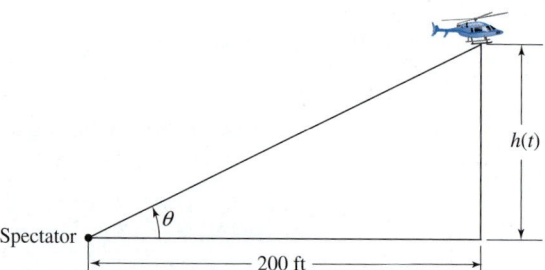

FIGURE 21
The helicopter attains a height of $h(t) = 4t^2$ after t sec.

Solution   The angle of elevation of the spectator's line of sight at time t is

$$\theta(t) = \tan^{-1}\left(\frac{h(t)}{200}\right) = \tan^{-1}\left(\frac{4t^2}{200}\right) = \tan^{-1}\left(\frac{t^2}{50}\right)$$

Therefore,

$$\frac{d\theta}{dt} = \frac{1}{1 + \left(\dfrac{t^2}{50}\right)^2} \cdot \frac{d}{dt}\left(\frac{t^2}{50}\right) = \frac{2500}{2500 + t^4} \cdot \frac{2t}{50}$$

$$= \frac{100t}{2500 + t^4}$$

To find when $d\theta/dt$ is maximal, we first compute

$$\frac{d^2\theta}{dt^2} = \frac{(2500 + t^4)100 - 100t(4t^3)}{(2500 + t^4)^2} = \frac{100(2500 - 3t^4)}{(2500 + t^4)^2}$$

Then, setting $d^2\theta/dt^2 = 0$ gives $t = (2500/3)^{1/4} \approx 5.37$ as the sole critical number of $d\theta/dt$. Using either the First or Second Derivative Test, we can show that this critical number gives rise to a maximum for $d\theta/dt$. The height of the helicopter at this instant of time is

$$h(\sqrt[4]{2500/3}) = 4(\sqrt[4]{2500/3})^2 = 4\sqrt{2500/3} \approx 115.47$$

or approximately 115 ft. ◼

Note   The point $(5.37, 115.47)$ in Example 8 is an inflection point of the graph of h.
◼

The Second Derivative Test is not useful if $f''(c) = 0$ at a critical number c. For example, each of the functions $f(x) = -x^4$, $g(x) = x^4$, and $h(x) = x^3$ has a critical number 0. Notice that $f''(0) = g''(0) = h''(0) = 0$; but as you can see from the graphs of these functions (Figure 22), f has a relative maximum at 0, g has a relative minimum at 0, and h has no extremum at 0.

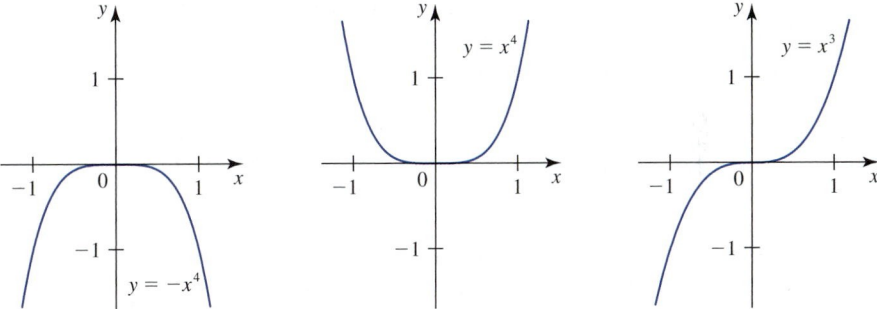

FIGURE 22
The Second Derivative Test is not useful when the second derivative is zero at a critical number c.

What are the pros and cons of using the First Derivative Test (FDT) and the Second Derivative Test (SDT) to determine the relative extrema of a function? First, because the SDT can be used only when f'' exists, it is less versatile than the FDT. For example, the SDT cannot be used to show that $f(x) = x^{2/3}$ has a relative minimum at 0. Furthermore, the SDT is inconclusive if f'' is equal to zero at a critical number of f, whereas the FDT always yields positive conclusions. The SDT is also inconvenient to use when f'' is difficult to compute. However, on the plus side, the SDT is easy to apply if f'' is easy to compute. (See Example 7.) Also, the conclusions of the SDT are often used in theoretical work.

◼ The Roles of f' and f'' in Determining the Shape of a Graph

Let's summarize our discussion of the properties of the graph of a function f that are determined by its first and second derivatives: The first derivative f' tells us where f is increasing and where f is decreasing, whereas the second derivative f'' tells us where the graph of f is concave upward and where it is concave downward. Each of these properties is determined by the signs of f' and f'' in the interval of interest and is reflected in the shape of the graph of f. Table 1 gives the characteristics of the graph of f for the various possible combinations of the signs of f' and f''.

TABLE 1

Signs of f' and f''	Properties of the graph of f	General shape of the graph of f
$f'(x) > 0$ $f''(x) > 0$	f increasing f concave upward	
$f'(x) > 0$ $f''(x) < 0$	f increasing f concave downward	
$f'(x) < 0$ $f''(x) > 0$	f decreasing f concave upward	
$f'(x) < 0$ $f''(x) < 0$	f decreasing f concave downward	

3.4 CONCEPT QUESTIONS

1. Explain what it means for the graph of a function f to be (a) concave upward and (b) concave downward on an open interval I. Given that f has a second derivative on I (except at isolated numbers), how do you determine where the graph of f is concave upward and where it is concave downward?

2. What is an inflection point of the graph of a function f? How do you find the inflection points of the graph of a function f whose rule is given?

3. State the Second Derivative Test. What are the pros and cons of using the First Derivative Test and the Second Derivative Test?

3.4 EXERCISES

In Exercises 1–6 you are given the graph of a function f. Determine the intervals where the graph of f is concave upward and where it is concave downward. Find all inflection points of f.

1.

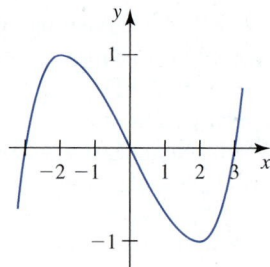

2.

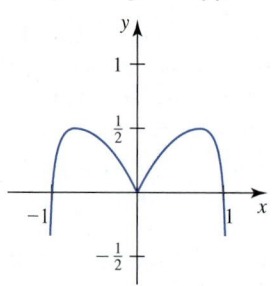

3.

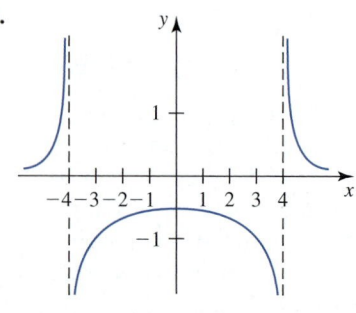

4.

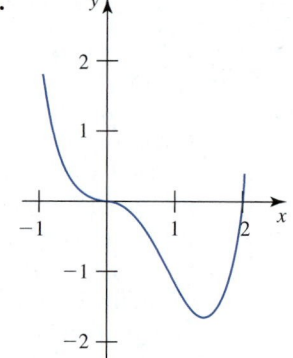

5.

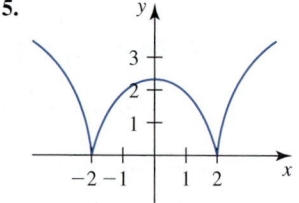

6.

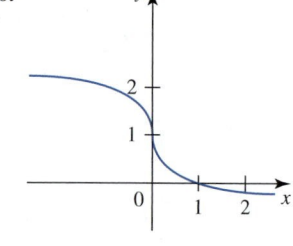

In Exercises 7 and 8 you are given the graph of the second derivative f″ of a function f. (a) Determine the intervals where the graph of f is concave upward and the intervals where it is concave downward. (b) Find the x-coordinates of the inflection points of f.

7.

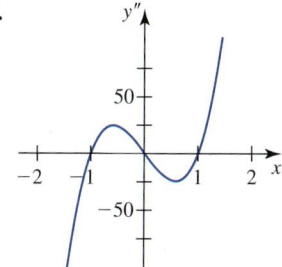

8.

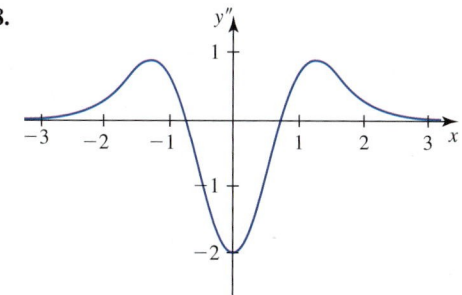

In Exercises 9–10, determine which graph—(a), (b), or (c)—is the graph of the function f with the specified properties. Explain.

9. $f'(0)$ is undefined, f is decreasing on $(-\infty, 0)$, the graph of f is concave downward on $(0, 3)$, and f has an inflection point at $x = 3$.

(a)

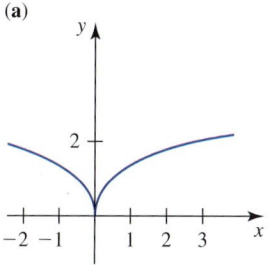

(b)

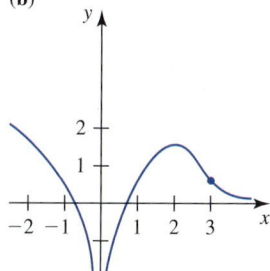

(c)

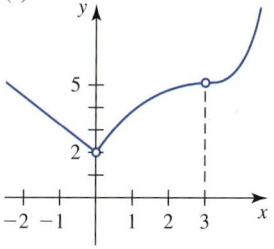

10. f is decreasing on $(-\infty, 2)$ and increasing on $(2, \infty)$, the graph of f is concave upward on $(1, \infty)$, and f has inflection points at $x = 0$ and $x = 1$.

(a)

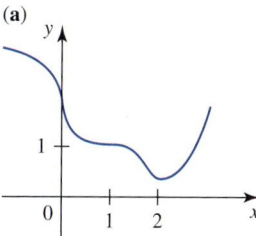

(b)

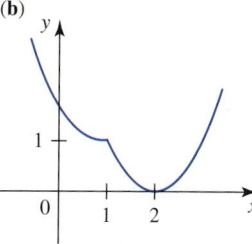

(c)

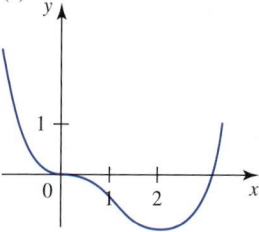

In Exercises 11–32, determine where the graph of the function is concave upward and where it is concave downward. Also, find all inflection points of the function.

11. $f(x) = x^3 - 6x$

12. $g(x) = x^3 - 6x^2 + 2x + 3$

13. $f(t) = t^4 - 2t^3$

14. $h(x) = 3x^4 + 4x^3 + 1$

15. $f(x) = 1 + 3x^{1/3}$

16. $g(x) = 2x - x^{1/3}$

17. $h(t) = \frac{1}{3}t^2 + \frac{3}{5}t^{5/3}$

18. $f(x) = x - \sqrt{1 - x^2}$

19. $h(x) = \sqrt{x^2 - x^4}$

20. $g(x) = x + \frac{1}{x}$

21. $h(x) = x^2 + \frac{1}{x^2}$

22. $f(x) = \frac{x}{x + 1}$

23. $f(u) = \frac{u}{u^2 - 1}$

24. $f(x) = \frac{x^2 - 9}{1 - x^2}$

25. $f(x) = \sin 2x, \quad 0 \le x \le \pi$

26. $g(x) = \cos^2 x, \quad 0 \le x \le 2\pi$

27. $h(t) = \sin t + \cos t, \quad 0 \le t \le 2\pi$

28. $f(x) = x - \sin x, \quad 0 \le x \le 4\pi$

29. $f(x) = \tan 2x, \quad -\pi \le x \le \pi$

30. $g(x) = x^2 e^{-x}$

31. $h(x) = \ln|x|$

32. $f(x) = e^{\sin x}, \quad -\frac{\pi}{2} \le x \le \frac{\pi}{2}$

In Exercises 33–36, *plot the graph of f, and find* (a) *the approximate intervals where the graph of f is concave upward and where it is concave downward and* (b) *the approximate coordinates of the point(s) of inflection accurate to 1 decimal place.*

33. $f(x) = x^5 - 2x^4 + 3x^2 - 5x + 4$

34. $f(x) = \dfrac{x^3 + x^2 - x + 1}{x^3 + 1}$ **35.** $f(x) = \dfrac{x}{\sqrt{x^2 + 1}}$

36. $f(x) = \cos(\sin x)$ $-2 < x < 2$

In Exercises 37–48, *find the relative extrema, if any, of the function. Use the Second Derivative Test, if applicable.*

37. $h(t) = \dfrac{1}{3}t^3 - 2t^2 - 5t - 10$

38. $h(x) = 2x^3 + 3x^2 - 12x - 2$

39. $f(x) = x^4 - 4x^3$ **40.** $f(x) = 2x^4 - 8x + 4$

41. $f(t) = 2t + \dfrac{1}{t}$ **42.** $h(t) = e^t - t - 1$

43. $g(t) = t - 2 \ln t$ **44.** $f(x) = x(\ln x)^2$

45. $f(x) = \sin x + \cos x$, $0 < x < \frac{\pi}{2}$

46. $f(x) = \sin^2 x$, $0 < x < \frac{3\pi}{2}$

47. $f(x) = 2 \sin x + \sin 2x$, $0 < x < \pi$

48. $f(x) = x^2 \ln x$

In Exercises 49 and 50, *find the point(s) of inflection of the graph of the function.*

49. $f(x) = \sin^{-1} x$ **50.** $f(x) = (\tan^{-1} x)^2$

In Exercises 51–54, *sketch the graph of a function having the given properties.*

51. $f(0) = 0, f'(0) = 0$
   $f'(x) < 0$ on $(-\infty, 0)$
   $f'(x) > 0$ on $(0, \infty)$
   $f''(x) > 0$ on $(-1, 1)$
   $f''(x) < 0$ on $(-\infty, -1) \cup (1, \infty)$

52. $f(0) = -1, f(-1) = f(1) = 0$
   $f'(0)$ does not exist
   $f'(x) < 0$ on $(-\infty, 0)$
   $f'(x) > 0$ on $(0, \infty)$
   $f''(x) < 0$ on $(-\infty, 0) \cup (0, \infty)$

53. $f(-1) = 0, f'(-1) = 0$
   $f(0) = 1, f'(0) = 0$
   $f'(x) < 0$ on $(-\infty, -1)$
   $f'(x) > 0$ on $(-1, \infty)$
   $f''(x) > 0$ on $\left(-\infty, -\frac{2}{3}\right) \cup (0, \infty)$
   $f''(x) < 0$ on $\left(-\frac{2}{3}, 0\right)$

54. $f(-1) = f(1) = 2, f'(-1) = f'(1) = 0$
   $f'(x) < 0$ on $(-\infty, -1) \cup (0, 1)$
   $f'(x) > 0$ on $(-1, 0) \cup (1, \infty)$
   $\lim\limits_{x \to 0} f(x) = \infty$
   $f''(x) > 0$ on $(-\infty, 0) \cup (0, \infty)$

55. Effect of Advertising on Bank Deposits The CEO of the Madison Savings Bank used the graphs on the following page to illustrate what effect a projected promotional campaign would have on its deposits over the next year. The functions D_1 and D_2 give the projected amount of money on deposit with the bank over the next 12 months with and without the proposed promotional campaign, respectively.
   a. Determine the signs of $D_1'(t), D_2'(t), D_1''(t),$ and $D_2''(t)$ on the interval $(0, 12)$.
   b. What can you conclude about the rate of change of the growth rate of the money on deposit with the bank with and without the proposed promotional campaign?

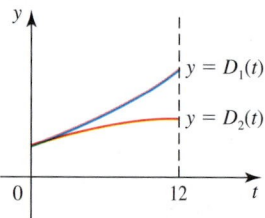

56. Assembly Time of a Worker In the following graph, $N(t)$ gives the number of satellite radios assembled by the average worker by the tth hour, where $t = 0$ corresponds to 8 A.M. and $0 \leq t \leq 4$. The point P is an inflection point of N.
   a. What can you say about the rate of change of the rate of the number of satellite radios assembled by the average worker between 8 A.M. and 10 A.M.? Between 10 A.M. and 12 P.M.?
   b. At what time is the rate at which the satellite radios are being assembled by the average worker greatest?

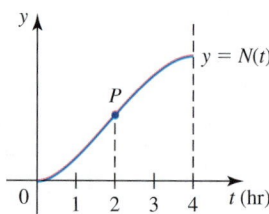

57. Water Pollution When organic waste is dumped into a pond, the oxidation process that takes place reduces the pond's oxygen content. However, given time, nature will restore the oxygen content to its natural level. In the following graph, $P(t)$ gives the oxygen content (as a percentage of its normal level) t days after organic waste has been dumped into the pond. Explain the significance of the inflection point Q.

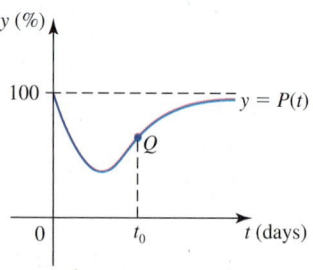

58. Effect of Budget Cuts on Drug-Related Crimes A police commissioner used the following graphs to illustrate what effect a budget cut would have on crime in the city. The number $N_1(t)$ gives the projected number of drug-related crimes in the next 12 months. The number $N_2(t)$ gives the projected number of drug-related crimes in the same time frame if next year's budget is cut.

a. Explain why $N_1'(t)$ and $N_2'(t)$ are both positive on the interval $(0, 12)$.

b. What are the signs of $N_1''(t)$ and $N_2''(t)$ on the interval $(0, 12)$?

c. Interpret the results of part (b).

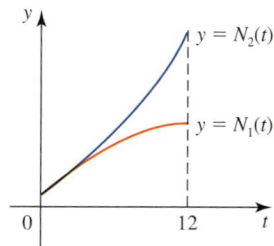

59. In the figure below, water is poured into the vase at a constant rate (in appropriate units), and the water level rises to a height of $f(t)$ units at time t as measured from the base of the vase. Sketch the graph of f, and explain its shape, indicating where it is concave upward and concave downward. Indicate the inflection point on the graph, and explain its significance.

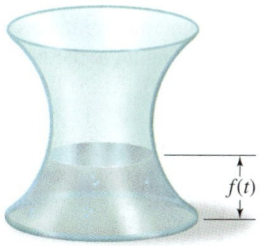

60. In the figure below, water is poured into an urn at a constant rate (in appropriate units), and the water level rises to a height of $f(t)$ units at time t as measured from the base of the urn. Sketch the graph of f, and explain its shape, indicating where it is concave upward and concave downward. Indicate the inflection point on the graph, and explain its significance.

61. Effect of Smoking Bans The sales (in billions of dollars) in restaurants and bars in California from the beginning of 1993 ($t = 0$) to the beginning of 2000 ($t = 7$) are approximated by the function

$$S(t) = 0.195t^2 + 0.32t + 23.7 \qquad 0 \le t \le 7$$

a. Show that the sales in restaurants and bars continued to rise after smoking bans were implemented in restaurants in 1995 and in bars in 1998.

Hint: Show that S is increasing on the interval $(2, 7)$.

b. What can you say about the rate at which the sales were rising after smoking bans were implemented?

Source: California Board of Equalization.

62. Global Warming The increase in carbon dioxide in the atmosphere is a major cause of global warming. Using data obtained by Charles David Keeling, professor at Scripps Institution of Oceanography, the average amount of CO_2 in the atmosphere from 1958 through 2007 is approximated by

$$A(t) = 0.010716t^2 + 0.8212t + 313.4$$

where $t = 1$ corresponds to the beginning of 1958 and $1 \le t \le 50$.

a. What can you say about the rate of change of the average amount of atmospheric CO_2 from 1958 through 2007?

b. What can you say about the rate of the rate of change of the average amount of atmospheric CO_2 from 1958 through 2007?

Source: Scripps Institution of Oceanography.

63. Population Growth in Clark County Clark County in Nevada, which is dominated by greater Las Vegas, is one of the fastest-growing metropolitan areas in the United States. The population of the county from 1970 through 2000 is approximated by the function

$$P(t) = 44{,}560t^3 - 89{,}394t^2 + 234{,}633t + 273{,}288$$
$$0 \le t \le 3$$

where t is measured in decades, with $t = 0$ corresponding to the beginning of 1970.

a. Show that the population of Clark County was always increasing over the time period in question.

b. Show that the population of Clark County was increasing at the slowest pace some time around the middle of August 1976.

Source: U.S. Census Bureau.

64. Air Pollution The level of ozone, an invisible gas that irritates and impairs breathing, that was present in the atmosphere on a certain day in May in the city of Riverside is approximated by

$$A(t) = 1.0974t^3 - 0.0915t^4 \qquad 0 \le t \le 11$$

where $A(t)$ is measured in pollutant standard index (PSI) and t is measured in hours, with $t = 0$ corresponding to 7 A.M. Use the Second Derivative Test to show that the function A has a relative maximum at approximately $t = 9$. Interpret your results.

Source: The Los Angeles Times.

65. Women's Soccer Starting with the youth movement that took hold in the 1970s and buoyed by the success of the U.S. national women's team in international competition in recent years, girls and women have taken to soccer in ever-growing numbers. The function

$$N(t) = -0.9307t^3 + 74.04t^2 + 46.8667t + 3967$$
$$0 \le t \le 16$$

gives the number of participants in women's soccer in year t with $t = 0$ corresponding to the beginning of 1985.
a. Verify that the number of participants in women's soccer has been increasing from 1985 through 2000.
b. Show that the number of participants in women's soccer has been growing at an increasing rate from 1985 through 2000.
Source: NCCA News.

66. Surveillance Cameras Research reports indicate that surveillance cameras at major intersections dramatically reduce the number of drivers who barrel through red lights. The cameras automatically photograph vehicles that drive into intersections after the light turns red. Vehicle owners are then mailed citations instructing them to pay a fine or sign an affidavit that they were not driving at the time. The function

$$N(t) = 6.08t^3 - 26.79t^2 + 53.06t + 69.5$$
$$0 \le t \le 4$$

gives the number, $N(t)$, of U.S. communities using surveillance cameras at intersections in year t with $t = 0$ corresponding to the beginning of 2003.
a. Show that N is increasing on $(0, 4)$.
b. When was the number of communities using surveillance cameras at intersections increasing least rapidly? What was the rate of increase?
Source: Insurance Institute for Highway Safety.

67. Measles Deaths Measles is still a leading cause of vaccine-preventable death among children, but because of improvements in immunizations, measles deaths have dropped globally. The function

$$N(t) = -2.42t^3 + 24.5t^2 - 123.3t + 506$$
$$0 \le t \le 6$$

gives the number of measles deaths (in thousands) in sub-Saharan Africa in year t with $t = 0$ corresponding to the beginning of 1999.
a. What was the number of measles deaths in 1999? In 2005?
b. Show that $N'(t) < 0$ on $(0, 6)$. What does this say about the number of measles deaths from 1999 through 2005?
c. When was the number of measles deaths decreasing most rapidly? What was the rate of measles death at that instant of time?
Source: Centers for Disease Control and World Health Organization.

68. Epidemic Growth During a flu epidemic the number of children in the Woodhaven Community School System who contracted influenza by the tth day is given by

$$N(t) = \frac{5000}{1 + 99e^{-0.8t}}$$

($t = 0$ corresponds to the date when data were first collected.)
a. How fast was the flu spreading on the third day ($t = 2$)?
b. When was the flu being spread at the fastest rate?

69. Von Bertalanffy Functions The mass $W(t)$ (in kilograms) of the average female African elephant at age t (in years) can be approximated by a *von Bertalanffy function*

$$W(t) = 2600(1 - 0.51e^{-0.075t})^3$$

a. How fast does a newborn female elephant gain weight? A 1600 kg female elephant?
b. At what age does a female elephant gain weight at the fastest rate?

70. Death Due to Strokes Before 1950, little was known about strokes. By 1960, however, risk factors such as hypertension had been identified. In recent years, CAT scans used as a diagnostic tool have helped to prevent strokes. As a result, deaths due to strokes have fallen dramatically. The function

$$N(t) = 130.7e^{-0.1155t^2} + 50 \qquad 0 \le t \le 6$$

gives the number of deaths per 100,000 people from the beginning of 1950 to the beginning of 2010, where t is measured in decades, with $t = 0$ corresponding to the beginning of 1950.
a. How fast was the number of deaths due to strokes per 100,000 people changing at the beginning of 1950? At the beginning of 1960? At the beginning of 1970? At the beginning of 1980?
b. When was the decline in the number of deaths due to strokes per 100,000 people greatest?
Source: American Heart Association, Centers for Disease Control and Prevention, and National Institutes of Health.

71. Oxygen Content of a Pond Refer to Exercise 57. When organic waste is dumped into a pond, the oxidation process that takes place reduces the pond's oxygen content. However, given time, nature will restore the oxygen content to its natural level. Suppose that the oxygen content t days after the organic waste has been dumped into the pond is given by

$$f(t) = 100\left(\frac{t^2 + 10t + 100}{t^2 + 20t + 100}\right)$$

percent of its normal level. Show that an inflection point of f occurs at $t = 20$.

72. Find the intervals where $f(x) = \log_3 |x|$ is concave upward or where it is concave downward.

73. a. Determine where the graph of $f(x) = 2 - |x^3 - 1|$ is concave upward and where it is concave downward.
   b. Does the graph of f have an inflection point at $x = 1$? Explain.
   c. Sketch the graph of f.

74. Show that the graph of the function $f(x) = x|x|$ has an inflection point at $(0, 0)$ but $f''(0)$ does not exist.

75. Find the values of c such that the graph of

$$f(x) = x^4 + 2x^3 + cx^2 + 2x + 2$$

is concave upward everywhere.

76. Find conditions on the coefficients a, b, and c such that the graph of $f(x) = ax^4 + bx^3 + cx^2 + dx + e$ has inflection points.

77. If the graph of a function f is concave upward on an open interval I, must the graph of the function f^2 also be concave upward on I?
   Hint: Study the function $f(x) = x^2 - 1$ on $(-1, 1)$. Plot the graphs of f and f^2 on the same set of axes.

78. Suppose f is twice differentiable on an open interval I. If f is positive and the graph of f is concave upward on I, show that the graph of the function f^2 is also concave upward. (Compare with Exercise 77.)

79. Show that a polynomial function of odd degree greater than or equal to three has at least one inflection point.

80. Show that the graph of a polynomial function of the form

$$f(x) = a_{2n}x^{2n} + a_{2n-2}x^{2n-2} + \cdots + a_2x^2 + a_0$$

where n is a positive integer and the coefficients $a_0, a_2, \ldots, a_{2n}$ are positive, is concave upward everywhere and that f has an absolute minimum.

81. Suppose that the point $(a, f(a))$ is a point of inflection of the graph of $y = f(x)$. Prove that the number a gives rise to a relative extremum of the function f'.

82. a. Suppose that f'' is continuous and $f'(a) = f''(a) = 0$, but $f'''(a) \neq 0$. Show that the graph of f has an inflection point at a.
   b. Find the relative maximum and minimum values of

$$f(x) = \cos x - 1 + \frac{x^2}{2} - \frac{x^3}{6}$$

   c. Verify the result of part (b) by plotting the graph of f using the viewing window $[-2, 2] \times [-1.5, 1.5]$.

In Exercises 83–86, determine whether the statement is true or false. If it is true, explain why it is true. If it is false, explain why or give an example to show why it is false.

83. If f has an inflection point at a, then $f'(a) = 0$.

84. If $f''(x)$ exists everywhere except at $x = a$ and $f''(x)$ changes sign as we move across a, then f has an inflection point at a.

85. A polynomial function of degree 3 has exactly one inflection point.

86. If the graph of a function f that has a second derivative is concave upward on an open interval I, then the graph of the function $-f$ is concave downward.

3.5 Limits Involving Infinity; Asymptotes

■ Infinite Limits

In Section 1.1 we were concerned primarily with whether or not the functional values of f approach a number L as x approaches a number a. Even if $f(x)$ does not approach a (finite) limit, there are situations in which it is useful to describe the behavior of $f(x)$ as x approaches a. Recall that the function $f(x) = 1/x^2$ does not have a limit as x approaches 0 because $f(x)$ becomes arbitrarily large as x gets arbitrarily close to 0. (See Example 7 in Section 1.1.) The graph of f is reproduced in Figure 1. We described this behavior by writing

$$\lim_{x \to 0} \frac{1}{x^2} = \infty$$

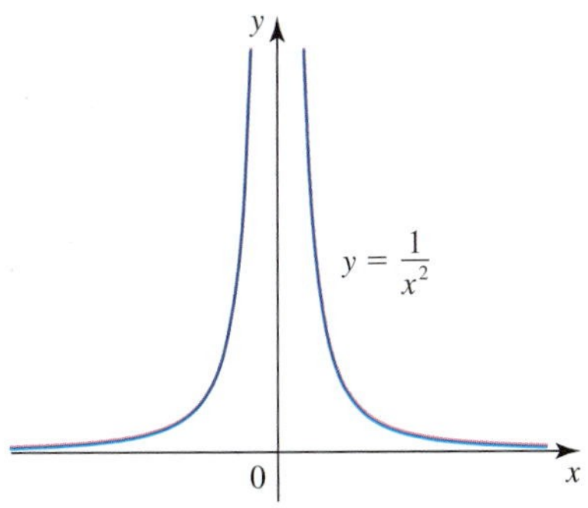

with the understanding that this is not a limit in the usual sense.

More generally, we have the following definitions concerning the behavior of functions whose values become unbounded as x approaches a.

FIGURE 1
$f(x)$ gets larger and larger without bound as x gets closer and closer to 0.

DEFINITIONS **Infinite Limits**

Let f be a function defined on an open interval containing a with the possible exception of a itself. Then

$$\lim_{x \to a} f(x) = \infty$$

if all the values of f can be made arbitrarily large (as large as we please) by taking x sufficiently close to but not equal to a. Similarly,

$$\lim_{x \to a} f(x) = -\infty$$

if all the values of f can be made as large in absolute value and negative as we please by taking x sufficiently close to but not equal to a.

These definitions are illustrated graphically in Figure 2.

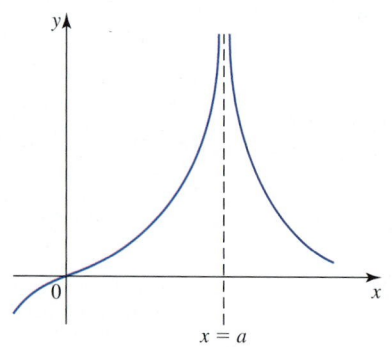

FIGURE 2
f has an infinite limit as x approaches a.

(a) $\lim_{x \to a} f(x) = \infty$

(b) $\lim_{x \to a} f(x) = -\infty$

Similar definitions can be given for the one-sided limits

$$\lim_{x \to a^-} f(x) = \infty \qquad \lim_{x \to a^+} f(x) = \infty$$

$$\lim_{x \to a^-} f(x) = -\infty \qquad \lim_{x \to a^+} f(x) = -\infty \tag{1}$$

(see Figure 3). The expression $\lim_{x \to a} f(x) = \infty$ is read "the limit of $f(x)$ as x approaches a is infinity." The expression $\lim_{x \to a} f(x) = -\infty$ is read "the limit of $f(x)$ as x approaches a is negative infinity."

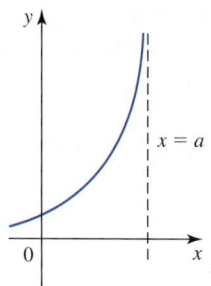
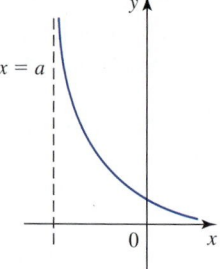
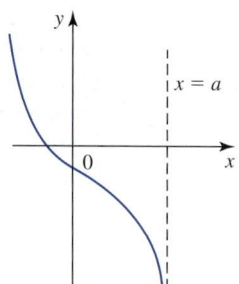
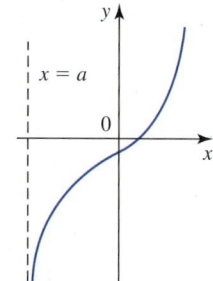

(a) $\lim_{x \to a^-} f(x) = \infty$

(b) $\lim_{x \to a^+} f(x) = \infty$

(c) $\lim_{x \to a^-} f(x) = -\infty$

(d) $\lim_{x \to a^+} f(x) = -\infty$

FIGURE 3
f has one-sided infinite limits as x approaches a.

⚠️ The "infinite limits" that are defined here are not limits in the sense defined in Section 1.1. They are simply expressions used to indicate the direction (positive or negative) taken by the unbounded values of $f(x)$ as x approaches a.

■ Vertical Asymptotes

Each vertical line $x = a$ shown in Figures 2a–b and 3a–d is called a *vertical asymptote* of the graph of f. Note that an asymptote does not constitute part of the graph of f, but it is a useful aid for sketching the graph of f.

> **DEFINITION** Vertical Asymptote
>
> The line $x = a$ is a **vertical asymptote** of the graph of a function f if at least one of the following statements is true:
>
> $$\lim_{x \to a} f(x) = \infty \quad (\text{or} -\infty); \quad \lim_{x \to a^+} f(x) = \infty \quad (\text{or} -\infty); \quad \lim_{x \to a} f(x) = \infty \quad (\text{or} -\infty)$$

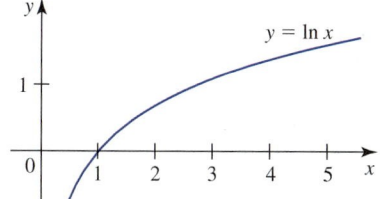

FIGURE 4
The graph of the natural logarithmic function $y = \ln x$

It follows from the above definition that $x = 0$ (the y-axis) is a vertical asymptote of the graph of $f(x) = 1/x^2$. (See Figure 1.) Another example of a function whose graph has a vertical asymptote is the natural logarithmic function $y = \ln x$. From the graph of $y = \ln x$ shown in Figure 4, we see that

$$\lim_{x \to 0^+} \ln x = -\infty$$

So $x = 0$ is a vertical asymptote of $f(x) = \ln x$. In fact, it is true that $x = 0$ is a vertical asymptote of $f(x) = \log_a x$ if $a > 1$. (See Figures 6 and 7 in Section 0.8.)

EXAMPLE 1 Find $\displaystyle\lim_{x \to 1^-} \frac{1}{x - 1}$ and $\displaystyle\lim_{x \to 1^+} \frac{1}{x - 1}$, and the vertical asymptote of the graph of $f(x) = \dfrac{1}{x - 1}$.

Solution From the graph of $f(x) = 1/(x - 1)$ shown in Figure 5, we see that

$$\lim_{x \to 1^-} \frac{1}{x - 1} = -\infty \quad \text{and} \quad \lim_{x \to 1^+} \frac{1}{x - 1} = \infty$$

The line $x = 1$ is a vertical asymptote of the graph of f.

x	$f(x)$
0.9	-10
0.99	-100
0.999	-1000

x	$f(x)$
1.1	10
1.01	100
1.001	1000

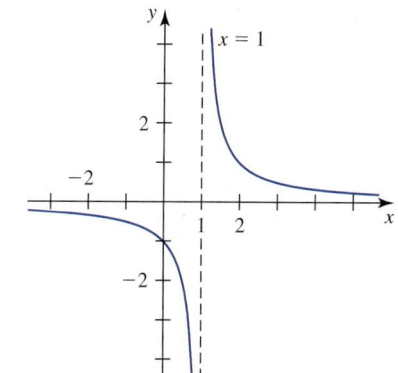

FIGURE 5

$$\lim_{x \to 1^-} \frac{1}{x - 1} = -\infty \quad \text{and} \quad \lim_{x \to 1^+} \frac{1}{x - 1} = \infty$$

Alternative Solution Observe that if x is close to but less than 1, then $(x - 1)$ is a small negative number. The numerator, however, remains constant with value 1. Therefore, $1/(x - 1)$ is a number that is large in absolute value and negative. Consequently, as x approaches 1 from the left, $1/(x - 1)$ becomes larger and larger in absolute value and negative; that is,

$$\lim_{x \to 1^-} \frac{1}{x - 1} = -\infty$$

Similarly, if x is close to but greater than 1, then $(x - 1)$ is a small positive number, and we see that $1/(x - 1)$ is a large positive number. Thus,

$$\lim_{x \to 1^+} \frac{1}{x - 1} = \infty$$

EXAMPLE 2 **Special Theory of Relativity** According to Einstein's special theory of relativity, the mass m of a particle moving with speed v is

$$m = f(v) = \frac{m_0}{\sqrt{1 - \dfrac{v^2}{c^2}}} \tag{2}$$

where c is the speed of light (approximately 3×10^8 m/sec) and m_0 is the rest mass.

a. Evaluate $\lim_{v \to c^-} f(v)$.

b. Sketch the graph of f, and interpret your result.

Solution

a. Observe that as v approaches c from the left, v^2/c^2 approaches 1 through values less than 1 and $1 - (v^2/c^2)$ approaches zero. Thus, the denominator of Equation (2) approaches zero through positive values, and the numerator remains constant, so $f(v)$ increases without bound. Thus, we have

$$\lim_{v \to c^-} f(v) = \lim_{v \to c^-} \frac{m_0}{\sqrt{1 - \dfrac{v^2}{c^2}}} = \infty$$

b. From the result of part (a) we see that $v = c$ is a vertical asymptote of the graph of f. The graph of f is shown in Figure 6. This mathematical model tells us that the mass of a particle grows without bound as its speed approaches the speed of light. This is why the speed of light is called the "ultimate speed."

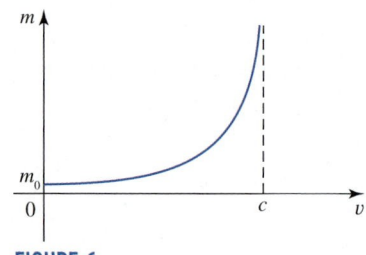

FIGURE 6

If a function f is the quotient of two functions, g and h, that is,

$$f(x) = \frac{g(x)}{h(x)}$$

then the zeros of the denominator $h(x)$ provide us with candidates for the vertical asymptotes of the graph of f, as the following example shows.

EXAMPLE 3 Find the vertical asymptotes of the graph of

$$f(x) = \frac{x}{x^2 - x - 2}$$

Solution By factoring the denominator, we can rewrite $f(x)$ in the form

$$f(x) = \frac{x}{(x+1)(x-2)}$$

Notice that the denominator of $f(x)$ is equal to zero when $x = -1$ or $x = 2$. The lines $x = -1$ and $x = 2$ are candidates for vertical asymptotes of the graph of f. To see whether $x = -1$ is, in fact, a vertical asymptote of the graph of f, let's evaluate

$$\lim_{x \to -1^-} f(x)$$

If x is close to but less than -1, then $(x+1)$ is a small negative number. Furthermore, $(x-2)$ is close to -3, so $[(x+1)(x-2)]$ is a small positive number. Also, the numerator of $f(x)$ is close to -1 when x is close to -1. Therefore, $x/[(x+1)(x-2)]$ is a number that is large in absolute value and negative. Thus,

$$\lim_{x \to -1^-} \frac{x}{(x+1)(x-2)} = -\infty$$

We conclude that $x = -1$ is a vertical asymptote of the graph of f. We leave it to you to show that

$$\lim_{x \to -1^+} \frac{x}{(x+1)(x-2)} = \infty$$

which also confirms that $x = -1$ is a vertical asymptote of the graph of f.

Next, notice that if x is close to but less than 2, then $(x-2)$ is a small negative number. Furthermore, $(x+1)$ is close to 3, so $[(x+1)(x-2)]$ is a small negative number. Also, the numerator of $f(x)$ is close to 2 when x is close to 2. Therefore,

$$\lim_{x \to 2^-} \frac{x}{(x+1)(x-2)} = -\infty$$

We conclude that $x = 2$ is also a vertical asymptote of the graph of f. We leave it to you to show that

$$\lim_{x \to 2^+} \frac{x}{(x+1)(x-2)} = \infty$$

The graph of f is shown in Figure 7. Don't worry about sketching it at this time. We will study curve sketching in Section 3.6.

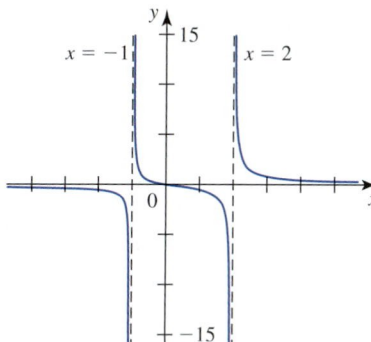

FIGURE 7
The graph of

$$y = \frac{x}{x^2 - x - 2}$$

has a vertical asymptote at $x = -1$ and another at $x = 2$.

EXAMPLE 4 Find the vertical asymptotes of the graph of $f(x) = \tan x$.

Solution We write

$$f(x) = \tan x = \frac{\sin x}{\cos x}$$

Since $\cos x = 0$ if $x = (2n+1)\pi/2$, where n is an integer, we see that the vertical lines $x = (2n+1)\pi/2$ are candidates for vertical asymptotes of the graph of f. Consider the line $x = \pi/2$, where $n = 0$. If x is close to but less than $\pi/2$, then $\sin x$ is close to 1, but $\cos x$ is positive and close to 0. Therefore, $(\sin x)/(\cos x)$ is positive and large. Thus,

$$\lim_{x \to (\pi/2)^-} \tan x = \infty$$

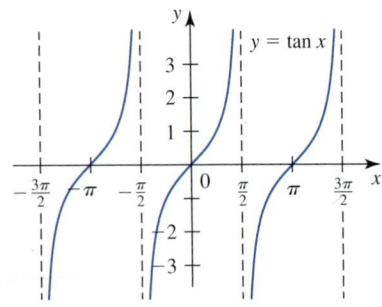

FIGURE 8
The lines $x = (2n + 1)\pi/2$ (n, an integer) are vertical asymptotes of the graph of f.

Next, if x is close to but greater than $\pi/2$, then $\sin x$ is close to 1, and $\cos x$ is negative and close to 0. Therefore, $(\sin x)/(\cos x)$ is negative and large in absolute value. Thus,

$$\lim_{x \to (\pi/2)^+} \tan x = -\infty$$

This shows that the line $x = \pi/2$ is a vertical asymptote of the graph of f. Similarly, you can show that the lines $x = (2n + 1)\pi/2$, where n is an integer, are vertical asymptotes of the graph of f (see Figure 8). ■

■ Limits at Infinity

Up to now we have studied the limit of a function as x approaches a finite number a. Sometimes we wish to know whether $f(x)$ approaches a unique number as x increases without bound. Consider, for example, the function P giving the number of fruit flies (*Drosophila melanogaster*) in a container under controlled laboratory conditions as a function of time t. The graph of P is shown in Figure 9. You can see from the graph of P that as t increases without bound (tends to infinity), $P(t)$ approaches the number 400. This number, called the *carrying capacity* of the environment, is determined by the amount of living space and food available, as well as other environmental factors.

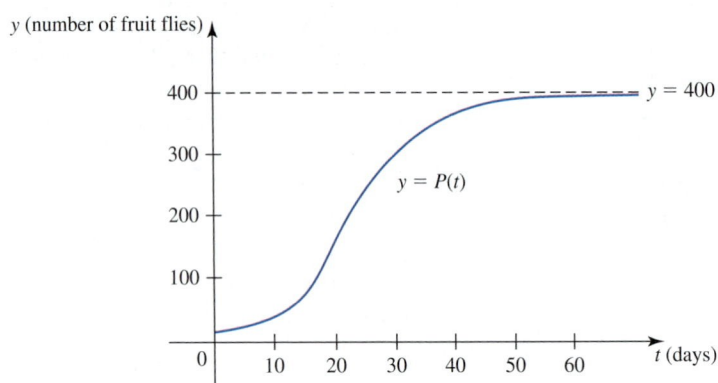

FIGURE 9
The graph of $P(t)$ gives the population of fruit flies in a laboratory experiment.

More generally, we have the following intuitive definition of the limit of a function at infinity.

DEFINITION **Limit of a Function at Infinity**

Let f be a function that is defined on an interval (a, ∞). Then the limit of $f(x)$ as x approaches infinity (increases without bound) is the number L, written

$$\lim_{x \to \infty} f(x) = L$$

if all the values of f can be made arbitrarily close to L by taking x to be sufficiently large.

This definition is illustrated graphically in Figure 10.

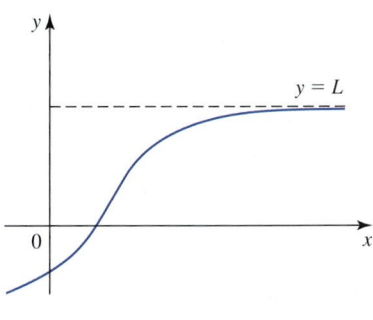

FIGURE 10   (a) $\lim\limits_{x \to \infty} f(x) = L$ (b) $\lim\limits_{x \to \infty} f(x) = L$

We define the limit at negative infinity in a similar manner.

> **DEFINITION**   **Limit of a Function at Negative Infinity**
>
> Let f be a function that is defined on an interval $(-\infty, a)$. Then the limit of $f(x)$ as x approaches negative infinity (decreases without bound) is the number L, written
>
> $$\lim_{x \to -\infty} f(x) = L$$
>
> if all the values of f can be made arbitrarily close to L by taking x to be sufficiently large in absolute value and negative. (See Figure 11.)

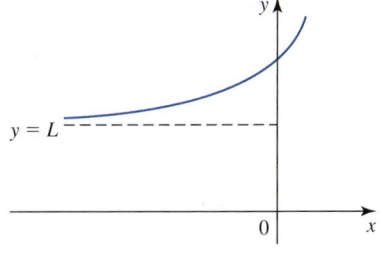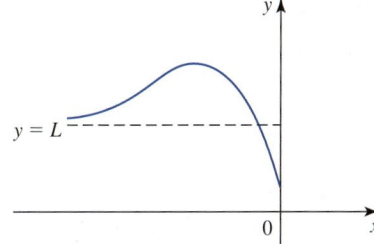

FIGURE 11   (a) $\lim\limits_{x \to -\infty} f(x) = L$ (b) $\lim\limits_{x \to -\infty} f(x) = L$

■ Horizontal Asymptotes

Each horizontal line $y = L$ shown in Figures 10a–b and 11a–b is called a *horizontal asymptote* of the graph of f.

> **DEFINITION**   **Horizontal Asymptote**
>
> The line $y = L$ is a **horizontal asymptote** of the graph of a function f if
>
> $$\lim_{x \to \infty} f(x) = L \qquad \text{or} \qquad \lim_{x \to -\infty} f(x) = L$$
>
> (or both).

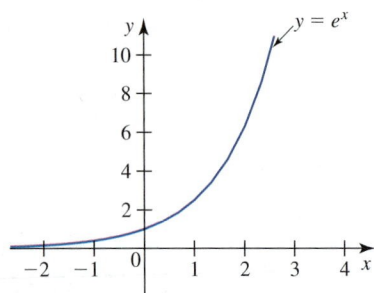

FIGURE 12
The graph of the natural exponential function $y = e^x$

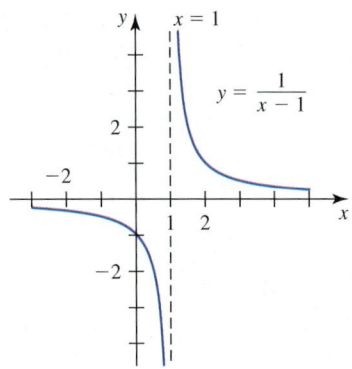

FIGURE 13
$\displaystyle\lim_{x\to\infty} \frac{1}{x-1} = 0$, $\displaystyle\lim_{x\to-\infty} \frac{1}{x-1} = 0$, and, therefore, $y = 0$ is a horizontal asymptote of the graph of f.

An example of a function whose graph has a horizontal asymptote is the natural exponential function $y = e^x$. From the graph of $y = e^x$ shown in Figure 12, we see that

$$\lim_{x\to-\infty} e^x = 0$$

and so $y = 0$ (the x-axis) is a horizontal asymptote of $f(x) = e^x$. (Also see Figure 5, Section 0.8.) In fact, it is true that $y = 0$ is a horizontal asymptote of $f(x) = a^x$ provided $a > 1$.

EXAMPLE 5   Find $\displaystyle\lim_{x\to\infty} \frac{1}{x-1}$, $\displaystyle\lim_{x\to-\infty} \frac{1}{x-1}$, and the horizontal asymptote of the graph of $f(x) = \dfrac{1}{x-1}$.

Solution   We have

$$\lim_{x\to\infty} \frac{1}{x-1} = 0 \qquad \text{and} \qquad \lim_{x\to-\infty} \frac{1}{x-1} = 0$$

We conclude that $y = 0$ is a horizontal asymptote of f (Figure 13). ■

The following theorem is useful for evaluating limits at infinity. We also point out that the laws of limits in Section 1.2 are valid if we replace $x \to a$ by $x \to -\infty$ or $x \to \infty$.

THEOREM 1

Let $r > 0$ be a rational number. Then

$$\lim_{x\to\infty} \frac{1}{x^r} = 0$$

Also, if x^r is defined for all x, then

$$\lim_{x\to-\infty} \frac{1}{x^r} = 0$$

EXAMPLE 6   Let $f(x) = \dfrac{2x^2 - x + 1}{3x^2 + 2x - 1}$. Find $\lim_{x\to\infty} f(x)$ and $\lim_{x\to-\infty} f(x)$, and find all horizontal asymptotes of the graph of f.

Solution   If we divide both the numerator and denominator by x^2, the highest power of x in the denominator, we obtain

$$\lim_{x\to\infty} \frac{2x^2 - x + 1}{3x^2 + 2x - 1} = \lim_{x\to\infty} \frac{2 - \dfrac{1}{x} + \dfrac{1}{x^2}}{3 + \dfrac{2}{x} - \dfrac{1}{x^2}} = \frac{\lim_{x\to\infty}\left(2 - \dfrac{1}{x} + \dfrac{1}{x^2}\right)}{\lim_{x\to\infty}\left(3 + \dfrac{2}{x} - \dfrac{1}{x^2}\right)}$$

$$= \frac{\lim_{x\to\infty} 2 - \lim_{x\to\infty}\dfrac{1}{x} + \lim_{x\to\infty}\dfrac{1}{x^2}}{\lim_{x\to\infty} 3 + \lim_{x\to\infty}\dfrac{2}{x} - \lim_{x\to\infty}\dfrac{1}{x^2}} = \frac{2 - 0 + 0}{3 + 0 - 0} = \frac{2}{3}$$

In a similar manner, we can show that

$$\lim_{x \to -\infty} \frac{2x^2 - x + 1}{3x^2 + 2x - 1} = \frac{2}{3}$$

We conclude that $y = \frac{2}{3}$ is a horizontal asymptote of the graph of f. ∎

EXAMPLE 7 Find the horizontal asymptotes of the graph of the function

$$f(x) = \frac{3x}{\sqrt{x^2 + 1}}$$

Solution First, let's investigate $\lim_{x \to \infty} f(x)$. We may assume that $x > 0$. In this case $\sqrt{x^2} = x$. Dividing the numerator and the denominator by x, the highest power of x in the denominator, we find

$$f(x) = \frac{\frac{1}{x}(3x)}{\frac{1}{x}\sqrt{x^2 + 1}} = \frac{3}{\frac{1}{\sqrt{x^2}}\sqrt{x^2 + 1}}$$

$$= \frac{3}{\sqrt{\frac{1}{x^2}(x^2 + 1)}} = \frac{3}{\sqrt{1 + \frac{1}{x^2}}}$$

Therefore,

$$\lim_{x \to \infty} f(x) = \lim_{x \to \infty} \frac{3}{\sqrt{1 + \frac{1}{x^2}}} = \frac{\lim_{x \to \infty} 3}{\lim_{x \to \infty} \sqrt{1 + \frac{1}{x^2}}}$$

$$= \frac{3}{\sqrt{\lim_{x \to \infty} 1 + \lim_{x \to \infty} \frac{1}{x^2}}} = \frac{3}{\sqrt{1 + 0}} = 3$$

We conclude that $y = 3$ is a horizontal asymptote of the graph of f. Next, we investigate $\lim_{x \to -\infty} f(x)$. In this case we may assume that $x < 0$. Then $\sqrt{x^2} = |x| = -x$. Dividing both the numerator and the denominator of $f(x)$ by $-x$, we obtain

$$f(x) = \frac{-\frac{1}{x}(3x)}{-\frac{1}{x}\sqrt{x^2 + 1}} = \frac{-3}{\frac{1}{\sqrt{x^2}}\sqrt{x^2 + 1}}$$

$$= \frac{-3}{\sqrt{\frac{1}{x^2}(x^2 + 1)}} = \frac{-3}{\sqrt{1 + \frac{1}{x^2}}}$$

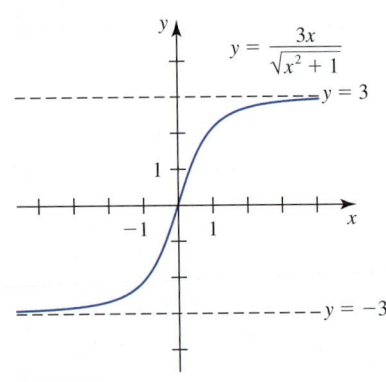

FIGURE 14
$y = 3$ and $y = -3$ are horizontal
asymptotes of the graph of f.

Therefore,

$$\lim_{x \to -\infty} f(x) = \lim_{x \to -\infty} \frac{-3}{\sqrt{1 + \dfrac{1}{x^2}}} = -3$$

and we see that $y = -3$ is also a horizontal asymptote of the graph of f. The graph of f is sketched in Figure 14.   ■

The Number e as a Limit at ∞

In Section 2.8 we saw that

$$e = \lim_{h \to 0}(1 + h)^{1/h} \tag{3}$$

We can obtain an alternative expression for e by putting $n = 1/h$ so that $h = 1/n$. Observe that $n \to \infty$ as $h \to 0$. Making this substitution in Equation (3), we have

$$e = \lim_{n \to \infty}\left(1 + \frac{1}{n}\right)^n \tag{4}$$

Equation (4) gives an expression for e as a limit at infinity. (See Exercise 69.)

Interest Compounded Continuously

In Section 0.8 we saw that when a principal of P dollars earns compound interest at the rate of r per year converted m times a year, then the accumulated amount at the end of t years is given by

$$A = P\left(1 + \frac{r}{m}\right)^{mt} \tag{5}$$

Also, the results of Example 5 in Section 0.8 suggest that as interest is converted more and more frequently, the accumulated amount over a fixed term seems to increase—but ever so slowly. This raises the question: Does the accumulated amount grow without bound, or does it approach a limit as interest is computed more and more frequently? To answer this question, we let m approach infinity in Equation (5), obtaining

$$A = \lim_{m \to \infty} P\left(1 + \frac{r}{m}\right)^{mt}$$

$$= \lim_{m \to \infty} P\left[\left(1 + \frac{r}{m}\right)^m\right]^t$$

If we make the substitution $u = m/r$ and observe that $u \to \infty$ as $m \to \infty$, then

$$A = \lim_{u \to \infty} P\left[\left(1 + \frac{1}{u}\right)^{ur}\right]^t = \lim_{u \to \infty} P\left[\left(1 + \frac{1}{u}\right)^u\right]^{rt}$$

$$= P\left[\lim_{u \to \infty}\left(1 + \frac{1}{u}\right)^u\right]^{rt}$$

But the limit in this expression is equal to the number e (see Equation (4)). Therefore,

$$A = Pe^{rt} \tag{6}$$

Equation (6) gives the accumulated amount of P dollars over a term of t years and earning interest at the rate of r per year **compounded continuously.**

EXAMPLE 8 Find the accumulated amount after 3 years if $1000 is invested at 8% per year compounded continuously.

Solution We use Equation (6) with $P = 1000$, $r = 0.08$, and $t = 3$, obtaining

$$A = 1000e^{(0.08)(3)} \approx 1271.25$$

or $1271.25.

■ Infinite Limits at Infinity

The notation

$$\lim_{x \to \infty} f(x) = \infty$$

is used to indicate that $f(x)$ becomes arbitrarily large as x increases without bound (approaches infinity). For example,

$$\lim_{x \to \infty} x^3 = \infty$$

(See Figure 15.) Similarly, we can define

$$\lim_{x \to \infty} f(x) = -\infty, \qquad \lim_{x \to -\infty} f(x) = \infty, \qquad \lim_{x \to -\infty} f(x) = -\infty$$

For example, an examination of Figure 15 once again will confirm that

$$\lim_{x \to -\infty} x^3 = -\infty$$

x	$f(x) = x^3$
-1	-1
-10	-1000
-100	-1000000
-1000	-1000000000

x	$f(x) = x^3$
1	1
10	1000
100	1000000
1000	1000000000

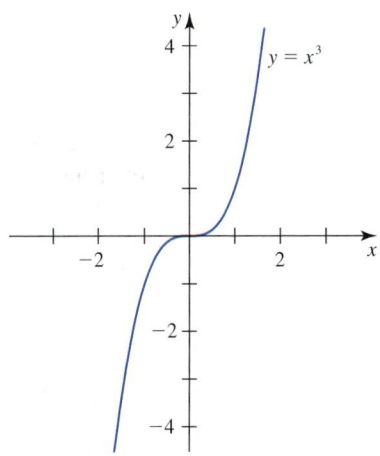

FIGURE 15

$\lim\limits_{x \to \infty} x^3 = \infty \quad$ and $\quad \lim\limits_{x \to -\infty} x^3 = -\infty$

EXAMPLE 9 Find $\lim_{x \to \infty}(2x^3 - x^2 + 1)$ and $\lim_{x \to -\infty}(2x^3 - x^2 + 1)$.

Solution We rewrite

$$2x^3 - x^2 + 1 = x^3\left(2 - \frac{1}{x} + \frac{1}{x^3}\right)$$

and note that if x is very large, then $\left(2 - \dfrac{1}{x} + \dfrac{1}{x^3}\right)$ is close to 2 and x^3 is very large. This shows that

$$\lim_{x \to \infty} (2x^3 - x^2 + 1) = \infty$$

Next, note that if x is large in absolute value and negative, so is x^3. Furthermore, $\left(2 - \dfrac{1}{x} + \dfrac{1}{x^3}\right)$ is close to 2. Therefore, $x^3\left(2 - \dfrac{1}{x} + \dfrac{1}{x^3}\right)$ is numerically very large and negative. So

$$\lim_{x \to -\infty} (2x^3 - x^2 + 1) = -\infty \qquad \blacksquare$$

EXAMPLE 10   Find $\displaystyle\lim_{x \to -\infty} \dfrac{x^2 + 1}{x - 2}$.

Solution   Dividing both the numerator and the denominator by x (the largest power of x in the denominator), we obtain

$$\lim_{x \to -\infty} \frac{x^2 + 1}{x - 2} = \lim_{x \to -\infty} \frac{x + \dfrac{1}{x}}{1 - \dfrac{2}{x}}$$

If x is very large in absolute value and negative, then the denominator of this last expression is close to 1, whereas the numerator is large in absolute value and negative. Thus, the quotient is large in absolute value and negative. We conclude that

$$\lim_{x \to -\infty} \frac{x^2 + 1}{x - 2} = -\infty \qquad \blacksquare$$

■ Precise Definitions

We begin by giving a precise definition of an infinite limit as x approaches a number a.

DEFINITION   **Infinite Limit**

Let f be a function defined on an open interval containing a, with the possible exception of a itself. We write

$$\lim_{x \to a} f(x) = \infty$$

if for every number $M > 0$ we can find a number $\delta > 0$ such that for all x satisfying

$$0 < |x - a| < \delta$$

then $f(x) > M$.

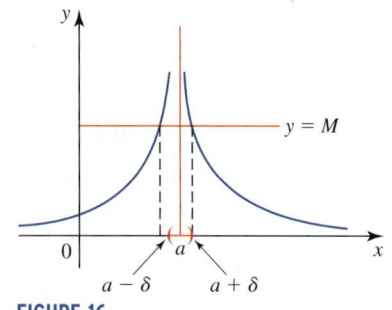

FIGURE 16
If $x \in (a - \delta, a) \cup (a, a + \delta)$, then $f(x) > M$.

For a geometric interpretation, let $M > 0$ be given. Draw the line $y = M$ shown in Figure 16. You can see that there exists a $\delta > 0$ such that whenever x lies in the interval $(a - \delta, a + \delta)$, the graph of $y = f(x)$ lies above the line $y = M$. You can also see from the figure that once you have found a number $\delta > 0$ called for in the definition, then any positive number smaller than δ will also satisfy the requirement in the definition.

EXAMPLE 11   Prove that $\lim\limits_{x \to 0} \dfrac{1}{x^2} = \infty$.

Solution   Let $M > 0$ be given. We want to show that there exists a $\delta > 0$ such that

$$\frac{1}{x^2} > M$$

whenever $0 < |x - 0| < \delta$. To find δ, consider

$$\frac{1}{x^2} > M$$

$$x^2 < \frac{1}{M}$$

or

$$|x| < \frac{1}{\sqrt{M}}$$

This suggests that we may take δ to be $1/\sqrt{M}$ or any positive number less than or equal to $1/\sqrt{M}$. Reversing the steps, we see that if $0 < |x| < \delta$, then

$$x^2 < \delta^2$$

so

$$\frac{1}{x^2} > \frac{1}{\delta^2} \geq M$$

Therefore,

$$\lim_{x \to 0} \frac{1}{x^2} = \infty$$ ∎

The precise definition of $\lim_{x \to a} f(x) = -\infty$ is similar to that of $\lim_{x \to a} f(x) = \infty$.

DEFINITION   Infinite Limit

Let f be a function defined on an open interval containing a, with the possible exception of a itself. We write

$$\lim_{x \to a} f(x) = -\infty$$

if for every number $N < 0$, we can find a number $\delta > 0$ such that for all x satisfying

$$0 < |x - a| < \delta$$

then $f(x) < N$.

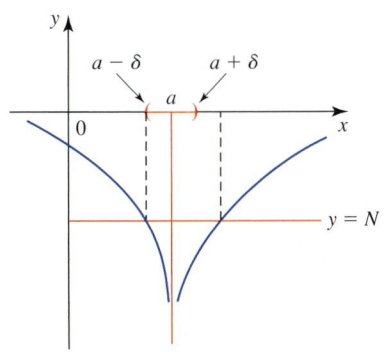

FIGURE 17
If $x \in (a - \delta, a) \cup (a, a + \delta)$, then $f(x) < N$.

(See Figure 17 for a geometric interpretation.)

The precise definitions for one-sided infinite limits are similar to the previous definitions. For example, in defining

$$\lim_{x \to a^-} f(x) = \infty$$

we must restrict x so that $x < a$. Otherwise, the definition is similar to that for

$$\lim_{x \to a} f(x) = \infty$$

We now turn our attention to the precise definition of the limit of a function at infinity.

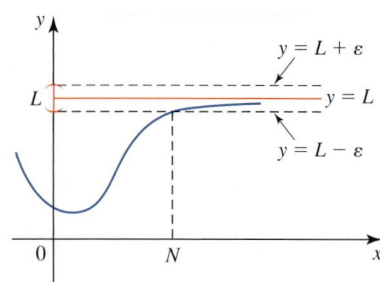

FIGURE 18
If $x > N$, then $f(x)$ lies in the band defined by $y = L - \varepsilon$ and $y = L + \varepsilon$.

> **DEFINITION   Limit at Infinity**
>
> Let f be a function defined on an interval (a, ∞). We write
>
> $$\lim_{x \to \infty} f(x) = L$$
>
> if for every number $\varepsilon > 0$ there exists a number N such that for all x satisfying $x > N$ then $|f(x) - L| < \varepsilon$.

As Figure 18 illustrates, the definition states that given any number $\varepsilon > 0$, we can find a number N such that $x > N$ implies that all the values of f lie inside the band of width 2ε determined by the lines $y = L - \varepsilon$ and $y = L + \varepsilon$.

Finally, infinite limits at infinity can also be defined precisely. For example, the precise definition of $\lim_{x \to \infty} f(x) = \infty$ follows.

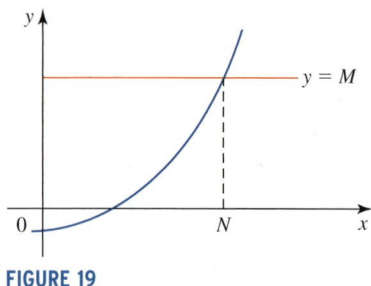

FIGURE 19
If $x > N$, then $f(x) > M$.

> **DEFINITION   Infinite Limit at Infinity**
>
> Let f be a function defined on an interval (a, ∞). We write
>
> $$\lim_{x \to \infty} f(x) = \infty$$
>
> if for every number $M > 0$ there exists a number N such that for all x satisfying $x > N$, then $f(x) > M$.

Figure 19 gives a geometric illustration of this definition. The precise definitions for $\lim_{x \to \infty} f(x) = -\infty$, $\lim_{x \to -\infty} f(x) = \infty$, and $\lim_{x \to -\infty} f(x) = -\infty$ are similar.

3.5 CONCEPT QUESTIONS

1. Explain what is meant by the statements
   (a) $\lim_{x \to 3} f(x) = \infty$ and (b) $\lim_{x \to 2^-} f(x) = -\infty$.
2. Explain what is meant by the statements
   (a) $\lim_{x \to -\infty} f(x) = 2$ and (b) $\lim_{x \to \infty} f(x) = -5$.
3. Explain the following terms in your own words:
   a. Vertical asymptote
   b. Horizontal asymptote

4. **a.** How many vertical asymptotes can the graph of a function f have? Explain using graphs.
   b. How many horizontal asymptotes can the graph of a function f have? Explain, using graphs.
5. State the precise definition of
   (a) $\lim_{x \to 2} \dfrac{3}{(x - 2)^2} = \infty$ and (b) $\lim_{x \to \infty} \dfrac{2x^2 + x + 1}{3x^2 + 4} = \dfrac{2}{3}$.

3.5   EXERCISES

In Exercises 1–6, use the graph of the function f to find the given limits.

1. a. $\lim\limits_{x \to 0^-} f(x)$ **b.** $\lim\limits_{x \to 0^+} f(x)$

   c. $\lim\limits_{x \to \infty} f(x)$ **d.** $\lim\limits_{x \to -\infty} f(x)$

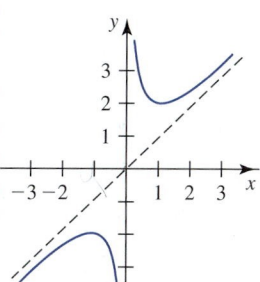

2. a. $\lim\limits_{x \to 0^-} f(x)$ **b.** $\lim\limits_{x \to 0^+} f(x)$

   c. $\lim\limits_{x \to \infty} f(x)$ **d.** $\lim\limits_{x \to -\infty} f(x)$

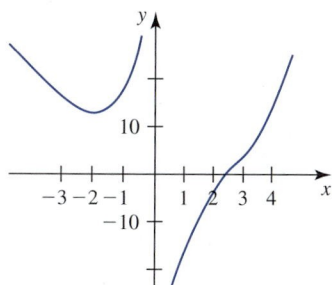

3. a. $\lim\limits_{x \to 0} f(x)$ **b.** $\lim\limits_{x \to -\infty} f(x)$ **c.** $\lim\limits_{x \to \infty} f(x)$

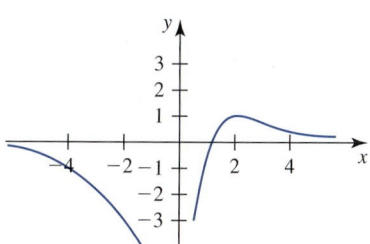

4. a. $\lim\limits_{x \to -\infty} f(x)$ **b.** $\lim\limits_{x \to \infty} f(x)$

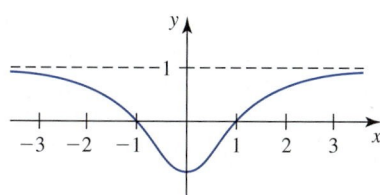

5. $\lim\limits_{x \to 2n\pi} f(x)$ for $n = 0, 1, 2, \ldots$

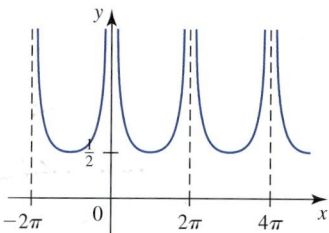

6. a. $\lim\limits_{x \to -\infty} f(x)$ **b.** $\lim\limits_{x \to \infty} f(x)$

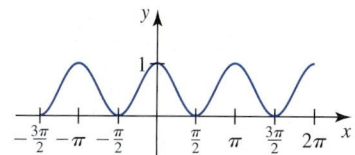

In Exercises 7–36, find the limit.

7. $\lim\limits_{x \to -1^-} \dfrac{1}{x + 1}$

8. $\lim\limits_{t \to -3^+} \dfrac{t}{t + 3}$

9. $\lim\limits_{x \to 1^-} \dfrac{1 + x}{1 - x}$

10. $\lim\limits_{x \to 1^+} \dfrac{x + 1}{1 - x}$

11. $\lim\limits_{u \to 4^+} \dfrac{u^2 + 1}{u - 4}$

12. $\lim\limits_{t \to 1} \dfrac{t^3}{(t^2 - 1)^2}$

13. $\lim\limits_{x \to 0^+} \dfrac{x + 1}{\sqrt{x}(x - 1)^2}$

14. $\lim\limits_{x \to -1^+} \left(\dfrac{1}{x} - \dfrac{1}{x + 1} \right)$

15. $\lim\limits_{x \to 0^+} \dfrac{1}{\sin x}$

16. $\lim\limits_{x \to 0^+} \cot 2x$

17. $\lim\limits_{x \to 0^+} \dfrac{1}{1 - e^{(1/\ln x)}}$

18. $\lim\limits_{x \to (\pi/2)^-} \dfrac{2e^{\tan x}}{2x - \pi}$

19. $\lim\limits_{t \to -(3/2)^-} \sec \pi t$

20. $\lim\limits_{x \to \infty} \dfrac{x + 1}{x - 5}$

21. $\lim\limits_{x \to -\infty} \dfrac{3x + 4}{2x - 3}$

22. $\lim\limits_{x \to \infty} \dfrac{2x^2 - 1}{4x^2 + 1}$

23. $\lim\limits_{x \to -\infty} \dfrac{1 - 2x^2}{x^3 + 1}$

24. $\lim\limits_{x \to -\infty} \dfrac{2x^3 + x^2 + 3}{x + 1}$

25. $\lim\limits_{x \to \infty} \left(\dfrac{x^3}{3x^2 - 2} - \dfrac{x^2}{3x + 1} \right)$

26. $\lim\limits_{x \to -\infty} \dfrac{x^4 + 1}{x^3 + 1}$

27. $\lim\limits_{x \to \infty} \dfrac{-2x^4}{3x^4 - 3x^2 + x + 1}$

28. $\lim\limits_{x \to \infty} \left(1 + \dfrac{1}{x} \right)\left(\dfrac{x^2 + 1}{x^2 - 1} \right)$

29. $\lim\limits_{t \to \infty} \left(\dfrac{t + 1}{2t - 1} + \dfrac{2t^2 - 1}{1 - 3t^2} \right)$

30. $\lim\limits_{s \to -\infty} \left(\dfrac{s}{s + 1} - \dfrac{s^2}{2s^2 + 1} \right)$

31. $\displaystyle\lim_{x\to\infty} \frac{2x}{\sqrt{3x^2 + 1}}$

32. $\displaystyle\lim_{t\to-\infty} \frac{2t^2}{\sqrt{t^4 + t^2}}$

33. $\displaystyle\lim_{x\to\infty} \frac{2e^x + 1}{3e^x + 2}$

34. $\displaystyle\lim_{t\to-\infty} \frac{e^{-t} - 2e^{2t}}{e^{-2t} + 3e^{2t}}$

35. $\displaystyle\lim_{t\to\infty} \left(\frac{3t^2 + 1}{2t^2 - 1}\right) e^{-0.1t}$

36. $\displaystyle\lim_{x\to\infty} \tan^{-1}(\ln x)$

37. Let

$$f(x) = \begin{cases} \dfrac{1}{x} & \text{if } x < 0 \\ 1 & \text{if } x > 0 \end{cases}$$

Find $\lim_{x\to 0^-} f(x)$, $\lim_{x\to 0^+} f(x)$, $\lim_{x\to-\infty} f(x)$, and $\lim_{x\to\infty} f(x)$.

38. Let

$$f(x) = \begin{cases} \dfrac{1}{\pi} x^2 + x & \text{if } x < 0 \\ \sin 2x & \text{if } x \geq 0 \end{cases}$$

(See the graph of f.) Find $\lim_{x\to-\infty} f(x)$ and $\lim_{x\to\infty} f(x)$.

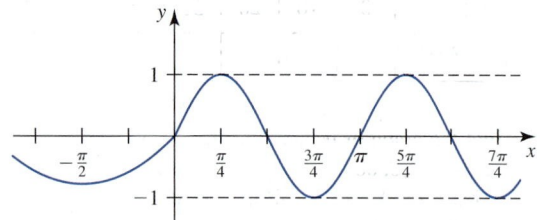

39. Let $f(x) = \dfrac{\sin x}{x}$.

a. Show that $-\dfrac{1}{x} \leq \dfrac{\sin x}{x} \leq \dfrac{1}{x}$, for $x > 0$.

b. Use the results of (a) and the Squeeze Theorem (which also holds for limits at infinity) to find $\displaystyle\lim_{x\to\infty} \frac{\sin x}{x}$.

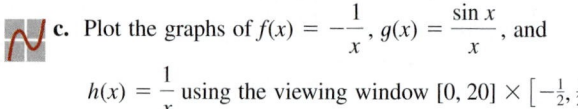 **c.** Plot the graphs of $f(x) = -\dfrac{1}{x}$, $g(x) = \dfrac{\sin x}{x}$, and

$h(x) = \dfrac{1}{x}$ using the viewing window $[0, 20] \times \left[-\frac{1}{2}, \frac{1}{2}\right]$.

40. Let $g(x) = \dfrac{\cos x^2}{\sqrt{x}}$. Find $\lim_{x\to\infty} g(x)$.

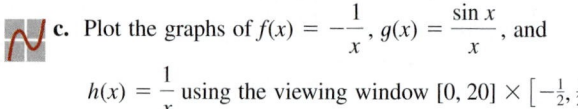 *In Exercises 41–44, (a) find an approximate value of the limit by plotting the graph of an appropriate function f, (b) find an approximate value of the limit by constructing a table of values of f, and (c) find the exact value of the limit.*

41. $\displaystyle\lim_{x\to\infty} x\left(\sqrt{x^2 + 1} - x\right)$

42. $\displaystyle\lim_{x\to-\infty} \left(x + \sqrt{x^2 + 5x}\right)$

43. $\displaystyle\lim_{x\to\infty} \left(\sqrt{2x^2 + 3x + 4} - \sqrt{2x^2 + x + 1}\right)$

44. $\displaystyle\lim_{x\to\infty} \frac{\sqrt{3x + 2} - \sqrt{3x}}{\sqrt{2x + 1} - \sqrt{2x}}$

In Exercises 45–48 you are given the graph of a function f. Find the horizontal and vertical asymptotes of the graph of f.

45.

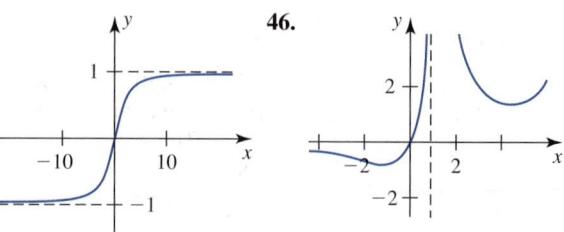

46.

47.

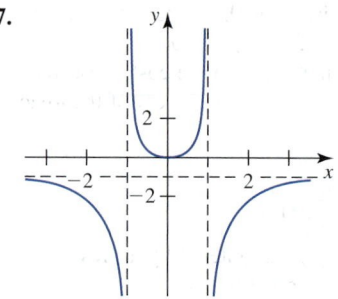

48.

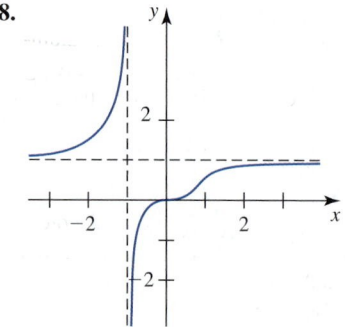

In Exercises 49–56, find the horizontal and vertical asymptotes of the graph of the function. Do not sketch the graph.

49. $f(x) = \dfrac{1}{x + 2}$

50. $g(x) = \dfrac{x}{x + 1}$

51. $h(x) = \dfrac{x - 1}{x + 1}$

52. $f(t) = \dfrac{t^2}{t^2 - 4}$

53. $f(x) = \dfrac{2x}{x^2 - x - 6}$

54. $h(x) = \dfrac{e^x}{e^x - 2}$

55. $f(t) = \dfrac{t^2 - 2}{t^2 - 4}$

56. $f(x) = \dfrac{2x^3}{\sqrt{3x^6 + 2}}$

In Exercises 57–60, sketch the graph of a function having the given properties.

57. $f(0) = 0$, $f'(0) = 1$, $f''(x) > 0$ on $(-\infty, 0)$, $f''(x) < 0$ on $(0, \infty)$, $\lim_{x\to-\infty} f(x) = -1$, $\lim_{x\to\infty} f(x) = 1$

58. $f(0) = \pi/2$, $f'(0)$ does not exist, $f(-1) = f(1) = 0$, $f''(x) > 0$ on $(-\infty, 0) \cup (0, \infty)$, $\lim_{x\to-\infty} f(x) = \lim_{x\to\infty} f(x) = -\pi/2$

59. Domain of f is $(-\infty, -1) \cup (1, \infty)$, $f(-2) = -1$,
$f'(-2) = 0$, $f''(x) < 0$ on $(-\infty, -1) \cup (1, \infty)$,
$\lim_{x \to -1^-} f(x) = -\infty$, $\lim_{x \to 1^+} f(x) = -\infty$,
$\lim_{x \to \infty} f(x) = -\infty$

60. $f(2) = 3$, $f'(2) = 0$, $f'(x) < 0$ on $(-\infty, 0) \cup (2, \infty)$,
$f'(x) > 0$ on $(0, 2)$, $\lim_{x \to 0^-} f(x) = -\infty$, $\lim_{x \to 0^+} f(x) = -\infty$,
$\lim_{x \to -\infty} f(x) = \lim_{x \to \infty} f(x) = 1$, $f''(x) < 0$ on
$(-\infty, 0) \cup (0, 3)$, $f''(x) > 0$ on $(3, \infty)$

61. Chemical Pollution As a result of an abandoned chemical dump leaching chemicals into the water, the main well of a town has been contaminated with trichloroethylene, a cancer-causing chemical. A proposal submitted by the town's board of health indicates that the cost, measured in millions of dollars, of removing x percent of the toxic pollutant is given by

$$C(x) = \frac{0.5x}{100 - x}$$

a. Evaluate $\lim_{x \to 100^-} C(x)$, and interpret your results.
b. Plot the graph of C using the viewing window
$[0, 100] \times [0, 10]$.

62. Driving Costs A study of driving costs of a 2008 medium-sized sedan found that the average cost (car payments, gas, insurance, upkeep, and depreciation) is given by the function

$$C(x) = \frac{1735.2}{x^{1.72}} + 38.6$$

where $C(x)$ is measured in cents per mile and x denotes the number of miles (in thousands) the car is driven in a year. Compute $\lim_{x \to \infty} C(x)$, and interpret your results.
Source: American Automobile Association.

63. City Planning A major developer is building a 5000-acre complex of homes, offices, stores, schools, and churches in the rural community of Marlboro. As a result of this development, the planners have estimated that Marlboro's population (in thousands) t years from now will be given by

$$P(t) = \frac{25t^2 + 125t + 200}{t^2 + 5t + 40}$$

a. What will the population of Marlboro be in the long run?
Hint: Find $\lim_{t \to \infty} P(t)$.
b. Plot the graph of P using the viewing window
$[0, 20] \times [0, 30]$.

64. Oxygen Content of a Pond When organic waste is dumped into a pond, the oxidation process that takes place reduces the pond's oxygen content. However, given time, nature will restore the oxygen content to its natural level. Suppose that the oxygen content t days after the organic waste has been dumped into the pond is given by

$$f(t) = 100\left(\frac{t^2 + 10t + 100}{t^2 + 20t + 100}\right)$$

percent of its normal level.

a. Evaluate $\lim_{t \to \infty} f(t)$ and interpret your result.
b. Plot the graph of f using the viewing window
$[0, 200] \times [70, 100]$.

65. Terminal Velocity A skydiver leaps from the gondola of a hot-air balloon. As she free-falls, air resistance, which is proportional to her velocity, builds up to a point at which it balances the force due to gravity. The resulting motion may be described in terms of her velocity as follows: Starting at rest (zero velocity), her velocity increases and approaches a constant velocity, called the *terminal velocity*. Sketch a graph of her velocity v versus time t.

66. Terminal Velocity A skydiver leaps from a helicopter hovering high above the ground. Her velocity t sec later and before deploying her parachute is given by

$$v(t) = 52[1 - (0.82)^t]$$

where $v(t)$ is measured in meters per second.
a. Complete the following table, giving her velocity at the indicated times.

t (sec)	0	10	20	30	40	50	60
$v(t)$ (m/sec)							

b. Plot the graph of v using the viewing window
$[0, 60] \times [0, 60]$.
c. What is her terminal velocity?
Hint: Evaluate $\lim_{t \to \infty} v(t)$.

67. Mass of a Moving Particle The mass m of a particle moving at a speed v is related to its rest mass m_0 by the equation

$$m = \frac{m_0}{\sqrt{1 - \dfrac{v^2}{c^2}}}$$

where c, a constant, is the speed of light. Show that

$$\lim_{v \to c^-} \frac{m_0}{\sqrt{1 - \dfrac{v^2}{c^2}}} = \infty$$

thus proving that the line $v = c$ is a vertical asymptote of the graph of m versus v. Make a sketch of the graph of m as a function of v.

68. Special Theory of Relativity According to the special theory of relativity

$$v = c\sqrt{1 - \left(\frac{E_0}{E}\right)^2}$$

where $E_0 = m_0 c^2$ is the rest energy and E is the total energy.
a. Find $\lim_{E \to \infty} v$.
b. Sketch the graph of v.
c. What do your results say about the speed of light?

69. Complete the following table to show that Equation (4),

$$e = \lim_{n \to \infty} \left(1 + \frac{1}{n}\right)^n$$

appears to be valid.

n	1	10	10^2	10^3	10^4	10^5	10^6
$\left(1 + \dfrac{1}{n}\right)^n$							

70. Find the accumulated amount after 5 years on an investment of $5000 earning interest at the rate of 10% per year compounded continuously.

71. Find the accumulated amount after 10 years on an investment of $10,000 earning interest at the rate of 12% per year compounded continuously.

72. Annual Return of an Investment A group of private investors purchased a condominium complex for $2.1 million and sold it 6 years later for $4.4 million. Find the annual rate of return (compounded continuously) on their investment.

73. Establishing a Trust Fund The parents of a child wish to establish a trust fund for the child's college education. If they need an estimated $96,000 8 years from now and they are able to invest the money at 8.5% compounded continuously in the interim, how much should they set aside in trust now?

74. Effect of Inflation on Salaries Mr. Gilbert's current annual salary is $75,000. Ten years from now, how much will he need to earn to retain his present purchasing power if the rate of inflation over that period is 5% per year? Assume that inflation is compounded continuously.

 75. Let $f(x) = \sqrt{3x + \sqrt{x}} - \sqrt{3x - \sqrt{x}}$.
   a. Plot the graph of f, and use it to estimate $\lim_{x \to \infty} f(x)$ to one decimal place.
   b. Use a table of values to estimate $\lim_{x \to \infty} f(x)$.
   c. Find the exact value of $\lim_{x \to \infty} f(x)$ analytically.

 76. Let

$$f(x) = \sqrt[3]{x^3 + 2x^2 + 3x - 1} - \sqrt[3]{x^3 - 3x^2 + x - 4}$$

   a. Plot the graph of f, and use it to estimate $\lim_{x \to \infty} f(x)$ to one decimal place.
   b. Use a table of values to estimate $\lim_{x \to \infty} f(x)$.
   c. Find the exact value of $\lim_{x \to \infty} f(x)$ analytically.

77. Escape Velocity An object is projected vertically upward from the earth's surface with an initial velocity v_0 of magnitude less than the *escape velocity* (the velocity that a projectile should have in order to break free of the earth forever). If

only the earth's influence is taken into consideration, then the maximum height reached by the rocket is

$$H = \frac{v_0^2 R}{2gR - v_0^2}$$

where R is the radius of the earth and g is the acceleration due to gravity.
   a. Show that the graph of H has a vertical asymptote at $v_0 = \sqrt{2gR}$, and interpret your result.
   b. Use the result of part (a) to find the escape velocity. Take the radius of the earth to be 4000 mi ($g = 32$ ft/sec^2).
   c. Sketch the graph of H as a function of v_0.

78. Determine the constants a and b such that

$$\lim_{x \to \infty} \left(\frac{2x^2 + 3}{x + 1} - ax - b\right) = 0$$

79. Let

$$P(x) = \frac{a_n x^n + a_{n-1} x^{n-1} + \cdots + a_0}{b_m x^m + b_{m-1} x^{m-1} + \cdots + b_0}$$

where $a_n \neq 0$, $b_m \neq 0$, and m, n, are positive integers. Show that

$$\lim_{x \to \infty} P(x) = \begin{cases} \pm\infty & \text{if } n > m \\ \dfrac{a_n}{b_m} & \text{if } n = m \\ 0 & \text{if } n < m \end{cases}$$

80. Prove that $\lim_{x \to \infty} f(x) = \lim_{t \to 0^+} f(1/t)$.

81. Use the result of Exercise 80 to find $\lim_{x \to \infty} x \sin(1/x)$.

82. a. Show that $\lim_{x \to \infty} (x^a/e^x) = 0$ for any fixed number a. Thus, e^x eventually grows faster than any power of x.
   Hint: Use the result of part (b) of Exercise 72, Section 3.3, to show that if $a = 1$, then $\lim_{x \to \infty} (x/e^x) = 0$. For the general case, introduce the variable y defined by $x = ay$ if $a > 0$.

 b. Plot the graph of $f(x) = (x^{10}/e^x)$ using the viewing window $[0, 40] \times [0, 460{,}000]$, thus verifying the result of part (a) for the special case in which $a = 10$.
   c. Find the value of x at which the graph of $f(x) = e^x$ eventually overtakes that of $g(x) = x^{10}$.

In Exercises 83–88, use the appropriate precise definition to prove the statement.

83. $\lim_{x \to 0} \dfrac{2}{x^4} = \infty$

84. $\lim_{x \to 0^+} \dfrac{1}{\sqrt{x}} = \infty$

85. $\lim_{x \to 0^-} \dfrac{1}{x} = -\infty$

86. $\lim_{x \to \infty} \dfrac{x}{x^2 + 1} = 0$

87. $\lim_{x \to -\infty} \dfrac{x}{x + 1} = 1$

88. $\lim_{x \to \infty} 3x = \infty$

In Exercises 89–94, determine whether the given statement is true or false. If it is true, explain why it is true. If it is false, explain why or give an example to show why it is false.

89. $\displaystyle\lim_{x \to 2} \frac{1}{x - 2} = \infty$

90. $\lim_{x \to \infty} c = c$ for any real number c.

91. If $y = L$ is a horizontal asymptote of the graph of the function f, then the graph of f cannot intersect $y = L$.

92. If the denominator of a rational function f is equal to zero at a, then $x = a$ is a vertical asymptote of the graph of f.

93. The graph of a function can have two distinct horizontal asymptotes.

94. If f is defined on $(0, \infty)$ and $\lim_{x \to 0^+} f(x) = L$, then $\lim_{x \to \infty} f(1/x) = 1/L$.

3.6 Curve Sketching

■ The Graph of a Function

We have seen on many occasions how the graph of a function can help us to visualize the properties of the function. From a practical point of view, the graph of a function also gives, at one glance, a complete summary of all the information captured by the function.

Consider, for example, the graph of the function giving the Dow-Jones Industrial Average (DJIA) on Black Monday: October 19, 1987 (Figure 1). Here, $t = 0$ corresponds to 9:30 A.M., when the market was open for business, and $t = 6.5$ corresponds to 4 P.M., the closing time. The following information can be gleaned from studying the graph.

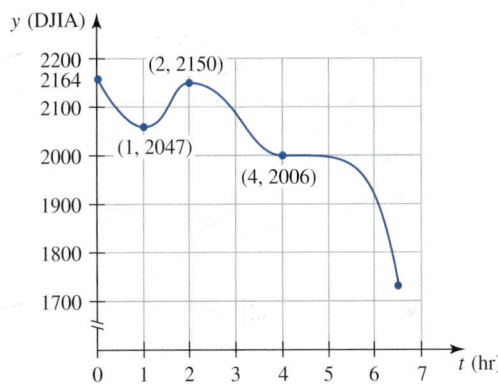

FIGURE 1

The Dow-Jones Industrial Average on Black Monday

Source: The Wall Street Journal.

The graph is *decreasing* rapidly from $t = 0$ to $t = 1$, reflecting the sharp drop in the index in the first hour of trading. The point $(1, 2047)$ is a *relative minimum* point of the function, and this turning point coincides with the start of an aborted recovery. The short-lived rally, represented by the portion of the graph that is *increasing* on the interval $(1, 2)$, quickly fizzled out at $t = 2$ (11:30 A.M.). The *relative maximum* point $(2, 2150)$ marks the highest point of the recovery. The function is decreasing on the rest of the interval. The point $(4, 2006)$ is an *inflection point* of the function; it shows that there was a temporary respite at $t = 4$ (1:30 P.M.). However, selling pressure continued unabated, and the DJIA continued to fall until the closing bell. Finally, the graph also shows that the index opened at the high of the day ($f(0) = 2164$ is the *absolute maximum* of the function) and closed at the low of the day ($f\left(\frac{13}{2}\right) = 1739$ is the *absolute minimum* of the function), a drop of 508 points, or approximately 23%, from the previous close.

Guide to Curve Sketching

A systematic approach to sketching the graph of a function f begins with an attempt to gather as much information as possible about f. The following guidelines provide us with a step-by-step procedure for doing this.

> **Guidelines for Curve Sketching**
>
> 1. Find the domain of f.
> 2. Find the x- and y-intercepts of f.
> 3. Determine whether the graph of f is symmetric with respect to the y-axis or the origin.
> 4. Determine the behavior of f for large absolute values of x.
> 5. Find the asymptotes of the graph of f.
> 6. Find the intervals where f is increasing and where f is decreasing.
> 7. Find the relative extrema of f.
> 8. Determine the concavity of the graph of f.
> 9. Find the inflection points of f.
> 10. Sketch the graph of f.

EXAMPLE 1 Sketch the graph of the function $f(x) = 2x^3 - 3x^2 - 12x + 12$.

Solution First, we obtain the following information about f.
1. Since f is a polynomial function of degree 3, the domain of f is $(-\infty, \infty)$.
2. By setting $x = 0$, we see that the y-intercept is 12. Since the cubic equation $2x^3 - 3x^2 - 12x + 12 = 0$ is not readily solved, we will not attempt to find the x-intercept.*
3. Since $f(-x) = -2x^3 - 3x^2 + 12x + 12$ is not equal to $f(x)$ or $-f(x)$, the graph of f is not symmetric with respect to the y-axis or the origin.
4. Since

$$\lim_{x \to -\infty} f(x) = -\infty \qquad \text{and} \qquad \lim_{x \to \infty} f(x) = \infty$$

we see that f decreases without bound as x decreases without bound and f increases without bound as x increases without bound.
5. Because f is a polynomial function (a rational function whose denominator is 1 and is therefore never zero), we see that the graph of f has no vertical asymptotes. From part (4), we see that the graph of f has no horizontal asymptotes.
6. $f'(x) = 6x^2 - 6x - 12 = 6(x^2 - x - 2) = 6(x + 1)(x - 2)$ and is continuous everywhere. Setting $f'(x) = 0$ gives -1 and 2 as critical numbers. The sign diagram for f' shows that f is increasing on $(-\infty, -1)$ and on $(2, \infty)$ and decreasing on $(-1, 2)$. (See Figure 2.)
7. From the results of part (6), we see that -1 and 2 are critical numbers of f. Furthermore, from the sign diagram of f', we see that f has a relative maximum at -1 with value

$$f(-1) = 2(-1)^3 - 3(-1)^2 - 12(-1) + 12 = 19$$

and a relative minimum at 2 with value

$$f(2) = 2(2)^3 - 3(2)^2 - 12(2) + 12 = -8$$

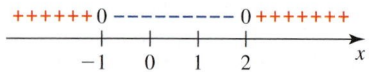

FIGURE 2
The sign diagram for f'

---

*If the equation $f(x) = 0$ is difficult to solve, disregard finding the x-intercepts.

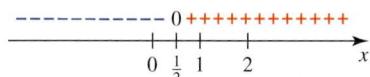

FIGURE 3
The sign diagram for f''

8. $f''(x) = 12x - 6 = 6(2x - 1)$

Setting $f''(x) = 0$ gives $x = \frac{1}{2}$. The sign diagram for f'' shows that the graph of f is concave downward on $\left(-\infty, \frac{1}{2}\right)$ and concave upward on $\left(\frac{1}{2}, \infty\right)$. (See Figure 3.)

9. From the results of part (8) we see that f has an inflection point when $x = \frac{1}{2}$. Next,

$$f\left(\tfrac{1}{2}\right) = 2\left(\tfrac{1}{2}\right)^3 - 3\left(\tfrac{1}{2}\right)^2 - 12\left(\tfrac{1}{2}\right) + 12 = \tfrac{11}{2}$$

so $\left(\frac{1}{2}, \frac{11}{2}\right)$ is the inflection point of f.

10. The following table summarizes this information.

Domain	$(-\infty, \infty)$
Intercepts	y-intercept: 12
Symmetry	None
End behavior	$\displaystyle\lim_{x \to -\infty} f(x) = -\infty$ and $\displaystyle\lim_{x \to \infty} f(x) = \infty$
Asymptotes	None
Intervals where f is $\nearrow$ or $\searrow$	$\nearrow$ on $(-\infty, -1)$ and on $(2, \infty)$; $\searrow$ on $(-1, 2)$
Relative extrema	Rel. max. at $(-1, 19)$; rel. min. at $(2, -8)$
Concavity	Downward on $\left(-\infty, \frac{1}{2}\right)$; upward on $\left(\frac{1}{2}, \infty\right)$
Point of inflection	$\left(\frac{1}{2}, \frac{11}{2}\right)$

We begin by plotting the intercepts, the inflection point, and the relative extrema of f as shown in Figure 4. Then, using the rest of the information, we complete the graph of f as shown in Figure 5.

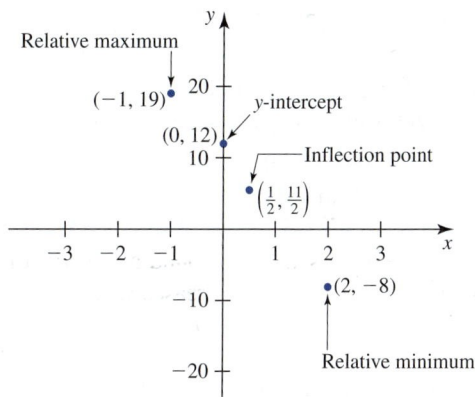

FIGURE 4
First plot the y-intercept, the relative extrema, and the inflection point.

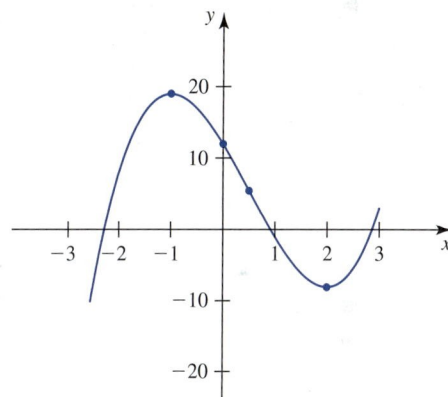

FIGURE 5
The graph of $y = 2x^3 - 3x^2 - 12x + 12$

EXAMPLE 2 Sketch the graph of the function $f(x) = \dfrac{x^2}{x^2 - 1}$.

Solution

1. The denominator of the rational function f is equal to zero if $x^2 - 1 = (x + 1)(x - 1) = 0$, that is, if $x = -1$ or $x = 1$. Therefore, the domain of f is $(-\infty, -1) \cup (-1, 1) \cup (1, \infty)$.

2. Setting $x = 0$ gives 0 as the y-intercept. Next, setting $f(x) = 0$ gives $x^2 = 0$, or $x = 0$. So the x-intercept is 0.

3. $f(-x) = \dfrac{(-x)^2}{(-x)^2 - 1} = \dfrac{x^2}{x^2 - 1} = f(x)$

and this shows that the graph of f is symmetric with respect to the y-axis.

4. $\displaystyle\lim_{x \to -\infty} \frac{x^2}{x^2 - 1} = \lim_{x \to \infty} \frac{x^2}{x^2 - 1} = 1$

5. Because the denominator of $f(x)$ is equal to zero at -1 and 1, the lines $x = -1$ and $x = 1$ are candidates for the vertical asymptotes of the graph of f. Since

$$\lim_{x \to -1^-} \frac{x^2}{x^2 - 1} = \infty \quad \text{and} \quad \lim_{x \to 1^-} \frac{x^2}{x^2 - 1} = -\infty$$

we see that $x = -1$ and $x = 1$ are indeed vertical asymptotes. From part (4) we see that $y = 1$ is a horizontal asymptote of the graph of f.

6. $f'(x) = \dfrac{(x^2 - 1)\dfrac{d}{dx}(x^2) - x^2 \dfrac{d}{dx}(x^2 - 1)}{(x^2 - 1)^2}$

$\qquad = \dfrac{(x^2 - 1)(2x) - x^2(2x)}{(x^2 - 1)^2} = -\dfrac{2x}{(x^2 - 1)^2}$

Notice that f' is continuous everywhere except at ± 1 and that it has a zero when $x = 0$. The sign diagram of f' is shown in Figure 6.

f' not defined here

$\begin{array}{c} +\!+\!+\!+\!+ \mid +\!+\!+\, 0 \,-\!-\!- \mid -\!-\!-\!-\!-\!-\!-\!-\!- \\[2pt] \underset{-1 \quad\; 0 \quad\; 1}{\rule{4cm}{0.4pt}} \to x \end{array}$

FIGURE 6
The sign diagram for f'

From the diagram we see that f is increasing on $(-\infty, -1)$ and on $(-1, 0)$ and decreasing on $(0, 1)$ and on $(1, \infty)$.

7. From the results of part (6) we see that 0 is a critical number of f. The numbers -1 and 1 are not in the domain of f and, therefore, are not critical numbers of f. Also, from Figure 6 we see that f has a relative maximum at $x = 0$. Its value is $f(0) = 0$.

8. $f''(x) = \dfrac{d}{dx}\left[\dfrac{-2x}{(x^2 - 1)^2}\right]$

$\qquad = \dfrac{(x^2 - 1)^2(-2) - (-2x)(2)(x^2 - 1)(2x)}{(x^2 - 1)^4}$

$\qquad = \dfrac{2(x^2 - 1)[-(x^2 - 1) + 4x^2]}{(x^2 - 1)^4} = \dfrac{2(3x^2 + 1)}{(x^2 - 1)^3}$

Notice that f'' is continuous everywhere except at ± 1 and that f'' has no zeros. From the sign diagram of f'' shown in Figure 7, we see that the graph of f is concave upward on $(-\infty, -1)$ and on $(1, \infty)$ and concave downward on $(-1, 1)$.

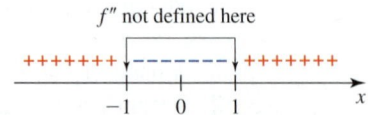

FIGURE 7
The sign diagram for f''

9. f has no inflection points. Remember that -1 and 1 are not in the domain of f.

10. The following table summarizes this information.

Domain	$(-\infty, -1) \cup (-1, 1) \cup (1, \infty)$
Intercepts	x- and y-intercepts: 0
Symmetry	With respect to the y-axis
Asymptotes	Vertical: $x = -1$ and $x = 1$
	Horizontal: $y = 1$
End behavior	$\displaystyle \lim_{x \to -\infty} \frac{x^2}{x^2 - 1} = \lim_{x \to \infty} \frac{x^2}{x^2 - 1} = 1$
Intervals where f is $\nearrow$ or $\searrow$	$\nearrow$ on $(-\infty, -1)$ and on $(-1, 0)$; $\searrow$ on $(0, 1)$ and on $(1, \infty)$
Relative extrema	Rel. max. at $(0, 0)$
Concavity	Downward on $(-1, 1)$; upward on $(-\infty, -1)$ and on $(1, \infty)$
Point of inflection	None

We begin by plotting the relative maximum of f and drawing the asymptotes of the graph of f as shown in Figure 8. In this case, plotting a few additional points will ensure a more accurate graph. For example, from the table

x	$\frac{1}{2}$	$\frac{3}{2}$	2
$f(x)$	$-\frac{1}{3}$	$\frac{9}{5}$	$\frac{4}{3}$

we see that the points $\left(\frac{1}{2}, -\frac{1}{3}\right)$, $\left(\frac{3}{2}, \frac{9}{5}\right)$, and $\left(2, \frac{4}{3}\right)$ and, by symmetry, $\left(-\frac{1}{2}, -\frac{1}{3}\right)$, $\left(-\frac{3}{2}, \frac{9}{5}\right)$, and $\left(-2, \frac{4}{3}\right)$ lie on the graph of f. Finally, using the rest of the information about f, we sketch its graph as shown in Figure 9.

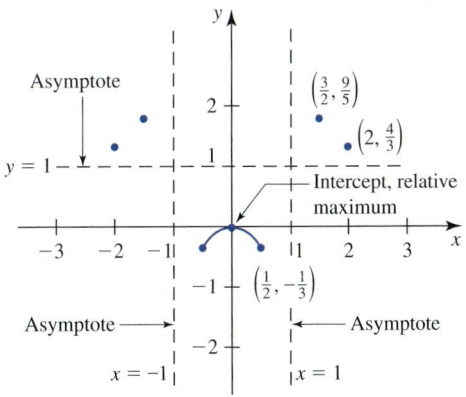

FIGURE 8
First plot the y-intercept, relative maximum, and asymptotes. Then plot a few additional points.

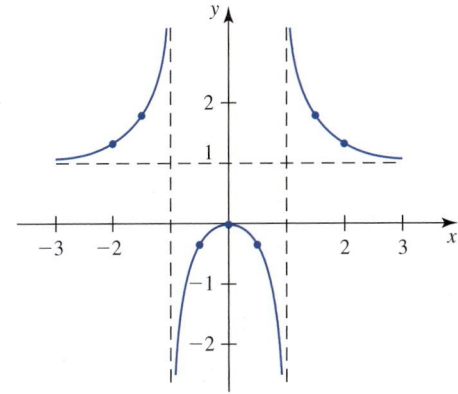

FIGURE 9
The graph of $f(x) = \dfrac{x^2}{x^2 - 1}$

EXAMPLE 3 Sketch the graph of the function $f(x) = \dfrac{1}{1 + \sin x}$.

Solution

1. The denominator of $f(x)$ is equal to zero if $1 + \sin x = 0$; that is, if $\sin x = -1$ or $x = (3\pi/2) + 2n\pi$ ($n = 0, \pm1, \pm2, \dots$). Therefore, the domain of f is $\dots \left(-\frac{\pi}{2}, \frac{3\pi}{2}\right) \cup \left(\frac{3\pi}{2}, \frac{7\pi}{2}\right) \cup \dots$.

2. Setting $x = 0$ gives 1 as the y-intercept. Since $y \neq 0$, there are no x-intercepts.

3. $f(-x) = \dfrac{1}{1 + \sin(-x)} = \dfrac{1}{1 - \sin x}$ $\sin(-x) = -\sin x$

   and is equal to neither $f(x)$ nor $-f(x)$. Therefore, f is not symmetric with respect to the y-axis or the origin.

4. $\displaystyle\lim_{x \to -\infty} \left[\dfrac{1}{1 + \sin x}\right]$ and $\displaystyle\lim_{x \to \infty} \left[\dfrac{1}{1 + \sin x}\right]$ do not exist.

5. The denominator of $f(x)$ is equal to zero when $1 + \sin x = 0$, that is, when $x = (3\pi/2) + 2n\pi$ ($n = 0, \pm1, \pm2, \dots$) (see part (1)). Since

$$\lim_{x \to (3\pi/2) + 2n\pi} \left[\dfrac{1}{1 + \sin x}\right] = \infty$$

we see that the lines $x = (3\pi/2) + 2n\pi$ ($n = 0, \pm1, \pm2, \dots$) are vertical asymptotes of the graph of f. From part (4) we see that there are no horizontal asymptotes.

6. $f'(x) = \dfrac{d}{dx}(1 + \sin x)^{-1}$

   $= -(1 + \sin x)^{-2}(\cos x)$ Use the Chain Rule.

   $= -\dfrac{\cos x}{(1 + \sin x)^2}$

   Notice that f' is continuous everywhere except at $x = (3\pi/2) + 2n\pi$ ($n = 0, \pm1, \pm2, \dots$) and has zeros at $x = (\pi/2) + 2n\pi$ ($n = 0, \pm1, \pm2, \dots$). The sign diagram of f' is shown in Figure 10. We see that f is increasing on $\dots \left(-\frac{3\pi}{2}, -\frac{\pi}{2}\right), \left(\frac{\pi}{2}, \frac{3\pi}{2}\right)$, and on $\left(\frac{5\pi}{2}, \frac{7\pi}{2}\right) \dots$ and decreasing on $\dots \left(-\frac{5\pi}{2}, -\frac{3\pi}{2}\right), \left(-\frac{\pi}{2}, \frac{\pi}{2}\right)$, and on $\left(\frac{3\pi}{2}, \frac{5\pi}{2}\right) \dots$.

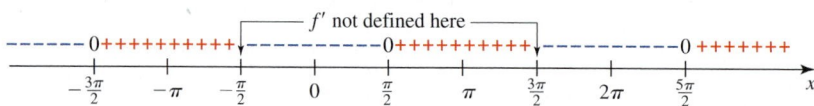

FIGURE 10
The sign diagram for f'

7. From the results of part (6) we see that $(\pi/2) + 2n\pi$ ($n = 0, \pm1, \pm2, \dots$) are critical numbers of f. From Figure 10 we see that these numbers give rise to the relative minima of f, each with value $\frac{1}{2}$, since

$$f\left(\tfrac{\pi}{2} + 2n\pi\right) = \dfrac{1}{1 + \sin\left(\frac{\pi}{2} + 2n\pi\right)} = \dfrac{1}{1 + \sin\frac{\pi}{2}} = \dfrac{1}{2}$$

8. $f''(x) = \dfrac{d}{dx}[-(\cos x)(1 + \sin x)^{-2}]$

$\qquad = (\sin x)(1 + \sin x)^{-2} - (\cos x)(-2)(1 + \sin x)^{-3}(\cos x)$

$\qquad = (1 + \sin x)^{-3}[(\sin x)(1 + \sin x) + 2\cos^2 x]$

$\qquad = \dfrac{\sin x + \sin^2 x + 2\cos^2 x}{(1 + \sin x)^3}$

$\qquad = \dfrac{\sin x + \sin^2 x + 2(1 - \sin^2 x)}{(1 + \sin x)^3} = -\dfrac{\sin^2 x - \sin x - 2}{(1 + \sin x)^3}$

$\qquad = -\dfrac{(\sin x - 2)(\sin x + 1)}{(1 + \sin x)^3} = \dfrac{2 - \sin x}{(1 + \sin x)^2}$

Because $|\sin x| \le 1$ for all values of x, we see that $f''(x) > 0$ whenever it is defined. From the sign diagram of f'' shown in Figure 11, we conclude that the graph of f is concave upward on $\cdots\left(-\frac{5\pi}{2}, -\frac{\pi}{2}\right), \left(-\frac{\pi}{2}, \frac{3\pi}{2}\right),$ and on $\left(\frac{3\pi}{2}, \frac{7\pi}{2}\right)\cdots$.

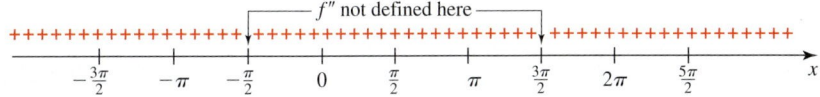

FIGURE 11
The sign diagram for f''

9. f has no inflection points.
10. The following table summarizes this information.

Domain	$\cdots\left(-\frac{\pi}{2}, \frac{3\pi}{2}\right) \cup \left(\frac{3\pi}{2}, \frac{7\pi}{2}\right) \cup \cdots$
Intercept	y-intercept: 1
Symmetry	None (with respect to the y-axis or the origin)
End behavior	$\displaystyle\lim_{x \to -\infty}\left[\dfrac{1}{1 + \sin x}\right]$ and $\displaystyle\lim_{x \to \infty}\left[\dfrac{1}{1 + \sin x}\right]$ do not exist.
Asymptotes	Vertical: $x = \frac{3\pi}{2} + 2n\pi$ $(n = 0, \pm 1, \pm 2, \dots)$
Intervals where f is ↗ or ↘	↗ on $\cdots\left(-\frac{3\pi}{2}, -\frac{\pi}{2}\right)$ and on $\left(\frac{\pi}{2}, \frac{3\pi}{2}\right)\cdots$ ↘ on $\cdots\left(-\frac{\pi}{2}, \frac{\pi}{2}\right)$ and on $\left(\frac{3\pi}{2}, \frac{5\pi}{2}\right)\cdots$
Relative extrema	Rel. min: $\cdots\left(-\frac{3\pi}{2}, \frac{1}{2}\right), \left(\frac{\pi}{2}, \frac{1}{2}\right), \left(\frac{5\pi}{2}, \frac{1}{2}\right)\cdots$
Concavity	Upward on $\cdots\left(-\frac{5\pi}{2}, -\frac{\pi}{2}\right)$ and on $\left(-\frac{\pi}{2}, \frac{3\pi}{2}\right)\cdots$
Point of inflection	None

The graph of f is shown in Figure 12.

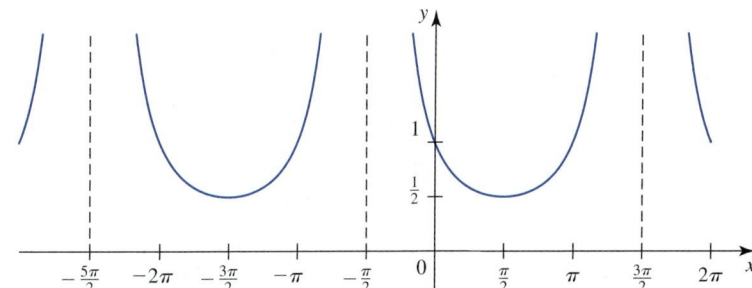

FIGURE 12
The graph of $f(x) = \dfrac{1}{1 + \sin x}$

EXAMPLE 4 Sketch the graph of the function $f(x) = e^{-x^2}$.

Solution First, we obtain the following information on the function f.

1. The domain of f is $(-\infty, \infty)$.
2. Setting $x = 0$ gives 1 as the y-intercept. Next, since $e^{-x^2} = 1/e^{x^2}$ is never zero, there are no x-intercepts.
3. Since

$$f(-x) = e^{-(-x)^2} = e^{-x^2} = f(x)$$

we see that the graph of f is symmetric with respect to the y-axis.
4. and 5. Since

$$\lim_{x \to -\infty} e^{-x^2} = \lim_{x \to -\infty} \frac{1}{e^{x^2}} = 0 = \lim_{x \to \infty} e^{-x^2}$$

we see that $y = 0$ is a horizontal asymptote of the graph of f.

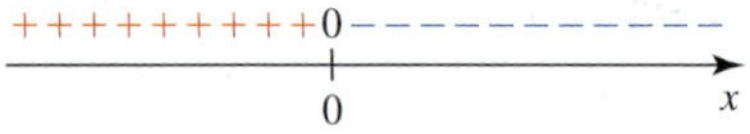

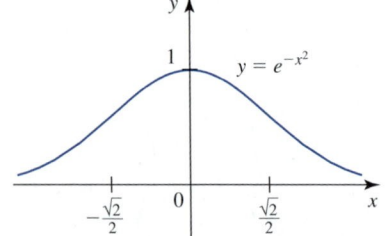

FIGURE 13
The sign diagram for f'

6. $f'(x) = \dfrac{d}{dx} e^{-x^2} = e^{-x^2} \dfrac{d}{dx}(-x^2) = -2xe^{-x^2}$

Setting $f'(x) = 0$ gives $x = 0$. The sign diagram of f' shows that f is increasing on $(-\infty, 0)$ and decreasing on $(0, \infty)$. (See Figure 13.)
7. From the results of Step 6 we see that 0 is the sole critical number of f. Furthermore, from the sign diagram of f', we see that f has a relative maximum at $x = 0$ with value $f(0) = e^0 = 1$.

8. $f''(x) = \dfrac{d}{dx}\left[-2xe^{-x^2}\right]$

$= -2e^{-x^2} - 2xe^{-x^2}(-2x)$ Use the Product Rule and the Chain Rule.

$= 2(2x^2 - 1)e^{-x^2}$

Setting $f''(x) = 0$ gives $2x^2 - 1 = 0$ or $x = \pm\sqrt{2}/2$. The sign diagram of f'' shows that f is concave upward on $\left(-\infty, -\frac{\sqrt{2}}{2}\right)$ and on $\left(\frac{\sqrt{2}}{2}, \infty\right)$ and concave downward on $\left(-\frac{\sqrt{2}}{2}, \frac{\sqrt{2}}{2}\right)$. (See Figure 14.)

FIGURE 14
The sign diagram for f''

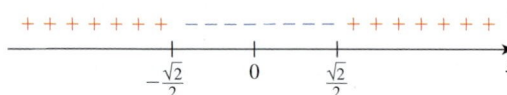

9. From the results of Step 8 we see that f has inflection points at $x = \pm\sqrt{2}/2$. Since $f\left(-\frac{\sqrt{2}}{2}\right) = e^{-1/2} = f\left(\frac{\sqrt{2}}{2}\right)$, we see that $\left(-\frac{\sqrt{2}}{2}, e^{-1/2}\right)$ and $\left(\frac{\sqrt{2}}{2}, e^{-1/2}\right)$ are inflection points of f.
10. The graph of $f(x) = e^{-x^2}$ is sketched in Figure 15. ∎

■ Slant Asymptotes

The graph of a function f may have an asymptote that is neither vertical nor horizontal but slanted. We call the line with equation $y = mx + b$ a **slant** or **oblique (right) asymptote** of the graph of f if

$$\lim_{x \to \infty} \frac{f(x)}{x} = m \qquad \text{and} \qquad \lim_{x \to \infty} [f(x) - mx] = b \qquad (1)$$

FIGURE 15
The graph of $y = e^{-x^2}$

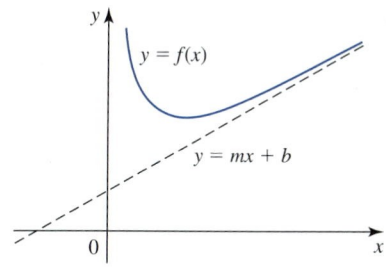

FIGURE 16
The graph of f has a slant asymptote.

Observe that the second equation in (1) is equivalent to the statement $\lim_{x \to \infty}[f(x) - mx - b] = 0$. Since $|f(x) - mx - b|$ measures the vertical distance between the graph of $f(x)$ and the line $y = mx + b$, the second equation in (1) simply states that the graph of f approaches the line with equation $y = mx + b$ as x approaches infinity. (See Figure 16.)

Similarly, if

$$\lim_{x \to -\infty} \frac{f(x)}{x} = m \qquad \text{and} \qquad \lim_{x \to -\infty} [f(x) - mx] = b \qquad (2)$$

then the line $y = mx + b$ is called a **slant (left) asymptote** of the graph of f. Note that a horizontal asymptote of the graph of f may be considered a special case of a slant asymptote where $m = 0$.

Before looking at the next example, we point out that the graph of a rational function has a slant asymptote if the degree of its numerator exceeds the degree of its denominator by 1 or more. In fact, if the degree of the numerator exceeds the degree of the denominator by 1, the slant asymptote is a straight line, as the next example shows; if it exceeds the denominator by 2, then the slant asymptote is parabolic, and so forth.

EXAMPLE 5 Find the slant asymptotes of the graph of $f(x) = \dfrac{2x^2 - 3}{x - 2}$.

Solution We compute

$$\lim_{x \to \infty} \frac{f(x)}{x} = \lim_{x \to \infty} \frac{\dfrac{2x^2 - 3}{x - 2}}{x} = \lim_{x \to \infty} \frac{2x - \dfrac{3}{x}}{x - 2}$$

$$= \lim_{x \to \infty} \frac{2 - \dfrac{3}{x^2}}{1 - \dfrac{2}{x}} \qquad \text{\color{brown}{Divide the numerator and the denominator by } } x.$$

$$= 2$$

Next, taking $m = 2$, we compute

$$\lim_{x \to \infty}[f(x) - mx] = \lim_{x \to \infty}\left(\frac{2x^2 - 3}{x - 2} - 2x\right)$$

$$= \lim_{x \to \infty} \frac{2x^2 - 3 - 2x^2 + 4x}{x - 2}$$

$$= \lim_{x \to \infty} \frac{4x - 3}{x - 2} = \lim_{x \to \infty} \frac{4 - \dfrac{3}{x}}{1 - \dfrac{2}{x}} = 4$$

So, taking $b = 4$, we see that the line with equation $y = 2x + 4$ is a slant asymptote of the graph of f. You can show that the computations using the equations in (2) lead to the same conclusion (see Exercise 43), so $y = 2x + 4$ is the only slant asymptote of the graph of f. The graph of f is sketched in Figure 17.

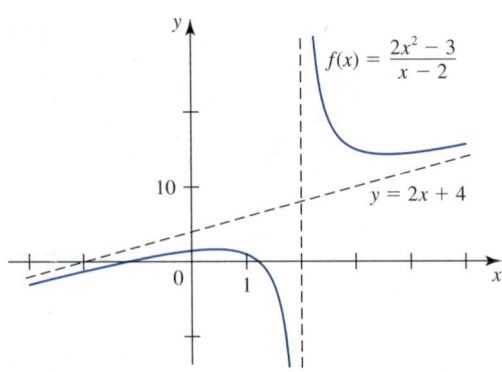

FIGURE 17
$y = 2x + 4$ is a slant asymptote
of the graph of f.

■ Finding Relative Extrema Using a Graphing Utility

Although we found the relative extrema of the functions in the previous examples analytically, these relative extrema can also be found with the aid of a graphing utility. For instance, the relative extrema of the graphs of the functions in Examples 3 and 5 are easily identified (see Figures 18 and 19).

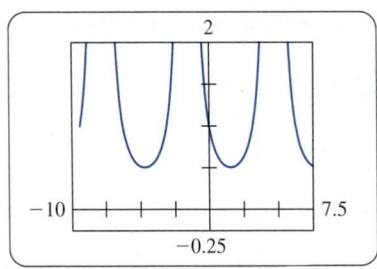

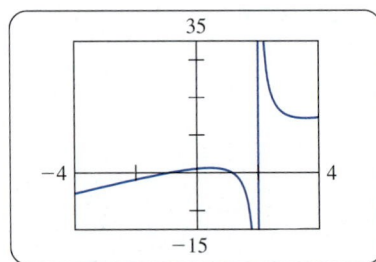

FIGURE 18
The graph of the function $f(x) = \dfrac{1}{1 + \sin x}$

FIGURE 19
The graph of the function $f(x) = \dfrac{2x^2 - 3}{x - 2}$

For more complicated functions, however, it could prove to be rather difficult to find their relative extrema by using only a graphing utility. Consider, for example, the function

$$f(x) = \frac{(x - 2)(x - 3)}{(x + 2)(x + 3)}$$

The graph of f in the viewing window $[-10, 10] \times [-10, 10]$ is shown in Figure 20.

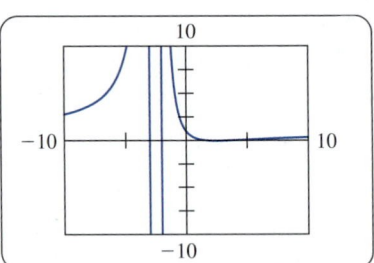

FIGURE 20

A cursory examination of the graph seems to indicate that f has no relative extrema, at least for x in the interval $(-10, 10)$.

Let's look at the problem analytically. We compute

$$f'(x) = \frac{d}{dx}\left[\frac{(x-2)(x-3)}{(x+2)(x+3)}\right] = \frac{d}{dx}\left(\frac{x^2 - 5x + 6}{x^2 + 5x + 6}\right)$$

$$= \frac{(x^2 + 5x + 6)(2x - 5) - (x^2 - 5x + 6)(2x + 5)}{(x + 2)^2(x + 3)^2} = \frac{10(x^2 - 6)}{(x + 2)^2(x + 3)^2}$$

$$= \frac{10(x - \sqrt{6})(x + \sqrt{6})}{(x + 2)^2(x + 3)^2}$$

We see that f has two critical numbers, $-\sqrt{6}$ and $\sqrt{6}$. Note that f' is discontinuous at -2 and -3, but because these numbers are not in the domain of f, they do not qualify as critical numbers. From the sign diagram of f' (Figure 21) we see that f has a relative maximum at $-\sqrt{6}$ with value $f(-\sqrt{6}) \approx -97.99$ and a relative minimum at $\sqrt{6}$ with value $f(\sqrt{6}) \approx -0.01$. These calculations tell us that we need to adjust the viewing window to see the relative maximum of f. Figure 22a shows the graph of f using the viewing window $[-5, 5] \times [-300, 100]$. A close-up of the relative maximum is shown in Figure 22b, where the viewing window $[-3, -2] \times [-300, 100]$ is used. The point $(-\sqrt{6}, f(-\sqrt{6}))$ can be estimated by using the function for finding the maximum on your graphing utility.

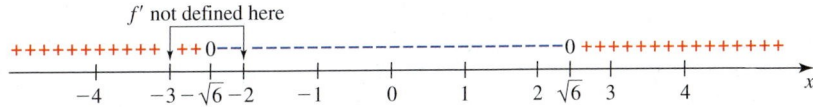

FIGURE 21
The sign diagram for f'

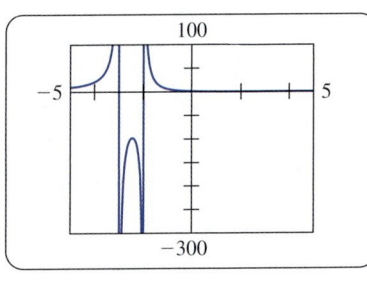

(a)

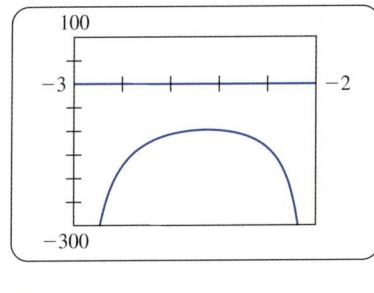

(b)

FIGURE 22
The relative maximum at $-\sqrt{6}$ an be seen in part (a). The same relative maximum is shown in close-up view in part (b).

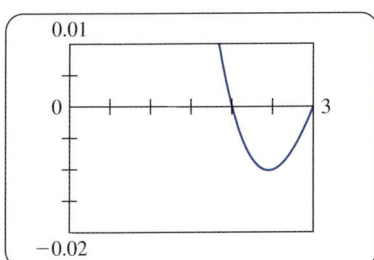

FIGURE 23
The relative minimum at $\sqrt{6}$ can be seen using the viewing window $[0, 3] \times [-0.02, 0.01]$.

Our calculations also indicate that there is a relative minimum at $\sqrt{6}$ with value $f(\sqrt{6}) \approx -0.01$. This relative minimum point $(\sqrt{6}, f(\sqrt{6}))$ shows up when we use the viewing window $[0, 3] \times [-0.02, 0.01]$. (See Figure 23.) You can use the graphing utility to approximate the point $(\sqrt{6}, f(\sqrt{6}))$.

This example shows that a combination of analytical and graphical techniques sometimes forms a powerful team when it comes to solving calculus problems.

3.6   CONCEPT QUESTIONS

1. Give the guidelines for sketching a curve.
2. Let $f(x) = x^2 + 1/x^2$.
   a. Show that if $|x|$ is very large, then $f(x)$ behaves like $g(x) = x^2$.

   b. Show that $\lim_{x \to 0} f(x) = \infty$.
   c. Use the guidelines for curve sketching and the results of parts (a) and (b) to sketch the graph of f.

3.6   EXERCISES

In Exercises 1–4, use the information summarized in the table to sketch the graph of f.

1. $f(x) = x^3 - 3x^2 + 1$

Domain	$(-\infty, \infty)$
Intercepts	y-intercept: 1
Symmetry	None
Asymptotes	None
Intervals where f is $\nearrow$ or $\searrow$	$\nearrow$ on $(-\infty, 0)$ and on $(2, \infty)$; $\searrow$ on $(0, 2)$
Relative extrema	Rel. max. at $(0, 1)$; rel. min. at $(2, -3)$
Concavity	Downward on $(-\infty, 1)$; upward on $(1, \infty)$
Point of inflection	$(1, -1)$

2. $f(x) = \dfrac{1}{9}(x^4 - 4x^3)$

Domain	$(-\infty, \infty)$
Intercepts	x-intercepts: 0, 4 y-intercept: 0
Symmetry	None
Asymptotes	None
Intervals where f is $\nearrow$ or $\searrow$	$\nearrow$ on $(3, \infty)$; $\searrow$ on $(-\infty, 3)$
Relative extrema	Rel. min. at $(3, -3)$
Concavity	Downward on $(0, 2)$; upward on $(-\infty, 0)$ and on $(2, \infty)$
Points of inflection	$(0, 0)$ and $\left(2, -\frac{16}{9}\right)$

3. $f(x) = \dfrac{4x - 4}{x^2}$

Domain	$(-\infty, 0) \cup (0, \infty)$
Intercepts	x-intercept: 1
Symmetry	None
Asymptotes	x-axis; y-axis
Intervals where f is $\nearrow$ or $\searrow$	$\nearrow$ on $(0, 2)$; $\searrow$ on $(-\infty, 0)$ and on $(2, \infty)$
Relative extrema	Rel. max. at $(2, 1)$
Concavity	Downward on $(-\infty, 0)$ and on $(0, 3)$; upward on $(3, \infty)$
Point of inflection	$\left(3, \frac{8}{9}\right)$

4. $f(x) = x - 3x^{1/3}$

Domain	$(-\infty, \infty)$
Intercepts	x-intercepts: $\pm 3\sqrt{3}$, 0; y-intercept: 0
Symmetry	With respect to the origin
Asymptotes	None
Intervals where f is $\nearrow$ or $\searrow$	$\nearrow$ on $(-\infty, -1)$ and on $(1, \infty)$; $\searrow$ on $(-1, 1)$
Relative extrema	Rel. max. at $(-1, 2)$; rel. min. at $(1, -2)$
Concavity	Downward on $(-\infty, 0)$ upward on $(0, \infty)$
Point of inflection	$(0, 0)$

In Exercises 5–38, sketch the graph of the function using the curve-sketching guidelines on page 348.

5. $f(x) = 4 - 3x - 2x^3$

6. $f(x) = x^3 - 3x^2 + 2$

7. $f(x) = x^3 - 6x^2 + 9x + 2$

8. $y = 2t^3 - 15t^2 + 36t - 20$

9. $f(x) = 2x^3 - 9x^2 + 12x - 3$

10. $f(t) = 3t^4 + 4t^3$

11. $g(x) = x^4 + 2x^3 - 2$

12. $f(x) = (x - 2)^4 + 1$

13. $f(x) = 4x^5 - 5x^4$

14. $g(x) = \dfrac{1}{2}x - \sqrt{x}$

15. $y = (x + 2)^{3/2} + 1$

16. $f(t) = \sqrt{t^2 - 4}$

17. $f(x) = \dfrac{x}{x + 1}$

18. $g(x) = \dfrac{x + 1}{x - 1}$

19. $h(x) = \dfrac{x}{x^2 - 9}$

20. $f(t) = \dfrac{t}{t^2 + 1}$

21. $f(x) = \dfrac{x^2}{x^2 + 1}$

22. $g(x) = \dfrac{x^2 - 9}{x^2 - 4}$

23. $h(x) = \dfrac{1}{x^2 - x - 2}$

24. $f(x) = x\sqrt{9 - x^2}$

25. $f(x) = x - \sin x, \quad 0 \le x \le 2\pi$

26. $g(x) = 2\sin x + \sin 2x, \quad 0 \le x \le 2\pi$

27. $f(x) = \dfrac{1}{1 - \cos x}, \quad -2\pi < x < 2\pi$

28. $y = \cos^2 x, \quad -\pi \le x \le \pi$

29. $g(x) = \dfrac{\sin x}{1 + \sin x}$

30. $f(x) = 2x - \tan x, \quad -\dfrac{\pi}{2} < x < \dfrac{\pi}{2}$

31. $f(x) = xe^{-x}$

32. $f(x) = xe^x$

33. $f(x) = \dfrac{e^x - e^{-x}}{2}$

34. $f(x) = e^x - x$

35. $f(x) = x + \ln x$

36. $f(x) = x \ln x - x$

37. $f(x) = \ln(x^2 + 1)$

38. $f(x) = \ln(\cos x), \quad -\dfrac{\pi}{2} < x < \dfrac{\pi}{2}$

In Exercises 39–42, find the slant asymptotes of the graphs of the function. Then sketch the graph of the function.

39. $g(u) = \dfrac{u^3 + 1}{u^2 - 1}$

40. $h(x) = \dfrac{x^3 + 1}{x(x + 1)}$

41. $f(x) = \dfrac{x^2 - 2x - 3}{2x - 2}$

42. $f(x) = e^{-x} + x$

43. Refer to Example 5. Show that

$$\lim_{x \to -\infty} \frac{f(x)}{x} = 2 \quad \text{and} \quad \lim_{x \to -\infty} [f(x) - 2x] = 4$$

so $y = 2x + 4$ is a (left) slant asymptote of the graph of $f(x) = \dfrac{2x^2 - 3}{x - 2}$.

44. Find the (right) slant asymptote and the (left) slant asymptote of the graph of the function $f(x) = \sqrt{1 + x^2} + 2x$. Plot the graph of f together with the slant asymptotes.

45. Worker Efficiency An efficiency study showed that the total number of cell phones assembled by the average worker at Alpha Communications t hours after starting work at 8 A.M. is given by

$$N(t) = -\frac{1}{2}t^3 + 3t^2 + 10t \qquad 0 \le t \le 4$$

Sketch the graph of the function N, and interpret your result.

46. Crime Rate The number of major crimes per 100,000 people committed in a city from the beginning of 2002 to the beginning of 2009 is approximated by the function

$$N(t) = -0.1t^3 + 1.5t^2 + 80 \qquad 0 \le t \le 7$$

where $N(t)$ denotes the number of crimes per 100,000 people committed in year t and $t = 0$ corresponds to the beginning of 2002. Enraged by the dramatic increase in the crime rate, the citizens, with the help of the local police, organized Neighborhood Crime Watch groups in early 2007 to combat this menace. Sketch the graph of the function N, and interpret your results. Is the Neighborhood Crime Watch program working?

47. Air Pollution The level of ozone, an invisible gas that irritates and impairs breathing, that is present in the atmosphere on a certain day in June in the city of Riverside is approximated by

$$S(t) = 1.0974t^3 - 0.0915t^4 \qquad 0 \le t \le 11$$

where $S(t)$ is measured in Pollutant Standard Index (PSI) and t is measured in hours with $t = 0$ corresponding to 7 A.M. Plot the graph of S, and interpret your results.

Source: The Los Angeles Times.

48. Production Costs The total daily cost in dollars incurred by the TKK Corporation in manufacturing x multipacks of DVDs is given by the function

$$f(x) = 0.000001x^3 - 0.003x^2 + 5x + 500$$

$$0 \le x \le 3000$$

Plot the graph of f, and interpret your results.

49. A Mixture Problem A tank initially contains 10 gal of brine with 2 lb of salt. Brine with 1.5 lb of salt per gallon enters the tank at the rate of 3 gal/min, and the well-stirred mixture leaves the tank at the rate of 4 gal/min. It can be shown that the amount of salt in the tank after t min is x lb, where

$$x = f(t) = 1.5(10 - t) - 0.0013(10 - t)^4 \qquad 0 \le t \le 10$$

Plot the graph of f, and interpret your result.

50. Traffic Flow Analysis The speed of traffic flow in miles per hour on a stretch of Route 123 between 6 A.M. and 10 A.M. on a typical workday is approximated by the function

$$f(t) = 20t - 40\sqrt{t} + 52 \qquad 0 \le t \le 4$$

where t is measured in hours and $t = 0$ corresponds to 6 A.M. Sketch the graph of f and interpret your results.

51. Einstein's Theory of Special Relativity The mass of a particle moving at a velocity v is related to its rest mass m_0 by the equation

$$m = f(v) = \frac{m_0}{\sqrt{1 - \dfrac{v^2}{c^2}}}$$

where c is the speed of light. Sketch the graph of the function f, and interpret your results.

52. Absorption of Drugs A liquid carries a drug into an organ of volume V cm^3 at the rate of a cm^3/sec and leaves at the same rate. The concentration of the drug in the entering liquid is c g/cm^3. Letting $x(t)$ denote the concentration of the drug in the organ at any time t, we have $x(t) = c(1 - e^{-at/V})$, where a is a positive constant that depends on the organ.
a. Show that x is an increasing function on $(0, \infty)$.
b. Sketch the graph of x.

53. Harbor Water Level The water level (in feet) at Boston Harbor during a certain 24-hour period is approximated by the function

$$H = f(t) = 4.8 \sin\left(\frac{\pi}{6}(t - 10)\right) + 7.6 \qquad 0 \le t \le 24$$

where $t = 0$ corresponds to 12 A.M. Plot the graph of f, and interpret your results.

54. Chemical Mixtures Two chemicals react to form another chemical. Suppose that the amount of the chemical formed in time t (in hours) is given by

$$x(t) = \frac{15\left[1 - \left(\frac{2}{3}\right)^{3t}\right]}{1 - \frac{1}{4}\left(\frac{2}{3}\right)^{3t}}$$

where $x(t)$ is measured in pounds.

a. Plot the graph of x using the viewing window $[0, 10] \times [0, 16]$.
b. Find the rate at which the chemical is formed when $t = 1$.
c. How many pounds of the chemical are formed eventually?

In Exercises 55–58, plot the graph of the function.

55. $f(t) = \dfrac{\sqrt{t^2 + 1}}{t - 1}$

56. $f(x) = \dfrac{x^2 + x}{3x^2 + x - 1}$

57. $g(t) = t^2 + 3 \sin 2t, \quad -2\pi < t < 2\pi$

58. $h(x) = 2 \sin x + 3 \cos 2x + \sin 3x, \quad -2\pi < x < 2\pi$

59. Snowfall Accumulation The snowfall accumulation at Logan Airport t hr after a 33-hr snowstorm in Boston in 1995 is given in the following table.

Hour	0	3	6	9	12	15
Inches	0.1	0.4	3.6	6.5	9.1	14.4

Hour	18	21	24	27	30	33
Inches	19.5	22	23.6	24.8	26.6	27

By using the logistic curve-fitting capability of a graphing calculator, it can be verified that a regression model for this data is given by

$$f(t) = \frac{26.71}{1 + 31.74e^{-0.24t}}$$

where t is measured in hours, $t = 0$ corresponds to noon of February 6, and $f(t)$ is measured in inches.
a. Plot the scatter diagram and the graph of the function f using the viewing window $[0, 36] \times [0, 30]$.
b. How fast was the snowfall accumulating at midnight on February 6? At noon on February 7?
c. At what time during the storm was the snowfall accumulating at the greatest rate? What was the rate of accumulation?
Source: The Boston Globe.

60. Worldwide PC Shipments The number of worldwide PC shipments (in millions of units) from 2005 through 2009, according to data from the International Data Corporation, are given in the following table.

Year	2005	2006	2007	2008	2009
PCs	207.1	226.2	252.9	283.3	302.4

By using the logistic curve-fitting capability of a graphing calculator, it can be verified that a regression model for this data is given by

$$f(t) = \frac{544.65}{1 + 1.65e^{-0.1846t}}$$

where t is measured in years and $t = 0$ corresponds to 2005.

a. Plot the scatter diagram and the graph of the function f using the viewing window $[0, 4] \times [200, 300]$.

b. How fast were the worldwide PC shipments increasing in 2006? In 2008?

Source: International Data Corporation.

61. Flight Path of a Plane The function

$$f(x) = \begin{cases} 0 & \text{if } 0 \leq x < 1 \\ -0.0411523x^3 + 0.679012x^2 \\ \quad - 1.23457x + 0.596708 & \text{if } 1 \leq x < 10 \\ 15 & \text{if } 10 \leq x \leq 11 \end{cases}$$

where both x and $f(x)$ are measured in units of 1000 ft, describes the flight path of a plane taking off from the origin and climbing to an altitude of 15,000 ft. Plot the graph of f to visualize the trajectory of the plane.

62. Let

$$f(x) = \frac{x^{2n} - 1}{x^{2n} + 1}$$

a. Plot the graphs of f for $n = 1, 5, 10, 100,$ and 1000. Do these graphs approach a "limiting" graph as n approaches infinity?

b. Can you prove this result analytically?

3.7 Optimization Problems

We first encountered optimization problems in Section 3.1. There, we solved certain problems by finding the absolute maximum value or the absolute minimum value of a continuous function on a *closed, bounded interval.* Thanks to the Extreme Value Theorem, we saw that these problems always have a solution.

In practice, however, there are optimization problems that are solved by finding the absolute extremum value of a continuous function on an *arbitrary interval.* If the interval is not closed, there is no guarantee that the function to be optimized has an absolute maximum value or an absolute minimum value on that interval (see Example 1 in Section 3.1). Thus, for these problems, a solution might not exist. But if the function to be maximized (minimized) has exactly one relative maximum (relative minimum) inside that interval, then there is a solution to the problem. In fact, as Figure 1 suggests, the relative extremum value at a critical number turns out to be the absolute extremum value of the function on the interval. Thus, the solutions to such problems are found by finding the relative extreme values of the function in that interval.

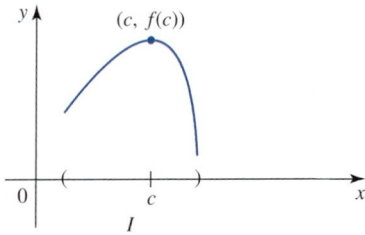

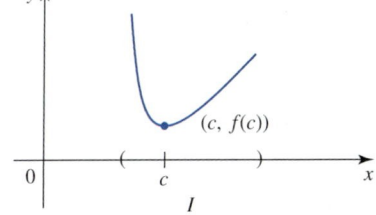

FIGURE 1
f has only one critical number on an interval I.

(a) The relative maximum value $f(c)$ is the absolute maximum value.

(b) The relative minimum value $f(c)$ is the absolute minimum value.

Before proceeding further, let us summarize this important observation.

> **Guidelines for Finding the Absolute Extrema of a Continuous Function f on an Arbitrary Interval**
>
> Suppose that a continuous function f has only one critical number c in an interval I.
>
> **1.** Use the First Derivative Test or the Second Derivative Test to ascertain whether f has a relative maximum (minimum) value at c.
> **2. a.** If f has a relative maximum value at c, then the number $f(c)$ is also the absolute maximum value of f on I.
>   **b.** If f has a relative minimum value at c, then the number $f(c)$ is also the absolute minimum value of f on I.

Armed with these guidelines and the guidelines for finding the absolute extrema of functions on closed intervals, we are ready to tackle a large class of optimization problems.

■ Formulating Optimization Problems

If you reexamine the optimization problems in Section 3.1, you will see that the functions to be optimized were given to you. More often than not, we first need to find an appropriate function and then optimize it. The following guidelines can be used to formulate these optimization problems.

> **Guidelines for Solving Optimization Problems**
>
> **1.** Assign a letter to each variable. Draw and label a figure (if appropriate).
> **2.** Find an expression for the quantity to be maximized or minimized.
> **3.** Use the conditions given in the problem to express the quantity to be optimized as a function f of one variable. Note any restrictions to be placed on the domain of f.
> **4.** Optimize the function f over its domain using the guidelines of Section 3.1 and the guidelines on this page.

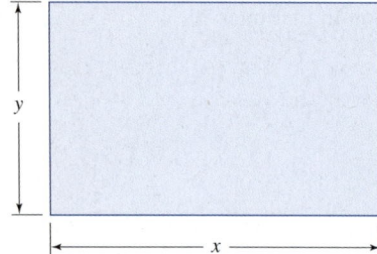

FIGURE 2
The area of the rectangle is $A = xy$.

EXAMPLE 1 **A Fencing Problem**   A man has 100 ft of fencing to enclose a rectangular garden in his backyard. Find the dimensions of the garden of largest area he can have if he uses all of the fencing.

Solution

Step 1   Let x and y denote the length and width of the garden (in feet) and let A denote its area (see Figure 2).

Step 2   The area of the rectangle is

$$A = xy \qquad (1)$$

and is the quantity to be maximized.

Step 3   The perimeter of the rectangle is $(2x + 2y)$ ft, and this must be equal to 100 ft. Therefore, we have the equation

$$2x + 2y = 100 \qquad (2)$$

relating the variables x and y. Solving Equation (2) for y in terms of x, we have

$$y = 50 - x \tag{3}$$

which, when substituted into Equation (1), yields

$$A = x(50 - x)$$
$$= -x^2 + 50x$$

(Remember, the function to be optimized must involve just one variable.) Because the sides of the rectangle must be positive, $x > 0$ and $y = 50 - x > 0$, giving us the inequality $0 < x < 50$. Thus, the problem is reduced to that of finding the value of x in $(0, 50)$ at which $f(x) = -x^2 + 50x$ attains the largest value.

Step 4 To find the critical number(s) of f, we compute

$$f'(x) = -2x + 50 = -2(x - 25)$$

Setting $f'(x) = 0$, yields 25 as the only critical number of f. Since $f''(x) = -2 < 0$, we see, by the Second Derivative Test, that f has a relative maximum at $x = 25$. But 25 is the only critical number in $(0, 50)$, so we conclude that f attains its largest value of $f(25) = 625$ at $x = 25$. From Equation (3) the corresponding value of y is 25. Thus, the man would have a garden of maximum area (625 ft^2) if it were in the form of a square with sides of length 25 ft. ■

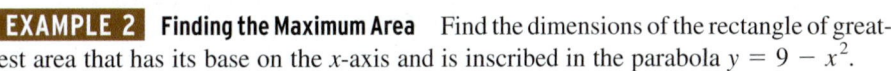

EXAMPLE 2 **Finding the Maximum Area** Find the dimensions of the rectangle of greatest area that has its base on the x-axis and is inscribed in the parabola $y = 9 - x^2$.

Solution
Step 1 Consider the rectangle of width $2x$ and height y as shown in Figure 3. Let A denote its area.
Step 2 The area of the rectangle is $A = 2xy$ and is the quantity to be maximized.
Step 3 Because the point (x, y) lies on the parabola, it must satisfy the equation of the parabola; that is, $y = 9 - x^2$. Therefore,

$$A = 2xy$$
$$= 2x(9 - x^2)$$
$$= -2x^3 + 18x$$

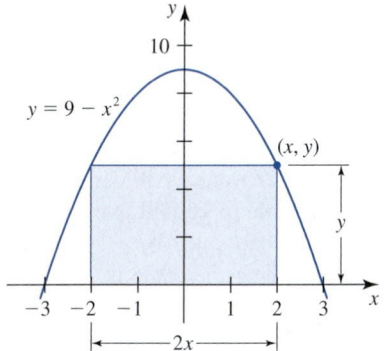

FIGURE 3
The area of the rectangle is $2xy = 2x(9 - x^2)$.

Furthermore, $y > 0$ implies that $9 - x^2 > 0$ or, equivalently, $-3 < x < 3$. Also, $x > 0$, since the side of a rectangle must be positive. Therefore, the problem is equivalent to the problem of finding the value of x in $(0, 3)$ for which $f(x) = -2x^3 + 18x$ attains the largest value.
Step 4 To find the critical numbers of f, we compute

$$f'(x) = -6x^2 + 18 = -6(x^2 - 3)$$

Setting $f'(x) = 0$ yields $x = \pm\sqrt{3}$. We consider only the critical number $\sqrt{3}$, since $-\sqrt{3}$ lies outside the interval $(0, 3)$. Since $f''(x) = -12x$ and $f''(\sqrt{3}) = -12\sqrt{3} < 0$, we see, by the Second Derivative Test, that f has a relative maximum at $x = \sqrt{3}$. Since f has only one critical number in $(0, 3)$, we see that f attains its largest value at $x = \sqrt{3}$. Substituting this value of x into $y = 9 - x^2$ gives $y = 6$. Thus, the dimensions of the desired rectangle are $2\sqrt{3}$ by 6 and its area is $12\sqrt{3}$. ■

EXAMPLE 3 **Minimizing the Cost of Laying Cable**   In Figure 4, the point S gives the location of a power relay station on a straight coast, and the point E gives the location of a marine biology experimental station on an island. The point Q is located 7 mi west of the point S, and the point Q is 3 mi south of the point E. A cable is to be laid connecting the relay station with the experimental station. If the cost of running the cable along the shoreline is \$10,000/mi and the cost of running the cable under water is \$30,000/mi, where should the point P be located to minimize the cost of laying the cable?

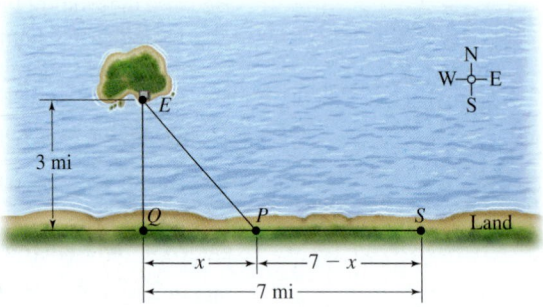

FIGURE 4
The cable connects the marine biology station at E to the power relay station at S. The cable from E to P will be laid under water, and the cable from P to S will be laid over land.

Solution

Step 1   It is clear that the point P should lie between Q and S, inclusive. Let x denote the distance between P and Q (in miles), and let C denote the cost of laying the cable (in thousands of dollars).

Step 2   The length of the cable to be laid under water is given by the distance between E and P. Using the Pythagorean Theorem, we find that this length is $\sqrt{x^2 + 9}$ mi. So the cost of laying the cable under water is $30\sqrt{x^2 + 9}$ thousand dollars. Next, we see that the length of cable to be laid over land is $(7 - x)$ mi. So the cost of laying this stretch of the cable is $10(7 - x)$ thousand dollars. Therefore, the total cost incurred in laying the cable is

$$C = 30\sqrt{x^2 + 9} + 10(7 - x)$$

thousand dollars, and this is the quantity to be minimized.

Step 3   Because the distance between Q and S is 7 mi, we see that x must satisfy the constraint $0 \le x \le 7$. So the problem is that of finding the value of x in $[0, 7]$ at which $f(x) = 30\sqrt{x^2 + 9} + 10(7 - x)$ attains the smallest value.

Step 4   Observe that f is continuous on the closed interval $[0, 7]$. So the absolute minimum value of f must be attained at an endpoint of $[0, 7]$ or at a critical number of f in the interval. To find the critical numbers of f, we compute

$$f'(x) = \frac{d}{dx}[30(x^2 + 9)^{1/2} + 10(7 - x)]$$

$$= (30)\left(\frac{1}{2}\right)(x^2 + 9)^{-1/2}(2x) - 10$$

$$= 10\left[\frac{3x}{\sqrt{x^2 + 9}} - 1\right]$$

Setting $f'(x) = 0$ gives

$$\frac{3x}{\sqrt{x^2 + 9}} - 1 = 0$$

$$3x = \sqrt{x^2 + 9}$$

$$9x^2 = x^2 + 9$$

$$8x^2 = 9$$

or

$$x = \pm\frac{3}{2\sqrt{2}} = \pm\frac{3\sqrt{2}}{4} \approx \pm 1.06$$

We reject the root $-3\sqrt{2}/4$ because it lies outside the interval $[0, 7]$. We are left with $x = 3\sqrt{2}/4$ as the only critical number of f. Finally, from the following table we see that $f(x)$ attains its smallest value of 154.85 at $x = 3\sqrt{2}/4 \approx 1.06$. We conclude that the cost of laying the cable will be minimized (approximately $155,000) if the point P is located at a distance of approximately 1.06 miles from Q.

$f(0)$	$f(3\sqrt{2}/4)$	$f(7)$
160	154.85	228.47

EXAMPLE 4 **Packaging** The Betty Moore Company requires that its beef stew containers have a capacity of 64 in.^{3}, have the shape of right circular cylinders, and be made of aluminum. Determine the radius and height of the container that requires the least amount of metal.

Solution

Step 1 Let r and h denote the radius and height, respectively, of a container (Figure 5). The amount of aluminum required to construct a container is given by the total surface area of the cylinder, which we denote by S.

Step 2 The area of the base or top of the cylinder is πr^2 in.^{2}, and the area of its lateral surface is $2\pi rh$ in.^{2}. Therefore,

$$S = 2\pi r^2 + 2\pi rh \tag{4}$$

and this is the quantity to be minimized.

Step 3 The requirement that the volume of the container be 64 in.^{3} translates into the equation

$$\pi r^2 h = 64 \tag{5}$$

Solving Equation (5) for h in terms of r, we obtain

$$h = \frac{64}{\pi r^2} \tag{6}$$

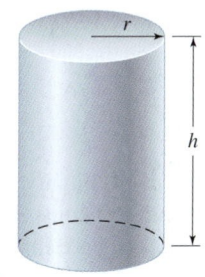

FIGURE 5
We want to minimize the amount of material used to construct the container.

which, when substituted into Equation (4), yields

$$S = 2\pi r^2 + 2\pi r \left(\frac{64}{\pi r^2} \right)$$

$$= 2\pi r^2 + \frac{128}{r}$$

The domain of S is $(0, \infty)$. The problem has been reduced to one of finding the value of r in $(0, \infty)$ at which $f(r) = 2\pi r^2 + (128/r)$ attains the smallest value.

Step 4 Observe that f is continuous on $(0, \infty)$. Following the guidelines given at the beginning of this section, we first find the critical number of f,

$$f'(r) = 4\pi r - \frac{128}{r^2}$$

Setting $f'(r) = 0$ gives

$$4\pi r - \frac{128}{r^2} = 0$$

$$4\pi r^3 - 128 = 0$$

$$r^3 = \frac{32}{\pi}$$

or

$$r = \left(\frac{32}{\pi} \right)^{1/3} \approx 2.17$$

as the only critical number of f.

To see whether this critical number gives rise to a relative extremum of f, we use the Second Derivative Test. Now

$$f''(r) = 4\pi + \frac{256}{r^3}$$

so

$$f''\left(\left(\frac{32}{\pi} \right)^{1/3} \right) = 4\pi + \frac{256}{\dfrac{32}{\pi}} = 12\pi > 0$$

Therefore, f has a relative minimum value at $r = (32/\pi)^{1/3}$. Finally, because f has only one critical number in $(0, \infty)$, we conclude that f attains the absolute minimum value at this number. Using Equation (6), we find that the corresponding value of h is

$$h = \frac{64}{\pi \left(\dfrac{32}{\pi} \right)^{2/3}} = \frac{64}{\pi \left(\dfrac{32}{\pi} \right)^{2/3}} \cdot \frac{\left(\dfrac{32}{\pi} \right)^{1/3}}{\left(\dfrac{32}{\pi} \right)^{1/3}}$$

$$= \frac{64}{\pi \left(\dfrac{32}{\pi} \right)} \cdot \left(\frac{32}{\pi} \right)^{1/3}$$

$$= 2r$$

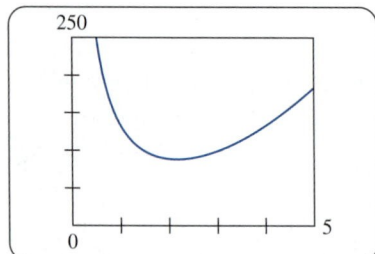

FIGURE 6

The graph of $S = 2\pi r^2 + \dfrac{128}{r}$

Thus, the required container has a radius of approximately 2.17 in. and a height twice the size of its radius, or approximately 4.34 in. The graph of S is shown in Figure 6. ■

EXAMPLE 5 **Finding the Minimum Distance** Figure 7 shows an aerial view of a race-track composed of two sides of a rectangle and two semicircles. It also shows the position P of a spectator watching a race from the roof of his car. Find the point Q on the track that is closest to the spectator. What is the distance between these two points?

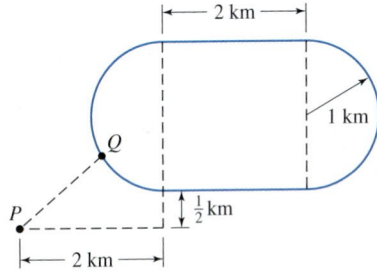

FIGURE 7
The diagram shows the position of a spectator, P, in relation to a racetrack.

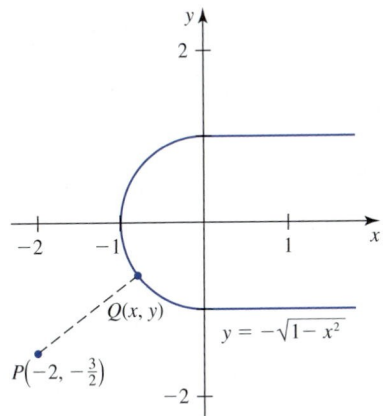

FIGURE 8
We want to minimize the distance between P and Q.

Solution

Step 1   Clearly, the required point must lie on the lower left semicircular stretch of the racetrack. Let us set up a rectangular coordinate system as shown in Figure 8. To find an equation describing this curve, begin with the equation $x^2 + y^2 = 1$ of the circle with center at the origin and radius 1. Solving for y in terms of x and observing that both x and y must be nonpositive, we are led to the following representation of the curve:

$$y = -\sqrt{1 - x^2} \qquad -1 \le x \le 0 \qquad (7)$$

Next, let D denote the distance between $P\left(-2, -\frac{3}{2}\right)$ and a point $Q(x, y)$ lying on the curve described by Equation (7).

Step 2   Using the distance formula, we see that the distance D between P and Q is given by

$$D = \sqrt{(x + 2)^2 + \left(y + \frac{3}{2}\right)^2}$$

Thus,

$$D^2 = (x + 2)^2 + \left(y + \frac{3}{2}\right)^2$$

$$= x^2 + 4x + 4 + y^2 + 3y + \frac{9}{4} \qquad (8)$$

Since D is minimal if and only if D^2 is minimal, we will minimize D^2 instead of D.

Step 3   Substituting Equation (7) into Equation (8), we obtain

$$D^2 = x^2 + 4x + 4 + (1 - x^2) - 3\sqrt{1 - x^2} + \frac{9}{4}$$

$$= 4x - 3\sqrt{1 - x^2} + \frac{29}{4}$$

So the problem is reduced to that of finding the value of x in $[-1, 0]$ at which $f(x) = 4x - 3\sqrt{1 - x^2} + (29/4)$ attains the smallest value.

Step 4   Observe that f is continuous on $[-1, 0]$. So the absolute minimum value of f must be attained at an endpoint of $[-1, 0]$ or at a critical number of f in that interval. To find the critical numbers of f, we compute

$$f'(x) = \frac{d}{dx}\left[4x - 3(1 - x^2)^{1/2} + \frac{29}{4}\right]$$

$$= 4 - 3\left(\frac{1}{2}\right)(1 - x^2)^{-1/2}(-2x) = 4 + \frac{3x}{\sqrt{1 - x^2}}$$

Setting $f'(x) = 0$ and solving for x, we obtain

$$4 + \frac{3x}{\sqrt{1 - x^2}} = 0$$

$$3x = -4\sqrt{1 - x^2}$$

$$9x^2 = 16(1 - x^2)$$

$$25x^2 = 16$$

or $x = \pm\frac{4}{5}$. Only $-\frac{4}{5}$ is a solution of $f'(x) = 0$; so it is the only critical number of interest. Finally, from the following table

$f(-1)$	$f\left(-\frac{4}{5}\right)$	$f(0)$
$\frac{13}{4} = 3.25$	$\frac{9}{4} = 2.25$	$\frac{17}{4} = 4.25$

we see that f attains its smallest value of 2.25 at $x = -\frac{4}{5}$. Using Equation (7), we find that the corresponding value of y is

$$y = -\sqrt{1 - \left(-\frac{4}{5}\right)^2} = -\frac{3}{5}$$

We conclude that the point $\left(-\frac{4}{5}, -\frac{3}{5}\right)$ is the point on the track closest to the spectator. The distance between the spectator and the point is

$$\sqrt{f\left(-\frac{4}{5}\right)} = \sqrt{4\left(-\frac{4}{5}\right) - 3\sqrt{1 - \left(-\frac{4}{5}\right)^2} + \frac{29}{4}} = \sqrt{\frac{9}{4}} = \frac{3}{2}$$

or 1.5 km.   ■

EXAMPLE 6   **Minimizing Length**   Figure 9a depicts a cross section of a high-rise building. A ladder from a fire engine to the front wall of the building must clear the canopy, which extends 12 ft from the building. Find the length of the shortest ladder that will enable the firefighters to accomplish this task.

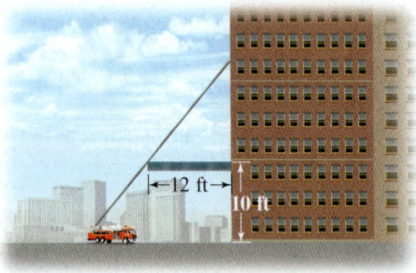
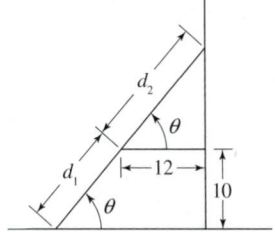

FIGURE 9   (**a**) The ladder touches the edge of the canopy.   (**b**) The length of the ladder is $L = d_1 + d_2$.

Solution

Step 1   Let L denote the length of the ladder, and let θ be the angle the ladder makes with the horizontal.

Step 2 From Figure 9b we see that

$$L = d_1 + d_2$$
$$= 10 \csc \theta + 12 \sec \theta \qquad \csc \theta = \frac{d_1}{10} \text{ and } \sec \theta = \frac{d_2}{12}$$

and this is the quantity to be minimized.

Step 3 The domain of L is $\left(0, \frac{\pi}{2}\right)$. So the problem is to find the value of θ in $\left(0, \frac{\pi}{2}\right)$ for which $f(\theta) = 10 \csc \theta + 12 \sec \theta$ has the smallest value.

Step 4 Observe that f is continuous on $\left(0, \frac{\pi}{2}\right)$. Following the guidelines given at the beginning of this section, we first find the critical numbers of f. Thus,

$$f'(\theta) = -10 \csc \theta \cot \theta + 12 \sec \theta \tan \theta$$

Setting $f'(\theta) = 0$ gives

$$12 \sec \theta \tan \theta = 10 \csc \theta \cot \theta$$

$$12\left(\frac{1}{\cos \theta}\right)\left(\frac{\sin \theta}{\cos \theta}\right) = 10\left(\frac{1}{\sin \theta}\right)\left(\frac{\cos \theta}{\sin \theta}\right)$$

$$\frac{\sin^3 \theta}{\cos^3 \theta} = \frac{10}{12}$$

$$\tan^3 \theta = \frac{5}{6}$$

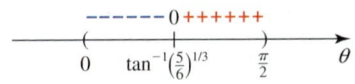

FIGURE 10
The sign diagram for f'

or $\theta = \tan^{-1} \sqrt[3]{5/6} \approx 0.76$. The sign diagram for f' shown in Figure 10 tells us that f has a relative minimum value at $\tan^{-1} \sqrt[3]{5/6}$. Since f has only one critical number in $\left(0, \frac{\pi}{2}\right)$, this value is also the absolute minimum value of f. Finally, $f(0.76) \approx 31.07$, so we conclude that the ladder must be at least 31.1 ft long.

3.7 CONCEPT QUESTIONS

1. Give the procedure for finding the absolute extrema of a continuous function f on (a) a closed interval and on (b) an arbitrary interval in which f possesses only one critical number at which an extremum occurs.

2. Give the guidelines for solving optimization problems.

3.7 EXERCISES

1. Find two positive numbers whose sum is 100 and whose product is a maximum.

2. Find two numbers whose difference is 50 and whose product is a minimum.

3. The product of two positive numbers is 54. Find the numbers if the sum of the first number plus the square of the second number is as small as possible.

4. The sum of a positive number and its reciprocal is to be as small as possible. What is the number?

5. Find the dimensions of a rectangle with a perimeter of 100 m that has the largest possible area.

6. Find the dimensions of a rectangle of area 144 ft^2 that has the smallest possible perimeter.

7. **A Fencing Problem** A rancher has 400 ft of fencing with which to enclose two adjacent rectangular parts of a corral. What are the dimensions of the parts if the area enclosed is to be as large as possible and she uses all of the fencing available?

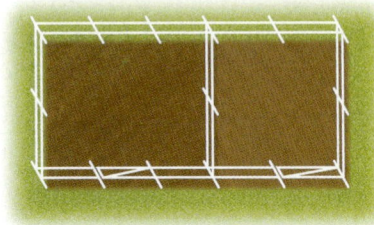

8. **A Fencing Problem** The owner of the Rancho Grande has 3000 yd of fencing with which to enclose a rectangular piece of grazing land situated along the straight portion of a river. If fencing is not required along the river, what are the dimensions of the largest area he can enclose? What is the area?

9. **Packaging** An open box is made from a rectangular piece of cardboard of dimensions 16×10 in. by cutting out identical squares from each corner and bending up the resulting flaps. Find the dimensions of the box with the largest volume that can be made.

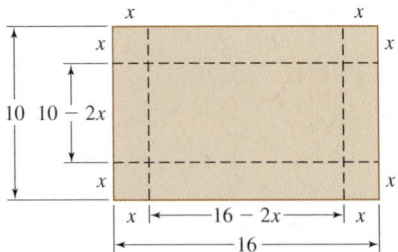

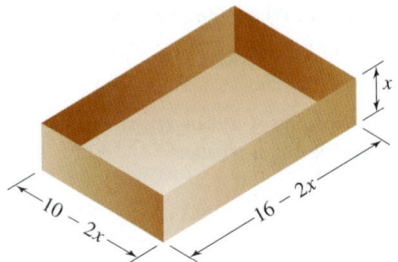

10. **Packaging** If an open box is made from a metal sheet 10 in. square by cutting out identical squares from each corner and bending up the resulting flaps, determine the dimensions of the box with the largest volume that can be made.

11. **Packaging** An open box constructed from a tin sheet has a square base and a volume of 216 in.³. Find the dimensions of the box, assuming that the minimum amount of material was used in its construction.

12. **Satisfying Postal Regulations** Postal regulations specify that a parcel sent by priority mail may have a combined length and girth of no more than 108 in. Find the dimensions of a rectangular package that has a square cross section and largest volume that may be sent by priority mail. What is the volume of such a package?

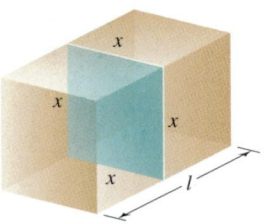

Hint: The length plus the girth is $4x + l$.

13. **Satisfying Postal Regulations** Postal regulations specify that a package sent by priority mail may have a combined length and girth of no more than 108 in. Find the dimensions of a cylindrical package with the greatest volume that may be sent by priority mail. What is the volume of such a package?

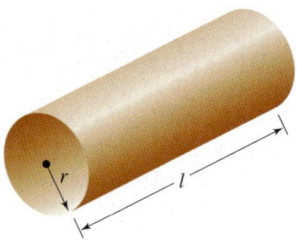

Hint: The length plus the girth is $2\pi r + l$.

14. **Packaging** A container for a soft drink is in the form of a right circular cylinder. If the container is to have a capacity of 12 fluid ounces (fl oz), find the dimensions of the container that can be constructed with a minimum of material. Hint: 1 fl oz ≈ 1.805 in.³.

15. **Designing a Loudspeaker** The rectangular enclosure for a loudspeaker system is to have an internal volume of 2.4 ft³. For aesthetic reasons the height of the enclosure is to be 1.5 times its width. If the top, bottom, and sides of the enclosure are to be constructed of veneer costing 80 cents per square foot and the front and rear are to be constructed of particle board cost-

ing 40 cents per square foot, find the dimensions of the enclosure that can be constructed at a minimum cost.

16. **Book Publishing** A production editor at Weston Publishers decided that the pages of a book should have a 1-in. margin at the top and the bottom, and a $\frac{1}{2}$-in. margin on each side of the page. She further stipulated that each page of the book should have an area of 50 in.^{2}. Determine the dimensions of the page that will result in the maximum printed area on the page.

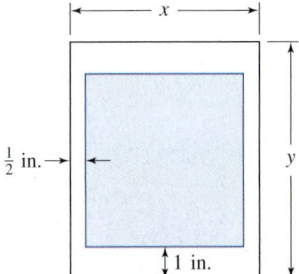

In Exercises 17–20, find the dimensions of the shaded region so that its area is maximized.

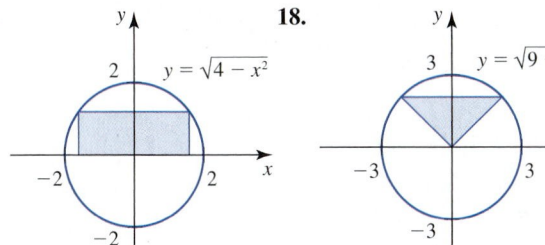

17. $y = \sqrt{4 - x^2}$

18. $y = \sqrt{9 - x^2}$

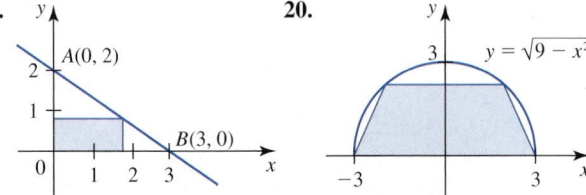

19. $A(0, 2)$ $B(3, 0)$

20. $y = \sqrt{9 - x^2}$

21. Find the point on the line $y = 2x + 5$ that is closest to the origin.

22. Find the points on the hyperbola $x^2/4 - y^2/9 = 1$ that are closest to the point $(0, 3)$.

 23. Find the approximate location of the points on the hyperbola $xy = 1$ that are closest to the point $(-1, 1)$.

24. Let P be a point lying on the axis of the parabola $y^2 = 2px$ at a distance a from its vertex. Find the x-coordinate(s) of the point(s) on the parabola that are closest to P.

25. Find the dimensions of the rectangle of maximum possible area that can be inscribed in a semicircle of radius 4.

26. Find the point on the graph of $x + y = 1$ that is closest to $(3, 2)$.

27. Find the point on the graph of the parabola $y = 4 - x^2$ that is closest to the point $(-3, -4)$.

28. **Optimal Driving Speed** A truck gets $600/x$ miles per gallon (mpg) when driven at a constant speed of x mph, where $40 \le x \le 80$. If the price of fuel is $2.80/gal and the driver is paid $12/hr, at what speed is it most economical for the trucker to drive?

29. **Maximizing Yield** An apple orchard has an average yield of 36 bushels of apples per tree if tree density is 22 trees per acre. For each unit increase in tree density, the yield decreases by 2 bushels per tree. How many trees per acre should be planted to maximize the yield?

30. **Packaging** A rectangular box is to have a square base and a volume of 20 ft^3. If the material for the base costs $0.30 per square foot, the material for the sides costs $0.10 per square foot, and the material for the top costs $0.20 per square foot, determine the dimensions of the box that can be constructed at minimum cost.

31. **Packaging** A rectangular box having a top and square base is to be constructed at a cost of $2. If the material for the bottom costs $0.30 per square foot, the material for the top costs $0.20 per square foot, and the material for the sides costs $0.15 per square foot, find the dimensions and volume of the box of maximum volume that can be constructed.

32. **Maximizing Revenue** If exactly 200 people sign up for a charter flight, the operators of a charter airline charge $300 for a round-trip ticket. However, if more than 200 people sign up for the flight, then each fare is reduced by $1 for each additional person. Assuming that more than 200 people sign up, determine how many passengers will result in a maximum revenue for the travel agency. What is the maximum revenue? What would the fare per person be in this case?

33. **Optimal Subway Fare** A city's Metropolitan Transit Authority (MTA) operates a subway line for commuters from a certain suburb to downtown. Currently, an average of 6000 passengers a day take the trains, paying a fare of $3.00 per ride. The board of the MTA, contemplating raising the fare to $3.50 per ride to generate a larger revenue, engages the services of a consulting firm. The firm's study reveals that for each $0.50 increase in fare, the ridership will be reduced by an average of 1000 passengers a day. Therefore, the consulting firm recommends that the MTA stick to the current fare of $3.00 per ride, which already yields a maximum revenue. Show that the consultants are correct.

34. **Strength of a Beam** A wooden beam has a rectangular cross section of height h and width w. The strength S of the beam is directly proportional to its width and the square of its

height. Find the dimensions of the cross section of such a beam of maximum strength that can be cut from a round log of diameter 24 in.

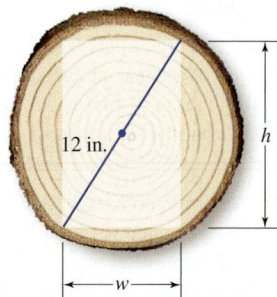

Hint: $S = kh^2w$, where k is the constant of proportionality.

35. **Stiffness of a Beam** The stiffness S of a wooden beam with a rectangular cross section is proportional to its width w and the cube of its height h. Find the dimensions of the cross section of the beam of maximum stiffness that can be cut from a round log of diameter 23 in.
    Hint: $S = kwh^3$, where k is the constant of proportionality.

36. **Maximizing Drainage Capacity** The cross section of a drain is a trapezoid as shown in the figure. The sides and the bottom of the trapezoid each have length 5 ft. Determine the angle θ such that the drain will have a maximal cross-sectional area.

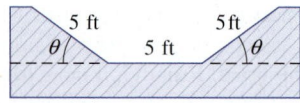

37. **Designing a Conical Figure** A cone is constructed by cutting out a sector of central angle θ from a circular sheet of radius 12 in. and then gluing the edges of the remaining piece together. Find the value of θ that will result in a cone of maximal volume. What is the maximal volume?

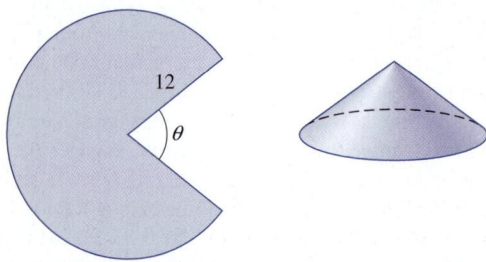

38. **A Norman Window** A Norman window has the shape of a rectangle surmounted by a semicircle. Find the dimensions of a Norman window of perimeter 28 ft that will admit the greatest possible amount of light.

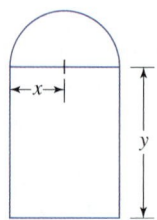

39. **Designing a Grain Silo** A grain silo has the shape of a right circular cylinder surmounted by a hemisphere. If the silo is to have a volume of 504π ft^3, determine the radius and height of the silo that requires the least amount of material to build.

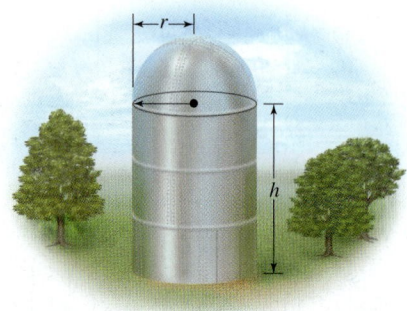

Hint: The volume of the silo is $\pi r^2 h + \frac{2}{3}\pi r^3$, and the surface area of the silo (including the floor) is $\pi(3r^2 + 2rh)$.

40. **Racetrack Design** The figure below depicts a racetrack with ends that are semicircular. The length of the track is 1760 ft $\left(\frac{1}{3}\text{ mi}\right)$. Find l and r so that the area of the rectangular portion of the region enclosed by the racetrack is as large as possible. What is the area enclosed by the track in this case?

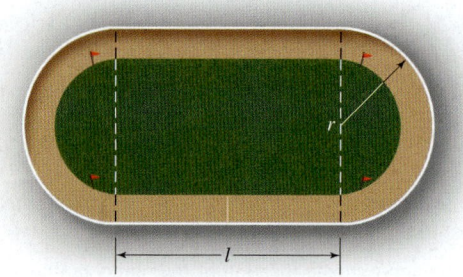

41. **Packaging** A container of capacity 64 in.^{3} is to be made in the form of a right circular cylinder. The top and the bottom of the can are to be cut from squares, whereas the side is to be made by bending a rectangular sheet so that the ends

match. Find the radius and height of the can that can be constructed with the least amount of material.

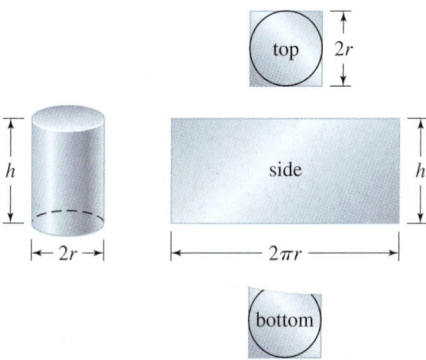

42. **A Fencing Problem** Joan has 50 ft of interlocking stone available for fencing off a flower bed in the form of a circular sector. Find the radius of the circle that will yield a flower bed with the largest area if Joan uses all of the stone.

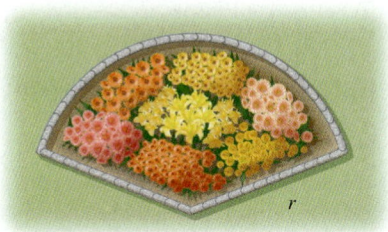

43. **Constructing a Marina** The figure below shows the position of two islands located off a straight stretch of coastal highway. A marina is to be constructed at the point M on the highway to serve both island communities. Determine the location of M if the total distance from both the islands to M is as small as possible.

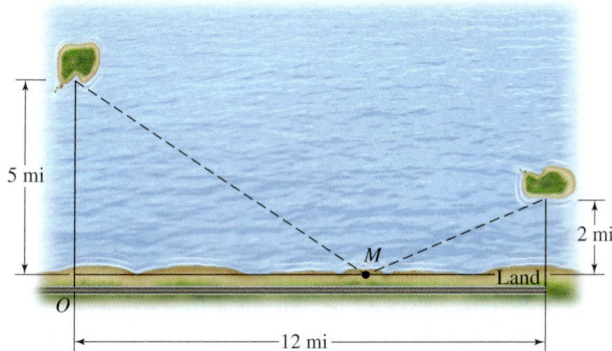

44. **Flights of Birds** During daylight hours some birds fly more slowly over water than over land because some of their

energy is expended in overcoming the downdrafts of air over open bodies of water. Suppose a bird that flies at a constant speed of 4 mph over water and 6 mph over land starts its journey at the point E on an island and ends at its nest N on the shore of the mainland, as shown in the figure. Find the location of the point P that allows the bird to complete its journey in the minimum time (solve for x).

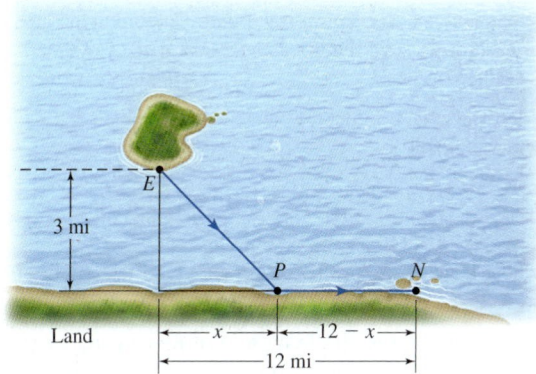

45. **Avoiding a Collision** Upon spotting a disabled and stationary boat, the driver of a speedboat took evasive action. Suppose that the disabled boat is located at the point $(0, 2)$ in an xy-coordinate system (both scales measured in miles) and the path of the speedboat is described by the graph of $f(x) = (x + 1)/\sqrt{x}$.
   a. Find an expression $D(x)$ that gives the distance between the speedboat and the disabled boat.
    b. Plot the graph of D, and use it to determine how close the speedboat came to the disabled boat before it changed its path.

46. **Minimizing Costs** Suppose that the cost incurred in operating a cruise ship for 1 hr is $a + bv^3$ dollars, where a and b are positive constants and v is the ship's speed in miles per hour. At what speed should the ship be operated between two ports to minimize the cost?

47. **Maximum Power Output** Suppose that the source of current in an electric circuit is a battery. Then the power output P (in watts) obtained if the circuit has a resistance of R ohms is given by

$$P = \frac{E^2 R}{(R + r)^2}$$

where E is the electromotive force in volts and r is the internal resistance of the battery in ohms. If E and r are constant, find the value of R that will result in the greatest power output. What is the maximum power output?

48. Optimal Inventory Control The equation

$$A(q) = \frac{km}{q} + cm + \frac{hq}{2}$$

gives the annual cost of ordering and storing (as yet unsold) merchandise. Here, q is the size of each order, k is the cost of placing each order, c is the unit cost of the product, m is the number of units of the product sold per year, and h is the annual cost for storing each unit. Determine the size of each order such that the annual cost $A(q)$ is as small as possible.

49. Velocity of a Wave In deep water a wave of length L travels with a velocity

$$v = k\sqrt{\frac{L}{C} + \frac{C}{L}}$$

where k and C are positive constants. Find the length of the wave that has a minimum velocity.

50. Show that the isosceles triangle of maximum area that can be inscribed in a circle of fixed radius a is equilateral.

51. Show that the rectangle of maximum area that can be inscribed in a circle of fixed radius a is a square.

52. Find the dimensions of the cylinder of largest volume that will fit inside a right circular cone of radius 3 in. and height 5 in. Assume that the axis of the cylinder coincides with the axis of the cone.

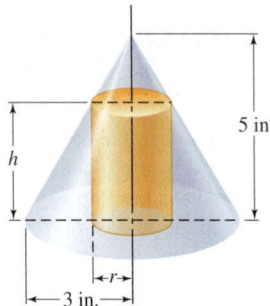

53. A right circular cylinder is inscribed in a cone of height H and base radius R so that the axis of the cylinder coincides with the axis of the cone. Determine the dimensions of the cylinder with the largest lateral surface area.

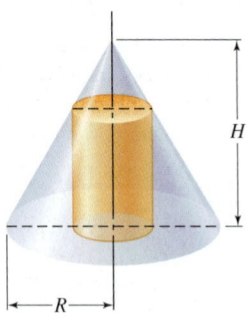

54. Find the radius and height of a right circular cylinder with the largest possible lateral surface area that can be inscribed in a sphere of radius a.

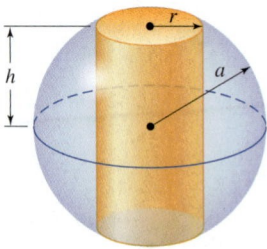

55. Find an equation of the line passing through the point $(1, 2)$ such that the area of the triangle formed by this line and the positive coordinate axes is as small as possible.

56. Range of a Projectile The range of an artillery shell fired at an angle of $\theta°$ with the horizontal is

$$R = \frac{v_0^2}{g} \sin 2\theta$$

feet, where v_0 is the muzzle velocity of the shell in feet per second, and g is the constant of acceleration due to gravity (32 ft/sec^2). Find the angle of elevation of the gun that will give it a maximum range.

57. Optimal Illumination A hobbyist has set up a railroad track on a circular table to display a recently acquired model railroad locomotive. The radius of the track is 5 ft, and the display is to be illuminated by a light source suspended from an 8-ft ceiling located directly above the center of the table (see the figure). How high above the table should the light source be placed in order to achieve maximum illumination on the railroad track?

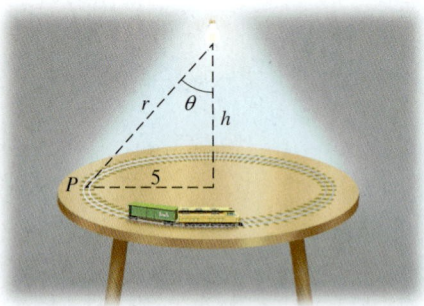

Hint: The intensity of light at P is proportional to the cosine of the angle that the incident light makes with the vertical and inversely proportional to the square of the distance r between P and the light source.

58. Cells of a Honeycomb The accompanying figure depicts a single prism-shaped cell in a honeycomb. The front end of the prism is a regular hexagon, and the back is formed by the sides of the cell coming together at a point. It can be shown that the surface area of a cell is given by

$$S(\theta) = 6ab + \frac{3}{2}b^2\left(\frac{\sqrt{3} - \cos\theta}{\sin\theta}\right) \qquad 0 < \theta < \frac{\pi}{2}$$

where θ is the angle between one of the (three) upper surfaces and the altitude. The lengths of the sides of the hexagon, b, and the altitude, a, are both constants.

a. Show that the surface area is minimized if $\cos\theta = 1/\sqrt{3}$, or $\theta \approx 54.7°$. (Measurements of actual honeycombs have confirmed that this is, in fact, the angle found in beehives.)

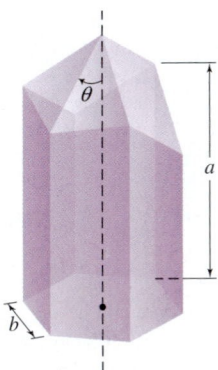

b. Using a graphing utility, verify the result of part (a) by finding the absolute minimum of

$$f(\theta) = \frac{\sqrt{3} - \cos\theta}{\sin\theta} \qquad 0 < \theta < \frac{\pi}{2}$$

59. **Maximizing Length** A metal pipe of length 16 ft is to be carried horizontally around a corner from a hallway 8 ft wide into a hallway 4 ft wide. Can this be done?

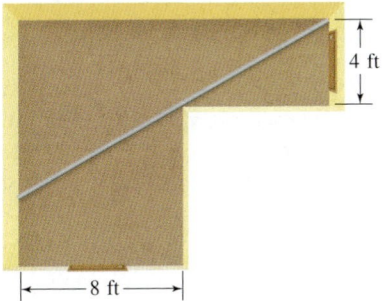

Hint: Find the length of the largest pipe that can be carried horizontally around the corner.

60. **Distance Between Two Aircraft** Two aircraft approach each other, each flying at a speed of 500 mph and at an altitude of 35,000 ft. Their paths are straight lines that intersect at an angle of 120°. At a certain instant of time, one aircraft is 200 mi from the point of intersection of their paths, while the other is 300 mi from it. At what time will the aircraft be closest to each other, and what will that distance be?

61. **Storing Radioactive Waste** A cylindrical container for storing radioactive waste is to be constructed from lead and have a thickness of 6 in. (see the figure). If the volume of the outside cylinder is to be 16π ft³, find the radius and the height of the inside cylinder that will result in a container of maximum storage capacity.

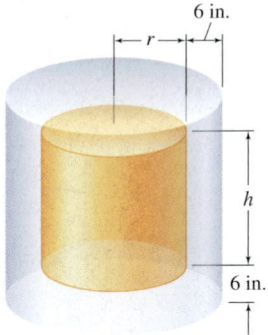

Hint: Show that the storage capacity (inside volume in ft³) is given by

$$V(r) = \pi r^2\left[\frac{16}{(r + \frac{1}{2})^2} - 1\right] \qquad 0 \le r \le \frac{7}{2}$$

62. **Electrical Force of a Conductor** A ring-shaped conductor of radius a carrying a total charge Q induces an electrical force of magnitude

$$F = \frac{Q}{4\pi\varepsilon_0} \cdot \frac{x}{(x^2 + a^2)^{3/2}}$$

where ε_0 is a constant called the permittivity of free space, at a point P, a distance x from the center, along the line perpendicular to the plane of the ring through its center. Find the value of x for which F is greatest.

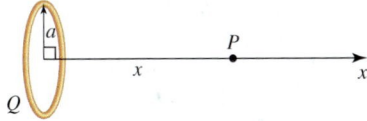

63. **Energy Expended by a Fish** It has been conjectured that the total energy expended by a fish swimming a distance of L ft at a speed of v ft/sec relative to the water and against a current flowing at the rate of u ft/sec ($u < v$) is given by

$$E(v) = \frac{aLv^3}{v - u}$$

where E is measured in foot-pounds (ft-lb) and a is a constant.

a. Find the speed at which the fish must swim to minimize the total energy expended.

b. Sketch the graph of E.

Note: This result has been verified by biologists.

64. A poster of height 36 in. is mounted on a wall so that its lower edge is 12 in. above the eye level of an observer. How far from the wall should the observer stand so that the viewing angle θ subtended at his eye by the poster is as large as possible?

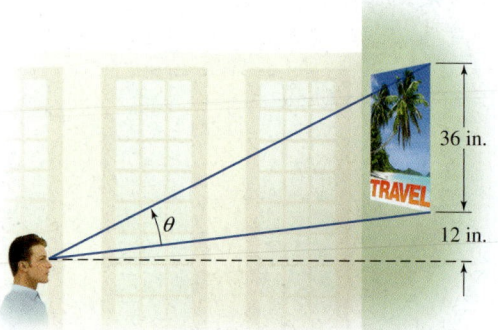

65. A restaurateur has a choice of a site for a restaurant to be constructed between two jetties. The two jetties lie along a straight stretch of a coastal highway and are 1000 ft apart. How far from the longer jetty should the restaurant be located in order to have the largest unobstructed view of the ocean?

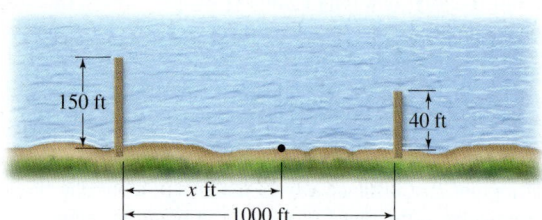

66. An observer stands on a straight path that is parallel to a straight test track. At $t = 0$ a Formula 1 car is directly opposite her and 200 ft away. As she watches, the car moves with a constant acceleration of 20 ft/sec², so it is at a distance of $10t^2$ ft from the starting position after t sec, where $0 \le t \le 15$. As the car moves, $d\theta/dt$ increases, slowly at first, then faster, and finally it slows down again. The observer perceives the car to be moving at the fastest speed when $d\theta/dt$ is maximal. Determine the position of the car at the moment she perceives it to be moving at the fastest speed.

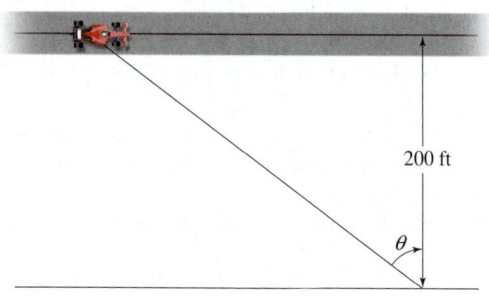

67. A woman is on a lake in a rowboat located one mile from the closest point P of a straight shoreline (see the figure). She wishes to get to a point Q, 10 miles along the shore from P, by rowing to a point R between P and Q and then walking the rest of the distance. If she can row at a speed of 3 mph and walk at a speed of 4 mph, how should she pick the point R to get to Q as quickly as possible? How much time does she require?

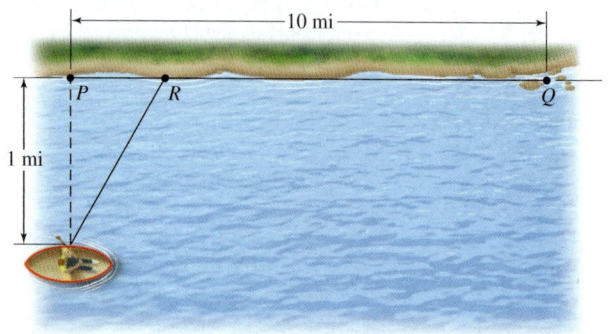

68. **Resonance** A spring system comprising a weight attached to a spring and a dashpot damping device (see the accompanying figure) is acted on by an oscillating external force. Its motion for large values of t is described by the equation

$$x(t) = \frac{F}{(\omega^2 - \gamma)^2 + 4\lambda^2\gamma^2} \sin(\gamma t + \theta)$$

where F, ω, λ, and θ are constants. (F is the amplitude of the external force, θ is a phase angle, γ is associated with the frequency of the external force, and ω and λ are associated with the stiffness of the spring and the degree of resistance of the dashpot damping device, respectively.) Show that the amplitude of the motion of the system

$$g(\gamma) = \frac{F}{\sqrt{(\omega^2 - \gamma^2)^2 + 4\lambda^2\gamma^2}}$$

has a maximum value at $\gamma_1 = \sqrt{\omega^2 - 2\lambda^2}$. When the frequency of the external force is $\sqrt{\omega^2 - 2\lambda^2}/2\pi$, the system is said to be in **resonance**. The figure below shows a typical **resonance curve**.

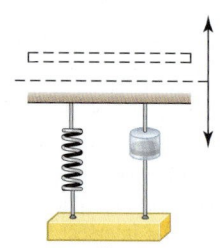
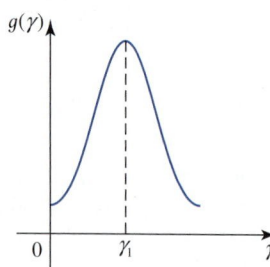

The external force imparts an oscillatory vertical motion on the support.

69. Snell's Law of Refraction The following figure shows the path of a ray of light traveling in air from the source A to the point C and then from C to the point B in water. Let v_1 denote the velocity of light in air, and let v_2 denote the velocity of light in water. Use Fermat's Principle, which states that a ray of light will travel from one point to another in the least time, to prove that

$$\frac{\sin \theta_1}{v_1} = \frac{\sin \theta_2}{v_2}$$

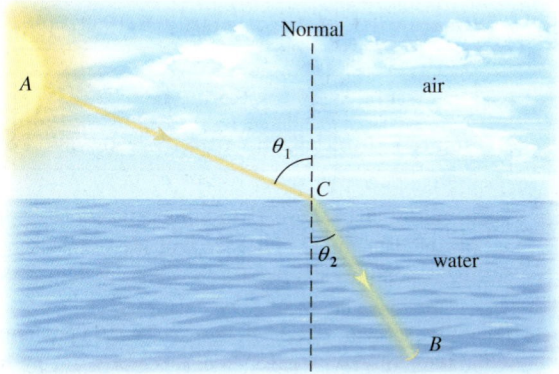

70. Flow of Blood Suppose that some of the fluid flowing along a pipe of radius R is diverted to a pipe of smaller radius r attached to the former at an angle θ (see the figure). Such is the case when blood flowing along an artery is pumped into an arteriole. What should the angle θ be so that the energy loss due to friction in moving the fluid is minimal? Solve the problem via the steps below.

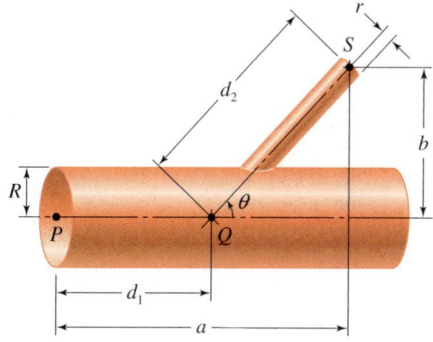

a. Use Poiseuille's Law, which states that the loss of energy due to friction in nonturbulent flow is proportional to the length of the path and inversely proportional to the fourth power of the radius, to show that the energy loss in moving the fluid from P to S via Q is

$$E = \frac{kd_1}{R^4} + \frac{kd_2}{r^4}$$

where k is a constant.

b. Suppose a and b are fixed. Find d_1 and d_2 in terms of a and b. Then use this result together with the result from part (a) to show that

$$E = k\left[\frac{a - b \cot \theta}{R^4} + \frac{b \csc \theta}{r^4}\right]$$

c. Using the technique of this section, show that E is minimized when

$$\theta = \cos^{-1}\frac{r^4}{R^4}$$

71. Least Squares Approximation Suppose we are given n data points

$$P_1(x_1, y_1), P_2(x_2, y_2), \ldots, P_n(x_n, y_n)$$

that are scattered about the graph of a straight line with equation $y = ax$ (see the figure). The error in approximating y_i by the value of the function $f(x) = ax$ at x_i is

$$[y_i - f(x_i)] \qquad 1 \le i \le n$$

a. Show that the sum of the squares of the errors in approximating y_i by $f(x_i)$ for $1 \le i \le n$ is

$$g(a) = (y_1 - ax_1)^2 + (y_2 - ax_2)^2 + \cdots + (y_n - ax_n)^2$$

b. Show that g is minimized if

$$a = \frac{x_1 y_1 + x_2 y_2 + \cdots + x_n y_n}{x_1^2 + x_2^2 + \cdots + x_n^2}$$

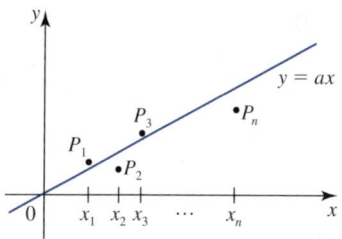

Note: The straight line with equation $y = ax$, where a is the number found in part (b), is called the least-squares, or regression, line associated with the given data and is a line that fits the data "best" in the sense of least squares.

 72. Calculating a Spring Constant The following table gives the force required to stretch a spring by an elongation x beyond its unstretched length.

x (ft)	0	0.1	0.2	0.3	0.4	0.5
Force, y (lb)	0	1.68	3.18	4.84	6.36	8.02

a. Use the result of Exercise 71 to find the straight line $y = ax$ that fits the data "best" in the sense of least squares.

b. Plot the data points and the least squares line found in part (a) on the same set of axes.

c. Using Hooke's Law, which states that $F = kx$, where k is the spring constant, what does the result of part (b) give as the spring constant?

73. Least Squares Approximation Suppose we are given n data points

$$P_1(x_1, y_1), P_2(x_2, y_2), \ldots, P_n(x_n, y_n)$$

that are scattered about the graph of a parabola with equation $y = ax^2$ (see the figure). The error in approximating y_i by the value of the function $f(x) = ax^2$ at x_i is

$$[y_i - f(x_i)] \qquad 1 \le i \le n$$

a. Show that the sum of the squares of the errors in approximating y_i by $f(x_i)$ for $1 \le i \le n$ is

$$g(a) = \left(y_1 - ax_1^2\right)^2 + \left(y_2 - ax_2^2\right)^2 + \cdots + \left(y_n - ax_n^2\right)^2$$

b. Show that g is minimized if

$$a = \frac{x_1^2 y_1 + x_2^2 y_2 + \cdots + x_n^2 y_n}{x_1^4 + x_2^4 + \cdots + x_n^4}$$

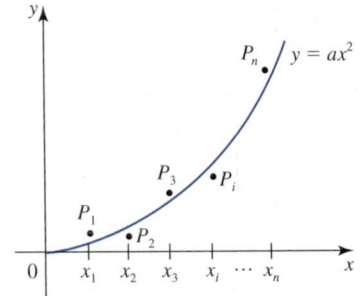

Note: The curve with equation $y = ax^2$, where a is the number found in part (b), is called the least-squares curve associated with the data given and is a curve that fits the data "best" in the sense of least squares.

74. Calculating the Constant of Acceleration A steel ball is dropped from a height of 10 ft. The distance covered by the ball at intervals of one tenth of a second is measured and recorded in the following table.

Time t (sec)	0.0	0.1	0.2	0.3
Distance y (ft)	0	0.1608	0.6416	1.4444

Time t (sec)	0.4	0.5	0.6	0.7
Distance y (ft)	2.5672	4.0108	5.7760	7.8614

a. Use the result of Exercise 73 to find the parabola $y = at^2$ that fits the data "best" in the sense of least squares.

b. Plot the data points and the least-squares curve found in part (a) on the same set of axes.

c. Using the fact that a free-falling object acted upon only by the force of gravity covers a distance of $s = \frac{1}{2}gt^2$ ft after t sec, what does the result of part (a) give as the constant of acceleration g?

3.8 Indeterminate Forms and l'Hôpital's Rule

In Section 1.1 we encountered the limit

$$\lim_{t \to 2} \frac{4(t^2 - 4)}{t - 2} \tag{1}$$

when we attempted to find the velocity of the maglev at time $t = 2$, and in Section 1.2 we studied the limit

$$\lim_{x \to 0} \frac{\sin x}{x} \tag{2}$$

Observe that both the numerator and the denominator of expression (1) approach zero as t approaches 2. Similarly, both the numerator and denominator of expression (2) also approach zero as x approaches zero.

More generally, if $\lim_{x \to a} f(x) = 0$ and $\lim_{x \to a} g(x) = 0$, then the limit

$$\lim_{x \to a} \frac{f(x)}{g(x)}$$

is called an **indeterminate form of the type 0/0**. As the name implies, the undefined expression $0/0$ does not provide us with a definitive answer concerning the existence of the limit or its value, if the limit exists.

Recall that we evaluated the limit in (1) through algebraic sleight of hand. Thus,

$$\lim_{t \to 2} \frac{4(t^2 - 4)}{t - 2} = \lim_{t \to 2} \frac{4(t + 2)(t - 2)}{t - 2} = \lim_{t \to 2} 4(t + 2) = 16$$

This method, however, will not work in evaluating the limit in (2). In Section 1.2 we used a geometric argument to show that

$$\lim_{x \to 0} \frac{\sin x}{x} = 1$$

These examples raise the following question: Given an indeterminate form of the type $0/0$, is there a more general and efficient method for resolving whether the limit

$$\lim_{x \to a} \frac{f(x)}{g(x)}$$

exists, and if so, what is the limit?

■ The Indeterminate Forms $0/0$ and ∞/∞

To gain insight into the nature of an indeterminate form of the type $0/0$, let's consider the following limits:

a. $\lim_{x \to 0^+} \frac{x^2}{x}$ **b.** $\lim_{x \to 0^+} \frac{2x}{3x}$ **c.** $\lim_{x \to 0^+} \frac{x}{x^2}$

Each of these limits is an indeterminate form of the type $0/0$. We can evaluate each limit as follows:

a. $\lim_{x \to 0^+} \frac{x^2}{x} = \lim_{x \to 0^+} x = 0$

b. $\lim_{x \to 0^+} \frac{2x}{3x} = \lim_{x \to 0^+} \frac{2}{3} = \frac{2}{3}$

c. $\lim_{x \to 0^+} \frac{x}{x^2} = \lim_{x \to 0^+} \frac{1}{x} = \infty$

Let's examine each limit in greater detail. In (a) the numerator $f_1(x) = x^2$ goes to zero faster than the denominator $g_1(x) = x$, when x is close to zero. So it is plausible that the ratio $f_1(x)/g_1(x)$ should approach 0 as x approaches 0. In (b) the numerator $f_2(x) = 2x$ goes to zero at $(2x)/(3x) = \frac{2}{3}$ the rate at which $g_2(x) = 3x$ goes to zero, so the answer seems reasonable. Finally, in (c) the denominator $g_3(x) = x^2$ goes to zero faster than the numerator $f_3(x) = x$, and consequently, we expect the ratio to "blow up."

These three examples show that the existence or nonexistence of the limit as well as the value of the limit depend on how fast the numerator $f(x)$ and the denominator $g(x)$ go to zero. This observation suggests the following technique for evaluating these indeterminate forms: Because both $f(x)$ and $g(x)$ go to 0 as x approaches 0, we cannot determine the limit of the quotient by using the Quotient Rule for limits. So we might consider the limit of the ratio of their *derivatives*, $f'(x)$ and $g'(x)$, since the derivatives measure how fast $f(x)$ and $g(x)$ change. In other words, it might be plausible that if both $f(x) \to 0$ and $g(x) \to 0$ as $x \to 0$, then

$$\lim_{x \to 0} \frac{f(x)}{g(x)} = \lim_{x \to 0} \frac{f'(x)}{g'(x)}$$

Let's try this on the limits in (1) and (2). For the limit in (1) we have

$$\lim_{t \to 2} \frac{4(t^2 - 4)}{t - 2} = \lim_{t \to 2} \frac{\frac{d}{dt}[4(t^2 - 4)]}{\frac{d}{dt}(t - 2)} = \lim_{t \to 2} \frac{8t}{1} = 16$$

Historical Biography

G. F. A. DE L'HÔPITAL
(1661-1704)

The wealthy Guillaume François Antoine de l'Hôpital (also spelled l'Hospital) commissioned mathematician Johann Bernoulli (page 624) to teach him differential and integral calculus and even to sell Bernoulli's own mathematical discoveries to l'Hôpital in exchange for a regular salary. Although it seems surprising to us, Bernoulli agreed to this arrangement, and l'Hôpital thereafter presented many of Bernoulli's results as his own. In fact l'Hôpital used one of Bernoulli's most impressive contributions in the first textbook to be written on differential calculus, *Analyse des infiniment petits pour l'intelligence des lignes courbes* (1696). l'Hôpital's writing style was exceptional, and his text appeared in numerous editions throughout the next century. l'Hôpital did acknowledge Bernoulli in the preface to his text, but he did not make clear the great extent to which the work was actually due to Bernoulli. Bernoulli kept silent during l'Hôpital's life, but after l'Hôpital's death Bernoulli accused him of plagiarism. Bernoulli's claims were not taken seriously at the time, and the rule on indeterminate forms has been known as l'Hôpital's Rule since 1696. Only after historical research and the publication of the correspondence between Bernoulli and l'Hôpital has it been substantiated that the rule is actually the result of Johann Bernoulli's insight.

which is the value we obtained before! For the limit in (2) we find

$$\lim_{x \to 0} \frac{\sin x}{x} = \lim_{x \to 0} \frac{\frac{d}{dx}(\sin x)}{\frac{d}{dx}(x)} = \lim_{x \to 0} \frac{\cos x}{1} = 1$$

which we demonstrated in Section 1.2.

This method, which we have arrived at intuitively, is given validity by the theorem known as l'Hôpital's Rule. The theorem is named after the French mathematician Guillaume Francois Antoine de l'Hôpital (1661–1704), who published the first calculus text in 1696. But before stating l'Hôpital's Rule, we need to define another type of indeterminate form.

If $\lim_{x \to a} f(x) = \pm\infty$ and $\lim_{x \to a} g(x) = \pm\infty$, then the limit

$$\lim_{x \to a} \frac{f(x)}{g(x)}$$

is said to be an indeterminate form of the type ∞/∞, $-\infty/\infty$, $\infty/-\infty$, or $-\infty/-\infty$. To see why this limit is an indeterminate form, we simply write

$$\lim_{x \to a} \frac{f(x)}{g(x)} = \lim_{x \to a} \frac{\frac{1}{g(x)}}{\frac{1}{f(x)}}$$

which has the form $0/0$ and, therefore, is indeterminate. We refer to each of these limits as an **indeterminate form of the type ∞/∞**, since the sign provides no useful information.

THEOREM 1 l'Hôpital's Rule

Suppose that f and g are differentiable on an open interval I that contains a, with the possible exception of a itself, and $g'(x) \neq 0$ for all x in I. If $\lim_{x \to a} \dfrac{f(x)}{g(x)}$ is an indeterminate form of the type $0/0$ or ∞/∞, then

$$\lim_{x \to a} \frac{f(x)}{g(x)} = \lim_{x \to a} \frac{f'(x)}{g'(x)}$$

provided that the limit on the right-hand side exists or is infinite.

We will prove this theorem in Appendix B.

 The expression $f'(x)/g'(x)$ is the *ratio* of the derivatives of $f(x)$ and $g(x)$—it is *not obtained* from f/g by using the Quotient Rule.

Notes

1. l'Hôpital's Rule is also valid for one-sided limits as well as limits at infinity or negative infinity; that is, we can replace "$x \to a$" by any of the symbols $x \to a^+$, $x \to a^-$, $x \to \infty$, or $x \to -\infty$.

2. Before applying l'Hôpital's Rule, check to see that the limit does have one of the indeterminate forms. For example, $\cos x \to 1$ as $x \to 0^+$, so

$$\lim_{x \to 0^+} \frac{\cos x}{x} = \infty$$

If we had applied l'Hôpital's Rule to evaluate the limit without first ascertaining that it had an indeterminate form, we would have obtained the erroneous result

$$\lim_{x \to 0^+} \frac{\cos x}{x} = \lim_{x \to 0^+} \frac{-\sin x}{1} = 0$$

FIGURE 1
The graph of $y = \dfrac{e^x - 1}{2x}$ gives a visual confirmation of the result of Example 1.

EXAMPLE 1 Evaluate $\lim\limits_{x \to 0} \dfrac{e^x - 1}{2x}$.

Solution We have an indeterminate form of the type $0/0$. Applying l'Hôpital's Rule, we obtain

$$\lim_{x \to 0} \frac{e^x - 1}{2x} = \lim_{x \to 0} \frac{\dfrac{d}{dx}(e^x - 1)}{\dfrac{d}{dx}(2x)} = \lim_{x \to 0} \frac{e^x}{2} = \frac{1}{2}$$

(See Figure 1.)

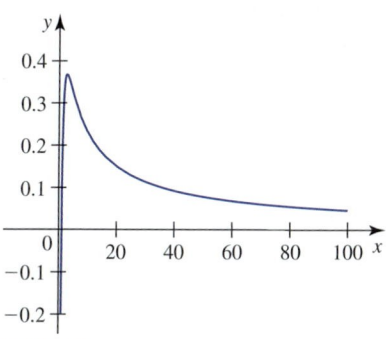

FIGURE 2
The graph of $y = \dfrac{\ln x}{x}$ shows that $y \to 0$ as $x \to \infty$.

EXAMPLE 2 Evaluate $\lim\limits_{x \to \infty} \dfrac{\ln x}{x}$.

Solution We have an indeterminate form of the type ∞/∞. Applying l'Hôpital's Rule, we obtain

$$\lim_{x \to \infty} \frac{\ln x}{x} = \lim_{x \to \infty} \frac{\dfrac{d}{dx}(\ln x)}{\dfrac{d}{dx}(x)} = \lim_{x \to \infty} \frac{1}{x} = 0$$

(See Figure 2.)

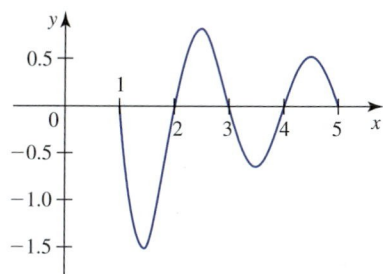

FIGURE 3
The graph of $y = \dfrac{\sin \pi x}{(x - 1)^{1/2}}$ shows that $\lim\limits_{x \to 1^+} \dfrac{\sin \pi x}{(x - 1)^{1/2}} = 0$.

EXAMPLE 3 Evaluate $\lim\limits_{x \to 1^+} \dfrac{\sin \pi x}{\sqrt{x - 1}}$.

Solution We have an indeterminate form of the type $0/0$. Applying l'Hôpital's Rule, we obtain

$$\lim_{x \to 1^+} \frac{\sin \pi x}{(x - 1)^{1/2}} = \lim_{x \to 1^+} \frac{\pi \cos \pi x}{\frac{1}{2}(x - 1)^{-1/2}}$$

$$= \lim_{x \to 1^+} 2\pi(\cos \pi x)\sqrt{x - 1}$$

$$= 0$$

(See Figure 3.)

Sometimes we need to apply l'Hôpital's Rule more than once to resolve a limit involving an indeterminate form. This is illustrated in the next two examples.

EXAMPLE 4 Evaluate $\lim\limits_{x\to\infty} \dfrac{x^3}{e^{2x}}$.

Solution Applying l'Hôpital's Rule (three times), we obtain

$$\lim_{x\to\infty} \frac{x^3}{e^{2x}} = \lim_{x\to\infty} \frac{3x^2}{2e^{2x}} \qquad \text{Type: } \infty/\infty$$

$$= \lim_{x\to\infty} \frac{6x}{4e^{2x}} \qquad \text{Type: } \infty/\infty$$

$$= \lim_{x\to\infty} \frac{6}{8e^{2x}} = 0$$

(See Figure 4.)

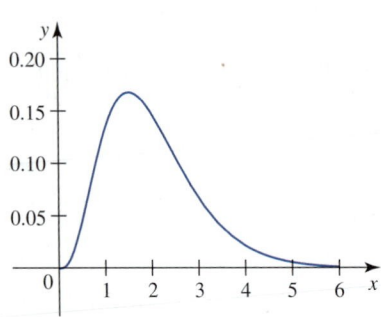

FIGURE 4
The graph of $y = \dfrac{x^3}{e^{2x}}$ shows that $y \to 0$ as $x \to \infty$.

EXAMPLE 5 Evaluate $\lim\limits_{x\to0} \dfrac{x^3}{x - \tan x}$.

Solution We have an indeterminate form of the type $0/0$. Using l'Hôpital's Rule, repeatedly, we obtain

$$\lim_{x\to0} \frac{x^3}{x - \tan x} = \lim_{x\to0} \frac{3x^2}{1 - \sec^2 x} \qquad \text{Type: } 0/0$$

$$= \lim_{x\to0} \frac{6x}{-2\sec^2 x \tan x} \qquad \text{Type: } 0/0$$

$$= \lim_{x\to0} \frac{6}{-4\sec^2 x \tan^2 x - 2\sec^4 x} = \frac{6}{-2} = -3$$

(See Figure 5.)

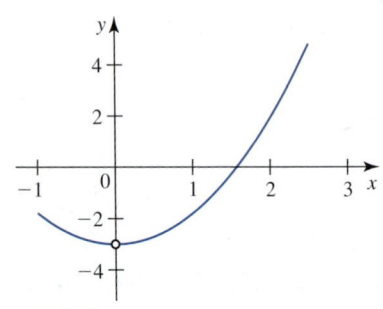

FIGURE 5
The graph of $y = \dfrac{x^3}{x - \tan x}$ shows that $y \to -3$ as $x \to 0$. Note that y is not defined at $x = 0$.

■ The Indeterminate Forms $\infty - \infty$ and $0 \cdot \infty$

If $\lim_{x\to a} f(x) = \infty$ and $\lim_{x\to a} g(x) = \infty$, then the limit

$$\lim_{x\to a}[f(x) - g(x)]$$

is said to be an **indeterminate form of the type $\infty - \infty$.** An indeterminate form of this type can often be expressed as one of the type $0/0$ or ∞/∞ by algebraic manipulation. This is illustrated in the following example.

EXAMPLE 6 Evaluate $\lim\limits_{x\to0^+}\left(\dfrac{1}{x} - \dfrac{1}{e^x - 1}\right)$.

Solution We have an indeterminate form of the type $\infty - \infty$. By writing the expression in parentheses as a single fraction, we obtain an indeterminate form of

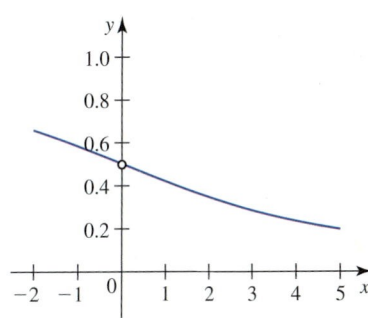

FIGURE 6

The graph of $y = \dfrac{1}{x} - \dfrac{1}{e^x - 1}$ shows

that $\lim\limits_{x \to 0^+} \left(\dfrac{1}{x} - \dfrac{1}{e^x - 1} \right) = \dfrac{1}{2}$.

the type $0/0$. This enables us to evaluate the resulting expression using l'Hôpital's Rule:

$$\lim_{x \to 0^+} \left(\frac{1}{x} - \frac{1}{e^x - 1} \right) = \lim_{x \to 0^+} \frac{e^x - x - 1}{x(e^x - 1)} \qquad \text{Type: } 0/0$$

$$= \lim_{x \to 0^+} \frac{e^x - 1}{e^x - 1 + xe^x} \qquad \text{Apply l'Hôpital's Rule.}$$

$$= \lim_{x \to 0^+} \frac{e^x}{(x + 2)e^x} = \frac{1}{2} \qquad \text{Apply l'Hôpital's Rule again.}$$

(See Figure 6.)　　　　　　　　　　　　　　　　　　　　■

If $\lim_{x \to a} f(x) = 0$ and $\lim_{x \to a} g(x) = \pm\infty$, then $\lim_{x \to a} f(x)g(x)$ is said to be an **indeterminate form of the type $0 \cdot \infty$.** An indeterminate form of this type also can be expressed as one of the type $0/0$ or ∞/∞ by algebraic manipulation, as illustrated in the following example.

EXAMPLE 7 Evaluate $\lim_{x \to 0^+} x \ln x$.

Solution We have an indeterminate form of the type $0 \cdot \infty$. By writing

$$x \ln x = \frac{\ln x}{\dfrac{1}{x}}$$

the given limit can be cast in an indeterminate form of the type ∞/∞. Then, applying l'Hôpital's Rule, we obtain

$$\lim_{x \to 0^+} x \ln x = \lim_{x \to 0^+} \frac{\ln x}{\dfrac{1}{x}} \qquad \text{Type: } \infty/\infty$$

$$= \lim_{x \to 0^+} \frac{\dfrac{1}{x}}{-\dfrac{1}{x^2}} = \lim_{x \to 0^+} (-x) = 0$$

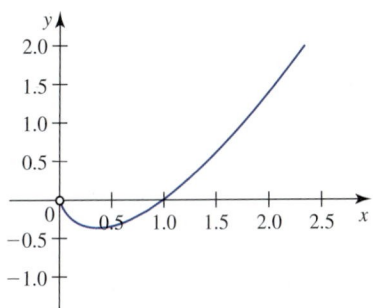

FIGURE 7

The graph of $y = x \ln x$ gives a visual verification of the result in Example 7.

(See Figure 7.)　　　　　　　　　　　　　　　　　　　　■

■ The Indeterminate Forms 0^0, ∞^0, and 1^∞

The limit

$$\lim_{x \to a} [f(x)]^{g(x)}$$

is said to be an **indeterminate form of the type**

0^0 if $\lim\limits_{x \to a} f(x) = 0$ and $\lim\limits_{x \to a} g(x) = 0$

∞^0 if $\lim\limits_{x \to a} f(x) = \infty$ and $\lim\limits_{x \to a} g(x) = 0$

1^∞ if $\lim\limits_{x \to a} f(x) = 1$ and $\lim\limits_{x \to a} g(x) = \pm\infty$

These indeterminate forms can usually be converted to indeterminate forms of the type $0 \cdot \infty$ by taking logarithms or by using the identity

$$[f(x)]^{g(x)} = e^{g(x)\ln f(x)}$$

EXAMPLE 8 Evaluate $\lim_{x \to 0^+} x^x$.

Solution We have an indeterminate form of the type 0^0. Let

$$y = x^x$$

Then

$$\ln y = \ln x^x = x \ln x$$

and using the result of Example 7, we obtain

$$\lim_{x \to 0^+} \ln y = \lim_{x \to 0^+} x \ln x = 0$$

Finally, using the identity $y = e^{\ln y}$ and the continuity of the exponential function, we have

$$\lim_{x \to 0^+} x^x = \lim_{x \to 0^+} y = \lim_{x \to 0^+} e^{\ln y} = e^{\lim_{x \to 0^+} \ln y} = e^0 = 1$$

(See Figure 8.)

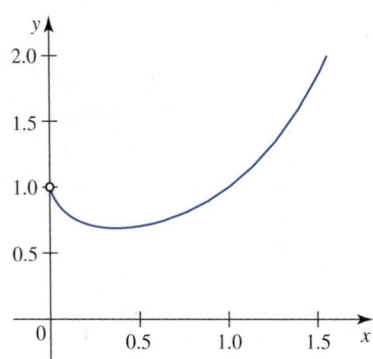

FIGURE 8
The graph of $y = x^x$
shows that $\lim_{x \to 0^+} x^x = 1$.

EXAMPLE 9 Evaluate $\lim_{x \to 0^+} \left(\dfrac{1}{x}\right)^{\sin x}$.

Solution We have an indeterminate form of the type ∞^0. Let

$$y = \left(\frac{1}{x}\right)^{\sin x}$$

Then

$$\ln y = \ln\left(\frac{1}{x}\right)^{\sin x} = (\sin x)\ln\frac{1}{x}$$

and

$$\lim_{x \to 0^+} \ln y = \lim_{x \to 0^+} (\sin x)\ln\frac{1}{x}$$

This last limit is an indeterminate form of the type $0 \cdot \infty$. By writing

$$(\sin x)\ln\left(\frac{1}{x}\right) = \frac{\ln\dfrac{1}{x}}{\dfrac{1}{\sin x}}$$

we can transform it into an indeterminate form of the type ∞/∞ and hence use l'Hôpital's Rule. We have

$$\lim_{x\to 0^+} \ln y = \lim_{x\to 0^+} \frac{\ln\dfrac{1}{x}}{\dfrac{1}{\sin x}} \qquad \text{Type: } \infty/\infty$$

$$= \lim_{x\to 0^+} -\frac{\ln x}{\dfrac{1}{\sin x}} \qquad \text{Rewrite } \ln\left(\dfrac{1}{x}\right).$$

$$= \lim_{x\to 0^+} \frac{-\dfrac{1}{x}}{-\dfrac{\cos x}{\sin^2 x}} = \lim_{x\to 0^+} \frac{\sin^2 x}{x\cos x} \qquad \text{Apply l'Hôpital's Rule.}$$

$$= \lim_{x\to 0^+} \left(\frac{\sin x}{x}\right)(\tan x) = 0 \qquad \lim_{x\to 0^+}\frac{\sin x}{x} = 1$$

Therefore,

$$\lim_{x\to 0^+}\left(\frac{1}{x}\right)^{\sin x} = \lim_{x\to 0^+} y = \lim_{x\to 0^+} e^{\ln y} = e^{\lim_{x\to 0^+}\ln y} = e^0 = 1$$

since the exponential function is continuous. (See Figure 9.)

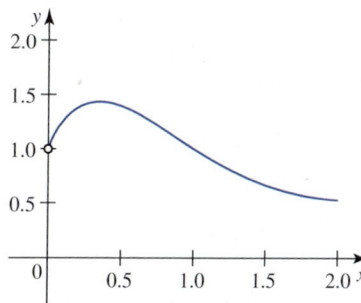

FIGURE 9
The graph of $y = \left(\dfrac{1}{x}\right)^{\sin x}$
shows that $\lim_{x\to 0^+} \ln y = 1$.

EXAMPLE 10 Evaluate $\displaystyle\lim_{x\to\infty}\left(1 + \frac{1}{x}\right)^x$.

Solution We have an indeterminate form of the type 1^∞. Let

$$y = \left(1 + \frac{1}{x}\right)^x$$

Then

$$\ln y = \ln\left(1 + \frac{1}{x}\right)^x = x\ln\left(1 + \frac{1}{x}\right)$$

so

$$\lim_{x\to\infty} \ln y = \lim_{x\to\infty} x \ln\left(1 + \frac{1}{x}\right)$$

has an indeterminate form of the type $0 \cdot \infty$. Rewriting and using l'Hôpital's Rule, we obtain

$$\lim_{x\to\infty} \ln y = \lim_{x\to\infty} x \ln\left(1 + \frac{1}{x}\right) \qquad \text{Type: } 0 \cdot \infty$$

$$= \lim_{x\to\infty} \frac{\ln\left(1 + \frac{1}{x}\right)}{\frac{1}{x}} \qquad \text{Type: } 0/0$$

$$= \lim_{x\to\infty} \left[\frac{\left(\frac{1}{1 + \frac{1}{x}}\right)\left(-\frac{1}{x^2}\right)}{-\frac{1}{x^2}}\right] \qquad \text{Apply l'Hôpital's Rule.}$$

$$= \lim_{x\to\infty} \frac{1}{1 + \frac{1}{x}} = 1$$

Therefore,

$$\lim_{x\to\infty}\left(1 + \frac{1}{x}\right)^x = \lim_{x\to\infty} y = \lim_{x\to\infty} e^{\ln y} = e^{\lim_{x\to\infty} \ln y} = e^1 = e$$

since the exponential function is continuous. (See Figure 10.)

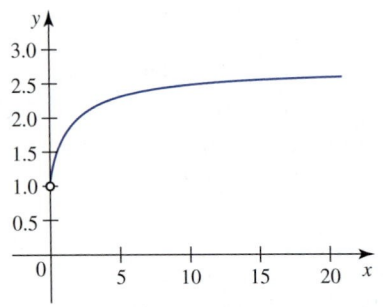

FIGURE 10
The graph of $y = \left(1 + \frac{1}{x}\right)^x$ shows that $y \to e \approx 2.718$ as $x \to \infty$.

3.8   CONCEPT QUESTIONS

In Exercises 1–8, evaluate the limit or classify the type of inde-terminate form to which it gives rise.

1. $\displaystyle\lim_{x\to a} \frac{f(x)}{g(x)}$ if $\displaystyle\lim_{x\to a} f(x) = 1$ and $\displaystyle\lim_{x\to a} g(x) = \infty$

2. $\displaystyle\lim_{x\to a} \frac{f(x)}{g(x)}$ if $\displaystyle\lim_{x\to a} f(x) = 4$ and $\displaystyle\lim_{x\to a} g(x) = 0$, where $g(x) > 0$

3. $\displaystyle\lim_{x\to a^+} \frac{f(x)}{x - a}$ if $\displaystyle\lim_{x\to a} f(x) = \infty$

4. $\displaystyle\lim_{x\to a}[f(x) - g(x)]$ if $\displaystyle\lim_{x\to a} f(x) = \infty$ and $\displaystyle\lim_{x\to a} g(x) = -\infty$

5. $\displaystyle\lim_{x\to 3}\left[\frac{f(x)}{(x - 3)^2} + \frac{2}{|x - 3|}\right]$ if $\displaystyle\lim_{x\to 3} f(x) = 8$

6. $\displaystyle\lim_{x\to\infty}[f(x)]^{g(x)}$ if $\displaystyle\lim_{x\to\infty} f(x) = 0$ and $\displaystyle\lim_{x\to\infty} g(x) = \infty$ where $f(x) \geq 0$

7. $\displaystyle\lim_{x\to\infty}\left[\frac{x}{\sqrt{x^2 + 5}}\right]^{f(x)}$ if $\displaystyle\lim_{x\to\infty} f(x) = \infty$

8. $\displaystyle\lim_{x\to a}\left[\frac{2}{f(x)}\right]^{g(x)}$ if $\displaystyle\lim_{x\to a} f(x) = 0$ and $\displaystyle\lim_{x\to a} g(x) = 0$ where $f(x) > 0$

9. a. State l'Hôpital's Rule.

   b. Explain how l'Hôpital's Rule can be used to evaluate

   (i) $\displaystyle\lim_{x\to a} f(x)g(x)$ if $\displaystyle\lim_{x\to a} f(x) = \infty$ and $\displaystyle\lim_{x\to a} g(x) = 0$

   (ii) $\displaystyle\lim_{x\to a}[f(x) - g(x)]$ if $\displaystyle\lim_{x\to a} f(x) = -\infty$ and $\displaystyle\lim_{x\to a} g(x) = -\infty$

   (iii) $\displaystyle\lim_{x\to a}[f(x)]^{g(x)}$ if $\displaystyle\lim_{x\to a} f(x) = 0$ and $\displaystyle\lim_{x\to a} g(x) = 0$, where $f(x) > 0$

3.8 EXERCISES

In Exercises 1–56, evaluate the limit using l'Hôpital's Rule if appropriate.

1. $\lim\limits_{x \to 1} \dfrac{x - 1}{x^2 - 1}$

2. $\lim\limits_{x \to -1} \dfrac{x^2 - 2x - 3}{x + 1}$

3. $\lim\limits_{x \to 2} \dfrac{x^3 - 8}{x - 2}$

4. $\lim\limits_{x \to 1} \dfrac{x^7 - 1}{x^4 - 1}$

5. $\lim\limits_{x \to 0} \dfrac{e^x - 1}{x^2 + x}$

6. $\lim\limits_{x \to 1} \dfrac{\ln x}{x - 1}$

7. $\lim\limits_{t \to \pi} \dfrac{\sin t}{\pi - t}$

8. $\lim\limits_{x \to 0} \dfrac{e^x - 1}{x + \sin x}$

9. $\lim\limits_{\theta \to 0} \dfrac{\tan 2\theta}{\theta}$

10. $\lim\limits_{x \to 0} \dfrac{\sin 2x}{x}$

11. $\lim\limits_{x \to \infty} \dfrac{x + \cos x}{2x + 1}$

12. $\lim\limits_{\theta \to 0} \dfrac{\theta + \sin \theta}{\tan \theta}$

13. $\lim\limits_{x \to 0} \dfrac{\sin x - x \cos x}{\tan^3 x}$

14. $\lim\limits_{u \to \pi} \dfrac{2 \sin^2 u}{1 + \cos u}$

15. $\lim\limits_{x \to \infty} \dfrac{\sqrt{x}}{\ln x}$

16. $\lim\limits_{x \to \infty} \dfrac{e^x}{x^4}$

17. $\lim\limits_{x \to \infty} \dfrac{(\ln x)^3}{x^2}$

18. $\lim\limits_{x \to 1} \dfrac{x^{1/2} - x^{1/3}}{x - 1}$

19. $\lim\limits_{x \to \infty} \dfrac{\ln(1 + e^x)}{x^2}$

20. $\lim\limits_{x \to 1} \dfrac{a^{\ln x} - x}{\ln x}$

21. $\lim\limits_{x \to -1} \dfrac{\sqrt{x + 2} + x}{\sqrt[3]{2x + 1} + 1}$

22. $\lim\limits_{x \to 0} \dfrac{\ln(x^2 + 1)}{\cos x - 1}$

23. $\lim\limits_{x \to 0^+} \dfrac{e^{x^2} + x - 1}{1 - \sqrt{1 - x^2}}$

24. $\lim\limits_{x \to 0} \dfrac{\ln(1 + x) - \tan x}{x^2}$

25. $\lim\limits_{x \to 0} \dfrac{\sin x - x}{e^x - e^{-x} - 2x}$

26. $\lim\limits_{x \to 0} \dfrac{e^{x^2} - 1}{1 - \cos x}$

27. $\lim\limits_{x \to 0} \dfrac{\sin^{-1}(2x)}{x}$

28. $\lim\limits_{x \to 0} \dfrac{2x}{\tan^{-1}(3x)}$

29. $\lim\limits_{x \to 0} \left(\cot x - \dfrac{1}{x} \right)$

30. $\lim\limits_{x \to 0} \dfrac{\sin^{-1} x - x}{\tan^{-1} x - x}$

31. $\lim\limits_{x \to 0} \dfrac{(\sin x)^2}{1 - \sec x}$

32. $\lim\limits_{t \to \pi/2} (\tan t - \sec t)$

33. $\lim\limits_{x \to 0^+} \left(\dfrac{1}{x} - \dfrac{1}{1 - \cos x} \right)$

34. $\lim\limits_{x \to 0} \left(\dfrac{e^x - \cos x + \tan x}{x + \tan x + \sin x} \right)$

35. $\lim\limits_{x \to 1} \left(\dfrac{1}{\ln x} - \dfrac{1}{x - 1} \right)$

36. $\lim\limits_{x \to \pi/2} [(\pi - 2x) \sec x]$

37. $\lim\limits_{x \to 0^+} [\csc x \cdot \ln(1 - \sin x)]$

38. $\lim\limits_{x \to 0^+} x^{\sin x}$

39. $\lim\limits_{x \to \infty} \left(\dfrac{1}{x} \right) e^{-x}$

40. $\lim\limits_{x \to 0^+} (x - \sin x)^{\sqrt{x}}$

41. $\lim\limits_{x \to 0^+} (1 - \cos x)^{\tan x}$

42. $\lim\limits_{x \to \infty} (\ln x)^{1/x}$

43. $\lim\limits_{x \to \infty} (e^{2x} + 1)^{1/x}$

44. $\lim\limits_{x \to \infty} (x^2 + e^x)^{1/x}$

45. $\lim\limits_{x \to \frac{\pi}{2}^-} (\tan x)^{\cos x}$

46. $\lim\limits_{x \to \infty} x^{\tan(1/x)}$

47. $\lim\limits_{x \to \infty} \left(1 + \dfrac{1}{x} \right)^{x^3}$

48. $\lim\limits_{x \to \infty} \left(1 - \dfrac{1}{x} \right)^x$

49. $\lim\limits_{x \to \infty} \left(\dfrac{2x + 1}{2x - 1} \right)^{\sqrt{x}}$

50. $\lim\limits_{x \to \frac{\pi}{2}^-} (\sin x)^{\tan x}$

51. $\lim\limits_{x \to \infty} (x - \sqrt{x^2 + 1})$

52. $\lim\limits_{x \to \infty} (2 \tan^{-1} x - \pi) \ln x$

53. $\lim\limits_{x \to \infty} \dfrac{2 \tan^{-1} x - \pi}{e^{1/x^2} - 1}$

54. $\lim\limits_{x \to 0} \dfrac{a^x - b^x}{x}, \quad a, b > 0$

55. $\lim\limits_{x \to 0^+} \dfrac{\ln x}{2 + 3 \ln(\sin x)}$

56. $\lim\limits_{x \to \infty} (a^{1/x} - 1)x$

In Exercises 57 and 58, l'Hôpital's Rule is used incorrectly. Find where the error is made, and give the correct solution.

57. $\lim\limits_{x \to 1} \dfrac{x^5 - 1}{x^2 - 1} = \lim\limits_{x \to 1} \dfrac{5x^4}{2x} = \lim\limits_{x \to 1} \dfrac{20x^3}{2} = 10$

58. $\lim\limits_{x \to 0} \dfrac{e^{3x} + x - 1}{e^x - 1} = \lim\limits_{x \to 0} \dfrac{3e^{3x} + 1}{e^x} = \lim\limits_{x \to 0} \dfrac{9e^{3x}}{e^x} = 9$

59. Continuous Compound Interest Formula See Section 3.5. Use l'Hôpital's Rule to derive the continuous compound interest formula

$$A = Pe^{rt}$$

where A is the accumulated amount, P is the principal, t is the time in years, and r is the nominal interest rate per year compounded continuously, from the compound interest formula

$$A = P \left(1 + \dfrac{r}{m} \right)^{mt}$$

where r is the nominal interest rate per year compounded m times per year.

60. Velocity of a Ballast A ballast of mass m slugs is dropped from a hot-air balloon with an initial velocity of v_0 ft/sec. If the ballast is subjected to air resistance that is directly proportional to its instantaneous velocity, then its velocity at time t is

$$v(t) = \dfrac{mg}{k} + \left(v_0 - \dfrac{mg}{k} \right) e^{-kt/m}$$

feet per second, where $k > 0$ is the constant of proportionality and g is the constant of acceleration. Find an expression for the velocity of the ballast at any time t, assuming that there is no air resistance.
Hint: Find $\lim\limits_{k \to 0} v(t)$.

61. Current in a Circuit A series *RL* circuit including a resistor *R* and inductance *L* is shown in the schematic. Suppose that the electromotive force $E(t)$ is V volts, the resistance is R ohms, and the inductance is L henries, where V, R, and L are positive constants. Then the current at time t is given by

$$I(t) = \frac{V}{R}\left(1 - e^{-Rt/L}\right)$$

amperes. Using l'Hôpital's Rule, evaluate $\lim_{R\to 0^+} I$ to find an expression for the current in a circuit in which the resistance is 0 ohms.

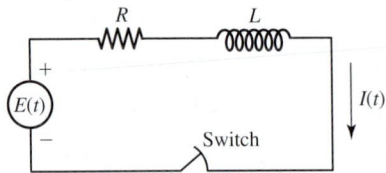

62. Resonance Refer to Section 2.5. A weight of mass *m* is attached to a spring suspended from a support. The weight is then set in motion by an oscillatory force $f(t) = F_0 \sin \omega t$ acting on the support. Here, F_0 and ω are positive constants, and *t* is time. In the absence of frictional and damping forces, the position of the weight from its equilibrium position at time *t* is given by

$$x(t) = \frac{F_0(-\omega \sin \omega_0 t + \omega_0 \sin \omega t)}{\omega_0(\omega_0^2 - \omega^2)}$$

with $\omega_0 = \sqrt{k/m}$, where *k* is the spring constant. Show that if ω approaches ω_0, the resulting oscillations of the mass increase without bound. This phenomenon is known as **pure resonance.**

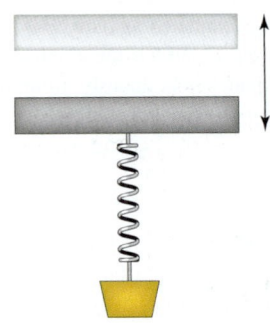

63. Bimolecular Reaction In a bimolecular reaction $A + B \to M$, *a* moles per liter of *A* and *b* moles per liter of *B* are combined. The number of moles per liter that have reacted after time *t* is given by

$$x = \frac{ab[1 - e^{(b-a)kt}]}{a - be^{(b-a)kt}} \qquad a \neq b$$

where the positive number *k* is called the *velocity constant*. Find an expression for *x* if $a = b$, and find $\lim_{t\to\infty} x$. Interpret your results.

64. a. Prove that $\lim_{x\to\infty} \dfrac{x^k}{e^x} = 0$ for every positive constant *k*. This shows that the natural exponential function approaches infinity faster than any power function.

   b. Prove that $\lim_{x\to\infty} \dfrac{\ln x}{x^k} = 0$ for every positive constant *k*. This shows that the natural logarithmic function approaches infinity slower than any power function.

65. Prove that $\lim_{x\to\pi/2} \dfrac{\tan x}{\sec x} = 1$. Can l'Hôpital's Rule be used to compute this limit?

66. Show that

$$\lim_{x\to 0} \frac{x^2 \sin\left(\dfrac{1}{x}\right)}{\sin x} = 0$$

Can l'Hôpital's Rule be used to compute this limit?

67. Use l'Hôpital's Rule to show that if f' is continuous, then

$$\lim_{h\to 0} \frac{f(x+h) - f(x-h)}{2h} = f'(x)$$

and that if f'' is continuous, then

$$\lim_{h\to 0} \frac{f(x+h) - 2f(x) + f(x-h)}{h^2} = f''(x)$$

68. Find the slant asymptotes of the graph of $f(x) = x \tan^{-1} x$. Then sketch the graph of the function.
Hint: See Section 3.6.

In Exercises 69–72, plot the graph the function and use it to guess at the limit. Verify your result using l'Hôpital's Rule.

69. $\displaystyle\lim_{x\to 0} \frac{e^x - e^{-2x}}{\ln(1+x)}$

70. $\displaystyle\lim_{x\to 0^+} \left(\frac{1}{x}\right)^{\tan x}$

71. $\displaystyle\lim_{x\to 1}\left(\frac{1}{\ln x} - \frac{1}{x-1}\right)$

72. $\displaystyle\lim_{x\to 0} \frac{1}{x}[(1+x)^{1/x} - e]$

In Exercises 73 and 74, determine whether the statement is true or false. If it is true, explain why it is true. If it is false, give an example to show why it is false.

73. If $\lim_{x\to a} f(x) = 0$ and $\lim_{x\to a} g(x) = 0$, then

$$\lim_{x\to a} \frac{f(x)}{g(x)} = \lim_{x\to a} \frac{d}{dx}\left[\frac{f(x)}{g(x)}\right].$$

74. $\displaystyle\lim_{x\to\pi^+} \frac{\sin x}{1 - \cos x} = \lim_{x\to\pi^+} \frac{\cos x}{\sin x} = \infty$

3.9 | Newton's Method

There are many occasions when we need to find one or more zeros of a function f. For example, the x-intercepts of a function f are precisely the values of x that satisfy $f(x) = 0$; the critical numbers of f include the roots of the equation $f'(x) = 0$; and the x-coordinates of the candidates for the inflection points of f include the roots of the equation $f''(x) = 0$.

If f is a linear or quadratic function, the zeros are easily found. But, if f is a polynomial function of degree three or higher, the task of finding its zeros is difficult unless the polynomial is easily factored.* We encounter similar difficulties when we try to solve transcendental equations such as

$$x - \tan x = 0 \qquad x > 0$$

In such situations, we have to settle for *approximations* of the roots of the equation.

Actually, in many practical applications we often have a rough idea as to where the zero(s) of interest are located. We can also determine the approximate location of a zero by making a rough sketch of the graph of f and noting where the graph crosses the x-axis. Now, once a crude approximation of the desired zero has been found, a procedure is needed that will yield an approximation of the zero to the desired accuracy. Newton's method (also called the Newton-Raphson method) provides us with one such procedure.

Newton's Method

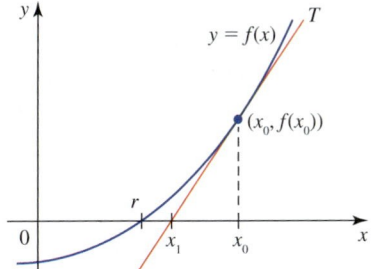

FIGURE 1
Starting with an initial estimate x_0, Newton's method gives a better approximation x_1 to the root r.

Suppose f has a zero at r that we want to approximate (see Figure 1). Let x_0 be an approximation of r obtained, for example, from a rough sketch of the graph of f or by using the Intermediate Value Theorem, and let T denote the tangent line to the graph of f at $(x_0, f(x_0))$.

Because of the proximity of the points on T to the graph of f near $(x_0, f(x_0))$, one may expect that if x_0 is close to r, then the x-intercept of T (call it x_1) will be close to r as well. As it turns out, x_1 often provides us with an even better approximation to r than x_0 does (such is the situation depicted in Figure 1).

To find a formula for x_1 in terms of x_0, we first find an equation for T. Since the slope of T is $f'(x_0)$ and its point of tangency is $(x_0, f(x_0))$, we see that such an equation is

$$y - f(x_0) = f'(x_0)(x - x_0) \tag{1}$$

Now, if $f'(x_0) \neq 0$, then setting $y = 0$ in Equation (1) and solving for x gives the x-intercept of T, x_1. Thus,

$$x_1 = x_0 - \frac{f(x_0)}{f'(x_0)} \tag{2}$$

---

*Although formulas exist for solving third- and fourth-degree polynomial equations, they are seldom used because of their complexity. No formula exists for finding the roots of a general polynomial equation of degree five or higher. This fact was demonstrated by the Norwegian mathematician Niels Henrik Abel (1802–1829).

If we repeat the process, this time letting x_1 play the role of x_0, we obtain yet another approximation of r:

$$x_2 = x_1 - \frac{f(x_1)}{f'(x_1)}$$

Continuing in this manner, we generate a sequence of approximations $x_1, x_2, \ldots x_n, x_{n+1}, \ldots$, with

$$x_{n+1} = x_n - \frac{f(x_n)}{f'(x_n)} \tag{3}$$

provided that $f'(x_n) \neq 0$.

The sequence obtained through the repetitive use, or *iteration*, of Equation (3) frequently converges to the root r of $f(x) = 0$. Our discussion leads us to the following algorithm for finding an approximation to the root r of $f(x) = 0$.

The Newton Algorithm

1. Pick an initial estimate x_0 of the root r.
2. Generate a sequence of estimates using the iterative formula

$$x_{n+1} = x_n - \frac{f(x_n)}{f'(x_n)} \qquad n = 0, 1, 2, \ldots \tag{4}$$

3. Compute $|x_n - x_{n+1}|$. If this number is less than a prescribed number, stop. The required approximation to the root r is x_{n+1}.

Note The initial estimate x_0 is normally taken to be a guess of the root r. For example, a rough sketch of the graph of f could reveal what a good choice of x_0 might be.

■ Applying Newton's Method

EXAMPLE 1 In Example 9 in Section 1.4, we saw that a zero of $f(x) = x^3 + x - 1$ lies in the interval $(0, 1)$. Use Newton's method with $x_0 = 0.5$ to obtain an approximation of this root. Stop the iteration when $|x_n - x_{n+1}| < 0.000001$.

Solution We have

$$f(x) = x^3 + x - 1$$

and

$$f'(x) = 3x^2 + 1$$

so the iterative formula (4) becomes

$$x_{n+1} = x_n - \frac{f(x_n)}{f'(x_n)} = x_n - \frac{x_n^3 + x_n - 1}{3x_n^2 + 1}$$

$$= \frac{2x_n^3 + 1}{3x_n^2 + 1} \tag{5}$$

Letting $n = 0$ in Equation (5) and using $x_0 = 0.5$, we obtain

$$x_1 = \frac{2(0.5)^3 + 1}{3(0.5)^2 + 1} \approx 0.714285714$$

Next, with $n = 1$ and the value of x_1 just obtained, we find

$$x_2 \approx \frac{2(0.714285714)^3 + 1}{3(0.714285714)^2 + 1} \approx 0.683179724$$

Continuing with the iteration, we obtain

$$x_3 \approx 0.682328423 \qquad \text{and} \qquad x_4 \approx 0.682327804$$

Since

$$|x_3 - x_4| \approx 0.000000619 < 0.000001$$

the process is terminated, and we find the required root to be approximately 0.682328. Note that $f(0.682328) \approx 4.7 \times 10^{-7}$ is very close to zero.

EXAMPLE 2 Use Newton's method to find an approximation to the root of the equation $\cos x - x = 0$ accurate to eight decimal places.

Solution By writing the given equation in the form

$$\cos x = x$$

we see that the root of the equation is just the x-coordinate of the point of intersection of the graphs of $y = \cos x$ and $y = x$. This observation enables us to obtain an initial estimate of the root of $\cos x - x = 0$ graphically (Figure 2).

From Figure 2 we see that $x_0 = 0.5$ is a reasonable approximation. Writing

$$f(x) = \cos x - x$$

we have

$$f'(x) = -\sin x - 1$$

and the required iterative formula is

$$x_{n+1} = x_n - \frac{f(x_n)}{f'(x_n)} = x_n - \frac{\cos x_n - x_n}{-\sin x_n - 1}$$

$$= \frac{x_n \sin x_n + \cos x_n}{\sin x_n + 1}$$

With $x_0 = 0.5$ we obtain the sequence

$$x_1 \approx 0.755222417$$

$$x_2 \approx 0.739141666$$

$$x_3 \approx 0.739085134$$

$$x_4 \approx 0.739085133$$

$$x_5 \approx 0.739085133$$

Therefore, the root of $\cos x - x = 0$ is approximately 0.73908513.

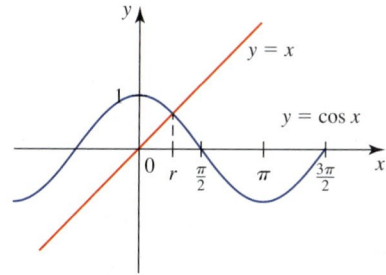

FIGURE 2
Our initial estimate of the root of the equation $\cos x = x$ is $x_0 = 0.5$.

EXAMPLE 3 **Approximating the Square Root of a Positive Number** Observe that the positive root of the positive number A can be found by solving the equation $x^2 - A = 0$. Thus, an approximation of $\sqrt{A}$ can be obtained by using Newton's method to solve the equation $f(x) = x^2 - A = 0$.

a. Find the iterative formula for solving the equation $x^2 - A = 0$.
b. Compute $\sqrt{2}$ using this formula with $x_0 = 1$, terminating the process when two successive approximations differ by less than 0.00001.

Solution
a. We have $f(x) = x^2 - A$ and $f'(x) = 2x$, so by Equation (4)

$$x_{n+1} = x_n - \frac{f(x_n)}{f'(x_n)} = x_n - \frac{x_n^2 - A}{2x_n} = \frac{x_n^2 + A}{2x_n}$$

b. With $A = 2$ and $x_0 = 1$, we find

$$x_1 = \frac{1^2 + 2}{2(1)} = 1.5$$

$$x_2 = \frac{(1.5)^2 + 2}{2(1.5)} \approx 1.416667$$

$$x_3 = \frac{(1.416667)^2 + 2}{2(1.416667)} \approx 1.414216$$

$$x_4 = \frac{(1.414216)^2 + 2}{2(1.414216)} \approx 1.414214$$

Since $|x_3 - x_4| = 0.000002 < 0.00001$, we terminate the process. The sequence that was generated converges to $\sqrt{2}$, which is one of the two roots of the equation $x^2 - 2 = 0$. Note that the value of $\sqrt{2}$ to six places is 1.414214. ∎

EXAMPLE 4 **A Mixture Problem** A tank initially contains 10 gal of brine with 2 lb of salt. Brine with 1.5 lb of salt per gallon enters the tank at the rate of 3 gal/min, and the well-stirred mixture leaves the tank at the rate of 4 gal/min (see Figure 3). It can be shown that the amount of salt in the tank after t min is x lb, where

$$x = f(t) = 1.5(10 - t) - 0.0013(10 - t)^4 \qquad 0 \le t \le 10$$

Find the time(s) t when the amount of salt in the tank is 5 lb.

Solution We solve the equation

$$1.5(10 - t) - 0.0013(10 - t)^4 = 5$$

or

$$(10 - t)^4 - 1153.846154(10 - t) + 3846.153846 = 0$$

Let's make the substitution $u = 10 - t$. Then the above equation becomes

$$u^4 - 1153.846154u + 3846.153846 = 0$$

Put

$$g(u) = u^4 - 1153.846154u + 3846.153846$$

FIGURE 3
Brine enters the tank at the rate of 3 gal/min, and the mixture exits at the rate of 4 gal/min.

Then

$$g'(u) = 4u^3 - 1153.846154$$

and the Newton iteration formula becomes

$$u_{n+1} = u_n - \frac{g(u_n)}{g'(u_n)} = u_n - \frac{u_n^4 - 1153.846154u_n + 3846.153846}{4u_n^3 - 1153.846154}$$

$$= \frac{3u_n^4 - 3846.153846}{4u_n^3 - 1153.846154}$$

A rough sketch of the graph of g on the interval $[0, 10]$ shows that g has zeros near $u = 3$ and $u = 9$. Taking $u_0 = 2$ as an initial guess, we obtain the following sequence:

$$u_1 \approx 3.38563, \qquad u_2 \approx 3.45677, \qquad u_3 \approx 3.45713, \qquad u_4 \approx 3.45713$$

Therefore, $t \approx 10 - 3.45713 = 6.54287 \approx 6.54$.

Next, taking $u_0 = 8$ as an initial guess, we obtain

$$u_1 = 9.44116, \quad u_2 = 9.03541, \quad u_3 = 8.98780, \quad u_4 = 8.98716, \quad u_5 = 8.98716$$

Therefore, $t = 10 - 8.98716 = 1.01284 \approx 1.01$.

So 5 lb of salt are in the tank approximately 1 min and approximately 6.5 min after the brine enters the tank.

■ When Newton's Method Does Not Work

Now that we have seen how effective Newton's method can be for finding the zeros of a function, we wish to point out that there are situations in which the method fails and that care must be exercised in applying it. Figure 4a illustrates a situation in which $f'(x_n) = 0$ for some n (in this case, $n = 2$). Since the iterative formula (4) involves division by $f'(x_n)$, it should be clear why the method fails to work in this case. However, we are sometimes able to salvage the situation by choosing a different initial estimate x_0 (see Figure 4b).

The situation shown in Figure 5 is more serious, and Newton's method will not work for any choice of the initial estimate x_0 other than the actual zero of the function $f(x)$. As you can see from the figure, the sequence $x_1, x_2, x_3, \ldots$ actually moves farther and farther away with each iteration, and thus the method fails.

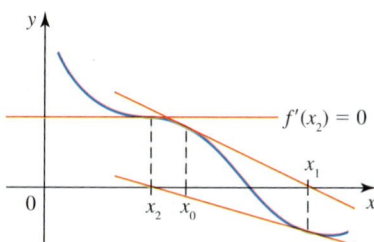

(a) Newton's method fails to work because $f'(x_2) = 0$.

FIGURE 4

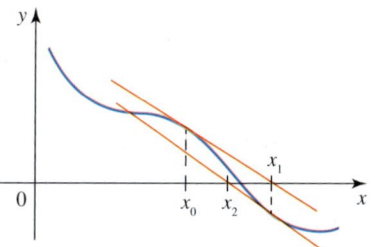

(b) The situation in part (a) is remedied by selecting a different initial estimate x_0.

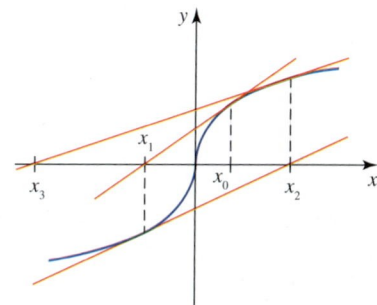

FIGURE 5

Newton's method fails here because the sequence of estimates diverges.

3.9 CONCEPT QUESTIONS

1. Give a geometric description of Newton's method for finding the zeros of a function f. Illustrate graphically.
2. Describe Newton's algorithm for finding the zero of a function f.

3. Does Newton's method always work for any choice of the initial estimate x_0 of the root of $f(x) = 0$? Explain graphically.

3.9 EXERCISES

In Exercises 1–4, use Newton's method to find the zero(s) of f to four decimal places by solving the equation $f(x) = 0$. Use the initial estimate(s) x_0.

1. $f(x) = -x^3 - 2x + 2$, $x_0 = 1$

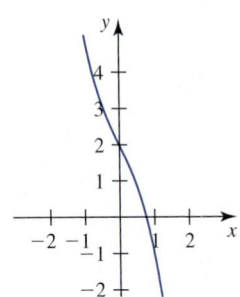

2. $f(x) = 2x^3 - 15x^2 + 36x - 20$, $x_0 = 1$

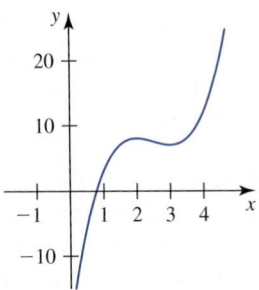

3. $f(x) = \dfrac{3}{2}x^4 - 2x^3 - 6x^2 + 8$, $x_0 = 1$ and $x_0 = 3$

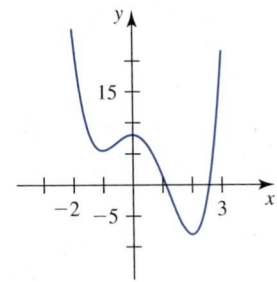

4. $f(x) = x - \sqrt{1 - x^2}$, $x_0 = 0.5$

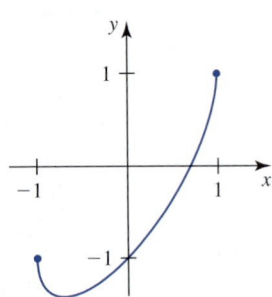

In Exercises 5–8, use Newton's method to find the point of intersection of the graphs to four decimal places of accuracy by solving the equation $f(x) - g(x) = 0$. Use the initial estimate x_0 for the x-coordinate.

5. $f(x) = x^2$, $g(x) = \sin x$, $x_0 = 1$

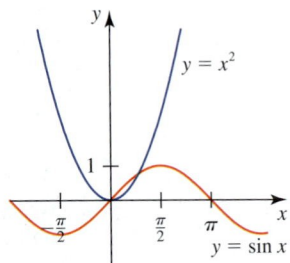

6. $f(x) = \tan x$, $g(x) = 1 - x$, $x_0 = 1$

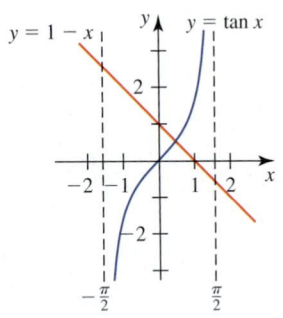

7. $f(x) = \dfrac{1}{2} \cos x,\ g(x) = x,\quad x_0 = 0.5$

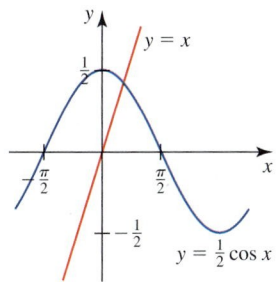

8. $f(x) = \sin x,\ g(x) = \dfrac{1}{5} x,\quad x_0 = 2$

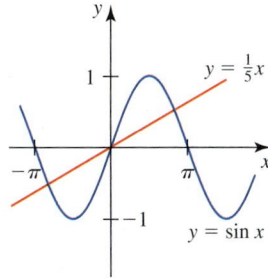

Hint: Use symmetry.

In Exercises 9–14, use Newton's method to approximate the indicated zero of the function. Continue with the iteration until two successive approximations differ by less than 0.0001.

9. The zero of $f(x) = x^3 + x - 4$ between $x = 0$ and $x = 2$. Take $x_0 = 1$.

10. The zero of $f(x) = x^3 + 2x^2 + x - 6$ between $x = 1$ and $x = 2$. Take $x_0 = 1.5$.

11. The zero of $f(x) = x^5 + x - 1$ between $x = 0$ and $x = 1$. Take $x_0 = 0.5$.

12. The zero of $f(x) = x^5 + 2x^4 + 2x - 4$ between $x = 0$ and $x = 1$. Take $x_0 = 0.5$.

13. The zero of $f(x) = 5x + \cos x - 5$ between $x = 0$ and $x = 1$. Take $x_0 = 0.5$.

14. The zero of $f(x) = x - \sin x - 0.5$ between $x = 0$ and $x = 2$. Take $x_0 = 1$.

 In Exercises 15–18, approximate the zero of the function in the indicated interval to six decimal places.

15. $f(x) = x^3 + 3x^2 - 3$ in $[-2, 0]$

16. $f(x) = x^3 - x - 1$ in $[1, 2]$

17. $f(x) = \cos x - x$ in $\left[0, \dfrac{\pi}{2}\right]$

18. $f(x) = 2x - \sin x - 2$ in $[0, \pi]$

In Exercises 19 and 20, use Newton's method to find the roots of the equation correct to five decimal places.

19. $x \ln x - 1 = 0$

20. $\ln x + x - 3 = 0$

21. What can you say about the sequence of approximations obtained using Newton's method if your initial estimate, through a stroke of luck, happens to be the root you are seeking?

22. Let $f(x) = 2x^3 - 9x^2 + 12x - 2$. Use the Intermediate Value Theorem to prove that f has a zero between $x = 0$ and $x = 1$, and then use Newton's method to find it.

23. Let $f(x) = x^3 - 3x - 1$. Use the Intermediate Value Theorem to prove that f has a zero between $x = 1$ and $x = 2$, and then use Newton's method to find it.

In Exercises 24–27, estimate the value of the radical accurate to four decimal places by using three iterations of Newton's method to solve the equation $f(x) = 0$ with initial estimate x_0.

24. $\sqrt{3}$; $f(x) = x^2 - 3$; $x_0 = 1.5$

25. $\sqrt{6}$; $f(x) = x^2 - 6$; $x_0 = 2.5$

26. $\sqrt[3]{7}$; $f(x) = x^3 - 7$; $x_0 = 2$

27. $\sqrt[4]{20}$; $f(x) = x^4 - 20$; $x_0 = 2.1$

28. Consider the equation $xe^x = 2$.
   a. Show that this equation has one positive root in the interval $(0, 1)$.
   b. Use Newton's method to compute the root accurate to five decimal places.

29. Use Newton's method to solve the equation
$$320(t + 10e^{-0.1t}) - 13{,}200 = 0$$
accurate to five decimal places.

30. Use Newton's method to obtain an approximation of the root of $\cos^{-1} x - x = 0$ accurate to three decimal places.

31. Use Newton's method to find the point of intersection of the graphs of $y = \tan^{-1} x$ and $y = \cos^{-1} x$ accurate to three decimal places.

32. Approximating the _k_th Root of a Positive Number
   a. Apply Newton's method to the solution of the equation $f(x) = x^k - A = 0$ to show that an approximation of $\sqrt[k]{A}$ can be found by using the iteration
$$x_{n+1} = \frac{1}{k}\left[(k - 1)x_n + \frac{A}{x_n^{k-1}}\right]$$
   b. Use this iteration to find $\sqrt[10]{50}$ accurate to four decimal places.

33 The graph of $f(x) = x - \sqrt{1 - x^2}$ accompanies Exercise 4. Explain why $x_0 = 1$ cannot be used as an initial estimate for solving the equation $f(x) = 0$ using Newton's method. Can you explain this analytically? How about the initial estimate $x_0 = 0$?

34. The temperature at 6 A.M. on a certain December day in Chicago was 15.6°F. As a cold front moved in gradually, the temperature in the next t hours was given by

$$T(t) = -0.05t^3 + 0.4t^2 + 3.8t + 15.6$$

degrees Fahrenheit, where $0 \leq t \leq 15$. At what time was the temperature 0°F?

35. Tracking a Submarine A submarine traveling along a path described by the equation $y = x^2 + 1$ (both x and y are measured in miles) is being tracked by a sonobuoy (sound detector) located at the point $(3, 0)$ in the figure.

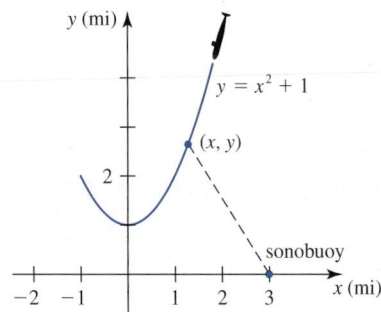

a. Show that the submarine is closest to the sonobuoy if x satisfies the equation $2x^3 + 3x - 3 = 0$.
Hint: Minimize the square of the distance between the points $(3, 0)$ and (x, y).

b. Use Newton's method to solve the equation $2x^3 + 3x - 3 = 0$.

c. What is the distance between the submarine and the sonobuoy at the closest point of approach?

 d. Plot the graph of the function $g(x)$ giving the distance between the submarine and the sonobuoy using the viewing window $[0, 1] \times [2, 4]$. Then use it to verify the result that you obtained in parts (b) and (c).

36. Finding the Position of a Planet As shown in the accompanying figure, the position of a planet that revolves about the sun with an elliptical orbit can be located by calculating the central angle θ. Suppose the central angle sustained by a planet on a certain day satisfies the equation $\theta - 0.5 \sin \theta = 1$. Using Newton's method, find an approximation of θ to five decimal places.
Hint: A rough sketch of the graphs of $y = x - 1$ and $y = \frac{1}{2} \sin x$ will show that an initial estimate of θ may be taken to be $\theta_0 = 1.5$ (radians).

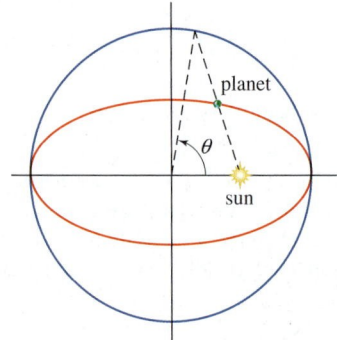

37. Loan Amortization The size of the monthly repayment k that amortizes a loan of A dollars in N years at an interest rate of r per year, compounded monthly, on the unpaid balance is given by

$$k = \frac{Ar}{12\left[1 - \left(1 + \dfrac{r}{12}\right)^{-12N}\right]}$$

Show that r can be found by performing the iteration

$$r_{n+1} = r_n - \frac{Ar_n + 12k\left[\left(1 + \dfrac{r_n}{12}\right)^{-12N} - 1\right]}{A - 12Nk\left(1 + \dfrac{r_n}{12}\right)^{-12N-1}}$$

Hint: Apply Newton's method to solve the equation

$$Ar + 12k\left[\left(1 + \frac{r}{12}\right)^{-12N} - 1\right] = 0$$

38. Financing a Home Refer to Exercise 37. The McCoys secured a loan of $360,000 from a bank to finance the purchase of a house. They have agreed to repay the loan in equal monthly installments of $2106 over 25 years. The bank charges interest at the rate of r per year, compounded monthly. Find r.

39. Financing a Home Refer to Exercise 37. The Wheatons borrowed a sum of $200,000 from a bank to help finance the purchase of a house. The bank charges interest at the rate of r per year on the unpaid balance, compounded monthly. The Wheatons have agreed to repay the loan in equal monthly installments of $1287.40 over 30 years. What is the rate of interest r charged by the bank?

40. a. Show that Newton's method fails when it is used to find the zero of $f(x) = x^{1/3}$ with any initial guess $x_0 \neq 0$.

 b. Illustrate the result graphically in the special case in which $x_0 = 1$ by plotting the graph of f at the points $(x_0, f(x_0))$ and $(x_1, f(x_1))$. Use the viewing window $[-5, 5] \times [-2, 2]$.

41. For a concrete interpretation of a situation similar to that depicted in Figure 4a, consider the function

$$f(x) = x^3 - 1.5x^2 - 6x + 2$$

a. Show that Newton's method fails to work if we choose $x_0 = -1$ or $x_0 = 2$ for an initial estimate.

b. Using the initial estimates $x_0 = -2.5$, $x_0 = 1$, and $x_0 = 2.5$, show that the three roots of $f(x) = 0$ are $-2, 0.313859$, and 3.186141, respectively.

 c. Plot the graph of f using the viewing window $[-3, 4] \times [-10, 7]$.

CHAPTER 3 REVIEW

CONCEPT REVIEW

In Exercises 1–9, fill in the blanks.

1. a. A function f has an absolute maximum at c if _____ for all x in the domain D of f. The number $f(c)$ is called the _____ _____ _____ of f on D.

b. A function f has a relative minimum at c if _____ for all values of x in some _____ _____ containing c.

2. a. A critical number of a function f is any number in the _____ of f at which $f'(c)$ _____ or $f'(c)$ _____ _____ _____.

b. If f has a relative extremum at c, then c must be a _____ _____ of f.

c. If c is a critical number of f, then f may or may not have a _____ _____ at c.

3. The Extreme Value Theorem states that if f is _____ on a closed interval $[a, b]$, then f attains an _____ _____ _____ $f(c)$ for some number c in $[a, b]$ and an _____ _____ _____ $f(d)$ for some number d in $[a, b]$.

4. Suppose that f is continuous on $[a, b]$ and differentiable on (a, b).

a. If $f(a) = f(b)$, then Rolle's Theorem states that there exists at least one number c in _____ such that _____.

b. The Mean Value Theorem states that there exists at least one number c in (a, b) such that $f'(c) =$ _____.

5. a. A function f is increasing on an interval I if for every pair of numbers x_1 and x_2 in I, $x_1 < x_2$ implies that _____.

b. A function f is monotonic if it is either _____ or _____ on I.

c. If f is differentiable on an open interval (a, b) and if $f'(x)$ _____ for all x in (a, b), then f is decreasing on (a, b).

6. a. The graph of a differentiable function f is concave upward on an interval I if _____ is increasing on I.

b. If f has a second derivative on an open interval I and $f''(x)$ _____ on I, then the graph of f is concave upward on I.

c. Suppose that the graph of f has a tangent line at $P(c, f(c))$ and the graph of f changes concavity from _____ to _____ or vice versa at P; then P is called an inflection point of the graph of f.

d. Suppose that f has a continuous second derivative on an interval (a, b), containing a critical number c of f. If $f''(c) < 0$, then f has a _____ _____ at c. If $f''(c) = 0$, then f may or may not have a _____ _____ at c.

7. a. The statement $\lim_{x \to a} f(x) = \infty$ means that all the _____ of f can be made _____ _____ by taking x sufficiently close but not equal to _____.

b. The statement $\lim_{x \to \infty} f(x) = L$ means that all the _____ of f can be made arbitrarily close _____ _____ by taking x _____ _____.

c. The statement $\lim_{x \to -\infty} f(x) = \infty$ means that all the values of f can be made _____ _____ as x _____ without bound.

8. a. The line $x = a$ is a vertical asymptote of the graph of f if at least one of the following is true: _____, _____, or _____.

b. The line $y = L$ is a horizontal asymptote of the graph of f if _____ or _____.

9. a. The precise definition of $\lim_{x \to a} f(x) = \infty$ is: Given any number _____, we can find a number _____ such that $0 < |x - a| < \delta$ implies that _____.

b. The precise definition of $\lim_{x \to -\infty} f(x) = \infty$ means for every number _____, there exists a number _____ such that _____ implies that _____.

REVIEW EXERCISES

In Exercises 1–14, find the absolute maximum value and the absolute minimum value, if any, of the function.

1. $f(x) = -x^2 + 4x - 3$ on $[-1, 3]$

2. $g(x) = \frac{1}{3}x^3 - x^2 + 1$ on $[0, 2]$

3. $h(x) = x^3 - 6x^2$ on $[2, 5]$

4. $f(t) = \dfrac{t}{t^2 + 1}$ on $[0, 5]$

5. $f(x) = 4x - \dfrac{1}{x^2}$ on $[1, 3]$

6. $g(x) = x\sqrt{1 - x^2}$ on $[-1, 1]$

7. $f(x) = -2x^3 + 9x^2 - 12x + 6$ on $(0, 3)$

8. $g(x) = x^3 - 2x^2 - 4x + 4$ on $(-1, 3)$

9. $f(x) = \cos x - \sin x$ on $[0, 2\pi]$

10. $f(x) = \sin 2x - 2 \sin x$ on $[-\pi, \pi]$

11. $f(x) = \dfrac{x}{2} - \sin x$ on $\left(0, \frac{\pi}{2}\right)$

12. $f(x) = x \tan x$ on $\left(-\frac{\pi}{2}, \frac{\pi}{2}\right)$

13. $f(x) = \dfrac{\ln x}{x}$ on $[1, 5]$

14. $f(x) = \tan^{-1} x - \frac{1}{2} \ln x$ on $\left[\frac{1}{\sqrt{3}}, \sqrt{3}\right]$

In Exercises 15–20, verify that the function satisfies the hypotheses of the Mean Value Theorem on the given interval, and find all values of c that satisfy the conclusion of the theorem.

15. $f(x) = x^3$; $[-2, 1]$

16. $g(x) = \sqrt{x}$; $[0, 4]$

17. $h(x) = x + \dfrac{1}{x}$; $[1, 3]$

18. $f(x) = \dfrac{1}{\sqrt{x + 1}}$; $[0, 3]$

19. $f(x) = x - \sin x$; $\left[0, \frac{\pi}{2}\right]$

20. $g(x) = \cos 2x$; $\left[0, \frac{\pi}{2}\right]$

21. Let $f(x) = \dfrac{x}{x + 1}$. Show that there is no value of c in $[-2, 0]$ such that

$$f'(c) = \frac{f(0) - f(-2)}{0 - (-2)}$$

Why doesn't this contradict the Mean Value Theorem?

22. Let $f(x) = |x - 1|$. Show that there is no value c in $[0, 2]$ such that

$$f'(c) = \frac{f(2) - f(0)}{2 - 0}$$

Why doesn't this contradict the Mean Value Theorem?

In Exercises 23–32, (a) find the intervals where the function f is increasing and where it is decreasing, (b) find the relative extrema of f, (c) find the intervals where the graph of f is concave upward and where it is concave downward, and (d) find the inflection points, if any, of f.

23. $f(x) = \dfrac{1}{3}x^3 - x^2 + x - 6$

24. $f(x) = (x - 2)^3$

25. $f(x) = x^4 - 2x^2$

26. $f(x) = x + \dfrac{4}{x}$

27. $f(x) = \dfrac{x^2}{x - 1}$

28. $f(x) = \sqrt{x - 1}$

29. $f(x) = (1 - x)^{1/3}$

30. $f(x) = x\sqrt{x - 1}$

31. $f(x) = \dfrac{2x}{x + 1}$

32. $f(x) = -\dfrac{1}{1 + x^2}$

33. Prove that the equation $x^5 + 5x - 2 = 0$ has only one real root.

34. Prove that the equation $x^5 + 3x^3 + x - 2 = 0$ has exactly one real root.

35. Find the intervals where $f(x) = 2x^2 - \ln x$ is increasing and where it is decreasing.

36. Find the intervals where $f(x) = x^2 e^{-x}$ is increasing and where it is decreasing.

In Exercises 37–46, find the limit.

37. $\displaystyle\lim_{x \to 2^-} \frac{x^2}{x - 2}$

38. $\displaystyle\lim_{x \to -2^+} \frac{1 - 2x}{x^2 - 4}$

39. $\displaystyle\lim_{x \to 3} \frac{2 - 3x}{(x - 3)^2}$

40. $\displaystyle\lim_{x \to 2^-} \frac{x^2 - x + 1}{x^2 - x - 2}$

41. $\displaystyle\lim_{x \to 0^+} \frac{\sqrt{x}}{\sin x}$

42. $\displaystyle\lim_{x \to (\pi/2)^-} \frac{2x + \sin x}{(\tan x)^2}$

43. $\displaystyle\lim_{x \to \infty} \frac{1 - 2x + 3x^2}{x^2 - 1}$

44. $\displaystyle\lim_{x \to \infty} \frac{x + 4}{\sqrt{x^2 - 1}}$

45. $\displaystyle\lim_{x \to -\infty} \frac{\sqrt{2 - x}}{x + 2}$

46. $\displaystyle\lim_{x \to -\infty} \frac{x}{\sin x - 2x}$

In Exercises 47 and 48, find the horizontal and vertical asymptotes of the graph of each function. Do not sketch the graph.

47. $f(x) = \dfrac{1}{2x + 3}$

48. $f(x) = \dfrac{x^2 + x}{x(x - 1)}$

In Exercises 49–64, use the guidelines on page 348 to sketch the graph of the function.

49. $f(x) = x^2 - 4x + 3$

50. $f(x) = 2x^3 - 6x^2 + 6x + 3$

51. $g(x) = 3x^4 - 4x^3$

52. $h(x) = 2x + \dfrac{3}{x}$

53. $f(x) = x^2 + \dfrac{1}{x}$

54. $f(x) = \dfrac{2}{1 + x^2}$

55. $f(x) = \dfrac{x^2}{x^2 - 1}$

56. $f(x) = \dfrac{x^2}{x^4 + 1}$

57. $f(x) = (1 - x)^{1/3}$

58. $g(x) = x + \cos x$, $0 \le x \le 2\pi$

59. $h(x) = 2 \sin x - \sin 2x$, $0 \le x \le 2\pi$

60. $f(x) = \dfrac{1}{1 - \sin x}$, $-\frac{3\pi}{2} < x < \frac{5\pi}{2}$

61. $f(x) = x \ln x$

62. $f(x) = 2 - e^{-x}$

63. $f(x) = \dfrac{3}{1 + e^{-x}}$

64. $f(x) = \sin^{-1}\left(\dfrac{1}{x}\right)$

In Exercises 65–74, evaluate the limit.

65. $\lim\limits_{x \to 1} \dfrac{x^3 - 2x^2 + x}{x^5 - 1}$

66. $\lim\limits_{x \to 2} \dfrac{\sqrt{x} - \sqrt{2}}{x - 2}$

67. $\lim\limits_{x \to 0} \dfrac{\sin 2x}{\sin 3x}$

68. $\lim\limits_{x \to \infty} \dfrac{e^{2x}}{e^x + x^2}$

69. $\lim\limits_{x \to \infty} e^{-x} \cos x$

70. $\lim\limits_{x \to (\pi/2)^-} (\cos x)^{\tan x}$

71. $\lim\limits_{x \to 0} \left(\csc x - \dfrac{1}{x} \right)$

72. $\lim\limits_{x \to (\pi/2)} (\sin x)^{\tan x}$

73. $\lim\limits_{x \to 0^+} \left(\dfrac{1}{x} - \dfrac{1}{e^x - 1} \right)$

74. $\lim\limits_{x \to 0^+} x^n \ln x$

75. Find the accumulated amount after 4 years if $5000 is invested at 8% per year compounded continuously.

76. Effect of Inflation on Salaries Omar's current annual salary is $70,000. How much will he need to earn 10 years from now in order to retain his present purchasing power if the rate of inflation over that period is 6% per year? Assume that inflation is continuously compounded.

77. Spread of an Epidemic The incidence (number of new cases per day) of a contagious disease that is spreading in a population of M people is given by

$$R(x) = kx(M - x)$$

where k is a positive constant and x denotes the number of people already infected. Show that the incidence R is greatest when half the population is infected.

78. Maximizing Profit The management of the company that makes Long Horn Barbecue Sauce estimates that the daily profit from the production and sale of x cases (each case contains 24 bottles of the sauce) is

$$P(x) = -0.000002x^3 + 6x - 350$$

dollars. Determine the largest possible daily profit the company can realize.

79. Senior Workforce The percent of women age 65 years and older in the workforce from the beginning of 1970 to the beginning of 2000 is approximated by the function

$$P(t) = -0.0002t^3 + 0.018t^2 - 0.36t + 10 \qquad 0 \le t \le 30$$

where t is measured in years, with $t = 0$ corresponding to the beginning of 1970.
   a. Find the interval where P is decreasing and the interval where P is increasing.
   b. Find the absolute minimum of P.
   c. Interpret the results of parts (a) and (b).
   Source: U.S. Census Bureau.

80. Find the point on the graph of $y = x^2 + 1$ that is closest to the point $(3, 1)$.

81. Find the dimensions of the rectangle of maximum area, with sides parallel to the coordinate axes, that can be inscribed in the ellipse with equation

$$\frac{x^2}{100} + \frac{y^2}{16} = 1$$

82. The sum of two numbers is 8. Find the numbers if the sum of their cubes is as small as possible.

83. Find the point on the parabola $y = (x - 3)^2$ that is closest to the origin.

84. Find the point on the parabola $y = x^2$ that is closest to the point $(5, -1)$.

85. Demand for Electricity The demand for electricity from 1 A.M. through 7 P.M. on August 1, 2006, in Boston is described by the function

$$D(t) = -11.3975t^3 + 285.991t^2 - 1467.73t + 23{,}755$$
$$0 \le t \le 18$$

where $D(t)$ is measured in megawatts (MW), with $t = 0$ corresponding to 1 A.M. Driven overwhelmingly by air-conditioning and refrigeration systems, the demand for electricity reached a new record high that day.

   a. Plot the graph of D in the viewing window $[0, 18] \times [20{,}000, 30{,}000]$.
   b. Show that the demand for electricity did not exceed the system capacity of 31,000 MW, thus negating the necessity for imposing rolling blackouts if electricity demand were to exceed supply.
   Source: ISO New England.

86. Sickouts In a sickout by pilots of American Airlines in February 1999, the number of canceled flights from February 6 ($t = 0$) through February 14 ($t = 8$) is approximated by the function

$$N(t) = 1.2576t^4 - 26.357t^3 + 127.98t^2 + 82.3t + 43$$
$$0 \le t \le 8$$

where t is measured in days. The sickout ended after the union was threatened with millions of dollars in fines.

   a. Plot the graph of N in the viewing window $[0, 8] \times [0, 1200]$.
   b. Show that the number of canceled flights was increasing at the fastest rate on February 8.
   c. Estimate the maximum number of canceled flights in a day during the sickout.
   Source: Associated Press.

87. Air Inhaled During Respiration Suppose that the volume of air inhaled by a person during respiration is given by

$$V(t) = \frac{6}{5\pi} \left(1 - \cos \frac{\pi t}{2} \right)$$

liters at time t (in seconds). At what time is the rate of flow of air at a maximum? At a minimum?

88. Path of an Acrobatic Plane In a fly-by, the path of an acrobatic plane may be described by the equation

$$y = 200(e^{0.01x} + e^{-0.02x})$$

where x and y are both measured in feet.

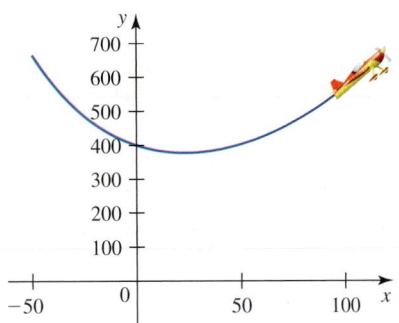

How close to the ground does the plane get?

89. Absorption of Drugs Jane took 100 mg of a drug in the morning and another 100 mg of the same drug at the same time the following morning. The amount of the drug in her body t days after the first dosage was taken is given by

$$A(t) = \begin{cases} 100e^{-1.4t} & \text{if } 0 \le t < 1 \\ 100(1 + e^{1.4})e^{-1.4t} & \text{if } t \ge 1 \end{cases}$$

a. How fast was the amount of drug in Jane's body changing after 12 hr $\left(t = \frac{1}{2}\right)$? After 2 days?

b. When was the amount of drug in Jane's body a maximum?

c. What was the maximum amount of drug in Jane's body?

90. A box with an open top is to be constructed from a square piece of cardboard, 8 in. wide, by cutting out a square from each of the four corners and bending up the sides. What is the largest volume of such a box?

91. Packaging A closed rectangular box with a volume of 4 ft³ is to be constructed. The length of the base of the box will be twice as long as its width. The material for the top and bottom of the box costs 30 cents per square foot, and the material for the sides of the box costs 20 cents per square foot. Find the dimensions of the least expensive box that can be constructed.

92. Minimizing Construction Costs A man wishes to construct a cylindrical barrel with a capacity of 32π ft³. The cost per square foot of the material for the side of the barrel is half that of the cost per square foot for the top and bottom. Help him find the dimensions of the barrel that can be constructed at a minimum cost in terms of material used.

93. Maximizing Light Intensity A light is suspended over the center of a 2 ft by 2 ft table. The intensity of the light striking a point on the table is directly proportional to the sine of the angle the path of the light makes with the table and inversely proportional to the square of the distance between

the point and the light. How high above the table should the light be positioned to maximize the light intensity at the corners of the table?

94. Minimizing Length Two towers, one 120 ft high, the other 300 ft high, and standing 500 ft apart, are to be stayed by two wires (among others) running from the top of the towers to the ground between them. Where should the stake on the ground be placed if we want to minimize the length of wire used?

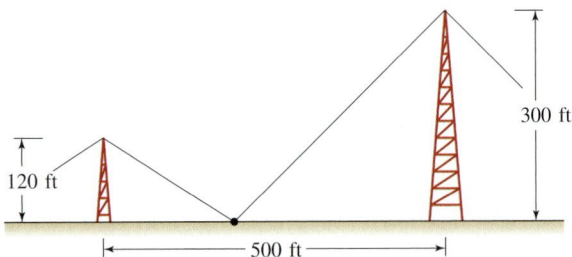

In Exercises 95–97, use three iterations of Newton's method to approximate the indicated zero of the given function accurate to 4 decimal places.

95. The zero of $f(x) = x^3 + 2x + 2$ between $x = -1$ and $x = 0$. Take $x_0 = -0.5$.

96. The zero of $f(x) = x^4 + x - 4$ between $x = 0$ and $x = 2$. Take $x_0 = 1$.

97. The zero of $f(x) = x^2 - \sin x$ between $x = \pi/6$ and $x = \pi/2$. Take $x_0 = \pi/4$.

98. On what interval is the quadratic function

$$f(x) = ax^2 + bx + c \qquad a \ne 0$$

increasing? On what interval is f decreasing?

99. Let $f(x) = x^2 + ax + b$. Determine the constants a and b such that f has a relative minimum at $x = 2$ and the relative minimum value is 7.

100. Find the values of c such that the graph of

$$f(x) = x^4 + 2x^3 + cx^2 + 2x + 2$$

is concave upward everywhere.

101. Let $f(x) = ax^6 + bx^4 + cx^2 + d$, where a, b, c, and d are positive constants. Can the graph of f have any inflection points? Explain.

102. Let

$$f(x) = \begin{cases} x^3 + 1 & \text{if } x \ne 0 \\ 2 & \text{if } x = 0 \end{cases}$$

a. Compute $f'(x)$, and show that it does not change sign as we move across $x = 0$.

b. Show that f has a relative maximum at $x = 0$. Does this contradict the first derivative test? Explain your answer.

PROBLEM-SOLVING TECHNIQUES

As a first step in analyzing functions involving absolute values, we often rewrite the function as an equivalent piecewise defined function. This is illustrated in the following example.

EXAMPLE   Find the absolute minimum value of the function $f(x) = |x|^3 + |x - 2|^3$.

Solution   We write

$$f(x) = |x|^3 + |x - 2|^3 = \begin{cases} -x^3 - (x - 2)^3 & \text{if } x < 0 \\ x^3 - (x - 2)^3 & \text{if } 0 \le x < 2 \\ x^3 + (x - 2)^3 & \text{if } x \ge 2 \end{cases}$$

Differentiating, we obtain

$$f'(x) = \begin{cases} -3x^2 - 3(x - 2)^2 & \text{if } x < 0 \\ 3x^2 - 3(x - 2)^2 & \text{if } 0 \le x < 2 \\ 3x^2 + 3(x - 2)^2 & \text{if } x \ge 2 \end{cases}$$

Since $\lim_{x \to 0^-} f'(x) = -12 = \lim_{x \to 0^+} f'(x)$, we see that f' exists at 0. Similarly, you can show that f' exists at 2. We see that $f'(x) < 0$ for $x < 0$ and $f'(x) > 0$ for $x > 2$. Next, observe that for $0 < x < 2, f'(x) = 0$ if

$$3x^2 - 3(x - 2)^2 = 3x^2 - 3x^2 + 12x - 12 = 12(x - 1) = 0$$

so $x = 1$ is a critical number of f. The sign diagram of f' follows.

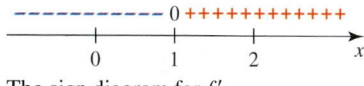

The sign diagram for f'

From the sign diagram for f' we see that f has a relative minimum at $x = 1$, where f takes on the value $f(1) = 2$. Therefore, the absolute minimum value of f is 2.   ■

CHALLENGE PROBLEMS

1. Find the absolute minimum value of $f(x) = |x|^3 + |x - 1|^3$.

2. Find the relative extrema of

$$f(x) = \frac{(x + a)(x + b)}{(x - a)(x - b)} \qquad a + b \neq 0$$

3. Find the highest and the lowest points on the graph of the equation $x^2 + xy + y^2 = 3$.

4. Use the Mean Value Theorem to prove the inequalities

$$nx^{n-1}(y - x) \le y^n - x^n \le ny^{n-1}(y - x)$$

for $0 \le x \le y$, where n is a natural number.

5. a. Show that if $0 \le p \le 1$ and $x \ge 0$, then
$(1 + x)^p \le 1 + x^p$.

b. Use the result of part (a) to show that $(a + b)^p < a^p + b^p$ for all positive numbers a and b and $0 \le p \le 1$.

c. Use the result of part (b) to show that
$\sqrt[n]{a + b} \le \sqrt[n]{a} + \sqrt[n]{b}$ if $n \ge 1$.

6. Find the maximum of $x + y$ defined on the part of the *eight curve* $x^4 = x^2 - y^2$ that lies in the first quadrant.

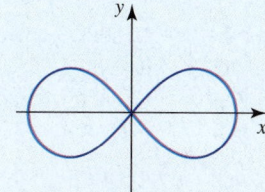

7. Show that the equation $x^5 - 5x + c = 0$ does not have two roots between 0 and 1 for any c.

8. The equation

$$v^3 - \frac{Tv^2}{2\alpha} + 1 = 0$$

is called *Rivlin's equation* and arises in the study of incompressible material. Specifically, v gives the factor by which a rubber cube is stretched in two directions and the factor by which it is contracted in the other when the material is pulled on all faces with a force T. The positive constant α is analogous to the spring constant for a spring. Show that Rivlin's equation has no positive root if $T < 3\sqrt[3]{2}\alpha$ and two positive roots if $T > 3\sqrt[3]{2}\alpha$.

9. Prove that between any two zeros of the polynomial function

$$f(x) = a_n x^n + a_{n-1} x^{n-1} + \cdots + a_1 x + a_0$$

there exists a zero of the polynomial function

$$g(x) = n a_n x^{n-1} + \cdots + 2 a_2 x + a_1$$

10. A generalization of the Mean Value Theorem is Cauchy's Theorem: If f and g are continuous on $[a, b]$ and differentiable on (a, b), $g'(x) \neq 0$ for all x in (a, b), and $g(a) \neq g(b)$, then there exists a number c in (a, b) such that

$$\frac{f(b) - f(a)}{g(b) - g(a)} = \frac{f'(c)}{g'(c)}$$

   a. Explain what is wrong with the following proof of Cauchy's Theorem: Since all the conditions of the Mean Value Theorem are satisfied by f and g, there exists a number c in (a, b) such that

$$\frac{f(b) - f(a)}{b - a} = f'(c) \quad \text{and} \quad \frac{g(b) - g(a)}{b - a} = g'(c)$$

   Therefore, dividing the first expression by the second gives the desired result.

   b. Prove Cauchy's Theorem by applying Rolle's Theorem to the function

$$h(x) = f(x) - f(a) - \frac{f(b) - f(a)}{g(b) - g(a)} [g(x) - g(a)]$$

11. Prove that if $a_0, a_1, a_2, \ldots, a_n$ are real numbers such that

$$\frac{a_0}{1} + \frac{a_1}{2} + \frac{a_2}{3} + \cdots + \frac{a_n}{n + 1} = 0$$

then the polynomial function $f(x) = a_0 + a_1 x + a_2 x^2 + \cdots + a_n x^n$ has at least one zero in the interval $(0, 1)$.

12. Suppose that F is continuous on $[a, b]$ and that it has a derivative f for all x in (a, b), and suppose that x_i, where $0 \leq i \leq n$, are real numbers satisfying

$$x_0 = a < x_1 < x_2 < \cdots < x_n = b$$

Show that there exist numbers c_1 in $[x_0, x_1]$, c_2 in $[x_1, x_2], \ldots, c_n$ in $[x_{n-1}, x_n]$ such that

$$F(b) - F(a) = f(c_1)(x_1 - x_0) + f(c_2)(x_2 - x_1)$$
$$+ \cdots + f(c_n)(x_n - x_{n-1})$$

13. Let f be defined and have a continuous derivative of order $(n - 1)$, $f^{(n-1)}(x)$, on an interval $[a, b]$ and a derivative of order n, $f^{(n)}(x)$, on the interval (a, b). Furthermore, let

$$f(x_0) = f(x_1) = \cdots = f(x_n)$$

$$x_0 = a < x_1 < x_2 < \cdots < x_n = b$$

Show that there exists at least one number c in (a, b) such that $f^{(n)}(c) = 0$.

14. Prove the inequality $x/(x + 1) \leq \ln(x + 1) \leq x$ for $x \geq 0$.

15. Prove the inequality $\dfrac{x}{1 + x^2} < \tan^{-1} x < x$ for $x > 0$.

16. Let $f(x) = x^x (1 - x)^{1-x}$ for $0 < x < 1$.
   a. Evaluate $\lim_{x \to 0^+} f(x)$ and $\lim_{x \to 1^-} f(x)$.
   b. Find the absolute extrema of f on $(0, 1)$.

17. Evaluate $\lim_{x \to \infty} \left(\sqrt[n]{x^n + a_{n-1} x^{n-1} + \cdots + a_0} - \sqrt[n]{x^n + b_{n-1} x^{n-1} + \cdots + b_0} \right)$.

In Chapter 2, we saw how the derivative of a function enabled us to calculate the velocity of the maglev knowing only its position function. In this chapter, we will see how the knowledge of the velocity of the maglev at time t will enable us to calculate its position at any time t. The tool used here is the antiderivative of a function. As it turns out, the derivative of a function and the antiderivative of the function are intimately related—one of the fundamental results of this chapter.

Liu Jin/Getty Images

4

Integration

IN THIS CHAPTER we begin the study of the other major branch of calculus, known as *integral calculus*. Historically, the development of integral calculus, like the development of differential calculus, was motivated by a geometric problem. In this case the problem is that of finding the area of a region in the plane.

The principal tool in the study of integral calculus is the *definite integral*, which, as in the case of the derivative, is defined by using the notion of a limit. As we shall see in the ensuing chapters, the concept of the integral allows us to solve not only the area problem, but also other geometric problems, such as finding the lengths of curves and the volumes and surface areas of solids. The integral also proves to be an all-important tool in solving problems in physics, chemistry, biology, engineering, economics, and other fields.

Although the two branches of calculus seem at first sight to be unconnected, they are, in fact, intimately related. This relationship is established via the Fundamental Theorem of Calculus, which is the main result of this chapter. This theorem also simplifies the calculations involved in solving many problems.

V This symbol indicates that one of the following video types is available for enhanced student learning at **www.academic.cengage.com/login:**
- Chapter lecture videos
- Solutions to selected exercises

4.1 Indefinite Integrals

■ Antiderivatives

Let's return to the example involving the motion of the maglev (see Figure 1). In Chapter 2 we discussed the following problem: *If we know the position of the maglev at all times t, can we find its velocity at any time t?* As it turns out, if the position of the maglev is described by the position function f, then its velocity at any time t is given by $f'(t)$. Here f', the velocity function of the maglev, is just the derivative of f.

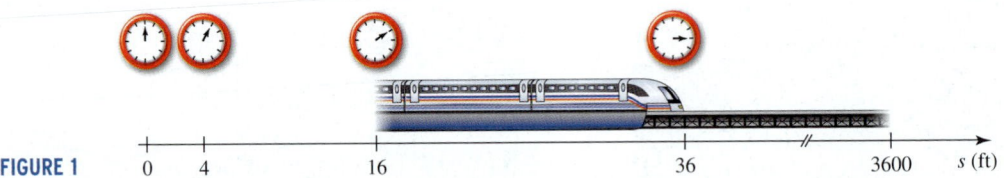

FIGURE 1

Now, in Chapters 4 and 5 we will consider precisely the opposite problem: *If we know the velocity of the maglev at all times t, can we find its position at any time t?* Stated another way, if we know the velocity function f' of the maglev, can we find its position function f? To solve this problem, we need the concept of an *antiderivative* of a function.

DEFINITION Antiderivative

A function F is an **antiderivative** of a function f on an interval I if $F'(x) = f(x)$ for all x in I.

Thus, an antiderivative of a function f is a function F whose derivative is f.

EXAMPLE 1 Show that $F_1(x) = x^3$, $F_2(x) = x^3 + 1$, and $F_3(x) = x^3 - \pi$ are antiderivatives of $f(x) = 3x^2$. How about the function $G(x) = x^3 + C$, where C is any constant?

Solution You can easily verify that $F_1'(x) = F_2'(x) = F_3'(x) = 3x^2 = f(x)$ for all x in $(-\infty, \infty)$. Therefore, by the definition of an antiderivative, F_1, F_2, and F_3 are all antiderivatives of f, as was asserted.

Next, we find

$$G'(x) = \frac{d}{dx}(x^3 + C)$$

$$= 3x^2 + 0 = 3x^2 = f(x)$$

so G is also an antiderivative of f.

Example 1 suggests the following more general result: If F is an antiderivative of f on I, then so is every function of the form $G(x) = F(x) + C$, where C is an arbitrary constant. To prove this, we find

$$G'(x) = \frac{d}{dx}[F(x) + C] = \frac{d}{dx}[F(x)] + \frac{d}{dx}(C) = F'(x) + 0 = F'(x) = f(x)$$

Are there any antiderivatives of f other than those that are obtained in this manner? To answer this question, suppose that H is any other antiderivative of f on I. Then

$$F'(x) = H'(x) = f(x)$$

Since two functions having the same derivative on an interval differ only by a constant, (by the corollary to the Mean Value Theorem, page 298), we have $H(x) - F(x) = C$, where C is a constant. Equivalently, $H(x) = F(x) + C$.

THEOREM 1

If F is an antiderivative of f on an interval I, then *every* antiderivative of f on I has the form

$$G(x) = F(x) + C$$

where C is a constant.

EXAMPLE 2 Let $f(x) = 1$.

a. Show that $F(x) = x$ is an antiderivative of f on $(-\infty, \infty)$.
b. Find all antiderivatives of f on $(-\infty, \infty)$.

Solution
a. $F'(x) = 1 = f(x)$, and this proves that F is an antiderivative of f.
b. By Theorem 1 the antiderivatives of f have the form $G(x) = x + C$, where C is an arbitrary constant.

Figure 2 shows the graphs of some antiderivatives of $f(x) = 1$. These graphs constitute part of a family of infinitely many parallel lines, each having slope 1. This result is expected, because an antiderivative G of f satisfies $G'(x) = f(x) = 1$, and there are infinitely many straight lines that have slope 1. The antiderivatives $G(x) = x + C$, where C is a constant, are precisely the functions representing this family of straight lines.

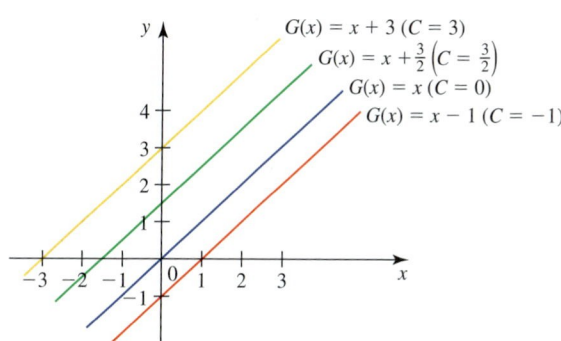

FIGURE 2
The graphs of the antiderivatives
G of $f(x) = 1$ constitute a family
of straight lines, each with slope 1.

■ The Indefinite Integral

The process of finding all antiderivatives of a function is called **antidifferentiation** or **integration.** We can view this process as an operation on a function f to produce the entire family of antiderivatives of f. The integral operator is denoted by the integral sign $\int$, and the process of integration is indicated by the expression

$$\int f(x)\, dx = F(x) + C$$

which is read "the **indefinite integral** of $f(x)$ with respect to x equals $F(x)$ plus C." The function f to be integrated is called the **integrand.** The differential dx reminds us that the integration is performed with respect to the variable x. The function F is an antiderivative of f, and the constant C is called a **constant of integration.** Using this notation, the result of Example 2 is written

$$\int 1\, dx = x + C$$

■ Basic Rules of Integration

Because integration and differentiation are, in a sense, reverse operations, we can discover many of the rules of integration by guessing at an antiderivative F of an integrand f and then verifying that F is an antiderivative of f by demonstrating that $F'(x) = f(x)$. For example, to find the indefinite integral of $f(x) = x^n$, we first recall the Power Rule for differentiating $f(x) = x^n$. Thus,

$$f'(x) = \frac{d}{dx}(x^n) = nx^{n-1}$$

In writing the derivative nx^{n-1}, we followed these steps:

Step 1 Diminish the power of x^n by 1 to obtain x^{n-1}.
Step 2 Multiply x^{n-1} by the "old" power n to obtain nx^{n-1}.

Now, if we reverse the operation in each step, we have

Step 1 Increase the power of x^n by 1 to obtain x^{n+1}.
Step 2 Divide x^{n+1} by the "new" power $n + 1$ to obtain $\dfrac{x^{n+1}}{n+1}$.

This argument suggests that

$$\int x^n\, dx = \frac{x^{n+1}}{n+1} + C \qquad n \neq -1$$

To verify that this formula is correct, we compute

$$\frac{d}{dx}\left[\frac{x^{n+1}}{n+1} + C\right] = \frac{n+1}{n+1}x^{(n+1)-1} = x^n$$

In a similar manner we obtain the following integration formulas by studying the corresponding differentiation formulas.

Basic Integration Formulas

Differentiation Formula	Integration Formula
1. $\dfrac{d}{dx}(C) = 0$	$\displaystyle\int 0\,dx = C$
2. $\dfrac{d}{dx}(x^n) = nx^{n-1}$	$\displaystyle\int x^n\,dx = \dfrac{x^{n+1}}{n+1} + C \qquad n \neq -1$
3. $\dfrac{d}{dx}(\sin x) = \cos x$	$\displaystyle\int \cos x\,dx = \sin x + C$
4. $\dfrac{d}{dx}(\cos x) = -\sin x$	$\displaystyle\int \sin x\,dx = -\cos x + C$
5. $\dfrac{d}{dx}(\tan x) = \sec^2 x$	$\displaystyle\int \sec^2 x\,dx = \tan x + C$
6. $\dfrac{d}{dx}(\sec x) = \sec x \tan x$	$\displaystyle\int \sec x \tan x\,dx = \sec x + C$
7. $\dfrac{d}{dx}(\csc x) = -\csc x \cot x$	$\displaystyle\int \csc x \cot x\,dx = -\csc x + C$
8. $\dfrac{d}{dx}(\cot x) = -\csc^2 x$	$\displaystyle\int \csc^2 x\,dx = -\cot x + C$
9. $\dfrac{d}{dx}(e^x) = e^x$	$\displaystyle\int e^x\,dx = e^x + C$
10. $\dfrac{d}{dx}(\ln|x|) = \dfrac{1}{x}$	$\displaystyle\int \dfrac{dx}{x} = \ln|x| + C$
11. $\dfrac{d}{dx}(a^x) = a^x \ln a$	$\displaystyle\int a^x\,dx = \dfrac{a^x}{\ln a} + C \qquad a > 0, a \neq 1$

Note The formulas for integrals such as $\int \tan x\,dx$ and $\int \sec x\,dx$ are not as easily found. We will learn how to find formulas for such integrals later on. ◼

EXAMPLE 3 Using Formula 2 for integration, we see that

a. $\displaystyle\int 1\,dx = \int x^0\,dx = \dfrac{x^{0+1}}{0+1} + C = x + C$ $\qquad$ Here, $n = 0$.

b. $\displaystyle\int x^2\,dx = \dfrac{x^{2+1}}{2+1} + C = \dfrac{1}{3}x^3 + C$

c. $\displaystyle\int \dfrac{1}{x^3}\,dx = \int x^{-3}\,dx = \dfrac{x^{-3+1}}{-3+1} + C = -\dfrac{1}{2x^2} + C$ $\qquad$ Here, $n = -3$.

d. $\displaystyle\int x^{1/4}\,dx = \dfrac{x^{1/4+1}}{\frac{1}{4}+1} + C = \dfrac{4}{5}x^{5/4} + C$ $\qquad$ Here, $n = \frac{1}{4}$. ◼

Note We can check our answers by differentiating each indefinite integral and showing that the result is equal to the integrand. Thus, to verify the result of Example 3d, we compute

$$\dfrac{d}{dx}\left(\dfrac{4}{5}x^{5/4} + C\right) = \dfrac{4}{5} \cdot \dfrac{5}{4}x^{5/4-1} = x^{1/4}$$ ◼

Rules of Integration

1. $\displaystyle\int c\,f(x)\,dx = c\int f(x)\,dx,$ where c is a constant

2. $\displaystyle\int [f(x) \pm g(x)]\,dx = \int f(x)\,dx \pm \int g(x)\,dx$

Thus, the indefinite integral of a constant multiple of a function is equal to the constant multiple of the indefinite integral, and the indefinite integral of the sum (difference) of two functions is equal to the sum (difference) of their indefinite integrals.

Also, Rule 2 is valid for any finite number of functions; that is,

$$\int [f_1(x) \pm \cdots \pm f_n(x)]\,dx = \int f_1(x)\,dx \pm \int f_2(x)\,dx \pm \cdots \pm \int f_n(x)\,dx$$

EXAMPLE 4 Find

a. $\displaystyle\int 2x^3\,dx$ **b.** $\displaystyle\int (2x + 3\sin x)\,dx$ **c.** $\displaystyle\int \left(3e^x - \frac{2}{x}\right)dx$

Solution

a. $\displaystyle\int 2x^3\,dx = 2\int x^3\,dx$ Rule 1

$$= 2\left(\frac{1}{4}x^4 + C_1\right)$$ Formula 2

$$= \frac{1}{2}x^4 + 2C_1$$

where C_1 is a constant of integration. Since C_1 is arbitrary, so is $2C_1$, and we can write $2C_1 = C$, where C is an arbitrary number. Therefore,

$$\int 2x^3\,dx = \frac{1}{2}x^4 + C$$

b. $\displaystyle\int (2x + 3\sin x)\,dx = \int 2x\,dx + \int 3\sin x\,dx$ Rule 2

$$= 2\int x\,dx + 3\int \sin x\,dx$$ Rule 1

$$= 2\left(\frac{1}{2}x^2 + C_1\right) + 3(-\cos x + C_2)$$ Formulas 2 and 4

$$= x^2 + 2C_1 - 3\cos x + 3C_2$$

$$= x^2 - 3\cos x + C$$

where $C = 2C_1 + 3C_2$ is an arbitrary constant.

c. $\displaystyle\int \left(3e^x - \frac{2}{x}\right)dx = \int 3e^x\,dx - \int \frac{2}{x}\,dx$ Rule 2

$$= 3\int e^x\,dx - 2\int \frac{1}{x}\,dx$$ Rule 1

$$= 3(e^x + C_1) - 2(\ln|x| + C_2)$$
$$= 3e^x + 3C_1 - 2\ln|x| - 2C_2$$
$$= 3e^x - \ln x^2 + C$$

where $C = 3C_1 - 2C_2$.

From now on, we will use a single letter C to represent any combination of constants of integration.

EXAMPLE 5 Find $\int (3x^5 - 2x^3 + 2 - 3x^{-1/3})\, dx$.

Solution Using the generalized sum rule, we have

$$\int (3x^5 - 2x^3 + 2 - 3x^{-1/3})\, dx = 3\int x^5\, dx - 2\int x^3\, dx + 2\int 1\, dx - 3\int x^{-1/3}\, dx$$
$$= 3\left(\frac{x^6}{6}\right) - 2\left(\frac{x^4}{4}\right) + 2x - 3\left(\frac{x^{2/3}}{\frac{2}{3}}\right) + C$$
$$= \frac{1}{2}x^6 - \frac{1}{2}x^4 + 2x - \frac{9}{2}x^{2/3} + C$$

Sometimes we need to rewrite the integrand in a different form before integrating, as is illustrated in the next example.

EXAMPLE 6 Find

a. $\int (x + 1)(x^2 - 2)\, dx$ **b.** $\int \frac{2x^2 - 1}{x^2}\, dx$ **c.** $\int \frac{\sin t}{\cos^2 t}\, dt$

Solution

a. $\int (x + 1)(x^2 - 2)\, dx = \int (x^3 + x^2 - 2x - 2)\, dx = \frac{1}{4}x^4 + \frac{1}{3}x^3 - x^2 - 2x + C$

b. $\int \frac{2x^2 - 1}{x^2}\, dx = \int \left(2 - \frac{1}{x^2}\right) dx = \int (2 - x^{-2})\, dx = 2x + x^{-1} + C = 2x + \frac{1}{x} + C$

c. $\int \frac{\sin t}{\cos^2 t}\, dt = \int \frac{1}{\cos t}\cdot\frac{\sin t}{\cos t}\, dt = \int \sec t \tan t\, dt = \sec t + C$

■ Differential Equations

Let's return to the problem posed at the beginning of the section: Given the derivative of a function, f', can we find the function f? As an example, suppose that we are given the function

$$f'(x) = 2x - 1 \tag{1}$$

and we wish to find $f(x)$. From what we now know, we can find f by integrating Equation (1). Thus,

$$f(x) = \int f'(x)\, dx = \int (2x - 1)\, dx = x^2 - x + C \tag{2}$$

where C is an arbitrary constant. So there are infinitely many functions having the derivative f'; these functions differ from each other by a constant.

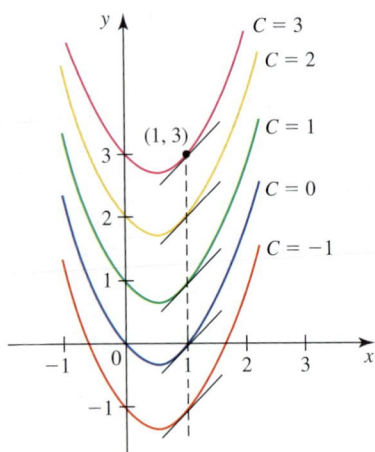

FIGURE 3
The graphs of some functions having the derivative $f'(x) = 2x - 1$

Equation (1) is called a *differential equation.* In general, a **differential equation** is an equation that involves the derivative or differential of an unknown function. (In Equation (1) the unknown function is f.) A **solution** of a differential equation on an interval I is any function that satisfies the differential equation on I. Thus, Equation (2) gives all solutions of the differential equation (1) on $(-\infty, \infty)$ and is, accordingly, called the **general solution** of the differential equation $f'(x) = 2x - 1$.

The graphs of $f(x) = x^2 - x + C$ for selected values of C are shown in Figure 3. These graphs have one property in common: For any fixed value of x, the tangent lines to these graphs have the same slope. This follows because any member of the family $f(x) = x^2 - x + C$ must have the same slope at x, namely, $2x - 1$. (We will study differential equations in greater depth in Chapter 8.)

Although there are infinitely many solutions to the differential equation $f'(x) = 2x - 1$, we can obtain a **particular solution** by specifying the value that the function must assume at a certain value of x. For example, suppose we stipulate that the solution $f(x) = x^2 - x + C$ must satisfy the condition $f(1) = 3$. Then, we find that

$$f(1) = 1 - 1 + C = 3$$

and $C = 3$. Thus, the particular solution is $f(x) = x^2 - x + 3$ (see Figure 3). The condition $f(1) = 3$ is an example of an *initial condition.* More generally, an **initial condition** is a condition that is imposed on the value of f at a number $x = a$. Geometrically, this means that the graph of the particular solution passes through the point $(a, f(a))$.

■ Initial Value Problems

In an **initial value problem** we are required to find a function satisfying (1) a differential equation and (2) one or more initial conditions. The following are examples.

EXAMPLE 7 **Finding the Position of a Maglev** In a test run of a maglev along a straight elevated monorail track, data obtained from reading its speedometer indicated that the velocity of the maglev at time t can be described by the velocity function

$$v(t) = 8t \qquad 0 \le t \le 30$$

Find the position function of the maglev. Assume that the maglev is initially located at the origin of a coordinate line.

Solution Let $s(t)$ denote the position of the maglev at time t, where $0 \le t \le 30$. Then $s'(t) = v(t)$. So we have the initial value problem

$$\begin{cases} s'(t) = 8t \\ s(0) = 0 \end{cases}$$

Integrating both sides of the differential equation $s'(t) = 8t$, we obtain

$$s(t) = \int s'(t)\, dt = \int 8t\, dt = 4t^2 + C$$

where C is an arbitrary constant. To evaluate C, we use the initial condition $s(0) = 0$ to write

$$s(0) = 4(0) + C = 0 \qquad \text{or} \qquad C = 0$$

Therefore, the required position function is $s(t) = 4t^2$, where $0 \le t \le 30$. ■

EXAMPLE 8   **Describing the Path of a Pop-Up**   In a baseball game, one of the batters hit a pop-up. Suppose that the initial velocity of the ball was 96 ft/sec and the initial height of the ball was 4 ft from the ground.

a. Find the position function giving the height of the ball at any time t.
b. How high did the ball go?
c. How long did the ball stay in the air after being struck?

Solution

a. Let $s(t)$ denote the position of the ball at time t, and let $t = 0$ represent the (initial) time when the ball was struck. The only force acting on the ball during the motion is the force of gravity; taking the acceleration due to this force as -32 ft/sec^2, we see that s must satisfy

$$s''(t) = -32$$

When $t = 0$,

$$s(0) = 4 \qquad \text{Initial height was 4 ft.}$$

and

$$s'(0) = 96 \qquad \text{Initial velocity was 96 ft/sec.}$$

To solve this initial value problem, we integrate the differential equation $s''(t) = -32$ with respect to t, obtaining

$$s'(t) = \int s''(t)\, dt = \int -32\, dt = -32t + C_1$$

where C_1 is an arbitrary constant. To determine the value of C_1, we use the initial condition $s'(0) = 96$. We find

$$s'(0) = -32(0) + C_1 = 96$$

which gives $C_1 = 96$. Therefore,

$$s'(t) = -32t + 96$$

Integrating again, we have

$$s(t) = \int s'(t)\, dt = \int (-32t + 96)\, dt = -16t^2 + 96t + C_2$$

where C_2 is an arbitrary constant. To evaluate C_2, we use the initial condition $s(0) = 4$ to obtain

$$s(0) = -16(0) + 96(0) + C_2 = 4 \qquad \text{or} \qquad C_2 = 4$$

Therefore, the required position function is

$$s(t) = -16t^2 + 96t + 4$$

b. At the highest point, the velocity of the ball is zero. But from part (a) the velocity of the ball at any time t is $v(t) = s'(t) = -32t + 96$. So setting $v(t) = 0$, we obtain $-32t + 96 = 0$, or $t = 3$. Substituting this value of t into the position function gives

$$s(3) = -16(3^2) + 96(3) + 4$$

or 148 ft as the maximum height attained by the ball. (See Figure 4.)

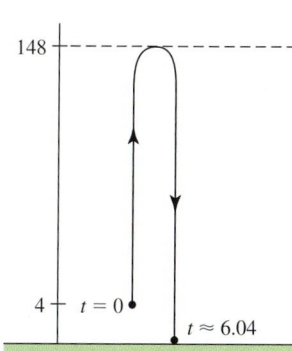

FIGURE 4
The ball attains a maximum height of 148 ft and stays in the air approximately 6 sec.

c. The ball hits the ground when $s(t) = 0$. Solving this equation, we have

$$-16t^2 + 96t + 4 = 0$$

or

$$4t^2 - 24t - 1 = 0$$

Next, using the quadratic formula, we obtain

$$t = \frac{24 \pm \sqrt{576 + 16}}{8} = \frac{24 \pm 4\sqrt{37}}{8} \approx -0.04 \quad \text{or} \quad 6.04$$

Since t must be positive, we see that the ball hit the ground when $t \approx 6.04$. Therefore, after the ball was struck, it remained in the air for approximately 6 sec.

4.1 CONCEPT QUESTIONS

1. What is an antiderivative of a function f? Give an example.
2. If $f'(x) = g'(x)$ for all x in an interval I, what is the relationship between f and g?
3. What is the difference between an antiderivative of f and the indefinite integral of f?

4. Define each of the following:
   a. A differential equation
   b. A solution of a differential equation
   c. A general solution of a differential equation
   d. A particular solution of a differential equation
   e. An initial value problem

4.1 EXERCISES

In Exercises 1–30, find the indefinite integral, and check your answer by differentiation.

1. $\displaystyle\int (x + 2)\, dx$

2. $\displaystyle\int (6x^2 - 2x + 1)\, dx$

3. $\displaystyle\int (3 - 2x + x^2)\, dx$

4. $\displaystyle\int (x^3 - 2x^2 + x + 1)\, dx$

5. $\displaystyle\int (2x^9 + 3e^x + 4)\, dx$

6. $\displaystyle\int (2x^{2/3} - 4x^{1/3} + 4)\, dx$

7. $\displaystyle\int \left(\sqrt{x} + \frac{3}{\sqrt{x}}\right) dx$

8. $\displaystyle\int x^{2/3}(x - 1)\, dx$

9. $\displaystyle\int (e^t + t^e)\, dt$

10. $\displaystyle\int \left(\frac{2}{1 + x^2} + \frac{1}{x}\right) dx$

11. $\displaystyle\int \frac{3}{\sqrt{u}}\, du$

12. $\displaystyle\int \frac{x^2 + 1}{x^2}\, dx$

13. $\displaystyle\int \frac{3x^4 - 2x^2 + 1}{x^4}\, dx$

14. $\displaystyle\int \frac{t^2 - 2\sqrt{t} + 1}{t^2}\, dt$

15. $\displaystyle\int \frac{x^2 - 2x + 3}{\sqrt{x}}\, dx$

16. $\displaystyle\int 10^x\, dx$

17. $\displaystyle\int \frac{x^2 2^x + x}{x^2}\, dx$

18. $\displaystyle\int (2t + 3\cos t)\, dt$

19. $\displaystyle\int (3\sin x - 4\cos x)\, dx$

20. $\displaystyle\int (\csc \theta \cot \theta - 3\sec^2 \theta)\, d\theta$

21. $\displaystyle\int (\csc^2 x + \sqrt{x})\, dx$

22. $\displaystyle\int \sec u\, (\tan u + \sec u)\, du$

23. $\displaystyle\int \frac{\cos x}{1 - \cos^2 x}\, dx$

24. $\displaystyle\int \frac{\sin 2x}{\cos x}\, dx$

25. $\displaystyle\int \frac{1 - 2\cot^2 x}{\cos^2 x}\, dx$

26. $\displaystyle\int \frac{\cos 2x}{\cos x - \sin x}\, dx$

27. $\displaystyle\int \frac{1}{\sin^2 x \cos^2 x}\, dx$

   Hint: Use the identity $\sin^2 x + \cos^2 x = 1$.

28. $\displaystyle\int \tan^2 x\, dx$

   Hint: Rewrite the integrand.

29. $\displaystyle\int \frac{dx}{1 - \sin x}$

30. $\displaystyle\int \cot^2 x\, dx$

In Exercises 31–34, (a) *find the indefinite integral, and* (b) *plot the graphs of the antiderivatives corresponding to* $C = -2, -1, 0, 1,$ *and* 2 *(C is the constant of integration).*

31. $\displaystyle\int (x - 3)\, dx$

32. $\displaystyle\int (3x^2 + x - 1)\, dx$

33. $\displaystyle\int (2x + \sin x)\, dx$

34. $\displaystyle\int (1 + \sec^2 x)\, dx$

In Exercises 35–46, *find f by solving the initial value problem.*

35. $f'(x) = 2x + 1, \quad f(1) = 3$

36. $f'(x) = 3x^2 - 6x, \quad f(2) = 4$

37. $f'(x) = \dfrac{1}{\sqrt{x}}, \quad f(4) = 2$

38. $f'(x) = 1 + \dfrac{1}{x^2}, \quad f(1) = 2$

39. $f'(x) = e^x + \sin x, \quad f(0) = 0$

40. $f'(t) = \sec^2 t + 2\cos t, \quad f\!\left(\dfrac{\pi}{4}\right) = \sqrt{2}$

41. $f''(x) = 6; \quad f(1) = 4, \quad f'(1) = 2$

42. $f''(x) = 2x + 1; \quad f(0) = 5, \quad f'(0) = 1$

43. $f''(x) = 6x^2 + 6x + 2; \quad f(-1) = \dfrac{1}{2}, \quad f'(-1) = 2$

44. $f''(x) = \dfrac{1}{x^3}; \quad f(1) = 1, \quad f'(1) = \dfrac{1}{2}$

45. $f''(t) = t^{-3/2}; \quad f(4) = 1, \quad f'(4) = 3$

46. $f''(t) = 2\sin t + 3\cos t; \quad f\!\left(\dfrac{\pi}{2}\right) = 1, \quad f'\!\left(\dfrac{\pi}{2}\right) = 2$

47. Find the function f given that the slope of the tangent line to the graph of f at any point $(x, f(x))$ is $x^2 - 2x + 3$ and the graph of f passes through the point $(1, 2)$.

48. Find the function f given that it satisfies $f''(x) = 36x^2 + 24x$ and its graph has a horizontal tangent line at the point $(0, 1)$.

In Exercises 49–50, *identify which of the two graphs 1 and 2 is the graph of the function f and the graph of its antiderivative. Give a reason for your choice.*

49.

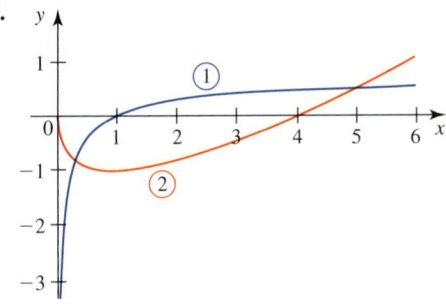

50.

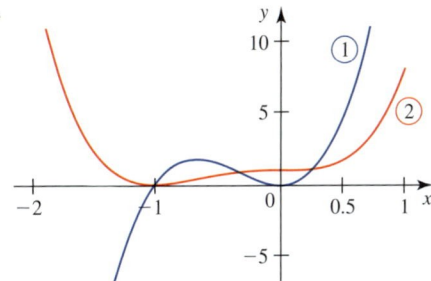

In Exercises 51–56, *find the position function of a particle moving along a coordinate line that satisfies the given condition(s).*

51. $v(t) = 6t^2 - 4t + 1, \quad s(1) = -1$

52. $v(t) = 2\sin t - 3\cos t, \quad s(0) = -2$

53. $a(t) = 6t - 4, \quad s(0) = -2, \quad v(0) = 4$

54. $a(t) = -6t^2 + 4t + 8, \quad s(1) = \dfrac{7}{6}, \quad v(1) = 4$

55. $a(t) = \sin t - 2\cos t, \quad s(0) = 3, \quad v(0) = 0$

56. $a(t) = 6\sin t, \quad s(\pi) = \pi, \quad v(0) = -4$

57. Velocity of a Maglev The velocity of a maglev is $v(t) = 0.2t + 3$ (ft/sec), where $0 \le t \le 120$. At $t = 0$ the maglev is at the station. Find the function that gives the position of the maglev at time t assuming that the motion takes place along a straight stretch of the track.

58. A ball is thrown straight up from a height of 3 ft with an initial velocity of 40 ft/sec. How high will the ball go? (Take $g = 32$ ft/sec².)

59. Ballast Dropped from a Balloon A ballast is dropped from a stationary hot-air balloon that is at an altitude of 400 ft. Find (a) an expression for the altitude of the ballast after t seconds, (b) the time when it strikes the ground, and (c) its velocity when it strikes the ground. (Disregard air resistance and take $g = 32$ ft/sec².)

ballast

60. Ballast Dropped from a Balloon Refer to Exercise 59. Suppose that the hot-air balloon is rising vertically with a velocity of 16 ft/sec at an altitude of 128 ft when the ballast is dropped. How long will it take for the ballast to strike the ground? What will its impact velocity be?

61. A particle located at the point $x = x_0$ on a coordinate line is given an initial velocity of v_0 ft/sec and a constant acceleration of a ft/sec^2. Show that its position at any time t is

$$x = x_0 + v_0 t + \frac{1}{2} at^2$$

62. Refer to Exercise 61. Show that the velocity v of the particle at any time t satisfies

$$v^2 = v_0^2 + 2a(x - x_0)$$

63. Flight of a Model Rocket A model rocket is fired vertically upward from a height of s_0 ft above the ground with a velocity of v_0 ft/sec. If air resistance is negligible, show that its height (in feet) after t seconds is given by

$$s(t) = -16t^2 + v_0 t + s_0$$

(Take $g = 32$ ft/sec^2.)

64. Kaitlyn drops a stone into a well. Approximately 4.22 sec later, she hears the splash made by the impact of the stone in the water. How deep is the well? (The speed of sound is approximately 1128 ft/sec.)

65. Jumping While on Mars The acceleration due to gravity on Mars is approximately 3.72 m/sec^2. If an astronaut jumps straight up on the surface of the planet with an initial velocity of 4 m/sec, what height will she attain? Find the comparable height that she would jump on the earth. (The constant of acceleration due to gravity on the earth is 9.8 m/sec^2.)

66. Acceleration of a Car A car traveling along a straight road at 66 ft/sec accelerated to a speed of 88 ft/sec over a distance of 440 ft. What was the acceleration of the car, assuming that the acceleration was constant?

67. Stopping Distance of a Car To what constant deceleration would a car moving along a straight road be subjected if the car were brought to rest from a speed of 88 ft/sec in 9 sec? What would the stopping distance be?

68. Acceleration of a Car A car traveling along a straight road at a constant speed was subjected to a constant acceleration of 12 ft/sec^2. It reached a speed of 60 mph after traveling 242 ft. What was the speed of the car just prior to the acceleration?

69. Crossing the Finish Line After rounding the final turn in the bell lap, two runners emerged ahead of the pack. When runner A is 200 ft from the finish line, his speed is 22 ft/sec, a speed that he maintains until he crosses the line. At that instant of time, runner B, who is 20 ft behind runner A and running at a speed of 20 ft/sec, begins to spurt. Assuming that runner B sprints with a constant acceleration, what minimum acceleration will enable him to cross the finish line ahead of runner A?

70. Velocity of a Car Two cars, side by side, start from rest and travel along a straight road. The velocity of car A is given by $v = f(t)$, and the velocity of car B is given by $v = g(t)$. The graphs of f and g are shown in the following figure. Are the cars still side by side after T sec? If not, which car is ahead of the other? Justify your answer.

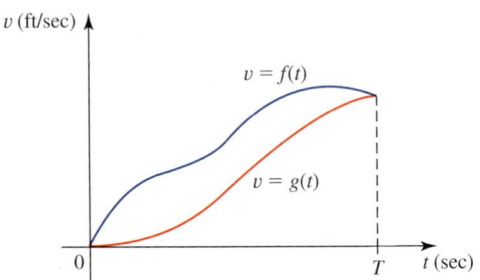

71. Bank Deposits Madison Finance opened two branches on September 1 ($t = 0$). Branch A is located in an established industrial park, and branch B is located in a fast-growing new development. The net rates at which money was deposited into branch A and branch B in the first 180 business days are given by the graphs of f and g, respectively. Which branch has a larger amount on deposit at the end of 180 business days? Justify your answer.

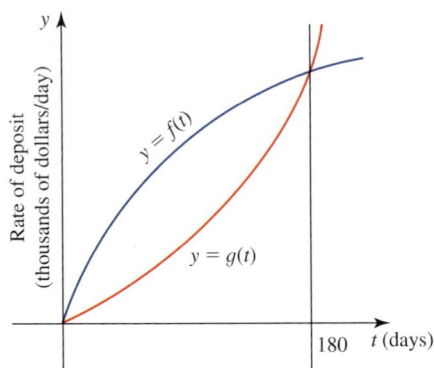

72. Collision of Two Particles Two points A and B are located 100 ft apart on a straight line. A particle moves from A toward B with an initial velocity of 10 ft/sec and an acceleration of $\frac{1}{2}$ ft/sec^2. Simultaneously, a particle moves from B toward A with an initial velocity of 5 ft/sec and an acceleration of $\frac{3}{4}$ ft/sec^2. When will the two particles collide? At what distance from A will the collision take place?

73. Revenue The monthly marginal revenue of Commuter Air Service is $R'(x) = 10{,}000 - 200x$ dollars per passenger, where x stands for the fare per passenger. Find the monthly total revenue function R if $R(0) = 0$.

74. Total Cost Function The weekly marginal cost of the Electra Electronics Company in producing its Zephyr laser jet printers is given by

$$C'(x) = 0.000006x^2 - 0.04x + 1000$$

dollars per printer, where x stands for the number of printers manufactured. Find the weekly total cost function C if the fixed cost of the company is $120,000 per week.

75. Risk of Down Syndrome The rate at which the risk of Down syndrome is changing is approximated by the function

$$r(x) = 0.004641x^2 - 0.3012x + 4.9 \qquad 20 \le x \le 45$$

where $r(x)$ is measured in percent of all births per year and x is the maternal age at delivery.
  a. Find a function f giving the risk as a percentage of all births when the maternal age at delivery is x years, given that the risk of Down syndrome at age 30 is 0.14% of all births.
  b. What is the risk of Down syndrome when the maternal age at delivery is 40 years? 45 years?
  Source: New England Journal of Medicine.

76. Online Ad Sales In a study conducted in 2004, it was found that the share of online advertisement worldwide, as a percentage of the total ad market, was expected to grow at the rate of

$$R(t) = -0.033t^2 + 0.3428t + 0.07 \qquad 0 \le t \le 6$$

percent per year at time t (in years), with $t = 0$ corresponding to the beginning of 2000. The online ad market at the beginning of 2000 was 2.9% of the total ad market.
  a. What is the projected online ad market share at any time t?
  b. What was the projected online ad market share at the beginning of 2006?
  Source: Jupiter Media Metrix, Inc.

77. Ozone Pollution The rate of change of the level of ozone, an invisible gas that is an irritant and impairs breathing, present in the atmosphere on a certain May day in the city of Riverside is given by

$$R(t) = 3.2922t^2 - 0.366t^3 \qquad 0 < t < 11$$

(measured in pollutant standard index per hour). Here, t is measured in hours, with $t = 0$ corresponding to 7 A.M. Find the ozone level $A(t)$ at any time t, assuming that at 7 A.M. it is 34.
  Source: The Los Angeles Times.

78. U.S. Sales of Organic Milk The sales of organic milk from 1999 through 2004 grew at the rate of approximately

$$R(t) = 3t^3 - 17.9445t^2 + 28.7222t + 26.632$$
$$0 \le t \le 5$$

million dollars per year, where t is measured in years with $t = 0$ corresponding to 1999. Sales of organic milk in 1999 totaled $108 million.
  a. Find an expression giving the total sales of organic milk by year t, where $0 \le t \le 5$.
  b. According to this model, what were the total sales of organic milk in 2004?
  Source: Resource, Inc.

79. Water Level of a Tank A tank has a constant cross-sectional area of 50 ft^2 and an orifice of constant cross-sectional area of $\frac{1}{2}$ ft^2 located at the bottom of the tank. If the tank is filled with water to a height of h ft and allowed to drain, then the height of the water decreases at a rate that is described by the equation

$$\frac{dh}{dt} = -\frac{1}{25}\left(\sqrt{20} - \frac{t}{50}\right) \qquad 0 \le t \le 50\sqrt{20}$$

Find an expression for the height of the water at any time t if its height initially is 20 ft.

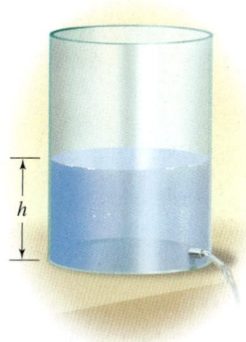

80. The Elastic Curve of a Beam A horizontal, uniform beam of length L, supported at its ends, bends under its own weight, w per unit length. The *elastic curve* of the beam (the shape that it assumes) has equation $y = f(x)$ satisfying

$$EIy'' = \frac{wx^2}{2} - \frac{wLx}{2}$$

where E and I are positive constants that depend on the material and the cross section of the beam.

(a) The distorted beam

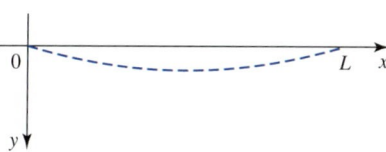

(b) The elastic curve in the xy-plane (The positive direction of the y-axis is downward.)

  a. Find an equation of the elastic curve.
    Hint: $y = 0$ at $x = 0$ and at $x = L$.
  b. Show that the maximum deflection of the beam occurs at $x = L/2$.

In Exercises 81–86, determine whether the statement is true or false. If it is true, explain why it is true. If it is false, explain why or give an example to show why it is false.

81. $\int f'(x)\, dx = f(x) + C$

82. $\int f(x)g(x)\, dx = F(x)G(x) + C,$ where $F' = f$ and $G' = g$

83. $\int x\, f(x)\, dx = x \int f(x)\, dx = x\, F(x) + C,$ where $F' = f$

84. If F and G are antiderivatives of f and g, respectively, then

$$\int [2f(x) - 3g(x)]\, dx = 2F(x) - 3G(x) + C$$

85. If $R(x) = P(x)/Q(x)$ is a rational function, then

$$\int R(x)\, dx = \frac{\displaystyle\int P(x)\, dx}{\displaystyle\int Q(x)\, dx}$$

86. $\int \left[\int f(x)\, dx \right] dx = G(x) + C_1 x + C_2,$

where $G' = F$ and $F' = f$

87. $\int e^{\ln x}\, dx = \frac{1}{2}\, xe^{\ln x} + C$

4.2 Integration by Substitution

In this section we introduce a technique of integration that will enable us to integrate a large class of functions. This method of integration, like the integration formulas of Section 4.1, is obtained by reversing a differentiation rule—in this case the Chain Rule.

■ How the Method of Substitution Works

Consider the indefinite integral

$$\int 2x\sqrt{x^2 + 3}\, dx \tag{1}$$

You can convince yourself that this integral cannot be evaluated as it stands by using any of the integration formulas that we are now familiar with.

Let's try to simplify the indefinite integral (1) by making a change of variable from x to a new variable u as follows. Write

$$u = x^2 + 3$$

with differential

$$du = 2x\, dx$$

If we *formally* substitute these quantities into the indefinite integral (1), we obtain

$$\int 2x\sqrt{x^2 + 3}\, dx = \int \sqrt{x^2 + 3}\, (2x\, dx) = \int \sqrt{u}\, du$$

Rewriting $\qquad\qquad \begin{cases} u = x^2 + 3 \\ du = 2x\, dx \end{cases}$

Now the integral is in a form that is easily integrated by using Formula (2) of Section 4.1. Thus,

$$\int \sqrt{u}\, du = \int u^{1/2}\, du = \frac{2}{3}\, u^{3/2} + C$$

Finally, replacing u by $x^2 + 3$, we see that

$$\int 2x\sqrt{x^2 + 3}\, dx = \frac{2}{3}(x^2 + 3)^{3/2} + C$$

To verify that this solution is correct, we compute

$$\frac{d}{dx}\left[\frac{2}{3}(x^2 + 3)^{3/2} + C\right] = \frac{2}{3} \cdot \frac{3}{2}(x^2 + 3)^{1/2}(2x) \qquad \text{Use the Chain Rule.}$$

$$= 2x\sqrt{x^2 + 3}$$

which is the integrand. This proves the assertion.

■ The Technique of Integration by Substitution

As was shown in the preceding example, it is sometimes possible to transform one indefinite integral into another that is easier to integrate by using a suitable change of variable. Now, by letting $f(x) = \sqrt{x}$ and $g(x) = x^2 + 3$, so that $g'(x) = 2x$, we can see that the indefinite integral (1) is a special case of an indefinite integral of the form

$$\int f(g(x))g'(x)\, dx \tag{2}$$

Let's show that the integral (2) can always be rewritten in the form

$$\int f(u)\, du \tag{3}$$

where u is differentiable on an interval (a, b) and f is continuous on the range of u. Suppose that F is an antiderivative of f. By the Chain Rule we have

$$\frac{d}{dx}[F(g(x))] = F'(g(x))g'(x)$$

or, equivalently,

$$\int F'(g(x))g'(x)\, dx = F(g(x)) + C$$

Writing $F' = f$ and making the substitution $u = g(x)$, we have

$$\int f(g(x))g'(x)\, dx = F(u) + C = \int F'(u)\, du = \int f(u)\, du$$

as was to be shown.

Before looking at an example, let's summarize the steps used in this method.

Integration by Substitution: Evaluating $\int f(g(x))g'(x)\, dx$

Step 1 Let $u = g(x)$, where $g(x)$ is part of the integrand, usually, the "inside function" of the composite function $f(g(x))$.

Step 2 Compute $du = g'(x)\, dx$.

Step 3 Use the substitution $u = g(x)$ and $du = g'(x)\, dx$ to transform the integral into one that involves *only* u: $\int f(u)\, du$.

Step 4 Find the resulting integral.

Step 5 Replace u by $g(x)$ so that the final solution is in terms of x.

EXAMPLE 1 Find $\int x^2(x^3 + 2)^4\, dx$.

Solution

Step 1 If you examine the integrand, you will see that it involves the composite function $(x^3 + 2)^4$, with "inside function" $g(x) = x^3 + 2$. So let us choose $u = x^3 + 2$.

Step 2 We compute $du = 3x^2\, dx$.

Step 3 Making the substitution $u = x^3 + 2$ and $du = 3x^2\, dx$ or $x^2 dx = \frac{1}{3}\, du$, we obtain

$$\int x^2(x^3 + 2)^4\, dx = \int (x^3 + 2)^4\, x^2\, dx = \int u^4 \left(\frac{1}{3}\, du\right) = \frac{1}{3} \int u^4\, du$$

↑
Rewriting

an integral involving only the variable u.

Step 4 We find

$$\frac{1}{3} \int u^4\, du = \frac{1}{3} \left(\frac{u^5}{5}\right) + C = \frac{1}{15} u^5 + C$$

Step 5 Replacing u by $x^3 + 2$, we find

$$\int x^2(x^3 + 2)^4\, dx = \frac{1}{15} (x^3 + 2)^5 + C$$ ■

EXAMPLE 2 Find $\int \dfrac{dx}{(2x - 4)^3}$.

Solution First rewrite the integral in the form

$$\int (2x - 4)^{-3}\, dx$$

Step 1 In the composite function $(2x - 4)^{-3}$, the "inside function" is $g(x) = 2x - 4$. So let $u = 2x - 4$.

Step 2 We find $du = 2\, dx$.

Step 3 Substituting $u = 2x - 4$ and $du = 2\, dx$ or $dx = \frac{1}{2}\, du$ into the integral yields

$$\int (2x - 4)^{-3}\, dx = \int u^{-3} \left(\frac{1}{2}\, du\right) = \frac{1}{2} \int u^{-3}\, du$$

an integral involving only the variable u.

Step 4 We integrate

$$\frac{1}{2} \int u^{-3}\, du = \frac{1}{2} \left(\frac{u^{-2}}{-2}\right) + C = -\frac{1}{4u^2} + C$$

Step 5 Replacing u by $2x - 4$ gives

$$\int \frac{dx}{(2x - 4)^3} = -\frac{1}{4(2x - 4)^2} + C$$ ■

EXAMPLE 3   Find $\int (x + 1)\sqrt{2x - 1}\, dx$.

Solution   First, rewrite the integral in the form

$$\int (x + 1)(2x - 1)^{1/2}\, dx$$

Step 1   Examining the integrand, we spot the composite function $(2x - 1)^{1/2}$, which has the "inside function" $g(x) = 2x - 1$. So let $u = 2x - 1$.

Step 2   We find $du = 2\, dx$.

Step 3   We use the substitution $u = 2x - 1$ and $du = 2\, dx$ or $dx = \frac{1}{2}\, du$. Because of the factor $x + 1$ in the integrand, we need to solve $u = 2x - 1$ for x, obtaining $x = \frac{1}{2}u + \frac{1}{2}$. Therefore,

$$\int (x + 1)(2x - 1)^{1/2}\, dx = \int \left(\frac{1}{2}u + \frac{1}{2} + 1\right)u^{1/2}\left(\frac{1}{2}\, du\right) = \frac{1}{2}\int \left(\frac{1}{2}u^{3/2} + \frac{3}{2}u^{1/2}\right) du$$

$$= \frac{1}{4}\int (u^{3/2} + 3u^{1/2})\, du$$

an integral involving only the variable u.

Step 4   Integrating, we find

$$\frac{1}{4}\int (u^{3/2} + 3u^{1/2})\, du = \frac{1}{4}\left(\frac{2}{5}u^{5/2} + 2u^{3/2}\right) + C$$

$$= \frac{1}{10}u^{5/2} + \frac{1}{2}u^{3/2} + C$$

Step 5   Replacing u by $2x - 1$, we have

$$\int (x + 1)\sqrt{2x - 1}\, dx = \frac{1}{10}(2x - 1)^{5/2} + \frac{1}{2}(2x - 1)^{3/2} + C \quad \blacksquare$$

Now that we are familiar with this procedure, we will drop the practice of labeling the steps as we work through the next several examples.

EXAMPLE 4   Find

a. $\int \sin 5x\, dx$ **b.** $\int \dfrac{\cos \sqrt{x}}{\sqrt{x}}\, dx$

Solution

a. Let $u = 5x$, so that $du = 5\, dx$ or $dx = \frac{1}{5}\, du$. Substituting these quantities into the integral gives

$$\int (\sin u)\frac{1}{5}\, du = \frac{1}{5}\int \sin u\, du = -\frac{1}{5}\cos u + C = -\frac{1}{5}\cos 5x + C$$

b. Let $u = \sqrt{x} = x^{1/2}$. (Here, $\sqrt{x}$ is the "inside function" of the composite function $y = \cos \sqrt{x}$.) Then

$$du = \frac{1}{2}x^{-1/2}\, dx = \frac{dx}{2\sqrt{x}} \quad \text{or} \quad \frac{dx}{\sqrt{x}} = 2\, du$$

Substituting these quantities into the integral yields

$$\int (\cos u)\, 2\, du = 2\int \cos u\, du = 2\sin u + C = 2\sin \sqrt{x} + C \quad \blacksquare$$

EXAMPLE 5 Find $\int \sin^3 x \cos x \, dx$.

Solution The integrand contains the composite function $y = \sin^3 x = (\sin x)^3$. So let's put $u = \sin x$. Then $du = \cos x \, dx$, so

$$\int \sin^3 x \cos x \, dx = \int u^3 \, du = \frac{1}{4} u^4 + C = \frac{1}{4} \sin^4 x + C$$ ∎

EXAMPLE 6 Find

a. $\int \dfrac{e^{2/x}}{x^2} \, dx$ **b.** $\int \dfrac{1}{2x + 1} \, dx$

Solution
a. Let $u = 2/x$, so that

$$du = -\frac{2}{x^2} \, dx \qquad \text{or} \qquad \frac{dx}{x^2} = -\frac{1}{2} \, du$$

Making these substitutions, we obtain

$$\int \frac{e^{2/x}}{x^2} \, dx = -\frac{1}{2} \int e^u \, du = -\frac{1}{2} e^u + C = -\frac{1}{2} e^{2/x} + C$$

b. Let $u = 2x + 1$, so that $du = 2 \, dx$ or $dx = \frac{1}{2} du$. Making these substitutions, we have

$$\int \frac{1}{2x + 1} \, dx = \frac{1}{2} \int \frac{1}{u} \, du = \frac{1}{2} \ln|u| + C$$

$$= \frac{1}{2} \ln|2x + 1| + C$$ ∎

EXAMPLE 7 Find $\int \tan x \, dx$.

Solution We first rewrite

$$\int \tan x \, dx = \int \frac{\sin x}{\cos x} \, dx$$

Then we use the substitution $u = \cos x$, so that $du = -\sin x \, dx$ or $\sin x \, dx = -du$, giving

$$\int \frac{\sin x}{\cos x} \, dx = -\int \frac{1}{u} \, du = -\ln|u| + C$$

Therefore,

$$\int \tan x \, dx = -\ln|\cos x| + C \quad \text{or} \quad \ln|\sec x| + C$$ ∎

We can obtain the formula for $\int \cot x \, dx$ in a similar manner by observing that $\cot x = (\cos x)/(\sin x)$.

EXAMPLE 8   Find $\int \sec x \, dx$.

Solution   Multiplying both the numerator and denominator of the integrand by $\sec x + \tan x$ gives

$$\int \sec x \, dx = \int \sec x \frac{\sec x + \tan x}{\sec x + \tan x} \, dx = \int \frac{\sec^2 x + \sec x \tan x}{\sec x + \tan x} \, dx$$

Next, use the substitution

$$u = \sec x + \tan x \qquad \text{so that} \qquad du = (\sec x \tan x + \sec^2 x) \, dx$$

This gives

$$\int \sec x \, dx = \int \frac{1}{u} \, du = \ln|u| + C = \ln|\sec x + \tan x| + C \qquad ■$$

We can find the formula for $\int \csc x \, dx$ by using the same technique. The results of Examples 7 and 8 are summarized below.

Integrals of Trigonometric Functions

$$\int \tan u \, du = \ln|\sec u| + C \qquad\qquad (4a)$$

$$\int \cot u \, du = \ln|\sin u| + C \qquad\qquad (4b)$$

$$\int \sec u \, du = \ln|\sec u + \tan u| + C \qquad\qquad (4c)$$

$$\int \csc u \, du = \ln|\csc u - \cot u| + C \qquad\qquad (4d)$$

EXAMPLE 9   Find $\int x \sec x^2 \, dx$.

Solution   Let $u = x^2$, so that $du = 2x \, dx$ or $x \, dx = \frac{1}{2} \, du$. Making these substitutions, we obtain

$$\int x \sec x^2 \, dx = \frac{1}{2} \int \sec u \, du = \frac{1}{2} \ln|\sec u + \tan u| + C$$

$$= \frac{1}{2} \ln|\sec x^2 + \tan x^2| + C \qquad ■$$

By reversing the rules of differentiation for inverse trigonometric functions, we obtain the following formulas.

Integrals Involving Inverse Trigonometric Functions

$$\int \frac{1}{\sqrt{1 - u^2}}\, du = \sin^{-1} u + C \tag{5a}$$

$$\int \frac{1}{1 + u^2}\, du = \tan^{-1} u + C \tag{5b}$$

$$\int \frac{1}{|u|\sqrt{u^2 - 1}}\, du = \sec^{-1}|u| + C \tag{5c}$$

EXAMPLE 10 Find

a. $\displaystyle\int \frac{1}{\sqrt{1 - 9x^2}}\, dx$ **b.** $\displaystyle\int \frac{1}{\sqrt{4 - x^2}}\, dx$

Solution

a. Comparing the integral with Formula (5a) suggests the substitution $u = 3x$. Then $du = 3\, dx$ or $dx = \frac{1}{3}\, du$. Therefore,

$$\int \frac{1}{\sqrt{1 - 9x^2}}\, dx = \int \frac{1}{\sqrt{1 - (3x)^2}}\, dx$$

$$= \frac{1}{3} \int \frac{1}{\sqrt{1 - u^2}}\, du$$

$$= \frac{1}{3} \sin^{-1} u + C = \frac{1}{3} \sin^{-1}(3x) + C$$

b. Once again, comparing the integral with Formula (5a) suggests that we write

$$\int \frac{1}{\sqrt{4 - x^2}}\, dx = \int \frac{1}{2\sqrt{1 - \left(\dfrac{x}{2}\right)^2}}\, dx$$

Next, we let $u = x/2$ so that $du = \frac{1}{2}\, dx$ or $dx = 2\, du$. Then

$$\int \frac{1}{\sqrt{4 - x^2}}\, dx = \frac{1}{2}\,(2) \int \frac{1}{\sqrt{1 - u^2}}\, du$$

$$= \sin^{-1} u + C = \sin^{-1}\left(\frac{x}{2}\right) + C \qquad \blacksquare$$

EXAMPLE 11 Find

a. $\displaystyle\int \frac{e^x}{e^{2x} + 1}\, dx$ **b.** $\displaystyle\int \frac{1}{x\sqrt{x^4 - 16}}\, dx$

Solution

a. Let $u = e^x$ so that $du = e^x\, dx$. Then

$$\int \frac{e^x}{e^{2x} + 1}\, dx = \int \frac{1}{u^2 + 1}\, du$$

$$= \tan^{-1} u + C = \tan^{-1} e^x + C$$

b. Let $u = x^2$. Then $du = 2x\, dx$, or $dx = \dfrac{du}{2x}$. Making these substitutions, we find

$$\int \frac{1}{x\sqrt{x^4 - 16}}\, dx = \frac{1}{2} \int \frac{1}{x^2\sqrt{u^2 - 16}}\, du$$

$$= \frac{1}{2} \int \frac{1}{u\sqrt{u^2 - 16}}\, du \qquad \text{Replace } x^2 \text{ by } u.$$

$$= \frac{1}{2} \cdot \frac{1}{4} \int \frac{1}{u\sqrt{\left(\dfrac{u}{4}\right)^2 - 1}}\, du$$

Next, we let $v = u/4$ so that $dv = \frac{1}{4}\, du$ or $du = 4\, dv$. Then

$$\int \frac{1}{x\sqrt{x^4 - 16}}\, dx = \left(\frac{1}{8}\right) 4 \int \frac{1}{4v\sqrt{v^2 - 1}}\, dv$$

$$= \frac{1}{8} \int \frac{1}{v\sqrt{v^2 - 1}}\, dv$$

$$= \frac{1}{8} \sec^{-1} v + C = \frac{1}{8} \sec^{-1}\left(\frac{x^2}{4}\right) + C \qquad \blacksquare$$

EXAMPLE 12 **A Falling Ballast** A ballast is dropped from a balloon at a height of 10,000 ft. The velocity at any time t (until it reaches the ground) is given by

$$v(t) = 320(e^{-0.1t} - 1)$$

where the velocity is measured in feet per second and t is measured in seconds.

a. Find an expression for the height $s(t)$ of the ballast at any time t. ($s(t)$ is measured from the ground.)
b. What is the terminal velocity of the ballast?
c. Estimate the time it takes for the ballast to hit the ground.

Solution
a. The velocity of the ballast is

$$s'(t) = v(t) = 320(e^{-0.1t} - 1)$$

Therefore, its height at any time t is

$$s(t) = \int v(t)\, dt = \int 320(e^{-0.1t} - 1)\, dt$$

$$= 320\left(-\frac{1}{0.1} e^{-0.1t} - t\right) + C \qquad \text{Use integration by substitution with } u = -0.1t.$$

$$= -320(t + 10e^{-0.1t}) + C$$

To determine C, we use the initial condition $s(0) = 10,000$, obtaining

$$s(0) = -3200 + C = 10,000 \qquad \text{or} \qquad C = 13,200$$

Therefore,

$$s(t) = -320(t + 10e^{-0.1t}) + 13,200$$

b. The terminal velocity is given by

$$\lim_{t \to \infty} v(t) = \lim_{t \to \infty} 320(e^{-0.1t} - 1)$$

$$= 320 \lim_{t \to \infty} (e^{-0.1t} - 1) = -320$$

or −320 ft/sec.

c. The ballast hits the ground when $s(t) = 0$. Using the result of part (a), we have

$$-320(t + 10e^{-0.1t}) + 13,200 = 0 \qquad (6)$$

This equation is not easily solved for t, but we can obtain an approximation of t by observing that if t is large, then the term $10e^{-0.1t}$ is relatively small in comparison to t. So, dropping this term, we solve the equation

$$-320t + 13,200 = 0$$

getting $t \approx 41$. Therefore, the ballast hits the ground approximately 41 sec after it has been jettisoned. The graph of $s(t) = -320(t + 10e^{-0.1t}) + 13,200$ is shown in Figure 1.

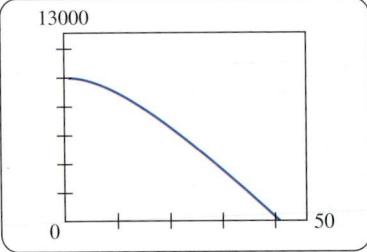

FIGURE 1
The graph of
$s(t) = -320(t + 10e^{-0.1t}) + 13,200$

Notes

1. Let's show that the approximation obtained in part (c) of Example 12 is reasonably accurate by computing the position of the ballast 41 sec after it was jettisoned. Thus,

$$s(41) = -320(41 + 10e^{-4.1}) + 13,200 \approx 27$$

or 27 ft above the ground. If greater accuracy is required, one can use Newton's method to solve the equation $s(t) = 0$. (See Exercise 29, Section 3.9.)

2. We can also estimate the time of impact of the ballast by using a graphing utility to find the zero of f. We find $t \approx 41.086$, accurate to three decimal places.

EXAMPLE 13 **Flight of a Projectile** Suppose that a projectile is launched vertically upward from the earth's surface with an initial velocity equal to the *escape velocity*. (The escape velocity is the velocity the projectile must attain to escape from the earth's gravitational pull.) Then, if we neglect the gravitational influence of the sun and other planets, the rotation of the earth, and air resistance, it can be shown that the differential equation governing the motion of the projectile is

$$\frac{dt}{dx} = \frac{\sqrt{x + R}}{R\sqrt{2g}}$$

where t is the time in seconds, x is the distance of the projectile from the surface of the earth, R is the radius of the earth (approximately 4000 mi), and g is the gravitational constant of acceleration (approximately 32 ft/sec^2). (See Figure 2.) Find the time it takes for the projectile to travel a distance of x miles. How long would it take the projectile to cover 100,000 mi?

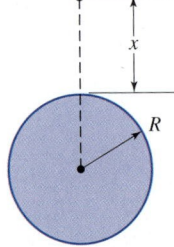

FIGURE 2
The projectile is launched vertically upward from the earth's surface.

Solution Integrating the given equation with respect to x, we obtain

$$t = \int \frac{dt}{dx} \, dx = \int \frac{\sqrt{x + R}}{R\sqrt{2g}} \, dx = \frac{1}{R\sqrt{2g}} \int (x + R)^{1/2} \, dx$$

Let $u = x + R$, so that $du = dx$. Then

$$t = \frac{1}{R\sqrt{2g}} \int u^{1/2}\, du = \frac{2u^{3/2}}{3R\sqrt{2g}} + C = \frac{2(x + R)^{3/2}}{3R\sqrt{2g}} + C$$

where C is the constant of integration. To determine the value of C, we use the condition that $t = 0$ when $x = 0$. This gives

$$0 = \frac{2R^{3/2}}{3R\sqrt{2g}} + C \quad \text{or} \quad C = -\frac{2R^{3/2}}{3R\sqrt{2g}}$$

Therefore,

$$t = \frac{2(x + R)^{3/2}}{3R\sqrt{2g}} - \frac{2R^{3/2}}{3R\sqrt{2g}} = \frac{2}{3R\sqrt{2g}}[(x + R)^{3/2} - R^{3/2}]$$

and this is the time it takes for the projectile to cover a distance x. Finally, to find the time it takes for the projectile to travel 100,000 mi, we substitute

$$R = 4000(5280) = 21,120,000 \quad \text{(ft)}$$

and

$$x = 100,000(5280) = 528,000,000 \quad \text{(ft)}$$

into the expression for t, obtaining

$$t = \frac{2}{3(21,120,000)\sqrt{64}}(549,120,000^{3/2} - 21,120,000^{3/2}) \approx 50,389.2$$

So it takes approximately 50,000 sec, or 14 hr, to cover 100,000 mi.

4.2 CONCEPT QUESTIONS

1. Explain how the method of integration by substitution works by showing the steps that are used to find $\int f(g(x))g'(x)\, dx$.

2. Explain why the method of substitution works for the integral $\int x^2 \cos(x^3 + 1)\, dx$. Does it work for $\int x \cos(x^3 + 1)\, dx$?

4.2 EXERCISES

In Exercises 1–6, find the integral using the indicated substitution.

1. $\displaystyle \int (2x + 3)^5\, dx, \quad u = 2x + 3$

2. $\displaystyle \int x^2\sqrt{x^3 + 2}\, dx, \quad u = x^3 + 2$

3. $\displaystyle \int \frac{x}{\sqrt{x^2 + 1}}\, dx, \quad u = x^2 + 1$

4. $\displaystyle \int e^{-3x}\, dx, \quad u = -3x$

5. $\displaystyle \int \tan^3 x \sec^2 x\, dx, \quad u = \tan x$

6. $\displaystyle \int \frac{\sin x}{\cos^2 x}\, dx, \quad u = \cos x$

In Exercises 7–72, find the indefinite integral.

7. $\displaystyle \int 2x(x^2 + 1)^4\, dx$

8. $\displaystyle \int x^2(2x^3 - 1)^4\, dx$

9. $\displaystyle \int (2x - 4)^{3/5}\, dx$

10. $\displaystyle \int (1 - 3x)^{1.4}\, dx$

— 11. $\displaystyle \int 3x(2x^2 + 3)^5\, dx$

12. $\displaystyle \int x^2(2x^3 - 1)^{-4}\, dx$

13. $\displaystyle \int \sqrt{1 - 2x}\, dx$

14. $\displaystyle \int 2x\sqrt[3]{1 - 4x^2}\, dx$

15. $\displaystyle \int s^3(s^4 - 1)^{3/2}\, ds$

16. $\displaystyle \int x^{-1/3}\sqrt{x^{2/3} - 1}\, dx$

— 17. $\displaystyle \int xe^{-x^2}\, dx$

18. $\displaystyle \int (x^2 - \tfrac{1}{3})e^{x^3 - x}\, dx$

19. $\int (e^x + e^{-x})^2 \, dx$

20. $\int \sqrt{1 + 2e^x} e^x \, dx$

21. $\int 2^{-x} \, dx$

22. $\int (x + 1)3^{x^2 + 2x} \, dx$

23. $\int \dfrac{e^{-1/x}}{x^2} \, dx$

24. $\int \dfrac{\sqrt{1 + u^{-1}}}{u^2} \, du$

25. $\int 2 \cos \dfrac{x}{2} \, dx$

26. $\int x \sin x^2 \, dx$

27. $\int x \cos \pi x^2 \, dx$

28. $\int x^2 \sec^2 x^3 \, dx$

29. $\int \sin \pi x \cos \pi x \, dx$

30. $\int \cot^3 x \csc^2 x \, dx$

31. $\int \tan 3x \sec 3x \, dx$

32. $\int \sqrt{\sin \theta} \cos \theta \, d\theta$

33. $\int \dfrac{\sin u^{-1}}{u^2} \, du$

34. $\int \dfrac{\sin x}{(1 + \cos x)^3} \, dx$

35. $\int \dfrac{\csc^2 3x}{\cot^3 3x} \, dx$

36. $\int \dfrac{\sin \sqrt{x}}{\sqrt{x}} \, dx$

37. $\int \dfrac{\cos 2t}{\sqrt{2 + \sin 2t}} \, dt$

38. $\int (\csc^2 x)(\cot x - 1)^3 \, dx$

39. $\int \dfrac{\sec x \tan x}{(\sec x - 1)^2} \, dx$

40. $\int \sec^2(x + 1)\sqrt{1 + \tan(x + 1)} \, dx$

41. $\int \sin^2 \pi x \, dx$

Hint: $\sin^2 \theta = \dfrac{1 - \cos 2\theta}{2}$

42. $\int \dfrac{1 + \sin x}{\cos^2 x} \, dx$

43. $\int \dfrac{e^{-x}}{1 + e^{-x}} \, dx$

44. $\int \dfrac{3^x}{1 + 3^x} \, dx$

45. $\int \dfrac{\sqrt{\log x}}{x} \, dx$

46. $\int e^x \sin e^x \, dx$

47. $\int \dfrac{e^x \ln(e^x + 1)}{e^x + 1} \, dx$

48. $\int e^{\sin x} \cos x \, dx$

49. $\int \dfrac{1}{2x + 3} \, dx$

50. $\int \dfrac{1}{x^{2/3}(x^{1/3} + 1)} \, dx$

51. $\int \dfrac{1}{x \ln x} \, dx$

52. $\int \dfrac{\sqrt{1 + \ln x}}{x} \, dx$

53. $\int \dfrac{\cos x}{1 + \sin x} \, dx$

54. $\int \dfrac{\sec^2 3x}{4 - \tan 3x} \, dx$

55. $\int (\sec \theta + \cos \theta) \, d\theta$

56. $\int \dfrac{\sin 2x}{1 + \sin^2 x} \, dx$

57. $\int \dfrac{1 + \ln x}{2 + x \ln x} \, dx$

58. $\int \dfrac{(\ln x)\sqrt{1 + \ln x}}{x} \, dx$

59. $\int \dfrac{1}{\sqrt{16 - x^2}} \, dx$

60. $\int \dfrac{1}{x\sqrt{9x^2 - 1}} \, dx$

61. $\int \dfrac{1}{x\sqrt{x^4 - 81}} \, dx$

62. $\int \dfrac{1}{t\sqrt{t^6 - 16}} \, dt$

63. $\int \dfrac{x^2 - 1}{x^2 + 1} \, dx$

64. $\int \dfrac{\cos 3x}{1 + \sin^2 3x} \, dx$

65. $\int \dfrac{\sin x}{\sqrt{4 - \cos^2 x}} \, dx$

66. $\int \dfrac{dx}{|x|(\sec^{-1} x)^3 \sqrt{x^2 - 1}}$

67. $\int \dfrac{e^{\tan^{-1} x}}{1 + x^2} \, dx$

68. $\int \dfrac{1}{(x + 1)\sqrt{(x + 1)^2 - 9}} \, dx$

69. $\int \dfrac{1}{\sqrt{x}\,(4 + x)} \, dx$

70. $\int \dfrac{e^x}{\sqrt{1 - e^{2x}}} \, dx$

71. $\int \dfrac{1}{4 + (x - 2)^2} \, dx$

72. $\int \dfrac{1}{x[9 + (\ln x)^2]} \, dx$

In Exercises 73–78, find the indefinite integral.

73. $\int x\sqrt{x - 4} \, dx$

74. $\int x^2(1 - x)^7 \, dx$

75. $\int x^3(x^2 + 1)^{5/2} \, dx$

76. $\int \dfrac{x + 1}{(\sqrt{x} - 1)^{3/2}} \, dx$

77. $\int \dfrac{dx}{\sqrt{x} + \sqrt{x + 1}}$

Hint: First rationalize the denominator of the integrand.

78. $\int \dfrac{\sqrt{a^2 - x^2}}{x^4} \, dx$

Hint: Let $x = 1/t$.

79. Find the function f given that its derivative is $f'(x) = x\sqrt{1 + x^2}$ and that its graph passes through the point $(0, 1)$.

80. The slope of the tangent line at any point on the graph of f is $\dfrac{x}{(2x^2 + 1)^{3/2}}$, and the graph of f passes through the point $\left(2, -\frac{1}{6}\right)$. Find f.

81. Rectilinear Motion A body moves along a coordinate line in such a way that its velocity at any time t, where $0 \le t \le 4$, is given by

$$v(t) = t\sqrt{16 - t^2}$$

Find its position function if the body is initially located at the origin.

82. Population Growth The population of a certain city is projected to grow at the rate of

$$r(t) = 400\left(1 + \dfrac{2t}{\sqrt{24 + t^2}}\right) \qquad 0 \le t \le 5$$

people per year t years from now. The current population is 60,000. What will be the population 5 years from now?

83. Life Expectancy of a Female Suppose that in a certain country the life expectancy at birth of a female is changing at the rate of

$$g'(t) = \frac{5.45218}{(1 + 1.09t)^{0.9}}$$

years per year. Here, t is measured in years, with $t = 0$ corresponding to the beginning of 1900. Find an expression $g(t)$ giving the life expectancy at birth (in years) of a female in that country if the life expectancy at the beginning of 1900 is 50.02 years. What is the life expectancy at birth of a female born at the beginning of 2000 in that country?

84. Revenue The weekly marginal revenue of a company selling x units (in lots of 100) of a portable hair dryer is given by

$$R'(x) = \frac{225 - 10x}{\sqrt{225 - 5x}}$$

dollars per lot. Find the weekly total revenue function.
Hint: $R(0) = 0$.

85. Respiratory Cycle Suppose that the rate at which air is inhaled by a person during respiration is

$$r(t) = \frac{3}{5} \sin \frac{\pi t}{2}$$

liters per second, at time t. Find $V(t)$, the volume of inhaled air in the lungs at any time t. Assume that $V(0) = 0$.

86. Revenue The total revenue of McMenamy's Fish Shanty at a popular summer resort is changing at the rate of approximately

$$R'(t) = 2\left(5 - 4 \cos \frac{\pi t}{6}\right) \qquad 0 \le t \le 12$$

thousand dollars per week, where t is measured in weeks, with $t = 0$ corresponding to the beginning of June. Find the total revenue R of the Shanty at the end of t weeks after its opening on June 1.

87. Simple Harmonic Motion The acceleration function of a body moving along a coordinate line is

$$a(t) = -4 \cos 2t - 3 \sin 2t \qquad t \ge 0$$

Find its velocity and position functions at any time t if the body is located at the origin and has an initial velocity of $\frac{3}{2}$ m/sec.

88. Special Theory of Relativity According to Einstein's special theory of relativity, the mass of a particle is given by

$$m = \frac{m_0}{\sqrt{1 - \dfrac{v^2}{c^2}}}$$

where m_0 is the rest mass of the particle, v is its velocity, and c is the speed of light. Suppose that a particle starts from rest at $t = 0$ and moves along a straight line under the action of a constant force F. Then, according to Newton's second law of motion, the equation of motion is

$$F = m_0 \frac{d}{dt}\left(\frac{v}{\sqrt{1 - \dfrac{v^2}{c^2}}}\right)$$

Find the velocity and position functions of the particle. What happens to the velocity of the particle as time goes by?

In Exercises 89 and 90, determine whether the statement is true or false. If it is true, explain why it is true. If it is false, explain why or give an example to show why it is false.

89. If f is continuous, then $\int x\, f(x^2)\, dx = \frac{1}{2}\int f(u)\, du$, where $u = x^2$.

90. If f is continuous, then $\int f(ax + b)\, dx = \int f(x)\, dx$.

4.3 Area

◼ An Intuitive Look

Consider a car moving on a straight road with a velocity function given by

$$v(t) = 44 \qquad 0 \le t \le 10$$

where t is measured in seconds and $v(t)$ in feet per second. Since $v(t) > 0$, it also gives the speed of the car over this time interval. The distance traveled by the car between $t = 1$ and $t = 5$ is

$$(44)(5 - 1) \qquad \text{constant speed · time elapsed}$$

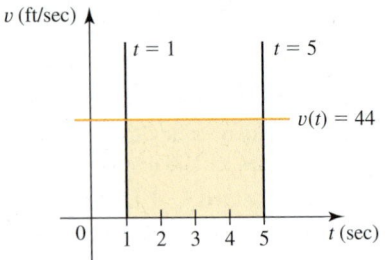

FIGURE 1
The distance traveled by the car can be represented by the area of the rectangular region.

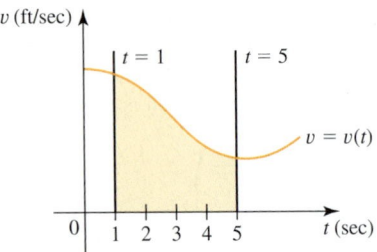

FIGURE 2
The distance covered by the car is given by the "area" of the shaded region.

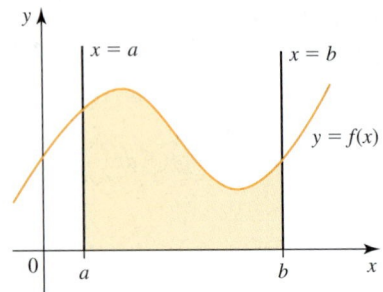

FIGURE 3
The shaded region is the area under the graph of f on $[a, b]$.

or 176 ft. If you examine the graph of v shown in Figure 1, you will see that this distance is just the area of the rectangular region bounded above by the graph of v, below by the t-axis, and to the left and right by the vertical lines $t = 1$ and $t = 5$, respectively.

Suppose that the same car moves along a straight road but this time with a velocity function v that is positive but not necessarily constant over an interval of time. What is the distance traveled by the car between $t = 1$ and $t = 5$? We might be tempted to conjecture that it is given by the "area" of the region bounded above by the graph of v, below by the t-axis, and to the left and right by the vertical lines $t = 1$ and $t = 5$, respectively (see Figure 2). Later, we will show that this is indeed the case.

This example raises two questions:

1. What do we mean by the "area" of a region such as the one shown in Figure 2?
2. How do we find the area of such a region?

■ The Area Problem

Here, we have touched upon the second fundamental problem in calculus: How do we find the area of the region bounded above by the graph of a *nonnegative* function f, below by the x-axis, and to the left and right by the vertical lines $x = a$ and $x = b$, as shown in Figure 3? We refer to the area of this region as the **area under the graph of f** on the interval $[a, b]$.

■ Defining the Area of the Region Under the Graph of a Function

When we defined the slope of the tangent line to the graph of a function at a point on the graph, we first approximated it with the slopes of secant lines (quantities that we could compute). We then took the limit of these approximations to give us the slope of the tangent line. We will now adopt a parallel approach to define the area of the region under the graph of a function.

The idea here is to approximate the area of a region by using the sums of the areas of rectangles (quantities that we can compute).* We can then find the desired area by taking the limit of these sums. Let's begin by looking at a specific example.

EXAMPLE 1 Consider the region S bounded above by the parabola $f(x) = x^2$, below by the x-axis, and to the left and right by the vertical lines $x = 0$ and $x = 1$, respectively (see Figure 4). As you can see, the area A of the region S can be approximated by the area A_1 of the rectangle R_1 with base lying on the interval $[0, 1]$ and height given by the value of $f(x) = x^2$ evaluated at the midpoint of $[0, 1]$. Thus,

$$A \approx A_1 = 1 \cdot f\left(\frac{1}{2}\right) = (1)\left(\frac{1}{2}\right)^2 = \frac{1}{4}$$

FIGURE 4
The area of the region S in part (a) is approximated by the area of the rectangle R_1 in part (b).

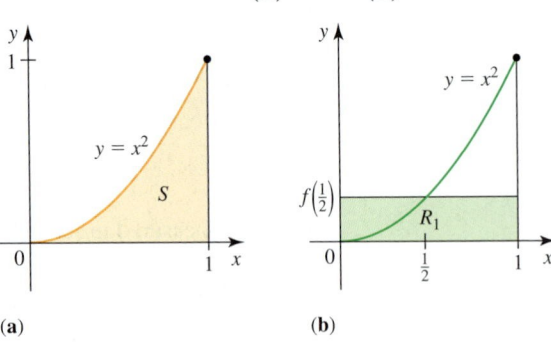

(a)   **(b)**

---

*Until a formal definition of area is given, the term *area* will refer to our intuitive notion of area.

We have used the midpoint of the interval [0, 1] to compute the height of the approximating rectangle because it seems to be a logical choice. But you should convince yourself that any other point in the interval, including the endpoints, would also serve our purpose. Of course, the approximation obtained will be different depending on your choice.

Can we do better? Let's divide the interval [0, 1] into two subintervals $\left[0, \frac{1}{2}\right]$ and $\left[\frac{1}{2}, 1\right]$, each of (equal) length $\frac{1}{2}$. Figure 5a shows the region S expressed as the union of two nonoverlapping subregions S_1 and S_2 with bases lying on the subintervals $\left[0, \frac{1}{2}\right]$ and $\left[\frac{1}{2}, 1\right]$, respectively. Figure 5b shows the rectangle R_1 with base lying on $\left[0, \frac{1}{2}\right]$ and height $f\left(\frac{1}{4}\right)$, the value of f evaluated at the midpoint of $\left[0, \frac{1}{2}\right]$, and the rectangle R_2 with base lying on $\left[\frac{1}{2}, 1\right]$ and height $f\left(\frac{3}{4}\right)$, where $x = \frac{3}{4}$ is the midpoint of $\left[\frac{1}{2}, 1\right]$. If we approximate the area of S_1 by the area of R_1 and the area of S_2 by the area of R_2 and denote the sum of the areas of the two rectangles by A_2, we obtain

$$A \approx A_2 = \frac{1}{2}f\left(\frac{1}{4}\right) + \frac{1}{2}f\left(\frac{3}{4}\right)$$

$$= \frac{1}{2}\left(\frac{1}{4}\right)^2 + \frac{1}{2}\left(\frac{3}{4}\right)^2$$

$$= \frac{1}{2}\left(\frac{1}{16} + \frac{9}{16}\right) = \frac{5}{16}$$

or 0.3125. Continuing with this process, we divide the interval [0, 1] into four subintervals of equal length $\frac{1}{4}$ using the five points

$$x_0 = 0, \qquad x_1 = \frac{1}{4}, \qquad x_2 = \frac{1}{2}, \qquad x_3 = \frac{3}{4}, \qquad \text{and} \qquad x_4 = 1$$

FIGURE 5

The subregions S_1 and S_2 in part (a) are approximated by the rectangles R_1 and R_2 in part (b).

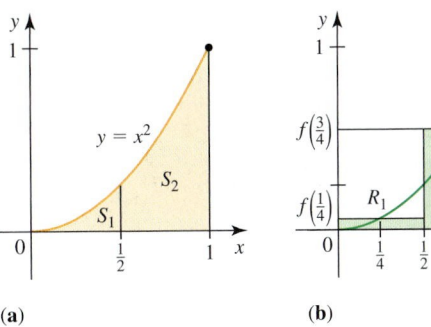

(a)   (b)

The resulting subintervals are

$$\left[0, \tfrac{1}{4}\right], \qquad \left[\tfrac{1}{4}, \tfrac{1}{2}\right], \qquad \left[\tfrac{1}{2}, \tfrac{3}{4}\right], \qquad \text{and} \qquad \left[\tfrac{3}{4}, 1\right]$$

Figure 6a shows the region S expressed as the union of four nonoverlapping subregions S_1, S_2, S_3, and S_4 with bases lying on these subintervals. The midpoints of the subintervals are

$$c_1 = \frac{1}{8}, \qquad c_2 = \frac{3}{8}, \qquad c_3 = \frac{5}{8}, \qquad \text{and} \qquad c_4 = \frac{7}{8}$$

respectively. The rectangles R_1, R_2, R_3, and R_4 with bases lying on these subintervals and having heights evaluated at their respective midpoints are shown in Figure 6b.

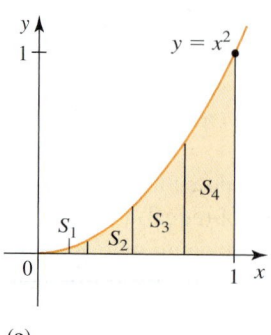

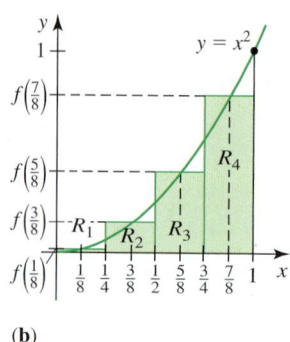

(a)   **(b)**

FIGURE 6

The area of subregion S_i in part (a) is approximated by the area of rectangle R_i for $1 \le i \le 4$.

Historical Biography

SONYA KOVALEVSKAYA
(1850-1891)

Karl Weierstrass's (page 645) favorite student, Sonya Kovalevskaya was born January 15, 1850, in Moscow. A room of her family's estate was wallpapered with sheets of Mikhailo Ostrogradsky's lithographed lectures on differential and integral calculus, and as a child Kovalevskaya spent hours trying to decipher the formulas. At the age of 15 she astonished her tutor with how easily she understood calculus and its foundation. Kovalevskaya desperately wanted to attend a university and secure a degree in mathematics, but her father would not allow it, so she entered a marriage of convenience to a man who allowed her to travel and continue her studies. After three semesters at the University of Heidelberg, Germany, she traveled to Berlin in search of the renowned mathematician Weierstrass. Weierstrass agreed to teach Kovalevskaya privately, as the university would not allow women to attend his lectures. By 1874 she had written three degree-worthy dissertations, and Weierstrass submitted the most profound of these to the University of Göttingen. Kovalevskaya was awarded her doctorial *summa cum laude* in 1874, becoming the first woman to earn a doctorate in mathematics. Unfortunately, as a woman, she could not find a university position until 1884. Kovalevskaya died of influenza in 1891 at the height of her mathematical career.

Approximating the area of the subregion S_i by the area of the rectangle R_i, where $1 \le i \le 4$, and letting A_4 denote the sum of the areas of the four rectangles, we obtain yet another approximation of the area A of S:

$$A \approx A_4 = \frac{1}{4}f\left(\frac{1}{8}\right) + \frac{1}{4}f\left(\frac{3}{8}\right) + \frac{1}{4}f\left(\frac{5}{8}\right) + \frac{1}{4}f\left(\frac{7}{8}\right)$$

$$= \frac{1}{4}\left(\frac{1}{8}\right)^2 + \frac{1}{4}\left(\frac{3}{8}\right)^2 + \frac{1}{4}\left(\frac{5}{8}\right)^2 + \frac{1}{4}\left(\frac{7}{8}\right)^2$$

$$= \frac{1}{4}\left(\frac{1}{64} + \frac{9}{64} + \frac{25}{64} + \frac{49}{64}\right) = \frac{21}{64}$$

or 0.328125.

We can keep going. Figure 7a shows what happens if we use eight rectangles to approximate the area of the region S, and Figure 7b shows the situation if sixteen rectangles are used.

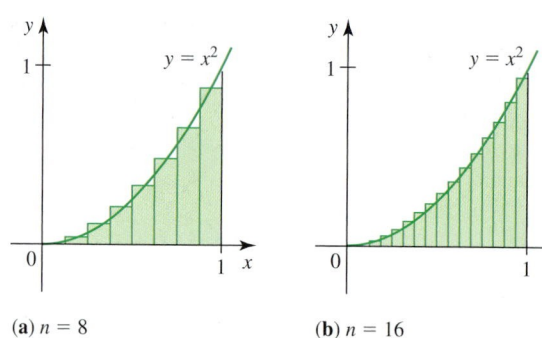

(a) $n = 8$ **(b)** $n = 16$

FIGURE 7

As the number of rectangles used increases, the approximation of the area of the region S seems to improve.

With the aid of a computer we can find the approximations of the area A of the region S using n approximating rectangles. In the following table, A_n denotes the approximation of A when n rectangles are used. The results include approximations obtained earlier ($n = 1, 2, 4$) and are rounded off to seven decimal places.

n	A_n
1	0.25
2	0.3125
4	0.328125
8	0.3320313
16	0.3330078
50	0.3333000
100	0.3333250
500	0.3333330
1000	0.3333333

These results seem to suggest that A_n approaches $\frac{1}{3}$ as n gets larger and larger; that is, $\lim_{n \to \infty} A_n = \frac{1}{3}$. This in turn suggests that we could take the area of the region S to be $\frac{1}{3}$.

■ Sigma Notation

Before confirming this result, we will digress a little to introduce a notation that will provide us with a shorthand method for writing sums involving a large number of terms. The notation uses the uppercase Greek letter sigma Σ and is accordingly called *sigma notation*.

> **DEFINITION   Sigma Notation**
>
> The sum of the n terms $a_1, a_2, a_3, \ldots, a_n$ is abbreviated $\sum_{k=1}^{n} a_k$. Thus,
>
> $$\sum_{k=1}^{n} a_k = a_1 + a_2 + a_3 + \cdots + a_k + \cdots + a_n$$
>
> The variable k is called the **index of summation,** the term a_k is called the **kth term** of the sum, and the numbers n and 1 are called the **upper** and **lower limits of summation,** respectively.

The sum $\sum_{k=1}^{n} a_k$ is read "the sum of a_k where k runs from 1 to n."

EXAMPLE 2 Write each of the following sums in expanded form:

a. $\displaystyle\sum_{k=1}^{5} k$ **b.** $\displaystyle\sum_{k=1}^{10} k^2$ **c.** $\displaystyle\sum_{k=1}^{20} \frac{1}{(k+1)^2}$ **d.** $\displaystyle\sum_{k=1}^{15} (-1)^k k^3$ **e.** $\displaystyle\sum_{k=1}^{10} \sin\left(\frac{k\pi}{4}\right)$

Solution

a. $\displaystyle\sum_{k=1}^{5} k = 1 + 2 + 3 + 4 + 5$ Here, $a_k = k$, so $a_1 = 1$, $a_2 = 2$, $a_3 = 3$, $a_4 = 4$, and $a_5 = 5$.

b. $\displaystyle\sum_{k=1}^{10} k^2 = 1^2 + 2^2 + 3^2 + \cdots + 10^2$ Here, $a_k = k^2$, so $a_1 = 1^2$, $a_2 = 2^2$, ...

c. $\displaystyle\sum_{k=1}^{20} \frac{1}{(k+1)^2} = \frac{1}{2^2} + \frac{1}{3^2} + \frac{1}{4^2} + \cdots + \frac{1}{21^2}$ Here, $a_k = \dfrac{1}{(k+1)^2}$.

d. $\displaystyle\sum_{k=1}^{15} (-1)^k k^3 = (-1)^1 1^3 + (-1)^2 2^3 + (-1)^3 3^3 + \cdots + (-1)^{15} 15^3$

$$= -1 + 2^3 - 3^3 + \cdots - 15^3$$

e. $\displaystyle\sum_{k=1}^{10} \sin\left(\frac{k\pi}{4}\right) = \sin\frac{\pi}{4} + \sin\frac{2\pi}{4} + \sin\frac{3\pi}{4} + \cdots + \sin\frac{10\pi}{4}$ ■

So far, we have used k as the index of summation, but any letter will do. For example, each of the following

$$\sum_{k=1}^{5} a_k, \qquad \sum_{i=1}^{5} a_i, \qquad \text{and} \qquad \sum_{j=1}^{5} a_j$$

represents the sum $a_1 + a_2 + a_3 + a_4 + a_5$. Sometimes it is more convenient to use a lower limit of summation other than 1. For example, we can write

$$\sum_{k=2}^{6} (2k+1) = 5 + 7 + 9 + 11 + 13$$

which is equivalent to

$$\sum_{k=1}^{5} (2k+3) = 5 + 7 + 9 + 11 + 13$$

Also, if the upper and lower indices of summation are the same, then the sum consists of just one term. For example,

$$\sum_{k=1}^{1} \frac{1}{k} = \frac{1}{1} = 1 \qquad \text{k runs from 1 to 1.}$$

In the next example, keep in mind that the upper limit of summation n is constant *with respect to the summation.*

EXAMPLE 3   Write each of the following sums in expanded form:

a. $\displaystyle\sum_{k=1}^{n} \frac{1}{n}(2k-1)$ **b.** $\displaystyle\sum_{k=1}^{n} \left(1 + \frac{k}{n}\right)^3 \left(\frac{1}{n}\right)$ **c.** $\displaystyle\sum_{k=1}^{n-1} \sin\left(\frac{k\pi}{n}\right)$

Solution

a. $\displaystyle\sum_{k=1}^{n} \frac{1}{n}(2k-1) = \frac{1}{n}(2-1) + \frac{1}{n}(4-1) + \frac{1}{n}(6-1) + \cdots + \frac{1}{n}(2n-1)$

$$= \frac{1}{n} + \frac{3}{n} + \frac{5}{n} + \cdots + \frac{2n-1}{n}$$

b. $\displaystyle\sum_{k=1}^{n}\left(1+\frac{k}{n}\right)^3\left(\frac{1}{n}\right) = \left(1+\frac{1}{n}\right)^3\left(\frac{1}{n}\right) + \left(1+\frac{2}{n}\right)^3\left(\frac{1}{n}\right)$

$$+ \left(1+\frac{3}{n}\right)^3\left(\frac{1}{n}\right) + \cdots + \left(1+\frac{n}{n}\right)^3\left(\frac{1}{n}\right)$$

$$= \left(\frac{1}{n}\right)\left[\left(1+\frac{1}{n}\right)^3 + \left(1+\frac{2}{n}\right)^3 + \left(1+\frac{3}{n}\right)^3 + \cdots + 2^3\right]$$

c. $\displaystyle\sum_{k=1}^{n-1}\sin\left(\frac{k\pi}{n}\right) = \sin\frac{\pi}{n} + \sin\frac{2\pi}{n} + \sin\frac{3\pi}{n} + \cdots + \sin\frac{(n-1)\pi}{n}$

■ Summation Formulas

The following rules are useful in manipulating sums written using sigma notation.

Rules of Summation

1. $\displaystyle\sum_{k=1}^{n} ca_k = c\sum_{k=1}^{n} a_k,$ where c is a constant

2. $\displaystyle\sum_{k=1}^{n}(a_k + b_k) = \sum_{k=1}^{n} a_k + \sum_{k=1}^{n} b_k$

3. $\displaystyle\sum_{k=1}^{n}(a_k - b_k) = \sum_{k=1}^{n} a_k - \sum_{k=1}^{n} b_k$

PROOF All three rules can be proved by writing the respective sums in expanded form. For example, to prove Rule 1, we write

$$\sum_{k=1}^{n} ca_k = ca_1 + ca_2 + \cdots + ca_n = c(a_1 + a_2 + \cdots + a_n)$$

Use the distributive property.

$$= c\sum_{k=1}^{n} a_k$$

The proof of Rules 2 and 3 are left as exercises.

EXAMPLE 4 Use the rules of summation to expand each sum:

a. $\displaystyle\sum_{k=1}^{10} 3k^2$ **b.** $\displaystyle\sum_{k=2}^{8}(k + 3k^3)$

Solution

a. $\displaystyle\sum_{k=1}^{10} 3k^2 = 3\sum_{k=1}^{10} k^2 = 3(1^2 + 2^2 + 3^2 + \cdots + 10^2)$

b. $\displaystyle\sum_{k=2}^{8}(k + 3k^3) = \sum_{k=2}^{8} k + \sum_{k=2}^{8} 3k^3 = \sum_{k=2}^{8} k + 3\sum_{k=2}^{8} k^3$

$$= (2 + 3 + \cdots + 8) + 3(2^3 + 3^3 + \cdots + 8^3)$$

The following summation formulas will be used later.

THEOREM 1 **Summation Formulas**

a. $\displaystyle\sum_{k=1}^{n} c = nc$, c a constant

b. $\displaystyle\sum_{k=1}^{n} k = \frac{n(n+1)}{2}$

c. $\displaystyle\sum_{k=1}^{n} k^2 = \frac{n(n+1)(2n+1)}{6}$

d. $\displaystyle\sum_{k=1}^{n} k^3 = \left[\frac{n(n+1)}{2}\right]^2$

We will omit the proofs.

EXAMPLE 5 Use Theorem 1 to evaluate each sum:

a. $\displaystyle\sum_{k=1}^{10} 3$ **b.** $\displaystyle\sum_{k=1}^{20} k$ **c.** $\displaystyle\sum_{k=1}^{50} k^2$

Solution

a. $\displaystyle\sum_{k=1}^{10} 3 = \underbrace{3 + 3 + 3 + \cdots + 3}_{10 \text{ terms}} = 10(3) = 30$ Use Theorem 1a.

b. $\displaystyle\sum_{k=1}^{20} k = 1 + 2 + 3 + \cdots + 20 = \frac{20(20 + 1)}{2} = 210$ Use Theorem 1b.

c. $\displaystyle\sum_{k=1}^{50} k^2 = 1^2 + 2^2 + 3^2 + \cdots + 50^2$

$$= \frac{50(50 + 1)(2 \cdot 50 + 1)}{6} = 42{,}925$$ Use Theorem 1c. ■

EXAMPLE 6 Evaluate $\displaystyle\sum_{k=1}^{10} 3k^2(2k + 1)$.

Solution

$$\sum_{k=1}^{10} 3k^2(2k + 1) = \sum_{k=1}^{10} (6k^3 + 3k^2)$$

$$= 6\sum_{k=1}^{10} k^3 + 3\sum_{k=1}^{10} k^2$$

$$= 6\left[\frac{10(10 + 1)}{2}\right]^2 + 3\left[\frac{10(10 + 1)(20 + 1)}{6}\right]$$

$$= 18{,}150 + 1155 = 19{,}305$$ ■

EXAMPLE 7 Evaluate

$$\lim_{n\to\infty} \sum_{k=1}^{n} \left[\left(\frac{k}{n}\right)^2 + 2 \right]\left(\frac{4}{n}\right)$$

Solution

$$\lim_{n\to\infty} \sum_{k=1}^{n} \left[\left(\frac{k}{n}\right)^2 + 2 \right]\left(\frac{4}{n}\right)$$

$$= \lim_{n\to\infty} \sum_{k=1}^{n} \left(\frac{4k^2}{n^3} + \frac{8}{n} \right)$$

$$= \lim_{n\to\infty} \left[\frac{4}{n^3} \sum_{k=1}^{n} k^2 + \frac{8}{n} \sum_{k=1}^{n} 1 \right] \qquad \text{Remember that } n \text{ is constant with respect to the summations.}$$

$$= \lim_{n\to\infty} \left[\frac{4}{n^3} \cdot \frac{n(n+1)(2n+1)}{6} + \frac{8}{n} \cdot n \right] \qquad \text{Use Theorems 1a and 1c.}$$

$$= \lim_{n\to\infty} \left[\frac{2}{3}\left(\frac{n}{n}\right)\left(\frac{n+1}{n}\right)\left(\frac{2n+1}{n}\right) + 8 \right]$$

$$= \lim_{n\to\infty} \left[\frac{2}{3}\left(1 + \frac{1}{n}\right)\left(2 + \frac{1}{n}\right) + 8 \right]$$

$$= \frac{2}{3}(1)(2) + 8 = \frac{28}{3} \qquad \blacksquare$$

■ An Intuitive Look at Area (Continued)

We are now ready to resume the discussion of the area concept.

EXAMPLE 8 Following the procedure of Example 1, we can obtain an expression, A_n, for approximating the area of the region under the graph of $f(x) = x^2$ on the interval $[0, 1]$ using n rectangles. Then, by letting n take on increasingly larger values, we will show that

$$\lim_{n\to\infty} A_n = \frac{1}{3}$$

To find such an expression, let's divide the interval $[0, 1]$ into n subintervals of equal length $1/n$ using the $(n + 1)$ points

$$x_0 = 0, \quad x_1 = \frac{1}{n}, \quad x_2 = \frac{2}{n}, \quad x_3 = \frac{3}{n}, \quad \dots, \quad x_k = \frac{k}{n}, \quad \dots, \quad x_n = 1$$

The subintervals are

$$\left[0, \frac{1}{n}\right], \qquad \left[\frac{1}{n}, \frac{2}{n}\right], \qquad \left[\frac{2}{n}, \frac{3}{n}\right], \quad \dots, \quad \left[\frac{k-1}{n}, \frac{k}{n}\right], \quad \dots, \quad \left[\frac{n-1}{n}, 1\right]$$

1st subinterval   2nd subinterval   3rd subinterval kth subinterval nth subinterval

Next, we note that the midpoints of these subintervals are

$$c_1 = \frac{1}{2n}, \quad c_2 = \frac{3}{2n}, \quad c_3 = \frac{5}{2n}, \quad \dots, \quad c_k = \frac{2k-1}{2n}, \quad \dots, \quad c_n = \frac{2n-1}{2n}$$

so the heights of the n corresponding rectangles are

$$f\left(\frac{1}{2n}\right), \quad f\left(\frac{3}{2n}\right), \quad f\left(\frac{5}{2n}\right), \quad \cdots, \quad f\left(\frac{2k-1}{2n}\right), \quad \cdots, \quad f\left(\frac{2n-1}{2n}\right)$$

(See Figure 8.) Letting A_n denote the sum of the areas of the n rectangles, we have

$$A_n = \frac{1}{n}f\left(\frac{1}{2n}\right) + \frac{1}{n}f\left(\frac{3}{2n}\right) + \frac{1}{n}f\left(\frac{5}{2n}\right) + \cdots + \frac{1}{n}f\left(\frac{2k-1}{2n}\right) + \cdots + \frac{1}{n}f\left(\frac{2n-1}{2n}\right)$$

$$= \frac{1}{n}\left[f\left(\frac{1}{2n}\right) + f\left(\frac{3}{2n}\right) + f\left(\frac{5}{2n}\right) + \cdots + f\left(\frac{2k-1}{2n}\right) + \cdots + f\left(\frac{2n-1}{2n}\right)\right] \qquad \text{Factor out } \frac{1}{n}.$$

$$= \frac{1}{n}\sum_{k=1}^{n} f\left(\frac{2k-1}{2n}\right) \qquad \text{Use sigma notation.}$$

$$= \frac{1}{n}\sum_{k=1}^{n}\left(\frac{2k-1}{2n}\right)^2 \qquad f(x) = x^2$$

$$= \frac{1}{n}\sum_{k=1}^{n}\left(\frac{4k^2 - 4k + 1}{4n^2}\right) \qquad \text{Expand the expression following the summation sign.}$$

$$= \frac{1}{4n^3}\sum_{k=1}^{n}(4k^2 - 4k + 1) \qquad n \text{ is constant with respect to summation.}$$

$$= \frac{1}{4n^3}\left[4\sum_{k=1}^{n}k^2 - 4\sum_{k=1}^{n}k + \sum_{k=1}^{n}1\right] \qquad \text{Use the rules of summation.}$$

$$= \frac{1}{4n^3}\left[\frac{4n(n+1)(2n+1)}{6} - \frac{4n(n+1)}{2} + n\right] \qquad \text{Use Theorems 1a, 1b, and 1c.}$$

$$= \frac{1}{4n^3}\cdot\frac{n(4n^2 - 1)}{3}$$

$$= \frac{4n^2 - 1}{12n^2}$$

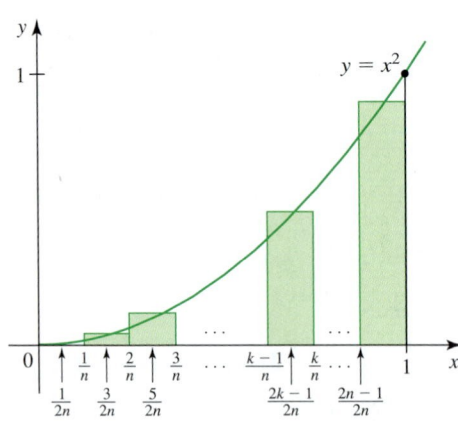

FIGURE 8
The area of the first rectangle is $\frac{1}{n}\cdot f\left(\frac{1}{2n}\right)$, the area of the second rectangle is $\frac{1}{n}\cdot f\left(\frac{3}{2n}\right),\ldots$, and the area of the nth rectangle is $\frac{1}{n}\cdot f\left(\frac{2n-1}{2n}\right)$.

By letting n take on the values 4, 10, and 100, for example, we see that

$$A_4 = \frac{4(4)^2 - 1}{12(4)^2} = 0.328125 \qquad \text{Compare this with Example 1.}$$

$$A_{10} = \frac{4(10)^2 - 1}{12(10)^2} = 0.3325$$

and

$$A_{100} = \frac{4(100)^2 - 1}{12(100)^2} = 0.333325$$

Our computations seem to show that A_n approaches $\frac{1}{3}$ as n gets larger and larger. This result is confirmed by the following calculation:

$$\lim_{n \to \infty} A_n = \lim_{n \to \infty} \frac{4n^2 - 1}{12n^2}$$

$$= \lim_{n \to \infty} \left(\frac{1}{3} - \frac{1}{12n^2} \right)$$

$$= \frac{1}{3}$$

The results of Example 8 suggest that we *define* the area of the region S under the graph of $f(x) = x^2$ on the interval $[0, 1]$ to be $\frac{1}{3}$.

■ Defining the Area of the Region Under the Graph of a Function

Example 8 paves the way to defining the area of the region under the graph of a continuous nonnegative function f on an interval $[a, b]$. (See Figure 9.) We begin by partitioning the interval $[a, b]$ using $n + 1$ equally spaced points

$$a = x_0 < x_1 < x_2 < x_3 < \cdots < x_{n-1} < x_n = b$$

This is called a **regular partition** of $[a, b]$. The resulting subintervals are

$$[x_0, x_1], \qquad [x_1, x_2], \qquad [x_2, x_3], \qquad \ldots, \qquad [x_{n-1}, x_n]$$

with $x_0 = a$ and $x_n = b$. The width of each subinterval is

$$\Delta x = \frac{b - a}{n}$$

This partitioning leads to the subdivision of the region S into n nonoverlapping subregions $S_1, S_2, S_3, \ldots, S_n$, where S_1 is the subregion under the graph of f on $[x_0, x_1]$, S_2 is the subregion under the graph of f on $[x_1, x_2]$, and so on. (See Figure 10.)

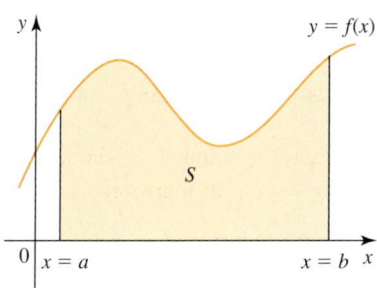

FIGURE 9
The region S under the graph of f on $[a, b]$.

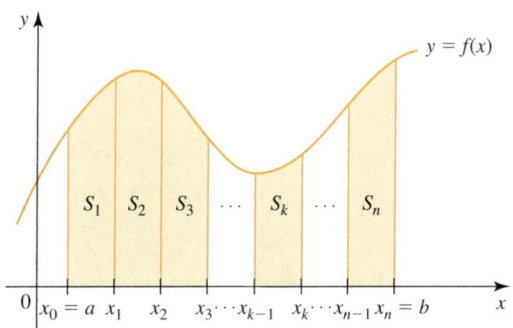

FIGURE 10
The region S is the union of n nonoverlapping subregions.

Next, we approximate the area of the subregion S_1 by the area of the rectangle R_1 with base $[x_0, x_1]$ and height $f(c_1)$, where c_1 is an arbitrarily chosen point in the subinterval $[x_0, x_1]$. (See Figure 11.) Thus,

$$\text{area of } S_1 \approx \text{area of } R_1 = f(c_1)\Delta x$$

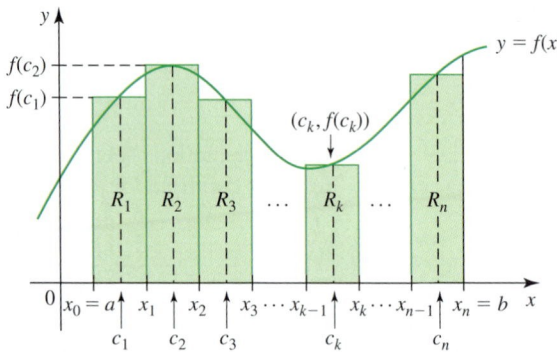

FIGURE 11
The area of the subregion S_k in Figure 10 is approximated by the area of the rectangle R_k.

Similarly, we approximate the area of the subregion S_2 by the area of the rectangle R_2 with base $[x_1, x_2]$ and height $f(c_2)$, where c_2 is an arbitrary point in $[x_1, x_2]$. Thus,

$$\text{area of } S_2 \approx \text{area of } R_2 = f(c_2)\Delta x$$

In general, we approximate the area of the subregion S_k by the area of the rectangle R_k with base $[x_{k-1}, x_k]$ and height $f(c_k)$, where c_k is an arbitrary point in $[x_{k-1}, x_k]$. Thus,

$$\text{area of } S_k \approx \text{area of } R_k = f(c_k)\Delta x$$

If we denote the area of the region S by A and the sum of the areas of the n rectangles by A_n, then, intuitively, we see that

$$A \approx A_n = f(c_1)\Delta x + f(c_2)\Delta x + \cdots + f(c_n)\Delta x = \sum_{k=1}^{n} f(c_k)\Delta x$$

If we let the number of partition points, n, increase, then the number of subregions increases, and, as shown in Figure 12, the approximations seem to improve.

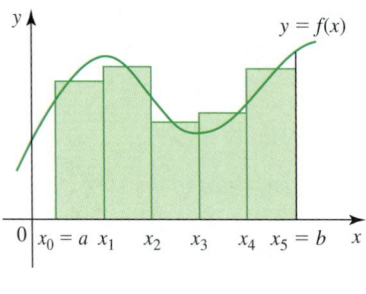

FIGURE 12
As n increases the approximation seems to improve.

(a) $n = 5$

(b) $n = 10$

This observation suggests that we define the area A of the region S as follows.

DEFINITION Area of the Region S Under the Graph of a Function

Let f be a continuous, nonnegative function defined on an interval $[a, b]$. Suppose that $[a, b]$ is divided into n subintervals of equal length $\Delta x = (b - a)/n$ by means of $(n + 1)$ equally spaced points

$$a = x_0 < x_1 < x_2 < \cdots < x_n = b$$

Then the area of the region S that lies under the graph of f on $[a, b]$ is

$$A = \lim_{n \to \infty} A_n = \lim_{n \to \infty} \sum_{k=1}^{n} f(c_k)\Delta x \qquad (1)$$

where c_k lies in the kth subinterval $[x_{k-1}, x_k]$.

Because of the supposition that f is continuous, it can be shown that the limit (1) in the definition always exists, regardless of how the points c_k in $[x_{k-1}, x_k]$, where $1 \le k \le n$, are chosen. In Exercises 47 and 48 you will be asked to compute the area of the region under the graph of $f(x) = x^2$ on $[0, 1]$ choosing c_k to be (1) the left endpoint of the subinterval $[x_{k-1}, x_k]$, that is, $c_k = x_{k-1}$, for $1 \le k \le n$, and (2) the right endpoint of the subinterval $[x_{k-1}, x_k]$, that is, $c_k = x_k$, for $1 \le k \le n$. You will see that the results are indeed the same as those obtained in Example 8, where c_k was chosen to be the midpoint of the subinterval $[x_{k-1}, x_k]$.

EXAMPLE 9 Find the area of the region under the graph of $f(x) = 4 - x^2$ on the interval $[-2, 1]$.

Solution Observe that f is continuous and nonnegative on $[-2, 1]$. The region under consideration is shown in Figure 13. If we partition the interval $[-2, 1]$ into n subintervals of equal length by means of $(n + 1)$ points, then the width of each subinterval is

$$\Delta x = \frac{b - a}{n} = \frac{1 - (-2)}{n} = \frac{3}{n}$$

and the partition points are

$$x_0 = -2, \qquad x_1 = -2 + \frac{3}{n}, \qquad x_2 = -2 + 2\left(\frac{3}{n}\right), \qquad \dots,$$

$$x_k = -2 + k\left(\frac{3}{n}\right), \qquad \dots, \qquad x_n = 1$$

Since f is continuous, we have a free hand at picking c_k in $[x_{k-1}, x_k]$. So let's pick c_k to be the right endpoint of the subinterval; that is,

$$c_k = x_k = -2 + \frac{3k}{n}$$

Using the definition of the area of the region under a graph of a function, we find the required area to be

$$A = \lim_{n \to \infty} \sum_{k=1}^{n} f(c_k)\Delta x$$

$$= \lim_{n \to \infty} \sum_{k=1}^{n} f\left(-2 + \frac{3k}{n}\right)\left(\frac{3}{n}\right)$$

$$= \lim_{n \to \infty} \sum_{k=1}^{n} \left[4 - \left(-2 + \frac{3k}{n}\right)^2\right]\left(\frac{3}{n}\right) \qquad f(x) = 4 - x^2$$

$$= \lim_{n \to \infty} \frac{3}{n} \sum_{k=1}^{n} \left(4 - 4 + \frac{12k}{n} - \frac{9k^2}{n^2}\right)$$

$$= \lim_{n \to \infty} \frac{3}{n} \left[\frac{12}{n} \sum_{k=1}^{n} k - \frac{9}{n^2} \sum_{k=1}^{n} k^2\right]$$

$$= \lim_{n \to \infty} \frac{3}{n} \left[\frac{12}{n} \cdot \frac{n(n + 1)}{2} - \frac{9}{n^2} \cdot \frac{n(n + 1)(2n + 1)}{6}\right] \qquad \text{Use Theorems 1b and 1c.}$$

$$= \lim_{n \to \infty} \left[18\left(1 + \frac{1}{n}\right) - \frac{9}{2}\left(1 + \frac{1}{n}\right)\left(2 + \frac{1}{n}\right)\right]$$

$$= 18 - \frac{9}{2}(2) = 9 \qquad \blacksquare$$

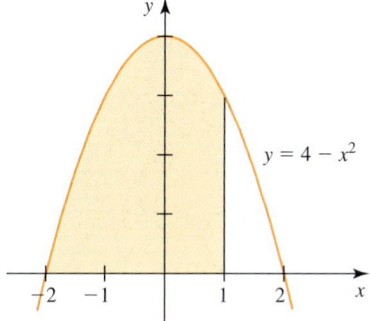

FIGURE 13
The region under the graph of $f(x) = 4 - x^2$ on $[-2, 1]$

Note In Exercise 49 you will be asked to solve Example 9 again, this time choosing the midpoint for c_k. Of course, you should obtain the same answer. ■

■ Area and Distance

We now show that if v is a (continuous) velocity function of a car traveling in a straight line and $v(t) \geq 0$ on $[a, b]$, then the distance covered by the car between $t = a$ and $t = b$ is numerically equal to the area of the region under the graph of the velocity function on $[a, b]$. (See Figure 14.) Let's divide the time interval $[a, b]$ into n subintervals each of equal length $\Delta t = (b - a)/n$ by means of $(n + 1)$ equally spaced points

$$t_0 = a, \quad t_1 = a + \Delta t, \quad t_2 = a + 2(\Delta t), \quad \ldots, \quad t_k = a + k(\Delta t), \quad \ldots, \quad t_n = b$$

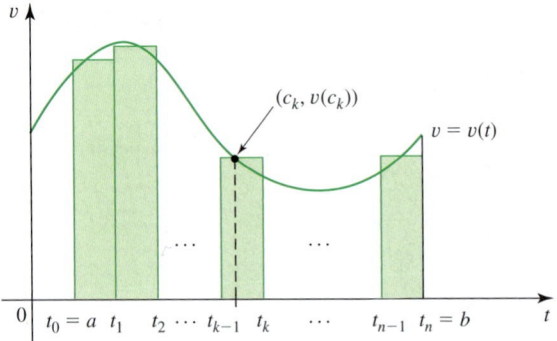

FIGURE 14
v is the velocity function on $[a, b]$.

Observe that if n is large, then the time intervals $[t_0, t_1], [t_1, t_2], \ldots, [t_{n-1}, t_n]$ are uniformly small.

Let's focus our attention on the first subinterval $[t_0, t_1]$. Because v is continuous, we see that the speed of the car does not vary appreciably in that interval and can be approximated by the *constant* speed $v(c_1)$, where c_1 is an arbitrary point in $[t_0, t_1]$.* Therefore, the distance covered by the car from $t = t_0$ to $t = t_1$ may be approximated by

$$v(c_1)\Delta t \qquad \text{distance = constant speed · time elapsed}$$

In a similar manner we see that the distance covered by the car from t_1 to t_2 is approximately

$$v(c_2)\Delta t$$

where c_2 is an arbitrary point in $[t_1, t_2]$. Continuing, we see that the distance covered by the car from t_{k-1} to t_k is approximately

$$v(c_k)\Delta t$$

where c_k is an arbitrary point in $[t_{k-1}, t_k]$. Therefore, the distance traveled by the car from $t = a$ to $t = b$ is approximately

$$v(c_1)\Delta t + v(c_2)\Delta t + \cdots + v(c_n)\Delta t = \sum_{k=1}^{n} v(c_k)\Delta t \tag{2}$$

*Recall that if a function f is continuous at t, then a small change in t implies a small change in $f(t)$.

As n gets larger and larger, the length of the time subintervals gets smaller and smaller. Intuitively, we expect that the approximations will improve. It seems reasonable, therefore, to define the distance covered by the car to be

$$\lim_{n \to \infty} \sum_{k=1}^{n} v(c_k) \Delta t$$

But as you can see from Figure 14, this quantity also gives the area of the region under the graph of v on $[a, b]$.

EXAMPLE 10 **Distance Covered by a Cyclist** The speed of a cyclist is measured at 4-sec intervals over a 32-sec time span and recorded in the following table.

Time (sec)	0	4	8	12	16	20	24	28	32
Speed (ft/sec)	2	4	6	10	12	14	10	8	6

If we let v denote the velocity function associated with the motion of the cyclist over the time interval $[0, 32]$, then the values of v are available to us only at a discrete set of numbers, even though v is clearly a continuous function defined on the interval. Using Equation (2), find the approximate distance D covered by the cyclist from $t = 0$ to $t = 32$ using

a. Eight ($n = 8$) rectangles and choosing c_k to be the left endpoint of the kth subinterval

b. Eight ($n = 8$) rectangles and choosing c_k to be the right endpoint of the kth subinterval

c. Four ($n = 4$) rectangles and choosing c_k to be the midpoint of the kth subinterval.

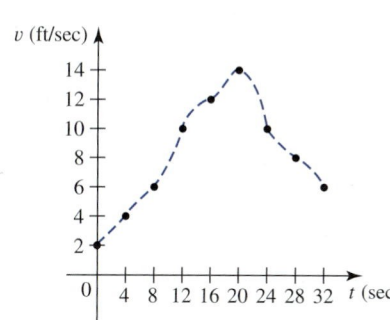

FIGURE 15
An approximation of the graph of v on $[0, 32]$

Solution An approximation to the graph of v is shown in Figure 15. (Remember that we know the values of v at only $t = 0, 4, 8, \ldots, 32$.)

a. Using eight rectangles with $t_0 = 0$, $t_1 = 4$, $t_2 = 8$, $\ldots$, $t_8 = 32$ and $c_1 = 0$, $c_2 = 4$, $c_3 = 8$, $\ldots$, $c_8 = 28$, we see that the required approximate distance is

$$D = \sum_{k=1}^{8} v(c_k) \Delta t = \sum_{k=1}^{8} v(t_{k-1}) \Delta t = v(t_0) \cdot 4 + v(t_1) \cdot 4 + \cdots + v(t_7) \cdot 4$$

$$= v(0) \cdot 4 + v(4) \cdot 4 + v(8) \cdot 4 + \cdots + v(28) \cdot 4$$

$$= 2 \cdot 4 + 4 \cdot 4 + 6 \cdot 4 + 10 \cdot 4 + 12 \cdot 4 + 14 \cdot 4 + 10 \cdot 4 + 8 \cdot 4$$

$$= 264$$

or 264 ft.

b. Using the same partition as in part (a) and

$$c_1 = 4, \qquad c_2 = 8, \qquad c_3 = 12, \qquad \ldots, \qquad c_8 = 32$$

we find

$$D = \sum_{k=1}^{8} v(c_k) \Delta t = \sum_{k=1}^{8} v(t_k) \Delta t = v(t_1) \cdot 4 + v(t_2) \cdot 4 + \cdots + v(t_8) \cdot 4$$

$$= v(4) \cdot 4 + v(8) \cdot 4 + v(12) \cdot 4 + \cdots + v(32) \cdot 4$$

$$= 4 \cdot 4 + 6 \cdot 4 + 10 \cdot 4 + 12 \cdot 4 + 14 \cdot 4 + 10 \cdot 4 + 8 \cdot 4 + 6 \cdot 4$$

$$= 280$$

or 280 ft.

c. We use four rectangles with $t_0 = 0$, $t_1 = 8$, $t_2 = 16$, $t_3 = 24$, and $t_4 = 32$ and

$$c_1 = 4, \qquad c_2 = 12, \qquad c_3 = 20, \qquad \text{and} \qquad c_4 = 28$$

obtaining

$$
\begin{aligned}
D &= \sum_{k=1}^{4} v(c_k)\Delta t = v(c_1) \cdot 8 + v(c_2) \cdot 8 + v(c_3) \cdot 8 + v(c_4) \cdot 8 \\
&= v(4) \cdot 8 + v(12) \cdot 8 + v(20) \cdot 8 + v(28) \cdot 8 \\
&= 4 \cdot 8 + 10 \cdot 8 + 14 \cdot 8 + 8 \cdot 8 \\
&= 288
\end{aligned}
$$

or 288 ft.

4.3 CONCEPT QUESTIONS

1. Let $f(x) = x + 1$ on the interval $[1, 5]$. Divide the interval $[1, 5]$ into four subintervals of length 1 using the points

$$x_0 = 1, \quad x_1 = 2, \quad x_2 = 3, \quad x_3 = 4, \quad \text{and} \quad x_4 = 5$$

Write the sum $\sum_{k=1}^{4} f(c_k)\Delta x$ to approximate the area of the region S under the graph of f on $[1, 5]$, choosing c_k in the subinterval $[x_{k-1}, x_k]$, where $1 \le k \le 4$, to be (a) the left endpoint of the subinterval, $c_k = x_{k-1}$; (b) the right endpoint of the subinterval; $c_k = x_k$; and (c) the midpoint of the

subinterval, $c_k = \frac{1}{2}(x_{k-1} + x_k) = x_{k-1} + \frac{1}{2}\Delta x$. Sketch the graph of f and the approximating rectangles for parts (a)–(c).

2. Refer to Exercise 1. Find the area of the region S under the graph of f in $[1, 5]$ by calculating $\lim_{n \to \infty} \sum_{k=1}^{n} f(c_k)\Delta x$, where c_k is chosen to be (a) the left endpoint, (b) the right endpoint, and (c) the midpoint of the subinterval $[x_{k-1}, x_k]$, where $1 \le k \le n$. Verify your result by using elementary geometry to find the area of the region S.

4.3 EXERCISES

In Exercises 1–12 you are given a function f, an interval $[a, b]$, the number n of subintervals into which $[a, b]$ is divided (each of length $\Delta x = (b - a)/n$), and the point c_k in $[x_{k-1}, x_k]$, where $1 \le k \le n$. (a) Sketch the graph of f and the rectangles with base on $[x_{k-1}, x_k]$ and height $f(c_k)$, and (b) find the approximation $\sum_{k=1}^{n} f(c_k)\Delta x$ of the area of the region S under the graph of f on $[a, b]$.

1. $f(x) = x$, $[0, 1]$, $n = 5$, c_k is the left endpoint

2. $f(x) = x$, $[1, 4]$, $n = 6$, c_k is the midpoint

3. $f(x) = 2x + 3$, $[0, 4]$, $n = 5$, c_k is the right endpoint

4. $f(x) = 3 - 2x$, $[0, 1]$, $n = 5$, c_k is the left endpoint

5. $f(x) = 8 - 2x$, $[1, 3]$, $n = 4$, c_k is the midpoint

6. $f(x) = x^2$, $[0, 1]$, $n = 5$, c_k is the right endpoint

7. $f(x) = x^2$, $[1, 3]$, $n = 4$, c_k is the midpoint

8. $f(x) = 4 - x^2$, $[0, 2]$, $n = 8$, c_k is the left endpoint

9. $f(x) = 16 - x^2$, $[1, 3]$, $n = 5$, c_k is the right endpoint

10. $f(x) = \sqrt{x}$, $[0, 4]$, $n = 8$, c_k is the left endpoint

11. $f(x) = \dfrac{1}{x}$, $[1, 2]$, $n = 10$, c_k is the left endpoint

12. $f(x) = \cos x$, $\left[0, \frac{\pi}{2}\right]$, $n = 4$, c_k is the midpoint

In Exercises 13–20, expand and then evaluate the sum.

13. $\sum_{k=1}^{10} 1$

14. $\sum_{k=1}^{5} 2k$

15. $\sum_{k=1}^{5} (2k - 1)$

16. $\sum_{k=1}^{5} k(k + 1)$

17. $\sum_{k=1}^{5} k^2$

18. $\sum_{k=1}^{5} \frac{1}{k}$

19. $\sum_{k=1}^{4} \sqrt{k}$

20. $\sum_{k=1}^{4} k \sin \frac{k\pi}{2}$

In Exercises 21–30, rewrite the sum using sigma notation. Do not evaluate.

21. $2 + 4 + 6 + 8 + \cdots + 60$

22. $2 \cdot 1 + 2 \cdot 2 + 2 \cdot 3 + \cdots + 2 \cdot 10$

23. $3 + 5 + 7 + 9 + \cdots + 23$

24. $\dfrac{1}{5} + \dfrac{2}{5} + \dfrac{3}{5} + \dfrac{4}{5} + \cdots + \dfrac{8}{5}$

25. $\left[2\left(\dfrac{1}{5}\right) + 1\right] + \left[2\left(\dfrac{2}{5}\right) + 1\right] + \left[2\left(\dfrac{3}{5}\right) + 1\right]$

$+ \left[2\left(\dfrac{4}{5}\right) + 1\right] + \left[2\left(\dfrac{5}{5}\right) + 1\right]$

26. $\left[\left(\dfrac{1}{4}\right)^2 - 1\right]\left(\dfrac{1}{4}\right) + \left[\left(\dfrac{2}{4}\right)^2 - 1\right]\left(\dfrac{1}{4}\right)$

$+ \left[\left(\dfrac{3}{4}\right)^2 - 1\right]\left(\dfrac{1}{4}\right) + \left[\left(\dfrac{4}{4}\right)^2 - 1\right]\left(\dfrac{1}{4}\right)$

27. $\left[2\left(\dfrac{1}{n}\right)^3 - 1\right]\left(\dfrac{1}{n}\right) + \left[2\left(\dfrac{2}{n}\right)^3 - 1\right]\left(\dfrac{1}{n}\right)$

$+ \left[2\left(\dfrac{3}{n}\right)^3 - 1\right]\left(\dfrac{1}{n}\right) + \cdots + \left[2\left(\dfrac{n}{n}\right)^3 - 1\right]\left(\dfrac{1}{n}\right)$

28. $\left[\sqrt{\dfrac{0}{n}} + 1\right]\left(\dfrac{1}{n}\right) + \left[\sqrt{\dfrac{1}{n}} + 1\right]\left(\dfrac{1}{n}\right)$

$+ \left[\sqrt{\dfrac{2}{n}} + 1\right]\left(\dfrac{1}{n}\right) + \cdots + \left[\sqrt{\dfrac{n-1}{n}} + 1\right]\left(\dfrac{1}{n}\right)$

29. $\dfrac{1}{n}\sin\left(1 + \dfrac{1}{n}\right) + \dfrac{1}{n}\sin\left(1 + \dfrac{2}{n}\right)$

$+ \dfrac{1}{n}\sin\left(1 + \dfrac{3}{n}\right) + \cdots + \dfrac{1}{n}\sin\left(1 + \dfrac{n}{n}\right)$

30. $\dfrac{1}{n}\sec^2\left(1 + \dfrac{1}{n}\right) + \dfrac{1}{n}\sec^2\left(1 + \dfrac{2}{n}\right)$

$+ \dfrac{1}{n}\sec^2\left(1 + \dfrac{3}{n}\right) + \cdots + \dfrac{1}{n}\sec^2\left(1 + \dfrac{n}{n}\right)$

In Exercises 31–38, use the rules of summation and the summation formulas to evaluate the sum.

31. $\displaystyle\sum_{k=1}^{10}(2k + 1)$
32. $\displaystyle\sum_{k=1}^{8}(3 - k^2)$

33. $\displaystyle\sum_{k=1}^{10}k(k - 2)$
34. $\displaystyle\sum_{k=1}^{40}k(k^2 - k)$

35. $\displaystyle\sum_{k=1}^{10}k(2k + 1)^2$
36. $\displaystyle\sum_{k=1}^{n}\dfrac{1}{n^2}(2k + 1)$

37. $\displaystyle\sum_{k=1}^{n}(2k + 1)^2$
38. $\displaystyle\sum_{k=1}^{n}\dfrac{1}{n}\left(1 + \dfrac{k}{n}\right)^2$

In Exercises 39–44, evaluate the limit after first finding the sum (as a function of n) using the summation formulas.

39. $\displaystyle\lim_{n\to\infty}\sum_{k=1}^{n}\dfrac{2k}{n^2}$
40. $\displaystyle\lim_{n\to\infty}\sum_{k=1}^{n}\dfrac{1}{n^3}(2k + 1)^2$

41. $\displaystyle\lim_{n\to\infty}\sum_{k=1}^{n}\left(\dfrac{k}{n} + 2\right)\left(\dfrac{3}{n}\right)$
42. $\displaystyle\lim_{n\to\infty}\sum_{k=1}^{n}\left[1 + 2\left(\dfrac{k}{n}\right)^2\right]\left(\dfrac{2}{n}\right)$

43. $\displaystyle\lim_{n\to\infty}\sum_{k=1}^{n}\left(1 + \dfrac{2k}{n}\right)^2\left(\dfrac{1}{n}\right)$
44. $\displaystyle\lim_{n\to\infty}\sum_{k=1}^{n}\left(1 + \dfrac{2k-1}{2n}\right)\left(\dfrac{1}{n}\right)$

In Exercises 45–52, use the definition of area (page 438) to find the area of the region under the graph of f on [a, b] using the indicated choice of c_k.

45. $f(x) = 2x + 1$, $[0, 2]$, c_k is the left endpoint
46. $f(x) = 3x - 1$, $[1, 3]$, c_k is the midpoint
47. $f(x) = x^2$, $[0, 1]$, c_k is the left endpoint
48. $f(x) = x^2$, $[0, 1]$, c_k is the right endpoint
49. $f(x) = 4 - x^2$, $[-2, 1]$, c_k is the midpoint
50. $f(x) = x - x^2$, $[-2, 1]$, c_k is the right endpoint
51. $f(x) = x^2 + 2x + 2$, $[-1, 1]$, c_k is the right endpoint
52. $f(x) = 2x - x^3$, $[0, 1]$, c_k is the right endpoint

cas *In Exercises 53–56, (a) express the area of the region under the graph of the function f over the interval as the limit of a sum (use the right endpoints), (b) use a computer algebra system (CAS) to find the sum obtained in part (a) in compact form, and (c) evaluate the limit of the sum found in part (b) to obtain the exact area of the region.*

53. $f(x) = x^4$; $[0, 2]$
54. $f(x) = x^5$; $[0, 2]$
55. $f(x) = x^4 + 2x^2 + x$; $[2, 5]$
56. $f(x) = \sin x$; $\left[0, \dfrac{\pi}{2}\right]$

57. A regular n-sided polygon is inscribed in a circle of radius r as shown in the figure with $n = 6$.

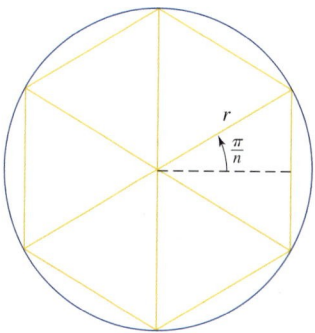

a. Show that the area of the polygon is $A_n = \frac{1}{2}nr^2\sin(2\pi/n)$.

b. Evaluate $\lim_{n\to\infty} A_n$ to obtain the area of the circle $A = \pi r^2$.

  Hint: Use the result $\displaystyle\lim_{x\to 0}\dfrac{\sin x}{x} = 1$.

58. Refer to Exercise 57.

a. Show that the perimeter of the polygon is $C_n = 2nr\sin(\pi/n)$.

b. Evaluate $\lim_{n\to\infty} C_n$ to obtain the circumference of the circle $C = 2\pi r$.

59. Real Estate Figure (a) shows a vacant lot with a 100-ft frontage in a development. To estimate its area, we introduce a coordinate system so that the x-axis coincides with the edge of the straight road forming the lower boundary of the property, as shown in Figure (b). Then, thinking of the upper boundary of the property as the graph of a continuous function f over the interval $[0, 100]$, we see that the problem is mathematically equivalent to that of finding the area of the region under the graph of f on $[0, 100]$. To estimate the area of the lot using the sum of the areas of rectangles, we divide the interval $[0, 100]$ into five equal subintervals of length 20 ft. Then, using surveyor's equipment, we measure the distance from the midpoint of each of these subintervals to the upper boundary of the property. These measurements give the values of $f(x)$ at $x = 10, 30, 50, 70,$ and 90. What is the approximate area of the lot?

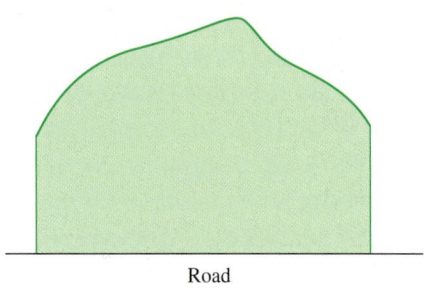

Road

(a)

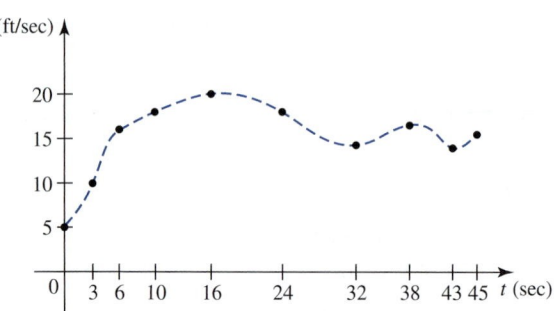

Time (sec)	0	3	6	10	16	24	32	38	43	45
Velocity (ft/sec)	5	10	16	18	20	18	14	17	14	15

Using nine rectangles determined by the 10 points

$$t_0 = 0, \qquad t_1 = 3, \qquad t_2 = 6, \qquad \dots, \qquad t_9 = 45$$

and choosing c_k to be the left endpoint of the kth subinterval, estimate the total height gained by the balloon over the time period from $t = 0$ to $t = 45$.

Note: Here, the partition points are not spaced equally apart, so the subintervals are not of equal length.

61. Prove that $\sum_{k=1}^{n}(a_k + b_k) = \sum_{k=1}^{n} a_k + \sum_{k=1}^{n} b_k$.

62. Prove that $\sum_{k=1}^{n}(a_k - b_k) = \sum_{k=1}^{n} a_k - \sum_{k=1}^{n} b_k$.

In Exercises 63–66, determine whether the statement is true or false. If it is true, explain why it is true. If it is false, explain why or give an example to show why it is false.

63. If f is a nonnegative function such that $f(x)$ is strictly positive for some value of x in $[a, b]$, $[a, b]$ is partitioned into n subintervals of equal length, and c_k lies in the kth subinterval $[x_{k-1}, x_k]$, then $\sum_{k=1}^{n} f(c_k)\Delta x$ must be strictly positive.
Hint: Study the Dirichlet function of Exercise 92 in Section 1.2.

64. $\displaystyle\sum_{k=1}^{n} (ca_k - db_k) = c \sum_{k=1}^{n} a_k - d \sum_{k=1}^{n} b_k,$
where c and d are constants

65. $\displaystyle\left(\sum_{k=1}^{n} a_k \right)\left(\sum_{k=1}^{n} b_k \right) = \sum_{k=1}^{n} a_k b_k$

66. $\displaystyle\sum_{k=1}^{n} (a_k - b_k)^2 = \sum_{k=1}^{n} a_k^2 - \sum_{k=1}^{n} b_k^2$

y (ft)

80   100   110   100   80

10 20 30 40 50 60 70 80 90 100   x (ft)

(b)

60. Hot-Air Balloon The rate of ascent or descent of a hot-air balloon is measured at certain instants of time from $t = 0$ to $t = 45$ as summarized in the following table. The dashed curve in the figure is an estimate of the graph of the velocity function on the time interval $[0, 45]$.

4.4   The Definite Integral

Historical Biography

SPL/Photo Researchers, Inc.

BERNHARD RIEMANN
(1826-1866)

Bernhard Riemann was one of the few mathematicians to impress his contemporary Carl Friedrich Gauss, and his work continues to deeply influence modern mathematics. Born the son of a poor country pastor in Northern Germany, Riemann was raised without family money to support his education. Nevertheless, he was able to secure a solid education and showed exceptional mathematical insight at an early age. While still in secondary school, he studied the works of Euler (page 19) and Legendre, mastering Legendre's treatise on number theory in less than a week. He obtained his doctorate in 1851 from the University of Göttingen after writing a thesis involving the theory of functions of a complex variable. In 1854, upon his appointment as Privatdozent (unpaid lecturer), Riemann was required to give a lecture to the current professors. He submitted three topics to then department chair Gauss, who, in past situations, had chosen whichever topic was listed first. But Riemann had submitted the foundations of geometry as his third topic, one that so interested Gauss that it was the topic chosen. After two months of preparation, Riemann presented his lecture, and that work is now considered one of the great classical masterpieces of mathematics. It was documented that even Gauss was impressed. Riemann's famous conjecture, the Riemann Hypothesis, remains unresolved to this day, and the search for a solution to that problem is still very active. The problem has been designated one of seven Prize Problems by the Clay Mathematics Institute, and $1,000,000 will be awarded to the person who finds a solution.

■ Definition of the Definite Integral

In Section 4.3 we saw that the area of the region under the graph of a continuous, nonnegative function f on an interval $[a, b]$ is defined by a limit of the form

$$\lim_{n \to \infty} \sum_{k=1}^{n} f(c_k)\Delta x = \lim_{n \to \infty}[f(c_1)\Delta x + f(c_2)\Delta x + \cdots + f(c_n)\Delta x] \quad (1)$$

where $\Delta x = (b - a)/n$ and c_k is in $[x_{k-1}, x_k]$. We also saw that the distance covered by an object moving along a straight line with a positive velocity is found by evaluating a similar limit.

In this section we will look at limits defined by Equation (1) in which f may take on both positive and negative values. We will give a geometric interpretation for this general case later on. We will also interpret such limits in terms of the position of an object that moves with both positive and negative velocities. Looking ahead, we will see that limits of this type arise when we try to find the length and mass of a curved wire, the center of mass of a body, the volume of a solid, the area of a surface, the pressure exerted by a fluid against the wall of a container, the amount of oil consumed over a certain period of time, the net sales of a department store over a certain period, and the total number of AIDS cases diagnosed over a certain period of time, just to name a few applications.

In the following definition we will assume, as before, that f is continuous. This allows for a relatively simple development of the material ahead of us.

DEFINITION   Definite Integral

Let f be a continuous function defined on an interval $[a, b]$. Suppose that $[a, b]$ is divided into n subintervals of equal length $\Delta x = (b - a)/n$ by means of $(n + 1)$ equally spaced points

$$a = x_0 < x_1 < x_2 < \cdots < x_n = b$$

Let $c_1, c_2, \ldots, c_n$ be arbitrary points in the respective subintervals with c_k lying in the kth subinterval $[x_{k-1}, x_k]$. Then the **definite integral of f on $[a, b]$,** denoted by $\int_a^b f(x)\, dx$, is

$$\int_a^b f(x)\, dx = \lim_{n \to \infty} \sum_{k=1}^{n} f(c_k)\Delta x \quad (2)$$

We also say that f is **integrable on $[a, b]$** if the limit (2) exists. The process of evaluating a definite integral is called **integration.** The number a in the definition is called the **lower limit of integration,** and the number b is called the **upper limit of integration.** Together, the numbers a and b are referred to as the **limits of integration.** As in the case of the indefinite integral, the function f to be integrated is called the **integrand.**

The sum $\sum_{k=1}^{n} f(c_k)\Delta x$ in the definition is called a **Riemann sum** in honor of the German mathematician Bernhard Riemann (1826–1866). Actually, this sum is a special case of a more general form of a Riemann sum in which no assumption is made requiring that f be continuous on $[a, b]$ or that the interval be partitioned in such a way that the resulting subintervals have equal length. For completeness we will discuss this general case at the end of this section.

Notes

1. The assumption that f is continuous on $[a, b]$ guarantees that the definite integral always exists. In other words, the limit in Equation (2) exists and is unique for all choices of the *evaluation* points c_k. Furthermore, if f is nonnegative, then the definite integral gives the area of the region under the graph of f on $[a, b]$ since the limit in Equation (2) reduces to the limit in Equation (1), page 438, in Section 4.3.

2. The symbol $\int$ in the definition of the definite integral is the same as that used to denote the indefinite integral of a function. (Remember that the definite integral is a number, in contrast to the indefinite integral, which is a family of functions (the antiderivatives of f).)

EXAMPLE 1 Compute the Riemann sum for $f(x) = 4 - x^2$ on $[-1, 3]$ using five subintervals ($n = 5$) and choosing the evaluation points to be the midpoints of the subintervals.

Solution   Here, $a = -1$, $b = 3$, and $n = 5$. So the length of each subinterval is

$$\Delta x = \frac{b - a}{n} = \frac{3 - (-1)}{5} = \frac{4}{5}$$

The partition points are

$$x_0 = -1, \qquad x_1 = -1 + \frac{4}{5} = -\frac{1}{5}, \qquad x_2 = -1 + 2\left(\frac{4}{5}\right) = \frac{3}{5},$$

$$x_3 = \frac{7}{5}, \qquad x_4 = \frac{11}{5}, \qquad \text{and} \qquad x_5 = 3$$

The midpoints of the subintervals are given by $c_k = \frac{1}{2}(x_k + x_{k-1})$, or

$$c_1 = -\frac{3}{5}, \qquad c_2 = \frac{1}{5}, \qquad c_3 = 1, \qquad c_4 = \frac{9}{5}, \qquad \text{and} \qquad c_5 = \frac{13}{5}$$

(See Figure 1.)

Therefore, the required Riemann sum is

$$\sum_{k=1}^{5} f(c_k)\Delta x = f(c_1)\Delta x + f(c_2)\Delta x + f(c_3)\Delta x + f(c_4)\Delta x + f(c_5)\Delta x$$

$$= \left[f\left(-\frac{3}{5}\right) + f\left(\frac{1}{5}\right) + f(1) + f\left(\frac{9}{5}\right) + f\left(\frac{13}{5}\right) \right]\Delta x$$

$$= \left(\frac{4}{5}\right)\left\{ \left[4 - \left(-\frac{3}{5}\right)^2 \right] + \left[4 - \left(\frac{1}{5}\right)^2 \right] + [4 - (1)^2] \right.$$

$$\left. + \left[4 - \left(\frac{9}{5}\right)^2 \right] + \left[4 - \left(\frac{13}{5}\right)^2 \right] \right\}$$

$$= \left(\frac{4}{5}\right)(3.64 + 3.96 + 3 + 0.76 - 2.76)$$

$$= 6.88$$

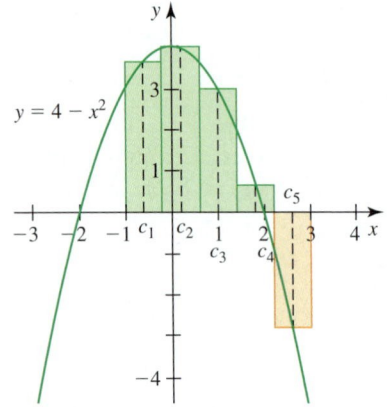

FIGURE 1
The positive terms of the Riemann sum are associated with the rectangles that lie above the x-axis; the negative term is associated with the rectangle that lies below the x-axis.

The Riemann sum computed in Example 1 is the sum of five terms. As you can see in Figure 1, these terms are associated with the areas of the five rectangles shown. The positive terms give the areas of the rectangles that lie above the x-axis, while the negative term is the negative of the area of the rectangle that lies below the x-axis.

EXAMPLE 2   Evaluate $\displaystyle\int_{-1}^{3} (4 - x^2)\, dx$.

Solution   Here, $a = -1$ and $b = 3$. Furthermore, $f(x) = 4 - x^2$ is continuous on $[-1, 3]$, so f is integrable on $[-1, 3]$. To evaluate the given definite integral, let's subdivide the interval $[-1, 3]$ into n equal subintervals of length

$$\Delta x = \frac{b - a}{n} = \frac{3 - (-1)}{n} = \frac{4}{n}$$

The partition points are

$$x_0 = -1, \qquad x_1 = -1 + \frac{4}{n}, \qquad x_2 = -1 + 2\left(\frac{4}{n}\right), \qquad \ldots,$$

$$x_{k-1} = -1 + (k - 1)\left(\frac{4}{n}\right), \qquad x_k = -1 + k\left(\frac{4}{n}\right), \qquad \ldots, \qquad x_n = 3$$

Next, we pick c_k to be the right endpoint of the subinterval $[x_{k-1}, x_k]$ so that

$$c_k = x_k = -1 + k\left(\frac{4}{n}\right) = -1 + \frac{4k}{n}$$

Then

$$\int_{-1}^{3} (4 - x^2)\,dx = \int_{-1}^{3} f(x)\, dx = \lim_{n \to \infty} \sum_{k=1}^{n} f(c_k)\Delta x$$

$$= \lim_{n \to \infty} \sum_{k=1}^{n} f\left(-1 + \frac{4k}{n}\right)\left(\frac{4}{n}\right)$$

$$= \lim_{n \to \infty} \sum_{k=1}^{n} \left[4 - \left(-1 + \frac{4k}{n}\right)^2\right]\left(\frac{4}{n}\right) \qquad {\color{red} f(x) = 4 - x^2}$$

$$= \lim_{n \to \infty} \left(\frac{4}{n}\right) \sum_{k=1}^{n} \left(3 + \frac{8k}{n} - \frac{16k^2}{n^2}\right)$$

$$= \lim_{n \to \infty} \left[\frac{4}{n} \sum_{k=1}^{n} 3 + \frac{32}{n^2} \sum_{k=1}^{n} k - \frac{64}{n^3} \sum_{k=1}^{n} k^2\right]$$

$$= \lim_{n \to \infty} \left[\frac{4}{n}(3n) + \frac{32}{n^2} \cdot \frac{n(n + 1)}{2} - \frac{64}{n^3} \cdot \frac{n(n + 1)(2n + 1)}{6}\right]$$

$$= \lim_{n \to \infty} \left[12 + 16\left(1 + \frac{1}{n}\right) - \frac{32}{3}\left(1 + \frac{1}{n}\right)\left(2 + \frac{1}{n}\right)\right]$$

$$= 12 + 16 - \frac{64}{3} = \frac{20}{3} = 6\frac{2}{3}$$

(Compare this with the approximate value of $\int_{-1}^{3} (4 - x^2)\, dx$ obtained in Example 1.)

EXAMPLE 3   Show that $\displaystyle\int_{a}^{b} x\, dx = \frac{1}{2}(b^2 - a^2)$.

Solution   Let's subdivide the interval $[a, b]$ into n subintervals of length

$$\Delta x = \frac{b - a}{n}$$

The partition points are

$$x_0 = a, \qquad x_1 = a + \frac{b-a}{n}, \qquad x_2 = a + 2\left(\frac{b-a}{n}\right), \qquad \ldots,$$

$$x_k = a + k\left(\frac{b-a}{n}\right), \qquad \ldots, \qquad x_n = b$$

Next we choose the evaluation point c_k to be the right endpoint of the subinterval $[x_{k-1}, x_k]$, where $1 \le k \le n$; that is, we pick $c_k = x_k$ for each $1 \le k \le n$. Then

$$\begin{aligned}
\int_a^b x \, dx &= \lim_{n\to\infty} \sum_{k=1}^n f(c_k)\Delta x \\
&= \lim_{n\to\infty} \sum_{k=1}^n \left[a + \left(\frac{b-a}{n}\right)k\right]\left(\frac{b-a}{n}\right) \\
&= (b-a)\lim_{n\to\infty} \frac{1}{n} \sum_{k=1}^n \left[a + \left(\frac{b-a}{n}\right)k\right] \\
&= (b-a)\lim_{n\to\infty} \frac{1}{n} \left[\sum_{k=1}^n a + \left(\frac{b-a}{n}\right)\sum_{k=1}^n k\right] \\
&= (b-a)\lim_{n\to\infty} \frac{1}{n} \left[na + \left(\frac{b-a}{n}\right)\cdot\frac{n(n+1)}{2}\right] \\
&= (b-a)\lim_{n\to\infty} \left[a + \left(\frac{b-a}{2}\right)\cdot\frac{n(n+1)}{n^2}\right] \\
&= (b-a)\left[a + \left(\frac{b-a}{2}\right)\lim_{n\to\infty}\frac{n+1}{n}\right] \\
&= (b-a)\left(a + \frac{b-a}{2}\right) = (b-a)\left(\frac{2a+b-a}{2}\right) \\
&= \frac{1}{2}(b-a)(b+a) = \frac{1}{2}(b^2 - a^2)
\end{aligned}$$

EXAMPLE 4 Divide the interval $[2, 5]$ into n subintervals of equal length, and let c_k be any point in $[x_{k-1}, x_k]$. Write

$$\lim_{n\to\infty} \sum_{k=1}^n \sqrt{1 + (c_k)^2}\, \Delta x$$

as an integral.

Solution Comparing the given expression with Equation (2), we see that it is the limit of a Riemann sum of the function $f(x) = \sqrt{1 + x^2}$ on the interval $[2, 5]$. Next, since f is continuous on $[2, 5]$, the limit exists, so by Equation (2),

$$\lim_{n\to\infty} \sum_{k=1}^n \sqrt{1 + (c_k)^2}\, \Delta x = \int_2^5 \sqrt{1 + x^2}\, dx$$

■ Geometric Interpretation of the Definite Integral

As was pointed out earlier, if f is a continuous, nonnegative function on $[a, b]$, then the definite integral $\int_a^b f(x)\, dx$ gives the area of the region under the graph of f on $[a, b]$. (See Figure 2.)

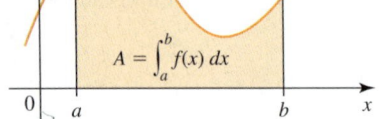

FIGURE 2
If $f(x) \ge 0$ on $[a, b]$, then $\int_a^b f(x)\, dx$ gives the area of the region under the graph of f on $[a, b]$.

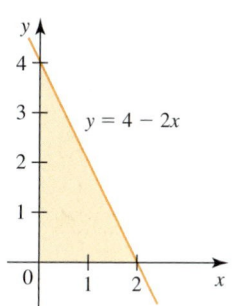

FIGURE 3
$\int_0^2 (4 - 2x)\, dx =$ area of the triangle.

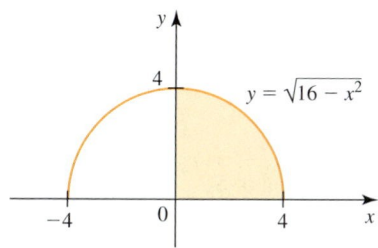

FIGURE 4
$f(x) = \sqrt{16 - x^2}$ represents the upper semicircle.

EXAMPLE 5 Evaluate the definite integral by interpreting it geometrically:

a. $\displaystyle\int_0^2 (4 - 2x)\, dx$ **b.** $\displaystyle\int_0^4 \sqrt{16 - x^2}\, dx$

Solution

a. The graph of the integrand $f(x) = 4 - 2x$ on $[0, 2]$ is the straight line segment shown in Figure 3. Since $f(x) \geq 0$ on $[0, 2]$, we can interpret the integral as the area of the triangle shown. Thus,

$$\int_0^2 (4 - 2x)\, dx = \frac{1}{2}(2)(4) = 4 \qquad \text{area} = \frac{1}{2}\,\text{base} \cdot \text{height}$$

b. The integrand $f(x) = \sqrt{16 - x^2}$ is the positive root obtained by solving the equation $x^2 + y^2 = 16$ for y, which represents the circle of radius 4 centered at the origin; therefore, it represents the upper semicircle shown in Figure 4. Since $f(x) \geq 0$ on $[0, 4]$, we can interpret the integral as the area of that part of the circle lying in the first quadrant. Since this area is $\frac{1}{4}\pi(4^2) = 4\pi$, we see that

$$\int_0^4 \sqrt{16 - x^2}\, dx = 4\pi$$

Next we look at a geometric interpretation of the definite integral for the case in which f assumes both positive and negative values on $[a, b]$. Consider a typical Riemann sum of the function f,

$$\sum_{k=1}^n f(c_k)\Delta x$$

corresponding to a partition P with points of subdivision

$$a = x_0 < x_1 < x_2 < \cdots < x_{k-1} < x_k < \cdots < x_{n-1} < x_n = b$$

and evaluation points c_k in $[x_{k-1}, x_k]$. The sum consists of n terms in which a positive term corresponds to the area of a rectangle of height $f(c_k)$ lying above the x-axis, and a negative term corresponds to the area of a rectangle of height $-f(c_k)$ lying below the x-axis. (See Figure 5, where $n = 6$.)

FIGURE 5
The positive (negative) terms in the Riemann sum are associated with the areas of the rectangles that lie above (below) the x-axis.

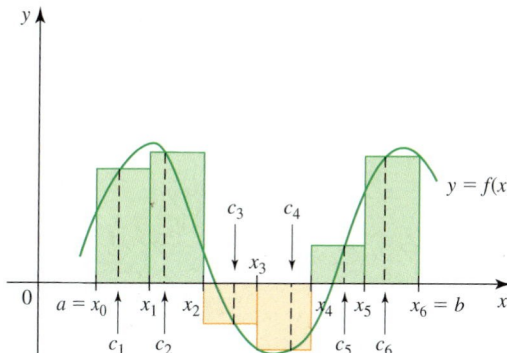

As n gets larger and larger, the sums of the areas of the rectangles lying above the x-axis seem to give a better and better approximation of the area of the region lying above the x-axis. Similarly, the sums of the area of the rectangles lying below the *x*-axis seem to give a better and better approximation of the area of the region lying below the *x*-axis. (See Figure 6, where $n = 12$.)

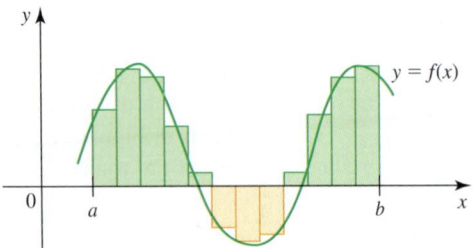

FIGURE 6
Approximating $\int_a^b f(x)\,dx$
with 12 rectangles

This observation suggests that we interpret the definite integral

$$\int_a^b f(x)\,dx = \lim_{n\to\infty} \sum_{k=1}^{n} f(c_k)\Delta x$$

as a difference of areas. Specifically,

$$\int_a^b f(x)\,dx = \text{area of } S_1 - \text{area of } S_2 + \text{area of } S_3$$

where S_2 is the region lying *above* the graph of f and below the *x*-axis. (See Figure 7.) More generally,

$$\int_a^b f(x)\,dx = \text{areas of the regions above } [a, b] - \text{areas of the regions below } [a, b]$$

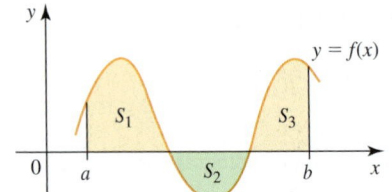

FIGURE 7
$\int_a^b f(x)\,dx = \text{area of } S_1 -$
area of S_2 + area of S_3

The Definite Integral and Displacement

In Section 4.3 we showed that if $v(t)$ is a nonnegative velocity function of a car traveling in a straight line, then the distance covered by the car between $t = a$ and $t = b$ is given by the area of the region under the graph of the velocity function on the time interval $[a, b]$. Since the area of the region under the graph of a nonnegative function $v(t)$ on $[a, b]$ is just the definite integral of v on $[a, b]$, we can write

$$\int_a^b v(t)\,dt = \text{displacement of the car between } t = a \text{ and } t = b$$

If we denote the position of the car at any time t by $s(t)$, then its position at $t = a$ is $s(a)$. So we can then write its final position at $t = b$ as

$$s(b) = s(a) + \int_a^b v(t)\,dt$$

(See Figure 8.)

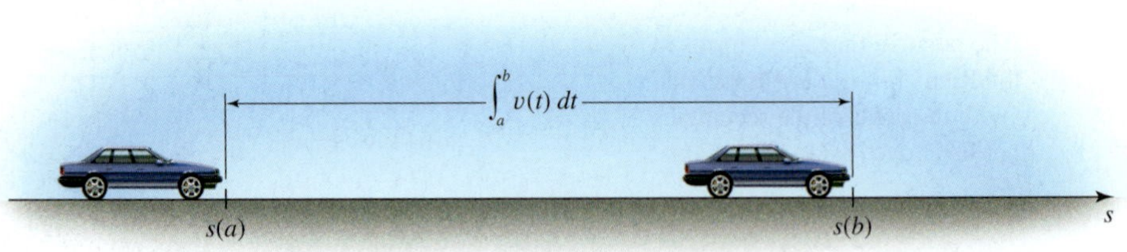

FIGURE 8
The position of the car at $t = b$
is $s(b) = s(a) + \int_a^b v(t)\,dt$.

Now suppose that $v(t)$ assumes both positive and negative values on $[a, b]$. (See Figure 9.)

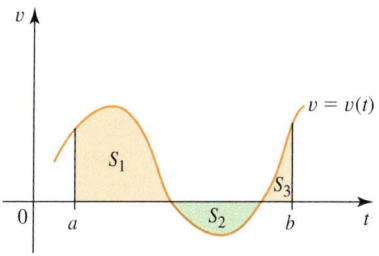

FIGURE 9
The area of S_1 and the area of S_3 give the distance the car moves in the positive direction, whereas the area of S_2 gives the distance it moves in the negative direction.

Then

$$\int_a^b v(t)\, dt = \text{area of the regions above } [a, b] - \text{area of the region below } [a, b]$$

$$= \text{distance covered by the car in the positive direction} - \text{distance covered by the car in the negative direction}$$

$$= \text{displacement of the car between } t = a \text{ and } t = b$$

In other words, the final position of the car at $t = b$ is

$$s(b) = s(a) + \int_a^b v(t)\, dt$$

as before.

EXAMPLE 6 The velocity function of a car moving along a straight road is given by $v(t) = t - 20$ for $0 \le t \le 40$, where $v(t)$ is measured in feet per second and t in seconds. Show that at $t = 40$ the car will be in the same position as it was initially.

Solution The graph of v is shown in Figure 10. We have

$$\int_0^{40} v(t)\, dt = \text{area of } S_2 - \text{area of } S_1$$

$$= \frac{1}{2}(20)(20) - \frac{1}{2}(20)(20)$$

$$= 200 - 200 = 0$$

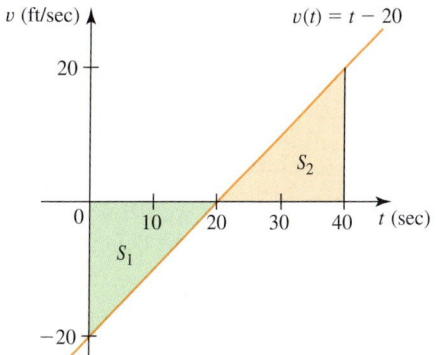

FIGURE 10
The area of S_1 is equal to the area of S_2.

Therefore,

$$s(40) = s(0) + \int_0^{40} v(t)\, dt = s(0)$$

so the net change in the position of the car is zero, as was to be shown.

We interpret this result as follows: The car moves a total of 200 ft in the negative direction in the first 20 sec and then moves a total of 200 ft in the positive direction in the next 20 sec, resulting in no net change in its position.

Alternative Solution Let $s(t)$ denote the position of the car at any time t. Then

$$\frac{ds}{dt} = v(t)$$

But $v(t) = t - 20$, so

$$\frac{ds}{dt} = t - 20$$

Integrating with respect to t, we have

$$s(t) = \int (t - 20)\, dt$$

$$= \frac{1}{2} t^2 - 20t + C \qquad \text{\textcolor{orange}{\textit{C} an arbitrary constant}}$$

The position of the car at $t = 0$ is $s(0)$, and this condition gives

$$s(0) = \frac{1}{2}(0) - 20(0) + C \qquad \text{or} \qquad C = s(0)$$

Therefore, the position of the car at any time t is

$$s(t) = \frac{1}{2} t^2 - 20t + s(0)$$

In particular, the position of the car at $t = 40$ is

$$s(40) = \frac{1}{2}(40^2) - 20(40) + s(0) = s(0)$$

its position at $t = 0$, as was to be shown.

Note The method used in the alternative solution of the problem in Example 6 hints at the relationship between the definite integral of a function and the indefinite integral of the function. We will exploit this relationship in the next section.

■ Properties of the Definite Integral

When we defined the definite integral $\int_a^b f(x)\, dx$, we assumed that $a < b$. We now extend the definition to cover the cases $a = b$ and $a > b$.

DEFINITIONS Two Special Definite Integrals

1. $\displaystyle\int_a^a f(x)\, dx = 0$

2. $\displaystyle\int_a^b f(x)\, dx = -\int_b^a f(x)\, dx, \quad \text{if } a > b$

The first definition is compatible with the definition of the definite integral if we observe that here,

$$\Delta x = \frac{b - a}{n} = \frac{a - a}{n} = 0$$

The second definition is also compatible with the definition by observing that if we interchange a and b, then the sign of the resulting Riemann sum changes because

$$\Delta x = \frac{b - a}{n} = -\frac{a - b}{n}$$

EXAMPLE 7 Evaluate the definite integral:

a. $\displaystyle\int_2^2 (x^2 - 2x + 4)\, dx$ **b.** $\displaystyle\int_3^{-1} (4 - x^2)\, dx$

Solution

a. $\displaystyle\int_2^2 (x^2 - 2x + 4)\, dx = 0$

b. $\displaystyle\int_3^{-1} (4 - x^2)\, dx = -\int_{-1}^3 (4 - x^2)\, dx = -6\frac{2}{3}$ using the result of Example 2. ■

In the expression $\int_a^b f(x)\, dx$ the variable of integration, x, is a *dummy variable* in the sense that it may be replaced by any other letter without changing the value of the integral. As an illustration, the results of Example 2 may be written

$$\int_{-1}^3 (4 - x^2)\, dx = \int_{-1}^3 (4 - u^2)\, du = \int_{-1}^3 (4 - s^2)\, ds = 6\frac{2}{3}$$

Suppose that $c > 0$. Interpreting $\int_a^b c\, dx$ as the area of the region under the graph of $f(x) = c$ on $[a, b]$ gives

$$\int_a^b c\, dx = c(b - a)$$

(See Figure 11.)

We will now look at some properties of the definite integral that will prove helpful later on when we evaluate integrals. Here we assume, as we did earlier, that all of the functions under consideration are continuous.

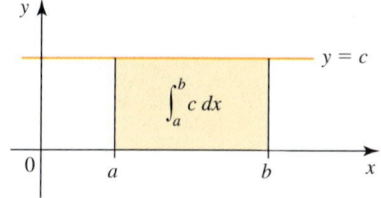

FIGURE 11

If $c > 0$, then interpreting $\int_a^b c\, dx$ as the area of the region under the graph of $f(x) = c$ on $[a, b]$ gives $\int_a^b f(x)\, dx = c(b - a)$.

The Definite Integral of a Constant Function

If c is a real number, then

$$\int_a^b c\, dx = c(b - a) \tag{3}$$

The special case where $c > 0$ was discussed earlier.

EXAMPLE 8 Evaluate $\displaystyle\int_2^7 3\, dx$.

Solution We use Equation (3) with $c = 3$, $a = 2$, and $b = 7$, obtaining

$$\int_2^7 3\, dx = 3(7 - 2) = 15$$

■

The next two properties of the definite integral are analogous to the rules of integration for indefinite integrals (see Section 4.1).

Properties of the Definite Integral

1. Sum (Difference)

$$\int_a^b [f(x) \pm g(x)]\, dx = \int_a^b f(x)\, dx \pm \int_a^b g(x)\, dx$$

2. Constant Multiple

$$\int_a^b cf(x)\, dx = c\int_a^b f(x)\, dx, \quad \text{where } c \text{ is any constant}$$

Property 1 states that the integral of the sum (difference) is the sum (difference) of the integrals. Property 2 states that the integral of a constant times a function is equal to the constant times the integral of the function. Thus, a constant (and only a constant!) can be moved in front of the integral sign. These properties are derived by using the corresponding limit laws. For example, to prove Property 2, we use the definition of the definite integral to write

$$\int_a^b cf(x)\, dx = \lim_{n \to \infty} \sum_{k=1}^{n} cf(c_k)\Delta x$$

$$= c \lim_{n \to \infty} \sum_{k=1}^{n} f(c_k)\Delta x \qquad \text{\color{orange}{Constant Multiple Law for limits}}$$

$$= c \int_a^b f(x)\, dx$$

EXAMPLE 9 Use the result $\int_0^1 x^2\, dx = \dfrac{1}{3}$ of Example 8 in Section 4.3 to evaluate

a. $\displaystyle\int_0^1 (x^2 - 4)\, dx$ **b.** $\displaystyle\int_0^1 5x^2\, dx$

Solution

a. $\displaystyle\int_0^1 (x^2 - 4)\, dx = \int_0^1 x^2\, dx - \int_0^1 4\, dx$ {\color{orange}Property 1}

$$= \frac{1}{3} - 4(1)$$

$$= -\frac{11}{3}$$

b. $\displaystyle\int_0^1 5x^2\, dx = 5\int_0^1 x^2\, dx$ {\color{orange}Property 2}

$$= 5\left(\frac{1}{3}\right) = \frac{5}{3}$$

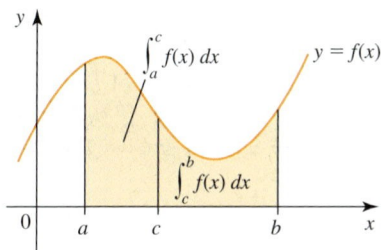

FIGURE 12

$$\int_a^b f(x) \, dx = \int_a^c f(x) \, dx + \int_c^b f(x) \, dx$$

Suppose that f is continuous and nonnegative on $[a, b]$. Then $\int_a^b f(x) \, dx$ gives the area of the region under the graph of f on $[a, b]$. Next, if $a < c < b$, then $\int_a^c f(x) \, dx$ and $\int_c^b f(x) \, dx$ give the area of the region under the graph of f on $[a, c]$ and $[c, b]$, respectively. Therefore, as you can see in Figure 12,

$$\int_a^b f(x) \, dx = \int_a^c f(x) \, dx + \int_c^b f(x) \, dx$$

This observation suggests the following property of definite integrals.

Property of the Definite Integral

3. If c is any number in $[a, b]$, then

$$\int_a^b f(x) \, dx = \int_a^c f(x) \, dx + \int_c^b f(x) \, dx$$

Note   The conclusion of Property 3 holds for *any* three numbers a, b, and c. ■

EXAMPLE 10   Suppose that $\int_1^6 f(x) \, dx = 8$ and $\int_4^6 f(x) \, dx = 5$. What is $\int_1^4 f(x) \, dx$?

Solution   Using Property 3, we have

$$\int_1^6 f(x) \, dx = \int_1^4 f(x) \, dx + \int_4^6 f(x) \, dx$$

from which we see that

$$\int_1^4 f(x) \, dx = \int_1^6 f(x) \, dx - \int_4^6 f(x) \, dx = 8 - 5 = 3$$ ■

The next three properties of the definite integral involve inequalities.

Properties of the Definite Integral

4. If $f(x) \geq 0$ on $[a, b]$, then

$$\int_a^b f(x) \, dx \geq 0$$

5. If $f(x) \geq g(x)$ on $[a, b]$, then

$$\int_a^b f(x) \, dx \geq \int_a^b g(x) \, dx$$

6. If $m \leq f(x) \leq M$ on $[a, b]$, then

$$m(b - a) \leq \int_a^b f(x) \, dx \leq M(b - a)$$

The plausibility of Property 4 stems from the observation that the area of the region under the graph of a nonnegative function is nonnegative. Also, if we assume that g and therefore f are both nonnegative on $[a, b]$, then Property 5 is a statement that the area of the region under the graph of f is larger than the area of the region under the graph of g. (See Figure 13.) The plausibility of Property 6 is suggested by Figure 14, where m and M are the absolute minimum and absolute maximum values of f, respectively, on $[a, b]$: The area of the region under the graph of f on $[a, b]$, $\int_a^b f(x)\,dx$, is greater than the area of the rectangle with height m, $m(b - a)$, and smaller than the area of the rectangle with height M, $M(b - a)$.

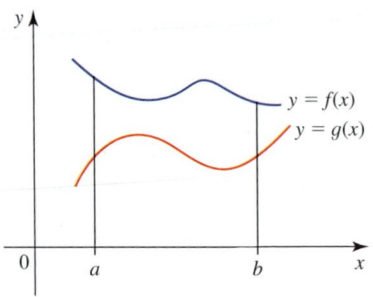

FIGURE 13
If $f(x) \geq g(x)$ on $[a, b]$, then the area of the region under the graph of f is greater than the area of the region under the graph of g.

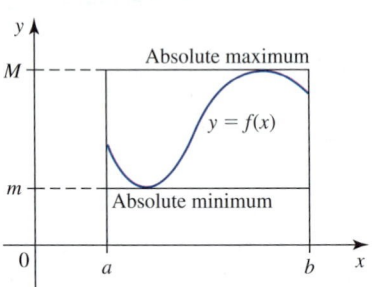

FIGURE 14
The area of the region under the graph of f is greater than or equal to $m(b - a)$ and less than or equal to $M(b - a)$.

It should be mentioned that all of the properties of the definite integral can be proved with mathematical rigor and without any assumption regarding the sign of $f(x)$ (see Exercise 62).

EXAMPLE 11 Use Property 6 to estimate $\displaystyle\int_0^1 e^{-\sqrt{x}}\,dx$.

Solution The integrand $f(x) = e^{-\sqrt{x}}$ is decreasing on $[0, 1]$. Therefore, its absolute maximum value occurs at $x = 0$ (the left endpoint of the interval), and its absolute minimum value occurs at $x = 1$ (the right endpoint of the interval). If we take $m = f(1) = e^{-1}$, $M = f(0) = 1$, $a = 0$, and $b = 1$, then Property 6 gives

$$e^{-1}(1 - 0) \leq \int_0^1 e^{-\sqrt{x}}\,dx \leq 1(1 - 0)$$

$$e^{-1} \leq \int_0^1 e^{-\sqrt{x}}\,dx \leq 1$$

Since $e^{-1} \approx 0.3679$, we have the estimate

$$0.367 \leq \int_0^1 e^{-\sqrt{x}}\,dx \leq 1$$

More General Definition of the Definite Integral

As was pointed out earlier, the points that make up a partition of an interval $[a, b]$ need not be chosen to be equally spaced. In general, a **partition of $[a, b]$** is any set $P = \{x_0, x_1, \ldots, x_n\}$ satisfying

$$a = x_0 < x_1 < x_2 < \cdots < x_{n-1} < x_n = b$$

The subintervals corresponding to this partition of $[a, b]$ are

$$[x_0, x_1], \qquad [x_1, x_2], \qquad \ldots, \qquad [x_{k-1}, x_k], \qquad \ldots, \qquad [x_{n-1}, x_n]$$

The length of the kth subinterval is

$$\Delta x_k = x_k - x_{k-1}$$

Figure 15 shows one possible partition of $[a, b]$.

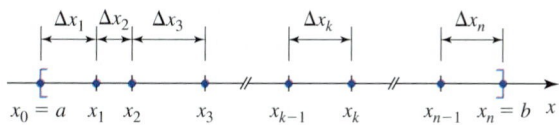

FIGURE 15
A possible partition of $[a, b]$

The length of the largest subinterval, denoted by $\|P\|$, is called the **norm** of P. For example, in the partition shown in Figure 16,

$$\Delta x_1 = \frac{1}{4}, \qquad \Delta x_2 = \frac{1}{4}, \qquad \Delta x_3 = \frac{1}{2}, \qquad \Delta x_4 = \frac{1}{4},$$

$$\Delta x_5 = \frac{1}{8}, \qquad \Delta x_6 = \frac{1}{8}, \qquad \Delta x_7 = \frac{1}{4}, \qquad \text{and} \qquad \Delta x_8 = \frac{1}{4}$$

so its norm is $\frac{1}{2}$.

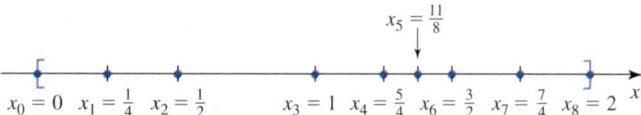

FIGURE 16
A possible partition of $[0, 2]$

If the $(n + 1)$ points of a partition of $[a, b]$ are chosen to be equally spaced so that the resulting n subintervals have equal length, then the partition is **regular.** In a regular partition, the norm satisfies

$$\|P\| = \Delta x = \frac{b - a}{n}$$

For a general partition P,

$$\|P\| \ge \frac{b - a}{n} \qquad \text{or} \qquad n \ge \frac{b - a}{\|P\|}$$

FIGURE 17
As the number of subintervals approach infinity, $\|P\|$ does not approach 0.

From this inequality we see that as the norm of a partition approaches 0, the number of subintervals approach infinity. The converse, however, is false. For example, the partition P of the interval $[0, 1]$ in Figure 17 is given by

$$0 < \frac{1}{2} < \frac{3}{4} < \frac{7}{8} < \cdots < 1 - \frac{1}{2^{n-1}} < 1 - \frac{1}{2^n} < 1$$

has norm $\frac{1}{2}$ for any positive integer n. Therefore, $n \to \infty$ does not imply that $\|P\| \to 0$. But for a regular partition,

$$\|P\| \to 0 \quad \text{if and only if} \quad n \to \infty$$

a fact that we will use shortly.

We are now in a position to give a more general definition of the definite integral, but first we observe that a function f is **bounded** on an interval $[a, b]$ if there exists some positive real number M such that $|f(x)| \le M$ for all x in $[a, b]$.

> **DEFINITION   Definite Integral (General Definition)**
>
> Let f be a bounded function defined on an interval $[a, b]$. Then the **definite integral of f on $[a, b]$,** denoted by $\displaystyle\int_a^b f(x)\, dx$, is
>
> $$\int_a^b f(x)\, dx = \lim_{\|P\| \to 0} \sum_{k=1}^{n} f(c_k)\Delta x \qquad (4)$$
>
> if the limit exists for *all* partitions P of $[a, b]$ and *all* choices of c_k in $[x_{k-1}, x_k]$.

It can be shown that if f is continuous on $[a, b]$, then the definite integral of f on $[a, b]$ always exists. Therefore, the limit (4) exists for all choices of P and c_k. In particular, the limit exists if we choose a regular partition, as was done in our earlier presentation. In fact, for regular partitions, $\|P\| \to 0$ if and only if $n \to \infty$. So the limit (4) is equivalent to

$$\int_a^b f(x)\, dx = \lim_{n \to \infty} \sum_{k=1}^{n} f(c_k)\Delta x$$

which is the definition of the definite integral given earlier.

Finally, we note the following precise definition of the definite integral.

> **DEFINITION   Precise Definition of the Definite Integral**
>
> The definite integral of f on $[a, b]$ is
>
> $$\int_a^b f(x)\, dx$$
>
> if for every number $\varepsilon > 0$ there exists a number $\delta > 0$ such that for every partition P of $[a, b]$ with $\|P\| < \delta$ and every choice of points c_k in $[x_{k-1}, x_k]$, the inequality
>
> $$\left| \sum_{k=1}^{n} f(c_k)\Delta x_k - \int_a^b f(x)\, dx \right| < \varepsilon$$
>
> holds.

4.4   CONCEPT QUESTIONS

1. What is a Riemann sum of a continuous function f on an interval $[a, b]$? Illustrate graphically the case in which f assumes both positive and negative values on $[a, b]$.

2. Define the definite integral of a continuous function on the interval $[a, b]$. Give a geometric interpretation of $\int_a^b f(x)\, dx$ for the case in which (a) f is nonnegative on $[a, b]$ and (b) f assumes both positive and negative values on $[a, b]$. Illustrate your answers graphically.

3. The following figure depicts the graph of the velocity function v of an object traveling along a coordinate line over the time interval $[a, b]$. The numbers c, d, and e satisfy $a < c < d < e < b$. The areas of the regions S_1, S_2, S_3, and S_4 are A_1, A_2, A_3, and A_4 respectively. Assume that the object is located at the origin at $t = a$.

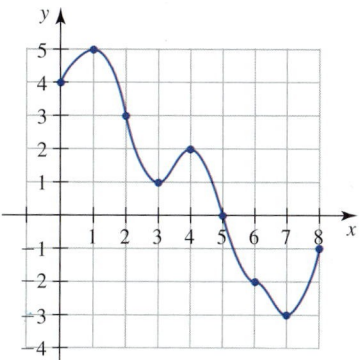

v (ft/sec)

S_1

S_2

S_3

S_4

0   *a*   *c*   *d*   *e*   *b*   *t* (sec)

a. Write the displacement of the object at $t = c$, $t = d$, $t = e$, and $t = b$ (i) in terms of A_1, A_2, A_3, and A_4 and (ii) in terms of definite integrals.

b. Write the distances covered by the object over the time intervals $[a, d]$ and $[a, b]$. Express your answer using A_1, A_2, A_3, and A_4 and also using definite integrals.

4.4 EXERCISES

1. The graph of a function f on the interval $[0, 8]$ is shown in the figure. Compute the Riemann sum for f on $[0, 8]$ using four subintervals of equal length and choosing the evaluation points to be (a) the left endpoints, (b) the right endpoints, and (c) the midpoints of the subintervals.

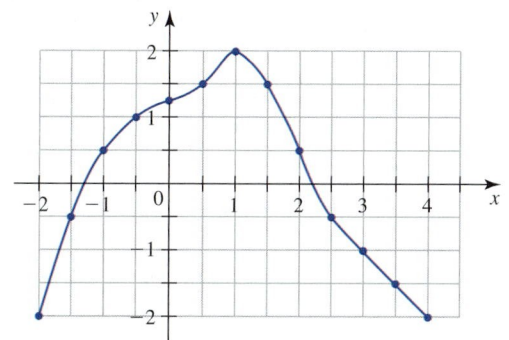

2. The graph of a function g on the interval $[-2, 4]$ is shown in the figure. Compute the Riemann sum for g on $[-2, 4]$ using six subintervals of equal length and choosing the evaluation points to be (a) the left endpoints, (b) the right endpoints, and (c) the midpoints of the subintervals.

In Exercises 3–6 you are given a function f defined on an interval [a, b], the number n of subintervals of equal length $\Delta x = (b - a)/n$, and the evaluation points c_k in $[x_{k-1}, x_k]$. (a) Sketch the graph of f and the rectangles associated with the Riemann sum for f on [a, b], and (b) find the Riemann sum.

3. $f(x) = 2x - 3$, $[0, 2]$, $n = 4$, c_k is the midpoint

4. $f(x) = -2x + 1$, $[-1, 2]$, $n = 6$, c_k is the left endpoint

5. $f(x) = \sqrt{x} - 1$, $[0, 3]$, $n = 6$, c_k is the right endpoint

6. $f(x) = 2 \sin x$, $\left[0, \frac{5\pi}{4}\right]$, $n = 5$, c_k is the right endpoint

In Exercises 7–12, use Equation (2) to evaluate the integral.

7. $\displaystyle\int_0^2 x \, dx$

8. $\displaystyle\int_{-1}^2 x^2 \, dx$

9. $\displaystyle\int_{-1}^3 (x - 2) \, dx$

10. $\displaystyle\int_{-1}^1 (2x + 1) \, dx$

11. $\displaystyle\int_1^2 (3 - 2x^2) \, dx$

12. $\displaystyle\int_{-2}^1 (x^3 + 2x) \, dx$

In Exercises 13–16, the given expression is the limit of a Riemann sum of a function f on [a, b]. Write this expression as a definite integral on [a, b].

13. $\displaystyle\lim_{n \to \infty} \sum_{k=1}^n (4c_k - 3)\Delta x$, $[-3, -1]$

14. $\displaystyle\lim_{n \to \infty} \sum_{k=1}^n 2c_k(1 - c_k)^2 \Delta x$, $[0, 3]$

15. $\displaystyle\lim_{n \to \infty} \sum_{k=1}^n \frac{2c_k}{c_k^2 + 1} \Delta x$, $[1, 2]$

16. $\displaystyle\lim_{n \to \infty} \sum_{k=1}^n c_k(\cos c_k)\Delta x$, $\left[0, \frac{\pi}{2}\right]$

In Exercises 17 and 18, express the integral as a limit of a Riemann sum using a regular partition. Do not evaluate the limit.

17. $\displaystyle\int_{-2}^1 (1 + x^3)^{1/3} \, dx$

18. $\displaystyle\int_1^4 \left(5 \ln x - \frac{1}{2}x^2\right) dx$

19. Use the graph of f shown in the figure to evaluate the integral by interpreting it geometrically.

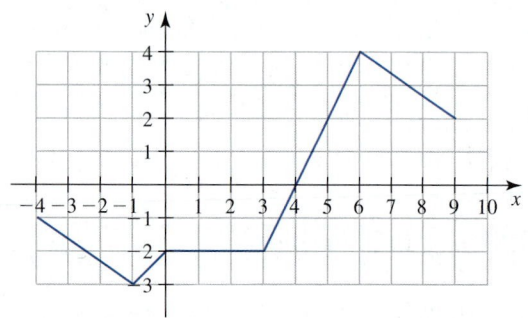

a. $\displaystyle\int_{-4}^{-1} f(x)\,dx$ **b.** $\displaystyle\int_{-1}^{4} f(x)\,dx$

c. $\displaystyle\int_{4}^{9} f(x)\,dx$ **d.** $\displaystyle\int_{-4}^{9} f(x)\,dx$

20. The graph of f shown in the figure consists of straight line segments and a semicircle. Evaluate each integral by interpreting it geometrically.

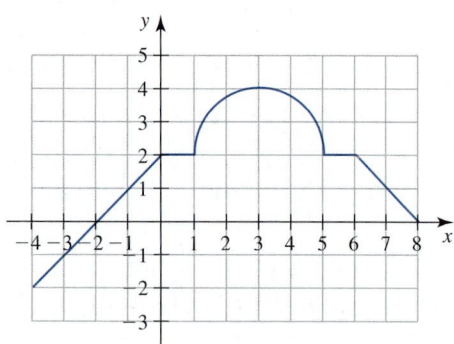

a. $\displaystyle\int_{-4}^{1} f(x)\,dx$ **b.** $\displaystyle\int_{1}^{5} f(x)\,dx$

c. $\displaystyle\int_{5}^{8} f(x)\,dx$ **d.** $\displaystyle\int_{-4}^{8} f(x)\,dx$

In Exercises 21–28 you are given a definite integral $\int_a^b f(x)\,dx$. Make a sketch of f on $[a, b]$. Then use the geometric interpretation of the integral to evaluate it.

21. $\displaystyle\int_{-2}^{4} 3\,dx$ **22.** $\displaystyle\int_{-2}^{3} (2x + 1)\,dx$

23. $\displaystyle\int_{0}^{3} (-3x + 6)\,dx$ **24.** $\displaystyle\int_{-1}^{2} |x|\,dx$

25. $\displaystyle\int_{-1}^{2} |x - 1|\,dx$ **26.** $\displaystyle\int_{-2}^{2} \sqrt{4 - x^2}\,dx$

27. $\displaystyle\int_{0}^{3} -\sqrt{9 - x^2}\,dx$ **28.** $\displaystyle\int_{0}^{2} \sqrt{-x^2 + 2x}\,dx$

29. Given that $\int_0^2 f(x)\,dx = 3$ and $\int_2^5 f(x)\,dx = -1$, evaluate the following integrals.

a. $\displaystyle\int_{0}^{5} f(x)\,dx$ **b.** $\displaystyle\int_{5}^{2} f(x)\,dx$

c. $\displaystyle\int_{0}^{2} 2f(x)\,dx$ **d.** $\displaystyle\int_{2}^{5} [f(x) - 4]\,dx$

30. Given that $\int_1^3 f(x)\,dx = 4$ and $\int_3^6 f(x)\,dx = 2$, evaluate the following integrals.

a. $\displaystyle\int_{3}^{1} 2f(x)\,dx$ **b.** $\displaystyle\int_{6}^{1} f(x)\,dx$

c. $\displaystyle\int_{3}^{1} -2f(x)\,dx$ **d.** $\displaystyle\int_{2}^{2} 3f(x)\,dx$

31. Given that $\int_{-1}^3 f(x)\,dx = 5$ and $\int_{-1}^3 g(x)\,dx = -2$, evaluate the following integrals.

a. $\displaystyle\int_{-1}^{3} [f(x) + g(x)]\,dx$

b. $\displaystyle\int_{-1}^{3} [g(x) - f(x)]\,dx$

c. $\displaystyle\int_{-1}^{3} [3f(x) - 2g(x)]\,dx$

32. Given that $\int_{-2}^2 f(x)\,dx = 3$ and $\int_0^2 f(x)\,dx = 2$, evaluate the following integrals.

a. $\displaystyle\int_{2}^{0} f(x)\,dx$ **b.** $\displaystyle\int_{-2}^{0} [f(x) + 3]\,dx$

c. $\displaystyle\int_{2}^{0} 3f(x)\,dx - \int_{0}^{-2} 2f(x)\,dx$

33. Evaluate $\int_2^2 \sqrt[3]{x^2 + x + 1}\,dx$.

34. Evaluate $\int_2^5 f(x)\,dx$ if it is known that $\int_5^2 f(x)\,dx = -10$.

35. Show that $\int_0^\pi \cos 2x\,dx = 0$ by interpreting the definite integral geometrically.

36. Show that

$$\int_0^x \sqrt{a^2 - t^2}\,dt = \frac{1}{2}x\sqrt{a^2 - x^2} + \frac{a^2}{2}\sin^{-1}\left(\frac{x}{a}\right)$$

$$0 < x \le a$$

by interpreting the definite integral geometrically.

In Exercises 37–40, use the properties of the integral to prove the inequality without evaluating the integral.

37. $\displaystyle\int_{0}^{1} \frac{\sqrt{x^3 + x}}{x^2 + 1}\,dx \ge 0$ **38.** $\displaystyle\int_{0}^{1} x^2\,dx \le \int_{0}^{1} \sqrt{x}\,dx$

39. $\displaystyle\int_{0}^{\pi/4} \sin^2 x \cos x\,dx \le \int_{0}^{\pi/4} \sin^2 x\,dx$

40. $\displaystyle\int_0^{\pi/2} \cos x \, dx \le \int_0^{\pi/2} (x^2 + 1) \, dx$

In Exercises 41–46, use Property 6 of the definite integral to estimate the definite integral.

41. $\displaystyle\int_1^2 \sqrt{1 + 2x^3} \, dx$ **42.** $\displaystyle\int_1^3 \frac{1}{x} \, dx$

43. $\displaystyle\int_{-1}^2 (x^2 - 2x + 2) \, dx$ **44.** $\displaystyle\int_0^2 \frac{x^2 + 5}{x^2 + 2} \, dx$

45. $\displaystyle\int_0^1 e^{-x^2} \, dx$ **46.** $\displaystyle\int_{\pi/4}^{\pi/2} x \sin x \, dx$

 47. a. Plot the graph of $f(x) = x\sqrt{x^4 + 1}$ on the interval $[-1, 1]$.
  b. Prove that the area of the region above the x-axis is equal to the area of the region below the x-axis.
  c. Use the result of part (b) to show that
  $\displaystyle\int_{-1}^1 x\sqrt{x^4 + 1} \, dx = 0$.

 48. a. Plot the graph of $f(x) = \sin^3 x$ on the interval $[0, 2\pi]$.
  b. Prove that the area of the region above the x-axis is equal to the area of the region below the x-axis.
  Hint: Look at $f(\pi + t)$ for $0 \le t \le \pi$.
  c. Use the result of part (b) to show that $\displaystyle\int_0^{2\pi} \sin^3 x \, dx = 0$.

49. Suppose that f is continuous on $[a, b]$ and $f(x) \le 0$ on $[a, b]$. Prove that $\int_a^b f(x) \, dx \le 0$.

50. Suppose that f is continuous on $[a, b]$. Prove that
$$\left| \int_a^b f(x) \, dx \right| \le \int_a^b |f(x)| \, dx$$
Hint: $-|f(x)| \le f(x) \le |f(x)|$.

51. Use the result of Exercise 50 to show that $\left| \int_a^b x \sin 2x \, dx \right| \le \frac{1}{2}(b^2 - a^2)$, where $0 \le a < b$.
Hint: Use the result of Example 3.

52. Suppose that f is continuous and increasing and its graph is concave upward on the interval $[a, b]$. Give a geometric argument to show that
$$(b - a)f(a) \le \int_a^b f(x) \, dx \le \frac{1}{2}(b - a)[f(a) + f(b)]$$

 53. a. Plot the graphs of $f(x) = \dfrac{x}{\sqrt{1 + x^5}}$ and $g(x) = x$ using the viewing window $[0, 1] \times [0, 1]$.
  b. Prove that $0 \le f(x) \le g(x)$.
  c. Use the result of part (b) and Property 5 to show that
  $$0 \le \int_0^1 \frac{x}{\sqrt{1 + x^5}} \, dx \le \frac{1}{2}$$
  Hint: Use the result of Example 3.

 54. a. Plot the graphs of $f(x) = \sin x$ and $g(x) = x$ using the viewing window $[0, \frac{\pi}{2}] \times [0, 2]$.
  b. Prove that $0 \le f(x) \le g(x)$.

c. Use the result of part (b) and Property 5 to show that
$$0 \le \int_{\pi/6}^{\pi/4} \sin x \, dx \le \frac{5\pi^2}{288}$$
Hint: Use the result of Example 3.

In Exercises 55 and 56, use Property 5 to prove the inequality.

55. $\displaystyle\int_2^4 \sqrt{x^4 + x} \, dx \ge \frac{56}{3}$ **56.** $\displaystyle\int_0^{\pi/4} x \sin x \, dx \le \frac{\pi^3}{192}$

57. Estimate the integral $\int_0^1 \sqrt{1 + x^2} \, dx$ using (a) Property 6 of the definite integral and (b) the result of Exercise 52. Which estimate is better? Explain.

58. Show that $\int_a^b x^2 \, dx = \frac{1}{3}(b^3 - a^3)$.

59. Find the constant b such that $\int_0^b (2\sqrt{x} - x) \, dx$ is as large as possible. Explain your answer.

60. Define the function F by $F(x) = \int_{-1}^x (t^4 - 2t^3) \, dt$ for x in $[-1, 2]$.
 a. Plot the graph of $f(t) = t^4 - 2t^3$ on $[-1, 2]$.
  b. Use the result of part (a) to find the interval where F is increasing and where F is decreasing on $(-1, 2)$.

61. Determine whether the Dirichlet function
$$f(x) = \begin{cases} 1 & \text{if } x \text{ is rational} \\ 0 & \text{if } x \text{ is irrational} \end{cases}$$
is integrable on the interval $[0, 1]$. Explain.

62. Prove Properties 4, 5, and 6 of the definite integral.

In Exercises 63–70, determine whether the statement is true or false. If it is true, explain why it is true. If it is false, explain why or give an example to show why it is false.

63. If f and g are continuous on $[a, b]$ and c is constant, then
$$\int_a^b [f(x) + cg(x)] \, dx = \int_a^b f(x) \, dx + c \int_a^b g(x) \, dx$$

64. If f and g are continuous on $[a, b]$, then
$$\int_a^b f(x) \, g(x) \, dx = \left[\int_a^b f(x) \, dx \right] \left[\int_a^b g(x) \, dx \right]$$

65. If f is continuous on $[a, b]$, then $\int_a^b xf(x) \, dx = x \int_a^b f(x) \, dx$.

66. If f is continuous on $[a, b]$ and $\int_a^b f(x) \, dx > 0$, then f must be positive on $[a, b]$.

67. If f is continuous and decreasing on $[a, b]$, then $(b - a)f(b) \le \int_a^b f(x) \, dx \le (b - a)f(a)$.

68. If f is nonnegative and continuous on $[a, b]$ and $a < c < d < b$, then $\int_c^d f(x) \, dx \le \int_a^b f(x) \, dx$.

69. $\displaystyle\int_1^3 \frac{dx}{x - 2} = -\int_3^1 \frac{dx}{x - 2}$

70. $\displaystyle\int_{-2}^2 \frac{dx}{x} = \ln|x| \Big|_{-2}^2 = \ln|2| - \ln|-2| = \ln 2 - \ln 2 = 0$

4.5 The Fundamental Theorem of Calculus

■ How Are Differentiation and Integration Related?

In Section 4.4 we defined the definite integral of a function by taking the limit of its Riemann sums. But as we saw, the actual process of finding the definite integral of a function based on this definition turned out to be rather tedious even for simple functions. This is reminiscent of the process of finding the derivative of a function by finding the limit of the difference quotient of the function. Fortunately, there are better and easier ways of evaluating definite integrals.

In this section we will look at what is undoubtedly the most important theorem in calculus. Because it establishes the relationship between differentiation and integration, it is called the **Fundamental Theorem of Calculus.** It was discovered independently by Sir Isaac Newton (1643–1727) in England and by Gottfried Wilhelm Leibniz (1646–1716) in Germany. Before looking at this theorem, we need the results of the following theorem.

■ The Mean Value Theorem for Definite Integrals

Suppose that the velocity of a maglev traveling along a straight track is $v(t)$ ft/sec for t between $t = a$ and $t = b$, where t is measured in seconds. What is the average velocity of the maglev over the time interval $[a, b]$?

To answer this question, let's assume that v is continuous on $[a, b]$. We begin by partitioning the interval $[a, b]$ into n equal subintervals of length

$$\Delta t = \frac{b - a}{n}$$

by means of equally spaced points

$$a = t_0 < t_1 < t_2 < \cdots < t_n = b$$

Next, we choose the evaluation points $c_1, c_2, \ldots, c_n$ lying in the subintervals $[t_0, t_1]$, $[t_1, t_2], \ldots, [t_{n-1}, t_n]$, respectively, and compute the velocities of the maglev at these points:

$$v(c_1), \quad v(c_2), \quad \ldots, \quad v(c_n)$$

The average of these n numbers

$$\frac{v(c_1) + v(c_2) + \cdots + v(c_n)}{n} = \frac{1}{n} \sum_{k=1}^{n} v(c_k)$$

gives an approximation of the average velocity of the maglev over $[a, b]$. Since

$$n = \frac{b - a}{\Delta t}$$

we can rewrite the expression in the form

$$\frac{1}{n} \sum_{k=1}^{n} v(c_k) = \frac{1}{\dfrac{b - a}{\Delta t}} \sum_{k=1}^{n} v(c_k) = \frac{1}{b - a} \sum_{k=1}^{n} v(c_k) \Delta t$$

By letting n get larger and larger, we are approximating the average velocity of the maglev using measurements of its velocity at more and more points over smaller and smaller time intervals. Intuitively, the approximations should improve with increasing n. This suggests that we define the average velocity of the maglev over the time interval $[a, b]$ to be

$$\lim_{n \to \infty} \frac{1}{b-a} \sum_{k=1}^{n} v(c_k) \Delta t$$

But by the definition of the definite integral, we have

$$\lim_{n \to \infty} \frac{1}{b-a} \sum_{k=1}^{n} v(c_k) \Delta t = \frac{1}{b-a} \lim_{n \to \infty} \sum_{k=1}^{n} v(c_k) \Delta t$$

$$= \frac{1}{b-a} \int_{a}^{b} v(t) \, dt$$

Thus, we are led to define the **average velocity** of the maglev over the time interval $[a, b]$ to be

$$\frac{1}{b-a} \int_{a}^{b} v(t) \, dt$$

More generally, we have the following definition of the average value of a function f over an interval $[a, b]$.

DEFINITION   Average Value of a Function

If f is integrable on $[a, b]$, then the **average value of f** over $[a, b]$ is the number

$$f_{\text{av}} = \frac{1}{b-a} \int_{a}^{b} f(x) \, dx \tag{1}$$

If we assume that f is nonnegative, then we have the following geometric interpretation for the average value of a function over $[a, b]$. Referring to Figure 1, we see that f_{av} is the height of the rectangle with base lying on the interval $[a, b]$ and having the same area as the area of the region under the graph of f on $[a, b]$.

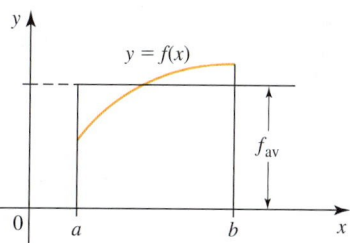

FIGURE 1
The area of the rectangle is
$(b - a)f_{\text{av}} = \int_{a}^{b} f(x) \, dx$ = area
of the region under the graph of f.

Returning to the example involving the motion of the maglev, we see that if we assume that $v(t) \geq 0$ on $[a, b]$, then the distance covered by the maglev over the time period $[a, b]$ is $\int_{a}^{b} v(t) \, dt$, the area of the region under the graph of v on $[a, b]$. But this area is equal to $(b - a)v_{\text{av}}$, where v_{av} is the average value of the velocity function v. Thus, we can cover the distance traveled by the maglev at a speed of $v(t)$ ft/sec from $t = a$ to $t = b$ by traveling at a *constant* speed, namely, at the average speed v_{av} ft/sec over the same time interval.

EXAMPLE 1 Find the average value of $f(x) = 4 - x^2$ over the interval $[-1, 3]$.

Solution Using Equation (1) with $a = -1$, $b = 3$, and $f(x) = 4 - x^2$, we find

$$f_{av} = \frac{1}{b - a} \int_a^b f(x)\, dx$$

$$= \frac{1}{3 - (-1)} \int_{-1}^3 (4 - x^2)\, dx$$

$$= \frac{1}{4}\left(\frac{20}{3}\right) \qquad \text{Use the result of Example 2 in Section 4.4.}$$

$$= \frac{5}{3}$$

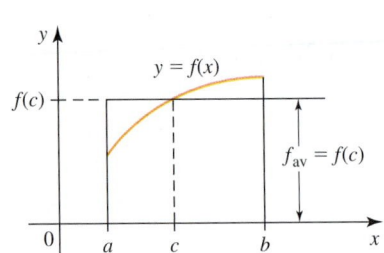

FIGURE 2

$$f_{av} = \frac{1}{b - a} \int_a^b f(x)\, dx$$

If you look at Figure 1 again, you will see that there is a number c on $[a, b]$ such that $f(c) = f_{av}$. (See Figure 2.)

The following theorem guarantees that f_{av} is always attained at (at least) one number in an interval $[a, b]$ if f is continuous.

THEOREM 1   The Mean Value Theorem for Integrals

If f is continuous on $[a, b]$, then there exists a number c in $[a, b]$ such that

$$f(c) = \frac{1}{b - a} \int_a^b f(x)\, dx$$

PROOF Since f is continuous on the interval $[a, b]$, the Extreme Value Theorem tells us that f attains an absolute minimum value m at some number in $[a, b]$ and an absolute maximum value M at some number in $[a, b]$. So $m \le f(x) \le M$ for all x in $[a, b]$.

By Property 6 of integrals we have

$$m(b - a) \le \int_a^b f(x)\, dx \le M(b - a)$$

If $b > a$, then, upon dividing by $(b - a)$, we obtain

$$m \le \frac{1}{b - a} \int_a^b f(x)\, dx \le M$$

Because the number

$$\frac{1}{b - a} \int_a^b f(x)\, dx$$

lies between m and M, the Intermediate Value Theorem guarantees the existence of at least one number c in $[a, b]$ such that

$$f(c) = \frac{1}{b - a} \int_a^b f(x)\, dx$$

as was to be shown.

EXAMPLE 2 Find the value of c guaranteed by the Mean Value Theorem for Integrals for $f(x) = 4 - 2x$ on the interval $[0, 2]$.

Solution The function $f(x) = 4 - 2x$ is continuous on the interval $[0, 2]$. Therefore, the Mean Value Theorem for Integrals states that there is a number c in $[0, 2]$ such that

$$\frac{1}{b - a} \int_a^b f(x) \, dx = f(c)$$

where $a = 0$ and $b = 2$. Thus,

$$\frac{1}{2 - 0} \int_0^2 (4 - 2x) \, dx = 4 - 2c$$

but

$$\int_0^2 (4 - 2x) \, dx = 4 \qquad \text{See Example 5a in Section 4.4.}$$

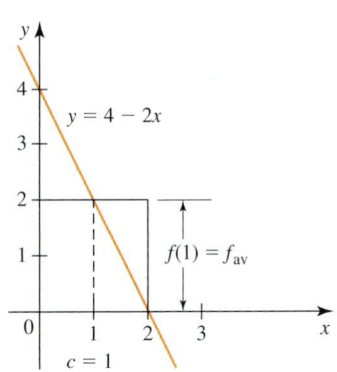

FIGURE 3
The number $c = 1$ in $[0, 2]$ gives $f(c) = f_{\text{av}}$ as guaranteed by the Mean Value Theorem for Integrals.

So we have

$$\frac{1}{2} (4) = 4 - 2c$$

or $c = 1$. (See Figure 3.) ∎

The Fundamental Theorem of Calculus, Part I

Suppose that f is a continuous, nonnegative function defined on the interval $[a, b]$. If x is any number in $[a, b]$, let us put

$$A(x) = \int_a^x f(t) \, dt$$

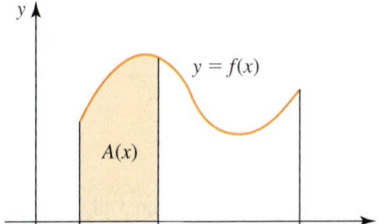

FIGURE 4
$A(x) = \int_a^x f(t) \, dt$ gives the area of the region under the graph of f on $[a, x]$.

(We use the dummy variable t because we are using x to denote the upper limit of integration.) Since f is nonnegative, we can interpret $A(x)$ to be the area of the region under the graph of f on $[a, x]$, as shown in Figure 4. Since the number $A(x)$ is unique for each x in $[a, b]$, we see that A is a function of x with domain $[a, b]$.

Let's look at a specific example. Suppose that $f(x) = x$ on the interval $[0, 1]$. If we use the result of Example 3 in Section 4.4, with $a = 0$ and $b = x$, we obtain

$$A(x) = \int_0^x t \, dt = \frac{1}{2} x^2 \qquad 0 \le x \le 1$$

This result is also evident if you refer to Figure 5 and interpret the integral $\int_0^x t \, dt$ as the area of the shaded triangle. Observe that

$$A'(x) = \frac{d}{dx} \int_0^x t \, dt = \frac{d}{dx} \left(\frac{1}{2} x^2 \right) = x = f(x)$$

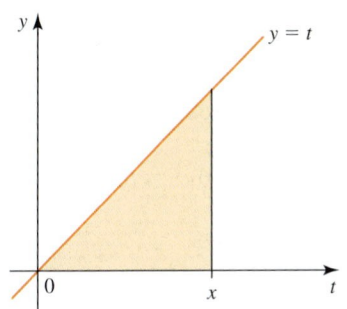

FIGURE 5
The area of the triangle is $\frac{1}{2}(x)(x) = \frac{1}{2} x^2$.

so $A(x)$ is an antiderivative of $f(x) = x$. Now if this result,

$$\frac{d}{dx} \int_a^x f(t) \, dt = f(x)$$

is true for all continuous functions f, then it is quite astounding because it provides a link between the processes of differentiation and integration. Roughly speaking, this equation says that differentiation undoes what integration does: The two operations are inverses of one another. Thus, the two seemingly unrelated problems of differential calculus (that of finding the slope of a tangent line to a curve) and integral calculus (that of finding the area of the region bounded by a curve) are indeed intimately related.

As it turns out, the result is true. Because of its importance, it is called the Fundamental Theorem of Calculus.

THEOREM 2 **The Fundamental Theorem of Calculus, Part 1**

If f is continuous on $[a, b]$, then the function F defined by

$$F(x) = \int_a^x f(t)\, dt \qquad a \leq x \leq b$$

is differentiable on (a, b), and

$$F'(x) = \frac{d}{dx} \int_a^x f(t)\, dt = f(x) \tag{2}$$

PROOF Fix x in (a, b), and suppose that $x + h$ is in (a, b), where $h \neq 0$. Then

$$F(x + h) - F(x) = \int_a^{x+h} f(t)\, dt - \int_a^x f(t)\, dt$$

$$= \int_a^x f(t)\, dt + \int_x^{x+h} f(t)\, dt - \int_a^x f(t)\, dt \qquad \text{\textcolor{orange}{By Property 3}}$$

$$= \int_x^{x+h} f(t)\, dt$$

By the Mean Value Theorem for Integrals there exists a number c between x and $x + h$ such that

$$\int_x^{x+h} f(t)\, dt = f(c) \cdot h$$

Therefore,

$$\frac{F(x + h) - F(x)}{h} = \frac{1}{h} \int_x^{x+h} f(t)\, dt = \frac{f(c) \cdot h}{h} = f(c)$$

Next, observe that as h approaches 0, the number c, which is squeezed between x and $x + h$, approaches x, and by continuity, $f(c)$ approaches $f(x)$. Therefore,

$$F'(x) = \lim_{h \to 0} \frac{F(x + h) - F(x)}{h} = \lim_{h \to 0} \frac{1}{h} \int_x^{x+h} f(t)\, dt = \lim_{h \to 0} f(c) = f(x)$$

which is the desired result.

EXAMPLE 3 Find the derivative of the function:

a. $F(x) = \int_{-1}^x \frac{1}{1 + t^2}\, dt$ \qquad **b.** $G(x) = \int_x^3 \sqrt{1 + t^2}\, dt$

Solution

a. The integrand

$$f(t) = \frac{1}{1 + t^2}$$

is continuous everywhere. Using the Fundamental Theorem of Calculus, Part 1, we find

$$F'(x) = \frac{d}{dx} \int_{-1}^{x} \frac{1}{1 + t^2} \, dt = f(x) = \frac{1}{1 + x^2}$$

b. The integrand $\sqrt{1 + t^2}$ is continuous everywhere. Therefore,

$$G'(x) = \frac{d}{dx} \int_{x}^{3} \sqrt{1 + t^2} \, dt = \frac{d}{dx} \left[-\int_{3}^{x} \sqrt{1 + t^2} \, dt \right] \qquad \int_{a}^{b} f(x)\,dx = -\int_{b}^{a} f(x)\,dx$$

$$= -\frac{d}{dx} \int_{3}^{x} \sqrt{1 + t^2} \, dt$$

$$= -\sqrt{1 + x^2} \qquad\blacksquare$$

EXAMPLE 4 If $y = \displaystyle\int_{0}^{x^3} \cos t^2 \, dt$, what is $\dfrac{dy}{dx}$?

Solution Notice that the upper limit of integration is not x, so the Fundamental Theorem of Calculus, Part 1, is not applicable as the problem now stands. Let's put

$$u = x^3 \qquad \text{so} \qquad \frac{du}{dx} = 3x^2$$

Using the Chain Rule and the Fundamental Theorem of Calculus, Part 1, we have

$$\frac{dy}{dx} = \frac{dy}{du} \cdot \frac{du}{dx} = \left[\frac{d}{du} \int_{0}^{u} \cos t^2 \, dt \right] \cdot \frac{du}{dx}$$

$$= (\cos u^2)(3x^2) = 3x^2 \cos x^6 \qquad\blacksquare$$

■ Fundamental Theorem of Calculus, Part 2

The following theorem, which is a consequence of Part 1 of the Fundamental Theorem of Calculus, shows how to evaluate a definite integral by finding an antiderivative of the integrand, rather than relying on evaluating the limit of a Riemann sum, thus simplifying the task greatly.

THEOREM 3 **The Fundamental Theorem of Calculus, Part 2**

If f is continuous on $[a, b]$, then

$$\int_{a}^{b} f(x) \, dx = F(b) - F(a) \qquad (3)$$

where F is any antiderivative of f, that is, $F' = f$.

PROOF Let $G(x) = \int_a^x f(t)\, dt$. By Theorem 2 we know that G is an antiderivative of f. If F is any other antiderivative of f, then Theorem 1 in Section 4.1 tells us that F and G differ by a constant. In other words, $F(x) = G(x) + C$. To determine C, we put $x = a$ to obtain

$$F(a) = G(a) + C = \int_a^a f(t)\, dt + C = C \qquad \int_a^a f(x)\, dx = 0$$

Therefore, evaluating F at b, we have

$$F(b) = G(b) + C = \int_a^b f(t)\, dt + F(a)$$

from which we conclude that

$$F(b) - F(a) = \int_a^b f(x)\, dx \qquad \blacksquare$$

When applying the Fundamental Theorem of Calculus, it is convenient to use the notation

$$\left[F(x) \right]_a^b = F(b) - F(a) \qquad \text{``}F(x) \text{ evaluated at } b \text{ minus } F(x) \text{ evaluated at } a.\text{''}$$

For example, by using this notation, Equation (3) is written

$$\int_a^b f(x)\, dx = \left[F(x) \right]_a^b = F(b) - F(a)$$

Also, by the Fundamental Theorem of Calculus, if $F(x) + C$ is any antiderivative of f, then

$$\int_a^b f(x)\, dx = \left[F(x) + C \right]_a^b$$
$$= [F(b) + C] - [F(a) + C]$$
$$= F(b) - F(a) = \left[F(x) \right]_a^b$$

This result shows that we can drop the constant of integration when we use the Fundamental Theorem of Calculus.

From now on, thanks to the Fundamental Theorem of Calculus, Part 2, we can use our knowledge for finding antiderivatives to help us evaluate definite integrals.

EXAMPLE 5 Evaluate

a. $\displaystyle\int_1^2 (x^3 - 2x^2 + 1)\, dx$ **b.** $\displaystyle\int_0^4 2\sqrt{x}\, dx$ **c.** $\displaystyle\int_0^{\pi/2} \cos x\, dx$

Solution

a. $\displaystyle\int_1^2 (x^3 - 2x^2 + 1)\, dx = \left[\frac{1}{4}x^4 - \frac{2}{3}x^3 + x \right]_1^2$

$$= \left(4 - \frac{16}{3} + 2 \right) - \left(\frac{1}{4} - \frac{2}{3} + 1 \right) = \frac{1}{12}$$

b. $\displaystyle\int_0^4 2\sqrt{x}\, dx = \int_0^4 2x^{1/2}\, dx = \left[\frac{4}{3}x^{3/2} \right]_0^4 = \frac{4}{3}(4)^{3/2} - \frac{4}{3}(0) = \frac{32}{3}$

c. $\displaystyle\int_0^{\pi/2} \cos x \, dx = \left[\sin x\right]_0^{\pi/2} = 1 - 0 = 1$ ■

The next example shows how to evaluate the definite integral of a function that is defined piecewise.

EXAMPLE 6 Evaluate $\displaystyle\int_{-2}^{2} f(x) \, dx$, where

$$f(x) = \begin{cases} -x^2 + 1 & \text{if } x < 0 \\ x^3 + 1 & \text{if } x \ge 0 \end{cases}$$

Solution The graph of f is shown in Figure 6. Observe that f is continuous on $[-2, 2]$. Since f is defined by different rules for x in the two subintervals $[-2, 0)$ and $[0, 2]$, we use Property 3 of definite integrals to write

$$\int_{-2}^{2} f(x) \, dx = \int_{-2}^{0} f(x) \, dx + \int_{0}^{2} f(x) \, dx$$

$$= \int_{-2}^{0} (-x^2 + 1) \, dx + \int_{0}^{2} (x^3 + 1) \, dx$$

$$= \left[-\frac{1}{3} x^3 + x\right]_{-2}^{0} + \left[\frac{1}{4} x^4 + x\right]_{0}^{2}$$

$$= 0 - \left(\frac{8}{3} - 2\right) + (4 + 2) - 0 = \frac{16}{3}$$ ■

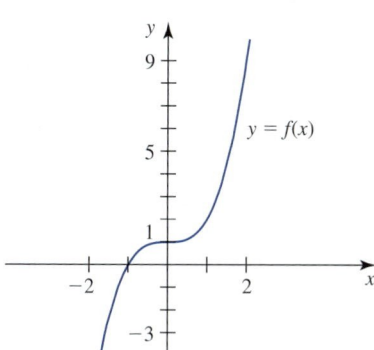

FIGURE 6
$\displaystyle\int_{-2}^{2} f(x) \, dx = \int_{-2}^{0} f(x) \, dx + \int_{0}^{2} f(x) \, dx$

Evaluating Definite Integrals Using Substitution

The next two examples show how the method of substitution can be used to help us evaluate definite integrals.

EXAMPLE 7 Evaluate $\displaystyle\int_0^2 x\sqrt{x^2 + 4} \, dx$.

Solution *Method I:* Consider the corresponding indefinite integral

$$I = \int x\sqrt{x^2 + 4} \, dx = \int x(x^2 + 4)^{1/2} \, dx$$

Let $u = x^2 + 4$, so that $du = 2x \, dx$ or $x \, dx = \frac{1}{2} \, du$. Substituting these quantities into the integral gives

$$I = \int \frac{1}{2} u^{1/2} \, du = \frac{1}{3} u^{3/2} + C = \frac{1}{3} (x^2 + 4)^{3/2} + C$$

Armed with the knowledge of the antiderivative of the function $f(x) = x\sqrt{x^2 + 4}$, we can evaluate the given integral as follows:

$$\int_0^2 x\sqrt{x^2 + 4} \, dx = \left[\frac{1}{3} (x^2 + 4)^{3/2}\right]_0^2 = \frac{1}{3} (8)^{3/2} - \frac{1}{3} (4)^{3/2} = \frac{8}{3} (2\sqrt{2} - 1)$$

Solution *Method II: Changing the Limits of Integration* As before, we make the substitution $u = x^2 + 4$, so that $du = 2x \, dx$ or $x \, dx = \frac{1}{2} \, du$. Next, we make the following intuitive observation: The given integral has lower and upper limits of integration 0 and 2, respectively, and hence a *range of integration* given by the interval $[0, 2]$. In making the substitution $u = x^2 + 4$, the original integral is transformed into another integral in which the integration is carried out with respect to the new variable u.

To obtain the new limits of integration, we note that if $x = 0$, then $u = 0 + 4 = 4$. This gives the lower limit of integration when integrating with respect to u. Similarly, if $x = 2$, then $u = 4 + 4 = 8$, and this gives the upper limit of integration. Thus, the range of integration when the integration is performed with respect to u is $[4, 8]$. In view of this, we can write

$$\int_0^2 x(x^2 + 4)^{1/2} \, dx = \int_4^8 \frac{1}{2} u^{1/2} \, du = \left[\frac{1}{3} u^{3/2}\right]_4^8$$

$$= \frac{1}{3}(8)^{3/2} - \frac{1}{3}(4)^{3/2} = \frac{8}{3}(2\sqrt{2} - 1)$$

as was obtained earlier.

EXAMPLE 8 Evaluate $\displaystyle\int_0^{\pi/4} \cos^3 2x \sin 2x \, dx$.

Solution Let $u = \cos 2x$, so that $du = -2 \sin 2x \, dx$ or $\sin 2x \, dx = -\frac{1}{2} du$. Also, if $x = 0$, then $u = 1$, and if $x = \pi/4$, then $u = 0$, giving 1 and 0 as the lower and upper limits of integration with respect to u. Making these substitutions, we obtain

$$\int_0^{\pi/4} \cos^3 2x \sin 2x \, dx = \int_1^0 u^3 \left(-\frac{1}{2} du\right)$$

$$= -\frac{1}{8} u^4 \Big|_1^0$$

$$= 0 - \left(-\frac{1}{8}\right) = \frac{1}{8}$$

Note Do not let the fact that the limits of integration with respect to u run from 1 to 0 alarm you. This is not uncommon when we integrate using the method of substitution. Of course,

$$\int_1^0 u^3 \left(-\frac{1}{2} du\right) = -\int_0^1 u^3 \left(-\frac{1}{2} du\right) \qquad \int_a^b f(x) \, dx = -\int_b^a f(x) \, dx$$

as you can verify.

EXAMPLE 9 **Drug Concentration in the Bloodstream** The concentration of a certain drug (in mg/cc) in a patient's bloodstream t hr after injection is

$$C(t) = \frac{0.2t}{t^2 + 1}$$

Determine the average concentration of the drug in the patient's bloodstream over the first 4 hr after the drug is injected.

Solution The average concentration of the drug over the time interval $[0, 4]$ is given by

$$A = \frac{1}{4 - 0} \int_0^4 C(t) \, dt = \frac{1}{4} \int_0^4 \frac{0.2t}{t^2 + 1} \, dt$$

To evaluate this definite integral, we make the substitution

$$u = t^2 + 1 \qquad \text{so that} \qquad du = 2t \, dt \qquad \text{or} \qquad t \, dt = \frac{du}{2}$$

Observe that when $t = 0$, $u = 0^2 + 1 = 1$, and when $t = 4$, $u = 4^2 + 1 = 17$, giving $u = 1$ and $u = 17$ as the lower and upper limits of integration with respect to u, respectively. We have

$$A = \frac{1}{20} \int_0^4 \frac{t}{t^2 + 1} \, dt$$

$$= \frac{1}{20} \left(\frac{1}{2}\right) \int_1^{17} \frac{1}{u} \, du = \left[\frac{1}{40} \ln u\right]_1^{17} = \frac{1}{40}(\ln 17 - \ln 1)$$

or approximately 0.071 mg/cc. ■

■ Definite Integrals of Odd and Even Functions

The following theorem makes use of the symmetry properties of the integrand to help us evaluate a definite integral.

THEOREM 4 Integrals of Odd and Even Functions

Suppose that f is continuous on $[-a, a]$.

a. If f is even, then $\displaystyle\int_{-a}^{a} f(x) \, dx = 2\int_0^a f(x) \, dx$.

b. If f is odd, then $\displaystyle\int_{-a}^{a} f(x) \, dx = 0$.

PROOF We write

$$\int_{-a}^{a} f(x) \, dx = \int_{-a}^{0} f(x) \, dx + \int_0^a f(x) \, dx = -\int_0^{-a} f(x) \, dx + \int_0^a f(x) \, dx \qquad (4)$$

For the integral

$$\int_0^{-a} f(x) \, dx$$

let's make the substitution $u = -x$, so that $du = -dx$. Also, if $x = 0$, then $u = 0$, and if $x = -a$, then $u = a$. So

$$\int_0^{-a} f(x) \, dx = \int_0^a f(-u)(-du) = -\int_0^a f(-x) \, dx$$

Therefore, Equation (4) can be written as

$$\int_{-a}^{a} f(x) \, dx = \int_0^a f(-x) \, dx + \int_0^a f(x) \, dx = \int_0^a [f(-x) + f(x)] \, dx \qquad (5)$$

If f is even, then $f(-x) = f(x)$, so, using Equation (5), we have

$$\int_{-a}^{a} f(x) \, dx = \int_0^a [f(x) + f(x)] \, dx = 2\int_0^a f(x) \, dx$$

If f is odd, then $f(-x) = -f(x)$, so Equation (5) gives

$$\int_{-a}^{a} f(x)\, dx = \int_{0}^{a} [-f(x) + f(x)]\, dx = 0$$ ■

Figure 7 gives a geometric interpretation of Theorem 4. In Figure 7a the area of the region under the graph of the nonnegative function f from $-a$ to 0 is the same as that under the graph of f from 0 to a, so the area of the region under the graph of f from $-a$ to a is equal to twice that from 0 to a. But each of these areas is given by an appropriate integral, leading to the first result in the theorem. In Figure 7b the area of the region above the graph of f and under the x-axis from $-a$ to 0 is equal to the area of the region under the graph of f from 0 to a; the former is given by the *negative* of the integral from 0 to a.

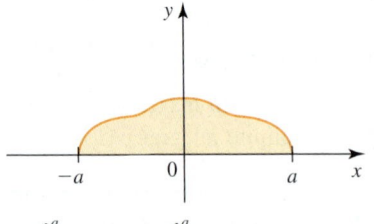

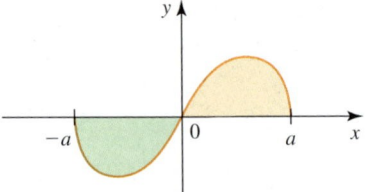

FIGURE 7
The integral of (a) an even function and (b) an odd function

(a) $\displaystyle\int_{-a}^{a} f(x)\, dx = 2\int_{0}^{a} f(x)\, dx$ (b) $\displaystyle\int_{-a}^{a} f(x)\, dx = 0$

EXAMPLE 10   Evaluate

a. $\displaystyle\int_{-1}^{1} (x^2 + 2)\, dx$ **b.** $\displaystyle\int_{-2}^{2} \frac{\sin x}{\sqrt{1 + x^2}}\, dx$

Solution

a. Here, $f(-x) = (-x)^2 + 2 = x^2 + 2 = f(x)$, so f is even. Therefore, by Theorem 4,

$$\int_{-1}^{1} (x^2 + 2)\, dx = 2\int_{0}^{1} (x^2 + 2)\, dx = 2\left(\frac{1}{3}x^3 + 2x\right)\Big|_{0}^{1} = 2\left(\frac{1}{3} + 2\right) = \frac{14}{3}$$

b. Here,

$$f(-x) = \frac{\sin(-x)}{\sqrt{1 + (-x)^2}} = -\frac{\sin x}{\sqrt{1 + x^2}} = -f(x)$$

so f is odd. Therefore, by Theorem 4,

$$\int_{-2}^{2} \frac{\sin x}{\sqrt{1 + x^2}}\, dx = 0$$ ■

The Definite Integral as a Measure of Net Change

In real-world applications we are often interested in the net change of a quantity over a period of time. For example, suppose that P is a function giving the population, $P(t)$, of a city at time t. Then the *net change* in the population over the period from $t = a$ to $t = b$ is given by

$$P(b) - P(a) \qquad \text{Population at } t = b \text{ minus population at } t = a$$

If P has a continuous derivative P' on $[a, b]$, then we can invoke the Fundamental Theorem of Calculus, Part 2, to write

$$P(b) - P(a) = \int_a^b P'(t)\, dt \qquad \text{\textit{P} is an antiderivative of \textit{P}'.}$$

Thus, if we know the *rate of change* of the population at any time t, then we can calculate the net change in the population from $t = a$ to $t = b$ by evaluating an appropriate definite integral.

EXAMPLE 11 **Population Growth in Clark County** Clark County in Nevada, dominated by Las Vegas, was one of the fastest-growing metropolitan areas in the United States. From 1970 through 2000 the population was growing at the rate of

$$R(t) = 133{,}680t^2 - 178{,}788t + 234{,}633 \qquad 0 \le t \le 4$$

people per decade, where $t = 0$ corresponds to the beginning of 1970. What was the net change in the population over the decade from the beginning of 1980 to the beginning of 1990?
Source: U.S. Census Bureau.

Solution The net change in the population over the decade from the beginning of 1980 ($t = 1$) to the beginning of 1990 ($t = 2$) is given by $P(2) - P(1)$, where P denotes the population in the county at time t. But $P' = R$, so

$$
\begin{aligned}
P(2) - P(1) &= \int_1^2 P'(t)\, dt = \int_1^2 R(t)\, dt \\
&= \int_1^2 (133{,}680t^2 - 178{,}788t + 234{,}633)\, dt \\
&= \left[44{,}560t^3 - 89{,}394t^2 + 234{,}633t \right]_1^2 \\
&= [44{,}560(2^3) - 89{,}394(2^2) + 234{,}633(2)] \\
&\quad - [44{,}560 - 89{,}394 + 234{,}633] \\
&= 278{,}371
\end{aligned}
$$

so the net change is 278,371 people.

More generally, we have the following result. We assume that f has a continuous derivative, even though the integrability of f' is sufficient.

Net Change Formula

The net change in a function f over an interval $[a, b]$ is given by

$$f(b) - f(a) = \int_a^b f'(x)\, dx \qquad \text{(6)}$$

provided that f' is continuous on $[a, b]$.

As another example of the net change of a function, let's consider the motion of an object along a straight line. Suppose that the position function and the velocity function of the object are s and v, respectively. Since $s'(t) = v(t)$, Equation (6) gives

$$s(b) - s(a) = \int_a^b s'(t)\,dt = \int_a^b v(t)\,dt$$

the net change in the position of the object over the time interval $[a, b]$. This net change of position is the *displacement* of the object between $t = a$ and $t = b$. (Recall that this result was also discussed in Section 4.4.)

To calculate the distance covered by the object between $t = a$ and $t = b$, we observe that if $v(t) \ge 0$ on an interval $[c, d]$, then the distance covered by the object between $t = c$ and $t = d$ is given by its displacement $\int_c^d v(t)\,dt$. On the other hand, if $v(t) \le 0$ on an interval $[c, d]$, then the distance covered by the object between $t = c$ and $t = d$ is given by the negative of its displacement, that is, by $-\int_c^d v(t)\,dt$. But $-\int_c^d v(t)\,dt = \int_c^d -v(t)\,dt$. Since

$$|v(t)| = \begin{cases} v(t) & \text{if } v(t) \ge 0 \\ -v(t) & \text{if } v(t) < 0 \end{cases}$$

we see that in either case the distance covered by the object is obtained by integrating the *speed* $|v(t)|$ of the object. Therefore, the distance covered by an object between $t = a$ and $t = b$ is

$$\int_a^b |v(t)|\,dt \tag{7}$$

Figure 8 gives a geometric interpretation of the displacement of an object and the distance covered by an object.

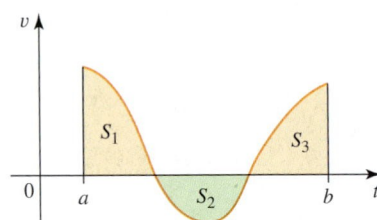

FIGURE 8
Displacement is $\int_a^b v(t)\,dt = $ area of $S_1 - $ area of $S_2 + $ area of S_3, and distance covered is $\int_a^b |v(t)|\,dt = $ area of $S_1 + $ area of $S_2 + $ area of S_3.

EXAMPLE 12 A car moves along a straight road with velocity function

$$v(t) = t^2 + t - 6 \qquad 0 \le t \le 10$$

where $v(t)$ is measured in feet per second.

a. Find the displacement of the car between $t = 1$ and $t = 4$.
b. Find the distance covered by the car during this period of time.

Solution
a. Using Equation (6), we see that the displacement is

$$s(4) - s(1) = \int_1^4 v(t)\,dt = \int_1^4 (t^2 + t - 6)\,dt$$

$$= \left[\frac{1}{3}t^3 + \frac{1}{2}t^2 - 6t\right]_1^4 = 10\frac{1}{2}$$

That is, at $t = 4$ the car is $10\frac{1}{2}$ ft to the right of its position at $t = 1$.

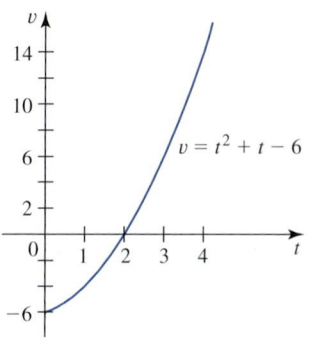

FIGURE 9
$v(t) \leq 0$ if $t \in [1, 2]$, and
$v(t) \geq 0$ if $t \in [2, 4]$.

b. Writing $v(t) = t^2 + t - 6 = (t - 2)(t + 3)$, we see that $v(t) \leq 0$ on $[1, 2]$ and $v(t) \geq 0$ on $[2, 4]$. (See Figure 9.) Using the integral in (7), we see that the distance covered by the car between $t = 1$ and $t = 4$ is given by

$$\int_1^4 |v(t)| \, dt = \int_1^2 (-v(t)) \, dt + \int_2^4 v(t) \, dt$$

$$= \int_1^2 (-t^2 - t + 6) \, dt + \int_2^4 (t^2 + t - 6) \, dt$$

$$= \left[-\frac{1}{3}t^3 - \frac{1}{2}t^2 + 6t \right]_1^2 + \left[\frac{1}{3}t^3 + \frac{1}{2}t^2 - 6t \right]_2^4$$

$$= 14\frac{5}{6}$$

or $14\frac{5}{6}$ ft.

4.5 CONCEPT QUESTIONS

1. Define the average value of a function f over an interval $[a, b]$. Give a geometric interpretation.
2. State the Mean Value Theorem for Integrals. Give a geometric interpretation.
3. State both parts of the Fundamental Theorem of Calculus.
4. State the Net Change Formula, and use it to answer the following:
   a. If water is flowing through a pipe at the rate of R ft^3/min, what does $\int_{t_1}^{t_2} R(t) \, dt$ measure, where t_1 and t_2 are measured in minutes with $t_1 < t_2$?

   b. If an object is moving along a straight line with an acceleration of $a(t)$ ft/sec^2, what does $\int_{t_1}^{t_2} a(t) \, dt$ measure if $t_1 < t_2$?
5. Suppose that a particle moves along a coordinate line with a velocity of $v(t)$ ft/sec. Explain the difference between $\int_a^b v(t) \, dt$ and $\int_a^b |v(t)| \, dt$.

4.5 EXERCISES

1. Let $F(x) = \int_2^x t^2 \, dt$.
   a. Use Part 1 of the Fundamental Theorem of Calculus to find $F'(x)$.
   b. Use Part 2 of the Fundamental Theorem of Calculus to integrate $\int_2^x t^2 \, dt$ to obtain an alternative expression for $F(x)$.
   c. Differentiate the expression for $F(x)$ found in part (b), and compare the result with that obtained in part (a). Comment on your result.

2. Repeat Exercise 1 with $G(x) = \int_0^x \sqrt{3t + 1} \, dt$.

In Exercises 3–14, find the derivative of the function.

3. $F(x) = \int_0^x \sqrt{3t + 5} \, dt$

4. $G(x) = \int_{-1}^x t \sqrt{t^2 + 1} \, dt$

5. $g(x) = \int_2^x \frac{1}{t^2 + 1} \, dt$

6. $h(x) = \int_x^3 \frac{t}{\sqrt{t + 1}} \, dt$

7. $F(x) = \int_x^\pi \sin 2t \, dt$

8. $G(x) = \int_0^{x^2} t \sin t \, dt$

9. $g(x) = \int_2^{\sqrt{x}} \frac{\sin t}{t} \, dt$

10. $h(x) = \int_0^{x^2} \sin t^2 \, dt$

11. $F(x) = \int_1^{\cos x} \frac{t^2}{t + 1} \, dt$

12. $G(x) = \int_{\sqrt{x}}^5 \frac{\sin t^2}{t} \, dt$

13. $f(x) = \int_{x^2}^{x^3} \ln t \, dt; \quad x > 0$

14. $g(x) = \int_{1/x}^{e^x} \tan^{-1} t \, dt$

In Exercises 15–36, evaluate the integral.

15. $\int_{-3}^2 4 \, dx$

16. $\int_{-2}^0 (2x - 3) \, dx$

17. $\int_{-1}^1 (t^2 - 4) \, dt$

18. $\int_0^2 (2 - 4u + u^2) \, du$

19. $\int_{-2}^1 (3t + 2)^2 \, dt$

20. $\int_1^2 \frac{3}{x^3} \, dx$

21. $\int_1^3 \frac{x^2 - x + 3}{x} \, dx$

22. $\int_1^3 \frac{\ln x}{x} \, dx$

23. $\displaystyle\int_1^4 \frac{1}{\sqrt{x}}\,dx$

24. $\displaystyle\int_1^2 \frac{3x^4 - 2x^2 + 1}{2x^2}\,dx$

25. $\displaystyle\int_4^9 \frac{x-1}{\sqrt{x}}\,dx$

26. $\displaystyle\int_1^0 (t^{1/2} - t^{5/2})\,dt$

27. $\displaystyle\int_2^0 \sqrt{x}(x+1)(x-2)\,dx$

28. $\displaystyle\int_0^{\pi/2} (\sin x + 1)\,dx$

29. $\displaystyle\int_{\pi/6}^{\pi/4} \sec^2 t\,dt$

30. $\displaystyle\int_{\pi/6}^{\pi/4} \csc\theta \cot\theta\,d\theta$

31. $\displaystyle\int_0^{\pi} \sin 2x \cos x\,dx$

32. $\displaystyle\int_0^{\pi} |\cos x|\,dx$

33. $\displaystyle\int_{\pi/4}^{\pi/3} \frac{dx}{\sin^2 x \cos^2 x}$

34. $\displaystyle\int_0^{\pi} \sqrt{\sin x - \sin^3 x}\,dx$

35. $\displaystyle\int_{-1}^1 f(x)\,dx$ where $f(x) = \begin{cases} -x+1 & \text{if } x \le 0 \\ 2x^2 + 1 & \text{if } x > 0 \end{cases}$

36. $\displaystyle\int_{-\pi}^{\pi/2} f(x)\,dx$ where $f(x) = \begin{cases} x^2 + 1 & \text{if } x \le 0 \\ \cos x & \text{if } x > 0 \end{cases}$

In Exercises 37–62, evaluate the integral.

37. $\displaystyle\int_0^1 (3 - 2x)^4\,dx$

38. $\displaystyle\int_0^2 (t+1)^{0.2}\,dt$

39. $\displaystyle\int_1^2 8t(t^2 - 1)^7\,dt$

40. $\displaystyle\int_1^5 \sqrt{2x - 1}\,dx$

41. $\displaystyle\int_1^4 \sqrt[3]{5 - u}\,du$

42. $\displaystyle\int_1^2 \frac{x}{2x^2 - 1}\,dx$

43. $\displaystyle\int_1^4 \frac{1}{\sqrt{x}(\sqrt{x}+1)^2}\,dx$

44. $\displaystyle\int_{\pi/4}^{\pi/2} \sin 2x\,dx$

45. $\displaystyle\int_{-1}^0 \frac{1}{1 + e^{-2x}}\,dx$

46. $\displaystyle\int_0^{\pi/4} \frac{e^{\tan x}}{\cos^2 x}\,dx$

47. $\displaystyle\int_{\pi/2}^{\pi} \cos\left(\frac{1}{2}x\right)dx$

48. $\displaystyle\int_0^{\pi/2} \sqrt{\cos\theta}\,\sin\theta\,d\theta$

49. $\displaystyle\int_{-1/4}^{1/4} \sec\pi t \tan\pi t\,dt$

50. $\displaystyle\int_{\pi/6}^{\pi/2} \csc^2\theta \cot\theta\,d\theta$

51. $\displaystyle\int_{1/\pi}^{2/\pi} \frac{\sin\frac{1}{x}}{x^2}\,dx$

52. $\displaystyle\int_{-\pi}^{\pi/2} \frac{x^2 \sin x}{\sqrt{1 + x^2}}\,dx$

53. $\displaystyle\int_0^1 xe^{x^2}e^{e^{x^2}}\,dx$

54. $\displaystyle\int_1^e \frac{\ln x}{x}e^{(\ln x)^2}\,dx$

55. $\displaystyle\int_0^1 (3^t + t^3)\,dt$

56. $\displaystyle\int_1^4 \frac{3^{\sqrt{x}}}{\sqrt{x}}\,dx$

57. $\displaystyle\int_0^{1/4} \frac{1}{\sqrt{1 - 4x^2}}\,dx$

58. $\displaystyle\int_0^{1/2} \frac{1}{1 + 4x^2}\,dx$

59. $\displaystyle\int_0^{4\sqrt{3}} \frac{1}{x^2 + 16}\,dx$

60. $\displaystyle\int_0^1 \frac{e^{2x}}{1 + e^{4x}}\,dx$

61. $\displaystyle\int_0^1 \frac{x^3}{1 + x^8}\,dx$

62. $\displaystyle\int_0^{\sqrt{3}/2} \frac{\sin^{-1}x}{\sqrt{1 - x^2}}\,dx$

63. a. Prove that $0 \le \displaystyle\int_0^1 \frac{x^5}{\sqrt[3]{1 + x^4}}\,dx \le \frac{1}{6}$.

   cas b. Use a calculator or a computer to find the value of the integral accurate to five decimal places.

64. a. Prove that $0 \le \displaystyle\int_0^1 \frac{dx}{\sqrt{4 - 3x + x^2}} \le \frac{2}{3}$.

   cas b. Use a calculator or a computer to find the value of the integral accurate to five decimal places.

In Exercises 65–70, find the area of the region under the graph of f on $[a, b]$.

65. $f(x) = x^2 - 2x + 2$; $[-1, 2]$

66. $f(x) = \dfrac{1}{x^2}$; $[1, 2]$

67. $f(x) = 2 + \sqrt{x + 1}$; $[0, 3]$

68. $f(x) = \sec^2 x$; $\left[0, \frac{\pi}{4}\right]$

69. $f(x) = e^{-x/2}$; $[-1, 2]$

70. $f(x) = \dfrac{1}{4 + x^2}$; $[0, 1]$

71. Let $f(x) = -2x^4 + x^2 + 2x$.

  a. Plot the graph of f.
     b. Find the x-intercepts of f accurate to three decimal places.
     c. Use the results of parts (a) and (b) to find the area of the region under the graph of f and above the x-axis.

72. Let $f(x) = \dfrac{e^x - 1}{e^x + 1}$.

  a. Plot the graph of f using the viewing window $[-5, 5] \times [-1, 1]$.
     b. Find the area of the region under the graph of f over the interval $[0, \ln 3]$.
     c. Verify your answer to part (b) using a calculator or a computer.

In Exercises 73–76, evaluate the limit by interpreting it as the limit of a Riemann sum of a function on the interval $[a, b]$.

73. $\displaystyle\lim_{n \to \infty} \frac{1}{n^5} \sum_{k=1}^n k^4$; $[0, 1]$

74. $\displaystyle\lim_{n \to \infty} \frac{1}{n} \sum_{k=1}^n \left(\frac{k}{n}\right)^{1/3}$; $[0, 1]$

75. $\displaystyle\lim_{n \to \infty} \frac{2}{n} \sum_{k=1}^n \left(2 + \frac{2k}{n}\right)^2$; $[2, 4]$

76. $\displaystyle\lim_{n \to \infty} \frac{\pi}{2n} \sum_{k=1}^n \cos\left(\frac{k\pi}{2n}\right)$; $\left[0, \frac{\pi}{2}\right]$

In Exercises 77–80, find the average value f_{av} of the function over the indicated interval.

77. $f(x) = 2x^2 - 3x$; $[-1, 2]$

78. $f(x) = 1 + \sqrt{x}$; $[0, 4]$

79. $f(x) = \dfrac{x}{\sqrt{x^2 + 1}}$; $[0, 3]$

80. $f(x) = \sin x$; $[0, \pi]$

In Exercises 81–84, (a) find the number c whose existence is guaranteed by the Mean Value Theorem for Integrals for the function f on [a, b], and (b) sketch the graph of f on [a, b] and the rectangle with base on [a, b] that has the same area as that of the region under the graph of f.

81. $f(x) = x^2 + 2x$; $[0, 1]$

82. $f(x) = x^3$; $[0, 2]$

83. $f(x) = \sqrt{x + 3}$; $[1, 6]$

84. $f(x) = \cos x$; $\left[-\frac{\pi}{3}, \frac{\pi}{3}\right]$

85. Distance Covered by a Car A car moves along a straight road with velocity function

$$v(t) = 2t^2 + t - 6 \qquad 0 \le t \le 8$$

where $v(t)$ is measured in feet per second.
a. Find the displacement of the car between $t = 0$ and $t = 3$.
b. Find the distance covered by the car during this period of time.

86. Projected U.S. Gasoline Use The White House wants to cut gasoline use from 140 billion gallons per year in 2007 to 128 billion gallons per year in 2017. But estimates by the Department of Energy's Energy Information Agency suggest that this will not happen. In fact, the agency's projection of gasoline use from the beginning of 2007 to the beginning of 2017 is given by

$$A(t) = 0.014t^2 + 1.93t + 140 \qquad 0 \le t \le 10$$

where $A(t)$ is measured in billions of gallons per year and t is in years with $t = 0$ corresponding to the beginning of 2007.
a. According to the agency's projection, what will be the gasoline consumption at the beginning of 2017?
b. What will be the average consumption per year from the beginning of 2007 to the beginning of 2017?
Source: U.S. Department of Energy, Energy Information Agency.

87. Air Purification To test air purifiers, engineers ran a purifier in a smoke-filled 10-ft × 20-ft room. While conducting a test for a certain brand of air purifier, it was determined that the amount of smoke in the room was decreasing at the rate of $R(t)$ percent of the (original) amount of smoke per minute, t min after the start of the test, where R is given by

$$R(t) = 0.00032t^4 - 0.01872t^3 + 0.3948t^2 - 3.83t + 17.63$$
$$0 \le t \le 20$$

How much smoke was left in the room 5 min after the start of the test? How much smoke was left in the room 10 min after the start of the test?
Source: Consumer Reports

88. Voltage in AC Circuits The voltage in an AC circuit is given by

$$V = V_0 \sin \omega t$$

a. Show that the average (mean) voltage from $t = 0$ to $t = \pi/\omega$ (a half-cycle) is $V_{av} = (2/\pi)V_0$, which is $2/\pi$ $\left(\text{about } \frac{2}{3}\right)$ times the maximum voltage V_0.
b. Show that the average voltage over a complete cycle is 0. Explain.

89. If a feet of fencing are used to enclose a rectangular garden, show that the average area of such a garden is $a^2/24$ ft^2.

90. Average Acceleration of a Car A car moves along a straight road with velocity function $v(t)$ and acceleration function $a(t)$. The average acceleration of the car over the time interval $[t_1, t_2]$ is

$$\bar{a} = \frac{v(t_2) - v(t_1)}{t_2 - t_1}$$

Show that $\bar{a}$ is equal to the average value of $a(t)$ on $[t_1, t_2]$.

91. Velocity of a Falling Hammer During the construction of a high-rise apartment building, a construction worker accidentally drops a hammer that falls vertically a distance of h ft. The velocity of the hammer after falling a distance of x ft is $v = \sqrt{2gx}$ ft/sec, where $0 \le x \le h$. Show that the average velocity of the hammer over this path is $\bar{v} = \frac{2}{3}\sqrt{2gh}$.

92. Flow of Water in a Canal Water at a depth of x ft in a wide rectangular canal flows at a velocity of

$$v = v_0 - 20\sqrt{hs}\left(\frac{x}{h}\right)^2$$

feet per second, where v_0 is the velocity of the water on the surface, h is the depth of the canal, and s is its gradient. Find the average velocity of flow in a cross section of the canal.

93. Flow of Blood in an Artery The velocity (in centimeters per second) of blood r cm from the central axis of an artery is given by $v(r) = k(R^2 - r^2)$, where k is a constant and R is the radius of the artery. Suppose that $k = 1000$ and $R = 0.2$. Find the average velocity of the blood across a cross section of the artery.

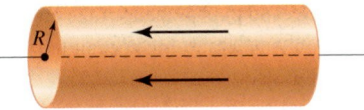

94. Newton's Law of Cooling A bottle of white wine at room temperature (68°F) is placed in a refrigerator at 4 P.M. Its temperature after t hr is changing at the rate of $-18e^{-0.6t}$ °F/hr. By how many degrees will the temperature of the wine have dropped by 7 P.M.? What will the temperature of the wine be at 7 P.M.?

95. Air Pollution According to the South Coast Air Quality Management District, the level of nitrogen dioxide, a brown gas that impairs breathing, present in the atmosphere on a certain June day in downtown Los Angeles is approximated by

$$A(t) = 0.03t^3(t - 7)^4 + 62.7 \qquad 0 \le t \le 7$$

where $A(t)$ is measured in pollutant standard index and t is measured in hours with $t = 0$ corresponding to 7 A.M. What is the average level of nitrogen dioxide present in the atmosphere from 7 A.M. to 2 P.M. on that day?
Source: The Los Angeles Times.

96. Water Level in Boston Harbor The water level (in feet) in Boston Harbor during a certain 24-hr period is approximated by the formula

$$H = 4.8 \sin\left[\frac{\pi}{6}(t - 10)\right] + 7.6 \qquad 0 \le t \le 24$$

where $t = 0$ corresponds to 12 A.M. What is the average water level in Boston Harbor over the 24-hr period on that day?

97 Predator-Prey Populations The wolf and caribou populations in a certain northern region are given by

$$P_1(t) = 8000 + 1000 \sin\frac{\pi t}{24}$$

and

$$P_2(t) = 40,000 + 12,000 \cos\frac{\pi t}{24}$$

respectively, at time t, where t is measured in months. What are the average wolf and caribou populations over the time interval $[0, 6]$?

98. Daylight Hours in Chicago The number of hours of daylight at any time t in Chicago is approximated by

$$L(t) = 2.8 \sin\left[\frac{2\pi}{365}(t - 79)\right] + 12$$

where t is measured in days and $t = 0$ corresponds to January 1. What is the daily average number of hours of daylight in Chicago over the year? Over the summer months from June 21 ($t = 171$) through September 20 ($t = 262$)?

99. Global Warming The increase in carbon dioxide in the atmosphere is a major cause of global warming. Using data obtained by Dr. Charles David Keeling, professor at Scripps Institution of Oceanography, the average amount of carbon dioxide in the atmosphere from 1958 through 2007 is approximated by

$$A(t) = 0.010716t^2 + 0.8212t + 313.4 \qquad 1 \le t \le 50$$

where $A(t)$ is measured in parts per million volume (ppmv) and t in years with $t = 1$ corresponding to the beginning of 1958. Find the average amount of carbon dioxide in the atmosphere from 1958 through 2007.
Source: Scripps Institution of Oceanography.

100. Depreciation: Double Declining Balance Method Suppose that a tractor purchased at a price of $60,000 is to be depreciated by the *double declining balance method* over a 10-year period. It can be shown that the rate at which the book value will be decreasing is given by

$$R(t) = 13,388.61e^{-0.22314t} \qquad 0 \le t \le 10$$

dollars per year at year t. Find the amount by which the book value of the tractor will depreciate over the first 5 years of its life.

101. Canadian Oil-Sands Production The production of oil (in millions of barrels per day) extracted from oil sands in Canada is projected to be

$$P(t) = \frac{4.76}{1 + 4.11e^{-0.22t}} \qquad 0 \le t \le 15$$

where t is measured in years, with $t = 0$ corresponding to the beginning of 2005. What will the total oil production of oil from oil sands be over the years from the beginning of 2005 until the beginning of 2020 ($t = 15$)?
Source: Canadian Association of Petroleum Producers.

102. Average Temperature A homogenous hollow metallic ball of inner radius r_1 and outer radius r_2 is in thermal equilibrium. The temperature T at a distance r from the center of the ball is given by

$$T = T_1 + \frac{r_1 r_2(T_2 - T_1)}{(r_1 - r_2)}\left(\frac{1}{r} - \frac{1}{r_1}\right) \qquad r_1 \le r \le r_2$$

where T_1 is the temperature on the inner surface and T_2 is the temperature on the outer surface. Find the average temperature of the ball in a radial direction between $r = r_1$ and $r = r_2$.

103. Motion of a Submersible A submersible moving in a straight line through water is subjected to a resistance R that is proportional to its velocity. Suppose that the submersible travels with its engine shut off. Then the time it takes for the submersible to slow down from a velocity of v_1 to a velocity of v_2 is

$$T = -\int_{v_1}^{v_2} \frac{m}{kv}\, dv$$

where m is the mass of the submersible and k is a constant. Find the time it takes the submersible to slow down from a velocity of 16 ft/sec to 8 ft/sec if its mass is 1250 slugs and $k = 20$ (slug/sec).

104. Lengths of Infants Medical records of infants delivered at Kaiser Memorial Hospital show that the percentage of infants whose length at birth is between 19 and 21 in. is given by

$$P = 100\int_{19}^{21} \frac{1}{2.6\sqrt{2\pi}} e^{-(1/2)[(x-20)/2.6]^2}\, dx$$

Use a calculator or computer to estimate P.

105. Serum Cholesterol Levels The percentage of a current Mediterranean population with serum cholesterol levels between 160 and 180 mg/dL is estimated to be

$$P = \sqrt{\frac{2}{\pi}} \int_{160}^{180} e^{-(1/2)[(x-160)/50]^2} \, dx$$

Estimate P.

 106. Absorption of Drugs The concentration of a drug in an organ at any time t, in seconds) is given by

$$C(t) = \begin{cases} 0.3t - 18(1 - e^{-t/60}) & \text{if } 0 \le t \le 20 \\ 18e^{-t/60} - 12e^{-(t-20)/60} & \text{if } t > 20 \end{cases}$$

where $C(t)$ is measured in grams per cubic centimeter (g/cm³). Find the average concentration of the drug in the organ over the first 30 sec after it is administered.

107. Prove that

$$\int_{-1/2}^{1/2} 2^{\cos x} \, dx = 2 \int_0^{1/2} 2^{\cos x} \, dx$$

108. Find dx/dy if

$$\int_0^x \sqrt{3 + 2 \cos t} \, dt + \int_0^y \sin t \, dt = 0$$

109. Find the x-coordinates of the relative extrema of the function

$$F(x) = \int_0^x \frac{\sin t}{t} \, dt \qquad x > 0$$

110. Find all functions f on $[0, 1]$ such that f is continuous on $[0, 1]$ and

$$\int_0^x f(t) \, dt = \int_x^1 f(t) \, dt \quad \text{for every } x \in (0, 1)$$

111. If $f(x) = \int_2^x \frac{dt}{\sqrt{1 + t^3}}$, where $x > -1$, what is $(f^{-1})'(0)$?

112. Let

$$f(x) = \begin{cases} 1 - x & \text{if } 0 \le x \le 1 \\ x - 1 & \text{if } 1 < x \le 3 \end{cases}$$

a. Find $F(x) = \int_0^x f(t) \, dt$.

 b. Plot the graph of F, and show that it is continuous on $[0, 3]$.

c. Where is f differentiable? Where is F differentiable?

113. Evaluate $\lim_{h \to 0} \frac{1}{h} \int_2^{2+h} \sqrt{5 + t^2} \, dt$.

114. Evaluate $\int_{-1}^{1} \frac{2x^5 + x^4 - 3x^3 + 2x^2 + 8x + 1}{x^2 + 1} \, dx$.

115. Evaluate $\int_{-\pi/4}^{\pi/4} (\cos x + 1) \tan^3 x \, dx$.

116. Show that

$$\int_{-1}^{1} \sqrt{x^2 + 1} \sec x \, dx = 2 \int_0^1 \sqrt{x^2 + 1} \sec x \, dx$$

117. a. Show that $\int_0^\pi x f(\sin x) \, dx = (\pi/2) \int_0^\pi f(\sin x) \, dx$.
   Hint: Use the substitution $x = \pi - u$.
   b. Use the result of part (a) to evaluate $\int_0^\pi x \sin x \, dx$.

118. a. If f is even, what can you say about $\int_{-\pi}^{\pi} f(x) \cos nx \, dx$ and $\int_{-\pi}^{\pi} f(x) \sin nx \, dx$ if n is an integer? Explain.
   b. If f is odd, what can you say about $\int_{-\pi}^{\pi} f(x) \cos nx \, dx$ and $\int_{-\pi}^{\pi} f(x) \sin nx \, dx$? Explain.

119. Use the identity

$$\frac{\sin(n + \frac{1}{2})x}{2 \sin \frac{x}{2}} = \frac{1}{2} + \cos x + \cos 2x + \cdots + \cos nx$$

to show that

$$\int_0^\pi \frac{\sin(n + \frac{1}{2})x}{\sin \frac{x}{2}} \, dx = \pi$$

120. a. Show that if f is a continuous function, then

$$\int_0^a f(x) \, dx = \int_0^a f(a - x) \, dx$$

and give a geometric interpretation of this result.
   b. Use the result of part (a) to prove that

$$\int_0^\pi \frac{\sin 2kx}{\sin x} \, dx = 0$$

where k is an integer.

 c. Plot the graph of

$$f(x) = \frac{\sin 2kx}{\sin x}$$

for $k = 1, 2, 3,$ and 4. Do these graphs support the result of part (b)?
   d. Prove that the graph of

$$f(x) = \frac{\sin 2kx}{\sin x}$$

on $[0, \pi]$ is antisymmetric with respect to the line $x = \pi/2$ by showing that $f(x + \frac{\pi}{2}) = -f(x - \frac{\pi}{2})$ for $0 \le x \le \frac{\pi}{2}$, and use this result to explain part (b).

121. A car travels along a straight road in such a way that the average velocity over *any* time interval $[a, b]$ is equal to the average of its velocities at a and at b.
   a. Show that its velocity $v(t)$ satisfies

$$\int_a^b v(t) \, dt = \frac{1}{2} [v(a) + v(b)](b - a) \tag{1}$$

   b. Show that $v(t) = ct + d$ for some constants c and d.
   Hint: Differentiate Equation (1) with respect to a and with respect to b.

122. Let f be continuous on $(-\infty, \infty)$. Show that

$$\int_a^b f(x + h)\, dx = \int_{a+h}^{b+h} f(x)\, dx$$

123. Let f be continuous on $(-\infty, \infty)$, and let c be a constant. Show that

$$\int_{ca}^{cb} f(x)\, dx = c \int_a^b f(cx)\, dx$$

In Exercises 124–128, determine whether the statement is true or false. If it is true, explain why it is true. If it is false, explain why or give an example to show why it is false.

124. Assuming that the integral exists, then
$\int_{-a}^a f(x^2)\, dx = 2 \int_0^a f(x^2)\, dx.$

125. Assuming that the integral exists and that f is even, then $\int_{-a}^a f(x^3)\, dx = 2 \int_0^a f(x^3)\, dx.$

126. Assuming that the integral exists and that f is even and g is odd, then

$$\int_{-a}^a f(x)[g(x)]^2\, dx = 2 \int_0^a f(x)[g(x)]^2\, dx$$

127. If F is defined by $F(x) = \int_0^x \sqrt[3]{1 + t^2}\, dt$, then F' has an inverse on $(0, \infty)$.

128. $\int_1^2 \dfrac{e^{-x}}{x}\, dx < 0$

4.6 Numerical Integration

■ Approximating Definite Integrals

Table 1 gives the daily consumption of oil in the United States in millions of barrels, in two-year intervals from 1987 through the year 2007. Suppose that we want to determine the average daily consumption of oil over the period in question. From our earlier work, we know that the solution is obtained by computing

$$\frac{1}{20} \int_0^{20} f(t)\, dt$$

where $f(t)$ is the oil consumption in year t and $t = 0$ corresponds to 1987. But the problem here is that we do not know the algebraic rule defining the integrand f for all values of t in $[0, 20]$. We are given its values only at a discrete set of points in that interval! Here, the Fundamental Theorem of Calculus cannot be used to help us evaluate the integral, since we cannot find an antiderivative of f. Other situations also arise (for example, $f(t) = \sin t^2$) in which, although the integrand of a definite integral is defined algebraically, we are not able to find its antiderivative in terms of elementary functions. In each of these situations the best we can do is to obtain an approximation to the definite integral. (We will return to the problem of finding the average daily consumption of petroleum in Example 5.)

TABLE 1

Year	1987	1989	1991	1993	1995	1997	1999	2001	2003	2005	2007
Consumption	16.7	17.3	16.7	17.2	17.7	18.6	19.5	19.6	20.0	20.8	20.7

Source: U.S. Energy Information Administration.

A Riemann sum gives us a good approximation of a definite integral of an integrable function if the norm of the partition is sufficiently small. But there are better methods for finding approximate values of definite integrals. In this section we will look at two such methods.

■ The Trapezoidal Rule

The Trapezoidal Rule uses the sum of the areas of trapezoids to approximate the definite integral $\int_a^b f(x)\,dx$. To derive this rule, let's assume that f is continuous and nonnegative on $[a, b]$.* We begin by subdividing the interval $[a, b]$ into n subintervals, each of equal length $\Delta x = (b - a)/n$. Because f is nonnegative, the definite integral $\int_a^b f(x)\,dx$ gives the area of the region R under the graph of f on $[a, b]$. (See Figure 1.) This area is given by the sum of the areas of the nonoverlapping subregions $R_1, R_2, \ldots, R_n$, where R_1 represents the region under the graph of f on $[x_0, x_1]$, R_2 represents the region under the graph of f on $[x_1, x_2], \ldots$, and R_n represents the region under the graph of f on $[x_{n-1}, x_n]$.

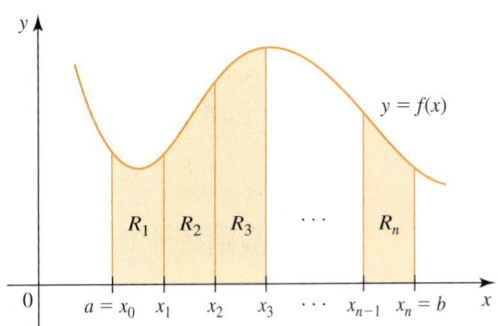

FIGURE 1
$\int_a^b f(x)\,dx =$ the sum of the areas of the subregions $R_1, R_2, \ldots, R_n$

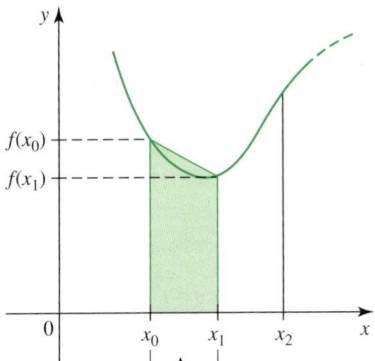

FIGURE 2
The area of R_1 is approximated by the area of a trapezoid.

The basis for the Trapezoidal Rule lies in the approximation of each of the subregions $R_1, R_2, \ldots, R_n$ by a suitable trapezoid. For example, consider the subregion R_1 reproduced in Figure 2. You can see that the area of R_1 may be approximated by the area of the trapezoid having width $\Delta x = x_1 - x_0$ and parallel sides of lengths $f(x_0)$ and $f(x_1)$. The area of this trapezoid is

$$\left[\frac{f(x_0) + f(x_1)}{2}\right]\Delta x \qquad \text{average of the lengths of the parallel sides} \cdot \text{the width}$$

Similarly, the area of the subregion R_2 may be approximated by the area of the trapezoid having width Δx and sides of length $f(x_1)$ and $f(x_2)$:

$$\left[\frac{f(x_1) + f(x_2)}{2}\right]\Delta x$$

Finally, the area of the subregion R_n is approximately

$$\left[\frac{f(x_{n-1}) + f(x_n)}{2}\right]\Delta x$$

---

*Actually, the nonnegativity condition is not necessary, but this assumption will simplify the derivation of the Trapezoidal Rule.

Therefore, the area of the region R is approximated by the sum of the areas of the n trapezoids:

$$\int_a^b f(x)\,dx \approx \Delta x\left[\frac{f(x_0)+f(x_1)}{2}+\frac{f(x_1)+f(x_2)}{2}+\cdots+\frac{f(x_{n-1})+f(x_n)}{2}\right]$$

$$=\frac{\Delta x}{2}[f(x_0)+2f(x_1)+2f(x_2)+\cdots+2f(x_{n-1})+f(x_n)]$$

where $\Delta x = (b-a)/n$.

The Trapezoidal Rule

$$\int_a^b f(x)\,dx \approx \frac{\Delta x}{2}[f(x_0)+2f(x_1)+2f(x_2)+\cdots+2f(x_{n-1})+f(x_n)] \quad \textbf{(1)}$$

where $\Delta x = (b-a)/n$ and $x_i = a + i\Delta x$, for $0 \le i \le n$.

EXAMPLE 1 Use the Trapezoidal Rule with $n = 10$ to approximate $\int_1^2 \dfrac{dx}{x}$.

Solution Here, $a = 1$, $b = 2$, and $n = 10$, so

$$\Delta x = \frac{b-a}{n} = \frac{2-1}{10} = 0.1$$

Also,

$$x_0 = 1, \quad x_1 = 1.1, \quad x_2 = 1.2, \quad x_3 = 1.3, \quad \ldots, \quad x_9 = 1.9, \quad x_{10} = 2$$

The Trapezoidal Rule yields

$$\int_1^2 \frac{dx}{x} \approx \frac{0.1}{2}\left[1 + 2\left(\frac{1}{1.1}\right) + 2\left(\frac{1}{1.2}\right) + 2\left(\frac{1}{1.3}\right) + \cdots + 2\left(\frac{1}{1.9}\right) + \frac{1}{2}\right]$$

$$\approx 0.693771$$

If we use Formula (10) of Section 4.1, we see that

$$\int_1^2 \frac{dx}{x} = \ln 2 \qquad \text{Natural logarithm of 2}$$

$$\approx 0.693147$$

Thus, the Trapezoidal Rule with $n = 10$ yields an approximation with an error of approximately 0.000624.

The Error in the Trapezoidal Rule

The **error** in the approximation of $I = \int_a^b f(x)\,dx$ by the Trapezoidal Rule is defined to be $E_n = I - T_n$, where

$$T_n = \frac{\Delta x}{2}[f(x_0)+2f(x_1)+2f(x_2)+\cdots+2f(x_{n-1})+f(x_n)]$$

An upper bound for this error follows.

Error Bound for the Trapezoidal Rule

If f'' is continuous on $[a, b]$, then the error E_n in approximating $\int_a^b f(x)\, dx$ by the Trapezoidal Rule satisfies

$$|E_n| \leq \frac{M(b-a)^3}{12n^2} \tag{2}$$

where M is a positive number such that $|f''(x)| \leq M$ for all x in $[a, b]$.

Note   Observe that as $n \to \infty$, $E_n \to 0$, as our intuition tells us. ■

EXAMPLE 2   Find an upper bound for the error in the approximation of $\int_1^2 \dfrac{dx}{x}$ using the Trapezoidal Rule with $n = 10$ (see Example 1).

Solution   Here $f(x) = 1/x$. So

$$f'(x) = -\frac{1}{x^2} \qquad \text{and} \qquad f''(x) = \frac{2}{x^3}$$

Since f'' is positive and decreasing on $(1, 2)$, it attains its maximum value at the left endpoint of the interval. So if we take $M = f''(1) = 2$, then $|f''(x)| \leq 2$ for all x in $[1, 2]$. Finally, using Inequality (2) with this value of M and $a = 1$, $b = 2$, and $n = 10$, we obtain

$$|E_{10}| \leq \frac{2(2-1)^3}{12(10)^2} = \frac{1}{600} \approx 0.0016667$$

The actual error is approximately 0.000624 (as we computed in Example 1), and this is less than the upper bound that we just found. ■

EXAMPLE 3   Use the Trapezoidal Rule to approximate $\int_0^1 \sin x^2\, dx$ with an error that is less than 0.01.

Solution   First, we determine the number of subintervals, n, required in the Trapezoidal Rule to ensure that the error will be less than 0.01. To find the value of M called for in Inequality (2), we compute the second derivative of $f(x) = \sin x^2$. Thus,

$$f'(x) = 2x \cos x^2$$

and

$$f''(x) = 2(\cos x^2 - 2x^2 \sin x^2)$$

Using the triangle inequality, we have

$$|f''(x)| \leq 2|\cos x^2| + 4x^2|\sin x^2| \leq 2 + 4 = 6$$

because $0 \leq x \leq 1$ and both $|\cos x^2|$ and $|\sin x^2|$ cannot exceed 1. Therefore, we can take $M = 6$.

Using Inequality (2) with $a = 0$, $b = 1$, and $M = 6$ and observing the requirement that the error in the approximation be less than 0.01, we have

$$\frac{6(1-0)^3}{12n^2} < 0.01, \qquad n^2 > \frac{100}{2}, \qquad \text{or} \qquad n > \sqrt{50} \approx 7.07$$

So by taking $n = 8$, the smallest integer exceeding 7.07, we are guaranteed that the error in the approximation will be smaller than that prescribed.

To obtain the approximation, we compute

$$\Delta x = \frac{b - a}{n} = \frac{1 - 0}{8} = \frac{1}{8}$$

Then with

$$x_0 = 0, \qquad x_1 = \frac{1}{8}, \qquad x_2 = \frac{1}{4}, \qquad x_3 = \frac{3}{8},$$

$$x_4 = \frac{1}{2}, \qquad x_5 = \frac{5}{8}, \qquad x_6 = \frac{3}{4}, \qquad x_7 = \frac{7}{8}, \qquad \text{and} \qquad x_8 = 1$$

the Trapezoidal Rule yields

$$\int_0^1 \sin x^2 \, dx \approx \frac{\frac{1}{8}}{2}\left[f(0) + 2f\left(\frac{1}{8}\right) + 2f\left(\frac{1}{4}\right) + 2f\left(\frac{3}{8}\right) + 2f\left(\frac{1}{2}\right) \right.$$

$$\left. + 2f\left(\frac{5}{8}\right) + 2f\left(\frac{3}{4}\right) + 2f\left(\frac{7}{8}\right) + f(1) \right]$$

$$= \frac{1}{16}\left(\sin 0 + 2 \sin \frac{1}{64} + 2 \sin \frac{1}{16} + 2 \sin \frac{9}{64} + 2 \sin \frac{1}{4} \right.$$

$$\left. + 2 \sin \frac{25}{64} + 2 \sin \frac{9}{16} + 2 \sin \frac{49}{64} + \sin 1 \right)$$

$$\approx 0.3117$$

■ Simpson's Rule

Before deriving Simpson's Rule, let's look at the two methods that we currently have for approximating the definite integral $\int_a^b f(x) \, dx$ from a fresh point of view. Let f be a continuous, nonnegative function on $[a, b]$.* Suppose that the interval $[a, b]$ is partitioned into n subintervals of equal length by means of $n + 1$ equally spaced points $x_0 = a, x_1, x_2, \ldots, x_n = b$, where n is a positive integer, so that the length of each subinterval is $\Delta x = (b - a)/n$. (See Figure 3.)

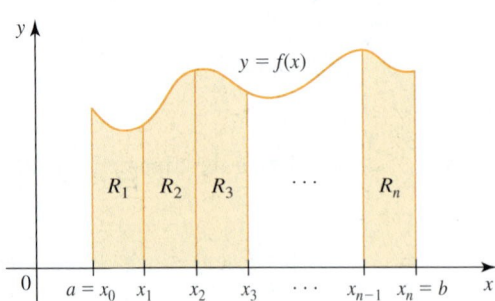

FIGURE 3

The area of the region under the graph of f on $[a, b]$ is given by the sum of the areas of the subregions $R_1, R_2, \ldots, R_n$.

---

*Again, the nonnegativity condition is not necessary for our results but will be assumed here to simplify the ensuing discussion.

Figure 4 shows the approximation of $\int_a^b f(x)\,dx$ using a Riemann sum consisting of n terms that are just the areas of the n rectangles shown shaded. Here is another view of that method: Approximate $f(x)$ on $[x_0, x_1]$ by the *constant* function $y = f(p_1)$, where p_1 is any point in $[x_0, x_1]$; approximate $f(x)$ on $[x_1, x_2]$ by the constant function $y = f(p_2)$, where p_2 is any point in $[x_1, x_2]$; and so on. Then the area of the region under the graph of f on $[a, b]$, $\int_a^b f(x)\,dx$, is approximated by the area of the region under the graph of the approximating "step function" $S(x)$ on $[a, b]$.

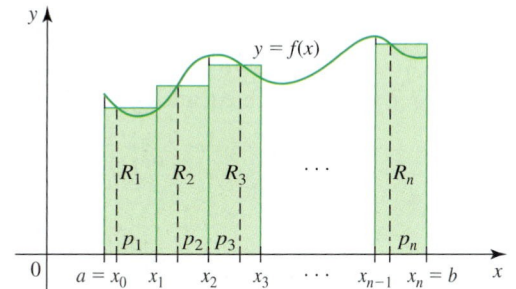

$$S(x) = \begin{cases} f(p_1) & \text{for } x_0 \le x < x_1 \\ f(p_2) & \text{for } x_1 \le x < x_2 \\ \vdots & \\ f(p_n) & \text{for } x_{n-1} \le x \le x_n \end{cases}$$

FIGURE 4
f is approximated by the step function S.

Next, Figure 5 shows the approximation of $\int_a^b f(x)\,dx$ using the Trapezoidal Rule. Another view of this method follows: Approximate $f(x)$ on $[x_0, x_1]$ by the *linear* function whose graph is the line passing through the two points $(x_0, f(x_0))$ and $(x_1, f(x_1))$; approximate $f(x)$ on $[x_1, x_2]$ by the linear function whose graph is the line passing through the points $(x_1, f(x_1))$ and $(x_2, f(x_2))$; and so on. Then the area of the region under the graph of f on $[a, b]$, $\int_a^b f(x)\,dx$, is approximated by the area of the region under the graph of the approximating "polygonal function" $P(x)$ on $[a, b]$ whose graph is a *polygonal* curve.

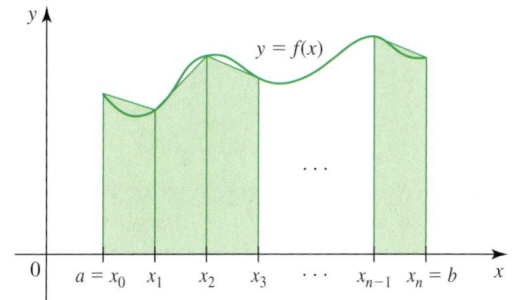

$$P(x) = \begin{cases} f(x_0) + \dfrac{f(x_1) - f(x_0)}{x_1 - x_0}(x - x_0) & \text{for } x_0 \le x < x_1 \\[2ex] f(x_1) + \dfrac{f(x_2) - f(x_1)}{x_2 - x_1}(x - x_1) & \text{for } x_1 \le x < x_2 \\ \vdots & \\ f(x_{n-1}) + \dfrac{f(x_n) - f(x_{n-1})}{x_n - x_{n-1}}(x - x_{n-1}) & \text{for } x_{n-1} \le x \le x_n \end{cases}$$

FIGURE 5
f is approximated by the "polygonal function" P.

A natural extension of the method used to approximate $\int_a^b f(x)\,dx$ is to approximate sections of the graph of f by sections of the graphs of second-degree polynomials (parts of parabolas). We begin by showing that the area of the region under the parabola $y = ax^2 + bx + c$ on $[-h, h]$ is

$$A = \frac{h}{3}(y_0 + 4y_1 + y_2) \tag{3}$$

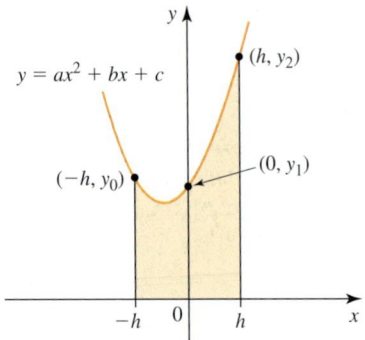

FIGURE 6

The area of the region under the parabola $y = ax^2 + bx + c$ on $[-h, h]$ is $A = \frac{h}{3}(y_0 + 4y_1 + y_2)$.

where y_0, y_1, and y_2 are the y-coordinates of the points lying on the parabola with x-coordinates $-h$, 0, and h, respectively. (See Figure 6.)

The area under the parabola $y = ax^2 + bx + c$ on $[-h, h]$ is

$$A = \int_{-h}^{h} (ax^2 + bx + c)\, dx = \left[\frac{a}{3}x^3 + \frac{b}{2}x^2 + cx \right]_{-h}^{h}$$

$$= \frac{2ah^3}{3} + 2ch = \frac{h}{3}(2ah^2 + 6c)$$

Since the parabola passes through the three points $(-h, y_0)$, $(0, y_1)$, and (h, y_2), the equation $y = ax^2 + bx + c$ must be satisfied at each of these points. This yields the following system of three equations:

$$ah^2 - bh + c = y_0$$

$$c = y_1$$

$$ah^2 + bh + c = y_2$$

from which we obtain

$$c = y_1$$

$$ah^2 - bh = y_0 - y_1$$

$$ah^2 + bh = y_2 - y_1$$

Adding the last two equations gives

$$2ah^2 = y_0 - 2y_1 + y_2$$

These expressions for $2ah^2$ and c enable us to express A in terms of y_0, y_1, and y_2 as follows:

$$A = \frac{h}{3}(2ah^2 + 6c) = \frac{h}{3}(y_0 - 2y_1 + y_2 + 6y_1) = \frac{h}{3}(y_0 + 4y_1 + y_2)$$

as was to be shown.

To derive Simpson's Rule for approximating $\int_a^b f(x)\, dx$, we divide the interval $[a, b]$ into an *even* number of subintervals of width $\Delta x = (b - a)/n$. If we approximate the area under the region of the graph of f on $[x_0, x_2]$ by the area of the region under the parabola passing through the three points $(x_0, f(x_0))$, $(x_1, f(x_1))$, and $(x_2, f(x_2))$ on $[x_0, x_2]$ (Figure 7) and use Equation (3) with $h = \Delta x$, $y_0 = f(x_0)$, $y_1 = f(x_1)$, and $y_2 = f(x_2)$, we obtain

$$\int_{x_0}^{x_2} f(x)\, dx \approx \frac{\Delta x}{3}[f(x_0) + 4f(x_1) + f(x_2)]$$

In a similar manner we approximate the area of the region under the graph of f on $[x_2, x_4]$ by the area of the region under the parabola passing through the three points $(x_2, f(x_2))$, $(x_3, f(x_3))$, and $(x_4, f(x_4))$ to obtain

$$\int_{x_2}^{x_4} f(x)\, dx \approx \frac{\Delta x}{3}[f(x_2) + 4f(x_3) + f(x_4)]$$

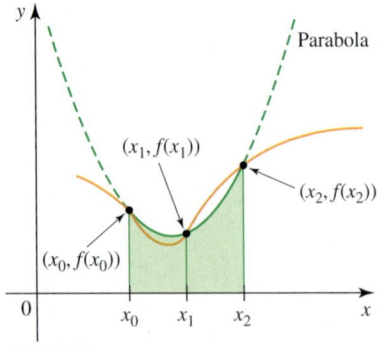

FIGURE 7

Simpson's Rule approximates portions of the area of the region under the curve by the area of the regions under parabolas.

Continuing, we approximate the area of the region under the graph of f on $[a, b]$ using the sum of the areas of the regions under $n/2$ parabolas. Thus,

$$\int_a^b f(x)\, dx = \int_{x_0}^{x_2} f(x)\, dx + \int_{x_2}^{x_4} f(x)\, dx + \cdots + \int_{x_{n-2}}^{x_n} f(x)\, dx$$

$$\approx \frac{\Delta x}{3}[f(x_0) + 4f(x_1) + f(x_2)] + \frac{\Delta x}{3}[f(x_2) + 4f(x_3) + f(x_4)] + \cdots$$

$$+ \frac{\Delta x}{3}[f(x_{n-2}) + 4f(x_{n-1}) + f(x_n)]$$

$$= \frac{\Delta x}{3}[f(x_0) + 4f(x_1) + 2f(x_2) + 4f(x_3) + \cdots$$

$$+ 2f(x_{n-2}) + 4f(x_{n-1}) + f(x_n)]$$

Let's summarize this result.

Simpson's Rule

$$\int_a^b f(x)\, dx \approx \frac{\Delta x}{3}[f(x_0) + 4f(x_1) + 2f(x_2) + 4f(x_3) + \cdots$$

$$+ 2f(x_{n-2}) + 4f(x_{n-1}) + f(x_n)] \tag{4}$$

where $\Delta x = (b - a)/n$, and n is even.

Simpson's rule is named after the English mathematician Thomas Simpson (1710–1761).

The Error in Simpson's Rule

If we denote the expression on the right of the approximation in (4) by S_n, then the error in approximating $I = \int_a^b f(x)\, dx$ by Simpson's Rule is defined to be $I - S_n$. An upper bound for this error follows. We omit the proof, which can be found in more advanced textbooks.

Error Bound for Simpson's Rule

If $f^{(4)}$ is continuous on $[a, b]$, then the error E_n in approximating $\int_a^b f(x)\, dx$ by Simpson's Rule satisfies

$$|E_n| \le \frac{M(b - a)^5}{180 n^4} \tag{5}$$

where M is a positive number such that $|f^4(x)| \le M$ for all x in $[a, b]$.

EXAMPLE 4 Use Simpson's Rule to approximate

$$\int_0^2 \frac{1}{\sqrt{x + 1}}\, dx$$

with an error that is less than 0.001.

Solution We first determine the number of subintervals, n, required in Simpson's Rule to guarantee that the error will be less than 0.001. To find the value of M that is called for in Inequality (5), we compute the fourth derivative of $f(x) = (x + 1)^{-1/2}$. Thus,

$$f'(x) = -\frac{1}{2}(x + 1)^{-3/2}, \qquad f''(x) = \frac{3}{4}(x + 1)^{-5/2}, \qquad f'''(x) = -\frac{15}{8}(x + 1)^{-7/2}$$

and

$$f^{(4)}(x) = \frac{105}{16}(x + 1)^{-9/2} = \frac{105}{16(x + 1)^{9/2}}$$

Because $f^{(4)}$ is decreasing on $(0, 2)$, we see that $|f^{(4)}(x)| \leq |f^{(4)}(0)| = \frac{105}{16}$. So we may take $M = \frac{105}{16}$. Using Inequality (5) with $a = 0$, $b = 2$, and $M = \frac{105}{16}$ and observing the requirement that the error in the approximation be less than 0.001, we have

$$\frac{105(2 - 0)^5}{16(180n^4)} < 0.001, \qquad n^4 > \frac{3500}{3}, \qquad \text{or} \qquad n > \left(\frac{3500}{3}\right)^{1/4} \approx 5.84$$

So by taking $n = 6$ (which is even), we are guaranteed that the error in the approximation will be less than 0.001.

To obtain the approximation, we compute

$$\Delta x = \frac{2 - 0}{6} = \frac{1}{3}$$

Then, with

$$x_0 = 0, \qquad x_1 = \frac{1}{3}, \qquad x_2 = \frac{2}{3}, \qquad x_3 = 1, \qquad x_4 = \frac{4}{3}, \qquad x_5 = \frac{5}{3}, \qquad \text{and} \qquad x_6 = 2$$

Simpson's Rule yields

$$\int_0^2 \frac{dx}{\sqrt{x + 1}} \approx \frac{\frac{1}{3}}{3}\left[f(0) + 4f\left(\frac{1}{3}\right) + 2f\left(\frac{2}{3}\right) + 4f(1) + 2f\left(\frac{4}{3}\right) + 4f\left(\frac{5}{3}\right) + f(2)\right]$$

$$= \frac{1}{9}\left(1 + \frac{4}{\sqrt{\frac{4}{3}}} + \frac{2}{\sqrt{\frac{5}{3}}} + \frac{4}{\sqrt{2}} + \frac{2}{\sqrt{\frac{7}{3}}} + \frac{4}{\sqrt{\frac{8}{3}}} + \frac{1}{\sqrt{3}}\right)$$

$$\approx 1.46421$$

The actual value of the definite integral to six decimal places is 1.464102 and can be found by using the method of substitution. ■

In the next example we solve the oil consumption problem posed at the beginning of this section.

EXAMPLE 5 **U.S. DAILY OIL CONSUMPTION** Table 2 gives the daily consumption of oil in the United States, in millions of barrels, measured in two-year intervals, from the beginning of 1987 to the beginning of 2007. Use Simpson's Rule to estimate the average daily consumption of oil over the period in question.

TABLE 2

Year	1987	1989	1991	1993	1995	1997	1999	2001	2003	2005	2007
Consumption	16.7	17.3	16.7	17.2	17.7	18.6	19.5	19.6	20.0	20.8	20.7

Source: U.S. Energy Information Administration

Solution The average daily consumption from the beginning of 1987 to the beginning of 2007 is given by

$$\frac{1}{20} \int_0^{20} f(t) \, dt$$

where t is measured in years, with $t = 0$ corresponding to 1987. Using Simpson's Rule with $a = 0$, $b = 20$, and $n = 10$, so that $\Delta t = (20 - 0)/10 = 2$, we have $t_0 = 0$, $t_1 = 2$, $t_3 = 4, \ldots, t_{10} = 20$. Therefore,

$$\frac{1}{20} \int_0^{20} f(t) \, dt \approx \left(\frac{1}{20}\right)\left(\frac{2}{3}\right)[f(0) + 4f(2) + 2f(4) + 4f(6) + \cdots + 4f(18) + f(20)]$$

$$= \frac{1}{30}[16.7 + 4(17.3) + 2(16.7) + 4(17.2) + 2(17.7) + 4(18.6)$$

$$+ 2(19.5) + 4(19.6) + 2(20.0) + 4(20.8) + 20.7]$$

$$= 18.64$$

and we conclude that the average daily oil consumption in the United States from 1987 through 2007 is approximately 18.6 million barrels per day. ◼

4.6 CONCEPT QUESTIONS

1. Describe (a) the Trapezoidal Rule and (b) Simpson's Rule. What are the main differences in approximating $\int_a^b f(x) \, dx$ using a Riemann sum, using the Trapezoidal Rule, and using Simpson's Rule?

2. Explain why n can be odd or even in the Trapezoidal Rule, but it must be even in Simpson's Rule.

3. Explain, without alluding to the error formulas, why the Trapezoidal Rule gives the exact value of $\int_a^b f(x) \, dx$ if f is a linear function and why Simpson's Rule gives the exact value of the integral if f is a quadratic function.

4.6 EXERCISES

In Exercises 1–8, use (a) the Trapezoidal Rule and (b) Simpson's Rule to approximate the integral. Compare your results with the exact value of the integral.

1. $\int_0^2 x^2 \, dx$; $\quad n = 4$

2. $\int_1^3 (x^2 - 1) \, dx$; $\quad n = 6$

3. $\int_1^2 x^3 \, dx$; $\quad n = 6$

4. $\int_1^2 \frac{1}{x^2} \, dx$; $\quad n = 4$

5. $\int_0^2 x\sqrt{2x^2 + 1} \, dx$; $\quad n = 6$

6. $\int_0^1 e^{-x} \, dx$; $\quad n = 6$

7. $\int_0^1 xe^{-x^2} \, dx$; $\quad n = 6$

8. $\int_0^{\pi/2} \cos 2x \, dx$; $\quad n = 6$

In Exercises 9–14, use the Trapezoidal Rule to approximate the integral with answers rounded to four decimal places.

9. $\int_0^1 \frac{dx}{2x + 1}$; $\quad n = 7$

10. $\int_1^3 \sqrt{x^2 + 1} \, dx$; $\quad n = 5$

11. $\int_0^2 e^{-x^2} \, dx$; $\quad n = 4$

12. $\int_0^1 \cos x^2 \, dx$; $\quad n = 6$

13. $\int_1^2 \sqrt{x} \sin x \, dx$; $\quad n = 5$

14. $\int_2^4 \frac{dx}{\ln x}$; $\quad n = 6$

In Exercises 15–20, use Simpson's Rule to approximate the integral with answers rounded to four decimal places.

15. $\int_0^2 \sqrt{x^3 + 1} \, dx$; $\quad n = 6$

16. $\int_0^1 \frac{dx}{x^2 + 1}$; $\quad n = 4$

17. $\int_{-1}^1 \sqrt{x^2 + 1} \, dx$; $\quad n = 6$

18. $\int_1^2 x^{-1/2} e^x \, dx$; $\quad n = 4$

19. $\int_0^{\pi/2} \sqrt{1 + \sin^2 x}\, dx; \quad n = 6$

20. $\int_2^4 \frac{dx}{\ln x}; \quad n = 6$

In Exercises 21–28, find a bound on the error in approximating the integral using (a) the Trapezoidal Rule and (b) Simpson's Rule with n subintervals.

21. $\int_{-1}^2 x^3\, dx; \quad n = 6$

22. $\int_0^1 \frac{dx}{x + 1}; \quad n = 8$

23. $\int_1^4 \sqrt{x}\, dx; \quad n = 8$

24. $\int_0^2 \frac{dx}{\sqrt{x + 1}}; \quad n = 6$

25. $\int_0^{\pi/2} x \sin x\, dx; \quad n = 6$

26. $\int_0^1 \cos x^2\, dx; \quad n = 6$

27. $\int_1^2 \ln x^2\, dx; \quad n = 8$

28. $\int_0^1 \cot^{-1} x\, dx; \quad n = 10$

cas *In Exercises 29–34, use a calculator or a computer and the error formula for the Trapezoidal Rule to find n such that the error in the approximation of the integral using the Trapezoidal Rule is less than 0.001.*

29. $\int_1^2 \frac{dx}{x}$

30. $\int_0^2 \frac{dx}{x^2 + 1}$

31. $\int_1^2 e^{1/x}\, dx$

32. $\int_0^2 \sqrt{x^2 + 1}\, dx$

33. $\int_0^{\pi/2} x \cos x\, dx$

34. $\int_0^1 \cos x^2\, dx$

cas *In Exercises 35–40, use a calculator or a computer and the error formula for Simpson's Rule to find n such that the error in the approximation of the integral using Simpson's Rule is less than 0.001.*

35. $\int_1^4 2x^{3/2}\, dx$

36. $\int_1^3 \frac{dx}{x}$

37. $\int_0^2 \frac{dx}{x^2 + 4}$

38. $\int_1^2 e^{x^2}\, dx$

39. $\int_0^{\pi/2} x \sin x\, dx$

40. $\int_0^1 \sin x^2\, dx$

41. Velocity of a Sports Car The velocity function for a sports car traveling on a straight road is given by

$$v(t) = \frac{80t^3}{t^3 + 100} \qquad 0 \le t \le 16$$

where t is measured in seconds and $v(t)$ in feet per second. Use Simpson's Rule with $n = 8$ to estimate the average velocity of the car over the time interval $[0, 16]$.

42. Air Pollution The amount of nitrogen dioxide, a brown gas that impairs breathing, present in the atmosphere on a certain May day in the city of Long Beach is approximated by

$$A(t) = \frac{136}{1 + 0.25(t - 4.5)^2} + 28 \qquad 0 \le t \le 11$$

where $A(t)$ is measured in pollutant standard index and t is measured in hours, with $t = 0$ corresponding to 7 A.M. Use the Trapezoidal Rule with $n = 11$ to find the approximate average level of nitrogen dioxide present in the atmosphere from 7 A.M. to 6 P.M. on that day.

Source: The Los Angeles Times.

43. U.S. Strategic Petroleum Reserves According to data from the American Petroleum Institute, the U.S. Strategic Petroleum Reserves from the beginning of 1981 to the beginning of 1990 can be approximated by the function

$$S(t) = \frac{613.7t^2 + 1449.1}{t^2 + 6.3} \qquad 0 \le t \le 9$$

where $S(t)$ is measured in millions of barrels and t in years, with $t = 0$ corresponding to the beginning of 1981. Using the Trapezoidal Rule with $n = 9$, estimate the average petroleum reserves from the beginning of 1981 to the beginning of 1990.

Source: American Petroleum Institute.

44. Velocity of an Attack Submarine The following data give the velocity of an attack submarine taken at 10-min intervals during a submerged trial run.

Time t (hr)	0	$\frac{1}{6}$	$\frac{1}{3}$	$\frac{1}{2}$	$\frac{2}{3}$	$\frac{5}{6}$	1
Velocity v (mph)	14.2	24.3	40.2	45.0	38.5	27.6	12.8

Use Simpson's Rule to estimate the distance traveled by the submarine during the 1-hr submerged trial run.

45. Flow of Water in a River At a certain point, a river is 78 ft wide and its depth, measured at 6-ft intervals across the river, is recorded in the following table.

x	0	6	12	18	24	30	36
y	0.8	2.6	5.8	6.2	8.2	10.1	10.8

x	42	48	54	60	66	72	78
y	9.8	7.6	6.4	5.2	3.9	2.4	1.4

Here, x denotes the distance (in feet) from one bank of the river, and y (in feet) is the corresponding depth. If the average rate of flow through this section of the river is 4 ft/sec, estimate the rate of the volume of flow of water in the river. Use the Trapezoidal Rule with $n = 13$.

46. **Measuring Cardiac Output** Eight milligrams of a dye are injected into a vein leading to an individual's heart. The concentration of the dye in the aorta (in milligrams per liter) measured at 2-sec intervals is shown in the accompanying table. Use Simpson's Rule with $n = 12$ and the formula

$$R = \frac{60D}{\int_0^{24} C(t)\,dt}$$

to estimate the person's cardiac output, where D is the quantity of dye injected in milligrams, $C(t)$ is the concentration of the dye in the aorta, and R is measured in liters per minute.

t	0	2	4	6	8	10	12
$C(t)$	0	0	2.8	6.1	9.7	7.6	4.8

t	14	16	18	20	22	24
$C(t)$	3.7	1.9	0.8	0.3	0.1	0

In Exercises 47–48, determine whether the statement is true or false. If it is true, explain why it is true. If it is false, explain why or give an example to show why it is false.

47. If f is a polynomial of degree greater than one, then the error E_n in approximating $\int_a^b f(x)\,dx$ by the Trapezoidal Rule must be nonzero.

48. If f is nonnegative and concave upward on $[a, b]$ and A is an approximation of $\int_a^b f(x)\,dx$ using the Trapezoidal Rule, then $A > \int_a^b f(x)\,dx$.

CHAPTER 4 REVIEW

CONCEPT REVIEW

In Exercises 1–10, fill in the blanks.

1. **a.** A function F is an antiderivative of f on an interval if _____ for all x in I.
   b. If F is an antiderivative of f on an interval I, then every antiderivative of f on I has the form _____.

2. **a.** $\int c f(x)\,dx =$ _____.
   b. $\int [f(x) \pm g(x)]\,dx =$ _____.

3. **a.** A differential equation is an equation that involves the derivative or differential of an _____ function.
   b. A solution of a differential equation on an interval I is any _____ that satisfies the differential equation.

4. If we let $u = g(x)$, then $du =$ _____, and the substitution transforms the integral $\int f(g(x))g'(x)\,dx$ into the integral _____ involving only u.

5. **a.** If f is continuous and nonnegative on an interval $[a, b]$, then the area of the region under the graph of f on $[a, b]$ is given by _____.
   b. If f is continuous on an interval $[a, b]$, then $\int_a^b f(x)\,dx$ is equal to the area(s) of the regions lying above the x-axis and bounded by the graph of f on $[a, b]$ _____ the

area(s) of the regions lying below the x-axis and bounded by the graph of f on $[a, b]$.

6. **a.** If f is continuous on $[a, b]$, then the average value of f over $[a, b]$ is the number $f_{av} =$ _____.
   b. If f is a continuous and nonnegative function on $[a, b]$, then the number f_{av} may be thought of as the _____ of the rectangle with base lying on the interval $[a, b]$ and having the same _____ as the area of the region under the graph of f on $[a, b]$.

7. **a.** The Fundamental Theorem of Calculus, Part 1, states that if f is continuous on (a, b), then $F(x) = \int_a^x f(t)\,dt$ is differentiable on (a, b), and $F'(x) =$ _____.
   b. The Fundamental Theorem of Calculus, Part 2, states that if f is continuous on $[a, b]$, then $\int_a^b f(x)\,dx =$ _____, where F is an _____ of f.
   c. The net change in a function f over an interval $[a, b]$ is given by $f(b) - f(a) =$ _____, provided that f' is continuous on $[a, b]$.

8. Let f be continuous on $[-a, a]$. If f is even, then $\int_{-a}^a f(x)\,dx =$ _____, and if f is odd, then $\int_{-a}^a f(x)\,dx =$ _____.

9. The Mean Value Theorem for Integrals states that if f is continuous on $[a, b]$, then there exists at least one point c in $[a, b]$ such that _____.

10. a. The Trapezoidal Rule states that $\int_a^b f(x)\, dx \approx$ _____, where $\Delta x = (b - a)/n$. The error E_n in approximating $\int_a^b f(x)\, dx$ by the Trapezoidal Rule satisfies $|E_n| \le$

_____, where M is a positive number such that $|f''(x)| \le M$ for all x in $[a, b]$.

b. Simpson's Rule states that $\int_a^b f(x)\, dx \approx$ _____, where $\Delta x = (b - a)/n$ and n is _____. The error E_n in approximating $\int_a^b f(x)\, dx$ by Simpson's Rule satisfies $|E_n| \le$ _____, where M is a positive number such that $|f^{(4)}(x)| \le M$ for all x in $[a, b]$.

REVIEW EXERCISES

In Exercises 1–32, find the indefinite integral.

1. $\displaystyle\int (2x^3 - 4x^2 + 3x + 4)\, dx$

2. $\displaystyle\int \frac{x^5 - 3x + 2}{x^3}\, dx$

3. $\displaystyle\int (x^{5/3} - 2x^{2/5})\, dx$

4. $\displaystyle\int x^{1/3}(2x^2 - 3x + 1)\, dx$

5. $\displaystyle\int \left(x^{2/3} - \frac{2}{x^4} + 3\right) dx$

6. $\displaystyle\int \frac{x^2 - x + \sqrt{x}}{\sqrt[3]{x}}\, dx$

7. $\displaystyle\int (1 + 2t)^3\, dt$

8. $\displaystyle\int \frac{(1 + x)^2}{\sqrt{x}}\, dx$

9. $\displaystyle\int (3t - 4)^8\, dt$

10. $\displaystyle\int \sqrt[3]{2u + 1}\, du$

11. $\displaystyle\int (x + x^{-1})^2\, dx$

12. $\displaystyle\int 2x^3\sqrt{x^4 + 1}\, dx$

13. $\displaystyle\int \frac{3x + 1}{(3x^2 + 2x)^3}\, dx$

14. $\displaystyle\int (\pi^2 + \sqrt{x} + 1)\, dx$

15. $\displaystyle\int \cos^4 t \sin t\, dt$

16. $\displaystyle\int \frac{\sec^2 x}{\sqrt{\tan x}}\, dx$

17. $\displaystyle\int \frac{\cos \theta}{\sqrt{1 - \sin \theta}}\, d\theta$

18. $\displaystyle\int \frac{\cos^3 \theta + 1}{\cos^2 \theta}\, d\theta$

19. $\displaystyle\int x \csc x^2 \cot x^2\, dx$

20. $\displaystyle\int \sec 3x \tan 3x\, dx$

21. $\displaystyle\int \frac{1}{5x - 3}\, dx$

22. $\displaystyle\int \frac{\cos x}{2 + 3\sin x}\, dx$

23. $\displaystyle\int \frac{2x + 1}{3x + 2}\, dx$

24. $\displaystyle\int \frac{(\ln x)^3}{x}\, dx$

25. $\displaystyle\int t \cdot 2^{t^2}\, dt$

26. $\displaystyle\int \frac{e^x}{e^x - 1}\, dx$

27. $\displaystyle\int \frac{\sin(\ln x)}{x}\, dx$

28. $\displaystyle\int \frac{e^{1/x}}{x^2}\, dx$

29. $\displaystyle\int \frac{\tan \sqrt{x}}{\sqrt{x}}\, dx$

30. $\displaystyle\int \frac{\sin^{-1} x}{\sqrt{1 - x^2}}\, dx$

31. $\displaystyle\int \frac{\tan^{-1} 2x}{1 + 4x^2}\, dx$

32. $\displaystyle\int \frac{\sec t \tan t}{1 + \sec t}\, dt$

In Exercises 33–46, evaluate the definite integral.

33. $\displaystyle\int_0^2 (3x + 5)\, dx$

34. $\displaystyle\int_{-1}^1 \sqrt[3]{8x}\, dx$

35. $\displaystyle\int_1^2 \left(\frac{1}{x^2} - \frac{1}{x^3}\right) dx$

36. $\displaystyle\int_0^2 t^2\sqrt{t^3 + 1}\, dt$

37. $\displaystyle\int_0^4 \frac{1}{\sqrt{1 + 2x}}\, dx$

38. $\displaystyle\int_0^{\sqrt{2}} \frac{x}{\sqrt{x^2 + 2}}\, dx$

39. $\displaystyle\int_0^1 (x + 1)(2x + 3)^2\, dx$

40. $\displaystyle\int_1^4 \frac{(\sqrt{x} + 1)^5}{\sqrt{x}}\, dx$

41. $\displaystyle\int_0^{\pi/8} \frac{\sin 2x}{\cos^2 2x}\, dx$

42. $\displaystyle\int_0^{\pi/2} \sin \theta\sqrt{2 + 7\cos \theta}\, d\theta$

43. $\displaystyle\int_{\pi/6}^{\pi/4} (\csc \theta + \cot \theta)(1 - \cos \theta)\, d\theta$

44. $\displaystyle\int_0^1 \frac{x}{2x^2 + 1}\, dx$

45. $\displaystyle\int_1^2 \frac{x^3 - 2x + 1}{x^2}\, dx$

46. $\displaystyle\int_0^1 \frac{e^{3x}}{1 + e^{3x}}\, dx$

In Exercises 47 and 48, use the properties of integrals to prove the inequality.

47. $\displaystyle\int_0^1 \sqrt{1 + e^{4x}}\, dx \ge \frac{1}{2}(e^2 - 1)$

48. $\displaystyle 1 < \int_0^1 e^{x^2}\, dx < e$

In Exercises 49–52, find the average value of the function over the given interval.

49. $f(x) = x^3$; $[0, 2]$

50. $f(x) = \dfrac{1}{\sqrt{x} + 1}$; $[1, 3]$

51. $f(x) = \dfrac{\ln x^2}{x}$; $[1, 2]$

52. $f(x) = \sin^2 x \cos x$; $\left[0, \frac{\pi}{2}\right]$

In Exercises 53 and 54, find f'(x).

53. $f(x) = \displaystyle\int_0^{x^2} \dfrac{e^t}{t^2 + 1}\, dt$

54. $f(x) = \displaystyle\int_{\ln x}^{\sqrt{x}} \sin t^2\, dt,$ where $x > 0$

55. Find the derivative of $F(x) = \displaystyle\int_{x^2}^{x^3} \sin t\, dt.$

56. Find the area of the region under the graph of $y = xe^{-x^2}$ on $[0, 4]$.

57. Evaluate $\displaystyle\int_0^1 \sqrt{1 + x^2}\, dx$ by using the Trapezoidal Rule with $n = 5$.

58. Evaluate $\displaystyle\int_1^2 \dfrac{\sqrt{x}}{1 + x^2}\, dx$ by using Simpson's Rule with $n = 4$.

In Exercises 59 and 60, find a bound on the error in approximating each definite integral using (a) the Trapezoidal Rule and (b) Simpson's Rule with n subintervals.

59. $\displaystyle\int_0^2 \dfrac{1}{\sqrt{1 + x}}\, dx;\quad n = 8$

60. $\displaystyle\int_1^3 \ln x\, dx;\quad n = 10$

61. Find the function f given that its derivative is $f'(x) = \sqrt{x} + \sin x$ and that its graph passes through the point $(0, 2)$.

62. A ball is thrown straight up from the ground with an initial velocity of 64 ft/sec. How long will it take for the ball to reach its highest point, and what will its height be at this point?

63. An electric drill rolls off the edge of a 128-ft-tall structure under construction.
   a. Find the position of the electric drill after t sec.
   b. Determine when it strikes the ground.
   c. What is its velocity at impact?

64. A stone is dropped from a height of h ft above the ground. Show that the speed at which the stone strikes the ground is $\sqrt{2gh}$ ft/sec, where g is the gravitational constant.

65. A car traveling along a straight road undergoes constant deceleration that reduces its speed from 44 ft/sec to 22 ft/sec in 8 sec. How far will the car travel if it is brought to rest from 44 ft/sec at that rate of deceleration?

66. Traffic Flow The traffic department of a certain city estimates that t years from now the number of vehicles (in thousands) in the city will be

$$0.2t^4 + 4t + 84$$

Find the estimated average number of motor vehicles in the city over the next 5 years.

67. An electromotive force (emf), E, is given by

$$E = E_0 \sin \dfrac{2\pi t}{T}$$

where T is the period in seconds and E_0 is the amplitude of the emf. Find the average value of the emf over one period (the time interval $[0, T]$).

68. Total Cost The weekly marginal cost of the Advanced Visuals Systems Corporation in manufacturing its 42-in. plasma television sets is given by

$$C'(x) = 0.000006x^2 - 0.04x + 120$$

dollars per set, where x stands for the number of sets manufactured. Find the weekly total cost function C if the fixed cost of the company is $70,000 per week.

69. Total Profit The weekly marginal profit of the Advanced Visuals Systems Corporation is given by

$$P'(x) = -0.000006x^2 - 0.04x + 200$$

dollars per set where x stands for the number of 42-in. plasma television sets sold. Find the weekly total profit function P if $P(0) = -80,000$.

70. Hotel Occupancy Rate The occupancy rate at a large hotel in Maui in month t is described by the function

$$R(t) = 60 + 37 \sin^2\left(\dfrac{\pi t}{12}\right) \qquad 0 \le t \le 12$$

where t is measured in months, $v(t)$ is measured in percent, and $t = 0$ corresponds to the beginning of June. Find the average occupancy rate of the hotel over a 1-year period.

Hint: $\sin^2 x = \dfrac{1 - \cos 2x}{2}$

71. TV on Mobile Phones The number of people watching TV on mobile phones is expected to grow at the rate of

$$N'(t) = \dfrac{5.4145}{\sqrt{1 + 0.91t}}\, dt \qquad 0 \le t \le 4$$

million per year. The number of people watching TV on mobile phones at the beginning of 2007 ($t = 0$) was 11.9 million.
   a. Find an expression giving the number of people watching TV on mobile phones in year t.
   b. According to this projection, how many people will be watching TV on mobile phones at the beginning of 2011?
   Source: International Data Corporation, U.S. forecast.

72. Net Investment Flow The net investment flow of a giant conglomerate, which is the rate of capital formation, is projected to be $t\sqrt{2t^2 + 1}$ million dollars per year t years from now. Find the accruement on the company's capital stock after 2 years; that is, compute

$$\int_0^2 t\sqrt{2t^2 + 1}\, dt$$

73. Respiratory Cycles The volume of air inhaled by a person during respiration is given by

$$V(t) = \frac{6}{5\pi}\left(1 - \cos\frac{\pi t}{2}\right)$$

liters at time t (in seconds). What is the average volume of air inhaled by a person over one cycle from $t = 0$ to $t = 4$?

74. Average Temperature The average daily temperature (in degrees Fahrenheit) on the tth day at a tourist resort in Cameron Highlands is approximated by

$$T = 62 - 18\cos\frac{2\pi(t - 23)}{365}$$

($t = 0$ corresponds to the beginning of the year). What is the average daily temperature in Cameron Highlands over the year?

75. Average Temperature The following graph shows the daily mean temperatures recorded during a certain month at Frazer's Hill. Using (a) the Trapezoidal Rule and

(b) Simpson's Rule with $n = 10$, estimate the average temperature during that month.

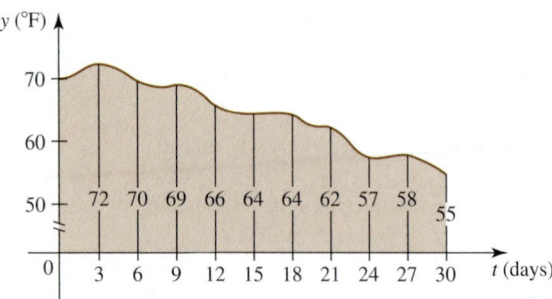

76. Show that $\dfrac{d}{dx}\displaystyle\int_x^c f(t)\,dt = -f(x)$.

77. a. Prove that

$$0.5 < \int_0^{1/2} \frac{dx}{\sqrt{1 - x^{2n}}} < 0.524 \qquad n > 1$$

 b. Use a computer or a calculator to find the value of the integral in part (a) with $n = 6$ accurate to six decimal places.

PROBLEM-SOLVING TECHNIQUES

The following example introduces a technique for transforming a definite integral into another having the same value as the original integral. This technique is then used to evaluate a definite integral without the need to find an antiderivative associated with the integral.

EXAMPLE

a. Show that if f is a continuous function, then

$$\int_0^a f(x)\,dx = \int_0^a f(a - x)\,dx$$

b. Use the result of part (a) to show that $\int_0^{\pi/2} \sin^m x\,dx = \int_0^{\pi/2} \cos^m x\,dx$.

c. Use the result of part (b) to evaluate $\int_0^{\pi/2} \sin^2 x\,dx$ and $\int_0^{\pi/2} \cos^2 x\,dx$.

Solution

a. Let us evaluate the integral on the right-hand side by using the substitution $u = a - x$, so that $du = -dx$. To obtain the limits of integration, observe that if $x = 0$, then $u = a$, and if $x = a$, then $u = 0$. Substituting, we obtain

$$\int_0^a f(a - x)\,dx = -\int_a^0 f(u)\,du = \int_0^a f(u)\,du = \int_0^a f(x)\,dx$$

This proves the assertion.

There is a simple geometric explanation for this result. It stems from the fact that the graph of f on the interval $[0, a]$ is the mirror image of the graph of $f(a - x)$ on the same interval with respect to the vertical line $x = a/2$. In fact, if the point A lies on the x-axis and has x-coordinate x, then the point A' that is

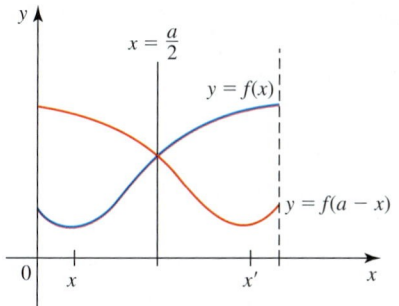

FIGURE 1
The graphs of $f(x)$ and $f(a - x)$ are mirror images with respect to $x = a/2$.

its mirror image with respect to the line $x = a/2$ has x-coordinate $x' = a - x$. Therefore, $f(a - x') = f(a - (a - x)) = f(x)$. (See Figure 1.) Since congruent figures have equal areas, the result follows from interpreting definite integrals as areas.

b. Using the result of part (a), we see that

$$\int_0^{\pi/2} \sin^m x \, dx = \int_0^{\pi/2} \sin^m \left(\frac{\pi}{2} - x \right) dx$$

$$= \int_0^{\pi/2} \left(\sin \frac{\pi}{2} \cos x - \cos \frac{\pi}{2} \sin x \right)^m dx$$

$$= \int_0^{\pi/2} \cos^m x \, dx$$

c. Using the result of part (b) with $m = 2$, we have

$$I = \int_0^{\pi/2} \sin^2 x \, dx = \int_0^{\pi/2} \cos^2 x \, dx$$

Therefore,

$$2I = \int_0^{\pi/2} \sin^2 x \, dx + \int_0^{\pi/2} \cos^2 x \, dx$$

$$= \int_0^{\pi/2} (\sin^2 x + \cos^2 x) \, dx = \int_0^{\pi/2} dx = \frac{\pi}{2}$$

and hence $I = \pi/4$.

CHALLENGE PROBLEMS

1. Evaluate $\displaystyle\int_a^b \frac{|x|}{x} \, dx$, where $a < b$.

2. Show that $\displaystyle\int_0^x [\![t]\!] \, dt = \frac{[\![x]\!]([\![x]\!] - 1)}{2} + [\![x]\!](x - [\![x]\!])$, where $[\![x]\!]$ is the greatest integer function.

3. Evaluate $\displaystyle\int_0^{10\pi} \sqrt{1 - \cos 2x} \, dx$.

4. By interpreting the integral geometrically, evaluate

$$\int_{\sqrt{2}/2}^{\sqrt{3}/2} \sqrt{1 - x^2} \, dx$$

5. Evaluate $\displaystyle\int_{-1}^1 \frac{3x^6 - 2x^5 + 4x^3 - 3x^2 + 5x}{x^2 + 1} \, dx$.

6. Find $\displaystyle\int \frac{1}{\sin^2 x \cos^4 x} \, dx$.

7. Find $\displaystyle\int \frac{dx}{1 + 2x \cos a + x^2}$, where $0 < |a| < \pi$.

Hint: Use the substitution $u = \dfrac{x + \cos a}{\sin a}$.

8. Evaluate

$$\lim_{x \to 0} \frac{\displaystyle\int_0^{x^3} \tan t^{1/3} \, dt}{\displaystyle\int_0^{2x^2} t \, dt}$$

9. Evaluate $\displaystyle\lim_{b \to a} \frac{1}{b - a} \int_a^b f(x) \, dx$, where f is a continuous function.

10. Evaluate $\displaystyle\lim_{n \to \infty} \sum_{k=1}^n \frac{n}{n^2 + k^2}$.
Hint: Relate the limit to the limit of a Riemann sum of an appropriate function.

11. a. Show that $\int_a^b f(x) \, dx = \int_a^b f(a + b - x) \, dx$, and give a geometric interpretation of the result.
b. Use the result of part (a) to show that $\int_0^\pi f(\sin x) \cos x \, dx = 0$.

12. Show that $\int_0^t f(x)g(t - x) \, dx = \int_0^t g(x)f(t - x) \, dx$.

13. a. Suppose that f is continuous and g and h are differentiable. Show that

$$\frac{d}{dx} \int_{g(x)}^{h(x)} f(t)dt = f[h(x)]h'(x) - f[g(x)]g'(x)$$

b. Use the result of part (a) to find $g'(x)$ if

$$g(x) = \int_{1/x}^{\sqrt{x}} \sin t^2 \, dt \qquad x > 0$$

14. Prove that if f and g are continuous functions on $[a, b]$, then

$$\left| \int_a^b f(x)g(x) \, dx \right| \leq \sqrt{\int_a^b [f(x)]^2 \, dx \int_a^b [g(x)]^2 \, dx}$$

This is known as Schwarz's inequality.
Hint: Consider the function $F(x) = [f(x) - tg(x)]^2$, where t is a real number.

15. a. Use Schwarz's inequality (see Exercise 14) to prove that

$$\int_0^1 \sqrt{1 + x^3} \, dx < \frac{\sqrt{5}}{2}$$

b. Is this estimate better than the one obtained by using the Mean Value Theorem for Integrals?

16. Find the values of x at which

$$F(x) = \int_0^{x^2} \frac{t^2 - 5t + 4}{t^2 + 1} \, dt$$

has relative extrema.

17. Suppose that f is continuous on an interval $[a, b]$. Show that

$$\lim_{n \to \infty} \frac{1}{n} \sum_{k=1}^n f\left[a + \frac{k(b-a)}{n}\right] = \frac{1}{b-a} \int_a^b f(x) \, dx$$

18. a. Prove that

$$\int_a^b f(x) \, dx = (b - a) \int_0^1 f[(b - a)t + a] \, dt$$

Thus, an integral with interval of integration $[a, b]$ can be transformed into one with interval of integration $[0, 1]$ by means of the substitution $x = (b - a)t + a$.

b. Use the result of part (a) to evaluate

$$\int_{-3}^{-4} \cos(x + 4)^2 \, dx + 3 \int_{1/3}^{2/3} \cos\left[9\left(x - \frac{2}{3}\right)^2\right] dx$$

19. Suppose that f is a continuous periodic function with period p.
a. Prove that if a is any real number, then

$$\int_0^a f(x) \, dx = \int_p^{a+p} f(x) \, dx$$

b. Use the result of part (a) to show that if a is any real number, then

$$\int_0^p f(x) \, dx = \int_a^{a+p} f(x) \, dx$$

20. Let f be continuous on an interval $[-a, a]$.
a. Show that $\int_{-a}^a f(x^2) \, dx = 2 \int_0^a f(x^2) \, dx$.
b. What can you say about $\int_{-a}^a f(x^2)\sin x \, dx$?

21. Let f be continuous on an interval $[a, b]$ and satisfy $\int_a^x f(t) \, dt = \int_x^b f(t) \, dt$ for all x in $[a, b]$. Show that $f(x) = 0$ on $[a, b]$.

22. The Fresnel function S is defined by the integral

$$S(x) = \int_0^x \sin\left(\frac{\pi t^2}{2}\right) dt$$

a. Sketch the graphs of $f(x) = \sin(\pi x^2/2)$ and $S(x)$ on the same set of axes for $0 \leq x \leq 3$. Interpret your results.
b. Sketch the graph of S on the interval $[-10, 10]$.

23. Find all continuous, nonnegative functions f defined on $[0, b]$, where $b > 0$, satisfying the equation $[f(x)]^2 = 2 \int_0^x f(t) \, dt$.

24. a. Prove that $e^{-R \sin x} < e^{(-2R/\pi)x}$ on the interval $\left(0, \frac{\pi}{2}\right)$, where $R > 0$.
Hint: Show that $f(x) = \sin x/x$ is decreasing on $\left(0, \frac{\pi}{2}\right)$.
b. Use the result of part (a) to prove the inequality

$$\int_0^{\pi/2} e^{-R \sin x} \, dx < \frac{\pi}{2R} (1 - e^{-R}) \qquad R > 0$$

Ambient Images/Alamy

The photograph shows the Jacqueline Kennedy Onassis Reservoir (formerly the Central Park Reservoir). Built between 1858 and 1862, it is located between 86th Street and 96th Street in the borough of Manhattan in New York City. In this chapter we will use calculus to help us estimate the surface area of the reservoir.

5

Applications of the Definite Integral

IN THIS CHAPTER we continue to exploit the integral as a tool for solving a variety of problems. More specifically, we will use the techniques of integration to find the areas of regions between curves, the volumes of solids, the arc lengths of plane curves, and the areas of surfaces. We will also show how the integral is used to compute the work done by a force acting on an object and the force exerted on an object by a hydrostatic force. Finally, we will use integration to find the moments and the centers of mass of thin plates.

V This symbol indicates that one of the following video types is available for enhanced student learning at **www.academic.cengage.com/login:**
• Chapter lecture videos   • Solutions to selected exercises

5.1　Areas Between Curves

■ A Real-Life Interpretation

Two cars are traveling in adjacent lanes along a straight stretch of a highway. The velocity functions for Car A and Car B are $v = f(t)$ and $v = g(t)$, respectively. The graphs of these functions are shown in Figure 1.

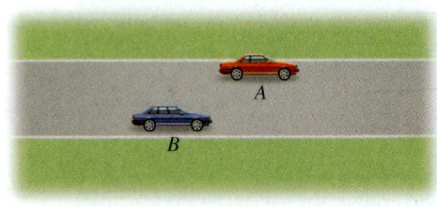

FIGURE 1

The shaded area S gives the distance that Car A is ahead of Car B at time $t = b$.

The area of the region under the graph of f from $t = 0$ to $t = b$ gives the total distance covered by Car A in b seconds over the time interval $[0, b]$. The distance covered by Car B over the same period of time is given by the area under the graph of g on the interval $[0, b]$. Intuitively, we see that the area of the (shaded) region S between the graphs of f and g on the interval $[0, b]$ gives the distance that Car A will be ahead of Car B at time $t = b$.

Since the area of the region under the graph of f on $[0, b]$ is

$$\int_0^b f(t)\, dt$$

and the area of the region under the graph of g on $[0, b]$ is

$$\int_0^b g(t)\, dt$$

we see that the area of the region S is given by

$$\int_0^b f(t)\, dt - \int_0^b g(t)\, dt = \int_0^b [f(t) - g(t)]\, dt$$

Therefore, the distance that Car A will be ahead of Car B at $t = b$ is

$$\int_0^b [f(t) - g(t)]\, dt$$

This example suggests that some applied problems can be solved by finding the area of a region between two curves, which in turn can be found by evaluating an appropriate definite integral. Let's make this notion more precise.

■ The Area Between Two Curves

Suppose f and g are continuous functions with $f(x) \geq g(x)$ for all x in $[a, b]$, so that the graph of f lies on or above that of g on $[a, b]$. Let's consider the region S bounded by the graphs of f and g between the vertical lines $x = a$ and $x = b$ as shown in Figure 2. To define the *area* of S, we take a regular partition of $[a, b]$,

$$a = x_0 < x_1 < x_2 < x_3 < \cdots < x_n = b$$

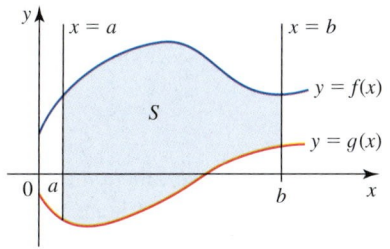

FIGURE 2
The region S between the graphs of f and g on $[a, b]$

and form the Riemann sum of the function $f - g$ over $[a, b]$ with respect to this partition:

$$\sum_{k=1}^{n} [f(c_k) - g(c_k)]\Delta x$$

where c_k is an evaluation point in the subinterval $[x_{k-1}, x_k]$ and $\Delta x = (b - a)/n$. The kth term of this sum gives the area of a rectangle with height $[f(c_k) - g(c_k)]$ and width Δx. As you can see in Figure 3, this area is an approximation of the area of the subregion of S that lies between the graphs of f and g on $[x_{k-1}, x_k]$.

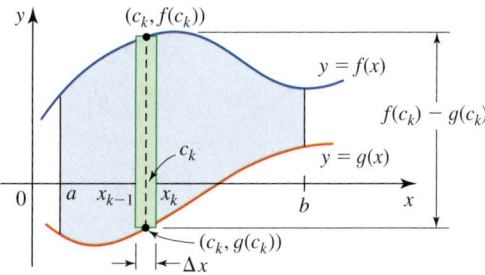

FIGURE 3
The kth term of the Riemann sum of $f - g$ gives the area of the kth rectangle of width Δx.

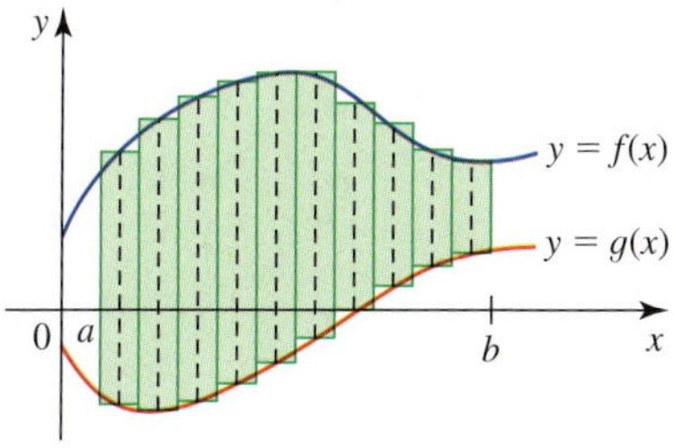

FIGURE 4
The Riemann sum of $f - g$ approximates the area of S.

Therefore, the Riemann sum provides us with an approximation of what we might intuitively think of as the area of S (see Figure 4). As n gets larger and larger, we might expect the approximation to get better and better. This suggests that we *define* the area A of S by

$$A = \lim_{n \to \infty} \sum_{k=1}^{n} [f(c_k) - g(c_k)]\Delta x \qquad (1)$$

Since $f - g$ is continuous on $[a, b]$, the limit in Equation (1) exists and is equal to the definite integral of $f - g$ from a to b. This leads us to the following definition of the area A of S.

DEFINITION Area of a Region Between Two Curves
Let f and g be continuous on $[a, b]$, and suppose that $f(x) \geq g(x)$ for all x in $[a, b]$. Then the area of the region between the graphs of f and g and the vertical lines $x = a$ and $x = b$ is

$$A = \int_a^b [f(x) - g(x)] \, dx \qquad (2)$$

Notes
1. If $g(x) = 0$ for all x in $[a, b]$, then the region S is just the region under the graph of f on $[a, b]$, and its area is

$$\int_a^b [f(x) - 0] \, dx = \int_a^b f(x) \, dx$$

as expected (see Figure 5a).

2. If $f(x) = 0$ for all x in $[a, b]$, then the region S lies on or below the x-axis, and its area is

$$\int_a^b [0 - g(x)]\, dx = -\int_a^b g(x)\, dx$$

This shows that we can interpret the definite integral of a negative function as the *negative* of the area of the region *above* the graph of g on $[a, b]$ (see Figure 5b).

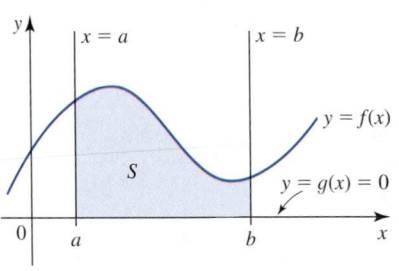

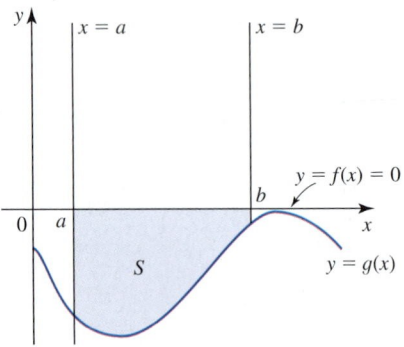

(a) If $g(x) = 0$ on $[a, b]$, then $\int_a^b f(x)\, dx$ gives the area of S.

(b) If $f(x) = 0$ on $[a, b]$, then $-\int_a^b g(x)\, dx$ gives the area of S.

FIGURE 5

The following guidelines are useful in setting up the integral in Equation (2).

> **Finding the Area Between Two Curves**
>
> **1.** Sketch the region between the graphs of f and g on $[a, b]$.
> **2.** Draw a representative rectangle with height $[f(x) - g(x)]$ and width Δx and note that its area is
>
> $$\Delta A = [f(x) - g(x)]\Delta x$$
>
> **3.** Observe that the height of the rectangle, $[f(x) - g(x)]$, is the integrand in Equation (2). The width Δx reminds us to integrate with respect to x. Thus,
>
> $$A = \int_a^b [f(x) - g(x)]\, dx$$
>
> (See Figure 6.)

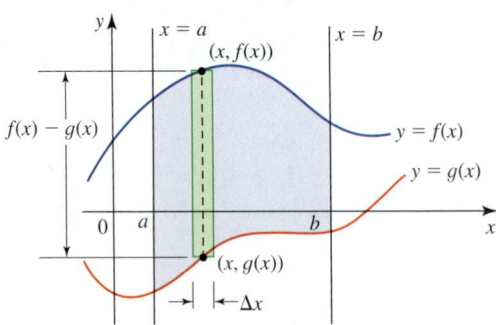

FIGURE 6

The area of the vertical rectangle is $\Delta A = [f(x) - g(x)]\Delta x$.

EXAMPLE 1 Find the area of the region between the graphs of $y = e^x$ and $y = x$ and the vertical lines $x = 0$ and $x = 1$.

Solution First, we make a sketch of the region and draw a representative rectangle. (See Figure 7.) Observe that the graph of $y = e^x$ lies above that of $y = x$. Therefore, if we let $f(x) = e^x$ and $g(x) = x$, then $f(x) \geq g(x)$ on $[0, 1]$. Also, from the figure we see that the area of the vertical rectangle is

$$\Delta A = [f(x) - g(x)]\Delta x = (e^x - x)\Delta x \qquad \text{(upper function } - \text{ lower function)}\Delta x$$

So the area of the required region is

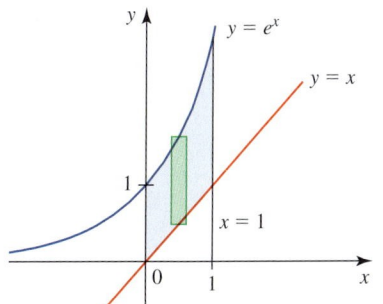

FIGURE 7
The graph of $y = e^x$ lies above that of $y = x$ on $[0, 1]$.

$$A = \int_a^b [f(x) - g(x)]\, dx = \int_0^1 (e^x - x)\, dx$$

$$= \left[e^x - \frac{1}{2}x^2 \right]_0^1 = \left(e - \frac{1}{2} \right) - (1 - 0)$$

$$= e - \frac{3}{2} \approx 1.22 \qquad \blacksquare$$

EXAMPLE 2 Find the area of the region bounded by the graphs of $y = 2 - x^2$ and $y = -x$.

Solution We first make a sketch of the desired region and draw a representative rectangle. (See Figure 8.) The points of intersection of the two graphs are found by solving the equations $y = 2 - x^2$ and $y = -x$ simultaneously. Substituting the second equation into the first yields

$$-x = 2 - x^2$$

$$x^2 - x - 2 = 0$$

$$(x + 1)(x - 2) = 0$$

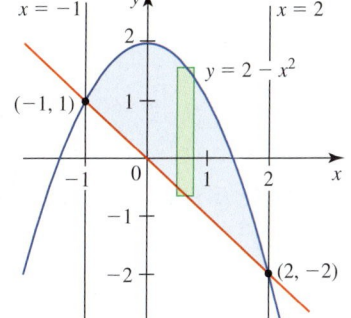

FIGURE 8
The graph of $f(x) = 2 - x^2$ lies above that of $g(x) = -x$ on $[-1, 2]$.

giving $x = -1$ and $x = 2$ as the x-coordinates of the points of intersection. We can think of the region in question as being bounded by the vertical lines $x = -1$ and $x = 2$. This gives the limits of integration as $a = -1$ and $b = 2$ in Equation (2). Next, if we let $f(x) = 2 - x^2$ and $g(x) = -x$, then $f(x) \geq g(x)$ on $[-1, 2]$, and the representative rectangle has area

$$\Delta A = [f(x) - g(x)]\Delta x = [(2 - x^2) - (-x)]\Delta x = (-x^2 + x + 2)\Delta x$$

Therefore, the area of the required region is

$$A = \int_a^b [f(x) - g(x)]\, dx = \int_{-1}^2 (-x^2 + x + 2)\, dx$$

$$= \left[-\frac{1}{3}x^3 + \frac{1}{2}x^2 + 2x \right]_{-1}^2$$

$$= \left(-\frac{8}{3} + 2 + 4 \right) - \left(\frac{1}{3} + \frac{1}{2} - 2 \right) = \frac{27}{6} \quad \text{or} \quad 4\frac{1}{2} \qquad \blacksquare$$

EXAMPLE 3 Refer to Figure 9. Find the area of the region enclosed by the graphs of

$$y = \frac{1}{4}x^2 \quad \text{and} \quad y = \frac{8}{x^2 + 4}$$

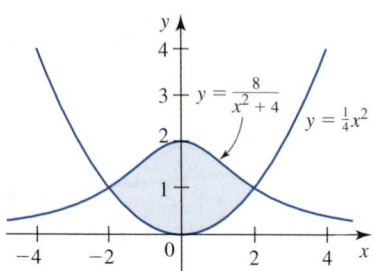

FIGURE 9
The region bounded by the graphs
of $y = \dfrac{8}{x^2 + 4}$ and $y = \dfrac{x^2}{4}$

Solution We first find the x-coordinates of the points of intersection of the two graphs by solving the system

$$\begin{cases} y = \frac{1}{4}x^2 \\ y = \dfrac{8}{x^2 + 4} \end{cases}$$

simultaneously. We have

$$\frac{1}{4}x^2 = \frac{8}{x^2 + 4}$$

$$x^4 + 4x^2 - 32 = 0$$

$$(x^2 + 8)(x^2 - 4) = 0$$

giving $x = \pm 2$. Next, observing that the graph of $f(x) = 8/(x^2 + 4)$ lies above that of $g(x) = x^2/4$ on $[-2, 2]$, we find the required area to be

$$A = \int_{-2}^{2} \left(\frac{8}{x^2 + 4} - \frac{x^2}{4} \right) dx$$

$$= 2 \int_{0}^{2} \left(\frac{8}{x^2 + 4} - \frac{x^2}{4} \right) dx \qquad \text{The integrand is even.}$$

$$= 2 \left[4 \tan^{-1} \frac{x}{2} - \frac{1}{12} x^3 \right]_{0}^{2}$$

$$= 2 \left(4 \tan^{-1} 1 - \frac{8}{12} \right) = 2\pi - \frac{4}{3} \qquad \blacksquare$$

■ Integrating with Respect to y

Sometimes it is easier to find the area of a region by integrating with respect to y rather than with respect to x. Consider, for example, the region S bounded by the graphs of $x = f(y)$ and $x = g(y)$, where $f(y) \geq g(y)$, and the horizontal lines $y = c$ and $y = d$, where $c \leq d$, as shown in Figure 10.

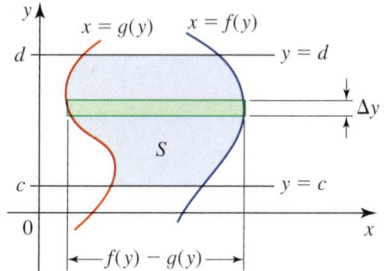

FIGURE 10
The region S is bounded on the left by the graph of $x = g(y)$ and on the right by that of $x = f(y)$ on $[c, d]$.

Observe that the condition $f(y) \geq g(y)$ implies that the graph of f lies to the right of the graph of g. Considering the horizontal rectangle of length $[f(y) - g(y)]$ and width Δy, we see that its area is

$$\Delta A = [f(y) - g(y)]\Delta y$$

This suggests that the area of S is

$$A = \int_c^d [f(y) - g(y)]\, dy \tag{3}$$

Since a rigorous derivation of Equation (3) proceeds along lines that are virtually identical to that of Equation (2), it will be omitted here.

EXAMPLE 4 Find the area of the region of Example 3 by integrating with respect to y.

Solution We view the region S as being bounded by the graphs of the functions $f(y) = y + 2$ (solve $y = x - 2$ for x), $g(y) = y^2$, and the horizontal lines $y = -1$ and $y = 2$. See Figure 11. Observe that $f(y) \geq g(y)$ for y in $[-1, 2]$. The area of the representative horizontal rectangle is

$$\Delta A = [f(y) - g(y)]\Delta y = [(y + 2) - y^2]\Delta y = (y + 2 - y^2)\Delta y$$

(right function − left function)Δy

This implies that

$$A = \int_{-1}^{2} (y + 2 - y^2)\, dy = \left[\frac{1}{2}y^2 + 2y - \frac{1}{3}y^3\right]_{-1}^{2}$$

$$= \left(2 + 4 - \frac{8}{3}\right) - \left(\frac{1}{2} - 2 + \frac{1}{3}\right) = \frac{9}{2} \quad \text{or} \quad 4\frac{1}{2} \quad \blacksquare$$

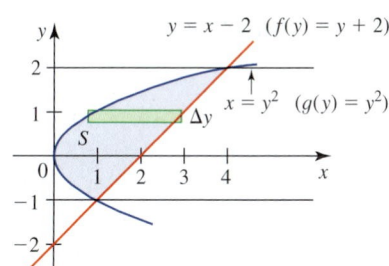

FIGURE 11
The horizontal rectangle has area $[f(y) - g(y)]\Delta y$.

Note Sometimes we prefer to use Equation (3) instead of Equation (2) or vice versa. In general, the choice of the formula depends on the shape of the region. Often one would integrate with respect to the variable that results in the minimal splitting of the region. But sometimes the use of one formula leads to an integral(s) that is difficult to evaluate, in which case the other formula should be used. $\blacksquare$

What Happens When the Curves Intertwine?

Sometimes we are required to find the area of a region S between two curves in which the graph of one function f lies above that of another function g for some values of x ($f(x) \geq g(x)$) and lies below it for other values of x ($f(x) \leq g(x)$). You will be asked to give a physical interpretation of a problem involving precisely such a situation in Exercise 48.

To find the area of the region S, we divide it into subregions $S_1, S_2, \ldots, S_n$, each of which is described by the sole condition $f(x) \geq g(x)$ or $f(x) \leq g(x)$. Figure 12 illustrates the case in which $n = 3$. We then use the guidelines developed earlier to calculate the area of each subregion. Adding up these areas gives the area of S. Thus, the area of the region S shown in Figure 12 between the graphs of f and g and between the vertical lines $x = a$ and $x = b$ is

$$A = \int_a^c [f(x) - g(x)]\, dx + \int_c^d [g(x) - f(x)]\, dx + \int_d^b [f(x) - g(x)]\, dx$$

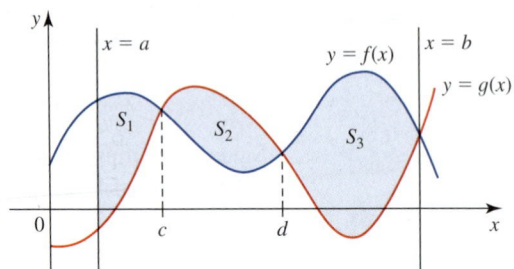

FIGURE 12

The region S is the union of S_1, where $f(x) \geq g(x)$; S_2, where $f(x) \leq g(x)$, and S_3, where $f(x) \geq g(x)$.

Since

$$|f(x) - g(x)| = \begin{cases} f(x) - g(x) & \text{if } f(x) \geq g(x) \\ g(x) - f(x) & \text{if } f(x) \leq g(x) \end{cases}$$

we can also write A in the abbreviated form

$$A = \int_a^b |f(x) - g(x)|\, dx \tag{4}$$

When using Equation (4), however, we still need to determine the subintervals of $[a, b]$ where $f(x) \geq g(x)$ and/or where $g(x) \geq f(x)$ and write A as the sum of integrals giving the areas of the subregions on these subintervals.

EXAMPLE 5 Find the area of the region S bounded by the graphs of $y = \cos x$ and $y = (2/\pi)x - 1$ and the vertical lines $x = 0$ and $x = \pi$.

Solution The region S is shown in Figure 13. To find the points of intersection of the graphs of $y = \cos x$ and $y = (2/\pi)x - 1$, we solve the two equations simultaneously. Substituting the first equation into the second, we obtain

$$\cos x = \frac{2}{\pi} x - 1$$

By inspecting the graphs, we see that $x = \pi/2$ is the only solution of the equation. Therefore, the point of intersection is $\left(\frac{\pi}{2}, 0\right)$. Let $f(x) = \cos x$ and $g(x) = (2/\pi)x - 1$. Referring to Figure 13, we see that the areas A_1 and A_2 of the subregions S_1 and S_2 are

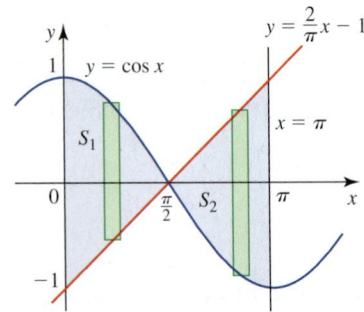

FIGURE 13

The area of S is the sum of the areas of S_1 and S_2.

$$A_1 = \int_0^{\pi/2} [f(x) - g(x)]\, dx \qquad f(x) \geq g(x)$$

$$= \int_0^{\pi/2} \left[\cos x - \left(\frac{2}{\pi}x - 1\right)\right] dx = \int_0^{\pi/2} \left(\cos x - \frac{2}{\pi}x + 1\right) dx$$

$$= \left[\sin x - \frac{1}{\pi}x^2 + x\right]_0^{\pi/2} = 1 - \frac{1}{\pi}\left(\frac{\pi}{2}\right)^2 + \frac{\pi}{2} = \frac{4\pi - \pi^2 + 2\pi^2}{4\pi} = \frac{4 + \pi}{4}$$

and

$$A_2 = \int_{\pi/2}^{\pi} [g(x) - f(x)]\, dx \qquad g(x) \geq f(x)$$

$$= \int_{\pi/2}^{\pi} \left(\frac{2}{\pi} x - 1 - \cos x \right) dx$$

$$= \left[\frac{1}{\pi} x^2 - x - \sin x \right]_{\pi/2}^{\pi} = \left[\frac{1}{\pi}(\pi^2) - \pi - 0 \right] - \left[\frac{1}{\pi} \left(\frac{\pi}{2} \right)^2 - \frac{\pi}{2} - 1 \right]$$

$$= \pi - \pi - \frac{\pi}{4} + \frac{\pi}{2} + 1 = \frac{4 + \pi}{4}$$

Therefore, the required area is

$$A = A_1 + A_2 = \frac{4 + \pi}{4} + \frac{4 + \pi}{4} = \frac{4 + \pi}{2} = 2 + \frac{\pi}{2} \qquad \blacksquare$$

The following example, drawn from the field of study known as the *theory of elasticity,* gives yet another physical interpretation of the area between two curves.

EXAMPLE 6   Elastic Hysteresis   Figure 14 shows a stress–strain curve for a sample of vulcanized rubber that has been stretched to seven times its original length. The function *f* whose graph is the upper curve gives the relationship between the stress and the strain as the load (the stress) is applied to the material. Because the material is elastic, the rubber returns to its original length when the load is removed. However, when the load is decreased, the graph of *f* is not retraced. Instead, the stress–strain curve given by the graph of the function *g* is obtained.

FIGURE 14

A stress–strain curve for a sample of vulcanized rubber: The upper curve shows what happens when the load is applied, and the lower curve shows what happens when the load is decreased.

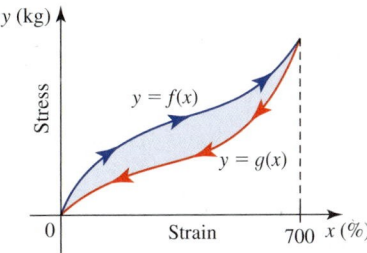

The lack of coincidence of the curves for increasing and decreasing stress is known as *elastic hysteresis.* The graphs of *f* and *g* on the interval [0, 700] form the *hysteresis loop* for the material. It can be shown that the area of the region enclosed by the hysteresis loop is proportional to the energy dissipated within the rubber. Thus, the elastic hysteresis of the rubber is given by

$$\int_0^{700} [f(x) - g(x)]\, dx \qquad \text{Since } f(x) \geq g(x) \text{ on } [0, 700]$$

Certain types of rubber have large hysteresis, and these materials are often used as vibration absorbers. Most of the internal energy is dissipated in the form of heat, thereby minimizing the transmission of the energy of vibration to the mediums to which the machinery is mounted. $\blacksquare$

5.1 CONCEPT QUESTIONS

1. Write an expression that gives the area of the region completely enclosed by the graphs of f and g in Figures 1 and 2 in terms of (a) a single integral and (b) two integrals.

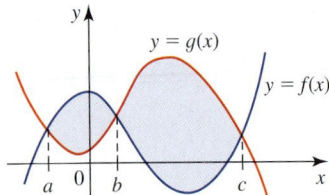

FIGURE 1

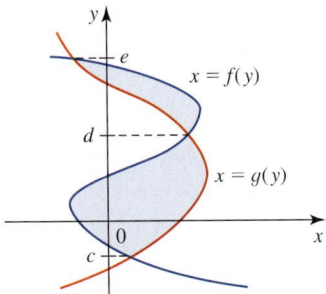

FIGURE 2

2. Two cars start out side by side moving down a straight road. The velocity functions for Car A and Car B are f and g, respectively. Their graphs are shown in Figure 3. Suppose that $f(t)$ and $g(t)$ are measured in feet per second and t in seconds, where t lies in the interval $[0, 10]$. Answer the following questions using definite integral(s) if appropriate.

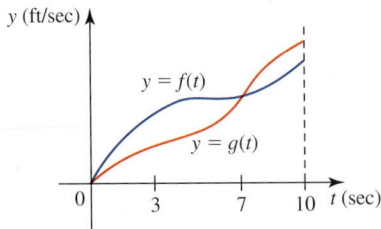

FIGURE 3

a. By what distance is Car A ahead of Car B after 3 sec? After 7 sec? After 10 sec?
b. Is one car always ahead of the other after the start of motion?
c. What is the greatest distance between the two cars over the 10-sec interval?

5.1 EXERCISES

In Exercises 1–6, find the area of the shaded region.

1.

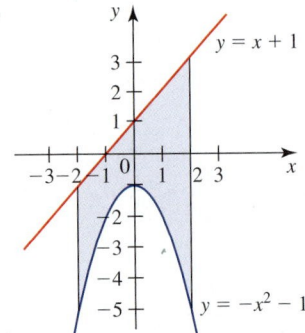

2.

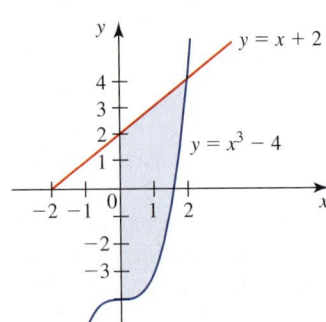

3.

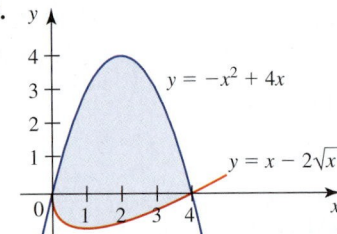

4.

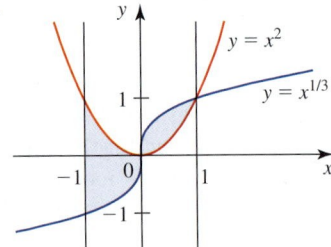

5.

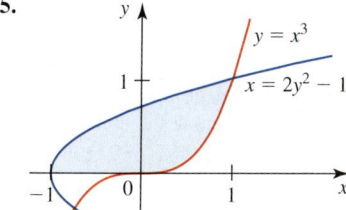

6.

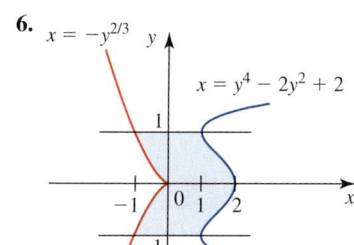

7. Oil Production Shortfall Energy experts disagree about when global oil production will begin to decline. In the following figure, the function f gives the annual world oil production in billions of barrels from 1980 to 2050 according to the U.S. Department of Energy projection. The function g gives the world oil production in billions of barrels per year over the same period according to longtime petroleum geologist Colin Campbell. Find an expression in terms of definite integrals involving f and g giving the shortfall in the total oil production over the period in question heeding Campbell's dire warnings.

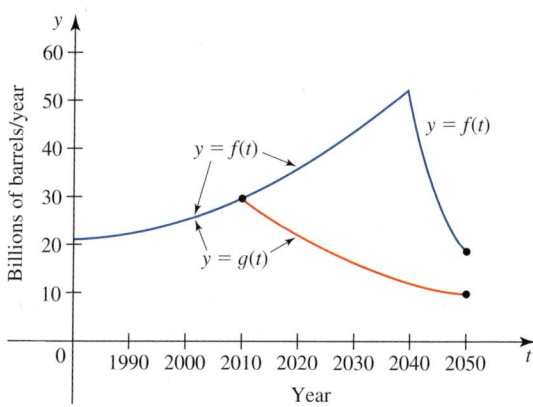

Source: U.S. Department of Energy and Colin Campbell.

8. Rate of Change of Revenue The rate of change of the revenue of Company A over the (time) interval $[0, T]$ is $f(t)$ dollars per week, whereas the rate of change of the revenue of Company B over the same period is $g(t)$ dollars per week. Suppose the graphs of f and g are as depicted in the following figure. Find an expression in terms of definite integrals involving f and g giving the additional revenue that Company B will have over Company A in the period $[0, T]$.

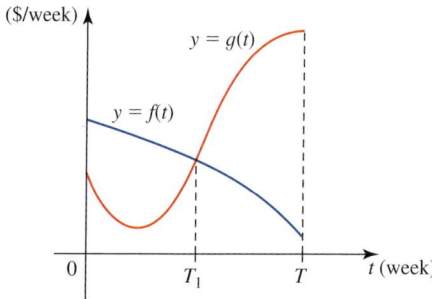

In Exercises 9–40, sketch the region bounded by the graphs of the given equations and find the area of that region.

9. $y = x^2 + 3$, $\quad y = x + 1$, $\quad x = -1$, $\quad x = 1$

10. $y = x^3 + 1$, $\quad y = x - 1$, $\quad x = -1$, $\quad x = 1$

11. $y = x^2 - 2x - 1$, $\quad y = -e^x - 1$, $\quad x = -1$, $\quad x = 1$

12. $y = \dfrac{1}{\sqrt{1 - x^2}}$, $\quad y = e^{-x}$, $\quad x = 0$, $\quad x = \dfrac{1}{2}$

13. $y = -x^2 + 4$, $\quad y = 3x + 4$

14. $y = x^2 - 4x$, $\quad y = -x + 4$

15. $y = x^2 - 4x + 3$, $\quad y = -x^2 + 2x + 3$

16. $y = (x - 2)^2$, $\quad y = 4 - x^2$

17. $y = x$, $\quad y = x^3$

18. $y = x^2$, $\quad y = x^4$

19. $y = \dfrac{x}{x^2 + 1}$, $\quad y = -\dfrac{1}{2}x^2$, $\quad x = 1$

20. $y = x^3 - 6x^2 + 9x$, $\quad y = x^2 - 3x$

21. $y = \sqrt{x}$, $\quad y = -\dfrac{1}{2}x + 1$, $\quad x = 1$, $\quad x = 4$

22. $y = 2\sqrt{x} - x$, $\quad y = -\sqrt{x}$

23. $y = \dfrac{1}{x^2}$, $\quad y = x^2$, $\quad x = 3$

24. $y = 2x$, $\quad y = x\sqrt{x + 1}$

25. $y = -x^2 + 6x + 5$, $\quad y = x^2 + 5$

26. $y = x\sqrt{4 - x^2}$, $\quad y = 0$

27. $y = \dfrac{x}{\sqrt{16 - x^2}}$, $\quad y = 0$, $\quad x = 3$

28. $x = y^2 + 1$, $\quad x = 0$, $\quad y = -1$, $\quad y = 2$

29. $x = y^2$, $\quad x = y - 3$, $\quad y = -1$, $\quad y = 2$

30. $x = y^2$, $\quad x = 2y + 3$

31. $y = -x^3 + x$, $\quad y = x^4 - 1$

32. $\sqrt{x} + \sqrt{y} = 1$, $\quad x + y = 1$

33. $y = |x|$, $y = x^2 - 2$

34. $y = 2^x$, $y = 2^{-x}$, $x = -2$, and $x = 2$.

35. $y = \sin 2x$, $y = \cos x$, $x = \dfrac{\pi}{6}$, $x = \dfrac{\pi}{2}$

36. $y = \cos 2x$, $y = \sin x$, $x = 0$, $x = \dfrac{3\pi}{2}$

37. $y = \sec^2 x$, $y = 2$, $x = -\dfrac{\pi}{4}$, $x = \dfrac{\pi}{4}$

38. $y = \sec^2 x$, $y = \cos x$, $x = -\dfrac{\pi}{3}$, $x = \dfrac{\pi}{3}$

39. $y = 2\sin x + \sin 2x$, $y = 0$, $x = 0$, $x = \pi$

40. $x = \sin y + \cos 2y$, $x = 0$, $y = 0$, $y = \dfrac{\pi}{2}$

41. Find the area of the region in the first quadrant bounded by the parabolas $y = x^2$ and $y = \frac{1}{4}x^2$ and the line $y = 2$.

42. Find the area of the region enclosed by the curve $y^2 = x^2(1 - x^2)$.

In Exercises 43 and 44, use integration to find the area of the triangle with the given vertices.

43. $(0, 0)$, $(1, 6)$, $(4, 2)$

44. $(-2, 4)$, $(0, -2)$, $(6, 2)$

In Exercises 45 and 46, find the area of the region bounded by the given curves (a) using integration with respect to x and (b) using integration with respect to y.

45. $y = x^3$, $y = 2x + 4$, $x = 0$

46. $y = \sqrt{x}$, $y = \dfrac{1}{2}x$, $y = 1$, $y = 2$

47. Effect of Advertising on Revenue In the accompanying figure, the function f gives the rate of change of Odyssey Travel's revenue with respect to the amount x it spends on advertising with its current advertising agency. By engaging the services of a different advertising agency, Odyssey expects its revenue to grow at the rate given by the function g. Give an interpretation of the area A of the region S, and find an expression for A in terms of a definite integral involving f and g.

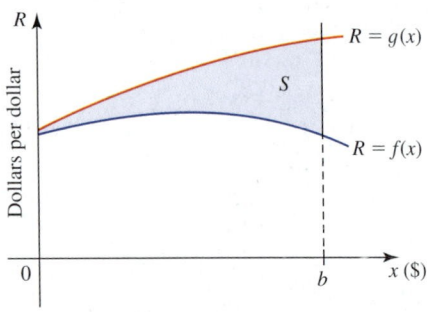

48. Two cars start out side by side and travel along a straight road. The velocity of Car A is $f(t)$ ft/sec, and the velocity of Car B is $g(t)$ ft/sec over the interval $[0, T]$, where $0 < T_1 < T$. Furthermore, suppose that the graphs of f and g are as depicted in the figure. Let A_1 and A_2 denote the areas of the regions shown shaded.

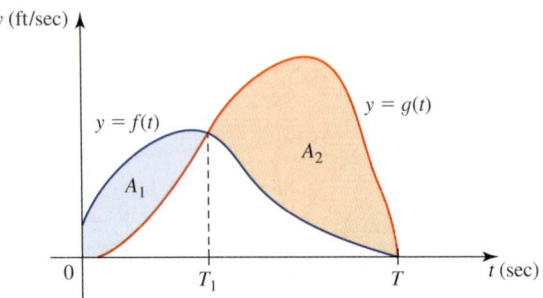

a. Write the number

$$\int_{T_1}^{T} [g(t) - f(t)]\, dt - \int_{0}^{T_1} [f(t) - g(t)]\, dt$$

in terms of A_1 and A_2.

b. What does the number obtained in part (a) represent?

 In Exercises 49–54, use a graphing utility to (a) plot the graphs of the given functions and (b) find the x-coordinates of the points of intersection of the curves. Then find an approximation of the area of the region bounded by the curves using the integration capabilities of the graphing utility.

49. $y = x^2$, $y = 4 - x^4$

50. $y = x^3 - 3x^2 + 1$, $y = x^2 - 4$

51. $y = x^3 - 4x^2$, $y = x^3 - 9x$

52. $y = x^4 - 2x^2 + 2$, $y = 4 - x^2$

53. $y = x^2$, $y = \sin x$

54. $y = \cos x$, $y = |x|$

55. Turbocharged Engine Versus Standard Engine In tests conducted by *Auto Test Magazine* on two identical models of the Phoenix Elite, one equipped with a standard engine and the other with a turbocharger, it was found that the acceleration of the former (in ft/sec^2) is given by

$$a = f(t) = 4 + 0.8t \qquad 0 \le t \le 12$$

t sec after starting from rest at full throttle, whereas the acceleration of the latter (in ft/sec^2) is given by

$$a = g(t) = 4 + 1.2t + 0.03t^2 \qquad 0 \le t \le 12$$

How much faster is the turbocharged model moving than the model with the standard engine at the end of a 10-sec test run at full throttle?

56. Velocity of Dragsters Two dragsters start out side by side. The velocity of Dragster A, V_A, and the velocity of Dragster B,

V_B, for the first 8 sec of the race are shown in the following table, where V_A and V_B are measured in feet per second. Use Simpson's Rule with $n = 8$ to estimate how far Dragster A is ahead of Dragster B 8 sec after the start of the race.

t (sec)	0	1	2	3	4	5	6	7	8
V_A (ft/sec)	0	22	46	70	94	118	142	166	190
V_B (ft/sec)	0	20	44	66	88	112	138	160	182

57. Surface Area of the Jacqueline Kennedy Onassis Reservoir The reservoir located in Central Park in New York City has the shape depicted in the figure below. The measurements shown were taken at 206-ft intervals. Use Simpson's Rule with $n = 10$ to estimate the surface area of the reservoir.

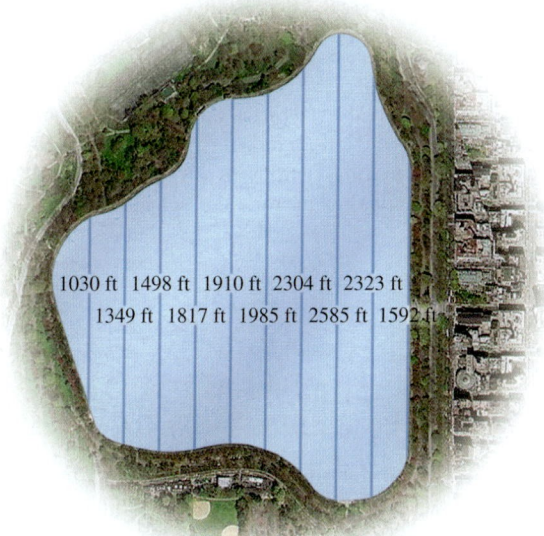

1030 ft 1498 ft 1910 ft 2304 ft 2323 ft
1349 ft 1817 ft 1985 ft 2585 ft 1592 ft

Source: The Boston Globe

58. Estimating the Rate of Flow of a River A stream is 120 ft wide. The following table gives the depths of the river measured across a section of the river in intervals of 6 ft. Here, x denotes the distance from one bank of the river, and y denotes the corresponding depth (in feet). The average rate of flow of the river across this section of the river is 4.2 ft/sec. Use Simpson's Rule to estimate the rate of flow of the river.

x (ft)	0	6	12	18	24	30	36	42	48	54	60
y (ft)	0.8	1.2	3.0	4.1	5.8	6.6	6.8	7.0	7.2	7.4	7.8

x (ft)	66	72	78	84	90	96	102	108	114	120
y (ft)	7.6	7.4	7.0	6.6	6.0	5.1	4.3	3.2	2.2	1.1

59. Profit Functions The weekly total marginal cost incurred by the Advance Visuals Systems Corporation in manufacturing x 19-inch LCD HDTVs is

$$C'(x) = 0.000006x^2 - 0.04x + 120$$

dollars per set. The weekly marginal revenue realized by the company from the sale of x sets is

$$R'(x) = -0.008x + 200$$

dollars per set.

a. Plot the graphs of C' and R' using the viewing window $[0, 10{,}000] \times [0, 300]$.

b. Find the area of the region bounded by the graphs of C' and R' and the vertical lines $x = 2000$ and $x = 5000$. Interpret your result.

60. Find the area of the region bounded by the curve $y^2 = x^3 - x^2$ and the line $x = 2$.

61. Find the area of the region bounded by the graph of $f(x) = \sqrt{x}$, the y-axis, and the tangent line to the graph of f at $(1, 1)$.

62. Find the number a such that the area of the region bounded by the graph of $x = (y - 1)^2$ and the line $x = a$ is $\frac{9}{2}$.

63. Find the area of the region bounded by the x-axis and the graph of $f(x) = x^4 - 2x^3$ and to the right of the vertical line that passes through the point at which f attains its absolute minimum.

64. The area of the region in the right half plane bounded by the y-axis, the parabola $y = -x^2 - 2x + 3$, and a line tangent to the parabola is $\frac{8}{3}$. Find the coordinates of the point of tangency.

65. The region S is bounded by the graphs of $y = \sqrt{x}$, the x-axis, and the line $x = 4$.
a. Find a such that the line $x = a$ divides S into two subregions of equal area.
b. Find b such that the line $y = b$ divides S into two subregions of equal area.

66. Find the value of c such that the parabola $y = cx^2$ divides the region bounded by the parabola $y = \frac{1}{9}x^2$, and the lines $y = 2$, and $x = 0$ into two subregions of equal area.

67. Let $A(x)$ denote the area of the region in the first quadrant completely enclosed by the graphs of $f(x) = x^m$ and $g(x) = x^{1/m}$, where m is a positive integer.
a. Find an expression for $A(m)$.
b. Evaluate $\lim_{m \to 1} A(m)$ and $\lim_{m \to \infty} A(m)$. Give a geometric interpretation.

c. Verify your observations in part (b) by plotting the graphs of f and g.

68. Let $f(x) = \dfrac{1}{x^2 + 1}$ and $g(x) = |x|$.

   a. Plot the graphs of f and g using the viewing window $[-1, 1] \times [0, 1.5]$. Find the points of intersection of the graphs of f and g accurate to three decimal places.

   b. Use a calculator or computer and the result of part (a) to find the area of the region bounded by the graphs of f and g.

69. The curve with equation $y^2 - 4x^3 + 4x^4 = 0$ is called a **piriform.**

   a. Plot the curve using the viewing window $[-1, 1] \times [-1, 1]$.

   b. Find the area of the region enclosed by the curve accurate to five decimal places.

70. The curve with equation $4y^2 - 4xy^2 - x^2 - x^3 = 0$ is called a **right strophoid.**

   a. Plot the curve using the viewing window $[-1.5, 1.5] \times [-0.5, 0.5]$.

   b. Find the area of the region enclosed by the loop of the curve.

In Exercises 71–74, determine whether the statement is true or false. If it is true, explain why it is true. If it is false, explain why or give an example to show why it is false.

71. If A denotes the area bounded by the graphs of f and g on $[a, b]$, then

$$A^2 = \int_a^b [f(x) - g(x)]^2 \, dx$$

72. If f and g are continuous on $[a, b]$ and $\int_a^b [f(t) - g(t)] \, dt > 0$, then $f(t) \geq g(t)$ for all t in $[a, b]$.

73. Two cars start out traveling side by side along a straight road at $t = 0$. Twenty seconds later, Car A is 30 ft behind Car B. If v_1 and v_2 are continuous velocity functions for Car A and Car B, respectively, where $v_1(t)$ and $v_2(t)$ are measured in feet per second, then

$$\int_0^{20} v_2(t) \, dt = \int_0^{20} v_1(t) \, dt + 30$$

74. Suppose that the acceleration of Car A and Car B along a straight road are $a_1(t)$ ft/sec^2 and $a_2(t)$ ft/sec^2, respectively, over the time interval $[t_1, t_2]$, where a_1 and a_2 are continuous functions with $a_1(t) \geq a_2(t)$ on $[t_1, t_2]$. Then at time $t = t_2$, Car A will be traveling $\int_{t_1}^{t_2} [a_1(t) - a_2(t)] \, dt$ ft/sec faster than Car B. (Assume that t is measured in seconds.)

5.2   Volumes: Disks, Washers, and Cross Sections

In Section 5.1 we saw the role played by the definite integral in finding the area of plane regions. In the next two sections we will see how the definite integral can be used to help us find the volumes of solids such as those shown in Figure 1.

FIGURE 1   **(a)** Wine barrel   **(b)** Pyramid   **(c)** Pontoon

Figure 1c depicts a pontoon for a seaplane. In designing a pontoon, the engineer needs to know the volume of water displaced by the part of the pontoon that lies below the waterline in order to determine the buoyancy of the pontoon (Archimedes' Principle).

Solids of Revolution

A **solid of revolution** is a solid obtained by revolving a region in the plane about a line in the plane. The line is called the **axis of revolution.** For example, if the region R under the graph of f on the interval $[a, b]$ shown in Figure 2a is revolved about the x-axis, we obtain the solid of revolution S shown in Figure 2b. Here, the axis of revolution of the solid is the x-axis.

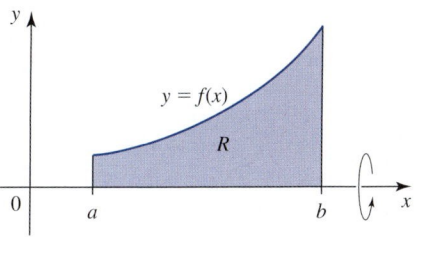
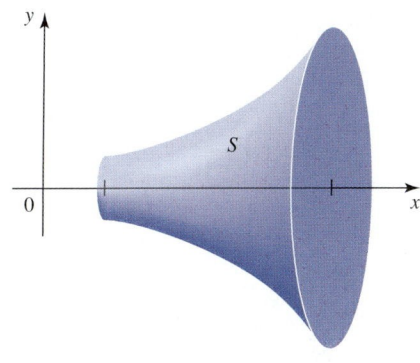

FIGURE 2   (**a**) Region R under the graph of f (**b**) Solid obtained by revolving R about the x-axis

The Disk Method

To define the volume of a solid of revolution and to devise a method for computing it, let's consider the solid S generated by the region R shown in Figure 3a. Let $P = \{x_0, x_1, \ldots, x_n\}$ be a regular partition of $[a, b]$. This partition divides the region R into n nonoverlapping subregions $R_1, R_2, \ldots, R_n$. When these regions are revolved about the x-axis, they give rise to the n nonoverlapping solids $S_1, S_2, \ldots, S_n$, whose union is S. (See Figure 3b.)

FIGURE 3

A partition of $[a, b]$ produces n subregions $R_1, R_2, \ldots, R_n$ that are revolved about the x-axis to obtain the n solids $S_1, S_2, \ldots, S_n$ that together form S. (Here $n = 8$.)

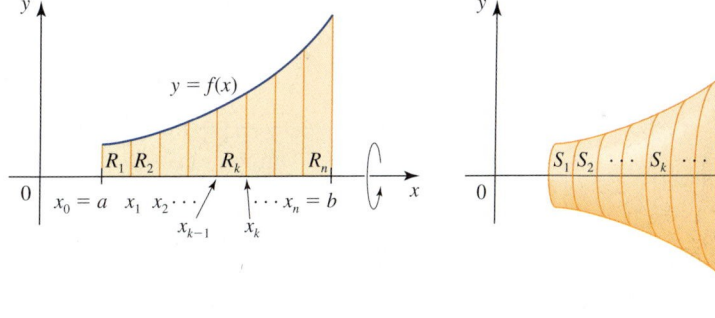

(**a**) The region R (**b**) The solid S

Let's concentrate on the part of the solid of revolution that is generated by the region R_k under the graph of f on the interval $[x_{k-1}, x_k]$. This region is shown enlarged for the sake of clarity in Figure 4. If c_k is an evaluation point in $[x_{k-1}, x_k]$, then the area of R_k is approximated by the rectangle of height $f(c_k)$ and width $\Delta x = (b - a)/n$. If this rectangle is revolved about the x-axis, it generates the disk D_k having radius $f(c_k)$ and width Δx; therefore, its volume is

$$\Delta V_k = \pi[f(c_k)]^2\, \Delta x \qquad \pi(\text{radius})^2 \cdot \text{width}$$

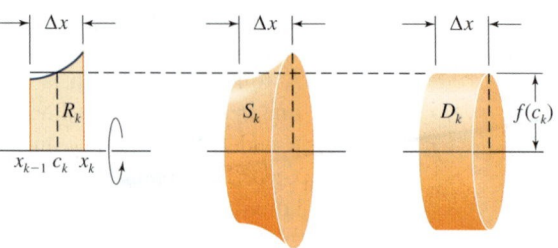

FIGURE 4

The region R_k, shown shaded, is approximated by the rectangle. The volume of S_k is approximated by the volume of the disk D_k.

The kth region and the approximating rectangle	The kth solid of revolution	The kth disk

The volume of D_k provides us with an approximation of the volume of S_k. Therefore, by approximating the volume of each solid $S_1, S_2, \ldots, S_n$ with the volume of a corresponding disk $D_1, D_2, \ldots, D_n$, we see that the volume V of S is approximated by the sum of the volumes of these disks. (See Figure 5.) Thus,

$$V \approx \sum_{k=1}^{n} \Delta V_k = \sum_{k=1}^{n} \pi [f(c_k)]^2 \, \Delta x$$

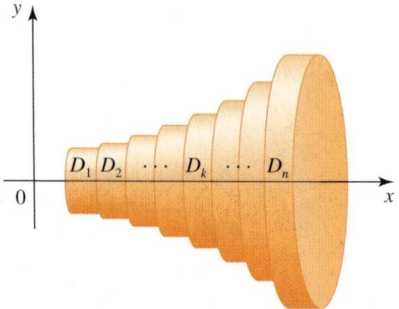

FIGURE 5

The volume V of the solid of revolution S is approximated by the sum of the volume of the n disks $D_1, \ldots, D_n$.

Recognizing this sum to be the Riemann sum of the function πf^2 on the interval $[a, b]$, we see that

$$\lim_{n \to \infty} \sum_{k=1}^{n} \pi [f(c_k)]^2 \, \Delta x = \int_a^b \pi [f(x)]^2 \, dx$$

DEFINITION **Volume of a Solid of Revolution**
(Region revolved about the x-axis)

Let f be a continuous nonnegative function on $[a, b]$, and let R be the region under the graph of f on the interval $[a, b]$. The volume of the solid of revolution generated by revolving R about the x-axis is

$$V = \lim_{n \to \infty} \sum_{k=1}^{n} \pi [f(c_k)]^2 \, \Delta x = \int_a^b \pi [f(x)]^2 \, dx \qquad (1)$$

Just as we were able to recall the formulas for finding the area under a curve by looking at the area of a representative rectangle, so can we recall Formula (1) by looking at the volume of the disk obtained by revolving a representative rectangle about the x-axis.

We proceed as follows: Having made a sketch of the region R under the graph of $y = f(x)$ on $[a, b]$, draw a representative vertical rectangle of height $f(x)$, or y, corresponding to a value of x in $[a, b]$, and width Δx. (See Figure 6.) We can regard this disk with volume

$$\Delta V = \pi[f(x)]^2 \, \Delta x = \pi y^2 \, \Delta x \qquad \pi(\text{radius})^2 \cdot \text{width}$$

as representing an element of volume of a solid. Now observe that the expression next to Δx, πy^2, is *the integrand* in Formula (1).

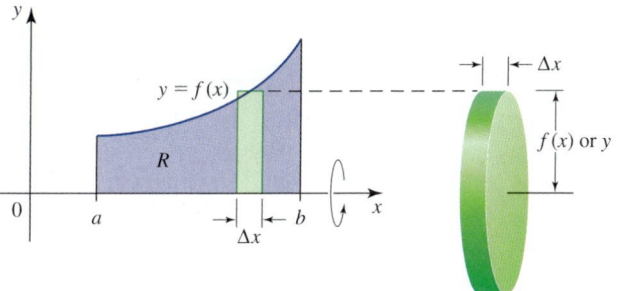

FIGURE 6
If a representative vertical rectangle is revolved about the x-axis, it generates a disk of radius $f(x)$, or y, and width Δx.

Volume by Disk Method (Region revolved about the x-axis)

$$V = \pi \int_a^b [f(x)]^2 \, dx = \pi \int_a^b y^2 \, dx \qquad f \geq 0$$

From now on, when we introduce a notion and/or derive a formula through the use of Riemann sums, we will often use the heuristic approach of looking at a representative element associated with the general term of the Riemann sum (without the subscripts) to help us recall the appropriate formula.

EXAMPLE 1 Find the volume of the solid obtained by revolving the region under the graph of $y = \sqrt{x}$ on $[0, 2]$ about the x-axis.

Solution From the graph of $y = \sqrt{x}$ sketched in Figure 7a, we see that the radius of the representative disk corresponding to a particular value of x in $[0, 2]$ (the height of the representative rectangle) is y, or $\sqrt{x}$. Therefore, the volume of the disk is

$$\Delta V = \pi y^2 \, \Delta x \qquad \text{Here } y = f(x) = \sqrt{x}.$$
$$= \pi(\sqrt{x})^2 \, \Delta x = \pi x (\Delta x)$$

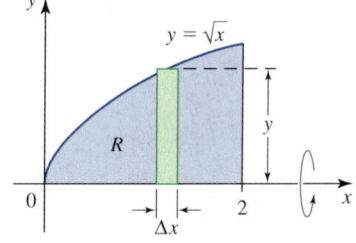

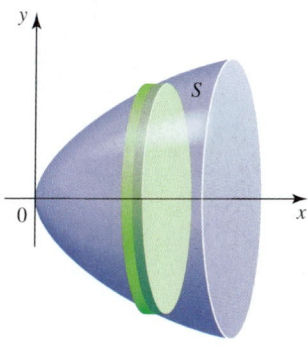

FIGURE 7
If R is revolved about the x-axis, we obtain the solid of revolution S.

(a) The region R

(b) The solid S

Summing the volumes of the disks and taking the limit, we find that the volume of the solid is

$$V = \int_0^2 \pi x \, dx = \pi \int_0^2 x \, dx$$

$$= \frac{1}{2} \pi x^2 \Big|_0^2 = \frac{1}{2} \pi (4 - 0) \quad \text{or} \quad 2\pi$$

EXAMPLE 2 By revolving the region under the graph of $y = \sqrt{r^2 - x^2}$ on $[-r, r]$, show that the volume of a sphere of radius r is $V = \frac{4}{3} \pi r^3$.

Solution The graph of $y = \sqrt{r^2 - x^2}$ is a semicircle, as shown in Figure 8a. We can see that the radius of a representative disk is y, the height of the vertical rectangle. Therefore, the volume of the disk is

$$\Delta V = \pi y^2 \Delta x$$

$$= \pi (r^2 - x^2) \Delta x \qquad \text{Since } y = \sqrt{r^2 - x^2}$$

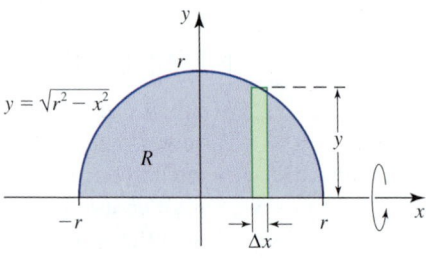

FIGURE 8
By revolving the region R about the x-axis, we obtain the sphere of radius r.

(**a**) The region R (**b**) The solid S

Summing the volumes of the disks and taking the limit, we obtain the required volume as

$$V = \int_{-r}^r \pi (r^2 - x^2) \, dx$$

$$= \pi \int_{-r}^r (r^2 - x^2) \, dx$$

$$= 2\pi \int_0^r (r^2 - x^2) \, dx \qquad \text{Use the symmetry of the region.}$$

$$= 2\pi \left[r^2 x - \frac{1}{3} x^3 \right]_0^r$$

$$= 2\pi \left(r^3 - \frac{1}{3} r^3 \right)$$

$$= \frac{4}{3} \pi r^3$$

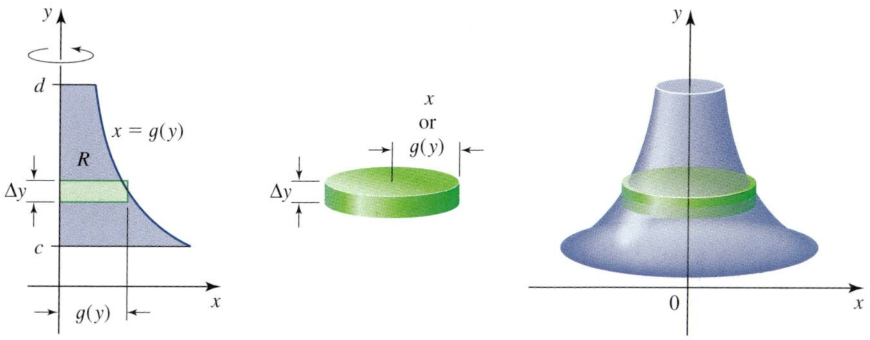

FIGURE 9

If a representative horizontal rectangle is revolved about the y-axis, it generates a disk of radius $g(y)$, or x, and width Δy and hence volume $\Delta V = \pi x^2 \Delta y$.

Formula (1) is used to find the volume of a solid of revolution when the axis of revolution is the x-axis. To derive a formula for the volume V of a solid of revolution obtained by revolving a region about the y-axis, consider the region R bounded by the graphs of $x = g(y)$, $x = 0$, $y = c$, and $y = d$ as shown in Figure 9.

If R is revolved about the y-axis, then a representative horizontal rectangle (perpendicular to the axis of revolution) with length x, or $g(y)$, and width Δy generates a disk with volume

$$\Delta V = \pi [g(y)]^2 \Delta y = \pi x^2 \Delta y$$

Summing the volumes of the disks and taking the limit, we obtain the following formula.

Volume by Disk Method (Region revolved about the y-axis)

$$V = \pi \int_c^d [g(y)]^2 \, dy = \pi \int_c^d x^2 \, dy \qquad g \geq 0$$

EXAMPLE 3   Find the volume of the solid obtained by revolving the region bounded by the graphs of $y = x^3$, $y = 8$, and $x = 0$ about the y-axis.

Solution   The region R in question together with the solid generated by revolving that region about the y-axis is shown in Figure 10. A representative horizontal rectangle sweeps out a disk of radius x and width Δy. Therefore, its volume is

$$
\begin{aligned}
\Delta V &= \pi x^2 \Delta y \\
&= \pi (y^{1/3})^2 \Delta y \qquad \text{Solve } y = x^3 \text{ for } x. \\
&= \pi y^{2/3} \Delta y
\end{aligned}
$$

FIGURE 10

If a horizontal rectangle is revolved about the y-axis, it generates a disk of radius $g(y) = y^{1/3}$, or x, and width Δy.

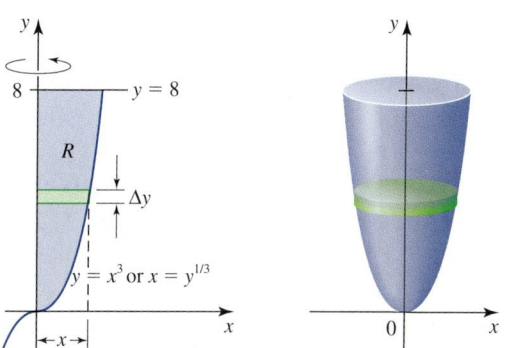

If we sum the volume of these disks and take the limit, we find that the required volume is

$$V = \pi \int_0^8 y^{2/3} \, dy$$

$$= \frac{3}{5} \pi y^{5/3} \Big|_0^8 = \frac{3}{5} \pi (8^{5/3}) \quad \text{or} \quad \frac{96\pi}{5}$$

■ The Washer Method

Let R be the region between the graphs of the functions f and g and between the vertical lines $x = a$ and $x = b$, where $f(x) \geq g(x) \geq 0$ on $[a, b]$. If R is revolved about the x-axis, we obtain a solid of revolution with a hole in it. (See Figure 11.) Observe that when a representative vertical rectangle between the curves is revolved about the x-axis, the resultant element of volume of the solid has the shape of a washer with outer radius $f(x)$ and inner radius $g(x)$. Therefore, the volume of this element is

$$\Delta V = \pi [f(x)]^2 \, \Delta x - \pi [g(x)]^2 \, \Delta x$$

$$\pi (\text{outer radius})^2 \cdot \text{width} - \pi (\text{inner radius})^2 \cdot \text{width}$$

$$= \pi \{ [f(x)]^2 - [g(x)]^2 \} \Delta x$$

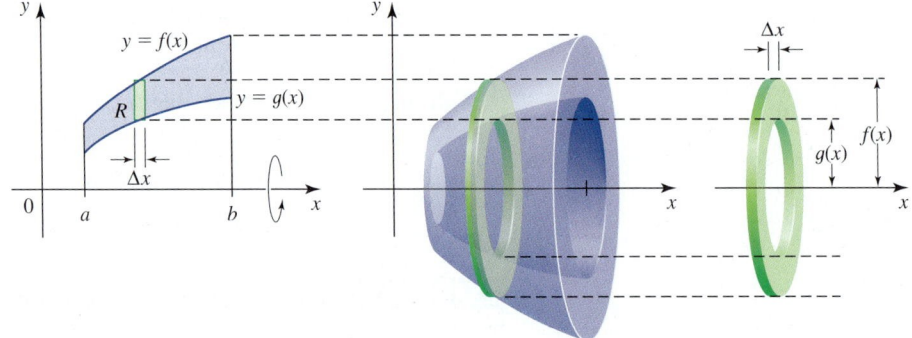

FIGURE 11
When a vertical rectangle is revolved about the x-axis, it generates a washer of outer radius $f(x)$, inner radius $g(x)$, and width Δx.

Summing the volumes of the washers and taking the limit, we see that the volume V of the solid S is given by the following.

Volume by Washer Method (Region revolved about the x-axis)

$$V = \pi \int_a^b \{ [f(x)]^2 - [g(x)]^2 \} \, dx \qquad f \geq g \geq 0$$

EXAMPLE 4   Find the volume of the solid obtained by revolving the region bounded by $y = \sqrt{x}$ and $y = x$ about the x-axis.

Solution   The region bounded by $y = \sqrt{x}$ and $y = x$ is shown in Figure 12. The curves $y = \sqrt{x}$ and $y = x$ intersect at $(0, 0)$ and $(1, 1)$, as may be verified by solving the equations simultaneously. The outer and inner radius of the washer generated by the representative vertical rectangle shown are $f(x) = \sqrt{x}$ and $g(x) = x$, respectively. Therefore, its volume is

$$\Delta V = \pi \{ [f(x)]^2 - [g(x)]^2 \} \Delta x$$

$$= \pi (x - x^2) \Delta x$$

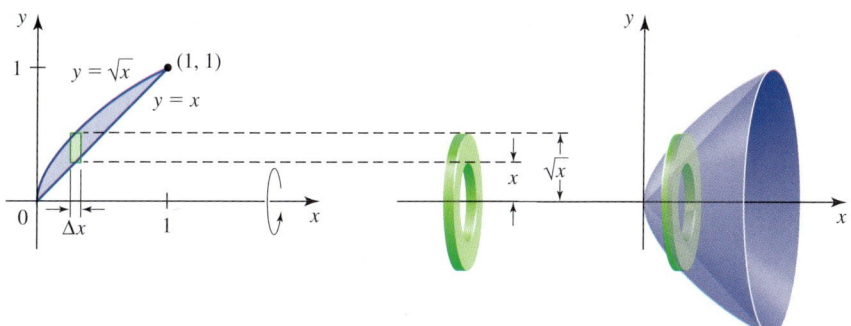

FIGURE 12
If a vertical rectangle is revolved about the *x*-axis, it generates a washer with outer radius $\sqrt{x}$, inner radius *x*, and width Δx.

Summing the volumes of the washers and taking the limit, we find that the required volume is

$$V = \int_0^1 \pi(x - x^2)\, dx$$

$$= \pi \int_0^1 (x - x^2)\, dx$$

$$= \pi\left[\frac{1}{2}x^2 - \frac{1}{3}x^3\right]_0^1 = \pi\left(\frac{1}{2} - \frac{1}{3}\right) \quad \text{or} \quad \frac{\pi}{6}$$

EXAMPLE 5 Find the volume of the solid generated by revolving the region of Example 4 about the line *y* = 2.

Solution The region and the resulting solid of revolution are shown in Figure 13. If a representative vertical rectangle is revolved about the line *y* = 2, the resultant solid is a washer with outer radius $2 - x$, inner radius $2 - \sqrt{x}$, and width Δx. Therefore, its volume is

$$\Delta V = \pi[(2 - x)^2 - (2 - \sqrt{x})^2]\Delta x$$

$$\pi[(\text{outer radius})^2 - (\text{inner radius})^2]\Delta x$$

$$= \pi(x^2 - 5x + 4\sqrt{x})\Delta x$$

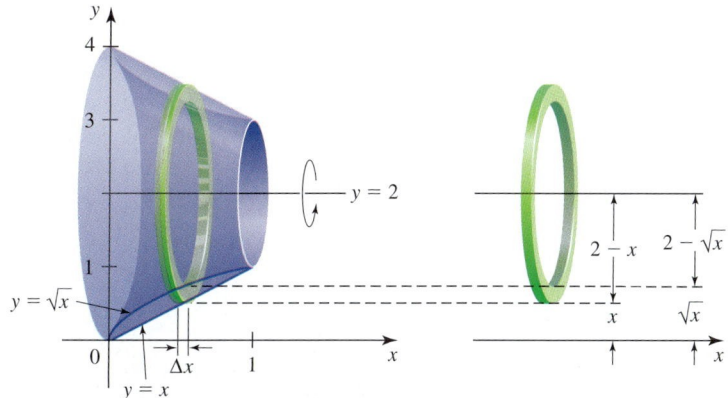

FIGURE 13
If a vertical rectangle is revolved about the line *y* = 2, it generates a washer with outer radius $2 - x$, inner radius $2 - \sqrt{x}$, and width Δx.

Summing the volumes of the washers and taking the limit, we find that the required volume is

$$V = \int_0^1 \pi(x^2 - 5x + 4\sqrt{x})\, dx$$

$$= \pi\left[\frac{1}{3}x^3 - \frac{5}{2}x^2 + \frac{8}{3}x^{3/2}\right]_0^1$$

$$= \pi\left(\frac{1}{3} - \frac{5}{2} + \frac{8}{3}\right) \quad \text{or} \quad \frac{\pi}{2} \qquad \blacksquare$$

EXAMPLE 6 Find the volume of the solid generated by revolving the region of Example 4 about the y-axis.

Solution   The region together with the solid of revolution is shown in Figure 14. When a horizontal rectangle is revolved about the y-axis, the resultant solid is a washer with outer radius y, inner radius y^2, and width Δy. Therefore, the volume of the solid is

$$\Delta V = \pi(y^2 - y^4)\Delta y$$

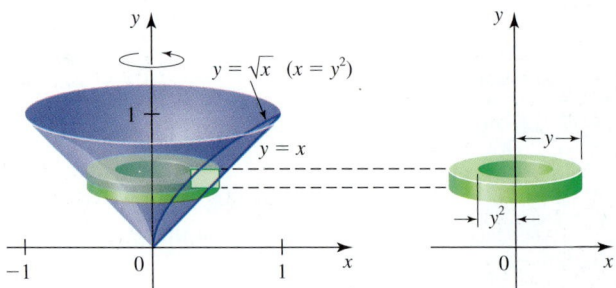

FIGURE 14

If a horizontal rectangle is revolved about the y-axis, it generates a washer with outer radius y, inner radius y^2, and width Δy.

Summing the volumes of the washers and taking the limit, we find that the volume of the solid is

$$V = \int_0^1 \pi(y^2 - y^4)\, dy = \pi\int_0^1 (y^2 - y^4)\, dy$$

$$= \pi\left[\frac{1}{3}y^3 - \frac{1}{5}y^5\right]_0^1 = \pi\left(\frac{1}{3} - \frac{1}{5}\right) \quad \text{or} \quad \frac{2\pi}{15} \qquad \blacksquare$$

■ The Method of Cross Sections

We now turn to the more general problem of defining the volume of an irregularly shaped object. Consider, for example, the solid that is the part of a pontoon that lies below the waterline. The side view of one such pontoon is shown in Figure 15. A cross section of the pontoon (by a plane perpendicular to the x-axis) at the point x is shown on the right.

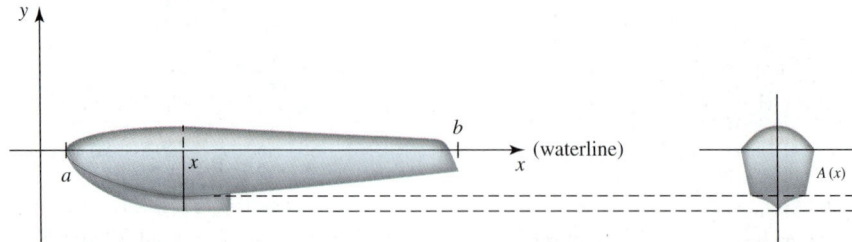

FIGURE 15

$A(x)$ is the area of a cross section of a pontoon at x.

To find the volume of the pontoon, let's take a regular partition $P = \{x_0, x_1, \ldots, x_n\}$ of the interval $[a, b]$. The planes that are perpendicular to the x-axis at the partition points will slice the pontoon into "slabs" much like the way one slices a loaf of bread. The volume ΔV of the kth slab between $x = x_{k-1}$ and $x = x_k$ is approximated by the volume of the cylinder with *constant* cross-sectional area $A(c_k)$ and height Δx, where c_k lies in $[x_{k-1}, x_k]$. (See Figure 16.) Thus,

$$\Delta V \approx A(c_k)\Delta x$$

FIGURE 16
The volume of the kth "slab" is approximately $A(c_k)\Delta x$.

If we add up these n terms, we obtain an approximation to the volume V of the pontoon. We can expect the approximations to get better and better as $n \to \infty$. Recognizing this sum to be the Riemann sum of the function $A(x)$ on the interval $[a, b]$, we are led to the following definition.

DEFINITION Volume of a Solid with Known Cross Section

Let S be a solid bounded by planes that are perpendicular to the x-axis at $x = a$ and $x = b$. If the cross-sectional area of S at any point x in $[a, b]$ is $A(x)$, where A is continuous on $[a, b]$, then the **volume** of S is

$$V = \lim_{n \to \infty} \sum_{k=1}^{n} A(c_k)\Delta x = \int_a^b A(x)\, dx \qquad (2)$$

EXAMPLE 7 A solid has a circular base of radius 2. Parallel cross sections of the solid perpendicular to its base are equilateral triangles. What is the volume of the solid?

Solution Suppose that the base of the solid is bounded by the circle with equation $x^2 + y^2 = 4$. The solid is shown in Figure 17a, where we have highlighted a typical cross section. To find the area of the cross section, observe that the base of the triangular cross section is $2y$, as shown in Figure 17b. Using the Pythagorean Theorem, we see that the height of the cross section is $\sqrt{3}y$ (see Figure 17c). Therefore, the area $A(x)$ of a typical cross section is

$$A(x) = \frac{1}{2}(2y)(\sqrt{3}y) = \sqrt{3}y^2 = \sqrt{3}(4 - x^2) \qquad \color{teal}{y^2 = 4 - x^2}$$

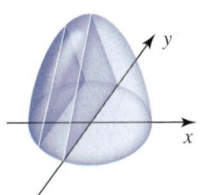 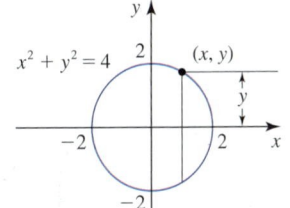 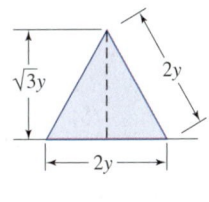

FIGURE 17   (a) The solid   (b) The base of a cross section   (c) A cross section

Using Formula (2), we see that the volume of the solid is

$$V = \int_{-2}^{2} A(x)\,dx = \int_{-2}^{2} \sqrt{3}(4 - x^2)\,dx = 2\int_{0}^{2} \sqrt{3}(4 - x^2)\,dx$$

The integrand is even.

$$= 2\sqrt{3}\left[4x - \frac{1}{3}x^3\right]_{0}^{2} = \frac{32\sqrt{3}}{3}$$

◾

EXAMPLE 8   Find the volume of a right pyramid with a square base of side b and height h.

Solution   Let's place the center of the base of the pyramid at the origin as shown in Figure 18a. A typical cross section of the pyramid perpendicular to the y-axis is a square of dimension $2x$ by $2x$. From Figure 18b we see by similar triangles that

$$\frac{x}{\dfrac{b}{2}} = \frac{h - y}{h}$$

or

$$x = \frac{b}{2h}(h - y)$$

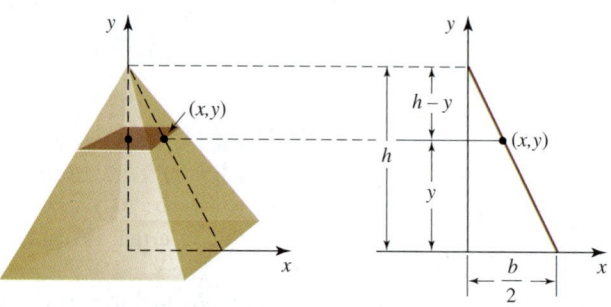

FIGURE 18   (a) A right pyramid   (b) A side view of the pyramid

Therefore, the area of the cross section is

$$A(y) = (2x)(2x) = 4x^2 = \frac{b^2}{h^2}(h - y)^2$$

The pyramid lies between $y = 0$ and $y = h$. Therefore, its volume is

$$V = \int_{0}^{h} A(y)\,dy = \int_{0}^{h} \frac{b^2}{h^2}(h - y)^2\,dy$$

$$= \left[-\frac{b^2}{3h^2}(h - y)^3\right]_{0}^{h} = 0 - \left(-\frac{b^2}{3h^2}\right)(h^3) = \frac{1}{3}b^2h$$

◾

EXAMPLE 9 The external fuel tank for a space shuttle has a shape that may be obtained by revolving the region under the curve

$$f(x) = \begin{cases} 4\sqrt{10} & \text{if } -120 \le x \le 10 \\ \dfrac{1}{5}\sqrt{x}(30 - x) & \text{if } 10 < x \le 30 \end{cases}$$

from $x = -120$ to $x = 30$ about the x-axis (Figure 19) where all measurements are given in feet. The tank carries liquid hydrogen for fueling the shuttle's three main engines. Estimate the capacity of the tank (231 cubic inches = 1 gallon).

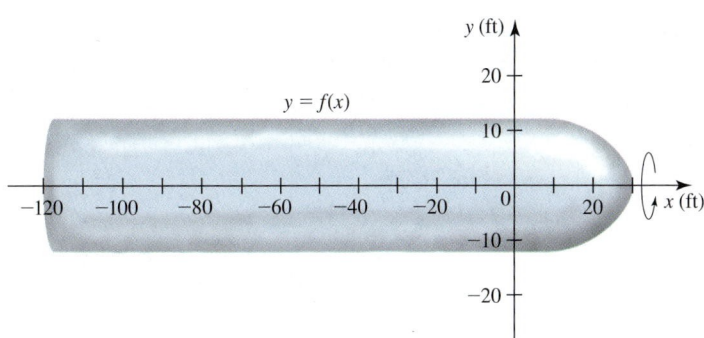

FIGURE 19
The solid of revolution obtained by revolving the region under the curve $y = f(x)$ about the x-axis

Solution   The volume of the tank is given by

$$V = \pi \int_{-120}^{30} [f(x)]^2 \, dx$$

$$= \pi \int_{-120}^{10} (4\sqrt{10})^2 \, dx + \pi \int_{10}^{30} \left[\frac{1}{5}\sqrt{x}(30 - x) \right]^2 dx$$

$$= 160\pi \int_{-120}^{10} dx + \frac{\pi}{25} \int_{10}^{30} x(30 - x)^2 \, dx$$

$$= 160\pi x \Big|_{-120}^{10} + \frac{\pi}{25} \int_{10}^{30} (900x - 60x^2 + x^3) \, dx$$

$$= 160\pi(130) + \frac{\pi}{25} \left(450x^2 - 20x^3 + \frac{1}{4}x^4 \right) \Big|_{10}^{30}$$

$$= 20{,}800\pi + \frac{\pi}{25} \left\{ \left[450(30)^2 - 20(30)^3 + \frac{1}{4}(30)^4 \right] \right.$$

$$\left. - \left[450(10)^2 - 20(10)^3 + \frac{1}{4}(10)^4 \right] \right\}$$

$$= 20{,}800\pi + 1600\pi = 22{,}400\pi$$

or approximately 70,372 cubic feet. Therefore, its capacity is approximately $(70{,}372)(12^3)/231$, or 526,419, gallons.   ▪

5.2 CONCEPT QUESTIONS

1. Write the integral that gives the volume of a solid of revolution using (a) the disk method and (b) the washer method. Illustrate each case graphically by drawing the region R, indicating the axis of revolution, and drawing a representative rectangle that helps you to derive the formula.

2. Write the integral that gives the volume of a solid using the method of cross sections.

5.2 EXERCISES

In Exercises 1–12, find the volume of the solid that is obtained by revolving the region about the indicated axis or line.

1.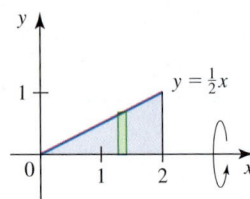

2.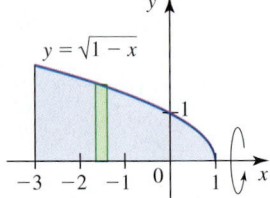

3.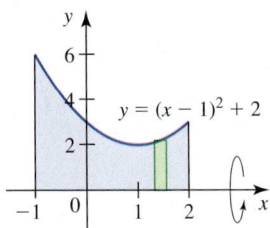

4.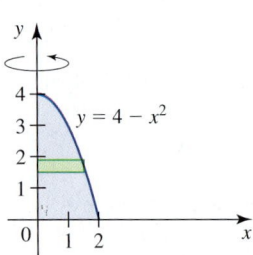

5.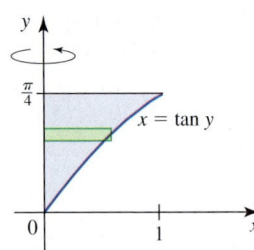

6.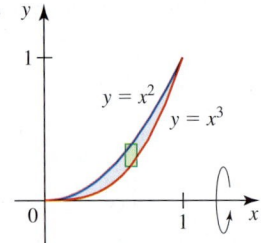

7.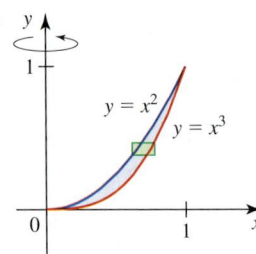

8.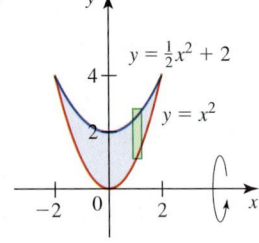

9.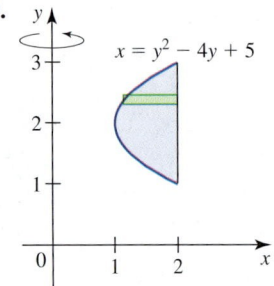

10.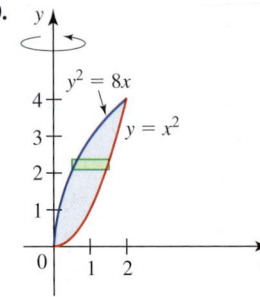

11.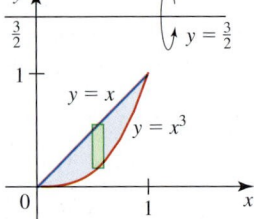

12.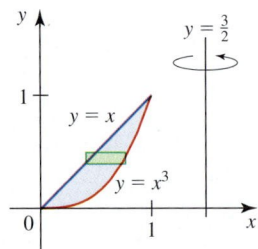

In Exercises 13–34, find the volume of the solid generated by revolving the region bounded by the graphs of the equations and/or inequalities about the indicated axis.

13. $y = x^2$, $y = 0$, $x = 2$; the x-axis

14. $y = x^3$, $y = 0$, $x = 1$; the x-axis

15. $y = -x^2 + 2x$, $y = 0$; the x-axis

16. $y = \sqrt{x - 1}$, $y = 0$, $x = 2$, $x = 5$; the x-axis

17. $y = e^x$, $y = 0$, $x = 0$, $x = 1$; the x-axis

18. $y = \dfrac{e^{x/2}}{(1 + e^x)^{3/2}}$, $y = 0$, $x = 0$, $x = 2$; the x-axis

19. $x = \dfrac{1}{y}$, $x = 0$, $y = 1$, $y = 2$; the y-axis

20. $x = y^{3/2}$, $x = 0$, $y = 1$; the y-axis

21. $x = \sqrt{4 - y^2}$, $x = 0$, $y = 0$; the y-axis

22. $x = -y^2 + 2y$, $x = 0$; the y-axis

23. $x^2 - y^2 = 4$, $x \geq 0$, $y = -2$, $y = 2$; the y-axis

24. $y = \dfrac{1}{\sqrt{x}(x^2 - 4)^{1/4}}$, $y = 0$, $x = 3$, $x = 4$; the y-axis

25. $x = y\sqrt{4 - y^2}$, $x = 0$; the y-axis

26. $y = \sqrt{\sin x}$, $y = 0$, $x \leq \dfrac{\pi}{2}$; the x-axis

27. $y = \cos x$, $x = 0$, $y = 0$, $x = \dfrac{\pi}{2}$; the x-axis

28. $y = x^2$, $y = x$; the x-axis

29. $y = x^2$, $y = \sqrt{x}$; the x-axis

30. $y = x^2$, $y = 2 - x^2$; the x-axis

31. $y = \ln x$, $y = 0$, $y = 1$, $x = 0$; the y-axis

32. $y = \sin^{-1} x$, $y = 0$, $y = \dfrac{\pi}{2}$, $x = 0$; the y-axis

33. $x^2 + y^2 = 1$, $y^2 = \dfrac{3}{2}x$, $y \geq 0$; the x-axis;

   (the smaller region)

34. $x^2 + y^2 = 1$, $y^2 = \dfrac{3}{2}x$; the y-axis; (the smaller region)

 In Exercises 35 and 36 use a graphing utility to (a) *plot the graphs of the given functions and* (b) *find the approximate x-coordinates of the points of intersection of the graphs. Then find an approximation of the volume of the solid obtained by revolving the region bounded by the graphs of the functions about the x-axis.*

35. $y = \dfrac{1}{2}x^5$, $y = 2x^2 - x^3$ 36. $y = x^5$, $y = \sin(x^2)$

In Exercises 37–42, find the volume of the solid generated by revolving the region bounded by the graphs of the equations about the indicated line.

37. $y = -x^2 + 2x$, $y = 0$; the line $y = 2$

38. $y = x$, $y = x^2$; the line $y = 2$

39. $y = 4 - x^2$, $y = 0$; the line $y = 5$

40. $y = x^2$, $y = \dfrac{1}{2}x^2 + 2$; the line $y = 5$

41. $x = y^2 - 4y + 5$, $x = 2$; the line $x = -1$

42. $y = x^2$, $y^2 = 8x$; the line $x = 2$

In Exercises 43–46, sketch a plane region, and indicate the axis about which it is revolved so that the resulting solid of revolution has the volume given by the integral. (The answer is not unique.)

43. $\pi \displaystyle\int_0^{\pi/2} \sin^2 x \, dx$ 44. $\pi \displaystyle\int_0^1 y^{2/3} \, dy$

45. $\pi \displaystyle\int_0^1 (x^2 - x^4) \, dx$ 46. $\pi \displaystyle\int_0^1 [(-1)^2 - (x^2 - 1)^2] \, dx$

47. Find the volume of the solid generated by revolving the region enclosed by the graph of $x^{1/2} + y^{1/2} = a^{1/2}$ and the coordinate axes about the x-axis.

48. **a.** Find the volume of the solid (a prolate spheroid) generated by revolving the upper half of the ellipse $9x^2 + 25y^2 = 225$ about the x-axis.
    b. Find the volume of the solid (an oblate spheroid) generated by revolving the right half of the ellipse $9x^2 + 25y^2 = 225$ about the y-axis.

49. Find the volume of the solid obtained by revolving the region enclosed by the curve $y^2 = \frac{1}{4}(2x^3 - x^4)$, where $y \geq 0$, about the x-axis.

50. Find the volume of the solid generated by revolving the region enclosed by the astroid $x^{2/3} + y^{2/3} = a^{2/3}$ about the x-axis.

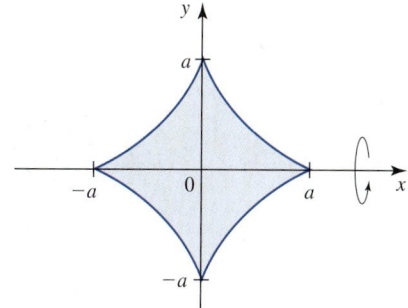

51. The function f is defined by
$$f = \begin{cases} \sqrt{x} & \text{if } 0 \leq x \leq 1 \\ x^2 - 2x + 2 & \text{if } 1 < x \leq 2 \end{cases}$$

Find the volume of the solid generated by revolving the region under the graph of f on $[0, 2]$ about the x-axis.

52. Verify the formula for the volume of a right circular cone by finding the volume of the solid obtained by revolving the triangular region with vertices $(0, 0)$, $(0, r)$, and $(h, 0)$ about the x-axis.

53. Find the volume of a frustum of a right circular cone with height h, lower base radius R, and upper radius r.

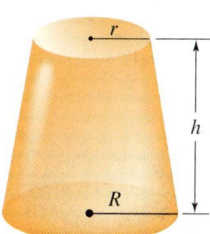

54. Verify the formula for the volume of a sphere of radius r by finding the volume of the solid obtained by revolving the region bounded by the graph of $x^2 + y^2 = r^2$, $x \geq 0$, and the y-axis about the y-axis.

55. Find the volume of a cap of height h formed from a sphere of radius r.

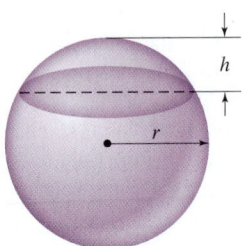

56. Newton's Wine Barrel Find the capacity of a wine barrel with the shape of a solid that is obtained by revolving the region bounded by the graphs of $x = R - ky^2$, $x = 0$, $y = -h/2$, and $y = h/2$ about the y-axis.

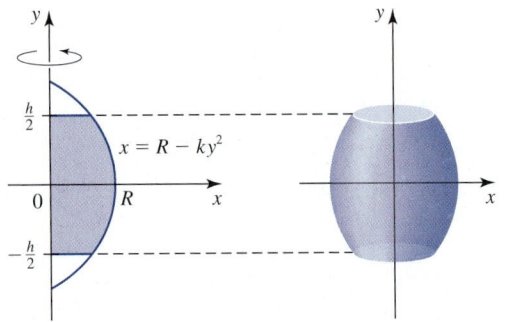

In Exercises 57–60, find the volume of the solid with the given base R and the indicated shape of every cross section taken perpendicular to the x-axis.

57. Cross section: a square

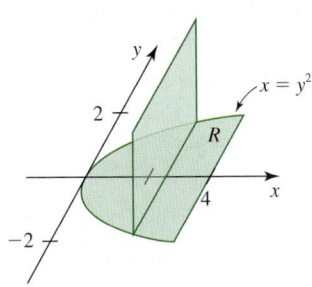

58. Cross section: a semicircle

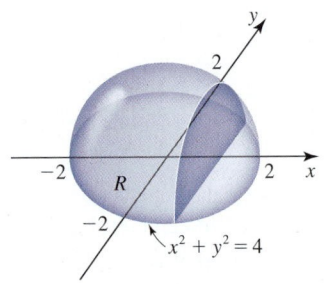

59. Cross section: an equilateral triangle

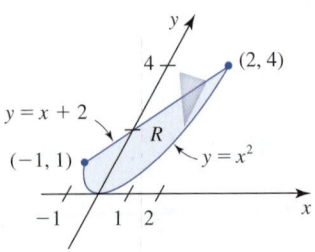

60. Cross section: a quarter circle

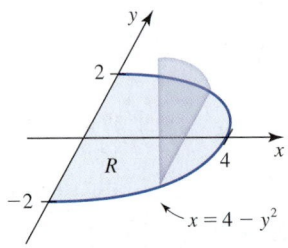

61. The curve defined by $y^4 = 1 - |x/2|^4$ is called a **hyperellipse.**

 a. Plot the curve using the viewing window $[-3, 3] \times [-2, 2]$.

b. Estimate the volume V of the solid obtained by revolving the region enclosed by the hyperellipse for $y \geq 0$ about the x-axis.

c. Use a calculator or computer to find V accurate to four decimal places.

Hint: The hyperellipse is almost rectangular in shape.

62. A solid has a circular base of radius 2, and its parallel cross sections perpendicular to its base are rectangles of height 2. Find the volume of the solid.

63. The curve defined by $2y^2 - x^3 - x^2 = 0$ is called a **Tschirnhausen's cubic.**

 a. Plot the curve using the viewing window $[-1.5, 1.5] \times [-1.5, 1.5]$.

b. Find the volume of the solid obtained by revolving the region enclosed by the loop of the curve about the x-axis.

64. Find the volume of the solid obtained by revolving the region bounded by the bullet-nosed curve $y = \dfrac{|x|}{\sqrt{2 - x^2}}$ about the y-axis for $0 \leq y \leq 12$.

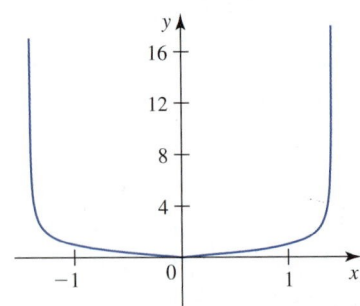

65. A solid has a circular base of radius 2, and its parallel cross sections perpendicular to its base are isosceles right triangles oriented so that the endpoints of the hypotenuse of a triangle lie on the circle. Find the volume of the solid.

66. The base of a solid is the region bounded by the graphs of $y = 4 - x^2$ and $y = 0$. The cross sections perpendicular to the y-axis are equilateral triangles. Find the volume of the solid.

67. The base of a wooden wedge is in the form of a semicircle with radius a, and its top is a plane that passes through the diameter of the base and makes a 45° angle with the plane of the base. Find the volume of the wedge.

68. The axes of two right cylinders, each of radius r, intersect at right angles. Find the volume of the resulting solid that is common to both cylinders. (The figure shows one eighth of the solid.)

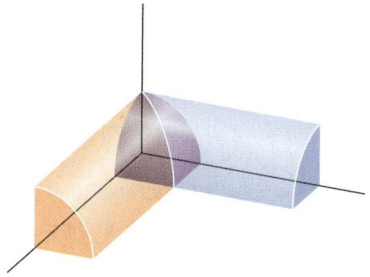

69. Cavalieri's Theorem Cavalieri's Theorem states that if two solids have equal altitudes and all cross sections parallel to their bases and at equal distance from their bases have the same area, then the solids have the same volume.

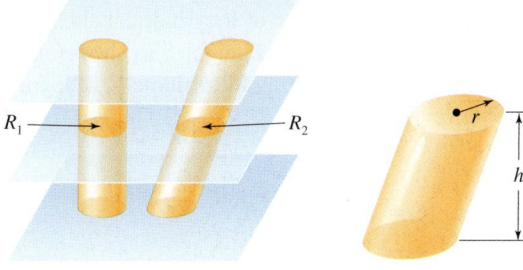

(**a**) area of R_1 = area of R_2 (**b**) An oblique circular cylinder

  a. Prove Cavalieri's Theorem.
  b. Use Cavalieri's Theorem to find the volume of the oblique circular cylinder shown in part (b) of the figure.

70. Capacity of a Fuel Tank The external fuel tank for a fighter aircraft is 8 m long. The areas of the cross sections in square meters measured from the front to the back of the tank at 1-m intervals are summarized in the following table.

x (distance from front)	0	1	2	3	4
$A(x)$	0	0.3041	0.6206	0.8937	0.8937

x (distance from front)	5	6	7	8
$A(x)$	0.8937	0.6206	0.3041	0

Use Simpson's Rule to estimate the capacity (in liters) of the fuel tank.

71. The Volume of a Pontoon A pontoon is 12 ft long. The areas of the cross sections in square feet measured from the blueprint at intervals of 2 ft from the front to the back of the part of the pontoon that is under the waterline are summarized in the following table.

x	0	2	4	6	8	10	12
$A(x)$	0	3.82	4.78	3.24	2.64	1.80	0

Use Simpson's Rule to estimate the volume of the pontoon.

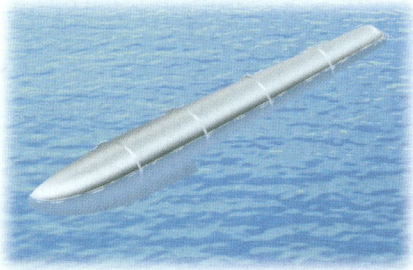

72. a. Let S be a solid bounded by planes that are perpendicular to the x-axis at $x = 0$ and $x = h$. If the cross-sectional area of S at any point x in $[0, h]$ is $A(x)$, where A is a polynomial of degree less than or equal to three, show that the volume of the solid is

$$V = \frac{h}{6}\left[A(0) + 4A\left(\frac{h}{2}\right) + A(h)\right]$$

  b. Use the result of part (a) to verify the result of Exercise 54.

5.3 | Volumes Using Cylindrical Shells

In Section 5.2 we saw how the volume of a solid of revolution can be found by using the method of disks or the method of washers. Sometimes these methods are difficult or inconvenient to use. For example, suppose that we want to find the volume of the solid generated by revolving the region R bounded by the graphs of the equations $y = -x^3 + 3x^2$, $y = 0$, $x = 0$, and $x = 3$ about the y-axis. (See Figure 1.)

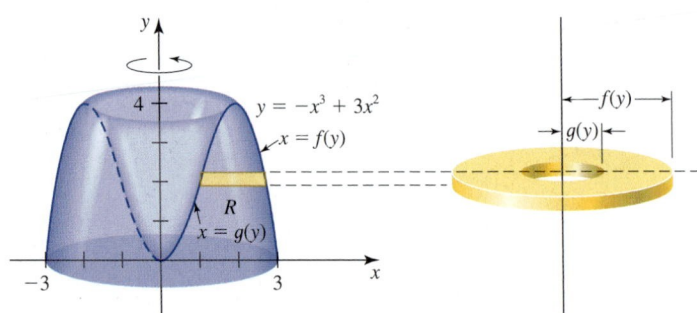

FIGURE 1
The washer generated by revolving the representative horizontal rectangle about the y-axis has outer radius $f(y)$ and inner radius $g(y)$.

As you can see from the figure, $f(y)$ is the outer radius and $g(y)$ is the inner radius of the washer generated by revolving a representative horizontal rectangle about the y-axis. Therefore, the volume of the solid is given by

$$\pi \int_0^b \{[f(y)]^2 - [g(y)]^2\}\, dy$$

where b is the maximum value of $F(x) = -x^3 + 3x^2$ on $[0, 3]$. Using the techniques of Section 3.1, we can show that $b = 4$. So finding the interval of integration does not, at least in this case, pose much difficulty. But finding the functions f and g is an entirely different matter. Here, we need to solve the cubic equation $x^3 - 3x^2 + y = 0$ for x, a far more complicated task. Fortunately, there is another method that will allow us to find the volumes of such solids with relative ease. We will complete the solution of this problem in Example 1 after introducing the *method of cylindrical shells*.

■ The Method of Cylindrical Shells

As the name suggests, the method of cylindrical shells makes use of the volumes of cylindrical shells (or tubes) to approximate the volume of a solid of revolution. We begin with the derivation of an expression for the volume of a cylindrical shell.

Suppose a shell has outer radius r_2, inner radius r_1, and height h as shown in Figure 2. The volume V of the shell can be found by subtracting the volume V_1 of the inner cylinder from the volume V_2 of the outer cylinder. Thus,

$$V = V_2 - V_1$$
$$= \pi r_2^2 h - \pi r_1^2 h = \pi (r_2^2 - r_1^2)h$$
$$= \pi (r_2 + r_1)(r_2 - r_1)h$$
$$= 2\pi \left(\frac{r_2 + r_1}{2}\right)(r_2 - r_1)h$$

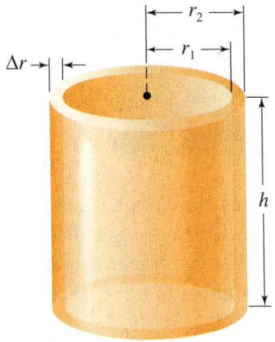

FIGURE 2
A cylindrical shell of outer radius r_2, inner radius r_1, and height h

This last equation can be written in the form

$$V = 2\pi r h \, \Delta r \qquad \qquad \textbf{(1)}$$

where $r = (r_1 + r_2)/2$ is the average radius of the shell and $\Delta r = r_2 - r_1$ is the thickness of the shell. Formula (1) may also be written in the following form.

Volume of a Cylindrical Shell

$$V = 2\pi(\text{average radius})(\text{height})(\text{thickness})$$

Now let R be the region under the graph of f on the interval $[a, b]$, where $a \geq 0$, shown in Figure 3a. If this region R is revolved about the y-axis, we obtain the solid shown in Figure 3b.

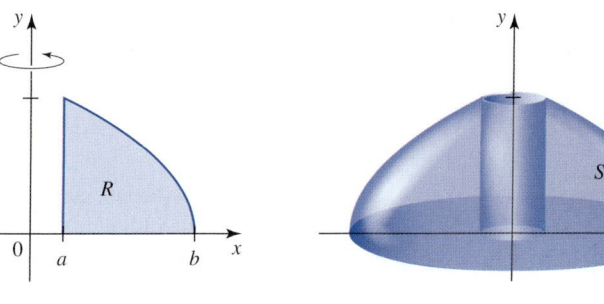

FIGURE 3   (**a**) The region R (**b**) The solid S

Let $P = \{x_0, x_1, x_2, \ldots, x_n\}$ be a regular partition of the interval $[a, b]$, and let c_k be the midpoint of the subinterval $[x_{k-1}, x_k]$; that is,

$$c_k = \frac{1}{2}(x_k + x_{k-1})$$

If the vertical rectangle with base $[x_{k-1}, x_k]$ and height $f(c_k)$ is revolved about the y-axis, we obtain a cylindrical shell with average radius c_k, height $f(c_k)$, and thickness $\Delta x = (b - a)/n$. (See Figure 4.) Therefore, by Formula (1) the volume of the shell is

$$\Delta V_k = 2\pi c_k f(c_k) \Delta x$$

FIGURE 4
If the vertical rectangle in (a) is revolved about the y-axis, we obtain the cylindrical shell (b). The volume of S is approximated by the volumes of the nested shells (c).

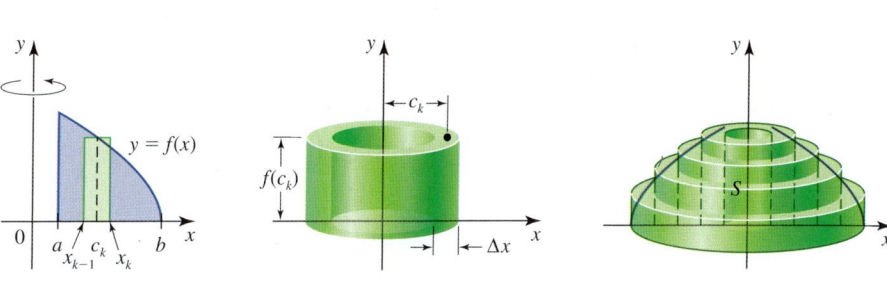

(**a**) A representative rectangle   (**b**) A cylindrical shell   (**c**) Nested shells

The volume V of S is approximated by the sum of the volumes of these shells (Figure 4c). Thus,

$$V \approx \sum_{k=1}^{n} \Delta V_k = \sum_{k=1}^{n} 2\pi c_k f(c_k)\Delta x$$

Recognizing this sum to be the Riemann sum of the function $2\pi x f(x)$ on the interval $[a, b]$, we see that

$$\lim_{n\to\infty} \sum_{k=1}^{n} 2\pi c_k f(c_k)\Delta x = \int_{a}^{b} 2\pi x f(x)\, dx = 2\pi \int_{a}^{b} x f(x)\, dx$$

This discussion leads to the following definition.

Method of Cylindrical Shells (Region revolved about the y-axis)

Let f be a continuous nonnegative function on $[a, b]$, where $0 \le a \le b$, and let R be the region under the graph of f on the interval $[a, b]$. The volume V of the solid of revolution generated by revolving R about the y-axis is

$$V = \lim_{n\to\infty} \sum_{k=1}^{n} 2\pi c_k f(c_k)\Delta x = \int_{a}^{b} 2\pi x f(x)\, dx \qquad (2)$$

As before, there is a convenient aid to help us recall this method. Draw a representative vertical rectangle of height $f(x)$, or y, and width Δx. Here we pick x to be the midpoint of the base of the rectangle. Observe that this rectangle is parallel to the axis of revolution. When this rectangle is revolved about the y-axis, it generates a cylindrical shell of radius x, height $f(x)$, and thickness Δx. (See Figure 5.) Therefore, its volume is

$$\Delta V = 2\pi x f(x)\Delta x$$

Summing the volume of the shells and taking the limit, we obtain

$$V = \int_{a}^{b} 2\pi x f(x)\, dx$$

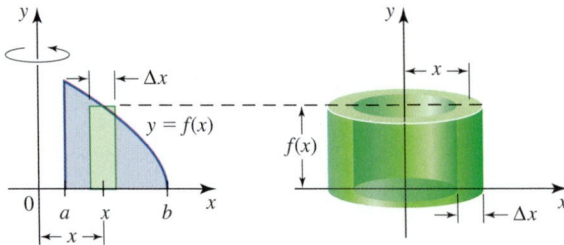

FIGURE 5
If a representative vertical rectangle is revolved about the y-axis, it generates a cylindrical shell of radius x, height $f(x)$, and thickness Δx and hence volume $\Delta V = 2\pi x f(x)\Delta x$.

■ Applying the Method of Cylindrical Shells

EXAMPLE 1   The region under the graph of $y = -x^3 + 3x^2$ on $[0, 3]$ is revolved about the y-axis. Find the volume of the resulting solid.

Solution   The region and the resulting solid of revolution are shown in Figure 6. If a representative vertical rectangle is revolved about the y-axis, the resulting cylindrical shell has an average radius x, height $-x^3 + 3x^2$, and thickness Δx. Therefore, its volume is

$$\Delta V = 2\pi x(-x^3 + 3x^2)\Delta x$$
$$= 2\pi(-x^4 + 3x^3)\Delta x$$

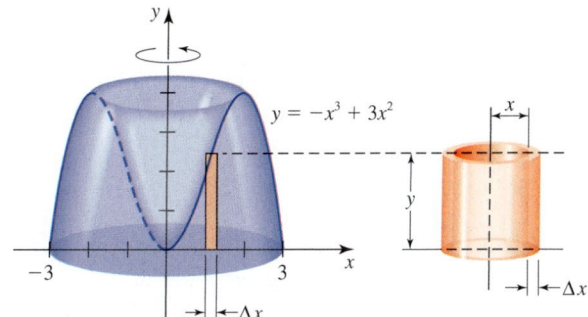

FIGURE 6

If a representative rectangle is revolved about the y-axis, it generates a cylindrical shell of volume $\Delta V = 2\pi xy\,\Delta x$.

Summing the volumes of the cylindrical shells and taking the limit, we find that the volume of the solid is

$$V = \int_0^3 2\pi(-x^4 + 3x^3)\,dx = 2\pi\int_0^3 (-x^4 + 3x^3)\,dx$$

$$= 2\pi\left[-\frac{1}{5}x^5 + \frac{3}{4}x^4\right]_0^3$$

$$= 2\pi\left(-\frac{243}{5} + \frac{243}{4}\right) \quad \text{or} \quad \frac{243\pi}{10}$$

Sometimes one method is preferable to another. In the next example the method of cylindrical shells is more convenient to use than the method of washers.

EXAMPLE 2   Let R be the region bounded by the graphs of $y = x^2 + 1$, $y = -x + 1$, and $x = 1$. Find the volume of the solid that is obtained by revolving R about the y-axis using (a) the method of washers and (b) the method of cylindrical shells.

Solution   The region R is shown in Figure 7a.

FIGURE 7

If each of the horizontal rectangles in part (b) is revolved about the y-axis, the resulting solid is a washer. If the vertical rectangle in part (c) is revolved about the y-axis, the resulting solid is a cylindrical shell.

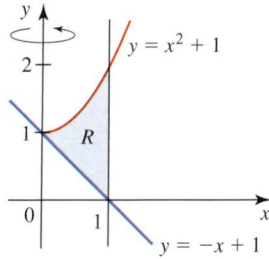

(a) The region R (b) The method of washers   (c) The method of shells

a. To use the method of washers, we regard the region R as being made up of two subregions R_1 and R_2. (See Figure 7b.) Observe that if a representative horizontal rectangle lying in R_1 is revolved about the y-axis, we obtain a washer with outer radius $x = 1$ and inner radius $x = 1 - y$ (obtained by solving the equation $y = -x + 1$ for x). Therefore, its volume is

$$\Delta V_1 = \pi\{[f(y)]^2 - [g(y)]^2\}\Delta y$$

$$= \pi[1 - (1 - y)^2]\Delta y \quad \text{Here,} f(y) = 1 \text{ and } g(y) = 1 - y.$$

$$= \pi(2y - y^2)\Delta y$$

Summing the volumes of the washers and taking the limit, we see that the volume of the solid obtained by revolving the subregion R_1 about the y-axis is

$$V_1 = \int_0^1 \pi(2y - y^2)\, dy$$

Similarly, we see that if a representative horizontal rectangle lying in R_2 is revolved about the y-axis, we obtain a washer with outer radius $x = 1$ and inner radius $x = \sqrt{y - 1}$ (obtained by solving the equation $y = x^2 + 1$ for x), where $x \geq 0$. Therefore, its volume is

$$\Delta V_2 = \pi\{[f(y)]^2 - [g(y)]^2\}\Delta y$$

$$= \pi[1 - (\sqrt{y - 1})^2]\Delta y \quad \text{Here,} f(y) = 1 \text{ and } g(y) = \sqrt{y - 1}.$$

$$= \pi(2 - y)\Delta y$$

Summing the volumes of the washers and taking the limit, we see that the volume of the solid obtained by revolving the subregion R_2 about the y-axis is

$$V_2 = \int_1^2 \pi(2 - y)\, dy$$

Therefore, the required volume is

$$V_1 + V_2 = \int_0^1 \pi(2y - y^2)\, dy + \int_1^2 \pi(2 - y)\, dy$$

$$= \pi\left[y^2 - \frac{1}{3}y^3\right]_0^1 + \pi\left[2y - \frac{1}{2}y^2\right]_1^2$$

$$= \pi\left(1 - \frac{1}{3}\right) + \pi\left\{\left[2(2) - \frac{1}{2}(4)\right] - \left[2 - \frac{1}{2}\right]\right\} = \frac{7\pi}{6}$$

b. If a representative vertical rectangle is revolved about the y-axis, the resulting cylindrical shell has an average radius of x, height $[(x^2 + 1) - (-x + 1)]$, or $x^2 + x$, and thickness Δx (Figure 7c). Therefore, its volume is

$$\Delta V = 2\pi x(x^2 + x)\Delta x$$

$$= 2\pi(x^3 + x^2)\Delta x$$

Summing the volumes of the cylindrical shells and taking the limit, we find that the volume of the solid is

$$V = \int_0^1 2\pi(x^3 + x^2)\, dx = 2\pi \int_0^1 (x^3 + x^2)\, dx$$

$$= 2\pi\left[\frac{1}{4}x^4 + \frac{1}{3}x^3\right]_0^1 = 2\pi\left(\frac{1}{4} + \frac{1}{3}\right) \quad \text{or} \quad \frac{7\pi}{6} \quad \blacksquare$$

Note Figure 7 reveals, once again, an intrinsic difference between the method of washers and the method of cylindrical shells. In the method of washers a representative rectangle is always *perpendicular* to the axis of revolution of the solid. In the method of cylindrical shells a representative rectangle is always *parallel* to the axis of revolution.

◾ Shells Generated by Revolving a Region About the *x*-axis

The method of cylindrical shells can also be used to find the volume of a solid obtained by revolving a region about the *x*-axis. For example, suppose that the region R bounded by the graphs of $x = f(y)$, $x = 0$, $y = c$, and $y = d$, where $f \geq 0$ and $c \leq d$, is revolved about the *x*-axis. (See Figure 8.) Then the volume of the resulting solid is given by the following formula.

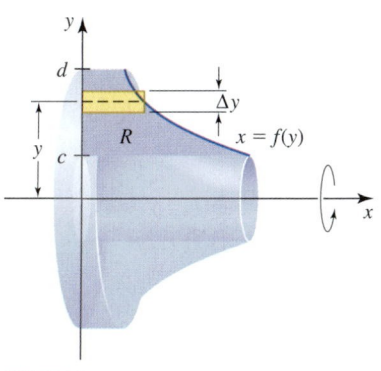

FIGURE 8
If a horizontal rectangle is revolved about the *x*-axis, it generates a cylindrical shell of volume $\Delta V = 2\pi y f(y) \Delta y$.

Volume by Cylindrical Shells (Region revolved about the *x*-axis)

$$V = \int_c^d 2\pi y f(y)\, dy \tag{3}$$

Equation (3) follows from Equation (2) if we interchange the roles of x and y. It also follows from this observation: The solid generated by revolving the representative rectangle shown in Figure 8 is a cylindrical shell of average radius y, height $f(y)$, thickness Δy, and therefore volume $\Delta V = 2\pi y f(y) \Delta y$. Summing the volumes of the shells and taking the limit, we obtain

$$V = \int_c^d 2\pi y f(y)\, dy$$

EXAMPLE 3 Let R be the region bounded by the graphs of $x = -y^2 + 6y$ and $x = 0$. Find the volume of the solid obtained by revolving R about the *x*-axis.

Solution The region R is shown in Figure 9. If a representative horizontal rectangle is revolved about the *x*-axis, the resulting cylindrical shell has an average radius of y, a height of x, or $-y^2 + 6y$, and a thickness Δy. Therefore, its volume is

$$\Delta V = 2\pi y(-y^2 + 6y)\Delta y$$
$$= 2\pi(-y^3 + 6y^2)\Delta y$$

Summing the volumes of the cylindrical shells and taking the limit, we find the volume of the solid to be

$$V = \int_0^6 2\pi(-y^3 + 6y^2)\, dy = 2\pi \int_0^6 (-y^3 + 6y^2)\, dy$$

$$= 2\pi\left[-\frac{1}{4}y^4 + 2y^3 \right]_0^6 = 2\pi(-324 + 432) = 216\pi$$

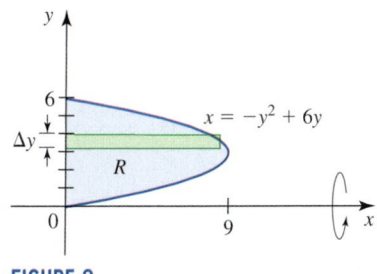

FIGURE 9
If a horizontal rectangle is revolved about the *x*-axis, it generates a cylindrical shell of volume $\Delta V = 2\pi y x \, \Delta y$.

EXAMPLE 4 Let R be the region bounded by the graphs of the equations $y = 4 - x^2$ and $y = -x + 2$. Find the volume of the solid obtained by revolving R about the line $x = 4$.

Solution The region R is shown in Figure 10. If a representative vertical rectangle is revolved about the line $x = 4$, it generates a cylindrical shell of average radius $4 - x$, height $(4 - x^2) - (-x + 2)$ or $-x^2 + x + 2$, and thickness Δx. Therefore, its volume is $\Delta V = 2\pi(4 - x)(-x^2 + x + 2)\Delta x$. Summing the volumes of the cylindrical shells and taking the limit, we find the volume of the solid to be

$$V = \int_{-1}^{2} 2\pi(4 - x)(-x^2 + x + 2)\, dx = 2\pi \int_{-1}^{2} (x^3 - 5x^2 + 2x + 8)\, dx$$

$$= 2\pi \left[\frac{1}{4}x^4 - \frac{5}{3}x^3 + x^2 + 8x \right]_{-1}^{2}$$

$$= 2\pi \left[\left(4 - \frac{40}{3} + 4 + 16 \right) - \left(\frac{1}{4} + \frac{5}{3} + 1 - 8 \right) \right] = \frac{63\pi}{2}$$

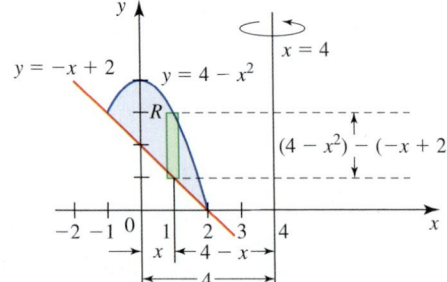

FIGURE 10
If a vertical rectangle is revolved about the line $x = 4$, it generates a cylindrical shell with average radius $4 - x$, height $(4 - x^2) - (-x + 2)$, and thickness Δx.

5.3 CONCEPT QUESTIONS

1. Let S be the solid obtained by revolving the region shown in the figure about the y-axis.
   a. Sketch representative horizontal rectangles, and use them to help you set up the integrals giving the volume of S using the disk and/or washer method.
   b. Sketch a representative vertical rectangle, and use it to help you set up an integral giving the volume of S using the shell method.
   c. Find the volume of S. Which method is easier?

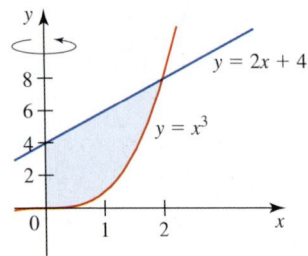

2. Let S be the solid that is obtained by revolving the region shown in the figure about the y-axis.
   a. Is it desirable to use the disk method to find the volume of S? Explain.
   b. Use the shell method to find the volume of S.

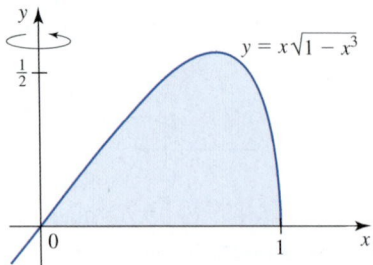

5.3 EXERCISES

In Exercises 1–6, use the method of cylindrical shells to find the volume of the solid generated by revolving the region about the indicated axis or line.

1.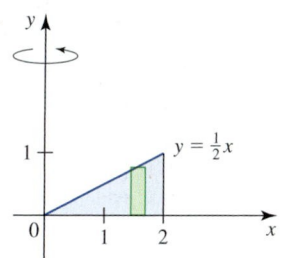
$y = \frac{1}{2}x$

2.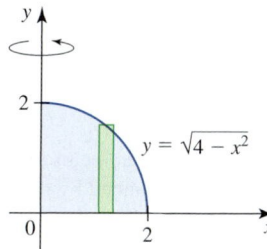
$y = \sqrt{4 - x^2}$

3.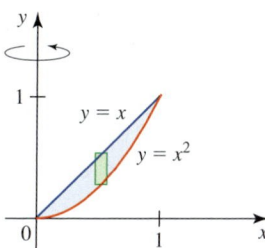
$y = x$
$y = x^2$

4.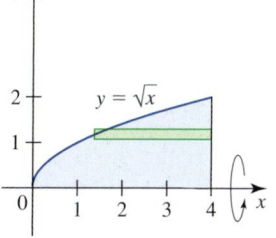
$y = \sqrt{x}$

5.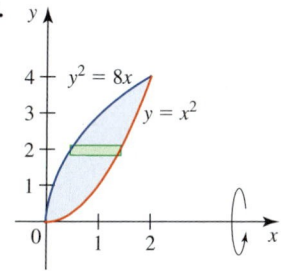
$y^2 = 8x$
$y = x^2$

6.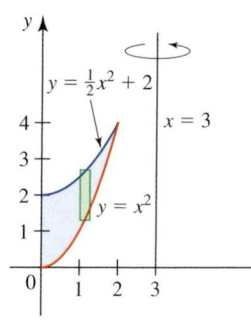
$y = \frac{1}{2}x^2 + 2$
$x = 3$
$y = x^2$

In Exercises 7–22, use the method of cylindrical shells to find the volume of the solid generated by revolving the region bounded by the graphs of the equations and/or inequalities about the indicated axis. Sketch the region and a representative rectangle.

7. $y = x^2$, $y = 0$, $x = 2$; the y-axis

8. $y = x^3$, $y = 0$, $x = 1$; the y-axis

9. $y = -x^2 + 2x$, $y = 0$; the y-axis

10. $y = \sqrt{x - 1}$, $y = 0$, $x = 5$; the y-axis

11. $y = \frac{1}{x}$, $y = 0$, $x = 1$, $x = 2$; the y-axis

12. $y = e^{-x^2}$, $y = 0$, $x = 0$, $x = 1$; the y-axis

13. $y = \frac{1}{x^2 + 1}$, $x = 0$, $x = 2$; the y-axis

14. $y = \frac{1}{x(x^2 + 4)}$, $x = 1$, $x = 2$; the y-axis

15. $x = \sqrt{9 - y^2}$, $x = 0$, $y = 0$; the x-axis

16. $y = 3^{x^2}$, $y = 0$, $x = 0$, $x = 1$; the y-axis

17. $y = x^2 + 1$, $x \geq 0$, $y = 5$; the y-axis

18. $y = x$, $y = \frac{1}{2}x^2$; the y-axis

19. $y = \sqrt{1 - x^2}$, $y = -x + 1$; the y-axis

20. $y = \sin x^2$, $y = 0$, $x = 0$, $x = \sqrt{\pi}$; the y-axis

21. $y^2 = \frac{3}{2}x$; $x^2 + y^2 = 1$, $y \geq 0$; the x-axis;
(the smaller region)

22. $y = \sqrt{x - 1}$, $y = x - 1$; the y-axis

In Exercises 23 and 24, use a graphing utility to (a) plot the graphs of the given functions, (b) find the approximate x-coordinates of the points of intersection of the graphs, and (c) find an approximation of the volume of the solid obtained by revolving the region bounded by the graphs of the functions about the y-axis.

23. $y = x$, $y = x^5 - x^2$; $x \geq 0$

24. $y = \sin x$, $y = x^2$

In Exercises 25–30, use the method of disks or washers, or the method of cylindrical shells to find the volume of the solid generated by revolving the region bounded by the graphs of the equations about the indicated axis. Sketch the region and a representative rectangle.

25. $y = \sqrt{x}$, $y = x - 2$, $y = 0$; the x-axis

26. $y = (x - 1)^2$, $y = x + 1$; the x-axis

27. $y = x^2$, $y = 2x - 1$, $y = 4$; the y-axis

28. $y = \sqrt{1 - x^2}$, $y = -x + 1$; the x-axis

29. $y = 2x^2$, $y = x + 1$, $y = 0$; the x-axis

30. $y = \sqrt{9 - x^2}$, $y = \frac{2}{3}\sqrt{9 - x^2}$, $x \geq 0$; the y-axis

In Exercises 31–36, find the volume of the solid generated by revolving the region bounded by the graphs of the equations about the indicated line. Sketch the region and a representative rectangle.

31. $y = x$, $y = 0$, $x = 2$; the line $x = 4$

32. $y = x^2 + 1$, $y = 0$, $x = 0$, $x = 2$; the line $x = 3$

33. $y = 4 - x^2$, $y = 0$; the line $x = -2$

34. $y = \sqrt{x}$, $y = 0$, $x = 4$; the line $y = 2$

35. $y = \sqrt{x - 1}$, $y = x - 1$; the line $x = 3$

36. $y = x$, $y = x^2$; the line $y = 2$

In Exercises 37–40, sketch a plane region and indicate the axis about which it is revolved so that the resulting solid of revolution (found using the shell method) is given by the integral. (Answers may not be unique.)

37. $2\pi \displaystyle\int_0^\pi x \sin x \, dx$

38. $2\pi \displaystyle\int_0^1 y^{4/3} \, dy$

39. $2\pi \displaystyle\int_0^1 y(y^{1/3} - y) \, dy$

40. $2\pi \displaystyle\int_0^1 (x + 1)x^2 \, dx$

41. Verify the formula for the volume of a right circular cone by applying the method of cylindrical shells to find the volume of the solid obtained by revolving the triangular region with vertices $(0, 0)$, $(h, 0)$, and $(0, r)$ about the x-axis.

42. Verify the formula for the volume of a sphere of radius r by applying the method of cylindrical shells to find the volume of the solid obtained by revolving the semicircular region $x^2 + y^2 = r^2$, where $x \geq 0$, about the y-axis.

43. Use the method of cylindrical shells to find the volume of the ellipsoid obtained by revolving the elliptical region enclosed by the graph of

$$\frac{x^2}{a^2} + \frac{y^2}{b^2} = 1 \qquad x \geq 0$$

about the y-axis.

44. Find the volume of the solid that remains after a circular hole of radius a is bored through the center of a solid sphere of radius $r > a$.

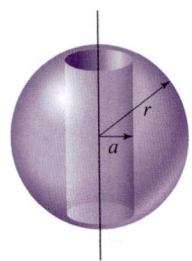

45. A torus (a doughnut-shaped object) is formed by revolving the circle $x^2 + y^2 = a^2$ about the vertical line $x = b$, where $0 < a < b$. Find its volume.

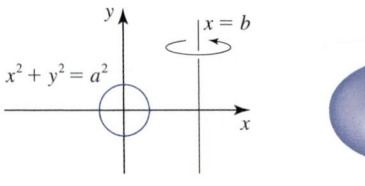
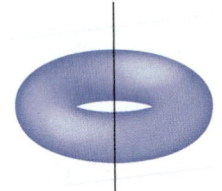

46. Find the volume of the solid obtained by revolving the region bounded by the graphs of $y = \sin x^2$ and $y = \cos x^2$ on $\left[0, \frac{\pi}{2}\right]$ about the y-axis.

47. Volume of Liquid in a Rotating Container A cylindrical container of radius 2 ft and height 4 ft is partially filled with a liquid. When the container is rotated about its axis of symmetry at a constant angular speed ω, the surface of the liquid assumes a parabolic cross section. Suppose that the parabola is given by

$$y = 2 + \frac{\omega^2 x^2}{2g}$$

Find the volume of the liquid in the rotating container if the liquid reaches a height of 3 ft on the side of the container.

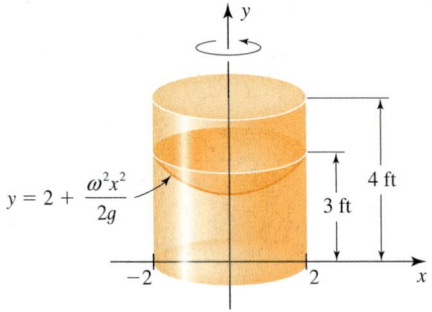

48. The Clepsydra or Water Clock A container having the shape of a solid of revolution obtained by revolving the graph of $y = kx^4$, $k > 0$, about the y-axis is made with a transparent material (see the following figure). A small hole is drilled in the bottom of the container to allow water to flow out.
   a. Find the volume $V(h)$ of water in the container as a function of h, the height of the water at time t.
   b. Use the Chain Rule and Torricelli's Law—which states that the rate of flow of water is $dV/dt = kA\sqrt{h}$, where k is a negative constant, A is the area of the hole at the bottom of the container, and h is the height of the water—to show that the water level in the container drops at a *constant* rate.
   c. Explain why this property allows us to construct a water clock.

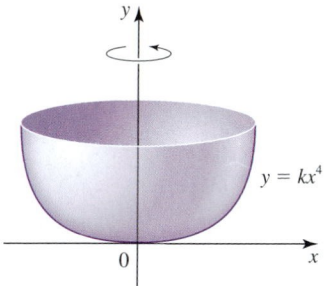

$y = kx^4$

49. Capacity of an Artificial Lake A circular artificial lake has a diameter of 4000 ft. The following figure gives the depth of the water measured in 200-ft intervals starting from the center of the lake. Assume that readings taken along other radial directions produce similar data, so that the capacity of the lake can be approximated by the volume of the solid obtained by revolving the region bounded above by the x-axis and below by the graph of $y = f(x)$ about the y-axis. Use Simpson's Rule to approximate this capacity.

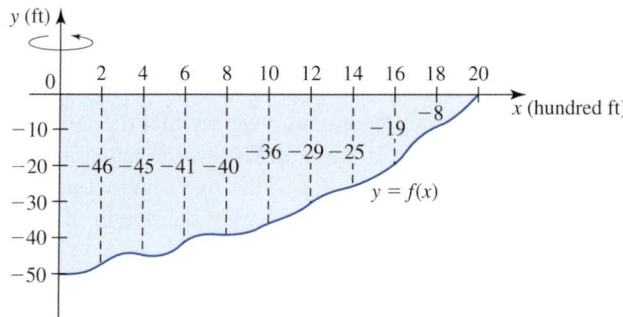

50. Land Reclamation A hill will be leveled, and the earth will be used in a land reclamation project that includes the construction of additional landing strips for an existing airport. The hill resembles the solid of revolution obtained by revolving the region under the graph of the function f on [0, 240] about the y-axis. Use Simpson's Rule to approximate the amount of earth that can be recovered.

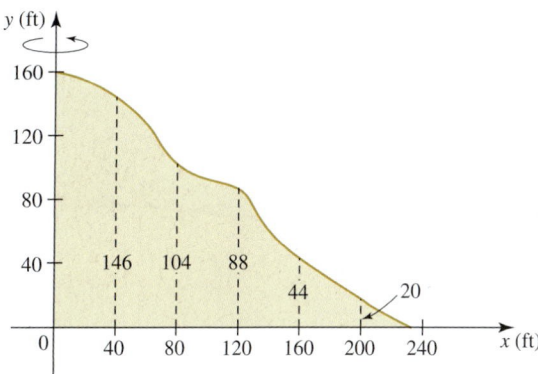

5.4 | **Arc Length and Areas of Surfaces of Revolution**

Upon leaving port, an oil tanker sails along a course given by the curve C shown in Figure 1, where the port is taken to be located at the origin of a coordinate system. What is the distance traveled by the tanker when it reaches a point on the course that is located 4 mi to the east and 2 mi to the north of the port?

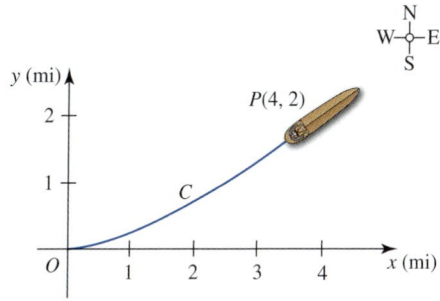

FIGURE 1
The curve C gives the course taken by an oil tanker.

Intuitively, we see that this distance is given by the length of the curve C between the points O and P. So to answer this question, we must (a) define what we mean by the length of a curve and (b) devise a way of computing it. (We will solve this problem in Example 1.)

■ Definition of Arc Length

Suppose that C is the graph of a continuous function f on a closed interval $[a, b]$. Let $P = \{x_0, x_1, \ldots, x_n\}$ be a regular partition of $[a, b]$. If $y_k = f(x_k)$, then the points $P_k(x_k, y_k)$ divide C into n arcs that we denote by $\overparen{P_0 P_1}, \overparen{P_1 P_2}, \ldots, \overparen{P_{n-1} P_n}$. (See Figure 2.)

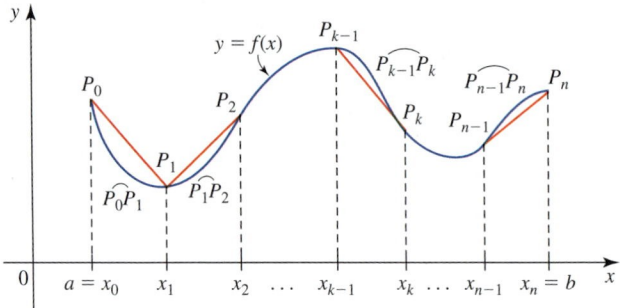

FIGURE 2
The graph of f on $[a, b]$ is the union of the arcs $\overparen{P_0 P_1}, \overparen{P_1 P_2}, \ldots, \overparen{P_{n-1} P_n}$.

Since these arcs are disjoint (except for their endpoints), we see that the length L of C from P_0 to P_n is just the sum of the lengths of these arcs. Now the length of the arc $\overparen{P_{k-1} P_k}$ can be approximated by the length $d(P_{k-1} P_k)$ of the line segment joining P_{k-1} and P_k (shown in red in Figure 2). Therefore, approximating the length of each arc with the length of the corresponding line segment, we see that

$$L \approx \sum_{k=1}^{n} d(P_{k-1} P_k)$$

This approximation improves as n gets larger and larger. This observation suggests that we define the length of C as follows.

DEFINITION   **Arc Length of a Curve**
Let f be a continuous function defined on $[a, b]$, and let $P = \{x_0, x_1, \ldots, x_n\}$ be a regular partition of $[a, b]$. The **arc length of the graph** of f from $P(a, f(a))$ to $Q(b, f(b))$ is

$$L = \lim_{n \to \infty} \sum_{k=1}^{n} d(P_{k-1} P_k) \tag{1}$$

if the limit exists.

Note L is also called the arc length of the graph of f on the interval $[a, b]$.   ■

■ Length of a Smooth Curve

A function f is **smooth** on an interval if its derivative f' is continuous on that interval. The continuity of f' implies that a small change in x produces a small change in the slope $f'(x)$ of the tangent line to the graph of f at any point $(x, f(x))$. Consequently, the graph of f cannot have an abrupt change in direction. In other words, the graph of f has no cusps or corners and is a **smooth curve.** (See Figure 3.)

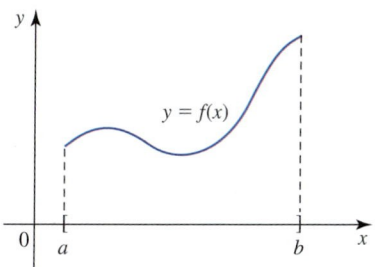

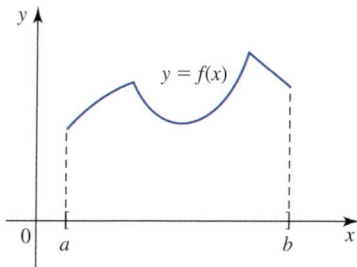

FIGURE 3   **(a)** The function f is smooth.   **(b)** The function f is not smooth.

The length of the graph of a smooth function can be found by integration. To derive a formula for finding the length L of such a graph, suppose f is a smooth function defined on the closed interval $[a, b]$ and $P = \{x_0, x_1, \ldots, x_n\}$ is a regular partition of $[a, b]$. Then by Equation (1),

$$L = \lim_{n \to \infty} \sum_{k=1}^{n} d(P_{k-1}P_k)$$

Using the distance formula, we have

$$\begin{aligned}
d(P_{k-1}P_k) &= \sqrt{(x_k - x_{k-1})^2 + (y_k - y_{k-1})^2} \\
&= \sqrt{(x_k - x_{k-1})^2 + [f(x_k) - f(x_{k-1})]^2} \qquad \textcolor{brown}{y_k = f(x_k)}
\end{aligned}$$

Applying the Mean Value Theorem to f on the interval $[x_{k-1}, x_k]$, we see that

$$f(x_k) - f(x_{k-1}) = f'(c_k)(x_k - x_{k-1})$$

where c_k is a number in the interval (x_{k-1}, x_k). Therefore,

$$\begin{aligned}
d(P_{k-1}P_k) &= \sqrt{(x_k - x_{k-1})^2 + [f'(c_k)(x_k - x_{k-1})]^2} \\
&= \sqrt{\{1 + [f'(c_k)]^2\}(x_k - x_{k-1})^2} \\
&= \sqrt{1 + [f'(c_k)]^2}\, \Delta x \qquad \textcolor{brown}{\Delta x = (b - a)/n}
\end{aligned}$$

So

$$\begin{aligned}
L &= \lim_{n \to \infty} \sum_{k=1}^{n} d(P_{k-1}P_k) \\
&= \lim_{n \to \infty} \sum_{k=1}^{n} \sqrt{1 + [f'(c_k)]^2}\, \Delta x
\end{aligned}$$

Recognizing this expression as the Riemann sum of the continuous function $g(x) = \sqrt{1 + [f'(x)]^2}$ leads to the following result.

The Arc Length Formula

Let f be smooth on $[a, b]$. Then the arc length of the graph of f from $P(a, f(a))$ to $Q(b, f(b))$ is

$$L = \int_a^b \sqrt{1 + [f'(x)]^2}\, dx \qquad (2)$$

Note   If the equation defining the function f is expressed in the form $y = f(x)$, then Equation (2) is sometimes written

$$L = \int_a^b \sqrt{1 + \left(\frac{dy}{dx}\right)^2}\, dx \qquad (3)$$

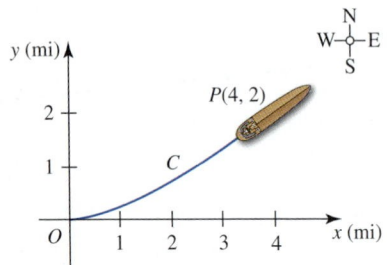

y (mi)

$P(4, 2)$

2

1

C

O 1   2   3   4 x (mi)

FIGURE 4
The course taken by the oil tanker

EXAMPLE 1   **Distance Traveled by a Tanker**   The graph C of the equation $y = \frac{1}{4}x^{3/2}$ gives the course taken by an oil tanker after leaving port, which is taken to be located at the origin of a coordinate system. (See Figure 4.) Find the distance traveled by the tanker when it reaches a point on the course that is located 4 mi to the east and 2 mi to the north of the port.

Solution   The required distance is given by the length L of the curve C from $x = 0$ to $x = 4$. To use Equation (3), we first find

$$\frac{dy}{dx} = \frac{d}{dx}\left(\frac{1}{4}x^{3/2}\right) = \frac{3}{8}x^{1/2}$$

and

$$1 + \left(\frac{dy}{dx}\right)^2 = 1 + \left(\frac{3}{8}x^{1/2}\right)^2 = 1 + \frac{9}{64}x$$

Then

$$L = \int_0^4 \sqrt{1 + \left(\frac{dy}{dx}\right)^2}\, dx = \int_0^4 \sqrt{1 + \frac{9}{64}x}\, dx$$

$$= \left(\frac{64}{9}\right)\left(\frac{2}{3}\right)\left(1 + \frac{9}{64}x\right)^{3/2}\Bigg|_0^4$$

$$= \frac{128}{27}\left[\left(1 + \frac{9}{16}\right)^{3/2} - 1\right] = \frac{128}{27}\left(\frac{125}{64} - 1\right) = \frac{122}{27} \approx 4.52$$

So the oil tanker will have traveled approximately 4.52 mi when it reaches the point in question.

EXAMPLE 2   Find the length of the graph $f(x) = \frac{1}{3}x^3 + \frac{1}{4x}$ on the interval $[1, 3]$.

Solution   The graph of f is sketched in Figure 5.
    We first find

$$f'(x) = \frac{d}{dx}\left[\frac{1}{3}x^3 + \frac{1}{4}x^{-1}\right] = x^2 - \frac{1}{4x^2}$$

Using Equation (2) with

$$1 + [f'(x)]^2 = 1 + \left(x^2 - \frac{1}{4x^2}\right)^2 = 1 + \left(\frac{4x^4 - 1}{4x^2}\right)^2$$

$$= 1 + \frac{16x^8 - 8x^4 + 1}{16x^4} = \frac{16x^8 + 8x^4 + 1}{16x^4}$$

$$= \frac{(4x^4 + 1)^2}{16x^4}$$

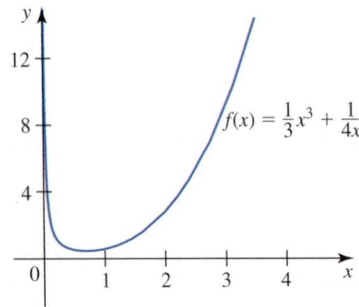

y

12

8    $f(x) = \frac{1}{3}x^3 + \frac{1}{4x}$

4

0   1   2   3   4 x

FIGURE 5
The graph of $f(x) = \frac{1}{3}x^3 + \frac{1}{4x}$

we see that the required length is

$$L = \int_1^3 \sqrt{1 + [f'(x)]^2}\, dx = \int_1^3 \sqrt{\frac{(4x^4 + 1)^2}{16x^4}}\, dx = \int_1^3 \frac{4x^4 + 1}{4x^2}\, dx$$

$$= \int_1^3 \left(x^2 + \frac{1}{4}x^{-2}\right) dx = \left[\frac{1}{3}x^3 - \frac{1}{4x}\right]_1^3 = \left(9 - \frac{1}{12}\right) - \left(\frac{1}{3} - \frac{1}{4}\right) = \frac{53}{6} \quad \blacksquare$$

EXAMPLE 3 Find the arc length of the graph of $f(x) = \ln(2\cos x)$ between the adjacent points of the intersection of the graph with the x-axis.

Solution To find the x-coordinates of the points of intersection of the graph of f with the x-axis, we set $f(x) = 0$ or

$$\ln(2\cos x) = 0$$

Solving this equation we find $2\cos x = 1$, $\cos x = \dfrac{1}{2}$, or $x = \pm\dfrac{\pi}{3} \pm 2n\pi$ ($n = 0, 1, 2, \ldots$). The graph of f is shown in Figure 6.

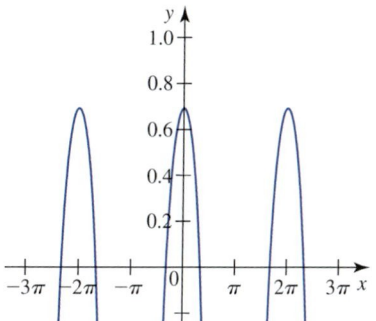

FIGURE 6
The graph of $f(x) = \ln(2\cos x)$

Making use of symmetry, we see that the required arc length is

$$L = 2\int_0^{\pi/3} \sqrt{1 + \left(\frac{dy}{dx}\right)^2}\, dx$$

But

$$\frac{dy}{dx} = \frac{d}{dx}[\ln(2\cos x)] = \frac{\dfrac{d}{dx}(2\cos x)}{2\cos x} = \frac{-2\sin x}{2\cos x} = -\tan x$$

and so

$$L = 2\int_0^{\pi/3} \sqrt{1 + (-\tan x)^2}\, dx = 2\int_0^{\pi/3} \sqrt{\sec^2 x}\, dx$$

$$= 2\int_0^{\pi/3} \sec x\, dx \qquad \text{\textcolor{orange}{Since } sec\,x > 0 \text{ on } \left[0, \frac{\pi}{3}\right]}$$

Using Formula (4c), Section 4.2, we have

$$L = 2\ln|\sec x + \tan x|\,\Big|_0^{\pi/3} = 2\ln(2 + \sqrt{3}) - 2\ln 1$$

$$= 2\ln(2 + \sqrt{3}) \qquad\qquad \blacksquare$$

By interchanging the roles of x and y in Equation (2), we obtain the following formula for finding the arc length of the graph of a smooth function defined by $x = g(y)$ on the interval $[c, d]$. (See Figure 7.)

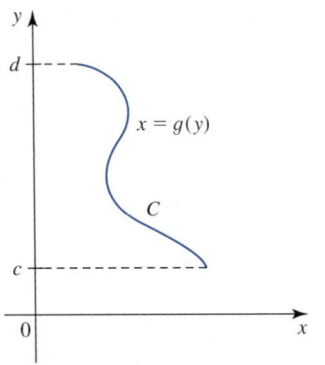

FIGURE 7
The curve C is the graph of $x = g(y)$
for $c \le y \le d$.

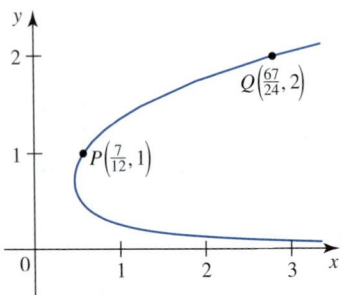

FIGURE 8
The graph of the function
$x = g(y) = \dfrac{1}{3}y^3 + \dfrac{1}{4y}$

Arc Length: Integrating with Respect to y

$$L = \int_c^d \sqrt{1 + [g'(y)]^2}\, dy = \int_c^d \sqrt{1 + \left(\frac{dx}{dy}\right)^2}\, dy \qquad (4)$$

EXAMPLE 4 Find the length of the graph of $x = \dfrac{1}{3}y^3 + \dfrac{1}{4y}$ from $P\left(\frac{7}{12}, 1\right)$ to $Q\left(\frac{67}{24}, 2\right)$.

Solution The graph of $x = g(y)$ is shown in Figure 8. Here x is a function of y, so we use Equation (4). First, we compute

$$1 + \left(\frac{dx}{dy}\right)^2 = 1 + \left(y^2 - \frac{1}{4y^2}\right)^2$$

$$= 1 + y^4 - \frac{1}{2} + \frac{1}{16y^4} = y^4 + \frac{1}{2} + \frac{1}{16y^4}$$

$$= \left(y^2 + \frac{1}{4y^2}\right)^2$$

Then observing that y runs from $y = 1$ to $y = 2$ and using Equation (4), we find that the required length is

$$L = \int_1^2 \sqrt{\left(y^2 + \frac{1}{4y^2}\right)^2}\, dy = \int_1^2 \left(y^2 + \frac{1}{4y^2}\right) dy$$

$$= \left[\frac{1}{3}y^3 - \frac{1}{4y}\right]_1^2 = \left[\left(\frac{8}{3} - \frac{1}{8}\right) - \left(\frac{1}{3} - \frac{1}{4}\right)\right] = \frac{59}{24} \qquad ∎$$

The Arc Length Function

Suppose that C is the graph of a smooth function f defined by $y = f(x)$ on the closed interval $[a, b]$. If x is a point in $[a, b]$, we can use Equation (2) to express the length of the arc of the graph of f from $P(a, f(a))$ to $Q(x, f(x))$. (See Figure 9.) Denoting this length by $s(x)$ (since it depends on x), we have

$$s(x) = \int_a^x \sqrt{1 + [f'(t)]^2}\, dt$$

This equation enables us to define the following function.

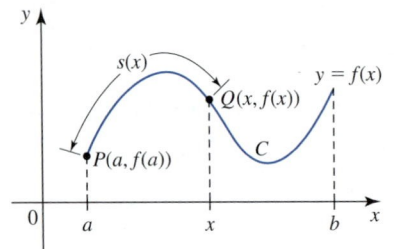

FIGURE 9
$s(x)$ is the length of the arc of the graph
of f from $P(a, f(a))$ to $Q(x, f(x))$.

DEFINITION **Arc Length Function**

Let f be smooth on $[a, b]$. The **arc length function** s for the graph of f is defined by

$$s(x) = \int_a^x \sqrt{1 + [f'(t)]^2}\, dt \qquad (5)$$

with domain $[a, b]$.

If we use the Fundamental Theorem of Calculus, Part 1, to differentiate Equation (5), we obtain

$$s'(x) = \frac{d}{dx} \int_a^x \sqrt{1 + [f'(t)]^2}\, dt = \sqrt{1 + [f'(x)]^2} = \sqrt{1 + \left(\frac{dy}{dx}\right)^2} \tag{6}$$

The quantity $ds = s'(x)\, dx$ is the *differential of arc length*. In view of Equation (6) we can express ds in the following forms.

Arc Length Differentials

$$ds = \sqrt{1 + \left(\frac{dy}{dx}\right)^2}\, dx \tag{7}$$

or, equivalently,

$$(ds)^2 = (dy)^2 + (dx)^2 \tag{8}$$

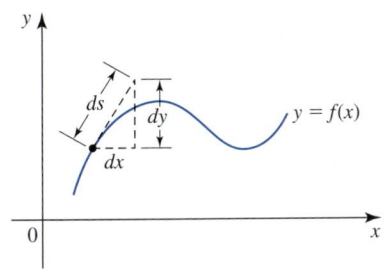

FIGURE 10
The relationship between ds, dy, and dx follows from the Pythagorean Theorem.

Figure 10 gives a geometric interpretation of the differential of arc length in terms of the differentials dx and dy. Observe that if $dx = \Delta x$ is small, then ds affords a good approximation of the arc length of the graph of f corresponding to the change, Δx, in x.

EXAMPLE 5 Use differentials to obtain an approximation of the arc length of the graph of $y = 2x^2 + x$ from $P(1, 3)$ to $Q(1.1, 3.52)$.

Solution   Using Equation (7), we find

$$ds = \sqrt{1 + \left(\frac{dy}{dx}\right)^2}\, dx = \sqrt{1 + (4x + 1)^2}\, dx$$

Letting $x = 1$ and $dx = 0.1$, we obtain the approximation

$$ds \approx \sqrt{1 + 5^2}(0.1) = 0.1\sqrt{26} \approx 0.51 \qquad\blacksquare$$

Note   The expression in Equation (8) provides us with an easy way of recalling the formula for the arc length L of the graph of a function f on $[a, b]$. From

$$(ds)^2 = (dy)^2 + (dx)^2$$

we see that

$$ds = \sqrt{1 + \left(\frac{dy}{dx}\right)^2}\, dx$$

and

$$s(x) = \int_a^x ds$$

Therefore,

$$L = s(b) = \int_a^b ds = \int_a^b \sqrt{1 + \left(\frac{dy}{dx}\right)^2}\, dx = \int_a^b \sqrt{1 + [f'(x)]^2}\, dx \qquad\blacksquare$$

■ Surfaces of Revolution

A **surface of revolution** is a surface that is obtained by revolving the graph of a continuous function about a line. For example, if the graph C of the function f on the interval $[a, b]$ shown in Figure 11a is revolved about the x-axis, we obtain the surface of revolution S shown in Figure 11b.

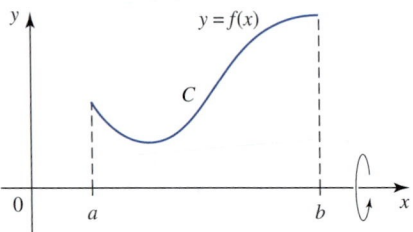

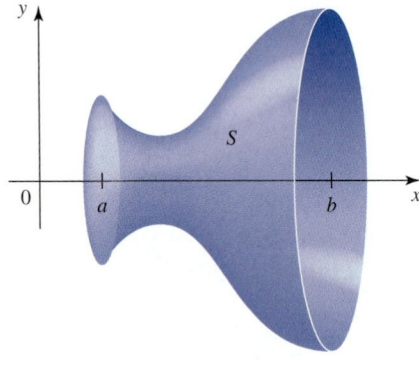

FIGURE 11

If we revolve the graph of f about the x-axis in (a), we obtain the surface S in (b). **(a)**

(b)

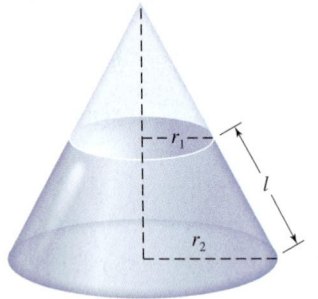

FIGURE 12

The frustum of a cone obtained by cutting off its top using a plane parallel to its base

Our immediate objective is to devise a formula for finding the surface area of S. To do this, we need the formula for the lateral surface area of a frustum of a right circular cone (Figure 12).

If the upper and lower radii of a frustum are r_1 and r_2, respectively, and its slant height is l, then the surface area S of the frustum is

$$S = 2\pi r l \tag{9}$$

where $r = \frac{1}{2}(r_1 + r_2)$ is the average radius of the frustum. (You will be asked to establish this formula in Exercise 59.)

Next, consider the surface S generated by revolving the graph C of a smooth non-negative function f about the x-axis from $x = a$ to $x = b$. Let $P = \{x_0, x_1, \ldots, x_n\}$ be a regular partition of $[a, b]$. If $y_k = f(x_k)$, then the points $P_k(x_k, y_k)$ divide C into n disjoint (except at their endpoints) arcs $\overparen{P_0P_1}, \overparen{P_1P_2}, \ldots, \overparen{P_{n-1}P_n}$ whose union is C (Figure 13a). The surface S is the union of the surfaces $S_1, S_2, \ldots, S_n$ obtained by revolving these arcs about the x-axis (Figure 13b).

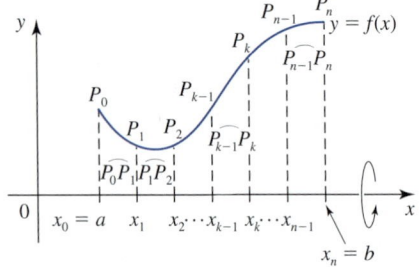

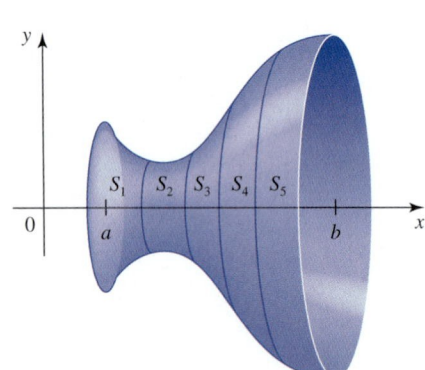

FIGURE 13

(a) A partition of $[a, b]$ produces n arcs $\overparen{P_0P_1}, \overparen{P_1P_2}, \ldots, \overparen{P_{n-1}P_n}$, which, when revolved about the x-axis, give n surfaces $S_1, S_2, \ldots, S_n$, which together form S. (b) Here, $n = 5$. **(a)**

(b)

Let's concentrate on the part of the surface generated by the arc of the graph of f on the interval $[x_{k-1}, x_k]$. This arc is shown in Figure 14. If Δx is small, then the arc $\overset{\frown}{P_{k-1}P_k}$ may be approximated by the line segment joining P_{k-1} and P_k. This suggests that the surface area of the frustum that is generated by revolving this line segment about the x-axis will provide us with a good approximation of the surface area of S_k. (See Figure 14.)

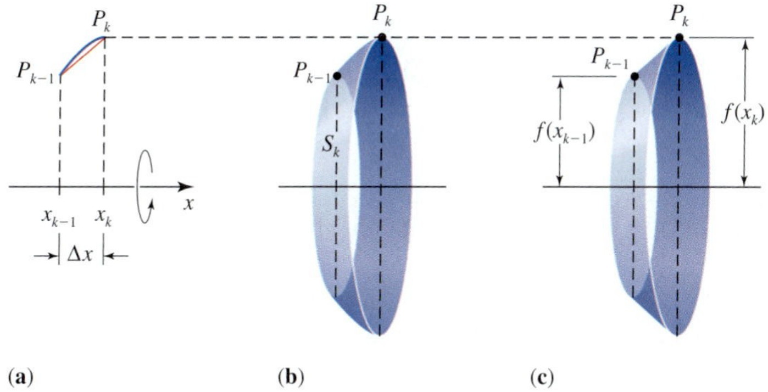

FIGURE 14
In part (a) the arc $\overset{\frown}{P_{k-1}P_k}$ is approximated by the line segment joining P_{k-1} to P_k. So the area of the surface generated by $\overset{\frown}{P_{k-1}P_k}$ in (b) is approximated by the lateral surface area of the frustum generated by the line segment in (c).

(a) (b) (c)

Since the frustum has an average radius of $r = \frac{1}{2}[f(x_{k-1}) + f(x_k)]$ and a slant height of $l = d(P_{k-1}P_k)$, Formula (9) tells us that its surface area is

$$\Delta S = 2\pi \left[\frac{f(x_{k-1}) + f(x_k)}{2}\right] d(P_{k-1}P_k)$$

But as in the computations leading to Equation (2), we have

$$d(P_{k-1}P_k) = \sqrt{1 + [f'(c_k)]^2}\, \Delta x$$

where c_k is a number in the interval (x_{k-1}, x_k). Also, if Δx is small, the continuity of f implies that $f(x_{k-1}) \approx f(c_k)$ and $f(x_k) \approx f(c_k)$. Therefore,

$$\Delta S = 2\pi \left[\frac{f(c_k) + f(c_k)}{2}\right]\sqrt{1 + [f'(c_k)]^2}\, \Delta x$$

Approximating the area of each surface S_k by the area of the corresponding frustum, we see that*

$$S \approx \sum_{k=1}^{n} 2\pi f(c_k)\sqrt{1 + [f'(c_k)]^2}\, \Delta x$$

This approximation can be expected to improve as n gets larger and larger. Finally, recognizing this sum to be the Riemann sum of the function $g(x) = 2\pi f(x)\sqrt{1 + [f'(x)]^2}$ on the interval $[a, b]$, we see that

$$\lim_{n \to \infty} \sum_{k=1}^{n} 2\pi f(c_k)\sqrt{1 + [f'(c_k)]^2}\, \Delta x = \int_a^b 2\pi f(x)\sqrt{1 + [f'(x)]^2}\, dx$$

This discussion leads to the following definition.

---

*It is conventional to denote the area of a surface by S, and we will do so here even though we have used this very letter to denote the surface of revolution itself.

> **DEFINITION** Surface Area of a Surface of Revolution
>
> Let f be a nonnegative smooth function on $[a, b]$. The **surface area** of the surface obtained by revolving the graph of f about the x-axis is
>
> $$S = 2\pi \int_a^b f(x)\sqrt{1 + [f'(x)]^2}\, dx \qquad (10)$$

Note If we use Equation (7), then we can write Equation (10) in the form

$$S = 2\pi \int_a^b y\, ds \qquad (11)$$

which is the *arc length differential form* of Equation (10). ◼

This formula can be remembered as follows: If Δx is small, the differential of arc length, ds, gives an approximation of the slant height of the frustum of a cone of average radius approximated by y (or $f(x)$). So $2\pi y\, ds$ represents an element of area of the surface (Figure 15). By summing and taking the limit, we then obtain Equation (11).

FIGURE 15

If Δx is small, ds approximates the slant height of the frustum, and $f(x)$ approximates the average height of the frustum.

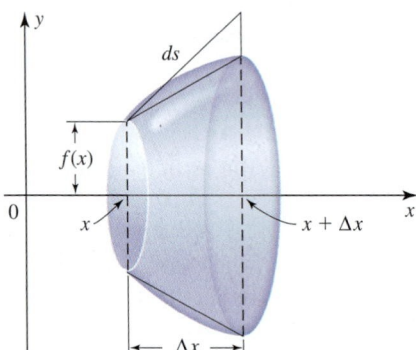

EXAMPLE 6 Find the area of the surface obtained by revolving the graph of $f(x) = \sqrt{x}$ on the interval $[0, 2]$ about the x-axis.

Solution The graph of f and the resulting surface of revolution are shown in Figure 16. We have

$$f'(x) = \frac{1}{2\sqrt{x}}$$

Using Equation (10), we find that the required area is given by

$$S = 2\pi \int_0^2 f(x)\sqrt{1 + [f'(x)]^2}\, dx$$

$$= 2\pi \int_0^2 \sqrt{x}\,\sqrt{1 + \left(\frac{1}{2\sqrt{x}}\right)^2}\, dx = 2\pi \int_0^2 \sqrt{x}\,\sqrt{1 + \frac{1}{4x}}\, dx$$

$$= \pi \int_0^2 \sqrt{4x + 1}\, dx$$

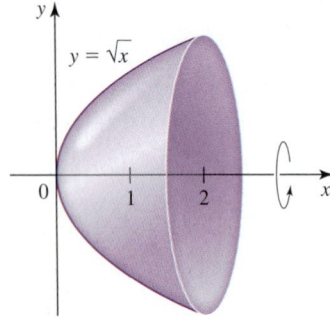

FIGURE 16

The graph of $y = \sqrt{x}$ on $[0, 2]$ and the resulting surface of revolution obtained by revolving the graph about the x-axis

We evaluate this integral using the method of substitution with $u = 4x + 1$, so that $du = 4\,dx$ or $dx = \frac{1}{4}\,du$. The lower and upper limits of integration with respect to u are 1 and 9, respectively. We obtain

$$S = \frac{\pi}{4}\int_1^9 \sqrt{u}\,du = \frac{\pi}{4}\left[\frac{2}{3}u^{3/2}\right]_1^9 = \frac{\pi}{4}\left(18 - \frac{2}{3}\right) = \frac{13\pi}{3}$$

By interchanging the roles of x and y in Equation (10), we obtain the following formula for finding the area of the surface obtained by revolving the graph of a smooth function defined by $x = g(y)$ on the interval $[c, d]$ about the y-axis.

Surface Area: Integrating with Respect to y

$$S = 2\pi\int_c^d g(y)\sqrt{1 + [g'(y)]^2}\,dy = 2\pi\int_c^d x\,ds \qquad \textbf{(12)}$$

EXAMPLE 7 Find the area of the surface obtained by revolving the graph of $x = y^3$ on the interval $[0, 1]$ about the y-axis.

Solution Here, $x = g(y) = y^3$ and so $g'(y) = 3y^2$. Therefore, Equation(12) gives the required surface area as

$$S = 2\pi\int_0^1 g(y)\sqrt{1 + [g'(y)]^2}\,dy$$

$$= 2\pi\int_0^1 y^3\sqrt{1 + (3y^2)^2}\,dy = 2\pi\int_0^1 y^3\sqrt{1 + 9y^4}\,dy$$

To evaluate the integral, we use the method of substitution with $u = 1 + 9y^4$ so that $du = 36y^3\,dy$. The lower and upper limits of integration are 1 and 10, respectively. We obtain

$$S = \frac{2\pi}{36}\int_1^{10}\sqrt{u}\,du = \frac{\pi}{18}\left[\frac{2}{3}u^{3/2}\right]_1^{10} = \frac{\pi}{18}\left(\frac{2}{3}10^{3/2} - \frac{2}{3}\right) = \frac{\pi}{27}(10\sqrt{10} - 1)$$

The surface is shown in Figure 17.

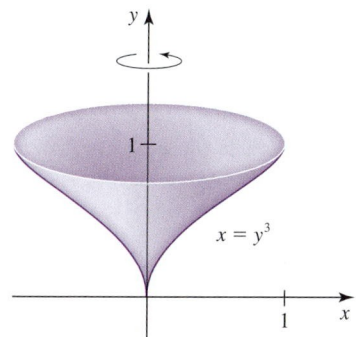

FIGURE 17
The graph of $x = y^3$ on $[0, 1]$ and the surface obtained by revolving it about the y-axis

5.4 CONCEPT QUESTIONS

1. **a.** Write an integral that gives the arc length of (1) a smooth function $y = f(x)$ on the interval $[a, b]$ and (2) a smooth function $x = g(y)$ on the interval $[c, d]$.
   b. Write two different integrals that give the arc length L of the curve defined by the equation $y = x^{2/3} - 1$ from the point $P(0, -1)$ to the point $Q(5\sqrt{5}, 4)$. Which integral would you choose to compute L? Explain. Then use your choice of integral to compute L.

2. Write the formulas for finding the surface area of a surface of revolution obtained by (a) revolving the graph of a non-negative smooth function $y = f(x)$ on the interval $[a, b]$ about the x-axis and (b) revolving the graph of a smooth function $x = g(y)$ on the interval $[c, d]$ about the y-axis.

5.4 EXERCISES

In Exercises 1–4, find the arc length of the graph from A to B.

1.
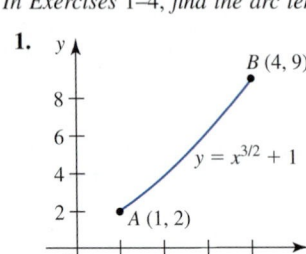

2.
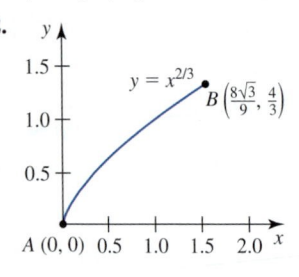

3.
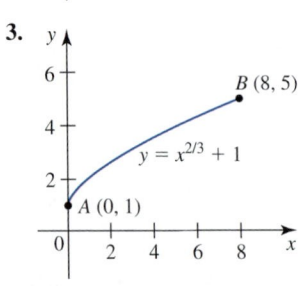

4.
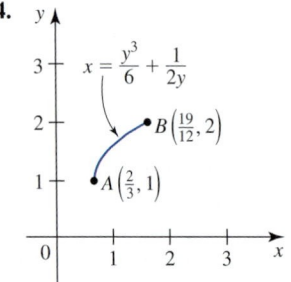

In Exercises 5 and 6, find the length of the line segment joining the two given points by finding the equation of the line and using Equation (2). Then check your answer by using the distance formula.

5. $(0, 0)$ and $(3, 8)$ **6.** $(-1, -2)$ and $(3, 6)$

In Exercises 7–18, find the arc length of the graph of the given equation from P to Q or on the specified interval.

7. $y = -2x + 3$; $P(-1, 5), Q(2, -1)$

8. $y = \frac{2}{3}x^{3/2} - 1$; $P(4, \frac{13}{3}), Q(9, 17)$

9. $y = 2(x - 1)^{3/2}$; $P(1, 0), Q(5, 16)$

10. $x = \frac{1}{4}y^4 + \frac{1}{8y^2}$; $P(\frac{3}{8}, 1), Q(\frac{129}{32}, 2)$

11. $y = \frac{2}{3}(x^2 + 1)^{3/2}$; $[1, 4]$

12. $y = (2 - x^{2/3})^{3/2}$; $[1, 2]$

13. $(y + 3)^2 = 4(x + 2)^3$; $P(-2, -3), Q(2, 13)$

14. $y = \frac{x^3}{3} + \frac{1}{4x}$; $[1, 3]$

15. $y = \frac{1}{2}(e^x + e^{-x})$; $[0, \ln 2]$

16. $y = \frac{1}{2}[x\sqrt{x^2 - 1} - \ln(x + \sqrt{x^2 - 1})]$; $[1, 3]$

17. $y = \ln \cos x$; $[0, \frac{\pi}{4}]$

18. $y = \sqrt{4 - x^2}$; $[0, 2]$

In Exercises 19–24, write an integral giving the arc length of the graph of the equation from P to Q or over the indicated interval. (Do not evaluate the integral.)

19. $y = x^2$; $P(-1, 1), Q(2, 4)$

20. $y = x^3 - 1$; $[0, 1]$

21. $y = \dfrac{1}{x^2 + 1}$; $P(-1, \frac{1}{2}), Q(2, \frac{1}{5})$

22. $y = \cos x$; $[0, \pi]$

23. $y = \tan x$; $P(0, 0), Q(\frac{\pi}{4}, 1)$

24. $x = \sec y$; $P(\sqrt{2}, -\frac{\pi}{4}), Q(1, 0)$

In Exercises 25–28, (a) plot the graph of the function f, (b) write an integral giving the arc length of the graph of the function over the indicated interval, and (c) find the arc length of the curve accurate to four decimal places.

25. $f(x) = 2x^3 - x^4$; $[0, 2]$ **26.** $f(x) = \sqrt{x^2 - x^4}$; $[0, 1]$

27. $f(x) = x - 2\sqrt{x}$; $[0, 4]$ **28.** $f(x) = \dfrac{x^2}{1 + x^4}$; $[0, 1]$

29. The graph of the equation $x^{2/3} + y^{2/3} = a^{2/3}$, where $a > 0$, shown in the following figure, is called an astroid. Find the arc length of the astroid.

Hint: By symmetry the arc length is equal to 8 times the length of the curve joining P to $Q(a, 0)$. To find the coordinates of P, find the point of intersection of the astroid with the line $y = x$.

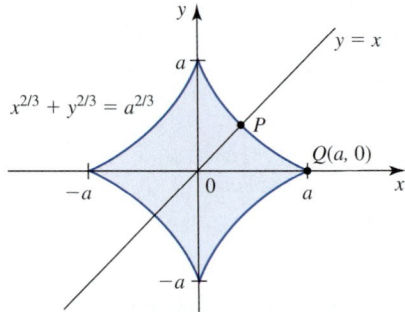

30. Use the fact that the circumference of a circle of radius 1 is 2π to evaluate the integral

$$\int_0^{\sqrt{2}/2} \frac{dx}{\sqrt{1 - x^2}}$$

Hint: Interpret $\int_0^{\sqrt{2}/2}\sqrt{1 + (y')^2}\, dx$, where $y = \sqrt{1 - x^2}$.

In Exercises 31 and 32, use differentials to approximate the arc length of the graph of the equation from P to Q.

31. $y = x^3 + 1$; $P(1, 2), Q(1.2, 2.728)$

32. $y = \sqrt{x} + 1$; $P(4, 3), Q(4.3, 3.074)$

In Exercises 33–43, find the area of the surface obtained by revolving the given curve about the indicated axis.

33. $y = \dfrac{1}{2}x + 2$ for $0 \le x \le 2$; *x*-axis

34. $y = \sqrt{x}$ on $[4, 9]$; *x*-axis

35. $y = x^3$ on $[0, 1]$; *x*-axis

36. $y = x^{1/3}$ on $[1, 8]$; *y*-axis

37. $y = 4 - x^2$ on $[0, 2]$; *y*-axis

38. $x = \dfrac{1}{6}y^3 + \dfrac{1}{2y}$ for $1 \le y \le 2$; *y*-axis

39. $2x + 3y = 6$ for $-2 \le y \le 1$; *y*-axis

40. $y = \dfrac{1}{4}x^4 + \dfrac{1}{8x^2}$ on $[1, 2]$; *x*-axis

41. $y = \dfrac{1}{2\sqrt{2}}\sqrt{x^2 - x^4}$ on $[0, 1]$; *x*-axis

42. $x = \dfrac{1}{3}\sqrt{y(3 - y)^2}$ on $0 \le y \le 3$; *y*-axis

43. $y = \dfrac{1}{2}(e^x + e^{-x})$ on $[0, \ln 2]$; *x*-axis.

In Exercises 44 and 45, write an integral giving the area of the surface obtained by revolving the curve about the x-axis. (Do not evaluate the integral.)

44. $y = \dfrac{1}{x}$ on $[1, 2]$ **45.** $y = \sin x$ on $\left[0, \frac{\pi}{2}\right]$

 46. a. Plot the graph of $f(x) = \tan^{-1} x$ and the graph of the secant line passing through $(0, 0)$ and $\left(1, \frac{\pi}{4}\right)$.
   b. Use the Pythagorean Theorem to estimate the arc length of the graph of f on the interval $[0, 1]$.
   c. Use a calculator or a computer to find the arc length of the graph of $f(x) = \tan^{-1} x$.

 47. Refer to Exercise 25. Use a calculator or computer to find the area of the surface formed by revolving the graph of $f(x) = 2x^3 - x^4$, where $0 \le x \le 2$, about the *x*-axis, accurate to four decimal places.

 48. Refer to Exercise 26. Use a calculator or computer to find the area of the surface formed by revolving the graph of $f(x) = \sqrt{x^2 - x^4}$, where $0 \le x \le 1$, about the *x*-axis, accurate to four decimal places.

49. Verify that the lateral surface area of a right circular cone of height h and base radius r is $S = \pi r\sqrt{r^2 + h^2}$ by evaluating a definite integral.
   Hint: The cone is generated by revolving the region bounded by $y = (h/r)x$, $y = h$, and $x = 0$ about the *y*-axis.

50. Verify that the surface area of a sphere of radius r is $S = 4\pi r^2$ by evaluating a definite integral.
   Hint: Generate this sphere by revolving the semicircle $x^2 + y^2 = r^2$, where $y \ge 0$, about the *x*-axis.

51. Find the area of the surface obtained by revolving the graph of $y = \sqrt{4 - x^2}$ on $[0, 1]$ about the *x*-axis. This surface is called a **spherical zone.**

52. Find the area of the spherical zone formed by revolving the graph of $y = \sqrt{r^2 - x^2}$ on $[a, b]$, where $0 < a < b < r$, about the *x*-axis.

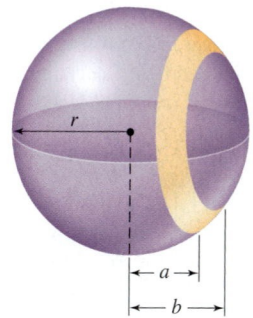

53. A Pursuit Curve The graph C of the function

$$y = \frac{2}{3}\left(1 - \frac{x}{2}\right)^{3/2} - 2\left(1 - \frac{x}{2}\right)^{1/2} + \frac{4}{3}$$

gives the path taken by Boat A as it pursues and eventually intercepts Boat B ($x = 2$). Initially, Boat A was at the origin, and Boat B was at the point $(2, 0)$, heading due north. Find the distance traveled by Boat A during the pursuit.

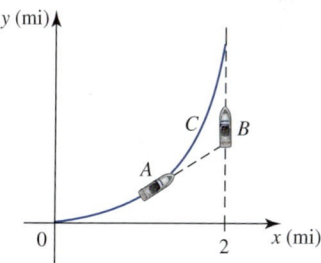

 54. Motion of a Projectile Refer to Exercise 29 in Section 2.4. A projectile is fired from a cannon located on a horizontal plane. If we think of the cannon as being located at the origin O of an *xy*-coordinate system, then the path of the projectile is

$$y = \sqrt{3}x - \frac{x^2}{400}$$

where x and y are measured in feet. Estimate the distance traveled by the projectile in the air.

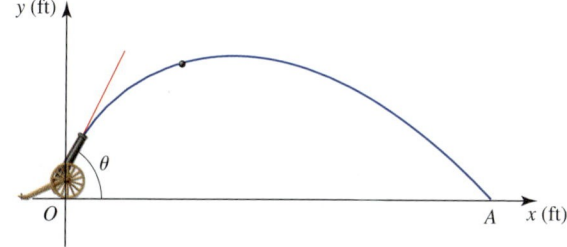

55. Flight Path of an Airplane The path of an airplane on its final approach to landing is described by the equation $y = f(x)$ with

$$f(x) = 4.3403 \times 10^{-10}x^3 - 1.5625 \times 10^{-5}x^2 + 3000$$
$$0 \le x \le 24{,}000$$

where x and y are both measured in feet. Estimate the distance traveled by the airplane during the landing approach.

56. Area of a Roof A hangar is 100 ft long and has a uniform cross section that is described by the equation $y = 10 - 0.0001x^4$, where both x and y are measured in feet. Estimate the area of the roof of the hangar.

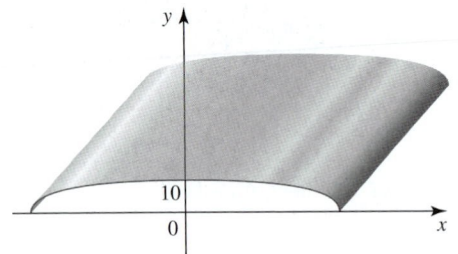

57. Manufacturing Corrugated Sheets A manufacturer of aluminum roofing products makes corrugated sheets as shown in the figure. The cross section of the corrugated sheets can be described by the equation

$$y = \sin\left(\frac{\pi x}{10}\right) \qquad 0 \le x \le 30$$

where x and y are measured in inches. If the corrugated sheets are made from flat sheets of aluminum using a stamping machine that does not stretch the metal, find the width w of a flat aluminum sheet that is needed to make a 30-in. panel.

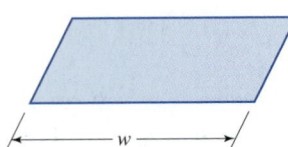
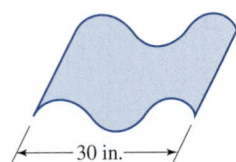

58. Let f be a smooth nonnegative function on $[a, b]$. Show that the area of the surface obtained by revolving the graph of f about the line $y = L$ is given by

$$S = 2\pi \int_a^b |f(x) - L|\sqrt{1 + [f'(x)]^2}\, dx$$

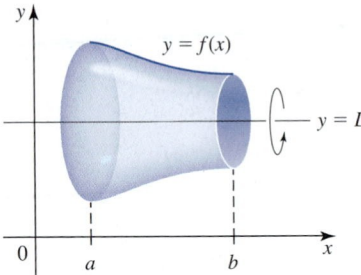

59. Show that the lateral surface area of a frustrum of a right circular cone of upper and lower radii r_1 and r_2, respectively, and slant height l is $S = 2\pi r l$, where $r = \frac{1}{2}(r_1 + r_2)$.

60. Let L denote the length of the graph of $y = f(x)$ connecting the points $(0, 0)$ and $(l, 0)$, and let $D = L - l$ (see the figure). Show that

$$\frac{1}{2}\int_0^l \frac{(y')^2}{\sqrt{1 + (y')^2}}\, dx \le D \le \frac{1}{2}\int_0^l (y')^2\, dx$$

assuming that y' is continuous on $(0, l)$.

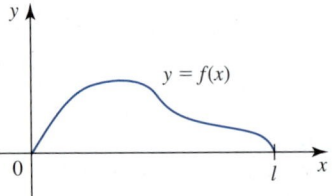

5.5 Work

The term *work,* as used in physics and engineering, is the transference of energy that results when the application of a force causes a body to move. Scientists and engineers need to know precisely how much energy is required to perform certain tasks. For example, a rocket scientist needs to know the amount of energy required to put an artificial satellite into an orbit around the earth, and a power engineer needs to know the amount of energy derived from water flowing through a dam.

■ Work Done by a Constant Force

We begin by defining the work done by a *constant* force in moving an object along a straight line.

DEFINITION Work Done by a Constant Force

The work W done by a constant force F in moving a body a distance d in the direction of the force is

$$W = Fd \qquad \text{work} = \text{force} \cdot \text{distance}$$

The unit of work in any system is the unit of force times the unit of distance. In the English system the unit of force is the pound (lb), the unit of distance is the foot (ft), and so the unit of work is the foot-pound (ft-lb). In the International System of Units, abbreviated SI (for Système international d'unités), the unit of force is the newton (N), the unit of distance is the meter (m), and so the unit of work is the newton-meter (N-m). A newton-meter is also called a *joule* (J).

EXAMPLE 1

a. Find the work done in lifting a 25-lb object 4 ft off the ground.
b. Find the work done in lifting a 2.4-kg package 0.8 m off the ground. (Take $g = 9.8$ m/sec^2.)

Solution

a. The force F required to do the job is 25 lb (the weight of the object). Therefore, the work done by the force is

$$W = Fd = 25(4) = 100$$

or 100 ft-lb.
b. The magnitude of the force required is $F = mg = (2.4)(9.8) = 23.52$, or 23.52 N. So the work done is

$$W = Fd = (23.52)(0.8) \approx 18.8$$

or 18.8 J. ◼

■ Work Done by a Variable Force

Suppose that a body moves along the x-axis in the positive direction from $x = a$ to $x = b$ under the action of a force $F(x)$ that depends on x. Suppose also that the function F is continuous on the interval $[a, b]$ with the graph depicted in Figure 1. Next, let $P = \{x_0, x_1, \ldots, x_n\}$ be a regular partition of $[a, b]$.

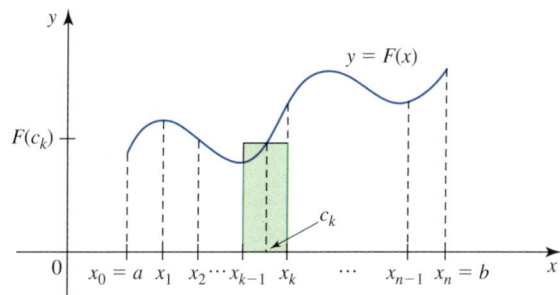

FIGURE 1
The graph of a variable force defined by the function F

Let's concentrate on the subinterval $[x_{k-1}, x_k]$. If $\Delta x = (b - a)/n$ is small, then the continuity of F guarantees that the values of $F(x)$ at any two points in $[x_{k-1}, x_k]$

do not differ by much. Therefore, if c_k is any point in $[x_{k-1}, x_k]$, we can approximate $F(x)$ by $F(c_k)$ for all x in $[x_{k-1}, x_k]$. Physically, we are saying that the force $F(x)$ is approximately constant when measured over a small distance. So if we assume that $F(x) = F(c_k)$ in $[x_{k-1}, x_k]$, then the work done by F in moving the body along the x-axis from $x = x_{k-1}$ to $x = x_k$ is

$$\Delta W_k \approx F(c_k)\Delta x \qquad \text{constant force} \cdot \text{distance}$$

(This is the area of a rectangle of height $F(c_k)$ and width Δx.) It follows that the work W done by F in moving the body from $x = a$ to $x = b$ is

$$W \approx \sum_{k=1}^{n} F(c_k)\Delta x$$

Intuitively, we see that the approximation improves as n gets larger and larger. This suggests that we define the work done by F by taking the limit of the sum

$$\sum_{k=1}^{n} F(c_k)\Delta x$$

as $n \to \infty$. But this sum is just the Riemann sum of F on the interval $[a, b]$. Therefore, our discussion leads to the following definition.

DEFINITION   Work Done by a Variable Force
Suppose that a force F, where F is continuous on $[a, b]$, acts on a body moving it along the x-axis. Then the **work** done by the force in moving the body from $x = a$ to $x = b$ is

$$W = \lim_{n \to \infty} \sum_{k=1}^{n} F(c_k)\Delta x = \int_a^b F(x)\, dx \qquad (1)$$

Note   When we derived Equation (1), we assumed that $b > a$. This condition is not necessary and may be dropped.   ∎

EXAMPLE 2   Find the work done by the force $F(x) = 3x^2 + x$ (measured in pounds) in moving a particle along the x-axis from $x = 2$ to $x = 4$ (measured in feet).

Solution   Here, $F(x) = 3x^2 + x$, so the work done by F in moving the body from $x = 2$ to $x = 4$ is

$$W = \int_2^4 F(x)\, dx = \int_2^4 (3x^2 + x)\, dx = \left[x^3 + \frac{1}{2}x^2 \right]_2^4 = 72 - 10 = 62$$

or 62 ft-lb.   ∎

■ Hooke's Law

As another application of Equation (1), let's find the work done in stretching or compressing a spring. Recall that **Hooke's Law** states that the force F required to stretch or compress a spring x units past its natural length is proportional to x. That is,

$$F(x) = kx$$

where k, the constant of proportionality, is called the **spring constant,** or the **stiffness.** Hooke's Law is valid provided that $|x|$ is not too large.

EXAMPLE 3 A force of 30 N is required to stretch a spring 4 cm beyond its natural length of 18 cm. Find the work required to stretch the spring from a length of 20 cm to a length of 24 cm.

Solution   Suppose that the spring is placed on the x-axis with the free end at the origin as shown in Figure 2. According to Hooke's Law, the force $F(x)$ required to stretch the spring x meters beyond its natural length is $F(x) = kx$. Since a 30-N force is required to stretch the spring 4 cm, or 0.04 m, beyond its natural length, we see that

$$30 = k(0.04) \qquad \text{or} \qquad k = 750$$

that is, 750 N/m. Therefore, $F(x) = 750x$ for this spring. Using Equation (1), we find that the work required to stretch the spring from 20 cm to 24 cm is

$$W = \int_{0.02}^{0.06} 750x \, dx = 750 \left[\frac{1}{2} x^2 \right]_{0.02}^{0.06}$$

$$= 375[(0.06)^2 - (0.02)^2] = 1.2$$

or 1.2 J.

(a) The unstretched spring

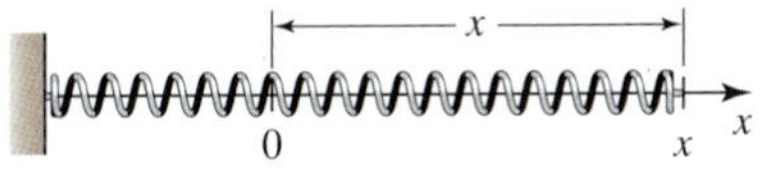

(b) The spring stretched x units beyond its natural length

FIGURE 2

■ Moving Nonrigid Matter

The next two examples involve the computation of the work involved in moving nonrigid matter, such as the evacuation of fluid from a container and the hoisting of an object.

EXAMPLE 4 A tank has the shape of an inverted right circular cone with a base of radius 5 ft and a height of 12 ft. If the tank is filled with water to a height of 8 ft, find the work required to empty the tank by pumping the water over the top of the tank. (Water weighs 62.4 lb/ft^3.)

Solution   We think of the tank as being placed on a coordinate system with its vertex at the origin and its axis along the y-axis as shown in Figure 3a. Think of the water as being subdivided into slabs by planes perpendicular to the y-axis from $y = 0$ to $y = 8$. The volume ΔV of a representative slab is approximated by a disk of radius x and width Δy, that is,

$$\Delta V \approx \pi x^2 \Delta y$$

You can see how to express x in terms of y by referring to Figure 3b. By similar triangles,

$$\frac{x}{5} = \frac{y}{12} \qquad \text{or} \qquad x = \frac{5}{12} y$$

FIGURE 3

We wish to find the amount of work required to pump all of the water out of the top of the conical tank.

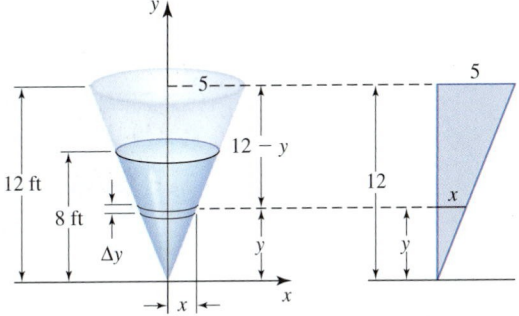

(a)   (b)

so

$$\Delta V \approx \pi \left(\frac{5}{12} y \right)^2 \Delta y = \frac{25\pi}{144} y^2 \Delta y$$

Since water weighs 62.4 lb/ft³, or $62\frac{2}{5}$ lb/ft³, the weight of a representative slab (the force required to lift this slab) is

$$\Delta F \approx 62\frac{2}{5} \cdot \frac{25\pi}{144} y^2 \Delta y = \frac{65\pi}{6} y^2 \Delta y$$

Since this slab is transported a distance of approximately $(12 - y)$ ft, the work done by the force is

$$\Delta W \approx \Delta F (12 - y)$$

$$= \frac{65\pi}{6} y^2 (12 - y) \Delta y$$

Finally, summing the work done in lifting each slab to the top of the tank and taking the limit, we see that the work required to empty the tank is

$$W = \int_0^8 \frac{65\pi}{6} y^2 (12 - y) \, dy$$

$$= \frac{65\pi}{6} \int_0^8 (12y^2 - y^3) \, dy$$

$$= \frac{65\pi}{6} \left[4y^3 - \frac{1}{4} y^4 \right]_0^8 = \frac{65\pi}{6} (1024)$$

or approximately 34,851 ft-lb. ∎

EXAMPLE 5 A ship's anchor, weighing 800 lb, is attached to a chain that weighs 10 lb per running foot. Find the work done by the winch if the anchor is pulled in from a height of 20 ft. (See Figure 4.)

Solution The work done by the winch is $W = W_A + W_C$, where W_A is the work required to hoist the anchor to the top of the ship and W_C is the work required to pull the cable to the top of the ship. To find W_A, observe that the force required to lift the anchor is 800 lb and that it will be applied over a distance of 20 ft. Therefore,

$$W_A = (800)(20) = 16,000$$

or 16,000 ft-lb.

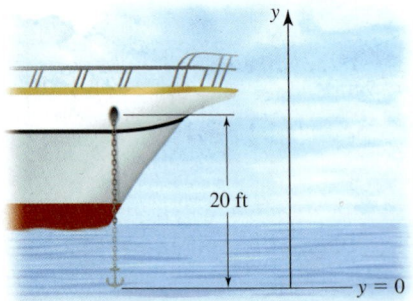 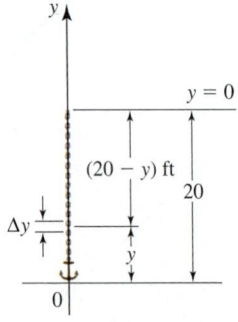

FIGURE 4
The anchor is hoisted to the top of the ship by means of a cable chain.

(a)   (b)

To find W_C, think of the chain as being subdivided into pieces. The length of a representative piece is Δy ft, and its weight is $10\Delta y$ lb (weight per running foot times length). This element is to be lifted a distance of approximately $(20 - y)$ ft, so the work required is

$$\Delta W_C \approx 10\,\Delta y(20 - y)$$

Summing the work done in lifting each piece of the chain to the top and taking the limit, we see that the work required is

$$W_C = \int_0^{20} 10(20 - y)\,dy$$

$$= 10\left[20y - \frac{1}{2}y^2\right]_0^{20} = 2000$$

or 2000 ft-lb. So the work required to pull in the anchor from a height of 20 ft is

$$W = W_A + W_C = 16{,}000 + 2{,}000 = 18{,}000$$

or 18,000 ft-lb.

Note   A 2-horsepower winch with a capacity of 1100 ft-lb/sec can pull in this anchor in approximately 16 sec.

■ Work Done by an Expanding Gas

EXAMPLE 6   Figure 5 shows the cross section of a cylindrical casing of internal radius r. When the confined gas expands, the resulting increase in pressure exerts a force against the piston, moving it and thus causing work to be done. If the confined gas has a pressure of p lb/in.^{2} and the gas expands from a volume of V_0 in.^{3} to V_1 in.^{3}, show that the work done by the expanding gas is

$$W = \int_{V_0}^{V_1} p\,dV$$

FIGURE 5
Work is done by the expanding gas against the piston.

Solution   Draw the x-axis parallel to the side of the casing as shown in Figure 6, and suppose that the piston has initial and final positions $x = a$ and $x = b$, respectively.

FIGURE 6
The work done by the expanding gas in moving the piston from x to $x + \Delta x$ is ΔW, which is approximately $p(x)(\pi r^2)\Delta x$.

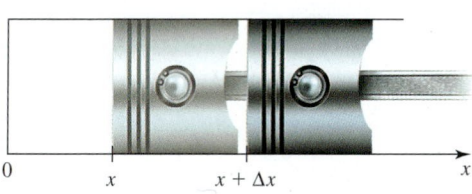

The force exerted by the expanding gas against the piston head at x ($a < x < b$) is

$$F(x) = p(x)(\pi r^2) \qquad \text{pressure} \cdot \text{area}$$

so the work done by the force in moving the piston a distance of Δx from x to $x + \Delta x$ is

$$\Delta W \approx p(x)(\pi r^2)\Delta x \qquad \text{constant force} \cdot \text{distance}$$

Summing the work done by the force in moving the piston over each of the subintervals in the interval $[a, b]$ and taking the limit, we see that the work done is

$$W = \int_a^b p(x)\pi r^2 \, dx$$

To express this integral in terms of the volume of the gas, observe that the volume V of the gas is related to x by $V = \pi r^2 x$, so $dV = \pi r^2 \, dx$. Furthermore, observe that when $x = a$, $V = V_0$, and when $x = b$, $V = V_1$. Therefore,

$$W = \int_{V_0}^{V_1} p \, dV \qquad \blacksquare$$

5.5 CONCEPT QUESTIONS

1. **a.** A force of 3 lb moves an object along a coordinate line from $x = 0$ to $x = 10$ (x is measured in feet). What is the work done by the force on the object?
   b. A force of magnitude 3 lb acts on an object in the negative direction with respect to a coordinate line as the object moves from $x = 0$ to $x = 10$ (x is measured in feet). What is the work done by the force on the object?
   c. As an object moves in the coordinate plane from the point $A(0, 0)$ to the point $B(10, 0)$ along the x-axis, a

force of magnitude 5 lb acts on the body in the positive y-direction. What is the work done by the force on the object? Explain.

2. **a.** Can the work done on a body by a force be negative? Explain with an example.
   b. A force acts on an object situated on a coordinate line. If the work done by the force on the object is 0 ft-lb, does this mean that the force has magnitude 0 and/or the distance moved by the object is 0 ft? Explain.

5.5 EXERCISES

1. Find the work done in lifting a 50-lb sack of potatoes to a height of 4 ft above the ground.

2. How much work is done in lifting a 4-kg bag of rice to a height of 1.5 m above the ground?

3. A particle moves a distance of 100 ft along a straight line. As it moves, it is acted upon by a constant force of magnitude 5 lb in a direction opposite to that of the motion. What is the work done by the force?

4. An engine crane is used to raise a 400-lb engine a vertical distance of 2 ft so that it can be placed in an engine dolly. Find the work done by the crane.

5. Find the work done by the force $F(x) = 2x - 1$ (measured in pounds) in moving an object along the x-axis from $x = -2$ to $x = 4$ (x is measured in feet).

6. Find the work done by the force $f(x) = 4/x^2$ (measured in pounds) in moving a particle along the x-axis from $x = 1$ to $x = 6$ (x is measured in feet).

7. When a particle is at the point x on the x-axis, it is acted upon by a force of $x^2 + 2x$ newtons. Find the work done by the force in moving the particle from the origin to the point $x = 3$ (x is measured in meters).

8. A particle moves along the x-axis from $x = 1$ to $x = 3$. As it moves, it is acted upon by a force $F(x) = -3x^2 + x$. If x is measured in meters and $F(x)$ is measured in newtons, find the work done by the force.

9. When a particle is at the point x on the x-axis, it is acted upon by a force of $\sin \pi x$ newtons. Find the work done by the force in moving the particle from $x = 1$ to $x = 2$ (x is measured in meters).

10. A force of 8 lb is required to stretch a spring 2 in. beyond its natural length. Find the work required to stretch the spring 3 in. beyond its natural length.

11. A force of 20 N is required to stretch a spring 3 cm beyond its natural length of 24 cm. Find the work required to stretch the spring from 30 to 35 cm.

12. Suppose that it takes 3 J of work to stretch a spring 5 cm beyond its natural length. How much work is required to stretch the spring from 2 cm beyond its natural length to 4 cm beyond its natural length?

13. A spring has a natural length of 8 in. If it takes a force of 14 lb to compress the spring to a length of 6 in., how much work is required to compress the spring from its natural length to 7 in.?

14. A chain with length 5 m and mass 30 kg is lying on the ground. Find the work done in pulling one end of the chain vertically upward to a height of 2 m.

15. A chain weighing 5 lb/ft hangs vertically from a winch located 12 ft above the ground, and the free end of the chain is just touching the ground. Find the work done by the winch in pulling in the whole chain.

16. A chain weighing 5 lb/ft hangs vertically from a winch located 16 ft above the ground, and the free end of the chain is 3 ft from the ground. Find the work done by the winch in pulling in 4 ft of the chain.

17. A steel girder weighing 200 lb is hoisted from ground level to the roof of a 60-ft building using a chain that weighs 2 lb/running foot. Find the work done.

18. An aquarium has the shape of a rectangular tank of length 4 ft, width 2 ft, and height 3 ft. If the tank is filled with water weighing 62.4 lb/ft^3, find the work required to empty the tank by pumping the water over the top of the tank.

19. A tank having the shape of a right-circular cylinder with a radius of 4 ft and a height of 6 ft is filled with water weighing 62.4 lb/ft^3. Find the work required to empty the tank by pumping the water over the top of the tank.

20. Leaking Bucket A bucket weighing 4 lb when empty and attached to a rope of negligible weight is used to draw water from a well that is 30 ft deep. Initially, the bucket contains 40 lb of water, but as it is pulled up at a constant rate of 2 ft/sec, the water leaks out of the bucket at the rate of 0.2 lb/sec. Find the work done in pulling the bucket to the top of the well.

21. Leaking Bucket A bucket weighing 4 lb when empty and attached to a rope of negligible weight is used to draw water from a well that is 40 ft deep. Initially, the bucket contains 40 lb of water and is pulled up at a constant rate of 2 ft/sec.

Halfway up, the bucket springs a leak and begins to lose water at the rate of 0.2 lb/sec. Find the work done in pulling the bucket to the top of the well.

22. A tank having the shape of a right circular cylinder with a radius of 5 ft and a height of 6 ft is filled with water weighing 62.4 lb/ft^3. Find the work required to empty the tank by pumping the water out of the tank through a pipe that extends to a height of 2 ft beyond the top of the tank.

23. A tank has the shape of an inverted right circular cone with a base radius of 2 m and a height of 5 m. If the tank is filled with water to a height of 3 m, find the work required to empty the tank by pumping the water over the top of the tank. (The mass of water is 1000 kg/m^3.)

24. Consider the tank described in Exercise 23. If water is pumped in through the bottom of the tank, find the work required to fill the empty tank to a depth of 2 m.

25. Emptying a Storage Tank A gasoline storage tank in the shape of a right cylinder of radius 3 ft and length 12 ft is buried in the ground in a horizontal position. If the top of the tank is 4 ft below the surface, find the work required to empty a full tank of gasoline weighing 42 lb/ft^3 by pumping it through a pipe that extends to a height of 2 ft above the ground.

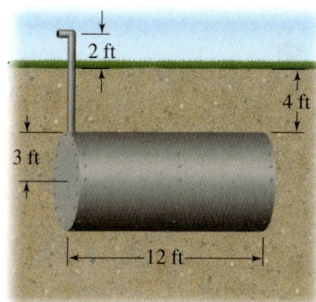

26. Emptying a Trough An 8-ft-long trough has ends that are equilateral triangles with sides that are 2 ft long. If the trough is full of water weighing 62.4 lb/ft^3, find the work required to empty it by pumping the water through a pipe that extends 1 ft above the top of the trough.

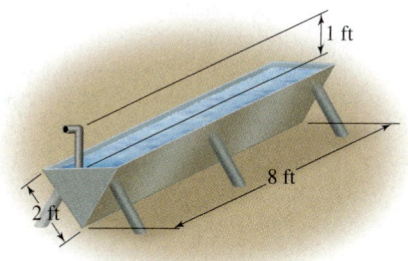

27. Emptying a Trough An 8-ft-long trough has ends that are semi-circles of radius 2 ft. If the trough is full of water weighing 62.4 lb/ft^3, find the work required to empty it by pumping the water through a pipe that extends 1 ft above the top of the trough.

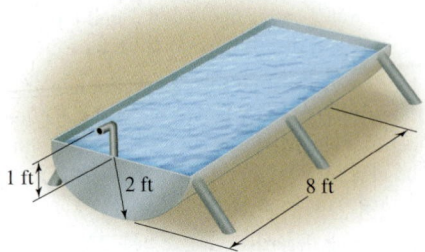

28. A boiler has the shape of a (lower) hemisphere of radius 5 ft. If it is filled with water weighing 62.4 lb/ft^3, find the work required to empty the boiler by pumping the water over the top of the boiler.

29. Refer to Example 6. Suppose that the pressure P and volume V of the steam in a steam engine are related by the law $PV^{1.4} = 100,000$, where P is measured in pounds per square inch and V is measured in cubic inches. Find the work done by the steam as it expands from a volume of 100 in.^{3} to a volume of 400 in.^{3}.

30. Refer to Example 6. The pressure P and volume V of the steam in a steam engine are related by the equation $PV^{1.2} = k$, where k is a constant. If the initial pressure of the steam is P_0 lb/in.^{2} and its initial volume is V_0 in.^{3}, find an expression for the work done by the steam as it expands to a volume of four times its initial volume.

31. Launching a Rocket Newton's Law of Gravitation states that two bodies having masses m_1 and m_2 attract each other with a force

$$F = G\frac{m_1 m_2}{r^2}$$

where G is the gravitational constant and r is the distance between the two bodies. Assume that the mass of the earth is 5.97×10^{24} kg and is concentrated at the center of the earth, the radius of the earth is 6.37×10^6 m, and $G = 6.67 \times 10^{-11}$ N-m^2/kg^2. Find the work required to launch a rocket of mass 500,000 kg vertically upwards to a height of 10,000 km.

32. Launching a Rocket Show that the work W required to launch a rocket of mass m from the ground vertically upward to a height h is given by the formula

$$W = \frac{mgRh}{R + h}$$

where R is the radius of the earth.

Hint: Use Newton's Law of Gravitation given in Exercise 31 and follow these steps: (i) Let m and M denote the mass of the rocket and the earth, respectively, so that $F = GmM/r^2$, where $R \leq r \leq R + h$. At $r = R$ the force will be the weight of the rocket, that is, $mg = GmM/R^2$. Therefore, $G = gR^2/M$, so $F = mgR^2/r^2$. (ii) $W = \int_R^{R+h} F\, dr$.

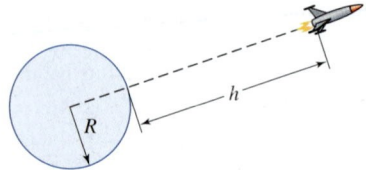

33. Launching a Lunar Landing Module A lunar landing module with a weight of 20,000 lb, as measured on the earth, is to be launched vertically upward from the surface of the moon to a height of 20 mi. Taking the radius of the moon to be 1100 mi and its gravitational force to be one sixth that of the earth, find the work required to accomplish the task. **Hint:** See Exercise 32.

34. Work Done by a Repulsive Charge Coulomb's Law states that the force exerted on two point charges q_1 and q_2 separated by a distance r is given by

$$F = \frac{1}{4\pi\varepsilon_0}\frac{q_1 q_2}{r^2}$$

where ε_0 is a constant known as the permittivity of free space. Suppose that an electrical charge q_1 is concentrated at the origin of the coordinate line and that it repulses a like charge q_2 from the point $x = a$ to the point $x = b$. Show that the work W done by the repulsive force is given by

$$W = \frac{q_1 q_2}{4\pi\varepsilon_0}\left(\frac{1}{a} - \frac{1}{b}\right)$$

35. Work Done by a Repulsive Charge An electric charge Q distributed uniformly along a ring-shaped conductor of radius a repulses a like charge q along the line perpendicular to the plane of the ring, through its center. The magnitude of the force acting on the charge q when it is at the point x is given by

$$F = \frac{1}{4\pi\varepsilon_0} \cdot \frac{qQx}{(x^2 + R^2)^{3/2}}$$

and the force acts in the direction of the positive x-axis. Find the work done by the force of repulsion in moving the charge q from $x = a$ to $x = b$.

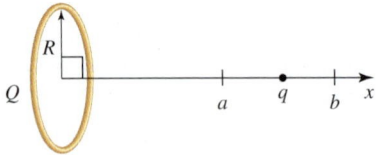

36. Work Done by an Expanding Gas In Example 6 we showed that the work done by an expanding gas against a piston as its volume expands from V_0 to V_1 is given by

$$W = \int_{V_0}^{V_1} p \, dV$$

where p is the pressure of the gas. If the pressure and volume of a gas are related by the equation $pV = k$, where k is a positive constant, show that $W = k \ln(V_1/V_0)$.

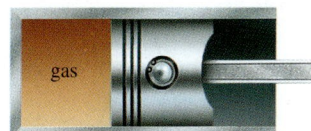

As the gas expands, work is done by the expanding gas against the piston.

37. Work Done by an Expanding Gas Refer to Exercise 36. At high pressure, the relationship between the volume V and pressure P of gases is approximated by the van der Waals equation:

$$\left(P + \frac{an^2}{V^2}\right)(V - nb) = nRT$$

where R is the gas constant, n is the number of moles, and a and b are constants having different values for different gases. (In the special case in which $a = b = 0$, we have the ideal gas equation.) Calculate the work done by a van der Waals gas when it undergoes isothermal expansion ($T =$ constant) from a volume of V_0 to a volume of V_1. Reconcile your result with that of Exercise 36 when $a = b = 0$ (that is, when expansion occurs under normal pressure).

38. The following table shows the force $F(x)$ (in pounds) exerted on an object as it is moved along a coordinate axis from $x = 0$ to $x = 10$ (x is measured in feet). Use Simpson's Rule to estimate the work done by the force.

x (ft)	0	1	2	3	4	5
$F(x)$ (lb)	0	0.69	1.61	2.28	2.88	3.20

x (ft)	6	7	8	9	10
$F(x)$ (lb)	3.58	3.95	4.20	4.38	4.64

39. Work and Kinetic Energy A force $F(x)$ acts on a body of mass m moving it along a coordinate axis. Show that the work done by the force in moving the body from $x = x_1$ to $x = x_2$ is

$$W = \int_{x_1}^{x_2} F(x) \, dx = \frac{1}{2} m v_2^2 - \frac{1}{2} m v_1^2$$

where v_1 and v_2 are the velocities of the body when it is at $x = x_1$ and $x = x_2$, respectively.
Hint: Use Newton's Second Law of Motion $\left(F = m\dfrac{dv}{dt}\right)$ and the Chain Rule to write

$$\frac{dv}{dt} = \frac{dv}{dx} \cdot \frac{dx}{dt} = v\frac{dv}{dx}$$

The quantity $\frac{1}{2} m v^2$ is the *kinetic energy* of a body of mass m moving with a velocity v. Thus, the work done by the force is equal to the net change in the kinetic energy of the body.

40. Refer to Exercise 39. A 4-kg block is attached to a horizontal spring with a spring constant of 400 N/m. The spring is compressed 5 cm from equilibrium and released from rest. Find the speed of the block when the spring is at its equilibrium position.

5.6 Fluid Pressure and Force

Whether designing a hydroelectric dam, an aquarium, or a submarine, an engineer must consider the *pressure* exerted by the water on the walls or surfaces of the object. (See Figure 1.)

FIGURE 1

■ Fluid Pressure

Consider a thin horizontal plate of area A ft^2 submerged to a depth of h ft in a liquid of weight density δ lb/ft^3 (Figure 2a). The force acting on the surface of the plate is just the weight of the column of liquid above it (Figure 2b). Since the volume of this column of liquid is Ah ft^3 and its weight density is δ lb/ft^3, we see that the force exerted on the plate by the liquid is given by

$$F = \delta Ah \qquad \text{weight density} \cdot \text{volume}$$

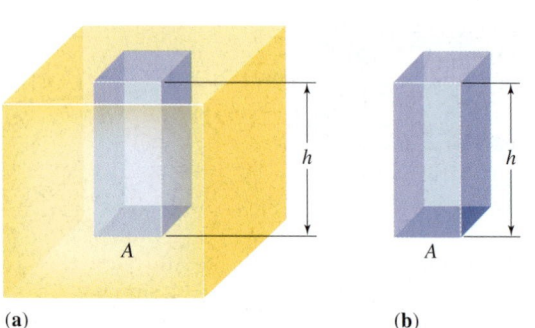

FIGURE 2

The force exerted by the fluid on the horizontal plate (a) is the weight of the column of liquid above it.

(a)   **(b)**

The pressure exerted by the liquid on the horizontal plate is

$$P = \delta h \qquad \text{force divided by area}$$

in lb/ft^2.

EXAMPLE 1   A rectangular fish aquarium has a base measuring 2 ft by 4 ft. (See Figure 3.) Find the pressure and the force exerted on the base of the tank when the tank is filled with water to a height of $1\frac{1}{2}$ ft. (The weight density of water is 62.4 lb/ft^3.)

FIGURE 3

A rectangular fish aquarium with base 2 ft × 4 ft

Solution   The pressure exerted by the water on the base of the tank is

$$P = \delta h = (62.4)(1.5) = 93.6$$

or 93.6 lb/ft^2.

Since the area of the base of the tank is $(4)(2)$ or 8 ft^2, we see that the force exerted on the base of the tank is

$$F = PA = (93.6)(8) \qquad \text{pressure} \cdot \text{area}$$

$$= 748.8$$

or 748.8 lb.

■

In the study of hydrostatics we are guided by the following important physical law: *The pressure at any point in a liquid is the same in all directions.* Thus, the water pressure at a point on the wall of a swimming pool h ft from the surface of the water is the same as that at a point located away from the sides of the pool and h ft from the surface of the water. (See Figure 4.) The pressure is δh lb/ft^2 and is the same in every direction. This physical law, known as Pascal's Principle, is named after the French mathematician Blaise Pascal (1623–1662).

FIGURE 4

A cross section of a swimming pool

Referring once again to Figure 4, you can see that as we move vertically downward along the wall of the swimming pool, the depth of the water increases, and, therefore, the water pressure on the wall increases as well. Thus, unlike the case of a thin *horizontal* plate, in which the pressure is constant at every point on the plate, we have here a situation in which the pressure varies as we proceed down the vertical wall. How do we find the force exerted by the water against the wall?

To answer this question, let's consider the more general situation in which a thin vertical plate is submerged in a liquid with weight density δ lb/ft^3 as shown in Figure 5.

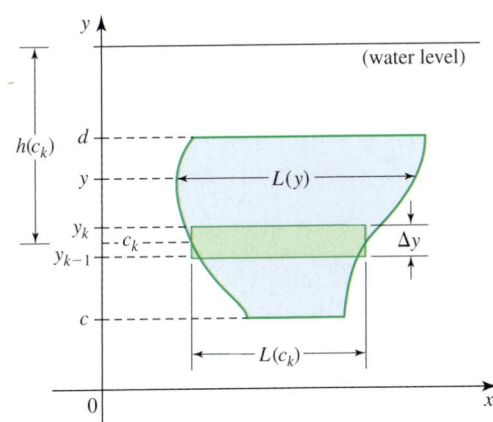

FIGURE 5

$L(y)$ gives the length of the vertical plate at y.

Let $P = \{y_0, y_1, \ldots, y_n\}$ be a regular partition of $[c, d]$, and let c_k be any point in $[y_{k-1}, y_k]$. If $\Delta y = (d - c)/n$ is small, then the depth of the kth (representative) rectangular strip, shown shaded in Figure 5, is approximately $h(c_k)$. Its length is approximately $L(c_k)$, where $L(y)$ is the horizontal length of the plate at y. Therefore, the force exerted by the liquid on this representative rectangular strip is

$$\Delta F_k \approx \delta h(c_k)L(c_k)\Delta y \qquad \text{pressure} \cdot \text{area}$$

and so the sum

$$\sum_{k=1}^{n} \Delta F_k \approx \sum_{k=1}^{n} \delta h(c_k)L(c_k)\Delta y$$

provides us with an approximation of the force F exerted by the liquid on the vertical plate. Recognizing this sum to be a Riemann sum of the function $g(y) = \delta h(y)L(y)$ on $[c, d]$, we have the following definition.

DEFINITION Force Exerted by a Fluid

The **force F exerted by a fluid** of constant weight density δ on one side of a submerged vertical plate from $y = c$ to $y = d$, where $c \le d$, is given by

$$F = \lim_{n \to \infty} \sum_{k=1}^{n} \delta h(c_k)L(c_k)\Delta y = \delta \int_{c}^{d} h(y)L(y)\, dy \qquad (1)$$

where $h(y)$ is the depth of the fluid at y and $L(y)$ is the horizontal length of the plate at y.

EXAMPLE 2 **Fluid Pressure** The vertical wall on the deep end of a rectangular swimming pool is 20 ft wide and 8 ft high. If water in the swimming pool is filled to a height of 7 ft as measured from the bottom of the wall, find the force exerted on the wall by the water. (The weight density of water is 62.4 lb/ft^3.)

Solution Imagine that the wall is placed on a coordinate system with the bottom of the wall lying along the x-axis, as shown in Figure 6. Here, the width of the wall is constant, so the length of the thin horizontal strip at y is $L(y) = 20$. The depth of the fluid at y is $h(y) = 7 - y$. Therefore, the force exerted by the water on the wall is

$$F = \delta \int_{c}^{d} h(y)L(y)\, dy$$

$$= 62.4 \int_{0}^{7} (7 - y)(20)\, dy = 1248 \int_{0}^{7} (7 - y)\, dy$$

$$= 1248 \left[7y - \frac{1}{2}y^2 \right]_{0}^{7} = 30{,}576$$

or 30,576 lb.

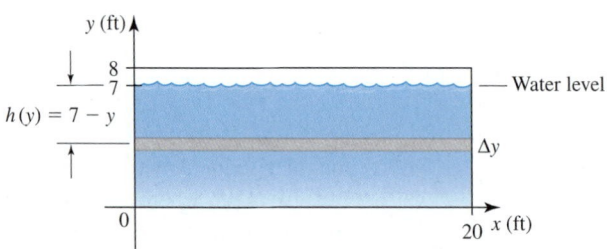

FIGURE 6
We want to find the force exerted on the wall of a swimming pool by the water. Here, the width of the wall is constant.

EXAMPLE 3 **Fluid Pressure** The vertical gate of a dam has the shape of a trapezoid as shown in Figure 7. What is the force on the gate when the surface of the water is 2 ft above the top of the gate?

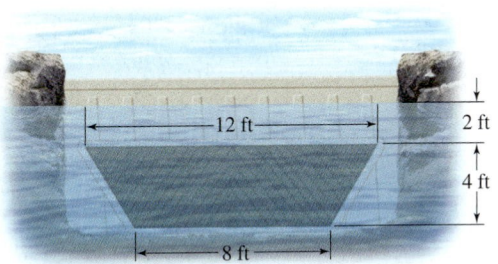

FIGURE 7
We want to find the force exerted on the gate of the dam by the water. Here, the width of the gate varies.

Solution Let's introduce a coordinate system so that the x-axis coincides with the water level, as shown in Figure 8. The length of the horizontal strip is $L(y) = 8 + 2t$. To find t, refer to Figure 8b. By similar triangles we have

$$\frac{t}{2} = \frac{6 + y}{4} \quad \text{or} \quad t = \frac{1}{2}(6 + y)$$

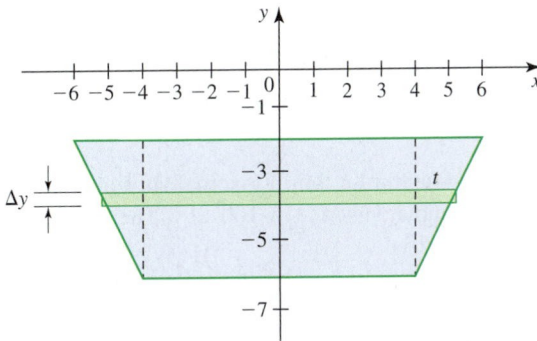

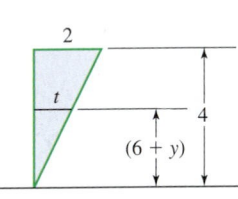

FIGURE 8   (a) The length of the strip is $8 + 2t$.   (b) We use similar triangles to find t.

So

$$L(y) = 8 + 2t = 8 + 6 + y = 14 + y$$

The depth of the fluid at y is $h(y) = -y$. Therefore, the force exerted by the water on the gate is

$$F = \delta \int_c^d h(y)L(y)\, dy = 62.4 \int_{-6}^{-2} (-y)(14 + y)\, dy$$

$$= 62.4 \left[-7y^2 - \frac{1}{3}y^3 \right]_{-6}^{-2}$$

$$= 62.4 \left[\left(-28 + \frac{8}{3} \right) - (-252 + 72) \right] = 9651.2$$

or 9651.2 lb.

EXAMPLE 4 The viewing port of a modern submersible used in oceanographic research has a radius of 1 ft. If the vertical viewing port is 100 ft under water as measured from its center, find the force exerted on it by the water.

Solution   Let's choose a coordinate system so that its origin coincides with the center of the viewing port. Then the viewing port is described by the equation $x^2 + y^2 = 1$. (See Figure 9.)

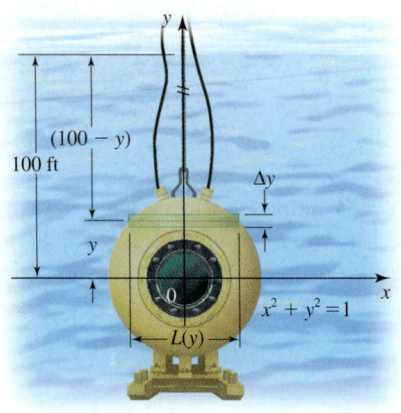

FIGURE 9
The viewing port of a modern submersible

The length of a thin horizontal strip at y is $L(y) = 2x = 2\sqrt{1 - y^2}$, and the depth of the fluid at y is $h(y) = 100 - y$. Therefore, the force exerted by the water on the viewing port is

$$F = \delta \int_c^d h(y)L(y)\, dy = 62.4 \int_{-1}^{1} (100 - y)(2)\sqrt{1 - y^2}\, dy$$

$$= 12{,}480 \int_{-1}^{1} \sqrt{1 - y^2}\, dy - 124.8 \int_{-1}^{1} y\sqrt{1 - y^2}\, dy$$

The second integral on the right is zero because the integrand is an odd function (see Theorem 4 in Section 4.5). To evaluate the first integral, observe that it represents the area of a semicircular disk with radius 1. Therefore,

$$F = 12{,}480 \int_{-1}^{1} \sqrt{1 - y^2}\, dy$$

$$= 12{,}480 \left(\frac{1}{2}\pi\right)(1)^2 = 6240\pi \approx 19{,}604$$

or 19,604 lb.

5.6 CONCEPT QUESTIONS

1. Explain Pascal's Principle.
2. **a.** A thin vertical plate is submerged in a fluid of constant weight density δ. Its length at a depth of y is $L(y)$ for $c \le y \le d$. Write an integral giving the force exerted on one side of the plate.

   b. If the plate described in part (a) is submerged in the fluid so that the plate is parallel to the surface of the liquid at a depth of h feet, what is the force exerted on the plate?

5.6 EXERCISES

1. An aquarium is 3 ft long, 1 ft wide, and 1 ft deep. If the aquarium is filled with water, find the force exerted by the water (a) on the bottom of the aquarium, (b) on the longer side of the aquarium, and (c) on the shorter side of the aquarium.

2. A rectangular swimming pool is 40 ft long, 15 ft wide, and 9 ft deep. If the pool is filled with water to a depth of 8 ft, find the force exerted by the water (a) on the bottom of the pool and (b) on one end of the pool.

In Exercises 3–10, you are given the shape of the vertical ends of a trough that is completely filled with water. Find the force exerted by the water on one end of the trough.

3.
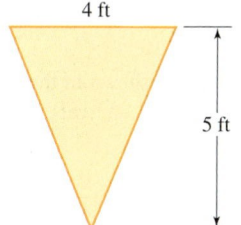
4 ft
5 ft

4.
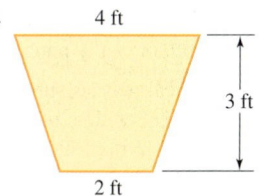
4 ft
3 ft
2 ft

5.
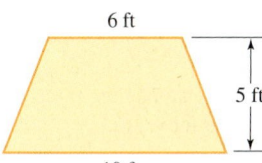
6 ft
5 ft
10 ft

6.
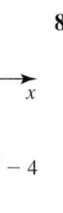
3 ft

7.

y (ft)
-2 0 2 x
$y = x^2 - 4$
-4

8.
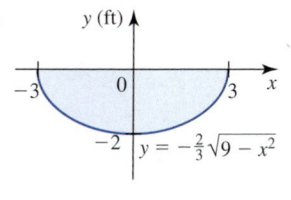
y (ft)
-3 0 3 x
-2 $y = -\frac{2}{3}\sqrt{9 - x^2}$

9.
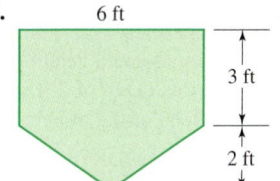
6 ft
3 ft
2 ft

10.
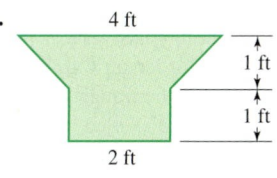
4 ft
1 ft
1 ft
2 ft

In Exercises 11–14, a vertical plate is submerged in water (the surface of the water coincides with the x-axis). Find the force exerted by the water on the plate.

11.
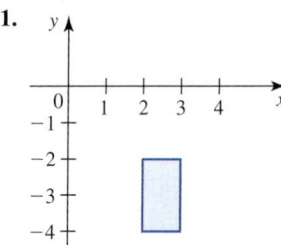
y
0 1 2 3 4 x
-1
-2
-3
-4

12.
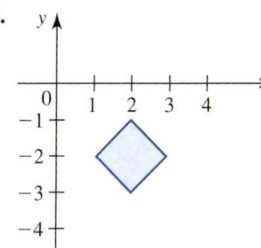
y
0 1 2 3 4 x
-1
-2
-3
-4

13.
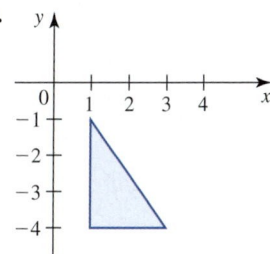
y
0 1 2 3 4 x
-1
-2
-3
-4

14.

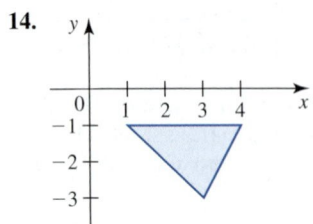

15. A trough has vertical ends that are equilateral triangles with sides of length 2 ft. If the trough is filled with water to a depth of 1 ft, find the force exerted by the water on one end of the trough.

16. A trough has vertical ends that are trapezoids with parallel sides of length 4 ft (top) and 2 ft (bottom) and a height of 3 ft. If the trough is filled with water to a depth of 2 ft, find the force exerted by the water on one end of the trough.

17. A cylindrical drum of diameter 4 ft and length 8 ft is lying on its side, submerged in water 12 ft deep. Find the force exerted by the water on one end of the drum.

18. A cylindrical oil storage tank of diameter 4 ft and length 8 ft is lying on its side. If the tank is half full of oil that weighs 50 lb/ft^3, find the force exerted by the oil on one end of the tank.

19. The first figure shows a vertical dam with a parabolic gate, and the second figure shows an enlargement of the parabolic gate.

20 ft

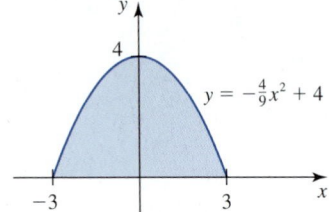

$y = -\frac{4}{9}x^2 + 4$

a. Find the force exerted by the water on the gate when the water is 10 ft deep.
b. The gate is designed to withstand twice the force that the water will exert on it under flood conditions (when the water level is 20 ft deep). What is this force?

20. Redo Exercise 19 for a semicircular gate as shown in the figure.

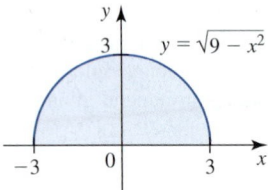

$y = \sqrt{9 - x^2}$

21. A rectangular tank has width 2 ft, height 3 ft, and length 6 ft. It is filled with equal volumes of water and oil. The oil has a weight density of 50 lb/ft^3 and floats on the water. Find the force exerted by the mixture on one end of the tank.

22. A rectangular swimming pool is 25 ft wide, 60 ft long, and 4 ft deep at the shallow end and 9 ft deep at the deep end. Its bottom is an inclined plane. If the pool is completely filled with water, find the force exerted by the water on each side of the pool.

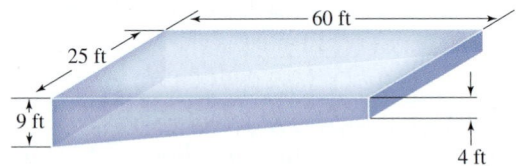

60 ft

25 ft

9 ft

4 ft

23. Refer to Exercise 22. Find the force exerted by the water on the bottom of the pool.

24. A vertical plate is submerged in water as shown in the figure below. The widths of the plate taken at $\frac{1}{2}$-ft intervals are recorded in the following table.

Depth (ft)	2.0	2.5	3.0	3.5	4.0	4.5	5.0
Width of plate (ft)	0	2.9	3.4	3.0	2.2	1.7	0

Use Simpson's Rule with $n = 6$ to estimate the force exerted by the water on one side of the plate.

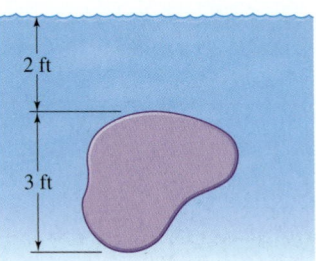

2 ft

3 ft

5.7 | Moments and Center of Mass

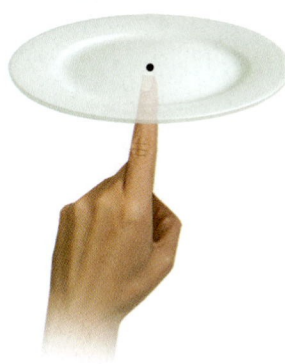

FIGURE 1
The center of mass of a circular plate is located at the center of the plate.

As every juggler knows, many objects will remain in equilibrium if supported at a certain point. As an example, for a homogeneous circular plate, this point is located at the center of the plate and is called the center of mass of the plate. (See Figure 1.)

The knowledge of the location of the center of mass of a body or a system of bodies is important in physics and engineering. As a matter of practical interest, every motorist knows that a car wheel must be balanced when a new tire is installed. Because of defects in the tire-manufacturing process, a wheel is seldom balanced when a new tire is installed; that is, the center of mass of the wheel is not located at "dead center." An unbalanced wheel causes the car to shimmy.

Before we learn how to find the center of mass of plane regions, we need to recall some basic notions from physics.

■ Measures of Mass

The **mass** of a body is the quantity of matter in the body. In the English system the unit of mass is the **slug;** in the international system (SI) the unit of mass is the **kilogram;** and in the centimeter-gram-second system (cgs) the unit of mass is the **gram.** On the surface of the earth, where the constant of acceleration due to gravity, g, is approximately 32 ft/sec^2 in the English system, 9.8 m/sec^2 in the international system, and 980 cm/sec^2 in the cgs system, Newton's Second Law of Motion ($F = ma$) tells us that

a body of mass m slugs has a weight of $32m$ pounds,

a body of mass m kilograms has a weight of $9.8m$ newtons, and

a body of mass m grams has a weight of $980m$ dynes.

■ Center of Mass of a System on a Line

Consider a simple system consisting of two particles of mass m_1 and m_2 connected by a rod of negligible mass. If we place this system on a fulcrum as shown in Figure 2, then equilibrium is achieved if

$$m_1 d_1 = m_2 d_2 \tag{1}$$

where d_1 and d_2 are the distances (called *moment arms*) between the particles and the fulcrum. The quantity $m_1 d_1$, called the *moment* of m_1 about the fulcrum, is a measure of the tendency of m_1 to rotate the system about the fulcrum (in this case in the counterclockwise direction). On the other hand, the moment $m_2 d_2$ is a measure of the tendency of m_2 to rotate the system about the fulcrum (in a clockwise direction). Balance is achieved when these moments are equal, that is, when Equation (1) holds.

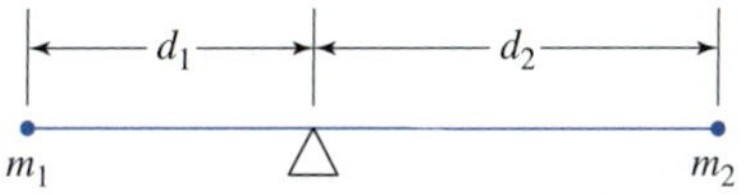

FIGURE 2
The condition for equilibrium of the system is $m_1 d_1 = m_2 d_2$.

We can use Equation (1) to derive a formula for calculating the center of mass of the system. Place the system on a coordinate line, and suppose that the coordinates of m_1, m_2, and the fulcrum are x_1, x_2, and $\bar{x}$, respectively, as shown in Figure 3. You can see immediately that the distance between m_1 and the fulcrum is $d_1 = \bar{x} - x_1$ and that the distance between m_2 and the fulcrum is $d_2 = x_2 - \bar{x}$. Therefore, Equation (1) gives

$$m_1(\bar{x} - x_1) = m_2(x_2 - \bar{x})$$

$$m_1\bar{x} + m_2\bar{x} = m_1x_1 + m_2x_2$$

and

$$\bar{x} = \frac{m_1x_1 + m_2x_2}{m_1 + m_2} \tag{2}$$

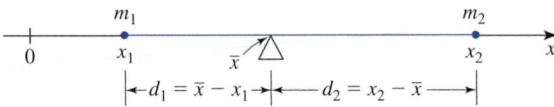

FIGURE 3
The system placed on a coordinate line

The numbers m_1x_1 and m_2x_2 in Equation (2) are called the *moments* of the masses m_1 and m_2 about the origin.

In general, if m is a mass located at the point x on a coordinate line, then mx is called the **moment of the mass m about the origin.** If you think of the mass m as being connected to the origin by a rod of negligible mass, then mx measures the tendency of m to rotate the rod about the origin. Observe that Equation (2) says that to find the coordinate of the center of mass of a system comprising two masses, m_1 and m_2, add the moments of the masses about the origin and divide the sum by the total mass $m_1 + m_2$. A similar analysis of a system comprising n particles located on a coordinate line, as shown in Figure 4, leads to the definition of the center of mass of that system.

FIGURE 4

A system of n masses connected by a rod of negligible mass on a coordinate line

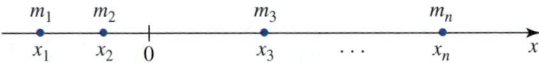

DEFINITION   The Center of Mass of a System of n Masses on a Line

Let S denote a system of n masses $m_1, m_2, \ldots, m_n$ located at $x_1, x_2, \ldots, x_n$, lying on a line, respectively, and let $m = \sum_{k=1}^{n} m_k$ denote the total mass of the system.

1. The **moment of S about the origin** is

$$M = \sum_{k=1}^{n} m_k x_k \tag{3a}$$

2. The **center of mass of S** is located at

$$\bar{x} = \frac{M}{m} = \frac{1}{m} \sum_{k=1}^{n} m_k x_k \tag{3b}$$

Note   If we write Equation (3b) in the form $m\bar{x} = M$, we obtain the following interpretation of the center of mass: Think of the total mass of the system as being concentrated at the center of mass $\bar{x}$. Then the moment of this mass about the origin will be the same as the moment of the system about the origin.

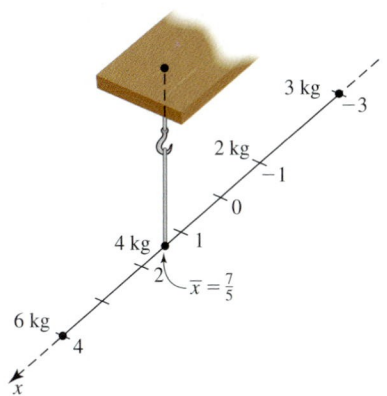

FIGURE 5
If the system is suspended at $\bar{x} = \frac{7}{5}$, it will hang in equilibrium horizontally.

EXAMPLE 1 Find the center of mass of a system of four objects located at the points -3, -1, 2, and 4, on the x-axis (x in meters), with masses 3, 2, 4, and 6 kilograms, respectively.

Solution Using Equation (3b) with $m_1 = 3$, $m_2 = 2$, $m_3 = 4$, and $m_4 = 6$ and $x_1 = -3$, $x_2 = -1$, $x_3 = 2$, and $x_4 = 4$ gives the coordinate of the center of mass of the system as

$$\bar{x} = \frac{3(-3) + 2(-1) + 4(2) + 6(4)}{3 + 2 + 4 + 6} = \frac{21}{15} = \frac{7}{5} \quad \text{or} \quad 1.4 \text{ m} \quad \blacksquare$$

Interpreting Our Results

Think of the four masses as being connected by a rod of length 7 m and of mass very small in comparison to the mass of the four given objects. If the system is suspended by a string at $\bar{x}$, then the system will hang in equilibrium horizontally. (See Figure 5.)

Center of Mass of a System in the Plane

Consider a particle of mass m located at the point $P(x, y)$ in a coordinate plane. (See Figure 6.) If you think of this mass as being connected to the x-axis by a rod perpendicular to the axis and of negligible mass, then the quantity my measures the tendency of the mass m to rotate the system about the x-axis. This quantity is called the **moment of the mass m about the x-axis** and is denoted by M_x. Similarly, we define the **moment of the mass m about the y-axis** to be $M_y = mx$. To find the moments about the x- and y-axes of a system comprising n particles in the plane, we simply add the respective moments of each mass. (See Figure 7.)

This leads to the following definition.

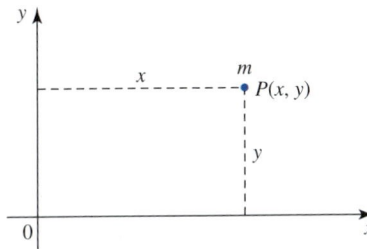

FIGURE 6
A particle of mass m located at the point $P(x, y)$

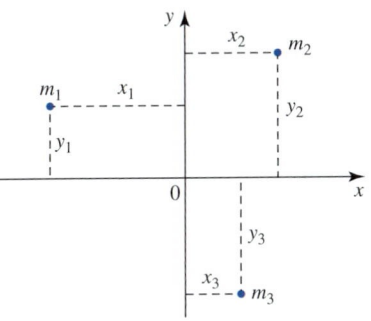

FIGURE 7
A system with three masses ($n = 3$)

DEFINITION **The Center of Mass of a System of n Particles in a Plane**

Let S denote a system of n particles with masses $m_1, m_2, \ldots, m_n$ located at the points (x_1, y_1), (x_2, y_2), ..., (x_n, y_n), respectively, and let $m = \sum_{k=1}^{n} m_k$ denote the total mass of the system.

1. The **moment of S about the x-axis** is

$$M_x = \sum_{k=1}^{n} m_k y_k \qquad \text{(4a)}$$

2. The **moment of S about the y-axis** is

$$M_y = \sum_{k=1}^{n} m_k x_k \qquad \text{(4b)}$$

3. The **center of mass of S** is located at the point $(\bar{x}, \bar{y})$ where

$$\bar{x} = \frac{M_y}{m} = \frac{1}{m} \sum_{k=1}^{n} m_k x_k \quad \text{and} \quad \bar{y} = \frac{M_x}{m} = \frac{1}{m} \sum_{k=1}^{n} m_k y_k \qquad \text{(4c)}$$

Note The center of mass of a system of n particles in the plane is the point at which, if the total mass m of the system were concentrated, it would generate the same moments as the system. This can be seen by writing Equation (4c) in the form $m\bar{x} = M_y$ and $m\bar{y} = M_x$. $\quad \blacksquare$

EXAMPLE 2 Find the center of mass of a system comprising three particles with masses 2, 3, and 5 slugs, located at the points $(-2, 2)$, $(4, 6)$, and $(2, -3)$, respectively. (Assume that all distances are measured in feet.)

Solution We first compute the moments

$$M_y = 2(-2) + 3(4) + 5(2) = 18$$

or 18 slug-ft, and

$$M_x = 2(2) + 3(6) + 5(-3) = 7$$

or 7 slug-ft. Since $m = 2 + 3 + 5 = 10$, an application of Equation (4c) yields

$$\bar{x} = \frac{M_y}{m} = \frac{18}{10} = \frac{9}{5} \quad \text{and} \quad \bar{y} = \frac{M_x}{m} = \frac{7}{10}$$

feet. Therefore, the center of mass of the system is located at $\left(\frac{9}{5}, \frac{7}{10}\right)$. ■

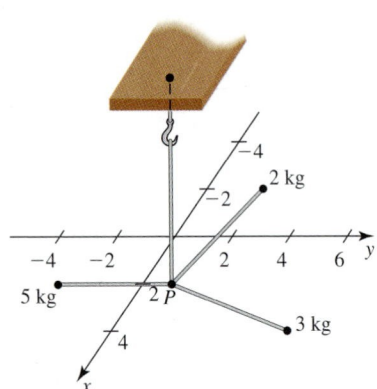

FIGURE 8
When the system is suspended by a string at P, it will rest in a horizontal position.

Interpreting Our Results

Think of the three particles as being connected by rods of negligible mass to the center of mass, P. If the system is suspended by a string at P, then it will rest in a horizontal position, much like a mobile. (See Figure 8.)

■ Center of Mass of Laminas

We now turn our attention to the problem of finding the center of mass of a lamina (a thin, flat plate). We will assume that the laminas we consider are homogeneous, that is, that they have uniform mass density ρ (the Greek letter rho), where ρ is a positive constant.

Let's begin by assuming that the lamina L has the shape of the region R under the graph of a continuous nonnegative function f on the interval $[a, b]$, as shown in Figure 9.

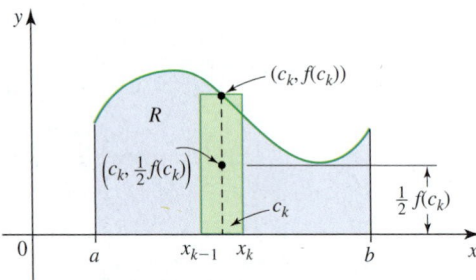

FIGURE 9
The lamina has the shape of the region R. The kth lamina is highlighted.

Let $P = \{x_0, x_1, \ldots, x_n\}$ be a regular partition of $[a, b]$. This partition divides R into n nonoverlapping subregions $R_1, \ldots, R_n$, each of which is again a lamina. The kth subregion is approximated by the kth rectangle of width $\Delta x = (b - a)/n$ and height $f(c_k)$, where c_k is the midpoint of the kth subinterval $[x_{k-1}, x_k]$, that is, $c_k = (x_k + x_{k-1})/2$. The area of the kth rectangle is $f(c_k)\Delta x$, so the mass of the kth lamina is approximately

$$\rho f(c_k)\Delta x \qquad \text{density} \cdot \text{area}$$

Next, since the center of mass of a rectangular lamina is located at its center, we conclude that the center of mass of the kth lamina is located at the point $\left(c_k, \frac{1}{2}f(c_k)\right)$.

This tells us that the moment arm of the kth lamina with respect to the y-axis is c_k and, therefore, that the moment of the kth lamina about the y-axis is

$$[\rho f(c_k)\Delta x]c_k = \rho c_k f(c_k)\Delta x \qquad \text{mass} \cdot \text{moment arm}$$

Adding the moments of the n laminas and taking the limit of the associated Riemann sum as $n \to \infty$ lead to the following definition of the moment of L about the y-axis:

$$M_y = \lim_{n \to \infty} \sum_{k=1}^{n} \rho c_k f(c_k)\Delta x = \rho \int_a^b x f(x) \, dx$$

Similarly, by observing that the moment arm of the kth rectangle about the x-axis is $\frac{1}{2}f(c_k)$, we see that the moment of L about the x-axis may be defined as

$$M_x = \lim_{n \to \infty} \sum_{k=1}^{n} \rho \cdot \frac{1}{2}[f(c_k)]^2 \Delta x = \rho \int_a^b \frac{1}{2}[f(x)]^2 \, dx$$

(See Figure 9.) Finally, the mass of L may be defined as

$$m = \lim_{n \to \infty} \sum_{k=1}^{n} \rho f(c_k)\Delta x = \rho \int_a^b f(x) \, dx$$

DEFINITION Moments and Center of Mass of a Lamina

Let L denote a lamina of constant mass density ρ, and suppose that L has the shape of the region R under the graph of a nonnegative continuous function f on $[a, b]$.

1. The **mass** of L is

$$m = \rho \int_a^b f(x) \, dx = \rho A \qquad \text{(5a)}$$

where $A = \int_a^b f(x) \, dx$ is the area of R.

2. The **moments of L about the x- and the y-axis** are

$$M_x = \rho \int_a^b \frac{1}{2}[f(x)]^2 \, dx \qquad \text{(5b)}$$

and

$$M_y = \rho \int_a^b x f(x) \, dx \qquad \text{(5c)}$$

3. The **center of mass** of L is located at $(\bar{x}, \bar{y})$, where

$$\bar{x} = \frac{M_y}{m} = \frac{1}{A}\int_a^b x f(x) \, dx \qquad \text{and} \qquad \bar{y} = \frac{M_x}{m} = \frac{1}{A}\int_a^b \frac{1}{2}[f(x)]^2 \, dx \quad \text{(5d)}$$

EXAMPLE 3 A lamina L of uniform area density ρ has the shape of the region R under the graph of $f(x) = x^2$ on $[0, 2]$. (See Figure 10.) Find the mass of L, the moments of L about each of the coordinate axes, and the center of mass of L.

Solution Using Equation (5a), we find that the mass of the lamina is

$$m = \rho \int_0^2 x^2 \, dx = \rho \left[\frac{1}{3}x^3\right]_0^2 = \frac{8\rho}{3}$$

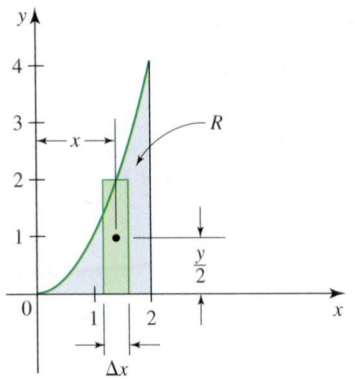

FIGURE 10
The center of mass of a representative rectangle is $\left(x, \frac{y}{2}\right)$.

To find the moment M_x of L about the x-axis, we can use Equation (5b), or we can proceed as follows: Draw a representative rectangle of width Δx and height y (Figure 10). The moment arm of this rectangle with respect to the x-axis is

$$\frac{y}{2} = \frac{f(x)}{2} = \frac{x^2}{2}$$

and the mass of the representative rectangle is

$$\rho y \, \Delta x = \rho f(x) \Delta x = \rho x^2 \, \Delta x \qquad \text{density} \cdot \text{area}$$

So the moment of this element about the y-axis is

$$\left(\frac{x^2}{2}\right) \rho x^2 \, \Delta x \qquad \text{moment arm} \cdot \text{mass}$$

Summing and taking the limit of the Riemann sum, we have

$$M_x = \int_0^2 \left(\frac{x^2}{2}\right) \rho x^2 \, dx = \frac{\rho}{2} \int_0^2 x^4 \, dx = \left[\frac{\rho}{10} x^5\right]_0^2 = \frac{16\rho}{5}$$

To find M_y, we can use Equation (5c) or proceed as before, observing that the moment arm of the representative rectangle is x. Thus,

$$M_y = \int_0^2 x\rho x^2 \, dx = \rho \int_0^2 x^3 \, dx = \left[\frac{\rho}{4} x^4\right]_0^2 = 4\rho$$

Finally, using Equation (5d), we see that the coordinates of the center of mass of L are

$$\bar{x} = \frac{M_y}{m} = \frac{4\rho}{\dfrac{8\rho}{3}} = \frac{3}{2} \qquad \text{and} \qquad \bar{y} = \frac{M_x}{m} = \frac{\dfrac{16\rho}{5}}{\dfrac{8\rho}{3}} = \frac{6}{5} \qquad \blacksquare$$

Interpreting Our Results

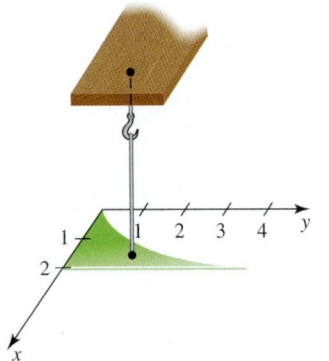

FIGURE 11
If the lamina is suspended at $\left(\frac{3}{2}, \frac{6}{5}\right)$, it will hang horizontally.

If the lamina is suspended by a string at $\left(\frac{3}{2}, \frac{6}{5}\right)$, it will hang in equilibrium horizontally. (See Figure 11.)

Observe that Equation (5d) does not involve ρ, the density of the lamina. This is always true when the lamina has *uniform* density. In other words, the center of mass of such a lamina depends only on the shape of the region R that it occupies in the plane and not on its density. The point $(\bar{x}, \bar{y})$ at which the center of mass of such a lamina is located is called the **centroid** of the region R.

EXAMPLE 4 Find the centroid of the region R under the graph of $y = \sqrt{x}$ on the interval $[0, 4]$. (See Figure 12.)

Solution The area of the region R is

$$A = \int_0^4 \sqrt{x} \, dx = \left[\frac{2}{3} x^{3/2}\right]_0^4 = \frac{16}{3}$$

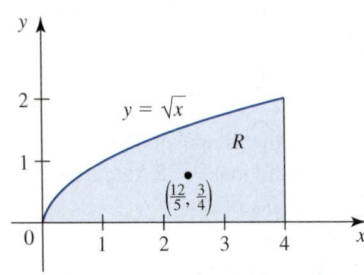

FIGURE 12
The centroid of the region R is $\left(\frac{12}{5}, \frac{3}{4}\right)$.

Using Equation (5d), we have

$$\bar{x} = \frac{1}{A} \int_0^4 x f(x) \, dx = \frac{3}{16} \int_0^4 x^{3/2} \, dx = \frac{3}{16} \left[\frac{2}{5} x^{5/2}\right]_0^4 = \left(\frac{3}{16}\right)\left(\frac{64}{5}\right) = \frac{12}{5}$$

and

$$\bar{y} = \frac{1}{A} \int_0^4 \frac{1}{2} [f(x)]^2 \, dx = \frac{3}{16} \int_0^4 \frac{1}{2} x \, dx = \frac{3}{16} \left[\frac{1}{4} x^2 \right]_0^4 = \left(\frac{3}{16} \right)(4) = \frac{3}{4}$$

Therefore, the centroid of R is $\left(\frac{12}{5}, \frac{3}{4} \right)$. ∎

We can use the following heuristic argument to derive the formulas for the centroid of a region R between the graphs of two functions. Suppose that R is bounded by the graphs of two continuous functions f and g, where $f(x) \geq g(x)$ on an interval $[a, b]$, and to the left and right by the lines $x = a$ and $x = b$. (See Figure 13.) Then let L be a lamina of uniform density that has the shape of R. Draw a representative rectangle of width Δx and height $[f(x) - g(x)]$ and, therefore, area $\Delta A = [f(x) - g(x)]\Delta x$. Its mass is

$$\rho \, \Delta A = \rho[f(x) - g(x)]\Delta x \qquad \text{density} \cdot \text{area}$$

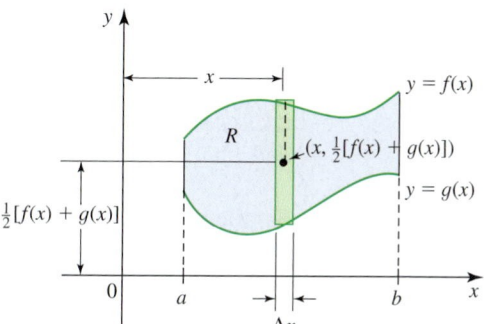

FIGURE 13

The region R lies between the graphs of f and g on $[a, b]$.

Therefore, the mass of R is

$$m = \rho \int_a^b [f(x) - g(x)] \, dx = \rho A$$

where A is the area of R.

Next, the moment of the rectangle about the x-axis is

$$\frac{1}{2} [f(x) + g(x)]\rho[f(x) - g(x)]\Delta x \qquad \text{moment arm} \cdot \text{mass}$$

and this gives

$$M_x = \rho \int_a^b \left[\frac{f(x) + g(x)}{2} \right] [f(x) - g(x)] \, dx = \frac{\rho}{2} \int_a^b \{ [f(x)]^2 - [g(x)]^2 \} \, dx$$

The moment of the rectangle about the y-axis is

$$x\rho[f(x) - g(x)]\Delta x$$

and this gives

$$M_y = \rho \int_a^b x[f(x) - g(x)] \, dx$$

Since the center of mass of L (also called the **centroid of R**) is given by $(\bar{x}, \bar{y})$, where $\bar{x} = M_y/m$ and $\bar{y} = M_x/m$, we have the following result.

DEFINITION **The Centroid of a Region Between Two Curves**

Let R be a region bounded by the graphs of two continuous functions f and g on $[a, b]$, where $f(x) \geq g(x)$ for all x in $[a, b]$. Then the centroid $(\bar{x}, \bar{y})$ of R is given by

$$\bar{x} = \frac{1}{A} \int_a^b x[f(x) - g(x)]\, dx \tag{6a}$$

and

$$\bar{y} = \frac{1}{A} \int_a^b \frac{1}{2}\{[f(x)]^2 - [g(x)]^2\}\, dx \tag{6b}$$

where

$$A = \int_a^b [f(x) - g(x)]\, dx$$

EXAMPLE 5 Find the centroid of the region bounded by the graphs of $y = x^2 - 3$ and $y = -x^2 + 2x + 1$.

Solution The region R in question is shown in Figure 14. The points of intersection of the two graphs are $(-1, -2)$ and $(2, 1)$. If we let $f(x) = -x^2 + 2x + 1$ and $g(x) = x^2 - 3$, then $f(x) \geq g(x)$ on $[-1, 2]$, so the area A of R is

$$A = \int_{-1}^2 [f(x) - g(x)]\, dx$$

$$= \int_{-1}^2 (-2x^2 + 2x + 4)\, dx = \left[-\frac{2}{3}x^3 + x^2 + 4x \right]_{-1}^2 = 9$$

Next, using Equations (6a) and (6b), we have

$$\bar{x} = \frac{1}{A} \int_{-1}^2 x[f(x) - g(x)]\, dx = \frac{1}{9} \int_{-1}^2 (-2x^3 + 2x^2 + 4x)\, dx$$

$$= \frac{1}{9} \left[-\frac{1}{2}x^4 + \frac{2}{3}x^3 + 2x^2 \right]_{-1}^2 = \frac{1}{2}$$

$$\bar{y} = \frac{1}{A} \int_{-1}^2 \frac{1}{2}[(-x^2 + 2x + 1)^2 - (x^2 - 3)^2]\, dx = \frac{1}{9} \int_{-1}^2 (-2x^3 + 4x^2 + 2x - 4)\, dx$$

$$= \frac{1}{9} \left[-\frac{1}{2}x^4 + \frac{4}{3}x^3 + x^2 - 4x \right]_{-1}^2 = -\frac{1}{2} \qquad \blacksquare$$

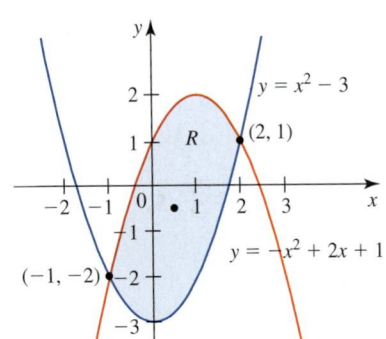

FIGURE 14
The region R bounded by the graphs of $y = x^2 - 3$ and $y = -x^2 + 2x + 1$ has centroid $\left(\frac{1}{2}, -\frac{1}{2}\right)$.

■ The Theorem of Pappus

Suppose that a solid of revolution is obtained by revolving a plane region R about a line. The Theorem of Pappus enables us to find the volume of the solid in terms of the centroid of the region. (See Figure 15.) The theorem is named after the Greek mathematician Pappus of Alexandria, who lived in the fourth century A.D.

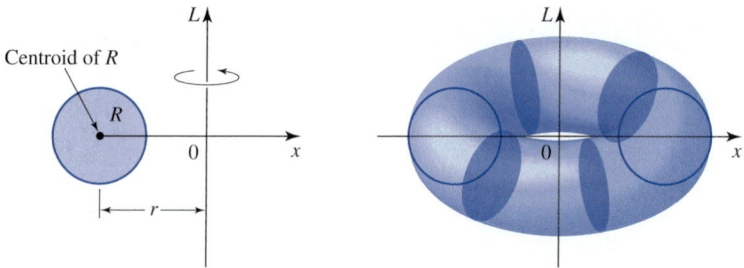

FIGURE 15
The volume of the solid generated by revolving R about L is $(2\pi r)$(area of R).

The Theorem of Pappus

Let R be a plane region that lies entirely on one side of a line L in the same plane. If r is the distance between the centroid of R and the line L, then the volume V of the solid of revolution obtained by revolving R about L is given by

$$V = 2\pi rA$$

where A is the area of R.

Note that $2\pi r$ is the distance traveled by the centroid as the region R is revolved about the line L.

EXAMPLE 6 **Volume of a Torus** A torus (a doughnut-shaped solid) is formed by revolving a circular region of radius a about a line lying in the same plane as the circle and at a distance b $(b > a)$ from the center of the circle. (See Figure 16.) Find the volume of the torus.

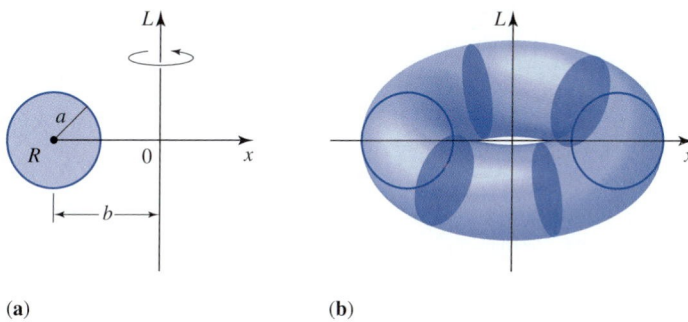

(**a**)   (**b**)

FIGURE 16
If the circular region R in part (a) is revolved about L, the resulting solid of revolution is a torus (b).

Solution The centroid of the circular region is the center of the circle. So the distance traveled by the centroid during one revolution of the circular region is $2\pi b$. Since the area of the region is πa^2, the Theorem of Pappus says that the volume of the torus is

$$V = 2\pi bA = (2\pi b)(\pi a^2) = 2\pi^2 a^2 b$$

5.7 CONCEPT QUESTIONS

1. **a.** Let S be a system of n masses $m_1, m_2, \ldots, m_n$ located at $x_1, x_2, \ldots, x_n$ on a coordinate line, respectively. What is the center of mass of S?

   b. Let S be a system of n masses $m_1, m_2, \ldots, m_n$ located at $(x_1, y_1), (x_2, y_2), \ldots, (x_n, y_n)$ in the plane. What is the moment of S about the x-axis? About the y-axis? What is the center of mass of S?

2. Let L denote a lamina having the shape of a region R under the graph of a nonnegative continuous function f on $[a, b]$ and having uniform mass density ρ. What is the center of mass of L?

3. Let R be a region bounded by the graphs of two continuous functions f and g on $[a, b]$, where $f(x) \geq g(x)$. What is the centroid of R?

4. State the Theorem of Pappus.

5.7 EXERCISES

In Exercises 1–4, find the center of mass of the system comprising masses m_k located at the points x_k on a coordinate line. Assume that mass is measured in kilograms and distance is measured in meters.

1. $m_1 = 2$, $m_2 = 4$, $m_3 = 6$; $x_1 = -3$, $x_2 = -1$, $x_3 = 4$

2. $m_1 = 3$, $m_2 = 1$, $m_3 = 5$, $m_4 = 6$; $x_1 = -4$, $x_2 = -1$, $x_3 = 1$, $x_4 = 3$

3. $m_1 = 4$, $m_2 = 3$, $m_3 = 2$, $m_4 = 4$, $m_5 = 8$; $x_1 = -5$, $x_2 = -3$, $x_3 = -2$, $x_4 = 2$, $x_5 = 4$

4. $m_1 = 6$, $m_2 = 4$, $m_3 = 5$, $m_4 = 8$, $m_5 = 4$; $x_1 = -4$, $x_2 = -2$, $x_3 = 0$, $x_4 = 3$, $x_5 = 6$

In Exercises 5–8, find the center of mass of the system comprising masses m_k located at the points P_k in a coordinate plane. Assume that mass is measured in grams and distance is measured in centimeters.

5. $m_1 = 4$, $m_2 = 3$, $m_3 = 5$; $P_1(-3, -2)$, $P_2(-1, 2)$, $P_3(2, 4)$

6. $m_1 = 2$, $m_2 = 4$, $m_3 = 1$; $P_1(-2, 2)$, $P_2(2, 1)$, $P_3(3, -1)$

7. $m_1 = 3$, $m_2 = 4$, $m_3 = 6$, $m_4 = 5$; $P_1(-3, -2)$, $P_2(-2, 3)$, $P_3(2, 3)$, $P_4(4, -2)$

8. $m_1 = 4$, $m_2 = 1$, $m_3 = 2$, $m_4 = 5$; $P_1(-2, 3)$, $P_2(-1, 4)$, $P_3(1, 4)$, $P_4(4, -3)$

In Exercises 9–24, find the centroid of the region bounded by the graphs of the given equations.

9. $y = -\dfrac{2}{3}x + 2$, $y = 0$, $x = 0$

10. $y = x^2$, $y = 0$, $x = 1$, $x = 2$

11. $y = 4 - x^2$, $y = 0$

12. $y = \sqrt{x}$, $y = 0$, $x = 1$, $x = 4$

13. $y = |x|\sqrt{1 - x^2}$, $y = 0$, $x = -1$, $x = 1$

14. $y = x^3$, $y = 0$, $x = 3$

15. $y = 2x - x^2$, $y = 0$

16. $y = \sqrt{1 - x^2}$, $y = 0$

17. $y = x^{2/3}$, $y = 0$, $x = 8$

18. $y = x^{2/3}$, $y = 4$, $x = 0$

19. $y = x^2$, $y = \sqrt{x}$

20. $y = x^3$, $y = x$, $x = 0$, $x = 1$

21. $y = x^3$, $y = \sqrt[3]{x}$, $x = 0$, $x = 1$

22. $y = \dfrac{1}{x^3}$, $y = 0$, $x = 1$, $x = 2$

23. $y = 6 - x^2$, $y = 3 - 2x$

24. $y = -x^2 + 3$, $y = x^2 - 2x - 1$

In Exercises 25–28, find the centroid of the region shown in the figure.

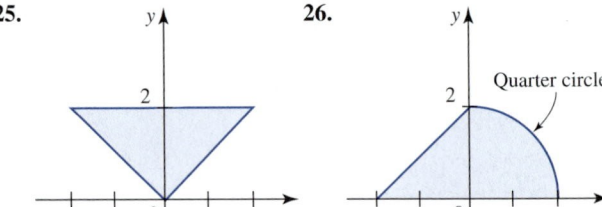

25.

26.

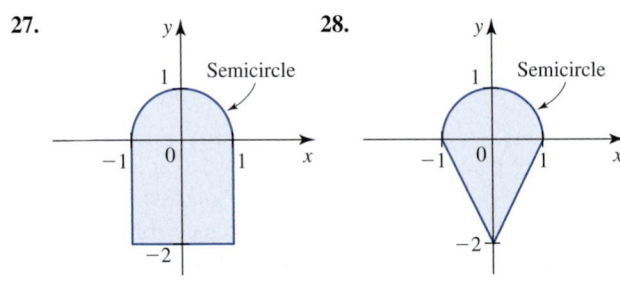

27.

28.

In Exercises 29–32, find the centroid of the region shown in the figure. (You can solve the problem without using integration.)

29.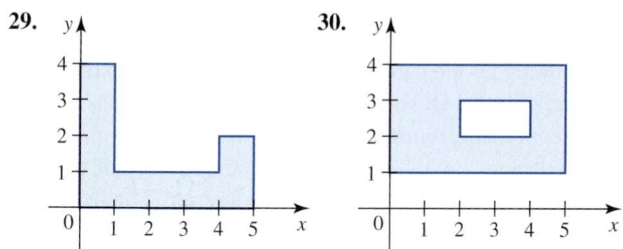

30.

31.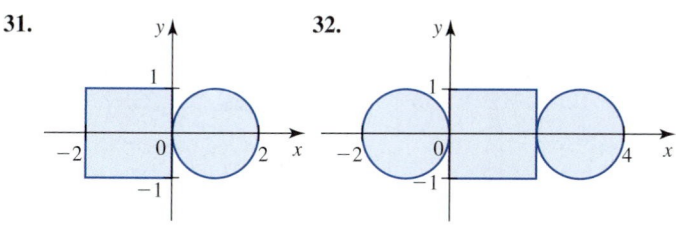

32.

33. Find the center of mass of the lamina of Exercise 31 if the density of the circular lamina is twice that of the square lamina.

34. Find the center of mass of the lamina of Exercise 32 if the density of the circular laminae is 3 times that of the square lamina.

35. Find the centroid of the region bounded by the graphs of $x/a + y/b = 1$, $x = 0$, and $y = 0$.

36. Find the centroid of the region bounded by the graph of the equation $x^{1/2} + y^{1/2} = a^{1/2}$ and the coordinate axes.

37. Prove that the centroid of a triangular region is located at the point of intersection of the medians of the triangle.
Hint: Suppose that the vertices of the triangle are located at $(0, 0)$, $(a, 0)$, and (b, h).

38. Find the centroid of the region bounded by the graphs of $y = \sqrt{1 - x^2}$ and $y = 1 - x$.

39. Find the centroid of the region bounded by the graphs of $y = 1/x$, $y = 0$, $x = 1$, and $x = 2$.

 40. Use a calculator or computer to find the centroid of the region R under the graph of $y = \tan^{-1} x$ on $[0, 1]$ accurate to three decimal places.

In Exercises 41–44, use the Theorem of Pappus to find the volume of the given solid.

41. The torus formed by revolving the region bounded by the circle $(x - 4)^2 + y^2 = 9$ about the y-axis

42. A cone of radius r and height h

43. The solid obtained by revolving the region bounded by the graphs of $y = 4 - x^2$, $y = 4$, and $x = 2$ about the y-axis

44. The solid obtained by revolving the region bounded by the graphs of $y = \sqrt{x - 2}$, $y = 0$, and $x = 6$ about the y-axis

45. Use the Theorem of Pappus to find the centroid of the region bounded by the upper semicircle $y = \sqrt{R^2 - x^2}$ and the x-axis.

46. Use the Theorem of Pappus to show that the y-coordinate of the centroid of a triangular region is located at the point that is one third of the distance along the altitude from the base of the triangle.
Hint: Suppose the vertices of the triangle are located at $(0, 0)$, $(a, 0)$, and (b, h).

In Exercises 47 and 48, C is a curve that is the graph of a continuous function $y = f(x)$ on the interval $[a, b]$, and the moments M_x and M_y of C about the x- and y-axis are defined by $M_x = \int_a^b y \, ds$ and $M_y = \int_a^b x \, ds$, respectively, where $ds = \sqrt{1 + (y')^2} \, dx$ is the element of arc length. The coordinates of the centroid of C are $\bar{x} = M_y/L$ and $\bar{y} = M_x/L$, where L is the arc length of C. Find the centroid of C.

47. $C: y = \sqrt{a^2 - x^2}$, $\quad -a \le x \le a$ (upper semicircle)

48. $C: x^{2/3} + y^{2/3} = a^{2/3}$, $\quad 0 \le x \le a, y \ge 0$ (astroid in the first quadrant)

 49. Find the centroid of the region under the graph of $y = \sin \pi x$ on the interval $[0, 1]$. Find the exact values of $\bar{x}$ and $\bar{y}$.

 50. Find the centroid of the region under the graph of $y = 1/(1 + x^2)$ on the interval $[-1, 1]$.

5.8 Hyperbolic Functions

Figure 1 depicts a uniform flexible cable, such as a telephone or power line, suspended between two poles. The shape assumed by the cable is called a *catenary,* from the Latin word *catena,* which means "chain." Figure 1b shows the path taken by a heat-seeking missile as it locks onto and intercepts an aircraft. We assume here that the aircraft is flying along a straight line at a constant height and at a constant speed and that the missile, also flying at a constant speed is always pointed at the aircraft. The trajectory of the missile is called a *pursuit curve.*

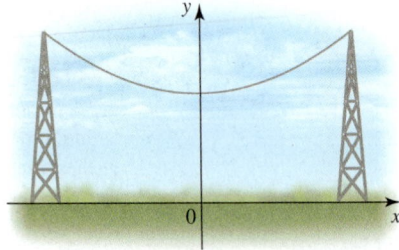
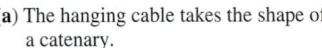
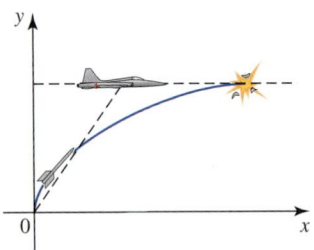

(**a**) The hanging cable takes the shape of a catenary.

(**b**) The trajectory of the missile is called a pursuit curve.

FIGURE 1

Historical Biography

SPL/Photo Researchers, Inc.

JOHANN HEINRICH LAMBERT
(1728-1777)

Not being from a wealthy family, Johann Lambert had to leave school at the age of 12 to work with his father as a tailor to help support his family. Lambert continued to study in the evenings and took a variety of jobs over the next several years, including clerk at an ironworks and secretary to the editor of a Basel newspaper. His father died in 1747. Shortly thereafter, Lambert was hired by the von Salis family to tutor their children and, with more time to devote to study, his scientific career took off. The scientific community eventually noticed Lambert's work in astronomy, and he was given a series of increasingly prestigious academic positions. He became a member of the Prussian Academy of Sciences in 1761 and so became a colleague of Leonhard Euler (page 19) and Joseph-Louis Lagrange (page 316). While he was with the Prussian Academy, Lambert wrote more than 150 papers. In 1776, he published a book on non-Euclidean geometry, and many of his results are still of interest today. Among Lambert's many important contributions is the first systematic development of the hyperbolic functions sinh x, cosh x, and tanh x.

The analysis of problems such as these involves combinations of exponential functions of the form e^{-cx} and e^{cx}, where c is a constant. Because combinations of these functions arise so frequently in mathematics and its applications, they have been given special names. These combinations—the **hyperbolic sine,** the **hyperbolic cosine,** the **hyperbolic tangent,** and so on—are referred to as **hyperbolic functions** and are so called because they have many properties in common with the trigonometric functions.

DEFINITIONS **The Hyperbolic Functions**

$$\sinh x = \frac{e^x - e^{-x}}{2} \qquad \cosh x = \frac{e^x + e^{-x}}{2} \qquad \tanh x = \frac{\sinh x}{\cosh x}$$

$$\operatorname{csch} x = \frac{1}{\sinh x}, \quad x \neq 0 \qquad \operatorname{sech} x = \frac{1}{\cosh x} \qquad \coth x = \frac{\cosh x}{\sinh x}, \quad x \neq 0$$

Note The expression sinh x is pronounced "cinch x," and cosh x is pronounced "kosh x," which rhymes with "gosh x." ▪

■ The Graphs of the Hyperbolic Functions

The graph of $y = \sinh x$ can be drawn by first sketching the graphs of $y = \frac{1}{2} e^x$ and $y = -\frac{1}{2} e^{-x}$ and then adding the y-coordinates of the points on these graphs corresponding to each x to obtain the y-coordinates of the points on $y = \sinh x$ (Figure 2a). Similarly, the graph of $y = \cosh x$ can be drawn by first sketching the graphs of $y = \frac{1}{2} e^x$ and $y = \frac{1}{2} e^{-x}$ and then adding the y-coordinates of the points on these graphs corresponding to each x to obtain the y-coordinates of the points on $y = \cosh x$ (Figure 2b).

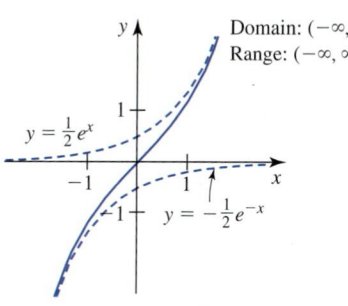

Domain: $(-\infty, \infty)$
Range: $(-\infty, \infty)$

$y = \frac{1}{2}e^x$

$y = -\frac{1}{2}e^{-x}$

(**a**) $y = \sinh x = \dfrac{e^x - e^{-x}}{2}$

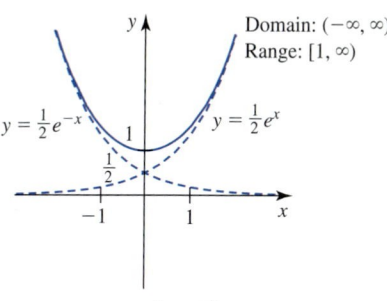

Domain: $(-\infty, \infty)$
Range: $[1, \infty)$

$y = \frac{1}{2}e^{-x}$

$y = \frac{1}{2}e^x$

(**b**) $y = \cosh x = \dfrac{e^x + e^{-x}}{2}$

FIGURE 2
The graphs of the hyperbolic
sine and cosine functions

The graphs of the other four hyperbolic functions are shown in Figure 3.

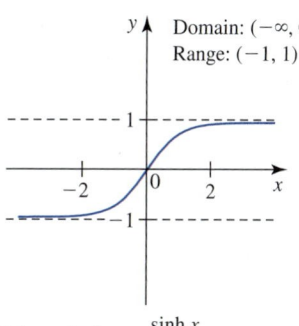

Domain: $(-\infty, \infty)$
Range: $(-1, 1)$

(**a**) $y = \tanh x = \dfrac{\sinh x}{\cosh x}$

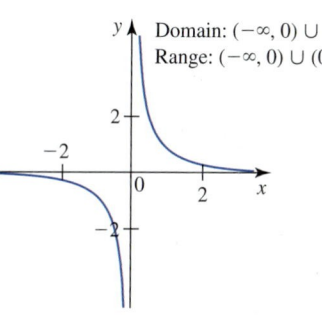

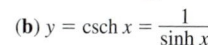

Domain: $(-\infty, 0) \cup (0, \infty)$
Range: $(-\infty, 0) \cup (0, \infty)$

(**b**) $y = \operatorname{csch} x = \dfrac{1}{\sinh x}$

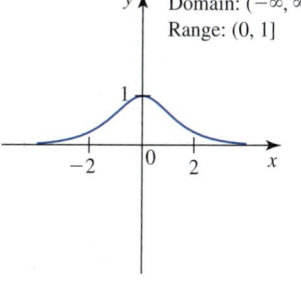

Domain: $(-\infty, \infty)$
Range: $(0, 1]$

(**c**) $y = \operatorname{sech} x = \dfrac{1}{\cosh x}$

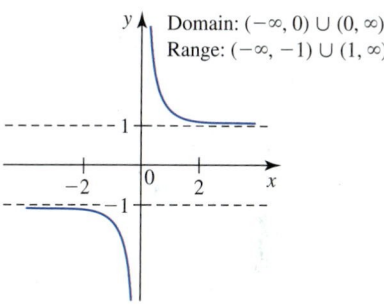

Domain: $(-\infty, 0) \cup (0, \infty)$
Range: $(-\infty, -1) \cup (1, \infty)$

(**d**) $y = \coth x = \dfrac{1}{\tanh x}$

FIGURE 3
The graphs of the hyperbolic
tangent, cosecant, secant,
and cotangent functions

■ Hyperbolic Identities

The hyperbolic functions satisfy certain identities that look very much like those sat-isfied by trigonometric functions. For example, the analog of $\sin(-x) = -\sin x$ is $\sinh(-x) = -\sinh x$. To prove this identity, we simply compute

$$\sinh(-x) = \frac{e^{(-x)} - e^{-(-x)}}{2} = \frac{e^{-x} - e^x}{2} = -\frac{e^x - e^{-x}}{2} = -\sinh x$$

A list of frequently used hyperbolic identities is given in Table 1.

TABLE 1 Hyperbolic Identities

$\sinh(-x) = -\sinh x$	$\cosh(-x) = \cosh x$
$\cosh^2 x - \sinh^2 x = 1$	$\operatorname{sech}^2 x = 1 - \tanh^2 x$
$\sinh(x + y) = \sinh x \cosh y + \cosh x \sinh y$	$\cosh(x + y) = \cosh x \cosh y + \sinh x \sinh y$
$\sinh 2x = 2 \sinh x \cosh x$	$\cosh 2x = \cosh^2 x + \sinh^2 x$
$\cosh^2 x = \frac{1}{2}(1 + \cosh 2x)$	$\sinh^2 x = \frac{1}{2}(-1 + \cosh 2x)$

We will prove the identity $\cosh^2 x - \sinh^2 x = 1$ in Example 1. The proofs of the others will be left as exercises.

EXAMPLE 1 Prove the identity $\cosh^2 x - \sinh^2 x = 1$.

Solution We compute

$$\cosh^2 x - \sinh^2 x = \left(\frac{e^x + e^{-x}}{2}\right)^2 - \left(\frac{e^x - e^{-x}}{2}\right)^2$$

$$= \frac{e^{2x} + 2 + e^{-2x}}{4} - \frac{e^{2x} - 2 + e^{-2x}}{4}$$

$$= \frac{4}{4} = 1$$

and this establishes the identity. ■

■ Derivatives and Integrals of Hyperbolic Functions

Since the hyperbolic functions are defined in terms of e^x and e^{-x}, their derivatives are easily computed. For example,

$$\frac{d}{dx}(\sinh x) = \frac{d}{dx}\left(\frac{e^x - e^{-x}}{2}\right) = \frac{e^x + e^{-x}}{2} = \cosh x$$

Similarly, we can show that

$$\frac{d}{dx}(\cosh x) = \sinh x$$

Then, using these results, we can compute

$$\frac{d}{dx}(\tanh x) = \frac{d}{dx}\frac{\sinh x}{\cosh x} = \frac{\cosh x \dfrac{d}{dx}(\sinh x) - \sinh x \dfrac{d}{dx}(\cosh x)}{\cosh^2 x}$$

$$= \frac{\cosh^2 x - \sinh^2 x}{\cosh^2 x} = \frac{1}{\cosh^2 x} = \operatorname{sech}^2 x$$

Following are the differentiation formulas together with the corresponding integration formulas for the six hyperbolic functions. We have assumed that $u = g(x)$, where g is a differentiable function, and we have used the Chain Rule. The proofs of these formulas are left as exercises.

Derivatives and Integrals of Hyperbolic Functions

$$\frac{d}{dx}(\sinh u) = (\cosh u)\frac{du}{dx} \qquad \int \cosh u\, du = \sinh u + C$$

$$\frac{d}{dx}(\cosh u) = (\sinh u)\frac{du}{dx} \qquad \int \sinh u\, du = \cosh u + C$$

$$\frac{d}{dx}(\tanh u) = (\operatorname{sech}^2 u)\frac{du}{dx} \qquad \int \operatorname{sech}^2 u\, du = \tanh u + C$$

$$\frac{d}{dx}(\operatorname{csch} u) = -(\operatorname{csch} u \coth u)\frac{du}{dx} \qquad \int \operatorname{csch} u \coth u\, du = -\operatorname{csch} u + C$$

$$\frac{d}{dx}(\operatorname{sech} u) = -(\operatorname{sech} u \tanh u)\frac{du}{dx} \qquad \int \operatorname{sech} u \tanh u\, du = -\operatorname{sech} u + C$$

$$\frac{d}{dx}(\coth u) = -(\operatorname{csch}^2 u)\frac{du}{dx} \qquad \int \operatorname{csch}^2 u\, du = -\coth u + C$$

EXAMPLE 2

a. $\dfrac{d}{dx}\sinh(x^2 + 1) = \cosh(x^2 + 1)\dfrac{d}{dx}(x^2 + 1) = 2x\cosh(x^2 + 1)$

b. $\dfrac{d}{dx}\cosh^2(\ln 2x) = 2\cosh(\ln 2x)\dfrac{d}{dx}\cosh(\ln 2x)$

$$= 2\cosh(\ln 2x)\sinh(\ln 2x)\frac{d}{dx}\ln 2x$$

$$= \frac{2}{x}\cosh(\ln 2x)\sinh(\ln 2x) \qquad \blacksquare$$

EXAMPLE 3 Find $\int \cosh^2 3x \sinh 3x\, dx$.

Solution Let $u = 3x$ so that $du = 3\, dx$ or $dx = \frac{1}{3}\, du$. Then

$$\int \cosh^2 3x \sinh 3x\, dx = \frac{1}{3}\int \cosh^2 u \sinh u\, du$$

Next, let $v = \cosh u$ so that $dv = \sinh u\, du$. Then

$$\frac{1}{3}\int \cosh^2 u \sinh u\, du = \frac{1}{3}\int v^2\, dv = \frac{1}{9}v^3 + C$$

So

$$\int \cosh^2 3x \sinh 3x\, dx = \frac{1}{9}\cosh^3 3x + C \qquad \blacksquare$$

■ Inverse Hyperbolic Functions

If you examine Figures 2a and 3a, you will notice that both $\sinh x$ and $\tanh x$ are *one-to-one* on $(-\infty, \infty)$ and hence have inverse functions that we denote by $\sinh^{-1} x$ and $\tanh^{-1} x$, respectively. Also, an examination of Figure 2b shows that $\cosh x$ is one-to-one on $[0, \infty)$, so, if restricted to this domain, it has an inverse, $\cosh^{-1} x$. By examining the graphs of the other hyperbolic functions and making the necessary restrictions on their domains, we are able to define the other inverse hyperbolic functions.

DEFINITIONS **Inverse Hyperbolic Functions**

			Domain
$y = \sinh^{-1} x$	if and only if	$x = \sinh y$	$(-\infty, \infty)$
$y = \cosh^{-1} x$	if and only if	$x = \cosh y$	$[1, \infty)$
$y = \tanh^{-1} x$	if and only if	$x = \tanh y$	$(-1, 1)$
$y = \operatorname{csch}^{-1} x$	if and only if	$x = \operatorname{csch} y$	$(-\infty, 0) \cup (0, \infty)$
$y = \operatorname{sech}^{-1} x$	if and only if	$x = \operatorname{sech} y$	$(0, 1]$
$y = \coth^{-1} x$	if and only if	$x = \coth y$	$(-\infty, -1) \cup (1, \infty)$

The graphs of $y = \sinh^{-1} x$, $y = \cosh^{-1} x$, and $y = \tanh^{-1} x$ are shown in Figure 4.

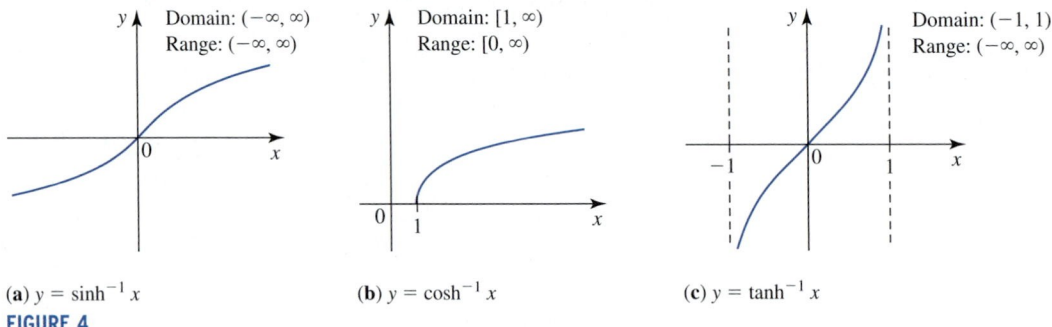

(a) $y = \sinh^{-1} x$ (b) $y = \cosh^{-1} x$ (c) $y = \tanh^{-1} x$

FIGURE 4

Since the hyperbolic functions are defined in terms of exponential functions, it seems natural that the inverse hyperbolic functions should be expressible in terms of logarithmic functions.

EXAMPLE 4 Show that $\sinh^{-1} x = \ln\left(x + \sqrt{x^2 + 1}\right)$.

Solution Let $y = \sinh^{-1} x$. Then

$$x = \sinh y = \frac{e^y - e^{-y}}{2}$$

or

$$e^y - 2x - e^{-y} = 0$$

On multiplying both sides of this equation by e^y, we obtain

$$e^{2y} - 2xe^y - 1 = 0$$

which is a quadratic in e^y. Using the quadratic formula, we have

$$e^y = \frac{2x \pm \sqrt{4x^2 + 4}}{2} = x \pm \sqrt{x^2 + 1}$$

Only the root $x + \sqrt{x^2 + 1}$ is admissible. To see why, observe that $e^y > 0$, but $x - \sqrt{x^2 + 1} < 0$, since $x < \sqrt{x^2 + 1}$. Therefore, we have

$$e^y = x + \sqrt{x^2 + 1}$$

so

$$y = \ln\!\left(x + \sqrt{x^2 + 1}\right)$$

that is,

$$\sinh^{-1} x = \ln\!\left(x + \sqrt{x^2 + 1}\right)$$

Proceeding in a similar manner, we can obtain the representations of the other five inverse hyperbolic functions in terms of logarithmic functions. Three such representations follow.

Representations of Inverse Hyperbolic Functions in Terms of Logarithmic Functions

	Domain
$\sinh^{-1} x = \ln\!\left(x + \sqrt{x^2 + 1}\right)$	$(-\infty, \infty)$
$\cosh^{-1} x = \ln\!\left(x + \sqrt{x^2 - 1}\right)$	$[1, \infty)$
$\tanh^{-1} x = \dfrac{1}{2}\ln\!\left(\dfrac{1 + x}{1 - x}\right)$	$(-1, 1)$

Derivatives of Inverse Hyperbolic Functions

The derivatives of the inverse hyperbolic functions can be found by differentiating the function in question directly. For example,

$$\frac{d}{dx}\sinh^{-1} x = \frac{d}{dx}\ln\!\left(x + \sqrt{x^2 + 1}\right)$$

$$= \frac{1}{x + \sqrt{x^2 + 1}}\left[1 + \frac{1}{2}(x^2 + 1)^{-1/2}(2x)\right]$$

$$= \frac{1}{x + \sqrt{x^2 + 1}} \cdot \frac{\sqrt{x^2 + 1} + x}{\sqrt{x^2 + 1}}$$

$$= \frac{1}{\sqrt{x^2 + 1}}$$

Alternatively, we may proceed as follows:

$$y = \sinh^{-1} x \qquad \text{if and only if} \qquad x = \sinh y$$

Differentiating this last equation implicitly with respect to x, we obtain

$$\frac{d}{dx}(x) = \frac{d}{dx}(\sinh y)$$

$$1 = (\cosh y)\frac{dy}{dx}$$

or

$$\frac{dy}{dx} = \frac{1}{\cosh y} = \frac{1}{\sqrt{\sinh^2 y + 1}} = \frac{1}{\sqrt{x^2 + 1}}$$

as before.

Using techniques such as these, we obtain the following formulas for differentiating the inverse hyperbolic functions (once again, $u = g(x)$, where g is a differentiable function).

Derivatives of Inverse Hyperbolic Functions

$$\frac{d}{dx}\sinh^{-1} u = \frac{1}{\sqrt{u^2 + 1}}\frac{du}{dx} \qquad \frac{d}{dx}\cosh^{-1} u = \frac{1}{\sqrt{u^2 - 1}}\frac{du}{dx}$$

$$\frac{d}{dx}\tanh^{-1} u = \frac{1}{1 - u^2}\frac{du}{dx} \qquad \frac{d}{dx}\operatorname{csch}^{-1} u = -\frac{1}{|u|\sqrt{u^2 + 1}}\frac{du}{dx}$$

$$\frac{d}{dx}\operatorname{sech}^{-1} u = -\frac{1}{u\sqrt{1 - u^2}}\frac{du}{dx} \qquad \frac{d}{dx}\coth^{-1} u = \frac{1}{1 - u^2}\frac{du}{dx}$$

EXAMPLE 5   Find the derivative of $y = x^2 \operatorname{sech}^{-1} 3x$.

Solution   We have

$$\frac{dy}{dx} = \operatorname{sech}^{-1} 3x \cdot \frac{d}{dx}(x^2) + x^2 \frac{d}{dx}\operatorname{sech}^{-1} 3x \qquad \text{\textcolor{orange}{Use the Product Rule.}}$$

$$= 2x \operatorname{sech}^{-1} 3x - x^2\left[\frac{1}{3x\sqrt{1 - 9x^2}}\right]\frac{d}{dx}(3x)$$

$$= 2x \operatorname{sech}^{-1} 3x - \frac{x}{\sqrt{1 - 9x^2}}$$

■ An Application

EXAMPLE 6   **Length of a Power Line**   A power line is suspended between two towers as depicted in Figure 5. The shape of the cable is a *catenary* with equation

$$y = 80 \cosh\frac{x}{80} \qquad -100 \le x \le 100$$

where x is measured in feet. Find the length of the cable.

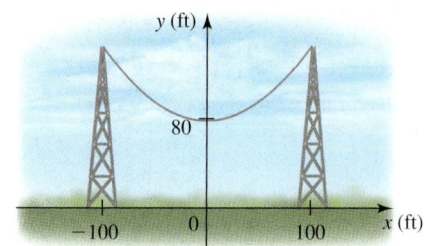

FIGURE 5
The shape of the hanging
cable is a catenary.

Solution   Taking advantage of the symmetry of the situation, we see that the required length is given by

$$L = 2\int_0^{100} \sqrt{1 + \left(\frac{dy}{dx}\right)^2}\, dx$$

But

$$\frac{dy}{dx} = \frac{d}{dx}\left[80 \cosh \frac{x}{80}\right] = 80 \sinh \frac{x}{80} \cdot \frac{d}{dx}\left(\frac{x}{80}\right) = \sinh \frac{x}{80}$$

So

$$\sqrt{1 + \left(\frac{dy}{dx}\right)^2} = \sqrt{1 + \sinh^2\left(\frac{x}{80}\right)} = \sqrt{1 + \cosh^2\left(\frac{x}{80}\right) - 1}$$

$$= \sqrt{\cosh^2\left(\frac{x}{80}\right)} = \cosh \frac{x}{80}$$

Therefore,

$$L = 2\int_0^{100} \cosh \frac{x}{80}\, dx$$

$$= 2\left[80 \sinh \frac{x}{80}\right]_0^{100} \qquad \text{Use the substitution } u = \frac{x}{80}.$$

$$= 160 \sinh \frac{100}{80} = 160 \sinh \frac{5}{4}$$

or approximately 256 ft.   ∎

5.8   CONCEPT QUESTIONS

1. Define (a) $\sinh x$, (b) $\cosh x$, and (c) $\tanh x$.
2. State the derivative of (a) $\sinh x$, (b) $\cosh x$, and (c) $\tanh x$.
3. Write an antiderivative of (a) $\operatorname{sech}^2 u$, (b) $\operatorname{sech} u \tanh u$, and (c) $\operatorname{csch}^2 u$.

4. Define (a) $\operatorname{csch}^{-1} x$, (b) $\operatorname{sech}^{-1} x$, and (c) $\coth^{-1} x$.
5. Write (a) $\sinh^{-1} x$, (b) $\cosh^{-1} x$, and (c) $\tanh^{-1} x$ in terms of logarithmic functions.
6. State the derivative of (a) $\sinh^{-1} u$, (b) $\cosh^{-1} u$, and (c) $\tanh^{-1} u$ with respect to x.

5.8 EXERCISES

In Exercises 1–6, find the value of the expression accurate to four decimal places.

1. a. $\sinh 2$ **b.** $\cosh 4$ **c.** $\operatorname{sech} 3$

2. a. $\operatorname{csch} 3$ **b.** $\tanh(-2)$ **c.** $\coth 5$

3. a. $\cosh 0$ **b.** $\operatorname{sech}(-1)$ **c.** $\operatorname{csch}(\ln 2)$

4. a. $\sinh^{-1} 1$ **b.** $\cosh^{-1} 2$ **c.** $\operatorname{sech}^{-1} \dfrac{1}{3}$

5. a. $\operatorname{csch}^{-1} 2$ **b.** $\operatorname{csch}^{-1}(-2)$ **c.** $\coth^{-1} \dfrac{3}{2}$

6. a. $\tanh^{-1}\left(-\dfrac{1}{2}\right)$ **b.** $\cosh^{-1}(\ln 5)$ **c.** $\tanh^{-1}(\sinh 0)$

In Exercises 7–16, prove the identity.

7. $\cosh(-x) = \cosh x$

8. $\tanh(-x) = -\tanh x$

9. $\operatorname{sech}^2 x + \tanh^2 x = 1$

10. $\sinh^2 x = \dfrac{\cosh 2x - 1}{2}$

11. $\cosh^2 x = \dfrac{1 + \cosh 2x}{2}$

12. $\sinh 2x = 2 \sinh x \cosh x$

13. $\cosh 2x = \cosh^2 x + \sinh^2 x$

14. $\sinh(x + y) = \sinh x \cosh y + \cosh x \sinh y$

15. $\cosh(x + y) = \cosh x \cosh y + \sinh x \sinh y$

16. $\tanh(x + y) = \dfrac{\tanh x + \tanh y}{1 + \tanh x \tanh y}$

17. If $\sinh x = \frac{4}{3}$, find the values of the other hyperbolic functions at x.

18. If $\cosh x = \frac{5}{4}$, find the values of the other hyperbolic functions at x.

In Exercises 19–54, find the derivative of the function.

19. $f(x) = \sinh 3x$

20. $f(x) = \cosh(2x + 1)$

21. $g(x) = \tanh(1 - 3x)$

22. $h(x) = \operatorname{sech}(x^2)$

23. $f(t) = e^t \sinh t$

24. $y = \coth \dfrac{1}{x}$

25. $F(x) = \ln(\cosh x)$

26. $y = \ln(\sinh 3x)$

27. $g(u) = \tanh(\cosh u^2)$

28. $h(s) = \coth(\cosh 2s)$

29. $f(t) = \cosh^2(3t^2 + 1)$

30. $f(x) = \sinh 2x \cosh 4x$

31. $g(v) = v \sinh v^2$

32. $F(t) = \cosh \sqrt{2t^2 + 1}$

33. $f(x) = \tanh(e^{2x} + 1)$

34. $y = \cosh \sqrt[3]{x^2 + 1}$

35. $f(x) = (\cosh x - \sinh x)^{2/3}$

36. $y = e^{\sinh 2t}$

37. $g(x) = \tanh^{-1}(\cosh x)$

38. $f(x) = \sqrt{2 + \coth 3x}$

39. $f(x) = \dfrac{\sinh x}{1 + \cosh x}$

40. $g(x) = \dfrac{\sinh x}{x}$

41. $y = \dfrac{\cosh^{-1} t}{1 + \tanh 2t}$

42. $f(x) = e^{-x} \operatorname{sech} 2x$

43. $f(x) = \sinh^{-1} 3x$

44. $g(x) = \tanh^{-1} \dfrac{x}{2}$

45. $y = \sqrt{\cosh^{-1} 2x}$

46. $f(x) = \operatorname{sech}^{-1} x^3$

47. $f(x) = \operatorname{sech}^{-1} \sqrt{2x + 1}$

48. $y = e^x \operatorname{sech}^{-1} x$

49. $y = x \cosh^{-1} x^2$

50. $g(x) = \ln(\tanh^{-1} x)$

51. $f(x) = \operatorname{sech}^{-1} \sqrt{x}$

52. $h(x) = \cosh^{-1}(\sinh x)$

53. $y = \sqrt{9x^2 - 1} - 3 \cosh^{-1} 3x$

54. $y = 2x \coth^{-1} 2x - \ln \sqrt{1 - 4x^2}$

In Exercises 55–62, find the given integral.

55. $\displaystyle\int \cosh(2x + 3)\, dx$

56. $\displaystyle\int \dfrac{\sinh \sqrt{x}}{\sqrt{x}}\, dx$

57. $\displaystyle\int \sqrt{\sinh x}\, \cosh x\, dx$

58. $\displaystyle\int \tanh x\, dx$

59. $\displaystyle\int \coth 3x\, dx$

60. $\displaystyle\int \operatorname{sech}^2(3x - 1)\, dx$

61. $\displaystyle\int \dfrac{\sinh x}{1 + \cosh x}\, dx$

62. $\displaystyle\int \dfrac{\operatorname{sech}\left(\dfrac{1}{x}\right) \tanh\left(\dfrac{1}{x}\right)}{x^2}\, dx$

63. Find the volume of the solid obtained by revolving the region under the graph of the catenary $y = a \cosh(x/a)$ on the interval $[-b, b]$ $(b > 0)$ about the x-axis.

64. The arc of the catenary $y = a \cosh(x/a)$ for x between $x = 0$ and $x = b$ is revolved about the x-axis. Show that the surface area S and the volume V of the resulting solid of revolution are related by the formula $S = 2V/a$.

65. Refer to Figure 5. Suppose that the cable has a constant weight density of W lb/ft. Then the tension on the cable is

$$T = T_0 \cosh \dfrac{Wx}{T_0} \qquad -b \leq x \leq b$$

where T_0 is the tension at the lowest point. Find the average tension on the cable.

66. The velocity of a body of mass m falling from rest through a viscous medium is given by

$$v(t) = \sqrt{\frac{mg}{k}} \tanh\left(\sqrt{\frac{gk}{m}}\, t\right)$$

where g is the acceleration of gravity and k is a positive constant that depends on the viscosity of the medium.

a. Find $\lim_{t\to\infty} v(t)$.

b. Plot the graph of v taking $m = 2$, $g = 32$, and $k = 8$.

Note: This limiting velocity of the body is called the *terminal velocity*.

 67. Damped Harmonic Motion The equation of motion of a weight attached to a spring and a dashpot damping device is

$$x(t) = -\frac{1}{\sqrt{2}}\, e^{-4t} \sinh 2\sqrt{2}t$$

where $x(t)$, measured in feet, is the displacement from the equilibrium position of the spring system and t is measured in seconds.

a. Find the initial position and the initial velocity of the weight.

b. Plot the graph of $x(t)$.

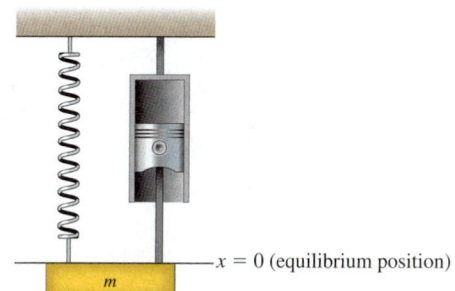

$x = 0$ (equilibrium position)

m

The system in equilibrium (The positive direction is downward.)

 68. Heat-Seeking Missiles In a test conducted on a heat-seeking Missile A, the target missile B, which is initially at a distance of b miles from Missile A, is launched vertically upward. Assume that Missile A travels at a constant speed v_A, that Missile B travels at a constant speed v_B ($v_A > v_B$), and that Missile A, which is launched from the origin, is always pointed at Missile B. Then the trajectory of Missile A is

$$y = \frac{b}{2}\left[\frac{\left(1 - \dfrac{x}{b}\right)^{1+c}}{1 + c} - \frac{\left(1 - \dfrac{x}{b}\right)^{1-c}}{1 - c}\right] + \frac{bc}{1 - c^2}$$

where $c = v_B/v_A$.

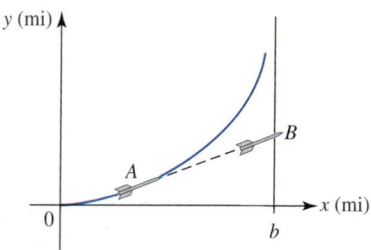

The trajectory of Missile A is a pursuit curve.

a. Find the point at which Missile A intercepts Missile B.

b. Show that

$$\frac{dy}{dx} = -\sinh\left[c \ln\left(1 - \frac{x}{b}\right)\right]$$

c. Suppose that $b = 1$ and $c = \frac{1}{2}$. Show that the distance D traveled by Missile A for the intercept is $1\frac{1}{3}$ mi.

Hint: $D = \displaystyle\int_0^1 \sqrt{1 + \left(\frac{dy}{dx}\right)^2}\, dx$

d. Plot the graph of the trajectory of the heat-seeking missile taking $b = 1$ and $c = \frac{1}{2}$.

69. The minimum-surface-of-revolution problem may be stated as follows: Of all curves joining two fixed points, find the one that, when revolved about the x-axis, will generate a surface of minimum area. It can be shown that the solution to the problem is a catenary. The resulting surface of revolution is called a *catenoid*. Suppose a catenary described by the equation

$$y = \cosh x \qquad a \le x \le b$$

is revolved about the x-axis. Find the surface area of the resulting catenoid.

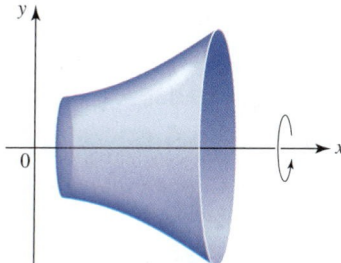

Hint: Use the identity $\cosh^2 x = \dfrac{1 + \cosh 2x}{2}$.

Note: A soap bubble formed by two parallel circular rings that are close to each other is an example of a catenoid.

70. Find the volume of the solid of revolution that is obtained by revolving the region bounded by the graph of $y = (x^2 - 1)^{3/4}$, the x-axis, and the lines $x = 1$ and $x = 2$, about the x-axis.
Hint: Use the substitution $x = \cosh u$.

71. Find the centroid of the region under the graph of $f(x) = \cosh x$ on $[-a, a]$.

72. A power line is suspended between two towers that are 200 ft apart, as shown in the figure. The shape of the cable is a *catenary* with equation

$$y = 80 \cosh \frac{x}{80} \qquad -100 \leq x \leq 100$$

where x is measured in feet. What is the angle θ that the line makes with the pole?

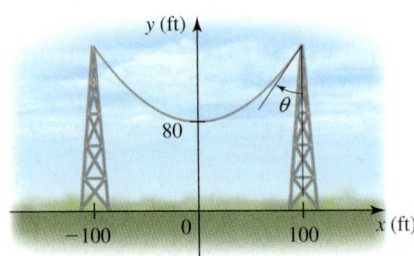

73. Prove that $\dfrac{d}{dx} \cosh u = (\sinh u) \dfrac{du}{dx}$.

74. Prove that $\dfrac{d}{dx} \operatorname{csch} u = -(\operatorname{csch} u \coth u) \dfrac{du}{dx}$.

75. Prove that $\dfrac{d}{dx} \operatorname{sech} u = -(\operatorname{sech} u \tanh u) \dfrac{du}{dx}$.

76. Prove that $\dfrac{d}{dx} \coth u = -(\operatorname{csch}^2 u) \dfrac{du}{dx}$.

In Exercises 77–80, determine whether the statement is true or false. If it is true, explain why it is true. If it is false, explain why or give an example to show why it is false.

77. $(\sinh x + \cosh x)^3 > 0$ for all x in $(-\infty, \infty)$.

78. $\dfrac{d}{dx} (\coth^2 x - \operatorname{csch}^2 x)^5 = 0$

79. $\displaystyle\int_{-\pi}^{\pi} (\cos x)\sinh x \, dx = 0$

80. $\displaystyle\int_{-3}^{3} x^2 \operatorname{sech} x \, dx = 2\int_{0}^{3} x^2 \operatorname{sech} x \, dx$

CHAPTER 5   REVIEW

CONCEPT REVIEW

In Exercises 1–12, fill in the blanks.

1. a. If f and g are continuous on $[a, b]$ and $f(x) \geq g(x)$ for all x in $[a, b]$, then the area of the region between the graphs of f and g and the vertical lines $x = a$ and $x = b$ is $A = $ _____.

   b. If f and g are continuous on $[a, b]$, then the area of the region bounded by the graphs of f and g and the vertical lines $x = a$ and $x = b$ is $A = $ _____.

2. a. If f is a continuous nonnegative function on $[a, b]$, then the volume of the solid obtained by revolving the region R under the graph of f on $[a, b]$ about the x-axis is $V = $ _____.

   b. If g is a continuous nonnegative function on $[c, d]$, then the volume of the solid obtained by revolving the region R between the graph of $x = g(y)$ and the y-axis on $[c, d]$ about the y-axis is $V = $ _____.

3. If f and g are continuous on $[a, b]$ and $f(x) \geq g(x) \geq 0$ for all x in $[a, b]$, then the volume of the solid obtained by revolving the region between the graphs of f and g on $[a, b]$ about the x-axis is $V = $ _____.

4. If S is a solid bounded by planes that are perpendicular to the x-axis at $x = a$ and $x = b$ and the cross sectional area of S at any point x in $[a, b]$ is $A(x)$, where A is continuous on $[a, b]$, then the volume of S is $V = $ _____.

5. a. If f is a continuous nonnegative function on $[a, b]$ and R is the region under the graph of f on $[a, b]$, then the volume of the solid obtained by revolving R about the y-axis is $V = $ _____.

   b. If the region R bounded by the graphs of $x = f(y)$, where f is nonnegative, $x = 0$, $y = c$, and $y = d$, is revolved about the x-axis, where $0 \leq c \leq d$, then the volume of the resulting solid is $V = $ _____.

6. a. If f is smooth on $[a, b]$, then the arc length of the graph of f from $P(a, f(a))$ to $Q(b, f(b))$ is $L = $ _____.

   b. If g is smooth on $[c, d]$, then the arc length of the graph of $x = g(y)$ from $P(g(c), c)$ to $Q(g(d), d)$ is $L = $ _____.

7. If f is smooth on $[a, b]$, then the arc length function s for the graph of f is defined by $s(x) = $ _____ with domain _____. The arc length differential is $ds = $ _____ or $(ds)^2 = $ _____.

8. a. If f is a nonnegative smooth function on $[a, b]$, then the surface area of the surface obtained by revolving the graph of f about the x-axis is $S = $ _____.

b. If g is a smooth function on $[c, d]$ with $g(y) \geq 0$, then the surface area of the surface obtained by revolving the graph of g about the y-axis is $S = $ _____.

9. If F is continuous on $[a, b]$, then the work done by the force $F(x)$ in moving a body from $x = a$ to $x = b$ is $W = $ _____.

10. The force F exerted by a fluid of constant weight density δ on one side of a submerged vertical plate from $y = c$ to $y = d$, where $c \leq d$, is given by $F = $ _____, where $L(y)$ is the horizontal length of the plate at y and $h(y)$ is the depth of the fluid at y.

11. If L denotes a lamina of constant mass density ρ and L has the shape of the region R under the graph of a nonnegative continuous function f on $[a, b]$, then

a. The mass of L is _____.

b. The moments of L about the x-axis and y-axis are $M_x = $ _____ and $M_y = $ _____.

c. The center of mass of L is located at $(\bar{x}, \bar{y})$, where $\bar{x} = $ _____ and $\bar{y} = $ _____.

12. If R is the region bounded by the graphs of two continuous functions f and g on $[a, b]$, where $f(x) \geq g(x)$ for all x in $[a, b]$, then the centroid $(\bar{x}, \bar{y})$ of R is given by $\bar{x} = $ _____, and $\bar{y} = $ _____, where $A = $ _____.

REVIEW EXERCISES

In Exercises 1–16, sketch the region bounded by the graphs of the equations and find the area of the region.

1. $y = \dfrac{1}{x^3}$, $y = 0$, $x = 1$, $x = 2$

2. $y = \dfrac{1}{\sqrt{x}}$, $y = 0$, $x = 1$, $x = 2$

3. $y = x^2 + 2$, $y = x + 1$, $x = 0$, $x = 1$

4. $y = 2x^3 - 1$, $y = x - 1$, $x = 1$, $x = 2$

5. $y = 2x^2 + 2x - 3$, $y = 3x^2 + 2x - 4$

6. $y = x^3$, $x = y^3$

7. $y = \sqrt{x - 1}$, $y = 2$, $y = 0$, $x = 0$

8. $y = (x + 1)^3$, $y = x + 1$

9. $x = (y - 1)^2$, $x = 1$

10. $x = \sqrt{y - 1}$, $x = y$, $x = 0$, $y = 2$

11. $x = y^2 - 1$, $x = 1 - y$

12. $y = \cos x$, $y = 1 - \dfrac{2}{\pi}x$, $x = -\dfrac{\pi}{2}$, $x = \dfrac{\pi}{2}$

13. $y = \cos x$, $y = -\sin x$, $x = 0$, $x = \dfrac{\pi}{2}$

14. $y = \sec^2 x$, $y = \sin x$, $x = -\dfrac{\pi}{4}$, $x = \dfrac{\pi}{4}$

15. $\sqrt{x} + \sqrt{y} = 1$, $x + y = 1$

16. $(y - x)^2 = x^3$, $x = 1$

17. $y = e^{2x}$, $y = -\dfrac{1}{x}$, $x = 1$; $x = 2$

In Exercises 18–26, find the volume of the solid generated by revolving the region bounded by the graphs of the equations about the indicated line.

18. $y = x - x^2$, $y = 0$; the x-axis

19. $y = \sqrt{x + 1}$, $y = 0$, $x = 3$; the x-axis

20. $y = x^2$, $y = x^3$; the line $y = 2$

21. $y = x^{1/3}$, $y = x$; the y-axis

22. $y = \dfrac{1}{\sqrt{1 + x^2}}$, $y = 0$, $x = 0$, $x = 1$; the y-axis

23. $y = x^2$, $y = x^3$; the line $x = 2$

24. $y = 1 - x^2$, $y = -x + 1$; the line $x = -2$

25. $y = \cos x^2$, $x \geq 0$, $y = 0$; the y-axis

26. $y = \dfrac{\ln x}{x^2}$, $y = 0$, $x = e$; the y-axis

In Exercises 27–32, find dy/dx.

27. $y = \ln(\sinh 2x)$

28. $y = e^{\tanh 3x}$

29. $y = \sinh^{-1}(\tanh x)$

30. $y = \sin^{-1}\left(\dfrac{x + 1}{x + 2}\right)$

31. $y = \tan^{-1}(\cos^{-1}\sqrt{x})$

32. $x \cosh y + e^{\sinh y} = 10$

33. Find $\displaystyle\int \sinh^2 x \, dx$.

34. Find $\displaystyle\int \cosh^3 x \, dx$.

35. Find $\displaystyle\int \sinh 2t \, dt$.

36. If $f(x) = \displaystyle\int_{\ln x}^{\sqrt{x}} \sinh t \, dt$, where $x > 0$, find $f'(x)$.

37. Find the area of the region completely enclosed by the parabola $y = x^2 - 6x + 11$ and the line passing through the point $(1, 0)$ and the vertex of the parabola.

38. The base of a solid is a circular disk of radius 2, and the cross sections perpendicular to the base are isosceles right triangles with the hypotenuse lying on the base. Find the volume of the solid.

39. A monument stands 50 m high. A horizontal cross section x m from the top is an equilateral triangle with sides $x/5$ m. What is the volume of the monument?

40. Write an integral giving the arc length of the graph of the function $f(x) = x^2 + x^3$ on $[0, 1]$. Do not evaluate the integral.

41. Find the length of the graph of $y = (9 - x^{2/3})^{3/2}$ on the interval $[1, 27]$.

42. Find the length of the graph of

$$y = \int_1^x \sqrt{t^{3/2} - 1}\, dt \qquad 1 \le x \le 16$$

43. Find the arc length of the graph of $y = \frac{1}{4}x^2 - \frac{1}{2}\ln x$ on $[1, 2]$.

44. Find the arc length of the graph of $y = \ln \dfrac{e^x + 1}{e^x - 1}$ on $[1, 2]$.

45. Find the area of the surface obtained by revolving the portion of the graph of $y = \frac{1}{2}x^2$ that lies below $y = \frac{3}{2}$ about the y-axis.

 46. Find the area of the surface formed by revolving the graph of $y = 2\sqrt{x} - x$, $0 \le x \le 2$, about the x-axis, accurate to four decimal places.

47. Write an integral giving the area of the surface obtained by revolving the graph of $f(x) = x^2 + \dfrac{1}{x}$ on $[1, 2]$ about the x-axis. Do not evaluate the integral.

48. The region under the graph of $y = x^2/(1 + x^4)$ on the interval $[0, 1]$ is revolved about the y-axis. Find the volume of the resulting solid.

49. Consider the portion of the unit circle lying in the first quadrant.

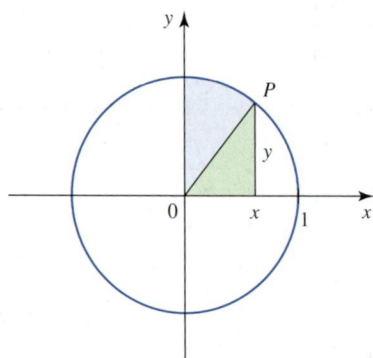

a. By considering the area of the shaded region, show that

$$\int_0^x \sqrt{1 - t^2}\, dt = \frac{1}{2}\sin^{-1} x + \frac{x}{2}\sqrt{1 - x^2}$$

and hence

$$\sin^{-1} x = 2\int_0^x \sqrt{1 - t^2}\, dt - x\sqrt{1 - x^2}$$

b. By differentiating the last equation in part (a) with respect to x, show that

$$\frac{d}{dx}(\sin^{-1} x) = \frac{1}{\sqrt{1 - x^2}}$$

50. When a particle is at the point x on the x-axis, it is acted upon by a force of $x + \cos \pi x$ dynes. Find the work done by the force in moving the particle from $x = 1$ to $x = 2$, where x is measured in centimeters.

51. A force of 6 lb is required to stretch a spring $1\frac{1}{2}$ in. beyond its natural length. Find the work required to stretch the spring 2 in. beyond its natural length.

52. A 1200-lb elevator at a construction site is suspended by a cable that weighs 10 lb/ft. How much work is done in raising the elevator from the ground to a height of 20 ft?

53. A tank having the shape of an inverted right circular cone with a base radius of 10 ft and a height of 15 ft is filled with water weighing 62.4 lb/ft^3. Find the work required to empty the tank by pumping the water over the rim of the tank.

54. A semicircular plate of radius r is submerged vertically in a liquid weighing 50 lb/ft^3 in such a way that the diameter of the plate is flush with the surface of the liquid. Find the force exerted by the liquid on one side of the plate.

55. A rectangular swimming pool is 50 ft long, 20 ft wide, and 3 ft deep at the shallow end and 7 ft deep at the deep end. Its bottom is an inclined plane. If the pool is completely filled with water, find the force exerted by the water on each vertical wall of the swimming pool.

In Exercises 56–59, find the centroid of the region bounded by the graphs of the equations.

56. $y = \sqrt{x}, \quad y = \dfrac{1}{2}x$

57. $y = x^2 - 2x, \quad y = 0$

58. $y = \sqrt{9 - x^2}, \quad y = 0$

59. $y = 2x^2 - 4x, \quad y = 2x - x^2$

PROBLEM-SOLVING TECHNIQUES

In the following example we invoke a rule of differentiation to solve a problem involving integration.

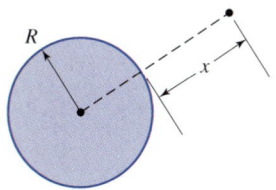

EXAMPLE **Height Reached by a Projectile** A projectile is launched vertically upward from the earth's surface with an initial velocity v_0 of magnitude less than the escape velocity. If only the earth's influence is taken into consideration, then the differential equation governing its motion is

$$\frac{d^2x}{dt^2} = -\frac{gR^2}{(x+R)^2}$$

where t is in seconds, x is the distance of the projectile from the surface of the earth in miles, R is the radius of the earth, and g is the constant of acceleration due to gravity. Show that the maximum height reached by the projectile is $v_0^2 R / (2gR - v_0^2)$.

Solution At first glance, we are tempted to integrate the given equation with respect to t. But the right-hand side of the equation is a function of the unknown variable x! The trick here is to use the Chain Rule to rewrite the left-hand side of the differential equation. Thus,

$$\frac{d^2x}{dt^2} = \frac{d}{dt}\left(\frac{dx}{dt}\right) = \frac{dv}{dt} = \frac{dv}{dx}\cdot\frac{dx}{dt} = \frac{dv}{dx}v$$

The given equation now reads

$$v\frac{dv}{dx} = -\frac{gR^2}{(x+R)^2}$$

Integrating both sides with respect to x, we obtain

$$\int v\frac{dv}{dx}\,dx = -gR^2\int(x+R)^{-2}dx$$

$$\frac{1}{2}v^2 = \frac{gR^2}{x+R} + C$$

or

$$v^2 = \frac{2gR^2}{x+R} + C$$

We have used the substitution $u = x + R$ to find the integral on the right-hand side. To find C, we use the initial condition $v = v_0$ if $x = 0$, obtaining

$$v_0^2 = \frac{2gR^2}{R} + C \quad \text{or} \quad C = v_0^2 - 2gR$$

Therefore,

$$v^2 = \frac{2gR^2}{x+R} + v_0^2 - 2gR$$

At the maximum height, $v = 0$, so we have

$$0 = \frac{2gR^2}{x + R} + v_0^2 - 2gR$$

$$\frac{2gR^2}{x + R} = 2gR - v_0^2$$

$$x + R = \frac{2gR^2}{2gR - v_0^2}$$

$$x = \frac{2gR^2}{2gR - v_0^2} - R = \frac{2gR^2 - 2gR^2 + v_0^2 R}{2gR - v_0^2}$$

$$= \frac{v_0^2 R}{2gR - v_0^2}$$

as was to be shown.

CHALLENGE PROBLEMS

1. The figure shows the region bounded by the parabola $y = x^2 - 2x + 2$, the line tangent to it at the point $P(a, b)$, and the y-axis. If the area of the region is 9 square units, what are the values of a and b?

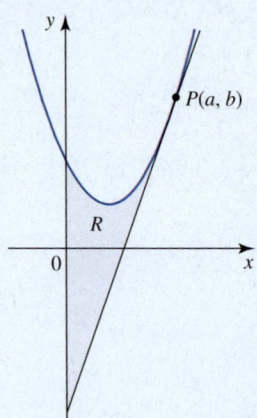

2. Find the area of the region completely enclosed by the parabolas $x = y^2$ and $x = \frac{3}{4} y^2 + 1$.

3. Let R be the region bounded by the graph of the function $f(x) = x\sqrt{1 - x^2}$ and the positive x-axis. Find the parabola $y = cx^2$ that divides R into two subregions of equal area.

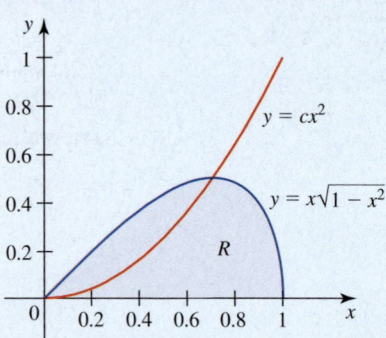

4. A trough has a cross section in the form of a parabola with base a ft and depth h ft. What is the average depth of the trough?

5. Let f be a nonnegative, continuous function on the interval $[0, \infty)$. Suppose that the arc length of the graph of f from $x = 0$ to $x = b$ is $b + \frac{2}{3} b^3$ and that $f(0) = \frac{2}{3}$. Find the function f.

6. A solid has a circular base of radius R, and its parallel cross sections perpendicular to its base are parabolas of height h. Find the volume of the solid.

7. Find the area of the region bounded by the graph of the function $f(x) = x^3 - 6x^2$, the x-axis, and the vertical lines passing through the inflection point and the relative minimum of f.

8. A semicircle of radius a is revolved about an axis parallel to the straight edge of the semicircle and located at a distance $b > a$ from the edge. Find the volume of the resulting solid of revolution.

9. The region bounded by the hyperbola $x^2/a^2 - y^2/b^2 = 1$, the line $bx - 2ay = 0$, and the x-axis is revolved about the x-axis. What is the volume of the resulting solid of revolution?

10. A solid ball of mass M and radius R rotates with an angular velocity ω about an axis through its center. Calculate the work required to stop the ball.
Hint: Calculate the kinetic energy $\left(\frac{1}{2}mv^2\right)$ of the ball.

11. The rate at which water evaporates from a pond is proportional to its surface area. Show that the depth of the water decreases at a constant rate and is not dependent on the shape of the pond.
Hint: If $V(t)$ denotes the volume of water in the pond at time t and $A(x)$ the surface area of the pond when the water has depth x, then $dV/dt = -kA(x)$, where k is the constant of proportionality.

12. Buffon's Needle Problem A needle of length l is dropped onto a board that is covered with parallel lines spaced at a distance w units apart where $w > l$. What is the probability that the needle will intersect one of the lines? To solve the problem, refer to the figure and observe that the needle intersects one of the lines if and only if $|(l/2)\cos\theta| > y$, where y is the distance from the center of the needle to the nearest line.

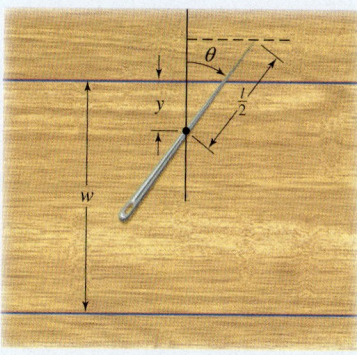

We can think of the set of all positions assumed by the needle as being associated with the rectangular region $S = \{(y, \theta) \mid 0 \leq y \leq \frac{w}{2} \text{ and } 0 \leq \theta \leq 2\pi\}$. The set of all positions assumed by the needle when it intersects a line can then be associated with the region $R = \{(y, \theta) \mid 0 \leq y < |(l/2)\cos\theta| \text{ and } 0 \leq \theta \leq 2\pi\}$. We define the probability p that the needle intersects a line to be the ratio of all "favorable" outcomes to "all" outcomes. Thus,

$$p = \frac{\text{area of } R}{\text{area of } S}$$

Show that $p = 2l/(\pi w)$.

Brand X/Alamy

There is a minimum speed that a rocket must attain in order to escape from the gravitational field of a planet. This speed is called the *escape velocity* for the planet. In this chapter, we will learn how to calculate the escape velocity for the earth.

6

Techniques of Integration

UP TO NOW we have relied on the basic integration formulas and the method of substitution to help us evaluate integrals. In this chapter we will look at some techniques of integration that will enable us to evaluate the integrals of more complicated functions.

We begin by introducing the method of integration by parts, which, like the method of substitution, is a general technique of integration. We then look at special methods for integrating trigonometric functions and rational functions. We also see how a table of integrals and a computer algebra system can help us to evaluate very general integrals.

Finally, we look at *improper integrals*, integrals in which the interval of integration is infinite or the integrand is unbounded (or both).

V This symbol indicates that one of the following video types is available for enhanced student learning at **www.academic.cengage.com/login:**
- Chapter lecture videos
- Solutions to selected exercises

6.1 Integration by Parts

As we have seen, a rule of integration can often be found by reversing a corresponding rule of differentiation. In this section we will look at a method of integration that is obtained by reversing the Product Rule for differentiation.

The Method of Integration by Parts

Recall that the Product Rule states that if f and g are differentiable functions, then

$$\frac{d}{dx}[f(x)g(x)] = f(x)g'(x) + g(x)f'(x)$$

If we integrate both sides of this equation with respect to x, we obtain

$$\int \frac{d}{dx}[f(x)g(x)]\, dx = \int [f(x)g'(x) + g(x)f'(x)]\, dx$$

or

$$f(x)g(x) = \int f(x)g'(x)\, dx + \int g(x)f'(x)\, dx$$

which may be written in the form

$$\int f(x)g'(x)\, dx = f(x)g(x) - \int g(x)f'(x)\, dx \tag{1}$$

Formula (1) is called the **formula for integration by parts.** We use this formula to express one integral in terms of another that is easier to integrate.

Formula (1) can be simplified by using differentials. Let $u = f(x)$ and $v = g(x)$ so that $du = f'(x)\, dx$ and $dv = g'(x)\, dx$. Substituting these quantities into Formula (1) leads to the following version of the formula for integration by parts.

Integration by Parts Formula

$$\int u\, dv = uv - \int v\, du \tag{2}$$

EXAMPLE 1 Find $\int xe^x\, dx$.

Solution Let's use Formula (2) by choosing

$$u = x \qquad \text{and} \qquad dv = e^x\, dx$$

so that

$$du = dx \qquad \text{and} \qquad v = \int e^x\, dx = e^x$$

Any antiderivative will do—see the *Note* following the example.

This gives

$$\int xe^x \, dx = uv - \int v \, du$$

$$= xe^x - \int e^x \, dx$$

$$= xe^x - e^x + C = (x - 1)e^x + C$$ ▪

Notes

1. In finding v from the expression for dv, we don't need to include the constant of integration (that is, we may take $C = 0$). To see why, suppose that we replace v in Formula (2) by $v + C$. Then we obtain

$$\int u \, dv = u(v + C) - \int (v + C) \, du = uv + Cu - \int v \, du - \int C \, du$$

$$= uv + Cu - \int v \, du - Cu = uv - \int v \, du$$

In other words, the constant C "drops out."

2. The success of the method of integration by parts depends on a judicious choice of u and dv. For instance, had we chosen $u = e^x$ and $dv = x \, dx$ in Example 1, then $du = e^x \, dx$ and $v = x^2/2$ and Formula (2) would have yielded

$$\int xe^x \, dx = uv - \int v \, du$$

$$= \frac{1}{2}x^2 e^x - \int \frac{1}{2}x^2 e^x \, dx$$

Since the integral on the right-hand side of this equation is more complicated than the original integral, we have not made a good choice of u and dv. ▪

Our original choice of u and dv in Example 1 suggests the following general guidelines.

> **Guidelines for Choosing u and dv**
>
> Choose u and dv so that
>
> 1. du is simpler than u (if possible).
> 2. dv is easily integrated.

EXAMPLE 2   Find $\int x \ln x \, dx$.

Solution   Let

$$u = \ln x \qquad \text{and} \qquad dv = x \, dx$$

so that

$$du = \frac{1}{x} \, dx \qquad \text{and} \qquad v = \int x \, dx = \frac{1}{2}x^2$$

Then Formula (2) gives

$$\int x \ln x \, dx = \frac{1}{2} x^2 \ln x - \int \frac{1}{2} x^2 \left(\frac{1}{x} \right) dx$$

$$= \frac{1}{2} x^2 \ln x - \frac{1}{4} x^2 + C = \frac{1}{4} x^2 (2 \ln x - 1) + C$$ ∎

Sometimes we need to apply the integration by parts formula more than once to find an integral, as illustrated in the next two examples.

EXAMPLE 3   Find $\int x^2 \sin x \, dx$.

Solution   Let

$$u = x^2 \qquad \text{and} \qquad dv = \sin x \, dx$$

so that

$$du = 2x \, dx \qquad \text{and} \qquad v = \int \sin x \, dx = -\cos x$$

Then Formula (2) yields

$$\int x^2 \sin x \, dx = -x^2 \cos x + \int 2x \cos x \, dx \tag{3}$$

Observe that the integral on the right-hand side, although not readily integrable, is simpler than the original integral. In fact, the power in x in the integrand is 1 instead of 2. This suggests that integrating by parts again might be a move in the right direction. So let's apply the formula once again to evaluate $\int 2x \cos x \, dx$. Let

$$u = 2x \qquad \text{and} \qquad dv = \cos x \, dx$$

so that

$$du = 2 \, dx \qquad \text{and} \qquad v = \int \cos x \, dx = \sin x$$

It follows from Formula (2) that

$$\int 2x \cos x \, dx = 2x \sin x - \int 2 \sin x \, dx = 2x \sin x + 2 \cos x + C \tag{4}$$

Finally, substituting Equation (4) into Equation (3) gives

$$\int x^2 \sin x \, dx = -x^2 \cos x + 2x \sin x + 2 \cos x + C$$ ∎

EXAMPLE 4   Find $\int e^x \sin 2x \, dx$.

Solution   Let

$$u = e^x \qquad \text{and} \qquad dv = \sin 2x \, dx$$

so that

$$du = e^x \, dx \qquad \text{and} \qquad v = \int \sin 2x \, dx = -\frac{1}{2} \cos 2x$$

(In this case the choice $u = \sin 2x$ and $dv = e^x\, dx$ will work equally well.) Substituting this value into Formula (2) yields

$$\int e^x \sin 2x\, dx = -\frac{1}{2} e^x \cos 2x + \frac{1}{2} \int e^x \cos 2x\, dx \tag{5}$$

The integral on the right-hand side is not readily integrable. But notice that it is certainly no more complicated than the original integral. So let's integrate by parts again and see where this leads us. Let

$$u = e^x \quad \text{and} \quad dv = \cos 2x\, dx$$

so that

$$du = e^x\, dx \quad \text{and} \quad v = \int \cos 2x\, dx = \frac{1}{2} \sin 2x$$

On using Formula (2), we find that

$$\int e^x \cos 2x\, dx = \frac{1}{2} e^x \sin 2x - \frac{1}{2} \int e^x \sin 2x\, dx \tag{6}$$

Substituting Equation (6) into Equation (5) yields

$$\int e^x \sin 2x\, dx = -\frac{1}{2} e^x \cos 2x + \frac{1}{4} e^x \sin 2x - \frac{1}{4} \int e^x \sin 2x\, dx$$

Since the integral on the right-hand side is, except for the constant of integration, a constant multiple of the (original) integral on the left side, we can combine them to yield

$$\frac{5}{4} \int e^x \sin 2x\, dx = -\frac{1}{2} e^x \cos 2x + \frac{1}{4} e^x \sin 2x + C_1$$

so the required result is

$$\int e^x \sin 2x\, dx = -\frac{2}{5} e^x \cos 2x + \frac{1}{5} e^x \sin 2x + C \qquad C = \frac{4}{5}C_1$$

$$= \frac{1}{5} e^x(\sin 2x - 2\cos 2x) + C \qquad\blacksquare$$

Note We leave it for you to show that if we had chosen $u = \cos 2x$ and $dv = e^x\, dx$ in finding the integral on the right-hand side of Equation (5), then our final result would have been $\int e^x \sin 2x\, dx = \int e^x \sin 2x\, dx$. This is certainly a true statement, but it is of no help to us in evaluating the given integral. ■

EXAMPLE 5 Evaluate $\displaystyle\int_0^{\pi/4} \sec^3 x\, dx$.

Solution We first find the indefinite integral

$$\int \sec^3 x\, dx = \int \sec x \cdot \sec^2 x\, dx$$

Let

$$u = \sec x \quad \text{and} \quad dv = \sec^2 x\, dx$$

so that

$$du = \sec x \tan x\, dx \quad \text{and} \quad v = \int \sec^2 x\, dx = \tan x$$

Using Formula (2), we obtain

$$\int \sec^3 x \, dx = \sec x \tan x - \int \tan^2 x \sec x \, dx$$

$$= \sec x \tan x - \int (\sec^2 x - 1)\sec x \, dx \qquad \sec^2 x = 1 + \tan^2 x$$

$$= \sec x \tan x - \int \sec^3 x \, dx + \int \sec x \, dx$$

$$= \sec x \tan x + \ln|\sec x + \tan x| - \int \sec^3 x \, dx$$

Combining the integrals, we obtain

$$2\int \sec^3 x \, dx = \sec x \tan x + \ln|\sec x + \tan x| + C_1$$

or

$$\int \sec^3 x \, dx = \frac{1}{2} \sec x \tan x + \frac{1}{2} \ln|\sec x + \tan x| + C \qquad C = \frac{1}{2} C_1$$

Finally, using this result, we find

$$\int_0^{\pi/4} \sec^3 x \, dx = \left[\frac{1}{2} \sec x \tan x + \frac{1}{2} \ln|\sec x + \tan x| \right]_0^{\pi/4}$$

$$= \left[\frac{1}{2} (\sqrt{2})(1) + \frac{1}{2} \ln|\sqrt{2} + 1| \right] - \left[\frac{1}{2} (1)(0) + \frac{1}{2} \ln 1 \right]$$

$$= \frac{1}{2} [\sqrt{2} + \ln(\sqrt{2} + 1)] \approx 1.148$$

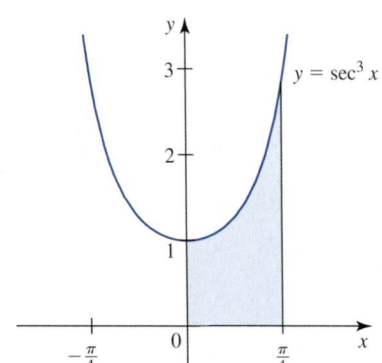

FIGURE 1
The area of the shaded region is given by $\int_0^{\pi/4} \sec^3 x \, dx$.

Since $f(x) = \sec^3 x$ is positive on $\left[0, \frac{\pi}{4}\right]$, we can interpret the integral in this example as the area of the region under the graph of f on $\left[0, \frac{\pi}{4}\right]$. (See Figure 1.)

An alternative method for evaluating a definite integral using integration by parts is based on the following formula. Here we assume that both f' and g' are continuous. Then the Fundamental Theorem of Calculus, Part 2, gives

$$\int_a^b f(x)g'(x) \, dx = f(x)g(x) \Big|_a^b - \int_a^b g(x)f'(x) \, dx$$

Letting $u = f(x)$ and $v = g(x)$ and keeping in mind that the limits of integration are stated for x, we have the following.

Integration by Parts Formula for a Definite Integral

$$\int_a^b u \, dv = \left[uv \right]_a^b - \int_a^b v \, du \qquad \qquad \textbf{(7)}$$

We illustrate the use of this formula in the next example.

EXAMPLE 6 Find the centroid of the region under the graph of $f(x) = \ln x$ on $[1, e]$.

Solution The region R under consideration is shown in Figure 2. The area of R is given by

$$A = \int_1^e \ln x \, dx$$

We integrate by parts, letting

$$u = \ln x \qquad \text{and} \qquad dv = dx$$

so that

$$du = \frac{1}{x} \, dx \qquad \text{and} \qquad v = x$$

Using Formula (7), we obtain

$$A = \left[x \ln x \right]_1^e - \int_1^e dx$$

$$= (e \ln e - \ln 1) - x \Big|_1^e$$

$$= e - (e - 1) \qquad \text{\color{red}{ln } } e = 1 \quad \text{and} \quad \ln 1 = 0$$

$$= 1$$

Then, using Equation (6) of Section 5.7, we find

$$\bar{x} = \frac{1}{A} \int_a^b x f(x) \, dx = \frac{1}{1} \int_1^e x \ln x \, dx$$

$$= \left[\frac{1}{4} x^2 (2 \ln x - 1) \right]_1^e \qquad \text{\color{red}{Use the result of Example 2.}}$$

$$= \frac{1}{4} e^2 (2 \ln e - 1) - \frac{1}{4} (2 \ln 1 - 1) = \frac{1}{4} (e^2 + 1)$$

and

$$\bar{y} = \frac{1}{A} \int_a^b \frac{1}{2} [f(x)]^2 \, dx = \frac{1}{1} \int_1^e \frac{1}{2} (\ln x)^2 \, dx$$

$$= \frac{1}{2} \int_1^e (\ln x)^2 \, dx$$

We integrate by parts, letting

$$u = (\ln x)^2 \qquad \text{and} \qquad dv = dx$$

so that

$$du = \frac{2 \ln x}{x} \, dx \qquad \text{and} \qquad v = x$$

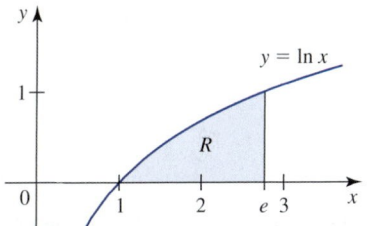

FIGURE 2
The region R

We obtain

$$\bar{y} = \frac{1}{2}\left\{\left[x(\ln x)^2\right]_1^e - \int_1^e 2\ln x \, dx\right\}$$

$$= \frac{1}{2}\left\{\left[e(\ln e)^2 - 1(\ln 1)^2\right] - 2\int_1^e \ln x \, dx\right\}$$

$$= \frac{1}{2}e - 1$$

because $\int_1^e \ln x \, dx = 1$, as we saw earlier. Therefore, the centroid of R is located at the point $\left(\frac{1}{4}(e^2 + 1), \frac{1}{2}e - 1\right)$.

■ **Reduction Formulas**

We can use the integration by parts formula to derive **reduction formulas** for evaluating certain integrals. These formulas enable us to express such integrals in terms of integrals whose integrands involve lower powers. In the next two examples we derive these formulas and then show how they are used.

EXAMPLE 7 Find a reduction formula for $\int \sin^n x \, dx$, where $n \geq 2$ is an integer.

Solution We first rewrite

$$\int \sin^n x \, dx = \int \sin^{n-1} x \sin x \, dx$$

and then integrate by parts, letting

$$u = \sin^{n-1} x \qquad \text{and} \qquad dv = \sin x \, dx$$

so that

$$du = (n - 1)\sin^{n-2} x \cos x \, dx \qquad \text{and} \qquad v = -\cos x$$

This gives

$$\int \sin^n x \, dx = uv - \int v \, du$$

$$= -\sin^{n-1} x \cos x + (n - 1)\int \sin^{n-2} x \cos^2 x \, dx$$

Since $\cos^2 x = 1 - \sin^2 x$, we can write

$$\int \sin^n x \, dx = -\sin^{n-1} x \cos x + (n - 1)\int \sin^{n-2} x \, dx - (n - 1)\int \sin^n x \, dx$$

Transposing the last term on the right to the left-hand side gives

$$n\int \sin^n x \, dx = -\sin^{n-1} x \cos x + (n - 1)\int \sin^{n-2} x \, dx$$

or

$$\int \sin^n dx = -\frac{1}{n}\sin^{n-1} x \cos x + \frac{n-1}{n}\int \sin^{n-2} x \, dx$$

EXAMPLE 8 Use the reduction formula obtained in Example 7 to find $\int \sin^4 x \, dx$.

Solution We use the reduction formula with $n = 4$ to obtain

$$\int \sin^4 x \, dx = -\frac{1}{4} \sin^3 x \cos x + \frac{3}{4} \int \sin^2 x \, dx$$

Applying the reduction formula once more to the integral on the right-hand side with $n = 2$, we have

$$\int \sin^2 x \, dx = -\frac{1}{2} \sin x \cos x + \frac{1}{2} \int dx$$

$$= -\frac{1}{2} \sin x \cos x + \frac{1}{2} x + C_1$$

Therefore,

$$\int \sin^4 x \, dx = -\frac{1}{4} \sin^3 x \cos x + \frac{3}{4} \left(-\frac{1}{2} \sin x \cos x + \frac{1}{2} x + C_1 \right)$$

$$= -\frac{1}{4} \sin^3 x \cos x - \frac{3}{8} \sin x \cos x + \frac{3}{8} x + C$$

where $C = \frac{3}{4} C_1$.

6.1 CONCEPT QUESTIONS

1. Write the formula for integration by parts for (a) indefinite integrals and (b) definite integrals.

2. Explain how you would choose u and dv when using the integration by parts formula. Illustrate your answer with the integral $\int xe^{-x} \, dx$.

6.1 EXERCISES

In Exercises 1–44, find or evaluate the integral.

1. $\int xe^{2x} \, dx$

2. $\int xe^{-x} \, dx$

3. $\int x \sin x \, dx$

4. $\int x \cos 2x \, dx$

5. $\int x \ln 2x \, dx$

6. $\int x^3 \ln x \, dx$

7. $\int x^2 e^{-x} \, dx$

8. $\int t^3 e^t \, dt$

9. $\int x^2 \cos x \, dx$

10. $\int x^2 \sin 2x \, dx$

11. $\int \tan^{-1} x \, dx$

12. $\int \sin^{-1} x \, dx$

13. $\int \sqrt{t} \ln t \, dt$

14. $\int \frac{\ln t}{\sqrt{t}} \, dt$

15. $\int x \sec^2 x \, dx$

16. $\int e^{-x} \sin x \, dx$

17. $\int e^{2x} \cos 3x \, dx$

18. $\int_0^1 x \tan^{-1} x \, dx$

19. $\int u \sin(2u + 1) \, du$

20. $\int \theta \csc^2 \theta \, d\theta$

21. $\int x \tan^2 x \, dx$

22. $\int \cos(\ln x) \, dx$

23. $\int \sqrt{x} \cos \sqrt{x} \, dx$

24. $\int \csc^3 \theta \, d\theta$

25. $\int \sec^5 \theta \, d\theta$

26. $\int x \sinh x \, dx$

27. $\displaystyle\int x^3 \sinh x \, dx$

28. $\displaystyle\int x \cosh 2x \, dx$

29. $\displaystyle\int e^{-x} \ln(e^x + 1) \, dx$

30. $\displaystyle\int (x^2 - 1)\cos x \, dx$

31. $\displaystyle\int \frac{\ln x}{\sqrt{1-x}} \, dx$

32. $\displaystyle\int_0^1 (t-1)e^{-2t} \, dt$

33. $\displaystyle\int_1^e x^2 \ln x \, dx$

34. $\displaystyle\int_0^2 \ln(x+1) \, dx$

35. $\displaystyle\int_0^{1/2} \cos^{-1} x \, dx$

36. $\displaystyle\int_0^{\pi} x \sin 2x \, dx$

37. $\displaystyle\int_{\sqrt{e}}^e x^{-2} \ln x \, dx$

38. $\displaystyle\int_0^{\pi/2} e^{2x} \cos x \, dx$

39. $\displaystyle\int_1^2 x \sec^{-1} x \, dx$

40. $\displaystyle\int_0^{\pi/2} (x + x \cos x) \, dx$

41. $\displaystyle\int_0^1 \ln(1 + t^2) \, dt$

42. $\displaystyle\int_0^{\pi^2/4} \sin\sqrt{x} \, dx$

43. $\displaystyle\int_{\pi/4}^{\pi/3} \frac{\theta}{\sin^2 \theta} \, d\theta$

44. $\displaystyle\int_0^1 \tan^{-1} \sqrt{x} \, dx$

45. Find the area of the region under the graph of $y = (\ln x)^2$ on the interval $[1, e]$.

46. Find the area of the region under the graph of $y = \dfrac{xe^x}{(1 + x)^2}$ on the interval $[0, 1]$.

47. Find the area of the region bounded by the graphs of $y = \tan^{-1} x$ and $y = (\pi/4)x$.
Hint: The graphs intersect at $(0, 0)$ and $\left(1, \frac{\pi}{4}\right)$.

48. Find the area of the region bounded by the graph of $y = e^{-x} \cos x$ and the x-axis for x in the interval $\left[0, \frac{3\pi}{2}\right]$.

49. Let $f(x) = x\sqrt{x+1}$ and $g(x) = 1 - x^2$.

   a. Plot the graphs of f and g using the viewing window $[-1.2, 1] \times [-1, 1.5]$. Find the x-coordinates of the points of intersection of the graphs of f and g accurate to three decimal places.
   b. Use the result of part (a) and integration by parts to find the approximate area of the region bounded by the graphs of f and g.

50. Let $f(x) = e^{-x} \sin x$ and $g(x) = -\sqrt{x} \cos \sqrt{x}$.
   a. Plot the graphs of f and g using the viewing window $[0, 3] \times [-0.7, 0.5]$. Find the x-coordinates of the points of intersection of the graphs of f and g accurate to three decimal places.
   b. Use the result of part (a) and integration by parts to find the approximate area of the region bounded by the graphs of f and g.

51. The region under the graph of $y = \sqrt{\cos^{-1} x}$ on the interval $[0, 1]$ is revolved about the x-axis. Find the volume of the solid generated.

52. The region bounded by the graphs of $y = \ln x$, $y = 0$, $x = 1$, and $x = e$ is revolved about the x-axis. Find the volume of the solid generated.

53. The region bounded by the graphs of $y = e^{x/2} \cos x$, $x = 0$, $y = 0$, and $x = \pi/2$ is revolved about the x-axis. Find the volume of the solid generated.

54. Find the volume of the solid generated by revolving the region enclosed by the graphs of $y = \sin x$, $x = 0$, and $y = 1$ about the y-axis.

55. The region bounded by the graphs of $y = \sin x$, $y = 0$, $x = 0$, and $x = \pi$ is revolved about the line $x = -1$. Find the volume of the solid generated.

56. Find the centroid of the region R bounded by the graphs of $y = \sin x$ and $y = (2/\pi)x$ on $\left[0, \frac{\pi}{2}\right]$.

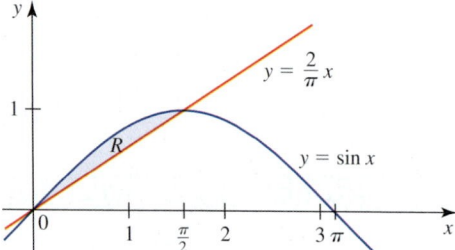

57. Energy Production To satisfy increased worldwide demand, the Metro Mining Company plans to increase its production of steam coal, the boiler-firing fuel used for generating electricity. Currently 20 million metric tons are produced per year; however, the company plans to increase production by $2te^{-0.05t}$ million metric tons per year, where t is measured in years, for the next 10 years. Find a function that describes the company's total production of steam coal at the end of t years. How much coal will the company have produced over the next 10 years if its production goals are met?

58. Alcohol-Related Traffic Accidents The number of alcohol-related accidents in a certain state, t months after the passage of a series of strict anti-drunk-driving laws, has been decreasing at the rate of $R(t) = 20 + te^{0.1t}$ accidents per month. There were 882 alcohol-related accidents for the year before enactment of the laws. Determine how many alcohol-related accidents were expected to occur during the first year after passage of the laws.

59. Damped Harmonic Motion Consider the system shown in the accompanying figure. Here, the weight is attached to a spring and a dashpot damping device. Suppose that at $t = 0$, the weight is set in motion from its equilibrium position so that its velocity at any time t is

$$v(t) = 3e^{-4t}(1 - 4t)$$

Find the position function $x(t)$ of the body.

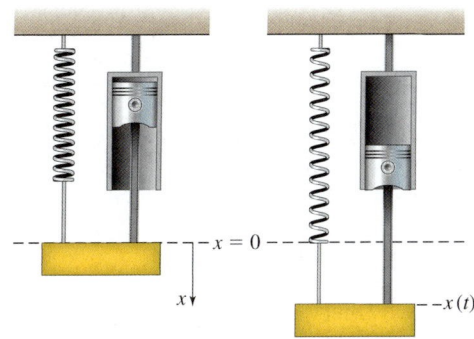

(a) System in equilibrium position   **(b)** System in motion

60. A Mixture Problem Two tanks are connected in tandem as shown in the figure. Each tank contains 60 gal of water. Starting at time $t = 0$, brine containing 3 lb/gal of salt flows into Tank 1 at the rate of 2 gal/min. The mixture then enters and leaves Tank 2 at the same rate. The mixtures in both tanks are stirred uniformly. It can be shown that the amount of salt in Tank 2 after t min is given by

$$A(t) = 180(1 - e^{-t/30}) - 6te^{-t/30}$$

where $A(t)$ is measured in pounds.
  a. What is the initial amount of salt in Tank 2?
  b. What is the amount of salt in Tank 2 after 3 hr?
  c. What is the average amount of salt in Tank 2 over the first 3 hr?

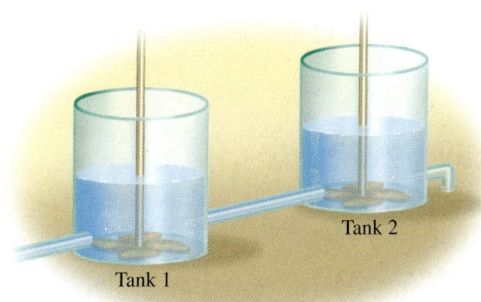

Tank 2

Tank 1

61. The Charge in an Electric Current The following figure shows an LRC series electrical circuit comprising an inductor, a resistor, and a capacitor with inductance L in henries, resistance R in ohms, and capacitance C in farads, respectively.

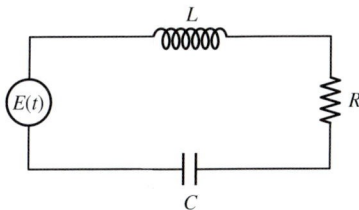

Here $E(t)$ is the electromotive force in volts. Suppose that the current in the system at time t is

$$i(t) = -\frac{200}{3}e^{-20t}\sin 60t$$

amperes. Find the charge $q(t)$ on the capacitor at any time t if the initial charge in the capacitor is q_0 coulombs.

Hint: $i = \dfrac{dq}{dt}$

62. Damped Harmonic Motion Refer to Exercise 59. Suppose that $t = 0$. The weight is set in motion from a point $\frac{1}{2}$ ft below the equilibrium position so that its velocity at any time t is

$$v(t) = e^{-2t}(\cos 4t - 3\sin 4t)$$

Find the position function of the body.

63. A Rocket Launch A rocket with a mass m (including fuel) is launched vertically upward from the surface of the earth ($t = 0$). If the fuel is consumed at a constant rate r during the interval $0 \le t \le T$ and the speed of the exhaust gas relative to the rocket is a constant s, then the velocity of the rocket at time t is given by

$$v(t) = v_0 - gt - s\ln\left(1 - \frac{r}{m}t\right)$$

where v_0 is its initial velocity and g is the gravitational constant. Find the height $h(t)$ of the rocket at any time t before burnout ($t = T$).

64. Growth of HMOs The membership of the Cambridge Community Health Plan, a health maintenance organization, is projected to grow at the rate of $9\sqrt{t+1}\,\ln\sqrt{t+1}$ thousand people per year t years from now. If the HMO's current membership is 50,000, what will the membership be 5 years from now?

65. Diffusion A cylindrical membrane with inner radius r_1 cm and outer radius r_2 cm containing a chemical solution is introduced into a salt bath with constant concentration c_2 moles/L. If the concentration of the chemical inside the membrane is kept constant at a different concentration of c_1 moles/L, then the concentration of chemical across the membrane will be given by

$$c(r) = \left(\frac{c_1 - c_2}{\ln r_1 - \ln r_2}\right)(\ln r - \ln r_2) + c_2 \qquad r_1 < r < r_2$$

moles/L. Find the average concentration of the chemical across the membrane from $r = r_1$ to $r = r_2$.

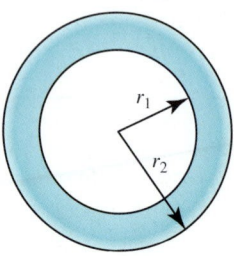

 66. Mechanical Resonance Refer to Example 5 of Section 2.5. Suppose that an external force is applied to the spring so that the velocity of the weight at time t is $v = 8t \sin 2t$.
   a. Find the position function of the weight if it is at the equilibrium position at $t = 0$.
   b. Plot the graph of the position function found in part (a).
   c. Show that $\lim_{t \to \infty} |s(t)| = \infty$ and hence the motion is one of resonance (see Exercise 44 of Section 2.5).

67. Suppose that f'' is continuous on $[1, 3]$ and $f(1) = 2$, $f(3) = -1$, $f'(1) = 2$, and $f'(3) = 5$. Evaluate $\int_1^3 x f''(x)\, dx$.

68. Consider the following "proof" that $0 = 1$. Integrate $\int (dx)/x$ by parts by letting $u = 1/x$ and $dv = dx$ so that $du = (-1/x^2)\, dx$ and $v = x$. This gives

$$\int \frac{dx}{x} = uv - \int v\, du$$

$$= \left(\frac{1}{x}\right)x - \int x\left(-\frac{1}{x^2}\right) dx = 1 + \int \frac{dx}{x}$$

Therefore, $0 = 1$. What is wrong with this argument?

In Exercises 69 and 70, determine whether the statement is true or false. If it is true, explain why it is true. If it is false, explain why or give an example to show why it is false.

69. $\displaystyle\int e^x f'(x)\, dx = e^x f(x) - \int e^x f(x)\, dx$

70. $\displaystyle\int uv\, dw = uvw - \int uw\, dv - \int vw\, du$

6.2 | Trigonometric Integrals

In this section we develop techniques for evaluating integrals involving combinations of trigonometric functions. Examples of such integrals are

$$\int \sin^5 x \cos^2 x\, dx, \qquad \int \csc 4x \cot^4 x\, dx, \qquad \text{and} \qquad \int \sin 5x \cos 4x\, dx$$

As you will see, these techniques rely on the use of the appropriate trigonometric identities.

■ Integrals of the Form $\int \sin^m x \cos^n x\, dx$

We begin by looking at integrals of the form

$$\int \sin^m x \cos^n x\, dx \tag{1}$$

where

1. m and/or n is an odd positive integer.
2. m and n are both even nonnegative integers.

Examples 1 and 2 illustrate how an integral belonging to category 1 is evaluated, and Example 3 shows how to evaluate an integral belonging to category 2.

EXAMPLE 1 Find $\int \sin^5 x \cos^2 x \, dx$.

Solution Here m (the power of $\sin x$) is an odd positive integer. Let's write

$$\sin^5 x = (\sin^4 x)(\sin x) \qquad \text{Retain a factor of } \sin x.$$

$$= (\sin^2 x)^2 \sin x$$

$$= (1 - \cos^2 x)^2 \sin x \qquad \begin{array}{l} \text{Use the identity } \sin^2 x + \cos^2 x = 1 \text{ to convert} \\ \text{the other factor to a function of } \cos x. \end{array}$$

$$= (1 - 2\cos^2 x + \cos^4 x)\sin x$$

Then

$$\int \sin^5 x \cos^2 x \, dx = \int \cos^2 x (1 - 2\cos^2 x + \cos^4 x)\sin x \, dx$$

$$= \int (\cos^2 x - 2\cos^4 x + \cos^6 x)\sin x \, dx$$

If we make the substitution $u = \cos x$, then $du = -\sin x \, dx$, so

$$\int \sin^5 x \cos^2 x \, dx = \int (u^2 - 2u^4 + u^6)(-du)$$

$$= -\int (u^2 - 2u^4 + u^6) \, du$$

$$= -\left(\frac{1}{3} u^3 - \frac{2}{5} u^5 + \frac{1}{7} u^7 \right) + C$$

$$= -\frac{1}{3} \cos^3 x + \frac{2}{5} \cos^5 x - \frac{1}{7} \cos^7 x + C \qquad \blacksquare$$

EXAMPLE 2 Find $\int \sin^4 x \cos^3 x \, dx$.

Solution Here n (the power of $\cos x$) is an odd positive integer. Let's write

$$\cos^3 x = (\cos^2 x)(\cos x) \qquad \text{Retain a factor of } \cos x.$$

$$= (1 - \sin^2 x)\cos x \qquad \begin{array}{l} \text{Use the identity } \sin^2 x + \cos^2 x = 1 \text{ to con-} \\ \text{vert the other factor to a function of } \sin x. \end{array}$$

Then

$$\int \sin^4 x \cos^3 x \, dx = \int \sin^4 x (1 - \sin^2 x)\cos x \, dx$$

$$= \int (\sin^4 x - \sin^6 x)\cos x \, dx$$

Let $u = \sin x$ so that $du = \cos x \, dx$. Then

$$\int \sin^4 x \cos^3 x \, dx = \int (u^4 - u^6) \, du$$

$$= \frac{1}{5} u^5 - \frac{1}{7} u^7 + C$$

$$= \frac{1}{5} \sin^5 x - \frac{1}{7} \sin^7 x + C \qquad \blacksquare$$

EXAMPLE 3 Find $\int \sin^4 x \, dx$.

Solution Here, $m = 4$ and $n = 0$. So both m and n are even nonnegative integers. In this case we use the half-angle formula for $\sin^2 x$:

$$\sin^2 x = \frac{1}{2}(1 - \cos 2x)$$

to write

$$\sin^4 x = (\sin^2 x)^2$$

$$= \left[\frac{1}{2}(1 - \cos 2x)\right]^2$$

$$= \frac{1}{4}(1 - 2\cos 2x + \cos^2 2x)$$

Applying the half-angle formula,

$$\cos^2 x = \frac{1}{2}(1 + \cos 2x)$$

to $\cos^2 2x$ in the last equation leads to

$$\sin^4 x = \frac{1}{4}\left(1 - 2\cos 2x + \frac{1}{2} + \frac{1}{2}\cos 4x\right)$$

$$= \frac{1}{4}\left(\frac{3}{2} - 2\cos 2x + \frac{1}{2}\cos 4x\right)$$

Therefore,

$$\int \sin^4 x \, dx = \frac{1}{4}\int \left(\frac{3}{2} - 2\cos 2x + \frac{1}{2}\cos 4x\right) dx$$

$$= \frac{3}{8}x - \frac{1}{4}\sin 2x + \frac{1}{32}\sin 4x + C \qquad \blacksquare$$

In general, we have the following guidelines for evaluating integrals of the form $\int \sin^m x \cos^n x \, dx$.

Guidelines for Evaluating $\int \sin^m x \cos^n x \, dx$

1. *If the power of* $\sin x$ *is odd and positive* ($m = 2k + 1$), retain a factor of $\sin x$, and use the identity $\sin^2 x = 1 - \cos^2 x$ to write

$$\int \sin^{2k+1} x \cos^n x \, dx = \int (\sin^2 x)^k \cos^n x \sin x \, dx$$

$$= \int (1 - \cos^2 x)^k \cos^n x \sin x \, dx$$

Then integrate using the substitution $u = \cos x$.

2. *If the power of* cos *x is odd and positive* ($n = 2k + 1$), *retain a factor of* cos *x, and use the identity* $\cos^2 x = 1 - \sin^2 x$ *to write*

$$\int \sin^m x \cos^{2k+1} x \, dx = \int \sin^m x \, (\cos^2 x)^k \cos x \, dx$$

$$= \int \sin^m x \, (1 - \sin^2 x)^k \cos x \, dx$$

Then integrate using the substitution $u = \sin x$.

3. *If the powers of* sin *x and* cos *x are both even and nonnegative,* use the half-angle formulas (repeatedly, if necessary) to write

$$\sin^2 x = \frac{1}{2}(1 - \cos 2x) \qquad \text{and} \qquad \cos^2 x = \frac{1}{2}(1 + \cos 2x)$$

EXAMPLE 4   Evaluate $\displaystyle\int_0^{\pi/2} \sin^3 x \cos^{1/2} x \, dx$.

Solution   The power of sin x is odd and positive. So we retain a factor of sin x. Thus,

$$\int_0^{\pi/2} \sin^3 x \cos^{1/2} x \, dx = \int_0^{\pi/2} \sin^2 x \cos^{1/2} x \sin x \, dx$$

$$= \int_0^{\pi/2} (1 - \cos^2 x)\cos^{1/2} x \sin x \, dx$$

$$= \int_0^{\pi/2} (\cos^{1/2} x - \cos^{5/2} x)\sin x \, dx$$

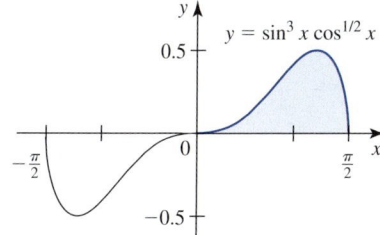

FIGURE 1
The area of the shaded region is given by $\int_0^{\pi/2} \sin^3 x \cos^{1/2} x \, dx$.

Let $u = \cos x$ so that $du = -\sin x \, dx$. Note that when $x = 0$, $u = \cos 0 = 1$, the lower limit of integration with respect to u; when $x = \pi/2$, $u = \cos(\pi/2) = 0$, the upper limit of integration with respect to u. Making these substitutions, we obtain

$$\int_0^{\pi/2} \sin^3 x \cos^{1/2} x \, dx = \int_1^0 (u^{1/2} - u^{5/2})(-du) = -\int_1^0 (u^{1/2} - u^{5/2}) \, du$$

$$= \left[-\frac{2}{3} u^{3/2} + \frac{2}{7} u^{7/2} \right]_1^0 = \left[0 - \left(-\frac{2}{3} + \frac{2}{7} \right) \right] = \frac{8}{21}$$

Since $f(x) = \sin^3 x \cos^{1/2} x$ is nonnegative on $\left[0, \frac{\pi}{2} \right]$, we can interpret the integral in this example as the area of the region under the graph of f on $\left[0, \frac{\pi}{2} \right]$. (See Figure 1.)

■ Integrals of the Form $\int \tan^m x \sec^n x \, dx$ and $\int \cot^m x \csc^n x \, dx$

The techniques for evaluating integrals of the form

$$\int \tan^m x \sec^n x \, dx$$

are developed in a similar manner. We have the following guidelines for evaluating such integrals.

Guidelines for Evaluating $\int \tan^m x \sec^n x \, dx$

1. *If the power of* $\tan x$ *is odd and positive* ($m = 2k + 1$), retain a factor of $\sec x \tan x$ and use the identity $\tan^2 x = \sec^2 x - 1$ to write

$$\int \tan^{2k+1} x \sec^n x \, dx = \int (\tan^2 x)^k \sec^{n-1} x \sec x \tan x \, dx$$

$$= \int (\sec^2 x - 1)^k \sec^{n-1} x \sec x \tan x \, dx$$

Then integrate using the substitution $u = \sec x$.

2. *If the power of* $\sec x$ *is even and positive* ($n = 2k, k \geq 2$), retain a factor of $\sec^2 x$ and use the identity $\sec^2 x = 1 + \tan^2 x$ to write

$$\int \tan^m x \sec^{2k} x \, dx = \int \tan^m x (\sec^2 x)^{k-1} \sec^2 x \, dx$$

$$= \int \tan^m x (1 + \tan^2 x)^{k-1} \sec^2 x \, dx$$

Then integrate using the substitution $u = \tan x$.

The guidelines for evaluating $\int \cot^m x \csc^n x \, dx$ are similar to those for evaluating $\int \tan^m x \sec^n x \, dx$.

EXAMPLE 5 Find $\int \tan^3 x \sec^7 x \, dx$.

Solution Here, m (the power of $\tan x$) is an odd positive integer. Let's retain the factor $\sec x \tan x$ from the integrand and write

$\tan^3 x \sec^7 x = \tan^2 x \sec^6 x (\sec x \tan x)$

$\qquad = (\sec^2 x - 1)\sec^6 x(\sec x \tan x)$ Use the identity $\tan^2 x = \sec^2 x - 1$ to convert the other factors to a function of $\sec x$.

$\qquad = (\sec^8 x - \sec^6 x)\sec x \tan x$

Then

$$\int \tan^3 x \sec^7 x \, dx = \int (\sec^8 x - \sec^6 x)\sec x \tan x \, dx$$

Let $u = \sec x$ so that $du = \sec x \tan x \, dx$. Then

$$\int \tan^3 x \sec^7 x \, dx = \int (u^8 - u^6) \, du$$

$$= \frac{1}{9}u^9 - \frac{1}{7}u^7 + C$$

$$= \frac{1}{9}\sec^9 x - \frac{1}{7}\sec^7 x + C$$

EXAMPLE 6   Evaluate $\displaystyle\int_0^{\pi/4} \sqrt{\tan x}\, \sec^6 x \, dx$.

Solution   Here, n (the power of $\sec x$) is an even positive integer. So let's retain a factor of $\sec^2 x$. Thus,

$$\int_0^{\pi/4} \sqrt{\tan x}\, \sec^6 x \, dx = \int_0^{\pi/4} \tan^{1/2} x \, \sec^4 x \, \sec^2 x \, dx$$

$$= \int_0^{\pi/4} (\tan^{1/2} x)(1 + \tan^2 x)^2 \sec^2 x \, dx \qquad \sec^2 x = 1 + \tan^2 x$$

$$= \int_0^{\pi/4} (\tan^{1/2} x + 2 \tan^{5/2} x + \tan^{9/2} x)\sec^2 x \, dx$$

Let $u = \tan x$ so that $du = \sec^2 x \, dx$. The lower and upper limits of integration with respect to u are $u = 0$ (set $x = 0$) and $u = 1$ (set $x = \pi/4$), respectively. Making these substitutions, we obtain

$$\int_0^{\pi/4} \sqrt{\tan x}\, \sec^6 x \, dx = \int_0^1 (u^{1/2} + 2u^{5/2} + u^{9/2}) \, du$$

$$= \left[\frac{2}{3} u^{3/2} + \frac{4}{7} u^{7/2} + \frac{2}{11} u^{11/2} \right]_0^1$$

$$= \frac{2}{3} + \frac{4}{7} + \frac{2}{11} = \frac{328}{231} \qquad \blacksquare$$

Integrals of the form $\int \cot^m x \, \csc^n x \, dx$ may be evaluated in a similar manner, as the following example illustrates.

EXAMPLE 7   Evaluate $\int \cot^5 x \, \csc^5 x \, dx$.

Solution   Here, the power of $\cot x$ is an odd positive integer. So we retain the factor $\csc x \cot x$ from the integrand. Thus,

$$\int \cot^5 x \, \csc^5 x \, dx = \int \cot^4 x \, (\csc^4 x)(\csc x \cot x) \, dx$$

$$= \int (\csc^2 x - 1)^2 \csc^4 x \, \csc x \cot x \, dx \qquad \cot^2 x = \csc^2 x - 1$$

$$= \int (\csc^8 x - 2 \csc^6 x + \csc^4 x) \csc x \cot x \, dx$$

Let $u = \csc x$ so that $du = -\csc x \cot x \, dx$. Then

$$\int \cot^5 x \, \csc^5 x \, dx = -\int (u^8 - 2u^6 + u^4) \, du$$

$$= -\frac{1}{9} u^9 + \frac{2}{7} u^7 - \frac{1}{5} u^5 + C$$

$$= -\frac{1}{9} \csc^9 x + \frac{2}{7} \csc^7 x - \frac{1}{5} \csc^5 x + C \qquad \blacksquare$$

■ Converting to Sines and Cosines

For integrals involving powers of trigonometric functions that are not covered by the formulas just considered, we are sometimes able to evaluate the integral by converting the integrand to an expression involving sines and cosines, as the following example illustrates.

EXAMPLE 8 Find $\displaystyle\int \frac{\tan x}{\sec^2 x} \, dx$.

Solution We have

$$\int \frac{\tan x}{\sec^2 x} \, dx = \int \left(\frac{\sin x}{\cos x} \right) \cos^2 x \, dx = \int \sin x \cos x \, dx$$

$$= \frac{1}{2} \sin^2 x + C \qquad \text{Let } u = \sin x. \qquad ■$$

■ Integrals of the Form $\int \sin mx \sin nx \, dx$, $\int \sin mx \cos nx \, dx$, and $\int \cos mx \cos nx \, dx$

Integrals in which the integrand is a product of sines and cosines of two *different* angles can be evaluated with the help of the following identities.

Trigonometric Identities

$$\sin mx \sin nx = \frac{1}{2} [\cos(m - n)x - \cos(m + n)x] \qquad \text{(2a)}$$

$$\sin mx \cos nx = \frac{1}{2} [\sin(m - n)x + \sin(m + n)x] \qquad \text{(2b)}$$

$$\cos mx \cos nx = \frac{1}{2} [\cos(m - n)x + \cos(m + n)x] \qquad \text{(2c)}$$

EXAMPLE 9 Find $\int \sin 4x \cos 5x \, dx$.

Solution Using Formula (2b), we have

$$\int \sin 4x \cos 5x \, dx = \int \frac{1}{2} [\sin(-x) + \sin 9x] \, dx$$

$$= \frac{1}{2} \int (-\sin x + \sin 9x) \, dx$$

$$= \frac{1}{2} \left(\cos x - \frac{1}{9} \cos 9x \right) + C \qquad ■$$

6.2 CONCEPT QUESTIONS

1. Explain how you would find $\int \sin^m x \cos^n x \, dx$ if (a) m is odd and positive, (b) n is odd and positive, and (c) m and n are both nonnegative and even.

2. Explain how you would find $\int \tan^m x \sec^n x \, dx$ if (a) m is odd and positive and (b) n is even and positive.

3. Explain how you would find $\int \cot^m x \csc^n x \, dx$ if (a) m is odd and positive and (b) n is even and positive.

4. Explain how you would find (a) $\int \sin mx \cos nx \, dx$, (b) $\int \sin mx \sin nx \, dx$, and (c) $\int \cos mx \cos nx \, dx$.

6.2 EXERCISES

In Exercises 1–48, find or evaluate the integral.

1. $\displaystyle\int \sin^3 x \cos x \, dx$

2. $\displaystyle\int \sin^3 x \cos^2 x \, dx$

3. $\displaystyle\int \cos^3 2x \sin^5 2x \, dx$

4. $\displaystyle\int_0^{\pi/2} \sqrt{\cos x} \, \sin^3 x \, dx$

5. $\displaystyle\int \sin^3 x \, dx$

6. $\displaystyle\int \cos^3 2x \, dx$

7. $\displaystyle\int_0^{\pi} \cos^2 \frac{x}{2} \, dx$

8. $\displaystyle\int \cos^4 x \, dx$

9. $\displaystyle\int_0^1 \sin^4 \pi x \, dx$

10. $\displaystyle\int \sin^2 2x \cos^4 2x \, dx$

11. $\displaystyle\int \sin^2\left(\frac{x}{2}\right)\cos^2\left(\frac{x}{2}\right) dx$

12. $\displaystyle\int \cos^4 x \sin^4 x \, dx$

13. $\displaystyle\int_0^{\pi} \sin^2 x \cos^4 x \, dx$

14. $\displaystyle\int \sin^6 u \, du$

15. $\displaystyle\int x \cos^4(x^2) \, dx$

16. $\displaystyle\int \theta \sin^2(\theta^2)\cos^2(\theta^2) \, d\theta$

17. $\displaystyle\int x \sin^2 x \, dx$

18. $\displaystyle\int x \cos^2 x \, dx$

Hint: Integrate by parts.

Hint: Integrate by parts.

19. $\displaystyle\int_0^{\pi/4} \tan^2 x \, dx$

20. $\displaystyle\int \tan^3(\pi - x) \, dx$

21. $\displaystyle\int \tan^5 \frac{x}{2} \, dx$

22. $\displaystyle\int \tan^5 x \sec^3 x \, dx$

23. $\displaystyle\int \sec^2(\pi x)\tan^3(\pi x) \, dx$

24. $\displaystyle\int_0^{\pi/4} \sec^2 x \tan^2 x \, dx$

25. $\displaystyle\int \sec^4 3x \tan^2 3x \, dx$

26. $\displaystyle\int \sec^4(\pi - x)\tan(\pi - x) \, dx$

27. $\displaystyle\int \sec^4 \theta \sqrt{\tan \theta} \, d\theta$

28. $\displaystyle\int \sec^4\left(\frac{x}{2}\right)\tan^4\left(\frac{x}{2}\right) dx$

29. $\displaystyle\int \cot^2 2x \, dx$

30. $\displaystyle\int_{\pi/4}^{\pi/2} \cot^3 x \, dx$

31. $\displaystyle\int \csc^3 x \, dx$

32. $\displaystyle\int \csc^5 x \, dx$

Hint: Integrate by parts.

Hint: Integrate by parts.

33. $\displaystyle\int \csc^4 t \, dt$

34. $\displaystyle\int \csc^4 \theta \cot^4 \theta \, d\theta$

35. $\displaystyle\int \cot^6 t \, dt$

36. $\displaystyle\int \cot^3 x \csc^4 x \, dx$

37. $\displaystyle\int_0^{\pi/4} \frac{1}{\cos^4 x} \, dx$

38. $\displaystyle\int (1 + \cot x)^2 \csc x \, dx$

39. $\displaystyle\int_0^{\pi/2} \sin x \cos 2x \, dx$

40. $\displaystyle\int \sin 3\theta \sin 4\theta \, d\theta$

41. $\displaystyle\int \cos 2\theta \cos 4\theta \, d\theta$

42. $\displaystyle\int \frac{\sin^3 x}{\sec^2 x} \, dx$

43. $\displaystyle\int \cos^2 2\theta \cot 2\theta \, d\theta$

44. $\displaystyle\int_0^{\pi/2} \frac{\sin t}{1 + \cos t} \, dt$

45. $\displaystyle\int \frac{\tan^3 \sqrt{t} \sec^2 \sqrt{t}}{\sqrt{t}} \, dt$

46. $\displaystyle\int \frac{1}{\csc x \cot^2 x} \, dx$

47. $\displaystyle\int \frac{1 - \tan^2 x}{\sec^2 x} \, dx$

48. $\displaystyle\int \frac{\cos 2\theta}{\cos \theta + \sin \theta} \, d\theta$

49. Find the average value of $f(x) = \cos^2 x$ over the interval $[0, 2\pi]$.

50. Find the average value of $f(x) = \cos^2 x \sin^3 x$ over the interval $[0, \pi]$.

51. Find the area of the region under the graph of $y = \sin^2 \pi x$ on the interval $[0, 1]$.

52. Find the area of the region bounded by the graphs of $y = \sin^4 x$, $y = \cos^4 x$, $x = 0$, and $x = \pi/4$.

53. Let $f(x) = \sin^4 x$ and $g(x) = 1 - x^2$.

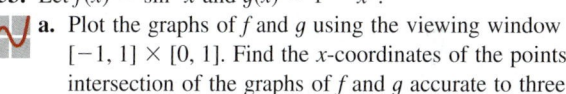

    a. Plot the graphs of f and g using the viewing window $[-1, 1] \times [0, 1]$. Find the x-coordinates of the points of intersection of the graphs of f and g accurate to three decimal places.

    b. Use the result of part (a) and the method of this section to find the approximate area of the region bounded by the graphs of f and g.

54. Let $f(x) = \cos 2x \cos 4x$ and $g(x) = \sqrt{x}$.

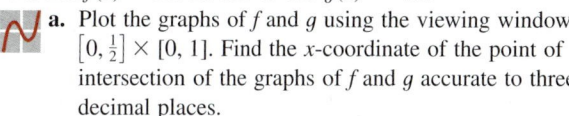

    a. Plot the graphs of f and g using the viewing window $[0, \frac{1}{2}] \times [0, 1]$. Find the x-coordinate of the point of intersection of the graphs of f and g accurate to three decimal places.

    b. Use the result of part (a) and the method of this section to find the approximate area of the region bounded by the graphs of f and g and the y-axis.

55. The region under the graph of $y = \tan^2 x$ on the interval $\left[0, \frac{\pi}{4}\right]$ is revolved about the x-axis. Find the volume of the solid generated.

56. The region under the graph of $y = \sin^3 x$ on the interval $[0, \pi]$ is revolved about the y-axis. Find the volume of the solid generated.

57. Find the centroid of the region R under the graph of $y = \cos x$ on the interval $\left[0, \frac{\pi}{2}\right]$.

58. Find the centroid of the region R bounded by the graphs of $y = \cos x$ and $y = 1 - (2/\pi)x$ on $\left[0, \frac{\pi}{2}\right]$.

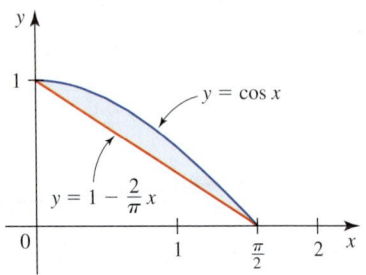

59. The velocity of a particle t sec after leaving the origin, moving along a coordinate line, is $v(t) = \sin^3 \pi t$ ft/sec. What is the distance traveled by the particle during the first 6 sec? What is its position at $t = 6$?

60. Find the volume of the solid generated by revolving the region bounded by the graphs of $y = \sin x$ and $y = (2/\pi)x$ about the x-axis.

61. Find the volume of the solid generated by revolving the region under the graph of $f(x) = (\sin x)/(\cos^3 x)$ on $\left[0, \frac{\pi}{4}\right]$ about the x-axis.

62. Interval Training As part of her speed training, a long-distance runner runs in spurts for a minute. Suppose that she runs along a straight line so that her velocity t sec after passing a marker is given by the velocity function

$$v(t) = 5 + 100 \sin^2\left(\frac{\pi}{16}t\right)\cos^2\left(\frac{\pi}{16}t\right) \qquad 0 \le t \le 60$$

where $v(t)$ is measured in feet per second. Find the position function as measured from the marker.

63. Alternating Current Intensity Find the average value of the alternating current intensity described by

$$I = I_0 \cos(\omega t + \alpha)$$

where I_0, ω, and α are constants, over the time interval $\left[0, \frac{\pi}{\omega}\right]$.

64. Electromotive Force An electromotive force (emf), E, is given by

$$E = E_0 \sin \frac{2\pi t}{T}$$

where T is the period in seconds and E_0 is the amplitude of the emf. Find the average value of the square of the emf, E^2, over the time interval $[0, T]$.

65. Heat Generated by an Alternating Current According to the Joule-Lenz Law, the amount of heat generated by an alternating current

$$I = I_0 \sin\left(\frac{2\pi t}{T} - \phi\right)$$

flowing in a conductor with resistance R ohms from $t = T_1$ to $t = T_2$ is given by

$$Q = 0.24R \int_{T_1}^{T_2} I^2 \, dt$$

joules. Find the amount of heat generated during a cycle (from $t = 0$ to $t = T$).

66. Fabricating Corrugated Metal Sheets A certain brand of corrugated metal sheets comes in 20 in. $\times$ 48 in. sizes. The cross section of the sheets can be described by the graph of

$$y = \sin \frac{\pi}{2} t \qquad 0 \le t \le 20$$

Use a calculator or computer to find the approximate length of the flat metal sheet before fabrication.

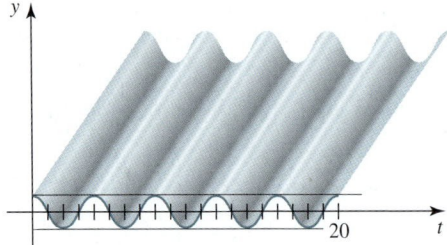

67. Plot the graph of $f(x) = e^{-x} \sin(\pi x/2)$ for $0 \le x \le 2$. Then use a calculator or computer to approximate the volume of the solid generated by revolving the region under the graph of f on $[0, 2]$ about (a) the x-axis and (b) the y-axis.

68. Refer to Exercise 62. Use a calculator or computer to approximate the distance traveled by the athlete in her 60-sec speed exercise.

69. Prove that if m and n are positive integers, then

$$\int_{-\pi}^{\pi} \sin mx \cos nx \, dx = 0$$

70. Prove that if m and n are positive integers, then

$$\int_{-\pi}^{\pi} \sin mx \sin nx \, dx = \begin{cases} 0 & \text{if } m \ne n \\ \pi & \text{if } m = n \end{cases}$$

71. Prove that if m and n are positive integers, then

$$\int_{-\pi}^{\pi} \cos mx \cos nx \, dx = \begin{cases} 0 & \text{if } m \ne n \\ \pi & \text{if } m = n \end{cases}$$

72. Finite Fourier Series Prove that if

$$f(x) = \frac{a_0}{2} + \sum_{k=1}^{n} (a_k \cos kx + b_k \sin kx)$$

$$= \frac{a_0}{2} + a_1 \cos x + b_1 \sin x + a_2 \cos 2x$$

$$+ b_2 \sin 2x + \cdots + a_n \cos nx + b_n \sin nx$$

then

$$a_0 = \frac{1}{\pi} \int_{-\pi}^{\pi} f(x)\, dx, \qquad a_k = \frac{1}{\pi} \int_{-\pi}^{\pi} f(x) \cos kx\, dx$$

and

$$b_k = \frac{1}{\pi} \int_{-\pi}^{\pi} f(x) \sin kx\, dx \qquad k = 1, 2, \dots, n$$

Hint: Use the results of Exercises 69–71.

6.3   Trigonometric Substitutions

Figure 1a depicts an aerial view of a scenic drive along the coast. We can approximate the part of the road between Point A and Point B by the graph of $f(x) = \frac{1}{2}x^2$ on the interval $[0, 1]$, where both x and $f(x)$ are measured in miles (see Figure 1b). How far is the drive between A and B?

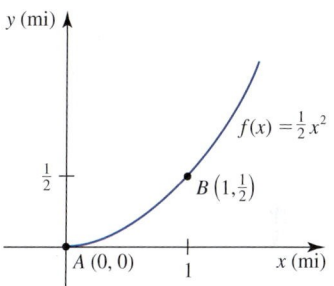

FIGURE 1
(a) A stretch of coastal road   (b) The graph of the function describing the road

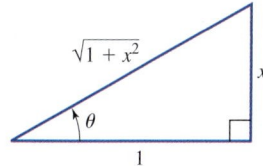

FIGURE 2
The right triangle associated with the integrand of Equation (1)

To answer this question, we need to find the arc length L of the graph of f from $A(0, 0)$ to $B\left(1, \frac{1}{2}\right)$. Now, from Section 5.4 we know that the required arc length is given by

$$L = \int_0^1 \sqrt{1 + [f'(x)]^2}\, dx = \int_0^1 \sqrt{1 + x^2}\, dx \tag{1}$$

To evaluate this integral, observe that the integrand can be written in the form $\sqrt{1^2 + x^2}$ —the square root of the sum of two squares. This brings to mind the Pythagorean Theorem. In fact, we can think of this quantity as being associated with the length of the hypotenuse of the right triangle shown in Figure 2. This suggests that we try the substitution

$$\tan \theta = \frac{x}{1} \qquad \text{or} \qquad x = \tan \theta$$

We then have

$$\sqrt{1 + x^2} = \sqrt{1 + \tan^2 \theta} = \sqrt{\sec^2 \theta}$$
$$= \sec \theta$$

provided that $\frac{-\pi}{2} < \theta < \frac{\pi}{2}$. Proceeding with the substitution $x = \tan\theta$ so that $dx = \sec^2\theta\, d\theta$, we see that the indefinite integral

$$\int \sqrt{1 + x^2}\, dx = \int \sec\theta\,(\sec^2\theta)\, d\theta$$

$$= \int \sec^3\theta\, d\theta$$

$$= \frac{1}{2}\left(\sec\theta\tan\theta + \ln|\sec\theta + \tan\theta|\right) + C$$

See Example 5 in Section 6.1.

Finally, referring to Figure 2 again, we see that

$$\sec\theta = \sqrt{1 + x^2} \qquad \text{hypotenuse/adjacent}$$

so

$$\int \sqrt{1 + x^2}\, dx = \frac{1}{2}\left(\sqrt{1 + x^2} \cdot x + \ln\left|\sqrt{1 + x^2} + x\right|\right) + C$$

Therefore, the distance of the drive between A and B is given by

$$L = \int_0^1 \sqrt{1 + x^2}\, dx = \frac{1}{2}\left[x\sqrt{1 + x^2} + \ln\left(\sqrt{1 + x^2} + x\right)\right]_0^1$$

$$= \frac{1}{2}\left[\sqrt{2} + \ln(\sqrt{2} + 1)\right] \approx 1.148$$

or approximately 1.15 mi.

■ Trigonometric Substitution

The techniques that we used in solving our introductory example involve **trigonometric substitution.** In general, this method can be used to evaluate integrals involving the radicals

$$\sqrt{a^2 - x^2}, \qquad \sqrt{a^2 + x^2}, \qquad \text{and} \qquad \sqrt{x^2 - a^2}$$

where $a > 0$.

The key to this technique lies in making an appropriate trigonometric substitution using one of the trigonometric identities

$$\cos^2\theta = 1 - \sin^2\theta$$

and

$$\sec^2\theta = 1 + \tan^2\theta$$

to transform the given integral into one that is radical free. The resulting *trigonometric integral* can then be evaluated by using the techniques developed earlier. Finally, the answer is written in terms of the original variable by converting from θ's to x's.

The trigonometric substitutions for evaluating integrals involving the indicated radicals are listed in Table 1.

Note that in each case the restriction on θ ensures that the function g in the substitution $x = g(\theta)$ is *one-to-one* and, therefore, has an inverse. This enables us to solve for θ in terms of x and, hence, to express the answer in terms of the original variable x.

TABLE 1 Trigonometric Substitutions

For integrals involving	Use the substitution	Use the identity	Right triangle associated with the substitution
$\sqrt{a^2 - x^2}$, $a > 0$	$x = a \sin \theta$, $-\frac{\pi}{2} \le \theta \le \frac{\pi}{2}$	$1 - \sin^2 \theta = \cos^2 \theta$	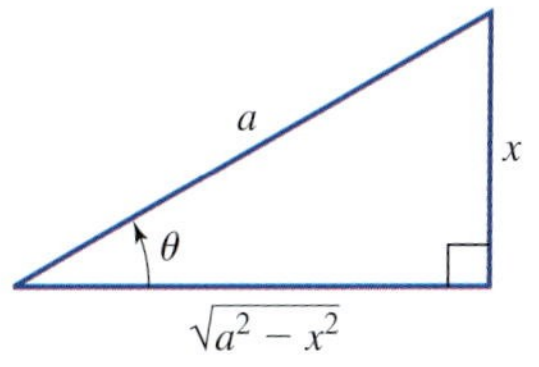
$\sqrt{a^2 + x^2}$, $a > 0$	$x = a \tan \theta$, $-\frac{\pi}{2} < \theta < \frac{\pi}{2}$	$1 + \tan^2 \theta = \sec^2 \theta$	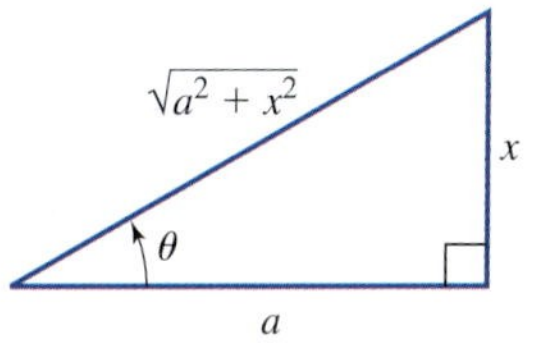
$\sqrt{x^2 - a^2}$, $a > 0$	$x = a \sec \theta$, $0 \le \theta < \frac{\pi}{2}$ or $\frac{\pi}{2} < \theta \le \pi$	$\sec^2 \theta - 1 = \tan^2 \theta$	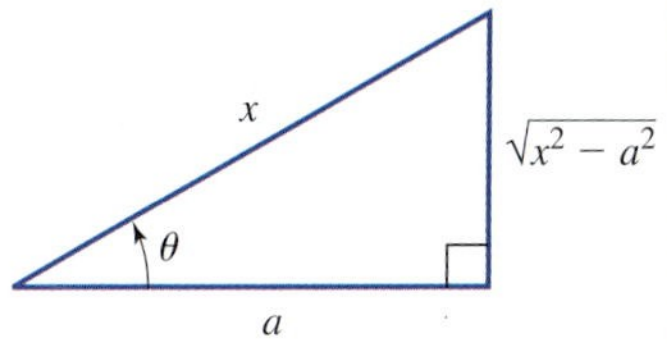

EXAMPLE 1 Find $\displaystyle\int \frac{x^2}{\sqrt{9 - x^2}} \, dx$.

Solution Note that the integrand involves a radical of the form $\sqrt{a^2 - x^2}$, where $a = 3$. This suggests that we use the trigonometric substitution

$$x = 3 \sin \theta \qquad \text{so that} \qquad dx = 3 \cos \theta \, d\theta$$

where $-\frac{\pi}{2} < \theta < \frac{\pi}{2}$. In this example we have the further restriction $\theta \ne \pm\pi/2$ to ensure that $x \ne \pm 3$ (the integrand is not defined at these points). Making these substitutions, we have

$$\int \frac{x^2}{\sqrt{9 - x^2}} \, dx = \int \frac{9 \sin^2 \theta}{\sqrt{9 - 9 \sin^2 \theta}} \, (3 \cos \theta \, d\theta)$$

$$= 9 \int \sin^2 \theta \, d\theta$$

$$= \frac{9}{2} \int (1 - \cos 2\theta) \, d\theta \qquad \text{Use a half-angle formula.}$$

$$= \frac{9}{2} \left(\theta - \frac{1}{2} \sin 2\theta \right) + C$$

To express this result in terms of the original variable x, observe that $\sin \theta = x/3$ implies that $\theta = \sin^{-1}(x/3)$. Next, observe that $\sin 2\theta = 2 \sin \theta \cos \theta$. With the help of Figure 3, we find

$$\sin 2\theta = 2(\sin \theta)(\cos \theta) = 2\left(\frac{x}{3}\right)\left(\frac{\sqrt{9 - x^2}}{3}\right)$$

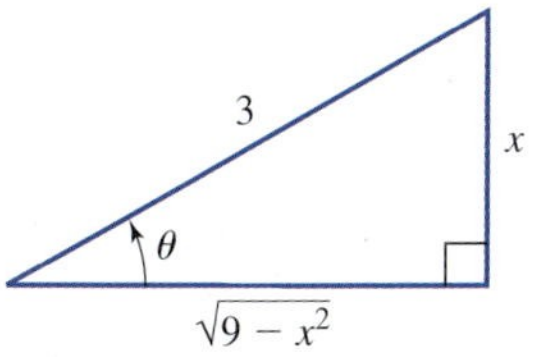

FIGURE 3
The right triangle associated with the substitution $x = 3 \sin \theta$

Therefore,

$$\int \frac{x^2}{\sqrt{9 - x^2}} \, dx = \frac{9}{2}\left[\sin^{-1}\left(\frac{x}{3}\right) - \frac{1}{9}x\sqrt{9 - x^2}\right] + C$$

EXAMPLE 2 Find the area enclosed by the ellipse $\dfrac{x^2}{a^2} + \dfrac{y^2}{b^2} = 1$.

Solution The ellipse is shown in Figure 4. By symmetry we see that the area A enclosed by the ellipse is just four times its area in the first quadrant. Next, to find the function describing the ellipse in this quadrant, we solve the given equation for y. Thus,

$$\frac{y^2}{b^2} = 1 - \frac{x^2}{a^2} = \frac{a^2 - x^2}{a^2} \qquad \text{or} \qquad y = \pm\frac{b}{a}\sqrt{a^2 - x^2}$$

Since $y > 0$ in this quadrant, the required function is $f(x) = (b/a)\sqrt{a^2 - x^2}$ for x in the interval $[0, a]$. Therefore, the desired area A is given by

$$A = 4\int_0^a \frac{b}{a}\sqrt{a^2 - x^2} \, dx = \frac{4b}{a}\int_0^a \sqrt{a^2 - x^2} \, dx$$

To evaluate this integral, we let

$$x = a \sin \theta \qquad \text{so that} \qquad dx = a \cos \theta \, d\theta$$

Note that when $x = 0$, $\sin \theta = 0$, so $\theta = 0$ is the lower limit of integration with respect to θ; when $x = a$, $\sin \theta = 1$, giving $\theta = \pi/2$ as the upper limit of integration. Also,

$$\sqrt{a^2 - x^2} = \sqrt{a^2 - a^2 \sin^2 \theta} = a\sqrt{1 - \sin^2 \theta} = a\sqrt{\cos^2 \theta} = a|\cos \theta| = a \cos \theta$$

since $0 \le \theta \le \frac{\pi}{2}$. Therefore,

$$A = \frac{4b}{a}\int_0^a \sqrt{a^2 - x^2} \, dx = \frac{4b}{a}\int_0^{\pi/2} a \cos \theta \cdot a \cos \theta \, d\theta$$

$$= 4ab\int_0^{\pi/2} \cos^2 \theta \, d\theta$$

$$= 4ab\int_0^{\pi/2} \frac{1}{2}(1 + \cos 2\theta) \, d\theta \qquad \text{Use a half-angle formula.}$$

$$= 2ab\left[\theta + \frac{1}{2}\sin 2\theta\right]_0^{\pi/2} = 2ab\left[\left(\frac{\pi}{2} + 0\right) - 0\right]$$

or πab.

Note For a circle, $a = b = r$, where r is the radius of the circle, and the result of Example 2 gives πr^2 as the area of the circle, as expected.

EXAMPLE 3 Find $\displaystyle\int \frac{1}{(4 + x^2)^{3/2}} \, dx$.

Solution Observe that the denominator of the integrand can be written as $\left(\sqrt{4 + x^2}\right)^3$ and thus involves a radical of the form $\sqrt{a^2 + x^2}$ with $a = 2$. Hence, we make the substitution

$$x = 2 \tan \theta \qquad \text{so that} \qquad dx = 2 \sec^2 \theta \, d\theta$$

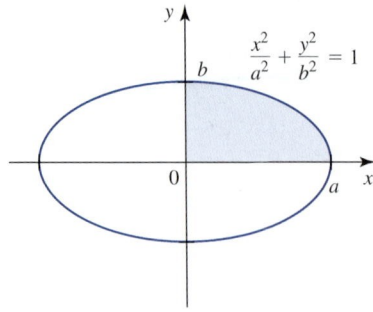

FIGURE 4
The area enclosed by the ellipse is four times its area in the first quadrant.

Then

$$\sqrt{4 + x^2} = \sqrt{4 + 4\tan^2\theta} = 2\sqrt{1 + \tan^2\theta} = 2\sqrt{\sec^2\theta} = 2\sec\theta$$

Therefore,

$$\int \frac{1}{(4 + x^2)^{3/2}}\, dx = \int \frac{1}{(2\sec\theta)^3} \cdot 2\sec^2\theta\, d\theta$$

$$= \frac{1}{4}\int \cos\theta\, d\theta$$

$$= \frac{1}{4}\sin\theta + C$$

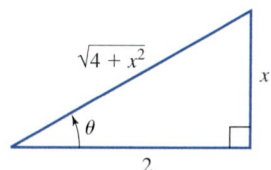

FIGURE 5
The right triangle associated
with the substitution $x = 2\tan\theta$

Finally, from the right triangle associated with the substitution $x = 2\tan\theta$, we see that $\sin\theta = x/\sqrt{4 + x^2}$. (See Figure 5.) Therefore,

$$\int \frac{1}{(4 + x^2)^{3/2}}\, dx = \frac{x}{4\sqrt{4 + x^2}} + C \qquad \blacksquare$$

EXAMPLE 4 Find $\displaystyle\int \frac{\sqrt{x^2 - 16}}{x}\, dx$.

Solution   Here, the integrand involves a radical of the form $\sqrt{x^2 - a^2}$, where $a = 4$. So we make the substitution

$$x = 4\sec\theta \qquad \text{so that} \qquad dx = 4\sec\theta\tan\theta\, d\theta$$

Then

$$\sqrt{x^2 - 16} = \sqrt{16\sec^2\theta - 16} = 4\sqrt{\sec^2\theta - 1} = 4\sqrt{\tan^2\theta} = 4\tan\theta$$

Therefore,

$$\int \frac{\sqrt{x^2 - 16}}{x}\, dx = \int \frac{4\tan\theta}{4\sec\theta} \cdot 4\sec\theta\tan\theta\, d\theta$$

$$= 4\int \tan^2\theta\, d\theta$$

$$= 4\int (\sec^2\theta - 1)\, d\theta = 4\int \sec^2\theta\, d\theta - 4\int d\theta$$

$$= 4\tan\theta - 4\theta + C$$

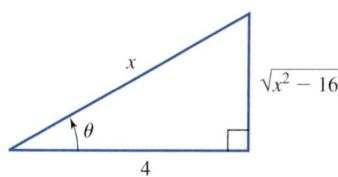

FIGURE 6
The right triangle associated with
the substitution $x = 4\sec\theta$

Since $x = 4\sec\theta$ or $\sec\theta = x/4$, we see that $\theta = \sec^{-1}(x/4)$. Furthermore, by inspecting the right triangle associated with the substitution, we see that

$$\tan\theta = \frac{\sqrt{x^2 - 16}}{4}$$

(See Figure 6.) Therefore,

$$\int \frac{\sqrt{x^2 - 16}}{x}\, dx = \sqrt{x^2 - 16} - 4\sec^{-1}\left(\frac{x}{4}\right) + C \qquad \blacksquare$$

Sometimes we can use the technique of completing the square to rewrite an integrand that involves a quadratic expression in the appropriate form before making a trigonometric substitution. This is illustrated in the next example.

EXAMPLE 5 Find $\displaystyle\int \frac{dx}{\sqrt{x^2 + 4x + 7}}$.

Solution By completing the square for the expression under the radical sign, we obtain

$$x^2 + 4x + 7 = [x^2 + 4x + (2)^2] + 7 - 4 = (x + 2)^2 + 3$$

If we let $u = x + 2$, then

$$x^2 + 4x + 7 = u^2 + 3 \qquad \text{and} \qquad du = dx$$

so we can write

$$\int \frac{dx}{\sqrt{x^2 + 4x + 7}} = \int \frac{du}{\sqrt{u^2 + 3}}$$

Observe that the integrand $\sqrt{u^2 + 3}$ has the form $\sqrt{u^2 + a^2}$, where $a = \sqrt{3}$. This suggests that we make the substitution

$$u = \sqrt{3}\tan\theta \qquad \text{so that} \qquad du = \sqrt{3}\sec^2\theta\, d\theta$$

Then

$$\sqrt{u^2 + 3} = \sqrt{3\tan^2\theta + 3} = \sqrt{3}\sqrt{\tan^2\theta + 1} = \sqrt{3}\sqrt{\sec^2\theta} = \sqrt{3}\sec\theta$$

Therefore,

$$\int \frac{du}{\sqrt{u^2 + 3}} = \int \frac{\sqrt{3}\sec^2\theta}{\sqrt{3}\sec\theta}\, d\theta = \int \sec\theta\, d\theta$$

$$= \ln|\sec\theta + \tan\theta| + C$$

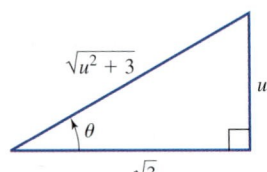

FIGURE 7
The right triangle associated with the substitution $u = \sqrt{3}\tan\theta$

From $u = \sqrt{3}\tan\theta$ we see that $\tan\theta = u/\sqrt{3}$. Also, from the right triangle associated with the substitution $u = \sqrt{3}\tan\theta$, we see that $\sec\theta = \sqrt{u^2 + 3}/\sqrt{3}$. (See Figure 7.) Therefore,

$$\int \frac{du}{\sqrt{u^2 + 3}} = \ln\left|\frac{\sqrt{u^2 + 3}}{\sqrt{3}} + \frac{u}{\sqrt{3}}\right| + \ln C_1$$

so

$$\int \frac{dx}{\sqrt{x^2 + 4x + 7}} = \ln|\sqrt{x^2 + 4x + 7} + x + 2| + C \qquad C = \ln\left(\frac{C_1}{\sqrt{3}}\right) \quad ■$$

EXAMPLE 6 **Hydrostatic Force on a Window** A cylindrical oil storage tank of radius 3 ft and length 10 ft is lying on its side. If the tank is filled to a height of 5 ft with oil having a weight density of 50 lb/ft³, find the force exerted by the oil on one end of the tank.

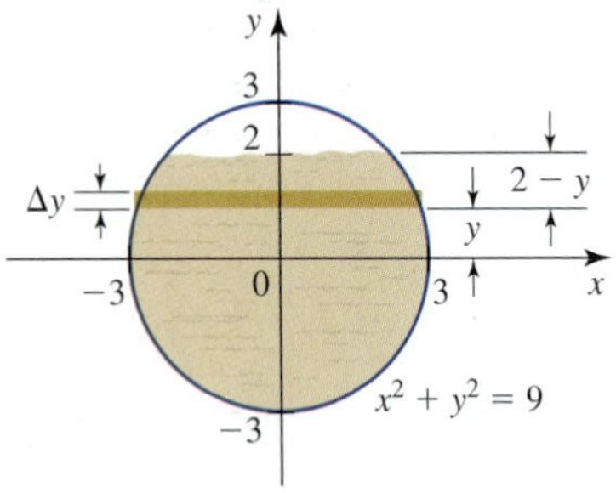

Solution Let's introduce a coordinate system in such a way that the end of the tank (a disk of radius 3 ft) has its center at the origin. Then the circle in question is described by the equation $x^2 + y^2 = 9$. (See Figure 8.) The length of a representative horizontal rectangle is $L(y) = 2x = 2\sqrt{9 - y^2}$. Therefore, the area of a horizontal rectangle is

$$\Delta A = L(y)\Delta y = 2\sqrt{9 - y^2}\, \Delta y$$

FIGURE 8
One end of a cylindrical oil storage tank

The pressure exerted by the oil on the rectangle is

$$\delta(2 - y) = 50(2 - y) \qquad \text{density} \cdot \text{depth}$$

Therefore, the force exerted by the oil on the rectangle is

$$\delta(2 - y)\Delta A = 50(2 - y)(2)\sqrt{9 - y^2}\,\Delta y \qquad \text{pressure} \cdot \text{area}$$
$$= 100(2 - y)\sqrt{9 - y^2}\,\Delta y$$

Summing the forces on these rectangles and taking the limit, we find that the force exerted by the oil on the end of the storage tank is

$$F = \int_{-3}^{2} 100(2 - y)\sqrt{9 - y^2}\,dy$$
$$= 200\int_{-3}^{2} \sqrt{9 - y^2}\,dy - 100\int_{-3}^{2} y\sqrt{9 - y^2}\,dy$$

The second integral can be evaluated by making the substitution $u = 9 - y^2$, whereas the first integral can be evaluated by using the trigonometric substitution $y = 3\sin\theta$. We will leave the evaluation of these integrals as an exercise (Exercise 58). You will find that the force is approximately 2890 lb. ■

6.3 CONCEPT QUESTIONS

1. What substitution would you use to find an integral whose integrand involves the expression (a) $\sqrt{a^2 - x^2}$, $a > 0$; (b) $\sqrt{a^2 + x^2}$, $a > 0$; and (c) $\sqrt{x^2 - a^2}$, $a > 0$?

2. How would you find an integral whose integrand involves the expression $\sqrt{ax^2 + bx + c}$?

6.3 EXERCISES

In Exercises 1–32, find or evaluate the integral using an appropriate trigonometric substitution.

1. $\displaystyle\int \frac{x}{\sqrt{9 - x^2}}\,dx$

2. $\displaystyle\int \frac{\sqrt{4 - x^2}}{x^2}\,dx$

3. $\displaystyle\int x\sqrt{4 - x^2}\,dx$

4. $\displaystyle\int \frac{1}{x^2\sqrt{1 - x^2}}\,dx$

5. $\displaystyle\int \frac{1}{x\sqrt{4 + x^2}}\,dx$

6. $\displaystyle\int x^3\sqrt{1 + x^2}\,dx$

7. $\displaystyle\int \frac{1}{x^2\sqrt{x^2 + 4}}\,dx$

8. $\displaystyle\int \frac{1}{x^3\sqrt{x^2 - 4}}\,dx$

9. $\displaystyle\int x^3\sqrt{1 - x^2}\,dx$

10. $\displaystyle\int x^3\sqrt{4 - x^2}\,dx$

11. $\displaystyle\int \frac{x^3}{\sqrt{x^2 + 9}}\,dx$

12. $\displaystyle\int_{0}^{3/4} \frac{x^2}{\sqrt{9 - 4x^2}}\,dx$

13. $\displaystyle\int \frac{1}{(x^2 - 9)^{3/2}}\,dx$

14. $\displaystyle\int (4 - x^2)^{3/2}\,dx$

15. $\displaystyle\int \frac{\sqrt{16x^2 - 9}}{x}\,dx$

16. $\displaystyle\int \frac{x^2}{\sqrt{3 - x^2}}\,dx$

17. $\displaystyle\int \frac{\sqrt{1 - x^2}}{x^4}\,dx$

18. $\displaystyle\int \frac{\sqrt{9x^2 + 4}}{x^4}\,dx$

19. $\displaystyle\int \frac{1}{x\sqrt{9x^2 + 4}}\,dx$

20. $\displaystyle\int \frac{1}{(9 + x^2)^2}\,dx$

21. $\displaystyle\int_{-\sqrt{3}}^{\sqrt{3}} \sqrt{4 - x^2}\,dx$

22. $\displaystyle\int_{2}^{4} \frac{\sqrt{x^2 - 4}}{x^4}\,dx$

23. $\displaystyle\int_{1}^{\sqrt{3}} \frac{1}{(1 + x^2)^{3/2}}\,dx$

24. $\displaystyle\int \frac{2x + 3}{\sqrt{1 - x^2}}\,dx$

25. $\displaystyle\int e^x\sqrt{4 - e^{2x}}\,dx$

26. $\displaystyle\int e^t\sqrt{1 + e^{2t}}\,dt$

27. $\displaystyle\int_{1/2}^{\sqrt{3}/2} \frac{dx}{x\sqrt{1 - x^2}}$

28. $\displaystyle\int \sqrt{4x - x^2}\,dx$

29. $\displaystyle\int \frac{1}{\sqrt{2t-t^2}}\,dt$

30. $\displaystyle\int \frac{t^2}{\sqrt{4t-t^2}}\,dt$

31. $\displaystyle\int \frac{1}{(x^2+4x+8)^2}\,dx$

32. $\displaystyle\int \frac{1}{(3-2x-x^2)^{5/2}}\,dx$

33. Find the area of the region under the graph of
$$y = \frac{1}{x\sqrt{4-x^2}} \quad \text{on the interval } [1, \sqrt{2}].$$

34. Find the area of the region enclosed by the hyperbola $16x^2 - 9y^2 = 144$ and the line $x = 5$.

35. Find the average value of the positive y-coordinates of the ellipse $\dfrac{x^2}{a^2} + \dfrac{y^2}{b^2} = 1$.

36. The region under the graph of $y = \dfrac{x}{(9+x^2)^{1/4}}$ on the interval $[0, 4]$ is revolved about the x-axis. Find the volume of the resulting solid.

37. The region under the graph of $y = \dfrac{x}{\sqrt{16-x^2}}$ on the interval $[0, 2]$ is revolved about the y-axis. Find the volume of the resulting solid.

38. The graph of $y = e^x$ between $x = 0$ and $x = 1$ is revolved about the x-axis. Find the surface area of the resulting solid.

39. Find the arc length of the graph of $y = \ln 2x$ on the interval $[1, \sqrt{3}]$.

40. Find the arc length of the graph of $y = -\frac{1}{2}x^2 + 2x$ from $P(0, 0)$ to $Q(2, 2)$.

41. Force Exerted on a Viewing Port The circular viewing port of a modern submersible used in oceanographic research has a radius of 1 ft. If the viewing port is partially submerged so that three fourths of it is under water, find the force exerted on it by the water. Note that the density of sea water is 64 lb/ft³.

42. Find the area of the region enclosed by the parabola $y = \frac{1}{4}x^2$ and the witch of Agnesi: $y = \dfrac{8}{x^2+4}$.

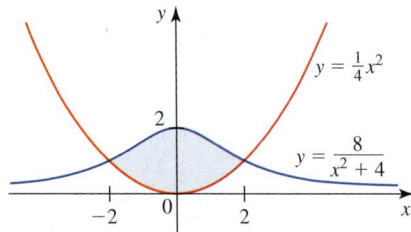

43. Let $f(x) = \dfrac{1}{(4+x^2)^{3/2}}$ and $g(x) = 1 - x^2$.

 a. Plot the graphs of f and g using the viewing window $[-1.5, 1.5] \times [0, 1.2]$. Find the x-coordinates of the points of intersection of the graphs of f and g accurate to three decimal places.

b. Use the result of part (a) and trigonometric substitution to find the approximate area of the region bounded by the graphs of f and g.

44. Let $f(x) = x^3\sqrt{1-x^2}$ and $g(x) = \dfrac{0.2}{(1+x^2)^{3/2}}$.

a. Plot the graphs of f and g using the viewing window $[0, 1.2] \times [0, 0.4]$. Find the x-coordinates of the points of intersection of the graphs of f and g accurate to three decimal places.

b. Use the result of part (a) and trigonometric substitution to find the approximate area of the region bounded by the graphs of f and g.

45. Find the surface area of the ellipsoid formed by revolving the ellipse $\dfrac{x^2}{a^2} + \dfrac{y^2}{b^2} = 1$, $a > b$, about the x-axis.

46. Find the force exerted by a liquid of constant weight density δ on a vertical ellipse $\dfrac{x^2}{a^2} + \dfrac{y^2}{b^2} = 1$ whose center is submerged in the liquid to a depth h, where $h \geq b$.

47. Air Pollution The amount of nitrogen dioxide, a brown gas that impairs breathing, present in the atmosphere on a certain day in May in the city of Long Beach is approximated by
$$A(t) = \frac{544}{4+(t-4.5)^2} + 28 \qquad 0 \leq t \leq 11$$
where $A(t)$ is measured in pollutant standard index (PSI) and t is measured in hours with $t = 0$ corresponding to 7 A.M. What is the average amount of the pollutant present in the atmosphere between 7 A.M. and noon on that day in the city?
Source: The Los Angeles Times.

48. Work Done by a Repulsive Charge An electric charge Q distributed uniformly along a line of length $2c$ lying along the y-axis repulses a like charge q from the point $x = a$ $(a > 0)$ to the point $x = b$, where $b > a$. The magnitude of the force acting on the charge q when it is at the point x is given by
$$F(x) = \frac{1}{4\pi\varepsilon_0}\frac{qQ}{x\sqrt{x^2+c^2}}$$
and the force acts in the direction of the positive x-axis. Find the work done by the force of repulsion.

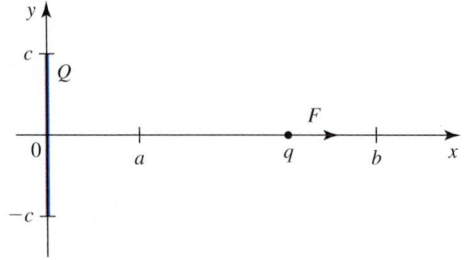

49. Work Done by a Magnetic Field The force of a circular electric current acting on a small magnet with its axis perpendicular to the plane of the circle and passing through its center has magnitude given by

$$F = \frac{k}{(x^2 + a^2)^{3/2}} \qquad 0 < x < \infty$$

where a is the radius of the circle, k is a constant, and x is the distance from the center of the circle to the magnet in the direction along the axis. Find the work done by the magnetic field in moving the magnet along the axis from $x = 0$ to $x = 2a$.

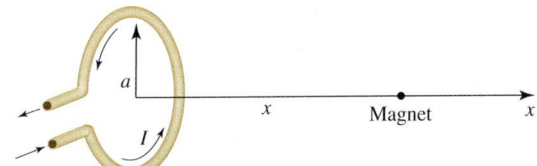

The electric current I flowing in the loop establishes a magnetic field that acts on the magnet.

50. Average Illumination Two lamp posts, each h ft tall, are located d ft apart. (See the figure.) If the intensity of each light is I lumens, find the average intensity of light along the straight line connecting the bases of the lamp posts.

Hint: The intensity of light at a point P from the base of the lamp post is proportional to the cosine of the angle that the incident light makes with the vertical and inversely proportional to the square of the distance between P and the light source.

In Exercises 51–54, use a trigonometric substitution to derive the formula.

51. $\displaystyle \int \sqrt{a^2 - u^2}\, du = \frac{u}{2}\sqrt{a^2 - u^2} + \frac{a^2}{2}\sin^{-1}\frac{u}{a} + C$

52. $\displaystyle \int \frac{\sqrt{a^2 + u^2}}{u}\, du = \sqrt{a^2 + u^2} - a\ln\left|\frac{a + \sqrt{a^2 + u^2}}{u}\right| + C$

53. $\displaystyle \int \frac{du}{u\sqrt{a^2 + u^2}} = -\frac{1}{a}\ln\left|\frac{\sqrt{a^2 + u^2} + a}{u}\right| + C$

54. $\displaystyle \int \frac{\sqrt{u^2 - a^2}}{u^2}\, du = -\frac{\sqrt{u^2 - a^2}}{u} + \ln\left|u + \sqrt{u^2 - a^2}\right| + C$

55. a. Use trigonometric substitution to show that

$$\int \frac{dx}{\sqrt{x^2 + a^2}} = \ln\left(x + \sqrt{x^2 + a^2}\right) + C$$

b. Use integration by parts and the result of part (a) to find

$$\int \frac{x\tan^{-1} x}{\sqrt{1 + x^2}}\, dx$$

56. Evaluate

$$\int_0^{\pi/4} \frac{dx}{a^2\cos^2 x + b^2\sin^2 x} \qquad a > 0, \quad b > 0$$

Hint: Use the substitution $u = \tan x$.

57. Prove that

$$\int_0^x \sqrt{a^2 - u^2}\, du = \frac{1}{2}x\sqrt{a^2 - x^2} + \frac{a^2}{2}\sin^{-1}\frac{x}{a} + C$$

where $0 \le x \le a$, by interpreting the integral geometrically.

58. Refer to Example 6. Show that

$$F = 200\int_{-3}^{2} \sqrt{9 - y^2}\, dy - 100\int_{-3}^{2} y\sqrt{9 - y^2}\, dy \approx 2890$$

6.4 The Method of Partial Fractions

■ Partial Fractions

In algebra we learned how to combine two or more rational expressions (fractions) into a single expression by putting them together over a common denominator. For example,

$$\frac{2}{x - 3} - \frac{1}{x + 1} = \frac{2(x + 1) - (x - 3)}{(x - 3)(x + 1)} = \frac{x + 5}{(x - 3)(x + 1)} \tag{1}$$

Sometimes, however, it is advantageous to reverse the process, that is, to express a complicated expression as a sum or difference of simpler ones. As an example, suppose that we wish to evaluate the integral

$$\int \frac{x + 5}{x^2 - 2x - 3}\, dx \tag{2}$$

Thanks to Equation (1), we can write the integrand in the form

$$\frac{x + 5}{x^2 - 2x - 3} = \frac{x + 5}{(x - 3)(x + 1)} = \frac{2}{x - 3} - \frac{1}{x + 1} \tag{3}$$

so that upon integrating both sides with respect to x, we obtain

$$\int \frac{x + 5}{x^2 - 2x - 3}\, dx = \int \left(\frac{2}{x - 3} - \frac{1}{x + 1} \right) dx$$

$$= 2 \ln|x - 3| - \ln|x + 1| + C$$

The expression on the right-hand side of Equation (3) is called the *partial fraction decomposition* of $(x + 5)/(x^2 - 2x - 3)$, and each of the terms is called a *partial fraction*. The technique of integration that we have used to evaluate the integrand in (2) is called the **method of partial fractions** and can be used to integrate any rational function.

Suppose that f is a rational function defined by

$$f(x) = \frac{P(x)}{Q(x)}$$

where P and Q are polynomials. If the degree of P is greater than or equal to the degree of Q, we can use long division to express $f(x)$ in the form

$$f(x) = S(x) + \frac{R(x)}{Q(x)} \tag{4}$$

where S is a *polynomial* and the degree of R is *less than* that of Q. For example, if

$$f(x) = \frac{x^3 - 4x^2 + 3x - 5}{x^2 - 1}$$

then using long division, we can write

$$f(x) = x - 4 + \frac{4x - 9}{x^2 - 1}$$

Now suppose that we want to integrate f. Using Equation (4), we have

$$\int f(x)\, dx = \int S(x)\, dx + \int \frac{R(x)}{Q(x)}\, dx$$

The first integral on the right is easily evaluated since its integrand is a polynomial. To evaluate the second integral, we decompose $R(x)/Q(x)$ into a sum of partial fractions and integrate the resulting expression term by term. That $R(x)/Q(x)$ can be so decomposed is guaranteed by the following results from algebra, which we state without proof:

1. Every polynomial Q can be factored into a product of linear factors (of the form $ax + b$) and irreducible quadratic factors (of the form $ax^2 + bx + c$ where $b^2 - 4ac < 0$).

2. Every rational function $R(x)/Q(x)$ where the degree of R is *less than* the degree of Q can be decomposed into a sum of partial fractions of the form

$$\frac{A}{(ax + b)^k} \quad \text{or} \quad \frac{Ax + B}{(ax^2 + bx + c)^k}$$

The form the partial fraction decomposition of the rational function $R(x)/Q(x)$ takes depends on the form of $Q(x)$ and can be illustrated through examining four cases.

Case 1: Distinct Linear Factors

If

$$\frac{R(x)}{Q(x)} = \frac{R(x)}{(a_1 x + b_1)(a_2 x + b_2) \cdots (a_n x + b_n)}$$

where all the factors $a_k x + b_k$, $k = 1, 2, \ldots, n$, are distinct, then there exist constants $A_1, A_2, \ldots, A_n$ such that

$$\frac{R(x)}{Q(x)} = \frac{A_1}{a_1 x + b_1} + \frac{A_2}{a_2 x + b_2} + \cdots + \frac{A_n}{a_n x + b_n}$$

EXAMPLE 1 Find $\displaystyle\int \frac{4x^2 - 4x + 6}{x^3 - x^2 - 6x}\, dx.$

Solution The degree of the numerator of the integrand is less than that of the denominator, and no long division is required in this case. The denominator can be written in the form $x(x - 3)(x + 2)$, a product of three distinct linear factors. Therefore, a partial fraction decomposition of the form

$$\frac{4x^2 - 4x + 6}{x(x - 3)(x + 2)} = \frac{A}{x} + \frac{B}{x - 3} + \frac{C}{x + 2}$$

exists. To determine A, B, and C, we multiply both sides of the equation by $x(x - 3)(x + 2)$, obtaining

$$4x^2 - 4x + 6 = A(x - 3)(x + 2) + Bx(x + 2) + Cx(x - 3)$$

If we expand the terms on the right and collect like powers of x, the equation can be written in the form

$$4x^2 - 4x + 6 = (A + B + C)x^2 + (-A + 2B - 3C)x - 6A$$

Because the two polynomials are equal, the coefficients of like powers of x must be equal. Equating, in turn, the coefficients of x^2, x^1, and x^0 leads to the following system of linear equations in A, B, and C:

$$A + B + C = 4$$
$$-A + 2B - 3C = -4$$
$$-6A \qquad\qquad = 6$$

Solving this system, we find $A = -1$, $B = 2$, and $C = 3$. Therefore, the partial fraction decomposition of the integrand is

$$\frac{4x^2 - 4x + 6}{x^3 - x^2 - 6x} = -\frac{1}{x} + \frac{2}{x - 3} + \frac{3}{x + 2}$$

Finally, integrating both sides of this equation gives

$$\int \frac{4x^2 - 4x + 6}{x^3 - x^2 - 6x}\, dx = \int \left(-\frac{1}{x} + \frac{2}{x-3} + \frac{3}{x+2} \right) dx$$

$$= -\ln|x| + 2\ln|x-3| + 3\ln|x+2| + k$$

where k is the constant of integration.

Note There is another way of finding the coefficients A, B, and C in Example 1. Our starting point is the equation

$$4x^2 - 4x + 6 = A(x-3)(x+2) + Bx(x+2) + Cx(x-3)$$

which holds for *all* values of x. If we let $x = 0$, then the second and third terms on the right are equal to zero, giving $6 = -6A$ or $A = -1$. Next, letting $x = 3$, so that the first and third terms are equal to zero, we find that $30 = 15B$, giving $B = 2$. Finally, letting $x = -2$ gives $30 = 10C$ or $C = 3$.

EXAMPLE 2 Find $\displaystyle\int \frac{4x^3 + x}{2x^2 + x - 3}\, dx$.

Solution Since the degree of the numerator of the integrand is greater than that of the denominator, we use long division to write

$$\frac{4x^3 + x}{2x^2 + x - 3} = 2x - 1 + \frac{8x - 3}{2x^2 + x - 3} \tag{5}$$

Next, we decompose $(8x - 3)/(2x^2 + x - 3)$ into a sum of partial fractions. Factoring, we see that $(2x^2 + x - 3) = (2x + 3)(x - 1)$ is a product of two distinct linear factors. Therefore,

$$\frac{8x - 3}{2x^2 + x - 3} = \frac{8x - 3}{(2x + 3)(x - 1)} = \frac{A}{2x + 3} + \frac{B}{x - 1}$$

Multiplying through by $(2x + 3)(x - 1)$ gives

$$8x - 3 = A(x - 1) + B(2x + 3)$$

If we let $x = 1$, then $5 = 5B$ or $B = 1$. Next, letting $x = -\frac{3}{2}$ yields $-15 = -\frac{5}{2}A$, or $A = 6$. Therefore,

$$\frac{8x - 3}{2x^2 + x - 3} = \frac{6}{2x + 3} + \frac{1}{x - 1}$$

Substituting the right-hand side of this equation into Equation (5) and integrating both sides of the resulting expression with respect to x, we get the desired result:

$$\int \frac{4x^3 + x}{2x^2 + x - 3}\, dx = \int \left(2x - 1 + \frac{6}{2x + 3} + \frac{1}{x - 1} \right) dx$$

$$= x^2 - x + 3\ln|2x + 3| + \ln|x - 1| + k$$

Observe that we have used the substitution $u = 2x + 3$ to evaluate the integral of the third term on the right and the substitution $u = x - 1$ to evaluate the last term on the right.

Case 2: Repeated Linear Factors

If $Q(x)$ contains a factor $(ax + b)^r$ with $r > 1$, then the partial fraction decomposition of $R(x)/Q(x)$ contains a sum of r partial fractions of the form

$$\frac{A_1}{ax + b} + \frac{A_2}{(ax + b)^2} + \cdots + \frac{A_r}{(ax + b)^r}$$

where each A_k is a real number.

For example,

$$\frac{2x^4 - 3x^2 + x - 4}{x(x - 1)(2x + 3)^3} = \frac{A}{x} + \frac{B}{x - 1} + \frac{C}{2x + 3} + \frac{D}{(2x + 3)^2} + \frac{E}{(2x + 3)^3}$$

EXAMPLE 3 Find $\displaystyle\int \frac{2x^2 + 3x + 7}{x^3 + x^2 - x - 1}\,dx$.

Solution The degree of the numerator of the integrand is less than that of the denominator, and no long division is necessary. Note that

$$Q(x) = x^3 + x^2 - x - 1 = x^2(x + 1) - (x + 1) = (x + 1)(x^2 - 1)$$

$$= (x - 1)(x + 1)^2$$

Since -1 is a zero of multiplicity 2 (here, $r = 2$), the partial fraction decomposition of the integrand has the form

$$\frac{2x^2 + 3x + 7}{(x + 1)^2(x - 1)} = \frac{A}{x + 1} + \frac{B}{(x + 1)^2} + \frac{C}{x - 1}$$

Multiplying both sides of this equation by $(x + 1)^2(x - 1)$, we obtain

$$2x^2 + 3x + 7 = A(x + 1)(x - 1) + B(x - 1) + C(x + 1)^2$$

If we let $x = 1$, then we obtain $12 = 4C$, which yields $C = 3$. Next, letting $x = -1$, we have $6 = -2B$, so $B = -3$. Finally, to determine A, we let $x = 0$ (which is the most convenient choice), obtaining $7 = -A - B + C$. Using the values of B and C that we obtained earlier, we see that $A = -B + C - 7 = -1$. Therefore,

$$\int \frac{2x^2 + 3x + 7}{x^3 + x^2 - x - 1}\,dx = \int \left(-\frac{1}{x + 1} - \frac{3}{(x + 1)^2} + \frac{3}{x - 1} \right) dx$$

$$= -\ln|x + 1| + \frac{3}{x + 1} + 3\ln|x - 1| + k$$

$$= \frac{3}{x + 1} + \ln\left| \frac{(x - 1)^3}{x + 1} \right| + k \qquad \blacksquare$$

Recall that a quadratic expression $ax^2 + bx + c$ is **irreducible** if it cannot be written as a product of linear factors with real roots. For example, $3x^2 + x + 1$ is irreducible.

> **Case 3: Distinct Irreducible Quadratic Factors**
> If
> $$\frac{R(x)}{Q(x)} = \frac{R(x)}{(a_1 x^2 + b_1 x + c_1)(a_2 x^2 + b_2 x + c_2) \cdots (a_n x^2 + b_n x + c_n)}$$
> where all the factors $a_k x^2 + b_k x + c_k$, $k = 1, 2, \ldots, n$, are distinct and irreducible, then there exist constants $A_1, A_2, \ldots, A_n, B_1, B_2, \ldots, B_n$ such that
> $$\frac{R(x)}{Q(x)} = \frac{A_1 x + B_1}{a_1 x^2 + b_1 x + c_1} + \frac{A_2 x + B_2}{a_2 x^2 + b_2 x + c_2} + \cdots + \frac{A_n x + B_n}{a_n x^2 + b_n x + c_n}$$

For example,

$$\frac{3x^3 + 8x^2 + 7x + 5}{(x^2 + 1)(x^2 + 2x + 2)} = \frac{Ax + B}{x^2 + 1} + \frac{Cx + D}{x^2 + 2x + 2}$$

EXAMPLE 4   Find $\displaystyle\int \frac{x^4 + 3x^3 + 14x^2 + 14x + 41}{(x^2 + 4)(x^2 + 2x + 5)}\, dx.$

Solution   Since the degree of the numerator is not less than the degree of the denominator, we use long division to write

$$\frac{x^4 + 3x^3 + 14x^2 + 14x + 41}{(x^2 + 4)(x^2 + 2x + 5)} = \frac{x^4 + 3x^3 + 14x^2 + 14x + 41}{x^4 + 2x^3 + 9x^2 + 8x + 20}$$

$$= 1 + \frac{x^3 + 5x^2 + 6x + 21}{(x^2 + 4)(x^2 + 2x + 5)}$$

Notice that the quadratic $x^2 + 2x + 5$ is irreducible because its discriminant

$$b^2 - 4ac = 2^2 - 4(1)(5) = -16 < 0$$

Since the quadratic factors are distinct, we can write

$$\frac{x^3 + 5x^2 + 6x + 21}{(x^2 + 4)(x^2 + 2x + 5)} = \frac{Ax + B}{x^2 + 4} + \frac{Cx + D}{x^2 + 2x + 5}$$

Multiplying both sides of the equation by $(x^2 + 4)(x^2 + 2x + 5)$ gives

$$x^3 + 5x^2 + 6x + 21 = (Ax + B)(x^2 + 2x + 5) + (Cx + D)(x^2 + 4)$$

$$= (A + C)x^3 + (2A + B + D)x^2$$

$$+ (5A + 2B + 4C)x + (5B + 4D)$$

Equating the coefficients of like powers of x yields the system

$$
\begin{aligned}
A +   C &= 1 \\
2A + B + D &= 5 \\
5A + 2B + 4C &= 6 \\
5B + 4D &= 21
\end{aligned}
$$

The solution of the system is $A = 0$, $B = 1$, $C = 1$, and $D = 4$. Therefore,

$$\int \frac{x^4 + 3x^3 + 14x^2 + 14x + 41}{(x^2 + 4)(x^2 + 2x + 5)}\, dx = \int \left(1 + \frac{1}{x^2 + 4} + \frac{x + 4}{x^2 + 2x + 5}\right) dx$$

$$= x + \frac{1}{2}\tan^{-1}\left(\frac{x}{2}\right) + \int \frac{x + 4}{x^2 + 2x + 5}\, dx$$

To evaluate the integral on the right, we complete the square in the denominator of the integrand. Thus, $x^2 + 2x + 5 = (x + 1)^2 + 4$. Next, using the substitution $u = x + 1$ so that $du = dx$ and $x = u - 1$, we obtain

$$\int \frac{x + 4}{x^2 + 2x + 5}\, dx = \int \frac{x + 4}{(x + 1)^2 + 4}\, dx = \int \frac{(u - 1) + 4}{u^2 + 4}\, du = \int \frac{u + 3}{u^2 + 4}\, du$$

$$= \int \frac{u}{u^2 + 4}\, du + \int \frac{3}{u^2 + 4}\, du$$

$$= \frac{1}{2}\ln(u^2 + 4) + \frac{3}{2}\tan^{-1}\left(\frac{u}{2}\right) + C_1$$

$$= \frac{1}{2}\ln(x^2 + 2x + 5) + \frac{3}{2}\tan^{-1}\left(\frac{x + 1}{2}\right) + C_1$$

So

$$\int \frac{x^4 + 3x^3 + 14x^2 + 14x + 41}{(x^2 + 4)(x^2 + 2x + 5)}\, dx$$

$$= x + \frac{1}{2}\tan^{-1}\left(\frac{x}{2}\right) + \frac{1}{2}\ln(x^2 + 2x + 5) + \frac{3}{2}\tan^{-1}\left(\frac{x + 1}{2}\right) + C \quad ∎$$

Case 4: Repeated Irreducible Quadratic Factors

If $Q(x)$ contains a factor $(ax^2 + bx + c)^r$ with $r > 1$, where $ax^2 + bx + c$ is irreducible, then the partial fraction decomposition of $R(x)/Q(x)$ contains a sum of r partial fractions of the form

$$\frac{A_1 x + B_1}{ax^2 + bx + c} + \frac{A_2 x + B_2}{(ax^2 + bx + c)^2} + \cdots + \frac{A_r x + B_r}{(ax^2 + bx + c)^r}$$

where A_k and B_k are real numbers.

For example,

$$\frac{x^4 - 3x^3 + x + 1}{x(x - 1)^2(x^2 + 1)(x^2 + x + 1)^2}$$

$$= \frac{A}{x} + \frac{B}{x - 1} + \frac{C}{(x - 1)^2} + \frac{Dx + E}{x^2 + 1} + \frac{Fx + G}{(x^2 + x + 1)} + \frac{Hx + I}{(x^2 + x + 1)^2}$$

EXAMPLE 5 Find $\displaystyle\int \frac{x^3 - 2x^2 + 3x + 2}{x(x^2 + 1)^2}\, dx$.

Solution The partial fraction decomposition of the integrand has the form

$$\frac{x^3 - 2x^2 + 3x + 2}{x(x^2 + 1)^2} = \frac{A}{x} + \frac{Bx + C}{x^2 + 1} + \frac{Dx + E}{(x^2 + 1)^2}$$

Multiplying both sides of the equation by $x(x^2 + 1)^2$ gives

$$\begin{aligned}
x^3 - 2x^2 + 3x + 2 &= A(x^2 + 1)^2 + (Bx + C)x(x^2 + 1) + (Dx + E)x \\
&= A(x^4 + 2x^2 + 1) + B(x^4 + x^2) + C(x^3 + x) + Dx^2 + Ex \\
&= (A + B)x^4 + Cx^3 + (2A + B + D)x^2 + (C + E)x + A
\end{aligned}$$

Equating the coefficients of like powers of x yields the system

$$\begin{aligned}
A + B && &= 0 \\
C && &= 1 \\
2A + B + \quad D && &= -2 \\
C \quad + E &= 3 \\
A && &= 2
\end{aligned}$$

The solution of this system is $A = 2, B = -2, C = 1, D = -4$, and $E = 2$. Therefore,

$$\begin{aligned}
\int \frac{x^3 - 2x^2 + 3x + 2}{x(x^2 + 1)^2}\, dx \\
= \int \left(\frac{2}{x} + \frac{-2x + 1}{x^2 + 1} + \frac{-4x + 2}{(x^2 + 1)^2} \right) dx \\
= 2 \int \frac{dx}{x} - 2 \int \frac{x}{x^2 + 1}\, dx + \int \frac{dx}{x^2 + 1} - 4 \int \frac{x}{(x^2 + 1)^2}\, dx + 2 \int \frac{1}{(x^2 + 1)^2}\, dx \\
= 2 \ln|x| - \ln(x^2 + 1) + \tan^{-1} x + \frac{2}{x^2 + 1} + 2 \int \frac{1}{(x^2 + 1)^2}\, dx
\end{aligned}$$

To find the integral on the right-hand side, we make the substitution

$$x = \tan \theta \qquad \text{so that} \qquad dx = \sec^2 \theta\, d\theta$$

Also, $x^2 + 1 = \tan^2 \theta + 1 = \sec^2 \theta$. So

$$\begin{aligned}
\int \frac{1}{(x^2 + 1)^2}\, dx &= \int \frac{1}{\sec^4 \theta} \cdot \sec^2 \theta\, d\theta = \int \cos^2 \theta\, d\theta \\
&= \frac{1}{2} \int (1 + \cos 2\theta)\, d\theta = \frac{1}{2}\left(\theta + \frac{1}{2}\sin 2\theta \right) + k \\
&= \frac{1}{2}(\theta + \sin \theta \cos \theta) + k \qquad \text{\textcolor{orange}{$\sin 2\theta = 2\sin\theta\cos\theta$}} \\
&= \frac{1}{2}\left(\tan^{-1} x + \frac{x}{\sqrt{x^2 + 1}} \cdot \frac{1}{\sqrt{x^2 + 1}} \right) + k \qquad \text{\textcolor{orange}{See Figure 1.}} \\
&= \frac{1}{2}\left(\tan^{-1} x + \frac{x}{x^2 + 1} \right) + k
\end{aligned}$$

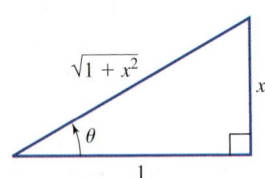

FIGURE 1
The right triangle associated
with the substitution $x = \tan \theta$

Therefore, the desired result is

$$\int \frac{x^3 - 2x^2 + 3x + 2}{x(x^2 + 1)^2} dx = \ln \frac{x^2}{x^2 + 1} + \tan^{-1} x + \frac{2}{x^2 + 1} + \tan^{-1} x + \frac{x}{x^2 + 1} + K$$

$$= \ln \frac{x^2}{x^2 + 1} + 2 \tan^{-1} x + \frac{x + 2}{x^2 + 1} + K \qquad \blacksquare$$

Note Certain integrals involving rational functions can be more easily evaluated by using the method of substitution. For example, the integral

$$\int \frac{6x^2 + 4x + 2}{x(x^2 + x + 1)} dx$$

can be evaluated by letting

$$u = x(x^2 + x + 1) = x^3 + x^2 + x$$

Then $du = (3x^2 + 2x + 1) \, dx$, so

$$\int \frac{6x^2 + 4x + 2}{x(x^2 + x + 1)} dx = \int \frac{2}{u} du$$

$$= 2 \ln|u| + C$$

$$= \ln(x^3 + x^2 + x)^2 + C$$

However, such integrals rarely occur in practice. $\qquad \blacksquare$

EXAMPLE 6 **Waste Disposal** When organic waste is dumped into a pond, the oxidization process that takes place reduces the pond's oxygen content. However, in time, nature will restore the oxygen content to its natural level. Suppose that the oxygen content t days after organic waste has been dumped into a pond is given by

$$f(t) = 100 \left(\frac{t^2 + 10t + 100}{t^2 + 20t + 100} \right)$$

percent of its normal level. Find the average content of oxygen in the pond over the first 10 days after organic waste has been dumped into it.

Solution The average content is given by

$$C = \frac{1}{10} \int_0^{10} f(t) \, dt = 10 \int_0^{10} \frac{t^2 + 10t + 100}{t^2 + 20t + 100} dt$$

Note that the degree of the numerator in the integrand is the same as that of the denominator, so long division is required here. Carrying through with the division, we find

$$\frac{t^2 + 10t + 100}{t^2 + 20t + 100} = 1 - \frac{10t}{t^2 + 20t + 100}$$

Next, observe that $t^2 + 20t + 100 = (t + 10)^2$. Therefore, we can write

$$\frac{10t}{t^2 + 20t + 100} = \frac{10t}{(t + 10)^2} = \frac{A}{t + 10} + \frac{B}{(t + 10)^2}$$

where A and B are real numbers to be determined. Multiplying both sides of this equation by $(t + 10)^2$, we obtain

$$10t = A(t + 10) + B$$

Equating the coefficients of like powers of x yields $A = 10$ and $10A + B = 0$. Substituting $A = 10$ into the second equation then gives $B = -100$. Therefore,

$$C = 10 \int_0^{10} \left(1 - \frac{10}{t + 10} + \frac{100}{(t + 10)^2} \right) dt$$

$$= 10 \left[t - 10 \ln(t + 10) - \frac{100}{t + 10} \right]_0^{10}$$

$$= 10[(10 - 10 \ln 20 - 5) - (0 - 10 \ln 10 - 10)] = 80.69$$

or approximately 81%.

6.4 CONCEPT QUESTIONS

Let $f(x) = P(x)/Q(x)$ be a rational function in which the degree of P is less than the degree of Q. What is the form of the partial fraction decomposition of f:

1. If Q has only distinct linear factors?

2. If Q contains a factor $(ax + b)^r$ with $r > 1$ that is repeated?

3. If Q contains a factor $(ax^2 + bx + c)^r$ with $r = 1$ that is not repeated?

6.4 EXERCISES

In Exercises 1–6, write the form of the partial fraction decomposition of the rational expression. Do not find the numerical values of the constants.

1. a. $\dfrac{3}{x(x - 5)}$ **b.** $\dfrac{2x}{(x + 1)(3x - 2)}$

2. a. $\dfrac{2x + 1}{x^2 - x - 2}$ **b.** $\dfrac{x - 4}{x^2 + 4x + 3}$

3. a. $\dfrac{2x^2 - 1}{x^3 + x^2}$ **b.** $\dfrac{7}{x^2 + 3x - 4}$

4. a. $\dfrac{2x + 1}{x^3 + x}$ **b.** $\dfrac{8x}{x^3 - 5x^2}$

5. a. $\dfrac{x^3 - 2x + 1}{x^4 - 16}$ **b.** $\dfrac{x^2 - x - 27}{2x^3 - x^2 + 8x - 4}$

6. a. $\dfrac{2x^3 - 3x - 5}{x^2(x + 1)^3}$ **b.** $\dfrac{2x^4 - 3x^2 + 8x + 1}{(x - 1)^2(x^2 + x + 1)^3}$

In Exercises 7–51, find or evaluate the integral.

7. $\displaystyle\int \frac{dx}{x(x - 4)}$

8. $\displaystyle\int \frac{3x + 2}{x(x - 2)} \, dx$

9. $\displaystyle\int \frac{t + 3}{t(t + 1)} \, dt$

10. $\displaystyle\int \frac{2x - 1}{2x^2 - x} \, dx$

11. $\displaystyle\int_3^4 \frac{1}{x^2 - 4} \, dx$

12. $\displaystyle\int \frac{1}{4x^2 - 9} \, dx$

13. $\displaystyle\int \frac{x - 1}{x^2 - x - 2} \, dx$

14. $\displaystyle\int_0^1 \frac{2u + 3}{u^2 + 4u + 3} \, du$

15. $\displaystyle\int \frac{2x^2 + 3x + 6}{(x + 3)(x^2 - 4)} \, dx$

16. $\displaystyle\int \frac{x^2 + 2x + 8}{x^3 - 4x} \, dx$

17. $\displaystyle\int \frac{2x^2 + x - 1}{x^2 - x} \, dx$

18. $\displaystyle\int_2^3 \frac{x^3 - 2x + 7}{x^2 + x - 2} \, dx$

19. $\displaystyle\int \frac{2x^2 - 3x + 3}{x^3 - 2x^2 + x} \, dx$

20. $\displaystyle\int \frac{x^4 - 3x^2 - 3x - 2}{x^3 - x^2 - 2x} \, dx$

21. $\displaystyle\int \frac{4x^2 + 3x + 2}{x^3 + x^2} \, dx$

22. $\displaystyle\int_2^4 \frac{3x - 5}{(x - 1)^2} \, dx$

23. $\displaystyle\int \frac{v^3 + 1}{v(v - 1)^3} \, dv$

24. $\displaystyle\int_1^2 \frac{x^2 + 10x - 36}{x(x - 3)^2} \, dx$

25. $\displaystyle\int \frac{x^3 - x + 2}{x^3 + 2x^2 + x} \, dx$

26. $\displaystyle\int \frac{4x^2}{(x^2 - 4)^2} \, dx$

27. $\displaystyle\int \frac{dx}{x(x^2 - 1)^2}$

28. $\displaystyle\int \frac{x^2}{(x^2 + 4x + 3)^2} \, dx$

29. $\displaystyle\int \frac{6x^2 + 28x + 28}{x^3 + 4x^2 + x - 6} \, dx$

30. $\displaystyle\int \frac{x^2 + 16x + 7}{x^3 - x^2 + x + 3} \, dx$

31. $\displaystyle\int \frac{x^3 + 3}{(x + 1)(x^2 + 1)}\, dx$ **32.** $\displaystyle\int \frac{2r^2 - 3r + 4}{(r^2 + 2)^2}\, dr$

33. $\displaystyle\int \frac{5x^3 - 3x^2 + 7x - 3}{(x^2 + 1)^2}\, dx$

34. $\displaystyle\int \frac{13x + 4}{(x - 2)(x^2 + 2x + 2)}\, dx$

35. $\displaystyle\int \frac{8 - 3x}{(x + 1)(x^2 - 4x + 6)}\, dx$

36. $\displaystyle\int \frac{x^2 + 1}{x^3 - 1}\, dx$ **37.** $\displaystyle\int \frac{x}{x^3 + 1}\, dx$

38. $\displaystyle\int \frac{x^2 - x - 21}{2x^3 - x^2 + 8x - 4}\, dx$ **39.** $\displaystyle\int_0^1 \frac{3x^3 + 5x^2 + 5x + 1}{(x + 1)^2(x^2 + 1)}\, dx$

40. $\displaystyle\int \frac{3x - x^2}{(x^2 + 1)(x^2 + 2)}\, dx$ **41.** $\displaystyle\int \frac{3x^2 + x + 2}{(x^2 + x + 1)^2}\, dx$

42. $\displaystyle\int \frac{t^4}{t^4 - 1}\, dt$ **43.** $\displaystyle\int \frac{3x^2 - x + 4}{(2x^3 - x^2 + 8x + 4)^2}\, dx$

44. $\displaystyle\int \frac{\cos x}{\sin^2 x - \sin x - 6}\, dx$ **45.** $\displaystyle\int \frac{\sin x}{\cos^3 x + \cos^2 x}\, dx$

46. $\displaystyle\int \frac{\sec^2 \theta}{\tan \theta\,(\tan \theta - 1)}\, d\theta$ **47.** $\displaystyle\int \frac{e^t}{(e^t - 1)(e^t + 2)}\, dt$

48. $\displaystyle\int \frac{e^x}{e^{2x} + 2e^x - 8}\, dx$ **49.** $\displaystyle\int \frac{e^{4t}}{(e^t + 2)(e^{2t} - 1)}\, dt$

50. $\displaystyle\int \frac{dx}{e^x(1 + e^{2x})}$ **51.** $\displaystyle\int \frac{x^{1/3}}{1 + x}\, dx$

Hint: Let $u = x^{1/3}$.

52. An integral of the form $\int R(\sin x, \cos x)\, dx$, where R is a rational function of $\sin x$ and $\cos x$, can be converted into an integral involving an ordinary rational function of u by means of the substitution $u = \tan (x/2)$. Prove this by showing that if $u = \tan (x/2)$, where $-\pi < x < \pi$, then

$$\cos x = \frac{1 - u^2}{1 + u^2}, \quad \sin x = \frac{2u}{1 + u^2}, \quad dx = \frac{2}{1 + u^2}\, du$$

Hint: Sketch a right triangle.

In Exercises 53–60, use the result of Exercise 52 to find the integral.

53. $\displaystyle\int \frac{1}{1 + \cos x}\, dx$

54. $\displaystyle\int \frac{1}{3 \sin x - 4 \cos x}\, dx$

55. $\displaystyle\int \frac{1}{5 + \sin x - 3 \cos x}\, dx$

56. $\displaystyle\int \frac{1}{\sin x\,(2 + \cos x - 2 \sin x)}\, dx$

57. $\displaystyle\int_0^{\pi/2} \frac{1}{1 + \cos x + \sin x}\, dx$

58. $\displaystyle\int_0^{\pi/4} \frac{\tan x}{1 + \cos x}\, dx$

59. $\displaystyle\int \frac{1}{1 + \tan x}\, dx$

60. $\displaystyle\int_0^{\pi/6} \frac{\sin x}{\cos x\,(1 + \sin x)}\, dx$

61. Find the area of the region under the graph of $y = \dfrac{1}{x(x + 1)}$ on the interval $[1, 2]$.

62. Find the area of the region under the graph of $y = \dfrac{x^3}{x^3 + 1}$ on the interval $[0, 2]$.

63. Let $f(x) = \dfrac{x + 3}{x(x + 1)}$ and $g(x) = \ln x$.

  a. Plot the graphs of f and g using the viewing window $[0, 3] \times [-1, 3]$. Find the x-coordinate of the point of intersection of the graphs of f and g accurate to three decimal places.

  b. Use the result of part (a) to find the approximate area of the region bounded by the graphs of f and g and the vertical line $x = 1$.

64. Let $f(x) = \dfrac{x}{x^3 + 1}$ and $g(x) = \dfrac{1}{3}x^3$.

  a. Plot the graphs of f and g using the viewing window $[0, 2] \times \left[-\frac{1}{2}, 1\right]$. Find the x-coordinate of the point of intersection of the graphs of f and g accurate to three decimal places.

  b. Use the result of part (a) to find the approximate area of the region enclosed by the graphs of f and g.

65. The region under the graph of $y = \dfrac{1}{x(x + 1)}$ on the interval $[1, 2]$ is revolved about the x-axis. Find the volume of the resulting solid.

66. The region under the graph of $y = \dfrac{2x}{x^2 + 1}$ on the interval $[0, 2]$ is revolved about the x-axis. Find the volume of the resulting solid.

67. Find the length of the graph of $y = 2 \ln\left(\dfrac{4}{4 - x^2}\right)$ from $A(0, 0)$ to $B\left(1, 2 \ln \frac{4}{3}\right)$.

68. Find the centroid of the region under the graph of
$$y = \frac{2x}{x^2 + 1} \quad \text{on } [0, 2].$$

69. Let $I = \displaystyle\int \frac{x^2 + 1}{x^4 + 6x^3 + 12x^2 + 11x + 6}\, dx$.

  a. Find I.

  Hint: One root of $x^4 + 6x^3 + 12x^2 + 11x + 6 = 0$ is -2.

cas **b.** Use a CAS to find the partial fraction decomposition of

$$f(x) = \frac{x^2 + 1}{x^4 + 6x^3 + 12x^2 + 11x + 6}$$

  c. Use a CAS to find I.

cas **70.** Let

$$I = \int \frac{8x^5 - 3x^4 + 2x^2 - 1}{36x^6 - 108x^5 + 105x^4 - 72x^3 + 58x^2 - 12x + 9}\, dx$$

  a. Use a CAS to find the partial fraction decomposition of

$$f(x) = \frac{8x^5 - 3x^4 + 2x^2 - 1}{36x^6 - 108x^5 + 105x^4 - 72x^3 + 58x^2 - 12x + 9}$$

  b. Use a CAS to find I.

71. City Planning A major corporation is building a 4325-acre complex of homes, offices, stores, schools, and churches in the rural community of Glen Cove. As a result of this development the planners have estimated that Glen Cove's population (in thousands) t years from now will be given by

$$P(t) = \frac{3t^2 + 130t + 270}{t^2 + 6t + 45}$$

What will the average population of Glen Cove be over the next 10 years?

72. Work Done in Moving a Charged Particle Suppose that a particle of charge $+1$ is placed on a coordinate line between two particles, each of charge -1, as shown in the figure. Then, according to Coulomb's Law, there is an electrical force acting on the particle of charge $+1$ given by

$$F(x) = k\left[\frac{2x - 3}{x^2(x - 3)^2} \right]$$

where k is a positive constant. Find the work required to move the particle of charge $+1$ along the coordinate line from $x = 1$ to $x = 2$.

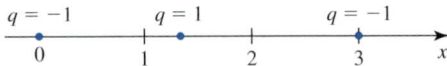

In Exercises 73–76, determine whether the statement is true or false. If it is true, explain why it is true. If it is false, explain why or give an example to show why it is false.

73. $\dfrac{x^3 + 2x}{(x + 1)(x - 2)}$ can be written in the form $\dfrac{A}{x + 1} + \dfrac{B}{x - 2}$.

74. $\dfrac{4x^2 - 15x - 1}{x(x^2 - 4x - 5)}$ can be written in the form

$$\frac{A}{x} + \frac{B}{x - 5} + \frac{C}{x + 1}.$$

75. $\dfrac{1}{x(x - 1)^2}$ can be written in the form $\dfrac{A}{x} + \dfrac{B}{(x - 1)^2}$.

76. $\dfrac{4x^3 - x^2 + 4x + 2}{(x^2 + 1)^2}$ can be written in the form

$$\frac{A}{x^2 + 1} + \frac{B}{(x^2 + 1)^2}.$$

6.5   Integration Using Tables of Integrals and a CAS; a Summary of Techniques

The techniques of integration that we have developed so far enable us to integrate a wide variety of functions. But in practice, there are many functions for which these techniques will not work or, if they do, work inefficiently. Other techniques have been developed that enable us to integrate many complicated functions. By using these techniques, extensive lists of integration formulas have been compiled. A small sample of such formulas can be found in the Table of Integrals on the reference pages at the back of this book. These formulas are grouped according to the following basic forms of the integrand: $a + bu$, $\sqrt{a + bu}$, $\sqrt{a^2 \pm u^2}$, $\sqrt{u^2 - a^2}$, $\sqrt{2au - u^2}$, trigonometric, inverse trigonometric, exponential, logarithmic, and hyperbolic functions.

■ Using a Table of Integrals

The Table of Integrals provides us with a quick and convenient way of integrating complicated functions. The idea is to match the integrand of the integral to be found with the integrand of an appropriate integral appearing in the table (whose antiderivative is known). Sometimes we need to recast the given integral by making an appropriate substitution or by using the integration by parts formula before we can use the Table of Integrals.

EXAMPLE 1 Use the Table of Integrals to find $\displaystyle\int \frac{3x}{\sqrt{2+x}}\, dx$.

Solution We first write

$$\int \frac{3x}{\sqrt{2+x}}\, dx = 3\int \frac{x}{\sqrt{2+x}}\, dx$$

Scanning the Table of Integrals for integrands involving $\sqrt{a+bu}$, we see that Formula 28,

$$\int \frac{u}{\sqrt{a+bu}}\, du = \frac{2}{3b^2}(bu - 2a)\sqrt{a+bu} + C$$

is the proper choice. With $a = 2$, $b = 1$, and $u = x$ we obtain

$$\int \frac{3x}{\sqrt{2+x}}\, dx = 3\left[\frac{2}{3}(x-4)\sqrt{2+x}\right] + C$$

$$= 2(x-4)\sqrt{2+x} + C \qquad \blacksquare$$

EXAMPLE 2 Use the Table of Integrals to find $\displaystyle\int \frac{\sqrt{3-4x^2}}{x^2}\, dx$.

Solution Looking at the Table of Integrals for integrands involving $\sqrt{a^2 - u^2}$, we find that Formula 49,

$$\int \frac{\sqrt{a^2 - u^2}}{u^2}\, du = -\frac{1}{u}\sqrt{a^2 - u^2} - \sin^{-1}\frac{u}{a} + C$$

is closest to the form of the given integral. Comparison of the two integrands suggests that we make the substitution $u = 2x$ and $du = 2\, dx$, obtaining

$$\int \frac{\sqrt{3-4x^2}}{x^2}\, dx = \int \frac{\sqrt{3-u^2}}{(u/2)^2}\left(\frac{du}{2}\right) = 2\int \frac{\sqrt{3-u^2}}{u^2}\, du$$

Then, using Formula 49 with $a = \sqrt{3}$, we obtain

$$\int \frac{\sqrt{3-4x^2}}{x^2}\, dx = 2\int \frac{\sqrt{3-u^2}}{u^2}\, du = 2\left[-\frac{1}{u}\sqrt{3-u^2} - \sin^{-1}\frac{u}{\sqrt{3}}\right] + C$$

$$= -\frac{\sqrt{3-4x^2}}{x} - 2\sin^{-1}\left(\frac{2x}{\sqrt{3}}\right) + C \qquad \blacksquare$$

EXAMPLE 3 Use the Table of Integrals to find $\int x^3 \cos x\, dx$.

Solution Looking in the section of the Table of Integrals for integrands involving trigonometric functions, we find Formula 78, a reduction formula.

$$\int u^n \cos u\, du = u^n \sin u - n\int u^{n-1}\sin u\, du$$

Using the formula with $n = 3$, we obtain

$$\int x^3 \cos x\, dx = x^3 \sin x - 3\int x^2 \sin x\, dx$$

Next, using Formulas 77 and 76, we obtain

$$\int x^3 \cos x \, dx = x^3 \sin x - 3\left[-x^2 \cos x + 2\int x \cos x \, dx \right]$$

$$= x^3 \sin x + 3x^2 \cos x - 6(\cos x + x \sin x) + C$$

$$= x^3 \sin x + 3x^2 \cos x - 6x \sin x - 6 \cos x + C \qquad \blacksquare$$

Using Tables of Integrals to Evaluate Definite Integrals

EXAMPLE 4 Use the Table of Integrals to evaluate $\displaystyle\int_0^{\pi/2} \frac{\sin 2x}{\sqrt{3 - 2\cos x}} \, dx$.

Solution Let's begin by evaluating the corresponding indefinite integral, which can also be rewritten as

$$\int \frac{\sin 2x}{\sqrt{3 - 2\cos x}} \, dx = \int \frac{2\sin x \cos x}{\sqrt{3 - 2\cos x}} \, dx$$

No formula in the Table of Integrals has either of these forms, so let's consider making a substitution. Letting $u = \cos x$, so that $du = -\sin x \, dx$, we find

$$\int \frac{\sin 2x}{\sqrt{3 - 2\cos x}} \, dx = -2\int \frac{\cos x \, (-\sin x)}{\sqrt{3 - 2\cos x}} \, dx = -2\int \frac{u}{\sqrt{3 - 2u}} \, du$$

Looking in the Table of Integrals for integrands involving $\sqrt{a + bu}$ leads to Formula 28,

$$\int \frac{u}{\sqrt{a + bu}} \, du = \frac{2}{3b^2} (bu - 2a)\sqrt{a + bu} + C$$

with $a = 3$ and $b = -2$. We obtain

$$-2\int \frac{u}{\sqrt{3 - 2u}} \, du = -2\left(\frac{2}{12}\right)(-2u - 6)\sqrt{3 - 2u} + C = \frac{2}{3} (u + 3)\sqrt{3 - 2u} + C$$

Therefore,

$$\int \frac{\sin 2x}{\sqrt{3 - 2\cos x}} \, dx = \frac{2}{3} (u + 3)\sqrt{3 - 2u} + C$$

$$= \frac{2}{3} (\cos x + 3)\sqrt{3 - 2\cos x} + C \qquad \text{Since } u = \cos x$$

So

$$\int_0^{\pi/2} \frac{\sin 2x}{\sqrt{3 - 2\cos x}} \, dx = \left[\frac{2}{3} (\cos x + 3)\sqrt{3 - 2\cos x} \right]_0^{\pi/2}$$

$$= \frac{2}{3} (3)\sqrt{3 - 0} - \frac{2}{3} (1 + 3)\sqrt{3 - 2} = 2\sqrt{3} - \frac{8}{3}$$

$$= \frac{6\sqrt{3} - 8}{3} \qquad \blacksquare$$

EXAMPLE 5 The region R under the graph of $y = \cos^{-1} x$ on the interval $[0, 1]$ is revolved about the y-axis. Find the volume of the resulting solid.

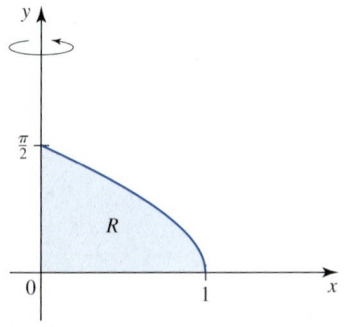

FIGURE 1
The region R under $y = \cos^{-1} x$ on $[0, 1]$

Solution   The region R is shown in Figure 1. Using the method of cylindrical shells, we see that the required volume is

$$V = 2\pi \int_0^1 x \cos^{-1} x \, dx$$

We use Formula 91 from the Table of Integrals to evaluate this integral, obtaining

$$V = 2\pi \int_0^1 x \cos^{-1} x \, dx = 2\pi \left[\frac{2x^2 - 1}{4} \cos^{-1} x - \frac{x\sqrt{1 - x^2}}{4} \right]_0^1$$

$$= 2\pi \left[\frac{1}{4} \cos^{-1} 1 - \left(-\frac{1}{4} \cos^{-1} 0 \right) \right] = 2\pi \left[\frac{1}{4} (0) + \frac{1}{4} \left(\frac{\pi}{2} \right) \right]$$

$$= \frac{\pi^2}{4}$$

Graphing Calculators and CAS

Most of the graphing calculators that are available today will perform numerical integration; that is, the calculator will give a numerical approximation to the value of a definite integral. The more sophisticated graphing calculators, such as the TI-89 or TI-92, will even do symbolic integration; that is, the calculator will find an antiderivative of a given function.

A computer equipped with the appropriate software, such as *Mathematica*, *Maple*, or *Derive*, can be used to perform both tasks. If you use these programs, bear in mind that the commands are different for different programs and, more important, the answers may appear in different forms even though they are equivalent.

EXAMPLE 6   Find $\int x(x^2 + 3)^6 \, dx$ using (a) the TI-89 and (b) CAS with *Maple* and *Mathematica*.

Solution
a. Using the TI-89:

$$\int x(x^2 + 3)^6 \, dx = \frac{(x^2 + 3)^7}{14}$$

b. Using *Mathematica*:

$$\int x(x^2 + 3)^6 \, dx = \frac{729x^2}{2} + \frac{729x^4}{2} + \frac{405x^6}{2} + \frac{135x^8}{2} + \frac{27x^{10}}{2} + \frac{3x^{12}}{2} + \frac{x^{14}}{14}$$

Using *Maple*:

$$\int x(x^2 + 3)^6 \, dx = \frac{1}{14}x^{14} + \frac{3}{2}x^{12} + \frac{27}{2}x^{10} + \frac{135}{2}x^8 + \frac{405}{2}x^6 + \frac{729}{2}x^4 + \frac{729}{2}x^2$$

Note that none of these programs include the constant of integration in their answers. Of course, we can find the integral under consideration by using the method of substitution. You can easily verify that your result is the same as that obtained by using the TI-89. The output obtained by using both *Maple* and *Mathematica* is in a more cumbersome form, but it is equivalent to the more compact answer obtained by using the TI-89. You can see this by expanding the latter using the Binomial Theorem.

EXAMPLE 7 Find $\displaystyle\int \frac{\cos^4 x}{\sin^3 x}\,dx$ by using (a) the TI-89 and (b) CAS with *Maple* and *Mathematica*.

Solution

a. Using the TI-89:

$$\int \frac{\cos^4 x}{\sin^3 x}\,dx$$

$$= \frac{3(\sin^2 x)\cdot\ln(|\cos x + 1|) - 3(\sin^2 x\cdot\ln(|\sin x|)) - (2(\sin^2 x)+1)\cdot\cos x}{2\sin^2 x}$$

b. Using *Mathematica*:

$$\int \frac{\cos^4 x}{\sin^3 x}\,dx = -\cos x - \frac{1}{8}\csc\!\left(\frac{x}{2}\right)^2 + \frac{3}{2}\ln\!\left(\cos\frac{x}{2}\right) - \frac{3}{2}\ln\!\left(\sin\frac{x}{2}\right) + \frac{1}{8}\sec\!\left(\frac{x}{2}\right)^2$$

Using *Maple*:

$$\int \frac{\cos^4 x}{\sin^3 x}\,dx = -\frac{\cos(x)^5}{2\sin(x)^2} - \frac{1}{2}\cos(x)^3 - \frac{3}{2}\cos(x) - \frac{3}{2}\ln(\csc(x) - \cot(x)) \quad\blacksquare$$

■ Summary of Integration Techniques

Our first table gives a summary of the basic integration formulas that were covered in this and the previous chapters.

BASIC INTEGRATION FORMULAS

1. $\displaystyle\int u^n\,du = \frac{u^{n+1}}{n+1} + C, \quad n \neq -1$	**9.** $\displaystyle\int \sec u \tan u\,du = \sec u + C$	**17.** $\displaystyle\int \frac{du}{1+u^2} = \tan^{-1} u + C$
2. $\displaystyle\int \frac{1}{u}\,du = \ln|u| + C$	**10.** $\displaystyle\int \csc u \cot u\,du = -\csc u + C$	**18.** $\displaystyle\int \sinh u\,du = \cosh u + C$
3. $\displaystyle\int e^u\,du = e^u + C$	**11.** $\displaystyle\int \sec u\,du = \ln|\sec u + \tan u| + C$	**19.** $\displaystyle\int \cosh u\,du = \sinh u + C$
4. $\displaystyle\int a^u\,du = \frac{a^u}{\ln a} + C$	**12.** $\displaystyle\int \csc u\,du = -\ln|\csc u + \cot u| + C$	**20.** $\displaystyle\int \text{sech}^2 u\,du = \tanh u + C$
5. $\displaystyle\int \sin u\,du = -\cos u + C$	**13.** $\displaystyle\int \tan u\,du = \ln|\sec u| + C$	**21.** $\displaystyle\int \text{csch } u \coth u\,du = -\text{csch } u + C$
6. $\displaystyle\int \cos u\,du = \sin u + C$	**14.** $\displaystyle\int \cot u\,du = \ln|\sin u| + C$	**22.** $\displaystyle\int \text{sech } u \tanh u\,du = -\text{sech } u + C$
7. $\displaystyle\int \sec^2 u\,du = \tan u + C$	**15.** $\displaystyle\int \frac{du}{\sqrt{1-u^2}} = \sin^{-1} u + C$	**23.** $\displaystyle\int \text{csch}^2 u\,du = -\coth u + C$
8. $\displaystyle\int \csc^2 u\,du = -\cot u + C$	**16.** $\displaystyle\int \frac{du}{u\sqrt{u^2-1}} = \sec^{-1}|u| + C$	

The next table lists the methods of integration developed in Chapter 4 (Integration by Substitution) and this chapter.

METHODS OF INTEGRATION

Integration	Method of integration	Section
1. $\displaystyle\int f(g(x))g'(x)\,dx$	Use the substitution $u = g(x)$.	Section 4.2
2. $\displaystyle\int f(x)g'(x)\,dx$	Use the integration by parts formula: $$\int f(x)g'(x)\,dx = f(x)g(x) - \int g(x)f'(x)\,dx \quad \text{or} \quad \int u\,dv = uv - \int v\,du$$ **Note:** Apply the method to integrals of the form $\int P(x)e^{ax}\,dx$, $\int P(x)\sin ax\,dx$, $\int P(x)\cos ax\,dx$, where $P(x)$ is a polynomial, $\int \ln x\,dx$, $\int \sin^{-1} x\,dx$, $\int \tan^{-1} x\,dx$, $\int \sec^m x\,dx$ ($m > 0$ and m odd), $\int e^{ax}\cos bx\,dx$, $\int e^{ax}\sin bx$, and so on.	Section 6.1
3. a. $\displaystyle\int \sin^m x \cos^n x\,dx$, where m or n is a positive integer	**a.** If m is odd and positive, use the substitution $u = \cos x$. **b.** If n is odd and positive, use the substitution $u = \sin x$. **c.** If m and n are even and nonnegative use the formulas $$\sin^2 x = \frac{1 - \cos 2x}{2}, \quad \cos^2 x = \frac{1 + \cos 2x}{2}$$	
b. $\displaystyle\int \tan^m x \sec^n x\,dx$, where m or n is a positive integer.	**a.** If m is odd and positive, use the substitution $u = \sec x$. **b.** If n is even and positive, use the substitution $u = \tan x$.	
Note: Also try converting the integrand to one involving sines and cosines.	Use the identities:	
c. $\displaystyle\int \sin mx \sin nx\,dx$	$\sin mx \sin nx = \frac{1}{2}[\cos(m-n)x - \cos(m+n)x]$	
$\displaystyle\int \sin mx \cos nx\,dx$	$\sin mx \cos nx = \frac{1}{2}[\sin(m-n)x + \sin(m+n)x]$	
$\displaystyle\int \cos mx \cos nx\,dx$	$\cos mx \cos nx = \frac{1}{2}[\cos(m-n)x + \cos(m+n)x]$	
4. $\displaystyle\int f(x)\,dx$, where f involves $\sqrt{a^2 - x^2}$ $\sqrt{a^2 + x^2}$ $\sqrt{x^2 - a^2}$	Use the substitution $x = a\sin\theta$, where $-\frac{\pi}{2} \le \theta \le \frac{\pi}{2}$. Use the substitution $x = a\tan\theta$, where $-\frac{\pi}{2} < \theta < \frac{\pi}{2}$. Use the substitution $x = a\sec\theta$, where $0 \le \theta < \frac{\pi}{2}$ or $\frac{\pi}{2} < \theta \le \pi$.	Section 6.3
5. $\displaystyle\int \frac{P(x)}{Q(x)}\,dx$, where $\deg P < \deg Q$ and $Q(x) = (p_1 x + q_1)^k (p_2 x + q_2)^l$ $\cdots (ax^2 + bx + c)^m \cdots$	Write the integrand as a sum of partial fractions: $$\frac{P(x)}{Q(x)} = \frac{A_1}{p_1 x + q_1} + \frac{A_2}{(p_1 x + q_1)^2} + \cdots + \frac{A_k}{(p_1 x + q_1)^k}$$ $$+ \frac{B_1}{p_2 x + q_2} + \frac{B_2}{(p_2 x + q_2)^2} + \cdots + \frac{B_l}{(p_2 x + q_2)^l}$$ $$+ \cdots + \frac{M_1 x + N_1}{ax^2 + bx + c} + \frac{M_2 x + N_2}{(ax^2 + bx + c)^2} + \cdots$$ $$+ \frac{M_m x + N_m}{(ax^2 + bx + c)^m} + \cdots$$	Section 6.4

EXAMPLE 8 Indicate the method of integration that you would use to find the integral. Explain how you arrive at your choice.

a. $\displaystyle\int x^2(1-x)^{30}\,dx$ **b.** $\displaystyle\int \frac{x\sin^{-1}x}{\sqrt{1-x^2}}\,dx$ **c.** $\displaystyle\int \sin x \sin 2x \cos 3x\,dx$

d. $\displaystyle\int \frac{\cos^4 x}{\sin^3 x}\,dx$ **e.** $\displaystyle\int \frac{x+4}{(x-1)(x^2+1)^2}\,dx$

Solution

a. We use the substitution $u = 1 - x$ so that $du = -dx$. Then

$$\int x^2(1-x)^{30}\,dx = -\int (1-u)^2 u^{30}\,du = -\int (1-2u+u^2)u^{30}\,du$$

$$= -\int (u^{30} - 2u^{31} + u^{32})\,du$$

which is easily integrated.

b. The integrand involves $\sin^{-1} x$, so we try the method of integration by parts with

$$u = \sin^{-1} x \qquad \text{and} \qquad dv = \frac{x}{\sqrt{1-x^2}}\,dx$$

so that

$$du = \frac{1}{\sqrt{1-x^2}}\,dx \qquad \text{and} \qquad v = \int \frac{x}{\sqrt{1-x^2}}\,dx = -\sqrt{1-x^2}$$

We obtain

$$\int \frac{x\sin^{-1}x}{\sqrt{1-x^2}}\,dx = -(\sin^{-1}x)\sqrt{1-x^2} + \int \frac{\sqrt{1-x^2}}{\sqrt{1-x^2}}\,dx$$

$$= -(\sin^{-1}x)\sqrt{1-x^2} + x + C$$

c. We use the trigonometric identities of Section 6.2. Thus,

$$\sin x \sin 2x \cos 3x = [(\sin x)(\sin 2x)]\cos 3x$$

$$= \frac{1}{2}(\cos x - \cos 3x)\cos 3x$$

$$= \frac{1}{2}[(\cos x)(\cos 3x) - (\cos 3x)(\cos 3x)]$$

$$= \frac{1}{4}(\cos 2x + \cos 4x - 1 - \cos 6x)$$

So

$$\int \sin x \sin 2x \cos 3x\,dx = \frac{1}{4}\int (\cos 2x + \cos 4x - \cos 6x - 1)\,dx$$

which is readily integrated.

d. We rewrite

$$I = \int \frac{\cos^4 x}{\sin^3 x}\, dx = \int \frac{\cos^4 x \sin x}{\sin^4 x}\, dx$$

Letting $u = \cos x$, we have $du = -\sin x\, dx$, and this gives

$$I = \int \frac{\cos^4 x \sin x}{(\sin^2 x)^2}\, dx = \int \frac{\cos^4 x \sin x}{(1 - \cos^2 x)^2}\, dx$$

$$= -\int \frac{u^4}{(1 - u^2)^2}\, du$$

To complete the solution, we first perform long division and then use the method of partial fractions.

e. The integrand is a rational function whose numerator has degree less than that of the denominator. So we use the method of partial fractions. The form of the decomposition is

$$\frac{A}{x - 1} + \frac{Bx + C}{x^2 + 1} + \frac{Dx + E}{(x^2 + 1)^2}$$

6.5 EXERCISES

In Exercises 1–36, use the Table of Integrals to evaluate the integral.

1. $\displaystyle\int x\sqrt{1 + 2x}\, dx$

2. $\displaystyle\int \frac{x}{\sqrt{2 + 3x}}\, dx$

3. $\displaystyle\int \frac{x^2}{(1 + 2x)^2}\, dx$

4. $\displaystyle\int \frac{1}{x\sqrt{4 + x}}\, dx$

5. $\displaystyle\int \frac{\sqrt{3 + 2x}}{x^2}\, dx$

6. $\displaystyle\int \frac{x^2}{\sqrt{9 + 4x^2}}\, dx$

7. $\displaystyle\int \frac{1}{x\sqrt{3 + 2x^2}}\, dx$

8. $\displaystyle\int x^2\sqrt{4 - 3x^2}\, dx$

9. $\displaystyle\int \frac{\sqrt{2 - x^2}}{x}\, dx$

10. $\displaystyle\int \frac{\sqrt{9 - 2x^2}}{x^2}\, dx$

11. $\displaystyle\int \frac{\sqrt{x^2 - 3}}{x}\, dx$

12. $\displaystyle\int \frac{1}{x^2\sqrt{x^2 - 5}}\, dx$

13. $\displaystyle\int \frac{e^x}{(1 - e^{2x})^{3/2}}\, dx$

14. $\displaystyle\int x^4 \sin x\, dx$

15. $\displaystyle\int x \cos^{-1} 2x\, dx$

16. $\displaystyle\int \csc^5 \theta\, d\theta$

17. $\displaystyle\int x^3 \sin(x^2 + 1)\, dx$

18. $\displaystyle\int x^2 \tan^{-1} 3x\, dx$

19. $\displaystyle\int_0^1 \sin^{-1}\sqrt{x}\, dx$

Hint: Let $u = \sqrt{x}$.

20. $\displaystyle\int \frac{\cos^{-1}\sqrt{x}}{\sqrt{x}}\, dx$

Hint: Let $u = \sqrt{x}$.

21. $\displaystyle\int e^{-2x} \sin 3x\, dx$

22. $\displaystyle\int e^{2x} \sin^{-1} e^x\, dx$

23. $\displaystyle\int x^3 e^{-2x}\, dx$

24. $\displaystyle\int \frac{1}{\sqrt{1 + e^{2x}}}\, dx$

25. $\displaystyle\int \frac{\sin x}{1 + \cos^2 x}\, dx$

26. $\displaystyle\int \frac{\sec^3\sqrt{x}}{\sqrt{x}}\, dx$

27. $\displaystyle\int x^3 \ln 5x\, dx$

28. $\displaystyle\int \frac{1}{x \ln\sqrt{x}}\, dx$

29. $\displaystyle\int \sqrt{6x - x^2}\, dx$

30. $\displaystyle\int \frac{\sqrt{4x - 2x^2}}{x}\, dx$

31. $\displaystyle\int \frac{x^2}{\sqrt{8x - 3x^2}}\, dx$

32. $\displaystyle\int e^{2x} \ln(1 + e^{2x})\, dx$

33. $\displaystyle\int_1^{e^2} \frac{\ln t}{t\sqrt{1 + \ln t}}\, dt$

34. $\displaystyle\int_0^{\pi/4} \frac{1}{a^2 \cos^2 x + b^2 \sin^2 x}\, dx \quad a > 0, b > 0$

Hint: Let $u = \tan x$.

35. $\displaystyle\int e^{\cos x} \sin 2x\, dx$

36. $\displaystyle\int e^{2x} \ln(1 + e^x)\, dx$

37. Find the area of the region under the graph of $y = x^2 \ln x$ on the interval $[1, e]$.

38. Find the length of the graph of $f(x) = \ln x$ from $A(1, 0)$ to $B(e, 1)$.

39. The region under the graph of $y = \cos^2 x$ on $\left[0, \frac{\pi}{2}\right]$ is revolved about the x-axis. Find the volume of the resulting solid of revolution.

40. The region under the graph of $y = \sin^{-1} x$ on $[0, 1]$ is revolved about the y-axis. Find the volume of the resulting solid of revolution.

41. Find the centroid of the region under the graph of $y = \cos^2 x$ on $\left[0, \frac{\pi}{2}\right]$.

42. Find the work done by the force $F(x) = x^2/(1 + 2x)^2$ (measured in pounds) in moving a particle along the x-axis from $x = 0$ to $x = 4$ (measured in feet).

43. Theme Park Attendance The management of Astro World ("The Amusement Park of the Future") estimates that visitors enter the park t hours after opening time at 8 A.M. at the rate of

$$R(t) = \frac{60}{(2 + t^2)^{3/2}}$$

thousand people per hour. Determine the number of visitors admitted by noon.

44. Voter Registration The number of voters in a certain district of a city is expected to grow at the rate of

$$R(t) = \frac{3000}{\sqrt{4 + t^2}}$$

people per year t years from now. If the number of voters at present is 20,000, how many voters will be in the district 5 years from now?

45. Growth of Fruit Flies On the basis of data collected during an experiment, a biologist found that the number of fruit flies (*Drosophila melanogaster*) with a limited food supply could be approximated by the exponential model

$$N(t) = \frac{1000}{1 + 24e^{-0.02t}}$$

where t denotes the number of days since the beginning of the experiment. Find the average number of fruit flies in the colony in the first 10 days of the experiment and in the first 20 days.

46. Average Mass of an Electron According to the special theory of relativity, the mass m of a particle moving at a velocity v is given by

$$m = \frac{m_0}{\sqrt{1 - \dfrac{v^2}{c^2}}}$$

where m_0 is the mass of the body at rest and $c = 3 \times 10^8$ m/sec is the speed of light. If an electron is accelerated from a speed of v_1 m/sec to a speed of v_2 m/sec, find an expression for the average mass of the electron between $v = v_1$ and $v = v_2$.

47. Find the area of the surface generated by revolving the graph of $y = x^2$ for $0 \le x \le 1$ about the x-axis.

48. Find the centroid of the region enclosed by the graph of $y^2 = x^3 - x^4$.

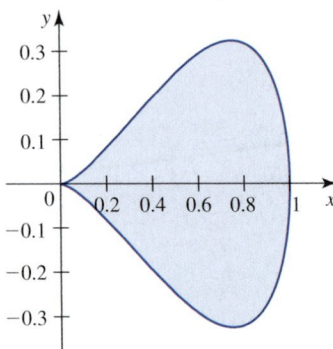

In Exercises 49–52, verify the integration formula.

49. $\displaystyle\int \frac{\sqrt{a^2 - u^2}}{u^2}\, du = -\frac{1}{u}\sqrt{a^2 - u^2} - \sin^{-1}\frac{u}{a} + C$

50. $\displaystyle\int \frac{1}{u^2(a + bu)}\, du = -\frac{1}{au} + \frac{b}{a^2}\ln\left|\frac{a + bu}{u}\right| + C$

51. $\displaystyle\int u^n \tan^{-1} u\, du = \frac{1}{n+1}\left[u^{n+1}\tan^{-1} u - \int \frac{u^{n+1}}{1 + u^2}\, du\right],$ $\quad n \ne -1$

52. $\displaystyle\int u^n \ln u\, du = \frac{u^{n+1}}{(n+1)^2}[(n+1)\ln u - 1] + C, \quad n \ne -1$

cas *In Exercises 53–62, use a CAS to find the integral.*

53. $\displaystyle\int x\sqrt{x + 2}\, dx$

54. $\displaystyle\int \frac{x}{\sqrt{1 + 2x}}\, dx$

55. $\displaystyle\int \frac{x + 1}{x\sqrt{x + 2}}\, dx$

56. $\displaystyle\int \frac{x^2 + x + 1}{x^3 + 1}\, dx$

57. $\displaystyle\int \cos^4 x\, dx$

58. $\displaystyle\int \tan^5 x\, dx$

59. $\displaystyle\int x^5 e^x\, dx$

60. $\displaystyle\int x^3 e^{-2x}\, dx$

61. $\displaystyle\int \frac{e^{2x}}{\sqrt{e^x + 1}}\, dx$

62. $\displaystyle\int x\sin^{-1} x\, dx$

In Exercises 63–100, find or evaluate the integral.

63. $\displaystyle\int \frac{x}{\sqrt[3]{2 - x}}\, dx$

64. $\displaystyle\int \frac{t^3}{\sqrt{1 - t^2}}\, dt$

65. $\displaystyle\int \frac{\cos\dfrac{1}{x}}{x^2}\, dx$

66. $\displaystyle\int \sqrt{1 + 2\cos^2 x}\, \sin 2x\, dx$

67. $\displaystyle\int_0^{1/2} \frac{x+1}{\sqrt{1-x^2}}\, dx$

68. $\displaystyle\int \frac{x+3}{\sqrt{5-4x-x^2}}\, dx$

69. $\displaystyle\int_0^1 \frac{x}{x^4+3}\, dx$

70. $\displaystyle\int_0^{1/\sqrt{2}} \frac{(\sin^{-1} x)^2}{\sqrt{1-x^2}}\, dx$

71. $\displaystyle\int \frac{dx}{x\sqrt{1+(\ln x)^2}}$

72. $\displaystyle\int e^{\sqrt{x}}\, dx$

73. $\displaystyle\int_1^e \frac{\sqrt{\ln x + 3}}{x}\, dx$

74. $\displaystyle\int \frac{e^x}{\sqrt{1-e^x}}\, dx$

75. $\displaystyle\int x^2(3^{x^3+1})\, dx$

76. $\displaystyle\int \tan^{-1} x\, dx$

77. $\displaystyle\int x \sin^{-1} x\, dx$

78. $\displaystyle\int_1^e \sin(\ln x)\, dx$

79. $\displaystyle\int_2^{\sqrt{5}} \sqrt{x^2-4}\, dx$

80. $\displaystyle\int x^2 e^{3x}\, dx$

81. $\displaystyle\int_1^2 \frac{\ln x}{x^2}\, dx$

82. $\displaystyle\int \sin^2 x \cos^5 x\, dx$

83. $\displaystyle\int x \tan^2 x\, dx$

84. $\displaystyle\int e^x \sin^2 x\, dx$

85. $\displaystyle\int_0^{\pi/3} \sqrt{1-\cos x}\, dx$

86. $\displaystyle\int \sin 3x \cos 4x\, dx$

87. $\displaystyle\int \frac{dx}{x+1+\sqrt{x+1}}$

88. $\displaystyle\int \frac{dx}{1+\tan x}$

89. $\displaystyle\int \frac{dx}{\sqrt{x^2-6x}}$

90. $\displaystyle\int_0^{\pi/2} \frac{\sin x}{1+\sqrt{\cos x}}\, dx$

91. $\displaystyle\int \cot^4(2x)\, dx$

92. $\displaystyle\int \frac{\sqrt{9-4x^2}}{x}\, dx$

93. $\displaystyle\int \frac{\sqrt{x^2+9}}{x}\, dx$

94. $\displaystyle\int_0^{\pi/4} \tan^{3/2} x \sec^4 x\, dx$

95. $\displaystyle\int \frac{dx}{x^3-1}$

96. $\displaystyle\int_0^1 \frac{x^3}{(x+1)^2(x^2+x+1)}\, dx$

97. $\displaystyle\int \frac{dx}{x^4+x^2+1}$

98. $\displaystyle\int \frac{x^4}{(1-x)^3}\, dx$

99. $\displaystyle\int xe^{x^2+e^{x^2}}\, dx$

100. $\displaystyle\int_0^{\pi/4} \frac{\sin x}{1-4\cos^2 x}\, dx$

6.6   Improper Integrals

In defining the definite integral $\int_a^b f(x)\, dx$, we required that the interval of integration $[a, b]$ be finite and that f be bounded. In many applications, one or both of these conditions do not hold. In this section we will extend the concept of the definite integral to include these cases:

1. The interval of integration is infinite (Figure 1a).
2. f is unbounded (Figure 1b).

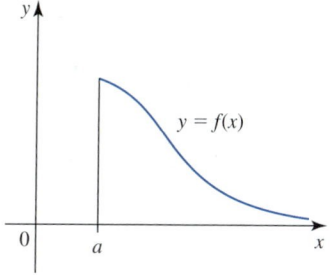

(a) The interval of integration $[a, \infty)$ is infinite.

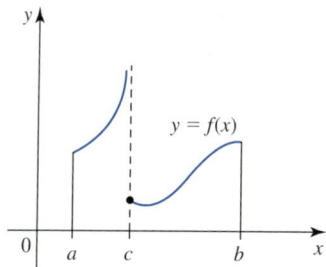

(b) f is unbounded on $[a, b]$ because it has an infinite discontinuity at $c\colon f(x) \to \infty$ as $x \to c^-$.

FIGURE 1

Integrals that have infinite intervals of integration or unbounded integrands are called **improper integrals.**

■ Infinite Intervals of Integration

Suppose that we want to find the area A of the unbounded region under the graph of $f(x) = 1/x^2$ on the interval $[1, \infty)$ as shown in Figure 2a. Because the interval $[1, \infty)$ is infinite, the definition of the integral that we have used thus far is not applicable, and a new approach to solving the problem is required. But observe that if $b > 1$, then A can be approximated by the area $A(b)$ of the region under the graph of f on $[1, b]$ (Figure 2b).

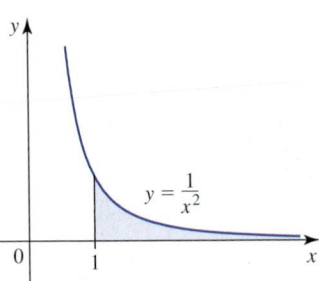

FIGURE 2
The shaded area in part (a) is approximated by the shaded area in part (b).

(a) The area of A of the region under the graph of $y = 1/x^2$ on $[1, \infty)$.

(b) The area $A(b)$ of the region under the graph of $y = 1/x^2$ on $[1, b]$.

The approximation seems to get better and better as b gets larger and larger (see Figure 3). Since $[1, b]$ is finite, we see that

$$A(b) = \int_1^b f(x)\, dx = \int_1^b \frac{1}{x^2}\, dx = -\frac{1}{x}\Big|_1^b = -\frac{1}{b} + 1$$

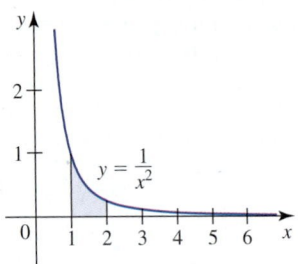
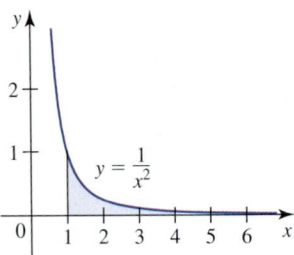

(a) Area of region under the graph of f on $[1, 2]$

(b) Area of region under the graph of f on $[1, 3]$

(c) Area of region under the graph of f on $[1, 4]$

FIGURE 3
As b increases, the approximation of A by the definite integral improves.

Letting $b \to \infty$, we obtain

$$\lim_{b \to \infty} A(b) = \lim_{b \to \infty}\left(-\frac{1}{b} + 1\right) = 1$$

This suggests that we *define* the area A to be 1 and write

$$A = \int_1^\infty \frac{1}{x^2}\, dx = \lim_{b \to \infty} \int_1^b \frac{1}{x^2}\, dx = 1$$

This example shows how we can define an integral over an infinite interval as the limit of integrals over finite intervals. More precisely, we have the following definitions. (Note that f need not be positive in the interval under consideration.)

DEFINITIONS **Improper Integrals with Infinite Limits of Integration**

1. If f is continuous on $[a, \infty)$, then

$$\int_a^\infty f(x)\, dx = \lim_{b \to \infty} \int_a^b f(x)\, dx \tag{1}$$

provided that the limit exists.

2. If f is continuous on $(-\infty, b]$, then

$$\int_{-\infty}^b f(x)\, dx = \lim_{a \to -\infty} \int_a^b f(x)\, dx \tag{2}$$

provided that the limit exists.

3. If f is continuous on $(-\infty, \infty)$, then

$$\int_{-\infty}^\infty f(x)\, dx = \int_{-\infty}^c f(x)\, dx + \int_c^\infty f(x)\, dx \tag{3}$$

where c is any real number, provided that both improper integrals on the right-hand side exist.

Convergence and Divergence

Each improper integral in Equation (1) and Equation (2) is **convergent** if the limit exists and **divergent** if the limit does not exist. The improper integral on the left-hand side in Equation (3) is **convergent** if both improper integrals on the right are convergent and **divergent** if one or both of the improper integrals on the right is divergent.

EXAMPLE 1 Evaluate $\displaystyle\int_1^\infty \frac{1}{x}\, dx$.

Solution By Equation (1) we have

$$\int_1^\infty \frac{1}{x}\, dx = \lim_{b \to \infty} \int_1^b \frac{1}{x}\, dx = \lim_{b \to \infty} \Big[\ln x\Big]_1^b$$

$$= \lim_{b \to \infty} (\ln b - \ln 1) = \infty$$

Therefore, the given improper integral is divergent. ■

Let's compare the integral $\int_1^\infty (1/x)\, dx$ of Example 1 with the integral $\int_1^\infty (1/x^2)\, dx$ that we considered earlier. If we interpret each integral as the area of the region under the graph of a function on the infinite interval $[1, \infty)$, then the result $\int_1^\infty (1/x^2)\, dx = 1$ tells us that the area under the graph of $y = 1/x^2$ is equal to 1 and hence finite, whereas the result $\int_1^\infty (1/x)\, dx = \infty$ tells us that the area under the graph of $y = 1/x$ is infinite. Observe that the graphs of $y = 1/x^2$ and $y = 1/x$ are similar. (See Figure 4.) Both $1/x^2$ and $1/x$ approach zero as x approaches infinity, but $1/x^2$ approaches zero faster than $1/x$ does.

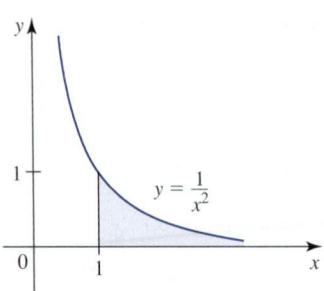
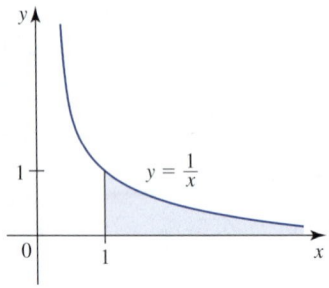

FIGURE 4   (**a**) The unbounded region has *finite* area.   (**b**) The unbounded region has *infinite* area.

These examples reveal the fine line between convergence and divergence of an improper integral. But a word of caution: It is not even necessary for $f(x)$ to approach zero as x approaches infinity for an integral $\int_a^\infty f(x)\,dx$ to converge (see Challenge Problem 16 at the end of this chapter).

EXAMPLE 2   Find the values of p for which $\displaystyle\int_1^\infty \frac{1}{x^p}\,dx$ is convergent.

Solution   From the result of Example 1 we see that the integral is divergent if $p = 1$. So let's assume that $p \neq 1$. We have

$$\int_1^\infty \frac{1}{x^p}\,dx = \lim_{b \to \infty} \int_1^b x^{-p}\,dx$$

$$= \lim_{b \to \infty} \left[\frac{x^{-p+1}}{-p+1} \right]_1^b$$

$$= \frac{1}{1-p} \lim_{b \to \infty} \left[\frac{1}{b^{p-1}} - 1 \right]$$

If $p < 1$, then $1 - p > 0$, so

$$\lim_{b \to \infty} \frac{1}{b^{p-1}} = \lim_{b \to \infty} b^{1-p} = \infty$$

Therefore, the integral diverges. If $p > 1$, then $p - 1 > 0$, so

$$\lim_{b \to \infty} \frac{1}{b^{p-1}} = 0$$

Therefore, the integral converges to $1/(p - 1)$. To summarize

$$\int_1^\infty \frac{1}{x^p}\,dx = \begin{cases} \dfrac{1}{p-1} & \text{if } p > 1 \\ \text{diverges} & \text{if } p \leq 1 \end{cases}$$

EXAMPLE 3   Evaluate

a. $\displaystyle\int_{-1}^\infty e^{-x}\,dx$ **b.** $\displaystyle\int_0^\infty \cos x\,dx$

Solution

a. $\int_{-1}^{\infty} e^{-x}\, dx = \lim_{b\to\infty} \int_{-1}^{b} e^{-x}\, dx = \lim_{b\to\infty}\left[-e^{-x}\right]_{-1}^{b} = \lim_{b\to\infty}(-e^{-b} + e^{1}) = e$

b. $\int_{0}^{\infty} \cos x\, dx = \lim_{b\to\infty} \int_{0}^{b} \cos x\, dx = \lim_{b\to\infty}\left[\sin x\right]_{0}^{b} = \lim_{b\to\infty}(\sin b - 0)$

Since $\lim_{b\to\infty}\sin b$ does not exist, we conclude that the given integral is divergent. (To see why, just examine the graph of $y = \sin x$.)

EXAMPLE 4 Evaluate $\int_{-\infty}^{0} xe^{x}\, dx$.

Solution By Equation (2) we have

$$\int_{-\infty}^{0} xe^{x}\, dx = \lim_{a\to-\infty} \int_{a}^{0} xe^{x}\, dx$$

From the result of Example 1 in Section 6.1 we have

$$\int xe^{x}\, dx = xe^{x} - \int e^{x}\, dx = (x - 1)e^{x} + C$$

Therefore,

$$\int_{-\infty}^{0} xe^{x}\, dx = \lim_{a\to-\infty} \int_{a}^{0} xe^{x}\, dx = \lim_{a\to-\infty}\left[(x - 1)e^{x}\right]_{a}^{0}$$

$$= \lim_{a\to-\infty}\left[-1 - (a - 1)e^{a}\right]$$

To evaluate the limit on the right-hand side, note that

$$\lim_{a\to-\infty} e^{a} = 0$$

and, by l'Hôpital's Rule,

$$\lim_{a\to-\infty} ae^{a} = \lim_{a\to-\infty} \frac{a}{e^{-a}} \qquad \text{Indeterminate form: } -\infty/\infty$$

$$= \lim_{a\to-\infty} \frac{1}{-e^{-a}} = 0$$

Therefore,

$$\int_{-\infty}^{0} xe^{x}\, dx = \lim_{a\to-\infty}(-1 - ae^{a} + e^{a})$$

$$= \lim_{a\to-\infty}(-1) - \lim_{a\to-\infty} ae^{a} + \lim_{a\to-\infty} e^{a}$$

$$= -1 - 0 + 0 = -1$$

EXAMPLE 5 Evaluate $\int_{-\infty}^{\infty} \frac{1}{1 + x^{2}}\, dx$, and interpret your result geometrically.

Solution By Equation (3) we have

$$\int_{-\infty}^{\infty} \frac{1}{1 + x^2} \, dx = \int_{-\infty}^{0} \frac{1}{1 + x^2} \, dx + \int_{0}^{\infty} \frac{1}{1 + x^2} \, dx$$

For convenience we have chosen $c = 0$.

$$= \lim_{a \to -\infty} \int_{a}^{0} \frac{1}{1 + x^2} \, dx + \lim_{b \to \infty} \int_{0}^{b} \frac{1}{1 + x^2} \, dx$$

$$= \lim_{a \to -\infty} \left[\tan^{-1} x \right]_{a}^{0} + \lim_{b \to \infty} \left[\tan^{-1} x \right]_{0}^{b}$$

$$= \lim_{a \to -\infty} (\tan^{-1} 0 - \tan^{-1} a) + \lim_{b \to \infty} (\tan^{-1} b - \tan^{-1} 0)$$

$$= \left[0 - \left(-\frac{\pi}{2} \right) \right] + \left(\frac{\pi}{2} - 0 \right) = \pi$$

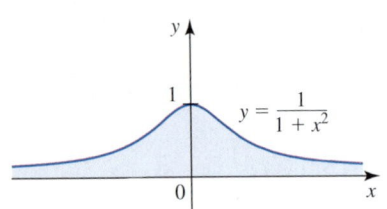

FIGURE 5
The area of the region under the graph of $y = \dfrac{1}{1 + x^2}$ on $(-\infty, \infty)$ is π.

Because the integrand $f(x) = 1/(1 + x^2)$ is nonnegative on $(-\infty, \infty)$, we can interpret the value of the improper integral as the area (π) of the region under the graph of f on $(-\infty, \infty)$. (See Figure 5.)

EXAMPLE 6 **A Rocket Launch** Find the work done in launching a rocket weighing P pounds, vertically upward from the surface of the earth so that the rocket completely escapes the earth's gravitational field.

Solution According to Newton's Law of Gravitation, the rocket is attracted to the earth by a force $F(x)$ given by

$$F(x) = \frac{GmM}{x^2}$$

where m is the mass of the rocket, M is the mass of the earth, x is the distance between the rocket and the center of the earth, and G is the universal gravitational constant. Writing $k = GmM$, we have

$$F(x) = \frac{k}{x^2} \qquad R \le x < \infty$$

where R is the radius of the earth. Since the rocket weighs P pounds on the surface of the earth, we have

$$F(R) = \frac{k}{R^2} = P$$

This gives $k = PR^2$, and therefore

$$F(x) = \frac{PR^2}{x^2}$$

(See Figure 6.) Therefore, the work required to propel the rocket to an infinite height (to escape the earth's gravitational field) is

$$W = \int_{R}^{\infty} F(x) \, dx = \int_{R}^{\infty} \frac{PR^2}{x^2} \, dx$$

$$= \lim_{b \to \infty} \int_{R}^{b} \frac{PR^2}{x^2} \, dx = \lim_{b \to \infty} \left[-\frac{PR^2}{x} \right]_{R}^{b}$$

$$= \lim_{b \to \infty} \left(-\frac{PR^2}{b} + \frac{PR^2}{R} \right) = PR$$

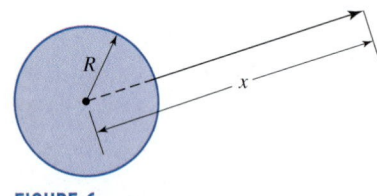

FIGURE 6
The force attracting the rocket to the earth when it is at a distance x is $F = PR^2/x^2$, where $R \le x < \infty$.

For example, if the rocket weighs 20 tons (40,000 lb) on the ground and the radius of the earth is approximately 4000 mi (21,120,000 ft), then the work required is $W \approx 40,000 \times 21,120,000$ or 8.448×10^{11} ft-lb. ◼

◼ Improper Integrals with Infinite Discontinuities

As we mentioned earlier, there is another kind of improper integral: those having integrands that are unbounded on the interval of integration (Figure 1b). To see how we define this type of integral, consider the problem of finding the area A of the unbounded region under the graph of $f(x) = 1/\sqrt{x}$ on the interval $(0, 4]$ shown in Figure 7a.

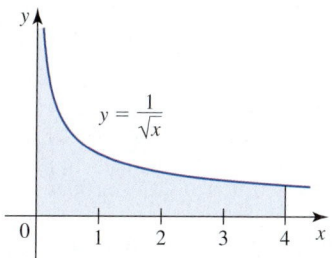

(**a**) The area A of the region under the graph of $y = 1/\sqrt{x}$ on $(0, 4]$

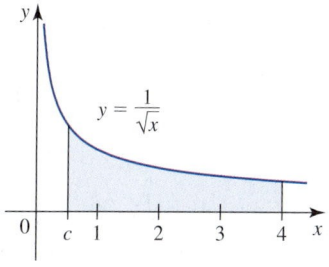

(**b**) The area $A(c)$ of the region under the graph of $y = 1/\sqrt{x}$ on $[c, 4]$

FIGURE 7

The area of the shaded region in part (a) is approximated by the area of the shaded region in part (b).

Because the integrand is unbounded on the interval $(0, 4]$ (that is, $1/\sqrt{x} \to \infty$ as $x \to 0^+$), the definition of the integral given in Chapter 4 cannot be used to find A. But observe that if c is any number such that $0 < c < 4$, then A can be approximated by the area $A(c)$ of the region under the graph of f on $[c, 4]$ (Figure 7b). Observe that the approximation appears to get better and better as c approaches 0 from the right. Since $f(x) = 1/\sqrt{x}$ is bounded on the finite interval $[c, 4]$, we see that

$$A(c) = \int_c^4 f(x)\, dx = \int_c^4 \frac{1}{\sqrt{x}}\, dx = 2\sqrt{x}\,\Big|_c^4 = 4 - 2\sqrt{c}$$

Letting $c \to 0^+$, we obtain

$$\lim_{c \to 0^+} A(c) = \lim_{c \to 0^+} (4 - 2\sqrt{c}) = 4$$

This suggests that we *define* the area A to be 4 and write

$$A = \int_0^4 \frac{1}{\sqrt{x}}\, dx = \lim_{c \to 0^+} \int_c^4 \frac{1}{\sqrt{x}}\, dx = 4$$

This example shows how we can define an integral whose integrand has an infinite discontinuity at a point as the limit of integrals whose integrands are bounded. More precisely, we have the following definitions. (Again, note that f need not be positive in the interval under consideration.)

> **DEFINITIONS** **Improper Integrals Whose Integrands Have Infinite Discontinuities**
>
> 1. If f is continuous on $[a, b)$ and f has an infinite discontinuity at b, then
>
> $$\int_a^b f(x)\, dx = \lim_{c \to b^-} \int_a^c f(x)\, dx \qquad (4)$$
>
> provided that the limit exists (Figure 8a).
> 2. If f is continuous on $(a, b]$ and f has an infinite discontinuity at a, then
>
> $$\int_a^b f(x)\, dx = \lim_{c \to a^+} \int_c^b f(x)\, dx \qquad (5)$$
>
> provided that the limit exists (Figure 8b).
> 3. If f has an infinite discontinuity at c, where $a < c < b$, but f is continuous elsewhere on $[a, b]$, then
>
> $$\int_a^b f(x)\, dx = \int_a^c f(x)\, dx + \int_c^b f(x)\, dx \qquad (6)$$
>
> provided that both improper integrals on the right exist (Figure 8c).
>
> **Convergence and Divergence**
>
> Each improper integral in Equations (4) and (5) is **convergent** if the limit exists and **divergent** if the limit does not exist. The improper integral on the left in Equation (6) is **convergent** if both improper integrals on the right are convergent and **divergent** if one or both improper integrals on the right is divergent.

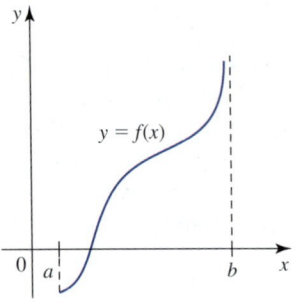

(a) f has an infinte discontinuity at b.

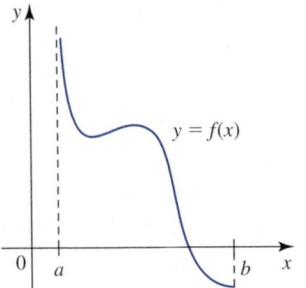

(b) f has an infinite discontinuity at a.

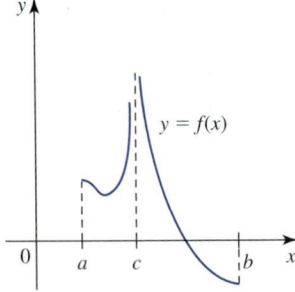

(c) f has an infinite discontinuity at c.

FIGURE 8

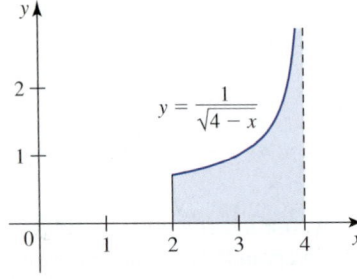

FIGURE 9
The area of the region under the graph of $y = 1/\sqrt{4 - x}$ on $[2, 4)$ is $2\sqrt{2}$.

EXAMPLE 7 Evaluate $\displaystyle\int_2^4 \frac{1}{\sqrt{4 - x}}\, dx$, and interpret your result geometrically.

Solution The integrand $f(x) = 1/\sqrt{4 - x}$ has an infinite discontinuity at $x = 4$, as shown in Figure 9. Using Equation (4), we have

$$\int_2^4 \frac{1}{\sqrt{4 - x}}\, dx = \lim_{c \to 4^-} \int_2^c \frac{1}{\sqrt{4 - x}}\, dx$$

$$= \lim_{c \to 4^-} \left[-2\sqrt{4 - x} \right]_2^c \qquad \text{Integrate using the substitution } u = 4 - x.$$

$$= \lim_{c \to 4^-} \left(-2\sqrt{4 - c} + 2\sqrt{2} \right) = 2\sqrt{2}$$

Since the integrand is positive on $[2, 4)$, we can interpret the value of the improper integral as the area of the region under the graph of f on $[2, 4)$. ∎

EXAMPLE 8   Evaluate $\displaystyle\int_0^1 \frac{dx}{x^2}$.

Solution   The integrand $1/x^2$ has an infinite discontinuity at $x = 0$. Using Equation (5), we have

$$\int_0^1 \frac{dx}{x^2} = \lim_{a \to 0^+} \int_a^1 \frac{dx}{x^2} = \lim_{a \to 0^+} \left[-\frac{1}{x}\right]_a^1 = \lim_{a \to 0^+} \left(-1 + \frac{1}{a}\right) = \infty$$

and we conclude that the given improper integral is divergent. ∎

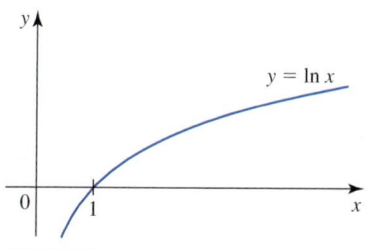

$y = \ln x$

FIGURE 10
The integrand $f(x) = \ln x$ approaches $-\infty$ as x approaches 0 from the right.

EXAMPLE 9   Evaluate $\displaystyle\int_0^1 \ln x \, dx$.

Solution   The integrand has an infinite discontinuity at $x = 0$. (See Figure 10.) Therefore, we write

$$\int_0^1 \ln x \, dx = \lim_{a \to 0^+} \int_a^1 \ln x \, dx$$

$$= \lim_{a \to 0^+} \left[x \ln x - x\right]_a^1 \qquad \text{\color{brown}Integrate by parts with } u = \ln x \text{ and } dv = dx.$$

$$= \lim_{a \to 0^+} (0 - 1 - a \ln a + a)$$

To evaluate the limit on the right, we apply l'Hôpital's Rule, obtaining

$$\lim_{a \to 0^+} a \ln a = \lim_{a \to 0^+} \frac{\ln a}{\dfrac{1}{a}} = \lim_{a \to 0^+} \frac{\dfrac{1}{a}}{-\dfrac{1}{a^2}} = \lim_{a \to 0^+} (-a) = 0$$

Therefore,

$$\int_0^1 \ln x \, dx = \lim_{a \to 0^+} (-1 - a \ln a + a) = -1 - 0 + 0 = -1 \qquad ∎$$

EXAMPLE 10   Evaluate $\displaystyle\int_{-1}^1 \frac{dx}{x^2}$.

Solution   The integrand $f(x) = 1/x^2$ has an infinite discontinuity at $x = 0$. (See Figure 11.) Using Equation (6), we have

$$\int_{-1}^1 \frac{dx}{x^2} = \int_{-1}^0 \frac{dx}{x^2} + \int_0^1 \frac{dx}{x^2}$$

Now, using the result of Example 8, we see that the second integral on the right is divergent; that is,

$$\int_0^1 \frac{dx}{x^2} = \infty$$

FIGURE 11
The integrand $f(x) = 1/x^2$ approaches ∞ as x approaches 0.

Therefore, the given improper integral is divergent. Note that it is not necessary to evaluate the first integral on the right. ∎

Note   If we had not realized that $f(x) = 1/x^2$ has an infinite discontinuity at $x = 0$, then we might have proceeded as follows:

$$\int_{-1}^{1} \frac{dx}{x^2} = -\frac{1}{x}\Big|_{-1}^{1} = -1 + (-1) = -2$$

giving a *wrong* answer. After all, a positive integrand could not possibly yield an integral whose value is negative! ▪

EXAMPLE 11   **Length of a Pursuit Curve**   The graph C of the equation

$$y = \frac{1}{3}\sqrt{x}(x - 3) + \frac{2}{3}$$

gives the path taken by a coast guard patrol boat (Boat A) as it pursued and eventually intercepted boat B that was suspected of carrying contraband. (See Figure 12.) Initially, the patrol boat was at point P, and Boat B was at the origin, heading north. At the time of interception both boats were at point Q. Find the distance traveled by the patrol boat during the pursuit.

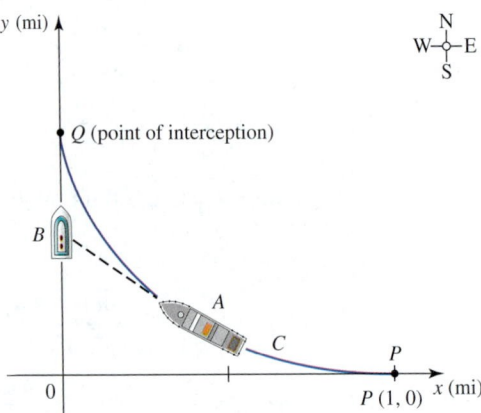

FIGURE 12
The pursuit curve C gives the path taken by patrol boat A.

Solution   The distance traveled by the patrol boat is given by the length L of the curve C from $x = 0$ to $x = 1$. To use Equation (5), we first compute

$$\frac{dy}{dx} = \frac{d}{dx}\left[\frac{1}{3}x^{3/2} - x^{1/2} + \frac{2}{3}\right]$$

$$= \frac{1}{2}x^{1/2} - \frac{1}{2}x^{-1/2} = \frac{1}{2}(x^{1/2} - x^{-1/2})$$

and

$$1 + \left(\frac{dy}{dx}\right)^2 = 1 + \frac{1}{4}(x^{1/2} - x^{-1/2})^2 = 1 + \frac{1}{4}(x - 2 + x^{-1})$$

$$= \frac{4x + x^2 - 2x + 1}{4x} = \frac{x^2 + 2x + 1}{4x} = \frac{(x + 1)^2}{4x}$$

Then

$$L = \int_0^1 \sqrt{1 + \left(\frac{dy}{dx}\right)^2} \, dx = \int_0^1 \sqrt{\frac{(x+1)^2}{4x}} \, dx = \frac{1}{2}\int_0^1 \frac{x+1}{\sqrt{x}} \, dx$$

$$= \frac{1}{2}\int_0^1 x^{1/2} \, dx + \frac{1}{2}\int_0^1 x^{-1/2} \, dx$$

The second integral on the right has an infinite discontinuity at $x = 0$. So we write

$$L = \frac{1}{2}\int_0^1 x^{1/2} \, dx + \frac{1}{2}\lim_{t \to 0^+}\int_t^1 x^{-1/2} \, dx$$

$$= \left(\frac{1}{2}\right)\left(\frac{2}{3}x^{3/2}\right)\Big|_0^1 + \frac{1}{2}\lim_{t \to 0^+}\left[2x^{1/2}\right]_t^1$$

$$= \frac{1}{3} + \frac{1}{2}\lim_{t \to 0^+}(2 - 2t^{1/2}) = \frac{1}{3} + 1 = \frac{4}{3}$$

Therefore, the patrol boat traveled $\frac{4}{3}$ miles from the time Boat B was spotted until the time it was intercepted. ■

The next example involves both an infinite limit of integration and an infinite discontinuity.

EXAMPLE 12 Evaluate $\displaystyle\int_0^\infty \frac{e^{-\sqrt{x}}}{\sqrt{x}} \, dx$.

Solution   We write

$$\int_0^\infty \frac{e^{-\sqrt{x}}}{\sqrt{x}} \, dx = \int_0^1 \frac{e^{-\sqrt{x}}}{\sqrt{x}} \, dx + \int_1^\infty \frac{e^{-\sqrt{x}}}{\sqrt{x}} \, dx$$

$$= \lim_{t \to 0^+}\int_t^1 \frac{e^{-\sqrt{x}}}{\sqrt{x}} \, dx + \lim_{b \to \infty}\int_1^b \frac{e^{-\sqrt{x}}}{\sqrt{x}} \, dx$$

$$= \lim_{t \to 0^+}\left[-2e^{-\sqrt{x}}\right]_t^1 + \lim_{b \to \infty}\left[-2e^{-\sqrt{x}}\right]_1^b$$

$$= \lim_{t \to 0^+}\left(-2e^{-1} + 2e^{-\sqrt{t}}\right) + \lim_{b \to \infty}\left(-2e^{-\sqrt{b}} + 2e^{-1}\right)$$

$$= -2e^{-1} + 2 + 2e^{-1} = 2$$ ■

■ A Comparison Test for Improper Integrals

Sometimes it is impossible to find the exact value of an improper integral. In such instances we need to determine whether the integral is convergent or divergent. If we can ascertain that the improper integral is convergent, then we can proceed to obtain a sufficiently accurate approximation of its value, which, in practice, is all that is required. The following theorem is stated without proof, but its plausibility should be evident by examining Figure 13.

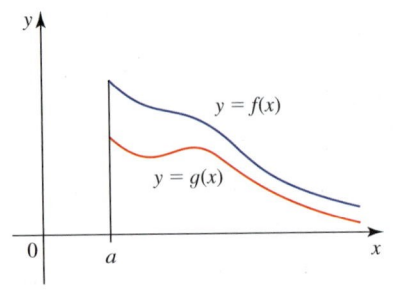

FIGURE 13
The function f dominates the function g on $[a, \infty)$.

> **THEOREM 1 A Comparison Test for Improper Integrals**
>
> Let f and g be continuous, and suppose that $f(x) \geq g(x) \geq 0$ for all $x \geq a$; that is, f dominates g on $[a, \infty)$.
>
> **a.** If $\displaystyle\int_a^\infty f(x)\, dx$ is convergent, then so is $\displaystyle\int_a^\infty g(x)\, dx$.
>
> **b.** If $\displaystyle\int_a^\infty g(x)\, dx$ is divergent, then so is $\displaystyle\int_a^\infty f(x)\, dx$.

Before looking at the next example, let's note that the functions that we have dealt with up until now have been functions such as polynomial, rational, power, exponential, logarithmic, trigonometric, and inverse trigonometric functions or functions obtained from this list by combining them using the operations of addition, subtraction, multiplication, division, and composition. Such functions are called *elementary functions*.

EXAMPLE 13 Show that $\displaystyle\int_0^\infty e^{-x^2}\, dx$ is convergent.

Solution   We cannot evaluate the integral directly because it turns out that the antiderivative of e^{-x^2} is not an elementary function. To show that this integral is convergent, let's write

$$\int_0^\infty e^{-x^2}\, dx = \int_0^1 e^{-x^2}\, dx + \int_1^\infty e^{-x^2}\, dx$$

Observe that the first integral on the right is a proper integral, and therefore, it has a finite value, even though we don't know what that value is. For the second integral we note that $x^2 \geq x$ for $x \geq 1$, so $e^{-x^2} \leq e^{-x}$ on $[1, \infty)$. (See Figure 14.) Now

$$\int_1^\infty e^{-x}\, dx = \lim_{b \to \infty} \int_1^b e^{-x}\, dx = \lim_{b \to \infty}\left[-e^{-x}\right]_1^b = \lim_{b \to \infty}(-e^{-b} + e^{-1}) = \frac{1}{e}$$

So if we take $f(x) = e^{-x}$ and $g(x) = e^{-x^2}$, the Comparison Test tells us that $\int_1^\infty e^{-x^2}\, dx$ is convergent. Therefore, $\int_0^\infty e^{-x^2}\, dx$ is convergent.

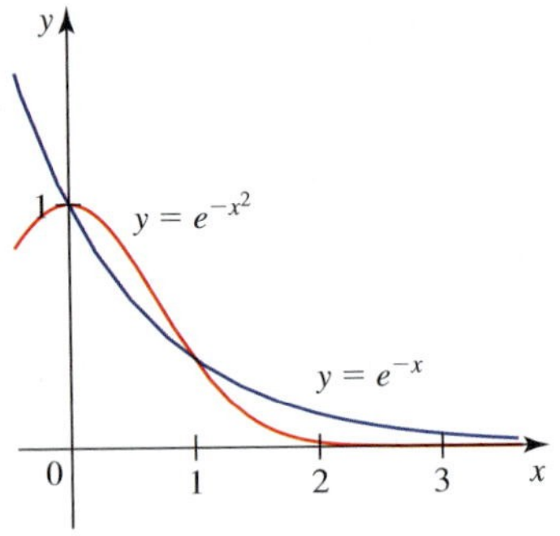

FIGURE 14
We use the Comparison Test to show that
$$\int_0^\infty e^{-x^2}\, dx = \int_0^1 e^{-x^2}\, dx + \int_1^\infty e^{-x^2}\, dx$$
is convergent.

6.6 CONCEPT QUESTIONS

1. Define the following improper integrals:

   a. $\displaystyle\int_{-\infty}^b f(x)\, dx$ **b.** $\displaystyle\int_a^\infty f(x)\, dx$ **c.** $\displaystyle\int_{-\infty}^\infty f(x)\, dx$

2. Define the improper integral $\displaystyle\int_a^b f(x)\, dx$ if
   a. f has an infinite discontinuity at a.
   b. f has an infinite discontinuity at b.
   c. f has an infinite discontinuity at c, where $a < c < b$.

3. State the Comparison Test for improper integrals.

6.6 EXERCISES

In Exercises 1–6, find the area of the shaded region, if it exists.

1.

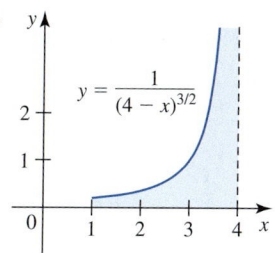

$y = \dfrac{1}{\sqrt{x-1}}$

2.

$y = \dfrac{1}{(4-x)^{3/2}}$

3.

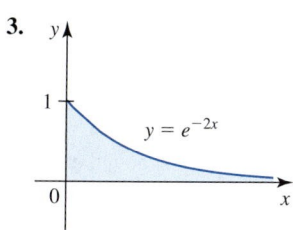

$y = e^{-2x}$

4.

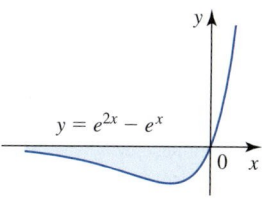

$y = e^{2x} - e^{x}$

5.

$y = 1$

$y = \dfrac{x^2}{1+x^2}$

6.

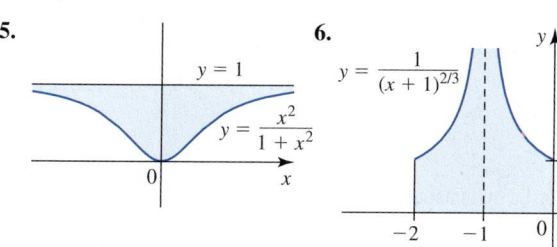

$y = \dfrac{1}{(x+1)^{2/3}}$

In Exercises 7–42, determine whether the improper integral converges or diverges, and if it converges, find its value.

7. $\displaystyle \int_1^\infty \frac{1}{x^3}\, dx$

8. $\displaystyle \int_1^\infty \frac{1}{x^{0.99}}\, dx$

9. $\displaystyle \int_1^\infty \frac{1}{x^{1.01}}\, dx$

10. $\displaystyle \int_0^\infty \frac{1}{(x+1)^2}\, dx$

11. $\displaystyle \int_1^\infty \frac{1}{(x+2)^{3/2}}\, dx$

12. $\displaystyle \int_2^\infty \frac{1}{\sqrt[3]{x-1}}\, dx$

13. $\displaystyle \int_1^\infty e^{-2x}\, dx$

14. $\displaystyle \int_e^\infty \frac{1}{x \ln^2 x}\, dx$

15. $\displaystyle \int_0^\infty \sin x\, dx$

16. $\displaystyle \int_0^\infty e^{-x} \sin x\, dx$

17. $\displaystyle \int_0^\infty \frac{x}{1+x^2}\, dx$

18. $\displaystyle \int_{-\infty}^0 \frac{1}{x^2+2x+5}\, dx$

19. $\displaystyle \int_{-\infty}^\infty \frac{1}{x^2+4}\, dx$

20. $\displaystyle \int_{-\infty}^\infty x e^{-x^2}\, dx$

21. $\displaystyle \int_{-\infty}^\infty \frac{e^x}{1+e^{2x}}\, dx$

22. $\displaystyle \int_{-\infty}^\infty \cos^2 x\, dx$

23. $\displaystyle \int_{-\infty}^\infty \frac{x}{(x^2+1)^{3/2}}\, dx$

24. $\displaystyle \int_{-\infty}^\infty e^{-|x|}\, dx$

25. $\displaystyle \int_0^1 \frac{1}{x^{2/3}}\, dx$

26. $\displaystyle \int_{-2}^1 \frac{1}{x^2}\, dx$

27. $\displaystyle \int_{-8}^1 \frac{1}{\sqrt[3]{x}}\, dx$

28. $\displaystyle \int_0^2 \frac{1}{2x-3}\, dx$

29. $\displaystyle \int_1^4 \frac{1}{(4-x)^{2/3}}\, dx$

30. $\displaystyle \int_0^2 \frac{1}{x^2-2x}\, dx$

31. $\displaystyle \int_0^4 \frac{1}{\sqrt{x-1}}\, dx$

32. $\displaystyle \int_0^e \ln x\, dx$

33. $\displaystyle \int_0^1 x \ln x\, dx$

34. $\displaystyle \int_0^\pi \sec^2 x\, dx$

35. $\displaystyle \int_{\pi/6}^{\pi/2} \frac{\cos x}{\sqrt{1-\sin x}}\, dx$

36. $\displaystyle \int_0^{\pi/2} \tan^2 x\, dx$

37. $\displaystyle \int_1^\infty \frac{\ln x}{x^{3/2}}\, dx$

38. $\displaystyle \int_0^\infty \frac{\sqrt{\tan^{-1} x}}{1+x^2}\, dx$

39. $\displaystyle \int_{-\infty}^\infty \frac{1}{x^{4/3}}\, dx$

40. $\displaystyle \int_0^\infty \frac{1}{e^x-1}\, dx$

41. $\displaystyle \int_0^1 \frac{\ln x}{\sqrt{x}}\, dx$

42. $\displaystyle \int_1^\infty \frac{dx}{x\sqrt{x^2-1}}$

In Exercises 43–48, use the Comparison Test to determine whether the integral is convergent or divergent by comparing it with the second integral.

43. $\displaystyle \int_1^\infty \frac{1}{1+x^2}\, dx; \quad \int_1^\infty \frac{1}{x^2}\, dx$

44. $\displaystyle \int_1^\infty \frac{1}{\sqrt{x^3+1}}\, dx; \quad \int_1^\infty \frac{1}{x^{3/2}}\, dx$

45. $\displaystyle \int_1^\infty \frac{\cos^2 x}{x^2}\, dx; \quad \int_1^\infty \frac{1}{x^2}\, dx$

46. $\displaystyle \int_1^\infty \frac{dx}{x+\sin^2 x}; \quad \int_1^\infty \frac{1}{1+x}\, dx$

47. $\displaystyle \int_1^\infty \frac{2+\cos x}{\sqrt{x}}\, dx; \quad \int_1^\infty \frac{1}{\sqrt{x}}\, dx$

48. $\displaystyle \int_1^\infty \frac{1}{\sqrt{1+x^2+x^4}}\, dx; \quad \int_1^\infty \frac{1}{x^2}\, dx$

49. Evaluate $\int_0^\infty x^5 e^{-x^2}\, dx$.

50. Find the area of the region bounded by the graph of $y = 1/\left(\frac{1}{2}x^2 - x + 1\right)$ and the *x*-axis.

51. Find the volume of the solid obtained by revolving the region under the graph of $y = 2\left(\dfrac{1}{x^2} - \dfrac{1}{x^4}\right)$ on $[1, \infty)$ about the x-axis.

52. Find the area of the surface obtained by revolving the graph of $y = e^{-x}$ on $[0, \infty)$ about the x-axis.

53. Find the volume of the solid obtained by revolving the region under the graph of $y = e^{-x}$ on $[0, \infty)$ about the x-axis.

54. Find the volume of the solid obtained by revolving the region under the graph of $y = e^{-x}$ on $[0, \infty)$ about the y-axis.

55. Find the area of the region bounded by the graphs of $y = 1/\sqrt{1 - x^2}$, $y = 0$, $x = 0$, and $x = 1$.

56. Gabriel's Horn The solid obtained by revolving the unbounded region under the graph of $f(x) = 1/x$ on the interval $[1, \infty)$ about the x-axis is called *Gabriel's Horn*. Show that this solid has a finite volume but an infinite surface area. Thus, Gabriel's Horn describes a can that does not hold enough paint to cover its outside surface!

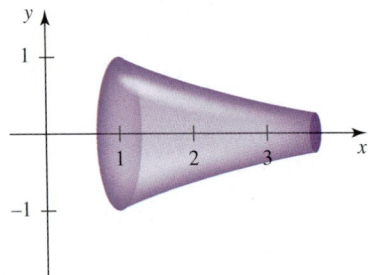

Hint: The surface area is

$$S = 2\pi \int_1^\infty \frac{\sqrt{1 + x^4}}{x^3}\, dx$$

Use the substitution $u = x^2$, and integrate using Formula 40 from the Table of Integrals.

57. Cissoid of Diocles Find the area of the region bounded by the *cissoid of Diocles*

$$y = \pm \frac{x^{3/2}}{\sqrt{1 - x}}\, dx$$

and its asymptote $x = 1$.

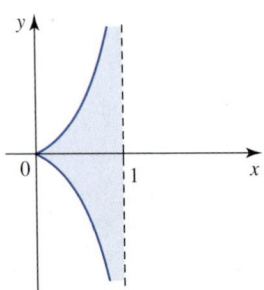

Hint: The area is

$$2 \int_0^1 \frac{x^{3/2}}{\sqrt{1 - x}}\, dx$$

Use the substitution $u = \sqrt{x}$ followed by the substitution $u = \sin \theta$.

58. Find the length of the astroid $x^{2/3} + y^{2/3} = a^{2/3}$, where $a > 0$.

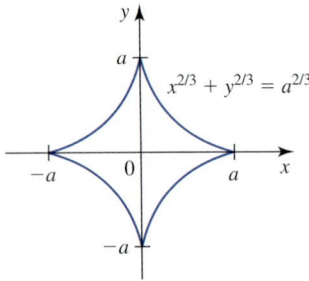

59. Work Done by a Repulsive Force An electric charge Q located at the origin of a coordinate line repulses a like charge q from the point $x = a$, where $a > 0$, an infinite distance to the right. Find the work done by the force of repulsion.
Hint: The magnitude of force acting on the charge q when it is at the point x is given by

$$F(x) = \frac{1}{4\pi\varepsilon_0} \frac{qQ}{x^2}$$

60. Elastic Deformation of a Long Beam The graph C of the function

$$y = \frac{P\alpha}{2k} e^{-\alpha|x|}(\cos \alpha x + \sin \alpha|x|)$$

where α and k are constants, gives the shape of a beam of infinite length lying on an elastic foundation and acted upon by a concentrated load P applied to the beam at the origin. Before application of the force, the beam lies on the x-axis. Find the potential energy of elastic deformation W using the formula

$$W = Ee \int_0^\infty (y'')^2\, dx$$

where E and e are constants.

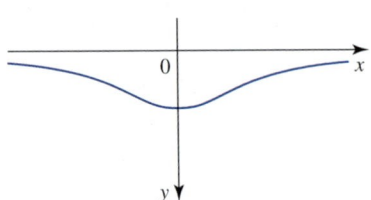

Note: This model provides a good approximation in working with long beams.

61. Work Done by a Repulsive Charge An electric charge Q distributed uniformly along a line of length $2c$ lying along the y-axis repulses a like charge q from the point $x = a$, where $a > 0$, an infinite distance to the right. The magnitude of the force acting on the charge q when it is at the point x is given by

$$F(x) = \frac{1}{4\pi\varepsilon_0} \frac{qQ}{x\sqrt{x^2 + c^2}}$$

and the force acts in the direction of the positive x-axis. Find the work done by the force of repulsion.

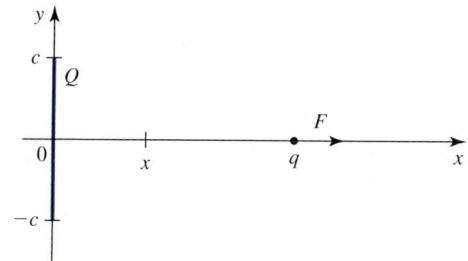

62. Escape Velocity of a Rocket The *escape velocity* v_0 is the minimum speed a rocket must attain in order to escape from the gravitational field of a planet. Use Newton's Law of Gravitation to find the escape velocity for the earth (see Exercise 32 in Section 5.5).

Hint: The work required to launch a rocket from the surface of the earth upward to escape from the earth's gravitational field is

$$W = \int_R^\infty \frac{mgR^2}{r^2} \, dr$$

Equate W with the initial kinetic energy $\frac{1}{2} m v_0^2$ of the rocket.

63. Capital Value of Property The *capital value* (present sale value) CV of a property that can be rented on a perpetual basis of R dollars annually is given by

$$CV \approx \int_0^\infty R e^{-it} \, dt$$

where i is the prevailing interest rate per year compounded continuously.

a. Show that $CV \approx R/i$.
b. Find the capital value of a property that can be rented out for \$10,000 annually when the prevailing interest rate is 10% per year.

64. Average Power in AC Circuits If f is defined on $[0, \infty)$, then the average value of f over $[0, \infty)$ is defined to be

$$f_{av} = \lim_{b \to \infty} \frac{1}{b} \int_0^b f(x) \, dx$$

Suppose that the voltage and current in an AC circuit are

$$V = V_0 \cos \omega t \qquad \text{and} \qquad I = I_0 \cos(\omega t + \phi)$$

so that the voltage and current differ by an angle ϕ. Then the power output is $P = VI$. Show that the average power output is $P_{av} = \frac{1}{2} I_0 V_0 \cos \phi$.

Note: The factor $\cos \phi$ is called the *power factor*. When V and I are in phase ($\phi = 0°$), the average power output is $\frac{1}{2} IV$, but when V and I are out of phase ($\phi = 90°$), then the average power output is zero.

cas 65. Serum Cholesterol Population Study The percentage of a current Mediterranean population with serum cholesterol levels at or above 200 mg/dL is estimated to be

$$P = \frac{1}{20\sqrt{2\pi}} \int_{200}^\infty e^{(-1/2)[(x - 160)/20]^2} \, dx$$

Use a CAS to find P.

66. Find the arc length of the loop defined by $3y^2 = x(x - 1)^2$ from $x = 0$ to $x = 1$.

67. Find the value of the constant C for which

$$\int_1^\infty \left(\frac{1}{\sqrt{x}} - \frac{C}{\sqrt{x + 1}} \right) dx$$

converges. Then evaluate the integral for this value of C.

68. Let $I = \int_0^\infty \frac{x^2}{x^4 + 1} \, dx$.

a. Use the substitution $u = 1/x$ to show that

$$I = \frac{1}{2} \int_0^\infty \frac{x^2 + 1}{x^4 + 1} \, dx = \frac{1}{2} \int_0^\infty \frac{1 + \frac{1}{x^2}}{x^2 + \frac{1}{x^2}} \, dx$$

b. Use the substitution $v = x - \frac{1}{x}$ to show that

$$I = \frac{1}{2} \int_{-\infty}^\infty \frac{dv}{v^2 + 2}.$$

c. Use the result of part (b) to show that

$$\int_0^\infty \frac{x^2}{x^4 + 1} \, dx = \frac{\sqrt{2} \, \pi}{4}$$

69. Find the values of p for which the integral $\int_0^1 1/x^p \, dx$ converges and the values of p for which it diverges.

70. Consider the integral $I = \int_3^\infty \frac{dx}{\sqrt{x(x - 1)(x - 2)}}$.

a. Plot the graphs of $f(x) = \frac{1}{\sqrt{x(x - 1)(x - 2)}}$ and $g(x) = \frac{3\sqrt{2}}{2x^{3/2}}$ using the viewing window $[0, 6] \times [0, 1.8]$ to see that $f(x) \le g(x)$ for all x in $(3, \infty)$.

b. Prove the assertion in part (a).
c. Prove that I converges.

71. Prove that $\int_0^1 \frac{\sin \frac{1}{\sqrt{x}}}{\sqrt{x}} \, dx$ converges.

72. Observe that $\int_0^\infty e^{-x^2} \, dx = \int_0^4 e^{-x^2} \, dx + \int_4^\infty e^{-x^2} \, dx$.

a. Show that $\int_4^\infty e^{-x^2} \, dx \le 10^{-7}$ so that $\int_0^\infty e^{-x^2} \, dx \approx \int_0^4 e^{-x^2} \, dx$.

b. Use a calculator or computer to obtain an estimate for $\int_0^\infty e^{-x^2} \, dx$.

In Exercises 73 and 74, (a) *find a "test integral" to be used in determining the convergence or divergence of the improper integral,* (b) *verify the result of part (a) by plotting the graphs of both integrands in the same viewing window, and* (c) *determine the convergence or divergence of the integral.*

73. $\displaystyle\int_1^\infty \frac{\sqrt{t^3 - t^2 + 1}}{t^5 + t + 2}\, dt$ **74.** $\displaystyle\int_1^\infty \frac{1 - 4\sin 2x}{x^3 + x^{1/3}}\, dx$

*Let $f(t)$ be continuous for $t > 0$. The **Laplace transform** of f is the function F defined by*

$$F(s) = \int_0^\infty f(t)e^{-st}\, dt$$

provided that the integral exists. In Exercises 75–79, use this definition.

75. Find the Laplace transform of $f(t) = 1$.

76. Find the Laplace transform of $f(t) = e^{at}$, where a is a constant.

77. Find the Laplace transform of $f(t) = t$.

78. Show that the Laplace transform of $f(t) = \cos \omega t$ is
$$F(s) = \frac{s}{s^2 + \omega^2}.$$

79. Suppose that f' is continuous for $t > 0$ and f satisfies the condition $\lim_{t\to\infty} e^{-st}f(t) = 0$. Show that the Laplace transform of $f'(t)$ for $t > 0$, denoted by G, satisfies $G(s) = sF(s) - f(0)$, where $s > 0$ and F is the Laplace transform of f.

In Exercises 80–87, *determine whether the statement is true or false. If it is true, explain why it is true. If it is false, explain why or give an example to show why it is false.*

80. If f is continuous on $(-\infty, \infty)$, then $\int_{-\infty}^\infty f(x)\, dx = \lim_{t\to\infty} \int_{-t}^t f(x)\, dx$.

81. If f is continuous on $[0, \infty)$ and $\lim_{x\to\infty} f(x) = 0$, then $\int_0^\infty f(x)\, dx$ is convergent.

82. If $\int_a^\infty [f(x) + g(x)]\, dx$ is convergent, then $\int_a^\infty f(x)\, dx$ and $\int_a^\infty g(x)\, dx$ must both be convergent.

83. If $\int_a^\infty f(x)\, dx$ and $\int_a^\infty g(x)\, dx$ are both convergent, then $\int_a^\infty [f(x) + g(x)]\, dx$ is convergent.

84. If both $\int_a^\infty f(x)\, dx$ and $\int_a^\infty g(x)\, dx$ are divergent, then $\int_a^\infty [f(x) + g(x)]\, dx$ must also be divergent.

85. If $f(x) \le g(x)$ for all x in $[a, \infty)$ and $\int_a^\infty g(x)\, dx$ diverges, then $\int_a^\infty f(x)\, dx$ may converge.

86. If $f(x) \le g(x)$ for all x in $[a, \infty)$ and $\int_a^\infty f(x)\, dx$ converges, then $\int_a^\infty g(x)\, dx$ also converges.

87. Suppose that f is continuous on $[a, b)$ and f has an infinite discontinuity at b. Furthermore, suppose that $\int_c^b f(x)\, dx$ is convergent, where c is a number between a and b. Then $\int_a^b f(x)\, dx$ is convergent.

CHAPTER 6   REVIEW

CONCEPT REVIEW

In Exercises 1–8, *fill in the blanks.*

1. The integration by parts formula is obtained by reversing the _____ Rule. The formula for indefinite integrals is $\int u\, dv = $ _____. In choosing u and dv, we want du to be simpler than _____ and dv to be _____ _____. The formula for definite integrals is $\int_a^b f(x)g'(x)\, dx = $ _____.

2. To integrate $\sin^m x \cos^n x$, where m and n are positive integers, we use the substitution (a) $u = $ _____ if m is _____ and (b) $u = $ _____ if n is _____. If m and n are both even and nonnegative, we use the half-angle formulas $\sin^2 x = $ _____ and $\cos^2 x = $ _____.

3. To integrate $\tan^m x \sec^n x$, where m and n are positive integers, we use the substitution (a) $u = $ _____ if m is _____ and (b) $u = $ _____ if n is _____.

4. To integrate $\sin mx \sin nx$, we use the identity $\sin mx \sin nx = $ _____; to integrate $\sin mx \cos nx$, we use the identity $\sin mx \cos nx = $ _____; to integrate $\cos mx \cos nx$, we use the identity $\cos mx \cos nx = $ _____.

5. a. If an integral involves $\sqrt{a^2 - x^2}$, we use the substitution $x = $ _____.
b. If an integral involves $\sqrt{a^2 + x^2}$, we use the substitution $x = $ _____.
c. If an integral involves $\sqrt{x^2 - a^2}$, we use the substitution $x = $ _____.

6. The method of partial fractions is used to integrate _____ functions. As a first step, the integrand $f(x) = P(x)/Q(x)$ should be written as $f(x) = S(x) + R(x)/Q(x)$ where the degree of R is _____ than the degree of Q. $R(x)/Q(x)$ is decomposed into a sum of partial fractions involving _____ and irreducible _____ factors. As an example, the form of the decomposition for $\dfrac{2x^4 + 3x^2 + 8x + 5}{(x-1)^3(x^2 + x + 1)^2}$ is _____. The integral of f is then found by _____ this last expression.

7. The improper integrals are defined by $\int_{-\infty}^{b} f(x)\,dx =$ _____; $\int_{a}^{\infty} f(x)\,dx =$ _____; $\int_{-\infty}^{\infty} f(x)\,dx =$ _____. If $\lim_{x \to b^-} f(x) = \pm\infty$, where $a < b$, then the improper integral $\int_{a}^{b} f(x)\,dx =$ _____. If f is continuous on $[a, b]$ except that f has an infinite discontinuity at c, where $a < c < b$, then $\int_{a}^{b} f(x)\,dx =$ _____.

8. If f and g are continuous and $f(x) \geq g(x) \geq 0$ for all $x \geq a$, then if $\int_{a}^{\infty} f(x)\,dx$ converges, _____ converges and if $\int_{a}^{\infty} g(x)\,dx$ diverges, _____ diverges.

REVIEW EXERCISES

In Exercises 1–42, evaluate or find the integral.

1. $\displaystyle\int \frac{2x}{x+1}\,dx$

2. $\displaystyle\int x^2 \cos 3x\,dx$

3. $\displaystyle\int \frac{x^3}{\sqrt{9-x^2}}\,dx$

4. $\displaystyle\int \frac{\cos^3 x}{\sin x}\,dx$

5. $\displaystyle\int x^2 \ln x\,dx$

6. $\displaystyle\int \frac{2x-1}{x(x^2-4)}\,dx$

7. $\displaystyle\int \frac{1}{1-\cos\theta}\,d\theta$

8. $\displaystyle\int \frac{1}{1-\sin x}\,dx$

9. $\displaystyle\int \frac{x+1}{x^4+6x^3+9x^2}\,dx$

10. $\displaystyle\int \sec^3 x \tan^5 x\,dx$

11. $\displaystyle\int \sqrt{x^2-4}\,dx$

12. $\displaystyle\int_0^1 \cos^4 \pi x\,dx$

13. $\displaystyle\int \theta \sin^{-1}\theta\,d\theta$

14. $\displaystyle\int \sqrt{4+x^2}\,dx$

15. $\displaystyle\int_1^{e^\pi} \cos(\ln x)\,dx$

16. $\displaystyle\int \frac{x^2+4x}{x^3-x^2+x-1}\,dx$

17. $\displaystyle\int \frac{x+2}{(x^2+x)(x^2+1)}\,dx$

18. $\displaystyle\int_1^\infty \sin(\ln x)\,dx$

19. $\displaystyle\int \sec^4 2x \tan^6 2x\,dx$

20. $\displaystyle\int_0^2 \sqrt{4x-x^2}\,dx$

21. $\displaystyle\int \frac{\cos x}{1+\cos x}\,dx$

22. $\displaystyle\int e^x \cos 2x\,dx$

23. $\displaystyle\int \frac{(\ln x)^3}{x}\,dx$

24. $\displaystyle\int \frac{1}{3\sin\theta + 4\cos\theta}\,d\theta$

25. $\displaystyle\int \frac{1}{x\sqrt{4x-1}}\,dx$

26. $\displaystyle\int \tan^3 x \sec^3 x\,dx$

27. $\displaystyle\int \sec^2 x \ln(\tan x)\,dx$

28. $\displaystyle\int \frac{1}{x^2+4x+20}\,dx$

29. $\displaystyle\int \sin x \cos 3x\,dx$

30. $\displaystyle\int \cosh^{-1} x\,dx$

31. $\displaystyle\int \frac{1}{\sqrt{1-(2x+3)^2}}\,dx$

32. $\displaystyle\int \frac{\sqrt{x^2+4}}{x^2}\,dx$

33. $\displaystyle\int \sin^2 t \cos^4 t\,dt$

34. $\displaystyle\int x \cos^2 x\,dx$

35. $\displaystyle\int \frac{\sqrt{x}}{\sqrt{x}-1}\,dx$

36. $\displaystyle\int (x+1)e^{2x}\,dx$

37. $\displaystyle\int x \cos^{-1} 2x\,dx$

38. $\displaystyle\int_1^4 \frac{e^{1/x}}{x^2}\,dx$

39. $\displaystyle\int e^{-x} \cosh x\,dx$

40. $\displaystyle\int \csc^4 2x\,dx$

41. $\displaystyle\int \frac{1}{\sqrt{4x^2+4x+10}}\,dx$

42. $\displaystyle\int_0^1 \frac{(\sin^{-1} x)^{3/2}}{\sqrt{1-x^2}}\,dx$

In Exercises 43–50, evaluate the integral or show that it is divergent.

43. $\displaystyle\int_{-\infty}^0 e^x\,dx$

44. $\displaystyle\int_0^\infty \frac{1}{(x+1)^{3/2}}\,dx$

45. $\displaystyle\int_{-\infty}^\infty \frac{x}{1+x^2}\,dx$

46. $\displaystyle\int_0^3 \frac{1}{\sqrt{3-x}}\,dx$

47. $\displaystyle\int_{-8}^1 \frac{1}{\sqrt[3]{x}}\,dx$

48. $\displaystyle\int_e^\infty \frac{1}{x\ln^4 x}\,dx$

49. $\displaystyle\int_1^e \frac{1}{x(\ln x)^{1/3}}\,dx$

50. $\displaystyle\int_0^2 \frac{x}{\sqrt{4-x^2}}\,dx$

In Exercises 51–56, use the Table of Integrals to find the integral.

51. $\displaystyle\int x^2\sqrt{3+x^2}\,dx$

52. $\displaystyle\int e^{2x}\sqrt{5+2e^x}\,dx$

53. $\displaystyle\int \frac{dx}{(x+1)\ln(1+x)}$

54. $\displaystyle\int (\ln x)^3\,dx$

55. $\displaystyle\int \sec^4 x\,dx$

56. $\displaystyle\int \frac{\tan x}{\sqrt{1+2\cos x}}\,dx$

57. Find $\int e^x f(x)\, dx + \int f'(x)\, e^x\, dx$, where f' is continuous.

58. Find the area of the region under the graph of

$$y = \frac{\sqrt{4 + x^2}}{x^2}$$

on the interval $[1, 2]$.

59. Find the area of the region bounded by the graphs of $y = 1/x^{2/3}$, $y = 0$, $x = -1$, and $x = 1$.

60. Find the area of the region bounded by the graphs of $y = \sin^2 x$, $y = \sin^3 x$, $x = 0$, and $x = \pi$.

61. Find the area of the region enclosed by the ellipse $9x^2 + 4y^2 = 36$.

62. Let $I = \displaystyle\int_{-4}^{4} \frac{x^2}{\sqrt{16 - x^2}}\, dx$.

   a. Plot the graph of $f(x) = \dfrac{x^2}{\sqrt{16 - x^2}}$ using the viewing window $[-5, 5] \times [0, 20]$.

    b. Evaluate I using the Table of Integrals.

63. Consider the integral $I = \displaystyle\int_{1}^{\infty} \left(1 - \cos\frac{2}{x}\right) dx$.

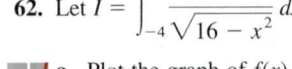   **a.** Plot the graphs of $f(x) = 1 - \cos(2/x)$ and $g(x) = (2/x^2)$ using the viewing window $[0, 5] \times [-1, 3]$ to see that $f(x) \le g(x)$ for all x in $(0, \infty)$.

    b. Prove the assertion in part (a).

    c. Prove that I converges.

64. The region under the graph of $y = \tan x$ on the interval $\left[0, \frac{\pi}{4}\right]$ is revolved about the x-axis. Find the volume of the resulting solid.

65. The region under the graph of $y = x \ln x$ on the interval $[1, e]$ is revolved about the y-axis. Find the volume of the resulting solid.

66. The region under the graph of $y = \tan^{-1} x$ on the interval $[0, 1]$ is revolved about the y-axis. Find the volume of the resulting solid.

67. Find the length of the graph of $y = \frac{1}{2}x^2$ from $(0, 0)$ to $\left(\sqrt{3}, \frac{3}{2}\right)$.

68. Use the Comparison Test to determine whether the integral

$$\int_{1}^{\infty} \frac{1 + 2 \cos x}{x^3 + \sqrt{x}}\, dx$$

is convergent.

69. **Velocity of a Dragster** The velocity of a dragster t sec after leaving the starting line is $v(t) = 80te^{-0.2t}$ ft/sec. What is the distance traveled by the dragster during the first 10 sec?

70. **Drug Concentration in the Bloodstream** The concentration of a certain drug (in mg/mL) in the bloodstream of a patient t hours after it has been administered is given by

$$C(t) = 2te^{-t/3}$$

What is the average concentration of the drug in the patient's bloodstream over the first 12 hr after administration of the drug?

PROBLEM-SOLVING TECHNIQUES

The following example shows that by making a suitable substitution, we can sometimes evaluate a definite integral whose indefinite integral cannot be expressed in terms of elementary functions, that is, as a sum, difference, product, quotient, or composition of the functions we have studied thus far.

EXAMPLE 1 Evaluate $I = \displaystyle\int_{0}^{\pi} \frac{x \sin x}{1 + \cos^2 x}\, dx$.

Solution Let $u = \pi - x$ or $x = \pi - u$. Then $du = -dx$. Furthermore, if $x = 0$, then $u = \pi$, and if $x = \pi$, then $u = 0$. Making these substitutions, we have

$$I = \int_{0}^{\pi} \frac{x \sin x}{1 + \cos^2 x}\, dx = -\int_{\pi}^{0} \frac{(\pi - u)\sin(\pi - u)}{1 + \cos^2(\pi - u)}\, du$$

$$= \int_{0}^{\pi} \frac{(\pi - u)\sin(\pi - u)}{1 + \cos^2(\pi - u)}\, du$$

Next, we observe that

$$\sin(\pi - u) = \sin \pi \cos u - \cos \pi \sin u = \sin u$$

$$\cos(\pi - u) = \cos \pi \cos u + \sin \pi \sin u = -\cos u$$

and this leads to

$$I = \int_0^\pi \frac{(\pi - u)\sin u}{1 + \cos^2 u} \, du = \pi \int_0^\pi \frac{\sin u}{1 + \cos^2 u} \, du - \int_0^\pi \frac{u \sin u}{1 + \cos^2 u} \, du$$

But the second integral on the right-hand side is the same as I. Therefore, we have

$$2I = \pi \int_0^\pi \frac{\sin u}{1 + \cos^2 u} \, du$$

or

$$I = \frac{\pi}{2} \int_0^\pi \frac{\sin u}{1 + \cos^2 u} \, du$$

In this form, I is easily evaluated. In fact, letting $t = \cos u$ so that $dt = -\sin u \, du$ and observing that if $u = 0$, then $t = 1$, and if $u = \pi$, then $t = -1$, we have

$$I = -\frac{\pi}{2} \int_1^{-1} \frac{dt}{1 + t^2} = -\frac{\pi}{2} \tan^{-1} t \Big|_1^{-1} = -\frac{\pi}{2} [\tan^{-1}(-1) - \tan^{-1}(1)]$$

$$= -\frac{\pi}{2} \left(-\frac{\pi}{4} - \frac{\pi}{4} \right) = \frac{\pi^2}{4}$$

If you look at the result of finding the integral of the form $\int P(x)e^{ax} \, dx$, where P is a polynomial function and a is a constant, you will see that $\int P(x)e^{ax} \, dx = Q(x)e^{ax} + C$, where Q is a polynomial having the same degree as that of P. A similar observation reveals that using the integration by parts formula gives

$$\int P(x)\sin ax \, dx = P_1(x)\sin ax + Q_1(x)\cos ax + C$$

and

$$\int P(x)\cos ax \, dx = P_2(x)\cos ax + Q_2(x)\sin ax + C$$

where P_1, P_2, Q_1, and Q_2 are polynomial functions having the same degree as that of P.

These observations reduce the problem of finding integrals of the aforementioned form to one of solving an algebraic problem: that of solving a system of equations for the "undetermined" coefficients of a polynomial or polynomials.

EXAMPLE 2 Find $I = \int (2x^3 - 3x^2 + 8)e^{2x} \, dx$.

Solution $\int (2x^3 - 3x^2 + 8)e^{2x} \, dx = (Ax^3 + Bx^2 + Dx + E)e^{2x} + C$. Differentiating both sides of the equation with respect to x yields

$$(2x^3 - 3x^2 + 8)e^{2x} = (3Ax^2 + 2Bx + D)e^{2x} + 2(Ax^3 + Bx^2 + Dx + E)e^{2x}$$

$$2x^3 - 3x^2 + 8 = 2Ax^3 + (3A + 2B)x^2 + (2B + 2D)x + (D + 2E)$$

Since this equation holds for all values of x, the coefficients of like terms must be equal. This observation leads to the system

$$
\begin{aligned}
2A &= 2 \\
3A + 2B &= -3 \\
2B + 2D &= 0 \\
D + 2E &= 8
\end{aligned}
$$

Solving this system, we find $A = 1$, $B = -3$, $D = 3$, and $E = \frac{5}{2}$. So

$$\int (2x^3 - 3x^2 + 8)\, e^{2x}\, dx = \left(x^3 - 3x^2 + 3x + \frac{5}{2}\right)e^{2x} + C$$

CHALLENGE PROBLEMS

1. Evaluate $\lim\limits_{n \to \infty} \dfrac{\sqrt[n]{n!}}{n}$.

Hint: Take logarithms and interpret the sum as an integral.

2. Find $\int \ln(\sqrt{1 - x} + \sqrt{1 + x})\, dx$.

3. a. Show that

$$f(x) = \begin{cases} x \ln\left(1 + \dfrac{1}{x}\right) & \text{if } 0 < x \le 1 \\ 0 & \text{if } x = 0 \end{cases}$$

is continuous on $[0, 1]$.

  b. Evaluate $\int_0^1 f(x)\, dx$.

4. Let $I_n = \int (a^2 - x^2)^n\, dx$, where $n > 0$.

  a. Show that $I_n = \dfrac{x(a^2 - x^2)^n}{2n + 1} + \dfrac{2na^2}{2n + 1} I_{n-1}$.

  b. Use the result of part (a) to show that

$$\int \sqrt{a^2 - x^2}\, dx = \frac{x}{2}\sqrt{a^2 - x^2} + \frac{a^2}{2} \arcsin\frac{x}{a} + C$$

5. Find

$$\lim_{n \to \infty}\left(\frac{1}{\sqrt{4n^2 - 1}} + \frac{1}{\sqrt{4n^2 - 2^2}} + \cdots + \frac{1}{\sqrt{4n^2 - n^2}}\right)$$

Hint: Interpret the sum as an integral.

6. Find

$$\int \frac{dx}{(1 + \sqrt{x})\sqrt{x - x^2}}$$

Hint: Use the substitution $u = \sin^{-1}\sqrt{x}$.

7. Prove that $\left|\displaystyle\int_0^a \frac{\cos bx}{1 + x^2}\, dx\right| < \dfrac{\pi}{2}$.

8. Find the area of the region lying between the cissoid

$$y^2 = \frac{x^3}{2a - x} \quad \text{and its asymptote.}$$

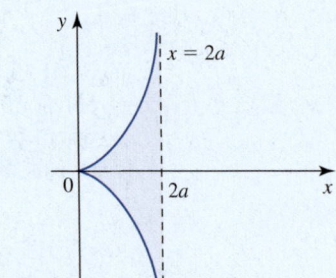

9. Prove that

$$\int_0^a (a^2 - x^2)^n\, dx = \frac{2^{2n} a^{2n+1}(n!)^2}{(2n + 1)!}$$

where n is a positive integer.

Hint: Denote the integral by I_n, and show that

$$I_n = a^2 \frac{2n}{2n + 1} I_{n-1}$$

10. Find the area enclosed by the ellipse with equation $2x^2 + \sqrt{3}xy + y^2 = 20$ as shown in the figure.

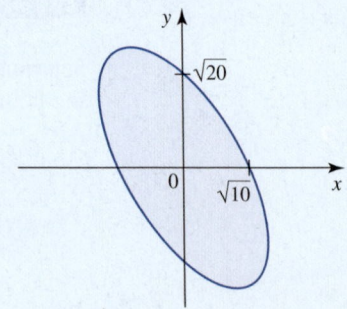

11. Show that

$$\int_0^1 x^m (\ln x)^n \, dx = \frac{(-1)^n n!}{(m+1)^{n+1}}$$

where n is a positive integer and $m > -1$.

12. Show that $\displaystyle\int_0^\infty \frac{\ln x}{1 + x^2} \, dx = 0$.

Hint: Write

$$\int_0^\infty \frac{\ln x}{1 + x^2} \, dx = \int_0^1 \frac{\ln x}{1 + x^2} \, dx + \int_1^\infty \frac{\ln x}{1 + x^2} \, dx$$

and use the substitution $u = 1/x$ on the second integral on the right.

13. Suppose that f is continuous on $(-\infty, \infty)$ and $\int_{-\infty}^\infty |f(x)| \, dx$ exists.
  a. Show that $\int_{-\infty}^\infty f(x) \, dx$ exists.
  b. If g is continuous and bounded on $(-\infty, \infty)$, that is, there exists a positive number M such that $|g(x)| \le M$ for all x in $(-\infty, \infty)$, show that $\int_{-\infty}^\infty f(x) \, g(x) \, dx$ exists.
  c. Use the result of part (b) to show that $\displaystyle\int_{-\infty}^\infty \frac{\sin x}{x^2 + 1} \, dx = 0$.

14. Find the area of the surface obtained by revolving the ellipse $4x^2 + y^2 = 4$ about the y-axis.

15. Consider the Dirichlet integral $\displaystyle\int_0^\infty \frac{\sin x}{x} \, dx$, which may be written as the sum of the integrals

$$I_1 = \int_0^{\pi/2} \frac{\sin x}{x} \, dx \qquad \text{and} \qquad I_2 = \int_{\pi/2}^\infty \frac{\sin x}{x} \, dx$$

  a. Show that I_1 exists and is finite.
  b. Show that I_2 converges.
  c. Conclude that the Dirichlet integral converges.

16. The integrals $\int_0^\infty \sin x^2 \, dx$ and $\int_0^\infty \cos x^2 \, dx$ are called Fresnel integrals and are used in the study of light diffraction. Show that a Fresnel integral is convergent.
Hint: Use the substitution $u = x^2$ and write the resulting integral as the sum of two integrals as in Exercise 15.
Note: A Fresnel integral shows that an improper integral can converge even though the integrand does not approach zero as x approaches infinity.

17. a. Use the result of Exercise 16 to show that the integral $\int_0^\infty 2x \cos x^4 \, dx$ converges.
    Hint: Use the substitution $x^2 = u$.
  b. Show that the integrand $f(x) = 2x \cos x^4$ is unbounded.
    Note: This integral shows that an improper integral can converge even though the integrand is unbounded.

18. The Path of a Water Skier A water skier is pulled along by means of a 40-ft tow rope attached to a boat. Initially, the boat is located at the origin and the skier is located at the point $(40, 0)$. As the boat moves along the y-axis, the tow rope is kept taut at all times. The path followed by the skier is a curve called a *tractrix* and has the property that the rope is tangent to the curve.

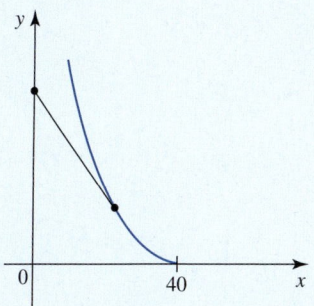

  a. Show that the path followed by the skier is the graph of $y = f(x)$, where y satisfies the equation

$$\frac{dy}{dx} = -\frac{\sqrt{1600 - x^2}}{x}$$

  b. Solve the equation in part (a) to show that the path followed by the skier is

$$y = -\sqrt{1600 - x^2} + 40 \ln\left[\frac{40 + \sqrt{1600 - x^2}}{x}\right]$$

19. Refer to Exercise 18. Find the distance covered by the water skier after the boat has traveled 100 ft from its starting point, which is located at the origin.

Once a skydiver jumps out of a plane, the force of gravity acts on the skydiver, accelerating her fall to earth. But air resistance builds up quickly as she falls and soon matches the force due to gravity. The result is that her rate of fall approaches a constant (maximum) rate, called the *terminal velocity*. We will see how the motion of the skydiver is described by the solution of a differential equation.

Steve Fitchett/Getty Images

7

Differential Equations

A DIFFERENTIAL EQUATION is one that involves the derivative, or differential, of one or more unknown functions. In this chapter we give a brief introduction to the all-important field of differential equations by looking at *first-order* differential equations and their applications. The applications of differential equations are many and varied and appear in virtually every field of study. Examples are the study of motion, population growth, radioactive decay, calculations involving compound interest, electrostatic and electromagnetic fields, carbon dating, chemical reactions, concentration of a drug in the bloodstream, and the spread of a disease. We also take a brief look at an important application involving a system of differential equations: the predator-prey problem, which deals with how one population (predator) affects another (prey).

V This symbol indicates that one of the following video types is available for enhanced student learning at **www.academic.cengage.com/login:**
- Chapter lecture videos
- Solutions to selected exercises

7.1 | Differential Equations: Separable Equations

We first encountered differential equations in Section 4.1, and you might want to review the material there before proceeding.

■ First-order Differential Equations and Solutions

Recall that a **differential equation** is an equation that involves the derivative or differential of an unknown function. The **order** of a differential equation is the order of the highest derivative that occurs in the equation. Thus, a **first-order differential equation** is one that involves only derivatives of order one. If we solve the equation for the derivative, it can be written in the form

$$\frac{dy}{dx} = f(x, y) \tag{1}$$

In Section 4.1 we considered first-order differential equations of the form

$$\frac{dy}{dx} = f(x)$$

where f is a function of x alone and, therefore, can be solved by integration. In fact, as we have seen, the general solution of this equation is

$$y = \int f(x)\, dx + C$$

where C is an arbitrary constant.

In the general case in which f involves both x and y, the solution is not so easily obtained, and more sophisticated methods of solution are needed. Before going further, let's recall that a **solution** of the differential equation (1) is a differentiable function $y = y(x)$ defined on an open interval I such that y satisfies the equation on I. Thus, for Equation (1) a **solution** is a function $y(x)$ that satisfies

$$\frac{d}{dx}[y(x)] = f(x, y(x))$$

for all x in the interval.

EXAMPLE 1 Show that the function $y = x + 1 + Ce^x$, where C is an arbitrary constant, is a solution of the differential equation $y' = y - x$.

Solution The function $y = x + 1 + Ce^x$ is defined and differentiable on the interval $(-\infty, \infty)$. To show that the differential equation is satisfied, we compute

$$y' = \frac{d}{dx}(x + 1 + Ce^x) = 1 + Ce^x$$

Next, substituting the expression for y into the right-hand side of the differential equation gives

$$y - x = x + 1 + Ce^x - x = 1 + Ce^x$$

which is the same as the left-hand side of the differential equation for all values of x in $(-\infty, \infty)$. Therefore, $y = x + 1 + Ce^x$ is a solution of the given differential equation on $(-\infty, \infty)$. ■

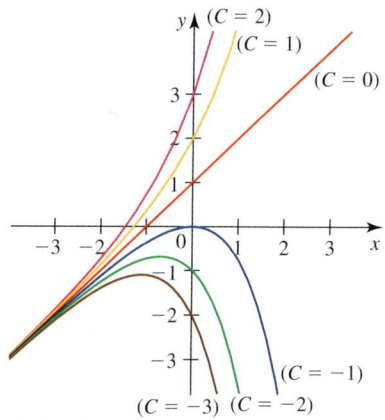

FIGURE 1
Some solution curves of $y' = y - x$

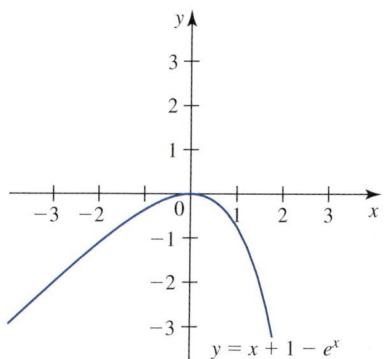

FIGURE 2
The particular solution of $y' = y - x$
satisfying $y(0) = 0$

In general, a first-order differential equation will have a solution involving one arbitrary constant. Such a solution is called the *general solution* of the differential equation. For example, the solution $y = x + 1 + Ce^x$ is called the general solution of the equation $y' = y - x$. Graphically, the general solution of the differential equation represents a family of curves called the **solution** or **integral curves** of the differential equation. Figure 1 shows six solution curves of the differential equation for selected values of the parameter C.

We can obtain a *particular solution* of a differential equation by choosing a particular value of the arbitrary constant. This is usually done by requiring that the differential equation satisfy a side condition $y(x_0) = y_0$. Geometrically, the solution of the **initial-value problem**

$$\begin{cases} \dfrac{dy}{dx} = f(x, y) \\ y(x_0) = y_0 \end{cases}$$

is the solution curve of the differential equation that passes through the point (x_0, y_0).

EXAMPLE 2 Solve the initial-value problem

$$\begin{cases} y' = y - x \\ y(0) = 0 \end{cases}$$

Solution In Example 1 we saw that $y = x + 1 + Ce^x$ is a one-parameter solution of the equation $y' = y - x$. To determine C, we use the initial condition $y(0) = 0$, or $y = 0$, when $x = 0$. We obtain

$$0 = 0 + 1 + Ce^0 \qquad \text{or} \qquad C = -1$$

Therefore, the required solution is $y = x + 1 - e^x$. The graph of this function is shown in Figure 2. If you compare this with the family of solution curves of this differential equation shown in Figure 1, you will see that the particular solution obtained here is the solution curve that passes through the origin $(0, 0)$.

The Laws of Natural Growth and Decay

When free of constraints, certain quantities in nature grow or decay at a rate that is proportional to their current size. Examples of such phenomena are

- the growth of a population of bacteria under ideal conditions,
- the decay of a radioactive substance, and
- the discharge of an electrical condenser.

Even quantities that occur outside the realm of nature sometimes exhibit this type of growth or decay. For example, the accumulated amount of money on deposit with a bank, earning interest at a fixed rate compounded continuously, grows in this manner (see Example 9).

We can describe these phenomena mathematically. Suppose that the size or magnitude of a quantity y at any time t is given by $y(t)$.* Since the rate of change of y with respect to t, dy/dt, is proportional to its size y at any time t, we have

$$\frac{dy}{dt} = ky \qquad (2)$$

---

*We use the letter y to denote the function in question.

where k is a constant. If $k > 0$, this equation is called the **law of natural growth;** and if $k < 0$, it is called the **law of natural decay.** Also, since the equation involves the derivative of an unknown function, it is a differential equation. If the amount present initially is y_0, then we have the initial-value problem

$$\begin{cases} \dfrac{dy}{dt} = ky \\ y(0) = y_0 \end{cases} \tag{3}$$

The differential equation $dy/dt = ky$ is an example of a **first-order separable differential equation.** It is separable because dy/dt can be written as a function of t times a function of y. In general, a first-order separable differential equation in x and y is one that can be written in the form

$$\frac{dy}{dx} = g(x)h(y) \tag{4}$$

where g is a function of x alone and h is a function of y alone. Equivalently, a separable equation is one that can be written in differential form as

$$G(x)\,dx + H(y)\,dy = 0 \tag{5}$$

where G is a function of x alone and H is a function of y alone. For example, the differential equation in system (3) is easily seen to be separable if we put $g(t) = k$ and $h(y) = y$.

As another example, the equation

$$(x^2 + 1)\,dx + \frac{1}{y}\,dy = 0$$

has the form of the differential equation in (5) with $G(x) = x^2 + 1$ and $H(y) = 1/y$, so it is separable. On the other hand, the differential equation

$$\frac{dy}{dx} = xy^2 + 2$$

is *not* separable, nor is the equation

$$(x + y)\,dx + xy\,dy = 1$$

We will return to the solution of the initial-value problem (3) later on.

■ The Method of Separation of Variables

First-order separable differentiable equations can be solved by using the *method of separation of variables.* If $h(y) \neq 0$, we write differential equation (4) in the form

$$\frac{1}{h(y)}\frac{dy}{dx} = g(x)$$

When it is written in this form, the variables x and y are said to be separated. Integrating both sides of the equation with respect to x then gives

$$\int \frac{1}{h(y)}\frac{dy}{dx}\,dx = \int g(x)\,dx$$

or

$$\int \frac{1}{h(y)}\,dy = \int g(x)\,dx \tag{6}$$

Carrying out the integration on each side of Equation (6) with respect to the appropriate variable gives the solution to the differential equation expressed implicitly by an equation in x and y. In some cases we may be able to solve for y explicitly in terms of x.

To justify the method of separation of variables, let's consider the separable Equation (4) in the general form:

$$\frac{dy}{dx} = g(x)h(y)$$

If $h(y) \neq 0$, we may rewrite the equation in the form

$$\frac{1}{h(y)}\frac{dy}{dx} - g(x) = 0$$

Now, suppose that H is an antiderivative of $1/h$ and G is an antiderivative of g. Using the chain rule, we see that

$$\frac{d}{dx}[H(y) - G(x)] = H'(y)\frac{dy}{dx} - G'(x) = \frac{1}{h(y)}\frac{dy}{dx} - g(x)$$

Therefore,

$$\frac{d}{dx}[H(y) - G(x)] = 0$$

and so

$$H(y) - G(x) = C \qquad \textcolor{orange}{C, \text{ a constant}}$$

But the last equation is equivalent to

$$H(y) = G(x) + C \qquad \text{or} \qquad \int \frac{dy}{h(y)} = \int g(x)\,dx$$

which is precisely Equation (6).

EXAMPLE 3   Solve the differential equation $\dfrac{dy}{dx} = \dfrac{y}{x}$.

Solution   First, observe that $y = 0$ is a solution of the separable equation. To find other solutions, assume that $y \neq 0$, separate variables, and integrate each side of the resulting equation with respect to the appropriate variable to obtain

$$\int \frac{dy}{y} = \int \frac{dx}{x}$$

or

$$\ln|y| = \ln|x| + \ln|C_1|$$

where C_1 is a nonzero but otherwise arbitrary constant. We write $\ln|C_1|$ rather than an arbitrary constant C_2 because we can then more readily apply the laws of logarithms. Proceeding, we have

$$\ln|y| = \ln\big(|C_1| \cdot |x|\big) = \ln|C_1 x|$$

or, upon exponentiating using the base e,

$$|y| = |C_1 x| \qquad C_1 \neq 0$$

$$y = \pm C_1 x$$

Since $y = 0$ is also a solution, we see that the general solution can be written in the form $y = Cx$, where C is an arbitrary constant. ∎

EXAMPLE 4 Solve the differential equation $y' = \dfrac{xy}{x^2 + 1}$.

Solution First, observe that $y = 0$ is a solution of the differential equation. Next, write the given equation in the form

$$\frac{dy}{dx} = \left(\frac{x}{x^2 + 1}\right) y = g(x)h(y)$$

which is separable. Separating variables and integrating, we have

$$\int \frac{dy}{y} = \int \frac{x}{x^2 + 1}\, dx$$

$$\ln|y| = \frac{1}{2}\ln(x^2 + 1) + \ln|C_1| \qquad C_1 \neq 0$$

where $\ln|C_1|$ represents the combined constants of integration. We have

$$\ln|y| = \ln(x^2 + 1)^{1/2} + \ln|C_1|$$
$$= \ln|C_1\sqrt{x^2 + 1}|$$

or

$$y = \pm C_1\sqrt{x^2 + 1}$$

Since $y = 0$ is also a solution of the differential equation, we conclude that the general solution is

$$y = C\sqrt{x^2 + 1}$$

where C is an arbitrary constant. Figure 3 shows six solution curves of the differential equation for selected values of C. ∎

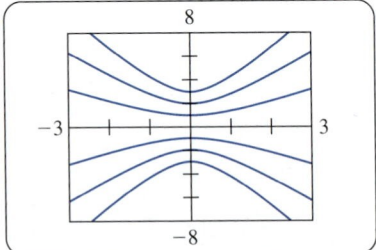

FIGURE 3
Solution curves for $y' = \dfrac{xy}{x^2 + 1}$

for $C = \pm 1, \pm 2, \pm 3$

EXAMPLE 5 Find the particular solution of the differential equation

$$ye^x\, dx + (y^2 - 1)\, dy = 0$$

that satisfies the condition $y(0) = 1$.

Solution The equation is separable. By inspection we see that $y = 0$ is a solution of the differential equation. But this solution does not satisfy the initial condition $y(0) = 1$ and is rejected. Next, suppose that $y \neq 0$. Separating variables and integrating, we have

$$\int \frac{y^2 - 1}{y}\, dy = -\int e^x\, dx$$

$$\int \left(y - \frac{1}{y}\right) dy = -\int e^x\, dx$$

$$\frac{1}{2}y^2 - \ln|y| = -e^x + C_1$$

$$y^2 - \ln y^2 = -2e^x + C_2 \qquad C_2 = 2C_1$$

Using the condition $y(0) = 1$, we have

$$1 - \ln 1 = -2 + C_2 \qquad \text{or} \qquad C_2 = 3$$

Therefore, the required solution is

$$y^2 - \ln y^2 = -2e^x + 3$$

Note that the solution of the differential equation in Example 3 is obtained in the form of an implicit equation in x and y.

We now turn to the solution of the initial-value problem (3) posed earlier:

$$\begin{cases} \dfrac{dy}{dt} = ky \\ y(0) = y_0 \end{cases}$$

The differential equation $dy/dt = ky$ is separable. Separating variables and integrating, we obtain

$$\int \frac{dy}{y} = \int k \, dt$$

$$\ln|y| = kt + C_1$$

$$|y| = e^{kt+C_1} = C_2 e^{kt} \qquad C_2 = e^{C_1}$$

$$y = \pm C_2 e^{kt}$$

$$= C e^{kt} \qquad C = \pm C_2$$

Using the initial condition $y(0) = y_0$ gives $y_0 = C$. Therefore, the solution is $y = y_0 e^{kt}$.

THEOREM 1 **Natural Law of Exponential Growth (Decay)**

The initial-value problem

$$\begin{cases} \dfrac{dy}{dt} = ky \qquad k, \text{ a constant} \\ y(0) = y_0 \end{cases}$$

has the unique solution $y = y_0 e^{kt}$.

The solution curves for the initial-value problem are shown in Figure 4.

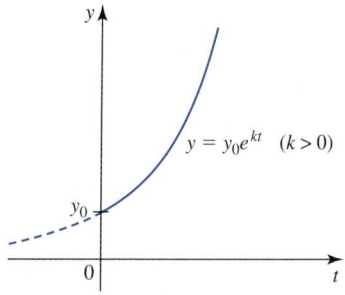

(a) Exponential growth $(k > 0)$

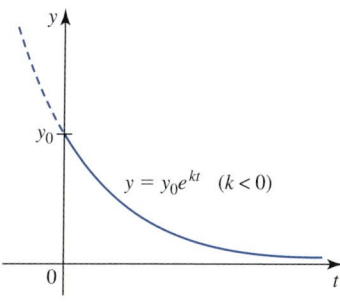

(b) Exponential decay $(k < 0)$

FIGURE 4
Graphs of $y = y_0 e^{kt}$

EXAMPLE 6 **Growth of Bacteria** The population of bacteria in a culture grows at a rate that is proportional to the number present. Suppose that there are 1000 bacteria present initially in a culture and 3000 present 2 hr later. How many bacteria will there be in the culture after 4 hr?

Solution Let $y = y(t)$ denote the number of bacteria present in the culture after t hr. Then $y(0) = 1000$ and $y(2) = 3000$. Since the rate of growth of bacteria in the culture is proportional to the number present, the quantity y satisfies the initial-value problem

$$\begin{cases} \dfrac{dy}{dt} = ky \\ y(0) = 1000 \end{cases}$$

By Theorem 1,

$$y(t) = y_0 e^{kt} = 1000 e^{kt}$$

Next, using the condition $y(2) = 3000$, we have

$$y(2) = 1000 e^{2k} = 3000$$

$$e^{2k} = \frac{3000}{1000} = 3$$

or

$$e^k = (e^{2k})^{1/2} = 3^{1/2}$$

Therefore, the number of bacteria present after t hr is

$$y(t) = 1000 e^{kt} = 1000(e^k)^t = 1000(3^{t/2})$$

In particular, the number of bacteria present in the culture after 4 hr is

$$y(4) = 1000(3^{4/2}) = 9000$$

Notice that it is not necessary to determine the value of k, the *growth constant,* which depends on the strain of the bacteria. However, if desired, its value can be found by solving the equation

$$e^{2k} = 3$$

Thus,

$$\ln e^{2k} = \ln 3$$

$$2k \ln e = \ln 3$$

$$2k = \ln 3$$

$$k = \frac{1}{2} \ln 3 \approx 0.5493$$

The graph of $y = 1000(3^{t/2})$, or $y = 1000 e^{0.5493t}$, is shown in Figure 5. ■

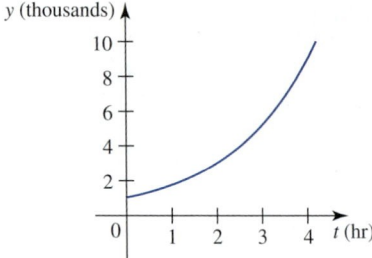

y (thousands)

FIGURE 5
The graph of $y = 1000 e^{0.5493t}$ shows the population of bacteria at time t.

EXAMPLE 7 **Radioactive Decay** Radioactive substances decay at a rate that is proportional to the amount present. The *half-life* of a substance is the time required for a given amount to be reduced by one-half. It is known that the half-life of radium-226 is approximately 1602 years. Suppose that initially there are 100 mg of radium in a sample.

a. Find a formula that gives the amount of radium-226 present after t years.
b. Find the amount of radium-226 present after 1000 years.
c. How long will it take for the radium-226 to be reduced to 40 mg?

Solution

a. Let $y(t)$ denote the amount of radium-226 present in the sample after t years. Then $y(0) = 100$ and $y(1602) = 50$. Since the rate of decay of the radium is proportional to the amount present, the quantity y satisfies the initial-value problem

$$\begin{cases} \dfrac{dy}{dt} = ky \\ y(0) = 100 \end{cases}$$

By Theorem 1,

$$y(t) = y_0 e^{kt} = 100 e^{kt}$$

Next, we use the condition $y(1602) = 50$ to write

$$y(1602) = 100 e^{1602k} = 50$$

$$e^{1602k} = \frac{1}{2}$$

or

$$e^k = \left(\frac{1}{2}\right)^{1/1602}$$

Therefore, the amount of radium-226 present after t years is

$$y(t) = 100 e^{kt} = 100(e^k)^t = 100\left(\frac{1}{2}\right)^{t/1602}$$

b. The amount of radium-226 present after 1000 years is given by

$$y(1000) = 100\left(\frac{1}{2}\right)^{1000/1602} \approx 64.88$$

or approximately 64.9 mg.

c. We want to find the value of t such that $y(t) = 40$; that is,

$$100\left(\frac{1}{2}\right)^{t/1602} = 40$$

$$\left(\frac{1}{2}\right)^{t/1602} = \frac{40}{100} = \frac{2}{5}$$

Taking the natural logarithm on both sides, we obtain

$$\ln\left(\frac{1}{2}\right)^{t/1602} = \ln\frac{2}{5}$$

$$\frac{t}{1602}\ln\left(\frac{1}{2}\right) = \ln\frac{2}{5}$$

$$t = 1602\left(\frac{\ln\frac{2}{5}}{\ln\frac{1}{2}}\right) \approx 2118$$

So it will take approximately 2118 years for the radium-226 to decay to 40 mg. The graph of $y = 100\left(\frac{1}{2}\right)^{t/1602}$ is shown in Figure 6.

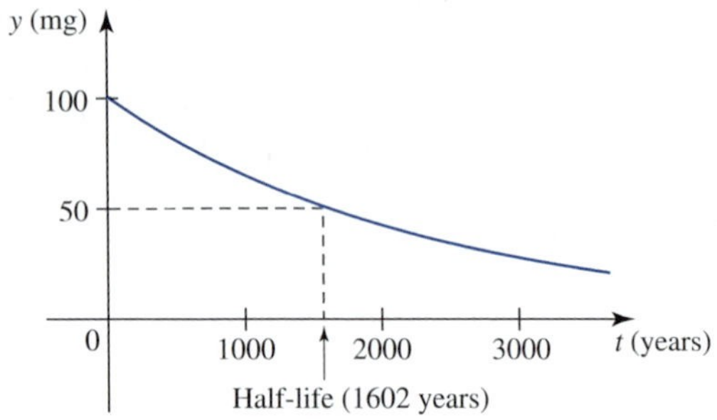

FIGURE 6
The graph of $y = 100\left(\frac{1}{2}\right)^{t/1602}$ shows
how the radium-226 decays.

EXAMPLE 8 **Newton's Law of Cooling** Newton's Law of Cooling states that the temperature of a body drops at a rate that is proportional to the difference between the temperature of the body and the temperature of the surrounding medium. An apple pie is taken out of an oven at a temperature of 200°F and placed on the counter in a room where the temperature is 70°F. The temperature of the pie is 160°F after 15 min.

a. What is the temperature of the pie after 30 min?
b. How long will it take for the pie to cool to 120°F?

Solution
a. Let $y(t)$ denote the temperature of the apple pie t min after it was placed on the counter. Then Newton's Law of Cooling gives

$$\frac{dy}{dt} = k(y - 70)$$

where k is the constant of proportionality. The initial temperature of the pie is 200°F, and this translates into the condition $y(0) = 200$. So we have the initial-value problem

$$\begin{cases} \dfrac{dy}{dt} = k(y - 70) \\ y(0) = 200 \end{cases}$$

Observe that the differential equation here does not have the same form as that in Theorem 1. Let's define the function u by $u(t) = y(t) - 70$. Then

$$\frac{du}{dt} = \frac{dy}{dt}$$

Making these substitutions, we obtain

$$\frac{du}{dt} = ku$$

Using Theorem 1, we obtain the solution

$$u(t) = u(0)e^{kt}$$

$$y(t) - 70 = [y(0) - 70]e^{kt} \qquad \text{Recall } u(t) = y(t) - 70.$$

or

$$y(t) = 70 + (200 - 70)e^{kt} = 70 + 130e^{kt}$$

Next, we use the condition $y(15) = 160$ to write

$$y(15) = 70 + 130e^{15k} = 160$$

$$130e^{15k} = 90$$

or

$$e^k = \left(\frac{9}{13}\right)^{1/15}$$

Therefore, the temperature of the pie after t min is

$$y(t) = 70 + 130\left(\frac{9}{13}\right)^{t/15}$$

In particular, the temperature of the pie after 30 min is

$$y(30) = 70 + 130\left(\frac{9}{13}\right)^{30/15} = 70 + 130\left(\frac{9}{13}\right)^2 \approx 132.3$$

or approximately 132°F.

b. We need to find the value of t for which $y(t) = 120$, that is,

$$70 + 130e^{kt} = 120$$

$$130e^{kt} = 50$$

$$(e^k)^t = \frac{5}{13}$$

$$\left(\frac{9}{13}\right)^{t/15} = \frac{5}{13} \qquad \text{\color{orange}Use the value of } e^k \text{ from part (a).}$$

Taking the natural logarithm on both sides, we obtain

$$\frac{t}{15}\ln\frac{9}{13} = \ln\frac{5}{13}$$

or

$$t = \frac{15\ln\frac{5}{13}}{\ln\frac{9}{13}} \approx 39$$

So it will take approximately 39 min for the pie to cool to 120°F. The graph of $y(t) = 70 + 130\left(\frac{9}{13}\right)^{t/15}$ is shown in Figure 7.

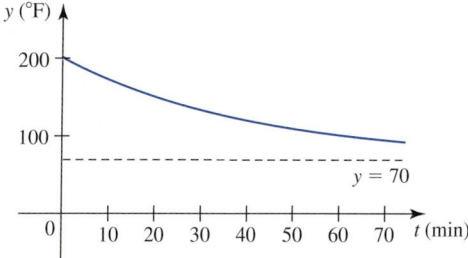

FIGURE 7
The graph of $y(t) = 70 + 130\left(\frac{9}{13}\right)^{t/15}$
gives the temperature of the
pie as a function of time.

Note   The differential equation $dy/dt = k(y - 70)$ in Example 8 is separable and can be solved directly without using Theorem 1.

EXAMPLE 9 **Continuously Compounded Interest** Suppose that money deposited into a bank grows at a rate that is proportional to the amount accumulated. If the amount on deposit initially is P dollars, find an expression for the accumulated amount A after t years. Reconcile your result with the continuous compound interest formula, $A = Pe^{rt}$.

Solution Since the rate of growth of the money is proportional to the amount present, we have the initial-value problem

$$\begin{cases} \dfrac{dA}{dt} = kA \\ A(0) = P \end{cases}$$

By Theorem 1,

$$A(t) = A(0)e^{kt} = Pe^{kt}$$

Therefore, the accumulated amount after t years is given by

$$A(t) = Pe^{kt}$$

dollars.

If we compare this result with the formula $A = Pe^{rt}$, we see that the formulas are identical when the growth constant k is taken to be equal to r, the nominal interest rate. This shows that money deposited into a bank with interest compounded continuously grows according to the law of natural growth. ■

Suppose all the curves of one family intersect all the curves of another family at right angles. Then the curves of the first family are said to be **orthogonal trajectories** of the other family, and vice versa. For example, the straight lines passing through the origin are the orthogonal trajectories of the concentric circles with center at the origin and vice versa. (See Figure 8.) Orthogonal trajectories occur in physics and engineering. For example, in electrostatics and electromagnetics the field lines are orthogonal to the equipotential curves.

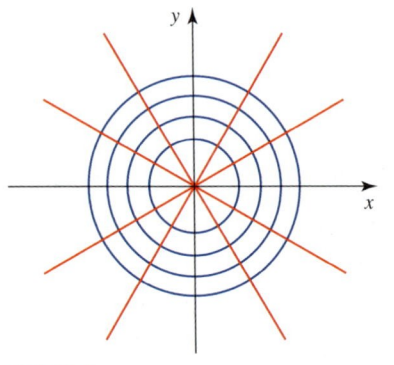

FIGURE 8
The concentric circles are the orthogonal trajectories of the lines, and vice versa.

EXAMPLE 10 Find the orthogonal trajectories of the family of curves given by $y = Cx^2$, where C is an arbitrary constant.

Solution First, recall that the slope of the tangent line to a curve in the given family at a point (x, y) is given by dy/dx. To find dy/dx, we differentiate the given equation to obtain

$$\frac{dy}{dx} = 2Cx$$

Next, we eliminate C from this equation. Solving for C in the given equation gives $C = y/x^2$. Substituting this value of C into the equation for dy/dx, we obtain

$$\frac{dy}{dx} = 2\left(\frac{y}{x^2}\right)x = \frac{2y}{x}$$

Since the required family is orthogonal to the given family, the slope of the tangent line to each member of the required family is given by the negative reciprocal of $2y/x$. Therefore, the orthogonal trajectories satisfy the differential equation

$$\frac{dy}{dx} = -\frac{x}{2y}$$

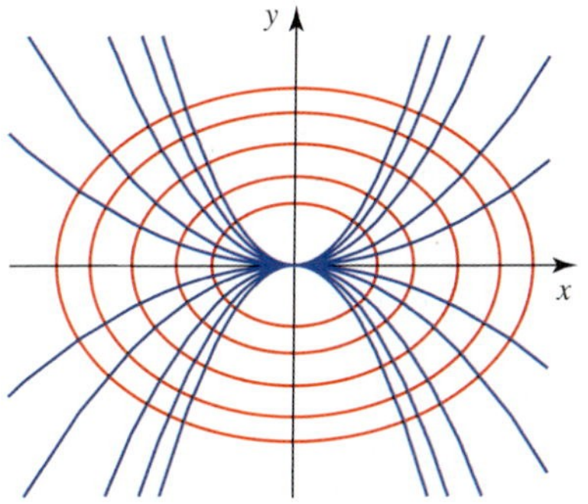

Separating variables and integrating with respect to the appropriate variable, we obtain

$$\int 2y \, dy + \int x \, dx = 0$$

$$y^2 + \frac{1}{2}x^2 = k$$

where k is an arbitrary constant. We recognize this family to be a family of ellipses. (See Figure 9.) ■

FIGURE 9
Each parabola is orthogonal to each ellipse, and vice versa.

7.1 CONCEPT QUESTIONS

1. **a.** What is a first-order separable differential equation?
   b. Is the equation $f(x)g(y)dx + F(x)G(y)dy = 0$ separable? Explain.

2. **a.** Is $x\dfrac{dy}{dx} + y = e^y$ a separable differential equation? Explain.
   b. Is $(x^2 + y^2)dx + xy^3 dy = 4$ a separable differential equation? Explain.

3. Describe a method for solving a separable differential equation.

4. **a.** Write a differential equation to describe the natural law of exponential growth (decay).
   b. Write a differential equation to describe Newton's Law of Cooling.

7.1 EXERCISES

1. Show that $y = \dfrac{1}{2} + \dfrac{3}{x^2}$ is a solution of the differential equation $xy' + 2y = 1$ on any interval that does not contain $x = 0$.

2. Show that $y = Ce^{-2x} + e^x$ is a solution of the differential equation $y' + 2y = 3e^x$ on $(-\infty, \infty)$.

3. Show that $y = x^4 + 3x^3$ is a solution of the initial-value problem $xy' - 3y = x^4$, $y(1) = 4$ on $(-\infty, \infty)$.

4. Show that $y = \sin x - \cos x$ is a solution of the initial-value problem $\cos x \dfrac{dy}{dx} + y \sin x = 1$, $y(0) = -1$ on the interval $(-\infty, \infty)$.

 5. Assume that the differential equation $y' = 3y$ has a solution of the form $y = Ce^{mx}$.
   a. Find the value of m.
   b. Plot the solution curves of $y' = 3y$ on the same set of axes for $C = -3, -2, -1, 0, 1, 2$, and 3.
   c. Find the solution of $y' = 3y$ that satisfies the initial condition $y(0) = 2$. Is the solution curve for this solution among those in part (b)?

6. Suppose that a solution of the second-order differential equation $y'' - y' - 2y = 0$ has the form $y = e^{mx}$.
   a. Find an equation that m must satisfy.
   b. Solve the equation found in part (a).
   c. Write two solutions of the differential equation.
   d. Verify the results of part (c) directly.

In Exercises 7 and 8, the general solution of a differential equation is given. (a) Find the particular solution that satisfies the given initial condition. (b) Plot the solution curves corresponding to the given values of C. Indicate the solution curve that corresponds to the solution found in part (a).

7. $x\dfrac{dy}{dx} + y = x^3$, $y = \dfrac{C}{x} + \dfrac{x^3}{4}$; $y(1) = \dfrac{5}{4}$;
   $C = -2, -1, 0, 1, 2$

8. $y\dfrac{dy}{dx} - e^{2x} = 0$, $y^2 = e^{2x} + C$; $y(0) = 1$;
   $C = -2, -1, 0, 1, 2$

In Exercises 9–18, solve the differential equation.

9. $\dfrac{dy}{dx} = \dfrac{2y}{x}$

10. $\dfrac{dy}{dx} = \dfrac{x+1}{y^2}$

11. $\dfrac{dy}{dx} = x^2 y$

12. $\dfrac{dy}{dx} = -\dfrac{xy}{x+1}$

13. $y' = \dfrac{2y+3}{x^2}$

14. $y' = e^{x-y}$

15. $\cos y \dfrac{dy}{dx} = \sec^2 x$

16. $(1-\cos\theta)\dfrac{dr}{d\theta} = r\sin\theta$

17. $xy' = y^2 \ln x$

18. $\dfrac{dy}{dt} = 1 + t + y + ty$

In Exercises 19–26, solve the initial-value problem.

19. $\dfrac{dy}{dx} = 3xy - 2x, \quad y(0) = 1$

20. $\dfrac{dy}{dx} = xe^{-y}, \quad y(0) = 1$

21. $y' = x^2 y^{-1/2}, \quad y(1) = 1$

22. $\dfrac{dy}{dx} = \dfrac{y^2}{x-2}, \quad y(3) = 1$

23. $y' = 3x^2 e^{-y}, \quad y(0) = 1$

24. $\sin^2 y\,dx + \cos^2 x\,dy = 0, \quad y\left(\dfrac{\pi}{4}\right) = \dfrac{\pi}{4}$

25. $\dfrac{dI}{dt} + 2I = 4, \quad I(0) = 0$

26. $\cos\theta \dfrac{du}{d\theta} = u\tan\theta, \quad u\left(\dfrac{\pi}{3}\right) = 2$

27. Find an equation defining a function f given that (a) the slope of the tangent line to the graph of f at any point $P(x, y)$ on the graph is given by

$$\dfrac{dy}{dx} = \dfrac{3x^2}{2y}$$

and (b) the graph of f passes through the point $(1, 3)$.

28. Find an equation of a curve given that it passes through the point $\left(2, \dfrac{2\sqrt{5}}{3}\right)$ and that the slope of the tangent line to the curve at any point $P(x, y)$ is given by

$$\dfrac{dy}{dx} = -\dfrac{4x}{9y}$$

 In Exercises 29–32, find the orthogonal trajectories of the family of curves. Use a graphing utility to draw several members of each family on the same set of axes.

29. $xy = c$

30. $y^2 = cx^3$

31. $y = ce^x$

32. $y = \ln(cx)$

33. Find the constant a such that the curves $x^2 + ay^2 = C_1$ and $y^3 = C_2 x$ are orthogonal trajectories of each other.

34. **Growth of Bacteria** The population of bacteria in a culture grows at a rate that is proportional to the number present. Initially, there are 600 bacteria, and after 3 hr there are 10,000 bacteria.
    a. What is the number of bacteria after t hr?
    b. What is the number of bacteria after 5 hr?
    c. When will the number of bacteria reach 24,000?

35. **Growth of Bacteria** The population of bacteria in a culture grows at a rate that is proportional to the number present. After 2 hr there are 800 bacteria present. After 4 hr there are 3200 bacteria present. How many bacteria were there initially?

36. **Growth of Bacteria** The population of bacteria in a certain culture grows at a rate that is proportional to the number present. If the original population increases by 50% in $\frac{1}{2}$ hr, how long will it take for the population to triple in size?

37. **Lambert's Law of Absorption** According to Lambert's Law of Absorption, the percentage of incident light L, absorbed in passing through a thin layer of material x, is proportional to the thickness of the material. For a certain material, if $\frac{1}{2}$ in. of the material reduces the light to half of its intensity, how much additional material is needed to reduce the intensity to one fourth of its initial value?

38. **Savings Accounts** An amount of money deposited in a savings account grows at a rate proportional to the amount present. (Thus it earns interest compounded continuously (see Example 9).) Suppose that $10,000 is deposited in a fixed account earning interest at the rate of 10% compounded continuously.
    a. What is the accumulated amount after 5 years?
    b. How long does it take for the original deposit to double in value?

39. **Chemical Reactions** In a certain chemical reaction a substance is converted into another substance at a rate proportional to the square of the amount of the first substance present at any time t. Initially ($t = 0$), 50 g of the first substance was present; 1 hr later, only 10 g of it remained. Find an expression that gives the amount of the first substance present at any time t. What is the amount present after 2 hr?

40. **Radioactive Decay** Phosphorus-32 has a half-life of 14.3 days. If 100 g of this substance is present initially, find the amount present after t days. What amount will be left after 7.1 days? How fast is the phosphorus-32 decaying when $t = 7.1$?

41. **Nuclear Fallout** Strontium-90, a radioactive isotope of strontium, is present in the fallout resulting from nuclear explosions. It is especially hazardous to animal life, including humans, because when contaminated food is ingested, the strontium-90 is absorbed into the bone structure. Its half-life is 28.9 years. If the amount of strontium-90 in a certain area is found to be four times the "safe" level, find how much time must elapse before an "acceptable" level is reached.

42. Carbon-14 Dating Wood deposits recovered from an archeological site contain 20% of the carbon-14 they originally contained. How long ago did the tree from which the wood was obtained die? (The half-life of carbon C-14 is 5730 years.)

43. Carbon-14 Dating Skeletal remains of the so-called Pittsburgh Man unearthed in Pennsylvania had lost 82% of the carbon-14 they originally contained. Determine the approximate age of the bones. (The half-life of carbon C-14 is 5730 years.)

44. Newton's Law of Cooling Newton's Law of Cooling states that the rate at which the temperature of an object changes is directly proportional to the difference in temperature between the object and that of the surrounding medium. A horseshoe that has been heated to a temperature of 600°C is immersed in a large tank of water at a (constant) temperature of 30°C at time $t = 0$. Two minutes later, the temperature of the horseshoe is reduced to 70°C. Derive an expression that gives the temperature of the horseshoe at any time t. What is the temperature of the horseshoe 3 min after it has been immersed in the water?

45. Newton's Law of Cooling Newton's Law of Cooling states that the rate at which the temperature of an object changes is directly proportional to the difference in temperature of the object and that of the surrounding medium. A cup of coffee is prepared with boiling water (212°F) and left to cool on a counter in a room where the temperature is 72°F. If the temperature of the coffee is 140°F after 2 min, when will the coffee will be cool enough to drink (say, 110°F)?

46. Newton's Law of Heating A thermometer is taken from a room where the temperature is 70°F to a patio. After 1 min the thermometer reads 50°F, and after 2 min it reads 40°F. What is the outdoor temperature?

47. Motion of a Motorboat A motorboat is traveling at a speed of 12 mph in calm water when its motor is cut off. Twenty seconds later, the boat's speed drops to 8 mph. Assuming that the water resistance on the boat is directly proportional to the speed of the boat, what will its speed be 2 min after the motor was cut off?

48. Learning Curves The American Court Reporter Institute finds that the average student taking elementary machine shorthand will progress at a rate given by

$$\frac{dQ}{dt} = k(80 - Q)$$

in a 20-week course, where k is a positive constant and $Q(t)$ measures the number of words of dictation a student can take per minute after t weeks in the course. If the average student can take 50 words of dictation per minute after 10 weeks in the course, how many words per minute can the average student take after completing the course?

49. Effect of Immigration on Population Growth Suppose that a country's population at any time t grows in accordance with the rule

$$\frac{dP}{dt} = kP + I$$

where P denotes the population at any time t, k is a positive constant reflecting the natural growth rate of the population, and I is a constant giving the (constant) rate of immigration into the country.

a. If the total population of the country at time $t = 0$ is P_0, find an expression for the population at any time t.

b. The population of the United States in the year 1980 ($t = 0$) was 226.5 million. Suppose that the natural growth rate is 0.8% annually ($k = 0.008$) and that net immigration is allowed at the rate of 0.5 million people per year ($I = 0.5$). What will the U.S. population be in 2010?

50. Chemical Reaction Rates Two chemical solutions, one containing N molecules of chemical A and another containing M molecules of chemical B, are mixed together at time $t = 0$. The molecules from the two chemicals combine to form another chemical solution containing y AB molecules. The rate at which the AB molecules are formed, dy/dt, is called the *reaction rate* and is jointly proportional to $(N - y)$ and $(M - y)$. Thus,

$$\frac{dy}{dt} = k(N - y)(M - y)$$

where k is a constant. (We assume that the temperature of the chemical mixture remains constant during the interaction.) Solve this differential equation with the side condition $y(0) = 0$, assuming that $N - y > 0$ and $M - y > 0$.
Hint: Use the identity

$$\frac{1}{(N - y)(M - y)} = \frac{1}{M - N}\left(\frac{1}{N - y} - \frac{1}{M - y}\right)$$

51. A Falling Raindrop As a raindrop falls, it picks up more moisture, and as a result, its mass increases. Suppose that the rate of change of its mass is directly proportional to its current mass.

a. Using Newton's Law of Motion, $\dfrac{d}{dt}(mv) = F = mg$, where $m(t)$ is the mass of the raindrop at time t, v is its velocity (positive direction is downward), and g is the acceleration due to gravity, derive the (differential) equation of motion of the raindrop.

b. Solve the differential equation of part (a) to find the velocity of the raindrop at time t. Assume that $v(0) = 0$.

c. Find the *terminal velocity* of the raindrop, that is, find $\lim_{t \to \infty} v(t)$.

52. Discharging Water from a Tank A container that has a constant cross section A is filled with water to height H. The water is discharged through an opening of cross section B at the base of the container. By using Torricelli's Law, it can be shown that the height h of the water at time t satisfies the initial-value problem

$$\frac{dh}{dt} = -\frac{B}{A}\sqrt{2gh} \qquad h(0) = H$$

a. Find an expression for h.

b. Find the time T it takes for the tank to empty.

c. Find T if $A = 4$ (ft^2), $B = 1$ (in.^{2}), $H = 16$ (ft), and $g = 32$ (ft/sec^2).

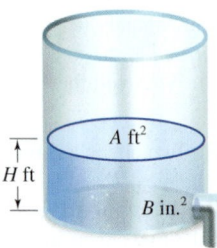

53. Doomsday Equation Suppose that the population P satisfies the differential equation $dP/dt = kP^{1.01}$, where k is a positive constant and $P(0) = 1$.

a. Solve the initial-value problem.

cas **b.** Suppose that $k = 0.1$. Plot the graph of $P(t)$.

c. Why is $dP/dt = kP^{1.01}$ called the "doomsday equation"?

54. Stefan's Law Stefan's Law states that the rate of change of the temperature T of a body is directly proportional to the difference of the fourth power of T and the fourth power of the temperature of the surrounding medium T_m. Thus,

$$\frac{dT}{dt} = k(T^4 - T_m^4)$$

where k is the constant of proportionality. Stefan's Law holds over a greater temperature range than does Newton's Law of Cooling. Show that T is given by the implicit equation

$$\ln\left(\frac{T + T_m}{T - T_m}\right) + 2\tan^{-1}\left(\frac{T}{T_m}\right) = -4T_m^3 kt + C \qquad T > T_m$$

55. Concentration of a Drug in the Bloodstream Suppose that the rate at which the concentration of a drug in the bloodstream decreases is proportional to the concentration at time t. Initially, there is no drug in the bloodstream. At time $t = 0$ a drug having a concentration of C_0 g/mL is introduced into the bloodstream.

a. What is the concentration of drug in the bloodstream at the end of T hr?

b. If at time T another dosage having the concentration of C_0 g/mL is infused into the bloodstream, what is the concentration of the drug at the end of $2T$ hr?

c. If the process were continued, what would the concentration of the drug be at the end of NT hr?

d. Find the concentration of the drug in the bloodstream in the long run.

Hint: Evaluate $\lim_{N \to \infty} y(NT)$, where $y(NT)$ denotes the concentration of the drug at the end of NT hr.

56. Spread of Disease A simple mathematical model in epidemiology for the spread of a disease assumes that the rate at which the disease spreads is jointly proportional to the number of infected people and the number of uninfected people. Suppose that there are a total of N people in the population, of whom N_0 are infected initially. Show that the number of infected people after t weeks, $x(t)$, is given by

$$x(t) = \frac{N}{1 + \left(\dfrac{N - N_0}{N_0}\right)e^{-kNt}}$$

where k is a positive constant.

57. Spread of Disease Refer to Exercise 56. Suppose that there are 8000 students in a college and 400 students had contracted the flu at the beginning of the week.

a. If 1200 had contracted the flu at the end of the week, how many will have contracted the flu at the end of 2, 3, and 4 weeks?

b. How long does it take for 80% of the student population to become infected?

 c. Plot the graph of the function $x(t)$.

58. Von Bertalanffy Growth Model The von Bertalanffy growth model is used to predict the length of commercial fish. The model is described by the differential equation

$$\frac{dx}{dt} = k(L - x)$$

where $x(t)$ is the length of the fish at time t, k is a positive constant called the von Bertalanffy growth rate, and L is the maximum length of the fish.

a. Find $x(t)$ given that the length of the fish at $t = 0$ is x_0.

b. At the time the larvae hatch, the North Sea haddock are about 0.4 cm long, and the average haddock grows to a length of 10 cm after 1 year. Find an expression for the length of the North Sea haddock at time t.

 c. Plot the graph of x. Take $L = 100$ (cm).

d. On average, the haddock that are caught today are between 40 cm and 60 cm long. What are the ages of the haddock that are caught?

In Exercises 59–62, determine whether the statement is true or false. If it is true, explain why. If it is false, explain why or give an example that shows it is false.

59. If f is a solution of a first-order differential equation, then so is cf, where c is a constant.

60. The differential equation $y' = x^2 - y^2$ is separable.

61. The differential equation $y' = xy + 2x - y - 2$ is separable.

62. The curves $cx^2 + y^2 = 1$ and $x^2 + y^2 - \ln y^2 = k$ are orthogonal to each other.

In Section 7.1 we considered differential equations with solutions that could be found analytically. Armed with these solutions, we were able to draw the solution curves of these first-order differential equations.

In this section we will describe a way to visualize the general solution of a differential equation without actually solving the equation. This is especially useful when we are unable to find an exact solution of the equation. We will also describe a method for constructing an approximation for the solution curve of an initial-value problem.

■ Direction Fields

Suppose that we are given a first-order differential equation of the form

$$y' = f(x, y) \tag{1}$$

If (a, b) is any point on a solution curve of Equation (1), then the slope of the tangent line to this curve at (a, b) is given by

$$y'|_{(a, b)} = f(a, b)$$

(See Figure 1a.) If we retain a small portion of the tangent line at (a, b), then we have a small line segment called a **lineal element** that indicates the direction of the solution curve at that point (Figure 1b).

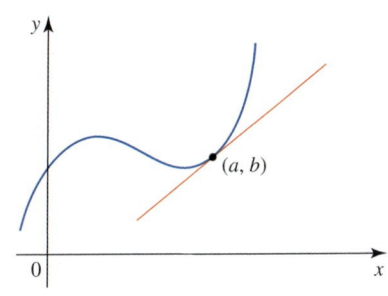

FIGURE 1
A short tangent line at (a, b) gives the direction of the solution curve at (a, b).

(**a**) The tangent line at (a, b) (**b**) A lineal element at (a, b)

If f is defined in a region in the xy-plane, then we can draw a lineal element at each point (x, y) in the region. A set of lineal elements drawn at various points is called a **slope field** or **direction field** of the differential equation. For example, the direction field for the differential equation $y' = y - x$ (see Example 1 in Section 7.1) is shown in Figure 2.

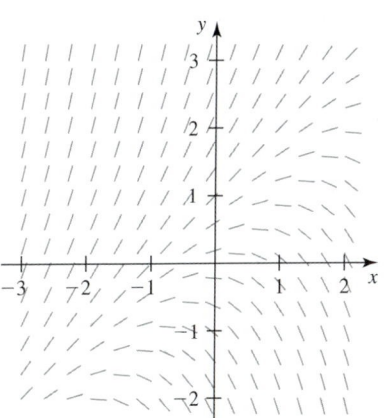

FIGURE 2
A direction field for $y' = y - x$

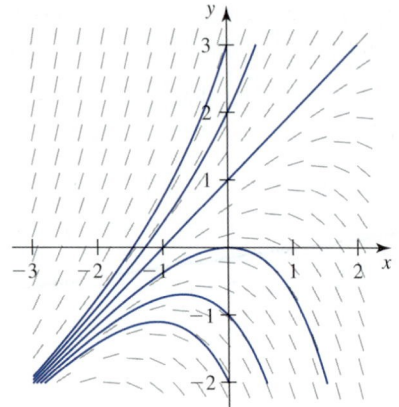

FIGURE 3
A few solution curves of $y' = y - x$
superimposed over its direction field

Since the small line segments represent tangent lines to the solution curves of the differential equation at these points, we see that a direction field indicates the general shape of the solution curves. Figure 3 shows a few solution curves of the differential equation $y' = y - x$ superimposed over its direction field.

EXAMPLE 1 Consider the differential equation $y' = x + 2y$.

a. Sketch the lineal elements at $(-1, 0)$, $(-1, 1)$, $(0, 0)$, $(1, -1)$, $(1, 1)$, $(1, 2)$, $(2, -1)$, and $(2, 1)$.
b. Use a calculator or computer to draw a direction field with more lineal elements.
c. Use the slope field to sketch the solution that passes through the point $(0, 1)$.

Solution
a. Here, $f(x, y) = x + 2y$. The slope of the lineal element at $(-1, 0)$ is

$$f(-1, 0) = -1 + 2(0) = -1$$

We summarize the results of the other calculations in the table.

(x, y)	$(-1, 0)$	$(-1, 1)$	$(0, 0)$	$(1, -1)$	$(1, 1)$	$(1, 2)$	$(2, -1)$	$(2, 1)$
$y' = f(x, y)$	-1	1	0	-1	3	5	0	4

The lineal elements for the given ordered pairs are shown in Figure 4.
b. The required direction field is shown in Figure 5. Note that the lineal elements that we obtained in part (a) are contained in the direction field, as expected.
c. To sketch the solution curve passing through the point $(0, 1)$, we draw a curve starting out at $(0, 1)$ and extend it first to the right, then to the left, always requiring that it be parallel to nearby lineal elements as we proceed. The resulting approximating solution curve is shown in Figure 6 superimposed over the direction field of the differential equation.

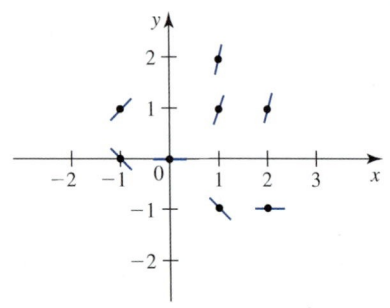

FIGURE 4
The lineal elements at selected points

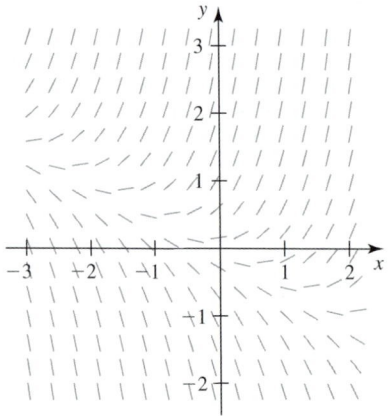

FIGURE 5
The direction field for $y' = x + 2y$

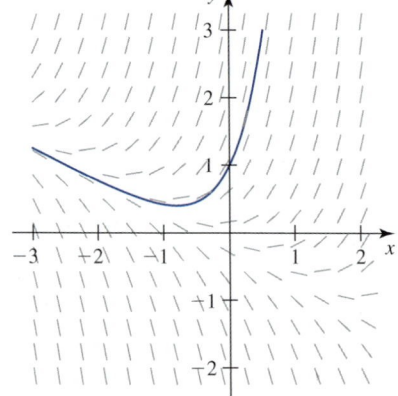

FIGURE 6
The solution curve of $y' = x + 2y$
passing through $(0, 1)$

In physical applications involving differential equations, the direction fields of the equations can shed much light on the nature of their solution. Suppose that an object of constant mass m falls vertically downward under the influence of gravity. Assum-

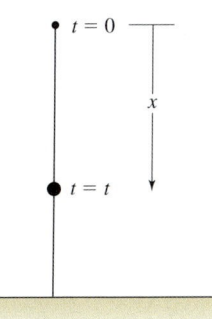

FIGURE 7
Here, $x(t)$ is the position of the object, $v = dx/dt$, and the positive direction is downward.

ing that air resistance is proportional to the velocity of the object at any instant during the fall, then according to Newton's Second Law of Motion,

$$F = ma = m\frac{dv}{dt}$$

where F is the net force acting on the object in the positive (downward) direction. (See Figure 7.) But F is given by the weight of the object mg (acting downward) minus the air resistance kv (acting upward), where $k > 0$ is the constant of proportionality; that is, $F = mg - kv$. Therefore, the equation of motion is

$$m\frac{dv}{dt} = mg - kv \tag{2}$$

EXAMPLE 2 **A Parachute Jump** A paratrooper and his equipment have a combined weight of 192 lb. At the instant that the parachute is deployed, he is traveling vertically downward at a speed of 30 ft/sec. Assume that air resistance is proportional to the instantaneous velocity with constant of proportionality $k = 4$ and that $g = 32$ (ft/sec^2).

a. Write an equation of motion, and draw a direction field associated with it.
b. Sketch the solution curve superimposed over the direction field.
c. What velocity does the paratrooper approach as t increases without bound?

(The velocity found in part (c) is called the terminal velocity.)

Solution
a. Here, $m = \frac{192}{32}$, or 6 slugs. Therefore, using Equation (2), we have the equation of motion

$$6\frac{dv}{dt} = 192 - 4v \qquad \text{or} \qquad \frac{dv}{dt} = 32 - \frac{2}{3}v$$

At $t = 0$, $v = 30$, so we have the initial-value problem

$$\begin{cases} \dfrac{dv}{dt} = 32 - \dfrac{2}{3}v \\ v(0) = 30 \end{cases}$$

A direction field for the differential equation $v' = 32 - \frac{2}{3}v$ is shown in Figure 8.

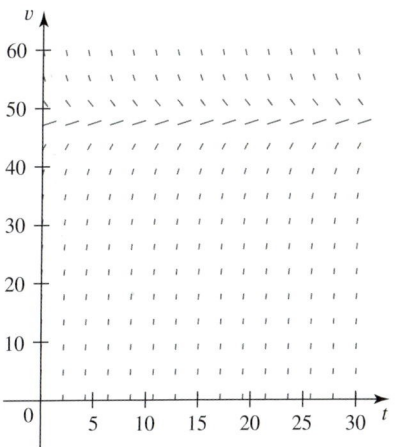

FIGURE 8
A direction field for $v' = 32 - \frac{2}{3}v$

b. The solution curve of the initial-value problem superimposed over the direction field is shown in Figure 9.

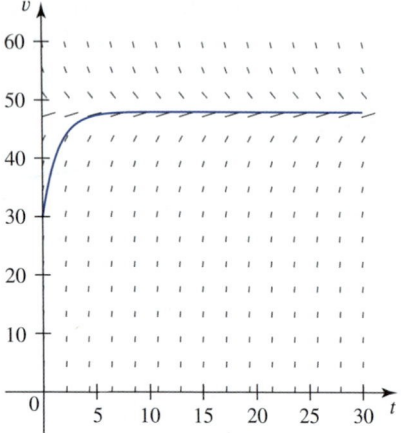

FIGURE 9
The solution curve passing
through the point (0, 30)

c. From the solution curve of part (b), we see that as $t \to \infty$, $v(t)$ seems to approach
48, and we conclude that the terminal velocity of the paratrooper is 48 ft/sec. ■

■ Euler's Method

The technique that we used to sketch the solution curves of a differential equation
helps to reveal the nature of the solution of the equation, but it does not provide us
with an accurate solution of the problem. We now turn our attention to a method that
provides us with a more accurate approximation to the solution of the initial-value
problem

$$\frac{dy}{dx} = F(x, y) \qquad y(x_0) = y_0 \tag{3}$$

Euler's method, named after the Swiss mathematician Leonard Euler (1707–1783),
calls for approximating the actual solution $y = f(x)$ at certain selected values of x. The
values of f between two adjacent values of x are then found by linear interpolation.
This situation is depicted geometrically in Figure 10. As you can see, the actual solution of the differential equation is approximated by a suitable polygonal curve.

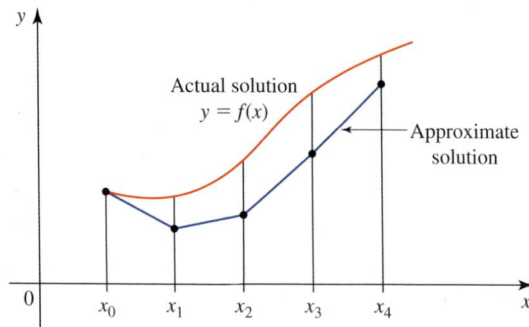

FIGURE 10
In using Euler's method, the
actual solution curve of the differential equation is approximated by a polygonal curve.

To describe the method, let h be a small positive number, and let

$$x_n = x_0 + nh \qquad n = 1, 2, 3, \dots$$

That is,

$$x_1 = x_0 + h, \qquad x_2 = x_0 + 2h, \qquad x_3 = x_0 + 3h, \qquad \dots$$

Observe that the points x_0, x_1, x_2, x_3, ... are spaced evenly apart, and the distance between any two adjacent points is h units.

We begin by finding an approximation y_1 to the value of the actual solution, $f(x_1)$, at $x = x_1$. Observe that the *initial* condition $y(x_0) = y_0$ tells us that the point (x_0, y_0) lies on the solution curve. Euler's method calls for approximating the part of the graph of f on the interval $[x_0, x_1]$ by the straight-line segment that is tangent to the graph of f at (x_0, y_0). (See Figure 11.) To find an equation of this straight-line segment, observe that the slope of this line segment is equal to $F(x_0, y_0)$. So using the point-slope form of an equation of a line, we see that the required equation is

$$y - y_0 = F(x_0, y_0)(x - x_0)$$

or

$$y = y_0 + F(x_0, y_0)(x - x_0)$$

Therefore, the approximation y_1 to $f(x_1)$ is obtained by replacing x by x_1. Thus,

$$y_1 = y_0 + F(x_0, y_0)(x_1 - x_0)$$
$$= y_0 + F(x_0, y_0)h \qquad \text{Since } x_1 - x_0 = h$$

This situation is depicted in Figure 11.

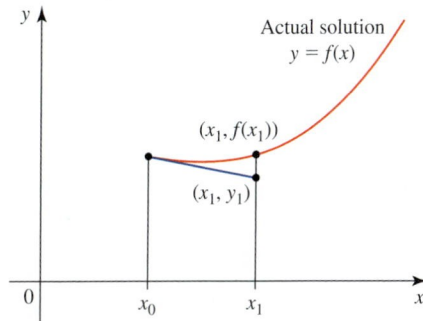

FIGURE 11

$y_1 = y_0 + hF(x_0, y_0)$ is an approximation of $f(x_1)$.

Next, to find an approximation y_2 to the value of the actual solution, $f(x_2)$, at $x = x_2$, we repeat the above procedure, this time taking the slope of the straight-line segment on $[x_1, x_2]$ to be $F(x_1, y_1)$. We obtain

$$y_2 = y_1 + hF(x_1, y_1)$$

(See Figure 12.)

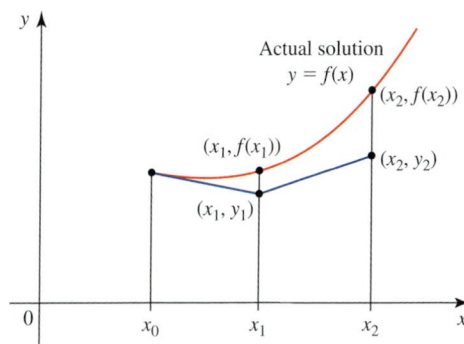

FIGURE 12

$y_2 = y_1 + hF(x_1, y_1)$ is the number used to approximate $f(x_2)$.

Continuing in this manner, we see that $y_1, y_2, \ldots, y_n$ can be found by the general formula

$$y_n = y_{n-1} + hF(x_{n-1}, y_{n-1}) \qquad n = 1, 2, \ldots$$

We now summarize this procedure.

Euler's Method

Suppose that we are given the differential equation

$$\frac{dy}{dx} = F(x, y)$$

subject to the initial condition $y(x_0) = y_0$, and we wish to find an approximation of $y(b)$, where b is a number greater than x_0 and n is a positive integer. Compute

$$h = \frac{b - x_0}{n}$$

$$x_1 = x_0 + h \qquad \text{and} \qquad y_0 = y(x_0)$$
$$x_2 = x_0 + 2h \qquad\qquad\qquad y_1 = y_0 + hF(x_0, y_0)$$
$$x_3 = x_0 + 3h \qquad\qquad\qquad y_2 = y_1 + hF(x_1, y_1)$$
$$\vdots \qquad\qquad\qquad\qquad\qquad \vdots$$
$$x_n = x_0 + nh = b \qquad\qquad y_n = y_{n-1} + hF(x_{n-1}, y_{n-1})$$

Then y_n gives an approximation of the true value $y(b)$ of the solution to the initial-value problem at $x = b$.

EXAMPLE 3 Use Euler's method with (a) $n = 5$ and (b) $n = 10$ to approximate the solution of the initial-value problem

$$y' = -2xy^2 \qquad y(0) = 1$$

on the interval $[0, 0.5]$. Find the actual solution of the initial-value problem. Finally, sketch the graphs of the approximate solutions and the actual solution for $0 \le x \le 0.5$ on the same set of axes.

Solution

a. Here, $x_0 = 0$ and $b = 0.5$. Taking $n = 5$, we find

$$h = \frac{0.5 - 0}{5} = 0.1$$

and $x_0 = 0$, $x_1 = 0.1$, $x_2 = 0.2$, $x_3 = 0.3$, $x_4 = 0.4$, and $x_5 = b = 0.5$. Also,

$$F(x, y) = -2xy^2 \qquad \text{and} \qquad y_0 = y(0) = 1$$

Therefore

$$y_0 = y(0) = 1$$

$$y_1 = y_0 + hF(x_0, y_0) = 1 + 0.1(-2)(0)(1)^2 = 1$$

$$y_2 = y_1 + hF(x_1, y_1) = 1 + 0.1(-2)(0.1)(1)^2 = 0.98$$

$$y_3 = y_2 + hF(x_2, y_2) = 0.98 + 0.1(-2)(0.2)(0.98)^2 \approx 0.9416$$

$$y_4 = y_3 + hF(x_3, y_3) = 0.9416 + 0.1(-2)(0.3)(0.9416)^2 \approx 0.8884$$

$$y_5 = y_4 + hF(x_4, y_4) = 0.8884 + 0.1(-2)(0.4)(0.8884)^2 \approx 0.8253$$

b. Here, $x_0 = 0$ and $b = 0.5$. Taking $n = 10$, we find

$$h = \frac{0.5 - 0}{10} = 0.05$$

and $x_0 = 0$, $x_1 = 0.05$, $x_2 = 0.10$, ..., $x_9 = 0.45$, and $x_{10} = 0.5 = b$. Proceeding as in part (a), we obtain the approximate solutions listed in the following table.

x_n	0.00	0.05	0.10	0.15	0.20	0.25	0.30	0.35	0.40	0.45	0.50
y_n	1.0000	1.0000	0.9950	0.9851	0.9705	0.9517	0.9291	0.9032	0.8746	0.8440	0.8119

To obtain the actual solution of the differential equation, we separate variables, obtaining

$$\frac{dy}{y^2} = -2x \, dx$$

Integrating each side of the last equation with respect to the appropriate variable, we have

$$\int \frac{dy}{y^2} = \int -2x \, dx$$

or

$$-\frac{1}{y} = -x^2 + C_1$$

$$\frac{1}{y} = x^2 + C \qquad C = -C_1$$

$$y = \frac{1}{x^2 + C}$$

Using the condition $y(0) = 1$, we have

$$\frac{1}{0 + C} = 1 \qquad \text{or} \qquad C = 1$$

Therefore, the required solution is given by

$$y = \frac{1}{x^2 + 1}$$

The graphs of the approximate solutions and the actual solution are sketched in Figure 13.

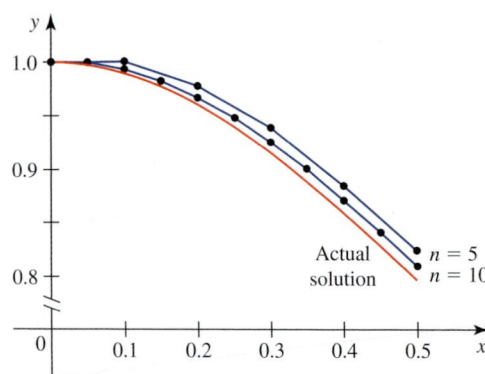

FIGURE 13

The approximate solutions
and the actual solution to
the initial-value problem

EXAMPLE 4 **Parachute Jump**   Refer to Example 2. There, we showed that the differential equation describing the motion of a paratrooper was

$$\frac{dv}{dt} = 32 - \frac{2}{3}v \qquad v(0) = 30$$

where v is the velocity of the paratrooper at time t. Use Euler's method with $h = 0.2$ to estimate the velocity of the paratrooper 2 sec after his parachute was deployed.

Solution   Here, $F(t, v) = 32 - \frac{2}{3}v$. With a step size of 0.2 ($n = 10$), we find

$$t_0 = 0, \qquad t_1 = 0.2, \qquad t_2 = 0.4, \qquad t_3 = 0.6, \qquad t_4 = 0.8, \qquad t_5 = 1.0,$$

$$t_6 = 1.2, \qquad t_7 = 1.4, \qquad t_8 = 1.6, \qquad t_9 = 1.8, \qquad t_{10} = 2.0$$

Therefore,

$$v_0 = 30$$

$$v_1 = v_0 + hF(t_0, v_0) = 30 + 0.2\left[32 - \frac{2}{3}(30)\right] = 32.4$$

$$v_2 = v_1 + hF(t_1, v_1) = 32.4 + 0.2\left[32 - \frac{2}{3}(32.4)\right] = 34.48$$

$$v_3 = v_2 + hF(t_2, v_2) = 34.48 + 0.2\left[32 - \frac{2}{3}(34.48)\right] \approx 36.28267$$

$$v_4 = v_3 + hF(t_3, v_3) = 36.28267 + 0.2\left[32 - \frac{2}{3}(36.28267)\right] \approx 37.84498$$

$$v_5 = v_4 + hF(t_4, v_4) = 37.84498 + 0.2\left[32 - \frac{2}{3}(37.84498)\right] \approx 39.19898$$

$$v_6 = v_5 + hF(t_5, v_5) = 39.19898 + 0.2\left[32 - \frac{2}{3}(39.19898)\right] \approx 40.37245$$

Carrying on, we find

$$v_7 \approx 41.38946, \qquad v_8 \approx 42.27087, \qquad v_9 \approx 43.03475, \qquad \text{and} \qquad v_{10} \approx 43.69678$$

So his velocity 2 sec after deployment of the parachute is $v_{10} \approx 43.70$ ft/sec.

7.2   CONCEPT QUESTIONS

1. **a.** What is the direction field of the differential equation
   $y' = f(x, y)$?
   b. Explain how you would use the direction field of part (a)
   to sketch a solution curve passing through the point
   (x_0, y_0).

2. Explain how Euler's method is used to approximate the
   solution of an initial-value problem.

7.2   EXERCISES

In Exercises 1–4, match the differential equation with the direction field labeled (a)–(d). Give a reason for your choice.

1. $y' = 1 - \dfrac{y}{2}$

2. $y' = 1 + xy$

3. $y' = 2x + y$

4. $y' = \sin x \cos y$

(a)

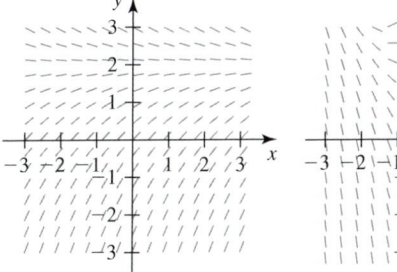

(b)

(c)

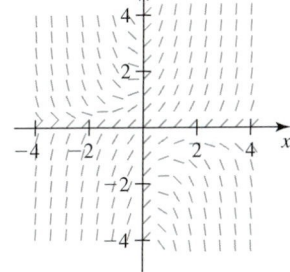

(d)

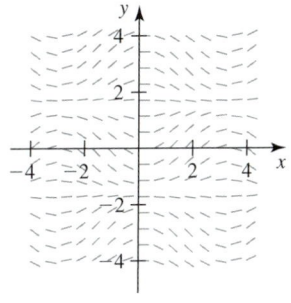

In Exercises 5–8 a direction field for the differential equation is given. Sketch the solution curves that satisfy the initial condition.

5. $y' = x^2 - y$
   a. $y(-2) = 0$ **b.** $y(0) = 0$ **c.** $y(0) = 1$ **d.** $y(1) = 0$

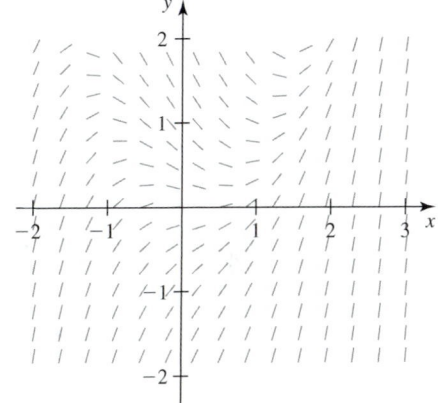

6. $y' = 1 - \dfrac{1}{4}y$
   a. $y(0) = 1$ **b.** $y(0) = 4$ **c.** $y(0) = 6$

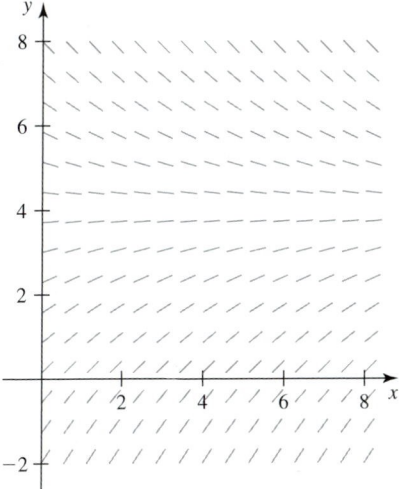

7. $y' = x^2 + y^2$

   a. $y(0) = 0$ **b.** $y(0) = 1$ **c.** $y(0) = 2$

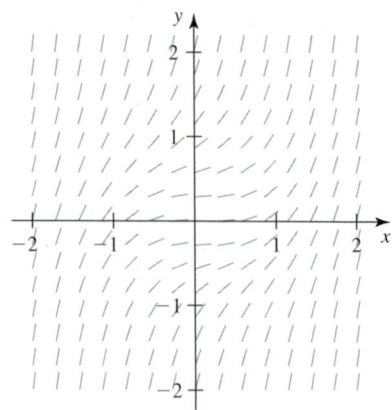

8. $y' = \sin x \sin y$

   a. $y(0) = -1$ **b.** $y(0) = 0$ **c.** $y(0) = 1$

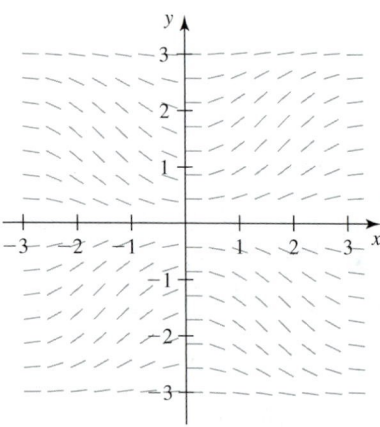

cas *In Exercises 9–16, use a computer algebra system (CAS) to draw a direction field for the differential equation. Then sketch approximate solution curves passing through the given points by hand superimposed over the direction field. Compare your sketch with the solution curve obtained by using a CAS.*

9. $y' = y$

   a. $(0, -1)$ **b.** $(0, 0)$ **c.** $(0, 1)$

10. $y' = y - 2$

   a. $(0, 1)$ **b.** $(0, 2)$ **c.** $(0, 4)$

11. $y' = x + y + 1$

   a. $(0, -2)$ **b.** $(0, 0)$ **c.** $(0, 1)$

12. $y' = \dfrac{1}{4}x^2 + y$

   a. $(0, -2)$ **b.** $(0, 1)$ **c.** $(1, 3)$

13. $y' = -\dfrac{x}{y}$

   a. $(-1, 1)$ **b.** $(2, 0)$ **c.** $(0, 4)$

14. $y' = x(2 - y)$

   a. $(0, -1)$ **b.** $(0, 2)$ **c.** $(0, 4)$

15. $y' = \cos x - y \tan x$

   a. $(0, -1)$ **b.** $(0, 0)$ **c.** $(0, 1)$

16. $y' = e^{x-y}$

   a. $(0, 0)$ **b.** $(0, 1)$ **c.** $(2, 1)$

In Exercises 17–26, use Euler's method with (a) $n = 4$ and (b) $n = 6$ to estimate $y(b)$, where y is the solution of the initial-value problem (accurate to two decimal places).

17. $y' = x + y, \quad y(0) = 1; \quad b = 1$

18. $y' = x - 2y, \quad y(0) = 1; \quad b = 2$

19. $y' = 2x - y + 1, \quad y(0) = 2; \quad b = 2$

20. $y' = 2xy, \quad y(0) = 1; \quad b = 0.5$

21. $y' = -2xy^2, \quad y(0) = 1; \quad b = 0.5$

22. $y' = 1 + xy^2, \quad y(0) = 1; \quad b = 0.8$

23. $y' = \sqrt{x + y}, \quad y(0) = 1; \quad b = 1.5$

24. $y' = (x^2 + y^2)^{-1}, \quad y(0) = 1; \quad b = 1$

25. $y' = \dfrac{x}{y}, \quad y(0) = 1; \quad b = 1$

26. $y' = xy^{1/3}, \quad y(0) = 1; \quad b = 1$

In Exercises 27 and 28, (a) sketch a few solution curves of the differential equation on the direction field, (b) solve the initial-value problem, and (c) sketch the solution curve found in part (b) on the direction field.

27. $\dfrac{dy}{dx} = -\dfrac{x}{y}, \quad y(2) = 2\sqrt{3}$

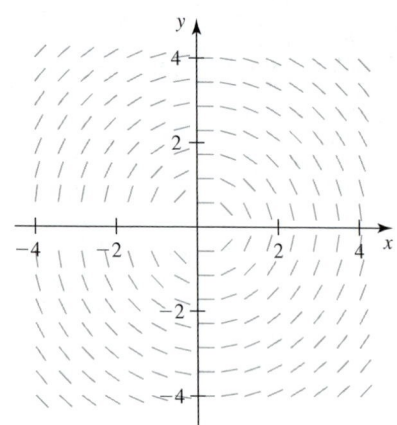

28. $\dfrac{dy}{dx} = y + xy, \quad y(0) = 1$

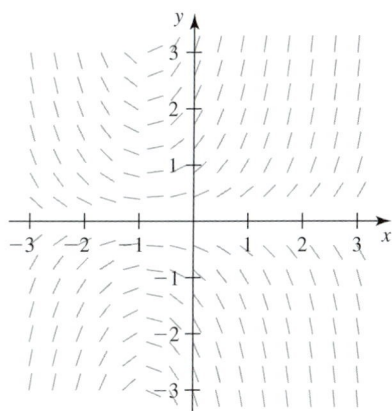

cas **29. Gompertz Growth Curves** The differential equation
$P' = P(a - b \ln P)$, where a and b are constants,
is called a **Gompertz differential equation.** This
differential equation occurs in the study of population
growth and the growth of tumors.
 a. Take $a = b = 1$ in the Gompertz differential equation,
 and use a CAS to draw a direction field for the differential equation.
 b. Use the direction field of part (a) to sketch the approximate curves for solutions satisfying the initial conditions
 $P(0) = 1$ and $P(0) = 4$.
 c. What can you say about $P(t)$ as t tends to infinity? If the
 limit exists, what is its approximate value?

cas **30. Restricted Population Growth** The differential equation
$P' = P(a - bP)$, where a and b are constants, is called a
logistic equation. This differential equation is used in the
study of restricted population growth.
 a. Take $a = 2$ and $b = 1$, and use a CAS to draw a direction field for the differential equation.
 b. Use the direction field of part (a) to sketch the approximate solution curves passing through the points $(0, -1)$,
 $(0, 1)$, and $(0, 3)$.
 c. Suppose that $P(0) = c$. For what values of c does
 $\lim_{t\to\infty} P(t)$ exist? What is the value of the limit?

31. Parachute Jump A skydiver, together with her parachute and
equipment, have a combined weight of 160 lb. At the instant
of deployment of the parachute, she is falling vertically
downward at a speed of 30 ft/sec. Suppose that the air
resistance varies directly as the instantaneous velocity and
that the air resistance is 30 lb when her velocity is 30 ft/sec.
 a. Use Euler's method with $n = 10$ to estimate her velocity
 2 sec after deployment of her parachute.
 b. Find the exact solution of the separable differential equation, and compute $v(2)$. Compare the answers obtained in
 parts (a) and (b).

32. R-C Series Circuit The figure shows an R-C series circuit containing a resistor with a resistance of R ohms, and a capacitor with a capacitance of C farads. The voltage drop across
the capacitor is $Q(t)/C$, where Q is the charge (in coulombs)
in the capacitor. Using Kirchhoff's Second Law, we have

$$RI + \frac{Q}{C} = E(t)$$

where $E(t)$ is the electromotive force (emf) in volts. But
$I = dQ/dt$, and this gives

$$R\frac{dQ}{dt} + \frac{1}{C}Q = E(t)$$

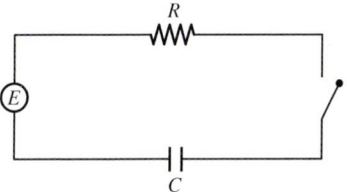

Suppose that an emf of 100 volts is applied to an R-C series
circuit in which the resistance is 50 ohms and the capacitance is 0.001 farad.
 a. Draw a direction field for this differential equation.
 b. Sketch the solution curve passing through the point
 $(0, 0)$.
 c. Using Euler's method with $n = 10$, estimate the charge
 0.1 sec after the switch is closed.
 d. Find the charge $Q(t)$ at time t by solving the separable
 differential equation if the initial charge is 0 coulomb.
 Sketch the graph of Q, and compare this result with that
 obtained in part (b).

*In Exercises 33–36, determine whether the statement is true or
false. If it is true, explain why. If it is false, explain why or give
an example that shows it is false.*

33. At each point (x, y) on a solution curve of the differential
equation $y' = f(x, y)$, a small line segment that contains the
point (x, y) and has slope $f(x, y)$ is drawn. The result is a
direction field of the differential equation.

34. The lineal elements in the direction field of a differential
equation constitute parts of the solution curve of the differential equation.

35. The lineal elements in the direction field of a differential
equation of the form $y' = f(y)$ at the point (x, y_0) are parallel to each other for all values of x and each fixed y_0.

36. The lineal elements in the direction field of a differential
equation of the form $y' = f(x)$ at the point (x_0, y) are parallel to each other for all values of y and each fixed x_0.

7.3 | The Logistic Equation

In Section 7.1 we considered a model for population growth in which the rate of change of the population at any time is proportional to the current population. Thus,

$$\frac{dP}{dt} = kP \tag{1}$$

where $P(t)$ is the population at time t, and k, the positive constant of proportionality, is the *growth constant*. Unfortunately, this model describing *unrestricted growth* is not very realistic. In the real world, one might expect that the population would grow rapidly at first and then eventually slow down because of overcrowding, scarcity of food, and other environmental factors. Indeed, one might expect that the population would eventually stabilize at a level that is compatible with the life-support capacity of the environment. In this section we will study a population model that exhibits precisely these characteristics.

■ The Logistic Model

We can rewrite Equation (1) in the form

$$\frac{\dfrac{dP}{dt}}{P} = k$$

This tells us that the *relative growth* rate of the population in the unrestricted growth model is a (positive) constant k. Suppose that the population cannot exceed a number L, called the **carrying capacity** of the environment. Then it is reasonable to assume that the relative growth rate of the population starts out at k when P is small and approaches zero when P is close to L. In other words, we want a model of the form

$$\frac{\dfrac{dP}{dt}}{P} = f(P)$$

where f satisfies $f(0) = k$ and $f(L) = 0$. The simplest function f satisfying these conditions is the linear function whose graph is the straight line passing through the points $(0, k)$ and $(L, 0)$. (See Figure 1.) You can verify that the desired function is

$$f(P) = k\left(1 - \frac{P}{L}\right)$$

This discussion leads to the following model for restricted population growth, known as the **logistic differential equation:**

$$\frac{dP}{dt} = kP\left(1 - \frac{P}{L}\right) \tag{2}$$

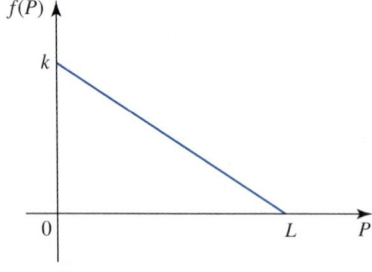

FIGURE 1
The graph of the linear function f satisfying $f(0) = k$ and $f(L) = 0$

Observe that if P is small relative to L, then P/L is small and $dP/dt \approx kP$; that is, the logistic model behaves like the unrestricted growth model. But as P approaches L, then P/L approaches 1, and the rate of growth of P, dP/dt, approaches 0. Thus, the logistic differential equation exhibits both the property of rapid growth initially and that of saturation eventually. Also, note that if the (initial) population P exceeds the carrying capacity L, then $1 - (P/L)$ is negative and $dP/dt < 0$, so the population decreases.

The following example of the logistic differential equation verifies these properties graphically.

EXAMPLE 1   **Logistic Growth Function**   Sketch the direction field for the logistic differential equation with $k = 0.05$ and $L = 1000$. Then draw the approximate solution curves of the equation satisfying the initial conditions $P(0) = 100$, $P(0) = 1400$, and $P(0) = 1000$ superimposed upon the direction field.

Solution   The logistic differential equation under consideration is

$$\frac{dP}{dt} = 0.05P\left(1 - \frac{P}{1000}\right)$$

Using a graphing utility, we obtain the direction field for this equation shown in Figure 2a. Note that the slopes are the same along any horizontal line. This occurs because the logistic differential equation is *autonomous;* that is, P' depends on P alone. The solution curves satisfying the initial conditions $P(0) = 100$, $P(0) = 1400$, and $P(0) = 1000$ are shown in Figure 2b–d.

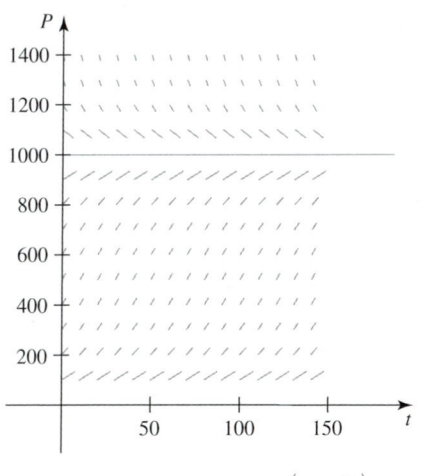

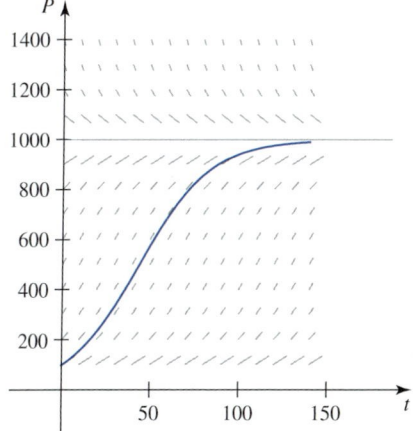

(a) Direction field for $P' = 0.05P\left(1 - \frac{P}{1000}\right)$

(b) Solution curve with $P(0) = 100$

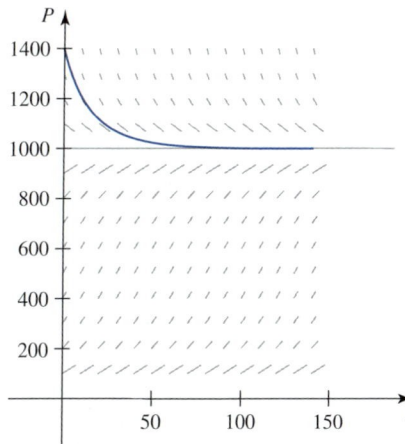

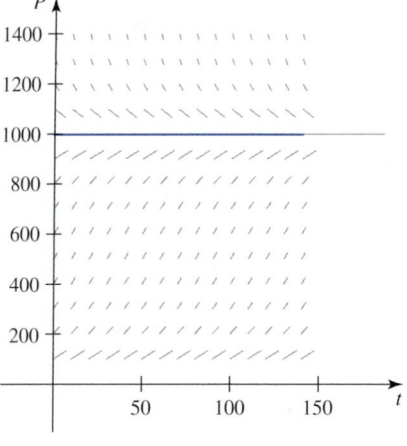

FIGURE 2   **(c)** Solution curve with $P(0) = 1400$

(d) Solution curve with $P(0) = 1000$

Note that in the two cases in which the initial populations do not begin at 1000, the carrying capacity of the environment, both populations tend to 1000 as t increases without bound. But in the case in which the initial population is 1000, the population remains steady at that level for all values of t. ■

■ Analytic Solution of the Logistic Differential Equation

The logistic differential equation (2) is separable and can be solved by using the method of Section 7.1.

EXAMPLE 2 Solve the logistic differential equation

$$\frac{dP}{dt} = kP\left(1 - \frac{P}{L}\right) \qquad P(0) = P_0$$

Solution First, $P = 0$ and $P = L$ are solutions, as you can verify by substituting these values into the differential equation. Next, suppose that $P \neq 0$ and $P \neq L$. Observe that the equation is separable. Separating variables leads to

$$\frac{dP}{P\left(1 - \dfrac{P}{L}\right)} = k\,dt$$

Integrating each side of this equation with respect to the appropriate variable, we obtain

$$\int \frac{dP}{P\left(1 - \dfrac{P}{L}\right)} = \int k\,dt \qquad \text{or} \qquad \int \frac{L}{P(L - P)}\,dP = k\int dt \tag{3}$$

To find the integral on the left-hand side, we use partial fraction decomposition (see Section 6.4) to write

$$\frac{L}{P(L - P)} = \frac{1}{P} + \frac{1}{L - P}$$

This leads to

$$\int \left(\frac{1}{P} + \frac{1}{L - P}\right) dP = \int k\,dt$$

$$\ln|P| - \ln|L - P| = kt + C_1$$

$$\ln|L - P| - \ln|P| = -kt - C_1 \qquad \text{\color{orange}Multiply each side by } -1.$$

$$\ln\left|\frac{L - P}{P}\right| = -kt - C_1$$

$$\left|\frac{L - P}{P}\right| = e^{-kt-C_1} = e^{-C_1}e^{-kt} = C_2 e^{-kt} \qquad \color{orange}C_2 = e^{-C_1}$$

or

$$\frac{L - P}{P} = Ce^{-kt} \tag{4}$$

where $C = \pm C_2$. We can solve for P in Equation (4) as follows:

$$\frac{L}{P} - 1 = Ce^{-kt}$$

$$\frac{L}{P} = 1 + Ce^{-kt}$$

and

$$P = \frac{L}{1 + Ce^{-kt}}$$

To determine C, we use the initial condition $P(0) = P_0$, where P_0 is the initial population. Putting $t = 0$ and $P = P_0$ in Equation (4) yields

$$\frac{L - P_0}{P_0} = Ce^0 = C$$

Thus, the solution of the initial-value problem is

$$P(t) = \frac{L}{1 + \left(\dfrac{L}{P_0} - 1\right)e^{-kt}} \tag{5}$$

Note that

$$\lim_{t \to \infty} P(t) = \lim_{t \to \infty} \frac{L}{1 + \left(\dfrac{L}{P_0} - 1\right)e^{-kt}} = L \tag{6}$$

as expected. The graph of Equation (5) is called the **logistic curve.**

Logistic Curve

Example 1 suggests the shape of the logistic curve. We are now in the position to confirm this observation analytically.

We begin by determining the intervals where P is increasing and where it is decreasing. To do this, we could compute $P'(t)$ from Equation (5), but this would be tedious and unnecessary. Instead, we will work with Equation (2), which expresses P' in terms of P. Observe that

$$P' = kP\left(1 - \frac{P}{L}\right) \tag{7}$$

is a continuous function of P on $(-\infty, \infty)$ and has zeros at $P = 0$ and $P = L$. The sign diagram for P' is shown in Figure 3.

FIGURE 3
The sign diagram for P'

$$0 +\!+\!+\!+\!+\!+\!+ 0 -\!-\!-\!-\!-\!-\!-\!-\!-\!-\!-$$

Observe that this sign diagram is not the same as the sign diagrams that we encountered in Chapter 3. Here, P is the *dependent variable,* and it lies on the *vertical* axis.

From the sign diagram for P' we conclude that P is increasing for $0 < P < L$ and P is decreasing for $P > L$. Next, we compute

$$P'' = \frac{d}{dt}\left[kP\left(1 - \frac{P}{L}\right)\right] = k\frac{d}{dt}\left(P - \frac{P^2}{L}\right) \qquad \text{Use Equation (7).}$$

$$= k\left(1 - \frac{2P}{L}\right)P' = k^2P\left(1 - \frac{2P}{L}\right)\left(1 - \frac{P}{L}\right)$$

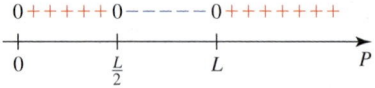

FIGURE 4

The sign diagram for P''

Observe that P'', as a function of P, is continuous on $(-\infty, \infty)$ and has zeros at $P = 0$, $L/2$, and L. The sign diagram of P'' is shown in Figure 4. From the sign diagram for P'' we conclude that the graph of P is concave upward for $0 < P < \frac{L}{2}$ and $P > L$ and is concave downward for $\frac{L}{2} < P < L$. Also, P has an inflection point at $P = L/2$. We have two cases. Referring to the sign diagrams for P' and P'', we see the following:

1. If $0 < P < L$, then P is increasing, concave upward for $0 < P < \frac{L}{2}$, and concave downward for $\frac{L}{2} < P < L$. Also, the graph of P has an inflection point at $P = L/2$. (See Figure 5.)
2. If $P > L$, then P is decreasing and concave upward.

It can be shown, although we will not do so here, that none of these curves can cross the horizontal lines $P = 0$ and $P = L$.

Let's summarize our results.

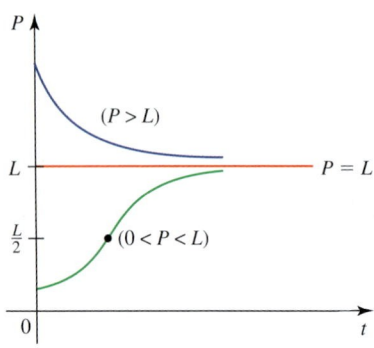

FIGURE 5

Two possible logistic curves

Suppose that a population at any time t satisfies the logistic differential equation (2) and that the initial population at $t = 0$ is P_0.

1. If $P_0 = 0$, the population stays at zero at all times
2. If $0 < P_0 < L$, then the population increases and approaches the limiting value L, called the carrying capacity of the environment, asymptotically. The population increases most rapidly at the instant of time when it reaches $\frac{1}{2}L$.
3. If $P_0 = L$, the population at any later time remains at L.
4. If $P_0 > L$, then the population decreases and approaches the carrying capacity L asymptotically.

In Exercise 9 you will be asked to show that the time referred to in part (2) when the population increases most rapidly is given by

$$T = \frac{\ln\left(\dfrac{L}{P_0} - 1\right)}{k} \tag{8}$$

Note The constant solutions $P = 0$ and $P = L$ are called **equilibrium solutions.** ■

EXAMPLE 3 **Logistic Growth Function** Refer to Example 1. Suppose that the population $P(t)$ satisfies the logistic differential equation

$$\frac{dP}{dt} = 0.05P\left(1 - \frac{P}{1000}\right)$$

a. What is the population at any time t if the initial population is 1000?
b. What is the population at any time t if the initial population is 1400?
c. What is the population at any time t if the initial population is 100?

Solution Using Equation (5) with $k = 0.05$ and $L = 1000$, we see that the population at time t is

$$P(t) = \frac{L}{1 + \left(\dfrac{L}{P_0} - 1\right)e^{-kt}} = \frac{1000}{1 + \left(\dfrac{1000}{P_0} - 1\right)e^{-0.05t}} \tag{9}$$

a. Here, $P_0 = 1000$, so Equation (9) gives

$$P(t) = \frac{1000}{1 + \left(\frac{1000}{1000} - 1\right)e^{-0.05t}} = 1000$$

That is, the population stays at the equilibrium level for all t. (See Figure 6a.)
b. Here, $P_0 = 1400$, so Equation (9) gives

$$P(t) = \frac{1000}{1 + \left(\frac{1000}{1400} - 1\right)e^{-0.05t}} = \frac{1000}{1 - \frac{2}{7}e^{-0.05t}}$$

(See Figure 6b.) The population decreases to 1000 asymptotically.
c. Here, $P_0 = 100$, so Equation (9) gives

$$P(t) = \frac{1000}{1 + \left(\frac{1000}{100} - 1\right)e^{-0.05t}} = \frac{1000}{1 + 9e^{-0.05t}}$$

(See Figure 6c.) The population increases to 1000 asymptotically.

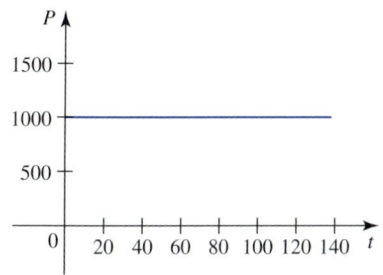

(a) $P(0) = 1000$

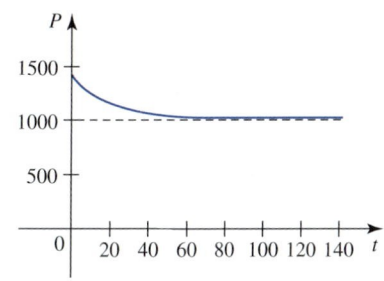

(b) $P(0) = 1400$

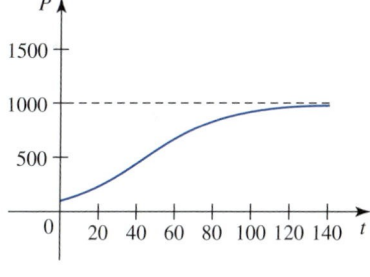

(c) $P(0) = 100$

FIGURE 6
Logistic curves for $\dfrac{dP}{dt} = 0.05P\left(1 - \dfrac{P}{1000}\right)$

EXAMPLE 4 **Logistic Growth Function** Refer to Example 3. Suppose that the population $P(t)$ satisfies the logistic differential equation

$$\frac{dP}{dt} = 0.05P\left(1 - \frac{P}{1000}\right)$$

where t is measured in days and the initial population is 100.

a. What is the population at $t = 30$? At $t = 50$?
b. At what time is the population increasing most rapidly?
c. At what time is the population equal to 800?

Solution

a. The population at any time t was obtained in part (c) of Example 3. We have

$$P(t) = \frac{1000}{1 + 9e^{-0.05t}}$$

The population at $t = 30$ is

$$P(30) = \frac{1000}{1 + 9e^{-0.05(30)}} \approx 332$$

The population at $t = 50$ is

$$P(50) = \frac{1000}{1 + 9e^{-0.05(50)}} \approx 575$$

b. The population increases most rapidly at the instant it reaches half the carrying capacity of the environment. Therefore, we can find the required time by solving the equation

$$500 = \frac{1000}{1 + 9e^{-0.05t}}$$

for t. Alternatively, we can use Equation (8) to find

$$T = \frac{\ln\left(\dfrac{L}{P_0} - 1\right)}{k} = \frac{\ln\left(\dfrac{1000}{100} - 1\right)}{0.05} \approx 43.944$$

So the population increases most rapidly on approximately the 44th day.

c. The required time is found by solving

$$800 = \frac{1000}{1 + 9e^{-0.05t}}$$

for t. We have

$$1 + 9e^{-0.05t} = \frac{1000}{800} = \frac{5}{4}$$

$$9e^{-0.05t} = \frac{1}{4}$$

$$e^{-0.05t} = \frac{1}{36}$$

$$-0.05t = \ln\frac{1}{36} = -\ln 36 \qquad \text{Take the natural logarithms.}$$

$$t = \frac{\ln 36}{0.05} \approx 71.67$$

So the population reaches 800 when t is approximately 72, or approximately after 72 days.

The graph of P is shown in Figure 7.

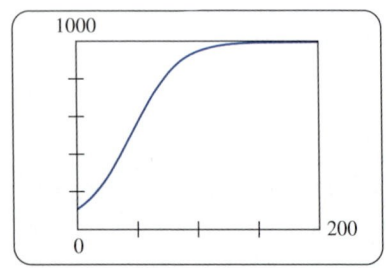

FIGURE 7
The graph of $P(t) = \dfrac{1000}{1 + 9e^{-0.05t}}$

EXAMPLE 5 **Rate of Growth of a Fish Population** A fish farm is stocked with 100 fish. Suppose that the fish population satisfies the logistic equation and that the carrying capacity of the pond is 2000.

a. Find an expression for the fish population after t years if the number of fish increased to 250 in the first year.
b. How long will it take for the fish population to reach 1000?

Solution

a. Using Equation (5), we see that the population after t years is

$$P(t) = \frac{L}{1 + \left(\dfrac{L}{P_0} - 1\right)e^{-kt}}$$

Here, $L = 2000$ and $P_0 = P(0) = 100$, so

$$P(t) = \frac{2000}{1 + \left(\dfrac{2000}{100} - 1\right)e^{-kt}} = \frac{2000}{1 + 19e^{-kt}}$$

To determine k, we use the condition $P(1) = 250$. This leads to

$$P(1) = \frac{2000}{1 + 19e^{-k}} = 250$$

$$1 + 19e^{-k} = \frac{2000}{250} = 8$$

$$19e^{-k} = 7$$

$$e^{-k} = \frac{7}{19}$$

$$k = -\ln\frac{7}{19} = 0.9985$$

so

$$P(t) = \frac{2000}{1 + 19e^{-0.9985t}}$$

b. We solve the equation $P(t) = 1000$ for t; that is,

$$1000 = \frac{2000}{1 + 19e^{-0.9985t}}$$

$$1 + 19e^{-0.9985t} = 2$$

$$e^{-0.9985t} = \frac{1}{19}$$

$$-0.9985t = \ln\left(\frac{1}{19}\right)$$

$$t \approx 2.949$$

So it takes approximately 2.9 years for the fish population to reach 1000.

The graph of P is shown in Figure 8.

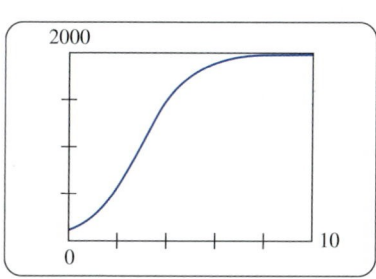

FIGURE 8
The graph of $P(t) = \dfrac{2000}{1 + 19e^{-0.9985t}}$

7.3   CONCEPT QUESTIONS

1. Consider the logistic differential equation $\dfrac{dP}{dt} = kP\left(1 - \dfrac{P}{L}\right)$.

   a. What does k represent? What does L represent?

   b. Write two constant solutions of the equation, and explain their meaning.

   c. What can you say about the rate of change of the population if the initial population is greater than L? If it is greater than zero but less than L? Interpret your answers.

2. a. Verify by direct computation that

   $$P(t) = \dfrac{L}{1 + \left(\dfrac{L}{P_0} - 1\right)e^{-kt}}$$

   is a solution of the initial-value problem

   $$\dfrac{dP}{dt} = kP\left(1 - \dfrac{P}{L}\right) \qquad P(0) = P_0$$

   b. Describe the solution of the logistic differential equation of part (a) with the aid of a graph. Assume that $P_0 > 0$.

7.3   EXERCISES

In Exercises 1–4, a logistic differential equation describing population growth is given. Use the equation to find (a) the growth constant and (b) the carrying capacity of the environment.

1. $\dfrac{dP}{dt} = 0.02P\left(1 - \dfrac{P}{1000}\right)$

2. $\dfrac{dP}{dt} = 0.03P - 0.000006P^2$

3. $\dfrac{dP}{dt} = P\left(0.5 - \dfrac{P}{1000}\right)$

4. $150{,}000\dfrac{dP}{dt} = 3P(2000 - P)$

5. A direction field of a logistic differential equation describing population growth is shown in the figure.

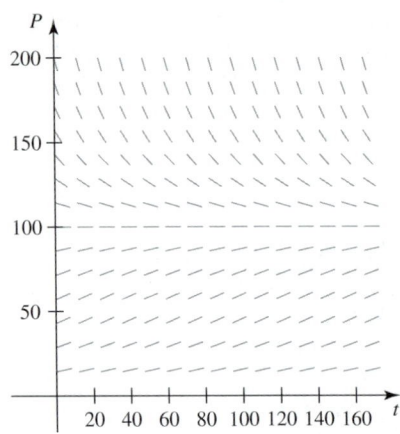

   a. What is the carrying capacity of the environment?

   b. What are the constant solutions?

   c. Sketch the solution curve with an initial population of 200.

   d. Sketch the solution curve with an initial population of 100.

   e. Sketch the solution curve with an initial population of 10.

6. A direction field of a modified logistic differential equation describing population growth is shown in the figure.

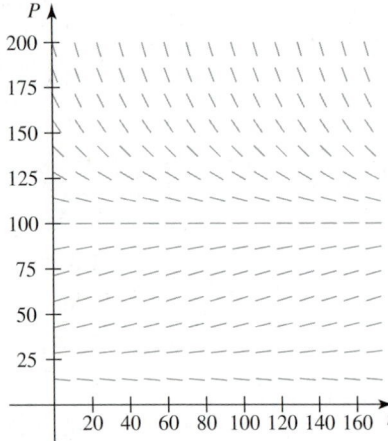

   a. What is the carrying capacity of the environment?

   b. What are the constant solutions?

   c. Sketch the solution curve with an initial population of 120.

   d. Sketch the solution curve with an initial population of 60.

   e. Sketch the solution curve with an initial population of 10.

In Exercises 7 and 8, use the given logistic equation to find (a) the growth constant, (b) the carrying capacity of the environment, and (c) the initial population.

7. $P(t) = \dfrac{8000}{2 + 798e^{-0.02t}}$

8. $P(t) = \dfrac{100e^{0.2t}}{e^{0.2t} + 19}$

9. Consider the logistic differential equation $\dfrac{dP}{dt} = kP\left(1 - \dfrac{P}{L}\right)$.

   a. Show that $P(t)$ grows most rapidly when $P = L/2$.

   b. Show that $P(t)$ grows most rapidly at time

   $$T = \frac{\ln\left(\dfrac{L}{P_0} - 1\right)}{k}$$

   where P_0 is the initial population.

10. **Spread of an Epidemic** During a flu epidemic the number of children in the Woodbridge Community School System who contracted influenza after t days was given by

    $$Q(t) = \frac{1000}{1 + 199e^{-0.8t}}$$

    a. How many children were stricken by the flu after the first day?

    b. How many children had the flu after 10 days?

    c. How many children eventually contracted the disease?

11. **Lay Teachers at Roman Catholic Schools** The change from religious to lay teachers at Roman Catholic schools has been partly attributed to the decline in the number of women and men entering religious orders. The percentage of teachers who are lay teachers is given by

    $$f(t) = \frac{98}{1 + 2.77e^{-t}} \qquad 0 \le t \le 4$$

    where t is measured in decades, with $t = 0$ corresponding to the beginning of 1960.

    a. What percentage of teachers were lay teachers at the beginning of 1990?

    b. Find the year when the percentage of lay teachers was increasing most rapidly.

    Source: National Catholic Education Association and the Department of Education.

12. **People Living with HIV** On the basis of data compiled by the World Health Organization, it is estimated that the number of people living with HIV worldwide from 1985 through 2006 is

    $$N(t) = \frac{39.88}{1 + 18.94e^{-0.2957t}} \qquad 0 \le t \le 21$$

    where $N(t)$ is measured in millions and t in years with $t = 0$ corresponding to the beginning of 1985.

    a. How many people were living with HIV worldwide at the beginning of 1985? At the beginning of 2005?

    b. Assuming that the trend continued, how many people were living with HIV worldwide at the beginning of 2008?

    Source: World Health Organization.

13. **Growth of a Fruit Fly Population** Initially, there were 10 fruit flies (*Drosophila melanogaster*) in an experiment. Because of a limit to be placed on the amount of food available, the maximum population of fruit flies was estimated to be 100.

    a. Suppose that the pattern of growth followed the logistic curve and that the population was 34 after 30 days. Find an expression for the fruit fly population t days after the start of the experiment.

    b. How long did it take the population to reach 80?

14. **Rate of Growth for a Plant** The rate of growth of a certain type of plant is described by a logistic differential equation. Botanists have estimated the maximum theoretical height of such plants to be 30 in. At the beginning of an experiment, the height of a plant was 5 in., and the plant grew to 12 in. after 20 days.

    a. Find an expression for the height of the plant after t days.

    b. What was the height of the plant after 30 days?

    c. How long did it take for the plant to reach 80% of its maximum theoretical height?

15. **Logistic Growth Function** Consider the logistic growth function

    $$P(t) = \frac{L}{1 + \left(\dfrac{L}{P_0} - 1\right)e^{-kt}}$$

    Suppose that the population is P_1 when $t = t_1$ and P_2 when $t = t_2$. Show that the value of k is

    $$k = \frac{1}{t_2 - t_1} \ln\left[\frac{P_2(L - P_1)}{P_1(L - P_2)}\right]$$

16. **Logistic Growth Function** The carrying capacity of a colony of fruit flies (*Drosophila melanogaster*) is 600. The population of fruit flies after 14 days is 76, and the population after 21 days is 167. What is the value of the growth constant k? **Hint:** Use the result of Exercise 15.

17. **Rate of Growth of a Fish Population** Let $P(t)$ denote the population of a certain species of fish in a lake, where t is measured in weeks. Then P can be described by the modified logistic differential equation

    $$\frac{dP}{dt} = kP\left(1 - \frac{P}{L}\right) - c$$

    where k is the growth rate, L is the carrying capacity of the environment, and c is the constant rate at which fish are being removed because of fishing. Suppose that $k = 0.2$, $L = 800$, and $c = 30$.

    a. Draw a direction field for the resulting differential equation.

    b. Use the direction field that was obtained in part (a) to find the equilibrium solutions. Verify your results algebraically.

    c. Sketch the solution curves for the solutions with initial populations of 100, 300, and 700, and describe what happens to the fish population in each case.

18. Gompertz Growth Curves The **Gompertz differential equation,** a model for restricted population growth, is obtained by modifying the logistic differential equation and is given by

$$\frac{dP}{dt} = cP \ln\left(\frac{L}{P}\right)$$

where c is a constant and L is the carrying capacity of the environment.

a. Find the equilibrium solution of the differential equation.

b. Illustrate graphically the solutions of the equation with initial conditions $P(0) = P_0$, where (i) $P_0 > L$, (ii) $P_0 = L$, and (iii) $0 < P_0 < L$.

19. Gompertz Growth Curves Refer to Exercise 18. Consider the Gompertz differential equation with $L = 1000$ and $c = 0.02$.

a. Draw a direction field for the differential equation.

b. Identify the equilibrium solution.

c. Plot the solution curve with initial conditions $P(0) = 1200$ and $P(0) = 100$.

20. Gompertz Growth Curves Refer to Exercises 18 and 19. Consider the Gompertz differential equation

$$\frac{dP}{dt} = cP \ln\left(\frac{L}{P}\right)$$

where c is a positive constant and L is the carrying capacity of the environment.

a. Solve the differential equation.

b. Find $\lim_{t \to \infty} P(t)$.

c. Show that $P(t)$ is increasing most rapidly when $P = L/e$.

d. Show that $P(t)$ is increasing most rapidly when

$$t = \frac{\ln \ln\left(\dfrac{L}{P_0}\right)}{c}$$

21. A Goldfish Population Refer to Exercise 20. A population of 20 goldfish was introduced into a pond that has an estimated carrying capacity of 200 fish. After 1 month, the population

of goldfish had grown to 80. If the pattern of growth of the population followed the Gompertz curve, how many goldfish were in the pond after 3 months?

22. Cyclical Models Some populations are subject to seasonal fluctuations. The population in a vacation resort serves as one example. A model for describing such situations is the differential equation

$$\frac{dP}{dt} = (k \cos t)P$$

where k is a constant and t is measured in months.

a. Find the solution of the differential equation subject to $P(0) = P_0$.

b. Let $k = 0.2$, and plot the graphs of P for $P_0 = 400$, 500, and 600.

c. What happens to $P(t)$ for large values of t?

In Exercises 23–26, determine whether the statement is true or false. If it is true, explain why. If it is false, explain why or give an example that shows it is false.

23. If P is the solution of the initial-value problem

$$P' = 0.2P\left(1 - \frac{P}{100}\right), \quad P(0) = 150, \text{ then } P(t) \text{ is}$$

decreasing on the interval $(0, \infty)$.

24. If P is the solution of the initial-value problem

$$P' = 0.3P\left(1 - \frac{P}{20}\right), \quad P(0) = 0, \text{ then } P(t) \text{ is}$$

increasing on the interval $(0, \infty)$.

25. If P is the solution of the initial-value problem

$$P' = 0.5P\left(1 - \frac{P}{50}\right), \quad P(0) = 10, \text{ then the graph}$$

of P has an inflection point.

26. If P is the solution of the initial-value problem

$$P' = 0.02P\left(1 - \frac{P}{1000}\right), \quad P(0) = 1000, \text{ then}$$

$\lim_{t \to \infty} P(t) = 1000$.

7.4   First-Order Linear Differential Equations

We now consider another class of first-order differential equations. A **first-order linear differential equation** is one that can be written in the form

$$\frac{dy}{dx} + P(x)y = Q(x) \tag{1}$$

where P and Q are continuous functions of x on a given interval. The equation is so named because it is *linear* in the unknown function and its derivative. A linear equation written in the form of Equation (1) is said to be in **standard form.** For example, the differential equation

$$x\frac{dy}{dx} - y - x^3 = 0$$

is a linear equation, since it is linear in both y and dy/dx. By dividing through by x and rearranging terms, we obtain the equation in standard form, namely,

$$\frac{dy}{dx} - \frac{1}{x}y = x^2$$

Here, $P(x) = -\dfrac{1}{x}$ and $Q(x) = x^2$. On the other hand, the equations

$$y\frac{dy}{dx} + 2y = e^x \qquad \text{and} \qquad \frac{dy}{dx} + 2\cos y = x^3$$

are not linear because of the nonlinear term $y(dy/dx)$ in the first equation and the nonlinear term $\cos y$ in the second equation.

■ Method of Solution

First-order linear differential equations can be solved by multiplying both sides of the equation

$$\frac{dy}{dx} + P(x)y = Q(x)$$

by a suitable function $u(x)$ that transforms the equation into one that can be solved by integration. To find $u(x)$, let's consider the equation that is obtained by putting $Q(x) = 0$. The resulting equation

$$\frac{dy}{dx} + P(x)y = 0 \tag{2}$$

is called a *homogeneous* linear equation. Observe that Equation (2) is a separable equation and, therefore, can be solved using the method of separation of variables. We find

$$\int \frac{dy}{y} = -\int P(x)\,dx$$

$$\ln|y| = -\int P(x)\,dx + C_1$$

$$|y| = e^{C_1}e^{-\int P(x)\,dx}$$

$$y = Ce^{-\int P(x)\,dx} \qquad C = \pm e^{C_1} \tag{3}$$

The solution of the homogeneous equation associated with Equation (1) points the way to solving the nonhomogeneous equation (1) itself. We rewrite Equation (3) in the form

$$ye^{\int P(x)\,dx} = C$$

Let's differentiate this last equation using the Fundamental Theorem of Calculus, Part 1. Thus,

$$\frac{d}{dx}\left[ye^{\int^x P(t)\,dt}\right] = \frac{d}{dx}(C) \qquad \text{\color{brown}Rewrite the integral using the dummy variable t.*}$$

$$\frac{dy}{dx}e^{\int^x P(t)\,dt} + ye^{\int^x P(t)\,dt}\cdot\frac{d}{dx}\int^x P(t)\,dt = 0 \qquad \text{\color{brown}Use the Product Rule and the Chain Rule.}$$

$$\frac{dy}{dx}e^{\int^x P(t)\,dt} + ye^{\int^x P(t)\,dt}\cdot P(x) = 0 \qquad \text{\color{brown}Use the Fundamental Theorem of Calculus, Part 1.}$$

---

*Henceforth, we will usually write $e^{\int^x P(t)\,dt}$ in the form $e^{\int P(x)\,dx}$ to conform with the more standard practice.

or

$$e^{\int^x P(t)\,dt}\left[\frac{dy}{dx} + P(x)y\right] = 0$$

Observe that the expression within the square brackets is just the expression on the left-hand side of Equation (1). This suggests that by multiplying both sides of Equation (1) by $e^{\int P(x)\,dx}$, we have

$$e^{\int P(x)\,dx}\left[\frac{dy}{dx} + P(x)y\right] = e^{\int P(x)\,dx}Q(x)$$

which in turn can be written in the form

$$\frac{d}{dx}\left[ye^{\int P(x)\,dx}\right] = e^{\int P(x)\,dx}Q(x) \tag{4}$$

The expression on the left of Equation (4) is easily integrated because it is the derivative of a function. Since the function on the right does not involve the unknown function, it can also be integrated. Thus, the function u that we are seeking is

$$u(x) = e^{\int P(x)\,dx}$$

This function is called an **integrating factor** because multiplying Equation (1) by u, as we have just seen, enables us to solve the problem by integration. Before looking at an example, let's summarize the steps in solving a linear differential equation.

Solving a First-Order Linear Differential Equation

1. Rewrite the equation in standard form $y' + P(x)y = Q(x)$ if necessary.
2. Find an integrating factor $u(x) = e^{\int P(x)\,dx}$.
3. Multiply both sides of the equation $y' + P(x)y = Q(x)$ by $u(x)$. The resulting equation can be written in the form

$$\frac{d}{dx}(yu) = uQ$$

Unknown function   Integrating factor

which can then be integrated.

EXAMPLE 1 Solve the equation $x\dfrac{dy}{dx} - y - x^3 = 0$, where $x > 0$.

Solution

Step 1 We rewrite the equation in standard form

$$\frac{dy}{dx} - \frac{1}{x}y = x^2 \tag{5}$$

and identify $P(x) = -1/x$ and $Q(x) = x^2$

Step 2 We find

$$u(x) = e^{\int P(x)\,dx} = e^{-\int (1/x)\,dx} = e^{-\ln x} = e^{\ln x^{-1}} = \frac{1}{x}$$

Step 3 Multiplying both sides of Equation (5) by $u(x) = 1/x$, we obtain

$$\frac{1}{x}\frac{dy}{dx} - \frac{1}{x^2}y = x$$

which can be written in the form

$$\frac{d}{dx}\left[\frac{1}{x}y\right] = x$$

Integrating
factor

Unknown
function

Integrating both sides with respect to x, we obtain

$$\frac{1}{x}y = \int x\,dx = \frac{1}{2}x^2 + C$$

so $y = \frac{1}{2}x^3 + Cx$. ∎

Note If we integrate Equation (4), we can obtain a formula for the solution to the problem. Thus,

$$ye^{\int^x P(t)\,dt} = \int^x e^{\int^u P(t)\,dt}Q(u)\,du + C$$

or

$$y = e^{-\int^x P(t)\,dt}\int^x e^{\int^u P(t)\,dt}Q(u)\,du + Ce^{-\int^x P(t)\,dt}$$

But it is preferable to use the method of solution just described. ∎

Historical Biography

Science Photo Library/Photo Researchers, Inc.

JACOB BERNOULLI
(1654-1705)

Going against his father's wish that he enter the ministry, Jacob Bernoulli followed his personal interests and studied mathematics and astronomy. He eventually founded a school for science and mathematics, where he lectured on mathematics and mechanics and carried out experiments in physics. He was the first in his mathematically talented family to pursue a career in mathematics. He was followed by his younger brother Johann and three of Johann's sons. Jacob's own two children did not pursue careers in mathematics or the sciences. In 1687, Jacob Bernoulli was named professor of mathematics at the University of Basel, a seat that he held until his death in 1705. He was among the first mathematicians of his time to fully understand the newly developing calculus. One of his many contributions to mathematics was a method to solve differential equations of the form $y' + P(x)y = Q(x)y^n$, a type of differential equation that is now known as Bernoulli's differential equation (see Exercise 23 in this section). Bernoulli also made important contributions to the theory of probability and to the study of mechanics.

EXAMPLE 2 Solve the initial-value problem

$$\begin{cases} x^2y' + 3xy = e^{-x^2} & x > 0 \\ y(1) = 0 \end{cases}$$

Solution First, we rewrite the differential equation in standard form by dividing both sides by x^2, obtaining

$$y' + \frac{3}{x}y = \frac{e^{-x^2}}{x^2} \tag{6}$$

An integrating factor is

$$u(x) = e^{\int (3/x)\,dx} = e^{3\ln x} = e^{\ln x^3} = x^3$$

Multiplying both sides of Equation (6) by x^3 gives

$$x^3y' + 3x^2y = xe^{-x^2}$$

which can be rewritten in the form

$$\frac{d}{dx}(x^3y) = xe^{-x^2}$$

Integrating both sides with respect to x yields

$$x^3 y = \int xe^{-x^2}\, dx = -\frac{1}{2} e^{-x^2} + C$$

$$y = -\frac{e^{-x^2}}{2x^3} + \frac{C}{x^3}$$

Since $y(1) = 0$, we have

$$-\frac{e^{-1}}{2} + C = 0 \qquad \text{or} \qquad C = \frac{1}{2e}$$

Therefore, the required solution is

$$y = -\frac{e^{-x^2}}{2x^3} + \frac{1}{2ex^3}$$

Our first application of first-order linear differential equations is a *mixture problem.*

FIGURE 1

EXAMPLE 3   Mixture Problem   A tank initially contains 400 gal of water in which 50 lb of salt has been dissolved. Brine containing 2 lb of salt per gallon enters the tank at the rate of 3 gal/min. The well-stirred solution drains from the tank at the rate of 5 gal/min. How much salt is in the tank after 30 min? (See Figure 1.)

Solution   Let $y(t)$ denote the amount of salt in the tank (in pounds) at time t (in minutes). Then the rate of change of the amount of salt at time t is

$$\frac{dy}{dt} = (\text{rate of salt entering}) - (\text{rate of salt exiting})$$

$$\text{rate of salt entering} = (\text{concentration of brine entering}) \cdot (\text{rate of flow in})$$

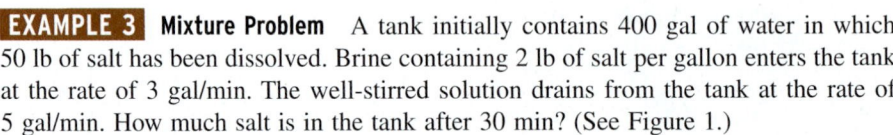

$$= (2)(3) = 6 \text{ lb/min} \qquad \frac{\text{lb}}{\text{gal}} \cdot \frac{\text{gal}}{\text{min}} = \frac{\text{lb}}{\text{min}}$$

$$\text{rate of salt exiting} = (\text{concentration of brine in tank}) \cdot (\text{rate of flow out})$$

$$= \left(\frac{\text{amount of salt in the tank at time } t}{\text{volume of brine in the tank at time } t} \right) \cdot (\text{rate of flow out})$$

But the volume of brine at time t is given by

$$(\text{initial volume}) + (\text{net change in volume})$$

$$= (\text{initial volume}) + (\text{rate of flow in minus rate of flow out})t$$

$$= 400 + (3 - 5)t = 400 - 2t$$

So the rate of salt exiting is

$$\frac{y}{400 - 2t} \cdot 5 = \frac{5y}{400 - 2t} \qquad 400 - 2t > 0$$

Therefore,

$$\frac{dy}{dt} = 6 - \frac{5y}{400 - 2t}$$

The condition that there are 50 lb of salt in the tank initially translates into the initial condition $y(0) = 50$. The mathematical formulation has led to the initial-value problem

$$\begin{cases} \dfrac{dy}{dt} = 6 - \dfrac{5y}{400 - 2t} & 0 \le t < 200 \\ y(0) = 50 \end{cases}$$

To solve the first-order linear differential equation, we first write it in standard form

$$\frac{dy}{dt} + \frac{5}{400 - 2t} y = 6 \tag{7}$$

and identify $P(t) = 5/(400 - 2t)$. An integrating factor is

$$u(t) = e^{\int [5/(400 - 2t)] \, dt} = e^{-(5/2)\ln(400 - 2t)} = (400 - 2t)^{-5/2}$$

Multiplying both sides of Equation (7) by $u(t)$, we obtain

$$(400 - 2t)^{-5/2} \frac{dy}{dt} + (400 - 2t)^{-5/2} \left(\frac{5}{400 - 2t} \right) y = 6(400 - 2t)^{-5/2}$$

which we can write as

$$\frac{d}{dt} [(400 - 2t)^{-5/2} y] = 6(400 - 2t)^{-5/2}$$

Integrating both sides with respect to t yields

$$(400 - 2t)^{-5/2} y = 6 \int (400 - 2t)^{-5/2} \, dt$$

$$= 6 \left(-\frac{1}{2} \right) \left(-\frac{2}{3} \right) (400 - 2t)^{-3/2} + C$$

or

$$y = 4(200 - t) + C(400 - 2t)^{5/2}$$

To determine the value of C, we use the initial condition $y = 50$ when $t = 0$, giving

$$4(200) + C(400)^{5/2} = 50$$

or

$$C = -\frac{750}{400^{5/2}}$$

Therefore,

$$y = 4(200 - t) - \frac{750}{400^{5/2}} (400 - 2t)^{5/2}$$

The amount of salt in the tank after 30 min is given by

$$y(30) = 4(200 - 30) - \frac{750}{400^{5/2}} (400 - 60)^{5/2} \approx 180.42$$

or approximately 180 lb.

First-order linear differential equations also arise in the analysis of electrical circuits. For example, suppose that we are given an electric circuit consisting of a battery

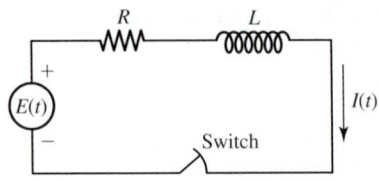

FIGURE 2
A single-loop series circuit

or generator having an electromotive force (emf) of $E(t)$ volts in series with a resistor having a resistance of R ohms and an inductor having an inductance of L henries. (See Figure 2.) According to Kirchhoff's Second Law, the emf that is supplied (E) is equal to the voltage drop across the inductor ($L\,dI/dt$) plus the voltage drop across the resistor (RI), where $I(t)$ is the current in amperes at time t. Thus, the differential equation for the circuit is a first-order linear equation

$$L\frac{dI}{dt} + RI = E \tag{8}$$

EXAMPLE 4 **Electric Circuits** A 12-volt battery is connected in series with a 10-ohm resistor and an inductor of 2 henries. If the switch is closed at time $t = 0$, determine (a) the current at time t, (b) the current $\frac{1}{10}$ second after the switch is closed, and (c) the current after a long time.

Solution
a. We put $L = 2$, $R = 10$, and $E = 12$ in Equation (8) and let $I(0) = 0$ to obtain the initial-value problem

$$\begin{cases} 2\dfrac{dI}{dt} + 10I = 12 \\ I(0) = 0 \end{cases}$$

Rewriting the first-order linear equation in standard form, we obtain

$$\frac{dI}{dt} + 5I = 6$$

An integrating factor for this equation is

$$u(t) = e^{\int 5\,dt} = e^{5t}$$

Multiplying the differential equation by $u(t)$ gives

$$e^{5t}\frac{dI}{dt} + 5e^{5t}I = 6e^{5t}$$

$$\frac{d}{dt}(e^{5t}I) = 6e^{5t}$$

$$e^{5t}I = \int 6e^{5t}dt = \frac{6}{5}e^{5t} + C$$

$$I(t) = \frac{6}{5} + Ce^{-5t}$$

Since $I(0) = 0$, we have $\frac{6}{5} + C = 0$, or $C = -\frac{6}{5}$. Therefore,

$$I(t) = \frac{6}{5}(1 - e^{-5t})$$

b. The current $\frac{1}{10}$ second after the switch is closed is given by

$$I\left(\frac{1}{10}\right) = \frac{6}{5}(1 - e^{-5/10}) \approx 0.47$$

or approximately 0.5 amp.

c. The current after a long time (called the **steady-state current**) is given by

$$\lim_{t \to \infty} I(t) = \lim_{t \to \infty} \frac{6}{5} (1 - e^{-5t})$$

$$= \frac{6}{5} - \frac{6}{5} \lim_{t \to \infty} e^{-5t}$$

$$= \frac{6}{5} - 0 = \frac{6}{5}$$

or 1.2 amp.

The graph of I is shown in Figure 3.

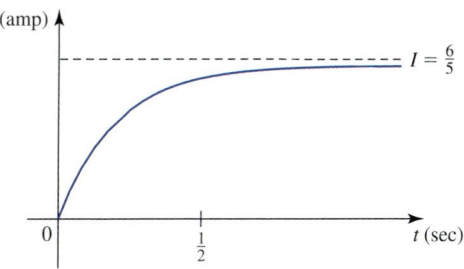

FIGURE 3
The current $I(t)$ approaches the steady-state current as $t \to \infty$.

EXAMPLE 5 **Electric Circuits** Suppose that the battery in the circuit of Example 4 is replaced by a generator having an emf of $E(t) = 20e^{-2t}$ volts. Find $I(t)$. What is the maximum current in the circuit?

Solution The only difference between this problem and that of Example 4 is that $E = 12$ is replaced by $E = 20e^{-2t}$. We have

$$2 \frac{dI}{dt} + 10I = 20e^{-2t} \qquad \text{or} \qquad \frac{dI}{dt} + 5I = 10e^{-2t}$$

We use the same integrating factor as before, obtaining

$$e^{5t} \frac{dI}{dt} + 5e^{5t}I = e^{5t} \cdot 10e^{-2t} = 10e^{3t}$$

$$\frac{d}{dt} (e^{5t}I) = 10e^{3t}$$

$$e^{5t}I = \int 10e^{3t} \, dt = \frac{10}{3} e^{3t} + C$$

$$I(t) = \frac{10}{3} e^{-2t} + Ce^{-5t}$$

Since $I(0) = 0$, we have $\frac{10}{3} + C = 0$ or $C = -\frac{10}{3}$. Therefore,

$$I(t) = \frac{10}{3} e^{-2t} - \frac{10}{3} e^{-5t}$$

To find the maximum current in the circuit, we set

$$I'(t) = -\frac{20}{3} e^{-2t} + \frac{50}{3} e^{-5t} = 0$$

obtaining

$$\frac{20}{3}e^{-2t} = \frac{50}{3}e^{-5t}$$

$$e^{3t} = \frac{50}{20}$$

$$\ln e^{3t} = \ln \frac{5}{2}$$

$$3t = \ln \frac{5}{2}$$

$$t \approx \frac{1}{3}\ln \frac{5}{2} \approx 0.3054$$

This is the only critical value of I on the interval $[0, \infty)$. Since $I(0) = 0$ and

$$\lim_{t\to\infty} I(t) = \lim_{t\to\infty}\left(\frac{10}{3}e^{-2t} - \frac{10}{3}e^{-5t}\right) = 0$$

we see that the maximum current occurs at $t \approx 0.3054$ and has a value of approximately

$$I(0.3054) \approx \frac{10}{3}e^{-2(0.3054)} - \frac{10}{3}e^{-5(0.3054)} \approx 1.09$$

that is, approximately 1.1 amp. The graph of I is shown in Figure 4.

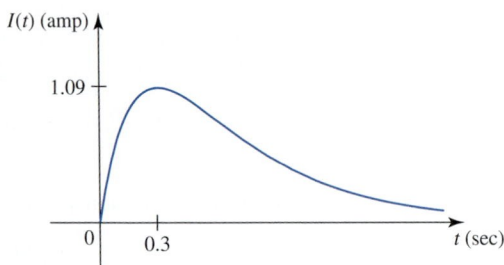

FIGURE 4
The graph of $I(t) = \dfrac{10}{3}e^{-2t} - \dfrac{10}{3}e^{-5t}$

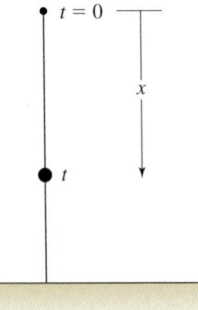

Our final example looks at an application of first-order linear differential equations to the motion of an object. Suppose that an object of constant mass m falls vertically downward under the influence of gravity. If we assume that air resistance is proportional to the speed of the object at any instant during the fall, then according to Newton's second law of motion, $F = ma = m(dv/dt)$, where F is the net force acting on the object in the positive (downward) direction. (See Figure 5.) But F is given by the weight of the object mg (acting downward) minus the air resistance kv (acting upward), where $k > 0$ is the constant of proportionality; that is, $F = mg - kv$. Therefore, the equation of motion is

FIGURE 5
The positive direction is downward.

$$m\frac{dv}{dt} = mg - kv \tag{9}$$

EXAMPLE 6 **Parachute Jump** A paratrooper and his equipment have a combined weight of 192 lb. At the instant that the parachute is deployed, he is traveling vertically downward at a speed of 30 ft/sec. Assume that air resistance is proportional to the instantaneous velocity with constant of proportionality $k = 4$.

a. Determine the velocity and position of the paratrooper at any time.
b. Find his limiting velocity by evaluating $\lim_{t \to \infty} v(t)$.

Solution
a. Here, $m = \frac{192}{32}$, or 6 slugs. Therefore, using Equation (9) we have the equation of motion

$$6 \frac{dv}{dt} = 192 - 4v \qquad \text{or} \qquad \frac{dv}{dt} = 32 - \frac{2}{3}v$$

At $t = 0$, $v = 30$, so we have the initial-value problem

$$\begin{cases} \dfrac{dv}{dt} + \dfrac{2}{3}v = 32 \\ v(0) = 30 \end{cases}$$

An integrating factor for this equation is

$$u(t) = e^{\int (2/3)\,dt} = e^{(2/3)t}$$

Multiplying the differential equation by $u(t)$ gives

$$e^{(2/3)t}\frac{dv}{dt} + \frac{2}{3}e^{(2/3)t}v = 32e^{(2/3)t}$$

$$\frac{d}{dt}(e^{(2/3)t}v) = 32e^{(2/3)t}$$

$$e^{(2/3)t}v = 32\int e^{(2/3)t}\,dt = 48e^{(2/3)t} + C$$

$$v(t) = 48 + Ce^{-(2/3)t}$$

Since $v(0) = 30$, we have $v(0) = 48 + C = 30$, or $C = -18$. Therefore,

$$v(t) = 48 - 18e^{-(2/3)t}$$

The position of the paratrooper at time t is

$$x(t) = \int v(t)\,dt = \int (48 - 18e^{-(2/3)t})\,dt = 48t + 27e^{-(2/3)t} + C_1$$

Since $x(0) = 0$, we have $27 + C_1 = 0$, or $C_1 = -27$. Therefore,

$$x(t) = 48t + 27e^{-(2/3)t} - 27$$

(See Figure 6 on the following page.)
b. The paratrooper's limiting (terminal) velocity is

$$\lim_{t \to \infty} v(t) = \lim_{t \to \infty}(48 - 18e^{-(2/3)t}) = 48 - \lim_{t \to \infty} 18e^{-(2/3)t} = 48$$

or 48 ft/sec. (See Figure 7 on the following page.)

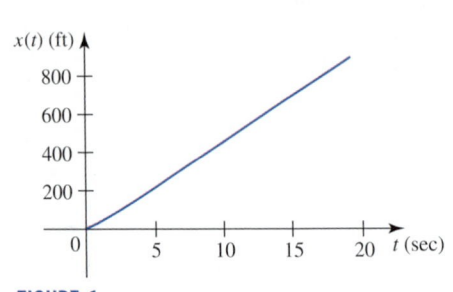

FIGURE 6
The position of the paratrooper at time t

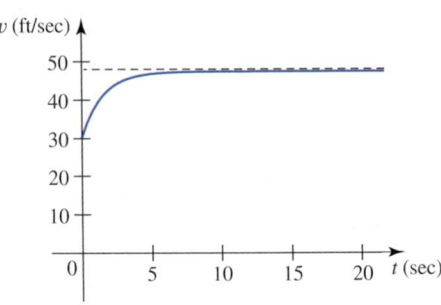

FIGURE 7
The velocity of the paratrooper at time t ▪

7.4   CONCEPT QUESTIONS

1. a. Write a first-order linear differential equation in standard form.
   b. Is the differential equation $a_0(x)y' + a_1(x)y = g(x)$, where $a_0(x) \neq 0$, a first-order linear differential equation?

2. a. What is an integrating factor for a first-order linear differential equation?
   b. Describe a method for solving a first-order linear differential equation.

7.4   EXERCISES

In Exercises 1–4, determine whether the differential equation is linear.

1. $\dfrac{dy}{dx} + xy^2 = \cos x$ **2.** $x^2 y' + e^x y = 4$

3. $y \cos y + \dfrac{1}{x}\dfrac{dy}{dx} - \ln x = 0$

4. $y^2 \dfrac{dx}{dy} + 3x = \tan y$

In Exercises 5–16, solve the differential equation.

5. $\dfrac{dy}{dx} + 2y = e^{2x}$ **6.** $x\dfrac{dy}{dx} + 3y = 2$

7. $xy' + y = x^3$ **8.** $y \sin x + y' \cos x = 1$

9. $\dfrac{dy}{dx} - \dfrac{2y}{x} = x^2 \cos 3x$ **10.** $\dfrac{dy}{dx} + y \cot x = \cos x$

11. $xy' - y = 2x(\ln x)^2$ **12.** $(\cos y - xe^y)\,dy = e^y\,dx$
                          Hint: Consider $x = f(y)$.

13. $(t + 1)\dfrac{dy}{dt} + y = t, \quad t > -1$

14. $xy' + (1 + x)y = e^{-x}(1 + \cos 2x)$

15. $xy' + (2x + 1)y = xe^{-2x}$

16. $\dfrac{dy}{dx} = \dfrac{y}{x + y^3}$
   Hint: Consider $x = f(y)$.

In Exercises 17–22, solve the initial-value problem.

17. $\dfrac{dy}{dx} + y = 1, \quad y(0) = -1$

18. $xy' - 3y = x^4, \quad y(1) = 5$

19. $\dfrac{dy}{dx} + 2xy = x, \quad y(0) = 1$

20. $\dfrac{dr}{d\theta} = \theta - \dfrac{r}{3\theta}, \quad r(1) = 1$

21. $x(x + 1)y' + xy = \ln x, \quad y(1) = \dfrac{1}{2}$

22. $(1 + e^x)\dfrac{dy}{dx} + e^x y = \sin x, \quad y(0) = \dfrac{1}{2}$

23. The equation

$$\frac{dy}{dx} + P(x)y = Q(x)y^n$$

where n is a constant, is called *Bernoulli's differential equation.*
   a. Show that the Bernoulli equation reduces to a linear equation if $n = 0$ or 1.
   b. Show that if $n \neq 0$ or 1, then changing the dependent variable from y to v using the transformation $v = y^{1-n}$ reduces the Bernoulli equation to the linear equation

$$\frac{dv}{dx} + (1 - n)P(x)v = (1 - n)Q(x)$$

24. Use the method of Exercise 23 to solve $y' - y = xy^2$.

25. Use the method of Exercise 23 to solve the initial-value problem

$$x^2y' - 2xy = 4y^3 \qquad y(1) = \sqrt{3}$$

26. a. Show that the differential equation

$$\frac{dy}{dx} + P(x)y = Q(x)y \ln y$$

can be solved by using the transformation $y = e^v$.
    b. Use the result of part (a) to solve $xy' - 2x^2y = y \ln y$.

27. The slope of the tangent line to the graph of a function $y = f(x)$ at the point (x, y) is $1 + y/x$. If the graph passes through the point $(1, 1)$, find f.

28. Mixture Problem A tank initially holds 16 gal of water in which 4 lb of salt has been dissolved. Brine that contains 6 lb of salt per gallon enters the tank at the rate of 2 gal/min, and the well-stirred mixture leaves at the same rate.
    a. Find a function that gives the amount of salt in the tank at time t.
    b. Find the amount of salt in the tank after 5 min.
    c. How much salt is in the tank after a long time?

29. Mixture Problem A tank initially holds 30 gal of pure water. Brine that contains 3 lb of salt per gallon enters the tank at the rate of 2 gal/min, and the well-stirred mixture leaves at the same rate.
    a. How much salt is in the tank at any time t?
    b. When will the tank hold 80 lb of salt?

30. Mixture Problem A tank initially holds 10 gal of water in which 2 lb of salt has been dissolved. Brine containing 1.5 lb of salt per gallon enters the tank at the rate of 2 gal/min, and the well-stirred mixture leaves at the rate of 3 gal/min.
    a. Find the amount of salt $y(t)$ in the tank at time t.
    b. Find the amount of salt in the tank after 10 min.
    c. Plot the graph of y.
    d. At what time is the amount of salt in the tank greatest? How much salt is in the tank at that time?

31. Mixture Problem A tank initially holds 40 gal of pure water. Brine that contains 2 lb of salt per gallon enters the tank at the rate of 1.5 gal/min, and the well-stirred mixture leaves at the rate of 2 gal/min.
    a. Find the amount of salt in the tank at time t.
    b. Find the amount of salt in the tank after 20 min.
    c. Find the amount of salt when the tank holds 20 gal of brine.
    d. Find the maximum amount of salt present.

32. Electric Circuit A 24-volt battery is connected in series with a 20-ohm resistor and an inductor of 4 henries. If the switch is closed at time $t = 0$, determine (a) the current at time t, (b) the current after 0.2 sec, and (c) the current after a long time.

33. Electric Circuit The figure shows an electric circuit consisting of a battery or generator of E volts in series with a resistor of R ohms and a capacitor of C farads. The voltage drop across the capacitor is Q/C where Q is the charge (in coulombs), so by Kirchhoff's Second Law,

$$RI + \frac{1}{C}Q = E$$

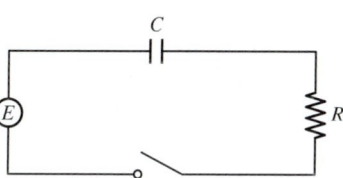

But $I = dQ/dt$, so we have the differential equation

$$R\frac{dQ}{dt} + \frac{1}{C}Q = E$$

Suppose that a circuit consists of a battery having a constant emf of 12 volts in series with a resistor of 10 ohms and a capacitor of 0.02 farad. The charge on the capacitor at $t = 0$ is 0.05 coulomb. Find the charge and the current at time t after the switch is closed.

34. Electric Circuit Suppose that the battery in the electric circuit of Exercise 33 is replaced by a generator having an emf of $E(t) = 30e^{-2t} + 10e^{-6t}$ volts. If the charge in the capacitor is 0 coulomb at $t = 0$, find the maximum charge on the capacitor.

35. Falling Weight An 8-lb weight is dropped from rest from a cliff. Assume that air resistance is equal to the weight's instantaneous velocity.
    a. Find the velocity of the weight at time t.
    b. What is the velocity of the weight after 1 sec?
    c. How long does it take for the weight to reach a speed of 4 ft/sec?

36. Parachute Jump A skydiver and his equipment have a combined weight of 192 lb. At the instant that his parachute is deployed, he is traveling vertically downward at a speed of 112 ft/sec. Assume that air resistance is proportional to the instantaneous velocity with a constant of proportionality of $k = 12$. Determine the position and velocity of the skydiver t sec after his parachute is deployed. What is his limiting velocity?

37. Sinking Boat As a boat weighing 1000 lb sinks in water from rest, it is acted upon by a buoyant force of 200 lb and a force of water resistance in pounds that is numerically equal to $100v$, where v is in feet per second. Find the distance traveled by the boat after 4 sec. What is its limiting velocity?

38. An object of mass m is thrown vertically upward with an initial velocity of v_0. Air resistance is proportional to its instantaneous velocity with constant of proportionality k. Show that the maximum height attained by the object is

$$\frac{mv_0}{k} - \frac{m^2 g}{k^2} \ln\left(1 + \frac{kv_0}{mg}\right)$$

39. Electric Circuit An electromotive force of $E_0 \cos \omega t$ volts, where E_0 and ω are constants, is applied to a series circuit consisting of a resistor of constant resistance R ohms and an inductor of constant inductance L henries. If we use Ohm's Law, the current $I(t)$, where t is time in seconds, satisfies the first-order linear differential equation

$$L\frac{dI}{dt} + RI = E_0 \cos \omega t$$

If the current I is 0 ampere initially, find an expression for the current at any time t.

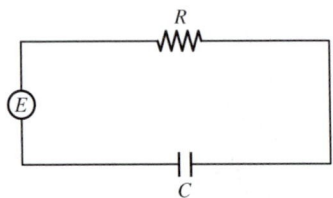

40. Market Equilibrium The quantity demanded of a certain commodity, $d(t)$, is related to its unit price $p(t)$, in dollars, by the *demand equation* $d(t) = 40 - p(t) + 2p'(t)$, where t denotes time. The quantity of the commodity made available by the supplier, $s(t)$, is related to the unit price $p(t)$, in dollars, by the *supply equation* $s(t) = 22 + 2p(t) + 3p'(t)$. Both $d(t)$ and $s(t)$ are measured in units of a thousand. *Market equilibrium* prevails when the demand is equal to the supply.
 a. If market equilibrium prevails, find the *equilibrium price* at time t if the price of the commodity is 10 dollars at $t = 0$.
 b. What happens to the price as $t \to \infty$?
 Note: If $\lim_{t \to \infty} p(t)$ exists, we say that there is price stability.

41. Suppose that y_1 is a solution of $y' + P(x)y = f(x)$ and y_2 is a solution of $y' + P(x)y = g(x)$. Show that $c_1 y_1 + c_2 y_2$ is a solution of $y' + P(x)y = c_1 f(x) + c_2 g(x)$ for all constants c_1 and c_2.

42. a. Find the general solution of

$$y' + \frac{2}{x}y = \frac{e^x}{x} \quad \text{and} \quad y' + \frac{2}{x}y = \frac{e^{-x}}{x}$$

 b. Use the result of Exercise 41 to write down the general solution of

$$y' + \frac{2}{x}y = \frac{2e^x}{x} - \frac{3e^{-x}}{x}$$

In Exercises 43–47, determine whether the statement is true or false. If it is true, explain why it is true. If it is false, explain why or give an example to show why it is false.

43. $y^2 \dfrac{dx}{dy} + e^y x = y \cos y$ is a first-order linear differential equation.

44. A first-order differential equation can be both separable and linear.

45. An integrating factor for the equation $a_0(x)y' + a_1(x)y = f(x)$ is $e^{\int [a_1(x)/a_0(x)]\, dx}$.

46. If y_1 is a solution of the homogeneous equation $y' + Py = 0$ associated with the nonhomogeneous equation $y' + Py = f$ and y_2 is a solution of the nonhomogeneous equation, then $y = cy_1 + y_2$ is a solution of the nonhomogeneous equation, where c is any constant.

47. The function $f(x) = 2e^x - \frac{1}{2}(\cos x + \sin x)$ is a solution of the differential equation $y' - y = \sin x$.

7.5   Predator-Prey Models

Up to now, we have dealt only with population models involving a single species. But there are many instances in nature in which one species of animals feeds on another species of animals that in turn feeds on other food that is readily available. For example, wolves hunt caribou, which feed on an unlimited supply of vegetation, and sharks feed on small fish, which in turn feed on plankton. The first species is called the *predator* and the second species is called the *prey*.

Let $x(t)$ denote the number of prey and let $y(t)$ denote the number of predators at time t. If there are no predators and there is an unlimited supply of food, then the prey population will grow at a rate that is proportional to the current population; that is,

$$\frac{dx}{dt} = ax \qquad a > 0 \tag{1}$$

In the absence of prey, the predator population will decline at a rate that is proportional to the current population; that is,

$$\frac{dy}{dt} = -ry \qquad r > 0 \tag{2}$$

When both predators and prey are present, however, we must modify both Equations (1) and (2) to take into account the interactions of the species. It seems reasonable to assume that the number of encounters between these two species is jointly proportional to their populations, that is, the number is proportional to the product xy. Since these encounters are detrimental to the prey population, the rate at which the prey population changes is *decreased* by the term bxy, where b is a positive constant. Similarly, these encounters are beneficial to the predator population, so the rate at which the predator population changes is *increased* by the term sxy, where s is a positive constant. Thus, we are led to the following:

$$\frac{dx}{dt} = ax - bxy$$
$$\frac{dy}{dt} = -ry + sxy \tag{3}$$

where a, b, r and s are positive constants.

The equations in system (3) are called *predator-prey equations*. They are also called **Lotka-Volterra equations** after the mathematicians Alfred Lotka (1880–1949) and Vito Volterra (1860–1940), who independently developed mathematical models to study how two species interact. The equations are autonomous since the expression on the right-hand side of each equation does not depend explicitly on the time t. A **solution** of system (3) is an ordered pair of functions $(x(t), y(t))$, where $x(t)$ and $y(t)$ give the populations of prey and predators at time t, respectively. Although the system of equations looks simple, no exact solutions have yet been found. Nevertheless, much insight into the nature of the solutions of the system of equations can be obtained without solving them.

Observe that system (3) can be written in the form

$$\frac{dx}{dt} = x(a - by)$$
$$\frac{dy}{dt} = y(-r + sx)$$

from which we see that $x = 0$ and $y = 0$, or $(0, 0)$, and $x = r/s$ and $y = a/b$, or $\left(\frac{r}{s}, \frac{a}{b}\right)$, are solutions of system (3). These points are called **critical points** or **equilibrium points** of the system.

The solution $(0, 0)$ merely represents the fact that if there aren't any predators and prey at some point in time, then this situation will remain so forever. The solution $\left(\frac{r}{s}, \frac{a}{b}\right)$ reflects the case in which the number of predators and prey are in a state of equilibrium. The number of prey, r/s, is at exactly the level that will sustain the number of predators, a/b.

What about other, less obvious solutions? To shed some light on the nature of these solutions, let's look at the problem graphically in an example.

EXAMPLE 1   Suppose that the population of rabbits (prey) in hundreds, $x(t)$, and the population of foxes (predators) in tens, $y(t)$, are described by the Lotka-Volterra equations with $a = 4$, $b = 1$, $r = 1$, and $s = 0.2$.

a. Write the Lotka-Volterra equations for this case.
b. Find the equilibrium points of the system.
c. Find an expression for dy/dx, and use it to draw a direction field for the resulting differential equation in the xy-plane.
d. Sketch some solution curves for the differential equation found in part (c).

Solution

a. The required equations are

$$\frac{dx}{dt} = 4x - xy$$

$$\frac{dy}{dt} = -y + 0.2xy$$

b. To find the equilibrium points of the system, we set $dx/dt = 0$ and $dy/dt = 0$, simultaneously, obtaining

$$4x - \quad xy = x(4 - y) \qquad = 0$$

$$-y + 0.2xy = y(-1 + 0.2x) = 0$$

from which we obtain the constant solutions (equilibrium points)

$$x = 0 \qquad \text{and} \qquad y = 0$$

or $(0, 0)$, and

$$x = \frac{1}{0.2} = 5 \qquad \text{and} \qquad y = 4$$

or $(5, 4)$. So, if at any moment in time there are no foxes and no rabbits, then there will be no predators, or prey at any time later. The other equilibrium point tells us that 500 rabbits is the exact number needed to support 40 foxes.

c. Using the Chain Rule, we have

$$\frac{dy}{dx} = \frac{\dfrac{dy}{dt}}{\dfrac{dx}{dt}} = \frac{-y + 0.2xy}{4x - xy}$$

Next, we draw a direction field for this equation using a graphing utility (see Figure 1).

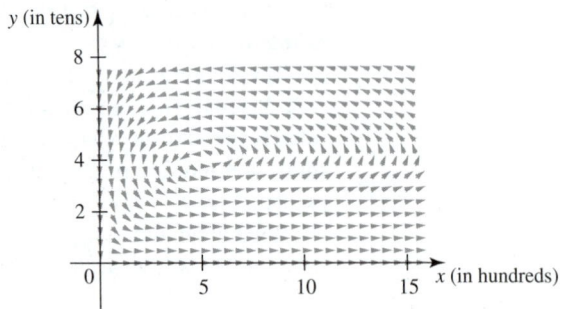

FIGURE 1
A direction field for Equation (1)

d. A few solution curves for the differential equation in part (c) are sketched in Figure 2. We have included the equilibrium points $(0, 0)$ and $(5, 4)$ in the figure. Note that the equilibrium point $(5, 4)$ lies inside the solution curves. Also, note that the solution curves are closed (a fact that we will not prove here).

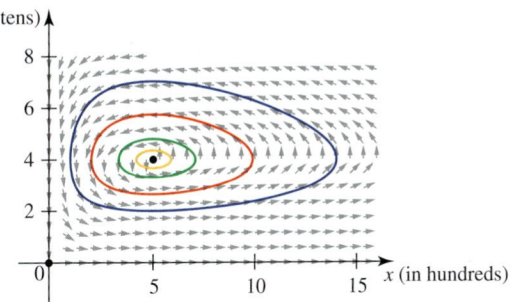

FIGURE 2

A few solution curves for system (3)

The solution curves shown in Figure 2 are called **phase curves,** or **phase trajectories,** of system (3) and the *xy*-plane in which the phase curves lie is called the **phase plane.** Recall that each point on the phase curve is a solution of the system (3). The next example shows that the solution points other than the equilibrium points "move" along a phase curve. A figure that consists of equilibrium points and typical phase curves is called a **phase portrait.**

EXAMPLE 2 Refer to the system of differential equations in Example 1:

$$\frac{dx}{dt} = 4x - xy$$

$$\frac{dy}{dt} = -y + 0.2xy$$

where $x(t)$ and $y(t)$ denote the number of rabbits (in hundreds) and foxes (in tens) at time t.

a. Suppose that at some time $t = 0$ there are 500 rabbits and 20 foxes. Draw the phase curve that corresponds to this situation.
b. What happens to the solution point (x, y) as t increases?
c. Use the results of parts (a) and (b) to sketch the graphs of $x(t)$ and $y(t)$.

Solution
a. We use the direction field obtained in the solution to Example 1c to help us draw the phase curve passing through the point $(5, 2)$. (See Figure 3.)

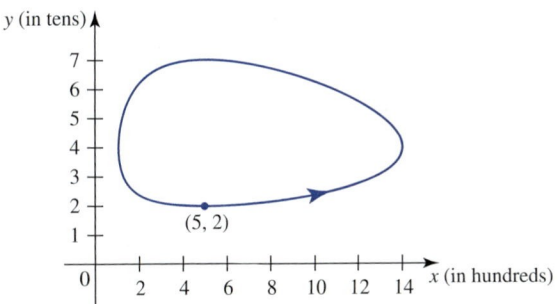

FIGURE 3

b. The phase curve is reproduced in Figure 4. At $t = 0$, $x = 5$, and $y = 2$. So

$$\left.\frac{dx}{dt}\right|_{t=0} = \left. 4x - xy\right|_{(5,\,2)}$$

$$= 4(5) - (5)(2) = 10$$

Since $dx/dt > 0$, we see that the solution point (x, y) moves counterclockwise around the phase curve.

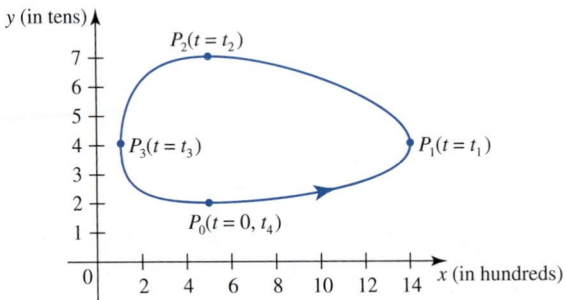

FIGURE 4

c. Refer to Figure 4. We begin at $P_0(5, 2)$ when $t = 0$ and move in a counterclockwise direction reaching $P_1(14, 4)$ at some time t_1, $P_2(5, 7)$ at some time t_2, and $P_3(1, 4)$ at some time t_3, and returning to $P_0(5, 2)$ at some time t_4. From this, we obtain the following table of values for x and y at different times t. For values of $t \geq t_4$, we simply replicate the graph obtained since the phase portrait is a closed curve.

t	0	t_1	t_2	t_3	t_4
x	5	14	5	1	5
y	2	4	7	4	2

Using this table of values, we obtain the graphs of x and t as shown in Figure 5.

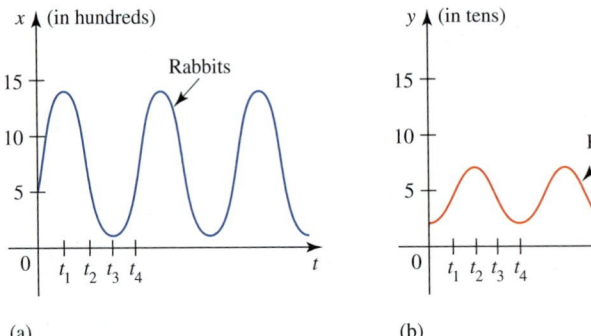

FIGURE 5

The graphs of the populations of (a) rabbits and (b) foxes as a function of time

Initially, there are 500 rabbits and 20 foxes. As t increases, the rabbit population increases rapidly. With a plentiful supply of food available, the fox population also increases rapidly. But this rapid increase of predators starts to take a toll on the prey and the rate of increase of the population of rabbits soon slows down, the population reaching a maximum population of approximately 1400 at time t_1. The fox population appears to be increasing most rapidly at this time. (The graph of y has an inflection point at $t = t_1$.)

From $t = t_1$ to $t = t_3$, the population of rabbits declines, the rate of decline being most rapid at $t = t_2$ (where x also has an inflection point). With the decline of the rabbit population, the rate of increase of the fox population soon begins to slow down. The fox population reaches a maximum of approximately 70 at $t = t_2$, then declines rapidly until $t = t_3$, then declines less rapidly from $t = t_3$ to $t = t_4$. Finally, as the fox population declines, the rabbit population once again begins to increase (from $t = t_3$ to $t = t_4$).

The cycle then repeats over and over again. ∎

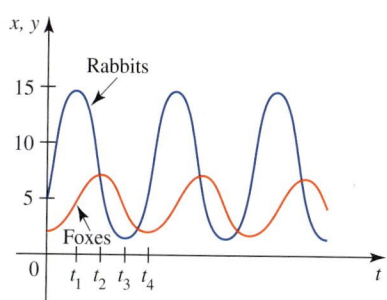

FIGURE 6

Graphs showing the population of rabbits, $x(t)$, in hundreds, and the population of foxes, $y(t)$, in tens

The relationship between the predator and prey populations in Examples 1 and 2 is illustrated in Figure 6, where the graphs of both x and y are plotted on the same set of axes. Observe that the fox population lags behind the rabbit population as both populations oscillate between their maximum and minimum values.

7.5 CONCEPT QUESTION

1. a. Let $x(t)$ and $y(t)$ denote the populations of prey and predators at time t, respectively. Write the Lotka-Volterra equations to model these populations.

  b. How are the equations modified if there are no predators? What do the resulting equation(s) say about the prey population?

  c. How are the equations modified if there are no prey? What do the resulting equation(s) say about the predator population?

7.5 EXERCISES

In Exercises 1 and 2 you are given the Lotka-Volterra equations describing the relationship between the prey population (in hundreds) at time t, x(t), and the predator population (in tens) at time t, y(t). (a) Find the equilibrium points of the system. (b) Find an expression for dy/dx and use it to draw a direction field for the resulting differential equation in the xy-plane. (c) Sketch some solution curves for the differential equation found in part (b).

1. $\dfrac{dx}{dt} = 2.4x - 1.2xy$

$\dfrac{dy}{dt} = -y + 0.8xy$

2. $\dfrac{dx}{dt} = 5x - 2xy$

$\dfrac{dy}{dt} = -0.6y + 0.2xy$

In Exercises 3 and 4 you are given the phase curve associated with a system of predator-prey equations, where x(t) denotes the prey (caribou) population, in hundreds, and y(t) denotes the predator (wolves) population, in tens, at time t. (a) Describe how each population changes over time t starting from t = 0. (b) Make a rough sketch of the graphs of x and y as a function of t on the same set of axes.

3.

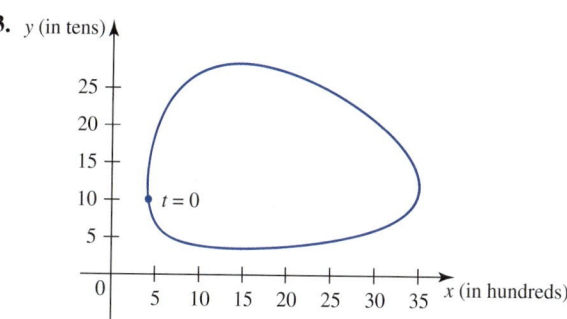

4.

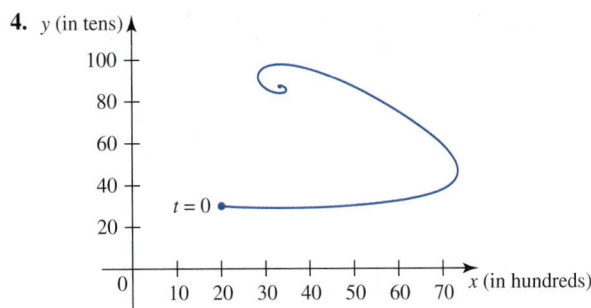

In Exercises 5 and 6 you are given the graphs of x and y as a function of time t, where x(t) denotes the prey population (in thousands) and y(t) denotes the predator population (in hundreds) at time t. Use them to sketch the associated phase curve.

5.

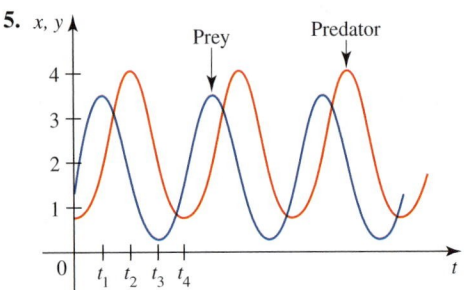

6.

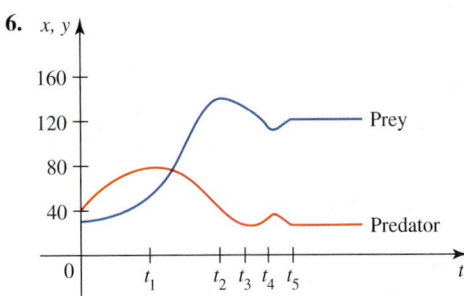

cas **7.** Consider the predator-prey equations

$$\frac{dx}{dt} = ax - bxy$$

$$\frac{dy}{dt} = -ry + sxy$$

where a, b, r, and s are positive constants.

a. Show that

$$\frac{dy}{dx} = \frac{y(-r + sx)}{x(a - by)}$$

b. Show that an implicit solution of the separable differential equation in part (a) is

$$\frac{x^r y^a}{e^{sx} e^{by}} = C$$

where C is a constant.

c. Find the equation of the phase curve passing through the point $(5, 2)$. Then use a CAS to plot the curve of this implicit equation. Compare this curve with the phase curve shown in Figure 4 of this section.

8. In nature, the population of aphids (small insects that suck plant juices) is held in check by ladybugs. Assume that the population of aphids (in thousands), $x(t)$, and the population of ladybugs (in hundreds), $y(t)$, satisfy the equations

$$\frac{dx}{dt} = 2x - 1.2xy$$

$$\frac{dy}{dt} = -y + 0.8xy$$

a. Find the equilibrium points and interpret your results.

b. A direction field for the differential equation

$$\frac{dy}{dx} = \frac{-y + 0.8xy}{2x - 1.2xy}$$

is shown. Sketch a phase portrait superimposed over the direction field.

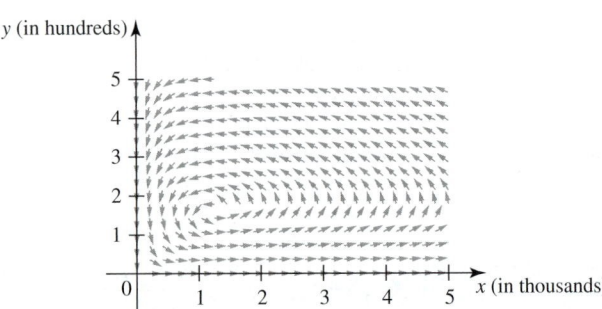

c. Suppose that initially there are 1000 aphids and 60 ladybugs. Draw the phase trajectory that satisfies this initial condition and use it to describe the behavior of both populations over time.

d. Use the result of part (c) to obtain the graphs of $x(t)$ and $y(t)$, and explain how they are related.

9. In the Lotka-Volterra model it was assumed that an unlimited amount of food was available to the prey. In a situation in which there is a finite amount of natural resources available to the prey, the Lotka-Volterra model can be modified to reflect this situation. Consider the following system of differential equations:

$$\frac{dx}{dt} = kx\left(1 - \frac{x}{L}\right) - axy$$

$$\frac{dy}{dt} = -by + cxy$$

where $x(t)$ and $y(t)$ represent the populations of prey and predators, respectively, and a, b, c, k, and L are positive constants.

a. Describe what happens to the prey population in the absence of predators.

b. Describe what happens to the predator population in the absence of prey.

c. Find all the equilibrium points and explain their significance.

10. Refer to Exercise 9. Consider the modified Lotka-Volterra equations

$$\frac{dx}{dt} = 0.5x\left(1 - \frac{x}{150}\right) - 0.005xy$$

$$\frac{dy}{dt} = -0.2y + 0.004xy$$

where x denotes the number of rabbits in tens and y denotes the number of foxes at time t.

a. Using the result of Exercise 9, or otherwise, find the equilibrium points and interpret your results.

b. The figure shows the phase curve that starts at the point $(40, 30)$. Describe what happens to the rabbit and fox populations as t increases.

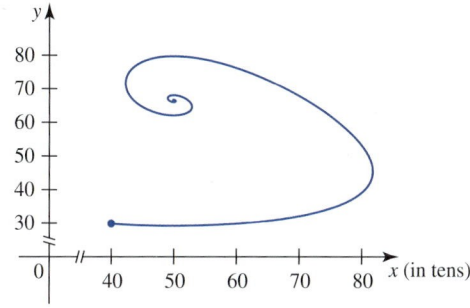

c. Sketch the graphs of the rabbit and fox populations as a function of time.

11. Consider the system of equations

$$\frac{dx}{dt} = k_1 x\left(1 - \frac{x}{L_1}\right) - axy$$

$$\frac{dy}{dt} = k_2 y\left(1 - \frac{y}{L_2}\right) - bxy$$

where $x(t)$ and $y(t)$ give the populations of two species A and B, respectively, and k_1, k_2, L_1, L_2, a, and b are positive constants.

a. Describe what happens to the population of A in the absence of B.

b. Describe what happens to the population of B in the absence of A.

c. Give a physical interpretation of the roles played by the terms axy and bxy, and explain why the equations are called **competing species equations.** (Examples of competing species are trout and bass.)

d. Find the equilibrium points and interpret your results.

12. A model for the populations of trout, x, and bass, y, that compete for food and space is given by

$$\frac{dx}{dt} = 0.6x\left(1 - \frac{x}{4}\right) - 0.01xy$$

$$\frac{dy}{dt} = 0.1y\left(1 - \frac{y}{2}\right) - 0.01xy$$

where x and y are in thousands.

a. Find the equilibrium points of the system.

b. Plot the direction field for dy/dx.

c. Plot the phase curve that satisfies the initial condition $(6, 6)$ superimposed upon the direction field found in part (b). Does this agree with the result of part (a)?

d. Interpret your result.

CHAPTER 7 REVIEW

CONCEPT REVIEW

In Exercises 1–8, fill in the blanks.

1. a. A differential equation is one that involves the _____ or _____ of a(n) _____ function.

b. The order of a differential equation is the order of the _____ derivative that occurs in the equation.

2. a. A solution of the differential equation $dy/dx = f(x, y)$ is a function $y = y(x)$ defined on an _____ interval I such that y satisfies the _____ on I.

b. Graphically, the general solution of a differential equation represents a family of _____ called the _____ or _____ _____ of the differential equation.

3. A first-order separable differential equation is one that can be written in the form $dy/dx =$ _____; the differential equation $G(x)\,dx + H(y)\,dy = 0$ is _____.

4. a. The initial-value problem $dy/dt = ky$, $y(0) = y_0$ has the unique solution $y =$ _____.

   b. The initial-value problem in part (a) is a model for exponential growth or decay; it is an exponential growth model if k _____ and is an exponential decay model if k _____.

5. a. A lineal element is a small portion of the _____ _____ at a point (a, b) on a solution curve of the differential equation $y' = f(x, y)$.

   b. A set of lineal elements drawn at various points is called a _____ field or _____ field of the differential equation.

6. a. The logistic differential equation has the form _____.

   b. The solution of the logistic differential equation with initial condition $P(0) = P_0$ is _____.

   c. If $P(0) = 0$, the population stays at _____ at all times; if $0 < P_0 < L$, then the population _____ and _____ the limiting value _____ asymptotically; if $P_0 = L$, the population at any time later remains at _____; if $P_0 > L$, then the population _____ and approaches _____ asymptotically.

7. a. A first-order linear differential equation is one that can be written in the form _____.

   b. An integrating factor for the first-order linear differential equation in part (a) is _____.

8. To solve a first-order linear differential equation, (a) rewrite the equation in _____ form, (b) multiply both sides of the resulting equation by the _____ _____, then (c) _____ both sides of the resulting equation.

REVIEW EXERCISES

1. Determine whether $y = C \cos x + \sin x$ is a solution of the differential equation $(\cos x)y' + (\sin x)y = 1$.

2. Determine whether $y = -2 + 3e^{3t}$ is a solution of the initial-value problem $y' - 3y = 6$, $y(0) = 1$.

In Exercises 3–6, solve the differential equation.

3. $\dfrac{dy}{dx} = 2xy^2$

4. $\dfrac{dy}{dx} = \dfrac{x}{y^2}$

5. $\dfrac{dy}{dx} = e^{y-x}$

6. $x\dfrac{dy}{dx} = y^2 + 1$

In Exercises 7–10, solve the initial-value problem.

7. $\dfrac{dy}{dx} = x^2y^3$, $y(0) = \dfrac{1}{2}$

8. $\dfrac{dy}{dx} + 3y = 6$, $y(0) = 0$

9. $\dfrac{dy}{dx} = -\dfrac{x}{y}$, $y(0) = 2$

10. $\dfrac{dy}{dx} = \dfrac{1+x}{e^y}$, $y(1) = 0$

11. Find the equation of a curve given that it passes through the point $\left(\frac{\pi}{4}, 1\right)$ and that the slope of the tangent line to the curve at any point $P(x, y)$ is given by $dy/dx = 4(y^2 + 1)$.

 12. Find the orthogonal trajectories of the family of curves given by $y^2 = x + C$. Use a graphing utility to draw several members of each family on the same set of axes.

13. Bacteria Growth A certain culture of bacteria grows at a rate that is proportional to the number present. If there are 1000 bacteria present initially and 4000 after 3 hr, find (a) an expression giving the number of bacteria in the culture after t hr, (b) the number of bacteria in the culture after 6 hr, and (c) the time it takes for the number of bacteria to reach 400,000.

14. Radioactivity If 4 g of a radioactive substance is present at time $t = 1$ (years) and 1 g at $t = 6$, how much was present initially? What is the half-life of the substance?

15. Radioactivity The radioactive element radium-226 has a half-life of 1602 years. What is its decay constant?

16. Rate of Return A conglomerate purchased a hotel for $4.5 million and sold it five years later for $8.2 million. Find the annual rate of return (compounded continuously).

17. Cost of Housing The Brennans are planning to buy a house 4 years from now. Housing experts in their area have estimated that the cost of a home will increase at a rate of 3% per year compounded continuously over that 4-year period. If their predictions are correct, how much can the Brennans expect to pay for a house that currently costs $300,000?

18. Newton's Law of Cooling Newton's Law of Cooling (heating) states that the rate at which the temperature of an object changes is directly proportional to the difference in the temperature of the object and that of the surrounding medium. A thermometer is taken from the patio, where the temperature is 40°F, into a room where the temperature is 70°F. After 1 min, the thermometer read 52°F. How long did it take for the thermometer to reach 64°F?

19. Newton's Law of Cooling Refer to Exercise 18. A cup of coffee had a temperature of 200°F when it was removed from a microwave oven and placed on a counter in a room that was kept at a temperature of 70°F. The temperature of the coffee was 180°F after 5 min.

   a. What was the temperature of the coffee after 10 min?

   b. How long did it take for the coffee to cool to 120°F?

20. A motorboat is traveling in calm water when its motor is suddenly cut off. Ten seconds later, the boat's speed is 10 mph; and another 10 sec later, its speed is 4 mph. What was its speed at the instant of time when the motor was cut off if the resistance of the water is proportional to the speed of the boat?

21. Future Value of an Annuity The future value S of an annuity (a stream of payments made continuously) satisfies the equation

$$\frac{dS}{dt} = rS + d$$

where r denotes the interest rate compounded continuously and d is a positive constant giving the rate at which payments are made into the account.
  a. If the future value of an annuity at time $t = 0$ is $\$S_0$, find an expression for the future value of the annuity at any time t.
  b. If the future value of an annuity at $t = 0$ is $\$10,000$, the interest rate is 6% compounded continuously, and a constant stream of payments of $\$2000$ per year are made into the account, what is the future value of the annuity after 5 years?

22. A direction field for the differential equation $y' = y(1 - y)$ follows. Sketch the solution curves that satisfy the given initial conditions.
  a. $y(-1) = -0.5$
  b. $y(0) = 0.4$
  c. $y(0) = 1$
  d. $y(-1) = 1.4$

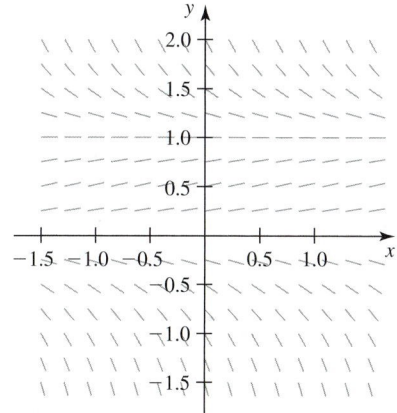

23. A direction field for the differential equation $y' = x^2 - y$ is shown in the figure.
  a. Sketch the solution curve for the initial-value problem

$$y' = x^2 - y \qquad y(0) = 1$$

b. Use the graph of part (a) to estimate the value of y when $x = 1$.

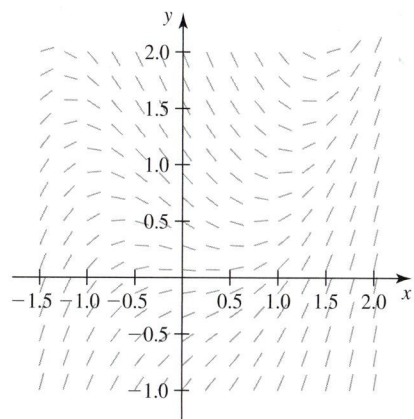

24. Use Euler's method with $n = 6$ to estimate $y(0.6)$, where y is the solution of the initial-value problem

$$y' = y^2 - x \qquad y(0) = 1$$

25. Use Euler's method with $n = 5$ to estimate $y(0.5)$, where y is the solution of the initial-value problem

$$y' = 2x^2 y \qquad y(0) = 1$$

26. Dissemination of Information Three hundred students attended the dedication ceremony of a new building on a college campus. The president of the college announced a new expansion program that included plans to make the college coeducational. The number of students who learned of the new program t hr later is given by the function

$$f(t) = \frac{3000}{1 + Be^{-kt}}$$

If 600 students on campus had heard about the new program 2 hr after the ceremony, how many students had heard about the policy after 4 hr? How fast was the news spreading 4 hr after the announcement?

In Exercises 27–30, solve the differential equation.

27. $xy' + 2y = 4x^2$ **28.** $y' + 2xy = 3x$

29. $\dfrac{dy}{dx} + y = e^{-x} \cos 2x$ **30.** $y \dfrac{dx}{dy} + x = 2y$

In Exercises 31–34, solve the initial-value problem.

31. $3y' - y = 0, \quad y(-1) = 2$

32. $y' - y = 3x^2 e^x, \quad y(0) = 0$

33. $xy' - y = x^2 \cos x, \quad y\left(\dfrac{\pi}{2}\right) = 0$

34. $xy' - y = x^2 e^x, \quad y(1) = e$

35. Trout Population Marine biologists released 400 trout into a lake that has an estimated carrying capacity of 10,000. The trout population after the first year was 1000. Suppose that the pattern of growth of the trout population follows a logistic curve.
   a. Find an equation giving the trout population after t years.
   b. What was the trout population after 6 years?
   c. How long did it take for the population to reach 8000?

36. Mixture Problem A tank initially holds 200 gal of water in which 20 lb of salt has been dissolved. Brine containing 2 lb of salt per gallon enters the tank at the rate of 2 gal/min. The well-stirred solution drains from the tank at the rate of 3 gal/min. How much salt is in the tank after 20 min?

37. The rabbit population x (in thousands) and the fox population y (in hundreds) are modeled by the equations

$$\frac{dx}{dt} = 2x - 0.4xy$$

$$\frac{dy}{dt} = -1.2y + 0.3xy$$

   a. What happens to the rabbit population in the absence of foxes? What happens to the fox population in the absence of rabbits?
   b. Find the equilibrium points and interpret your results.
   c. Find an expression for dy/dx and use it to plot a phase portrait.
   d. Use the phase curve corresponding to the initial populations of 4000 rabbits and 200 foxes to make a rough sketch of the graphs of x and y on the same set of axes.

38. The caribou population x (in tens) and the wolf population y are modeled by the equations

$$\frac{dx}{dt} = 0.5x\left(1 - \frac{x}{180}\right) - 0.005xy$$

$$\frac{dy}{dt} = -0.2y + 0.004xy$$

   a. What can you say about the caribou population in the absence of wolves?
   b. Find the equilibrium points and interpret your results.
   c. The phase curve with initial point (40, 30) is shown in the following figure. Use the phase curve to sketch the graphs of x and y. What happens to the populations eventually?

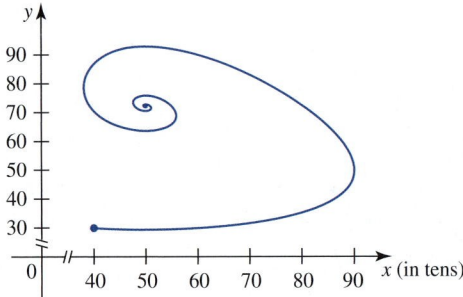

CHALLENGE PROBLEMS

1. As a boat crosses the finish line in a regatta, the athlete stops rowing and allows the boat to coast to a stop. Assuming that the race takes place in a calm lake and that water resistance to the boat is directly proportional to its velocity, with constant of proportionality k, and that the velocity of the boat at the instant it crosses the finish line is v_0, show that the distance covered by the boat before it comes to a stop is mv_0/k, where m is the combined mass of the boat and the athlete.
   Hint: Find an expression for $s(t)$, the distance covered by the boat at time t after crossing the finish line, and evaluate $\lim_{t \to \infty} s(t)$.

2. A *first-order homogeneous differential equation* is one of the form

$$\frac{dy}{dx} = f\left(\frac{y}{x}\right)$$

   a. Show that the substitution $u = y/x$ reduces a homogeneous equation to a separable equation in the variables u and x.
   b. Solve $\dfrac{dy}{dx} = \dfrac{y - x}{y + x}$.

3. Population Growth Consider the logistic function

$$P(t) = \frac{L}{1 + \left(\dfrac{L}{P_0} - 1\right)e^{-kt}}$$

giving the population at time t (see Example 2 in Section 7.3). Here, k is the growth constant, L is the carrying capacity of the environment, and P_0 is the initial population. Suppose that $P(1) = P_1$ and $P(2) = P_2$, where P_1 and P_2 are constants.

a. Show that

$$e^{-k} = \frac{P_0(P_2 - P_1)}{P_2(P_1 - P_0)} \quad \text{and} \quad L = \frac{P_1(P_0P_1 - 2P_0P_2 + P_1P_2)}{P_1^2 - P_0P_2}$$

b. The following table gives the population of the United States from the year 1900 through the year 2000.

Year	1900	1910	1920	1930	1940	1950
Population (millions)	76.21	92.23	106.02	123.20	132.16	151.33

Year	1960	1970	1980	1990	2000
Population (millions)	179.32	203.30	226.54	248.71	281.42

By taking $P_0 = 76.21$, $P_1 = 151.33$, and $P_2 = 281.42$, find an expression giving the population of the United States in year t, where t is measured in 50-year intervals and $t = 0$ corresponds to 1900.

 c. Plot the graph of $P(t)$.

d. Use the result of part (b) or part (c) to estimate the population of the United States in 2020.

Source: U.S. Census Bureau.

4. Find the orthogonal trajectories of the family of curves $x^2 + y^2 = 2ax$. Sketch a few members of each family.

5. a. Suppose that y_1 and y_2 are two different solutions of

$$\frac{dy}{dx} + P(x)y = Q(x)$$

Show that

$$y_2 = y_1\left(1 + Ce^{-\int [Q(x)/y_1(x)]\,dx}\right)$$

b. Use the result of part (a) to solve $y' + xy = x^2 + 1$ by observing that $y = x$ is a solution.

c. Use the technique of Section 7.4 to verify the solution that you obtained in part (b).

6. The differential equation

$$\frac{dy}{dx} = P(x)y^2 + Q(x)y + R(x)$$

is called the *Riccati equation*. This equation occurs in electromagnetic theory and the study of optics. Suppose that one solution, $y_1(x)$, of the Riccati equation is known.

a. Show that the substitution $y = y_1 + 1/u$ reduces the Riccati equation to the first-order linear differential equation

$$\frac{du}{dx} + (2Py_1 + Q)u = -P$$

b. Verify that $y_1 = 1/x$ is a solution of

$$\frac{dy}{dx} = y^2 - \frac{2}{x^2}$$

Then use the result of part (a) to find the general solution.

7. Radioactive Decay A radioactive substance A decays into another substance B at a rate that is proportional to the amount present at time t. The new substance B in turn decays into yet another substance C at a rate that is proportional to the amount present at time t. If the amount of substance A present initially is A_0 and there is no substance B present initially, show that the amount of substance B present at time t is given by

$$B(t) = \frac{aA_0}{b - a}(e^{-at} - e^{-bt})$$

where a and b are the decay constants for substance A and substance B, respectively, and $a \neq b$.

8. Show that the function defined by

$$f(x) = \frac{1}{2}\int_0^x \frac{e^{x-t}}{t}\,dt - \frac{1}{2}\int_0^x \frac{e^{t-x}}{t}\,dt$$

satisfies the differential equation

$$y'' - y = \frac{1}{x}$$

Horizon International Images Limited/Alamy

What percentage of the non-farm workforce in a country will be in the service industries one decade from now? In this chapter, we will see how a Taylor polynomial can be used to help answer this question.

8

Infinite Sequences and Series

IF WE ALLOW the number of terms of a sequence of real numbers to grow indefinitely, we obtain an *infinite sequence*. Infinite sequences that are *convergent* are of practical and theoretical interest. Indeed, it is the concept of a convergent sequence that allows us to define the *sum* of an *infinite series* (a series that is obtained by letting the number of terms of a series grow indefinitely). In this chapter we will see how a special type of infinite series called a *power series* affords us yet another way of representing a function. By representing a function in this manner, we are able to solve problems that we might otherwise not be able to solve.

 This symbol indicates that one of the following video types is available for enhanced student learning at **www.academic.cengage.com/login:**
- Chapter lecture videos
- Solutions to selected exercises

8.1 | Sequences

An idealized superball is dropped from a height of 1 m onto a flat surface. Suppose that each time the ball hits the surface, it rebounds to two thirds of its previous height. If we let a_1 denote the initial height of the ball, a_2 denote the maximum height attained on the first rebound, a_3 denote the maximum height attained on the second rebound, and so on, then we have

$$a_1 = 1, \qquad a_2 = \frac{2}{3}, \qquad a_3 = \frac{4}{9}, \qquad a_4 = \frac{8}{27}, \qquad \dots$$

(See Figure 1.) This array of numbers, $a_1, a_2, a_3, \dots$, is an example of an *infinite sequence*, or simply a *sequence*. If we define the function f by $f(x) = \left(\frac{2}{3}\right)^{x-1}$ and allow x to take on the positive integral values $x = 1, 2, 3, \dots, n, \dots$, then we see that the sequence $a_1, a_2, a_3, \dots$, may be viewed as the functional values of f at these numbers. Thus,

$$f(1) = 1 \qquad f(2) = \frac{2}{3} \qquad f(3) = \frac{4}{9} \qquad \dots \qquad f(n) = \left(\frac{2}{3}\right)^{n-1} \qquad \dots$$

$$\downarrow \qquad\qquad \downarrow \qquad\qquad \downarrow \qquad\qquad\qquad \downarrow$$

$$a_1 \qquad\qquad a_2 \qquad\qquad a_3 \qquad\qquad\qquad a_n \qquad \dots$$

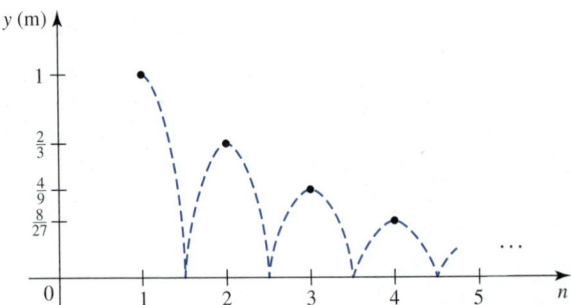

FIGURE 1
The ball rebounds to two thirds of its previous height upon hitting the surface.

This discussion motivates the following definition.

DEFINITION   **Sequence**

A **sequence** $\{a_n\}$ is a function whose domain is the set of positive integers. The functional values $a_1, a_2, a_3, \dots, a_n, \dots$ are the **terms** of the sequence, and the term a_n is called the ***n*th term** of the sequence.

Notes

1. The sequence $\{a_n\}$ is also denoted by $\{a_n\}_{n=1}^{\infty}$.
2. Sometimes it is convenient to begin a sequence with a_k. In this case the sequence is $\{a_n\}_{n=k}^{\infty}$, and its terms are $a_k, a_{k+1}, a_{k+2}, \dots, a_n, \dots$.

EXAMPLE 1   List the terms of the sequence.

a. $\left\{\dfrac{n}{n+1}\right\}$ **b.** $\left\{\dfrac{\sqrt{n}}{2^{n-1}}\right\}$ **c.** $\{(-1)^n \sqrt{n-2}\}_{n=2}^{\infty}$ **d.** $\left\{\sin \dfrac{n\pi}{3}\right\}_{n=0}^{\infty}$

Solution

a. Here, $a_n = f(n) = \dfrac{n}{n+1}$. Thus,

$$a_1 = f(1) = \frac{1}{1+1} = \frac{1}{2}, \quad a_2 = f(2) = \frac{2}{2+1} = \frac{2}{3}, \quad a_3 = f(3) = \frac{3}{3+1} = \frac{3}{4}, \quad \ldots$$

and we see that the given sequence can be written as

$$\left\{\frac{n}{n+1}\right\} = \left\{\frac{1}{2}, \frac{2}{3}, \frac{3}{4}, \frac{4}{5}, \ldots, \frac{n}{n+1}, \ldots\right\}$$

b. $\left\{\dfrac{\sqrt{n}}{2^{n-1}}\right\} = \left\{\dfrac{\sqrt{1}}{2^0}, \dfrac{\sqrt{2}}{2^1}, \dfrac{\sqrt{3}}{2^2}, \dfrac{\sqrt{4}}{2^3}, \ldots, \dfrac{\sqrt{n}}{2^{n-1}}, \ldots\right\}$

c. $\{(-1)^n\sqrt{n-2}\}_{n=2}^{\infty} = \{(-1)^2\sqrt{0}, (-1)^3\sqrt{1}, (-1)^4\sqrt{2},$

$$(-1)^5\sqrt{3}, \ldots, (-1)^n\sqrt{n-2}, \ldots\}$$

$$= \{0, -\sqrt{1}, \sqrt{2}, -\sqrt{3}, \ldots, (-1)^n\sqrt{n-2}, \ldots\}$$

Notice that n starts from 2 in this example. (See Note 2 on page 726.)

d. $\left\{\sin\dfrac{n\pi}{3}\right\}_{n=0}^{\infty} = \left\{\sin 0, \sin\dfrac{\pi}{3}, \sin\dfrac{2\pi}{3}, \sin\dfrac{3\pi}{3}, \sin\dfrac{4\pi}{3}, \sin\dfrac{5\pi}{3}, \ldots, \sin\dfrac{n\pi}{3}, \ldots\right\}$

$$= \left\{0, \frac{\sqrt{3}}{2}, \frac{\sqrt{3}}{2}, 0, -\frac{\sqrt{3}}{2}, -\frac{\sqrt{3}}{2}, \ldots, \sin\frac{n\pi}{3}, \ldots\right\}$$

Once again, refer to Note 2. ■

We can often determine the nth term of a sequence by studying the first few terms of the sequence and recognizing the pattern that emerges.

EXAMPLE 2 Find an expression for the nth term of each sequence.

a. $\left\{2, \dfrac{3}{\sqrt{2}}, \dfrac{4}{\sqrt{3}}, \dfrac{5}{\sqrt{4}}, \ldots\right\}$ **b.** $\left\{1, \dfrac{1}{8}, \dfrac{1}{27}, \dfrac{1}{64}, \ldots\right\}$ **c.** $\left\{1, -\dfrac{1}{2}, \dfrac{1}{3}, -\dfrac{1}{4}, \ldots\right\}$

Solution

a. The terms of the sequence may be written in the form

$$a_1 = \frac{1+1}{\sqrt{1}}, \qquad a_2 = \frac{2+1}{\sqrt{2}}, \qquad a_3 = \frac{3+1}{\sqrt{3}}, \qquad a_4 = \frac{4+1}{\sqrt{4}}, \qquad \ldots$$

from which we see that $a_n = \dfrac{n+1}{\sqrt{n}}$.

b. Here,

$$a_1 = \frac{1}{1^3}, \qquad a_2 = \frac{1}{2^3}, \qquad a_3 = \frac{1}{3^3}, \qquad a_4 = \frac{1}{4^3}, \qquad \ldots$$

so $a_n = \dfrac{1}{n^3}$.

c. Note that $(-1)^r$ is equal to 1 if r is an even integer and -1 if r is an odd integer. Using this result, we obtain

$$a_1 = \frac{(-1)^0}{1}, \qquad a_2 = \frac{(-1)^1}{2}, \qquad a_3 = \frac{(-1)^2}{3}, \qquad a_4 = \frac{(-1)^3}{4}, \qquad \ldots$$

We conclude that the nth term is $a_n = (-1)^{n-1}/n$. ■

Some sequences are defined **recursively;** that is, the sequence is defined by specifying the first term or the first few terms of the sequence and a rule for calculating any other term of the sequence from the preceding term(s).

EXAMPLE 3 List the first five terms of the recursively defined sequence $a_1 = 2$, $a_2 = 4$, and $a_{n+1} = 2a_n - a_{n-1}$ for $n \geq 2$.

Solution The first two terms of the sequence are given as $a_1 = 2$ and $a_2 = 4$. To find the third term of the sequence, we put $n = 2$ in the recursion formula to obtain

$$a_3 = 2a_2 - a_1 = 2 \cdot 4 - 2 = 6$$

Next, putting $n = 3$ and $n = 4$ in succession in the recursive formula gives

$$a_4 = 2a_3 - a_2 = 2 \cdot 6 - 4 = 8 \qquad \text{and} \qquad a_5 = 2a_4 - a_3 = 2 \cdot 8 - 6 = 10 \quad \blacksquare$$

Since a sequence is a function, we can draw its graph. The graphs of the sequences $\{n/(n + 1)\}$ and $\{(-1)^n\}$ are shown in Figure 2. They are just the graphs of the functions $f(n) = n/(n + 1)$ and $g(n) = (-1)^n$ for $n = 1, 2, 3, \ldots$.

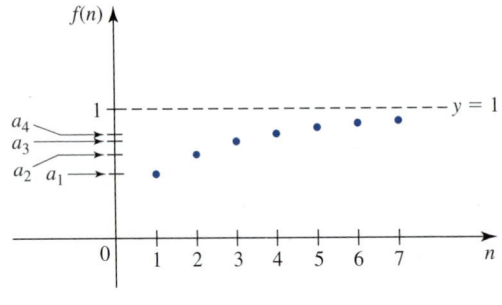

(a) The graph of $\left\{\dfrac{n}{n + 1}\right\}$

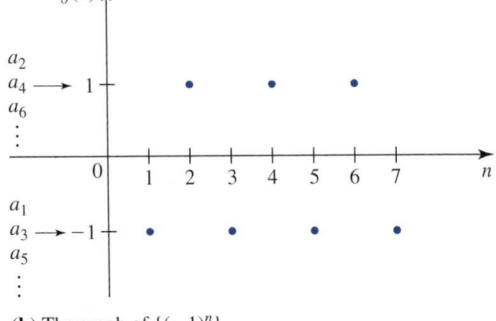

(b) The graph of $\{(-1)^n\}$

FIGURE 2

■ Limit of a Sequence

If you examine the graph of the sequence $\{n/(n + 1)\}$ sketched in Figure 2a, you will see that the terms of the sequence seem to get closer and closer to 1 as n gets larger and larger. In this situation we say that the sequence $\{n/(n + 1)\}$ converges to the *limit* 1, written

$$\lim_{n \to \infty} \frac{n}{n + 1} = 1$$

In general, we have the following informal definition of the limit of a sequence.

DEFINITION **Limit of a Sequence**

A sequence $\{a_n\}$ has the **limit** L, written

$$\lim_{n \to \infty} a_n = L$$

if a_n can be made as close to L as we please by taking n sufficiently large. If $\lim_{n \to \infty} a_n$ exists, we say that the sequence **converges.** Otherwise, we say that the sequence **diverges.**

A more precise definition of the limit of a sequence follows.

DEFINITION (Precise)   Limit of a Sequence

A sequence $\{a_n\}$ **converges** and has the **limit** L, written

$$\lim_{n \to \infty} a_n = L$$

if for every $\varepsilon > 0$ there exists a positive integer N such that $|a_n - L| < \varepsilon$ whenever $n > N$.

To illustrate this definition, suppose that a challenger selects an $\varepsilon > 0$. Then we must show that there exists a positive integer N such that all points (n, a_n) on the graph of $\{a_n\}$, where $n > N$, lie inside a band of width 2ε about the line $y = L$. (See Figure 3.)

FIGURE 3
If $n > N$, then $L - \varepsilon < a_n < L + \varepsilon$
or, equivalently, $|a_n - L| < \varepsilon$.

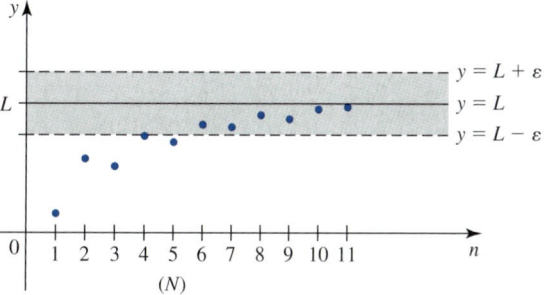

To reconcile this definition with the intuitive definition of a limit, recall that ε is arbitrary. Therefore, by choosing ε very small, the challenger ensures that a_n is "close" to L. Furthermore, if corresponding to *each* choice of ε, we can produce an N such that $n > N$ implies that $|a_n - L| < \varepsilon$, then we have shown that a_n can be made as close to L as we please by taking n sufficiently large.

Notice that the definition of the limit of a sequence is very similar to the definition of the limit of a function at infinity given in Section 3.5. This is expected, since the only difference between a function f defined by $y = f(x)$ on the interval $(0, \infty)$ and the sequence $\{a_n\}$ defined by $a_n = f(n)$ is that n is an integer. (See Figure 4.) This observation tells us that we can often evaluate $\lim_{n \to \infty} a_n$ by evaluating $\lim_{x \to \infty} f(x)$, where f is defined on $(0, \infty)$ and $a_n = f(n)$.

FIGURE 4
The graph of $\{a_n\}$ comprises
the points $(n, f(n))$ that lie
on the graph of $y = f(x)$.

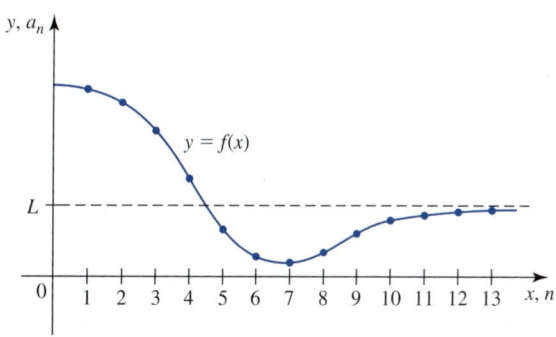

THEOREM 1

If $\lim_{x \to \infty} f(x) = L$ and $\{a_n\}$ is a sequence defined by $a_n = f(n)$, where n is a positive integer, then $\lim_{n \to \infty} a_n = L$.

You will be asked to prove Theorem 1 in Exercise 75.

EXAMPLE 4 Find $\lim_{n \to \infty} \dfrac{1}{n^r}$ if $r > 0$.

Solution Since $a_n = 1/n^r$, we choose $f(x) = 1/x^r$, where $x > 0$. By Theorem 1 in Section 3.5 we have

$$\lim_{x \to \infty} \frac{1}{x^r} = 0$$

Using Theorem 1 of this section, we conclude that

$$\lim_{n \to \infty} \frac{1}{n^r} = 0$$

(See Figure 5.)

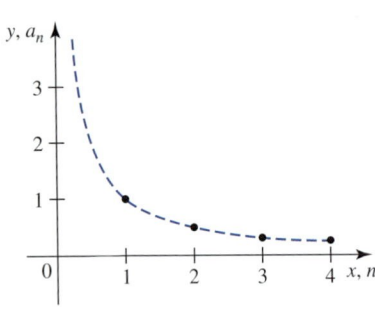

FIGURE 5
The graph of $\{1/n^r\}$ for $n = 1, 2, 3, 4$, and $r = 1$. The graph of $f(x) = 1/x^r$ is shown with a dashed curve.

The converse of Theorem 1 is false. Consider, for example, the sequence $\{\sin n\pi\} = \{0\}$. This sequence evidently converges to 0, since every term of the sequence is 0. But $\lim_{x \to \infty} \sin \pi x$ does not exist. (See Figure 6.)

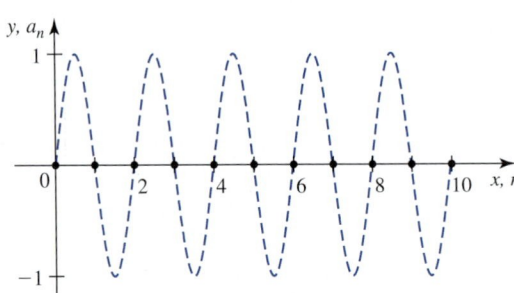

FIGURE 6
The graph of $\{\sin n\pi\}$ for $n = 0, 1, 2, \ldots, 10$. The graph of $f(x) = \sin \pi x$ is shown with a dashed curve.

The following limit laws for sequences are the analogs of the limit laws for functions studied in Section 1.2 and are proved in a similar manner.

THEOREM 2   Limit Laws for Sequences

Suppose that $\lim_{n \to \infty} a_n = L$ and $\lim_{n \to \infty} b_n = M$ and that c is a constant. Then

1. $\lim_{n \to \infty} c a_n = cL$

2. $\lim_{n \to \infty} (a_n \pm b_n) = L \pm M$

3. $\lim_{n \to \infty} a_n b_n = LM$

4. $\lim_{n \to \infty} \dfrac{a_n}{b_n} = \dfrac{L}{M}$,   provided that $b_n \neq 0$ and $M \neq 0$

5. $\lim_{n \to \infty} a_n^p = L^p$,   if $p > 0$ and $a_n > 0$

EXAMPLE 5 Determine whether the sequence converges or diverges.

a. $\left\{ \dfrac{n}{n+1} \right\}$ **b.** $\{(-1)^n\}$

Solution

a. Both the numerator and the denominator of $n/(n+1)$ approach infinity as n approaches infinity. So their limits do not exist, and we cannot use Law 4 of Theorem 2. But we can divide the numerator and denominator by n and then apply Law 4 to evaluate the resulting limit. Thus,

$$\lim_{n\to\infty} \frac{n}{n+1} = \lim_{n\to\infty} \frac{1}{1+\dfrac{1}{n}} = 1$$

and we conclude that the sequence converges to 1. (See Figure 2a.)

b. The terms of the sequence are

$$-1, 1, -1, 1, \ldots$$

The sequence evidently does not approach a unique number, and we conclude that it diverges. (See Figure 2b.) ∎

EXAMPLE 6 Find

a. $\lim_{n\to\infty} \dfrac{\ln n}{n}$ **b.** $\lim_{n\to\infty} \dfrac{e^n}{n^2}$

Solution

a. Observe that both the numerator and the denominator of $(\ln n)/n$ approach infinity as $n \to \infty$. Therefore, we may not use Law 4 of Theorem 2 directly. Since $a_n = f(n) = (\ln n)/n$, we consider the function $f(x) = (\ln x)/x$. Using l'Hôpital's Rule, we find

$$\lim_{x\to\infty} \frac{\ln x}{x} = \lim_{x\to\infty} \frac{1/x}{1} = \lim_{x\to\infty} \frac{1}{x} = 0$$

Therefore, by Theorem 1 we conclude that

$$\lim_{n\to\infty} \frac{\ln n}{n} = 0$$

(See Figure 7.)

b. Once again both e^n and n^2 approach infinity as $n \to \infty$. Choose $f(x) = e^x/x^2$, and use l'Hôpital's Rule twice to find that

$$\lim_{x\to\infty} \frac{e^x}{x^2} = \lim_{x\to\infty} \frac{e^x}{2x} = \lim_{x\to\infty} \frac{e^x}{2} = \infty$$

from which we see that

$$\lim_{n\to\infty} \frac{e^n}{n^2} = \infty$$

and we conclude that the sequence $\{e^n/n^2\}$ is divergent. (See Figure 8.) ∎

The Squeeze Theorem has the following counterpart for sequences. (The proof is similar to that of the Squeeze Theorem and will be omitted.)

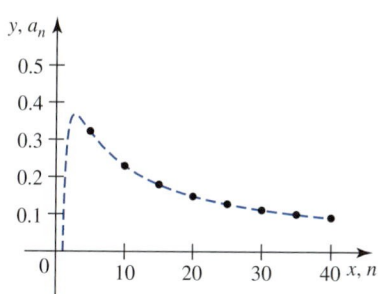

FIGURE 7
The graph of $\{(\ln n)/n\}$ for $n = 5, 10, 15, \ldots, 40$. The graph of $f(x) = (\ln x)/x$ is shown with a dashed curve.

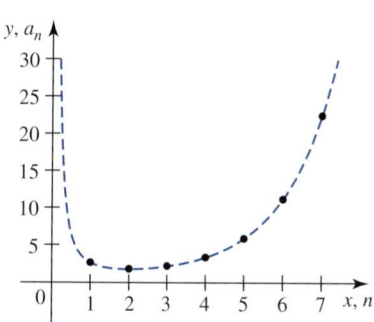

FIGURE 8
The graph of $\{e^n/n^2\}$. The graph of $f(x) = e^x/x^2$ is shown with a dashed curve.

> **THEOREM 3** **Squeeze Theorem for Sequences**
>
> If there exists some integer N such that $a_n \le b_n \le c_n$ for all $n \ge N$ and $\lim_{n \to \infty} a_n = \lim_{n \to \infty} c_n = L$, then $\lim_{n \to \infty} b_n = L$.

(See Figure 9.)

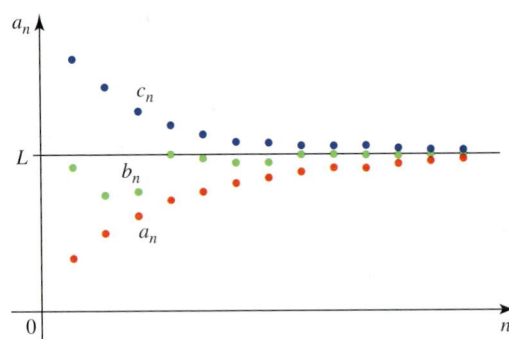

FIGURE 9

The sequence $\{b_n\}$ is squeezed between the sequences $\{a_n\}$ and $\{c_n\}$.

EXAMPLE 7 Find $\lim\limits_{n \to \infty} \dfrac{n!}{n^n}$, where $n!$ (read "n factorial") is defined by

$$n! = n(n-1)(n-2) \cdots 1$$

Solution Let $a_n = n!/n^n$. The first few terms of $\{a_n\}$ are

$$a_1 = \frac{1!}{1} = 1, \qquad a_2 = \frac{2!}{2^2} = \frac{2 \cdot 1}{2 \cdot 2}, \qquad a_3 = \frac{3!}{3^3} = \frac{3 \cdot 2 \cdot 1}{3 \cdot 3 \cdot 3}$$

and its nth term is

$$a_n = \frac{n!}{n^n} = \frac{n(n-1) \cdot \cdots \cdot 3 \cdot 2 \cdot 1}{n \cdot n \cdot \cdots \cdot n \cdot n \cdot n} = \left(\frac{n}{n}\right)\left(\frac{n-1}{n}\right) \cdot \cdots \cdot \left(\frac{3}{n}\right)\left(\frac{2}{n}\right)\left(\frac{1}{n}\right) \le \frac{1}{n}$$

Therefore,

$$0 < a_n \le \frac{1}{n}$$

Since $\lim_{n \to \infty} 1/n = 0$, the Squeeze Theorem implies that

$$\lim_{n \to \infty} a_n = \lim_{n \to \infty} \frac{n!}{n^n} = 0$$

The next theorem is an immediate consequence of the Squeeze Theorem.

> **THEOREM 4**
>
> If $\lim_{n \to \infty} |a_n| = 0$, then $\lim_{n \to \infty} a_n = 0$.

You are asked to prove Theorem 4 in Exercise 76.

EXAMPLE 8 Find $\lim\limits_{n\to\infty} \dfrac{(-1)^n}{n}$.

Solution Since

$$\lim_{n\to\infty} \left| \frac{(-1)^n}{n} \right| = \lim_{n\to\infty} \frac{1}{n} = 0$$

we conclude by Theorem 4 that

$$\lim_{n\to\infty} \frac{(-1)^n}{n} = 0$$

The graph of the sequence $\{(-1)^n/n\}$ confirms this result. (See Figure 10.)

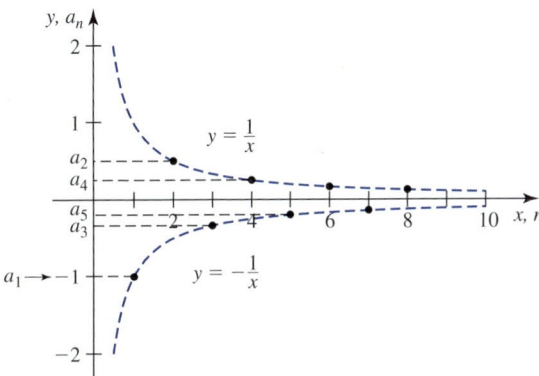

FIGURE 10
The terms of the sequence $\{(-1)^n/n\}$ oscillate between the graphs of $y = 1/x$ and $y = -1/x$.

If we take the composition of a function f with a sequence $\{a_n\}$, we obtain another sequence $\{f(a_n)\}$. The following theorem shows how to compute the limit of the latter. The proof will be given in Appendix B.

> **THEOREM 5**
> If $\lim_{n\to\infty} a_n = L$ and the function f is continuous at L, then
> $$\lim_{n\to\infty} f(a_n) = f(\lim_{n\to\infty} a_n) = f(L)$$

Note Compare this theorem with Theorem 5 in Section 1.4.

EXAMPLE 9 Find $\lim_{n\to\infty} e^{\sin(1/n)}$.

Solution Observe that $e^{\sin(1/n)} = f(a_n)$, where $f(x) = e^x$ and $a_n = \sin(1/n)$. Since

$$\lim_{n\to\infty} \sin \frac{1}{n} = 0$$

and f is continuous at 0, Theorem 5 gives $\lim_{n\to\infty} e^{\sin(1/n)} = e^{\lim\limits_{n\to\infty} \sin(1/n)} = e^0 = 1$.

■ Bounded Monotonic Sequences

Up to now, the convergent sequences that we have dealt with had limits that are readily found. Sometimes, however, we need to show that a sequence is convergent even if its precise limit is not readily found. Our immediate goal here is to find conditions that will guarantee that a sequence converges. To do this, we need to make use of two further properties of sequences.

DEFINITION **Monotonic Sequence**

A sequence $\{a_n\}$ is **increasing** if

$$a_1 < a_2 < a_3 < \cdots < a_n < a_{n+1} < \cdots$$

and **decreasing** if

$$a_1 > a_2 > a_3 > \cdots > a_n > a_{n+1} > \cdots$$

A sequence is **monotonic** if it is either increasing or decreasing.

EXAMPLE 10 Show that the sequence $\left\{\dfrac{n}{n+1}\right\}$ is increasing.

Solution Let $a_n = n/(n+1)$. We must show that $a_n \le a_{n+1}$ for all $n \ge 1$; that is,

$$\frac{n}{n+1} \le \frac{n+1}{(n+1)+1}$$

or

$$\frac{n}{n+1} \le \frac{n+1}{n+2}$$

To show that this inequality is true, we obtain the following equivalent inequalities:

$$n(n+2) \le (n+1)(n+1) \qquad \text{Cross-multiply.}$$
$$n^2 + 2n \le n^2 + 2n + 1$$
$$0 \le 1$$

which is true for $n \ge 1$. Therefore, $a_n \le a_{n+1}$, so $\{a_n\}$ is increasing.

Alternative Solution Here, $a_n = f(n) = n/(n+1)$. So consider the function $f(x) = x/(x+1)$. Since

$$f'(x) = \frac{(x+1)(1) - x(1)}{(x+1)^2} = \frac{1}{(x+1)^2} > 0 \qquad \text{if} \quad x > 0$$

we see that f is increasing on $(0, \infty)$. Therefore, the given sequence is increasing. ■

EXAMPLE 11 Show that the sequence $\left\{\dfrac{n}{e^n}\right\}$ is decreasing.

Solution   We must show that $a_n \geq a_{n+1}$ for $n \geq 1$; that is,

$$\frac{n}{e^n} \geq \frac{n+1}{e^{n+1}}$$

$$ne^{n+1} \geq (n+1)e^n$$

$$ne \geq n + 1 \qquad \text{Divide both sides by } e^n.$$

$$n(e-1) \geq 1$$

which is true for all $n \geq 1$, so $\{n/e^n\}$ is decreasing. ∎

Next, we explain what is meant by a *bounded* sequence.

DEFINITION   Bounded Sequence

A sequence $\{a_n\}$ is **bounded above** if there exists a number M such that

$$a_n \leq M \qquad \text{for all } n \geq 1$$

A sequence is **bounded below** if there exists a number m such that

$$m \leq a_n \qquad \text{for all } n \geq 1$$

A sequence is **bounded** if it is both bounded above and bounded below.

For example, the sequence $\{n\}$ is bounded below by 0, but it is not bounded above. The sequence $\{n/(n+1)\}$ is bounded below by $\frac{1}{2}$ and above by 1 and is therefore bounded. (See Figure 2a.)

A bounded sequence need not be convergent. For example, the sequence $\{(-1)^n\}$ is bounded, since $-1 \leq (-1)^n \leq 1$; but it is evidently divergent. (See Figure 2b.) Also, a monotonic sequence need not be convergent. For example, the sequence $\{n\}$ is increasing and evidently divergent. However, if a sequence is *both* bounded and monotonic, then it must be convergent.

THEOREM 6   Monotone Convergence Theorem for Sequences

Every bounded, monotonic sequence is convergent.

The plausibility of Theorem 6 is suggested by the sequence $\{n/(n+1)\}$ whose graph is reproduced in Figure 11. This sequence is increasing and bounded above by

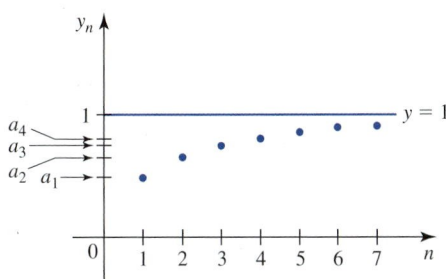

FIGURE 11
The increasing, bounded sequence $\{n/(n+1)\}$ is convergent.

any number $M \geq 1$. Therefore, as n increases, the terms a_n approach a number (which is no larger than M) from below. In this case the number is 1, which is also the limit of this sequence. (A proof of Theorem 6 is given at the end of this section.)

Theorem 6 can be used to find the limit of a convergent sequence indirectly, as the next example shows. It will also play an important role in infinite series (Sections 8.2–8.9).

EXAMPLE 12 Show that $\left\{\dfrac{2^n}{n!}\right\}$ is convergent and find its limit.

Solution Here, $a_n = 2^n/n!$. The first few terms of the sequence are

$$a_1 = 2, \qquad a_2 = 2, \qquad a_3 \approx 1.333333, \qquad a_4 \approx 0.666667, \qquad a_5 \approx 0.266667,$$

$$a_6 \approx 0.088889, \qquad \ldots, \qquad a_{10} \approx 0.000282$$

These terms suggest that the sequence is decreasing from $n = 2$ onward. To prove this, we compute

$$\frac{a_{n+1}}{a_n} = \frac{\dfrac{2^{n+1}}{(n+1)!}}{\dfrac{2^n}{n!}} = \frac{2^{n+1} n!}{2^n (n+1)!} = \frac{2n!}{(n+1)n!} = \frac{2}{n+1} \tag{1}$$

So

$$\frac{a_{n+1}}{a_n} \leq 1 \quad \text{if} \quad n \geq 1$$

Thus, $a_{n+1} \leq a_n$ if $n \geq 1$, and this proves the assertion. Since all of the terms of the sequence are positive, $\{a_n\}$ is bounded below by 0. Therefore, the sequence is decreasing and bounded below, and Theorem 6 guarantees that it converges to a nonnegative limit L.

To find L, we first use Equation (1) to write

$$a_{n+1} = \frac{2}{n+1} a_n \tag{2}$$

Since $\lim_{n\to\infty} a_n = L$, we also have $\lim_{n\to\infty} a_{n+1} = L$. Taking the limit on both sides of Equation (2) and using Law (3) for limits of sequences, we obtain

$$L = \lim_{n\to\infty} a_{n+1} = \lim_{n\to\infty}\left(\frac{2}{n+1} a_n\right) = \lim_{n\to\infty} \frac{2}{n+1} \cdot \lim_{n\to\infty} a_n = 0 \cdot L = 0$$

We conclude that $\lim_{n\to\infty} 2^n/n! = 0$.

Alternative Solution Observe that

$$a_2 = \frac{2 \cdot 2}{2 \cdot 1} = 2, \; a_3 = \frac{2 \cdot 2 \cdot 2}{3 \cdot 2 \cdot 1} = 2\left(\frac{2}{3}\right), \; a_4 = \frac{2 \cdot 2 \cdot 2 \cdot 2}{4 \cdot 3 \cdot 2 \cdot 1} = \left(\frac{2}{4}\right)\left(\frac{2}{3}\right)2 < 2\left(\frac{2}{3}\right)^2,$$

$$a_5 = \frac{2 \cdot 2 \cdot 2 \cdot 2 \cdot 2}{5 \cdot 4 \cdot 3 \cdot 2 \cdot 1} = \left(\frac{2}{5}\right)\left(\frac{2}{4}\right)\left(\frac{2}{3}\right)2 < 2\left(\frac{2}{3}\right)^3$$

and

$$a_n = \frac{2 \cdot 2 \cdot 2 \cdots \cdot 2}{n \cdot (n-1) \cdot (n-2) \cdots \cdot 1} < 2\left(\frac{2}{3}\right)^{n-2}$$

Therefore

$$0 < a_n < 2\left(\frac{2}{3}\right)^{n-2}$$

Since $\lim_{n \to \infty}\left(\frac{2}{3}\right)^{n-2} = 0$, the Squeeze Theorem gives the desired result.   ■

The next example contains some important results that we will derive here using the Squeeze Theorem. (You will also be asked to demonstrate their validity using the properties of exponential functions in Exercise 77.)

EXAMPLE 13   Show that $\lim_{n \to \infty} r^n = 0$ if $|r| < 1$.

Solution   If $r = 0$, then each term of the sequence $\{r^n\}$ is 0, and the sequence converges to 0. Now suppose that $0 < |r| < 1$. Then $1/|r|$ is greater than 1. So there exists a positive number p such that

$$\frac{1}{|r|} = 1 + p$$

Using the Binomial Theorem, we have

$$(1 + p)^n = 1 + np + \frac{n(n-1)}{2!}p^2 + \cdots + p^n > np$$

Thus,

$$0 < |r|^n = \frac{1}{(1 + p)^n} < \frac{1}{np}$$

But

$$\lim_{n \to \infty} \frac{1}{np} = 0$$

so by the Squeeze Theorem

$$\lim_{n \to \infty} |r|^n = 0$$

Finally, using Theorem 4, we conclude that $\lim_{n \to \infty} r^n = 0$.   ■

If $r = 1$, then $r^n = 1$ for all n, and the sequence $\{r^n\}$ evidently converges to 1. If $r = -1$, then the sequence $\{r^n\} = \{(-1)^n\}$ is divergent. (See Example 5b.) If $|r| > 1$, then $|r| = 1 + p$ for some positive number p. Using the Binomial Theorem again, we have

$$|r|^n = (1 + p)^n > np$$

Since $p > 0$, $\lim_{n \to \infty} np = \infty$. This shows that $\{r^n\}$ diverges if $|r| > 1$.

A summary of these results follows.

Properties of the Sequence $\{r^n\}$

The sequence $\{r^n\}$ converges if $-1 < r \leq 1$ and

$$\lim_{n \to \infty} r^n = \begin{cases} 0 & \text{if } -1 < r < 1 \\ 1 & \text{if } r = 1 \end{cases}$$

It diverges for all other values of r.

■ Proof of Theorem 6

The proof of Theorem 6 depends on the **Completeness Axiom** for the real number system, which states that *every nonempty set S of real numbers that is bounded above has a least upper bound.* Thus, if $x \leq M$ for all x in S, then there must be a real number b such that b is an upper bound of S ($x \leq b$ for all $x \in S$), and if N is any upper bound of S, then $N \geq b$. For example, if S is the interval $(-2, 3)$, then the number 4 (or any number greater than 3) is an upper bound of S and 3 is the least upper bound of S. As a consequence of this axiom, it can be shown that every nonempty set of real numbers that is bounded below has a greatest lower bound. The Completeness Axiom merely states that the real number line has no gaps or holes.

PROOF Suppose that $\{a_n\}$ is an increasing sequence. Since $\{a_n\}$ is bounded, the set $S = \{a_n | n \geq 1\}$ is bounded above, and by the Completeness Axiom it has a least upper bound L. (See Figure 12.) We claim that L is the limit of the sequence.

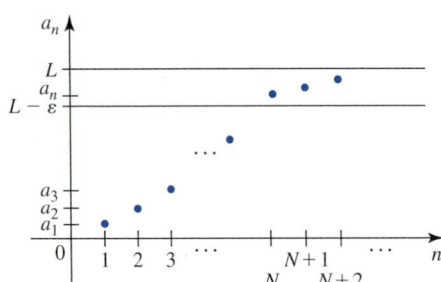

FIGURE 12

An increasing sequence bounded above must converge to its least upper bound.

To show this, let $\varepsilon > 0$ be given. Then $L - \varepsilon$ is *not* an upper bound of S (since L is the least upper bound of S). Therefore, there exists an integer N such that $a_N > L - \varepsilon$. But the sequence is increasing, so $a_n \geq a_N$ for every $n > N$. In other words, if $n > N$, we have $a_n > L - \varepsilon$. Since $a_n \leq L$,

$$0 \leq L - a_n < \varepsilon$$

This shows that

$$|L - a_n| < \varepsilon$$

whenever $n > N$, so $\lim_{n \to \infty} a_n = L$.

The proof is similar for the case in which $\{a_n\}$ is decreasing, except that we use the greatest lower bound instead. ■

8.1 CONCEPT QUESTIONS

1. Explain each of the following terms in your own words, and give an example of each.
   a. Sequence
   b. Convergent sequence
   c. Divergent sequence
   d. Limit of a sequence

2. Explain each of the following terms in your own words, and give an example of each.
   a. Bounded sequence
   b. Monotonic sequence

8.1 EXERCISES

In Exercises 1–6, write the first five terms of the sequence $\{a_n\}$ whose nth term is given.

1. $a_n = \dfrac{n+1}{2n-1}$

2. $a_n = \dfrac{(-1)^{n+1}2^n}{n+1}$

3. $a_n = \sin \dfrac{n\pi}{2}$

4. $a_n = \dfrac{1 \cdot 3 \cdot 5 \cdot \cdots \cdot (2n-1)}{n!}$

5. $a_n = \dfrac{2^n}{(2n)!}$

6. $a_1 = 2, \quad a_{n+1} = 3a_n + 1$

In Exercises 7–12, find an expression for the nth term of the sequence. (Assume that the pattern continues.)

7. $\left\{\dfrac{1}{2}, \dfrac{2}{3}, \dfrac{3}{4}, \dfrac{4}{5}, \dfrac{5}{6}, \cdots\right\}$

8. $\left\{\dfrac{3}{4}, \dfrac{4}{9}, \dfrac{5}{16}, \dfrac{6}{25}, \dfrac{7}{36}, \cdots\right\}$

9. $\left\{-1, \dfrac{1}{2}, -\dfrac{1}{6}, \dfrac{1}{24}, -\dfrac{1}{120}, \cdots\right\}$

10. $\{0, 2, 0, 2, 0, \ldots\}$

11. $\left\{\dfrac{1}{1 \cdot 2}, \dfrac{2}{2 \cdot 3}, \dfrac{3}{3 \cdot 4}, \dfrac{4}{4 \cdot 5}, \dfrac{5}{5 \cdot 6}, \cdots\right\}$

12. $\left\{\dfrac{1}{2}, \dfrac{1 \cdot 3}{2 \cdot 4}, \dfrac{1 \cdot 3 \cdot 5}{2 \cdot 4 \cdot 6}, \dfrac{1 \cdot 3 \cdot 5 \cdot 7}{2 \cdot 4 \cdot 6 \cdot 8}, \dfrac{1 \cdot 3 \cdot 5 \cdot 7 \cdot 9}{2 \cdot 4 \cdot 6 \cdot 8 \cdot 10}, \cdots\right\}$

In Exercises 13–42, determine whether the sequence $\{a_n\}$ converges or diverges. If it converges, find its limit.

13. $a_n = \dfrac{2n}{n+1}$

14. $a_n = \sqrt{n+1}$

15. $a_n = 1 + 2(-1)^n$

16. $a_n = 1 + \dfrac{(-1)^n}{n^{3/2}}$

17. $a_n = \dfrac{n-1}{n} - \dfrac{2n+1}{n^2}$

18. $a_n = \dfrac{n^2-1}{2n^2+1}$

19. $a_n = \dfrac{2n^2 - 3n + 4}{3n^2 + 1}$

20. $a_n = (-1)^n \dfrac{n+2}{3n+1}$

21. $a_n = \dfrac{2 + (-1)^n}{n}$

22. $a_n = \dfrac{\sqrt{2n^2+1}}{n}$

23. $a_n = \dfrac{2n}{\sqrt{n+1}}$

24. $a_n = 1 + \left(-\dfrac{2}{e}\right)^n$

25. $a_n = \dfrac{n^{1/2} + n^{1/3}}{n + 2n^{2/3}}$

26. $a_n = \cos n\pi + 2$

27. $a_n = \sin \dfrac{n\pi}{2}$

28. $a_n = \sin\left(\dfrac{n\pi}{2n+1}\right)$

29. $a_n = \dfrac{\sin \sqrt{n}}{\sqrt{n}}$

30. $a_n = \tan^{-1} n^2$

31. $a_n = \tanh n$

32. $a_n = \dfrac{\ln n^2}{\sqrt{n}}$

33. $a_n = \dfrac{2^n}{3^n + 1}$

34. $a_n = \dfrac{2^n + 1}{e^n}$

35. $a_n = \sqrt{n+1} - \sqrt{n}$

36. $a_n = \dfrac{n^p}{e^n}, \quad p > 0$

37. $a_n = \left(1 + \dfrac{2}{n}\right)^{1/n}$

38. $a_n = \dfrac{(-2)^n}{n!}$

39. $a_n = \dfrac{\sin^2 n}{\sqrt{n}}$

40. $a_n = \dfrac{1 \cdot 3 \cdot 5 \cdot \cdots \cdot (2n-1)}{n!}$

41. $a_n = \dfrac{1}{n^2} + \dfrac{2}{n^2} + \dfrac{3}{n^2} + \cdots + \dfrac{n}{n^2}$

42. $a_n = \dfrac{1 + 2 + 3 + \cdots + n}{n+2} - \dfrac{n}{2}$

In Exercises 43–48, (a) graph the sequence $\{a_n\}$ with a graphing utility, (b) use your graph to guess at the convergence or divergence of the sequence, and (c) use the properties of limits to verify your guess and to find the limit of the sequence if it converges.

43. $a_n = \dfrac{n-1}{n+2}$

44. $a_n = (-1)^n \dfrac{2n+1}{n+3}$

45. $a_n = \dfrac{n!}{n^n}$

46. $a_n = 2 \tan^{-1}\left(\dfrac{n+1}{n+3}\right)$

47. $a_n = n \sin \dfrac{1}{n}$

48. $a_n = \left(1 - \dfrac{2}{n}\right)^n$

49. Evaluate

$$\lim_{n \to \infty} \dfrac{1 - \left(1 - \dfrac{1}{n}\right)^9}{1 - \left(1 - \dfrac{1}{n}\right)}$$

Hint: Use Theorem 1.

50. Evaluate

$$\lim_{n \to \infty} n\left(1 - \sqrt[7]{1 - \dfrac{1}{n}}\right)$$

Hint: Use Theorem 1.

In Exercises 51–58, determine whether the sequence $\{a_n\}$ is monotonic. Is the sequence bounded?

51. $a_n = \dfrac{3}{2n+5}$

52. $a_n = \dfrac{2n}{n+1}$

53. $a_n = 3 - \dfrac{1}{n}$

54. $a_n = 2 + \dfrac{(-1)^n}{n}$

Videos for selected exercises are available online at **www.academic.cengage.com/login**.

55. $a_n = \dfrac{\sin n}{n}$

56. $a_n = \tan^{-1} n$

57. $a_n = \dfrac{n}{2^n}$

58. $a_n = \dfrac{\ln n}{n}$

59. Compound Interest If a principal of P dollars is invested in an account earning interest at the rate of r per year compounded monthly, then the accumulated amount A_n at the end of n months is

$$A_n = P\left(1 + \frac{r}{12}\right)^n$$

a. Write the first six terms of the sequence $\{A_n\}$ if $P = 10,000$ and $r = 0.105$. Interpret your results.

b. Does the sequence $\{A_n\}$ converge or diverge?

60. Quality Control Half a percent of the microprocessors manufactured by Alpha Corporation for use in regulating fuel consumption in automobiles are defective. It can be shown that the probability of finding at least one defective microprocessor in a random sample of n microprocessors is $f(n) = 1 - (0.995)^n$. Consider the sequence $\{a_n\}$ defined by $a_n = f(n)$.

a. Write the terms a_{10}, a_{100}, and a_{1000}.

b. Evaluate $\lim_{n \to \infty} a_n$, and interpret your result.

61. Annuities An annuity is a sequence of payments made at regular intervals. Suppose that a sum of $200 is deposited at the end of each month into an account earning interest at the rate of 12% per year compounded monthly. Then the amount on deposit (called the future value of the annuity) at the end of the nth month is $f(n) = 20,000[(1.01)^n - 1]$. Consider the sequence $\{a_n\}$ defined by $a_n = f(n)$.

a. Find the 24th term of the sequence $\{a_n\}$, and interpret your result.

b. Evaluate $\lim_{n \to \infty} a_n$, and interpret your result.

62. Continuously Compounded Interest If P dollars is invested in an account paying interest at the rate of r per year compounded m times per year, then the accumulated amount at the end of t years is

$$A_m = P\left(1 + \frac{r}{m}\right)^{mt} \qquad m = 1, 2, 3, \ldots$$

a. Find the limit of the sequence $\{A_m\}$.

b. Interpret the result in part (a).

Note: In this situation, interest is said to be *compounded continuously.*

c. What is the accumulated amount at the end of 3 years if $1000 is invested in an account paying interest at the rate of 10% per year compounded continuously?

 63. Find the limit of the sequence $\left\{\left(1 + \dfrac{2}{n}\right)^{3n}\right\}$. Confirm your results visually by plotting the graph of

$$f(x) = \left(1 + \frac{2}{x}\right)^{3x}$$

 64. Define the sequence $\{a_n\}$ recursively by $a_0 = 2$ and $a_{n+1} = \sqrt{a_n}$ for $n \geq 1$.

a. Show that $a_n = 2^{1/2^n}$.

b. Evaluate $\lim_{n \to \infty} a_n$.

c. Verify the result of part (b) graphically.

 65. Newton's Method Suppose that $A > 0$. Applying Newton's method to the solution of the equation $x^2 - A = 0$ leads to the sequence $\{x_n\}$ defined by

$$x_{n+1} = \frac{1}{2}\left(x_n + \frac{A}{x_n}\right) \qquad x_0 > 0$$

a. Show that if $L = \lim_{n \to \infty} x_n$ exists, then $L = \sqrt{A}$.

Hint: $\lim_{n \to \infty} x_{n+1} = L$

b. Find $\sqrt{5}$ accurate to four decimal places.

 66. Finding the Roots of an Equation Suppose that we want to find a root of $f(x) = 0$. Newton's method provides one way of finding it. Here is another method that works under suitable conditions.

a. Write $f(x) = 0$ in the form $x = g(x)$, where g is continuous. Then generate the sequence $\{x_n\}$ by the recursive formula $x_{n+1} = g(x_n)$, where x_0 is arbitrary.

b. Show that if the sequence $\{x_n\}$ converges to a number r, then r is a solution of $f(x) = 0$.

Hint: $\lim_{n \to \infty} x_{n+1} = r$

c. Use this method to find the root of $f(x) = 3x^3 - 9x + 2$ (accurate to four decimal places) that lies in the interval $(0, 1)$.

Hint: Write $3x^3 - 9x + 2 = 0$ in the form $x = \frac{1}{9}(3x^3 + 2)$. Take $x_0 = 0$.

67. A Floating Object A sphere of radius 1 ft is made of wood that has a specific gravity of $\frac{2}{3}$. If the sphere is placed in water, it sinks to a depth of h ft. It can be shown that h satisfies the equation

$$h^3 - 3h^2 + \frac{8}{3} = 0$$

Use the method described in Exercise 66 to find h accurate to three decimal places.

Hint: Show that the equation can be written in the form $h = \frac{1}{3}\sqrt{3h^3 + 8}$.

68. Find the limit of the sequence

$$\left\{\sqrt{2}, \sqrt{2\sqrt{2}}, \sqrt{2\sqrt{2\sqrt{2}}}, \ldots\right\}$$

Hint: Show that $a_n = 2^{(2^n - 1)/2^n} = 2^{1 - 1/2^n}$.

69. Consider the sequence $\{a_n\}$ defined by $a_1 = \sqrt{2}$ and $a_n = \sqrt{2 + a_{n-1}}$ for $n \geq 2$. Assuming that the sequence converges, find its limit.

Note: Using the principle of mathematical induction, it can be shown that $\{a_n\}$ is increasing and bounded by 2 and, hence, by Theorem 6 is convergent.

70. Show that if $\lim_{n \to \infty} a_{2n} = L$ and $\lim_{n \to \infty} a_{2n+1} = L$, then $\lim_{n \to \infty} a_n = L$.

71. Let the sequence $\{a_n\}$ be defined by

$$a_n = 1 + \frac{1}{2^2} + \frac{1}{3^2} + \cdots + \frac{1}{n^2}$$

a. Show that $\{a_n\}$ is increasing.

b. Show that $\{a_n\}$ is bounded above by establishing that $a_n < 2 - 1/n$ for $n \geq 2$.

Hint: $\dfrac{1}{n^2} < \dfrac{1}{n(n-1)} = \dfrac{1}{n-1} - \dfrac{1}{n}$, for $n \geq 2$

c. Using the results of parts (a) and (b), what can you deduce about the convergence of $\{a_n\}$?

72. Let the sequence $\{a_n\}$ be defined by

$$a_n = \frac{1}{2+1} + \frac{1}{2^2+2} + \cdots + \frac{1}{2^n+n}$$

a. Show that $\{a_n\}$ is increasing.

b. Show that $\{a_n\}$ is bounded above.

c. Using the results of parts (a) and (b), what can you deduce about the convergence of $\{a_n\}$?

73. Let the sequence $\{a_n\}$ be defined by

$$a_1 = \frac{a_0}{2+a_0}, \qquad a_2 = \frac{a_1}{2+a_1},$$

$$a_3 = \frac{a_2}{2+a_2}, \qquad \ldots, \qquad a_n = \frac{a_{n-1}}{2+a_{n-1}}, \qquad \ldots$$

where $a_n > 0$.

a. Show that $\{a_n\}$ is convergent.

b. Find the limit of $\{a_n\}$.

74. Use the Squeeze Theorem for Sequences to prove that

$$\lim_{n \to \infty} \sqrt[n]{a} = 1 \qquad a > 0$$

Hint: For n sufficiently large, $1/n < a < n$.

75. Prove Theorem 1: If $\lim_{x \to \infty} f(x) = L$ and $\{a_n\}$ is a sequence defined by $a_n = f(n)$, where n is a positive integer, then $\lim_{n \to \infty} a_n = L$.

76. Prove Theorem 4: If $\lim_{n \to \infty} |a_n| = 0$, then $\lim_{n \to \infty} a_n = 0$.

77. Prove the properties of the sequence $\{r^n\}$ given on page 737 using the results $\lim_{x \to \infty} a^x = 0$ if $0 < a < 1$ and $\lim_{x \to \infty} a^x = \infty$ if $a > 1$.

78. Fibonacci Sequence The Fibonacci sequence $\{F_n\}$ is defined by $F_1 = 1$, $F_2 = 1$, and $F_{n+1} = F_n + F_{n-1}$ for $n \geq 2$. Let $a_n = F_{n+1}/F_n$. Assuming that $\{a_n\}$ is convergent, show that

$$\lim_{n \to \infty} a_n = \frac{1}{2}(1 + \sqrt{5})$$

Hint: First, show that $a_{n-1} = 1 + 1/a_{n-2}$. Then use the fact that if $\lim_{n \to \infty} a_n = L$, then $\lim_{n \to \infty} a_{n-2} = \lim_{n \to \infty} a_{n-1} = L$.

Note: The number $\frac{1}{2}(1 + \sqrt{5})$, which is approximately 1.6, has the following special property: A picture with a ratio of width to height equal to this number is especially pleasing to the eye. The ancient Greeks used this "golden" ratio in designing their beautiful temples and public buildings, such as the Parthenon.

The front of the Parthenon has a ratio of width to height that is approximately 1.6 to 1.

In Exercises 79–86, determine whether the statement is true or false. If it is true, explain why it is true. If it is false, explain why or give an example to show why it is false.

79. If $\{a_n\}$ and $\{b_n\}$ are divergent, then $\{a_n + b_n\}$ is divergent.

80. If $\{a_n\}$ is divergent, then $\{|a_n|\}$ is divergent.

81. If $\{a_n\}$ converges to L and $\{b_n\}$ converges to 0, then $\{a_n b_n\}$ converges to 0.

82. If $\{a_n\}$ converges and $\{b_n\}$ converges, then $\{a_n/b_n\}$ converges.

83. If $\{a_n\}$ is bounded and $\{b_n\}$ converges, then $\{a_n b_n\}$ converges.

84. If $\{a_n\}$ is bounded, then $\{a_n/n\}$ converges to 0.

85. If $\lim_{n \to \infty} a_n b_n$ exists, then both $\lim_{n \to \infty} a_n$ and $\lim_{n \to \infty} b_n$ exist.

86. If $\lim_{n \to \infty} |a_n|$ exists, then $\lim_{n \to \infty} a_n$ exists.

8.2 Series

Consider again the example involving the bouncing ball. Earlier we found a sequence describing the maximum height attained by the ball on each rebound after hitting a surface. The question that follows naturally is: How do we find the total distance traveled by the ball? To answer this question, recall that the initial height and the heights attained on each subsequent rebound are

$$1, \quad \frac{2}{3}, \quad \left(\frac{2}{3}\right)^2, \quad \left(\frac{2}{3}\right)^3, \quad \cdots$$

meters, respectively. (See Figure 1.) Observe that the distance traveled by the ball when it first hits the surface is 1 m. When it hits the surface the second time, it will have traveled a total distance of

$$1 + 2\left(\frac{2}{3}\right) \qquad \text{or} \qquad 1 + \frac{4}{3}$$

meters. When it hits the surface the third time, it will have traveled a distance of

$$1 + 2\left(\frac{2}{3}\right) + 2\left(\frac{2}{3}\right)^2 \qquad \text{or} \qquad 1 + \frac{4}{3} + \frac{8}{9}$$

meters. Continuing in this fashion, we see that the total distance traveled by the ball is

$$1 + 2\left(\frac{2}{3}\right) + 2\left(\frac{2}{3}\right)^2 + 2\left(\frac{2}{3}\right)^3 + \cdots \tag{1}$$

meters. Observe that this last expression involves the sum of infinitely many terms.

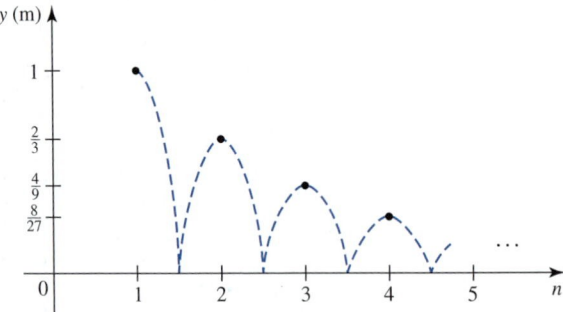

FIGURE 1

In general, an expression of the form

$$a_1 + a_2 + a_3 + \cdots + a_n + \cdots$$

is called an **infinite series** or, more simply, a **series.** The numbers $a_1, a_2, a_3, \ldots$ are called the **terms** of the series; a_n is called the ***n*th term**, or **general term**, of the series; and the series itself is denoted by the symbol

$$\sum_{n=1}^{\infty} a_n$$

or simply $\sum a_n$.

How do we define the "sum" of an infinite series, if it exists? To answer this question, we use the same technique that we have employed several times before: using quantities that we can compute to help us define new ones. For example, in defining the slope of the tangent line to the graph of a function, we take the limit of the slope of secant lines (quantities that we can compute); and in defining the area under the graph of a function, we take the limit of the sum of the area of rectangles (again, quantities that we can compute). Here, we define the sum of an infinite series as the limit of a sequence of *finite* sums (quantities that we can compute).

We can get an inkling of how this may be done from examining the series (1) giving the total distance traveled by the ball. Define the sequence $\{S_n\}$ by

$$S_1 = 1$$

$$S_2 = 1 + 2\left(\frac{2}{3}\right)$$

$$S_3 = 1 + 2\left(\frac{2}{3}\right) + 2\left(\frac{2}{3}\right)^2$$

$$\vdots$$

$$S_n = 1 + 2\left(\frac{2}{3}\right) + 2\left(\frac{2}{3}\right)^2 + \cdots + 2\left(\frac{2}{3}\right)^{n-1}$$

giving the total vertical distance traveled by the ball when it hits the surface the first time, the second time, the third time, ... , and the nth time, respectively. If the series (1) has a sum S (the total distance traveled by the ball), then the terms of the sequence $\{S_n\}$ form a sequence of increasingly accurate approximations to S. This suggests that we define

$$S = \lim_{n \to \infty} S_n$$

We will complete the solution to this problem in Example 5.

Motivated by this discussion, we define the sum of an infinite series.

DEFINITION **Convergence of Infinite Series**

Given an infinite series

$$\sum_{n=1}^{\infty} a_n = a_1 + a_2 + a_3 + \cdots + a_n + \cdots$$

the **nth partial sum** of the series is

$$S_n = \sum_{k=1}^{n} a_k = a_1 + a_2 + a_3 + \cdots + a_n$$

If the sequence of partial sums $\{S_n\}$ **converges** to the number S, that is, if $\lim_{n \to \infty} S_n = S$, then the series $\Sigma \, a_n$ **converges** and has **sum S**, written

$$\sum_{n=1}^{\infty} a_n = a_1 + a_2 + a_3 + \cdots + a_n + \cdots = S$$

If $\{S_n\}$ diverges, then the series $\Sigma \, a_n$ **diverges.**

> ⚠️ Be sure to note the difference between a sequence and a series. A sequence is a *succession* of terms, whereas a series is a *sum* of terms.

EXAMPLE 1 Determine whether the series converges. If the series converges, find its sum.

a. $\displaystyle\sum_{n=1}^{\infty} n$ **b.** $\displaystyle\sum_{n=1}^{\infty} \left(\frac{1}{n} - \frac{1}{n+1} \right)$

Solution

a. The nth partial sum of the series is

$$S_n = 1 + 2 + 3 + \cdots + n = \frac{n(n+1)}{2}$$

Since

$$\lim_{n\to\infty} S_n = \lim_{n\to\infty} \frac{n(n+1)}{2} = \infty$$

we conclude that the limit does not exist and $\sum_{n=1}^{\infty} n$ diverges.

b. The nth partial sum of the series is

$$S_n = \left(1 - \frac{1}{2}\right) + \left(\frac{1}{2} - \frac{1}{3}\right) + \left(\frac{1}{3} - \frac{1}{4}\right) + \cdots + \left(\frac{1}{n-1} - \frac{1}{n}\right) + \left(\frac{1}{n} - \frac{1}{n+1}\right)$$

Removing the parentheses, we see that all the terms of S_n, except for the first and last, cancel out. So

$$S_n = 1 - \frac{1}{n+1}$$

Since

$$\lim_{n\to\infty} S_n = \lim_{n\to\infty}\left(1 - \frac{1}{n+1}\right) = 1$$

we conclude that the series converges and has sum 1, that is,

$$\sum_{n=1}^{\infty} \left(\frac{1}{n} - \frac{1}{n+1} \right) = 1$$ ∎

The series in Example 1b is called a **telescoping series.**

EXAMPLE 2 Show that the series $\displaystyle\sum_{n=1}^{\infty} \frac{4}{4n^2 - 1}$ is convergent, and find its sum.

Solution First, we use partial fraction decomposition to rewrite the general term $a_n = 4/(4n^2 - 1)$:

$$a_n = \frac{4}{4n^2 - 1} = \frac{4}{(2n-1)(2n+1)} = \frac{2}{2n-1} - \frac{2}{2n+1}$$

Then we write the nth partial sum of the series as

$$S_n = \sum_{k=1}^{n} \frac{4}{4k^2 - 1} = \sum_{k=1}^{n} \left(\frac{2}{2k - 1} - \frac{2}{2k + 1} \right)$$

$$= \left(\frac{2}{1} - \frac{2}{3} \right) + \left(\frac{2}{3} - \frac{2}{5} \right) + \left(\frac{2}{5} - \frac{2}{7} \right) + \cdots + \left(\frac{2}{2n - 1} - \frac{2}{2n + 1} \right)$$

$$= 2 - \frac{2}{2n + 1} \qquad \text{This is a telescoping series.}$$

Since

$$\lim_{n \to \infty} S_n = \lim_{n \to \infty} \left(2 - \frac{2}{2n + 1} \right) = 2$$

we conclude that the given series is convergent and has sum 2; that is,

$$\sum_{n=1}^{\infty} \frac{4}{4n^2 - 1} = 2 \qquad \blacksquare$$

■ Geometric Series

Geometric series play an important role in mathematical analysis. They also arise frequently in the field of finance. The convergence or divergence of a geometric series is easily established.

> **DEFINITION** Geometric Series
>
> A series of the form
>
> $$\sum_{n=1}^{\infty} ar^{n-1} = a + ar + ar^2 + \cdots + ar^{n-1} + \cdots \qquad a \neq 0$$
>
> is called a **geometric series** with common ratio r.

The following theorem tells us the conditions under which a geometric series is convergent.

> **THEOREM 1**
>
> If $|r| < 1$, then the geometric series
>
> $$\sum_{n=1}^{\infty} ar^{n-1} = a + ar + ar^2 + \cdots + ar^{n-1} + \cdots$$
>
> converges, and its sum is $\displaystyle\sum_{n=1}^{\infty} ar^{n-1} = \frac{a}{1 - r}$. The series diverges if $|r| \geq 1$.

PROOF The nth partial sum of the series is

$$S_n = a + ar + ar^2 + \cdots + ar^{n-1}$$

Multiplying both sides of this equation by r gives

$$rS_n = ar + ar^2 + ar^3 + \cdots + ar^n$$

Subtracting the second equation from the first then yields

$$(1 - r)S_n = a - ar^n = a(1 - r^n)$$

If $r \neq 1$, we can solve the last equation for S_n, obtaining

$$S_n = \frac{a(1 - r^n)}{1 - r}$$

From Example 13 on page 737 we know that $\lim_{n \to \infty} r^n = 0$ if $|r| < 1$, so

$$\lim_{n \to \infty} S_n = \lim_{n \to \infty} \frac{a(1 - r^n)}{1 - r} = \frac{a}{1 - r}$$

This implies that

$$\sum_{n=1}^{\infty} ar^{n-1} = \frac{a}{1 - r} \qquad |r| < 1$$

If $|r| > 1$, then the sequence $\{r^n\}$ diverges, so $\lim_{n \to \infty} S_n$ does not exist. This means that the geometric series diverges. We leave it as an exercise to show that $\{S_n\}$ diverges if $r = \pm 1$, so the series also diverges for these values of r. ■

EXAMPLE 3 Determine whether the series converges or diverges. If it converges, find its sum.

a. $\displaystyle\sum_{n=1}^{\infty} 3\left(-\frac{1}{2}\right)^{n-1} = 3 - \frac{3}{2} + \frac{3}{4} - \frac{3}{8} + \cdots$

b. $\displaystyle\sum_{n=1}^{\infty} 5\left(\frac{4}{3}\right)^{n-1} = 5 + \frac{20}{3} + \frac{80}{9} + \frac{320}{27} + \cdots$

Solution

a. This is a geometric series with $a = 3$ and common ratio $r = -\frac{1}{2}$. Since $\left|-\frac{1}{2}\right| < 1$, Theorem 1 tells us that the series converges and has sum

$$\sum_{n=1}^{\infty} 3\left(-\frac{1}{2}\right)^{n-1} = \frac{3}{1 - \left(-\frac{1}{2}\right)} = 2$$

The graphs of $\{a_n\}$ and $\{S_n\}$ for this series are shown in Figure 2a.

b. This is a geometric series with $a = 5$ and common ratio $r = \frac{4}{3}$. Since $\frac{4}{3} > 1$, Theorem 1 tells us that the series is divergent. The graphs of $\{a_n\}$ and $\{S_n\}$ for this series are shown in Figure 2b.

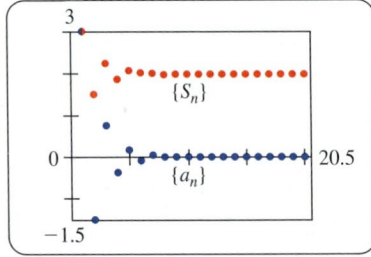

(a) The geometric series converges because $|r| < 1$.

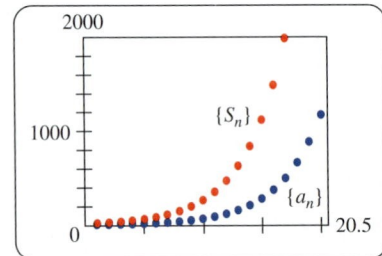

(b) The geometric series diverges because $|r| > 1$.

FIGURE 2

EXAMPLE 4 Express the number $3.2\overline{14} = 3.2141414\ldots$ as a rational number.

Solution We rewrite the number as

$$3.2141414\ldots = 3.2 + \frac{14}{10^3} + \frac{14}{10^5} + \frac{14}{10^7} + \cdots$$

$$= \frac{32}{10} + \frac{14}{10^3}\left[1 + \frac{1}{10^2} + \frac{1}{10^4} + \cdots\right]$$

$$= \frac{32}{10} + \sum_{n=1}^{\infty}\left(\frac{14}{10^3}\right)\left(\frac{1}{10^2}\right)^{n-1}$$

The expression after the first term is a geometric series with $a = \frac{14}{1000}$ and $r = \frac{1}{100}$. Using Theorem 1, we have

$$3.2141414\ldots = \frac{32}{10} + \frac{\frac{14}{1000}}{1 - \frac{1}{100}}$$

$$= \frac{32}{10} + \frac{14}{990} = \frac{3182}{990} \qquad \blacksquare$$

EXAMPLE 5 Complete the solution of the bouncing ball problem that was introduced at the beginning of this section. Recall that the total vertical distance traveled by the ball is given by

$$1 + 2\left(\frac{2}{3}\right) + 2\left(\frac{2}{3}\right)^2 + 2\left(\frac{2}{3}\right)^3 + \cdots$$

meters.

Solution If we let d denote the total vertical distance traveled by the ball, then

$$d = 1 + \sum_{n=1}^{\infty}\left(\frac{4}{3}\right)\left(\frac{2}{3}\right)^{n-1}$$

The expression after the first term is a geometric series with $a = \frac{4}{3}$ and $r = \frac{2}{3}$. Using Theorem 1, we obtain

$$d = 1 + \frac{\frac{4}{3}}{1 - \frac{2}{3}} = 1 + 4 = 5$$

and conclude that the total distance traveled by the ball is 5 m. $\qquad \blacksquare$

■ The Harmonic Series

The series

$$\sum_{n=1}^{\infty} \frac{1}{n} = 1 + \frac{1}{2} + \frac{1}{3} + \frac{1}{4} + \cdots$$

is called the **harmonic series.** Before showing that this series is divergent, we make this observation: If a sequence $\{b_n\}$ is convergent, then any *subsequence* obtained by deleting any number of terms from the parent sequence $\{b_n\}$ must also converge to the same limit. Therefore, to show that a sequence is divergent, it suffices to produce a subsequence of the parent sequence that is divergent.

In keeping with this strategy, let us show that the subsequence

$$S_2, S_4, S_8, S_{16}, \ldots, S_{2^n}, \ldots$$

of the sequence $\{S_n\}$ of partial sums of the harmonic series is divergent. We have

$$S_2 = 1 + \frac{1}{2}$$

$$S_4 = 1 + \frac{1}{2} + \left(\frac{1}{3} + \frac{1}{4}\right) > 1 + \frac{1}{2} + \left(\frac{1}{4} + \frac{1}{4}\right) = 1 + 2\left(\frac{1}{2}\right)$$

$$S_8 = 1 + \frac{1}{2} + \left(\frac{1}{3} + \frac{1}{4}\right) + \left(\frac{1}{5} + \frac{1}{6} + \frac{1}{7} + \frac{1}{8}\right)$$

$$> 1 + \frac{1}{2} + \left(\frac{1}{4} + \frac{1}{4}\right) + \left(\frac{1}{8} + \frac{1}{8} + \frac{1}{8} + \frac{1}{8}\right) = 1 + \frac{1}{2} + \frac{1}{2} + \frac{1}{2} = 1 + 3\left(\frac{1}{2}\right)$$

$$S_{16} = 1 + \frac{1}{2} + \left(\frac{1}{3} + \frac{1}{4}\right) + \left(\frac{1}{5} + \cdots + \frac{1}{8}\right) + \left(\frac{1}{9} + \cdots + \frac{1}{16}\right)$$

$$> 1 + \frac{1}{2} + \left(\frac{1}{4} + \frac{1}{4}\right) + \underbrace{\left(\frac{1}{8} + \cdots + \frac{1}{8}\right)}_{\text{4 terms}} + \underbrace{\left(\frac{1}{16} + \cdots + \frac{1}{16}\right)}_{\text{8 terms}}$$

$$= 1 + \frac{1}{2} + \frac{1}{2} + \frac{1}{2} + \frac{1}{2} = 1 + 4\left(\frac{1}{2}\right)$$

and, in general, $S_{2^n} > 1 + n\left(\frac{1}{2}\right)$. Therefore,

$$\lim_{n \to \infty} S_{2^n} = \infty$$

so $\{S_n\}$ is divergent. This proves that the harmonic series is divergent.

■ The Divergence Test

The next theorem tells us that the terms of a convergent series must ultimately approach zero.

THEOREM 2

If $\sum_{n=1}^{\infty} a_n$ converges, then $\lim_{n \to \infty} a_n = 0$.

PROOF   We have $S_n = a_1 + a_2 + \cdots + a_{n-1} + a_n = S_{n-1} + a_n$, so $a_n = S_n - S_{n-1}$. Since $\sum_{n=1}^{\infty} a_n$ is convergent, the sequence $\{S_n\}$ is convergent. Let $\lim_{n \to \infty} S_n = S$. Then

$$\lim_{n \to \infty} a_n = \lim_{n \to \infty}(S_n - S_{n-1}) = \lim_{n \to \infty} S_n - \lim_{n \to \infty} S_{n-1} = S - S = 0 \qquad ■$$

The *Divergence Test* is an important consequence of Theorem 2.

THEOREM 3   **The Divergence Test**

If $\lim_{n \to \infty} a_n$ does not exist or $\lim_{n \to \infty} a_n \neq 0$, then $\sum_{n=1}^{\infty} a_n$ diverges.

The Divergence Test does *not* say that if $\lim_{n\to\infty} a_n = 0$, then $\sum_{n=1}^{\infty} a_n$ must converge. In other words, the converse of Theorem 2 is not true in general. For example, $\lim_{n\to\infty} 1/n = 0$, yet the harmonic series $\sum_{n=1}^{\infty} 1/n$ is divergent. In short, the Divergence Test rules out convergence for a series whose nth term does not approach zero but yields no information if a_n does approach zero—that is, the series might or might not converge.

EXAMPLE 6 Show that the following series are divergent.

a. $\displaystyle\sum_{n=1}^{\infty} (-1)^{n-1}$ **b.** $\displaystyle\sum_{n=1}^{\infty} \frac{2n^2 + 1}{3n^2 - 1}$

Solution

a. Here, $a_n = (-1)^{n-1}$, and

$$\lim_{n\to\infty} a_n = \lim_{n\to\infty} (-1)^{n-1}$$

does not exist. We conclude by the Divergence Test that the series diverges.

b. Here, $a_n = \dfrac{2n^2 + 1}{3n^2 - 1}$, and

$$\lim_{n\to\infty} a_n = \lim_{n\to\infty} \frac{2n^2 + 1}{3n^2 - 1} = \lim_{n\to\infty} \frac{2 + \dfrac{1}{n^2}}{3 - \dfrac{1}{n^2}} = \frac{2}{3} \neq 0$$

so by the Divergence Test, the series diverges. ■

Properties of Convergent Series

The following properties of series are immediate consequences of the corresponding properties of the limits of sequences. We omit the proofs.

THEOREM 4   Properties of Convergent Series

If $\sum_{n=1}^{\infty} a_n = A$ and $\sum_{n=1}^{\infty} b_n = B$ are convergent and c is any real number, then $\sum_{n=1}^{\infty} ca_n$ and $\sum_{n=1}^{\infty} (a_n \pm b_n)$ are also convergent, and

a. $\displaystyle\sum_{n=1}^{\infty} ca_n = c\sum_{n=1}^{\infty} a_n = cA$ **b.** $\displaystyle\sum_{n=1}^{\infty} (a_n \pm b_n) = \sum_{n=1}^{\infty} a_n \pm \sum_{n=1}^{\infty} b_n = A \pm B$

EXAMPLE 7 Show that the series $\displaystyle\sum_{n=1}^{\infty} \left[\frac{2}{n(n + 1)} - \frac{4}{3^n} \right]$ is convergent, and find its sum.

Solution First, consider the series $\sum_{n=1}^{\infty} 1/[n(n + 1)]$. Using partial fraction decomposition, we can write this series in the form

$$\sum_{n=1}^{\infty} \frac{1}{n(n + 1)} = \sum_{n=1}^{\infty} \left(\frac{1}{n} - \frac{1}{n + 1} \right)$$

Using the result of Example 1, we see that

$$\sum_{n=1}^{\infty} \frac{1}{n(n+1)} = 1$$

Next, observe that $\sum_{n=1}^{\infty} \frac{4}{3^n}$ is a geometric series with $a = \frac{4}{3}$ and $r = \frac{1}{3}$, so

$$\sum_{n=1}^{\infty} \frac{4}{3^n} = \frac{\frac{4}{3}}{1 - \frac{1}{3}} = 2$$

Therefore, by Theorem 4 the given series is convergent, and

$$\sum_{n=1}^{\infty} \left[\frac{2}{n(n+1)} - \frac{4}{3^n} \right] = 2 \sum_{n=1}^{\infty} \frac{1}{n(n+1)} - \sum_{n=1}^{\infty} \frac{4}{3^n}$$
$$= 2 \cdot 1 - 2 = 0 \qquad \blacksquare$$

8.2 CONCEPT QUESTIONS

1. Explain the difference between
   a. A sequence and a series
   b. A convergent sequence and a convergent series
   c. A divergent sequence and a divergent series
   d. The limit of a sequence and the sum of a series

2. Suppose that $\sum_{n=1}^{\infty} a_n = 6$.
   a. Evaluate $\lim_{n \to \infty} S_n$, where S_n is the nth partial sum of $\sum_{n=1}^{\infty} a_n$.
   b. Find $\sum_{n=2}^{\infty} a_n$ if it is known that $a_1 = \frac{1}{2}$.

8.2 EXERCISES

In Exercises 1–6, find the nth partial sum S_n of the telescoping series, and use it to determine whether the series converges or diverges. If it converges, find its sum.

1. $\sum_{n=2}^{\infty} \left(\frac{1}{n-1} - \frac{1}{n} \right)$ **2.** $\sum_{n=1}^{\infty} \left(\frac{1}{2n+3} - \frac{1}{2n+1} \right)$

3. $\sum_{n=1}^{\infty} \frac{4}{(2n+3)(2n+5)}$ **4.** $\sum_{n=1}^{\infty} \left(\frac{-8}{4n^2 + 4n - 3} \right)$

5. $\sum_{n=2}^{\infty} \left(\frac{1}{\ln n} - \frac{1}{\ln(n+1)} \right)$ **6.** $\sum_{n=1}^{\infty} \frac{2}{\sqrt{n+1} + \sqrt{n}}$

In Exercises 7–14, determine whether the geometric series converges or diverges. If it converges, find its sum.

7. $4 + \frac{8}{3} + \frac{16}{9} + \frac{32}{27} + \cdots$ **8.** $-\frac{1}{2} + \frac{1}{4} - \frac{1}{8} + \frac{1}{16} - \cdots$

9. $\frac{5}{3} - \frac{5}{9} + \frac{5}{27} - \frac{5}{81} + \cdots$ **10.** $1 + \frac{4}{3} + \frac{16}{9} + \frac{64}{27} + \cdots$

11. $\sum_{n=0}^{\infty} 2\left(-\frac{1}{\sqrt{2}} \right)^n$ **12.** $\sum_{n=1}^{\infty} \frac{e^n}{3^{n+1}}$

13. $\sum_{n=0}^{\infty} 2^n 3^{-n+1}$ **14.** $\sum_{n=1}^{\infty} (-1)^{n-1} 3^n 2^{1-n}$

In Exercises 15–22, show that the series diverges.

15. $\frac{1}{2} + \frac{2}{3} + \frac{3}{4} + \cdots$ **16.** $1 - \frac{3}{2} + \frac{9}{4} - \frac{27}{8} + \cdots$

17. $\sum_{n=1}^{\infty} \frac{2n}{3n+1}$ **18.** $\sum_{n=1}^{\infty} \frac{n^2}{2n^2 + 1}$

19. $\sum_{n=1}^{\infty} 2(1.5)^n$ **20.** $\sum_{n=0}^{\infty} \frac{(-1)^n 3^n}{2^{n-1}}$

21. $\sum_{n=1}^{\infty} \frac{1}{2 + 3^{-n}}$ **22.** $\sum_{n=1}^{\infty} \frac{n}{\sqrt{2n^2 + 1}}$

 In Exercises 23–28, (a) compute as many terms of the sequence of partial sums, S_n, as is necessary to convince yourself that the series converges or diverges. If it converges, estimate its sum. (b) Plot $\{S_n\}$ to give a visual confirmation of your observation in part (a). (c). If the series converges, find the exact sum. If it diverges, prove it, using the Divergence Theorem.

23. $\sum_{n=1}^{\infty} \frac{6}{n(n+1)}$ **24.** $\sum_{n=1}^{\infty} \frac{2n}{\sqrt{n^2 + 1}}$

25. $\sum_{n=1}^{\infty} 3\left(\frac{7}{8} \right)^{n-1}$ **26.** $\sum_{n=1}^{\infty} 5\left(-\frac{2}{3} \right)^{n-1}$

27. $\sum_{n=1}^{\infty} \sin n^2$ **28.** $\sum_{n=1}^{\infty} \left(\frac{1}{2^n} - \frac{1}{3^n} \right)$

In Exercises 29–54, determine whether the given series converges or diverges. If it converges, find its sum.

29. $\displaystyle\sum_{n=1}^{\infty} \frac{1}{n(n+2)}$

30. $\displaystyle\sum_{n=2}^{\infty} \frac{1}{n^2-1}$

31. $\displaystyle\sum_{n=0}^{\infty} \frac{2^n}{5^n}$

32. $\displaystyle\sum_{n=0}^{\infty} \frac{3^{n+1}}{5^n}$

33. $\displaystyle\sum_{n=0}^{\infty} \frac{(-3)^n}{2^{n+1}}$

34. $\displaystyle\sum_{n=1}^{\infty} 2^{-n}5^{n+1}$

35. $\displaystyle\sum_{n=1}^{\infty} \frac{2n-1}{3n+1}$

36. $\displaystyle\sum_{n=0}^{\infty} \frac{2n^2+n+1}{3n^2+2}$

37. $\displaystyle\sum_{n=0}^{\infty} 3(1.01)^n$

38. $\displaystyle\sum_{n=1}^{\infty} \frac{3^n-1}{3^{n+1}}$

39. $\displaystyle\sum_{n=1}^{\infty} \left[\frac{1}{2^n} - \frac{1}{n(n+1)}\right]$

40. $\displaystyle\sum_{n=1}^{\infty} \left[\frac{2^n}{3^{n-1}} + \frac{(-1)^{n-1}2^n}{3^{n+1}}\right]$

41. $\displaystyle\sum_{n=1}^{\infty} \frac{2}{1+(0.2)^n}$

42. $\displaystyle\sum_{n=1}^{\infty} \ln\left(\frac{n}{n+1}\right)$

43. $\displaystyle\sum_{n=1}^{\infty} \left[\cos\left(\frac{1}{n}\right) - \cos\left(\frac{1}{n+1}\right)\right]$

44. $\displaystyle\sum_{n=1}^{\infty} \frac{n!}{2^n}$

45. $\displaystyle\sum_{n=1}^{\infty} [2(0.1)^n + 3(-1)^n(0.2)^n]$

46. $\displaystyle\sum_{n=0}^{\infty} \left[\left(-\frac{3}{\pi}\right)^n + \left(\frac{e}{3}\right)^{n+1}\right]$

47. $\displaystyle\sum_{n=0}^{\infty} \left(\frac{2^n+3^n}{6^n}\right)$

48. $\displaystyle\sum_{n=1}^{\infty} \left(\frac{2^n-5^n}{3^n}\right)$

49. $\displaystyle\sum_{n=1}^{\infty} \tan^{-1} n$

50. $\displaystyle\sum_{n=1}^{\infty} \sin^2 n$

51. $\displaystyle\sum_{n=1}^{\infty} n \sin\frac{1}{n}$

52. $\displaystyle\sum_{n=1}^{\infty} \frac{\sin n}{1+e^{-n}}$

53. $\displaystyle\sum_{n=2}^{\infty} \frac{n}{\ln n}$

54. $\displaystyle\sum_{n=1}^{\infty} \left(1+\frac{2}{n}\right)^n$

In Exercises 55–58, express each number as a rational number.

55. $0.\overline{4} = 0.444\ldots$

56. $-0.\overline{23} = -0.232323\ldots$

57. $1.\overline{213} = 1.213213213\ldots$

58. $3.14\overline{234} = 3.142343434\ldots$

In Exercises 59–62, find the values of x for which the series converges, and find the sum of the series. (Hint: First show that the series is a geometric series.)

59. $\displaystyle\sum_{n=0}^{\infty} (-x)^n$

60. $\displaystyle\sum_{n=0}^{\infty} (x-2)^n$

61. $\displaystyle\sum_{n=1}^{\infty} 2^n(x-1)^n$

62. $\displaystyle\sum_{n=0}^{\infty} \frac{x^{2n}}{3^n}$

63. Distance Traveled by a Bouncing Ball A rubber ball is dropped from a height of 2 m onto a flat surface. Each time the ball hits the surface, it rebounds to half its previous height. Find the total distance the ball travels.

64. Finding the Coefficient of Restitution The *coefficient of restitution* for steel onto steel is measured by dropping a steel ball onto a steel plate. If the ball is dropped from a height H and rebounds to a height h, then the coefficient of restitution is $\sqrt{h/H}$. Suppose that a steel ball is dropped from a height of 1 m onto a steel plate. Each time the ball hits the plate, it rebounds to r times it previous height ($0 < r < 1$). If the ball travels a total distance of 2 m, find the coefficient of restitution for steel on steel.

65. Probability of Winning a Dice Toss Peter and Paul take turns tossing a pair of dice. The first person to throw a 7 wins. If Peter starts the game, then it can be shown that his chances of winning are

$$p = \frac{1}{6} + \left(\frac{1}{6}\right)\left(\frac{5}{6}\right)^2 + \left(\frac{1}{6}\right)\left(\frac{5}{6}\right)^4 + \cdots$$

Find p.

66. Multiplier Effect of a Tax Cut Suppose that the average wage earner saves 9% of his or her take-home pay and spends the other 91%. What is the estimated impact that a proposed $30 billion tax cut will have on the economy over the long run because of the additional spending generated by the proposed tax cut?

Note: This phenomenon in economics is known as the *multiplier effect.*

67. Perpetuities An *annuity* is a sequence of payments that are made at regular time intervals. If the payments are allowed to continue indefinitely, then it is a *perpetuity*.

a. Suppose that P dollars is paid into an account at the beginning of each month and that the account earns interest at the rate of r per year compounded monthly. Then the present value V of the perpetuity (that is, the value of the perpetuity in today's dollars) is

$$V = P\left(1+\frac{r}{12}\right)^{-1} + P\left(1+\frac{r}{12}\right)^{-2} + \cdots + P\left(1+\frac{r}{12}\right)^{-n} + \cdots$$

Show that $V = 12P/r$.

b. Mrs. Thompson wishes to establish a fund to provide a university medical center with a monthly research grant of $150,000. If the fund will earn interest at the rate of 9% per year compounded monthly, use the result of part (a) to find the amount of the endowment she is required to make now.

68. Residual Concentration of a Drug in the Bloodstream Suppose that a dose of C units of a certain drug is administered to a patient and that the fraction of the dose remaining in the patient's bloodstream t hr after the dose is administered is given by Ce^{-kt}, where k is a positive constant.

a. Show that the residual concentration of the drug in the bloodstream after extended treatment when a dose of C units is administered at intervals of t hr is given by

$$R = \frac{Ce^{-kt}}{1-e^{-kt}}$$

b. If the highest concentration of this particular drug that is considered safe is S units, find the minimal time that must exist between doses.
Hint: $C + R \leq S$

69. Capital Value of a Perpetuity The capital value of a perpetuity involving payments of P dollars paid at the end of each investment period into a fund that earns interest at the rate of r per year compounded continuously is given by

$$A = Pe^{-r} + Pe^{-2r} + Pe^{-3r} + \cdots$$

Find an expression for A that does not involve an infinite series.

70. Sum of Areas of Nested Squares An infinite sequence of nested squares is constructed as follows: Starting with a square with a side of length 2, each square in the sequence is constructed from the preceding square by drawing line segments connecting the midpoints of the sides of the square. Find the sum of the areas of all the squares in the sequence.

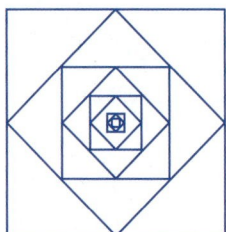

71. Sum of Areas of Nested Triangles and Circles An infinite sequence of nested equilateral triangles and circles is constructed as follows: Beginning with an equilateral triangle with a side of length 1, inscribe a circle followed by a triangle, followed by a circle, and so on, ad infinitum. Find the total area of the shaded regions.

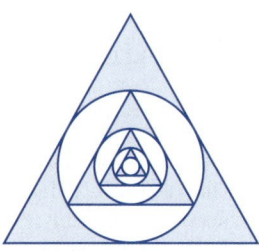

72. Prove or disprove: If $\Sigma \, a_n$ and $\Sigma \, b_n$ are both divergent, then $\Sigma(a_n + b_n)$ is divergent.

73. Suppose that $\Sigma \, a_n$ ($a_n \neq 0$) is convergent. Prove that $\Sigma \, 1/a_n$ is divergent.

74. Suppose that $\Sigma \, a_n$ is convergent and $\Sigma \, b_n$ is divergent. Prove that $\Sigma(a_n + b_n)$ is divergent.
Hint: Prove by contradiction, using Theorem 4.

75. Suppose that $\Sigma \, a_n$ is divergent and $c \neq 0$. Prove that $\Sigma \, ca_n$ is divergent.
Hint: Prove by contradiction, using Theorem 4.

76. Prove that if the sequence $\{a_n\}$ converges, then the series $\Sigma(a_{n+1} - a_n)$ converges. Conversely, prove that if $\Sigma(a_{n+1} - a_n)$ converges, then $\{a_n\}$ converges.

77. Show that $\displaystyle\sum_{n=1}^{\infty} \frac{1}{n^2}$ converges and $\dfrac{3}{2} \leq \displaystyle\sum_{n=1}^{\infty} \frac{1}{n^2} \leq 2$.
Hint: See Exercise 71 in Section 8.1.

78. Prove that $\displaystyle\sum_{n=1}^{\infty} \frac{1}{2^n + 1}$ converges by showing that $\{S_n\}$ is increasing and bounded above, where S_n is the nth partial sum of the series.

In Exercises 79–84, determine whether the statement is true or false. If it is true, explain why it is true. If it is false, explain why or give an example to show why it is false.

79. If $\lim_{n \to \infty} a_n = 0$, then $\sum_{n=1}^{\infty} a_n$ converges.

80. If $\lim_{n \to \infty} a_n = L$, then the telescoping series $\sum_{n=1}^{\infty}(a_{n+1} - a_n)$ converges and has sum $L - a_1$.

81. $\sum_{n=1}^{\infty} \sin^n x$ converges for all x in $[0, 2\pi]$.

82. $\displaystyle\sum_{n=p}^{\infty} ar^n = \frac{ar^p}{1 - r}$ provided that $|r| < 1$.

83. If the sequence of partial sums of a series $\Sigma \, a_n$ is bounded above, then $\Sigma \, a_n$ must converge.

84. If $\Sigma(a_n + b_n)$ converges, then both $\Sigma \, a_n$ and $\Sigma \, b_n$ must converge.

8.3 The Integral Test

The convergence or divergence of a telescoping or geometric series is relatively easy to determine because we are able to find a simple formula involving a finite number of terms for the nth partial sum S_n of these series. As we saw in Section 8.2, we can find the actual sum of a convergent series in this case by simply evaluating $\lim_{n \to \infty} S_n$. However, it is often very difficult or impossible to obtain a simple formula for the nth partial sum of an infinite series, and we are forced to look for alternative ways to investigate the convergence or divergence of the series.

In this and the next two sections we will develop several tests for determining the convergence or divergence of an infinite series by examining the nth term a_n of the series. These tests will confirm the convergence of a series without yielding a value for its sum. From the practical point of view, however, this is all that is required. Once it has been ascertained that a series is convergent, we can approximate its sum to any degree of accuracy desired by adding up the terms of its nth partial sum S_n, provided that n is chosen large enough. The convergence tests that are given here and in Section 8.4 apply only to series with positive terms.

■ The Integral Test

The Integral Test ties the convergence or divergence of an infinite series $\sum_{n=1}^{\infty} a_n$ to the convergence or divergence of the improper integral $\int_1^\infty f(x)\,dx$, where $f(n) = a_n$.

THEOREM 1   The Integral Test

Suppose that f is a continuous, positive, and decreasing function on $[1, \infty)$. If $f(n) = a_n$ for $n \ge 1$, then

$$\sum_{n=1}^{\infty} a_n \qquad \text{and} \qquad \int_1^\infty f(x)\,dx$$

either both converge or both diverge.

PROOF   If you examine Figure 1a, you will see that the height of the first rectangle is $a_2 = f(2)$. Since this rectangle has width 1, the area of the rectangle is also $a_2 = f(2)$. Similarly, the area of the second rectangle is a_3, and so on. Comparing the sum of the areas of the first $(n - 1)$ inscribed rectangles with the area of the region under the graph of f over the interval $[1, n]$, we see that

$$a_2 + a_3 + \cdots + a_n \le \int_1^n f(x)\,dx$$

which implies that

$$S_n = a_1 + a_2 + a_3 + \cdots + a_n \le a_1 + \int_1^n f(x)\,dx \tag{1}$$

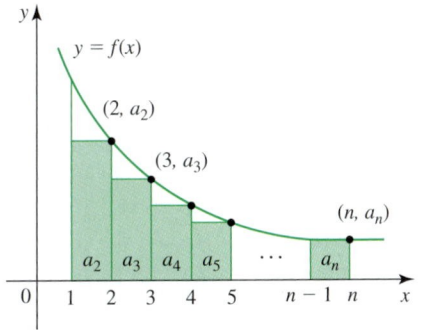

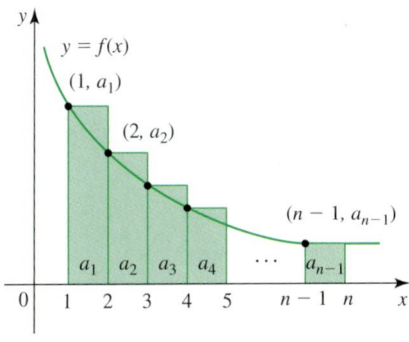

FIGURE 1   (a) $a_2 + a_3 + \cdots + a_n \le \int_1^n f(x)\,dx$ (b) $\int_1^n f(x)\,dx \le a_1 + a_2 + \cdots + a_{n-1}$

If $\int_1^\infty f(x)\,dx$ is convergent and has value L, then

$$S_n \le a_1 + \int_1^n f(x)\,dx \le a_1 + L$$

This shows that $\{S_n\}$ is bounded above. Also,

$$S_{n+1} = S_n + a_{n+1} \ge S_n \qquad \text{Because } a_{n+1} = f(n+1) \ge 0$$

shows that $\{S_n\}$ is increasing as well. Therefore, by Theorem 6, Section 8.1, $\{S_n\}$ is convergent. In other words, $\sum_{n=1}^\infty a_n$ is convergent.

Next, by examining Figure 1b, we can see that

$$\int_1^n f(x)\,dx \le a_1 + a_2 + \cdots + a_{n-1} = S_{n-1} \tag{2}$$

So if $\int_1^\infty f(x)\,dx$ diverges (to infinity because $f(x) \ge 0$), then $\lim_{n\to\infty} S_{n-1} = \lim_{n\to\infty} S_n = \infty$, and $\sum_{n=1}^\infty a_n$ is divergent. ∎

Notes

1. The Integral Test simply tells us whether a series converges or diverges. If it indicates that a series converges, we may not conclude that the (finite) value of the improper integral used in conjunction with the test is the sum of the convergent series (see Exercise 54).

2. Since the convergence of an infinite series is not affected by adding or subtracting a finite number of terms to the series, we sometimes study the series $\sum_{n=N}^\infty a_n = a_N + a_{N+1} + \cdots$ rather than the series $\sum_{n=1}^\infty a_n$. In this case the series is compared with the improper integral $\int_N^\infty f(x)\,dx$, as we will see in Example 2. ∎

EXAMPLE 1 Use the Integral Test to determine whether $\displaystyle\sum_{n=1}^\infty \frac{1}{n^2 + 1}$ converges or diverges.

Solution Here, $a_n = f(n) = 1/(n^2 + 1)$, so we consider the function $f(x) = 1/(x^2 + 1)$. Since f is continuous, positive, and decreasing on $[1, \infty)$, we may use the Integral Test. Next,

$$\int_1^\infty \frac{1}{x^2 + 1}\,dx = \lim_{b\to\infty} \int_1^b \frac{1}{x^2 + 1}\,dx = \lim_{b\to\infty}\left[\tan^{-1} x\right]_1^b$$

$$= \lim_{b\to\infty}(\tan^{-1} b - \tan^{-1} 1) = \frac{\pi}{2} - \frac{\pi}{4} = \frac{\pi}{4}$$

Since $\int_1^\infty 1/(x^2 + 1)\,dx$ converges, we conclude that $\sum_{n=1}^\infty 1/(n^2 + 1)$ converges as well. ∎

EXAMPLE 2 Use the Integral Test to determine whether $\displaystyle\sum_{n=1}^\infty \frac{\ln n}{n}$ converges or diverges.

Solution Here, $a_n = (\ln n)/n$, so we consider the function $f(x) = (\ln x)/x$. Observe that f is continuous and positive on $[1, \infty)$. Next, we compute

$$f'(x) = \frac{x\left(\dfrac{1}{x}\right) - \ln x}{x^2} = \frac{1 - \ln x}{x^2}$$

Note that $f'(x) < 0$ if $\ln x > 1$, that is, if $x > e$. This shows that f is decreasing on $[3, \infty)$. Therefore, we may use the Integral Test:

$$\int_3^\infty \frac{\ln x}{x}\, dx = \lim_{b \to \infty} \int_3^b \frac{\ln x}{x}\, dx = \lim_{b \to \infty} \left[\frac{1}{2}(\ln x)^2\right]_3^b$$

$$= \lim_{b \to \infty} \frac{1}{2}[(\ln b)^2 - (\ln 3)^2] = \infty$$

and we conclude that $\sum_{n=1}^\infty (\ln n)/n$ diverges.

■ The *p*-Series

The following series will play an important role in our work later on.

> **DEFINITION** *p*-Series
>
> A **_p_-series** is a series of the form
>
> $$\sum_{n=1}^\infty \frac{1}{n^p} = 1 + \frac{1}{2^p} + \frac{1}{3^p} + \cdots$$
>
> where p is a constant.

Observe that if $p = 1$, the p-series is just the harmonic series $\sum_{n=1}^\infty 1/n$.

The conditions for the convergence or divergence of the p-series can be found by applying the Integral Test to the series.

> **THEOREM 2** **Convergence of the *p*-Series**
>
> The p-series $\displaystyle\sum_{n=1}^\infty \frac{1}{n^p}$ converges if $p > 1$ and diverges if $p \le 1$.

PROOF If $p < 0$, then $\lim_{n \to \infty}(1/n^p) = \infty$. If $p = 0$, then $\lim_{n \to \infty}(1/n^p) = 1$. In either case, $\lim_{n \to \infty}(1/n^p) \ne 0$, so the p-series diverges by the Divergence Test.

If $p > 0$, then the function $f(x) = 1/x^p$ is continuous, positive, and decreasing on $[1, \infty)$. In Example 2 in Section 6.6 we found that $\int_1^\infty 1/x^p\, dx$ converges if $p > 1$ and diverges if $p \le 1$. Using this result and the Integral Test, we conclude that $\sum_{n=1}^\infty 1/n^p$ converges if $p > 1$ and diverges if $0 < p \le 1$. Therefore, $\sum_{n=1}^\infty 1/n^p$ converges if $p > 1$ and diverges if $p \le 1$.

EXAMPLE 3 Determine whether the given series converges or diverges.

a. $\displaystyle\sum_{n=1}^\infty \frac{1}{n^2}$ **b.** $\displaystyle\sum_{n=1}^\infty \frac{1}{\sqrt{n}}$ **c.** $\displaystyle\sum_{n=1}^\infty n^{-1.001}$

Solution
a. This is a p-series with $p = 2 > 1$, and hence it converges by Theorem 2.
b. Rewriting the series in the form $\sum_{n=1}^\infty 1/n^{1/2}$, we see that the series is a p-series with $p = \frac{1}{2} < 1$, and hence it diverges by Theorem 2.
c. We rewrite the series in the form $\sum_{n=1}^\infty 1/n^{1.001}$, which we recognize to be a p-series with $p = 1.001 > 1$ and conclude that the series converges.

8.3 CONCEPT QUESTIONS

1. Consider the series $\displaystyle\sum_{n=1}^{\infty} \frac{1}{n^2} = \frac{1}{1^2} + \frac{1}{2^2} + \frac{1}{3^2} + \frac{1}{4^2} + \frac{1}{5^2} + \cdots$.
   Let $f(x) = \dfrac{1}{x^2}$.

   a. Sketch a figure similar to Figure 1a for this series and function, and compute $a_1 = f(1)$, $a_2 = f(2)$, $a_3 = f(3)$, $\ldots$, $a_n = f(n)$.

   b. Explain why $\displaystyle S_n = \frac{1}{1^2} + \frac{1}{2^2} + \frac{1}{3^2} + \cdots + \frac{1}{n^2}$
   $$\leq \frac{1}{1^2} + \int_1^n \frac{1}{x^2}\,dx < 1 + \int_1^\infty \frac{1}{x^2}\,dx.$$

   c. By evaluating the improper integral in part (b), show that $S_n \leq 2$ for each $n = 1, 2, 3, \ldots$. Then use the Monotone Convergence Theorem (Section 8.1) to show that $\displaystyle\sum_{n=1}^{\infty} \frac{1}{n^2}$ converges.

   Note: The Swiss mathematician Leonhard Euler showed that the sum of this series is $\pi^2/6$.

2. Consider the series $\displaystyle\sum_{n=1}^{\infty} \frac{1}{n} = \frac{1}{1} + \frac{1}{2} + \frac{1}{3} + \frac{1}{4} + \frac{1}{5} + \cdots$.
   Let $f(x) = \dfrac{1}{x}$.

   a. Sketch a figure similar to Figure 1b for this series and function, and compute $a_1 = f(1)$, $a_2 = f(2)$, $a_3 = f(3)$, $\ldots$, $a_n = f(n)$.

   b. Explain why $\displaystyle S_{n-1} = \frac{1}{1} + \frac{1}{2} + \frac{1}{3} + \cdots + \frac{1}{n-1} \geq \int_1^n \frac{1}{x}\,dx$.

   c. Show that $\displaystyle\int_1^\infty \frac{1}{x}\,dx$ is divergent, and conclude that $\displaystyle\sum_{n=1}^{\infty} \frac{1}{n}$ diverges.

   Note: This is the harmonic series that was shown to be divergent in Section 8.2.

8.3 EXERCISES

In Exercises 1–8, use the Integral Test to determine whether the series is convergent or divergent.

1. $\displaystyle\sum_{n=1}^{\infty} \frac{1}{n^4}$

2. $\displaystyle\sum_{n=1}^{\infty} \frac{3}{2n-1}$

3. $\displaystyle\sum_{n=1}^{\infty} e^{-n}$

4. $\displaystyle\sum_{n=1}^{\infty} ne^{-n}$

5. $\dfrac{1}{2} + \dfrac{1}{5} + \dfrac{1}{10} + \dfrac{1}{17} + \dfrac{1}{26} + \cdots$

6. $\dfrac{1}{3} + \dfrac{1}{7} + \dfrac{1}{11} + \dfrac{1}{15} + \dfrac{1}{19} + \cdots$

7. $\displaystyle\sum_{n=1}^{\infty} \frac{n}{(n^2+1)^{3/2}}$

8. $\displaystyle\sum_{n=2}^{\infty} \frac{1}{n\sqrt{\ln n}}$

In Exercises 9–14, determine whether the p-series is convergent or divergent.

9. $\displaystyle\sum_{n=1}^{\infty} \frac{1}{n^3}$

10. $\displaystyle\sum_{n=1}^{\infty} \frac{1}{n^{2/3}}$

11. $\displaystyle\sum_{n=1}^{\infty} \frac{1}{n^{1.01}}$

12. $\displaystyle\sum_{n=1}^{\infty} \frac{1}{n^e}$

13. $\displaystyle\sum_{n=1}^{\infty} n^{-\pi}$

14. $\displaystyle\sum_{n=1}^{\infty} n^{-0.98}$

In Exercises 15–32 determine whether the given series is convergent or divergent.

15. $\displaystyle\sum_{n=0}^{\infty} \frac{1}{\sqrt{n+1}}$

16. $\displaystyle\sum_{n=1}^{\infty} \frac{n}{\sqrt{2n^2+1}}$

17. $\displaystyle\sum_{n=1}^{\infty} \frac{1}{n\sqrt{n}}$

18. $\displaystyle\sum_{n=1}^{\infty} n^{-0.75}$

19. $\displaystyle\sum_{n=1}^{\infty} \left(\frac{1}{n\sqrt{n}} + \frac{2}{n^2} \right)$

20. $\displaystyle\sum_{n=1}^{\infty} \left[\left(\frac{2}{3}\right)^n + \frac{1}{n^{3/2}} \right]$

21. $\displaystyle\sum_{n=2}^{\infty} \frac{\ln n}{n}$

22. $\displaystyle\sum_{n=2}^{\infty} \frac{\ln n}{n^2}$

23. $\displaystyle\sum_{n=2}^{\infty} \frac{1}{n(\ln n)^2}$

24. $\displaystyle\sum_{n=1}^{\infty} \frac{e^{1/n}}{n^2}$

25. $\displaystyle\sum_{n=1}^{\infty} \frac{\sin\left(\frac{1}{n}\right)}{n^2}$

26. $\displaystyle\sum_{n=1}^{\infty} \frac{1}{\sqrt{n}+4}$

27. $\displaystyle\sum_{n=1}^{\infty} \frac{1}{4n^2-1}$

28. $\displaystyle\sum_{n=1}^{\infty} \frac{n}{2^n}$

29. $\displaystyle\sum_{n=1}^{\infty} \frac{\tan^{-1} n}{n^2+1}$

30. $\displaystyle\sum_{n=1}^{\infty} \frac{1}{e^{-n}+1}$

31. $\displaystyle\sum_{n=1}^{\infty} \frac{1}{n^2+2n+5}$

32. $\displaystyle\sum_{n=1}^{\infty} \frac{1}{2n^2+7n+3}$

In Exercises 33 and 34, find the values of p for which the series is convergent.

33. $\displaystyle\sum_{n=2}^{\infty} \frac{1}{n(\ln n)^p}$

34. $\displaystyle\sum_{n=1}^{\infty} \frac{\ln n}{n^p}$

35. Find the value(s) of a for which the series
$$\sum_{n=1}^{\infty} \left[\frac{a}{n+1} - \frac{1}{n+2} \right]$$ converges. Justify your answer.

V Videos for selected exercises are available online at **www.academic.cengage.com/login**.

36. a. Show that if S_n is the nth partial sum of the harmonic series, then $S_n \le 1 + \ln n$.
  Hint: Use Inequality (1), page 753, with $f(x) = 1/x$.
  b. Use part (a) to show that the sum of the first 1,000,000 terms of the harmonic series is less than 15. The harmonic series diverges very slowly!

37. Euler's Constant
  a. Show that

$$\ln(n + 1) \le 1 + \frac{1}{2} + \cdots + \frac{1}{n}$$

  and therefore,

$$0 < \ln(n + 1) - \ln n \le 1 + \frac{1}{2} + \cdots + \frac{1}{n} - \ln n$$

  Hence, deduce that the sequence $\{a_n\}$ defined by

$$a_n = 1 + \frac{1}{2} + \cdots + \frac{1}{n} - \ln n$$

  is bounded below.
  Hint: Use Inequality (2), page 754, with $f(x) = 1/x$.
  b. Show that

$$\frac{1}{n + 1} < \int_n^{n+1} \frac{1}{x}\, dx = \ln(n + 1) - \ln n$$

  and use this result to show that the sequence $\{a_n\}$ defined in part (a) is decreasing.
  Hint: Draw a figure similar to Figure 1.
  c. Use the Monotone Convergence Theorem to show that $\{a_n\}$ is convergent.
  Note: The number

$$\gamma = \lim_{n \to \infty} a_n = \lim_{n \to \infty}\left(1 + \frac{1}{2} + \cdots + \frac{1}{n} - \ln n\right)$$

  whose value is $0.5772\ldots$, is called Euler's constant.

38. Riemann Zeta Function The *Riemann zeta function* for real numbers is defined by

$$\xi(x) = \sum_{n=1}^{\infty} n^{-x}$$

  What is the domain of the function?

39. Let $a_k = f(k)$, where f is a continuous, positive, and decreasing function on $[n, \infty)$, and suppose that $\sum_{n=1}^{\infty} a_n$ is convergent.
  a. Show, by sketching appropriate figures, that if $R_n = S - S_n$, where $S = \sum_{n=1}^{\infty} a_n$ and $S_n = \sum_{k=1}^{n} a_k$, then

$$\int_{n+1}^{\infty} f(x)\, dx \le R_n \le \int_n^{\infty} f(x)\, dx$$

  Note: R_n is the error estimate for the Integral Test.
  b. Use the result of part (a) to deduce that

$$S_n + \int_{n+1}^{\infty} f(x)\, dx \le S \le S_n + \int_n^{\infty} f(x)\, dx$$

40. Consider the series $\sum_{n=1}^{\infty} \frac{1}{n^2}$, which is a convergent p-series $(p = 2)$.
  a. Use the result of Exercise 39b to show that

$$S_n + \frac{1}{n + 1} \le \sum_{n=1}^{\infty} \frac{1}{n^2} \le S_n + \frac{1}{n}$$

  where $S_n = \sum_{k=1}^{n} \frac{1}{k^2}$ is the nth partial sum of $\sum_{n=1}^{\infty} \frac{1}{n^2}$.
  b. In Exercise 77 in Section 8.2 you were asked to show that

$$\frac{3}{2} \le \sum_{n=1}^{\infty} \frac{1}{n^2} \le 2$$

  Confirm this result, using the result of part (a).
  c. Use the result of Exercise 39a to find the upper and lower bounds on the error incurred in approximating

$$\sum_{n=1}^{\infty} \frac{1}{n^2}$$ using the 100th partial sum of the series.

  cas d. It can be shown that $\sum_{n=1}^{\infty} \frac{1}{n^2} = \frac{\pi^2}{6}$. Use a calculator or computer to verify this.

In Exercises 41–44, use the result of Exercise 39 to find the maximum error if the sum of the series is approximated by S_n.

41. $\sum_{n=1}^{\infty} \frac{2}{n^2}$; S_{40}

42. $\sum_{n=1}^{\infty} \frac{1}{n^{5/2}}$; S_{20}

43. $\sum_{n=1}^{\infty} \frac{1}{n^2 + 1}$; S_{50}

44. $\sum_{n=1}^{\infty} ne^{-n^2}$; S_3

In Exercises 45–48, use the result of Exercise 39 to find the number of terms of the series that is sufficient to obtain an approximation of the sum of the series accurate to two decimal places.

45. $\sum_{n=1}^{\infty} \frac{1}{n^2}$

46. $\sum_{n=1}^{\infty} \frac{1}{n^3}$

47. $\sum_{n=1}^{\infty} \frac{\tan^{-1} n}{1 + n^2}$

48. $\sum_{n=2}^{\infty} \frac{1}{n(\ln n)^2}$

In Exercises 49 and 50, use the result of Exercise 39 to find the sum of the series accurate to three decimal places using the nth partial sum of the series.

49. $\sum_{n=1}^{\infty} \frac{1}{n^4}$

50. $\sum_{n=1}^{\infty} \frac{1}{n^{9/2}}$

51. a. Show that

$$\sum_{n=1}^{\infty} \frac{1}{n(n + 1)(n + 2)} = \sum_{n=1}^{\infty}\left[\frac{1}{2n(n + 1)} - \frac{1}{2(n + 1)(n + 2)}\right]$$

  b. Use the results of part (a) to evaluate

$$\sum_{n=1}^{\infty} \frac{1}{n(n + 1)(n + 2)}$$

52. Evaluate $\displaystyle\sum_{n=1}^{\infty} \frac{1}{n^3}$ accurate to four decimal places by establishing parts (a) and (b) and using the results of Exercise 51.

a. $\displaystyle\sum_{n=1}^{\infty} \frac{1}{n^3} = 1 + \sum_{n=2}^{\infty} \frac{1}{(n-1)n(n+1)} - \sum_{n=2}^{\infty} \frac{1}{n^3(n^2-1)}$

b. $\displaystyle\sum_{n=2}^{\infty} \frac{1}{n^3(n^2-1)}$ can be approximated with an accuracy of four decimal places by using six terms of the series.
Hint: Show that

$$\frac{1}{n^3(n^2-1)} \le \frac{2}{n^5}$$

if $n \ge 2$, and use the result of Exercise 39.

53. Use the Integral Test to show that $\displaystyle\sum_{n=3}^{\infty} \frac{1}{n(\ln n)[\ln(\ln n)]^p}$ converges if $p > 1$ and diverges if $p \le 1$.

54. Consider the series $\sum_{n=0}^{\infty} e^{-n}$.
a. Evaluate $\int_0^{\infty} e^{-x}\, dx$, and deduce from the Integral Test that the given series is convergent.

b. Show that the given series is a geometric series, and find its sum.

c. Conclude that although the convergence of $\int_0^{\infty} e^{-x}\, dx$ implies convergence of the infinite series, its value does not give the sum of the infinite series.

In Exercises 55–58, determine whether the statement is true or false. If it is true, explain why it is true. If it is false, explain why or give an example to show why it is false.

55. Suppose that f is a continuous, positive, and decreasing function on $[1, \infty)$. If $f(n) = a_n$ for $n \ge 1$ and $\sum_{n=1}^{\infty} a_n$ is convergent, then $\sum_{n=1}^{\infty} a_n \le a_1 + \int_1^{\infty} f(x)\, dx$.

56. Suppose that f is a continuous, positive, and decreasing function on $[1, \infty)$. If $f(n) = a_n$ for $n \ge 1$ and $\int_N^{\infty} f(x)\, dx = \infty$, where N is a positive integer, then $\sum_{n=1}^{\infty} a_n$ diverges.

57. $\displaystyle\int_1^{\infty} \frac{dx}{x(x+1)} < \infty$

58. If $\sum_{n=1}^{\infty} a_n$ is a convergent series with positive terms, then $\sum_{n=1}^{\infty} \sqrt{a_n}$ must also converge.

8.4 **The Comparison Tests**

The rationale for the comparison tests is that the convergence or divergence of a given series $\Sigma\, a_n$ can be determined by comparing its terms with the corresponding terms of a *test series* whose convergence or divergence is known. The series that we will consider in this section have positive terms.

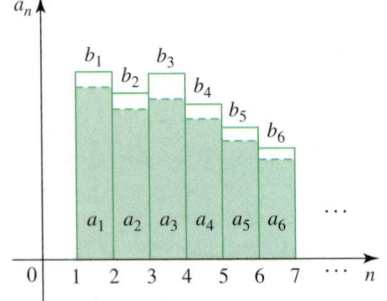

FIGURE 1
Each rectangle representing a_n is contained in the rectangle representing b_n.

■ The Comparison Test

Suppose that the terms of a series $\Sigma\, a_n$ are smaller than the corresponding terms of a series $\Sigma\, b_n$. This situation is illustrated in Figure 1, where the respective terms of each series are represented by rectangles, each of width 1 and having the appropriate height.

If $\Sigma\, b_n$ is convergent, the total area of the rectangles representing this series is finite. Since each rectangle representing the series $\Sigma\, a_n$ is contained in a corresponding rectangle representing the terms of $\Sigma\, b_n$, the total area of the rectangles representing $\Sigma\, a_n$ must also be finite; that is, the series $\Sigma\, a_n$ must be convergent. A similar argument suggests that if all the terms of a series $\Sigma\, a_n$ are larger than the corresponding terms of a series $\Sigma\, b_n$ that is known to be divergent, then $\Sigma\, a_n$ must itself be divergent. These observations lead to the following theorem.

> **THEOREM 1** **The Comparison Test**
>
> Suppose that $\Sigma\, a_n$ and $\Sigma\, b_n$ are series with positive terms.
>
> **a.** If $\Sigma\, b_n$ is convergent and $a_n \le b_n$ for all n, then $\Sigma\, a_n$ is also convergent.
> **b.** If $\Sigma\, b_n$ is divergent and $a_n \ge b_n$ for all n, then $\Sigma\, a_n$ is also divergent.

PROOF Let

$$S_n = \sum_{k=1}^{n} a_k \quad \text{and} \quad T_n = \sum_{k=1}^{n} b_k$$

be the nth terms of the sequence of partial sums of $\Sigma\, a_n$ and $\Sigma\, b_n$, respectively. Since both series have positive terms, $\{S_n\}$ and $\{T_n\}$ are increasing.

a. If $\sum_{n=1}^{\infty} b_n$ is convergent, then there exists a number L such that $\lim_{n\to\infty} T_n = L$ and $T_n \leq L$ for all n. Since $a_n \leq b_n$ for all n, we have $S_n \leq T_n$, and this implies that $S_n \leq L$ for all n. We have shown that $\{S_n\}$ is increasing and bounded above, so by the Monotone Convergence Theorem for Sequences of Section 8.1, $\Sigma\, a_n$ converges.

b. If $\sum_{n=1}^{\infty} b_n$ is divergent, then $\lim_{n\to\infty} T_n = \infty$, since $\{T_n\}$ is increasing. But $a_n \geq b_n$ for all n, and this implies that $S_n \geq T_n$, which in turn implies that $\lim_{n\to\infty} S_n = \infty$. Therefore, $\Sigma\, a_n$ diverges. ■

To use the Comparison Test, we need a catalog of test series whose convergence or divergence is known. For the moment we can use the geometric series and the p-series as test series.

EXAMPLE 1 Determine whether the series $\displaystyle\sum_{n=1}^{\infty} \frac{1}{n^2 + 2}$ converges or diverges.

Solution Let

$$a_n = \frac{1}{n^2 + 2}$$

If n is large, $n^2 + 2$ behaves like n^2, so a_n behaves like

$$b_n = \frac{1}{n^2}$$

This observation suggests that we compare $\Sigma\, a_n$ with the test series $\Sigma\, b_n$, which is a convergent p-series with $p = 2$. Now,

$$0 < \frac{1}{n^2 + 2} < \frac{1}{n^2} \qquad n \geq 1$$

and the given series is indeed "smaller" than the test series $\Sigma\, 1/n^2$. Since the test series converges, we conclude by the Comparison Test that $\Sigma\, 1/(n^2 + 2)$ also converges. ■

EXAMPLE 2 Determine whether the series $\displaystyle\sum_{n=1}^{\infty} \frac{1}{3 + 2^n}$ converges or diverges.

Solution Let

$$a_n = \frac{1}{3 + 2^n}$$

If n is large, $3 + 2^n$ behaves like 2^n, so a_n behaves like $b_n = \left(\frac{1}{2}\right)^n$. This observation suggests that we compare $\Sigma\, a_n$ with $\Sigma\, b_n$. Now the series $\Sigma\, \frac{1}{2^n} = \Sigma\, \left(\frac{1}{2}\right)^n$ is a geometric series with $r = \frac{1}{2} < 1$, so it is convergent. Since

$$a_n = \frac{1}{3 + 2^n} < \frac{1}{2^n} = b_n \qquad n \geq 1$$

the Comparison Test tells us that the given series is convergent. ■

Note Since the convergence or divergence of a series is not affected by the omission of a finite number of terms of the series, the condition $a_n \le b_n$ (or $a_n \ge b_n$) for all n can be replaced by the condition that these inequalities hold for all $n \ge N$ for some integer N.

EXAMPLE 3 Determine whether the series $\displaystyle\sum_{n=2}^{\infty} \frac{1}{\sqrt{n} - 1}$ is convergent or divergent.

Solution Let

$$a_n = \frac{1}{\sqrt{n} - 1}$$

If n is large, $\sqrt{n} - 1$ behaves like $\sqrt{n}$, so a_n behaves like

$$b_n = \frac{1}{\sqrt{n}}$$

Now the series

$$\sum_{n=2}^{\infty} b_n = \sum_{n=2}^{\infty} \frac{1}{\sqrt{n}} = \sum_{n=2}^{\infty} \frac{1}{n^{1/2}}$$

is a p-series with $p = \frac{1}{2} < 1$, so it is divergent. Since

$$a_n = \frac{1}{\sqrt{n} - 1} > \frac{1}{\sqrt{n}} = b_n \qquad \text{for } n \ge 2$$

the Comparison Test implies that the given series is divergent.

■ The Limit Comparison Test

Consider the series

$$\sum_{n=1}^{\infty} \frac{1}{\sqrt{n} + 1}$$

If n is large, $\sqrt{n} + 1$ behaves like $\sqrt{n}$, so the nth term of the given series

$$a_n = \frac{1}{\sqrt{n} + 1}$$

behaves like

$$b_n = \frac{1}{\sqrt{n}}$$

Since the series $\sum_{n=1}^{\infty} b_n = \sum_{n=1}^{\infty} 1/\sqrt{n}$ is a divergent p-series with $p = \frac{1}{2}$, we expect the series $\sum_{n=1}^{\infty} 1/(\sqrt{n} + 1)$ to be divergent as well. But the inequality

$$a_n = \frac{1}{\sqrt{n} + 1} < \frac{1}{\sqrt{n}} = b_n \qquad n \ge 1$$

tells us that $\sum_{n=1}^{\infty} a_n$ is "smaller" than a divergent series, and this is of no help if we try to use the Comparison Test!

In situations like this, the *Limit Comparison Test* might be applicable. The rationale for this test follows: Suppose that $\sum a_n$ and $\sum b_n$ are series with positive terms and suppose that $\lim_{n\to\infty}(a_n/b_n) = L$, where L is a positive constant. If n is large, $a_n/b_n \approx L$ or $a_n \approx Lb_n$. It is reasonable to conjecture that the series $\Sigma\, a_n$ and $\Sigma\, b_n$ must both converge or both diverge.

THEOREM 2   The Limit Comparison Test

Suppose that $\Sigma\, a_n$ and $\Sigma\, b_n$ are series with positive terms and

$$\lim_{n\to\infty} \frac{a_n}{b_n} = L$$

where L is a positive number. Then either both series converge or both diverge.

PROOF   Since $\lim_{n\to\infty}(a_n/b_n) = L > 0$, there exists an integer N such that $n \geq N$ implies that

$$\left| \frac{a_n}{b_n} - L \right| < \frac{1}{2}L$$

$$\frac{1}{2}L < \frac{a_n}{b_n} < \frac{3}{2}L$$

or

$$\frac{1}{2}Lb_n < a_n < \frac{3}{2}Lb_n$$

If $\Sigma\, b_n$ converges, so does $\Sigma\, \frac{3}{2}Lb_n$. Therefore, the right side of the last inequality implies that $\Sigma\, a_n$ converges by the Comparison Test. On the other hand, if $\Sigma\, b_n$ diverges, so does $\Sigma\, \frac{1}{2}Lb_n$, and the left side of the last inequality implies by the Comparison Test that $\Sigma\, a_n$ diverges as well.   ∎

EXAMPLE 4   Show that the series $\displaystyle\sum_{n=1}^{\infty} \frac{1}{\sqrt{n}+1}$ is divergent.

Solution   As we saw earlier, $1/(\sqrt{n}+1)$ behaves like $1/\sqrt{n}$ if n is large. This suggests that we use the Limit Comparison Test with $a_n = 1/(\sqrt{n}+1)$ and $b_n = 1/\sqrt{n}$. Thus,

$$\lim_{n\to\infty} \frac{a_n}{b_n} = \lim_{n\to\infty} \frac{\dfrac{1}{\sqrt{n}+1}}{\dfrac{1}{\sqrt{n}}} = \lim_{n\to\infty} \frac{\sqrt{n}}{\sqrt{n}+1} = \lim_{n\to\infty} \frac{1}{1 + \dfrac{1}{\sqrt{n}}} = 1$$

Since $\sum_{n=1}^{\infty} 1/\sqrt{n}$ is divergent $\left(\text{it is a } p\text{-series with } p = \frac{1}{2}\right)$, we conclude that the given series is divergent as well.   ∎

Note   You can still use the Comparison Test to solve the problem. Simply observe that

$$\frac{1}{\sqrt{n}+1} \geq \frac{1}{\sqrt{n}+\sqrt{n}} = \frac{1}{2\sqrt{n}} \qquad \text{for } n \geq 1$$

This suggests picking $\Sigma\, b_n$, where $b_n = 1/(2\sqrt{n})$, for the test series.   ∎

EXAMPLE 5   Determine whether the series $\displaystyle\sum_{n=1}^{\infty} \frac{2n^2 + n}{\sqrt{4n^7 + 3}}$ converges or diverges.

Solution   If n is large, $2n^2 + n$ behaves like $2n^2$, and $4n^7 + 3$ behaves like $4n^7$. Therefore,

$$a_n = \frac{2n^2 + n}{\sqrt{4n^7 + 3}}$$

behaves like

$$\frac{2n^2}{\sqrt{4n^7}} = \frac{2n^2}{2n^{7/2}} = \frac{1}{n^{3/2}} = b_n$$

Now

$$\lim_{n \to \infty} \frac{a_n}{b_n} = \lim_{n \to \infty} \frac{2n^2 + n}{(4n^7 + 3)^{1/2}} \cdot \frac{n^{3/2}}{1}$$

$$= \lim_{n \to \infty} \frac{2n^{7/2} + n^{5/2}}{(4n^7 + 3)^{1/2}}$$

$$= \lim_{n \to \infty} \frac{2 + \dfrac{1}{n}}{\left(4 + \dfrac{3}{n^7}\right)^{1/2}} \qquad \text{\color{brown}Divide numerator and denominator by $n^{7/2}$.}$$

$$= 1$$

Since $\Sigma\, 1/n^{3/2}$ converges $\left(\text{it is a } p\text{-series with } p = \tfrac{3}{2}\right)$, the given series converges, by the Limit Comparison Test.   ■

EXAMPLE 6   Determine whether the series $\displaystyle\sum_{n=1}^{\infty} \frac{\sqrt{n} + \ln n}{n^2 + 1}$ converges or diverges.

Solution   If n is large, $\sqrt{n} + \ln n$ behaves like $\sqrt{n}$. You can see this by comparing the derivatives of $f(x) = \sqrt{x}$ and $g(x) = \ln x$:

$$f'(x) = \frac{1}{2\sqrt{x}} \qquad \text{and} \qquad g'(x) = \frac{1}{x}$$

Observe that $g'(x)$ approaches zero faster than $f'(x)$ approaches zero, as $x \to \infty$. This shows that $\sqrt{x}$ grows faster than $\ln x$. Also, if n is large, $n^2 + 1$ behaves like n^2. Therefore,

$$a_n = \frac{\sqrt{n} + \ln n}{n^2 + 1}$$

behaves like

$$\frac{\sqrt{n}}{n^2} = \frac{1}{n^{3/2}} = b_n$$

Next, we compute

$$\lim_{n\to\infty}\frac{a_n}{b_n} = \lim_{n\to\infty}\frac{n^{1/2}+\ln n}{n^2+1}\cdot\frac{n^{3/2}}{1}$$

$$= \lim_{n\to\infty}\frac{n^2+n^{3/2}\ln n}{n^2+1}$$

$$= \lim_{n\to\infty}\frac{1+\dfrac{\ln n}{n^{1/2}}}{1+\dfrac{1}{n^2}} \qquad \text{\color{red}Divide the numerator and denominator by } n^2.$$

In evaluating this limit, we need to compute

$$\lim_{x\to\infty}\frac{\ln x}{x^{1/2}} = \lim_{x\to\infty}\frac{\dfrac{1}{x}}{\frac{1}{2}x^{-1/2}} = \lim_{x\to\infty}\frac{2}{\sqrt{x}} = 0 \qquad \text{\color{red}Use l'Hôpital's Rule.}$$

(Incidentally, this result supports the observation made earlier that $\sqrt{x}$ grows faster than $\ln x$.) Using this result, we find

$$\lim_{n\to\infty}\frac{a_n}{b_n} = \lim_{n\to\infty}\frac{1+\dfrac{\ln n}{n^{1/2}}}{1+\dfrac{1}{n^2}} = 1$$

Since $\Sigma\, 1/n^{3/2}$ converges $\left(\text{it is a } p\text{-series with } p = \tfrac{3}{2}\right)$, the given series converges, by the Limit Comparison Test. ∎

8.4 CONCEPT QUESTIONS

1. **a.** State the Comparison Test and the Limit Comparison Test.
   b. When is the Comparison Test used? When is the Limit Comparison Test used?
2. Let $\Sigma\, a_n$ and $\Sigma\, b_n$ be series with positive terms.
   a. If $\Sigma\, b_n$ is convergent and $a_n \geq b_n$ for all n, what can you say about the convergence or divergence of $\Sigma\, a_n$? Give examples.
   b. If $\Sigma\, b_n$ is divergent and $a_n \leq b_n$ for all n, what can you say about the convergence or divergence of $\Sigma\, a_n$? Give examples.

In Exercises 3 and 4, let $\Sigma\, a_n$, $\Sigma\, b_n$, and $\Sigma\, c_n$ be series with positive terms.

3. If $\Sigma\, a_n$ is convergent and $b_n + c_n \leq a_n$ for all n, what can you say about the convergence or divergence of $\Sigma\, b_n$ and $\Sigma\, c_n$?
4. If $\Sigma\, a_n$ is divergent and $b_n + c_n \geq a_n$ for all n, what can you say about the convergence or divergence of $\Sigma\, b_n$ and $\Sigma\, c_n$?

8.4 EXERCISES

In Exercises 1–12, use the Comparison Test to determine whether the series is convergent or divergent.

1. $\displaystyle\sum_{n=1}^{\infty}\frac{1}{2n^2+1}$

2. $\displaystyle\sum_{n=1}^{\infty}\frac{1}{n^2+2n}$

3. $\displaystyle\sum_{n=3}^{\infty}\frac{1}{n-2}$

4. $\displaystyle\sum_{n=2}^{\infty}\frac{1}{n^{2/3}-1}$

5. $\displaystyle\sum_{n=2}^{\infty}\frac{1}{\sqrt{n^2-1}}$

6. $\displaystyle\sum_{n=0}^{\infty}\frac{1}{\sqrt{n^3+1}}$

7. $\displaystyle\sum_{n=0}^{\infty} \frac{2^n}{3^n + 1}$

8. $\displaystyle\sum_{n=3}^{\infty} \frac{3^n}{2^n - 4}$

9. $\displaystyle\sum_{n=2}^{\infty} \frac{\ln n}{n}$

10. $\displaystyle\sum_{n=1}^{\infty} \frac{\cos^2 n}{n^2}$

11. $\displaystyle\sum_{n=1}^{\infty} \frac{2 + \sin n}{3^n}$

12. $\displaystyle\sum_{n=1}^{\infty} \frac{1}{n^n}$

In Exercises 13–24, use the Limit Comparison Test to determine whether the series is convergent or divergent.

13. $\displaystyle\sum_{n=2}^{\infty} \frac{n}{n^2 + 1}$

14. $\displaystyle\sum_{n=1}^{\infty} \frac{1}{\sqrt{n} + 2}$

15. $\displaystyle\sum_{n=2}^{\infty} \frac{n}{\sqrt{n^5 - 1}}$

16. $\displaystyle\sum_{n=1}^{\infty} \frac{2n + 1}{3n^2 - n + 1}$

17. $\displaystyle\sum_{n=1}^{\infty} \frac{3n^2 + 1}{2n^5 + n + 2}$

18. $\displaystyle\sum_{n=1}^{\infty} \frac{n^2 + 1}{n^2(n + 3)}$

19. $\displaystyle\sum_{n=2}^{\infty} \frac{1}{\sqrt{n^3 - n - 1}}$

20. $\displaystyle\sum_{n=2}^{\infty} \frac{1}{2^n - 3}$

21. $\displaystyle\sum_{n=1}^{\infty} \frac{n}{2^n - 1}$

22. $\displaystyle\sum_{n=2}^{\infty} \frac{\ln n}{n^3 - 1}$

23. $\displaystyle\sum_{n=1}^{\infty} \sin \frac{1}{n}$

24. $\displaystyle\sum_{n=1}^{\infty} \tan \frac{1}{n}$

In Exercises 25–40, determine whether the series is convergent or divergent.

25. $\displaystyle\sum_{n=1}^{\infty} \frac{n + 1}{(n + 2)(2n^2 + 1)}$

26. $\displaystyle\sum_{n=1}^{\infty} \frac{n}{\sqrt{n^5 + n}}$

27. $\displaystyle\sum_{n=1}^{\infty} \frac{n - 1}{n^3 + 2}$

28. $\displaystyle\sum_{n=1}^{\infty} \frac{n + 1}{2n^3 + 1}$

29. $\displaystyle\sum_{n=1}^{\infty} \frac{2^{n-1}}{n^2 + n}$

30. $\displaystyle\sum_{n=1}^{\infty} \frac{1}{n + \sqrt{n^2 - 1}}$

31. $\displaystyle\sum_{n=1}^{\infty} \frac{\sin^2 n}{n\sqrt{n + 1}}$

32. $\displaystyle\sum_{n=1}^{\infty} \frac{\tan^{-1} n}{n^3 + 1}$

33. $\displaystyle\sum_{n=2}^{\infty} \frac{1}{\ln n}$

34. $\displaystyle\sum_{n=1}^{\infty} \frac{\ln n}{n + 2}$

35. $\displaystyle\sum_{n=0}^{\infty} \frac{1}{n!}$

36. $\displaystyle\sum_{n=1}^{\infty} \frac{n^2}{n!}$

37. $\displaystyle\sum_{n=1}^{\infty} \frac{n!}{n^n}$

38. $\displaystyle\sum_{n=1}^{\infty} \frac{1}{1 + 2 + 3 + \cdots + n}$

39. $\displaystyle\sum_{n=1}^{\infty} \frac{\sqrt{n} + \ln n}{2n^2 + 3}$

40. $\displaystyle\sum_{n=1}^{\infty} \frac{2n^2 + n}{\sqrt{3n^7 + \ln n}}$

41. Let $\Sigma\, a_n$ and $\Sigma\, b_n$ be series with $0 \le a_n \le b_n$, and suppose that $\Sigma\, b_n$ is convergent with sum T. Then the Comparison Test implies that $\Sigma\, a_n$ also converges, say, with sum S. Put $R_n = S - S_n$ and $T_n = T - U_n$, where U_n is the nth-partial sum of $\Sigma\, b_n$. Show that the remainders R_n and T_n satisfy $R_n \le T_n$.

In Exercises 42–45, use the result of Exercise 41 to find an approximation of the sum of the series using its partial sum, accurate to two decimal places.

42. $\displaystyle\sum_{n=1}^{\infty} \frac{1}{n^3 + 2n}$

43. $\displaystyle\sum_{n=1}^{\infty} \frac{\sin n + 2}{n^4}$

44. $\displaystyle\sum_{n=1}^{\infty} \frac{1}{3^n + 1}$

45. $\displaystyle\sum_{n=1}^{\infty} \frac{\tan^{-1} n}{2^n}$

46. Suppose that $\Sigma\, a_n$ is a convergent series with positive terms. Show that $\Sigma(a_n/n)$ is also convergent.

47. Suppose that $\Sigma\, a_n$ and $\Sigma\, b_n$ are convergent series with positive terms. Show that $\Sigma\, a_n b_n$ is convergent.
    Hint: There exists an integer N such that $n \ge N$ implies that $b_n \le 1$, and therefore, $a_n b_n \le a_n$ for $n \ge N$.

48. Suppose that $\Sigma\, a_n$ is a convergent series with positive terms and $\{c_n\}$ is a sequence of positive numbers that converges to zero. Prove that $\Sigma\, a_n c_n$ is convergent.
    Hint: There exists an integer N such that $n \ge N$ implies that $c_n < L$, where L is a positive number, and therefore, $a_n c_n < L a_n$ for $n \ge N$.

49. Prove that if $a_n \ge 0$ and $\Sigma\, a_n$ converges, then $\Sigma\, a_n^2$ also converges. Is the converse true? Explain.

50. Using the result of Exercise 48 or otherwise, show that $\sum_{n=2}^{\infty} 1/(n^p \ln n)$ is convergent if $p > 1$.

51. a. Suppose that $\Sigma\, a_n$ and $\Sigma\, b_n$ are series with positive terms and $\Sigma\, b_n$ is convergent. Show that if $\lim_{n\to\infty} a_n/b_n = 0$, then $\Sigma\, a_n$ is convergent.

    b. Use part (a) to show that $\displaystyle\sum_{n=1}^{\infty} \frac{\ln n}{n^2}$ is convergent.

52. Give an example of a pair of series $\Sigma\, a_n$ and $\Sigma\, b_n$ with positive terms such that $\lim_{n\to\infty} a_n/b_n = 0$, $\Sigma\, b_n$ is divergent, but $\Sigma\, a_n$ is convergent. (Compare this with the result of Exercise 51.)

53. a. Show that if $\Sigma\, a_n$ is a convergent series with positive terms, then $\Sigma \sin a_n$ is also convergent.

    b. If $\Sigma\, a_n$ diverges can $\Sigma \sin a_n$ converge? Explain.

54. Prove that (a) $\displaystyle\int_1^{\infty} \frac{1}{\sqrt{x(x + 1)(x + 2)}}\, dx$ converges and

    (b) $\displaystyle\int_1^{\infty} \frac{1}{\sqrt{x(x + 1)}}\, dx$ diverges.

In Exercises 55–58, determine whether the statement is true or false. If it is true, explain why it is true. If it is false, explain why or give an example to show why it is false.

55. If $0 \le a_n \le b_n$ and $\Sigma\, a_n$ converges, then $\Sigma\, b_n$ diverges.

56. If $0 \le a_n \le b_n$ and $\Sigma\, b_n$ diverges, then $\Sigma\, a_n$ diverges.

57. If $a_n > 0$ and $b_n > 0$ and $\Sigma\, a_n b_n$ converges, then $\Sigma\, a_n$ and $\Sigma\, b_n$ both converge.

58. If $a_n > 0$ and $b_n > 0$ and $\Sigma \sqrt{a_n^2 + b_n^2}$ converges, then $\Sigma\, a_n$ and $\Sigma\, b_n$ both converge.

8.5 Alternating Series

Up to now, we have dealt mainly with series that have positive terms, and the convergence tests that we have developed are applicable only to these series. In this section and Section 8.6 we will consider series that contain both positive and negative terms. Series whose terms alternate in sign are called *alternating series*.

Examples are the *alternating harmonic series*

$$\sum_{n=1}^{\infty} \frac{(-1)^{n-1}}{n} = 1 - \frac{1}{2} + \frac{1}{3} - \frac{1}{4} + \frac{1}{5} - \frac{1}{6} + \cdots$$

and the series

$$\sum_{n=1}^{\infty} \frac{(-1)^n n^2}{(n+1)!} = -\frac{1}{2!} + \frac{4}{3!} - \frac{9}{4!} + \frac{16}{5!} - \frac{25}{6!} + \cdots$$

More generally, an **alternating series** is a series of the form

$$\sum_{n=1}^{\infty} (-1)^{n-1} a_n \qquad \text{or} \qquad \sum_{n=1}^{\infty} (-1)^n a_n$$

where a_n is a positive number. We use the *Alternating Series Test* to determine convergence for these series.

THEOREM 1 The Alternating Series Test

If the alternating series

$$\sum_{n=1}^{\infty} (-1)^{n-1} a_n = a_1 - a_2 + a_3 - a_4 + a_5 - a_6 + \cdots \qquad a_n > 0$$

satisfies the conditions

1. $a_{n+1} \leq a_n$ for all n
2. $\displaystyle\lim_{n \to \infty} a_n = 0$

then the series converges.

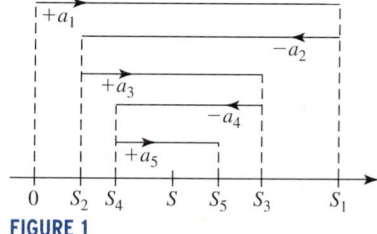

FIGURE 1
The terms of $\{S_n\}$ oscillate in smaller and smaller steps, and this suggests that $\lim_{n \to \infty} S_n = S$.

The plausibility of Theorem 1 is suggested by Figure 1, which shows the first few terms of the sequence of partial sums $\{S_n\}$ of the alternating series

$$\sum_{n=1}^{\infty} (-1)^{n-1} a_n = a_1 - a_2 + a_3 - \cdots$$

plotted on the number line. The point $S_2 = a_1 - a_2$ lies to the left of the number $S_1 = a_1$, since it is obtained by subtracting the positive number a_2 from S_1. But the number S_2 also lies to the right of the origin because $a_2 \leq a_1$. The number $S_3 = a_1 - a_2 + a_3 = S_2 + a_3$ is obtained by adding a_3 to S_2, and hence it lies to the right of S_2. But because $a_3 < a_2$, S_3 lies to the left of S_1. Continuing in this fashion, we see that numbers corresponding to the partial sums $\{S_n\}$ oscillate. Because $\lim_{n \to \infty} a_n = 0$, the steps get smaller and smaller. Thus, it appears that the sequence $\{S_n\}$ will approach a limit. In particular, observe that the even terms of the sequence $\{S_n\}$ are increasing, whereas the odd terms of the sequence are decreasing. This suggests that the subsequence $\{S_{2n}\}$ will approach the limit S from below and the subsequence $\{S_{2n+1}\}$ will approach S from above. These observations form the basis of the proof of Theorem 1.

PROOF OF THEOREM 1   We first consider the subsequence $\{S_{2n}\}$ comprising the even terms of $\{S_n\}$. Now,

$$S_2 = a_1 - a_2 \geq 0 \qquad \text{Since } a_1 \geq a_2$$

$$S_4 = S_2 + (a_3 - a_4) \geq S_2 \qquad \text{Since } a_3 \geq a_4$$

and, in general,

$$S_{2n+2} = S_{2n} + (a_{2n+1} - a_{2n+2}) \geq S_{2n} \qquad \text{Since } a_{2n+1} \geq a_{2n+2}$$

This shows that

$$0 \leq S_2 \leq S_4 \leq \cdots \leq S_{2n} \leq \cdots$$

that is, $\{S_{2n}\}$ is increasing. Next, we write S_{2n} in the form

$$S_{2n} = a_1 - (a_2 - a_3) - (a_4 - a_5) - \cdots - (a_{2n-2} - a_{2n-1}) - a_{2n}$$

and observe that every expression within the parenthesis is nonnegative (again, because $a_{n+1} \leq a_n$). Thus, we see that $S_{2n} \leq a_1$ for all n. This shows that the sequence $\{S_{2n}\}$ is bounded above as well. Therefore, by the Monotone Convergence Theorem for Sequences of Section 8.1, the sequence $\{S_{2n}\}$ is convergent; that is, there exists a number S such that $\lim_{n \to \infty} S_{2n} = S$.

Next, we consider the subsequence $\{S_{2n+1}\}$ comprising the odd terms of $\{S_n\}$. Since $S_{2n+1} = S_{2n} + a_{2n+1}$ and $\lim_{n \to \infty} a_{2n+1} = 0$ by assumption, we have

$$\lim_{n \to \infty} S_{2n+1} = \lim_{n \to \infty} (S_{2n} + a_{2n+1})$$

$$= \lim_{n \to \infty} S_{2n} + \lim_{n \to \infty} a_{2n+1}$$

$$= S$$

Since the subsequences $\{S_{2n}\}$ and $\{S_{2n+1}\}$ of the sequence of partial sums $\{S_n\}$ both converge to S, we have $\lim_{n \to \infty} S_n = S$, so the series converges.   ◼

EXAMPLE 1   Show that the alternating harmonic series

$$\sum_{n=1}^{\infty} \frac{(-1)^{n-1}}{n} = 1 - \frac{1}{2} + \frac{1}{3} - \frac{1}{4} + \cdots$$

converges.

Solution   This is an alternating series with $a_n = 1/n$, so we use the Alternating Series Test. We need to verify that (1) $a_{n+1} \leq a_n$ and (2) $\lim_{n \to \infty} a_n = 0$. But the first condition follows from the computation

$$a_{n+1} = \frac{1}{n+1} < \frac{1}{n} = a_n$$

while the second condition follows from

$$\lim_{n \to \infty} a_n = \lim_{n \to \infty} \frac{1}{n} = 0$$

Therefore, by the Alternating Series Test, the given series converges.   ◼

EXAMPLE 2 Determine whether the series converges or diverges.

a. $\displaystyle\sum_{n=1}^{\infty}(-1)^n\frac{2n}{4n-1}$ **b.** $\displaystyle\sum_{n=1}^{\infty}(-1)^{n-1}\frac{3n}{4n^2-1}$

Solution Since both series are alternating series we use the Alternating Series Test.
a. Here, $a_n = 2n/(4n-1)$. Because

$$\lim_{n\to\infty}\frac{2n}{4n-1}=\frac{1}{2}\neq 0$$

we see that condition (2) in the Alternating Series Test is not satisfied. In fact, this computation shows that

$$\lim_{n\to\infty}(-1)^n\frac{2n}{4n-1}$$

does not exist, and the divergence of the series follows from the Divergence Test.
b. Here $a_n = 3n/(4n^2-1)$. First we show that $a_n \geq a_{n+1}$ for all n. We can do this by showing that $f(x) = 3x/(4x^2-1)$ is decreasing for $x \geq 0$. We compute

$$f'(x) = \frac{(4x^2-1)(3)-(3x)(8x)}{(4x^2-1)^2}$$

$$= \frac{-12x^2-3}{(4x^2-1)^2} < 0$$

and the desired conclusion follows. Next, we compute

$$\lim_{n\to\infty}a_n = \lim_{n\to\infty}\frac{3n}{4n^2-1}=\lim_{n\to\infty}\frac{\dfrac{3}{n}}{4-\dfrac{1}{n^2}}=0$$

Since both conditions of the Alternating Series Test are satisfied, we conclude that the series is convergent. ■

Notes

1. Example 2a reminds us once again that it is a good idea to begin investigating the convergence of a series by checking for divergence using the Divergence Test.
2. Because the behavior of a finite number of terms will not affect the convergence or divergence of a series, the first condition in the Alternating Series Test can be replaced by the condition $a_{n+1} \leq a_n$ for $n \geq N$, where N is some positive integer. ■

■ Approximating the Sum of an Alternating Series by S_n

Suppose that we can show that the series $\sum a_n$ is convergent so that it has a sum S. If $\{S_n\}$ is the sequence of partial sums of $\sum a_n$, then $\lim_{n\to\infty}S_n = S$ or, equivalently,

$$\lim_{n\to\infty}(S - S_n) = 0$$

Thus, the sum of a convergent series can be approximated to any degree of accuracy by its nth partial sum S_n, provided that n is taken large enough. To measure the accuracy of the approximation, we introduce the quantity

$$R_n = S - S_n = \sum_{k=1}^{\infty} a_k - \sum_{k=1}^{n} a_k = \sum_{k=n+1}^{\infty} a_k = a_{n+1} + a_{n+2} + a_{n+3} + \cdots$$

called the **remainder after n terms** of the series $\sum_{n=1}^{\infty} a_n$. The remainder measures the error incurred when S is approximated by S_n.

In general, it is difficult to determine the accuracy of such an approximation, but for alternating series the following theorem gives us a simple way of estimating the error.

THEOREM 2 **Error Estimate in Approximating an Alternating Series**

Suppose $\sum_{n=1}^{\infty} (-1)^{n-1} a_n$ is an alternating series satisfying

1. $0 \le a_{n+1} \le a_n$ for all n

2. $\lim\limits_{n\to\infty} a_n = 0$

If S is the sum of the series, then

$$|R_n| = |S - S_n| \le a_{n+1}$$

In other words, the absolute value of the error incurred in approximating S by S_n is no larger than a_{n+1}, the first term omitted.

PROOF We have

$$S - S_n = \sum_{k=1}^{\infty} (-1)^{k-1} a_k - \sum_{k=1}^{n} (-1)^{k-1} a_k = \sum_{k=n+1}^{\infty} (-1)^{k-1} a_k$$

$$= (-1)^n a_{n+1} + (-1)^{n+1} a_{n+2} + (-1)^{n+2} a_{n+3} + \cdots$$

$$= (-1)^n (a_{n+1} - a_{n+2} + a_{n+3} - \cdots)$$

Next,

$$a_{n+1} - a_{n+2} + a_{n+3} - a_{n+4} + \cdots$$

$$= (a_{n+1} - a_{n+2}) + (a_{n+3} - a_{n+4}) + \cdots$$

$$\ge 0 \qquad \text{Since } a_{n+1} \le a_n \text{ for all } n$$

So

$$|S - S_n| = a_{n+1} - a_{n+2} + a_{n+3} - a_{n+4} + a_{n+5} - \cdots$$

$$= a_{n+1} - (a_{n+2} - a_{n+3}) - (a_{n+4} - a_{n+5}) - \cdots$$

Since every expression within each parenthesis is nonnegative, we see that $|S - S_n| \le a_{n+1}$.

⚠ This error estimate holds only for alternating series.

EXAMPLE 3 Show that the series $\sum_{n=0}^{\infty} (-1)^n \dfrac{1}{n!}$ is convergent, and find its sum correct to three decimal places.

Solution Since

$$a_{n+1} = \frac{1}{(n+1)!} = \frac{1}{n!(n+1)} < \frac{1}{n!} = a_n$$

for all n and

$$\lim_{n\to\infty} a_n = \lim_{n\to\infty} \frac{1}{n!} = 0$$

we conclude that the series converges by the Alternating Series Test.

To see how many terms of the series are needed to ensure the specified accuracy of the approximation, we turn to Theorem 2. It tells us that

$$|R_n| = |S - S_n| \le a_{n+1} = \frac{1}{(n+1)!}$$

We require that $|R_n| < 0.0005$, which is satisfied if

$$\frac{1}{(n+1)!} < 0.0005 \quad \text{or} \quad (n+1)! > \frac{1}{0.0005} = 2000$$

The smallest positive integer that satisfies the last inequality is $n = 6$. Hence, the required approximation is

$$S \approx S_6 = \frac{1}{0!} - \frac{1}{1!} + \frac{1}{2!} - \frac{1}{3!} + \frac{1}{4!} - \frac{1}{5!} + \frac{1}{6!}$$

$$= 1 - 1 + \frac{1}{2} - \frac{1}{6} + \frac{1}{24} - \frac{1}{120} + \frac{1}{720}$$

$$\approx 0.368$$

8.5 CONCEPT QUESTIONS

1. a. What is an alternating series? Give an example.
   b. State the Alternating Series Test, and use it to determine whether the series in your example converges or diverges.
   c. What is the maximum error that can occur if you approximate the sum of a convergent alternating series by its nth partial sum?

8.5 EXERCISES

In Exercises 1–24, determine whether the series converges or diverges.

1. $\displaystyle\sum_{n=1}^{\infty} \frac{(-1)^{n-1}}{n+2}$

2. $\displaystyle\sum_{n=1}^{\infty} \frac{(-1)^n n}{3n-1}$

3. $\displaystyle\sum_{n=1}^{\infty} \frac{(-1)^{n+1}}{n^2}$

4. $\displaystyle\sum_{n=1}^{\infty} \frac{(-1)^{n-1} n^2}{2n^2-1}$

5. $\displaystyle\sum_{n=1}^{\infty} \frac{(-1)^{n-1}}{\sqrt{n}}$

6. $\displaystyle\sum_{n=1}^{\infty} \frac{(-1)^{n+1} n}{\sqrt{n^2+1}}$

7. $\displaystyle\sum_{n=2}^{\infty} \frac{(-1)^{n-1} \sqrt{n+1}}{n-1}$

8. $\displaystyle\sum_{n=2}^{\infty} \frac{(-1)^{n-1}}{\ln n}$

9. $\displaystyle\sum_{n=2}^{\infty} \frac{(-1)^n n}{\ln n}$

10. $\displaystyle\sum_{n=1}^{\infty} \frac{(-1)^n \ln(n+1)}{n+2}$

Ⓥ Videos for selected exercises are available online at **www.academic.cengage.com/login**.

11. $\displaystyle\sum_{n=1}^{\infty} \frac{(-1)^n n}{2^n}$

12. $\displaystyle\sum_{n=1}^{\infty} \frac{(-1)^{n-1}}{ne^{-n}}$

13. $\displaystyle\sum_{n=0}^{\infty} \frac{(-1)^{n+1} e^n}{\pi^{n+1}}$

14. $\displaystyle\sum_{n=1}^{\infty} \frac{\cos n\pi}{n}$

15. $\displaystyle\sum_{n=1}^{\infty} \frac{1}{\sqrt{n}} \sin \frac{(2n-1)\pi}{2}$

16. $\displaystyle\sum_{n=2}^{\infty} (\ln n) \sin \frac{(2n-1)\pi}{2}$

17. $\displaystyle\sum_{n=1}^{\infty} \frac{\sin\left(\dfrac{n\pi}{2}\right)}{\sqrt{n^3+1}}$

18. $\displaystyle\sum_{n=1}^{\infty} (-1)^n \cos\left(\frac{\pi}{n}\right)$

19. $\displaystyle\sum_{n=1}^{\infty} (-1)^n n \sin\left(\frac{\pi}{n}\right)$

20. $\displaystyle\sum_{n=1}^{\infty} \frac{(-1)^n n!}{n^n}$

21. $\displaystyle\sum_{n=2}^{\infty} \frac{(-1)^n \ln n}{e^n}$

22. $\displaystyle\sum_{n=2}^{\infty} \frac{(-1)^{n-1}\sqrt{\ln n}}{n}$

23. $\displaystyle\sum_{n=1}^{\infty} \frac{(-1)^n}{\sqrt{n}+\sqrt{n+1}}$

24. $\displaystyle\sum_{n=1}^{\infty} \frac{(-1)^{n-1}}{\sqrt[n]{n}}$

In Exercises 25 and 26, find the values of p for which the series is convergent.

25. $\displaystyle\sum_{n=2}^{\infty} \frac{(-1)^n}{(\ln n)^p}$

26. $\displaystyle\sum_{n=2}^{\infty} (-1)^{n-1} \frac{(\ln n)^p}{n}$

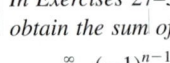 *In Exercises 27–30, determine the number of terms sufficient to obtain the sum of the series accurate to three decimal places.*

27. $\displaystyle\sum_{n=1}^{\infty} \frac{(-1)^{n-1}}{n^2+1}$

28. $\displaystyle\sum_{n=1}^{\infty} \frac{(-1)^{n-1}}{\sqrt{n}}$

29. $\displaystyle\sum_{n=0}^{\infty} \frac{(-2)^{n+3}}{(n+1)!}$

30. $\displaystyle\sum_{n=2}^{\infty} \frac{(-1)^{n-1}}{n \ln n}$

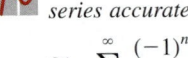 *In Exercises 31–34, find an approximation of the sum of the series accurate to two decimal places.*

31. $\displaystyle\sum_{n=1}^{\infty} \frac{(-1)^n}{n^3}$

32. $\displaystyle\sum_{n=0}^{\infty} \frac{(-1)^n}{(2n)!}$

33. $\displaystyle\sum_{n=1}^{\infty} \frac{(-1)^{n-1}(n+1)}{2^n}$

34. $\displaystyle\sum_{n=1}^{\infty} \frac{(-1)^{n-1}}{n \cdot 2^n}$

35. Show that the series

$$\frac{1}{2} - \frac{1}{3} + \frac{1}{4} - \frac{1}{9} + \frac{1}{8} - \frac{1}{27} + \cdots + \frac{1}{2^n} - \frac{1}{3^n} + \cdots$$

converges, and find its sum. Why isn't the Alternating Series Test applicable?

36. Show that the series

$$1 - \frac{1}{4} + \frac{1}{3} - \frac{1}{16} + \frac{1}{5} - \frac{1}{36} + \cdots + \frac{1}{2n-1} - \frac{1}{(2n)^2} + \cdots$$

diverges. Why isn't the Alternating Series Test applicable?

37. a. Suppose that $\Sigma\, a_n$ and $\Sigma\, b_n$ are both convergent. Does it follow that $\Sigma\, a_n b_n$ must be convergent? Justify your answer.

   b. Suppose that $\Sigma\, a_n$ and $\Sigma\, b_n$ are both divergent. Does it follow that $\Sigma\, a_n b_n$ must be divergent? Justify your answer.

38. Find all values of s for which $\displaystyle\sum_{n=1}^{\infty} \frac{(-1)^n}{n^s}$ converges.

39. a. Show that $\displaystyle\sum_{n=1}^{\infty} \frac{(-1)^n(2n+1)}{n(n+1)}$ converges.

   b. Find the sum of the series of part (a).

40. a. Show that $\displaystyle\sum_{n=0}^{\infty} \frac{(-1)^n}{n!}$ converges.

   b. Denote the sum of the infinite series in part (a) by S. Show that S is irrational.

   Hint: Use Theorem 2.

In Exercises 41–44, determine whether the statement is true or false. If it is true, explain why it is true. If it is false, explain why or give an example to show why it is false.

41. If the alternating series $\Sigma_{n=1}^{\infty}(-1)^{n-1} a_n$, where $a_n > 0$, is divergent, then the series $\Sigma_{n=1}^{\infty} a_n$ is also divergent.

42. Let $\Sigma_{n=1}^{\infty}(-1)^{n+1} a_n$ be an alternating series, where $a_n > 0$. If $\lim_{n\to\infty} a_n = 0$, then $\Sigma_{n=1}^{\infty}(-1)^{n+1} a_n$ converges.

43. If the alternating series $\Sigma_{n=1}^{\infty}(-1)^{n+1} a_n$, where $a_n > 0$, converges, then both the series $\Sigma_{n=1}^{\infty} a_{2n-1}$ and $\Sigma_{n=1}^{\infty} a_{2n}$ converge.

44. Let $\Sigma_{n=1}^{\infty}(-1)^{n+1} a_n$ be an alternating series, where $a_n > 0$. If $a_{n+1} \le a_n$ for all n, then $\Sigma_{n=1}^{\infty}(-1)^{n+1} a_n$ converges.

8.6 Absolute Convergence; the Ratio and Root Tests

■ Absolute Convergence

Up to now, we have considered series whose terms are all positive and series whose terms alternate between being positive and negative. Now, consider the series

$$\sum_{n=1}^{\infty} \frac{\sin 2n}{n^2} = \sin 2 + \frac{\sin 4}{2^2} + \frac{\sin 6}{3^2} + \cdots$$

With the aid of a calculator you can verify that the first term of this series is positive, the next two terms are negative, and the next term is positive. Therefore, this series is neither a series with positive terms nor an alternating series. To study the convergence of such series, we introduce the notion of *absolute convergence.*

Suppose that $\sum_{n=1}^{\infty} a_n$ is any series. Then we can form the series

$$\sum_{n=1}^{\infty} |a_n| = |a_1| + |a_2| + |a_3| + \cdots$$

by taking the absolute value of each term of the given series. Since this series contains only positive terms, we can use the tests developed in Sections 8.3 and 8.4 to determine its convergence or divergence.

DEFINITION **Absolutely Convergent Series**

A series $\sum a_n$ is **absolutely convergent** if the series $\sum |a_n|$ is convergent.

Notice that if the terms of the series $\sum a_n$ are positive, then $|a_n| = a_n$. In this case absolute convergence is the same as convergence.

EXAMPLE 1 Show that the series

$$\sum_{n=1}^{\infty} \frac{(-1)^{n-1}}{n^2} = 1 - \frac{1}{2^2} + \frac{1}{3^2} - \frac{1}{4^2} + \cdots$$

is absolutely convergent.

Solution Taking the absolute value of each term of the series, we obtain

$$\sum_{n=1}^{\infty} \left| \frac{(-1)^{n-1}}{n^2} \right| = \sum_{n=1}^{\infty} \frac{1}{n^2} = 1 + \frac{1}{2^2} + \frac{1}{3^2} + \frac{1}{4^2} + \cdots$$

which is a convergent p-series ($p = 2$). Hence the series is absolutely convergent. ∎

EXAMPLE 2 Show that the alternating harmonic series

$$\sum_{n=1}^{\infty} \frac{(-1)^{n-1}}{n} = 1 - \frac{1}{2} + \frac{1}{3} - \frac{1}{4} + \cdots$$

is not absolutely convergent.

Solution Taking the absolute value of each term of the series leads to

$$\sum_{n=1}^{\infty} \left| \frac{(-1)^{n-1}}{n} \right| = \sum_{n=1}^{\infty} \frac{1}{n} = 1 + \frac{1}{2} + \frac{1}{3} + \cdots$$

which is the divergent harmonic series. This shows that the series is not absolutely convergent. ∎

In Example 2 we saw that the alternating harmonic series is not absolutely convergent; but as we proved earlier, it is convergent. Such a series is said to be *conditionally convergent.*

DEFINITION   **Conditionally Convergent Series**

A series $\Sigma\, a_n$ is **conditionally convergent** if it is convergent but not absolutely convergent.

The following theorem tells us that absolute convergence is, loosely speaking, stronger than convergence.

THEOREM 1

If a series $\Sigma\, a_n$ is absolutely convergent, then it is convergent.

PROOF   Using an absolute value property, we have

$$-|a_n| \le a_n \le |a_n|$$

Adding $|a_n|$ to both sides of this inequality yields

$$0 \le a_n + |a_n| \le 2|a_n|$$

If we let $b_n = a_n + |a_n|$, then the last inequality becomes $0 \le b_n \le 2|a_n|$. If $\Sigma\, a_n$ is absolutely convergent, then $\Sigma\, |a_n|$ is convergent, which in turn implies, by Theorem 4a of Section 8.2, that $\Sigma\, 2|a_n|$ is convergent. Therefore, $\Sigma\, b_n$ is convergent by the Comparison Test. Finally, since $a_n = b_n - |a_n|$, we see that $\Sigma\, a_n = \Sigma\, b_n - \Sigma\, |a_n|$ is convergent by Theorem 4b of Section 8.2. ∎

As an illustration, the series $\Sigma (-1)^{n-1}/n^2$ of Example 1 is an alternating series that can be shown to be convergent by the Alternating Series Test. Alternatively, we can show that the series is absolutely convergent (as was done in Example 1) and conclude by Theorem 1 that it must be convergent.

EXAMPLE 3   Determine whether the series

$$\sum_{n=1}^{\infty} \frac{\sin 2n}{n^2} = \sin 2 + \frac{\sin 4}{2^2} + \frac{\sin 6}{3^2} + \cdots$$

converges or diverges.

Solution   As was pointed out at the beginning of this section, this series contains both positive and negative terms, but it is not an alternating series because the first term is positive, the next two terms are negative, and the next term is positive.

Let's show that the series is absolutely convergent. To do this, we consider the series

$$\sum_{n=1}^{\infty} \left| \frac{\sin 2n}{n^2} \right| = \sum_{n=1}^{\infty} \frac{|\sin 2n|}{n^2}$$

Since $|\sin 2n| \le 1$ for all n, we see that

$$\frac{|\sin 2n|}{n^2} \le \frac{1}{n^2}$$

Now, because $\Sigma\, 1/n^2$ is a convergent p-series, the Comparison Test tells us that $\sum_{n=1}^{\infty} |\sin 2n|/n^2$ is convergent. This shows that the given series is absolutely convergent, and we conclude by Theorem 1 that it is convergent. ∎

◼ The Ratio Test

The *Ratio Test* is a test for determining whether a series is absolutely convergent. Of course, for series that contain only positive terms, the Ratio Test will just be yet another test for convergence. To gain insight into why the Ratio Test works, consider the ratios of the consecutive terms of the series $\Sigma\,|a_n|$:

$$\frac{|a_2|}{|a_1|}, \quad \frac{|a_3|}{|a_2|}, \quad \frac{|a_4|}{|a_3|}, \quad \cdots$$

If the terms of this sequence are ultimately less than 1, then the terms of the series $\Sigma\,|a_n|$ ultimately behave roughly like the terms of a geometric series $\Sigma\,ar^n$ with $0 < r < 1$, and we can expect the series to be convergent. On the other hand, if the terms of the series are ultimately greater than 1, then we can expect the series to be divergent.

THEOREM 2 The Ratio Test

Let $\Sigma\,a_n$ be a series with nonzero terms.

a. If $\displaystyle\lim_{n\to\infty}\left|\frac{a_{n+1}}{a_n}\right| = L < 1$, then $\displaystyle\sum_{n=1}^{\infty} a_n$ converges absolutely.

b. If $\displaystyle\lim_{n\to\infty}\left|\frac{a_{n+1}}{a_n}\right| = L > 1$, or $\displaystyle\lim_{n\to\infty}\left|\frac{a_{n+1}}{a_n}\right| = \infty$, then $\displaystyle\sum_{n=1}^{\infty} a_n$ diverges.

c. If $\displaystyle\lim_{n\to\infty}\left|\frac{a_{n+1}}{a_n}\right| = 1$, the test is inconclusive, and another test should be used.

PROOF

a. Suppose that

$$\lim_{n\to\infty}\left|\frac{a_{n+1}}{a_n}\right| = L < 1$$

Let r be any number such that $0 \le L < r < 1$. Then there exists an integer N such that

$$\left|\frac{a_{n+1}}{a_n}\right| < r$$

whenever $n \ge N$ or, equivalently,

$$|a_{n+1}| < |a_n|\,r$$

whenever $n \ge N$. Letting n take on the values $N, N+1, N+2, \ldots$, successively, we obtain

$$|a_{N+1}| < |a_N|\,r$$
$$|a_{N+2}| < |a_{N+1}|\,r < |a_N|\,r^2$$
$$|a_{N+3}| < |a_{N+2}|\,r < |a_N|\,r^3$$

and, in general,

$$|a_{N+k}| < |a_N|\,r^k \qquad \text{for all } k \ge 1$$

Since the series

$$\sum_{k=1}^{\infty} |a_N|\,r^k = |a_N|\,r + |a_N|\,r^2 + |a_N|\,r^3 + \cdots \tag{1}$$

is a convergent geometric series with $0 < r < 1$ and each term of the series

$$\sum_{k=1}^{\infty} |a_{N+k}| = |a_{N+1}| + |a_{N+2}| + |a_{N+3}| + \cdots \tag{2}$$

is less than the corresponding term of the geometric series (1), the Comparison Test then implies that series (2) is convergent. Since convergence or divergence is unaffected by the omission of a finite number of terms, we see that the series $\sum_{n=1}^{\infty} |a_n|$ is also convergent.

b. Suppose that

$$\lim_{n \to \infty} \left| \frac{a_{n+1}}{a_n} \right| = L > 1$$

Let r be any number such that $L > r > 1$. Then there exists an integer N such that

$$\left| \frac{a_{n+1}}{a_n} \right| > r > 1$$

whenever $n \geq N$. This implies that $|a_{n+1}| > |a_n|$ when $n \geq N$. Thus, $\lim_{n \to \infty} a_n \neq 0$, and $\Sigma\, a_n$ is divergent by the Divergence Test.

c. Consider the series $\sum_{n=1}^{\infty} 1/n$ and $\sum_{n=1}^{\infty} 1/n^2$. For the first series we have

$$\lim_{n \to \infty} \left| \frac{a_{n+1}}{a_n} \right| = \lim_{n \to \infty} \frac{1}{n+1} \cdot \frac{n}{1} = \lim_{n \to \infty} \frac{1}{1 + \dfrac{1}{n}} = 1$$

and for the second series we have

$$\lim_{n \to \infty} \left| \frac{a_{n+1}}{a_n} \right| = \lim_{n \to \infty} \frac{1}{(n+1)^2} \cdot \frac{n^2}{1} = \lim_{n \to \infty} \frac{1}{\left(1 + \dfrac{1}{n}\right)^2} = 1$$

Thus,

$$\lim_{n \to \infty} \left| \frac{a_{n+1}}{a_n} \right| = 1$$

for both series. The first series is the divergent harmonic series, whereas the second series is a convergent p-series with $p = 2$. Thus, if $L = 1$, the series may converge or diverge, and the Ratio Test is inconclusive.

EXAMPLE 4 Determine whether the series $\displaystyle\sum_{n=1}^{\infty} (-1)^{n-1} \frac{n^2 + 1}{2^n}$ is absolutely convergent, conditionally convergent, or divergent.

Solution We use the Ratio Test with $a_n = (-1)^{n-1}(n^2 + 1)/2^n$. We have

$$\lim_{n \to \infty} \left| \frac{a_{n+1}}{a_n} \right| = \lim_{n \to \infty} \left| \frac{(-1)^n[(n+1)^2 + 1]}{2^{n+1}} \cdot \frac{2^n}{(-1)^{n-1}(n^2 + 1)} \right|$$

$$= \lim_{n \to \infty} \frac{1}{2}\left(\frac{n^2 + 2n + 2}{n^2 + 1} \right) = \frac{1}{2} < 1$$

Therefore, by the Ratio Test, the series is absolutely convergent.

EXAMPLE 5 Determine whether the series $\displaystyle\sum_{n=1}^{\infty} \frac{n!}{n^n}$ is convergent or divergent.

Solution Let $a_n = n!/n^n$. Then

$$\lim_{n\to\infty} \left| \frac{a_{n+1}}{a_n} \right| = \lim_{n\to\infty} \frac{a_{n+1}}{a_n} \qquad \text{\color{red}Since } a_n \text{ and } a_{n+1} \text{ are positive}$$

$$= \lim_{n\to\infty} \frac{(n+1)!}{(n+1)^{n+1}} \cdot \frac{n^n}{n!}$$

$$= \lim_{n\to\infty} \frac{(n+1)n!}{(n+1)(n+1)^n} \cdot \frac{n^n}{n!}$$

$$= \lim_{n\to\infty} \left(\frac{n}{n+1} \right)^n$$

$$= \lim_{n\to\infty} \frac{1}{\left(\dfrac{n+1}{n} \right)^n} = \lim_{n\to\infty} \frac{1}{\left(1 + \dfrac{1}{n} \right)^n} = \frac{1}{\lim_{n\to\infty} \left(1 + \dfrac{1}{n} \right)^n} = \frac{1}{e} < 1$$

Therefore, the series converges, by the Ratio Test. ∎

EXAMPLE 6 Determine whether the series $\displaystyle\sum_{n=1}^{\infty} (-1)^n \frac{n!}{3^n}$ is absolutely convergent, conditionally convergent, or divergent.

Solution Let $a_n = (-1)^n n!/3^n$. Then

$$\lim_{n\to\infty} \left| \frac{a_{n+1}}{a_n} \right| = \lim_{n\to\infty} \left| \frac{(-1)^{n+1}(n+1)!}{3^{n+1}} \cdot \frac{3^n}{(-1)^n n!} \right|$$

$$= \lim_{n\to\infty} \frac{n+1}{3} = \infty$$

and we conclude that the given series is divergent by the Ratio Test.

Alternative Solution Observe that for $n \ge 2$,

$$\frac{n!}{3^n} = \frac{n \cdot (n-1) \cdot \cdots \cdot 3 \cdot 2 \cdot 1}{3 \cdot 3 \cdot \cdots \cdot 3 \cdot 3 \cdot 3} \ge \frac{2 \cdot 1}{3 \cdot 3} = \frac{2}{9} \ne 0$$

Therefore, $\lim_{n\to\infty} a_n = \lim_{n\to\infty} (-1)^n n!/3^n$ does not exist, so the Divergence Test implies that the series must diverge. ∎

■ The Root Test

The following test is especially useful when the nth term of a series involves the nth power. Since the proof is similar to that of the Ratio Test, it will be omitted.

> **THEOREM 3** **The Root Test**
>
> Let $\sum_{n=1}^{\infty} a_n$ be a series.
>
> **a.** If $\lim_{n\to\infty} \sqrt[n]{|a_n|} = L < 1$, then $\sum_{n=1}^{\infty} a_n$ converges absolutely.
> **b.** If $\lim_{n\to\infty} \sqrt[n]{|a_n|} = L > 1$ or $\lim_{n\to\infty} \sqrt[n]{|a_n|} = \infty$, then $\sum_{n=1}^{\infty} a_n$ diverges.
> **c.** If $\lim_{n\to\infty} \sqrt[n]{|a_n|} = 1$, the test is inconclusive, and another test should be used.

EXAMPLE 7 Determine whether the series $\sum_{n=1}^{\infty} (-1)^{n-1} \dfrac{2^{n+3}}{(n+1)^n}$ is absolutely convergent, conditionally convergent, or divergent.

Solution We apply the Root Test with $a_n = (-1)^{n-1} 2^{n+3}/(n+1)^n$. We have

$$\lim_{n \to \infty} \sqrt[n]{|a_n|} = \lim_{n \to \infty} \sqrt[n]{\left| (-1)^{n-1} \frac{2^{n+3}}{(n+1)^n} \right|} = \lim_{n \to \infty} \left| \frac{2^{n+3}}{(n+1)^n} \right|^{1/n}$$

$$= \lim_{n \to \infty} \frac{2^{1+3/n}}{n+1} = 0 < 1$$

and conclude that the series is absolutely convergent.

Summary of Tests for Convergence and Divergence of Series

We have developed several ways of determining whether a series is convergent or divergent. Next, we give a summary of the available tests and suggest when it might be advantageous to use each test.

> **Summary of the Convergence and Divergence Tests for Series**
>
> 1. The **Divergence Test** often settles the question of convergence or divergence of a series $\Sigma \, a_n$ simply and quickly:
>
> $$\text{If } \lim_{n \to \infty} a_n \neq 0, \text{ then the series diverges.}$$
>
> 2. If you recognize that the series is
>   a. a **geometric series** $\sum_{n=1}^{\infty} ar^{n-1}$, then it converges with sum $a/(1-r)$ if $|r| < 1$. If $|r| \geq 1$, the series diverges.
>   b. a **telescoping series,** then use partial fraction decomposition (if necessary) to find its nth partial sum S_n. Next determine convergence or divergence by evaluating $\lim_{n \to \infty} S_n$.
>   c. a **p-series** $\sum_{n=1}^{\infty} 1/n^p$, then the series converges if $p > 1$ and diverges if $p \leq 1$.
>   Sometimes a little algebraic manipulation might be required to cast the series into one of these forms. Also, a series might involve a combination (for example, a sum or difference) of these series.
> 3. If $f(n) = a_n$ for $n \geq 1$, where f is a continuous, positive, decreasing function on $[1, \infty)$ and readily integrable, then we may use the **Integral Test:**
>
> $\sum_{n=1}^{\infty} a_n$ converges if $\int_1^{\infty} f(x) \, dx$ converges and diverges if $\int_1^{\infty} f(x) \, dx$ diverges.
> 4. If a_n is positive and behaves like the nth term of a geometric or p-series for large values of n, then the Comparison Test or Limit Comparison Test may be used. The tests and conclusions follow:
>   a. If $a_n \leq b_n$ for all n and $\Sigma \, b_n$ converges, then $\Sigma \, a_n$ converges.
>   b. If $a_n \geq b_n \geq 0$ for all n and $\Sigma \, b_n \geq 0$ diverges, then $\Sigma \, a_n$ diverges.
>   c. If b_n is positive and $\lim_{n \to \infty}(a_n/b_n) = L > 0$, then both series converge or both diverge.
>   The comparison tests can also be used on $\Sigma \, |a_n|$ to test for absolute convergence.

5. If the series is an **alternating series**, $\Sigma_{n=1}^{\infty}(-1)^n a_n$ or $\Sigma_{n=1}^{\infty}(-1)^{n-1} a_n$, then the Alternating Series Test should be considered:

   If $a_n \geq a_{n+1}$ for all n and $\lim_{n\to\infty} a_n = 0$, then the series converges.

6. The **Ratio Test** is useful if a_n involves factorials or nth powers. The series

   a. converges absolutely if $\lim\limits_{n\to\infty}\left|\dfrac{a_{n+1}}{a_n}\right| < 1$.

   b. diverges if $\lim\limits_{n\to\infty}\left|\dfrac{a_{n+1}}{a_n}\right| > 1$ or $\lim\limits_{n\to\infty}\left|\dfrac{a_{n+1}}{a_n}\right| = \infty$.

   The test is inconclusive if $\lim\limits_{n\to\infty}\left|\dfrac{a_{n+1}}{a_n}\right| = 1$.

7. The **Root Test** is useful if a_n involves nth powers. The series

   a. converges absolutely if $\lim_{n\to\infty}\sqrt[n]{|a_n|} < 1$.

   b. diverges if $\lim_{n\to\infty}\sqrt[n]{|a_n|} > 1$ or $\lim_{n\to\infty}\sqrt[n]{|a_n|} = \infty$.

   The test is inconclusive if $\lim_{n\to\infty}\sqrt[n]{|a_n|} = 1$.

8. If the series $\Sigma\, a_n$ involves terms that are both positive and negative but it is not alternating, then one sometimes can prove convergence of the series by proving that $\Sigma\, |a_n|$ is convergent.

■ Rearrangement of Series

A series with a finite number of terms has the same sum regardless of how the terms of the series are rearranged. The situation gets a little more complicated, however, when we deal with infinite series. The following example shows that a rearrangement of a convergent series could result in a series with a different sum!

EXAMPLE 8 Consider the alternating harmonic series that converges to ln 2 (see Problem 57 in Exercises 8.8):

$$1 - \frac{1}{2} + \frac{1}{3} - \frac{1}{4} + \frac{1}{5} - \frac{1}{6} + \frac{1}{7} - \frac{1}{8} + \cdots = \ln 2$$

If we rearrange the series so that every positive term is followed by two negative terms, we obtain

$$1 - \frac{1}{2} - \frac{1}{4} + \frac{1}{3} - \frac{1}{6} - \frac{1}{8} + \frac{1}{5} - \frac{1}{10} - \frac{1}{12} + \cdots$$

$$= \left(1 - \frac{1}{2}\right) - \frac{1}{4} + \left(\frac{1}{3} - \frac{1}{6}\right) - \frac{1}{8} + \left(\frac{1}{5} - \frac{1}{10}\right) - \frac{1}{12} + \cdots$$

$$= \frac{1}{2} - \frac{1}{4} + \frac{1}{6} - \frac{1}{8} + \frac{1}{10} - \frac{1}{12} + \cdots$$

$$= \frac{1}{2}\left(1 - \frac{1}{2} + \frac{1}{3} - \frac{1}{4} + \frac{1}{5} - \frac{1}{6} + \cdots\right) = \frac{1}{2}\ln 2$$

Thus, rearrangement of the alternating harmonic series has a sum that is one half that of the original series! ∎

You might have noticed that the alternating harmonic series in Example 8 is *conditionally convergent*. In fact, for such series, Riemann proved the following result:

If x is any real number and $\sum_{n=1}^{\infty} a_n$ is conditionally convergent, then there is a rearrangement of $\sum_{n=1}^{\infty} a_n$ that converges to x.

A proof of this result can be found in more advanced textbooks.

Riemann's result tells us that for conditionally convergent series, we may not rearrange their terms, lest we end up with a totally different series, that is, a series with a different sum. Actually, for conditionally convergent series, one can find rearrangements of the series that diverge to infinity, diverge to minus infinity, or oscillate between any two prescribed real numbers!

So what kind of convergent series will have rearrangements that converge to the same sum as the original series? The answer is found in the following result, which we state without proof:

If $\sum_{n=1}^{\infty} a_n$ converges absolutely and $\sum_{n=1}^{\infty} b_n$ is any rearrangement of $\sum_{n=1}^{\infty} a_n$, then $\sum_{n=1}^{\infty} b_n$ converges and $\sum_{n=1}^{\infty} a_n = \sum_{n=1}^{\infty} b_n$.

Finally, since a convergent series with positive terms is absolutely convergent, its terms can be written in any order, and the resultant series will converge and have the same sum as the original series.

EXAMPLE 9   Indicate the test(s) that you would use to determine whether the series converges or diverges. Explain how you arrived at your choice.

a. $\displaystyle\sum_{n=1}^{\infty} \frac{2n-1}{3n+1}$ **b.** $\displaystyle\sum_{n=1}^{\infty} \left[\frac{2}{3^n} - \frac{1}{n(n+1)}\right]$ **c.** $\displaystyle\sum_{n=1}^{\infty} \left(\frac{1}{n}\right)^e$

d. $\displaystyle\sum_{n=3}^{\infty} \frac{1}{n\sqrt{\ln n}}$ **e.** $\displaystyle\sum_{n=3}^{\infty} \frac{\ln n}{n^2}$ **f.** $\displaystyle\sum_{n=1}^{\infty} \frac{\sqrt{n^3+2}}{n^4+3n^2+1}$

g. $\displaystyle\sum_{n=1}^{\infty} (-1)^n \frac{\sqrt{n}}{n^2+1}$ **h.** $\displaystyle\sum_{n=1}^{\infty} \frac{n}{2^n}$ **i.** $\displaystyle\sum_{n=1}^{\infty} \frac{\sin n}{\sqrt{n^3+1}}$

Solution
a. Since
$$\lim_{n\to\infty} a_n = \lim_{n\to\infty} \frac{2n-1}{3n+1} = \frac{2}{3} \neq 0$$
we use the Divergence Test.
b. The series is the difference of a geometric series and a telescoping series, so we use the properties of these series to determine convergence.
c. Here, $a_n = \left(\dfrac{1}{n}\right)^e = \dfrac{1}{n^e}$ is a *p*-series, so we use the properties of a *p*-series to study its convergence.
d. The function $f(x) = \dfrac{1}{x\sqrt{\ln x}}$ is continuous, positive, and decreasing on $[3, \infty)$ and is integrable, so we choose the Integral Test.
e. Here,
$$a_n = \frac{\ln n}{n^2} < \frac{\sqrt{n}}{n^2} = \frac{1}{n^{3/2}} = b_n$$
and we use the Comparison Test with the test series $\sum b_n$.

f. $a_n = \dfrac{(n^3 + 2)^{1/2}}{n^4 + 3n^2 + 1}$ is positive and behaves like

$$b_n = \frac{(n^3)^{1/2}}{n^4} = \frac{n^{3/2}}{n^4} = \frac{1}{n^{5/2}}$$

for large values of n, so we use the Limit Comparison Test with test series $\sum_{n=1}^{\infty} 1/n^{5/2}$.

g. This is an alternating series, and we use the Alternating Series Test.

h. Here, $a_n = \dfrac{n}{2^n} = \left(\dfrac{n^{1/n}}{2}\right)^n$ involves the nth power, so the Root Test is a candidate.

In fact, here $\lim_{n\to\infty} \sqrt[n]{|a_n|} = \lim_{n\to\infty} \dfrac{\sqrt[n]{n}}{2} = \dfrac{1}{2} < 1$ and the series converges.

i. The series involves both positive and negative terms and is not an alternating series, so we use the test for absolute convergence.

8.6 CONCEPT QUESTIONS

1. a. What is an absolutely convergent series? Give an example.
   b. What is a conditionally convergent series? Give an example.

2. a. State the Ratio Test and the Root Test.
   b. Give an example of a convergent series and an example of a divergent series for which the Ratio Test is inconclusive.
   c. Give an example of a convergent series and an example of a divergent series for which the Root Test is inconclusive.

8.6 EXERCISES

In Exercises 1–34, determine whether the series is convergent, absolutely convergent, conditionally convergent, or divergent.

1. $\displaystyle\sum_{n=1}^{\infty} \frac{(-1)^{n-1}}{\sqrt{n}}$

2. $\displaystyle\sum_{n=1}^{\infty} \frac{(-1)^n}{n\sqrt{n}}$

3. $\displaystyle\sum_{n=1}^{\infty} \frac{(-2)^{n-1}}{n^2}$

4. $\displaystyle\sum_{n=1}^{\infty} \frac{(-2)^n}{n!}$

5. $\displaystyle\sum_{n=1}^{\infty} \frac{(-1)^{n+1}}{n+1}$

6. $\displaystyle\sum_{n=1}^{\infty} \frac{(-1)^n n}{n^2 + 1}$

7. $\displaystyle\sum_{n=1}^{\infty} \frac{(-1)^n n^2}{n^2 + 3}$

8. $\displaystyle\sum_{n=1}^{\infty} \frac{(-1)^{n-1} n}{\sqrt{2n^2 + 1}}$

9. $\displaystyle\sum_{n=2}^{\infty} \frac{(-1)^n}{n \ln n}$

10. $\displaystyle\sum_{n=3}^{\infty} \frac{(-1)^n}{n\sqrt{\ln n}}$

11. $\displaystyle\sum_{n=1}^{\infty} \frac{n!}{e^n}$

12. $\displaystyle\sum_{n=1}^{\infty} \frac{\cos(n+1)}{n\sqrt{n}}$

13. $\displaystyle\sum_{n=1}^{\infty} (-1)^{n-1} \sin\left(\frac{1}{n}\right)$

14. $\displaystyle\sum_{n=1}^{\infty} \frac{(-1)^{n-1} \tan^{-1} n}{n^2}$

15. $\displaystyle\sum_{n=1}^{\infty} \frac{2^n}{n! \, n}$

16. $\displaystyle\sum_{n=1}^{\infty} \frac{(-5)^{n-1}}{n^2 \cdot 3^n}$

17. $\displaystyle\sum_{n=1}^{\infty} \frac{(-2)^n n}{(n+1)3^{n-1}}$

18. $\displaystyle\sum_{n=2}^{\infty} \frac{(-1)^{n+1} \ln n}{n^2 + 1}$

19. $\displaystyle\sum_{n=2}^{\infty} \frac{(-1)^n \ln n}{2^n}$

20. $\displaystyle\sum_{n=0}^{\infty} \frac{\cos n\pi}{n!}$

21. $\displaystyle\sum_{n=2}^{\infty} \frac{\sin\left(\dfrac{n\pi}{4}\right)}{n(\ln n)^2}$

22. $\displaystyle\sum_{n=1}^{\infty} \frac{(-1)^{n-1} n^5}{e^n}$

23. $\displaystyle\sum_{n=1}^{\infty} \frac{(-1)^{n+1} n^n}{n!}$

24. $\displaystyle\sum_{n=2}^{\infty} \left(\frac{\ln n}{n}\right)^n$

25. $\displaystyle\sum_{n=2}^{\infty} \frac{(-1)^n}{(\ln n)^n}$

26. $\displaystyle\sum_{n=1}^{\infty} \left(\frac{n}{2n+1}\right)^n$

27. $\displaystyle\sum_{n=1}^{\infty} (-1)^n \tan\left(\frac{1}{n}\right)$

28. $\displaystyle\sum_{n=1}^{\infty} (-1)^n \, n \sin\left(\frac{\pi}{n}\right)$

29. $\displaystyle\sum_{n=1}^{\infty} \frac{(-n)^n}{[(n+1)\tan^{-1} n]^n}$

30. $\displaystyle\sum_{n=2}^{\infty} \left(\sqrt[n]{n} - 1\right)^n$

V Videos for selected exercises are available online at **www.academic.cengage.com/login**.

31. $\displaystyle\sum_{n=1}^{\infty} (-1)^{n-1} \frac{3 \cdot 5 \cdot 7 \cdot \cdots \cdot (2n+1)}{1 \cdot 4 \cdot 7 \cdot \cdots \cdot (3n-2)}$

32. $\displaystyle\sum_{n=1}^{\infty} (-1)^n \frac{2^n}{3 \cdot 5 \cdot 7 \cdot \cdots \cdot (2n+1)}$

33. $\displaystyle\sum_{n=1}^{\infty} \frac{4 \cdot 7 \cdot 10 \cdot \cdots \cdot (3n+1)}{4^n(n+1)!}$

34. $\displaystyle\sum_{n=1}^{\infty} \frac{(n!)^2}{(3n)!}$

35. Find all values of x for which the series $\displaystyle\sum_{n=1}^{\infty} \frac{x^n}{n}$ (a) converges absolutely and (b) converges conditionally.

36. Show that the Ratio Test is inconclusive for the p-series.

37. Show that the Root Test is inconclusive for the p-series.

38. a. Show that if $\Sigma\, a_n$ converges absolutely, then $\Sigma\, a_n^2$ converges.
   b. Show that the converse of the result in part (a) is false by finding a series $\Sigma\, a_n$ for which $\Sigma\, a_n^2$ converges, but $\Sigma\, |a_n|$ diverges.

39. Show that if $\Sigma\, a_n$ diverges, then $\Sigma\, |a_n|$ diverges.

40. Show that if $\Sigma\, a_n$ converges absolutely, then $\Sigma\, a_n \leq \Sigma\, |a_n|$.

41. Suppose that $\Sigma\, a_n^2$ and $\Sigma\, b_n^2$ are convergent. Show that $\Sigma\, a_n b_n$ is absolutely convergent.
   Hint: Show that $2|ab| \leq a^2 + b^2$ by looking at $(a+b)^2$ and $(a-b)^2$.

42. Prove that $\displaystyle\lim_{n\to\infty} \frac{2^n n!}{n^n} = 0$.

   Hint: Show that $\displaystyle\sum_{n=1}^{\infty} \frac{2^n n!}{n^n}$ is convergent.

43. a. Show that the series $\Sigma_{n=1}^{\infty}\, np^n$, where $0 < p < 1$, is convergent.
   b. Show that its sum is $S = \dfrac{p}{(1-p)^2}$.
   Hint: Find an expression for $S_n - pS_n$.

44. Average Number of Coin Tosses An unbiased coin is tossed until the coin lands heads and the number of throws in the experiment is recorded. As more and more experiments are performed, the average number of tosses obtained from these experiments approaches $\Sigma_{n=1}^{\infty}\, n\left(\frac{1}{2}\right)^n$. Use the result of Exercise 43 to find this number.

45. Show that if $\Sigma_{n=1}^{\infty}\, |a_n|$ converges, then so does $\Sigma_{n=2}^{\infty}\, |a_n - a_{n-1}|$.

46. Show that if $\Sigma_{n=1}^{\infty}\, a_n$ is absolutely convergent, then $|\Sigma_{n=1}^{\infty}\, a_n| \leq \Sigma_{n=1}^{\infty}\, |a_n|$.

In Exercises 47–50, determine whether the statement is true or false. If it is true, explain why it is true. If it is false, explain why or give an example to show why it is false.

47. If $\Sigma_{n=1}^{\infty}\, a_n$ and $\Sigma_{n=1}^{\infty}\, b_n$ converge absolutely, then $\Sigma_{n=1}^{\infty}(a_n + b_n)$ converges absolutely.

48. If $a_n > 0$ for $n \geq 1$ and $\Sigma_{n=1}^{\infty}\, a_n$ converges, then $\Sigma_{n=1}^{\infty}(-1)^n a_n$ converges.

49. If $\Sigma_{n=1}^{\infty}\sqrt{a_n^2 + b_n^2}$ converges, then $\Sigma_{n=1}^{\infty}\, a_n$ and $\Sigma_{n=1}^{\infty}\, b_n$ converge absolutely.

50. If $a_n \neq 0$ for any $n \geq 1$ and $\displaystyle\sum_{n=1}^{\infty} a_n$ converges absolutely,

   then $\displaystyle\sum_{n=1}^{\infty} \frac{1}{|a_n|}$ diverges.

8.7 Power Series

Power Series

Until now, we have dealt with series with constant terms. In this section we will study infinite series of the form

$$\sum_{n=0}^{\infty} a_n x^n = a_0 + a_1 x + a_2 x^2 + a_3 x^3 + \cdots + a_n x^n + \cdots$$

where x is a variable. More generally, we will consider series of the form

$$\sum_{n=0}^{\infty} a_n(x-c)^n = a_0 + a_1(x-c) + a_2(x-c)^2 + a_3(x-c)^3 + \cdots + a_n(x-c)^n + \cdots$$

from which $\Sigma_{n=0}^{\infty}\, a_n x^n$ may be obtained as a special case by putting $c = 0$. We may view such series as generalizations of the notion of a polynomial to an infinite series.
Examples of power series are

$$\sum_{n=0}^{\infty} x^n = 1 + x + x^2 + x^3 + \cdots$$

$$\sum_{n=0}^{\infty} \frac{(-1)^n x^n}{n!} = 1 - x + \frac{x^2}{2!} - \frac{x^3}{3!} + \cdots$$

and

$$\sum_{n=0}^{\infty} \frac{(-1)^n \left(x - \frac{\pi}{4}\right)^{2n+1}}{(2n+1)!} = \left(x - \frac{\pi}{4}\right) - \frac{\left(x - \frac{\pi}{4}\right)^3}{3!} + \frac{\left(x - \frac{\pi}{4}\right)^5}{5!} - \cdots$$

Observe that if we truncate each of these series, we obtain a polynomial.

DEFINITION Power Series

Let x be a variable. A **power series in x** is a series of the form

$$\sum_{n=0}^{\infty} a_n x^n = a_0 + a_1 x + a_2 x^2 + a_3 x^3 + \cdots + a_n x^n + \cdots$$

where the a_n's are constants and are called the **coefficients** of the series. More generally, a **power series in $(x - c)$**, where c is a constant, is a series of the form

$$\sum_{n=0}^{\infty} a_n (x - c)^n = a_0 + a_1 (x - c) + a_2 (x - c)^2$$
$$+ a_3 (x - c)^3 + \cdots + a_n (x - c)^n + \cdots$$

Notes

1. A power series in $(x - c)$ is also called a **power series centered at c** or a **power series about c.** Thus, a power series in x is just a series centered at the origin.
2. To simplify the notation used for a power series, we have adopted the convention that $(x - c)^0 = 1$, even when $x = c$. ▪

We can view a power series as a function f defined by the rule

$$f(x) = \sum_{n=0}^{\infty} a_n (x - c)^n$$

The *domain* of f is the set of all x for which the power series converges, and the *range of f* comprises the sums of the series obtained by allowing x to take on all values in the domain of f. If a function f is defined in this manner, we say that **f is represented by the power series $\sum_{n=0}^{\infty} a_n (x - c)^n$**.

EXAMPLE 1 As an example, consider the power series

$$\sum_{n=0}^{\infty} x^n = 1 + x + x^2 + x^3 + \cdots + x^n + \cdots \tag{1}$$

Recognizing that this is a geometric series with common ratio x, we see that it converges for $-1 < x < 1$. Thus, the power series (1) is a rule for a function f with interval $(-1, 1)$ as its domain; that is,

$$f(x) = \sum_{n=0}^{\infty} x^n = 1 + x + x^2 + x^3 + \cdots + x^n + \cdots$$

There is a simple formula for the sum of the geometric series (1), namely, $1/(1 - x)$,

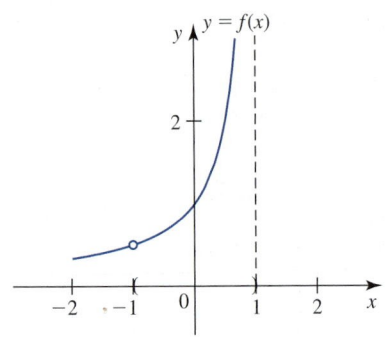

FIGURE 1
The function $f(x) = \sum_{n=0}^{\infty} x^n$
represents the function

$g(x) = \dfrac{1}{1 - x}$ on $(-1, 1)$ only.

FIGURE 2
Observe that $S_n(x) = \sum_{k=0}^{n} x^k$
approximates $g(x)$ better and
better as $n \to \infty$ for $-1 < x < 1$.

and we see that the function *represented* by the series is the function

$$f(x) = \frac{1}{1 - x} \qquad -1 < x < 1$$

Even though the domain of the function $g(x) = 1/(1 - x)$ is the set of all real numbers except $x = 1$, the power series (1) represents the function $g(x) = 1/(1 - x)$ only in the interval of convergence $(-1, 1)$ of the series. (See Figure 1.) Observe that the nth partial sum $S_n(x) = 1 + x + x^2 + \cdots + x^n$ of $\sum_{n=0}^{\infty} x^n$ approximates $g(x)$ better and better as n increases for $-1 < x < 1$. But outside this interval, $S_n(x)$ diverges from $g(x)$. (See Figure 2.)

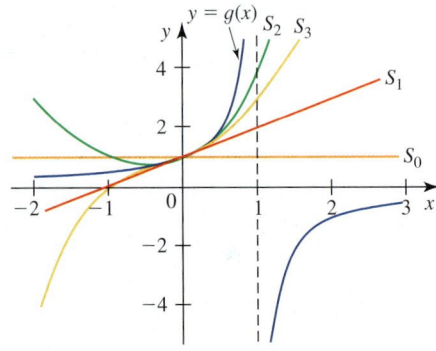

Example 1 reveals one shortcoming in representing a function by a power series. But as we will see later on, the advantages far outweigh the disadvantages.

■ Interval of Convergence

How do we find the domain of a function represented by a power series? Suppose that f is the function represented by the power series

$$f(x) = \sum_{n=0}^{\infty} a_n(x - c)^n = a_0 + a_1(x - c) + a_2(x - c)^2 \qquad \text{(2)}$$
$$+ a_3(x - c)^3 + \cdots + a_n(x - c)^n + \cdots$$

Since $f(c) = a_0$, we see that the domain of f always contains at least one number (the center of the power series) and is therefore nonempty. The following theorem, which we state without proof, tells us that the domain of a power series is always an interval with $x = c$ as its center. In the extreme cases the domain consists of the infinite interval $(-\infty, \infty)$ or just the point $x = c$, which may be regarded as a degenerate interval.

THEOREM 1   Convergence of a Power Series

Given a power series $\sum_{n=0}^{\infty} a_n(x - c)^n$, exactly one of the following is true:

a. The series converges only at $x = c$.
b. The series converges for all x.
c. There is a number $R > 0$ such that the series converges for $|x - c| < R$ and diverges for $|x - c| > R$.

A proof of Theorem 1 is given in Appendix B.

The number R referred to in Theorem 1 is called the **radius of convergence** of the power series. The radius of convergence is $R = 0$ in case (a) and $R = \infty$ in case (b). The set of all values for which the power series converges is called the **interval of convergence** of the power series. Thus, Theorem 1 tells us that the interval of convergence of a power series centered at c is (a) just the single point c, (b) the interval $(-\infty, \infty)$, or (c) the interval $(c - R, c + R)$. (See Figure 3.) But in the last case, Theorem 1 does not tell us whether the endpoints $x = c - R$ and $x = c + R$ are included in the interval of convergence. To determine whether they are included, we simply replace x in the power series (2) by $c - R$ and $c + R$ in succession and use a convergence test on the resultant series.

FIGURE 3
The power series $\sum_{n=0}^{\infty} a_n(x - c)^n$
converges for $|x - c| < R$
and diverges for $|x - c| > R$.

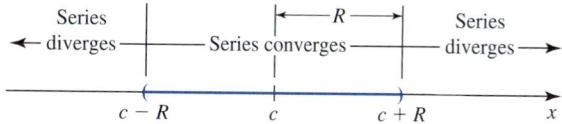

EXAMPLE 2 Find the radius of convergence and the interval of convergence of $\sum_{n=0}^{\infty} n! \, x^n$.

Solution We can think of the given series as $\sum_{n=0}^{\infty} u_n$, where $u_n = n! \, x^n$. Applying the Ratio Test, we have

$$\lim_{n \to \infty} \left| \frac{u_{n+1}}{u_n} \right| = \lim_{n \to \infty} \left| \frac{(n + 1)! \, x^{n+1}}{n! \, x^n} \right| = \lim_{n \to \infty} (n + 1)|x| = \infty$$

whenever $x \neq 0$, and we conclude that the series diverges whenever $x \neq 0$. Therefore, the series converges only when $x = 0$, and its radius of convergence is accordingly $R = 0$.

EXAMPLE 3 Find the radius of convergence and the interval of convergence of

$$\sum_{n=0}^{\infty} \frac{(-1)^n x^{2n}}{(2n)!}$$

Solution Let

$$u_n = \frac{(-1)^n x^{2n}}{(2n)!}$$

Then

$$\lim_{n \to \infty} \left| \frac{u_{n+1}}{u_n} \right| = \lim_{n \to \infty} \left| \frac{(-1)^{n+1} x^{2n+2}}{(2n + 2)!} \cdot \frac{(2n)!}{(-1)^n x^{2n}} \right|$$

$$= \lim_{n \to \infty} \frac{x^2}{(2n + 1)(2n + 2)} = 0 < 1$$

for each fixed value of x, so by the Ratio Test, the given series converges for all values of x. Therefore, the radius of convergence of the series is $R = \infty$, and its interval of convergence is $(-\infty, \infty)$.

EXAMPLE 4   Find the radius of convergence and the interval of convergence of $\displaystyle\sum_{n=1}^{\infty} \frac{x^n}{n}$.

Solution   Let $u_n = x^n/n$. Then

$$\lim_{n\to\infty} \left| \frac{u_{n+1}}{u_n} \right| = \lim_{n\to\infty} \left| \frac{x^{n+1}}{n+1} \cdot \frac{n}{x^n} \right| = \lim_{n\to\infty} \left(\frac{n}{n+1} \right) |x| = |x|$$

By the Ratio Test, the series converges if $|x| < 1$, that is, if $-1 < x < 1$. Therefore, the radius of convergence of the series is $R = 1$. To determine the interval of convergence of the power series, we need to examine the behavior of the series at the endpoints $x = -1$ and $x = 1$. Now, if $x = -1$, the series becomes

$$\sum_{n=1}^{\infty} \frac{(-1)^n}{n}$$

which is the convergent alternating harmonic series, and we see that $x = -1$ is in the interval of convergence of the power series. If $x = 1$, we obtain the harmonic series $\sum_{n=1}^{\infty} 1/n$, which is divergent, so $x = 1$ is not in the interval of convergence. We conclude that the interval of convergence of the given power series is $[-1, 1)$, as shown in Figure 4.

FIGURE 4
The interval of convergence of $\sum_{n=1}^{\infty} x^n/n$ is the interval $[-1, 1)$ with center $c = 0$ and radius $R = 1$.

EXAMPLE 5   Find the radius of convergence and the interval of convergence of $\displaystyle\sum_{n=1}^{\infty} \frac{(x-2)^n}{n^2 \cdot 3^n}$.

Solution   Letting

$$u_n = \frac{(x-2)^n}{n^2 \cdot 3^n}$$

we have

$$\lim_{n\to\infty} \left| \frac{u_{n+1}}{u_n} \right| = \lim_{n\to\infty} \left| \frac{(x-2)^{n+1}}{(n+1)^2 3^{n+1}} \cdot \frac{n^2 \cdot 3^n}{(x-2)^n} \right|$$

$$= \lim_{n\to\infty} \left(\frac{n}{n+1} \right)^2 \frac{|x-2|}{3} = \frac{|x-2|}{3}$$

By the Ratio Test, the series converges if $|x-2|/3 < 1$ or $|x-2| < 3$. The last inequality tells us that the radius of convergence of the given series is $R = 3$ and that the power series converges for x in the interval $(-1, 5)$.

Next, we check the endpoints $x = -1$ and $x = 5$. If $x = -1$, the power series becomes

$$\sum_{n=1}^{\infty} \frac{(-3)^n}{n^2 \cdot 3^n} = \sum_{n=1}^{\infty} \frac{(-1)^n}{n^2}$$

which is a convergent alternating series. Therefore, $x = -1$ is in the interval of convergence. Next, if $x = 5$, we obtain

$$\sum_{n=1}^{\infty} \frac{3^n}{n^2 \cdot 3^n} = \sum_{n=1}^{\infty} \frac{1}{n^2}$$

Figure 5 number line shown with marks at −1, 0, 2, 5, x

FIGURE 5
The interval of convergence of
$$\sum_{n=1}^{\infty} \frac{(x-2)^n}{n^2 \cdot 3^n}$$ is the interval $[-1, 5]$
with center $c = 2$ and radius $R = 3$.

which is a convergent p-series. Therefore, $x = 5$ is also in the interval of convergence. We conclude, accordingly, that the interval of convergence of the given power series is $[-1, 5]$, as shown in Figure 5.

EXAMPLE 6 Find the radius of convergence and the interval of convergence of
$$\sum_{n=0}^{\infty} \frac{(-1)^n 2^n x^n}{\sqrt{n+1}}.$$

Solution Let

$$u_n = \frac{(-1)^n 2^n x^n}{\sqrt{n+1}}$$

Then

$$\lim_{n \to \infty} \left| \frac{u_{n+1}}{u_n} \right| = \lim_{n \to \infty} \left| \frac{(-1)^{n+1} 2^{n+1} x^{n+1}}{\sqrt{n+2}} \cdot \frac{\sqrt{n+1}}{(-1)^n 2^n x^n} \right|$$

$$= \lim_{n \to \infty} 2\sqrt{\frac{n+1}{n+2}} |x| = 2|x| \lim_{n \to \infty} \sqrt{\frac{1 + (1/n)}{1 + (2/n)}} = 2|x|$$

By the Ratio Test, the series converges if $2|x| < 1$ or $|x| < \frac{1}{2}$. The last inequality tells us that the radius of convergence of the power series is $R = \frac{1}{2}$, and the series converges in the interval $\left(-\frac{1}{2}, \frac{1}{2}\right)$.

Next, we check the endpoints $x = -\frac{1}{2}$ and $x = \frac{1}{2}$. If $x = -\frac{1}{2}$, the power series becomes

$$\sum_{n=0}^{\infty} \frac{(-1)^n 2^n \left(-\frac{1}{2}\right)^n}{\sqrt{n+1}} = \sum_{n=0}^{\infty} \frac{1}{\sqrt{n+1}}$$

which can be shown to be divergent by the Limit Comparison Test. (Compare it with the p-series $\sum_{n=1}^{\infty} 1/n^{1/2}$.) Next, if $x = \frac{1}{2}$, we have

$$\sum_{n=0}^{\infty} \frac{(-1)^n 2^n \left(\frac{1}{2}\right)^n}{\sqrt{n+1}} = \sum_{n=0}^{\infty} \frac{(-1)^n}{\sqrt{n+1}}$$

which converges, by the Alternating Series Test. Therefore, the interval of convergence of the power series is $\left(-\frac{1}{2}, \frac{1}{2}\right]$.

Differentiation and Integration of Power Series

Suppose that f is a function represented by a power series centered at c, that is,

$$f(x) = \sum_{n=0}^{\infty} a_n (x - c)^n$$

where x lies in the interval of convergence of the series (domain of f). The following question arises naturally: Can we differentiate and integrate f, and if so, what are the series representations of the derivative and integral of f? The next theorem answers this question in the affirmative and tells us that the series representations of the derivative and integral of f are found by differentiating and integrating the power series representation of f term by term. (We omit its proof.)

Historical Biography

Mary Evans Picture Library/
The Image Works

FRIEDRICH WILHELM BESSEL
(1784-1846)

The first person to use the term *light-years* as a way to express extreme distances, Friedrich Bessel astounded other astronomers when he proved that one of the nearest stars to the earth, 61 Cygni, was about 10 light-years (more that 60 trillion miles) away. He was able to apply his method of computation to compile a catalog of the positions of 50,000 stars. This was no small task for a mathematician and astronomer who did not have a university education. Born in Minden, Germany, on July 22, 1784, Bessel began to study navigation, geography, and foreign languages after he decided to enter foreign trade. From the study of navigation, his interest of astronomy blossomed, and his mathematical and analytical talents quickly surfaced. He made many discoveries in astronomy, and his mathematical discoveries were often derived from his work in that field. The so-called Bessel functions (see Exercise 32 in this section) were studied extensively by Bessel and are still important in mathematics, with important applications in areas such as geology, physics, and engineering.

THEOREM 2 **Differentiation and Integration of Power Series**

Suppose that the power series $\sum_{n=0}^{\infty} a_n(x - c)^n$ has a radius of convergence $R > 0$. Then the function f defined by

$$f(x) = \sum_{n=0}^{\infty} a_n(x - c)^n = a_0 + a_1(x - c) + a_2(x - c)^2 + a_3(x - c)^3 + \cdots$$

for all x in $(c - R, c + R)$ is both differentiable and integrable on $(c - R, c + R)$. Moreover, the derivative of f and the indefinite integral of f are

a. $f'(x) = a_1 + 2a_2(x - c) + 3a_3(x - c)^2 + \cdots = \sum_{n=1}^{\infty} na_n(x - c)^{n-1}$

b. $\displaystyle\int f(x)\,dx = C + a_0(x - c) + a_1\frac{(x - c)^2}{2} + a_2\frac{(x - c)^3}{3} + \cdots$

$$= \sum_{n=0}^{\infty} a_n\frac{(x - c)^{n+1}}{n + 1} + C$$

Notes

1. The series in parts (a) and (b) of Theorem 2 have the same radius of convergence, R, as the series $\sum_{n=0}^{\infty} a_n(x - c)^n$. But the interval of convergence may change. More specifically, you may lose convergence at the endpoints when you differentiate (Exercise 38) and gain convergence there when you integrate (Example 9).
2. Theorem 2 implies that a function that is represented by a power series in an interval $(c - R, c + R)$ is continuous on that interval. This follows from Theorem 1 in Section 2.1. ∎

EXAMPLE 7 Find a power series representation for $1/(1 - x)^2$ on $(-1, 1)$ by differentiating a power series representation of $f(x) = 1/(1 - x)$.

Solution Recalling that $1/(1 - x)$ is the sum of a geometric series, we have

$$f(x) = \frac{1}{1 - x} = 1 + x + x^2 + x^3 + \cdots = \sum_{n=0}^{\infty} x^n \qquad |x| < 1$$

Differentiating both sides of this equation with respect to x and using Theorem 2, we obtain

$$f'(x) = \frac{1}{(1 - x)^2} = 1 + 2x + 3x^2 + \cdots = \sum_{n=1}^{\infty} nx^{n-1} \qquad \blacksquare$$

EXAMPLE 8 Find a power series representation for $\ln(1 - x)$ on $(-1, 1)$.

Solution We start with the equation

$$\frac{1}{1 - x} = 1 + x + x^2 + x^3 + \cdots = \sum_{n=0}^{\infty} x^n \qquad |x| < 1$$

Integrating both sides of this equation with respect to x and using Theorem 2, we obtain

$$\int \frac{1}{1 - x}\,dx = \int (1 + x + x^2 + x^3 + \cdots)\,dx$$

or

$$-\ln(1 - x) = x + \frac{1}{2}x^2 + \frac{1}{3}x^3 + \cdots + C$$

To determine the value of C, we set $x = 0$ in this equation to obtain $-\ln 1 = 0 = C$. Using this value of C, we see that

$$\ln(1 - x) = -x - \frac{1}{2}x^2 - \frac{1}{3}x^3 - \cdots = -\sum_{n=1}^{\infty} \frac{x^n}{n} \qquad |x| < 1 \qquad \blacksquare$$

EXAMPLE 9   Find a power series representation for $\tan^{-1} x$ by integrating a power series representation of $f(x) = 1/(1 + x^2)$.

Solution   Observe that we can obtain a power series representation of f by replacing x with $-x^2$ in the equation

$$\frac{1}{1 - x} = 1 + x + x^2 + \cdots \qquad |x| < 1$$

Thus,

$$\frac{1}{1 + x^2} = \frac{1}{1 - (-x^2)} = 1 + (-x^2) + (-x^2)^2 + (-x^2)^3 + \cdots$$

$$= 1 - x^2 + x^4 - x^6 + \cdots = \sum_{n=0}^{\infty} (-1)^n x^{2n}$$

Since the geometric series converges for $|x| < 1$, we see that this series converges for $|-x^2| < 1$, that is, $x^2 < 1$ or $|x| < 1$. Finally, integrating this equation, we have, by Theorem 2,

$$\tan^{-1} x = \int \frac{1}{1 + x^2} \, dx = \int (1 - x^2 + x^4 - x^6 + \cdots) \, dx$$

$$= C + x - \frac{x^3}{3} + \frac{x^5}{5} - \frac{x^7}{7} + \cdots$$

To find C, we use the condition $\tan^{-1} 0 = 0$ to obtain $0 = C$. Therefore,

$$\tan^{-1} x = x - \frac{x^3}{3} + \frac{x^5}{5} - \frac{x^7}{7} + \cdots = \sum_{n=0}^{\infty} (-1)^n \frac{x^{2n+1}}{2n + 1}$$

We leave it for you to show that the interval of convergence of the series is $[-1, 1]$. $\blacksquare$

8.7   CONCEPT QUESTIONS

1. **a.** Define a power series in x.
   b. Define a power series in $(x - c)$.
2. **a.** What is the radius of convergence of a power series?
   b. What is the interval of convergence of a power series?
   c. How do you find the radius and the interval of convergence of a power series?

3. Suppose that $\sum_{n=0}^{\infty} a_n x^n$ has radius of convergence 2. What can you say about the convergence or divergence of $\sum_{n=0}^{\infty} a_n \left(\frac{3}{2}\right)^n$?
4. Suppose that $\sum_{n=0}^{\infty} a_n(x - 2)^n$ diverges for $x = 0$. What can you say about the convergence or divergence of $\sum_{n=0}^{\infty} a_n 5^n$? What about $\sum_{n=0}^{\infty} a_n 2^n$?

8.7   EXERCISES

In Exercises 1–30, find the radius of convergence and the interval of convergence of the power series.

1. $\displaystyle\sum_{n=0}^{\infty} \frac{x^n}{n+1}$

2. $\displaystyle\sum_{n=1}^{\infty} (-1)^{n-1}\, nx^n$

3. $\displaystyle\sum_{n=1}^{\infty} \frac{x^n}{\sqrt{n}}$

4. $\displaystyle\sum_{n=1}^{\infty} \frac{x^n}{n^2}$

5. $\displaystyle\sum_{n=0}^{\infty} \frac{(2x)^n}{n!}$

6. $\displaystyle\sum_{n=1}^{\infty} \frac{(-1)^n x^n}{n \cdot 3^n}$

7. $\displaystyle\sum_{n=1}^{\infty} (nx)^n$

8. $\displaystyle\sum_{n=0}^{\infty} \frac{n!\, x^n}{(2n)!}$

9. $\displaystyle\sum_{n=2}^{\infty} \frac{x^n}{\ln n}$

10. $\displaystyle\sum_{n=2}^{\infty} (x \ln n)^n$

11. $\displaystyle\sum_{n=1}^{\infty} \frac{e^n x^n}{n}$

12. $\displaystyle\sum_{n=0}^{\infty} \frac{(-1)^n n!\, x^n}{2^n}$

13. $\displaystyle\sum_{n=1}^{\infty} \frac{(-1)^n (x-3)^n}{\sqrt{n}}$

14. $\displaystyle\sum_{n=1}^{\infty} \sqrt{n}\,(2x+3)^n$

15. $\displaystyle\sum_{n=1}^{\infty} \frac{(-1)^{n-1}(x-2)^n}{n \cdot 3^n}$

16. $\displaystyle\sum_{n=1}^{\infty} \frac{n(2x+1)^n}{2^n}$

17. $\displaystyle\sum_{n=0}^{\infty} \frac{(-1)^n n(x-1)^n}{n^2+1}$

18. $\displaystyle\sum_{n=0}^{\infty} \frac{n(x+2)^n}{(n^2+1)2^n}$

19. $\displaystyle\sum_{n=0}^{\infty} \frac{(-1)^n (x+2)^{2n+1}}{(2n+1)!}$

20. $\displaystyle\sum_{n=0}^{\infty} \frac{(-1)^n (3x+2)^{2n}}{(2n)!}$

21. $\displaystyle\sum_{n=1}^{\infty} \frac{2^n (x+2)^n}{n^n}$

22. $\displaystyle\sum_{n=1}^{\infty} \frac{(3x-1)^n}{n^3+n}$

23. $\displaystyle\sum_{n=2}^{\infty} \frac{(-1)^n (3x+5)^n}{n \ln n}$

24. $\displaystyle\sum_{n=2}^{\infty} \frac{(-1)^n (x+2)^n}{(\ln n)^n}$

25. $\displaystyle\sum_{n=2}^{\infty} \frac{x^n}{n(\ln n)^2}$

26. $\displaystyle\sum_{n=1}^{\infty} \frac{n^n (3x+5)^n}{(2n)!}$

27. $\displaystyle\sum_{n=1}^{\infty} \frac{2 \cdot 4 \cdot 6 \cdot \cdots \cdot 2n}{3 \cdot 5 \cdot 7 \cdot \cdots \cdot (2n+1)}\, x^{2n+1}$

28. $\displaystyle\sum_{n=1}^{\infty} \frac{(-1)^n n!\, (x-1)^n}{1 \cdot 3 \cdot 5 \cdot \cdots \cdot (2n-1)}$

29. $\displaystyle\sum_{n=1}^{\infty} \frac{(-1)^n 2^n n!\, x^n}{5 \cdot 8 \cdot 11 \cdot \cdots \cdot (3n+2)}$

30. $\displaystyle\sum_{n=1}^{\infty} \frac{(-1)^n 2 \cdot 4 \cdot 6 \cdot \cdots \cdot 2n(x-\pi)^n}{n!}$

 31. Consider the series $\sum_{n=0}^{\infty} x^n$ and the (sum) function $f(x) = 1/(1-x)$ represented by the series for $-1 < x < 1$.
   a. Find the remainder $R_n(x) = f(x) - S_n(x)$, where $S_n(x) = \sum_{k=0}^{n} x^k$ is the nth partial sum of $\sum_{n=0}^{\infty} x^n$ and x is fixed.

   b. Evaluate $\lim_{n\to\infty} R_n(x)$ for each fixed x in the interval $(-1, 1)$. What happens to $\lim_{n\to\infty} R_n(x)$ for $|x| > 1$?
   c. Plot the graphs of $R_n(x)$ for $n = 1, 2, 3, \ldots, 5$, and 20 using the viewing window $[-2, 2] \times [-10, 5]$.

 32. A Bessel Function The function J_0 defined by

$$J_0(x) = \sum_{n=0}^{\infty} \frac{(-1)^n x^{2n}}{2^{2n}(n!)^2}$$

is called the *Bessel function of order* 0.
   a. What is the domain of J_0?
   b. Plot the graph of J_0 in the viewing window $[-10, 10] \times [-0.5, 1.2]$, and plot the graphs of $S_n(x)$ for $n = 0, 1, 2, 3$, and 4 in the viewing window $[-8, 8] \times [-2, 2]$.

33. A Bessel Function The function J_1 defined by

$$J_1(x) = \sum_{n=0}^{\infty} \frac{(-1)^n x^{2n+1}}{n!(n+1)!\, 2^{2n+1}}$$

is called the *Bessel function of order* 1. What is its domain?

34. If a is a constant, find the radius and the interval of convergence of the power series $\sum_{n=0}^{\infty} a^n (x-c)^n$.

35. If the radius of convergence of the power series $\sum a_n x^n$ is R, what is the radius of convergence of the power series $\sum a_n x^{2n}$?

36. Suppose that $\lim_{n\to\infty} |a_{n+1}/a_n| = L$ and $L \neq 0$. Show that the radius of convergence of the power series $\sum a_n x^n$ is $1/L$.

37. Suppose that $\lim_{n\to\infty} \sqrt[n]{|a_n|} = L$ and $L \neq 0$. What is the radius of convergence of the power series $\sum a_n x^n$?

38. Let $f(x) = \displaystyle\sum_{n=1}^{\infty} \frac{(x-2)^n}{n^2 3^n}$. Show that the domain of f is $[-1, 5]$ but the domain of f' is $[-1, 5)$.

39. Let $f(x) = \displaystyle\sum_{n=1}^{\infty} \frac{x^n}{n^2}$. Find $f'(x)$ and $f''(x)$. What are the intervals of convergence of f, f', and f''?

40. Show that the series $\displaystyle\sum_{n=1}^{\infty} \frac{\sin(n^3 x)}{n^2}$ converges for all values of x, but $\displaystyle\sum_{n=1}^{\infty} \frac{d}{dx}\!\left[\frac{\sin(n^3 x)}{n^2}\right]$ diverges for all values of x. Does this contradict Theorem 2? Explain your answer.

41. Find the sum of the series $\sum_{n=1}^{\infty} nx^{n-1}$, $|x| < 1$.
   Hint: Differentiate the geometric series $\sum_{n=0}^{\infty} x^n$.

42. a. Find the sum of the series $\sum_{n=1}^{\infty} nx^n$, $|x| < 1$.
   Hint: See the hint for Exercise 41.

   b. Use the result of part (a) to find the sum of $\displaystyle\sum_{n=1}^{\infty} \frac{n}{2^n}$.

43. Suppose that the interval of convergence of the series $\sum_{n=0}^{\infty} a_n (x-c)^n$ is $(c-R, c+R]$. Prove that the series is conditionally convergent at $c+R$.

44. Suppose that the series $\sum_{n=0}^{\infty} a_n(x - c)^n$ is absolutely convergent at one endpoint of its interval of convergence. Prove that the series is also absolutely convergent at the other endpoint.

45. a. Find a power series representation for $1/(1 - t^2)$.
   b. Use the result of part (a) to find a power series representation of $\tanh^{-1} x$ using the relationship

$$\tanh^{-1} x = \int_0^x \frac{1}{1 - t^2} \, dt$$

   What is the radius of convergence of the series?

46. Use the result of Example 8

$$\ln(1 - x) = -\sum_{n=1}^{\infty} \frac{x^n}{n}$$

to obtain an approximation of $\ln 1.2$ accurate to five decimal places.
Hint: Use Theorem 2 in Section 8.5.

47. Use the result of Example 9,

$$\tan^{-1} x = \sum_{n=0}^{\infty} (-1)^n \frac{x^{2n+1}}{2n + 1}$$

to obtain an approximation of π accurate to five decimal places.
Hint: Use Theorem 2 in Section 8.5.

48. Motion Along an Inclined Plane An object of mass m is thrown up an inclined plane that makes an angle of α with the horizontal. If air resistance proportional to the instantaneous

velocity is taken into consideration, then the object reaches a maximum distance up the incline given by

$$D = \frac{mv_0}{k} - \frac{m^2 g}{k^2} (\sin \alpha) \ln\left(1 + \frac{kv_0}{mg \sin \alpha} \right)$$

where k is the constant of proportionality and g is the constant of acceleration due to gravity.
   a. Show that D can be written as

$$D = \frac{1}{2} \frac{v_0^2}{g \sin \alpha} - \frac{1}{3} \frac{v_0^3}{m(g \sin \alpha)^2} k + \frac{1}{4} \frac{v_0^4}{m^2(g \sin \alpha)^3} k^2 - \cdots$$

   Hint: Use the result of Example 8.
   b. Use the result of part (a) to show that in the absence of air resistance the object reaches a maximum distance of $v_0^2/(2g \sin \alpha)$ up the incline.

In Exercises 49–52, determine whether the statement is true or false. If it is true, explain why it is true. If it is false, explain why or give an example to show why it is false.

49. If the power series $\sum_{n=0}^{\infty} a_n x^n$ converges for $x = 3$, then it converges for $x = -2$.

50. If the power series $\sum_{n=0}^{\infty} a_n x^n$ converges for x in $(-1, 1)$, then $f(x) = \sum_{n=0}^{\infty} a_n x^n$ is continuous on $(-1, 1)$.

51. If the interval of convergence of $\sum_{n=0}^{\infty} a_n x^n$ is $[-2, 2)$, then the interval of convergence of $\sum_{n=0}^{\infty} a_n(x - 3)^n$ is $[1, 5)$.

52. If the radius of convergence of $\sum_{n=0}^{\infty} a_n x^n$ is $R > 0$, then the radius of convergence of the power series in $\frac{1}{x}$, $\sum_{n=0}^{\infty} \frac{a_n}{x^n}$, is $\frac{1}{R}$.

8.8 **Taylor and Maclaurin Series**

In Section 8.7 we saw that every power series represents a function whose domain is precisely the interval of convergence of the series. We also touched upon the converse problem: Given a function f defined on an interval containing a point c, is there a power series centered at c that represents f, and if so, how do we find it? There, we were able to look only at functions whose power series representations are obtained by manipulating the geometric series.

We now look at the general problem of finding power series representations for functions. The problem centers on finding the answers to two questions:

1. What form does the power series representation of the function f take? (In other words, what does a_n look like?)
2. What conditions will guarantee that such a power series will represent f?

We will consider the first question here and leave the second for Section 8.9.

■ **Taylor and Maclaurin Series**

Suppose that f is a function that can be represented by a power series that is centered at c and has a radius of convergence $R > 0$. If $|x - c| < R$, we have

$$f(x) = a_0 + a_1(x - c) + a_2(x - c)^2 + a_3(x - c)^3 + a_4(x - c)^4 + \cdots + a_n(x - c)^n + \cdots$$

Applying Theorem 2 of Section 8.7 repeatedly, we obtain

$$f'(x) = a_1 + 2a_2(x - c) + 3a_3(x - c)^2 + 4a_4(x - c)^3 + \cdots + na_n(x - c)^{n-1} + \cdots$$

$$f''(x) = 2a_2 + 3 \cdot 2a_3(x - c) + 4 \cdot 3a_4(x - c)^2 + \cdots + n(n - 1)a_n(x - c)^{n-2} + \cdots$$

$$f'''(x) = 3 \cdot 2a_3 + 4 \cdot 3 \cdot 2a_4(x - c) + \cdots + n(n - 1)(n - 2)a_n(x - c)^{n-3} + \cdots$$

$$\vdots$$

$$f^{(n)}(x) = n(n - 1)(n - 2)(n - 3) \cdots \cdot 2a_n + \cdots$$

$$\vdots$$

Each of these series is valid for x satisfying $|x - c| < R$. Substituting $x = c$ in each of the above expressions, we obtain

$$f(c) = a_0, \qquad f'(c) = a_1, \qquad f''(c) = 2a_2,$$

$$f'''(c) = 3! \, a_3, \qquad \ldots, \qquad f^{(n)}(c) = n! \, a_n, \qquad \ldots$$

from which we find

$$a_0 = f(c), \qquad a_1 = f'(c), \qquad a_2 = \frac{f''(c)}{2!},$$

$$a_3 = \frac{f'''(c)}{3!}, \qquad \ldots, \qquad a_n = \frac{f^{(n)}(c)}{n!}, \qquad \ldots$$

We have proved that if f has a power series representation, then the series must have the form given in the following theorem.

THEOREM 1   **Taylor Series of f at c**

If f has a power series representation at c, that is, if

$$f(x) = \sum_{n=0}^{\infty} a_n(x - c)^n \qquad |x - c| < R$$

then $f^{(n)}(c)$ exists for every positive integer n and

$$a_n = \frac{f^{(n)}(c)}{n!}$$

Thus,

$$f(x) = \sum_{n=0}^{\infty} \frac{f^{(n)}(c)}{n!} (x - c)^n$$

$$= f(c) + f'(c)(x - c) + \frac{f''(c)}{2!} (x - c)^2 + \frac{f'''(c)}{3!} (x - c)^3 + \cdots \tag{1}$$

A series of this form is called the **Taylor series of the function f at c** after the English mathematician Brook Taylor (1685–1731).

In the special case in which $c = 0$, the Taylor series becomes

$$f(x) = \sum_{n=0}^{\infty} \frac{f^{(n)}(0)}{n!} x^n = f(0) + f'(0)x + \frac{f''(0)}{2!} x^2 + \frac{f'''(0)}{3!} x^3 + \cdots \tag{2}$$

This series is just the Taylor series of f centered at the origin. It is called the **Maclaurin series of f** in honor of the Scottish mathematician Colin Maclaurin (1698–1746).

Note   Theorem 1 states that if a function f has a power series representation at c, then the (unique) series must be the Taylor series at c. The converse is not necessarily true. Given a function f with derivatives of *all* orders at c, we can compute the Taylor coefficients of f at c,

$$\frac{f^{(n)}(c)}{n!} \qquad n = 0, 1, 2, \ldots$$

and, therefore, the Taylor series of f at c (Equation (1)). But the series that is obtained *formally* in this fashion need *not* represent f. Situations such as these, however, are rare. (We give an example of such a function in Exercise 75.) In view of this *we will assume*, in the rest of this section, *that the Taylor series of a function does represent the function, unless otherwise noted.*

EXAMPLE 1   Let $f(x) = e^x$. Find the Maclaurin series of f, and determine its radius of convergence.

Solution   The derivatives of $f(x) = e^x$ are $f'(x) = e^x$, $f''(x) = e^x$, and, in general, $f^{(n)}(x) = e^x$, where $n \geq 1$. So

$$f(0) = 1, \qquad f'(0) = 1, \qquad f''(0) = 1, \qquad \ldots, \qquad f^{(n)}(0) = 1, \qquad \ldots$$

Therefore, if we use Equation (2), the Maclaurin series of f (the Taylor series of f at 0) is

$$\sum_{n=0}^{\infty} \frac{f^{(n)}(0)}{n!} x^n = \sum_{n=0}^{\infty} \frac{1}{n!} x^n = 1 + x + \frac{x^2}{2!} + \frac{x^3}{3!} + \cdots + \frac{x^n}{n!} + \cdots$$

To determine the radius of convergence of the power series, we use the ratio test with $u_n = x^n/n!$. Since

$$\lim_{n \to \infty} \left| \frac{u_{n+1}}{u_n} \right| = \lim_{n \to \infty} \left| \frac{x^{n+1}}{(n+1)!} \cdot \frac{n!}{x^n} \right| = \lim_{n \to \infty} \frac{|x|}{n+1} = 0$$

we conclude that the radius of convergence of the series is $R = \infty$.

EXAMPLE 2   Find the Taylor series for $f(x) = \ln x$ at 1, and determine its interval of convergence.

Solution   We compute the values of f and its derivatives at 1. Thus,

$$f(x) = \ln x \qquad\qquad\qquad f(1) = \ln 1 = 0$$

$$f'(x) = \frac{1}{x} = x^{-1} \qquad\qquad\qquad f'(1) = 1$$

$$f''(x) = -x^{-2} \qquad\qquad\qquad f''(1) = -1$$

$$f'''(x) = 2x^{-3} \qquad\qquad\qquad f'''(1) = 2$$

$$f^{(4)}(x) = -3 \cdot 2x^{-4} \qquad\qquad\qquad f^{(4)}(1) = -3 \cdot 2$$

$$\vdots \qquad\qquad\qquad\qquad\qquad \vdots$$

$$f^{(n)}(x) = (-1)^{n-1}(n-1)! \, x^{-n} \qquad\qquad f^{(n)}(1) = (-1)^{n-1}(n-1)!$$

Then using Equation (1), we obtain the Taylor series of $f(x) = \ln x$:

$$\sum_{n=0}^{\infty} \frac{f^{(n)}(1)}{n!}(x-1)^n = f(1) + f'(1)(x-1) + \frac{f''(1)}{2!}(x-1)^2 + \frac{f'''(1)}{3!}(x-1)^3 + \cdots$$

$$= (x-1) - \frac{1}{2!}(x-1)^2 + \frac{2}{3!}(x-1)^3 - \frac{3!}{4!}(x-1)^4 + \cdots$$

$$= (x-1) - \frac{(x-1)^2}{2} + \frac{(x-1)^3}{3} - \frac{(x-1)^4}{4} + \cdots$$

$$= \sum_{n=1}^{\infty} (-1)^{n-1} \frac{(x-1)^n}{n}$$

To find the interval of convergence of the series, we use the Ratio Test with $u_n = (-1)^{n-1}(x-1)^n/n$. Since

$$\lim_{n\to\infty} \left| \frac{u_{n+1}}{u_n} \right| = \lim_{n\to\infty} \left| \frac{(-1)^n(x-1)^{n+1}}{n+1} \cdot \frac{n}{(-1)^{n-1}(x-1)^n} \right|$$

$$= \lim_{n\to\infty} |x-1| \left(\frac{n}{n+1} \right) = |x-1| \lim_{n\to\infty} \frac{1}{1 + \dfrac{1}{n}} = |x-1|$$

we see that the series converges for x in the interval $(0, 2)$. Next, we notice that if $x = 0$, the series becomes

$$\sum_{n=1}^{\infty} \frac{(-1)^{2n-1}}{n} = -\sum_{n=1}^{\infty} \frac{1}{n}$$

Since this is the negative of the harmonic series, it is divergent. If $x = 2$, the series becomes

$$\sum_{n=1}^{\infty} \frac{(-1)^{n-1}}{n}$$

This is the alternating harmonic series and, hence, is convergent. Therefore, the Taylor series for $f(x) = \ln x$ at 1 has interval of convergence $(0, 2]$.   ■

EXAMPLE 3   Find the Maclaurin series of $f(x) = \sin x$, and determine its interval of convergence.

Solution   To find the Maclaurin series of $f(x) = \sin x$, we compute the values of f and its derivatives at $x = 0$. We obtain

$$f(x) = \sin x \qquad\qquad f(0) = 0$$
$$f'(x) = \cos x \qquad\qquad f'(0) = 1$$
$$f''(x) = -\sin x \qquad\qquad f''(0) = 0$$
$$f'''(x) = -\cos x \qquad\qquad f'''(0) = -1$$
$$f^{(4)}(x) = \sin x \qquad\qquad f^{(4)}(0) = 0$$

We need not go further, since it is clear that successive derivatives of f follow this same pattern. Then, using Equation (2), we obtain the Maclaurin series of $f(x) = \sin x$:

$$\sum_{n=0}^{\infty} \frac{f^{(n)}(0)}{n!} x^n = f(0) + f'(0)x + \frac{f''(0)}{2!} x^2 + \frac{f'''(0)}{3!} x^3 + \frac{f^{(4)}(0)}{4!} x^4 + \cdots$$

$$= x - \frac{x^3}{3!} + \frac{x^5}{5!} - \frac{x^7}{7!} + \cdots$$

$$= \sum_{n=0}^{\infty} \frac{(-1)^n}{(2n+1)!} x^{2n+1}$$

To find the interval of convergence of the series, we use the Ratio Test with $u_n = (-1)^n x^{2n+1}/(2n+1)!$. Since

$$\lim_{n \to \infty} \left| \frac{u_{n+1}}{u_n} \right| = \lim_{n \to \infty} \left| \frac{(-1)^{n+1} x^{2n+3}}{(2n+3)!} \cdot \frac{(2n+1)!}{(-1)^n x^{2n+1}} \right|$$

$$= \lim_{n \to \infty} \frac{|x|^2}{(2n+2)(2n+3)} = 0 < 1$$

we conclude that the interval of convergence of the series is $(-\infty, \infty)$. ■

EXAMPLE 4 Find the Maclaurin series of $f(x) = \cos x$.

Solution We could proceed as in Example 3, but it is easier to make use of Theorem 2 of Section 8.7 to differentiate the expression for $\sin x$ that we obtained in Example 3. Thus,

$$f(x) = \cos x = \frac{d}{dx} (\sin x) = \frac{d}{dx} \left(x - \frac{x^3}{3!} + \frac{x^5}{5!} - \frac{x^7}{7!} + \cdots \right)$$

$$= 1 - \frac{x^2}{2!} + \frac{x^4}{4!} - \frac{x^6}{6!} + \cdots$$

$$= \sum_{n=0}^{\infty} \frac{(-1)^n}{(2n)!} x^{2n}$$

Since the Maclaurin series for $\sin x$ converges for all x, Theorem 2 of Section 8.7 tells us that this series converges in $(-\infty, \infty)$ as well. ■

EXAMPLE 5 Find the Maclaurin series for $f(x) = (1 + x)^k$, where k is a real number.

Solution We compute the values of f and its derivatives at $x = 0$, obtaining

$$f(x) = (1 + x)^k \qquad\qquad f(0) = 1$$

$$f'(x) = k(1 + x)^{k-1} \qquad\qquad f'(0) = k$$

$$f''(x) = k(k - 1)(1 + x)^{k-2} \qquad\qquad f''(0) = k(k - 1)$$

$$f'''(x) = k(k - 1)(k - 2)(1 + x)^{k-3} \qquad\qquad f'''(0) = k(k - 1)(k - 2)$$

$$\vdots \qquad\qquad\qquad\qquad \vdots$$

$$f^{(n)}(x) = k(k - 1) \cdots (k - n + 1)(1 + x)^{k-n} \qquad f^{(n)}(0) = k(k - 1) \cdots (k - n + 1)$$

So the Maclaurin series of $f(x) = (1 + x)^k$ is

$$\sum_{n=0}^{\infty} \frac{f^{(n)}(0)}{n!} x^n = f(0) + f'(0)x + \frac{f''(0)}{2!} x^2 + \frac{f'''(0)}{3!} x^3 + \cdots$$

$$= 1 + kx + \frac{k(k-1)}{2!} x^2 + \frac{k(k-1)(k-2)}{3!} x^3 + \cdots$$

$$= \sum_{n=0}^{\infty} \frac{k(k-1)(k-2) \cdots (k-n+1)}{n!} x^n$$

Observe that if k is a positive integer, then the series is infinite (by the Binomial Theorem), and so it converges for all x.

If k is not a positive integer, then we use the Ratio Test to find the interval of convergence. Denoting the nth term of the series by u_n, we find

$$\lim_{n \to \infty} \left| \frac{u_{n+1}}{u_n} \right| = \lim_{n \to \infty} \left| \frac{k(k-1) \cdots (k-n+1)(k-n)x^{n+1}}{(n+1)!} \cdot \frac{n!}{k(k-1) \cdots (k-n+1)x^n} \right|$$

$$= \lim_{n \to \infty} \frac{|k-n|}{n+1} |x| = \lim_{n \to \infty} \frac{\left| \dfrac{k}{n} - 1 \right|}{1 + \dfrac{1}{n}} |x| = |x|$$

and we see that the series converges for x in the interval $(-1, 1)$. ■

The series in Example 5 is called the *binomial series*.

The Binomial Series

If k is any real number and $|x| < 1$, then

$$(1 + x)^k = 1 + kx + \frac{k(k-1)}{2!} x^2 + \frac{k(k-1)(k-2)}{3!} x^3 + \cdots = \sum_{n=0}^{\infty} \binom{k}{n} x^n \quad (3)$$

Notes

1. The coefficients in the binomial series are referred to as **binomial coefficients** and are denoted by

$$\binom{k}{n} = \frac{k(k-1) \cdots (k-n+1)}{n!} \quad n \geq 1, \qquad \binom{k}{0} = 1$$

2. If k is a positive integer and $n > k$, then the binomial coefficient contains a factor $(k - k)$, so $\binom{k}{n} = 0$ for $n > k$. The binomial series then reduces to a polynomial of degree k:

$$(1 + x)^k = 1 + kx + \frac{k(k-1)}{2!} x^2 + \cdots + x^k = \sum_{n=0}^{k} \binom{k}{n} x^n$$

In other words, the expression $(1 + x)^k$ can be represented by a finite sum if k is a positive integer and by an infinite series if k is not a positive integer. Thus, we can view the binomial series as an extension of the Binomial Theorem to the case in which k is not a positive integer.

3. Even though the binomial series always converges for $-1 < x < 1$, its convergence at the endpoints $x = -1$ or $x = 1$ depends on the value of k. It can be shown that the series converges at $x = 1$ if $-1 < k < 0$ and at both endpoints $x = \pm 1$ if $k \geq 0$.

4. We have derived Equation (3) under the assumption that $(1 + x)^k$ has a power series representation. In Exercise 78 we outline a procedure for deriving Equation (3) without this assumption. ■

EXAMPLE 6 Find a power series representation for the function $f(x) = \sqrt{1 + x}$.

Solution   Using Equation (3) with $k = \frac{1}{2}$, we obtain

$$f(x) = (1 + x)^{1/2} = 1 + \frac{1}{2}x + \frac{\frac{1}{2}\left(\frac{1}{2} - 1\right)}{2!}x^2 + \frac{\frac{1}{2}\left(\frac{1}{2} - 1\right)\left(\frac{1}{2} - 2\right)}{3!}x^3 + \cdots$$

$$+ \frac{\frac{1}{2}\left(\frac{1}{2} - 1\right)\cdots\left(\frac{1}{2} - n + 1\right)}{n!}x^n + \cdots$$

$$= 1 + \frac{1}{2}x - \frac{1}{2 \cdot 2^2}x^2 + \frac{1 \cdot 3}{3! \, 2^3}x^3 + \cdots$$

$$+ (-1)^{n+1}\frac{1 \cdot 3 \cdot 5 \cdot \cdots \cdot (2n - 3)}{n! \, 2^n}x^n + \cdots$$

$$= 1 + \frac{1}{2}x + \sum_{n=2}^{\infty}(-1)^{n+1}\frac{1 \cdot 3 \cdot 5 \cdot \cdots \cdot (2n - 3)}{n! \, 2^n}x^n$$

This representation is valid for $|x| \leq 1$.

The graph of f and the first three partial sums $P_1(x) = 1$, $P_2(x) = 1 + \frac{1}{2}x$, and $P_3(x) = 1 + \frac{1}{2}x - \frac{1}{8}x^2$ are shown in Figure 1. Observe that the partial sums of f, $P_n(x)$, approximate f better and better in the interval of convergence of the series as n increases. ■

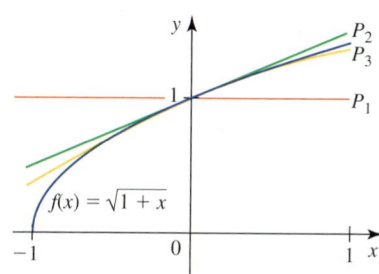

FIGURE 1
The graphs of $f(x) = \sqrt{1 + x}$ and the first three partial sums of the binomial series

Techniques for Finding Taylor Series

The Taylor series of a function can always be found by using Equation (1). But as Examples 7, 8, and 9 of Section 8.7 and Example 4 of this section show, it is often easier to find the series by algebraic manipulation, differentiation, or integration of some well-known series. We now elaborate further on such techniques. First, we list some common functions and their power series representations in Table 1.

TABLE 1

Maclaurin Series	Interval of Convergence
1. $\dfrac{1}{1 - x} = 1 + x + x^2 + x^3 + \cdots = \displaystyle\sum_{n=0}^{\infty} x^n$	$(-1, 1)$
2. $e^x = 1 + x + \dfrac{x^2}{2!} + \dfrac{x^3}{3!} + \cdots = \displaystyle\sum_{n=0}^{\infty} \dfrac{x^n}{n!}$	$(-\infty, \infty)$
3. $\sin x = x - \dfrac{x^3}{3!} + \dfrac{x^5}{5!} - \dfrac{x^7}{7!} + \cdots = \displaystyle\sum_{n=0}^{\infty} (-1)^n \dfrac{x^{2n+1}}{(2n + 1)!}$	$(-\infty, \infty)$
4. $\cos x = 1 - \dfrac{x^2}{2!} + \dfrac{x^4}{4!} - \dfrac{x^6}{6!} + \cdots = \displaystyle\sum_{n=0}^{\infty} (-1)^n \dfrac{x^{2n}}{(2n)!}$	$(-\infty, \infty)$

(continued)

TABLE 1 (*continued*)

Maclaurin Series	Interval of Convergence
5. $\ln(1 + x) = x - \dfrac{x^2}{2} + \dfrac{x^3}{3} - \dfrac{x^4}{4} + \cdots = \displaystyle\sum_{n=1}^{\infty} (-1)^{n-1} \dfrac{x^n}{n}$	$(-1, 1]$
6. $\sin^{-1} x = x + \dfrac{x^3}{2 \cdot 3} + \dfrac{1 \cdot 3x^5}{2 \cdot 4 \cdot 5} + \cdots = \displaystyle\sum_{n=0}^{\infty} \dfrac{(2n)!\, x^{2n+1}}{(2^n\, n!)^2 (2n + 1)}$	$[-1, 1]$
7. $\tan^{-1} x = x - \dfrac{x^3}{3} + \dfrac{x^5}{5} - \dfrac{x^7}{7} + \cdots = \displaystyle\sum_{n=0}^{\infty} (-1)^n \dfrac{x^{2n+1}}{2n + 1}$	$[-1, 1]$
8. $(1 + x)^k = \displaystyle\sum_{n=0}^{\infty} \binom{k}{n} x^n = 1 + kx + \dfrac{k(k - 1)}{2!} x^2 + \dfrac{k(k - 1)(k - 2)}{3!} x^3 + \cdots$	$(-1, 1)$

All of the formulas in the table except Formulas (5) and (6) have been derived in this and the previous sections. (See note on page 791.) Formula (5) follows from the result of Example 8 in Section 8.7 by replacing $-x$ by x. Formula (6) will be derived in Example 14.

EXAMPLE 7 Find the Taylor series representation of $f(x) = \dfrac{1}{1 + x}$ at $x = 2$.

Solution We first rewrite $f(x)$ so that it includes the expression $(x - 2)$. Thus,

$$f(x) = \frac{1}{1 + x} = \frac{1}{3 + (x - 2)} = \frac{1}{3\left[1 + \left(\dfrac{x - 2}{3}\right)\right]} = \frac{1}{3} \cdot \frac{1}{1 + \left(\dfrac{x - 2}{3}\right)}$$

Then, using Formula (1) in Table 1 with x replaced by $-(x - 2)/3$, we obtain

$$f(x) = \frac{1}{3}\left\{\frac{1}{1 - \left[-\left(\dfrac{x - 2}{3}\right)\right]}\right\}$$

$$= \frac{1}{3}\left\{1 + \left[-\left(\frac{x - 2}{3}\right)\right] + \left[-\left(\frac{x - 2}{3}\right)\right]^2 + \left[-\left(\frac{x - 2}{3}\right)\right]^3 + \cdots\right\}$$

$$= \frac{1}{3}\left[1 - \left(\frac{x - 2}{3}\right) + \left(\frac{x - 2}{3}\right)^2 - \left(\frac{x - 2}{3}\right)^3 + \cdots\right]$$

$$= \frac{1}{3} - \frac{1}{3^2}(x - 2) + \frac{1}{3^3}(x - 2)^2 - \frac{1}{3^4}(x - 2)^3 + \cdots = \sum_{n=0}^{\infty}(-1)^n \frac{(x - 2)^n}{3^{n+1}}$$

The series converges for $|(x - 2)/3| < 1$, that is, $|x - 2| < 3$ or $-1 < x < 5$. You can verify that the series diverges at both endpoints. ∎

EXAMPLE 8 Find the Maclaurin series for $f(x) = x^2 \sin 2x$.

Solution If we replace x by $2x$ in Formula (3) in Table 1, we obtain

$$\sin 2x = (2x) - \frac{(2x)^3}{3!} + \frac{(2x)^5}{5!} - \frac{(2x)^7}{7!} + \cdots$$

$$= 2x - \frac{2^3 x^3}{3!} + \frac{2^5 x^5}{5!} - \frac{2^7 x^7}{7!} + \cdots = \sum_{n=0}^{\infty}(-1)^n \frac{2^{2n+1} x^{2n+1}}{(2n + 1)!}$$

which is valid for all x in $(-\infty, \infty)$. Therefore, using Theorem 4a of Section 8.2, we obtain

$$f(x) = x^2 \sin 2x = x^2\left(2x - \frac{2^3 x^3}{3!} + \frac{2^5 x^5}{5!} - \frac{2^7 x^7}{7!} + \cdots\right)$$

$$= 2x^3 - \frac{2^3 x^5}{3!} + \frac{2^5 x^7}{5!} - \frac{2^7 x^9}{7!} + \cdots$$

$$= \sum_{n=0}^{\infty} (-1)^n \frac{2^{2n+1} x^{2n+3}}{(2n+1)!}$$

which converges for all x in $(-\infty, \infty)$. ∎

The next example shows how the use of trigonometric identities can help us find the Taylor series of a trigonometric function.

EXAMPLE 9 Find the Taylor series for $f(x) = \sin x$ at $x = \pi/6$.

Solution We write

$$f(x) = \sin x = \sin\left[\left(x - \frac{\pi}{6}\right) + \frac{\pi}{6}\right]$$

$$= \sin\left(x - \frac{\pi}{6}\right)\cos\frac{\pi}{6} + \cos\left(x - \frac{\pi}{6}\right)\sin\frac{\pi}{6}$$

$$= \frac{\sqrt{3}}{2}\sin\left(x - \frac{\pi}{6}\right) + \frac{1}{2}\cos\left(x - \frac{\pi}{6}\right)$$

Then using Formulas 3 and 4 with $x - (\pi/6)$ in place of x, we obtain

$$f(x) = \frac{\sqrt{3}}{2}\sum_{n=0}^{\infty} \frac{(-1)^n}{(2n+1)!}\left(x - \frac{\pi}{6}\right)^{2n+1} + \frac{1}{2}\sum_{n=0}^{\infty} \frac{(-1)^n}{(2n)!}\left(x - \frac{\pi}{6}\right)^{2n}$$

which converges for all x in $(-\infty, \infty)$. ∎

The power series representations of certain functions can also be found by adding, multiplying, or dividing the Maclaurin or Taylor series of some familiar functions as the following examples show.

EXAMPLE 10 Find the Maclaurin series representation for $f(x) = \sinh x$.

Solution We have

$$f(x) = \sinh x = \frac{1}{2}(e^x - e^{-x}) = \frac{1}{2}e^x - \frac{1}{2}e^{-x}$$

$$= \frac{1}{2}\left(1 + x + \frac{x^2}{2!} + \frac{x^3}{3!} + \cdots\right) - \frac{1}{2}\left(1 - x + \frac{x^2}{2!} - \frac{x^3}{3!} + \cdots\right)$$

$$= x + \frac{x^3}{3!} + \frac{x^5}{5!} + \cdots = \sum_{n=0}^{\infty} \frac{x^{2n+1}}{(2n+1)!}$$

Since the Maclaurin series of both e^x and e^{-x} converge for x in $(-\infty, \infty)$, we see that this representation of $\sinh x$ is also valid for all values of x. ∎

EXAMPLE 11 Find the first three terms of the Maclaurin series representation for $f(x) = e^x \cos x$.

Solution Using Formulas (2) and (4) in Table 1, we can write

$$f(x) = e^x \cos x = \left(1 + x + \frac{x^2}{2} + \frac{x^3}{6} + \cdots \right)\left(1 - \frac{x^2}{2} + \frac{x^4}{24} - \cdots \right)$$

Multiplying and collecting like terms, we obtain

$$f(x) = (1)\left(1 - \frac{x^2}{2} + \frac{x^4}{24} - \cdots \right) + x\left(1 - \frac{x^2}{2} + \frac{x^4}{24} - \cdots \right)$$

$$+ \frac{x^2}{2}\left(1 - \frac{x^2}{2} + \frac{x^4}{24} - \cdots \right) + \frac{x^3}{6}\left(1 - \frac{x^2}{2} + \cdots \right) + \cdots$$

$$= 1 - \frac{x^2}{2} + \frac{x^4}{24} - \cdots + x - \frac{x^3}{2} + \frac{x^5}{24} - \cdots + \frac{x^2}{2} - \frac{x^4}{4}$$

$$+ \frac{x^6}{48} - \cdots + \frac{x^3}{6} - \frac{x^5}{12} + \cdots$$

$$= 1 + x - \frac{x^3}{3} + \cdots$$

■

EXAMPLE 12 Find the first three terms of the Maclaurin series representation for $f(x) = \tan x$.

Solution Using Formulas (3) and (4) in Table 1, we have

$$f(x) = \tan x = \frac{\sin x}{\cos x} = \frac{x - \dfrac{x^3}{3!} + \dfrac{x^5}{5!} - \cdots}{1 - \dfrac{x^2}{2!} + \dfrac{x^4}{4!} - \cdots}$$

By long division we find

$$\begin{array}{r} x + \frac{1}{3}x^3 + \frac{2}{15}x^5 + \cdots \\ 1 - \frac{1}{2}x^2 + \frac{1}{24}x^4 - \cdots \overline{\smash{\big)}\, x - \frac{1}{6}x^3 + \frac{1}{120}x^5 - \cdots} \\ \underline{x - \frac{1}{2}x^3 + \frac{1}{24}x^5 - \cdots} \\ \frac{1}{3}x^3 - \frac{1}{30}x^5 + \cdots \\ \underline{\frac{1}{3}x^3 - \frac{1}{6}x^5 + \cdots} \\ \frac{2}{15}x^5 + \cdots \end{array}$$

Therefore,

$$f(x) = \tan x = x + \frac{1}{3}x^3 + \frac{2}{15}x^5 + \cdots$$

■

In both Examples 11 and 12 we computed only the first three terms of each series. In practice, the retention of just the first few terms of a series is sufficient to obtain an acceptable approximation to the solution of a problem.

We can also use Taylor series to integrate functions whose antiderivatives cannot be found in terms of elementary functions (see page 814). Examples of such functions are e^{-x^2} and $\sin x^2$. In particular, the use of Taylor series enables us to obtain approximations to definite integrals involving such functions, as illustrated in the following example.

EXAMPLE 13

a. Find $\displaystyle\int e^{-x^2}\, dx$.

b. Find an approximation of $\displaystyle\int_0^{0.5} e^{-x^2}\, dx$ accurate to four decimal places.

Solution
a. Replacing x in Formula (2) in Table 1 by $-x^2$ gives

$$e^{-x^2} = 1 - x^2 + \frac{x^4}{2!} - \frac{x^6}{3!} + \cdots = \sum_{n=0}^{\infty} (-1)^n \frac{x^{2n}}{n!}$$

Integrating both sides of this equation with respect to x, we obtain, by Theorem 2,

$$\int e^{-x^2}\, dx = \int \left(1 - x^2 + \frac{x^4}{2!} - \frac{x^6}{3!} + \cdots\right) dx$$

$$= C + x - \frac{1}{3}x^3 + \frac{1}{5\cdot 2!}x^5 - \frac{1}{7\cdot 3!}x^7 + \cdots$$

$$= C + \sum_{n=0}^{\infty} (-1)^n \frac{1}{(2n+1)\cdot n!} x^{2n+1}$$

Since the power series representation of e^{-x^2} converges for x in $(-\infty, \infty)$, this result is valid for all values of x.

b. Using the result from part (a), we obtain

$$\int_0^{0.5} e^{-x^2}\, dx = \left[x - \frac{1}{3}x^3 + \frac{1}{5\cdot 2!}x^5 - \frac{1}{7\cdot 3!}x^7 + \frac{1}{9\cdot 4!}x^9 - \frac{1}{11\cdot 5!}x^{11} + \cdots\right]_0^{1/2}$$

$$= \frac{1}{2} - \frac{1}{3}\left(\frac{1}{2}\right)^3 + \frac{1}{5\cdot 2!}\left(\frac{1}{2}\right)^5 - \frac{1}{7\cdot 3!}\left(\frac{1}{2}\right)^7 + \frac{1}{9\cdot 4!}\left(\frac{1}{2}\right)^9 - \cdots$$

$$= \frac{1}{2} - \frac{1}{24} + \frac{1}{320} - \frac{1}{5376} + \frac{1}{110592} - \cdots$$

$$\approx 0.4613$$

Since this series is alternating and its terms decrease to 0, we know, by Theorem 2 of Section 8.5, that the error incurred in the approximation does not exceed

$$\frac{1}{9\cdot 4!}\left(\frac{1}{2}\right)^9 = \frac{1}{110592} \approx 0.000009 < 0.00005$$

So the result is accurate to within four decimal places, as desired. ■

EXAMPLE 14 Find a power series representation for $\sin^{-1} x$.

Solution Observe that

$$\sin^{-1} x = \int_0^x \frac{1}{\sqrt{1 - t^2}} \, dt$$

Using Equation (3) with $k = -\frac{1}{2}$ and $x = -t^2$, we have

$$\frac{1}{\sqrt{1 - t^2}} = (1 - t^2)^{-1/2} = 1 + \left(-\frac{1}{2} \right)(-t^2) + \frac{-\frac{1}{2}\left(-\frac{1}{2} - 1 \right)}{2!}(-t^2)^2 + \cdots$$

$$+ \frac{-\frac{1}{2}\left(-\frac{1}{2} - 1 \right) \cdots \left(-\frac{1}{2} - n + 1 \right)}{n!}(-t^2)^n + \cdots$$

$$= 1 + \frac{1}{2}t^2 + \frac{1 \cdot 3}{2! \, 2^2}t^4 + \cdots + \frac{1 \cdot 3 \cdot 5 \cdot \, \cdots \, (2n - 1)}{n! \, 2^n}t^{2n} + \cdots$$

Therefore,

$$\sin^{-1} x = \int_0^x \frac{1}{\sqrt{1 - t^2}} \, dt = x + \frac{1}{2 \cdot 3}x^3 + \frac{1 \cdot 3}{2! \, 2^2 \cdot 5}x^5 + \cdots$$

$$= x + \sum_{n=1}^{\infty} \frac{1 \cdot 3 \cdot 5 \cdot \, \cdots \, (2n - 1)}{2 \cdot 4 \cdot 6 \cdot \, \cdots \, (2n)} \cdot \frac{x^{2n+1}}{2n + 1}$$

$$= \sum_{n=0}^{\infty} \frac{(2n)! \, x^{2n+1}}{(2^n \, n!)^2 (2n + 1)}$$

It can be shown that the series converges in $[-1, 1]$. ◼

EXAMPLE 15 **Einstein's Special Theory of Relativity**   According to Einstein's special theory of relativity, a body of mass m_0 at rest has a *rest energy* $E_0 = m_0 c^2$ due to the *mass* itself. (Here, the constant c denotes the speed of light.) The same body, moving at a speed v, has *total energy*

$$E = \frac{m_0 c^2}{\sqrt{1 - \dfrac{v^2}{c^2}}}$$

The kinetic energy, the energy of motion, is the difference between the total energy and the rest energy and is therefore given by

$$K = E - E_0 = \frac{m_0 c^2}{\sqrt{1 - \dfrac{v^2}{c^2}}} - m_0 c^2$$

Show that if v is very small in comparison to c, the kinetic energy of the body assumes the classical form $K = \frac{1}{2} m_0 v^2$.

Solution We have

$$K = \frac{m_0 c^2}{\sqrt{1 - \dfrac{v^2}{c^2}}} - m_0 c^2 = m_0 c^2 \left[\left(1 - \frac{v^2}{c^2} \right)^{-1/2} - 1 \right]$$

Using Equation (3) with $k = -\frac{1}{2}$ and $x = -v^2/c^2$, we obtain

$$\left(1 - \frac{v^2}{c^2}\right)^{-1/2} = 1 - \frac{1}{2}\left(-\frac{v^2}{c^2}\right) + \frac{\left(-\frac{1}{2}\right)\left(-\frac{1}{2} - 1\right)}{2!}\left(-\frac{v^2}{c^2}\right)^2 + \cdots$$

$$= 1 + \frac{1}{2} \cdot \frac{v^2}{c^2} + \frac{3}{8} \cdot \frac{v^4}{c^4} + \cdots$$

Therefore, the kinetic energy is

$$K = m_0 c^2\left[\left(1 + \frac{1}{2} \cdot \frac{v^2}{c^2} + \frac{3}{8} \cdot \frac{v^4}{c^4} + \cdots\right) - 1\right]$$

$$= m_0 c^2\left(\frac{1}{2} \cdot \frac{v^2}{c^2} + \frac{3}{8} \cdot \frac{v^4}{c^4} + \cdots\right)$$

For speeds much less than the speed of light (v much smaller than c), all the terms after the first are very small in comparison to the first and may be neglected, leading to

$$K = \frac{1}{2} m_0 v^2$$

the classical expression for kinetic energy.

8.8 CONCEPT QUESTIONS

1. **a.** What is a Taylor series? What is a Maclaurin series?
   b. What is the difference between a Taylor series and a Maclaurin series?
2. **a.** Suppose $f(x) = \sum_{n=0}^{\infty} a_n(x - c)^n$ for x in $(-R, R)$, where $R > 0$. What is $f^{(n)}(c)$?
   b. The Taylor series of $f(x)$ at $x = 1$ is
   $$\sum_{n=0}^{\infty} (-1)^n \frac{(x - 1)^{n+1}}{n + 1}$$
   What is $f^{(5)}(1)$?
3. **a.** What is a binomial series?
   b. What is the nth term of a binomial series?
   c. What is the radius of convergence of a binomial series if the exponent is a nonnegative integer?

4. **a.** Consider the function $f(x) = (1 + x)^k$, where k is negative. What can you say about $f(-1)$?
   b. Consider the function $f(x) = (1 + x)^k$, where k is positive but not an integer. What can you say about the derivative or higher derivatives of f at -1? Illustrate with an example.
   c. Use the results of parts (a) and (b) to explain why we can only assert, in general, that the binomial series converges for $|x| < 1$.

8.8 EXERCISES

Note: *In this exercise set, assume that all the functions have power series representations.*

In Exercises 1–10, *use Equation (1) to find the Taylor series of f at the given value of c. Then find the radius of convergence of the series.*

1. $f(x) = e^{2x}$, $c = 0$
2. $f(x) = e^{-3x}$, $c = 0$
3. $f(x) = e^x$, $c = 2$
4. $f(x) = e^{-2x}$, $c = 3$
5. $f(x) = \sin 2x$, $c = 0$
6. $f(x) = \sin x$, $c = \frac{\pi}{4}$
7. $f(x) = \cos x$, $c = -\frac{\pi}{6}$
8. $f(x) = \frac{1}{x}$, $c = -1$
9. $f(x) = \ln x$, $c = 2$
10. $f(x) = \sinh x$, $c = 0$

In Exercises 11–28, use the power series representations of functions established in this section to find the Taylor series of f at the given value of c. Then find the radius of convergence of the series.

11. $f(x) = \dfrac{1}{1+x}$, $c = 1$

12. $f(x) = \dfrac{1}{1+x}$, $c = -2$

13. $f(x) = \dfrac{1}{1-2x}$, $c = 1$

14. $f(x) = \dfrac{1}{1+3x}$, $c = 2$

15. $f(x) = \dfrac{x^2}{x^2-1}$, $c = 0$

16. $f(x) = \dfrac{1}{4+x^2}$, $c = 0$

17. $f(x) = xe^{-x}$, $c = 0$

18. $f(x) = e^{2x}$, $c = -1$

19. $f(x) = x^2 \cos x$, $c = 0$

20. $f(x) = x \cos 3x$, $c = 0$

21. $f(x) = \cos^2 x$, $c = 0$

  Hint: $\cos^2 x = \dfrac{1}{2}(1 + \cos 2x)$

22. $f(x) = \sin^2 x$, $c = 0$

  Hint: $\sin^2 x = \dfrac{1}{2}(1 - \cos 2x)$

23. $f(x) = \sin x$, $c = \dfrac{\pi}{3}$

24. $f(x) = \cos x$, $c = \dfrac{\pi}{6}$

25. $f(x) = \sqrt{x} \sin^{-1} x$, $c = 0$

26. $f(x) = (1 + x^2)\tan^{-1} x$, $c = 0$

27. $f(x) = \ln(1 + x^2)$, $c = 0$

28. $f(x) = \ln\left(\dfrac{1+x}{1-x}\right)$, $c = 0$

In Exercises 29–34, use the binomial series to find the power series representation of the function. Then find the radius of convergence of the series.

29. $f(x) = \dfrac{1}{(1+x)^2}$

30. $f(x) = \sqrt[3]{1+x}$

31. $f(x) = \sqrt{1-x^2}$

32. $f(x) = \dfrac{1}{\sqrt[3]{8+x}}$

33. $f(x) = (1-x)^{3/5}$

34. $f(x) = \dfrac{x}{(1+x)^2}$

In Exercises 35–40, find the first three terms of the Taylor series of f at the given value of c.

35. $f(x) = \tan x$, $c = \dfrac{\pi}{4}$

36. $f(x) = \sec x$, $c = 0$

37. $f(x) = \sin^{-1} x$, $c = \dfrac{1}{2}$

38. $f(x) = \tan^{-1} x$, $c = 1$

39. $f(x) = e^{-x} \sin x$, $c = 0$

40. $f(x) = e^x \tan x$, $c = 0$

In Exercises 41 and 42, (a) find the power series representation for the function; (b) write the first three partial sums P_1, P_2, and P_3; and (c) plot the graphs of f and P_1, P_2, and P_3 using a viewing window that includes the interval of convergence of the power series.

41. $f(x) = \sqrt[3]{1+x}$

42. $f(x) = \dfrac{1}{\sqrt{9-x}}$

43. Use the Maclaurin series for e^{-x^2} to calculate $e^{-0.01}$ accurate to five decimal places.

44. Use the Maclaurin series for $\cos x$ to calculate $\cos 3°$ accurate to five decimal places.

In Exercises 45–50, find a power series representation for the indefinite integral.

45. $\displaystyle\int \dfrac{1}{1+x^3}\, dx$

46. $\displaystyle\int e^{-\sqrt{x}}\, dx$

47. $\displaystyle\int \sin x^2\, dx$

48. $\displaystyle\int x \tan^{-1} x\, dx$

49. $\displaystyle\int \dfrac{\ln(1+x)}{x}\, dx$

50. $\displaystyle\int \dfrac{\sin x}{x}\, dx$

In Exercises 51–56, use a power series to obtain an approximation of the definite integral to four decimal places of accuracy.

51. $\displaystyle\int_0^1 e^{-x^2}\, dx$

52. $\displaystyle\int_0^{0.5} x^2 e^{-x^2}\, dx$

53. $\displaystyle\int_0^{0.5} \cos x^2\, dx$

54. $\displaystyle\int_0^1 \sin x^2\, dx$

55. $\displaystyle\int_0^{0.5} x \cos x^3\, dx$

56. $\displaystyle\int_0^{0.5} \tan^{-1} x^3\, dx$

In Exercises 57–62, find the sum of the given series. (Hint: Each series is the Maclaurin series of a function evaluated at an appropriate point.)

57. $\displaystyle\sum_{n=1}^{\infty} (-1)^{n-1}\dfrac{1}{n}$

58. $\displaystyle\sum_{n=0}^{\infty} \dfrac{(-1)^n}{n!\, 2^n}$

59. $\displaystyle\sum_{n=0}^{\infty} (-1)^n \dfrac{\pi^{2n}}{(2n)!}$

60. $\displaystyle\sum_{n=0}^{\infty} (-1)^n \dfrac{\pi^{2n+1}}{(2n+1)!\, 2^{2n+1}}$

61. $\displaystyle\sum_{n=1}^{\infty} (-1)^{n-1} \frac{1}{n2^n}$

62. $\displaystyle\sum_{n=0}^{\infty} \frac{(-1)^n}{2n+1}$

63. Evaluate $\displaystyle\lim_{x\to 0} \frac{\sin x - x + \frac{1}{6}x^3}{x^5}$.

Hint: Use the Maclaurin series representation of $\sin x$.

64. Evaluate $\displaystyle\lim_{x\to 0} \frac{\cos x^2 - 1 + \frac{1}{2}x^4}{x^8}$.

Hint: Use the Maclaurin series representation of $\cos x^2$.

65. Evaluate $\displaystyle\lim_{x\to 0} \frac{\tan x - x - \frac{1}{3}x^3}{x^5}$.

Hint: Use the result of Example 12.

66. Evaluate $\displaystyle\lim_{x\to 1} \frac{\ln x}{x-1}$.

Hint: Use the Taylor series representation of $\ln x$ at 1.

67. a. Find the power series representation for $\dfrac{1}{\sqrt{1-u^2}}$.

b. Use the result of part (a) to find a power series representation of

$$\sin^{-1} x = \int_0^x \frac{1}{\sqrt{1-t^2}} \, dt$$

What is the radius of convergence of the series?

68. a. Find a power series representation of $f(x) = \sqrt[3]{1+x^2}$.

b. Use the result of part (a) to find $f^{(6)}(0)$.

69. Force Exerted by a Charge Distribution Suppose that a charge Q is distributed uniformly along the positive x-axis from $x = 0$ to $x = a$ and that a negative charge $-Q$ is distributed uniformly along the negative x-axis from $x = -a$ to $x = 0$. If a positive charge q is placed on the positive x-axis a distance of x units ($x > a$) from the origin, then the force exerted by the charge distribution on q has magnitude

$$F = \frac{qQ}{4\pi\varepsilon_0 a}\left[\frac{1}{x-a} + \frac{1}{x+a} - \frac{2}{x}\right]$$

and direction along the positive x-axis.

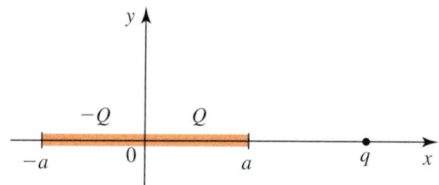

Show that if x is large, then

$$F \approx \frac{qQa}{2\pi\varepsilon_0 x^3}$$

70. Speed of a Wave A wave of water of length L travels across a body of water of depth d, as illustrated in the figure below. The speed of the wave is given by

$$v = \left(\frac{gL}{2\pi}\tanh\frac{2\pi d}{L}\right)^{1/2}$$

where g is the constant of acceleration due to gravity.

a. Show that in deep water, $v \approx \sqrt{gL/(2\pi)}$, and hence the speed of the wave is independent of the depth of the body of water.

b. Show that in shallow water, $v \approx \sqrt{gd}$, so the speed of the wave is independent of the length of the wave.

Hint: Find the first three nonzero terms of the Maclaurin series for $f(x) = \tanh x$.

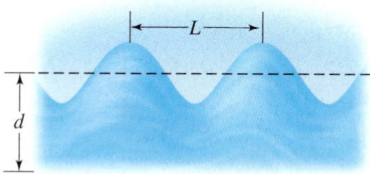

71. Gravitational Force Between Two Masses Suppose that a mass M is distributed uniformly over a disk of radius a. Then it can be shown that the attractive gravitational force between the disk-shaped mass and a point mass m located a distance of x units above the center of the disk has magnitude

$$F = \frac{2GmM}{a^2}\left[1 - \frac{x}{\sqrt{x^2+a^2}}\right]$$

Here, g is the gravitational constant. Show that if x is large, then

$$F \approx \frac{GmM}{x^2}$$

Thus, from this distance the disk "looks" like a point mass.

72. Force of Attraction of a Cylinder on a Point It can be shown that the magnitude of the force of attraction of a homogeneous right-circular cylinder upon a point P on its axis is

$$F = 2\pi\sigma\left[h + \sqrt{R^2 + a^2} - \sqrt{(R+h)^2 + a^2}\right]$$

where h and a are the height and radius of the cylinder, R is the distance between P and the top of the cylinder, and σ is the (constant) density of the solid. Show that if R is large in comparison to a and h, then

$$F \approx \frac{M}{R^2}$$

where $M = \pi a^2 h \sigma$ is the mass of the cylinder.

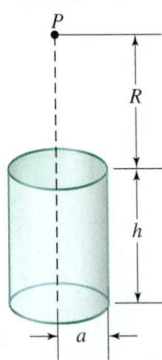

73. Volume of Water in a Trough A trough of length L feet has a cross section in the shape of a semicircle with radius r feet. When the trough is filled with water to a level that is h feet as measured from the top of the trough, the volume of the water is

$$V = L\left[\frac{1}{2}\pi r^2 - r^2 \sin^{-1}\left(\frac{h}{r}\right) - h\sqrt{r^2 - h^2}\right]$$

Show that if h is small in comparison to r (that is, h/r is small), then

$$V \approx L\left[\frac{1}{2}\pi r^2 - 2rh + \frac{1}{2}\cdot\frac{h^3}{r}\right]$$

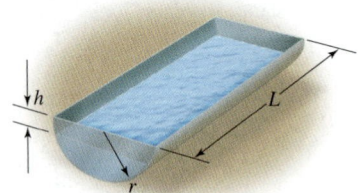

74. Formula (5) in Table 1 can be used to compute the value of $\ln x$ for $-1 < x \le 1$. However, the restriction on x and the slow convergence of the series limit its effectiveness from the computational point of view. A more effective formula, first obtained by the Scottish mathematician James Gregory (1638–1675), follows.
a. Use Formula (5) to show that

$$\ln\left(\frac{1+x}{1-x}\right) = 2\left(x + \frac{x^3}{3} + \frac{x^5}{5} + \frac{x^7}{7} + \cdots\right) \qquad -1 < x < 1$$

b. To compute the natural logarithm of a positive number p, let $p = (1 + x)/(1 - x)$ and show that

$$x = \frac{p-1}{p+1} \qquad -1 < x < 1$$

c. Use parts (a) and (b) to find $\ln 2$ accurate to four decimal places.

75. Let f be the function defined by

$$f(x) = \begin{cases} e^{-1/x^2} & \text{if } x \ne 0 \\ 0 & \text{if } x = 0 \end{cases}$$

Show that f cannot be represented by a Maclaurin series.

76. a. Find the Taylor series for $f(x) = 2x^3 + 3x^2 + 1$ at $x = 1$.
b. Show that the Taylor series and $f(x)$ are equal.
c. What can you say about a Taylor series for a polynomial function? Justify your answer.

77. Show that $(1 + x)^n > 1 + nx$ for all $x > 0$ and $n > 1$.

78. Prove that $(1 + x)^k = \displaystyle\sum_{n=0}^{\infty} \binom{k}{n} x^n = \sum_{n=0}^{\infty} \frac{k!}{n!(k-n)!} x^n$, where k is any real number and $|x| < 1$, by verifying the following steps.
a. Let

$$f(x) = \sum_{n=0}^{\infty} \binom{k}{n} x^n = 1 + kx + \frac{k(k-1)}{2!} x^2 + \cdots$$
$$+ \frac{k(k-1)\cdots(k-n+1)}{n!} x^n + \cdots$$

Differentiate the equation with respect to x to show that

$$f'(x)(1 + x) - kf(x) = 0$$

b. Define the function g by $g(x) = f(x)/(1 + x)^k$ and show that $g'(x) = 0$.
c. Deduce that $f(x) = (1 + x)^k$.

In Exercises 79–84, determine whether the statement is true or false. If it is true, explain why it is true. If it is false, explain why or give an example to show why it is false.

79. If $P(x)$ is a polynomial function of degree n, then the Maclaurin series for P is P.

80. Suppose that $f(x) = \sum_{n=0}^{\infty} a_n x^n$ for x in $(-R, R)$, where $R > 0$ and f is odd. Then $a_{2n} = 0$ for $n \ge 0$.

81. The function $f(x) = x^{5/3}$ has a Maclaurin series.

82. The Taylor series of $f(x) = (1 - x)^7$ at $x = 1$ is $(x - 1)^7$.

83. The Maclaurin series for $f(x) = (2 + x)^k$ is $\displaystyle\sum_{n=0}^{\infty} \binom{k}{n} 2^{k-n} x^n$.

84. If k is a positive integer, then the Maclaurin series for $f(x) = (1 + x)^k$ is a polynomial of degree k.

In Section 8.8 we saw how Maclaurin and Taylor series of functions can be used to help us find the values of the functions they represent. We also saw how we can use these series to find the antiderivatives as well as the values of definite integrals of functions that we could not otherwise evaluate. You will recall that in each instance we were able to obtain satisfactory approximations to the actual values of these quantities by retaining just the first few terms of the series. These truncated series—the nth partial sums of the power series representations of the functions—are polynomials. The nth partial sum of the Taylor series of f centered at c,

$$P_n(x) = \sum_{k=0}^{n} \frac{f^{(k)}(c)}{k!}(x - c)^k$$

$$= f(c) + \frac{f'(c)}{1!}(x - c) + \frac{f''(c)}{2!}(x - c)^2 + \cdots + \frac{f^{(n)}(c)}{n!}(x - c)^n$$

(1)

is called the **nth-degree Taylor polynomial of f at c.** If $c = 0$, we have the **nth-degree Maclaurin polynomial of f.**

The accuracy with which the Taylor polynomials approximate a function f in a neighborhood of c is demonstrated graphically in Figure 1. Here, the function $f(x) = e^x$ with Maclaurin series

$$f(x) = 1 + x + \frac{x^2}{2!} + \frac{x^3}{3!} + \cdots + \frac{x^n}{n!} + \cdots$$

is approximated by the Maclaurin polynomials of degrees 1, 2, and 3:

$$P_1(x) = 1 + x, \quad P_2(x) = 1 + x + \frac{1}{2}x^2, \quad \text{and} \quad P_3(x) = 1 + x + \frac{1}{2}x^2 + \frac{1}{6}x^3$$

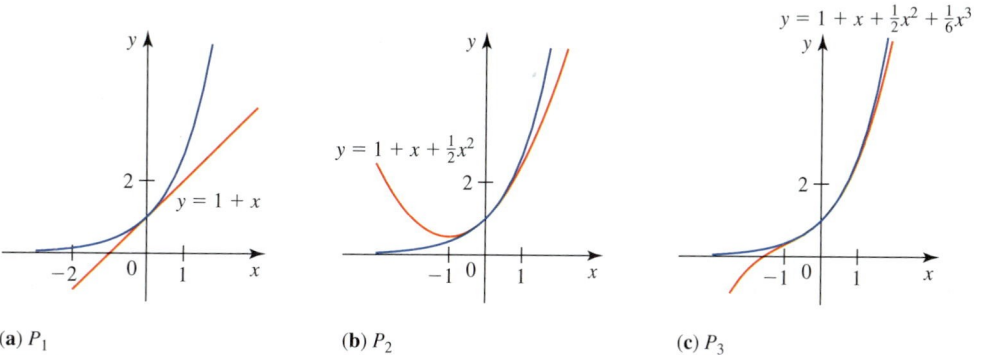

(a) P_1 (b) P_2 (c) P_3

FIGURE 1

As n increases, $P_n(x)$ gives a better and better approximation of $f(x)$ in a neighborhood of $x = 0$.

Observe that the graph of

$$P_1(x) = 1 + x$$

is a straight line that is tangent to the graph of f at $(0, 1)$. [$P_1(0) = f(0)$ and $P_1'(0) = f'(0)$]. The graph of

$$P_2(x) = 1 + x + \frac{1}{2}x^2$$

is a parabola that passes through $(0, 1)$ $[P_2(0) = f(0)]$, has a tangent line that coincides with that of f at $(0, 1)$ $[P_2'(0) = f'(0)]$, and has concavity that matches that of the graph of f at $(0, 1)$ $[P_2''(0) = f''(0)]$. The graph of

$$P_3(x) = 1 + x + \frac{1}{2}x^2 + \frac{1}{6}x^3$$

provides an even better approximation to the graph of f than that of $P_2(x)$ near $(0, 1)$. Not only does it have the same tangent line and concavity as that of f at $(0, 1)$ $[P_3'(0) = f'(0)$ and $P_3''(0) = f''(0)]$, but both P_3 and f satisfy the condition $P_3'''(0) = f'''(0)$.

In general, you can show that if P_n is the nth-degree Taylor polynomial of f at c, then the derivatives of P_n at c agree with the derivatives of f at c up to and including those of order n (see Exercise 48). This explains why the graph of P_n more closely conforms to the graph of f near $x = c$ as n gets larger and larger.

Figure 2 illustrates how the Maclaurin polynomials P_2, P_4, P_6, and P_8 of $f(x) = \cos x$ approximate with increasing accuracy the function f. For fixed n the accuracy in the approximation decreases as we move away from the center c.

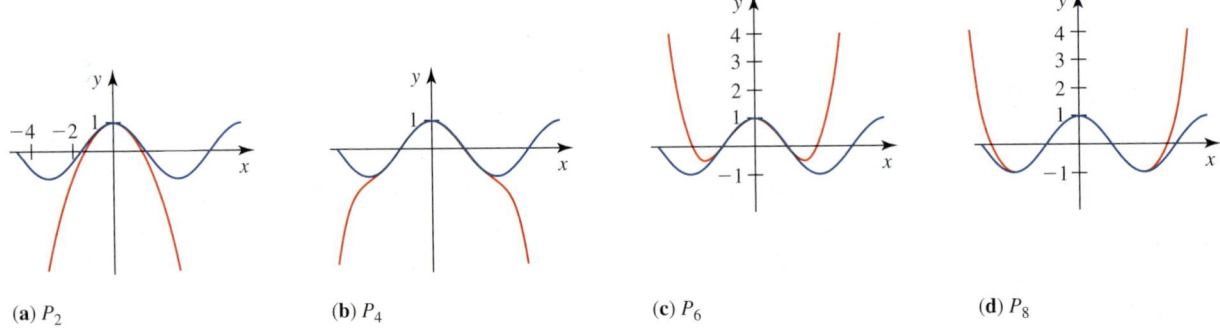

(a) P_2 (b) P_4 (c) P_6 (d) P_8

FIGURE 2
As n increases, $P_{2n}(x)$ approximates $f(x)$ with greater and greater accuracy.

To obtain the same degree of accuracy for x farther away from the center, we need to use an approximating polynomial of higher degree. Figure 3 shows the approximation of $f(x) = \cos x$ using a Maclaurin polynomial $P_{24}(x)$ of degree 24.

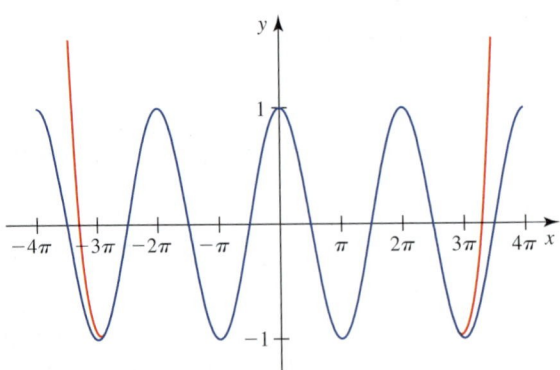

FIGURE 3
Approximating $f(x) = \cos x$ using a Maclaurin polynomial of degree 24

■ Taylor's Formula with Remainder

Two important questions arise when a function f is approximated by a Taylor polynomial P_n:

1. How good is the approximation?
2. How large should n be taken to ensure that a specified degree of accuracy is obtained?

To answer these questions, we need the following theorem, which gives the relationship between f and P_n.

THEOREM 1 **Taylor's Theorem**

If f has derivatives up to order $n + 1$ in an interval I containing c, then for each x in I, there exists a number z between x and c such that

$$f(x) = f(c) + f'(c)(x - c) + \frac{f''(c)}{2!}(x - c)^2 + \cdots + \frac{f^{(n)}(c)}{n!}(x - c)^n + R_n(x)$$

$$= P_n(x) + R_n(x)$$

where

$$R_n(x) = \frac{f^{(n+1)}(z)}{(n+1)!}(x - c)^{n+1} \tag{2}$$

PROOF Let x be any point in I that is different from c and define

$$R_n(x) = f(x) - P_n(x)$$

where $P_n(x)$ is the nth-degree Taylor polynomial of f at c. For any point t in I, define the function g by

$$g(t) = f(x) - f(t) - f'(t)(x - t) - \cdots - \frac{f^{(n)}(t)}{n!}(x - t)^n - R_n(x)\frac{(x - t)^{n+1}}{(x - c)^{n+1}}$$

If we differentiate both sides of this equation with respect to t, then the expression on the right side of the resulting equation will be a telescoping finite series (to see this, just write out the first several terms). Canceling like terms, we obtain the following expression for $g'(t)$:

$$g'(t) = -\frac{f^{(n+1)}(t)}{n!}(x - t)^n + (n + 1)R_n(x)\frac{(x - t)^n}{(x - c)^{n+1}}$$

We now apply Rolle's Theorem to the function g defined on the interval $[c, x]$ or $[x, c]$, depending on whether $c < x$ or $c > x$. In either case we see that $g(x) = 0$. Furthermore,

$$g(c) = f(x) - f(c) - f'(c)(x - c) - \cdots - \frac{f^{(n)}(c)}{n!}(x - c)^n - R_n(x)\frac{(x - c)^{n+1}}{(x - c)^{n+1}}$$

$$= f(x) - P_n(x) - R_n(x)$$

$$= f(x) - P_n(x) - [f(x) - P_n(x)] \qquad \text{By the definition of } R_n(x)$$

$$= 0$$

Therefore, g satisfies the conditions of Rolle's Theorem, so there exists a number z between c and x such that $g'(z) = 0$. Using the expression for g' obtained earlier, we have

$$g'(z) = -\frac{f^{(n+1)}(z)}{n!}(x-z)^n + (n+1)R_n(x)\frac{(x-z)^n}{(x-c)^{n+1}} = 0$$

Solving for $R_n(x)$, we obtain

$$R_n(x) = \frac{f^{(n+1)}(z)}{(n+1)!}(x-c)^{n+1}$$

Finally, since $g(c) = 0$, we have

$$0 = f(x) - f(c) - f'(c)(x-c) - \cdots - \frac{f^{(n)}(c)}{n!}(x-c)^n - R_n(x)$$

or

$$f(x) = f(c) + f'(c)(x-c) + \cdots + \frac{f^{(n)}(c)}{n!}(x-c)^n + R_n(x) \qquad \blacksquare$$

The expression $R_n(x)$ is called the **Taylor remainder of f at c.** If $c = 0$, $R_n(x)$ is called the **Maclaurin remainder of f.** We can regard

$$R_n(x) = f(x) - P_n(x)$$

as the error that is incurred when $f(x)$ is approximated by the nth-degree Taylor polynomial of f at c. Since we usually don't know the value of z in Equation (2)—all we know is that it lies between x and c—we often use Equation (2) to find a *bound* for the error in the approximation rather than attempting to find the actual error itself. Incidentally, the presence of the factor $(x-c)^{n+1}$ in Equation (2) explains why (for fixed n) $P_n(x)$ gives a better approximation when x is closer to c.

EXAMPLE 1   Let $f(x) = \ln x$.

a. Find the fourth-degree Taylor polynomial of f at $c = 1$, and use it to approximate $\ln 1.1$.

b. Estimate the accuracy of the approximation that you obtained in part (a).

Solution
a. The first five derivatives of $f(x) = \ln x$ are

$$f'(x) = \frac{1}{x}, \ f''(x) = -\frac{1}{x^2}, \ f'''(x) = \frac{2}{x^3}, \ f^{(4)}(x) = -\frac{3!}{x^4}, \ \text{and} \ f^{(5)}(x) = \frac{4!}{x^5}$$

and the values of $f(x)$ and its first four derivatives at $x = 1$ are

$$f(1) = 0, \ f'(1) = 1, \ f''(1) = -1, \ f'''(1) = 2, \ \text{and} \ f^{(4)}(1) = -3!$$

Using Equation (1) with $n = 4$ and $c = 1$, we obtain

$$P_4(x) = f(1) + f'(1)(x-1) + \frac{f''(1)}{2!}(x-1)^2 + \frac{f'''(1)}{3!}(x-1)^3 + \frac{f^{(4)}(1)}{4!}(x-1)^4$$

$$= (x-1) - \frac{1}{2}(x-1)^2 + \frac{1}{3}(x-1)^3 - \frac{1}{4}(x-1)^4$$

Replacing x by 1.1 then gives the required approximation

$$\ln 1.1 \approx 0.1 - \frac{1}{2}(0.1)^2 + \frac{1}{3}(0.1)^3 - \frac{1}{4}(0.1)^4$$

$$\approx 0.09530833$$

b. The error in the approximation is found by using Equation (2) with $n = 4$, $c = 1$, and $x = 1.1$. Thus,

$$R_4(1.1) = \frac{f^{(5)}(z)}{5!}(1.1 - 1)^5 = \frac{(0.1)^5}{5z^5}$$

where $1 < z < 1.1$. The largest possible value of $R_4(1.1)$ is obtained when $z = 1$. (The denominator of $R_4(1.1)$ is smallest for this value of z in the interval $[1, 1.1]$.) Therefore,

$$R_4(1.1) < \frac{(0.1)^5}{5} = 0.000002$$

so the error in the approximation is less than 0.000002.

Alternative Solution By replacing x by $x - 1$ in Formula (5) in Section 8.8, we obtain the following power series representation of $f(x)$:

$$\ln x = (x - 1) - \frac{1}{2}(x - 1)^2 + \frac{1}{3}(x - 1)^3 - \frac{1}{4}(x - 4)^4 + \frac{1}{5}(x - 1)^5 - \cdots \qquad 0 < x \le 2$$

Therefore,

$$\ln 1.1 = 0.1 - \frac{1}{2}(0.1)^2 + \frac{1}{3}(0.1)^3 - \frac{1}{4}(0.1)^4 + \frac{1}{5}(0.1)^5 - \cdots$$

If we use just the first four terms of the series on the right to approximate $\ln 1.1$, we obtain the approximation of $\ln 1.1$ by $P_4(1.1)$. Next, since the series is an *alternating series* with terms decreasing to 0, the error in this approximation is no larger than $\frac{1}{5}(0.1)^5$, the first term that is omitted—that is, no larger than 0.000002, which is in agreement with the result that we obtained earlier. ◼

EXAMPLE 2 Let $f(x) = \sqrt{x}$.

a. Find the Taylor polynomial $P_2(x)$ of degree 2 at $c = 4$.
b. What is the maximum error incurred if f is approximated by $P_2(x)$ on the interval $[3, 5]$?

Solution
a. The first two derivatives of $f(x) = \sqrt{x}$ are

$$f'(x) = \frac{1}{2}x^{-1/2} \qquad \text{and} \qquad f''(x) = -\frac{1}{4}x^{-3/2}$$

and the values of $f(x)$ and its first two derivatives at $x = 4$ are

$$f(4) = 2, \qquad f'(4) = \frac{1}{4}, \qquad \text{and} \qquad f''(4) = -\frac{1}{32}$$

Therefore, the required Taylor polynomial is

$$P_2(x) = f(4) + f'(4)(x - 4) + \frac{f''(4)}{2!}(x - 4)^2$$

$$= 2 + \frac{1}{4}(x - 4) - \frac{1}{64}(x - 4)^2$$

b. The Taylor remainder is

$$R_2(x) = \frac{f'''(z)}{3!}(x - 4)^3$$

where z lies between 4 and x. But

$$f'''(x) = \frac{3}{8}x^{-5/2}$$

so

$$R_2(x) = \frac{3}{8}z^{-5/2}\frac{(x - 4)^3}{3!} = \frac{(x - 4)^3}{16z^{5/2}}$$

Now, if x lies in the interval $[3, 5]$, then $3 \le x \le 5$, so $-1 \le x - 4 \le 1$, or $|x - 4| \le 1$. Furthermore, since $z > 3$, we see that

$$z^{5/2} > 3^{5/2} > 15$$

so a bound on the error incurred in approximating f by P_2 on the interval $[3, 5]$ is

$$|R_2(x)| = \frac{|x - 4|^3}{16z^{5/2}} < \frac{1}{16 \cdot 15} < 0.0042 \qquad \blacksquare$$

EXAMPLE 3 Determine the degree of the Maclaurin polynomial of $f(x) = e^x$ that allows us to find the value of $\sqrt{e}$ to within an accuracy of 0.0001. Then use the polynomial to obtain the approximation.

Solution We are required to estimate the value of $\sqrt{e} = e^{1/2} = f\left(\frac{1}{2}\right)$. Since $f^{(n)}(x) = e^x$ for all n, we see that the error in approximating $f(x)$ by $P_n(x)$ is

$$R_n(x) = \frac{f^{(n+1)}(z)}{(n + 1)!}x^{n+1} = \frac{e^z}{(n + 1)!}x^{n+1}$$

where z lies between $c = 0$ and x. We are interested in approximating $e^{1/2}$, so we take $x = \frac{1}{2}$. Then, $0 < z < \frac{1}{2}$. Because $g(z) = e^z$ is an increasing function of z, we see that

$$e^z < e^{1/2} < 4^{1/2} = 2$$

Therefore,

$$R_n\left(\frac{1}{2}\right) < \frac{e^{1/2}}{(n + 1)!}\left(\frac{1}{2}\right)^{n+1} < \frac{2}{(n + 1)! \, 2^{n+1}} = \frac{1}{(n + 1)! \, 2^n}$$

Let's try $n = 4$. We obtain

$$R_4\left(\frac{1}{2}\right) < \frac{1}{5! \, 2^4} \approx 0.0005$$

Since this bound is not within the specified error bound of 0.0001, we next try $n = 5$, obtaining

$$R_5\left(\frac{1}{2}\right) < \frac{1}{6! \, 2^5} \approx 0.00004$$

This bound is less than the prescribed error bound, so we can use

$$P_5(x) = 1 + x + \frac{x^2}{2!} + \frac{x^3}{3!} + \frac{x^4}{4!} + \frac{x^5}{5!}$$

for the approximation, obtaining

$$e^{1/2} \approx P_5\left(\frac{1}{2}\right) = 1 + \frac{1}{2} + \frac{1}{2!}\left(\frac{1}{2}\right)^2 + \frac{1}{3!}\left(\frac{1}{2}\right)^3 + \frac{1}{4!}\left(\frac{1}{2}\right)^4 + \frac{1}{5!}\left(\frac{1}{2}\right)^5$$

$$\approx 1.64870 \qquad \blacksquare$$

As we saw earlier, the approximation of a function f by the Taylor polynomial P_n of f centered at c diminishes in accuracy as we move away from the center. Therefore, in approximating $f(x_0)$ by Taylor polynomials of f, it is best to pick the center c as close to x_0 as possible. This is illustrated in the following example.

EXAMPLE 4 Suppose that we want to approximate $\cos 50°$ using the second-order Taylor polynomial $P_2(x)$ of $f(x) = \cos x$ with center at $x = \pi/4$. (Note that $\pi/4$ rad $= 45°$ is close to $50°$.)

a. Find $P_2(x)$.
b. Find the maximum error in the approximation $f(x) \approx P_2(x)$ when $|x - (\pi/4)| < 0.1$.
c. Use the results of parts (a) and (b) to find $\cos 50°$. How accurate is your estimate?

Solution
a. Since

$$f(x) = \cos x, \qquad f'(x) = -\sin x, \qquad f''(x) = -\cos x, \qquad \text{and} \qquad f'''(x) = \sin x$$

we find

$$f\left(\frac{\pi}{4}\right) = \frac{1}{\sqrt{2}}, \qquad f'\left(\frac{\pi}{4}\right) = -\frac{1}{\sqrt{2}}, \qquad \text{and} \qquad f''\left(\frac{\pi}{4}\right) = -\frac{1}{\sqrt{2}}$$

Therefore, the required Taylor polynomial is

$$P_2(x) = f\left(\frac{\pi}{4}\right) + f'\left(\frac{\pi}{4}\right)\left(x - \frac{\pi}{4}\right) + \frac{1}{2}f''\left(\frac{\pi}{4}\right)\left(x - \frac{\pi}{4}\right)^2$$

$$= \frac{1}{\sqrt{2}} - \frac{1}{\sqrt{2}}\left(x - \frac{\pi}{4}\right) - \frac{1}{2\sqrt{2}}\left(x - \frac{\pi}{4}\right)^2$$

$$= \frac{1}{\sqrt{2}}\left[1 - \left(x - \frac{\pi}{4}\right) - \frac{1}{2}\left(x - \frac{\pi}{4}\right)^2\right]$$

b. The error in the approximation $f(x) \approx P_2(x)$ is

$$R_2(x) = \frac{f'''(z)}{3!}\left(x - \frac{\pi}{4}\right)^3 = \frac{\sin z}{6}\left(x - \frac{\pi}{4}\right)^3$$

where z lies between $\pi/4$ and x. Now $|\sin z| \le 1$ for any z, and if $|x - (\pi/4)| < 0.1$, then

$$|R_2(x)| = \frac{|\sin z|}{6}\left|x - \frac{\pi}{4}\right|^3 < \frac{(0.1)^3}{6} \approx 0.000167 < 0.0002$$

Therefore, the maximum error in approximating $f(x)$ by $P_2(x)$ for x satisfying $|x - (\pi/4)| < 0.1$ is less than 0.0002.

c. Since

$$5° = \frac{5\pi}{180} = \frac{\pi}{36}$$

we see that

$$50° = 45° + 5° = \frac{\pi}{4} + \frac{\pi}{36}$$

Therefore, using the result of part (a), we have

$$\cos 50° = \cos\left(\frac{\pi}{4} + \frac{\pi}{36}\right) = \frac{1}{\sqrt{2}}\left[1 - \frac{\pi}{36} - \frac{1}{2}\left(\frac{\pi}{36}\right)^2\right] \approx 0.643$$

Since $\pi/36 \approx 0.0873 < 0.1$, the results of part (b) guarantee that the error in the approximation that we just obtained is accurate to three decimal places. The true value of $\cos 50°$ is approximately 0.642787610. ■

Representing a Function by a Series

We now turn our attention to finding the conditions under which the function f has a power series representation. These conditions are spelled out in the following theorem.

THEOREM 2

Suppose that f has derivatives of all order on an interval I containing c and that $R_n(x)$ is the Taylor remainder of f at c. If

$$\lim_{n\to\infty} R_n(x) = 0$$

for every x in I, then $f(x)$ is represented by the Taylor series of f at c; that is,

$$f(x) = \sum_{n=0}^{\infty} \frac{f^{(n)}(c)}{n!}(x - c)^n$$

PROOF As was noted earlier, $P_n(x)$ is the nth partial sum of the Taylor series of f at c. By Taylor's Theorem, $P_n(x) = f(x) - R_n(x)$, and therefore,

$$\lim_{n\to\infty} P_n(x) = \lim_{n\to\infty}[f(x) - R_n(x)] = \lim_{n\to\infty} f(x) - \lim_{n\to\infty} R_n(x)$$
$$= f(x) - \lim_{n\to\infty} R_n(x) = f(x) - 0 = f(x)$$

for all x in I. Thus, the sequence of partial sums converges to $f(x)$ for each x in I, and the theorem is proved. ■

Before looking at an application of Theorem 2, we state the following result, which will be used in the solution.

THEOREM 3

If x is any real number, then

$$\lim_{n \to \infty} \frac{|x|^n}{n!} = 0$$

PROOF In Example 1 in Section 8.8 we proved that the power series $\sum_{n=0}^{\infty} x^n/n!$ is absolutely convergent for every real number x. Since the nth term of a convergent series must approach zero as n approaches infinity (Theorem 2 in Section 8.2), we conclude that

$$\lim_{n \to \infty} \frac{|x|^n}{n!} = 0$$

EXAMPLE 5 Show that the Maclaurin series $\sum_{n=0}^{\infty} x^n/n!$ of the function $f(x) = e^x$ does represent f.

Solution We use Taylor's Theorem with $c = 0$. Since $f^{(n+1)}(x) = e^x$, we see that

$$R_n(x) = \frac{f^{(n+1)}(z)}{(n+1)!} x^{n+1} = \frac{e^z}{(n+1)!} x^{n+1}$$

where z is a number between 0 and x. If $x > 0$, then $e^z < e^x$, since the function $f(x) = e^x$ is increasing. Therefore,

$$0 < R_n(x) < \frac{e^x}{(n+1)!} x^{n+1}$$

By Theorem 3,

$$\lim_{n \to \infty} \frac{e^x}{(n+1)!} x^{n+1} = e^x \lim_{n \to \infty} \frac{x^{n+1}}{(n+1)!} = 0$$

so the Squeeze Theorem implies that

$$\lim_{n \to \infty} R_n(x) = 0$$

If $x < 0$, then $z < 0$, and hence $e^z < e^0 = 1$. Therefore,

$$0 < |R_n(x)| < \left| \frac{x^{n+1}}{(n+1)!} \right| = \frac{|x|^{n+1}}{(n+1)!}$$

and once again, the Squeeze Theorem implies that

$$\lim_{n \to \infty} R_n(x) = 0$$

It follows from Theorem 2 that the Maclaurin series of $f(x) = e^x$ represents the function f for all $x \neq 0$. Finally, the series represents f at $x = 0$, since $f(0) = e^0 = 1$, and this is also the value of the sum of the series at 0.

EXAMPLE 6 Show that the Maclaurin series $\displaystyle\sum_{n=0}^{\infty}(-1)^n \frac{x^{2n+1}}{(2n+1)!}$ of the function $f(x) = \sin x$ does represent f.

Solution   Using Taylor's Theorem with $c = 0$, we have

$$R_n(x) = \frac{f^{(n+1)}(z)}{(n+1)!}x^{n+1}$$

where z is a number between 0 and x. But $f^{(n+1)}(x)$ is either $\pm\sin x$ or $\pm\cos x$ for any n ($n = 0, 1, 2, \ldots$). Therefore, $|f^{(n+1)}(z)| \le 1$, so

$$|R_n(x)| = \frac{|f^{(n+1)}(z)|}{(n+1)!}|x|^{n+1} \le \frac{|x|^{n+1}}{(n+1)!}$$

By Theorem 3,

$$\lim_{n\to\infty} \frac{|x|^{n+1}}{(n+1)!} = 0$$

so the Squeeze Theorem implies that

$$\lim_{n\to\infty} R_n(x) = 0$$

It follows from Theorem 2 that

$$f(x) = \sum_{n=0}^{\infty}(-1)^n \frac{x^{2n+1}}{(2n+1)!}$$

as was to be shown.   ■

The next example shows how a Taylor polynomial can be used to approximate an integral that involves an integrand whose antiderivative cannot be expressed as an elementary function.

EXAMPLE 7 **Growth of the Service Industries**   It has been estimated that service industries, which currently make up 30% of the nonfarm workforce in a certain country, will continue to grow at the rate of

$$R(t) = 5e^{1/(t+1)}$$

percent per decade, t decades from now. Estimate the percentage of the nonfarm workforce in the service industries one decade from now.

Solution   The percentage of the nonfarm workforce in the service industries t decades from now will be given by

$$P(t) = \int 5e^{1/(t+1)}\,dt \qquad P(0) = 30$$

This integral cannot be expressed in terms of an elementary function. To obtain an approximate solution to the problem at hand, let's first make the substitution

$$u = \frac{1}{t+1}$$

So that

$$t + 1 = \frac{1}{u} \quad \text{and} \quad t = \frac{1}{u} - 1$$

giving

$$dt = -\frac{1}{u^2}\, du$$

The integral becomes

$$F(u) = 5 \int e^u \left(-\frac{du}{u^2} \right) = -5 \int \frac{e^u}{u^2}\, du$$

Next, let's approximate e^u at $u = 0$ by a fourth-degree Taylor polynomial. Using Formula 2 in Section 8.8, we have

$$e^u \approx 1 + u + \frac{u^2}{2!} + \frac{u^3}{3!} + \frac{u^4}{4!}$$

Thus,

$$F(u) \approx -5 \int \frac{1}{u^2} \left(1 + u + \frac{u^2}{2} + \frac{u^3}{6} + \frac{u^4}{24} \right) du$$

$$= -5 \int \left(\frac{1}{u^2} + \frac{1}{u} + \frac{1}{2} + \frac{u}{6} + \frac{u^2}{24} \right) du$$

$$= -5 \left(-\frac{1}{u} + \ln u + \frac{1}{2} u + \frac{u^2}{12} + \frac{u^3}{72} \right) + C$$

Therefore,

$$P(t) \approx -5 \left[-(t + 1) + \ln\left(\frac{1}{t + 1} \right) + \frac{1}{2(t + 1)} + \frac{1}{12(t + 1)^2} + \frac{1}{72(t + 1)^3} \right] + C$$

Using the condition $P(0) = 30$, we find

$$30 = P(0) \approx -5 \left(-1 + \ln 1 + \frac{1}{2} + \frac{1}{12} + \frac{1}{72} \right) + C$$

or $C \approx 27.99$. So

$$P(t) \approx -5 \left[-(t + 1) + \ln\left(\frac{1}{t + 1} \right) + \frac{1}{2(t + 1)} + \frac{1}{12(t + 1)^2} + \frac{1}{72(t + 1)^3} \right] + 27.99$$

In particular, the percentage of the nonfarm workforce in the service industries one decade from now will be given by

$$P(1) \approx -5 \left[-2 + \ln\left(\frac{1}{2} \right) + \frac{1}{4} + \frac{1}{48} + \frac{1}{576} \right] + 27.99 \approx 40.09$$

or approximately 40.1%.

8.9 CONCEPT QUESTIONS

1. What is the nth-degree Taylor polynomial of f at c? What is the nth-degree Maclaurin polynomial of f?
2. Match each of the Taylor polynomials with the graph of the appropriate function f, g, or h.

   a. $1 + \dfrac{2}{3}(x - 1) - \dfrac{1}{9}(x - 1)^2 + \dfrac{4}{81}(x - 1)^3$

   b. $1 + \dfrac{1}{8}(x - 1)^2 - \dfrac{1}{8}(x - 1)^3$

   c. $1 - \dfrac{1}{4}(x - 1)^2 + \dfrac{1}{8}(x - 1)^3$

   (1)

   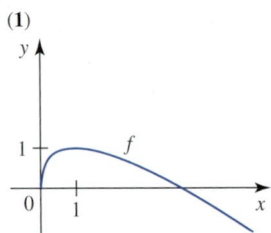

(2)

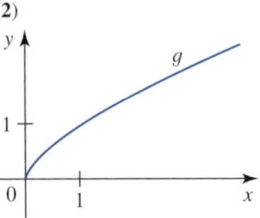

(3)

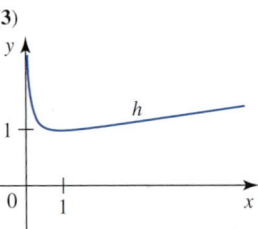

3. Write the expression for the Taylor remainder $R_n(x)$ of f at c.
4. State the conditions under which a function f will have a power series representation at c.

8.9 EXERCISES

In Exercises 1 and 2, find the nth-order Taylor polynomial $P_n(x)$ at c for the function f and the values of n. Then plot the graphs of f and the approximating polynomials on the same set of axes.

1. $f(x) = e^{-x}$, $c = 0$, $n = 1, 2, 3$

2. $f(x) = \sin x$, $c = 0$, $n = 1, 2, 3, 4$

In Exercises 3–16, find the Taylor polynomial $P_n(x)$ and the Taylor remainder $R_n(x)$ for the function f and the values of c and n.

3. $f(x) = 2x^3 + 3x^2 + x + 1$, $c = 1$, $n = 4$

4. $f(x) = x^4 + 3x^3 + 2x + 3$, $c = -1$, $n = 4$

5. $f(x) = \sin x$, $c = \dfrac{\pi}{2}$, $n = 3$

6. $f(x) = \cos x$, $c = \dfrac{\pi}{6}$, $n = 3$

7. $f(x) = \tan x$, $c = \dfrac{\pi}{4}$, $n = 2$

8. $f(x) = \sqrt{x}$, $c = 4$, $n = 3$

9. $f(x) = \sqrt[3]{x}$, $c = -8$, $n = 3$

10. $f(x) = \dfrac{1}{x}$, $c = -1$, $n = 5$

11. $f(x) = \tan^{-1} x$, $c = 1$, $n = 2$

12. $f(x) = \ln x$, $c = 4$, $n = 3$

13. $f(x) = xe^x$, $c = -1$, $n = 3$

14. $f(x) = e^x \sin x$, $c = 0$, $n = 2$

15. $f(x) = e^x \cos 2x$, $c = \dfrac{\pi}{6}$, $n = 2$

16. $f(x) = \ln \sin x$, $c = \dfrac{\pi}{6}$, $n = 3$

In Exercises 17–28, find the Taylor or Maclaurin polynomial $P_n(x)$ for the function f with the given values of c and n. Then give a bound on the error that is incurred if $P_n(x)$ is used to approximate $f(x)$ on the given interval.

17. $f(x) = x^4 - 1$, $c = 1$, $n = 2$, $[0.8, 1.2]$

18. $f(x) = \sin x$, $c = \dfrac{\pi}{6}$, $n = 5$, $\left[0, \dfrac{\pi}{3}\right]$

19. $f(x) = \cos x$, $\quad c = \dfrac{\pi}{4}$, $\quad n = 4$, $\quad \left[0, \dfrac{\pi}{2}\right]$

20. $f(x) = x^{1/3}$, $\quad c = 1$, $\quad n = 3$, $\quad [0.8, 1.2]$

21. $f(x) = e^{2x}$, $\quad c = 1$, $\quad n = 4$, $\quad [1, 1.1]$

22. $f(x) = e^{-x^2}$, $\quad c = 0$, $\quad n = 2$, $\quad [0, 0.1]$

23. $f(x) = \sqrt{x}$, $\quad c = 9$, $\quad n = 3$, $\quad [8, 10]$

24. $f(x) = \dfrac{1}{x}$, $\quad c = 1$, $\quad n = 5$, $\quad [0.9, 1.1]$

25. $f(x) = \tan x$, $\quad c = 0$, $\quad n = 3$, $\quad \left[0, \dfrac{\pi}{4}\right]$

26. $f(x) = \sec x$, $\quad c = 0$, $\quad n = 2$, $\quad \left[0, \dfrac{\pi}{6}\right]$

27. $f(x) = \ln(x + 1)$, $\quad c = 3$, $\quad n = 3$, $\quad [2, 4]$

28. $f(x) = \cosh x$, $\quad c = 0$, $\quad n = 5$, $\quad [-1, 1]$

In Exercises 29–38, find the Taylor polynomial of smallest degree of an appropriate function about a suitable point to approximate the given number to within the indicated accuracy.

29. $e^{0.2}$, $\quad 0.0001$

30. $e^{-1/2}$, $\quad 0.0002$

31. $\sqrt{9.01}$, $\quad 0.00005$

32. $\sqrt[3]{-8.2}$, $\quad 0.000005$

33. $-\dfrac{1}{2.1}$, $\quad 0.0005$

34. $\ln 1.2$, $\quad 0.0001$

35. $\sin 0.1$, $\quad 0.00001$

36. $\cos 0.5$, $\quad 0.0005$

37. $\cos 32°$, $\quad 0.0001$

38. $\sin 69°$, $\quad 0.0001$

In Exercises 39–44, prove that the given Taylor (Maclaurin) series does represent the function.

39. $\displaystyle\sum_{n=0}^{\infty} (-1)^n \dfrac{x^n}{n!}$, $\quad f(x) = e^{-x}$

40. $\displaystyle\sum_{n=0}^{\infty} (-1)^n \dfrac{x^{2n}}{(2n)!}$, $\quad f(x) = \cos x$

41. $\dfrac{1}{\sqrt{2}} \displaystyle\sum_{n=0}^{\infty} (-1)^{n(n-1)/2} \dfrac{\left(x - \frac{\pi}{4}\right)^n}{n!}$, $\quad f(x) = \sin x$

42. $\displaystyle\sum_{n=0}^{\infty} \dfrac{x^{2n+1}}{(2n + 1)!}$, $\quad f(x) = \sinh x$

43. $\displaystyle\sum_{n=0}^{\infty} \dfrac{x^{2n}}{(2n)!}$, $\quad f(x) = \cosh x$

44. $\dfrac{\sqrt{3}}{2} \displaystyle\sum_{n=0}^{\infty} (-1)^n \dfrac{\left(x - \frac{\pi}{6}\right)^{2n+1}}{(2n + 1)!} + \dfrac{1}{2} \displaystyle\sum_{n=0}^{\infty} (-1)^n \dfrac{\left(x - \frac{\pi}{6}\right)^{2n}}{(2n)!}$,

$\qquad\qquad\qquad\qquad\qquad f(x) = \sin x$

45. Growth of Service Industries It has been estimated that service industries, which currently make up 30% of the nonfarm workforce in a certain country, will continue to grow at the rate of

$$R(t) = 6e^{1/(2t+1)}$$

percent per decade, t decades from now. Estimate the percentage of the nonfarm workforce in service industries two decades from now.

46. Concentration of Carbon Monoxide in the Air According to a joint study conducted by a certain city's Environmental Management Department and a state government agency, the concentration of carbon monoxide (CO) in the air due to automobile exhaust t years from now is given by

$$C(t) = 0.01(0.2t^2 + 4t + 64)^{2/3}$$

parts per million. Use the second Taylor polynomial of C at $t = 0$ to obtain an approximation of the average level of concentration of CO in the air between $t = 0$ and $t = 2$.

47. Show that $y = P_1(x)$, where P_1 is the first-order Taylor polynomial of f at c, is an equation of the tangent line to the graph of f at the point $(c, f(c))$.

48. Let $P_n(x)$ be the nth-order Taylor polynomial of f at c. Show that P_n and f have the same derivatives at c up to order n.

49. Prove that

$$x - \dfrac{x^2}{2} < \ln(1 + x) < x$$

if $x > 0$.

Hint: Use Taylor's Theorem with $c = 0$ and $n = 1$ and $n = 2$.

50. Show that the error that is incurred in approximating $\sin(c + h)$ by $\sin c + h \cos c$ does not exceed $h^2/2$.

In Exercises 51–54, determine whether the statement is true or false. If it is true, explain why it is true. If it is false, explain why or give an example to show why it is false.

51. If f is a polynomial function of degree n, then the Maclaurin polynomial of degree n of f is f itself.

52. If $f(x) = e^x$ and $P_n(x)$ is the nth-degree Maclaurin polynomial of f, then $P_n(0.1) = e^{0.1}$ for some positive integer n.

53. The inequality $1 + x \le e^x$ holds for all real values of x.

54. The binomial expansion in the Binomial Theorem is the Maclaurin polynomial of $f(x) = (1 + x)^n$, where n is a positive integer.

CHAPTER 8 REVIEW

CONCEPT REVIEW

In Exercises 1–10, fill in the blanks.

1. a. A sequence is a _____ whose domain is the set of positive _____. The term a_n is called the _____ _____ of the sequence.

b. If a_n can be made as close to the number L as we please by taking n sufficiently large, then $\{a_n\}$ is said to _____ to L.

c. The precise definition of a limit states that $\lim_{n \to \infty} a_n = L$ if _____ _____ $\varepsilon > 0$ there exists a _____ _____ N such that $|a_n - L| < \varepsilon$ whenever _____.

2. a. If $\lim_{n \to \infty} a_n = L$, $\lim_{n \to \infty} b_n = M$, and c is any real number, then $\lim_{n \to \infty} ca_n =$ _____, $\lim_{n \to \infty}(a_n + b_n) =$ _____, $\lim_{n \to \infty} a_n b_n =$ _____, and $\lim_{n \to \infty} \dfrac{a_n}{b_n} =$ _____, provided that _____.

b. If there exists some integer N such that $a_n \le b_n \le c_n$ for all $n \ge N$ and $\lim_{n \to \infty} a_n =$ _____ $= L$, then $\lim_{n \to \infty} b_n = L$.

3. a. A series $\sum_{n=1}^{\infty} a_n$ converges and has sum S if its sequence of _____ _____ converges to S.

b. The series $\sum_{n=1}^{\infty} ar^{n-1}$, $a \ne 0$, is called a _____ series. It converges if $|r| <$ _____ and diverges if $|r| \ge$ _____.

4. a. If $\sum_{n=1}^{\infty} a_n$ converges, then $\lim_{n \to \infty} a_n =$ _____. If $\lim_{n \to \infty} a_n$ does not exist or $\lim_{n \to \infty} a_n \ne 0$, then $\sum_{n=1}^{\infty} a_n$ _____.

b. If $\sum_{n=1}^{\infty} a_n = A$, $\sum_{n=1}^{\infty} b_n = B$, and c is any real number, then $\sum_{n=1}^{\infty}(ca_n + b_n) =$ _____.

5. a. If f is positive, continuous, and decreasing, and if $a_n = f(n)$, then $\sum_{n=1}^{\infty} a_n$ and $\int_1^{\infty} f(x)\, dx$ are either both _____ or _____.

b. The p-series has the form _____ and converges if _____ and diverges if _____.

6. a. If $\sum a_n$ and $\sum b_n$ are series with positive terms with $a_n \le b_n$ for all n, then $\sum b_n$ converges implies that $\sum a_n$ _____. If $\sum b_n$ diverges and _____ for all n, then $\sum a_n$ also diverges.

b. If $\sum a_n$ and $\sum b_n$ are series with positive terms and $\lim_{n \to \infty} \dfrac{a_n}{b_n} = L$, where L is _____ _____ _____, then either both series _____ or both _____.

7. a. The series $\sum_{n=1}^{\infty}(-1)^{n-1} a_n$ is called _____ _____ series. It converges if both the conditions a_{n+1} _____ a_n for all n and $\lim_{n \to \infty} a_n =$ _____ are satisfied.

b. If both the conditions in part (a) are satisfied, then the error that is incurred in approximating the sum of the alternating series by S_n is no larger than _____.

8. a. A series $\sum a_n$ is absolutely convergent if the series _____ converges.

b. A series $\sum a_n$ is _____ convergent if it is convergent but not _____ convergent.

c. A(n) _____ convergent series is convergent.

d. Suppose that $\lim_{n \to \infty} |a_{n+1}/a_n| = L$. Then if $L < 1$, the series _____; if $L > 1$ or $L = \infty$, the series _____; and if $L = 1$, the Ratio Test is _____.

e. Suppose that $\lim_{n \to \infty} \sqrt[n]{|a_n|} = L$. Then if $L < 1$, the series _____; if $L > 1$ or $L = \infty$, the series _____; and if $L = 1$, the Root Test is _____.

9. a. A power series in $(x - c)$ is a series of the form _____.

b. For a power series in $(x - c)$, exactly one of the following is true: It converges only at _____, it converges for all _____, or it converges for $|x - c| < R$, where R is the radius of convergence of the series. In the last case the series diverges for $|x - c| > R$.

10. a. The Taylor series of a function f at c has the form _____.

b. The nth-degree Taylor polynomial P_n of f at c is the nth _____ _____ of the Taylor series at c.

c. Taylor's Theorem states that if f has derivatives up to order $n + 1$ in an interval I containing _____, then for each x in _____ there exists a number z between _____ _____ _____ such that $f(x) = P_n(x) + R_n(x)$ where $R_n(x) =$ _____.

d. If f has derivatives of all order in I and $\lim_{n \to \infty} R_n(x) = 0$, then f is represented by the _____ _____ of f at c.

REVIEW EXERCISES

In Exercises 1–8, determine whether the sequence with the given nth term converges or diverges. If it converges, find the limit.

1. $a_n = \dfrac{n}{3n-2}$

2. $a_n = \dfrac{n+1}{2n^2}$

3. $a_n = 2 + 3(0.9)^n$

4. $a_n = 10(-1.01)^n$

5. $a_n = \dfrac{n}{\ln n}$

6. $a_n = \dfrac{\ln(n^2+1)}{n}$

7. $a_n = \dfrac{\cos n}{n}$

8. $a_n = \left(1 + \dfrac{3}{n}\right)^{2n}$

In Exercises 9–12, find the sum of the series.

9. $\displaystyle\sum_{n=0}^{\infty} \left(\dfrac{2}{3}\right)^n$

10. $\displaystyle\sum_{n=0}^{\infty} \left(\dfrac{1}{3^n} - \dfrac{1}{4^{n+1}}\right)$

11. $\displaystyle\sum_{n=1}^{\infty} \dfrac{1}{n(n+3)}$

12. $\displaystyle\sum_{n=1}^{\infty} \left[\left(\dfrac{3}{5}\right)^n - \dfrac{1}{n(n+1)}\right]$

In Exercises 13–26, determine whether the series is convergent or divergent.

13. $\displaystyle\sum_{n=1}^{\infty} \dfrac{n}{2n^3+1}$

14. $\displaystyle\sum_{n=1}^{\infty} \dfrac{n^3+1}{2n^3-1}$

15. $\displaystyle\sum_{n=1}^{\infty} \dfrac{n^3}{2^n}$

16. $\displaystyle\sum_{n=1}^{\infty} \dfrac{(-1)^n}{\sqrt[3]{n+1}}$

17. $\displaystyle\sum_{n=1}^{\infty} \dfrac{1}{\sqrt{n^3+n}}$

18. $\displaystyle\sum_{n=1}^{\infty} \dfrac{\sin n}{n^2+1}$

19. $\displaystyle\sum_{n=1}^{\infty} \dfrac{n+\cos n}{n^3+1}$

20. $\displaystyle\sum_{n=2}^{\infty} \dfrac{(-1)^n \ln n}{n}$

21. $\displaystyle\sum_{n=1}^{\infty} \dfrac{(-1)^n 3^n}{n \cdot 2^n}$

22. $\displaystyle\sum_{n=1}^{\infty} \dfrac{e^n}{(n+1)^{2n}}$

23. $\displaystyle\sum_{n=2}^{\infty} \dfrac{1}{n(\ln n)^2}$

24. $\displaystyle\sum_{n=1}^{\infty} \dfrac{\tan^{-1} n}{\sqrt{n^2+1}}$

25. $\displaystyle\sum_{n=1}^{\infty} \dfrac{1 \cdot 3 \cdot 5 \cdot \cdots \cdot (2n-1)}{2 \cdot 5 \cdot 8 \cdot \cdots \cdot (3n-1)}$

26. $\displaystyle\sum_{n=1}^{\infty} \dfrac{1 \cdot 3 \cdot 5 \cdot \cdots \cdot (2n-1)}{n! \, 3^n}$

In Exercises 27–32, determine whether the series is absolutely convergent, conditionally convergent, or divergent.

27. $\displaystyle\sum_{n=1}^{\infty} \dfrac{(-1)^{n-1}}{2n+1}$

28. $\displaystyle\sum_{n=2}^{\infty} \dfrac{(-1)^n}{n \ln n}$

29. $\displaystyle\sum_{n=2}^{\infty} \dfrac{(-1)^n}{(\ln n)^{n/2}}$

30. $\displaystyle\sum_{n=1}^{\infty} \dfrac{(-1)^n \tan^{-1} n}{n^2+1}$

31. $\displaystyle\sum_{n=1}^{\infty} \dfrac{(-1)^n \sqrt{n}}{2n+1}$

32. $\displaystyle\sum_{n=1}^{\infty} (-1)^n \dfrac{1 \cdot 3 \cdot 5 \cdot \cdots \cdot (2n+1)}{2 \cdot 5 \cdot 8 \cdot \cdots \cdot (3n+2)}$

33. Express $1.3\overline{617} = 1.3617617617\ldots$ as a rational number.

34. Find an approximation of the sum of the series $\displaystyle\sum_{n=1}^{\infty} \dfrac{(-1)^{n-1}}{n^3}$ accurate to three decimal places.

35. True or false? If $\lim_{n\to\infty} a_n \neq 0$, then $\Sigma \, a_n$ may converge conditionally but not absolutely.

36. True or false? If $0 \leq a_n \leq b_n$ and $\Sigma \, b_n$ diverges, then $\Sigma \, a_n$ diverges.

37. True or false? If $\Sigma \, a_n$ diverges, then $\Sigma |a_n|$ also diverges.

38. True or false? If $\sum_{n=1}^{\infty} a_n$ diverges and $a_n \geq 0$ for every n, then $\lim_{n\to\infty} S_n = \lim_{n\to\infty} \sum_{k=1}^{n} a_k = \infty$.

39. Find all values of x for which the series $\sum_{n=1}^{\infty}(\cos x)^n$ converges.

40. Show that $\lim_{n\to\infty} nx^n = 0$ if $|x| < 1$.
Hint: Show that $\sum_{n=1}^{\infty} nx^n$ is convergent.

41. Show that $\sum_{n=1}^{\infty} a_n \; (a_n > 0)$ is convergent if and only if the sequence of partial sums of $\Sigma \, a_n$, S_n, are bounded for $n \geq 1$.

42. If c is a nonzero constant and $\sum_{n=1}^{\infty} a_n$ diverges, prove that $\sum_{n=1}^{\infty} ca_n$ diverges.

In Exercises 43–48, find the radius of convergence and the interval of convergence of the power series.

43. $\displaystyle\sum_{n=0}^{\infty} \dfrac{(-1)^n x^n}{n+1}$

44. $\displaystyle\sum_{n=1}^{\infty} \dfrac{n^2(x-2)^n}{2^n}$

45. $\displaystyle\sum_{n=0}^{\infty} \dfrac{(-2x)^n}{n^2+1}$

46. $\displaystyle\sum_{n=1}^{\infty} \dfrac{(x+2)^n}{n^n}$

47. $\displaystyle\sum_{n=2}^{\infty} \dfrac{x^n}{n(\ln n)^2}$

48. $\displaystyle\sum_{n=1}^{\infty} \dfrac{(-1)^n n! (x-1)^n}{2 \cdot 4 \cdot 6 \cdot \cdots \cdot (2n)}$

In Exercises 49–54, find the Taylor series of f at the given value of c.

49. $f(x) = \dfrac{x^3}{1+x}$, $\quad c = 0$

50. $f(x) = xe^{-x^2}$, $\quad c = 0$

51. $f(x) = \cos x^2$, $\quad c = 0$

52. $f(x) = \ln x$, $\quad c = 3$

53. $f(x) = \sqrt{1+x^2}$, $\quad c = 0$

54. $f(x) = \cos x$, $\quad c = \dfrac{\pi}{6}$

55. Find the radius of convergence of the series $\displaystyle\sum_{n=1}^{\infty} \dfrac{(2n)!}{(n!)^2} x^n$.

56. Find the radius of convergence of the series

$$\sum_{n=1}^{\infty} \frac{n^n}{(2n)!} (x-1)^n.$$

57. Find a power series representation of $\displaystyle\int \frac{e^{-x}}{x} dx.$

58. Find a power series representation of $\displaystyle\int \frac{e^x - 1}{x} dx.$

59. Approximate $\displaystyle\int_0^{0.2} \sqrt{1 - x^2} \, dx$ to three decimal places of accuracy.

60. Approximate $\displaystyle\int_0^{0.1} \cos \sqrt{x} \, dx$ to three decimal places of accuracy.

In Exercises 61 and 62, use a Taylor polynomial to approximate the number with an error of less than 0.001.

61. $e^{-0.25}$ **62.** $\sin 2°$

In Exercises 63–66, find the Taylor polynomial $P_n(x)$ and the Taylor remainder $R_n(x)$ for the given function f and the given values of c and n.

63. $f(x) = \sqrt{x}, \quad c = 1, \quad n = 3$

64. $f(x) = \cos x, \quad c = \dfrac{\pi}{3}, \quad n = 3$

65. $f(x) = \csc x, \quad c = \dfrac{\pi}{2}, \quad n = 2$

66. $f(x) = \ln \cos x, \quad c = \dfrac{\pi}{6}, \quad n = 2$

67. Suppose that $\sum_{n=1}^{\infty} a_n$ and $\sum_{n=1}^{\infty} b_n$ are both convergent series with positive terms. Show that $\sum_{n=1}^{\infty} \sqrt{a_n b_n}$ also converges. **Hint:** $(\sqrt{a_n} - \sqrt{b_n})^2 \geq 0$

68. Show that the series

$$\frac{1}{\sqrt{2} - 1} - \frac{1}{\sqrt{2} + 1} + \frac{1}{\sqrt{3} - 1} - \frac{1}{\sqrt{3} + 1} + \cdots$$

diverges. Is the Alternating Series Test applicable? Explain.

69. It can be shown that the magnetic field at a point P a distance y above the center of a circular loop of radius R and carrying a steady current I is directed upward and has magnitude

$$B = \frac{\mu_0 I}{2} \cdot \frac{R^2}{(R^2 + y^2)^{3/2}}$$

where μ_0 is a constant called the permeability of free space (see the figure). Show that

$$B \approx \frac{\mu_0 I R^2}{2y^3}$$

if y is large in comparison to R.

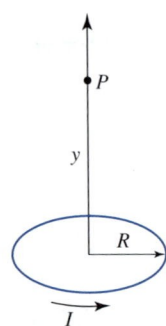

70. Electric Field Induced by a Line Charge The figure below shows a straight line segment of length $2L$ that carries a uniform line charge λ. It can be shown that the electric field induced by this line charge at a distance y above the origin is directed along the y-axis and has magnitude

$$E = \frac{1}{4\pi\varepsilon_0} \cdot \frac{2\lambda L}{y\sqrt{y^2 + L^2}}$$

where ε_0 is a constant called the permittivity of free space. Show that the formula becomes

$$E \approx \frac{1}{4\pi\varepsilon_0} \cdot \frac{2\lambda L}{y^2}$$

if y is large in comparison to L. This suggests that the line charge "looks" like a point charge $q = 2\lambda L$ from this distance, so the field reduces to that induced by a point charge q, namely, $q/(4\pi\varepsilon_0)y^2$.

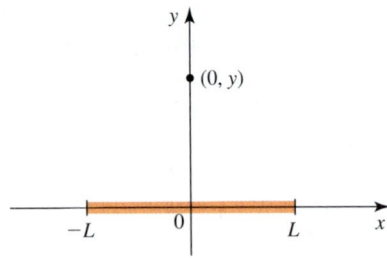

PROBLEM-SOLVING TECHNIQUES

Although l'Hôpital's Rule is a powerful tool for evaluating limits involving an indeterminate form, it is not always the ideal choice. The following technique illustrates the usefulness of the Taylor series in solving such problems.

EXAMPLE Find $\displaystyle\lim_{x\to 0}\frac{e^{x^2}\sin x - x\left(1+\frac{5}{6}x^2\right)}{x^5}$.

Solution Evaluating the limit, we are led to the indeterminate form 0/0. An obvious approach is to use l'Hôpital's Rule to find the limit. But we would soon be deterred by the number of calculations (of derivatives) involved in the process. Alternatively, we can solve the problem with the aid of the power series representations of functions.

Displaying terms up to those of degree five, we find

$$\lim_{x\to 0}\frac{e^{x^2}\sin x - x\left(1+\frac{5}{6}x^2\right)}{x^5}$$

$$= \lim_{x\to 0}\frac{\left(1+x^2+\dfrac{x^4}{2!}+\cdots\right)\left(x-\dfrac{x^3}{6}+\dfrac{x^5}{120}-\cdots\right)-x-\dfrac{5}{6}x^3}{x^5}$$

$$= \lim_{x\to 0}\frac{x-\dfrac{x^3}{6}+\dfrac{x^5}{120}+x^3-\dfrac{x^5}{6}+\dfrac{x^5}{2}+\cdots-x-\dfrac{5}{6}x^3}{x^5}$$

$$= \lim_{x\to 0}\frac{\dfrac{41}{120}x^5+\cdots}{x^5}=\frac{41}{120}\approx 0.342 \qquad\blacksquare$$

CHALLENGE PROBLEMS

1. Let $x_n = \dfrac{1}{\sqrt{n^2+1}}+\dfrac{1}{\sqrt{n^2+2}}+\cdots+\dfrac{1}{\sqrt{n^2+n}}$. Find $\lim_{n\to\infty} x_n$.
Hint: Show that $y_n < x_n < z_n$, where $y_n = \dfrac{n}{\sqrt{n^2+n}}$ and $z_n = \dfrac{n}{\sqrt{n^2+1}}$, and use the Squeeze Theorem.

2. Define f by
$$f(x) = \lim_{n\to\infty}\frac{x^{2n}-1}{x^{2n}+1}$$
  a. Find a rule for f that does not involve a limit.
  b. Sketch the graph of f.

3. Cantor Set Start with the interval $[0, 1]$, and remove the open middle third $\left(\frac{1}{3},\frac{2}{3}\right)$. Next, remove the open middle third from each of the two remaining closed intervals, then the open middle third from each of the four remaining closed intervals, and so on.
  a. Find an expression for c_n, the sum of the lengths of the intervals remaining after n steps.
  b. Show that $\lim_{n\to\infty} c_n = 0$. The set c remaining after all the deletions is called the **Cantor middle-third set** and can be said to have total length zero.

4. Find $\lim_{n\to\infty} 4n\left(\sqrt{n^2+1}-n\right)$.

5. Let $a > 0$, and define the sequence $\{x_n\}$ by
$$x_1 = \sqrt{a},\ x_2 = \sqrt{a+\sqrt{a}},\ x_3 = \sqrt{a+\sqrt{a+\sqrt{a}}},\ldots$$
$$x_n = \underbrace{\sqrt{a+\sqrt{a+\cdots+\sqrt{a}}}}_{n\text{ radicals}}$$
Assuming that $\lim_{n\to\infty} x_n$ exists, what is the limit?

6. Find the largest term in the sequence $\{a_n\}$, where
$$a_n = \frac{2n^2}{3n^3+400}.$$

7. Show that $\displaystyle\lim_{n\to\infty}\frac{1}{n^2}\sum_{k=1}^{n}\llbracket kx \rrbracket = \frac{x}{2}$.

8. Let $\displaystyle\sum_{n=1}^{\infty} a_n$ be a convergent series with positive terms.
Show that $\displaystyle\sum_{n=1}^{\infty}\frac{\sqrt{a_n}}{n}$ converges.
Hint: Consider $\displaystyle\sum_{n=1}^{N}\left(\sqrt{a_n}-\frac{1}{n}\right)^2$.

9. Show that if $a > 1$ and k is any positive integer, then
$$\lim_{n\to\infty}\frac{n^k}{a^n}=0.$$
Hint: Show that the series $\displaystyle\sum_{n=1}^{\infty}\frac{n^k}{a^n}$ converges.

10. Let $A_0, A_1, A_2, A_3, \ldots$ denote the areas of the regions $R_0, R_1, R_2, R_3, \ldots$ bounded by the x-axis and the graph of $f(x) = e^{-ax} \sin bx \ (a > 0)$ for $x \geq 0$.

    a. Show that

    $$A_n = (-1)^n \int_{n\pi/b}^{(n+1)\pi/b} e^{-ax} \sin bx \, dx \qquad n = 0, 1, 2, 3, \ldots$$

    b. Integrate by parts to show that

    $$\int e^{-ax} \sin bx \, dx = -\frac{e^{-ax}}{a^2 + b^2} (a \sin bx + b \cos bx) + C$$

    c. Using the results of parts (a) and (b), show that

    $$A_n = \frac{b}{a^2 + b^2} e^{-na\pi/b} (1 + e^{-a\pi/b}) \qquad n = 0, 1, 2, \ldots$$

    d. Using the results of part (c), find the sum of the areas of the regions $A_0, A_1, A_2, A_3, \ldots$.
       Hint: $\Sigma_{n=0}^{\infty} A_n$ is a geometric series.

11. Evaluate $\displaystyle\lim_{x \to 0} \frac{1 - \sqrt{1 + x^2} \cos x}{x^4}$.

12. Evaluate $\displaystyle\lim_{n \to \infty} \left(\frac{1 - 2 + 3 - 4 + \cdots - 2n}{\sqrt{4n^2 + 1}} \right)$.

13. Find the Maclaurin series of $f(x) = \dfrac{1}{1 + x + x^2 + x^3}$.

14. Determine whether $\Sigma_{n=1}^{\infty} n^{100} e^{-n} \sin n$ is convergent or divergent.

15. Find the values of x for which the series $\displaystyle\sum_{n=1}^{\infty} \frac{(-1)^{n-1}}{n + x^2}$ converges.

16. Suppose that f, f', and f'' are continuous in an interval containing x. Show that

    $$\lim_{h \to 0} \frac{f(x + 2h) - 2f(x + h) + f(x)}{h^2} = f''(x)$$

17. Find the Maclaurin series of $f(x) = (1 + x)^x$ up to the x^3 term.

The shape assumed by the cable in a suspension bridge is a parabola. The parabola, like the ellipse and hyperbola, is a curve that is obtained as a result of the intersection of a plane and a double-napped cone, and is, accordingly, called a conic section. Conic sections appear in various fields of study such as astronomy, physics, engineering, and navigation.

James Osmond/Alamy

9 Conic Sections, Plane Curves, and Polar Coordinates

CONIC SECTIONS ARE curves that can be obtained by intersecting a double-napped right circular cone with a plane. Our immediate goal is to describe conics using algebraic equations. We then turn to applications of conics, which range from the design of suspension bridges to the design of satellite-signal receiving dishes and to the design of whispering galleries, in which a person standing at one spot in a gallery can hear a whisper coming from another spot in the gallery. The orbits of celestial bodies and human-made satellites can also be described by using conics.

Parametric equations afford a way of describing curves in the plane and in space. We will study these representations and use them to describe the motion of projectiles and the motion of other objects.

Polar coordinates provide an alternative way of representing points in the plane. We will see that certain curves have simpler representations with polar equations than with rectangular equations. We will also make use of polar equations to help us find the arc length of a curve, the area of a region bounded by a curve, and the area of a surface obtained by revolving a curve about a given line.

V This symbol indicates that one of the following video types is available for enhanced student learning at **www.academic.cengage.com/login**:
- Chapter lecture videos
- Solutions to selected exercises

9.1 Conic Sections

FIGURE 1

The reflector of a radio telescope

Figure 1 shows the reflector of a radio telescope. The shape of the surface of the reflector is obtained by revolving a plane curve called a *parabola* about its axis of symmetry. (See Figure 2a.) Figure 2b depicts the orbit of a planet P around the sun, S. This curve is called an *ellipse*. Figure 2c depicts the trajectory of an incoming alpha particle heading toward and then repulsed by a massive atomic nucleus located at the point F. The trajectory is one of two branches of a *hyperbola*.

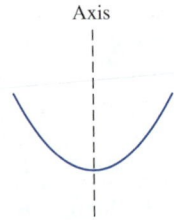

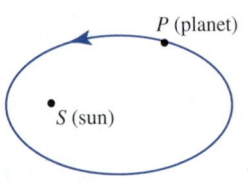

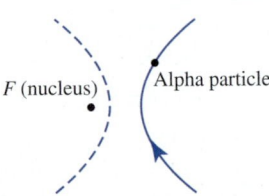

FIGURE 2

(**a**) The cross section of a radio telescope is part of a parabola.

(**b**) The orbit of a planet around the sun is an ellipse.

(**c**) The trajectory of an alpha particle in a Rutherford scattering is part of a branch of a hyperbola.

These curves—parabolas, ellipses, and hyperbolas—are called *conic sections* or, more simply, *conics* because they result from the intersection of a plane and a double-napped cone, as shown in Figure 3.

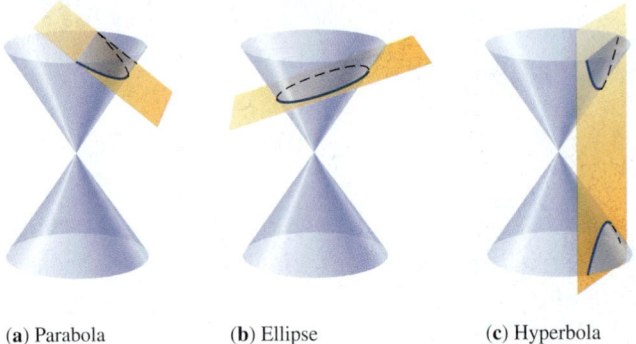

FIGURE 3

The conic sections

(**a**) Parabola   (**b**) Ellipse   (**c**) Hyperbola

In this section we give the geometric definition of each conic section, and we derive an equation for describing each conic section algebraically.

■ Parabola

We first consider a conic section called a *parabola*.

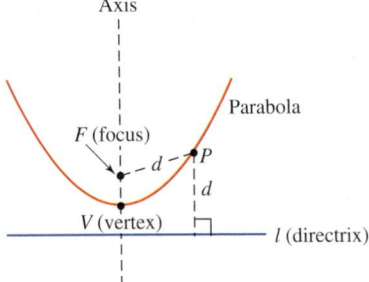

FIGURE 4

The distance between a point P on a parabola and its focus F is the same as the distance between P and the directrix l of the parabola.

DEFINITION   Parabola

A **parabola** is the set of all points in a plane that are equidistant from a fixed point (called the **focus**) and a fixed line (called the **directrix**). (See Figure 4.)

By definition the point halfway between the focus and directrix lies on the parabola. This point V is called the **vertex** of the parabola. The line passing through the focus and perpendicular to the directrix is called the **axis** of the parabola. Observe that the parabola is symmetric with respect to its axis.

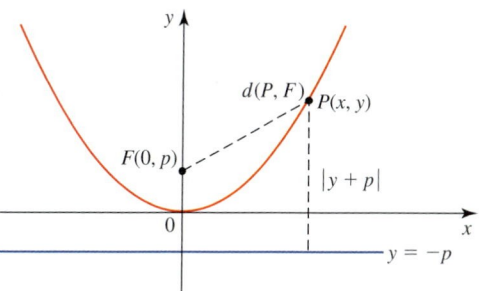

FIGURE 5
The parabola with focus $F(0, p)$ and directrix $y = -p$, where $p > 0$

To find an equation of a parabola, suppose that the parabola is placed so that its vertex is at the origin and its axis is along the y-axis, as shown in Figure 5. Further, suppose that its focus F is at $(0, p)$, and its directrix is the line with equation $y = -p$. If $P(x, y)$ is any point on the parabola, then the distance between P and F is

$$d(P, F) = \sqrt{x^2 + (y - p)^2}$$

whereas the distance between P and the directrix is $|y + p|$. By definition these distances are equal, so

$$\sqrt{x^2 + (y - p)^2} = |y + p|$$

Squaring both sides and simplifying, we obtain

$$x^2 + (y - p)^2 = |y + p|^2 = (y + p)^2$$
$$x^2 + y^2 - 2py + p^2 = y^2 + 2py + p^2$$
$$x^2 = 4py$$

Standard Equation of a Parabola

An equation of the parabola with focus $(0, p)$ and directrix $y = -p$ is

$$x^2 = 4py \qquad \qquad \textbf{(1)}$$

If we write $a = 1/(4p)$, then Equation (1) becomes $y = ax^2$. Observe that the parabola opens upward if $p > 0$ and opens downward if $p < 0$. (See Figure 6.) Also, the parabola is symmetric with respect to the y-axis (that is, the axis of the parabola coincides with the y-axis), since Equation (1) remains unchanged if we replace x by $-x$.

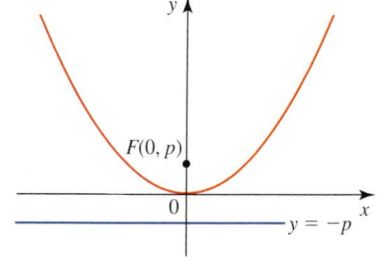

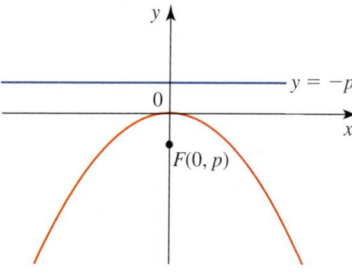

FIGURE 6
The parabola $x^2 = 4py$ opens upward if $p > 0$ and downward if $p < 0$.

(a) $p > 0$

(b) $p < 0$

Interchanging x and y in Equation (1) gives

$$y^2 = 4px \qquad (2)$$

which is an equation of the parabola with focus $F(p, 0)$ and directrix $x = -p$. The parabola opens to the right if $p > 0$ and opens to the left if $p < 0$. (See Figure 7.) In both cases the axis of the parabola coincides with the x-axis.

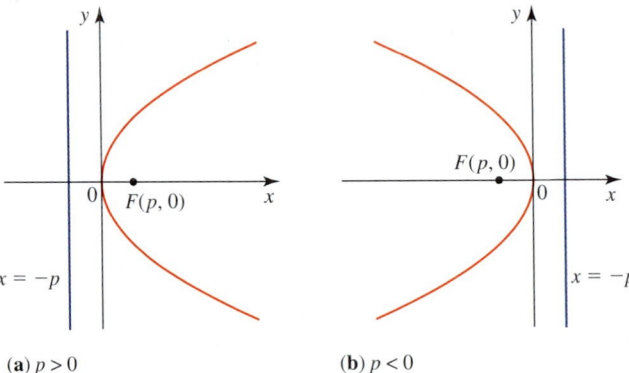

FIGURE 7
The parabola $y^2 = 4px$ opens to the right if $p > 0$ and to the left if $p < 0$.

(a) $p > 0$ **(b)** $p < 0$

Note A parabola with vertex at the origin and axis of symmetry lying on the x-axis or y-axis is said to be in **standard position.** (See Figures 6 and 7.)

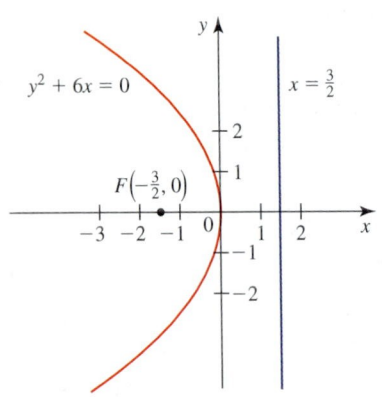

FIGURE 8
The parabola $y^2 + 6x = 0$

EXAMPLE 1 Find the focus and directrix of the parabola $y^2 + 6x = 0$, and make a sketch of the parabola.

Solution Rewriting the given equation in the form $y^2 = -6x$ and comparing it with Equation (2), we see that $4p = -6$ or $p = -\frac{3}{2}$. Therefore, the focus of the parabola is $F\left(-\frac{3}{2}, 0\right)$, and its directrix is $x = \frac{3}{2}$. The parabola is sketched in Figure 8.

EXAMPLE 2 Find an equation of the parabola that has its vertex at the origin with axis of symmetry lying on the y-axis, and passes through the point $P(3, -4)$. What are the focus and directrix of the parabola?

Solution An equation of the parabola has the form $y = ax^2$. To determine the value of a, we use the condition that the point $P(3, -4)$ lies on the parabola to obtain the equation $-4 = a(3)^2$ giving $a = -\frac{4}{9}$. Therefore, an equation of the parabola is

$$y = -\frac{4}{9}x^2$$

To find the focus of the parabola, observe that it has the form $F(0, p)$. Now

$$p = \frac{1}{4a} = \frac{1}{4\left(-\frac{4}{9}\right)} = -\frac{9}{16}$$

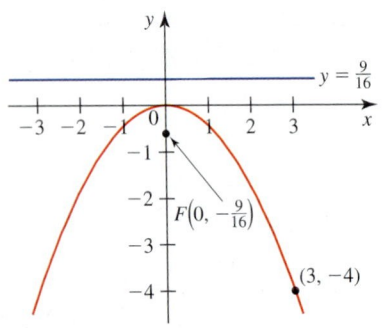

FIGURE 9
The parabola $y = -\frac{4}{9}x^2$

Therefore, the focus is $F\left(0, -\frac{9}{16}\right)$. Its directrix is $y = -p = -\left(-\frac{9}{16}\right)$, or $y = \frac{9}{16}$. The graph of the parabola is sketched in Figure 9.

The parabola has many applications. For example, the cables of certain suspension bridges assume shapes that are parabolic.

EXAMPLE 3 **Suspension Bridge Cables** Figure 10 depicts a bridge, suspended by a flexible cable. If we assume that the weight of the cable is negligible in comparison to the weight of the bridge, then it can be shown that the shape of the cable is described by the equation

$$y = \frac{Wx^2}{2H}$$

where W is the weight of the bridge in pounds per foot and H is the tension at the lowest point of the cable in pounds (the origin). (See Exercise 89.) Suppose that the *span* of the cable is $2a$ ft and the *sag* is h ft.

a. Find an equation describing the shape assumed by the cable in terms of a and h.
b. Find the length of the cable if the span of the cable is 400 ft and the sag is 80 ft.

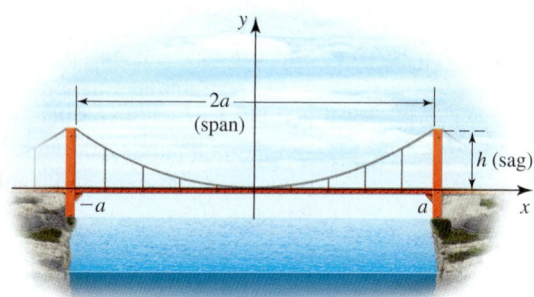

FIGURE 10
A bridge of length $2a$
suspended by a flexible cable

Solution

a. We can write the given equation in the form $y = kx^2$, where $k = W/(2H)$. Since the point (a, h) lies on the parabola $y = kx^2$, we have

$$h = ka^2$$

or $k = h/a^2$, so the required equation is $y = hx^2/a^2$.

b. With $a = 200$ and $h = 80$ an equation that describes the shape of the cable is

$$y = \frac{80x^2}{200^2} = \frac{x^2}{500}$$

Next, the length of the cable is given by

$$s = 2\int_0^{200} \sqrt{1 + (y')^2} \, dx$$

But $y' = x/250$, so

$$s = 2\int_0^{200} \sqrt{1 + \left(\frac{x}{250}\right)^2} \, dx = \frac{1}{125}\int_0^{200} \sqrt{250^2 + x^2} \, dx$$

The easiest way to evaluate this integral is to use Formula 37 from the Table of Integrals found on the reference pages of this book:

$$\int \sqrt{a^2 + u^2} \, du = \frac{u}{2}\sqrt{a^2 + u^2} + \frac{a^2}{2} \ln|u + \sqrt{a^2 + u^2}| + C$$

If we let $a = 250$ and $u = x$, then

$$s = \frac{1}{125}\left[\frac{x}{2}\sqrt{250^2 + x^2} + \frac{250^2}{2}\ln|x + \sqrt{250^2 + x^2}|\right]_0^{200}$$

$$= \frac{1}{125}\left[100\sqrt{62500 + 40000} + 31250\ln|200 + \sqrt{62500 + 40000}| - 31250\ln 250\right]$$

$$= \frac{4}{5}\sqrt{102500} + 250\ln\left(\frac{200 + \sqrt{102500}}{250}\right) \approx 439$$

or 439 ft.

Other applications of the parabola include the trajectory of a projectile in the absence of air resistance.

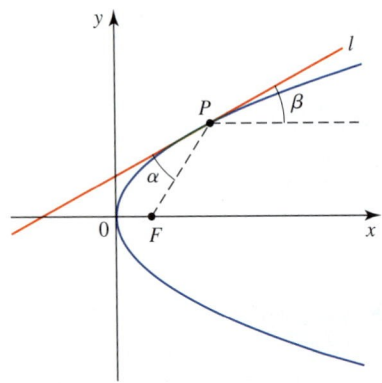

FIGURE 11

The reflective property states that $\alpha = \beta$.

■ Reflective Property of the Parabola

Suppose that P is any point on a parabola with focus F, and let l be the tangent line to the parabola at P. (See Figure 11.) The reflective property states that the angle α that lies between l and the line segment FP is equal to the angle β that lies between l and the line passing through P and parallel to the axis of the parabola. This property is the basis for many applications. (An outline of the proof of this property is given in Exercise 105.)

As was mentioned earlier, the reflector of a radio telescope has a shape that is obtained by revolving a parabola about its axis. Figure 12a shows a cross section of such a reflector. A radio wave coming in from a great distance may be assumed to be parallel to the axis of the parabola. This wave will strike the surface of the reflector and be reflected toward the focus F, where a collector is located. (The angle of incidence is equal to the angle of reflection.)

FIGURE 12

Applications of the reflective property of a parabola

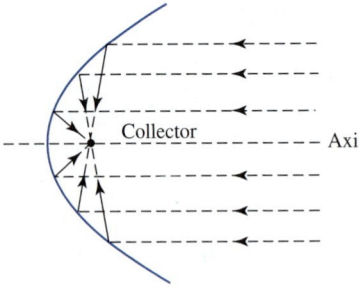

(**a**) A cross section of a radio telescope

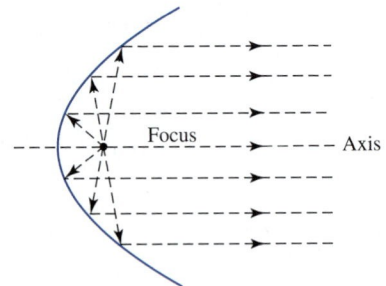

(**b**) A cross section of a headlight

The reflective property of the parabola is also used in the design of headlights of automobiles. Here, a light bulb is placed at the focus of the parabola. A ray of light emanating from the light bulb will strike the surface of the reflector and be reflected outward along a direction parallel to the axis of the parabola (see Figure 12b).

■ Ellipses

Next, we consider a conic section called an ellipse.

> **DEFINITION** **Ellipse**
>
> An ellipse is the set of all points in a plane the sum of whose distances from two fixed points (called the **foci**) is a constant.

Figure 13 shows an ellipse with foci F_1 and F_2. The line passing through the foci intersects the ellipse at two points, V_1 and V_2, called the **vertices** of the ellipse. The chord joining the vertices is called the **major axis,** and its midpoint is called the **center** of the ellipse. The chord passing through the center of the ellipse and perpendicular to the major axis is called the **minor axis** of the ellipse.

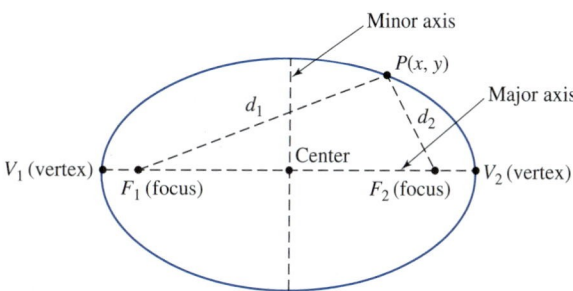

FIGURE 13

An ellipse with foci F_1 and F_2. A point $P(x, y)$ is on the ellipse if and only if $d_1 + d_2 =$ a constant.

Note We can construct an ellipse on paper in the following way: Place a piece of paper on a flat wooden board. Next, secure the ends of a piece of string to two points (the foci of the ellipse) with thumbtacks. Then trace the required ellipse with a pencil pushed against the string, as shown in Figure 14, making sure that the string is kept taut at all times. ■

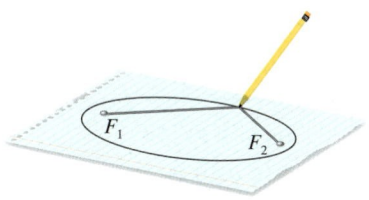

FIGURE 14

Drawing an ellipse on paper using thumbtacks, a string, and a pencil

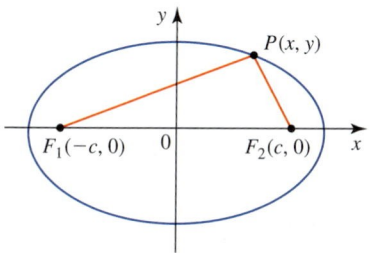

FIGURE 15

The ellipse with foci $F_1(-c, 0)$ and $F_2(c, 0)$

To find an equation for an ellipse, suppose that the ellipse is placed so that its major axis lies along the x-axis and its center is at the origin, as shown in Figure 15. Then its foci F_1 and F_2 are at the points $(-c, 0)$ and $(c, 0)$, respectively. Let the sum of the distances between any point $P(x, y)$ on the ellipse and its foci be $2a > 2c > 0$. Then, by the definition of an ellipse we have

$$d(P, F_1) + d(P, F_2) = 2a$$

that is,

$$\sqrt{(x + c)^2 + y^2} + \sqrt{(x - c)^2 + y^2} = 2a$$

or

$$\sqrt{(x - c)^2 + y^2} = 2a - \sqrt{(x + c)^2 + y^2}$$

Squaring both sides of this equation, we obtain

$$x^2 - 2cx + c^2 + y^2 = 4a^2 - 4a\sqrt{(x + c)^2 + y^2} + x^2 + 2cx + c^2 + y^2$$

or, upon simplification,

$$a\sqrt{(x + c)^2 + y^2} = a^2 + cx$$

Squaring both sides again, we have

$$a^2(x^2 + 2cx + c^2 + y^2) = a^4 + 2a^2cx + c^2x^2$$

which yields

$$(a^2 - c^2)x^2 + a^2y^2 = a^2(a^2 - c^2)$$

Recall that $a > c$, so $a^2 - c^2 > 0$. Let $b^2 = a^2 - c^2$ with $b > 0$. Then the equation of the ellipse becomes

$$b^2x^2 + a^2y^2 = a^2b^2$$

or, upon dividing both sides by a^2b^2, we obtain

$$\frac{x^2}{a^2} + \frac{y^2}{b^2} = 1$$

By setting $y = 0$, we obtain $x = \pm a$, which gives $(-a, 0)$ and $(a, 0)$ as the vertices of the ellipse. Similarly, by setting $x = 0$, we see that the ellipse intersects the y-axis at the points $(0, -b)$ and $(0, b)$. Since the equation remains unchanged if x is replaced by $-x$ and y is replaced by $-y$, we see that the ellipse is symmetric with respect to both axes.

Observe, too, that $b < a$, since

$$b^2 = a^2 - c^2 < a^2$$

So as the name implies, the length of the major axis, $2a$, is greater than the length of the minor axis, $2b$. Finally, observe that if the foci coincide, then $c = 0$ and $a = b$, so the ellipse is a circle with radius $r = a = b$.

Placing the ellipse so that its major axis lies along the y-axis and its center is at the origin leads to an equation in which the roles of x and y are reversed. To summarize, we have the following.

Standard Equation of an Ellipse

An equation of the ellipse with foci $(\pm c, 0)$ and vertices $(\pm a, 0)$ is

$$\frac{x^2}{a^2} + \frac{y^2}{b^2} = 1 \qquad a \geq b > 0 \tag{3}$$

and an equation of the ellipse with foci $(0, \pm c)$ and vertices $(0, \pm a)$ is

$$\frac{x^2}{b^2} + \frac{y^2}{a^2} = 1 \qquad a \geq b > 0 \tag{4}$$

where $c^2 = a^2 - b^2$. (See Figure 16.)

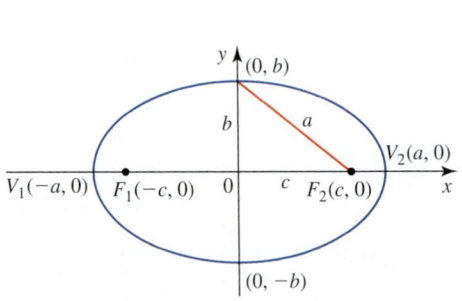

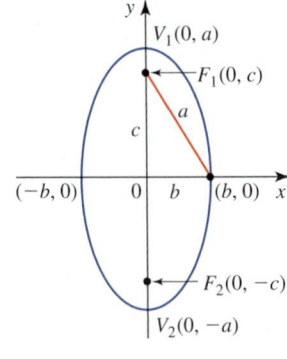

FIGURE 16
Two ellipses in standard position
with center at the origin   (a) The major axis is along the x-axis.

(b) The major axis is along the y-axis.

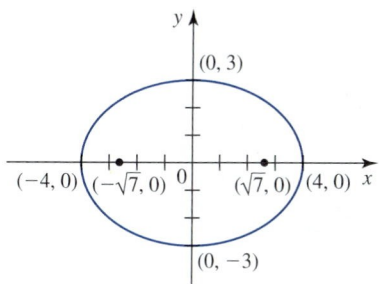

FIGURE 17

The ellipse $\dfrac{x^2}{16} + \dfrac{y^2}{9} = 1$

Note   An ellipse with center at the origin and foci lying along the x-axis or the y-axis is said to be in standard position. (See Figure 16.)   ■

EXAMPLE 4   Sketch the ellipse $\dfrac{x^2}{16} + \dfrac{y^2}{9} = 1$. What are the foci and vertices?

Solution   Here, $a^2 = 16$ and $b^2 = 9$, so $a = 4$ and $b = 3$. Setting $y = 0$ and $x = 0$ in succession gives the x- and y-intercepts as ± 4 and ± 3, respectively. Also, from

$$c^2 = a^2 - b^2 = 16 - 9 = 7$$

we obtain $c = \sqrt{7}$ and conclude that the foci of the ellipse are $(\pm\sqrt{7}, 0)$. Its vertices are $(\pm 4, 0)$. The ellipse is sketched in Figure 17.   ■

EXAMPLE 5   Find an equation of the ellipse with foci $(0, \pm 2)$ and vertices $(0, \pm 4)$.

Solution   Since the foci and therefore the major axis of the ellipse lie along the y-axis, we use Equation (4). Here, $c = 2$ and $a = 4$, so

$$b^2 = a^2 - c^2 = 16 - 4 = 12$$

Therefore, the standard form of the equation for the ellipse is

$$\frac{x^2}{12} + \frac{y^2}{16} = 1$$

or

$$4x^2 + 3y^2 = 48$$ ■

■ Reflective Property of the Ellipse

The ellipse, like the parabola, has a reflective property. To describe this property, consider an ellipse with foci F_1 and F_2 as shown in Figure 18. Let P be a point on the ellipse, and let l be the tangent line to the ellipse at P. Then the angle α between the line segment F_1P and l is equal to the angle β between the line segment F_2P and l. You will be asked to establish this property in Exercise 106.

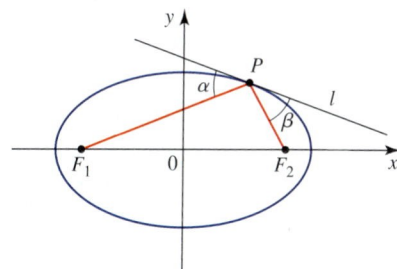

FIGURE 18
The reflective property
states that $\alpha = \beta$.

The reflective property of the ellipse is used to design *whispering galleries*—rooms with elliptical-shaped ceilings, in which a person standing at one focus can hear the whisper of another person standing at the other focus. A whispering gallery can be found in the rotunda of the Capitol Building in Washington, D.C. Also, Paris subway tunnels are almost elliptical, and because of the reflective property of the ellipse, whispering on one platform can be heard on the other. (See Figure 19.)

FIGURE 19
A cross section of a Paris subway
tunnel is almost elliptical.

Yet another application of the reflective property of the ellipse can be found in the field of medicine in a procedure for removing kidney stones called *shock wave lithotripsy*. In this procedure an ellipsoidal reflector is positioned so that a transducer is at one focus and a kidney stone is at the other focus. Shock waves emanating from the transducer are reflected according to the reflective property of the ellipse onto the kidney stone, pulverizing it. This procedure obviates the necessity for surgery.

Eccentricity of an Ellipse

To measure the ovalness of an ellipse, we introduce the notion of eccentricity.

> **DEFINITION   Eccentricity of an Ellipse**
> The **eccentricity** of an ellipse is given by the ratio $e = c/a$.

The eccentricity of an ellipse satisfies $0 < e < 1$, since $0 < c < a$. The closer e is to zero, the more circular is the ellipse.

Hyperbolas

The definition of a hyperbola is similar to that of an ellipse. The *sum* of the distances between the foci and a point on an ellipse is fixed, whereas the *difference* of these distances is fixed for a hyperbola.

> **DEFINITION   Hyperbola**
> A **hyperbola** is the set of all points in a plane the difference of whose distances from two fixed points (called the **foci**) is a constant.

FIGURE 20
A hyperbola with foci F_1 and F_2. A point $P(x, y)$ is on the hyperbola if and only if $|d_1 - d_2|$ is a constant.

Figure 20 shows a hyperbola with foci F_1 and F_2. The line passing through the foci intersects the hyperbola at two points, V_1 and V_2, called the **vertices** of the hyperbola. The line segment joining the vertices is called the **transverse axis** of the hyperbola, and the midpoint of the transverse axis is called the **center** of the hyperbola. Observe that a hyperbola, in contrast to a parabola or an ellipse, has two separate branches.

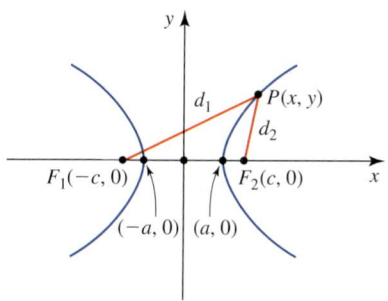

FIGURE 21

An equation of the hyperbola with center $(0, 0)$ and foci $(-c, 0)$ and $(c, 0)$ is $\dfrac{x^2}{a^2} - \dfrac{y^2}{b^2} = 1$

The derivation of an equation of a hyperbola is similar to that of an ellipse. Consider, for example, the hyperbola with center at the origin and foci $F_1(-c, 0)$ and $F_2(c, 0)$ on the x-axis. (See Figure 21.) Using the condition $d(P, F_1) - d(P, F_2) = 2a$, where a is a positive constant, it can be shown that if $P(x, y)$ is any point on the hyperbola, then an equation of the hyperbola is

$$\frac{x^2}{a^2} - \frac{y^2}{b^2} = 1$$

where $b = \sqrt{c^2 - a^2}$ or $c = \sqrt{a^2 + b^2}$.

Observe that the x-intercepts of the hyperbola are $x = \pm a$, giving $(-a, 0)$ and $(a, 0)$ as its vertices. But there are no y-intercepts, since setting $x = 0$ gives $y^2 = -b^2$, which has no real solution. Also, observe that the hyperbola is symmetric with respect to both axes.

If we solve the equation

$$\frac{x^2}{a^2} - \frac{y^2}{b^2} = 1$$

for y, we obtain

$$y = \pm\frac{b}{a}\sqrt{x^2 - a^2}$$

Since $x^2 - a^2 \geq 0$ or, equivalently, $x \leq -a$ or $x \geq a$, we see that the hyperbola actually consists of two separate branches, as was noted earlier. Also, observe that if x is large in magnitude, then $x^2 - a^2 \approx x^2$, so $y = \pm(b/a)x$. This heuristic argument suggests that both branches of the hyperbola approach the slant asymptotes $y = \pm(b/a)x$ as x increases or decreases without bound. (See Figure 22.) You will be asked in Exercise 101 to demonstrate that this is true.

Finally, if the foci of a hyperbola are on the y-axis, then by reversing the roles of x and y, we obtain

$$\frac{y^2}{a^2} - \frac{x^2}{b^2} = 1$$

as an equation of the hyperbola.

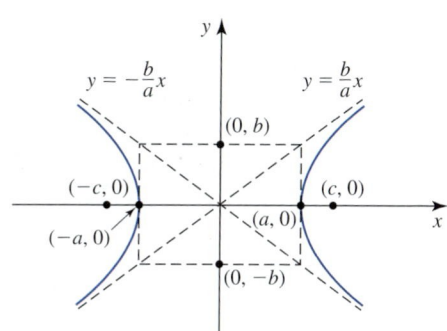

(a) $\dfrac{x^2}{a^2} - \dfrac{y^2}{b^2} = 1$ (The transverse axis is along the x-axis.)

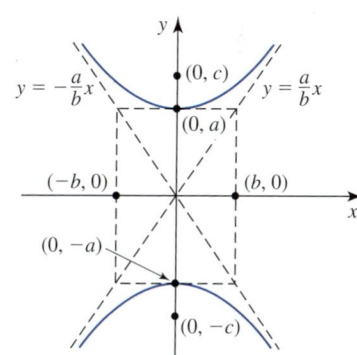

(b) $\dfrac{y^2}{a^2} - \dfrac{x^2}{b^2} = 1$ (The transverse axis is along the y-axis.)

FIGURE 22

Two hyperbolas in standard position with center at the origin

> **Standard Equation of a Hyperbola**
>
> An equation of the hyperbola with foci $(\pm c, 0)$ and vertices $(\pm a, 0)$ is
>
> $$\frac{x^2}{a^2} - \frac{y^2}{b^2} = 1 \qquad (5)$$
>
> where $c = \sqrt{a^2 + b^2}$. The hyperbola has asymptotes $y = \pm(b/a)x$. An equation of the hyperbola with foci $(0, \pm c)$ and vertices $(0, \pm a)$ is
>
> $$\frac{y^2}{a^2} - \frac{x^2}{b^2} = 1 \qquad (6)$$
>
> where $c = \sqrt{a^2 + b^2}$. The hyperbola has asymptotes $y = \pm(a/b)x$.

The line segment of length $2b$ joining the points $(0, -b)$ and $(0, b)$ or $(-b, 0)$ and $(b, 0)$ is called the **conjugate axis** of the hyperbola.

EXAMPLE 6 Find the foci, vertices, and asymptotes of the hyperbola $4x^2 - 9y^2 = 36$.

Solution Dividing both sides of the given equation by 36 leads to the standard equation

$$\frac{x^2}{9} - \frac{y^2}{4} = 1$$

of a hyperbola. Here, $a^2 = 9$ and $b^2 = 4$, so $a = 3$ and $b = 2$. Setting $y = 0$ gives ± 3 as the x-intercepts, so $(\pm 3, 0)$ are the vertices of the hyperbola. Also, we have $c = \sqrt{a^2 + b^2} = \sqrt{13}$, and conclude that the foci of the hyperbola are $(\pm\sqrt{13}, 0)$. Finally, the asymptotes of the hyperbola are

$$y = \pm\frac{b}{a}x = \pm\frac{2}{3}x$$

When you sketch this hyperbola, draw the asymptotes first so that you can then use them as guides for sketching the hyperbola itself. (See Figure 23.) ∎

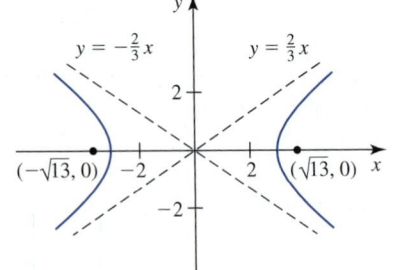

FIGURE 23
The graph of the hyperbola
$4x^2 - 9y^2 = 36$

EXAMPLE 7 A hyperbola has vertices $(0, \pm 3)$ and passes through the point $(2, 5)$. Find an equation of the hyperbola. What are its foci and asymptotes?

Solution Here, the foci lie along the y-axis, so the standard equation of the hyperbola has the form

$$\frac{y^2}{9} - \frac{x^2}{b^2} = 1 \qquad \text{Note that } a = 3.$$

To determine b, we use the condition that the hyperbola passes through the point $(2, 5)$ to write

$$\frac{25}{9} - \frac{4}{b^2} = 1$$

$$\frac{4}{b^2} = \frac{25}{9} - 1 = \frac{16}{9}$$

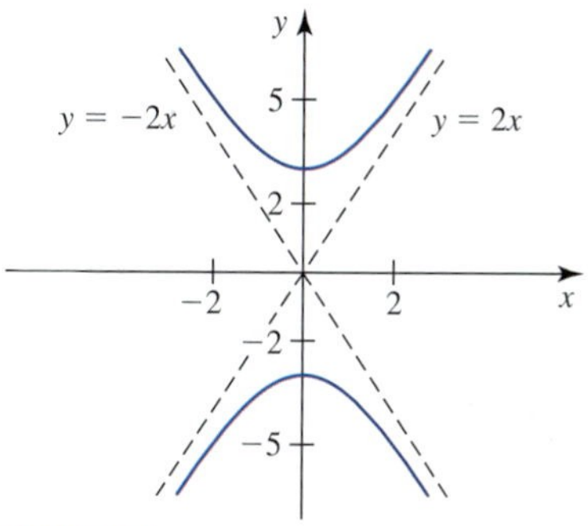

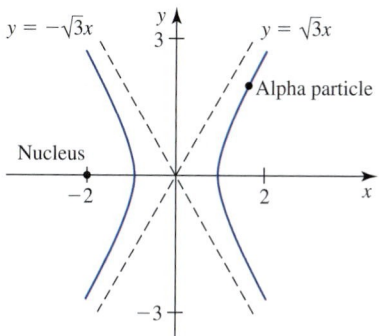

FIGURE 24
The graph of the hyperbola
$y^2 - 4x^2 = 9$

or $b^2 = \frac{9}{4}$. Therefore, a required equation of the hyperbola is

$$\frac{y^2}{9} - \frac{x^2}{\frac{9}{4}} = 1$$

or, equivalently, $y^2 - 4x^2 = 9$. To find the foci of the hyperbola, we compute

$$c^2 = a^2 + b^2 = 9 + \frac{9}{4} = \frac{45}{4}$$

or $c = \pm\sqrt{45/4} = \pm 3\sqrt{5}/2$, from which we see that the foci are $\left(0, \pm\frac{3\sqrt{5}}{2}\right)$. Finally, the asymptotes are obtained by substituting $a = 3$ and $b = \frac{3}{2}$ into the equations $y = \pm(a/b)x$, giving $y = \pm 2x$. The graph of the hyperbola is shown in Figure 24. ∎

EXAMPLE 8 **A Rutherford Scattering** A massive atomic nucleus used as a target for incoming alpha particles is located at the point $(-2, 0)$, as shown in Figure 25. Suppose that an alpha particle approaching the nucleus has a trajectory that is a branch of the hyperbola shown with asymptotes $y = \pm\sqrt{3}x$ and foci $(\pm 2, 0)$. Find an equation of the trajectory.

Solution The asymptotes of a hyperbola with center at the origin and foci lying on the x-axis have equations of the form $y = \pm(b/a)x$. Since the asymptotes of the trajectory are $y = \pm\sqrt{3}x$, we see that

$$\frac{b}{a} = \sqrt{3} \qquad \text{or} \qquad b = \sqrt{3}a$$

Next, since the foci of the hyperbola are $(\pm 2, 0)$, we know that $c = 2$. But $c^2 = a^2 + b^2$, and this gives

$$4 = a^2 + (\sqrt{3}a)^2 = a^2 + 3a^2 = 4a^2$$

or $a = 1$, so $b = \sqrt{3}$. Therefore, an equation of the trajectory is

$$\frac{x^2}{1} - \frac{y^2}{3} = 1$$

or $3x^2 - y^2 = 3$, where $x > 0$. ∎

FIGURE 25
The trajectory of an alpha particle in a Rutherford scattering is a branch of a hyperbola.

DEFINITION **Eccentricity of a Hyperbola**

The **eccentricity** of a hyperbola is given by the ratio $e = c/a$.

Since $c > a$, the eccentricity of a hyperbola satisfies $e > 1$. The larger the eccentricity is, the flatter are the branches of the hyperbola.

■ Shifted Conics

By using the techniques in Section 0.4, we can obtain the equations of conics that are translated from their standard positions. In fact, by replacing x by $x - h$ and y by $y - k$ in their standard equations, we obtain the equation of a parabola whose vertex is translated from the origin to the point (h, k) and the equation of an ellipse (or hyperbola) whose center is translated from the origin to the point (h, k).

We summarize these results in Table 1. Figure 26 shows the graphs of these conics.

TABLE 1

Conic	Orientation of axis	Equation of conic		
Parabola	Axis horizontal	$(y - k)^2 = 4p(x - h)$	**(7)**	(See Figure 26a.)
Parabola	Axis vertical	$(x - h)^2 = 4p(y - k)$	**(8)**	(See Figure 26b.)
Ellipse	Major axis horizontal	$\dfrac{(x - h)^2}{a^2} + \dfrac{(y - k)^2}{b^2} = 1$	**(9)**	(See Figure 26c.)
Ellipse	Major axis vertical	$\dfrac{(x - h)^2}{b^2} + \dfrac{(y - k)^2}{a^2} = 1$	**(10)**	(See Figure 26d.)
Hyperbola	Transverse axis horizontal	$\dfrac{(x - h)^2}{a^2} - \dfrac{(y - k)^2}{b^2} = 1$	**(11)**	(See Figure 26e.)
Hyperbola	Transverse axis vertical	$\dfrac{(y - k)^2}{a^2} - \dfrac{(x - h)^2}{b^2} = 1$	**(12)**	(See Figure 26f.)

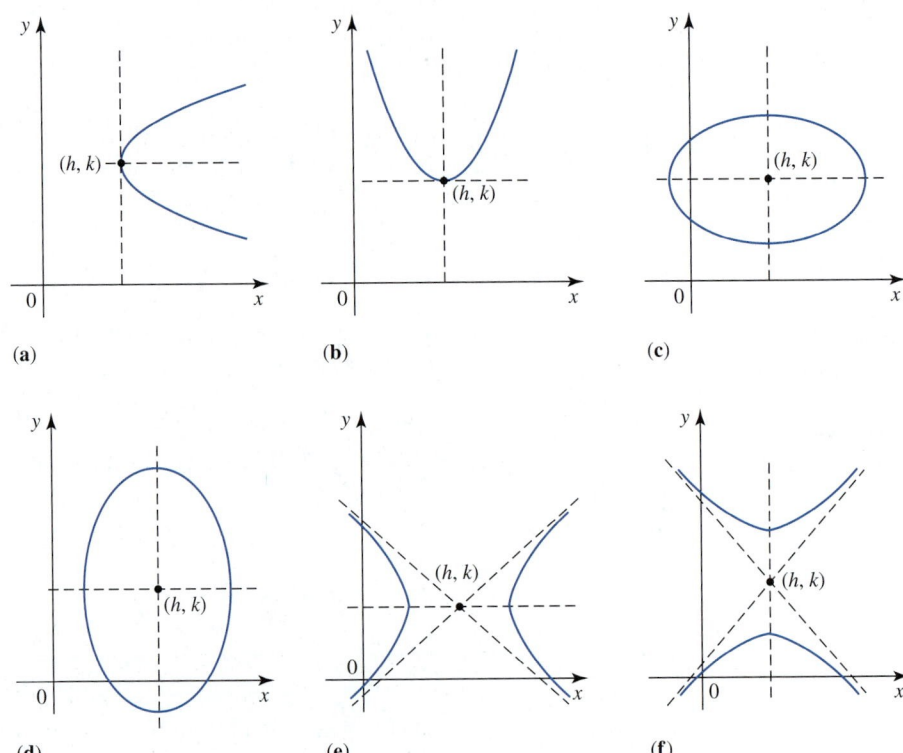

FIGURE 26
Shifted conics with centers at (h, k)

(a) (b) (c)

(d) (e) (f)

Observe that if $h = k = 0$, then each of the equations listed in Table 1 reduces to the corresponding standard equation of a conic centered at the origin, as expected.

EXAMPLE 9 Find the standard equation of the ellipse with foci at $(1, 2)$ and $(5, 2)$ and major axis of length 6. Sketch the ellipse.

Solution   Since the foci $(1, 2)$ and $(5, 2)$ have the same y-coordinate, we see that they lie along the line $y = 2$ parallel to the x-axis. The midpoint of the line segment joining $(1, 2)$ to $(5, 2)$ is $(3, 2)$, and this is the center of the ellipse. From this we can see that the distance from the center of the ellipse to each of the foci is 2, so $c = 2$. Next,

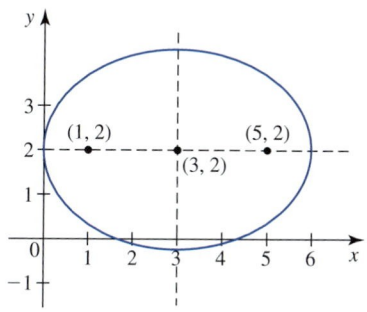

FIGURE 27

The ellipse $\dfrac{(x-3)^2}{9} + \dfrac{(y-2)^2}{5} = 1$

since the major axis of the ellipse is known to have length 6, we have $2a = 6$, or $a = 3$. Finally, from the relation $c^2 = a^2 - b^2$, we obtain $4 = 9 - b^2$, or $b^2 = 5$. Therefore, using Equation (9) from Table 1 with $h = 3$, $k = 2$, $a = 3$, and $b = \sqrt{5}$, we obtain the desired equation:

$$\frac{(x-3)^2}{9} + \frac{(y-2)^2}{5} = 1$$

The ellipse is sketched in Figure 27.

If you expand and simplify each equation in Table 1, you will see that these equations have the general form

$$Ax^2 + By^2 + Dx + Ey + F = 0$$

where the coefficients are real numbers. Conversely, given such an equation, we can obtain an equivalent equation in the form listed in Table 1 by using the technique of completing the square. The latter can then be analyzed readily to obtain the properties of the conic that it represents.

EXAMPLE 10 Find the standard equation of the hyperbola

$$3x^2 - 4y^2 + 6x + 16y - 25 = 0$$

Find its foci, vertices, and asymptotes, and sketch its graph.

Solution We complete the squares in x and y:

$$3(x^2 + 2x) - 4(y^2 - 4y) = 25$$

$$3[x^2 + 2x + (1)^2] - 4[y^2 - 4y + (-2)^2] = 25 + 3 - 16$$

$$3(x+1)^2 - 4(y-2)^2 = 12$$

Then, dividing both sides of this equation by 12 gives the desired equation

$$\frac{(x+1)^2}{4} - \frac{(y-2)^2}{3} = 1$$

Comparing this equation with Equation (11) in Table 1, we see that it is an equation of a hyperbola with center $(-1, 2)$ and transverse axis parallel to the x-axis. We also see that $a^2 = 4$ and $b^2 = 3$, from which it follows that $c^2 = a^2 + b^2 = 4 + 3 = 7$. We can think of this hyperbola as one that is obtained by shifting a similar hyperbola, centered at the origin, one unit to the left and two units upward. Then the required foci, vertices, and asymptotes are obtained by shifting the foci, vertices, and asymptotes of this latter hyperbola accordingly. The results are as follows:

Foci	$(-\sqrt{7} - 1, 2)$ and $(\sqrt{7} - 1, 2)$
Vertices	$(-3, 2)$ and $(1, 2)$
Asymptotes	$y - 2 = \pm\frac{\sqrt{3}}{2}(x + 1)$

The hyperbola is sketched in Figure 28.

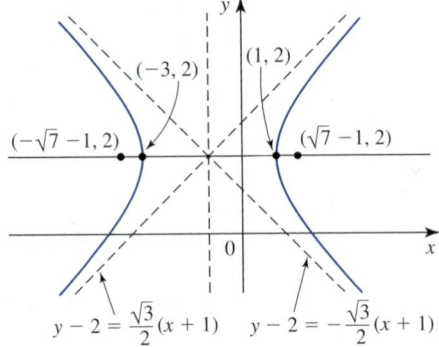

FIGURE 28
The hyperbola
$$3x^2 - 4y^2 + 6x + 16y - 25 = 0$$

The properties of the hyperbola are exploited in the navigational system LORAN (Long Range Navigation). This system utilizes two sets of transmitters: one set located at F_1 and F_2 and another set located at G_1 and G_2. (See Figure 29.) Suppose that synchronized signals sent out by the transmitters located at F_1 and F_2 reach a ship that is located at P. The difference in the times of arrival of the signals are converted by an onboard computer into the difference in the distance $d(P, F_1) - d(P, F_2)$. Using the definition of the hyperbola, we see that this places the ship on a branch of a hyperbola with foci F_1 and F_2 (Figure 29). Similarly, we see that the ship must also lie on a branch of a hyperbola with foci G_1 and G_2. Thus, the position of P is given by the intersection of these two branches of the hyperbolas.

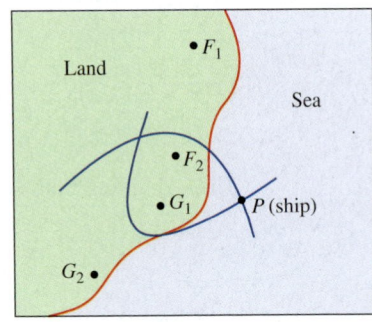

FIGURE 29
In the LORAN navigational system the position of a ship is the point of intersection of two branches of hyperbolas.

9.1 CONCEPT QUESTIONS

1. a. Give the definition of a parabola. What are the focus, directrix, vertex, and axis of a parabola? Illustrate with a sketch.

   b. Write the standard equation of (i) a parabola whose axis lies on the y-axis and (ii) a parabola whose axis lies on the x-axis. Illustrate with sketches.

2. a. Give the definition of an ellipse. What are the foci, vertices, center, major axis, and minor axis of an ellipse? Illustrate with a sketch.

   b. Write the standard equation of (i) an ellipse with foci $(\pm c, 0)$ and vertices $(\pm a, 0)$ and (ii) an ellipse with foci $(0, \pm c)$ and vertices $(0, \pm a)$. Illustrate with sketches.

3. a. Give the definition of a hyperbola. What are the center, the foci, and the transverse axis of the hyperbola? Illustrate with a sketch.

   b. Write the standard equation of (i) a hyperbola with foci $(\pm c, 0)$ and vertices $(\pm a, 0)$ and (ii) a hyperbola with foci $(0, \pm c)$ and vertices $(0, \pm a)$. Illustrate with sketches.

9.1 EXERCISES

In Exercises 1–8, match the equation with one of the conics labeled (a)–(h). If the conic is a parabola, find its vertex, focus and directrix. If it is an ellipse or a hyperbola, find its vertices, foci, and eccentricity.

1. $x^2 = -4y$

2. $y = \dfrac{x^2}{8}$

3. $y^2 = 8x$

4. $x = -\dfrac{1}{4}y^2$

5. $\dfrac{x^2}{9} + \dfrac{y^2}{4} = 1$

6. $x^2 + \dfrac{y^2}{4} = 1$

7. $\dfrac{x^2}{16} - \dfrac{y^2}{9} = 1$

8. $y^2 - \dfrac{x^2}{4} = 1$

(a)

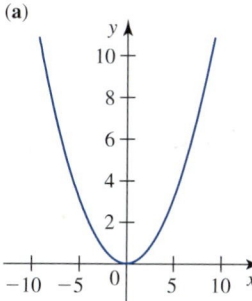

(b)

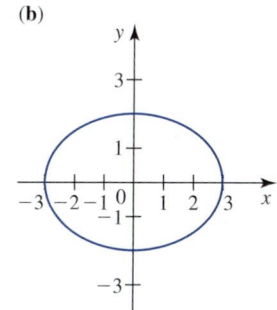

(c)

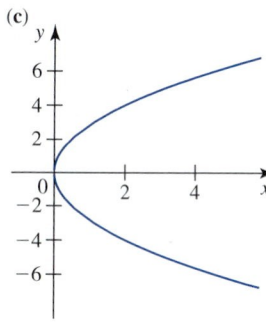

(d)

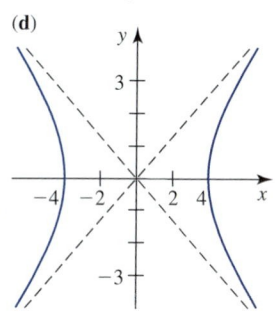

(e)

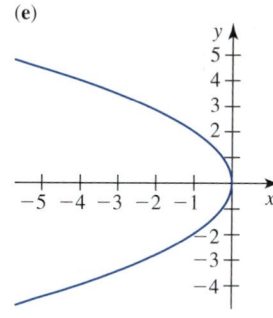

(f)

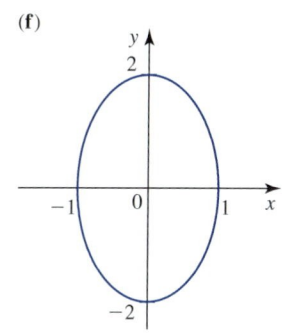

(g)

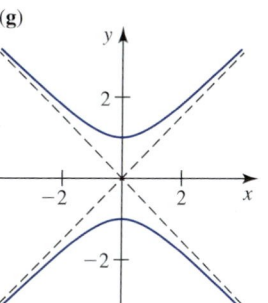

(h)

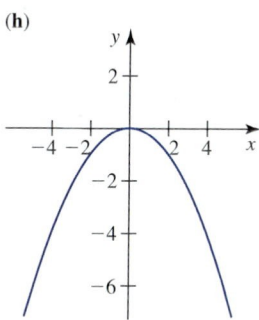

In Exercises 9–14, find the vertex, focus, and directrix of the parabola with the given equation, and sketch the parabola.

9. $y = 2x^2$

10. $x^2 = -12y$

11. $x = 2y^2$

12. $y^2 = -8x$

13. $5y^2 = 12x$

14. $y^2 = -40x$

In Exercises 15–20, find the foci and vertices of the ellipse, and sketch its graph.

15. $\dfrac{x^2}{4} + \dfrac{y^2}{25} = 1$

16. $\dfrac{x^2}{16} + \dfrac{y^2}{9} = 1$

17. $4x^2 + 9y^2 = 36$

18. $25x^2 + 16y^2 = 400$

19. $x^2 + 4y^2 = 4$

20. $2x^2 + y^2 = 4$

In Exercises 21–26, find the vertices, foci, and asymptotes of the hyperbola, and sketch its graph using its asymptotes as an aid.

21. $\dfrac{x^2}{25} - \dfrac{y^2}{144} = 1$

22. $\dfrac{y^2}{16} - \dfrac{x^2}{81} = 1$

23. $x^2 - y^2 = 1$

24. $4y^2 - x^2 = 4$

25. $y^2 - 5x^2 = 25$

26. $x^2 - 2y^2 = 8$

In Exercises 27–30, find an equation of the parabola that satisfies the conditions.

27. Focus $(3, 0)$, directrix $x = -3$

28. Focus $(0, -2)$, directrix $y = 2$

29. Focus $\left(-\dfrac{5}{2}, 0\right)$, directrix $x = \dfrac{5}{2}$

30. Focus $\left(0, \dfrac{3}{2}\right)$, directrix $y = -\dfrac{3}{2}$

In Exercises 31–38, find an equation of the ellipse that satisfies the given conditions.

31. Foci $(\pm 1, 0)$, vertices $(\pm 3, 0)$

32. Foci $(0, \pm 3)$, vertices $(0, \pm 5)$

33. Foci $(0, \pm 1)$, length of major axis 6

34. Vertices $(0, \pm 5)$, length of minor axis 5

35. Vertices $(\pm 3, 0)$, passing through $(1, \sqrt{2})$

36. Passes through $(1, 5)$ and $(2, 4)$ and its center is at $(0, 0)$

37. Passes through $\left(2, \frac{3\sqrt{3}}{2}\right)$ with vertices at $(0, \pm 5)$

38. x-intercepts ± 3, y-intercepts $\pm \frac{1}{2}$

In Exercises 39–44, find an equation of the hyperbola centered at the origin that satisfies the given conditions.

39. foci $(\pm 5, 0)$, vertices $(\pm 3, 0)$

40. foci $(0, \pm 8)$, vertices $(0, \pm 4)$

41. foci $(0, \pm 5)$, conjugate axis of length 4

42. vertices $(\pm 4, 0)$ passing through $\left(5, \frac{9}{4}\right)$

43. vertices $(\pm 2, 0)$, asymptotes $y = \pm \frac{3}{2}x$

44. y-intercepts ± 1, asymptotes $y = \pm \frac{1}{2\sqrt{2}}x$

In Exercises 45–48, match the equation with one of the conic sections labeled (a)–(d).

45. $(x + 3)^2 = -2(y - 4)$ **46.** $\dfrac{(x - 2)^2}{16} + \dfrac{(y + 3)^2}{4} = 1$

47. $\dfrac{(y - 3)^2}{16} - \dfrac{(x + 1)^2}{9} = 1$ **48.** $(y - 1)^2 = -4(x - 2)$

(a)

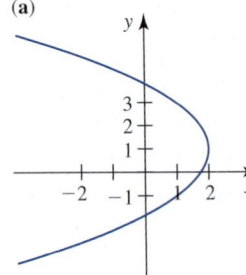

(b)

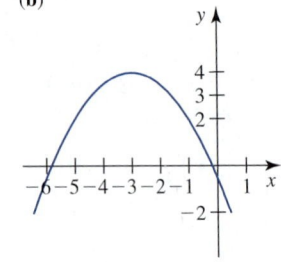

(c)

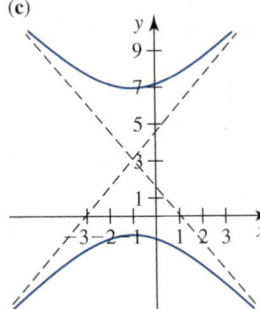

(d)

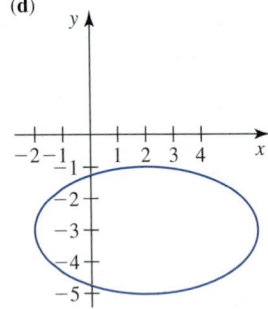

In Exercises 49–66, find an equation of the conic satisfying the given conditions.

49. Parabola, focus $(3, 1)$, directrix $x = 1$

50. Parabola, focus $(-2, 3)$, directrix $y = 5$

51. Parabola, vertex $(2, 2)$, focus $\left(\frac{3}{2}, 2\right)$

52. Parabola, vertex $(1, -2)$, directrix $y = 1$

53. Parabola, axis parallel to the y-axis, passes through $(-3, 2)$, $\left(0, -\frac{5}{2}\right)$, and $(1, -6)$

54. Parabola, axis parallel to the x-axis, passes through $(-6, 6)$, $(0, 0)$, and $(2, 2)$

55. Ellipse, foci $(\pm 1, 3)$, vertices $(\pm 3, 3)$

56. Ellipse, foci $(0, 2)$ and $(4, 2)$, vertices $(-1, 2)$ and $(5, 2)$

57. Ellipse, foci $(\pm 1, 2)$, length of major axis 8

58. Ellipse, foci $(1, \pm 3)$, length of minor axis 2

59. Ellipse, center $(2, 1)$, focus $(0, 1)$, vertex $(5, 1)$

60. Ellipse, foci $(2 - \sqrt{3}, -1)$ and $(2 + \sqrt{3}, -1)$, passes through $(2, 0)$

61. Hyperbola, foci $(-2, 2)$ and $(8, 2)$, vertices $(0, 2)$ and $(6, 2)$

62. Hyperbola, foci $(-4, 5)$ and $(-4, -15)$, vertices $(-4, -3)$ and $(-4, -7)$

63. Hyperbola, foci $(6, -3)$ and $(-4, -3)$, asymptotes $y + 3 = \pm \frac{4}{3}(x - 1)$

64. Hyperbola, foci $(2, 2)$ and $(2, 6)$, asymptotes $x = -2 + y$ and $x = 6 - y$

65. Hyperbola, vertices $(4, -2)$ and $(4, 4)$, asymptotes $y - 1 = \pm \frac{3}{2}(x - 4)$

66. Hyperbola, vertices $(0, -2)$ and $(4, -2)$, asymptotes $x = -y$ and $x = y + 4$

In Exercises 67–72, find the vertex, focus, and directrix of the parabola, and sketch its graph.

67. $y^2 - 2y - 4x + 9 = 0$ **68.** $y^2 - 4y - 2x - 4 = 0$

69. $x^2 + 6x - y + 11 = 0$ **70.** $2x^2 - 8x - y + 5 = 0$

71. $4y^2 - 4y - 32x - 31 = 0$

72. $9x^2 + 6x + 9y - 8 = 0$

In Exercises 73–78, find the center, foci, and vertices of the ellipse, and sketch its graph.

73. $(x - 1)^2 + 4(y + 2)^2 = 1$

74. $2x^2 + y^2 - 20x + 2y + 43 = 0$

75. $x^2 + 4y^2 - 2x + 16y + 13 = 0$

76. $2x^2 + y^2 + 12x - 6y + 25 = 0$

77. $4x^2 + 9y^2 - 18x - 27 = 0$

78. $9x^2 + 36y^2 - 36x + 48y + 43 = 0$

In Exercises 79–84, find the center, foci, vertices, and equations of the asymptotes of the hyperbola with the given equation, and sketch its graph using its asymptotes as an aid.

79. $3x^2 - 4y^2 - 8y - 16 = 0$

80. $4x^2 - 9y^2 - 16x - 54y + 79 = 0$

81. $2x^2 - 3y^2 - 4x + 12y + 8 = 0$

82. $4y^2 - 9x^2 + 18x + 16y + 43 = 0$

83. $4x^2 - 2y^2 + 8x + 8y - 12 = 0$

84. $4x^2 - 3y^2 - 12y - 3 = 0$

85. Parabolic Reflectors The following figure shows the cross section of a parabolic reflector. If the reflector is 2 ft wide at the opening and 1 ft deep, how far from the vertex should the light source be placed along the axis of symmetry of the parabola?

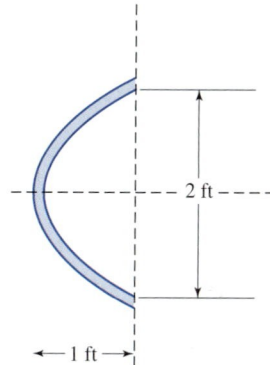

86. Length of Suspension Bridge Cable The figure below depicts a suspension bridge. The shape of the cable is described by the equation

$$y = \frac{hx^2}{a^2}$$

where $2a$ ft is the span of the bridge and h ft is the sag. (See Example 3.) Assuming that the sag is small in comparison to the span (that is, h/a is small), show that the length of the cable is

$$s \approx 2a\left(1 + \frac{2h^2}{3a^2}\right)$$

Use this approximation to estimate the length of the cable in Example 3, where $a = 200$ and $h = 80$. Compare your result with that obtained in the example.

Hint: Retain the first two terms of a binomial series.

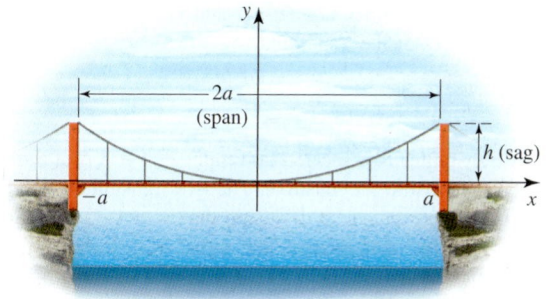

87. Length of Suspension Bridge Cable The cable of the suspension bridge shown in the figure below has the shape of a parabola. If the span of the bridge is 600 ft and the sag is 60 ft, what is the length of the cable?

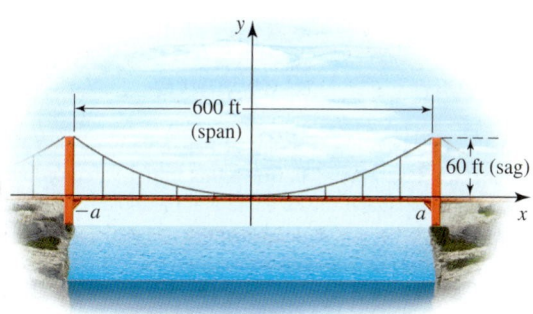

88. Surface Area of a Satellite Dish An 18-in. satellite dish is obtained by revolving the parabola with equation $y = \frac{4}{81}x^2$ about the y-axis. Find the surface area of the dish.

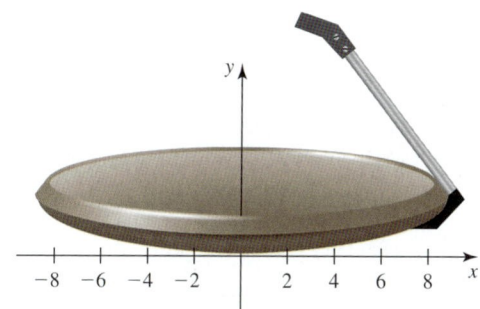

89. Shape of a Suspension Bridge Cable Consider a bridge of weight W lb/ft suspended by a flexible cable. Assume that the weight of the cable is negligible in comparison to the weight of the bridge. The following figure shows a portion of such a structure with the lowest point of the cable located at the origin. Let P be any point on the cable, and suppose that the tension of the cable at P is T lb and lies along the tangent at P (this is the case with flexible cables).

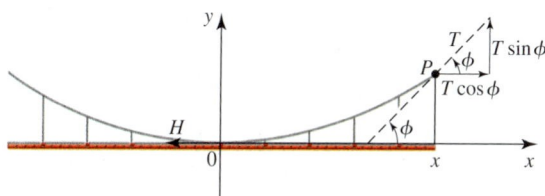

Referring to the figure, we see that

$$\frac{dy}{dx} = \tan \phi = \frac{\sin \phi}{\cos \phi} = \frac{T \sin \phi}{T \cos \phi}$$

But since the bridge is in equilibrium, the horizontal component of T must be equal to H, the tension at the lowest point of the cable (the origin); that is, $T \cos \phi = H$. Similarly, the vertical component of T must be equal to Wx, the load carried over that section of the cable from 0 to P; that is, $T \sin \phi = Wx$. Therefore,

$$\frac{dy}{dx} = \frac{Wx}{H}$$

Finally, since the lowest point of the cable is located at the origin, we have $y(0) = 0$. Solve this initial-value problem to show that the shape of the cable is a parabola.

Note: Observe that in a suspension bridge, the cable supports a load that is *uniformly distributed horizontally*. A cable supporting a load distributed *uniformly along its length* (for example, a cable supporting its own weight) assumes the shape of a catenary, as we saw in Section 5.8.

90. Shape of a Suspension Bridge Cable Refer to Exercise 89. Suppose that the span of the cable is $2a$ ft and the sag is h ft. (See Figure 10.) Show that the tension (in pounds) at the highest point of the cable has magnitude

$$T = \frac{Wa\sqrt{a^2 + 4h^2}}{2h}$$

91. Arch of a Bridge A bridge spanning the Charles River has three arches that are semielliptical in shape. The base of the center arch is 24 ft across, and the maximum height of the arch is 8 ft. What is the height of the arch 6 ft from the center of the base?

8 ft (maximum height of arch)

←24 ft→
(maximum width of arch)

92. a. Find an equation of the tangent line to the parabola $y = ax^2$ at the point where $x = x_0$.
  b. Use the result of part (a) to show that the x-intercept of this tangent line is $x_0/2$.
  c. Use the result of part (b) to draw the tangent line.

93. Prove that any two distinct tangent lines to a parabola must intersect at one and only one point.

94. a. Show that an equation of the tangent line to the parabola $y^2 = 4px$ at the point (x_0, y_0) can be written in the form
$$y_0 y = 2p(x + x_0)$$
  b. Use the result of part (a) to show that the x-intercept of this tangent line is $-x_0$.
  c. Use the result of part (b) to draw the tangent line for $p > 0$.

95. Show that an equation of the tangent line to the ellipse
$$\frac{x^2}{a^2} + \frac{y^2}{b^2} = 1$$
at the point (x_0, y_0) can be written in the form
$$\frac{xx_0}{a^2} + \frac{yy_0}{b^2} = 1$$

96. Use the result of Exercise 95 to find an equation of the tangent line to the ellipse
$$\frac{x^2}{4} + \frac{y^2}{25} = 1$$
at the point $\left(1, \frac{5\sqrt{3}}{2}\right)$.

97. Show that an equation of the tangent line to the hyperbola
$$\frac{x^2}{a^2} - \frac{y^2}{b^2} = 1$$
at the point (x_0, y_0) can be written in the form
$$\frac{xx_0}{a^2} - \frac{yy_0}{b^2} = 1$$

98. Use the result of Exercise 97 to find an equation of the tangent line to the hyperbola
$$\frac{x^2}{4} - \frac{y^2}{9} = 1$$
at the point $(4, 3\sqrt{3})$.

99. Show that the ellipse
$$\frac{x^2}{a^2} + \frac{2y^2}{b^2} = 1$$
and the hyperbola
$$\frac{x^2}{a^2 - b^2} - \frac{2y^2}{b^2} = 1$$
intersect at right angles.

100. Use the definition of a hyperbola to derive Equation (5) for a hyperbola with foci $F_1(-c, 0)$ and $F_2(c, 0)$ and vertices $V_1(a, 0)$ and $V_2(-a, 0)$.

101. Show that the lines $y = (b/a)x$ and $y = -(b/a)x$ are slant asymptotes of the hyperbola
$$\frac{x^2}{a^2} - \frac{y^2}{b^2} = 1$$

102. A transmitter B is located 200 miles due east of a transmitter A on a straight coastline. The two transmitters send out signals simultaneously to a ship that is located at P. Suppose that the ship receives the signal from B, 800 microseconds (μ sec) before it receives the signal from A.
  a. Assuming that radio waves travel at a speed of 980 ft/μ sec, find an equation of the hyperbola on which the ship lies (see page 838).
    Hint: $d(P, A) - d(P, B) = 2a$.

b. If the ship is sailing in a direction parallel to and 20 mi north of the coastline, locate the position of the ship at that instant of time.

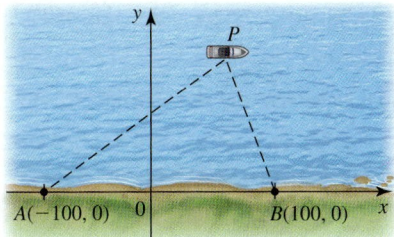

cas **103.** Use a computer algebra system (CAS) to find an approximation of the circumference of the ellipse

$$4x^2 + 25y^2 = 100$$

cas **104.** The dwarf planet Pluto has an elliptical orbit with the sun at one focus. The length of the major axis of the ellipse is 7.33×10^9 miles, and the length of the minor axis is 7.08×10^9 miles. Use a CAS to approximate the distance traveled by the planet during one complete orbit around the sun.

105. The Reflective Property of the Parabola The figure shows a parabola with equation $y^2 = 4px$. The line l is tangent to the parabola at the point $P(x_0, y_0)$.

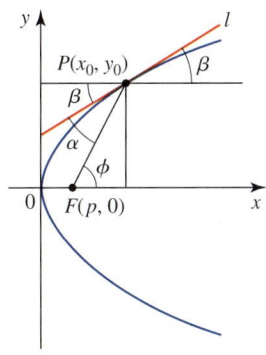

Show that $\alpha = \beta$ by establishing the following:

a. $\tan \beta = \dfrac{2p}{y_0}$ **b.** $\tan \phi = \dfrac{y_0}{x_0 - p}$

c. $\tan \alpha = \dfrac{2p}{y_0}$

Hint: $\tan \alpha = \dfrac{\tan \phi - \tan \beta}{1 + \tan \phi \tan \beta}$

106. The Reflective Property of the Ellipse Establish the reflective property of the ellipse by showing that $\alpha = \beta$ in Figure 18, page 831.

Hint: Use the trigonometric formula

$$\tan(\theta_1 - \theta_2) = \dfrac{\tan \theta_1 - \tan \theta_2}{1 + \tan \theta_1 \tan \theta_2}$$

107. Reflective Property of the Hyperbola The hyperbola also has the reflective property that the other two conics enjoy. Consider a mirror that has the shape of one branch of a hyperbola as shown in Figure (a). A ray of light aimed at a focus F_2 will be reflected toward the other focus, F_1. To establish the reflective property of the hyperbola, let $P(x_0, y_0)$ be a point on the hyperbola $x^2/a^2 - y^2/b^2 = 1$ with foci F_1 and F_2, and let α and β be the angles between the lines PF_1 and PF_2, as shown in Figure (b). Show that $\alpha = \beta$.

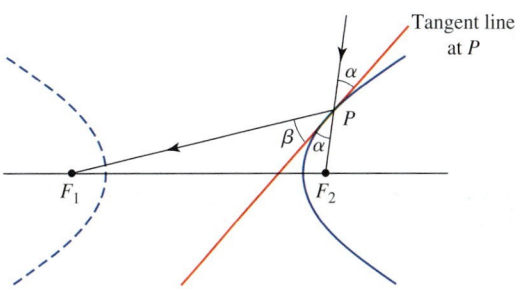

(a)

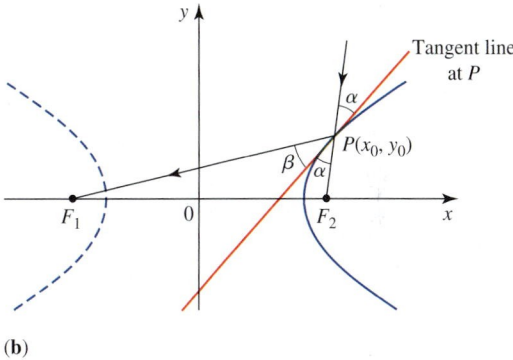

(b)

108. Reflecting Telescopes The reflective properties of the parabola and the hyperbola are exploited in designing a reflecting telescope. A hyperbolic mirror and a parabolic mirror are placed so that one focus of the hyperbola coincides with the focus F of the parabola, as shown in the figure. Use the reflective properties of the two conics to explain why rays of light coming from great distances are finally focused at the eyepiece placed at F', the other focus of the hyperbola.

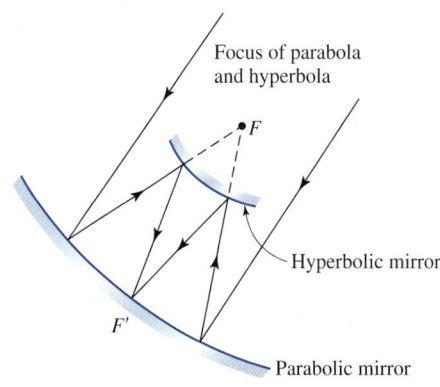

In Exercises 109–114, determine whether the statement is true or false. If it is true, explain why it is true. If it is false, give an example to show why it is false.

109. The graph of $2x^2 - y^2 + F = 0$ is a hyperbola, provided that $F \neq 0$.

110. The graph of $y^4 = 16ax^2$, where $a > 0$, is a parabola.

111. The ellipse $b^2x^2 + a^2y^2 = a^2b^2$, where $a > b > 0$, is contained in the circle $x^2 + y^2 = a^2$ and contains the circle $x^2 + y^2 = b^2$.

112. The asymptotes of the hyperbola $x^2/a^2 - y^2/b^2 = 1$ are perpendicular to each other if and only if $a = b$.

113. If A and C are both positive constants, then
$$Ax^2 + Cy^2 + Dx + Ey + F = 0$$
is an ellipse.

114. If A and C have opposite signs, then
$$Ax^2 + Cy^2 + Dx + Ey + F = 0$$
is a hyperbola.

9.2 Plane Curves and Parametric Equations

■ Why We Use Parametric Equations

Figure 1a gives a bird's-eye view of a proposed training course for a yacht. In Figure 1b we have introduced an *xy*-coordinate system in the plane to describe the position of the yacht. With respect to this coordinate system the position of the yacht is given by the point $P(x, y)$, and the course itself is the graph of the rectangular equation $4x^4 - 4x^2 + y^2 = 0$, which is called a *lemniscate*. But representing the lemniscate in terms of a rectangular equation in this instance has three major drawbacks.

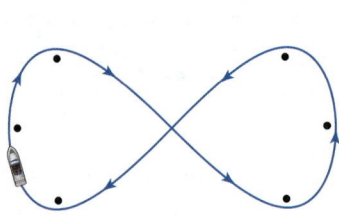

FIGURE 1   **(a)** The dots give the position of markers.   **(b)** An equation of the curve C is $4x^4 - 4x^2 + y^2 = 0$.

First, the equation does not define y explicitly as a function of x or x as a function of y. You can also convince yourself that this is not the graph of a function by applying the vertical and horizontal line tests to the curve in Figure 1b (see Section 0.2). Because of this, we cannot make direct use of many of the results for functions developed earlier. Second, the equation does not tell us when the yacht is at a given point (x, y). Third, the equation gives no inkling as to the direction of motion of the yacht.

To overcome these drawbacks when we consider the motion of an object in the plane or plane curves that are not graphs of functions, we turn to the following representation. If (x, y) is a point on a curve in the *xy*-plane, we write

$$x = f(t) \qquad y = g(t)$$

where f and g are functions of an auxiliary variable t with (common) domain some interval I. These equations are called **parametric equations,** t is called a **parameter,** and the interval I is called a **parameter interval.**

If we think of t on the closed interval $[a, b]$ as representing time, then we can interpret the parametric equations in terms of the motion of a particle as follows: At $t = a$ the particle is at the **initial point** $(f(a), g(a))$ of the curve or **trajectory** C. As t increases from $t = a$ to $t = b$, the particle traverses the curve in a specific direction called the **orientation** of the curve, eventually ending up at the **terminal point** $(f(b), g(b))$ of the curve. (See Figure 2.)

FIGURE 2
As t increases from a to b, the particle traces the curve from $(f(a), g(a))$ to $(f(b), g(b))$ in a specific direction.

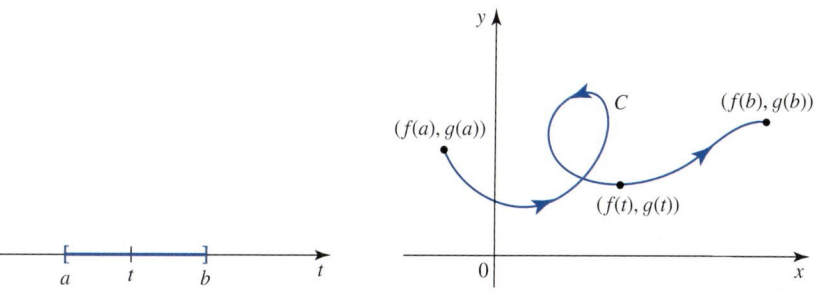

Parameter interval is $[a, b]$.

We can also interpret the parametric equations in geometric terms as follows: We take the line segment $[a, b]$ and, by a process of stretching, bending, and twisting, make it conform geometrically to the curve C.

Sketching Curves Defined by Parametric Equations

Before looking at some examples, let's define the following term.

DEFINITION **Plane Curve**

A plane curve is a set C of ordered pairs (x, y) defined by the parametric equations

$$x = f(t) \qquad \text{and} \qquad y = g(t)$$

where f and g are continuous functions on a parameter interval I.

EXAMPLE 1 Sketch the curve described by the parametric equations

$$x = t^2 - 4 \qquad \text{and} \qquad y = 2t \qquad -1 \leq t \leq 2$$

Solution By plotting and connecting the points (x, y) for selected values of t (Table 1), we obtain the curve shown in Figure 3.

TABLE 1

t	-1	$-\frac{1}{2}$	0	$\frac{1}{2}$	1	2
(x, y)	$(-3, -2)$	$\left(-\frac{15}{4}, -1\right)$	$(-4, 0)$	$\left(-\frac{15}{4}, 1\right)$	$(-3, 2)$	$(0, 4)$

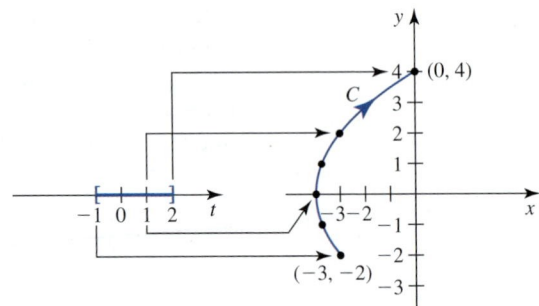

FIGURE 3
As t increases from -1 to 2, the curve
C is traced from the initial point
$(-3, -2)$ to the terminal point $(0, 4)$.

Alternative Solution We eliminate the parameter t by solving the second of the two given parametric equations for t, obtaining $t = \frac{1}{2}y$. We then substitute this value of t into the first equation to obtain

$$x = \left(\frac{1}{2}y\right)^2 - 4 \qquad \text{or} \qquad x = \frac{1}{4}y^2 - 4$$

This is an equation of a parabola that has the x-axis as its axis of symmetry and its vertex at $(-4, 0)$. Now observe that $t = -1$ gives $(-3, -2)$ as the initial point of the curve and that $t = 2$ gives $(0, 4)$ as the terminal point of the curve. So tracing the graph from the initial point to the terminal point gives the desired curve, as obtained earlier. ∎

We will adopt the convention here, just as we did with the domain of a function, that the parameter interval for $x = f(t)$ and $y = g(t)$ will consist of all values of t for which $f(t)$ and $g(t)$ are real numbers, unless otherwise noted.

EXAMPLE 2 Sketch the curves represented by

a. $x = \sqrt{t}$ and $y = t$
b. $x = t$ and $y = t^2$

Solution
a. We eliminate the parameter t by squaring the first equation to obtain $x^2 = t$. Substituting this value of t into the second equation, we obtain $y = x^2$, which is an equation of a parabola. But note that the first parametric equation implies that $t \geq 0$, so $x \geq 0$. Therefore, the desired curve is the right portion of the parabola shown in Figure 4. Finally, note that the parameter interval is $[0, \infty)$, and as t increases from 0, the desired curve starts at the initial point $(0, 0)$ and moves away from it along the parabola.

FIGURE 4
As t increases from 0, the curve
starts out at $(0, 0)$ and follows
the right portion of the parabola
with indicated orientation.

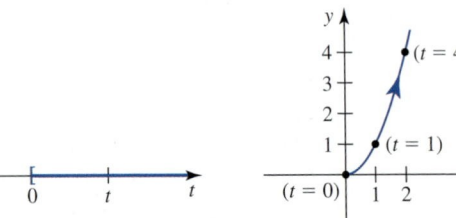

Parameter interval is $[0, \infty)$.

t	(x, y)
0	$(0, 0)$
1	$(1, 1)$
2	$(\sqrt{2}, 2)$
4	$(2, 4)$

b. Substituting the first equation into the second yields $y = x^2$. Although the rectangular equation is the same as that in part (a), the curve described by the parametric equations here is different from that of part (a), as we will now see. In this instance the parameter interval is $(-\infty, \infty)$. Furthermore, as t increases from $-\infty$ to ∞, the curve runs along the parabola $y = x^2$ from left to right, as you can see by plotting the points corresponding to, say, $t = -1, 0$, and 1. You can also see this by examining the parametric equation $x = t$, which tells us that x increases as t increases. (See Figure 5.)

FIGURE 5
As t increases from $-\infty$ to ∞, the entire parabola is traced out, from left to right.

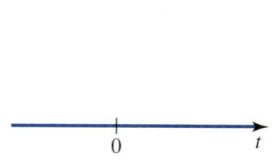

Parameter interval is $(-\infty, \infty)$.

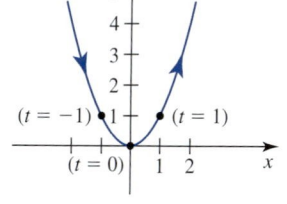

t	(x, y)
-1	$(-1, 1)$
0	$(0, 0)$
1	$(1, 1)$

For problems involving motion, it is natural to use the parameter t to represent time. But other situations call for different representations or interpretations of the parameters, as the next two examples show. Here, we use an *angle* as a parameter.

EXAMPLE 3 Describe the curves represented by the parametric equations

$$x = a \cos \theta \qquad \text{and} \qquad y = a \sin \theta \qquad a > 0$$

with parameter intervals

a. $[0, \pi]$
b. $[0, 2\pi]$
c. $[0, 4\pi]$

Solution We have $\cos \theta = x/a$ and $\sin \theta = y/a$. So

$$1 = \cos^2 \theta + \sin^2 \theta = \left(\frac{x}{a}\right)^2 + \left(\frac{y}{a}\right)^2$$

giving us

$$x^2 + y^2 = a^2$$

This tells us that each of the curves under consideration is contained in a circle of radius a, centered at the origin.

a. If $\theta = 0$, then $x = a$ and $y = 0$, giving $(a, 0)$ as the initial point on the curve. As θ increases from 0 to π, the required curve is traced out in a counterclockwise direction, terminating at the point $(-a, 0)$. (See Figure 6a.)
b. Here, the curve is a complete circle that is traced out in a counterclockwise direction, starting at $(a, 0)$ and terminating at the same point (see Figure 6b).
c. The curve here is a circle that is traced out *twice* in a counterclockwise direction starting at $(a, 0)$ and terminating at the same point (see Figure 6c).

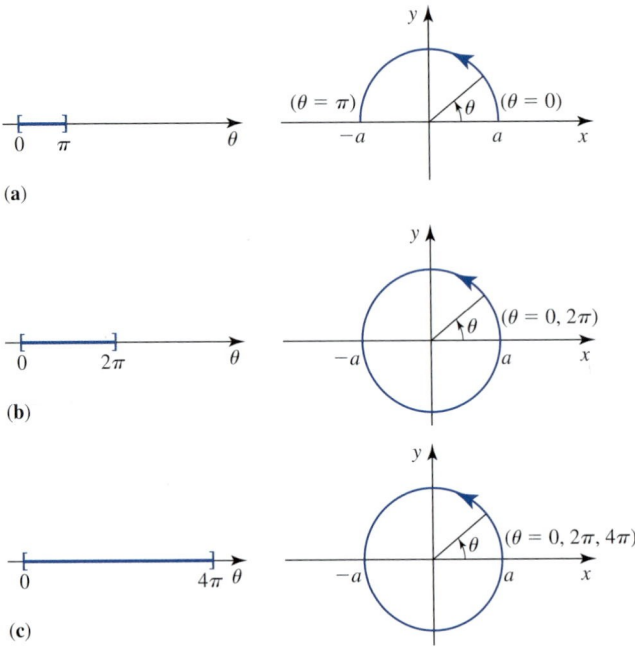

(a)

(b)

(c)

Parameter interval

FIGURE 6

The curve is (a) a semicircle, (b) a complete circle, and (c) a complete circle traced out twice. All curves are traced in a counterclockwise direction.

EXAMPLE 4 Describe the curve represented by

$$x = 4 \cos \theta \qquad \text{and} \qquad y = 3 \sin \theta \qquad 0 \le \theta \le 2\pi$$

Solution Solving the first equation for $\cos \theta$ and the second equation for $\sin \theta$ gives

$$\cos \theta = \frac{x}{4}$$

and

$$\sin \theta = \frac{y}{3}$$

Squaring each equation and adding the resulting equations, we obtain

$$\cos^2 \theta + \sin^2 \theta = \left(\frac{x}{4}\right)^2 + \left(\frac{y}{3}\right)^2$$

Since $\cos^2 \theta + \sin^2 \theta = 1$, we end up with the rectangular equation

$$\frac{x^2}{16} + \frac{y^2}{9} = 1$$

From this we see that the curve is contained in an ellipse centered at the origin. If $\theta = 0$, then $x = 4$ and $y = 0$, giving $(4, 0)$ as the initial point of the curve. As θ increases from 0 to 2π, the elliptical curve is traced out in a counterclockwise direction, terminating at $(4, 0)$. (See Figure 7.)

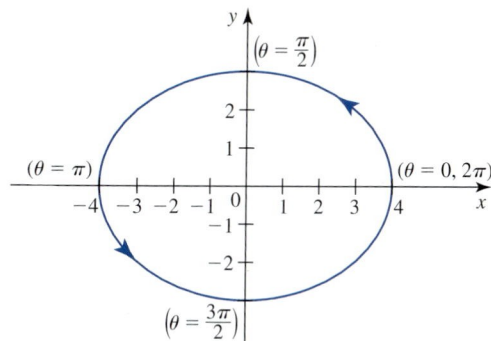

FIGURE 7
As θ increases from 0 to 2π, the curve that is traced out in a counterclockwise direction beginning and ending at (4, 0) is an ellipse.

EXAMPLE 5 A proposed training course for a yacht is represented by the parametric equations

$$x = \sin t \quad \text{and} \quad y = \sin 2t \quad 0 \le t \le 2\pi$$

where x and y are measured in miles.

a. Show that the rectangular equation of the course is $4x^4 - 4x^2 + y^2 = 0$.
b. Describe the course.

Solution

a. Using the trigonometric identity $\sin 2t = 2 \sin t \cos t$, we rewrite the second of the parametric equations in the form

$$y = 2 \sin t \cos t = 2x \cos t \qquad \text{Since } x = \sin t$$

Solving for $\cos t$, we have

$$\cos t = \frac{y}{2x}$$

Then, using the identity $\sin^2 t + \cos^2 t = 1$, we obtain

$$x^2 + \left(\frac{y}{2x}\right)^2 = 1$$

$$x^2 + \frac{y^2}{4x^2} = 1$$

or

$$4x^4 - 4x^2 + y^2 = 0$$

b. From the results of part (a) we see that the required curve is symmetric with respect to the x-axis, the y-axis, and the origin. Therefore, it suffices to concentrate first on drawing the part of the curve that lies in the first quadrant and then make use of symmetry to complete the curve. Since both $\sin t$ and $\sin 2t$ are nonnegative only for $0 \le t \le \frac{\pi}{2}$, we first sketch the curve corresponding to values of t in $\left[0, \frac{\pi}{2}\right]$. With the help of the following table we obtain the curve shown in Figure 8. The direction of the yacht is indicated by the arrows.

t	0	$\frac{\pi}{6}$	$\frac{\pi}{4}$	$\frac{\pi}{3}$	$\frac{\pi}{2}$
(x, y)	$(0, 0)$	$\left(\frac{1}{2}, \frac{\sqrt{3}}{2}\right)$	$\left(\frac{\sqrt{2}}{2}, 1\right)$	$\left(\frac{\sqrt{3}}{2}, \frac{\sqrt{3}}{2}\right)$	$(1, 0)$

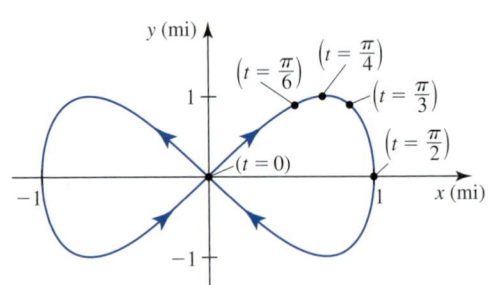

FIGURE 8
The training course for the yacht

EXAMPLE 6 **Cycloids**   Let P be a fixed point on the rim of a wheel. If the wheel is allowed to roll along a straight line without slipping, then the point P traces out a curve called a **cycloid** (see Figure 9). Suppose that the wheel has radius a and rolls along the x-axis. Find parametric equations for the cycloid.

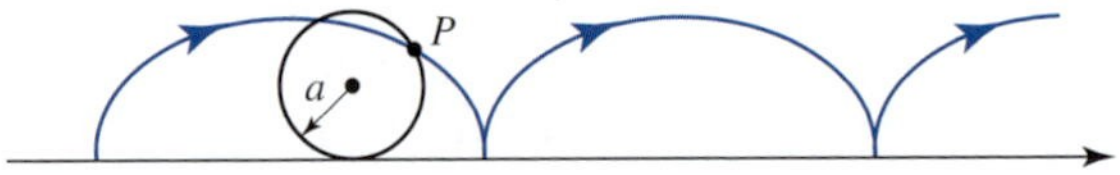

FIGURE 9
The cycloid is the curve
traced out by a fixed point P
on the rim of a rolling wheel.

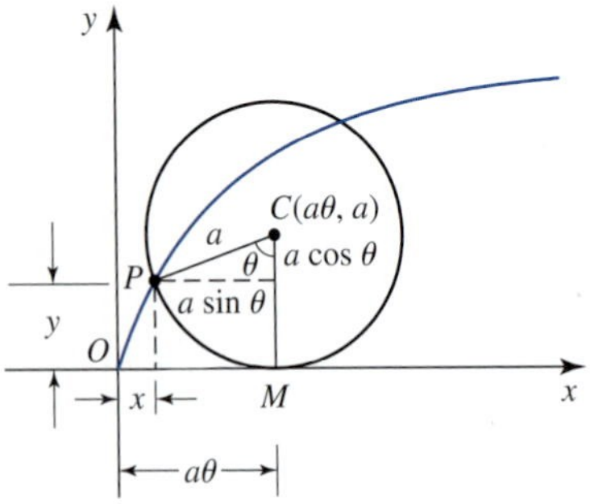

Solution   Suppose that the wheel rolls in a positive direction with the point P initially at the origin of the coordinate system. Figure 10 shows the position of the wheel after it has rotated through θ radians. Because there is no slippage, the distance the wheel has rolled from the origin is

$$d(O, M) = \text{length of arc } PM = a\theta$$

giving its center as $C(a\theta, a)$. Also, from Figure 10 we see that the coordinates of $P(x, y)$ satisfy

$$x = d(O, M) - a \sin \theta = a\theta - a \sin \theta = a(\theta - \sin \theta)$$

and

$$y = d(C, M) - a \cos \theta = a - a \cos \theta = a(1 - \cos \theta)$$

Although these results are derived under the tacit assumption that $0 < \theta < \frac{\pi}{2}$, it can be demonstrated that they are valid for other values of θ. Therefore, the required parametric equations of the cycloid are

$$x = a(\theta - \sin \theta) \qquad \text{and} \qquad y = a(1 - \cos \theta) \qquad -\infty < \theta < \infty$$

FIGURE 10
The position of the wheel after it has
rotated through θ radians

The cycloid provides the solution to two famous problems in mathematics:

1. *The brachistochrone problem:* Find the curve along which a moving particle (under the influence of gravity) will slide from a point A to another point B, not directly beneath A, in the shortest time (see Figure 11a).
2. *The tautochrone problem:* Find the curve having the property that it takes the same time for a particle to slide to the bottom of the curve no matter where the particle is placed on the curve (see Figure 11b).

The brachistochrone problem—the problem of finding the curve of quickest descent—was advanced in 1696 by the Swiss mathematician Johann Bernoulli. Offhand, one might conjecture that such a curve should be a straight line, since it yields

FIGURE 11
The cycloid provides the solution
to both the brachistochrone
and the tautochrone problem.

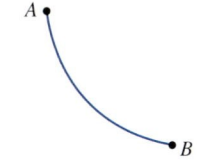

(**a**) The brachistochrone problem

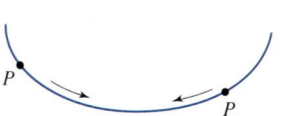

(**b**) The tautochrone problem

the shortest distance between the two points. But the velocity of the particle moving on the straight line will build up comparatively slowly, whereas if we take a curve that is steeper near A, even though the path becomes longer, the particle will cover a large portion of the distance at a greater speed. The problem was solved by Johann Bernoulli, his older brother Jacob Bernoulli, Leibniz, Newton, and l'Hôpital. They found that the curve of quickest descent is an inverted arc of a cycloid (Figure 11a). As it turns out, this same curve is also the solution to the tautochrone problem.

9.2 CONCEPT QUESTIONS

1. What is a plane curve? Give an example of a plane curve that is not the graph of a function.

2. What is the difference between the curve C_1 with parametric representation $x = \cos t$ and $y = \sin t$, where $0 \le t \le 2\pi$, and the curve C_2 with parametric representation $x = \sin t$ and $y = \cos t$, where $0 \le t \le 2\pi$?

3. Describe the relationship between the curve C_1 with parametric equations $x = f(t)$ and $y = g(t)$, where $0 \le t \le 1$, and the curve C_2 with parametric equations $x = f(1 - t)$ and $y = g(1 - t)$, where $0 \le t \le 1$.

9.2 EXERCISES

In Exercises 1–28, (a) find a rectangular equation whose graph contains the curve C with the given parametric equations, and (b) sketch the curve C and indicate its orientation.

1. $x = 2t + 1$, $y = t - 3$

2. $x = t - 2$, $y = 2t - 1$; $-1 \le t \le 5$

3. $x = \sqrt{t}$, $y = 9 - t$

4. $x = t^2$, $y = t - 1$; $0 \le t \le 3$

5. $x = t^2 + 1$, $y = 2t^2 - 1$; $-2 \le t \le 2$

6. $x = t^3$, $y = 2t + 1$

7. $x = t^2$, $y = t^3$; $-2 \le t \le 2$

8. $x = 1 + \dfrac{1}{t}$, $y = t + 1$

9. $x = 2 \sin \theta$, $y = 2 \cos \theta$; $0 \le \theta \le 2\pi$

10. $x = \cos 2\theta$, $y = 3 \sin \theta$; $0 \le \theta \le 2\pi$

11. $x = 2 \sin \theta$, $y = 3 \cos \theta$; $0 \le \theta \le 2\pi$

12. $x = \cos \theta + 1$, $y = \sin \theta - 2$; $0 \le \theta \le 2\pi$

13. $x = 2 \cos \theta + 2$, $y = 3 \sin \theta - 1$; $0 \le \theta \le 2\pi$

14. $x = \sin \theta + 3$, $y = 3 \cos \theta + 1$; $0 \le \theta \le 2\pi$

15. $x = \cos \theta$, $y = \cos 2\theta$

16. $x = \sec \theta$, $y = \cos \theta$

17. $x = \sec \theta$, $y = \tan \theta$; $-\dfrac{\pi}{2} < \theta < \dfrac{\pi}{2}$

18. $x = \cos^3 \theta$, $y = \sin^3 \theta$

19. $x = \sin^2 \theta$, $y = \sin^4 \theta$; $0 \le \theta \le \dfrac{\pi}{2}$

20. $x = e^t$, $y = e^{-t}$

21. $x = -e^t$, $y = e^{2t}$

22. $x = t^3$, $y = 3 \ln t$

23. $x = \ln 2t$, $y = t^2$

24. $x = e^t$, $y = \ln t$

25. $x = \cosh t$, $y = \sinh t$

26. $x = 3 \sinh t$, $y = 2 \cosh t$

27. $x = (t - 1)^2$, $y = (t - 1)^3$; $1 \le t \le 2$

28. $x = \dfrac{2t}{1 + t^2}$, $y = \dfrac{1 - t^2}{1 + t^2}$

ⓥ Videos for selected exercises are available online at **www.academic.cengage.com/login**.

In Exercises 29–34, the position of a particle at time t is (x, y). Describe the motion of the particle as t varies over the time interval [a, b].

29. $x = t + 1$, $y = \sqrt{t}$; $[0, 4]$

30. $x = \sin \pi t$, $y = \cos \pi t$; $[0, 6]$

31. $x = 1 + \cos t$, $y = 2 + \sin t$; $[0, 2\pi]$

32. $x = 1 + 2 \sin 2t$, $y = 2 + 4 \sin 2t$; $[0, 2\pi]$

33. $x = \sin t$, $y = \sin^2 t$; $[0, 3\pi]$

34. $x = e^{-t}$, $y = e^{2t-1}$; $[0, \infty)$

35. Flight Path of an Aircraft The position (x, y) of an aircraft flying in a fixed direction t seconds after takeoff is given by $x = \tan(0.025\pi t)$ and $y = \sec(0.025\pi t) - 1$, where x and y are measured in miles. Sketch the flight path of the aircraft for $0 \le t \le \frac{40}{3}$.

36. Trajectory of a Shell A shell is fired from a howitzer with a muzzle speed of v_0 ft/sec. If the angle of elevation of the howitzer is α, then the position of the shell after t sec is described by the parametric equations

$$x = (v_0 \cos \alpha)t \quad \text{and} \quad y = (v_0 \sin \alpha)t - \frac{1}{2}gt^2$$

where g is the acceleration due to gravity (32 ft/sec^2).
   a. Find the range of the shell.
   b. Find the maximum height attained by the shell.
   c. Show that the trajectory of the shell is a parabola by eliminating the parameter t.

37. Let $P_1(x_1, y_1)$ and $P_2(x_2, y_2)$ be two distinct points in the plane. Show that the parametric equations

$$x = x_1 + (x_2 - x_1)t \quad \text{and} \quad y = y_1 + (y_2 - y_1)t$$

describe (a) the line passing through P_1 and P_2 if $-\infty < t < \infty$ and (b) the line segment joining P_1 and P_2 if $0 \le t \le 1$.

38. Show that

$$x = a \cos t + h \quad \text{and} \quad y = b \sin t + k \quad 0 \le t \le 2\pi$$

are parametric equations of an ellipse with center at (h, k) and axes of lengths $2a$ and $2b$.

39. Show that

$$x = a \sec t + h \quad \text{and} \quad y = b \tan t + k$$
$$t \in \left(-\frac{\pi}{2}, \frac{\pi}{2}\right) \cup \left(\frac{\pi}{2}, \frac{3\pi}{2}\right)$$

are parametric equations of a hyperbola with center at (h, k) and transverse and conjugate axes of lengths $2a$ and $2b$, respectively.

40. Let P be a point located a distance d from the center of a circle of radius r. The curve traced out by P as the circle

rolls without slipping along a straight line is called a **trochoid**. (The cycloid is the special case of a trochoid with $d = r$.) Suppose that the circle rolls along the x-axis in the positive direction with $\theta = 0$ when the point P is at one of the lowest points on the trochoid. Show that the parametric equations of the trochoid are

$$x = r\theta - d \sin \theta \quad \text{and} \quad y = r - d \cos \theta$$

where θ is the same parameter as that for the cycloid. Sketch the trochoid for the cases in which $d < r$ and $d > r$.

41. The **witch of Agnesi** is the curve shown in the following figure. Show that the parametric equations of this curve are

$$x = 2a \cot \theta \quad \text{and} \quad y = 2a \sin^2 \theta$$

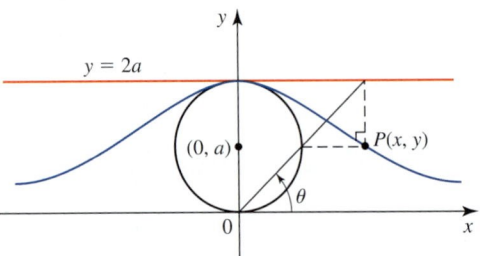

42. If a string is unwound from a circle of radius a in such a way that it is held taut in the plane of the circle, then its end P will trace a curve called the **involute of the circle**. Referring to the following figure, show that the parametric equations of the involute are

$$x = a(\cos \theta + \theta \sin \theta) \quad \text{and} \quad y = a(\sin \theta - \theta \cos \theta)$$

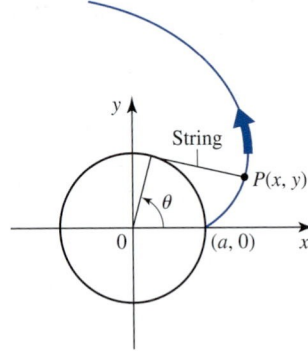

In Exercises 43–46, use a graphing utility to plot the curve with the given parametric equations.

43. $x = 2 \sin 3t$, $y = 3 \sin 1.5t$; $t \ge 0$

44. $x = \cos t + 5 \cos 3t$, $y = 6 \cos t - 5 \sin 3t$; $0 \le t \le 2\pi$

45. $x = 2 \cos t + \cos 2t$, $y = 2 \sin t - \sin 2t$; $0 \le t \le 2\pi$

46. $x = 3 \cos t + \cos 3t$, $y = 3 \sin t - \sin 3t$; $0 \le t \le 2\pi$

47. The **butterfly catastrophe curve,** which is described by the parametric equations

$$x = c(8at^3 + 24t^5) \quad \text{and} \quad y = c(-6at^2 - 15t^4)$$

occurs in the study of catastrophe theory. Plot the curve with $a = -7$ and $c = 0.03$ for t in the parameter interval $[-1.629, 1.629]$.

 48. The **swallowtail catastrophe curve,** which is described by the parametric equations

$$x = c(-2at - 4t^3) \quad \text{and} \quad y = c(at^2 + 3t^4)$$

occurs in the study of catastrophe theory. Plot the curve with $a = -2$ and $c = 0.5$ for t in the parameter interval $[-1.25, 1.25]$.

 49. The **Lissajous curves,** also known as **Bowditch curves,** have applications in physics, astronomy, and other sciences. They are described by the parametric equations

$$x = \sin(at + b\pi), \quad a \text{ a rational number,} \quad \text{and} \quad y = \sin t$$

Plot the curve with $a = 0.75$ and $b = 0$ for t in the parameter interval $[0, 8\pi]$.

 50. The **prolate cycloid** is the path traced out by a fixed point at a distance $b > a$ from the center of a rolling circle, where a

is the radius of the circle. The prolate cycloid is described by the parametric equations

$$x = a(t - b \sin t) \quad \text{and} \quad y = c(1 - d \cos t)$$

Plot the curve with $a = 0.1$, $b = 2$, $c = 0.25$, and $d = 2$ for t in the parameter interval $[-10, 10]$.

In Exercises 51–54, determine whether the statement is true or false. If it is true, explain why it is true. If it is false, give an example to show why it is false.

51. The parametric equations $x = \cos^2 t$ and $y = \sin^2 t$, where $-\infty < t < \infty$, have the same graph as $x + y = 1$.

52. The graph of a function $y = f(x)$ can always be represented by a pair of parametric equations.

53. The curve with parametric equations $x = f(t) + a$ and $y = g(t) + b$ is obtained from the curve C with parametric equations $x = f(t)$ and $y = g(t)$ by shifting the latter horizontally and vertically.

54. The ellipse with center at the origin and major and minor axes a and b, respectively, can be obtained from the circle with equations $x = f(t) = \cos t$ and $y = g(t) = \sin t$ by multiplying $f(t)$ and $g(t)$ by appropriate nonzero constants.

9.3   The Calculus of Parametric Equations

■ Tangent Lines to Curves Defined by Parametric Equations

Suppose that C is a smooth curve that is parametrized by the equations $x = f(t)$ and $y = g(t)$ with parameter interval I and we wish to find the slope of the tangent line to the curve at the point P. (See Figure 1.) Let t_0 be the point in I that corresponds to P, and let (a, b) be the subinterval of I containing t_0 corresponding to the highlighted portion of the curve C in Figure 1. This subset of C is the graph of a function of x, as you can verify using the Vertical Line Test. (The general conditions that f and g must satisfy for this to be true are given in Exercise 66.)

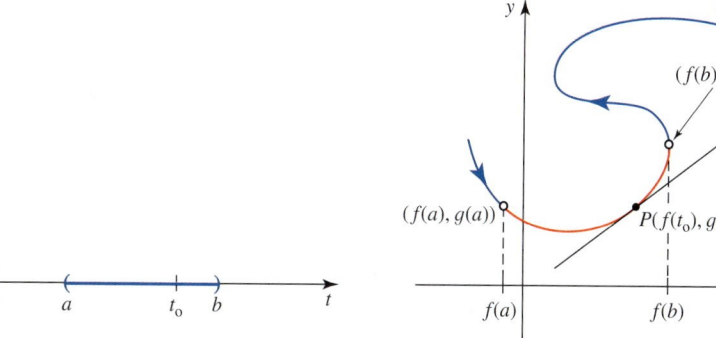

FIGURE 1
We want to find the slope of the tangent line to the curve at the point P.

Parameter interval

Let's denote this function by F so that $y = F(x)$, where $f(a) < x < f(b)$. Since $x = f(t)$ and $y = g(t)$, we may rewrite this equation in the form

$$g(t) = F[f(t)]$$

Using the Chain Rule, we obtain

$$g'(t) = F'[f(t)]f'(t)$$

$$= F'(x)f'(t) \qquad \text{Replace } f(t) \text{ by } x.$$

If $f'(t) \neq 0$, we can solve for $F'(x)$, obtaining

$$F'(x) = \frac{g'(t)}{f'(t)}$$

which can also be written

$$\frac{dy}{dx} = \frac{\dfrac{dy}{dt}}{\dfrac{dx}{dt}} \qquad \text{if} \quad \frac{dx}{dt} \neq 0 \tag{1}$$

The required slope of the tangent line at P is then found by evaluating Equation (1) at t_0. Observe that Equation (1) enables us to solve the problem without eliminating t.

EXAMPLE 1 Find an equation of the tangent line to the curve

$$x = \sec t \qquad y = \tan t \qquad -\frac{\pi}{2} < t < \frac{\pi}{2}$$

at the point where $t = \pi/4$. (See Figure 2.)

Solution The slope of the tangent line at any point (x, y) on the curve is

$$\frac{dy}{dx} = \frac{\dfrac{dy}{dt}}{\dfrac{dx}{dt}}$$

$$= \frac{\sec^2 t}{\sec t \tan t} = \frac{\sec t}{\tan t}$$

In particular, the slope of the tangent line at the point where $t = \pi/4$ is

$$\frac{dy}{dx}\bigg|_{t=\pi/4} = \frac{\sec \dfrac{\pi}{4}}{\tan \dfrac{\pi}{4}} = \frac{\sqrt{2}}{1} = \sqrt{2}$$

Also, when $t = \pi/4$, we have $x = \sec(\pi/4) = \sqrt{2}$ and $y = \tan(\pi/4) = 1$ giving $(\sqrt{2}, 1)$ as the point of tangency. Finally, using the point-slope form of the equation of a line, we obtain the required equation:

$$y - 1 = \sqrt{2}(x - \sqrt{2}) \qquad \text{or} \qquad y = \sqrt{2}x - 1 \qquad \blacksquare$$

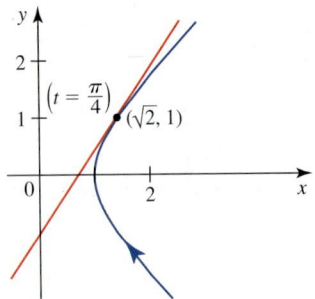

FIGURE 2
The tangent line to the curve at $(\sqrt{2}, 1)$

Horizontal and Vertical Tangents

A curve C represented by the parametric equations $x = f(t)$ and $y = g(t)$ has a **horizontal** tangent at a point (x, y) on C where $dy/dt = 0$ and $dx/dt \neq 0$ and a **vertical** tangent where $dx/dt = 0$ and $dy/dt \neq 0$, so that dy/dx is undefined there. Points where both dy/dt and dx/dt are equal to zero are candidates for horizontal or vertical tangents and may be investigated by using l'Hôpital's Rule.

EXAMPLE 2 A curve C is defined by the parametric equations $x = t^2$ and $y = t^3 - 3t$.

a. Find the points on C where the tangent lines are horizontal or vertical.
b. Find the x- and y-intercepts of C.
c. Sketch the graph of C.

Solution
a. Setting $dy/dt = 0$ gives $3t^2 - 3 = 0$, or $t = \pm 1$. Since $dx/dt = 2t \neq 0$ at these values of t, we conclude that C has horizontal tangents at the points on C corresponding to $t = \pm 1$, that is, at $(1, -2)$ and $(1, 2)$. Next, setting $dx/dt = 0$ gives $2t = 0$, or $t = 0$. Since $dy/dt \neq 0$ for this value of t, we conclude that C has a vertical tangent at the point corresponding to $t = 0$, or at $(0, 0)$.
b. To find the x-intercepts, we set $y = 0$, which gives $t^3 - 3t = t(t^2 - 3) = 0$, or $t = -\sqrt{3}, 0$, and $\sqrt{3}$. Substituting these values of t into the expression for x gives 0 and 3 as the x-intercepts. Next, setting $x = 0$ gives $t = 0$, which, when substituted into the expression for y, gives 0 as the y-intercept.
c. Using the information obtained in parts (a) and (b), we obtain the graph of C shown in Figure 3. ∎

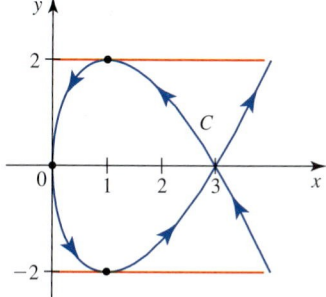

FIGURE 3
The graph of $x = t^2$, $y = t^3 - 3t$, and the tangent lines at $t = \pm 1$

Finding d^2y/dx^2 from Parametric Equations

Suppose that the parametric equations $x = f(t)$ and $y = g(t)$ define y as a twice-differentiable function of x over some suitable interval. Then d^2y/dx^2 may be found from Equation (1) with another application of the Chain Rule.

$$\frac{d^2y}{dx^2} = \frac{d}{dx}\left(\frac{dy}{dx}\right) = \frac{\dfrac{d}{dt}\left(\dfrac{dy}{dx}\right)}{\dfrac{dx}{dt}} \qquad \text{if} \quad \frac{dx}{dt} \neq 0 \tag{2}$$

Higher-order derivatives are found in a similar manner.

EXAMPLE 3 Find $\dfrac{d^2y}{dx^2}$ if $x = t^2 - 4$ and $y = t^3 - 3t$.

Solution First, we use Equation (1) to compute

$$\frac{dy}{dx} = \frac{\dfrac{dy}{dt}}{\dfrac{dx}{dt}} = \frac{3t^2 - 3}{2t}$$

Then, using Equation (2), we obtain

$$\frac{d^2y}{dx^2} = \frac{\dfrac{d}{dt}\left(\dfrac{dy}{dx}\right)}{\dfrac{dx}{dt}} = \frac{\dfrac{d}{dt}\left(\dfrac{3t^2-3}{2t}\right)}{2t}$$

$$= \frac{\dfrac{(2t)(6t)-(3t^2-3)(2)}{4t^2}}{2t} \qquad \text{Use the Quotient Rule.}$$

$$= \frac{6t^2+6}{8t^3} = \frac{3(t^2+1)}{4t^3}$$

■ The Length of a Smooth Curve

In Section 5.4 we showed that the length L of the graph of a smooth function f on an interval $[a, b]$ can be found by using the formula

$$L = \int_a^b \sqrt{1 + [f'(x)]^2}\, dx \qquad (3)$$

We now generalize this result to include curves defined by parametric equations. We begin by explaining what is meant by a *smooth* curve defined parametrically. Suppose that C is represented by $x = f(t)$ and $y = g(t)$ on a parameter interval I. Then C is **smooth** if f' and g' are continuous on I and are not simultaneously zero, except possibly at the endpoints of I. A smooth curve is devoid of corners or cusps. For example, the cycloid that we discussed in Section 9.2 (see Figure 9 in that section) has sharp corners at the values $x = 2n\pi a$ and, therefore, is not smooth. However, it is smooth between these points.

Now let $P = \{t_0, t_1, \ldots, t_n\}$ be a regular partition of the parameter interval $[a, b]$. Then the point $P_k(f(t_k), g(t_k))$ lies on C, and the length of C is approximated by the length of the polygonal curve with vertices $P_0, P_1, \ldots, P_n$. (See Figure 4.) Thus,

$$L \approx \sum_{k=1}^n d(P_{k-1}, P_k) \qquad (4)$$

where

$$d(P_{k-1}, P_k) = \sqrt{[f(t_k) - f(t_{k-1})]^2 + [g(t_k) - g(t_{k-1})]^2}$$

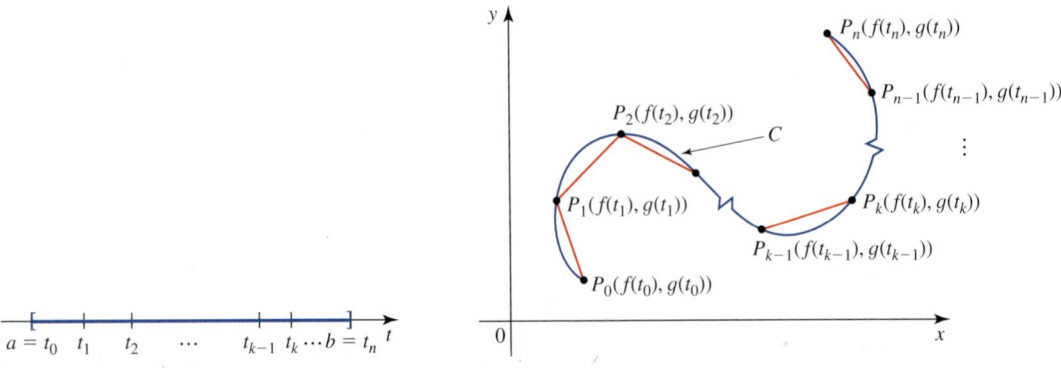

Parameter interval

FIGURE 4
The length of C is approximated by the length of the polygonal curve (the red lines).

Now, since f and g both have continuous derivatives, we can use the Mean Value Theorem to write

$$f(t_k) - f(t_{k-1}) = f'(t_k^*)(t_k - t_{k-1})$$

and

$$g(t_k) - g(t_{k-1}) = g'(t_k^{**})(t_k - t_{k-1})$$

where t_k^* and t_k^{**} are numbers in (t_{k-1}, t_k). Substituting these expressions into Equation (4) gives

$$L \approx \sum_{k=1}^{n} d(P_{k-1}, P_k) = \sum_{k=1}^{n} \sqrt{[f'(t_k^*)]^2 + [g'(t_k^{**})]^2} \, \Delta t \tag{5}$$

As in Section 5.4, we define

$$L = \lim_{n \to \infty} \sum_{k=1}^{n} d(P_{k-1}, P_k)$$

$$= \lim_{n \to \infty} \sum_{k=1}^{n} \sqrt{[f'(t_k^*)]^2 + [g'(t_k^{**})]^2} \, \Delta t \tag{6}$$

The sum in Equation (6) looks like a Riemann sum of the function $\sqrt{[f']^2 + [g']^2}$, but it is not, because t_k^* is not necessarily equal to t_k^{**}. But it can be shown that the limit in Equation (6) is the same as that of an expression in which $t_k^* = t_k^{**}$. Therefore,

$$L = \int_a^b \sqrt{[f'(t)]^2 + [g'(t)]^2} \, dt$$

and we have the following result.

> **THEOREM 1   Length of a Smooth Curve**
>
> Let C be a smooth curve represented by the parametric equations $x = f(t)$ and $y = g(t)$ with parameter interval $[a, b]$. If C does not intersect itself, except possibly for $t = a$ and $t = b$, then the length of C is
>
> $$L = \int_a^b \sqrt{[f'(t)]^2 + [g'(t)]^2} \, dt = \int_a^b \sqrt{\left(\frac{dx}{dt}\right)^2 + \left(\frac{dy}{dt}\right)^2} \, dt \tag{7}$$

Note   Equation (7) is consistent with Equations (3) and (4) of Section 5.4. Both have the form $L = \int ds$, where $(ds)^2 = (dx)^2 + (dy)^2$.   ∎

EXAMPLE 4   Find the length of one arch of the cycloid

$$x = a(\theta - \sin \theta) \qquad y = a(1 - \cos \theta)$$

(See Example 6 in Section 9.2.)

Solution   One arch of the cycloid is traced out by letting θ run from 0 to 2π. Now

$$\frac{dx}{d\theta} = a(1 - \cos \theta) \qquad \text{and} \qquad \frac{dy}{d\theta} = a \sin \theta$$

Therefore, using Equation (7), we find the required length to be

$$L = \int_0^{2\pi} \sqrt{\left(\frac{dx}{d\theta}\right)^2 + \left(\frac{dy}{d\theta}\right)^2}\, d\theta = \int_0^{2\pi} \sqrt{a^2(1 - \cos\theta)^2 + a^2 \sin^2\theta}\, d\theta$$

$$= \int_0^{2\pi} \sqrt{a^2 - 2a^2\cos\theta + a^2\cos^2\theta + a^2\sin^2\theta}\, d\theta$$

$$= a \int_0^{2\pi} \sqrt{2(1 - \cos\theta)}\, d\theta \qquad \textcolor{orange}{\sin^2\theta + \cos^2\theta = 1}$$

To evaluate this integral, we use the identity $\sin^2 x = \frac{1}{2}(1 - \cos 2x)$ with $\theta = 2x$. This gives $1 - \cos\theta = 2\sin^2(\theta/2)$, so

$$L = a \int_0^{2\pi} \sqrt{4 \sin^2 \frac{\theta}{2}}\, d\theta$$

$$= 2a \int_0^{2\pi} \sin \frac{\theta}{2}\, d\theta \qquad \textcolor{orange}{\sin\frac{\theta}{2} \ge 0 \text{ on } [0, 2\pi]}$$

$$= -4a \left[\cos \frac{\theta}{2} \right]_0^{2\pi}$$

$$= -4a(-1 - 1) = 8a \qquad \blacksquare$$

■ The Area of a Surface of Revolution

Recall that the formulas $S = 2\pi \int y\, ds$ and $S = 2\pi \int x\, ds$ (Formulas 11 and 12 of Section 5.4) give the area of the surface of revolution that is obtained by revolving the graph of a function about the x- and y-axes, respectively. These formulas are valid for finding the area of the surface of revolution that is obtained by revolving a curve described by parametric equations about the x- and the y-axes, provided that we replace the element of arc length ds by the appropriate expression. These results, which may be derived by using the method used to derive Equation (7), are stated in the next theorem.

THEOREM 2   Area of a Surface of Revolution

Let C be a smooth curve represented by the parametric equations $x = f(t)$ and $y = g(t)$ with parameter interval $[a, b]$, and suppose that C does not intersect itself, except possibly for $t = a$ and $t = b$. If $g(t) \ge 0$ for all t in $[a, b]$, then the area S of the surface obtained by revolving C about the x-axis is

$$S = 2\pi \int_a^b y\sqrt{[f'(t)]^2 + [g'(t)]^2}\, dt = 2\pi \int_a^b y\sqrt{\left(\frac{dx}{dt}\right)^2 + \left(\frac{dy}{dt}\right)^2}\, dt \qquad (8)$$

If $f(t) \ge 0$ for all t in $[a, b]$, then the area S of the surface that is obtained by revolving C about the y-axis is

$$S = 2\pi \int_a^b x\sqrt{[f'(t)]^2 + [g'(t)]^2}\, dt = 2\pi \int_a^b x\sqrt{\left(\frac{dx}{dt}\right)^2 + \left(\frac{dy}{dt}\right)^2}\, dt \qquad (9)$$

EXAMPLE 5   Show that the surface area of a sphere of radius r is $4\pi r^2$.

Solution   We obtain this sphere by revolving the semicircle

$$x = r\cos t \qquad y = r\sin t \qquad 0 \le t \le \pi$$

about the *x*-axis. Using Equation (8), the surface area of the sphere is

$$S = 2\pi \int_0^\pi r \sin t \sqrt{(-r \sin t)^2 + (r \cos t)^2} \, dt$$

$$= 2\pi r \int_0^\pi \sin t \sqrt{r^2(\sin^2 t + \cos^2 t)} \, dt$$

$$= 2\pi r \int_0^\pi r \sin t \, dt \qquad \sin^2 t + \cos^2 t = 1$$

$$= 2\pi r^2 \left[-\cos t \right]_0^\pi = 2\pi r^2 [-(-1) + 1] = 4\pi r^2$$

9.3 CONCEPT QUESTIONS

1. Suppose that *C* is a smooth curve with parametric equations $x = f(t)$ and $y = g(t)$ and parameter interval *I*. Write an expression for the slope of the tangent line to *C* at the point (x_0, y_0) corresponding to t_0 in *I*.
2. Suppose that *C* is a smooth curve with parametric equations $x = f(t)$ and $y = g(t)$ and parameter interval $[a, b]$. Furthermore, suppose that *C* does not cross itself, except possibly for $t = a$. Write an expression giving the length of *C*.
3. Suppose that *C* is a smooth curve with parametric equations $x = f(t)$ and $y = g(t)$ and parameter interval $[a, b]$. Suppose,

further, that *C* does not intersect itself, except possibly for $t = a$ and $t = b$.
 a. Write an integral giving the area of the surface obtained by revolving *C* about the *x*-axis assuming that $g(t) \geq 0$ for all *t* in $[a, b]$.
 b. Write an integral giving the area of the surface obtained by revolving *C* about the *y*-axis assuming that $f(t) \geq 0$ for all *t* in $[a, b]$.

9.3 EXERCISES

In Exercises 1–6, find the slope of the tangent line to the curve at the point corresponding to the value of the parameter.

1. $x = t^2 + 1$, $y = t^2 - t$; $t = 1$
2. $x = t^3 - t$, $y = t^2 - 2t + 2$; $t = 2$
3. $x = \sqrt{t}$, $y = \dfrac{1}{t}$; $t = 1$
4. $x = e^{2t}$, $y = \ln t$; $t = 1$
5. $x = 2 \sin \theta$, $y = 3 \cos \theta$; $\theta = \dfrac{\pi}{4}$
6. $x = 2(\theta - \sin \theta)$, $y = 2(1 - \cos \theta)$; $\theta = \dfrac{\pi}{6}$

In Exercises 7 and 8, find an equation of the tangent line to the curve at the point corresponding to the value of the parameter.

7. $x = 2t - 1$, $y = t^3 - t^2$; $t = 1$
8. $x = \theta \cos \theta$, $y = \theta \sin \theta$; $\theta = \dfrac{\pi}{2}$

In Exercises 9 and 10, find an equation of the tangent line to the curve at the given point. Then sketch the curve and the tangent line(s).

9. $x = t^2 + t$, $y = t^2 - t^3$; $(0, 2)$
10. $x = e^t$, $y = e^{-t}$; $(1, 1)$

In Exercises 11 and 12, find the points on the curve at which the slope of the tangent line is m.

11. $x = 2t^2 - 1$, $y = t^3$; $m = 3$
12. $x = t^3$, $y = t^2 + t$; $m = 1$

In Exercises 13–16, find the points on the curve at which the tangent line is either horizontal or vertical. Sketch the curve.

13. $x = t^2 - 4$, $y = t^3 - 3t$
14. $x = t^3 - 3t$, $y = t^2$
15. $x = 1 + 3 \cos t$, $y = 2 - 2 \sin t$
16. $x = \sin t$, $y = \sin 2t$

In Exercises 17–24, find dy/dx and d²y/dx².

17. $x = 3t^2 + 1, \quad y = 2t^3$ **18.** $x = t^3 - t, \quad y = t^3 + 2t^2$

19. $x = \sqrt{t}, \quad y = \dfrac{1}{t}$ **20.** $x = \sin 2t, \quad y = \cos 2t$

21. $x = \theta + \cos \theta, \quad y = \theta - \sin \theta$

22. $x = e^{-t}, \quad y = e^{2t}$

23. $x = \cosh t, \quad y = \sinh t$

24. $x = \sqrt{t^2 + 1}, \quad y = t \ln t$

25. Let C be the curve defined by the parametric equations $x = t^2$ and $y = t^3 - 3t$ (see Example 2). Find d^2y/dx^2, and use this result to determine the intervals where C is concave upward and where it is concave downward.

26. Show that the curve defined by the parametric equations $x = t^2$ and $y = t^3 - 3t$ crosses itself. Find equations of the tangent lines to the curve at that point (see Example 2).

27. The parametric equations of the astroid $x^{2/3} + y^{2/3} = a^{2/3}$ are $x = a \cos^3 t$ and $y = a \sin^3 t$. (Verify this!) Find an expression for the slope of the tangent line to the astroid in terms of t. At what points on the astroid is the slope of the tangent line equal to -1? Equal to 1?

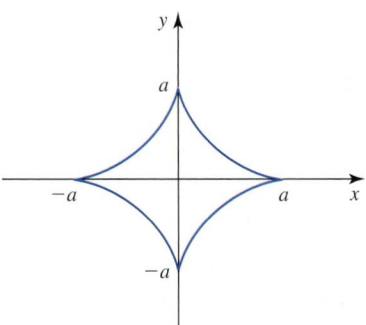

28. Find dy/dx and d^2y/dx^2 if

$$x = \int_1^t \frac{\sin u}{u}\, du \quad \text{and} \quad y = \int_2^{\ln t} e^u \, du$$

29. The function $y = f(x)$ is defined by the parametric equations

$$x = t^5 + 5t^3 + 10t + 2 \quad \text{and} \quad y = 2t^3 - 3t^2 - 12t + 1$$
$$-2 \le t \le 2$$

Find the absolute maximum and the absolute minimum values of f.

30. Find the points on the curve with parametric equations $x = t^3 - t$ and $y = t^2$ at which the tangent line is parallel to the line with parametric equations $x = 2t$ and $y = 2t + 4$.

In Exercises 31–36, find the length of the curve defined by the parametric equations.

31. $x = 2t^2, \quad y = 3t^3; \quad 0 \le t \le 1$

32. $x = 2t^{3/2}, \quad y = 3t + 1; \quad 0 \le t \le 4$

33. $x = \sin^2 t, \quad y = \cos 2t; \quad 0 \le t \le \pi$

34. $x = e^t \cos t, \quad y = e^t \sin t; \quad 0 \le t \le \pi$

35. $x = a(\cos t + t \sin t), \quad y = a(\sin t - t \cos t); \quad 0 \le t \le \frac{\pi}{2}$

36. $x = (t^2 - 2)\sin t + 2t \cos t, \quad y = (2 - t^2)\cos t + 2t \sin t;$ $0 \le t \le \pi$

37. Find the length of the cardioid with parametric equations

$$x = a(2 \cos t - \cos 2t) \quad \text{and} \quad y = a(2 \sin t - \sin 2t)$$

38. Find the length of the astroid with parametric equations

$$x = a \cos^3 t \quad \text{and} \quad y = a \sin^3 t$$

(See the figure for Exercise 27. Compare with Exercise 29 in Section 5.4.)

39. The position of an object at any time t is (x, y), where $x = \cos^2 t$ and $y = \sin^2 t$, $0 \le t \le 2\pi$. Find the distance covered by the object as t runs from $t = 0$ to $t = 2\pi$.

40. The following figure shows the course taken by a yacht during a practice run. The parametric equations of the course are

$$x = 4\sqrt{2} \sin t \qquad y = \sin 2t \qquad 0 \le t \le 2\pi$$

where x and y are measured in miles. Find the length of the course.

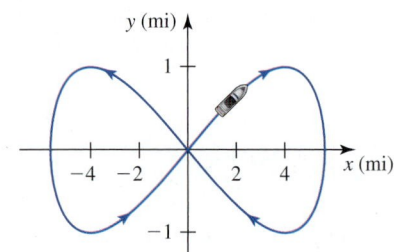

41. Path of a Boat Two towns, A and B, are located directly opposite each other on the banks of a river that is 1600 ft wide and flows east with a constant speed of 4 ft/sec. A boat leaving Town A travels with a constant speed of 18 ft/sec always aimed toward Town B. It can be shown that the path of the boat is given by the parametric equations

$$x = 800(t^{7/9} - t^{11/9}) \qquad y = 1600t \qquad 0 \le t \le 1$$

Find the distance covered by the boat in traveling from A to B.

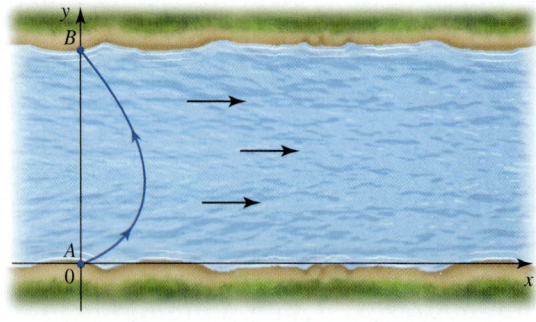

42. Trajectory of an Electron An electron initially located at the origin of a coordinate system is projected horizontally into a uniform electric field with magnitude E and directed upward. If the initial speed of the electron is v_0, then its trajectory is

$$x = v_0 t \qquad y = -\frac{1}{2}\left(\frac{eE}{m}\right)t^2$$

where e is the charge of the electron and m is its mass. Show that the trajectory of the electron is a parabola.

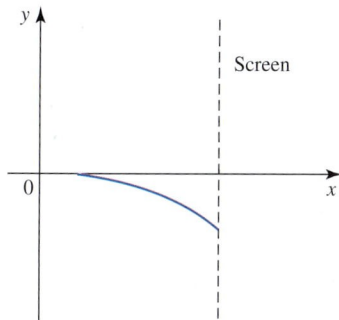

Note: The deflection of electrons by an electric field is used to control the direction of an electron beam in an electron gun.

43. Refer to Exercise 42. If a screen is placed along the vertical line $x = a$, at what point will the electron beam hit the screen?

 44. Find the point that is located one quarter of the way along the arch of the cycloid

$$x = a(t - \sin t) \qquad y = a(1 - \cos t) \qquad 0 \le t \le 2\pi$$

as measured from the origin. What is the slope of the tangent line to the cycloid at that point? Plot the arch of the cycloid and the tangent line on the same set of axes.

 45. The **cornu spiral** is a curve defined by the parametric equations

$$x = C(t) = \int_0^t \cos(\pi u^2/2)\, du \qquad y = S(t) = \int_0^t \sin(\pi u^2/2)\, du$$

where C and S are called Fresnel integrals. They are used to explain the phenomenon of light diffraction.
a. Plot the spiral. Describe the behavior of the curve as $t \to \infty$ and as $t \to -\infty$.
b. Find the length of the spiral from $t = 0$ to $t = a$.

46. Suppose that the graph of a nonnegative function F on an interval $[a, b]$ is represented by the parametric equations $x = f(t)$ and $y = g(t)$ for t in $[\alpha, \beta]$. Show that the area of the region under the graph of F is given by

$$\int_\alpha^\beta g(t)f'(t)\, dt \qquad \text{or} \qquad \int_\beta^\alpha g(t)f'(t)\, dt$$

47. Use the result of Exercise 46 to find the area of the region under one arch of the cycloid $x = a(\theta - \sin \theta)$, $y = a(1 - \cos \theta)$.

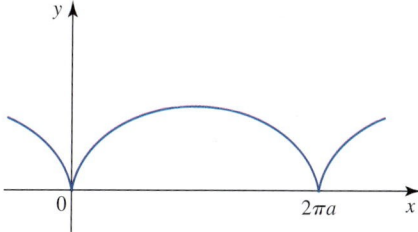

48. Use the result of Exercise 46 to find the area of the region enclosed by the ellipse with parametric equations $x = a \cos \theta$, $y = b \sin \theta$, where $0 \le \theta \le 2\pi$.

49. Use the result of Exercise 46 to find the area of the region enclosed by the astroid $x = a \cos^3 \theta$, $y = a \sin^3 \theta$. (See the figure for Exercise 27.)

50. Use the result of Exercise 46 to find the area of the region enclosed by the curve $x = a \sin t$, $y = b \sin 2t$.

51. Use the result of Exercise 46 to find the area of the region lying inside the course taken by the yacht of Exercise 40.

In Exercises 52–57, find the area of the surface obtained by revolving the curve about the x-axis.

52. $x = t$, $y = 2 - t$; $0 \le t \le 2$

53. $x = t^3$, $y = t^2$; $0 \le t \le 1$

54. $x = \sqrt{3}t^2$, $y = t - t^3$; $0 \le t \le 1$

55. $x = \dfrac{1}{3}t^3$, $y = 4 - \dfrac{1}{2}t^2$; $0 \le t \le 2\sqrt{2}$

56. $x = e^t \sin t$, $y = e^t \cos t$; $0 \le t \le \frac{\pi}{2}$

57. $x = t - \sin t$, $y = 1 - \cos t$; $0 \le t \le 2\pi$

In Exercises 58–61, find the area of the surface obtained by rotating the curve about the y-axis.

58. $x = t$, $y = 2t$; $0 \le t \le 4$

59. $x = 3t^2$, $y = 2t^3$; $0 \le t \le 1$

60. $x = a \cos t$, $y = b \sin t$; $-\frac{\pi}{2} \le t \le \frac{\pi}{2}$

61. $x = e^t - t$, $y = 4e^{t/2}$; $0 \le t \le 1$

62. Find the area of the surface obtained by revolving the cardioid

$$x = a(2 \cos t - \cos 2t) \qquad y = a(2 \sin t - \sin 2t)$$

about the x-axis.

63. Find the area of the surface obtained by revolving the astroid

$$x = a \cos^3 t \qquad y = a \sin^3 t$$

about the x-axis.

64. Find the areas of the surface obtained by revolving one arch of the cycloid $x = a(\theta - \sin \theta)$, $y = a(1 - \cos \theta)$ about the x- and y-axes.

65. Find the surface area of the torus obtained by revolving the circle $x^2 + (y - b)^2 = r^2$ $(0 < r < b)$ about the x-axis. **Hint:** Represent the equation of the circle in parametric form: $x = r \cos t$, $y = b + r \sin t$, $0 \le t \le 2\pi$.

66. Show that if f' is continuous and $f'(t) \ne 0$ for $a \le t \le b$, then the parametric curve defined by $x = f(t)$ and $y = g(t)$ for $a \le t \le b$ can be put in the form $y = F(x)$.

cas *In Exercises 67–70, (a) plot the curve defined by the parametric equations and (b) estimate the arc length of the curve accurate to four decimal places.*

67. $x = 2t^2$, $y = t - t^3$; $0 \le t \le 1$

68. $x = \sin(0.5t + 0.4\pi)$, $y = \sin t$; $0 < t \le 4\pi$

69. $x = 0.2(6 \cos t - \cos 6t)$, $y = 0.2(6 \sin t - \sin 6t)$; $0 \le t \le 2\pi$

70. $x = 2t(1 - t^2)$, $y = -t^2\left(1 - \dfrac{3}{2}t^2\right)$; $-2 < t < 2$

(swallowtail castastrophe)

cas 71. Use a calculator or computer to approximate the area of the surface obtained by revolving the curve

$$x = 4 \sin 2t \qquad y = 2 \cos 3t \qquad 0 \le t \le \tfrac{\pi}{6}$$

about the x-axis.

72. a. Find an expression for the arc length of the curve defined by the parametric equations

$$x = f''(t)\cos t + f'(t)\sin t \qquad y = -f''(t)\sin t + f'(t)\cos t$$

where $a \le t \le b$ and f has continuous third-order derivatives.

b. Use the result of part (a) to find the arc length of the curve $x = 6t \cos t + 3t^2 \sin t$ and $y = -6t \sin t + 3t^2 \cos t$, where $0 \le t \le 1$.

73. Show that

$$x = \frac{2at}{1 + t^2} \qquad y = \frac{a(1 - t^2)}{1 + t^2}$$

where $a > 0$ and $-\infty < t < \infty$, are parametric equations of a circle. What are its center and radius?

74. Use the parametric representation of a circle in Exercise 73 to show that the circumference of a circle of radius a is $2\pi a$.

75. Find parametric equations for the *Folium of Descartes*, $x^3 + y^3 = 3axy$ with parameter $t = y/x$.

cas 76. Use the parametric representation of the *Folium of Descartes* to estimate the length of the loop.

cas 77. Show that the length of the ellipse $x = a \cos t$, $y = b \sin t$, $0 \le t \le 2\pi$, where $a > b > 0$, is given by

$$L = 4a\int_0^{\pi/2} \sqrt{1 - e^2 \sin^2 t}\, dt$$

where

$$e = \frac{c}{a} = \frac{\sqrt{a^2 - b^2}}{a}$$

is the eccentricity of the ellipse.

Note: The integral is called an *elliptical integral of the second kind.*

cas 78. Use a computer or calculator and the result of Exercise 77 to estimate the circumference of the ellipse

$$\frac{x^2}{100} + \frac{y^2}{36} = 1$$

accurate to three decimal places.

In Exercises 79–80, determine whether the statement is true or false. If it is true, explain why it is true. If it is false, give an example to show why it is false.

79. If $x = f(t)$ and $y = g(t)$, f and g have second-order derivatives, and $f'(t) \ne 0$, then

$$\frac{d^2y}{dx^2} = \frac{f'(t)g''(t) - g'(t)f''(t)}{[f'(t)]^2}$$

80. The curve with parametric equations $x = f(t)$ and $y = g(t)$ is a line if and only if f and g are both linear functions of t.

9.4 Polar Coordinates

The curve shown in Figure 1a is a lemniscate, and the one shown in Figure 1b is called a cardioid. The rectangular equations of these curves are

$$(x^2 + y^2)^2 = 4(x^2 - y^2) \qquad \text{and} \qquad x^4 - 2x^3 + 2x^2y^2 - 2xy^2 - y^2 + y^4 = 0$$

respectively. As you can see, these equations are somewhat complicated. For example, they will not prove very helpful if we want to calculate the area enclosed by the two loops of the lemniscate shown in Figure 1a or the length of the cardioid shown in Figure 1b.

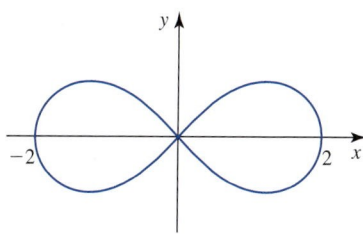
(**a**) A lemniscate

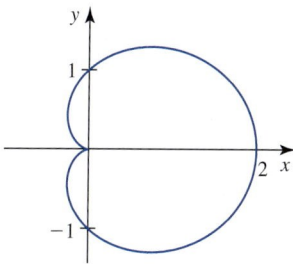
(**b**) A cardioid

FIGURE 1

A rectangular equation of the lemniscate in part (a) is $(x^2 + y^2)^2 = 4(x^2 - y^2)$, and an equation of the cardioid in part (b) is $x^4 - 2x^3 + 2x^2y^2 - 2xy^2 - y^2 + y^4 = 0$.

A question that arises naturally is: Is there a coordinate system other than the rectangular system that we can use to give a simpler representation for curves such as the lemniscate and cardioid? One such system is the *polar coordinate system.*

■ The Polar Coordinate System

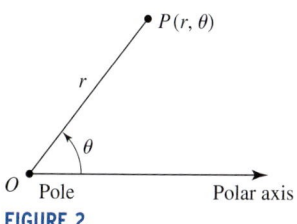

FIGURE 2

To construct the polar coordinate system, we fix a point O called the **pole** (or **origin**) and draw a ray (half-line) emanating from O called the **polar axis.** Suppose that P is any point in the plane, let r denote the distance from O to P, and let θ denote the angle (in degrees or radians) between the polar axis and the line segment OP. (See Figure 2.) Then the point P is represented by the ordered pair (r, θ), also written $P(r, \theta)$, where the numbers r and θ are called the **polar coordinates** of P.

The **angular coordinate** θ is positive if it is measured in the counterclockwise direction from the polar axis and negative if it is measured in the clockwise direction. The **radial coordinate** r may assume positive as well as negative values. If $r > 0$, then $P(r, \theta)$ is on the terminal side of θ and at a distance r from the origin. If $r < 0$, then $P(r, \theta)$ lies on the ray that is opposite the terminal side of θ and at a distance of $|r| = -r$ from the pole. (See Figure 3.) Also, by convention the pole O is represented by the ordered pair $(0, \theta)$ for *any* value of θ. Finally, a plane that is endowed with a polar coordinate system is referred to as an $r\theta$-plane.

FIGURE 3

(**a**) $r > 0$

(**b**) $r < 0$

EXAMPLE 1 Plot the following points in the $r\theta$-plane.

a. $\left(1, \frac{2\pi}{3}\right)$ **b.** $\left(2, -\frac{\pi}{4}\right)$ **c.** $\left(-2, \frac{\pi}{3}\right)$ **d.** $(2, -3\pi)$

Solution The points are plotted in Figure 4.

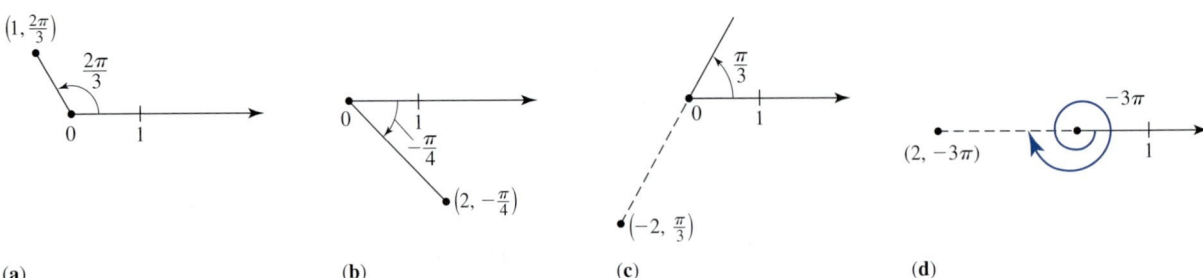

(a)   **(b)**   **(c)**   **(d)**

FIGURE 4
The points in Example 1

Unlike the representation of points in the rectangular system, the representation of points using polar coordinates is *not* unique. For example, the point (r, θ) can also be written as $(r, \theta + 2n\pi)$ or $(-r, \theta + (2n + 1)\pi)$, where n is any integer. Figures 5a and 5b illustrate this for the case $n = 1$ and $n = 0$, respectively.

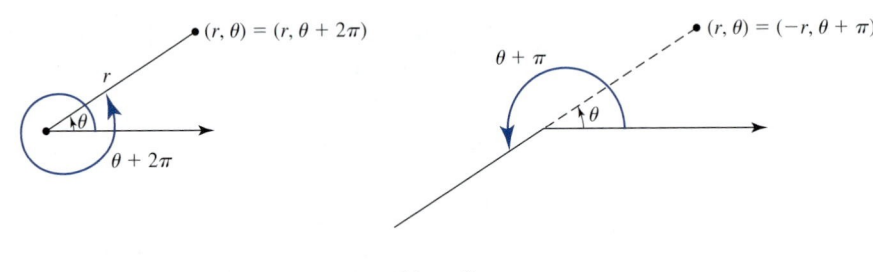

FIGURE 5
Representation of points using polar coordinates is not unique.

(a) $n = 1$ **(b)** $n = 0$

■ Relationship Between Polar and Rectangular Coordinates

To establish the relationship between polar and rectangular coordinates, let's superimpose an xy-plane on an $r\theta$-plane in such a way that the origins coincide and the positive x-axis coincides with the polar axis. Let P be any point in the plane other than the origin with rectangular representation (x, y) and polar representation (r, θ). Figure 6a shows a situation in which $r > 0$, and Figure 6b shows a situation in which $r < 0$. If $r > 0$, we see immediately from the figure that

$$\cos \theta = \frac{x}{r} \qquad \sin \theta = \frac{y}{r}$$

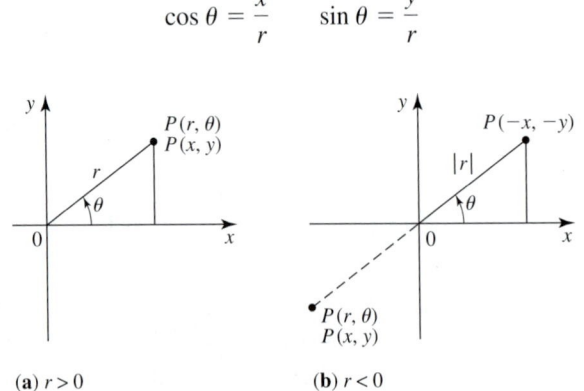

FIGURE 6
The relationship between polar and rectangular coordinates

(a) $r > 0$ **(b)** $r < 0$

so $x = r \cos \theta$ and $y = r \sin \theta$. If $r < 0$, we see by referring to Figure 6b that

$$\cos \theta = \frac{-x}{|r|} = \frac{-x}{-r} = \frac{x}{r} \qquad \sin \theta = \frac{-y}{|r|} = \frac{-y}{-r} = \frac{y}{r}$$

so again $x = r \cos \theta$ and $y = r \sin \theta$. Finally, in either case we have

$$x^2 + y^2 = r^2 \qquad \text{and} \qquad \tan \theta = \frac{y}{x} \qquad \text{if } x \neq 0$$

Relationship Between Rectangular and Polar Coordinates

Suppose that a point P (other than the origin) has representation (r, θ) in polar coordinates and (x, y) in rectangular coordinates. Then

$$x = r \cos \theta \qquad \text{and} \qquad y = r \sin \theta \tag{1}$$

$$r^2 = x^2 + y^2 \qquad \text{and} \qquad \tan \theta = \frac{y}{x} \qquad \text{if } x \neq 0 \tag{2}$$

EXAMPLE 2 The point $\left(4, \frac{\pi}{6}\right)$ is given in polar coordinates. Find its representation in rectangular coordinates.

Solution   Here, $r = 4$ and $\theta = \pi/6$. Using Equation (1), we obtain

$$x = r \cos \theta = 4 \cos \frac{\pi}{6} = 4 \cdot \frac{\sqrt{3}}{2} = 2\sqrt{3}$$

$$y = r \sin \theta = 4 \sin \frac{\pi}{6} = 4 \cdot \frac{1}{2} = 2$$

Therefore, the given point has rectangular representation $(2\sqrt{3}, 2)$. ■

EXAMPLE 3 The point $(-1, 1)$ is given in rectangular coordinates. Find its representation in polar coordinates.

Solution   Here, $x = -1$ and $y = 1$. Using Equation (2), we have

$$r^2 = x^2 + y^2 = (-1)^2 + 1^2 = 2$$

and

$$\tan \theta = \frac{y}{x} = -1$$

Let's choose r to be positive; that is, $r = \sqrt{2}$. Next, observe that the point $(-1, 1)$ lies in the second quadrant and so we choose $\theta = 3\pi/4$ (other choices are $\theta = (3\pi/4) \pm 2n\pi$, where n is an integer). Therefore, one representation of the given point is $\left(\sqrt{2}, \frac{3\pi}{4}\right)$. ■

■ Graphs of Polar Equations

The graph of a **polar equation** $r = f(\theta)$ or, more generally, $F(r, \theta) = 0$ is the set of all points (r, θ) whose coordinates satisfy the equation.

EXAMPLE 4 Sketch the graphs of the polar equations, and reconcile your results by finding the corresponding rectangular equations.

a. $r = 2$ **b.** $\theta = \dfrac{2\pi}{3}$

Solution

a. The graph of $r = 2$ consists of all points $P(r, \theta)$ where $r = 2$ and θ can assume *any* value. Since r gives the distance between P and the pole O, we see that the graph consists of all points that are located a distance of 2 units from the pole; in other words, the graph of $r = 2$ is the circle of radius 2 centered at the pole. (See Figure 7a.) To find the corresponding rectangular equation, square both sides of the given equation obtaining $r^2 = 4$. But by Equation (2), $r^2 = x^2 + y^2$, and this gives the desired equation $x^2 + y^2 = 4$. Since this is a rectangular equation of a circle with center at the origin and radius 2, the result obtained earlier has been confirmed.

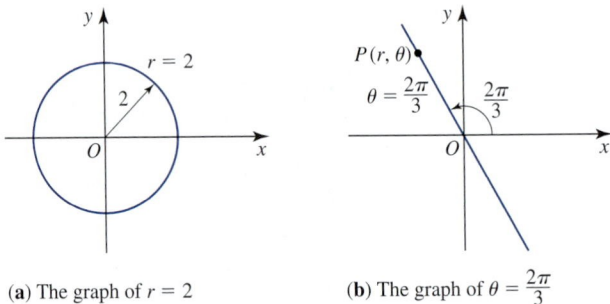

FIGURE 7   (a) The graph of $r = 2$ (b) The graph of $\theta = \dfrac{2\pi}{3}$

b. The graph of $\theta = 2\pi/3$ consists of all points $P(r, \theta)$ where $\theta = 2\pi/3$ and r can assume *any* value. Since θ measures the angle the line segment OP makes with the polar axis, we see that the graph consists of all points that are located on the straight line passing through the pole O and making an angle of $2\pi/3$ radians with the polar axis. (See Figure 7b.) Observe that the half-line in the second quadrant consists of points for which $r > 0$, whereas the half-line in the fourth quadrant consists of points for which $r < 0$. To find the corresponding rectangular equation, we use Equation (2), $\tan \theta = y/x$, to obtain

$$\tan \frac{2\pi}{3} = \frac{y}{x} \qquad \text{or} \qquad \frac{y}{x} = -\sqrt{3}$$

or $y = -\sqrt{3}x$. This equation confirms that the graph of $\theta = 2\pi/3$ is a straight line with slope $-\sqrt{3}$. ∎

As in the case with rectangular equations, we can often obtain a sketch of the graph of a simple polar equation by plotting and connecting some points that lie on the graph.

EXAMPLE 5 Sketch the graph of the polar equation $r = 2 \sin \theta$. Find a corresponding rectangular equation and reconcile your results.

Solution   The following table shows the values of r corresponding to some convenient values of θ. It suffices to restrict the values of θ to those lying between 0 and π, since values of θ beyond π will give the same points (r, θ) again.

θ	0	$\frac{\pi}{6}$	$\frac{\pi}{4}$	$\frac{\pi}{3}$	$\frac{\pi}{2}$	$\frac{2\pi}{3}$	$\frac{3\pi}{4}$	$\frac{5\pi}{6}$	π
r	0	1	$\sqrt{2} \approx 1.4$	$\sqrt{3} \approx 1.7$	2	$\sqrt{3} \approx 1.7$	$\sqrt{2} \approx 1.4$	1	0

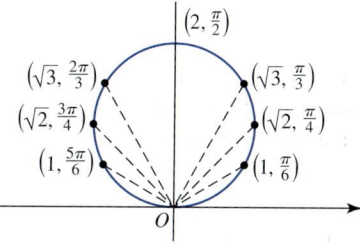

FIGURE 8

The graph of $r = 2 \sin \theta$ is a circle. To plot the points, first draw the ray with the desired angle, then locate the point by measuring off the required distance from the pole.

The graph of $r = 2 \sin \theta$ is sketched in Figure 8. To find a corresponding rectangular equation, we multiply both sides of $r = 2 \sin \theta$ by r to obtain $r^2 = 2r \sin \theta$ and then use the relationships $r^2 = x^2 + y^2$ (Equation (2)) and $y = r \sin \theta$ (Equation (1)), to obtain the desired equation

$$x^2 + y^2 = 2y \qquad \text{or} \qquad x^2 + y^2 - 2y = 0$$

Finally, completing the square in y, we have

$$x^2 + y^2 - 2y + (-1)^2 = 1$$

or

$$x^2 + (y - 1)^2 = 1$$

which is an equation of the circle with center $(0, 1)$ and radius 1, as obtained earlier.

It might have occurred to you that in the last several examples we could have obtained the graphs of the polar equations by first converting them to the corresponding rectangular equations. But as you will see, some curves are easier to graph using polar coordinates.

■ Symmetry

Just as the use of symmetry is helpful in graphing rectangular equations, its use is equally helpful in graphing polar equations. Three types of symmetry are illustrated in Figure 9. The test for each type of symmetry follows.

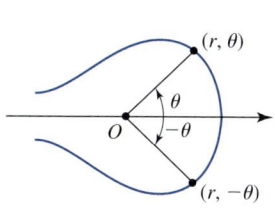 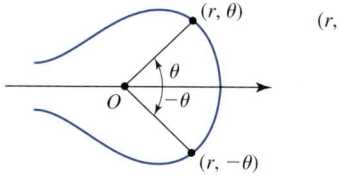 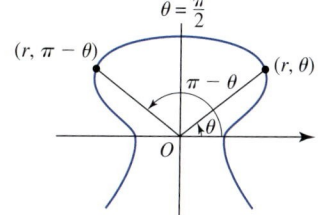 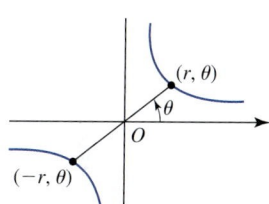

FIGURE 9
Symmetries of graphs of polar equations

(a) Symmetry with respect to the polar axis

(b) Symmetry with respect to the line $\theta = \frac{\pi}{2}$

(c) Symmetry with respect to the pole

Tests for Symmetry

a. The graph of $r = f(\theta)$ is **symmetric with respect to the polar axis** if the equation is unchanged when θ is replaced by $-\theta$.

b. The graph of $r = f(\theta)$ is **symmetric with respect to the vertical line** $\theta = \pi/2$ if the equation is unchanged when θ is replaced by $\pi - \theta$.

c. The graph of $r = f(\theta)$ is **symmetric with respect to the pole** if the equation is unchanged when r is replaced by $-r$ or when θ is replaced by $\theta + \pi$.

To illustrate the use of the tests for symmetry, consider the equation $r = 2 \sin \theta$ of Example 5. Here, $f(\theta) = 2 \sin \theta$, and since

$$f(\pi - \theta) = 2 \sin(\pi - \theta) = 2(\sin \pi \cos \theta - \cos \pi \sin \theta) = 2 \sin \theta = f(\theta)$$

we conclude that the graph of $r = 2 \sin \theta$ is symmetric with respect to the vertical line $\theta = \pi/2$ (Figure 8).

EXAMPLE 6 Sketch the graph of the polar equation $r = 1 + \cos \theta$. This is the polar form of the rectangular equation $x^4 - 2x^3 + 2x^2 y^2 - 2xy^2 - y^2 + y^4 = 0$ of the cardioid that was mentioned at the beginning of this section (Figure 1b).

Solution Writing $f(\theta) = 1 + \cos \theta$ and observing that

$$f(-\theta) = 1 + \cos(-\theta) = 1 + \cos \theta = f(\theta)$$

we conclude that the graph of $r = 1 + \cos \theta$ is symmetric with respect to the polar axis. In view of this, we need only to obtain that part of the graph between $\theta = 0$ and $\theta = \pi$. We can then complete the graph using symmetry.

To sketch the graph of $r = 1 + \cos \theta$ for $0 \le \theta \le \pi$, we can proceed as we did in Example 5 by first plotting some points lying on that part of the graph, or we may proceed as follows: Treat r and θ as *rectangular* coordinates, and make use of our knowledge of graphing rectangular equations to obtain the graph of $r = f(\theta) = 1 + \cos \theta$ on the interval $[0, \pi]$. (See Figure 10a.) Then recalling that θ is the angular coordinate and r is the radial coordinate, we see that as θ increases from 0 to π, the points on the respective rays shrink to 0. (See Figure 10b, where the corresponding points are shown.)

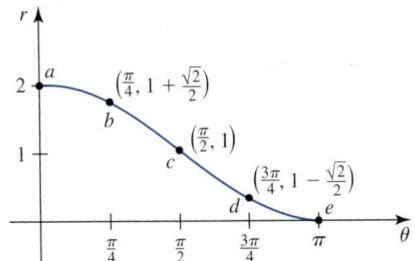

(a) $r = f(\theta)$, treating r and θ as rectangular coordinates

FIGURE 10
Two steps in sketching the graph of the polar equation $r = 1 + \cos \theta$

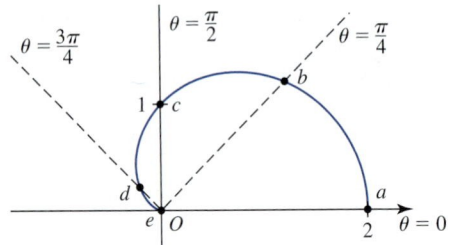

(b) $r = f(\theta)$, treating r and θ as polar coordinates

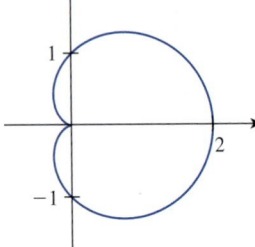

FIGURE 11
The graph of $r = 1 + \cos \theta$ is a cardioid.

Finally, using symmetry, we complete the graph of $r = 1 + \cos \theta$, as shown in Figure 11. It is called a **cardioid** because it is heart-shaped. ∎

EXAMPLE 7 Sketch the graph of the polar equation $r = 2 \cos 2\theta$.

Solution Write $f(\theta) = 2 \cos 2\theta$, and observe that

$$f(-\theta) = 2 \cos 2(-\theta) = 2 \cos 2\theta = f(\theta)$$

and

$$f(\pi - \theta) = 2 \cos 2(\pi - \theta) = 2 \cos(2\pi - 2\theta)$$
$$= 2[\cos 2\pi \cos 2\theta + \sin 2\pi \sin 2\theta] = 2 \cos 2\theta = f(\theta)$$

Therefore, the graph of the given equation is symmetric with respect to both the polar axis and the vertical line $\theta = \pi/2$. It suffices, therefore, to obtain an accurate sketch of that part of the graph for $0 \le \theta \le \frac{\pi}{2}$ and then complete the sketch of the graph using symmetry. Proceeding as in Example 6, we first sketch the graph of $r = 2 \cos 2\theta$ for $0 \le \theta \le \frac{\pi}{2}$ treating r and θ as rectangular coordinates (Figure 12a), and then transcribe the information contained in this graph onto the graph in the $r\theta$-plane for $0 \le \theta \le \frac{\pi}{2}$. (See Figure 12b.)

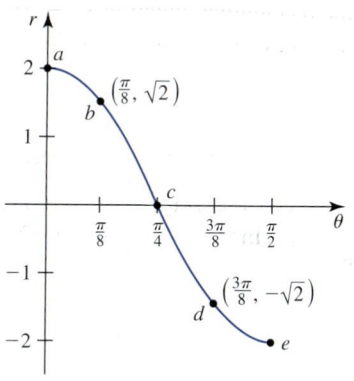

FIGURE 12

Two steps in sketching the graph of $r = 2 \cos 2\theta$

(a) $r = f(\theta)$ treating r and θ as rectangular coordinates

(b) $r = f(\theta)$, treating r and θ as polar coordinates

Finally, using the symmetry that was established earlier (Figure 13a), we complete the graph of $r = 2 \cos 2\theta$ as shown in Figure 13b. This graph is called a **four-leaved rose.**

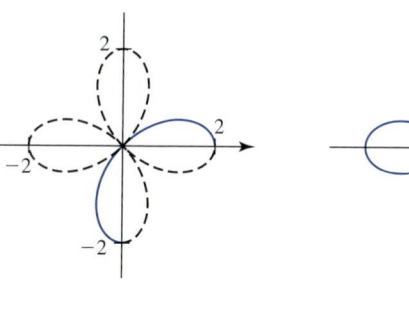

FIGURE 13

The graph of $r = 2 \cos 2\theta$ is a four-leaved rose.

(a)

(b)

The next example shows how the graph of a rectangular equation can be sketched more easily by first converting it to polar form.

EXAMPLE 8   Sketch the graph of the equation $(x^2 + y^2)^2 = 4(x^2 - y^2)$ by first converting it to polar form. This is an equation of the lemniscate that was mentioned at the beginning of this section.

Solution   To convert the given equation to polar form, we use Equations (1) and (2), obtaining

$$(r^2)^2 = 4(r^2 \cos^2 \theta - r^2 \sin^2 \theta)$$

$$= 4r^2(\cos^2 \theta - \sin^2 \theta)$$

$$r^4 = 4r^2 \cos 2\theta$$

or

$$r^2 = 4 \cos 2\theta$$

Observe that $f(\theta) = 2\sqrt{\cos 2\theta}$ is defined for $-\frac{\pi}{4} \le \theta \le \frac{\pi}{4}$ and $\frac{3\pi}{4} \le \theta \le \frac{5\pi}{4}$. Also, observe that $f(-\theta) = f(\theta)$ and $f(\pi - \theta) = f(\theta)$. (These computations are similar to those in Example 7.) So the graph of $r = 2\sqrt{\cos 2\theta}$ is symmetric with respect to the polar axis and the line $\theta = \pi/2$. The graph of $r = f(\theta)$ for $0 \le \theta \le \frac{\pi}{4}$, where r and θ are treated as rectangular coordinates, is shown in Figure 14a. This leads to the part of the required graph for $0 \le \theta \le \frac{\pi}{4}$ shown in Figure 14b. Then, using symmetry, we obtain the graph of $r = 2\sqrt{\cos 2\theta}$ and, therefore, that of $(x^2 + y^2)^2 = 4(x^2 - y^2)$, as shown in Figure 15.

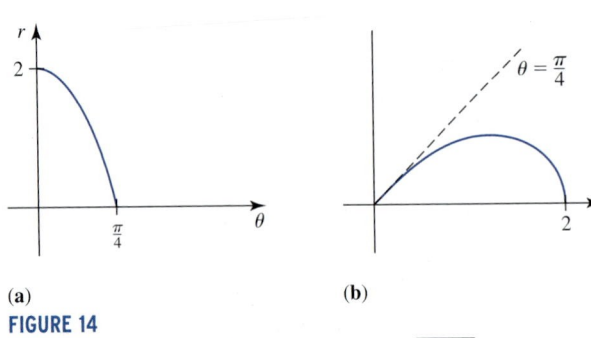

(a)

(b)

FIGURE 14

Two steps in sketching the graph of $r = 2\sqrt{\cos 2\theta}$

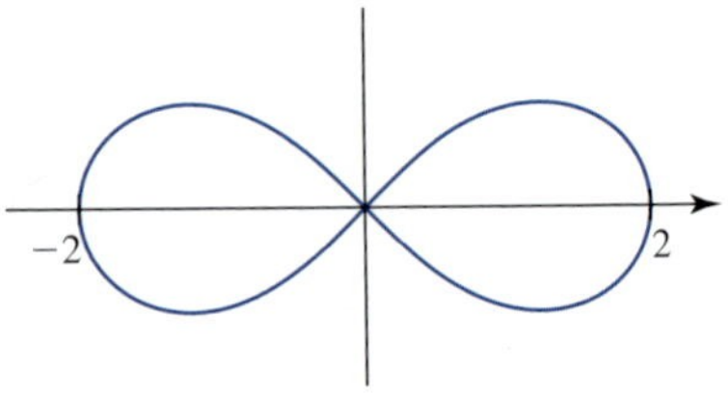

FIGURE 15

The graph of $r = 2\sqrt{\cos 2\theta}$ is a lemniscate.

■ Tangent Lines to Graphs of Polar Equations

To find the slope of the tangent line to the graph of $r = f(\theta)$ at the point $P(r, \theta)$, let $P(x, y)$ be the rectangular representation of P. Then

$$x = r \cos \theta = f(\theta) \cos \theta$$

$$y = r \sin \theta = f(\theta) \sin \theta$$

We can view these equations as parametric equations for the graph of $r = f(\theta)$ with parameter θ. Then, using Equation (1) of Section 9.3, we have

$$\frac{dy}{dx} = \frac{\dfrac{dy}{d\theta}}{\dfrac{dx}{d\theta}} = \frac{\dfrac{dr}{d\theta} \sin \theta + r \cos \theta}{\dfrac{dr}{d\theta} \cos \theta - r \sin \theta} \qquad \text{if} \quad \frac{dx}{d\theta} \ne 0 \tag{3}$$

and this gives the slope of the tangent line to the graph of $r = f(\theta)$ at any point $P(r, \theta)$.

The horizontal tangent lines to the graph of $r = f(\theta)$ are located at the points where $dy/d\theta = 0$ and $dx/d\theta \ne 0$. The vertical tangent lines are located at the points where $dx/d\theta = 0$ and $dy/d\theta \ne 0$ (so that dy/dx is undefined). Also, points where both $dy/d\theta$ and $dx/d\theta$ are equal to zero are candidates for horizontal or vertical tangent lines, respectively, and may be investigated using l'Hôpital's Rule.

Equation (3) can be used to help us find the tangent lines to the graph of $r = f(\theta)$ at the pole. To see this, suppose that the graph of f passes through the pole when $\theta = \theta_0$. Then $f(\theta_0) = 0$. If $f'(\theta_0) \ne 0$, then Equation (3) reduces to

$$\frac{dy}{dx} = \frac{f'(\theta_0) \sin \theta_0 + f(\theta_0) \cos \theta_0}{f'(\theta_0) \cos \theta_0 - f(\theta_0) \sin \theta_0} = \frac{\sin \theta_0}{\cos \theta_0} = \tan \theta_0$$

This shows that $\theta = \theta_0$ is a tangent line to the graph of $r = f(\theta)$ at the pole $(0, \theta_0)$. The following summarizes this discussion.

$\theta = \theta_0$ is a tangent line to the graph of $r = f(\theta)$ at the pole if $f(\theta_0) = 0$ and $f'(\theta_0) \neq 0$.

EXAMPLE 9 Consider the cardioid $r = 1 + \cos\theta$ of Example 6.

a. Find the slope of the tangent line to the cardioid at the point where $\theta = \pi/6$.
b. Find the points on the cardioid where the tangent lines are horizontal and where the tangent lines are vertical.

Solution

a. The slope of the tangent line to the cardioid $r = 1 + \cos\theta$ at any point $P(r, \theta)$ is given by

$$\frac{dy}{dx} = \frac{\dfrac{dr}{d\theta}\sin\theta + r\cos\theta}{\dfrac{dr}{d\theta}\cos\theta - r\sin\theta} = \frac{(-\sin\theta)(\sin\theta) + (1 + \cos\theta)\cos\theta}{(-\sin\theta)(\cos\theta) - (1 + \cos\theta)\sin\theta}$$

$$= \frac{(\cos^2\theta - \sin^2\theta) + \cos\theta}{-2\sin\theta\cos\theta - \sin\theta} = -\frac{\cos 2\theta + \cos\theta}{\sin 2\theta + \sin\theta}$$

At the point on the cardioid where $\theta = \pi/6$, the slope of the tangent line is

$$\frac{dy}{dx}\bigg|_{\theta=\pi/6} = -\frac{\cos\left(\dfrac{\pi}{3}\right) + \cos\left(\dfrac{\pi}{6}\right)}{\sin\left(\dfrac{\pi}{3}\right) + \sin\left(\dfrac{\pi}{6}\right)} = -\frac{\dfrac{1}{2} + \dfrac{\sqrt{3}}{2}}{\dfrac{\sqrt{3}}{2} + \dfrac{1}{2}} = -1$$

b. Observe that $dy/d\theta = 0$ if

$$\cos 2\theta + \cos\theta = 0$$

$$2\cos^2\theta + \cos\theta - 1 = 0$$

$$(2\cos\theta - 1)(\cos\theta + 1) = 0$$

that is, if $\cos\theta = \frac{1}{2}$ or $\cos\theta = -1$. This gives

$$\theta = \frac{\pi}{3}, \quad \pi, \quad \text{or} \quad \frac{5\pi}{3}$$

Next, $dx/d\theta = 0$ if

$$\sin 2\theta + \sin\theta = 0$$

$$2\sin\theta\cos\theta + \sin\theta = 0$$

$$\sin\theta\,(2\cos\theta + 1) = 0$$

that is, if $\sin\theta = 0$ or $\cos\theta = -\frac{1}{2}$. This gives

$$\theta = 0, \quad \pi, \quad \frac{2\pi}{3}, \quad \text{or} \quad \frac{4\pi}{3}$$

In view of the remarks following Equation (3), we see that $\theta = \pi/3$ and $\theta = 5\pi/3$ give rise to horizontal tangents. To investigate the candidate $\theta = \pi$, where both $dy/d\theta$ and $dx/d\theta$ are equal to zero, we use l'Hôpital's Rule. Thus,

$$\lim_{\theta \to \pi^-} \frac{dy}{dx} = -\lim_{\theta \to \pi^-} \frac{\cos 2\theta + \cos \theta}{\sin 2\theta + \sin \theta}$$

$$= -\lim_{\theta \to \pi^-} \frac{-2 \sin 2\theta - \sin \theta}{2 \cos 2\theta + \cos \theta} = 0$$

Similarly, we see that

$$\lim_{\theta \to \pi^+} \frac{dy}{dx} = 0$$

Therefore, $\theta = \pi$ also gives rise to a horizontal tangent. Thus, the horizontal tangent lines occur at

$$\left(\tfrac{3}{2}, \tfrac{\pi}{3}\right), \quad (0, \pi), \quad \text{and} \quad \left(\tfrac{3}{2}, \tfrac{5\pi}{3}\right)$$

The vertical tangent lines occur at $\theta = 0$, $2\pi/3$, and $4\pi/3$. The points are $(2, 0)$, $\left(\tfrac{1}{2}, \tfrac{2\pi}{3}\right)$, and $\left(\tfrac{1}{2}, \tfrac{4\pi}{3}\right)$. These tangent lines are shown in Figure 16. ◼

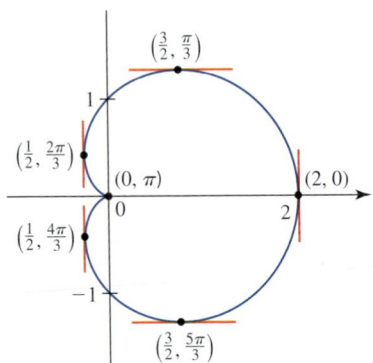

FIGURE 16
The horizontal and vertical tangents to the graph of $r = 1 + \cos \theta$

EXAMPLE 10   Find the tangent lines of $r = \cos 2\theta$ at the origin.

Solution   Setting $f(\theta) = \cos 2\theta = 0$, we find that

$$2\theta = \frac{\pi}{2}, \quad \frac{3\pi}{2}, \quad \frac{5\pi}{2}, \quad \text{or} \quad \frac{7\pi}{2}$$

or

$$\theta = \frac{\pi}{4}, \quad \frac{3\pi}{4}, \quad \frac{5\pi}{4}, \quad \text{or} \quad \frac{7\pi}{4}$$

Next, we compute $f'(\theta) = -2 \sin 2\theta$. Since $f'(\theta) \neq 0$ for each of these values of θ, we see that $\theta = \pi/4$ and $\theta = 3\pi/4$ (that is, $y = x$ and $y = -x$) are tangent lines to the graph of $r = \cos 2\theta$ at the pole (see Figure 17). ◼

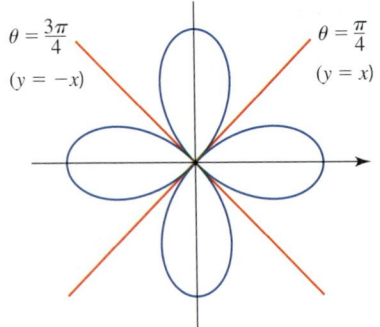

FIGURE 17
The tangent lines to the graph of $r = \cos 2\theta$ at the origin

9.4   CONCEPT QUESTIONS

1. Let $P(r, \theta)$ be a point in the plane with polar coordinates r and θ. Find all possible representations of $P(r, \theta)$.
2. Suppose that P has representation (r, θ) in polar coordinates and (x, y) in rectangular coordinates. Express (a) x and y in terms of r and θ and (b) r and θ in terms of x and y.
3. Explain how you would determine whether the graph of $r = f(\theta)$ is symmetric with respect to (a) the polar axis, (b) the vertical line $\theta = \pi/2$, and (c) the pole.

4. Suppose that $r = f(\theta)$, where f is differentiable.
   a. Write an expression for dy/dx.
   b. How do you find the points on the graph of $r = f(\theta)$ where the tangent lines are horizontal and where the tangent lines are vertical?
   c. How do you find the tangent lines to the graph of $r = f(\theta)$ (if they exist) at the pole?

9.4 EXERCISES

In Exercises 1–8, plot the point with the polar coordinates. Then find the rectangular coordinates of the point.

1. $\left(4, \frac{\pi}{4}\right)$

2. $\left(2, \frac{\pi}{6}\right)$

3. $\left(4, \frac{3\pi}{2}\right)$

4. $(6, 3\pi)$

5. $\left(-\sqrt{2}, \frac{\pi}{4}\right)$

6. $\left(-1, \frac{\pi}{3}\right)$

7. $\left(-4, -\frac{3\pi}{4}\right)$

8. $\left(5, -\frac{5\pi}{6}\right)$

In Exercises 9–16, plot the point with the rectangular coordinates. Then find the polar coordinates of the point taking $r > 0$ and $0 \le \theta < 2\pi$.

9. $(2, 2)$

10. $(1, -1)$

11. $(0, 5)$

12. $(3, -4)$

13. $(-\sqrt{3}, -\sqrt{3})$

14. $(2\sqrt{3}, -2)$

15. $(5, -12)$

16. $(3, -1)$

In Exercises 17–24, sketch the region comprising points whose polar coordinates satisfy the given conditions.

17. $r \ge 1$

18. $r > 1$

19. $0 \le r \le 2$

20. $1 \le r < 2$

21. $0 \le \theta \le \frac{\pi}{4}$

22. $0 \le r \le 3, \quad 0 \le \theta \le \frac{\pi}{3}$

23. $1 \le r \le 3, \quad -\frac{\pi}{6} \le \theta \le \frac{\pi}{6}$

24. $2 < r < 4, \quad -\frac{\pi}{2} < \theta < \frac{\pi}{2}$

In Exercises 25–32, convert the polar equation to a rectangular equation.

25. $r \cos \theta = 2$

26. $r \sin \theta = -3$

27. $2r \cos \theta + 3r \sin \theta = 6$

28. $r \sin \theta = 2r \cos \theta$

29. $r^2 = 4r \cos \theta$

30. $r^2 = \sin 2\theta$

31. $r = \dfrac{1}{1 - \sin \theta}$

32. $r = \dfrac{3}{4 - 5 \cos \theta}$

In Exercises 33–38, convert the rectangular equation to a polar equation.

33. $x = 4$

34. $x + 2y = 3$

35. $x^2 + y^2 = 9$

36. $x^2 - y^2 = 1$

37. $xy = 4$

38. $y^2 - x^2 = 4\sqrt{x^2 + y^2}$

In Exercises 39–64, sketch the curve with the polar equation.

39. $r = 3$

40. $r = -2$

41. $\theta = \dfrac{\pi}{3}$

42. $\theta = -\dfrac{\pi}{6}$

43. $r = 3 \cos \theta$

44. $r = -4 \sin \theta$

45. $r = 3 \cos \theta - 2 \sin \theta$

46. $r = 2 \sin \theta + 4 \cos \theta$

47. $r = 1 + \cos \theta$

48. $r = 1 + \sin \theta$

49. $r = 4(1 - \sin \theta)$

50. $r = 3 - 3 \cos \theta$

51. $r = 2 \csc \theta$

52. $r = -3 \sec \theta$

53. $r = \theta, \quad \theta \ge 0$ (spiral)

54. $r = \dfrac{1}{\theta}$ (spiral)

55. $r = e^\theta, \quad \theta \ge 0$ (logarithmic spiral)

56. $r^2 = \dfrac{1}{\theta}$ (lituus)

57. $r^2 = 4 \sin 2\theta$ (lemniscate)

58. $r = 1 - 2 \cos \theta$ (limaçon)

59. $r = 3 + 2 \sin \theta$ (limaçon)

60. $r = \sin 2\theta$ (four-leaved rose)

61. $r = \sin 3\theta$ (three-leaved rose)

62. $r = 2 \cos 4\theta$ (eight-leaved rose)

63. $r = 4 \sin 4\theta$ (eight-leaved rose)

64. $r = 2 \sin 5\theta$ (five-leaved rose)

In Exercises 65–72, find the slope of the tangent line to the curve with the polar equation at the point corresponding to the given value of θ.

65. $r = 4 \cos \theta, \quad \theta = \dfrac{\pi}{3}$

66. $r = 3 \sin \theta, \quad \theta = \dfrac{\pi}{4}$

67. $r = \sin \theta + \cos \theta, \quad \theta = \dfrac{\pi}{4}$

68. $r = 1 + 3 \cos \theta, \quad \theta = \dfrac{\pi}{2}$

69. $r = \theta, \quad \theta = \pi$

70. $r = \sin 3\theta, \quad \theta = \dfrac{\pi}{3}$

71. $r^2 = 4 \cos 2\theta, \quad \theta = \dfrac{\pi}{6}$

72. $r = 2 \sec \theta, \quad \theta = \dfrac{\pi}{4}$

In Exercises 73–78, find the points on the curve with the given polar equation where the tangent line is horizontal or vertical.

73. $r = 4 \cos \theta$

74. $r = \sin \theta + \cos \theta$

75. $r = \sin 2\theta$

76. $r^2 = 4 \cos 2\theta$

77. $r = 1 + 2 \cos \theta$

78. $r = 1 + \sin \theta$

79. Show that the rectangular equation

$$x^4 - 2x^3 + 2x^2y^2 - 2xy^2 - y^2 + y^4 = 0$$

is an equation of the cardioid with polar equation $r = 1 + \cos \theta$.

80. Show that the polar equation $r = a \sin \theta + b \cos \theta$, where a and b are nonzero, represents a circle. What are the center and radius of the circle?

81. **a.** Show that the distance between the points with polar coordinates (r_1, θ_1) and (r_2, θ_2) is given by

$$d = \sqrt{r_1^2 + r_2^2 - 2r_1 r_2 \cos(\theta_1 - \theta_2)}$$

   b. Find the distance between the points with polar coordinates $\left(4, \frac{2\pi}{3}\right)$ and $\left(2, \frac{\pi}{3}\right)$.

82. Show that the curves with polar equations $r = a \sin \theta$ and $r = a \cos \theta$ intersect at right angles.

 83. **a.** Plot the graphs of the cardioids $r = a(1 + \cos \theta)$ and $r = a(1 - \cos \theta)$.

   b. Show that the cardioids intersect at right angles except at the pole.

84. Let ψ be the angle between the radial line OP and the tangent line to the curve with polar equation $r = f(\theta)$ at P (see the figure). Show that

$$\tan \psi = r \frac{d\theta}{dr}$$

Hint: Observe that $\psi = \phi - \theta$. Then use the trigonometric identity

$$\tan(a - b) = \frac{\tan a - \tan b}{1 + \tan a \tan b}$$

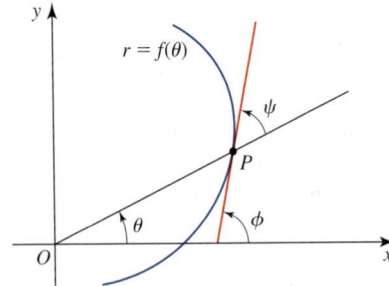

 In Exercises 85–92, use a graphing utility to plot the curve with the polar equation.

85. $r = \cos \theta(4 \sin^2 \theta - 1)$, $0 \le \theta < 2\pi$

86. $r = 3 \sin \theta \cos^2 \theta$, $0 \le \theta < 2\pi$

87. $r = 0.3\left[1 + 2 \sin\left(\frac{\theta}{2}\right)\right]$, $0 \le \theta < 4\pi$
   (nephroid of Freeth)

88. $r = \dfrac{1 - 10 \cos \theta}{1 + 10 \cos \theta}$, $0 \le \theta < 2\pi$

89. $r^2 = 0.8(1 - 0.8 \sin^2 \theta)$, $0 \le \theta < 2\pi$ (hippopede curve)

90. $r^2 = \dfrac{\frac{1}{4} \sin^2 \theta - 3.6 \cos^2 \theta}{\sin^2 \theta - \cos^2 \theta}$, $0 \le \theta < 2\pi$ (devil's curve)

91. $r = \dfrac{0.1}{\cos 3\theta}$, $0 \le \theta < \pi$ (epi-spiral)

92. $r = \dfrac{\sin \theta}{\theta}$, $-6\pi \le \theta < 6\pi$ (cochleoid)

In Exercises 93–95, determine whether the statement is true or false. If it is true, explain why it is true. If it is false, give an example to show why it is false.

93. If $P(r_1, \theta_1)$ and $P(r_2, \theta_2)$ represent the same point in polar coordinates, then $r_1 = r_2$.

94. If $P(r_1, \theta_1)$ and $P(r_2, \theta_2)$ represent the same point in polar coordinates, then $\theta_1 = \theta_2$.

95. The graph of $r = f(\theta)$ has a horizontal tangent line at a point on the graph if $dy/d\theta = 0$, where $y = f(\theta)\sin \theta$.

9.5   Areas and Arc Lengths in Polar Coordinates

In this section we see how the use of polar equations to represent curves such as lemniscates and cardioids will simplify the task of finding the areas of the regions enclosed by these curves as well as the lengths of these curves.

■ Areas in Polar Coordinates

To develop a formula for finding the area of a region bounded by a curve defined by a polar equation, we need the formula for the area of a sector of a circle

$$A = \frac{1}{2} r^2 \theta \tag{1}$$

where r is the radius of the circle and θ is the central angle measured in radians. (See Figure 1.) This formula follows by observing that the area of a sector is $\theta/(2\pi)$ times that of the area of a circle; that is,

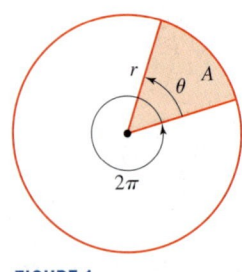

FIGURE 1
The area of a sector of a circle is $A = \frac{1}{2} r^2 \theta$.

$$A = \frac{\theta}{2\pi} \cdot \pi r^2 = \frac{1}{2} r^2 \theta$$

Now let R be a region bounded by the graph of the polar equation $r = f(\theta)$ and the rays $\theta = \alpha$ and $\theta = \beta$, where f is a nonnegative continuous function and $0 \le \beta - \alpha < 2\pi$, as shown in Figure 2a. Let P be a regular partition of the interval $[\alpha, \beta]$:

$$\alpha = \theta_0 < \theta_1 < \theta_2 < \cdots < \theta_n = \beta$$

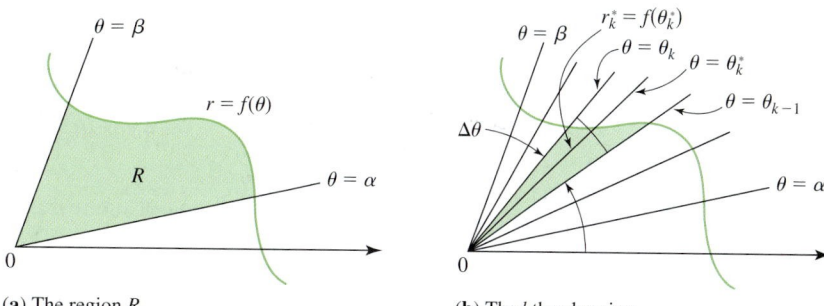

FIGURE 2   (**a**) The region R (**b**) The kth subregion

The rays $\theta = \theta_k$ divide R into n subregions $R_1, R_2, \ldots, R_n$ of area $\Delta A_1, \Delta A_2, \ldots, \Delta A_n$, respectively. If we choose θ_k^* in the interval $[\theta_{k-1}, \theta_k]$, then the area of ΔA_k of the kth subregion bounded by the rays $\theta = \theta_{k-1}$ and $\theta = \theta_k$ is approximated by the sector of a circle with central angle

$$\Delta \theta = \frac{\beta - \alpha}{n}$$

and radius $f(\theta_k^*)$ (highlighted in Figure 2b). Using Equation (1), we have

$$\Delta A_k \approx \frac{1}{2} [f(\theta_k^*)]^2 \Delta \theta$$

Therefore, an approximation of the area A of R is

$$A = \sum_{k=1}^{n} \Delta A_k \approx \sum_{k=1}^{n} \frac{1}{2} [f(\theta_k^*)]^2 \Delta \theta \tag{2}$$

But the sum in Equation (2) is a Riemann sum of the continuous function $\frac{1}{2} f^2$ over the interval $[\alpha, \beta]$. Therefore, it is true, although we will not prove it here, that

$$A = \lim_{n \to \infty} \sum_{k=1}^{n} \frac{1}{2} [f(\theta_k^*)]^2 \Delta \theta = \int_{\alpha}^{\beta} \frac{1}{2} [f(\theta)]^2 \, d\theta$$

THEOREM 1   Area Bounded by a Polar Curve

Let f be a continuous, nonnegative function on $[\alpha, \beta]$ where $0 \le \beta - \alpha < 2\pi$. Then the area A of the region bounded by the graphs of $r = f(\theta)$, $\theta = \alpha$, and $\theta = \beta$ is given by

$$A = \int_{\alpha}^{\beta} \frac{1}{2} [f(\theta)]^2 \, d\theta = \int_{\alpha}^{\beta} \frac{1}{2} r^2 \, d\theta$$

Note   When you determine the limits of integration, keep in mind that the region R is swept out in a counterclockwise direction by the ray emanating from the origin, starting at the angle α and terminating at the angle β.

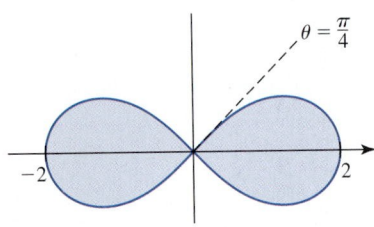

FIGURE 3

The region enclosed by the lemniscate $r^2 = 4 \cos 2\theta$

EXAMPLE 1   Find the area of the region enclosed by the lemniscate $r^2 = 4 \cos 2\theta$. This lemniscate has rectangular equation $x^4 + 2x^2y^2 - 4x^2 + 4y^2 + y^4 = 0$, as you can verify.

Solution   The lemniscate is shown in Figure 3. Making use of symmetry, we see that the required area A is four times that of the area swept out by the ray emanating from the origin as θ increases from 0 to $\pi/4$. In other words,

$$A = 4 \int_0^{\pi/4} \frac{1}{2} r^2 \, d\theta = 8 \int_0^{\pi/4} \cos 2\theta \, d\theta$$

$$= \Big[4 \sin 2\theta \Big]_0^{\pi/4} = 4$$

EXAMPLE 2   Find the area of the region enclosed by the cardioid $r = 1 + \cos \theta$.

Solution   The graph of the cardioid $r = 1 + \cos \theta$, sketched previously in Example 6 in Section 9.4, is reproduced in Figure 4. Observe that the ray emanating from the origin sweeps out the required region exactly once as θ increases from 0 to 2π. Therefore, the required area A is

$$A = \int_0^{2\pi} \frac{1}{2} r^2 \, d\theta = \int_0^{2\pi} \frac{1}{2} (1 + \cos \theta)^2 \, d\theta$$

$$= \frac{1}{2} \int_0^{2\pi} (1 + 2 \cos \theta + \cos^2 \theta) \, d\theta$$

$$= \frac{1}{2} \int_0^{2\pi} \left(1 + 2 \cos \theta + \frac{1 + \cos 2\theta}{2} \right) d\theta$$

$$= \frac{1}{2} \int_0^{2\pi} \left(\frac{3}{2} + 2 \cos \theta + \frac{1}{2} \cos 2\theta \right) d\theta$$

$$= \frac{1}{2} \left[\frac{3}{2} \theta + 2 \sin \theta + \frac{1}{4} \sin 2\theta \right]_0^{2\pi} = \frac{3}{2} \pi$$

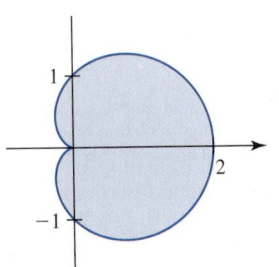

FIGURE 4

The region enclosed by the cardioid $r = 1 + \cos \theta$

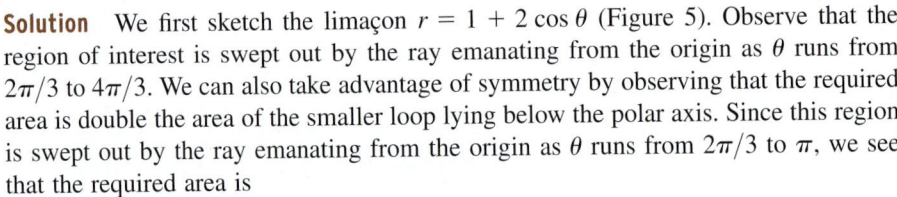

EXAMPLE 3   Find the area inside the smaller loop of the limaçon $r = 1 + 2 \cos \theta$.

Solution   We first sketch the limaçon $r = 1 + 2 \cos \theta$ (Figure 5). Observe that the region of interest is swept out by the ray emanating from the origin as θ runs from $2\pi/3$ to $4\pi/3$. We can also take advantage of symmetry by observing that the required area is double the area of the smaller loop lying below the polar axis. Since this region is swept out by the ray emanating from the origin as θ runs from $2\pi/3$ to π, we see that the required area is

$$A = 2 \int_{2\pi/3}^{\pi} \frac{1}{2} r^2 \, d\theta = \int_{2\pi/3}^{\pi} r^2 \, d\theta$$

$$= \int_{2\pi/3}^{\pi} (1 + 2 \cos \theta)^2 \, d\theta$$

$$= \int_{2\pi/3}^{\pi} (1 + 4 \cos \theta + 4 \cos^2 \theta) \, d\theta$$

$$= \int_{2\pi/3}^{\pi} \left[1 + 4 \cos \theta + 4 \left(\frac{1 + \cos 2\theta}{2} \right) \right] d\theta$$

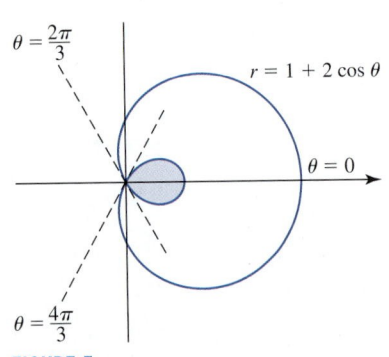

FIGURE 5

The limaçon $r = 1 + 2 \cos \theta$

$$= \int_{2\pi/3}^{\pi} (3 + 4 \cos \theta + 2 \cos 2\theta) \, d\theta$$

$$= \left[3\theta + 4 \sin \theta + \sin 2\theta \right]_{2\pi/3}^{\pi}$$

$$= 3\pi - \left(2\pi + 4 \cdot \frac{\sqrt{3}}{2} - \frac{\sqrt{3}}{2} \right) = \pi - \frac{3\sqrt{3}}{2}$$

Area Bounded by Two Graphs

Consider the region R bounded by the graphs of the polar equations $r = f(\theta)$ and $r = g(\theta)$, and the rays $\theta = \alpha$ and $\theta = \beta$, where $f(\theta) \geq g(\theta) \geq 0$ and $0 \leq \beta - \alpha < 2\pi$. (See Figure 6.) From the figure we can see that the area A of R is found by subtracting the area of the region inside $r = g(\theta)$ from the area of the region inside $r = f(\theta)$. Using Theorem 1, we obtain the following theorem.

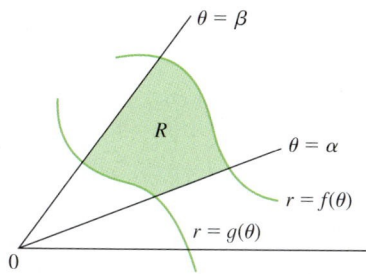

FIGURE 6
R is the region bounded by the graphs of $r = f(\theta)$ and $r = g(\theta)$ for $\alpha \leq \theta \leq \beta$.

THEOREM 2   Area Bounded by Two Polar Curves

Let f and g be continuous on $[\alpha, \beta]$, where $0 \leq g(\theta) \leq f(\theta)$ and $0 \leq \beta - \alpha < 2\pi$. Then the area A of the region bounded by the graphs of $r = g(\theta)$, $r = f(\theta)$, $\theta = \alpha$, and $\theta = \beta$ is given by

$$A = \frac{1}{2} \int_{\alpha}^{\beta} \{ [f(\theta)]^2 - [g(\theta)]^2 \} \, d\theta$$

EXAMPLE 4   Find the area of the region that lies outside the circle $r = 3$ and inside the cardioid $r = 2 + 2 \cos \theta$.

Solution   We first sketch the circle $r = 3$ and the cardioid $r = 2 + 2 \cos \theta$. The required region is shown shaded in Figure 7.

To find the points of intersection of the two curves, we solve the two equations simultaneously. We have $2 + 2 \cos \theta = 3$ or $\cos \theta = \frac{1}{2}$, which gives $\theta = \pm \pi/3$. Since the region of interest is swept out by the ray emanating from the origin as θ varies from $-\pi/3$ to $\pi/3$, we see that the required area is, by Theorem 2,

$$A = \frac{1}{2} \int_{\alpha}^{\beta} \{ [f(\theta)]^2 - [g(\theta)]^2 \} \, d\theta$$

where $f(\theta) = 2 + 2 \cos \theta = 2(1 + \cos \theta)$, $g(\theta) = 3$, $\alpha = -\pi/3$, and $\beta = \pi/3$. If we take advantage of symmetry, we can write

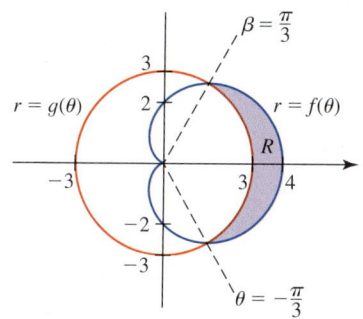

FIGURE 7
R is the region outside the circle $r = 3$ and inside the cardioid $r = 2 + 2 \cos \theta$.

$$A = 2 \left(\frac{1}{2} \right) \int_0^{\pi/3} \{ [2(1 + \cos \theta)]^2 - 3^2 \} \, d\theta$$

$$= \int_0^{\pi/3} (4 + 8 \cos \theta + 4 \cos^2 \theta - 9) \, d\theta$$

$$= \int_0^{\pi/3} \left(-5 + 8 \cos \theta + 4 \cdot \frac{1 + \cos 2\theta}{2} \right) d\theta$$

$$= \int_0^{\pi/3} (-3 + 8 \cos \theta + 2 \cos 2\theta) \, d\theta$$

$$= \left[-3\theta + 8 \sin \theta + \sin 2\theta \right]_0^{\pi/3}$$

$$= \left(-\pi + 8 \left(\frac{\sqrt{3}}{2} \right) + \frac{\sqrt{3}}{2} \right) = \frac{9\sqrt{3}}{2} - \pi$$

■ Arc Length in Polar Coordinates

To find the length of a curve C defined by a polar equation $r = f(\theta)$ for $\alpha \le \theta \le \beta$, we use Equation (1) in Section 9.4 to write the parametric equations

$$x = r \cos \theta = f(\theta) \cos \theta \qquad \text{and} \qquad y = r \sin \theta = f(\theta) \sin \theta \qquad \alpha \le \theta \le \beta$$

for the curve, regarding θ as the parameter. Then

$$\frac{dx}{d\theta} = f'(\theta) \cos \theta - f(\theta) \sin \theta \qquad \text{and} \qquad \frac{dy}{d\theta} = f'(\theta) \sin \theta + f(\theta) \cos \theta$$

Therefore,

$$\begin{aligned}
\left(\frac{dx}{d\theta}\right)^2 + \left(\frac{dy}{d\theta}\right)^2 &= [f'(\theta)]^2 \cos^2 \theta - 2f'(\theta)f(\theta) \cos \theta \sin \theta + [f(\theta)]^2 \sin^2 \theta \\
&\quad + [f'(\theta)]^2 \sin^2 \theta + 2f'(\theta)f(\theta) \cos \theta \sin \theta + [f(\theta)]^2 \cos^2 \theta \\
&= [f'(\theta)]^2 + [f(\theta)]^2 \qquad \sin^2 \theta + \cos^2 \theta = 1
\end{aligned}$$

Consequently, if f' is continuous, then Theorem 1 in Section 9.3 gives the arc length of C as

$$L = \int_\alpha^\beta \sqrt{\left(\frac{dx}{d\theta}\right)^2 + \left(\frac{dy}{d\theta}\right)^2} \, d\theta = \int_\alpha^\beta \sqrt{[f'(\theta)]^2 + [f(\theta)]^2} \, d\theta$$

THEOREM 3 **Arc Length**

Let f be a function with a continuous derivative on an interval $[\alpha, \beta]$. If the graph C of $r = f(\theta)$ is traced exactly once as θ increases from α to β, then the length L of C is given by

$$L = \int_\alpha^\beta \sqrt{[f'(\theta)]^2 + [f(\theta)]^2} \, d\theta = \int_\alpha^\beta \sqrt{\left(\frac{dr}{d\theta}\right)^2 + r^2} \, d\theta$$

EXAMPLE 5 Find the length of the cardioid $r = 1 + \cos \theta$.

Solution The cardioid is shown in Figure 8. Observe that the cardioid is traced exactly once as θ runs from θ to 2π. However, we can also take advantage of symmetry to see that the required length is twice that of the length of the cardioid lying above the polar axis. Thus,

$$L = 2 \int_0^\pi \sqrt{\left(\frac{dr}{d\theta}\right)^2 + r^2} \, d\theta$$

But $r = 1 + \cos \theta$, so

$$\frac{dr}{d\theta} = -\sin \theta$$

Therefore,

$$\begin{aligned}
L &= 2 \int_0^\pi \sqrt{(-\sin \theta)^2 + (1 + \cos \theta)^2} \, d\theta \\
&= 2 \int_0^\pi \sqrt{\sin^2 \theta + 1 + 2 \cos \theta + \cos^2 \theta} \, d\theta
\end{aligned}$$

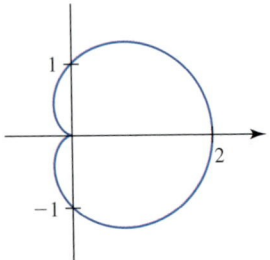

FIGURE 8
The cardioid $r = 1 + \cos \theta$

$$= 2 \int_0^{\pi} \sqrt{2 + 2 \cos \theta} \, d\theta \qquad \sin^2 \theta + \cos^2 \theta = 1$$

$$= 2\sqrt{2} \int_0^{\pi} \sqrt{1 + \cos \theta} \, d\theta = 2\sqrt{2} \int_0^{\pi} \sqrt{2 \cos^2 \frac{\theta}{2}} \, d\theta$$

$$= 4 \int_0^{\pi} \left| \cos \frac{\theta}{2} \right| d\theta = 4 \int_0^{\pi} \cos \frac{\theta}{2} \, d\theta \qquad \cos \frac{\theta}{2} \geq 0 \text{ on } [0, \pi]$$

$$= \left[4(2) \sin \frac{\theta}{2} \right]_0^{\pi} = 8$$

Area of a Surface of Revolution

The formulas for finding the area of a surface obtained by revolving a curve defined by a polar equation about the polar axis or about the line $\theta = \pi/2$ can be derived by using Equations (8) and (9) of Section 9.3 and the equations $x = r \cos \theta$ and $y = r \sin \theta$.

> **THEOREM 4   Area of a Surface of a Revolution**
>
> Let f be a function with a continuous derivative on an interval $[\alpha, \beta]$. If the graph C of $r = f(\theta)$ is traced exactly once as θ increases from α to β, then the area of the surface obtained by revolving C about the indicated line is given by
>
> **a.** $\displaystyle S = 2\pi \int_\alpha^\beta r \sin \theta \sqrt{\left(\frac{dr}{d\theta} \right)^2 + r^2} \, d\theta$ (about the polar axis)
>
> **b.** $\displaystyle S = 2\pi \int_\alpha^\beta r \cos \theta \sqrt{\left(\frac{dr}{d\theta} \right)^2 + r^2} \, d\theta$ (about the line $\theta = \pi/2$)

Note   In using Theorem 4, we must choose $[\alpha, \beta]$ so that the surface is only traced once when C is revolved about the line.

EXAMPLE 6   Find the area S of the surface obtained by revolving the circle $r = \cos \theta$ about the line $\theta = \pi/2$. (See Figure 9.)

Solution   Observe that the circle is traced exactly once as θ increases from 0 to π. Therefore, using Theorem 4 with $r = \cos \theta$, $\alpha = 0$, and $\beta = \pi$, we obtain

$$S = 2\pi \int_\alpha^\beta f(\theta) \cos \theta \sqrt{\left(\frac{dr}{d\theta} \right)^2 + r^2} \, d\theta$$

$$= 2\pi \int_0^{\pi} \cos \theta (\cos \theta) \sqrt{(-\sin \theta)^2 + (\cos \theta)^2} \, d\theta$$

$$= 2\pi \int_0^{\pi} \cos^2 \theta \, d\theta = \pi \int_0^{\pi} (1 + \cos 2\theta) \, d\theta$$

$$= \pi \left[\theta + \frac{\sin 2\theta}{2} \right]_0^{\pi} = \pi^2$$

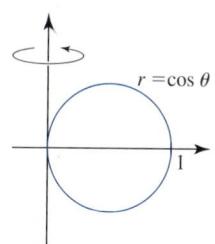

(a)

(b)

FIGURE 9
The solid obtained by revolving the circle $r = \cos \theta$ (a) about the line $\theta = \pi/2$ is a torus (b).

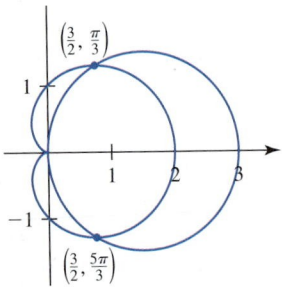

FIGURE 10
The graphs of the cardioid
$r = 1 + \cos\theta$ and the circle
$r = 3\cos\theta$

Points of Intersection of Graphs in Polar Coordinates

In Example 4 we were able to find the points of intersection of two curves with representations in polar coordinates by solving a system of two equations simultaneously. This is not always the case. Consider for example, the graphs of the cardioid $r = 1 + \cos\theta$ and the circle $r = 3\cos\theta$ shown in Figure 10. Solving the two equations simultaneously, we obtain

$$3\cos\theta = 1 + \cos\theta$$

$$\cos\theta = \frac{1}{2}$$

(3)

or $\theta = \pi/3$ and $5\pi/3$. Therefore, the points of intersection are $\left(\frac{3}{2}, \frac{\pi}{3}\right)$ and $\left(\frac{3}{2}, \frac{5\pi}{3}\right)$. But one glance at Figure 10 shows the pole as a third point of intersection that is not revealed in our calculation. To see how this can happen, think of the cardioid as being traced by the point (r, θ) satisfying

$$r = f(\theta) = 1 + \cos\theta \qquad 0 \le \theta \le 2\pi$$

with θ as a parameter. If we think of θ as representing time, then as θ runs from $\theta = 0$ through $\theta = 2\pi$, the point (r, θ) starts at $(2, 0)$ and traverses the cardioid in a counterclockwise direction, eventually returning to the point $(2, 0)$. (See Figure 11a.) Similarly, the circle is traced *twice* in the counterclockwise direction, by the point (r, θ), where

$$r = g(\theta) = 3\cos\theta \qquad 0 \le \theta \le 2\pi$$

and the parameter θ, once again representing time, runs from $\theta = 0$ through $\theta = 2\pi$ (see Figure 11b).

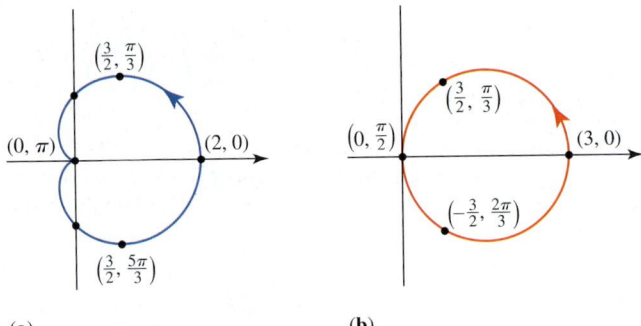

FIGURE 11   **(a)**   **(b)**

Observe that the point tracing the cardioid arrives at the point $\left(\frac{3}{2}, \frac{\pi}{3}\right)$ on the cardioid at precisely the same time that the point tracing the circle arrives at the point $\left(\frac{3}{2}, \frac{\pi}{3}\right)$ on the circle. A similar observation holds at the point $\left(\frac{3}{2}, \frac{5\pi}{3}\right)$ on each of the two curves. These are the points of intersection found earlier.

Next, observe that the point tracing the cardioid arrives at the origin when $\theta = \pi$. But the point tracing the circle first arrives at the origin when $\theta = \pi/2$ and then again when $\theta = 3\pi/2$. In other words, these two points arrive at the origin at *different* times, so there is no (common) value of θ corresponding to the origin that satisfies both Equations (3) simultaneously. Thus, although the origin is a point of intersection of the two curves, this fact will not show up in the solution of the system of equations. For this reason it is recommended that we sketch the graphs of polar equations when finding their points of intersection.

EXAMPLE 7 Find the points of intersection of $r = \cos\theta$ and $r = \cos 2\theta$.

Solution We solve the system of equations

$$r = \cos\theta$$

$$r = \cos 2\theta$$

We set $\cos\theta = \cos 2\theta$ and use the identity $\cos 2\theta = 2\cos^2\theta - 1$. We obtain

$$2\cos^2\theta - \cos\theta - 1 = 0$$

$$(2\cos\theta + 1)(\cos\theta - 1) = 0$$

So

$$\cos\theta = -\frac{1}{2} \qquad \text{or} \qquad \cos\theta = 1$$

that is,

$$\theta = \frac{2\pi}{3}, \qquad \frac{4\pi}{3}, \qquad \text{or} \qquad 0$$

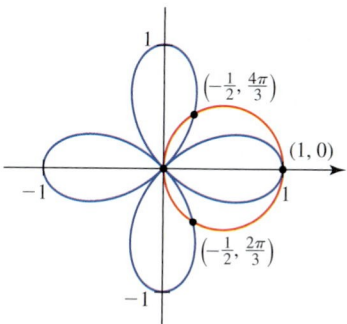

FIGURE 12

These values of θ give $\left(-\frac{1}{2}, \frac{2\pi}{3}\right)$, $\left(-\frac{1}{2}, \frac{4\pi}{3}\right)$, and $(1, 0)$ as the points of intersection. Since both graphs also pass through the pole, we conclude that the pole is also a point of intersection. (See Figure 12.)

9.5 CONCEPT QUESTIONS

1. a. Let f be nonnegative and continuous on $[\alpha, \beta]$, where $0 \le \beta - \alpha < 2\pi$. Write an integral giving the area of the region bounded by the graphs of $r = f(\theta)$, $\theta = \alpha$, and $\theta = \beta$. Make a sketch of the region.

    b. If f and g are continuous on $[\alpha, \beta]$ and $0 \le g(\theta) \le f(\theta)$, where $0 \le \alpha \le \beta \le 2\pi$, write an integral giving the area of the region bounded by the graphs of $r = g(\theta)$, $r = f(\theta)$, $\theta = \alpha$, and $\theta = \beta$. Make a sketch of the region.

2. Suppose that f has a continuous derivative on an interval $[\alpha, \beta]$. If the graph C of $r = f(\theta)$ is traced exactly once as θ increases from α to β, write an integral giving the length of C.

3. Suppose that f is a function with a continuous derivative on $[\alpha, \beta]$ and the graph C of $r = f(\theta)$ is traced exactly once as θ increases from α to β. Write an integral giving the area of the surface obtained by revolving C about (a) the polar axis, $y \ge 0$, and (b) the line $\theta = \pi/2$, $x \ge 0$.

9.5 EXERCISES

1. a. Find a rectangular equation of the circle $r = 4\cos\theta$, and use it to find its area.

    b. Find the area of the circle of part (a) by integration.

2. a. By finding a rectangular equation, show that the polar equation $r = 2\cos\theta - 2\sin\theta$ represents a circle. Then find the area of the circle.

    b. Find the area of the circle of part (a) by integration.

In Exercises 3–8, find the area of the region bounded by the curve and the rays.

3. $r = \theta$, $\theta = 0$, $\theta = \pi$

4. $r = \dfrac{1}{\theta}$, $\theta = \dfrac{\pi}{6}$, $\theta = \dfrac{\pi}{3}$

5. $r = e^\theta$, $\theta = -\dfrac{\pi}{2}$, $\theta = 0$

6. $r = e^{-2\theta}$, $\theta = 0$, $\theta = \dfrac{\pi}{4}$

7. $r = \sqrt{\cos\theta}$, $\theta = 0$, $\theta = \dfrac{\pi}{2}$

8. $r = \cos 2\theta$, $\theta = 0$, $\theta = \dfrac{\pi}{16}$

In Exercises 9–12, find the area of the shaded region.

9.

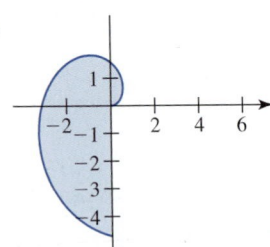

$$r = \theta$$

10.

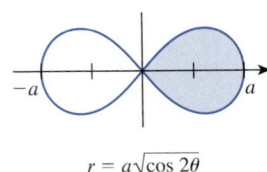

$$r = a\sqrt{\cos 2\theta}$$

11.

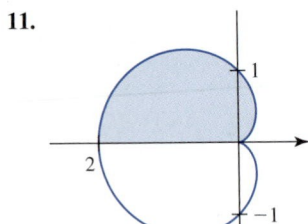

$$r = 1 - \cos \theta$$

12.

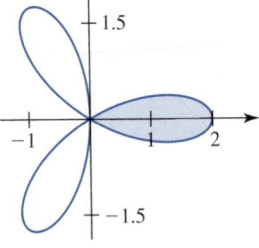

$$r = 2 \cos 3\theta$$

In Exercises 13–18, sketch the curve, and find the area of the region enclosed by it.

13. $r = 3 \sin \theta$

14. $r = 2(1 - \cos \theta)$

15. $r^2 = \sin \theta$

16. $r^2 = 3 \sin 2\theta$

17. $r = 2 \sin 2\theta$

18. $r = 2 \sin 3\theta$

In Exercises 19–22, find the area of the region enclosed by one loop of the curve.

19. $r = \cos 2\theta$

20. $r = 2 \cos 3\theta$

21. $r = \sin 4\theta$

22. $r = 2 \cos 4\theta$

In Exercises 23–24, find the area of the region described.

23. The inner loop of the limaçon $r = 1 + 2 \cos \theta$

24. Between the loops of the limaçon $r = 1 + 2 \sin \theta$

In Exercises 25–28, find the area of the shaded region.

25.

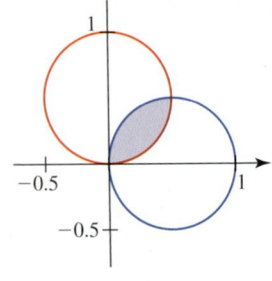

$$r = \sin \theta, r = \cos \theta$$

26.

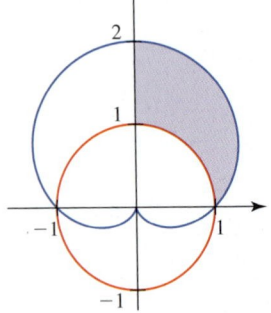

$$r = 1, r = 1 + \sin \theta$$

27.

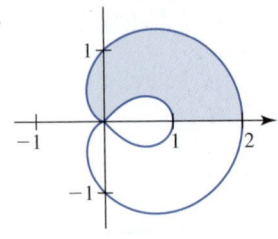

$$r = 1 + \cos \theta, r = \sqrt{\cos 2\theta}$$

28.

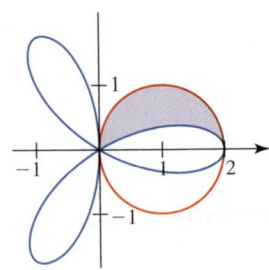

$$r = 2 \cos 3\theta, r = 2 \cos \theta$$

In Exercises 29–34, find all points of intersection of the given curves.

29. $r = 1$ and $r = 1 + \cos \theta$

30. $r = 3$ and $r = 2 + 2 \cos \theta$

31. $r = 2$ and $r = 4 \cos 2\theta$

32. $r = 1$ and $r^2 = 2 \cos 2\theta$

33. $r = \sin \theta$ and $r = \sin 2\theta$

34. $r = \cos \theta$ and $r = \cos 2\theta$

In Exercises 35–40, find the area of the region that lies outside the first curve and inside the second curve.

35. $r = 1 + \cos \theta$, $r = 3 \cos \theta$

36. $r = 1 - \sin \theta$, $r = 1$

37. $r = 4 \cos \theta$, $r = 2$

38. $r = 3 \sin \theta$, $r = 2 - \sin \theta$

39. $r = 1 - \cos \theta$, $r = \dfrac{3}{2}$

40. $r = 2 \cos 3\theta$, $r = 1$

In Exercises 41–46, find the area of the region that is enclosed by both of the curves.

41. $r = 1$, $r = 2 \sin \theta$

42. $r = \cos \theta$, $r = \sqrt{3} \sin \theta$

43. $r = \sin \theta$, $r = 1 - \sin \theta$

44. $r = \cos \theta$, $r = 1 - \cos \theta$

45. $r^2 = 4 \cos 2\theta$, $r = \sqrt{2}$

46. $r = \sqrt{3} \sin \theta$, $r = 1 + \cos \theta$

In Exercises 47–54, find the length of the given curve.

47. $r = 5 \sin \theta$

48. $r = 2\theta$; $0 \le \theta \le 2\pi$

49. $r = e^{-\theta}$; $0 \le \theta \le 4\pi$

50. $r = 1 + \sin \theta$; $0 \le \theta \le 2\pi$

51. $r = \sin^3 \dfrac{\theta}{3}$; $0 \le \theta \le \pi$

52. $r = \cos^2 \dfrac{\theta}{2}$

53. $r = a \sin^4 \dfrac{\theta}{4}$

54. $r = \sec \theta; \quad 0 \le \theta \le \frac{\pi}{3}$

In Exercises 55–60, find the area of the surface obtained by revolving the given curve about the given line.

55. $r = 4 \cos \theta$ about the polar axis

56. $r = 2 \cos \theta$ about the line $\theta = \dfrac{\pi}{2}$

57. $r = 2 + 2 \cos \theta$ about the polar axis

58. $r^2 = \cos 2\theta$ about the polar axis

59. $r^2 = \cos 2\theta$ about the line $\theta = \dfrac{\pi}{2}$

60. $r = e^{a\theta}, \quad 0 \le \theta \le \frac{\pi}{2}$ about the line $\theta = \dfrac{\pi}{2}$

In Exercises 61 and 62, find the area of the region enclosed by the given curve. (Hint: Convert the rectangular equation to a polar equation.)

61. $(x^2 + y^2)^3 = 16x^2y^2$ **62.** $x^4 + y^4 = 4(x^2 + y^2)$

63. Let P be a point other than the origin lying on the curve $r = f(\theta)$. If ψ is the angle between the tangent line to the curve at P and the radial line OP, then $\tan \psi = \dfrac{r}{dr/d\theta}$. (See Section 9.4, Exercise 84.)

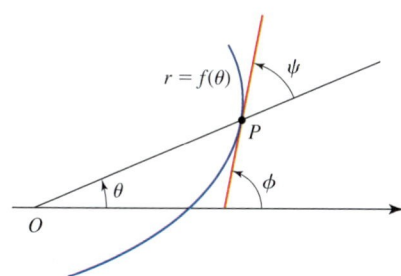

a. Show that the angle between the tangent line to the logarithmic spiral $r = e^{m\theta}$ and the radial line at the point of tangency is a constant.

b. Suppose the curve with polar equation $r = f(\theta)$ has the property that at any point on the curve, the angle ψ between the tangent line to the curve at that point and the radial line from the origin to that point is a constant. Show that $f(\theta) = Ce^{m\theta}$, where C and m are constants.

64. Find the length of the logarithmic spiral $r = ae^{m\theta}$ between the point (r_0, θ_0) and the point (r, θ), and use this result to deduce that the length of a logarithmic spiral is proportional to the difference between the radial coordinates of the points.

65. Show that the length of the parabola $y = (1/2p)x^2$ on the interval $[0, a]$ is the same as the length of the spiral $r = p\theta$ for $0 \le r \le a$.

cas **66.** Plot the curve $r = \sin(3 \cos \theta)$, and find an approximation of the area enclosed by the curve accurate to four decimal places.

cas *In Exercises 67–69, (a) plot the curve, and (b) find an approximation of its length accurate to two decimal places.*

67. $r = \sqrt{1 + \theta^2}$, where $0 \le \theta \le 2\pi$ (involute of a circle)

68. $r = 0.2\sqrt{\theta} + 1$, where $0 \le \theta \le 6\pi$ (parabolic spiral)

69. $r = 3 \sin \theta \cos^2 \theta$, where $0 \le \theta \le \pi$ (bifolia)

70. a. Let f be a function with a continuous derivative in an interval $[\alpha, \beta]$. If the graph C of $r = f(\theta)$ is traced exactly once as θ increases from α to β, show that the rectangular coordinates of the centroid of C are

$$\bar{x} = \dfrac{\displaystyle\int_\alpha^\beta r \cos \theta \sqrt{(r')^2 + r^2}\, d\theta}{\displaystyle\int_\alpha^\beta \sqrt{(r')^2 + r^2}\, d\theta}$$

and

$$\bar{y} = \dfrac{\displaystyle\int_\alpha^\beta r \sin \theta \sqrt{(r')^2 + r^2}\, d\theta}{\displaystyle\int_\alpha^\beta \sqrt{(r')^2 + r^2}\, d\theta}$$

Hint: See the directions for Exercises 47 and 48 in Section 5.7.

b. Use the result of part (a) to find the centroid of the upper semicircle $r = a$, where $a > 0$ and $0 \le \theta \le \pi$.

cas **71. a.** Plot the curve with polar equation $r = 2 \cos^3 \theta$ where $-\frac{\pi}{2} \le \theta \le \frac{\pi}{2}$.

b. Find the Cartesian coordinates of the centroid of the region bounded by the curve of part (a).

cas **72. a.** Plot the graphs of $r = 1 + \cos \theta$ and $r = 3 \cos \theta$ for $0 \le \theta \le 2\pi$, treating r and θ as rectangular coordinates.

b. Refer to page 880. Reconcile your results with the discussion of finding the points of intersection of graphs in polar coordinates.

cas *In Exercises 73 and 74, (a) find the polar representation of the curve given in rectangular coordinates, (b) plot the curve, and (c) find the area of the region enclosed by a loop (or loops) of the curve.*

73. $x^3 - 3xy + y^3 = 0$ (folium of Descartes)

74. $(x^2 + y^2)^{1/2} - \cos\left[4 \tan^{-1}\left(\dfrac{y}{x}\right)\right] = 0$ (rhodenea)

In Exercises 75 and 76, determine whether the statement is true or false. If it is true, explain why it is true. If it is false, give an example to show why it is false.

75. If there exists a θ_0 such that $f(\theta_0) = g(\theta_0)$, then the graphs of $r = f(\theta)$ and $r = g(\theta)$ have at least one point of intersection.

76. If P is a point of intersection of the graphs of $r = f(\theta)$ and $r = g(\theta)$, then there must exist a θ_0 such that $f(\theta_0) = g(\theta_0)$.

9.6   Conic Sections in Polar Coordinates

In Section 9.1 we obtained representations of the conic sections—the parabola, the ellipse, and the hyperbola—in terms of rectangular equations. In this section we will show that all three types of conic sections can be represented by a single polar equation. As you saw in the preceding sections, some problems can be solved more easily using polar coordinates rather than rectangular coordinates.

We begin by proving the following theorem, which gives an equivalent definition of each conic section in terms of its focus and directrix. As a corollary, we will obtain the desired representation of the conic sections in polar form.

THEOREM 1

Let F be a fixed point, let l be a fixed line in the plane, and let e be a fixed positive number. Then the set of all points P in the plane satisfying

$$\frac{d(P, F)}{d(P, l)} = e$$

is a conic section. The point F is the **focus** of the conic section, and the line l is its **directrix**. The number e, which is the ratio of the distance between P and F and the distance between P and l, is called the **eccentricity** of the conic. The conic is an ellipse if $e < 1$, a parabola if $e = 1$, or a hyperbola if $e > 1$.

The three types of conics are illustrated in Figure 1.

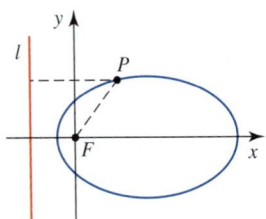

(a) $\dfrac{d(P, F)}{d(P, l)} = e < 1$ (ellipse)   (b) $\dfrac{d(P, F)}{d(P, l)} = e = 1$ (parabola)   (c) $\dfrac{d(P, F)}{d(P, l)} = e > 1$ (hyperbola)

FIGURE 1

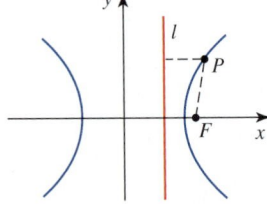

FIGURE 2

PROOF   Observe that if $e = 1$, then $d(P, F) = d(P, l)$. That is, the distance between a point on the curve and the focus is equal to the distance between the point and the directrix. But this is just the definition of a parabola, so the curve is a conic section.

In what follows, we will assume that $e \neq 1$. Refer to Figure 2, where we have placed the focus F at the origin and the directrix l parallel to and d units to the left of the y-axis. Therefore, the directrix has equation $x = -d$, where $d > 0$. If $P(r, \theta)$ is any point lying on the curve, then you can see from Figure 2 that

$$d(P, F) = r \qquad \text{and} \qquad d(P, l) = d + r \cos \theta$$

Therefore, the condition $d(P, F)/d(P, l) = e$ or, equivalently, $d(P, F) = e \cdot d(P, l)$, implies that

$$r = e(d + r \cos \theta) \tag{1}$$

Converting this equation to rectangular coordinates gives

$$\sqrt{x^2 + y^2} = e(d + x)$$

which, upon squaring, yields

$$x^2 + y^2 = e^2(d + x)^2 = e^2(d^2 + 2dx + x^2)$$

$$(1 - e^2)x^2 - 2de^2x + y^2 = e^2d^2$$

or

$$x^2 - \left(\frac{2e^2d}{1 - e^2}\right)x + \frac{y^2}{1 - e^2} = \frac{e^2d^2}{1 - e^2}$$

Completing the square in x, we obtain

$$\left(x - \frac{e^2d}{1 - e^2}\right)^2 + \frac{y^2}{1 - e^2} = \frac{e^2d^2}{1 - e^2} + \frac{e^4d^2}{(1 - e^2)^2} = \frac{e^2d^2}{(1 - e^2)^2} \tag{2}$$

Now, if $e < 1$, then $1 - e^2 > 0$. Dividing both sides by $e^2d^2/(1 - e^2)^2$, we can write Equation (2) in the form

$$\frac{(x - h)^2}{a^2} + \frac{y^2}{b^2} = 1$$

where

$$h = \frac{e^2d}{1 - e^2}, \qquad a^2 = \frac{e^2d^2}{(1 - e^2)^2}, \qquad \text{and} \qquad b^2 = \frac{e^2d^2}{1 - e^2} \tag{3}$$

This is an equation of an ellipse centered at the point $(h, 0)$ on the x-axis.

Next, we compute

$$c^2 = a^2 - b^2 = \frac{e^4d^2}{(1 - e^2)^2} \tag{4}$$

from which we obtain

$$c = \frac{e^2d}{1 - e^2} = h$$

Recalling that the foci of an ellipse are located at a distance c from its center, we have shown that F is indeed the focus of the ellipse. It also follows from Equations (3) and (4) that the eccentricity of the ellipse is given by

$$e = \frac{c}{a} \tag{5}$$

where $c^2 = a^2 - b^2$.

If $e > 1$, then $1 - e^2 < 0$. Proceeding in a similar manner as before, we can write Equation (2) in the form

$$\frac{(x - h)^2}{a^2} - \frac{y^2}{b^2} = 1$$

Historical Biography

JOHANNES KEPLER
(1571–1630)

Born in 1571 in Weil der Stadt, Germany, Johannes Kepler was introduced to the wonders of the universe at a young age when his mother took him to observe the comet of 1577. Kepler completed a master's degree in theology at the University of Tübingen and established a reputation as a talented mathematician and astronomer, but an unorthodox Lutheran. Given the volatile religious situation of the time, Kepler was advised not to pursue a career in the ministry and was instead recommended for a position teaching mathematics and astronomy at a school in Graz. He took that position in 1594, and a year later he published his first major work, *Mysterium cosmographicum* ("The Mystery of the Cosmos"). In this work, he explained his discovery of the three laws of planetary motion that now bear his name (see page 889), and he became the first person to correctly explain planetary orbits within our solar system. Kepler's First Law is that every planet's orbit is an ellipse with the sun at one focus.

While Kepler was writing his famous work *The Harmony of the World* (1619), his mother was charged with witchcraft. Kepler enlisted the help of the legal faculty at the University of Tübingen in his effort to prevent his mother from being convicted. Katharina Kepler was eventually released as the result of technical objections on the part of the defense, arising from the authorities' failure to follow the legal procedures of the time regarding the use of torture.

which is an equation of a hyperbola. We also see that the eccentricity of the hyperbola is

$$e = \frac{c}{a} \tag{6}$$

where $c^2 = a^2 + b^2$.

If we solve Equation (1) for r, we obtain the polar equation

$$r = \frac{ed}{1 - e \cos \theta}$$

of the conic shown in Figure 2. If the directrix is chosen so that it lies to the right of the focus, say, $x = d$, where $d > 0$, then the polar equation of the conic is

$$r = \frac{ed}{1 + e \cos \theta}$$

Similarly, we can show that if the directrix $y = \pm d$ is chosen to be parallel to the polar axis, then the polar equation of the conic is

$$r = \frac{ed}{1 \pm e \sin \theta}$$

(See Exercises 28–30.)

The conics are illustrated in Figure 3.

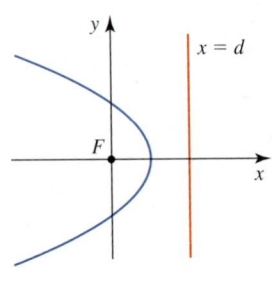
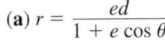
(a) $r = \dfrac{ed}{1 + e \cos \theta}$

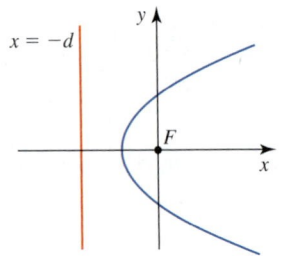
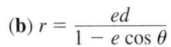
(b) $r = \dfrac{ed}{1 - e \cos \theta}$

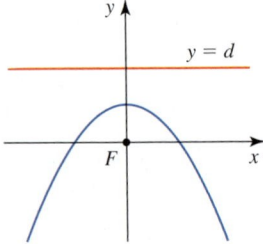
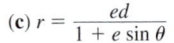
(c) $r = \dfrac{ed}{1 + e \sin \theta}$

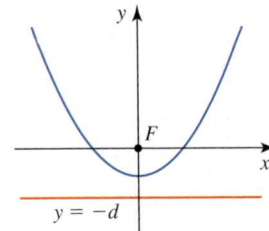
(d) $r = \dfrac{ed}{1 - e \sin \theta}$

FIGURE 3
Polar equations of conics

THEOREM 2

A polar equation of the form

$$r = \frac{ed}{1 \pm e \cos \theta} \qquad \text{or} \qquad r = \frac{ed}{1 \pm e \sin \theta}$$

represents a conic section with eccentricity e. The conic is a parabola if $e = 1$, an ellipse if $e < 1$, and a hyperbola if $e > 1$.

EXAMPLE 1 Find a polar equation of a parabola that has its focus at the pole and the line $y = 2$ as its directrix.

Solution Since this conic section is a parabola, we see that $e = 1$. Next, observe that its directrix, $y = 2$, is parallel to and lies above the polar axis. So letting $d = 2$ and referring to Figure 3c, we see that a required equation of the parabola is

$$r = \frac{2}{1 + \sin \theta}$$ ∎

EXAMPLE 2 A conic has polar equation

$$r = \frac{15}{3 + 2 \cos \theta}$$

Find the eccentricity and the directrix of the conic section, and sketch the conic section.

Solution We begin by rewriting the given equation in standard form by dividing both its numerator and denominator by 3, obtaining

$$r = \frac{5}{1 + \frac{2}{3} \cos \theta}$$

Then using Theorem 2, we see that $e = \frac{2}{3}$. Since $ed = 5$, we have

$$d = \frac{5}{e} = \frac{5}{\frac{2}{3}} = \frac{15}{2}$$

Since $e < 1$, we conclude that the conic section is an ellipse with focus at the pole and major axis lying along the polar axis. Its directrix has rectangular equation $x = \frac{15}{2}$. Setting $\theta = 0$ and $\theta = \pi$ successively gives $r = 3$ and $r = 15$, giving the vertices of the ellipse in polar coordinates as $(3, 0)$ and $(15, \pi)$. The center of the ellipse is the midpoint $(6, \pi)$ in polar coordinates of the line segment joining the vertices. Since the length of the major axes of the ellipse is 18, we have $2a = 18$, or $a = 9$. Finally, since $e = c/a$, we find that

$$c = ae = 9\left(\frac{2}{3}\right) = 6$$

So

$$b^2 = a^2 - c^2 = 81 - 36 = 45$$

or

$$b = 3\sqrt{5}$$

The graph of the conic is sketched in Figure 4. ∎

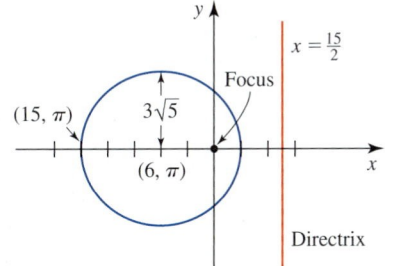

FIGURE 4
The graph of $r = \dfrac{15}{3 + 2 \cos \theta}$

EXAMPLE 3 Sketch the graph of the polar equation

$$r = \frac{20}{2 + 3 \sin \theta}$$

Solution By dividing the numerator and the denominator of the given equation by 2, we obtain the equation

$$r = \frac{10}{1 + \frac{3}{2} \sin \theta}$$

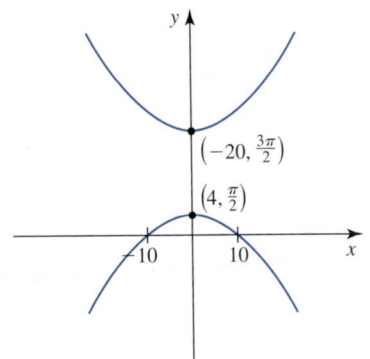

FIGURE 5
The graph of $r = \dfrac{20}{2 + 3 \sin \theta}$

in standard form. We see that $e = \frac{3}{2}$, so the equation represents a hyperbola with one focus at the pole. Comparing this equation with the equation associated with Figure 3c, we see that the transverse axis of the hyperbola lies along the line $\theta = \pi/2$. To find the vertices of the hyperbola, we set $\theta = \pi/2$ and $\theta = 3\pi/2$ successively, giving $\left(4, \frac{\pi}{2}\right)$ and $\left(-20, \frac{3\pi}{2}\right)$ as the required vertices in polar coordinates. The center of the hyperbola in polar coordinates is the midpoint $\left(12, \frac{\pi}{2}\right)$ of the line segment joining the vertices. The x-intercepts (we superimpose the Cartesian system over the polar system) are found by setting $\theta = 0$ and $\theta = \pi$, giving the x-intercepts as 10 and -10. The required graph may be sketched in two steps; first, we sketch the lower branch of the hyperbola, making use of the x-intercepts that we just found. Then, using symmetry, we sketch the upper branch of the hyperbola. (See Figure 5.) ◼

◼ Eccentricity of a Conic

As we saw in Theorem 1, the nature of a conic section is determined by its eccentricity e. To see in greater detail the role that is played by the eccentricity of a conic, let's first consider the case in which $e < 1$, so that the conic under consideration is an ellipse. Now by Equation (5) we have

$$e = \frac{c}{a} = \frac{\sqrt{a^2 - b^2}}{a}$$

If e is close to 0, then $\sqrt{a^2 - b^2}$ is close to 0, or a is close to b. This means that the ellipse is almost circular (see Figure 6a). On the other hand, if e is close to 1, then $\sqrt{a^2 - b^2} \approx a$, $a^2 - b^2 \approx a^2$, or b is small. This means that the ellipse is very flat (see Figure 6b).

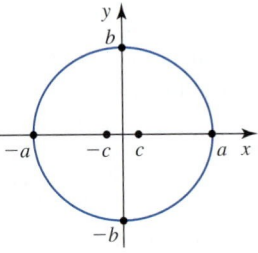

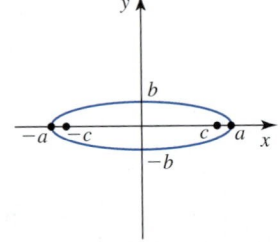

FIGURE 6
The ellipse is almost circular if e is close to 0 and is very flat if e is close to 1.

(a) e is close to 0.

(b) e is close to 1.

If $e = 1$, then the conic is a parabola. We leave it to you to perform a similar analysis in the case in which $e > 1$ (so that the conic is a hyperbola).

In Figure 7 we show two hyperbolas: In part (a) the eccentricity e is close to but greater than 1. In part (b) the eccentricity e is much larger than 1.

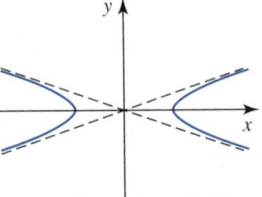

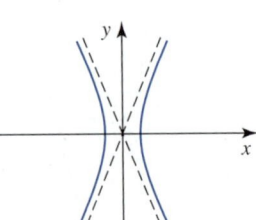

FIGURE 7
The eccentricity of the hyperbola in part (a) is close to 1, whereas the eccentricity of the hyperbola in part (b) is much larger than 1.

(a)

(b)

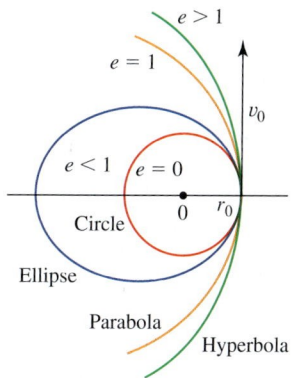

FIGURE 8
The speed v_0 determines
the orbit of the body.

Motion of Celestial Bodies

In the last few sections we have seen numerous applications of conics. Yet another important application of the conics arises in the motion of celestial bodies.

Figure 8 shows a body a distance r_0 from the origin 0 moving with a speed v_0 and in a direction perpendicular to the line passing through 0 and v_0. It can be shown, although we will not do so here, that the orbit of the body about the origin depends on the magnitude of v_0. For the planets in the solar system (with the sun at the origin) and for certain comets such as Halley's comet, the speed v_0 is such that they remain captive and will never leave the system; their orbits are ellipses. However, if the speed v_0 of a body is sufficiently large, then its orbit about the sun is a parabola ($e = 1$) or a branch of a hyperbola ($e > 1$). In both these cases the body makes but a single pass about the sun!

The orbits of the planets about the sun, moreover, are described by Kepler's Laws.

Kepler's Laws

1. Planets move in orbits that are ellipses with the sun at one focus.
2. The line from the sun to a planet sweeps out equal areas in equal times. (See Figure 9.)
3. The square of a planet's period is proportional to the cube of the length of the semimajor axis of its orbit.

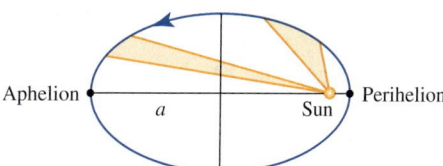

FIGURE 9
Equal areas are swept out in equal times, $T^2 \propto a^3$, where T is the period.

The positions of a planet that are closest to and farthest from the sun are called the *perihelion* and *aphelion*, respectively.

EXAMPLE 4 **The Orbit of Halley's Comet** Halley's comet has an elliptical orbit with an eccentricity of 0.967. Its perihelion distance (shortest distance from the sun) is 8.9×10^7 km.

a. Find a polar equation for the orbit.
b. Find the distance of the comet from the sun when it is at the aphelion.

Solution
a. Suppose that the axis is horizontal as shown in Figure 10. Then the polar equation can be chosen to have the form

$$r = \frac{ed}{1 + e \cos \theta}$$

The distance of the comet from the sun when it is at the perihelion is given by

$$a - c = a - ea = a(1 - e)$$

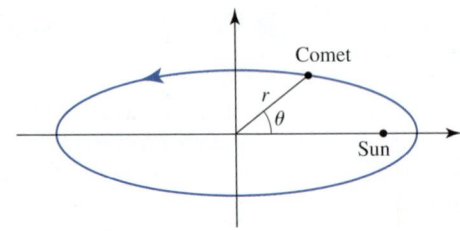

FIGURE 10
In actuality the trajectory is much flatter.

But we are given that at the perihelion the distance from the sun is 8.9×10^7 km and $e = 0.967$. So

$$a(1 - 0.967) = 8.9 \times 10^7$$

or

$$a = \frac{8.9 \times 10^7}{1 - 0.967} \approx 2.697 \times 10^9$$

Next, from Equation (3) we see that

$$ed = a(1 - e^2)$$
$$= (2.697 \times 10^9)(1 - 0.967^2) = 1.75 \times 10^8$$

So the required equation is

$$r = \frac{1.75 \times 10^8}{1 + 0.967 \cos \theta}$$

b. The aphelion distance (farthest distance from the sun) is

$$a + c = a + ea = a(1 + e) \approx (2.697 \times 10^9)(1 + 0.967) \approx 5.305 \times 10^9$$

kilometers.

9.6 CONCEPT QUESTIONS

1. Consider the polar equations

$$r = \frac{ed}{1 \pm e \cos \theta} \quad \text{and} \quad r = \frac{ed}{1 \pm e \sin \theta}$$

Explain the role of the numbers d and e. Illustrate each with a sketch.

2. Give a classification of the conics in terms of their eccentricities.

3. Identify the conic:

a. $r = \dfrac{3}{1 + 2 \sin \theta}$
b. $r = \dfrac{6}{3 + \cos \theta}$

c. $r = \dfrac{2}{3(1 + \cos \theta)}$
d. $r = \dfrac{5}{3 - 2 \sin \theta}$

9.6 EXERCISES

In Exercises 1–8, write a polar equation of the conic that has a focus at the origin and the given properties. Identify the conic.

1. Eccentricity 1, directrix $x = -2$

2. Eccentricity $\frac{1}{3}$, directrix $x = 3$

3. Eccentricity $\frac{1}{2}$, directrix $y = -2$

4. Eccentricity 1, directrix $y = -3$

5. Eccentricity $\frac{3}{2}$, directrix $x = 1$

6. Eccentricity $\frac{5}{4}$, directrix $y = -2$

7. Eccentricity 0.4, directrix $y = 0.4$

8. Eccentricity $\frac{1}{2}$, directrix $r = -2 \sec \theta$

In Exercises 9–20, (a) *find the eccentricity and an equation of the directrix of the conic,* (b) *identify the conic, and* (c) *sketch the curve.*

9. $r = \dfrac{8}{6 + 2 \sin \theta}$

10. $r = \dfrac{8}{6 - 2 \sin \theta}$

11. $r = \dfrac{10}{4 + 6 \cos \theta}$

12. $r = \dfrac{10}{4 - 6 \cos \theta}$

13. $r = \dfrac{5}{2 + 2 \cos \theta}$

14. $r = \dfrac{5}{2 - 2 \sin \theta}$

15. $r = \dfrac{1}{3 - 2 \cos \theta}$

16. $r = \dfrac{12}{3 + \cos \theta}$

17. $r = \dfrac{1}{1 - \sin \theta}$

18. $r = \dfrac{1}{1 + \cos \theta}$

19. $r = -\dfrac{6}{\sin \theta - 2}$

20. $r = -\dfrac{2}{\cos \theta - 3}$

In Exercises 21–26, use Equation (5) or Equation (6) to find the eccentricity of the conic with the given rectangular equation.

21. $\dfrac{x^2}{9} + \dfrac{y^2}{16} = 1$

22. $\dfrac{x^2}{5} - \dfrac{y^2}{3} = 1$

23. $x^2 - y^2 = 1$

24. $9x^2 + 25y^2 = 225$

25. $x^2 - 9y^2 + 2x - 54y = 105$

26. $2x^2 + y^2 + 4x - 6y + 7 = 0$

27. Show that the parabolas with polar equations

$$r = \frac{c}{1 + \sin \theta} \quad \text{and} \quad r = \frac{d}{1 - \sin \theta}$$

intersect at right angles.

28. Show that a conic with focus at the origin, eccentricity e, and directrix $x = d$ has polar equation

$$r = \frac{ed}{1 + e \cos \theta}$$

29. Show that a conic with focus at the origin, eccentricity e, and directrix $y = d$ has polar equation

$$r = \frac{ed}{1 + e \sin \theta}$$

30. Show that a conic with focus at the origin, eccentricity e, and directrix $y = -d$ has polar equation

$$r = \frac{ed}{1 - e \sin \theta}$$

31. a. Show that the polar equation of an ellipse with one focus at the pole and major axis lying along the polar axis is given by

$$r = \frac{a(1 - e^2)}{1 - e \cos \theta}$$

where e is the eccentricity of the ellipse and $2a$ is the length of its major axis.

b. The planets revolve about the sun in elliptical orbits with the sun at one focus. The points on the orbit where a planet is nearest to and farthest from the sun are called the *perihelion* and the *aphelion* of the orbit, respectively. Use the result of part (a) to show that the perihelion distance (minimum distance from the planet to the sun) is $a(1 - e)$.

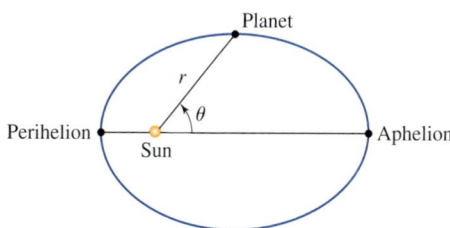

In Exercises 32 and 33, use the results of Exercise 31 to find a polar equation describing the approximate orbit of the given planet and to find the perihelion and aphelion distances.

32. Earth: $e = 0.017$, $a = 92.957 \times 10^6$ mi

33. Saturn: $e = 0.056$, $a = 1.427 \times 10^9$ km

34. The dwarf planet Pluto revolves about the sun in an elliptical orbit. The eccentricity of the orbit is 0.249, and its perihelion distance is 4.43×10^9 km. Use the results of Exercise 31 to find a polar equation for the orbit of Pluto and find its aphelion distance.

35. The planet Mercury revolves about the sun in an elliptical orbit. Its perihelion distance is approximately 4.6×10^7 km, and its aphelion distance is approximately 7.0×10^7 km. Use the results of Exercise 31 to estimate the eccentricity of Mercury's orbit.

CHAPTER 9 REVIEW

CONCEPT REVIEW

In Exercises 1–16, fill in the blanks.

1. a. A parabola is the set of all points in the plane that are _____ from a fixed _____ and a fixed _____. The fixed _____ is called the _____, and the fixed _____ is called the _____.

b. The point halfway between the focus and the directrix of a parabola is called its _____. The axis of the parabola is the line passing through the _____ and perpendicular to the _____.

2. a. An equation of a parabola with focus $(0, p)$ and directrix $y = -p$ is _____.

b. An equation of a parabola with focus _____ and directrix _____ is $y^2 = 4px$.

3. a. An ellipse is the set of all points in a plane the _____ of whose distances from two fixed points (called the _____) is a _____.

b. The vertices of an ellipse are the points of intersection of the line passing through the _____ and the ellipse. The chord joining the vertices is called the _____ _____, and its midpoint is called the _____ of the ellipse. The chord passing through the center of the ellipse and perpendicular to the major axis is called the _____ _____ of the ellipse.

4. a. An equation of the ellipse with foci $(\pm c, 0)$ and vertices $(\pm a, 0)$ is _____, where $c^2 = $ _____.

b. An equation of the ellipse with foci _____ and vertices _____ is $x^2/b^2 + y^2/a^2 = 1$.

5. a. A hyperbola is the set of all points in a plane the _____ of whose distances from two fixed points (called the _____) is a _____.

b. The line passing through the foci intersects the hyperbola at two points called the _____ of the hyperbola. The line segment joining the vertices is called the _____ axis of the hyperbola, and the midpoint of the _____ axis is called the _____ of the hyperbola. A hyperbola has _____ _____ branches.

6. a. An equation of a hyperbola with foci $(\pm c, 0)$ and vertices $(\pm a, 0)$ is _____, where $c^2 = $ _____. The hyperbola has asymptotes _____.

b. An equation of the hyperbola with foci _____ and vertices _____ is $y^2/a^2 - x^2/b^2 = 1$, where $c^2 = $ _____. The hyperbola has asymptotes _____.

7. a. A plane curve is a set C of ordered pairs (x, y) defined by the parametric equations _____, where f and g are continuous functions on an interval I; I is called the _____ interval.

8. a. If $x = f(t)$ and $y = g(t)$, where f and g are differentiable and $f'(t) \neq 0$, then $dy/dx = $ _____.

b. If $x = f(t)$ and $y = g(t)$ define y as a twice-differentiable function of x over some suitable interval, then $d^2y/dx^2 = $ _____.

9. a. A curve C represented by $x = f(t)$ and $y = g(t)$ on a parameter interval I is smooth if _____ and _____ are continuous on I and are not _____ _____, except possibly at the _____ of I.

b. If C is a smooth curve represented by $x = f(t)$ and $y = g(t)$ with parameter interval $[a, b]$, then the length of C is $L = $ _____.

10. If C is a smooth curve as described in Question 9b, C does not intersect itself, except possibly at _____, and $g(t) \geq 0$, then the area of the surface obtained by revolving C about the x-axis is $S = $ _____. If $f(t) \geq 0$ for all t in $[a, b]$, then the area of the surface obtained by revolving C about the y-axis is $S = $ _____.

11. a. The rectangular coordinates (x, y) of a point P are related to the polar coordinates of P by the equations $x = $ _____ and $y = $ _____.

b. The polar coordinates (r, θ) of a point P are related to the rectangular coordinates of P by the equations $r^2 = $ _____ and $\tan \theta = $ _____.

12. The horizontal tangent lines to the graph of $r = f(\theta)$ are located at the points where $dy/d\theta$ _____ and $dx/d\theta$ _____. The vertical tangent lines are located at the points where $dx/d\theta$ _____ and $dy/d\theta$ _____. Horizontal and vertical tangent lines may also be located at points where $dy/d\theta$ and $dx/d\theta$ are both equal to _____.

13. a. If f is nonnegative and continuous on $[\alpha, \beta]$, where $0 \leq \alpha < \beta \leq 2\pi$, then the area of the region bounded by the graphs of $r = f(\theta)$, $\theta = \alpha$, and $\theta = \beta$ is given by _____.

b. Let f and g be continuous on $[\alpha, \beta]$, where $0 \leq g(\theta) \leq f(\theta)$ and $0 \leq \alpha < \beta \leq 2\pi$. Then the area of the region bounded by the graphs of $r = g(\theta)$, $r = f(\theta)$, $\theta = \alpha$, and $\theta = \beta$ is given by _____.

14. Suppose that f has a continuous derivative on $[\alpha, \beta]$. If the graph C of $r = f(\theta)$ is traced exactly _____ as θ increases from α to β, then the length of C is given by _____.

15. Let F be a fixed point, let l be a fixed line in the plane, and let e be a fixed positive number. Then a conic section defined by the equation _____ is an ellipse if e satisfies

_____, a parabola if e satisfies _____, and a hyperbola if e satisfies _____.

16. A conic section can be represented by a polar equation of the form _____ or _____. It is an ellipse, a parabola, or a hyperbola depending on whether e satisfies _____, _____, or _____, respectively.

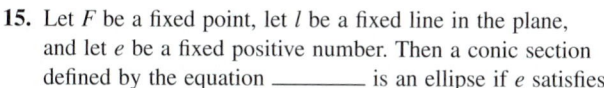

REVIEW EXERCISES

In Exercises 1–6, find the vertices and the foci of the conic and sketch its graph.

1. $\dfrac{x^2}{4} + \dfrac{y^2}{9} = 1$

2. $\dfrac{(x - 1)^2}{2} + \dfrac{(y + 1)^2}{4} = 1$

3. $x^2 - 9y^2 = 9$

4. $y^2 - 2y - 8x - 15 = 0$

5. $y^2 - 9x^2 + 8y + 7 = 0$

6. $4x^2 + 25y^2 - 16x + 50y - 59 = 0$

In Exercises 7–12, find a rectangular equation of the conic satisfying the given conditions.

7. parabola, focus $(-2, 0)$, directrix $x = 2$

8. parabola, vertex $(-2, 2)$, directrix $y = 4$

9. ellipse, vertices $(\pm 7, 0)$, foci $(\pm 2, 0)$

10. ellipse, foci $(\pm 2, 3)$, major axis has length 8

11. hyperbola, foci $\left(0, \pm \frac{3}{2}\sqrt{5}\right)$, vertices $(0, \pm 3)$

12. hyperbola, vertices $(-2, 0)$ and $(2, 0)$, asymptotes $y = \pm \dfrac{3}{2} x$

13. Show that if m is any real number, then there is exactly one line of slope m that is tangent to the parabola $x^2 = 4py$ and its equation is $y = mx - pm^2$.

14. Show that if m is any real number, then there are exactly two lines of slope m that are tangent to the ellipse $x^2/a^2 + y^2/b^2 = 1$ and their equations are $y = mx \pm \sqrt{a^2 m^2 + b^2}$.

In Exercises 15–18, (a) find a rectangular equation whose graph contains the curve C with the given parametric equations, and (b) sketch the curve C and indicate its orientation.

15. $x = 1 + 2t, \quad y = 3 - 2t$

16. $x = e^t, \quad y = e^{-2t}$

17. $x = 1 + 2 \sin t, \quad y = 3 + 2 \cos t$

18. $x = \cos^3 t, \quad y = 4 \sin^3 t$

In Exercises 19–22, find the slope of the tangent line to the curve at the point corresponding to the value of the parameter.

19. $x = t^3 + 1, \quad y = 2t^2 - 1; \quad t = 1$

20. $x = \sqrt{t + 1}, \quad y = \sqrt{16 - t}; \quad t = 0$

21. $x = te^{-t}, \quad y = \dfrac{1}{t^2 + 1}; \quad t = 0$

22. $x = 1 - \sin^2 t, \quad y = \cos^3 t; \quad t = \dfrac{\pi}{4}$

In Exercises 23 and 24, find dy/dx and d^2y/dx^2.

23. $x = t^3 + 1, \quad y = t^4 + 2t^2$

24. $x = e^t \sin t, \quad y = e^t \cos t$

In Exercises 25 and 26, find the points on the curve with the given parametric equations at which the tangent lines are vertical or horizontal.

25. $x = t^3 - 4t, \quad y = t^2 + 2$

26. $x = 1 - 2 \cos t, \quad y = 1 - 2 \sin t$

In Exercises 27 and 28, find the length of the curve defined by the given parametric equations.

27. $x = \dfrac{1}{6} t^6, \quad y = 2 - \dfrac{1}{4} t^4; \quad 0 \le t \le \sqrt[4]{8}$

28. $x = \sqrt{3} t^2, \quad y = t - t^3; \quad -1 \le t \le 1$

29. The position of a body at time t is (x, y), where $x = e^{-t} \cos t$ and $y = e^{-t} \sin t$. Find the distance covered by the body as t runs from 0 to $\pi/2$.

30. The course taken by an oceangoing racing boat during a practice run is described by the parametric equations
$$x = \sqrt{3}(t - 1)^2 \quad \text{and} \quad y = (t - 1) - (t - 1)^3$$
$$0 \le t \le 2$$
where x and y are measured in miles. Sketch the path of the boat, and find the length of the course.

In Exercises 31 and 32, find the area of the surface obtained by revolving the given curve about the x-axis.

31. $x = t^2, \quad y = \dfrac{t}{3}(3 - t^2); \quad 0 \le t \le \sqrt{3}$

32. $x = \ln(\sec t + \tan t) - \sin t, \quad y = \cos t; \quad 0 \le t \le \frac{\pi}{3}$

In Exercises 33–38, sketch the curve with the given polar equation.

33. $r = 2 \sin \theta$

34. $r = 3 - 4 \cos \theta$

35. $r = 2 \cos 5\theta$

36. $r = e^{-\theta}$

37. $r^2 = \cos 2\theta$

38. $r = 2 \sin \theta \cos^2 \theta$

In Exercises 39 and 40, find the slope of the tangent line to the curve with the given polar equation at the point corresponding to the given value of θ.

39. $r = e^{2\theta}$, $\theta = \dfrac{\pi}{2}$

40. $r = 2 - \sin \theta$, $\theta = \dfrac{\pi}{2}$

In Exercises 41 and 42, find the points of intersection of the given curves.

41. $r = \sin \theta$, $r = 1 - \sin \theta$

42. $r = \cos \theta$, $r = \cos 2\theta$

43. Find the area of the region enclosed by the curve with polar equation $r = 2 + \cos \theta$.

44. Find the area of the region enclosed by the curve with polar equation $r = 1 + \sin \theta$.

45. Find the area of the region that is enclosed between the petals of the curves with polar equations $r = 2 \sin 2\theta$ and $r = 2 \cos 2\theta$.

46. Find the area of the region that is enclosed between the curves with polar equations $r = 3 + 2 \sin \theta$ and $r = 2$.

In Exercises 47 and 48, find the length of the given curve.

47. $r = \theta^2$, $0 \le \theta \le 2\pi$

48. $r = 2(\sin \theta + \cos \theta)$, $0 \le \theta \le 2\pi$

In Exercises 49 and 50, find the area of the surface obtained by revolving the curve about the given line.

49. $r = 2 \sin \theta$ about the polar axis

50. $r = \sqrt{\cos 2\theta}$, $0 \le \theta \le \frac{\pi}{4}$, about the line $\theta = \dfrac{\pi}{2}$

In Exercises 51 and 52, sketch the curve with the given equation.

51. $r = \dfrac{1}{1 + \sin \theta}$

52. $r = \dfrac{16}{3 - 5 \cos \theta}$

In Exercises 53–54, plot the curve with the parametric equations.

 53. $x = 0.15(2 \cos t + 3 \cos 2t)$, $y = 0.15(2 \sin t - 3 \sin 2t)$; $0 \le t \le 2\pi$ (hypotrochoid)

 54. $x = 0.24(-7t^3 + 3t^5)$, $y = 0.09(14t^2 - 5t^4)$; $-1.629 \le t \le 1.629$ (butterfly catastrophe)

In Exercises 55–57, plot the curve with the polar equation.

 55. $r = 0.1e^{0.1\theta}$ where $0 \le \theta \le \frac{15\pi}{2}$

 56. $r = \dfrac{0.1}{\cos 4\theta}$ where $0 \le \theta \le 2\pi$ (epi-spiral)

57. $r = \dfrac{1}{2 \sin \theta}$ where $-2\pi \le \theta \le 2\pi$ (spiral of Poinsot)

58. An ant crawls along the curve $x = \frac{1}{2}t^2$, $y = \frac{1}{3}(2t + 1)^{3/2}$ starting at the point $\left(0, \frac{1}{3}\right)$ and ending at the point $\left(\frac{1}{2}, \sqrt{3}\right)$, where x and y are measured in feet. Find the distance traveled by the ant.

59. An egg has the shape of a solid obtained by revolving the upper half of the ellipse $x^2 + 2y^2 = 2$ about the x-axis. What is the surface area of the egg?

60. A piston is attached to a crankshaft by means of a connecting rod of length L, as shown in the figure. If the disk is of radius r, find the parametric equations giving the position of the point P using the angle θ as a parameter.

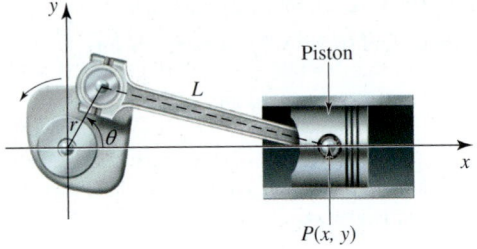

CHALLENGE PROBLEMS

1. a. In the following figure, the axes of an xy-coordinate system have been rotated about the origin through an angle of θ to produce a new $x'y'$-coordinate system.

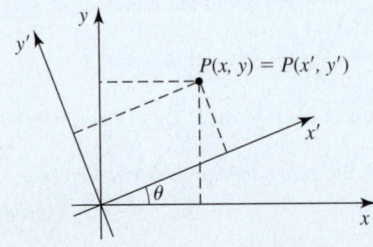

Show that

$$x = x' \cos \theta - y' \sin \theta \qquad y = x' \sin \theta + y' \cos \theta$$

b. Show that the equation $Ax^2 + Bxy + Cy^2 + F = 0$, where $B \ne 0$, will have the form $(A'x')^2 + (C'y')^2 + F = 0$ in the $x'y'$-coordinate system obtained by rotating the xy-system through an angle θ given by

$$\cot 2\theta = \frac{A - C}{B}$$

c. Sketch the ellipse $2x^2 + \sqrt{3}xy + y^2 - 20 = 0$.

2. a. Show that the area of the ellipse

$$Ax^2 + Bxy + Cy^2 + F = 0$$

where $B^2 - 4AC < 0$, is given by

$$S = -\frac{2\pi F}{\sqrt{4AC - B^2}}$$

b. Find the area of the ellipse $2x^2 + \sqrt{3}xy + y^2 = 20$.

3. Find the length of the curve with parametric equations

$$x = \int_1^t \frac{\cos u}{u}\, du \qquad y = \int_1^t \frac{\sin u}{u}\, du$$

between the origin and the nearest point from the vertical tangent line.

4. Find the area of the surface obtained by revolving one branch of the lemniscate $r = a\sqrt{\cos 2\theta}$ about the line $\theta = \pi/4$.

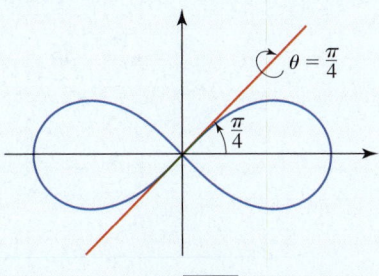

$$r = a\sqrt{\cos 2\theta}$$

5. The curve with equation $x^3 + y^3 = 3axy$, where a is a nonzero constant, is called the **folium of Descartes**.

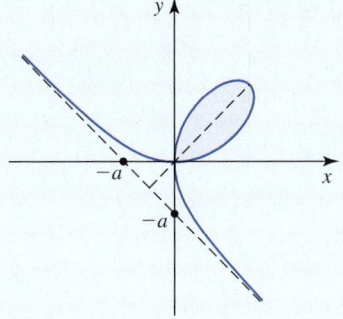

a. Show that the polar equation of the curve is

$$r = \frac{3a \sec\theta \tan\theta}{1 + \tan^3\theta}$$

b. Find the area of the region enclosed by the loop of the curve.

6. Find the rectangular coordinates of the centroid of the region that is completely enclosed by the curve $r = 3\cos^3\theta$.

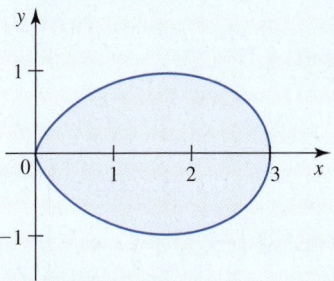

7. An ant is placed at each corner of a square with sides of length a. Starting at the same instant of time, all four ants begin to move counterclockwise at the same speed and in such a way that each ant moves toward the next at all times. The resulting path of each ant is a spiral curve that converges to the center of the square.

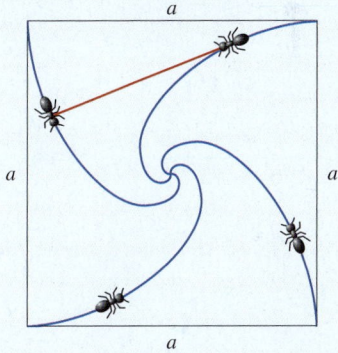

a. Taking the pole to be the center of the square, find the polar equation describing the path taken by one of the ants.
Hint: The line passing through the position of two adjacent ants is tangent to the path of one of them.

b. Find the distance traveled by an ant as it moves from a corner of the square to its center.

The Real Number Line, Inequalities, and Absolute Value

■ The Real Number Line

The real number system is made up of the set of real numbers together with the usual operations of addition, subtraction, multiplication, and division.

We can represent real numbers geometrically by points on a **real number,** or **coordinate, line.** This line can be constructed as follows. Arbitrarily select a point on a straight line to represent the number 0. This point is called the **origin.** If the line is horizontal, then a point at a convenient distance to the right of the origin is chosen to represent the number 1. This determines the scale for the number line. Each positive real number lies at an appropriate distance to the right of the origin, and each negative real number lies at an appropriate distance to the left of the origin (see Figure 1).

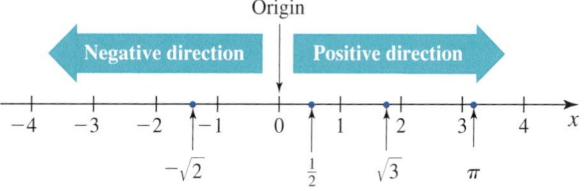

FIGURE 1
The real number line

A *one-to-one correspondence* is set up between the set of all real numbers and the set of points on the number line; that is, exactly one point on the line is associated with each real number. Conversely, exactly one real number is associated with each point on the line. The real number that is associated with a point on the real number line is called the **coordinate** of that point.

■ Intervals

Throughout this book we often restrict our attention to subsets of the set of real numbers. For example, if x denotes the number of cars rolling off a plant assembly line each day, then x must be nonnegative—that is, $x \geq 0$. Further, suppose that management decides that the daily production must not exceed 200 cars. Then, x must satisfy the inequality $0 \leq x \leq 200$.

More generally, we will be interested in the following subsets of real numbers: open intervals, closed intervals, and half-open intervals. The set of all real numbers that lie *strictly* between two fixed numbers a and b is called an **open interval** (a, b). It consists of all real numbers x that satisfy the inequalities $a < x < b$, and it is called "open" because neither of its endpoints is included in the interval. A **closed interval** contains *both* of its endpoints. Thus, the set of all real numbers x that satisfy the inequalities

$a \leq x \leq b$ is the closed interval $[a, b]$. Notice that square brackets are used to indicate that the endpoints are included in this interval. **Half-open intervals** contain only *one* of their endpoints. Thus, the interval $[a, b)$ is the set of all real numbers x that satisfy $a \leq x < b$, whereas the interval $(a, b]$ is described by the inequalities $a < x \leq b$. Examples of these **finite intervals** are illustrated in Table 1.

TABLE 1 Finite Intervals

Interval	Graph	Example
Open: (a, b)		$(-2, 1)$
Closed: $[a, b]$		$[-1, 2]$
Half-open: $(a, b]$		$\left(\frac{1}{2}, 3\right]$
Half-open: $[a, b)$		$\left[-\frac{1}{2}, 3\right)$

In addition to finite intervals, we will encounter **infinite intervals.** Examples of infinite intervals are the half lines (a, ∞), $[a, \infty)$, $(-\infty, a)$, and $(-\infty, a]$ defined by the set of all real numbers that satisfy $x > a$, $x \geq a$, $x < a$, and $x \leq a$, respectively. The symbol ∞, called *infinity,* is not a real number. It is used here only for notational purposes. The notation $(-\infty, \infty)$ is used for the set of all real numbers x, since by definition the inequalities $-\infty < x < \infty$ hold for any real number x. Infinite intervals are illustrated in Table 2.

TABLE 2 Infinite Intervals

Interval	Graph	Example
(a, ∞)		$(2, \infty)$
$[a, \infty)$		$[-1, \infty)$
$(-\infty, a)$		$(-\infty, 1)$
$(-\infty, a]$		$\left(-\infty, -\frac{1}{2}\right]$

■ Inequalities

The following properties may be used to solve one or more inequalities involving a variable.

Properties of Inequalities

If a, b, and c, are any real numbers, then

		Example
Property 1	If $a < b$ and $b < c$, then $a < c$.	$2 < 3$ and $3 < 8$, so $2 < 8$.
Property 2	If $a < b$, then $a + c < b + c$.	$-5 < -3$, so $-5 + 2 < -3 + 2$; that is, $-3 < -1$.
Property 3	If $a < b$ and $c > 0$, then $ac < bc$.	$-5 < -3$, and since $2 > 0$, we have $(-5)(2) < (-3)(2)$; that is, $-10 < -6$.
Property 4	If $a < b$ and $c < 0$, then $ac > bc$.	$-2 < 4$, and since $-3 < 0$, we have $(-2)(-3) > (4)(-3)$; that is, $6 > -12$.

Similar properties hold if each inequality sign, $<$, between a and b and between b and c is replaced by $\geq$, $>$, or $\leq$. Note that Property 4 says that an inequality sign is reversed if the inequality is multiplied by a negative number.

A real number is a *solution of an inequality* involving a variable if a true statement is obtained when the variable is replaced by that number. The set of all real numbers satisfying the inequality is called the *solution set*. We often use interval notation to describe the solution set.

EXAMPLE 1   Find the set of real numbers that satisfy $-1 \leq 2x - 5 < 7$.

Solution   Add 5 to each member of the given double inequality, obtaining

$$4 \leq 2x < 12$$

Next, multiply each member of the resulting double inequality by $\frac{1}{2}$, yielding

$$2 \leq x < 6$$

Thus, the solution is the set of all values of x lying in the interval $[2, 6)$.   ■

EXAMPLE 2   Solve the inequality $x^2 + 2x - 8 < 0$.

Solution   Observe that $x^2 + 2x - 8 = (x + 4)(x - 2)$, so the given inequality is equivalent to the inequality $(x + 4)(x - 2) < 0$. Since the product of two real numbers is negative if and only if the two numbers have opposite signs, we solve the inequality $(x + 4)(x - 2) < 0$ by studying the signs of the two factors $x + 4$ and $x - 2$. Now, $x + 4 > 0$ if $x > -4$, and $x + 4 < 0$ if $x < -4$. Similarly, $x - 2 > 0$ if $x > 2$, and $x - 2 < 0$ if $x < 2$. These results are summarized graphically in Figure 2.

FIGURE 2
Sign diagram for $(x + 4)(x - 2)$

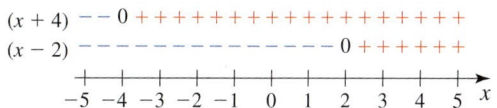

From Figure 2 we see that the two factors $x + 4$ and $x - 2$ have opposite signs if and only if x lies strictly between -4 and 2. Therefore, the required solution is the interval $(-4, 2)$. ∎

EXAMPLE 3 Solve the inequality $\dfrac{x + 1}{x - 1} \geq 0$.

Solution   The quotient $(x + 1)/(x - 1)$ is strictly positive if and only if both the numerator and the denominator have the same sign. The signs of $x + 1$ and $x - 1$ are shown in Figure 3.

FIGURE 3
Sign diagram for $\dfrac{x + 1}{x - 1}$

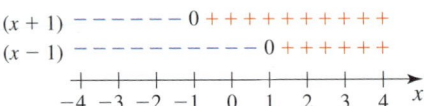

From Figure 3 we see that $x + 1$ and $x - 1$ have the same sign if and only if $x < -1$ or $x > 1$. The quotient $(x + 1)/(x - 1)$ is equal to zero if and only if $x = -1$. Therefore, the required solution is the set of all x in the intervals $(-\infty, -1]$ and $(1, \infty)$. ∎

■ Absolute Value

> **DEFINITION**   **Absolute Value**
>
> The **absolute value** of a number a is denoted by $|a|$ and is defined by
>
> $$|a| = \begin{cases} a & \text{if } a \geq 0 \\ -a & \text{if } a < 0 \end{cases}$$

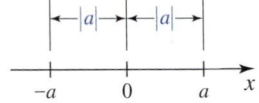

FIGURE 4
The absolute value of a number

Since $-a$ is a positive number when a is negative, it follows that the absolute value of a number is always nonnegative. For example, $|5| = 5$ and $|-5| = -(-5) = 5$. Geometrically, $|a|$ is the distance between the origin and the point on the number line that represents the number a. (See Figure 4.)

Absolute Value Properties

If a and b are any real numbers, then

		Example
Property 5	$\lvert -a \rvert = \lvert a \rvert$	$\lvert -3 \rvert = -(-3) = 3 = \lvert 3 \rvert$
Property 6	$\lvert ab \rvert = \lvert a \rvert \lvert b \rvert$	$\lvert (2)(-3) \rvert = \lvert -6 \rvert = 6 = (2)(3)$ $= \lvert 2 \rvert \lvert -3 \rvert$
Property 7	$\left\lvert \dfrac{a}{b} \right\rvert = \dfrac{\lvert a \rvert}{\lvert b \rvert}$ $(b \neq 0)$	$\left\lvert \dfrac{(-3)}{(-4)} \right\rvert = \left\lvert \dfrac{3}{4} \right\rvert = \dfrac{3}{4} = \dfrac{\lvert -3 \rvert}{\lvert -4 \rvert}$
Property 8	$\lvert a + b \rvert \leq \lvert a \rvert + \lvert b \rvert$	$\lvert 8 + (-5) \rvert = \lvert 3 \rvert = 3$ $\leq \lvert 8 \rvert + \lvert -5 \rvert = 13$

Property 8 is called the **triangle inequality.** To prove the triangle inequality, note that

$$-|a| < a < |a| \qquad \text{and} \qquad -|b| < b < |b|$$

Adding the respective numbers in these inequalities, we have

$$-(|a| + |b|) \leq a + b \leq |a| + |b|$$

which is equivalent to

$$|a + b| \leq |a| + |b|$$

as was to be shown.

EXAMPLE 4   Evaluate each of the following expressions:

a. $|\pi - 5| + 3$ **b.** $|\sqrt{3} - 2| + |2 - \sqrt{3}|$

Solution
a. Since $\pi - 5 < 0$, we see that $|\pi - 5| = -(\pi - 5)$. Therefore,

$$|\pi - 5| + 3 = -(\pi - 5) + 3 = 8 - \pi$$

b. Since $\sqrt{3} - 2 < 0$, we see that $|\sqrt{3} - 2| = -(\sqrt{3} - 2)$. Next, observe that $2 - \sqrt{3} > 0$, so $|2 - \sqrt{3}| = 2 - \sqrt{3}$. Therefore,

$$|\sqrt{3} - 2| + |2 - \sqrt{3}| = -(\sqrt{3} - 2) + (2 - \sqrt{3})$$
$$= 4 - 2\sqrt{3} = 2(2 - \sqrt{3}) \qquad \blacksquare$$

EXAMPLE 5   Solve the inequalities $|x| \leq 5$ and $|x| \geq 5$.

Solution   First, we consider the inequality $|x| \leq 5$. If $x \geq 0$, then $|x| = x$, so $|x| \leq 5$ implies $x \leq 5$ in this case. On the other hand, if $x < 0$, then $|x| = -x$, so $|x| \leq 5$ implies $-x \leq 5$ or $x \geq -5$. Thus, $|x| \leq 5$ means $-5 \leq x \leq 5$. (See Figure 5a.) To obtain an alternative solution, observe that $|x|$ is the distance from the point x to zero, so the inequality $|x| \leq 5$ implies immediately that $-5 \leq x \leq 5$.

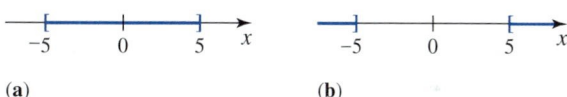

FIGURE 5   (a)   (b)

Next, the inequality $|x| \geq 5$ states that the distance from x to zero is greater than or equal to 5. This observation yields the result $x \geq 5$ or $x \leq -5$. (See Figure 5b.) $\blacksquare$

EXAMPLE 6   Solve the inequality $|2x - 3| \leq 1$.

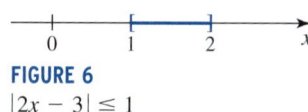

FIGURE 6
$|2x - 3| \leq 1$

Solution   The inequality $|2x - 3| \leq 1$ is equivalent to the inequalities $-1 \leq 2x - 3 \leq 1$. (See Example 5.) Thus, $2 \leq 2x \leq 4$ and $1 \leq x \leq 2$. The solution is therefore given by the set of all x in the interval $[1, 2]$. (See Figure 6.) $\blacksquare$

EXAMPLE 7   Solve $|2x + 3| \geq 5$.

Solution   The inequality $|2x + 3| \geq 5$ is equivalent to $2x + 3 \geq 5$ or $2x + 3 \leq -5$. (See Example 5 with x replaced by $2x + 3$.) The first inequality gives $x \geq 1$, and the second gives $x \leq -4$. So the solution is $\{x | x \leq -4 \quad \text{or} \quad x \geq 1\} = (-\infty, -4] \cup [1, \infty)$.

$\blacksquare$

EXAMPLE 8 If $|x - 2| < 0.1$ and $|y - 3| < 0.2$, find an upper bound for $|x + y - 5|$.

Solution We have

$$|x + y - 5| = |(x - 2) + (y - 3)|$$
$$\leq |x - 2| + |y - 3| \qquad \text{Use the Triangle Inequality.}$$
$$< 0.1 + 0.2 = 0.3$$

So $|x + y - 5| < 0.3$.

EXERCISES

In Exercises 1–6, show the interval on a number line.

1. $(3, 6)$

2. $(-2, 5]$

3. $[-1, 4)$

4. $\left[-\dfrac{6}{5}, -\dfrac{1}{2}\right]$

5. $(0, \infty)$

6. $(-\infty, 5]$

In Exercises 7–10, determine whether the statement is true or false.

7. $-3 < -20$

8. $-5 \leq -5$

9. $\dfrac{2}{3} > \dfrac{5}{6}$

10. $-\dfrac{5}{6} < -\dfrac{11}{12}$

In Exercises 11–28, find the values of x that satisfy the inequality (inequalities).

11. $2x + 4 < 8$

12. $-6 > 4 + 5x$

13. $-4x \geq 20$

14. $-12 \leq -3x$

15. $-6 < x - 2 < 4$

16. $0 \leq x + 1 \leq 4$

17. $x + 1 > 4$ or $x + 2 < -1$

18. $x + 1 > 2$ or $x - 1 < -2$

19. $x + 3 > 1$ and $x - 2 < 1$

20. $x - 4 \leq 1$ and $x + 3 > 2$

21. $(x + 3)(x - 5) \leq 0$

22. $(2x - 4)(x + 2) \geq 0$

23. $(2x - 3)(x - 1) \geq 0$

24. $(3x - 4)(2x + 2) \leq 0$

25. $\dfrac{x + 3}{x - 2} \geq 0$

26. $\dfrac{2x - 3}{x + 1} \geq 4$

27. $\dfrac{x - 2}{x - 1} \leq 2$

28. $\dfrac{2x - 1}{x + 2} \leq 4$

In Exercises 29–38, evaluate the expression.

29. $|-6 + 2|$

30. $4 + |-4|$

31. $\dfrac{|-12 + 4|}{|16 - 12|}$

32. $\left|\dfrac{0.2 - 1.4}{1.6 - 2.4}\right|$

33. $\sqrt{3}|-2| + 3|-\sqrt{3}|$

34. $|-1| + \sqrt{2}|-2|$

35. $|\pi - 1| + 2$

36. $|\pi - 6| - 3$

37. $|\sqrt{2} - 1| + |3 - \sqrt{2}|$

38. $|2\sqrt{3} - 3| - |\sqrt{3} - 4|$

In Exercises 39–44, suppose that a and b are real numbers other than zero and that $a > b$. State whether the inequality is true or false.

39. $b - a > 0$

40. $\dfrac{a}{b} > 1$

41. $a^2 > b^2$

42. $\dfrac{1}{a} > \dfrac{1}{b}$

43. $a^3 > b^3$

44. $-a < -b$

In Exercises 45–50, determine whether the statement is true for all real numbers a and b.

45. $|-a| = a$

46. $|b^2| = b^2$

47. $|a - 4| = |4 - a|$

48. $|a + 1| = |a| + 1$

49. $|a + b| = |a| + |b|$

50. $|a - b| = |a| - |b|$

In Exercises 51–54, solve the equation for x.

51. $|3x| = 4$

52. $|2x + 4| = 1$

53. $|x + 2| = |2x + 3|$

54. $\left|\dfrac{3x + 1}{x + 2}\right| = 3$

In Exercises 55–64, solve the inequality.

55. $|x| < 4$

56. $|x| > 3$

57. $|x - 2| < 1$

58. $|x - 4| < 0.1$

59. $|x + 3| \geq 2$

60. $|3x - 2| > 1$

61. $|2x + 3| \leq 0.2$

62. $|3x - 2| < 4$

63. $1 \leq |x| \leq 3$

64. $0 < |x - 2| < \frac{1}{3}$

65. If $|x - 1| < 0.2$ and $|y - 4| < 0.2$, find an upper bound for $|x + y - 5|$.

66. Prove that $|x - y| \geq |x| - |y|$.
Hint: Write $x = (x - y) + y$ and use the Triangle Inequality.

Proofs of Theorems

In this appendix we give the proofs of some of the theorems that appear in the body of the text.

■ Chapter 1

Theorem 1   Limit Laws   (Section 1.2)

PRODUCT LAW

$$\text{If } \lim_{x \to a} f(x) = L \text{ and } \lim_{x \to a} g(x) = M, \text{ then } \lim_{x \to a} [f(x)g(x)] = LM.$$

PROOF   Let $\varepsilon > 0$ be given. We want to show that there exists a $\delta > 0$ such that

$$0 < |x - a| < \delta \Rightarrow |f(x)g(x) - LM| < \varepsilon.$$

We begin by considering

$$
\begin{aligned}
|f(x)g(x) - LM| &= |f(x)g(x) - Lg(x) + Lg(x) - LM| \\
&= |[f(x) - L]g(x) + L[g(x) - M]| \\
&\leq |[f(x) - L]g(x)| + |L[g(x) - M]| \qquad \text{\color{orange}Use the Triangle Inequality.} \\
&= |f(x) - L||g(x)| + |L||g(x) - M| \qquad\qquad\qquad\qquad \textbf{(1)}
\end{aligned}
$$

Thus, our goal will be achieved if we can show that the expression on the right of Equation (1) is less than ε whenever $0 < |x - a| < \delta$. Since $\lim_{x \to a} g(x) = M$, there exists a $\delta_1 > 0$ such that

$$0 < |x - a| < \delta_1 \Rightarrow |g(x) - M| < 1$$

and therefore,

$$|g(x)| = |g(x) - M + M| = |[g(x) - M] + M| \leq |g(x) - M| + |M| < 1 + |M| \quad \textbf{(2)}$$

Also, because $\lim_{x \to a} g(x) = M$, there exists a $\delta_2 > 0$ such that

$$0 < |x - a| < \delta_2 \Rightarrow |g(x) - M| < \frac{\varepsilon}{2(1 + |L|)} \qquad\qquad \textbf{(3)}$$

Next, since $\lim_{x \to a} f(x) = L$, there exists a $\delta_3 > 0$ such that

$$0 < |x - a| < \delta_3 \Rightarrow |f(x) - L| < \frac{\varepsilon}{2(1 + |M|)} \qquad\qquad \textbf{(4)}$$

If we let δ denote the smallest of the three numbers δ_1, δ_2, and δ_3, then $\delta > 0$, and if $0 < |x - a| < \delta$, then we have $0 < |x - a| < \delta_1$, $0 < |x - a| < \delta_2$, and

$0 < |x - a| < \delta_3$, so all three Inequalities (2)–(4) hold simultaneously. Therefore, if $0 < |x - a| < \delta$, then by Equation (1),

$$|f(x)g(x) - LM| \leq \frac{\varepsilon}{2(1 + |M|)} \cdot (1 + |M|) + |L| \cdot \frac{\varepsilon}{2(1 + |L|)}$$

$$< \frac{\varepsilon}{2} + \frac{\varepsilon}{2} = \varepsilon$$

This completes our proof.   ■

CONSTANT MULTIPLE LAW

$$\lim_{x \to a} [cf(x)] = c \lim_{x \to a} f(x) \quad \text{for every } c$$

PROOF   Put $g(x) = c$ in the Product Law, obtaining

$$\lim_{x \to a} [cf(x)] = \lim_{x \to a} [g(x)f(x)] = [\lim_{x \to a} g(x)][\lim_{x \to a} f(x)]$$

$$= c \lim_{x \to a} f(x) \qquad \text{Use Law 1 in Section 1.2.}$$

as was to be shown.   ■

QUOTIENT LAW

$$\lim_{x \to a} \frac{f(x)}{g(x)} = \frac{L}{M} \quad \text{provided that } M \neq 0$$

PROOF   We can write

$$\frac{f(x)}{g(x)} = f(x) \cdot \frac{1}{g(x)}$$

and thus use the Product Law established earlier to help us with the proof. Let's first show that

$$\lim_{x \to a} \frac{1}{g(x)} = \frac{1}{M}$$

Let $\varepsilon > 0$ be given. We need to show that there exists a $\delta > 0$ such that

$$0 < |x - a| < \delta \Rightarrow \left| \frac{1}{g(x)} - \frac{1}{M} \right| < \varepsilon$$

We consider

$$\left| \frac{1}{g(x)} - \frac{1}{M} \right| = \left| \frac{M - g(x)}{Mg(x)} \right| = \frac{|M - g(x)|}{|M||g(x)|} \tag{5}$$

We want to show that the denominator of the last expression is bounded away from 0 when x is close to a. To do this, observe that if $\lim_{x \to a} g(x) = M$, then there exists a $\delta_1 > 0$ such that

$$0 < |x - a| < \delta_1 \Rightarrow |g(x) - M| < \frac{|M|}{2} \qquad \text{Remember that } M \neq 0, \text{ so } |M| > 0.$$

Then

$$|M| = |M - g(x) + g(x)| \leq |M - g(x)| + |g(x)|$$

$$< \frac{|M|}{2} + |g(x)|$$

and this implies that if $0 < |x - a| < \delta_1$, then

$$|g(x)| > \frac{|M|}{2} \tag{6}$$

Thus, if x satisfies $0 < |x - a| < \delta_1$ then Inequality (6) implies that

$$\frac{1}{|M| \, |g(x)|} < \frac{1}{|M|} \cdot \frac{2}{|M|} = \frac{2}{|M|^2}$$

Also, since $\lim_{x \to a} g(x) = M$, there exists a $\delta_2 > 0$ such that

$$0 < |x - a| < \delta_2 \Rightarrow |g(x) - M| < \frac{|M|^2 \varepsilon}{2} \tag{7}$$

If we let δ denote the smaller of the two numbers δ_1 and δ_2, then $\delta > 0$ and if $0 < |x - a| < \delta$, then $0 < |x - a| < \delta_1$ and $0 < |x - a| < \delta_2$, so both Inequalities (6) and (7) hold simultaneously. Therefore, if $0 < |x - a| < \delta$, then Equation (5) implies that

$$\left| \frac{1}{g(x)} - \frac{1}{M} \right| = \frac{|M - g(x)|}{|M| \, |g(x)|} < \frac{2}{|M|^2} \cdot \frac{|M|^2 \varepsilon}{2} = \varepsilon$$

This establishes that

$$\lim_{x \to a} \frac{1}{g(x)} = \frac{1}{M}$$

Thus, using the Product Law proved earlier, we have

$$\lim_{x \to a} \frac{f(x)}{g(x)} = \lim_{x \to a} f(x) \cdot \frac{1}{g(x)} = L\left(\frac{1}{M} \right) = \frac{L}{M}$$

and the desired result follows.

ROOT LAW

If $\lim_{x \to a} f(x) = L$, then $\lim_{x \to a} \sqrt[n]{f(x)} = \sqrt[n]{L}$, provided that $L > 0$ if n is even.

PROOF   First, we prove that

$$\lim_{x \to a} \sqrt{x} = \sqrt{a} \qquad a > 0$$

Let $\varepsilon > 0$ be given. We have

$$|\sqrt{x} - \sqrt{a}| = \left| \frac{\sqrt{x} - \sqrt{a}}{1} \cdot \frac{\sqrt{x} + \sqrt{a}}{\sqrt{x} + \sqrt{a}} \right| = \frac{|x - a|}{\sqrt{x} + \sqrt{a}}$$

Let's agree to pick $\delta \le \frac{3}{4}a$. Then $\delta > 0$. If $|x - a| < \delta$, then $-\frac{3}{4}a < x - a < \frac{3}{4}a$ or $\frac{1}{4}a < x < \frac{7}{4}a$, and so $\sqrt{x} + \sqrt{a} > \dfrac{\sqrt{a}}{2} + \sqrt{a} = \dfrac{3\sqrt{a}}{2}$. Let's pick δ to be the smaller of $\dfrac{3}{4}a$ and $\dfrac{3\sqrt{a}}{2}\varepsilon$. Then $0 < |x - a| < \delta$ implies

$$|\sqrt{x} - \sqrt{a}| = \frac{|x - a|}{\sqrt{x} + \sqrt{a}} < \frac{\dfrac{3\sqrt{a}}{2}\varepsilon}{\dfrac{3\sqrt{a}}{2}} = \varepsilon$$

Since ε is arbitrary, the result follows.

To complete the proof of the Root Law, we apply Theorem 4 (Section 1.4) to the special case where $f(x) = \sqrt[n]{x}$, with n a positive integer. Then, assuming that all roots exist, we have

$$f(g(x)) = \sqrt[n]{g(x)}$$

and

$$f\left[\lim_{x \to a} g(x)\right] = \sqrt[n]{\lim_{x \to a} g(x)}$$

So, Theorem 4 gives

$$\lim_{x \to a} \sqrt[n]{g(x)} = \sqrt[n]{\lim_{x \to a} g(x)}$$

and this completes the proof of the Root Law. ■

Theorem 3   The Squeeze Theorem   (Section 1.2)

Suppose that $f(x) \le g(x) \le h(x)$ for all x in an open interval containing a, except possibly at a and that

$$\lim_{x \to a} f(x) = L = \lim_{x \to a} h(x)$$

Then $\lim_{x \to a} g(x) = L$.

PROOF Let $\varepsilon > 0$ be given. Since $\lim_{x \to a} f(x) = L$, there exists a $\delta_1 > 0$ such that

$$0 < |x - a| < \delta_1 \Rightarrow |f(x) - L| < \varepsilon$$

or, equivalently,

$$L - \varepsilon < f(x) < L + \varepsilon \tag{8}$$

Next, since $\lim_{x \to a} h(x) = L$, there exists a $\delta_2 > 0$ such that

$$0 < |x - a| < \delta_2 \Rightarrow |h(x) - L| < \varepsilon$$

or, equivalently,

$$L - \varepsilon < h(x) < L + \varepsilon \tag{9}$$

If we let δ denote the smaller of the two numbers δ_1 and δ_2, then $\delta > 0$, and if $0 < |x - a| < \delta$, then $0 < |x - a| < \delta_1$ and $0 < |x - a| < \delta_2$, so both Inequalities (8) and (9) hold simultaneously. Therefore, if $0 < |x - a| < \delta$, then

$$L - \varepsilon < f(x) \le g(x) \le h(x) < L + \varepsilon$$

which implies that

$$L - \varepsilon < g(x) < L + \varepsilon$$

or, equivalently, $|g(x) - L| < \varepsilon$. Therefore, $\lim_{x \to a} g(x) = L$. ■

Theorem 4   (Section 1.2)

Suppose that $f(x) \leq g(x)$ for all x in an open interval containing a, except possibly at a and that

$$\lim_{x \to a} f(x) = L \qquad \text{and} \qquad \lim_{x \to a} g(x) = M$$

Then $L \leq M$.

PROOF   Suppose, to the contrary, that $L > M$. Now, since $\lim_{x \to a} f(x) = L$ and $\lim_{x \to a} g(x) = M$, by the Sum Law (Theorem 1a),

$$\lim_{x \to a} [g(x) - f(x)] = \lim_{x \to a} g(x) - \lim_{x \to a} f(x) = M - L$$

Since $L - M > 0$, we can take $\varepsilon = L - M$ and find a $\delta > 0$ such that

$$|[g(x) - f(x)] - (M - L)| < L - M$$

or

$$g(x) - f(x) - M + L < L - M$$

or

$$g(x) < f(x)$$

But this contradicts the condition in the hypothesis, which states that $g(x) \geq f(x)$. Therefore, the assumption $L > M$ must be false, and we conclude that $L \leq M$, as was to be shown. ■

Theorem 4   Limit of a Composite Function   (Section 1.4)

If the function f is continuous at L and $\lim_{x \to a} g(x) = L$, then

$$\lim_{x \to a} f(g(x)) = f(L)$$

PROOF   Let $\varepsilon > 0$ be given. We want to show that there exists a $\delta > 0$ such that

$$0 < |x - a| < \delta \Rightarrow |f(g(x)) - f(L)| < \varepsilon$$

Now, since f is continuous at L, we have

$$\lim_{y \to L} f(y) = f(L) \tag{10}$$

so there exists a $\delta_1 > 0$ such that

$$0 < |y - L| < \delta_1 \Rightarrow |f(y) - f(L)| < \varepsilon$$

Next, since $\lim_{x \to a} g(x) = L$, there exists a $\delta > 0$ such that

$$0 < |x - a| < \delta \Rightarrow |g(x) - L| < \delta_1 \tag{11}$$

Combining (10) and (11), we see that

$$0 < |x - a| < \delta \Rightarrow |g(x) - L| < \delta_1 \Rightarrow |f(g(x)) - f(L)| < \varepsilon$$

as was to be shown. ■

■ Chapter 2

Theorem 1   The Chain Rule

(Section 2.6)

If f is differentiable at x and g is differentiable at $f(x)$, then the composition $h = g \circ f$ defined by $h(x) = g[f(x)]$ is differentiable, and

$$h'(x) = g'[f(x)]f'(x)$$

Also, if we write $u = f(x)$ and $y = g(u) = g[f(x)]$, then

$$\frac{dy}{dx} = \frac{dy}{du}\frac{du}{dx}$$

PROOF   According to the definition of the derivative, we have, for fixed a,

$$f'(a) = \lim_{\Delta x \to 0} \frac{f(a + \Delta x) - f(a)}{\Delta x} = \lim_{\Delta x \to 0} \frac{\Delta y}{\Delta x} \qquad (12)$$

where $\Delta y = f(a + \Delta x) - f(a)$. Let us define a function of Δx by

$$\varepsilon(\Delta x) = \begin{cases} \dfrac{\Delta y}{\Delta x} - f'(a) & \text{if } \Delta x \neq 0 \\ 0 & \text{if } \Delta x = 0 \end{cases}$$

Then

$$\lim_{\Delta x \to 0} \varepsilon(\Delta x) = \lim_{\Delta x \to 0} \left(\frac{\Delta y}{\Delta x} - f'(a) \right) = 0$$

by Equation (12). So ε is a continuous function of Δx. Therefore, for a differentiable function f, we can write

$$\Delta y = f'(a)\,\Delta x + \varepsilon(\Delta x)\,\Delta x \qquad (13)$$

where $\lim_{\Delta x \to 0} \varepsilon(\Delta x) = 0$.

Now suppose that $u = f(x)$, where f is differentiable at a and $y = g(u)$ is differentiable at $b = f(a)$. If Δx is an increment in x and Δu and Δy are the corresponding increments in u and y, then using Equation (13), we have

$$\Delta u = f'(a)\,\Delta x + \varepsilon_1(\Delta x)\,\Delta x = [f'(a) + \varepsilon_1(\Delta x)]\,\Delta x \qquad (14)$$

where $\varepsilon_1(\Delta x) \to 0$ as $\Delta x \to 0$. Similarly,

$$\Delta y = g'(b)\,\Delta u + \varepsilon_2(\Delta u)\,\Delta u = [g'(b) + \varepsilon_2(\Delta u)]\,\Delta u \qquad (15)$$

where $\varepsilon_2(\Delta u) \to 0$ as $\Delta u \to 0$. Substituting the expression for Δu in Equation (14) into Equation (15), we obtain

$$\Delta y = [g'(b) + \varepsilon_2(\Delta u)][f'(a) + \varepsilon_1(\Delta x)]\,\Delta x$$

or

$$\frac{\Delta y}{\Delta x} = [g'(b) + \varepsilon_2(\Delta u)][f'(a) + \varepsilon_1(\Delta x)] \qquad (16)$$

Now, if $\Delta x \to 0$, then Equation (14) implies that $\Delta u \to 0$. So both $\varepsilon_1(\Delta x)$ and $\varepsilon_2(\Delta u)$ approach 0 as $\Delta x \to 0$. Therefore, from Equation (16) we have

$$\frac{dy}{dx} = \lim_{\Delta x \to 0} \frac{\Delta y}{\Delta x} = \lim_{\Delta x \to 0} [g'(b) + \varepsilon_2(\Delta u)][f'(a) + \varepsilon_1(\Delta x)]$$

$$= g'(b)f'(a) = g'(f(a))f'(a)$$

Since a is arbitrary, the proof is complete.   ■

Theorem 1   Continuity and Differentiability of Inverse Functions

(Sections 1.4 and 2.7)

Let f be one-to-one so that it has an inverse f^{-1}.

1. If f is continuous on its domain, then f^{-1} is continuous on its domain.
2. If f is differentiable at c and $f'(c) \neq 0$, then f^{-1} is differentiable at $f(c)$.

PROOF OF (1)   We first show that f is monotonic. Suppose, on the contrary, that f is neither increasing nor decreasing. Then there exists numbers x_1, x_2, x_3 in (a, b) with $x_1 < x_2 < x_3$ such that $f(x_2)$ does not lie between $f(x_1)$ and $f(x_3)$.

There are two cases: (i) $f(x_3)$ lies between $f(x_1)$ and $f(x_2)$ or (ii) $f(x_1)$ lies between $f(x_2)$ and $f(x_3)$. In case (i) the Intermediate Value Theorem implies that there exists a c satisfying $x_1 < c < x_2$ such that $f(c) = f(x_3)$. In case (ii) the Intermediate Value Theorem implies that there exists a c satisfying $x_2 < c < x_3$ such that $f(c) = f(x_1)$. In either case we have shown that f is not one-to-one. This is a contradiction. Thus, f is indeed monotonic.

Without loss of generality, let us assume that f is increasing on (a, b). Let y_0 be any number in the domain of f^{-1}. Then $f^{-1}(y_0) = x_0$, where x_0 is a number in (a, b) such that $f(x_0) = y_0$. We want to show that f^{-1} is continuous at y_0. So let $\varepsilon > 0$ be given and be sufficiently small so that the interval $(x_0 - \varepsilon, x_0 + \varepsilon)$ is contained in the interval (a, b). Since f is increasing, we see that the interval $I = (x_0 - \varepsilon, x_0 + \varepsilon)$ is mapped onto the interval $J = (f(x_0 - \varepsilon), f(x_0 + \varepsilon))$.

The function f^{-1} maps J onto I. Now let δ denote the smaller of the numbers $\delta_1 = y_0 - f(x_0 - \varepsilon)$ and $\delta_2 = f(x_0 + \varepsilon) - y_0$. Then the interval $(y_0 - \delta, y_0 + \delta)$ is contained in J and so is mapped onto the interval I by f^{-1}. (See Figure 1.) Thus, we have found a $\delta > 0$ such that $0 < |y - y_0| < \delta$ implies that $|f^{-1}(y) - f^{-1}(y_0)| < \varepsilon$. This shows that f^{-1} is continuous at y_0. Since y_0 is an arbitrary number in the domain of f^{-1}, we have shown that f^{-1} is continuous in its domain.

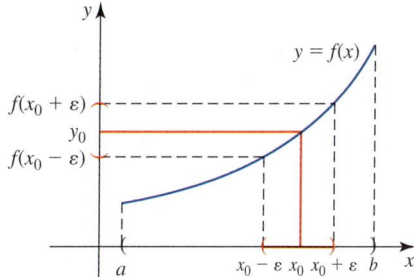

FIGURE 1

■ Chapter 3

Note on the Definition of Concavity of a Function   (Section 3.4)

If the graph of f is concave upward on an open interval I, then it lies above its tangent lines, and if the graph is concave downward on I, then it lies below all of its tangent lines.

PROOF   Suppose that the graph of f is concave upward on an interval $I = (a, b)$. Then f' is increasing on I. If c is a number in I, then an equation of the tangent line to the graph of f at $(c, f(c))$ is

$$L(x) = f(c) + f'(c)(x - c)$$

If x is any number in the interval (a, c), then the directed distance from the point $(x, f(x))$ on the graph of f to the point $(x, L(x))$ on the tangent line is given by

$$D(x) = f(x) - L(x) = f(x) - [f(c) + f'(c)(x - c)] \tag{17}$$
$$= f(x) - f(c) - f'(c)(x - c)$$

Now, by the Mean Value Theorem there exists a number z in (x, c) such that

$$f'(z) = \frac{f(x) - f(c)}{x - c} \tag{18}$$

Substituting Equation (18) into Equation (17) gives

$$D(x) = f'(z)(x - c) - f'(c)(x - c) = [f'(z) - f'(c)](x - c)$$

Now, $[f'(z) - f'(c)] < 0$ because f' is increasing. Furthermore, $(x - c) < 0$ because x lies in (a, c). So $D(x) > 0$ for all x in (a, c), and this tells us that the graph of f lies above the tangent line at x for all x in (a, c). In a similar manner you can show that the graph of f also lies above the tangent line at x for all x on (c, b). This completes the proof for the case in which f is concave upward. The proof for the case in which f is concave downward is similar. ∎

Theorem 1   l'Hôpital's Rule   (Section 3.8)

Suppose that f and g are differentiable on an open interval I that contains a, with the possible exception of a itself, and $g'(a) \neq 0$ for all x in I. If $\lim\limits_{x \to a} \dfrac{f(x)}{g(x)}$ has an indeterminate form of type $0/0$ or ∞/∞, then

$$\lim_{x \to a} \frac{f(x)}{g(x)} = \lim_{x \to a} \frac{f'(x)}{g'(x)}$$

provided that the limit on the right exists or is infinite.

PROOF for the Case $\lim\limits_{x \to a} \dfrac{f(x)}{g(x)} = \dfrac{0}{0}$ First, we need to show that if f and g are continuous on $[a, b]$ and differentiable on (a, b) and $g'(x) \neq 0$ for all x in (a, b), then there exists a number c in (a, b) such that

$$\frac{f'(c)}{g'(c)} = \frac{f(b) - f(a)}{g(b) - g(a)} \tag{19}$$

This is called Cauchy's Mean Value Theorem.

To prove this, first observe that $g(a) \neq g(b)$; otherwise, an application of Rolle's Theorem implies that there exists a c in (a, b) such that $g'(c) = 0$, contradicting the assumption that $g'(x) \neq 0$ for all x in (a, b). Put

$$h(x) = f(x) - f(a) - \frac{f(b) - f(a)}{g(b) - g(a)} [g(x) - g(a)] \tag{20}$$

Then Equation (19) follows by applying Rolle's Theorem to Equation (20). To prove l'Hôpital's Rule, suppose that $\lim_{x \to a} f(x) = 0$ and $\lim_{x \to a} g(x) = 0$. Define

$$F(x) = \begin{cases} f(x) & \text{if } x \neq a \\ 0 & \text{if } x = a \end{cases} \quad \text{and} \quad G(x) = \begin{cases} g(x) & \text{if } x \neq a \\ 0 & \text{if } x = a \end{cases}$$

Then F is continuous on I, since f is continuous on $\{x \in I \mid x \neq a\}$ and $\lim_{x \to a} F(x) = \lim_{x \to a} f(x) = 0 = F(a)$. Similarly, we see that G is also continuous on I.

For any x in I and $x > a$, F and G are continuous on $[a, x]$ and differentiable on (a, x), and $G'(x) \neq 0$, since $G'(x) = g'(x)$. So, by Cauchy's Mean Value Theorem there exists a z such that $a < z < x$, and

$$\frac{F'(z)}{G'(z)} = \frac{F(x) - F(a)}{G(x) - G(a)} = \frac{F(x)}{G(x)}$$

because, by definition, $F(a) = 0$ and $G(a) = 0$. Now, if we let x approach a from the right, then $z \to a^+$ because $a < z < x$. Therefore,

$$\lim_{x \to a^+} \frac{f(x)}{g(x)} = \lim_{x \to a^+} \frac{F(x)}{G(x)} = \lim_{z \to a^+} \frac{F'(z)}{G'(z)} = \lim_{z \to a^+} \frac{f'(z)}{g'(z)} = L$$

where

$$L = \lim_{x \to a^+} \frac{f'(x)}{g'(x)}$$

which we assume to exist. In a similar manner we can show that

$$\lim_{x \to a^-} \frac{f(x)}{g(x)} = L$$

So

$$\lim_{x \to a} \frac{f(x)}{g(x)} = L = \lim_{x \to a} \frac{f'(x)}{g'(x)}$$

as was to be shown. ■

■ Chapter 8

Theorem 5 (Section 8.1)

If $\lim_{n \to \infty} a_n = L$ and the function f is continuous at L, then

$$\lim_{n \to \infty} f(a_n) = f(\lim_{n \to \infty} a_n) = f(L)$$

PROOF   Let $\varepsilon > 0$ be given. We want to show that there exists a positive integer N such that $n > N$ implies that $|f(a_n) - f(L)| < \varepsilon$. Since f is continuous at L, there exists a $\delta > 0$ such that $0 < |x - L| < \delta$ implies that $|f(x) - f(L)| < \varepsilon$. Next, since $\lim_{n \to \infty} a_n = L$, there exists a positive integer N such that $n > N \Rightarrow |a_n - L| < \delta$. Now suppose that $n > N$. Then $0 < |a_n - L| < \delta$, which implies that $|f(a_n) - f(L)| < \varepsilon$ and thus completes the proof. ■

Theorem 1   Convergence of Power Series (Section 8.7)

Given a power series $\sum_{n=0}^{\infty} a_n(x - c)^n$, exactly one of the following is true:

a. The series converges only at $x = c$.
b. The series converges for all x.
c. There is a number $R > 0$ such that the series converges for $|x - c| < R$ and diverges for $|x - c| > R$.

PROOF   It suffices to prove the theorem for the special case where $c = 0$. The general case then follows if we replace x by $x - c$. So, let us prove the following result.
Given a power series $\sum_{n=0}^{\infty} a_n x^n$, exactly one of the following is true:

a. The series converges only at $x = 0$.
b. The series converges for all x.
c. There is a number $R > 0$ such that the series converges for $|x| < R$ and diverges for $|x| > R$.

We begin by establishing the following results:

1. If a power series $\sum_{n=0}^{\infty} a_n x^n$ converges at $x = b$, where $b \neq 0$, then it converges for all x satisfying $|x| < |b|$.

2. If a power series $\sum_{n=0}^{\infty} a_n x^n$ diverges at $x = c$, where $c \neq 0$, then it diverges for all x satisfying $|x| > |c|$.

PROOF OF (1) AND (2)   Suppose $\sum_{n=0}^{\infty} a_n x^n$ converges at b. Then $\lim_{n \to \infty} a_n b^n = 0$. So there exists a positive integer N such that $n \geq N$ implies that $|a_n b^n| < 1$. Therefore, if $n \geq N$, then we have

$$|a_n x^n| = \left| a_n x^n \frac{b^n}{b^n} \right| = |a_n b^n| \left| \frac{x}{b} \right|^n < \left| \frac{x}{b} \right|^n$$

If $|x| < |b|$, then $|x/b| < 1$ and so $\sum_{n=0}^{\infty} |x/b|^n$ is a convergent geometric series. Therefore, using the Comparison Test, we see that $\sum_{n=0}^{\infty} |a_n x^n|$ is convergent, and this shows that $\sum_{n=0}^{\infty} a_n x^n$ is absolutely convergent and, therefore, convergent. Next, suppose $\sum_{n=0}^{\infty} a_n x^n$ diverges at $x = c$. If x is any number satisfying $|x| > |c|$, then part (1) of the theorem shows that $\sum_{n=0}^{\infty} a_n x^n$ cannot converge because, otherwise, $\sum_{n=0}^{\infty} a_n c^n$ would converge, a contradiction. Therefore, $\sum_{n=0}^{\infty} a_n x^n$ diverges if $|x| > |c|$.
We are now in the position to prove the theorem. Suppose that neither case (a) nor case (b) is true. Then there exists nonzero numbers b and d such that $\sum_{n=0}^{\infty} a_n x^n$ converges at $x = b$ and diverges at $x = d$. Let $S = \{x \mid \sum_{n=0}^{\infty} a_n x^n \text{ converges}\}$. Then S is nonempty since $b \in S$. Furthermore, our earlier result shows that $\sum_{n=0}^{\infty} a_n x^n$ diverges if $|x| > |d|$ and so $|x| \leq |d|$ for all x in S. Thus $|d|$ is an upper bound for S. By the Completeness Axiom (see Section 8.1), S has a least upper bound R. Now, if $|x| > R$, then $x \notin S$ and so $\sum_{n=0}^{\infty} a_n x^n$ diverges. If $|x| < R$, then $|x|$ is not an upper bound for S and so there exists a b in S satisfying $|b| > |x|$. Since $b \in S$, $\sum_{n=0}^{\infty} a_n b^n$ converges, and this implies that $\sum_{n=0}^{\infty} a_n x^n$ converges.   ■

The Definition of the Logarithm as an Integral

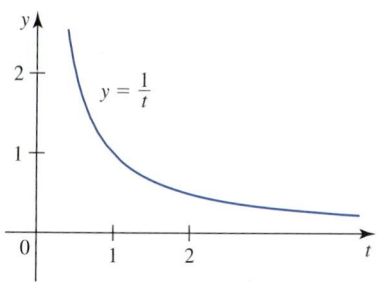

FIGURE 1
The function $f(t) = 1/t$ is continuous on $(0, \infty)$.

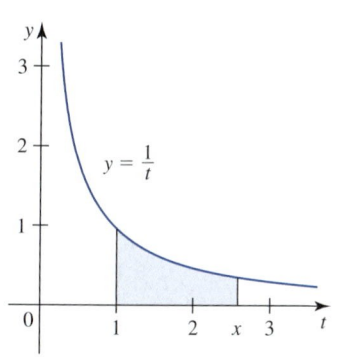

(a) If $x > 1$, $\ln x = \int_1^x \frac{1}{t} dt$

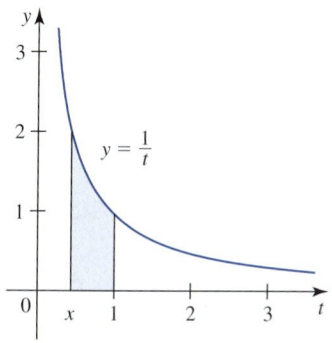

(b) If $0 < x < 1$, $\ln x = -\int_x^1 \frac{1}{t} dt$

FIGURE 2
$\ln x$ interpreted in terms of area

In Section 0.8, we gave an intuitive introduction of the exponential and logarithmic functions based on a numerical and graphical approach. In this appendix, we use the Fundamental Theorem of Calculus to help us define the logarithmic function. Then we use this definition to define the exponential function. This alternative approach, independent of the intuitive approach used earlier, puts the definition of these functions on a firm mathematical footing.

■ The Natural Logarithm Function

Recall that the Fundamental Theorem of Calculus, Part 1, states that if f is a continuous function on an open interval I and if a is any number in I, then we can define a differentiable function F by

$$F(x) = \int_a^x f(t) \, dt \qquad x \in I$$

Now consider the function f defined by $f(t) = 1/t$ on the interval $(0, \infty)$. (See Figure 1.) Since f is continuous on $(0, \infty)$, the Fundamental Theorem of Calculus, Part 1, guarantees that we can define a differentiable function on $(0, \infty)$ as follows.

DEFINITION   The Natural Logarithmic Function

The **natural logarithmic function,** denoted by **ln,** is the function defined by

$$\ln x = \int_1^x \frac{1}{t} \, dt \qquad (1)$$

for all $x > 0$.

The expression $\ln x$, read "ell-en of x," is called the **natural logarithm of x** because it has all the properties of logarithmic functions, as we shall see.

If $x > 1$, we can interpret $\ln x$ as the area of the region under the graph of $y = 1/t$ on the interval $[1, x]$. (See Figure 2.) For $x = 1$ we have

$$\ln 1 = \int_1^1 \frac{1}{t} \, dt = 0$$

If $0 < x < 1$, then

$$\ln x = \int_1^x \frac{1}{t} \, dt = -\int_x^1 \frac{1}{t} \, dt < 0$$

so $\ln x$ can be interpreted as the *negative* of the area of the region under the graph of $y = 1/t$ on the interval $[x, 1]$ (Figure 2b).

■ The Derivative of $\ln x$

Recall that the Fundamental Theorem of Calculus, Part 1, states that if f is continuous on an open interval I and the function F is defined by

$$F(x) = \int_a^x f(t)\, dt \qquad a \in I$$

then $F'(x) = f(x)$. Applying this theorem to the function $f(t) = 1/t$ gives

$$\frac{d}{dx} \ln x = \frac{d}{dx} \int_1^x \frac{1}{t}\, dt = \frac{1}{x} \qquad x > 0 \tag{2}$$

Next, using the Chain Rule, we see that if u is a differentiable function of x, then

$$\frac{d}{dx} \ln u = \frac{1}{u} \frac{du}{dx} \qquad u > 0 \tag{3}$$

■ Laws of Logarithms

The laws for differentiating the logarithmic function can be used to prove the following familiar laws of logarithms.

THEOREM 1   Laws of Logarithms

Let x and y be positive numbers and let r be a rational number. Then

a. $\ln 1 = 0$ **b.** $\ln xy = \ln x + \ln y$

c. $\ln \dfrac{x}{y} = \ln x - \ln y$ **d.** $\ln x^r = r \ln x$

PROOF

a. Law (a) was proved on page A 17.

b. Define the function $F(x) = \ln ax$, where a is a positive constant. Then, using Equation (3), we have

$$F'(x) = \frac{d}{dx} (\ln ax) = \frac{1}{ax} \frac{d}{dx} (ax) = \frac{a}{ax} = \frac{1}{x}$$

But by Equation (2) we have

$$\frac{d}{dx} \ln x = \frac{1}{x}$$

Therefore, $F(x)$ and $\ln x$ have the same derivative and, by Theorem 1 of Section 4.1, must differ by a constant; that is,

$$F(x) = \ln ax = \ln x + C$$

Letting $x = 1$ in this equation and recalling that $\ln 1 = 0$, we have

$$\ln a = \ln 1 + C = C$$

Therefore,

$$\ln ax = \ln x + \ln a$$

Since a can be any positive number, we have shown that

$$\ln xy = \ln x + \ln y$$

c. Using the result of part (b) with $x = 1/y$, we have

$$\ln \frac{1}{y} + \ln y = \ln\left(\frac{1}{y} \cdot y\right) = \ln 1 = 0$$

so

$$\ln \frac{1}{y} = -\ln y$$

Using the result of part (b) once again, we obtain

$$\ln \frac{x}{y} = \ln\left(x \cdot \frac{1}{y}\right) = \ln x + \ln \frac{1}{y} = \ln x - \ln y$$

as desired.

d. Define the functions F and G by $F(x) = \ln x^r$ and $G(x) = r \ln x$, respectively. Then using Equation (3), we have

$$F'(x) = \frac{1}{x^r} \cdot rx^{r-1} = \frac{r}{x}$$

Next, using Equation (2), we find

$$G'(x) = \frac{r}{x}$$

Therefore, F and G must differ by a constant; that is,

$$\ln x^r = r \ln x + C$$

Letting $x = 1$ in this equation gives

$$\ln 1 = r \ln 1 + C$$

or $C = 0$, so

$$\ln x^r = r \ln x$$

as was to be shown.

■ The Graph of the Natural Logarithmic Function

To help us draw the graph of the natural logarithmic function, we first note that $f(x) = \ln x$ has the following properties:

1. The domain of f is $(0, \infty)$, by definition.
2. f is continuous on $(0, \infty)$, since it is differentiable there.
3. f is increasing on $(0, \infty)$, since $f'(x) = \dfrac{1}{x} > 0$ on $(0, \infty)$.
4. The graph of f is concave downward on $(0, \infty)$ since $f''(x) = -\dfrac{1}{x^2} < 0$ on $(0, \infty)$.

Next, using the Trapezoidal Rule or Simpson's Rule, we have

$$f(2) = \ln 2 = \int_1^2 \frac{1}{t}\, dt \approx 0.693$$

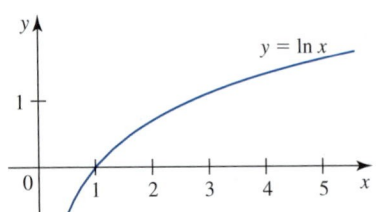

FIGURE 3
The graph of the natural logarithmic function $y = \ln x$

Then, using Theorem 1d, we obtain the following table of values.

x	4	8	$\frac{1}{2}$	$\frac{1}{4}$	$\frac{1}{8}$
$f(x)$	1.386	2.079	-0.693	-1.386	-2.079

Using the properties of $f(x) = \ln x$, the sample values just obtained, and the results

$$\lim_{x \to 0^+} \ln x = -\infty \qquad \text{and} \qquad \lim_{x \to \infty} \ln x = \infty \qquad (4)$$

which we will establish at the end of this section, we sketch the graph of $f(x) = \ln x$, as shown in Figure 3.

The Natural Exponential Function

Since the natural logarithmic function defined by $y = \ln x$ is continuous and increasing on the interval $(0, \infty)$, it is one-to-one there and, hence, has an inverse function. This inverse function is called the *natural exponential function* and is defined as follows.

> **DEFINITION**   **The Natural Exponential Function**
>
> The **natural exponential function,** denoted by **exp,** is the function satisfying the conditions:
>
> **1.** $\ln(\exp x) = x$ for all x in $(-\infty, \infty)$.
> **2.** $\exp(\ln x) = x$ for all x in $(0, \infty)$.
>
> Equivalently,
>
> $$\exp(x) = y \qquad \text{if and only if} \qquad \ln y = x$$

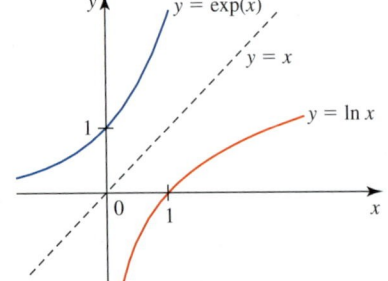

FIGURE 4
The graph of $y = \exp(x)$ is obtained by reflecting the graph of $y = \ln x$ with respect to the line $y = x$.

That the domain of exp is $(-\infty, \infty)$ and its range is $(0, \infty)$ follows because the range of ln is $(-\infty, \infty)$ and its domain is $(0, \infty)$. The graph of $y = \exp(x)$ can be obtained by reflecting the graph of $y = \ln x$ about the line $y = x$. (See Figure 4.)

The Number e

We begin by recalling that the natural logarithmic function ln is continuous and one-to-one and that its range is $(-\infty, \infty)$. Therefore, by the Intermediate Value Theorem there must be a unique real number x_0 such that $\ln x_0 = 1$. Let's denote x_0 by e. In view of the definition of ln, the number e can be defined as follows.

> **DEFINITION**   **The Number e**
>
> The number e is the number such that
>
> $$\ln e = \int_1^e \frac{1}{t}\, dt = 1$$

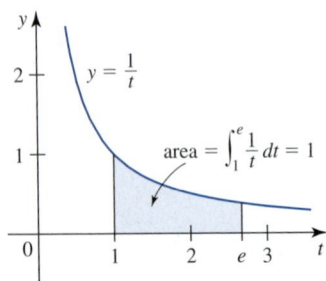

FIGURE 5
The area of the region under the graph of $f(t) = 1/t$ on $[1, e]$ is 1.

Figure 5 gives a geometric interpretation of the number e. It can be shown that the number e is irrational and has the following approximation:

$$e \approx 2.718281828$$

You can verify this using a graphing calculator. Plot the graphs of the functions $y_1 = \ln x$ and $y_2 = 1$. Then use the function for finding the intersection of two curves to estimate the x-coordinate of the point of intersection.

◼ Defining the Natural Exponential Function

Using Law (c) of logarithms, we see that if r is a *rational* number, then

$$\ln e^r = r \ln e = r(1) = r$$

Equivalently, $e^r = y$ if and only if $\ln y = r$. The equation $\ln e^r = r$ can be used to motivate the definition of e^x for every *real* number x.

DEFINITION e^x

If x is any real number, then

$$e^x = y \qquad \text{if and only if} \qquad \ln y = x$$

Now, by definition of the natural exponential function we have

$$\exp(x) = y \qquad \text{if and only if} \qquad \ln y = x$$

Comparing this definition with the definition of e^x gives the following rule for defining the natural exponential function.

DEFINITION   The Natural Exponential Function

The natural exponential function, exp, is defined by the rule

$$\exp(x) = e^x$$

In view of this, we have the following theorem, which gives us another way of expressing the fact that exp and ln are inverse functions.

THEOREM 1

a. $\ln e^x = x$,   for $x \in (-\infty, \infty)$ **b.** $e^{\ln x} = x$,   for $x \in (0, \infty)$

The graph of the natural exponential function $y = e^x$ was sketched earlier (Figure 4). We summarize the important properties of this function.

Properties of the Natural Exponential Function

1. The domain of $f(x) = e^x$ is $(-\infty, \infty)$, and its range is $(0, \infty)$.
2. The function $f(x) = e^x$ is continuous and increasing on $(-\infty, \infty)$.
3. The graph of $f(x) = e^x$ is concave upward on $(-\infty, \infty)$.
4. $\lim_{x \to -\infty} e^x = 0$ and $\lim_{x \to \infty} e^x = \infty$.

■ The Laws of Exponents

The following laws of exponents are useful when working with exponential functions.

THEOREM 2   Laws of Exponents

Let x and y be real numbers and r be a rational number. Then

a. $e^x e^y = e^{x+y}$ **b.** $\dfrac{e^x}{e^y} = e^{x-y}$ **c.** $(e^x)^r = e^{rx}$

PROOF We will prove Law (a). The proofs of the other two laws are similar and will be omitted. We have

$$\ln(e^x e^y) = \ln e^x + \ln e^y = x + y = \ln e^{x+y}$$

Since the natural logarithmic function is one-to-one, we see that

$$e^x e^y = e^{x+y}$$

■ The Derivatives of Exponential Functions

Since the inverse function of a differentiable function is itself differentiable, we see that the natural exponential function is differentiable. In fact, as the following theorem shows, the natural exponential function is its own derivative!

THEOREM 3   The Derivatives of Exponential Functions

Let u be a differentiable function of x. Then

a. $\dfrac{d}{dx} e^x = e^x$ **b.** $\dfrac{d}{dx} e^u = e^u \dfrac{du}{dx}$

PROOF

a. Let $y = e^x$, so that $\ln y = x$. Differentiating both sides of the last equation implicitly with respect to x gives

$$\frac{1}{y}\frac{dy}{dx} = 1 \qquad \text{or} \qquad \frac{dy}{dx} = y = e^x$$

b. This follows from part (a) by using the Chain Rule.

■ Logarithmic Functions with Base a

If a is a positive real number with $a \neq 1$, then the function f defined by $f(x) = a^x$ is one-to-one on $(-\infty, \infty)$, and its range is $(0, \infty)$. Therefore, it has an inverse on $(0, \infty)$. This function is called the logarithmic function with base a and is denoted by $\log_a$.

DEFINITION   Logarithmic Function with Base a

The **logarithmic function with base a,** denoted by $\log_a$, is the function satisfying the relationship

$$y = \log_a x \qquad \text{if and only if} \qquad x = a^y$$

Observe that if $a = e$, then this definition reduces to the relationship between the natural logarithmic function ln and the natural exponential function exp.

■ The Definition of the Number e as a Limit

If we use the definition of the derivative as a limit to compute $f'(1)$, where $f(x) = \ln x$, we obtain

$$f'(1) = \lim_{h \to 0} \frac{f(1 + h) - f(1)}{h}$$

$$= \lim_{h \to 0} \frac{\ln(1 + h) - \ln 1}{h} = \lim_{h \to 0} \frac{\ln(1 + h)}{h} \qquad \text{ln } 1 = 0$$

$$= \lim_{h \to 0} \ln(1 + h)^{1/h}$$

$$= \ln\left[\lim_{h \to 0} (1 + h)^{1/h}\right] \qquad \text{Use the continuity of ln.}$$

But

$$f'(1) = \left[\frac{d}{dx} \ln x\right]_{x=1} = \left[\frac{1}{x}\right]_{x=1} = 1$$

so

$$\ln\left[\lim_{h \to 0} (1 + h)^{1/h}\right] = 1$$

or

$$\lim_{h \to 0} (1 + h)^{1/h} = e$$

We now prove Equation (4). First, using Law (d) of logarithms with $x = 2$, we have $\ln 2^r = r \ln 2$, where r is any rational number. If we pick $r = n$, where n is a positive integer, then

$$\lim_{n \to \infty} \ln 2^n = \lim_{n \to \infty} n(\ln 2) = \infty$$

since $\ln 2 > 0$. But, as we established earlier, $\ln x$ is an increasing function, and this allows us to conclude that

$$\lim_{x \to \infty} \ln x = \infty$$

Next, let's put $u = 1/x$. Then $u \to \infty$ as $x \to 0^+$. Therefore,

$$\lim_{x \to 0^+} \ln x = \lim_{u \to \infty} \ln \frac{1}{u} = \lim_{u \to \infty} (-\ln u) = -\infty$$

and (4) is proved.

ANSWERS TO SELECTED EXERCISES

CHAPTER 0

0.1 Self-Check Diagnostic Test • page 2

1. $y = -\frac{7}{3}x + \frac{2}{3}$ **2.** $y = \frac{3}{2}x - \frac{13}{2}$ **3.** No, they do not.

4. $y = -\frac{3}{2}x + 6$ **5.** $y = -\frac{3}{4}x - \frac{7}{4}$

Exercises 0.1 • page 13

1. 6   **3.** -1.1

5. a. $m_1 > 0, m_2 > 0, m_3 = 0, m_4 < 0$ **b.** L_4, L_3, L_2, L_1

7. -22 **9.** 1   **11.** $\sqrt{3}/3$ **13.** $\sqrt{3}$ **15.** $135°$

17. $60°$ **19.** $150°$

21. 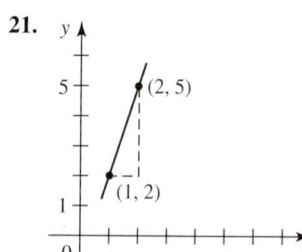   **23.**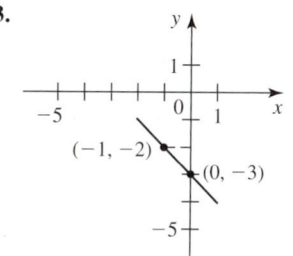

25. Parallel   **27.** Perpendicular   **29.** -6 **31.** 5

35. No   **37.** $y = \frac{3}{4}x + 2, m = \frac{3}{4}, b = 2$

39. $y = -\frac{A}{B}x + \frac{C}{B}, m = -\frac{A}{B}, b = \frac{C}{B}$

41. $y = \frac{\sqrt{6}}{3}x - \frac{4\sqrt{3}}{3}, m = \frac{\sqrt{6}}{3}, b = -\frac{4\sqrt{3}}{3}$

43. $60°$ **45.** $135°$ **47.** $y = 2x - 11$

49. $y = 4x - 4$ **51.** $x = 2$ **53.** $y = 3x - 5$

55. $y = -\frac{2}{3}x - 3$ **57.** $y = -\frac{1}{3}x - \frac{14}{3}$

59. $y = \frac{\sqrt{3}}{3}x + \left(3 - \frac{2\sqrt{3}}{3}\right)$ **61.** Perpendicular

63. Perpendicular   **65.** $(2, 1)$

67. a. $(x - 2)^2 + (y + 3)^2 = 25$ **b.** $x^2 + y^2 = 13$

c. $(x - 2)^2 + (y + 3)^2 = 34$

d. $(x + a)^2 + (y - a)^2 = 4a^2$

71. $\frac{x}{-4} + \frac{y}{-1} = 1$ **73.** $2\sqrt{5}$ **75.** $\arctan\left(\frac{1}{6}\right) \approx 9.5°$

77. a. $y = \frac{12}{7}x$ **b.** $y = -\frac{12}{7}x + 6$

c. $y = 1.2x - 4.2$ **d.** $y = -\frac{6}{7}x + \frac{57}{7}$

79. a.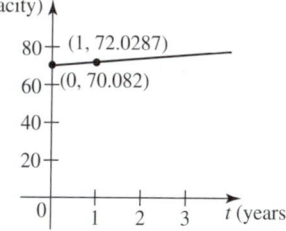

b. $1.9467, 70.082$ **d.** 2005

81. a. 88.8 tons

b.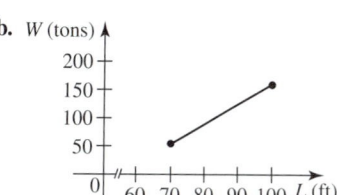

85. True   **87.** True

0.2 Self-Check Diagnostic Test • page 16

1. $2; 1; 4$ **2.** $2x + h + 2$ **3.** $\left[-\frac{1}{2}, 1\right) \cup (1, \infty)$

4. $(-\infty, \infty); [-1, \infty)$

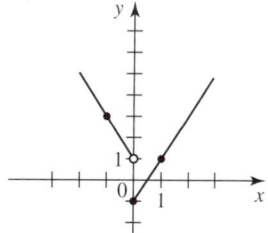

5. Odd

Exercises 0.2 • page 23

Abbreviations: D, domain; R, range.

1. $f(0) = 4, f(-4) = -8, f(a) = 3a + 4, f(-a) = -3a + 4,$
$f(a + 1) = 3a + 7, f(2a) = 6a + 4, f(\sqrt{a}) = 3\sqrt{a} + 4,$
$f(x + 1) = 3x + 7$

3. $g(-2) = -8, g(\sqrt{3}) = -3 + 2\sqrt{3}, g(a^2) = -a^4 + 2a^2,$
$g(a + h) = -a^2 - 2ah - h^2 + 2a + 2h, 1/(g(3)) = -\frac{1}{3}$

5. $f(-1) = -1, f(0) = 0, f(x^2) = 2x^6 - x^2,$
$f(\sqrt{x}) = 2x^{3/2} - \sqrt{x}, f\left(\frac{1}{x}\right) = \frac{2}{x^3} - \frac{1}{x}$

7. $f(-2) = 5, f(0) = 1, f(1) = 1$ **9.** $x + 1$

11. $1 - 2x - h$ **13. a.** 4   **b.** -2

15. $(-\infty, 0) \cup (0, \infty)$ **17.** $\left(-\infty, -\frac{1}{2}\right) \cup \left(-\frac{1}{2}, 1\right) \cup (1, \infty)$

19. $[-3, 3]$ **21.** $[2, 4]$

23. $[-2, -1) \cup (-1, 0) \cup (0, 1) \cup (1, 2]$

25. $(-\infty, -1) \cup (1, \infty)$

27. a. -2 **b. (i)** 2   **(ii)** 1   **c.** $[0, 6]$ **d.** $[-2, 6]$ **29.** Yes

31. D: $(-\infty, \infty)$, R: $(-\infty, \infty)$ **33.** D: $[1, \infty)$, R: $[0, \infty)$

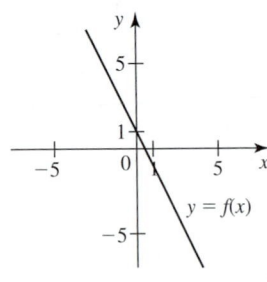

35. D: $(-\infty, -1] \cup [1, \infty)$, R: $[0, \infty)$

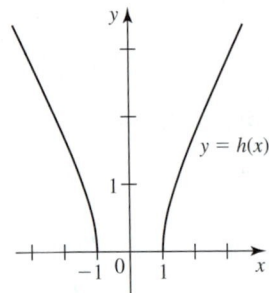

37. D: $(-\infty, \infty)$, R: $[0, \infty)$

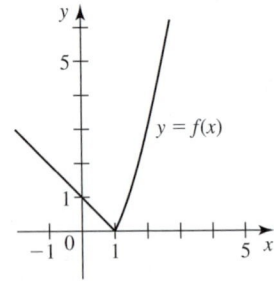

39. No   **41.** No   **43.** Yes   **45.** Even   **47.** Odd

49. Even   **51.** Neither   **53.** Even

55. a.   **b.**

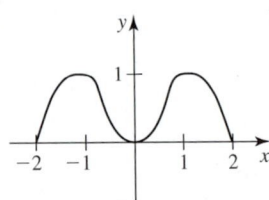

57. They stopped from 9:00 to 9:15 A.M. and from 12:00 to 1:00 P.M. They traveled at constant rates between stops.

59. $P = f(t)$

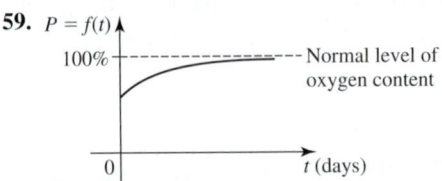

61. 4.9623%, 70.0923%

63. a. $82.95   **b.** $85.65, $90.25, $96.75

c. $f(t)$ (dollars)

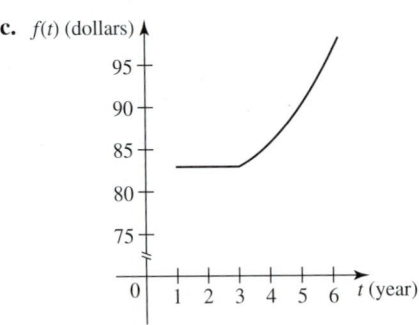

65. a. 130 tons, 100 tons, 40 tons

b. y (tons/day)

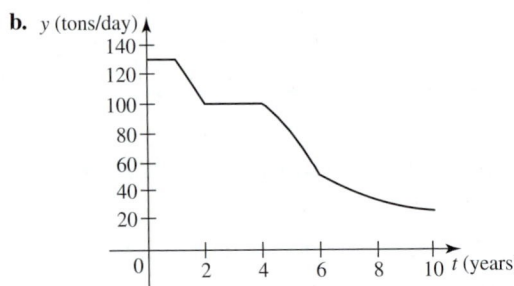

69. True   **71.** False

0.3 Self-Check Diagnostic Test • page 27

1. $\frac{4}{3}$ **2.** Odd   **4.** $a^2 \cos^2 \theta$ **5.** $\theta = \pi/3, \pi$, and $5\pi/3$

Exercises 0.3 • page 37

Abbreviations: D, domain; R, range; A, amplitude; P, period.

1. $5\pi/6$ **3.** $11\pi/6$ **5.** $-2\pi/3$ **7.** $-5\pi/12$

9. $60°$ **11.** $150°$ **13.** $-90°$ **15.** $-585°$

17. $\sqrt{3}/2, \frac{1}{2}, \sqrt{3}$ **19.** $-\frac{1}{2}, -\sqrt{3}, -2$

21. 0, 0, undefined   **23.** $2, -2\sqrt{3}/3, -\sqrt{3}$

25. $\cos \theta = -\frac{4}{5}, \tan \theta = -\frac{3}{4}, \cot \theta = -\frac{4}{3}, \sec \theta = -\frac{5}{4}, \csc \theta = \frac{5}{3}$

27. $0, \sqrt{2}/2, -\sqrt{3}/2, 0, \cos a$

29. $\left\{\frac{\pi}{2} + 2n\pi \,|\, n \text{ is an integer}\right\}$

31. a. Odd   **b.** Odd   **c.** Odd

43. $D = (-\infty, \infty), R = [-2, 2]$

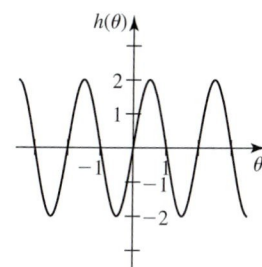

45. $A = 1, P = 2\pi$ **47.** $A = 1, P = 2\pi$

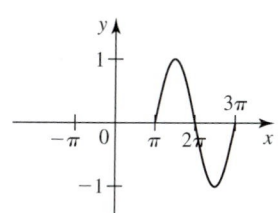   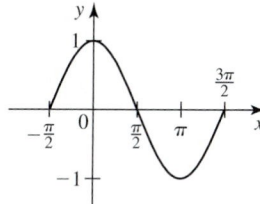

49. $A = 1, P = 2\pi$ **51.** $A = 2, P = \pi$

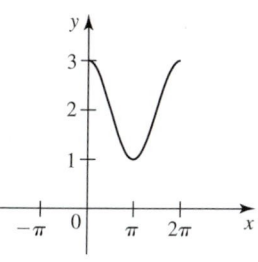   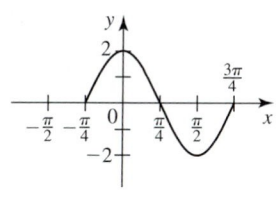

53. $A = 1, P = \pi$ **55.** $A = 2, P = 2\pi/3$

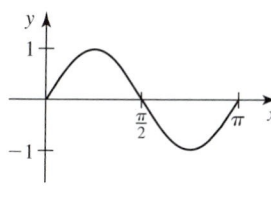   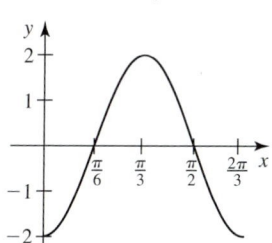

57. $A = 3, P = \pi$

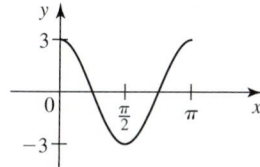

59. $x = \pi/4, 5\pi/4$ **61.** $t = \pi$

63. $x = \pi/4, \pi/2, 5\pi/4, 3\pi/2$ **65.** $x = 0, \pi/3, 5\pi/3$

67. About 2.4 min

69. True

71. True

73. False

0.4 Self-Check Diagnostic Test • page 39

1. $2x + \dfrac{1}{x+1}, (-\infty, -1) \cup (-1, \infty); 2x - \dfrac{1}{x+1},$

$(-\infty, -1) \cup (-1, \infty); \dfrac{2x}{x+1}, (-\infty, -1) \cup (-1, \infty);$

$2x^2 + 2x, (-\infty, -1) \cup (-1, \infty)$

2. $\dfrac{\sqrt{x+1}}{\sqrt{x+1}+1}; [-1, \infty)$ **3.** $f(x) = 3x^2 + 1; g(x) = \dfrac{10}{\sqrt{x}}$

4. $g(x) = 3\sqrt{x-2} - 5$ **5.** $f(x) = 2x^2 + x - 1$

Exercises 0.4 • page 48

Abbreviations: D, domain; R, range.

1. a. $x^2 + 3x - 1, D = (-\infty, \infty)$

  b. $-x^2 + 3x + 1, D = (-\infty, \infty)$

  c. $3x^3 - 3x, D = (-\infty, \infty)$

  d. $\dfrac{3x}{x^2 - 1}, D = (-\infty, -1) \cup (-1, 1) \cup (1, \infty)$

3. a. $\sqrt{x+1} + \sqrt{x-1}, D = [1, \infty)$

  b. $\sqrt{x+1} - \sqrt{x-1}, D = [1, \infty)$

  c. $\sqrt{x^2 - 1}, D = [1, \infty)$ **d.** $\dfrac{\sqrt{x+1}}{\sqrt{x-1}}, D = (1, \infty)$

5. $4x^2 + 12x + 9, D = (-\infty, \infty); 2x^2 + 3, D = (-\infty, \infty)$

7. $\dfrac{x-1}{x+1}, D = (-\infty, -1) \cup (-1, 1) \cup (1, \infty); \dfrac{1+x}{1-x},$

$D = (-\infty, 0) \cup (0, 1) \cup (1, \infty)$

9. 10

11. a. $g \circ f = \begin{cases} x^2 + 2x + 1 & \text{if } x < 0 \\ x^2 - 2x + 1 & \text{if } x \geq 0 \end{cases}$

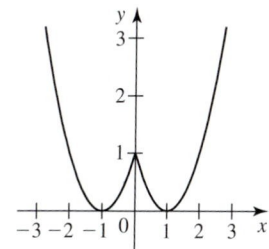

  b. $f \circ g = \begin{cases} x^2 + 1 & \text{if } x < 0 \\ x^2 - 1 & \text{if } x \geq 0 \end{cases}$

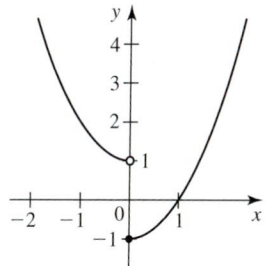

13. a. $8 + \sqrt{2}$ **b.** 34   **c.** 36   **d.** $18 + \sqrt{3}$

15. $\sqrt{2x^2 - 1}$ **17.** $f(x) = 3x^2 + 4$, $g(x) = x^{3/2}$

19. $f(x) = \sqrt{x^2 - 4}$, $g(x) = 1/x$

21. $f(t) = t^2$, $g(t) = \sin t$

23. a. $f(x) = \sqrt{x}$, $g(x) = 1 - x$, $h(x) = \sqrt{x}$

  b. $f(x) = x^3$, $g(x) = \sin x$, $h(x) = 2x + 3$

25. a. 4   **b.** 5   **c.** 1   **d.** 3   **e.** 2   **f.** 6

27. Curve 1: $y = f(x) + 1$; curve 2: $y = f(x) - 1$

29. Curve 1: $y = f\left(\frac{x}{2}\right)$; curve 2: $y = f(2x)$

31. $g(x) = x^3 + x + 2$ **33.** $g(x) = x + 3 + \dfrac{1}{\sqrt{x + 3}}$

35. $g(x) = \dfrac{3\sqrt{x}}{x^2 + 1}$ **37.** $g(x) = \dfrac{x}{2} \sin \dfrac{x}{2}$

39. $g(x) = \sqrt{8x - 4x^2} + 1$

41. a.

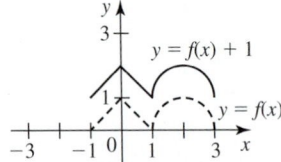

b.

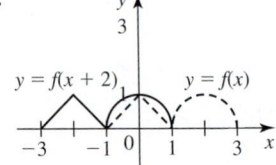

c.

d.

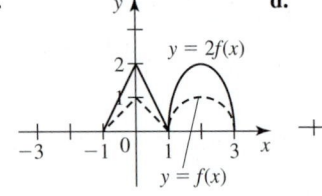

e.

f.

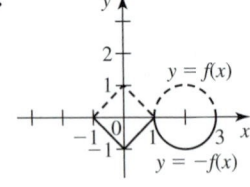
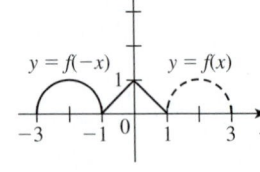

g.

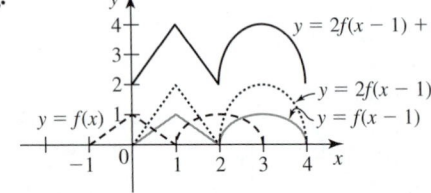

h.

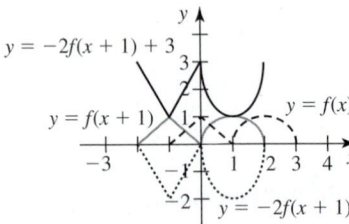

43.

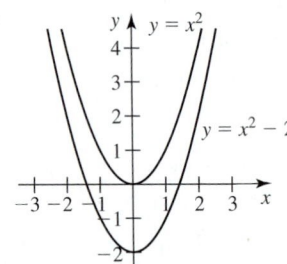

45.

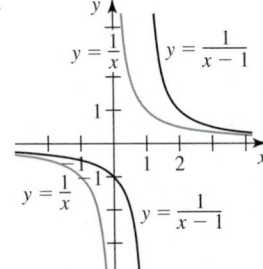

47.

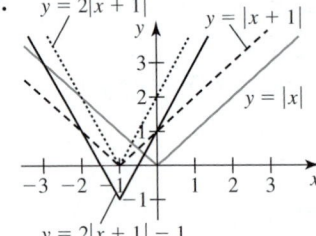

49.

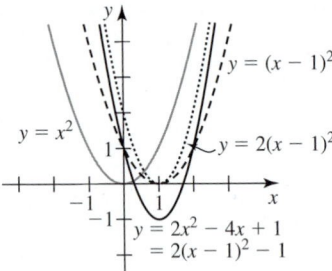

51.

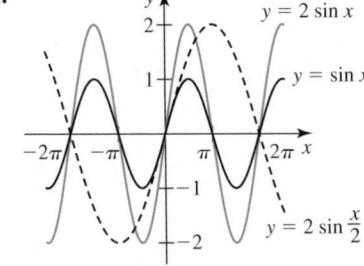

53.

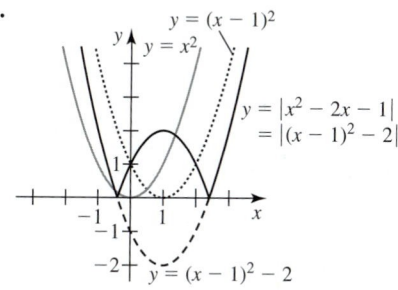

55. a. Graph $f(x)$ for $x \geq 0$ and $f(-x)$ for $x < 0$.

b.

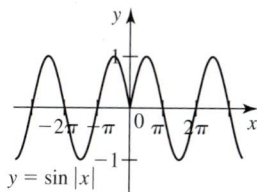

$y = \sin |x|$

57. a. $2x + 5$ **b.** $\frac{4}{3}x - 4$ **59.** $3x - 1$

61. a. $\begin{cases} \sqrt{-x} - \sin x & \text{if } -2\pi \leq x < 0 \\ \sqrt{x} + \sin x & \text{if } 0 \leq x \leq 2\pi \end{cases}$

b. $\begin{cases} -\sqrt{-x} + \sin x & \text{if } -2\pi \leq x < 0 \\ \sqrt{x} + \sin x & \text{if } 0 \leq x \leq 2\pi \end{cases}$

63. a. $D(t) = 0.33t^2 + 1.1t + 16.9, D(4) = 26.58$

b. $P(t) = \dfrac{1.21t^2 + 6t + 14.5}{1.54t^2 + 7.1t + 31.4}, P(4) \approx 0.69$

65. a. $(g \circ f)(0) = 26$ **b.** $(g \circ f)(6) = 42$

67. a. $3.5t^2 + 2.4t + 71.2$ **b.** $71,200; 109,900$

69. True   **71.** True   **73.** True

Exercises 0.5 • page 56

1. a.

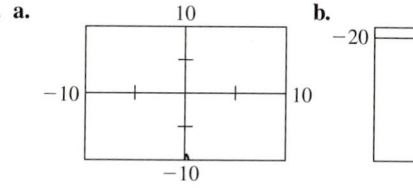

b.

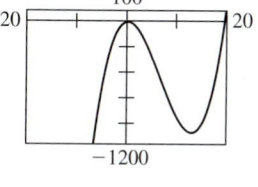

3. a.

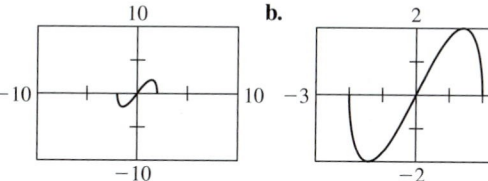

b.

5.

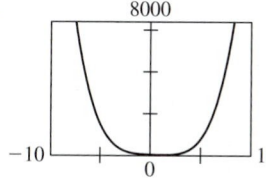

7.

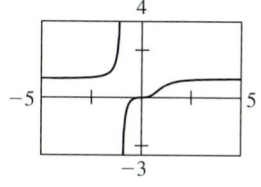

9.

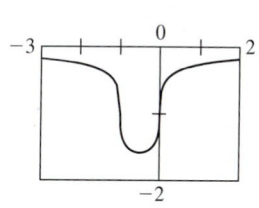

11.

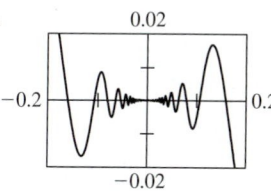

13. 1.25

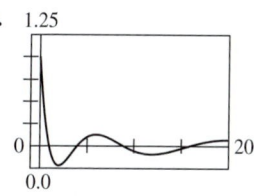

15. 0.3

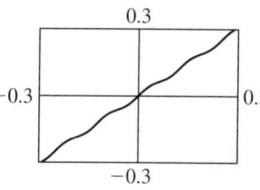

17. -1.47569 **19.** $-1.11769, 0.35855$

21. $-0.45662, 1.25873$

23. $(-2.33712, 2.41174), (6.05141, -2.50154)$

25. $(-1.02193, -6.34606), (1.2414, -1.59306),$
$(5.78053, 7.93912)$

27. $(-2.51746, -1.16879), (0.91325, 1.58299)$

29. a.

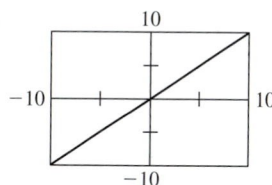

b.

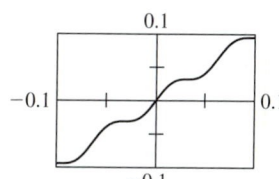

31. a.

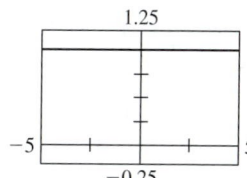

b. No; f is not defined at $x = 0$.

33. a.

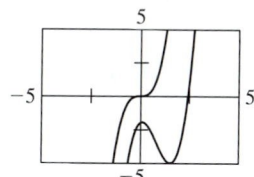

b.

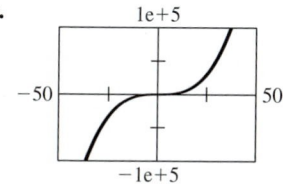

0.6 Self-Check Diagnostic Test • page 57

1. a. $f(x) = 3x - 1$ **b.** $f(x) = 2x^4 - 3x^2 + 7$

  c. $f(x) = \dfrac{x^2}{x^2 - 9}$ **d.** $f(x) = x^{5/7}$

  e. $f(x) = \sqrt{x - 1}$ **f.** $f(x) = 2 \sin x$

2. a. $V(0) = C$; the initial book value of the asset

  b. $\dfrac{C - S}{n}$ dollars/year

3. $V(x) = 4x^3 - 48x^2 + 144x$

Exercises 0.6 • page 69

1. a. Polynomial, 3   **b.** Power   **c.** Rational

  d. Rational   **e.** Algebraic   **f.** Trigonometric

3. a. 59.7 million   **b.** 152.54 million

5. 648,000; 902,000; 1,345,200; 1,762,800

7. 4.6%, 8.51%, 15.91%

11. a. $\dfrac{100x}{7960 + x}$ **b.** 74.8%

13. a.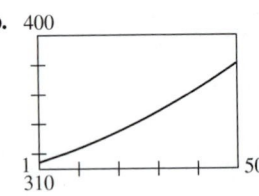

  b. 157 million

15. a. $A(t) = 0.010716t^2 + 0.8212t + 313.4$

  b.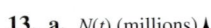

  c. 338 ppmv   **d.** 387 ppmv

17. a. $f(t) = -0.425t^3 + 3.65714t^2 + 4.01786t + 43.6643$

  b.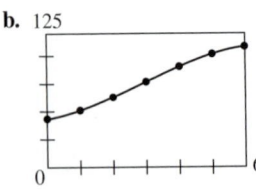

  c. $43.66 million; $77.16 million; $107.63 million

19. a. $f(t) = 0.00125t^4 - 0.005093t^3 - 0.024306t^2 + 0.128624t + 1.70992$

  b.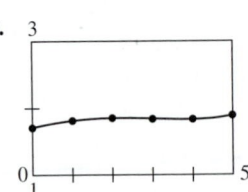

  c. 1.71, 1.81, 1.85, 1.84, 1.83, 1.89

21. $f(x) = 40x - x^2$, $(0, 40)$

23. $V = 4x^3 - 46x^2 + 120x$; $(0, 4)$

25. $f(x) = 28x - \left(\dfrac{\pi}{2} + 2\right)x^2$; $\left(0, \dfrac{56}{4 + \pi}\right)$

27. $f(x) = -2x + 52 - \dfrac{50}{x}$, $(1, 25)$

29. a. $R(x) = -4x^2 + 520x + 12{,}000$

  b. $26,400   **c.** $28,800

0.7 Self-Check Diagnostic Test • page 73

1. No   **2.** 1

3. $\dfrac{x + 2}{3}$

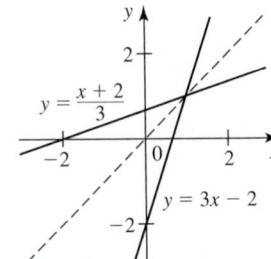

4. a. $-\dfrac{\pi}{4}$ **b.** $\dfrac{5\pi}{6}$ **5.** $\dfrac{x}{\sqrt{1 - x^2}}$

Exercises 0.7 • page 82

7. Yes   **9.** No   **11.** Yes   **13.** Yes

15. 2   **17.** 0   **19.** $\pi/6$

21.

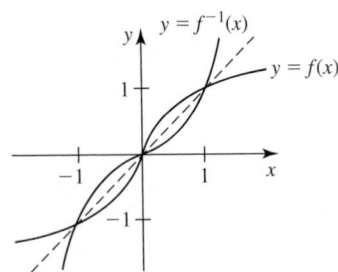

23. $\sqrt[3]{x-1}$

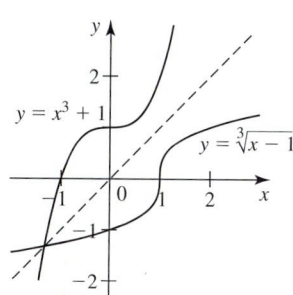

25. $\sqrt{9-x^2}$, $x \geq 0$

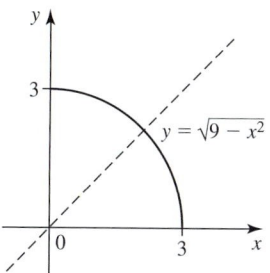

27. $\frac{1}{2}(1 + \sin^{-1} x)$

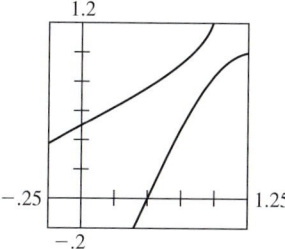

29. $x^3 + 1$

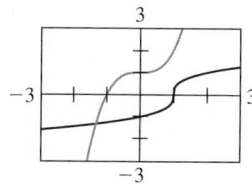

31. $\dfrac{1 - \sqrt{1 - 4x^2}}{2x}$, $-\frac{2}{5} \leq x \leq \frac{2}{5}$

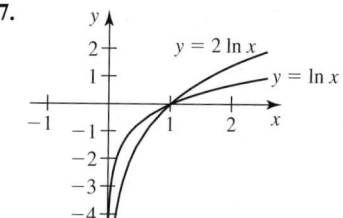

33. $\begin{cases} \dfrac{x+1}{2} & \text{if } x < 1 \\ x^2 & \text{if } 1 \leq x < 2 \\ \sqrt{2x + 12} & \text{if } x \geq 2 \end{cases}$ Domain: $(-\infty, \infty)$

35. 0   **37.** $\pi/6$ **39.** $\pi/3$ **41.** $\pi/3$ **43.** $\pi/3$

45. $-\pi/6$ **47.** $\sqrt{2}/2$ **49.** $\sqrt{1 - x^2}$ **51.** x

53. $\dfrac{1}{\sqrt{x^2 - 1}}$

55. a. $f^{-1}(F) = \frac{5}{9}(F - 32)$ **b.** $[-459.67, \infty)$

57. a. $f^{-1}(p) = \dfrac{10}{9}\left[\left(\dfrac{p}{10.72}\right)^{10/3} - 10\right]$ **b.** 7.58

61. True   **63.** True   **65.** False

0.8 Self-Check Diagnostic Test • page 84

1. $3x$ **2.** 3.6620   **3.** $\left(0, \dfrac{1}{e}\right) \cup \left(\dfrac{1}{e}, \infty\right)$

4. $e^{x/2} - 1$, $(-\infty, \infty)$ **5.** $\ln \dfrac{x\sqrt{x + 1}}{\cos x}$

Exercises 0.8 • page 93

1. a. 1.7917 **b.** 0.4055   **3. a.** 3.4011 **b.** 2.0149

5. $\ln 2 + \frac{1}{2}\ln 3 - \ln 5$ **7.** $\frac{1}{3}\ln x + \frac{2}{3}\ln y - \frac{1}{2}\ln z$

9. $\frac{1}{3}\ln(x + 1) - \frac{1}{3}\ln(x - 1)$ **11.** $\ln 2$

13. $\ln \dfrac{8}{\sqrt{x + 1}}$ **15. a.** 3 **b.** x^2 **17. a.** 9 **b.** $\sqrt{x}$

19. $(-\infty, \infty)$ **21.** $[0, \infty)$

23. $\left(-\frac{1}{2}, \infty\right)$ **25.** $\left(2k\pi - \frac{\pi}{2}, 2k\pi + \frac{\pi}{2}\right)$, $k = 0, \pm 1, \pm 2, \ldots$

27. a. $x = 2$ **b.** $x = -\frac{3}{2}$

29. a. $x = \ln \frac{5}{2} - 2$ **b.** $x = e^2 - 1$

31. a. $x = 5\ln\frac{3}{8}$ **b.** $x = \ln 2$ or $x = \ln 3$

33. Odd   **35.** Odd

37.

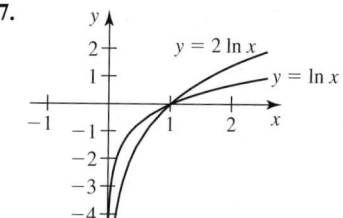

39. 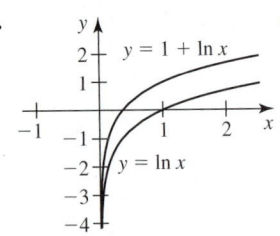   **41.**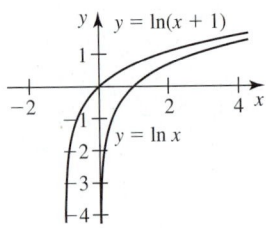

43. a. 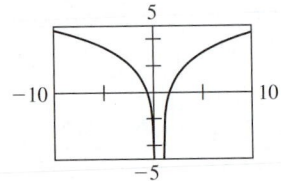   **b.** $x > 1$

The graph of f is the right branch
and the graph of g consists of
both branches.

45. 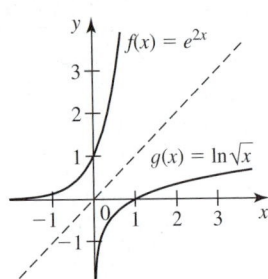   **47.**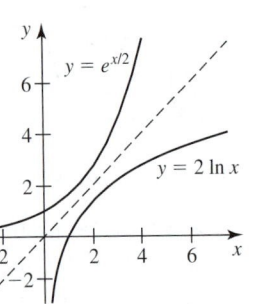

49. $\ln(x - 1)$ **51.** $\ln\left(\dfrac{x+1}{x-1}\right)$

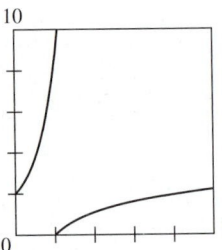   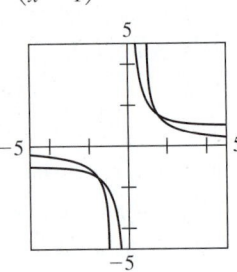

53. a. 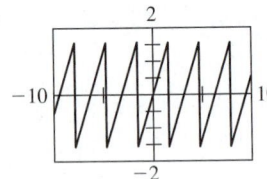   **b.** Yes

55. No   **57.** 0.83% per decade

59. 0.094 percent; 0.075 percent

61. a. 180.7 per decade   **b.** 52

63. a.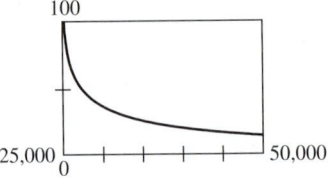

b. $35,038.78 per year

65. $0.4732p_0$; mountaineers experience difficulty in breathing at
very high altitudes because the air is more rarefied owing to
a decrease in atmospheric pressure.

67. 153,024; 235,181

69. a. 50   **b.**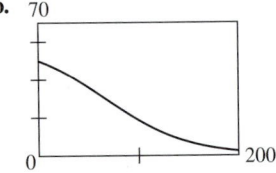

71. a. $8052.55   **b.** $8144.47   **c.** $8193.08

d. $8226.54   **e.** $8243.04

73. $2.58 million   **75.** True   **77.** True   **79.** False

Chapter O Review Exercises • page 96

1. $-\dfrac{7}{3}$ **3.** -2 **5.** $y = -4$ **7.** $y = -\dfrac{4}{3}x + \dfrac{1}{3}$

9. $y = -\dfrac{3}{5}x + \dfrac{12}{5}$ **11.** $y = -1$

13. a–b.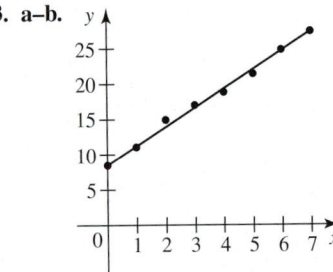

c. $y = 2.7x + 8.5$ **d.** 30.1 million

15. $f(0) = 0, f\left(\dfrac{\pi}{6}\right) = \sqrt{3}/3, f\left(\dfrac{\pi}{4}\right) = 1, f\left(\dfrac{\pi}{3}\right) = \sqrt{3}, f(\pi) = 0$

17. $(-\infty, -2) \cup (-2, 2) \cup (2, \infty)$

19. $[1, 2) \cup (2, \infty)$ **21.** $(-\infty, \infty)$ **23.** $(0, \infty)$

25. Domain $(-\infty, \infty)$, range $[1, 2]$

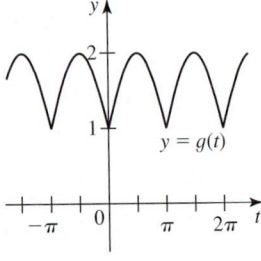

27. Even   **29.** Even   **31. a.** $330°$ **b.** $-450°$ **c.** $-315°$

33. a. $\theta = \pi/3, 5\pi/3$ **b.** $\theta = 5\pi/6, 11\pi/6$

35. a. $\pi/4, \pi/2, 5\pi/4, 3\pi/2$ **b.** $0, 2\pi/3, \pi, 4\pi/3$

37. $|x|, (-\infty, \infty)$

39. $f(x) = x^2, g(x) = \cos x, h(x) = 1 + \sqrt{x+2}$

41. $e^{2/5}$ **43.** 9   **45.** $(\ln 4)^2$ **47.** $\ln \frac{3}{4}$ **49.** $1 + \sqrt{2}$

51. 1, 2   **53.** $\tan 1$ **55.** $\frac{1}{2}\ln(y-2)$

57. $3 \ln x + \frac{1}{2} \ln y - \ln z$ **59.** $\ln \dfrac{x^5}{y^2(x+y)^2}$

61.

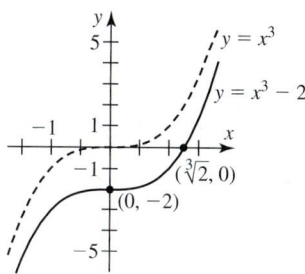

63.

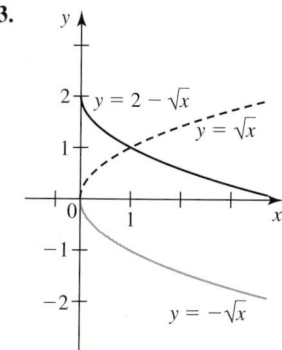

65.

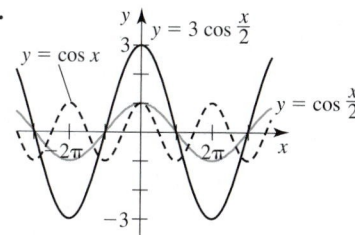

67.

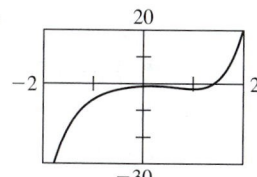

69. $-2.18271, 0.59237, 1.89858$

71. $-1.08659, -1, 1.26456$ **73.** $116\frac{2}{3}$ mg

75. $f(4) = 0, f(5) = 20, f(6) \approx 34, f(7) \approx 48, f(8) = 64,$
$f(9) \approx 80, f(10) \approx 98, f(11) \approx 116, f(12) \approx 136$

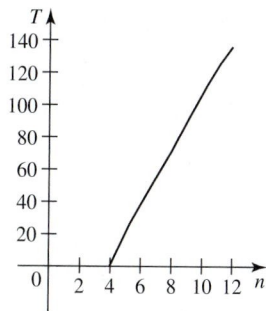

77. a. $f(r) = \pi r^2$ **b.** $g(t) = 2t$

c. $h(t) = f[g(t)] = \pi(2t)^2 = 4\pi t^2$ **d.** 3600π ft^2

79. a.

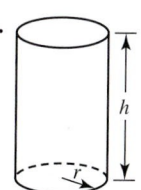

b. $V(x) = 4x^3 - 40x^2 + 100x$ **c.** 64 in.^{3}

81. a.

b. $C(r) = 16\pi r^2 + \dfrac{256\pi}{r}$ **c.** \$603.19   **83.** \$337,653

CHAPTER 1

Exercises 1.1 • page 109

1. a. 2 **b.** -1 **c.** Does not exist   **3. a.** 2 **b.** 2 **c.** 2

5. a. ∞ **b.** 1 **c.** Does not exist

7. a. True **b.** True **c.** False **d.** True **e.** True **f.** False

9.

x	$f(x)$
0.9	-0.90909
0.99	-0.99010
0.999	-0.99900
1.001	-1.00100
1.01	-1.01010
1.1	-1.11111

-1

11.

x	$f(x)$
1.9	0.25158
1.99	0.25016
1.999	0.25002
2.001	0.24998
2.01	0.24984
2.1	0.24846

$\frac{1}{4}$

13.

x	$f(x)$
1.9	-0.06370
1.99	-0.06262
1.999	-0.06251
2.001	-0.06249
2.01	-0.06238
2.1	-0.06135

-0.0625

15.

x	$f(x)$
-0.1	0.95163
-0.01	0.99502
-0.001	0.99950
0.001	1.00050
0.01	1.00502
0.1	1.05171

1

17.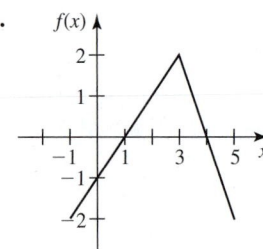

   a. 2 **b.** 2 **c.** 2

19.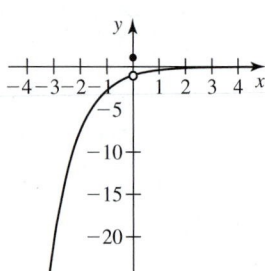

   a. -1 **b.** -1 **c.** -1

21.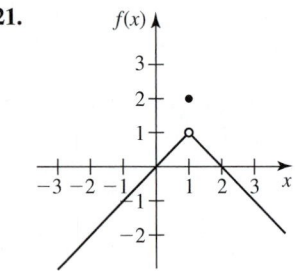

   a. 1 **b.** 1 **c.** 1

23. 2   **25.** -1 **27.** 3

29.

x	$f(x)$
-0.01	0
-0.001	0
-0.0001	0
0	0
0.0001	-0.3056
0.001	0.8269
0.01	-0.5064

Does not exist; 0;
does not exist

31. a.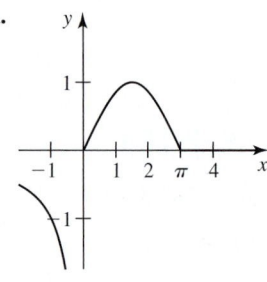

   b. $(-\infty, 0) \cup (0, \infty)$
   c. $(-\infty, 0) \cup (0, \infty)$
   d. $(-\infty, \infty)$

35. 2.7183

37.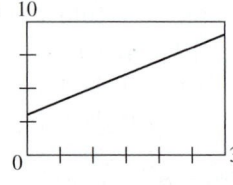

7

39.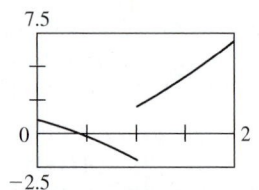

Does not exist

41.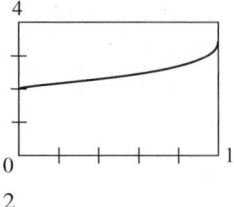

43. False   **45.** False

Exercises 1.2 • page 123

1. 10 **3.** 0 **5.** 1 **7.** $-\frac{1}{3}$ **9.** $4 - 2\sqrt{2}$ **11.** 4
13. $\frac{1}{2}$ **15.** $-\frac{2}{3}$ **17.** 1 **19.** $2\sqrt{2}/\pi$ **21.** 1
23. 16 **25.** 1 **27.** -7 **29.** 11 **31.** 2 **33.** 1
35. 0 **37.** Incorrect **39.** $f(x) = 1$, $g(x) = x$, $a = 0$
41. 4 **43.** ∞ **45.** 2 **47.** $\frac{\sqrt{6}}{2}$ **49.** $-\frac{3}{2}$ **51.** 3
53. $\frac{1}{2}$ **55.** Does not exist **57.** $\sqrt{3}/6$ **59.** $\frac{1}{2}$ **61.** 6
63. 1 **65.** $\frac{1}{3}$ **67.** $\sqrt{2}$ **69.** 0 **71.** $-\frac{1}{2}$
73. $-\sqrt{2}/2$ **75.** 1 **77.** -1

79. a. 12   **b.** 12

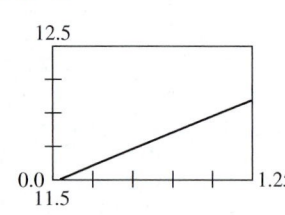

x	$f(x)$
0.99	11.995
0.999	11.999
1	undefined
1.001	12.001
1.01	12.005

  c. 12

81. a. $[0, c)$ **b.** 0   **83.** 0
85. a. 1, 2 **b.** No, $\lim_{x \to -1^-} f(x) \neq \lim_{x \to -1^+} f(x)$
87. 1, 1, yes   **89.** Yes, 1   **93.** No   **95.** No
99. False   **101.** False

Exercises 1.3 • page 133

1. 0.003 **3.** 0.005 **5.** 0.02 **7.** 0.0007
9. 0.01 **31.** False **33.** True

Exercises 1.4 • page 145

1. Nowhere **3.** At ± 1 **5.** At 0 **7.** None
9. 2 **11.** ± 2 **13.** 0, 2 **15.** -2, 0
17. $0, \pm 1, \pm 2, \ldots$ **19.** None **21.** -1 **23.** 0
25. $\pm \pi/4, \pm 3\pi/4, \pm 5\pi/4, \ldots$ **27.** 3
29. $a = \frac{8}{3}$, $b = \frac{4}{3}$ **31.** 2 **33.** Yes **35.** No
37. $(-\infty, \infty)$ **39.** $(-\infty, \infty)$ **41.** $[-3, 3]$
43. $(-3, 0)$ and $(0, 3)$ **45.** $[0, \infty)$ **47.** $(2, \infty)$

49. 5   **51.** $f(0) = -\frac{2}{5}$ **53.** $f(0) = \frac{1}{2}$ **55.** $f(0) = 1$

57. No   **59.** 3   **61.** 3

69. b. 0.57926   **71.** No   **73.** 1.34

75. v_A (m/sec)

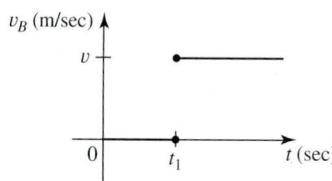

77. c. $\frac{1}{2}$ sec and $\frac{7}{2}$ sec

83. a. No   **b.** No

89.

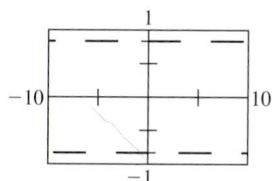

f is continuous at all numbers except $n\pi$, n an integer

97. False   **99.** True

Exercises 1.5 • page 155

1. -0.15 mph per thousand cars, -0.3 mph per thousand cars

3. Rising at 3.08%/hr; falling at 21.15%/hr

5. a. At t_1, the velocity of Car A is greater, but the acceleration of Car B is greater.
   b. At t_2, both cars have the same velocity, but Car B has greater acceleration.

7. a. 0   **b.** 0   **c.** $y = 5$

9. a. $2h + 8$ **b.** 8   **c.** $y = 8x - 9$

11. a. $h^2 + 6h + 12$ **b.** 12   **c.** $y = 12x - 16$

13. a. $-\dfrac{1}{1 + h}$ **b.** -1 **c.** $y = -x + 2$

15. 4   **17.** 13   **19.** -1

21. a. 1.25 ft/sec, 1.125 ft/sec, 1.025 ft/sec, 1.0025 ft/sec, 1.00025 ft/sec
   b. 1 ft/sec

23. a. 48 ft/sec, 56 ft/sec, 62.4 ft/sec
   b. 64 ft/sec
   c. -32 ft/sec, falling
   d. $t = 8$ sec

25. a. 820 ft   **b.** 20.5 ft/sec   **c.** 40.5 ft/sec

27. a. 3π units²/unit   **b.** 4π units²/unit

29. a. $-\$2.005$ per thousand tents, $-\$2.001$ per thousand tents
   b. Decrease of \$2 per thousand tents

31. 30.8 ft/sec

33. a.

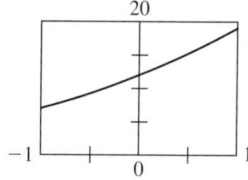

   b. 12

35. $f(x) = x^5$, $a = 1$ **37.** $f(x) = x^2 + \sqrt{x}$, $a = 4$

39. $f(x) = x^4$, $a = 1$ **41.** True   **43.** False

Chapter 1 Concept Review • page 158

1. a. L, f, L, a **b.** right   **c.** exist, L **d.** $\varepsilon > 0, \delta > 0$

3. $\lim_{x \to a} g(x) = L$

5. a. $(-\infty, \infty)$ **b.** its domain   **c.** continuous

7. a. $m_{\tan} = \lim\limits_{h \to 0} \dfrac{f(a + h) - f(a)}{h}$ **b.** $y - f(a) = m_{\tan}(x - a)$

Chapter 1 Review Exercises • page 159

1. a. 0   **b.** 0   **c.** 0

3. 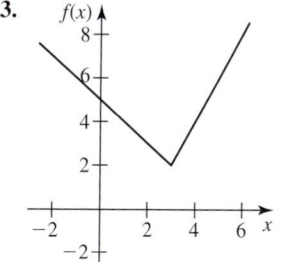   **5.**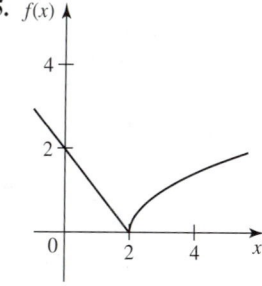

   a. 2   **b.** 2   **c.** 2   **a.** 0   **b.** 0   **c.** 0

7. 34   **9.** $2\sqrt{3}$ **11.** 9   **13.** $\frac{1}{2}$ **15.** $\frac{7}{8}$ **17.** $-\frac{1}{16}$

19. 0   **21.** 6   **23.** ∞ **25.** 0   **27.** 1

31. At b and c **33. a.** Yes   **b.** Yes   **c.** No   **35.** None

37. None   **39.** $n\pi$, n an integer   **41.** $x \le 0$ and 1

43. $\frac{1}{2}$ **49.** No   **53. a.** $[0, c)$ **b.** 0

55. a. 7 ft/sec; 6 ft/sec; 5.2 ft/sec; 5.02 ft/sec   **b.** 5 ft/sec

Chapter 1 Challenge Problems • page 162

1. $\frac{1}{3}$ **3.** $\pi/4$ **7.** 2   **11.** At 0, 1, and $\frac{3}{2}$

CHAPTER 2

2.1 Exercises • page 172

1. $0, (-\infty, \infty)$ **3.** $3, (-\infty, \infty)$ **5.** $6x - 1, (-\infty, \infty)$

7. $6x^2 + 1, (-\infty, \infty)$ **9.** $\dfrac{1}{2\sqrt{x + 1}}, (-1, \infty)$

11. $-\dfrac{1}{(x + 2)^2}, (-\infty, -2) \cup (-2, \infty)$

13. $-\dfrac{6}{(2x + 1)^2}, \left(-\infty, -\tfrac{1}{2}\right) \cup \left(-\tfrac{1}{2}, \infty\right)$

15. $y = 4x - 3$ **17.** $y = 6x - 4$ **19.** $y = \dfrac{\sqrt{3}}{6}x + \dfrac{\sqrt{3}}{3}$

21. a. $y = -x + 2$

b.

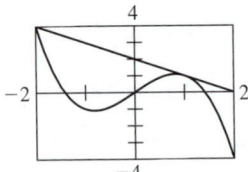

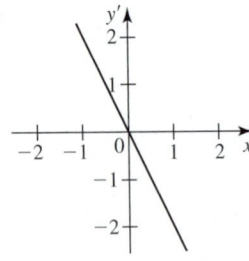

23. -3 **25.** $\tfrac{1}{2}$ **27.** (c)   **29.** (b)

31.

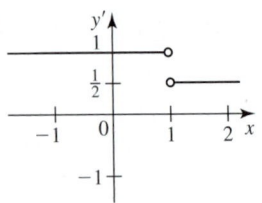

33.

35.

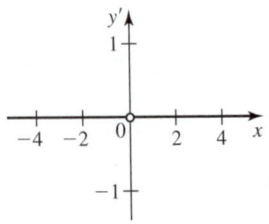

37. a. $f'(h)$ is measured in degrees Fahrenheit per foot and gives the instantaneous rate of change of the temperature at a given height h.

b. Negative

c. $-0.05°F$

39. a. $C'(x)$, measured in dollars per unit, gives the instantaneous rate of change of the total manufacturing cost C when x units of a certain product are produced.

b. Positive

c. $\$20$

41. a. $2x - 2$ **b.** $(1, 0)$

c.

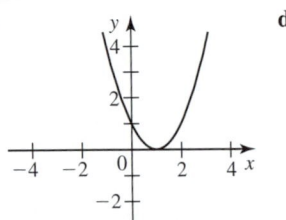

d. 0

43. 0 **45.** ± 2 **47.** ± 2

53.

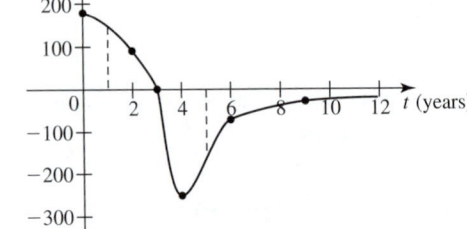

$\$150$ million per year, $-\$160$ million per year

55. a. The average rate of change of $f(x) = x^3$ over the interval $[x, x + h]$

b. $f'(x) = 3x^2$

c.

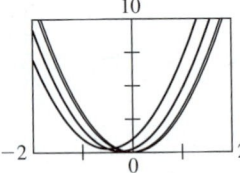

57. a.

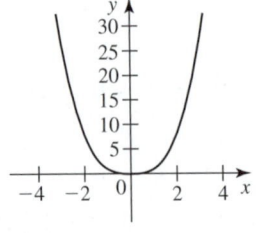

b. $x \in (-\infty, \infty)$

c. $f'(x) = \begin{cases} -3x^2 & \text{if } x < 0 \\ 3x^2 & \text{if } x \geq 0 \end{cases}$

61. a. 0 **b.**

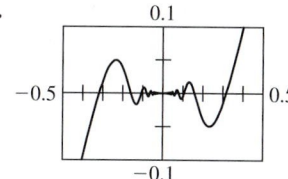

65. True **67.** False **69.** False

2.2 Exercises • page 183

1. 0 **3.** $6x$ **5.** $2.1x^{1.1}$ **7.** $\dfrac{3}{2\sqrt{x}} + 2e^x$ **9.** $-\dfrac{84}{x^{13}}$

11. $2x - 2$ **13.** $2\pi r + 2\pi$ **15.** $0.06x - 0.4$

17. $10x^4 - 12x^3 + 3x^2 + 8x$ **19.** $2x - 4 - \dfrac{3}{x^2}$

21. $16x^3 - \frac{15}{2}x^{3/2}$ **23.** $-\dfrac{3}{x^2} - \dfrac{8}{x^3}$

25. $-\dfrac{16}{t^5} + \dfrac{9}{t^4} - \dfrac{2}{t^2}$ **27.** $0.002x - 0.4 - \dfrac{200}{x^2}$

29. $2 - \dfrac{5}{2\sqrt{x}} + e^{x+1}$ **31.** $\dfrac{1}{3x^{2/3}} - \dfrac{1}{2x^{3/2}}$

33. a. 20 **b.** -4 **c.** 20

35. a. $y = 5x - 4$ **b.**

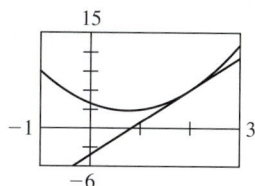

37. a. $y = -x + 1$ **b.**

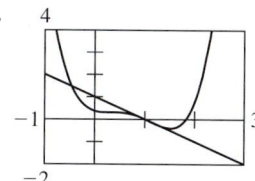

39. $(-2, 10)$ and $(1, -17)$ **41.** $\left(-\frac{1}{2}, -3\right)$ and $\left(\frac{1}{2}, 3\right)$

43. a. $y = 9x - 15$, $y = -\frac{1}{9}x + \frac{29}{9}$

b.

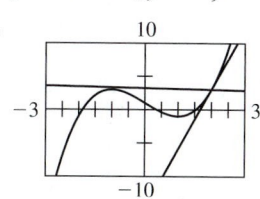

45. ± 3

47. a. $\left(1, -\frac{13}{12}\right)$ and $(0, 0)$

b. $(0, 0)$, $\left(2, -\frac{8}{3}\right)$, and $\left(-1, -\frac{5}{12}\right)$

c. $(0, 0)$, $\left(4, \frac{80}{3}\right)$, and $\left(-3, \frac{81}{4}\right)$

49. $\left(-1, \frac{20}{3}\right)$ and $\left(1, \frac{10}{3}\right)$ **51.** $y = 2x - 4$ and $y = 6x - 16$

53. 3 **55.** 11 **57.** $2x(7x^5 - 1)$ **59.** 1 P.M.

61. a. 1.94% per decade, 2.48% per decade

b. 3.87%, 6.08%

63. a. $3.6t^2 - 10.62t + 80$

b. $-\$3.42$ per person per year, $\$32.58$ per person per year

65. a.

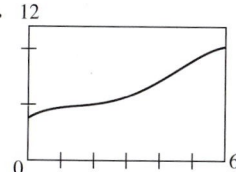

b. 0.3835, 1.0489, 1.7311

c. The number of Alzheimer's patients will be increasing at the rate of approximately 0.3835 million patients per decade at the beginning of 2010, 1.0489 million patients per decade in 2020, and 1.7311 million patients per decade in 2030.

67. a. $-0.0123t^4 + 0.2639t^3 - 2.2601t^2 + 5.6608t + 143.6$

b.

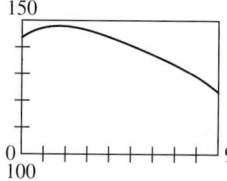

c. 5.6608, -0.6064, -6.7602

69. $\dfrac{3\pi}{R}\sqrt{\dfrac{r}{g}}$ **71.** $A = \frac{1}{4}$, $B = \frac{1}{2}$, $C = \frac{1}{4}$

75. False **77.** True

2.3 Exercises • page 193

1. $xe^x(x + 2)$ **3.** $\dfrac{(2t^2 + 7t + 2)e^t}{2\sqrt{t}}$ **5.** $2e^x(e^x + 1)$

7. $-\dfrac{1}{(x - 1)^2}$ **9.** $-\dfrac{7}{(3x - 2)^2}$ **11.** $\dfrac{1 - x^2 e^x}{(1 + xe^x)^2}$

13. $3x^2 + 2x - 1$ **15.** $2x + xe^x + e^x - 4e^{2x}$

17. $\dfrac{-3x^2 + 1}{\sqrt{x}(x^2 + 1)^2}$ **19.** $\dfrac{2x(x + 2)}{(x^2 + x + 1)^2}$

21. $\dfrac{(x^2 - x - 5)e^x}{(x - 2)^2}$ **23.** $-\dfrac{x^2 + 2x + 2}{x^2(x + 2)^2}$

25. $-\dfrac{3\sqrt{3}x + 2\sqrt{x} + \sqrt{3}}{2\sqrt{x}(3x - 1)^2}$ **27.** $\dfrac{ad - bc}{(cx + d)^2}$

29. $\dfrac{1 + e^x + x^2e^x + e^{2x}}{(1 - xe^x)^2}$ **31.** 10

33. 29   **35.** $\left(1, \dfrac{1}{e}\right)$ **37.** $(1, e + 1)$

39. a. $y = \frac{1}{4} ex + \frac{1}{4} e$

b.

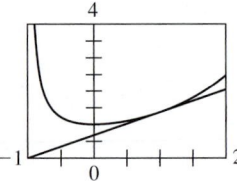

41. a. $y = -\frac{13}{3} x - \frac{14}{3}$

b.

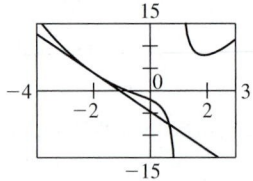

43. a. $y = 7x - 5, y = -\frac{1}{7}x + \frac{15}{7}$

b.

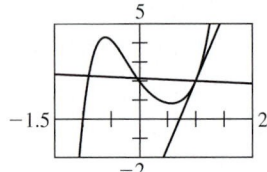

45. 8   **47.** -9 **49.** -2 **51.** $56x^6 - 12x^2 + 4$

53. $\dfrac{e^x(x^2 - 2x + 2)}{x^3}$ **55.** $6x - 4$

57. $\dfrac{\sqrt{x}(4x^2 - 20x + 15)}{4e^x}$ **59. a.** 44   **b.** 10

61. $f'(x) = 8x^3 - 8x, f''(x) = 24x^2 - 8, f'''(x) = 48x,$
$f^{(4)}(x) = 48, f^{(n)}(x) = 0$ for $n \geq 5$

63. 0.0375 part per million per year, 0.006 part per million per year

65. a. 120

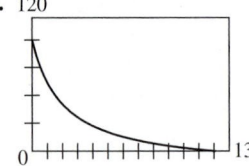

b. $f'(0) \approx -57.5266, f'(2) = -14.6165$

67. h **69.** $f''(0) = 0$, yes   **71.** False   **73.** True

75. True

2.4 Exercises • page 204

1. 7.44 ft, 7.44 ft/sec, 7.44 ft/sec

3. -2 ft, -8 ft/sec, 8 ft/sec

5. $\frac{4}{5}$ ft, $-\frac{6}{25}$ ft/sec, $\frac{6}{25}$ ft/sec

7. 0 ft, 0 ft/sec, 0 ft/sec

9. a. Never

b. Always positive

c.

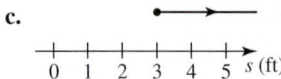

11. a. $s(1) = 9$ ft

b. Positive when $0 < t < 1$, negative when $t > 1$

c.

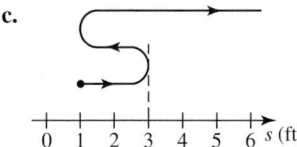

13. a. $s(0) = 1$ ft, $s(1) = 3$ ft, $s(2) = 1$ ft

b. Positive when $0 < t < 1$ and when $t > 2$, negative when $1 < t < 2$

c.

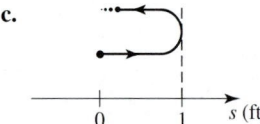

15. a. $s(1) = 1$ ft

b. Positive when $0 < t < 1$, negative when $t > 1$

c.

17. a. $6(2t - 3)$

b. $a\left(\frac{3}{2}\right) = 0, a(t) < 0$ if $0 < t < \frac{3}{2}, a(t) > 0$ if $t > \frac{3}{2}$

19. a. $\dfrac{4t(t^2 - 3)}{(t^2 + 1)^3}$

b. $a(0) = a(\sqrt{3}) = 0, a(t) < 0$ if $0 < t < \sqrt{3}, a(t) > 0$ if $t > \sqrt{3}$

21. a. 100 (kg · m)/sec   **b.** 250 J

23. a. Ascending at $t = t_0$, stationary at $t = t_1$, descending at $t = t_2$

b. Positive at $t = t_0$, 0 at $t = t_1$, positive at $t = t_2$

25. a. 1.65 sec   **b.** -14.14 m/sec

27. a. $\frac{1}{16} t^3 - \frac{3}{2} t^2 + 8t$ **b.** $v(0) = 0, v(8) = 0, v(16) = 0$

c. 64 ft

29. a. 60°   **b.** $(200\sqrt{3}, 300)$ **c.** 300 ft

d. $400\sqrt{3}$ ft   **e.** 120°

31. a. 3,000   **b.** 10.6°

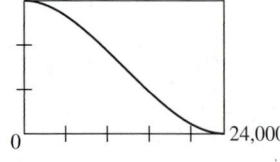

33. 22 ft/sec^2, 0.88 ft/sec^2 **35.** \$120, \$120.06

37. a. $10{,}000 - 200x$ **b.** 200, 0, -200

39. a. 2.38 cm **b.** 0.00227 cm/cm

2.5 Exercises • page 213

1. $-4 \sin x - 2$ **3.** $(\sec t)(3 \sec t - 4 \tan t)$

5. $e^u(\cot u - \csc^2 u)$ **7.** $\cos 2x$ **9.** $-\sin \theta$

11. $e^{-x}(\cos x - \sin x)$ **13.** $\dfrac{1 + \sec x - x \sec x \tan x}{(1 + \sec x)^2}$

15. $\dfrac{e^{2x} \sec x(\tan x + 1) - \csc^2 x - \cot x}{e^x}$

17. $\dfrac{\cos x - \sin x - 1}{(1 - \cos x)^2}$ **19.** $-\dfrac{2}{(\sin \theta - \cos \theta)^2}$

21. $e^x \sin x(2 \cos x + \sin x)$ **23.** $2e^x \cos x$

25. $-5 \cos x + x \sin x$ **27.** $-\dfrac{4x^2 \cos x + 4x \sin x + \cos x}{4x\sqrt{x}}$

29. a. $y = \dfrac{\sqrt{3}}{2}x + \dfrac{6 - \sqrt{3}\pi}{12}$ **b.**

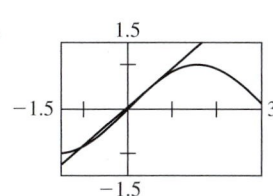

31. a. $y = 2\sqrt{3}x + \dfrac{6 - 2\sqrt{3}\pi}{3}$

b.

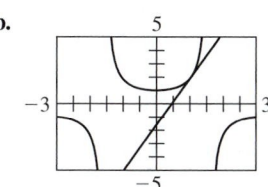

33. $\dfrac{\sqrt{2}\pi(8 + \pi)}{16}$ **35.** -1 **37.** $2k\pi, k = 0, \pm 1, \pm 2, \ldots$

39. $\dfrac{(2k + 1)\pi}{2}, k = 0, \pm 1, \pm 2, \ldots$

41. $f'(x) = \cos x, f''(x) = -\sin x, f'''(x) = -\cos x,$
$f^{(4)}(x) = \sin x, \ldots$

43. 2 ft, -3 ft/sec, 3 ft/sec, -2 ft/sec^2

45. $-\csc x \cot x$ **51.** True

2.6 Exercises • page 226

1. $6(2x + 4)^2$ **3.** $-\dfrac{2x}{3(x^2 + 1)^{4/3}}$

5. $\dfrac{e^x - \sin x}{2\sqrt{e^x + \cos x}}$ **7.** $10(2x + 1)^4$

9. $(2x - 1)e^{x^2-x}$ **11.** $6\left(t + \dfrac{2}{t}\right)^5\left(1 - \dfrac{2}{t^2}\right)$

13. $u^2(2u^2 - 1)^3(22u^2 - 3)$ **15.** $2xe^{-2x}(1 - x)$

17. $\dfrac{1 - 2u^2}{\sqrt{1 - u^2}}$ **19.** $-\dfrac{20t^2 + 40t + 9}{t^4(1 + 2t)^4}$

21. $\dfrac{5s(1 + \sqrt{1 + s^2})^4}{\sqrt{1 + s^2}}$ **23.** $\dfrac{e^x(2e^x + 3)}{(1 + e^{-x})^2}$ **25.** $3 \cos 3x$

27. $\pi \sec^2(\pi t - 1)$ **29.** $3 \cos x \sin^2 x$

31. $2 \cos 2x + \dfrac{\sec^2 \sqrt{x}}{2\sqrt{x}}$ **33.** $3 \sin x \cos x(\sin x - \cos x)$

35. $\dfrac{4 \sin 3x \cos 3x}{\sqrt[3]{1 + \sin^2 3x}}$ **37.** $-\dfrac{3(2x - \pi \sec \pi x \tan \pi x)}{(x^2 - \sec \pi x)^4}$

39. $-\dfrac{\sin x}{\sqrt{1 + 2 \cos x}}$ **41.** $-\dfrac{6 \sin 3x}{(1 - \cos 3x)^2}$

43. $-e^{\cos x} \sin x$ **45.** $\dfrac{\cos 2x + \sin 2x}{\sqrt{\sin 2x - \cos 2x}}$

47. $\dfrac{4}{(1 - x)^2} \sin\left(\dfrac{1 + x}{1 - x}\right) \cos\left(\dfrac{1 + x}{1 - x}\right)$

49. $-\dfrac{2(1 + x^2) \sin 2x + x \cos 2x}{(1 + x^2)^{3/2}}$

51. $\sec^2 x \, (3 \sec^2 3x + 2 \tan x \tan 3x)$ **53.** $\cos x \cos(\sin x)$

55. $-3\pi \cos \pi x \cos^2(\sin \pi x) \sin(\sin \pi x)$

57. $[1 + (3 \ln 5)x]5^{3x}$ **59.** $6(2^x + 3^{-x})^5(2^x \ln 2 - 3^{-x} \ln 3)$

61. $\left(\dfrac{e}{x} + 1\right)x^e e^x$ **63.** $-\ln 2(\csc^2 x)2^{\cot x}$

65. $48x(6x^2 - 1)(2x^2 - 1)^2$

67. $2(\cos 2t + 2t^2 \sin t^2 - \cos t^2)$

69. a. $y = \dfrac{3\sqrt{2}}{2}x - \dfrac{\sqrt{2}}{2}$ **b.**

71. a. $y = \dfrac{1}{e}$ **b.**

73. 300   **75.** No   **77.** $y = (\ln 2)x + 2$

79. $-4, \frac{1}{2}$, does not exist

81. $a \cdot \cos x \cdot f'(\sin x) - b \cdot \sin x \cdot g'(\cos x)$

83. $2x[f'(x^2 + 1) + g'(x^2 - 1)]$ **85.** -4

87. a. $f'(x) = \begin{cases} -\dfrac{2}{(2 - x^2)^{3/2}} & \text{if } x < 0 \\[2mm] \dfrac{2}{(2 - x^2)^{3/2}} & \text{if } x > 0 \end{cases}$

$y = -2x - 1, \; y = 2x - 1$

b.

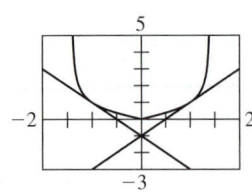

89. 0.6% per year, 0.4% per year, 25.9%

91. a. 0.83% per decade **b.** -0.18% per decade per decade

93. a. 80 mg **b.** $-\dfrac{4\ln 2}{7} \cdot 2^{-t/140}$ mg/day

95. $\frac{3}{4}$ ft, -1 ft/sec, 1 ft/sec, -3 ft/sec^2

97. $0 per share per day, $28 per share

99. $\dfrac{6u_0\sigma^6}{r^7}\left[2\left(\dfrac{\sigma}{r}\right)^6 - 1\right]$ **101.** $\sqrt{3}/3$ ft/sec, $-10\sqrt{3}/9$ ft/sec^2

103. a. 1.94 sec **b.** 7.57 m/sec

105. $\dfrac{k_1(EC - q_0)}{C(k_1 + k_2 t)^2}\left(\dfrac{k_1}{k_1 + k_2 t}\right)^{(1 - Ck_2)/(Ck_2)}$

107. $-\dfrac{\sigma}{2\varepsilon_0}\left(\dfrac{r}{\sqrt{r^2 + R^2}} - 1\right)$

109. b. 6.1 ft/sec, 12.3 ft/sec **c.** ∞

d. 100

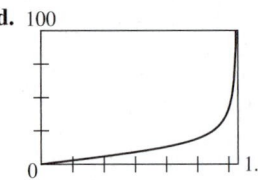

111. $\left(2 - \dfrac{1}{x^2}\right)\sin\dfrac{1}{x} - \dfrac{2}{x}\cos\dfrac{1}{x}$ $(x \neq 0)$, no

113. $\dfrac{x + 1}{|x + 1|}, \; x \neq -1$ **115.** $\dfrac{\sin 2x}{2|\sin x|}, \; x \neq k\pi, \; k$ an integer

117. a. $\dfrac{(2x - 3)(x - 1)(x - 2)}{2|x^2 - 3x + 2|^{3/2}}, \; x \neq 1, 2$

b.

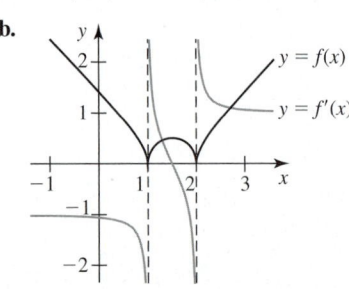

119. False   **121.** True

2.7 Exercises • page 240

1. $-\dfrac{2x}{y}$ **3.** $-\dfrac{y(y + 2x)}{x(2y + x)}$ **5.** $\dfrac{3x^2 - 1}{6y^2 + 1}$ **7.** $-\dfrac{y^2}{x^2}$

9. $\dfrac{x^4 + 2x^2y^2 + x^2y - y^3 + y^4}{x(x^2 - y^2)}$ or $\dfrac{3x^2 + 2x + y^2 - y}{x - 2xy - 2y}$

11. $\dfrac{x + 1}{2 - y}$ **13.** $-\sqrt{y/x}$ **15.** $\dfrac{\cos(x + y)}{2y - \cos(x + y)}$

17. $\dfrac{y - 6x^2\tan(x^3 + y^3)\sec^2(x^3 + y^3)}{6y^2\tan(x^3 + y^3)\sec^2(x^3 + y^3) - x}$

19. $\dfrac{-y\sqrt{1 + \cos^2 y}}{\cos y \sin y + x\sqrt{1 + \cos^2 y}}$ **21.** $\dfrac{3x^2 - e^{2y}}{2xe^{2y} + 2}$

23. $y = \dfrac{\sqrt{3}}{6}x - \dfrac{2\sqrt{3}}{3}$ **25.** $y = x - 2$ **27.** 1

29. $\sqrt{3}/3$ **31.** $y = -\dfrac{1}{e}x + 1$ **33.** $2y/x^2$

35. $-\dfrac{\sin x \sin^2 y + \cos y \cos^2 x}{\sin^3 y}$ **37. a.** $\dfrac{1}{2\sqrt{x}}$ **b.** $\dfrac{1}{2\sqrt{x}}$

39. b. $\frac{1}{2}$ **41. b.** 1   **43. b.** $\frac{1}{36}$ **45. b.** $-\frac{25}{4}$ **47.** $\frac{1}{3}$

49. $\dfrac{3}{\sqrt{1 - 9x^2}}$ **51.** $\dfrac{2x}{1 + x^4}$ **53.** $\tan^{-1} 3t + \dfrac{3t}{1 + 9t^2}$

55. $\dfrac{1}{|u|\sqrt{4u^2 - 1}}$ **57.** $-\dfrac{1}{\sqrt{1 - x^2}}$ **59.** $\dfrac{1 - x}{1 + x^2} + \cot^{-1} x$

61. $2x\tan^{-1} x + 1$ **63.** $\dfrac{1}{t^2 + 1}$ **65.** $\dfrac{2\cos 2x}{1 + \sin^2 2x}$

67. $\dfrac{2e^{2x}}{\sqrt{1 - e^{4x}}}$ **69.** $\dfrac{2x}{(1 - x^4)^{3/2}}$

71. $-\dfrac{1}{|\theta|(\sec^{-1}\theta)^2\sqrt{\theta^2 - 1}}$ **73.** $y = \dfrac{\sqrt{3}}{2}x + 2\sqrt{3}$

75. $y = 2x - \frac{1}{2}$ **77.** $y = -\frac{9}{13}x + \frac{40}{13}$

79. $y = \dfrac{\pi + 2\sqrt{3}}{6}x - \dfrac{\sqrt{3}}{6}$

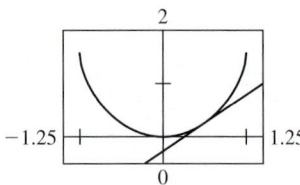

81. a. $y = -\frac{1}{4}x + \frac{3}{2}, \; y = 4x - \frac{45}{4}$

b.

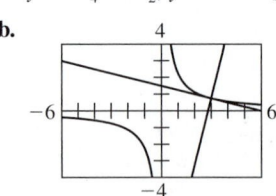

83. a. $y = \frac{15}{2}x + 18$, $y = -\frac{2}{15}x + \frac{41}{15}$

b.
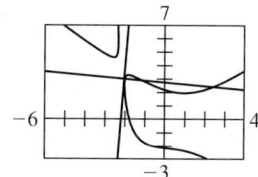

85. a. $\dfrac{y - x^2}{y^2 - x}$ **b.** $y = -x + 3$ **c.** $(0,0), \left(\sqrt[3]{2}, \sqrt[3]{4}\right)$

87. $\frac{3}{25}$ ft/sec

93. b.
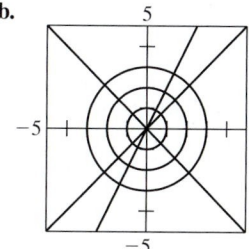

95. b.
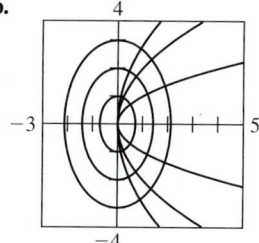

97. b.
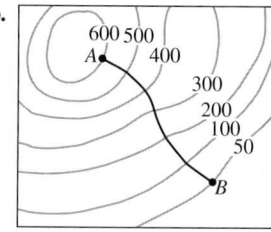

99. 17.9°/sec   **103.** True

2.8 Exercises • page 250

1. $\dfrac{2}{2x + 3}$ **3.** $\dfrac{1}{2x}$ **5.** $\dfrac{1}{u(u + 1)}$

7. $(\ln x)^2 + 2\ln x$ **9.** $\dfrac{x(1 - \ln x) + 1}{x(x + 1)^2}$ **11.** $\dfrac{1}{x \ln x}$

13. $\dfrac{\ln x + 1}{x \ln x}$ **15.** $\dfrac{\cos(\ln x)}{x}$ **17.** $2x \ln \cos x - x^2 \tan x$

19. $\tan u$ **21.** $\dfrac{2\cos t + \sin t + 1}{(\sin t + 1)(\cos t + 2)}$

23. $\dfrac{2x + 1}{(x^2 + x + 1)\ln 2}$ **25.** $\dfrac{t}{(t^2 + 1)\ln 10}$ **27.** $y = x - 1$

29. $2(2x + 1)(3x^2 - 4)^2(24x^2 + 9x - 8)$

31. $-\dfrac{x^2 - 2x - 1}{3(x - 1)^{2/3}(x^2 + 1)^{4/3}}$ **33.** $[x(\ln x + 1)^2 + 1]x^{x-1}$

35. $3^x \ln 3$ **37.** $\left[\dfrac{1}{x(x + 2)} - \dfrac{\ln(x + 2)}{x^2}\right](x + 2)^{1/x}$

39. $\dfrac{\cos x \ln(\cos x) - x \sin x}{2 \cos x}(\sqrt{\cos x})^x$

41. $y(\ln x + 1)$ **43.** $\dfrac{x + y}{x - y}$

45. a. $\dfrac{ab}{M + m - at} - g$ **b.** $-b \ln \dfrac{M}{M + m} - \dfrac{gm}{a}$; $\dfrac{ab}{M} - g$

47. 0.0580%/kg; 0.0133%/kg

51. a. $0.3307\, p_0$ **b.** $-0.0000455\, p_0$ atm/m

53. False   **55.** False

2.9 Exercises • page 257

1. $\frac{3}{2}$ **3.** -6 **5.** 1

7. a. $\dfrac{dV}{dt} = 3x^2 \dfrac{dx}{dt}$ **b.** 150 in.³/sec   **9.** -3 units/sec

11. a. $-2 < x < 2$ **b.** $x = \pm 2$ **c.** $x < -2$ or $x > 2$

13. 188.5 ft²/sec   **17.** 19.2 ft/sec   **19.** 29 packs per week

21. -6.96 ft/sec   **23.** 10 ft/sec   **25.** 20.1 in.³/sec

27. 17 ft/sec   **29.** -23 km/h   **31.** $\frac{1}{7}$ L/sec

33. 9.1×10^{-32} kg/sec   **35.** 196.8 ft/sec

37. a. $x^2 - (6\cos\theta)x - 40 = 0$ **b.** -1205.5 m/sec

39. -80.15 ft/sec   **41.** 0.04 in.³/sec

2.10 Exercises • page 270

1. a. 0.02, 0.0804 **b.** 0.08 **c.** 0.0004

3. a. 0.1, 0.033150 **b.** 0.033333 **c.** -0.000183

5. dx **7.** $-dx$ **9.** $\frac{57}{4}dx$ **11.** $-\dfrac{\sqrt{2}}{2}dx$ **13.** $-dx$

15. $2\,dx$ **17.** $2\,dx$ **19.** $7x - 4$ **21.** $x - 1$

23. 1.975, 2.025

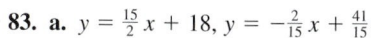

25. 1.006   **27.** 1.9885   **29.** 6%   **31.** $2000

33. -40% **35.** 3%   **37.** 15%   **39.** 1.75 ft

41. 18.1%   **43.** 0.68 hr   **45.** $\phi = 45°$ **47.** 2.78 kg

49. True   **51.** False

Chapter 2 Concept Review • page 273

1. a. $f'(x) = \lim\limits_{h \to 0} \dfrac{f(x + h) - f(x)}{h}$ **b.** limit

   c. tangent line; $(a, f(a))$ **d.** $f(x)$; x; $(a, f(a))$

   e. $y = f'(a)(x - a) + f(a)$

3. a. 0   **b.** nx^{n-1}

5. dy/dx **7.** $C'; R'; P'; \overline{C}'$

9. a. $n[f(x)]^{n-1} \cdot f'(x)$

   b. $\cos f(x) \cdot f'(x); \; -\sin f(x) \cdot f'(x); \; \sec^2 f(x) \cdot f'(x);$
   $\sec f(x) \tan f(x) \cdot f'(x); \; -\csc f(x) \cot f(x) \cdot f'(x);$
   $-\csc^2 f(x) \cdot f'(x)$

11. $\dfrac{1}{u \ln a} \dfrac{du}{dx}$ **13.** $y; \; dy/dt; \; a$

15. a. $x_2 - x_1$ **b.** $f(x + \Delta x) - f(x)$

Chapter 2 Review Exercises • page 274

1. $2x - 2$

3.

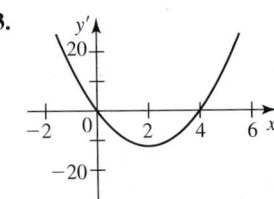

5. a. The rate of change of the amount of money on deposit with respect to the interest rate, measured in dollars per unit change in interest.

   b. Positive   **c.** \$607.75

7. $f(x) = 3x^{3/2}, \; a = 4$ **9.** $2x^5 - 8x^3 + 2x$

11. $4t + \dfrac{4}{t^2} - \dfrac{1}{t\sqrt{t}}$ **13.** $\dfrac{3}{(2t + 1)^2}$ **15.** $\dfrac{1 - 3u^2}{2\sqrt{u}(u^2 + 1)^2}$

17. $-\sin\theta - 2\cos\theta$ **19.** $(1 - x^2)\sin x + 3x\cos x$

21. $\dfrac{\cos t - t\sin t + \sin t - t\sin t \tan t - t\sec t}{(1 + \tan t)^2}$

23. $-\frac{3}{2}(3t^2 + 2)(t^3 + 2t + 1)^{-5/2}$

25. $\frac{1}{2}\sqrt{s^3 + s + 1}(11s^3 + 5s + 2)$

27. $\dfrac{t + 2}{(\sqrt{t} + 1)^3}$ **29.** $-2\sin(2x + 1)$

31. $2x + \dfrac{2x\cos 2x - \sin 2x}{x^2}$ **33.** $-\dfrac{2}{x^2}\sec^2\dfrac{2}{x}$

35. $-3\cot^2 x \csc^2 x$ **37.** $-\dfrac{\theta\sin\theta + 2\cos\theta}{\theta^3}$

39. $\dfrac{1}{2(x + 1)}$ **41.** $\dfrac{\ln x + 2}{2\sqrt{x}}$ **43.** $5e^{-x}(\cos 2x - \sin 2x)$

45. $-\dfrac{y(y + x\ln y)}{x(x + y\ln x)}$ **47.** $\dfrac{2(1 - x)}{x}$ **49.** $\dfrac{2e^x + 3}{2(1 + e^{-x})^{3/2}}$

51. $-\csc x \cot x \cdot e^{\csc x}$ **53.** $e^{e^x + x}$ **55.** $\dfrac{ye^{-x} - e^{y^2}}{e^{-x} + 2xye^{y^2}}$

57. $3^{x\cot x}(\cot x - x\csc^2 x)\ln 3$ **59.** $\sec^{-1} x + \dfrac{x}{|x|\sqrt{x^2 - 1}}$

61. $\dfrac{x}{(x^2 + 2)\sqrt{x^2 + 1}}$ **63.** $\dfrac{-1}{2\sqrt{x - x^2}\,[1 + (\cos^{-1}\sqrt{x})^2]}$

65. $-\frac{47}{1156}$ **67.** $6x + 2 - \dfrac{2}{x^3}$ **69.** $\dfrac{3x - 2}{(2x - 1)^{3/2}}$

71. $-2\cos 2x$ **73.** $\dfrac{1}{2}\csc^2\dfrac{\theta}{2}\cot\dfrac{\theta}{2}$

75. $-2(\csc^2 t)(1 - t\cot t)$ **77.** $-\frac{1}{27}$ **79.** 10

81. $\dfrac{1}{3}\left[\dfrac{f(x)}{g(x)}\right]^{-2/3}\dfrac{g(x)f'(x) - f(x)g'(x)}{[g(x)]^2}$ **83.** $\dfrac{3x}{2y}$ **85.** $-\dfrac{y^3}{x^3}$

87. $-\dfrac{2x^2 + 2xy + y^2}{x^2 + 2xy + 2y^2}$ **89.** $\dfrac{\sin y - \sin(x + y)}{\sin(x + y) - x\cos y}$

91. $-\dfrac{y}{x}$ **93.** $10x^4 - 3x^2$ **95.** $\frac{3}{16}\, dx$

97. $\frac{7}{3}\, dx$ **99.** $\dfrac{6 - \sqrt{3}\pi}{12}\, dx$ **101.** $y = 2e$

103. $y = -x + 2, \; y = x$ **105.** $-\dfrac{2x}{y^5}$

107. $-\dfrac{\sqrt{3}}{2}x + \dfrac{1}{12}(\sqrt{3}\pi + 9)$ **109. a.** $(1, 2)$ **b.** $y = 2x$

111. 0   **113.** $\frac{20}{3}t^{-1/3} - \frac{5}{3}t^{2/3}, \; -\frac{20}{9}t^{-4/3} - \frac{10}{9}t^{-1/3}$

115. 30 cm/sec, -200 cm/sec/cm

117. 0.0107 m²/kg   **119.** 0.05164

121. a. $\dfrac{\pi r^2}{3}$ **b.** $\dfrac{2\pi rh}{3}$ **125.** $\frac{15}{4}$ **127.** 7.7 ft/sec

Chapter 2 Challenge Problems • page 279

1. 64

3. b. $(-1)^n n!\left[\dfrac{1}{(x + 2)^{n+1}} + \dfrac{1}{(x - 1)^{n+1}}\right]$

5. $\dfrac{1 \cdot 3 \cdot 5 \cdot \cdots \cdot 17}{2^{10}}(1 - x)^{-21/2}(39 - x)$

9. $\dfrac{xf'(\sqrt{1 + x^2})}{\sqrt{1 + x^2}}$ **13.** -10 **17.** No

CHAPTER 3

Exercises 3.1 • page 291

Abbreviations: abs. max., absolute maximum; abs. min., absolute minimum.

1. Abs. max. $f(1) = 3$

3. Abs. max. $f(2n) = 1$, abs. min. $f(2n + 1) = 0$, n, an integer

5. Abs. max. $f(1) = 37$, abs. min. $f(5) = -5$

7.

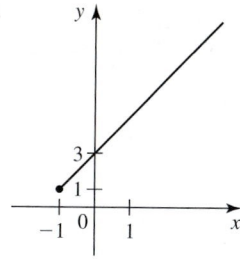

Abs. min. $f(-1) = 1$

9.

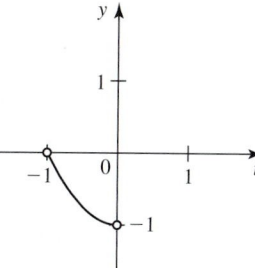

None

11.

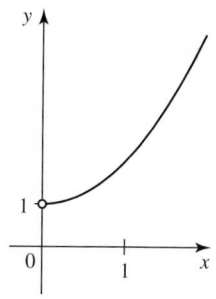

None

13.

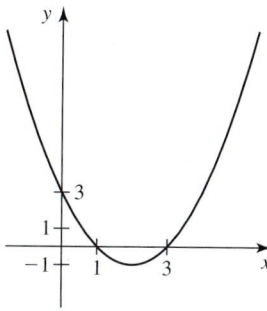

Abs. min. $f(2) = -1$

15.

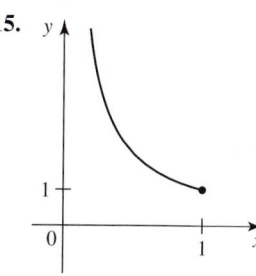

Abs. min. $f(1) = 1$

17.

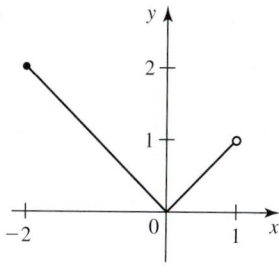

Abs. max. $f(-2) = 2$,
abs. min. $f(0) = 0$

19.

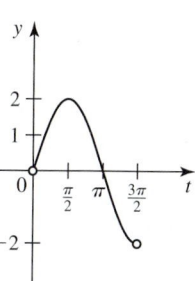

Abs. max. $f\left(\frac{\pi}{2}\right) = 2$

21.

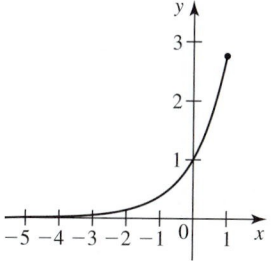

Abs. max. $f(1) = e$

23.

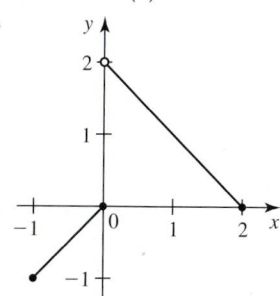

Abs. min. $f(-1) = -1$

25. None   **27.** -1 **29.** $\pm\sqrt{2}$ **31.** $0, 3$ **33.** 0

35. ± 1 **37.** $n\pi/4, n = 0, \pm 1, \pm 2, \ldots$ **39.** 0

41. Abs. max. $f(2) = 0$, abs. min. $f\left(\frac{1}{2}\right) = -\frac{9}{4}$

43. Abs. max. $h(2) = 21$, abs. min. $h(-3) = h(0) = 1$

45. Abs. max. $g(-2) = 17$, abs. min. $g(-1) = 0$

47. Abs. max. $f(1) = \frac{1}{2}$, abs. min. $f(-1) = -\frac{1}{2}$

49. Abs. max. $g(2) = 2$, abs. min. $g(4) = \frac{4}{3}$

51. Abs. max. $f(9) = 3$, abs. min. $f(1) = -1$

53. Abs. max. $f(0) = f(2) = 0$, abs. min. $f(\pm 1) = -3$

55. Abs. max. $f\left(\frac{\pi}{4}\right) = 5$, abs. min. $f(0) = f\left(\frac{\pi}{2}\right) = 2$

57. Abs. max. $f(1) = 1/e$, abs. min. $f(-1) = -e$

59. Abs. max. $f(2) = -0.61$, abs. min. $f(1) = -1$

61. 1667 dozen

63. Highest at 6 A.M. and 10 A.M., lowest at 7 A.M.

67. 10,000   **71.** 7.4 lb   **73.** $52.79/ft^2

75. a. 40

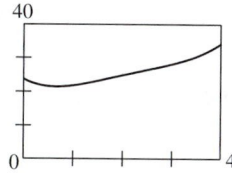

b. 21.5%

77. Halfway up the side of the cylinder

79. $\approx$93 ft   **81.** $0.85a, 0.53a$

83. a. $\tan^{-1}\mu$ **b.** $\dfrac{\mu W}{\sqrt{\mu^2 + 1}}$ lb   **c.** 60

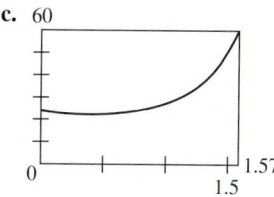

85. No   **89.** No abs. max., abs. min. 0

91.

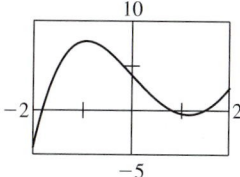

Abs. max. $f(-0.9) \approx 7.8$, abs. min. $f(-2) = -4.16$

93.

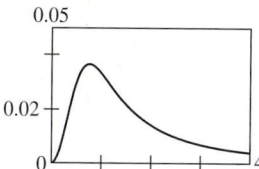

Abs. max. $f(0.8) \approx 0.037$, abs. min. $f(0) = 0$

95. a.

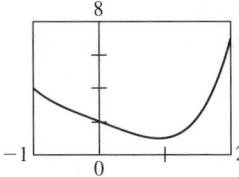

b. Abs. max. 7, abs. min. $2 - \frac{9}{8}\sqrt[3]{3/4} \approx 0.978$

97. a.

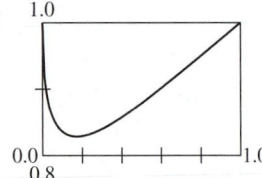

b. Abs. max. 1, abs. min. $\sqrt{2}(2 - \sqrt{2}) \approx 0.828$

99. False   **101.** False

Exercises 3.2 • page 302

1. 2   **3.** $-\dfrac{1 + \sqrt{7}}{3}$ **5.** $\pm\sqrt{2}/2$ **7.** $\pi/2$ **9.** 1

11. $\sqrt{3}$ **13.** $\dfrac{2 + 2\sqrt{3}}{3}$ **15.** $2\ln\left(\dfrac{2e^2}{e^2 - 1}\right)$

17. There is at least one instant during the 30-min flight when the plane is neither climbing nor descending.

19. $c = 4$; the aircraft attains the highest altitude at $t = 4$.

21. f is not differentiable on $(-1, 1)$.

23. No; f is not differentiable at $x = 1$.

31. b.

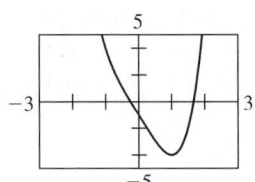

39. $(3, 3)$ **41. b.** $\frac{1}{2}$

43. a. 1.325   **b.**

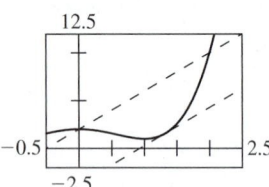

45. a. 0.691   **b.** 1.25

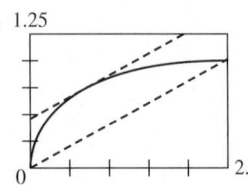

47. False   **49.** True   **51.** True

Exercises 3.3 • page 311

Abbreviations: rel. max., relative maximum; rel. min., relative minimum.

1. a. Increasing on $(-\infty, -2)$, constant on $(-2, 2)$, decreasing on $(2, \infty)$

  b. Rel. max. $f(x) = 2$, $x \in [-2, 2]$

3. a. Increasing on $(-1, \infty)$, decreasing on $(-\infty, -1)$

  b. Rel. min. $f(-1) = 0$

5. a. Increasing on $(-\infty, -1)$ and $(-1, \infty)$ **b.** None

7. a. Increasing on $(-2.5, 2.5)$, decreasing on $(-\infty, -2.5)$ and $(2.5, \infty)$

  b. Rel. max. at 2.5, rel. min. at -2.5

9. a. Increasing on $(1, \infty)$, decreasing on $(-\infty, 1)$

  b. Rel. min. $f(1) = -1$

11. a. Increasing on $(-\infty, -\sqrt{2})$ and $(\sqrt{2}, \infty)$, decreasing on $(-\sqrt{2}, \sqrt{2})$

  b. Rel. max. $f(-\sqrt{2}) = 1 + 4\sqrt{2}$, rel. min. $f(\sqrt{2}) = 1 - 4\sqrt{2}$

13. a. Increasing on $(-\infty, -2)$ and $(1, \infty)$, decreasing on $(-2, 1)$

  b. Rel. max. $f(-2) = 25$, rel. min. $f(1) = -2$

15. a. Increasing on $(3, \infty)$, decreasing on $(-\infty, 3)$

  b. Rel. min. $f(3) = -21$

17. a. Increasing on $(-\infty, \infty)$ **b.** None

19. a. Increasing on $(-\infty, 0)$ and $\left(\frac{4}{5}, \infty\right)$, decreasing on $\left(0, \frac{4}{5}\right)$

  b. Rel. max. $f(0) = 0$, rel. min. $f\left(\frac{4}{5}\right) = -1.10592$

21. a. Increasing on $(-\infty, -1)$ and $(1, \infty)$, decreasing on $(-1, 0)$ and $(0, 1)$

  b. Rel. max. $f(-1) = -2$, rel. min. $f(1) = 2$

23. a. Increasing on $(-\infty, 0)$ and $(2, \infty)$, decreasing on $(0, 1)$ and $(1, 2)$

  b. Rel. max. $f(0) = 0$, rel. min. $f(2) = 4$

25. a. Decreasing on $(-\infty, -2)$, $(-2, 2)$, and $(2, \infty)$

  b. None

27. a. Increasing on $(-\infty, 0)$ and $\left(\frac{6}{5}, \infty\right)$, decreasing on $\left(0, \frac{6}{5}\right)$

  b. Rel. max. $f(0) = 0$, rel. min. $f\left(\frac{6}{5}\right) \approx -2.03$

29. a. Increasing on $\left(0, \frac{3}{4}\right)$, decreasing on $\left(\frac{3}{4}, 1\right)$

  b. Rel. max. $f\left(\frac{3}{4}\right) = 3\sqrt{3}/16$

31. a. Increasing on $\left(\frac{\pi}{3}, \frac{5\pi}{3}\right)$, decreasing on $\left(0, \frac{\pi}{3}\right)$ and $\left(\frac{5\pi}{3}, 2\pi\right)$

  b. Rel. max. $f\left(\frac{5\pi}{3}\right) \approx 6.97$, rel. min. $f\left(\frac{\pi}{3}\right) \approx -0.68$

33. a. Increasing on $\left(\frac{\pi}{2}, \pi\right)$ and $\left(\frac{3\pi}{2}, 2\pi\right)$, decreasing on $\left(0, \frac{\pi}{2}\right)$ and $\left(\pi, \frac{3\pi}{2}\right)$

  b. Rel. max. $f(\pi) = 1$, rel. min. $f\left(\frac{\pi}{2}\right) = f\left(\frac{3\pi}{2}\right) = 0$

35. a. Increasing on $(0, 2)$, decreasing on $(-\infty, 0)$ and $(2, \infty)$

    b. Rel. max. $f(2) = 4e^{-2}$, rel. min. $f(0) = 0$

37. a. Increasing on (e, ∞), decreasing on $(0, 1)$ and $(1, e)$

    b. Rel. min. $f(e) = 2e$

39. Rel. max. $f\left(-\dfrac{\sqrt{3}}{2}\right) = \sqrt{3} - \dfrac{\pi}{3}$,

    rel. min. $f\left(\dfrac{\sqrt{3}}{2}\right) = \dfrac{\pi}{3} - \sqrt{3}$

41. a. Increasing on $(20.2, 20.6)$ and $(21.7, 21.8)$, constant on $(19.6, 20.2)$ and $(20.6, 21.1)$, decreasing on $(21.1, 21.7)$ and $(21.8, 22.7)$

43. Increasing on $(1, 4)$, decreasing on $(0, 1)$, rel. min. $f(1) = 32$

45. Decreasing on $(0, 500)$, increasing on $(500, \infty)$, $\overline{C}(500) = 35$

47. Rising on $(0, 1)$, $(7, 13)$, and $(19, 24)$; falling on $(1, 7)$ and $(13, 19)$; rel. max. $(1, 12.4)$ and $(13, 12.4)$, rel. min. $(7, 2.8)$ and $(19, 2.8)$

49. a.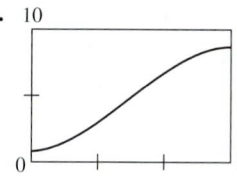

51. a.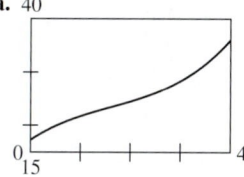

53. b. 4505 cases per year, 272 cases per year

55. Increasing on $(0, e)$, decreasing on (e, ∞)

57. a. 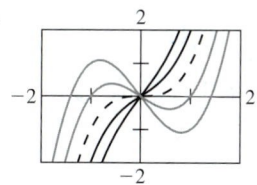   **b.** $a \leq 0$

63. $a = -4$, $b = 24$

65. No. f is not continuous on an interval containing $x = 0$.

69. a.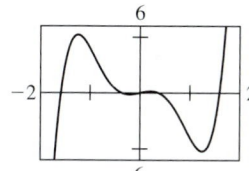

    b.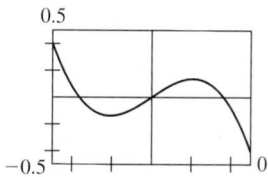

    Increasing on $(-\infty, -1.2)$, $(-0.2, 0.2)$, and $(1.2, \infty)$; decreasing on $(-1.2, -0.2)$ and $(0.2, 1.2)$

71. a.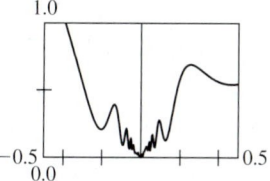

73. True   **75.** False   **77.** False

Exercises 3.4 • page 324

Abbreviations: CU, concave upward; CD, concave downward; IP, inflection point; rel. max., relative maximum; rel. min., relative minimum.

1. CU on $(0, \infty)$, CD on $(-\infty, 0)$, IP $(0, 0)$

3. CU on $(-\infty, -4)$ and $(4, \infty)$, CD on $(-4, 4)$

5. CD on $(-\infty, -2)$, $(-2, 2)$, and $(2, \infty)$

7. CU on $(-1, 0)$ and $(1, \infty)$, CD on $(-\infty, -1)$ and $(0, 1)$, IP at $x = -1, 0, 1$

9. (b)   **11.** CU on $(0, \infty)$, CD on $(-\infty, 0)$, IP $(0, 0)$

13. CU on $(-\infty, 0)$ and $(1, \infty)$, CD on $(0, 1)$, IP $(0, 0)$ and $(1, -1)$

15. CU on $(-\infty, 0)$, CD on $(0, \infty)$, IP $(0, 1)$

17. CU on $(-\infty, -1)$ and $(0, \infty)$, CD on $(-1, 0)$, IP $\left(-1, -\frac{4}{15}\right)$ and $(0, 0)$

19. CD on $(-1, 0)$ and $(0, 1)$

21. CU on $(-\infty, 0)$ and $(0, \infty)$

23. CU on $(-1, 0)$ and $(1, \infty)$, CD on $(-\infty, -1)$ and $(0, 1)$, IP $(0, 0)$

25. CU on $\left(\frac{\pi}{2}, \pi\right)$, CD on $\left(0, \frac{\pi}{2}\right)$, IP $\left(\frac{\pi}{2}, 0\right)$

27. CU on $\left(\frac{3\pi}{4}, \frac{7\pi}{4}\right)$, CD on $\left(0, \frac{3\pi}{4}\right)$ and $\left(\frac{7\pi}{4}, 2\pi\right)$, IP $\left(\frac{3\pi}{4}, 0\right)$ and $\left(\frac{7\pi}{4}, 0\right)$

29. CU on $\left(-\pi, -\frac{3\pi}{4}\right)$, $\left(-\frac{\pi}{2}, -\frac{\pi}{4}\right)$, $\left(0, \frac{\pi}{4}\right)$, and $\left(\frac{\pi}{2}, \frac{3\pi}{4}\right)$, CD on $\left(-\frac{3\pi}{4}, -\frac{\pi}{2}\right)$, $\left(-\frac{\pi}{4}, 0\right)$, $\left(\frac{\pi}{4}, \frac{\pi}{2}\right)$, and $\left(\frac{3\pi}{4}, \pi\right)$, IP $\left(-\frac{\pi}{2}, 0\right)$, $(0, 0)$, and $\left(\frac{\pi}{2}, 0\right)$

31. CD on $(-\infty, 0)$ and $(0, \infty)$

33.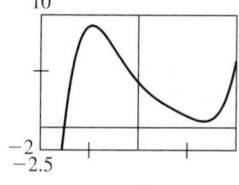

    a. CU on $(-0.4, \infty)$, CD on $(-\infty, -0.4)$ **b.** IP $(-0.4, 6.4)$

35.

    a. CU on $(-\infty, 0)$, CD on $(0, \infty)$ **b.** IP $(0, 0)$

37. Rel. max. $h(-1) = -\frac{22}{3}$, rel. min. $h(5) = -\frac{130}{3}$

39. Rel. min. $f(3) = -27$

41. Rel. max. $f\left(-\frac{\sqrt{2}}{2}\right) = -2\sqrt{2}$, rel. min. $f\left(\frac{\sqrt{2}}{2}\right) = 2\sqrt{2}$

43. Rel. min. $g(2) = 2 - 2\ln 2$

45. Rel. max. $f\left(\frac{\pi}{4}\right) = \sqrt{2}$ **47.** Rel. max. $f\left(\frac{\pi}{3}\right) = 3\sqrt{3}/2$

49. IP $(0, 0)$

51. 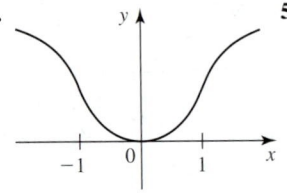   **53.**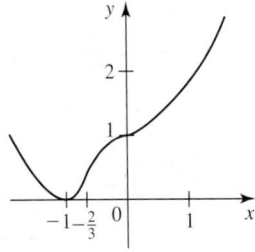

55. a. $D_1'(t) > 0$, $D_2'(t) > 0$, $D_1''(t) > 0$, $D_2''(t) < 0$

b. With the promotion the deposits will increase at an increasing rate; without it they will increase at a decreasing rate.

57. The restoration process is working at its peak at the time t_0 corresponding to the t-coordinate of Q.

59.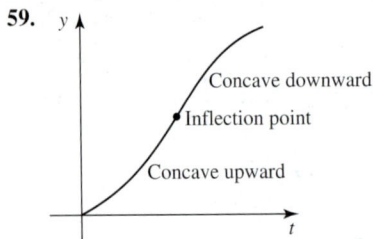

$f(t)$ increases at an increasing rate before the IP and a decreasing rate after the IP.

61. b. Sales continued to accelerate.

67. a. 506,000 in 1999, 125,480 in 2005

b. The number of measles deaths was dropping.

c. April 2002; decreasing at about 41,000 deaths annually

69. a. 71.6 kg/yr, 63.1 kg/yr   **b.** 5.7 years

73. a. CU on $(0, 1)$, CD on $(-\infty, 0)$ and $(1, \infty)$

b. No; the graph has no tangent line at $(1, 2)$.

c.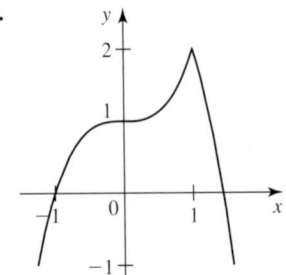

75. $c \geq \frac{3}{2}$ **77.** No   **83.** False   **85.** True

---

Exercises 3.5 • page 343

Abbreviations: HA, horizontal asymptote(s); VA, vertical asymptote(s).

1. a. $-\infty$ **b.** ∞ **c.** ∞ **d.** $-\infty$ **3. a.** $-\infty$ **b.** 0   **c.** 0

5. ∞ **7.** $-\infty$ **9.** ∞ **11.** ∞ **13.** ∞ **15.** ∞

17. ∞ **19.** ∞ **21.** $\frac{3}{2}$ **23.** 0   **25.** $\frac{1}{9}$ **27.** $-\frac{2}{3}$

29. $-\frac{1}{6}$ **31.** $2\sqrt{3}/3$ **33.** $\frac{2}{3}$ **35.** 0   **37.** $-\infty, 1, 0, 1$

39. b. 0   **c.** 0.5

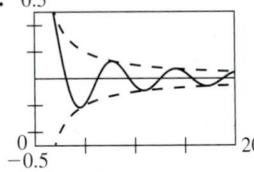

41. a. $\frac{1}{2}$ **b.** $\frac{1}{2}$ **c.** $\frac{1}{2}$

43. a. 0.7   **b.** 0.707   **c.** $\sqrt{2}/2$

45. HA $y = \pm 1$ **47.** HA $y = -1$, VA $x = \pm 1$

49. HA $y = 0$, VA $x = -2$ **51.** HA $y = 1$, VA $x = -1$

53. HA $y = 0$, VA $x = -2, 3$ **55.** HA $y = 1$, VA $t = \pm 2$

57. 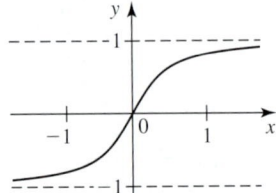   **59.**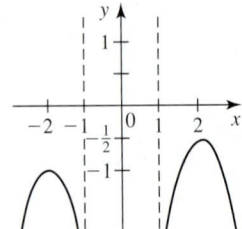

61. a. $\lim_{x \to 100^-} C(x) = \infty$; the cost increases dramatically as the amount of pollutant removed approaches 100%.

b.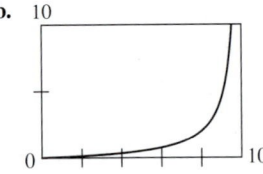

63. a. 25,000   **b.** 30

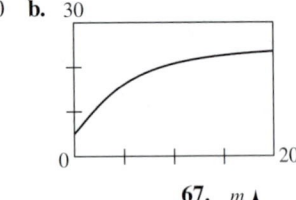

65. 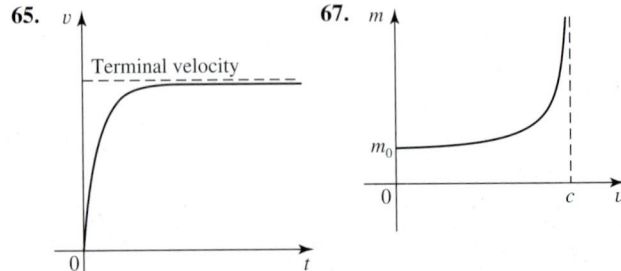   **67.**

69.

n	1	10	10^2	10^3	10^4	10^5	10^6
$\left(1 + \dfrac{1}{n}\right)^n$	2	2.59374	2.70481	2.71692	2.71815	2.71827	2.71828

71. $33,201.17   **73.** $48,635.23

75. a. 0.6   **b.** 0.577   **c.** $\sqrt{3}/3$

77. b. $\approx$25,067 mph   **c.**
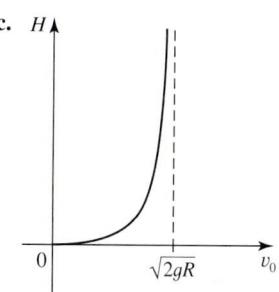

81. 1   **89.** False   **91.** False   **93.** True

Exercises 3.6 • page 358

1.
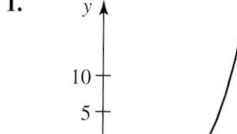

3.
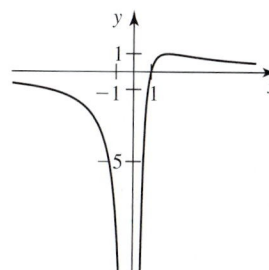

5.
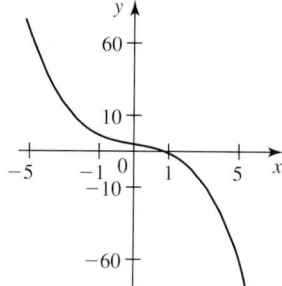

7.
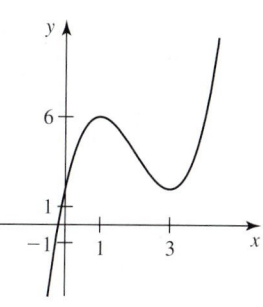

9.
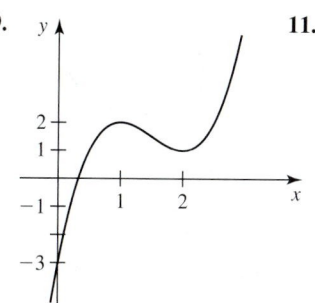

11.
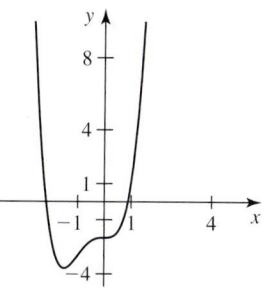

13.
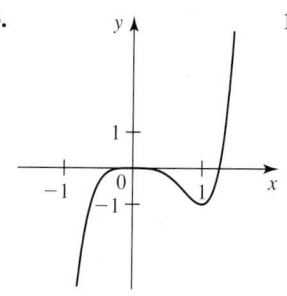

15.
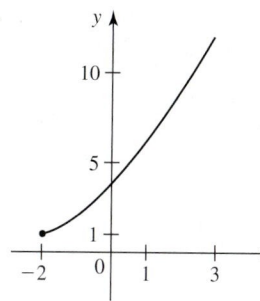

17.
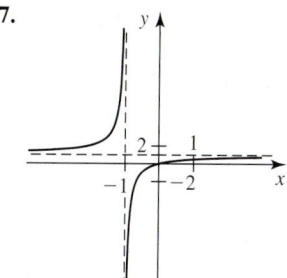

19.
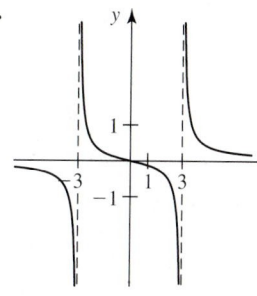

21.
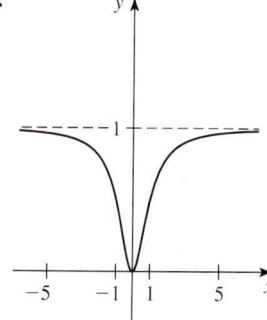

23.
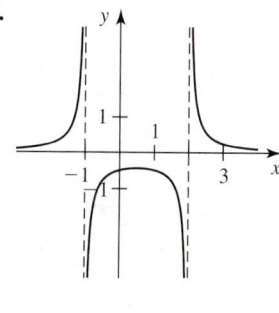

25.
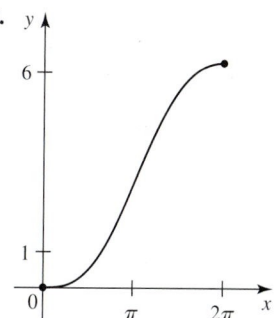

27.
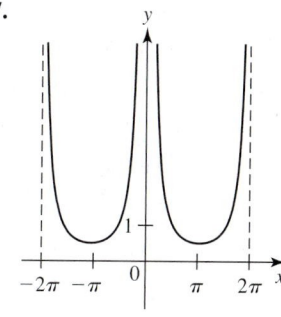

29.
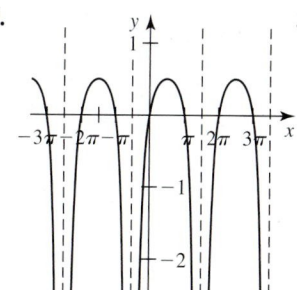

31.
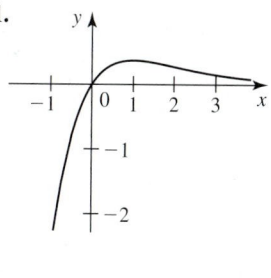

33.

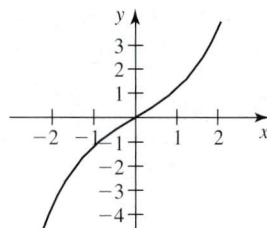

35.

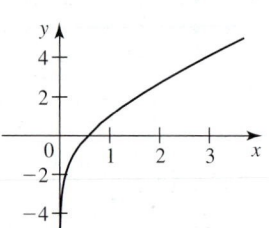

37.

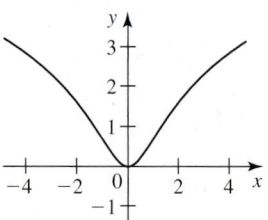

39. $y = u$ **41.** $y = \frac{1}{2}x - \frac{1}{2}$

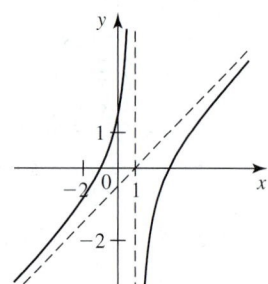

45.

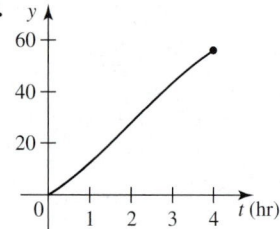

The average worker's efficiency increases until 10 A.M. and then begins to decrease.

47.

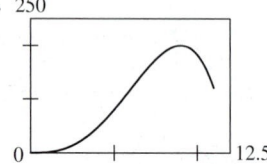

The rate of increase of the ozone level is maximum at about 1 P.M.

49.

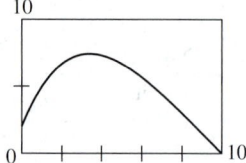

The amount of salt increases to a peak level of approximately 7.43 lb after approximately 3.4 min, and then it declines.

51.

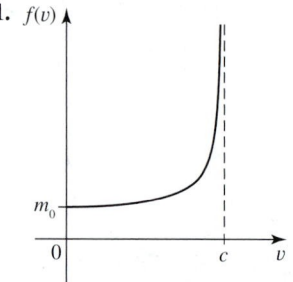

A particle's mass increases as its speed increases.

53.

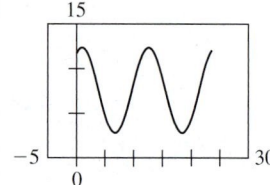

At 12 A.M. the water level is ≈11.8 ft. At 7 A.M. and 7 P.M. it has a rel. min. of ≈2.8 ft; at 1 A.M. and 1 P.M. it has a rel. max. of ≈12.4 ft.

55.

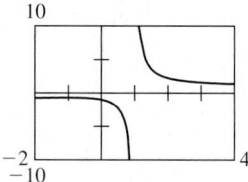

57.

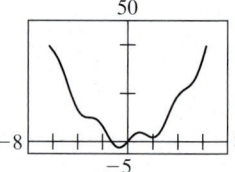

59. a.

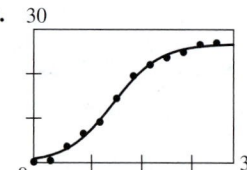

b. 1.5 in./hr, 0.5 in./hr

c. 2:24 A.M. on February 7, 1.6 in./hr

61.

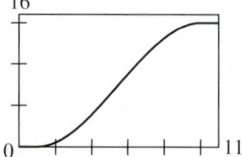

Exercises 3.7 • page 369

1. 50, 50   **3.** 18, 3   **5.** 25 m × 25 m

7. 100 ft × $66\frac{2}{3}$ ft   **9.** 12 in. × 6 in. × 2 in.

11. 7.6 in. × 7.6 in. × 3.8 in.

13. Radius $36/\pi$ in., length 36 in., volume $46{,}656/\pi$ in.³

15. 1.39 ft × 2.08 ft × 0.83 ft   **17.** $2\sqrt{2} \times \sqrt{2}$ **19.** $\frac{3}{2} \times 1$

21. $(-2, 1)$ **23.** $(-1.6180, -0.6180)$, $(0.6180, 1.6180)$

25. $4\sqrt{2} \times 2\sqrt{2}$ **27.** $(-2.834, -4.032)$ **29.** 20 trees/acre

31. $\dfrac{2\sqrt{3}}{3}$ ft × $\dfrac{2\sqrt{3}}{3}$ ft × $\dfrac{10\sqrt{3}}{9}$ ft; $\dfrac{40\sqrt{3}}{27}$ ft³

35. Width 11.5 in., height 19.9 in.

37. $\dfrac{2\pi(3 - \sqrt{6})}{3}$ rad, $128\pi\sqrt{3}$ in.³

39. ≈6.7 ft, ≈6.7 ft   **41.** $\sqrt[3]{32/\pi}$, $2\sqrt[3]{32/\pi}$

43. ≈8.6 mi from O

45. a. $D(x) = \sqrt{x^2 + \dfrac{(x + 1)^2}{x} - \dfrac{4(x + 1)}{\sqrt{x}} + 4}$

b. ≈0.45 mi

47. r, $\dfrac{E^2}{4r}$ watts   **49.** C **53.** Radius $R/2$, height $H/2$

55. $y = -2x + 4$ **57.** $5\sqrt{2}/2$ ft   **59.** Yes

61. Radius 1.5 ft, height 3 ft

63. a. $\frac{3}{2}u$ ft/sec

b. E

65. 655 ft   **67.** $\overline{PR} = \dfrac{3\sqrt{7}}{7}$ mi, $\dfrac{30 + \sqrt{7}}{12}$ hr

Exercises 3.8 • page 387

1. $\frac{1}{2}$ **3.** 12   **5.** 1   **7.** 1   **9.** 2   **11.** $\frac{1}{2}$ **13.** $\frac{1}{3}$

15. ∞   **17.** 0   **19.** 0   **21.** $\frac{9}{4}$ **23.** ∞   **25.** $-\frac{1}{2}$

27. 2   **29.** 0   **31.** -2 **33.** $-\infty$ **35.** $\frac{1}{2}$ **37.** -1

39. 0   **41.** 1   **43.** e^2 **45.** 1   **47.** ∞   **49.** 1

51. 0   **53.** $-\infty$ **55.** $\frac{1}{3}$ **57.** $\frac{5}{2}$ **61.** $\dfrac{Vt}{L}$ **63.** a

69. 3   **71.** $\frac{1}{2}$ **73.** False

Exercises 3.9 • page 394

1. 0.7709   **3.** 1.1219, 2.5745   **5.** (0.8767, 0.7686)

7. (0.4502, 0.4502)   **9.** 1.37880   **11.** 0.75488

13. 0.87122   **15.** -1.347296 **17.** 0.739085

19. 1.76322   **21.** All terms are x_0.   **23.** 1.8794

25. 2.4495   **27.** 2.1147   **29.** $t \approx 41.08569$

31. (0.786, 0.666)

33. $f'(1)$ is not defined; using $x_0 = 0$ leads to $x_1 = 1$.

35. b. 0.7351   **c.** ≈2.74 mi

d.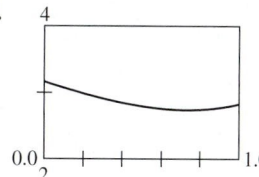

39. 6.7%/year

41. c.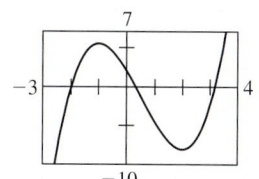

Chapter 3 Concept Review • page 397

1. a. $f(x) \le f(c)$; absolute maximum value

b. $f(c) \le f(x)$; open interval

3. Continuous; absolute maximum value; absolute minimum value

5. a. $f(x_1) < f(x_2)$

b. increasing; decreasing

c. < 0

7. a. values; arbitrarily large; a

b. values; to L; sufficiently large

c. arbitrarily large; decreases

9. a. $M > 0$; $\delta > 0$; $f(x) > M$

b. $M > 0$; N; $x < N$; $f(x) > M$

Chapter 3 Review Exercises • page 397

Abbreviations: abs. max., absolute maximum; abs. min., absolute minimum; CU, concave upward; CD, concave downward; IP, inflection point; HA, horizontal asymptote(s); VA, vertical asymptote(s).

1. Abs. max. $f(2) = 1$, abs. min. $f(-1) = -8$

3. Abs. max. $h(2) = -16$, abs. min. $h(4) = -32$

5. Abs. max. $f(3) = \frac{107}{9}$, abs. min. $f(1) = 3$

7. Abs. min. $f(3) = -3$

9. Abs. max. $f\left(\frac{7\pi}{4}\right) = \sqrt{2}$, abs. min. $f\left(\frac{3\pi}{4}\right) = -\sqrt{2}$

11. Abs. min. $f\left(\frac{\pi}{3}\right) = \dfrac{-3\sqrt{3} + \pi}{6}$

13. Absolute minimum $f(1) = 0$, absolute maximum $f(e) = 1/e$

15. -1 **17.** $\sqrt{3}$ **19.** $\cos^{-1}\left(\frac{2}{\pi}\right)$

21. f is not continuous on $[-2, 0]$.

23. a. Increasing on $(-\infty, \infty)$ **b.** None
  c. CU on $(1, \infty)$, CD on $(-\infty, 1)$ **d.** $\left(1, -\frac{17}{3}\right)$

25. a. Increasing on $(-1, 0)$ and on $(1, \infty)$, decreasing on $(-\infty, -1)$ and on $(0, 1)$
  b. Rel. max. $(0, 0)$, rel. min. $(-1, -1)$ and $(1, -1)$
  c. CU on $\left(-\infty, -\frac{\sqrt{3}}{3}\right)$ and on $\left(\frac{\sqrt{3}}{3}, \infty\right)$, CD on $\left(-\frac{\sqrt{3}}{3}, \frac{\sqrt{3}}{3}\right)$
  d. $\left(-\frac{\sqrt{3}}{3}, -\frac{5}{9}\right)$, $\left(\frac{\sqrt{3}}{3}, -\frac{5}{9}\right)$

27. a. Increasing on $(-\infty, 0)$ and on $(2, \infty)$, decreasing on $(0, 1)$ and on $(1, 2)$
  b. Rel. max. $(0, 0)$, rel. min. $(2, 4)$
  c. CU on $(1, \infty)$, CD on $(-\infty, 1)$
  d. None

29. a. Decreasing on $(-\infty, \infty)$ **b.** None
  c. CU on $(1, \infty)$, CD on $(-\infty, 1)$ **d.** $(1, 0)$

31. a. Increasing on $(-\infty, -1)$ and on $(-1, \infty)$ **b.** None
  c. CU on $(-\infty, -1)$, CD on $(-1, \infty)$ **d.** None

35. Decreasing on $\left(0, \frac{1}{2}\right)$, increasing on $\left(\frac{1}{2}, \infty\right)$

37. $-\infty$ **39.** $-\infty$ **41.** ∞ **43.** 3   **45.** 0

47. HA $y = 0$, VA $x = -\frac{3}{2}$

49.

51.

53.

55.
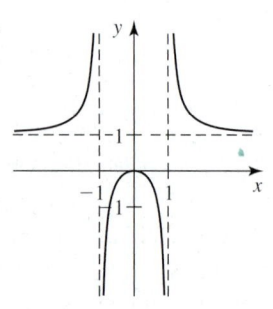

57.
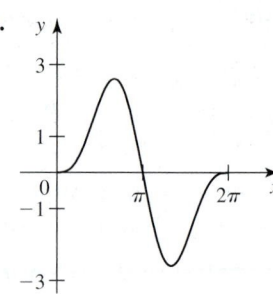

59.

61.

63.
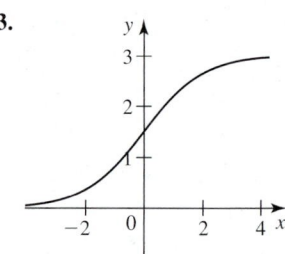

65. 0   **67.** $\frac{2}{3}$ **69.** 0   **71.** 0

73. $\frac{1}{2}$ **75.** \$6885.64

79. a. Increasing on $(12.7, 30)$, decreasing on $(0, 12.7)$
  b. $P(12.7) \approx 7.9$
  c. The smallest percentage of women over 65 in the workforce was about 7.9% in late 1982.

81. $10\sqrt{2} \times 4\sqrt{2}$ **83.** $(2, 1)$

85. a. 30,000
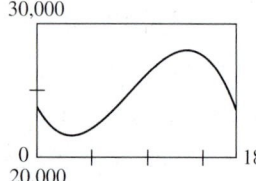

87. Maximum when $t = 2, 6, 10, 14, \ldots$; minimum when $t = 0, 4, 8, 12, \ldots$

89. a. 43 mg/day   **b.** $t = 1$ **c.** 125 mg

91. 1 ft $\times$ 2 ft $\times$ 2 ft   **93.** 1 ft

95. -0.7709 **97.** 0.8767

99. $a = -4, b = 11$ **101.** No

Chapter 3 Challenge Problems • page 401

1. $f\left(\frac{1}{2}\right) = \frac{1}{4}$

3. Highest point: $(-1, 2)$, lowest point: $(1, -2)$

17. $\dfrac{a_{n-1} - b_{n-1}}{n}$

CHAPTER 4

4.1 Exercises • page 412

1. $\frac{1}{2}x^2 + 2x + C$ **3.** $\frac{1}{3}x^3 - x^2 + 3x + C$

5. $\frac{1}{5}x^{10} + 3e^x + 4x + C$ **7.** $\frac{2}{3}x^{3/2} + 6\sqrt{x} + C$

9. $e^t + \dfrac{t^{e+1}}{e+1} + C$ **11.** $6\sqrt{u} + C$

13. $3x + \dfrac{2}{x} - \dfrac{1}{3x^3} + C$ **15.** $\frac{2}{5}x^{5/2} - \frac{4}{3}x^{3/2} + 6x^{1/2} + C$

17. $\dfrac{2^x}{\ln 2} + \ln|x| + C$ **19.** $-3\cos x - 4\sin x + C$

21. $-\cot x + \frac{2}{3}x^{3/2} + C$ **23.** $-\csc x + C$

25. $\tan x + 2\cot x + C$ **27.** $\tan x - \cot x + C$

29. $\tan x + \sec x + C$

31. a. $\frac{1}{2}x^2 - 3x + C$ **b.**

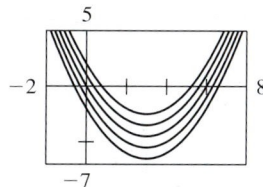

33. a. $x^2 - \cos x + C$ **b.**

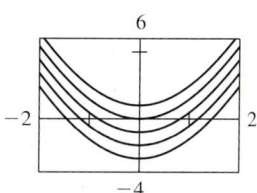

35. $x^2 + x + 1$ **37.** $2\sqrt{x} - 2$ **39.** $e^x - \cos x$

41. $3x^2 - 4x + 5$ **43.** $\frac{1}{2}x^4 + x^3 + x^2 + 3x + 3$

45. $-4t^{1/2} + 4t - 7$ **47.** $\frac{1}{3}x^3 - x^2 + 3x - \frac{1}{3}$

49. Graph 1 is f.   **51.** $2t^3 - 2t^2 + t - 2$

53. $t^3 - 2t^2 + 4t - 2$ **55.** $-\sin t + 2\cos t + t + 1$

57. $0.1t^2 + 3t$

59. a. $-16t^2 + 400$ **b.** 5 sec   **c.** -160 ft/sec

65. 2.15 m, 0.82 m   **67.** $-\frac{88}{9}$ ft/sec², 396 ft

69. 0.924 ft/sec²   **71.** Branch A **73.** $10{,}000x - 100x^2$

75. a. $0.001547x^3 - 0.1506x^2 + 4.9x - 53.09$

  b. 0.96%, 3.42%

77. $1.0974t^3 - 0.0915t^4 + 34$

79. $-\frac{1}{25}\left(2\sqrt{5}t - \frac{1}{100}t^2\right) + 20$

81. True   **83.** False   **85.** False   **87.** True

4.2 Exercises • page 425

1. $\frac{1}{12}(2x + 3)^6 + C$ **3.** $\sqrt{x^2 + 1} + C$ **5.** $\frac{1}{4}\tan^4 x + C$

7. $\frac{1}{5}(x^2 + 1)^5 + C$ **9.** $\frac{5}{16}(2x - 4)^{8/5} + C$

11. $\frac{1}{8}(2x^2 + 3)^6 + C$ **13.** $-\frac{1}{3}\sqrt{(1 - 2x)^3} + C$

15. $\frac{1}{10}(s^4 - 1)^{5/2} + C$ **17.** $-\frac{1}{2}e^{-x^2} + C$

19. $\frac{1}{2}(e^{2x} + 4x - e^{-2x}) + C$ **21.** $-\dfrac{1}{2^x \ln 2} + C$

23. $e^{-1/x} + C$ **25.** $4\sin\dfrac{x}{2} + C$ **27.** $\dfrac{1}{2\pi}\sin \pi x^2 + C$

29. $-\dfrac{1}{2\pi}\cos^2 \pi x + C$ or $\dfrac{1}{2\pi}\sin^2 \pi x + C$ **31.** $\frac{1}{3}\sec 3x + C$

33. $\cos u^{-1} + C$ **35.** $\frac{1}{6}\tan^2 3x + C$

37. $\sqrt{2 + \sin 2t} + C$ **39.** $\dfrac{1}{1 - \sec x} + C$

41. $\dfrac{1}{2}x - \dfrac{1}{4\pi}\sin 2\pi x + C$ **43.** $-\ln(1 + e^{-x}) + C$

45. $\dfrac{2(\ln x)^{3/2}}{3\sqrt{\ln 10}} + C$ **47.** $\frac{1}{2}[\ln(e^x + 1)]^2 + C$

49. $\frac{1}{2}\ln|2x + 3| + C$ **51.** $\ln|\ln x| + C$

53. $\ln|1 + \sin x| + C$ **55.** $\ln|\sec \theta + \tan \theta| + \sin \theta + C$

57. $\ln|2 + x\ln x| + C$ **59.** $\sin^{-1}\left(\frac{1}{4}x\right) + C$

61. $\dfrac{1}{18}\sec^{-1}\left(\dfrac{x^2}{9}\right) + C$ **63.** $x - 2\tan^{-1}x + C$

65. $-\sin^{-1}\left(\dfrac{\cos x}{2}\right) + C$ **67.** $e^{\tan^{-1}x} + C$

69. $\tan^{-1}\left(\dfrac{\sqrt{x}}{2}\right) + C$ **71.** $\dfrac{1}{2}\tan^{-1}\left(\dfrac{x - 2}{2}\right) + C$

73. $\frac{2}{15}(3x + 8)\sqrt{(x - 4)^3} + C$

75. $\frac{1}{9}(x^2 + 1)^{9/2} - \frac{1}{7}(x^2 + 1)^{7/2} + C$

77. $\frac{2}{3}(x + 1)^{3/2} - \frac{2}{3}x^{3/2} + C$ **79.** $\frac{1}{3}(1 + x^2)^{3/2} + \frac{2}{3}$

81. $-\frac{1}{3}(16 - t^2)^{3/2} + \frac{64}{3}$ **83.** 80.04 years

85. $\dfrac{1.2}{\pi}\left(1 - \cos\dfrac{\pi t}{2}\right)$

87. $v(t) = -2\sin 2t + \frac{3}{2}\cos 2t$, $s(t) = \cos 2t + \frac{3}{4}\sin 2t - 1$

89. True

4.3 Exercises • page 442

1. a.   **b.** $\frac{2}{5}$

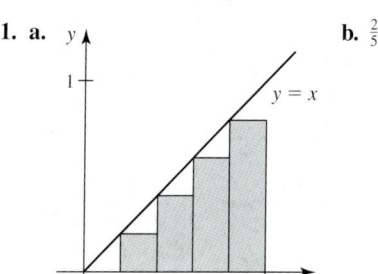

3. a. 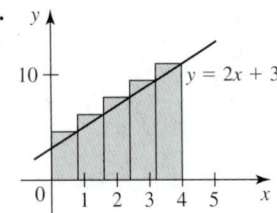 **b.** $\frac{156}{5}$

5. a. 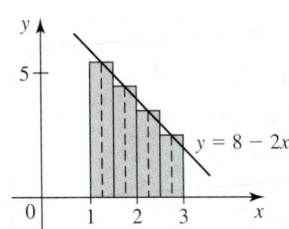 **b.** 8

7. a. 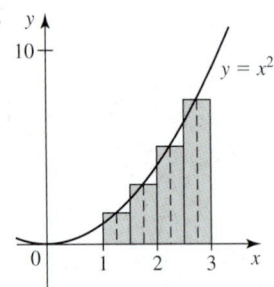 **b.** $\frac{69}{8} = 8.625$

9. a. 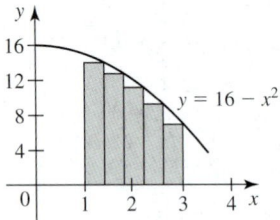 **b.** 21.68

11. a. 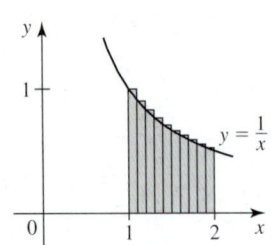 **b.** 0.72

13. 10 **15.** 25 **17.** 55 **19.** 6.15

21. $\sum_{k=1}^{30} 2k$ **23.** $\sum_{k=1}^{11} (2k + 1)$ **25.** $\sum_{k=1}^{5} \left(\frac{2k}{5} + 1 \right)$

27. $\frac{1}{n} \sum_{k=1}^{n} \left[2\left(\frac{k}{n} \right)^3 - 1 \right]$ **29.** $\frac{1}{n} \sum_{k=1}^{n} \sin\left(1 + \frac{k}{n} \right)$

31. 120 **33.** 275 **35.** 13,695

37. $\dfrac{4n^3 + 12n^2 + 11n}{3}$ **39.** 1 **41.** $\frac{15}{2}$ **43.** $\frac{13}{3}$

45. 6 **47.** $\frac{1}{3}$ **49.** 9 **51.** $\frac{14}{3}$

53. a. $\displaystyle\lim_{n\to\infty} \frac{32}{n^5} \sum_{k=1}^{n} k^4$

b. $\dfrac{16(n + 1)(2n + 1)(3n^2 + 3n - 1)}{15n^4}$ **c.** $\frac{32}{5}$

55. a. $\displaystyle\lim_{n\to\infty} \sum_{k=1}^{n} \left[\left(2 + \frac{3k}{n} \right)^4 + 2\left(2 + \frac{3k}{n} \right)^2 + \left(2 + \frac{3k}{n} \right) \right]\left(\frac{3}{n} \right)$

b. $\dfrac{3(2357n^4 + 3270n^3 + 1200n^2 - 27)}{10n^4}$ **c.** $\frac{7071}{10}$

59. 9400 ft^2 **63.** False **65.** False

4.4 Exercises • page 459

1. a. 14 **b.** 4 **c.** 6

3. a. 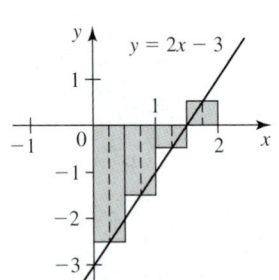 **b.** −2

5. a. 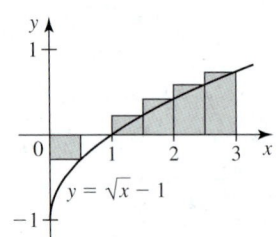 **b.** 0.83

7. 2 **9.** −4 **11.** $-\frac{5}{3}$ **13.** $\displaystyle\int_{-3}^{-1} (4x - 3)\, dx$

15. $\displaystyle\int_{1}^{2} \frac{2x}{x^2 + 1}\, dx$ **17.** $\displaystyle\lim_{n\to\infty} \sum_{k=1}^{n} \left(1 + c_k^3 \right)^{1/3} \Delta x$

19. a. −6 **b.** $-\frac{19}{2}$ **c.** 13 **d.** $-\frac{5}{2}$

21.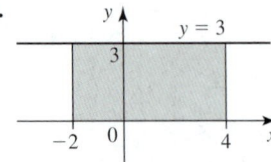

18

23.

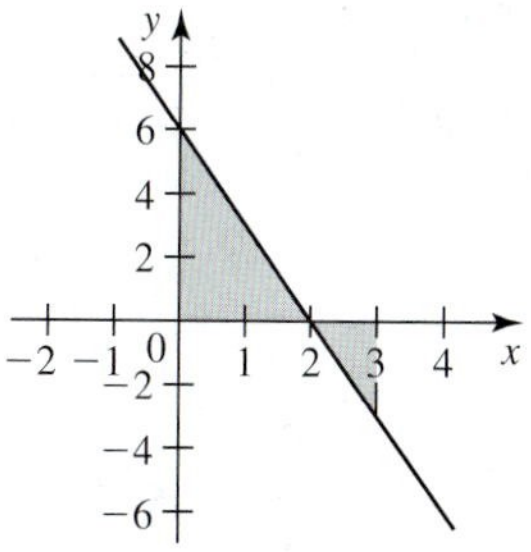

$\frac{9}{2}$

25.

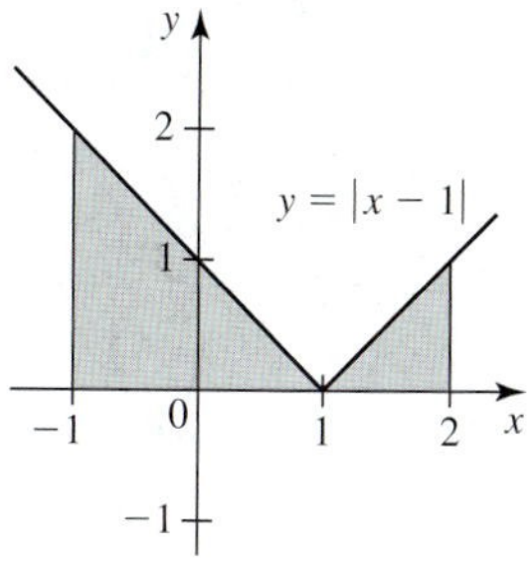

$\frac{5}{2}$

27.

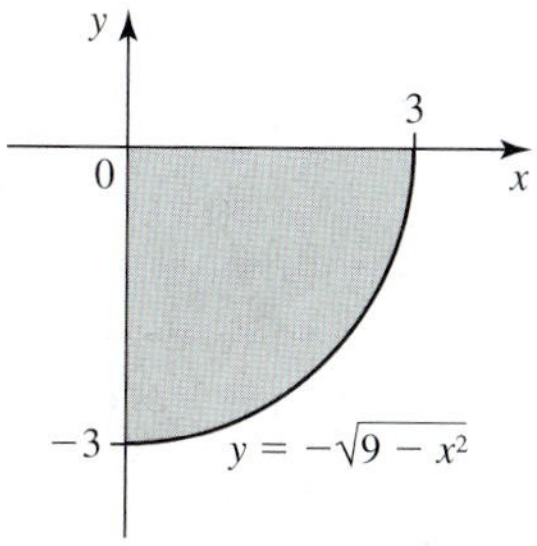

$-\frac{9\pi}{4}$

29. a. 2 **b.** 1 **c.** 6 **d.** −13

31. a. 3 **b.** −7 **c.** 19 **33.** 0

41. $\sqrt{3} \le \int_1^2 \sqrt{1 + 2x^3}\, dx \le \sqrt{17}$

43. $3 \le \int_{-1}^2 (x^2 - 2x + 2)\, dx \le 15$

45. $0.367 \le \int_0^1 e^{-x^2}\, dx \le 1$

47. a.

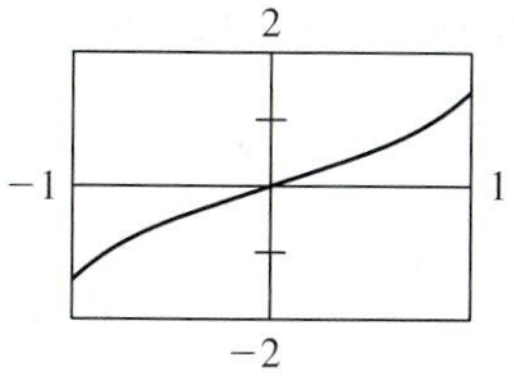

53. a.

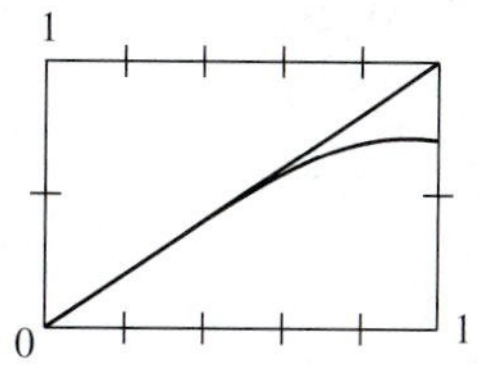

57. a. $1 \le \int_0^1 \sqrt{1 + x^2}\, dx \le \sqrt{2}$

b. $1 \le \int_0^1 \sqrt{1 + x^2}\, dx \le \dfrac{1 + \sqrt{2}}{2}$

59. 4 **61.** No **63.** True

65. False **67.** True **69.** False

4.5 Exercises • page 475

1. a. x^2 **b.** $\frac{1}{3}x^3 - \frac{8}{3}$ **c.** x^2 **3.** $\sqrt{3x + 5}$

5. $\dfrac{1}{x^2 + 1}$ **7.** $-\sin 2x$ **9.** $\dfrac{\sin\sqrt{x}}{2x}$

11. $-\dfrac{\sin x \cos^2 x}{\cos x + 1}$ **13.** $x(9x - 4)\ln x$

15. 20 **17.** $-\frac{22}{3}$ **19.** 21

21. $2 + 3 \ln 3$ **23.** 2 **25.** $\frac{32}{3}$

27. $208\sqrt{2}/105$ **29.** $\dfrac{3 - \sqrt{3}}{3}$

31. $\frac{4}{3}$ **33.** $2\sqrt{3}/3$ **35.** $\frac{19}{6}$ **37.** $\frac{121}{5}$

39. $\frac{6561}{2}$ **41.** $\frac{3}{4}(4\sqrt[3]{4} - 1)$ **43.** $\frac{1}{3}$

45. $\frac{1}{2}[\ln 2 - \ln(e^{-2} + 1)]$ **47.** $2 - \sqrt{2}$

49. 0 **51.** 1 **53.** $\frac{1}{2}e(e^{e-1} - 1)$

55. $\dfrac{2}{\ln 3} + \dfrac{1}{4}$ **57.** $\dfrac{\pi}{12}$ **59.** $\dfrac{\pi}{12}$

61. $\dfrac{\pi}{16}$ **63. b.** 0.14342 **65.** 6

67. $\frac{32}{3}$ **69.** $2\left(\sqrt{e} - \dfrac{1}{e}\right)$

71. a.

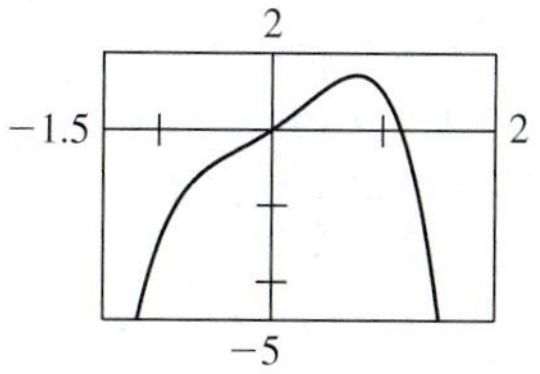

b. 0, 1.165 **c.** 1.026

73. $\frac{1}{5}$ **75.** $\frac{56}{3}$ **77.** $\frac{1}{2}$ **79.** $\dfrac{\sqrt{10} - 1}{3}$

81. a. $\dfrac{\sqrt{21} - 3}{3}$ **b.** y

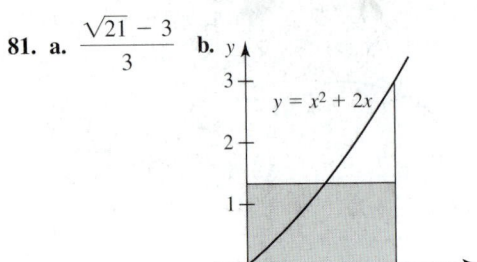

83. a. $\frac{769}{225}$ **b.** y

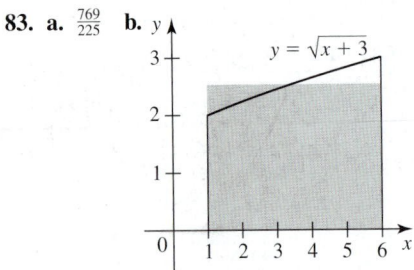

85. a. 4.5 ft **b.** 15.75 ft

87. 46%; 24% **93.** $\frac{80}{3}$ cm/sec

95. 150.937 pollutant standard index

97. 8373 wolves, 50,804 caribou

99. 343.45 ppmv/year

101. 39.16 million barrels

103. 43.3 sec **105.** 15.54

109. Maxima at $(2n - 1)\pi$, minima at $2n\pi$, $n = 1, 2, \ldots$

111. 3 **113.** 3 **115.** 0 **117. b.** π

125. True **127.** True

4.6 Exercises • page 489

1. a. 2.75 **b.** $\frac{8}{3}$ Exact value: $\frac{8}{3}$

3. a. 3.7708 **b.** $\frac{15}{4}$ Exact value: $\frac{15}{4}$

5. a. 4.3766 **b.** 4.3328 Exact value: $\frac{13}{3}$

7. a. 0.3129 **b.** 0.3161 Exact value: 0.316060

9. 0.5523 **11.** 0.8806 **13.** 1.1643

15. 3.2411 **17.** 2.2955 **19.** 1.9101

21. a. $\frac{3}{4}$ **b.** 0 **23. a.** $\frac{9}{1024}$ **b.** $\frac{81}{262,144}$

25. a. $\pi^3/1728$ **b.** $\pi^5/1,866,240$

27. 0.0026; 1.63×10^{-5} **29.** 13

31. 27 **33.** 28 **35.** 8

37. 4 **39.** 4 **41.** 52.82 ft/sec

43. 474.77 million barrels

45. 1922.4 ft³/sec **47.** False

Chapter 4 Concept Review • page 491

1. a. $F' = f$ **b.** $F(x) + C$ **3. a.** unknown **b.** function

5. a. $\displaystyle\int_a^b f(x)\, dx$ **b.** minus

7. a. $f(x)$ **b.** $F(b) - F(a)$; antiderivative **c.** $\displaystyle\int_a^b f'(x)\, dx$

9. $f(c) = \dfrac{1}{b - a}\displaystyle\int_a^b f(x)\, dx$

Chapter 4 Review Exercises • page 492

1. $\frac{1}{2}x^4 - \frac{4}{3}x^3 + \frac{3}{2}x^2 + 4x + C$ **3.** $\frac{3}{8}x^{8/3} - \frac{10}{7}x^{7/5} + C$

5. $\frac{3}{5}x^{5/3} + 3x + \frac{2}{3}x^{-3} + C$ **7.** $\frac{1}{8}(1 + 2t)^4 + C$

9. $\frac{1}{27}(3t - 4)^9 + C$ **11.** $\frac{1}{3}x^3 + 2x - \dfrac{1}{x} + C$

13. $-\dfrac{1}{4(3x^2 + 2x)^2} + C$ **15.** $-\frac{1}{5}\cos^5 t + C$

17. $-2\sqrt{1 - \sin\theta} + C$ **19.** $-\frac{1}{2}\csc x^2 + C$

21. $\frac{1}{5}\ln|5x - 3| + C$ **23.** $\frac{2}{3}x - \frac{1}{9}\ln|3x + 2| + C$

25. $\dfrac{2^{t^2}}{2\ln 2} + C$ **27.** $-\cos\ln x + C$

29. $-2\ln|\cos\sqrt{x}| + C$ **31.** $\frac{1}{4}(\tan^{-1} 2x)^2 + C$

33. 16 **35.** $\frac{1}{8}$ **37.** 2 **39.** $\frac{155}{6}$

41. $\frac{1}{2}(\sqrt{2} - 1)$ **43.** $\dfrac{\sqrt{3} - \sqrt{2}}{2}$

45. $2(1 - \ln 2)$ **47.** $\frac{1}{2}(e^2 - 1)$

49. 2 **51.** $(\ln 2)^2$ **53.** $\dfrac{2xe^{x^2}}{x^4 + 1}$

55. $3x^2 \sin x^3 - 2x \sin x^2$ **57.** 1.1502

59. a. $\frac{1}{128}$ **b.** $\frac{7}{24,576}$ **61.** $\frac{2}{3}x^{3/2} - \cos x + 3$

63. a. $-16t^2 + 128$ **b.** $2\sqrt{2}$ sec **c.** $-64\sqrt{2}$ ft/sec

65. 352 ft **67.** 0

69. $-0.000002x^3 - 0.02x^2 + 200x - 80,000$

71. a. $11.9\sqrt{1 + 0.91t}$ **b.** 25.6 million **73.** $6/(5\pi)$ L/sec

75. a. 64.45°F **b.** 64.63°F **77. b.** 0.500005

Chapter 4 Challenge Problems • page 495

1. $|b| - |a|$ **3.** $20\sqrt{2}$ **5.** $-\frac{4}{5}$

7. $\dfrac{1}{\sin a}\tan^{-1}\left(\dfrac{x + \cos a}{\sin a}\right) + C$ **9.** $f(a)$ **11. b.** 0

13. b. $\dfrac{\sin x}{2\sqrt{x}} + \dfrac{\sin(1/x^2)}{x^2}$ **15. b.** Yes **23.** $f(x) = x$

CHAPTER 5

Exercises 5.1 • page 506

1. $\frac{40}{3}$ **3.** $\frac{40}{3}$ **5.** $\frac{13}{12}$

7. $\displaystyle\int_{2010}^{2050} [f(t) - g(t)]\, dt$ billion barrels

9.

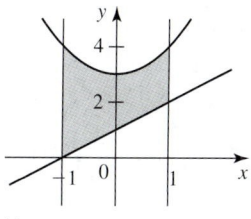

$\frac{14}{3}$

11.

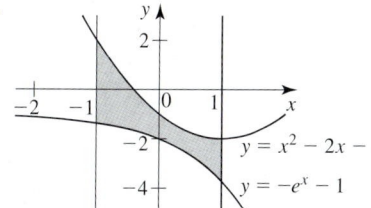

$y = x^2 - 2x - 1$

$y = -e^x - 1$

$\dfrac{3e^2 + 2e - 3}{3e}$

13.

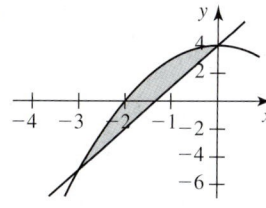

$\frac{9}{2}$

15.

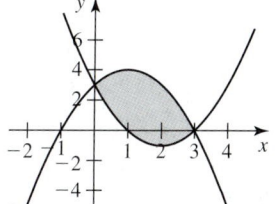

9

17.

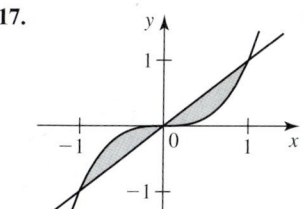

$\frac{1}{2}$

19.

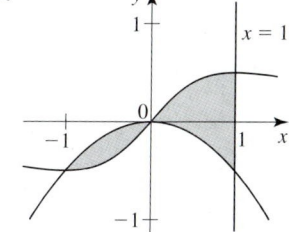

$x = 1$

$\ln 2$

21.

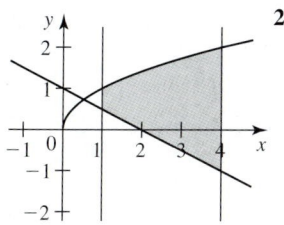

$\frac{65}{12}$

23.

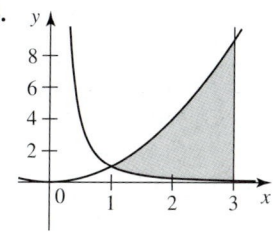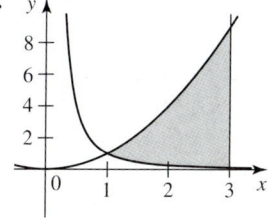

8

25.

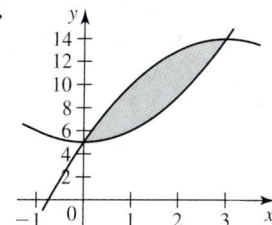

9

27.

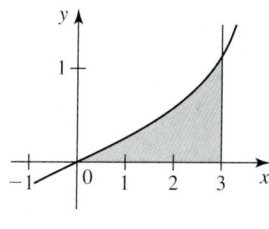

$4 - \sqrt{7}$

29.

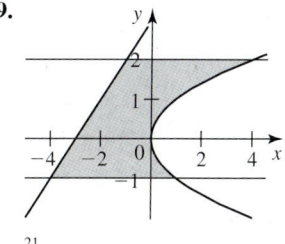

$\frac{21}{2}$

31.

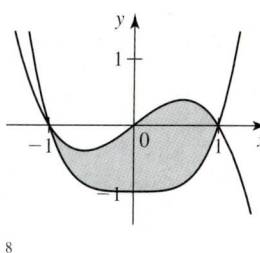

$\frac{8}{5}$

33.

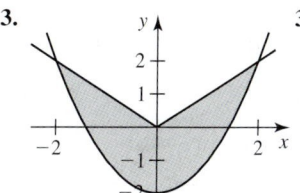

$\frac{20}{3}$

35.

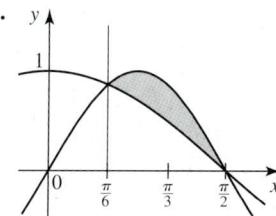

$\frac{1}{4}$

37.

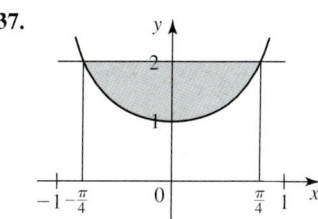

$\pi - 2$

39.

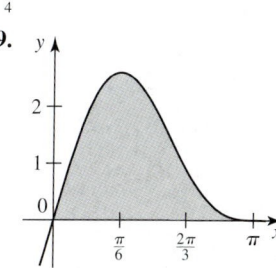

4

41. $4\sqrt{2}/3$ **43.** 11   **45. a.** 8   **b.** 8

47. $A = \displaystyle\int_0^b [g(x) - f(x)]\, dx$

49. a.

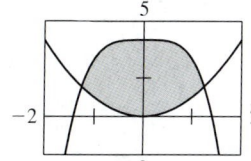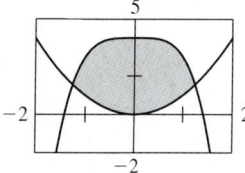

b. -1.25 and $1.25,\ 7.48$

51. a.

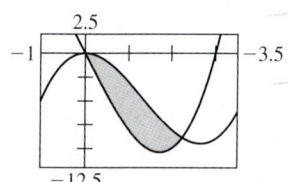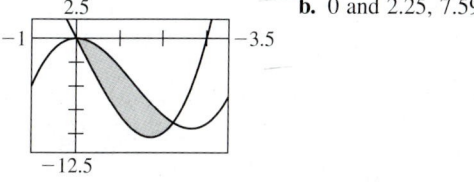

b. 0 and $2.25,\ 7.59$

53. a. 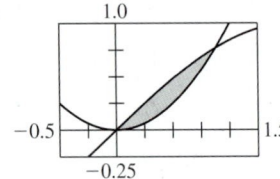   **b.** 0 and 0.88, 0.14

55. 30 ft/sec   **57.** 3,661,581 ft^2

59. a. 300 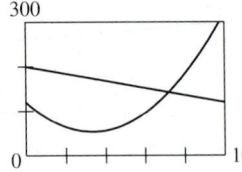   **b.** 342,000

61. $\frac{1}{12}$ **63.** 0.5875   **65. a.** $4^{2/3}$ **b.** 0.69

67. a. $(m-1)/(m+1)$ **b.** 0, 1

69. a. 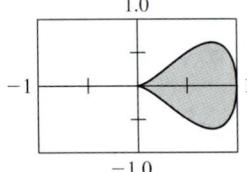   **b.** 0.78540

71. False   **73.** True

Exercises 5.2 • page 522

1. $2\pi/3$ **3.** $153\pi/5$ **5.** $\pi(1-(\pi/4))$ **7.** $\pi/10$

9. $64\pi/15$ **11.** $47\pi/84$ **13.** $32\pi/5$ **15.** $16\pi/15$

17. $(\pi/2)(e^2-1)$ **19.** $\pi/2$ **21.** $16\pi/3$ **23.** $64\pi/3$

25. $128\pi/15$ **27.** $\pi^2/4$ **29.** $3\pi/10$

31. $(\pi/2)(e^2-1)$ **33.** $19\pi/48$

35. a.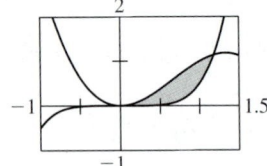

b. $(0,0)$ and $(1.18, 1.14)$, 1.08

37. $64\pi/15$ **39.** $1088\pi/15$ **41.** $104\pi/15$

43. 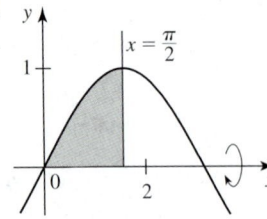   **45.**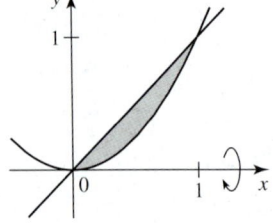

47. $(\pi/15)a^3$ **49.** $2\pi/5$ **51.** $71\pi/30$

53. $(\pi h/3)(R^2+rR+r^2)$ **55.** $(\pi h^2/3)(3r-h)$

57. 32   **59.** $81\sqrt{3}/40$

61. a.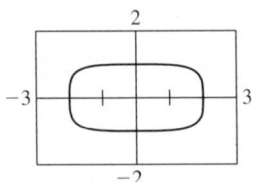

b. 12.6   **c.** 10.9832

63. a. 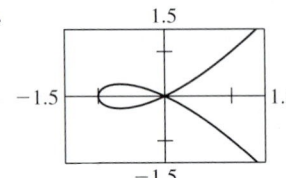   **b.** $\pi/24$

65. $\frac{32}{2}$ **67.** $\frac{2}{3}a^3$ **69. b.** $\pi r^2 h$ **71.** 33.52 ft^3

Exercises 5.3 • page 533

1. $8\pi/3$ **3.** $\pi/6$ **5.** $48\pi/5$

7. 8π **9.** $8\pi/3$

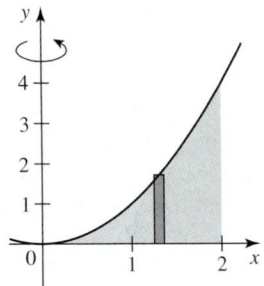

11. 2π **13.** $\pi \ln 5$

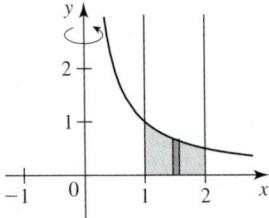   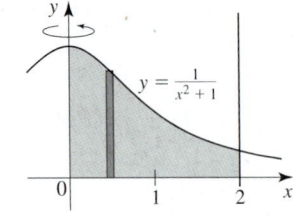

15. 18π **17.** 8π

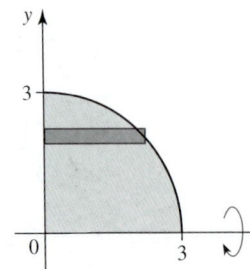   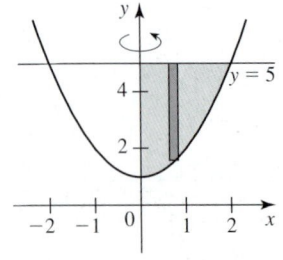

19. $\pi/3$

21. $19\pi/48$

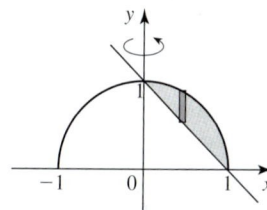

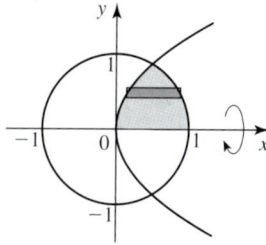

23. a.

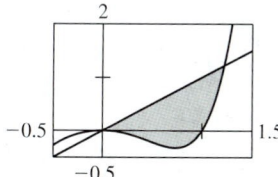

b. $(0, 0)$ and $(1.22, 1.22)$ **c.** 3.67

25. $16\pi/3$

27. $9\pi/4$

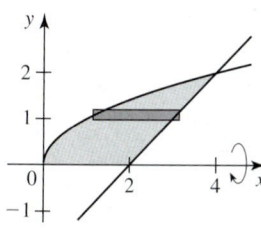

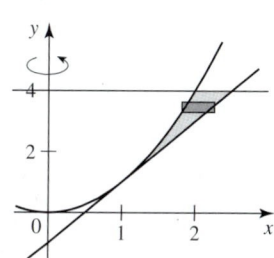

29. $\pi/15$

31. $32\pi/3$

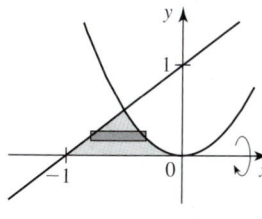

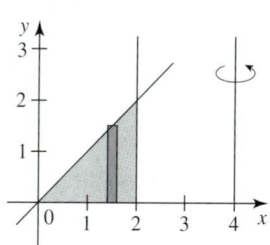

33. $128\pi/3$

35. $8\pi/15$

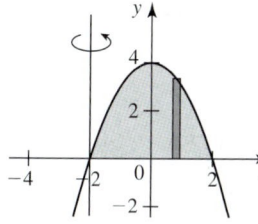

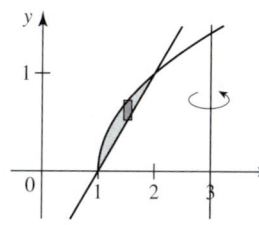

37.

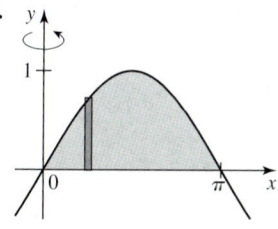

39.

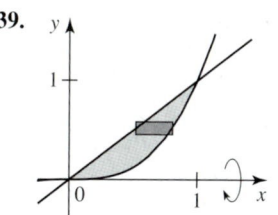

43. $\frac{4}{3}\pi a^2 b$ **45.** $2\pi^2 a^2 b$

47. 10π ft^3 **49.** $296{,}231{,}243$ ft^3

Exercises 5.4 • page 546

1. $\frac{1}{27}(80\sqrt{10} - 13\sqrt{13})$ **3.** $\frac{8}{27}(10\sqrt{10} - 1)$

5. $\sqrt{73}$ **7.** $3\sqrt{5}$ **9.** $\frac{2}{27}(37\sqrt{37} - 1)$ **11.** 45

13. $\frac{2}{27}(37\sqrt{37} - 1)$ **15.** $\frac{3}{4}$ **17.** 0.8814

19. $\displaystyle\int_{-1}^{2} \sqrt{1 + 4x^2}\, dx$ **21.** $\displaystyle\int_{-1}^{2} \sqrt{1 + \frac{4x^2}{(x^2 + 1)^4}}\, dx$

23. $\displaystyle\int_{0}^{\pi/4} \sqrt{1 + \sec^4 x}\, dx$

25. a.

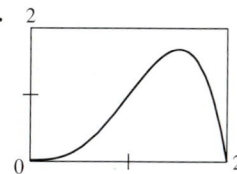

b. $\displaystyle\int_{0}^{2} \sqrt{1 + 4x^4(3 - 2x)^2}\, dx$ **c.** 4.2008

27. a.

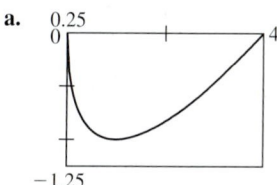

b. $\displaystyle\int_{0}^{4} \sqrt{1 + \left(1 - \frac{1}{\sqrt{x}}\right)^2}\, dx$ **c.** 4.8086

29. $6a$ **31.** 0.6325 **33.** $5\sqrt{5}\pi$

35. $(\pi/27)(10\sqrt{10} - 1)$ **37.** $(\pi/6)(17\sqrt{17} - 1)$

39. $\dfrac{45\sqrt{13}\pi}{4}$ **41.** $\pi/4$ **43.** $(\pi/16)(16 \ln 2 + 15)$

45. $2\pi \displaystyle\int_{0}^{\pi/2} \sin x \sqrt{1 + \cos^2 x}\, dx$ **47.** 21.4018

51. 4π **53.** $\frac{8}{3}$ mi **55.** $24{,}223.5$ ft **57.** 30.73 in.

Exercises 5.5 • page 554

1. 200 ft-lb **3.** -500 ft-lb

5. 6 ft-lb **7.** 18 J **9.** $-\dfrac{2}{\pi}$ J

11. $\frac{17}{6}$ J **13.** $\frac{7}{24}$ ft-lb **15.** 360 ft-lb

17. $15{,}600$ ft-lb **19.** $56{,}458$ ft-lb

21. 1740 ft-lb **23.** $121{,}919$ J **25.** $128{,}252$ ft-lb

27. 5799 ft-lb **29.** 1405 ft-lb **31.** 1.91×10^{13} J

33. 3.46×10^{8} ft-lb

35. $\dfrac{qQ}{4\pi\varepsilon_0}\left(\dfrac{1}{\sqrt{a^2+R^2}}-\dfrac{1}{\sqrt{b^2+R^2}}\right)$

37. $nRT\ln\dfrac{V_1-nb}{V_0-nb}+an^2\dfrac{V_0-V_1}{V_0V_1}$

Exercises 5.6 • page 563

1. a. 187.2 lb   **b.** 93.6 lb   **c.** 31.2 lb

3. 1040 lb   **5.** 6760 lb   **7.** 1064.96 lb   **9.** 3057.6 lb

11. 374.4 lb   **13.** 561.6 lb   **15.** 12.0 lb   **17.** 7841 lb

19. a. 8387 lb   **b.** 36,741 lb

21. 477.9 lb   **23.** 610,509 lb

Exercises 5.7 • page 574

1. $\frac{7}{6}$ m   **3.** $\frac{1}{3}$ m   **5.** $\left(-\frac{5}{12},\frac{3}{2}\right)$ **7.** $\left(\frac{5}{6},\frac{7}{9}\right)$ **9.** $\left(1,\frac{2}{3}\right)$

11. $\left(0,\frac{8}{5}\right)$ **13.** $\left(0,\frac{1}{5}\right)$ **15.** $\left(1,\frac{2}{5}\right)$ **17.** $\left(5,\frac{10}{7}\right)$

19. $\left(\frac{9}{20},\frac{9}{20}\right)$ **21.** $\left(\frac{16}{35},\frac{16}{35}\right)$ **23.** $\left(1,\frac{13}{5}\right)$ **25.** $\left(0,\frac{4}{3}\right)$

27. $\left(0,-\dfrac{20}{3(8+\pi)}\right)$ **29.** $\left(\frac{37}{18},\frac{23}{18}\right)$ **31.** $\left(\dfrac{\pi-4}{\pi+4},0\right)$

33. $\left(\dfrac{\pi-2}{\pi+2},0\right)$ **35.** $\left(\dfrac{a}{3},\dfrac{b}{3}\right)$ **39.** $(1.44,0.36)$

41. $72\pi^2$ **43.** 8π **45.** $\left(0,\dfrac{4R}{3\pi}\right)$

47. $\left(0,\dfrac{2a}{\pi}\right)$ **49.** $\left(\dfrac{1}{2},\dfrac{\pi}{8}\right)$

Exercises 5.8 • page 584

1. a. 3.6269   **b.** 27.3082   **c.** 0.0993

3. a. 1   **b.** 0.6481   **c.** 1.3333

5. a. 0.4812   **b.** −0.4812   **c.** 0.8047

17. $\operatorname{csch} x=\frac{3}{4}$, $\cosh x=\frac{5}{3}$, $\operatorname{sech} x=\frac{3}{5}$, $\tanh x=\frac{4}{5}$, $\coth x=\frac{5}{4}$

19. $3\cosh 3x$ **21.** $-3\operatorname{sech}^2(1-3x)$ **23.** e^{2t}

25. $\tanh x$ **27.** $\dfrac{2u\sinh u^2}{\cosh^2(\cosh u^2)}$

29. $12t\cosh(3t^2+1)\sinh(3t^2+1)$

31. $\sinh v^2+2v^2\cosh v^2$ **33.** $2e^{2x}\operatorname{sech}^2(e^{2x}+1)$

35. $-\frac{2}{3}(\cosh x-\sinh x)^{2/3}$ **37.** $-\operatorname{csch} x$

39. $\dfrac{1}{1+\cosh x}$

41. $\dfrac{(1+\tanh 2t)\cdot\dfrac{1}{\sqrt{t^2-1}}-2\cosh^{-1}t\operatorname{sech}^2 2t}{(1+\tanh 2t)^2}$

43. $\dfrac{3}{\sqrt{1+9x^2}}$ **45.** $\dfrac{1}{\sqrt{\cosh^{-1}2x}\sqrt{4x^2-1}}$

47. $-\dfrac{1}{(2x+1)\sqrt{-2x}}$ **49.** $\cosh^{-1}x^2+\dfrac{2x^2}{\sqrt{x^4-1}}$

51. $-\dfrac{1}{2x\sqrt{1-x}}$ **53.** $\dfrac{9(x-1)}{\sqrt{9x^2-1}}$

55. $\frac{1}{2}\sinh(2x+3)+C$ **57.** $\frac{2}{3}(\sinh x)^{3/2}+C$

59. $\frac{1}{3}\ln|\sinh 3x|+C$ **61.** $\ln(1+\cosh x)+C$

63. $\pi a^3\left(\dfrac{b}{a}+\dfrac{1}{2}\sinh\dfrac{2b}{a}\right)$ **65.** $\dfrac{T_0^2}{bW}\sinh\dfrac{Wb}{T_0}$

67. a. $0,-2$ ft/sec

b.

69. $\pi[(b-a)+\frac{1}{2}(\sinh 2b-\sinh 2a)]$

71. $\left(0,\dfrac{e^{2a}+4a-e^{-2a}}{8(e^a-e^{-a})}\right)$

77. True

79. True

Chapter 5 Concept Review • page 586

1. a. $\displaystyle\int_a^b [f(x)-g(x)]\,dx$

b. $\displaystyle\int_a^b |f(x)-g(x)|\,dx$

3. $\displaystyle\pi\int_a^b \{[f(x)]^2-[g(x)]^2\}\,dx$

5. a. $\displaystyle 2\pi\int_a^b xf(x)\,dx$

b. $\displaystyle 2\pi\int_c^d yf(y)\,dy$

7. $\displaystyle\int_a^x \sqrt{1+[f'(t)]^2}\,dt$; $[a,b]$; $\sqrt{1+(y')^2}\,dx$; $(dx)^2+(dy)^2$

9. $\displaystyle\int_a^b F(x)\,dx$

11. a. $\displaystyle\int_a^b \rho f(x)\,dx$

b. $\displaystyle\frac{1}{2}\int_a^b \rho[f(x)]^2\,dx$; $\displaystyle\int_a^b \rho xf(x)\,dx$

c. $\dfrac{M_y}{m};\dfrac{M_x}{m}$

Chapter 5 Review Exercises • page 587

1.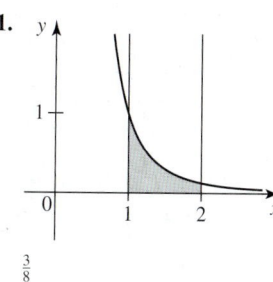

$\frac{3}{8}$

3.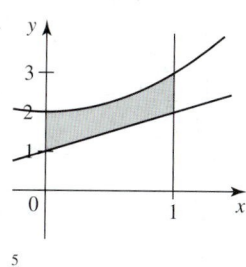

$\frac{5}{6}$

5.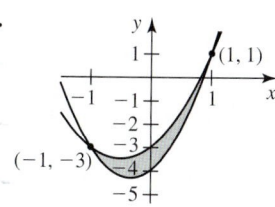

$\frac{4}{3}$

7.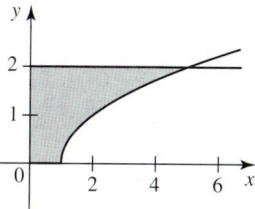

$\frac{14}{3}$

9.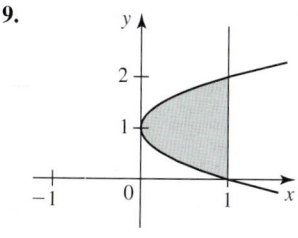

$\frac{4}{3}$

11.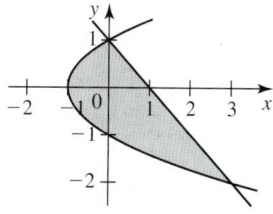

$\frac{9}{2}$

13.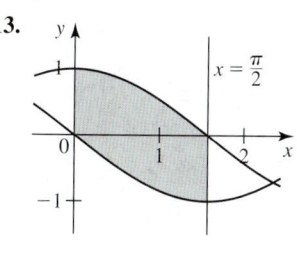

2

15.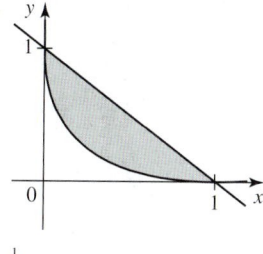

$\frac{1}{3}$

17. $\frac{1}{2}(e^4 + 2 \ln 2 - e^2)$ **19.** 8π **21.** $8\pi/21$

23. $7\pi/30$ **25.** π **27.** $2 \coth 2x$ **29.** $\dfrac{\operatorname{sech}^2 x}{\sqrt{2 - \operatorname{sech}^2 x}}$

31. $\dfrac{-1}{2\sqrt{x - x^2}\left[1 + (\cos^{-1}\sqrt{x})^2\right]}$ **33.** $-\frac{1}{2}x + \frac{1}{4}\sinh 2x + C$

35. $\frac{1}{2}\cosh 2t + C$ **37.** $\frac{1}{6}$ **39.** 722 m^3

41. 36   **43.** $\frac{1}{4}(3 + 2 \ln 2)$ **45.** $14\pi/3$

47. $2\pi \displaystyle\int_1^2 \dfrac{(x^3 + 1)\sqrt{4x^6 + x^4 - 4x^3 + 1}}{x^3}\,dx$

51. $\frac{2}{3}$ ft-lb   **53.** 367,566 ft-lb

55. deep end: 30,576 lb; shallow end: 5616 lb; other sides: 41,080 lb

57. $\left(1, -\frac{2}{5}\right)$ **59.** $\left(1, -\frac{2}{5}\right)$

Chapter 5 Challenge Problems • page 590

1. $a = 3, b = 5$ **3.** $y = \dfrac{\sqrt{3}}{3}x^2$ **5.** $f(x) = \frac{2}{3}(1 + x^2)^{3/2}$

7. 52   **9.** $\dfrac{4\pi ab^2(2\sqrt{3} - 3)}{9}$

CHAPTER 6

Exercises 6.1 • page 601

1. $\frac{1}{4}(2x - 1)e^{2x} + C$ **3.** $\sin x - x \cos x + C$

5. $\frac{1}{4}x^2(2 \ln 2x - 1) + C$ **7.** $-(x^2 + 2x + 2)e^{-x} + C$

9. $(x^2 - 2)\sin x + 2x \cos x + C$

11. $x \tan^{-1} x - \frac{1}{2}\ln(1 + x^2) + C$

13. $\frac{2}{9}t\sqrt{t}(3 \ln t - 2) + C$

15. $x \tan x + \ln|\cos x| + C$

17. $\frac{1}{13}e^{2x}(3 \sin 3x + 2 \cos 3x) + C$

19. $\frac{1}{4}[\sin(2u + 1) - 2u \cos(2u + 1)] + C$

21. $x \tan x + \ln|\cos x| - \frac{1}{2}x^2 + C$

23. $2[(x - 2)\sin \sqrt{x} + 2\sqrt{x} \cos \sqrt{x}] + C$

25. $\frac{1}{4}\sec^3 \theta \tan \theta + \frac{3}{8}(\sec \theta \tan \theta + \ln|\sec \theta + \tan \theta|) + C$

27. $x^3 \cosh x - 3x^2 \sinh x + 6x \cosh x - 6 \sinh x + C$

29. $-e^{-x} \ln(e^x + 1) - \ln(1 + e^{-x}) + C$

31. $4\sqrt{1 - x} - 2 \ln\left|\dfrac{\sqrt{1 - x} + 1}{\sqrt{1 - x} - 1}\right| - 2\sqrt{1 - x}\,\ln x + C$

33. $\frac{1}{9}(2e^3 + 1)$ **35.** $\frac{1}{6}(\pi + 6 - 3\sqrt{3})$

37. $\dfrac{3\sqrt{e} - 4}{2e}$ **39.** $\dfrac{4\pi - 3\sqrt{3}}{6}$ **41.** $\dfrac{2 \ln 2 + \pi - 4}{2}$

43. $\frac{1}{2}\ln\frac{3}{2} + \frac{\pi}{36}(9 - 4\sqrt{3})$ **45.** $e - 2$ **47.** $\frac{1}{8}(\pi - 4 \ln 2)$

49. a.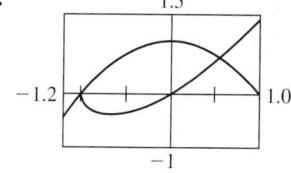

1.5

−1.2   1.0

−1

−1, 0.555

b. 1.251

51. π **53.** $\frac{\pi}{5}(2e^{\pi/2} - 3)$ **55.** $2\pi(\pi + 2)$

57. $P(t) = 820 - 40(t + 20)e^{-0.05t}$; 92.2 million metric tons

59. $3te^{-4t}$ **61.** $\frac{1}{3}e^{-20t}(\sin 60t + 3 \cos 60t) + q_0 - 1$

63. $v_0 t - \dfrac{1}{2}gt^2 + \dfrac{ms}{r}\left\{\left(1 - \dfrac{r}{m}t\right)\left[\ln\left(1 - \dfrac{r}{m}t\right) - 1\right] + 1\right\}$

65. $(c_2 - c_1)\left(\dfrac{r_1}{r_2 - r_1} + \dfrac{1}{\ln r_1 - \ln r_2}\right) + c_2$

67. 16   **69.** True

Exercises 6.2 • page 611

1. $\frac{1}{4}\sin^4 x + C$ **3.** $\frac{1}{4}\left(\frac{1}{3}\sin^6 2x - \frac{1}{4}\sin^8 2x\right) + C$

5. $\frac{1}{3}\cos^3 x - \cos x + C$ **7.** $\pi/2$ **9.** $\frac{3}{8}$

11. $\frac{1}{16}(2x - \sin 2x) + C$ **13.** $\pi/16$

15. $\frac{1}{64}\left[12x^2 + 8\sin(2x^2) + \sin(4x^2)\right] + C$

17. $\frac{1}{8}(2x^2 - \cos 2x - 2x\sin 2x) + C$

19. $\frac{4 - \pi}{4}$ **21.** $\frac{1}{2}\sec^4\frac{x}{2} - 2\tan^2\frac{x}{2} - \ln\cos^2\frac{x}{2} + C$

23. $\frac{1}{4\pi}\tan^4(\pi x) + C$ **25.** $\frac{1}{9}\tan^3 3x + \frac{1}{15}\tan^5 3x + C$

27. $\frac{2}{3}\tan^{3/2}\theta + \frac{2}{7}\tan^{7/2}\theta + C$ **29.** $-\frac{1}{2}\cot 2x - x + C$

31. $-\frac{1}{2}(\cot x \csc x + \ln|\csc x + \cot x|) + C$

33. $-\left(\cot t + \frac{1}{3}\cot^3 t\right) + C$

35. $-\frac{1}{5}\cot^5 t + \frac{1}{3}\cot^3 t - \cot t - t + C$

37. $\frac{4}{3}$ **39.** $-\frac{1}{3}$ **41.** $\frac{1}{4}\sin 2\theta + \frac{1}{12}\sin 6\theta + C$

43. $\frac{\ln|\sin 2\theta|}{2} - \frac{\sin^2 2\theta}{4} + C$ **45.** $\frac{1}{2}\tan^4\sqrt{t} + C$

47. $\frac{1}{2}\sin 2x + C$ **49.** $\frac{1}{2}$ **51.** $\frac{1}{2}$

53. a.

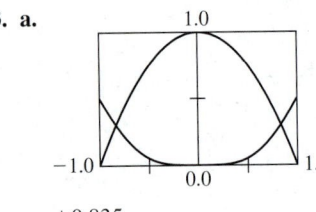

±0.835

b. 1.165

55. $\frac{\pi(3\pi - 8)}{12}$ **57.** $\left(\frac{1}{2}(\pi - 2), \frac{\pi}{8}\right)$ **59.** $8/\pi, 0$

61. $8\pi/15$ **63.** $-\frac{2I_0 \sin \alpha}{\pi}$ **65.** $0.12RI_0^2 T$

67.

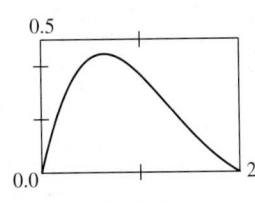

a. 0.55   **b.** 2.63

Exercises 6.3 • page 619

1. $-\sqrt{9 - x^2} + C$ **3.** $-\frac{(4 - x^2)^{3/2}}{3} + C$

5. $-\frac{1}{2}\ln\left|\frac{\sqrt{4 + x^2} + 2}{x}\right| + C$ **7.** $-\frac{\sqrt{x^2 + 4}}{4x} + C$

9. $-\frac{1}{15}(3x^2 + 2)(1 - x^2)^{3/2} + C$

11. $\frac{1}{3}(x^2 - 18)\sqrt{x^2 + 9} + C$ **13.** $-\frac{x}{9\sqrt{x^2 - 9}} + C$

15. $\sqrt{16x^2 - 9} - 3\sec^{-1}\left(\frac{4}{3}x\right) + C$

17. $-\frac{(1 - x^2)^{3/2}}{3x^3} + C$ **19.** $-\frac{1}{2}\ln\left|\frac{\sqrt{9x^2 + 4} + 2}{x}\right| + C$

21. $\frac{1}{3}(4\pi + 3\sqrt{3})$ **23.** $\frac{1}{2}(\sqrt{3} - \sqrt{2})$

25. $2\sin^{-1}\left(\frac{1}{2}e^x\right) + \frac{1}{2}e^x\sqrt{4 - e^{2x}} + C$ **27.** $\ln\frac{2\sqrt{3} + 3}{3}$

29. $\sin^{-1}(t - 1) + C$

31. $\frac{1}{16}\tan^{-1}\left(\frac{x + 2}{2}\right) + \frac{x + 2}{8(x^2 + 4x + 8)} + C$

33. $-\frac{1}{2}\ln\frac{\sqrt{2} + 1}{\sqrt{3} + 2}$ **35.** $\frac{\pi}{4}b$ **37.** $\frac{4\pi}{3}(2\pi - 3\sqrt{3})$

39. $2 + \ln\left(\frac{2\sqrt{3}}{3} - \frac{\sqrt{3}}{3}\right) - \sqrt{2} - \ln(\sqrt{2} - 1)$

41. $\frac{8(8\pi + 9\sqrt{3})}{3}$ or ≈ 108.59 lb

43. a.

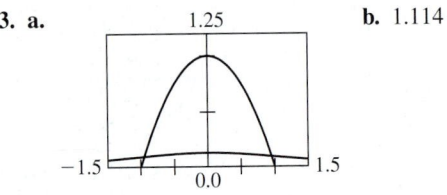

±0.953

b. 1.114

45. $\frac{2\pi a^2 b}{\sqrt{a^2 - b^2}}\left[\sin^{-1}\left(\frac{\sqrt{a^2 - b^2}}{a}\right) + \frac{b\sqrt{a^2 - b^2}}{a^2}\right]$

47. $\left\{\frac{544}{10}[\tan^{-1}(0.25) - \tan^{-1}(-2.25)] + 28\right\}$ or ≈ 104 PSI

49. $\frac{2\sqrt{5}k}{5\,a^2}$

55. b. $\sqrt{1 + x^2}\tan^{-1} x - \ln(\sqrt{1 + x^2} + x) + C$

Exercises 6.4 • page 630

1. a. $\frac{A}{x} + \frac{B}{x - 5}$ **b.** $\frac{A}{x + 1} + \frac{B}{3x - 2}$

3. a. $\frac{A}{x} + \frac{B}{x^2} + \frac{C}{x + 1}$ **b.** $\frac{A}{x + 4} + \frac{B}{x - 1}$

5. a. $\frac{A}{x + 2} + \frac{B}{x - 2} + \frac{Cx + D}{x^2 + 4}$ **b.** $\frac{A}{2x - 1} + \frac{Bx + C}{x^2 + 4}$

7. $\frac{1}{4}\ln\left|\frac{x - 4}{x}\right| + C$ **9.** $\ln\frac{|t|^3}{(t + 1)^2} + C$ **11.** $\frac{1}{4}\ln\frac{5}{3}$

13. $\frac{1}{3}\ln|(x - 2)(x + 1)^2| + C$ **15.** $\ln\left|\frac{(x + 3)^3(x - 2)}{(x + 2)^2}\right| + C$

17. $2x + \ln|x(x - 1)^2| + C$ **19.** $\ln\left|\frac{x^3}{x - 1}\right| - \frac{2}{x - 1} + C$

21. $\ln|x(x+1)^3| - \dfrac{2}{x} + C$

23. $\ln\left|\dfrac{(v-1)^2}{v}\right| - \dfrac{v}{(v-1)^2} + C$

25. $2\ln\left|\dfrac{x}{(x+1)^2}\right| + \dfrac{2}{x+1} + x + C$

27. $\ln|x| - \dfrac{1}{2}\ln|x^2 - 1| - \dfrac{1}{2(x^2-1)} + C$

29. $\frac{1}{6}(31\ln|x-1| + 8\ln|x+2| - 3\ln|x+3|) + C$

31. $x + \tan^{-1}x + \ln\left|\dfrac{x+1}{x^2+1}\right| + C$

33. $\dfrac{5}{2}\ln(x^2+1) - 3\tan^{-1}x - \dfrac{1}{x^2+1} + C$

35. $\dfrac{1}{2}\ln\dfrac{(x+1)^2}{x^2-4x+6} + C$

37. $\dfrac{1}{6}\left[\ln\dfrac{x^2-x+1}{(x+1)^2} + 2\sqrt{3}\tan^{-1}\dfrac{\sqrt{3}(2x-1)}{3}\right] + C$

39. $2\ln 2 + \dfrac{\pi}{4} - \dfrac{1}{2}$

41. $2\sqrt{3}\tan^{-1}\left[\dfrac{\sqrt{3}}{3}(2x+1)\right] + \dfrac{1}{x^2+x+1} + C$

43. $-\dfrac{1}{2(2x^3 - x^2 + 8x + 4)} + C$

45. $\ln\left|\dfrac{\cos x}{\cos x + 1}\right| + \sec x + C$

47. $\dfrac{1}{3}\ln\left|\dfrac{e^t - 1}{e^t + 2}\right| + C$

49. $e^t + \dfrac{1}{6}\ln\left|\dfrac{(e^t+1)^3(e^t-1)}{(e^t+2)^{16}}\right| + C$

51. $3x^{1/3} + \dfrac{1}{2}\ln\dfrac{x^{2/3} - x^{1/3} + 1}{(x^{1/3}+1)^2}$
$\qquad - \sqrt{3}\tan^{-1}\left[\dfrac{\sqrt{3}}{3}(2x^{1/3} - 1)\right] + C$

53. $\tan\dfrac{x}{2} + C$

55. $\dfrac{2\sqrt{15}}{15}\tan^{-1}\dfrac{\sqrt{15}(8\sin x + \cos x + 1)}{15(1+\cos x)} + C$

57. $\ln 2$ **59.** $\frac{1}{2}(\ln|\sin x + \cos x| + x) + C$ **61.** $\ln\frac{4}{3}$

63. a. 3   **b.** 0.892

2.144

65. $\dfrac{2\pi}{3}(1 + 3\ln 3 - 6\ln 2)$ **67.** $2\ln 3 - 1$

69. a. $\dfrac{5}{3}\ln|x+2| - \dfrac{10}{7}\ln|x+3|$
$\qquad - \dfrac{1}{42}\left[5\ln(x^2 + x + 1) + 2\sqrt{3}\tan^{-1}\dfrac{2x+1}{\sqrt{3}}\right] + C$

b. $\dfrac{5}{3(x+2)} - \dfrac{10}{7(x+3)} - \dfrac{5x+4}{21(x^2+x+1)}$

c. $\dfrac{1}{42}\left\{-2\sqrt{3}\tan^{-1}\left(\dfrac{2x+1}{\sqrt{3}}\right) + 70\ln(x+2)\right.$
$\qquad \left. - 5[12\ln(x+3) + \ln(x^2+x+1)]\right\}$

71. 9231   **73.** False   **75.** False

Exercises 6.5 • page 639

1. $\frac{1}{15}(3x-1)(1+2x)^{3/2} + C$

3. $\dfrac{1}{8}\left(1 + 2x - \dfrac{1}{1+2x} - 2\ln|1+2x|\right) + C$

5. $-\dfrac{\sqrt{3+2x}}{x} + \dfrac{\sqrt{3}}{3}\ln\left|\dfrac{\sqrt{3+2x}-\sqrt{3}}{\sqrt{3+2x}+\sqrt{3}}\right| + C$

7. $-\dfrac{\sqrt{3}}{3}\ln\left|\dfrac{\sqrt{3+2x^2}+\sqrt{3}}{\sqrt{2}x}\right| + C$

9. $\sqrt{2-x^2} - \sqrt{2}\ln\left|\dfrac{\sqrt{2}+\sqrt{2-x^2}}{x}\right| + C$

11. $\sqrt{x^2-3} - \sqrt{3}\cos^{-1}\dfrac{\sqrt{3}}{|x|} + C$

13. $\dfrac{e^x}{\sqrt{1-e^{2x}}} + C$

15. $\frac{1}{16}[(8x^2-1)\cos^{-1}2x - 2x\sqrt{1-4x^2}] + C$

17. $\frac{1}{2}\sin(x^2+1) - \frac{1}{2}x^2\cos(x^2+1) + C$ **19.** $\pi/4$

21. $-\frac{1}{13}e^{-2x}(2\sin 3x + 3\cos 3x) + C$

23. $-\frac{1}{8}e^{-2x}(4x^3 + 6x^2 + 6x + 3) + C$

25. $-\tan^{-1}(\cos x) + C$

27. $\dfrac{x^4}{16}(4\ln 5x - 1) + C$

29. $\dfrac{x-3}{2}\sqrt{6x-x^2} + \dfrac{9}{2}\cos^{-1}\left(\dfrac{3-x}{3}\right) + C$

31. $\dfrac{8\sqrt{3}}{9}\cos^{-1}\left(\dfrac{4-3x}{4}\right) - \dfrac{x+4}{6}\sqrt{8x-3x^2} + C$ **33.** $\frac{4}{3}$

35. $-2(\cos x - 1)e^{\cos x} + C$ **37.** $\frac{1}{9}(2e^3 + 1)$ **39.** $\frac{3}{16}\pi^2$

41. $\left(\dfrac{\pi^2 - 4}{4\pi}, \dfrac{3}{8}\right)$ **43.** 28,284   **45.** 44; 49

47. $\dfrac{\pi}{32}[18\sqrt{5} - \ln(2+\sqrt{5})]$

53. *Maple:* $\dfrac{2(x+2)^{5/2}}{5} - \dfrac{4(x+2)^{3/2}}{3}$;

Mathematica: $\frac{2}{15}(2+x)^{3/2}(-4+3x)$;

TI-89: $\dfrac{2(x+2)^{3/2}(3x-4)}{15}$

55. *Maple:* $2\sqrt{x+2} - \sqrt{2}\,\text{arctanh}\left(\dfrac{\sqrt{x+2}\sqrt{2}}{2}\right)$;

Mathematica: $2\sqrt{2+x} - \sqrt{2}\,\text{arctanh}\,\dfrac{\sqrt{2+x}}{\sqrt{2}}$;

TI-89:

$\dfrac{\sqrt{2}\,\ln(|\sqrt{x+2}-\sqrt{2}|) - \sqrt{2}\,\ln(|\sqrt{x+2}+\sqrt{2}|) + 4\sqrt{x+2}}{2}$

57. *Maple and TI-89:* $\frac{1}{4}\cos^3 x\sin x + \frac{3}{8}\cos x\sin x + \frac{3}{8}x$;

Mathematica: $\frac{3}{8}x + \frac{1}{4}\sin 2x + \frac{1}{32}\sin 4x$

59. *Maple:* $x^5 e^x - 5x^4 e^x + 20x^3 e^x - 60x^2 e^x + 120xe^x - 120e^x$;

Mathematica: $e^x(-120 + 120x - 60x^2 + 20x^3 - 5x^4 + x^5)$;

TI-89: $(x^5 - 5x^4 + 20x^3 - 60x^2 + 120x - 120)\,e^x$

61. *Maple:* $\frac{2}{3}(e^x+1)^{3/2} - 2\sqrt{e^x+1}$;

Mathematica: $\frac{2}{3}(-2+e^x)\sqrt{1+e^x}$;

TI-89: $\dfrac{2(e^x-2)\sqrt{e^x+1}}{3}$

63. $-\frac{3}{5}(x+3)(2-x)^{2/3} + C$　　**65.** $-\sin(1/x) + C$

67. $1 + \dfrac{\pi}{6} - \dfrac{\sqrt{3}}{2}$　　**69.** $\dfrac{\sqrt{3}}{36}\pi$

71. $\ln\left(\ln x + \sqrt{1+(\ln x)^2}\right) + C$　　**73.** $\dfrac{2(8-3\sqrt{3})}{3}$

75. $\dfrac{3^{x^3}}{\ln 3} + C$　　**77.** $\frac{1}{4}\left[(2x^2-1)\sin^{-1}x + x\sqrt{1-x^2}\right] + C$

79. $2\ln\dfrac{\sqrt{5}-1}{2} + \dfrac{\sqrt{5}}{2}$　　**81.** $\frac{1}{2}(1-\ln 2)$

83. $x\tan x + \ln|\cos x| - \frac{1}{2}x^2 + C$　　**85.** $\sqrt{2}(2-\sqrt{3})$

87. $2\ln(\sqrt{x+1}+1) + C$　　**89.** $\ln\left|x-3+\sqrt{x^2-6x}\right| + C$

91. $-\frac{1}{6}\cot^3 2x + \frac{1}{2}\cot 2x + x + C$

93. $\sqrt{x^2+9} - 3\ln\left|\dfrac{3+\sqrt{x^2+9}}{x}\right| + C$

95. $\frac{1}{3}\ln|x-1| - \frac{1}{6}\ln(x^2+x+1)$

$\qquad - \dfrac{\sqrt{3}}{3}\tan^{-1}\left[\dfrac{\sqrt{3}}{3}(2x+1)\right] + C$

97. $\frac{1}{4}\ln\left|\dfrac{x^2+x+1}{x^2-x+1}\right|$

$\qquad + \dfrac{\sqrt{3}}{6}\left[\tan^{-1}\dfrac{\sqrt{3}(2x+1)}{3} + \tan^{-1}\dfrac{\sqrt{3}(2x-1)}{3}\right] + C$

99. $\frac{1}{2}e^{e^{x^2}} + C$

Exercises 6.6 • page 653

1. $2\sqrt{3}$　**3.** $\frac{1}{2}$　**5.** π　**7.** $\frac{1}{2}$　**9.** 100　**11.** $2\sqrt{3}/3$

13. $1/(2e^2)$　**15.** Diverges　**17.** Diverges　**19.** $\pi/2$

21. $\pi/2$　**23.** 0　**25.** 3　**27.** $-\frac{9}{2}$　**29.** $3\sqrt[3]{3}$

31. Diverges　**33.** $-\frac{1}{4}$　**35.** $\sqrt{2}$　**37.** 4

39. Diverges　**41.** -4　**43.** Convergent

45. Convergent　**47.** Divergent　**49.** 1　**51.** $\frac{32}{105}\pi$

53. $\pi/2$　**55.** $\pi/2$　**57.** $3\pi/4$　**59.** $\dfrac{qQ}{4\pi a\varepsilon_0}$

61. $\dfrac{qQ}{4\pi\varepsilon_0 c}\ln\left|\dfrac{\sqrt{a^2+c^2}+c}{a}\right|$　　**63. b.** \$100,000

65. 2.3%　　**67.** 1; $2(\sqrt{2}-1)$

69. Converges if $p > 1$, diverges if $p \le 1$

73. a. $\displaystyle\int_1^\infty t^{-7/2}dt$　**c.** Converges

75. $\dfrac{1}{s}$, $s > 0$　　**77.** $\dfrac{1}{s^2}$, $s > 0$

81. False　**83.** True　**85.** True　**87.** True

Chapter 6 Concept Review • page 656

1. Product; $uv - \displaystyle\int v\,du$; u; easily integrated;

$\qquad f(x)g(x)\Big|_a^b - \displaystyle\int_a^b g(x)f'(x)\,dx$

3. a. $\sec x$; odd　**b.** $\tan x$; even

5. a. $a\sin\theta$　**b.** $a\tan\theta$　**c.** $a\sec\theta$

7. $\displaystyle\lim_{a\to-\infty}\int_a^b f(x)\,dx$; $\displaystyle\lim_{b\to\infty}\int_a^b f(x)\,dx$; $\displaystyle\int_{-\infty}^c f(x)\,dx + \int_c^\infty f(x)\,dx$;

$\qquad \displaystyle\lim_{c\to b-}\int_a^c f(x)\,dx$; $\displaystyle\int_a^c f(x)\,dx + \int_c^b f(x)\,dx$

Chapter 6 Review Exercises • page 657

1. $2(x - \ln|x+1|) + C$　　**3.** $-\frac{1}{3}(x^2+18)\sqrt{9-x^2} + C$

5. $\frac{1}{9}x^3(3\ln x - 1) + C$　　**7.** $-\cot\theta - \csc\theta + C$

9. $\dfrac{1}{27}\left(\ln\left|\dfrac{x}{x+3}\right| + \dfrac{6}{x+3} - \dfrac{3}{x}\right) + C$

11. $\dfrac{x}{2}\sqrt{x^2-4} - 2\ln\left|x+\sqrt{x^2-4}\right| + C$

13. $\frac{1}{4}\left[(2\theta^2-1)\sin^{-1}\theta + \theta\sqrt{1-\theta^2}\right] + C$

15. $-\frac{1}{2}(e^\pi + 1)$

17. $\dfrac{1}{4}\left[\ln\dfrac{x^8}{(x+1)^2(x^2+1)^3} - 2\tan^{-1}x\right] + C$

19. $\frac{1}{2}\left(\frac{1}{7}\tan^7 2x + \frac{1}{9}\tan^9 2x\right) + C$

21. $\dfrac{x \sin x + \cos x - 1}{\sin x} + C$ **23.** $\frac{1}{4}(\ln x)^4 + C$

25. $2 \tan^{-1}\sqrt{4x - 1} + C$ **27.** $\tan x\big(\ln|\tan x| - 1\big) + C$

29. $\frac{1}{8}(2 \cos 2x - \cos 4x) + C$

31. $\frac{1}{2}\sin^{-1}(2x + 3) + C$

33. $-\frac{1}{6}\sin t \cos^5 t + \frac{1}{24}\sin t \cos^3 t + \frac{1}{16}\sin t \cos t + \frac{1}{16}t + C$

35. $x + 2\sqrt{x} + 2 \ln|\sqrt{x} - 1| + C$

37. $\frac{1}{2}x^2 \cos^{-1} 2x - \frac{1}{8}x\sqrt{1 - 4x^2} + 16 \sin^{-1} 2x + C$

39. $\dfrac{x}{2} - \dfrac{1}{4}e^{-2x} + C$

41. $\frac{1}{2}\ln\big(2x + 1 + \sqrt{4x^2 + 4x + 10}\big) + C$ **43.** 1

45. Divergent   **47.** $-\frac{9}{2}$ **49.** $\frac{3}{2}$

51. $\dfrac{x}{8}(3 + 2x^2)\sqrt{3 + x^2} - \dfrac{9}{8}\ln|x + \sqrt{3 + x^2}| + C$

53. $\ln|\ln(x + 1)| + C$ **55.** $\frac{1}{3}\tan x \sec^2 x + \frac{2}{3}\tan x + C$

57. $e^x f(x) + C$ **59.** 6   **61.** 6π

63. a. 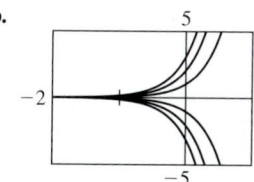   **65.** $\dfrac{2\pi}{9}(2e^3 + 1)$

67. $\frac{1}{2}\ln(\sqrt{3} + 2) + \sqrt{3}$ **69.** 1187.99 ft

Chapter 6 Challenge Problems • page 660

1. $1/e$ **3. b.** $\frac{1}{2}$ **5.** $\pi/6$ **19.** 72.6 ft

CHAPTER 7

Exercises 7.1 • page 675

5. a. 3

b. 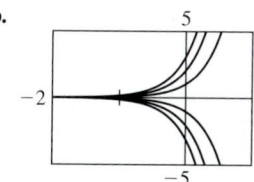   **c.** $y = 2e^{3x}$, yes

7. a. $y = \dfrac{1}{x} + \dfrac{x^3}{4}$ **b.**

9. $y = Cx^2$ **11.** $y = Ce^{x^3/3}$ **13.** $y = Ce^{-2/x} - \frac{3}{2}$

15. $y = \sin^{-1}(\tan x + C)$ **17.** $y = -\dfrac{2}{(\ln x)^2 + C}$

19. $y = \frac{2}{3} + \frac{1}{3}e^{3x^2/2}$ **21.** $y^{3/2} = \frac{1}{2}(x^3 + 1)$

23. $y = \ln(x^3 + e)$ **25.** $I = 2(1 - e^{-2t})$

27. $f(x) = \sqrt{x^3 + 8}$

29. $y^2 - x^2 = C$ **31.** $y^2 + 2x = C$

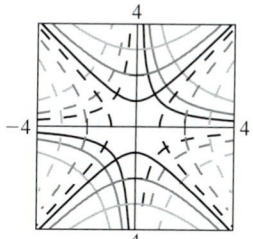 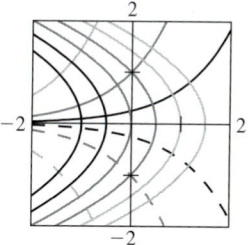

33. $\frac{1}{3}$ **35.** 200   **37.** $\frac{1}{2}$ in.   **39.** $y = \dfrac{50}{4t + 1}; \dfrac{50}{9}$ g

41. 57.8 years   **43.** 14,176 years old   **45.** After 3.6 min

47. 1.05 mph

49. a. $P(t) = (P_0 + I/k)e^{kt} - I/k$ **b.** 304.9 million

51. a. $\dfrac{dv}{dt} = g - kv$ **b.** $v(t) = \dfrac{g}{k}(1 - e^{-kt})$ **c.** $\dfrac{g}{k}$

53. a. $P(t) = \dfrac{1}{(1 - 0.01kt)^{100}}$

b.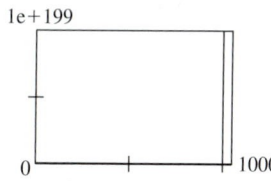

c. The population grows without bound after a finite period of time.

55. a. $C_0 e^{-kT}$ **b.** $C_0(e^{-kT} + e^{-2kT})$

c. $C_0(e^{-kT} + e^{-2kT} + \cdots + e^{-NkT})$ **d.** $\dfrac{C_0 e^{-kT}}{1 - e^{-kT}}$ g/mL

57. a. 2974, 5319, 6955   **b.** 3.6 weeks

c. 10,000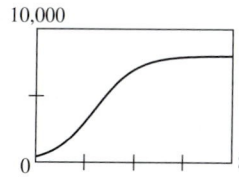

59. False   **61.** True

Exercises 7.2 • page 687

1. (a)

3. (b)

5. 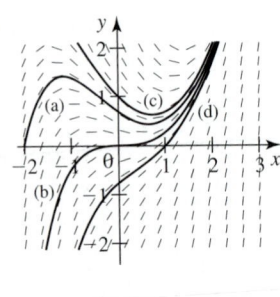 **7.**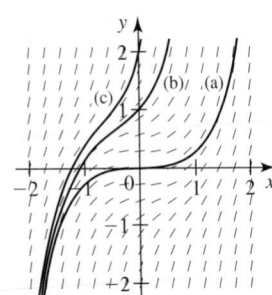

9. 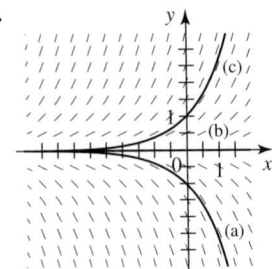 **11.**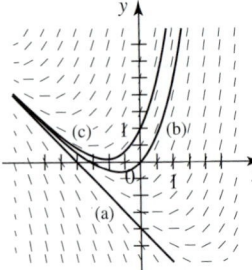

13. 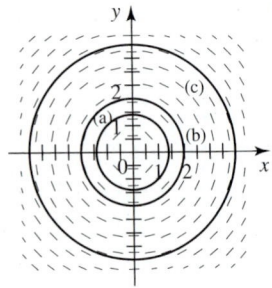 **15.**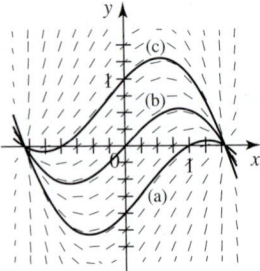

17. a. 2.88 **b.** 3.04

19. a. 3.19 **b.** 3.26

21. a. 0.83 **b.** 0.82

23. a. 1.78 **b.** 1.79

25. a. 1.34 **b.** 1.37

27. a, c.

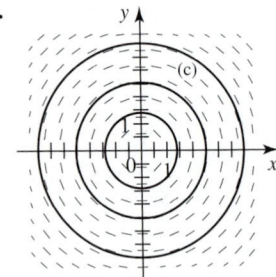

b. $x^2 + y^2 = 16$

29. a, b.

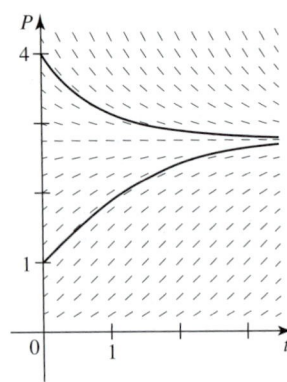

c. $\lim_{t \to \infty} P(t) \approx 2.72$

31. a. 73.6 ft/sec **b.** $v(t) = 160 - 30e^{-t/5}$; $v(2) \approx 72.9$ ft/sec

33. False **35.** True

Exercises 7.3 • page 698

1. a. 0.02 **b.** 1000 **3. a.** 0.5 **b.** 500

5. a. 100 **b.** $P(t) = 0$, $P(t) = 100$

c–e.

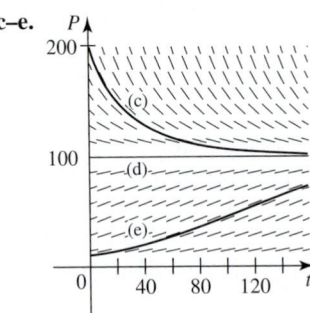

7. a. 0.02 **b.** 4000 **c.** 10

11. a. 86.12% **b.** 1970

13. a. $P(t) = \dfrac{100}{1 + 9\left(\frac{11}{51}\right)^{t/30}}$ **b.** 70 days

17. a, c.

If the initial fish population is 100, then it will be gone after 5 weeks. If the initial population is 300, then it increases to 600 over time. If it is 700, then it decreases to 600 over time.

b. $P(t) = 200$, $P(t) = 600$

19. a, c.

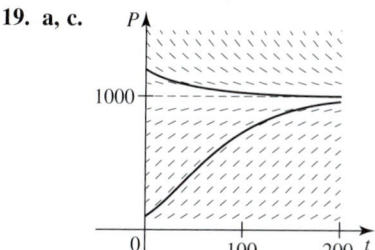

b. $P(t) = 1000$

21. 173   **23.** True   **25.** True

Exercises 7.4 • page 710

1. No   **3.** No   **5.** $y = \frac{1}{4}e^{2x} + Ce^{-2x}$

7. $y = \frac{1}{4}x^3 + \frac{C}{x}$ **9.** $y = \frac{1}{3}x^2 \sin 3x + Cx^2$

11. $y = \frac{2}{3}x(\ln x)^3 + Cx$ **13.** $y = \frac{t^2}{2(t+1)} + \frac{C}{t+1}$

15. $y = \frac{x}{2}e^{-2x} + \frac{Ce^{-2x}}{x}$ **17.** $y = 1 - 2e^{-x}$

19. $y = \frac{1}{2}\left(1 + e^{-x^2}\right)$ **21.** $y = \frac{(\ln x)^2 + 2}{2(x+1)}$

25. $y = \frac{\sqrt{3}x^2}{\sqrt{9 - 8x^3}}$ **27.** $f(x) = x \ln x + x$

29. a. $y(t) = 90(1 - e^{-t/15})$ **b.** 33 min

31. a. $y(t) = 80 - t - \frac{(80 - t)^4}{512{,}000}$ **b.** 34.7 lb

   c. 35 lb   **d.** 37.8 lb

33. $Q(t) = 0.24 - 0.19e^{-5t}$, $I(t) = 0.95e^{-5t}$

35. a. $8(1 - e^{-4t})$ **b.** 7.85 ft/sec   **c.** 0.17 sec

37. 29.5 ft; 8 ft/sec

39. $I = \frac{E_0 R}{R^2 + \omega^2 L^2}\left[\cos \omega t + \frac{L\omega}{R} \sin \omega t - e^{-(R/L)t}\right]$

43. True   **45.** True   **47.** True

Exercises 7.5 • page 717

1. a. $(0, 0)$ and $\left(\frac{5}{4}, 2\right)$ **b.** $\dfrac{-y + 0.8xy}{2.4x - 1.2xy}$

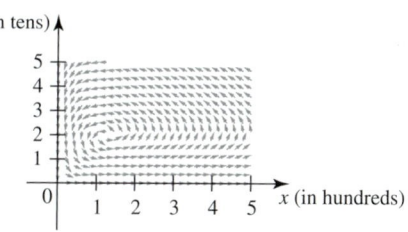

c.

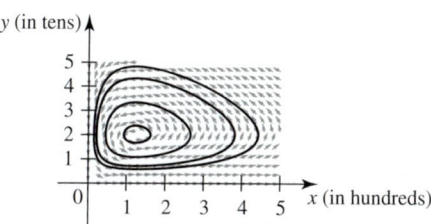

3. a. At the start there are approximately 480 caribou and 100 wolves. The caribou population fluctuates between a minimum of $\approx$480 and a maximum of $\approx$3500; the wolf population fluctuates between $\approx$30 and $\approx$290. The caribou population leads the wolf population.

b.

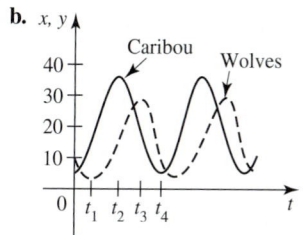

5.

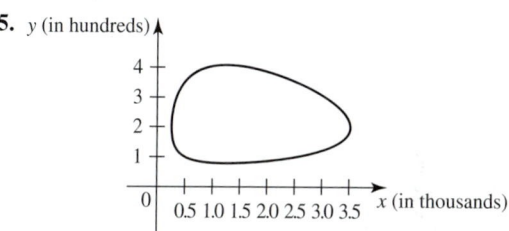

7. c. $xy^4 - 3.9830e^{0.2x}e^y = 0$

9. a. The system of differential equations reduces to the single differential equation $\dfrac{dx}{dt} = kx\left(1 - \dfrac{x}{L}\right)$. If $x(0) = 0$, then the prey population stays at zero at all times. If $0 < x(0) < L$, then the prey population approaches the carrying capacity L of the environment. If $x(0) = L$, the prey population at any time $t > 0$ remains at L. If $x(0) > L$, then the prey population decreases and approaches L.

b. The system of differential equations reduces to $\dfrac{dy}{dt} = -by$ with solution $y = y_0 e^{-bt}$, where $y_0 = y(0)$ is the initial predator population. Since $y \to 0$ as $t \to \infty$, the predator population eventually dies out.

c. $(0, 0)$, $(L, 0)$, and $\left(\dfrac{b}{c}, \dfrac{k(cL - b)}{acL}\right)$; If there are no predators or prey at any time t, the situation will persist forever. If there are exactly L prey and no predators at some time, then the prey population will remain at L forever. If at least one of the populations is not equal to zero, then we conclude that a prey population of b/c is exactly the number that will support a stable predator population of $\dfrac{k(cL - b)}{acL}$.

11. a. The system reduces to $\dfrac{dx}{dt} = k_1 x\left(1 - \dfrac{x}{L_1}\right)$. If $x(0) = 0$,

then the population of species A stays at zero at all times. If $0 < x(0) < L_1$, then the population of species A approaches L_1. If $x(0) = L_1$, the population remains at L_1. If $x(0) > L_1$, then the population decreases and approaches L_1.

b. The system reduces to $\dfrac{dy}{dt} = k_2 y\left(1 - \dfrac{y}{L_2}\right)$. The popula-

tion of species B behaves in a manner similar to that of the population of species A described in part (a).

c. The terms axy and bxy arise from the interaction of the species. The rates of change of the populations decline as a result of their competition for resources.

d. $(0, 0)$, $(0, L_2)$, $(L_1, 0)$, and

$$\left(\dfrac{k_2 L_1(aL_2 - k_1)}{abL_1 L_2 - k_1 k_2}, \dfrac{k_1 L_2(bL_1 - k_2)}{abL_1 L_2 - k_1 k_2}\right)$$

Chapter 7 Concept Review • page 719

1. a. derivative; differential; unknown   **b.** highest

3. $g(x)h(y)$; separable

5. a. tangent line   **b.** slope; direction

7. a. $\dfrac{dy}{dx} + P(x)y = Q(x)$ **b.** $u(x) = e^{\int P(x)\,dx}$

Chapter 7 Review Exercises • page 720

1. Yes   **3.** $y = -\dfrac{1}{x^2 + C}$ **5.** $e^y = \dfrac{e^x}{1 + Ce^x}$

7. $y^2 = \dfrac{3}{12 - 2x^3}$ **9.** $x^2 + y^2 = 4$ **11.** $y = \dfrac{1 + \tan 4x}{1 - \tan 4x}$

13. a. $x(t) = 1000 \cdot 4^{t/3}$ **b.** 16,000   **c.** 13 hr

15. -0.000433 **17.** \$338,249

19. a. 163°F   **b.** 28.6 min

21. a. $S(t) = \dfrac{1}{r}[(rS_0 + d)e^{rt} - d]$ **b.** \$25,160.55

23. a.

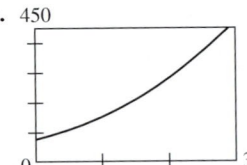

b. 0.6

25. 1.061   **27.** $y = x^2 + (C/x^2)$

29. $y = \frac{1}{2}e^{-x}\sin 2x + Ce^{-x}$ **31.** $y = 2e^{(x+1)/3}$

33. $y = x\sin x - x$

35. a. $P(t) = \dfrac{10{,}000}{1 + 24\left(\frac{3}{8}\right)^t}$ **b.** 9374   **c.** 4.7 years

37. a. The fox population dies out; the rabbit population grows indefinitely.

b. $(0, 0)$, $(4, 5)$. There are no rabbits and no foxes or 4000 rabbits will support a stable population of 500 foxes.

c. $\dfrac{-1.2y + 0.3xy}{2x - 0.4xy}$

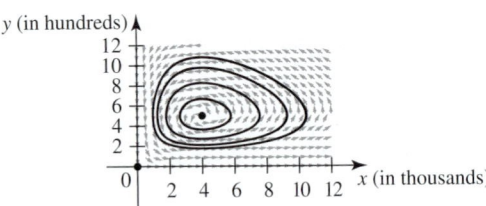

d.

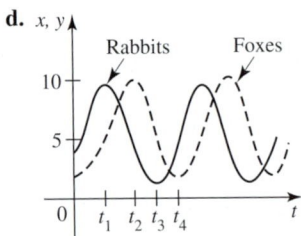

Chapter 7 Challenge Problems • page 722

3. b. $P(t) \approx \dfrac{1168.60}{1 + 14.334e^{-0.7572t}}$

c. 450

d. 351.06 million

5. b. $y_1 = x$, $y_2 = x + Ce^{-x^2/2}$

CHAPTER 8

Abbreviations: C, convergent; AC, absolutely convergent; CC, conditionally convergent; D, divergent; R, radius of convergence; I, interval of convergence

Exercises 8.1 • page 739

1. $2, 1, \frac{4}{5}, \frac{5}{7}, \frac{2}{3}$ **3.** $1, 0, -1, 0, 1$ **5.** $1, \frac{1}{6}, \frac{1}{90}, \frac{1}{2520}, \frac{1}{113,400}$

7. $a_n = \dfrac{n}{n+1}$ **9.** $a_n = \dfrac{(-1)^n}{n!}$ **11.** $a_n = \dfrac{1}{n+1}$

13. 2   **15.** D   **17.** 1   **19.** $\frac{2}{3}$ **21.** 0

23. D   **25.** 0   **27.** D   **29.** 0   **31.** 1

33. 0   **35.** 0   **37.** 1   **39.** 0   **41.** $\frac{1}{2}$

43. a. 1.0

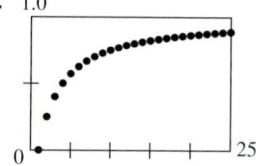

   b. 1   **c.** 1

45. a. 1.25

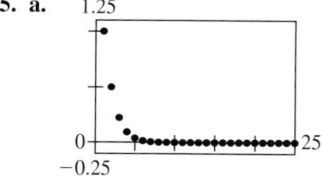

   b. 0   **c.** 0

47. a. 1.0

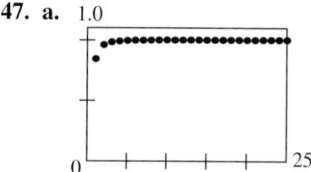

   b. 1   **c.** 1

49. 9   **51.** Monotonic, bounded   **53.** Monotonic, bounded

55. Not monotonic, bounded   **57.** Monotonic, bounded

59. a. 10088, 10176, 10265, 10355, 10445, 10537

   b. Diverges

61. a. 5394.69   **b.** ∞

63. e^6 **65. b.** 2.2361   **67.** 1.226   **69.** 2

71. c. $\{a_n\}$ converges   **73. b.** 0   **79.** False

81. True   **83.** False   **85.** False

Exercises 8.2 • page 750

1. 1   **3.** $\frac{2}{5}$ **5.** 1/ln 2   **7.** 12   **9.** $\frac{5}{4}$

11. $2(2 - \sqrt{2})$ **13.** 9

23. a. 6

   b. 8

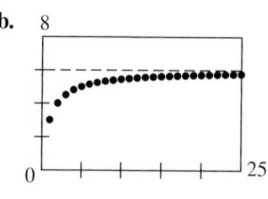

   c. 6

25. a. 24

   b. 30

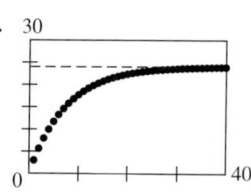

   c. 24

27. a. diverges

   b. 5

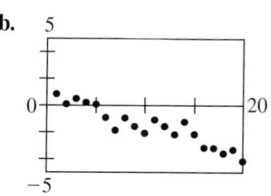

   c. D

29. $\frac{3}{4}$ **31.** $\frac{5}{3}$ **33.** D   **35.** D   **37.** D   **39.** 0

41. D   **43.** cos 1 − 1   **45.** $-\frac{5}{18}$ **47.** $\frac{7}{2}$ **49.** D

51. D   **53.** D   **55.** $\frac{4}{9}$ **57.** $\frac{404}{333}$

59. $|x| < 1, \dfrac{1}{1 + x}$ **61.** $\dfrac{1}{2} < x < \dfrac{3}{2}, \dfrac{2(x - 1)}{3 - 2x}$

63. 6 m   **65.** $\frac{6}{11}$ **67. b.** \$20 million

69. $A = \dfrac{P}{e^r - 1}$ **71.** $\dfrac{3\sqrt{3} - \pi}{9}$

79. False   **81.** False   **83.** False

Exercises 8.3 • page 756

1. C   **3.** C   **5.** C   **7.** C   **9.** C   **11.** C

13. C   **15.** D   **17.** C   **19.** C   **21.** D   **23.** C

25. C   **27.** C   **29.** C   **31.** C   **33.** $p > 1$

35. $a = 1$ **41.** 0.05   **43.** $(\pi/2) - \tan^{-1} 50$

45. $n \geq 200$ **47.** $n \geq 314$ **49.** 1.082

51. b. $\frac{1}{4}$ **55.** True   **57.** True

Exercises 8.4 • page 763

1. C   **3.** D   **5.** D   **7.** C   **9.** D   **11.** C

13. D   **15.** C   **17.** C   **19.** C   **21.** C   **23.** D

25. C   **27.** C   **29.** D   **31.** C   **33.** D   **35.** C

37. C   **39.** C   **43.** 3.06   **45.** 0.99

49. The converse is false.   **53. b.** Yes

55. False   **57.** False

Exercises 8.5 • page 769

1. C   **3.** C   **5.** C   **7.** C   **9.** D   **11.** C
13. C   **15.** C   **17.** C   **19.** D   **21.** C   **23.** C
25. $p > 0$ **27.** 44   **29.** 10   **31.** -0.90 **33.** 0.56
35. $\frac{1}{2}$; $a_{n+1} \le a_n$ is not satisfied.   **37. a.** No   **b.** No
39. b. -1 **41.** True   **43.** False

Exercises 8.6 • page 779

1. CC   **3.** D   **5.** CC   **7.** D   **9.** CC   **11.** D
13. CC   **15.** C   **17.** AC   **19.** AC   **21.** AC
23. D   **25.** AC   **27.** CC   **29.** AC   **31.** AC
33. C   **35. a.** $-1 < x < 1$ **b.** $x = -1$
47. True   **49.** True

Exercises 8.7 • page 788

1. $R = 1, I = [-1, 1)$ **3.** $R = 1, I = [-1, 1)$
5. $R = \infty, I = (-\infty, \infty)$ **7.** $R = 0, I = \{0\}$
9. $R = 1, I = [-1, 1)$ **11.** $R = \frac{1}{e}, I = \left[-\frac{1}{e}, \frac{1}{e}\right)$
13. $R = 1, I = (2, 4]$ **15.** $R = 3, I = (-1, 5]$
17. $R = 1, I = (0, 2]$ **19.** $R = \infty, I = (-\infty, \infty)$
21. $R = \infty, I = (-\infty, \infty)$
23. $R = \frac{1}{3}, I = \left(-2, -\frac{4}{3}\right]$
25. $R = 1, I = [-1, 1]$
27. $R = 1, I = (-1, 1)$
29. $R = \frac{3}{2}, I = \left(-\frac{3}{2}, \frac{3}{2}\right]$

31. a. $\dfrac{x^{n+1}}{1-x}$ **b.** 0; the limit does not exist.

c.
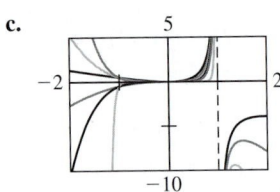

33. $(-\infty, \infty)$ **35.** $\sqrt{R}$ **37.** $1/L$
39. $\displaystyle\sum_{n=1}^{\infty} \frac{x^{n-1}}{n}$; $\displaystyle\sum_{n=2}^{\infty} \frac{(n-1)x^{n-2}}{n}$; $[-1, 1], [-1, 1), (-1, 1)$

41. $\dfrac{1}{(1-x)^2}$ **45. a.** $\displaystyle\sum_{n=0}^{\infty} t^{2n}$ **b.** $\displaystyle\sum_{n=0}^{\infty} \frac{x^{2n+1}}{2n+1}$, 1

47. 3.14159   **49.** True   **51.** True

Exercises 8.8 • page 801

1. $\displaystyle\sum_{n=0}^{\infty} \frac{2^n x^n}{n!}$, ∞ **3.** $\displaystyle\sum_{n=0}^{\infty} \frac{e^2}{n!}(x - 2)^n$, ∞

5. $2x - \dfrac{2^3}{3!}x^3 + \dfrac{2^5}{5!}x^5 - \cdots + \dfrac{(-1)^k 2^{2k+1}}{(2k+1)!}x^{2k+1} + \cdots$, ∞

7. $\dfrac{\sqrt{3}}{2}\displaystyle\sum_{n=0}^{\infty} \frac{(-1)^n}{(2n)!}\left(x + \frac{\pi}{6}\right)^{2n}$
$\quad + \dfrac{1}{2}\displaystyle\sum_{n=0}^{\infty} \frac{(-1)^n}{(2n+1)!}\left(x + \frac{\pi}{6}\right)^{2n+1}$, ∞

9. $\ln 2 + \displaystyle\sum_{n=1}^{\infty} \frac{(-1)^{n-1}}{n2^n}(x - 2)^n$, 2

11. $\displaystyle\sum_{n=0}^{\infty} \frac{(-1)^n}{2^{n+1}}(x - 1)^n$, 2   **13.** $\displaystyle\sum_{n=0}^{\infty}(-1)^{n+1} 2^n(x - 1)^n$, $\frac{1}{2}$

15. $-\displaystyle\sum_{n=0}^{\infty} x^{2n+2}$, 1   **17.** $\displaystyle\sum_{n=0}^{\infty} \frac{(-1)^n x^{n+1}}{n!}$, ∞

19. $\displaystyle\sum_{n=0}^{\infty} \frac{(-1)^n x^{2n+2}}{(2n)!}$, ∞ **21.** $1 + \displaystyle\sum_{n=1}^{\infty} \frac{(-1)^n 2^{2n-1} x^{2n}}{(2n)!}$, ∞

23. $\dfrac{1}{2}\displaystyle\sum_{n=0}^{\infty} \frac{(-1)^n\left(x - \frac{\pi}{3}\right)^{2n+1}}{(2n+1)!} + \dfrac{\sqrt{3}}{2}\displaystyle\sum_{n=0}^{\infty} \frac{(-1)^n\left(x - \frac{\pi}{3}\right)^{2n}}{(2n)!}$, ∞

25. $\displaystyle\sum_{n=0}^{\infty} \frac{(2n)! x^{3/2} x^{2n}}{(2^n n!)^2(2n+1)}$, 1   **27.** $\displaystyle\sum_{n=1}^{\infty}(-1)^{n-1} \frac{x^{2n}}{n}$, 1

29. $\displaystyle\sum_{n=0}^{\infty}(-1)^n(n+1)x^n$, 1

31. $1 - \dfrac{1}{2}x^2 - \displaystyle\sum_{n=2}^{\infty} \frac{1 \cdot 3 \cdot 5 \cdot \cdots \cdot (2n - 3)}{n!\, 2^n}x^{2n}$, 1

33. $1 - \dfrac{3}{5}x - 3\displaystyle\sum_{n=2}^{\infty} \frac{2 \cdot 7 \cdot 12 \cdot \cdots \cdot (5n - 8)}{n!\, 5^n}x^n$, 1

35. $f(x) = 1 + 2\left(x - \dfrac{\pi}{4}\right) + 2\left(x - \dfrac{\pi}{4}\right)^2 + \cdots$

37. $f(x) = \dfrac{\pi}{6} + \dfrac{2\sqrt{3}}{3}\left(x - \dfrac{1}{2}\right) + \dfrac{2\sqrt{3}}{9}\left(x - \dfrac{1}{2}\right)^2 + \cdots$

39. $f(x) = x - x^2 + \frac{1}{3}x^3 - \cdots$

41. a. $f(x) = 1 + \dfrac{1}{3}x$
$\quad + \displaystyle\sum_{n=2}^{\infty}(-1)^{n-1} \frac{2 \cdot 5 \cdot 8 \cdot \cdots \cdot (3n - 4)}{n!\, 3^n}x^n$, $R = 1$
b. $P_1(x) = 1, P_2(x) = 1 + \frac{1}{3}x, P_3(x) = 1 + \frac{1}{3}x - \frac{1}{9}x^2$
c.
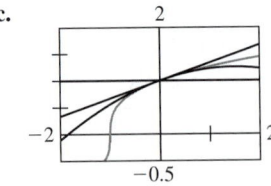

43. 0.99005

45. $\displaystyle\sum_{n=0}^{\infty} (-1)^n \frac{x^{3n+1}}{3n+1} + C$ **47.** $\displaystyle\sum_{n=0}^{\infty} \frac{(-1)^n x^{4n+3}}{(2n+1)!\,(4n+3)} + C$

49. $\displaystyle\sum_{n=1}^{\infty} \frac{(-1)^{n-1} x^n}{n^2} + C$ **51.** 0.7468   **53.** 0.4969

55. 0.1248   **57.** $\ln 2$ **59.** $\cos \pi = -1$

61. $\ln 3 - \ln 2$ **63.** $\frac{1}{120}$ **65.** $\frac{2}{15}$

67. a. $\displaystyle\frac{1}{\sqrt{1-u^2}} = 1 + \frac{1}{2}u^2 + \frac{1 \cdot 3}{2!\,2^2}u^4 + \cdots$
$$+ \frac{1 \cdot 3 \cdot 5 \cdots (2n-1)}{n!\,2^n}u^{2n} + \cdots$$

b. $\displaystyle\int_0^x \frac{dt}{\sqrt{1-t^2}} = \sum_{n=0}^{\infty} \frac{(2n)!}{(n!\,2^n)^2(2n+1)}x^{2n+1},\ R = 1$

79. True   **81.** False   **83.** True

Exercises 8.9 • page 816

1. $P_1(x) = 1 - x$, $P_2(x) = 1 - x + \frac{1}{2}x^2$,
$P_3(x) = 1 - x + \frac{1}{2}x^2 - \frac{1}{6}x^3$

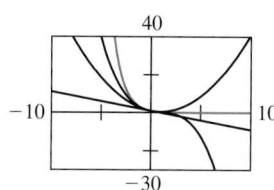

3. $P_4(x) = 7 + 13(x-1) + 9(x-1)^2 + 2(x-1)^3$, $R_4(x) = 0$

5. $P_3(x) = 1 - \frac{1}{2}\left(x - \frac{\pi}{2}\right)^2$, $R_3(x) = \frac{\sin z}{24}\left(x - \frac{\pi}{2}\right)^4$,
z lies between x and $\frac{\pi}{2}$

7. $P_2(x) = 1 + 2\left(x - \frac{\pi}{4}\right) + 2\left(x - \frac{\pi}{4}\right)^2$,
$R_2(x) = \frac{\sec^2 z(2\tan^2 z + \sec^2 z)}{3}\left(x - \frac{\pi}{4}\right)^3$,
z lies between x and $\frac{\pi}{4}$

9. $P_3(x) = -2 + \frac{1}{12}(x+8) + \frac{1}{288}(x+8)^2$
$\quad + \frac{5}{20{,}736}(x+8)^3$,
$R_3(x) = -\frac{10}{243z^{11/3}}(x+8)^4$, z lies between x and -8

11. $P_2(x) = \frac{\pi}{4} + \frac{1}{2}(x-1) - \frac{1}{4}(x-1)^2$,
$R_2(x) = \frac{3z^2 - 1}{3(1+z^2)^3}(x-1)^3$, z lies between x and 1

13. $P_3(x) = -\frac{1}{e} + \frac{1}{2e}(x+1)^2 + \frac{1}{3e}(x+1)^3$,
$R_3(x) = \frac{(z+4)e^z}{24}(x+1)^4$, z lies between x and -1

15. $P_2(x) = \frac{1}{2}e^{\pi/6} + \left(\frac{1}{2} - \sqrt{3}\right)e^{\pi/6}\left(x - \frac{\pi}{6}\right)$
$\quad - \left(\frac{3}{4} + \sqrt{3}\right)e^{\pi/6}\left(x - \frac{\pi}{6}\right)^2$,
$R_2(x) = \frac{e^z(2\sin 2z - 11\cos 2z)}{6}\left(x - \frac{\pi}{6}\right)^3$,
z lies between x and $\frac{\pi}{6}$

17. $4(x-1) + 6(x-1)^2$, $4(1.2)(1.2-1)^3$

19. $\frac{\sqrt{2}}{2} - \frac{\sqrt{2}}{2}\left(x - \frac{\pi}{4}\right) - \frac{\sqrt{2}}{4}\left(x - \frac{\pi}{4}\right)^2$
$\quad + \frac{\sqrt{2}}{12}\left(x - \frac{\pi}{4}\right)^3 + \frac{\sqrt{2}}{48}\left(x - \frac{\pi}{4}\right)^4$, $\frac{\sin\frac{\pi}{2}}{120}\left(\frac{\pi}{4}\right)^5$

21. $e^2 + 2e^2(x-1) + 2e^2(x-1)^2 + \frac{4}{3}e^2(x-1)^3$
$\quad + \frac{2}{3}e^2(x-1)^4$, $\frac{32e^{2(1.1)}}{5!}(0.1)^5$

23. $3 + \frac{1}{6}(x-9) - \frac{1}{216}(x-9)^2 + \frac{1}{3888}(x-9)^3$,
$\frac{5}{128} \cdot \frac{1}{8^{7/2}}(1)^4$

25. $x + \frac{1}{3}x^3$, $\frac{10\left(\frac{\pi}{4}\right)^4}{3}$

27. $\ln 4 + \frac{1}{4}(x-3) - \frac{1}{32}(x-3)^2 + \frac{1}{192}(x-3)^3$, $\frac{6(1)^4}{3^4 4!}$

29. $f(x) = e^x$, $c = 0$, $P_3(0.2) \approx 1.2213$

31. $f(x) = \sqrt{x}$, $c = 9$, $P_1(9.01) \approx 3.00167$

33. $f(x) = \frac{1}{x}$, $c = -2$, $P_2(-2.1) \approx -0.476$

35. $f(x) = \sin x$, $c = 0$, $P_3(0.1) \approx 0.09983$

37. $f(x) = \cos x$, $c = \frac{\pi}{6}$, $P_2\left(\frac{8\pi}{45}\right) \approx 0.8480$ **45.** 48%

51. True   **53.** True

Chapter 8 Concept Review • page 818

1. a. function; integers; nth term
   b. converge
   c. for every; positive integer; $n > N$

3. a. partial sums $\{S_n\}$
   b. geometric; 1; 1

5. a. convergent; divergent **b.** $\displaystyle\sum_{n=1}^{\infty} \frac{1}{n^p}; p > 1; p \le 1$

7. a. an alternating; $\le$; 0 **b.** a_{n+1}

9. a. $\displaystyle\sum_{n=0}^{\infty} a_n(x-c)^n$ **b.** $x = c; x$

Chapter 8 Review Exercises • page 819

1. $\frac{1}{3}$ **3.** 2 **5.** D **7.** 0 **9.** 3 **11.** $\frac{11}{18}$ **13.** C

15. C **17.** C **19.** C **21.** D **23.** C **25.** C

27. CC **29.** AC **31.** CC **33.** $\frac{6802}{4995}$ **35.** False

37. True **39.** $x \ne k\pi, k = 0, \pm 1, \pm 2, \ldots$

43. $R = 1, I = (-1, 1]$ **45.** $R = \frac{1}{2}, I = \left[-\frac{1}{2}, \frac{1}{2}\right]$

47. $R = 1, I = [-1, 1]$ **49.** $\displaystyle\sum_{n=0}^{\infty} (-1)^n x^{n+3}$

51. $\displaystyle\sum_{n=0}^{\infty} (-1)^n \frac{x^{4n}}{(2n)!}$

53. $1 + \frac{1}{2}x^2 + \displaystyle\sum_{n=2}^{\infty} (-1)^{n-1} \frac{1 \cdot 3 \cdot 5 \cdots (2n-3)}{n! \, 2^n} x^{2n}$

55. $\frac{1}{4}$ **57.** $\ln|x| + \displaystyle\sum_{n=1}^{\infty} (-1)^n \frac{x^n}{n! \, n} + C$

59. 0.199 **61.** 0.779

63. $P_3(x) = 1 + \frac{1}{2}(x-1) - \frac{1}{8}(x-1)^2 + \frac{1}{16}(x-1)^3$,

$R_3(x) = \dfrac{-5(x-1)^4}{128 z^{7/2}}$, z lies between x and 1

65. $P_2(x) = 1 + \frac{1}{2}\left(x - \frac{\pi}{2}\right)^2$,

$R_2(x) = -\dfrac{\csc z \cot z (\cot^2 z + 5 \csc^2 z)}{6}\left(x - \frac{\pi}{2}\right)^3$,

z lies between x and $\dfrac{\pi}{2}$

Chapter 8 Challenge Problems • page 821

1. 1 **3. a.** $c_n = \left(\frac{2}{3}\right)^n$ **5.** $\frac{1}{2}(1 + \sqrt{1 + 4a})$ **11.** $\frac{1}{3}$

13. $\displaystyle\sum_{n=0}^{\infty} x^{4n} - \sum_{n=0}^{\infty} x^{4n+1}$ **15.** $(-\infty, \infty)$

17. $P_3 = 1 + x^2 - \frac{1}{2}x^3, |x| < 1$

CHAPTER 9

Exercises 9.1 • page 839

1. Parabola (h), vertex $(0, 0)$, focus $(0, -1)$, directrix $y = 1$

3. Parabola (c), vertex $(0, 0)$, focus $(2, 0)$, directrix $x = -2$

5. Ellipse (b), vertices $(\pm 3, 0)$, foci $(\pm\sqrt{5}, 0)$, eccentricity $\frac{\sqrt{5}}{3}$

7. Hyperbola (d), vertices $(\pm 4, 0)$, foci $(\pm 5, 0)$, eccentricity $\frac{5}{4}$

9. Vertex $(0, 0)$, focus $\left(0, \frac{1}{8}\right)$, directrix $y = -\frac{1}{8}$

11. Vertex $(0, 0)$, focus $\left(\frac{1}{8}, 0\right)$, directrix $x = -\frac{1}{8}$

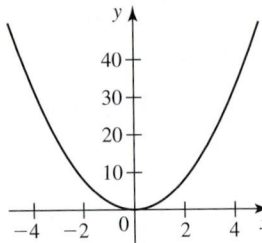

13. Vertex $(0, 0)$, focus $\left(\frac{3}{5}, 0\right)$, directrix $x = -\frac{3}{5}$

15. Foci $(0, \pm\sqrt{21})$, vertices $(0, \pm 5)$

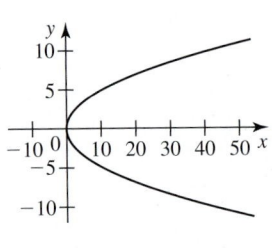
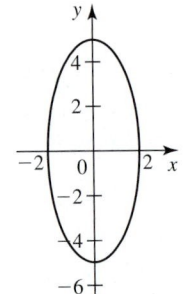

17. Foci $(\pm\sqrt{5}, 0)$, vertices $(\pm 3, 0)$

19. Foci $(\pm\sqrt{3}, 0)$, vertices $(\pm 2, 0)$

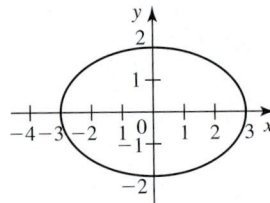
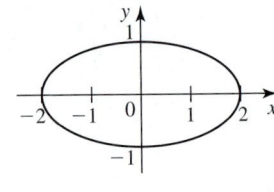

21. Foci $(\pm 13, 0)$, vertices $(\pm 5, 0)$, asymptotes $y = \pm\frac{12}{5}x$

23. Foci $(\pm\sqrt{2}, 0)$, vertices $(\pm 1, 0)$, asymptotes $y = \pm x$

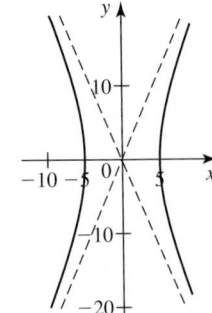
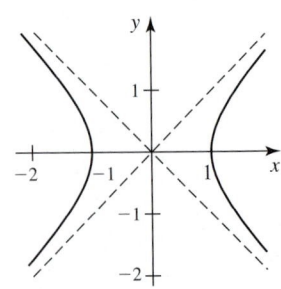

25. Foci $(0, \pm\sqrt{30})$, vertices $(0, \pm 5)$, asymptotes $y = \pm\sqrt{5}x$

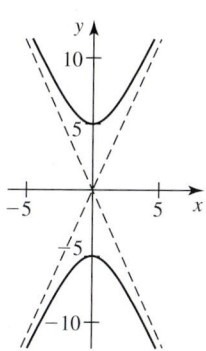

27. $y^2 = 12x$ **29.** $y^2 = -10x$ **31.** $\dfrac{x^2}{9} + \dfrac{y^2}{8} = 1$

33. $\dfrac{x^2}{8} + \dfrac{y^2}{9} = 1$ **35.** $\dfrac{x^2}{9} + \dfrac{4y^2}{9} = 1$ **37.** $\dfrac{73x^2}{400} + \dfrac{y^2}{25} = 1$

39. $\dfrac{x^2}{9} - \dfrac{y^2}{16} = 1$ **41.** $\dfrac{y^2}{21} - \dfrac{x^2}{4} = 1$ **43.** $\dfrac{x^2}{4} - \dfrac{y^2}{9} = 1$

45. (b) **47.** (c) **49.** $(y-1)^2 = 4(x-2)$

51. $(y-2)^2 = -2(x-2)$ **53.** $(x+3)^2 = -2(y-2)$

55. $\dfrac{x^2}{9} + \dfrac{(y-3)^2}{8} = 1$ **57.** $\dfrac{x^2}{16} + \dfrac{(y-2)^2}{15} = 1$

59. $\dfrac{(x-2)^2}{9} + \dfrac{(y-1)^2}{5} = 1$ **61.** $\dfrac{(x-3)^2}{9} - \dfrac{(y-2)^2}{16} = 1$

63. $\dfrac{(x-1)^2}{9} - \dfrac{(y+3)^2}{16} = 1$ **65.** $\dfrac{(y-1)^2}{9} - \dfrac{(x-4)^2}{4} = 1$

67. Vertex $(2, 1)$,   **69.** Vertex $(-3, 2)$,
focus $(3, 1)$,   focus $\left(-3, \dfrac{9}{4}\right)$,
directrix $x = 1$ directrix $y = \dfrac{7}{4}$

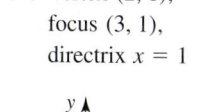

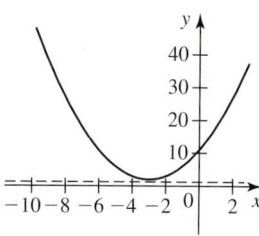

71. Vertex $\left(-1, \dfrac{1}{2}\right)$,   **73.** Center $(1, -2)$,
focus $\left(1, \dfrac{1}{2}\right)$,   foci $\left(1 \pm \dfrac{\sqrt{3}}{2}, -2\right)$,
directrix $x = -3$ vertices $(0, -2)$ and $(2, -2)$

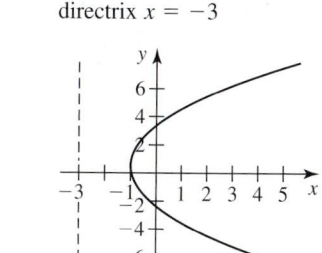

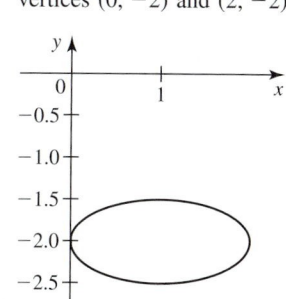

75. Center $(1, -2)$, foci $(1 \pm \sqrt{3}, -2)$, vertices $(-1, -2)$ and $(3, -2)$

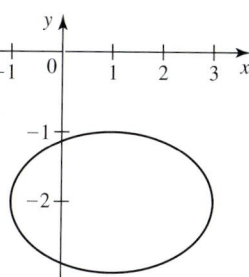

77. Center $\left(\dfrac{9}{4}, 0\right)$, foci $\left(\dfrac{9}{4} \pm \dfrac{\sqrt{105}}{4}, 0\right)$, vertices $\left(\dfrac{9}{4} \pm \dfrac{3\sqrt{21}}{4}, 0\right)$

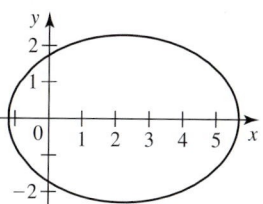

79. Center $(0, -1)$, foci $(\pm\sqrt{7}, -1)$, vertices $(\pm 2, -1)$, asymptotes $y = \pm\dfrac{\sqrt{3}}{2}x - 1$

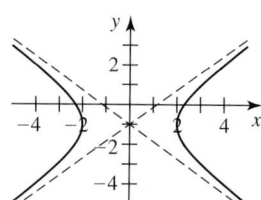

81. Center $(1, 2)$, foci $(1, 2 \pm \sqrt{15})$, vertices $(1, 2 \pm \sqrt{6})$, asymptotes $y - 2 = \pm\dfrac{\sqrt{6}}{3}(x - 1)$

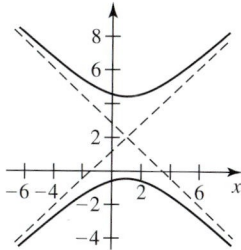

83. Center $(-1, 2)$, foci $(-1 \pm \sqrt{6}, 2)$, vertices $(-1 \pm \sqrt{2}, 2)$, asymptotes $y - 2 = \pm\sqrt{2}(x + 1)$

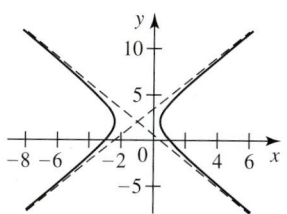

85. $\dfrac{1}{4}$ ft **87.** 616 ft **91.** 6.93 ft **103.** 23.013

109. True **111.** True **113.** False

Exercises 9.2 • page 851

1. a. $x - 2y - 7 = 0$
b.
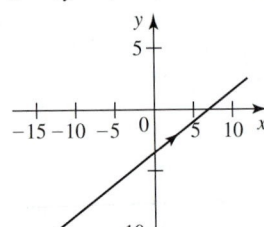

3. a. $y = 9 - x^2, x \geq 0$
b.
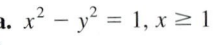

5. a. $y = 2x - 3, 1 \leq x \leq 5$
b.
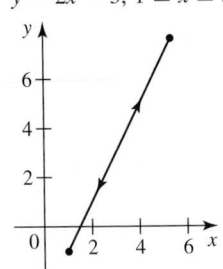

7. a. $x = y^{2/3}, -8 \leq y \leq 8$
b.
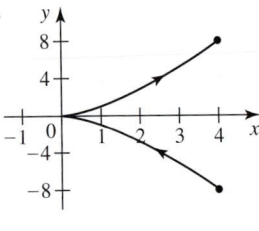

9. a. $x^2 + y^2 = 4$
b.
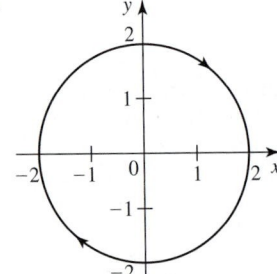

11. a. $\frac{1}{4}x^2 + \frac{1}{9}y^2 = 1$
b.
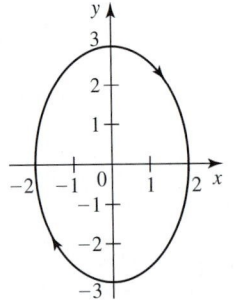

13. a. $\frac{1}{4}(x - 2)^2 + \frac{1}{9}(y + 1)^2 = 1$
b.
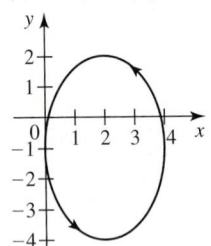

15. a. $y = 2x^2 - 1, -1 \leq x \leq 1$
b.
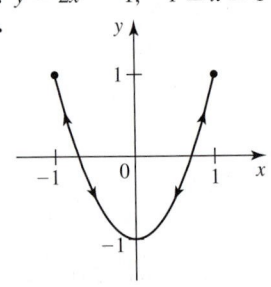

17. a. $x^2 - y^2 = 1, x \geq 1$
b.
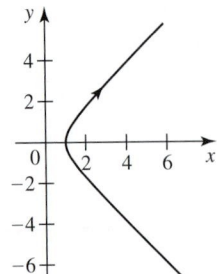

19. a. $y = x^2, 0 \leq x \leq 1$
b.
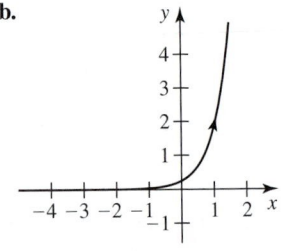

21. a. $y = x^2, x < 0$
b.
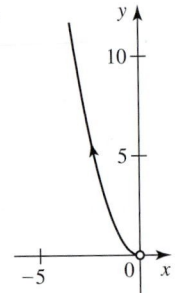

23. a. $y = \frac{1}{4}e^{2x}$
b.

25. a. $x^2 - y^2 = 1, x \geq 1$
b.
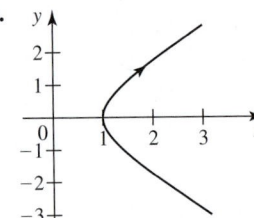

27. a. $y = x^{3/2}, 0 \leq x \leq 1$
b.

29. As t increases, the particle moves along the parabola $y = \sqrt{x - 1}$ from $(1, 0)$ to $(5, 2)$.

31. The particle starts out at $(2, 2)$ and travels once counterclockwise along the circle of radius 1 centered at $(1, 2)$.

33. The particle starts at $(0, 0)$ and moves to the right along the parabola $y = x^2$ to $(1, 1)$, then back to $(-1, 1)$, then again to $(1, 1)$, and finally back to $(0, 0)$.

35.
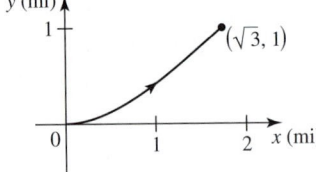

43.
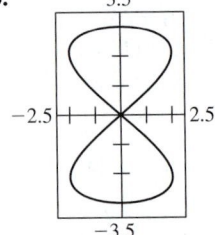

45.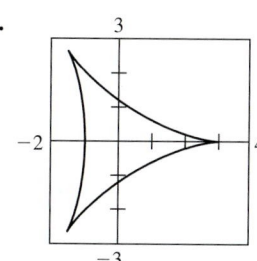

47.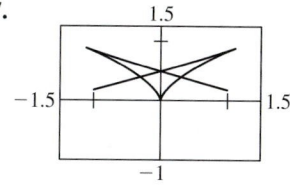

49. 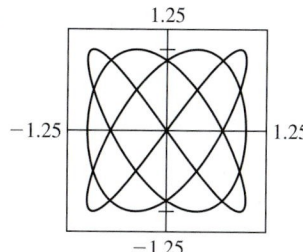   **51.** False   **53.** True

Exercises 9.3 • page 859

1. $\frac{1}{2}$ **3.** -2 **5.** $-\frac{3}{2}$ **7.** $y = \frac{1}{2}x - \frac{1}{2}$

9. $y = 5x + 2$

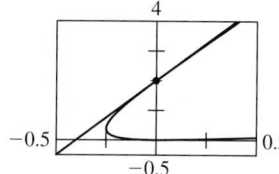

11. $(31, 64)$

13. Horizontal at $(-3, \pm 2)$, vertical at $(-4, 0)$

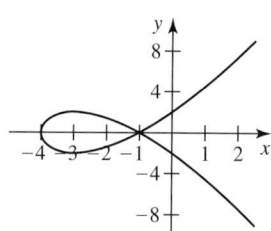

15. Horizontal at $(1, 0)$ and $(1, 4)$, vertical at $(4, 2)$ and $(-2, 2)$

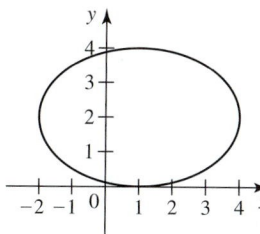

17. $\frac{dy}{dx} = t, \frac{d^2y}{dx^2} = \frac{1}{6t}$ **19.** $\frac{dy}{dx} = -\frac{2}{t^{3/2}}, \frac{d^2y}{dx^2} = \frac{6}{t^2}$

21. $\frac{dy}{dx} = \frac{1 - \cos\theta}{1 - \sin\theta}, \frac{d^2y}{dx^2} = \frac{\sin\theta + \cos\theta - 1}{(1 - \sin\theta)^3}$

23. $\frac{dy}{dx} = \coth t, \frac{d^2y}{dx^2} = -\frac{1}{\sinh^3 t}$

25. Concave downward on $(-\infty, 0)$, concave upward on $(0, \infty)$

27. $dy/dx = -\tan t, m = -1$ at $\left(\frac{\sqrt{2}}{4}a, \frac{\sqrt{2}}{4}a\right)$ and $\left(-\frac{\sqrt{2}}{4}a, -\frac{\sqrt{2}}{4}a\right)$, $m = 1$ at $\left(-\frac{\sqrt{2}}{4}a, \frac{\sqrt{2}}{4}a\right)$ and $\left(\frac{\sqrt{2}}{4}a, -\frac{\sqrt{2}}{4}a\right)$

29. Absolute maximum $f(-14) = 8$, absolute minimum $f(94) = -19$

31. $\frac{1}{243}(97^{3/2} - 64)$ **33.** $2\sqrt{5}$ **35.** $\frac{1}{8}a\pi^2$ **37.** $16a$

39. $4\sqrt{2}$ **41.** Approximately 1639 ft

43. $\left(a, -\frac{1}{2}\left(\frac{eE}{mv_0^2}\right)a^2\right)$

45. a.

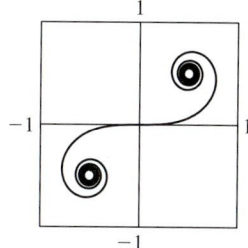

As $t \to \infty$, the curve spirals about and converges to the point $\left(\frac{1}{2}, \frac{1}{2}\right)$. As $t \to -\infty$, the curve spirals about and converges to the point $\left(-\frac{1}{2}, -\frac{1}{2}\right)$.

b. a

47. $3\pi a^2$ **49.** $3\pi a^2/8$ **51.** $32\sqrt{2}/3$

53. $\frac{2(247\sqrt{13} + 64)\pi}{1215}$ **55.** $148\pi/5$ **57.** $64\pi/3$

59. $\frac{24(\sqrt{2} + 1)\pi}{5}$ **61.** $\pi(e^2 + 2e - 6)$ **63.** $12\pi a^2/5$

65. $4\pi^2 rb$

67. 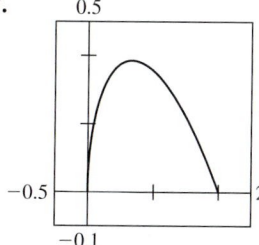   **69.**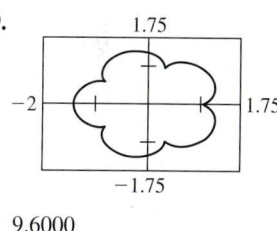

9.6000

2.2469

71. 33.66   **73.** Center $(0, 0)$, radius a

75. $x = \frac{3at}{t^3 + 1}, y = \frac{3at^2}{t^3 + 1}$ **79.** False

Exercises 9.4 • page 873

1.

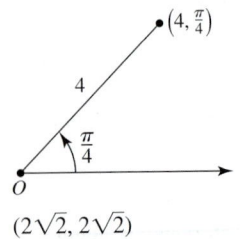

$(2\sqrt{2}, 2\sqrt{2})$

3. $\frac{3\pi}{2}$

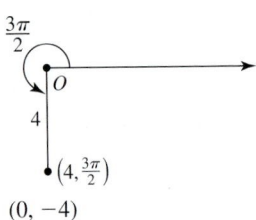

$(0, -4)$

5.

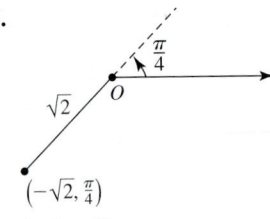

$(-\sqrt{2}, \frac{\pi}{4})$
$(-1, -1)$

7.

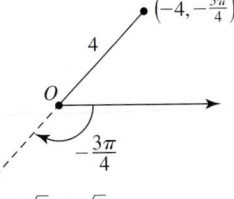

$(2\sqrt{2}, 2\sqrt{2})$

9.

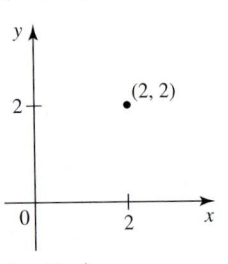

$(2\sqrt{2}, \frac{\pi}{4})$

11.

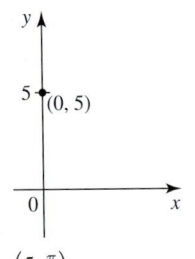

$(5, \frac{\pi}{2})$

13.

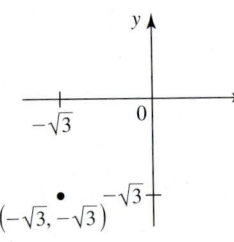

$(\sqrt{6}, \frac{5\pi}{4})$

15.

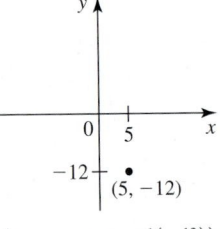

$(13, 2\pi + \tan^{-1}(-\frac{12}{5}))$

17.

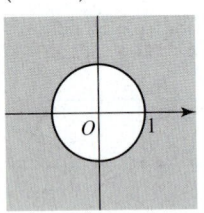

19.

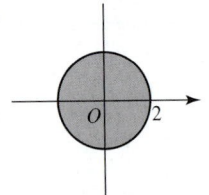

21.

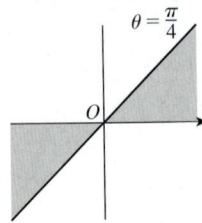

23.

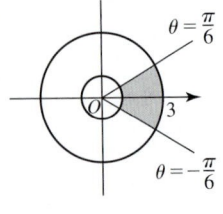

25. $x = 2$ **27.** $2x + 3y = 6$ **29.** $x^2 + y^2 - 4x = 0$

31. $x^2 - 2y - 1 = 0$ **33.** $r = 4 \sec \theta$ **35.** $r = 3$

37. $r^2 = 8 \csc 2\theta$

39.

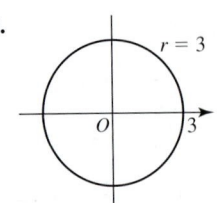

41.

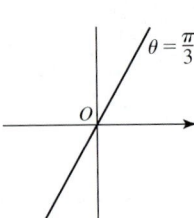

43.

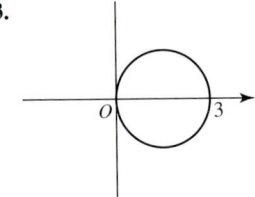

45.

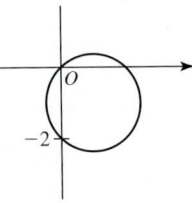

47.

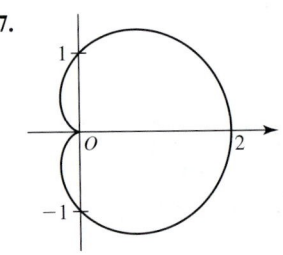

49.

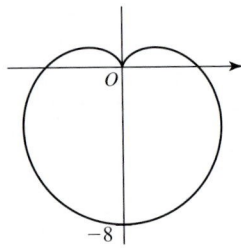

51.

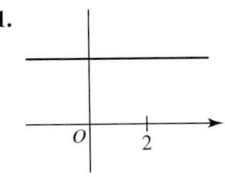

53.

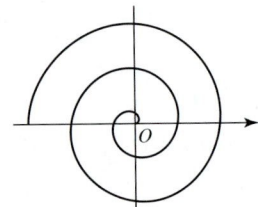

55.

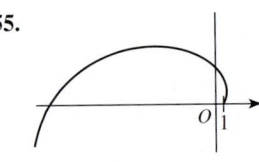

57.

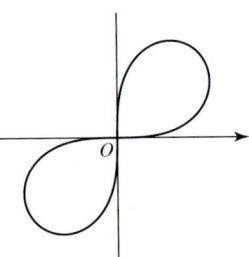

59.

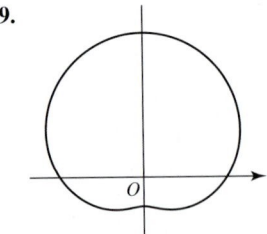

61.

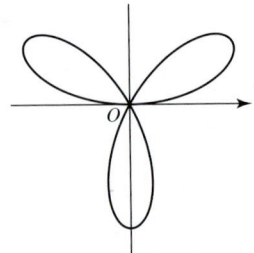

63.

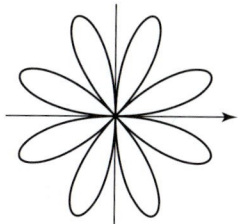

65. $\sqrt{3}/3$ **67.** -1 **69.** π **71.** 0

73. Horizontal at $\left(2\sqrt{2}, \frac{\pi}{4}\right)$ and $\left(-2\sqrt{2}, \frac{3\pi}{4}\right)$, vertical at the pole and $(4, 0)$

75. Horizontal at $(0, 0)$, $(\sin(2\tan^{-1}\sqrt{2}), \tan^{-1}\sqrt{2})$, $(\sin(2\tan^{-1}(-\sqrt{2})), \pi + \tan^{-1}(-\sqrt{2})), (0, \pi)$, $(\sin(2\tan^{-1}\sqrt{2}), \pi + \tan^{-1}\sqrt{2})$, and $(\sin(2\tan^{-1}(-\sqrt{2})), 2\pi + \tan^{-1}(-\sqrt{2}))$; vertical at $\left(\sin\left(2\sin^{-1}\frac{\sqrt{3}}{3}\right), \sin^{-1}\frac{\sqrt{3}}{3}\right), \left(0, \frac{\pi}{2}\right)$, $\left(\sin\left(2\sin^{-1}\left(-\frac{\sqrt{3}}{3}\right)\right), \pi + \sin^{-1}\left(-\frac{\sqrt{3}}{3}\right)\right)$, $\left(\sin\left(2\sin^{-1}\frac{\sqrt{3}}{3}\right), \pi + \sin^{-1}\frac{\sqrt{3}}{3}\right)$, $\left(0, \frac{3\pi}{2}\right)$, and $\left(\sin\left(2\sin^{-1}\left(-\frac{\sqrt{3}}{3}\right)\right), 2\pi + \sin^{-1}\left(-\frac{\sqrt{3}}{3}\right)\right)$

77. Horizontal at $\left(1 + \frac{\sqrt{33} - 1}{4}, \cos^{-1}\frac{\sqrt{33} - 1}{8}\right)$, $\left(1 - \frac{\sqrt{33} + 1}{4}, \cos^{-1}\frac{-\sqrt{33} - 1}{8}\right)$, $\left(1 - \frac{\sqrt{33} + 1}{4}, 2\pi - \cos^{-1}\left(\frac{-1 - \sqrt{33}}{8}\right)\right)$, and $\left(1 + \frac{\sqrt{33} - 1}{4}, 2\pi - \cos^{-1}\frac{\sqrt{33} - 1}{8}\right)$, vertical at $(3, 0)$, $\left(\frac{1}{2}, \cos^{-1}\left(-\frac{1}{4}\right)\right)$, $(-1, \pi)$, and $\left(\frac{1}{2}, 2\pi - \cos^{-1}\left(-\frac{1}{4}\right)\right)$

81. b. $2\sqrt{3}$

83. a.

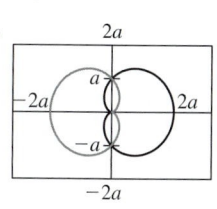

85.

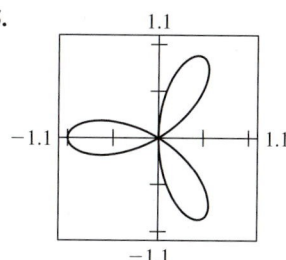

87.

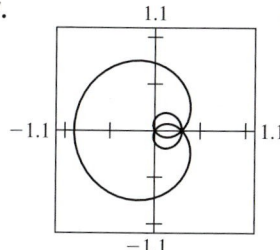

89.

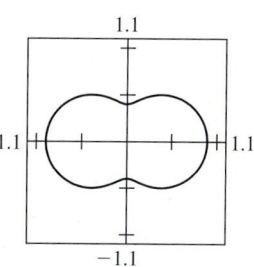

91.

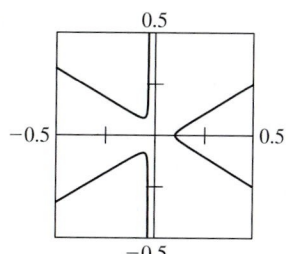

93. False **95.** False

Exercises 9.5 • page 881

1. a. 4π **b.** 4π **3.** $\frac{1}{6}\pi^3$

5. $\frac{e^\pi - 1}{4e^\pi}$ **7.** $\frac{1}{2}$ **9.** $9\pi^3/16$ **11.** $3\pi/4$

13.

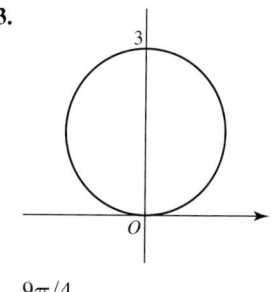

$9\pi/4$

15.

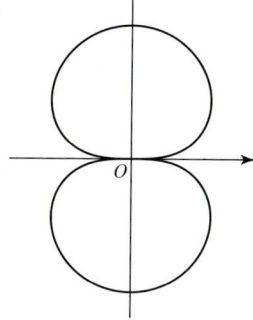

2

17.

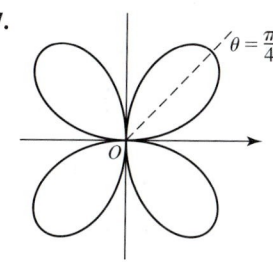

2π

19. $\pi/8$ **21.** $\pi/16$ **23.** $\pi - \dfrac{3\sqrt{3}}{2}$ **25.** $\dfrac{\pi - 2}{8}$

27. $\dfrac{3\pi - 1}{4}$ **29.** $\left(1, \frac{\pi}{2}\right)$ and $\left(1, \frac{3\pi}{2}\right)$

31. $\left(2, \frac{\pi}{6}\right), \left(2, \frac{\pi}{3}\right), \left(2, \frac{2\pi}{3}\right), \left(2, \frac{5\pi}{6}\right), \left(2, \frac{7\pi}{6}\right), \left(2, \frac{4\pi}{3}\right), \left(2, \frac{5\pi}{3}\right)$, and $\left(2, \frac{11\pi}{6}\right)$

33. $\left(\frac{\sqrt{3}}{2}, \frac{\pi}{3}\right), \left(-\frac{\sqrt{3}}{2}, \frac{5\pi}{3}\right)$, and the pole **35.** π

37. $\dfrac{4\pi + 6\sqrt{3}}{3}$ **39.** $\dfrac{4\pi + 9\sqrt{3}}{8}$ **41.** $\dfrac{4\pi - 3\sqrt{3}}{6}$

43. $\dfrac{7\pi - 12\sqrt{3}}{12}$ **45.** $\dfrac{2\pi + 12 - 6\sqrt{3}}{3}$ **47.** 5π

49. $\sqrt{2}(1 - e^{-4\pi})$ **51.** $\dfrac{4\pi - 3\sqrt{3}}{8}$ **53.** $16a/3$

55. 16π **57.** $128\pi/5$ **59.** $2\sqrt{2}\pi$ **61.** 2π

67. a. 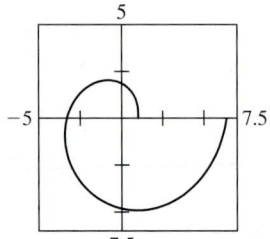   **b.** 22.01

69. a. 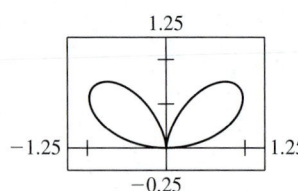   **b.** 5.37

71. a. 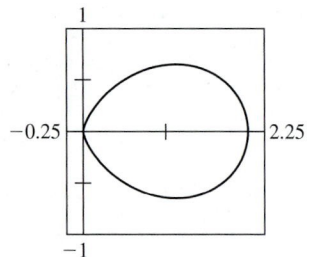   **b.** $\left(\frac{21}{20}, 0\right)$

73. a. $r = \dfrac{3\cos\theta\sin\theta}{\cos^3\theta + \sin^3\theta}$, $-\dfrac{\pi}{4} < \theta < \dfrac{3\pi}{4}$

   b. 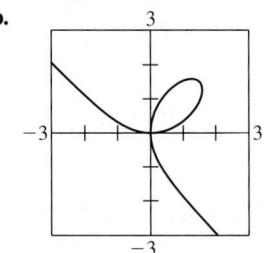   **c.** $\frac{3}{2}$

75. True

Exercises 9.6 • page 890

1. $r = \dfrac{2}{1 - \cos\theta}$, parabola

3. $r = \dfrac{2}{2 - \sin\theta}$, ellipse

5. $r = \dfrac{3}{2 + 3\cos\theta}$, hyperbola

7. $r = \dfrac{4}{25 + 10\sin\theta}$, ellipse

9. a. $e = \frac{1}{3}$, $y = 4$
   b. Ellipse
   c.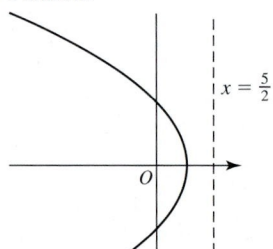

11. a. $e = \frac{3}{2}$, $x = \frac{5}{3}$
   b. Hyperbola
   c.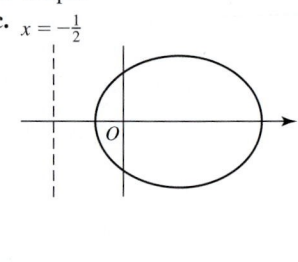

13. a. $e = 1$, $x = \frac{5}{2}$
   b. Parabola
   c.

15. a. $e = \frac{2}{3}$, $x = -\frac{1}{2}$
   b. Ellipse
   c. $x = -\frac{1}{2}$

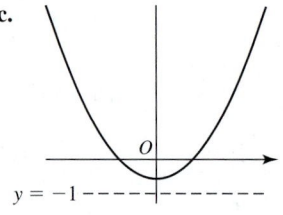

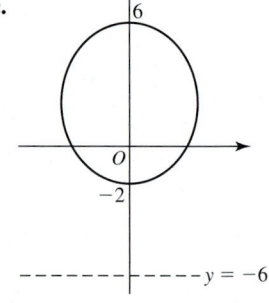

17. a. $e = 1$, $y = -1$
   b. Parabola
   c.

19. a. $e = \frac{1}{2}$, $y = -6$
   b. Ellipse
   c.

21. $\sqrt{7}/4$ **23.** $\sqrt{2}$ **25.** $\sqrt{10}/3$

33. $r = \dfrac{1.423 \times 10^9}{1 - 0.056\cos\theta}$, perihelion 1.347×10^9 km, aphelion 1.507×10^9 km

35. 0.207

Chapter 9 Concept Review • page 892

1. a. equidistant; point; line; point; focus; line; directrix
   b. vertex; focus (or vertex); directrix

3. a. sum; foci; constant
   b. foci; major axis; center; minor axis

5. a. difference; foci; constant
   b. vertices; transverse; transverse; center; two separate

7. $x = f(t)$, $y = g(t)$; parameter

9. a. $f'(t)$; $g'(t)$; simultaneously zero; endpoints

b. $\int_a^b \sqrt{[f'(t)]^2 + [g'(t)]^2}\, dt = \int_a^b \sqrt{\left(\frac{dx}{dt}\right)^2 + \left(\frac{dy}{dt}\right)^2}\, dt$

11. a. $r\cos\theta$; $r\sin\theta$

b. $x^2 + y^2$; $\dfrac{y}{x}$ $(x \ne 0)$

13. a. $A = \dfrac{1}{2}\int_\alpha^\beta r^2\, d\theta = \dfrac{1}{2}\int_\alpha^\beta [f(\theta)]^2\, d\theta$

b. $A = \dfrac{1}{2}\int_\alpha^\beta \{[f(\theta)]^2 - [g(\theta)]^2\}\, d\theta$

15. $\dfrac{d(P, F)}{d(P, l)} = e$; $0 < e < 1$; $e = 1$; $e > 1$

Chapter 9 Review Exercises • page 893

1. Vertices $(0, \pm 3)$, foci $(0, \pm\sqrt{5})$

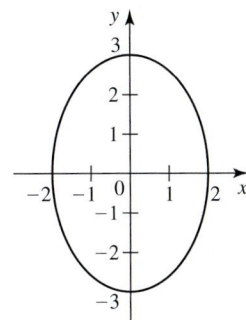

3. Vertices $(\pm 3, 0)$, foci $(\pm\sqrt{10}, 0)$

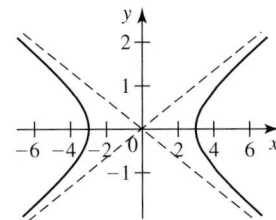

5. Vertices $(0, -1)$ and $(0, -7)$, foci $(0, -4 \pm \sqrt{10})$

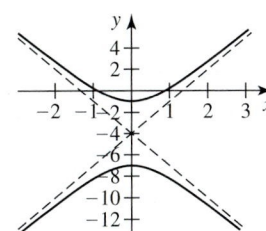

7. $y^2 = -8x$ **9.** $\dfrac{x^2}{49} + \dfrac{y^2}{45} = 1$ **11.** $y^2 - 4x^2 = 9$

15. a. $y = 4 - x$

b.
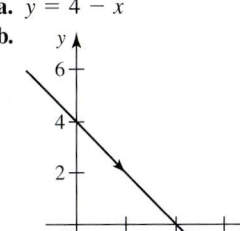

17. a. $(x - 1)^2 + (y - 3)^2 = 4$

b.
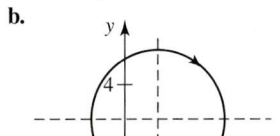

19. $\frac{4}{3}$ **21.** 0 **23.** $\dfrac{dy}{dx} = \dfrac{4(t^2 + 1)}{3t}$, $\dfrac{d^2y}{dx^2} = \dfrac{4(t^2 - 1)}{9t^4}$

25. Vertical at $\left(\pm\frac{16\sqrt{3}}{9}, \frac{10}{3}\right)$, horizontal at $(0, 2)$

27. $\frac{13}{3}$ **29.** $\sqrt{2}(1 - e^{-\pi/2})$ **31.** 3π

33.   **35.**

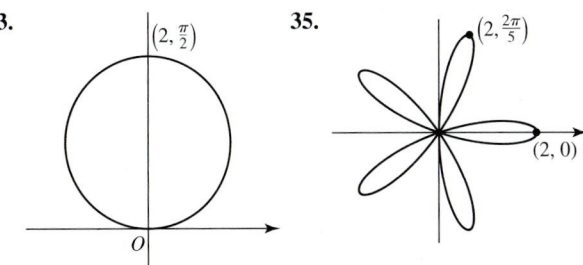

37.

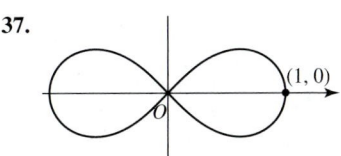

39. -2 **41.** $\left(\frac{1}{2}, \frac{\pi}{6}\right)$, $\left(\frac{1}{2}, \frac{5\pi}{6}\right)$, and the pole   **43.** $9\pi/2$

45. $2(\pi - 2)$ **47.** $\frac{8}{3}[(\pi^2 + 1)^{3/2} - 1]$ **49.** $4\pi^2$

51.   **53.**

55.   **57.**

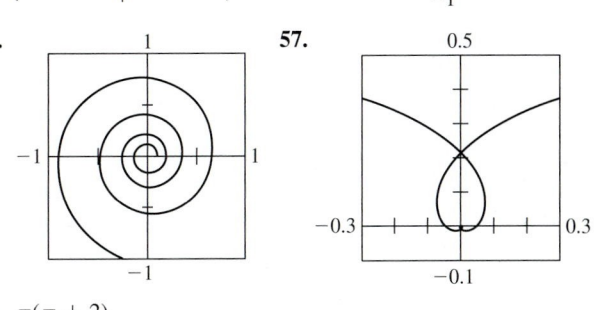

59. $\pi(\pi + 2)$

Chapter 9 Challenge Problems • page 894

1. c.

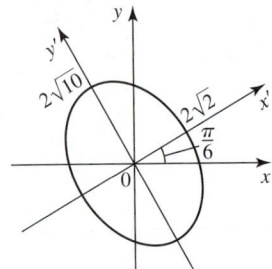

3. $\ln \dfrac{\pi}{2}$ **5. b.** $3a^2/2$

7. a. $r = \dfrac{\sqrt{2}}{2} \, ae^{(\pi/4)-\theta}$ (for an ant starting at the northeast corner)

  b. a

APPENDIX A

Exercises • page A 6

1.

3.

5.

7. False   **9.** False   **11.** $(-\infty, 2)$ **13.** $(-\infty, -5]$

15. $(-4, 6)$ **17.** $(-\infty, -3) \cup (3, \infty)$ **19.** $(-2, 3)$

21. $[-3, 5]$ **23.** $(-\infty, 1] \cup \left[\frac{3}{2}, \infty\right)$

25. $(-\infty, -3] \cup (2, \infty)$ **27.** $(-\infty, 0] \cup (1, \infty)$ **29.** 4

31. 2   **33.** $5\sqrt{3}$ **35.** $\pi + 1$ **37.** 2   **39.** False

41. False   **43.** True   **45.** False   **47.** True

49. False   **51.** $x = \pm\frac{4}{3}$ **53.** $x = -1$ or $x = -\frac{5}{3}$

55. $(-4, 4)$ **57.** $(1, 3)$ **59.** $(-\infty, -5] \cup [-1, \infty)$

61. $[-1.6, -1.4]$ **63.** $[-3, -1] \cup [1, 3]$ **65.** 0.4

INDEX

ENGLISH
ROMANTIC
WRITERS

EDITED BY

DAVID PERKINS

Harvard University

HARCOURT BRACE JOVANOVICH, INC.

New York | Chicago | San Francisco | Atlanta

COPYRIGHTS AND ACKNOWLEDGMENTS

WILLIAM BLAKE. Selections from the Nonesuch Compendium Edition of the *Poetry and Prose of William Blake*, Fourth Edition, ed. by Geoffrey Keynes, 1939, by permission of the Bodley Head Ltd.

GEORGE GORDON, LORD BYRON. Letters from *Lord Byron's Correspondence*, 2 vols., ed. by John Murray, 1922, and letters from *Byron: A Self-Portrait*, 2 vols., ed. by Peter Quennell, 1950, by permission of John Murray Ltd. Letters from *The Letters and Journals of Lord Byron*, 6 vols., ed. by Rowland E. Prothero, 1904: John Murray Ltd.

JOHN CLARE. Poems from *The Poems of John Clare*, 2 vols., ed. by J. W. Tibble, 1935, by permission of J. M. Dent & Sons Ltd., and poems from *John Clare, Poems Chiefly from Manuscript*, ed. by Edmund Blunden and Alan Porter, 1920, by permission of Putnam & Co. Ltd.

SAMUEL TAYLOR COLERIDGE. Letters from *The Collected Letters of Samuel Taylor Coleridge*, 4 vols., ed. by E. L. Griggs, by permission of the Clarendon Press, Oxford, and letter from *Unpublished Letters of Samuel Taylor Coleridge*, ed. by E. L. Griggs, 1932, by permission of Constable and Company Ltd. Selections from the book *Shakespearean Criticism* by Samuel Taylor Coleridge, ed. by Professor T. M. Raysor, Everyman's Library, reprinted by permission of E. P. Dutton & Co., Inc., and J. M. Dent & Sons Ltd. Selections from *Criticism: The Major Texts*, ed. by Walter Jackson Bate, copyright, 1952, by Harcourt Brace Jovanovich, Inc., and reprinted with their permission.

JOHN KEATS. Letters reprinted by permission of the publishers from *The Keats Circle: Letters and Papers, 1816–1878*, ed. by Hyder Edward Rollins, Cambridge, Mass.: Harvard University Press, Copyright, 1948, by the President and Fellows of Harvard College, and *The Letters of John Keats, 1814–1821*, ed. by Hyder Edward Rollins, Cambridge, Mass.: Harvard University Press, Copyright, 1958, by the President and Fellows of Harvard College.

CHARLES LAMB. Letters from *The Letters of Charles Lamb*, ed. by E. V. Lucas, 1935, by permission of J. M. Dent & Sons Ltd.

PERCY BYSSHE SHELLEY. Letters from *The Complete Works of Percy Bysshe Shelley*, ed. by Roger Ingpen and Walter E. Peck, 1926–30: Ernest Benn Ltd. Poems from *The Complete Works of Percy Bysshe Shelley*, ed. by Thomas Hutchinson: Oxford University Press.

WILLIAM WORDSWORTH. Letters from *Letters of William and Dorothy Wordsworth*, 6 vols., ed. by Ernest de Selincourt, 1935–39, by permission of the Clarendon Press, Oxford. Selections from *Lake Country Sketches* by H. D. Rawnsley, 1903, by permission of Macmillan & Co. Ltd. Selections from *Journals of Dorothy Wordsworth*, 2 vols., ed. by Ernest de Selincourt, 1941, by permission of The Macmillan Company and Macmillan & Co. Ltd.

A NOTE ON THE ILLUSTRATIONS: The illustrations of "The Lamb," "The Blossom," and "A Poison Tree" in the Blake section are from the Harry Elkins Widener Collection in the Harvard College Library and are reproduced with permission. The illustration of "The Tyger" is taken from a facsimile.

A NOTE ON THE COVER: The coat of arms reproduced on the cover of this book is that of the Worshipful Company of Stationers and Papermakers of London. This ancient City Company was originally a guild of stationers and craftsmen who made and dealt in parchment, paper, quill pens, and materials for binding books. In 1557, eighty-one years after printing was first introduced into England, the Stationers were granted a charter that gave them the monopoly of the "Art or Mystery of Printing." At the same time they received their coat of arms from the College of Heralds. For many years it was the custom to enter the titles of books to be printed by members of the Company in the Stationers' Register, which is thus a record of the first printing of many of the famous works known to students of English literature.

ISBN 0-15-522660-6

Library of Congress Catalog Card Number: 67–10010

PRINTED IN THE UNITED STATES OF AMERICA

PREFACE

MODERN LITERATURE begins with what we conventionally call the "Romantic Period," and on a large, dramatic scale. One of the principal considerations in selecting works and preparing the editorial commentary for this collection was the relevance of the English Romantic writers to our own generation. The book therefore concentrates on major figures, but without omitting or too sharply cutting works that have rightly been standard in anthologies during the last fifty years. The amount of intellectual prose usually presented has been considerably expanded, especially that of Coleridge and Hazlitt. One of the long prophetic books of Blake, *Milton*, is included for perhaps the first time in anthologies of the period, and there is also an unusually large representation of ancillary materials—letters, journals, reviews, and reminiscences by or about the major figures.

The main reason for including the ancillary materials is the quality of the writing; for if some of these selections are less famous than those regularly published, it is only because they are less accessible, being either out of print or available only in large, expensive editions. A second reason is the formative example and perennial fascination of genius: One has a rare opportunity to watch these writers in the welter of daily life, or to see them as they appeared to contemporaries who knew them. (For example, I have printed the accounts of Keats by his friends Charles Cowden Clarke, Benjamin Bailey, and Benjamin Robert Haydon.) Finally, such material often speaks of the personal experiences that were transformed into poems.

The focus, however, is on works of art. So far as possible, entire works are printed rather than extracts. I have departed from the precedent of some of the earlier anthologies by not providing selections from novels, since the excerpting of novels is unsatisfactory at best and unnecessary when paperback editions are available.

In the annotation and introduction to each writer the prevailing interests have been five: style and form, ideas and themes, biography, the critical and general intellectual background, and the writer's relation to English literary tradition. Accordingly the introductions attempt more than is usual in anthologies. They include a biographical narrative and one or more sections devoted to critical commentary and explanation; further discussion is presented in headnotes to particular works. The General Introduction seeks to convey a sense of the period as a whole—the historical background, literary and intellectual milieu, dominant literary themes and techniques. Through the commentary I have tried to free instructors to pursue more advanced or special topics in the classroom with the assurance that essential facts or basic critical themes will not have been ignored.

The bibliography at the end of each introduction is, for the major writers, something more than a list of books; it traces the history of critical and popular views of the writer from his own day to the present. (A bibliography of discussions of the period as a whole or of groups of writers concludes the General Introduction.)

In order to present this amount and variety of material, it was necessary to restrict the anthology to writers whose principal work was written within the conventional limits of the period, 1798–1832. (The sole exception is Clare, whose affinities are so obviously with the Romantic rather than the Victorian poets.) But this still allowed considerable leeway, and the general span of writing is about seventy-five years. The works of each author are printed in chronological order of composition, so far as this has been determined. The date of composition, as closely as is known, appears after each work, with the date of first publication in parentheses. Where much time elapsed between composition and publication, the work will ordinarily have been revised, and, of course, the texts used here may include revisions made after the first publication. I have not always printed from the last edition in the author's lifetime but have preferred the best texts that contemporary editorial scholarship can supply.

For very helpful advice in selecting works I am especially indebted to four men: Professor Earl Wasserman of Johns Hopkins, Professor Northrop Frye of the University of Toronto, and Professors W. J. Bate and Douglas Bush of Harvard.

DAVID PERKINS

CONTENTS

WILLIAM WORDSWORTH

THOMAS CAMPBELL

WILLIAM HAZLITT

THOMAS LOVE PEACOCK

BENJAMIN ROBERT HAYDON

GEORGE GORDON, LORD BYRON

EDWARD JOHN TRELAWNY

PERCY BYSSHE SHELLEY

EDWARD JOHN TRELAWNY

JOHN CLARE

JOHN KEATS

CHARLES COWDEN CLARKE

BENJAMIN BAILEY

RICHARD WOODHOUSE

J. G. LOCKHART

THOMAS LOVELL BEDDOES

ENGLISH ROMANTIC WRITERS

GENERAL INTRODUCTION

THE HALF-CENTURY from approximately 1775 to 1830 saw the American Revolution and the emergence of the United States, the French Revolution and Napoleon, the spread throughout Europe and America of democratic and egalitarian ideals, the origin or intensification in every European country of a sentiment of national identity, and, especially in England, the first important development of the industrial system. At the same time, virtually every realm of thought and art underwent a profound modification of which we still feel the impact. Three famous names in Germany alone—Kant, Goethe, Beethoven—may briefly suggest something of the scope of this cultural achievement and transition.

The term "Romantic," which is conventionally applied to the last thirty years of this period (1800–30), is far from adequate. All historical labels are, of course, simplifications and fail to suggest the diversity of life, thought, and art in the period to which they are applied. And this particular era, which in many respects ushered in the modern world, was at least as diverse as any previous thirty-year period. As a qualitative or descriptive term, the word "Romantic" —in its traditional and popular sense—is strictly applicable to only some aspects of the intellectual and cultural character of these thirty years. To begin with, the word still carries overtones it acquired in the early and middle eighteenth century, when it was connected with one of several reactions against neoclassical taste. In contrast to the rational order, regularity, and generalization associated with neoclassical art, "Romanticism"—largely because it was associated with the art and literature of the Middle Ages—suggested the "irregular," "picturesque," "wild," and distant. Of course interest in these qualities was strong in the early nineteenth century. But if the literature of the "Romantic" era was concerned with the remote or the distant, it was also concerned, in a new and vital way, with

I

the concrete and the directly familiar; and in it we find the beginnings of modern naturalism and realism. If it was idealistic, it was also at times directly empirical, and it flourished contemporaneously with a rapid development of historical study and psychological analysis. If it was attracted to the Middle Ages, it was equally drawn to classical antiquity; and its critical values both encouraged and profited from a resurgent and more informed study of the classical.

Beginning in 1798 the word "Romantic" was caught up by the influential German critics Friedrich and August Wilhelm von Schlegel, who gave it a deeper and more specific group of meanings; and throughout the nineteenth century the term became more widely used for the simple reason that nothing better appeared. The Schlegel brothers were interested in defining a contrast between the art and literature of the classical world and that of the Middle Ages and Renaissance (which they called "modern" in antithesis to "ancient"). According to the Schlegels, the "modern" is relatively indifferent to artistic form and seeks instead "fullness and life"—a complete expression of all life in its dynamism and its endless variety and particularity. Because this ideal is infinite, the spiritual quality of "modern" or "Romantic" art differs totally from the classical. The "Romantic" refuses to recognize restraints in subject matter or form and so is free to represent the abnormal, grotesque, and monstrous and to mingle standpoints, *genres*, modes of expression (such as philosophy and poetry), and even the separate arts in a single work. Ultimately it mirrors the struggle of genius against all limitation, and it leads to a glorification of yearning, striving, and becoming and of the personality of the artist as larger and more significant than the necessarily incomplete expression of it in his work. In this antithesis the Schlegels exalted the Romantic over the classical, but, it may be repeated, they applied the term to medieval literature and to such figures as Cervantes and Shakespeare and did not, on the whole, have in mind the writers of their own day. Nevertheless, they defined an ideal with which men of their time could identify, and it is especially as a result of their writings that the word "Romantic," as Goethe said in 1830, "goes over the whole world and causes so many quarrels and divisions . . . everyone talks about classicism and romanticism—of which nobody thought fifty years ago."

Even this more comprehensive use of the term is frequently attacked. It is argued that English writers did not think of themselves as "Romantic" or as constituting a "movement"; even in Germany the group that called itself "Romantic" did not include most of the figures now embraced by that generalization. Furthermore, as a critical term it tends to equivocate between the Romantic considered as a recurring type of personality and as a particular historical era. Moreover, even as the name of a cultural epoch the term constantly shifts meaning. Romanticism was not the same phenomenon in literature, fine arts, music, philosophy, historiography, and science; nor did it fall within the same span of time in the separate lines of endeavor and in the several nations. Hence, the argument goes, Romanticism is a name without a corresponding object, and it should either be used in the plural ("Romanticisms") or scrapped. On the other hand, it can be urged that the era under consideration, however variously delimited, was relatively short, and that within it one can identify widespread, though not necessarily harmonious, ideals, concepts, tastes, interests, and feelings. The term will be used here to refer to leading aspects of this cultural era, keeping in mind the era's variousness and distinctness.

THE HISTORICAL BACKGROUND

In discussing the historical background of the brilliant achievement of this period, one must begin with the French Revolution. For us, after nearly two centuries of revolution in various parts of the world, it is almost impossible to imagine the shock and awakening that swept over Europe when in 1789 the Bastille, a prison that symbolized royal power, was stormed by a Paris mob and a great popular uprising broke out in France. With the possible exception of the Reformation, which had developed much more slowly, there was nothing like it within historical memory. Two of its myriad effects may be noted at once. With the dissolution of social and class barriers—or even the hope of it—individual energies were released and it seemed that no limit needed to be set to personal ambition. A touching illustration is given by Crane Brinton in his *Decade of Revolution, 1789–1799* (1935), where he quotes the recollection of the scientist and philosopher Henrik Steffens:

> his father came home one night in Copenhagen, gathered his sons about him, and with tears of joy told them that the Bastille had fallen, that a new era had begun, that if they were failures in life they must blame themselves, for henceforth "poverty would vanish, the lowliest would begin the struggles of life on equal terms with the mightiest, with equal arms, on equal ground."

A few years later the prime example was, of course, the rise of Napoleon, but well before this the Revolution had an effect throughout Europe in stimulating the soaring aspiration that is so marked in the major figures of the age. In the second place, to sympathizers everywhere it appeared that society was about to be established on a rational and democratic basis, or even on a basis of fraternal love. "Nothing," said Southey long afterward, "was dreamt of but the regeneration of the human race." "Bliss was it in that dawn to be alive," remembered Wordsworth, "France standing on the top of golden hours / And human nature seeming born again." Throughout Europe the example of France intensified the already widespread questioning of social and political institutions, and, with the possibility of a transformation before their eyes, men set themselves with a new excitement and determination to seek ultimate principles—what is justice? human nature? the rights of man? whence derived?—and the quickening challenge of fundamental rethinking touched intellectual life at all points. There was, said Hazlitt, "a mighty ferment in the heads of statesmen and poets, kings and people." A "new impulse had been given to men's minds" and "philosophy took a higher, poetry could afford a deeper range."

The general causes of the Revolution in France may be found, first, in institutions and conditions that made social conflict inevitable and its peaceful solution difficult and, second, in the writings of intellectuals throughout the eighteenth century. The absolute monarchy of Louis XIV had weakened in the seventy years after his death, and by the time of the Revolution was incapable either of enforcing its will or of renouncing its claims to unlimited rule. Feudal privileges of the nobles and clerics were deeply resented, and the nobles, though willing to restrict the power of the king, would not surrender their own immunities and rights. The thriving bourgeoisie was excluded from power; the peasants and artisans were restless. Meanwhile, such writers as Voltaire, Diderot, and Rousseau waged a pamphlet war against abuses and argued that social, moral, and religious questions should be decided by an appeal to free reason rather than by tradition or authority. It was Rousseau—whose *Social Contract* (1762) begins with the famous inflamatory sentence "Man is born free, and is everywhere in chains" —who especially popularized the doctrine that all men are by nature free and equal. Thus the ideology of the Revolution was rationalistic, free-thinking, and democratic. The democratic ideal, partly expressed in the slogan "liberty, equality, fraternity," had an immense popular appeal throughout Europe, and when war later broke out, France at first conquered by ideology as well as by arms.

During the Revolution and subsequent war with France, the English upper classes lived in terror of a similar outbreak at home. To what extent this was a likelihood is not easy to say. Certainly English political and social institutions were unlike those of France and most of Europe. In the first place, power, instead of being concentrated in the monarch and his appointees, was more widely distributed. England was an aristocracy in which the lords and gentry controlled most positions of influence—they sat in Parliament, officered the army and navy, held the high places in the Anglican Church, and, as Justices of the Peace, dispensed law throughout the countryside. But they were not a closed group; middle class persons of talent or wealth could be assimilated into the governing class. In the second place, the ideals of the Revolution did not have the same impact in England as in the rest of Europe, partly because— with their traditional insular complacence—Englishmen viewed foreign ideology with suspicion and partly because of the widespread religious revival that began with the Methodist movement in the later part of the eighteenth century. This revival both distracted and consoled the suffering poor and made the "atheist" Jacobin an object of horror. The feeling was reinforced a few years later by the patriotic sentiment (powerfully expressed in Wordsworth's sonnets of 1802) that united the nation against France during the most dangerous period of the war. Finally, there was the continuing, pervasive effect of the oratory and pamphlets of the Whig statesman Edmund Burke, who attacked the Revolution. Then, as since, Burke was able, as one writer put it, to "sway the intelligent as a demagogue sways a mob." Nevertheless, the Revolution and its principles had active partisans, such as Tom Paine, who published his *Rights of Man* in 1791–92. Most of the Romantic writers of the first generation—notably Blake, Wordsworth, Coleridge, Southey, Lamb, and Hazlitt—were at first sympathetic to the Revolution, but, except for Blake and Hazlitt, they gradually adopted a more conservative position as their hopes were disappointed by events: the fall from power in France of the more moderate revolutionary party, the Girondins, and the rise of the extreme radicals, or Jacobins; the massacre in September, 1792, of more than a thousand prisoners by the Paris mob; the English war against France, which produced an acute conflict of emotions in these writers, setting love of country against revolutionary hope for mankind; the Reign of Terror in Paris (May, 1793–July, 1794), when thousands of supposed counterrevolutionaries were guillotined by the government of Robespierre; the French aggression against republican Switzerland in 1798; and the *coup d'état* of November, 1799, by which Napoleon siezed power and established a military dictatorship.

Except for one brief peace, England was at war with France for twenty-two years (1793–1815), and for part of this time was without allies and in daily fear of invasion. Thanks to the national ardor and, later, to the genius of Napoleon the French armies usually won on land. England and her allies eventually triumphed because of superior financial strength, control of the sea (confirmed by Nelson's victory at Trafalgar, 1805), Napoleon's insatiable ambition that caused him to overextend himself, and the overwhelming numerical preponderance of the allied armies in the closing years of the war. After Napoleon's abdication the Bourbon kings were restored in France, and the monarchs of Prussia, Russia, and Austria formed an alliance which gradually became an instrument for the suppression of liberal sentiments throughout Europe. For several years English statesmen uneasily cooperated with this policy and thereby earned the hatred of liberals such as Byron and Shelley, who experienced Austrian rule firsthand in Italy. After 1822 English foreign policy wore a more liberal aspect, and the government supported the revolt of the South American republics against Spain and of Greece against Turkey.

Particularly during the early years of the war the Tory government was reactionary and repressive. Liberal and reform movements were blackened with the cry "Jacobin," and their proponents were in jeopardy of prison. Though supported by philosophic radicals, factory workers, wealthy industrialists, and Dissenters (Protestant sects outside the established Anglican Church), proposals to reform Parliament by making it more representative came to nothing. Collective bargaining was outlawed (1799) and mass public meetings of any kind could be held only with permission of the magistrates. Suspected persons were sometimes kept in prison indefinitely without being brought to trial, a procedure that became possible only by suspending *habeas corpus*. Writers in such journals as Francis Jeffrey's liberal *Edinburgh Review* or Leigh Hunt's radical *Examiner* had to be cautious for fear of prosecution under the libel or sedition laws. Government spies and informers—"Satan's Watch-fiends," Blake called them— were active. Coleridge tells how one such eavesdropper reported that he and Wordsworth were discoursing on "Spy Nozy" (actually the philosopher Spinoza). This atmosphere, in which the government feared revolution and the liberals the government, naturally intensified passions and helps to explain, for example, how Blake could be tried for sedition on the word of a soldier and why the Tory journals *Blackwood's Edinburgh Magazine* and *The Quarterly Review* vilified Keats's poetry simply because Keats was associated with Hunt. It also helps to explain the bitterness of liberals such as Hazlitt toward the later conservatism of Southey, Coleridge, and Wordsworth.

Meanwhile, profound social and economic changes were taking place. The Industrial Revolution—the transition from hand and home manufacture of goods to the machine and factory system—meant that people were moving from the country villages to the new factory towns that sprang up, especially in the midlands. Local governments made little provision for the needs of this population for law, health, education, religion, or diversion. Life in the towns was grim and often vicious. The hardships inevitable in such a dislocation were further exacerbated by the reigning economic doctrine of *laissez faire*. It was held that the cheapest production of goods and thus the largest sum of national wealth would be secured if the government did not interfere in the operation of economic laws through the free market. Thus, except to outlaw collective bargaining, the government did nothing to adjust relations between the new capitalists and their employees, and workers had no protection against a sudden loss of employment or reduction of wages. In the countryside other developments resulted in similar distress among the poor. The members of a village had traditionally used some land in common, and for the marginal rural class the opportunity to pasture animals or raise crops on the common land had often made the difference between a meager sufficiency and pauperism. But throughout the eighteenth century improvements brought about by scientific agriculture made it increasingly profitable for private owners to "enclose" and farm these common lands for themselves. The process of enclosure was further accelerated by shortages of food during the war. At the same time, the loss of cottage industries—notably spinning and weaving, which now migrated to the factories—further impoverished the rural poor. During the war the price of food, like all other prices, fluctuated wildly but generally rose because European grain could not be imported. The high price of bread was artificially continued by the Corn Law of 1815 that protected English farmers from foreign competition. Because of the Industrial and Agricultural Revolutions England's wealth and productivity increased enormously during the Romantic Period, and it is probable that even the poor enjoyed, on the whole, a higher standard of living than previously. The cost was a more distant relationship between rich and poor, employers and employees, and a new resentment in the poor, who both in industry and agriculture now felt themselves dependent and insecure.

THE LITERARY SCENE

A sketch of the literary milieu may begin with the publishing trade itself. Literacy in Britain had been growing steadily since the Renaissance, and acquaintance with current books was to some degree a social obligation. As a result a large public was willing to pay the relatively high prices of from six shillings to one pound for a volume. (A pound at that time was worth approximately five to seven pounds, or about fifteen to twenty dollars, in present-day purchasing power.) A popular writer could earn a fortune for himself and his publisher. John

Murray sold 10,000 copies of Byron's *The Corsair* on the day of publication, and Byron later demanded £2,625 for the fourth canto of *Childe Harold*. Sir Walter Scott may have received from writing as much as £80,000 in his lifetime. But, except for Byron, Scott, and Wordsworth in his later years, the major poets of the age sold very little.

The leading magazines and reviews also had a wide audience. The first of these, *The Edinburgh Review*, was founded in 1802 and became an organ of liberal thought. It included articles on history, politics, and science as well as on imaginative literature, and its influence was considerable. In fact, it was partly because of the strictures and ridicule of its editor, Francis Jeffrey, that Wordsworth's reputation advanced so slowly. *The Quarterly Review* was established in 1809 as a Tory counterpart to *The Edinburgh Review*, and two other prominent journals were the Tory *Blackwood's Edinburgh Magazine*, founded in 1817, and *The London Magazine*, which dates from 1820. *The London Magazine* printed some poems of Keats, Clare, and Hood and first published De Quincey's *Confessions of an English Opium Eater*, portions of Hazlitt's *Table Talk*, and Lamb's *Essays of Elia*. The journals paid contributors well, usually between ten and twenty guineas per sheet; and without the stimulus they provided, less nonfictional prose would have been written. Yet the familiar essays, criticism, and miscellaneous prose of Lamb, Hazlitt, Southey, and De Quincey—all written chiefly for the journals—are among the enduring achievements of the age.

Another outlet, especially for criticism, was public lectures, usually paid for by subscription and given as a series. Most of Coleridge's influential Shakespearean criticism was delivered as such lectures. He spoke from notes, trusting to inspiration, and what he said survives chiefly in fragments. A different method was adopted by Hazlitt, who wrote out his lectures and later published them as books: *Lectures on the English Poets* (1818), *The English Comic Writers* (1819), and *The Dramatic Literature of the Age of Elizabeth* (1820). The first of these series was attended by the young Keats and, with other writings of Hazlitt, helped to waken his moral and poetic ideals of disinterestedness and the sympathetic imagination.

The age produced much drama, but little of value. There are many explanations, including the intimidation the Romantics felt before the examples of Shakespeare and the Elizabethans and their effort to imitate too exclusively the more external aspects of the Elizabethan drama. To this we could add the cramping effect of censorship, which precluded serious discussion of religion or politics. But most important was the state of the acting profession, the theaters, and the audience. The star system prevailed and created a tendency to conceive plays with one or two overwhelming characters. Acting style was declamatory. The theaters were large, so that most playgoers were relatively far from the stage. (Covent Garden Theater had a proscenium approximately forty-three feet wide and thirty-five high and could seat 3,044.) Production costs were high. The audience was noisy and undiscriminating. In these circumstances the theater managers, striving to fill their capacious houses, vied with one another in producing spectacular scenic effects. One piece packed Drury Lane Theater for three weeks by offering such excitements as a burning forest and a lady on horseback riding up the cataract of the

Ganges. The possibility and prevalence of such effects may indicate that Shelley's *Prometheus Unbound,* or more especially Byron's *Manfred,* though closet drama, were not so remote from the stage as they now seem. Certainly they show the influence of the contemporary theater. When Byron requires that Arimanes be seated on a "Globe of Fire" or Shelley brings on stage a rushing, whirling sphere of "Ten thousand orbs involving and involved," they are only slightly straining the resources of the theater managers, and the opening of *Prometheus Unbound* in "A Ravine of Icy Rocks in the Indian Caucasus" was routine.

By the early years of the nineteenth century the novel at least rivaled poetry and certainly outstripped the drama as the most popular of literary forms. A swarm of hacks engendered yearly broods for the circulating libraries. At a higher level Ann Radcliffe specialized in gothic novels of terror such as *The Mysteries of Udolpho* (1794) and *The Italian* (1797), the latter presenting as a main attraction the sinister monk Schedoni. To Jane Austen the gothic and sentimental cast of mind was an object for parody, and her portraits of the life of the gentry were worked with a conscious artistry and irony that make her now the most highly esteemed novelist of the period. In the judgment of the age, however, the leading novelist by far was Sir Walter Scott. A mine of antiquarian and historical information, he virtually invented the historical novel in a series of works set in the Middle Ages and in different eras and locales of the Scottish past and present. The charm of local custom and manners was reinforced by picturesque landscape, memorable characters, and rapid adventure. His fame and popularity spread throughout Europe and is one of the important cultural phenomena of the period.

THE ROMANTIC MODE

Nevertheless, the supreme achievement of the age was in poetry. From the standpoint of literary history, many of the common themes and styles of the Romantic poets are a development of the efforts of forerunners in the eighteenth century, poets who came after Pope and who, despite their admiration of him, tried to explore new possibilities. From another point of view, the Romantic mode concretely expresses underlying premises as to the nature of reality (which are taken up in a separate section below).

Among the most obvious general features of Romantic poetry are persistent reference to nature and natural objects, intimate self-revelation of the poet, and direct expression of strong, personal emotion. These aspects are so obvious that they are often cited reductively, and some preliminary qualifications are in order. The prominence of landscape and natural objects in this poetry does not necessarily involve a flight from wider concerns to the simplicities of the countryside. For the greater poets, the description of landscape and natural objects was not only a theme but a vehicle, a medium in which intellection was expressed. As Schiller pointed out in his famous essay "On Naive and Sentimental Poetry" (1795–96), the poet in the modern world does not depict nature for its own sake but to convey the "ideal." Similarly, the *étalage du moi* ("display of the self") in this poetry need not reflect egoistic self-absorption but arises partly in the struggle of thinking men to ground speculation and belief in what seemed the most certain facts of experience. So also with the emphasis on feeling, manifested in frequent exclamatory interjections, weighted use of names of emotions together with descriptions of emotional states, and a tendency through the course of a poem to trace not so much the sequence of logical argument or of narration, but rather the evolution and turn of feeling. Wordsworth's formula that "poetry is the spontaneous overflow of powerful feelings" boldly sums up one general presupposition that influenced poets and critics throughout the period. But neither this nor countless similar statements justifies an easy antithesis between the classical and the Romantic by equating it with such antitheses as reason and passion or control and release. Nor does it inevitably imply a supposition that feeling is of itself a good. The essential meaning of the Romantic emphasis on feeling is not cultivation of one quality or power at the expense of others but the pursuit of an ideal of unity or completeness of being.

The imagery of the English Romantic poets reflects, in one of its aspects, a living attention to concrete particulars. "To Generalize," declared Blake in a marginal note, "is to be an Idiot. To Particularize is the Alone Distinction of Merit." The frequent result is a full, vivid, and exact realization of objects. In "This Lime-Tree Bower My Prison," for example, Coleridge devotes three lines to a detailed description of a "Broad and sunny leaf" dappled by the "shadow of the leaf and stem above." Through four lines of "Resolution and Independence" Wordsworth describes a hare running on a moor and raising a mist about her from the "plashy earth." From some of the enthusiastic comments in Keats's letters, it might be inferred that the principal aim of poetry is to catch and render each thing of this world in its full individuality, and his own work provides memorable examples: "The hare limped trembling through the frozen grass," in *The Eve of St. Agnes*, or the "wailful choir" of "small gnats" in the ode "To Autumn":

> Then in a wailful choir the small gnats mourn
> Among the river sallows, borne aloft
> Or sinking as the light wind lives or dies.

In each of these examples objects are seen in a context of other objects and circumstances, and the context gives a further particularity to the image. This vision of things in context informs much Romantic imagery. It has an ontological meaning, but for the moment a more purely literary effect may be noticed, namely, the impression it gives of actual observation and immediate experience.

Yet with this "purchase on matter," as Hazlitt called it, this keeping to "natural bones or substance," the Romantics seldom sought or achieved the detached, scientific objectivity that was held as an ideal in the "realism" of the later part of the century. Keats especially valued a responsiveness so keen and massive that it brings about a sympathetic identification. This state of mind, which Hazlitt characterized as "gusto," manifests itself in an imagery of densely intertwined impressions, often taking the form of synaesthesia (the interpretation of one sense by another, as in the phrase "soft incense") and often including a dynamic, even organic participation in the life of the thing portrayed. More commonly, Romantic poetry renders things not as they are "in themselves" (supposing that were possible) but as they are modified in perception by the thoughts and feelings of the poet. In this process objects and scenes, though present in thick, substantial being, also become outward expressions of human insight and feeling.

To this must be added the pervasive reliance on the quality of suggestion. For the whole context of any event or object might be specified to some degree, but it could not possibly be exhausted descriptively or known completely. Hence the Romantics prized a use of language that would remain open to all that could not be said and, at the same time, summon the imaginative energies of the reader. "The power of poetry," said Coleridge, lecturing on Shakespeare's *The Tempest*,

> is, by a single word perhaps, to instil that energy into the mind, which compels the imagination to produce the picture. Prospero tells Miranda,
>
> > "One midnight,
> > Fated to the purpose, did Antonio open
> > The gates of Milan; and i' the dead of darkness,
> > The ministers for the purpose hurried thence
> > Me, and thy *crying* self."
>
> Here, by introducing a single happy epithet, "crying," in the last line, a complete picture is presented to the mind, and in the production of such pictures the power of genius consists.

Most Romantic poets believed in the reality of a supersensuous or noumenal realm of being. The intuition of this could be expressed only by suggestion, and thus suggestive uses of language often appear in poems or passages that, for readers in the period, belonged to the widely recognized category of the "sublime." The term, popularized throughout the eighteenth century, had referred to objects of overwhelming vastness or power, or to the state of mind felt in their presence; in the Romantic age it came also to designate more particularly that state of mind arising in contact, either direct or through analogy, with the transcendent and infinite. Thus "the grandest efforts of poetry," Coleridge said,

> are when the imagination is called forth, not to produce a distinct form, but a strong working of the mind, still offering what is still repelled, and again creating what is again rejected; the result being what the poet wishes to impress, namely, the substitution of a sublime feeling of the unimaginable for a mere image.

Often the quest for the "unimaginable" through suggestion took another form in evoking the mysterious, occult, and supernatural. This became something of a fad and was often pursued for its own sake, but of such poems as Keats's "La Belle Dame sans Merci" and Coleridge's "The Ancient Mariner," "Christabel," and "Kubla Khan" with its famous lines

> A savage place! as holy and enchanted
> As e'er beneath a waning moon was haunted
> By woman wailing for her demon-lover!

the most general meaning lies in the reminder of what Wordsworth called "unknown modes of being," of a reality beyond the limited, empirical one that comes to us through our senses and understanding. It was from reading the *Arabian Nights*, said Coleridge, and similar tales of "Giants & Magicians, & Genii" that his mind "had been habituated *to the Vast*—& I never regarded *my senses* in any way as the criteria of my belief."

Some of the aspects of imagery already mentioned—the involvement of the mind in the object, the reliance on suggestion—approximate what would ordinarily be called symbolism. Certainly both the term and the way of thinking and writing it describes have a large place in poetic theory of the period, especially that of Coleridge and Shelley, and in the practice of all the major poets. There was, however, no one theory or practice, and perhaps the chief general statement that can be made is simply that the Romantics did not possess a traditional symbolism of which the meaning or reference is determined by convention. In them one finds instead what has become more explicit and conscious in contemporary poets such as Yeats and T. S. Eliot, who adopted it as the chief technique of the *Four Quartets*: images recur in the work of a writer and bring associations from the poems or passages in which they had been used before; these images thus acquire increasing depth of implication and in them the writer gradually develops a personal symbolism. This appears in its most systematic form in Blake's "prophetic" poems and is perhaps least apparent in Keats. However, Keats's odes are the finest examples in the period of what may be called the poetry of symbolic debate. In these poems a central symbolism—the nightingale, the Grecian urn—enables the poet to engage conflicting values and attitudes concretely and to achieve at least an *ad hoc* resolution. A special type of Romantic symbolism arises in what may be called visionary poetry. This is found in poems presenting superhuman beings in a dreamlike realm, as in Shelley's *Prometheus Unbound* or Keats's "The Fall of Hyperion," and also in the use of an imagery wrenched out of natural contexts and relations. For example, Blake's "The Tyger" shows a blacksmith creating a tiger, and Shelley for symbolic purposes in "The Witch of Atlas" sends an open boat beneath the surface of a river. Visionary symbolism often implies a belief that ultimate truth lies beyond nature rather than in it, and that to reflect the supernatural and eternal, a poet must articulate images in total freedom.

The diction and syntax of Romantic verse, like much else besides, were strongly influenced by ideals of spontaneity or naturalness. When in the Preface to the *Lyrical Ballads* Wordsworth asked himself the crucial question "What is a Poet?" he gave the revolutionary though not unprecedented answer that a poet is a "man speaking to men." The traditional view had been that a poet is a "maker," and this implied a more detached relation between the poet and his work and also between the poet and his readers. As a "speaker" the Romantic poet was likely to avoid the appearance and to some degree the fact of predeliberating artistry, partly because this might seem insincere and partly because spontaneity, immediacy, or naturalness were thought to have an inherent value in bringing us closer to the poet and to actual experience. Thus in "Hyperion" Keats produced the most brilliant of the many Miltonic imitations, yet he quickly abandoned the style: "Miltonic verse cannot be written but in an artful or rather artist's humour." One of his "Axioms" was that "if Poetry comes not as naturally as the Leaves to a tree it had better not come at all." In this spirit Wordsworth's Preface to the *Lyrical Ballads* specifically condemned poetic diction, periphrasis, and "personifications of abstract ideas"—conventions in the poetry of the previous hundred years—and affirmed that one "principal object" was "as far as is possible, to adopt the very language of men." Though this intention applied *in toto* only to some poems in the *Lyrical Ballads*, an ideal of easy naturalness modified the phrasing of Wordsworth and Coleridge throughout their careers—one may think of Wordsworth's "Michael," for example, where a child herding sheep is "Something between a hindrance and a help"—and this ideal was caught up and carried further by the second generation of poets: Keats, Shelley, and especially Byron. In syntactical arrangement the Romantics (except Byron in his satires) strove to avoid the balance and antithesis they especially associated with Pope and judged "mechanical." Instead their sentence structure is often unpremeditated, digressive, and accumulative as it enacts the drama of the mind's free movement. "My thoughts," said Coleridge, "bustle along like a Surinam toad, with little toads sprouting out of back, side and belly, vegetating while it crawls."

INTELLECTUAL BACKGROUND

British Empiricism

The most influential English philosopher throughout the eighteenth century was John Locke, two of whose assertions may be noticed here. In the first place, Locke put in a new form the old argument that only particular, concrete things exist. General terms such as "man," "freedom," and so forth name ideas that we have abstracted from experience, and such ideas have no counterpart in reality. (For example, there is no essence or substance "man" but only individual men.) In the second place, denying the existence of innate ("inborn") ideas, Locke held that the mind at birth may be compared to a *tabula rasa* (smoothed tablet) on which impressions are engraved as we grow up. These impressions are always sensory, and thus sensations are the ultimate source of all our ideas.

This analysis, or rather the whole tradition of British empirical psychology and philosophy that flowed from it, profoundly supported tendencies and premises in Romantic literature that were also developing from other sources. For one thing, if real things are always particular, it might easily be argued that they are always unique. Thus, especially in the criticism of Hazlitt, "truth" as the aim of art might shift its traditional meaning. No longer referring to permanent, universal forms or types, truth might now require that an object be presented with full concreteness and with a highlight on what peculiarly characterizes or differentiates it. An art thus bringing forth distinct individualities or identities was, for Hazlitt, "expressive," and in his estimates of books and paintings the criterion of "expressiveness" usually predominates.

In this sense, British empiricism strongly reinforced the particularist and circumstantial bias in Romantic poetry.

The same tradition also encouraged a literature of self-expression and the subjective exploration of the writer's mind and feelings. One stimulus was simply the attention given by Locke and his followers to facts and processes of mind. This of itself promoted introspection. There were, however, more urgent considerations. If all ideas, including general ones, come ultimately from concrete experience and refer back to it, that experience is by definition personal and differs from one individual to another. Moreover, psychological theory in the later eighteenth century usually explained the unity of the mind by the association of ideas. It was held that when two or more ideas are linked together, as by succession in time, simultaneity in space, and so forth, one idea will thereafter recall the other. Thus gradually a system of associative links is set up such that the whole mind is latently present at any moment. But in any particular mind what associative links are actually established is to some degree accidental. To this must be added the disturbing arguments of David Hume, in whose philosophy the empirical tradition led to a throughgoing scepticism. To Hume it seemed that since we have cognizance only of the contents of our own minds, we can never know that our sensations or ideas correspond to what exists outside us. For example, we can never know that one thing is the cause of another, for we experience only the succession of two ideas, not those realities in which the supposed principle of cause and effect may or may not operate. Hume thus denied the possibilities either of directly knowing the real world or of making any necessary inferences concerning it. While this extreme position was not generally adopted, there was a widespread assumption that one's hold upon objective truth is uncertain and that opinions and assertions can be spoken only from a personal point of view, that is, with reference to one's own experience and consciousness. It was perhaps with some such recognition that Coleridge resolved "to write my metaphysical works, as my Life, & in my Life, intermixed with all the other events or history of the mind & fortunes of S. T. Coleridge," a project partly fulfilled in *Biographia Literaria*.

A further inducement to self-expression developed from the premise of the unique character of each individual. Just as art, looking outward, might strive to disclose the particular identity of other objects or persons, so the poet might legitimately disclose his own identity, and readers might welcome intimacy with a being both unique and exceptionally gifted. "Poetry without egotism," Coleridge remarked, is "comparatively uninteresting." Thus originality, which had once implied merely novel inventiveness or had been applied as a cant term to "natural" or untaught poets such as Burns and Clare, came to be regarded as the inevitable result of successful self-expression, and at the same time it acquired the peculiar sanction it still to some degree retains. "Points have we all of us," said Wordsworth, "where all stand single,"— "Something within which yet is shared by none." The effort of the poet is to impart his special dower, "Heaven's gift . . . that fits him to perceive / Objects unseen before." In an extreme espousal of Romantic views, the development and expression of one's originality might seem almost a sacred duty. It was a dynamic cooperation with the creating, proliferating energies of the cosmos, in which the cosmos itself was enriched and at the same time reflected from a novel point of view.

Transcendentalism

Romantic transcendentalism arose as a direct reaction against the empirical tradition. In the philosophy of Kant "transcendental" refers to the *a priori* element in experience—that is, the way in which the mind determines and orders its own contents through its own laws—and the term "transcendent" refers to ideas, such as freedom of the will, God, and immortality, that cannot become objects of knowledge. More generally, transcendentalism is the belief in the existence of a timeless realm of being beyond the shifting, sensory world of common experience. It was nourished in the Romantic Period from many sources—Plato and the Neoplatonic philosophers, contemporary German idealism, occult and theosophical writings, and, at least in Blake, Wordsworth, and Shelley, from personal experiences sometimes called mystical. (Transcendentalism as an element in the Romantic philosophy of organicism will be discussed later.) In Shelley especially, however, the cosmos is conceived less as an organic whole and more as divided between the actual, concrete, present world and the ideal or transcendent that lies beyond. The first has an inferior degree of reality, while the second is the goal of aspiration and the true home of the human spirit. At moments one may intuit or, as Keats said, "guess at" the realm of timeless and perfect being, and to express or convey such moments, with their overwhelming impact on thought and feeling, is the highest aim of poetry.

Time, History, and Nostalgia

As a third influence on Romantic literature, in addition to empirical and transcendental premises, one may note the more modern consciousness of time and history that had been emerging throughout the eighteenth century. This may be described in brief as a shift toward a genetic and historically relative point of view. Increasingly the assumption was that both social institutions and personal character could be understood only by examining the historical circumstances that shaped their origin and growth. For example, Wordsworth's *Prelude* is a long, introspective, biographical account of his own development, the purpose of which was to interpret his present character and to determine, he says, his qualifications for writing a major philosophic poem. Human nature, which classical and neoclassical thinkers had presumed to be essentially the same in all times and places, now appeared to be molded by history and therefore susceptible of fundamental change. Supposedly universal standards could not so readily be invoked, for it could be argued that moral and literary rules lose validity and pertinence with the passage of time. The novels of Scott especially made this historical relativism concretely meaningful. They were virtually the first literary attempt, based on genuine study, to enter imaginatively into the alien character and feelings of men in the past. For this reason they had a profound impact on historiography and on all thinking readers.

The historical point of view induced a tendency to balance progress against loss. But however mindful of present advantages one might strive to be, it was not easy to think of the times forever gone without also sliding into regret. Unquestionably the consciousness that time brings fundamental change and that the past is irrecoverable was a principal source of Romantic nostalgia, a state of mind so pervasive that it is often said to be the essence of Romanticism. With respect to what may be called cultural nostalgia, however, it should be stressed that at least in England it rarely involved a genuine assumption that some past age or alien state was more to be valued than the present. Thus "primitivism," the glowing description of the virtues of the "state of nature" as it might be imagined in prehistoric man, American Indians (the "noble savage"), Scotch Highlanders, children, or (notably by Wordsworth) the peasantry of England, did not usually express a wish to go back or away but was rather a wistful sense of loss, a reaction against urban complexity and artifice, and a potent means of criticizing contemporary society. So also with the attraction to the Middle Ages, as in Scott's novels and poems or Keats's *The Eve of St. Agnes,* or to the Renaissance, as in Hunt's "Story of Rimini" or Keats's *Isabella,* or to ancient Greece dreamily conceived, as in Keats's *Endymion,* or to the contemporary Near East as with Byron. In all these instances the nostalgia was partly for a setting that contrasted with the present by allowing color, adventure, mystery, sincerity, simplicity, passion, and beauty.

Nostalgia toward one's own past was naturally more intense and less manageable. The changes that time brings in the self and its world are a major theme of Romantic literature, and, though Wordsworth tried to do so in such poems as "Intimations of Immortality" and "Elegiac Stanzas on . . . Peele Castle," it was not easy to find in adult maturity a compensation for the loss of youth. Most especially is this true with such writers as Wordsworth and Shelley, who felt that childhood or youth had been marked by moments of profound intuition or visionary insight no longer experienced. But the same thing was felt by writers who lay no claim to moments of special illumination. Nostalgia—the vivid, tender, descriptive cherishing and sorrow for the past—dominates the familiar essays of Lamb and Hazlitt, and, with the implication of a heart now jaded and guilt-hardened, it sweeps through much of Byron: "No more—no more—Oh! never more on me / The freshness of the heart can fall like dew."

Organicism

Romantic organicism was an attempt to reconcile in a profounder conception the several premises already discussed. The opposition of empiricism and transcendentalism was itself one ramification of questions that had been taken up in philosophy at least since Plato and that Descartes (1596–1650) had formulated in a way that challenged thinkers for the next hundred and fifty years. According to Descartes reality is comprised of three principles or substances—

matter, mind, and God—and the difficulty was to show how these are interconnected. By the early nineteenth century the issue was usually posed in terms of such dualisms as matter and mind, the real and the ideal, or nature and God. One might, with Spinoza, adopt a pantheist solution by arguing that matter and mind are both modifications of the one substance, God, or one might, as was often the case in British empiricism (especially as it was taken over and developed by French writers), strive to show that the mind is not a separate substance but a mode of physical or material activity. In Kant and his successors throughout the nineteenth century, on the other hand, the philosophy of idealism stressed the mental or ideal as a distinct component of reality and maintained that the mind, instead of depending on impressions from without, has the prime role in the creation of what we know as experience. The idealists argued, for example, that values such as beauty or goodness are real but not learned from experience. Instead they are either determined by the mind in an *a priori* way, or, as in Plato, they exist in a transcendent realm that is at least partially disclosed to human beings. Through his free will (the existence of which empiricists often denied) man then strives to express these values in his own acts—in other words, to bring value into the realm of concrete phenomena. Organicism abandons these dualisms by conceiving the cosmos (reality) as a process rather than as a substance, an activity in which the material world, the mental or ideal, and the Divine mutually involve or interpenetrate each other.

Among philosophers the concept of organicism was especially developed by the German idealists and, in an entirely different way in our own century, in the brilliant, difficult metaphysics of Alfred North Whitehead. In the English Romantic writers it is present as a basic conception of reality that is often expounded abstractly (especially by Wordsworth and Coleridge) but still more frequently manifests itself as it guides critical interpretation and poetic vision. This poetry, as Whitehead said, could be described as a "protest on behalf of the organic view of nature," and the protest based itself on "the concrete facts of our apprehension." That is, in immediate awareness we do not distinguish what the mind has contributed from what comes from without. Values, for example, such as beauty or goodness are felt as aspects of concrete experience, and it is only by a secondary act of abstraction that we say they have their source only in the mind or only in the transcendent or empirical realms. Unless, therefore, by analysis we chop an experience into separate aspects ("we murder to dissect," as Wordsworth put it), we will not argue that values are any more or less "real" than whatever other aspect or object we may discriminate in the wholeness of the immediate event. Objects themselves, moreover, emerge in actual perception with an infinite background of which we are always to some extent aware. So in Wordsworth, said Whitehead, it is

> the brooding presence of the hills which haunts him. His theme is nature *in solido*, that is to say, he dwells on that mysterious presence of surrounding things, which imposes itself on any separate element that we set up as an individual for its own sake. He always grasps the whole of nature as involved in the tonality of the particular instance.

And we should add that, for Wordsworth, the "whole of nature" involves also the Divine.

The organicist vision of nature thus had a profound impact on poetic theory and critical values. The implications were articulated chiefly by Coleridge in the concept of "organic form," which is, in many respects, a rationale of the working procedures of Romantic poetry. The argument was that a poem manifests a process that is at least analogous to that of reality itself. It fuses, Coleridge said, the universal with the particular, the idea with the image, the part with the whole; it exists both as a unity and as an activity that develops through time; it reconciles spontaneity with inevitability and law. A model or analogy often proposed was a growing tree. One may say that the universal principle or form of a tree is implicit in the seed, yet it declares itself only through time. The form does not exist independent of the concrete manifestation, and it expresses itself in each part, as also the existence of each part presupposes the whole. Coleridge often contrasted organic with "mechanical" form. The "mechanical," he said (paraphrasing A. W. Schlegel), is predetermined and subsequently impressed on whatever material we choose, as when "to a mass of wet clay we give whatever shape we wish it to retain when hardened." The organic form, on the other hand, "shapes as it develops itself from within, and the fullness of its development is one and the same with the perfection of its outward form." Each exterior thus becomes a "true image" of "the being within." The concept of organic form provided a fundamental justification for art, which could be regarded as a means of concretely knowing or even participating in reality as process. It also gave rise to an approach to art that stressed sympathetic identification rather than analysis from a critical distance. And it stimulated a criterion of evaluation that rests on the extent to which all the "parts" of a work of art (plot, character, emotion, theme, image, and the like) interconnect and sustain one another. This relationship was, however, to be sensed immediately, and it is doubtful whether Coleridge would have approved much of the criticism of our own time that applies this criterion by piecemeal analysis.

The Imagination

In the Romantic Period much of the speculation outlined above centers around the term "imagination." Until the later eighteenth century the term had usually denoted the faculty by which we form mental images of things not present to our senses or held in memory. This faculty could also combine ideas in ways not warranted by experience, as when the idea of supernatural power is joined to the shape and passions of man in order to produce the gods of Greek mythology. As the example suggests, the imagination thus described was thought to have a large role in poetry, but it was far from being the essential power of poetry simply because it was not necessarily or even probably directed to reality or truth. As conceptions of reality changed, however, a new way of knowing it had to be described, and the term "imagination" gradually extended its meaning. There was, however, no single interpretation of the imagination, for this varied with the general context of thought in the different writers. Hence it is more appropriately discussed in relation to particular thinkers (see the introductions to Blake, Wordsworth, Coleridge, Hazlitt, Shelley, and Keats) and is only briefly treated here.

In its most general significance, "imagination" denoted a working of the mind that is total, synthetic, immediate, and dynamic. In this sense, the theory of the imagination was a reaction not only against empirical analysis but also against the traditional faculty psychology, of which the vocabulary, at least, persisted throughout the nineteenth century. This psychology conceived the mind as exhibiting separate powers or functions (sensation, memory, reason, emotion, and so forth), and not all of these functions were necessarily engaged in any particular act of the mind. Thus mathematics was thought to involve a step-by-step proceeding of just one faculty, and similarly emotion might be disengaged from reason or sensation from emotion. Granting that the mind can indeed work this way, the Romantic writers held that in doing so it loses touch with the fullness of reality. If this reality inheres in concretely individuated particulars, then the imagination may be described, with Hazlitt, as the immediate response of all faculties at once to the particular object, and also the coalescence or synthesis of the impressions thus received into a unified mental construct that corresponds in every way to its object. One should stress the concrete richness of the image thus created, for it may be so distinct and vivid that one forgets one's own separate identity and lives in the object. Thus the imagination in its highest working becomes "sympathetic," freeing one from self-consciousness and self-interest and enabling one to enter into the experience and feelings of other persons and even into animals and inanimate objects. "If a sparrow come before my window," said Keats, "I take part in its existence and pick about the gravel," and Byron proclaimed, "I live not in myself, but I become / Portion of that around me." The same sympathy could also be given, Keats reminds us, to "creations of my own brain"; this further explains Keats's conception, derived from Hazlitt, of the poet of "Negative Capability" whose own identity never appears in his work, a type of poetic character of which Shakespeare was clearly the supreme example. Moreover, the same loss of self in sympathetic identification could be felt before the landscape as a whole, as is often the case in Wordsworth and Byron, and sometimes in Shelley.

However, the term could also designate a power able to commune with transcendent reality. In Shelley it is sometimes identified with the Platonic reason (*nous*). In connection with organicism, on the other hand, the imagination, as the active coalescing of all faculties of the mind, involves faculties empirically directed to the outer world, as the senses, faculties through which we have access to our own human nature, as the emotions, and a faculty of transcendent intuition (which Coleridge, at least, usually called reason). Thus only through the imagination can we apprehend reality in its organic wholeness and process. Especially in Wordsworth, however, the term is invested with so much significance that different implications are highlighted in different contexts. Sometimes it appears that when the imagination is most fully roused the senses fail ("the light of sense goes out"), and the imagination achieves a direct intuition of one's own nature and of transcendent truth inwardly possessed. More frequently, however, Wordsworth describes imagination as a "modifying" power by which the mind imposes itself in the act of perception, thereby immediately endowing the thing perceived with passion and with symbolic meaning.

THE FIGURE OF THE POET

"A man may write at any time," said Dr. Johnson, "if he will set himself doggedly to it." As for genius, it is nothing more than general and capacious power of mind. These opinions, like so many of Johnson's, express an instinctive recoil from what he regarded as cant, and they help measure the distance between the classicial position (for which Johnson is in many ways the greatest English spokesman) and the oncoming Romantics, standing in awe before what Coleridge called "the creative, productive life-power of inspired genius." The difference, however, lay not in reverence for poets (or rather, for the ideal figure of the poet, since living writers profited little from the emotion). This had always been felt. The point is that it was now felt with an intensity that itself distinguished the Romantic Period. And it was based on different concepts. Creation, for example, had earlier been the prerogative of God; artists aspired to reproduce or "imitate" His works, not His way of working, and thereby to express the general and permanent truths of nature and the moral law. Throughout the eighteenth century, however, there was a growing tendency to understand the artist's creative act by analogy with that of God. A poet, it seemed, might bring forth a new nature; and his works, or even a single work, might be regarded as his independent universe, complete in itself, harmonious, and obedient to its own laws. The analogy, which still influences criticism, naturally reorganized emotions along with conceptions.

In Plato the image of the inspired poet is that of a man suddenly possessed by a god who speaks through him. Coming after a hundred years of empirical, psychological speculation, the Romantics seldom claimed divine seizure (though there are versions of this view, especially in Shelley and Blake), but they nevertheless stressed that a poet is inspired. This was interpreted either as an invasion of the conscious mind from the unconscious or, more frequently, as a peculiarly rapid and total mental functioning, a sudden and more intense degree of what was described above in connection with the imagination. Inspiration implied that poetry was to some degree involuntary—according to Shelley, "A man cannot say, 'I will compose poetry' "—and also that poetry allows readers access to moments of peculiarly heightened intellection. To quote Shelley again, "Poetry is the record of the best and happiest moments of the happiest and best minds." A mind thus creative and inspired characterized "genius," which was often contrasted with mere talent. Where the man of talent may note only a few premises, proceed step-by-step, and stay within the boundaries of tradition and convention, the working of genius is original, comprehensive, immediate, and unerring. For Hazlitt, Coleridge was "the only person I ever knew who answered to the idea of a man of genius. . . . His thoughts did not seem to come with labour and effort; but as if borne on the gusts of genius, and as if the wings of his imagination lifted him from off his feet." In *The Prelude* Wordsworth looks back to the strength of imagination in childhood and youth with the same Romantic awe:

> Of genius, power,
> Creation and divinity itself
> I have been speaking, for my theme has been
> What passed within me
>
> [III. 173–76].

Inevitably the figure of the poet or artist was likely to be presented in strongest contrast with the rest of mankind, the "trembling throng," Shelley called them, "Whose sails were never to the tempest given." The poet is on a quest, and as such he is a dedicated and heroic being. For this reason, he may be lonely among other men who cannot share his insight or his quest. To the generality of mankind, Wordsworth acknowledged in *The Prelude*, the poet may appear mad. Even if he were, reverence would still be due him, for he would have been

> crazed
> By love and feeling, and internal thought
> Protracted among endless solitudes
>
> [V. 145–47].

But the truth is, Wordsworth adds, that even "in the blind and awful lair / Of such a madness" dwells a power of insight inconceivable to others. Moreover, the intense self-commitment of the poet to his quest justifies him in casting off the lesser responsibilities of littler men:

> Enow there are on earth to take in charge
> Their wives, their children, and their virgin loves,
> Or whatsoever else the heart holds dear;
> Enow to stir for these;
>
> [V. 153–56].

and Wordsworth, like most of the Romantic poets, could think of himself as a "Pilgrim of Eternity"—Shelley's phrase for Byron.

The quest pursues an infinite, perhaps unknown ideal—complete being, final truth—and therefore the poet's life is endless striving. In this, however, as in many other respects, he is not distinguished from other men so far as they are sensitive and aware. Perhaps at death, Shelley felt, "the pure spirit shall flow / Back to the burning fountain whence it came," but while we live in time we are trapped in a hopeless wish for the inaccessible—"the desire of the moth for the star." Or perhaps, as sometimes in Wordsworth, a man through his infinite striving participates in the open, endless process of becoming that characterizes reality itself:

> Our destiny, our being's heart and home,
> Is with infinitude, and only there;
> With hope it is, hope that can never die,
> Effort, and expectation, and desire,
> And something evermore about to be
>
> [XI. 604–08].

Or perhaps man's infinite desire is simply to undergo all experience. In this sense, greatness may be fatal, a fever, Byron said, that drives him ever onward without rest or satisfaction.

In the quest of the infinite the poet-hero might be represented not only as great but as guilty. Even in Keats's conception of the Shakespearean poet of "negative capability," the imagination, as it seeks to encompass the whole of reality, is not limited by moral allegiances and "has as much delight in conceiving an Iago as an Imogen." More generally, to seek ultimate truth implies that it is not already possessed, and thus the poet might figure as the rejecter or defier of tradition, convention, law, religion, and the collected social or even supernatural powers that seek to chain the human spirit. Thus the poet could be seen as creative in metaphysical insight and moral conduct, redeeming men from error and empty guilt and awakening in them a higher ideal. Or he could be seen as disclosing the existential loneliness of man in a vast cosmos without meaning or purpose. Or, finally, he could be seen as a guilty transgressor, stepping beyond the legitimate boundaries of knowledge and morality. Many of the major, recurrent figures of what one might call the Romantic legendry are heroic rebels—Prometheus, Napoleon, Satan, Don Juan, Cain—but the implications of this rebellion are often ambiguous, and they vary from one writer to another, reflecting tensions in the writers themselves and in the period.

BIBLIOGRAPHY

Critical and biographical works on individual writers are listed in the bibliographies appended to the introduction to each writer. The following list, concerned only with general discussions of English Romantic literature, is extremely condensed. For more detailed bibliographies, the student may consult: Ernest Bernbaum, *Guide Through the Romantic Movement*, 2nd ed., 1949; *The English Romantic Poets: A Review of Research*, ed. T. G. Raysor (Modern Language Association), rev. ed., 1956, which provides critical bibliographies of the Romantic movement as a whole and of five writers, Wordsworth, Coleridge, Byron, Shelley, and Keats; *The English Romantic Poets and Essayists: A Review of Research and Criticism*, eds. C. W. and L. H. Houtchens (Modern Language Association), rev. ed., 1966; and the annual bibliography of work concerned with the Romantic Period published through 1964 in the *Philological Quarterly* and thereafter in *English Language Notes*.

GENERAL HISTORIES AND DISCUSSIONS OF ENGLISH ROMANTICISM

Representative early histories of English Romanticism are C. H. Herford, *The Age of Wordsworth*, 1897; H. A. Beers, *A History of English Romanticism in the Eighteenth Century*, 1898, and *A History of English Romanticism in the Nineteenth Century*, 1901; and Oliver Elton, *A Survey of English Literature, 1780–1830*, 2 vols., 1912.

Three more recent histories include: Samuel Chew, *The Nineteenth Century and After, 1789–1939*, Pt. IV of *A Literary History of England*, ed. A. C. Baugh, 1948; W. L. Renwick's far from satisfactory history of the first half of the period, *English Literature, 1789–1815*, 1963,

Vol. IX of the *Oxford History of English Literature;* and the more informative study of the second half of the period, Ian Jack, *English Literature, 1815–1832,* 1963, Vol. X of the *Oxford History.*

Most of the perceptive discussions of English Romanticism that appeared during the last generation focused on particular subjects or themes as they relate to a large number of writers. These are listed separately below. But several are so broad in their approach that they may also be mentioned among general histories of the period. Notable examples are J. W. Beach, *The Concept of Nature in Nineteenth-Century English Poetry,* 1936; Douglas Bush, *Mythology and the Romantic Tradition in English Poetry,* 1937; and F. L. Lucas, *The Decline and Fall of the Romantic Ideal,* 1936. Of perennial fascination, if only because of the answers they provoked, are the marvellously readable onslaughts on Romanticism by Irving Babbitt, typified by his *Rousseau and Romanticism,* 1919, and his essays in *On Being Creative and Other Essays,* 1932. Concerned with the critical theory of the time but dealing with broad premises and concepts are W. J. Bate, *From Classic to Romantic,* 1946, which focuses on the general eighteenth-century transition to Romanticism, and M. H. Abrams, *The Mirror and the Lamp: Romantic Theory and the Critical Tradition,* 1953. Condensed answers to the attacks on Romanticism first by the "New Humanism," led by Paul Elmer More and Irving Babbitt, and later during the revival of "metaphysical poetry" may be represented respectively by J. W. Beach, *A Romantic View of Poetry,* 1944, and R. H. Fogle, "Romantic Bards and Metaphysical Reviewers," *ELH: A Journal of English Literary History,* XII (1945), 221–50.

Collections of critical essays and works relating a group of writers through a common theme include D. G. James, *Scepticism and Poetry,* 1937, and *The Romantic Comedy,* 1948; *The Major English Romantic Poets: A Symposium in Reappraisal,* ed. C. D. Thorpe and others, 1957; G. Wilson Knight, *The Starlit Dome,* 1941; Malcolm Elwin, *The First Romantics,* 1947; David Perkins, *The Quest for Permanence: The Symbolism of Wordsworth, Shelley, and Keats,* 1959; *English Romantic Poets: Modern Essays in Criticism,* ed. M. H. Abrams, 1960; Karl Kroeber, *Romantic Narrative Art,* 1960; Brian Wilkie, *Romantic Poets and Epic Tradition,* 1965; E. E. Bostetter, *The Romantic Ventriloquists,* 1963; and James Benziger, *Images of Eternity: Studies in the Poetry of Religious Vision,* 1962. Suggestive explications of individual poems are provided in Earl Wasserman, *The Subtler Language,* 1959, and Harold Bloom, *The Visionary Company,* 1961.

For the political and social background of the period, consult the well-known works of G. M. Trevelyan, *British History in the Nineteenth Century,* 1922, and *English Social History,* 1942. More detailed are two volumes of Élie Halévy's great history of nineteenth-century England, *England in 1815,* 1924, rev. ed., 1949, and *The Liberal Awakening, 1815–1830,* 1927, rev. ed., 1949; and *The New Cambridge Modern History,* Vol. VIII, *The American and French Revolutions, 1763–93,* ed. A. Goodwin, 1965, and Vol. IX, *War and Peace in an Age of Upheaval, 1793–1832,* ed. C. W. Crawley, 1965. Particularly readable discussions of the social background are found in Arthur Bryant, *The Age of Elegance, 1812–1822,* 1950, and R. J. White, *Waterloo to Peterloo,* 1957.

SPECIFIC SUBJECTS

Works dealing with the history of ideas in the eighteenth century that are especially valuable as intellectual background include A. O. Lovejoy's famous study of cosmology, *The Great Chain of Being*, 1936; Basil Willey, *The Eighteenth Century Background: Studies on the Idea of Nature in the Thought of the Period*, 1940; and t*v*o studies of eighteenth-century Romantic primitivism, H. N. Fairchild, *The Noble Savage: A Study in Romantic Naturalism*, 1928, and Lois Whitney, *Primitivism and the Idea of Progress*, 1934.

For literary forms other than poetry and criticism, see, for fiction, Vol. VI of E. A. Baker, *The History of the English Novel*, 10 vols., 1924–39, which is far from satisfactory but is at least comprehensive. Other works, on special topics, include Edith Birkhead, *The Tale of Terror: A Study of the Gothic Romance*, 1921; Montague Summers, *The Gothic Quest*, 1938; and J. M. S. Tompkins, *The Popular Novel in England, 1770–1800*, 1932. The most helpful account of the drama remains Allardyce Nicoll, *History of Early Nineteenth Century Drama, 1800–50*, 2 vols., 1930, rev. ed., 1955. For the essay, see M. H. Law, *The English Familiar Essay*, 1934. History is considered in T. P. Peardon, *The Transition in English Historical Writing, 1760–1830*, 1933. For an account of newspapers of the time, see H. R. Fox Bourne, *English Newspapers*, 2 vols., 1887, and A. Aspinall, *Politics and the Press, 1780–1850*, 1949; and, for magazines, Edmund Blunden, *Leigh Hunt's "Examiner" Examined*, 1928; John Clive, *Scotch Reviewers: The Edinburgh Review, 1802–15*, 1957; and H. and H. C. Shine, *The Quarterly Review Under Gifford, 1809–24*, 1949.

Much of the commentary on the relation of English Romantic writing to foreign literatures is concerned with German influence, for which see especially V. Stockley, *German Literature as Known in England, 1750–1830*, 1929; F. W. Stokoe, *German Influence in the English Romantic Period, 1788–1818*, 1926; René Wellek, *Immanuel Kant in England, 1793–1838*, 1931, the relevant chapters in his *A History of Modern Criticism, 1750–1950*, 1955, and *Confrontations*, 1965. Little is available on French influence. An exception is the influence of the French Revolution, treated in A. E. Hancock, *The French Revolution and the English Poets*, 1899, and C. Cestre, *La révolution française et les poètes anglais, 1789–1809*, 1906. For Italian influence, see C. P. Brand, *Italy and the English Romantics*, 1957, and R. Marshall, *Italy in English Literature, 1755–1815*, 1934.

The visual arts and their relation to romantic literature are dealt with in Edmund Blunden, *Romantic Poetry and the Fine Arts* (British Academy Lecture), 1942; Ronald Bradbury, *The Romantic Theories of Architecture*, 1934; S. A. Larrabee, *English Bards and Grecian Marbles: The Relationship Between Sculpture and Poetry*, 1943; and Sacheverell Sitwell, *British Architects and Craftsmen, 1600–1830*, 1946. Special emphasis on the eighteenth-century background is given in B. S. Allen, *Tides in English Taste, 1619–1800*, 2 vols., 1937; Sir Kenneth Clark, *The Gothic Revival*, 1928; Elizabeth Manwaring, *Italian Landscape in Eighteenth Century England*, 1925; and Christopher Hussey, *The Picturesque*, 1927.

Political attitudes are discussed in Crane Brinton, *The Political Ideas of the English Romanticists*, 1926, and Alfred Cobban, *Edmund Burke and the Revolt Against the Eighteenth Century*, 1929. A provocative interpretation is Jacques Barzun, *Romanticism and the Modern Ego*, 1943. A large literature is concerned with religious thought and institutions of the period. In addition to V. F. Storr's general work, *The Development of English Theology in the Nineteenth Century, 1800–1860*, 1913, and Vol. III of Horton Davies, *Worship and Theology in England*, 1962, mention should be made of Basil Willey, *Nineteenth Century Studies: Coleridge to Matthew Arnold*, 1949. More specific works include M. L. Edwards, *After Wesley: A Study of the Social and Political Influence of Methodism, 1791–1849*, 1935; F. C. Gill, *The Romantic Movement and Methodism*, 1937; and Dean W. R. Inge, *The Platonic Tradition in English Religious Thought*, 1926. For the religious views of the Romantic poets and their immediate predecessors see H. N. Fairchild, *Religious Trends in English Poetry*, Vols. II, 1942, and III, 1949. See also James Benziger, *Images of Eternity*, 1962, cited above.

Romantic critical theory during the late eighteenth and early nineteenth centuries is dealt with in the introductions by W. J. Bate in *Criticism: The Major Texts*, 1952, reprinted in *Prefaces to Criticism*, 1959; the somewhat outdated, but still helpful, A. E. Powell, *The Romantic Theory of Poetry*, 1926; René Wellek, *A History of Modern Criticism*, Vol. II, 1955, W. J. Bate, *From Classic to Romantic*, 1946, and M. H. Abrams, *The Mirror and the Lamp*, 1953, all cited above; and, on more specific topics, R. W. Babcock, *The Genesis of Shakespeare Idolatry*, 1931; S. H. Monk, *The Sublime: A Study of Critical Theories in Eighteenth-Century England*, 1935; Augustus Ralli, *A History of Shakespearian Criticism*, Vol. II, 1932; and Earl Wasserman, *Elizabethan Poetry in the Eighteenth Century*, 1947. Broad premises and distinctions, relevant to critical theory and to Romanticism as a whole, are discussed in two important articles: A. O. Lovejoy's famous "On the Discrimination of Romanticisms," *Publications of the Modern Language Association*, XXXIX (1924), reprinted in *English Romantic Poets*, ed. M. H. Abrams, 1960, cited above, and René Wellek, "The Concept of Romanticism," *Comparative Literature*, I (1949).

GEORGE CRABBE

1754–1832

THE WHOLE of English literature offers no better example than Crabbe of a poet so respected by other writers and by critics generally—beginning with his first major poems—and yet so completely neglected by the reading public throughout the revolutions in poetic taste that have taken place since he was midway in his career. The explanation lies in the extraordinarily novel combination of two qualities: a stark, uncompromising realism and, at the same time, a formal use of neoclassic versification and idiom (the closed couplet, stylized devices of rhythm and pause, the use of balance and antithesis). Neither of these qualities was popular in poetry during most of the nineteenth century, while, in the period since, if one has happened to be valued the other has not (not at least by the same kind of reader). But it was exactly this unusual combination that aroused the interest of Edmund Burke and Samuel Johnson when Crabbe began to write. A list of some of his other admirers, down to the end of the last century, quickly indicates that, if select, they have also been diverse: Wordsworth, Byron, the critic Francis Jeffrey, Scott (who, like the statesman Charles James Fox, found comfort in reading Crabbe in his last hours), and Jane Austen (who thought it would be pleasant to be Mrs. Crabbe). In a similar way Victorian admirers ranged from Cardinal Newman to the skeptical Sir Leslie Stephen, and, among poets, from Edward Fitzgerald, whose own *Rubáiyat of Omar Khayyám* is about as far from Crabbe as can be imagined, to Thomas Hardy, who eagerly acknowledged Crabbe's influence on his own novels.

Born in Suffolk, where his father was a collector of salt duties, he was apprenticed to a physician, began to practice medicine, and in his late twenties, after a few other poems, wrote *The Village* (published in 1783), with its somber presentation of the life of the rural poor. He then studied for the clergy and held various positions as a country curate. Twenty years passed before Crabbe brought out his next volume, *The Parish Register* (1807), in which his gift for narrative poetry first appeared. The volume is especially remembered for its powerful account of the decline of a patient in a madhouse ("Sir Eustace Grey"). Two other volumes of tales then followed: *The Borough* (1810), and *Tales in Verse* (1812) containing twenty-one stories. Here especially we find what Edwin Arlington Robinson has praised as Crabbe's "hard, human pulse . . . his plain excellence and stubborn skill." And there is considerable variety. Two examples, of very different kinds, are those printed below: "Abel Keene," a short sketch from *The Borough*, and the well-known comic tale "The Frank Courtship." After the death of his wife, Crabbe moved (1814) to Trowbridge, Wiltshire, where he served as vicar, and in 1819 published one more volume, *Tales of the Hall*. Additional stories in verse, written in his last years, were published posthumously with a collected edition (1834) of his works. He died in 1832.

Bibliography

In addition to the almost complete *Works*, 5 vols., 1823, *Poems*, ed. Adolphus W. Ward, 3 vols., 1905–07, and *Poetical Works*, ed. A. J. and R. M. Carlyle, 1914, there are selections by A. C. Deane, 1932, and F. L. Lucas, 1933. To these should be added the recent *New Poems by George Crabbe*, ed. Arthur Pollard, 1960.

    For biography consult *Life of George Crabbe* by his son, George Crabbe, with an introduction by E. M. Forster, 1932, and Alfred Ainger, *Crabbe*, 1903 (English Men of Letters Series). For general criticism see J. H. Evans, *The Poems of George Crabbe: A Literary and Historical Study*, 1933, and O. F. Sigworth, *Nature's Sternest Painter*, 1965.

FROM

THE BOROUGH

ABEL KEENE

Written as a series of twenty-four verse-letters, *The Borough* (1810) offers a comprehensive picture of rural and village life—the church, the law, trade, medicine, the schools, the prisons, the almshouse. The section entitled "The Poor of the Borough" includes four sketches, one of which follows.

Abel, a poor Man, Teacher of a School of the lower Order; is placed in the Office of a Merchant; is alarmed by Discourses of the Clerks; unable to reply; becomes a Convert; dresses, drinks, and ridicules his former Conduct—The Remonstrance of his Sister, a devout Maiden—Its Effect—The Merchant dies— Abel returns to Poverty unpitied; but relieved—His abject Condition—His Melancholy—He wanders about: is found—His own Account of himself, and the Revolutions in his Mind.

A quiet simple man was Abel Keene,
He meant no harm, nor did he often mean:
He kept a school of loud rebellious boys,
And growing old, grew nervous with the noise;
When a kind merchant hired his useful pen,
And made him happiest of accompting men,
With glee he rose to every easy day,
When half the labour brought him twice the pay.
  There were young clerks, and there the merchant's
    son,
Choice spirits all, who wish'd him to be one;   10

It must, no question, give them lively joy,
Hopes long indulged to combat and destroy;
At these they level'd all their skill and strength,—
He fell not quickly, but he fell at length:
They quoted books, to him both bold and new,
And scorn'd as fables all he held as true;
"Such monkish stories and such nursery lies,"
That he was struck with terror and surprise.

    "What! all his life had he the laws obey'd,
Which they broke through and were not once afraid?
Had he so long his evil passions check'd,   21
And yet at last had nothing to expect?
While they their lives in joy and pleasure led,
And then had nothing, at the end, to dread?
Was all his priest with so much zeal convey'd,
A part! a speech! for which the man was paid?
And were his pious books, his solemn prayers,
Not worth one tale of the admired Voltaire's?
Then was it time, while yet some years remain'd,
To drink untroubled and to think unchain'd,   30
And on all pleasures, which his purse could give,
Freely to seize, and while he lived, to live."

    Much time he passed in this important strife,
The bliss or bane of his remaining life;
For converts all are made with care and grief;
And pangs attend the birth of unbelief;
Nor pass they soon;—with awe and fear he took
The flow'ry way, and cast back many a look,

    The youths applauded much his wise design,
With weighty reasoning o'er their evening wine;   40
And much in private 'twould their mirth improve,
To hear how Abel spake of life and love;
To hear him own what grievous pains it cost,
Ere the old saint was in the sinner lost,
Ere his poor mind with every deed alarm'd,
By wit was settled, and by vice was charm'd.

    For Abel enter'd in his bold career,
Like boys on ice, with pleasure and with fear;
Lingering, yet longing for the joy, he went,
Repenting now, now dreading to repent: 50
With awkward pace, and with himself at war,
Far gone, yet frighten'd that he went so far;
Oft for his efforts he'd solicit praise,
And then proceed with blunders and delays:
The young more aptly passion's calls pursue,
But age and weakness start at scenes so new,
And tremble when they've done, for all they dared to
    do.
    At length example Abel's dread removed,
With small concern he sought the joys he loved;

Not resting here, he claim'd his share of fame,   60
And first their votary, then their wit became;
His jest was bitter and his satire bold,
When he his tales of formal brethren told;
What time with pious neighbours he discuss'd,
Their boasted treasure and their boundless trust:
"Such were our dreams," the jovial elder cried;
"Awake and live," his youthful friends replied.

    Now the gay clerk a modest drab despised,
And clad him smartly as his friends advised;
So fine a coat upon his back he threw,   70
That not an alley-boy old Abel knew;
Broad polish'd buttons blazed that coat upon,
And just beneath the watch's trinkets shone,—
A splendid watch, that pointed out the time,
To fly from business and make free with crime:
The crimson waistcoat and the silken hose
Rank'd the lean man among the Borough beaux:
His raven hair he cropp'd with fierce disdain,
And light elastic locks encased his brain:
More pliant pupil who could hope to find,   80
So deck'd in person and so changed in mind?

    When Abel walk'd the streets, with pleasant mien
He met his friends, delighted to be seen;
And when he rode along the public way,
No beau so gaudy and no youth so gay.

    His pious sister, now an ancient maid,
For Abel fearing, first in secret pray'd;
Then thus in love and scorn her notions she convey'd:
    "Alas! my brother! can I see thee pace
Hoodwink'd to hell, and not lament thy case,   90
Nor stretch my feeble hand to stop thy headlong race?
Lo! thou art bound; a slave in Satan's chain,
The righteous Abel turn'd the wretched Cain;
His brother's blood against the murderer cried,
Against thee thine, unhappy suicide!
Are all our pious nights and peaceful days,
Our evening readings and our morning praise,
Our spirits' comfort in the trials sent,
Our hearts' rejoicings in the blessings lent,
All that o'er grief a cheering influence shed,   100
Are these for ever and for ever fled?

    "When in the years gone by, the trying years,
When faith and hope had strife with wants and fears,
Thy nerves have trembled till thou couldst not eat
(Dress'd by this hand) thy mess of simple meat;
When, grieved by fastings, gall'd by fates severe,
Slow pass'd the days of the successless year;
Still in these gloomy hours, my brother then
Had glorious views, unseen by prosperous men;

And when thy heart has felt its wish denied,   110
What gracious texts has thou to grief applied;
Till thou hast enter'd in thine humble bed,
By lofty hopes and heavenly musings fed;
Then I have seen thy lively looks express
The spirit's comforts in the man's distress.
  "Then didst thou cry, exulting, 'Yes, 'tis fit,
'Tis meet and right, my heart! that we submit:'
And wilt thou, Abel, thy new pleasures weigh
Against such triumphs?—Oh! repent and pray.
  "What are thy pleasures?—with the gay to sit,   120
And thy poor brain torment for awkward wit;
All thy good thoughts (thou hat'st them) to restrain,
And give a wicked pleasure to the vain;
Thy long lean frame by fashion to attire,
That lads may laugh and wantons may admire;
To raise the mirth of boys, and not to see,
Unhappy maniac! that they laugh at thee.
  "These boyish follies, which alone the boy
Can idly act or gracefully enjoy,
Add new reproaches to thy fallen state,   130
And make men scorn what they would only hate.
  "What pains, my brother, dost thou take to prove
A taste for follies which thou canst not love?
Why do thy stiffening limbs the steed bestride—
That lads may laugh to see thou canst not ride?
And why (I feel the crimson tinge my cheek)
Dost thou by night in Diamond-Alley sneak?
  "Farewell! the parish will thy sister keep,
Where she in peace shall pray and sing and sleep,
Save when for thee she mourns, thou wicked, wander-
        ing sheep!   140
When youth is fall'n, there's hope the young may rise,
But fallen age for ever hopeless lies:
Torn up by storms and placed in earth once more,
The younger tree may sun and soil restore;
But when the old and sapless trunk lies low,
No care or soil can former life bestow;
Reserved for burning is the worthless tree;
And what, O Abel! is reserved for thee?"
  These angry words our hero deeply felt,
Though hard his heart, and indisposed to melt!   150
To gain relief he took a glass the more,
And then went on as careless as before;
Thenceforth, uncheck'd, amusements he partook,
And (save his ledger) saw no decent book;
Him found the merchant punctual at his task,
And that perform'd, he'd nothing more to ask;
He cared not how old Abel played the fool,
No master he, beyond the hours of school:

Thus they proceeding, had their wine and joke
Till merchant Dixon felt a warning stroke,   160
And, after struggling half a gloomy week,
Left his poor clerk another friend to seek.
  Alas! the son, who led the saint astray,
Forgot the man whose follies made him gay;
He cared no more for Abel in his need,
Than Abel cared about his hackney steed;
He now, alas! had all his earnings spent,
And thus was left to languish and repent;
No school nor clerkship found he in the place,
Now lost to fortune, as before to grace.   170
  For town-relief the grieving man applied,
And begg'd with tears what some with scorn denied;
Others look'd down upon the glowing vest,
And frowning, ask'd him at what price he dress'd?
Happy for him his country's laws are mild,
They must support him, though they still reviled;
Grieved, abject, scorn'd, insulted, and betray'd,
Of God unmindful, and of man afraid,—
No more he talk'd; 'twas pain, 'twas shame to speak,
His heart was sinking and his frame was weak.   180
His sister died with such serene delight,
He once again began to think her right;
Poor like himself, the happy spinster lay,
And sweet assurance bless'd her dying-day:
Poor like the spinster, he, when death was nigh,
Assured of nothing, felt afraid to die.
The cheerful clerks who sometimes pass'd the door,
Just mention'd "Abel!" and then thought no more.
So Abel, pondering on his state forlorn,
Look'd round for comfort, and was chased by scorn.
And now we saw him on the beach reclined,   191
Or causeless walking in the wint'ry wind;
And when it raised a loud and angry sea,
He stood and gazed, in wretched reverie:
He heeded not the frost, the rain, the snow;
Close by the sea he walk'd alone and slow:
Sometimes his frame through many an hour he spread
Upon a tombstone, moveless as the dead;
And was there found a sad and silent place,
There would he creep with slow and measured pace:
Then would he wander by the river's side,   201
And fix his eyes upon the falling tide;
The deep dry ditch, the rushes in the fen,
And mossy crag-pits were his lodgings then:
There, to his discontented thoughts a prey,
The melancholy mortal pined away.
  The neighb'ring poor at length began to speak
Of Abel's ramblings—he'd been gone a week;

They knew not where, and little care they took
For one so friendless and so poor to look;   210
At last a stranger, in a pedler's shed,
Beheld him hanging—he had long been dead.
He left a paper, penn'd at sundry times,
Intitled thus—"My Groanings and my Crimes!"
    "I was a Christian man, and none could lay
Aught to my charge; I walk'd the narrow way:
All then was simple faith, serene and pure,
My hope was steadfast and my prospects sure;
Then was I tried by want and sickness sore,
But these I clapp'd my shield of faith before,   220
And cares and wants and man's rebukes I bore:
Alas! new foes assail'd me; I was vain,
They stung my pride and they confused my brain:
Oh! these deluders! with what glee they saw
Their simple dupe transgress the righteous law;
'Twas joy to them to view that dreadful strife,
When faith and frailty warr'd for more than life;
So with their pleasures they beguiled the heart,
Then with their logic they allay'd the smart;
They proved (so thought I then) with reasons strong,
That no man's feelings ever led him wrong: 231
And thus I went, as on the varnish'd ice,
The smooth career of unbelief and vice.
Oft would the youths, with sprightly speech and bold,
Their witty tales of naughty priests unfold;
''Twas all a craft,' they said, 'a cunning trade,
Not she the priests, but priests religion made:'
So I believed:"—No, Abel! to thy grief,
So thou relinquish'dst all that was belief:—
    "I grew as very flint, and when the rest   240
Laugh'd at devotion, I enjoy'd the jest;
But this all vanish'd like the morning-dew,
When unemploy'd, and poor again I grew;
Yea! I was doubly poor, for I was wicked too.
    "The mouse that trespass'd and the treasure stole,
Found his lean body fitted to the hole;
Till having fatted, he was forced to stay,
And, fasting, starve his stolen bulk away:
Ah! worse for me—grown poor, I yet remain
In sinful bonds, and pray and fast in vain.   250
    "At length I thought, although these friends of sin
Have spread their net and caught their prey therein;
Though my hard heart could not for mercy call,
Because, though great my grief, my faith was small;
Yet, as the sick on skilful men rely,
The soul diseased may to a doctor fly.
    "A famous one there was, whose skill had wrought
Cures past belief, and him the sinners sought;

Numbers there were defiled by mire and filth,
Whom he recover'd by his goodly tilth:—   260
'Come then,' I said, 'let me the man behold,
And tell my case'—I saw him and I told.
    "With trembling voice, 'Oh! reverend sir,' I said,
'I once believed, and I was then misled;
And now such doubts my sinful soul beset,
I dare not say that I'm a Christian yet;
Canst thou, good sir, by thy superior skill,
Inform my judgment and direct my will?
Ah! give thy cordial; let my soul have rest,
And be the outward man alone distress'd;   270
For at my state I tremble.'—'Tremble more,'
Said the good man, 'and then rejoice therefore;
'Tis good to tremble; prospects then are fair,
When the lost soul is plunged in deep despair:
Once thou wert simply honest, just and pure,
Whole, as thou thought'st, and never wish'd a cure:
Now thou hast plunged in folly, shame, disgrace;
Now thou'rt an object meet for healing grace;
No merit thine, no virtue, hope, belief,
Nothing hast thou, but misery, sin, and grief,   280
The best, the only titles to relief.'
    "'What must I do,' I said, 'my soul to free?'
'—Do nothing, man; it will be done for thee.'
'But must I not, my reverend guide, believe?'
'—If thou art call'd, thou wilt the faith receive:'—
'But I repent not.'—Angry he replied,
'If thou art call'd, thou needest nought beside:
Attend on us, and if 'tis Heaven's decree,
The call will come,—if not, ah! wo for thee.'
    "There then I waited, ever on the watch,   290
A spark of hope, a ray of light to catch;
His words fell softly like the flakes of snow,
But I could never find my heart o'erflow:
He cried aloud, till in the flock began
The sigh, the tear, as caught from man to man;
They wept and they rejoiced, and there was I,
Hard as a flint, and as the desert dry:
To me no tokens of the call would come,
I felt my sentence and received my doom;
But I complain'd—'Let thy repinings cease,   300
Oh! man of sin, for they thy guilt increase;
It bloweth where it listeth;—die in peace.'
—'In peace, and perish?' I replied; 'impart
Some better comfort to a burthen'd heart.'—
'Alas!' the priest return'd, 'can I direct
The heavenly call?—Do I proclaim th' elect?
Raise not thy voice against th' Eternal will,
But take thy part with sinners and be still.'

"Alas! for me, no more the times of peace
Are mine on earth—in death my pains may cease. 310
  "Foes to my soul! ye young seducers, know,
What serious ills from your amusements flow;
Opinions, you with so much ease profess,
O'erwhelm the simple and their minds oppress:
Let such be happy, nor with reasons strong,
That make them wretched, prove their notions wrong;
Let them proceed in that they deem the way,
Fast when they will, and at their pleasure pray:
Yes, I have pity for my brethren's lot,
And so had Dives, but it help'd him not: 320
And is it thus?—I'm full of doubts:—Adieu!
Perhaps his reverence is mistaken too."

FROM

TALES IN VERSE

THE FRANK COURTSHIP

∽⚬∾

Francis Jeffrey, reviewing the poem, described it as
follows: "*The Frank Courtship* is rather in the merry
vein. . . . The whole of the story is, that the daughter of
a rigid Quaker, having been educated from home,
conceives a slight prejudice against the ungallant manners
of the sect, and is prepared to be very contemptuous and
uncomplying when her father proposes a sober youth
of the persuasion for a husband; but is so much struck
with the beauty of his person, and the cheerful reasonableness of his deportment, at their first interview, that
she instantly yields her consent. There is an excellent
description of the father, and the unbending elders of his
tribe; and some fine traits of natural coquetry."

∽⚬∾

Grave Jonas Kindred, Sybil Kindred's sire,
Was six feet high, and look'd six inches higher;
Erect, morose, determined, solemn, slow,
Who knew the man, could never cease to know;
His faithful spouse, when Jonas was not by,
Had a firm presence and a steady eye;
But with her husband dropp'd her look and tone,
And Jonas ruled unquestion'd and alone.
  He read, and oft would quote the sacred words,
How pious husbands of their wives were lords; 10
Sarah called Abraham lord! and who could be,
So Jonas thought, a greater man than he?

Himself view'd with undisguised respect,
And never pardon'd freedom or neglect.
  They had one daughter, and this favourite child
Had oft the father of his spleen beguiled;
Soothed by attention from her early years,
She gain'd all wishes by her smiles or tears:
But Sybil then was in that playful time,
When contradiction is not held a crime; 20
When parents yield their children idle praise
For faults corrected in their after days.
  Peace in the sober house of Jonas dwelt,
Where each his duty and his station felt:
Yet not that peace some favour'd mortals find,
In equal views and harmony of mind;
Not the soft peace that blesses those who love,
Where all with one consent in union move;
But it was that which one superior will
Commands, by making all inferiors still; 30
Who bids all murmurs, all objections cease,
And with imperious voice announces—Peace!
  They were, to wit, a remnant of that crew,
Who, as their foes maintain, their sovereign slew;
An independent race, precise, correct,
Who ever married in the kindred sect:
No son or daughter of their order wed
A friend to England's king who lost his head;
Cromwell was still their saint, and when they met,
They mourn'd that saints were not our rulers yet. 40
  Fix'd were their habits; they arose betimes,
Then pray'd their hour, and sang their party-rhymes:
Their meals were plenteous, regular, and plain;
The trade of Jonas brought him constant gain;
Vendor of hops and malt, of coals and corn—
And, like his father, he was merchant born:
Neat was their house; each table, chair, and stool,
Stood in its place, or moving moved by rule;
No lively print or picture graced the room;
A plain brown paper lent its decent gloom; 50
But here the eye, in glancing round, survey'd
A small recess that seem'd for china made;
Such pleasing pictures seem'd this pencill'd ware,
That few would search for nobler objects there—
Yet, turn'd by chosen friends, and there appear'd
His stern, strong features, whom they all revered;
For there in lofty air was seen to stand
The bold protector of the conquer'd land;
Drawn in that look with which he wept and swore,
Turn'd out the members, and made fast the door, 60
Ridding the house of every knave and drone,
Forced, though it grieved his soul, to rule alone.

The stern still smile each friend approving gave,
Then turn'd the view, and all again were grave.

There stood a clock, though small the owner's need,
For habit told when all things should proceed;
Few their amusements, but when friends appear'd,
They with the world's distress their spirits cheer'd;
The nation's guilt, that would not long endure
The reign of men so modest and so pure;   70
Their town was large, and seldom pass'd a day
But some had fail'd, and others gone astray;
Clerks had absconded, wives eloped, girls flown
To Gretna-Green, or sons rebellious grown;
Quarrels and fires arose;—and it was plain
The times were bad; the saints had ceased to reign!
A few yet lived to languish and to mourn
For good old manners never to return.

Jonas had sisters, and of these was one
Who lost a husband and an only son: 80
Twelve months her sables she in sorrow wore,
And mourn'd so long that she could mourn no more.
Distant from Jonas, and from all her race,
She now resided in a lively place;
There, by the sect unseen, at whist she play'd,
Nor was of churchmen or their church afraid:
If much of this the graver brother heard,
He something censured, but he little fear'd;
He knew her rich and frugal; for the rest,
He felt no care, or, if he felt, suppress'd: 90
Nor for companion when she ask'd her niece,
Had he suspicions that disturb'd his peace;
Frugal and rich, these virtues as a charm
Preserved the thoughtful man from all alarm;
An infant yet, she soon would home return,
Nor stay the manners of the world to learn;
Meantime his boys would all his care engross,
And be his comforts if he felt the loss.

The sprightly Sybil, pleased and unconfined,
Felt the pure pleasure of the op'ning mind: 100
All here was gay and cheerful—all at home
Unvaried quiet and unruffled gloom:
There were no changes, and amusements few;
Here, all was varied, wonderful, and new;
There were plain meals, plain dresses, and grave looks—
Here, gay companions and amusing books;
And the young beauty soon began to taste
The light vocations of the scene she graced.

A man of business feels it as a crime
On calls domestic to consume his time;   110
Yet this grave man had not so cold a heart,
But with his daughter he was grieved to part:

And he demanded that in every year
The aunt and niece should at his house appear.

"Yes! we must go, my child, and by our dress
A grave conformity of mind express;
Must sing at meeting, and from cards refrain,
The more t' enjoy when we return again."

Thus spake the aunt, and the discerning child
Was pleased to learn how fathers are beguiled.   120
Her artful part the young dissembler took,
And from the matron caught th' approving look:
When thrice the friends had met, excuse was sent
For more delay, and Jonas was content;
Till a tall maiden by her sire was seen,
In all the bloom and beauty of sixteen;
He gazed admiring;—she, with visage prim,
Glanced an arch look of gravity on him;
For she was gay at heart, but wore disguise,
And stood a vestal in her father's eyes: 130
Pure, pensive, simple, sad; the damsel's heart,
When Jonas praised, reproved her for the part;
For Sybil, fond of pleasure, gay and light,
Had still a secret bias to the right;
Vain as she was—and flattery made her vain—
Her simulation gave her bosom pain.

Again return'd, the matron and the niece
Found the late quiet gave their joy increase;
The aunt infirm, no more her visits paid,
But still with her sojourn'd the favourite maid.   140
Letters were sent when franks could be procured,
And when they could not, silence was endured;
All were in health, and if they older grew,
It seem'd a fact that none among them knew;
The aunt and niece still led a pleasant life,
And quiet days had Jonas and his wife.

Near him a widow dwelt of worthy fame,
Like his her manners, and her creed the same;
The wealth her husband left, her care retain'd
For one tall youth, and widow she remain'd;   150
His love respectful, all her care repaid,
Her wishes watch'd, and her commands obey'd.

Sober he was and grave from early youth,
Mindful of forms, but more intent on truth;
In a light drab he uniformly dress'd,
And look serene th' unruffled mind express'd;
A hat with ample verge his brows o'erspread,
And his brown locks curl'd graceful on his head;
Yet might observers in his speaking eye
Some observation, some acuteness spy;   160
The friendly thought it keen, the treacherous deem'd
   it sly;

Yet not a crime could foe or friend detect,
His actions all were, like his speech, correct;
And they who jested on a mind so sound,
Upon his virtues must their laughter found;
Chaste, sober, solemn, and devout they named
Him who was thus, and not of *this* ashamed.

 Such were the virtues Jonas found in one
In whom he warmly wish'd to find a son:
Three years had pass'd since he had Sybil seen;   170
But she was doubtless what she once had been,
Lovely and mild, obedient and discreet;
The pair must love whenever they should meet;
Then ere the widow or her son should choose
Some happier maid, he would explain his views;
Now she, like him, was politic and shrewd,
With strong desire of lawful gain embued;
To all he said, she bow'd with much respect,
Pleased to comply, yet seeming to reject;
Cool and yet eager, each admired the strength   180
Of the opponent, and agreed at length:
As a drawn battle shows to each a force,
Powerful as his, he honours it of course;
So in these neighbours, each the power discern'd,
And gave the praise that was to each return'd,
 Jonas now ask'd his daughter—and the aunt,
Though loth to lose her, was obliged to grant:—
But would not Sybil to the matron cling,
And fear to leave the shelter of her wing?
No! in the young there lives a love of change,   190
And to the easy they prefer the strange!
Then too the joys she once pursued with zeal,
From whist and visits sprung, she ceased to feel;
When with the matrons Sybil first sat down,
To cut for partners and to stake her crown,
This to the youthful maid preferment seem'd,
Who thought what woman she was then esteem'd;
But in few years, when she perceived, indeed,
The real woman to the girl succeed,
No longer tricks and honours fill'd her mind,   200
But other feelings, not so well defined;
She then reluctant grew, and thought it hard,
To sit and ponder o'er an ugly card;
Rather the nut-tree shade the nymph preferr'd,
Pleased with the pensive gloom and evening bird;
Thither, from company retired, she took
The silent walk, or read the fav'rite book.

 The father's letter, sudden, short, and kind,
Awaked her wonder, and disturb'd her mind;
She found new dreams upon her fancy seize,   210
 Wild roving thoughts and endless reveries:

The parting came;—and when the aunt perceived
The tears of Sybil, and how much she grieved—
To love for her that tender grief she laid,
That various, soft, contending passions made.

 When Sybil rested in her father's arms,
His pride exulted in a daughter's charms;
A maid accomplish'd he was pleased to find,
Nor seem'd the form more lovely than the mind:
But when the fit of pride and fondness fled,   220
He saw his judgment by his hopes misled;
High were the lady's spirits, far more free
Her mode of speaking than a maid's should be;
Too much, as Jonas thought, she seem'd to know,
And all her knowledge was disposed to show;
"Too gay her dress, like theirs who idly dote
On a young coxcomb, or a coxcomb's coat;
In foolish spirits when our friends appear,
And vainly grave when not a man is near."

 Thus Jonas, adding to his sorrow blame,   230
And terms disdainful to his sister's name:—
"The sinful wretch has by her arts defiled
The ductile spirit of my darling child."

 "The maid is virtuous," said the dame—Quoth he,
 "Let her give proof, by acting virtuously:
Is it in gaping when the elders pray?
In reading nonsense half a summer's day?
In those mock forms that she delights to trace,
Or her loud laughs in Hezekiah's face?
She—O Susannah!—to the world belongs;   240
She loves the follies of its idle throngs,
And reads soft tales of love, and sings love's soft'ning
   songs.
But, as our friend is yet delay'd in town,
We must prepare her till the youth comes down;
You shall advise the maiden; I will threat;
Her fears and hopes may yield us comfort yet."

 Now the grave father took the lass aside,
Demanding sternly, "Wilt thou be a bride?"
She answer'd, calling up an air sedate,
"I have not vow'd against the holy state."   250

 "No folly, Sybil," said the parent; "know
What to their parents virtuous maidens owe:
A worthy, wealthy youth, whom I approve,
Must thou prepare to honour and to love.
Formal to thee his air and dress may seem,
But the good youth is worthy of esteem;
Shouldst thou with rudeness treat him; of disdain
Should he with justice or of slight complain,
Or of one taunting speech give certain proof,
Girl! I reject thee from my sober roof."   260

"My aunt," said Sybil, "will with pride protect
One whom a father can for this reject;
Nor shall a formal, rigid, soul-less boy
My manners alter, or my views destroy!"

Jonas then lifted up his hands on high,
And utt'ring something 'twixt a groan and sigh,
Left the determined maid, her doubtful mother by.

"Hear me," she said; "incline thy heart, my child,
And fix thy fancy on a man so mild:
Thy father, Sybil, never could be moved   270
By one who loved him, or by one he loved.
Union like ours is but a bargain made
By slave and tyrant—he will be obey'd;
Then calls the quiet, comfort—but thy youth
Is mild by nature, and as frank as truth."

"But will he love?" said Sybil; "I am told
That these mild creatures are by nature cold."

"Alas!" the matron answer'd, "much I dread
That dangerous love by which the young are led!
That love is earthy; you the creature prize,   280
And trust your feelings and believe your eyes:
Can eyes and feelings inward worth descry?
No! my fair daughter, on our choice rely!
Your love, like that display'd upon the stage,
Indulged is folly, and opposed is rage;—
More prudent love our sober couples show,
All that to mortal beings, mortals owe;
All flesh is grass—before you give a heart,
Remember, Sybil, that in death you part;
And should your husband die before your love,   290
What needless anguish must a widow prove!
No! my fair child, let all such visions cease;
Yield but esteem, and only try for peace."

"I must be loved," said Sybil; "I must see
The man in terrors who aspires to me;
At my forbidding frown, his heart must ache,
His tongue must falter, and his frame must shake:
And if I grant him at my feet to kneel,
What trembling, fearful pleasure must he feel;
Nay, such the raptures that my smiles inspire,   300
That reason's self must for a time retire."

"Alas! for good Josiah," said the dame,
"These wicked thoughts would fill his soul with
       shame;
He kneel and tremble at a thing of dust!
He cannot, child:"—the child replied, "He must."

They ceased: the matron left her with a frown;
So Jonas met her when the youth came down:
"Behold," said he, "thy future spouse attends;
Receive him, daughter, as the best of friends;

Observe, respect him—humble be each word,   310
That welcomes home thy husband and thy lord."

Forewarn'd, thought Sybil, with a bitter smile,
I shall prepare my manner and my style.

Ere yet Josiah enter'd on his task,
The father met him—"Deign to wear a mask
A few dull days, Josiah—but a few—
It is our duty, and the sex's due;
I wore it once, and every grateful wife
Repays it with obedience through her life:
Have no regard to Sybil's dress, have none   320
To her pert language, to her flippant tone:
Henceforward thou shalt rule unquestioned and alone;
And she thy pleasure in thy looks shall seek—
How she shall dress, and whether she may speak."

A sober smile return'd the youth, and said,
"Can I cause fear, who am myself afraid?"

Sybil, meantime, sat thoughtful in her room,
And often wonder'd—"Will the creature come?
Nothing shall tempt, shall force me to bestow
My hand upon him—yet I wish to know."   330

The door unclosed, and she beheld her sire
Lead in the youth, then hasten to retire;
"Daughter, my friend—my daughter, friend"—he
       cried,
And gave a meaning look, and stepp'd aside;
That look contain'd a mingled threat and prayer,
"Do take him, child—offend him, if you dare."

The couple gazed—were silent, and the maid
Look'd in his face, to make the man afraid;
The man, unmoved, upon the maiden cast
A steady view—so salutation pass'd: 340
But in this instant Sybil's eye had seen
The tall fair person, and the still staid mien;
The glow that temp'rance o'er the cheek had spread,
Where the soft down half veil'd the purest red;
And the serene deportment that proclaim'd
A heart unspotted, and a life unblamed:
But then with these she saw attire too plain,
The pale brown coat, though worn without a stain;
The formal air, and something of the pride
That indicates the wealth it seems to hide;   350
And looks that were not, she conceived, exempt
From a proud pity, or a sly contempt.

Josiah's eyes had their employment too,
Engaged and soften'd by so bright a view;
A fair and meaning face, an eye of fire,
That check'd the bold, and made the free retire:
But then with these he mark'd the studied dress
And lofty air, that scorn or pride express;

With that insidious look, that seem'd to hide
In an affected smile the scorn and pride;   360
And if his mind the virgin's meaning caught,
He saw a foe with treacherous purpose fraught—
Captive the heart to take, and to reject it caught.

  Silent they sate—thought Sybil, that he seeks
Something, no doubt; I wonder if he speaks:
Scarcely she wonder'd, when these accents fell
Slow in her ear—"Fair maiden, art thou well?"
"Art thou physician?" she replied; "my hand,
My pulse, at least, shall be at thy command."

  She said—and saw, surprised, Josiah kneel,   370
And gave his lips the offer'd pulse to feel;
The rosy colour rising in her cheek,
Seem'd that surprise unmix'd with wrath to speak;
Then sternness she assumed, and—"Doctor, tell,
Thy words cannot alarm me—am I well?"

  "Thou art," said he; "and yet thy dress so light,
I do conceive, some danger must excite:"
"In whom?" said Sybil, with a look demure:
"In more," said he, "than I expect to cure.
I, in thy light luxuriant robe, behold   380
Want and excess, abounding and yet cold;
Here needed, there display'd, in many a wanton fold:
Both health and beauty, learned authors show,
From a just medium in our clothing flow."

  "Proceed, good doctor; if so great my need,
What is thy fee? Good doctor! pray proceed."

  "Large is my fee, fair lady, but I take
None till some progress in my cure I make:
Thou hast disease, fair maiden; thou art vain;
Within that face sit insult and disdain;   390
Thou art enamour'd of thyself; my art
Can see the naughty malice of thy heart:
With a strong pleasure would thy bosom move,
Were I to own thy power, and ask thy love;
And such thy beauty, damsel, that I might,
But for thy pride, feel danger in thy sight,
And lose my present peace in dreams of vain de-
    light."

  "And can thy patients," said the nymph, "endure
Physic like this? and will it work a cure?"

  "Such is my hope, fair damsel; thou, I find,   400
Hast the true tokens of a noble mind;
But the world wins thee, Sybil, and thy joys
Are placed in trifles, fashions, follies, toys;
Thou hast sought pleasure in the world around,
That in thine own bosom should be found:
Did all that world admire thee, praise and love,
Could it the least of nature's pains remove?

Could it for errors, follies, sins atone,
Or give thee comfort, thoughtful and alone?
It has, believe me, maid, no power to charm   410
Thy soul from sorrow, or thy flesh from harm:
Turn then, fair creature, from a world of sin,
And seek the jewel happiness within."

  "Speak'st thou at meeting?" said the nymph; "thy
    speech
Is that of mortal very prone to teach;
But wouldst thou, doctor, from the patient learn
Thine own disease?—The cure is thy concern."

  "Yea, with good will."—"Then know, 'tis thy com-
    plaint,
That, for a sinner, thou'rt too much a saint;
Hast too much show of the sedate and pure,   420
And without cause art formal and demure:
This makes a man unsocial, unpolite;
Odious when wrong, and insolent if right.
Thou may'st be good, but why should goodness be
Wrapt in a garb of such formality?
Thy person well might please a damsel's eye,
In decent habit with a scarlet dye;
But, jest apart—what virtue canst thou trace
In that broad brim that hides thy sober face?
Does that long-skirted drab, that over-nice   430
And formal clothing, prove a scorn of vice?
Then for thine accent—what in sound can be
So void of grace as dull monotony?
Love has a thousand varied notes to move
The human heart;—thou may'st not speak of love
Till thou hast cast thy formal ways aside,
And those becoming youth and nature tried;
Not till exterior freedom, spirit, ease,
Prove it thy study and delight to please;
Not till these follies meet thy just disdain,   440
While yet thy virtues and thy worth remain."

  "This is severe!—Oh! maiden, wilt not thou
Something for habits, manners, modes, allow?"—
"Yes! but allowing much, I much require
In my behalf, for manners, modes, attire!"

  "True, lovely Sybil; and, this point agreed,
Let me to those of greater weight proceed:
Thy father!"—"Nay," she quickly interposed,
"Good doctor, here our conference is closed!"

  Then left the youth, who, lost in his retreat,   450
Pass'd the good matron on her garden-seat;
His looks were troubled, and his air, once mild
And calm, was hurried:—"My audacious child!"
Exclaim'd the dame, "I read what she has done
In thy displeasure—Ah! the thoughtless one;

But yet, Josiah, to my stern good man
Speak of the maid as mildly as you can:
Can you not seem to woo a little while
The daughter's will, the father to beguile?
So that his wrath in time may wear away;   460
Will you preserve our peace, Josiah? say."

  "Yes! my good neighbour," said the gentle youth,
"Rely securely on my care and truth;
And should thy comfort with my efforts cease,
And only then—perpetual is thy peace."

  The dame had doubts: she well his virtues knew,
His deeds were friendly, and his words were true;
"But to address this vixen is a task
He is ashamed to take, and I to ask."
Soon as the father from Josiah learn'd   470
What pass'd with Sybil, he the truth discern'd.
"He loves," the man exclaim'd, "he loves, 'tis plain,
The thoughtless girl, and shall he love in vain?
She may be stubborn, but she shall be tried,
Born as she is of wilfulness and pride."

  With anger fraught, but willing to persuade,
The wrathful father met the smiling maid:
"Sybil," said he, "I long, and yet I dread
To know thy conduct—hath Josiah fled?
And, grieved and fretted by thy scornful air,   480
For his lost peace betaken him to prayer?
Couldst thou his pure and modest mind distress,
By vile remarks upon his speech, address,
Attire, and voice?"—"All this I must confess."—
"Unhappy child! what labour will it cost
To win him back!"—"I do not think him lost."
"Courts he then, trifler! insult and disdain?"—
"No: but from these he courts me to refrain."
"Then hear me, Sybil—should Josiah leave
Thy father's house?"—"My father's child would
    grieve:"   490
"That is of grace, and if he come again
To speak of love?"—"I might from grief refrain."—
"Then wilt thou, daughter, our design embrace?"—
"Can I resist it, if it be of grace?"
"Dear child! in three plain words thy mind express—
Wilt thou have this good youth?" "Dear father! yes."

WILLIAM BLAKE

1757–1827

BORN in London on November 28, 1757, Blake was the second son of a well-to-do hosier. From his childhood the mundane earth was to him visionary. As he walked through the streets of the city or the green fields around he saw angels and talked with spirits, and in later years he always spoke of his paintings as "copied" from the visionary world and of his poetry as having been "dictated" to him. In the earliest of these visitations, when he was four years old, God looked through the window at him and set him screaming. When he was eight or ten he saw a tree filled with angels, bright wings shining from every bough. He naturally reported this to his father, and the honest hosier threatened to spank him for lying. In later years the world would think him mad, so far as his work and existence were known at all. But from Blake's point of view the world was mad, with its anxiety and bloodshed, its cruelty, repressive morality, and selfish lovelessness. In later years one of his symbols of this society was the ancient civilization of the Druids; these priests had practiced human sacrifice, and he believed human sacrifice to be the basis of social organization—turning little children into chimney sweeps, for example. A root cause of our warped civilization is a failure of imagination, an inability even to conceive nature and society other than as they now appear. To Blake such passivity seemed a slavish bowing down so unnecessary and desperately mistaken that one could hardly understand it. His poetry and painting are a lifetime's effort to explain how this passivity comes about and how it can be changed. As he put it, "My Work . . . is an Endeavour to Restore what the Ancients call'd the Golden Age."

He showed himself a born artist from the time he could hold a pencil. At the age of ten he was sent to a drawing school and subsequently apprenticed to an engraver. He studied with unflagging assiduity in order to become a skillful craftsman, and afterward earned his living—so far as it can be called a living—mainly from booksellers' commissions to reproduce the designs of artists far inferior to himself. His taste even during his apprenticeship was startlingly independent. The idols of the age were landscape painters such as Salvator Rosa and Claude Lorrain and the Flemish and Venetian schools of colorists in which the leading names were Rubens and Titian. Blake had no interest in landscape and scorned the colorists partly because he felt they appealed to the senses rather than the intellect. All these so-called masters were hired, Blake bursts out, to depress art by portraying "Mortal & Perishing Nature" rather than eternal truth. His more technical expression of this was that they worked with shadings of color ("blurs and blots") rather than with forms definitely outlined, and "the want of determinate and bounding form evidences the want of idea in the artist's mind."

When his apprenticeship ended he enrolled as a student in the Royal Academy, but he also set up shop as an engraver. In these years he seemed to those who knew him to be a youth of hopeful promise clouded only by eccentric ideas that time would disperse. He had formed friendships with some of the rising artists—James Barry, Thomas Stothard, John Flaxman, Henry Fuseli. He had a character of remarkable sweetness, simplicity, and charm.

Blake was astonishingly gifted. Besides his work as an engraver and artist he had been writing poems—many of them afterward printed in *Poetical Sketches* (1783)—and for these he had composed melodies "sometimes most singularly beautiful." In appearance he was stocky, with a big head and "fiery eye." He was "not handsome," but had "a noble countenance, full of expression and animation." The description comes from his devoted wife Catherine, who almost fainted when she first met him and felt it necessary to quit his presence until she recovered. They married in 1782. The marriage was childless.

In 1787 occurred an event of utmost importance in Blake's life, the death of his youngest brother, Robert, to whom he was devoted. (See the letter to William Hayley, May 6, 1800, p. 164.) His reaction was characteristic. He had sat by his brother without sleep for a fortnight and saw Robert's spirit ascend in joy from the deathbed. Henceforth the gates of Eternity were always open. He talked with Robert "daily and hourly." The happiness of this heightened communion with the heavenly world is apparent in the *Songs of Innocence* (1789). He now associated himself with followers of the Swedish religious teacher Emanuel Swedenborg and began a study of his works. Though they were to have a permanent influence on him, he quickly fell into vehement dissent, particularly as he encountered the "Cursed Folly" of belief in predestination. Swedenborg, he decided, was just a vanity-smitten inaugurator of another sanctimonious sect, and all such groups were publicly abjured in *The Marriage of Heaven and Hell* (1790–93).

Meanwhile the French Revolution erupted. It had for a while no more avid partisan than Blake, who, like Wordsworth, Coleridge, Southey, and Hazlitt, saw in it the onset of the millennium. He met many of London's leading radicals—William Godwin, author of *Political Justice* (1793), Dr. Price, the irritant whose sermon had provoked Burke's *Reflections on the Revolution in France,* Tom Paine, author of the *Rights of Man* (1791–92), and others associated with Joseph Johnson, the bookseller who first published Wordsworth. Their protest against contemporary society is immortally caught up in the *Songs of Experience* (1789–94), and the same radical milieu reflects itself in *The French Revolution, America,* and the *Visions of the Daughters of Albion,* the latter of which seems to attack the institution of marriage. (Blake was a faithful husband, but his theoretical standpoint matched Milton's indignation at having to "grind in the Mill of an undelighted and servile copulation.") He also wrote the delicately beautiful *Book of Thel,* in which a soul sees the world of experience and refuses to be born into it. All of these works are included in a Prospectus of 1793, and in the next two years, with undiminished vigor, he began to elaborate his mythology in *The Book of Urizen, The Song of Los, The Book of Ahania,* and *The Book of Los.*

Hoping to find a patron for Blake, the sculptor Flaxman introduced him to a blandly sentimental, gentleman poet, William Hayley, who had already befriended Cowper. Hayley invited Blake to settle near him in the rural village of Felpham and prepare engravings for his forthcoming *Life of Cowper.* He also planned to find Blake commissions among well-to-do friends, thus enabling him to acquire a reputation and assure his future livelihood. The Blakes, for their part, jumped at the chance to leave London—Mrs. Blake was "like a flame of many colours of precious jewels" at the prospect—and, on arriving, Blake heard a ploughboy say, "Father, the Gate is Open," a manifest sign that great works were now to be performed. But Hayley, though kindness itself by any usual judgment, proved a trial to genius. His own poetry was shallow and facile. He utterly lacked Blake's sense of high purpose. He did not understand the man with whom he was dealing. His letters indicate that he thought his protégé a simple, good soul, and accordingly he found him commissions that included painting miniatures and decorating ladies' fans. Blake bore it as patiently as he could, but this was far from what he meant by "Mental Fight" or "Spiritual Life," ideals to which Hayley was opaque, most especially when they resulted in visionary poems and drawings. (See the letter to Thomas Butts, July 6, 1803, p. 166.) There was mounting tension, a quarrel, and an outward reconciliation. But at the end of three years Blake returned to London. He had heard a voice saying,

> If you, who are organized by Divine Providence for Spiritual communion, Refuse, & bury your Talent in the Earth, even tho' you should want Natural Bread, Sorrow & Desperation pursues you thro' life, & after death shame & confusion of face to eternity. . . . You will be call'd the base Judas who betray'd his Friend!
>
> [*Letters,* ed. Keynes, p. 71].

Before leaving Felpham, however, he was involved in a nasty crisis through the villainy of a soldier named Scholfield, whom Blake had forcibly ejected from his cottage garden. In revenge the soldier accused him of treasonable utterances, for which Blake was later tried and acquitted, but not without having suffered much anxiety. Hayley contributed all possible help on this occasion.

At Felpham (1800–03) Blake had got through a quantity of work incredible in anyone else. Besides the commissions executed for Hayley and others, *The Four Zoas* had been composed and probably most of *Milton*. Sometime between 1804 and 1820, in London, he wrote the third of his major "prophetic" works, *Jerusalem*. (The "prophetic" works are a group of poems intended by Blake to be analogous in form to the Bible; for Blake read the Bible as visionary narrative embodying spiritual truth.) He also tried to draw public attention to himself as an artist and engraver. He entered into a verbal agreement with a publisher named Cromek to produce from his own designs forty illustrations of Robert Blair's *The Grave* (1743). But Cromek, taking the designs for one guinea apiece, gave the job of engraving to another man. Blake was furious at the cheat but doubtless still hoped for something from his designs. When the book was published, however, these apparently made little impression, except that Leigh and Robert Hunt's newspaper, the *Examiner*, vilified them as far-fetched, absurd, and libidinous. In 1809 he again tried to recoup his fortunes with a private exhibition, but this too failed, partly because of another attack in the *Examiner*, which this time openly described him as a lunatic. Little is known about his life during the next nine years. He was a forgotten artist, deemed mad by the few visitors, such as Southey, who recorded their impressions, and he was reduced for a while to earning his bread by drawing dishes and cream bowls for the catalogues of the Wedgwood china manufacturers. Except for *The Everlasting Gospel* (c. 1818) his literary production was finished with *Jerusalem*.

Beginning about 1818 he emerged from obscurity. Now sixty-one, he made friends in the younger generation of painters—such men as John Linnell, Samuel Palmer, John Varley, and George Richmond. They provided him with commissions, though he always remained poor, and, more important, with the only intelligent appreciation he ever experienced. In fact they were virtually disciples. Palmer jotted down and pondered his sayings. For John Varley he would draw from apparitions portraits of historical persons—such as William Wallace and Edward III—whom Varley wished to paint, the two friends sometimes staying up much of the night as the spirits came to sit for Blake. On one such occasion he saw and painted the ghost of a flea. He studied Italian in order to read Dante, and his art rose to new heights in the great illustrations for Dante, unfinished at his death, and for the book of Job. From 1824 he was ill with a wasting disease and, though he retained his power as an artist to the end, grew gradually feebler in body, at times suffering severe pain. He died August 12, 1827, like a figure from Bunyan's *Pilgrim's Progress:* "He said He was going to that Country he had all His life wished to see. . . . Just before he died His eyes Brighten'd and He burst out in Singing of the things he Saw in Heaven."

2

The idiom and imagery of poets in our time tend to be compressed, elliptical, allusive, recondite, and often deliberately remote from the conventional idea of the poetic. The Romantic writers used a plainer language and drew on images that in the aggregate created a splendid and beautiful poetic world. Their verse is difficult nonetheless, not so much in its style or way of speaking as in the conceptions and psychological states it explores—that is, in its complex interpretation of man and the cosmos. Blake's early lyrics are no exception. They seize us at once and are steadily the more meaningful the more they are read. His long poems, however, are likely to intimidate readers from the start, as they did his first biographer, Alexander Gilchrist, who though militant in 1863 for his neglected hero, could only let his "eyes wander, hopeless and dispirited," up and down the obscure pages.

The special difficulty of Blake's poetry for many readers is that while it fully shares in the inner complexity of Romantic verse, it also adopts habits of style that are at least analogous to those of poets in our own time—though, of course, the total effect is very different. His associations are at times almost unbelievably condensed. Secondly, being a painter he tended in an unusual degree to express conceptions visually—

> And the hapless Soldier's sigh
> Runs in blood down Palace walls
>
> ["London," ll. 11–12].

(In this connection it should also be remembered that all of his poems after the *Poetical Sketches* were conceived and engraved with designs and illustrations that make up an integral part of the text. To read him without them is somewhat like reading a song without hearing the music.) Moreover, like many Englishmen of his time he possessed a minute recollection of biblical texts, and allusions come casually and thickly, almost as a type of shorthand.

Finally, he interpreted traditional symbols (including the Bible) in a personal way and also invented a mythology uniquely his own. "I must," he once said, "create a system or be enslaved by another man's." A mythology is a language. It is not something one believes or disbelieves, but a medium in which one thinks and speaks. And it is not allegory. Blake once described his work as "allegory addressed to the intellectual powers," but the phrase can be misleading. For allegory usually implies doctrines or propositions translated with a one-to-one parallelism into characters or events (or vice versa). Blake's vocabulary of myth is much too connotative for this, and the best of translations inevitably becomes travesty. His vocabulary must be learned in context, and since each of his major works seems to presuppose the others, ideally the entire corpus should be read. As space forbids printing it here, the headnotes and footnotes below are often interpretive, but it may be emphasized that they are at best only fossils and footprints of the living suggestiveness of the poems. They are not intended to fetter the imagination.

3

As with several of the great Romantic poets, Blake's theme is the possibility of salvation—no lesser word can be used—and he affirms that it lies in the human imagination. The theory of the imagination in the Romantic period arose as a protest against the scientific and positivistic spirit that prevailed in English intellectual life throughout the eighteenth and nineteenth centuries and perhaps still does. Except at the radical fringe, the Romantics were not objecting to science as we now conceive it, but rather to a tendency to reduce all "reality" to material or at least quantifiable elements, and thus to approach and answer virtually all questions, including those in morality and psychology, by procedures derived from mathematics or mechanics or analogous to them. For Blake the leading exponents of this tendency were Bacon, Newton, and Locke, and he also included the Deists, who rejected biblical revelation and derived religious beliefs from observation of "Nature." Had he been much aware of them he would certainly have added the Utilitarians because of their attempt to deduce morality from a calculation of the pleasures and pains attendant upon proposed actions.

The reasons for picking out Bacon and Newton are evident. For Blake and other writers the root error of Locke was his assumption that "reality" is comprised in a world of objects external to the mind and absolutely independent of it. We learn, Locke taught, about these objects from our senses, collecting impressions out of which we form "ideas" of the things to which our senses have been exposed. Knowledge must, of course, be relative to our sensory structure, but, this limitation being granted, it remains true that "ideas" can be formed with greater accuracy to the extent that we carefully observe the world around us, avoiding the interferences that are only too likely to arise from emotion and imagination. These interferences may range from such simple confusions as calling a mountain "threatening" or "sublime" (it is only a mountain), to wild mythic inventions, such as Pegasus, the legendary winged horse whose hoof-blow was a source of poetic inspiration. All such divagations would have seemed to Locke a sort of daydream.

To the Romantics these views were false and dangerous. They were indefensible on Locke's own grounds because, as was quickly pointed out, the assumption that external objects exist at all is wholly gratuitous if we are aware only of impressions and "ideas," which might just as well come from the mind itself as from a supposed external world. From this striking assertion emerged every variety of philosophic idealism down to present-day existentialism. But Locke and his disciples could not be left merely to opponents in the stratosphere of philosophic abstraction. By calling into court the spontaneously creative and emotional life of the psyche, the poetry, as it were, unceasingly composed in each individual mind, they assailed, many persons thought, the prime source of human happiness and virtue.

As the Romantics tried to clarify their opposition they offered a more profound analysis of what takes place in imaginative activity. The imagination is the process by which, as we perceive objects, our thoughts and feelings blend with them and they become expressive of our inner life. From one point of view this is indeed a creative transformation, precisely what Lockeans deplored, but it is attackable only if it involves a retreat from "reality." If, however, we assume that as compared with images of sense the inner world of thought and feeling is as much or more in touch with "reality," the imagination becomes not a retreat but a power of concretely articulating. But for the moment, the points to be stressed are that this process goes on spontaneously and constantly, and that through it experience acquires a heightened vividness and significance that amounts to an increased fullness of life. Thus when a sky is felt to be "sullen" or an autumn wind is a "dirge" we are raising events into symbols, and a world thus seen has meaning for us because it emerges from us. Imaginative acts of this sort are essentially the same as Blake's when he sees a gray thistle and calls it an old man (see the letter to Butts, November 22, 1802, p. 164). As for outright "vision," that is, creation not reared on an "objective" source (Pegasus), it is simply a further extension of the same process, a further symbolic reflection of the inner life.

Clearly, to the extent that a man is imaginative he is independent of the outer world, not only because he radically transforms what he encounters, but because the excitement of life for him does not depend on external crises and shocks. He may find any common experience momentous because symbolic. That is one reason why Blake says that imaginative arousal is itself a form of personal salvation, the Last Judgment that may happen at any time in any life. It is at least salvation from the world of the unimaginative, what Coleridge calls the

> inanimate cold world allowed
> To the poor loveless ever-anxious crowd
>
> ["Dejection: An Ode," ll. 51–52].

Blake also speaks of it as a dwelling in "Eternity," meaning, among other things, that types and situations perennially recur in our own life and in history. Thus the imagination, creating a coherent myth out of individual experience, may comprehend all time and history at once in its "Eternal Form." Works of art, needless to say, are simply heightened, more organized instances of the imaginative processes taking place in us all the time.

Beyond the meaning it has for individual happiness, the imagination has a potential power of transforming society. The lack of it brings about the state of mind French writers in the nineteenth century called *ennui*, a form of damnation, the condition and explanation of the Devil as he appears in their poetry. It is a state of blasé, heartless apathy in which nothing seems very real or significant, and it leads to cruelty because of indifference or even because of boredom. The imagination, on the other hand, confers the possibility of sympathy. It is the process by which we identify ourselves with other people, and hence it could potentially be the salvation of society, redeeming it from "human sacrifice" and making of all one Man, as Blake puts it, whose life is "Mercy, Pity, Peace and Love."

The theory of the imagination is not, however, a uniform doctrine in the Romantic period. It has rather the character of a broad movement of thought in which individual writers differ and overlap. (Compare the discussions below in headnotes to Wordsworth, Coleridge, Hazlitt, and Shelley.) And of all the great Romantics Blake is the most extreme and assured. For others the ghosts of Bacon, Newton, and Locke were not easily or finally exorcised. Granted that the imagination creatively insinuates itself into all experience, the haunting question, forced especially upon Wordsworth and Keats, was whether the active indulgence of this did not finally lead one to live in a dream. To Blake, however, all reality is a mental construction, the world of experience just as much as the world of vision. We see what we believe. This being the case, he was forced to explain how it is that the world of experience appears at all, and he accounts for it as a product of psychic division and conflict. Instead of trusting and releasing imagination, we chain it under the usurped domination of reason, and thus create the world of experience, itself characterized by divisions and contradictions. For example, the intention of pitying and merciful actions is to make others happy, and yet,

> Pity would be no more
> If we did not make somebody Poor;
> And Mercy no more could be
> If all were as happy as we
>
> ["The Human Abstract," ll. 1–4].

The dilemma is clearly absurd and just as clearly unresolvable by reason, and it is typical of the situation in which reason, as the Romantics conceived it, continually finds itself. In fact, to the mere reason, in Blake's view, the world is a labyrinth in which we are trapped unless we are freed by the imagination.

Bibliography

To his contemporaries Blake was almost unknown. Coleridge read and admired the *Songs of Innocence and Experience*. Lamb was mildly appreciative, but relished mainly what he thought flavorful eccentricities. Wordsworth felt he was certainly mad but, even so, more interesting than Byron or Scott, an opinion that tells more about his opinion of Byron or Scott than it does about Blake. Except in his engraved volumes (and an edition in 1839 of the *Songs of Innocence and Experience*) his works were not in print and there was no significant discussion until 1863, when Alexander Gilchrist brought out the first biography, *The Life of William Blake, Pictor Ignotus*, for which D. G. Rossetti edited a selection of poems. Thereafter Swinburne, in 1868, and William Rossetti, in 1874, produced appreciative criticisms, to which may be added two essays by W. B. Yeats in *Ideas of Good and Evil*, 1903. As attention was gradually directed to Blake he came to seem one of the strangest instances in literary history. On the one hand, in his prophetic books he seemed an incoherent, utterly subjective, possibly mad writer. On the other hand, his early lyrics and a few later ones were obviously among the finest in English. Moreover, in his prophetic books he plainly wished to communicate, in fact he intended to shake the world. This was not esoteric amusement for a coterie, but

> long resounding, strong heroic Verse
> Marshall'd in order for the day of Intellectual Battle
>
> [*The Four Zoas*, "Night the First," ll. 5–6].

And anyone could see that in these long poems (*The Four Zoas*, *Milton*, and *Jerusalem*), there were passages of marvelous strength and beauty. The situation, therefore, seemed quite inexplicable, and around the turn of the century critics began seriously to assume that Blake was sane and therefore intelligible even in his prophetic works. They began, that is, to read them with concentrated attention and were rewarded by finding them poems rather than sibylline leaves. Any reader who makes the same effort may have the same experience. As with any unusually difficult writer, the reward may or may not justify the labor, but only those who have been through it are entitled to an opinion.

Of these early studies Pierre Berger's *William Blake: mysticisme et poésie*, 1907 (translated by D. H. Conner as *William Blake: Poet and Mystic*, 1914) is still valuable, but virtually all recent commentary descends from S. Foster Damon's monumental *William Blake: His Philosophy and Symbols*, 1924, still generally regarded as the most helpful single work. It achieves a coherent interpretation that embraces both the poetry and illustration, and it offers a "key" to the symbolism. Thereafter the most important general works are Northrop Frye, *Fearful Symmetry*, 1947, and David V. Erdman, *Blake: Prophet Against Empire*, 1954. The first brilliantly relates Blake's mythology and beliefs to other literary symbolism and myth as well as to the intellectual tradition against which he rebelled, and the second brings out allusions in the poetry to historical events in Blake's own age, showing him as a writer *engagé*. In this effort Erdman draws upon the still valuable studies of Jacob Bronowski, *A Man Without a Mask*, 1943, and Mark Schorer, *William Blake: The Politics of Vision*, 1946. For briefer introductions one may turn to three works, all entitled *William Blake*, by Philippe Soupault, 1928 (translated by Lewis May), J. M. Murry, 1933, and H. M. Margoliouth, 1951, and also to Peter Fisher, *The Valley of Vision*, 1961, a clear and powerful exposition of Blake's thought. There are also two collections of essays by several writers, *The Divine Vision*, ed. Vivian de Sola Pinto, 1957, and *Discussions of William Blake*, ed. John E. Grant, 1961.

Most writing on Blake has been devoted to elucidating particular aspects or particular poems. Among these one may mention M. Bottrall, *The Divine Image: A Study of Blake's Interpretation of Christianity*, 1950; Bernard Blackstone, *English Blake*, 1949, on Blake's relation to eighteenth-century culture and philosophy; George Mills Harper, *The Neoplatonism of William Blake*, 1961, a study of the Romantic interest in Plato; and Milton Percival, *William Blake's Circle of Destiny*, 1938, on his use of occult traditions and lore. Désirée Hirst, *Hidden Riches: Traditional Symbolism from the Renaissance to Blake*, 1964, is also valuable in this connection. Close explication was inaugurated by Joseph H. Wicksteed, *Blake's Innocence and Experience*, 1928, which also reproduces and interprets the illustrations to these songs. Hazard Adams, *William Blake: A Reading of the Shorter Poems*, 1963, takes them up in the light of the prophetic works. Robert F. Gleckner, *The Piper and the Bard*, 1959, concentrates on the *Songs of Innocence and Experience* and *Tiriel*, *The Book of Thel*, *Visions of the Daughters of Albion*, and *The Marriage of Heaven and Hell*. The latter poem is also the subject of Martin K. Nurmi, "Blake's *Marriage of Heaven and Hell*," *Kent University Bulletin*, XLV, April, 1957. The entire corpus is viewed poem by poem in Harold Bloom, *Blake's Apocalypse: A Study in Poetic Argument*, 1963.

Gilchrist's *Life* has been reprinted with notes by Ruthven Todd in Everyman's Library. More recent biographies are by Mona Wilson, *The Life of William Blake*, 1927, revised 1948, and

Thomas Wright, *The Life of William Blake*, 2 vols., 1929. One standard edition is *The Complete Writings of William Blake*, ed. Geoffrey Keynes, 1966. A second standard edition is *The Poetry and Prose of William Blake*, ed. David Erdman and Harold Bloom, 1965. It should be remembered that Blake's punctuation was scanty and erratic; the punctuation in all editions has been contributed by the editor.

FROM

POETICAL SKETCHES

Written between 1769 and 1777, the *Poetical Sketches* were printed (1783) but never offered for sale, presumably because Blake decided against it. They are apprentice exercises, but some of them are also complex and supremely felicitous poems in their own right. Moreover, as the youthful poet assimilates literary conventions of the time he also modifies them. Lyrics on spring, summer, autumn, and winter, for example, are and were stock items, and in the eighteenth century such poems usually proceeded by description and simple personification. Blake, however, evokes or envisions the seasons by a myth-making habit of mind, the organ and evidence of a greater imaginative freedom. In doing so he was also taking the first steps in the creation of a larger myth. The figures of summer and winter were to undergo a tremendous imaginative development, reappearing as Orc and Urizen respectively in the later prophetic books.

TO SPRING

O thou, with dewy locks, who lookest down
Thro' the clear windows of the morning, turn
Thine angel eyes upon our western isle,
Which in full choir hails thy approach, O Spring!

The hills tell each other, and the list'ning
Vallies hear; all our longing eyes are turned
Up to thy bright pavillions: issue forth,
And let thy holy feet visit our clime.

Come o'er the eastern hills, and let our winds
Kiss thy perfumed garments; let us taste   10
Thy morn and evening breath; scatter thy pearls
Upon our love-sick land that mourns for thee.

O deck her forth with thy fair fingers; pour
Thy soft kisses on her bosom; and put
Thy golden crown upon her languish'd head,
Whose modest tresses were bound up for thee.

                                    1769–77 (1783)

TO SUMMER

O thou who passest thro' our vallies in
Thy strength, curb thy fierce steeds, allay the heat
That flames from their large nostrils! thou, O Summer,
Oft pitched'st here thy golden tent, and oft
Beneath our oaks has slept, while we beheld
With joy thy ruddy limbs and flourishing hair.

Beneath our thickest shades we oft have heard
Thy voice, when noon upon his fervid car
Rode o'er the deep of heaven; beside our springs
Sit down, and in our mossy vallies, on   10
Some bank beside a river clear, throw thy
Silk draperies off, and rush into the stream:
Our vallies love the Summer in his pride.

Our bards are fam'd who strike the silver wire:
Our youth are bolder than the southern swains:
Our maidens fairer in the sprightly dance:
We lack not songs, nor instruments of joy,
Nor echoes sweet, nor waters clear as heaven,
Nor laurel wreaths against the sultry heat.

                                    1769–77 (1783)

TO AUTUMN

O Autumn, laden with fruit, and stained
With the blood of the grape, pass not, but sit
Beneath my shady roof; there thou may'st rest,
And tune thy jolly voice to my fresh pipe,
And all the daughters of the year shall dance!
Sing now the lusty song of fruits and flowers.

"The narrow bud opens her beauties to
The sun, and love runs in her thrilling veins;
Blossoms hang round the brows of morning, and
Flourish down the bright cheek of modest eve,   10
Till clustr'ing Summer breaks forth into singing,
And feather'd clouds strew flowers round her head.

"The spirits of the air live on the smells
Of fruit; and joy, with pinions light, roves round
The gardens, or sits singing in the trees."
Thus sang the jolly Autumn as he sat;
Then rose, girded himself, and o'er the bleak
Hills fled from our sight; but left his golden load.

1769–77 (1783)

TO WINTER

"O Winter! bar thine adamantine doors:
The north is thine; there hast thou built thy dark
Deep-founded habitation. Shake not thy roofs,
Nor bend thy pillars with thine iron car."

He hears me not, but o'er the yawning deep
Rides heavy; his storms are unchain'd, sheathed
In ribbed steel; I dare not lift mine eyes,
For he hath rear'd his sceptre o'er the world.

Lo! now the direful monster, whose skin clings
To his strong bones, strides o'er the groaning rocks:
He withers all in silence, and his hand   11
Unclothes the earth, and freezes up frail life.

He takes his seat upon the cliffs; the mariner
Cries in vain. Poor little wretch! that deal'st
With storms, till heaven smiles, and the monster
Is driv'n yelling to his caves beneath mount Hecla.

1769–77 (1783)

SONG

How sweet I roam'd from field to field,
  And tasted all the summer's pride,
'Till I the prince of love beheld,
  Who in the sunny beams did glide!

He shew'd me lillies for my hair,
  And blushing roses for my brow;
He led me through his gardens fair,
  Where all his golden pleasures grow.

With sweet May dews my wings were wet,
  And Phoebus fir'd my vocal rage;   10
He caught me in his silken net,
  And shut me in his golden cage.

He loves to sit and hear me sing,
  Then, laughing, sports and plays with me;
Then stretches out my golden wing,
  And mocks my loss of liberty.

1769–77 (1783)

SONG

My silks and fine array,
  My smiles and languish'd air,
By love are driv'n away;
  And mournful lean Despair
Brings me yew to deck my grave:
Such end true lovers have.

His face is fair as heav'n
  When springing buds unfold;
O why to him was't giv'n
  Whose heart is wintry cold?   10
His breast is love's all worship'd tomb,
Where all love's pilgrims come.

Bring me an axe and spade,
  Bring me a winding sheet;
When I my grave have made
  Let winds and tempests beat:
Then down I'll lie, as cold as clay.
True love doth pass away!

1769–77 (1783)

SONG

Love and harmony combine,
And around our souls intwine,
While thy branches mix with mine,
And our roots together join.

Joys upon our branches sit,
Chirping loud and singing sweet;
Like gentle streams beneath our feet,
Innocence and virtue meet.

Thou the golden fruit dost bear,
I am clad in flowers fair;   10
Thy sweet boughs perfume the air,
And the turtle buildeth there.

POETICAL SKETCHES. TO WINTER. **16. Hecla:** a volcano in
Iceland.

There she sits and feeds her young,
Sweet I hear her mournful song;
And thy lovely leaves among,
There is love: I hear his tongue.

There his charming nest doth lay,
There he sleeps the night away;
There he sports along the day,
And doth among our branches play.   20

1769–77 (1783)

SONG

Memory, hither come,
    And tune your merry notes;
And, while upon the wind
    Your music floats,
I'll pore upon the stream,
Where sighing lovers dream,
And fish for fancies as they pass
Within the watery glass.

I'll drink of the clear stream,
    And hear the linnet's song;   10
And there I'll lie and dream
    The day along:
And when night comes, I'll go
To places fit for woe,
Walking along the darken'd valley
With silent Melancholy.

1769–77 (1783)

MAD SONG

The wild winds weep,
    And the night is a-cold;
Come hither, Sleep,
    And my griefs infold:
But lo! the morning peeps
    Over the eastern steeps,
And the rustling birds of dawn
The earth do scorn.

Lo! to the vault
    Of paved heaven,   10
With sorrow fraught
    My notes are driven:

They strike the ear of night,
    Make weep the eyes of day;
They make mad the roaring winds,
    And with tempests play.

Like a fiend in a cloud,
    With howling woe,
After night I do croud,
    And with night will go;   20
I turn my back to the east
From whence comforts have increas'd;
For light doth seize my brain
With frantic pain.

1769–77 (1783)

TO THE MUSES

Whether on Ida's shady brow,
    Or in the chambers of the East,
The chambers of the sun, that now
    From antient melody have ceas'd;

Whether in Heav'n ye wander fair,
    Or the green corners of the earth,
Or the blue regions of the air,
    Where the melodious winds have birth;

Whether on chrystal rocks ye rove,
    Beneath the bosom of the sea   10
Wand'ring in many a coral grove,
    Fair Nine, forsaking Poetry!

How have you left the antient love
    That bards of old enjoy'd in you!
The languid strings do scarcely move!
    The sound is forc'd, the notes are few!

1769–77 (1783)

❦

Like all of Blake's poems except the *Poetical Sketches*, the series of assertions *There Is No Natural Religion* and *All Religions Are One* were not regularly printed in Blake's lifetime but issued by him in engraved and illustrated copies that are now quite valuable. These date from approximately 1788. Natural religion, espoused particularly by the Deists, comprised whatever religious beliefs a reasonable man would derive independently of the Bible and the Church.

❦

MAD SONG. **4. infold:** printed "unfold," altered by pen in some of the original copies. **7. birds:** printed "beds," altered by pen in some copies. "Beds of dawn" might refer to clouds, as George Saintsbury first suggested.

TO THE MUSES. **1. Ida:** mountain in Crete.

THERE IS NO NATURAL RELIGION

[First Series]

THE *Argument.* Man has no notion of moral fitness but from Education. Naturally he is only a natural organ subject to Sense.

I. Man cannot naturally Perceive but through his natural or bodily organs.

II. Man by his reasoning power can only compare & judge of what he has already perceiv'd.

III. From a perception of only 3 senses or 3 elements none could deduce a fourth or fifth.

IV. None could have other than natural or organic thoughts if he had none but organic perceptions.

V. Man's desires are limited by his perceptions, none can desire what he has not perceiv'd.

VI. The desires & perceptions of man, untaught by any thing but organs of sense, must be limited to objects of sense.

Conclusion. If it were not for the Poetic or Prophetic character the Philosophic & Experimental would soon be at the ratio[1] of all things, & stand still, unable to do other than repeat the same dull round over again.

(*1788*)

THERE IS NO NATURAL RELIGION

[Second Series]

I. Man's perceptions are not bounded by organs of perception; he perceives more than sense (tho' ever so acute) can discover.

II. Reason, or the ratio of all we have already known, is not the same that it shall be when we know more.

III. [*This proposition has been lost.*]

IV. The bounded is loathed by its possessor. The same dull round, even of a universe, would soon become a mill with complicated wheels.

V. If the many become the same as the few when possess'd, More! More! is the cry of a mistaken soul; less than All cannot satisfy Man.

THERE IS NO NATURAL RELIGION. FIRST SERIES. I. **ratio:** the sum of knowledge derived from experience. See *The Marriage of Heaven and Hell,* below, n. 11.

VI. If any could desire what he is incapable of possessing, despair must be his eternal lot.

VII. The desire of Man being Infinite, the possession is Infinite & himself Infinite.

Application. He who sees the Infinite in all things, sees God. He who sees the Ratio only, sees himself only.

Therefore God becomes as we are, that we may be as he is.

(c. *1788*)

ALL RELIGIONS ARE ONE

The Voice of one crying in the Wilderness

THE *Argument.* As the true method of knowledge is experiment, the true faculty of knowing must be the faculty which experiences. This faculty I treat of.

PRINCIPLE 1st. That the Poetic Genius is the true Man, and that the body or outward form of Man is derived from the Poetic Genius. Likewise that the forms of all things are derived from their Genius, which by the Ancients was call'd an Angel & Spirit & Demon.

PRINCIPLE 2d. As all men are alike in outward form, So (and with the same infinite variety) all are alike in the Poetic Genius.

PRINCIPLE 3d. No man can think, write, or speak from his heart, but he must intend truth. Thus all sects of Philosophy are from the Poetic Genius adapted to the weaknesses of every individual.

PRINCIPLE 4th. As none by traveling over known lands can find out the unknown, So from already acquired knowledge Man could not acquire more: therefore an universal Poetic Genius exists.

PRINCIPLE 5th. The Religions of all Nations are derived from each Nation's different reception of the Poetic Genius, which is every where call'd the Spirit of Prophecy.

PRINCIPLE 6th. The Jewish & Christian Testaments are An original derivation from the Poetic Genius; this is necessary from the confined nature of bodily sensation.

PRINCIPLE 7th. As all men are alike (tho' infinitely various), So all Religions &, as all similars, have one source.

The true Man is the source, he being the Poetic Genius.

(c. *1788*)

SONGS OF INNOCENCE

❧

Though they were conceived independently, the *Songs of Innocence* (1789) should be read in conjunction with the *Songs of Experience* written a few years later. The two groups of lyrics depict, as Blake said in 1794, "the Two Contrary States of the Human Soul," and in many instances they develop opposite points of view toward the same subject matter. The speakers in the *Songs of Innocence* are children or childlike, and the songs express their emotions and their vision. To them the cosmos is one of "Mercy, Pity, Peace, and Love" received and bestowed by all creatures and by their creator, who himself became a child and is called a Lamb. But if the speakers are innocent in their apprehension, the poems often are not. Interpretation of "Holy Thursday," for example, is complicated by such considerations as that children marshaled by "Grey-headed beadles" would usually suggest to Blake something negative and hateful, and the phrase "wise guardians of the poor" would usually be bitter irony. In "The Chimney Sweeper" the child has been "sold" by his father to be a sweep and he is treated harshly. But with a child's terrible readiness to accept the only world he knows, he does not realize he is wronged, and he is happy. To adult readers, however, this abuse of a child's trust is a searing indictment. From this point of view the religious teachings, to which the child gives full faith, sound vulgar and suspiciously useful to his exploiters:

> if he'd be a good boy,
> He'd have God for his father, & never want joy
>
> ["The Chimney Sweeper," ll. 19–20].

The point is not that adults with their longer experience of the world have a truer vision. As Blake once remarked, in an annotation to *Aphorisms on Man* by the Swiss mystic and poet Johan Kaspar Lavater, "the innocence of a child" can reproach us "with the errors of acquired folly," and one of these errors is the belief in the "reality" of the world presented to our senses as opposed to the world of imagination and vision. Thus "The Chimney Sweeper" may also be read as a celebration of the child's imaginative freedom to create the world in which he truly lives. But it may be noted that, as the pastoral imagery of shepherd and flock especially indicates, the imaginative world of the child is one in which he imagines himself to be dependent. While he is actually exercising his creative freedom he does not know it, and he may thus be contrasted with the artist who works in conscious liberty. For this among other reasons, the state of innocence does not represent Blake's final conception of man's highest joy. To attain this man must go through experience.

Blake issued many copies of the *Songs of Innocence* and they vary somewhat in the order of the poems and in what poems they include. Some of the songs were later usually transferred to the *Songs of Experience*. The order adopted here observes Blake's own instructions, as he wrote them in a memorandum, but in subsequent copies Blake himself did not always follow this order.

Four of Blake's engravings (two from the *Songs of Innocence* and two from the *Songs of Experience*) are reproduced here in black and white. These reproductions are the same size as the engravings.

❧

INTRODUCTION

Piping down the valleys wild,
Piping songs of pleasant glee,
On a cloud I saw a child,
And he laughing said to me:

"Pipe a song about a Lamb!"
So I piped with merry chear.
"Piper, pipe that song again";
So I piped: he wept to hear.

"Drop thy pipe, thy happy pipe;
Sing thy songs of happy chear":   10
So I sung the same again,
While he wept with joy to hear.

"Piper, sit thee down, and write
In a book, that all may read."
So he vanish'd from my sight,
And I pluck'd a hollow reed,

And I made a rural pen,
And I stain'd the water clear,
And I wrote my happy songs
Every child may joy to hear.   20

(*1789*)

THE ECCHOING GREEN

The Sun does arise,
And make happy the skies;
The merry bells ring
To welcome the Spring;
The skylark and thrush,
The birds of the bush,
Sing louder around
To the bells' chearful sound,
While our sports shall be seen
On the Ecchoing Green.   10

Old John, with white hair,
Does laugh away care,
Sitting under the oak,
Among the old folk.
They laugh at our play,
And soon they all say:
"Such, such were the joys
When we all, girls & boys,
In our youth time were seen
On the Ecchoing Green."   20

Till the little ones, weary,
No more can be merry;
The sun does descend,
And our sports have an end.
Round the laps of their mothers
Many sisters and brothers,
Like birds in their nest,

Are ready for rest,
And sport no more seen
On the darkening Green.   30

(*1789*)

THE LAMB

Little Lamb, who made thee?
  Dost thou know who made thee?
Gave thee life, & bid thee feed
By the stream & o'er the mead;
Gave thee clothing of delight,
Softest clothing, wooly, bright;
Gave thee such a tender voice,
Making all the vales rejoice?
  Little Lamb, who made thee?
  Dost thou know who made thee?   10

Little Lamb, I'll tell thee,
Little Lamb, I'll tell thee:
He is called by thy name,
For He calls himself a Lamb.
He is meek, & he is mild;
He became a little child.
I a child, & thou a lamb,
We are called by his name.
   Little Lamb, God bless thee!
   Little Lamb, God bless thee!   20

(1789)

THE SHEPHERD

How sweet is the Shepherd's sweet lot!
From the morn to the evening he strays;
He shall follow his sheep all the day,
And his tongue shall be filled with praise.

For he hears the lamb's innocent call,
And he hears the ewe's tender reply;
He is watchful while they are in peace,
For they know when their Shepherd is nigh.

(1789)

INFANT JOY

"I have no name:
I am but two days old."
What shall I call thee?
"I happy am,
Joy is my name."
Sweet joy befall thee!

Pretty joy!
Sweet joy but two days old.
Sweet joy I call thee:
Thou dost smile,   10
I sing the while,
Sweet joy befall thee!

(1789)

THE LITTLE BLACK BOY

My mother bore me in the southern wild,
And I am black, but O! my soul is white;

White as an angel is the English child,
But I am black, as if bereav'd of light.

My mother taught me underneath a tree,
And, sitting down before the heat of day,
She took me on her lap and kissed me,
And, pointing to the east, began to say:

"Look on the rising sun: there God does live,
And gives his light, and gives his heat away;   10
And flowers and trees and beasts and men receive
Comfort in morning, joy in the noonday.

"And we are put on earth a little space,
That we may learn to bear the beams of love;
And these black bodies and this sunburnt face
Is but a cloud, and like a shady grove.

"For when our souls have learn'd the heat to bear,
The cloud will vanish; we shall hear his voice,
Saying: 'Come out from the grove, my love & care,
And round my golden tent like lambs rejoice.'"   20

Thus did my mother say, and kissed me;
And thus I say to little English boy.
When I from black and he from white cloud free,
And round the tent of God like lambs we joy,

I'll shade him from the heat till he can bear
To lean in joy upon our father's knee;
And then I'll stand and stroke his silver hair,
And be like him, and he will then love me.

(1789)

LAUGHING SONG

When the green woods laugh with the voice of joy,
And the dimpling stream runs laughing by;
When the air does laugh with our merry wit,
And the green hill laughs with the noise of it;

When the meadows laugh with lively green,
And the grasshopper laughs in the merry scene,
When Mary and Susan and Emily
With their sweet round mouths sing "Ha, Ha, He!"

When the painted birds laugh in the shade,
Where our table with cherries and nuts is spread,
Come live & be merry, and join with me,   11
To sing the sweet chorus of "Ha, Ha, He!"

c. 1787 (1789)

SPRING

Sound the Flute!
Now it's mute.
Birds delight
Day and Night;
Nightingale
In the dale,
Lark in Sky,
Merrily,
Merrily, Merrily, to welcome in the Year.

Little Boy,   10
Full of joy;
Little Girl,
Sweet and small;
Cock does crow,
So do you;
Merry voice,
Infant noise,
Merrily, Merrily, to welcome in the Year.

Little Lamb,
Here I am;   20
Come and lick
My white neck;
Let me pull
Your soft Wool;
Let me kiss
Your soft face:
Merrily, Merrily, we welcome in the Year.

(*1789*)

A CRADLE SONG

Sweet dreams, form a shade
O'er my lovely infant's head;
Sweet dreams of pleasant streams
By happy, silent, moony beams.

Sweet sleep, with soft down
Weave thy brows an infant crown.
Sweet sleep, Angel mild,
Hover o'er my happy child.

Sweet smiles, in the night
Hover over my delight;   10
Sweet smiles, Mother's smiles,
All the livelong night beguiles.

Sweet moans, dovelike sighs,
Chase not slumber from thy eyes.
Sweet moans, sweeter smiles,
All the dovelike moans beguiles.

Sleep, sleep, happy child,
All creation slept and smil'd;
Sleep, sleep, happy sleep,
While o'er thee thy mother weep.   20

Sweet babe, in thy face
Holy image I can trace.
Sweet babe, once like thee,
Thy maker lay and wept for me,

Wept for me, for thee, for all,
When he was an infant small.
Thou his image ever see,
Heavenly face that smiles on thee,

Smiles on thee, on me, on all;
Who became an infant small.   30
Infant smiles are his own smiles;
Heaven & earth to peace beguiles.

(*1789*)

NURSE'S SONG

The song was first written out in Blake's burlesque
satire, *An Island in the Moon*, where it is sung by Mrs.
Nannicantipot. In that context it would be read as a
satire on namby-pamby sentimentalism. In *An Island in
the Moon* it comes immediately after "Holy Thursday"
(below), and it is followed by "A Little Boy Lost" and
then by the stark contrast of a dialogue of children at play
that is frankly realistic and even scatological.

When the voices of children are heard on the green,
And laughing is heard on the hill,
My heart is at rest within my breast,
   And everything else is still.

"Then come home, my children, the sun is gone down,
And the dews of night arise;
Come, come, leave off play, and let us away
Till the morning appears in the skies."

"No, no, let us play, for it is yet day,
And we cannot go to sleep;   10
Besides, in the sky the little birds fly
And the hills are all cover'd with sheep."

"Well, well, go & play till the light fades away,
And then go home to bed."
The little ones leaped & shouted & laugh'd
    And all the hills ecchoed.

                                    1784–85 (1789)

HOLY THURSDAY

Holy Thursday is Ascension Day, which celebrates the
ascension of Jesus forty days after Easter. Each year
thousands of children from the charity schools of
London were marched to a service in St. Paul's Cathedral.
An odd light is thrown on this poem by the fact that its
first draft appears in *An Island in the Moon* where it is
sung by Mr. Obtuse Angle, who "always understood
better when he shut his eyes."

'Twas on a Holy Thursday, their innocent faces clean,
The children walking two & two, in red & blue & green,
Grey-headed beadles walk'd before, with wands as
    white as snow,
Till into the high dome of Paul's they like Thames'
    waters flow.

O what a multitude they seem'd, these flowers of Lon-
    don town!
Seated in companies they sit with radiance all their own.
The hum of multitudes was there, but multitudes of
    lambs,
Thousands of little boys & girls raising their innocent
    hands.

Now like a mighty wind they raise to heaven the voice
    of song,
Or like harmonious thunderings the seats of Heaven
    among.   10
Beneath them sit the aged men, wise guardians of the
    poor;
Then cherish pity, lest you drive an angel from your
    door.

                                    1784–85 (1789)

SONGS OF INNOCENCE. HOLY THURSDAY. **2. red . . . green:**
Each school had its own uniform. **3. beadles:** functionaries
who serve as ushers and keep order during the church
service. **4. Thames' waters:** the "charter'd Thames" of
"London" in *Songs of Experience*?

THE BLOSSOM

Merry, Merry Sparrow!
Under leaves so green
A happy Blossom
Sees you swift as arrow
Seek your cradle narrow
Near my Bosom.

Pretty, Pretty Robin!
Under leaves so green
A happy Blossom
Hears you sobbing, sobbing,   10
Pretty, Pretty Robin,
Near my Bosom.

                                    (1789)

THE CHIMNEY SWEEPER

When my mother died I was very young,
And my father sold me while yet my tongue
Could scarcely cry "'weep! 'weep! 'weep! 'weep!"
So your chimneys I sweep, & in soot I sleep.

There's little Tom Dacre, who cried when his head,
That curl'd like a lamb's back, was shav'd: so I said
"Hush, Tom! never mind it, for when your head's bare
You know that the soot cannot spoil your white hair."

And so he was quiet, & that very night,
As Tom was a-sleeping, he had such a sight!—   10
That thousands of sweepers, Dick, Joe, Ned, & Jack,
Were all of them lock'd up in coffins of black.

And by came an Angel who had a bright key,
And he open'd the coffins & set them all free;
Then down a green plain leaping, laughing, they run,
And wash in a river, and shine in the Sun.

Then naked & white, all their bags left behind,
They rise upon clouds and sport in the wind;
And the Angel told Tom, if he'd be a good boy,
He'd have God for his father, & never want joy.   20

And so Tom awoke; and we rose in the dark,
And got with our bags & our brushes to work.
Tho' the morning was cold, Tom was happy & warm;
So if all do their duty they need not fear harm.

                                    (1789)

THE CHIMNEY SWEEPER. **3. 'weep:** the child's lisping way of
uttering his cry through the streets, "sweep, sweep."

THE DIVINE IMAGE

To Mercy, Pity, Peace, and Love
  All pray in their distress;
And to these virtues of delight
  Return their thankfulness.

For Mercy, Pity, Peace, and Love
  Is God, our father dear;
And Mercy, Pity, Peace, and Love
  Is Man, his child and care.

For Mercy has a human heart,
  Pity a human face,   10
And love, the human form divine;
  And Peace, the human dress.

Then every man, of every clime,
  That prays in his distress,
Prays to the human form divine,
  Love, Mercy, Pity, Peace.

And all must love the human form,
  In heathen, turk, or jew;
Where Mercy, Love, & Pity dwell
  There God is dwelling too.   20

          (*1789*)

NIGHT

The sun descending in the west,
The evening star does shine;
The birds are silent in their nest,
And I must seek for mine.
The moon like a flower
In heaven's high bower,
With silent delight
Sits and smiles on the night.

Farewell, green fields and happy groves,
Where flocks have took delight.   10
Where lambs have nibbled, silent moves
The feet of angels bright;
Unseen they pour blessing,
And joy without ceasing,
On each bud and blossom,
And each sleeping bosom.

They look in every thoughtless nest,
Where birds are cover'd warm;

They visit caves of every beast,
To keep them all from harm.   20
If they see any weeping
That should have been sleeping,
They pour sleep on their head,
And sit down by their bed.

When wolves and tygers howl for prey,
They pitying stand and weep;
Seeking to drive their thirst away,
And keep them from the sheep;
But if they rush dreadful,
The angels, most heedful,   30
Receive each mild spirit,
New worlds to inherit.

And there the lion's ruddy eyes
Shall flow with tears of gold,
And pitying the tender cries,
And walking round the fold,
Saying "Wrath, by his meekness,
And by his health, sickness
Is driven away
From our immortal day.   40

"And now beside thee, bleating lamb,
I can lie down and sleep;
Or think on him who bore thy name,
Graze after thee and weep.
For, wash'd in life's river,
My bright mane for ever
Shall shine like the gold
As I guard o'er the fold."

          (*1789*)

A DREAM

∾✵∾

In one copy this poem was printed with the *Songs of Experience.*

∾✵∾

Once a dream did weave a shade
O'er my Angel-guarded bed,
That an Emmet lost its way
Where on grass methought I lay.

Troubled, 'wilder'd, and forlorn,
Dark, benighted, travel-worn,

NIGHT. **33–36. And there . . . fold:** Cf. Isaiah 11:6.

Over many a tangled spray,
All heart-broke I heard her say:

"O, my children! do they cry?
Do they hear their father sigh?   10
Now they look abroad to see:
Now return and weep for me."

Pitying, I drop'd a tear;
But I saw a glow-worm near,
Who replied: "What wailing wight
Calls the watchman of the night?

"I am set to light the ground,
While the beetle goes his round:
Follow now the beetle's hum;
Little wanderer, hie thee home!"   20

(1789)

ON ANOTHER'S SORROW

Can I see another's woe,
And not be in sorrow too?
Can I see another's grief,
And not seek for kind relief?

Can I see a falling tear,
And not feel my sorrow's share?
Can a father see his child
Weep, nor be with sorrow fill'd?

Can a mother sit and hear
An infant groan, an infant fear?   10
No, no! never can it be!
Never, never can it be!

And can he who smiles on all
Hear the wren with sorrows small,
Hear the small bird's grief & care,
Hear the woes that infants bear,

And not sit beside the nest,
Pouring pity in their breast;
And not sit the cradle near,
Weeping tear on infant's tear;   20

And not sit both night & day,
Wiping all our tears away?
O, no! never can it be!
Never, never can it be!

He doth give his joy to all;
He becomes an infant small;

He becomes a man of woe;
He doth feel the sorrow too.

Think not thou canst sigh a sigh,
And thy maker is not by;   30
Think not thou canst weep a tear,
And thy maker is not near.

O! he gives to us his joy
That our grief he may destroy;
Till our grief is fled & gone
He doth sit by us and moan.

(1789)

THE LITTLE BOY LOST

"Father! father! where are you going?
O do not walk so fast.
Speak, father, speak to your little boy,
Or else I shall be lost."

The night was dark, no father was there;
The child was wet with dew;
The mire was deep, & the child did weep,
And away the vapour flew.

1784–85 (1789)

THE LITTLE BOY FOUND

The little boy lost in the lonely fen,
Led by the wand'ring light,
Began to cry; but God, ever nigh,
Appear'd like his father in white.

He kissed the child & by the hand led
And to his mother brought,
Who in sorrow pale, thro' the lonely dale,
Her little boy weeping sought.

(1789)

SONGS OF EXPERIENCE

❧

Most of these poems were first engraved in 1794 when they were bound with the *Songs of Innocence*. So far as is known, they were never issued separately by Blake, who intended the two sets of songs to be read together. First drafts of eighteen of these songs were written in a notebook of 1793 (printed in the Keynes edition) that repays

THE LITTLE BOY LOST. 8. **vapour:** the will-o'-the-wisp that the little boy has mistaken for his father.

careful study; the order, the details of composition, and the context of other lyrics never issued by Blake are extremely revealing. As with the *Songs of Innocence* the order adopted here follows Blake's later memorandum, but he frequently varied the arrangement of the plates.

The Bard of Experience is no more to be completely identified with Blake himself than is the Piper of Innocence. Each expresses a vision of life that is true from its own standpoint and at the same time limited. Looking back to the *Songs of Experience* from the perspective of the prophetic books of the next two years, we can see that they depict the world to which we are confined when our knowledge is merely sensory, not imaginative. It is the fallen world man creates and perceives because he is himself fallen. But it is doubtful how precisely Blake had this in mind when he wrote the songs. More probably the unresolved tension between the contrary standpoints of Innocence and Experience was one of the things forcing him to a more comprehensive vision in his prophetic books.

In chronological order the *Songs of Experience* come after *America*, below. They are here placed next to the *Songs of Innocence* because Blake intended the two volumes to be read together.

༃

INTRODUCTION

Hear the voice of the Bard!
Who Present, Past, & Future, sees;
Whose ears have heard
The Holy Word
That walk'd among the ancient trees,

Calling the lapsed Soul,
And weeping in the evening dew;
That might controll
The starry pole,
And fallen, fallen light renew!   10

"O Earth, O Earth, return!
Arise from out the dewy grass;
Night is worn,
And the morn
Rises from the slumberous mass.

"Turn away no more;
Why wilt thou turn away?
The starry floor,
The wat'ry shore,
Is giv'n thee till the break of day."   20
                                                  (*1794*)

SONGS OF EXPERIENCE. INTRODUCTION. **5. walk'd . . . trees:**
Cf. Genesis 3:8.

EARTH'S ANSWER

Earth rais'd up her head
From the darkness dread & drear.
Her light fled,
Stony dread!
And her locks cover'd with grey despair.

"Prison'd on wat'ry shore,
Starry Jealousy does keep my den:
Cold and hoar,
Weeping o'er,
I hear the father of the ancient men.   10

"Selfish father of men!
Cruel, jealous, selfish fear!
Can delight,
Chain'd in night,
The virgins of youth and morning bear?

"Does spring hide its joy
When buds and blossoms grow?
Does the sower
Sow by night,
Or the plowman in darkness plow?   20

"Break this heavy chain
That does freeze my bones around.
Selfish! vain!
Eternal bane!
That free Love with bondage bound."

                                              1793 (*1794*)

NURSE'S SONG

When the voices of children are heard on the green,
And whisp'rings are in the dale,
The days of my youth rise fresh in my mind,
My face turns green and pale.

Then come home, my children, the sun is gone down,
And the dews of night arise;
Your spring & your day are wasted in play,
And your winter and night in disguise.

                                              1793 (*1794*)

THE FLY

Little Fly,
Thy summer's play
My thoughtless hand
Has brush'd away.

Am not I
A fly like thee?
Or art not thou
A man like me?

For I dance,
And drink, & sing,   10
Till some blind hand
Shall brush my wing.

If thought is life
And strength & breath,
And the want
Of thought is death;

Then am I
A happy fly,
If I live
Or if I die.   20

1793 (1794)

THE TYGER

Tyger! Tyger! burning bright
In the forests of the night,
What immortal hand or eye
Could frame thy fearful symmetry?

In what distant deeps or skies
Burnt the fire of thine eyes?
On what wings dare he aspire?
What the hand dare sieze the fire?

And what shoulder, & what art,
Could twist the sinews of thy heart?   10
And when thy heart began to beat,
What dread hand? & what dread feet?

What the hammer? what the chain?
In what furnace was thy brain?
What the anvil? what dread grasp
Dare its deadly terrors clasp?

When the stars threw down their spears,
And water'd heaven with their tears,
Did he smile his work to see?
Did he who made the Lamb make thee?   20

Tyger! Tyger! burning bright
In the forests of the night,
What immortal hand or eye,
Dare frame thy fearful symmetry?

1793 (1794)

"The Little Girl Lost" and "The Little Girl Found" were
first engraved with the *Songs of Innocence* and transferred
when the *Songs of Experience* were engraved.

THE LITTLE GIRL LOST

In futurity
I prophetic see
That the earth from sleep
(Grave the sentence deep)

Shall arise and seek
For her maker meek;
And the desart wild
Become a garden mild.

In the southern clime,
Where the summer's prime   10
Never fades away,
Lovely Lyca lay.

Seven summers old
Lovely Lyca told;
She had wander'd long
Hearing wild birds' song.

"Sweet sleep, come to me
Underneath this tree.
Do father, mother, weep,
Where can Lyca sleep?   20

"Lost in desart wild
Is your little child.
How can Lyca sleep
If her mother weep?

"If her heart does ake
Then let Lyca wake;
If my mother sleep,
Lyca shall not weep.

"Frowning, frowning night,
O'er this desart bright   30
Let thy moon arise
While I close my eyes."

Sleeping Lyca lay
While the beasts of prey,
Come from caverns deep,
View'd the maid asleep.

The kingly lion stood
And the virgin view'd,
Then he gamboll'd round
O'er the hallow'd ground.   40

Leopards, tygers, play
Round her as she lay,
While the lion old
Bow'd his mane of gold

And her bosom lick,
And upon her neck
From his eyes of flame
Ruby tears there came;

While the lioness
Loos'd her slender dress,   50
And naked they convey'd
To caves the sleeping maid.

(1789)

THE LITTLE GIRL FOUND

All the night in woe
Lyca's parents go
Over vallies deep,
While the desarts weep.

Tired and woe-begone,
Hoarse with making moan,
Arm in arm seven days
They trac'd the desart ways.

Seven nights they sleep
Among shadows deep,   10
And dream they see their child
Starv'd in desert wild.

Pale, thro' pathless ways
The fancied image strays
Famish'd, weeping, weak,
With hollow piteous shriek.

Rising from unrest,
The trembling woman prest
With feet of weary woe:
She could no further go.   20

In his arms he bore
Her, arm'd with sorrow sore;
Till before their way
A couching lion lay.

Turning back was vain:
Soon his heavy mane
Bore them to the ground.
Then he stalk'd around,

Smelling to his prey;
But their fears allay   30
When he licks their hands,
And silent by them stands.

They look upon his eyes
Fill'd with deep surprise;
And wondering behold
A spirit arm'd in gold.

On his head a crown,
On his shoulders down
Flow'd his golden hair.
Gone was all their care.   40

"Follow me," he said;
"Weep not for the maid;
In my palace deep
Lyca lies asleep."

Then they followed
Where the vision led,
And saw their sleeping child
Among tygers wild.

To this day they dwell
In a lonely dell;   50
Nor fear the wolvish howl
Nor the lion's growl.

                           (1789)

THE CLOD AND THE PEBBLE

"Love seeketh not Itself to please,
Nor for itself hath any care,
But for another gives its ease,
And builds a Heaven in Hell's despair."

So sung a little Clod of Clay,
Trodden with the cattle's feet,
But a Pebble of the brook
Warbled out these metres meet:

"Love seeketh only Self to please,
To bind another to Its delight,   10
Joys in another's loss of ease,
And builds a Hell in Heaven's despite."

                           1793 (1794)

THE LITTLE VAGABOND

Dear Mother, dear Mother, the Church is cold,
But the Ale-house is healthy & pleasant & warm;
Besides I can tell where I am used well,
Such usage in Heaven will never do well.

But if at the Church they would give us some Ale,
And a pleasant fire our souls to regale,
We'd sing and we'd pray all the live-long day,
Nor ever once wish from the Church to stray.

Then the Parson might preach, & drink, & sing,
And we'd be as happy as birds in the spring;   10
And modest Dame Lurch, who is always at Church,
Would not have bandy children, nor fasting, nor birch.

And God, like a father rejoicing to see
His children as pleasant and happy as he,
Would have no more quarrel with the Devil or the
    Barrel,
But kiss him, & give him both drink and apparel.

                              1793 (*1794*)

HOLY THURSDAY

꧁ꕥ꧂

See headnote to "Holy Thursday" in the *Songs of Innocence*, above, p. 54.

꧁ꕥ꧂

Is this a holy thing to see
In a rich and fruitful land,
Babes reduc'd to misery,
Fed with cold and usurous hand?

Is that trembling cry a song?
Can it be a song of joy?
And so many children poor?
It is a land of poverty!

And their sun does never shine,
And their fields are bleak & bare,   10
And their ways are fill'd with thorns:
It is eternal winter there.

For where-e'er the sun does shine,
And where-e'er the rain does fall,
Babe can never hunger there,
Nor poverty the mind appall.

                              1793 (*1794*)

A POISON TREE

I was angry with my friend:
I told my wrath, my wrath did end.
I was angry with my foe:
I told it not, my wrath did grow.

And I water'd it in fears,
Night & morning with my tears;
And I sunned it with smiles,
And with soft deceitful wiles.

And it grew both day and night,
Till it bore an apple bright;   10
And my foe beheld it shine,
And he knew that it was mine,

And into my garden stole
When the night had veil'd the pole:
In the morning glad I see
My foe outstretch'd beneath the tree.

                              1793 (*1794*)

THE ANGEL

I dreamt a Dream! what can it mean?
And that I was a maiden Queen,
Guarded by an Angel mild:
Witless woe was ne'er beguil'd!

And I wept both night and day,
And he wip'd my tears away,
And I wept both day and night,
And hid from him my heart's delight.

So he took his wings and fled;
Then the morn blush'd rosy red;   10
I dried my tears, & arm'd my fears
With ten thousand shields and spears.

Soon my Angel came again:
I was arm'd, he came in vain;
For the time of youth was fled,
And grey hairs were on my head.

                              1793 (*1794*)

THE SICK ROSE

O Rose, thou art sick!
The invisible worm,
That flies in the night,
In the howling storm,

Has found out thy bed
Of crimson joy,
And his dark secret love
Does thy life destroy.

                              1793 (*1794*)

TO TIRZAH

This was composed later than the other *Songs of Experience*. Tirzah was the capital of Israel, the northern kingdom, in opposition to Jerusalem, the capital of the redeemed, the southern kingdom of Judah. In Blake's prophetic books Tirzah becomes a symbol of the five senses and of Nature, the "Mother of my Mortal part."

Whate'er is Born of Mortal Birth
Must be consumed with the Earth
To rise from Generation free:
Then what have I to do with thee?

The Sexes sprung from Shame & Pride,
Blow'd in the morn; in evening died;
But Mercy chang'd Death into Sleep;
The Sexes rose to work & weep.

Thou, Mother of my Mortal part,
With cruelty didst mould my Heart,   10
And with false self-decieving tears
Didst bind my Nostrils, Eyes, & Ears;

Didst close my Tongue in senseless clay,
And me to Mortal Life betray.
The Death of Jesus set me free:
Then what have I to do with thee?

                          c. 1801

THE VOICE OF THE ANCIENT BARD

This poem is generally included by Blake with the *Songs of Innocence*, often as the last poem in the volume to make a transition to the *Songs of Experience*.

Youth of delight, come hither,
And see the opening morn,
Image of truth new born.
Doubt is fled, & clouds of reason,
Dark disputes & artful teazing.
Folly is an endless maze,
Tangled roots perplex her ways.

TO TIRZAH. **4. what . . . thee:** Cf. John 2:4. **6. Blow'd:** blossomed.

How many have fallen there!
They stumble all night over bones of the dead,
And feel they know not what but care,   10
And wish to lead others, when they should be led.

                          (1789)

MY PRETTY ROSE TREE

A flower was offer'd to me,
Such a flower as May never bore;
But I said "I've a Pretty Rose-tree,"
And I passed the sweet flower o'er.

Then I went to my Pretty Rose-tree,
To tend her by day and by night;
But my Rose turn'd away with jealousy,
And her thorns were my only delight.

                          1793 (1794)

AH! SUN-FLOWER

Ah, Sun-flower! weary of time,
Who countest the steps of the Sun,
Seeking after that sweet golden clime
Where the traveller's journey is done:

Where the Youth pined away with desire,
And the pale Virgin shrouded in snow
Arise from their graves, and aspire
Where my Sun-flower wishes to go.

                          (1794)

THE LILLY

The modest Rose puts forth a thorn,
The humble Sheep a threat'ning horn;
While the Lilly white shall in Love delight,
Nor a thorn, nor a threat, stain her beauty bright.

                          1793 (1794)

THE GARDEN OF LOVE

I went to the Garden of Love,
And saw what I never had seen:
A Chapel was built in the midst,
Where I used to play on the green.

And the gates of this Chapel were shut,
And "Thou shalt not" writ over the door;
So I turn'd to the Garden of Love
That so many sweet flowers bore;

And I saw it was filled with graves,
And tomb-stones where flowers should be;   10
And Priests in black gowns were walking their rounds,
And binding with briars my joys & desires.

1793 (1794)

A LITTLE BOY LOST

"Nought loves another as itself,
Nor venerates another so,
Nor is it possible to Thought
A greater than itself to know:

"And Father, how can I love you
Or any of my brothers more?
I love you like the little bird
That picks up crumbs around the door."

The Priest sat by and heard the child,
In trembling zeal he siez'd his hair: 10
He led him by his little coat,
And all admir'd the Priestly care.

And standing on the altar high,
"Lo! what a fiend is here!" said he,
"One who sets reason up for judge
Of our most holy Mystery."

The weeping child could not be heard,
The weeping parents wept in vain;
They strip'd him to his little shirt,
And bound him in an iron chain;   20

And burn'd him in a holy place,
Where many had been burn'd before:
The weeping parents wept in vain.
Are such things done on Albion's shore?

1793 (1794)

INFANT SORROW

My mother groan'd! my father wept.
Into the dangerous world I leapt:
Helpless, naked, piping loud:
Like a fiend hid in a cloud.

Struggling in my father's hands,
Striving against my swaddling bands,
Bound and weary I thought best
To sulk upon my mother's breast.

1793 (1794)

THE SCHOOLBOY

❦

Originally one of the *Songs of Innocence*.

❦

I love to rise in a summer morn
When the birds sing on every tree;
The distant huntsman winds his horn,
And the sky-lark sings with me.
O! what sweet company.

But to go to school in a summer morn,
O! it drives all joy away;
Under a cruel eye outworn,
The little ones spend the day
In sighing and dismay.   10

Ah! then at times I drooping sit,
And spend many an anxious hour,
Nor in my book can I take delight,
Nor sit in learning's bower,
Worn thro' with the dreary shower.

How can the bird that is born for joy
Sit in a cage and sing?
How can a child, when fears annoy,
But droop his tender wing,
And forget his youthful spring?   20

O! father & mother, if buds are nip'd
And blossoms blown away,
And if the tender plants are strip'd
Of their joy in the springing day,
By sorrow and care's dismay,

How shall the summer arise in joy,
Or the summer fruits appear?
Or how shall we gather what griefs destroy,
Or bless the mellowing year,
When the blasts of winter appear?   30

(1789)

LONDON

I wander thro' each charter'd street,
Near where the charter'd Thames does flow,
And mark in every face I meet
Marks of weakness, marks of woe.

In every cry of every Man,
In every Infant's cry of fear,
In every voice, in every ban,
The mind-forg'd manacles I hear.

How the Chimney-sweeper's cry
Every black'ning Church appalls;　　　　10
And the hapless Soldier's sigh
Runs in blood down Palace walls.

But most thro' midnight streets I hear
How the youthful Harlot's curse
Blasts the new born Infant's tear,
And blights with plagues the Marriage hearse.

　　　　　　　　　　1793 (1794)

A LITTLE GIRL LOST

Children of the future Age
Reading this indignant page,
Know that in a former time
Love! sweet Love! was thought a crime.

In the Age of Gold,
Free from winter's cold,
Youth and maiden bright
To the holy light,
Naked in the sunny beams delight.

Once a youthful pair,　　　　　　10
Fill'd with softest care,
Met in garden bright
Where the holy light
Had just remov'd the curtains of the night.

There, in rising day,
On the grass they play;
Parents were afar,
Strangers came not near,
And the maiden soon forgot her fear.

Tired with kisses sweet,　　　　　20
They agree to meet
When the silent sleep
Waves o'er heaven's deep,
And the weary tired wanderers weep.

To her father white
Came the maiden bright;
But his loving look,
Like the holy book,
All her tender limbs with terror shook.

"Ona! pale and weak!　　　　　　30
To thy father speak:
O, the trembling fear!
O, the dismal care!
That shakes the blossoms of my hoary hair."

　　　　　　　　　　　　(1794)

THE CHIMNEY SWEEPER

A little black thing among the snow,
Crying "'weep! 'weep!" in notes of woe!
"Where are thy father & mother? say?"
"They are both gone up to the church to pray.

"Because I was happy upon the heath,
And smil'd among the winter's snow,
They clothed me in the clothes of death,
And taught me to sing the notes of woe.

"And because I am happy & dance & sing,
They think they have done me no injury,　　10
And are gone to praise God & his Priest & King,
Who make up a heaven of our misery."

　　　　　　　　　　1793 (1794)

THE HUMAN ABSTRACT

Pity would be no more
If we did not make somebody Poor;
And Mercy no more could be
If all were as happy as we.

And mutual fear brings peace,
Till the selfish loves increase:
Then Cruelty knits a snare,
And spreads his baits with care.

He sits down with holy fears,
And waters the ground with tears;　　　10
Then Humility takes its root
Underneath his foot.

Soon spreads the dismal shade
Of Mystery over his head;
And the Catterpiller and Fly
Feed on the Mystery.

And it bears the fruit of Deceit,
Ruddy and sweet to eat;
And the Raven his nest has made
In its thickest shade.　　　　　20

The Gods of the earth and sea
Sought thro' Nature to find this Tree;
But their search was all in vain:
There grows one in the Human Brain.

1793 (1794)

A DIVINE IMAGE

Engraved about 1794, this poem was never actually
bound by Blake with the *Songs of Experience*, presumably
because "The Human Abstract" served as a better
counterpart to "The Divine Image" in the *Songs of
Innocence*.

Cruelty has a Human Heart,
And Jealousy a Human Face;
Terror the Human Form Divine,
And Secrecy the Human Dress.

The Human Dress is forged Iron,
The Human Form a fiery Forge,
The Human Face a Furnace seal'd,
The Human Heart its hungry Gorge.

(1866)

LOVE'S SECRET

This and the following lyric come from a notebook and
were written in approximately 1793. The "Cradle Song"
was not included in the *Songs of Experience* though it was
obviously intended to match the poem with the same
title in the *Songs of Innocence*.

Never seek to tell thy love,
Love that never told can be;
For the gentle wind does move
Silently, invisibly.

I told my love, I told my love,
I told her all my heart;
Trembling, cold, in ghastly fears,
Ah! she doth depart.

Soon as she was gone from me
A traveller came by   10
Silently, invisibly—
He took her with a sigh.

1793 (1863)

A CRADLE SONG

Sleep! Sleep! beauty bright,
Dreaming o'er the joys of night.
Sleep! Sleep! in thy sleep
Little sorrows sit & weep.

Sweet Babe, in thy face
Soft desires I can trace,
Secret joys & secret smiles,
Little pretty infant wiles.

As thy softest limbs I feel,
Smiles as of the morning steal   10
O'er thy cheek, & o'er thy breast
Where thy little heart does rest.

O! the cunning wiles that creep
In thy little heart asleep.
When thy little heart does wake
Then the dreadful lightnings break.

From thy cheek & from thy eye
O'er the youthful harvests nigh
Infant wiles & infant smiles
Heaven & Earth of peace beguiles.   20

1793 (1863)

THE BOOK OF THEL

Composed probably in 1789, *The Book of Thel* transposes some aspects of the dialectic of Innocence and
Experience into mythical narrative.

Thel's Motto

Does the Eagle know what is in the pit,
Or wilt thou go ask the Mole?
Can Wisdom be put in a silver rod?
Or Love in a golden bowl?

THE BOOK OF THEL. **Title:** The name Thel is derived from a
Greek root meaning "desire." **3–4. silver . . . bowl:** Cf.
Ecclesiastes 12:6.

I

The Daughters of Mne Seraphim led round their sunny
flocks—
All but the youngest: she in paleness sought the secret
air,
To fade away like morning beauty from her mortal
day:
Down by the river of Adona her soft voice is heard,
And thus her gentle lamentation falls like morning
dew:

"O life of this our spring! why fades the lotus of the
water?   10
Why fade these children of the spring, born but to smile
& fall?
Ah! Thel is like a wat'ry bow, and like a parting cloud;
Like a reflection in a glass; like shadows in the water;
Like dreams of infants, like a smile upon an infant's
face;
Like the dove's voice; like transient day; like music in
the air.
Ah! gentle may I lay me down, and gentle rest my
head,
And gentle sleep the sleep of death, and gentle hear the
voice
Of him that walketh in the garden in the evening time."

The Lilly of the valley, breathing in the humble grass,
Answer'd the lovely maid and said: "I am a wat'ry
weed,   20
And I am very small and love to dwell in lowly vales;
So weak, the gilded butterfly scarce perches on my
head.
Yet I am visited from heaven, and he that smiles on all
Walks in the valley and each morn over me spreads his
hand,
Saying, 'Rejoice, thou humble grass, thou new-born
lilly flower,
Thou gentle maid of silent valleys and of modest
brooks;
For thou shalt be clothed in light, and fed with morning
manna,
Till summer's heat melts thee beside the fountains and
the springs
To flourish in eternal vales.' Then why should Thel
complain?

5. **Mne:** This was intended by Blake. Its meaning is not
known. It may be a corruption of Bne Seraphim, a name
found in occult lore. 8. **Adona:** derived from Adonis. Cf.
Paradise Lost, I. 450. 18. **him . . . time:** Cf. Genesis 3:8.

Why should the mistress of the vales of Har utter a
sigh?"   30

She ceas'd, & smil'd in tears, then sat down in her silver
shrine.

Thel answer'd: "O thou little virgin of the peaceful
valley,
Giving to those that cannot crave, the voiceless, the
o'ertired;
Thy breath doth nourish the innocent lamb, he smells
thy milky garments,
He crops thy flowers while thou sittest smiling in his
face,
Wiping his mild and meekin mouth from all contagi-
ous taints.
Thy wine doth purify the golden honey; thy perfume,
Which thou dost scatter on every little blade of grass
that springs,
Revives the milked cow, & tames the fire-breathing
steed.
But Thel is like a faint cloud kindled at the rising sun:
I vanish from my pearly throne, and who shall find
my place?"   41

"Queen of the vales," the Lilly answer'd, "ask the
tender cloud,
And it shall tell thee why it glitters in the morning sky,
And why it scatters its bright beauty thro' the humid
air.
Descend, O little Cloud, & hover before the eyes of
Thel."

The Cloud descended, and the Lilly bow'd her modest
head
And went to mind her numerous charge among the
verdant grass.

II

"O little Cloud," the virgin said, "I charge thee tell
to me
Why thou complainest not when in one hour thou
fade away:
Then we shall seek thee, but not find. Ah! Thel is like
to thee: 50
I pass away: yet I complain, and no one hears my
voice."

30. **Har:** the state of Innocence. 36. **meekin:** gentle.

The Cloud then shew'd his golden head & his bright
form emerg'd,
Hovering and glittering on the air before the face of
Thel.

"O virgin, know'st thou not our steeds drink of the
golden springs
Where Luvah doth renew his horses? Look'st thou on
my youth,
And fearest thou, because I vanish and am seen no
more,
Nothing remains? O maid, I tell thee, when I pass
away
It is to tenfold life, to love, to peace and raptures holy:
Unseen descending, weigh my light wings upon balmy
flowers,
And court the fair-eyed dew to take me to her shining
tent: 60
The weeping virgin, trembling, kneels before the risen
sun,
Till we arise link'd in a golden band and never part,
But walk united, bearing food to all our tender
flowers."

"Dost thou, O little Cloud? I fear that I am not like
thee,
For I walk thro' the vales of Har, and smell the sweetest
flowers,
But I feed not the little flowers; I hear the warbling
birds,
But I feed not the warbling birds; they fly and seek
their food:
But Thel delights in these no more, because I fade
away;
And all shall say, 'Without a use this shining woman
liv'd,
Or did she only live to be at death the food of worms?'"

The Cloud reclin'd upon his airy throne and answer'd
thus: 71

"Then if thou art the food of worms, O virgin of the
skies,
How great thy use, how great thy blessing! Every
thing that lives
Lives not alone nor for itself. Fear not, and I will call
The weak worm from its lowly bed, and thou shalt
hear its voice.
Come forth, worm of the silent valley, to thy pensive
queen."

The helpless worm arose, and sat upon the Lilly's leaf,
And the bright Cloud sail'd on, to find his partner in
the vale.

III

Then Thel astonish'd view'd the Worm upon its
dewy bed.

"Art thou a Worm? Image of weakness, art thou but
a Worm?   80
I see thee like an infant wrapped in the Lilly's leaf.
Ah! weep not, little voice, thou canst not speak, but
thou canst weep.
Is this a Worm? I see thee lay helpless & naked, weeping,
And none to answer, none to cherish thee with mother's
smiles."

The Clod of Clay heard the Worm's voice & rais'd her
pitying head:
She bow'd over the weeping infant, and her life
exhal'd
In milky fondness: then on Thel she fix'd her humble
eyes.

"O beauty of the vales of Har! we live not for our-
selves.
Thou seest me the meanest thing, and so I am indeed.
My bosom of itself is cold, and of itself is dark;   90
But he, that loves the lowly, pours his oil upon my
head,
And kisses me, and binds his nuptial bands around my
breast,
And says: 'Thou mother of my children, I have loved
thee,
And I have given thee a crown that none can take
away.'
But how this is, sweet maid, I know not, and I cannot
know;
I ponder, and I cannot ponder; yet I live and love."

---

52. Cloud . . . emerg'd: To the imaginative eye of Thel
in the paradise of Innocence everything appears in human
form. Cf. Blake's marginal annotation to Swedenborg's
Wisdom of Angels Concerning Divine Love and Divine Wisdom:
"Think of a white cloud as being holy, you cannot love it;
but think of a holy man within the cloud, love springs up in
your thoughts, for to think of holiness distinct from man is
impossible to the affections. Thought alone can make monsters, but the affections cannot." 55. Luvah: Later one of
Blake's four "zoas," Luvah here is perhaps derived from
"Lover."

The daughter of beauty wip'd her pitying tears with
    her white veil,
And said: "Alas! I knew not this, and therefore did I
    weep.
That God would love a Worm I knew, and punish the
    evil foot
That wilful bruis'd its helpless form; but that he
    cherish'd it   100
With milk and oil I never knew, and therefore did I
    weep;
And I complain'd in the mild air, because I fade away,
And lay me down in thy cold bed, and leave my
    shining lot."

"Queen of the vales," the matron Clay answer'd, "I
    heard thy sighs,
And all thy moans flew o'er my roof, but I have call'd
    them down.
Wilt thou, O Queen, enter my house? 'Tis given thee
    to enter
And to return: fear nothing, enter with thy virgin
    feet."

IV

The eternal gates' terrific porter lifted the northern
    bar:
Thel enter'd in & saw the secrets of the land unknown.
She saw the couches of the dead, & where the fibrous
    roots   110
Of every heart on earth infixes deep its restless twists:
A land of sorrows & of tears where never smile was
    seen.

She wander'd in the land of clouds thro' valleys dark,
    list'ning
Dolours & lamentations; waiting oft beside a dewy
    grave
She stood in silence, list'ning to the voices of the
    ground,
Till to her own grave plot she came, & there she sat
    down,
And heard this voice of sorrow breathed from the
    hollow pit.

"Why cannot the Ear be closed to its own destruction?
Or the glist'ning Eye to the poison of a smile?
Why are Eyelids stor'd with arrows ready drawn, 120
Where a thousand fighting men in ambush lie?

Or an Eye of gifts & graces show'ring fruits & coined
    gold?
Why a Tongue impress'd with honey from every
    wind?
Why an Ear, a whirlpool fierce to draw creations in?
Why a Nostril wide inhaling terror, trembling, &
    affright?
Why a tender curb upon the youthful burning boy?
Why a little curtain of flesh on the bed of our desire?"

The Virgin started from her seat, & with a shriek
Fled back unhinder'd till she came into the vales of Har.

                                        c. 1789 (1789)

THE MARRIAGE
OF HEAVEN AND HELL

Composed between 1790 and 1793, this brilliant work is
a satire—if it can even be classified. It ridicules the Swedish
religious writer Emanuel Swedenborg, 1688–1772, a
prophet Blake had once respected. Much more import-
antly, however, it attacks the sexual and social morality
that, backed by the apparatus of religion and law,
restrains energy, passion, and genius and, in Blake's
opinion, condemns us to the spectral half-existence that
is our common lot. The vehicle of this attack is an ironic
upset of stock associations. The Angels of Heaven sym-
bolize whatever limits and confines—priests, policemen,
prudence, logic, self-doubt, and, in metaphysics, belief in an objective world to which we must conform
ourselves. Blake speaks as one of the Devils, the inspired
creators who are in every respect the opposite of Angels.
  His title is partly ironic. The "marriage" is a dynamic
confrontation of contraries in which each must preserve
its own identity. For "without Contraries is no progression"; the Devourer and the Prolific must be "enemies;
whoever tries to reconcile them seeks to destroy existence." At the same time, there may be suggestions of a
"marriage" more in the sense of synthesis. When Blake
says that "Reason is the bound or outward circumference
of Energy," he speaks as a Devil subordinating reason
to energy. Reason is simply a boundary line that can be
changed and expanded. But he also means that a work of
art, for example, cannot be infinite. It must have shape or
form. "The Prolific would cease to be Prolific unless the
Devourer . . . received the excess of his delights." But
again he is also stressing that form cannot be imposed
from without. (Compare Coleridge's concept of organic
form.) The last "Memorable Fancy" is qualified by its
partly jocular tone, but there is a marriage in which the
Angel embraces the fire. He is consumed, but he is also
reborn as poetic inspiration or prophecy, the biblical
Elijah.

Blake is, of course, saying that much we call "evil" is really good. But it would be a wild misinterpretation to say that he has gone beyond good and evil into nihilism. He is simply asserting his own moral beliefs against conventional ones, and his beliefs give no endorsement to cruelty. From this point of view, the last of his marginal annotations to Lavater's *Aphorisms on Man* may be a helpful gloss:

There is a strong objection to Lavater's principles (as I understand them) & that is He makes every thing originate in its accident; he makes the vicious propensity not only a leading feature of the man, but the stamina on which all his virtues grow. But as I understand Vice it is a Negative. It does not signify what the laws of Kings & Priests have call'd Vice; we who are philosophers ought not to call the Staminal Virtues of Humanity by the same name that we call the omissions of intellect springing from poverty.

Every man's leading propensity ought to be call'd his leading Virtue & his good Angel. But the Philosophy of Causes & Consequences misled Lavater as it has all his Cotemporaries. Each thing is its own cause & its own effect. Accident is the omission of act in self & the hindering of act in another; This is Vice, but all Act is Virtue. To hinder another is not an act; it is the contrary; it is a restraint on action both in ourselves & in the person hinder'd, for he who hinders another omits his own duty at the same time.

Murder is Hindering Another.

Theft is Hindering Another.

Backbiting, Undermining, Circumventing, & whatever is Negative is Vice. But the origin of this mistake in Lavater & his cotemporaries is, They suppose that Woman's Love is Sin; in consequence all the Loves & Graces with them are Sins.

❧

THE ARGUMENT

Rintrah roars & shakes his fires in the burden'd air;
Hungry clouds swag on the deep.

Once meek, and in a perilous path,
The just man kept his course along
The vale of death.
Roses are planted where thorns grow,
And on the barren heath
Sing the honey bees.

Then the perilous path was planted,
And a river and a spring   10

THE MARRIAGE OF HEAVEN AND HELL. **1. Rintrah:** In Blake's prophetic works he is the angry prophet. Some of the imagery of the following lines occurs in Isaiah 35. **2. swag:** sway awkwardly or sink down, or both.

On every cliff and tomb,
And on the bleached bones
Red clay brought forth;

Till the villain left the paths of ease,
To walk in perilous paths, and drive
The just man into barren climes.

Now the sneaking serpent walks
In mild humility,
And the just man rages in the wilds
Where lions roam.   20

Rintrah roars & shakes his fires in the burden'd air;
Hungry clouds swag on the deep.

As a new heaven is begun, and it is now thirty-three years since its advent,[1] the Eternal Hell revives. And lo! Swedenborg is the Angel sitting at the tomb:[2] his writings are the linen clothes folded up.[3] Now is the dominion of Edom,[4] & the return of Adam into Paradise.[5] See Isaiah xxxiv & xxxv Chap.[6]

Without Contraries is no progression. Attraction and Repulsion, Reason and Energy, Love and Hate, are necessary to Human existence.

From these contraries spring what the religious call Good & Evil. Good is the passive that obeys Reason. Evil is the active springing from Energy.

Good is Heaven. Evil is Hell.

13. Red clay: "Adam" in Hebrew means "red clay." **1. advent:** According to Swedenborg the Last Judgment began in 1757, which was also the year of Blake's birth. Blake is writing in 1790, thirty-three years later. It was at the age of thirty-three that Christ rose from the tomb. The Eternal Hell revives as the contrary to the "new heaven." **2. Angel . . . tomb:** Matthew 28:1–7, and variously narrated in the other gospels. **3. linen clothes . . . up:** the now useless cloths that had been wrapped around the dead body of Jesus. **4. Edom:** identified with Esau, older brother of the crafty Jacob. By trickery Jacob stole from Esau the blessing of his dying father, Isaac. Isaac then prophesied to Esau that he would serve his brother until "it shall come to pass when thou shalt have the dominion, that thou shalt shake his yoke from off thy neck" (Genesis 28:40). Cf. also Isaiah 63:1–4. **5. Adam . . . Paradise:** Only according to Blake does Adam return into paradise. **6. Isaiah xxxiv & xxxv:** In these chapters the prophet foretells the destruction of the wicked and the eternal paradise when "The wilderness and the solitary place shall be glad for them; and the desert shall rejoice, and blossom as the rose . . . and the ransomed of the Lord shall return, and come to Zion with songs and everlasting joy upon their heads."

THE VOICE OF THE DEVIL

All Bibles or sacred codes have been the causes of the following Errors:

1. That Man has two real existing principles: Viz: a Body & a Soul.[7]

2. That Energy, call'd Evil, is alone from the Body; & that Reason, call'd Good, is alone from the Soul.

3. That God will torment Man in Eternity for following his Energies.

But the following Contraries to these are True:

1. Man has no Body distinct from his Soul; for that call'd Body is a portion of Soul discern'd by the five Senses, the chief inlets of Soul in this age.

2. Energy is the only life, and is from the Body; and Reason is the bound or outward circumference of Energy.

3. Energy is Eternal Delight.

Those who restrain desire, do so because theirs is weak enough to be restrained; and the restrainer or reason usurps its place & governs the unwilling.

And being restrain'd, it by degrees becomes passive, till it is only the shadow of desire.

The history of this is written in Paradise Lost, & the Governor or Reason is call'd Messiah.[8]

And the original Archangel, or possessor of the command of the heavenly host, is call'd the Devil or Satan, and his children are call'd Sin & Death.

But in the Book of Job, Milton's Messiah is call'd Satan.[9]

For this history has been adopted by both parties.

It indeed appear'd to Reason as if Desire was cast out; but the Devil's account is, that the Messiah fell, & formed a heaven of what he stole from the Abyss.

This is shewn in the Gospel,[10] where he prays to the Father to send the comforter, or Desire, that Reason may have Ideas to build on; the Jehovah of the Bible being no other than he who dwells in flaming fire.

Know that after Christ's death, he became Jehovah.

But in Milton, the Father is Destiny, the Son a Ratio[11] of the five senses, & the Holy-ghost Vacuum!

Note: The reason Milton wrote in fetters when he wrote of Angels & God, and at liberty when of Devils & Hell, is because he was a true Poet and of the Devil's party without knowing it.

A MEMORABLE FANCY[12]

As I was walking among the fires of hell, delighted with the enjoyments of Genius, which to Angels look like torment and insanity, I collected some of their Proverbs; thinking that as the sayings used in a nation mark its character, so the Proverbs of Hell show the nature of Infernal wisdom better than any description of buildings or garments.

When I came home: on the abyss of the five senses, where a flat sided steep frowns over the present world, I saw a mighty Devil folded in black clouds, hovering on the sides of the rock: with corroding fires he wrote the following sentence now percieved by the minds of men, & read by them on earth:

How do you know but ev'ry Bird that cuts the airy
    way,
Is an immense world of delight, clos'd by your
    senses five?

PROVERBS OF HELL

In seed time learn, in harvest teach, in winter enjoy.

Drive your cart and your plow over the bones of the dead.[13]

The road of excess leads to the palace of wisdom.

Prudence is a rich, ugly old maid courted by Incapacity.

He who desires but acts not, breeds pestilence.

The cut worm forgives the plow.

---

7. **Body . . . Soul:** Blake holds that the body is not a substance separate from the soul. **8. The history . . . Messiah:** In *Paradise Lost* Satan gradually loses the magnificence with which Milton endows him in the opening book. During the war in heaven of Book VI he is defeated by the Messiah. The birth of Sin and Death from Satan is narrated in Book II. **9. Messiah . . . Satan:** God allows Satan to afflict the just man, Job, with torments that include accusations of sin. **10. Gospel:** John 16:7?

11. **Ratio:** reason, the mind as described by Locke, who taught that all ideas are ultimately derived from sense impressions. To Blake and other Romantic writers this meant that the mind was not creative or that its creative productions could only be unreal. See Introduction and cf. Blake's use of the term "ratio" in *There Is No Natural Religion*, above. **12. A Memorable Fancy:** The title parodies Swedenborg's *Memorable Relations*, in which he describes visions of the spiritual world. **13. Drive . . . dead:** Cf. "The Voice of the Ancient Bard," above, l. 9, in *Songs of Experience*.

Dip him in the river who loves water.

A fool sees not the same tree that a wise man sees.[14]

He whose face gives no light, shall never become a star.

Eternity is in love with the productions of time.

The busy bee has no time for sorrow.

The hours of folly are measur'd by the clock; but of wisdom, no clock can measure.

All wholesome food is caught without a net or a trap.

Bring out number, weight & measure in a year of dearth.

No bird soars too high, if he soars with his own wings.

A dead body revenges not injuries.

The most sublime act is to set another before you.

If the fool would persist in his folly he would become wise.

Folly is the cloke of knavery.

Shame is Pride's cloke.

Prisons are built with stones of Law, Brothels with bricks of Religion.

The pride of the peacock is the glory of God.

The lust of the goat is the bounty of God.

The wrath of the lion is the wisdom of God.

The nakedness of woman is the work of God.

Excess of sorrow laughs. Excess of joy weeps.

The roaring of lions, the howling of wolves, the raging of the stormy sea, and the destructive sword, are portions of eternity too great for the eye of man.

The fox condemns the trap, not himself.

Joys impregnate. Sorrows bring forth.

Let man wear the fell of the lion, woman the fleece of the sheep.

The bird a nest, the spider a web, man friendship.

The selfish, smiling fool, & the sullen, frowning fool shall be both thought wise, that they may be a rod.

What is now proved was once only imagin'd.

The rat, the mouse, the fox, the rabbet watch the roots; the lion, the tyger, the horse, the elephant watch the fruits.

The cistern contains: the fountain overflows.

One thought fills immensity.

Always be ready to speak your mind, and a base man will avoid you.

Every thing possible to be believ'd is an image of truth.

The eagle never lost so much time as when he submitted to learn of the crow.

The fox provides for himself, but God provides for the lion.

Think in the morning. Act in the noon. Eat in the evening. Sleep in the night.

He who has suffer'd you to impose on him, knows you.

As the plow follows words, so God rewards prayers.

The tygers of wrath are wiser than the horses of instruction.

Expect poison from the standing water.

You never know what is enough unless you know what is more than enough.

Listen to the fool's reproach! it is a kingly title!

The eyes of fire, the nostrils of air, the mouth of water, the beard of earth.

The weak in courage is strong in cunning.

The apple tree never asks the beech how he shall grow; nor the lion, the horse, how he shall take his prey.

The thankful reciever bears a plentiful harvest.

If others had not been foolish, we should be so.

The soul of sweet delight can never be defil'd.

When thou seest an Eagle, thou seest a portion of Genius; lift up thy head!

As the caterpiller chooses the fairest leaves to lay her eggs on, so the priest lays his curse on the fairest joys.

To create a little flower is the labour of ages.

Damn braces. Bless relaxes.

The best wine is the oldest, the best water the newest.

Prayers plow not! Praises reap not!

Joys laugh not! Sorrows weep not!

The head Sublime, the heart Pathos, the genitals Beauty, the hands & feet Proportion.

As the air to a bird or the sea to a fish, so is contempt to the contemptible.

The crow wish'd every thing was black, the owl that every thing was white.

Exuberance is Beauty.

If the lion was advised by the fox, he would be cunning.

Improvement makes strait roads; but the crooked roads without Improvement are roads of Genius.

Sooner murder an infant in its cradle than nurse unacted desires.

Where man is not, nature is barren.

---

14. **A fool . . . sees:** Cf. the letter to the Revd. Dr. Trusler, below, August 23, 1799.

Truth can never be told so as to be understood, and not be believ'd.

Enough! or Too much.

The ancient Poets animated all sensible objects with Gods or Geniuses, calling them by the names and adorning them with the properties of woods, rivers, mountains, lakes, cities, nations, and whatever their enlarged & numerous senses could perceive.

And particularly they studied the genius of each city & country, placing it under its mental deity;

Till a system was formed, which some took advantage of, & enslav'd the vulgar by attempting to realize or abstract the mental deities from their objects: thus began Priesthood;

Choosing forms of worship from poetic tales.

And at length they pronounc'd that the Gods had order'd such things.

Thus men forgot that All deities reside in the human breast.

A MEMORABLE FANCY

The Prophets Isaiah and Ezekiel dined with me, and I asked them how they dared so roundly to assert that God spoke to them; and whether they did not think at the time that they would be misunderstood, & so be the cause of imposition.

Isaiah answer'd: "I saw no God, nor heard any, in a finite organical perception; but my senses discover'd the infinite in everything, and as I was then perswaded, & remain confirm'd, that the voice of honest indignation is the voice of God, I cared not for consequences, but wrote."

Then I asked: "does a firm perswasion that a thing is so, make it so?"

He replied: "All poets believe that it does, & in ages of imagination this firm perswasion removed mountains; but many are not capable of a firm perswasion of any thing."

Then Ezekiel said: "The philosophy of the east taught the first principles of human perception: some nations held one principle for the origin, and some another: we of Israel taught that the Poetic Genius (as you now call it) was the first principle and all the others merely derivative, which was the cause of our despising the Priests & Philosophers of other countries, and prophecying that all Gods would at last be proved

to originate in ours & to be the tributaries of the Poetic Genius; it was this that our great poet, King David, desired so fervently & invokes so pathetic'ly, saying by this he conquers enemies & governs kingdoms; and we so loved our God, that we cursed in his name all the deities of surrounding nations, and asserted that they had rebelled: from these opinions the vulgar came to think that all nations would at last be subject to the jews."

"This," said he, "like all firm perswasions, is come to pass; for all nations believe the jews' code and worship the jews' god, and what greater subjection can be?"

I heard this with some wonder, & must confess my own conviction. After dinner I ask'd Isaiah to favour the world with his lost works; he said none of equal value was lost. Ezekiel said the same of his.

I also asked Isaiah what made him go naked and barefoot three years? he answer'd: "the same that made our friend Diogenes, the Grecian." [15]

I then asked Ezekiel why he eat dung, & lay so long on his right & left side? [16] he answer'd, "the desire of raising other men into a perception of the infinite: this the North American tribes practise, & is he honest who resists his genius or conscience only for the sake of present ease or gratification?"

The ancient tradition that the world will be consumed in fire at the end of six thousand years is true, as I have heard from Hell.

For the cherub with his flaming sword is hereby commanded to leave his guard at tree of life; [17] and when he does, the whole creation will be consumed and appear infinite and holy, whereas it now appears finite & corrupt.

This will come to pass by an improvement of sensual enjoyment.

But first the notion that man has a body distinct from his soul is to be expunged; this I shall do by printing in the infernal method, by corrosives, [18] which in Hell

---

15. Isaiah . . . Grecian: Isaiah 20:3; Diogenes, 412?–323 B.C., Greek cynic philosopher. **16. Ezekiel . . . side:** Ezekiel 4:4–6, 12. **17. cherub . . . life:** Genesis 3:24. **18. printing . . . corrosives:** satire, but Blake also refers to the process of "Illuminated Printing" he invented and employed for his prophetic books. He would write and draw on a copper plate in some material impervious to acid and then apply acid to the plate so that his designs were left in relief. From these relief plates he printed. Cf. the phrase "corroding fires" in the first "Memorable Fancy," above.

are salutary and medicinal, melting apparent surfaces away, and displaying the infinite which was hid.

If the doors of perception were cleansed every thing would appear to man as it is, infinite.

For man has closed himself up, till he sees all things thro' narrow chinks of his cavern.

A MEMORABLE FANCY

I was in a Printing house in Hell, & saw the method in which knowledge is transmitted from generation to generation.

In the first chamber was a Dragon-Man, clearing away the rubbish from a cave's mouth; within, a number of Dragons were hollowing the cave.[19]

In the second chamber was a Viper folding round the rock & the cave, and others adorning it with gold, silver and precious stones.

In the third chamber was an Eagle with wings and feathers of air: he caused the inside of the cave to be infinite; around were numbers of Eagle-like men who built palaces in the immense cliffs.

In the fourth chamber were Lions of flaming fire, raging around & melting the metals into living fluids.

In the fifth chamber were Unnam'd forms, which cast the metals into the expanse.

There they were reciev'd by Men who occupied the sixth chamber, and took the forms of books & were arranged in libraries.

The Giants who formed this world into its sensual existence, and now seem to live in it in chains, are in truth the causes of its life & the sources of all activity; but the chains are the cunning of weak and tame minds which have power to resist energy; according to the proverb, the weak in courage is strong in cunning.

Thus one portion of being is the Prolific, the other the Devouring: to the Devourer it seems as if the producer was in his chains; but it is not so, he only takes portions of existence and fancies that the whole.

But the Prolific would cease to be Prolific unless the Devourer, as a sea, received the excess of his delights.

Some will say: "Is not God alone the Prolific?" I answer: "God only Acts & Is, in existing beings or Men."

These two classes of men are always upon earth, &

they should be enemies: whoever tries to reconcile them seeks to destroy existence.

Religion is an endeavour to reconcile the two.

Note: Jesus Christ did not wish to unite, but to seperate them, as in the Parable of sheep and goats! & he says: "I came not to send Peace, but a Sword."[20]

Messiah or Satan or Tempter was formerly thought to be one of the Antediluvians who are our Energies.

A MEMORABLE FANCY

An Angel came to me and said: "O pitiable foolish young man! O horrible! O dreadful state! consider the hot burning dungeon thou art preparing for thyself to all eternity, to which thou art going in such career."

I said: "Perhaps you will be willing to shew me my eternal lot, & we will contemplate together upon it, and see whether your lot or mine is most desirable."

So he took me thro' a stable & thro' a church & down into the church vault, at the end of which was a mill: thro' the mill we went, and came to a cave: down the winding cavern we groped our tedious way, till a void boundless as a nether sky appear'd beneath us, & we held by the roots of trees and hung over this immensity; but I said: "if you please, we will commit ourselves to this void, and see whether providence is here also: if you will not, I will": but he answer'd: "do not presume, O young man, but as we here remain, behold thy lot which will soon appear when the darkness passes away."

So I remain'd with him, sitting in the twisted root of an oak; he was suspended in a fungus, which hung with the head downward into the deep.

By degrees we beheld the infinite Abyss, fiery as the smoke of a burning city; beneath us, at an immense distance, was the sun, black but shining; round it were fiery tracks on which revolv'd vast spiders, crawling after their prey, which flew, or rather swum, in the infinite deep, in the most terrific shapes of animals sprung from corruption; & the air was full of them, & seem'd composed of them: these are Devils, and are called Powers of the air. I now asked my companion which was my eternal lot? he said: "between the black & white spiders."

But now, from between the black & white spiders, a cloud and fire burst and rolled thro' the deep, black'ning all beneath, so that the nether deep grew black as a sea, & rolled with a terrible noise; beneath us was

---

19. **In . . . cave:** the "improvement of sensual enjoyment" mentioned above.

20. **Parable . . . Sword:** Matthew 10:34 and 25:33.

nothing now to be seen but a black tempest, till looking east between the clouds & the waves, we saw a cataract of blood mixed with fire, and not many stones' throw from us appear'd and sunk again the scaly fold of a monstrous serpent; at last, to the east, distant about three degrees, appear'd a fiery crest above the waves; slowly it reared like a ridge of golden rocks, till we discover'd two globes of crimson fire, from which the sea fled away in clouds of smoke; and now we saw it was the head of Leviathan; his forehead was divided into streaks of green & purple like those on a tyger's forehead: soon we saw his mouth & red gills hang just above the raging foam, tinging the black deep with beams of blood, advancing toward us with all the fury of a spiritual existence.

My friend the Angel climb'd up from his station into the mill: I remain'd alone; & then this appearance was no more, but I found myself sitting on a pleasant bank beside a river by moonlight, hearing a harper, who sung to the harp; & his theme was: "The man who never alters his opinion is like standing water, & breeds reptiles of the mind."

But I arose and sought for the mill, & there I found my Angel, who, surprised, asked me how I escaped?

I answer'd: "All that we saw was owing to your metaphysics; for when you ran away, I found myself on a bank by moonlight hearing a harper. But now we have seen my eternal lot, shall I shew you yours?" he laugh'd at my proposal; but I by force suddenly caught him in my arms, & flew westerly thro' the night, till we were elevated above the earth's shadow; then I flung myself with him directly into the body of the sun; here I clothed myself in white, & taking in my hand Swedenborg's volumes, sunk from the glorious clime, and passed all the planets till we came to saturn: here I stay'd to rest, & then leap'd into the void between saturn & the fixed stars.

"Here," said I, "is your lot, in this space—if space it may be call'd." Soon we saw the stable and the church, & I took him to the altar and open'd the Bible, and lo! it was a deep pit, into which I descended, driving the Angel before me; soon we saw seven houses of brick;[21] one we enter'd; in it were a number of monkeys, baboons, & all of that species, chain'd by the middle, grinning and snatching at one another, but withheld by the shortness of their chains: however, I saw that they sometimes grew numerous, and then the weak were caught by the strong, and with a grinning aspect, first coupled with, & then devour'd, by plucking off first one limb and then another, till the body was left a helpless trunk; this, after grinning & kissing it with seeming fondness, they devour'd too; and here & there I saw one savourily picking the flesh off his own tail; as the stench terribly annoy'd us both, we went into the mill, & I in my hand brought the skeleton of a body, which in the mill was Aristotle's Analytics.

So the Angel said: "thy phantasy has imposed upon me, & thou oughtest to be ashamed."

I answer'd: "we impose on one another, & it is but lost time to converse with you whose works are only Analytics."

Opposition is true Friendship.

I have always found that Angels have the vanity to speak of themselves as the only wise; this they do with a confident insolence sprouting from systematic reasoning.

Thus Swedenborg boasts that what he writes is new: tho' it is only the Contents or Index of already publish'd books.

A man carried a monkey about for a shew, & because he was a little wiser than the monkey, grew vain, and conciev'd himself as much wiser than seven men. It is so with Swedenborg: he shews the folly of churches, & exposes hypocrites till he imagines that all are religious, & himself the single one on earth that ever broke a net.

Now hear a plain fact: Swedenborg has not written one new truth. Now hear another: he has written all the old falsehoods.

And now hear the reason. He conversed with Angels who are all religious, & conversed not with Devils who all hate religion, for he was incapable thro' his conceited notions.

Thus Swedenborg's writings are a recapitulation of all superficial opinions, and an analysis of the more sublime—but no further.

Have now another plain fact. Any man of mechanical talents may, from the writings of Paracelsus or Jacob Behmen,[22] produce ten thousand volumes of equal value with Swedenborg's, and from those of Dante or Shakespear an infinite number.

21. seven . . . brick: the seven churches in Asia addressed by St. John the Divine (Revelation 1). What follows is an image of the Angel's "eternal lot"—theological quarrel within one of the churches built in the "deep pit" of the Bible.

22. Paracelsus . . . Behmen: Paracelsus, c. 1490–1541, was a German physician, alchemist, and Neoplatonist; by "Behmen" is intended Jakob Böhme, 1575–1624, a mystical philosopher and theologian.

But when he has done this, let him not say that he knows better than his master, for he only holds a candle in sunshine.

A MEMORABLE FANCY

Once I saw a Devil in a flame of fire, who arose before an Angel that sat on a cloud, and the Devil utter'd these words:

"The worship of God is: Honouring his gifts in other men, each according to his genius, and loving the greatest men best: those who envy or calumniate great men hate God; for there is no other God."

The Angel hearing this became almost blue; but mastering himself he grew yellow, & at last white, pink, & smiling, and then replied:

"Thou Idolater! is not God One? & is not he visible in Jesus Christ? and has not Jesus Christ given his sanction to the law of ten commandments? and are not all other men fools, sinners, & nothings?"

The Devil answer'd: "bray a fool in a morter with wheat, yet shall not his folly be beaten out of him; if Jesus Christ is the greatest man, you ought to love him in the greatest degree; now hear how he has given his sanction to the law of ten commandments: did he not mock at the sabbath and so mock the sabbath's God? murder those who were murder'd because of him? turn away the law from the woman taken in adultery? steal the labor of others to support him? bear false witness when he omitted making a defence before Pilate? covet when he pray'd for his disciples, and when he bid them shake off the dust of their feet against such as refused to lodge them? I tell you, no virtue can exist without breaking these ten commandments. Jesus was all virtue, and acted from impulse, not from rules."

When he had so spoken, I beheld the Angel, who stretched out his arms, embracing the flame of fire, & he was consumed and arose as Elijah.

Note: This Angel, who is now become a Devil, is my particular friend; we often read the Bible together in its infernal or diabolical sense, which the world shall have if they behave well.

I have also The Bible of Hell, which the world shall have whether they will or no.

One Law for the Lion & Ox is Oppression.[23]

1790–93 (1790–93)

23. One . . . Oppression: the lion compelled to eat grass!

A SONG OF LIBERTY

Engraved in 1792, this poem was usually bound with *The Marriage of Heaven and Hell* as a coda. The verses were probably written as captions for illustrations.

1. The Eternal Female groan'd![1] it was heard over all the Earth.
2. Albion's[2] coast is sick, silent; the American meadows faint!
3. Shadows of Prophecy shiver along by the lakes and the rivers, and mutter across the ocean: France, rend down thy dungeon!
4. Golden Spain, burst the barriers of old Rome!
5. Cast thy keys, O Rome, into the deep down falling, even to eternity down falling,
6. And weep.
7. In her trembling hand she took the new born terror, howling.
8. On those infinite mountains of light, now barr'd out by the atlantic sea,[3] the new born fire stood before the starry king!
9. Flag'd with grey brow'd snows and thunderous visages, the jealous wings wav'd over the deep.
10. The speary hand burned aloft, unbuckled was the shield; forth went the hand of jealousy among the flaming hair, and hurl'd the new born wonder thro' the starry night.
11. The fire, the fire is falling!
12. Look up! look up! O citizen of London, enlarge thy countenance! O Jew, leave counting gold! return to thy oil and wine. O African! black African! (go, winged thought, widen his forehead.)
13. The fiery limbs, the flaming hair, shot like the sinking sun into the western sea.
14. Wak'd from his eternal sleep, the hoary element roaring fled away.
15. Down rush'd, beating his wings in vain, the jealous king; his grey brow'd councellors, thunderous warriors, curl'd veterans, among helms, and shields, and chariots, horses, elephants, banners, castles, slings, and rocks.

A SONG OF LIBERTY. 1. **The Eternal . . . groan'd:** in childbirth. 2. **Albion:** England. 3. **infinite . . . sea:** the now submerged continent of Atlantis.

16. Falling, rushing, ruining! buried in the ruins, on Urthona's dens; [4]

17. All night beneath the ruins; then, their sullen flames faded, emerge round the gloomy king.

18. With thunder and fire, leading his starry hosts thro' the waste wilderness, he promulgates his ten commands, glancing his beamy eyelids over the deep in dark dismay,

19. Where the son of fire in his eastern cloud, while the morning plumes her golden breast,

20. Spurning the clouds written with curses, stamps the stony law to dust, loosing the eternal horses from the dens of night, crying:

EMPIRE IS NO MORE! AND NOW THE LION & WOLF SHALL CEASE.

Chorus

Let the Priests of the Raven of dawn no longer, in deadly black, with hoarse note curse the sons of joy. Nor his accepted brethren—whom, tyrant, he calls free—lay the bound or build the roof. Nor pale religious letchery call that virginity that wishes but acts not!

For every thing that lives is Holy.

(*1792*)

AMERICA: A PROPHECY

Blake eagerly sympathized with the American Revolution. More than that, he later felt it had been a major fact in his own spiritual life as well as in world history. Composed and engraved in 1793, his poetic rendering of it is more immediately understandable than many of his longer works, primarily because it refers to familiar history. But it is one of the "prophetic books"—in other words, it sets forth spiritual truths in visionary form— and history is caught up into myth. The protagonists are George III and the colonial patriots, but they are at the same time Urizen and Orc (who are also Heaven and Hell as Blake continues the ironic vocabulary of *The Marriage of Heaven and Hell*). Some lines from a verse letter to John Flaxman, September 12, 1800, may be used as a brief summary of the action:

terrors appear'd in the Heavens above
And in Hell beneath, & a mighty & awful change
threatened the Earth.

4. Urthona's dens: In Blake's later poems Urthona becomes a symbol of creative or imaginative power. His "dens" would be where he is now confined. His name may be translated "Earth-owner." When he has power the earth is the primal paradise of Eden, and when he falls the earth falls also.

The American War began. All its dark horrors passed before my face
Across the Atlantic to France. Then the French Revolution commenc'd in thick clouds.

PRELUDIUM

The shadowy Daughter of Urthona stood before red Orc,
When fourteen suns had faintly journey'd o'er his dark abode:
His food she brought in iron baskets, his drink in cups of iron:
Crown'd with a helmet & dark hair the nameless female stood;
A quiver with its burning stores, a bow like that of night,
When pestilence is shot from heaven: no other arms she need!
Invulnerable tho' naked, save where clouds roll round her loins
Their awful folds in the dark air: silent she stood as night;
For never from her iron tongue could voice or sound arise,
But dumb till that dread day when Orc assay'd his fierce embrace.   10

"Dark Virgin," said the hairy youth, "thy father stern, abhorr'd,
Rivets my tenfold chains while still on high my spirit soars;
Sometimes an eagle screaming in the sky, sometimes a lion
Stalking upon the mountains, & sometimes a whale, I lash
The raging fathomless abyss; anon a serpent folding

AMERICA. PRELUDIUM. **1. Daughter of Urthona:** See *A Song of Liberty*, above, vs. 16. Urthona fallen keeps Orc (energy, libido, revolt; the name is from Latin *orcus*, hell) in chains. Even so Urthona can still create, but the world he makes is primitive and incomplete. Virtually every female figure in Blake symbolizes a form of Nature or, more precisely, a possible relation of Nature to mankind. All possible relations are products of man's mind. **13–14: eagle . . . whale:** Orc is fire; he has already escaped from his prison and is manifesting himself in the other three elements: air, earth, and water. These beasts are also symbols historically associated with revolution in the new world.

Around the pillars of Urthona, and round thy dark
limbs
On the Canadian wilds I fold; feeble my spirit folds,
For chain'd beneath I rend these caverns: when thou
bringest food
I howl my joy, and my red eyes seek to behold thy
face—
In vain! these clouds roll to & fro, & hide thee from
my sight."   20

Silent as despairing love, and strong as jealousy,
The hairy shoulders rend the links; free are the wrists
of fire;
Round the terrific loins he siez'd the panting, struggling womb;
It joy'd: she put aside her clouds & smiled her first-
born smile,
As when a black cloud shews its lightnings to the
silent deep.

Soon as she saw the terrible boy, then burst the virgin
cry:

"I know thee, I have found thee, & I will not let thee
go:
Thou art the image of God who dwells in darkness of
Africa,
And thou art fall'n to give me life in regions of dark
death.
On my American plains I feel the struggling afflic-
tions   30
Endur'd by roots that writhe their arms into the
nether deep.
I see a Serpent in Canada who courts me to his love,
In Mexico an Eagle, and a Lion in Peru;
I see a Whale in the South-sea, drinking my soul away.
O what limb rending pains I feel! thy fire & my frost
Mingle in howling pains, in furrows by thy lightnings
rent.
This is eternal death, and this the torment long fore-
told."

A PROPHECY

The Guardian Prince of Albion burns in his nightly
tent:
Sullen fires across the Atlantic glow to America's shore,

Piercing the souls of warlike men who rise in silent
night.
Washington, Franklin, Paine & Warren, Gates, Hancock & Green
Meet on the coast glowing with blood from Albion's
fiery Prince.

Washington spoke: "Friends of America! look over
the Atlantic sea;
A bended bow is lifted in heaven, & a heavy iron chain
Descends, link by link, from Albion's cliffs across the
sea, to bind
Brothers & sons of America till our faces pale and
yellow,
Heads deprest, voices weak, eyes downcast, hands
work-bruis'd,   10
Feet bleeding on the sultry sands, and the furrows of
the whip
Descend to generations that in future times forget."

The strong voice ceas'd, for a terrible blast swept over
the heaving sea:
The eastern cloud rent: on his cliffs stood Albion's
wrathful Prince,
A dragon form, clashing his scales: at midnight he
arose,
And flam'd red meteors round the land of Albion
beneath;
His voice, his locks, his awful shoulders, and his glowing eyes
Appear to the Americans upon the cloudy night.
Solemn heave the Atlantic waves between the gloomy
nations,
Swelling, belching from its deeps red clouds & raging
fires.   20
Albion is sick! America faints! enrag'd the Zenith
grew.
As human blood shooting its veins all round the orbed
heaven,
Red rose the clouds from the Atlantic in vast wheels
of blood,
And in the red clouds rose a Wonder o'er the Atlantic
sea,

28-34. **Africa . . . South-sea:** all places where men were
struggling for freedom. A PROPHECY. 1. **Guardian Prince:**
George III, also Albion's (England's) Angel.

4. **Paine . . . Green:** Thomas Paine, 1737-1809, author
of the *Rights of Man;* Joseph Warren, 1741-75, American
general killed at Bunker Hill; Horatio Gates, 1728-1806,
American general; John Hancock, 1737-93, statesman, signer
of the Declaration of Independence; Nathaniel Greene,
1742-86, American general. 24. **Wonder:** Orc, the spirit of
revolt.

Intense! naked! a Human fire, fierce glowing, as the
    wedge
Of iron heated in the furnace: his terrible limbs were
    fire
With myriads of cloudy terrors, banners dark &
    towers
Surrounded: heat but not light went thro' the murky
    atmosphere.

The King of England looking westward trembles at
    the vision.

Albion's Angel stood beside the Stone of night, and
    saw   30
The terror like a comet, or more like the planet red
That once enclos'd the terrible wandering comets in
    its sphere.
Then, Mars, thou wast our center, & the planets three
    flew round
Thy crimson disk: so e'er the Sun was rent from thy
    red sphere.
The Spectre glow'd his horrid length staining the
    temple long
With beams of blood; & thus a voice came forth, and
    shook the temple:

"The morning comes, the night decays, the watchmen
    leave their stations;
The grave is burst, the spices shed, the linen wrapped
    up;
The bones of death, the cov'ring clay, the sinews
    shrunk & dry'd
Reviving shake, inspiring move, breathing, awaken-
    ing,   40
Spring like redeemed captives when their bonds &
    bars are burst.
Let the slave grinding at the mill run out into the field,
Let him look up into the heavens & laugh in the bright
    air;
Let the inchained soul, shut up in darkness and in
    sighing,
Whose face has never seen a smile in thirty weary
    years,
Rise and look out; his chains are loose, his dungeon
    doors are open;

And let his wife and children return from the oppres-
    sor's scourge.
They look behind at every step & believe it is a dream,
Singing: 'The Sun has left his blackness & has found
    a fresher morning,
And the fair Moon rejoices in the clear & cloudless
    night;   50
For Empire is no more, and now the Lion & Wolf
    shall cease.' "

In thunders ends the voice. Then Albion's Angel
    wrathful burnt
Beside the Stone of Night, and like the Eternal Lion's
    howl
In famine & war, reply'd: "Art thou not Orc, who
    serpent-form'd
Stands at the gate of Enitharmon to devour her
    children?
Blasphemous Demon, Antichrist, hater of Dignities,
Lover of wild rebellion, and transgressor of God's
    Law,
Why dost thou come to Angel's eyes in this terrific
    form?"

The Terror answer'd: "I am Orc, wreath'd round the
    accursed tree:
The times are ended; shadows pass, the morning
    'gins to break;   60
The fiery joy, that Urizen perverted to ten commands,
What night he led the starry hosts thro' the wide
    wilderness,
That stony law I stamp to dust; and scatter religion
    abroad
To the four winds as a torn book, & none shall gather
    the leaves;
But they shall rot on desert sands, & consume in bot-
    tomless deeps,

---

55. Enitharmon: woman, Nature. **61. Urizen:** An important
figure in Blake's mythology, Urizen represents generally what,
in Blake's opinion, was the orthodox idea of the Divine Creator, the God of the universe of mathematical materialism conceived by Newton and Locke. As this God is an abstraction
he is associated with things that are wintry, skeletal, dark,
vague, and cloudy. He is the "starry king" of the *Song of
Liberty*, for the stars are symbols of mathematical order. At
times he is identified with the Jehovah of the Old Testament
promulgating the Ten Commandments from the darkness
of Mount Sinai. He is a tyrant and the God of tyrants. His
name has been variously derived from "*Your* reason,"
"horizon," and "Urian Zeus."

---

28. heat . . . light: Cf. *Paradise Lost*, I. 62–63. **30. Stone
of night:** the stone tablets on which the Ten Commandments
were written. **35. Spectre:** Albion's Angel. **36. a voice:**
Orc's.

To make the desarts blossom, & the deeps shrink to
    their fountains,
And to renew the fiery joy, and burst the stony roof;
That pale religious letchery, seeking Virginity,
May find it in a harlot, and in coarse-clad honesty
The undefil'd, tho' ravish'd in her cradle night and
    morn;   70
For everything that lives is holy, life delights in life;
Because the soul of sweet delight can never be defil'd.
Fires inwrap the earthly globe, yet man is not con-
    sum'd;
Amidst the lustful fires he walks; his feet become like
    brass,
His knees and thighs like silver, & his breast and head
    like gold."

"Sound! sound! my loud war-trumpets, & alarm my
    Thirteen Angels!
Loud howls the eternal Wolf! the eternal Lion lashes
    his tail!
America is darken'd; and my punishing Demons,
    terrified,
Crouch howling before their caverns deep, like skins
    dry'd in the wind.
They cannot smite the wheat, nor quench the fatness
    of the earth;   80
They cannot smite with sorrows, nor subdue the plow
    and spade;
They cannot wall the city, nor moat round the castle
    of princes;
They cannot bring the stubbed oak to overgrow the
    hills;
For terrible men stand on the shores, & in their robes
    I see
Children take shelter from the lightnings: there stands
    Washington
And Paine and Warren with their foreheads rear'd
    toward the east.
But clouds obscure my aged sight. A vision from afar!
Sound! sound! my loud war-trumpets, & alarm my
    thirteen Angels!
Ah vision from afar! Ah rebel form that rent the
    ancient
Heavens! Eternal Viper, self-renew'd, rolling in
    clouds,   90

I see thee in thick clouds and darkness on America's
    shore,
Writhing in pangs of abhorred birth; red flames the
    crest rebellious
And eyes of death; the harlot womb, oft opened in
    vain,
Heaves in enormous circles: now the times are return'd
    upon thee,
Devourer of thy parent, now thy unutterable torment
    renews.
Sound! sound! my loud war-trumpets, & alarm my
    thirteen Angels!
Ah terrible birth! a young one bursting! where is the
    weeping mouth,
And where the mother's milk? instead, those ever-
    hissing jaws
And parched lips drop with fresh gore: now roll thou
    in the clouds;
Thy mother lays her length outstretch'd upon the
    shore beneath.   100
Sound! sound! my loud war-trumpets, & alarm my
    thirteen Angels!
Loud howls the eternal Wolf! the eternal Lion lashes
    his tail!"

Thus wept the Angel voice, & as he wept, the terrible
    blasts
Of trumpets blew a loud alarm across the Atlantic
    deep.
No trumpets answer; no reply of clarions or of fifes:
Silent the Colonies remain and refuse the loud alarm.

On those vast shady hills between America & Albion's
    shore,
Now barr'd out by the Atlantic sea, call'd Atlantean
    hills,
Because from their bright summits you may pass to
    the Golden world,
An ancient palace, archetype of mighty Emperies, 110
Rears its immortal pinnacles, built in the forest of God
By Ariston, the king of beauty, for his stolen bride.

76. Sound . . . trumpets: The Angel of Albion speaks. His thirteen Angels are the thirteen states or the angelic powers in them.

108. Atlantean hills: The mythical, drowned continent of Atlantis, a utopia according to Sir Francis Bacon. Had it not sunk beneath the ocean it would unite England and America. Thus the ocean, together with other forms of moisture such as clouds and rain, protects against the fire of revolution. Cf. ll. 174–75, 207–16. **112. Ariston:** in Herodotus, VI. 61–66, a king of Sparta who steals the bride of his friend. His name means "best."

Here on their magic seats the thirteen Angels sat
    perturb'd,
For clouds from the Atlantic hover o'er the solemn
    roof.

Fiery the Angels rose, & as they rose deep thunder
    roll'd
Around their shores, indignant burning with the fires
    of Orc;
And Boston's Angel cried aloud as they flew thro' the
    dark night.

He cried: "Why trembles honesty, and like a murderer
Why seeks he refuge from the frowns of his immortal
    station?
Must the generous tremble & leave his joy to the idle,
    to the pestilence,   120
That mock him? who commanded this? what God?
    what Angel?
To keep the gen'rous from experience till the un-
    generous
Are unrestrain'd performers of the energies of nature;
Till pity is become a trade, and generosity a science
That men get rich by; & the sandy desert is giv'n to
    the strong?
What God is he writes laws of peace & clothes him in
    a tempest?
What pitying Angel lusts for tears and fans himself
    with sighs?
What crawling villain preaches abstinence & wraps
    himself
In fat of lambs? no more I follow, no more obedience
    pay!"

So cried he, rending off his robe & throwing down his
    scepter   130
In sight of Albion's Guardian; and all the thirteen
    Angels
Rent off their robes to the hungry wind, & threw
    their golden scepters
Down on the land of America; indignant they de-
    scended
Headlong from out their heav'nly heights, descending
    swift as fires
Over the land; naked & flaming are their lineaments
    seen
In the deep gloom; by Washington & Paine & Warren
    they stood;
And the flame folded, roaring fierce within the pitchy
    night

Before the Demon red, who burnt towards America,
In black smoke, thunders, and loud winds, rejoicing
    in its terror,
Breaking in smoky wreaths from the wild deep, &
    gath'ring thick   140
In flames as of a furnace on the land from North to
    South,
What time the thirteen Governors, that England sent,
    convene
In Bernard's house. The flames cover'd the land; they
    rouze; they cry;
Shaking their mental chains, they rush in fury to the
    sea
To quench their anguish; at the feet of Washington
    down fall'n
They grovel on the sand and writhing lie, while all
The British soldiers thro' the thirteen states sent up a
    howl
Of anguish, threw their swords & muskets to the earth,
    & ran
From their encampments and dark castles, seeking
    where to hide
From the grim flames, and from the visions of Orc, in
    sight   150
Of Albion's Angel; who, enrag'd, his secret clouds
    open'd
From north to south and burnt outstretch'd on wings
    of wrath, cov'ring
The eastern sky, spreading his awful wings across the
    heavens.
Beneath him roll'd his num'rous hosts, all Albion's
    Angels camp'd
Darken'd the Atlantic mountains; & their trumpets
    shook the valleys,
Arm'd with diseases of the earth to cast upon the
    Abyss,
Their numbers forty millions, must'ring in the eastern
    sky.

In the flames stood & view'd the armies drawn out in
    the sky
Washington, Franklin, Paine, & Warren, Allen, Gates,
    & Lee,
And heard the voice of Albion's Angel give the thunder-
    ous command;   160

---

143. Bernard: Sir Francis Bernard, governor of Massachusetts from 1760–69, detested by the Americans. Though he
had been recalled before the Revolution, Blake adopts him as
a symbol.

His plagues, obedient to his voice, flew forth out of
their clouds,
Falling upon America, as a storm to cut them off,
As a blight cuts the tender corn when it begins to
appear.
Dark is the heaven above, & cold & hard the earth
beneath:
And as a plague wind fill'd with insects cuts off man &
beast,
And as a sea o'erwhelms a land in the day of an earth-
quake,
Fury! rage! madness! in a wind swept through
America;
And the red flames of Orc, that folded roaring, fierce,
around
The angry shores; and the fierce rushing of th' in-
habitants together!
The citizens of New York close their books & lock
their chests;   170
The mariners of Boston drop their anchors and unlade;
The scribe of Pensylvania casts his pen upon the earth;
The builder of Virginia throws his hammer down in
fear.

Then had America been lost, o'erwhelm'd by the
Atlantic,
And Earth had lost another portion of the infinite,
But all rush together in the night in wrath and raging
fire.
The red fires rag'd! the plagues recoil'd! then roll'd
they back with fury
On Albion's Angels: then the Pestilence began in
streaks of red
Across the limbs of Albion's Guardian; the spotted
plague smote Bristol's
And the Leprosy London's Spirit, sickening all their
bands: 180
The millions sent up a howl of anguish and threw off
their hammer'd mail,
And cast their swords & spears to earth, & stood, a
naked multitude:
Albion's Guardian writhed in torment on the eastern
sky,
Pale, quiv'ring toward the brain his glimmering eyes,
teeth chattering,

Howling & shuddering, his legs quivering, convuls'd
each muscle & sinew:
Sick'ning lay London's Guardian, and the ancient
miterd York,
Their heads on snowy hills, their ensigns sick'ning in
the sky.
The plagues creep on the burning winds driven by
flames of Orc,
And by the fierce Americans rushing together in the
night,
Driven o'er the Guardians of Ireland, and Scotland and
Wales.   190
They, spotted with plagues, forsook the frontiers; &
their banners, sear'd
With fires of hell, deform their ancient heavens with
shame & woe.
Hid in his caves the Bard of Albion felt the enormous
plagues,
And a cowl of flesh grew o'er his head, & scales on his
back & ribs;
And, rough with black scales, all his Angels fright their
ancient heavens.
The doors of marriage are open, and the Priests in
rustling scales
Rush into reptile coverts, hiding from the fires of Orc,
That play around the golden roofs in wreaths of fierce
desire,
Leaving the females naked and glowing with the lusts
of youth.

For the female spirits of the dead, pining in bonds of
religion,   200
Run from their fetters reddening, & in long drawn
arches sitting,
They feel the nerves of youth renew, and desires of
ancient times
Over their pale limbs, as a vine when the tender grape
appears.

Over the hills, the vales, the cities, rage the red flames
fierce:
The Heavens melted from north to south; and Urizen,
who sat
Above all heavens, in thunders wrap'd, emerg'd his
leprous head
From out his holy shrine, his tears in deluge piteous

---

175. another portion: another Atlantis. 177. the plagues
recoil'd: The plagues Albion's Angel had sent against
America recoil upon him as the plague he fears most, the
spirit of revolutionary freedom.

186. ancient miterd York: The Bishop of York holds the
second oldest episcopal see in England. 193. Bard of Albion:
the poet laureate, at this time William Whitehead, 1715–85.

Falling into the deep sublime; flag'd with grey-brow'd
    snows
And thunderous visages, his jealous wings wav'd over
    the deep;
Weeping in dismal howling woe, he dark descended,
    howling   210
Around the smitten bands, clothed in tears & trembling,
    shudd'ring cold.
His stored snows he poured forth, and his icy magazines
He open'd on the deep, and on the Atlantic sea white
    shiv'ring
Leprous his limbs, all over white, and hoary was his
    visage,
Weeping in dismal howlings before the stern Ameri-
    cans,
Hiding the Demon red with clouds & cold mists from
    the earth;
Till Angels & weak men twelve years should govern
    o'er the strong;
And then their end should come, when France receiv'd
    the Demon's light.

Stiff shudderings shook the heav'nly thrones! France,
    Spain, & Italy
In terror view'd the bands of Albion, and the ancient
    Guardians,   220
Fainting upon the elements, smitten with their own
    plagues.
They slow advance to shut the five gates of their law-
    built heaven,
Filled with blasting fancies and with mildews of
    despair,
With fierce disease and lust, unable to stem the fires of
    Orc.
But the five gates were consum'd, & their bolts and
    hinges melted;
And the fierce flames burnt round the heavens & round
    the abodes of men.

                                        1793 (1793)

THE BOOK OF URIZEN

The Book of Urizen is a myth of the fall of man bringing
about the creation of the universe, at least of the universe
as we now perceive it. It was engraved in 1794.

The cosmos before the fall is imagined by Blake as a
vast man, a being not closed up in five senses and possessed
of eternal life. The persons of his mythology are most
easily explained as powers or faculties, not originally
divided or distinguished, of this man; his reason is or
becomes Urizen, imagination Los, libido Orc. What
happens to him is the story of the cosmos. At the same
time these persons are faculties of any man; both the fall
and the Last Judgment may take place in each of us.

The fall occurs as an act of separation. Urizen withdraws himself from the unified life of the whole, and
every further event in the gradual creation of the world
we live in repeats this original act. Separation creates
chaos, not only in Urizen but in the original man. Seeking
to restore unity, the fallen Urizen proclaims that all must
conform to one law, obey one supreme ruler, and this
attempted usurpation immediately causes a further
separation as the other Powers refuse and depart. In
the strife that follows Urizen succeeds in protectively
enclosing himself. But he is still a chaos, summoning, as
were, the imaginative power of Los. Los can work only
by binding the flux of Urizen into steadily more definite
form, and thus also confining and limiting him (specifically within the realm of time and the prison of the body).
But as all this is taking place within one (self-divided)
man, it is his own imaginative energy with which he sets
bounds to himself, and in doing so he enfeebles imagination, which therefore becomes pent up with Urizen and
sundered from Eternity. Thereafter Los himself is
divided, his feelings of pity manifesting themselves
separately in Enitharmon, on whom he begets Orc.
From one point of view the birth of Orc is a further
moment in the fall, imagination declining into mere
libido, but at the same time (as *America* has made clear)
Orc is the possibility of regeneration, and the fallen
imagination attempts to chain him. His birth brings
Urizen into activity and causes him to divide and limit
his world still further. In doing so, Urizen produces
existence as we find it, with its division, strife, and suffering. Looking about him, Urizen thinks he pities his own
creatures and, ever in character, automatically attempts
once more to impose unity and peace by restriction and
confinement, the net of religion. In a universe thus
persistently on the wrong track, Orc is the only hope,
though at the end of the poem Blake chooses to call him
Fuzon.

PRELUDIUM TO THE FIRST BOOK OF URIZEN

Of the primeval Priest's assum'd power,
When Eternals spurn'd back his religion
And gave him a place in the north,
Obscure, shadowy, void, solitary.

Eternals! I hear your call gladly.
Dictate swift winged words & fear not
To unfold your dark visions of torment.

Chap: I

1. Lo, a shadow of horror is risen
In Eternity! Unknown, unprolific,
Self-clos'd, all-repelling: what Demon
Hath form'd this abominable void,
This soul-shudd'ring vacuum? Some said
"It is Urizen." But unknown, abstracted,
Brooding, secret, the dark power hid.

2. Times on times he divided & measur'd
Space by space in his ninefold darkness,
Unseen, unknown; changes appear'd   10
Like desolate mountains, rifted furious
By the black winds of perturbation.

3. For he strove in battles dire,
In unseen conflictions with shapes
Bred from his forsaken wilderness
Of beast, bird, fish, serpent & element,
Combustion, blast, vapour and cloud.

4. Dark, revolving in silent activity:
Unseen in tormenting passions:
An activity unknown and horrible,
A self-contemplating shadow,   20
In enormous labours occupied.

5. But Eternals beheld his vast forests;
Age on ages he lay, clos'd, unknown,
Brooding shut in the deep; all avoid
The petrific, abominable chaos.

6. His cold horrors silent, dark Urizen
Prepar'd; his ten thousands of thunders,
Rang'd in gloom'd array, stretch out across
The dread world; & the rolling of wheels,   30
As of swelling seas, sound in his clouds,

In his hills of stor'd snows, in his mountains
Of hail & ice; voices of terror
Are heard, like thunders of autumn
When the cloud blazes over the harvests.

Chap: II

1. Earth was not: nor globes of attraction;
The will of the Immortal expanded
Or contracted his all flexible senses;
Death was not, but eternal life sprung.

2. The sound of a trumpet the heavens   40
Awoke, & vast clouds of blood roll'd
Round the dim rocks of Urizen, so nam'd
That solitary one in Immensity.

3. Shrill the trumpet: & myriads of Eternity
Muster around the bleak desarts,
Now fill'd with clouds, darkness, & waters,
That roll'd perplex'd, lab'ring; & utter'd
Words articulate bursting in thunders
That roll'd on the tops of his mountains:

4. "From the depths of dark solitude, From   50
The eternal abode in my holiness,
Hidden, set apart, in my stern counsels,
Reserv'd for the days of futurity,
I have sought for a joy without pain,
For a solid without fluctuation.
Why will you die, O Eternals?
Why live in unquenchable burnings?

5. "First I fought with the fire, consum'd
Inwards into a deep world within:
A void immense, wild, dark & deep,   60
Where nothing was: Nature's wide womb;
And self balanc'd, stretch'd o'er the void,
I alone, even I! the winds merciless
Bound; but condensing in torrents
They fall & fall; strong I repell'd
The vast waves, & arose on the waters
A wide world of solid obstruction.

6. "Here alone I, in books form'd of metals,
Have written the secrets of wisdom,
The secrets of dark contemplation,   70
By fighting and conflicts dire
With terrible monsters Sin-bred

PRELUDIUM. **6. Dictate:** The poem is being dictated to Blake
by the Eternals.

45–93. Muster . . . soul: These lines were omitted in four
copies of the book. **61. Nature's . . . womb:** chaos.
Cf. *Paradise Lost,* II. 911.

Which the bosoms of all inhabit,
Seven deadly Sins of the soul.

7. "Lo! I unfold my darkness, and on
This rock place with strong hand the Book
Of eternal brass, written in my solitude:

8. "Laws of peace, of love, of unity,
Of pity, compassion, forgiveness;
Let each chuse one habitation,   80
His ancient infinite mansion,
One command, one joy, one desire,
One curse, one weight, one measure,
One King, one God, one Law."

Chap: III

1. The voice ended: they saw his pale visage
Emerge from the darkness, his hand
On the rock of eternity unclasping
The Book of brass. Rage siez'd the strong,

2. Rage, fury, intense indignation,
In cataracts of fire, blood, & gall,   90
In whirlwinds of sulphurous smoke,
And enormous forms of energy,
All the seven deadly sins of the soul
In living creations appear'd,
In the flames of eternal fury.

3. Sund'ring, dark'ning, thund'ring,
Rent away with a terrible crash,
Eternity roll'd wide apart,
Wide asunder rolling;
Mountainous all around   100
Departing, departing, departing,
Leaving ruinous fragments of life
Hanging, frowning cliffs & all between,
An ocean of voidness unfathomable.

4. The roaring fires ran o'er the heav'ns
In whirlwinds & cataracts of blood,
And o'er the dark desarts of Urizen
Fires pour thro' the void on all sides
On Urizen's self-begotten armies.

5. But no light from the fires: all was darkness   110
In the flames of Eternal fury.

6. In fierce anguish & quenchless flames
To the desarts and rocks he ran raging
To hide; but he could not: combining,
He dug mountains & hills in vast strength,

110. no light . . . fires: Cf. *Paradise Lost*, I. 62–63.

He piled them in incessant labour,
In howling & pangs & fierce madness,
Long periods in burning fires labouring
Till hoary, and age-broke, and aged,
In despair and the shadows of death.   120

7. And a roof vast, petrific around
On all sides he fram'd, like a womb,
Where thousands of rivers in veins
Of blood pour down the mountains to cool
The eternal fires, beating without
From Eternals; & like a black globe,
View'd by sons of Eternity standing
On the shore of the infinite ocean,
Like a human heart, strugling & beating,
The vast world of Urizen appear'd.   130

8. And Los, round the dark globe of Urizen,
Kept watch for Eternals to confine
To obscure separation alone;
For Eternity stood wide apart,
As the stars are apart from the earth.

9. Los wept, howling around the dark Demon,
And cursing his lot; for in anguish
Urizen was rent from his side,
And a fathomless void for his feet,
And intense fires for his dwelling.   140

10. But Urizen laid in a stony sleep,
Unorganiz'd, rent from Eternity.

11. The Eternals said: "What is this? Death.
Urizen is a clod of clay."

12. Los howl'd in a dismal stupor,
Groaning, gnashing, groaning,
Till the wrenching apart was healed.

13. But the wrenching of Urizen heal'd not.
Cold, featureless, flesh or clay,
Rifted with direful changes,   150
He lay in a dreamless night,

14. Till Los rouz'd his fires, affrighted
At the formless, unmeasurable death.

Chap: IV [a]

1. Los, smitten with astonishment,
Frighten'd at the hurtling bones

Chap. IV (a): probably written as an after-thought when
IV (b) had already been engraved.

2. And at the surging, sulphureous,
Perturbed Immortal, mad raging

3. In whirlwinds & pitch & nitre
Round the furious limbs of Los.

4. And Los formed nets & gins   160
And threw the nets round about.

5. He watch'd in shudd'ring fear
The dark changes, & bound every change
With rivets of iron & brass.

6. And these were the changes of Urizen:

Chap: IV [b]

1. Ages on ages roll'd over him;
In stony sleep ages roll'd over him,
Like a dark waste stretching, chang'able,
By earthquakes riv'n, belching sullen fires:
On ages roll'd ages in ghastly   170
Sick torment; around him in whirlwinds
Of darkness the eternal Prophet howl'd,
Beating still on his rivets of iron,
Pouring sodor of iron; dividing
The horrible night into watches.

2. And Urizen (so his eternal name)
His prolific delight obscur'd more & more
In dark secresy, hiding in surgeing
Sulphureous fluid his phantasies,
The Eternal Prophet heav'd the dark bellows,   180
And turn'd restless the tongs, and the hammer
Incessant beat, forging chains new & new,
Numb'ring with links hours, days & years.

3. The Eternal mind, bounded, began to roll
Eddies of wrath ceaseless round & round,
And the sulphureous foam, surgeing thick,
Settled, a lake, bright & shining clear,
White as the snow on the mountains cold.

4. Forgetfulness, dumbness, necessity,
In chains of the mind locked up,   190
Like fetters of ice shrinking together,
Disorganiz'd, rent from Eternity,
Los beat on his fetters of iron,
And heated his furnaces, & pour'd
Iron sodor and sodor of brass.

5. Restless turn'd the Immortal inchain'd,
Heaving dolorous, anguish'd unbearable;
Till a roof, shaggy wild, inclos'd
In an orb his fountain of thought.

6. In a horrible, dreamful slumber,   200
Like the linked infernal chain,
A vast Spine writh'd in torment
Upon the winds, shooting pain'd
Ribs, like a bending cavern;
And bones of solidness froze
Over all his nerves of joy.
And a first Age passed over,
And a state of dismal woe.

7. From the caverns of his jointed Spine
Down sunk with fright a red   210
Round Globe, hot burning, deep,
Deep down into the Abyss;
Panting, Conglobing, Trembling,
Shooting out ten thousand branches
Around his solid bones.
And a second Age passed over,
And a state of dismal woe.

8. In harrowing fear rolling round,
His nervous brain shot branches
Round the branches of his heart,   **220**
On high into two little orbs,
And fixed in two little caves,
Hiding carefully from the wind,
His Eyes beheld the deep.
And a third Age passed over,
And a state of dismal woe.

9. The pangs of hope began.
In heavy pain, striving, struggling,
Two Ears in close volutions
From beneath his orbs of vision   230
Shot spiring out and petrified
As they grew. And a fourth Age passed,
And a state of dismal woe.

10. In ghastly torment sick,
Hanging upon the wind,
Two Nostrils bend down to the deep.
And a fifth Age passed over,
And a state of dismal woe.

11. In ghastly torment sick,
Within his ribs bloated round,   240

172. **Prophet:** Los, here imagined as a smith.

198. **roof:** the skull.

A craving Hungry Cavern;
Thence arose his channel'd Throat.
And, like a red flame, a Tongue
Of thirst & of hunger appear'd.
And a sixth Age passed over,
And a state of dismal woe.

12. Enraged & stifled with torment,
He threw his right Arm to the north,
His left Arm to the south
Shooting out in anguish deep,   250
And his feet stamp'd the nether Abyss
In trembling & howling & dismay.
And a seventh Age passed over,
And a state of dismal woe.

Chap: V

1. In terrors Los shrunk from his task:
His great hammer fell from his hand.
His fires beheld, and sickening
Hid their strong limbs in smoke;
For with noises, ruinous, loud,
With hurtlings & clashings & groans,   260
The Immortal endur'd his chains,
Tho' bound in a deadly sleep.

2. All the myriads of Eternity,
All the wisdom & joy of life
Roll like a sea around him,
Except what his little orbs
Of sight by degrees unfold.

3. And now his eternal life
Like a dream was obliterated.

4. Shudd'ring, the Eternal Prophet smote   270
With a stroke from his north to south region.
The bellows & hammer are silent now;
A nerveless silence his prophetic voice
Siez'd; a cold solitude & dark void
The Eternal Prophet & Urizen clos'd.

5. Ages on ages roll'd over them,
Cut off from life & light, frozen
Into horrible forms of deformity.
Los suffer'd his fires to decay;
Then he look'd back with anxious desire,   280
But the space, undivided by existence,
Struck horror into his soul.

6. Los wept obscur'd with mourning,
His bosom earthquak'd with sighs;

He saw Urizen deadly black
In his chains bound, & Pity began,

7. In anguish dividing & dividing,
For pity divides the soul
In pangs, eternity on eternity,
Life in cataracts pour'd down his cliffs.   290
The void shrunk the lymph into Nerves
Wand'ring wide on the bosom of night
And left a round globe of blood
Trembling upon the void.
Thus the Eternal Prophet was divided
Before the death image of Urizen;
For in changeable clouds and darkness,
In a winterly night beneath,
The Abyss of Los stretch'd immense;
And now seen, now obscur'd, to the eyes   300
Of Eternals the visions remote
Of the dark seperation appear'd:
As glasses discover Worlds
In the endless Abyss of space,
So the expanding eyes of Immortals
Beheld the dark visions of Los
And the globe of life trembling.

8. The globe of life blood trembled
Branching out into roots,
Fibrous, writhing upon the winds,   310
Fibres of blood, milk and tears,
In pangs, eternity on eternity.
At length in tears & cries imbodied,
A female form, trembling and pale,
Waves before his deathy face.

9. All Eternity shudder'd at sight
Of the first female now separate,
Pale as a cloud of snow
Waving before the face of Los.

10. Wonder, awe, fear, astonishment   320
Petrify the eternal myriads
At the first female form now separate.
They call'd her Pity, and fled.

11. "Spread a Tent with strong cutains around them.
Let cords & stakes bind in the Void,
That Eternals may no more behold them."

12. They began to weave curtains of darkness,
They erected large pillars round the Void,
With golden hooks fasten'd in the pillars;
With infinite labour the Eternals   330
A woof wove, and called it Science.

Chap: VI

1. But Los saw the Female & pitied;
He embrac'd her; she wept, she refus'd;
In perverse and cruel delight
She fled from his arms, yet he follow'd.

2. Eternity shudder'd when they saw
Man begetting his likeness
On his own divided image.

3. A time passed over: the Eternals
Began to erect the tent,   340
When Enitharmon, sick,
Felt a Worm within her Womb.

4. Yet helplessly it lay like a Worm
In the trembling womb
To be moulded into existence.

5. All day the worm lay on her bosom;
All night within her womb
The worm lay till it grew to a serpent,
With dolorous hissings & poisons
Round Enitharmon's loins folding.   350

6. Coil'd within Enitharmon's womb
The serpent grew, casting its scales;
With sharp pangs the hissings began
To change to a grating cry:
Many sorrows and dismal throes,
Many forms of fish, bird & beast
Brought forth an Infant form
Where was a worm before.

7. The Eternals their tent finished
Alarm'd with these gloomy visions,   360
When Enitharmon groaning
Produc'd a man Child to the light.

8. A shriek ran thro' Eternity,
And a paralytic stroke,
At the birth of the Human shadow.

9. Delving earth in his resistless way,
Howling, the Child with fierce flames
Issu'd from Enitharmon.

10. The Eternals closed the tent;
They beat down the stakes, the cords   370
Stretch'd for a work of eternity.
No more Los beheld Eternity.

11. In his hands he siez'd the infant,
He bathed him in springs of sorrow,
He gave him to Enitharmon.

Chap: VII

1. They named the child Orc; he grew,
Fed with milk of Enitharmon.

2. Los awoke her. O sorrow & pain!
A tight'ning girdle grew
Around his bosom. In sobbings   380
He burst the girdle in twain;
But still another girdle
Oppress'd his bosom. In sobbings
Again he burst it. Again
Another girdle succeeds.
The girdle was form'd by day,
By night was burst in twain.

3. These falling down on the rock
Into an iron Chain
In each other link by link lock'd.   390

4. They took Orc to the top of a mountain.
O how Enitharmon wept!
They chain'd his young limbs to the rock
With the Chain of Jealousy
Beneath Urizen's deathful shadow.

5. The dead heard the voice of the child
And began to awake from sleep;
All things heard the voice of the child
And began to awake to life.

6. And Urizen, craving with hunger,   400
Stung with the odours of Nature,
Explor'd his dens around.

7. He form'd a line & a plummet
To divide the Abyss beneath;
He form'd a dividing rule;

8. He formed scales to weigh,
He formed massy weights;
He formed a brazen quadrant;
He formed golden compasses,
And began to explore the Abyss;   410
And he planted a garden of fruits.

409. compasses: Cf. Proverbs 8:27; *Paradise Lost*, **VII.** 227.

9. But Los encircled Enitharmon
With fires of Prophecy
From the sight of Urizen & Orc.

10. And she bore an enormous race.

Chap: VIII

1. Urizen explor'd his dens,
Mountain, moor & wilderness,
With a globe of fire lighting his journey,
A fearful journey, annoy'd
By cruel enormities, forms   420
Of life on his forsaken mountains.

2. And his world teem'd vast enormities,
Fright'ning, faithless, fawning
Portions of life, similitudes
Of a foot, or a hand, or a head,
Or a heart, or an eye; they swam mischevous,
Dread terrors, delighting in blood.

3. Most Urizen sicken'd to see
His eternal creations appear,
Sons & daughters of sorrow on mountains   430
Weeping, wailing. First Thiriel appear'd,
Astonish'd at his own existence,
Like a man from a cloud born; & Utha,
From the waters emerging, laments;
Grodna rent the deep earth, howling
Amaz'd; his heavens immense cracks
Like the ground parch'd with heat, then Fuzon
Flam'd out, first begotten, last born;
All his Eternal sons in like manner;
His daughters from green herbs & cattle,   440
From monsters & worms of the pit.

4. He in darkness clos'd view'd all his race,
And his soul sicken'd! he curs'd
Both sons & daughters; for he saw
That no flesh nor spirit could keep
His iron laws one moment.

5. For he saw that life liv'd upon death:
The Ox in the slaughter house moans,
The Dog at the wintry door;
And he wept & he called it Pity,   450
And his tears flowed down on the winds.

431–37. Thiriel . . . Fuzon: The four elements appear.
446. iron laws: See ll. 78–79. Urizen would make these
virtues a matter of "iron law," thus inevitably making them
impossible.

6. Cold he wander'd on high, over their cities
In weeping & pain & woe;
And wherever he wander'd, in sorrows
Upon the aged heavens,
A cold shadow follow'd behind him
Like a spider's web, moist, cold & dim,
Drawing out from his sorrowing soul,
The dungeon-like heaven dividing,
Where ever the footsteps of Urizen   460
Walked over the cities in sorrow;

7. Till a Web, dark & cold, throughout all
The tormented element stretch'd
From the sorrows of Urizen's soul.
And the Web is a Female in embrio.
None could break the Web, no wings of fire,

8. So twisted the cords, & so knotted
The meshes, twisted like to the human brain.

9. And all call'd it The Net of Religion.

Chap: IX

1. Then the Inhabitants of those Cities   470
Felt their Nerves change into Marrow,
And hardening Bones began
In swift diseases and torments,
In throbbings & shootings & grindings
Thro' all the coasts; till weaken'd
The Senses inward rush'd, shrinking
Beneath the dark net of infection;

2. Till the shrunken eyes, clouded over,
Discern'd not the woven hipocrisy;
But the streaky slime in their heavens,   480
Brought together by narrowing perceptions,
Appear'd transparent air; for their eyes
Grew small like the eyes of a man,
And in reptile forms shrinking together,
Of seven feet stature they remain'd.

3. Six days they shrunk up from existence,
And on the seventh day they rested,
And they bless'd the seventh day, in sick hope,
And forgot their eternal life.

4. And their thirty cities divided   490
In form of a human heart.
No more could they rise at will

465. And . . . embrio: In most copies this line has been
erased.

In the infinite void, but bound down
To earth by their narrowing perceptions
They lived a period of years;
Then left a noisom body
To the jaws of devouring darkness.

5. And their children wept, & built
Tombs in the desolate places,
And form'd laws of prudence, and call'd them   500
The eternal laws of God.

6. And the thirty cities remain'd,
Surrounded by salt floods, now call'd
Africa: its name was then Egypt.

7. The remaining sons of Urizen
Beheld their brethren shrink together
Beneath the Net of Urizen.
Perswasion was in vain;
For the ears of the inhabitants
Were wither'd & deafen'd & cold,   510
And their eyes could not discern
Their brethren of other cities.

8. So Fuzon call'd all together
The remaining children of Urizen,
And they left the pendulous earth.
They called it Egypt, & left it.

9. And the salt Ocean rolled englob'd.

(1794)

THE BOOK OF AHANIA

∿⌘∿

Composed in 1795, this is a sequel to *The Book of Urizen*.
The two books may be regarded as Blake's Genesis and
Exodus respectively, the first chapter of "The Bible of
Hell" promised at the end of *The Marriage of Heaven and
Hell*. The action is relatively uncomplicated. Fuzon, a
variant of Orc, at first triumphs over Urizen, his passion
awaking Urizen's lust. Urizen's soul divides, and the
female Ahania (who is a representation of nature) emerges
as the object of his desire. But instead of openly rejoicing
in sexuality, Urizen, the "primal priest," calls her Sin and
banishes her, whereupon the outcast Ahania becomes the
mother of Pestilence. In his "dire Contemplations"
Urizen now produces a "lust-form'd monster" with
which he struggles, killing it and making a bow from its
dead bones and sinews. With this he hurls the rock of the
Ten Commandments, piercing the bosom of Fuzon,
who sinks down "smitten with darkness." He crucifies
Fuzon to the Tree of Mystery, a complex symbol that

516. **They . . . left it:** the Exodus.

evokes, among other things, the deliberate mystification
by which the Church keeps man in awe; the Tree of
Knowledge in Genesis, associated by Blake with a jealous
god, with prohibition, and with sexual shame; and, more
generally, the fallen world of Blake's *Songs of Experience*,
the labyrinth in which we find ourselves when we have
merely a sensory and rational awareness. At this point
the action is completed and the poem concludes with the
beautiful lament of the banished Ahania.

∿⌘∿

Chap: Ist

1. Fuzon on a chariot iron-wing'd
On spiked flames rose; his hot visage
Flam'd furious; sparkles his hair & beard
Shot down his wide bosom and shoulders.
On clouds of smoke rages his chariot
And his right hand burns red in its cloud
Moulding into a vast Globe his wrath,
As the thunder-stone is moulded.
Son of Urizen's silent burnings:

2. "Shall we worship this Demon of smoke,"   10
Said Fuzon, "this abstract non-entity,
This cloudy God seated on waters,
Now seen, now obscur'd, King of sorrow?"

3. So he spoke in a fiery flame,
On Urizen frowning indignant,
The Globe of wrath shaking on high;
Roaring with fury he threw
The howling Globe; burning it flew
Length'ning into a hungry beam. Swiftly

4. Oppos'd to the exulting flam'd beam,   20
The broad Disk of Urizen upheav'd
Across the Void many a mile.

5. It was forg'd in mills where the winter
Beats incessant: ten winters the disk
Unremitting endur'd the cold hammer.

6. But the strong arm that sent it remember'd
The sounding beam: laughing, it tore through
That beaten mass, keeping its direction,
The cold loins of Urizen dividing.

7. Dire shriek'd his invisible Lust;   30
Deep groan'd Urizen! stretching his awful hand,

THE BOOK OF AHANIA. **9. Son of Urizen:** See *The Book
of Urizen*, above, ll. 436–38. **12. cloudy . . . waters:** Cf.
Genesis 1:2.

Ahania (so name his parted soul)
He siez'd on his mountains of Jealousy.
He groan'd anguish'd, & called her Sin,
Kissing her and weeping over her;
Then hid her in darkness, in silence,
Jealous, tho' she was invisible.

8. She fell down a faint shadow wand'ring
In chaos and circling dark Urizen,
As the moon anguish'd circles the earth,   40
Hopeless! abhorr'd! a death-shadow,
Unseen, unbodied, unknown,
The mother of Pestilence.

9. But the fiery beam of Fuzon
Was a pillar of fire to Egypt
Five hundred years wand'ring on earth,
Till Los siez'd it and beat in a mass
With the body of the sun.

Chap: IId

1. But the forehead of Urizen gathering,
And his eyes pale with anguish, his lips   50
Blue & changing, in tears and bitter
Contrition he prepar'd his Bow,

2. Form'd of Ribs, that in his dark solitude,
When obscur'd in his forests, fell monsters
Arose. For his dire Contemplations
Rush'd down like floods from his mountains,
In torrents of mud settling thick,
With Eggs of unnatural production:
Forthwith hatching, some howl'd on his hills,
Some in vales, some aloft flew in air.   60

3. Of these, an enormous dread Serpent,
Scaled and poisonous horned,
Approach'd Urizen, even to his knees,
As he sat on his dark rooted Oak.

4. With his horns he push'd furious:
Great the conflict & great the jealousy
In cold poisons, but Urizen smote him.

5. First he poison'd the rocks with his blood,
Then polish'd his ribs, and his sinews
Dried, laid them apart till winter;   70
Then a Bow black prepar'd: on this Bow
A poisoned rock plac'd in silence.
He utter'd these words to the Bow:

72. **poisoned rock:** the Ten Commandments.

6. "O Bow of the clouds of secresy!
O nerve of that lust-form'd monster!
Send this rock swift, invisible thro'
The black clouds on the bosom of Fuzon."

7. So saying, In torment of his wounds
He bent the enormous ribs slowly,
A circle of darkness! then fixed   80
The sinew in its rest; then the Rock,
Poisonous source, plac'd with art, lifting difficult
Its weighty bulk; silent the rock lay,

8. While Fuzon, his tygers unloosing,
Thought Urizen slain by his wrath,
"I am God!" said he, "eldest of things."

9. Sudden sings the rock; swift & invisible
On Fuzon flew, enter'd his bosom;
His beautiful visage, his tresses
That gave light to the mornings of heaven,   90
Were smitten with darkness, deform'd
And outstretch'd on the edge of the forest.

10. But the Rock fell upon the Earth,
Mount Sinai in Arabia.

Chap: III

1. The Globe shook, and Urizen seated
On black clouds his sore wound anointed;
The ointment flow'd down on the void
Mix'd with blood—here the snake gets her poison.

2. With difficulty & great pain Urizen
Lifted on high the dead corse: 100
On his shoulders he bore it to where
A Tree hung over the Immensity.

3. For when Urizen shrunk away
From Eternals, he sat on a rock
Barren: a rock which himself
From redounding fancies had petrified.
Many tears fell on the rock,
Many sparks of vegetation.
Soon shot the pained root
Of Mystery under his heel: 110

94. **Mount . . . Arabia:** Cf. Galatians 4:24–25. **110–15.**
Mystery . . . tree: Cf. Revelation 17:5. Blake describes
the banyan tree, from which, according to *Paradise Lost*, IX.
1101–14, Adam and Eve wove garments to cover their
nakedness. For the Tree of Mystery see the headnote above
and also "The Human Abstract" in *Songs of Experience*, above.

It grew a thick tree: he wrote
In silence his book of iron,
Till the horrid plant bending its boughs
Grew to roots when it felt the earth,
And again sprung to many a tree.

4. Amaz'd started Urizen when
He beheld himself compassed round
And high roofed over with trees.
He arose, but the stems stood so thick
He with difficulty and great pain   120
Brought his Books, all but the Book
Of iron, from the dismal shade.

5. The Tree still grows over the Void
Enrooting itself all around,
An endless labyrinth of woe!

6. The corse of his first begotten
On the accursed Tree of Mystery,
On the topmost stem of this Tree,
Urizen nail'd Fuzon's corse.

Chap: IV

1. Forth flew the arrows of pestilence   130
Round the pale living Corse on the tree.

2. For in Urizen's slumbers of abstraction
In the infinite ages of Eternity,
When his Nerves of Joy melted & flow'd,
A white Lake on the dark blue air
In perturb'd pain and dismal torment
Now stretching out, now swift conglobing,

3. Effluvia vapor'd above
In noxious clouds; these hover'd thick
Over the disorganiz'd Immortal,   140
Till petrific pain scurf'd o'er the Lakes
As the bones of man, solid & dark.

4. The clouds of disease hover'd wide
Around the Immortal in torment,
Perching around the hurtling bones,
Disease on disease, shape on shape
Winged screaming in blood & torment.

5. The Eternal Prophet beat on his anvils;
Enrag'd in the desolate darkness
He forg'd nets of iron around   150
And Los threw them around the bones.

132–42. **For . . . dark:** Cf. *The Book of Urizen,* above, ll. 184–88. **148. Eternal Prophet:** Los, as in *The Book of Urizen.*

6. The shapes screaming flutter'd vain:
Some combin'd into muscles & glands,
Some organs for craving and lust;
Most remain'd on the tormented void,
Urizen's army of horrors.

7. Round the pale living Corse on the Tree
Forty years flew the arrows of pestilence.

8. Wailing and terror and woe
Ran thro' all his dismal world;   160
Forty years all his sons & daughters
Felt their skulls harden; then Asia
Arose in the pendulous deep.

9. They reptilize upon the Earth.

10. Fuzon groan'd on the Tree.

Chap: V

1. The lamenting voice of Ahania
Weeping upon the void!
And round the Tree of Fuzon,
Distant in solitary night,
Her voice was heard, but no form   170
Had she; but her tears from clouds
Eternal fell round the Tree.

2. And the voice cried: "Ah, Urizen! Love!
Flower of morning! I weep on the verge
Of Non-entity; how wide the Abyss
Between Ahania and thee!

3. "I lie on the verge of the deep;
I see thy dark clouds ascend;
I see thy black forests and floods,
A horrible waste to my eyes!   180

4. "Weeping I walk over rocks,
Over dens & thro' valleys of death.
Why didst thou despise Ahania
To cast me from thy bright presence
Into the World of Loneness?

158. Forty years: The Israelites wandered in the desert for forty years before they came to the promised land. **161–63. Forty . . . deep:** Cf. *The Book of Urizen,* above, ll. 505–16. After their captivity in Egypt (Africa) the Israelites finally settled in Asia. **173. Urizen! Love!:** In the primal man Urizen was himself the unfallen Prince of Light and lover of Ahania. See *The Four Zoas,* "Night the Ninth," below, ll. 122–34.

5. "I cannot touch his hand,
Nor weep on his knees, nor hear
His voice & bow, nor see his eyes
And joy, nor hear his footsteps and   190
My heart leap at the lovely sound!
I cannot kiss the place
Whereon his bright feet have trod,
But I wander on the rocks
With hard necessity.

6. "Where is my golden palace?
Where my ivory bed?
Where the joy of my morning hour?
Where the sons of eternity singing

7. "To awake bright Urizen, my king,
To arise to the mountain sport,   200
To the bliss of eternal valleys;

8. "To awake my king in the morn,
To embrace Ahania's joy
On the bredth of his open bosom?
From my soft cloud of dew to fall
In showers of life on his harvests,

9. "When he gave my happy soul
To the sons of eternal joy,
When he took the daughters of life
Into my chambers of love,   210

10. "When I found babes of bliss on my beds
And bosoms of milk in my chambers
Fill'd with eternal seed.
O eternal births sung round Ahania
In interchange sweet of their joys!

11. "Swell'd with ripeness & fat with fatness,
Bursting on winds, my odors,
My ripe figs and rich pomegranates
In infant joy at thy feet,
O Urizen, sported and sang.   220

12. "Then thou with thy lap full of seed,
With thy hand full of generous fire
Walked forth from the clouds of morning,
On the virgins of springing joy,
On the human soul to cast
The seed of eternal science.

13. "The sweat poured down thy temples;
To Ahania return'd in evening,
The moisture awoke to birth
My mothers-joys, sleeping in bliss.   230

14. "But now alone over rocks, mountains,
Cast out from thy lovely bosom,
Cruel jealousy! selfish fear!
Self-destroying, how can delight
Renew in these chains of darkness,
Where bones of beasts are strown
On the bleak and snowy mountains,
Where bones from the birth are buried
Before they see the light?"

1795 (1795)

FROM

THE FOUR ZOAS

NIGHT THE NINTH

BEING THE LAST JUDGMENT

The Four Zoas is an epic of 3,600 lines narrating the fall and redintegration of the eternal man (see headnote to *The Book of Urizen,* p. 82). It is divided into nine "Nights," the term having been derived from Edward Young's *Night Thoughts,* for which Blake was preparing illustrations. It was begun sometime between 1795 and 1797 and was abandoned in manuscript in 1804. Probably the greater part of it was written between 1800 and 1803. It is actually a revision and expansion (by about 900 lines) of another manuscript poem, *Vala.* The original *Vala* may be read in the edition of H. M. Margoliouth (1956), who recovered the text from the manuscript of *The Four Zoas.*

Blake derived his title from the four "beasts" (in Greek, *zoa*) of Revelation 4: 6, identified with the four "living creatures" beheld by Ezekiel in his visions of God by the river of Chebar. For Blake they are the "Four Mighty Ones . . . in every Man," powers or faculties whose disharmony brings about the fallen world of experience, and whose

Perfect Unity,
Cannot Exist but from the Universal Brotherhood of
    Eden,
The Universal Man

["Night the First," ll. 9–11].

These powers, that is, cannot be completely unified until the cosmos and society become once again perfected and paradisal. This great return is narrated in "Night the Ninth," a vision of the Last Judgment, the book of Revelation in Blake's "Bible of Hell." As in the Revelation of St. John the Divine (or in Isaiah 34–35) the Last

Judgment culminates in stages and includes terror and suffering. Yet the meaning of "Night the Ninth" is altogether different from that of the Bible. At the end no portion of existence has been cast out but has instead been transformed and restored.

The four powers or "Zoas" are named Tharmas, Urizen, Urthona, and Luvah. Or rather, these are their names in Eternity. In the fallen world these powers are themselves fallen and some of them are given different names, Urthona becoming Los and Luvah Orc. Moreover, each has a female counterpart, Enion, Ahania, Enitharmon, and Vala. The eternal Man is called Albion, and his bride is Jerusalem. Many of these names have been encountered in previous poems, but though the names are the same, one need not assume that the persons thus designated do not evolve through the course of Blake's writings. The consistency one finds in him is not mechanical but organic, not that of a system fixed and finished from the start but of development and growth.

On the versification the best comment is Blake's own:

> When this Verse was first dictated to me, I consider'd a Monotonous Cadence, like that used by Milton & Shakespeare & all writers of English Blank Verse, derived from the modern bondage of Rhyming, to be a necessary and indispensible part of Verse. But I soon found that in the mouth of a true Orator such monotony was not only awkward, but as much a bondage as rhyme itself. I therefore have produced a variety in every line, both of cadences & number of syllables. Every word and every letter is studied and put into its fit place; the terrific numbers are reserved for the terrific parts, the mild & gentle for the mild & gentle parts, and the prosaic for inferior parts; all are necessary to each other. Poetry Fetter'd Fetters the Human Race. Nations are Destroy'd or Flourish in proportion as Their Poetry, Painting and Music are Destroy'd or Flourish! The Primeval State of Man was Wisdom, Art and Science [from the Preface to *Jerusalem*].

∽∾∽

And Los & Enitharmon builded Jerusalem, weeping
Over the Sepulcher & over the Crucified body
Which, to their Phantom Eyes, appear'd still in the
    Sepulcher;
But Jesus stood beside them in the spirit, separating
Their spirit from their body. Terrified at Non Existence,
For such they deem'd the death of the body, Los his
    vegetable hands
Outstretch'd; his right hand, branching out in fibrous
    strength,

Siez'd the Sun; His left hand, like dark roots, cover'd
    the Moon,
And tore them down, cracking the heavens across from
    immense to immense.
Then fell the fires of Eternity with loud & shrill   10
Sound of Loud Trumpet thundering along from
    heaven to heaven
A mighty sound articulate: "Awake, ye dead, & come
To Judgment from the four winds! Awake & Come
    away!"
Folding like scrolls of the Enormous volume of
    Heaven & Earth,
With thunderous noise & dreadful shakings, rocking
    to & fro,
The heavens are shaken & the Earth removed from its
    place,
The foundations of the Eternal hills discover'd:
The thrones of Kings are shaken, they have lost their
    robes & crowns,
The poor smite their oppressors, they awake up to the
    harvest,
The naked warriors rush together down to the sea
    shore   20
Trembling before the multitudes of slaves now set at
    liberty:
They are become like wintry flocks, like forests strip'd
    of leaves:
The oppressed pursue like the wind; there is no room
    for escape.

The Spectre of Enitharmon, let loose on the troubled
    deep,
Wail'd shrill in the confusion, & the Spectre of
    Urthona
Reciev'd her in the darkening south; their bodies lost,
    they stood
Trembling & weak, a faint embrace, a fierce desire, as
    when
Two shadows mingle on a wall; they wail & shadowy
    tears
Fell down, & shadowy forms of joy mix'd with despair
    & grief—
Their bodies buried in the ruins of the Universe—   30
Mingled with the confusion. Who shall call them from
    the Grave?

THE FOUR ZOAS. NIGHT THE NINTH. **3. Phantom Eyes:** Los and Enitharmon are fallen beings, imprisoned in sensory limitations. Hence to them bodily death seems nonexistent (l. 5). **4. Jesus:** the risen man.

14–17. scrolls . . . discover'd: Cf. Revelation 6:14; Isaiah 25:3–4. **24. Spectre:** the shadowy and distorted way in which real being appears in the fallen world.

Rahab & Tirzah wail aloud in the wild flames; they
    give up themselves to Consummation.

The books of Urizen unroll with dreadful noise; the
    folding Serpent
Of Orc began to Consume in fierce raving fire; his
    fierce flames
Issu'd on all sides, gathering strength in animating
    volumes,
Roaming abroad on all the winds, raging intense,
    reddening
Into resistless pillars of fire rolling round & round,
    gathering
Strength from the Earths consum'd & heavens & all
    hidden abysses,
Where'er the Eagle has Explor'd, or Lion or Tyger
    trod,
Or where the Comets of the night or stars of asterial
    day   40
Have shot their arrows or long beamed spears in wrath
    & fury.

And all the while the trumpet sounds, "Awake, ye
    dead, & come
To Judgment!" From the clotted gore & from the
    hollow den
Start forth the trembling millions into flames of mental
    fire,
Bathing their limbs in the bright visions of Eternity.
Then, like the doves from pillars of Smoke, the trem-
    bling families
Of women & children throughout every nation under
    heaven
Cling round the men in bands of twenties & of fifties,
    pale

32. Rahab & Tirzah: In *Jerusalem* Blake defines Rahab as "the System of Moral Virtue." In Joshua 2, the harlot Rahab is preserved from destruction with all her household because she helps the Israelites conquer her own city of Jericho. In Dante and traditional commentaries on the Bible she is allegorized as the Church with its faithful saved at the Last Judgment. For Blake she might also be the Church thriving on its accusations of sin, and he elsewhere identifies her with the harlot of Revelation 17, who is described as Mystery and the whore of Babylon. He also identifies Rahab with Egypt keeping the Israelites in bondage, and with Leviathan. And she is a form of Nature—Nature conceived as the mysterious world of external things on which the mind, according to Locke, is passively dependent. Perhaps she may be described in summary as selfishness combined with a secret lust for power. For Tirzah, see "To Tirzah" in the *Songs of Experience*, above, and *Milton*, below, I. n. 177.

As snow that falls around a leafless tree upon the green.
Their oppressors are fall'n, they have stricken them,
    they awake to life.   50
Yet pale the just man stands erect & looking up to
    heav'n.
Trembling & strucken by the Universal stroke, the
    trees unroot,
The rocks groan horrible & run about; the mountains &
Their rivers cry with a dismal cry; the cattle gather
    together,
Lowing they kneel before the heavens; the wild beasts
    of the forests
Tremble; the Lion shuddering asks the Leopard:
    "Feelest thou
The dread I feel, unknown before? My voice refuses to
    roar,
And in weak moans I speak to thee. This night,
Before the morning's dawn, the Eagle call'd the
    Vulture,
The Raven call'd the hawk, I heard them from my
    forests black,   60
Saying: 'Let us go up far, for soon, I smell upon the
    wind,
A terror coming from the south.' The Eagle & Hawk
    fled away
At dawn, & e'er the sun arose, the raven & Vulture
    follow'd.
Let us flee also to the north." They fled. The Sons of
    Men
Saw them depart in dismal droves. The trumpet
    sounded loud
And all the Sons of Eternity Descended into Beulah.

In the fierce flames the limbs of Mystery lay consuming
    with howling
And deep despair. Rattling go up the flames around
    the Synagogue
Of Satan. Loud the Serpent Orc rag'd thro' his twenty
    seven
Folds. The tree of Mystery went up in folding flames.
Blood issu'd out in rushing volumes, pouring in whirl-
    pools fierce   71
From out the flood gates of the Sky. The Gates are
    burst; down pour
The torrents black upon the Earth; the blood pours
    down incessant.

66. Beulah: See Isaiah 62:4. Beulah represents Blake's state of Innocence. See *Milton*, below, I. n. 1.

Kings in their palaces lie drown'd. Shepherds, their
  flocks, their tents,
Roll down the mountains in black torrents. Cities,
  Villages,
High spires & Castles drown'd in the black deluge;
  shoal on shoal
Float the dead carcases of Men & Beasts, driven to &
  fro on waves
Of foaming blood beneath the black incessant sky, till
  all
Mystery's tyrants are cut off & not one left on Earth.

And when all Tyranny was cut off from the face of
  the Earth,   80
Around the dragon form of Urizen, & round his strong
  form,
The flames rolling intense thro' the wide Universe
Began to enter the Holy City. Ent'ring the dismal
  clouds
In furrow'd lightnings break their way, the wild flames
  licking up
The Bloody Deluge: living flames winged with in-
  tellect
And Reason, round the Earth they march in order,
  flame by flame.
From the clotted gore & from the hollow den
Start forth the trembling millions into flames of mental
  fire,
Bathing their limbs in the bright visions of Eternity.

Beyond this Universal Confusion, beyond the re-
  motest Pole   90
Where their vortexes began to operate, there stands
A Horrible rock far in the South; it was forsaken when
Urizen gave the horses of Light into the hands of
  Luvah.
On this rock lay the faded head of the Eternal Man
Enwrapped round with weeds of death, pale cold in
  sorrow & woe
He lifts the blue lamps of his Eyes & cries with heavenly
  voice:

90–95. Beyond . . . woe: Cf. *Milton*, below, I. 549–68.
The "Horrible rock" (l. 92) is the Rock of Ages as in
Milton, I. 564. 93. Luvah: the name of Orc before the
primal fall. In *The Book of Urizen* the fall of the primal man
takes place when Urizen separates himself from the whole.
In *The Four Zoas* Urizen appears both in his fallen form
and also as the unfallen Prince of Light (see *The Book of
Ahania*, above, ll. 173–239), and the fall is also an act of
Luvah usurping the function of Urizen.

Bowing his head over the consuming Universe, he
  cried:
"O weakness & O weariness! O war within my
  members!
My sons, exiled from my breast, pass to & fro before
  me.
My birds are silent on my hills, flocks die beneath my
  branches.   100
My tents are fallen, my trumpets & the sweet sound of
  my harp
Is silent on my clouded hills that belch forth storms &
  fire.
My milk of cows & honey of bees & fruit of golden
  harvest
Are gather'd in the scorching heat & in the driving rain.
My robe is turned to confusion, & my bright gold to
  stone.
Where once I sat, I weary walk in misery & pain,
For from within my wither'd breast grown narrow
  with my woes
The Corn is turned to thistles & the apples into poison,
The birds of song to murderous crows, My joys to
  bitter groans,
The voices of children in my tents to cries of helpless
  infants,   110
And all exiled from the face of light & shine of morning
In this dark world, a narrow house, I wander up &
  down.
I hear Mystery howling in these flames of Consumma-
  tion.
When shall the Man of future times become as in days
  of old?
O weary life! why sit I here & give up all my powers
To indolence, to the night of death, when indolence &
  mourning
Sit hovering over my dark threshold? tho' I arise, look
  out
And scorn the war within my members, yet my heart
  is weak
And my head faint. Yet will I look again into the
  morning.
Whence is this sound of rage of Men drinking each
  other's blood,   120
Drunk with the smoking gore, & red, but not with
  nourishing wine?"

The Eternal Man sat on the Rocks & cried with awful
  voice:
"O Prince of Light, where art thou? I behold thee not
  as once

In those Eternal fields, in clouds of morning stepping
forth
With harps & songs when bright Ahania sang before
thy face
And all thy sons & daughters gather'd round my ample
table.
See you not all this wracking furious confusion?
Come forth from slumbers of thy cold abstraction! Come forth,
Arise to Eternal births! Shake off thy cold repose,
Schoolmaster of souls, great opposer of change,
arise!   130
That the Eternal worlds may see thy face in peace &
joy,
That thou, dread form of Certainty, maist sit in town
& village
While little children play around thy feet in gentle
awe,
Fearing thy frown, loving thy smile, O Urizen, Prince
of Light."

He call'd; the deep buried his voice & answer none
return'd.
Then wrath burst round; the Eternal Man was wrath;
again he cried:
"Arise, O stony form of death! O dragon of the Deeps!
Lie down before my feet, O Dragon! let Urizen arise.
O how couldst thou deform those beautiful proportions
Of life & person; for as the Person, so is his life pro-
portion'd.
Let Luvah rage in the dark deep, even to Consumma-
tion,   141
For if thou feedest not his rage, it will subside in peace.
But if thou darest obstinate refuse my stern behest,
Thy crown & scepter I will sieze, & regulate all my
members
In stern severity, & cast thee out into the indefinite
Where nothing lives, there to wander; & if thou re-
turnest weary,
Weeping at the threshold of Existence, I will steel my
heart
Against thee to Eternity, & never recieve thee more.
Thy self-destroying, beast form'd Science shall be thy
eternal lot.
My anger against thee is greater than against this
Luvah,   150
For war is energy Enslav'd, but thy religion,
The first author of this war & the distracting of honest
minds
Into confused perturbation & strife & horrour & pride,

Is a deciet so detestable that I will cast thee out
If thou repentest not, & leave thee as a rotten branch to
be burn'd
With Mystery the Harlot & with Satan for Ever &
Ever.
Error can never be redeemed in all Eternity,
But Sin, Even Rahab, is redeem'd in blood & fury &
jealousy—
That line of blood that stretch'd across the windows of
the morning—
Redeem'd from Error's power. Wake, thou dragon of
the deeps!"   160

Urizen wept in the dark deep, anxious his scaly form
To reassume the human; & he wept in the dark deep,
Saying: "O that I had never drunk the wine nor eat
the bread
Of dark mortality, or cast my view into futurity, nor
turn'd
My back, dark'ning the present, clouding with a cloud,
And building arches high, & cities, turrets & towers &
domes
Whose smoke destroy'd the pleasant gardens, & whose
running kennels
Chok'd the bright rivers; burd'ning with my Ships the
angry deep;
Thro' Chaos seeking for delight, & in spaces remote
Seeking the Eternal which is always present to the wise;
Seeking for pleasure which unsought falls round the
infant's path   171
And on the fleeces of mild flocks who neither care nor
labour;
But I, the labourer of ages, whose unwearied hands
Are thus deform'd with hardness, with the sword &
with the spear
And with the chisel & the mallet, I, whose labours vast
Order the nations, separating family by family,
Alone enjoy not. I alone, in misery supreme,
Ungratified give all my joy unto this Luvah & Vala.
Then Go, O dark futurity! I will cast thee forth from
these
Heavens of my brain, nor will I look upon futurity
more.   180
I cast futurity away, & turn my back upon that void
Which I have made; for lo! futurity is in this moment.
Let Orc consume, let Tharmas rage, let dark Urthona
give

158. Rahab: See n. 32. Perhaps one may say that as Mystery
Rahab cannot be redeemed, as Sin she can. **159. line of
blood:** See Joshua 2:18.

All strength to Los & Enitharmon, & let Los self-curs'd
Rend down this fabric, as a wall ruin'd & family
    extinct.
Rage Orc! Rage Tharmas! Urizen no longer curbs
    your rage."

So Urizen spoke; he shook his snows from off his
    shoulders & arose
As on a Pyramid of mist, his white robes scattering
The fleecy white: renew'd, he shook his aged mantles
    off
Into the fires. Then, glorious bright, Exulting in his
    joy,   190
He sounding rose into the heavens in naked majesty,
In radiant Youth; when Lo! like garlands in the
    Eastern sky
When vocal may comes dancing from the East,
    Ahania came
Exulting in her flight, as when a bubble rises up
On the surface of a lake, Ahania rose in joy.
Excess of Joy is worse than grief; her heart beat high,
    her blood
Burst its bright vessels: she fell down dead at the feet
    of Urizen
Outstretch'd, a smiling corse: they buried her in a
    silent cave.
Urizen dropped a tear; the Eternal Man Darken'd
    with sorrow.

The three daughters of Urizen guard Ahania's death
    couch;   200
Rising from the confusion in tears & howlings &
    despair,
Calling upon their father's Name, upon their Rivers
    dark.

And the Eternal Man said: "Hear my words, O Prince
    of Light
Behold Jerusalem in whose bosom the Lamb of God
Is seen; tho' slain before her Gates, he self-renew'd
    remains
Eternal, & I thro' him awake from death's dark vale.
The times revolve; the time is coming when all these
    delights
Shall be renew'd, & all these Elements that now con-
    sume
Shall reflourish. Then bright Ahania shall awake from
    death,
A glorious Vision to thine Eyes, a Self-renewing
    Vision: 210

The spring, the summer, to be thine; then sleep the
    wintry days
In silken garments spun by her own hands against her
    funeral.
The winter thou shalt plow & lay thy stores into thy
    barns
Expecting to recieve Ahania in the spring with joy.
Immortal thou, Regenerate She, & all the lovely Sex
From her shall learn obedience & prepare for a wintry
    grave,
That spring may see them rise in tenfold joy & sweet
    delight.
Thus shall the male & female live the life of Eternity,
Because the Lamb of God Creates himself a bride &
    wife
That we his Children evermore may live in Jerusalem
Which now descendeth out of heaven, a City, yet a
    Woman,   221
Mother of myriads redeem'd & born in her spiritual
    palaces,
By a New Spiritual birth Regenerated from Death."

Urizen said: "I have Erred, & my Error remains with
    me.
What Chain encompasses? in what Lock is the river
    of light confin'd
That issues forth in the morning by measure & in the
    evening by carefulness?
Where shall we take our stand to view the infinite &
    unbounded?
Or where are human feet? for Lo, our eyes are in the
    heavens."

He ceas'd, for riv'n link from link, the bursting Uni-
    verse explodes.
All things revers'd flew from their centers: rattling
    bones   230
To bones Join: shaking convuls'd, the shivering clay
    breathes:
Each speck of dust to the Earth's center nestles round &
    round
In pangs of an Eternal Birth: in torment & awe &
    fear,
All spirits deceas'd, let loose from reptile prisons, come
    in shoals:
Wild furies from the tyger's brain & from the lion's
    eyes,

---

221. descendeth . . . Woman: Cf. Revelation, 21:2.
230-31. bones . . . Join: Cf. Ezekiel 38:7.

And from the ox & ass come moping terrors, from the eagle
And raven: numerous as the leaves of autumn, every species
Flock to the trumpet, mutt'ring over the sides of the grave & crying
In the fierce wind round heaving rocks & mountains fill'd with groans.
On rifted rocks, suspended in the air by inward fires,   240
Many a woful company & many on clouds & waters,
Fathers & friends, Mothers & Infants, Kings & Warriors,
Priests & chain'd Captives, met together in a horrible fear;
And every one of the dead appears as he had liv'd before,
And all the marks remain of the slave's scourge & tyrant's Crown,
And of the Priest's o'ergorged Abdomen, & of the merchant's thin
Sinewy deception, & of the warrior's outbraving & thoughtlessness
In lineaments too extended & in bones too strait & long.
They shew their wounds: they accuse: they sieze the opressor; howlings began
On the golden palace, songs & joy on the desart; the Cold babe   250
Stands in the furious air; he cries: "the children of six thousand years
Who died in infancy rage furious: a mighty multitude rage furious,
Naked & pale standing in the expecting air, to be deliver'd.
Rend limb from limb the warrior & the tyrant, re-uniting in pain."

The furious wind still rends around; they flee in sluggish effort;
They beg, they intreat in vain now; they listened not to intreaty;
They view the flames red rolling on thro' the wide universe
From the dark jaws of death beneath & desolate shores remote,
These covering vaults of heaven & these trembling globes of earth.
One Planet calls to another & one star enquires of another: 260
"What flames are these, coming from the South? what noise, what dreadful rout

As of a battle in the heavens? hark! heard you not the trumpet
As of fierce battle?" While they spoke, the flames come on intense roaring.
They see him whom they have pierc'd, they wail because of him,
They magnify themselves no more against Jerusalem, Nor
Against her little ones; the innocent, accused before the Judges,
Shines with immortal glory; trembling, the judge springs from his throne
Hiding his face in the dust beneath the prisoner's feet & saying:
"Brother of Jesus, what have I done? intreat thy lord for me:
Perhaps I may be forgiven." While he speaks the flames roll on,   270
And after the flames appears the Cloud of the Son of Man
Descending from Jerusalem with power and great Glory.
All nations look up to the Cloud & behold him who was crucified.
The Prisoner answers: "You scourg'd my father to death before my face
While I stood bound with cords & heavy chains. Your hipocrisy
Shall now avail you nought." So speaking, he dash'd him with his foot.

The Cloud is Blood, dazling upon the heavens, & in the cloud,
Above upon its volumes, is beheld a throne & a pavement
Of precious stones surrounded by twenty-four venerable patriarchs,
And these again surrounded by four Wonders of the Almighty,   280
Incomprehensible, pervading all, amidst & round about,
Fourfold, each in the other reflected; they are named Life's—in Eternity—
Four Starry Universes going forward from Eternity to Eternity.
And the Fall'n Man who was arisen upon the Rock of Ages

271. **Cloud . . . Man:** Cf. Revelation 1:7. **278–80.**
throne . . . Almighty: Cf. Revelation 4:2–6.

Beheld the Vision of God, & he arose up from the Rock,
And Urizen arose up with him, walking thro' the flames
To meet the Lord coming to Judgment; but the flames
   repell'd them
Still to the Rock; in vain they strove to Enter the
   Consummation
Together, for the Redeem'd Man could not enter the
   Consummation.

Then siez'd the sons of Urizen the Plow: they polish'd
   it   290
From rust of ages; all its ornaments of gold & silver &
   ivory
Reshone across the field immense where all the nations
Darken'd like Mould in the divided fallows where the
   weed
Triumphs in its own destruction; they took down the
   harness
From the blue walls of heaven, starry jingling, orna-
   mented
With beautiful art, the study of angels, the workman-
   ship of Demons
When Heaven & Hell in Emulation strove in sports of
   Glory.

The noise of rural works resounded thro' the heavens
   of heavens,
The horses neigh from the battle, the wild bulls from
   the sultry waste,
The tygers from the forests, & the lions from the sandy
   desarts.   300
They sing: they sieze the instruments of harmony;
   they throw away
The spear, the bow, the gun, the mortar; they level
   the fortifications.
They beat the iron engines of destruction into wedges;
They give them to Urthona's sons; ringing the
   hammers sound
In dens of death to forge the spade, the mattock &
   the ax,
The heavy roller to break the clods, to pass over the
   nations.

The Sons of Urizen shout. Their father rose. The
   Eternal horses

289. Redeem'd . . . Consummation: There must first be
the harvest and the vintage as in Revelation 14:15–20. **303.
beat . . . wedges:** Cf. Isaiah 2:4.

Harness'd, They call'd to Urizen; the heavens moved
   at their call.
The limbs of Urizen shone with ardor. He laid his
   hand on the Plow,
Thro' dismal darkness drave the Plow of ages over
   Cities   310
And all their Villages; over Mountains & all their
   Vallies;
Over the graves & caverns of the dead; Over the
   Planets
And over the void spaces; over sun & moon & star &
   constellation.

Then Urizen commanded & they brought the Seed
   of Men.
The trembling souls of All the dead stood before
   Urizen,
Weak wailing in the troubled air. East, west & north
   & south
He turn'd the horses loose & laid his Plow in the northern
   corner
Of the wide Universal field, then step'd forth into the
   immense.

Then he began to sow the seed; he girded round his
   loins
With a bright girdle, & his skirt fill'd with immortal
   souls.   320
Howling & Wailing fly the souls from Urizen's strong
   hand,
For from the hand of Urizen the myriads fall like
   stars
Into their own appointed places, driven back by the
   winds.
The naked warriors rush together down to the sea
   shores:
They are become like wintry flocks, like forests strip'd
   of leaves;
The Kings & Princes of the Earth cry with a feeble cry,
Driven on the unproducing sands & on the harden'd
   rocks;
And all the while the flames of Orc follow the vent'rous
   feet
Of Urizen, & all the while the Trump of Tharmas
   sounds.
Weeping & wailing fly the souls from Urizen's strong
   hands—   330
The daughters of Urizen stand with Cups & measures
   of foaming wine

Immense upon the heavens with bread & delicate
     repasts—
Then follows the golden harrow in the midst of Mental
     fires.
To ravishing melody of flutes & harps & softest voice
The seed is harrow'd in, while flames heat the black
     mould & cause
The human harvest to begin. Towards the south first
     sprang
The myriads, & in silent fear they look out from their
     graves.

Then Urizen sits down to rest, & all his wearied sons
Take their repose on beds; they drink, they sing, they
     view the flames
Of Orc; in joy they view the human harvest springing
     up.   340
A time they give to sweet repose, till all the harvest is
     ripe.
And Lo, like the harvest Moon, Ahania cast off her
     death clothes;
She folded them up in care, in silence, & her bright'ning
     limbs
Bath'd in the clear spring of the rock; then from her
     darksome cave
Issu'd in majesty divine. Urizen rose up from his couch
On wings of tenfold joy, clapping his hands, his feet,
     his radiant wings
In the immense: as when the Sun dances upon the
     mountains
A shout of jubilee in lovely notes responds from daugh-
     ter to daughter,
From son to son: as if the stars beaming innumerable
Thro' night should sing soft warbling, filling earth &
     heaven;   350
And bright Ahania took her seat by Urizen in songs &
     joy.

The Eternal Man also sat down upon the Couches of
     Beulah,
Sorrowful that he could not put off his new risen
     body
In mental flames; the flames refus'd, they drove him
     back to Beulah.
His body was redeem'd to be permanent thro' Mercy
     Divine.

And now fierce Orc had quite consum'd himself in
     Mental flames,
Expending all his energy against the fuel of fire.

The Regenerate Man stoop'd his head over the Uni-
     verse & in
His holy hands reciev'd the flaming Demon & Demon-
     ess of smoke
And gave them to Urizen's hands; the Immortal
     frown'd, saying,   360

"Luvah & Vala, henceforth you are Servants; obey &
     live.
You shall forget your former state; return, & Love in
     peace,
Into your place, the place of seed, not in the brain or
     heart.
If Gods combine against Man, setting their dominion
     above
The Human form Divine, Thrown down from their
     high station
In the Eternal heavens of Human Imagination, buried
     beneath
In dark Oblivion, with incessant pangs, ages on ages,
In enmity & war first weaken'd, then in stern repent-
     ance
They must renew their brightness, & their disorganiz'd
     functions
Again reorganize, till they resume the image of the
     human,   370
Co-operating in the bliss of Man, obeying his Will,
Servants to the infinite & Eternal of the Human form."

Luvah & Vala descended & enter'd the Gates of Dark
     Urthona,
And walk'd from the hands of Urizen in the shadows
     of Vala's Garden
Where the impressions of Despair & Hope for ever
     vegetate
In flowers, in fruits, in fishes, birds & beasts & clouds &
     waters,
The land of doubts & shadows, sweet delusions, un-
     form'd hopes.
They saw no more the terrible confusion of the wrack-
     ing universe.
They heard not, saw not, felt not all the terrible con-
     fusion,
For in their orbed senses, within clos'd up, they
     wander'd at will.   380
And those upon the Couches view'd them, in the
     dreams of Beulah,

361. Luvah & Vala: Orc has been regenerated as Luvah, the
Prince of Love. Vala is his female counterpart.

As they repos'd from the terrible wide universal
    harvest.
Invisible Luvah in bright clouds hover'd over Vala's
    head,
And thus their ancient golden age renew'd; for Luvah
    spoke
With voice mild from his golden Cloud upon the
    breath of morning:

    "Come forth, O Vala, from the grass & from the
      silent dew,
    Rise from the dews of death, for the Eternal Man is
      Risen."

She rises among flowers & looks toward the Eastern
    clearness,
She walks yea runs, her feet are wing'd, on the tops of
    the bending grass,
Her garments rejoice in the vocal wind & her hair
    glistens with dew.   390

She answer'd thus: "Whose voice is this, in the voice
    of the nourishing air,
In the spirit of the morning, awaking the Soul from
    its grassy bed?
Where dost thou dwell? for it is thee I seek, & but for
    thee
I must have slept Eternally, nor have felt the dew of
    thy morning.
Look how the opening dawn advances with vocal
    harmony!
Look how the beams foreshew the rising of some
    glorious power!
The sun is thine, he goeth forth in his majestic bright-
    ness.
O thou creating voice that callest! & who shall answer
    thee?"

"Where dost thou flee, O fair one? where doest thou
    seek thy happy place?"

"To yonder brightness, there I haste, for sure I came
    from thence   400
Or I must have slept eternally, nor have felt the dew of
    morning."

"Eternally thou must have slept, not have felt the
    morning dew,
But for yon nourishing sun; 'tis that by which thou
    art arisen.

The birds adore the sun: the beasts rise up & play in his
    beams,
And every flower & every leaf rejoices in his light.
Then, O thou fair one, sit thee down, for thou art as
    the grass,
Thou risest in the dew of morning & at night art folded
    up."

"Alas! am I but as a flower? then will I sit me down,
Then will I weep, then I'll complain & sigh for im-
    mortality,
And chide my maker, thee O sun, that raisedst me to
    fall."   410

So saying she sat down & wept beneath the apple trees.

"O be thou blotted out, thou Sun! that raisedst me to
    trouble,
That gavest me a heart to crave, & raisedst me, thy
    phantom,
To feel thy heat & see thy light & wander here alone,
Hopeless, if I am like the grass & so shall pass away."

"Rise, sluggish Soul, why sit'st thou here? why dost
    thou sit & weep?
Yon sun shall wax old & decay, but thou shalt ever
    flourish.
The fruit shall ripen & fall down, & the flowers con-
    sume away,
But thou shalt still survive; arise, O dry thy dewy tears."

"Hah! shall I still survive? whence came that sweet &
    comforting voice?   420
And whence that voice of sorrow? O sun! thou art
    nothing now to me.
Go on thy course rejoicing, & let us both rejoice
    together.
I walk among his flocks & hear the bleating of his
    lambs.
O that I could behold his face & follow his pure feet!
I walk by the footsteps of his flocks; come hither,
    tender flocks.
Can you converse with a pure soul that seeketh for her
    maker?
You answer not: then am I set your mistress in this
    garden.
I'll watch you & attend your footsteps; you are not like
    the birds

406–07. grass . . . up: Cf. Psalm 90:6.

That sing & fly in the bright air; but you do lick my feet
And let me touch your woolly backs; follow me as I
    sing,   430
For in my bosom a new song arises to my Lord:

"Rise up, O sun, most glorious minister & light of day.
Flow on, ye gentle airs, & bear the voice of my re-
    joicing.
Wave freshly, clear waters flowing around the tender
    grass;
And thou, sweet smelling ground, put forth thy life in
    fruits & flowers.
Follow me, O my flocks, & hear me sing my raptur-
    ous song.
I will cause my voice to be heard on the clouds that
    glitter in the sun.
I will call; & who shall answer me? I will sing; who shall
    reply?
For from my pleasant hills behold the living, living
    springs,
Running among my green pastures, delighting among
    my trees.   440
I am not here alone: my flocks, you are my brethren;
And you birds that sing & adorn the sky, you are my
    sisters.
I sing, & you reply to my song; I rejoice, & you are
    glad.
Follow me, O my flocks; we will now descend into the
    valley.
O how delicious are the grapes, flourishing in the sun!
How clear the spring of the rock, running among the
    golden sand!
How cool the breezes of the valley, & the arms of the
    branching trees!
Cover us from the sun; come & let us sit in the shade.
My Luvah here hath plac'd me in a sweet & pleasant
    land,
And given me fruits & pleasant waters, & warm hills &
    cool valleys.   450
Here will I build myself a house, & here I'll call on his
    name,
Here I'll return when I am weary & take my pleasant
    rest."

   So spoke the sinless soul, & laid her head on the downy
    fleece
Of a curl'd Ram who stretch'd himself in sleep beside
    his mistress,
And soft sleep fell upon her eyelids in the silent noon
    of day.

Then Luvah passed by, & saw the sinless soul,
And said: "Let a pleasant house arise to be the dwelling
    place
Of this immortal spirit growing in lower Paradise."
He spoke, & pillars were builded, & walls as white as
    ivory.
The grass she slept upon was pav'd with pavement as
    of pearl.   460
Beneath her rose a downy bed, & a cieling cover'd all.

Vala awoke. "When in the pleasant gates of sleep I
    enter'd,
I saw my Luvah like a spirit stand in the bright air.
Round him stood spirits like me, who rear'd me a
    bright house,
And here I see thee, house, remain in my most pleasant
    world.
My Luvah smil'd: I kneeled down: he laid his hand on
    my head,
And when he laid his hand upon me, from the gates of
    sleep I came
Into this bodily house to tend my flocks in my pleasant
    garden."

So saying, she arose & walked round her beautiful
    house,
And then from her white door she look'd to see her
    bleating lambs,   470
But her flocks were gone up from beneath the trees
    into the hills.

"I see the hand that leadeth me doth also lead my
    flocks."
She went up to her flocks & turned oft to see her
    shining house.
She stop'd to drink of the clear spring & eat the grapes
    & apples.
She bore the fruits in her lap; she gather'd flowers for
    her bosom.
She called to her flocks, saying, "Follow me, O my
    flocks!"

They follow'd her to the silent valley beneath the
    spreading trees.
And on the river's margin she ungirded her golden
    girdle;
She stood in the river & view'd herself within the
    wat'ry glass,
And her bright hair was wet with the waters: she rose
    up from the river,   480

And as she rose her eyes were open'd to the world
of waters:
She saw Tharmas sitting upon the rocks beside the
wavy sea.
He strok'd the water from his beard & mourn'd faint
thro' the summer vales.

And Vala stood on the rocks of Tharmas & heard his
mournful voice:

"O Enion, my weary head is in the bed of death,
For weeds of death have wrap'd around my limbs in
the hoary deeps.
I sit in the place of shells & mourn, & thou art clos'd in
clouds.
When will the time of Clouds be past, & the dismal
night of Tharmas?
Arise, O Enion! Arise & smile upon my head
As thou dost smile upon the barren mountains and
they rejoice.   490
When wilt thou smile on Tharmas, O thou bringer of
golden day?
Arise, O Enion, arise, for Lo, I have calm'd my seas."

So saying, his faint head he laid upon the Oozy rock,
And darkness cover'd all the deep; the light of Enion
faded
Like a faint flame quivering upon the surface of the
darkness.

Then Vala lifted up her hands to heaven to call on
Enion.
She call'd, but none could answer her & the eccho her
voice return'd:

"Where is the voice of God that call'd me from the
silent dew?
Where is the Lord of Vala? dost thou hide in clefts of
the rock?
Why shouldst thou hide thyself from Vala, from the
soul that wanders desolate?"   500

She ceas'd, & light beamed round her like the glory of
the morning,
And she arose out of the river & girded her golden
girdle.

482. Tharmas: Of the Four Zoas (see headnote) Tharmas is
described as the "Parent power" ("Night the First," l. 24).
In his fall he becomes the sea, a symbol of chaos. **485.**
Enion: the female counterpart of Tharmas.

And now her feet step on the grassy bosom of the ground
Among her flocks, & she turn'd her eyes toward her
pleasant house
And saw in the door way beneath the trees two little
children playing.
She drew near to her house & her flocks follow'd her
footsteps.
The children clung around her knees, she embrac'd
them & wept over them.

"Thou, little Boy, art Tharmas, & thou, bright Girl,
Enion.
How are ye thus renew'd & brought into the Gardens
of Vala?"

She embrac'd them in tears, till the sun descended the
western hills,   510
And then she enter'd her bright house, leading her
mighty children.
And when night came, the flocks laid round the house
beneath the trees.
She laid the children on the beds which she saw prepar'd
in the house,
Then last, herself laid down & clos'd her Eyelids in
soft slumbers.

And in the morning, when the sun arose in the crystal
sky,
Vala awoke & call'd the children from their gentle
slumbers:

"Awake, O Enion, awake & let thine innocent Eyes
Enlighten all the Crystal house of Vala! awake! awake!
Awake, Tharmas, awake, awake thou child of dewy
tears.
Open the orbs of thy blue eyes & smile upon my
gardens."   520

The Children woke & smil'd on Vala; she kneel'd by
the golden couch,
She pres'd them to her bosom & her pearly tears drop'd
down.
"O my sweet Children! Enion, let Tharmas kiss thy
Cheek.
Why dost thou turn thyself away from his sweet
wat'ry eyes?
Tharmas, henceforth in Vala's bosom thou shalt find
sweet peace.
O bless the lovely eyes of Tharmas & the Eyes of
Enion!"

They rose; they went out wand'ring, sometimes to-
gether, sometimes alone.
"Why weep'st thou, Tharmas, Child of tears, in the
bright house of joy?
Doth Enion avoid the sight of thy blue heavenly Eyes?
And dost thou wander with my lambs & wet their
innocent faces   530
With thy bright tears because the steps of Enion are in
the gardens?
Arise, sweet boy, & let us follow the path of Enion."

So saying, they went down into the garden among the
fruits.
And Enion sang among the flowers that grew among
the trees,
And Vala said: "Go, Tharmas; weep not. Go to
Enion."

He said: "O Vala, I am sick, & all this garden of
Pleasure
Swims like a dream before my eyes; but the sweet
smiling fruit
Revives me to new deaths. I fade, even as a water lilly
In the sun's heat, till in the night on the couch of Enion
I drink new life & feel the breath of sleeping Enion. 540
But in the morning she arises to avoid my Eyes,
Then my loins fade & in the house I sit me down &
weep."

"Chear up thy Countenance, bright boy, & go to
Enion.
Tell her that Vala waits her in the shadows of her
garden."

He went with timid steps, & Enion, like the ruddy morn
When infant spring appears in swelling buds & opening
flowers,
Behind her Veil withdraws; so Enion turn'd her modest
head.

But Tharmas spoke: "Vala seeks thee, sweet Enion, in
the shades.
Follow the steps of Tharmas, O thou brightness of the
gardens."
He took her hand reluctant; she follow'd in infant
doubts.   550
Thus in Eternal Childhood, straying among Vala's
flocks
In infant sorrow & joy alternate, Enion & Tharmas
play'd

Round Vala in the Gardens of Vala & by her river's
margin.
They are the shadows of Tharmas & of Enion in Vala's
world.

And the sleepers who rested from their harvest work
beheld these visions.
Thus were the sleepers entertain'd upon the Couches
of Beulah.

When Luvah & Vala were clos'd up in their world of
shadowy forms,
Darkness was all beneath the heavens: only a little
light
Such as glows out from sleeping spirits, appear'd in the
deeps beneath.
As when the wind sweeps over a corn field, the noise
of souls   560
Thro' all the immense, borne down by Clouds swag-
ging in autumnal heat,
Mutt'ring along from heaven to heaven, hoarse roll
the human forms
Beneath thick clouds, dreadful lightnings burst &
thunders roll,
Down pour the torrent floods of heaven on all the
human harvest.
Then Urizen, sitting at his repose on beds in the bright
South,
Cried, "Times are Ended!" he exulted; he arose in
joy; he exulted;
He pour'd his light, & all his sons & daughters pour'd
their light
To exhale the spirits of Luvah & Vala thro' the atmos-
phere.
And Luvah & Vala saw the Light; their spirits were
exhal'd
In all their ancient innocence; the floods depart; the
clouds   570
Dissipate or sink into the Seas of Tharmas. Luvah sat
Above on the bright heavens in peace; the Spirits of
Men beneath
Cried out to be deliver'd, & the spirit of Luvah wept
Over the human harvest & over Vala, the sweet
wanderer.
In pain the human harvest wav'd, in horrible groans of
woe.
The Universal Groan went up; the Eternal Man was
darken'd.

566. "Times are Ended!": Cf. Revelation 10:6.

Then Urizen arose & took his sickle in his hand.
There is a brazen sickle, & a scythe of iron hid
Deep in the South, guarded by a few solitary stars.
This sickle Urizen took; the scythe his sons embrac'd
And went forth & began to reap; & all his joyful
 sons   581
Reap'd the wide Universe & bound in sheaves a
 wondrous harvest.
They took them into the wide barns with loud re-
 joicings & triumph
Of flute & harp & drum & trumpet, horn & clarion.

The feast was spread in the bright South, & the Regener-
 ate Man
Sat at the feast rejoicing, & the wine of Eternity
Was serv'd round by the flames of Luvah all day & all
 the Night.
And when Morning began to dawn upon the distant
 hills,
A whirlwind rose up in the Center, & in the whirlwind
 a shriek,
And in the shriek a rattling of bones, & in the rattling
 of bones   590
A dolorous groan, & from the dolorous groan in tears
Rose Enion like a gentle light; & Enion spoke, saying:

"O Dreams of Death! the human form dissolving, com-
 panied
By beasts & worms & creeping things, & darkness &
 despair.
The clouds fall off from my wet brow, the dust from
 my cold limbs
Into the sea of Tharmas. Soon renew'd, a Golden
 Moth,
I shall cast off my death clothes & Embrace Tharmas
 again.
For Lo, the winter melted away upon the distant hills,
And all the black mould sings." She speaks to her infant
 race; her milk
Descends down on the sand; the thirsty sand drinks &
 rejoices   600
Wondering to behold the Emmet, the Grasshopper,
 the jointed worm.
The roots shoot thick thro' the solid rocks, bursting
 their way
They cry out in joys of existence; the broad stems
Rear on the mountains stem after stem; the scaly
 newt creeps

577–84. **Then . . . clarion:** Cf. Revelation 14:14–16.

From the stone, & the armed fly springs from the
 rocky crevice,
The spider, The bat burst from the harden'd slime,
 crying
To one another: "What are we, & whence is our joy &
 delight?
Lo, the little moss begins to spring, & the tender weed
Creeps round our secret nest." Flocks brighten the
 Mountains,
Herds throng up the Valley, wild beasts fill the forests.

Joy thrill'd thro' all the Furious forms of Tharmas
 humanizing.   611
Mild he Embrac'd her whom he sought; he rais'd her
 thro' the heavens,
Sounding his trumpet to awake the dead, on high he
 soar'd
Over the ruin'd worlds, the smoking tomb of the
 Eternal Prophet.

The Eternal Man arose. He welcom'd them to the Feast.
The feast was spread in the bright South, & the Eternal
 Man
Sat at the feast rejoicing, & the wine of Eternity
Was serv'd round by the flames of Luvah all day & all
 the night.

And Many Eternal Men sat at the golden feast to see
The female form now separate. They shudder'd at the
 horrible thing   620
Not born for the sport and amusement of Man, but
 born to drink up all his powers.
They wept to see their shadows; they said to one
 another: "This is Sin:
This is the Generative world;" they remember'd the
 days of old.

And One of the Eternals spoke. All was silent at the
 feast.

"Man is a Worm; wearied with joy, he seeks the caves
 of sleep
Among the Flowers of Beulah, in his selfish cold
 repose
Forsaking Brotherhood & Universal love, in selfish
 clay
Folding the pure wings of his mind, seeking the places
 dark
Abstracted from the roots of Science; then inclos'd
 around

In walls of Gold we cast him like a Seed into the
  Earth   630
Till times & spaces have pass'd over him; duly every
  morn
We visit him, covering with a Veil the immortal seed;
With windows from the inclement sky we cover him,
  & with walls
And hearths protect the selfish terror, till divided all
In families we see our shadows born, & thence we
  know
That Man subsists by Brotherhood & Universal Love.
We fall on one another's necks, more closely we em-
  brace.
Not for ourselves, but for the Eternal family we live.
Man liveth not by Self alone, but in his brother's face
Each shall behold the Eternal Father & love & joy
  abound."   640

So spoke the Eternal at the Feast; they embrac'd the
  New born Man,
Calling him Brother, image of the Eternal Father; they
  sat down
At the immortal tables, sounding loud their instru-
  ments of joy,
Calling the Morning into Beulah; the Eternal Man
  rejoic'd.

When Morning dawn'd, The Eternals rose to labour at
  the Vintage.
Beneath they saw their sons & daughters, wond'ring
  inconcievable
At the dark myriads in shadows in the worlds beneath.

The morning dawn'd. Urizen rose, & in his hand the
  Flail
Sounds on the Floor, heard terrible by all beneath
  the heavens.
Dismal loud redounding, the nether floor shakes with
  the sound,   650
And all Nations were threshed out, & the stars
  thresh'd from their husks.

  Then Tharmas took the Winnowing fan; the win-
    nowing wind furious
Above, veer'd round by violent whirlwind, driven
  west & south,
Tossed the Nations like chaff into the seas of Tharmas.

"O Mystery," Fierce Tharmas cries, "Behold thy end
  is come!
Art thou she that made the nations drunk with the cup
  of Religion?
Go down, ye Kings & Councellors & Giant Warriors,
Go down into the depths, go down & hide yourselves
  beneath,
Go down with horse & Chariots & Trumpets of hoarse
  war.

"Lo, how the Pomp of Mystery goes down into the
  Caves!   660
Her great men howl & throw the dust, & rend their
  hoary hair.
Her delicate women & children shriek upon the bitter
  wind,
Spoil'd of their beauty, their hair rent & their skin
  shrivel'd up.

"Lo, darkness covers the long pomp of banners on the
  wind,
And black horses & armed men & miserable bound
  captives.
Where shall the graves recieve them all, & where shall
  be their place?
And who shall mourn for Mystery who never loos'd
  her Captives?

"Let the slave, grinding at the mill, run out into the
  field;
Let him look up into the heavens & laugh in the bright
  air.
Let the inchained soul, shut up in darkness & in sighing,
Whose face has never seen a smile in thirty weary
  years,   671
Rise & look out: his chains are loose, his dungeon doors
  are open;
And let his wife & children return from the opressor's
  scourge.

"They look behind at every step & believe it is a dream.
Are these the slaves that groan'd along the streets of
  Mystery?
Where are your bonds & task masters? are these the
  prisoners?
Where are your chains? where are your tears? why do
  you look around?

635–37. In . . . embrace: Ephesians 3 : 10 (Blake's marginal
gloss).

668–74. Let . . . dream: repeated from *America*, ll. 42–48.

If you are thirsty, there is the river: go, bathe your
    parched limbs,
The good of all the Land is before you, for Mystery is
    no more."

Then All the Slaves from every Earth in the wide
    Universe   680
Sing a New Song, drowning confusion in its happy
    notes,
While the flail of Urizen sounded loud, & the winnow-
    ing wind of Tharmas
So loud, so clear in the wide heavens; & the song that
    they sung was this,
Composed by an African Black from the little Earth
    of Sotha:

"Aha! Aha! how came I here so soon in my sweet
    native land?
How came I here? Methinks I am as I was in my youth
When in my father's house I sat & heard his chearing
    voice.
Methinks I see his flocks & herds & feel my limbs
    renew'd,
And Lo, my Brethren in their tents, & their little ones
    arornd them!"

The song arose to the Golden feast; the Eternal Man
    rejoic'd.   690
Then the Eternal Man said: "Luvah, the Vintage is
    ripe: arise!
The sons of Urizen shall gather the vintage with sharp
    hooks,
And all thy sons, O Luvah! bear away the families of
    Earth.
I hear the flail of Urizen; his barns are full; no room
Remains, & in the Vineyards stand the abounding
    sheaves beneath
The falling Grapes that odorous burst upon the winds.
    Arise
My flocks & herds, trample the Corn! my cattle,
    browze upon
The ripe Clusters! The shepherds shout for Luvah,
    prince of Love.
Let the Bulls of Luvah tread the Corn & draw the
    loaded waggon
Into the Barn while children glean the Ears around the
    door.   700

Then shall they lift their innocent hands & stroke his
    furious nose,
And he shall lick the little girl's white neck & on her
    head
Scatter the perfume of his breath; while from his moun-
    tains high
The lion of terror shall come down, & bending his
    bright mane
And crouching at their side, shall eat from the curl'd
    boy's white lap
His golden food, and in the evening sleep before the
    door."

"Attempting to be more than Man We become less,"
    said Luvah
As he arose from the bright feast, drunk with the wine
    of ages.
His crown of thorns fell from his head, he hung his
    living Lyre
Behind the seat of the Eternal Man & took his
    way   710
Sounding the Song of Los, descending to the Vineyards
    bright.
His sons, arising from the feast with golden baskets,
    follow,
A fiery train, as when the Sun sings in the ripe vine-
    yards.
Then Luvah stood before the Wine press; all his fiery
    sons
Brought up the loaded Waggons with shoutings;
    ramping tygers play
In the jingling traces; furious lions sound the song of
    joy
To the golden wheels circling upon the pavement of
    heaven, & all
The Villages of Luvah ring; the golden tiles of the
    villages
Reply to violins & tabors, to the pipe, flute, lyre &
    cymbal.
Then fell the Legions of Mystery in madd'ning con-
    fusion,   720
Down, down thro' the immense, with outcry, fury &
    despair,
Into the wine presses of Luvah; howling fell the
    clusters
Of human families thro' the deep; the wine presses
    were fill'd;

691. Vintage: Cf. Revelation 14:17–19.

723. human families: Cf. *Jerusalem,* "To the Jews,"
below, ll. 77–80.

The blood of life flow'd plentiful. Odors of life arose
All round the heavenly arches, & the Odors rose singing
    this song:

"O terrible wine presses of Luvah! O caverns of the
    Grave!
How lovely the delights of those risen again from
    death!
O trembling joy! excess of joy is like Excess of grief."

So sang the Human Odors round the wine presses of
    Luvah;

But in the Wine presses is wailing, terror & despair. 730
Forsaken of their Elements they vanish & are no more,
No more but a desire of Being, a distracted, ravening
    desire,
Desiring like the hungry worm & like the gaping
    grave.
They plunge into the Elements; the Elements cast
    them forth
Or else consume their shadowy semblance. Yet they,
    obstinate
Tho' pained to distraction, cry, "O let us Exist! for
This dreadful Non Existence is worse than pains of
    Eternal Birth:
Eternal death who can Endure? let us consume in
    fires,
In waters stifling, or in air corroding, or in earth shut
    up.
The Pangs of Eternal birth are better than the Pangs of
    Eternal death."   740

How red the sons & daughters of Luvah! how they
    tread the Grapes!
Laughing & shouting, drunk with odors, many fall
    o'erwearied:
Drown'd in the wine is many a youth & maiden; those
    around
Lay them on skins of tygers or the spotted Leopard or
    wild Ass
Till they revive, or bury them in cool Grots making
    lamentation.

But in the Wine Presses the Human Grapes sing not
    nor dance,
They howl & writhe in shoals of torment, in fierce
    flames consuming,
In chains of iron & in dungeons circled with ceaseless
    fires,

In pits & dens & shades of death, in shapes of torment &
    woe;
The Plates, the Screws & Racks & Saws & cords &
    fires & floods,   750
The cruel joy of Luvah's daughters, lacerating with
    knives
And whips their Victims, & the deadly sport of Luvah's
    sons.
Timbrels & Violins sport round the Wine Presses. The
    little Seed,
The sportive root, the Earthworm, the small beetle,
    the wise Emmett,
Dance round the Wine Presses of Luvah; the Centipede
    is there,
The ground Spider with many eyes, the Mole clothed
    in Velvet,
The Earwig arm'd, the tender maggot, emblem of
    Immortality;
The slow slug, the grasshopper that sings & laughs &
    drinks:
The winter comes; he folds his slender bones without
    a murmur.
There is the Nettle that stings with soft down; &
    there
The indignant Thistle whose bitterness is bred in his
    milk   761
And who lives on the contempt of his neighbour;
    there all the idle weeds,
That creep about the obscure places, shew their various
    limbs
Naked in all their beauty, dancing round the Wine
    Presses.
They dance around the dying & they drink the howl &
    groan;
They catch the shrieks in cups of gold; they hand them
    to one another.
These are the sports of love & these the sweet delights
    of amorous play:
Tears of the grape, the death sweat of the Cluster, the
    last sigh
Of the mild youth who listens to the luring songs of
    Luvah.
The Eternal Man darken'd with sorrow & a wintry
    mantle   770
Cover'd the Hills. He said, "O Tharmas, rise! & O
    Urthona!"
Then Tharmas & Urthona rose from the Golden feast,
    satiated
With Mirth & Joy: Urthona, limping from his fall,
    on Tharmas lean'd,

In his right hand his hammer. Tharmas held his shepherd's crook

Beset with gold, gold were the ornaments form'd by
sons of Urizen.

Then Enion & Ahania & Vala & the wife of dark
Urthona

Rose from the feast, in joy ascending to their Golden
Looms.

There the wing'd shuttle sang, the spindle & the
distaff & the Reel

Rang sweet the praise of industry. Thro' all the golden
rooms

Heaven rang with winged Exultation. All beneath
howl'd loud;   780

With tenfold rout & desolation roar'd the Chasms beneath

Where the wide woof flow'd down & where the
Nations are gather'd together.

Tharmas went down to the Wine presses & beheld
the sons & daughters

Of Luvah quite exhausted with the labour & quite
fill'd

With new wine, that they began to torment one another
and to tread

The weak. Luvah & Vala slept on the floor, o'erwearied.

Urthona call'd his sons around him: Tharmas call'd
his sons

Numerous; they took the wine, they separated the
Lees,

And Luvah was put for dung on the ground by the
Sons of Tharmas & Urthona.

They formed heavens of sweetest woods, of gold &
silver & ivory,   790

Of glass & precious stones. They loaded all the waggons
of heaven

And took away the wine of ages with solemn songs &
joy.

Luvah & Vala woke, & all the sons & daughters of
Luvah

Awoke; they wept to one another & they reascended

To the Eternal Man in woe: he cast them wailing
into

The world of shadows, thro' the air; till winter is over
& gone;

But the Human Wine stood wondering; in all their
delightful Expanses

The elements subside; the heavens roll'd on with vocal
harmony.

Then Los, who is Urthona, rose in all his regenerate
power.

The Sea that roll'd & foam'd with darkness, & the
shadows of death   800

Vomited out & gave up all; the floods lift up their hands

Singing & shouting to the Man; they bow their hoary
heads

And murmuring in their channels flow & circle round
his feet.

Then Dark Urthona took the Corn out of the Stores
of Urizen;

He ground it in his rumbling Mills. Terrible the
distress

Of all the Nations of Earth, ground in the Mills of
Urthona.

In his hand Tharmas takes the Storms: he turns the
whirlwind loose

Upon the wheels; the stormy seas howl at his dread
command

And Eddying fierce rejoice in the fierce agitation of
the wheels

Of Dark Urthona. Thunders, Earthquakes, Fires,
Water floods,   810

Rejoice to one another; loud their voices shake the
Abyss,

Their dread forms tending the dire mills. The grey
hoar frost was there,

And his pale wife, the aged Snow; they watch over
the fires,

They build the Ovens of Urthona. Nature in darkness
groans

And Men are bound to sullen contemplation in the
night:

Restless they turn on beds of sorrow; in their inmost
brain

Feeling the crushing Wheels, they rise, they write the
bitter words

Of Stern Philosophy & knead the bread of knowledge
with tears & groans.

Such are the works of Dark Urthona. Tharmas sifts
the corn.

Urthona made the Bread of Ages, & he placed it,   820

In golden & in silver baskets, in heavens of precious
stone

And then took his repose in Winter, in the night of
Time.

800–01. The Sea . . . all: Cf. Revelation 20:13.

The Sun has left his blackness & has found a fresher
    morning,
And the mild moon rejoices in the clear & cloudless
    night,
And Man walks forth from midst of the fires: the evil
    is all consum'd.
His eyes behold the Angelic spheres arising night &
    day;
The stars consum'd like a lamp blown out, & in their
    stead, behold
The Expanding Eyes of Man behold the depths of
    wondrous worlds!
One Earth, one sea beneath; nor Erring Globes wander,
    but Stars
Of fire rise up nightly from the Ocean; & one Sun
Each morning, like a New born Man, issues with
    songs & joy   831
Calling the Plowman to his Labour & the Shepherd to
    his rest.
He walks upon the Eternal Mountains, raising his
    heavenly voice,
Conversing with the Animal forms of wisdom night
    & day,
That, risen from the Sea of fire, renew'd walk o'er the
    Earth;
For Tharmas brought his flocks upon the hills, & in
    the Vales
Around the Eternal Man's bright tent, the little Chil-
    dren play
Among the wooly flocks. The hammer of Urthona
    sounds
In the deep caves beneath; his limbs renew'd, his Lions
    roar
Around the Furnaces & in Evening sport upon the
    plains.   840
They raise their faces from the Earth, conversing with
    the Man:

"How is it we have walk'd thro' fires & yet are not
    consum'd?
How is it that all things are chang'd, even as in ancient
    times?"

The Sun arises from his dewy bed, & the fresh airs
Play in his smiling beams giving the seeds of life to
    grow,
And the fresh Earth beams forth ten thousand thousand
    springs of life.
Urthona is arisen in his strength, no longer now
Divided from Enitharmon, no longer the Spectre Los.

Where is the Spectre of Prophecy? where is the delu-
    sive Phantom?
Departed: & Urthona rises from the ruinous Walls 850
In all his ancient strength to form the golden armour of
    science
For intellectual War. The war of swords departed
    now,
The dark Religions are departed & sweet Science
    reigns.

c. 1797–1804 (1893)

MOCK ON, MOCK ON, VOLTAIRE, ROUSSEAU

From a notebook of 1800–03.

Mock on, Mock on, Voltaire, Rousseau:
Mock on, Mock on; 'tis all in vain!
You throw the sand against the wind,
And the wind blows it back again.

And every sand becomes a Gem
Reflected in the beams divine;
Blown back they blind the mocking Eye,
But still in Israel's paths they shine.

The Atoms of Democritus
And Newton's Particles of light   10
Are sands upon the Red sea shore,
Where Israel's tents do shine so bright.

1800–03 (1863)

853. **Science:** knowledge (*scientia*), not Newtonian science.
MOCK ON, MOCK ON, VOLTAIRE, ROUSSEAU. **1. Voltaire,
Rousseau:** Deists, in Blake's opinion mockers of faith.
9–10. Atoms . . . light: The Greek philosopher, Democritus, 460?–362? B.C., taught that all things are made up of
configurations of atoms; Newton advanced a corpuscular
theory of light as opposed to the wave theory that was
developed later. From Blake's point of view both men
were materialists.

THE MENTAL TRAVELLER

Composed probably in 1803, this extraordinarily compressed and allusive poem has been interpreted in many
different ways. It is an appalled conception of life in time
as a wheel on which man and his counterpart, woman or
Nature, move in opposite directions. Because they cannot
unite they are always proceeding toward another form of
suffering. The poem describes one completed cycle; at
the end the whole movement is beginning again.

I travel'd thro' a Land of Men,
A Land of Men & Women too,
And heard & saw such dreadful things
As cold Earth wanderers never knew.

For there the Babe is born in joy
That was begotten in dire woe;
Just as we Reap in joy the fruit
Which we in bitter tears did sow.

And if the Babe is born a Boy
He's given to a Woman Old,   10
Who nails him down upon a rock,
Catches his shrieks in cups of gold.

She binds iron thorns around his head,
She pierces both his hands & feet,
She cuts his heart out at his side
To make it feel both cold & heat.

Her fingers number every Nerve,
Just as a Miser counts his gold;
She lives upon his shrieks & cries,
And she grows young as he grows old.   20

Till he becomes a bleeding youth,
And she becomes a Virgin bright;
Then he rends up his Manacles
And binds her down for his delight.

He plants himself in all her Nerves,
Just as a Husbandman his mould;
And she becomes his dwelling place
And Garden fruitful seventy fold.

An aged Shadow, soon he fades,
Wand'ring round an Earthly Cot,   30
Full filled all with gems & gold
Which he by industry had got.

And these are the gems of the Human Soul,
The rubies & pearls of a lovesick eye,
The countless gold of the akeing heart,
The martyr's groan & the lover's sigh.

They are his meat, they are his drink;
He feeds the Beggar & the Poor
And the wayfaring Traveller:
For ever open is his door.   40

His grief is their eternal joy;
They make the roof & walls to ring;
Till from the fire on the hearth
A little Female Babe does spring.

And she is all of solid fire
And gems & gold, that none his hand
Dares stretch to touch her Baby form,
Or wrap her in his swaddling-band.

But she comes to the Man she loves,
If young or old, or rich or poor;   50
They soon drive out the aged Host,
A Beggar at another's door.

He wanders weeping far away,
Until some other take him in;
Oft blind & age-bent, sore distrest,
Untill he can a Maiden win.

And to allay his freezing Age
The Poor Man takes her in his arms;
The Cottage fades before his sight,
The Garden & its lovely Charms.   60

The Guests are scatter'd thro' the land,
For the Eye altering alters all;
The Senses roll themselves in fear,
And the flat Earth becomes a Ball;

The stars, sun, Moon, all shrink away,
A desart vast without a bound,
And nothing left to eat or drink,
And a dark desart all around.

THE MENTAL TRAVELLER. 1. 1: the mental traveler, a visitor
dismayed at what he sees in this strange world.

64. becomes a Ball: closed on itself.

The honey of her Infant lips,
The bread & wine of her sweet smile,   70
The wild game of her roving Eye,
Does him to Infancy beguile;

For as he eats & drinks he grows
Younger & younger every day;
And on the desart wild they both
Wander in terror & dismay.

Like the wild Stag she flees away,
Her fear plants many a thicket wild;
While he pursues her night & day,
By various arts of Love beguil'd,   80

By various arts of Love & Hate,
Till the wide desart planted o'er
With Labyrinths of wayward Love,
Where roam the Lion, Wolf & Boar.

Till he becomes a wayward Babe,
And she a weeping Woman Old.
Then many a Lover wanders here;
The Sun & Stars are nearer roll'd.

The trees bring forth sweet Extacy
To all who in the desart roam;   90
Till many a City there is Built,
And many a pleasant Shepherd's home.

But when they find the frowning Babe,
Terror strikes thro' the region wide:
They cry "The Babe! the Babe is Born!"
And flee away on Every side.

For who dare touch the frowning form,
His arm is wither'd to its root;
Lions, Boars, Wolves, all howling flee,
And every Tree does shed its fruit.   100

And none can touch that frowning form,
Except it be a Woman Old;
She nails him down upon the Rock,
And all is done as I have told.

                    1803 (1863)

THE CRYSTAL CABINET

"The Crystal Cabinet" was composed in 1803. As a poem about sexual love it is beautiful in its own right, but further implications emerge in the light of *Milton*. The word "threefold" (l. 15) is meaningful without a gloss, but in Blake it especially indicates that this happiness is not complete; if it were, it would be "fourfold." (Cf. *Milton*, I. 74, and the letter to Thomas Butts, November 22, 1802, below, ll. 83–88, pp. 165–66.)

The Maiden caught me in the Wild,
Where I was dancing merrily;
She put me into her Cabinet
And Lock'd me up with a golden Key.

This Cabinet is form'd of Gold
And Pearl & Crystal shining bright,
And within it opens into a World
And a little lovely Moony Night.

Another England there I saw,
Another London with its Tower,   10
Another Thames & other Hills,
And another pleasant Surrey Bower,

Another Maiden like herself,
Translucent, lovely, shining clear,
Threefold each in the other clos'd—
O, what a pleasant trembling fear!

O, what a smile! a threefold Smile
Fill'd me, that like a flame I burn'd;
I bent to Kiss the lovely Maid,
And found a Threefold Kiss return'd.   20

I strove to sieze the inmost Form
With ardor fierce & hands of flame,
But burst the Crystal Cabinet,
And like a Weeping Babe became—

A weeping Babe upon the wild,
And Weeping Woman pale reclin'd,
And in the outward air again
I fill'd with woes the passing Wind.

                    1803 (1863)

AUGURIES OF INNOCENCE

∽∾∽

This famous string of aphorisms comes from a notebook of 1803. The poem would undoubtedly have been much revised before being engraved. It is possible that the title refers only to the first four lines.

∽∾∽

To see a World in a Grain of Sand
And a Heaven in a Wild Flower,
Hold Infinity in the palm of your hand
And Eternity in an hour.
A Robin Red breast in a Cage
Puts all Heaven in a Rage.
A dove house fill'd with doves & Pigeons
Shudders Hell thro' all its regions.
A dog starv'd at his Master's Gate
Predicts the ruin of the State.   10
A Horse misus'd upon the Road
Calls to Heaven for Human blood.
Each outcry of the hunted Hare
A fibre from the Brain does tear.
A Skylark wounded in the wing,
A Cherubim does cease to sing.
The Game Cock clip'd & arm'd for fight
Does the Rising Sun affright.
Every Wolf's & Lion's howl
Raises from Hell a Human Soul.   20
The wild deer, wand'ring here & there,
Keeps the Human Soul from Care.
The Lamb misus'd breeds Public strife
And yet forgives the Butcher's Knife.
The Bat that flits at close of Eve
Has left the Brain that won't Believe.
The Owl that calls upon the Night
Speaks the Unbeliever's fright.
He who shall hurt the little Wren
Shall never be belov'd by Men.   30
He who the Ox to wrath has mov'd
Shall never be by Woman lov'd.
The wanton Boy that kills the Fly
Shall feel the Spider's enmity.
He who torments the Chafer's sprite
Weaves a Bower in endless Night.

The Catterpiller on the Leaf
Repeats to thee thy Mother's grief.
Kill not the Moth nor Butterfly,
For the Last Judgment draweth nigh.   40
He who shall train the Horse to War
Shall never pass the Polar Bar.
The Begger's Dog & Widow's Cat,
Feed them & thou wilt grow fat.
The Gnat that sings his Summer's song
Poison gets from Slander's tongue.
The poison of the Snake & Newt
Is the sweat of Envy's Foot.
The Poison of the Honey Bee
Is the Artist's Jealousy.   50
The Prince's Robes & Beggar's Rags
Are Toadstools on the Miser's Bags.
A truth that's told with bad intent
Beats all the Lies you can invent.
It is right it should be so;
Man was made for Joy & Woe;
And when this we rightly know
Thro' the World we safely go.
Joy & Woe are woven fine,
A Clothing for the Soul divine;   60
Under every grief & pine
Runs a joy with silken twine.
The Babe is more than swadling Bands;
Throughout all these Human Lands
Tools were made, & Born were hands,
Every Farmer Understands.
Every Tear from Every Eye
Becomes a Babe in Eternity;
This is caught by Females bright
And return'd to its own delight.   70
The Bleat, the Bark, Bellow & Roar
Are Waves that Beat on Heaven's Shore.
The Babe that weeps the Rod beneath
Writes Revenge in realms of death.
The Beggar's Rags, fluttering in Air,
Does to Rags the Heavens tear.
The Soldier, arm'd with Sword & Gun,
Palsied strikes the Summer's Sun.
The poor Man's Farthing is worth more
Than all the Gold on Afric's Shore.   80
One Mite wrung from the Labrer's hands
Shall buy & sell the Miser's Lands:
Or, if protected from on high,
Does that whole Nation sell & buy.
He who mocks the Infant's Faith
Shall be mock'd in Age & Death.

AUGURIES OF INNOCENCE. **17. Cock . . . fight:** in the sport of cockfighting. **35. Chafer:** beetle.

He who shall teach the Child to Doubt
The rotting Grave shall ne'er get out.
He who respects the Infant's faith
Triumphs over Hell & Death.   90
The Child's Toys & the Old Man's Reasons
Are the Fruits of the Two seasons.
The Questioner, who sits so sly,
Shall never know how to Reply.
He who replies to words of Doubt
Doth put the Light of Knowledge out.
The Strongest Poison ever known
Came from Caesar's Laurel Crown.
Nought can deform the Human Race
Like to the Armour's iron brace.   100
When Gold & Gems adorn the Plow
To peaceful Arts shall Envy Bow.
A Riddle or the Cricket's Cry
Is to Doubt a fit Reply.
The Emmet's Inch & Eagle's Mile
Make Lame Philosophy to smile.
He who Doubts from what he sees
Will ne'er Believe, do what you Please.
If the Sun & Moon should doubt,
They'd immediately Go out.   110
To be in a Passion you Good may do,
But no Good if a Passion is in you.
The Whore & Gambler, by the State
Licenc'd, build that Nation's Fate.
The Harlot's cry from Street to Street
Shall weave Old England's winding Sheet.
The Winner's Shout, the Loser's Curse,
Dance before dead England's Hearse.
Every Night & every Morn
Some to Misery are Born.   120
Every Morn & every Night
Some are Born to sweet delight.
Some are Born to sweet delight,
Some are Born to Endless Night.
We are led to Believe a Lie
When we see With not Thro' the Eye
Which was Born in a Night to perish in a Night
When the Soul Slept in Beams of Light.
God Appears & God is Light
To those poor Souls who dwell in Night,   130
But does a Human Form Display
To those who Dwell in Realms of day.

1803 (1863)

126. **With . . . Eye:** Cf. the concluding paragraph of "A Vision of the Last Judgment," below.

MILTON

A POEM IN 2 BOOKS
TO JUSTIFY THE WAYS OF GOD TO MEN

Composed probably between 1800 and 1804, with later additions, *Milton* is the shortest of the major prophetic works. To Blake, Milton was a supreme poet, but he had unaccountably celebrated the false god, Urizen, who struggles to keep human energy and imagination in chains, and he had thus divided and vitiated his own being. (See "The Voice of the Devil" at the start of *The Marriage of Heaven and Hell*, p. 70.) Hence for Milton to become his true self (rather than his specter) he would have to cast off his errors, and Blake's poem is the story of his doing so. It begins with the myth of creation, briefly recapitulated from *The Book of Urizen*, and then with a Bard singing the quarrel of Satan and Palamabron, who is, among other things, a representation of Blake himself. The song moves Milton to descend from Eternity into our world in order to redeem all that he has produced, his "emanation," Ololon. To redeem is to restore fullness of life by abolishing the usurped authority of Urizen (now also Satan). In narrative terms this means that Urizen tries to block the coming of Milton, and an epic struggle takes place. The sleeping Albion begins to waken, and at this point Milton enters into Blake, giving him a heightened power of imagination or prophecy. From now to the end of Book I the whole of time or history is seen in a moment of vision. In the second book Ololon descends. Seeking Milton, she appears in Blake's garden at Felpham. At this final crisis the error or Satan in Milton collects itself for its most tremendous struggle against him, but Satan is repudiated and Albion awakes (though still not completely). Gazing on Milton, Ololon begins to understand that she is a source of error, but under the guise of pity she hesitates to purge or sacrifice herself. Thus she shows herself as Rahab, as selfishness, mystery, coy virginity, and a hidden thirst for power. But Milton confronts and masters this manifestation. Ololon is redeemed, and, as the poem ends, the earth is now prepared to go forward to the Last Judgment.

The events of the bard's song in Book I are an imaginative transformation and use of the unpleasant situation between Blake (Palamabron) and his would-be patron, William Hayley (Satan). (See Introduction.)

The numbers (1–49) by which the poem is divided are of Blake's engraved plates, not all of which contain text. Blake could and sometimes did change the order of these plates.

MILTON. **Motto:** *Paradise Lost*, I. 26.

PREFACE

I

The Stolen and Perverted Writings[1] of Homer & Ovid, of Plato & Cicero, which all Men ought to contemn, are set up by artifice against the Sublime of the Bible; but when the New Age is at leisure to Pronounce, all will be set right, & those Grand Works of the more ancient & consciously & professedly Inspired Men[2] will hold their proper rank, & the Daughters of Memory shall become the Daughters of Inspiration. Shakespeare & Milton were both curb'd by the general malady & infection from the silly Greek & Latin slaves of the Sword.

Rouze up, O Young Men of the New Age! set your foreheads against the ignorant Hirelings! For we have Hirelings in the Camp, the Court & the University, who would, if they could, for ever depress Mental & prolong Corporeal War. Painters! on you I call. Sculptors! Architects! Suffer not the fashionable Fools to depress your powers by the prices they pretend to give for contemptible works, or the expensive advertizing boasts that they make of such works: believe Christ & His Apostles that there is a Class of Men whose whole delight is in Destroying. We do not want either Greek or Roman Models if we are but just & true to our own Imaginations, those Worlds of Eternity in which we shall live for ever in Jesus our Lord.

> And did those feet in ancient time
> Walk upon England's mountains green?
> And was the holy Lamb of God
> On England's pleasant pastures seen?
>
> And did the Countenance Divine
> Shine forth upon our clouded hills?
> And was Jerusalem builded here
> Among these dark Satanic Mills?
>
> Bring me my Bow of burning gold:
> Bring me my Arrows of desire: 10
> Bring me my Spear: O clouds, unfold!
> Bring me my Chariot of fire.

> I will not cease from Mental Fight,
> Nor shall my Sword sleep in my hand
> Till we have built Jerusalem
> In England's green & pleasant Land.

"Would to God that all the Lord's people were Prophets."[3]

Numbers, XI. 29.

BOOK THE FIRST

2

> Daughters of Beulah! Muses who inspire the Poet's Song,
> Record the journey of immortal Milton thro' your Realms
> Of terror & mild moony lustre in soft sexual delusions
> Of varied beauty, to delight the wanderer and repose
> His burning thirst & freezing hunger! Come into my hand,
> By your mild power descending down the Nerves of my right arm
> From out the portals of my Brain, where by your ministry
> The Eternal Great Humanity Divine planted his Paradise

3. all . . . Prophets: Cf. Milton's *Areopagitica*: "For now the time seems come, wherein Moses, the great prophet, may sit in heaven rejoicing to see that memorable and glorious wish of his fulfilled, when not only our seventy elders, but all the Lord's people, are become prophets." BOOK I. **1. Beulah:** Beulah is the "threefold" paradise of happy sexual love as opposed to the completed, "fourfold" paradise of art or imagination. It is also a psychological state, approximately that portrayed in the *Songs of Innocence*, where the mind is married to a lovingly responsive Nature and can be spontaneously joyful. To the Eternals Beulah is a place of relaxation from the "Mental Fight" of their higher paradise, but to men on earth Beulah is a source of poetic or prophetic inspiration. As a vision or aspiration it may be vague, but it still produces a continuing, potentially explosive or revolutionary dissatisfaction with the world as we find it. In Blake's cyclic interpretation of history, this revolutionary potential periodically breaks out, creates a new civilization (of which there have been seven), and gradually ages into a new tyranny. From this point of view Beulah is the reservoir of the recurrently renewed youth of time. But the Last Judgment that will finally annihilate these rotating cycles of history cannot come from Beulah. It may be noticed that despite Blake's animus against the classics, lines 1–20 are a classical invocation to the muses imitating the start of *Paradise Lost*.

PREFACE. **1. Stolen . . . Writings:** Greek art and religion were sometimes interpreted as a degenerate rendering of what had been originally derived from the ancient Hebrews. Cf. Milton's *Paradise Regained*, IV. 336–40. **2. Inspired Men:** the biblical prophets.

And in it caus'd the Spectres of the Dead to take sweet
    forms
In likeness of himself. Tell also of the False Tongue!
    vegetated   10
Beneath your land of shadows, of its sacrifices and
Its offerings: even till Jesus, the image of the Invisible
    God,
Became its prey, a curse, an offering and an atonement
For Death Eternal in the heavens of Albion & before
    the Gates
Of Jerusalem his Emanation, in the heavens beneath
    Beulah.

Say first! what mov'd Milton, who walk'd about in
    Eternity
One hundred years, pond'ring the intricate mazes of
    Providence,
Unhappy tho' in heav'n—he obey'd, he murmur'd not,
    he was silent
Viewing his Sixfold Emanation scatter'd thro' the
    deep
In torment—To go into the deep her to redeem & him-
    self perish?   20
That cause at length mov'd Milton to this unexampled
    deed.
A Bard's prophetic Song! for sitting at eternal tables,
Terrific among the Sons of Albion, in chorus solemn &
    loud
A Bard broke forth: all sat attentive to the awful man.

Mark well my words! they are of your eternal salvation.

Three Classes are Created by the Hammer of Los &
    Woven

3

By Enitharmon's Looms when Albion was slain upon
    his Mountains
And in his Tent, thro' envy of the Living Form, even
    of the Divine Vision,

And of the sports of Wisdom in the Human Imagina-
    tion,
Which is the Divine Body of the Lord Jesus, blessed
    for ever.   30
Mark well my words! they are of your eternal salvation.

Urizen lay in darkness & solitude, in chains of the mind
    lock'd up
Los siez'd his Hammer & Tongs; he labour'd at his
    resolute Anvil
Among indefinite Druid rocks & snows of doubt &
    reasoning.

Refusing all Definite Form, the Abstract Horror
    roof'd, stony hard;
And a first Age passed over, & a State of dismal woe.

Down sunk with fright a red round Globe, hot burning,
    deep,
Deep down into the Abyss, panting, conglobing,
    trembling;
And a second Age passed over, & a State of dismal woe.

Rolling round into two little Orbs, & closed in two
    little Caves,   40
The Eyes beheld the Abyss, lest bones of solidness
    freeze over all;
And a third Age passed over, & a State of dismal woe.

From beneath his Orbs of Vision, Two Ears in close
    volutions
Shot spiring out in the deep darkness & petrified as
    they grew;
And a fourth Age passed over, & a State of dismal woe.

Hanging upon the wind, Two Nostrils bent down into
    the Deep;
And a fifth Age passed over, & a state of dismal woe.

In ghastly torment sick, a Tongue of hunger & thirst
    flamed out;
And a sixth Age passed over, & a State of dismal woe.

---

10. False Tongue: Satan. Blake no longer uses the ironic
inversion of *The Marriage of Heaven and Hell.* **11–12. sacrifices . . . offerings:** characteristic of false religions. **15.
Emanation:** Blake's word for the totality of all that a man
loves and creates represented as his bride. **17. intricate . . .
Providence:** Cf. the Devils in *Paradise Lost*, II. 558–61.
19. Sixfold: Milton had three wives and three daughters.
26. Three Classes: See I. 176–80, 368–69, 1080–84. **27–66.
Albion . . . family:** Blake repeats in condensed form the
creation myth of *The Book of Urizen.* It is also retold in *The
Four Zoas.*

34. Druid: The Druids were the priestly class of the ancient
Celts. In Blake's mythology they are giant survivors of the
drowned continent of Atlantis and symbols of natural
religion. They are associated with reason as opposed to imaginative inspiration; hence to them Nature and its God are a
mystery that they attempt to propitiate with human sacrifice.

Enraged & stifled without & within, in terror & woe he
   threw his   50
Right Arm to the north, his left Arm to the south, &
   his feet
Stamp'd the nether Abyss in trembling & howling &
   dismay;
And a seventh Age passed over, & a State of dismal
   woe.

Terrified, Los stood in the Abyss, & his immortal
   limbs
Grew deadly pale: he became what he beheld, for a
   red
Round Globe sunk down from his Bosom into the
   Deep; in pangs
He hover'd over it trembling & weeping: suspended it
   shook
The nether Abyss; in tremblings he wept over it, he
   cherish'd it
In deadly, sickening pain, till separated into a Female
   pale
As the cloud that brings the snow; all the while from
   his Back   60
A blue fluid exuded in Sinews, hardening in the Abyss
Till it separated into a Male Form howling in Jealousy.

Within labouring, beholding Without, from Particu-
   lars to Generals
Subduing his Spectre, they Builded the Looms of
   Generation;
They builded Great Golgonooza Times on Times, Ages
   on Ages.
First Orc was Born, then the shadowy Female: then
   All Los's family.
At last Enitharmon brought forth Satan, Refusing
   Form in vain,
The Miller of Eternity made subservient to the Great
   Harvest
That he may go to his own Place, Prince of the Starry
   Wheels

4

Beneath the Plow of Rintrah & the Harrow of the
   Almighty   70
In the hands of Palamabron, Where the Starry Mills
   of Satan
Are built beneath the Earth & Waters of the Mundane
   Shell:
Here the Three Classes of Men take their Sexual texture,
   Woven;
The Sexual is Threefold, the Human is Fourfold.

"If you account it Wisdom when you are angry to be
   silent and
Not to shew it, I do not account that Wisdom, but
   Folly.
Every Man's Wisdom is peculiar to his own Individu-
   ality.
O Satan, my youngest born, art thou not Prince of the
   Starry Hosts
And of the Wheels of Heaven, to turn the Mills day
   & night?
Art thou not Newton's Pantocrator, weaving the
   Woof of Locke?   80
To Mortals thy Mills seem every thing, & the Harrow
   of Shaddai
A Scheme of Human conduct invisible & incompre-
   hensible.
Get to thy Labours at the Mills & leave me to my
   wrath."

Satan was going to reply, but Los roll'd his loud
   thunders.

"Anger me not! thou canst not drive the Harrow in
   pity's paths:
Thy work is Eternal Death with Mills & Ovens &
   Cauldrons.
Trouble me no more; thou canst not have Eternal
   Life."

64. Generation: the world of Experience. Lines 54–62 describe the fall of Los into this world, in which reality becomes "twofold," split into the subjective and objective, the perceiver and thing perceived. Nature, in other words, becomes a removed and tantalizing "Female pale," and man "a Male Form howling in Jealousy" to possess her. History in this realm of generation moves in the recurrent cyclic pattern ("Times on Times") summed up in lines 66–72. **65. Golgonooza:** the city of Art. See I. 142. **67. Satan:** In previous poems he has been the fallen Urizen, "Prince of the Starry Wheels" (l. 69). See I. 327.

70. Rintrah: See *The Marriage of Heaven and Hell*, above, n. 1. For the symbolism of the plow, see especially *The Four Zoas*, "Night the Ninth," above, ll. 290–313. The harrow follows the plow, further breaking up the earth for the sowing and harvest. **71. Palamabron:** "the Redeem'd" (l. 369). **72. Mundane Shell:** the physical universe of the world of generation. See I. 601–07. **80. Pantocrator:** the ruler of the universe as conceived by Newton and Locke. **81. Shaddai:** a Hebrew name for God.

So Los spoke. Satan trembling obey'd, weeping along
   the way.
Mark well my words! they are of your eternal Salvation.

Between South Molton Street & Stratford Place,
   Calvary's foot,   90
Where the Victims were preparing for Sacrifice their
   Cherubim;
Around their Loins pour'd forth their arrows, & their
   bosoms beam
With all colours of precious stones, & their inmost
   palaces
Resounded with preparation of animals wild & tame,
(Mark well my words: Corporeal Friends are Spiritual
   Enemies)
Mocking Druidical Mathematical Proportion of
   Length, Bredth, Highth:
Displaying Naked Beauty, with Flute & Harp & Song.

                          5

Palamabron with the fiery Harrow in morning return-
   ing
From breathing fields, Satan fainted beneath the
   artillery.
Christ took on Sin in the Virgin's Womb & put it off
   on the Cross.   100
All pitied the piteous & was wrath with the wrathful,
   & Los heard it.

And this is the manner of the Daughters of Albion in
   their beauty.
Every one is threefold in Head & Heart & Reins, &
   every one
Has three Gates into the Three Heavens of Beulah,
   which shine
Translucent in their Foreheads & their Bosoms & their
   Loins
Surrounded with fires unapproachable: but whom
   they please
They take up into their Heavens in intoxicating
   delight;
For the Elect cannot be Redeem'd, but Created con-
   tinually

By Offering & Atonement in the cruelties of Moral
   Law.
Hence the three Classes of Men take their fix'd destina-
   tions.   110
They are the Two Contraries & the Reasoning Nega-
   tive.

While the Females prepare the Victims, the Males at
   Furnaces
And Anvils dance the dance of tears & pain: loud
   lightnings
Lash on their limbs as they turn the whirlwinds loose
   upon
The Furnaces, lamenting around the Anvils, & this
   their Song:

"Ah weak & wide astray! Ah shut in narrow doleful
   form,
Creeping in reptile flesh upon the bosom of the
   ground!
The Eye of Man a little narrow orb, clos'd up & dark,
Scarcely beholding the great light, conversing with
   the Void;
The Ear a little shell, in small volutions shutting out 120
All melodies & comprehending only Discord and
   Harmony;
The Tongue a little moisture fills, a little food it cloys,
A little sound it utters & its cries are faintly heard,
Then brings forth Moral Virtue the cruel Virgin
   Babylon.

"Can such an Eye judge of the stars? & looking thro'
   its tubes
Measure the sunny rays that point their spears on
   Udanadan?
Can such an Ear, fill'd with the vapours of the yawning
   pit,
Judge of the pure melodious harp struck by a hand
   divine?
Can such closed Nostrils feel a joy? or tell of autumn
   fruits
When grapes & figs burst their covering to the joyful
   air?   130
Can such a Tongue boast of the living waters? or take
   in
Ought but the Vegetable Ratio & loathe the faint
   delight?

90. South Molton . . . Calvary: Blake lived at 17 South
Molton Street after returning to London from his three-year
stay in the village of Felpham. Christ was crucified at
Calvary. 91. Victims: prophetic artists such as Jesus preparing for mental strife.

126. Udanadan: the sea, chaos, the indefinite. 132. Vegetable
Ratio: See The Marriage of Heaven and Hell, above, n. 11.

Can such gross Lips percieve? alas, folded within themselves

They touch not ought, but pallid turn & tremble at
every wind."

Thus they sing Creating the Three Classes among
Druid Rocks.

Charles calls on Milton for Atonement. Cromwell is
ready.

James calls for fires in Golgonooza, for heaps of smoking ruins

In the night of prosperity and wantonness which he
himself Created

Among the Daughters of Albion, among the Rocks of
the Druids

When Satan fainted beneath the arrows of Elynittria,

And Mathematic Proportion was subdued by Living
Proportion.   141

6

From Golgonooza the spiritual Four-fold London
eternal,

In immense labours & sorrows, every building, ever
falling,

Thro' Albion's four Forests which overspread all the
Earth

From London Stone to Blackheath east: to Hounslow
west:

To Finchley north: to Norwood south: and the
weights

Of Enitharmon's Loom play lulling cadences on the
winds of Albion

From Caithness in the north to Lizard-point & Dover
in the south.

Loud sounds the Hammer of Los & loud his Bellows
is heard

Before London to Hampstead's breadths & Highgate's
heights, To   150

Stratford & old Bow & across to the Gardens of
Kensington

On Tyburn's Brook: loud groans Thames beneath the
iron Forge

Of Rintrah & Palamabron, of Theotorm & Bromion,
to forge the instruments

Of Harvest, the Plow & Harrow to pass over the
Nations.

The Surrey hills glow like the clinkers of the furnace;
Lambeth's Vale

Where Jerusalem's foundations began, where they
were laid in ruins,

Where they were laid in ruins from every Nation, &
Oak Groves rooted,

Dark gleams before the Furnace-mouth a heap of
burning ashes.

When shall Jerusalem return & overspread all the
Nations?

Return, return to Lambeth's Vale, O building of
human souls!   160

Thence stony Druid Temples overspread the Island
white,

And thence from Jerusalem's ruins, from her walls of
salvation

And praise, thro' the whole Earth were rear'd from
Ireland

To Mexico & Peru west, & east to China & Japan, till
Babel

The Spectre of Albion frown'd over the Nations in
glory & war.

All things begin & end in Albion's ancient Druid rocky
shore:

But now the Starry Heavens are fled from the mighty
limbs of Albion.

Loud sounds the Hammer of Los, loud turn the Wheels
of Enitharmon:

Her Looms vibrate with soft affections, weaving the
Web of Life,   169

Out from the ashes of the Dead; Los lifts his iron Ladles.

With molten ore: he heaves the iron cliffs in his rattling
chains

From Hyde Park to the Alms-houses of Mile-end & old
Bow.

Here the Three Classes of Mortal Men take their
fix'd destinations,

---

153. Rintrah ... Bromion: Blake enumerates four sons of
Los who vaguely correspond to the four Zoas or human
faculties. **155. Lambeth:** Blake wrote most of his shorter
prophetic works while living in Lambeth. **157. Oak Groves:**
Associated with the Druids, oak groves are symbols of ancient,
deep-rooted, labyrinthine error. **164. Babel:** See Genesis
11:1–9.

---

140. Elynittria: Palamabron's emanation. **144. Albion:**
the eternal man of *The Four Zoas,* also England. **150–52.**
Hampstead ... Tyburn: suburbs of London.

And hence they overspread the Nations of the whole
    Earth, & hence
The Web of Life is woven & the tender sinews of life
    created
And the Three Classes of Men regulated by Los's
    Hammers [and woven

7

By Enitharmon's Looms & Spun beneath the Spindle
    of Tirzah. *erased*]
The first, The Elect from before the foundation of the
    World:
The second, The Redeem'd: The Third, The Repro-
    bate & form'd
To destruction from the mother's womb: .   . 180
.   . [*words erased*] follow with me my plow.

Of the first class was Satan: with incomparable mild-
    ness,
His primitive tyrannical attempts on Los, with most
    endearing love
He soft intreated Los to give to him Palamabron's
    station,
For Palamabron return'd with labour wearied every
    evening.
Palamabron oft refus'd, and as often Satan offer'd
His service, till by repeated offers and repeated in-
    treaties
Los gave to him the Harrow of the Almighty; alas,
    blamable,
Palamabron fear'd to be angry lest Satan should accuse
    him of
Ingratitude & Los believe the accusation thro' Satan's
    extreme   190
Mildness. Satan labour'd all day: it was a thousand
    years:
In the evening returning terrified, overlabour'd &
    astonish'd,
Embrac'd soft with a brother's tears Palamabron, who
    also wept.

177. Tirzah: See "To Tirzah," above, in *Songs of Experience*. In Numbers 26:33, Tirzah is one of the five
daughters of Zelophehad, whom Blake identifies with the
five senses (often by synechdoche he uses Tirzah to stand for
the five). As Zelophehad had no sons his daughters were
granted his inheritance (Numbers 27:1–8), which may be
interpreted as a victory of the "female will," that is, Nature or
woman becoming coyly elusive, mysterious, and independent. See I. 54–62.

Mark well my words! they are of your eternal salva-
    tion.

Next morning Palamabron rose; the horses of the
    Harrow
Were madden'd with tormenting fury, & the servants
    of the Harrow,
The Gnomes, accus'd Satan with indignation, fury and
    fire.
Then Palamabron, reddening like the Moon in an
    Eclipse,
Spoke, saying: "You know Satan's mildness and his
    self-imposition,
Seeming a brother, being a tyrant, even thinking himself
    a brother   200
While he is murdering the just: prophetic I behold
His future course thro' darkness and despair to eternal
    death.
But we must not be tyrants also: he hath assum'd my
    place
For one whole day under pretence of pity and love to
    me.
My horses hath he madden'd and my fellow servants
    injur'd.
How should he, he, know the duties of another? O
    foolish forbearance!
Would I had told Los all my heart! but patience, O
    my friends,
All may be well: silent remain, while I call Los and
    Satan."

Loud as the wind of Beulah that unroots the rocks &
    hills
Palamabron call'd, and Los & Satan came before
    him,
And Palamabron shew'd the horses & the servants.
    Satan wept   211
And mildly cursing Palamabron, him accus'd of
    crimes
Himself had wrought. Los trembled: Satan's blandish-
    ments almost
Perswaded the Prophet of Eternity that Palamabron
Was Satan's enemy & that the Gnomes, being Palama-
    bron's friends,
Were leagued together against Satan thro' ancient
    enmity.
What could Los do? how could he judge, when
    Satan's self believ'd
That he had not oppres'd the horses of the Harrow nor
    the servants.

So Los said: "Henceforth, Palamabron, let each his
   own station
Keep: nor in pity false, nor in officious brotherhood,
   where   220
None needs, be active." Mean time Palamabron's horses
Rag'd with thick flames redundant, & the Harrow
   madden'd with fury.
Trembling Palamabron stood; the strongest of De-
   mons trembled,
Curbing his living creatures; many of the strongest
   Gnomes
They bit in their wild fury, who also madden'd like
   wildest beasts.

Mark well my words! they are of your eternal salvation.

8

Mean while wept Satan before Los accusing Palamabron
Himself exculpating with mildest speech, for himself
   believ'd
That he had not oppress'd nor injur'd the refractory
   servants.

But Satan returning to his Mills (for Palamabron had
   serv'd   230
The Mills of Satan as the easier task) found all confusion,
And back return'd to Los, not fill'd with vengeance
   but with tears,
Himself convinc'd of Palamabron's turpitude. Los beheld
The servants of the Mills drunken with wine and
   dancing wild
With shouts and Palamabron's songs, rending the
   forests green
With ecchoing confusion, tho' the Sun was risen on
   high.

Then Los took off his left sandal, placing it on his head,
Signal of solemn mourning: when the servants of the
   Mills
Beheld the signal they in silence stood, tho' drunk with
   wine.   239
Los wept! But Rintrah also came, and Enitharmon on
His arm lean'd tremblingly, observing all these things.

And Los said: "Ye Genii of the Mills! the Sun is on
   high,
Your labours call you: Palamabron is also in sad
   dilemma:

His horses are mad, his Harrow confounded, his companions enrag'd.
Mine is the fault! I should have remember'd that pity
   divides the soul
And man unmans; follow with me my Plow: this
   mournful day
Must be a blank in Nature: follow with me and to-
   morrow again
Resume your labours, & this day shall be a mournful
   day."

Wildly they follow'd Los and Rintrah, & the Mills
   were silent.
They mourn'd all day, this mournful day of Satan &
   Palamabron: 250
And all the Elect & all the Redeem'd mourn'd one
   toward another
Upon the mountains of Albion among the cliffs of the
   Dead.

They Plow'd in tears; incessant pour'd Jehovah's rain
   & Molech's
Thick fires contending with the rain thunder'd above,
   rolling
Terrible over their heads; Satan wept over Palamabron.
Theotormon & Bromion contended on the side of
   Satan,
Pitying his youth and beauty, trembling at eternal
   death.
Michael contended against Satan in the rolling thunder:
Thulloh the friend of Satan also reprov'd him: faint
   their reproof.

But Rintrah who is of the reprobate, of those form'd to
   destruction,   260
In indignation for Satan's soft dissimulation of friendship
Flam'd above all the plowed furrows, angry, red and
   furious,
Till Michael sat down in the furrow, weary, dissolv'd
   in tears.
Satan, who drave the team beside him, stood angry &
   red:
He smote Thulloh & slew him, & he stood terrible
   over Michael

---

253. Molech: One of the false gods of the Old Testament,
Moloch was a fire deity to whom children were sacrificed.
Cf. *Paradise Lost*, I. 392–93. **258. Michael . . . Satan:** Cf.
Revelation 12:7 and *Paradise Lost*, VI.

Urging him to arise: he wept: Enitharmon saw his
    tears.
But Los hid Thulloh from her sight, lest she should die
    of grief.
She wept, she trembled, she kissed Satan, she wept
    over Michael:
She form'd a Space for Satan & Michael & for the poor
    infected.
Trembling she wept over the Space & clos'd it with a
    tender Moon.   270

Los secret buried Thulloh, weeping disconsolate over
    the moony Space.

But Palamabron called down a Great Solemn Assembly,
That he who will not defend Truth, may be compelled
    to
Defend a Lie, that he may be snared & caught & taken.

                        9

And all Eden descended into Palamabron's tent
Among Albion's Druids & Bards in the caves beneath
    Albion's
Death Couch, in the caverns of death, in the corner of
    the Atlantic.
And in the midst of the Great Assembly Palamabron
    pray'd:
"O God, protect me from my friends, that they have
    not power over me.
Thou hast giv'n me power to protect myself from my
    bitterest enemies."   280

Mark well my words! they are of your eternal salvation.

Then rose the Two Witnesses, Rintrah & Palamabron:
And Palamabron appeal'd to all Eden and reciev'd
Judgment: and Lo! it fell on Rintrah and his rage,
Which now flam'd high & furious in Satan against
    Palamabron
Till it became a proverb in Eden: Satan is among the
    Reprobate.

Los in his wrath curs'd heaven & earth; he rent up
    Nations,
Standing on Albion's rocks among high-rear'd Druid
    temples
Which reach the stars of heaven & stretch from pole
    to pole.

282. Witnesses: Cf. Revelation 11.

He displac'd continents, the oceans fled before his
    face: 290
He alter'd the poles of the world, east, west & north &
    south,
But he clos'd up Enitharmon from the sight of all
    these things.

For Satan, flaming with Rintrah's fury hidden beneath
    his own mildness,
Accus'd Palamabron before the Assembly of ingrati-
    tude, of malice.
He created Seven deadly Sins, drawing out his infernal
    scroll
Of Moral laws and cruel punishments upon the clouds
    of Jehovah,
To pervert the Divine voice in its entrance to the earth
With thunder of war & trumpet's sound, with armies
    of disease,
Punishments & deaths muster'd & number'd, Saying:
    "I am God alone:
There is no other! let all obey my principles of moral
    individuality.   300
I have brought them from the uppermost, innermost
    recesses
Of my Eternal Mind: transgressors I will rend off for
    ever
As now I rend this accursed Family from my covering."

Thus Satan rag'd amidst the Assembly, and his
    bosom grew
Opake against the Divine Vision: the paved terraces of
His bosom inwards shone with fires, but the stones be-
    coming opake
Hid him from sight in an extreme blackness and dark-
    ness.
And there a World of deeper Ulro was open'd in the
    midst
Of the Assembly. In Satan's bosom, a vast unfathom-
    able Abyss.

Astonishment held the Assembly in an awful silence,
    and tears   310
Fell down as dews of night, & a loud solemn universal
    groan

299. I . . . alone: Cf. Ezekiel 28:2. 308. Ulro: hell,
the diametric opposite of paradise, or Eden. Psychologically
it is "single vision," a state of total isolation and self-absorption in which the mind broods among its own abstractions.

Was utter'd from the east & from the west & from the
south
And from the north; and Satan stood opake immeasurable,
Covering the east with solid blackness round his hidden
heart,
With thunders utter'd from his hidden wheels, accusing loud
The Divine Mercy for protecting Palamabron in his
tent.

Rintrah rear'd up walls of rocks and pour'd rivers &
moats
Of fire round the walls: columns of fire guard around
Between Satan and Palamabron in the terrible darkness.

And Satan not having the Science of Wrath, but only
of Pity,   320
Rent them asunder, and wrath was left to wrath, &
pity to pity.
He sunk down, a dreadful Death unlike the slumbers of
Beulah.

The Separation was terrible: the Dead was repos'd on
his Couch
Beneath the Couch of Albion, on the seven mountains
of Rome,
In the whole place of the Covering Cherub, Rome,
Babylon & Tyre.
His Spectre raging furious descended into its Space.

11

Then Los & Enitharmon knew that Satan is Urizen,
Drawn down by Orc & the Shadowy Female into
Generation.
Oft Enitharmon enter'd weeping into the Space, there
appearing
An aged Woman raving along the Streets (the Space is
named   330
Canaan): then she returned to Los, weary, frighted as
from dreams.

325. **Covering Cherub**: See Exodus 25:20 and Ezekiel
28:14–16. 331. **Canaan**: the land the Israelites conquered
and settled. It is mentioned several times in *Milton* and may
be considered in diverse aspects: the Promised Land; prior
to the occupation, the country of the enemies of the
Hebrews with their many false religions; the place of
warfare; and the land where the Israelites lived under the law.

The nature of a Female Space is this: it shrinks the
Organs
Of Life till they become Finite & Itself seems Infinite.

And Satan vibrated in the immensity of the Space,
Limited
To those without, but Infinite to those within: it fell
down and
Became Canaan, closing Los from Eternity in Albion's
Cliffs.
A mighty Fiend against the Divine Humanity, must'ring
to War.

"Satan, Ah me! is gone to his own place," said Los:
"their God
I will not worship in their Churches, nor King in their
Theatres.
Elynittria! whence is this Jealousy running along the
mountains?   340
British Women were not Jealous when Greek &
Roman were Jealous.
Every thing in Eternity shines by its own Internal light,
but thou
Darkenest every Internal light with the arrows of thy
quiver,
Bound up in the horns of Jealousy to a deadly fading
Moon,
And Ocalythron binds the Sun into a Jealous Globe,
That every thing is fix'd Opake without Internal
light."

So Los lamented over Satan who triumphant divided
the Nations.

12

He set his face against Jerusalem to destroy the Eon of
Albion.

But Los hid Enitharmon from the sight of all these
things
Upon the Thames whose lulling harmony repos'd her
soul,   350
Where Beulah lovely terminates in rocky Albion,
Terminating in Hyde Park on Tyburn's awful brook.

345. **Ocalythron**: Rintrah's emanation. 348. **He**: Satan.
Eon: emanation. 352. **Tyburn**: where the gallows stood.

And the Mills of Satan were separated into a moony
    Space
Among the rocks of Albion's Temples, and Satan's
    Druid sons
Offer the Human Victims throughout all the Earth,
    and Albion's
Dread Tomb, immortal on his Rock, overshadow'd the
    whole Earth,
Where Satan, making to himself Laws from his own
    identity,
Compell'd others to serve him in moral gratitude &
    submission,
Being call'd God, setting himself above all that is
    called God;
And all the Spectres of the Dead, calling themselves
    Sons of God,   360
In his Synagogues worship Satan under the Unutter-
    able Name.

    And it was enquir'd Why in a Great Solemn Assem-
    bly
The Innocent should be condemn'd for the Guilty.
    Then an Eternal rose,

Saying: "If the Guilty should be condemn'd he must be
    an Eternal Death,
And one must die for another throughout all Eternity.
Satan is fall'n from his station & never can be redeem'd,
But must be new Created continually moment by
    moment.
And therefore the Class of Satan shall be call'd the
    Elect, & those
Of Rintrah the Reprobate, & those of Palamabron the
    Redeem'd:
For he is redeem'd from Satan's Law, the wrath falling
    on Rintrah.   370
And therefore Palamabron dared not to call a solemn
    Assembly
Till Satan had assum'd Rintrah's wrath in the day of
    mourning,
In a feminine delusion of false pride self-deciev'd."

So spake the Eternal and confirm'd it with a thunder-
    ous oath.

But when Leutha (a Daughter of Beulah) beheld
    Satan's condemnation,

She down descended into the midst of the Great
    Solemn Assembly,
Offering herself a Ransom for Satan, taking on her his
    Sin.

Mark well my words! they are of your eternal salva-
    tion.

And Leutha stood glowing with varying colours, im-
    mortal, heart-piercing
And lovely, & her moth-like elegance shone over the
    Assembly.   380

At length, standing upon the golden floor of Palama-
    bron,
She spake: "I am the Author of this Sin! by my sugges-
    tion
My Parent power Satan has committed this transgres-
    sion.
I loved Palamabron & I sought to approach his Tent,
But beautiful Elynittria with her silver arrows repell'd
    me.

13

"For her light is terrible to me: I fade before her
    immortal beauty.
O wherefore doth a Dragon-form forth issue from my
    limbs
To sieze her new born son? Ah me! the wretched
    Leutha!
This to prevent, entering the doors of Satan's brain
    night after night
Like sweet perfumes, I stupified the masculine percep-
    tions   390
And kept only the feminine awake: hence rose his
    soft
Delusory love to Palamabron, admiration join'd with
    envy.
Cupidity unconquerable! my fault, when at noon of
    day
The Horses of Palamabron call'd for rest and pleasant
    death,
I sprang out of the breast of Satan, over the Harrow
    beaming
In all my beauty, that I might unloose the flaming
    steeds
As Elynittria used to do; but too well those living
    creatures

356. **Tomb . . . Rock:** See I. 564–67. **375. Leutha:**
Satan's emanation, sin.

397. **living creatures:** in Greek, *zoa.*

Knew that I was not Elynittria and they brake the traces.

But me the servants of the Harrow saw not but as a bow

Of varying colours on the hills; terribly rag'd the horses.   400

Satan astonish'd and with power above his own controll

Compell'd the Gnomes to curb the horses & to throw banks of sand

Around the fiery flaming Harrow in labyrinthine forms,

And brooks between to intersect the meadows in their course.

The Harrow cast thick flames: Jehovah thunder'd above

Chaos & ancient night fled from beneath the fiery Harrow:

The Harrow cast thick flames & orb'd us round in concave fires,

A Hell of our own making; see! its flames still gird me round.

Jehovah thunder'd above; Satan in pride of heart

Drove the fierce Harrow among the constellations of Jehovah,   410

Drawing a third part in the fires as stubble north & south

To devour Albion and Jerusalem, the Emanation of Albion,

Driving the Harrow in Pity's paths: 'twas then, with our dark fires

Which now gird round us (O eternal torment!) I form'd the Serpent

Of precious stones & gold, turn'd poisons on the sultry wastes.

The Gnomes in all that day spar'd not; they curs'd Satan bitterly

To do unkind things in kindness, with power arm'd to say

The most irritating things in the midst of tears and love:

These are the stings of the Serpent! thus did we by them till thus

They in return retaliated, and the Living Creatures madden'd.   420

The Gnomes labour'd. I weeping hid in Satan's inmost brain.

But when the Gnomes refus'd to labour more, with blandishments

I came forth from the head of Satan: back the Gnomes recoil'd

And called me Sin and for a sign portentous held me. Soon

Day sunk and Palamabron return'd; trembling I hid myself

In Satan's inmost Palace of his nervous fine wrought Brain:

For Elynittria met Satan with all her singing women,

Terrific in their joy & pouring wine of wildest power.

They gave Satan their wine; indignant at the burning wrath,

Wild with prophetic fury, his former life became like a dream.   430

Cloth'd in the Serpent's folds, in selfish holiness demanding purity,

Being most impure, self-condemn'd to eternal tears, he drove

Me from his inmost Brain & the doors clos'd with thunder's sound.

O Divine Vision who didst create the Female to repose

The Sleepers of Beulah, pity the repentant Leutha! My

14

Sick Couch bears the dark shades of Eternal Death infolding

The Spectre of Satan: he furious refuses to repose in sleep.

I humbly bow in all my Sin before the Throne Divine.

Not so the Sick-one. Alas, what shall be done him to restore

Who calls the Individual Law Holy and despises the Saviour,   440

Glorying to involve Albion's Body in fires of eternal War?"

Now Leutha ceas'd: tears flow'd, but the Divine Pity supported her.

"All is my fault! We are the Spectre of Luvah, the murderer

Of Albion. O Vala! O Luvah! O Albion! O lovely Jerusalem!

The Sin was begun in Eternity and will not rest to Eternity

Till two Eternitys meet together. Ah! lost, lost, lost for ever!"

---

424. Sin . . . me: Cf. *Paradise Lost*, II. 760–61. **443. murderer:** See *The Four Zoas*, "Night the Ninth," above, n. 93.

So Leutha spoke. But when she saw that Enitharmon
  had
Created a New Space to protect Satan from punish-
  ment,
She fled to Enitharmon's Tent & hid herself. Loud
  raging
Thunder'd the Assembly dark & clouded, and they
  ratify'd   450
The kind decision of Enitharmon & gave a Time to the
  Space,
Even Six Thousand years, and sent Lucifer for its
  Guard.
But Lucifer refus'd to die & in pride he forsook his
  charge:
And they elected Molech, and when Molech was im-
  patient
The Divine hand found the Two Limits, first of
  Opacity, then of Contraction.
Opacity was named Satan, Contraction was named
  Adam.
Triple Elohim came: Elohim wearied fainted: they
  elected Shaddai:
Shaddai angry, Pahad descended: Pahad terrified, they
  sent Jehovah,
And Jehovah was leprous; loud he call'd, stretching his
  hand to Eternity,
For then the Body of Death was perfected in hypocritic
  holiness,   460
Around the Lamb, a Female Tabernacle woven in
  Cathedron's Looms.
He died as a Reprobate, he was Punish'd as a Trans-
  gressor.
Glory! Glory! Glory! to the Holy Lamb of God!
I touch the heavens as an instrument to glorify the
  Lord!

The Elect shall meet the Redeem'd on Albion's
  rocks, they shall meet
Astonish'd at the Transgressor, in him beholding the
  Saviour.
And the Elect shall say to the Redeem'd: "We behold
  it is of Divine
Mercy alone, of Free Gift and Election that we live:
Our Virtues & Cruel Goodnesses have deserv'd Eternal
  Death."   469
Thus they weep upon the fatal Brook of Albion's River.

But Elynittria met Leutha in the place where she was
  hidden
And threw aside her arrows and laid down her sound-
  ing Bow.
She sooth'd her with soft words & brought her to
  Palamabron's bed
In moments new created for delusion, interwoven
  round about.
In dreams she bore the shadowy Spectre of Sleep &
  nam'd him Death:
In dreams she bore Rahab, the mother of Tirzah, &
  her sisters
In Lambeth's vales, in Cambridge & in Oxford, places
  of Thought,
Intricate labyrinths of Times and Spaces unknown, that
  Leutha lived
In Palamabron's Tent and Oothoon was her charming
  guard.

The Bard ceas'd. All consider'd and a loud resounding
  murmur   480
Continu'd round the Halls; and much they question'd
  the immortal
Loud voic'd Bard, and many condemn'd the high
  toned Song,
Saying: "Pity and Love are too venerable for the
  imputation
Of Guilt." Others said: "If it is true, if the acts have
  been perform'd,
Let the Bard himself witness. Where hadst thou this
  terrible Song?"

The Bard replied: "I am Inspired! I know it is
  Truth! for I Sing

452–58. Six . . . Jehovah: The seven cycles of history are
established (see I. n. 1), which Blake termed the "Seven
Eyes of God." In succession the cycles are those of Lucifer
(Isaiah 14:12), whose sin according to Christian doctrine was
pride (Blake's Selfhood or self-sufficiency); Moloch (see I.
n. 253), preeminently the age of the Druids; and then four
Hebrew names for God: Elohim, Shaddai, Pahad, and
Jehovah. During the Eye or cycle of Elohim the human
body assumes its present form (the limit of "Contraction" is
established; the body and senses will shrink no further), and
a limit is also set to the materiality of physical matter (the
limit of "Opacity"). The seventh and last cycle is that of
Jesus. 461. Cathedron: the place where Enitharmon sets up
her looms.

470. Brook: See I. 352. 476. Rahab . . . Tirzah: See
The Four Zoas, "Night the Ninth," above, n. 32, and I.
n. 177. 479. Oothoon: Heroine of Blake's Visions of the
Daughters of Albion (1793), she may represent an ideal of love.

15

According to the inspiration of the Poetic Genius
Who is the eternal all-protecting Divine Humanity,
To whom be Glory & Power & Dominion Evermore.
Amen."

Then there was great murmuring in the Heavens of
Albion   490
Concerning Generation & the Vegetative power & con-
cerning
The Lamb the Saviour. Albion trembled to Italy,
Greece & Egypt
To Tartary & Hindostan & China & to Great America,
Shaking the roots & fast foundations of the Earth in
doubtfulness.
The loud voic'd Bard terrify'd took refuge in Milton's
bosom.

Then Milton rose up from the heavens of Albion
ardorous.
The whole Assembly wept prophetic, seeing in Milton's
face
And in his lineaments divine the shades of Death &
Ulro:
He took off the robe of the promise & ungirded himself
from the oath of God.

And Milton said: "I go to Eternal Death! The Nations
still   500
Follow after the detestable Gods of Priam, in pomp
Of warlike selfhood contradicting and blaspheming.
When will the Resurrection come to deliver the sleep-
ing body
From corruptibility? O when, Lord Jesus, wilt thou
come?
Tarry no longer, for my soul lies at the gates of death.
I will arise and look forth for the morning of the grave:
I will go down to the sepulcher to see if morning breaks:
I will go down to self annihilation and eternal death,
Lest the Last Judgment come & find me unannihilate
And I be siez'd & giv'n into the hands of my own
Selfhood.   510
The Lamb of God is seen thro' mists & shadows,
hov'ring
Over the sepulchers in clouds of Jehovah & winds of
Elohim,

A disk of blood distant, & heav'ns & earths roll dark
between.
What do I here before the Judgment? without my
Emanation?
With the daughters of memory & not with the daughters
of inspiration?
I in my Selfhood am that Satan: I am that Evil One!
He is my Spectre! in my obedience to loose him from
my Hells,
To claim the Hells, my Furnaces, I go to Eternal
Death."

And Milton said: "I go to Eternal Death!" Eternity
shudder'd,
For he took the outside course among the graves of
the dead,   520
A mournful shade. Eternity shudder'd at the image of
eternal death.

Then on the verge of Beulah he beheld his own
Shadow,
A mournful form double, hermaphroditic, male &
female
In one wonderful body; and he enter'd into it
In direful pain, for the dread shadow twenty-seven fold
Reach'd to the depths of direst Hell & thence to Albion's
land,
Which is this earth of vegetation on which now I write.

The Seven Angels of the Presence wept over Milton's
Shadow.

17

As when a man dreams he reflects not that his body
sleeps,
Else he would wake, so seem'd he entering his Shadow:
but   530
With him the Spirits of the Seven Angels of the
Presence
Entering, they gave him still perceptions of his Sleeping
Body
Which now arose and walk'd with them in Eden, as an
Eighth

---

501. Gods of Priam: See Blake's *Descriptive Catalogue*,
below.

515. With . . . inspiration: See Blake's Preface to *Milton*,
above. 525. twenty-seven fold: Cf. I. 604–06, II. n. 352–60.
528. Seven . . . Presence: See I. n. 452–58.

Image Divine tho' darken'd and tho' walking as one walks
In sleep, and the Seven comforted and supported him.

Like as a Polypus that vegetates beneath the deep,
They saw his Shadow vegetated underneath the Couch
Of death: for when he enter'd into his Shadow, Himself,
His real and immortal Self, was, as appear'd to those
Who dwell in immortality, as One sleeping on a couch   540
Of gold, and those in immortality gave forth their Emanations
Like Females of sweet beauty to guard round him & to feed
His lips with food of Eden in his cold and dim repose:
But to himself he seem'd a wanderer lost in dreary night.

Onwards his Shadow kept its course among the Spectres call'd
Satan, but swift as lightning passing them, startled the shades
Of Hell beheld him in a trail of light as of a comet
That travels into Chaos: so Milton went guarded within.

The nature of infinity is this: That every thing has its
Own Vortex, and when once a traveller thro' Eternity
He pass'd that Vortex, he percieves it roll backward behind   551
His path, into a globe itself infolding like a sun,
Or like a moon, or like a universe of starry majesty,
While he keeps onwards in his wondrous journey on the earth,
Or like a human form, a friend with whom he liv'd benevolent.
As the eye of man views both the east & west encompassing
Its vortex, and the north & south with all their starry host,
Also the rising sun & setting moon he viewing surrounding
His corn-field and his valleys of five hundred acres square,
Thus is the earth one infinite plane, and not as apparent
To the weak traveller confin'd beneath the moony shade.   561

Thus is the heaven a vortex pass'd already, and the earth
A vortex not yet pass'd by the traveller thro' Eternity.

First Milton saw Albion upon the Rock of Ages,
Deadly pale outstretch'd and snowy cold, storm cover'd,
A Giant form of perfect beauty outstretch'd on the rock
In solemn death: the Sea of Time & Space thunder'd aloud
Against the rock, which was inwrapped with the weeds of death.
Hovering over the cold bosom in its vortex Milton bent down
To the bosom of death: what was underneath soon seem'd above: 570
A cloudy heaven mingled with stormy seas in loudest ruin;
But as a wintry globe descends precipitant thro' Beulah bursting
With thunders loud and terrible, so Milton's shadow fell
Precipitant, loud thunder'ing into the Sea of Time & Space.

Then first I saw him in the Zenith as a falling star
Descending perpendicular, swift as the swallow or swift:
And on my left foot falling on the tarsus, enter'd there:
But from my left foot a black cloud redounding spread over Europe.

Then Milton knew that the Three Heavens of Beulah were beheld
By him on earth in his bright pilgrimage of sixty years   580

19

In those three females whom his wives, & those three whom his Daughters
Had represented and contain'd, that they might be resum'd
By giving up Selfhood: & they distant view'd his journey
In their eternal spheres, now Human, tho' their Bodies remain clos'd

563. traveller thro' Eternity: Milton. 579. Three . . . Beulah: Cf. I. 104.

In the dark Ulro till the Judgment: also Milton knew
   they and
Himself was Human, tho' now wandering thro'
   Death's Vale
In conflict with those Female forms, which in blood &
   jealousy
Surrounded him, dividing & uniting without end or
   number.

He saw the Cruelties of Ulro and he wrote them down
In iron tablets; and his Wives' & Daughters' names
   were these: 590
Rahab and Tirzah, & Milcah & Malah & Noah &
   Hoglah.
They sat rang'd round him as the rocks of Horeb
   round the land
Of Canaan, and they wrote in thunder, smoke and
   fire
His dictate; and his body was the Rock Sinai, that
   body
Which was on earth born to corruption; & the six
   Females
Are Hor & Peor & Bashan & Abarim & Lebanon &
   Hermon,
Seven rocky masses terrible in the Desarts of Midian.

But Milton's Human Shadow continu'd journeying
   above
The rocky masses of The Mundane Shell, in the Lands
Of Edom & Aram & Moab & Midian & Amalek.   600

The Mundane Shell is a vast Concave Earth, an immense
Harden'd shadow of all things upon our Vegetated
   Earth,
Enlarg'd into dimension & deform'd into indefinite
   space,
In Twenty-seven Heavens and all their Hells, with
   Chaos
And Ancient Night & Purgatory. It is a cavernous
   Earth
Of labyrinthine intricacy, twenty-seven-folds of
   opakeness,

586. **Human:** i.e., they are in Eternity. **591. Tirzah . . .
Hoglah:** the five daughters of Zelophehad (Numbers
26:33). See I. n. 177. **596. Hor . . . Hermon:** names of
mountains surrounding Canaan. To say they are in "Desarts
of Midian" is not literal but spiritual geography. **600. Edom
. . . Amalek:** nations in and around Canaan. Since they are
"rocky masses" it might be that Milton, like the Israelites,
is wandering through the desert.

And finishes where the lark mounts; here Milton
   journeyed
In that Region call'd Midian among the Rocks of
   Horeb.
For travellers from Eternity pass outward to Satan's
   seat,
But travellers to Eternity pass inward to Golgonooza.

Los, the Vehicular terror, beheld him, & divine Enithar-
   mon   611
Call'd all her daughters, Saying: "Surely to unloose
   my bond
Is this Man come! O Satan shall be unloos'd upon
   Albion!"

Los heard in terror Enitharmon's words; in fibrous
   strength
His limbs shot forth like roots of trees against the for-
   ward path
Of Milton's journey. Urizen beheld the immortal
   Man

                         20

And Tharmas, Demon of the Waters, & Orc, who is
   Luvah.

The Shadowy Female seeing Milton, howl'd in her
   lamentation
Over the Deeps, outstretching her Twenty seven
   Heavens over Albion,

And thus the Shadowy Female howls in articulate
   howlings: 620

"I will lament over Milton in the lamentations of the
   afflicted:
My Garments shall be woven of sighs & heart broken
   lamentations:
The misery of unhappy Families shall be drawn out
   into its border,
Wrought with the needle with dire sufferings, poverty,
   pain & woe
Along the rocky Island & thence throughout the
   whole Earth;
There shall be the sick Father & his starving Family,
   there

617. **Tharmas . . . Waters:** See *The Four Zoas,* "Night
the Ninth," above, n. 482.

The Prisoner in the stone Dungeon & the Slave at the
    Mill.
I will have writings written all over it in Human
    Words
That every Infant that is born upon the Earth shall
    read
And get by rote as a hard task of a life of sixty years.
I will have Kings inwoven upon it & Councellors &
    Mighty Men: 631
The Famine shall clasp it together with buckles &
    Clasps,
And the Pestilence shall be its fringe & the War its
    girdle,
To divide into Rahab & Tirzah that Milton may
    come to our tents.
For I will put on the Human Form & take the Image of
    God,
Even Pity & Humanity, but my Clothing shall be
    Cruelty:
And I will put on Holiness as a breastplate & as a
    helmet,
And all my ornaments shall be of the gold of broken
    hearts,
And the precious stones of anxiety & care & desperation
    & death
And repentance for sin & sorrow & punishment &
    fear,   640
To defend me from thy terrors, O Orc, my only
    beloved!"

Orc answer'd: "Take not the Human Form, O
    loveliest, Take not
Terror upon thee! Behold how I am & tremble lest
    thou also
Consume in my Consummation; but thou maist
    take a Form
Female & lovely, that cannot consume in Man's con-
    summation.
Wherefore dost thou Create & Weave this Satan for a
    Covering?
When thou attemptest to put on the Human form, my
    wrath
Burns to the top of heaven against thee in Jealousy &
    Fear;
Then I rend thee asunder, then I howl over thy clay &
    ashes.
When wilt thou put on the Female Form as in times
    of old,   650
With a Garment of Pity & Compassion like the Gar-
    ment of God?

His Garments are long sufferings for the Children of
    Men;
Jerusalem is his Garment, & not thy Covering Cherub,
    O lovely
Shadow of my delight, who wanderest seeking for
    the prey."

So spoke Orc when Oothoon & Leutha hover'd over
    his Couch
Of fire, in interchange of Beauty & Perfection in the
    darkness
Opening interiorly into Jerusalem & Babylon, shining
    glorious
In the Shadowy Female's bosom. Jealous her darkness
    grew:
Howlings fill'd all the desolate places in accusations of
    Sin,
In Female beauty shining in the unform'd void; & Orc
    in vain   660
Stretch'd out his hands of fire & wooed: they triumph
    in his pain.

Thus darken'd the Shadowy Female tenfold, & Orc
    tenfold
Glow'd on his rocky Couch against the darkness: loud
    thunders
Told of the enormous conflict. Earthquake beneath,
    around,
Rent the Immortal Females limb from limb & joint
    from joint,
And moved the fast foundations of the Earth to wake
    the Dead.

Urizen emerged from his Rocky Form & from his
    Snows,

21

And he also darken'd his brows, freezing dark rocks
    between
The footsteps and infixing deep the feet in marble
    beds,
That Milton labour'd with his journey & his feet bled
    sore   670
Upon the clay now chang'd to marble; also Urizen
    rose
And met him on the shores of Arnon & by the streams
    of the brooks.

653. Covering Cherub: See I. n. 325. **672. Arnon:** the river
Arnon bounded the kingdom of the Ammonites, whose
god was Moloch. See *Paradise Lost*, I. 399.

Silent they met and silent strove among the streams
   of Arnon
Even to Mahanaim, when with cold hand Urizen
   stoop'd down
And took up water from the river Jordan, pouring on
To Milton's brain the icy fluid from his broad cold
   palm.
But Milton took of the red clay of Succoth, moulding
   it with care
Between his palms and filling up the furrows of many
   years,
Beginning at the feet of Urizen, and on the bones
Creating new flesh on the Demon cold and building
   him   680
As with new clay, a Human form in the Valley of
   Beth Peor.

Four Universes round the Mundane Egg remain
   Chaotic,
One to the North, named Urthona: One to the South,
   named Urizen:
One to the East, named Luvah: One to the West,
   named Tharmas;
They are the Four Zoas that stood around the Throne
   Divine.
But when Luvah assum'd the World of Urizen to the
   South
And Albion was slain upon his mountains & in his
   tent,
All fell towards the Center in dire ruin sinking down.
And in the South remains a burning fire: in the East,
   a void:
In the West, a world of raging waters: in the North,
   a solid,   690
Unfathomable, without end. But in the midst of
   these
Is built eternally the Universe of Los and Enitharmon,
Towards which Milton went, but Urizen oppos'd his
   path.

The Man and Demon strove many periods. Rahab
   beheld,
Standing on Carmel. Rahab and Tirzah trembled to
   behold
The enormous strife, one giving life, the other giving
   death

To his adversary, and they sent forth all their sons &
   daughters
In all their beauty to entice Milton across the river.

The Twofold form Hermaphroditic and the Double-
   sexed,
The Female-male & the Male-female, self-dividing
   stood   700
Before him in their beauty & in cruelties of holiness,
Shining in darkness, glorious upon the deeps of
   Entuthon,

Saying: "Come thou to Ephraim! behold the Kings of
   Canaan!
The beautiful Amalekites behold the fires of youth
Bound with the Chain of Jealousy by Los & Enitharmon.
The banks of Cam, cold learning's streams, London's
   dark frowning towers
Lament upon the winds of Europe in Rephaim's Vale,
Because Ahania, rent apart into a desolate night,
Laments, & Enion wanders like a weeping inarticulate
   voice,
And Vala labours for her bread & water among the
   Furnaces.   710
Therefore bright Tirzah triumphs, putting on all
   beauty
And all perfection in her cruel sports among the
   Victims.
Come, bring with thee Jerusalem with songs on the
   Grecian Lyre!
In Natural Religion, in experiments on Men
Let her be Offer'd up to Holiness! Tirzah numbers her:
She numbers with her fingers every fibre ere it grow.
Where is the Lamb of God? where is the promise of
   his coming?
Her shadowy Sisters form the bones, even the bones of
   Horeb
Around the marrow, and the orbed scull around the
   brain.
His Images are born for War, for Sacrifice to Tirzah,
To Natural Religion, to Tirzah, the Daughter of
   Rahab the Holy!   721

702. Entuthon: Entuthon Benython is a portion of Ulro,
nature as labyrinthine forest or wilderness, the Tree of
Mystery. **703. Come . . . Ephraim:** They are tempting
Milton to become an Old Testament king in Canaan. See I. n.
331. Ephraim was Joseph's second son of whom Jacob prophesied that "his seed shall become a multitude of nations"
(Genesis, 43:19). **706. Cam:** river at Cambridge University.
707. Rephaim: See II Samuel 5:18.

674. Mahanaim: See Genesis, 32:2. **677. Succoth:** See
Genesis 33:17. **681. Beth Peor:** where Moses was buried
Deuteronomy 34:6).

She ties the knot of nervous fibres into a white brain!
She ties the knot of bloody veins into a red hot heart!
Within her bosom Albion lies embalm'd, never to
    awake.
Hand is become a rock: Sinai & Horeb is Hyle & Coban:
Scofield is bound in iron armour before Reuben's Gate.
She ties the knot of milky seed into two lovely Heavens,

22

Two yet but one, each in the other sweet reflected;
    these
Are our Three Heavens beneath the shades of Beulah,
    land of rest.   729
Come then to Ephraim & Manasseh, O beloved-one!
Come to my ivory palaces, O beloved of thy mother!
And let us bind thee in the bands of War, & be thou
    King
Of Canaan and reign in Hazor where the Twelve
    Tribes meet."

So spoke they as in one voice. Silent Milton stood before
The darken'd Urizen, as the sculptor silent stands before
His forming image; he walks round it patient labouring.
Thus Milton stood forming bright Urizen, while his
    Mortal part
Sat frozen in the rock of Horeb, and his Redeemed
    portion
Thus form'd the Clay of Urizen; but within that
    portion
His real Human walk'd above in power and majesty,
Tho' darken'd, and the Seven Angels of the Presence
    attended him.   741

O how can I with my gross tongue that cleaveth to
    the dust
Tell of the Four-fold Man in starry numbers fitly
    order'd,
Or how can I with my cold hand of clay! But thou,
    O Lord,
Do with me as thou wilt! for I am nothing, and
    vanity.

725–26. Hand . . . Reuben: These persons figure more
prominently in *Jerusalem*. Hand, Hyle, Coban, and Scofield
are four of Albion's sons born after his fall. Scofield can be
identified as Scholfield, the soldier Blake evicted from his
garden at Felpham (see Introduction). Reuben, the eldest son
of Jacob, seems in *Jerusalem* to represent the natural man, the
opposite of the man of imagination. 730. Manasseh: Joseph's
first-born son. 733. Hazor: See Joshua, 11:1–11.

If thou chuse to elect a worm, it shall remove the
    mountains.
For that portion nam'd the Elect, the Spectrous body
    of Milton,
Redounding from my left foot into Los's Mundane
    Space,
Brooded over his Body in Horeb against the Resurrec-
    tion,
Preparing it for the Great Consummation; red the
    Cherub on Sinai   750
Glow'd, but in terrors folded round his clouds of
    blood.

Now Albion's sleeping Humanity began to turn
    upon his Couch,
Feeling the electric flame of Milton's awful precipitate
    descent.
Seest thou the little winged fly, smaller than a grain of
    sand?
It has a heart like thee, a brain open to heaven & hell,
Withinside wondrous & expansive: its gates are not
    clos'd:
I hope thine are not: hence it clothes itself in rich array:
Hence thou art cloth'd with human beauty, O thou
    mortal man.
Seek not thy heavenly father then beyond the skies,
There Chaos dwells & ancient Night & Og & Anak
    old.   760
For every human heart has gates of brass & bars of
    adamant
Which few dare unbar, because dread Og & Anak
    guard the gates
Terrific: and each mortal brain is wall'd and moated
    round
Within, and Og & Anak watch here: here is the Seat
Of Satan in its Webs: for in brain and heart and loins
Gates open behind Satan's Seat to the City of Golgo-
    nooza,
Which is the spiritual fourfold London in the loins of
    Albion.

Thus Milton fell thro' Albion's heart, travelling outside
    of Humanity
Beyond the Stars in Chaos, in Caverns of the Mundane
    Shell.

760. Og & Anak: giants, associates of Satan. A third giant,
Sihon, is often listed with them. (See I. 881). Their biblical
sources are Numbers 21:21–24, 33, and Joshua 11:21.

But many of the Eternals rose up from eternal tables
Drunk with the Spirit; burning round the Couch of
    death they stood   771
Looking down into Beulah; wrathful, fill'd with rage
They rend the heavens round the Watchers in a fiery
    circle
And round the Shadowy Eighth: the Eight close up
    the Couch
Into a tabernacle and flee with cries down to the
    Deeps,
Where Los opens his three wide gates surrounded by
    raging fires.
They soon find their own place & join the Watchers
    of the Ulro.

Los saw them and a cold pale horror cover'd o'er his
    limbs.
Pondering he knew that Rintrah & Palamabron might
    depart,
Even as Reuben & as Gad: gave up himself to tears,
He sat down on his anvil-stock and lean'd upon the
    trough,   781
Looking into the black water, mingling it with tears.

At last when desperation almost tore his heart in twain
He recollected an old Prophecy in Eden recorded
And often sung to the loud harp at the immortal
    feasts:
That Milton of the Land of Albion should up ascend
Forwards from Ulro from the Vale of Felpham, and
    set free
Orc from his Chain of Jealousy: he started at the
    thought

23

And down descended into Udan-Adan; it was night,
And Satan sat sleeping upon his Couch in Udan-
    Adan: 790
His Spectre slept, his Shadow woke; when one sleeps
    th'other wakes.

But Milton entering my Foot, I saw in the nether
Regions of the Imagination—also all men on Earth
And all in Heaven saw in the nether regions of the
    Imagination
In Ulro beneath Beulah—the vast breach of Milton's
    descent.

But I knew not that it was Milton, for man cannot
    know
What passes in his members till periods of Space &
    Time
Reveal the secrets of Eternity: for more extensive
Than any other earthly things are Man's earthly linea-
    ments.
And all this Vegetable World appear'd on my left
    Foot   800
As a bright sandal form'd immortal of precious stones
    & gold.
I stooped down & bound it on to walk forward thro'
    Eternity.

There is in Eden a sweet River of milk & liquid pearl
Nam'd Ololon, on whose mild banks dwelt those
    who Milton drove
Down into Ulro: and they wept in long resounding
    song
For seven days of eternity, and the river's living banks,
The mountains, wail'd, & every plant that grew, in
    solemn sighs lamented.

When Luvah's bulls each morning drag the sulphur
    Sun out of the Deep
Harness'd with starry harness, black & shining, kept
    by black slaves
That work all night at the starry harness, Strong and
    vigorous   810
They drag the unwilling Orb: at this time all the
    Family
Of Eden heard the lamentation and Providence
    began.
But when the clarions of day sounded, they drown'd
    the lamentations,
And when night came, all was silent in Ololon, & all
    refus'd to lament
In the still night, fearing lest they should others molest.

Seven mornings Los heard them, as the poor bird
    within the shell
Hears its impatient parent bird, and Enitharmon heard
    them
But saw them not, for the blue Mundane Shell inclos'd
    them in.

779–80. depart . . . Gad: Gad was the seventh son of
Jacob. See Joshua 22:9–34.

804. Ololon: Milton's emanation. Being sixfold she may be
either singular or plural, and Blake will use both forms in
referring to her.

And they lamented that they had in wrath & fury &
    fire
Driven Milton into the Ulro; for now they knew too
    late   820
That it was Milton the Awakener: they had not heard
    the Bard
Whose song call'd Milton to the attempt; and Los
    heard these laments.
He heard them call in prayer all the Divine Family,
And he beheld the Cloud of Milton stretching over
    Europe.

But all the Family Divine collected as Four Suns
In the Four Points of heaven, East, West & North &
    South,
Enlarging and enlarging till their Disks approach'd each
    other,
And when they touch'd, closed together Southward in
    One Sun
Over Ololon; and as One Man who weeps over his
    brother
In a dark tomb, so all the Family Divine wept over
    Ololon,   830

Saying: "Milton goes to Eternal Death!" so saying
    they groan'd in spirit
And were troubled; and again the Divine Family
    groaned in spirit.

And Ololon said: "Let us descend also, and let us
    give
Ourselves to death in Ulro among the Transgressors.
Is Virtue a Punisher? O no! how is this wondrous
    thing,
This World beneath, unseen before, this refuge from
    the wars
Of Great Eternity! unnatural refuge! unknown by us
    till now?
Or are these the pangs of repentance? let us enter into
    them."

Then the Divine Family said: "Six Thousand Years
    are now
Accomplished in this World of Sorrow. Milton's
    Angel knew   840
The Universal Dictate, and you also feel this Dictate.
And now you know this World of Sorrow and feel
    Pity. Obey
The Dictate! Watch over this World, and with your
    brooding wings

Renew it to Eternal Life. Lo! I am with you alway.
But you cannot renew Milton: he goes to Eternal
    Death."

So spake the Family Divine as One Man, even Jesus,
Uniting in One with Ololon, & the appearance of One
    Man,
Jesus the Saviour, appear'd coming in the Clouds of
    Ololon

24

Tho' driven away with the Seven Starry Ones into the
    Ulro,
Yet the Divine Vision remains Every-where For-ever.
    Amen.   850
And Ololon lamented for Milton with a great lamenta-
    tion.

While Los heard indistinct in fear, what time I bound
    my sandals
On to walk forward thro' Eternity, Los descended to
    me:
And Los behind me stood, a terrible flaming Sun, just
    close
Behind my back. I turned round in terror, and behold!
Los stood in that fierce glowing fire, & he also stoop'd
    down
And bound my sandals on in Udan-Adan; trembling
    I stood
Exceedingly with fear & terror, standing in the Vale
Of Lambeth; but he kissed me and wish'd me health,
And I became One Man with him arising in my
    strength.   860
'Twas too late now to recede. Los had enter'd into my
    soul:
His terrors now possess'd me whole! I arose in fury &
    strength.

"I am that Shadowy Prophet who Six Thousand Years
    ago
Fell from my station in the Eternal bosom. Six Thou-
    sand Years
Are finish'd. I return! both Time & Space obey my
    will.
I in Six Thousand Years walk up and down; for not
    one Moment
Of Time is lost, nor one Event of Space unpermanent,

844. Lo! . . . alway: Cf. Matthew 28:20.

But all remain: every fabric of Six Thousand Years
Remains permanent, tho' on the Earth where Satan
Fell and was cut off, all things vanish & are seen no
    more,   870
They vanish not from me & mine, we guard them
    first & last.
The generations of men run on in the tide of Time,
But leave their destin'd lineaments permanent for
    ever & ever."

So spoke Los as we went along to his supreme abodes.

Rintrah and Palamabron met us at the Gate of
    Golgonooza,
Clouded with discontent & brooding in their minds
    terrible things.

They said: "O Father most beloved! O merciful
    Parent
Pitying and permitting evil, tho' strong & mighty to
    destroy!
Whence is this Shadow terrible? wherefore dost thou
    refuse
To throw him into the Furnaces? knowest thou not
    that he   880
Will unchain Orc & let loose Satan, Og, Sihon & Anak
Upon the Body of Albion? for this he is come! behold
    it written
Upon his fibrous left Foot black, most dismal to our
    eyes.
The Shadowy Female shudders thro' heaven in torment
    inexpressible,
And all the Daughters of Los prophetic wail; yet in
    deceit
They weave a new Religion from new Jealousy of
    Theotormon.
Milton's Religion is the cause: there is no end to
    destruction.
Seeing the Churches at their Period in terror & despair,
Rahab created Voltaire, Tirzah created Rousseau,
Asserting the Self-righteousness against the Universal
    Saviour,   890
Mocking the Confessors & Martyrs, claiming Self-
    righteousness,
With cruel Virtue making War upon the Lamb's
    Redeemed

To perpetuate War & Glory, to perpetuate the Laws of
    Sin.
They perverted Swedenborg's Visions in Beulah &
    in Ulro
To destroy Jerusalem as a Harlot & her Sons as Repro-
    bates,
To raise up Mystery the Virgin Harlot, Mother of
    War,
Babylon the Great, the Abomination of Desolation.
O Swedenborg! strongest of men, the Samson shorn
    by the Churches,
Shewing the Transgressors in Hell, the proud Warriors
    in Heaven,
Heaven as a Punisher, & Hell as One under Punish-
    ment,   900
With Laws from Plato & his Greeks to renew the
    Trojan Gods
In Albion, & to deny the value of the Saviour's blood.
But then I rais'd up Whitefield, Palamabron rais'd up
    Westley,
And these are the cries of the Churches before the two
    Witnesses.
Faith in God the dear Saviour who took on the likeness
    of men,
Becoming obedient to death, even the death of the
    Cross.
The Witnesses lie dead in the Street of the Great City:
No Faith is in all the Earth: the Book of God is trodden
    under Foot.
He sent his two Servants, Whitefield & Westley: were
    they Prophets,
Or were they Idiots or Madmen? shew us Miracles! 910

25

"Can you have greater Miracles than these? Men who
    devote
Their life's whole comfort to intire scorn & injury &
    death?
Awake, thou sleeper on the Rock of Eternity! Albion
    awake!
The trumpet of Judgment hath twice sounded: all
    Nations are awake,
But thou art still heavy and dull. Awake, Albion awake!
Lo, Orc arises on the Atlantic. Lo, his blood and fire

---

881. Og . . . Anak: See I. n. 760. **889. Voltaire . . .
Rousseau:** See *Jerusalem*, "To the Deists," below.

894. Swedenborg: See headnote to *The Marriage of Heaven
and Hell*, above. **901. Trojan Gods:** See I. n. 501. **903.
Whitefield . . . Westley:** George Whitefield, 1714–70,
and John Wesley, 1703–91, founders of Methodism.

Glow on America's shore. Albion turns upon his
   Couch:
He listens to the sounds of War, astonished and con-
   founded:
He weeps into the Atlantic deep, yet still in dismal
   dreams
Unwaken'd, and the Covering Cherub advances from
   the East.   920
How long shall we lay dead in the Street of the great
   City?
How long beneath the Covering Cherub give our
   Emanations?
Milton will utterly consume us & thee our beloved
   Father.
He hath enter'd into the Covering Cherub, becoming
   one with
Albion's dread Sons: Hand, Hyle & Coban surround
   him as
A girdle, Gwendolen & Conwenna as a garment
   woven
Of War & Religion; let us descend & bring him
   chained
To Bowlahoola, O father most beloved! O mild
   Parent!
Cruel in thy mildness, pitying and permitting evil,
Tho' strong and mighty to destroy, O Los our beloved
   Father!"   930

Like the black storm, coming out of Chaos beyond
   the stars,
It issues thro' the dark & intricate caves of the Mundane
   Shell,
Passing the planetary visions & the well adorned
   Firmament.
The Sun rolls into Chaos & the stars into the Desarts,
And then the storms become visible, audible & terrible,
Covering the light of day & rolling down upon the
   mountains,
Deluge all the country round. Such is a vision of Los
When Rintrah & Palamabron spake, and such his
   stormy face
Appear'd as does the face of heaven when cover'd with
   thick storms,
Pitying and loving tho' in frowns of terrible perturba-
   tion.   940

926. Gwendolen & Conwenna: daughters of Albion. **928.**
Bowlahoola: Usually paired with Allamanda, the two together represent the necessary physiological basis of any higher
life (Golgonooza) in the awakened Albion, his digestive and
circulatory organs, i.e., commerce and law. See I. 1008–39.

But Los dispers'd the clouds even as the strong winds of
   Jehovah,
And Los thus spoke: "O noble Sons, be patient yet a
   little!
I have embrac'd the falling Death, he is become One
   with me:
O Sons, we live not by wrath, by mercy alone we live!
I recollect an old Prophecy in Eden recorded in gold
   and oft
Sung to the harp, That Milton of the land of Albion
Should up ascend forward from Felpham's Vale &
   break the Chain
Of Jealousy from all its roots; be patient therefore, O
   my Sons!
These lovely Females form sweet night and silence
   and secret
Obscurities to hide from Satan's Watch-Fiends
   Human loves   950
And graces, lest they write them in their Books & in
   the Scroll
Of mortal life to condemn the accused, who at Satan's
   Bar
Tremble in Spectrous Bodies continually day and
   night,
While on the Earth they live in sorrowful Vegetations.
O when shall we tread our Wine-presses in heaven and
   Reap
Our wheat with shoutings of joy, and leave the Earth
   in Peace?
Remember how Calvin and Luther in fury premature
Sow'd War and stern division between Papists &
   Protestants.
Let it not be so now! O go not forth in Martyrdoms &
   Wars!
We were plac'd here by the Universal Brotherhood &
   Mercy   960
With powers fitted to circumscribe this dark Satanic
   death,
And that the Seven Eyes of God may have space for
   Redemption.
But how this is as yet we know not, and we cannot
   know
Till Albion is arisen; then patient wait a little while.
Six Thousand years are pass'd away, the end approaches
   fast:
This mighty one is come from Eden, he is of the Elect
Who died from Earth & he is return'd before the
   Judgment. This thing

943. falling Death: Milton, who has entered into Blake.

Was never known, that one of the holy dead should
    willing return.
Then patient wait a little while till the Last Vintage is
    over,
Till we have quench'd the Sun of Salah in the Lake of
    Udan-Adan.   970
O my dear Sons, leave not your Father as your brethren
    left me!
Twelve Sons successive fled away in that thousand
    years of sorrow

26

Of Palamabron's Harrow & of Rintrah's wrath &
    fury:
Reuben & Manazzoth & Gad & Simeon & Levi
And Ephraim & Judah were Generated because
They left me, wandering with Tirzah. Enitharmon
    wept
One thousand years, and all the Earth was in a wat'ry
    deluge.
We call'd him Menassheh because of the Generations
    of Tirzah,
Because of Satan: & the Seven Eyes of God continually
Guard round them, but I, the Fourth Zoa, am also
    set   980
The Watchman of Eternity: the Three are not, & I
    am preserved.
Still my four mighty ones are left to me in Golgonooza,
Still Rintrah fierce, and Palamabron mild & piteous,
Theotormon fill'd with care, Bromion loving Science.
You, O my Sons, still guard round Los: O wander not
    & leave me!
Rintrah, thou well rememberest when Amalek &
    Canaan
Fled with their Sister Moab into that abhorred Void,
They became Nations in our sight beneath the hands
    of Tirzah.
And Palamabron, thou rememberest when Joseph, an
    infant,
Stolen from his nurse's cradle, wrap'd in needle-
    work   990
Of emblematic texture, was sold to the Amalekite

972. **Twelve Sons:** the twelve tribes of Israel, of which an
incomplete list comes in the next two lines. **978. Menassheh:**
Manasseh. See I. n. 730. **989–91. Joseph . . . Amalekite:**
an example of Blake revising the Bible as he adapts it to
his symbolic meaning. In Genesis Joseph was seventeen when
he was sold to the Ishmaelites, and his coat is simply described
as being of many colors.

Who carried him down into Egypt where Ephraim &
    Menassheh
Gather'd my Sons together in the Sands of Midian.
And if you also flee away and leave your Father's
    side
Following Milton into Ulro, altho' your power is
    great,
Surely you also shall become poor mortal vegetations
Beneath the Moon of Ulro: pity then your Father's
    tears.
When Jesus rais'd Lazarus from the Grave I stood &
    saw
Lazarus, who is the Vehicular Body of Albion the
    Redeem'd,
Arise into the Covering Cherub, who is the Spectre of
    Albion,   1000
By martyrdoms to suffer, to watch over the Sleeping
    Body
Upon his Rock beneath his Tomb. I saw the Covering
    Cherub
Divide Four-fold into Four Churches when Lazarus
    arose,
Paul, Constantine, Charlemaine, Luther; behold, they
    stand before us
Stretch'd over Europe & Asia! come O Sons, come,
    come away!
Arise, O Sons, give all your strength against Eternal
    Death,
Lest we are vegetated, for Cathedron's Looms weave
    only Death,
A Web of Death: & were it not for Bowlahoola &
    Allamanda
No Human Form but only a Fibrous Vegetation,
A Polypus of soft affections without Thought or
    Vision,   1010
Must tremble in the Heavens & Earths thro' all the
    Ulro space.
Throw all the Vegetated Mortals into Bowlahoola:
But as to this Elected Form who is return'd again,
He is the Signal that the Last Vintage now approaches,
Nor Vegetation may go on till all the Earth is
    reap'd."

So Los spoke. Furious they descended to Bowlahoola
    & Allamanda,
Indignant, unconvinc'd by Los's arguments & thunders
    rolling:
They saw that wrath now sway'd and now pity
    absorb'd him.
As it was so it remain'd & no hope of an end.

Bowlahoola is nam'd Law by mortals; Tharmas
    founded it   1020
Because of Satan, before Luban in the City of Golgo-
    nooza.
But Golgonooza is nam'd Art & Manufacture by
    mortal men.

In Bowlahoola Los's Anvils stand & his Furnaces rage;
Thundering the Hammers beat & the Bellows blow
    loud,
Living, self moving, mourning, lamenting & howling
    incessantly.
Bowlahoola thro' all its porches feels, tho' too fast
    founded
Its pillars & porticoes to tremble at the force
Of mortal or immortal arm: and softly lilling flutes,
Accordant with the horrid labours, make sweet
    melody.
The Bellows are the Animal Lungs: the Hammers the
    Animal Heart: 1030
The Furnaces the Stomach for digestion: terrible
    their fury.
Thousands & thousands labour, thousands play on
    instruments
Stringed or fluted to ameliorate the sorrows of slavery.
Loud sport the dancers in the dance of death, rejoicing
    in carnage.
The hard dentant Hammers are lull'd by the flutes'
    lula lula,
The bellowing Furnaces blare by the long sounding
    clarion,
The double drum drowns howls & groans, the shrill
    fife shrieks & cries,
The crooked horn mellows the hoarse raving serpent,
    terrible but harmonious:
Bowlahoola is the Stomach in every individual man.

Los is by mortals nam'd Time, Enitharmon is nam'd
    Space: 1040
But they depict him bald & aged who is in eternal
    youth
All powerful and his locks flourish like the brows of
    morning:
He is the Spirit of Prophecy, the ever apparent Elias.
Time is the mercy of Eternity; without Time's swift-
    ness,

Which is the swiftest of all things, all were eternal
    torment.
All the Gods of the Kingdoms of Earth labour in Los's
    Halls:
Every one is a fallen Son of the Spirit of Prophecy.
He is the Fourth Zoa that stood around the Throne
    Divine.

27

Loud shout the Sons of Luvah at the Wine-presses
    as Los descended
With Rintrah & Palamabron in his fires of resistless
    fury.   1050

The Wine-press on the Rhine groans loud, but all
    its central beams
Act more terrific in the central Cities of the Nations
Where Human Thought is crush'd beneath the iron
    hand of Power:
There Los puts all into the Press, the Opressor & the
    Opressed
Together, ripe for the Harvest & Vintage & ready for
    the Loom.

They sang at the Vintage: "This is the Last Vintage, &
    Seed
Shall no more be sown upon Earth till all the Vintage
    is over
And all gather'd in, till the Plow has pass'd over the
    Nations
And the Harrow & heavy thundering Roller upon the
    mountains."

And loud the Souls howl round the Porches of Golgo-
    nooza,   1060
Crying: "O God deliver us to the Heavens or to the
    Earths,
That we may preach righteousness & punish the sinner
    with death."
But Los refused, till all the Vintage of Earth was
    gathered in.

And Los stood & cried to the Labourers of the Vintage
    in voice of awe:

1021. **Luban:** the gate of Golgonooza. 1028. **lilling:**
either Blake's new coinage or a misprint. 1043. **Elias:**
Elijah.

1051. **Rhine:** where the battles of the Napoleonic Wars
were taking place.

"Fellow Labourers! The Great Vintage & Harvest is now upon Earth.

The whole extent of the Globe is explored. Every scatter'd Atom

Of Human Intellect now is flocking to the sound of the Trumpet.

All the Wisdom which was hidden in caves & dens from ancient

Time is now sought out from Animal & Vegetable & Mineral.

The Awakener is come outstretch'd over Europe: the Vision of God is fulfilled: 1070

The Ancient Man upon the Rock of Albion Awakes,

He listens to the sounds of War astonish'd & ashamed,

He sees his Children mock at Faith and deny Providence.

Therefore you must bind the Sheaves not by Nations or Families,

You shall bind them in Three Classes, according to their Classes

So shall you bind them, Separating What has been Mixed

Since Men began to be Wove into Nations by Rahab & Tirzah,

Since Albion's Death & Satan's Cutting off from our awful Fields,

When under pretence to benevolence the Elect Subdu'd All

From the Foundations of the World. The Elect is one Class: You   1080

Shall bind them separate: they cannot Believe in Eternal Life

Except by Miracle & a New Birth. The other two Classes,

The Reprobate who never cease to Believe, and the Redeem'd

Who live in doubts & fears perpetually tormented by the Elect,

These you shall bind in a twin-bundle for the Consummation:

But the Elect must be saved from fires of Eternal Death,

To be formed into the Churches of Beulah that they destroy not the Earth.

For in every Nation & every Family the Three Classes are born,

And in every Species of Earth, Metal, Tree, Fish, Bird & Beast.

We form the Mundane Egg, that Spectres coming by fury or amity,   1090

All is the same, & every one remains in his own energy.

Go forth Reapers with rejoicing; you sowed in tears,

But the time of your refreshing cometh: only a little moment

Still abstain from pleasure & rest in the labours of eternity,

And you shall Reap the whole Earth from Pole to Pole, from Sea to Sea,

Beginning at Jerusalem's Inner Court, Lambeth, ruin'd and given

To the detestable Gods of Priam, to Apollo, and at the Asylum

Given to Hercules, who labour in Tirzah's Looms for bread,

Who set Pleasure against Duty, who Create Olympic crowns

To make Learning a burden & the Work of the Holy Spirit, Strife: 1100

The Thor & cruel Odin who first rear'd the Polar Caves.

Lambeth mourns, calling Jerusalem: she weeps & looks abroad

For the Lord's coming, that Jerusalem may overspread all Nations.

Crave not for the mortal & perishing delights, but leave them

To the weak, and pity the weak as your infant care. Break not

Forth in your wrath, lest you also are vegetated by Tirzah.

Wait till the Judgement is past, till the Creation is consumed,

And then rush forward with me into the glorious spiritual

Vegetation, the Supper of the Lamb & his Bride, and the

Awaking of Albion our friend and ancient companion."   1110

So Los spoke. But lightnings of discontent broke on all sides round

And murmurs of thunder rolling heavy long & loud over the mountains,

While Los call'd his Sons around him to the Harvest & the Vintage.

1097. Asylum: the hospital of St. Mary of Bethlehem for the insane, popularly called Bedlam, in Lambeth. **1098. Tirzah's Looms:** Tirzah is here identified with Omphale, a queen of Lydia whom for three years Hercules was forced to serve, wearing female clothes and spinning among her maids.

Thou seest the Constellations in the deep & wondrous
Night:
They rise in order and continue their immortal
courses
Upon the mountains & in vales with harp & heavenly
song,
With flute & clarion, with cups & measures fill'd with
foaming wine.
Glitt'ring the streams reflect the Vision of beatitude,
And the calm Ocean joys beneath & smooths his
awful waves:

28

These are the Sons of Los, & these the Labourers of the
Vintage.   1120
Thou seest the gorgeous clothed Flies that dance &
sport in summer
Upon the sunny brooks & meadows: every one the
dance
Knows in its intricate mazes of delight artful to weave:
Each one to sound his instruments of music in the
dance,
To touch each other & recede, to cross & change &
return:
These are the Children of Los; thou seest the Trees
on mountains,
The wind blows heavy, loud they thunder thro' the
darksom sky,
Uttering prophecies & speaking instructive words to
the sons
Of men: These are the Sons of Los: These the Visions
of Eternity,
But we see only as it were the hem of their garments
When with our vegetable eyes we view these won-
drous Visions.   1131

There are Two Gates thro' which all Souls descend,
One Southward
From Dover Cliff to Lizard Point, the other toward
the North,
Caithness & rocky Durness, Pentland & John Groat's
House.

The Souls descending to the Body wail on the right
hand
Of Los, & those deliver'd from the Body on the left
hand.
For Los against the east his force continually bends
Along the Valleys of Middlesex from Hounslow to
Blackheath,

Lest those Three Heavens of Beulah should the Creation
destroy;
And lest they should descend before the north & south
Gates,   1140
Groaning with pity, he among the wailing Souls
laments.

And these the Labours of the Sons of Los in Allamanda
And in the City of Golgonooza & in Luban &
around
The Lake of Udan-Adan in the Forests of Entuthon
Benython,
Where Souls incessant wail, being piteous Passions &
Desires
With neither lineament nor form, but like to wat'ry
clouds
The Passions & Desires descend upon the hungry
winds,
For such alone Sleepers remain, meer passion & appetite.
The Sons of Los clothe them & feed & provide houses
& fields.

And every Generated Body in its inward form   1150
Is a garden of delight & a building of magnificence,
Built by the Sons of Los in Bowlahoola & Allamanda:
And the herbs & flowers & furniture & beds &
chambers
Continually woven in the Looms of Enitharmon's
Daughters,
In bright Cathedron's golden Dome with care & love
& tears.
For the various Classes of Men are all mark'd out
determinate
In Bowlahoola, & as the Spectres choose their affinities,
So they are born on Earth, & every Class is deter-
minate:
But not by Natural, but by Spiritual power alone,
Because
The Natural power continually seeks & tends to
Destruction,   1160
Ending in Death, which would of itself be Eternal
Death.
And all are Class'd by Spiritual & not by Natural
power.

And every Natural Effect has a Spiritual Cause, and
Not
A Natural; for a Natural Cause only seems: it is a
Delusion
Of Ulro & a ratio of the perishing Vegetable Memory.

29

But the Wine-press of Los is eastward of Golgonooza
    before the Seat
Of Satan: Luvah laid the foundation & Urizen finish'd
    it in howling woe.
How red the sons & daughters of Luvah! here they
    tread the grapes:
Laughing & shouting, drunk with odours many fall
    o'erwearied,
Drown'd in the wine is many a youth & maiden: those
    around   1170
Lay them on skins of Tygers & of the spotted Leopard
    & the Wild Ass
Till they revive, or bury them in cool grots, making
    lamentation.

This Wine-press is call'd War on Earth: it is the
    Printing-Press
Of Los, and here he lays his words in order above the
    mortal brain,
As cogs are form'd in a wheel to turn the cogs of the
    adverse wheel.

Timbrels & violins sport round the Wine-presses;
    the little Seed,
The sportive Root, the Earth-worm, the gold Beetle,
    the wise Emmet
Dance round the Wine-presses of Luvah: the Centipede
    is there,
The ground Spider with many eyes, the Mole clothed
    in velvet,
The ambitious Spider in his sullen web, the lucky
    golden Spinner,   1180
The Earwig arm'd, the tender Maggot, emblem of
    immortality,
The Flea, Louse, Bug, the Tape-Worm, all the Armies
    of Disease,
Visible or invisible to the slothful vegetating Man.
The slow Slug, the Grasshopper that sings & laughs &
    drinks:
Winter comes, he folds his slender bones without a
    murmur.
The cruel Scorpion is there, the Gnat, Wasp, Hornet &
    the Honey Bee,
The Toad & venomous Newt, the Serpent cloth'd in
    gems & gold.

1168–1206. **How . . . Luvah:** Cf. *The Four Zoas,* "Night
the Ninth," above, ll. 741–69.

They throw off their gorgeous raiment: they rejoice
    with loud jubilee
Around the Wine-presses of Luvah, naked & drunk
    with wine.

There is the Nettle that stings with soft down, and
    there   1190
The indignant Thistle whose bitterness is bred in his
    milk,
Who feeds on contempt of his neighbour: there all the
    idle Weeds
That creep around the obscure places shew their
    various limbs
Naked in all their beauty dancing round the Wine-
    presses.

But in the Wine-presses the Human grapes sing not
    nor dance:
They howl & writhe in shoals of torment, in fierce
    flames consuming,
In chains of iron & in dungeons circled with ceaseless
    fires,
In pits & dens & shades of death, in shapes of torment &
    woe:
The plates & screws & wracks & saws & cords & fires &
    cisterns,
The cruel joys of Luvah's Daughters, lacerating with
    knives   1200
And whips their Victims, & the deadly sport of Luvah's
    Sons.

They dance around the dying & they drink the howl &
    groan,
They catch the shrieks in cups of gold, they hand them
    to one another:
These are the sports of love, & these the sweet delights
    of amorous play,
Tears of the grape, the death sweat of the cluster, the
    last sigh
Of the mild youth who listens to the lureing songs of
    Luvah.

But Allamanda, call'd on Earth Commerce, is the
    Cultivated land
Around the City of Golgonooza in the Forests of
    Entuthon:
Here the Sons of Los labour against Death Eternal,
    through all
The Twenty-seven Heavens of Beulah in Ulro, Seat
    of Satan,   1210

Which is the False Tongue beneath Beulah: it is the
    Sense of Touch.
The Plow goes forth in tempest & lightnings, & the
    Harrow cruel
In blights of the east, the heavy Roller follows in
    howlings of woe.

Urizen's sons here labour also, & here are seen the Mills
Of Theotormon on the verge of the Lake of Udan-
    Adan.
These are the starry voids of night & the depths &
    caverns of earth.
These Mills are oceans, clouds & waters ungovernable
    in their fury:
Here are the stars created & the seeds of all things
    planted,
And here the Sun & Moon recieve their fixed destina-
    tions.

But in Eternity the Four Arts, Poetry, Painting,
    Music   1220
And Architecture, which is Science, are the Four Faces
    of Man.
Not so in Time & Space: there Three are shut out, and
    only
Science remains thro' Mercy, & by means of Science
    the Three
Become apparent in Time & Space in the Three Pro-
    fessions,
[Poetry in Religion: Music, Law: Painting, in Physic
    & Surgery: *erased*]
That Man may live upon Earth till the time of his
    awaking.
And from these Three Science derives every Occupa-
    tion of Men,
And Science is divided into Bowlahoola & Allamanda.

30

Some Sons of Los surround the Passions with porches
    of iron & silver,
Creating form & beauty around the dark regions of
    sorrow,   1230
Giving to airy nothing a name and a habitation
Delightful, with bounds to the Infinite putting off the
    Indefinite

Into most holy forms of Thought; such is the power
    of inspiration.
They labour incessant with many tears & afflictions,
Creating the beautiful House for the piteous sufferer.

Others Cabinets richly fabricate of gold & ivory
For Doubts & fears unform'd & wretched & melan-
    choly.
The little weeping Spectre stands on the threshold of
    Death
Eternal, and sometimes two Spectres like lamps
    quivering,
And often malignant they combat; heart-breaking
    sorrowful & piteous,   1240
Antamon takes them into his beautiful flexible hands:
As the Sower takes the seed or as the Artist his clay
Or fine wax, to mould artful a model for golden
    ornaments.
The soft hands of Antamon draw the indelible line,
Form immortal with golden pen, such as the Spectre
    admiring
Puts on the sweet form; then smiles Antamon bright
    thro' his windows.
The Daughters of beauty look up from their Loom &
    prepare
The integument soft for its clothing with joy & delight.

    But Theotormon & Sotha stand in the Gate of Luban
    anxious.
Their numbers are seven million & seven thousand &
    seven hundred.   1250
They contend with the weak Spectres, they fabricate
    soothing forms.
The Spectre refuses, he seeks cruelty: they create the
    crested Cock.
Terrified the Spectre screams & rushes in fear into their
    Net
Of kindness & compassion, & is born a weeping
    terror.
Or they create the Lion & Tyger in compassionate
    thunderings:
Howling the Spectres flee: they take refuge in Human
    lineaments.

The Sons of Ozoth within the Optic Nerve stand fiery
    glowing,
And the number of his Sons is eight millions & eight.

1231. Giving . . . habitation: quoted from *A Midsummer
Night's Dream,* V. i. 16–17.

1241. Antamon: another of the Sons of Los, as is Ozoth
(l. 1257).

They give delights to the man unknown; artificial riches
They give to scorn, & their possessors to trouble & sorrow & care,   1260
Shutting the sun & moon & stars & trees & clouds & waters
And hills out from the Optic Nerve, & hardening it into a bone
Opake and like the black pebble on the enraged beach,
While the poor indigent is like the diamond which, tho' cloth'd
In rugged covering in the mine, is open all within
And in his hallow'd center holds the heavens of bright eternity.
Ozoth here builds walls of rocks against the surging sea,
And timbers crampt with iron cramps bar in the joys of life
From fell destruction in the Spectrous cunning or rage. He Creates
The speckled Newt, the Spider & Beetle, the Rat & Mouse.   1270
The Badger & Fox: they worship before his feet in trembling fear.

But others of the Sons of Los build Moments & Minutes & Hours
And Days & Months & Years & Ages & Periods, wondrous buildings;
And every Moment has a Couch of gold for soft repose,
(A Moment equals a pulsation of the artery),
And between every two Moments stands a Daughter of Beulah
To feed the Sleepers on their Couches with maternal care.
And every Minute has an azure Tent with silken Veils:
And every Hour has a bright golden Gate carved with skill:
And every Day & Night has Walls of brass & Gates of adamant,   1280
Shining like precious Stones & ornamented with appropriate signs:
And every Month a silver paved Terrace builded high:
And every Year invulnerable Barriers with high Towers:
And every Age is Moated deep with Bridges of silver & gold:
And every Seven Ages is Incircled with a Flaming Fire.

Now Seven Ages is amounting to Two Hundred Years.
Each has its Guard, each Moment, Minute, Hour, Day, Month & Year.
All are the work of Fairy hands of the Four Elements:
The Guard are Angels of Providence on duty evermore.
Every Time less than a pulsation of the artery   1290
Is equal in its period & value to Six Thousand Years,

31

For in this Period the Poet's Work is Done, and all the Great
Events of Time start forth & are conciev'd in such a Period,
Within a Moment, a Pulsation of the Artery.

The Sky is an immortal Tent built by the Sons of Los:
And every Space that a Man views around his dwelling-place
Standing on his own roof or in his garden on a mount
Of twenty-five cubits in height, such space is his Universe:
And on its verge the Sun rises & sets, the Clouds bow
To meet the flat Earth & the Sea in such an order'd Space: 1300
The Starry heavens reach no further, but here bend and set
On all sides, & the two Poles turn on their valves of gold;
And if he move his dwelling-place, his heavens also move
Where'er he goes, & all his neighbourhood bewail his loss.
Such are the Spaces called Earth & such its dimension.
As to that false appearance which appears to the reasoner
As of a Globe rolling thro' Voidness, it is a delusion of Ulro.
The Microscope knows not of this nor the Telescope: they alter
The ratio of the Spectator's Organs, but leave Objects untouch'd.
For every Space larger than a red Globule of Man's blood   1310
Is visionary, and is created by the Hammer of Los:
And every Space smaller than a Globule of Man's blood opens
Into Eternity of which this vegetable Earth is but a shadow.

The red Globule is the unwearied Sun by Los created
To measure Time and Space to mortal Men every
    morning.
Bowlahoola & Allamanda are placed on each side
Of that Pulsation & that Globule, terrible their power.

But Rintrah & Palamabron govern over Day &
    Night
In Allamanda & Entuthon Benython where Souls
    wail,
Where Orc incessant howls, burning in fires of Eternal
    Youth,   1320
Within the vegetated mortal Nerves; for every Man
    born is joined
Within into One mighty Polypus, and this Polypus is
    Orc.

But in the Optic vegetative Nerves, Sleep was trans-
    formed
To Death in old time by Satan the father of Sin &
    Death:
And Satan is the Spectre of Orc, & Orc is the generate
    Luvah.

    But in the Nerves of the Nostrils, Accident being
        formed
Into Substance & Principle by the cruelties of Demon-
    stration
It became Opake & Indefinite, but the Divine Saviour
Formed it into a Solid by Los's Mathematic power.
He named the Opake, Satan: he named the Solid,
    Adam.   1330

And in the Nerves of the Ear (for the Nerves of the
    Tongue are closed)
On Albion's Rock Los stands creating the glorious
    Sun each morning,
And when unwearied in the evening, he creates the
    Moon,
Death to delude, who all in terror at their splendor
    leaves
His prey, while Los appoints & Rintrah & Palamabron
    guide
The Souls clear from the Rock of Death, that Death
    himself may wake
In his appointed season when the ends of heaven meet.

Then Los conducts the Spirits to be Vegetated into
Great Golgonooza, free from the four iron pillars of
    Satan's Throne,

(Temperance, Prudence, Justice, Fortitude, the four
    pillars of tyranny)   1340
That Satan's Watch-Fiends touch them not before
    they Vegetate.

But Enitharmon and her Daughters take the pleasant
    charge
To give them to their lovely heavens till the Great
    Judgment Day:
Such is their lovely charge. But Rahab & Tirzah
    pervert
Their mild influences; therefore the Seven Eyes of
    God walk round
The Three Heavens of Ulro where Tirzah & her Sisters
Weave the black Woof of Death upon Entuthon
    Benython,
In the Vale of Surrey where Horeb terminates in
    Rephaim.
The stamping feet of Zelophehad's Daughters are
    cover'd with Human gore
Upon the treddles of the Loom: they sing to the winged
    shuttle.   1350
The River rises above his banks to wash the Woof:
He takes it in his arms; he passes it in strength thro' his
    current;
The veil of human miseries is woven over the Ocean
From the Atlantic to the Great South Sea, the Erythrean.

Such is the World of Los, the labour of six thousand
    years.
Thus Nature is a Vision of the Science of the Elohim.

End of the First Book

BOOK THE SECOND

33

There is a place where Contrarieties are equally True:
This place is called Beulah. It is a pleasant lovely
    Shadow
Where no dispute can come, Because of those who
    Sleep.
Into this place the Sons & Daughters of Ololon de-
    scended
With solemn mourning, into Beulah's moony shades
    & hills
Weeping for Milton: mute wonder held the Daughters
    of Beulah,
Enraptur'd with affection sweet and mild benevolence.

Beulah is evermore Created around Eternity, appearing
To the Inhabitants of Eden around them on all sides.
But Beulah to its Inhabitants appears within each
    district   10
As the beloved infant in his mother's bosom round
    incircled
With arms of love & pity & sweet compassion. But to
The Sons of Eden the moony habitations of Beulah
Are from Great Eternity a mild & pleasant Rest.

And it is thus Created. Lo, the Eternal Great Humanity,
To whom be Glory & Dominion Evermore, Amen,
Walks among all his awful Family seen in every face:
As the breath of the Almighty such are the words of
    man to man
In the great Wars of Eternity, in fury of Poetic Inspira-
    tion,
To build the Universe stupendous, Mental forms
    Creating.   20

But the Emanations trembled exceedingly, nor could
    they
Live, because the life of Man was too exceeding un-
    bounded.
His joy became terrible to them; they trembled &
    wept,
Crying with one voice: "Give us a habitation & a place
In which we may be hidden under the shadow of
    wings:
For if we, who are but for a time & who pass away in
    winter
Behold these wonders of Eternity we shall consume:
But you, O our Fathers & Brothers, remain in Eternity.
But grant us a Temporal Habitation, do you speak
To us; we will obey your words as you obey Jesus 30
The Eternal who is blessed for ever & ever. Amen."

So spake the lovely Emanations, & there appear'd a
    pleasant
Mild Shadow above, beneath, & on all sides round.

34

Into this pleasant Shadow all the weak & weary
Like Women & Children were taken away as on wings
Of dovelike softness, & shadowy habitations prepared
    for them
But every Man return'd & went still going forward
    thro'
The Bosom of the Father in Eternity on Eternity,

Neither did any lack or fall into Error without
A Shadow to repose in all the Days of happy Eter-
    nity.   40

Into this pleasant Shadow, Beulah, all Ololon des-
    cended,
And when the Daughters of Beulah heard the lamen-
    tation
All Beulah wept, for they saw the Lord coming in the
    Clouds.
And the Shadows of Beulah terminate in rocky Albion.

And all Nations wept in affliction, Family by Family:
Germany wept towards France & Italy, England wept
    & trembled
Towards America, India rose up from his golden bed
As one awaken'd in the night; they saw the Lord
    coming
In the Clouds of Ololon with Power & Great Glory.

And all the Living Creatures of the Four Elements
    wail'd   50
With bitter wailing; these in the aggregate are named
    Satan
And Rahab: they know not of Regeneration, but only
    of Generation:
The Fairies, Nymphs, Gnomes & Genii of the Four
    Elements,
Unforgiving & unalterable, these cannot be Regener-
    ated
But must be Created, for they know only of Genera-
    tion:
These are the Gods of the Kingdoms of the Earth, in
    contrarious
And cruel opposition, Element against Element,
    opposed in War
Not Mental, as the Wars of Eternity, but a Corporeal
    Strife
In Los's Halls, continual labouring in the Furnaces of
    Golgonooza.
Orc howls on the Atlantic: Enitharmon trembles: All
    Beulah weeps.   60

Thou hearest the Nightingale begin the Song of
    Spring.
The Lark sitting upon his earthy bed, just as the
    morn
Appears, listens silent; then springing from the waving
    Cornfield, loud
He leads the Choir of Day: trill, trill, trill, trill,

Mounting upon the wings of light into the Great
    Expanse,
Reecchoing against the lovely blue & shining heavenly
    Shell,
His little throat labours with inspiration; every feather
On throat & breast & wings vibrates with the effluence
    Divine.
All Nature listens silent to him, & the awful Sun
Stands still upon the Mountain looking on this little
    Bird   70
With eyes of soft humility & wonder, love & awe,
Then loud from their green covert all the Birds begin
    their Song:
The Thrush, the Linnet & the Goldfinch, Robin &
    the Wren
Awake the Sun from his sweet reverie upon the
    Mountain.
The Nightingale again assays his song, & thro' the day
And thro' the night warbles luxuriant, every Bird of
    Song
Attending his loud harmony with admiration & love.
This is a Vision of the lamentation of Beulah over
    Ololon.

Thou percievest the Flowers put forth their precious
    Odours,
And none can tell how from so small a center comes
    such sweets,   80
Forgetting that within that Center Eternity expands
Its ever during doors that Og & Anak fiercely guard.
First, e'er the morning breaks, joy opens in the flowery
    bosoms,
Joy even to tears, which the Sun rising dries; first the
    Wild Thyme
And Meadow-sweet, downy & soft waving among the
    reeds,
Light springing on the air, lead the sweet Dance: they
    wake
The Honeysuckle sleeping on the Oak; the flaunting
    beauty
Revels along upon the wind; the White-thorn, lovely
    May,
Opens her many lovely eyes listening; the Rose still
    sleeps,
None dare to wake her; soon she bursts her crimson
    curtain'd bed   90
And comes forth in the majesty of beauty; every
    Flower,
The Pink, the Jessamine, the Wall-flower, the Carna-
    tion,

The Jonquil, the mild Lilly, opes her heavens; every
    Tree
And Flower & Herb soon fill the air with an innumer-
    able Dance,
Yet all in order sweet & lovely. Men are sick with Love,
Such is a Vision of the lamentation of Beulah over
    Ololon.

35

And Milton oft sat upon the Couch of Death & oft
    conversed
In vision & dream beatific with the Seven Angels of
    the Presence.

"I have turned my back upon these Heavens builded
    on cruelty;
My Spectre still wandering thro' them follows my
    Emanation,   100
He hunts her footsteps thro' the snow & the wintry
    hail & rain.
The idiot Reasoner laughs at the Man of Imagination,
And from laughter proceeds to murder by under-
    valuing calumny."

Then Hillel, who is Lucifer, replied over the Couch
    of Death,
And thus the Seven Angels instructed him, & thus they
    converse:

"We are not Individuals but States, Combinations of
    Individuals.
We were Angels of the Divine Presence, & were
    Druids in Annandale,
Compell'd to combine into Form by Satan, the Spectre
    of Albion,
Who made himself a God & destroyed the Human
    Form Divine.
But the Divine Humanity & Mercy gave us a Human
    Form   110
Because we were combin'd in Freedom & holy
    Brotherhood,
While those combin'd by Satan's Tyranny, first in the
    blood of War
And Sacrifice & next in Chains of imprisonment, are
    Shapeless Rocks
Retaining only Satan's Mathematic Holiness, Length,
    Bredth & Highth,
Calling the Human Imagination, which is the Divine
    Vision & Fruition

In which Man liveth eternally, madness & blasphemy
    against
Its own Qualities, which are Servants of Humanity, not
    Gods or Lords.
Distinguish therefore States from Individuals in those
    States.
States Change, but Individual Identities never change
    nor cease.
You cannot go to Eternal Death in that which can
    never Die.   120
Satan & Adam are States Created into Twenty-seven
    Churches,
And thou, O Milton, art a State about to be Created,
Called Eternal Annihilation, that none but the Living
    shall
Dare to enter, & they shall enter triumphant over
    Death
And Hell & the Grave: States that are not, but ah!
    Seem to be.

"Judge then of thy Own Self: thy Eternal Lineaments
    explore,
What is Eternal & what Changeable, & what Annihil-
    able.
The Imagination is not a State: it is the Human Exist-
    ence itself.
Affection or Love becomes a State when divided from
    Imagination.
The Memory is a State always, & the Reason is a
    State   130
Created to be Annihilated & a new Ratio Created.
Whatever can be Created can be Annihilated: Forms
    cannot:
The Oak is cut down by the Ax, the Lamb falls by
    the Knife,
But their Forms Eternal Exist For-ever. Amen.
    Hallelujah!"

Thus they converse with the Dead, watching round
    the Couch of Death;
For God himself enters Death's Door always with
    those that enter
And lays down in the Grave with them, in Visions of
    Eternity,
Till they awake & see Jesus & the Linen Clothes lying
That the Females had Woven for them, & the Gates
    of their Father's House.

BOOK II. **138. Jesus . . . Clothes:** See *The Marriage of Heaven and Hell*, above, n. 3.

36

And the Divine Voice was heard in the Songs of
    Beulah, Saying: 140

"When I first Married you, I gave you all my whole
    Soul.
I thought that you would love my loves & joy in my
    delights,
Seeking for pleasures in my pleasures, O Daughter of
    Babylon.
Then thou wast lovely, mild & gentle; now thou art
    terrible
In jealousy & unlovely in my sight, because thou hast
    cruelly
Cut off my loves in fury till I have no love left for
    thee.
Thy love depends on him thou lovest, & on his dear
    loves
Depend thy pleasures, which thou hast cut off by
    jealousy.
Therefore I shew my Jealousy & set before you
    Death.
Behold Milton descended to Redeem the Female
    Shade   150
From Death Eternal; such your lot, to be continually
    Redeem'd
By death & misery of those you love & by Annihila-
    tion.
When the Sixfold Female percieves that Milton an-
    nihilates
Himself, that seeing all his loves by her cut off, he
    leaves
Her also, intirely abstracting himself from Female
    loves,
She shall relent in fear of death; She shall begin to
    give
Her maidens to her husband, delighting in his delight.
And then & then alone begins the happy Female joy
As it is done in Beulah, & thou, O Virgin Babylon,
    Mother of Whoredoms,
Shall bring Jerusalem in thine arms in the night watches,
    and   160
No longer turning her a wandering Harlot in the
    streets,
Shalt give her into the arms of God your Lord &
    Husband."

Such are the Songs of Beulah in the Lamentations of
    Ololon.

38

And all the Songs of Beulah sounded comfortable
    notes
To comfort Ololon's lamentation, for they said:
"Are you the Fiery Circle that late drove in fury & fire
The Eight Immortal Starry-Ones down into Ulro dark
Rending the Heavens of Beulah with your thunders &
    lightnings?
And can you thus lament & can you pity & forgive?
Is terror chang'd to pity? O wonder of Eternity!" 170

And the Four States of Humanity in its Repose
Were shewed them. First of Beulah, a most pleasant
    Sleep
On Couches soft with mild music, tended by Flowers
    of Beulah,
Sweet Female forms, winged or floating in the air
    spontaneous:
The Second State is Alla, & the third State Al-Ulro:
But the Fourth State is dreadful, it is named Or-Ulro.
The First State is in the Head, the Second is in the
    Heart,
The Third in the Loins & Seminal Vessels, & the
    Fourth
In the Stomach & Intestines terrible, deadly, unutterable.
And he whose Gates are open'd in those Regions of
    his Body   180
Can from those Gates view all these wondrous Imagina-
    tions.

But Ololon sought the Or-Ulro & its fiery Gates
And the Couches of the Martyrs, & many Daughters of
    Beulah
Accompany them down to the Ulro with soft melo-
    dious tears,
A long journey & dark thro' Chaos in the track of
    Milton's course,
To where the Contraries of Beulah War beneath
    Negation's Banner.

Then view'd from Milton's Track they see the Ulro a
    vast Polypus
Of living fibres down into the Sea of Time & Space
    growing
A self-devouring monstrous Human Death Twenty
    seven fold.

175–76. **Alla; Al-Ulro; Or-Ulro:** Blake used these names
nowhere else.

Within it sit Five Females & the nameless Shadowy
    Mother,   190
Spinning it from their bowels with songs of amorous
    delight
And melting cadences that lure the Sleepers of Beulah
    down
The River Storge (which is Arnon) into the Dead Sea.
Around this Polypus Los continual builds the Mundane
    Shell.
Four Universes round the Universe of Los remain
    Chaotic,
Four intersecting Globes, & the Egg form'd World of
    Los
In midst, stretching from Zenith to Nadir in midst of
    Chaos.
One of these Ruin'd Universes is to the North, named
    Urthona:
One to the South, this was the glorious World of
    Urizen:
One to the East, of Luvah: One to the West, of
    Tharmas.   200
But when Luvah assumed the World of Urizen in the
    South
All fell towards the Center sinking downward in dire
    Ruin.

Here in these Chaoses the Sons of Ololon took their
    abode,
In Chasms of the Mundane Shell which open on all
    sides round,
Southward & by the East within the Breach of Milton's
    descent,
To watch the time, pitying, & gentle to awaken Urizen.
They stood in a dark land of death, of fiery corroding
    waters,
Where lie in evil death the Four Immortals pale and
    cold
And the Eternal Man, even Albion, upon the Rock of
    Ages.
Seeing Milton's Shadow, some Daughters of Beulah
    trembling   210
Return'd, but Ololon remain'd before the Gates of
    the Dead.

And Ololon looked down into the Heavens of Ulro in
    fear.
They said: "How are the Wars of man, which in
    Great Eternity

195–202. **Four . . . Ruin:** Cf. I. 682–88.

Appear around in the External Spheres of Visionary
  Life,
Here render'd Deadly within the Life & Interior
  Vision?
How are the Beasts & Birds & Fishes & Plants & Minerals
Here fix'd into a frozen bulk subject to decay & death?
Those Visions of Human Life & Shadows of Wisdom
  & Knowledge

39

"Are here frozen to unexpansive deadly destroying
  terrors,
And War & Hunting, the Two Fountains of the River
  of Life,   220
Are become Fountains of bitter Death & of corroding
  Hell,
Till Brotherhood is chang'd into a Curse & a Flattery
By Differences between Ideas, that Ideas themselves
  (which are
The Divine Members) may be slain in offerings for
  sin.
O dreadful Loom of Death! O piteous Female forms
  compell'd
To weave the Woof of Death! On Camberwell
  Tirzah's Courts,
Malah's on Blackheath, Rahab & Noah dwell on
  Windsor's heights:
Where once the Cherubs of Jerusalem spread to
  Lambeth's Vale
Milcah's Pillar's shine from Harrow to Hampstead,
  where Hoglah
On Highgate's heights magnificent Weaves over
  trembling Thames   230
To Shooters' Hill and thence to Blackheath, the dark
  Woof. Loud,
Loud roll the Weights & Spindles over the whole
  Earth, let down
On all sides round to the Four Quarters of the World,
  eastward on
Europe to Euphrates & Hindu to Nile, & back in
  Clouds
Of Death across the Atlantic to America North &
  South."

So spake Ololon in reminiscence astonish'd, but they
Could not behold Golgonooza without passing the
  Polypus,
A wondrous journey not passable by Immortal feet,
  & none

But the Divine Saviour can pass it without annihilation.
For Golgonooza cannot be seen till having pass'd the
  Polypus   240
It is viewed on all sides round by a Four-fold Vision,
Or till you become Mortal & Vegetable in Sexuality,
Then you behold its mighty Spires & Domes of ivory
  & gold.

And Ololon examined all the Couches of the Dead,
Even of Los & Enitharmon & all the Sons of Albion
And his Four Zoas terrified & on the verge of Death.
In midst of these was Milton's Couch, & when they
  saw Eight
Immortal Starry-Ones guarding the Couch in flaming
  fires,
They thunderous utter'd all a universal groan, falling
  down
Prostrate before the Starry Eight asking with tears
  forgiveness,   250
Confessing their crime with humiliation and sorrow.

O how the Starry Eight rejoic'd to see Ololon de-
  scended,
And now that a wide road was open to Eternity
By Ololon's descent thro' Beulah to Los & Enitharmon!
For mighty were the multitudes of Ololon, vast the
  extent
Of their great sway reaching from Ulro to Eternity,
Surrounding the Mundane Shell outside in its Caverns
And through Beulah, and all silent forbore to contend
With Ololon, for they saw the Lord in the Clouds of
  Ololon.

There is a Moment in each Day that Satan cannot
  find,   260
Nor can his Watch Fiends find it; but the Industrious
  find
This Moment & it multiply, & when it once is found
It renovates every Moment of the Day if rightly
  placed.
In this Moment Ololon descended to Los & Enitharmon
Unseen beyond the Mundane Shell, Southward in
  Milton's track.

Just in this Moment, when the morning odours rise
  abroad
And first from the Wild Thyme, stands a Fountain
  in a rock

260. Moment: Cf. I. 1290–94.

Of crystal flowing into two Streams: one flows thro'
    Golgonooza
And thro' Beulah to Eden beneath Los's western Wall:
The other flows thro' the Aerial Void & all the
    Churches,   270
Meeting again in Golgonooza beyond Satan's Seat.

The Wild Thyme is Los's Messenger to Eden, a mighty
    Demon,
Terrible, deadly & poisonous his presence in Ulro dark;
Therefore he appears only a small Root creeping in
    grass
Covering over the Rock of Odours his bright purple
    mantle
Beside the Fount above the Lark's nest in Golgonooza.
Luvah slept here in death & here is Luvah's empty
    Tomb.
Ololon sat beside this Fountain on the Rock of Odours.

Just at the place to where the Lark mounts is a Crystal
    Gate:
It is the enterance of the First Heaven, named Luther;
    for   280
The Lark is Los's Messenger thro' the Twenty-seven
    Churches,
That the Seven Eyes of God, who walk even to Satan's
    Seat
Thro' all the Twenty-seven Heavens, may not slumber
    nor sleep.
But the Lark's Nest is at the Gate of Los, at the eastern
Gate of wide Golgonooza, & the Lark is Los's Messenger.

40

When on the highest lift of his light pinions he arrives
At the bright Gate, another Lark meets him, & back
    to back
They touch their pinions, tip [to] tip, and each descend
To their respective Earths & there all night consult with
    Angels
Of Providence & with the eyes of God all night in
    slumbers   290
Inspired, & at the dawn of day send out another Lark
Into another Heaven to carry news upon his wings.
Thus are the Messengers dispatch'd till they reach the
    Earth again

280. Luther: Luther's was the last of the twenty-seven
churches and is therefore the first to be passed on the ascent
to Eternity.

In the East Gate of Golgonooza, & the Twenty-eighth
    bright
Lark met the Female Ololon descending into my
    Garden.
Thus it appears to Mortal eyes & those of the Ulro
    Heavens,
But not thus to Immortals: the Lark is a mighty
    Angel.

For Ololon step'd into the Polypus within the Mundane
    Shell.
They could not step into Vegetable Worlds without
    becoming
The enemies of Humanity, except in a Female
    Form,   300
And as One Female Ololon and all its mighty Hosts
Appear'd, a Virgin of twelve years: nor time nor
    space was
To the perception of the Virgin Ololon, but as the
Flash of lightning, but more quick the Virgin in my
    Garden
Before my Cottage stood, for the Satanic Space is
    delusion.

For when Los join'd with me he took me in his fi'ry
    whirlwind:
My Vegetated portion was hurried from Lambeth's
    shades,
He set me down in Felpham's Vale & prepar'd a
    beautiful
Cottage for me, that in three years I might write all
    these Visions
To display Nature's cruel holiness, the deceits of
    Natural Religion.   310
Walking in my Cottage Garden, sudden I beheld
The Virgin Ololon & address'd her as a Daughter of
    Beulah:

"Virgin of Providence, fear not to enter into my
    Cottage.
What is thy message to thy friend? What am I now
    to do?
Is it again to plunge into deeper affliction? behold me
Ready to obey, but pity thou my Shadow of
    Delight:
Enter my Cottage, comfort her, for she is sick with
    fatigue."

316. Shadow of Delight: his wife.

41

The Virgin answer'd: "Knowest thou of Milton who descended

Driven from Eternity? him I seek, terrified at my Act

In Great Eternity which thou knowest: I come him to seek."   320

So Ololon utter'd in words distinct the anxious thought:

Mild was the voice but more distinct than any earthly.

That Milton's Shadow heard, & condensing all his Fibres

Into a strength impregnable of majesty & beauty infinite,

I saw he was the Covering Cherub & within him Satan

And Rahab in an outside which is fallacious, within,

Beyond the outline of Identity, in the Selfhood deadly;

And he appear'd the Wicker Man of Scandinavia, in whom

Jerusalem's children consume in flames among the Stars.

Descending down into my Garden, a Human Wonder of God   330

Reaching from heaven to earth, a Cloud & Human Form,

I beheld Milton with astonishment & in him beheld

The Monstrous Churches of Beulah, the Gods of Ulro dark,

Twelve monstrous dishumaniz'd terrors, Synagogues of Satan,

A Double Twelve & Thrice Nine: such their divisions.

And these their Names & their Places within the Mundane Shell:

In Tyre & Sidon I saw Baal & Ashtaroth: In Moab Chemosh:

In Ammon Molech, loud his Furnaces rage among the Wheels

Of Og, & pealing loud the cries of the Victims of Fire,

And pale his Priestesses infolded in Veils of Pestilence border'd   340

With War, Woven in Looms of Tyre & Sidon by beautiful Ashtaroth:

In Palestine Dagon, Sea Monster, worship'd o'er the Sea:

Thammuz in Lebanon & Rimmon in Damascus curtain'd:

Osiris, Isis, Orus in Egypt, dark their Tabernacles on Nile

Floating with solemn songs & on the Lakes of Egypt nightly

With pomp even till morning break & Osiris appear in the sky:

But Belial of Sodom & Gomorrha, obscure Demon of Bribes

And secret Assasinations, not worship'd nor ador'd, but

With a finger on the lips & the back turn'd to the light:

And Saturn, Jove & Rhea of the Isles of the Sea remote.

These Twelve Gods are the Twelve Spectre Sons of the Druid Albion.   351

And these the names of the Twenty-seven Heavens & their Churches:

Adam, Seth, Enos, Cainan, Mahalaleel, Jared, Enoch,

Methuselah, Lamech, these are Giants Mighty, Hermaphroditic;

Noah, Shem, Arphaxad, Cainan the second, Salah, Heber,

Peleg, Reu, Serug, Nahor, Terah, these are the Female-Males,

A Male within a Female hid as in an Ark & Curtains;

Abraham, Moses, Solomon, Paul, Constantine, Charlemaine,

Luther, these seven are the Male-Females, the Dragon Forms,

Religion hid in War, a Dragon red & hidden Harlot. 360

---

328. **Wicker . . . Scandinavia:** According to Caesar's *Commentaries*, VI. 16, the Druids sacrificed their victims on wicker frames that they manufactured in human shape and set on fire. 337–49. **Baal . . . light:** the twelve Gods of Ulro. Blake takes the names from *Paradise Lost*, I. 392–521.

352–60. **Churches . . . Harlot:** These are not Gods but dogmas and ways of organizing religion. The churches follow each other in time, twenty-seven making up one cycle: "And where Luther ends, Adam begins again in Eternal Circle" (*Jerusalem*, 75:24). To Blake it often seemed that Adam was beginning again in the eighteenth century, but the return of Milton as poetic inspiration, casting out the error or Satan within him and uniting himself with Blake, shatters time and space and opens the way to Eternity.

All these are seen in Milton's Shadow, who is the
    Covering Cherub,
The Spectre of Albion in which the Spectre of Luvah
    inhabits
In the Newtonian Voids between the Substances of
    Creation.

For the Chaotic Voids outside of the Stars are measured
    by
The Stars, which are the boundaries of Kingdoms,
    Provinces
And Empire of Chaos invisible to the Vegetable Man.
The Kingdom of Og is in Orion: Sihon is in Ophiucus.
Og has Twenty-seven Districts: Sihon's Districts
    Twenty-one.
From Star to Star, Mountains & Valleys, terrible
    dimension
Stretch'd out, compose the Mundane Shell, a mighty
    Incrustation   370
Of Forty-eight deformed Human Wonders of the
    Almighty,
With Caverns whose remotest bottoms meet again
    beyond
The Mundane Shell in Golgonooza; but the Fires of
    Los rage
In the remotest bottoms of the Caves, that none can
    pass
Into Eternity that way, but all descend to Los,
To Bowlahoola & Allamanda & to Entuthon Benython.

The Heavens are the Cherub: the Twelve Gods are
    Satan,

43

And the Forty-eight Starry Regions are Cities of the
    Levites,
The Heads of the Great Polypus, Four-fold twelve
    enormity,
In mighty & mysterious comingling, enemy with
    enemy,   380
Woven by Urizen into Sexes from his mantle of
    years.
And Milton collecting all his fibres into impregnable
    strength
Descended down a Paved work of all kinds of precious
    stones
Out from the eastern sky; descending down into my
    Cottage
Garden, clothed in black, severe & silent he descended.

The Spectre of Satan stood upon the roaring sea &
    beheld
Milton within his sleeping Humanity; trembling &
    shudd'ring
He stood upon the waves a Twenty-seven fold mighty
    Demon
Gorgeous & beautiful; loud roll his thunders against
    Milton.
Loud Satan thunder'd, loud & dark upon mild Fel-
    pham shore   390
Not daring to touch one fibre he howl'd round upon
    the Sea.

I also stood in Satan's bosom & beheld its desolations:
A ruin'd Man, a ruin'd building of God, not made with
    hands:
Its plains of burning sand, its mountains of marble
    terrible:
Its pits & declivities flowing with molten ore & fountains
Of pitch & nitre: its ruin'd palaces & cities & mighty
    works:
Its furnaces of affliction, in which his Angels & Emana-
    tions
Labour with blacken'd visages among its stupendous
    ruins,
Arches & pyramids & porches, colonades & domes,
In which dwells Mystery, Babylon; here is her secret
    place,   400
From hence she comes forth on the Churches in delight;
Here is her Cup fill'd with its poisons in these horrid
    vales,
And here her scarlet Veil woven in pestilence & war;
Here is Jerusalem bound in chains in the Dens of
    Babylon.

In the Eastern porch of Satan's Universe Milton stood
    & said:

"Satan! my Spectre! I know my power thee to annihi-
    late
And be a greater in thy place & be thy Tabernacle,
A covering for thee to do thy will, till one greater comes
And smites me as I smote thee & becomes my covering.
Such are the Laws of thy false Heav'ns; but Laws of
    Eternity   410
Are not such; know thou, I come to Self Annihilation.
Such are the Laws of Eternity, that each shall mutually
Annihilate himself for others' good, as I for thee.
Thy purpose & the purpose of thy Priests & of thy
    Churches

Is to impress on men the fear of death, to teach
Trembling & fear, terror, constriction, abject selfishness.
Mine is to teach Men to despise death & to go on
In fearless majesty annihilating Self, laughing to scorn
Thy Laws & terrors, shaking down thy Synagogues as
  webs.
I come to discover before Heav'n & Hell the Self
  righteousness   420
In all its Hypocritic turpitude, opening to every eye
These wonders of Satan's holiness, shewing to the
  Earth
The Idol Virtues of the Natural Heart, & Satan's Seat
Explore in all its Selfish Natural Virtue, & put off
In Self annihilation all that is not of God alone,
To put off Self & all I have, ever & ever. Amen."

Satan heard, Coming in a cloud, with trumpets &
  flaming fire,
Saying: "I am God the judge of all, the living & the
  dead.
Fall therefore down & worship me, submit thy supreme
Dictate to my eternal Will, & to my dictate bow. 430
I hold the Balances of Right & Just & mine the Sword.
Seven Angels bear my Name & in those Seven I appear,
But I alone am God & I alone in Heav'n & Earth
Of all that live dare utter this, others tremble & bow,

44

"Till All Things become One Great Satan, in Holiness
Oppos'd to Mercy, and the Divine Delusion, Jesus, be
  no more."

Suddenly around Milton on my Path the Starry Seven
Burn'd terrible; my Path became a solid fire, as bright
As the clear Sun, & Milton silent came down on my
  Path.
And there went forth from the Starry limbs of the
  Seven, Forms   440
Human, with Trumpets innumerable, sounding articu-
  late
As the Seven spake; and they stood in a mighty
  Column of Fire
Surrounding Felpham's Vale, reaching to the Mundane
  Shell, Saying:

"Awake, Albion awake! reclaim thy Reasoning
  Spectre, Subdue
Him to the Divine Mercy. Cast him down into the
  Lake

Of Los that ever burneth with fire ever & ever, Amen!
Let the Four Zoas awake from Slumbers of Six
  Thousand Years."

Then loud the Furnaces of Los were heard, & seen as
  Seven Heavens
Stretching from south to north over the mountains of
  Albion.

Satan heard; trembling round his Body, he incircled
  it: 450
He trembled with exceeding great trembling &
  astonishment,
Howling in his Spectre round his Body, hung'ring to
  devour
But fearing for the pain, for if he touches a Vital
His torment is unendurable: therefore he cannot devour
But howls round it as a lion round his prey continually.
Loud Satan thunder'd, loud & dark upon mild Fel-
  pham's Shore,
Coming in a Cloud with Trumpets & with Fiery
  Flame,
An awful Form eastward from midst of a bright
  Paved-work
Of precious stones by Cherubim surrounded, so per-
  mitted
(Lest he should fall apart in his Eternal Death) to
  imitate   460
The Eternal Great Humanity Divine surrounded by
His Cherubim & Seraphim in ever happy Eternity.
Beneath sat Chaos: Sin on his right hand, Death on his
  left,
And Ancient Night spread over all the heav'n his
  Mantle of Laws.
He trembled with exceeding great trembling & aston-
  ishment.

Then Albion rose up in the Night of Beulah on his
  Couch
Of dread repose; seen by the visionary eye, his face is
  toward
The east, toward Jerusalem's Gates; groaning he sat
  above
His rocks. London & Bath & Legions & Edinburgh
Are the four pillars of his Throne: his left foot near
  London   470
Covers the shades of Tyburn: his instep from Windsor

457–59. **Coming . . . surrounded**: Cf. *Paradise Lost*, II.
511–17.

To Primrose Hill stretching to Highgate & Holloway.
London is between his knees, its basements fourfold;
His right foot stretches to the sea on Dover cliffs, his heel
On Canterbury's ruins; his right hand covers lofty Wales,
His left Scotland; his bosom girt with gold involves
York, Edinburgh, Durham & Carlisle, & on the front
Bath, Oxford, Cambridge, Norwich; his right elbow
Leans on the Rocks of Erin's Land, Ireland, ancient nation;
His head bends over London; he sees his embodied Spectre   480
Trembling before him with exceeding great trembling & fear.
He views Jerusalem & Babylon, his tears flow down.
He mov'd his right foot to Cornwall, his left to the Rocks of Bognor.
He strove to rise to walk into the Deep, but strength failing
Forbad, & down with dreadful groans he sunk upon his Couch
In moony Beulah. Los, his strong Guard, walks round beneath the Moon.

Urizen faints in terror striving among the Brooks of Arnon
With Milton's Spirit; as the Plowman or Artificer or Shepherd
While in the labours of his Calling sends his Thought abroad
To labour in the ocean or in the starry heaven, So Milton   490
Labour'd in Chasms of the Mundane Shell, tho' here before
My Cottage midst the Starry Seven where the Virgin Ololon
Stood trembling in the Porch; loud Satan thunder'd on the stormy Sea
Circling Albion's Cliffs, in which the Four-fold World resides,
Tho' seen in fallacy outside, a fallacy of Satan's Churches.

46

Before Ololon Milton stood & perciev'd the Eternal Form
Of that mild Vision; wondrous were their acts, by me unknown
Except remotely, and I heard Ololon say to Milton:

"I see thee strive upon the Brooks of Arnon: there a dread
And awful Man I see, o'ercover'd with the mantle of years.   500
I behold Los & Urizen, I behold Orc & Tharmas,
The Four Zoas of Albion, & thy Spirit with them striving,
In Self annihilation giving thy life to thy enemies.
Are those who contemn Religion & seek to annihilate it
Become in their Feminine portions the causes & promoters
Of these Religions? how is this thing, this Newtonian Phantasm,
This Voltaire & Rousseau, this Hume & Gibbon & Bolingbroke,
This Natural Religion, this impossible absurdity?
Is Ololon the cause of this? O where shall I hide my face?
These tears fall for the little ones, the Children of Jerusalem,   510
Lest they be annihilated in thy annihilation."

No sooner had she spoke but Rahab Babylon appear'd
Eastward upon the Paved work across Europe & Asia,
Glorious as the midday Sun in Satan's bosom glowing,
A Female hidden in a Male, Religion hidden in War,
Nam'd Moral Virtue, cruel two-fold Monster shining bright,
A Dragon red & hidden Harlot which John in Patmos saw.

507. Voltaire . . . Bolingbroke: To Blake these were Deists, teachers of natural religion and scoffers at faith. They are attacked again in *Jerusalem*, "To the Deists," below. David Hume, 1711–76, the great philosophical skeptic; Edward Gibbon, 1737–94, whose *Decline and Fall of the Roman Empire* gives a slightly satiric portrait of the early Christians and accounts for the spread of Christianity by natural causes; Henry Saint-John, Viscount Bolingbroke, 1678–1751, whose deistic writings were published posthumously (1754). **517. Dragon . . . Patmos:** Revelation 12:1–3.

485–86. down . . . Beulah: Albion's effort to rise is not completely successful. He attains only the "threefold" state of Beulah. But in this state he is guarded by the restored power of imagination (Los) that, as Milton, will continue to struggle with Urizen until the Last Judgment.

And all beneath the Nations innumerable of Ulro
Appear'd: the Seven Kingdoms of Canaan & Five
 Baalim 519
Of Philistea into Twelve divided, call'd after the Names
Of Israel, as they are in Eden, Mountain, River & Plain,
City & sandy Desert intermingled beyond mortal ken.

But turning toward Ololon in terrible majesty Milton
Replied: "Obey thou the Words of the Inspired Man.
All that can be annihilated must be annihilated
That the Children of Jerusalem may be saved from
 slavery.
There is a Negation, & there is a Contrary:
The Negation must be destroy'd to redeem the Con-
 traries.
The Negation is the Spectre, the Reasoning Power in
 Man:
This is a false Body, an Incrustation over my Immortal
Spirit, a Selfhood which must be put off & annihilated
 alway. 531
To cleanse the Face of my Spirit by Self-examination,

48

"To bathe in the Waters of Life, to wash off the Not
 Human,
I come in Self-annihilation & the grandeur of Inspira-
 tion,
To cast off Rational Demonstration by Faith in the
 Saviour,
To cast off the rotten rags of Memory by Inspiration,
To cast off Bacon, Locke & Newton from Albion's
 covering,
To take off his filthy garments & clothe him with
 Imagination,
To cast aside from Poetry all that is not Inspiration,
That it no longer shall dare to mock with the aspersion
 of Madness 540
Cast on the Inspired by the tame high finisher of paltry
 Blots
Indefinite, or paltry Rhymes, or paltry Harmonies,
Who creeps into State Government like a catterpiller
 to destroy;
To cast off the idiot Questioner who is always question-
 ing
But never capable of answering, who sits with a sly grin
Silent plotting when to question, like a thief in a cave,

525. All . . . annihilated: Cf. *The Four Zoas*, "Night the
Ninth," ll. 157-60.

Who publishes doubt & calls it knowledge, whose
 Science is Despair,
Whose pretence to knowledge is Envy, whose whole
 Science is
To destroy the wisdom of ages to gratify ravenous
 Envy
That rages round him like a Wolf day & night without
 rest: 550
He smiles with condescension, he talks of Benevolence
 & Virtue,
And those who act with Benevolence & Virtue they
 murder time on time.
These are the destroyers of Jerusalem, these are the
 murderers
Of Jesus, who deny the Faith & mock at Eternal Life,
Who pretend to Poetry that they may destroy Imagina-
 tion
By imitation of Nature's Images drawn from Remem-
 brance.
These are the Sexual Garments, the Abomination of
 Desolation,
Hiding the Human Lineaments as with an Ark &
 Curtains
Which Jesus rent & now shall wholly purge away with
 Fire
Till Generation is swallow'd up in Regeneration." 560

Then trembled the Virgin Ololon & reply'd in clouds
 of despair:

"Is this our Feminine Portion, the Six-fold Miltonic
 Female?
Terribly this Portion trembles before thee, O awful
 Man.
Altho' our Human Power can sustain the severe con-
 tentions
Of Friendship, our Sexual cannot, but flies into the Ulro.
Hence arose all our terrors in Eternity; & now remem-
 brance
Returns upon us; are we Contraries, O Milton, Thou
 & I?
O Immortal, how were we led to War the Wars of
 Death?
Is this the Void Outside of Existence, which if enter'd
 into

49

"Becomes a Womb? & is this the Death Couch of
 Albion? 570
Thou goest to Eternal Death & all must go with thee."

So saying, the Virgin divided Six-fold, & with a
　　shriek
Dolorous that ran thro' all Creation, a Double Six-fold
　　Wonder
Away from Ololon she divided & fled into the depths
Of Milton's Shadow, as a Dove upon the stormy Sea.

Then as a Moony Ark Ololon descended to Felpham's
　　Vale
In clouds of blood, in streams of gore, with dreadful
　　thunderings
Into the Fires of Intellect that rejoic'd in Felpham's
　　Vale
Around the Starry Eight; with one accord the Starry
　　Eight became
One Man, Jesus the Saviour, wonderful! round his
　　limbs　　　　　　　　　　　　　　　　　　580
The Clouds of Ololon folded as a Garment dipped in
　　blood,
Written within & without in woven letters, & the
　　Writing
Is the Divine Revelation in the Litteral expression,
A Garment of War. I heard it nam'd the Woof of
　　Six Thousand Years.

And I beheld the Twenty-four Cities of Albion
Arise upon their Thrones to Judge the Nations of the
　　Earth;
And the Immortal Four in whom the Twenty-four
　　appear Four-fold
Arose around Albion's body. Jesus wept & walked
　　forth
From Felpham's Vale clothed in Clouds of blood, to
　　enter into
Albion's Bosom, the bosom of death, & the Four
　　surrounded him　　　　　　　　　　　　　590
In the Column of Fire in Felpham's Vale; then to their
　　mouths the Four
Applied their Four Trumpets & them sounded to the
　　Four winds.

Terror struck in the Vale I stood at that immortal
　　sound.
My bones trembled, I fell outstretch'd upon the path
A moment, & my Soul return'd into its mortal state
To Resurrection & Judgment in the Vegetable Body,
And my sweet Shadow of Delight stood trembling by
　　my side.

574. she: "the Virgin" (l. 572).

Immediately the Lark mounted with a loud trill from
　　Felpham's Vale,
And the Wild Thyme from Wimbleton's green &
　　impurpled Hills,
And Los & Enitharmon rose over the Hills of Surrey:
Their clouds roll over London with a south wind; soft
　　Oothoon　　　　　　　　　　　　　　　601
Pants in the Vales of Lambeth, weeping o'er her
　　Human Harvest.
Los listens to the Cry of the Poor Man, his Cloud
Over London in volume terrific low bended in anger.

Rintrah & Palamabron view the Human Harvest
　　beneath.
Their Wine-presses & Barns stand open, the Ovens are
　　prepar'd,
The Waggons ready; terrific Lions & Tygers sport &
　　play.
All Animals upon the Earth are prepar'd in all their
　　strength

50

To go forth to the Great Harvest & Vintage of the
　　Nations.

1800–04 (1804–08)

FROM

JERUSALEM

The last of the major prophetic works, *Jerusalem* is more
than twice as long as *Milton*. Blake had begun composing
it at least by 1804. It may have been substantially finished
by 1809, though no surviving copy was engraved before
1818. The selections below are introductions to Chapters
2, 3, and 4. They do not typify the style of the poem,
which resembles *Milton*.

TO THE JEWS

The fields from Islington to Marybone,
To Primrose Hill and Saint John's Wood,
　Were builded over with pillars of gold,
　And there Jerusalem's pillars stood.

JERUSALEM. TO THE JEWS. 1–2. Islington . . . Saint
John's Wood: districts in and around London.

Her Little-ones ran on the fields,
The Lamb of God among them seen,
  And fair Jerusalem his Bride,
Among the little meadows green.

Pancrass & Kentish-town repose
Among her golden pillars high,   10
  Among her golden arches which
Shine upon the starry sky.

The Jew's-harp-house & the Green Man,
The Ponds where Boys to bathe delight,
  The fields of Cows by Willan's farm,
Shine in Jerusalem's pleasant sight.

She walks upon our meadows green,
The Lamb of God walks by her side,
  And every English Child is seen
Children of Jesus & his Bride.   20

Forgiving trespasses and sins
Lest Babylon with cruel Og
  With Moral & Self-righteous Law
Should Crucify in Satan's Synagogue!

What are those golden Builders doing
Near mournful ever-weeping Paddington,
  Standing above that mighty Ruin
Where Satan the first victory won,

Where Albion slept beneath the Fatal Tree,
And the Druid's golden Knife   30
  Rioted in human gore,
In Offerings of Human Life?

They groan'd aloud on London Stone,
They groan'd aloud on Tyburn's Brook,
  Albion gave his deadly groan,
And all the Atlantic Mountains shook.

Albion's Spectre from his Loins
Tore forth in all the pomp of War:
  Satan his name: in flames of fire
He stretch'd his Druid Pillars far.   40

Jerusalem fell from Lambeth's Vale
Down thro' Poplar & Old Bow,
  Thro' Malden & across the Sea,
In War & howling, death & woe.

The Rhine was red with human blood,
The Danube roll'd a purple tide,
  On the Euphrates Satan stood,
And over Asia stretch'd his pride.

He wither'd up sweet Zion's Hill
From every Nation of the Earth;   50
  He wither'd up Jerusalem's Gates,
And in a dark Land gave her birth.

He wither'd up the Human Form
By laws of sacrifice for sin,
  Till it became a Mortal Worm,
But O! translucent all within.

The Divine Vision still was seen,
Still was the Human Form Divine,
  Weeping in weak & mortal clay,
O Jesus, still the Form was thine.   60

And thine the Human Face, & thine
The Human Hands & Feet & Breath,
  Entering thro' the Gates of Birth
And passing thro' the Gates of Death.

And O thou Lamb of God, whom I
Slew in my dark self-righteous pride,
  Art thou return'd to Albion's Land?
And is Jerusalem thy Bride?

Come to my arms & never more
Depart, but dwell for ever here: 70
  Create my Spirit to thy Love:
Subdue my Spectre to thy Fear.

Spectre of Albion! warlike Fiend!
In clouds of blood & ruin roll'd,
  I here reclaim thee as my own.
My Selfhood! Satan! arm'd in gold.

Is this thy soft Family-Love,
Thy cruel Patriarchal pride,
  Planting thy Family alone,
Destroying all the World beside?   80

A man's worst enemies are those
Of his own house & family;
  And he who makes his law a curse,
By his own law shall surely die.

In my Exchanges every Land
Shall walk, & mine in every Land,
  Mutual shall build Jerusalem,
Both heart in heart & hand in hand.

---

13. The Jew's . . . Man: entertainments in London of the
sort now to be found in an amusement park.

TO THE DEISTS

He never can be a Friend to the Human Race who is the Preacher of Natural Morality or Natural Religion; he is a flatterer who means to betray, to perpetuate Tyrant Pride & the Laws of that Babylon which he Foresees shall shortly be destroyed, with the Spiritual and not the Natural Sword. He is in the State named Rahab, which State must be put off before he can be the Friend of Man.

You, O Deists, profess yourselves the Enemies of Christianity, and you are so: you are also the Enemies of the Human Race & of Universal Nature. Man is born a Spectre or Satan & is altogether an Evil, & requires a New Selfhood continually, & must continually be changed into his direct Contrary. But your Greek Philosophy (which is a remnant of Druidism) teaches that Man is Righteous in his Vegetated Spectre: an Opinion of fatal & accursed consequence to Man, as the Ancients saw plainly by Revelation, to the intire abrogation of Experimental Theory; and many believed what they saw and Prophecied of Jesus.

Man must & will have Some Religion: if he has not the Religion of Jesus, he will have the Religion of Satan & will erect the Synagogue of Satan, calling the Prince of this World, God, and destroying all who do not worship Satan under the Name of God. Will any one say, "Where are those who worship Satan under the Name of God?" Where are they? Listen! Every Religion that Preaches Vengeance for Sin is the Religion of the Enemy & Avenger and not of the Forgiver of Sin, and their God is Satan, Named by the Divine Name. Your Religion, O Deists! Deism, is the Worship of the God of this World by the means of what you call Natural Religion and Natural Philosophy, and of Natural Morality or Self-Righteousness, the Selfish Virtues of the Natural Heart. This was the Religion of the Pharisees who murder'd Jesus. Deism is the same & ends in the same.

Voltaire, Rousseau, Gibbon, Hume,[1] charge the Spiritually Religious with Hypocrisy; but how a Monk, or a Methodist either, can be a Hypocrite, I cannot concieve. We are Men of like passions with others & pretend not to be holier than others; therefore, when a Religious Man falls into Sin, he ought not be call'd a Hypocrite; this title is more properly to be given to a Player who falls into Sin, whose profession is Virtue & Morality & the making Men Self-Righteous. Foote in calling Whitefield,[2] Hypocrite, was himself one; for Whitefield pretended not to be holier than others, but confessed his Sins before all the World. Voltaire! Rousseau! You cannot escape my charge that you are Pharisees & Hypocrites, for you are constantly talking of the Virtues of the Human Heart and particularly of your own, that you may accuse others, & especially the Religious, whose errors you, by this display of pretended Virtue, chiefly design to expose. Rousseau thought Men Good by Nature: he found them Evil & found no friend. Friendship cannot exist without Forgiveness of Sins continually. The Book written by Rousseau call'd his Confessions, is an apology & cloke for his sin & not a confession.

But you also charge the poor Monks & Religious with being the causes of War, while you acquit & flatter the Alexanders & Caesars, the Lewis's & Fredericks,[3] who alone are its causes & its actors. But the Religion of Jesus, Forgiveness of Sin, can never be the cause of a War nor of a single Martyrdom.

Those who Martyr others or who cause War are Deists, but never can be Forgivers of Sin. The Glory of Christianity is To Conquer by Forgiveness. All the Destruction, therefore, in Christian Europe has arisen from Deism, which is Natural Religion.

I saw a Monk of Charlemaine
Arise before my sight:
I talk'd with the Grey Monk as we stood
In beams of infernal light.

Gibbon arose with a lash of steel,
And Voltaire with a wracking wheel:
The Schools, in clouds of learning roll'd,
Arose with War in iron & gold.

"Thou lazy Monk," they sound afar,
"In vain condemning glorious War;   10
And in your Cell you shall ever dwell:
Rise, War, & bind him in his Cell!"

The blood red ran from the Grey Monk's side,
His hands & feet were wounded wide,
His body bent, his arms & knees
Like to the roots of ancient trees.

TO THE DEISTS. **1. Voltaire . . . Hume:** See *Milton*, above, II. n. 507.

2. Whitefield: See *Milton*, above, I. n. 903. **3. Lewis's & Fredericks:** Louis XIV of France and Frederick the Great of Prussia.

When Satan first the black bow bent
And the Moral Law from the Gospel rent,
 He forg'd the Law into a Sword
And spill'd the blood of mercy's Lord.   20

 Titus! Constantine! Charlemaine!
O Voltaire! Rousseau! Gibbon! Vain
 Your Grecian Mocks & Romans Sword
Against this image of his Lord!

For a Tear is an Intellectual thing,
And a Sigh is the Sword of an Angel King,
 And the bitter groan of a Martyr's woe
Is an Arrow from the Almightie's Bow.

TO THE CHRISTIANS

I give you the end of a golden string,
 Only wind it into a ball,
It will lead you in at Heaven's gate,
 Built in Jerusalem's wall.

We are told to abstain from fleshly desires that we may lose no time from the Work of the Lord: Every moment lost is a moment that cannot be redeemed; every pleasure that intermingles with the duty of our station is a folly unredeemable, & is planted like the seed of a wild flower among our wheat: All the tortures of repentance are tortures of self-reproach on account of our leaving the Divine Harvest to the Enemy: the struggles of intanglement with incoherent roots. I know of no other Christianity and of no other Gospel than the liberty both of the body & mind to exercise the Divine Arts of Imagination, Imagination, the real & eternal World of which this Vegetable Universe is but a faint shadow, & in which we shall live in our Eternal or Imaginative Bodies when these Vegetable Mortal Bodies are no more. The Apostles knew of no other Gospel. What were all their spiritual gifts? What is the Divine Spirit? is the Holy Ghost any other than an Intellectual Fountain? What is the harvest of the Gospel & its Labours? What is that Talent which it is a curse to hide? What are the Treasures of Heaven which we are to lay up for ourselves, are they any other than Mental Studies & Performances? What are all the Gifts of the Gospel, are they not all Mental Gifts? Is God a Spirit who must be worshipped in Spirit & in Truth, and are not the Gifts of the Spirit Every-thing to Man? O ye Religious, discountenance every one among you who shall pretend to despise Art & Science! I call upon you in the Name of Jesus! What is the Life of Man but Art & Science? is it Meat & Drink? is not the Body more than Raiment? What is Mortality but the things relating to the Body which Dies? What is Immortality but the things relating to the Spirit which Lives Eternally? What is the Joy of Heaven but Improvement in the things of the Spirit? What are the Pains of Hell but Ignorance, Bodily Lust, Idleness & devastation of the things of the Spirit? Answer this to yourselves, & expel from among you those who pretend to despise the labours of Art & Science, which alone are the labours of the Gospel. Is not this plain & manifest to the thought? Can you think at all & not pronounce heartily That to Labour in Knowledge is to Build up Jerusalem, and to Despise Knowledge is to Despise Jerusalem & her Builders. And remember: He who despises & mocks a Mental Gift in another, calling it pride & selfishness & sin, mocks Jesus the giver of every Mental Gift, which always appear to the ignorance-loving Hypocrite as Sins; but that which is a Sin in the sight of cruel Man is not so in the sight of our kind God. Let every Christian, as much as in him lies, engage himself openly & publicly before all the World in some Mental pursuit for the Building up of Jerusalem.

I stood among my valleys of the south
And saw a flame of fire, even as a Wheel
Of fire surrounding all the heavens: it went
From west to east, against the current of
Creation, and devour'd all things in its loud
Fury & thundering course round heaven & earth.
By it the Sun was roll'd into an orb,
By it the Moon faded into a globe
Travelling thro' the night; for, from its dire
And restless fury, Man himself shrunk up   10
Into a little root a fathom long.
And I asked a Watcher & a Holy-One
Its Name; he answered: "It is the Wheel of Religion."
I wept & said: "Is this the law of Jesus,
This terrible devouring sword turning every way?"
He answer'd: "Jesus died because he strove
Against the current of this Wheel; its Name
Is Caiaphas, the dark preacher of Death,
Of sin, of sorrow & of punishment:
Opposing Nature! It is Natural Religion;   20
But Jesus is the bright Preacher of Life
Creating Nature from this fiery Law

TO THE CHRISTIANS. **18. Caiaphas:** the high priest who demanded the death of Christ.

By self-denial & forgiveness of Sin.
Go therefore, cast out devils in Christ's name,
Heal thou the sick of spiritual disease,
Pity the evil, for thou art not sent
To smite with terror & with punishments
Those that are sick, like to the Pharisees
Crucifying & encompassing sea & land
For proselytes to tyranny & wrath;   30
But to the Publicans & Harlots go,
Teach them True Happiness, but let no curse
Go forth out of thy mouth to blight their peace;
For Hell is open'd to Heaven: thine eyes beheld
The dungeons burst & the Prisoners set free."

England! awake! awake! awake!
    Jerusalem thy Sister calls!
Why wilt thou sleep the sleep of death
    And close her from thy ancient walls?

Thy hills & valleys felt her feet
    Gently upon their bosoms move:
Thy gates beheld sweet Zion's ways:
    Then was a time of joy and love.

And now the time returns again:
    Our souls exult, & London's towers   10
Recieve the Lamb of God to dwell
    In England's green & pleasant bowers.

                                    1804–09 (1818)

⌒⌒⌒

The following extracts from *A Descriptive Catalogue,
Public Address*, and *A Vision of the Last Judgment* are from
Blake's descriptions of his own pictures. The *Descriptive
Catalogue* was prepared in 1809 for an exhibition; the
numbers in brackets correspond to the numbers given on
Blake's paintings at the exhibition. The *Public Address*
and *A Vision of the Last Judgment*, referring to a painting of
that subject, come from a notebook of 1810.

⌒⌒⌒

from A DESCRIPTIVE CATALOGUE

Visions of these eternal principles[1] or characters of
human life appear to poets, in all ages; the Grecian gods

A DESCRIPTIVE CATALOGUE. **1. eternal principles:** Blake is
describing his picture of Chaucer's pilgrims on their way
to Canterbury. He has just asserted that "Chaucer has divided
the ancient character of Hercules between his Miller and his
Plowman."

were the ancient Cherubim of Phoenicia; but the
Greeks, and since then the Moderns, have neglected to
subdue the gods of Priam. These gods are visions of the
eternal attributes, or divine names, which, when erected
into gods, become destructive to humanity. They ought
to be the servants, and not the masters of man, or of
society. They ought to be made to sacrifice to Man, and
not man compelled to sacrifice to them; for when
separated from man or humanity, who is Jesus the
Saviour, the vine of eternity, they are thieves and rebels,
they are destroyers. [III]

The connoisseurs and artists who have made objections to Mr. B.'s mode of representing spirits with real
bodies, would do well to consider that the Venus, the
Minerva, the Jupiter, the Apollo, which they admire in
Greek statues are all of them representations of spiritual
existences, of Gods immortal, to the mortal perishing
organ of sight; and yet they are embodied and organized
in solid marble. Mr. B. requires the same latitude, and
all is well. The Prophets describe what they saw in
Vision as real and existing men, whom they saw with
their imaginative and immortal organs; the Apostles
the same; the clearer the organ the more distinct the
object. A Spirit and a Vision are not, as the modern
philosophy supposes, a cloudy vapour, or a nothing:
they are organized and minutely articulated beyond all
that the mortal and perishing nature can produce. He
who does not imagine in stronger and better lineaments,
and in stronger and better light than his perishing and
mortal eye can see, does not imagine at all. The painter
of this work asserts that all his imaginations appear to
him infinitely more perfect and more minutely organized than any thing seen by his mortal eye. Spirits are
organized men. Moderns wish to draw figures without
lines, and with great and heavy shadows; are not
shadows more unmeaning than lines, and more heavy?
O who can doubt this! [IV]

The distinction that is made in modern times between
a Painting and a Drawing proceeds from ignorance of
art. The merit of a Picture is the same as the merit of a
Drawing. The dawber dawbs his Drawings; he who
draws his Drawings draws his Pictures. There is no
difference between Rafael's Cartoons[2] and his Frescos,
or Pictures, except that the Frescos, or Pictures, are
more finished. When Mr. B. formerly painted in oil
colours his Pictures were shewn to certain painters and

2. Cartoons: full-sized drawings that can be copied in
making tapestries, mosaics, fresco paintings, and the like.

connoisseurs, who said that they were very admirable Drawings on canvass, but not Pictures; but they said the same of Rafael's Pictures. Mr. B. thought this the greatest of compliments, though it was meant otherwise. If losing and obliterating the outline constitutes a Picture, Mr. B. will never be so foolish as to do one. Such art of losing the outlines is the art of Venice and Flanders; it loses all character, and leaves what some people call expression; but this is a false notion of expression; expression cannot exist without character as its stamina; and neither character nor expression can exist without firm and determinate outline. Fresco Painting is susceptible of higher finishing than Drawing on Paper, or than any other method of Painting. But he must have a strange organization of sight who does not prefer a Drawing on Paper to a Dawbing in Oil by the same master, supposing both to be done with equal care.

The great and golden rule of art, as well as of life, is this: That the more distinct, sharp, and wirey the bounding line, the more perfect the work of art, and the less keen and sharp, the greater is the evidence of weak imitation, plagiarism, and bungling. Great inventors, in all ages, knew this: Protogenes and Apelles[3] knew each other by this line. Rafael and Michael Angelo and Albert Dürer are known by this and this alone. The want of this determinate and bounding form evidences the want of idea in the artist's mind, and the pretence of the plagiary in all its branches. How do we distinguish the oak from the beech, the horse from the ox, but by the bounding outline? How do we distinguish one face or countenance from another, but by the bounding line and its infinite inflexions and movements? What is it that builds a house and plants a garden, but the definite and determinate? What is it that distinguishes honesty from knavery, but the hard and wirey line of rectitude and certainty in the actions and intentions? Leave out this line, and you leave out life itself; all is chaos again, and the line of the almighty must be drawn out upon it before man or beast can exist. Talk no more then of Correggio, or Rembrandt, or any other of those plagiaries of Venice or Flanders. They were but the lame imitators of lines drawn by their predecessors, and their works prove themselves contemptible, disarranged imitations, and blundering, misapplied copies. [XV]

1809

from PUBLIC ADDRESS

I have heard many People say, "Give me the Ideas. It is no matter what Words you put them into," & others say, "Give me the Design, it is no matter for the Execution." These People know Enough of Artifice, but Nothing Of Art. Ideas cannot be Given but in their minutely Appropriate Words, nor Can a Design be made without its minutely Appropriate Execution. The unorganized Blots & Blurs of Rubens & Titian are not Art, nor can their Method ever express Ideas or Imaginations any more than Pope's Metaphysical Jargon of Rhyming. Unappropriate Execution is the Most nauseous of all affectation & foppery. He who copies does not Execute; he only Imitates what is already Executed. Execution is only the result of Invention.

1810

from A VISION
OF THE LAST JUDGMENT

The Last Judgment[1] is not Fable or Allegory, but Vision. Fable or Allegory are a totally distinct & inferior kind of Poetry. Vision or Imagination is a Representation of what Eternally Exists, Really & Unchangeably. Fable or Allegory is Form'd by the daughters of Memory. Imagination is surrounded by the daughters of Inspiration, who in the aggregate are call'd Jerusalem. Fable is allegory, but what Critics call The Fable,[2] is Vision itself. The Hebrew Bible & the Gospel of Jesus are not Allegory, but Eternal Vision or Imagination of All that Exists. Note here that Fable or Allegory is seldom without some Vision. Pilgrim's Progress is full of it, the Greek Poets the same; but Allegory & Vision ought to be known as Two Distinct Things, & so call'd for the Sake of Eternal Life. Plato has made Socrates say that Poets & Prophets do not know or Understand what they write or Utter; this is a most Pernicious Falshood. If they do not, pray is an inferior kind to be call'd Knowing? Plato confutes himself.

The Last Judgment is one of these Stupendous Visions. I have represented it as I saw it; to different People it appears differently as every thing else does; for tho' on

---

3. **Protogenes and Apelles:** Greek painters of *c.* 330 B.C.

A VISION OF THE LAST JUDGMENT. 1. **Last Judgment:** Blake's painting of that subject. 2. **Critics . . . Fable:** the plot.

Earth things seem Permanent, they are less permanent than a Shadow, as we all know too well.

The Nature of Visionary Fancy, or Imagination, is very little known, & the Eternal nature & permanence of its ever Existent Images is consider'd as less permanent than the things of Vegetative & Generative Nature; yet the Oak dies as well as the Lettuce, but Its Eternal Image & Individuality never dies, but renews by its seed; just so the Imaginative Image returns by the seed of Contemplative Thought; the Writings of the Prophets illustrate these conceptions of the Visionary Fancy by their various sublime & Divine Images as seen in the Worlds of Vision. . . . The Nature of my Work is Visionary or Imaginative; it is an Endeavour to Restore what the Ancients call'd the Golden Age.

This world of Imagination is the world of Eternity; it is the divine bosom into which we shall all go after the death of the Vegetated body. This World of Imagination is Infinite & Eternal, whereas the world of Generation, or Vegetation, is Finite & Temporal. There Exist in that Eternal World the Permanent Realities of Every Thing which we see reflected in this Vegetable Glass of Nature. All Things are comprehended in their Eternal Forms in the divine body of the Saviour, the True Vine of Eternity, The Human Imagination, who appear'd to Me as Coming to Judgment among his Saints & throwing off the Temporal that the Eternal might be Establish'd; around him were seen the Images of Existences according to a certain order Suited to my Imaginative Eye.

If the Spectator could enter into these Images in his Imagination, approaching them on the Fiery Chariot of his Contemplative Thought, if he could Enter into Noah's Rainbow or into his bosom, or could make a Friend & Companion of one of these Images of wonder, which always intreats him to leave mortal things (as he must know), then would he arise from his Grave, then would he meet the Lord in the Air & then he would be happy. General Knowledge is Remote Knowledge; it is in Particulars that Wisdom consists & Happiness too. Both in Art & in Life, General Masses are as Much Art as a Pasteboard Man is Human. Every Man has Eyes, Nose & Mouth; this Every Idiot knows, but he who enters into & discriminates most minutely the Manners & Intentions, the Characters in all their branches, is the alone Wise or Sensible Man, & on this discrimination

All Art is founded. I intreat, then, that the Spectator will attend to the Hands & Feet, to the Lineaments of the Countenances; they are all descriptive of Character, & not a line is drawn without intention, & that most discriminate & particular. As Poetry admits not a Letter that is Insignificant, so Painting admits not a Grain of Sand or a Blade of Grass Insignificant—much less an Insignificant Blur or Mark.

Men are admitted into Heaven not because they have curbed & govern'd their Passions or have No Passions, but because they have Cultivated their Understandings. The Treasures of Heaven are not Negations of Passion, but Realities of Intellect, from which all the Passions Emanate Uncurbed in their Eternal Glory. The Fool shall not enter into Heaven let him be ever so Holy. Holiness is not The Price of Enterance into Heaven. Those who are cast out are All Those who, having no Passions of their own because No Intellect, Have spent their lives in Curbing & Governing other People's by the Various arts of Poverty & Cruelty of all kinds. Wo, Wo, Wo to you Hypocrites. Even Murder, the Courts of Justice, more merciful than the Church, are compell'd to allow is not done in Passion, but in Cool Blooded designs & Intention.

The Modern Church Crucifies Christ with the Head Downwards.

The Last Judgment is an Overwhelming of Bad Art & Science. Mental Things are alone Real; what is call'd Corporeal, Nobody Knows of its Dwelling Place: it is in Fallacy, & its Existence an Imposture. Where is the Existence Out of Mind or Thought? Where is it but in the Mind of a Fool? Some People flatter themselves that there will be No Last Judgment & that Bad Art will be adopted & mixed with Good Art, That Error or Experiment will make a Part of Truth, & they Boast that it is its Foundation; these People flatter themselves: I will not Flatter them. Error is Created. Truth is Eternal. Error, or Creation, will be Burned up, & then, & not till Then, Truth or Eternity will appear. It is Burnt up the Moment Men cease to behold it. I assert for My Self that I do not behold the outward Creation & that to me it is hindrance & not Action; it is as the dirt upon my feet, No part of Me. "What," it will be Question'd, "When the Sun rises, do you not see a round disk of fire somewhat like a Guinea?" O no, no, I see an Innumerable company of the Heavenly host crying, "Holy, Holy, Holy is the Lord God Almighty." I question not my

Corporeal or Vegetative Eye any more than I would Question a Window concerning a Sight. I look thro' it & not with it.

1810

LETTERS

❧

Of Blake's correspondents the Rev. John Trusler was an author to whom Blake had been introduced as a possible illustrator. The two, however, could not please each other. "*Your Fancy*," Trusler pointed out, "seems to be in the other world, or the World of Spirits, which accords not with my Intentions, which, whilst living in This World, Wish to follow *the Nature of it*," a wish also indicated by the titles of his best-known books, *Hogarth Moralized* and *The Way To Be Rich and Respectable*. On receiving the beautiful letter printed below, Trusler endorsed it "Blake, dim'd with Superstition." For information about William Hayley see the Introduction. The letter to him printed here was written in the early period of their friendship. Thomas Butts was a friend who purchased much of Blake's work. George Cumberland, artist and engraver, met Blake in the early 1790's and was a lifelong friend.

❧

To the Revd. Dr. Trusler
13 Hercules Buildings, Lambeth,
August 23, 1799.

REVD. SIR,

I really am sorry that you are fall'n out with the Spiritual World, Especially if I should have to answer for it. I feel very sorry that your Ideas & Mine on Moral Painting differ so much as to have made you angry with my method of study. If I am wrong, I am wrong in good company. I had hoped your plan comprehended All Species of this Art, & Expecially that you would not regret that Species which gives Existence to Every other, namely, Visions of Eternity. You say that I want somebody to Elucidate my Ideas. But you ought to know that What is Grand is necessarily obscure to Weak men. That which can be made Explicit to the Idiot is not worth my care. The wisest of the Ancients consider'd what is not too Explicit as the fittest for Instruction, because it rouzes the faculties to act. I name Moses, Solomon, Esop, Homer, Plato.

But as you have favor'd me with your remarks on my Design, permit me in return to defend it against a mistaken one, which is, That I have supposed Malevolence

without a Cause.[1] Is not Merit in one a Cause of Envy in another, & Serenity & Happiness & Beauty a Cause of Malevolence? But Want of Money & the Distress of A Thief can never be alleged as the Cause of his Thieving, for many honest people endure greater hardships with Fortitude. We must therefore seek the Cause elsewhere than in want of Money, for that is the Miser's passion, not the Thief's.

I have therefore proved your Reasonings Ill proportion'd, which you can never prove my figures to be; they are those of Michael Angelo, Rafael & the Antique, & of the best living Models. I percieve that your Eye is perverted by Caricature Prints, which ought not to abound so much as they do. Fun I love, but too much Fun is of all things the most loathsom. Mirth is better than Fun, & Happiness is better than Mirth. I feel that a Man may be happy in This World. And I know that This World Is a World of Imagination & Vision. I see Every thing I paint In This World, but Every body does not see alike. To the Eyes of a Miser a Guinea is far more beautiful than the Sun, & a bag worn with the use of Money has more beautiful proportions than a Vine filled with Grapes. The tree which moves some to tears of joy is in the Eyes of others only a Green thing which stands in the way. Some see Nature all Ridicule & Deformity, & by these I shall not regulate my proportions; & some scarce see Nature at all. But to the Eyes of the Man of Imagination, Nature is Imagination itself. As a man is, so he sees. As the Eye is formed, such are its Powers. You certainly Mistake, when you say that the Visions of Fancy are not to be found in This World. To Me This World is all One continued Vision of Fancy or Imagination, & I feel Flatter'd when I am told so. What is it sets Homer, Virgil & Milton in so high a rank of Art? Why is the Bible more Entertaining & Instructive than any other book? Is it not because they are addressed to the Imagination, which is Spiritual Sensation, & but mediately to the Understanding or Reason? Such is True Painting, and such was alone valued by the Greeks & the best modern Artists. Consider what Lord Bacon says: "Sense sends over to Imagination before Reason have judged, & Reason sends over to Imagination before the Decree can be acted." See Advancemt. of Learning, Part 2 P. 47 of first Edition.

But I am happy to find a Great Majority of Fellow

LETTERS. 1. **Malevolence . . . Cause:** Blake refers to one of his designs for Trusler illustrating the subject of Malevolence: "A Father, taking leave of his Wife & Child, Is watch'd by Two Fiends incarnate, with intention that when his back is turned they will murder the mother & her infant."

Mortals who can Elucidate My Visions, & Particularly they have been Elucidated by Children, who have taken a greater delight in contemplating my Pictures than I even hoped. Neither Youth nor Childhood is Folly or Incapacity. Some Children are Fools & so are some Old Men. But There is a vast Majority on the side of Imagination or Spiritual Sensation.

To Engrave after another Painter is infinitely more laborious than to Engrave one's own Inventions. And of the size you require my price has been Thirty Guineas, & I cannot afford to do it for less. I had Twelve for the Head I sent you as a Specimen; but after my own designs I could do at least Six times the quantity of labour in the same time, which will account for the difference of price as also that Chalk Engraving is at least six times as laborious as Aqua tinta. I have no objection to Engraving after another Artist. Engraving is the profession I was apprenticed to, & should never have attempted to live by anything else, If orders had not come in for my Designs & Paintings, which I have the pleasure to tell you are Increasing Every Day. Thus If I am a Painter it is not to be attributed to Seeking after. But I am contented whether I live by Painting or Engraving.

I am, Revᵈ Sir, your very obedient servant,

WILLIAM BLAKE.

To William Hayley
Lambeth, May 6, 1800.

DEAR SIR,

I am very sorry for your immense loss,[2] which is a repetition of what all feel in this valley of misery and happiness mixed. I send the shadow of the departed angel,[3] and hope the likeness is improved. The lips I have again lessened as you advise, and done a good many other softenings to the whole. I know that our deceased friends are more really with us than when they were apparent to our mortal part. Thirteen years ago I lost a brother,[4] and with his spirit I converse daily and hourly in the spirit, and see him in my remembrance, in the regions of my imagination. I hear his advice, and even now write from his dictate. Forgive me for expressing to you my enthusiasm, which I wish all to partake of,

since it is to me a source of immortal joy, even in this world. By it I am the companion of angels. May you continue to be so more and more; and to be more and more persuaded that every mortal loss is an immortal gain. The ruins of Time build mansions in Eternity.

I have also sent a proof of Pericles[5] for your remarks, thanking you for the kindness with which you express them, and feeling heartily your grief with a brother's sympathy.

I remain,
Dear Sir,
Your humble servant,
WILLIAM BLAKE.

To Thomas Butts
[November 22, 1802.]

DEAR SIR,

After I had finish'd my Letter,[6] I found that I had not said half what I intended to say, & in particular I wish to ask you what subject you choose to be painted on the remaining Canvas which I brought down with me (for there were three), and to tell you that several of the Drawings were in great forwardness; you will see by the Inclosed Account that the remaining Number of Drawings which you gave me orders for is Eighteen. I will finish these with all possible Expedition, if indeed I have not tired you, or, as it is politely call'd, Bored you too much already; or, if you would rather cry out "Enough, Off, Off!", tell me in a Letter of forgiveness if you were offended, & of accustom'd friendship if you were not. But I will bore you more with some Verses which My Wife desires me to Copy out & send you with her kind love & Respect; they were Composed above a twelvemonth ago, while walking from Felpham to Lavant to meet my Sister:

With happiness stretch'd across the hills
In a cloud that dewy sweetness distills,
With a blue sky spread over with wings
And a mild sun that mounts & sings,
With trees & fields full of Fairy elves
And little devils who fight for themselves—
Rememb'ring the Verses that Hayley sung

---

2. **loss**: the death of his illegitimate son. 3. **shadow . . . angel**: Blake's engraving of the child after a medallion portrait. 4. **brother**: Robert, died February, 1787.

5. **proof of Pericles**: Blake was preparing engravings for Hayley's *Essay on Sculpture* (1800). 6. **Letter**: a previous letter on the same day.

When my heart knock'd against the root of my
    tongue—[7]
With Angels planted in Hawthorn bowers
And God himself in the passing hours,   10
With Silver Angels across my way
And Golden Demons that none can stay,
With my Father hovering upon the wind
And my Brother Robert just behind
And my Brother John, the evil one,[8]
In a black cloud making his mone;
Tho' dead, they appear upon my path,
Notwithstanding my terrible wrath:
They beg, they intreat, they drop their tears,
Fill'd full of hopes, fill'd full of fears—   20
With a thousand Angels upon the Wind
Pouring disconsolate from behind
To drive them off, & before my way
A frowning Thistle implores my stay.
What to others a trifle appears
Fills me full of smiles or tears;
For double the vision my Eyes do see,
And a double vision is always with me.
With my inward Eye 'tis an old Man grey;
With my outward, a Thistle across my way.   30
"If thou goest back," the thistle said,
"Thou art to endless woe betray'd;
For here does Theotormon lower
And here is Enitharmon's bower
And Los the terrible thus hath sworn,
Because thou backward dost return,
Poverty, Envy, old age & fear
Shall bring thy Wife upon a bier;
And Butts shall give what Fuseli[9] gave,
A dark black Rock & a gloomy Cave."   40

I struck the Thistle with my foot,
And broke him up from his delving root:
"Must the duties of life each other cross?
Must every joy be dung & dross?
Must my dear Butts feel cold neglect
Because I give Hayley his due respect?

Must Flaxman[10] look upon me as wild,
And all my friends be with doubts beguil'd?
Must my Wife live in my Sister's bane,[11]
Or my Sister survive on my Love's pain?   50
The curses of Los, the terrible shade
And his dismal terrors make me afraid."

So I spoke & struck in my wrath
The old man weltering upon my path.
Then Los appear'd in all his power:
In the Sun he appear'd, descending before
My face in fierce flames; in my double sight
'Twas outward a Sun, inward Los in his might.

"My hands are labour'd day & night,
And Ease comes never in my sight.   60
My Wife has no indulgence given
Except what comes to her from heaven.
We eat little, we drink less;
This Earth breeds not our happiness.
Another Sun feeds our life's streams,
We are not warmed with thy beams;
Thou measurest not the Time to me,
Nor yet the Space that I do see;
My Mind is not with thy light array'd,
Thy terrors shall not make me afraid."   70

When I had my Defiance given,
The Sun stood trembling in heaven;
The Moon that glow'd remote below,
Became leprous & white as snow;
And every soul of men on the Earth
Felt affliction & sorrow & sickness & dearth.
Los flam'd in my path, & the Sun was hot
With the bows of my Mind & the Arrows of
    Thought—
My bowstring fierce with Ardour breathes,
My arrows glow in their golden sheaves;   80
My brother & father march before;
The heavens drop with human gore.

Now I a fourfold vision see,
And a fourfold vision is given to me;
'Tis fourfold in my supreme delight
And threefold in soft Beulah's night

---

7. **Verses . . . tongue:** probably Hayley's translation
of the opening lines of Tasso's *Le Sette Giornale del Mundo
Creato* (*The Seven Days of the Created World*). 8. **John,
the evil one:** Blake's younger brother had been a soldier.
He died young and dissipated. 9. **Fuseli:** Henry Fuseli,
1742?–1825, English painter born in Switzerland. He was
actually a lifelong friend of Blake, but at this time there
had been a falling out, probably because Fuseli had tem-
porarily lost sympathy with Blake's art and ceased to
employ him as an engraver.

10. **Flaxman:** John Flaxman, 1755–1826, sculptor. He was
a friend and admirer, but he, like Hayley, felt that Blake
should stick to drawing and engraving without attempting
large paintings and visionary poems. 11. **Wife . . . bane:**
Blake's sister was living with him at Felpham and did not
get along with his wife.

And twofold Always. May God us keep
From Single vision & Newton's sleep!

I also inclose you some Ballads by Mr. Hayley, with prints to them by your Hble. Servt. I should have sent them before now, but could not get any thing done for you to please myself; for I do assure you that I have truly studied the two little pictures I now send, & do not repent of the time I have spent upon them.

God bless you.

Yours,

W. B.

To Thomas Butts
Felpham, July 6, 1803.

DEAR SIR,

I send you the Riposo, which I hope you will think my best Picture in many respects. It represents the Holy Family in Egypt, Guarded in their Repose from those Fiends, the Egyptian Gods, and tho' not directly taken from a Poem of Milton's (for till I had design'd it Milton's Poem did not come into my Thoughts), Yet it is very similar to his Hymn on the Nativity, which you will find among his smaller Poems, & will read with great delight. I have given, in the background, a building, which may be supposed the ruin of a Part of Nimrod's tower,[12] which I conjecture to have spread over many Countries; for he ought to be reckon'd of the Giant brood.

I have now on the Stocks the following drawings for you: 1. Jephthah sacrificing his Daughter; 2. Ruth & her mother in Law & Sister; 3. The three Maries at the Sepulcher; 4. The Death of Joseph; 5. The Death of the Virgin Mary; 6. St. Paul Preaching; & 7. The Angel of the Divine Presence clothing Adam & Eve with Coats of Skins.

These are all in great forwardness, & I am satisfied that I improve very much & shall continue to do so while I live, which is a blessing I can never be too thankful for both to God & Man.

We look forward every day with pleasure toward our meeting again in London with those whom we have learn'd to value by absence no less perhaps than we did by presence; for recollection often surpasses every thing, indeed, the prospect of returning to our friends is supremely delightful—Then, I am determined that Mrs. Butts shall have a good likeness of You, if I have

hands & eyes left; for I am become a likeness taker & succeed admirably well; but this is not to be atchiev'd without the original sitting before you for Every touch, all likenesses from memory being necessarily very, very defective; But Nature & Fancy are Two Things & can Never be join'd; neither ought any one to attempt it, for it is Idolatry & destroys the Soul.

I ought to tell you that Mr. H.[13] is quite agreeable to our return, & that there is all the appearance in the world of our being fully employ'd in Engraving for his projected Works, Particularly Cowper's Milton,[14] a Work now on foot by Subscription, & I understand that the Subscription goes on briskly. This work is to be a very Elegant one & to consist of All Milton's Poems, with Cowper's Notes and translations by Cowper from Milton's Latin & Italian Poems. These works will be ornamented with Engravings from Designs from Romney, Flaxman & Yr. hble Servt., & to be Engrav'd also by the last mention'd. The Profits of the work are intended to be appropriated to Erect a Monument to the Memory of Cowper in St. Paul's or Westminster Abbey. Such is the Project—& Mr. Addington[15] & Mr. Pitt are both among the Subscribers, which are already numerous & of the first rank; the price of the Work is Six Guineas—Thus I hope that all our three years' trouble Ends in Good Luck at last & shall be forgot by my affections & only remember'd by my Understanding; to be a Memento in time to come, & to speak to future generations by a Sublime Allegory, which is now perfectly completed into a Grand Poem. I may praise it, since I dare not pretend to be any other than the Secretary; the Authors are in Eternity. I consider it as the Grandest Poem that this World Contains. Allegory addressed to the Intellectual powers, while it is altogether hidden from the Corporeal Understanding, is My Definition of the Most Sublime Poetry; it is also somewhat in the same manner defin'd by Plato. This Poem shall, by Divine Assistance, be progressively Printed & Ornamented with Prints & given to the Public. But of this work I take care to say little to Mr. H., since he is as much averse to my poetry as he is to a Chapter in the Bible. He knows that I have writ it, for I have shewn it to him, & he has read Part by his own desire & has looked with sufficient contempt to enhance my opinion of it. But I do not wish to irritate by seeming

---

12. **Nimrod's tower:** the tower of Babel.

13. **Mr. H.:** Hayley. 14. **Cowper's Milton:** *Latin and Italian Poems of Milton* translated [by] . . . *William Cowper,* ed. William Hayley (1808). Blake was not employed on this work. 15. **Addington:** Henry Addington, 1757–1844, Speaker of the House of Commons, 1789–1801.

too obstinate in Poetic pursuits. But if all the World should set their faces against This, I have Orders to set my face like a flint (Ezekiel, III. 9) against their faces, & my forehead against their foreheads.

As to Mr. H., I feel myself at liberty to say as follows upon this ticklish subject: I regard Fashion in Poetry as little as I do in Painting; so, if both Poets & Painters should alternately dislike (but I know the majority of them will not), I am not to regard it at all, but Mr. H. approves of My Designs as little as he does of my Poems, and I have been forced to insist on his leaving me in both to my own Self Will; for I am determin'd to be no longer Pester'd with his Genteel Ignorance & Polite Disapprobation. I know myself both Poet & Painter, & it is not affected Contempt that can move me to any thing but a more assiduous pursuit of both Arts. Indeed, by my late Firmness I have brought down his affected Loftiness, & he begins to think I have some Genius: as if Genius & Assurance were the same thing! but his imbecile attempts to depress Me only deserve laughter. I say thus much to you, knowing that you will not make a bad use of it. But it is a Fact too true That, if I had only depended on Mortal Things, both myself & my wife must have been Lost. I shall leave every one in This Country astonish'd at my Patience & Forbearance of Injuries upon Injuries; & I do assure you that, if I could have return'd to London a Month after my arrival here, I should have done so, but I was commanded by my Spiritual friends to bear all, to be silent, & to go thro' all without murmuring, &, in fine, hope, till my three years should be almost accomplish'd; at which time I was set at liberty to remonstrate against former conduct & to demand Justice & Truth; which I have done in so effectual a manner that my antagonist is silenc'd completely, & I have compell'd what should have been of freedom—My Just Right as an Artist & as a Man; & if any attempt should be made to refuse me this, I am inflexible & will relinquish any engagement of Designing at all, unless altogether left to my own Judgment, As you, My dear Friend, have always left me; for which I shall never cease to honour & respect you.

When we meet, I will perfectly describe to you my Conduct & the Conduct of others toward me, & you will see that I have labour'd hard indeed, & have been borne on angel's wings. Till we meet I beg of God our Saviour to be with you & me, & yours & mine. Pray give my & my wife's love to Mrs. Butts & Family, & believe me to remain,

Yours in truth & sincerity,

WILL BLAKE.

To William Hayley
London, 7 October, 1803.

DEAR SIR,

Your generous & tender solicitude about your devoted rebel makes it absolutely necessary that he should trouble you with an account of his safe arrival, which will excuse his begging the favor of a few lines to inform him how you escaped the contagion of the Court of Justice—I fear that you have & must suffer more on my account than I shall ever be worth—Arrived safe in London, my wife in very poor health, still I resolve not to lose hope of seeing better days.

Art in London flourishes. Engravers in particular are wanted. Every Engraver turns away work that he cannot execute from his superabundant Employment. Yet no one brings work to me. I am content that it shall be so as long as God pleases. I know that many works of a lucrative nature are in want of hands; other Engravers are courted. I suppose that I must go a Courting, which I shall do awkwardly; in the meantime I lose no moment to complete Romney [16] to satisfaction.

How is it possible that a Man almost 50 years of Age, who has not lost any of his life since he was five years old without incessant labour & study, how is it possible that such a one with ordinary common sense can be inferior to a boy of twenty, who scarcely has taken or deigns to take pencil in hand, but who rides about the Parks or saunters about the Playhouses, who Eats & drinks for business not for need, how is it possible that such a fop can be superior to the studious lover of Art can scarcely be imagin'd. Yet such is somewhat like my fate & such it is likely to remain. Yet I laugh & sing, for if on Earth neglected I am in heaven a Prince among Princes, & even on Earth beloved by the Good as a Good Man; this I should be perfectly contented with, but at certain periods a blaze of reputation arises round me in which I am consider'd as one distinguish'd by some mental perfection, but the flame soon dies again & I am left stupified and astonish'd. O that I could live as others do in a regular succession of Employment, this wish I fear is not to be accomplish'd to me—Forgive this Dirge-like lamentation over a dead horse, & now I have lamented over the dead horse let me laugh & be merry with my friends till Christmas, for as Man liveth not by bread alone, I shall live altho I should want

16. **Romney:** Blake was engraving a head of Romney for Hayley.

bread—nothing is necessary to me but to do my Duty & to rejoice in the exceeding joy that is always poured out on my Spirit, to pray that my friends & you above the rest may be made partakers of the joy that the world cannot concieve, that you may still be replenish'd with the same & be as you always have been, a glorious & triumphant Dweller in immortality. Please to pay for me my best thanks to Miss Poole: tell her that I wish her a continued Excess of Happiness—some say that Happiness is not Good for Mortals, & they ought to be answer'd that Sorrow is not fit for Immortals & is utterly useless to any one; a blight never does good to a tree, & if a blight kill not a tree but it still bear fruit, let none say that the fruit was in consequence of the blight. When this Soldier-like danger[17] is over I will do double the work I do now, for it will hang heavy on my Devil who terribly resents it; but I soothe him to peace, & indeed he is a good natur'd Devil after all & certainly does not lead me into scrapes—he is not in the least to be blamed for the present scrape, as he was out of the way all the time on other employment seeking amusement in making Verses, to which he constantly leads me very much to my hurt & sometimes to the annoyance of my friends; as I percieve he is now doing the same work by my letter, I will finish it, wishing you health & joy in God our Saviour.

To Eternity yours,

WILL^M BLAKE.

To George Cumberland
N3, Fountain Court, Strand.
12 April, 1827.

I have been very near the gates of death, and have returned very weak and an old man, feeble and tottering, but not in spirit and life, not in the real man, the imagination, which liveth for ever. In that I am stronger and stronger, as this foolish body decays. I thank you for the pains you have taken with poor Job.[18] I know too well that the great majority of Englishmen are fond of the indefinite, which they measure by Newton's doctrine of the fluxions of an atom, a thing which does not exist. These are politicians, and think that Republican art is inimical to their atom, for a line or a lineament is not formed by chance. A line is a line in its

minutest subdivisions, straight or crooked. It is itself, not intermeasurable by anything else. Such is Job. But since the French Revolution Englishmen are all intermeasurable by one another: certainly a happy state of agreement, in which I for one do not agree. God keep you and me from the divinity of yes and no too—the yea, nay, creeping Jesus—from supposing up and down to be the same thing, as all experimentalists must suppose.

You are desirous, I know, to dispose of some of my works, but having none remaining of all I have printed, I cannot print more except at a great loss. I am now painting a set of the Songs of Innocence and Experience for a friend at ten guineas. The last work I produced is a poem entitled Jerusalem, the Emanation of the Giant Albion, but find that to print it will cost my time the amount of Twenty Guineas. One I have Finish'd. It contains 100 Plates, but it is not likely I shall get a Customer for it.

As you wish me to send you a list with the Prices[19] of these things, they are as follows:

	£	s.	d.
America	6	6	0
Europe	6	6	0
Visions, &c.	5	5	0
Thel	3	3	0
Songs of Inn. & Exp.	10	10	0
Urizen	6	6	0

The Little Card[20] I will do as soon as Possible, but when you Consider that I have been reduced to a Skeleton, from which I am slowly recovering, you will, I hope, have Patience with me.

Flaxman[21] is Gone, & we must All soon follow, every one to his Own Eternal House, Leaving the delusive Goddess Nature & her Laws, to get into Freedom from all Law of the Members, into The Mind, in which every one is King & Priest in his own House. God send it so on Earth, as it is in Heaven.

I am, dear Sir, Yours affectionately,

WILLIAM BLAKE.

17. **Soldier-like danger:** See Introduction, p. 39. 18. **Job:** Cumberland had been trying to sell a set of Blake's illustrations to the Book of Job.

19. **Prices:** The purchasing power of £1 was then equivalent to about £5 today. £6 6s. would thus come to approximately £35 (or $98). Blake did not regulate his prices by supply and demand but according to his need. If he was offered an unusually good price he would color his engraved books more elaborately. Thus they vary considerably in quality. 20. **Card:** a copperplate with the name "Mr Cumberland" and a design. This was Blake's last engraving. 21. **Flaxman:** See n. 10.

WILLIAM WORDSWORTH

1770–1850

TO MOST people Wordsworth did not look like a poet. He had nothing of Shelley's ethereal delicacy of feature. He could not, like Byron, be portrayed as a corsair. He lacked even the rapt expression and massive forehead of Coleridge, the outward symbols of genius and learning. Yet his countenance could be romanticized. There was something formidable about it—the wide slash of the mouth, the commanding nose, the fierce eyes, "half burning, half smouldering, with an acrid fixture of regard." Though capable of utmost delicacy in feeling and affection, his character was independent, craggy, intense, brooding, and inward. When Hazlitt first met him, he noticed a "severe, worn pressure of thought about his temples," and forty-two years later Carlyle saw the same thing: "His face bore marks of much, not always peaceful meditation; the look of it not bland or benevolent so much as close impregnable and hard." He was stubborn in effort, reflection, and the beliefs he won from reflection, proud of his work, and immensely high-minded in aspiration. Indeed, his lofty sense of poetry had an influence on his age quite apart from any particular doctrine or style he fostered. With the example of Wordsworth lodged in the conscience, a poet could not—as the development of Keats testifies—think of his art solely as reverie or amusement. To Wordsworth the "getting and spending" in which we consume time and ourselves are the escape, the evasion, the dream on which we float out of existence, having cheated ourselves of life. Poetry is a way of confronting ultimate questions, and hence of living with the depth and passion that come only in their presence.

Born April 7, 1770, he was the second of five children. His willful and emotional character must have declared itself early. When he was seven his mother told a friend that the only one of her children "about whose future life she was anxious, was William; and he, she said, would be remarkable either for good or evil." At the age of nine he was sent to school at Hawkshead, a small market village at the head of the lake of Esthwaite. Most of the episodes in *Prelude*, I and II refer to this period. Hawkshead lies in the heart of a region in Cumberland and Westmorland of bare hills, mountains, lakes, scattered villages, lonely upland valleys and tarns, and

> rocks and pools shut out from every star,
> All the green summer, [and] forlorn cascades
> Among the windings hid of mountain brooks
>
> [ll. 488–90].

Celebrated for its beauty ever since Europeans first learned to appreciate mountain scenery—the poet Gray was an early tourist and enthusiast—the Lake District was still rural and isolated in Wordsworth's youth.

Though his academic training had been excellent, he was otherwise hardly prepared for the bustling and fashionable world of Cambridge, which he entered in 1787 as a member of St. John's College. Except for a walking tour of the Alps (narrated in *Descriptive Sketches* and *Prelude*, VI), his college years were uneventful and are important mainly for a negative reason: he refused to compete for the academic honors that would have led to easy advancement and financial security. Partly he felt an aversion to mathematics, which with the renown of Sir Isaac Newton had become a central subject at Cambridge, but mainly he was declaring independence from the uncles who were now his guardians. After graduation Wordsworth would have no means of supporting himself, yet to the exasperation of his uncles he would suggest only vague and impracticable schemes. They hoped he would take clerical orders, but he refused, being convinced that "small certainties are the bane of great talents."

He sailed for France, having persuaded his guardians to finance a few months more. The French Revolution was now entering its third year. England was filled with radical agitation (Paine's *Rights of Man* appeared in 1791) and the aging Edmund Burke was pouring forth his magnificent counterblasts. Wordsworth had hitherto shown little interest in politics, but in France he became wholly committed to the revolutionary cause. He also entered into a liaison with Annette Vallon, who bore him a child. They were deeply in love and both then and for many years afterward intended to marry. But lack of money forced him to leave France in December, 1792, and the outbreak of war in 1793 made it impossible to return.

There followed a period of spiritual crisis (described in *Prelude*, XI. 173–370). He felt that in warring against France his country opposed human liberty itself, and like many radicals of the time he found his sympathies miserably divided. Also the Reign of Terror seemed to belie the hope roused by the Revolution. In the face of these shocks he tried to rethink completely his beliefs. (At this time he read William Godwin's *Political Justice* and was strongly influenced by Godwin's attempt to approach all questions in a severely rationalistic way. The influence lasted only for a short time.) The upshot of this intense effort was, he says, that he utterly despaired of any intellectual certainty. Meanwhile he had to live, and in 1795 a small legacy from a friend enabled him to set up housekeeping at Racedown in Dorset. There, with his sister Dorothy, he continued to write poetry—he had published two poems already—and gradually recovered a happier state of mind. His revolutionary faith waned with the rise of Napoleon's dictatorship, and particularly with the French invasion of Switzerland in 1798, and he became increasingly conservative in politics as he grew older.

His early poems had won the admiration of Coleridge, and in 1795 the two poets first met in Bristol. In order to be near Coleridge, Wordsworth later moved to Alfoxden House in the vicinity of Nether Stowey, where Coleridge lived. From the intimate, daily companionship of this marvelous year (1797–98) emerged Coleridge's "The Rime of the Ancient Mariner," Wordsworth's "Tintern Abbey," and the other poems that made up the *Lyrical Ballads* of 1798.

Wordsworth's days of *Sturm und Drang* were over. After a stay in Germany from 1798 to 1799, he resided in the Lake District, married, and had five children. In 1813 he became Distributor of Stamps for Westmorland, an office that paid between £400 and £600 a year. The drowning of his sailor brother, John, in 1805, was a deep sorrow, as was the later death in one year of his children Catherine and Thomas. Meanwhile the friendship with Coleridge had lapsed. The two were estranged and finally reconciled, but rarely met after 1810. Nevertheless Coleridge's brilliant criticism in *Biographia Literaria* (1817) did much to alert readers to Wordsworth's genius. His fame had been increasing slowly; it now soared and in the 1820's and thirties he became an institution. Tourists (Keats among them) thronged to his home on what Charles Lamb called "gaping missions." They might be charged for tea—Wordsworth was always parsimonious, and there were sometimes as many as thirty visitors a day—and "stunned with oratory," but they took much satisfaction in his obvious integrity, his simple manners, and his boundless self-confidence. Thomas Arnold and his family had a house not far away, and young Matthew early became a Wordsworthian, with important results for English poetry. In 1839 Wordsworth received an honorary degree at Oxford to "thunders of applause, repeated over and over." But these later years were saddened, first by the chronic illness of his sister—the aging poet would stand for a long time by her bed, saying nothing, comforting and comforted by being near—and then by the death of his beloved daughter, Dora. He lived on till 1850, a venerable and respected figure in an England that was rapidly changing.

2

When Wordsworth's verse ascends to impassioned meditation, no poetry in English is more swelling and stately. Yet at other times he dares in the other extreme. No one can appear more fresh and artless than Wordsworth when he looks around, delighted like Adam in the garden with

> bare trees, and mountains bare,
> And grass in the green field

["To My Sister," ll. 7–8].

Hence no description of him satisfies his admirers: each admires a different person. If we think we can isolate a distinctive quality in his poetry, we quickly feel that the opposite is equally distinctive. From introspection he could render subtlest shades and processes of thought and feeling—what Coleridge calls the "flux and reflux of [our] inmost nature . . . twilight realms of consciousness . . . modes of inmost being." But he also wrote wholly objective tales of pathos, such as "Michael" or the first book of *The Excursion*. These in turn bring to mind his strong bent toward realism, a literal accuracy in representation; yet, without ceasing to be realistic, description in his poetry becomes visionary symbolism. Coleridge was sure Wordsworth would write the first truly philosophic poem, and philosophies—or religions—have been reared on his work, yet he himself described poetry as the "spontaneous overflow of powerful feelings." And despite the many sides of his genius, Wordsworth is always entirely himself.

If, however, we sought stylistic affinities, we would go more to eighteenth-century predecessors than to Romantics of the next generation. The affinities, needless to say, are not with Pope, whom Byron sometimes tried to imitate, but with writers of the middle and latter part of the century, such as William Cowper and Robert Burns. Their influence helps account for his realism, his interest in ballads, the Miltonic cast of his blank verse, and, perhaps most important, for the descriptive-meditative technique he adopts and modifies in his longer poems. At the same time, this technique is a response to a bent and need of his personality. The general habit of his mind is to catch and hold for contemplation some particular image or scene. It may be the sudden glimpse of a shepherd through thick fog, or of cliffs that "wheeled by" when he was dizzy from skating, or of a blind man on the streets of London. In his poetry he makes such events marvelously vivid, fixing them before the mind, and goes on to draw forth a meaning in discursive terms. For these events are felt to be moments of revelation. The meaning they hold is precious for our moral and religious life. They must become explicit objects of meditation, and hence the typical movement of the longer poems—especially *The Prelude*—is from description to reflection, from concrete vignette to impassioned statement of belief.

Of these beliefs at least three worked powerfully throughout the nineteenth century, transforming the way many readers felt about their own lives and about the cosmos. They are doctrines of nature, the imagination, and the true source of human joy. In the great childhood episodes of the first book of *The Prelude*—snaring woodcocks, climbing the perilous ridge to the ravens' nests, rowing out at night on the lonely lake—Wordsworth presents the dawning upon his mind of a fundamental conviction: in the natural world there is not simply a congeries of objects but a Consciousness, a pervading Being, or, to use a favorite word of his, a Presence. Since he is a child, he does not grasp this philosophically, as an abstract proposition, but directly in experience. The mysterious Presence is felt in the "low breathings" and "steps" that seem to pursue him at night when he steals a bird from the snare of someone else, in the "strange utterance" of the "loud dry wind" as he hangs above the raven's nest, in the "grim shape" of the mountain that uprears its head and strides after him across the lake, and in the "huge and mighty forms" that move slowly through his mind for many days afterward. Once the awareness of this has been impressed upon him (and the reader), it persists through all subsequent childhood experience, or the poetry that describes it. Often it is suggested indirectly, but it is there, something remote and mysterious breaking in upon the thoughtless play. For example, it is in the "alien sound" from "far distant hills" as the children are ice-skating, or in the gleams of "moonlight from the sea" as the young horsemen "beat with thundering hoofs the level" shore.

Putting it another way, one can notice that in Wordsworth the natural world has an infinite depth. There is, so to speak, a clear foreground, often a scene of human sports and tasks. But the background extends and endlessly recedes in vast and tranquil forms of mountain, lake, and sky. It has a brooding dominance for the imagination. It engulfs our human bustle in space and stillness. Conceived as a living reality, or as the countenance or expression of an indwelling Presence, it might seem utterly indifferent to human life, and Wordsworth is haunted by the terror of this. He will not think it true, yet it persists in his imagination. On the other hand, to the child nature is not simply awe-inspiring, stunning in its grandeur and force; with its breezes, streams, and bowers it can also seem gentle and caressing. It is beautiful. Above all, it seems to seek the child out, and even its more terrifying interventions—the dread "steps," the "grim shape" of the mountain—can be felt as a warning and disciplining. The truths embodied in the direct experience of the child will be evolved into distinct propositions for belief by the mature man.

One need not suppose that this belief originated precisely as he says it did. Dwelling introspectively on his own life, he might inevitably have been influenced to modify the past in support of present convictions. What matters is that he could convince other men. We cannot, needless to say, understand the power of his beliefs to compel assent without encountering them in his surging, vivid language and prophetic tones. Nor can we understand without putting ourselves in the historical circumstances of the men who caught a faith from him. For he showed that nature was not the dead mechanism—matter pushed by matter in accordance with mathematics—that science presumed in the eighteenth and nineteenth centuries. In such a cosmos how can moral and aesthetic values be real? But when Wordsworth brought God into nature, he also brought the values science excluded. Moreover, to men who could not accept the God of narrow rationalism or biblical fundamentalism—the main forms of Christianity of the time— Wordsworth held out another Deity for worship, a God remote and awful, yet also omnipresent. To show what Wordsworth could mean to his readers, we can quote the testimony of the English novelist William Hale White, writing in 1881.

> Instead of an object of worship which was altogether artificial, remote, never coming into genuine contact with me, I had now one which I thought to be real, one in which literally I could live and move and have my being, an actual fact present before my eyes. God was brought from that heaven of the books, and dwelt on the downs in the far-away distances, and in every cloud-shadow which wandered across the valley. Wordsworth unconsciously did for me what every religious reformer has done—he recreated my Supreme Divinity; substituting a new and living spirit for the old deity, once alive, but gradually hardened into an idol.

The characteristic emotions of the religion of nature are awe and gratitude.

Yet intelligent men did not accept Wordsworth's faith simply because they wanted to. His religion of nature was anchored in further metaphysical speculation. He knew very well that no mountain literally chased him across a lake. What he saw was, he knew, the joint product of the actual configuration of things and his own emotions projected upon them, and the same would hold true for all experience. The mind is not "a mere pensioner / On outward forms." It endows or shapes experience as it occurs. This modifying or contributing activity Wordsworth calls creation. Sometimes he also names it imagination, but the word has a broader significance. For the power or function of mind through which experience comes to us in its *wholeness* is the imagination. Values and emotions are not less involved in an experience than, for example, motion and extension. We know the one as directly as the other. If we grant reality to the latter, we can deny it to the former only by an act of abstraction. Wordsworth preferred to take as real what we directly experience, that is, an event in which inner and outer are indissolubly blended.

The Divine Being dwells in "All thinking things" and "all objects of all thought," which is to say that all things are alive. The "great mass / Lay bedded in a quickening soul"; Wordsworth felt "the sentiment of Being spread / O'er all that moves and all that seemeth still." The intuition of God everywhere is "bliss ineffable." It is also the mystical ground for Wordsworth's philosophy of organicism, that is, the belief that the universe is a living whole. Parts or individual things are not related as cogs and levers of a machine, but are functions of one process working through all. The whole inheres in each individual thing and in all as one. In an accompanying intuition Wordsworth also felt that the Divine Life is joy. At this point, however, one becomes unsure of his meaning. He may hold that man can sometimes have knowledge of this Joy as it is immanent in things. Or he may believe that things are themselves also conscious of it, which is plainly a more difficult premise to accept. In some sense, however, he believes that existence is joy, and that this must also be true fundamentally for human beings, but that men have distinct aims and purposes of their own—"A wedding or a festival . . . dialogues of business, love, or strife"—which, as it were, usurp consciousness and distract them from the deep happiness that is their birthright. One must put aside the hot chase of ambition, fear, vanity, hope, and love to recover the natural or spontaneous joy within. That is one reason why children and peasants are more likely than others to be happy. As these beliefs arise in intuition, they cannot be fully conveyed in language. Wordsworth can talk in "Michael," for example, about "the pleasure that there is in life itself," but the phrase lies inert unless caught up by the meditating and realizing powers of the reader's mind. We should at least acknowledge, however, the depth and strangeness of his faith. When Matthew Arnold says that Wordsworth is "priest to us all / Of the wonder and bloom of the world," or that his poetry makes us feel "the joy offered to us in nature" and in "the simple primary affections and duties," the remarks are fine and true, but they skim lightly over profundities.

Perhaps it was necessary to skim in order not to founder. Even Coleridge, despite his transcendental buoys and air bladders of which Carlyle speaks, was lost in Wordsworth's assertion, in the great ode on "Intimations of Immortality," that a "little Child" possesses truths that "we are toiling all our lives to find," and Wordsworth himself could not continue to maintain all aspects of the faith we have sketched. The process of relinquishment is one of the most moving things in his personal life, a slow, unhappy revision caused by further experience and stubbornly honest thinking. The first stages can be seen in "Resolution and Independence," the "Ode to Duty," the "Elegiac Stanzas," and other poems, and *The Excursion* points toward the culmination in orthodox Christianity. With the growth of his later faith his genius as a poet declined, and there is probably a causal connection. But if there could be any conflict, far more than poetry he would always value integrity in thinking, the human struggle for truth.

Bibliography

The best single discussion of Wordsworth is still that of Coleridge in *Biographia Literaria*, 1817, especially Chapter XXII. In this pioneer work—one of the finest critical achievements in English—Coleridge laid out paths that discussion has followed ever since. He praises Wordsworth's power of mind and deeply felt moral concern, his brooding reflectiveness that is comprehensive, sane, and central. He calls attention to his realism or naturalism (the "perfect truth of nature in his images and descriptions") and also his gift of imagination. And though Coleridge notes "characteristic defects," he implies that Wordsworth is the greatest English poet since Milton. Matthew Arnold later echoed this evaluation and carried it further: Among English poets since the Elizabethans Wordsworth stands third after Shakespeare and Milton. Arnold's opinion was generally shared until the revival of metaphysical poetry in the 1920's. Now, however, readers and poets are once more sensing affinities with Wordsworth, who towers not only as a classic of our language but as a precursor of much in the literature and sensibility of our age.

Even before the *Biographia Literaria*, Hazlitt provided the chief grounds for a less appreciative view (mainly in his review of *The Excursion*, 1814, and the last of the *Lectures on the English Poets*, 1818). Though sensitively aware of Wordsworth's genius, he stresses, first, the unreality or deluded optimism of his view of man and the cosmos, and, second, the devouring "egotism" of his poetry—that is, his poetry does not enter empathically into natural objects or other human beings, but uses them to reflect his own identity. The second of these criticisms is now usually accepted as a description of the poetry, though not as a charge against it (but see Mark Van Doren, *The Noble Voice*, 1946). The first naturally remains controversial. Among twentieth-century restatements one may mention Aldous Huxley, "Wordsworth in the Tropics," *Holy Face and Other Essays*, 1929, and the much more sophisticated and persuasive argument of Douglas Bush, "Wordsworth: A Minority Report," in *Wordsworth Centenary Studies*, ed. G. T. Dunklin, 1951. In his few, brilliant remarks Keats sometimes echoes Hazlitt, but he was haunted by Wordsworth's weight of phrasing and his moral and metaphysical profundity, and he implicitly accepted Coleridge's ranking of him after Milton. But the main significance of Keats's comments, as W. J. Bate (*John Keats*, 1963, Ch. XIII) has shown, is that he saw in Wordsworth the preeminent example of what modern poetry must inevitably become, both in its limitations and in its achievement.

Matthew Arnold in *Essays in Criticism; Second Series*, 1888, like John Stuart Mill in his *Autobiography*, 1873, presents Wordsworth as a poet who makes us feel the joy offered to us in the common things of life. That is, his poetry is valued for its true and powerfully affecting moral inspiration, its "criticism of life." As for Wordsworth's metaphysical depth, Arnold denies it—the philosophy is an "illusion." Arnold's point of view recurs with variations in such contemporary critics as F. R. Leavis, *Revaluation*, 1936, and Lionel Trilling, "Wordsworth's 'Ode: Intimations of Immortality,'" in *English Institute Annual*, 1941, reprinted in *The Liberal Imagination*, 1950. In 1909, however, A. C. Bradley, *Oxford Lectures on Poetry*, had offered a deliberate correction of Arnold; he emphasized how much in Wordsworth is visionary and alien.

The opening up in the twenties of the general field of the history of ideas stimulated a renewed interest in Wordsworth's philosophy, and it gradually made possible a more precise understanding. An early effort was that of Arthur Beatty (*William Wordsworth*, 1922), who showed but exaggerated the influence on him of the eighteenth-century philosopher David Hartley. Beatty was influential, but the insights of Alfred North Whitehead in *Science and the Modern World*, 1925, are of lasting importance. Whitehead saw Wordsworth reacting against the eighteenth-century philosophies and science of mechanism and anticipating his own philosophy of organicism. J. W. Beach, *The Concept of Nature in Nineteenth Century English Poetry*, 1936, and Basil Willey, *The Eighteenth Century Background*, 1940, throw light on Wordsworth's beliefs while dissenting from them, and N. P. Stallknecht, *Strange Seas of Thought*, 1945, elaborates on the comments of Whitehead. More recent works are Thomas J. Rountree, *This Mighty Sum of Things: Wordsworth's Theme of Benevolent Necessity*, 1965, and Melvin Rader, *Wordsworth: A Philosophical Approach*, 1967.

Though Coleridge had been struck by the symbol-making habits of Wordsworth's imagination, the subject was not pursued until interest in symbolism generally revived in the twentieth century. Contemporary studies of his symbolism have usually followed Bradley in concentrating on what is dark, mysterious, irrational, and ineffable in his poetic vision. Among them may be mentioned Charles Williams, *The English Poetic Mind*, 1932; Bennett Weaver, "Forms and Images," in *Studies in Philology*, XXXV (1938), 433–45; G. Wilson Knight, *The Starlit Dome*, 1941; Kenneth MacLean, "The Water Symbol in *The Prelude*," in *University of Toronto Quarterly*, XVII (1948), 372–89; S. C. Wilcox, "Wordsworth's River Duddon Sonnets," *PMLA*, LXIX (1954), 131–41; Geoffrey Hartman, *The Unmediated Vision*, 1954; David Ferry, *The Limits of Mortality*, 1959; and David Perkins, *The Quest for Permanence: The Symbolism of Wordsworth, Shelley, and Keats*, 1959.

Critical discussion of Wordsworth's style also began with Coleridge, and most studies have made contributions. F. W. Bateson, *Wordsworth: A Re-interpretation*, 1954, is good on the lyrical ballads, though unreliable in other respects, and John Jones, *The Egotistical Sublime*, 1954, pays sympathetic attention to the style of the late poems. The latter work is difficult to classify, being a thoughtful and comprehensive treatment of Wordsworth as a whole. It might be used as a general introduction, and for the same purpose one might turn to Sir Walter Raleigh, *Wordsworth*, 1903, Helen Darbishire, *The Poet Wordsworth*, 1950, and David Perkins, *Wordsworth and the Poetry of Sincerity*, 1964, and Geoffrey Hartman, *Wordsworth's Poetry, 1787–1814*, 1964. R. D. Havens, *The Mind of a Poet*, 1941, is a close and massive study of the *Prelude;* it is still the most comprehensive discussion of the poem, though one should also mention Abbie Findlay Potts, *Wordsworth's Prelude*, 1953, and the far more sophisticated work of Herbert Lindenberger, *On Wordsworth's Prelude*, 1963. Where Miss Potts is concerned mainly with sources or possible sources, Lindenberger sensitively places *The Prelude* in literary history, showing both what it picks up from eighteenth-century poetry and what it anticipates in our own time. James Scoggins, *Imagination and Fancy*, 1966, applies these critical categories of Wordsworth to his poetry.

The standard biography—though less valuable as criticism—is that of Mary Moorman, *William Wordsworth*, 2 vols., 1957–65. George McL. Harper's *William Wordsworth*, 2 vols., 1916, revised and abridged, 1929, is still important but Harper is antipathetic to the conservative politics of Wordsworth's later years. As a result he is less than fair to the aging poet, and may be balanced by Edith Batho, *The Later Wordsworth*, 1933, which swings too far in the opposite direction. Still valuable for the early years is Emile H. Legouis, *The Early Life of William Wordsworth*, 1897. W. L. Sperry, *Wordsworth's Anti-Climax*, 1935, pays attention to his life during the years when his poetic powers were starting to fail. Special mention should be made of H. M. Margoliouth's wise and readable *Wordsworth and Coleridge, 1795–1834*, 1953, a brief biography of a friendship. Mark L. Reed, *Wordsworth: The Chronology of the Early Years, 1770–1799*, 1967, determines the date of all possible events and poetic compositions in Wordsworth's early career.

Perhaps the most notable event in twentieth-century scholarly writing on Wordsworth was the disclosure by Harper in 1916 of the affair with Annette Vallon, which had been a family secret till that time. Legouis printed the available documents in *William Wordsworth and Annette Vallon*, 1922. This revelation naturally inspired theories—no longer accepted—that Wordsworth was haunted by remorse, and the supposed psychological aftermath was invoked to explain both his poetic greatness and his decline. See, for example, Sir Herbert Read, *Wordsworth*, 1930, and H. I'Anson Fausset, *The Lost Leader*, 1933.

The standard edition is that of Ernest De Selincourt and Helen Darbishire, 5 vols., 1940–49. *The Prelude*, ed. De Selincourt, 1933, rev. by Helen Darbishire, 1959, prints both the 1805 and the 1849–50 text of the poem. The *Prose Works* are available in editions prepared by A. B. Grosart, 3 vols., 1876, and W. A. Knight, 2 vols., 1896. *The Letters of William and Dorothy Wordsworth* have been edited by De Selincourt, 6 vols., 1935–39, but the edition is not complete and must be supplemented from other works. Vol. I, *The Early Years, 1787–1805*, has been revised by Chester L. Shaver, 1967. De Selincourt also edited the *Journals of Dorothy Wordsworth*, 2 vols., 1941.

from DESCRIPTIVE SKETCHES

TAKEN DURING A PEDESTRIAN TOUR
IN THE ITALIAN, GRISON, SWISS,
AND SAVOYARD ALPS

This, with a companion piece entitled *An Evening Walk*,
was Wordsworth's first published poem. The text
reprinted is that of the first edition. The poem describes
his walking tour of the Alps in the summer of 1790, the
same material that was used again in *Prelude* VI. The
extract given here suggests the ardent republicanism of
his youth (for example, ll. 317–31), and also his feeling
for natural scenery, more particularly for what in his day
was termed the "sublime" in nature—that is, whatever
aroused feelings of awe, grandeur, or terror. In style, the
poem may seem a conventional example of eighteenth-
century verse in the "topographical" genre, but that is
partly because we have not read enough similar poems
to make distinctions. For Coleridge's highly favorable
reaction see *Biographia Literaria*, Chapter IV, below,
pp. 448–49, where he quotes and comments on lines
317–24 and 332–47.

Now, passing Urseren's open vale serene,
Her quiet streams, and hills of downy green,
Plunge with the Russ embrown'd by Terror's breath,
Where danger roofs the narrow walks of death;
By floods, that, thundering from their dizzy height,
Swell more gigantic on the stedfast sight;
Black drizzling craggs, that beaten by the din,
Vibrate, as if a voice complain'd within;   250
Bare steeps, where Desolation stalks, afraid,
Unstedfast, by a blasted yew upstay'd;
By cells whose image, trembling as he prays,
Awe-struck, the kneeling peasant scarce surveys;
Loose-hanging rocks the Day's bless'd eye that hide,
And crosses rear'd to Death on every side,

DESCRIPTIVE SKETCHES. **243. Urseren:** the valley of Ursern in
southeastern Switzerland. In the lines given here Wordsworth
follows the Reuss river from Ursern to the Lake of Uri
adjacent to Lake Lucerne. **249. Black . . . craggs:** Cf. *The
Prelude*, below, VI. 631. **253. cells:** "These cells are . . . very
common in Catholic countries, planted, like Roman tombs,
along the road side" (Wordsworth's note). **256. crosses:**
"Crosses commemorative of the deaths of travellers by the
fall of snow, and other accidents very common along this
dreadful road" (W. W.).

Which with cold kiss Devotion planted near,
And, bending, water'd with the human tear,
Soon fading "silent" from her upward eye,
Unmov'd with each rude form of Danger nigh,   260
Fix'd on the anchor left by him who saves
Alike in whelming snows and roaring waves.

    On as we move, a softer prospect opes,
Calm huts, and lawns between, and sylvan slopes.
While mists, suspended on th' expiring gale,
Moveless o'er-hang the deep secluded vale,
The beams of evening, slipping soft between,
Light up of tranquil joy a sober scene;
Winding it's dark-green wood and emerald glade,
The still vale lengthens underneath the shade;   270
While in soft gloom the scattering bowers recede,
Green dewy lights adorn the freshen'd mead,
Where solitary forms illumin'd stray
Turning with quiet touch the valley's hay,
On the low brown wood-huts delighted sleep
Along the brighten'd gloom reposing deep.
While pastoral pipes and streams the landscape lull,
And bells of passing mules that tinkle dull,
In solemn shapes before th' admiring eye
Dilated hang the misty pines on high,   280
Huge convent domes with pinnacles and tow'rs,
And antique castles seen thro' drizzling show'rs.

    From such romantic dreams my soul awake,
Lo! Fear looks silent down on Uri's lake,
By whose unpathway'd margin still and dread
Was never heard the plodding peasant's tread.
Tower like a wall the naked rocks, or reach
Far o'er the secret water dark with beech,
More high, to where creation seems to end,
Shade above shade the desert pines ascend,   290
And still, below, where mid the savage scene
Peeps out a little speck of smiling green,
There with his infants man undaunted creeps
And hangs his small wood-hut upon the steeps.
A garden-plot the desert air perfumes,
Mid the dark pines a little orchard blooms,
A zig-zag path from the domestic skiff
Threading the painful cragg surmounts the cliff.
—Before those hermit doors, that never know
The face of traveller passing to and fro,   300
No peasant leans upon his pole, to tell
For whom at morning toll'd the funeral bell,
Their watch-dog ne'er his angry bark forgoes,
Touch'd by the beggar's moan of human woes,
The grassy seat beneath their casement shade

The pilgrim's wistful eye hath never stay'd.
—There, did the iron Genius not disdain
The gentle Power that haunts the myrtle plain,
There might the love-sick maiden sit, and chide
Th' insuperable rocks and severing tide,   310
There watch at eve her lover's sun-gilt sail
Approaching, and upbraid the tardy gale,
There list at midnight till is heard no more,
Below, the echo of his parting oar,
There hang in fear, when growls the frozen stream,
To guide his dangerous tread the taper's gleam.

Mid stormy vapours ever driving by,
Where ospreys, cormorants, and herons cry,
Where hardly giv'n the hopeless waste to cheer,
Deny'd the bread of life the foodful ear,   320
Dwindles the pear on autumn's latest spray,
And apple sickens pale in summer's ray,
Ev'n here Content has fix'd her smiling reign
With Independance, child of high Disdain.
Exulting mid the winter of the skies,
Shy as the jealous chamois, Freedom flies,
And often grasps her sword, and often eyes,
Her crest a bough of Winter's bleakest pine,
Strange "weeds" and alpine plants her helm entwine,
And wildly-pausing oft she hangs aghast,   330
While thrills the "Spartan fife" between the blast.

'Tis storm; and hid in mist from hour to hour
All day the floods a deeper murmur pour,
And mournful sounds, as of a Spirit lost,
Pipe wild along the hollow-blustering coast,
'Till the Sun walking on his western field
Shakes from behind the clouds his flashing shield.

329. "weeds": Wordsworth alludes to William Cowper's
The Task, V. 446–48. 331. "Spartan fife": See William
Collins, "Ode to Liberty," l. 1. 336–47. 'Till . . . fire:
"I had once given to these sketches the title of Picturesque;
but the Alps are insulted in applying to them that term.
Whoever, in attempting to describe their sublime features,
should confine himself to the cold rules of painting would
give his reader but a very inperfect idea of those emotions
which they have the irresistible power of communicating
to the most impassive imaginations. The fact is, that con-
trouling influence, which distinguishes the Alps from all
other scenery, is derived from images which disdain the
pencil. Had I wished to make a picture of this scene I had
thrown much less light into it. But I consulted nature and
my feelings. The ideas excited by the stormy sunset I am here
describing owed their sublimity to that deluge of light, or
rather of fire, in which nature had wrapped the immense
forms around me; any intrusion of shade, by destroying the
unity of the impression, had necessarily diminished it's
grandeur" (W. W.).

Triumphant on the bosom of the storm,
Glances the fire-clad eagle's wheeling form;
Eastward, in long perspective glittering, shine   340
The wood-crown'd cliffs that o'er the lake recline;
Wide o'er the Alps a hundred streams unfold,
At once to pillars turn'd that flame with gold;
Behind his sail the peasant strives to shun
The west that burns like one dilated sun,
Where in a mighty crucible expire
The mountains, glowing hot, like coals of fire.

1791–92 (1793)

GUILT AND SORROW

This much revised poem never quite satisfied Wordsworth, partly because it reflected political beliefs he no
longer held, and partly because it presented weaknesses of
style and form he was unable wholly to correct. The final
text is given here, but even this retains suggestions—the
vision of the cruelty of war and social injustice—of the
angry radicalism in which the poem was begun. His
moving sympathy with the poor, outcast, and oppressed
recurs in the *Lyrical Ballads*. The tale is melodramatic and
sentimental—"addressed to coarse sympathies," as
Wordsworth put it—and he never managed successfully
to unite the two stories of the sailor and of the soldier's
widow. As he said in a note, "the incidents of this
attempt do only in a small degree produce each other,
and it deviates accordingly from the general rule by
which narrative pieces ought to be governed." But he
also felt that "it is not therefore wanting in continuous
hold upon the mind, or in unity, which is effected by the
identity of moral interest that places the two personages
upon the same footing in the reader's sympathies." For
Coleridge's impressions on hearing an early version of
the poem, see *Biographia Literaria*, Chapter IV, below,
p. 449.

A Traveller on the skirt of Sarum's Plain
Pursued his vagrant way, with feet half bare;
Stooping his gait, but not as if to gain
Help from the staff he bore; for mien and air
Were hardy, though his cheek seemed worn with care

GUILT AND SORROW. 1. Sarum's Plain: Salisbury Plain.
Cf. Wordsworth's later account of traveling on Salisbury
Plain, *The Prelude*, below, XIII. 312–49, where he uses again
many of the same images.

Both of the time to come, and time long fled:
Down fell in straggling locks his thin grey hair;
A coat he wore of military red
But faded, and stuck o'er with many a patch and shred.

While thus he journeyed, step by step led on,   10
He saw and passed a stately inn, full sure
That welcome in such house for him was none.
No board inscribed the needy to allure
Hung there, no bush proclaimed to old and poor
And desolate, "Here you will find a friend!"
The pendent grapes glittered above the door;—
On he must pace, perchance till night descend,
Where'er the dreary roads their bare white lines extend.

The gathering clouds grew red with stormy fire,
In streaks diverging wide and mounting high;   20
That inn he long had passed; the distant spire,
Which oft as he looked back had fixed his eye,
Was lost, though still he looked, in the blank sky.
Perplexed and comfortless he gazed around,
And scarce could any trace of man descry,
Save cornfields stretched and stretching without bound;
But where the sower dwelt was nowhere to be found.

No tree was there, no meadow's pleasant green,
No brook to wet his lip or soothe his ear;
Long files of corn-stacks here and there were seen,   30
But not one dwelling-place his heart to cheer.
Some labourer, thought he, may perchance be near;
And so he sent a feeble shout—in vain;
No voice made answer, he could only hear
Winds rustling over plots of unripe grain,
Or whistling thro' thin grass along the unfurrowed
    plain.

Long had he fancied each successive slope
Concealed some cottage, whither he might turn
And rest; but now along heaven's darkening cope
The crows rushed by in eddies, homeward borne.   40
Thus warned he sought some shepherd's spreading
    thorn
Or hovel from the storm to shield his head,
But sought in vain; for now, all wild, forlorn,
And vacant, a huge waste around him spread;
The wet cold ground, he feared, must be his only bed.

And be it so—for to the chill night shower
And the sharp wind his head he oft hath bared;
A Sailor he, who many a wretched hour
Hath told; for, landing after labour hard,
Full long endured in hope of just reward,   50

He to an armèd fleet was forced away
By seamen, who perhaps themselves had shared
Like fate; was hurried off, a helpless prey,
'Gainst all that in *his* heart, or theirs perhaps, said nay.

For years the work of carnage did not cease,
And death's dire aspect daily he surveyed,
Death's minister; then came his glad release,
And hope returned, and pleasure fondly made
Her dwelling in his dreams. By Fancy's aid
The happy husband flies, his arms to throw   60
Round his wife's neck; the prize of victory laid
In her full lap, he sees such sweet tears flow
As if thenceforth nor pain nor trouble she could know.

Vain hope! for fraud took all that he had earned.
The lion roars and gluts his tawny brood
Even in the desert's heart; but he, returned,
Bears not to those he loves their needful food.
His home approaching, but in such a mood
That from his sight his children might have run,
He met a traveller, robbed him, shed his blood;   70
And when the miserable work was done
He fled, a vagrant since, the murderer's fate to shun.

From that day forth no place to him could be
So lonely, but that thence might come a pang
Brought from without to inward misery.
Now, as he plodded on, with sullen clang
A sound of chains along the desert rang;
He looked, and saw upon a gibbet high
A human body that in irons swang,
Uplifted by the tempest whirling by;   80
And, hovering, round it often did a raven fly.

It was a spectacle which none might view,
In spot so savage, but with shuddering pain;
Nor only did for him at once renew
All he had feared from man, but roused a train
Of the mind's phantoms, horrible as vain.
The stones, as if to cover him from day,
Rolled at his back along the living plain;
He fell, and without sense or motion lay;
But, when the trance was gone, feebly pursued his
    way.   90

As one whose brain habitual frenzy fires
Owes to the fit in which his soul hath tossed
Profounder quiet, when the fit retires,
Even so the dire phantasma which had crossed

76–90. Now . . . way: Cf. *The Prelude*, below, XII. 235–47.

His sense, in sudden vacancy quite lost,
Left his mind still as a deep evening stream.
Nor, if accosted now, in thought engrossed,
Moody, or inly troubled, would he seem
To traveller who might talk of any casual theme.

Hurtle the clouds in deeper darkness piled,   100
Gone is the raven timely rest to seek;
He seemed the only creature in the wild
On whom the elements their rage might wreak;
Save that the bustard, of those regions bleak
Shy tenant, seeing by the uncertain light
A man there wandering, gave a mournful shriek,
And half upon the ground, with strange affright,
Forced hard against the wind a thick unwieldy flight.

All, all was cheerless to the horizon's bound;
The weary eye—which, wheresoe'er it strays,   110
Marks nothing but the red sun's setting round,
Or on the earth strange lines, in former days
Left by gigantic arms—at length surveys
What seems an antique castle spreading wide;
Hoary and naked are its walls, and raise
Their brow sublime: in shelter there to bide
He turned, while rain poured down smoking on every
     side.

Pile of Stone-henge! so proud to hint yet keep
Thy secrets, thou that lov'st to stand and hear
The Plain resounding to the whirlwind's sweep,   120
Inmate of lonesome Nature's endless year;
Even if thou saw'st the giant wicker rear
For sacrifice its throngs of living men,
Before thy face did ever wretch appear,
Who in his heart had groaned with deadlier pain
Than he who, tempest-driven, thy shelter now would
     gain?

Within that fabric of mysterious form
Winds met in conflict, each by turns supreme;
And, from the perilous ground dislodged, through
     storm
And rain he wildered on, no moon to stream   130
From gulf of parting clouds one friendly beam,
Nor any friendly sound his footsteps led;
Once did the lightning's faint disastrous gleam
Disclose a naked guide-post's double head,
Sight which, tho' lost at once, a gleam of pleasure shed.

No swinging sign-board creaked from cottage elm
To stay his steps with faintness overcome;
'Twas dark and void as ocean's watery realm

Roaring with storms beneath night's starless gloom;
No gipsy cower'd o'er fire of furze or broom;   140
No labourer watched his red kiln glaring bright,
Nor taper glimmered dim from sick man's room;
Along the waste no line of mournful light
From lamp of lonely toll-gate streamed athwart the
     night.

At length, though hid in clouds, the moon arose;
The downs were visible—and now revealed
A structure stands, which two bare slopes enclose.
It was a spot where, ancient vows fulfilled,
Kind pious hands did to the Virgin build
A lonely Spital, the belated swain   150
From the night terrors of that waste to shield:
But there no human being could remain,
And now the walls are named the "Dead House" of
     the plain.

Though he had little cause to love the abode
Of man, or covet sight of mortal face,
Yet when faint beams of light that ruin showed,
How glad he was at length to find some trace
Of human shelter in that dreary place.
Till to his flock the early shepherd goes,
Here shall much-needed sleep his frame embrace.   160
In a dry nook where fern the floor bestrows
He lays his stiffened limbs,—his eyes begin to close;

When hearing a deep sigh, that seemed to come
From one who mourned in sleep, he raised his head,
And saw a woman in the naked room
Outstretched, and turning on a restless bed:
The moon a wan dead light around her shed.
He waked her—spake in tone that would not fail,
He hoped, to calm her mind; but ill he sped,
For of that ruin she had heard a tale   170
Which now with freezing thoughts did all her powers
     assail;

Had heard of one who, forced from storms to shroud,
Felt the loose walls of this decayed Retreat
Rock to incessant neighings shrill and loud,
While his horse pawed the floor with furious heat;
Till on a stone, that sparkled to his feet,
Struck, and still struck again, the troubled horse:
The man half raised the stone with pain and sweat,
Half raised, for well his arm might lose its force
Disclosing the grim head of a late murdered corse.   180

150. Spital: hospital, here meaning a place for the shelter
or entertainment of travelers.

Such tale of this lone mansion she had learned,
And when that shape, with eyes in sleep half drowned,
By the moon's sullen lamp she first discerned,
Cold stony horror all her senses bound.
Her he addressed in words of cheering sound;
Recovering heart, like answer did she make;
And well it was that of the corse there found
In converse that ensued she nothing spake;
She knew not what dire pangs in him such tale could
    wake.

But soon his voice and words of kind intent   190
Banished that dismal thought; and now the wind
In fainter howlings told its *rage* was spent:
Meanwhile discourse ensued of various kind,
Which by degrees a confidence of mind
And mutual interest failed not to create.
And, to a natural sympathy resigned,
In that forsaken building where they sate
The Woman thus retraced her own untoward fate.

"By Derwent's side my father dwelt—a man
Of virtuous life, by pious parents bred;   200
And I believe that, soon as I began
To lisp, he made me kneel beside my bed,
And in his hearing there my prayers I said:
And afterwards, by my good father taught,
I read, and loved the books in which I read;
For books in every neighbouring house I sought,
And nothing to my mind a sweeter pleasure brought.

"A little croft we owned—a plot of corn,
A garden stored with peas, and mint, and thyme,
And flowers for posies, oft on Sunday morn   210
Plucked while the church bells rang their earliest chime.
Can I forget our freaks at shearing time!
My hen's rich nest through long grass scarce espied;
The cowslip-gathering in June's dewy prime;
The swans that with white chests upreared in pride
Rushing and racing came to meet me at the waterside!

"The staff I well remember which upbore
The bending body of my active sire;
His seat beneath the honied sycamore
Where the bees hummed, and chair by winter fire;
When market-morning came, the neat attire   221
With which, though bent on haste, myself I decked;
Our watchful house-dog, that would tease and tire
The stranger till its barking-fit I checked;
The red-breast, known for years, which at my casement
    pecked.

"The suns of twenty summers danced along,—
Too little marked how fast they rolled away:
But, through severe mischance and cruel wrong,
My father's substance fell into decay:
We toiled and struggled, hoping for a day   230
When Fortune might put on a kinder look;
But vain were wishes, efforts vain as they;
He from his old hereditary nook
Must part; the summons came;—our final leave we
    took.

"It was indeed a miserable hour
When, from the last hill-top, my sire surveyed,
Peering above the trees, the steeple tower
That on his marriage day sweet music made!
Till then he hoped his bones might there be laid
Close by my mother in their native bowers: 240
Bidding me trust in God, he stood and prayed;—
I could not pray:—through tears that fell in showers
Glimmered our dear-loved home, alas! no longer ours!

"There was a Youth whom I had loved so long,
That when I loved him not I cannot say:
'Mid the green mountains many a thoughtless song
We two had sung, like gladsome birds in May;
When we began to tire of childish play,
We seemed still more and more to prize each other;
We talked of marriage and our marriage day;   250
And I in truth did love him like a brother,
For never could I hope to meet with such another.

"Two years were passed since to a distant town
He had repaired to ply a gainful trade:
What tears of bitter grief, till then unknown,
What tender vows our last sad kiss delayed!
To him we turned:—we had no other aid:
Like one revived, upon his neck I wept;
And her whom he had loved in joy, he said,
He well could love in grief; his faith he kept;   260
And in a quiet home once more my father slept.

"We lived in peace and comfort; and were blest
With daily bread, by constant toil supplied.
Three lovely babes had lain upon my breast;
And often, viewing their sweet smiles, I sighed,
And knew not why. My happy father died,
When threatened war reduced the children's meal:
Thrice happy! that for him the grave could hide
The empty loom, cold hearth, and silent wheel,
And tears which flowed for ills which patience might
    not heal.   270

" 'Twas a hard change; an evil time was come;
We had no hope, and no relief could gain:
But soon, with proud parade, the noisy drum
Beat round to clear the streets of want and pain.
My husband's arms now only served to strain
Me and his children hungering in his view;
In such dismay my prayers and tears were vain:
To join those miserable men he flew,
And now to the sea-coast, with numbers more, we
    drew.

"There were we long neglected, and we bore   280
Much sorrow ere the fleet its anchor weighed;
Green fields before us, and our native shore,
We breathed a pestilential air, that made
Ravage for which no knell was heard. We prayed
For our departure; wished and wished—nor knew,
'Mid that long sickness and those hopes delayed,
That happier days we never more must view.
The parting signal streamed—at last the land withdrew.

"But the calm summer season now was past.
On as we drove, the equinoctial deep   290
Ran mountains high before the howling blast,
And many perished in the whirlwind's sweep.
We gazed with terror on their gloomy sleep,
Untaught that soon such anguish must ensue,
Our hopes such harvest of affliction reap,
That we the mercy of the waves should rue:
We reached the western world, a poor devoted crew.

"The pains and plagues that on our heads came down,
Disease and famine, agony and fear,
In wood or wilderness, in camp or town,   300
It would unman the firmest heart to hear.
All perished—all in one remorseless year,
Husband and children! one by one, by sword
And ravenous plague, all perished: every tear
Dried up, despairing, desolate, on board
A British ship I waked, as from a trance restored."

Here paused she, of all present thought forlorn,
Nor voice, nor sound, that moment's pain expressed,
Yet Nature, with excess of grief o'erborne,
From her full eyes their watery load released.   310
He too was mute: and, ere her weeping ceased,
He rose, and to the ruin's portal went,
And saw the dawn opening the silvery east
With rays of promise, north and southward sent;
And soon with crimson fire kindled the firmament.

"O come," he cried, "come, after weary night
Of such rough storm, this happy change to view."
So forth she came, and eastward looked; the sight
Over her brow like dawn of gladness threw;
Upon her cheek, to which its youthful hue   320
Seemed to return, dried the last lingering tear,
And from her grateful heart a fresh one drew:
The whilst her comrade to her pensive cheer
Tempered fit words of hope; and the lark warbled near.

They looked and saw a lengthening road, and wain
That rang down a bare slope not far remote:
The barrows glistered bright with drops of rain,
Whistled the waggoner with merry note,
The cock far off sounded his clarion throat;
But town, or farm, or hamlet, none they viewed, 330
Only were told there stood a lonely cot
A long mile thence. While thither they pursued
Their way, the Woman thus her mournful tale renewed.

"Peaceful as this immeasurable plain
Is now, by beams of dawning light imprest,
In the calm sunshine slept the glittering main;
The very ocean hath its hour of rest.
I too forgot the heavings of my breast.
How quiet 'round me ship and ocean were!
As quiet all within me. I was blest,   340
And looked, and fed upon the silent air
Until it seemed to bring a joy to my despair.

"Ah! how unlike those late terrific sleeps,
And groans that rage of racking famine spoke;
The unburied dead that lay in festering heaps,
The breathing pestilence that rose like smoke,
The shriek that from the distant battle broke,
The mine's dire earthquake, and the pallid host
Driven by the bomb's incessant thunder-stroke
To loathsome vaults, where heart-sick anguish tossed,
Hope died, and fear itself in agony was lost!   351

"Some mighty gulf of separation passed,
I seemed transported to another world;
A thought resigned with pain, when from the mast
The impatient mariner the sail unfurled,
And, whistling, called the wind that hardly curled
The silent sea. From the sweet thoughts of home
And from all hope I was for ever hurled.
For me—farthest from earthly port to roam
Was best, could I but shun the spot where man might
    come.   360

327. barrows: ancient burial mounds, tumuli.

"And oft I thought (my fancy was so strong)
That I, at last, a resting-place had found;
'Here will I dwell,' said I, 'my whole life long,
Roaming the illimitable waters round;
Here will I live, of all but heaven disowned,
And end my days upon the peaceful flood.'—
To break my dream the vessel reached its bound;
And homeless near a thousand homes I stood,
And near a thousand tables pined and wanted food.

"No help I sought; in sorrow turned adrift,   370
Was hopeless, as if cast on some bare rock;
Nor morsel to my mouth that day did lift,
Nor raised my hand at any door to knock.
I lay where, with his drowsy mates, the cock
From the cross-timber of an outhouse hung:
Dismally tolled, that night, the city clock!
At morn my sick heart hunger scarcely stung,
Nor to the beggar's language could I fit my tongue.

"So passed a second day; and, when the third
Was come, I tried in vain the crowd's resort.   380
—In deep despair, by frightful wishes stirred,
Near the sea-side I reached a ruined fort;
There, pains which nature could no more support,
With blindness linked, did on my vitals fall;
And, after many interruptions short
Of hideous sense, I sank, nor step could crawl:
Unsought for was the help that did my life recall.

"Borne to a hospital, I lay with brain
Drowsy and weak, and shattered memory;
I heard my neighbours in their beds complain   390
Of many things which never troubled me—
Of feet still bustling round with busy glee,
Of looks where common kindness had no part,
Of service done with cold formality,
Fretting the fever round the languid heart,
And groans which, as they said, might make a dead
    man start.

"These things just served to stir the slumbering sense,
Nor pain nor pity in my bosom raised.
With strength did memory return; and, thence
Dismissed, again on open day I gazed,   400
At houses, men, and common light, amazed.
The lanes I sought, and, as the sun retired,
Came where beneath the trees a faggot blazed;
The travellers saw me weep, my fate inquired,
And gave me food—and rest, more welcome, more
    desired.

"Rough potters seemed they, trading soberly
With panniered asses driven from door to door;
But life of happier sort set forth to me,
And other joys my fancy to allure—
The bag-pipe dinning on the midnight moor   410
In barn uplighted; and companions boon,
Well met from far with revelry secure
Among the forest glades, while jocund June
Rolled fast along the sky his warm and genial moon.

"But ill they suited me—those journeys dark
O'er moor and mountain, midnight theft to hatch!
To charm the surly house-dog's faithful bark,
Or hang on tip-toe at the lifted latch.
The gloomy lantern, and the dim blue match,
The black disguise, the warning whistle shrill,   420
And ear still busy on its nightly watch,
Were not for me, brought up in nothing ill:
Besides, on griefs so fresh my thoughts were brooding
    still.

"What could I do, unaided and unblest?
My father! gone was every friend of thine:
And kindred of dead husband are at best
Small help; and, after marriage such as mine,
With little kindness would to me incline.
Nor was I then for toil or service fit;
My deep-drawn sighs no effort could confine;   430
In open air forgetful would I sit
Whole hours, with idle arms in moping sorrow knit.

"The roads I paced, I loitered through the fields;
Contentedly, yet sometimes self-accused,
Trusted my life to what chance bounty yields,
Now coldly given, now utterly refused.
The ground I for my bed have often used:
But what afflicts my peace with keenest ruth,
Is that I have my inner self abused,
Forgone the home delight of constant truth,   440
And clear and open soul, so prized in fearless youth.

"Through tears the rising sun I oft have viewed,
Through tears have seen him towards that world
    descend
Where my poor heart lost all its fortitude:
Three years a wanderer now my course I bend—
Oh! tell me whither—for no earthly friend

---

406. potters: persons who traveled about selling crockery or
earthenware.

Have I."—She ceased, and weeping turned away;
As if because her tale was at an end,
She wept; because she had no more to say
Of that perpetual weight which on her spirit lay. 450

True sympathy the Sailor's looks expressed,
His looks—for pondering he was mute the while.
Of social Order's care for wretchedness,
Of Time's sure help to calm and reconcile,
Joy's second spring and Hope's long-treasured smile,
'Twas not for *him* to speak—a man so tried.
Yet, to relieve her heart, in friendly style
Proverbial words of comfort he applied,
And not in vain, while they went pacing side by side.

Ere long, from heaps of turf, before their sight, 460
Together smoking in the sun's slant beam,
Rise various wreaths that into one unite
Which high and higher mounts with silver gleam:
Fair spectacle,—but instantly a scream
Thence bursting shrill did all remark prevent;
They paused, and heard a hoarser voice blaspheme,
And female cries. Their course they thither bent,
And met a man who foamed with anger vehement.

A woman stood with quivering lips and pale,
And, pointing to a little child that lay 470
Stretched on the ground, began a piteous tale;
How in a simple freak of thoughtless play
He had provoked his father, who straightway,
As if each blow were deadlier than the last,
Struck the poor innocent. Pallid with dismay
The Soldier's Widow heard and stood aghast;
And stern looks on the man her grey-haired Comrade
cast.

His voice with indignation rising high
Such further deed in manhood's name forbade;
The peasant, wild in passion, made reply 480
With bitter insult and revilings sad;
Asked him in scorn what business there he had;
What kind of plunder he was hunting now;
The gallows would one day of him be glad;—
Though inward anguish damped the Sailor's brow,
Yet calm he seemed as thoughts so poignant would
allow.

Softly he stroked the child, who lay outstretched
With face to earth; and, as the boy turned round
His battered head, a groan the Sailor fetched
As if he saw—there and upon that ground— 490
Strange repetition of the deadly wound

He had himself inflicted. Through his brain
At once the griding iron passage found;
Deluge of tender thoughts then rushed amain,
Nor could his sunken eyes the starting tear restrain.

Within himself he said—What hearts have we!
The blessing this a father gives his child!
Yet happy thou, poor boy! compared with m ,
Suffering not doing ill—fate far more d.
The stranger's looks and tears of wrath beguiled 500
The father, and relenting thoughts awoke;
He kissed his son—so all was reconciled.
Then, with a voice which inward trouble broke
Ere to his lips it came, the Sailor them bespoke.

"Bad is the world, and hard is the world's law
Even for the man who wears the warmest fleece;
Much need have ye that time more closely draw
The bond of nature, all unkindness cease,
And that among so few there still be peace:
Else can ye hope but with such numerous foes 510
Your pains shall ever with your years increase?"—
While from his heart the appropriate lesson flows,
A correspondent calm stole gently o'er his woes.

Forthwith the pair passed on; and down they look
Into a narrow valley's pleasant scene
Where wreaths of vapour tracked a winding brook,
That babbled on through groves and meadows green;
A low-roofed house peeped out the trees between;
The dripping groves resound with cheerful lays,
And melancholy lowings intervene 520
Of scattered herds, that in the meadow graze,
Some amid lingering shade, some touched by the sun's
rays.

They saw and heard, and, winding with the road
Down a thick wood, they dropt into the vale;
Comfort by prouder mansions unbestowed
Their wearied frames, she hoped, would soon regale.
Ere long they reached that cottage in the dale:
It was a rustic inn;—the board was spread,
The milk-maid followed with her brimming pail,
And lustily the master carved the bread, 530
Kindly the housewife pressed, and they in comfort fed.

Their breakfast done, the pair, though loth, must part;
Wanderers whose course no longer now agrees.
She rose and bade farewell! and, while her heart
Struggled with tears nor could its sorrow ease,

493. griding: piercing.

She left him there; for, clustering round his knees,
With his oak-staff the cottage children played;
And soon she reached a spot o'erhung with trees
And banks of ragged earth; beneath the shade
Across the pebbly road a little runnel strayed.   540

A cart and horse beside the rivulet stood;
Chequering the canvas roof the sunbeams shone.
She saw the carman bend to scoop the flood
As the wain fronted her,—wherein lay one,
A pale-faced Woman, in disease far gone.
The carman wet her lips as well behoved;
Bed under her lean body there was none,
Though even to die near one she most had loved
She could not of herself those wasted limbs have
    moved.

The Soldier's Widow learned with honest pain   550
And homefelt force of sympathy sincere,
Why thus that worn-out wretch must there sustain
The jolting road and morning air severe.
The wain pursued its way; and following near
In pure compassion she her steps retraced
Far as the cottage. "A sad sight is here,"
She cried aloud; and forth ran out in haste
The friends whom she had left but a few minutes past.

While to the door with eager speed they ran,
From her bare straw the Woman half upraised   560
Her bony visage—gaunt and deadly wan;
No pity asking, on the group she gazed
With a dim eye, distracted and amazed;
Then sank upon her straw with feeble moan.
Fervently cried the housewife—"God be praised,
I have a house that I can call my own;
Nor shall she perish there, untended and alone!"

So in they bear her to the chimney seat,
And busily, though yet with fear, untie
Her garments, and, to warm her icy feet   570
And chafe her temples, careful hands apply.
Nature reviving, with a deep-drawn sigh
She strove, and not in vain, her head to rear;
Then said—"I thank you all; if I must die,
The God in heaven my prayers for you will hear;
Till now I did not think my end had been so near.

"Barred every comfort labour could procure,
Suffering what no endurance could assuage,
I was compelled to seek my father's door,
Though loth to be a burthen on his age.   580
But sickness stopped me in an early stage

Of my sad journey; and within the wain
They placed me—there to end life's pilgrimage,
Unless beneath your roof I may remain:
For I shall never see my father's door again.

"My life, Heaven knows, hath long been burthensome;
But, if I have not meekly suffered, meek
May my end be! Soon will this voice be dumb:
Should child of mine e'er wander hither, speak
Of me, say that the worm is on my cheek.—   590
Torn from our hut, that stood beside the sea
Near Portland lighthouse in a lonesome creek,
My husband served in sad captivity
On shipboard, bound till peace or death should set him
    free.

"A sailor's wife I knew a widow's cares,
Yet two sweet little ones partook my bed;
Hope cheered my dreams, and to my daily prayers
Our heavenly Father granted each day's bread;
Till one was found by stroke of violence dead,
Whose body near our cottage chanced to lie;   600
A dire suspicion drove us from our shed;
In vain to find a friendly face we try,
Nor could we live together those poor boys and I;

"For evil tongues made oath how on that day
My husband lurked about the neighbourhood;
Now he had fled, and whither none could say,
And he had done the deed in the dark wood—
Near his own home!—but he was mild and good;
Never on earth was gentler creature seen;
He'd not have robbed the raven of its food.   610
My husband's loving kindness stood between
Me and all worldly harms and wrongs however keen."

Alas! the thing she told with labouring breath
The Sailor knew too well. That wickedness
His hand had wrought; and when, in the hour of death,
He saw his Wife's lips move his name to bless
With her last words, unable to suppress
His anguish, with his heart he ceased to strive;
And, weeping loud in this extreme distress,
He cried—"Do pity me! That thou shouldst live   620
I neither ask nor wish—forgive me, but forgive!"

To tell the change that Voice within her wrought
Nature by sign or sound made no essay;
A sudden joy surprised expiring thought,
And every mortal pang dissolved away.
Borne gently to a bed, in death she lay;
Yet still, while over her the husband bent,

A look was in her face which seemed to say,
"Be blest: by sight of thee from heaven was sent
Peace to my parting soul, the fulness of content." 630

She slept in peace,—his pulses throbbed and stopped,
Breathless he gazed upon her face,—then took
Her hand in his, and raised it, but both dropped,
When on his own he cast a rueful look.
His ears were never silent; sleep forsook
His burning eyelids stretched and stiff as lead;
All night from time to time under him shook
The floor as he lay shuddering on his bed;
And oft he groaned aloud, "O God, that I were dead!"

The Soldier's Widow lingered in the cot;   640
And, when he rose, he thanked her pious care
Through which his Wife, to that kind shelter brought,
Died in his arms; and with those thanks a prayer
He breathed for her, and for that merciful pair.
The corse interred, not one hour he remained
Beneath their roof, but to the open air
A burthen, now with fortitude sustained,
He bore within a breast where dreadful quiet reigned.

Confirmed of purpose, fearlessly prepared
For act and suffering, to the city straight   650
He journeyed, and forthwith his crime declared:
"And from your doom," he added, "now I wait,
Nor let it linger long, the murderer's fate."
Not ineffectual was that piteous claim:
"O welcome sentence which will end though late,"
He said, "the pangs that to my conscience came
Out of that deed. My trust, Saviour! is in thy name!"

His fate was pitied. Him in iron case
(Reader, forgive the intolerable thought)
They hung not:—no one on *his* form or face   660
Could gaze, as on a show by idlers sought;
No kindred sufferer, to his death-place brought
By lawless curiosity or chance,
When into storm the evening sky is wrought,
Upon his swinging corse an eye can glance,
And drop, as he once dropped, in miserable trance.

                    1791–94 (1842)

658. iron case: See ll. 76–79.

THE OLD CUMBERLAND BEGGAR

"The class of Beggars, to which the Old Man here described belongs, will probably soon be extinct. It consisted of poor, and, mostly, old and infirm persons, who confined themselves to a stated round in their neighbourhood, and had certain fixed days, on which, at different houses, they regularly received alms, sometimes in money, but mostly in provisions" (W. W.).

I saw an aged Beggar in my walk;
And he was seated, by the highway side,
On a low structure of rude masonry
Built at the foot of a huge hill, that they
Who lead their horses down the steep rough road
May thence remount at ease. The aged Man
Had placed his staff across the broad smooth stone
That overlays the pile; and, from a bag
All white with flour, the dole of village dames,
He drew his scraps and fragments, one by one;   10
And scanned them with a fixed and serious look
Of idle computation. In the sun,
Upon the second step of that small pile,
Surrounded by those wild unpeopled hills,
He sat, and ate his food in solitude:
And ever, scattered from his palsied hand,
That, still attempting to prevent the waste,
Was baffled still, the crumbs in little showers
Fell on the ground; and the small mountain birds,
Not venturing yet to peck their destined meal,   20
Approached within the length of half his staff.

Him from my childhood have I known; and then
He was so old, he seems not older now;
He travels on, a solitary Man,
So helpless in appearance, that for him
The sauntering Horseman throws not with a slack
And careless hand his alms upon the ground,
But stops,—that he may safely lodge the coin
Within the old Man's hat; nor quits him so,
But still, when he has given his horse the rein, 30
Watches the aged Beggar with a look
Sidelong, and half-reverted. She who tends
The toll-gate, when in summer at her door
She turns her wheel, if on the road she sees

The aged Beggar coming, quits her work,
And lifts the latch for him that he may pass.
The post-boy, when his rattling wheels o'ertake
The aged Beggar in the woody lane,
Shouts to him from behind; and, if thus warned
The old man does not change his course, the boy   40
Turns with less noisy wheels to the roadside,
And passes gently by, without a curse
Upon his lips, or anger at his heart.

    He travels on, a solitary Man;
His age has no companion. On the ground
His eyes are turned, and, as he moves along,
They move along the ground; and, evermore,
Instead of common and habitual sight
Of fields with rural works, of hill and dale,
And the blue sky, one little span of earth   50
Is all his prospect. Thus, from day to day,
Bow-bent, his eyes for ever on the ground,
He plies his weary journey; seeing still,
And seldom knowing that he sees, some straw,
Some scattered leaf, or marks which, in one track,
The nails of cart or chariot-wheel have left
Impressed on the white road,—in the same line,
At distance still the same. Poor Traveller!
His staff trails with him; scarcely do his feet
Disturb the summer dust; he is so still   60
In look and motion, that the cottage curs,
Ere he has passed the door, will turn away,
Weary of barking at him. Boys and girls,
The vacant and the busy, maids and youths,
And urchins newly breeched—all pass him by:
Him even the slow-paced waggon leaves behind.

    But deem not this Man useless.—Statesmen! ye
Who are so restless in your wisdom, ye
Who have a broom still ready in your hands
To rid the world of nuisances; ye proud,   70
Heart-swoln, while in your pride ye contemplate
Your talents, power, or wisdom, deem him not
A burthen of the earth! 'Tis Nature's law
That none, the meanest of created things,
Of forms created the most vile and brute,
The dullest or most noxious, should exist
Divorced from good—a spirit and pulse of good,
A life and soul, to every mode of being
Inseparably linked. Then be assured
That least of all can aught—that ever owned   80
The heaven-regarding eye and front sublime
Which man is born to—sink, howe'er depressed,
So low as to be scorned without a sin;

Without offence to God cast out of view;
Like the dry remnant of a garden-flower
Whose seeds are shed, or as an implement
Worn out and worthless. While from door to door,
This old Man creeps, the villagers in him
Behold a record which together binds
Past deeds and offices of charity,   90
Else unremembered, and so keeps alive
The kindly mood in hearts which lapse of years,
And that half-wisdom half-experience gives,
Make slow to feel, and by sure steps resign
To selfishness and cold oblivious cares.
Among the farms and solitary huts,
Hamlets and thinly-scattered villages,
Where'er the aged Beggar takes his rounds,
The mild necessity of use compels
To acts of love; and habit does the work   100
Of reason; yet prepares that after-joy
Which reason cherishes. And thus the soul,
By that sweet taste of pleasure unpursued,
Doth find herself insensibly disposed
To virtue and true goodness.
                                    Some there are,
By their good works exalted, lofty minds,
And meditative, authors of delight
And happiness, which to the end of time
Will live, and spread, and kindle: even such minds
In childhood, from this solitary Being,   110
Or from like wanderer, haply have received
(A thing more precious far than all that books
Or the solicitudes of love can do!)
That first mild touch of sympathy and thought,
In which they found their kindred with a world
Where want and sorrow were. The easy man
Who sits at his own door,—and, like the pear
That overhangs his head from the green wall,
Feeds in the sunshine; the robust and young,
The prosperous and unthinking, they who live   120
Sheltered, and flourish in a little grove
Of their own kindred;—all behold in him
A silent monitor, which on their minds
Must needs impress a transitory thought
Of self-congratulation, to the heart
Of each recalling his peculiar boons,
His charters and exemptions; and, perchance,
Though he to no one give the fortitude
And circumspection needful to preserve
His present blessings, and to husband up   130
The respite of the season, he, at least,
And 'tis no vulgar service, makes them felt.

Yet further.——Many, I believe, there are
Who live a life of virtuous decency,
Men who can hear the Decalogue and feel
No self-reproach; who of the moral law
Established in the land where they abide
Are strict observers; and not negligent
In acts of love to those with whom they dwell,
Their kindred, and the children of their blood.   140
Praise be to such, and to their slumbers peace!
—But of the poor man ask, the abject poor;
Go, and demand of him, if there be here
In this cold abstinence from evil deeds,
And these inevitable charities,
Wherewith to satisfy the human soul?
No—man is dear to man; the poorest poor
Long for some moments in a weary life
When they can know and feel that they have been,
Themselves, the fathers and the dealers-out   150
Of some small blessings; have been kind to such
As needed kindness, for this single cause,
That we have all of us one human heart.
—Such pleasure is to one kind Being known,
My neighbour, when with punctual care, each week,
Duly as Friday comes, though pressed herself
By her own wants, she from her store of meal
Takes one unsparing handful for the scrip
Of this old Mendicant, and, from her door
Returning with exhilarated heart,   160
Sits by her fire, and builds her hope in heaven.

Then let him pass, a blessing on his head!
And while in that vast solitude to which
The tide of things has borne him, he appears
To breathe and live but for himself alone,
Unblamed, uninjured, let him bear about
The good which the benignant law of Heaven
Has hung around him: and, while life is his,
Still let him prompt the unlettered villagers
To tender offices and pensive thoughts.   170
—Then let him pass, a blessing on his head!
And, long as he can wander, let him breathe
The freshness of the valleys; let his blood
Struggle with frosty air and winter snows;
And let the chartered wind that sweeps the heath
Beat his grey locks against his withered face.
Reverence the hope whose vital anxiousness
Gives the last human interest to his heart.

May never HOUSE, misnamed of INDUSTRY,
Make him a captive!—for that pent-up din,   180
Those life-consuming sounds that clog the air,
Be his the natural silence of old age!
Let him be free of mountain solitudes;
And have around him, whether heard or not,
The pleasant melody of woodland birds.
Few are his pleasures: if his eyes have now
Been doomed so long to settle upon earth
That not without some effort they behold
The countenance of the horizontal sun,
Rising or setting, let the light at least   190
Find a free entrance to their languid orbs,
And let him, *where* and *when* he will, sit down
Beneath the trees, or on a grassy bank
Of highway side, and with the little birds
Share his chance-gathered meal; and, finally,
As in the eye of Nature he has lived,
So in the eye of Nature let him die!

1797 (1800)

THE REVERIE
OF POOR SUSAN

At the corner of Wood Street, when daylight appears,
Hangs a Thrush that sings loud, it has sung for three
            years:
Poor Susan has passed by the spot, and has heard
In the silence of morning the song of the Bird.

'Tis a note of enchantment; what ails her? She sees
A mountain ascending, a vision of trees;
Bright volumes of vapour through Lothbury glide,
And a river flows on through the vale of Cheapside.

Green pastures she views in the midst of the dale,
Down which she so often has tripped with her pail;10
And a single small cottage, a nest like a dove's,
The one only dwelling on earth that she loves.

She looks, and her heart is in heaven: but they fade,
The mist and the river, the hill and the shade:
The stream will not flow, and the hill will not rise,
And the colours have all passed away from her eyes!

1797 (1800)

THE OLD CUMBERLAND BEGGAR. 158. scrip: a small bag.

179. house . . . industry: a poorhouse. THE REVERIE OF
POOR SUSAN. 1. Wood Street: in London.

A NIGHT-PIECE

——The sky is overcast
With a continuous cloud of texture close,
Heavy and wan, all whitened by the Moon,
Which through that veil is indistinctly seen,
A dull, contracted circle, yielding light
So feebly spread that not a shadow falls,
Chequering the ground—from rock, plant, tree, or
    tower.
At length a pleasant instantaneous gleam
Startles the pensive traveller while he treads
His lonesome path, with unobserving eye   10
Bent earthwards; he looks up—the clouds are split
Asunder,—and above his head he sees
The clear Moon, and the glory of the heavens.
There, in a black-blue vault she sails along,
Followed by multitudes of stars, that, small
And sharp, and bright, along the dark abyss
Drive as she drives: how fast they wheel away,
Yet vanish not!—the wind is in the tree,
But they are silent;—still they roll along
Immeasurably distant; and the vault,   20
Built round by those white clouds, enormous clouds,
Still deepens its unfathomable depth.
At length the Vision closes; and the mind,
Not undisturbed by the delight it feels,
Which slowly settles into peaceful calm,
Is left to muse upon the solemn scene.

1798 (1815)

FROM

LYRICAL BALLADS (1798)

❧

From "Lines . . . Yew-tree" through "Tintern Abbey,"
poems included here are printed in the order in which
they appeared in the *Lyrical Ballads* of 1798. The purpose
of this famous volume is set forth by Wordsworth in the
Preface to the second edition, in the note to "We Are
Seven," and by Coleridge in *Biographia Literaria,*
Chapter XIV. The aims stated in the Preface do not apply
to all the lyrical ballads (e.g. "Tintern Abbey"), much
less to all his poems. They apply particularly to such
poems as "Goody Blake," "We Are Seven," "The
Thorn," and "The Idiot Boy," which were, as Wordsworth says, a deliberate "experiment."

Popular appreciation of the early English ballads had
been growing ever since Bishop Thomas Percy's
compilation in the *Reliques of Ancient English Poetry*
(1765). The phrase "lyrical ballads" may have been
intended to signify that the interest lay both in the
narrative or action, as with the true ballad, and in the
expression of feeling, as with a lyric. In the Preface
Wordsworth remarks, "the feeling therein developed
gives importance to the action and situation, and not the
action and situation to the feeling."

This means, among other things, that the poems are
designed to reveal the psychology of the human mind
and heart—to disclose, as Wordsworth says, "the primary
laws of our nature," "the essential passions of the heart."
For example, "We Are Seven" points out our utter
inability, as children, to accept or even entertain the idea
of death. For Wordsworth, this is an original fact of
human nature, and it evolves, as we grow older, into a
firm belief in immortality. "Goody Blake" illustrates the
psychological basis of superstition: Harry Gill is a victim
of what would now be called hysteria. Similarly, much
that the speaker describes in "The Thorn" might be
interpreted as a product of his imagination. At the same
time, these poems are charged with social protest and
humanitarian sympathy. "The Last of the Flock," by
suggesting that property is a natural source of satisfaction
closely twined with the domestic affections, may, as has
been argued, have been intended to contradict Godwin,
who held that love of property is a cause of evil. But the
poem also attacks the cruel operation of the Poor Laws,
which required that a man forfeit dignity and hope before
he could be aided. Like many great works of art, however, the lyrical ballads transcend the conscious intentions that went into their making.

Reactions to the poems have included all degrees of
praise and blame. Of the more extreme experiments, such
as "The Thorn," it may be said that most readers find
they improve steadily with longer acquaintance, and that
they provide a continual exercise in poetic discrimination
and taste, the verse ranging from lines such as

all have joined in one endeavour
To bury this poor Thorn for ever,

to the magic of,

And she is known to every star,
And every wind that blows.

❧

LINES

*Left upon a Seat in a Yew-tree, which stands near the lake
of Esthwaite, on a desolate part of the shore, commanding a
beautiful prospect.*

Nay, Traveller! rest. This lonely Yew-tree stands
Far from all human dwelling: what if here
No sparkling rivulet spread the verdant herb?
What if the bee love not these barren boughs?

Yet, if the wind breathe soft, the curling waves,
That break against the shore, shall lull thy mind
By one soft impulse saved from vacancy.
———————————— Who he was
That piled these stones and with the mossy sod
First covered, and here taught this aged Tree   10
With its dark arms to form a circling bower,
I well remember.—He was one who owned
No common soul. In youth by science nursed,
And led by nature into a wild scene
Of lofty hopes, he to the world went forth
A favoured Being, knowing no desire
Which genius did not hallow; 'gainst the taint
Of dissolute tongues, and jealousy, and hate,
And scorn,—against all enemies prepared,
All but neglect. The world, for so it thought,   20
Owed him no service; wherefore he at once
With indignation turned himself away,
And with the food of pride sustained his soul
In solitude.—Stranger! these gloomy boughs
Had charms for him; and here he loved to sit,
His only visitants a straggling sheep,
The stone-chat, or the glancing sand-piper:
And on these barren rocks, with fern and heath,
And juniper and thistle, sprinkled o'er,
Fixing his downcast eye, he many an hour   30
A morbid pleasure nourished, tracing here
An emblem of his own unfruitful life:
And, lifting up his head, he then would gaze
On the more distant scene,—how lovely 'tis
Thou seest,—and he would gaze till it became
Far lovelier, and his heart could not sustain
The beauty, still more beauteous! Nor, that time,
When nature had subdued him to herself,
Would he forget those Beings, to whose minds,
Warm from the labours of benevolence,   40
The world, and human life, appeared a scene
Of kindred loveliness: then he would sigh,
Inly disturbed, to think that others felt
What he must never feel: and so, lost Man!
On visionary views would fancy feed,
Till his eye streamed with tears. In this deep vale
He died,—this seat his only monument.

If Thou be one whose heart the holy forms
Of young imagination have kept pure,
Stranger! henceforth be warned; and know that
          pride,   50
Howe'er disguised in its own majesty,
Is littleness; that he who feels contempt

For any living thing, hath faculties
Which he has never used; that thought with him
Is in its infancy. The man whose eye
Is ever on himself doth look on one,
The least of Nature's works, one who might move
The wise man to that scorn which wisdom holds
Unlawful, ever. O be wiser, Thou!
Instructed that true knowledge leads to love;   60
True dignity abides with him alone
Who, in the silent hour of inward thought,
Can still suspect, and still revere himself,
In lowliness of heart.

1795–97 (1798)

GOODY BLAKE
AND HARRY GILL

A TRUE STORY

꩜

See Wordsworth's discussion in the Preface, below, p.
329 n.

꩜

Oh! what's the matter? what's the matter?
What is't that ails young Harry Gill?
That evermore his teeth they chatter,
Chatter, chatter, chatter still!
Of waistcoats Harry has no lack,
Good duffle grey, and flannel fine;
He has a blanket on his back,
And coats enough to smother nine.

In March, December, and in July,
'Tis all the same with Harry Gill;   10
The neighbours tell, and tell you truly,
His teeth they chatter, chatter still.
At night, at morning, and at noon,
'Tis all the same with Harry Gill;
Beneath the sun, beneath the moon,
His teeth they chatter, chatter still!

Young Harry was a lusty drover,
And who so stout of limb as he?

LYRICAL BALLADS. GOODY BLAKE AND HARRY GILL. **6. duffle:** a
coarse woolen cloth—named from the town of Duffel in
Belgium.

His cheeks were red as ruddy clover;
His voice was like the voice of three.   20
Old Goody Blake was old and poor;
Ill fed she was, and thinly clad;
And any man who passed her door
Might see how poor a hut she had.

All day she spun in her poor dwelling:
And then her three hours' work at night,
Alas! 'twas hardly worth the telling,
It would not pay for candle-light.
Remote from sheltered village-green,
On a hill's northern side she dwelt,   30
Where from sea-blasts the hawthorns lean,
And hoary dews are slow to melt.

By the same fire to boil their pottage,
Two poor old Dames, as I have known,
Will often live in one small cottage;
But she, poor Woman! housed alone.
'Twas well enough, when summer came,
The long, warm, lightsome summer-day,
Then at her door the *canty* Dame
Would sit, as any linnet, gay.   40

But when the ice our streams did fetter,
Oh then how her old bones would shake!
You would have said, if you had met her,
'Twas a hard time for Goody Blake.
Her evenings then were dull and dead:
Sad case it was, as you may think,
For very cold to go to bed;
And then for cold not sleep a wink.

O joy for her! whene'er in winter
The winds at night had made a rout;   50
And scattered many a lusty splinter
And many a rotten bough about.
Yet never had she, well or sick,
As every man who knew her says,
A pile beforehand, turf or stick,
Enough to warm her for three days.

Now, when the frost was past enduring,
And made her poor old bones to ache,
Could any thing be more alluring
Than an old hedge to Goody Blake?   60
And, now and then, it must be said,
When her old bones were cold and chill,
She left her fire, or left her bed,
To seek the hedge of Harry Gill.

39. canty: cheerful, merry.

Now Harry he had long suspected
This trespass of old Goody Blake;
And vowed that she should be detected—
That he on her would vengeance take.
And oft from his warm fire he'd go,
And to the fields his road would take;   70
And there, at night, in frost and snow,
He watched to seize old Goody Blake.

And once, behind a rick of barley,
Thus looking out did Harry stand:
The moon was full and shining clearly,
And crisp with frost the stubble land.
—He hears a noise—he's all awake—
Again?—on tip-toe down the hill
He softly creeps—'tis Goody Blake;
She's at the hedge of Harry Gill!   80

Right glad was he when he beheld her:
Stick after stick did Goody pull:
He stood behind a bush of elder,
Till she had filled her apron full.
When with her load she turned about,
The by-way back again to take;
He started forward, with a shout,
And sprang upon poor Goody Blake.

And fiercely by the arm he took her,
And by the arm he held her fast,   90
And fiercely by the arm he shook her,
And cried, "I've caught you then at last!"
Then Goody, who had nothing said,
Her bundle from her lap let fall;
And, kneeling on the sticks, she prayed
To God that is the judge of all.

She prayed, her withered hand uprearing,
While Harry held her by the arm—
"God! who art never out of hearing,
O may he never more be warm!"   100
The cold, cold moon above her head,
Thus on her knees did Goody pray;
Young Harry heard what she had said:
And icy cold he turned away.

He went complaining all the morrow
That he was cold and very chill:
His face was gloom, his heart was sorrow,
Alas! that day for Harry Gill!
That day he wore a riding-coat,
But not a whit the warmer he: 110
Another was on Thursday brought,
And ere the Sabbath he had three.

'Twas all in vain, a useless matter,
And blankets were about him pinned;
Yet still his jaws and teeth they clatter,
Like a loose casement in the wind.
And Harry's flesh it fell away;
And all who see him say, 'tis plain,
That, live as long as live he may,
He never will be warm again.   120

No word to any man he utters,
A-bed or up, to young or old;
But ever to himself he mutters,
"Poor Harry Gill is very cold."
A-bed or up, by night or day;
His teeth they chatter, chatter still.
Now think, ye farmers all, I pray,
Of Goody Blake and Harry Gill!

1798 (1798)

TO MY SISTER

It is the first mild day of March:
Each minute sweeter than before,
The redbreast sings from the tall larch
That stands beside our door.

There is a blessing in the air,
Which seems a sense of joy to yield
To the bare trees, and mountains bare,
And grass in the green field.

My sister! ('tis a wish of mine)
Now that our morning meal is done,   10
Make haste, your morning task resign;
Come forth and feel the sun.

Edward will come with you;—and, pray,
Put on with speed your woodland dress;
And bring no book: for this one day
We'll give to idleness.

No joyless forms shall regulate
Our living calendar:
We from to-day, my Friend, will date
The opening of the year.   20

Love, now a universal birth,
From heart to heart is stealing,
From earth to man, from man to earth:
—It is the hour of feeling.

TO MY SISTER. **13. Edward:** a child.

One moment now may give us more
Than years of toiling reason:
Our minds shall drink at every pore
The spirit of the season.

Some silent laws our hearts will make,
Which they shall long obey: 30
We for the year to come may take
Our temper from to-day.

And from the blessed power that rolls
About, below, above,
We'll frame the measure of our souls:
They shall be tuned to love.

Then come, my Sister! come, I pray,
With speed put on your woodland dress;
And bring no book: for this one day
We'll give to idleness.   40

1798 (1798)

SIMON LEE

THE OLD HUNTSMAN;

WITH AN INCIDENT IN WHICH HE WAS CONCERNED

In the sweet shire of Cardigan,
Not far from pleasant Ivor-hall,
An old Man dwells, a little man,—
'Tis said he once was tall.
Full five-and-thirty years he lived
A running huntsman merry;
And still the centre of his cheek
Is red as a ripe cherry.

No man like him the horn could sound,
And hill and valley rang with glee   10
When Echo bandied, round and round,
The halloo of Simon Lee.
In those proud days, he little cared
For husbandry or tillage;
To blither tasks did Simon rouse
The sleepers of the village.

He all the country could outrun,
Could leave both man and horse behind;
And often, ere the chase was done,
He reeled, and was stone-blind.   20

And still there's something in the world
At which his heart rejoices;
For when the chiming hounds are out,
He dearly loves their voices!

But, oh the heavy change!—bereft
Of health, strength, friends, and kindred, see!
Old Simon to the world is left
In liveried poverty.
His Master's dead,—and no one now
Dwells in the Hall of Ivor;   30
Men, dogs, and horses, all are dead;
He is the sole survivor.

And he is lean and he is sick;
His body, dwindled and awry,
Rests upon ankles swoln and thick;
His legs are thin and dry.
One prop he has, and only one,
His wife, an aged woman,
Lives with him, near the waterfall,
Upon the village Common.   40

Beside their moss-grown hut of clay,
Not twenty paces from the door,
A scrap of land they have, but they
Are poorest of the poor.
This scrap of land he from the heath
Enclosed when he was stronger;
But what to them avails the land
Which he can till no longer?

Oft, working by her Husband's side,
Ruth does what Simon cannot do;   50
For she, with scanty cause for pride,
Is stouter of the two.
And, though you with your utmost skill
From labour could not wean them,
'Tis little, very little—all
That they can do between them.

Few months of life has he in store
As he to you will tell,
For still, the more he works, the more
Do his weak ankles swell.   60
My gentle Reader, I perceive
How patiently you've waited,
And now I fear that you expect
Some tale will be related.

O Reader! had you in your mind
Such stores as silent thought can bring,

O gentle Reader! you would find
A tale in every thing.
What more I have to say is short,
And you must kindly take it: 70
It is no tale; but, should you think,
Perhaps a tale you'll make it.

One summer-day I chanced to see
This old Man doing all he could
To unearth the root of an old tree,
A stump of rotten wood.
The mattock tottered in his hand;
So vain was his endeavour,
That at the root of the old tree
He might have worked for ever.   80

"You're overtasked, good Simon Lee,
Give me your tool," to him I said;
And at the word right gladly he
Received my proffered aid.
I struck, and with a single blow
The tangled root I severed,
At which the poor old Man so long
And vainly had endeavoured.

The tears into his eyes were brought,
And thanks and praises seemed to run   90
So fast out of his heart, I thought
They never would have done.
—I've heard of hearts unkind, kind deeds
With coldness still returning;
Alas! the gratitude of men
Hath oftener left me mourning.

                                          1798 (1798)

ANECDOTE FOR FATHERS

〜✸〜

From 1800 to 1843 the title was "Anecdote for Fathers,
showing how the practice of lying may be taught." But it
would seem that the poem is a more complex moral
lesson, contrasting, among other things, the spontaneous
and unthinking motions of the child's mind with the
adult insistence on reasons.

〜✸〜

I have a boy of five years old;
His face is fair and fresh to see;
His limbs are cast in beauty's mould,
And dearly he loves me.

One morn we strolled on our dry walk,
Our quiet home all full in view,
And held such intermitted talk
As we are wont to do.

My thoughts on former pleasures ran;   10
I thought of Kilve's delightful shore,
Our pleasant home when spring began,
A long, long year before.

A day it was when I could bear
Some fond regrets to entertain;
With so much happiness to spare,
I could not feel a pain.

The green earth echoed to the feet
Of lambs that bounded through the glade,
From shade to sunshine, and as fleet
From sunshine back to shade.   20

Birds warbled round me—and each trace
Of inward sadness had its charm;
Kilve, thought I, was a favoured place,
And so is Liswyn farm.

My boy beside me tripped, so slim
And graceful in his rustic dress!
And, as we talked, I questioned him,
In very idleness.

"Now tell me, had you rather be,"
I said, and took him by the arm,   30
"On Kilve's smooth shore, by the green sea,
Or here at Liswyn farm?"

In careless mood he looked at me,
While still I held him by the arm,
And said, "At Kilve I'd rather be
Than here at Liswyn farm."

"Now, little Edward, say why so:
My little Edward, tell me why."—
"I cannot tell, I do not know."—
"Why, this is strange," said I;   40

"For here are woods, hills smooth and warm:
There surely must some reason be
Why you would change sweet Liswyn farm
For Kilve by the green sea."

At this my boy hung down his head,
He blushed with shame, nor made reply;
And three times to the child I said,
"Why, Edward, tell me why?"

His head he raised—there was in sight,
It caught his eye, he saw it plain—   50
Upon the house-top, glittering bright,
A broad and gilded vane.

Then did the boy his tongue unlock,
And eased his mind with this reply:
"At Kilve there was no weather-cock;
And that's the reason why."

O dearest, dearest boy! my heart
For better lore would seldom yearn,
Could I but teach the hundredth part
Of what from thee I learn.   60

1798 (1798)

WE ARE SEVEN

"In reference to this poem I will here mention one of the
most remarkable facts in my own poetic history and that
of Mr. Coleridge. In the spring of the year 1798, he, my
sister, and myself, started from Alfoxden, pretty late in
the afternoon, with a view to visit Lenton and the valley
of Stones near it; and as our united funds were very
small, we agreed to defray the expense of the tour by
writing a poem, to be sent to the new *Monthly Magazine*
set up by Phillips the bookseller, and edited by Dr. Aikin.
Accordingly we set off and proceeded along the Quantock Hills towards Watchet, and in the course of this
walk was planned the poem of *The Ancient Mariner*,
founded on a dream, as Mr. Coleridge said, of his friend,
Mr. Cruikshank. Much the greatest part of the story was
Mr. Coleridge's invention; but certain parts I myself
suggested:—for example, some crime was to be committed which should bring upon the old navigator, as
Coleridge afterwards delighted to call him, the spectral
persecution, as a consequence of that crime, and his own
wanderings. I had been reading in Shelvock's *Voyages* a
day or two before that while doubling Cape Horn they
frequently saw albatrosses in that latitude, the largest
sort of sea-fowl, some extending their wings twelve or
fifteen feet. 'Suppose,' said I, 'you represent him as
having killed one of these birds on entering the South
Sea, and that the tutelary Spirits of those regions take upon
them to avenge the crime.' The incident was thought fit
for the purpose and adopted accordingly. I also suggested
the navigation of the ship by the dead men, but do not
recollect that I had anything more to do with the scheme
of the poem. The Gloss with which it was subsequently
accompanied was not thought of by either of us at the
time; at least, not a hint of it was given to me, and I have
no doubt it was a gratuitous afterthought. We began the
composition together on that, to me, memorable

evening. I furnished two or three lines at the beginning
of the poem, in particular:—

    'And listened like a three years' child;
     The Mariner had his will.'

These trifling contributions, all but one (which Mr. C.
has with unnecessary scrupulosity recorded) slipt out of
his mind as they well might. As we endeavored to proceed conjointly (I speak of the same evening) our respective manners proved so widely different that it would
have been quite presumptuous in me to do anything but
separate from an undertaking upon which I could only
have been a clog. We returned after a few days from a
delightful tour, of which I have many pleasant, and some
of them droll-enough, recollections. We returned by
Dulverton to Alfoxden. *The Ancient Mariner* grew and
grew till it became too important for our first object,
which was limited to our expectation of five pounds, and
we began to talk of a volume, which was to consist, as
Mr. Coleridge has told the world, of poems chiefly on
supernatural subjects taken from common life, but
looked at, as much as might be, through an imaginative
medium. Accordingly I wrote *The Idiot Boy, Her Eyes
Are Wild,* etc., *We Are Seven, The Thorn,* and some others.
To return to *We Are Seven,* the piece that called forth this
note, I composed it while walking in the grove at
Alfoxden. My friends will not deem it too trifling to
relate that while walking to and fro I composed the last
stanza first, having begun with the last line. When it was
all but finished, I came in and recited it to Mr. Coleridge
and my sister, and said, 'A prefatory stanza must be
added, and I should sit down to our little tea-meal with
greater pleasure if my task were finished.' I mentioned in
substance what I wished to be expressed, and Coleridge
immediately threw off the stanza thus:

    'A little child, dear brother Jem'—

I objected to the rhyme, 'dear brother Jem,' as being
ludicrous, but we all enjoyed the joke of hitching-in our
friend, James Tobin's name, who was familiarly called
Jem. He was the brother of the dramatist, and this reminds me of an anecdote which it may be worth while
here to notice. The said Jem got a sight of the Lyrical
Ballads as it was going through the press at Bristol, during
which time I was residing in that city. One evening he
came to me with a grave face, and said, 'Wordsworth, I
have seen the volume that Coleridge and you are about to
publish. There is one poem in it, which I earnestly
entreat you will cancel, for, if published, it will make you
everlastingly ridiculous.' I answered that I felt much
obliged by the interest he took in my good name as a
writer, and begged to know what was the unfortunate
piece he alluded to. He said, 'It is called "We are seven."'
Nay! said I, that shall take its chance, however, and he
left me in despair" (W. W.).

   See Wordsworth's discussion of this poem in the
Preface to the *Lyrical Ballads,* p. 322 n.

                         ᘓᙨᕑ

————A simple Child,
That lightly draws its breath,
And feels its life in every limb,
What should it know of death?

I met a little cottage Girl:
She was eight years old, she said;
Her hair was thick with many a curl
That clustered round her head.

She had a rustic, woodland air,   10
And she was wildly clad:
Her eyes were fair, and very fair;
—Her beauty made me glad.

"Sisters and brothers, little maid,
How many may you be?"
"How many? Seven in all," she said,
And wondering looked at me.

"And where are they? I pray you tell."
She answered, "Seven are we;
And two of us at Conway dwell,
And two are gone to sea.   20

"Two of us in the church-yard lie,
My sister and my brother;
And, in the church-yard cottage, I
Dwell near them with my mother."

"You say that two at Conway dwell,
And two are gone to sea,
Yet ye are seven! I pray you tell,
Sweet Maid, how this may be."

Then did the little Maid reply,
"Seven boys and girls are we;   30
Two of us in the church-yard lie,
Beneath the church-yard tree."

"You run about, my little Maid,
Your limbs they are alive;
If two are in the church-yard laid,
Then ye are only five."

"Their graves are green, they may be seen,"
The little Maid replied,
"Twelve steps or more from my mother's door,
And they are side by side.   40

"My stockings there I often knit,
My kerchief there I hem;
And there upon the ground I sit,
And sing a song to them.

"And often after sun-set, Sir,
When it is light and fair,
I take my little porringer,
And eat my supper there.

"The first that died was sister Jane;
In bed she moaning lay,   50
Till God released her of her pain;
And then she went away.

"So in the church-yard she was laid;
And, when the grass was dry,
Together round her grave we played,
My brother John and I.

"And when the ground was white with snow,
And I could run and slide,
My brother John was forced to go,
And he lies by her side."   60

"How many are you, then," said I,
"If they two are in heaven?"
Quick was the little Maid's reply,
"O Master! we are seven."

"But they are dead; those two are dead!
Their spirits are in heaven!"
'Twas throwing words away; for still
The little Maid would have her will,
And said, "Nay, we are seven!"

                        1798 (1800)

LINES WRITTEN
IN EARLY SPRING

I heard a thousand blended notes,
While in a grove I sate reclined,
In that sweet mood when pleasant thoughts
Bring sad thoughts to the mind.

To her fair works did Nature link
The human soul that through me ran;
And much it grieved my heart to think
What man has made of man.

Through primrose tufts, in that green bower,
The periwinkle trailed its wreaths;   10
And 'tis my faith that every flower
Enjoys the air it breathes.

The birds around me hopped and played,
Their thoughts I cannot measure:—
But the least motion which they made,
It seemed a thrill of pleasure.

The budding twigs spread out their fan,
To catch the breezy air;
And I must think, do all I can,
That there was pleasure there.   20

If this belief from heaven be sent,
If such be Nature's holy plan,
Have I not reason to lament
What man has made of man?

                        1798 (1798)

THE THORN

⌒〜⌒

"Arose out of my observing, on the ridge of Quantock Hill, on a stormy day, a thorn which I had often past, in calm and bright weather, without noticing it. I said to myself, 'Cannot I by some invention do as much to make this thorn permanently an impressive object as the storm has made it to my eyes at this moment?' I began the poem accordingly, and composed it with great rapidity" (W. W.).

To the Preface to the second edition of the *Lyrical Ballads* Wordsworth added the following note:

"This poem ought to have been preceded by an introductory poem, which I have been prevented from writing by never having felt myself in a mood when it was probable that I should write it well. The character which I have here introduced speaking is sufficiently common. The reader will perhaps have a general notion of it, if he has ever known a man, a captain of a small trading vessel, for example, who being past the middle age of life, had retired upon an annuity or small independent income to some village or country town of which he was not a native, or in which he had not been accustomed to live. Such men, having little to do, become credulous and talkative from indolence; and from the same cause, and other predisposing causes by which it is probable that such men may have been affected, they are prone to superstition. On which account it appeared to me proper to select a character like this to exhibit some of the general laws by which superstition acts upon the mind. Superstitious men are almost always men of slow faculties and deep feelings; their minds are not loose, but adhesive; they have a reasonable share of imagination, by which word I mean the faculty which produces impressive effects out of simple elements; but they are utterly destitute of fancy, the power by which pleasure and surprise are excited by sudden varieties of situation and by accumulated imagery.

"It was my wish in this poem to show the manner in which such men cleave to the same ideas; and to follow the turns of passion, always different, yet not palpably different, by which their conversation is swayed. I had two objects to attain; first, to represent a picture which should not be unimpressive, yet consistent with the character that should describe it; secondly, while I adhered to the style in which such persons describe, to take care that words, which in their minds are impregnated with passion, should likewise convey passion to readers who are not accustomed to sympathize with men feeling in that manner or using such language. It seemed to me that this might be done by calling in the assistance of lyrical and rapid metre. It was necessary that the poem, to be natural, should in reality move slowly; yet I hoped that, by the aid of the metre, to those who should at all enter into the spirit of the poem, it would appear to move quickly. The reader will have the kindness to excuse this note, as I am sensible that an introductory poem is necessary to give the poem its full effect.

"Upon this occasion I will request permission to add a few words closely connected with *The Thorn* and many other poems in these volumes. There is a numerous class of readers who imagine that the same words cannot be repeated without tautology: this is a great error; virtual tautology is much oftener produced by using different words when the meaning is exactly the same. Words, a poet's words more particularly, ought to be weighed in the balance of feeling, and not measured by the space which they occupy upon paper. For the reader cannot be too often reminded that poetry is passion: it is the history or science of feelings. Now every man must know that an attempt is rarely made to communicate impassioned feelings without something of an accompanying consciousness of the inadequateness of our own powers, or the deficiences of language. During such efforts there will be a craving in the mind, and as long as it is unsatisfied the speaker will cling to the same words, or words of the same character. There are also various other reasons why repetition and apparent tautology are frequently beauties of the highest kind. Among the chief of these reasons is the interest which the mind attaches to words, not only as symbols of the passion, but as *things*, active and efficient, which are of themselves part of the passion. And further, from a spirit of fondness, exultation, and gratitude, the mind luxuriates in the repetition of words which appear successfully to communicate its feelings. The truth of these remarks might be shown by innumerable passages from the Bible, and from the impassioned poetry of every nation. 'Awake, awake, Deborah!' &c. Judges, chap. v., verses 12th, 27th, and part of 28th. See also the whole of that tumultuous and wonderful Poem."

Coleridge discusses the poem in *Biographia Literaria*, Chapter XVII, below, p. 462, and see Francis Jeffrey, below, p. 366.

✤

"There is a Thorn—it looks so old,
In truth, you'd find it hard to say
How it could ever have been young,
It looks so old and grey.
Not higher than a two years' child
It stands erect, this aged Thorn;
No leaves it has, no prickly points;
It is a mass of knotted joints,
A wretched thing forlorn.
It stands erect, and like a stone   10
With lichens is it overgrown.

"Like rock or stone, it is o'ergrown,
With lichens to the very top,
And hung with heavy tufts of moss,
A melancholy crop:
Up from the earth these mosses creep,
And this poor Thorn they clasp it round
So close, you'd say that they are bent
With plain and manifest intent
To drag it to the ground;   20
And all have joined in one endeavour
To bury this poor Thorn for ever.

"High on a mountain's highest ridge,
Where oft the stormy winter gale
Cuts like a scythe, while through the clouds
It sweeps from vale to vale;
Not five yards from the mountain path,
This Thorn you on your left espy;
And to the left, three yards beyond,
You see a little muddy pond   30
Of water—never dry,
Though but of compass small, and bare
To thirsty suns and parching air.

"And, close beside this aged Thorn,
There is a fresh and lovely sight,
A beauteous heap, a hill of moss,
Just half a foot in height.
All lovely colours there you see,
All colours that were ever seen;
And mossy network too is there,   40
As if by hand of lady fair
The work had woven been;
And cups, the darlings of the eye,
So deep is their vermilion dye.

"Ah me! what lovely tints are there
Of olive green and scarlet bright,
In spikes, in branches, and in stars,
Green, red, and pearly white!

This heap of earth o'ergrown with moss,
Which close beside the Thorn you see,   50
So fresh in all its beauteous dyes,
Is like an infant's grave in size,
As like as like can be:
But never, never any where,
An infant's grave was half so fair.

"Now would you see this aged Thorn,
This pond, and beauteous hill of moss,
You must take care and choose your time
The mountain when to cross.
For oft there sits between the heap,   60
So like an infant's grave in size,
And that same pond of which I spoke,
A Woman in a scarlet cloak,
And to herself she cries,
'Oh misery! oh misery!
Oh woe is me! oh misery!'

"At all times of the day and night
This wretched Woman thither goes;
And she is known to every star,
And every wind that blows;   70
And there, beside the Thorn, she sits
When the blue daylight's in the skies,
And when the whirlwind's on the hill,
Or frosty air is keen and still,
And to herself she cries,
'Oh misery! oh misery!
Oh woe is me! oh misery!'"

"Now wherefore, thus, by day and night,
In rain, in tempest, and in snow,
Thus to the dreary mountain-top   80
Does this poor Woman go?
And why sits she beside the Thorn
When the blue daylight's in the sky
Or when the whirlwind's on the hill,
Or frosty air is keen and still,
And wherefore does she cry?—
O wherefore? wherefore? tell me why
Does she repeat that doleful cry?"

"I cannot tell; I wish I could;
For the true reason no one knows:
But would you gladly view the spot,   90
The spot to which she goes;
The hillock like an infant's grave,
The pond—and Thorn, so old and grey;
Pass by her door—'tis seldom shut—
And if you see her in her hut—

Then to the spot away!
I never heard of such as dare
Approach the spot when she is there."

"But wherefore to the mountain-top   100
Can this unhappy Woman go,
Whatever star is in the skies,
Whatever wind may blow?"
"Full twenty years are past and gone
Since she (her name is Martha Ray)
Gave with a maiden's true good-will
Her company to Stephen Hill;
And she was blithe and gay,
While friends and kindred all approved
Of him whom tenderly she loved.   110

"And they had fixed the wedding day,
The morning that must wed them both;
But Stephen to another Maid
Had sworn another oath;
And, with this other Maid, to church
Unthinking Stephen went—
Poor Martha! on that woeful day
A pang of pitiless dismay
Into her soul was sent;
A fire was kindled in her breast,   120
Which might not burn itself to rest.

"They say, full six months after this,
While yet the summer leaves were green,
She to the mountain-top would go,
And there was often seen.
What could she seek?—or wish to hide?
Her state to any eye was plain;
She was with child, and she was mad;
Yet often was she sober sad
From her exceeding pain.   130
O guilty Father—would that death
Had saved him from that breach of faith!

"Sad case for such a brain to hold
Communion with a stirring child!
Sad case, as you may think, for one
Who had a brain so wild!
Last Christmas-eve we talked of this,
And grey-haired Wilfred of the glen
Held that the unborn infant wrought
About its mother's heart, and brought   140
Her senses back again:
And, when at last her time drew near,
Her looks were calm, her senses clear.

"More know I not, I wish I did,
And it should all be told to you;
For what became of this poor child
No mortal ever knew;
Nay—if a child to her was born
No earthly tongue could ever tell;
And if 'twas born alive or dead,   150
Far less could this with proof be said;
But some remember well,
That Martha Ray about this time
Would up the mountain often climb.

"And all that winter, when at night
The wind blew from the mountain-peak,
'Twas worth your while, though in the dark,
The churchyard path to seek:
For many a time and oft were heard
Cries coming from the mountain head: 160
Some plainly living voices were;
And others, I've heard many swear,
Were voices of the dead:
I cannot think, whate'er they say,
They had to do with Martha Ray.

"But that she goes to this old Thorn,
The Thorn which I described to you,
And there sits in a scarlet cloak,
I will be sworn is true.
For one day with my telescope,   170
To view the ocean wide and bright,
When to this country first I came,
Ere I had heard of Martha's name,
I climbed the mountain's height:—
A storm came on, and I could see
No object higher than my knee.

"'Twas mist and rain, and storm and rain:
No screen, no fence could I discover;
And then the wind! in sooth, it was
A wind full ten times over.   180
I looked around, I thought I saw
A jutting crag,—and off I ran,
Head-foremost, through the driving rain,
The shelter of the crag to gain;
And, as I am a man,
Instead of jutting crag, I found
A Woman seated on the ground.

"I did not speak—I saw her face;
Her face!—it was enough for me;
I turned about and heard her cry,   190
'Oh misery! oh misery!'

And there she sits, until the moon
Through half the clear blue sky will go;
And, when the little breezes make
The waters of the pond to shake,
As all the country know,
She shudders, and you hear her cry,
'Oh misery! oh misery!'"

"But what's the Thorn? and what the pond?
And what the hill of moss to her?   200
And what the creeping breeze that comes
The little pond to stir?"
"I cannot tell; but some will say
She hanged her baby on the tree;
Some say she drowned it in the pond,
Which is a little step beyond:
But all and each agree,
The little Babe was buried there,
Beneath that hill of moss so fair.

"I've heard, the moss is spotted red   210
With drops of that poor infant's blood;
But kill a new-born infant thus,
I do not think she could!
Some say if to the pond you go,
And fix on it a steady view,
The shadow of a babe you trace,
A baby and a baby's face,
And that it looks as you;
Whene'er you look on it, 'tis plain
The baby looks at you again.   220

"And some had sworn an oath that she
Should be to public justice brought;
And for the little infant's bones
With spades they would have sought.
But instantly the hill of moss
Before their eyes began to stir!
And, for full fifty yards around,
The grass—it shook upon the ground!
Yet all do still aver
The little Babe lies buried there,   230
Beneath that hill of moss so fair.

"I cannot tell how this may be,
But plain it is the Thorn is bound
With heavy tufts of moss that strive
To drag it to the ground;
And this I know, full many a time,
When she was on the mountain high,
By day, and in the silent night,
When all the stars shone clear and bright,

That I have heard her cry,   240
'Oh misery! oh misery!
Oh woe is me! oh misery!'"

                                        1798 (1798)

THE LAST OF THE FLOCK

In distant countries have I been,
And yet I have not often seen
A healthy man, a man full grown,
Weep in the public roads, alone.
But such a one, on English ground,
And in the broad highway, I met;
Along the broad highway he came,
His cheeks with tears were wet:
Sturdy he seemed, though he was sad;
And in his arms a Lamb he had.   10

He saw me, and he turned aside,
As if he wished himself to hide:
And with his coat did then essay
To wipe those briny tears away.
I followed him, and said, "My friend,
What ails you? wherefore weep you so?"
—"Shame on me, Sir! this lusty Lamb,
He makes my tears to flow.
To-day I fetched him from the rock;
He is the last of all my flock.   20

"When I was young, a single man,
And after youthful follies ran,
Though little given to care and thought,
Yet, so it was, an ewe I bought;
And other sheep from her I raised,
As healthy sheep as you might see;
And then I married, and was rich
As I could wish to be;
Of sheep I numbered a full score,
And every year increased my store.   30

"Year after year my stock it grew;
And from this one, this single ewe,
Full fifty comely sheep I raised,
As fine a flock as ever grazed!
Upon the Quantock hills they fed;
They throve, and we at home did thrive:
—This lusty Lamb of all my store
Is all that is alive;
And now I care not if we die,
And perish all of poverty.   40

"Six Children, Sir! had I to feed;
Hard labour in a time of need!
My pride was tamed, and in our grief
I of the Parish asked relief.
They said, I was a wealthy man;
My sheep upon the uplands fed,
And it was fit that thence I took
Whereof to buy us bread.
'Do this: how can we give to you,'
They cried, 'what to the poor is due?'   50

"I sold a sheep, as they had said,
And bought my little children bread,
And they were healthy with their food;
For me—it never did me good.
A woeful time it was for me,
To see the end of all my gains,
The pretty flock which I had reared
With all my care and pains,
To see it melt like snow away—
For me it was a woeful day.   60

"Another still! and still another!
A little lamb, and then its mother!
It was a vein that never stopped—
Like blood-drops from my heart they dropped.
Till thirty were not left alive
They dwindled, dwindled, one by one;
And I may say, that many a time
I wished they all were gone—
Reckless of what might come at last
Were but the bitter struggle past.   70

"To wicked deeds I was inclined,
And wicked fancies crossed my mind;
And every man I chanced to see,
I thought he knew some ill of me:
No peace, no comfort could I find,
No ease, within doors or without;
And crazily and wearily
I went my work about;
And oft was moved to flee from home,
And hide my head where wild beasts roam.   80

"Sir! 'twas a precious flock to me,
As dear as my own children be;
For daily with my growing store
I loved my children more and more.
Alas! it was an evil time;
God cursed me in my sore distress;
I prayed, yet every day I thought
I loved my children less;

And every week, and every day,
My flock it seemed to melt away.   90

"They dwindled, Sir, sad sight to see!
From ten to five, from five to three,
A lamb, a wether, and a ewe;—
And then at last from three to two;
And, of my fifty, yesterday
I had but only one:
And here it lies upon my arm,
Alas! and I have none;—
To-day I fetched it from the rock;
It is the last of all my flock."   100

                              1798 (1798)

HER EYES ARE WILD

Her eyes are wild, her head is bare,
The sun has burnt her coal-black hair;
Her eyebrows have a rusty stain,
And she came far from over the main.
She has a baby on her arm,
Or else she were alone:
And underneath the hay-stack warm,
And on the greenwood stone,
She talked and sung the woods among,
And it was in the English tongue.   10

"Sweet babe! they say that I am mad,
But nay, my heart is far too glad;
And I am happy when I sing
Full many a sad and doleful thing:
Then, lovely baby, do not fear!
I pray thee have no fear of me;
But safe as in a cradle, here
My lovely baby! thou shalt be:
To thee I know too much I owe;
I cannot work thee any woe.   20

"A fire was once within my brain;
And in my head a dull, dull pain;
And fiendish faces, one, two, three,
Hung at my breast, and pulled at me;
But then there came a sight of joy;
It came at once to do me good;
I waked, and saw my little boy,
My little boy of flesh and blood;
Oh joy for me that sight to see!
For he was here, and only he.   30

"Suck, little babe, oh suck again!
It cools my blood; it cools my brain;
Thy lips I feel them, baby! they
Draw from my heart the pain away.
Oh! press me with thy little hand;
It loosens something at my chest;
About that tight and deadly band
I feel thy little fingers prest.
The breeze I see is in the tree:
It comes to cool my babe and me.   40

"Oh! love me, love me, little boy!
Thou art thy mother's only joy;
And do not dread the waves below,
When o'er the sea-rock's edge we go;
The high crag cannot work me harm,
Nor leaping torrents when they howl;
The babe I carry on my arm,
He saves for me my precious soul;
Then happy lie; for blest am I;
Without me my sweet babe would die.   50

"Then do not fear, my boy! for thee
Bold as a lion will I be;
And I will always be thy guide,
Through hollow snows and rivers wide.
I'll build an Indian bower; I know
The leaves that make the softest bed:
And if from me thou wilt not go,
But still be true till I am dead,
My pretty thing! then thou shalt sing
As merry as the birds in spring.   60

"Thy father cares not for my breast,
'Tis thine, sweet baby, there to rest;
'Tis all thine own!—and if its hue
Be changed, that was so fair to view,
'Tis fair enough for thee, my dove!
My beauty, little child, is flown,
But thou wilt live with me in love;
And what if my poor cheek be brown?
'Tis well for me thou canst not see
How pale and wan it else would be.   70

"Dread not their taunts, my little Life;
I am thy father's wedded wife;
And underneath the spreading tree
We two will live in honesty.
If his sweet boy he could forsake,
With me he never would have stayed:
From him no harm my babe can take;
But he, poor man! is wretched made;

And every day we two will pray
For him that's gone and far away. 80

"I'll teach my boy the sweetest things:
I'll teach him how the owlet sings.
My little babe! thy lips are still,
And thou hast almost sucked thy fill.
—Where art thou gone, my own dear child?
What wicked looks are those I see?
Alas! Alas! that look so wild,
It never, never came from me:
If thou art mad, my pretty lad,
Then I must be for ever sad. 90

"Oh! smile on me, my little lamb!
For I thy own dear mother am:
My love for thee has well been tried:
I've sought thy father far and wide.
I know the poisons of the shade;
I know the earth-nuts fit for food:
Then, pretty dear, be not afraid:
We'll find thy father in the wood.
Now laugh and be gay, to the woods away!
And there, my babe, we'll live for aye." 100

1798 (1798)

THE IDIOT BOY

"This long poem was composed in the groves of Alfoxden, almost extempore; not a word, I believe, being
corrected, though one stanza was omitted. I mention this
in gratitude to those happy moments, for, in truth, I
never wrote anything with so much glee" (W. W.).
The word "glee" more or less defines the tone of the
poem and of Wordsworthian comedy in general. He is
almost never witty and he rarely makes one laugh, but he
has, at times, a high-spirited jollity. It manifests itself
in his obvious amusement here, or, for another example,
in the light-heartedness of "The Tables Turned," p. 209.
For discussion of the poem see the Preface to the *Lyrical
Ballads*, below, p. 322 n., Wordsworth's letter to John
Wilson, below, pp. 350–52, and Coleridge in *Biographia
Literaria*, Chapter XVII, below, pp. 461–62.

'Tis eight o'clock,—a clear March night,
The moon is up,—the sky is blue,
The owlet, in the moonlight air,
Shouts from nobody knows where;
He lengthens out his lonely shout,
Halloo! halloo! a long halloo!

—Why bustle thus about your door,
What means this bustle, Betty Foy?
Why are you in this mighty fret?
And why on horseback have you set 10
Him whom you love, your Idiot Boy?

Scarcely a soul is out of bed;
Good Betty, put him down again;
His lips with joy they burr at you;
But, Betty! what has he to do
With stirrup, saddle, or with rein?

But Betty's bent on her intent;
For her good neighbour Susan Gale,
Old Susan, she who dwells alone,
Is sick, and makes a piteous moan, 20
As if her very life would fail.

There's not a house within a mile,
No hand to help them in distress;
Old Susan lies a-bed in pain,
And sorely puzzled are the twain,
For what she ails they cannot guess.

And Betty's husband's at the wood,
Where by the week he doth abide,
A woodman in the distant vale;
There's none to help poor Susan Gale; 30
What must be done? what will betide?

And Betty from the lane has fetched
Her Pony, that is mild and good;
Whether he be in joy or pain,
Feeding at will along the lane,
Or bringing fagots from the wood.

And he is all in travelling trim,—
And, by the moonlight, Betty Foy
Has on the well-girt saddle set
(The like was never heard of yet) 40
Him whom she loves, her Idiot Boy.

And he must post without delay
Across the bridge and through the dale,
And by the church, and o'er the down,
To bring a Doctor from the town,
Or she will die, old Susan Gale.

There is no need of boot or spur,
There is no need of whip or wand;
For Johnny has his holly-bough,
And with a *hurly-burly* now 50
He shakes the green bough in his hand.

And Betty o'er and o'er has told
The Boy, who is her best delight,
Both what to follow, what to shun,
What do, and what to leave undone,
How turn to left, and how to right.

And Betty's most especial charge,
Was, "Johnny! Johnny! mind that you
Come home again, nor stop at all,—
Come home again, whate'er befall,   60
My Johnny, do, I pray you, do."

To this did Johnny answer make,
Both with his head and with his hand,
And proudly shook the bridle too;
And then! his words were not a few,
Which Betty well could understand.

And now that Johnny is just going,
Though Betty's in a mighty flurry,
She gently pats the Pony's side,
On which her Idiot Boy must ride,   70
And seems no longer in a hurry.

But when the Pony moved his legs,
Oh! then for the poor Idiot Boy!
For joy he cannot hold the bridle,
For joy his head and heels are idle,
He's idle all for very joy.

And, while the Pony moves his legs,
In Johnny's left hand you may see
The green bough motionless and dead:
The Moon that shines above his head   80
Is not more still and mute than he.

His heart it was so full of glee
That, till full fifty yards were gone,
He quite forgot his holly whip,
And all his skill in horsemanship:
Oh! happy, happy, happy John.

And while the Mother, at the door,
Stands fixed, her face with joy o'erflows,
Proud of herself, and proud of him,
She sees him in his travelling trim,   90
How quietly her Johnny goes.

The silence of her Idiot Boy,
What hopes it sends to Betty's heart!
He's at the guide-post—he turns right;
She watches till he's out of sight,
And Betty will not then depart.

Burr, burr—now Johnny's lips they burr,
As loud as any mill, or near it;
Meek as a lamb the Pony moves,
And Johnny makes the noise he loves,   100
And Betty listens, glad to hear it.

Away she hies to Susan Gale:
Her Messenger's in merry tune;
The owlets hoot, the owlets curr,
And Johnny's lips they burr, burr, burr,
As on he goes beneath the moon.

His steed and he right well agree;
For of this Pony there's a rumour
That, should he lose his eyes and ears,
And should he live a thousand years,   110
He never will be out of humour.

But then he is a horse that thinks!
And, when he thinks, his pace is slack;
Now, though he knows poor Johnny well,
Yet, for his life, he cannot tell
What he has got upon his back.

So through the moonlight lanes they go,
And far into the moonlight dale,
And by the church, and o'er the down,
To bring a Doctor from the town,   120
To comfort poor old Susan Gale.

And Betty, now at Susan's side,
Is in the middle of her story,
What speedy help her Boy will bring,
With many a most diverting thing,
Of Johnny's wit, and Johnny's glory.

And Betty, still at Susan's side,
By this time is not quite so flurried:
Demure with porringer and plate
She sits, as if in Susan's fate   130
Her life and soul were buried.

But Betty, poor good woman! she,
You plainly in her face may read it,
Could lend out of that moment's store
Five years of happiness or more
To any that might need it.

But yet I guess that now and then
With Betty all was not so well;
And to the road she turns her ears,
And thence full many a sound she hears,   140
Which she to Susan will not tell.

Poor Susan moans, poor Susan groans;
"As sure as there's a moon in heaven,"
Cries Betty, "he'll be back again;
They'll both be here—'tis almost ten—
Both will be here before eleven."

Poor Susan moans, poor Susan groans;
The clock gives warning for eleven;
'Tis on the stroke—"He must be near,"
Quoth Betty, "and will soon be here,   150
As sure as there's a moon in heaven."

The clock is on the stroke of twelve,
And Johnny is not yet in sight:
—The Moon's in heaven, as Betty sees,
But Betty is not quite at ease;
And Susan has a dreadful night.

And Betty, half an hour ago,
On Johnny vile reflections cast:
"A little idle sauntering Thing!"
With other names, an endless string;   160
But now that time is gone and past.

And Betty's drooping at the heart,
That happy time all past and gone,
"How can it be he is so late?
The Doctor, he has made him wait;
Susan! they'll both be here anon."

And Susan's growing worse and worse,
And Betty's in a sad *quandary;*
And then there's nobody to say
If she must go, or she must stay!   170
—She's in a sad *quandary.*

The clock is on the stroke of one;
But neither Doctor nor his Guide
Appears along the moonlight road;
There's neither horse nor man abroad,
And Betty's still at Susan's side.

And Susan now begins to fear
Of sad mischances not a few,
That Johnny may perhaps be drowned;
Or lost, perhaps, and never found;   180
Which they must both for ever rue.

She prefaced half a hint of this
With, "God forbid it should be true!"
At the first word that Susan said
Cried Betty, rising from the bed,
"Susan, I'd gladly stay with you.

"I must be gone, I must away:
Consider, Johnny's but half-wise;
Susan, we must take care of him,
If he is hurt in life or limb"—   190
"Oh God forbid!" poor Susan cries.

"What can I do?" says Betty, going,
"What can I do to ease your pain?
Good Susan tell me, and I'll stay;
I fear you're in a dreadful way,
But I shall soon be back again."

"Nay, Betty, go! good Betty, go!
There's nothing that can ease my pain."
Then off she hies; but with a prayer,
That God poor Susan's life would spare,   200
Till she comes back again.

So, through the moonlight lane she goes,
And far into the moonlight dale;
And how she ran, and how she walked,
And all that to herself she talked,
Would surely be a tedious tale.

In high and low, above, below,
In great and small, in round and square,
In tree and tower was Johnny seen,
In bush and brake, in black and green;   210
'Twas Johnny, Johnny, everywhere.

And while she crossed the bridge, there came
A thought with which her heart is sore—
Johnny perhaps his horse forsook,
To hunt the moon within the brook,
And never will be heard of more.

Now is she high upon the down,
Alone amid a prospect wide;
There's neither Johnny nor his Horse
Among the fern or in the gorse;   220
There's neither Doctor nor his Guide.

"Oh saints! what is become of him?
Perhaps he's climbed into an oak,
Where he will stay till he is dead;
Or sadly he has been misled,
And joined the wandering gipsy-folk.

"Or him that wicked Pony's carried
To the dark cave, the goblin's hall;
Or in the castle he's pursuing
Among the ghosts his own undoing;   230
Or playing with the waterfall."

At poor old Susan then she railed,
While to the town she posts away;
"If Susan had not been so ill,
Alas! I should have had him still,
My Johnny, till my dying day."

Poor Betty, in this sad distemper,
The Doctor's self could hardly spare:
Unworthy things she talked, and wild;
Even he, of cattle the most mild,   240
The Pony had his share.

But now she's fairly in the town,
And to the Doctor's door she hies;
'Tis silence all on every side;
The town so long, the town so wide,
Is silent as the skies.

And now she's at the Doctor's door,
She lifts the knocker, rap, rap, rap;
The Doctor at the casement shows
His glimmering eyes that peep and doze!   250
And one hand rubs his old night-cap.

"Oh Doctor! Doctor! where's my Johnny?"
"I'm here, what is't you want with me?"
"Oh Sir! you know I'm Betty Foy,
And I have lost my poor dear Boy,
You know him—him you often see;

"He's not so wise as some folks be:"
"The devil take his wisdom!" said
The Doctor, looking somewhat grim,
"What, Woman! should I know of him?"   260
And, grumbling, he went back to bed!

"O woe is me! O woe is me!
Here will I die; here will I die;
I thought to find my lost one here,
But he is neither far nor near,
Oh! what a wretched Mother I!"

She stops, she stands, she looks about;
Which way to turn she cannot tell.
Poor Betty! it would ease her pain
If she had heart to knock again;   270
—The clock strikes three—a dismal knell!

Then up along the town she hies,
No wonder if her senses fail;
This piteous news so much it shocked her,
She quite forgot to send the Doctor,
To comfort poor old Susan Gale.

And now she's high upon the down,
And she can see a mile of road:
"O cruel! I'm almost threescore;
Such night as this was ne'er before,   280
There's not a single soul abroad."

She listens, but she cannot hear
The foot of horse, the voice of man;
The streams with softest sound are flowing,
The grass you almost hear it growing,
You hear it now, if e'er you can.

The owlets through the long blue night
Are shouting to each other still:
Fond lovers! yet not quite hob nob,
They lengthen out the tremulous sob,   290
That echoes far from hill to hill.

Poor Betty now has lost all hope,
Her thoughts are bent on deadly sin,
A green-grown pond she just has past,
And from the brink she hurries fast,
Lest she should drown herself therein.

And now she sits her down and weeps;
Such tears she never shed before;
"Oh dear, dear Pony! my sweet joy!
Oh carry back my Idiot Boy!   300
And we will ne'er o'erload thee more."

A thought is come into her head:
The Pony he is mild and good,
And we have always used him well;
Perhaps he's gone along the dell,
And carried Johnny to the wood.

Then up she springs as if on wings;
She thinks no more of deadly sin;
If Betty fifty ponds should see,
The last of all her thoughts would be   310
To drown herself therein.

Oh Reader! now that I might tell
What Johnny and his Horse are doing!
What they've been doing all this time,
Oh could I put it into rhyme,
A most delightful tale pursuing!

Perhaps, and no unlikely thought!
He with his Pony now doth roam
The cliffs and peaks so high that are,
To lay his hands upon a star,   320
And in his pocket bring it home.

Perhaps he's turned himself about,
His face unto his horse's tail,
And, still and mute, in wonder lost,
All silent as a horseman-ghost,
He travels slowly down the vale.

And now, perhaps, is hunting sheep,
A fierce and dreadful hunter he;
Yon valley, now so trim and green,
In five months' time, should he be seen,   330
A desert wilderness will be!

Perhaps, with head and heels on fire,
And like the very soul of evil,
He's galloping away, away,
And so will gallop on for aye,
The bane of all that dread the devil!

I to the Muses have been bound
These fourteen years, by strong indentures:
O gentle Muses! let me tell
But half of what to him befell;   340
He surely met with strange adventures.

O gentle Muses! is this kind?
Why will ye thus my suit repel?
Why of your further aid bereave me?
And can ye thus unfriended leave me;
Ye Muses! whom I love so well?

Who's yon, that, near the waterfall,
Which thunders down with headlong force,
Beneath the moon, yet shining fair,
As careless as if nothing were,   350
Sits upright on a feeding horse?

Unto his horse—there feeding free,
He seems, I think, the rein to give;
Of moon or stars he takes no heed;
Of such we in romances read:
—'Tis Johnny! Johnny! as I live.

And that's the very Pony, too!
Where is she, where is Betty Foy?
She hardly can sustain her fears;
The roaring waterfall she hears,   360
And cannot find her Idiot Boy.

Your Pony's worth his weight in gold:
Then calm your terrors, Betty Foy!
She's coming from among the trees,
And now all full in view she sees
Him whom she loves, her Idiot Boy.

And Betty sees the Pony too:
Why stand you thus, good Betty Foy?
It is no goblin, 'tis no ghost,
'Tis he whom you so long have lost,   370
He whom you love, your Idiot Boy.

She looks again—her arms are up—
She screams—she cannot move for joy;
She darts, as with a torrent's force,
She almost has o'erturned the Horse,
And fast she holds her Idiot Boy.

And Johnny burrs, and laughs aloud;
Whether in cunning or in joy
I cannot tell; but, while he laughs,
Betty a drunken pleasure quaffs   380
To hear again her Idiot Boy.

And now she's at the Pony's tail,
And now is at the Pony's head,—
On that side now, and now on this;
And, almost stifled with her bliss,
A few sad tears does Betty shed.

She kisses o'er and o'er again
Him whom she loves, her Idiot Boy;
She's happy here, is happy there,
She is uneasy everywhere;   390
Her limbs are all alive with joy.

She pats the Pony, where or when
She knows not, happy Betty Foy!
The little Pony glad may be,
But he is milder far than she,
You hardly can perceive his joy.

"Oh! Johnny, never mind the Doctor;
You've done your best, and that is all:"
She took the reins, when this was said,
And gently turned the Pony's head   400
From the loud waterfall.

By this the stars were almost gone,
The moon was setting on the hill,
So pale you scarcely looked at her:
The little birds began to stir,
Though yet their tongues were still.

The Pony, Betty, and her Boy,
Wind slowly through the woody dale;
And who is she, betimes abroad,
That hobbles up the steep rough road?   410
Who is it, but old Susan Gale?

Long time lay Susan lost in thought;
And many dreadful fears beset her,
Both for her Messenger and Nurse;
And, as her mind grew worse and worse,
Her body—it grew better.

She turned, she tossed herself in bed,
On all sides doubts and terrors met her;
Point after point did she discuss;
And, while her mind was fighting thus,   420
Her body still grew better.

"Alas! what is become of them?
These fears can never be endured;
I'll to the wood."—The word scarce said,
Did Susan rise up from her bed,
As if by magic cured.

Away she goes up hill and down,
And to the wood at length is come;
She spies her Friends, she shouts a greeting;
Oh me! it is a merry meeting   430
As ever was in Christendom.

The owls have hardly sung their last,
While our four travellers homeward wend;
The owls have hooted all night long,
And with the owls began my song,
And with the owls must end.

For while they all were travelling home,
Cried Betty, "Tell us, Johnny, do,
Where all this long night you have been,
What you have heard, what you have seen: 440
And, Johnny, mind you tell us true."

Now Johnny all night long had heard
The owls in tuneful concert strive;
No doubt too he the moon had seen;
For in the moonlight he had been
From eight o'clock till five.

And thus, to Betty's question, he
Made answer, like a traveller bold,
(His very words I give to you,)
"The cocks did crow to-whoo, to-whoo,   450
And the sun did shine so cold!"
—Thus answered Johnny in his glory,
And that was all his travel's story.

1798 (1798)

EXPOSTULATION AND REPLY

"Why, William, on that old grey stone,
Thus for the length of half a day,
Why, William, sit you thus alone,
And dream your time away?

"Where are your books?—that light bequeathed
To Beings else forlorn and blind!
Up! up! and drink the spirit breathed
From dead men to their kind.

"You look round on your Mother Earth,
As if she for no purpose bore you;   10
As if you were her first-born birth,
And none had lived before you!"

One morning thus, by Esthwaite lake,
When life was sweet, I knew not why,
To me my good friend Matthew spake,
And thus I made reply:

"The eye—it cannot choose but see;
We cannot bid the ear be still;
Our bodies feel, where'er they be,
Against or with our will.   20

"Nor less I deem that there are Powers
Which of themselves our minds impress;
That we can feed this mind of ours
In a wise passiveness.

"Think you, 'mid all this mighty sum
Of things for ever speaking,
That nothing of itself will come,
But we must still be seeking?

"—Then ask not wherefore, here, alone,
Conversing as I may,   30
I sit upon this old grey stone,
And dream my time away."

1798 (1798)

EXPOSTULATION AND REPLY. **15. Matthew:** See the headnote to the "Matthew poems," below.

THE TABLES TURNED

AN EVENING SCENE
ON THE SAME SUBJECT

Up! up! my Friend, and quit your books;
Or surely you'll grow double:
Up! up! my Friend, and clear your looks;
Why all this toil and trouble?

The sun, above the mountain's head,
A freshening lustre mellow
Through all the long green fields has spread,
His first sweet evening yellow.

Books! 'tis a dull and endless strife:
Come, hear the woodland linnet,   10
How sweet his music! on my life,
There's more of wisdom in it.

And hark! how blithe the throstle sings!
He, too, is no mean preacher:
Come forth into the light of things,
Let Nature be your Teacher.

She has a world of ready wealth,
Our minds and hearts to bless—
Spontaneous wisdom breathed by health,
Truth breathed by cheerfulness.   20

One impulse from a vernal wood
May teach you more of man,
Of moral evil and of good,
Than all the sages can.

Sweet is the lore which Nature brings;
Our meddling intellect
Mis-shapes the beauteous forms of things:—
We murder to dissect.

Enough of Science and of Art;
Close up those barren leaves;   30
Come forth, and bring with you a heart
That watches and receives.

1798 (1798)

LINES

COMPOSED A FEW MILES ABOVE TINTERN ABBEY,
ON REVISITING THE BANKS OF THE WYE DURING A
TOUR. JULY 13, 1798

"I have not ventured to call this Poem an Ode; but it was written with a hope that in the transitions, and the impassioned music of the versification, would be found the principal requisites of that species of composition" (W. W.).

"No poem of mine was composed under circumstances more pleasant for me to remember than this. I began it upon leaving Tintern, after crossing the Wye, and concluded it just as I was entering Bristol in the evening, after a ramble of 4 or 5 days, with my sister. Not a line of it was altered, and not any part of it written down till I reached Bristol" (W. W.).

The poem illuminates what was to become an explicit theme of *The Prelude*, that is, the influence of Nature in forming and sustaining the mind and character. As elsewhere in Wordsworth's poetry, at least three meanings can be distinguished in the word Nature: (1) external nature, here the scenery of the Wye valley; (2) all existence, a harmonious, integrated, and living whole; (3) a "Presence" or Divine Life that informs the whole and every part. When Wordsworth speaks of Nature, he involves all these meanings at the same time; for in keeping with his philosophy of organicism he held that they interfused each other. Hence he can assert that "the language of the sense"—sensory awareness of nature—is

> The guide, the guardian of my heart, and soul
> Of all my moral being
>
> [ll. 110–11].

At the same time, he recognizes that in perception the senses are not passive, but can themselves "half-create." That is, the mind may have already modified or endowed the impressions it receives, and which it interprets as coming solely from without. Following Coleridge, Wordsworth soon begins to call this active and creative process "imagination."

Five years have past; five summers, with the length
Of five long winters! and again I hear
These waters, rolling from their mountain-springs
With a soft inland murmur.—Once again
Do I behold these steep and lofty cliffs,
That on a wild secluded scene impress
Thoughts of more deep seclusion; and connect
The landscape with the quiet of the sky.
The day is come when I again repose

Here, under this dark sycamore, and view   10
These plots of cottage-ground, these orchard-tufts,
Which at this season, with their unripe fruits,
Are clad in one green hue, and lose themselves
'Mid groves and copses. Once again I see
These hedge-rows, hardly hedge-rows, little lines
Of sportive wood run wild: these pastoral farms,
Green to the very door; and wreaths of smoke
Sent up, in silence, from among the trees!
With some uncertain notice, as might seem
Of vagrant dwellers in the houseless woods,   20
Or of some Hermit's cave, where by his fire
The Hermit sits alone.
                These beauteous forms,
Through a long absence, have not been to me
As is a landscape to a blind man's eye:
But oft, in lonely rooms, and 'mid the din
Of towns and cities, I have owed to them
In hours of weariness, sensations sweet,
Felt in the blood, and felt along the heart;
And passing even into my purer mind,
With tranquil restoration:—feelings too   30
Of unremembered pleasure: such, perhaps,
As have no slight or trivial influence
On that best portion of a good man's life,
His little, nameless, unremembered, acts
Of kindness and of love. Nor less, I trust,
To them I may have owed another gift,
Of aspect more sublime; that blessed mood,
In which the burthen of the mystery,
In which the heavy and the weary weight
Of all this unintelligible world,   40
Is lightened:—that serene and blessed mood,
In which the affections gently lead us on,—
Until, the breath of this corporeal frame
And even the motion of our human blood
Almost suspended, we are laid asleep
In body, and become a living soul:
While with an eye made quiet by the power
Of harmony, and the deep power of joy,
We see into the life of things.
                 If this
Be but a vain belief, yet, oh! how oft—   50
In darkness and amid the many shapes
Of joyless daylight; when the fretful stir
Unprofitable, and the fever of the world,
Have hung upon the beatings of my heart—
How oft, in spirit, have I turned to thee,
O sylvan Wye! thou wanderer thro' the woods,
How often has my spirit turned to thee!

And now, with gleams of half-extinguished thought
With many recognitions dim and faint,
And somewhat of a sad perplexity,   60
The picture of the mind revives again:
While here I stand, not only with the sense
Of present pleasure, but with pleasing thoughts
That in this moment there is life and food
For future years. And so I dare to hope,
Though changed, no doubt, from what I was when first
I came among these hills; when like a roe
I bounded o'er the mountains, by the sides
Of the deep rivers, and the lonely streams,
Wherever nature led: more like a man   70
Flying from something that he dreads, than one
Who sought the thing he loved. For nature then
(The coarser pleasures of my boyish days,
And their glad animal movements all gone by)
To me was all in all.—I cannot paint
What then I was. The sounding cataract
Haunted me like a passion: the tall rock,
The mountain, and the deep and gloomy wood,
Their colours and their forms, were then to me
An appetite; a feeling and a love,   80
That had no need of a remoter charm,
By thought supplied, nor any interest
Unborrowed from the eye.—That time is past,
And all its aching joys are now no more,
And all its dizzy raptures. Not for this
Faint I, nor mourn nor murmur; other gifts
Have followed; for such loss, I would believe,
Abundant recompense. For I have learned
To look on nature, not as in the hour
Of thoughtless youth; but hearing oftentimes   90
The still, sad music of humanity,
Nor harsh nor grating, though of ample power
To chasten and subdue. And I have felt
A presence that disturbs me with the joy
Of elevated thoughts; a sense sublime
Of something far more deeply interfused,
Whose dwelling is the light of setting suns,
And the round ocean and the living air,
And the blue sky, and in the mind of man:
A motion and a spirit, that impels   100
All thinking things, all objects of all thought,
And rolls through all things. Therefore am I still
A lover of the meadows and the woods,
And mountains; and of all that we behold
From this green earth; of all the mighty world
Of eye, and ear,—both what they half create,
And what perceive; well pleased to recognise

In nature and the language of the sense
The anchor of my purest thoughts, the nurse,
The guide, the guardian of my heart, and soul   110
Of all my moral being.

              Nor perchance,
If I were not thus taught, should I the more
Suffer my genial spirits to decay:
For thou art with me here upon the banks
Of this fair river; thou my dearest Friend,
My dear, dear Friend; and in thy voice I catch
The language of my former heart, and read
My former pleasures in the shooting lights
Of thy wild eyes. Oh! yet a little while
May I behold in thee what I was once,
My dear, dear Sister! and this prayer I make,   120
Knowing that Nature never did betray
The heart that loved her; 'tis her privilege,
Through all the years of this our life, to lead
From joy to joy: for she can so inform
The mind that is within us, so impress
With quietness and beauty, and so feed
With lofty thoughts, that neither evil tongues,
Rash judgments, nor the sneers of selfish men,
Nor greetings where no kindness is, nor all   130
The dreary intercourse of daily life,
Shall e'er prevail against us, or disturb
Our cheerful faith, that all which we behold
Is full of blessings. Therefore let the moon
Shine on thee in thy solitary walk;
And let the misty mountain-winds be free
To blow against thee: and, in after years,
When these wild ecstasies shall be matured
Into a sober pleasure; when thy mind
Shall be a mansion for all lovely forms,   140
Thy memory be as a dwelling-place
For all sweet sounds and harmonies; oh! then,
If solitude, or fear, or pain, or grief,
Should be thy portion, with what healing thoughts
Of tender joy wilt thou remember me,
And these my exhortations! Nor, perchance—
If I should be where I no more can hear
Thy voice, nor catch from thy wild eyes these gleams
Of past existence—wilt thou then forget
That on the banks of this delightful stream   150
We stood together; and that I, so long
A worshipper of Nature, hither came
Unwearied in that service: rather say
With warmer love—oh! with far deeper zeal

Of holier love. Nor wilt thou then forget
That after many wanderings, many years
Of absence, these steep woods and lofty cliffs,
And this green pastoral landscape, were to me
More dear, both for themselves and for thy sake!

                         1798 (1798)

THERE WAS A BOY

Both this and the following poem were written during a trip to Germany (fall to spring, 1798–99) together with much of *Prelude* I. "There Was a Boy" finally found its way into *The Prelude* (V. 364–97). "Nutting" did not. Both poems have affinities with the memorable episodes in *The Prelude*, such as I. 306–39, 357–400, 425–63. These poems and episodes begin with the child too intensely excited to be aware of the vast, natural scene in which he moves. But with a turn of feeling or a start of surprise the natural world or some aspect of or in it is strangely impressed on his consciousness. These happenings are presented as moments of vision or revelation, but the vision is of nature—here, its vastness, calm, and infinite depth. Of lines 24–25 Coleridge remarked in a letter, "I should have recognized them anywhere; and had I met these lines running wild in the deserts of Arabia, I should have instantly screamed out, 'Wordsworth!'"

There was a Boy; ye knew him well, ye cliffs
And islands of Winander!—many a time,
At evening, when the earliest stars began
To move along the edges of the hills,
Rising or setting, would he stand alone,
Beneath the trees, or by the glimmering lake;
And there, with fingers interwoven, both hands
Pressed closely palm to palm and to his mouth
Uplifted, he, as through an instrument,
Blew mimic hootings to the silent owls,   10
That they might answer him.—And they would shout
Across the watery vale, and shout again,
Responsive to his call,—with quivering peals,
And long halloos, and screams, and echoes loud
Redoubled and redoubled; concourse wild
Of jocund din! And, when there came a pause
Of silence such as baffled his best skill:
Then sometimes, in that silence, while he hung
Listening, a gentle shock of mild surprise

TINTERN ABBEY. 113. **genial**: inborn, native.

Has carried far into his heart the voice   20
Of mountain-torrents; or the visible scene
Would enter unawares into his mind
With all its solemn imagery, its rocks,
Its woods, and that uncertain heaven received
Into the bosom of the steady lake.

This boy was taken from his mates, and died
In childhood, ere he was full twelve years old.
Pre-eminent in beauty is the vale
Where he was born and bred: the churchyard hangs
Upon a slope above the village-school;   30
And through that churchyard when my way has led
On summer-evenings, I believe that there
A long half-hour together I have stood
Mute—looking at the grave in which he lies!

1798 (1800)

NUTTING

—————————————It seems a day
(I speak of one from many singled out)
One of those heavenly days that cannot die;
When, in the eagerness of boyish hope,
I left our cottage-threshold, sallying forth
With a huge wallet o'er my shoulders slung,
A nutting-crook in hand; and turned my steps
Tow'rd some far-distant wood, a Figure quaint,
Tricked out in proud disguise of cast-off weeds
Which for that service had been husbanded,   10
By exhortation of my frugal Dame—
Motley accoutrement, of power to smile
At thorns, and brakes, and brambles,—and, in truth,
More ragged than need was! O'er pathless rocks,
Through beds of matted fern, and tangled thickets,
Forcing my way, I came to one dear nook
Unvisited, where not a broken bough
Drooped with its withered leaves, ungracious sign
Of devastation; but the hazels rose
Tall and erect, with tempting clusters hung,   20
A virgin scene!—A little while I stood,
Breathing with such suppression of the heart
As joy delights in; and, with wise restraint
Voluptuous, fearless of a rival, eyed
The banquet;—or beneath the trees I sate
Among the flowers, and with the flowers I played;
A temper known to those who, after long
And weary expectation, have been blest
With sudden happiness beyond all hope.

Perhaps it was a bower beneath whose leaves   30
The violets of five seasons re-appear
And fade, unseen by any human eye;
Where fairy water-breaks do murmur on
For ever; and I saw the sparkling foam,
And—with my cheek on one of those green stones
That, fleeced with moss, under the shady trees,
Lay round me, scattered like a flock of sheep—
I heard the murmur and the murmuring sound,
In that sweet mood when pleasure loves to pay
Tribute to ease; and, of its joy secure,   40
The heart luxuriates with indifferent things,
Wasting its kindliness on stocks and stones,
And on the vacant air. Then up I rose,
And dragged to earth both branch and bough, with
     crash
And merciless ravage: and the shady nook
Of hazels, and the green and mossy bower,
Deformed and sullied, patiently gave up
Their quiet being: and unless I now
Confound my present feelings with the past,
Ere from the mutilated bower I turned   50
Exulting, rich beyond the wealth of kings,
I felt a sense of pain when I beheld
The silent trees, and saw the intruding sky.—
Then, dearest Maiden, move along these shades
In gentleness of heart; with gentle hand
Touch—for there is a spirit in the woods.

1798 (1800)

FROM

THE PRELUDE

OR, GROWTH OF A POET'S MIND

AN AUTOBIOGRAPHICAL POEM

The Prelude was begun in Germany in 1798, and Books I
and II were substantially finished by early 1800. The rest
of the poem was written mainly in 1804–05. Wordsworth
did not publish it, however. He continued to revise it as
long as he lived, and it was brought out after his death by
his nephew, Christopher Wordsworth, in 1850. The
1805 text has been published (in *The Prelude*, ed. Ernest
De Selincourt, second edition revised by Helen Darbishire,
1959), but the 1850 text is given here.

Personal reminiscence or autobiography was common
in the Romantic period. The chief model was Rousseau's
Confessions, and many of the familiar essays of Lamb and

Hazlitt are in a similar vein. Coleridge's *Biographia Literaria* was conceived (and subtitled) as *Biographical Sketches of My Literary Life and Opinions*. These works tend to be varied and unpredictable in their winding flow; they make their effects by rapid shifts of tone and style, by intense emotion directly expressed, by vividness of recollection and nostalgic yearning, and by the idiosyncrasy and charm of the speaking voice.

Wordsworth's is far from being a literal autobiography. He naturally omits or transposes much that happened to him. More important, the poem is subtitled "Growth of a Poet's Mind," which means that it describes a peculiarly favored and fortunate development, a history of how the mind can preserve and foster its highest potentialities, and of what it then becomes. As such the poem is the central statement of the philosophy Wordsworth held during the years of his greatest achievement.

BOOK FIRST

INTRODUCTION—CHILDHOOD AND SCHOOL-TIME

O there is blessing in this gentle breeze,
A visitant that while it fans my cheek
Doth seem half-conscious of the joy it brings
From the green fields, and from yon azure sky.
Whate'er its mission, the soft breeze can come
To none more grateful than to me; escaped
From the vast city, where I long had pined
A discontented sojourner: now free,

Free as a bird to settle where I will.
What dwelling shall receive me? in what vale   10
Shall be my harbour? underneath what grove
Shall I take up my home? and what clear stream
Shall with its murmur lull me into rest?
The earth is all before me. With a heart
Joyous, nor scared at its own liberty,
I look about; and should the chosen guide
Be nothing better than a wandering cloud,
I cannot miss my way. I breathe again!
Trances of thought and mountings of the mind
Come fast upon me: it is shaken off,   20
That burthen of my own unnatural self,
The heavy weight of many a weary day
Not mine, and such as were not made for me.
Long months of peace (if such bold word accord
With any promises of human life),
Long months of ease and undisturbed delight
Are mine in prospect; whither shall I turn,
By road or pathway, or through trackless field,
Up hill or down, or shall some floating thing
Upon the river point me out my course?   30

  Dear Liberty! Yet what would it avail
But for a gift that consecrates the joy?
For I, methought, while the sweet breath of heaven
Was blowing on my body, felt within
A correspondent breeze, that gently moved
With quickening virtue, but is now become
A tempest, a redundant energy,
Vexing its own creation. Thanks to both,
And their congenial powers, that, while they join
In breaking up a long-continued frost,   40
Bring with them vernal promises, the hope
Of active days urged on by flying hours,—
Days of sweet leisure, taxed with patient thought
Abstruse, nor wanting punctual service high,
Matins and vespers of harmonious verse!

  Thus far, O Friend! did I, not used to make
A present joy the matter of a song,
Pour forth that day my soul in measured strains
That would not be forgotten, and are here
Recorded: to the open fields I told   50
A prophecy: poetic numbers came
Spontaneously to clothe in priestly robe

THE PRELUDE. BOOK I. **1–45.** Lines 1–45, Wordsworth's "glad preamble" to *The Prelude*, were composed extempore (though later revised) in September, 1795, as he walked to Racedown, Dorset, where he was to stay with his sister until June, 1797. He had been sharing lodgings in London with a friend, and had just received the first installment of a small legacy which promised financial independence. The joy he feels arises from the shock of sudden freedom, and also from the unexpected chance to devote himself wholly to a great work of poetry, of which the present release of creative power is a hopeful augury. *The Prelude* thus begins with the sense of large anticipation that one finds at the start of the classical epic, except that Wordsworth characteristically transposes it into autobiography. The "breeze" seems "half-conscious," suggesting the divine life or being that Wordsworth intuited in nature, and quickens a "correspondent breeze" of creative inspiration within the poet. The symbolism allows Wordsworth to assert the divine inspiration traditionally claimed by epic poets and biblical prophets, and yet to assert it within the context of his own philosophy: as the breeze moves both within and without, the divine life animates both the human soul and the outer world of natural things.

46. Friend: Coleridge, to whom the entire poem is addressed. He was in Malta during most of the years *The Prelude* was composed. Wordsworth read the poem to him on January 7, 1807. For the impression made on Coleridge, see "To William Wordsworth," below.

A renovated spirit singled out,
Such hope was mine, for holy services.
My own voice cheered me, and, far more, the mind's
Internal echo of the imperfect sound;
To both I listened, drawing from them both
A cheerful confidence in things to come.

Content and not unwilling now to give
A respite to this passion, I paced on   60
With brisk and eager steps; and came, at length,
To a green shady place, where down I sate
Beneath a tree, slackening my thoughts by choice,
And settling into gentler happiness.
T'was autumn, and a clear and placid day,
With warmth, as much as needed, from a sun
Two hours declined towards the west; a day
With silver clouds, and sunshine on the grass,
And in the sheltered and the sheltering grove
A perfect stillness. Many were the thoughts   70
Encouraged and dismissed, till choice was made
Of a known Vale, whither my feet should turn,
Nor rest till they had reached the very door
Of the one cottage which methought I saw.
No picture of mere memory ever looked
So fair; and while upon the fancied scene
I gazed with growing love, a higher power
Than Fancy gave assurance of some work
Of glory there forthwith to be begun,
Perhaps too there performed. Thus long I mused,   80
Nor e'er lost sight of what I mused upon,
Save when, amid the stately grove of oaks,
Now here, now there, an acorn, from its cup
Dislodged, through sere leaves rustled, or at once
To the bare earth dropped with a startling sound.
From that soft couch I rose not, till the sun
Had almost touched the horizon; casting then
A backward glance upon the curling cloud
Of city smoke, by distance ruralised;
Keen as a Truant or a Fugitive,   90
But as a Pilgrim resolute, I took,
Even with the chance equipment of that hour,
The road that pointed toward the chosen Vale.
It was a splendid evening, and my soul
Once more made trial of her strength, nor lacked
Æolian visitations; but the harp
Was soon defrauded, and the banded host

Of harmony dispersed in straggling sounds,
And lastly utter silence! "Be it so;
Why think of anything but present good?"   100
So, like a home-bound labourer I pursued
My way beneath the mellowing sun, that shed
Mild influence; nor left in me one wish
Again to bend the Sabbath of that time
To a servile yoke. What need of many words?
A pleasant loitering journey, through three days
Continued, brought me to my hermitage.
I spare to tell of what ensued, the life
In common things—the endless store of things,
Rare, or at least so seeming, every day   110
Found all about me in one neighbourhood—
The self-congratulation, and, from morn
To night, unbroken cheerfulness serene.
But speedily an earnest longing rose
To brace myself to some determined aim,
Reading or thinking; either to lay up
New stores, or rescue from decay the old
By timely interference: and therewith
Came hopes still higher, that with outward life
I might endue some airy phantasies   120
That had been floating loose about for years,
And to such beings temperately deal forth
The many feelings that oppressed my heart.
That hope hath been discouraged; welcome light
Dawns from the east, but dawns to disappear
And mock me with a sky that ripens not
Into a steady morning: if my mind,
Remembering the bold promise of the past,
Would gladly grapple with some noble theme,
Vain is her wish; where'er she turns she finds   130
Impediments from day to day renewed.

And now it would content me to yield up
Those lofty hopes awhile, for present gifts
Of humbler industry. But, oh, dear Friend!
The Poet, gentle creature as he is,
Hath, like the Lover, his unruly times;
His fits when he is neither sick nor well,
Though no distress be near him but his own
Unmanageable thoughts: his mind, best pleased
While she as duteous as the mother dove   140
Sits brooding, lives not always to that end,

---

96. **Æolian . . . harp:** An Aeolian harp is a stringed instrument that produces musical sounds when touched by a current of air. Cf. Coleridge's "The Eolian Harp" and "Dejection: An Ode," below.

122. **temperately:** with control, without excess passion in the act of composition. 133–34. **gifts . . . industry:** less ambitious poems. 140–41. **dove . . . brooding:** Cf. *Paradise Lost,* I. 21: the creative power of God "Dove-like satst brooding on the vast Abyss."

But like the innocent bird, hath goadings on
That drive her as in trouble through the groves;
With me is now such passion, to be blamed
No otherwise than as it lasts too long.

When, as becomes a man who would prepare
For such an arduous work, I through myself
Make rigorous inquisition, the report
Is often cheering; for I neither seem
To lack that first great gift, the vital soul,   150
Nor general Truths, which are themselves a sort
Of Elements and Agents, Under-powers,
Subordinate helpers of the living mind:
Nor am I naked of external things,
Forms, images, nor numerous other aids
Of less regard, though won perhaps with toil
And needful to build up a Poet's praise.
Time, place, and manners do I seek, and these
Are found in plenteous store, but nowhere such
As may be singled out with steady choice;   160
No little band of yet remembered names
Whom I, in perfect confidence, might hope
To summon back from lonesome banishment,
And make them dwellers in the hearts of men
Now living, or to live in future years.
Sometimes the ambitious Power of choice, mistaking
Proud spring-tide swellings for a regular sea,
Will settle on some British theme, some old
Romantic tale by Milton left unsung;
More often turning to some gentle place   170
Within the groves of Chivalry, I pipe
To shepherd swains, or seated harp in hand,
Amid reposing knights by a river side
Or fountain, listen to the grave reports
Of dire enchantments faced and overcome
By the strong mind, and tales of warlike feats,
Where spear encountered spear, and sword with sword
Fought, as if conscious of the blazonry
That the shield bore, so glorious was the strife;
Whence inspiration for a song that winds   180
Through ever-changing scenes of votive quest
Wrongs to redress, harmonious tribute paid
To patient courage and unblemished truth,
To firm devotion, zeal unquenchable,
And Christian meekness hallowing faithful loves.
Sometimes, more sternly moved, I would relate

How vanquished Mithridates northward passed,
And, hidden in the cloud of years, became
Odin, the Father of a race by whom
Perished the Roman Empire: how the friends   190
And followers of Sertorius, out of Spain
Flying, found shelter in the Fortunate Isles,
And left their usages, their arts and laws,
To disappear by a slow gradual death,
To dwindle and to perish one by one,
Starved in those narrow bounds: but not the soul
Of Liberty, which fifteen hundred years
Survived, and, when the European came
With skill and power that might not be withstood,
Did, like a pestilence, maintain its hold   200
And wasted down by glorious death that race
Of natural heroes: or I would record
How, in tyrannic times, some high-souled man,
Unnamed among the chronicles of kings,
Suffered in silence for Truth's sake: or tell,
How that one Frenchman, through continued force
Of meditation on the inhuman deeds
Of those who conquered first the Indian Isles,
Went single in his ministry across
The Ocean; not to comfort the oppressed,   210
But, like a thirsty wind, to roam about
Withering the Oppressor: how Gustavus sought
Help at his need in Dalecarlia's mines:
How Wallace fought for Scotland; left the name

187. Mithridates: King of Pontus, an ancient country on the southern coast of the Black Sea; defeated by the Roman general Pompey in 66 B.C. **189. Odin:** according to legend, a barbarian chief in south Russia who marched with his tribe north into Sweden when threatened by the Roman armies. There he founded the Gothic people who later conquered Rome. Gibbon remarked that the story of Odin "might supply the noble groundwork of an Epic poem." **191. Sertorius:** a Roman general who fought against the Senatorial armies for eight years, until assassinated in 72 B.C. Like Mithridates and Odin, he could be regarded as a defender of liberty. Wordsworth adds that the followers of Sertorius fled to the Fortunate Isles (the Canary Islands), where they kept alive the "soul of Liberty" for fifteen hundred years—that is, until 1493 when the Spaniards invaded Teneriffe and met a heroic resistance, which was finally overcome. **206. Frenchman:** Dominique de Gourges, who went in 1567 to Florida and took vengeance upon the Spaniards for a massacre there of French Protestants. **212. Gustavus:** Gustavus I of Sweden, 1496-1560, liberated his country from Danish rule. During the war he sought refuge in Dalecarlia, a mining district in Sweden. **214. Wallace:** Sir William Wallace, 1272-1305, led the struggle against the armies of Edward I of England, who attempted to conquer Scotland.

158. manners: distinctive characters or varieties of character to be used in a literary work. Cf. IV. 302.

Of Wallace to be found, like a wild flower,
All over his dear Country; left the deeds
Of Wallace, like a family of Ghosts,
To people the steep rocks and river banks,
Her natural sanctuaries, with a local soul
Of independence and stern liberty.   220
Sometimes it suits me better to invent
A tale from my own heart, more near akin
To my own passions and habitual thoughts;
Some variegated story, in the main
Lofty, but the unsubstantial structure melts
Before the very sun that brightens it,
Mist into air dissolving! Then a wish,
My last and favourite aspiration, mounts
With yearning toward some philosophic song
Of Truth that cherishes our daily life;   230
With meditations passionate from deep
Recesses in man's heart, immortal verse
Thoughtfully fitted to the Orphean lyre;
But from this awful burthen I full soon
Take refuge and beguile myself with trust
That mellower years will bring a riper mind
And clearer insight. Thus my days are past
In contradiction; with no skill to part
Vague longing, haply bred by want of power,
From paramount impulse not to be withstood,   240
A timorous capacity from prudence,
From circumspection, infinite delay.
Humility and modest awe themselves
Betray me, serving often for a cloak
To a more subtle selfishness; that now
Locks every function up in blank reserve,
Now dupes me, trusting to an anxious eye
That with intrusive restlessness beats off
Simplicity and self-presented truth.
Ah! better far than this, to stray about   250
Voluptuously through fields and rural walks,
And ask no record of the hours, resigned
To vacant musing, unreproved neglect
Of all things, and deliberate holiday.
Far better never to have heard the name
Of zeal and just ambition, than to live
Baffled and plagued by a mind that every hour
Turns recreant to her task; takes heart again,
Then feels immediately some hollow thought
Hang like an interdict upon her hopes.   260
This is my lot; for either still I find
Some imperfection in the chosen theme,
Or see of absolute accomplishment
Much wanting, so much wanting, in myself,

That I recoil and droop, and seek repose
In listlessness from vain perplexity,
Unprofitably travelling toward the grave,
Like a false steward who hath much received
And renders nothing back.
                  Was it for this
That one, the fairest of all rivers, loved   270
To blend his murmurs with my nurse's song,
And, from his alder shades and rocky falls,
And from his fords and shallows, sent a voice
That flowed along my dreams? For this, didst thou,
O Derwent! winding among grassy holms
Where I was looking on, a babe in arms,
Make ceaseless music that composed my thoughts
To more than infant softness, giving me
Amid the fretful dwellings of mankind
A foretaste, a dim earnest, of the calm   280
That Nature breathes among the hills and groves.
   When he had left the mountains and received
On his smooth breast the shadow of those towers
That yet survive, a shattered monument
Of feudal sway, the bright blue river passed
Along the margin of our terrace walk;
A tempting playmate whom we dearly loved.
Oh, many a time have I, a five years' child,
In a small mill-race severed from his stream,
Made one long bathing of a summer's day;   290
Basked in the sun, and plunged and basked again
Alternate, all a summer's day, or scoured
The sandy fields, leaping through flowery groves
Of yellow ragwort; or, when rock and hill,
The woods, and distant Skiddaw's lofty height,
Were bronzed with deepest radiance, stood alone
Beneath the sky, as if I had been born
On Indian plains, and from my mother's hut
Had run abroad in wantonness, to sport
A naked savage, in the thunder shower.   300

   Fair seed-time had my soul, and I grew up
Fostered alike by beauty and by fear:

275. **Derwent:** the river that flows close behind the house
in Cockermouth where Wordsworth was born. **holms:**
low, flat lands by a river. 283. **towers:** the ruins of
Cockermouth Castle. "At the end of the garden of my
father's house at Cockermouth was a high terrace that commanded a fine view of the river Derwent and Cockermouth
Castle. This was our favorite playground" (W. W.'s note
to "The Sparrow's Nest"). 295. **Skiddaw:** Cockermouth
lies in open, relatively flat country. The peaks and massive
shoulders of Mount Skiddaw are plainly visible about nine
miles to the east.

Much favoured in my birth-place, and no less
In that beloved Vale to which erelong
We were transplanted—there were we let loose
For sports of wider range. Ere I had told
Ten birth-days, when among the mountain slopes
Frost, and the breath of frosty wind, had snapped
The last autumnal crocus, 'twas my joy
With store of springes o'er my shoulder hung   310
To range the open heights where woodcocks run
Along the smooth green turf. Through half the night,
Scudding away from snare to snare, I plied
That anxious visitation;—moon and stars
Were shining o'er my head. I was alone,
And seemed to be a trouble to the peace
That dwelt among them. Sometimes it befell
In these night wanderings, that a strong desire
O'erpowered my better reason, and the bird
Which was the captive of another's toil   320
Became my prey; and when the deed was done
I heard among the solitary hills
Low breathings coming after me, and sounds
Of undistinguishable motion, steps
Almost as silent as the turf they trod.

Nor less, when spring had warmed the cultured
  Vale,
Roved we as plunderers where the mother-bird
Had in high places built her lodge; though mean
Our object and inglorious, yet the end
Was not ignoble. Oh! when I have hung   330
Above the raven's nest, by knots of grass
And half-inch fissures in the slippery rock
But ill sustained, and almost (so it seemed)
Suspended by the blast that blew amain,
Shouldering the naked crag, oh, at that time
While on the perilous ridge I hung alone,
With what strange utterance did the loud dry wind
Blow through my ear! the sky seemed not a sky
Of earth—and with what motion moved the clouds!

Dust as we are, the immortal spirit grows   340
Like harmony in music; there is a dark
Inscrutable workmanship that reconciles
Discordant elements, makes them cling together
In one society. How strange that all
The terrors, pains, and early miseries,
Regrets, vexations, lassitudes interfused

Within my mind, should e'er have borne a part,
And that a needful part, in making up
The calm existence that is mine when I
Am worthy of myself! Praise to the end!   350
Thanks to the means which Nature deigned to employ;
Whether her fearless visiting, or those
That came with soft alarm, like hurtless light
Opening the peaceful clouds; or she may use
Severer interventions, ministry
More palpable, as best might suit her aim.

One summer evening (led by her) I found
A little boat tied to a willow tree
Within a rocky cave, its usual home.
Straight I unloosed her chain, and stepping in   360
Pushed from the shore. It was an act of stealth
And troubled pleasure, nor without the voice
Of mountain-echoes did my boat move on;
Leaving behind her still, on either side,
Small circles glittering idly in the moon,
Until they melted all into one track
Of sparkling light. But now, like one who rows,
Proud of his skill, to reach a chosen point
With an unswerving line, I fixed my view
Upon the summit of a craggy ridge,   370
The horizon's utmost boundary; for above
Was nothing but the stars and the grey sky.
She was an elfin pinnace; lustily
I dipped my oars into the silent lake,
And, as I rose upon the stroke, my boat
Went heaving through the water like a swan;
When, from behind that craggy steep till then
The horizon's bound, a huge peak, black and huge,
As if with voluntary power instinct
Upreared its head. I struck and struck again,   380
And growing still in stature the grim shape
Towered up between me and the stars, and still,
For so it seemed, with purpose of its own
And measured motion like a living thing,
Strode after me. With trembling oars I turned,
And through the silent water stole my way
Back to the covert of the willow tree;
There in her mooring-place I left my bark,—
And through the meadows homeward went, in grave

378–80. peak . . . Upreared: Wordsworth faces the shore
as he rows. While he is still close to shore, a "craggy steep"
hides the higher summits behind. But as he moves out into
the lake, the changing angle of vision allows him to see over
the "craggy steep." A "huge peak" seems suddenly to rise
up, and the farther he rows the larger it appears.

304. Vale: Esthwaite. At the head of the lake is the town of
Hawkshead where Wordsworth attended school, boarding
with the kindly matron Ann Tyson. **310. springes:** snares.

And serious mood; but after I had seen   390
That spectacle, for many days, my brain
Worked with a dim and undetermined sense
Of unknown modes of being; o'er my thoughts
There hung a darkness, call it solitude
Or blank desertion. No familiar shapes
Remained, no pleasant images of trees,
Of sea or sky, no colours of green fields;
But huge and mighty forms, that do not live
Like living men, moved slowly through the mind
By day, and were a trouble to my dreams.   400

   Wisdom and Spirit of the universe!
Thou Soul that art the eternity of thought,
That givest to forms and images a breath
And everlasting motion, not in vain
By day or star-light thus from my first dawn
Of childhood didst thou intertwine for me
The passions that build up our human soul;
Not with the mean and vulgar works of man,
But with high objects, with enduring things—
With life and nature—purifying thus   410
The elements of feeling and of thought,
And sanctifying, by such discipline,
Both pain and fear, until we recognise
A grandeur in the beatings of the heart.
Nor was this fellowship vouchsafed to me
With stinted kindness. In November days,
When vapours rolling down the valley made
A lonely scene more lonesome, among woods
At noon, and 'mid the calm of summer nights
When, by the margin of the trembling lake,   420
Beneath the gloomy hills homeward I went
In solitude, such intercourse was mine;
Mine was it in the fields both day and night,
And by the waters, all the summer long.

   And in the frosty season, when the sun
Was set, and visible for many a mile
The cottage windows blazed through twilight gloom,
I heeded not their summons: happy time
It was indeed for all of us—for me
It was a time of rapture! Clear and loud   430
The village clock tolled six,—I wheeled about,
Proud and exulting like an untired horse
That cares not for his home. All shod with steel,
We hissed along the polished ice in games
Confederate, imitative of the chase
And woodland pleasures,—the resounding horn,
The pack loud chiming, and the hunted hare.
So through the darkness and the cold we flew,

And not a voice was idle; with the din
Smitten, the precipices rang aloud;   440
The leafless trees and every icy crag
Tinkled like iron; while far distant hills
Into the tumult sent an alien sound
Of melancholy not unnoticed, while the stars
Eastward were sparkling clear, and in the west
The orange sky of evening died away.
Not seldom from the uproar I retired
Into a silent bay, or sportively
Glanced sideway, leaving the tumultuous throng,
To cut across the reflex of a star   450
That fled, and, flying still before me, gleamed
Upon the glassy plain; and oftentimes,
When we had given our bodies to the wind,
And all the shadowy banks on either side
Came sweeping through the darkness, spinning still
The rapid line of motion, then at once
Have I, reclining back upon my heels,
Stopped short; yet still the solitary cliffs
Wheeled by me—even as if the earth had rolled
With visible motion her diurnal round!   460
Behind me did they stretch in solemn train,
Feebler and feebler, and I stood and watched
Till all was tranquil as a dreamless sleep.

   Ye Presences of Nature in the sky
And on the earth! Ye Visions of the hills!
And Souls of lonely places! can I think
A vulgar hope was yours when ye employed
Such ministry, when ye, through many a year
Haunting me thus among my boyish sports,
On caves and trees, upon the woods and hills,   470
Impressed upon all forms the characters
Of danger or desire; and thus did make
The surface of the universal earth
With triumph and delight, with hope and fear,
Work like a sea?
                Not uselessly employed,
Might I pursue this theme through every change
Of exercise and play, to which the year
Did summon us in his delightful round.

   We were a noisy crew; the sun in heaven
Beheld not vales more beautiful than ours;   480
Nor saw a band in happiness and joy
Richer, or worthier of the ground they trod.
I could record with no reluctant voice
The woods of autumn, and their hazel bowers

450. reflex: reflection.

With milk-white clusters hung; the rod and line,
True symbol of hope's foolishness, whose strong
And unreproved enchantment led us on
By rocks and pools shut out from every star,
All the green summer, to forlorn cascades
Among the windings hid of mountain brooks.   490
—Unfading recollections! at this hour
The heart is almost mine with which I felt,
From some hill-top on sunny afternoons,
The paper kite high among fleecy clouds
Pull at her rein like an impetuous courser;
Or, from the meadows sent on gusty days,
Beheld her breast the wind, then suddenly
Dashed headlong, and rejected by the storm.

Ye lowly cottages wherein we dwelt,
A ministration of your own was yours;   500
Can I forget you, being as you were
So beautiful among the pleasant fields
In which ye stood? or can I here forget
The plain and seemly countenance with which
Ye dealt out your plain comforts? Yet had ye
Delights and exultations of your own.
Eager and never weary we pursued
Our home-amusements by the warm peat-fire
At evening, when with pencil, and smooth slate
In square divisions parcelled out and all   510
With crosses and with cyphers scribbled o'er,
We schemed and puzzled, head opposed to head
In strife too humble to be named in verse:
Or round the naked table, snow-white deal,
Cherry or maple, sate in close array,
And to the combat, Loo or Whist, led on
A thick-ribbed army; not, as in the world,
Neglected and ungratefully thrown by
Even for the very service they had wrought,
But husbanded through many a long campaign.   520
Uncouth assemblage was it, where no few
Had changed their functions: some, plebeian cards
Which Fate, beyond the promise of their birth
Had dignified, and called to represent
The persons of departed potentates.
Oh, with what echoes on the board they fell!
Ironic diamonds,—clubs, hearts, diamonds, spades,
A congregation piteously akin!
Cheap matter offered they to boyish wit,

Those sooty knaves, precipitated down   530
With scoffs and taunts, like Vulcan out of heaven:
The paramount ace, a moon in her eclipse,
Queens gleaming through their splendour's last decay,
And monarchs surly at the wrongs sustained
By royal visages. Meanwhile abroad
Incessant rain was falling, or the frost
Raged bitterly, with keen and silent tooth;
And, interrupting oft that eager game,
From under Esthwaite's splitting fields of ice
The pent-up air, struggling to free itself,   540
Gave out to meadow grounds and hills a loud
Protracted yelling, like the noise of wolves
Howling in troops along the Bothnic Main.

Nor, sedulous as I have been to trace
How Nature by extrinsic passion first
Peopled the mind with forms sublime or fair,
And made me love them, may I here omit
How other pleasures have been mine, and joys
Of subtler origin; how I have felt,
Not seldom even in that tempestuous time,   550
Those hallowed and pure motions of the sense
Which seem, in their simplicity, to own
An intellectual charm; that calm delight
Which, if I err not, surely must belong
To those first-born affinities that fit
Our new existence to existing things,
And, in our dawn of being, constitute
The bond of union between life and joy.

Yes, I remember when the changeful earth,
And twice five summers on my mind had stamped
The faces of the moving year, even then   561
I held unconscious intercourse with beauty
Old as creation, drinking in a pure
Organic pleasure from the silver wreaths
Of curling mist, or from the level plain
Of waters coloured by impending clouds.

---

510. square divisions: They were playing tic-tac-toe.
514–35. Or . . . visages: one of the many passages in
The Prelude of rather heavy, mock-heroic humor. **514. deal:**
plank or board of fir or pine.

530. knaves: jacks. **531. Vulcan:** god of fire, thrown from
Olympus ("heaven") by Zeus. **543. Bothnic Main:** at the
northern end of the Baltic Sea. **545. extrinsic passion:** a
passion directed not to nature itself, but to outdoor sports
and activities—trapping, boating, skating—in the course of
which natural objects and forms were unforgettably impressed
on his mind. Cf. ll. 581–612. **555–58. first-born . . . joy:** In
infancy we begin to discover the power of the natural world
to give us pleasure, to shape, express, and answer to our
instincts and feelings. Cf. "Prospectus" to *The Recluse*, ll.
62–71. Thus from our "dawn of infancy" we sense the truth,
which Wordsworth deeply believed, that existence is
fundamentally joy.

The sands of Westmoreland, the creeks and bays
Of Cumbria's rocky limits, they can tell
How, when the Sea threw off his evening shade,
And to the shepherd's hut on distant hills   570
Sent welcome notice of the rising moon,
How I have stood, to fancies such as these
A stranger, linking with the spectacle
No conscious memory of a kindred sight,
And bringing with me no peculiar sense
Of quietness or peace; yet have I stood,
Even while mine eye hath moved o'er many a league
Of shining water, gathering as it seemed,
Through every hair-breadth in that field of light,
New pleasure like a bee among the flowers.   580

    Thus oft amid those fits of vulgar joy
Which, through all seasons, on a child's pursuits
Are prompt attendants, 'mid that giddy bliss
Which, like a tempest, works along the blood
And is forgotten; even then I felt
Gleams like the flashing of a shield;—the earth
And common face of Nature spake to me
Rememberable things; sometimes, 'tis true,
By chance collisions and quaint accidents
(Like those ill-sorted unions, work supposed   590
Of evil-minded fairies), yet not vain
Nor profitless, if haply they impressed
Collateral objects and appearances,
Albeit lifeless then, and doomed to sleep
Until maturer seasons called them forth
To impregnate and to elevate the mind.
—And if the vulgar joy by its own weight
Wearied itself out of the memory,
The scenes which were a witness of that joy
Remained in their substantial lineaments   600
Depicted on the brain, and to the eye
Were visible, a daily sight; and thus
By the impressive discipline of fear,
By pleasure and repeated happiness,
So frequently repeated, and by force
Of obscure feelings representative
Of things forgotten, these same scenes so bright,
So beautiful, so majestic in themselves,
Though yet the day was distant, did become

Habitually dear, and all their forms   610
And changeful colours by invisible links
Were fastened to the affections.
                I began
My story early—not misled, I trust,
By an infirmity of love for days
Disowned by memory—fancying flowers where none
Not even the sweetest do or can survive
For him at least whose dawning day they cheered.
Nor will it seem to thee, O Friend! so prompt
In sympathy, that I have lengthened out
With fond and feeble tongue a tedious tale.   620
Meanwhile, my hope has been, that I might fetch
Invigorating thoughts from former years;
Might fix the wavering balance of my mind,
And haply meet reproaches too, whose power
May spur me on, in manhood now mature,
To honourable toil. Yet should these hopes
Prove vain and thus should neither I be taught
To understand myself, nor thou to know
With better knowledge how the heart was framed
Of him thou lovest; need I dread from thee   630
Harsh judgments, if the song be loth to quit
Those recollected hours that have the charm
Of visionary things, those lovely forms
And sweet sensations that throw back our life,
And almost make remotest infancy
A visible scene, on which the sun is shining?

    One end at least hath been attained; my mind
Hath been revived, and if this genial mood
Desert me not, forthwith shall be brought down
Through later years the story of my life.   640
The road lies plain before me;—'tis a theme
Single and of determined bounds; and hence
I choose it rather at this time, than work
Of ampler or more varied argument,
Where I might be discomfited and lost:
And certain hopes are with me, that to thee
This labour will be welcome, honoured Friend!

BOOK SECOND
SCHOOL-TIME (continued)

Thus far, O Friend! have we, though leaving much
Unvisited, endeavoured to retrace
The simple ways in which my childhood walked;
Those chiefly that first led me to the love

---

572. fancies . . . these: fancies such as those of lines 569–71, where the rising moon is felt to be "welcome." This interpretive activity of mind, like the conscious associating mentioned in line 573, involves something more than a purely sensuous response to the beauty of the natural world.
581. vulgar: commonplace.

Of rivers, woods, and fields. The passion yet
Was in its birth, sustained as might befall
By nourishment that came unsought; for still
From week to week, from month to month, we lived
A round of tumult. Duly were our games
Prolonged in summer till the daylight failed: 10
No chair remained before the doors; the bench
And threshold steps were empty; fast asleep
The labourer, and the old man who had sate
A later lingerer; yet the revelry
Continued and the loud uproar: at last,
When all the ground was dark, and twinkling stars
Edged the black clouds, home and to bed we went,
Feverish with weary joints and beating minds.
Ah! is there one who ever has been young,
Nor needs a warning voice to tame the pride   20
Of intellect and virtue's self-esteem?
One is there, though the wisest and the best
Of all mankind, who covets not at times
Union that cannot be;—who would not give,
If so he might, to duty and to truth
The eagerness of infantine desire?
A tranquillising spirit presses now
On my corporeal frame, so wide appears
The vacancy between me and those days
Which yet have such self-presence in my mind,   30
That, musing on them, often do I seem
Two consciousnesses, conscious of myself
And of some other Being. A rude mass
Of native rock, left midway in the square
Of our small market village, was the goal
Or centre of these sports; and when, returned
After long absence, thither I repaired,
Gone was the old grey stone, and in its place
A smart Assembly-room usurped the ground
That had been ours. There let the fiddle scream,   40
And be ye happy! Yet, my Friends! I know
That more than one of you will think with me
Of those soft starry nights, and that old Dame
From whom the stone was named, who there had sate,
And watched her table with its huckster's wares
Assiduous, through the length of sixty years.

We ran a boisterous course; the year span round
With giddy motion. But the time approached
That brought with it a regular desire
For calmer pleasures, when the winning forms   50
Of Nature were collaterally attached
To every scheme of holiday delight
And every boyish sport, less grateful else

And languidly pursued.
                          When summer came,
Our pastime was, on bright half-holidays,
To sweep along the plain of Windermere
With rival oars; and the selected bourne
Was now an Island musical with birds
That sang and ceased not; now a Sister Isle
Beneath the oaks' umbrageous covert, sown   60
With lilies of the valley like a field;
And now a third small Island, where survived
In solitude the ruins of a shrine
Once to Our Lady dedicate, and served
Daily with chaunted rites. In such a race
So ended, disappointment could be none,
Uneasiness, or pain, or jealousy:
We rested in the shade, all pleased alike,
Conquered and conqueror. Thus the pride of strength,
And the vain-glory of superior skill,   70
Were tempered; thus was gradually produced
A quiet independence of the heart;
And to my Friend who knows me I may add,
Fearless of blame, that hence for future days
Ensued a diffidence and modesty,
And I was taught to feel, perhaps too much,
The self-sufficing power of Solitude.

   Our daily meals were frugal, Sabine fare!
More than we wished we knew the blessing then
Of vigorous hunger—hence corporeal strength   80
Unsapped by delicate viands; for, exclude
A little weekly stipend, and we lived
Through three divisions of the quartered year
In penniless poverty. But now to school
From the half-yearly holidays returned,
We came with weightier purses, that sufficed
To furnish treats more costly than the Dame
Of the old grey stone, from her scant board, supplied.
Hence rustic dinners on the cool green ground,
Or in the woods, or by a river side   90
Or shady fountain, while among the leaves
Soft airs were stirring, and the mid-day sun
Unfelt shone brightly round us in our joy.
Nor is my aim neglected if I tell
How sometimes, in the length of those half-years,
We from our funds drew largely;—proud to curb,
And eager to spur on, the galloping steed;
And with the cautious inn-keeper, whose stud

BOOK II. **78. Sabine:** The Sabines were an ancient people of
austere life who lived in the Apennine Mountains northeast
of Rome.

Supplied our want, we haply might employ
Sly subterfuge, if the adventure's bound   100
Were distant: some famed temple where of yore
The Druids worshipped, or the antique walls
Of that large abbey, where within the Vale
Of Nightshade, to St. Mary's honour built,
Stands yet a mouldering pile with fractured arch,
Belfry, and images, and living trees,
A holy scene! Along the smooth green turf
Our horses grazed. To more than inland peace,
Left by the west wind sweeping overhead
From a tumultuous ocean, trees and towers   110
In that sequestered valley may be seen,
Both silent and both motionless alike;
Such the deep shelter that is there, and such
The safeguard for repose and quietness.

Our steeds remounted and the summons given,
With whip and spur we through the chauntry flew
In uncouth race, and left the cross-legged knight,
And the stone-abbot, and that single wren
Which one day sang so sweetly in the nave
Of the old church, that—though from recent showers
The earth was comfortless, and, touched by faint   121
Internal breezes, sobbings of the place
And respirations, from the roofless walls
The shuddering ivy dripped large drops—yet still
So sweetly 'mid the gloom the invisible bird
Sang to herself, that there I could have made
My dwelling-place, and lived for ever there
To hear such music. Through the walls we flew
And down the valley, and, a circuit made   129
In wantonness of heart, through rough and smooth
We scampered homewards. Oh, ye rocks and streams,
And that still spirit shed from evening air!
Even in this joyous time I sometimes felt
Your presence, when with slackened step we breathed
Along the sides of the steep hills, or when
Lighted by gleams of moonlight from the sea
We beat with thundering hoofs the level sand.

Midway on long Winander's eastern shore,
Within the crescent of a pleasant bay,
A tavern stood; no homely-featured house,   140
Primeval like its neighbouring cottages,
But 't was a splendid place, the door beset
With chaises, grooms, and liveries, and within
Decanters, glasses, and the blood-red wine.

In ancient times, and ere the Hall was built
On the large island, had this dwelling been
More worthy of a poet's love, a hut
Proud of its one bright fire and sycamore shade.
But—though the rhymes were gone that once inscribed
The threshold, and large golden characters,   150
Spread o'er the spangled sign-board, had dislodged
The old Lion and usurped his place, in slight
And mockery of the rustic painter's hand—
Yet, to this hour, the spot to me is dear
With all its foolish pomp. The garden lay
Upon a slope surmounted by a plain
Of a small bowling-green; beneath us stood
A grove, with gleams of water through the trees
And over the tree-tops; nor did we want
Refreshment, strawberries and mellow cream.   160
There, while through half an afternoon we played
On the smooth platform, whether skill prevailed
Or happy blunder triumphed, bursts of glee
Made all the mountains ring. But, ere night-fall,
When in our pinnace we returned at leisure
Over the shadowy lake, and to the beach
Of some small island steered our course with one,
The Minstrel of the Troop, and left him there,
And rowed off gently, while he blew his flute
Alone upon the rock—oh, then, the calm   170
And dead still water lay upon my mind
Even with a weight of pleasure, and the sky,
Never before so beautiful, sank down
Into my heart, and held me like a dream!
Thus were my sympathies enlarged, and thus
Daily the common range of visible things
Grew dear to me: already I began
To love the sun; a boy I loved the sun,
Not as I since have loved him, as a pledge
And surety of our earthly life, a light   180
Which we behold and feel we are alive;
Nor for his bounty to so many worlds—
But for this cause, that I had seen him lay
His beauty on the morning hills, had seen
The western mountain touch his setting orb,
In many a thoughtless hour, when, from excess
Of happiness, my blood appeared to flow
For its own pleasure, and I breathed with joy.
And, from like feelings, humble though intense,
To patriotic and domestic love   190

---

103. abbey: Furness Abbey, in ruins, twenty-one miles from Hawkshead. **116. chauntry:** chapel.

157. bowling-green: a greensward for playing bowls, a game in which balls are rolled as near as possible to a stationary ball.

Analogous, the moon to me was dear;
For I could dream away my purposes,
Standing to gaze upon her while she hung
Midway between the hills, as if she knew
No other region, but belonged to thee,
Yea, appertained by a peculiar right
To thee and thy grey huts, thou one dear Vale!

Those incidental charms which first attached
My heart to rural objects, day by day
Grew weaker, and I hasten on to tell   200
How Nature, intervenient till this time
And secondary, now at length was sought
For her own sake. But who shall parcel out
His intellect by geometric rules,
Split like a province into round and square?
Who knows the individual hour in which
His habits were first sown, even as a seed?
Who that shall point as with a wand and say
"This portion of the river of my mind
Came from yon fountain?" Thou, my Friend!
        art one   210
More deeply read in thy own thoughts; to thee
Science appears but what in truth she is,
Not as our glory and our absolute boast,
But as a succedaneum, and a prop
To our infirmity. No officious slave
Art thou of that false secondary power
By which we multiply distinctions, then
Deem that our puny boundaries are things
That we perceive, and not that we have made.
To thee, unblinded by these formal arts,   220
The unity of all hath been revealed,
And thou wilt doubt, with me less aptly skilled
Than many are to range the faculties
In scale and order, class the cabinet
Of their sensations, and in voluble phrase
Run through the history and birth of each
As of a single independent thing.

201. **intervenient:** coming in as something incidental or extraneous. 214. **succedaneum:** substitute. 223. **faculties:** According to faculty psychology, the mind was an aggregate of separate powers, such as reason, will, desire, instinct, etc., which interacted with each other. Wordsworth and Coleridge protested against this view of the mind and substituted an organic psychology, conceiving the mind as a complex but unified process in which, for example, intellectual and emotional processes take place simultaneously and modify each other. Faculty psychology is itself a product of the type of analysis Wordsworth deplores. 224–25. **class . . . sensations:** classify their sensations as though they were a collection (of shells, for example, or butterflies) in a cabinet.

Hard task, vain hope, to analyse the mind,
If each most obvious and particular thought,
Not in a mystical and idle sense,   230
But in the words of Reason deeply weighed,
Hath no beginning.
                Blest the infant Babe,
(For with my best conjecture I would trace
Our Being's earthly progress,) blest the Babe,
Nursed in his Mother's arms, who sinks to sleep
Rocked on his Mother's breast; who with his soul
Drinks in the feelings of his Mother's eye!
For him, in one dear Presence, there exists
A virtue which irradiates and exalts
Objects through widest intercourse of sense.   240
No outcast he, bewildered and depressed:
Along his infant veins are interfused
The gravitation and the filial bond
Of nature that connect him with the world.
Is there a flower, to which he points with hand
Too weak to gather it, already love
Drawn from love's purest earthly fount for him
Hath beautified that flower; already shades
Of pity cast from inward tenderness
Do fall around him upon aught that bears   250
Unsightly marks of violence or harm.
Emphatically such a Being lives,
Frail creature as he is, helpless as frail,
An inmate of this active universe.
For feeling has to him imparted power
That through the growing faculties of sense
Doth like an agent of the one great Mind
Create, creator and receiver both,
Working but in alliance with the works
Which it beholds.—Such, verily, is the first   260
Poetic spirit of our human life,
By uniform control of after years,
In most, abated or suppressed; in some,
Through every change of growth and of decay,
Pre-eminent till death.
                From early days,
Beginning not long after that first time

255–58. **power . . . both:** Like Coleridge, Wordsworth held that the mind partially creates the world it perceives. That is, in the act of perception the mind both endows its object with qualities and receives impressions from it. This process was described as imaginative or creative, and both Wordsworth and Coleridge wanted to believe that it produced valid and ultimate insight. Cf. "Tintern Abbey," above, ll. 105–07, and Coleridge's *Biographia Literaria*, below, Chap. XIII.

In which, a Babe, by intercourse of touch
I held mute dialogues with my Mother's heart,
I have endeavoured to display the means
Whereby this infant sensibility,   270
Great birthright of our being, was in me
Augmented and sustained. Yet is a path
More difficult before me; and I fear
That in its broken windings we shall need
The chamois' sinews, and the eagle's wing:
For now a trouble came into my mind
From unknown causes. I was left alone
Seeking the visible world, nor knowing why.
The props of my affections were removed,
And yet the building stood, as if sustained   280
By its own spirit! All that I beheld
Was dear, and hence to finer influxes
The mind lay open to a more exact
And close communion. Many are our joys
In youth, but oh! what happiness to live
When every hour brings palpable access
Of knowledge, when all knowledge is delight,
And sorrow is not there! The seasons came,
And every season wheresoe'er I moved
Unfolded transitory qualities,   290
Which, but for this most watchful power of love,
Had been neglected; left a register
Of permanent relations, else unknown.
Hence life, and change, and beauty, solitude
More active even than "best society"—
Society made sweet as solitude
By inward concords, silent, inobtrusive,
And gentle agitations of the mind
From manifold distinctions, difference
Perceived in things, where, to the unwatchful eye, 300
No difference is, and hence, from the same source,
Sublimer joy; for I would walk alone,
Under the quiet stars, and at that time
Have felt whate'er there is of power in sound
To breathe an elevated mood, by form
Or image unprofaned; and I would stand,
If the night blackened with a coming storm,
Beneath some rock, listening to notes that are
The ghostly language of the ancient earth,
Or make their dim abode in distant winds.   310
Thence did I drink the visionary power;
And deem not profitless those fleeting moods
Of shadowy exultation: not for this,
That they are kindred to our purer mind
And intellectual life; but that the soul,
Remembering how she felt, but what she felt

Remembering not, retains an obscure sense
Of possible sublimity, whereto
With growing faculties she doth aspire,
With faculties still growing, feeling still   320
That whatsoever point they gain, they yet
Have something to pursue.
                            And not alone,
'Mid gloom and tumult, but no less 'mid fair
And tranquil scenes, that universal power
And fitness in the latent qualities
And essences of things, by which the mind
Is moved with feelings of delight, to me
Came strengthened with a superadded soul,
A virtue not its own. My morning walks
Were early;—oft before the hours of school   330
I travelled round our little lake, five miles
Of pleasant wandering, Happy time! more dear
For this, that one was by my side, a Friend,
Then passionately loved; with heart how full
Would he peruse these lines! For many years
Have since flowed in between us, and, our minds
Both silent to each other, at this time
We live as if those hours had never been.
Nor seldom did I lift our cottage latch
Far earlier, ere one smoke-wreath had risen   340
From human dwelling, or the thrush, high-perched,
Piped to the woods his shrill reveillé, sate
Alone upon some jutting eminence,
At the first gleam of dawn-light, when the Vale,
Yet slumbering, lay in utter solitude.
How shall I seek the origin? where find
Faith in the marvellous things which then I felt?
Oft in these moments such a holy calm
Would overspread my soul, that bodily eyes
Were utterly forgotten, and what I saw   350
Appeared like something in myself, a dream,
A prospect in the mind.
                            'T were long to tell
What spring and autumn, what the winter snows,
And what the summer shade, what day and night,
Evening and morning, sleep and waking, thought
From sources inexhaustible, poured forth
To feed the spirit of religious love
In which I walked with Nature. But let this
Be not forgotten, that I still retained
My first creative sensibility;   360
That by the regular action of the world

---

330. hours of school: Work began at 6.00 A.M. in summer
and 7.00 A.M. in winter.

My soul was unsubdued. A plastic power
Abode with me; a forming hand, at times
Rebellious, acting in a devious mood;
A local spirit of his own, at war
With general tendency, but, for the most,
Subservient strictly to external things
With which it communed. An auxiliar light
Came from my mind, which on the setting sun
Bestowed new splendour; the melodious birds,   370
The fluttering breezes, fountains that run on
Murmuring so sweetly in themselves, obeyed
A like dominion, and the midnight storm
Grew darker in the presence of my eye:
Hence my obeisance, my devotion hence,
And hence my transport.
               Nor should this, perchance,
Pass unrecorded, that I still had loved
The exercise and produce of a toil,
Than analytic industry to me
More pleasing, and whose character I deem   380
Is more poetic as resembling more
Creative agency. The song would speak
Of that interminable building reared
By observation of affinities
In objects where no brotherhood exists
To passive minds. My seventeenth year was come;
And, whether from this habit rooted now
So deeply in my mind, or from excess
In the great social principle of life
Coercing all things into sympathy,   390
To unorganic natures were transferred
My own enjoyments; or the power of truth
Coming in revelation, did converse
With things that really are; I, at this time,
Saw blessings spread around me like a sea.
Thus while the days flew by, and years passed on,
From Nature overflowing in my soul,
I had received so much, that all my thoughts
Were steeped in feeling; I was only then
Contented, when with bliss ineffable   400
I felt the sentiment of Being spread
O'er all that moves and all that seemeth still;
O'er all that, lost beyond the reach of thought
And human knowledge, to the human eye
Invisible, yet liveth to the heart;
O'er all that leaps and runs, and shouts and sings,
Or beats the gladsome air; o'er all that glides
Beneath the wave, yea, in the wave itself,
And mighty depth of waters. Wonder not
If high the transport, great the joy I felt,   410

Communing in this sort through earth and heaven
With every form of creature, as it looked
Towards the Uncreated with a countenance
Of adoration, with an eye of love.
One song they sang, and it was audible,
Most audible, then, when the fleshly ear,
O'ercome by humblest prelude of that strain,
Forgot her functions, and slept undisturbed.

If this be error, and another faith
Find easier access to the pious mind,   420
Yet were I grossly destitute of all
Those human sentiments that make this earth
So dear, if I should fail with grateful voice
To speak of you, ye mountains, and ye lakes
And sounding cataracts, ye mists and winds
That dwell among the hills where I was born.
If in my youth I have been pure in heart,
If, mingling with the world, I am content
With my own modest pleasures, and have lived
With God and Nature communing, removed   430
From little enmities and low desires,
The gift is yours; if in these times of fear,
This melancholy waste of hopes o'erthrown,
If, 'mid indifference and apathy,
And wicked exultation when good men
On every side fall off, we know not how,
To selfishness, disguised in gentle names
Of peace and quiet and domestic love,
Yet mingled not unwillingly with sneers
On visionary minds; if, in this time   440
Of dereliction and dismay, I yet
Despair not of our nature, but retain
A more than Roman confidence, a faith
That fails not, in all sorrow my support,
The blessing of my life; the gift is yours,
Ye winds and sounding cataracts! 't is yours,
Ye mountains! thine, O Nature! Thou hast fed
My lofty speculations; and in thee,
For this uneasy heart of ours, I find
A never-failing principle of joy   450
And purest passion.
               Thou, my Friend! wert reared
In the great city, 'mid far other scenes;
But we, by different roads, at length have gained

432–41. **times . . . dismay:** 1799, by which time the ardent hope the French Revolution originally excited in Wordsworth, Coleridge, and many other liberals had been disappointed. **452. city:** London, where Coleridge had attended school. Cf. VI. 264–74.

The selfsame bourne. And for this cause to thee
I speak, unapprehensive of contempt,
The insinuated scoff of coward tongues,
And all that silent language which so oft
In conversation between man and man
Blots from the human countenance all trace
Of beauty and of love. For thou hast sought   460
The truth in solitude, and, since the days
That gave thee liberty, full long desired,
To serve in Nature's temple, thou hast been
The most assiduous of her ministers;
In many things my brother, chiefly here
In this our deep devotion.
                              Fare thee well!
Health and the quiet of a healthful mind
Attend thee! seeking oft the haunts of men,
And yet more often living with thyself,
And for thyself, so haply shall thy days   470
Be many, and a blessing to mankind.

from BOOK THIRD

RESIDENCE AT CAMBRIDGE

It was a dreary morning when the wheels
Rolled over a wide plain o'erhung with clouds,
And nothing cheered our way till first we saw
The long-roofed chapel of King's College lift
Turrets and pinnacles in answering files,
Extended high above a dusky grove.

Advancing, we espied upon the road
A student clothed in gown and tasselled cap,
Striding along as if o'ertasked by Time,
Or covetous of exercise and air;   10
He passed—nor was I master of my eyes
Till he was left an arrow's flight behind.
As near and nearer to the spot we drew,
It seemed to suck us in with an eddy's force.
Onward we drove beneath the Castle; caught,
While crossing Magdalene Bridge, a glimpse of Cam;
And at the *Hoop* alighted, famous Inn.

BOOK III. **16. Magdalene:** pronounced "Maudlin." **Cam:**
river that flows through Cambridge. **17. And . . . Inn:**
To Matthew Arnold, this line seemed pompous: Wordsworth
was trying to endow a simple matter with poetic elevation.
Other readers have taken it as mock-heroic, the deliberately
comic Miltonizing familiar in many poems of the eighteenth
century. The line is something of a test case for many passages
in *The Prelude:* do we laugh at Wordsworth or with him?

My spirit was up, my thoughts were full of hope;
Some friends I had, acquaintances who there
Seemed friends, poor simple schoolboys, now hung
    round   20
With honour and importance; in a world
Of welcome faces up and down I roved;
Questions, directions, warnings and advice,
Flowed in upon me from all sides; fresh day
Of pride and pleasure! to myself I seemed
A man of business and expense, and went
From shop to shop about my own affairs,
To Tutor or to Tailor, as befell,
From street to street with loose and careless mind.

I was the Dreamer, they the Dream; I roamed   30
Delighted through the motley spectacle;
Gowns grave, or gaudy, doctors, students, streets,
Courts, cloisters, flocks of churches, gateways, towers:
Migration strange for a stripling of the hills,
A northern villager.
                        As if the change
Had waited on some Fairy's wand, at once
Behold me rich in monies, and attired
In splendid garb, with hose of silk, and hair
Powdered like rimy trees, when frost is keen.
My lordly dressing-gown, I pass it by,   40
With other signs of manhood that supplied
The lack of beard.—The weeks went roundly on,
With invitations, suppers, wine and fruit,
Smooth housekeeping within, and all without
Liberal, and suiting gentleman's array,

The Evangelist St. John my patron was:
Three Gothic courts are his, and in the first
Was my abiding-place, a nook obscure;
Right underneath, the College kitchens made
A humming sound, less tuneable than bees,   50
But hardly less industrious; with shrill notes
Of sharp command and scolding intermixed.
Near me hung Trinity's loquacious clock,
Who never let the quarters, night or day,
Slip by him unproclaimed, and told the hours
Twice over with a male and female voice.
Her pealing organ was my neighbour too;
And from my pillow, looking forth by light
Of moon or favouring stars, I could behold

46. St. John my patron: Wordsworth was a student at
St. John's College from 1787 to 1791. **53. Trinity's . . .
clock:** Trinity College stands next to St. John's.

The antechapel where the statue stood   60
Of Newton with his prism and silent face,
The marble index of a mind for ever
Voyaging through strange seas of Thought, alone.

    Of College labours, of the Lecturer's room
All studded round, as thick as chairs could stand,
With loyal students, faithful to their books,
Half-and-half idlers, hardy recusants,
And honest dunces—of important days,
Examinations, when the man was weighed
As in a balance! of excessive hopes,   70
Trembling withal and commendable fears,
Small jealousies, and triumphs good or bad,
Let others that know more speak as they know.
Such glory was but little sought by me,
And little won. Yet from the first crude days
Of settling time in this untried abode,
I was disturbed at times by prudent thoughts,
Wishing to hope without a hope, some fears
About my future wordly maintenance,
And, more than all, a strangeness in the mind,   80
A feeling that I was not for that hour,
Nor for that place. But wherefore be cast down?
For (not to speak of Reason and her pure
Reflective acts to fix the moral law
Deep in the conscience, nor of Christian Hope,
Bowing her head before her sister Faith
As one far mightier), hither I had come,
Bear witness Truth, endowed with holy powers
And faculties, whether to work or feel.
Oft when the dazzling show no longer new   90
Had ceased to dazzle, ofttimes did I quit
My comrades, leave the crowd, buildings and groves,
And as I paced alone the level fields
Far from those lovely sights and sounds sublime
With which I had been conversant, the mind
Drooped not; but there into herself returning,
With prompt rebound seemed fresh as heretofore.
At least I more distinctly recognised

Her native instincts: let me dare to speak
A higher language, say that now I felt   100
What independent solaces were mine,
To mitigate the injurious sway of place
Or circumstance, how far soever changed
In youth, or *to* be changed in manhood's prime;
Or for the few who shall be called to look
On the long shadows in our evening years,
Ordained precursors to the night of death.
As if awakened, summoned, roused, constrained,
I looked for universal things; perused
The common countenance of earth and sky: 110
Earth, nowhere unembellished by some trace
Of that first Paradise whence man was driven;
And sky, whose beauty and bounty are expressed
By the proud name she bears—the name of Heaven.
I called on both to teach me what they might;
Or, turning the mind in upon herself,
Pored, watched, expected, listened, spread my thoughts
And spread them with a wider creeping; felt
Incumbencies more awful, visitings
Of the Upholder of the tranquil soul,   120
That tolerates the indignities of Time,
And, from the centre of Eternity
All finite motions overruling, lives
In glory immutable. But peace! enough
Here to record I had ascended now
To such community with highest truth—
A track pursuing, not untrod before,
From strict analogies by thought supplied
Or consciousnesses not to be subdued.
To every natural form, rock, fruit or flower,   130
Even the loose stones that cover the highway,
I gave a moral life: I saw them feel,
Or linked them to some feeling: the great mass
Lay bedded in a quickening soul, and all
That I beheld respired with inward meaning.
Add that whate'er of Terror or of Love
Or Beauty, Nature's daily face put on
From transitory passion, unto this
I was as sensitive as waters are
To the sky's influence; in a kindred mood   140
Of passion was obedient as a lute
That waits upon the touches of the wind.
Unknown, unthought of, yet I was most rich—
I had a world about me—'t was my own;
I made it, for it only lived to me,
And to the God who sees into the heart.
Such sympathies, though rarely, were betrayed

60–63. antechapel . . . alone: Roubiliac's famous statue of
Newton stands in the antechapel (a space separated from the
chapel and located at its west end) of Trinity College.
Wordsworth would not have seen the statue itself from his
window, but only the building. The face has something of
the shape of a "prism," and the word, like "index" (sign,
token, indication of) brings to mind geometry, optics, and
mathematics. The great lines 62–63 were a very late (*c.*
1832) addition to the poem.

By outward gestures and by visible looks:
Some called it madness—so indeed it was,
If child-like fruitfulness in passing joy,   150
If steady moods of thoughtfulness matured
To inspiration, sort with such a name;
If prophecy be madness; if things viewed
By poets in old time, and higher up
By the first men, earth's first inhabitants,
May in these tutored days no more be seen
With undisordered sight. But leaving this,
It was no madness, for the bodily eye
Amid my strongest workings evermore
Was searching out the lines of difference   160
As they lie hid in all external forms,
Near or remote, minute or vast; an eye
Which, from a tree, a stone, a withered leaf,
To the broad ocean and the azure heavens
Spangled with kindred multitudes of stars,
Could find no surface where its power might sleep;
Which spake perpetual logic to my soul,
And by an unrelenting agency
Did bind my feelings even as in a chain.

    And here, O Friend! have I retraced my life   170
Up to an eminence, and told a tale
Of matters which not falsely may be called
The glory of my youth. Of genius, power,
Creation and divinity itself
I have been speaking, for my theme has been
What passed within me. Not of outward things
Done visibly for other minds, words, signs,
Symbols or actions, but of my own heart
Have I been speaking, and my youthful mind.
O Heavens! how awful is the might of souls,   180
And what they do within themselves while yet
The yoke of earth is new to them, the world
Nothing but a wild field where they were sown.
This is, in truth, heroic argument,
This genuine prowess, which I wished to touch
With hand however weak, but in the main
It lies far hidden from the reach of words.
Points have we all of us within our souls
Where all stand single; this I feel, and make
Breathings for incommunicable powers;   190
But is not each a memory to himself,
And, therefore, now that we must quit this theme,
I am not heartless, for there's not a man
That lives who hath not known his god-like hours,
And feels not what an empire we inherit
As natural beings in the strength of Nature.

BOOK FOURTH
SUMMER VACATION

Bright was the summer's noon when quickening steps
Followed each other till a dreary moor
Was crossed, a bare ridge clomb, upon whose top
Standing alone, as from a rampart's edge,
I overlooked the bed of Windermere,
Like a vast river, stretching in the sun.
With exultation, at my feet I saw
Lake, islands, promontories, gleaming bays,
A universe of Nature's fairest forms
Proudly revealed with instantaneous burst,   10
Magnificent, and beautiful, and gay.
I bounded down the hill shouting amain
For the old Ferryman; to the shout the rocks
Replied, and when the Charon of the flood
Had staid his oars, and touched the jutting pier,
I did not step into the well-known boat
Without a cordial greeting. Thence with speed
Up the familiar hill I took my way
Towards that sweet Valley where I had been reared;
'Twas but a short hour's walk, ere veering round   20
I saw the snow-white church upon her hill
Sit like a thronèd Lady, sending out
A gracious look all over her domain.
Yon azure smoke betrays the lurking town;
With eager footsteps I advance and reach
The cottage threshold where my journey closed.
Glad welcome had I, with some tears, perhaps,
From my old Dame, so kind and motherly,
While she perused me with a parent's pride.
The thoughts of gratitude shall fall like dew   30
Upon thy grave, good creature! While my heart
Can beat never will I forget thy name.
Heaven's blessing be upon thee where thou liest
After thy innocent and busy stir
In narrow cares, thy little daily growth
Of calm enjoyments, after eighty years,
And more than eighty, of untroubled life,
Childless, yet by the strangers to thy blood
Honoured with little less than filial love.
What joy was mine to see thee once again,   40
Thee and thy dwelling, and a crowd of things
About its narrow precincts all beloved,

BOOK IV. **5–11. overlooked . . . gay:** Cf. Keats's response to the same scene in his letter to Tom Keats, June 26, 1818, below.

And many of them seeming yet my own!
Why should I speak of what a thousand hearts
Have felt, and every man alive can guess?
The rooms, the court, the garden were not left
Long unsaluted, nor the sunny seat
Round the stone table under the dark pine,
Friendly to studious or to festive hours;
Nor that unruly child of mountain birth,   50
The froward brook, who, soon as he was boxed
Within our garden, found himself at once,
As if by trick insidious and unkind,
Stripped of his voice and left to dimple down
(Without an effort and without a will)
A channel paved by man's officious care.
I looked at him and smiled, and smiled again,
And in the press of twenty thousand thoughts,
"Ha," quoth I, "pretty prisoner, are you there!"
Well might sarcastic Fancy then have whispered,   60
"An emblem here behold of thy own life;
In its late course of even days with all
Their smooth enthralment;" but the heart was full,
Too full for that reproach. My aged Dame
Walked proudly at my side; she guided me;
I willing, nay—nay, wishing to be led.
—The face of every neighbour whom I met
Was like a volume to me; some were hailed
Upon the road, some busy at their work,
Unceremonious greetings interchanged   70
With half the length of a long field between.
Among my schoolfellows I scattered round
Like recognitions, but with some constraint
Attended, doubtless, with a little pride,
But with more shame, for my habiliments,
The transformation wrought by gay attire.
Not less delighted did I take my place
At our domestic table: and, dear Friend!
In this endeavour simply to relate
A Poet's history, may I leave untold   80
The thankfulness with which I laid me down
In my accustomed bed, more welcome now
Perhaps than if it had been more desired
Or been more often thought of with regret;
That lowly bed whence I had heard the wind
Roar and the rain beat hard, where I so oft
Had lain awake on summer nights to watch
The moon in splendour couched among the leaves
Of a tall ash, that near our cottage stood;
Had watched her with fixed eyes while to and fro   90
In the dark summit of the waving tree
She rocked with every impulse of the breeze.

Among the favourites whom it pleased me well
To see again, was one by ancient right
Our inmate, a rough terrier of the hills;
By birth and call of nature pre-ordained
To hunt the badger and unearth the fox
Among the impervious crags, but having been
From youth our own adopted, he had passed
Into a gentler service. And when first   100
The boyish spirit flagged, and day by day
Along my veins I kindled with the stir,
The fermentation, and the vernal heat
Of poesy, affecting private shades
Like a sick Lover, then this dog was used
To watch me, an attendant and a friend,
Obsequious to my steps early and late,
Though often of such dilatory walk
Tired, and uneasy at the halts I made.
A hundred times when, roving high and low,   110
I have been harassed with the toil of verse,
Much pains and little progress, and at once
Some lovely Image in the song rose up
Full-formed, like Venus rising from the sea;
Then have I darted forwards to let loose
My hand upon his back with stormy joy,
Caressing him again and yet again.
And when at evening on the public way
I sauntered, like a river murmuring
And talking to itself when all things else   120
Are still, the creature trotted on before;
Such was his custom; but whene'er he met
A passenger approaching, he would turn
To give me timely notice, and straightway,
Grateful for that admonishment, I hushed
My voice, composed my gait, and, with the air
And mien of one whose thoughts are free, advanced
To give and take a greeting that might save
My name from piteous rumours, such as wait
On men suspected to be crazed in brain.   130

Those walks well worthy to be prized and loved—
Regretted!—that word, too, was on my tongue,
But they were richly laden with all good,
And cannot be remembered but with thanks
And gratitude, and perfect joy of heart—
Those walks in all their freshness now came back
Like a returning Spring. When first I made
Once more the circuit of our little lake,

119–20. **river . . . talking:** Wordsworth usually composed
out loud.

If ever happiness hath lodged with man,
That day consummate happiness was mine,   140
Wide-spreading, steady, calm, contemplative.
The sun was set, or setting, when I left
Our cottage door, and evening soon brought on
A sober hour, not winning or serene,
For cold and raw the air was, and untuned;
But as a face we love is sweetest then
When sorrow damps it, or, whatever look
It chance to wear, is sweetest if the heart
Have fulness in herself; even so with me
It fared that evening. Gently did my soul   150
Put off her veil, and, self-transmuted, stood
Naked, as in the presence of her God.
While on I walked, a comfort seemed to touch
A heart that had not been disconsolate:
Strength came where weakness was not known to be,
At least not felt; and restoration came
Like an intruder knocking at the door
Of unacknowledged weariness. I took
The balance, and with firm hand weighed myself.
—Of that external scene which round me lay,   160
Little, in this abstraction, did I see;
Remembered less; but I had inward hopes
And swellings of the spirit, was rapt and soothed,
Conversed with promises, had glimmering views
How life pervades the undecaying mind;
How the immortal soul with God-like power
Informs, creates, and thaws the deepest sleep
That time can lay upon her; how on earth,
Man, if he do but live within the light
Of high endeavours, daily spreads abroad   170
His being armed with strength that cannot fail.
Nor was there want of milder thoughts, of love,
Of innocence, and holiday repose;
And more than pastoral quiet, 'mid the stir
Of boldest projects, and a peaceful end
At last, or glorious, by endurance won.
Thus musing, in a wood I sate me down
Alone, continuing there to muse: the slopes
And heights meanwhile were slowly overspread
With darkness, and before a rippling breeze   180
The long lake lengthened out its hoary line,
And in the sheltered coppice where I sate,
Around me from among the hazel leaves,
Now here, now there, moved by the straggling wind,
Came ever and anon a breath-like sound,
Quick as the pantings of the faithful dog,
The off and on companion of my walk;
And such, at times, believing them to be,

I turned my head to look if he were there;
Then into solemn thought I passed once more.   190

     A freshness also found I at this time
In human Life, the daily life of those
Whose occupations really I loved;
The peaceful scene oft filled me with surprise,
Changed like a garden in the heat of spring
After an eight-days' absence. For (to omit
The things which were the same and yet appeared
Far otherwise) amid this rural solitude,
A narrow Vale where each was known to all,
'Twas not indifferent to a youthful mind   200
To mark some sheltering bower or sunny nook,
Where an old man had used to sit alone,
Now vacant; pale-faced babes whom I had left
In arms, now rosy prattlers at the feet
Of a pleased grandame tottering up and down;
And growing girls whose beauty, filched away
With all its pleasant promises, was gone
To deck some slighted playmate's homely cheek.

     Yes, I had something of a subtler sense,
And often looking round was moved to smiles   210
Such as a delicate work of humour breeds;
I read, without design, the opinions, thoughts,
Of those plain-living people now observed
With clearer knowledge; with another eye
I saw the quiet woodman in the woods,
The shepherd roam the hills. With new delight,
This chiefly, did I note my grey-haired Dame;
Saw her go forth to church or other work
Of state, equipped in monumental trim;
Short velvet cloak, (her bonnet of the like),   220
A mantle such as Spanish Cavaliers
Wore in old times. Her smooth domestic life,
Affectionate without disquietude,
Her talk, her business, pleased me; and no less
Her clear though shallow stream of piety
That ran on Sabbath days a fresher course;
With thoughts unfelt till now I saw her read
Her Bible on hot Sunday afternoons,
And loved the book, when she had dropped asleep
And made of it a pillow for her head.   230

     Nor less do I remember to have felt,
Distinctly manifested at this time,
A human-heartedness about my love
For objects hitherto the absolute wealth
Of my own private being and no more;
Which I had loved, even as a blessed spirit

Or Angel, if he were to dwell on earth,
Might love in individual happiness.
But now there opened on me other thoughts
Of change, congratulation or regret,   240
A pensive feeling! It spread far and wide;
The trees, the mountains shared it, and the brooks,
The stars of Heaven, now seen in their old haunts—
White Sirius glittering o'er the southern crags,
Orion with his belt, and those fair Seven,
Acquaintances of every little child,
And Jupiter, my own beloved star!
Whatever shadings of mortality,
Whatever imports from the world of death
Had come among these objects heretofore,   250
Were, in the main, of mood less tender: strong,
Deep, gloomy were they, and severe; the scatterings
Of awe or tremulous dread, that had given way
In later youth to yearnings of a love
Enthusiastic, to delight and hope.

As one who hangs down-bending from the side
Of a slow-moving boat, upon the breast
Of a still water, solacing himself
With such discoveries as his eye can make
Beneath him in the bottom of the deep,   260
Sees many beauteous sights—weeds, fishes, flowers,
Grots, pebbles, roots of trees, and fancies more,
Yet often is perplexed, and cannot part
The shadow from the substance, rocks and sky,
Mountains and clouds, reflected in the depth
Of the clear flood, from things which there abide
In their true dwelling; now is crossed by gleam
Of his own image, by a sunbeam now,
And wavering motions sent he knows not whence,
Impediments that make his task more sweet;   270
Such pleasant office have we long pursued
Incumbent o'er the surface of past time
With like success, nor often have appeared
Shapes fairer or less doubtfully discerned
Than these to which the Tale, indulgent Friend!
Would now direct thy notice. Yet in spite
Of pleasure won, and knowledge not withheld,
There was an inner falling off—I loved,
Loved deeply all that had been loved before,
More deeply even than ever: but a swarm   280
Of heady schemes jostling each other, gawds,
And feast and dance, and public revelry,
And sports and games (too grateful in themselves,
Yet in themselves less grateful, I believe,
Than as they were a badge glossy and fresh

Of manliness and freedom) all conspired
To lure my mind from firm habitual quest
Of feeding pleasures, to depress the zeal
And damp those daily yearnings which had once
          been mine—
A wild, unworldly-minded youth, given up   290
To his own eager thoughts. It would demand
Some skill, and longer time than may be spared,
To paint these vanities, and how they wrought
In haunts where they, till now, had been unknown.
It seemed the very garments that I wore
Preyed on my strength, and stopped the quiet stream
Of self-forgetfulness.
               Yes, that heartless chase
Of trivial pleasures was a poor exchange
For books and nature at that early age.
'Tis true, some casual knowledge might be gained 300
Of character or life; but at that time,
Of manners put to school I took small note,
And all my deeper passions lay elsewhere.
Far better had it been to exalt the mind
By solitary study, to uphold
Intense desire through meditative peace;
And yet, for chastisement of these regrets,
The memory of one particular hour
Doth here rise up against me. 'Mid a throng
Of maids and youths, old men, and matrons staid, 310
A medley of all tempers, I had passed
The night in dancing, gaiety, and mirth,
With din of instruments and shuffling feet,
And glancing forms, and tapers glittering,
And unaimed prattle flying up and down;
Spirits upon the stretch, and here and there
Slight shocks of young love-liking interspersed,
Whose transient pleasure mounted to the head,
And tingled through the veins. Ere we retired,
The cock had crowed, and now the eastern sky 320
Was kindling, not unseen, from humble copse
And open field, through which the pathway wound,
And homeward led my steps. Magnificent
The morning rose, in memorable pomp,
Glorious as e'er I had beheld—in front,
The sea lay laughing at a distance; near,
The solid mountains shone, bright as the clouds,
Grain-tinctured, drenched in empyrean light;
And in the meadows and the lower grounds
Was all the sweetness of a common dawn—   330

328. Grain-tinctured: crimson. ("Grain" is the name of the color.)

Dews, vapours, and the melody of birds,
And labourers going forth to till the fields.

Ah! need I say, dear Friend! that to the brim
My heart was full; I made no vows, but vows
Were then made for me; bond unknown to me
Was given, that I should be, else sinning greatly,
A dedicated Spirit. On I walked
In thankful blessedness, which yet survives.

Strange rendezvous my mind was at that time,
A parti-coloured show of grave and gay,   340
Solid and light, short-sighted and profound;
Of inconsiderate habits and sedate,
Consorting in one mansion unreproved.
The worth I knew of powers that I possessed,
Though slighted and too oft misused. Besides,
That summer, swarming as it did with thoughts
Transient and idle, lacked not intervals
When Folly from the frown of fleeting Time
Shrunk, and the mind experienced in herself
Conformity as just as that of old   350
To the end and written spirit of God's works,
Whether held forth in Nature or in Man,
Through pregnant vision, separate or conjoined.

When from our better selves we have too long
Been parted by the hurrying world, and droop,
Sick of its business, of its pleasures tired,
How gracious, how benign, is Solitude;
How potent a mere image of her sway;
Most potent when impressed upon the mind
With an appropriate human centre—hermit,   360
Deep in the bosom of the wilderness;
Votary (in vast cathedral, where no foot
Is treading, where no other face is seen)
Kneeling at prayers; or watchman on the top
Of lighthouse, beaten by Atlantic waves;
Or as the soul of that great Power is met
Sometimes embodied on a public road,
When, for the night deserted, it assumes
A character of quiet more profound   369
Than pathless wastes.
                        Once, when those summer months
Were flown, and autumn brought its annual show
Of oars with oars contending, sails with sails,

354–65. These lines were a very late addition (1832 or 1839),
doubtless intended to account for Wordsworth's strange
emotions in the encounter.

Upon Winander's spacious breast, it chanced
That—after I had left a flower-decked room
(Whose in-door pastime, lighted up, survived
To a late hour), and spirits overwrought
Were making night do penance for a day
Spent in a round of strenuous idleness—
My homeward course led up a long ascent,
Where the road's watery surface, to the top   380
Of that sharp rising, glittered to the moon
And bore the semblance of another stream
Stealing with silent lapse to join the brook
That murmured in the vale. All else was still;
No living thing appeared in earth or air,
And, save the flowing water's peaceful voice,
Sound there was none—but, lo! an uncouth shape,
Shown by a sudden turning of the road,
So near that, slipping back into the shade
Of a thick hawthorn, I could mark him well,   390
Myself unseen. He was of stature tall,
A span above man's common measure, tall,
Stiff, lank, and upright; a more meagre man
Was never seen before by night or day.
Long were his arms, pallid his hands; his mouth
Looked ghastly in the moonlight: from behind,
A mile-stone propped him; I could also ken
That he was clothed in military garb,
Though faded, yet entire. Companionless,
No dog attending, by no staff sustained,   400
He stood, and in his very dress appeared
A desolation, a simplicity,
To which the trappings of a gaudy world
Make a strange back-ground. From his lips, ere long,
Issued low muttered sounds, as if of pain
Or some uneasy thought; yet still his form
Kept the same awful steadiness—at his feet
His shadow lay, and moved not. From self-blame
Not wholly free, I watched him thus; at length
Subduing my heart's specious cowardice,   410
I left the shady nook where I had stood
And hailed him. Slowly from his resting-place
He rose, and with a lean and wasted arm
In measured gesture lifted to his head
Returned my salutation; then resumed
His station as before; and when I asked
His history, the veteran, in reply,
Was neither slow nor eager; but, unmoved,
And with a quiet uncomplaining voice,
A stately air of mild indifference,   420
He told in few plain words a soldier's tale—

That in the Tropic Islands he had served,
Whence he had landed scarcely three weeks past;
That on his landing he had been dismissed,
And now was travelling towards his native home.
This heard, I said, in pity, "Come with me."
He stooped, and straightway from the ground took up
An oaken staff by me yet unobserved—
A staff which must have dropped from his slack hand
And lay till now neglected in the grass.   430
Though weak his step and cautious, he appeared
To travel without pain, and I beheld,
With an astonishment but ill suppressed,
His ghostly figure moving at my side;
Nor could I, while we journeyed thus, forbear
To turn from present hardships to the past,
And speak of war, battle, and pestilence,
Sprinkling this talk with questions, better spared,
On what he might himself have seen or felt.
He all the while was in demeanour calm,   440
Concise in answer; solemn and sublime
He might have seemed, but that in all he said
There was a strange half-absence, as of one
Knowing too well the importance of his theme,
But feeling it no longer. Our discourse
Soon ended, and together on we passed
In silence through a wood gloomy and still.
Up-turning, then, along an open field,
We reached a cottage. At the door I knocked,
And earnestly to charitable care   450
Commended him as a poor friendless man,
Belated and by sickness overcome.
Assured that now the traveller would repose
In comfort, I entreated that henceforth
He would not linger in the public ways,
But ask for timely furtherance and help
Such as his state required. At this reproof,
With the same ghastly mildness in his look,
He said, "My trust is in the God of Heaven,
And in the eye of him who passes me!"   460

   The cottage door was speedily unbarred,
And now the soldier touched his hat once more
With his lean hand, and in a faltering voice,
Whose tone bespake reviving interests
Till then unfelt, he thanked me; I returned
The farewell blessing of the patient man,
And so we parted. Back I cast a look,
And lingered near the door a little space,
Then sought with quiet heart my distant home.

from BOOK FIFTH

BOOKS

When Contemplation, like the night-calm felt
Through earth and sky, spreads widely, and sends deep
Into the soul its tranquillizing power,
Even then I sometimes grieve for thee, O Man,
Earth's paramount Creature! not so much for woes
That thou endurest; heavy though that weight be,
Cloud-like it mounts, or touched with light divine
Doth melt away; but for those palms achieved
Through length of time, by patient exercise
Of study and hard thought; there, there, it is   10
That sadness finds its fuel. Hitherto,
In progress through this Verse, my mind hath looked
Upon the speaking face of earth and heaven
As her prime teacher, intercourse with man
Established by the sovereign Intellect,
Who through that bodily image hath diffused,
As might appear to the eye of fleeting time,
A deathless spirit. Thou also, man! hast wrought,
For commerce of thy nature with herself,
Things that aspire to unconquerable life;   20
And yet we feel—we cannot choose but feel—
That they must perish. Tremblings of the heart
It gives, to think that our immortal being
No more shall need such garments; and yet man,
As long as he shall be the child of earth,
Might almost "weep to have" what he may lose,
Nor be himself extinguished, but survive,
Abject, depressed, forlorn, disconsolate.
A thought is with me sometimes, and I say,—
Should the whole frame of earth by inward throes 30
Be wrenched, or fire come down from far to scorch
Her pleasant habitations, and dry up
Old Ocean, in his bed left singed and bare,
Yet would the living Presence still subsist
Victorious, and composure would ensue,
And kindlings like the morning—presage sure
Of day returning and of life revived.
But all the meditations of mankind,
Yea, all the adamantine holds of truth
By reason built, or passion, which itself   40
Is highest reason in a soul sublime;

BOOK V. **26. "weep to have"**: quoted from Shakespeare's
Sonnet LXIV.

The consecrated works of Bard and Sage,
Sensuous or intellectual, wrought by men,
Twin labourers and heirs of the same hopes;
Where would they be? Oh! why hath not the Mind
Some element to stamp her image on
In nature somewhat nearer to her own?
Why, gifted with such powers to send abroad
Her spirit, must it lodge in shrines so frail?

   One day, when from my lips a like complaint   50
Had fallen in presence of a studious friend,
He with a smile made answer, that in truth
'Twas going far to seek disquietude;
But on the front of his reproof confessed
That he himself had oftentimes given way
To kindred hauntings. Whereupon I told,
That once in the stillness of a summer's noon,
While I was seated in a rocky cave
By the sea-side, perusing, so it chanced,
The famous history of the errant knight   60
Recorded by Cervantes, these same thoughts
Beset me, and to height unusual rose,
While listlessly I sate, and, having closed
The book, had turned my eyes toward the wide sea.
On poetry and geometric truth,
And their high privilege of lasting life,
From all internal injury exempt,
I mused, upon these chiefly: and at length,
My senses yielding to the sultry air,
Sleep seized me, and I passed into a dream.   70
I saw before me stretched a boundless plain
Of sandy wilderness, all black and void,
And as I looked around, distress and fear
Came creeping over me, when at my side,
Close at my side, an uncouth shape appeared
Upon a dromedary, mounted high.
He seemed an Arab of the Bedouin tribes:
A lance he bore, and underneath one arm
A stone, and in the opposite hand a shell
Of a surpassing brightness. At the sight   80
Much I rejoiced, not doubting but a guide
Was present, one who with unerring skill
Would through the desert lead me; and while yet
I looked and looked, self-questioned what this freight
Which the new-comer carried through the waste
Could mean, the Arab told me that the stone
(To give it in the language of the dream)
Was "Euclid's Elements;" and "This," said he,
"Is something of more worth;" and at the word
Stretched forth the shell, so beautiful in shape,   90

In colour so resplendent, with command
That I should hold it to my ear. I did so,
And heard that instant in an unknown tongue,
Which yet I understood, articulate sounds,
A loud prophetic blast of harmony;
An Ode, in passion uttered, which foretold
Destruction to the children of the earth
By deluge, now at hand. No sooner ceased
The song, than the Arab with calm look declared
That all would come to pass of which the voice   100
Had given forewarning, and that he himself
Was going then to bury those two books:
The one that held acquaintance with the stars,
And wedded soul to soul in purest bond
Of reason, undisturbed by space or time;
The other that was a god, yea many gods,
Had voices more than all the winds, with power
To exhilarate the spirit, and to soothe,
Through every clime, the heart of human kind.
While this was uttering, strange as it may seem,   110
I wondered not, although I plainly saw
The one to be a stone, the other a shell;
Nor doubted once but that they both were books,
Having a perfect faith in all that passed.
Far stronger, now, grew the desire I felt
To cleave unto this man; but when I prayed
To share his enterprise, he hurried on
Reckless of me: I followed, not unseen,
For oftentimes he cast a backward look,
Grasping his twofold treasure.—Lance in rest,   120
He rode, I keeping pace with him; and now
He, to my fancy, had become the knight
Whose tale Cervantes tells; yet not the knight,
But was an Arab of the desert too;
Of these was neither, and was both at once.
His countenance, meanwhile, grew more disturbed;
And, looking backwards when he looked, mine eyes
Saw, over half the wilderness diffused,
A bed of glittering light: I asked the cause:
"It is," said he, "the waters of the deep   130
Gathering upon us;" quickening then the pace
Of the unwieldy creature he bestrode,
He left me: I called after him aloud;
He heeded not; but, with his twofold charge
Still in his grasp, before me, full in view,
Went hurrying o'er the illimitable waste,
With the fleet waters of a drowning world
In chase of him; whereat I waked in terror,
And saw the sea before me, and the book,
In which I had been reading, at my side.   140

Full often, taking from the world of sleep
This Arab phantom, which I thus beheld,
This semi-Quixote, I to him have given
A substance, fancied him a living man,
A gentle dweller in the desert, crazed
By love and feeling, and internal thought
Protracted among endless solitudes;
Have shaped him wandering upon this quest!
Nor have I pitied him; but rather felt
Reverence was due to a being thus employed;   150
And thought that, in the blind and awful lair
Of such a madness, reason did lie couched.
Enow there are on earth to take in charge
Their wives, their children, and their virgin loves,
Or whatsoever else the heart holds dear;
Enow to stir for these; yea, will I say,
Contemplating in soberness the approach
Of an event so dire, by signs in earth
Or heaven made manifest, that I could share
That maniac's fond anxiety, and go   160
Upon like errand. Oftentimes at least
Me hath such strong entrancement overcome,
When I have held a volume in my hand,
Poor earthly casket of immortal verse,
Shakespeare, or Milton, labourers divine!

. . .

Well do I call to mind the very week
When I was first intrusted to the care
Of that sweet Valley; when its paths, its shores,
And brooks were like a dream of novelty
To my half-infant thoughts; that very week,   430
While I was roving up and down alone,
Seeking I knew not what, I chanced to cross
One of those open fields, which, shaped like ears,
Make green peninsulas on Esthwaite's Lake:
Twilight was coming on, yet through the gloom
Appeared distinctly on the opposite shore
A heap of garments, as if left by one
Who might have there been bathing. Long I watched,
But no one owned them; meanwhile the calm lake
Grew dark with all the shadows on its breast,   440
And, now and then, a fish up-leaping snapped
The breathless stillness. The succeeding day,
Those unclaimed garments telling a plain tale
Drew to the spot an anxious crowd; some looked
In passive expectation from the shore,
While from a boat others hung o'er the deep,

428. **Valley:** Esthwaite.

Sounding with grappling irons and long poles.
At last, the dead man, 'mid that beauteous scene
Of trees and hills and water, bolt upright
Rose, with his ghastly face, a spectre shape   450
Of terror; yet no soul-debasing fear,
Young as I was, a child not nine years old,
Possessed me, for my inner eye had seen
Such sights before, among the shining streams
Of faery land, the forest of romance.
Their spirit hallowed the sad spectacle
With decoration of ideal grace;
A dignity, a smoothness, like the works
Of Grecian art, and purest poesy.

A precious treasure had I long possessed,   460
A little yellow, canvas-covered book,
A slender abstract of the Arabian tales;
And, from companions in a new abode,
When first I learnt, that this dear prize of mine
Was but a block hewn from a mighty quarry—
That there were four large volumes, laden all
With kindred matter, 'twas to me, in truth,
A promise scarcely earthly. Instantly,
With one not richer than myself, I made
A covenant that each should lay aside   470
The moneys he possessed, and hoard up more,
Till our joint savings had amassed enough
To make this book our own. Through several months
In spite of all temptation, we preserved
Religiously that vow; but firmness failed,
Nor were we ever masters of our wish.

And when thereafter to my father's house
The holidays returned me, there to find
That golden store of books which I had left,
What joy was mine! How often in the course   480
Of those glad respites, though a soft west wind
Ruffled the waters to the angler's wish,
For a whole day together, have I lain
Down by thy side, O Derwent! murmuring stream,
On the hot stones, and in the glaring sun,
And there have read, devouring as I read,
Defrauding the day's glory, desperate!
Till with a sudden bound of smart reproach,
Such as an idler deals with in his shame,
I to the sport betook myself again.   490

455. **faery land . . . romance:** Wordsworth here pays
tribute to the fairy tales and romances that, together with the
phenomena of nature, had excited and filled his childhood
imagination. 462. **Arabian tales:** the *Arabian Nights*.

A gracious spirit o'er this earth presides,
And o'er the heart of man: invisibly
It comes, to works of unreproved delight,
And tendency benign, directing those
Who care not, know not, think not what they do.
The tales that charm away the wakeful night
In Araby; romances; legends penned
For solace by dim light of monkish lamps;
Fictions, for ladies of their love, devised
By youthful squires; adventures endless, spun   500
By the dismantled warrior in old age,
Out of the bowels of those very schemes
In which his youth did first extravagate;
These spread like day, and something in the shape
Of these will live till man shall be no more.
Dumb yearnings, hidden appetites, are ours,
And *they must* have their food. Our childhood sits,
Our simple childhood, sits upon a throne
That hath more power than all the elements.
I guess not what this tells of Being past,   510
Nor what it augurs of the life to come;
But so it is, and, in that dubious hour,
That twilight when we first begin to see
This dawning earth, to recognise, expect,
And in the long probation that ensues,
The time of trial, ere we learn to live
In reconcilement with our stinted powers;
To endure this state of meagre vassalage,
Unwilling to forego, confess, submit,
Uneasy and unsettled, yoke-fellows   520
To custom, mettlesome, and not yet tamed
And humbled down; oh! then we feel, we feel,
We know where we have friends. Ye dreamers, then,
Forgers of daring tales! we bless you then,
Impostors, drivellers, dotards, as the ape
Philosophy will call you: *then* we feel
With what, and how great might ye are in league,
Who make our wish our power, our thought a deed,
An empire, a possession,—ye whom time
And seasons serve; all Faculties; to whom   530
Earth crouches, the elements are potter's clay,
Space like a heaven filled up with northern lights,
Here, nowhere, there, and everywhere at once.

   Relinquishing this lofty eminence
For ground, though humbler, not the less a tract
Of the same isthmus, which our spirits cross
In progress from their native continent

537. **native continent:** Cf. "Intimations of Immortality,"
below, especially sts. v and vi.

To earth and human life, the Song might dwell
On that delightful time of growing youth,
When craving for the marvellous gives way   540
To strengthening love for things that we have seen;
When sober truth and steady sympathies,
Offered to notice by less daring pens,
Take firmer hold of us, and words themselves
Move us with conscious pleasure.
               I am sad
At thought of raptures now for ever flown;
Almost to tears I sometimes could be sad
To think of, to read over, many a page,
Poems withal of name, which at that time
Did never fail to entrance me, and are now   550
Dead in my eyes, dead as a theatre
Fresh emptied of spectators. Twice five years
Or less I might have seen, when first my mind
With conscious pleasure opened to the charm
Of words in tuneful order, found them sweet
For their own *sakes*, a passion, and a power;
And phrases pleased me chosen for delight,
For pomp, or love. Oft, in the public roads
Yet unfrequented, while the morning light
Was yellowing the hill tops, I went abroad   560
With a dear friend, and for the better part
Of two delightful hours we strolled along
By the still borders of the misty lake,
Repeating favourite verses with one voice,
Or conning more, as happy as the birds
That round us chaunted. Well might we be glad,
Lifted above the ground by airy fancies,
More bright than madness or the dreams of wine;
And, though full oft the objects of our love
Were false, and in their splendour overwrought,   570
Yet was there surely then no vulgar power
Working within us,—nothing less, in truth,
Than that most noble attribute of man,
Though yet untutored and inordinate,
That wish for something loftier, more adorned,
Than is the common aspect, daily garb,
Of human life. What wonder, then, if sounds
Of exultation echoed through the groves!
For images, and sentiments, and words,
And everything encountered or pursued   580
In that delicious world of poesy,
Kept holiday, a never-ending show,
With music, incense, festival, and flowers!

   Here must we pause: this only let me add,
From heart-experience, and in humblest sense

Of modesty, that he, who in his youth
A daily wanderer among woods and fields
With living Nature hath been intimate,
Not only in that raw unpractised time
Is stirred to ecstasy, as others are,   590
By glittering verse; but further, doth receive,
In measure only dealt out to himself,
Knowledge and increase of enduring joy
From the great Nature that exists in works
Of mighty Poets. Visionary power
Attends the motions of the viewless winds,
Embodied in the mystery of words:
There, darkness makes abode, and all the host
Of shadowy things work endless changes,—there,
As in a mansion like their proper home,   600
Even forms and substances are circumfused
By that transparent veil with light divine,
And, through the turnings intricate of verse,
Present themselves as objects recognised,
In flashes, and with glory not their own.

from BOOK SIXTH

CAMBRIDGE AND THE ALPS

The leaves were fading when to Esthwaite's banks
And the simplicities of cottage life
I bade farewell; and, one among the youth
Who, summoned by that season, reunite
As scattered birds troop to the fowler's lure,
Went back to Granta's cloisters, not so prompt
Or eager, though as gay and undepressed
In mind, as when I thence had taken flight
A few short months before. I turned my face
Without repining from the coves and heights   10
Clothed in the sunshine of the withering fern;
Quitted, not loth, the mild magnificence
Of calmer lakes and louder streams; and you,
Frank-hearted maids of rocky Cumberland,
You and your not unwelcome days of mirth,
Relinquished, and your nights of revelry,
And in my own unlovely cell sate down
In lightsome mood—such privilege has youth
That cannot take long leave of pleasant thoughts.

The bonds of indolent society   20
Relaxing in their hold, henceforth I lived
More to myself. Two winters may be passed
Without a separate notice: many books
Were skimmed, devoured, or studiously perused,
But with no settled plan. I was detached
Internally from academic cares;
Yet independent study seemed a course
Of hardy disobedience toward friends
And kindred, proud rebellion and unkind.
This spurious virtue, rather let it bear   30
A name it more deserves, this cowardice,
Gave treacherous sanction to that over-love
Of freedom which encouraged me to turn
From regulations even of my own
As from restraints and bonds. Yet who can tell—
Who knows what thus may have been gained, both
     then
And at a later season, or preserved;
What love of nature, what original strength
Of contemplation, what intuitive truths
The deepest and the best, what keen research,   40
Unbiassed, unbewildered, and unawed?

The Poet's soul was with me at that time;
Sweet meditations, the still overflow
Of present happiness, while future years
Lacked not anticipations, tender dreams,
No few of which have since been realised;
And some remain, hopes for my future life.
Four years and thirty, told this very week,
Have I been now a sojourner on earth,
By sorrow not unsmitten; yet for me   50
Life's morning radiance hath not left the hills,
Her dew is on the flowers. Those were the days
Which also first emboldened me to trust
With firmness, hitherto but lightly touched
By such a daring thought, that I might leave
Some monument behind me which pure hearts
Should reverence. The instinctive humbleness,
Maintained even by the very name and thought
Of printed books and authorship, began
To melt away; and further, the dread awe   60

599. **shadowy . . . changes:** Cf. VIII. 560–89. BOOK VI.
6. **Granta's cloisters:** Cambridge. 14. **Cumberland:**
county.

25–29. **I . . . unkind:** His relatives hoped he would excel
in academic studies, obtain a fellowship at Cambridge,
and then be ordained an Anglican clergyman—a safe path to
advancement and financial security for an able youth in
Wordsworth's day. His brother Christopher fulfilled these
hopes.

Of mighty names was softened down and seemed
Approachable, admitting fellowship
Of modest sympathy. Such aspect now,
Though not familiarly, my mind put on,
Content to observe, to admire, and to enjoy.

All winter long, whenever free to choose,
Did I by night frequent the College groves
And tributary walks; the last, and oft
The only one, who had been lingering there
Through hours of silence, till the porter's bell,   70
A punctual follower on the stroke of nine,
Rang with its blunt unceremonious voice,
Inexorable summons! Lofty elms,
Inviting shades of opportune recess,
Bestowed composure on a neighbourhood
Unpeaceful in itself. A single tree
With sinuous trunk, boughs exquisitely wreathed,
Grew there; and ash which Winter for himself
Decked as in pride, and with outlandish grace:
Up from the ground, and almost to the top,   80
The trunk and every master branch were green
With clustering ivy, and the lightsome twigs
And outer spray profusely tipped with seeds
That hung in yellow tassels, while the air
Stirred them, not voiceless. Often have I stood
Foot-bound uplooking at this lovely tree
Beneath a frosty moon. The hemisphere
Of magic fiction, verse of mine perchance
May never tread; but scarcely Spenser's self
Could have more tranquil visions in his youth,   90
Or could more bright appearances create
Of human forms with superhuman powers,
Than I beheld, loitering on calm clear nights
Alone, beneath this fairy work of earth.

. . .

In summer, making quest for works of art,   190
Or scenes renowned for beauty, I explored
That streamlet whose blue current works its way
Between romantic Dovedale's spiry rocks;
Pried into Yorkshire dales, or hidden tracts
Of my own native region, and was blest
Between these sundry wanderings with a joy
Above all joys, that seemed another morn
Risen on mid noon; blest with the presence, Friend,
Of that sole Sister, she who hath been long
Dear to thee also, thy true friend and mine,   200
Now, after separation desolate,
Restored to me—such absence that she seemed

A gift then first bestowed. The varied banks
Of Emont, hitherto unnamed in song,
And that monastic castle, 'mid tall trees,
Low standing by the margin of the stream,
A mansion visited (as fame reports)
By Sidney, where, in sight of our Helvellyn,
Or stormy Cross-fell, snatches he might pen
Of his Arcadia, by fraternal love   210
Inspired;—that river and those mouldering towers
Have seen us side by side, when, having clomb
The darksome windings of a broken stair,
And crept along a ridge of fractured wall,
Not without trembling, we in safety looked
Forth, through some Gothic window's open space,
And gathered with one mind a rich reward
From the far-stretching landscape, by the light
Of morning beautified, or purple eve;
Or, not less pleased, lay on some turret's head,   220
Catching from tufts of grass and hare-bell flowers
Their faintest whisper to the passing breeze,
Given out while mid-day heat oppressed the plains.

Another maid there was, who also shed
A gladness o'er that season, then to me,
By her exulting outside look of youth
And placid under-countenance, first endeared;
That other spirit, Coleridge! who is now
So near to us, that meek confiding heart,
So reverenced by us both. O'er paths and fields   230
In all that neighbourhood, through narrow lanes
Of eglantine, and through the shady woods,
And o'er the Border Beacon, and the waste
Of naked pools, and common crags that lay
Exposed on the bare fell, were scattered love,
The spirit of pleasure, and youth's golden gleam.
O Friend! we had not seen thee at that time,
And yet a power is on me, and a strong
Confusion, and I seem to plant thee there.
Far art thou wandered now in search of health   240
And milder breezes,—melancholy lot!

203. gift: The name of Wordsworth's sister, Dorothy,
derives from Greek and means "gift of God." 203–10.
The erroneous legend was that Sidney visited Brougham
Castle on the Eamont River while his sister, the Countess
of Pembroke, dwelt there. 208–09. Helvellyn . . . Cross-
fell: mountains in the Lake District. 224. maid: Wordsworth's future wife, Mary. 227. under-countenance: a
neologism. He refers to the inner repose everyone noticed
in Mary Wordsworth. 233. Beacon: marker. Cf. XII.
253–66. 235. fell: a stretch of high, open, uncultivated land
in mountain country.

But thou art with us, with us in the past,
The present, with us in the times to come.
There is no grief, no sorrow, no despair,
No languor, no dejection, no dismay,
No absence scarcely can there be, for those
Who love as we do. Speed thee well! divide
With us thy pleasure; thy returning strength,
Receive it daily as a joy of ours;
Share with us thy fresh spirits, whether gift   250
Of gales Etesian or of tender thoughts.

  I, too, have been a wanderer; but, alas!
How different the fate of different men.
Though mutually unknown, yea nursed and reared
As if in several elements, we were framed
To bend at last to the same discipline,
Predestined, if two beings ever were,
To seek the same delights, and have one health,
One happiness. Throughout this narrative,
Else sooner ended, I have borne in mind   260
For whom it registers the birth, and marks the growth,
Of gentleness, simplicity, and truth,
And joyous loves, that hallow innocent days
Of peace and self-command. Of rivers, fields,
And groves I speak to thee, my Friend! to thee,
Who, yet a liveried schoolboy, in the depths
Of the huge city, on the leaded roof
Of that wide edifice, thy school and home,
Wert used to lie and gaze upon the clouds
Moving in heaven; or, of that pleasure tired,   270
To shut thine eyes, and by internal light
See trees, and meadows, and thy native stream,
Far distant, thus beheld from year to year
Of a long exile. Nor could I forget,
In this late portion of my argument,
That scarcely, as my term of pupilage
Ceased, had I left those academic bowers
When thou wert thither guided. From the heart
Of London, and from cloisters there, thou camest,
And didst sit down in temperance and peace,   280
A rigorous student. What a stormy course

Then followed. Oh! it is a pang that calls
For utterance, to think what easy change
Of circumstances might to thee have spared
A world of pain, ripened a thousand hopes,
For ever withered. Through this retrospect
Of my collegiate life I still have had
Thy after-sojourn in the self-same place
Present before my eyes, have played with times
And accidents as children do with cards,   290
Or as a man, who, when his house is built,
A frame locked up in wood and stone, doth still,
As impotent fancy prompts, by his fireside,
Rebuild it to his liking. I have thought
Of thee, thy learning, gorgeous eloquence,
And all the strength and plumage of thy youth,
Thy subtle speculations, toils abtruse
Among the schoolmen, and Platonic forms
Of wild ideal pageantry, shaped out
From things well-matched or ill, and words for things,
The self-created sustenance of a mind   301
Debarred from Nature's living images,
Compelled to be a life unto herself,
And unrelentingly possessed by thirst
Of greatness, love, and beauty. Not alone,
Ah! surely not in singleness of heart
Should I have seen the light of evening fade
From smooth Cam's silent waters: had we met,
Even at that early time, needs must I trust
In the belief, that my maturer age,   310
My calmer habits, and more steady voice,
Would with an influence benign have soothed,
Or chased away, the airy wretchedness
That battened on thy youth. But thou hast trod
A march of glory, which doth put to shame
These vain regrets: health suffers in thee, else
Such grief for thee would be the weakest thought
That ever harboured in the breast of man.

  A passing word erewhile did lightly touch
On wanderings of my own, that now embraced   320
With livelier hope a region wider far.

  When the third summer freed us from restraint,
A youthful friend, he too a mountaineer,
Not slow to share my wishes, took his staff,

---

251. **Etesian:** "The distinctive epithet of certain winds in the region of the Mediterranean, blowing from the NW for about 40 days annually in the summer" (Oxford English Dictionary). Coleridge was at Malta. **271–73. shut . . . distant:** an allusion to Coleridge's sonnet "To the River Otter." **281–86. stormy . . . withered:** Wordsworth may refer to Coleridge's financial troubles in college and desperate enlistment in the army, his disappointed hopes for "Pantisocracy" (see Introduction to Coleridge, below), his unhappy marriage, ill health, and habit of taking laudanum.

295–99. **thee . . . pageantry:** Cf. Lamb's description of Coleridge as a schoolboy in "Christ's Hospital Five and Thirty Years Ago," below. 298. **schoolmen:** medieval scholastic philosophers. 299. **ideal:** In Romantic literature "ideal" usually signifies "mental" more than "supremely excellent." 322. **third summer:** 1790.

And sallying forth, we journeyed side by side,
Bound to the distant Alps. A hardy slight
Did this unprecedented course imply
Of college studies and their set rewards;
Nor had, in truth, the scheme been formed by me
Without uneasy forethought of the pain,   330
The censures, and ill-omening of those
To whom my wordly interests were dear.
But Nature then was sovereign in my mind,
And mighty forms, seizing a youthful fancy,
Had given a charter to irregular hopes.
In any age of uneventful calm
Among the nations, surely would my heart
Have been possessed by similar desire;
But Europe at that time was thrilled with joy,
France standing on the top of golden hours,   340
And human nature seeming born again.

    Lightly equipped, and but a few brief looks
Cast on the white cliffs of our native shore
From the receding vessel's deck, we chanced
To land at Calais on the very eve
Of that great federal day; and there we saw,
In a mean city, and among a few,
How bright a face is worn when joy of one
Is joy for tens of millions. Southward thence
We held our way, direct through hamlets, towns,   350
Gaudy with reliques of that festival,
Flowers left to wither on triumphal arcs,
And window-garlands. On the public roads,
And, once, three days successively, through paths
By which our toilsome journey was abridged,
Among sequestered villages we walked
And found benevolence and blessedness
Spread like a fragrance everywhere, when spring
Hath left no corner of the land untouched:
Where elms for many and many a league in files   360
With their thin umbrage, on the stately roads
Of that great kingdom, rustled o'er our heads,
For ever near us as we paced along:
How sweet at such a time, with such delight
On every side, in prime of youthful strength,
To feed a Poet's tender melancholy
And fond conceit of sadness, with the sound

Of undulations varying as might please
The wind that swayed them; once, and more than
    once,
Unhoused beneath the evening star we saw   370
Dances of liberty, and, in late hours
Of darkness, dances in the open air
Deftly prolonged, though grey-haired lookers on
Might waste their breath in chiding.
                            Under hills—
The vine-clad hills and slopes of Burgundy,
Upon the bosom of the gentle Saone
We glided forward with the flowing stream.
Swift Rhone! thou wert the *wings* on which we cut
A winding passage with majestic ease
Between thy lofty rocks. Enchanting show   380
Those woods and farms and orchards did present,
And single cottages and lurking towns,
Reach after reach, succession without end
Of deep and stately vales! A lonely pair
Of strangers, till day closed, we sailed along,
Clustered together with a merry crowd
Of those emancipated, a blithe host
Of travellers, chiefly delegates returning
From the great spousals newly solemnised
At their chief city, in the sight of Heaven.   390
Like bees they swarmed, gaudy and gay as bees;
Some vapoured in the unruliness of joy,
And with their swords flourished as if to fight
The saucy air. In this proud company
We landed—took with them our evening meal,
Guests welcome almost as the angels were
To Abraham of old. The supper done,
With flowing cups elate and happy thoughts
We rose at signal given, and formed a ring
And, hand in hand, danced round and round the board;
All hearts were open, every tongue was loud   401
With amity and glee; we bore a name
Honoured in France, the name of Englishmen,
And hospitably did they give us hail,
As their forerunners in a glorious course;
And round and round the board we danced again.
With these blithe friends our voyage we renewed
At early dawn. The monastery bells
Made a sweet jingling in our youthful ears;
The rapid river flowing without noise,   410
And each uprising or receding spire
Spake with a sense of peace, at intervals

339-41. **Europe . . . again:** The French Revolution
had begun the year before with the fall of the Bastille, July
14, 1789. **346. federal day:** On the first anniversary of the
fall of the Bastille, a Festival of Confederation, at which
Louis XVI swore allegiance to the new constitution, was
held in Paris and celebrated throughout France.

389-90. **spousals . . . Heaven:** See n. 346. **405. course:**
of liberty.

Touching the heart amid the boisterous crew
By whom we were encompassed. Taking leave
Of this glad throng, foot-travellers side by side,
Measuring our steps in quiet, we pursued
Our journey, and ere twice the sun had set
Beheld the Convent of Chartreuse, and there
Rested within an awful *solitude:*
Yes, for even then no other than a place   420
Of soul-affecting *solitude* appeared
That far-famed region, though our eyes had seen,
As toward the sacred mansion we advanced,
Arms flashing, and a military glare
Of riotous men commissioned to expel
The blameless inmates, and belike subvert
That frame of social being, which so long
Had bodied forth the ghostliness of things
In silence visible and perpetual calm.
—"Stay, stay your sacrilegious hands!"—The voice
Was Nature's, uttered from her Alpine throne;   431
I heard it then and seem to hear it now—
"Your impious work forbear, perish what may,
Let this one temple last, be this one spot
Of earth devoted to eternity!"
She ceased to speak, but while St. Bruno's pines
Waved their dark tops, not silent as they waved,
And while below, along their several beds,
Murmured the sister streams of Life and Death,
Thus by conflicting passions pressed, my heart   440
Responded; "Honour to the patriot's zeal!
Glory and hope to new-born Liberty!
Hail to the mighty projects of the time!
Discerning sword that Justice wields, do thou
Go forth and prosper; and, ye purging fires,
Up to the loftiest towers of Pride ascend,
Fanned by the breath of angry Providence.
But oh! if Past and Future be the wings
On whose support harmoniously conjoined
Moves the great spirit of human knowledge, spare 450
These courts of mystery, where a step advanced
Between the portals of the shadowy rocks
Leaves far behind life's treacherous vanities,
For penitential tears and trembling hopes
Exchanged—to equalise in God's pure sight

Monarch and peasant: be the house redeemed
With its unworldly votaries, for the sake
Of conquest over sense, hourly achieved
Through faith and meditative reason, resting
Upon the word of heaven-imparted truth,   460
Calmly triumphant; and for humbler claim
Of that imaginative impulse sent
From these majestic floods, yon shining cliffs,
The untransmuted shapes of many worlds,
Cerulean ether's pure inhabitants,
These forests unapproachable by death,
That shall endure as long as man endures,
To think, to hope, to worship, and to feel,
To struggle, to be lost within himself
In trepidation, from the blank abyss   470
To look with bodily eyes, and be consoled."
Not seldom since that moment have I wished
That thou, O Friend! the trouble or the calm
Hadst shared, when, from profane regards apart,
In sympathetic reverence we trod
The floors of those dim cloisters, till that hour,
From their foundation, strangers to the presence
Of unrestricted and unthinking man.
Abroad, how cheeringly the sunshine lay
Upon the open lawns! Vallombre's groves   480
Entering, we fed the soul with darkness; thence
Issued, and with uplifted eyes beheld,
In different quarters of the bending sky,
The cross of Jesus stand erect, as if
Hands of angelic powers had fixed it there,
Memorial reverenced by a thousand storms;
Yet then, from the undiscriminating sweep
And rage of one State-whirlwind, insecure.

'Tis not my present purpose to retrace
That variegated journey step by step.   490
A march it was of military speed,
And Earth did change her images and forms
Before us, fast as clouds are changed in heaven.
Day after day, up early and down late,
From hill to vale we dropped, from vale to hill
Mounted—from province on to province swept,
Keen hunters in a chase of fourteen weeks,
Eager as birds of prey, or as a ship
Upon the stretch, when winds are blowing fair:

---

418. Chartreuse: La Grande Chartreuse, chief monastery of the Carthusian order, located in the mountains near Grenoble. The site is rocky, precipitous, and covered by snow much of the year. The Carthusian monks live a life of solitude, silence, and prayer. The order was founded by St. Bruno in 1084. **439. streams . . . Death:** rivers at the Chartreuse named the *Guiers vif* and the *Guiers mort.*

480. Vallombre: "name of one of the vallies of the Chartreuse" (W. W.'s note in the *Descriptive Sketches*). **484-85. cross . . . there:** "Crosses seen on the tops of the spiry rocks of the Chartreuse which have every appearance of being inaccessible" (W. W.'s note in the *Descriptive Sketches*).

Sweet coverts did we cross of pastoral life,   500
Enticing valleys, greeted them and left
Too soon, while yet the very flash and gleam
Of salutation were not passed away.
Oh! sorrow for the youth who could have seen,
Unchastened, unsubdued, unawed, unraised
To patriarchal dignity of mind,
And pure simplicity of wish and will,
Those sanctified abodes of peaceful man,
Pleased (though to hardship born, and compassed round
With danger, varying as the seasons change),   510
Pleased with his daily task, or, if not pleased,
Contented, from the moment that the dawn
(Ah! surely not without attendant gleams
Of soul-illumination) calls him forth
To industry, by glistenings flung on rocks,
Whose evening shadows lead him to repose.

   Well might a stranger look with bounding heart
Down on a green recess, the first I saw
Of those deep haunts, an aboriginal vale,
Quiet and lorded over and possessed   520
By naked huts, wood-built, and sown like tents
Or Indian cabins over the fresh lawns
And by the river side.
              That very day,
From a bare ridge we also first beheld
Unveiled the summit of Mont Blanc, and grieved
To have a soulless image on the eye
That had usurped upon a living thought
That never more could be. The wondrous Vale
Of Chamouny stretched far below, and soon
With its dumb cataracts and streams of ice,   530
A motionless array of mighty waves,
Five rivers broad and vast, made rich amends,
And reconciled us to realities;
There small birds warble from the leafy trees,
The eagle soars high in the element,
There doth the reaper bind the yellow sheaf,
The maiden spread the haycock in the sun,
While Winter like a well-tamed lion walks,
Descending from the mountain to make sport
Among the cottages by beds of flowers.   540

   Whate'er in this wide circuit we beheld,
Or heard, was fitted to our unripe state
Of intellect and heart. With such a book
Before our eyes, we could not choose but read

Lessons of genuine brotherhood, the plain
And universal reason of mankind,
The truths of young and old. Nor, side by side
Pacing, two social pilgrims, or alone
Each with his humour, could we fail to abound
In dreams and fictions, pensively composed: 550
Dejection taken up for pleasure's sake,
And gilded sympathies, the willow wreath,
And sober posies of funereal flowers,
Gathered among those solitudes sublime
From formal gardens of the lady Sorrow,
Did sweeten many a meditative hour.

   Yet still in me with those soft luxuries
Mixed something of stern mood, an under-thirst
Of vigour seldom utterly allayed.
And from that source how different a sadness   560
Would issue, let one incident make known.
When from the Vallais we had turned, and clomb
Along the Simplon's steep and rugged road,
Following a band of muleteers, we reached
A halting-place, where all together took
Their noon-tide meal. Hastily rose our guide,
Leaving us at the board; awhile we lingered,
Then paced the beaten downward way that led
Right to a rough stream's edge, and there broke off;
The only track now visible was one   570
That from the torrent's further brink held forth
Conspicuous invitation to ascend
A lofty mountain. After brief delay
Crossing the unbridged stream, that road we took,
And clomb with eagerness, till anxious fears
Intruded, for we failed to overtake
Our comrades gone before. By fortunate chance,
While every moment added doubt to doubt,
A peasant met us, from whose mouth we learned
That to the spot which had perplexed us first   580
We must descend, and there should find the road,
Which in the stony channel of the stream
Lay a few steps, and then along its banks;
And, that our future course, all plain to sight,
Was downwards, with the current of that stream.
Loth to believe what we so grieved to hear,
For still we had hopes that pointed to the clouds,
We questioned him again, and yet again;
But every word that from the peasant's lips
Came in reply, translated by our feelings,   590
Ended in this,—*that we had crossed the Alps.*

518–19. **first . . . haunts:** They have passed from France
to Switzerland. **521. huts:** Swiss chalets.

563. **Simplon's . . . road:** the Simplon Pass through the
Alps.

Imagination—here the Power so called
Through sad incompetence of human speech,
That awful Power rose from the mind's abyss
Like an unfathered vapour that enwraps,
At once, some lonely traveller. I was lost;
Halted without an effort to break through;
But to my conscious soul I now can say—
"I recognise thy glory:" in such strength
Of usurpation, when the light of sense   600
Goes out, but with a flash that has revealed
The invisible world, doth greatness make abode,
There harbours whether we be young or old.
Our destiny, our being's heart and home,
Is with infinitude, and only there;
With hope it is, hope that can never die,
Effort, and expectation, and desire,
And something evermore about to be.
Under such banners militant, the soul
Seeks for no trophies, struggles for no spoils   610
That may attest her prowess, blest in thoughts
That are their own perfection and reward,
Strong in herself and in beatitude
That hides her, like the mighty flood of Nile
Poured from his fount of Abyssinian clouds
To fertilise the whole Egyptian plain.

The melancholy slackening that ensued
Upon those tidings by the peasant given
Was soon dislodged. Downwards we hurried fast,
And, with the half-shaped road which we had missed,
Entered a narrow chasm. The brook and road   621
Were fellow-travellers in this gloomy strait,
And with them did we journey several hours
At a slow pace. The immeasurable height
Of woods decaying, never to be decayed,
The stationary blasts of waterfalls,
And in the narrow rent at every turn
Winds thwarting winds, bewildered and forlorn,
The torrents shooting from the clear blue sky,
The rocks that muttered close upon our ears,   630
Black drizzling crags that spake by the way-side
As if a voice were in them, the sick sight
And giddy prospect of the raving stream,
The unfettered clouds and region of the Heavens,
Tumult and peace, the darkness and the light—

Were all like workings of one mind, the features
Of the same face, blossoms upon one tree;
Characters of the great Apocalypse,
The types and symbols of Eternity,
Of first, and last, and midst, and without end.   640

That night our lodging was a house that stood
Alone within the valley, at a point
Where, tumbling from aloft, a torrent swelled
The rapid stream whose margin we had trod;
A dreary mansion, large beyond all need,
With high and spacious rooms, deafened and stunned
By noise of waters, making innocent sleep
Lie melancholy among weary bones.

Uprisen betimes, our journey we renewed,
Led by the stream, ere noon-day magnified   650
Into a lordly river, broad and deep,
Dimpling along in silent majesty,
With mountains for its neighbours, and in view
Of distant mountains and their snowy tops,
And thus proceeding to Locarno's Lake,
Fit resting-place for such a visitant.
Locarno! spreading out in width like Heaven,
How dost thou cleave to the poetic heart,
Bask in the sunshine of the memory;
And Como! thou, a treasure whom the earth   660
Keeps to herself, confined as in a depth
Of Abyssinian privacy. I spake
Of thee, thy chestnut woods, and garden plots
Of Indian corn tended by dark-eyed maids;
Thy lofty steeps, and pathways roofed with vines,
Winding from house to house, from town to town,
Sole link that binds them to each other; walks,
League after league, and cloistral avenues,
Where silence dwells if music be not there:
While yet a youth undisciplined in verse,   670
Through fond ambition of that hour, I strove
To chant your praise; nor can approach you now
Ungreeted by a more melodious Song,
Where tones of Nature smoothed by learned Art
May flow in lasting current. Like a breeze
Or sunbeam over your domain I passed
In motion without pause; but ye have left
Your beauty with me, a serene accord
Of forms and colours, passive, yet endowed
In their submissiveness with power as sweet   680
And gracious, almost might I dare to say,

592–616: That this imaginative arousal and insight came not while crossing the Alps but fourteen years later, while Wordsworth was writing the poem, is made clear in the 1805 text: "Imagination! lifting up itself / Before the eye and progress of my Song."

638. great Apocalypse: the Revelation of St. John the Divine, the last book of the Bible.

As virtue is, or goodness; sweet as love,
Or the remembrance of a generous deed,
Or mildest visitations of pure thought,
When God, the giver of all joy, is thanked
Religiously, in silent blessedness;
Sweet as this last herself, for such it is.

With those delightful pathways we advanced,
For two days' space in presence of the Lake,
That, stretching far among the Alps, assumed   690
A character more stern. The second night,
From sleep awakened, and misled by sound
Of the church clock telling the hours with strokes
Whose import then we had not learned, we rose
By moonlight, doubting not that day was nigh,
And that meanwhile, by no uncertain path,
Along the winding margin of the lake,
Led, as before, we should behold the scene
Hushed in profound repose. We left the town
Of Gravedona with this hope; but soon   700
Were lost, bewildered among woods immense,
And on a rock sate down, to wait for day.
An open place it was, and overlooked,
From high, the sullen water far beneath,
On which a dull red image of the moon
Lay bedded, changing oftentimes its form
Like an uneasy snake. From hour to hour
We sate and sate, wondering, as if the night
Had been ensnared by witchcraft. On the rock
At last we stretched our weary limbs for sleep,   710
But *could not* sleep, tormented by the stings
Of insects, which, with noise like that of noon,
Filled all the woods: the cry of unknown birds;
The mountains more by blackness visible
And their own size, than any outward light;
The breathless wilderness of clouds; the clock
That told, with unintelligible voice,
The widely parted hours; the noise of streams,
And sometimes rustling motions nigh at hand,
That did not leave us free from personal fear;   720
And, lastly, the withdrawing moon, that set
Before us, while she still was high in heaven;—
These were our food; and such a summer's night
Followed that pair of golden days that shed
On Como's Lake, and all that round it lay,
Their fairest, softest, happiest influence.

But here I must break off, and bid farewell
To days, each offering some new sight, or fraught
With some untried adventure, in a course
Prolonged till sprinklings of autumnal snow   730

Checked our unwearied steps. Let this alone
Be mentioned as a parting word, that not
In hollow exultation, dealing out
Hyperboles of praise comparative;
Not rich one moment to be poor for ever;
Not prostrate, overborne, as if the mind
Herself were nothing, a mere pensioner
On outward forms—did we in presence stand
Of that magnificent region. On the front
Of this whole Song is written that my heart   740
Must, in such Temple, needs have offered up
A different worship. Finally, whate'er
I saw, or heard, or felt, was but a stream
That flowed into a kindred stream; a gale,
Confederate with the current of the soul,
To speed my voyage; every sound or sight,
In its degree of power, administered
To grandeur or to tenderness,—to the one
Directly, but to tender thoughts by means
Less often instantaneous in effect;   750
Led me to these by paths that, in the main,
Were more circuitous, but not less sure
Duly to reach the point marked out by Heaven.

Oh, most belovèd Friend! a glorious time,
A happy time that was; triumphant looks
Were then the common language of all eyes;
As if awaked from sleep, the Nations hailed
Their great expectancy: the fife of war
Was then a spirit-stirring sound indeed,
A blackbird's whistle in a budding grove.   760
We left the Swiss exulting in the fate
Of their near neighbours; and, when shortening fast
Our pilgrimage, nor distant far from home,
We crossed the Brabant armies on the fret
For battle in the cause of Liberty.
A stripling, scarcely of the household then
Of social life, I looked upon these things
As from a distance; heard, and saw, and felt,
Was touched, but with no intimate concern;
I seemed to move along them, as a bird   770
Moves through the air, or as a fish pursues
Its sport, or feeds in its proper element;
I wanted not that joy, I did not need
Such help; the ever-living universe,
Turn where I might, was opening out its glories,
And the independent spirit of pure youth
Called forth, at every season, new delights
Spread round my steps like sunshine o'er green fields.

764. Brabant: Belgian.

from BOOK SEVENTH

RESIDENCE IN LONDON

Rise up, thou monstrous ant-hill on the plain
Of a too busy world! Before me flow,   150
Thou endless stream of men and moving things!
Thy every-day appearance, as it strikes—
With wonder heightened, or sublimed by awe—
On strangers, of all ages; the quick dance
Of colours, lights, and forms; the deafening din;
The comers and the goers face to face,
Face after face; the string of dazzling wares,
Shop after shop, with symbols, blazoned names,
And all the tradesman's honours overhead:
Here, fronts of houses, like a title-page,   160
With letters huge inscribed from top to toe,
Stationed above the door, like guardian saints;
There, allegoric shapes, female or male,
Or physiognomies of real men,
Land-warriors, kings, or admirals of the sea,
Boyle, Shakespeare, Newton, or the attractive head
Of some quack-doctor, famous in his day.

Meanwhile the roar continues, till at length,
Escaped as from an enemy, we turn
Abruptly into some sequestered nook,   170
Still as a sheltered place when winds blow loud!
At leisure, thence, through tracts of thin resort,
And sights and sounds that come at intervals,
We take our way. A raree-show is here,
With children gathered round; another street
Presents a company of dancing dogs,
Or dromedary, with an antic pair
Of monkeys on his back; a minstrel band
Of Savoyards; or, single and alone,
An English ballad-singer. Private courts,   180
Gloomy as coffins, and unsightly lanes
Thrilled by some female vendor's scream, belike
The very shrillest of all London cries,
May then entangle our impatient steps;
Conducted through those labyrinths, unawares,
To privileged regions and inviolate,
Where from their airy lodges studious lawyers
Look out on waters, walks, and gardens green.

. . .

Genius of Burke! forgive the pen seduced
By specious wonders, and too slow to tell
Of what the ingenuous, what bewildered men,
Beginning to mistrust their boastful guides,
And wise men, willing to grow wiser, caught,
Rapt auditors! from thy most eloquent tongue—
Now mute, for ever mute in the cold grave.
I see him,—old, but vigorous in age,—
Stand like an oak whose stag-horn branches start   520
Out of its leafy brow, the more to awe
The younger brethren of the grove. But some—
While he forewarns, denounces, launches forth,
Against all systems built on abstract rights,
Keen ridicule; the majesty proclaims
Of Institutes and Laws, hallowed by time;
Declares the vital power of social ties
Endeared by Custom; and with high disdain,
Exploding upstart Theory, insists
Upon the allegiance to which men are born—   530
Some—say at once a froward multitude—
Murmur (for truth is hated, where not loved)
As the winds fret within the Æolian cave,
Galled by their monarch's chain. The times were big
With ominous change, which, night by night, provoked
Keen struggles, and black clouds of passion raised;
But memorable moments intervened,
When Wisdom, like the Goddess from Jove's brain,
Broke forth in armour of resplendent words,
Startling the Synod. Could a youth, and one   540
In ancient story versed, whose breast had heaved
Under the weight of classic eloquence,
Sit, see, and hear, unthankful, uninspired?

. . .

But foolishness and madness in parade,
Though most at home in this their dear domain,
Are scattered everywhere, no rarities,
Even to the rudest novice of the Schools.
Me, rather, it employed, to note, and keep
In memory, those individual sights
Of courage, or integrity, or truth,   600
Or tenderness, which there, set off by foil,
Appeared more touching. One will I select—
A Father—for he bore that sacred name—
Him saw I, sitting in an open square,
Upon a corner-stone of that low wall,

BOOK VII. **166. Boyle:** Robert Boyle, the famous seventeenth-century physicist and chemist. **174. raree-show:** a peep-show. **179. Savoyards:** natives of Savoy, a former duchy located between southeast France and northwest Italy.

512–43. Genius . . . uninspired: A late addition (in the 1820's) to the poem. **533. Æolian cave:** Aeolus, god and father of the winds, keeps them chained in a vast cave. **595. domain:** London.

Wherein were fixed the iron pales that fenced
A spacious grass-plot; there, in silence, sate
This One Man, with a sickly babe outstretched
Upon his knee, whom he had thither brought
For sunshine, and to breathe the fresher air.   610
Of those who passed, and me who looked at him,
He took no heed; but in his brawny arms
(The Artificer was to the elbow bare,
And from his work this moment had been stolen)
He held the child, and, bending over it,
As if he were afraid both of the sun
And of the air, which he had come to seek,
Eyed the poor babe with love unutterable.

    As the black storm upon the mountain top
Sets off the sunbeam in the valley, so   620
That huge fermenting mass of human-kind
Serves as a solemn back-ground, or relief,
To single forms and objects, whence they draw,
For feeling and contemplative regard,
More than inherent liveliness and power.
How oft, amid those overflowing streets,
Have I gone forward with the crowd, and said
Unto myself, "The face of every one
That passes by me is a mystery!"
Thus have I looked, nor ceased to look, oppressed 630
By thoughts of what and whither, when and how,
Until the shapes before my eyes became
A second-sight procession, such as glides
Over still mountains, or appears in dreams;
And once, far-travelled in such mood, beyond
The reach of common indication, lost
Amid the moving pageant, I was smitten
Abruptly, with the view (a sight not rare)
Of a blind Beggar, who, with upright face,
Stood, propped against a wall, upon his chest   640
Wearing a written paper, to explain
His story, whence he came, and who he was.
Caught by the spectacle my mind turned round
As with the might of waters; an apt type
This label seemed of the utmost we can know,
Both of ourselves and of the universe;
And, on the shape of that unmoving man,
His steadfast face and sightless eyes, I gazed,
As if admonished from another world.

    Though reared upon the base of outward things, 650
Structures like these the excited spirit mainly
Builds for herself; scenes different there are,
Full-formed, that take, with small internal help,
Possession of the faculties,—the peace

That comes with night; the deep solemnity
Of nature's intermediate hours of rest,
When the great tide of human life stands still:
The business of the day to come, unborn,
Of that gone by, locked up, as in the grave;
The blended calmness of the heavens and earth,   660
Moonlight and stars, and empty streets, and sounds
Unfrequent as in deserts; at late hours
Of winter evenings, when unwholesome rains
Are falling hard, with people yet astir,
The feeble salutation from the voice
Of some unhappy woman, now and then
Heard as we pass, when no one looks about,
Nothing is listened to. But these, I fear,
Are falsely catalogued; things that are, are not,
As the mind answers to them, or the heart   670
Is prompt, or slow, to feel. What say you, then,
To times, when half the city shall break out
Full of one passion, vengeance, rage, or fear?
To executions, to a street on fire,
Mobs, riots, or rejoicings? From these sights
Take one,—that ancient festival, the Fair,
Holden where martyrs suffered in past time,
And named of St. Bartholomew; there, see
A work completed to our hands, that lays,
If any spectacle on earth can do,   680
The whole creative powers of man asleep!—
For once, the Muse's help will we implore,
And she shall lodge us, wafted on her wings,
Above the press and danger of the crowd,
Upon some showman's platform. What a shock
For eyes and ears! what anarchy and din,
Barbarian and infernal,—a phantasma,
Monstrous in colour, motion, shape, sight, sound!
Below, the open space, through every nook
Of the wide area, twinkles, is alive   690
With heads; the midway region, and above,
Is thronged with staring pictures and huge scrolls,
Dumb proclamations of the Prodigies;
With chattering monkeys dangling from their poles,
And children whirling in their roundabouts;
With those that stretch the neck and strain the eyes,
And crack the voice in rivalship, the crowd
Inviting; with buffoons against buffoons

676–78. Fair . . . Bartholomew: St. Bartholomew Fair
was held in Smithfield, where Protestants had been executed
in the reign of Queen Mary (1553–58). **687. phantasma:** fig-
ment or fantasy of a disordered mind. **695. roundabouts:**
merry-go-rounds.

Grimacing, writhing, screaming,—him who grinds
The hurdy-gurdy, at the fiddle weaves,   700
Rattles the salt-box, thumps the kettle-drum,
And him who at the trumpet puffs his cheeks,
The silver-collared Negro with his timbrel,
Equestrians, tumblers, women, girls, and boys,
Blue-breeched, pink-vested, with high-towering
    plumes.
All moveables of wonder, from all parts,
Are here—Albinos, painted Indians, Dwarfs,
The Horse of knowledge, and the learned Pig,
The Stone-eater, the man that swallows fire,
Giants, Ventriloquists, the Invisible Girl,   710
The Bust that speaks and moves its goggling eyes,
The Wax-work, Clock-work, all the marvellous craft
Of modern Merlins, Wild Beasts, Puppet-shows,
All out-o'-the-way, far-fetched, perverted things,
All freaks of nature, all Promethean thoughts
Of man, his dulness, madness, and their feats
All jumbled up together, to compose
A Parliament of Monsters. Tents and Booths
Meanwhile, as if the whole were one vast mill,
Are vomiting, receiving on all sides,   720
Men, Women, three-years' Children, Babes in arms.

   Oh, blank confusion! true epitome
Of what the mighty City is herself,
To thousands upon thousands of her sons,
Living amid the same perpetual whirl
Of trivial objects, melted and reduced
To one identity, by differences
That have no law, no meaning, and no end—
Oppression, under which even highest minds
Must labour, whence the strongest are not free.   730
But though the picture weary out the eye,
By nature an unmanageable sight,
It is not wholly so to him who looks
In steadiness, who hath among least things
An under-sense of greatest; sees the parts
As parts, but with a feeling of the whole.
This, of all acquisitions first, awaits
On sundry and most widely different modes
Of education, nor with least delight

708. **Horse of knowledge:** presumably a horse trained to
indicate answers to questions, pick out cards, etc. **713.**
Merlins: Merlin was the famous magician of the Arthurian
romances. **715. Promethean:** daring, inventive, creative.
Prometheus stole fire from the Olympian gods and gave it
to man, and hence has become a symbol of man's pride and
self-sufficiency in his creative power.

On that through which I passed. Attention springs, 740
And comprehensiveness and memory flow,
From early converse with the works of God
Among all regions; chiefly where appear
Most obviously simplicity and power.
Think, how the everlasting streams and woods,
Stretched and still stretching far and wide, exalt
The roving Indian. On his desert sands
What grandeur not unfelt, what pregnant show
Of beauty, meets the sun-burnt Arab's eye:
And, as the sea propels, from zone to zone,   750
Its currents; magnifies its shoals of life
Beyond all compass; spreads, and sends aloft
Armies of clouds,—even so, its powers and aspects
Shape for mankind, by principles as fixed,
The views and aspirations of the soul
To majesty. Like virtue have the forms
Perennial of the ancient hills; nor less
The changeful language of their countenances
Quickens the slumbering mind, and aids the thoughts,
However multitudinous, to move   760
With order and relation. This, if still,
As hitherto, in freedom I may speak,
Not violating any just restraint,
As may be hoped, of real modesty,—
This did I feel, in London's vast domain.
The Spirit of Nature was upon me there;
The soul of Beauty and enduring Life
Vouchsafed her inspiration, and diffused,
Through meagre lines and colours, and the press
Of self-destroying, transitory things,   770
Composure, and ennobling Harmony.

from BOOK EIGHTH

RETROSPECT—LOVE OF NATURE LEADING
TO LOVE OF MAN

What sounds are those, Helvellyn, that are heard
Up to thy summit, through the depth of air
Ascending, as if distance had the power
To make the sounds more audible? What crowd
Covers, or sprinkles o'er, yon village green?
Crowd seems it, solitary hill! to thee,
Though but a little family of men,
Shepherds and tillers of the ground—betimes
Assembled with their children and their wives,
And here and there a stranger interspersed.   10
They hold a rustic fair—a festival,

Such as, on this side now, and now on that,
Repeated through his tributary vales,
Helvellyn, in the silence of his rest,
Sees annually, if clouds towards either ocean
Blown from their favourite resting-place, or mists
Dissolved, have left him an unshrouded head.
Delightful day it is for all who dwell
In this secluded glen, and eagerly
They give it welcome. Long ere heat of noon,   20
From byre or field the kine were brought; the sheep
Are penned in cotes; the chaffering is begun.
The heifer lows, uneasy at the voice
Of a new master; bleat the flocks aloud.
Booths are there none; a stall or two is here;
A lame man or a blind, the one to beg,
The other to make music; hither, too,
From far, with basket, slung upon her arm,
Of hawker's wares—books, pictures, combs, and pins—
Some aged woman finds her way again,   30
Year after year, a punctual visitant!
There also stands a speech-maker by rote,
Pulling the strings of his boxed raree-show;
And in the lapse of many years may come
Prouder itinerant, mountebank, or he
Whose wonders in a covered wain lie hid.
But one there is, the loveliest of them all,
Some sweet lass of the valley, looking out
For gains, and who that sees her would not buy?
Fruits of her father's orchard are her wares,   40
And with the ruddy produce she walks round
Among the crowd, half pleased with, half ashamed
Of her new office, blushing restlessly.
The children now are rich, for the old to-day
Are generous as the young; and, if content
With looking on, some ancient wedded pair
Sit in the shade together, while they gaze,
"A cheerful smile unbends the wrinkled brow,
The days departed start again to life,
And all the scenes of childhood reappear,   50
Faint, but more tranquil, like the changing sun
To him who slept at noon and wakes at eve."
Thus gaiety and cheerfulness prevail,
Spreading from young to old, from old to young,
And no one seems to want his share.—Immense
Is the recess, the circumambient world
Magnificent, by which they are embraced:

BOOK VIII. **48–52.** "**A . . . eve**": "These lines are from a
descriptive poem—'Malvern Hills'—by one of Mr. Wordsworth's oldest friends, Mr. Joseph Cottle" (note in the first
edition of *The Prelude*).

They move about upon the soft green turf:
How little they, they and their doings, seem,
And all that they can further or obstruct!   60
Through utter weakness pitiably dear,
As tender infants are: and yet how great!
For all things serve them: them the morning light
Loves, as it glistens on the silent rocks;
And them the silent rocks, which now from high
Look down upon them; the reposing clouds;
The wild brooks prattling from invisible haunts;
And old Helvellyn, conscious of the stir
Which animates this day their calm abode.

With deep devotion, Nature, did I feel,   70
In that enormous City's turbulent world
Of men and things, what benefit I owed
To thee, and those domains of rural peace,
Where to the sense of beauty first my heart
Was opened.

                    . . .

                         Yet, hail to you
Moors, mountains, headlands, and ye hollow vales,
Ye long deep channels for the Atlantic's voice,
Powers of my native region! Ye that seize
The heart with firmer grasp! Your snows and streams
Ungovernable, and your terrifying winds,   220
That howl so dismally for him who treads
Companionless your awful solitudes!
There, 'tis the shepherd's task the winter long
To wait upon the storms: of their approach
Sagacious, into sheltering coves he drives
His flock, and thither from the homestead bears
A toilsome burden up the craggy ways,
And deals it out, their regular nourishment
Strewn on the frozen snow. And when the spring
Looks out, and all the pastures dance with lambs,   230
And when the flock, with warmer weather, climbs
Higher and higher, him his office leads
To watch their goings, whatsoever track
The wanderers choose. For this he quits his home
At day-spring, and no sooner doth the sun
Begin to strike him with a fire-like heat,
Than he lies down upon some shining rock,
And breakfasts with his dog. When they have stolen,
As is their wont, a pittance from strict time,
For rest not needed or exchange of love,   240
Then from his couch he starts; and now his feet
Crush out a livelier fragrance from the flowers
Of lowly thyme, by Nature's skill enwrought
In the wild turf: the lingering dews of morn

Smoke round him, as from hill to hill he hies,
His staff protending like a hunter's spear,
Or by its aid leaping from crag to crag,
And o'er the brawling beds of unbridged streams.
Philosophy, methinks, at Fancy's call,
Might deign to follow him through what he does 250
Or sees in his day's march; himself he feels,
In those vast regions where his service lies,
A freeman, wedded to his life of hope
And hazard, and hard labour interchanged
With that majestic indolence so dear
To native man. A rambling schoolboy, thus
I felt his presence in his own domain,
As of a lord and master, or a power,
Or genius, under Nature, under God,
Presiding; and severest solitude   260
Had more commanding looks when he was there.
When up the lonely brooks on rainy days
Angling I went, or trod the trackless hills
By mists bewildered, suddenly mine eyes
Have glanced upon him distant a few steps,
In size a giant, stalking through thick fog,
His sheep like Greenland bears; or, as he stepped
Beyond the boundary line of some hill-shadow,
His form hath flashed upon me, glorified
By the deep radiance of the setting sun: 270
Or him have I descried in distant sky,
A solitary object and sublime,
Above all height! like an aerial cross
Stationed alone upon a spiry rock
Of the Chartreuse, for worship. Thus was man
Ennobled outwardly before my sight,
And thus my heart was early introduced
To an unconscious love and reverence
Of human nature; hence the human form
To me became an index of delight,   280
Of grace and honour, power and worthiness.
Meanwhile this creature—spiritual almost
As those of books, but more exalted far;
Far more of an imaginative form
Than the gay Corin of the groves, who lives
For his own fancies, or to dance by the hour,
In coronal, with Phyllis in the midst—
Was, for the purposes of kind, a man
With the most common; husband, father; learned,
Could teach, admonish; suffered with the rest   290

246. **protending**: stretching forth. 267. **Greenland bears:**
polar bears. 273–75. **aerial . . . worship:** Cf. VI. 484–85.
285–87. **Corin; Phyllis:** stock names in pastoral literature.

From vice and folly, wretchedness and fear;
Of this I little saw, cared less for it,
But something must have felt.
                                Call ye these appearances—
Which I beheld of shepherds in my youth,
This sanctity of Nature given to man—
A shadow, a delusion, ye who pore
On the dead letter, miss the spirit of things;
Whose truth is not a motion or a shape
Instinct with vital functions, but a block
Or waxen image which yourselves have made,   300
And ye adore! But blessed be the God
Of Nature and of Man that this was so;
That men before my inexperienced eyes
Did first present themselves thus purified,
Removed, and to a distance that was fit:
And so we all of us in some degree
Are led to knowledge, wheresoever led,
And howsoever; were it otherwise,
And we found evil fast as we find good
In our first years, or think that it is found, 310
How could the innocent heart bear up and live!
But doubly fortunate my lot; not here
Alone, that something of a better life
Perhaps was round me than it is the privilege
Of most to move in, but that first I looked
At Man through objects that were great or fair;
First communed with him by their help. And thus
Was founded a sure safeguard and defence
Against the weight of meanness, selfish cares,
Coarse manners, vulgar passions, that beat in   320
On all sides from the ordinary world
In which we traffic. Starting from this point
I had my face turned toward the truth, began
With an advantage furnished by that kind
Of prepossession, without which the soul
Receives no knowledge that can bring forth good,
No genuine insight ever comes to her.

                    . . .

  Grave Teacher, stern Preceptress! for at times 530
Thou canst put on an aspect most severe;
London, to thee I willingly return.
Erewhile my verse played idly with the flowers
Enwrought upon thy mantle; satisfied
With that amusement, and a simple look
Of child-like inquisition now and then
Cast upwards on thy countenance, to detect
Some inner meanings which might harbour there.
But how could I in mood so light indulge,

Keeping such fresh remembrance of the day, 540
When, having thridded the long labyrinth
Of the suburban villages, I first
Entered thy vast dominion? On the roof
Of an itinerant vehicle I sate,
With vulgar men about me, trivial forms
Of houses, pavement, streets, of men and things,—
Mean shapes on every side: but, at the instant,
When to myself it fairly might be said,
The threshold now is overpast, (how strange
That aught external to the living mind 550
Should have such mighty sway! yet so it was),
A weight of ages did at once descend
Upon my heart; no thought embodied, no
Distinct remembrances, but weight and power,—
Power growing under weight: alas! I feel
That I am trifling: 'twas a moment's pause,—
All that took place within me came and went
As in a moment; yet with Time it dwells,
And grateful memory, as a thing divine.

The curious traveller, who, from open day, 560
Hath passed with torches into some huge cave,
The Grotto of Antiparos, or the Den
In old time haunted by that Danish Witch,
Yordas; he looks around and sees the vault
Widening on all sides; sees, or thinks he sees,
Erelong, the massy roof above his head,
That instantly unsettles and recedes,—
Substance and shadow, light and darkness, all
Commingled, making up a canopy
Of shapes and forms and tendencies to shape 570
That shift and vanish, change and interchange
Like spectres,—ferment silent and sublime!
That after a short space works less and less,
Till, every effort, every motion gone,
The scene before him stands in perfect view
Exposed, and lifeless as a written book!—
But let him pause awhile, and look again,
And a new quickening shall succeed, at first
Beginning timidly, then creeping fast,
Till the whole cave, so late a senseless mass, 580
Busies the eye with images and forms
Boldly assembled,—here is shadowed forth
From the projections, wrinkles, cavities,
A variegated landscape,—there the shape
Of some gigantic warrior clad in mail,
The ghostly semblance of a hooded monk,

Veiled nun, or pilgrim resting on his staff:
Strange congregation! yet not slow to meet
Eyes that perceive through minds that can inspire.

Even in such sort had I at first been moved, 590
Nor otherwise continued to be moved,
As I explored the vast metropolis,
Fount of my country's destiny and the world's;
That great emporium, chronicle at once
And burial-place of passions, and their home
Imperial, their chief living residence.

With strong sensations teeming as it did
Of past and present, such a place must needs
Have pleased me, seeking knowledge at that time
Far less than craving power; yet knowledge came, 600
Sought or unsought, and influxes of power
Came, of themselves, or at her call derived
In fits of kindliest apprehensiveness,
From all sides, when whate'er was in itself
Capacious found, or seemed to find, in me
A correspondent amplitude of mind;
Such is the strength and glory of our youth!
The human nature unto which I felt
That I belonged, and reverenced with love,
Was not a punctual presence, but a spirit 610
Diffused through time and space, with aid derived
Of evidence from monuments, erect,
Prostrate, or leaning towards their common rest
In earth, the widely scattered wreck sublime
Of vanished nations, or more clearly drawn
From books and what they picture and record.

'Tis true, the history of our native land—
With those of Greece compared and popular Rome,
And in our high-wrought modern narratives
Stript of their harmonising soul, the life 620
Of manners and familiar incidents,
Had never much delighted me. And less
Than other intellects had mine been used
To lean upon extrinsic circumstance
Of record or tradition; but a sense
Of what in the Great City had been done
And suffered, and was doing, suffering, still,
Weighed with me, could support the test of thought;
And, in despite of all that had gone by,
Or was departing never to return, 630
There I conversed with majesty and power

---

541. thridded: threaded. 562. Antiparos: an island in the Aegean Sea.

600. power: See De Quincey's famous distinction, which he owed to Wordsworth, of the "literature of knowledge and the literature of power."

Like independent natures. Hence the place
Was thronged with impregnations like the Wilds
In which my early feelings had been nursed—
Bare hills and valleys, full of caverns, rocks,
And audible seclusions, dashing lakes,
Echoes and waterfalls, and pointed crags
That into music touch the passing wind.
Here then my young imagination found
No uncongenial element; could here   640
Among new objects serve or give command,
Even as the heart's occasions might require,
To forward reason's else too-scrupulous march.
The effect was, still more elevated views
Of human nature. Neither vice nor guilt,
Debasement undergone by body or mind,
Nor all the misery forced upon my sight,
Misery not lightly passed, but sometimes scanned
Most feelingly, could overthrow my trust
In what we *may* become; induce belief   650
That I was ignorant, had been falsely taught,
A solitary, who with vain conceits
Had been inspired, and walked about in dreams.
From those sad scenes when meditation turned,
Lo! everything that was indeed divine
Retained its purity inviolate,
Nay brighter shone, by this portentous gloom
Set off; such opposition as aroused
The mind of Adam, yet in Paradise
Though fallen from bliss, when in the East he saw 660
Darkness ere day's mid course, and morning light
More orient in the western cloud, that drew
O'er the blue firmament a radiant white,
Descending slow with something heavenly fraught.

     Add also, that among the multitudes
Of that huge city, oftentimes was seen
Affectingly set forth, more than elsewhere
Is possible, the unity of man,
One spirit over ignorance and vice
Predominant, in good and evil hearts;   670
One sense for moral judgments, as one eye
For the sun's light. The soul when smitten thus
By a sublime *idea*, whencesoe'er
Vouchsafed for union or communion, feeds
On the pure bliss, and takes her rest with God.

     Thus from a very early age, O Friend!
My thoughts by slow gradations had been drawn

To human-kind, and to the good and ill
Of human life: Nature had led me on;
And oft amid the "busy hum" I seemed   680
To travel independent of her help,
As if I had forgotten her; but no,
The world of human-kind outweighed not hers
In my habitual thoughts; the scale of love,
Though filling daily, still was light, compared
With that in which *her* mighty objects lay.

from BOOK TENTH

RESIDENCE IN FRANCE

Most melancholy at that time, O Friend!
Were my day-thoughts,—my nights were miserable;
Through months, through years, long after the last
     beat
Of those atrocities, the hour of sleep   400
To me came rarely charged with natural gifts,
Such ghastly visions had I of despair
And tyranny, and implements of death;
And innocent victims sinking under fear,
And momentary hope, and worn-out prayer,
Each in his separate cell, or penned in crowds
For sacrifice, and struggling with forced mirth
And levity in dungeons, where the dust
Was laid with tears. Then suddenly the scene
Changed, and the unbroken dream entangled me 410
In long orations, which I strove to plead
Before unjust tribunals,—with a voice
Labouring, a brain confounded, and a sense,
Death-like, of treacherous desertion, felt
In the last place of refuge—my own soul.

from BOOK ELEVENTH

FRANCE

     It hath been told
That I was led to take an eager part
In arguments of civil polity,
Abruptly, and indeed before my time:

BOOK X. **397. time:** the Reign of Terror in France.
Wordsworth had been sympathetic to the French Revolution.
BOOK XI. **76–77. I . . . polity:** In France from November,
1791, to December, 1792, he naturally entered into the
political arguments going on everywhere.

660–64. **East . . . fraught:** quoted with slight changes
from *Paradise Lost*, XI. 203–07.

I had approached, like other youths, the shield
Of human nature from the golden side,   80
And would have fought, even to the death, to attest
The quality of the metal which I saw.
What there is best in individual man,
Of wise in passion, and sublime in power,
Benevolent in small societies,
And great in large ones, I had oft revolved,
Felt deeply, but not thoroughly understood
By reason: nay, far from it; they were yet,
As cause was given me afterwards to learn,
Not proof against the injuries of the day;   90
Lodged only at the sanctuary's door,
Not safe within its bosom. Thus prepared,
And with such general insight into evil,
And of the bounds which sever it from good,
As books and common intercourse with life
Must needs have given—to the inexperienced mind,
When the world travels in a beaten road,
Guide faithful as is needed—I began
To meditate with ardour on the rule
And management of nations; what it is   100
And ought to be; and strove to learn how far
Their power or weakness, wealth or poverty,
Their happiness or misery, depends
Upon their laws, and fashion of the State.

O pleasant exercise of hope and joy!
For mighty were the auxiliars which then stood
Upon our side, we who were strong in love!
Bliss was it in that dawn to be alive,
But to be young was very Heaven! O times,
In which the meagre, stale, forbidding ways   110
Of custom, law, and statute, took at once
The attraction of a country in romance!
When Reason seemed the most to assert her rights
When most intent on making of herself
A prime enchantress—to assist the work,
Which then was going forward in her name!
Not favoured spots alone, but the whole Earth,
The beauty wore of promise—that which sets
(As at some moments might not be unfelt
Among the bowers of Paradise itself)   120
The budding rose above the rose full blown.
What temper at the prospect did not wake
To happiness unthought of? The inert
Were roused, and lively natures rapt away!
They who had fed their childhood upon dreams,
The play-fellows of fancy, who had made
All powers of swiftness, subtilty, and strength
Their ministers,—who in lordly wise had stirred

Among the grandest objects of the sense,
And dealt with whatsoever they found there   130
As if they had within some lurking right
To wield it;—they, too, who of gentle mood
Had watched all gentle motions, and to these
Had fitted their own thoughts, schemers more mild,
And in the region of their peaceful selves;—
Now was it that *both* found, the meek and lofty
Did both find, helpers to their hearts' desire,
And stuff at hand, plastic as they could wish,—
Were called upon to exercise their skill,
Not in Utopia,—subterranean fields,—   140
Or some secreted island, Heaven knows where!
But in the very world, which is the world
Of all of us,—the place where, in the end,
We find our happiness, or not at all!

Why should I not confess that Earth was then
To me, what an inheritance, new-fallen,
Seems, when the first time visited, to one
Who thither comes to find in it his home?
He walks about and looks upon the spot
With cordial transport, moulds it and remoulds,   150
And is half-pleased with things that are amiss,
'Twill be such joy to see them disappear.

An active partisan, I thus convoked
From every object pleasant circumstance
To suit my ends; I moved among mankind
With genial feelings still predominant;
When erring, erring on the better part,
And in the kinder spirit; placable,
Indulgent, as not uninformed that men
See as they have been taught, and that Antiquity   160
Gives rights to error; and aware, no less,
That throwing off oppression must be work
As well of License as of Liberty;
And above all—for this was more than all—
Not caring if the wind did now and then
Blow keen upon an eminence that gave
Prospect so large into futurity;
In brief, a child of Nature, as at first,
Diffusing only those affections wider
That from the cradle had grown up with me,   170
And losing, in no other way than light
Is lost in light, the weak in the more strong.

In the main outline, such it might be said
Was my condition, till with open war

174. **war**: France declared war against England on February
1, 1793.

Britain opposed the liberties of France.
This threw me first out of the pale of love;
Soured and corrupted, upwards to the source,
My sentiments; was not, as hitherto,
A swallowing up of lesser things in great,
But change of them into their contraries;   180
And thus a way was opened for mistakes
And false conclusions, in degree as gross,
In kind more dangerous. What had been a pride,
Was now a shame; my likings and my loves
Ran in new channels, leaving old ones dry;
And hence a blow that, in maturer age,
Would but have touched the judgment, struck more
        deep
Into sensations near the heart: meantime,
As from the first, wild theories were afloat,
To whose pretensions, sedulously urged,   190
I had but lent a careless ear, assured
That time was ready to set all things right,
And that the multitude, so long oppressed,
Would be oppressed no more.
                              But when events
Brought less encouragement, and unto these
The immediate proof of principles no more
Could be entrusted, while the events themselves,
Worn out in greatness, stripped of novelty,
Less occupied the mind, and sentiments
Could through my understanding's natural growth 200
No longer keep their ground, by faith maintained
Of inward consciousness, and hope that laid
Her hand upon her object—evidence
Safer, of universal application, such
As could not be impeached, was sought elsewhere.

                    . . .

This was the time, when, all things tending fast
To depravation, speculative schemes—
That promised to abstract the hopes of Man
Out of his feelings, to be fixed thenceforth
For ever in a purer element—
Found ready welcome. Tempting region *that*
For Zeal to enter and refresh herself,
Where passions had the privilege to work,   230
And never hear the sound of their own names.
But, speaking more in charity, the dream

224–27. **speculative . . . element:** Wordsworth refers to
the rationalistic political philosophies of the time. In England
the chief example was William Godwin's *Political Justice*,
which was published in February, 1793.

Flattered the young, pleased with extremes, nor least
With that which makes our Reason's naked self
The object of its fervour. What delight!
How glorious! in self-knowledge and self-rule,
To look through all the frailties of the world,
And, with a resolute mastery shaking off
Infirmities of nature, time, and place,
Build social upon personal Liberty,   240
Which, to the blind restraints of general laws
Superior, magisterially adopts
One guide, the light of circumstances, flashed
Upon an independent intellect.
Thus expectation rose again; thus hope,
From her first ground expelled, grew proud once more.
Oft, as my thoughts were turned to human kind,
I scorned indifference; but, inflamed with thirst
Of a secure intelligence, and sick
Of other longing, I pursued what seemed   250
A more exalted nature; wished that Man
Should start out of his earthly, worm-like state,
And spread abroad the wings of Liberty,
Lord of himself, in undisturbed delight—
A noble aspiration! *yet* I feel
(Sustained by worthier as by wiser thoughts)
The aspiration, nor shall ever cease
To feel it;—but return we to our course.

    Enough, 'tis true—could such a plea excuse
Those aberrations—had the clamorous friends   260
Of ancient Institutions said and done
To bring disgrace upon their very names;
Disgrace, of which, custom and written law,
And sundry moral sentiments as props
Or emanations of those institutes,
Too justly bore a part. A veil had been
Uplifted; why deceive ourselves? in sooth,
'Twas even so; and sorrow for the man
Who either had not eyes wherewith to see,
Or, seeing, had forgotten! A strong shock   270
Was given to old opinions; all men's minds
Had felt its power, and mine was both let loose,
Let loose and goaded. After what hath been
Already said of patriotic love,
Suffice it here to add, that, somewhat stern
In temperament, withal a happy man,
And therefore bold to look on painful things,
Free likewise of the world, and thence more bold,
I summoned my best skill, and toiled, intent
To anatomise the frame of social life;   280
Yea, the whole body of society

Searched to its heart. Share with me, Friend! the
    wish
That some dramatic tale, endued with shapes
Livelier, and flinging out less guarded words
Than suit the work we fashion, might set forth
What then I learned, or think I learned, of truth,
And the errors into which I fell, betrayed
By present objects, and by reasonings false
From their beginnings, inasmuch as drawn
Out of a heart that had been turned aside   290
From Nature's way by outward accidents,
And which was thus confounded, more and more
Misguided, and misguiding. So I fared,
Dragging all precepts, judgments, maxims, creeds,
Like culprits to the bar; calling the mind,
Suspiciously, to establish in plain day
Her titles and her honours; now believing,
Now disbelieving; endlessly perplexed
With impulse, motive, right and wrong, the ground
Of obligation, what the rule and whence   300
The sanction; till, demanding formal *proof*,
And seeking it in every thing, I lost
All feeling of conviction, and, in fine,
Sick, wearied out with contrarieties,
Yielded up moral questions in despair.

    This was the crisis of that strong disease,
This the soul's last and lowest ebb; I drooped,
Deeming our blessed reason of least use
Where wanted most: "The lordly attributes
Of will and choice," I bitterly exclaimed,   310
"What are they but a mockery of a Being
Who hath in no concerns of his a test
Of good and evil; knows not what to fear
Or hope for, what to covet or to shun;
And who, if those could be discerned, would yet
Be little profited, would see, and ask
Where is the obligation to enforce?
And, to acknowledged law rebellious, still,
As selfish passion urged, would act amiss;
The dupe of folly, or the slave of crime."   320

    Depressed, bewildered thus, I did not walk
With scoffers, seeking light and gay revenge
From indiscriminate laughter, nor sate down
In reconcilement with an utter waste
Of intellect; such sloth I could not brook,
(Too well I loved, in that my spring of life,
Pains-taking thoughts, and truth, their dear reward)
But turned to abstract science, and there sought
Work for the reasoning faculty enthroned

Where the disturbances of space and time—   330
Whether in matter's various properties
Inherent, or from human will and power
Derived—find no admission. Then it was—
Thanks to the bounteous Giver of all good!—
That the beloved Sister in whose sight
Those days were passed, now speaking in a voice
Of sudden admonition—like a brook
That did but *cross* a lonely road, and now
Is seen, heard, felt, and caught at every turn,
Companion never lost through many a league—   340
Maintained for me a saving intercourse
With my true self; for, though bedimmed and changed
Much, as it seemed, I was no further changed
Than as a clouded and a waning moon:
She whispered still that brightness would return,
She, in the midst of all, preserved me still
A Poet, made me seek beneath that name,
And that alone, my office upon earth;
And, lastly, as hereafter will be shown,
If willing audience fail not, Nature's self,   350
By all varieties of human love
Assisted, led me back through opening day
To those sweet counsels between head and heart
Whence grew that genuine knowledge, fraught with
    peace,
Which, through the later sinkings of this cause,
Hath still upheld me, and upholds me now
In the catastrophe (for so they dream,
And nothing less), when, finally to close
And rivet down all the gains of France, a Pope
Is summoned in, to crown an Emperor—   360
This last opprobrium, when we see a people
That once looked up in faith, as if to Heaven
For manna, take a lesson from the dog
Returning to his vomit; when the sun
That rose in splendour, was alive, and moved
In exultation with a living pomp
Of clouds—his glory's natural retinue—
Hath dropped all functions by the gods bestowed,
And, turned into a gewgaw, a machine,
Sets like an Opera phantom.   370

---

335. **Sister:** Away at school and college, and then in France,
Wordsworth saw relatively little of his sister before September, 1795, when they lived together at Racedown. Thereafter
she was always a member of his household.

from BOOK TWELFTH
IMAGINATION AND TASTE, HOW IMPAIRED AND RESTORED

Long time have human ignorance and guilt
Detained us, on what spectacles of woe
Compelled to look, and inwardly oppressed
With sorrow, disappointment, vexing thoughts,
Confusion of the judgment, zeal decayed,
And, lastly, utter loss of hope itself
And things to hope for! Not with these began
Our song, and not with these our song must end.—
Ye motions of delight, that haunt the sides
Of the green hills; ye breezes and soft airs,⁣   10
Whose subtle intercourse with breathing flowers,
Feelingly watched, might teach Man's haughty race
How without injury to take, to give
Without offence; ye who, as if to show
The wondrous influence of power gently used,
Bend the complying heads of lordly pines,
And, with a touch, shift the stupendous clouds
Through the whole compass of the sky; ye brooks,
Muttering along the stones, a busy noise
By day, a quiet sound in silent night;⁣   20
Ye waves, that out of the great deep steal forth
In a calm hour to kiss the pebbly shore,
Not mute, and then retire, fearing no storm;
And you, ye groves, whose ministry it is
To interpose the covert of your shades,
Even as a sleep, between the heart of man
And outward troubles, between man himself,
Not seldom, and his own uneasy heart:
Oh! that I had a music and a voice
Harmonious as your own, that I might tell⁣   30
What ye have done for me. The morning shines,
Nor heedeth Man's perverseness; Spring returns,—
I saw the Spring return, and could rejoice,
In common with the children of her love,
Piping on boughs, or sporting on fresh fields,
Or boldly seeking pleasure nearer heaven
On wings that navigate cerulean skies.
So neither were complacency, nor peace,
Nor tender yearnings, wanting for my good
Through these distracted times; in Nature still⁣   40
Glorying, I found a counterpoise in her,
Which, when the spirit of evil reached its height,
Maintained for me a secret happiness.

. . . .

There are in our existence spots of time,
That with distinct pre-eminence retain
A renovating virtue, whence, depressed⁣   210
By false opinion and contentious thought,
Or aught of heavier or more deadly weight,
In trivial occupations, and the round
Of ordinary intercourse, our minds
Are nourished and invisibly repaired;
A virtue, by which pleasure is enhanced,
That penetrates, enables us to mount,
When high, more high, and lifts us up when fallen.
This efficacious spirit chiefly lurks
Among those passages of life that give⁣   220
Profoundest knowledge to what point, and how,
The mind is lord and master—outward sense
The obedient servant of her will. Such moments
Are scattered everywhere, taking their date
From our first childhood. I remember well,
That once, while yet my inexperienced hand
Could scarcely hold a bridle, with proud hopes
I mounted, and we journeyed towards the hills:
An ancient servant of my father's house
Was with me, my encourager and guide:⁣   230
We had not travelled long, ere some mischance
Disjoined me from my comrade; and, through fear
Dismounting, down the rough and stony moor
I led my horse, and, stumbling on, at length
Came to a bottom, where in former times
A murderer had been hung in iron chains.
The gibbet-mast had mouldered down, the bones
And iron case were gone; but on the turf,
Hard by, soon after that fell deed was wrought,
Some unknown hand had carved the murderer's name.
The monumental letters were inscribed⁣   241
In times long past; but still, from year to year
By superstition of the neighbourhood,
The grass is cleared away, and to this hour
The characters are fresh and visible:
A casual glance had shown them, and I fled,
Faltering and faint, and ignorant of the road:
Then, reascending the bare common, saw
A naked pool that lay beneath the hills,
The beacon on the summit, and, more near,⁣   250
A girl, who bore a pitcher on her head,
And seemed with difficult steps to force her way
Against the blowing wind. It was, in truth,
An ordinary sight; but I should need

BOOK XII. **250. beacon:** Cf. VI. 233–36.

Colours and words that are unknown to man,
To paint the visionary dreariness
Which, while I looked all round for my lost guide,
Invested moorland waste, and naked pool,
The beacon crowning the lone eminence,
The female and her garments vexed and tossed   260
By the strong wind. When, in the blessed hours
Of early love, the loved one at my side,
I roamed, in daily presence of this scene,
Upon the naked pool and dreary crags,
And on the melancholy beacon, fell
A spirit of pleasure and youth's golden gleam;
And think ye not with radiance more sublime
For these remembrances, and for the power
They had left behind? So feeling comes in aid
Of feeling, and diversity of strength   270
Attends us, if but once we have been strong.
Oh! mystery of man, from what a depth
Proceed thy honours. I am lost, but see
In simple childhood something of the base
On which thy greatness stands; but this I feel,
That from thyself it comes, that thou must give,
Else never canst receive. The days gone by
Return upon me almost from the dawn
Of life: the hiding-places of man's power
Open; I would approach them, but they close.   280
I see by glimpses now; when age comes on,
May scarcely see at all; and I would give,
While yet we may, as far as words can give,
Substance and life to what I feel, enshrining,
Such is my hope, the spirit of the Past
For future restoration.—Yet another
Of these memorials:—
                         One Christmas-time,
On the glad eve of its dear holidays,
Feverish, and tired, and restless, I went forth
Into the fields, impatient for the sight   290
Of those led palfreys that should bear us home;
My brothers and myself. There rose a crag,
That, from the meeting-point of two highways
Ascending, overlooked them both, far stretched;
Thither, uncertain on which road to fix
My expectation, thither I repaired,
Scout-like, and gained the summit; 'twas a day
Tempestuous, dark, and wild, and on the grass
I sate half-sheltered by a naked wall;
Upon my right hand couched a single sheep,   300

291. **home:** from school.

Upon my left a blasted hawthorn stood;
With those companions at my side, I watched,
Straining my eyes intensely, as the mist
Gave intermitting prospect of the copse
And plain beneath. Ere we to school returned,—
That dreary time,—ere we had been ten days
Sojourners in my father's house, he died,
And I and my three brothers, orphans then,
Followed his body to the grave. The event,
With all the sorrow that it brought, appeared   310
A chastisement; and when I called to mind
That day so lately past, when from the crag
I looked in such anxiety of hope;
With trite reflections of morality,
Yet in the deepest passion, I bowed low
To God, Who thus corrected my desires;
And, afterwards, the wind and sleety rain,
And all the business of the elements,
The single sheep, and the one blasted tree,
And the bleak music from that old stone wall,   320
The noise of wood and water, and the mist
That on the line of each of those two roads
Advanced in such indisputable shapes;
All these were kindred spectacles and sounds
To which I oft repaired, and thence would drink,
As at a fountain; and on winter nights,
Down to this very time, when storm and rain
Beat on my roof, or, haply, at noon-day,
While in a grove I walk, whose lofty trees,
Laden with summer's thickest foliage, rock   330
In a strong wind, some working of the spirit,
Some inward agitations thence are brought,
Whate'er their office, whether to beguile
Thoughts over busy in the course they took,
Or animate an hour of vacant ease.

311-16. **chastisement . . . desires:** The words are plain
enough, but it is not easy to say how or whom the child had
offended. A sensitive boy schooled in a Christian household
of the nineteenth century might conceivably take the death
of his father as a punishment for his own impatience, or more
generally for forgetting—despite the dreary, grim, and threatening aspect of the natural world—that earthly life is a time
of sorrow and probation. One may also recall Book I, ll. 306-
400, episodes in which the child is unresponsive to the tone
or mood of nature and is chastised.

from BOOK THIRTEENTH

IMAGINATION AND TASTE, HOW IMPAIRED
AND RESTORED (*concluded*)

From Nature doth emotion come, and moods
Of calmness equally are Nature's gift:
This is her glory; these two attributes
Are sister horns that constitute her strength.
Hence Genius, born to thrive by interchange
Of peace and excitation, finds in her
His best and purest friend; from her receives
That energy by which he seeks the truth,
From her that happy stillness of the mind
Which fits him to receive it when unsought.   10

Such benefit the humblest intellects
Partake of, each in their degree; 'tis mine
To speak, what I myself have known and felt;
Smooth task! for words find easy way, inspired
By gratitude, and confidence in truth.
Long time in search of knowledge did I range
The field of human life, in heart and mind
Benighted; but, the dawn beginning now
To re-appear, 'twas proved that not in vain
I had been taught to reverence a Power   20
That is the visible quality and shape
And image of right reason; that matures
Her processes by steadfast laws; gives birth
To no impatient or fallacious hopes,
No heat of passion or excessive zeal,
No vain conceits; provokes to no quick turns
Of self-applauding intellect; but trains
To meekness, and exalts by humble faith;
Holds up before the mind intoxicate
With present objects, and the busy dance   30
Of things that pass away, a temperate show
Of objects that endure; and by this course
Disposes her, when over-fondly set
On throwing off incumbrances, to seek
In man, and in the frame of social life,
Whate'er there is desirable and good
Of kindred permanence, unchanged in form
And function, or, through strict vicissitude,
Of life and death, revolving. Above all
Were re-established now those watchful thoughts   40
Which, seeing little worthy or sublime

In what the Historian's pen so much delights
To blazon—power and energy detached
From moral purpose—early tutored me
To look with feelings of fraternal love
Upon the unassuming things that hold
A silent station in this beauteous world.

Thus moderated, thus composed, I found
Once more in Man an object of delight,
Of pure imagination, and of love;   50
And, as the horizon of my mind enlarged,
Again I took the intellectual eye
For my instructor, studious more to see
Great truths, than touch and handle little ones.
Knowledge was given accordingly; my trust
Became more firm in feelings that had stood
The test of such a trial; clearer far
My sense of excellence—of right and wrong:
The promise of the present time retired
Into its true proportion; sanguine schemes,   60
Ambitious projects, pleased me less; I sought
For present good in life's familiar face,
And built thereon my hopes of good to come.

. . . .

Who doth not love to follow with his eye
The windings of a public way? the sight,
Familiar object as it is, hath wrought
On my imagination since the morn
Of childhood, when a disappearing line,
One daily present to my eyes, that crossed
The naked summit of a far-off hill
Beyond the limits that my feet had trod,
Was like an invitation into space   150
Boundless, or guide into eternity.
Yes, something of the granduer which invests
The mariner who sails the roaring sea
Through storm and darkness, early in my mind
Surrounded, too, the wanderers of the earth;
Grandeur as much, and loveliness far more.
Awed have I been by strolling Bedlamites;
From many other uncouth vagrants (passed
In fear) have walked with quicker step; but why
Take note of this? When I began to enquire,   160
To watch and question those I met, and speak
Without reserve to them, the lonely roads
Were open schools in which I daily read
With most delight the passions of mankind,

BOOK XIII. **22. right reason:** Cf. XIV. 192.

157. Bedlamites: lunatics.

Whether by words, looks, sighs, or tears, revealed;
There saw into the depth of human souls,
Souls that appear to have no depth at all
To careless eyes. And—now convinced at heart
How little those formalities, to which
With overweening trust alone we give   170
The name of Education, have to do
With real feeling and just sense; how vain
A correspondence with the talking world
Proves to the most; and called to make good search
If man's estate, by doom of Nature yoked
With toil, be therefore yoked with ignorance;
If virtue be indeed so hard to rear,
And intellectual strength so rare a boon—
I prized such walks still more, for there I found
Hope to my hope, and to my pleasure peace   180
And steadiness, and healing and repose
To every angry passion. There I heard,
From mouths of men obscure and lowly, truths
Replete with honour; sounds in unison
With loftiest promises of good and fair.

   There are who think that strong affection, love
Known by whatever name, is falsely deemed
A gift, to use a term which they would use,
Of vulgar nature; that its growth requires
Retirement, leisure, language purified   190
By manners studied and elaborate;
That whoso feels such passion in its strength
Must live within the very light and air
Of courteous usages refined by art.
True is it, where oppression worse than death
Salutes the being at his birth, where grace
Of culture hath been utterly unknown,
And poverty and labour in excess
From day to day pre-occupy the ground
Of the affections, and to Nature's self   200
Oppose a deeper nature; there, indeed,
Love cannot be; nor does it thrive with ease
Among the close and overcrowded haunts
Of cities, where the human heart is sick,
And the eye feeds it not, and cannot feed.
—Yes, in those wanderings deeply did I feel
How we mislead each other; above all,
How books mislead us, seeking their reward
From judgments of the wealthy Few, who see
By artificial lights; how they debase   210
The Many for the pleasure of those Few;
Effeminately level down the truth
To certain general notions, for the sake

Of being understood at once, or else
Through want of better knowledge in the heads
That framed them; flattering self-conceit with words,
That, while they most ambitiously set forth
Extrinsic differences, the outward marks
Whereby society has parted man
From man, neglect the universal heart.   220

   Here, calling up to mind what then I saw,
A youthful traveller, and see daily now
In the familiar circuit of my home,
Here might I pause, and bend in reverence
To Nature, and the power of human minds,
To men as they are men within themselves.
How oft high service is performed within,
When all the external man is rude in show,—
Not like a temple rich with pomp and gold,
But a mere mountain chapel, that protects   230
Its simple worshippers from sun and shower.
Of these, said I, shall be my song; of these,
If future years mature me for the task,
Will I record the praises, making verse
Deal boldly with substantial things; in truth
And sanctity of passion, speak of these,
That justice may be done, obeisance paid
Where it is due: thus haply shall I teach,
Inspire, through unadulterated ears
Pour rapture, tenderness, and hope,—my theme   240
No other than the very heart of man,
As found among the best of those who live,
Not unexalted by religious faith,
Nor uninformed by books, good books, though few,
In Nature's presence: thence may I select
Sorrow, that is not sorrow, but delight;
And miserable love, that is not pain
To hear of, for the glory that redounds
Therefrom to human kind, and what we are.
Be mine to follow with no timid step   250
Where knowledge leads me: it shall be my pride
That I have dared to tread this holy ground,
Speaking no dream, but things oracular;
Matter not lightly to be heard by those
Who to the letter of the outward promise
Do read the invisible soul; by men adroit
In speech, and for communion with the world
Accomplished; minds whose faculties are then
Most active when they are most eloquent,
And elevated most when most admired.   260
Men may be found of other mould than these,
Who are their own upholders, to themselves

Encouragement, and energy, and will,
Expressing liveliest thoughts in lively words
As native passion dictates. Others, too,
There are among the walks of homely life
Still higher, men for contemplation framed,
Shy, and unpractised in the strife of phrase;
Meek men, whose very souls perhaps would sink
Beneath them, summoned to such intercourse: 270
Theirs is the language of the heavens, the power,
The thought, the image, and the silent joy:
Words are but under-agents in their souls;
When they are grasping with their greatest strength,
They do not breathe among them: this I speak
In gratitude to God, Who feeds our hearts
For His own service; knoweth, loveth us,
When we are unregarded by the world.

  Also, about this time did I receive
Convictions still more strong than heretofore,   280
Not only that the inner frame is good,
And graciously composed, but that, no less,
Nature for all conditions wants not power
To consecrate, if we have eyes to see,
The outside of her creatures, and to breathe
Grandeur upon the very humblest face
Of human life. I felt that the array
Of act and circumstance, and visible form,
Is mainly to the pleasure of the mind
What passion makes them; that meanwhile the forms
Of Nature have a passion in themselves,   291
That intermingles with those works of man
To which she summons him, although the works
Be mean, have nothing lofty of their own;
And that the Genius of the Poet hence
May boldly take his way among mankind
Wherever Nature leads; that he hath stood
By Nature's side among the men of old,
And so shall stand for ever. Dearest Friend!
If thou partake the animating faith   300
That Poets, even as Prophets, each with each
Connected in a mighty scheme of truth,
Have each his own peculiar faculty,
Heaven's gift, a sense that fits him to perceive
Objects unseen before, thou wilt not blame
The humblest of this band who dares to hope
That unto him hath also been vouchsafed
An insight that in some sort he possesses,
A privilege whereby a work of his,
Proceeding from a source of untaught things,   310
Creative and enduring, may become

A power like one of Nature's. To a hope
Not less ambitious once among the wilds
Of Sarum's Plain, my youthful spirit was raised;
There, as I ranged at will the pastoral downs
Trackless and smooth, or paced the bare white roads
Lengthening in solitude their dreary line,
Time with his retinue of ages fled
Backwards, nor checked his flight until I saw
Our dim ancestral Past in vision clear;   320
Saw multitudes of men, and here and there,
A single Briton clothed in wolf-skin vest,
With shield and stone-axe, stride across the wold;
The voice of spears was heard, the rattling spear
Shaken by arms of mighty bone, in strength,
Long mouldered, of barbaric majesty.
I called on Darkness—but before the word
Was uttered, midnight darkness seemed to take
All objects from my sight; and lo! again
The Desert visible by dismal flames;   330
It is the sacrificial altar, fed
With living men—how deep the groans! the voice
Of those that crowd the giant wicker thrills
The monumental hillocks, and the pomp
Is for both worlds, the living and the dead.
At other moments (for through that wide waste
Three summer days I roamed) where'er the Plain
Was figured o'er with circles, lines, or mounds,
That yet survive, a work, as some divine,
Shaped by the Druids, so to represent   340
Their knowledge of the heavens, and image forth
The constellations; gently was I charmed
Into a waking dream, a reverie
That, with believing eyes, where'er I turned,
Beheld long-bearded teachers, with white wands
Uplifted, pointing to the starry sky,
Alternately, and plain below, while breath
Of music swayed their motions, and the waste
Rejoiced with them and me in those sweet sounds.

  This for the past, and things that may be viewed 350
Or fancied in the obscurity of years
From monumental hints: and thou, O Friend!

314. Sarum's Plain: Salisbury Plain in southern England;
it contains Stonehenge. **329-49.** According to Caesar's
Gallic War, the Druids were the learned and priestly caste
among the Britons and Gauls. He mentions that they performed human sacrifices by setting on fire vast images
made of wicker-work and filled with living men, and also
that they taught the young lore concerning the stars and
their motions.

Pleased with some unpremeditated strains
That served those wanderings to beguile, hast said
That then and there my mind had exercised
Upon the vulgar forms of present things,
The actual world of our familiar days,
Yet higher power; had caught from them a tone,
An image, and a character, by books
Not hitherto reflected. Call we this   360
A partial judgment—and yet why? for *then*
We were as strangers; and I may not speak
Thus wrongfully of verse, however rude,
Which on thy young imagination, trained
In the great City, broke like light from far.
Moreover, each man's Mind is to herself
Witness and judge; and I remember well
That in life's every-day appearances
I seemed about this time to gain clear sight
Of a new world—a world, too, that was fit   370
To be transmitted, and to other eyes
Made visible; as ruled by those fixed laws
Whence spiritual dignity originates,
Which do both give it being and maintain
A balance, an ennobling interchange
Of action from without and from within;
The excellence, pure function, and best power
Both of the object seen, and eye that sees.

from BOOK FOURTEENTH

CONCLUSION

In one of those excursions (may they ne'er
Fade from remembrance!) through the Northern tracts
Of Cambria ranging with a youthful friend,
I left Bethgelert's huts at couching-time,
And westward took my way, to see the sun
Rise, from the top of Snowdon. To the door
Of a rude cottage at the mountain's base
We came, and roused the shepherd who attends
The adventurous stranger's steps, a trusty guide;
Then, cheered by short refreshment, sallied forth.   10

It was a close, warm, breezeless summer night,
Wan, dull, and glaring, with a dripping fog
Low-hung and thick that covered all the sky;
But, undiscouraged, we began to climb
The mountain-side. The mist soon girt us round,
And, after ordinary travellers' talk
With our conductor, pensively we sank
Each into commerce with his private thoughts:
Thus did we breast the ascent, and by myself
Was nothing either seen or heard that checked   20
Those musings or diverted, save that once
The shepherd's lurcher, who, among the crags,
Had to his joy unearthed a hedgehog, teased
His coiled-up prey with barkings turbulent.
This small adventure, for even such it seemed
In that wild place and at the dead of night,
Being over and forgotten, on we wound
In silence as before. With forehead bent
Earthward, as if in opposition set
Against an enemy, I panted up   30
With eager pace, and no less eager thoughts.
Thus might we wear a midnight hour away,
Ascending at loose distance each from each,
And I, as chanced, the foremost of the band;
When at my feet the ground appeared to brighten,
And with a step or two seemed brighter still;
Nor was time given to ask or learn the cause,
For instantly a light upon the turf
Fell like a flash, and lo! as I looked up
The Moon hung naked in a firmament   40
Of azure without cloud, and at my feet
Rested a silent sea of hoary mist.
A hundred hills their dusky backs upheaved
All over this still ocean; and beyond,
Far, far beyond, the solid vapours stretched,
In headlands, tongues, and promontory shapes,
Into the main Atlantic, that appeared
To dwindle, and give up his majesty,
Usurped upon far as the sight could reach.
Not so the ethereal vault; encroachment none   50
Was there, nor loss; only the inferior stars
Had disappeared, or shed a fainter light
In the clear presence of the full-orbed Moon,
Who, from her sovereign elevation, gazed
Upon the billowy ocean, as it lay
All meek and silent, save that through a rift—
Not distant from the shore whereon we stood,
A fixed, abysmal, gloomy, breathing-place—

---

353. strains: an early version of "Guilt and Sorrow." See
Biographia Literaria, below, Chap. IV, where Coleridge
describes his impressions on first hearing this poem read
aloud. BOOK XIV. **1. one . . . excursions:** The climb actually
took place in 1791. Wordsworth deliberately displaces it in
chronological time in order to use it at the close of the poem.
3. Cambria: Wales. **4. Bethgelert:** a village. **6. Snowdon:**
highest mountain in England and Wales.

22. lurcher: crossbred hunting dog.

Mounted the roar of waters, torrents, streams
Innumerable, roaring with one voice!   60
Heard over earth and sea, and, in that hour,
For so it seemed, felt by the starry heavens.

When into air had partially dissolved
That vision, given to spirits of the night
And three chance human wanderers, in calm thought
Reflected, it appeared to me the type
Of a majestic intellect, its acts
And its possessions, what it has and craves,
What in itself it is, and would become.
There I beheld the emblem of a mind   70
That feeds upon infinity, that broods
Over the dark abyss, intent to hear
Its voices issuing forth to silent light
In one continuous stream; a mind sustained
By recognitions of transcendent power,
In sense conducting to ideal form,
In soul of more than mortal privilege.
One function, above all, of such a mind
Had Nature shadowed there, by putting forth,
'Mid circumstances awful and sublime,   80
That mutual domination which she loves
To exert upon the face of outward things,
So moulded, joined, abstracted, so endowed
With interchangeable supremacy,
That men, least sensitive, see, hear, perceive,
And cannot choose but feel. The power, which all
Acknowledge when thus moved, which Nature thus
To bodily sense exhibits, is the express
Resemblance of that glorious faculty
That higher minds bear with them as their own.   90
This is the very spirit in which they deal
With the whole compass of the universe:
They from their native selves can send abroad
Kindred mutations; for themselves create
A like existence; and, whene'er it dawns
Created for them, catch it, or are caught
By its inevitable mastery,
Like angels stopped upon the wing by sound
Of harmony from Heaven's remotest spheres.
Them the enduring and the transient both   100
Serve to exalt; they build up greatest things
From least suggestions; ever on the watch,
Willing to work and to be wrought upon,
They need not extraordinary calls
To rouse them; in a world of life they live,

By sensible impressions not enthralled,
But by their quickening impulse made more prompt
To hold fit converse with the spiritual world,
And with the generations of mankind
Spread over time, past, present, and to come,   110
Age after age, till Time shall be no more.
Such minds are truly from the Deity,
For they are Powers; and hence the highest bliss
That flesh can know is theirs—the consciousness
Of Whom they are, habitually infused
Through every image and through every thought,
And all affections by communion raised
From earth to heaven, from human to divine;
Hence endless occupation for the Soul,
Whether discursive or intuitive;   120
Hence cheerfulness for acts of daily life,
Emotions which best foresight need not fear,
Most worthy then of trust when most intense.
Hence, amid ills that vex and wrongs that crush
Our hearts—if here the words of Holy Writ
May with fit reverence be applied—that peace
Which passeth understanding, that repose
In moral judgments which from this pure source
Must come, or will by man be sought in vain.

Oh! who is he that hath his whole life long   130
Preserved, enlarged, this freedom in himself?
For this alone is genuine liberty:
Where is the favoured being who hath held
That course unchecked, unerring, and untired,
In one perpetual progress smooth and bright?—
A humbler destiny have we retraced,
And told of lapse and hesitating choice,
And backward wanderings along thorny ways:
Yet—compassed round by mountain solitudes,
Within whose solemn temple I received   140
My earliest visitations, careless then
Of what was given me; and which now I range,
A meditative, oft a suffering man—
Do I declare—in accents which, from truth
Deriving cheerful confidence, shall blend
Their modulation with these vocal streams—
That, whatsoever falls my better mind,
Revolving with the accidents of life,
May have sustained, that, howsoe'er misled,
Never did I, in quest of right and wrong,   150

75–76. **recognitions . . . form**: See ll. 105–08. **89. faculty**: imagination.

119–20. **occupation . . . intuitive**: Line 120 modifies "occupation." **discursive**: reasoning from premises to conclusions. Wordsworth quotes from *Paradise Lost*, V. 486–88.

Tamper with conscience from a private aim;
Nor was in any public hope the dupe
Of selfish passions; nor did ever yield
Wilfully to mean cares or low pursuits,
But shrunk with apprehensive jealousy
From every combination which might aid
The tendency, too potent in itself,
Of use and custom to bow down the soul
Under a growing weight of vulgar sense,
And substitute a universe of death   160
For that which moves with light and life informed,
Actual, divine, and true. To fear and love,
To love as prime and chief, for there fear ends,
Be this ascribed; to early intercourse,
In presence of sublime or beautiful forms,
With the adverse principles of pain and joy—
Evil as one is rashly named by men
Who know not what they speak. By love subsists
All lasting grandeur, by pervading love;
That gone, we are as dust.—Behold the fields   170
In balmy spring-time full of rising flowers
And joyous creatures; see that pair, the lamb
And the lamb's mother, and their tender ways
Shall touch thee to the heart; thou callest this love,
And not inaptly so, for love it is,
Far as it carries thee. In some green bower
Rest, and be not alone, but have thou there
The One who is thy choice of all the world:
There linger, listening, gazing, with delight
Impassioned, but delight how pitiable!   180
Unless this love by a still higher love
Be hallowed, love that breathes not without awe;
Love that adores, but on the knees of prayer,
By heaven inspired; that frees from chains the soul,
Bearing in union with the purest, best,
Of earth-born passions on the wings of praise
A mutual tribute to the Almighty's Throne.

    This spiritual Love acts not nor can exist
Without Imagination, which, in truth,
Is but another name for absolute power   190
And clearest insight, amplitude of mind,
And Reason in her most exalted mood.
This faculty hath been the feeding source
Of our long labour: we have traced the stream
From the blind cavern whence is faintly heard
Its natal murmur; followed it to light
And open day; accompanied its course

Among the ways of Nature, for a time
Lost sight of it bewildered and engulphed:
Then given it greeting as it rose once more   200
In strength, reflecting from its placid breast
The works of man and face of human life;
And lastly, from its progress have we drawn
Faith in life endless, the sustaining thought
Of human Being, Eternity, and God.

    Imagination having been our theme,
So also hath that intellectual Love,
For they are each in each, and cannot stand
Dividually.—Here must thou be, O Man!
Power to thyself; no Helper hast thou here;   210
Here keepest thou in singleness thy state:
No other can divide with thee this work:
No secondary hand can intervene
To fashion this ability; 'tis thine,
The prime and vital principle is thine
In the recesses of thy nature, far
From any reach of outward fellowship,
Else is not thine at all. But joy to him,
Oh, joy to him who here hath sown, hath laid
Here, the foundation of his future years!   220
For all that friendship, all that love can do,
All that a darling countenance can look
Or dear voice utter, to complete the man,
Perfect him, made imperfect in himself,
All shall be his: and he whose soul hath risen
Up to the height of feeling intellect
Shall want no humbler tenderness; his heart
Be tender as a nursing mother's heart;
Of female softness shall his life be full,
Of humble cares and delicate desires,   230
Mild interests and gentlest sympathies.

                    . . .

    And now, O Friend! this history is brought
To its appointed close: the discipline
And consummation of a Poet's mind,
In everything that stood most prominent,
Have faithfully been pictured; we have reached
The time (our guiding object from the first)
When we may, not presumptuously, I hope,
Suppose my powers so far confirmed, and such
My knowledge, as to make me capable   310
Of building up a Work that shall endure.

                    . . .

181–87. Unless . . . Throne: one of the many specifically
Christian revisions of the 1805 text.

309–11. powers . . . endure: See I. 146–237, 636–46.

Whether to me shall be allotted life,
And, with life, power to accomplish aught of worth,
That will be deemed no insufficient plea   390
For having given the story of myself,
Is all uncertain: but, beloved Friend!
When, looking back, thou seest, in clearer view
Than any liveliest sight of yesterday,
That summer, under whose indulgent skies,
Upon smooth Quantock's airy ridge we roved
Unchecked, or loitered 'mid her sylvan combs,
Thou in bewitching words, with happy heart,
Didst chaunt the vision of that Ancient Man,
The bright-eyed Mariner, and rueful woes   400
Didst utter of the Lady Christabel;
And I, associate with such labour, steeped
In soft forgetfulness the livelong hours,
Murmuring of him who, joyous hap, was found,
After the perils of his moonlight ride,
Near the loud waterfall; or her who sate
In misery near the miserable Thorn—
When thou dost to that summer turn thy thoughts,
And hast before thee all which then we were,
To thee, in memory of that happiness,   410
It will be known, by thee at least, my Friend!
Felt, that the history of a Poet's mind
Is labour not unworthy of regard;
To thee the work shall justify itself.

. . .

Oh! yet a few short years of useful life,   430
And all will be complete, thy race be run,
Thy monument of glory will be raised;
They, though (too weak to tread the ways of truth)
This age fall back to old idolatry,
Though men return to servitude as fast
As the tide ebbs, to ignominy and shame
By nations sink together, we shall still
Find solace—knowing what we have learnt to know,
Rich in true happiness if allowed to be
Faithful alike in forwarding a day   440
Of firmer trust, joint labourers in the work

(Should Providence such grace to us vouchsafe)
Of their deliverance, surely yet to come.
Prophets of Nature, we to them will speak
A lasting inspiration, sanctified
By reason, blest by faith: what we have loved,
Others will love, and we will teach them how;
Instruct them how the mind of man becomes
A thousand times more beautiful than the earth
On which he dwells, above this frame of things   450
(Which, 'mid all revolution in the hopes
And fears of men, doth still remain unchanged)
In beauty exalted, as it is itself
Of quality and fabric more divine.

⟨⟨⟩⟩

The next six are generally called the "Lucy" poems.
Nothing is known as to the identity of "Lucy" or
whether Wordsworth had any actual person in mind.
Speaking in a letter of "A Slumber Did My Spirit Seal,"
Coleridge remarked: "Some months ago Wordsworth
transmitted to me a most sublime epitaph. Whether it
had any reality I cannot say. Most probably, in some
gloomier moment he had fancied the moment in which
his sister might die." "I Travelled Among Unknown
Men" is displaced from chronological order and inserted
here because Wordsworth told Mary Hutchinson it
should be read after "She Dwelt Among the Untrodden
Ways." Francis Jeffrey was nonplused by "Strange Fits
of Passion" (see pp. 365–66).

⟨⟨⟩⟩

SHE DWELT AMONG
THE UNTRODDEN WAYS

She dwelt among the untrodden ways
  Beside the springs of Dove,
A Maid whom there were none to praise
  And very few to love:

A violet by a mossy stone
  Half hidden from the eye!
—Fair as a star, when only one
  Is shining in the sky.

She lived unknown, and few could know
  When Lucy ceased to be;   10
But she is in her grave, and, oh,
  The difference to me!

1799 (1800)

395–407. **That . . . Thorn:** From July, 1797, to June,
1798, William and Dorothy lived at Alfoxden House in
Somersetshire, on the edge of the Quantock Hills and about
four miles from the village of Nether Stowey, where
Coleridge lived. During this period Coleridge's writings
included "The Rime of the Ancient Mariner" and "Christabel," and Wordsworth's included "The Idiot Boy" and "The
Thorn." **431. thy:** Coleridge's.

I TRAVELLED
AMONG UNKNOWN MEN

I travelled among unknown men,
  In lands beyond the sea;
Nor, England! did I know till then
  What love I bore to thee.

'Tis past, that melancholy dream!
  Nor will I quit thy shore
A second time; for still I seem
  To love thee more and more.

Among thy mountains did I feel
  The joy of my desire;   10
And she I cherished turned her wheel
  Beside an English fire.

Thy mornings showed, thy nights concealed,
  The bowers where Lucy played;
And thine too is the last green field
  That Lucy's eyes surveyed.

                    1801 (1807)

STRANGE FITS OF
PASSION HAVE I KNOWN

Strange fits of passion have I known:
  And I will dare to tell,
But in the Lover's ear alone,
  What once to me befell.

When she I loved looked every day
  Fresh as a rose in June,
I to her cottage bent my way,
  Beneath an evening-moon.

Upon the moon I fixed my eye,
  All over the wide lea;   10
With quickening pace my horse drew nigh
  Those paths so dear to me.

And now we reached the orchard-plot;
  And, as we climbed the hill,
The sinking moon to Lucy's cot
  Came near, and nearer still.

In one of those sweet dreams I slept,
  Kind Nature's gentlest boon!

And all the while my eyes I kept
  On the descending moon.   20

My horse moved on; hoof after hoof
  He raised, and never stopped:
When down behind the cottage roof,
  At once, the bright moon dropped.

What fond and wayward thoughts will slide
  Into a Lover's head!
"O mercy!" to myself I cried,
  "If Lucy should be dead!"

                    1799 (1800)

THREE YEARS SHE
GREW IN SUN AND SHOWER

Three years she grew in sun and shower,
Then Nature said, "A lovelier flower
  On earth was never sown;
This Child I to myself will take;
She shall be mine, and I will make
  A Lady of my own.

"Myself will to my darling be
Both law and impulse: and with me
  The Girl, in rock and plain,
In earth and heaven, in glade and bower,   10
Shall feel an overseeing power
  To kindle or restrain.

"She shall be sportive as the fawn
That wild with glee across the lawn
  Or up the mountain springs;
And hers shall be the breathing balm,
And hers the silence and the calm
  Of mute insensate things.

"The floating clouds their state shall lend
To her; for her the willow bend;   20
  Nor shall she fail to see
Even in the motions of the Storm
Grace that shall mould the Maiden's form
  By silent sympathy.

"The stars of midnight shall be dear
To her; and she shall lean her ear
  In many a secret place
Where rivulets dance their wayward round,
And beauty born of murmuring sound
  Shall pass into her face.   30

"And vital feelings of delight
Shall rear her form to stately height,
Her virgin bosom swell;
Such thoughts to Lucy I will give
While she and I together live
Here in this happy dell."

Thus Nature spake—The work was done—
How soon my Lucy's race was run!
She died, and left to me
This heath, this calm, and quiet scene;   40
The memory of what has been,
And never more will be.

1799 (1800)

A SLUMBER
DID MY SPIRIT SEAL

A slumber did my spirit seal;
  I had no human fears:
She seemed a thing that could not feel
  The touch of earthly years.

No motion has she now, no force;
  She neither hears nor sees;
Rolled round in earth's diurnal course,
  With rocks, and stones, and trees.

1799 (1800)

LUCY GRAY

OR, SOLITUDE

Oft I had heard of Lucy Gray:
And, when I crossed the wild,
I chanced to see at break of day
The solitary child.

No mate, no comrade Lucy knew;
She dwelt on a wide moor,
—The sweetest thing that ever grew
Beside a human door!

You yet may spy the fawn at play,
The hare upon the green;   10
But the sweet face of Lucy Gray
Will never more be seen.

"To-night will be a stormy night—
You to the town must go;
And take a lantern, Child, to light
Your mother through the snow."

"That, Father! will I gladly do:
'Tis scarcely afternoon—
The minster-clock has just struck two,
And yonder is the moon!"   20

At this the Father raised his hook,
And snapped a faggot-band;
He plied his work;—and Lucy took
The lantern in her hand.

Not blither is the mountain roe:
With many a wanton stroke
Her feet disperse the powdery snow,
That rises up like smoke.

The storm came on before its time:
She wandered up and down;   30
And many a hill did Lucy climb:
But never reached the town.

The wretched parents all that night
Went shouting far and wide;
But there was neither sound nor sight
To serve them for a guide.

At day-break on a hill they stood
That overlooked the moor;
And thence they saw the bridge of wood,
A furlong from their door.   40

They wept—and, turning homeward, cried,
"In heaven we all shall meet;"
—When in the snow the mother spied
The print of Lucy's feet.

Then downwards from the steep hill's edge
They tracked the footmarks small;
And through the broken hawthorn hedge,
And by the long stone-wall;

And then an open field they crossed:
The marks were still the same;   50
They tracked them on, nor ever lost;
And to the bridge they came.

They followed from the snowy bank
Those footmarks, one by one,
Into the middle of the plank;
And further there were none!

—Yet some maintain that to this day
She is a living child;
That you may see sweet Lucy Gray
Upon the lonesome wild.   60

O'er rough and smooth she trips along,
And never looks behind;
And sings a solitary song
That whistles in the wind.

                                    1799 (1800)

❧❧❧

"In the School of [Hawkshead] is a tablet, on which are
inscribed, in gilt letters, the Names of the several persons
who have been Schoolmasters there since the foundation
of the School, with the time at which they entered upon
and quitted their office. Opposite to one of those Names
the Author wrote the following lines" (W. W.'s note to
"Matthew"). Matthew also figures in "The Two April
Mornings," "The Fountain," and "Expostulation and
Reply" (p. 208). Francis Jeffrey attacked these poems for
failing to preserve the neoclassical principle of decorum
of type of character (see p. 365). In other words, Jeffrey
felt that Matthew did not sufficiently resemble the typical
schoolmaster.

❧❧❧

MATTHEW

If Nature, for a favourite child,
In thee hath tempered so her clay,
That every hour thy heart runs wild,
Yet never once doth go astray,

Read o'er these lines; and then review
This tablet, that thus humbly rears
In such diversity of hue
Its history of two hundred years.

—When through this little wreck of fame,
Cipher and syllable! thine eye   10
Has travelled down to Matthew's name,
Pause with no common sympathy.

And, if a sleeping tear should wake,
Then be it neither checked nor stayed:
For Matthew a request I make
Which for himself he had not made.

Poor Matthew, all his frolics o'er,
Is silent as a standing pool;
Far from the chimney's merry roar,
And murmur of the village school.   20

The sighs which Matthew heaved were sighs
Of one tired out with fun and madness;
The tears which came to Matthew's eyes
Were tears of light, the dew of gladness.

Yet, sometimes, when the secret cup
Of still and serious thought went round,
It seemed as if he drank it up—
He felt with spirit so profound.

—Thou soul of God's best earthly mould!
Thou happy Soul! and can it be   30
That these two words of glittering gold
Are all that must remain of thee?

                                    1799 (1800)

THE TWO APRIL MORNINGS

We walked along, while bright and red
Uprose the morning sun;
And Matthew stopped, he looked, and said,
"The will of God be done!"

A village schoolmaster was he,
With hair of glittering grey;
As blithe a man as you could see
On a spring holiday.

And on that morning, through the grass,
And by the steaming rills,   10
We travelled merrily, to pass
A day among the hills.

"Our work," said I, "was well begun,
Then, from thy breast what thought,
Beneath so beautiful a sun,
So sad a sigh has brought?"

A second time did Matthew stop;
And fixing still his eye
Upon the eastern mountain-top,
To me he made reply: 20

"Yon cloud with that long purple cleft
Brings fresh into my mind
A day like this which I have left
Full thirty years behind.

"And just above yon slope of corn
Such colours, and no other,
Were in the sky, that April morn,
Of this the very brother.

"With rod and line I sued the sport
Which that sweet season gave,   30
And, to the churchyard come, stopped short
Beside my daughter's grave.

"Nine summers had she scarcely seen,
The pride of all the vale;
And then she sang;—she would have been
A very nightingale.

"Six feet in earth my Emma lay;
And yet I loved her more,
For so it seemed, than till that day
I e'er had loved before.   40

"And, turning from her grave, I met,
Beside the churchyard yew,
A blooming Girl, whose hair was wet
With points of morning dew.

"A basket on her head she bare;
Her brow was smooth and white:
To see a child so very fair,
It was a pure delight!

"No fountain from its rocky cave
E'er tripped with foot so free;   50
She seemed as happy as a wave
That dances on the sea.

"There came from me a sigh of pain
Which I could ill confine;
I looked at her, and looked again:
And did not wish her mine!"

Matthew is in his grave, yet now,
Methinks, I see him stand,
As at that moment, with a bough
Of wilding in his hand.   60

1799 (1800)

THE FOUNTAIN

A CONVERSATION

We talked with open heart, and tongue
Affectionate and true,
A pair of friends, though I was young,
And Matthew seventy-two.

We lay beneath a spreading oak,
Beside a mossy seat;
And from the turf a fountain broke,
And gurgled at our feet.

"Now, Matthew!" said I, "let us match
This water's pleasant tune   10
With some old border-song, or catch
That suits a summer's noon;

"Or of the church-clock and the chimes
Sing here beneath the shade,
That half-mad thing of witty rhymes
Which you last April made!"

In silence Matthew lay, and eyed
The spring beneath the tree;
And thus the dear old Man replied,
The grey-haired man of glee: 20

"No check, no stay, this Streamlet fears:
How merrily it goes!
'Twill murmur on a thousand years,
And flow as now it flows.

"And here, on this delightful day,
I cannot choose but think
How oft, a vigorous man, I lay
Beside this fountain's brink.

"My eyes are dim with childish tears,
My heart is idly stirred,   30
For the same sound is in my ears
Which in those days I heard.

"Thus fares it still in our decay:
And yet the wiser mind
Mourns less for what age takes away
Than what it leaves behind.

"The blackbird amid leafy trees,
The lark above the hill,
Let loose their carols when they please,
Are quiet when they will.   40

"With Nature never do *they* wage
A foolish strife; they see
A happy youth, and their old age
Is beautiful and free:

"But we are pressed by heavy laws;
And often, glad no more,
We wear a face of joy, because
We have been glad of yore.

"If there be one who need bemoan
His kindred laid in earth,   50
The household hearts that were his own;
It is the man of mirth.

"My days, my Friend, are almost gone,
My life has been approved,
And many love me; but by none
Am I enough beloved."

"Now both himself and me he wrongs,
The man who thus complains!
I live and sing my idle songs
Upon these happy plains;   60

"And, Matthew, for thy children dead
I'll be a son to thee!"
At this he grasped my hand, and said,
"Alas! that cannot be."

We rose up from the fountain-side;
And down the smooth descent
Of the green sheep-track did we glide;
And through the wood we went;

And, ere we came to Leonard's rock,
He sang those witty rhymes   70
About the crazy old church-clock,
And the bewildered chimes.

                                    1799 (1800)

A POET'S EPITAPH

Art thou a Statist in the van
Of public conflicts trained and bred?
—First learn to love one living man;
Then may'st thou think upon the dead.

A Lawyer art thou?—draw not nigh!
Go, carry to some fitter place
The keenness of that practised eye,
The hardness of that sallow face.

Art thou a Man of purple cheer?
A rosy Man, right plump to see?   10
Approach; yet, Doctor, not too near,
This grave no cushion is for thee.

Or art thou one of gallant pride,
A Soldier and no man of chaff?
Welcome!—but lay thy sword aside,
And lean upon a peasant's staff.

Physician art thou?—one, all eyes,
Philosopher!—a fingering slave,
One that would peep and botanize
Upon his mother's grave?   20

Wrapt closely in thy sensual fleece,
O turn aside,—and take, I pray,
That he below may rest in peace,
Thy ever-dwindling soul, away!

A Moralist perchance appears;
Led, Heaven knows how! to this poor sod:
And he has neither eyes nor ears;
Himself his world, and his own God;

One to whose smooth-rubbed soul can cling
Nor form, nor feeling, great or small;   30
A reasoning, self-sufficing thing,
An intellectual All-in-all!

Shut close the door; press down the latch;
Sleep in thy intellectual crust;
Nor lose ten tickings of thy watch
Near this unprofitable dust.

But who is He, with modest looks,
And clad in homely russet brown?
He murmurs near the running brooks
A music sweeter than their own.   40

He is retired as noontide dew,
Or fountain in a noon-day grove;
And you must love him, ere to you
He will seem worthy of your love.

The outward shows of sky and earth,
Of hill and valley, he has viewed;
And impulses of deeper birth
Have come to him in solitude.

In common things that round us lie
Some random truths he can impart,—   50
The harvest of a quiet eye
That broods and sleeps on his own heart.

But he is weak; both Man and Boy,
Hath been an idler in the land;
Contented if he might enjoy
The things which others understand.

—Come hither in thy hour of strength;
Come, weak as is a breaking wave!
Here stretch thy body at full length;
Or build thy house upon this grave.   60

                                    1799 (1800)

HART-LEAP WELL

PART FIRST

The Knight had ridden down from Wensley Moor
With the slow motion of a summer's cloud,
And now, as he approached a vassal's door,
"Bring forth another horse!" he cried aloud.

"Another horse!"—That shout the vassal heard
And saddled his best Steed, a comely grey;
Sir Walter mounted him; he was the third
Which he had mounted on that glorious day.

Joy sparkled in the prancing courser's eyes;
The horse and horseman are a happy pair;   10
But, though Sir Walter like a falcon flies,
There is a doleful silence in the air.

A rout this morning left Sir Walter's Hall,
That as they galloped made the echoes roar;
But horse and man are vanished, one and all;
Such race, I think, was never seen before.

Sir Walter, restless as a veering wind,
Calls to the few tired dogs that yet remain:
Blanch, Swift, and Music, noblest of their kind,
Follow, and up the weary mountain strain.   20

The Knight hallooed, he cheered and chid them on
With suppliant gestures and upbraidings stern;
But breath and eyesight fail; and, one by one,
The dogs are stretched among the mountain fern.

Where is the throng, the tumult of the race?
The bugles that so joyfully were blown?
—This chase it looks not like an earthly chase;
Sir Walter and the Hart are left alone.

The poor Hart toils along the mountain-side;
I will not stop to tell how far he fled,   30
Nor will I mention by what death he died;
But now the Knight beholds him lying dead.

Dismounting, then, he leaned against a thorn;
He had no follower, dog, nor man, nor boy:
He neither cracked his whip, nor blew his horn,
But gazed upon the spoil with silent joy.

Close to the thorn on which Sir Walter leaned
Stood his dumb partner in this glorious feat;
Weak as a lamb the hour that it is yeaned;
And white with foam as if with cleaving sleet.   40

Upon his side the Hart was lying stretched:
His nostril touched a spring beneath a hill,
And with the last deep groan his breath had fetched
The waters of the spring were trembling still.

And now, too happy for repose or rest,
(Never had living man such joyful lot!)
Sir Walter walked all round, north, south, and west,
And gazed and gazed upon that darling spot.

And climbing up the hill—(it was at least
Four roods of sheer ascent) Sir Walter found   50
Three several hoof-marks which the hunted Beast
Had left imprinted on the grassy ground.

Sir Walter wiped his face, and cried, "Till now
Such sight was never seen by human eyes:
Three leaps have borne him from this lofty brow
Down to the very fountain where he lies.

"I'll build a pleasure-house upon this spot,
And a small arbour, made for rural joy;
'Twill be the traveller's shed, the pilgrim's cot,
A place of love for damsels that are coy.   60

"A cunning artist will I have to frame
A basin for that fountain in the dell!
And they who do make mention of the same,
From this day forth, shall call it HART-LEAP WELL.

"And, gallant Stag! to make thy praises known,
Another monument shall here be raised;
Three several pillars, each a rough-hewn stone,
And planted where thy hoofs the turf have grazed.

"And in the summer-time, when days are long,
I will come hither with my Paramour;   70
And with the dancers and the minstrel's song
We will make merry in that pleasant bower.

"Till the foundations of the mountains fail
My mansion with its arbour shall endure;—
The joy of them who till the fields of Swale,
And them who dwell among the woods of Ure!"

Then home he went, and left the Hart stone-dead,
With breathless nostrils stretched above the spring.
—Soon did the Knight perform what he had said;
And far and wide the fame thereof did ring.   80

Ere thrice the Moon into her port had steered,
A cup of stone received the living well;
Three pillars of rude stone Sir Walter reared,
And built a house of pleasure in the dell.

And, near the fountain, flowers of stature tall
With trailing plants and trees were intertwined,—
Which soon composed a little sylvan hall,
A leafy shelter from the sun and wind.

And thither, when the summer days were long,
Sir Walter led his wondering Paramour;   90
And with the dancers and the minstrel's song
Made merriment within that pleasant bower.

The Knight, Sir Walter, died in course of time,
And his bones lie in his paternal vale.—
But there is matter for a second rhyme,
And I to this would add another tale.

PART SECOND

The moving accident is not my trade;
To freeze the blood I have no ready arts:
'Tis my delight, alone in summer shade,
To pipe a simple song for thinking hearts.   100

As I from Hawes to Richmond did repair,
It chanced that I saw standing in a dell
Three aspens at three corners of a square;
And one, not four yards distant, near a well.

What this imported I could ill divine:
And, pulling now the rein my horse to stop,
I saw three pillars standing in a line,—
The last stone-pillar on a dark hill-top.

The trees were grey, with neither arms nor head;
Half wasted the square mound of tawny green;   110
So that you just might say, as then I said,
"Here in old time the hand of man hath been."

I looked upon the hill both far and near,
More doleful place did never eye survey;
It seemed as if the spring-time came not here,
And Nature here were willing to decay.

I stood in various thoughts and fancies lost,
When one, who was in shepherd's garb attired,
Came up the hollow:—him did I accost,
And what this place might be I then enquired.   120

The Shepherd stopped, and that same story told
Which in my former rhyme I have rehearsed.
"A jolly place," said he, "in times of old!
But something ails it now: the spot is curst.

"You see these lifeless stumps of aspen wood—
Some say that they are beeches, others elms—

These were the bower; and here a mansion stood,
The finest palace of a hundred realms!

"The arbour does its own condition tell;
You see the stones, the fountain, and the stream;   130
But as to the great Lodge! you might as well
Hunt half a day for a forgotten dream.

"There's neither dog nor heifer, horse nor sheep,
Will wet his lips within that cup of stone;
And oftentimes, when all are fast asleep,
This water doth send forth a dolorous groan.

"Some say that here a murder has been done,
And blood cries out for blood: but, for my part,
I've guessed, when I've been sitting in the sun,
That it was all for that unhappy Hart.   140

"What thoughts must through the creature's brain
           have past!
Even from the topmost stone, upon the steep,
Are but three bounds—and look, Sir, at this last—
O Master! it has been a cruel leap.

"For thirteen hours he ran a desperate race;
And in my simple mind we cannot tell
What cause the Hart might have to love this place,
And come and make his death-bed near the well.

"Here on the grass perhaps asleep he sank,
Lulled by the fountain in the summer tide;   150
This water was perhaps the first he drank
When he had wandered from his mother's side.

"In April here beneath the flowering thorn
He heard the birds their morning carols sing;
And he, perhaps, for aught we know, was born
Not half a furlong from that self-same spring.

"Now, here is neither grass nor pleasant shade;
The sun on drearier hollow never shone;
So will it be, as I have often said,
Till trees, and stones, and fountain, all are gone."   160

"Grey-headed Shepherd, thou has spoken well;
Small difference lies between thy creed and mine:
This Beast not unobserved by Nature fell;
His death was mourned by sympathy divine.

"The Being that is in the clouds and air,
That is in the green leaves among the groves,
Maintains a deep and reverential care
For the unoffending creatures whom he loves.

"The pleasure-house is dust:—behind, before,
This is no common waste, no common gloom;   170
But Nature, in due course of time, once more
Shall here put on her beauty and her bloom.

"She leaves these objects to a slow decay,
That what we are, and have been, may be known;
But at the coming of the milder day
These monuments shall all be overgrown.

"One lesson, Shepherd, let us two divide,
Taught both by what she shows, and what conceals;
Never to blend our pleasure or our pride
With sorrow of the meanest thing that feels."   180

1800 (1800)

THE CHILDLESS FATHER

❧

This poem is discussed by Wordsworth in the Preface to
the *Lyrical Ballads*, below, p. 322 n.

❧

"Up, Timothy, up with your staff and away!
Not a soul in the village this morning will stay;
The hare has just started from Hamilton's grounds,
And Skiddaw is glad with the cry of the hounds."

—Of coats and of jackets grey, scarlet, and green,
On the slopes of the pastures all colours were seen;
With their comely blue aprons, and caps white as
    snow,
The girls on the hills made a holiday show.

Fresh sprigs of green box-wood, not six months
    before,
Filled the funeral basin at Timothy's door;   10
A coffin through Timothy's threshold had past;
One Child did it bear, and that Child was his last.

Now fast up the dell came the noise and the fray,
The horse, and the horn, and the hark! hark away!
Old Timothy took up his staff, and he shut
With a leisurely motion the door of his hut.

THE CHILDLESS FATHER. **9. box-wood:** "In several parts of
the North of England, when a funeral takes place, a basin
full of sprigs of box-wood is placed at the door of the house
from which the coffin is taken up, and each person who
attends the funeral ordinarily takes a sprig of this box-
wood, and throws it into the grave of the deceased" (W. W.).

Perhaps to himself at that moment he said;
"The key I must take, for my Ellen is dead."
But of this in my ears not a word did he speak;
And he went to the chase with a tear on his cheek.   20

1800 (1800)

MICHAEL

A PASTORAL POEM

❧

There is no finer example of Wordsworth's purity and
naturalness of diction. By calling the poem a pastoral he
perhaps wished to contrast it with the classical genre of
pastoral (imitated in Shelley's "Adonais"), with its
deliberate artifice and lack of realism. He discusses the
poem in his letter to Fox, below, p. 349. Compare the
description of shepherds in *Prelude*, VIII. 215–327.

❧

If from the public way you turn your steps
Up the tumultuous brook of Green-head Ghyll,
You will suppose that with an upright path
Your feet must struggle; in such bold ascent   5
The pastoral mountains front you, face to face.
But, courage! for around that boisterous brook
The mountains have all opened out themselves,
And made a hidden valley of their own.   40
No habitation can be seen; but they
Who journey thither find themselves alone   10
With a few sheep, with rocks and stones, and kites
That overhead are sailing in the sky.
It is in truth an utter solitude;
Nor should I have made mention of this Dell
But for one object which you might pass by,
Might see and notice not. Beside the brook
Appears a straggling heap of unhewn stones!
And to that simple object appertains
A story—unenriched with strange events,
Yet not unfit, I deem, for the fireside,   20
Or for the summer shade. It was the first
Of those domestic tales that spake to me
Of Shepherds, dwellers in the valleys, men
Whom I already loved;—not verily
For their own sakes, but for the fields and hills
Where was their occupation and abode.
And hence this Tale, while I was yet a Boy
Careless of books, yet having felt the power

Of Nature, by the gentle agency
Of natural objects, led me on to feel   30
For passions that were not my own, and think
(At random and imperfectly indeed)
On man, the heart of man, and human life.
Therefore, although it be a history
Homely and rude, I will relate the same
For the delight of a few natural hearts;
And, with yet fonder feeling, for the sake
Of youthful Poets, who among these hills
Will be my second self when I am gone.

Upon the forest-side in Grasmere Vale   40
There dwelt a Shepherd, Michael was his name;
An old man, stout of heart, and strong of limb.
His bodily frame had been from youth to age
Of an unusual strength: his mind was keen,
Intense, and frugal, apt for all affairs,
And in his shepherd's calling he was prompt
And watchful more than ordinary men.
Hence had he learned the meaning of all winds,
Of blasts of every tone; and oftentimes,
When others heeded not, He heard the South   50
Make subterraneous music, like the noise
Of bagpipers on distant Highland hills.
The Shepherd, at such warning, of his flock
Bethought him, and he to himself would say,
"The winds are now devising work for me!"
And, truly, at all times, the storm, that drives
The traveller to a shelter, summoned him
Up to the mountains: he had been alone
Amid the heart of many thousand mists,
That came to him, and left him, on the heights.   60
So lived he till his eightieth year was past.
And grossly that man errs, who should suppose
That the green valleys, and the streams and rocks,
Were things indifferent to the Shepherd's thoughts.
Fields, where with cheerful spirits he had breathed
The common air; hills, which with vigorous step
He had so often climbed; which had impressed
So many incidents upon his mind
Of hardship, skill or courage, joy or fear;
Which, like a book, preserved the memory   70
Of the dumb animals, whom he had saved,
Had fed or sheltered, linking to such acts
The certainty of honourable gain;
Those fields, those hills—what could they less? had laid
Strong hold on his affections, were to him
A pleasurable feeling of blind love,
The pleasure which there is in life itself.

His days had not been passed in singleness.
His Helpmate was a comely matron, old—
Though younger than himself full twenty years.   80
She was a woman of a stirring life,
Whose heart was in her house: two wheels she had
Of antique form; this large, for spinning wool;
That small, for flax; and, if one wheel had rest,
It was because the other was at work.
The Pair had but one inmate in their house,
An only Child, who had been born to them
When Michael, telling o'er his years, began
To deem that he was old,—in shepherd's phrase,
With one foot in the grave. This only Son,   90
With two brave sheep-dogs tried in many a storm,
The one of an inestimable worth,
Made all their household. I may truly say,
That they were as a proverb in the vale
For endless industry. When day was gone,
And from their occupations out of doors
The Son and Father were come home, even then,
Their labour did not cease; unless when all
Turned to the cleanly supper-board, and there,
Each with a mess of pottage and skimmed milk,   100
Sat round the basket piled with oaten cakes,
And their plain home-made cheese. Yet when the
    meal
Was ended, Luke (for so the Son was named)
And his old Father both betook themselves
To such convenient work as might employ
Their hands by the fire-side; perhaps to card
Wool for the Housewife's spindle, or repair
Some injury done to sickle, flail, or scythe,
Or other implement of house or field.

Down from the ceiling, by the chimney's edge,   110
That in our ancient uncouth country style
With huge and black projection overbrowed
Large space beneath, as duly as the light
Of day grew dim the Housewife hung a lamp;
An aged utensil, which had performed
Service beyond all others of its kind.
Early at evening did it burn—and late,
Surviving comrade of uncounted hours,
Which, going by from year to year, had found,
And left, the couple neither gay perhaps   120
Nor cheerful, yet with objects and with hopes,
Living a life of eager industry.
And now, when Luke had reached his eighteenth
    year,
There by the light of this old lamp they sate,

Father and Son, while far into the night
The Housewife plied her own peculiar work,
Making the cottage through the silent hours
Murmur as with the sound of summer flies.
This light was famous in its neighbourhood,
And was a public symbol of the life   130
That thrifty Pair had lived. For, as it chanced,
Their cottage on a plot of rising ground
Stood single, with large prospect, north and south,
High into Easedale, up to Dunmail-Raise,
And westward to the village near the lake;
And from this constant light, so regular
And so far seen, the House itself, by all
Who dwelt within the limits of the vale,
Both old and young, was named THE EVENING STAR.

   Thus living on through such a length of years,   140
The Shepherd, if he loved himself, must needs
Have loved his Helpmate; but to Michael's heart
This son of his old age was yet more dear—
Less from instinctive tenderness, the same
Fond spirit that blindly works in the blood of all—
Than that a child, more than all other gifts
That earth can offer to declining man,
Brings hope with it, and forward-looking thoughts,
And stirrings of inquietude, when they
By tendency of nature needs must fail.   150
Exceeding was the love he bare to him,
His heart and his heart's joy! For oftentimes
Old Michael, while he was a babe in arms,
Had done him female service, not alone
For pastime and delight, as is the use
Of fathers, but with patient mind enforced
To acts of tenderness; and he had rocked
His cradle, as with a woman's gentle hand.

   And in a later time, ere yet the Boy
Had put on boy's attire, did Michael love,   160
Albeit of a stern unbending mind,
To have the Young-one in his sight, when he
Wrought in the field, or on his shepherd's stool
Sate with a fettered sheep before him stretched
Under the large old oak, that near his door
Stood single, and, from matchless depth of shade,
Chosen for the Shearer's covert from the sun,
Thence in our rustic dialect was called
The CLIPPING TREE, a name which yet it bears.
There, while they two were sitting in the shade,   170
With others round them, earnest all and blithe,
Would Michael exercise his heart with looks
Of fond correction and reproof bestowed

Upon the Child, if he disturbed the sheep
By catching at their legs, or with his shouts
Scared them, while they lay still beneath the shears.

   And when by Heaven's good grace the boy grew
     up
A healthy Lad, and carried in his cheek
Two steady roses that were five years old;
Then Michael from a winter coppice cut   180
With his own hand a sapling, which he hooped
With iron, making it throughout in all
Due requisites a perfect shepherd's staff,
And gave it to the Boy; wherewith equipt
He as a watchman oftentimes was placed
At gate or gap, to stem or turn the flock;
And, to his office prematurely called,
There stood the urchin, as you will divine,
Something between a hindrance and a help;
And for this cause not always, I believe,   190
Receiving from his Father hire of praise;
Though nought was left undone which staff, or voice,
Or looks, or threatening gestures, could perform.

   But soon as Luke, full ten years old, could stand
Against the mountain blasts; and to the heights,
Not fearing toil, nor length of weary ways,
He with his Father daily went, and they
Were as companions, why should I relate
That objects which the Shepherd loved before
Were dearer now? that from the Boy there came   200
Feelings and emanations—things which were
Light to the sun and music to the wind;
And that the old Man's heart seemed born again?

   Thus in his Father's sight the Boy grew up:
And now, when he had reached his eighteenth year,
He was his comfort and his daily hope.

   While in this sort the simple household lived
From day to day, to Michael's ear there came
Distressful tidings. Long before the time
Of which I speak, the Shepherd had been bound   210
In surety for his brother's son, a man
Of an industrious life, and ample means;
But unforeseen misfortunes suddenly
Had prest upon him; and old Michael now
Was summoned to discharge the forfeiture,
A grievous penalty, but little less
Than half his substance. This unlooked-for claim,
At the first hearing, for a moment took
More hope out of his life than he supposed
That any old man ever could have lost.   220

As soon as he had armed himself with strength
To look his trouble in the face, it seemed
The Shepherd's sole resource to sell at once
A portion of his patrimonial fields.
Such was his first resolve; he thought again,
And his heart failed him. "Isabel," said he,
Two evenings after he had heard the news,
"I have been toiling more than seventy years,
And in the open sunshine of God's love
Have we all lived; yet, if these fields of ours   230
Should pass into a stranger's hand, I think
That I could not lie quiet in my grave.
Our lot is a hard lot; the sun himself
Has scarcely been more diligent than I;
And I have lived to be a fool at last
To my own family. An evil man
That was, and made an evil choice, if he
Were false to us; and, if he were not false,
There are ten thousand to whom loss like this
Had been no sorrow. I forgive him;—but   240
'Twere better to be dumb than to talk thus.

"When I began, my purpose was to speak
Of remedies and of a cheerful hope.
Our Luke shall leave us, Isabel; the land
Shall not go from us, and it shall be free;
He shall possess it, free as is the wind
That passes over it. We have, thou know'st,
Another kinsman—he will be our friend
In this distress. He is a prosperous man,
Thriving in trade—and Luke to him shall go,   250
And with his kinsman's help and his own thrift
He quickly will repair this loss, and then
He may return to us. If here he stay,
What can be done? Where every one is poor,
What can be gained?"
                    At this the old Man paused,
And Isabel sat silent, for her mind
Was busy, looking back into past times.
There's Richard Bateman, thought she to herself,
He was a parish-boy—at the church-door
They made a gathering for him, shillings, pence,   260
And halfpennies, wherewith the neighbours bought
A basket, which they filled with pedlar's wares;
And, with this basket on his arm, the lad
Went up to London, found a master there,
Who, out of many, chose the trusty boy
To go and overlook his merchandise
Beyond the seas; where he grew wondrous rich,
And left estates and monies to the poor,

And, at his birth-place, built a chapel floored
With marble, which he sent from foreign lands.   270
These thoughts, and many others of like sort,
Passed quickly through the mind of Isabel,
And her face brightened. The old Man was glad,
And thus resumed:—"Well, Isabel! this scheme
These two days has been meat and drink to me.
Far more than we have lost is left us yet.
—We have enough—I wish indeed that I
Were younger;—but this hope is a good hope.
Make ready Luke's best garments, of the best
Buy for him more, and let us send him forth   280
To-morrow, or the next day, or to-night:
—If he could go, the Boy should go to-night."

Here Michael ceased, and to the fields went forth
With a light heart. The Housewife for five days
Was restless morn and night, and all day long
Wrought on with her best fingers to prepare
Things needful for the journey of her son.
But Isabel was glad when Sunday came
To stop her in her work: for, when she lay
By Michael's side, she through the last two nights   290
Heard him, how he was troubled in his sleep:
And when they rose at morning she could see
That all his hopes were gone. That day at noon
She said to Luke, while they two by themselves
Were sitting at the door, "Thou must not go:
We have no other Child but thee to lose,
None to remember—do not go away,
For if thou leave thy Father he will die."
The Youth made answer with a jocund voice;
And Isabel, when she had told her fears,   300
Recovered heart. That evening her best fare
Did she bring forth, and all together sat
Like happy people round a Christmas fire.

With daylight Isabel resumed her work;
And all the ensuing week the house appeared
As cheerful as a grove in Spring: at length
The expected letter from their kinsman came,
With kind assurances that he would do
His utmost for the welfare of the Boy;
To which, requests were added, that forthwith   310
He might be sent to him. Ten times or more
The letter was read over; Isabel
Went forth to show it to the neighbours round;
Nor was there at that time on English land
A prouder heart than Luke's. When Isabel
Had to her house returned, the old Man said,
"He shall depart to-morrow." To this word

The Housewife answered, talking much of things
Which, if at such short notice he should go,
Would surely be forgotten. But at length   320
She gave consent, and Michael was at ease.

Near the tumultuous brook of Greenhead Ghyll,
In that deep valley, Michael had designed
To build a Sheep-fold; and, before he heard
The tidings of his melancholy loss,
For this same purpose he had gathered up
A heap of stones, which by the streamlet's edge
Lay thrown together, ready for the work.
With Luke that evening thitherward he walked:
And soon as they had reached the place he stopped,
And thus the old Man spake to him:—"My son,   331
To-morrow thou wilt leave me: with full heart
I look upon thee, for thou art the same
That wert a promise to me ere thy birth,
And all thy life has been my daily joy.
I will relate to thee some little part
Of our two histories; 'twill do thee good
When thou art from me, even if I should touch
On things thou canst not know of.—After thou
First cam'st into the world—as oft befalls   340
To new-born infants—thou didst sleep away
Two days, and blessings from thy Father's tongue
Then fell upon thee. Day by day passed on,
And still I loved thee with increasing love.
Never to living ear came sweeter sounds
Than when I heard thee by our own fire-side
First uttering, without words, a natural tune;
While thou, a feeding babe, didst in thy joy
Sing at thy Mother's breast. Month followed month,
And in the open fields my life was passed   350
And on the mountains; else I think that thou
Hadst been brought up upon thy Father's knees.
But we were playmates, Luke: among these hills,
As well thou knowest, in us the old and young
Have played together, nor with me didst thou
Lack any pleasure which a boy can know."
Luke had a manly heart; but at these words
He sobbed aloud. The old Man grasped his hand,
And said, "Nay, do not take it so—I see
That these are things of which I need not speak.   360
—Even to the utmost I have been to thee
A kind and a good Father: and herein
I but repay a gift which I myself
Received at others' hands; for, though now old
Beyond the common life of man, I still
Remember them who loved me in my youth.

Both of them sleep together: here they lived,
As all their Forefathers had done; and, when
At length their time was come, they were not loth
To give their bodies to the family mould.   370
I wished that thou shouldst live the life they lived,
But 'tis a long time to look back, my Son,
And see so little gain from threescore years.
These fields were burthened when they came to me;
Till I was forty years of age, not more
Than half of my inheritance was mine.
I toiled and toiled; God blessed me in my work,
And till these three weeks past the land was free.
—It looks as if it never could endure
Another Master. Heaven forgive me, Luke,   380
If I judge ill for thee, but it seems good
That thou shouldst go."
                              At this the old Man paused;
Then, pointing to the stones near which they stood,
Thus, after a short silence, he resumed:
"This was a work for us; and now, my Son,
It is a work for me. But, lay one stone—
Here, lay it for me, Luke, with thine own hands.
Nay, Boy, be of good hope;—we both may live
To see a better day. At eighty-four
I still am strong and hale;—do thou thy part;   390
I will do mine.—I will begin again
With many tasks that were resigned to thee:
Up to the heights, and in among the storms,
Will I without thee go again, and do
All works which I was wont to do alone,
Before I knew thy face.—Heaven bless thee, Boy!
Thy heart these two weeks has been beating fast
With many hopes; it should be so—yes—yes—
I knew that thou couldst never have a wish
To leave me, Luke: thou hast been bound to me   400
Only by links of love: when thou art gone,
What will be left to us!—But I forget
My purposes. Lay now the corner-stone,
As I requested; and hereafter, Luke,
When thou art gone away, should evil men
Be thy companions, think of me, my Son,
And of this moment; hither turn thy thoughts,
And God will strengthen thee: amid all fear
And all temptation, Luke, I pray that thou
May'st bear in mind the life thy Fathers lived,   410
Who, being innocent, did for that cause
Bestir them in good deeds. Now, fare thee well—
When thou return'st, thou in this place wilt see
A work which is not here: a covenant
'Twill be between us; but, whatever fate

Befall thee, I shall love thee to the last,
And bear thy memory with me to the grave."

The Shepherd ended here; and Luke stooped down,
And, as his Father had requested, laid
The first stone of the Sheep-fold. At the sight   420
The old Man's grief broke from him; to his heart
He pressed his Son, he kissèd him and wept;
And to the house together they returned.
—Hushed was that House in peace, or seeming peace,
Ere the night fell:—with morrow's dawn the Boy
Began his journey, and, when he had reached
The public way, he put on a bold face;
And all the neighbours, as he passed their doors,
Came forth with wishes and with farewell prayers,
That followed him till he was out of sight.   430

A good report did from their Kinsman come,
Of Luke and his well-doing: and the Boy
Wrote loving letters, full of wondrous news,
Which, as the Housewife phrased it, were throughout
"The prettiest letters that were ever seen."
Both parents read them with rejoicing hearts.
So, many months passed on: and once again
The Shepherd went about his daily work
With confident and cheerful thoughts; and now
Sometimes when he could find a leisure hour   440
He to that valley took his way, and there
Wrought at the Sheep-fold. Meantime Luke began
To slacken in his duty; and, at length,
He in the dissolute city gave himself
To evil courses: ignominy and shame
Fell on him, so that he was driven at last
To seek a hiding-place beyond the seas.

There is a comfort in the strength of love;
'Twill make a thing endurable, which else
Would overset the brain, or break the heart: 450
I have conversed with more than one who well
Remember the old Man, and what he was
Years after he had heard this heavy news.
His bodily frame had been from youth to age
Of an unusual strength. Among the rocks
He went, and still looked up to sun and cloud,
And listened to the wind; and, as before,
Performed all kinds of labour for his sheep,
And for the land, his small inheritance.
And to that hollow dell from time to time   460
Did he repair, to build the Fold of which
His flock had need. 'Tis not forgotten yet
The pity which was then in every heart

For the old Man—and 'tis believed by all
That many and many a day he thither went,
And never lifted up a single stone.

There, by the Sheep-fold, sometimes was he seen
Sitting alone, or with his faithful Dog,
Then old, beside him, lying at his feet.
The length of full seven years, from time to time,   470
He at the building of this Sheep-fold wrought,
And left the work unfinished when he died.
Three years, or little more, did Isabel
Survive her Husband: at her death the estate
Was sold, and went into a stranger's hand.
The Cottage which was named the EVENING STAR
Is gone—the ploughshare has been through the
  ground
On which it stood; great changes have been wrought
In all the neighbourhood:—yet the oak is left
That grew beside their door; and the remains   480
Of the unfinished Sheep-fold may be seen
Beside the boisterous brook of Greenhead Ghyll.

1800 (1800)

THE SPARROW'S NEST

Behold, within the leafy shade,
Those bright blue eggs together laid!
On me the chance-discovered sight
Gleamed like a vision of delight.
I started—seeming to espy
The home and sheltered bed,
The Sparrow's dwelling, which, hard by
My Father's house, in wet or dry
My sister Emmeline and I
  Together visited.   10

She looked at it and seemed to fear it;
Dreading, tho' wishing, to be near it:
Such heart was in her, being then
A little Prattler among men.
The Blessing of my later years
Was with me when a boy:
She gave me eyes, she gave me ears;
And humble cares, and delicate fears;
A heart, the fountain of sweet tears;
  And love, and thought, and joy.   20

1801 (1807)

Like the winter of 1798–99 in Germany, the spring of 1802 was a period of astonishing creative energy for Wordsworth. At this time he resided in Grasmere, and his way of life can be seen in Dorothy's *Grasmere Journal*, below. On March 11, he composed "The Sailor's Mother"; March 12, "Alice Fell"; March 14, "To a Butterfly"; March 23, "To the Cuckoo"; March 26, "My Heart Leaps Up"; March 27, the beginning of "Intimations of Immortality"; in April he composed "Written in March" and "To a Skylark"; and on May 3–7, "Resolution and Independence." Most of these poems were later revised, of course, and only the first four stanzas of "Intimations of Immortality" were written that spring. The three compositions from March 23 to 27 deal with more or less the same theme, and the four stanzas of "Intimations of Immortality" deny what is asserted at the end of "To a Cuckoo." Taken together, the three poems testify to uneasiness in Wordsworth's mind at this time. Fear that his imaginative power was failing alternates with joyful moods of reassurance. Coleridge was going through a similar crisis. See "Dejection: An Ode," below, p. 432.

THE SAILOR'S MOTHER

One morning (raw it was and wet—
A foggy day in winter time)
A Woman on the road I met,
Not old, though something past her prime:
Majestic in her person, tall and straight;
And like a Roman matron's was her mien and gait.

The ancient spirit is not dead;
Old times, thought I, are breathing there;
Proud was I that my country bred
Such strength, a dignity so fair: 10
She begged an alms, like one in poor estate;
I looked at her again, nor did my pride abate.

When from these lofty thoughts I woke,
"What is it," said I, "that you bear,
Beneath the covert of your Cloak,
Protected from this cold damp air?"
She answered, soon as she the question heard,
"A simple burthen, Sir, a little Singing-bird."

And, thus continuing, she said,
"I had a Son, who many a day   20
Sailed on the seas, but he is dead;
In Denmark he was cast away:
And I have travelled weary miles to see
If aught which he had owned might still remain for me.

"The bird and cage they both were his:
'Twas my Son's bird; and neat and trim
He kept it: many voyages
The singing-bird had gone with him;
When last he sailed, he left the bird behind;
From bodings, as might be, that hung upon his mind.

"He to a fellow-lodger's care   31
Had left it, to be watched and fed,
And pipe its song in safety;—there
I found it when my Son was dead;
And now, God help me for my little wit!
I bear it with me, Sir;—he took so much delight
    in it."

                                    1802 (1807)

ALICE FELL

OR, POVERTY

See Dorothy Wordsworth's *Grasmere Journal*, February 16, 1802, below, p. 361.

The post-boy drove with fierce career,
For threatening clouds the moon had drowned;
When, as we hurried on, my ear
Was smitten with a startling sound.

As if the wind blew many ways,
I heard the sound,—and more and more;
It seemed to follow with the chaise,
And still I heard it as before.

At length I to the boy called out;
He stopped his horses at the word,   10
But neither cry, nor voice, nor shout,
Nor aught else like it, could be heard.

The boy then smacked his whip, and fast
The horses scampered through the rain;
But, hearing soon upon the blast
The cry, I bade him halt again.

Forthwith alighting on the ground,
"Whence comes," said I, "this piteous moan?"
And there a little Girl I found,
Sitting behind the chaise, alone.   20

"My cloak!" no other word she spake,
But loud and bitterly she wept,
As if her innocent heart would break;
And down from off her seat she leapt.

"What ails you, child?"—she sobbed, "Look here!"
I saw it in the wheel entangled,
A weather-beaten rag as e'er
From any garden scare-crow dangled.

There, twisted between nave and spoke,
It hung, nor could at once be freed;   30
But our joint pains unloosed the cloak,
A miserable rag indeed!

"And whither are you going, child,
To-night along these lonesome ways?"
"To Durham," answered she, half wild—
"Then come with me into the chaise."

Insensible to all relief
Sat the poor girl, and forth did send
Sob after sob, as if her grief
Could never, never have an end.   40

"My child, in Durham do you dwell?"
She checked herself in her distress,
And said, "My name is Alice Fell;
I'm fatherless and motherless.

"And I to Durham, Sir, belong."
Again, as if the thought would choke
Her very heart, her grief grew strong;
And all was for her tattered cloak!

The chaise drove on; our journey's end
Was nigh; and, sitting by my side,   50
As if she had lost her only friend
She wept, nor would be pacified.

Up to the tavern-door we post;
Of Alice and her grief I told;
And I gave money to the host,
To buy a new cloak for the old.

"And let it be of duffil grey,
As warm a cloak as man can sell!"
Proud creature was she the next day,
The little orphan, Alice Fell!   60

1802 (1807)

TO A BUTTERFLY

See the *Grasmere Journal*, March 14, 1802, below, p. 362.

Stay near me—do not take thy flight!
A little longer stay in sight!
Much converse do I find in thee,
Historian of my infancy!
Float near me; do not yet depart!
Dead times revive in thee:
Thou bring'st, gay creature as thou art!
A solemn image to my heart,
My father's family!

Oh! pleasant, pleasant were the days,   10
The time, when in our childish plays,
My sister Emmeline and I
Together chased the butterfly!
A very hunter did I rush
Upon the prey;—with leaps and springs
I followed on from brake to bush;
But she, God love her! feared to brush
The dust from off its wings.

1802 (1807)

TO THE CUCKOO

See Wordsworth's discussion in the Preface of 1815, below, pp. 333–34.

O blithe New-comer! I have heard,
I hear thee and rejoice.
O Cuckoo! shall I call thee Bird,
Or but a wandering Voice?

While I am lying on the grass
Thy twofold shout I hear;
From hill to hill it seems to pass
At once far off, and near.

Though babbling only to the Vale,
Of sunshine and of flowers,   10
Thou bringest unto me a tale
Of visionary hours.

Thrice welcome, darling of the Spring!
Even yet thou art to me
No bird, but an invisible thing,
A voice, a mystery;

The same whom in my schoolboy days
I listened to; that Cry
Which made me look a thousand ways
In bush, and tree, and sky.   20

To seek thee did I often rove
Through woods and on the green;
And thou wert still a hope, a love;
Still longed for, never seen.

And I can listen to thee yet;
Can lie upon the plain
And listen, till I do beget
That golden time again.

O blessèd Bird! the earth we pace
Again appears to be   30
An unsubstantial, faery place;
That is fit home for Thee!

                              1802 (1807)

MY HEART LEAPS UP
WHEN I BEHOLD

My heart leaps up when I behold
    A rainbow in the sky:
So was it when my life began;
So is it now I am a man;
So be it when I shall grow old,
    Or let me die!
The Child is father of the Man;
And I could wish my days to be
Bound each to each by natural piety.

                              1802 (1807)

ODE

INTIMATIONS OF IMMORTALITY
FROM RECOLLECTIONS OF EARLY CHILDHOOD

"This was composed during my residence at Town End,
Grasmere; two years at least passed between the writing
of the four first stanzas and the remaining part. To the
attentive and competent reader the whole sufficiently
explains itself; but there may be no harm in adverting
here to particular feelings or *experiences* of my own mind
on which the structure of the poem partly rests. Nothing
was more difficult for me in childhood than to admit the
notion of death as a state applicable to my own being. I
have said elsewhere:—

        'A simple child,
        That lightly draws its breath,
        And feels its life in every limb,
        What should it know of death?'

                        [*We Are Seven*, ll. 1–4].

But it was not so much from feelings of animal vivacity
that *my* difficulty came as from a sense of the indomitable-
ness of the spirit within me. I used to brood over the
stories of Enoch and Elijah, and almost to persuade myself that, whatever might become of others, I should be
translated, in something of the same way, to heaven.
With a feeling congenial to this, I was often unable to
think of external things as having external existence, and
I communed with all that I saw as something not apart
from, but inherent in, my own immaterial nature. Many
times while going to school have I grasped at a wall or
tree to recall myself from this abyss of idealism to the
reality. At that time I was afraid of such processes. In
later periods of life I have deplored, as we have all reason
to do, a subjugation of an opposite character, and have
rejoiced over the remembrances, as is expressed in the
lines—

        'Obstinate questionings
        Of sense and outward things,
        Fallings from us, vanishings;' etc.

                        [ll. 141–43].

To that dream-like vividness and splendour which invest
objects of sight in childhood, every one, I believe, if he
would look back, could bear testimony, and I need not
dwell upon it here: but having in the poem regarded it as
presumptive evidence of a prior state of existence, I
think it right to protest against a conclusion, which has
given pain to some good and pious persons, that I
meant to inculcate such a belief. It is far too shadowy a
notion to be recommended to faith, as more than an
element in our instincts of immortality. But let us bear in
mind that, though the idea is not advanced in revelation,
there is nothing there to contradict it, and the fall of man

presents an analogy in its favor. Accordingly, a pre-existent state has entered into the popular creeds of many nations; and, among all persons acquainted with classic literature, is known as an ingredient in Platonic philosophy. Archimedes said that he could move the world if he had a point whereon to rest his machine. Who has not felt the same aspirations as regards the world of his own mind? Having to wield some of its elements when I was impelled to write this Poem on the 'Immortality of the Soul,' I took hold of the notion of pre-existence as having sufficient foundation in humanity for authorizing me to make for my purpose the best use of it I could as a Poet" (W. W.).

In a letter to Mrs. Clarkson, December, 1814, Wordsworth remarks, "The poem rests entirely upon two recollections of childhood, one that of a splendour in the objects of sense which is passed away, and the other an indisposition to bend to the law of death as applying to our particular case. A Reader who has not a vivid recollection of these feelings having existed in his mind cannot understand that poem."

Coleridge discusses the poem at some length in *Biographia Literaria*, Chapter XXII. His "Dejection: An Ode" may also be read as a personal counterpart and comment.

❦

The Child is father of the Man;
And I could wish my days to be
Bound each to each by natural piety.

I

There was a time when meadow, grove, and stream,
The earth, and every common sight,
        To me did seem
        Apparelled in celestial light,
The glory and the freshness of a dream.
It is not now as it hath been of yore;—
        Turn wheresoe'er I may,
        By night or day,
The things which I have seen I now can see no more.

II

        The Rainbow comes and goes,   10
        And lovely is the Rose,
        The Moon doth with delight
Look round her when the heavens are bare,
        Waters on a starry night
        Are beautiful and fair;
        The sunshine is a glorious birth;
        But yet I know, where'er I go,
That there hath past away a glory from the earth.

III

Now, while the birds thus sing a joyous song,
        And while the young lambs bound   20
                As to the tabor's sound,
To me alone there came a thought of grief:
A timely utterance gave that thought relief,
        And I again am strong:
The cataracts blow their trumpets from the steep;
No more shall grief of mine the season wrong;
I hear the Echoes through the mountains throng,
The Winds come to me from the fields of sleep,
        And all the earth is gay;
                Land and sea   30
        Give themselves up to jollity,
                And with the heart of May
        Doth every Beast keep holiday;—
                Thou Child of Joy,
Shout round me, let me hear thy shouts, thou happy
        Shepherd-boy!

IV

Ye blessèd Creatures, I have heard the call
        Ye to each other make; I see
The heavens laugh with you in your jubilee;
        My heart is at your festival,
        My head hath its coronal,   40
The fulness of your bliss, I feel—I feel it all.
        Oh evil day! if I were sullen
        While Earth herself is adorning,
                This sweet May-morning,
        And the Children are culling
                On every side,
        In a thousand valleys far and wide,
        Fresh flowers; while the sun shines warm,
And the Babe leaps up on his Mother's arm:—
        I hear, I hear, with joy I hear!   50
        —But there's a Tree, of many, one,
A single Field which I have looked upon,
Both of them speak of something that is gone:
        The Pansy at my feet
        Doth the same tale repeat:
Whither is fled the visionary gleam?
Where is it now, the glory and the dream?

ODE: INTIMATIONS OF IMMORTALITY. **21. tabor:** a small drum used chiefly to accompany the pipe or trumpet. **40. coronal:** a garland of flowers worn by shepherds in pastoral poetry.

V

Our birth is but a sleep and a forgetting:
The Soul that rises with us, our life's Star,
    Hath had elsewhere its setting,   60
    And cometh from afar:
    Not in entire forgetfulness,
    And not in utter nakedness,
But trailing clouds of glory do we come
    From God, who is our home:
Heaven lies about us in our infancy!
Shades of the prison-house begin to close
    Upon the growing Boy,
But He beholds the light, and whence it flows,
    He sees it in his joy;   70
The Youth, who daily farther from the east
    Must travel, still is Nature's Priest,
    And by the vision splendid
    Is on his way attended;
At length the Man perceives it die away,
And fade into the light of common day.

VI

Earth fills her lap with pleasures of her own;
Yearnings she hath in her own natural kind,
And, even with something of a Mother's mind,
    And no unworthy aim,   80
    The homely Nurse doth all she can
To make her Foster-child, her Inmate Man,
    Forget the glories he hath known,
And that imperial palace whence he came.

VII

Behold the Child among his new-born blisses,
A six years' Darling of a pigmy size!
See, where 'mid work of his own hand he lies
Fretted by sallies of his mother's kisses,
With light upon him from his father's eyes!
See, at his feet, some little plan or chart,   90
Some fragment from his dream of human life,
Shaped by himself with newly-learned art;
    A wedding or a festival,
    A mourning or a funeral;
    And this hath now his heart,
    And unto this he frames his song:
    Then will he fit his tongue
To dialogues of business, love, or strife;
    But it will not be long

Ere this be thrown aside,   100
    And with new joy and pride
The little Actor cons another part;
Filling from time to time his "humorous stage"
With all the Persons, down to palsied Age,
That life brings with her in her equipage;
    As if his whole vocation
    Were endless imitation.

VIII

Thou, whose exterior semblance doth belie
    Thy Soul's immensity;
Thou best Philosopher, who yet dost keep   110
Thy heritage, thou Eye among the blind,
That, deaf and silent, read'st the eternal deep,
Haunted for ever by the eternal mind,—
    Mighty Prophet! Seer blest!
    On whom those truths do rest,
Which we are toiling all our lives to find,
In darkness lost, the darkness of the grave;
Thou, over whom thy Immortality
Broods like the Day, a Master o'er a Slave,
A Presence which is not to be put by;   120
Thou little Child, yet glorious in the might
Of heaven-born freedom on thy being's height,
Why with such earnest pains dost thou provoke
The years to bring the inevitable yoke,
Thus blindly with thy blessedness at strife?
Full soon thy Soul shall have her earthly freight,
And custom lie upon thee with a weight,
Heavy as frost, and deep almost as life!

IX

    O joy! that in our embers
    Is something that doth live,   130
    That nature yet remembers
    What was so fugitive!
The thought of our past years in me doth breed
Perpetual benediction: not indeed
For that which is most worthy to be blest;
Delight and liberty, the simple creed
Of Childhood, whether busy or at rest,
With new-fledged hope still fluttering in his breast:—
    Not for these I raise
    The song of thanks and praise;   140

103. "humorous stage": quoted from the dedicatory sonnet of *Musophilus*, by Samuel Daniel, 1562–1619.

But for those obstinate questionings
Of sense and outward things,
Fallings from us, vanishings;
Blank misgivings of a Creature
Moving about in worlds not realised,
High instincts before which our mortal Nature
Did tremble like a guilty Thing surprised:
    But for those first affections,
    Those shadowy recollections,
    Which, be they what they may,   150
Are yet the fountain-light of all our day,
Are yet a master-light of all our seeing;
    Uphold us, cherish, and have power to make
Our noisy years seem moments in the being
Of the eternal Silence: truths that wake,
    To perish never:
Which neither listlessness, nor mad endeavour,
    Nor Man nor Boy,
Nor all that is at enmity with joy,
Can utterly abolish or destroy.   160
    Hence in a season of calm weather
    Though inland far we be,
Our Souls have sight of that immortal sea
    Which brought us hither,
    Can in a moment travel thither,
And see the Children sport upon the shore,
And hear the mighty waters rolling evermore.

X

Then sing, ye Birds, sing, sing a joyous song!
    And let the young Lambs bound
    As to the tabor's sound!   170
We in thought will join your throng,
    Ye that pipe and ye that play,
    Ye that through your hearts to-day
    Feel the gladness of the May!
What though the radiance which was once so bright
Be now for ever taken from my sight,
    Though nothing can bring back the hour
Of splendour in the grass, of glory in the flower;
    We will grieve not, rather find
    Strength in what remains behind;   180
    In the primal sympathy
    Which having been must ever be;
    In the soothing thoughts that spring
    Out of human suffering;
    In the faith that looks through death,
In years that bring the philosophic mind.

XI

And O, ye Fountains, Meadows, Hills, and Groves,
Forebode not any severing of our loves!
Yet in my heart of hearts I feel your might;
I only have relinquished one delight   190
To live beneath your more habitual sway.
I love the Brooks which down their channels fret,
Even more than when I tripped lightly as they;
The innocent brightness of a new-born Day
    Is lovely yet;
The Clouds that gather round the setting sun
Do take a sober colouring from an eye
That hath kept watch o'er man's mortality;
Another race hath been, and other palms are won.
Thanks to the human heart by which we live,   200
Thanks to its tenderness, its joys, and fears,
To me the meanest flower that blows can give
Thoughts that do often lie too deep for tears.

1802–04 (1807)

WRITTEN IN MARCH

WHILE RESTING ON THE BRIDGE
AT THE FOOT OF BROTHER'S WATER

See Dorothy's *Grasmere Journal*, April 16, 1802, below,
p. 363.

    The Cock is crowing,
    The stream is flowing,
    The small birds twitter,
    The lake doth glitter,
The green field sleeps in the sun;
    The oldest and youngest
    Are at work with the strongest;
    The cattle are grazing,
    Their heads never raising;
There are forty feeding like one!   10

    Like an army defeated
    The snow hath retreated,
    And now doth fare ill
    On the top of the bare hill;

WRITTEN IN MARCH. **Brother's Water:** the name of a lake
in the Lake District.

The Ploughboy is whooping—anon—anon:
  There's joy in the mountains;
  There's life in the fountains;
  Small clouds are sailing,
  Blue sky prevailing;
The rain is over and gone!   20

                                    1802 (1807)

TO A SKY-LARK

Up with me! up with me into the clouds!
  For thy song, Lark, is strong;
Up with me, up with me into the clouds!
  Singing, singing,
With clouds and sky about thee ringing,
  Lift me, guide me, till I find
That spot which seems so to thy mind!

I have walked through wildernesses dreary,
And to-day my heart is weary;
Had I now the wings of a Faery,   10
Up to thee would I fly.
There is madness about thee, and joy divine
In that song of thine;
Lift me, guide me, high and high
To thy banqueting place in the sky.

        Joyous as morning,
Thou art laughing and scorning;
Thou hast a nest for thy love and thy rest,
And, though little troubled with sloth,
Drunken Lark! thou wouldst be loth   20
To be such a traveller as I.
Happy, happy Liver,
With a soul as strong as a mountain river
Pouring out praise to the almighty Giver,
Joy and jollity be with us both!

Alas! my journey, rugged and uneven,
Through prickly moors or dusty ways must wind;
But hearing thee, or others of thy kind,
As full of gladness and as free of heaven,
I, with my fate contented, will plod on,   30
And hope for higher raptures, when life's day is done.

                              c. 1802 (1807)

TO A BUTTERFLY

I've watched you now a full half-hour,
Self-poised upon that yellow flower;
And, little Butterfly! indeed
I know not if you sleep or feed.
How motionless!—not frozen seas
More motionless! and then
What joy awaits you, when the breeze
Hath found you out among the trees,
And calls you forth again!

This plot of orchard-ground is ours;   10
My trees they are, my Sister's flowers;
Here rest your wings when they are weary;
Here lodge as in a sanctuary!
Come often to us, fear no wrong;
Sit near us on the bough!
We'll talk of sunshine and of song,
And summer days, when we were young;
Sweet childish days, that were as long
As twenty days are now.

                                    1802 (1807)

TO H. C.
SIX YEARS OLD

H. C. is Coleridge's child Hartley. Coleridge described
him as "a spirit that dances on an aspin leaf—the air . . .
is to my Babe a perpetual Nitrous Oxyde. Never was
more joyous creature born." "From morning to night he
whisks about and about, whisks, whirls, and eddies like a
blossom in a May-breeze."

O thou! whose fancies from afar are brought;
Who of thy words dost make a mock apparel,
And fittest to unutterable thought
The breeze-like motion and the self-born carol;
Thou faery voyager! that dost float
In such clear water, that thy boat
May rather seem
To brood on air than on an earthly stream;
Suspended in a stream as clear as sky,
Where earth and heaven do make one imagery;   10
O blessèd vision! happy child!
Thou art so exquisitely wild,

I think of thee with many fears
For what may be thy lot in future years.

  I thought of times when Pain might be thy guest,
Lord of thy house and hospitality;
And Grief, uneasy lover! never rest
But when she sate within the touch of thee.
O too industrious folly!
O vain and causeless melancholy!   20
Nature will either end thee quite;
Or, lengthening out thy season of delight,
Preserve for thee, by individual right,
A young lamb's heart among the full-grown flocks.
What hast thou to do with sorrow,
Of the injuries of to-morrow?
Thou art a dew-drop, which the morn brings forth,
Ill fitted to sustain unkindly shocks,
Or to be trailed along the soiling earth;
A gem that glitters while it lives,   30
And no forewarning gives;
But, at the touch of wrong, without a strife
Slips in a moment out of life.

                    1802 (1807)

RESOLUTION AND
INDEPENDENCE

∽✵∾

For discussion see Wordsworth's letter to Sara Hutchinson, June 14, 1802, below, pp. 352–53, the Preface of 1815, below, p. 334, and Coleridge in *Biographia Literaria*, Chapter XXII, below, p. 478.

∽✵∾

There was a roaring in the wind all night;
The rain came heavily and fell in floods;
But now the sun is rising calm and bright;
The birds are singing in the distant woods;
Over his own sweet voice the Stock-dove broods;
The Jay makes answer as the Magpie chatters;
And all the air is filled with pleasant noise of waters.

All things that love the sun are out of doors;
The sky rejoices in the morning's birth;
The grass is bright with rain-drops;—on the moors 10
The hare is running races in her mirth;
And with her feet she from the plashy earth
Raises a mist; that, glittering in the sun,
Runs with her all the way, wherever she doth run.

I was a Traveller then upon the moor;
I saw the hare that raced about with joy;
I heard the woods and distant waters roar;
Or heard them not, as happy as a boy:
The pleasant season did my heart employ:
My old remembrances went from me wholly;   20
And all the ways of men, so vain and melancholy.

But, as it sometimes chanceth, from the might
Of joy in minds that can no further go,
As high as we have mounted in delight
In our dejection do we sink as low;
To me that morning did it happen so;
And fears and fancies thick upon me came;
Dim sadness—and blind thoughts, I knew not, nor
    could name.

I heard the sky-lark warbling in the sky;
And I bethought me of the playful hare: 30
Even such a happy Child of earth am I;
Even as these blissful creatures do I fare;
Far from the world I walk, and from all care;
But there may come another day to me—
Solitude, pain of heart, distress, and poverty.

My whole life I have lived in pleasant thought,
As if life's business were a summer mood;
As if all needful things would come unsought
To genial faith, still rich in genial good;
But how can He expect that others should   40
Build for him, sow for him, and at his call
Love him, who for himself will take no heed at all?

I thought of Chatterton, the marvellous Boy,
The sleepless Soul that perished in his pride;
Of Him who walked in glory and in joy
Following his plough, along the mountain-side:
By our own spirits are we deified:
We Poets in our youth begin in gladness;
But thereof come in the end despondency and madness.

Now, whether it were by peculiar grace,   50
A leading from above, a something given,
Yet it befell that, in this lonely place,
When I with these untoward thoughts had striven,
Beside a pool bare to the eye of heaven

RESOLUTION AND INDEPENDENCE. **43. Chatterton:** Thomas Chatterton, 1752–70, a remarkably promising and gifted young poet. Lonely, impoverished, and despairing, he committed suicide at the age of seventeen. **45. Him:** Robert Burns, 1759–96, was to Wordsworth an example of a peasant poet. He died at the age of thirty-seven, broken by toil, dissipation, sickness, and poverty.

I saw a Man before me unawares:
The oldest man he seemed that ever wore grey hairs.

As a huge stone is sometimes seen to lie
Couched on the bald top of an eminence;
Wonder to all who do the same espy,
By what means it could thither come, and whence; 60
So that it seems a thing endued with sense:
Like a sea-beast crawled forth, that on a shelf
Of rock or sand reposeth, there to sun itself;

Such seemed this Man, not all alive nor dead,
Nor all asleep—in his extreme old age:
His body was bent double, feet and head
Coming together in life's pilgrimage;
As if some dire constraint of pain, or rage
Of sickness felt by him in times long past,
A more than human weight upon his frame had cast. 70

Himself he propped, limbs, body, and pale face,
Upon a long grey staff of shaven wood:
And, still as I drew near with gentle pace,
Upon the margin of that moorish flood
Motionless as a cloud the old Man stood,
That heareth not the loud winds when they call;
And moveth all together, if it move at all.

At length, himself unsettling, he the pond
Stirred with his staff, and fixedly did look
Upon the muddy water, which he conned,   80
As if he had been reading in a book:
And now a stranger's privilege I took;
And, drawing to his side, to him did say,
"This morning gives us promise of a glorious day."

A gentle answer did the old Man make,
In courteous speech which forth he slowly drew:
And him with further words I thus bespake,
"What occupation do you there pursue?
This is a lonesome place for one like you."
Ere he replied, a flash of mild surprise   90
Broke from the sable orbs of his yet-vivid eyes.

His words came feebly, from a feeble chest,
But each in solemn order followed each,
With something of a lofty utterance drest—
Choice word and measured phrase, above the reach
Of ordinary men; a stately speech;
Such as grave Livers do in Scotland use,
Religious men, who give to God and man their dues.

He told, that to these waters he had come
To gather leeches, being old and poor: 100
Employment hazardous and wearisome!
And he had many hardships to endure:
From pond to pond he roamed, from moor to moor;
Housing, with God's good help, by choice or chance;
And in this way he gained an honest maintenance.

The old Man still stood talking by my side;
But now his voice to me was like a stream
Scarce heard; nor word from word could I divide;
And the whole body of the Man did seem
Like one whom I had met with in a dream;   110
Or like a man from some far region sent,
To give me human strength, by apt admonishment.

My former thoughts returned: the fear that kills;
And hope that is unwilling to be fed;
Cold, pain, and labour, and all fleshly ills;
And mighty Poets in their misery dead.
—Perplexed, and longing to be comforted,
My question eagerly did I renew,
"How is it that you live, and what is it you do?"

He with a smile did then his words repeat;   120
And said that, gathering leeches, far and wide
He travelled; stirring thus about his feet
The waters of the pools where they abide.
"Once I could meet with them on every side;
But they have dwindled long by slow decay;
Yet still I persevere, and find them where I may."

While he was talking thus the lonely place,
The old Man's shape, and speech—all troubled me:
In my mind's eye I seemed to see him pace
About the weary moors continually,   130
Wandering about alone and silently.
While I these thoughts within myself pursued,
He, having made a pause, the same discourse renewed.

And soon with this he other matter blended,
Cheerfully uttered, with demeanour kind,
But stately in the main; and, when he ended,
I could have laughed myself to scorn to find
In that decrepit Man so firm a mind.
"God," said I, "be my help and stay secure;   139
I'll think of the Leech-gatherer on the lonely moor!"

1802 (1807)

100. **leeches:** commonly prescribed in the medical treatment
of the time. They were applied to suck blood as a relief for
various minor ailments.

Wordsworth became a prolific writer of sonnets, but it is a curious fact that most of the memorable ones were produced within a year or so after his first use of the form on May 21, 1802. About the composition of the sonnets of 1802, he wrote:

"In the cottage, Town-end, Grasmere, one afternoon in 1801, my sister read to me the sonnets of Milton. I had long been well acquainted with them, but I was particularly struck on that occasion with the dignified simplicity and majestic harmony that runs through most of them,—in character so totally different from the Italian, and still more so from Shakespeare's fine sonnets. I took fire, if I may be allowed to say so, and produced three sonnets the same afternoon, the first I ever wrote except an irregular one at school. Of these three, the only one I distinctly remember is—*I grieved for Buonaparté*. One was never written down: the third, which was, I believe, preserved, I cannot particularize."

The date of 1801 is incorrect. Dorothy's *Journal* reveals that "I grieved for Buonaparté" was composed May 21, 1802. Wordsworth's sonnets are Miltonic both in form and in emotion. He used the Petrarchan rhyme scheme (in which the octave rhymes *abbaabba*; the sestet has various patterns), and he caught from Milton a tendency not to break the sense at the close of the octave but to run on into the sestet. He also adopts Milton's rich play of long vowels, particularly in the rhymes, his frequent practice of opening the sonnet in direct address, and his use of a relatively high proportion of run-on lines. ("To Toussaint L'Ouverture" is an example.) Moreover, Wordsworth learned from Milton's sonnets that the form could be an effective vehicle for national or public themes, and he found in them an inspiration and echo of his own high-minded patriotism.

Of the eighteen sonnets printed below, the nine that follow the first were written in August and September, 1802. In August Wordsworth was in Calais during the brief peace of Amiens between England and France.

1801

I grieved for Buonaparté, with a vain
And an unthinking grief! The tenderest mood
Of that Man's mind—what can it be? what food
Fed his first hopes? what knowledge could *he* gain?
'Tis not in battles that from youth we train
The Governor who must be wise and good,
And temper with the sternness of the brain
Thoughts motherly, and meek as womanhood.
Wisdom doth live with children round her knees:
Books, leisure, perfect freedom, and the talk   10
Man holds with week-day man in the hourly walk

Of the mind's business: these are the degrees
By which true Sway doth mount; this is the stalk
True Power doth grow on; and her rights are these.

1802 (*1802*)

IT IS A BEAUTEOUS EVENING, CALM AND FREE

It is a beauteous evening, calm and free,
The holy time is quiet as a Nun
Breathless with adoration; the broad sun
Is sinking down in its tranquillity;
The gentleness of heaven broods o'er the Sea:
Listen! the mighty Being is awake,
And doth with his eternal motion make
A sound like thunder—everlastingly.
Dear Child! dear Girl! that walkest with me here,
If thou appear untouched by solemn thought,   10
Thy nature is not therefore less divine:
Thou liest in Abraham's bosom all the year;
And worshipp'st at the Temple's inner shrine,
God being with thee when we know it not.

1802 (*1807*)

COMPOSED BY THE SEA-SIDE, NEAR CALAIS, AUGUST, 1802

Fair Star of evening, Splendour of the west,
Star of my Country!—on the horizon's brink
Thou hangest, stooping, as might seem, to sink
On England's bosom; yet well pleased to rest,
Meanwhile, and be to her a glorious crest
Conspicuous to the Nations. Thou, I think,
Shouldst be my Country's emblem; and shouldst wink,
Bright Star! with laughter on her banners, drest
In thy fresh beauty. There! that dusky spot
Beneath thee, that is England; there she lies.   10

IT IS A BEAUTEOUS EVENING, CALM AND FREE. **12. Abraham's bosom:** an abode near heaven where the souls of the blessed go after death and prepare for the final vision of God. (See Luke 16:19–31.) Often used vaguely to suggest a place of bliss in the other world.

Blessings be on you both! one hope, one lot,
One life, one glory!—I, with many a fear
For my dear Country, many heartfelt sighs,
Among men who do not love her, linger here.

                                    1802 (1807)

CALAIS, AUGUST, 1802

Is it a reed that's shaken by the wind,
Or what is it that ye go forth to see?
Lords, lawyers, statesmen, squires of low degree,
Men known, and men unknown, sick, lame, and blind,
Post forward all, like creatures of one kind,
With first-fruit offerings crowd to bend the knee
In France, before the new-born Majesty.
'Tis ever thus. Ye men of prostrate mind,
A seemly reverence may be paid to power;
But that's a loyal virtue, never sown   10
In haste, nor springing with a transient shower:
When truth, when sense, when liberty were flown,
What hardship had it been to wait an hour?
Shame on you, feeble Heads, to slavery prone!

                                    1802 (1803)

ON THE EXTINCTION OF
THE VENETIAN REPUBLIC

Once did She hold the gorgeous east in fee;
And was the safeguard of the west: the worth
Of Venice did not fall below her birth,
Venice, the eldest Child of Liberty.
She was a maiden City, bright and free;
No guile seduced, no force could violate;
And, when she took unto herself a Mate,
She must espouse the everlasting Sea.

CALAIS, AUGUST, 1802. **7. new-born Majesty:** Napoleon
had lately been named Consul for Life, and constitutional
changes of August, 1802, made him absolute ruler of France.
ON THE EXTINCTION OF THE VENETIAN REPUBLIC. **4. eldest . . .
Liberty:** According to Gibbon, Venice was founded by
refugees from Attila's invasion of Italy (452). "In the midst
of the waters, free, indigent, laborious, and inaccessible,
they gradually coalesced into a republic . . . the Venetians
exult in the belief of primitive and perpetual independence."
8. She . . . Sea: Each Ascension Day at a solemn ceremony
a ring was cast into the sea, symbolizing the wedding of the
Doge to the Adriatic.

And what if she had seen those glories fade,
Those titles vanish, and that strength decay;   10
Yet shall some tribute of regret be paid
When her long life hath reached its final day:
Men are we, and must grieve when even the Shade
Of that which once was great, is passed away.

                                    1802 (1807)

TO TOUSSAINT L'OUVERTURE

Toussaint, the most unhappy man of men!
Whether the whistling Rustic tend his plough
Within thy hearing, or thy head be now
Pillowed in some deep dungeon's earless den;—
O miserable Chieftain! where and when
Wilt thou find patience! Yet die not; do thou
Wear rather in thy bonds a cheerful brow:
Though fallen thyself, never to rise again,
Live, and take comfort. Thou hast left behind
Powers that will work for thee; air, earth, and skies; 10
There's not a breathing of the common wind
That will forget thee; thou hast great allies;
Thy friends are exultations, agonies,
And love, and man's unconquerable mind.

                                    1802 (1803)

SEPTEMBER, 1802.
NEAR DOVER

Inland, within a hollow vale, I stood;
And saw, while sea was calm and air was clear,
The coast of France—the coast of France how near!
Drawn almost into frightful neighbourhood.
I shrunk; for verily the barrier flood
Was like a lake, or river bright and fair,
A span of waters; yet what power is there!
What mightiness for evil and for good!

12. final day: Venice was taken by Napoleon, who proclaimed the end of the Republic on May 16, 1797. TO
TOUSSAINT L'OUVERTURE. **1–4. Toussaint . . . den:** A leader
in the Haitian struggle for independence from France,
Toussaint l'Ouverture was captured by treachery and died
in a French prison on April 27, 1803.

Even so doth God protect us if we be
Virtuous and wise. Winds blow, and waters roll,   10
Strength to the brave, and Power, and Deity;
Yet in themselves are nothing! One decree
Spake laws to *them*, and said that by the soul
Only, the Nations shall be great and free.

1802 (1807)

COMPOSED UPON WESTMINSTER BRIDGE, SEPTEMBER 3, 1802

Earth has not anything to show more fair:
Dull would he be of soul who could pass by
A sight so touching in its majesty:
This City now doth, like a garment, wear
The beauty of the morning; silent, bare,
Ships, towers, domes, theatres, and temples lie
Open unto the fields, and to the sky;
All bright and glittering in the smokeless air.
Never did sun more beautifully steep
In his first splendour, valley, rock, or hill;   10
Ne'er saw I, never felt, a calm so deep!
The river glideth at his own sweet will:
Dear God! the very houses seem asleep;
And all that mighty heart is lying still!

1802 (1807)

WRITTEN IN LONDON, SEPTEMBER, 1802

O friend! I know not which way I must look
For comfort, being, as I am, opprest,
To think that now our life is only drest
For show; mean handy-work of craftsman, cook,
Or groom!—We must run glittering like a brook
In the open sunshine, or we are unblest:
The wealthiest man among us is the best:
No grandeur now in nature or in book
Delights us. Rapine, avarice, expense,
This is idolatry; and these we adore: 10
Plain living and high thinking are no more:
The homely beauty of the good old cause
Is gone; our peace, our fearful innocence,
And pure religion breathing household laws.

1802 (1807)

LONDON, 1802

Milton! thou shouldst be living at this hour:
England hath need of thee: she is a fen
Of stagnant waters: altar, sword, and pen,
Fireside, the heroic wealth of hall and bower,
Have forfeited their ancient English dower
Of inward happiness. We are selfish men;
Oh! raise us up, return to us again;
And give us manners, virtue, freedom, power.
Thy soul was like a Star, and dwelt apart;
Thou hadst a voice whose sound was like the sea: 10
Pure as the naked heavens, majestic, free,
So didst thou travel on life's common way,
In cheerful godliness; and yet thy heart
The lowliest duties on herself did lay.

1802 (1807)

GREAT MEN HAVE BEEN AMONG US; HANDS THAT PENNED

Great men have been among us; hands that penned
And tongues that uttered wisdom—better none:
The later Sidney, Marvel, Harrington,
Young Vane, and others who called Milton friend.
These moralists could act and comprehend:
They knew how genuine glory was put on;
Taught us how rightfully a nation shone
In spendour: what strength was, that would not bend
But in magnanimous meekness. France, 'tis strange,
Hath brought forth no such souls as we had then.   10
Perpetual emptiness! unceasing change!
No single volume paramount, no code,
No master spirit, no determined road;
But equally a want of books and men!

1802 (1807)

GREAT MEN HAVE BEEN AMONG US; HANDS THAT PENNED. 3-4.
Sidney . . . Vane: Algernon Sidney, the poet Andrew
Marvell, James Harrington, and Sir Henry Vane, to whom
Milton addressed a sonnet, all figured in the Puritan and
republican cause in the seventeenth century.

WHEN I HAVE BORNE
IN MEMORY WHAT HAS TAMED

When I have borne in memory what has tamed
Great Nations, how ennobling thoughts depart
When men change swords for ledgers, and desert
The student's bower for gold, some fears unnamed
I had, my Country!—am I to be blamed?
Now, when I think of thee, and what thou art,
Verily, in the bottom of my heart,
Of those unfilial fears I am ashamed.
For dearly must we prize thee; we who find
In thee a bulwark for the cause of men;   10
And I by my affection was beguiled:
What wonder if a Poet now and then,
Among the many movements of his mind,
Felt for thee as a lover or a child!

1802 or 1803 (*1803*)

IT IS NOT
TO BE THOUGHT OF
THAT THE FLOOD

It is not to be thought of that the Flood
Of British freedom, which, to the open sea
Of the world's praise, from dark antiquity
Hath flowed, "with pomp of waters, unwithstood,"
Roused though it be full often to a mood
Which spurns the check of salutary bands,
That this most famous Stream in bogs and sands
Should perish; and to evil and to good
Be lost for ever. In our halls is hung
Armoury of the invincible Knights of old: 10
We must be free or die, who speak the tongue
That Shakspeare spake; the faith and morals hold
Which Milton held.—In every thing we are sprung
Of Earth's first blood, have titles manifold.

1802 or 1803 (*1803*)

IT IS NOT TO BE THOUGHT OF THAT THE FLOOD. **4. "with . . . unwithstood"**: quoted from Samuel Daniel's *Civil Wars* (1595), II. 7.

WITH SHIPS THE SEA WAS
SPRINKLED FAR AND NIGH

See Wordsworth's analysis of this sonnet in his letter to Lady Beaumont, May 21, 1807, below, pp. 356–57.

With Ships the sea was sprinkled far and nigh,
Like stars in heaven, and joyously it showed;
Some lying fast at anchor in the road,
Some veering up and down, one knew not why.
A goodly Vessel did I then espy
Come like a giant from a haven broad;
And lustily along the bay she strode,
Her tackling rich, and of apparel high.
This Ship was nought to me, nor I to her,
Yet I pursued her with a Lover's look;   10
This Ship to all the rest did I prefer:
When will she turn, and whither? She will brook
No tarrying; where She comes the winds must stir:
On went She, and due north her journey took.

1802–04 (*1807*)

THE WORLD IS TOO MUCH
WITH US; LATE AND SOON

The world is too much with us; late and soon,
Getting and spending, we lay waste our powers:
Little we see in Nature that is ours;
We have given our hearts away, a sordid boon!
This Sea that bares her bosom to the moon;
The winds that will be howling at all hours,
And are up-gathered now like sleeping flowers;
For this, for everything, we are out of tune;
It moves us not.—Great God! I'd rather be
A Pagan suckled in a creed outworn;   10
So might I, standing on this pleasant lea,
Have glimpses that would make me less forlorn;
Have sight of Proteus rising from the sea;
Or hear old Triton blow his wreathèd horn.

1802–04 (*1807*)

THE WORLD IS TOO MUCH WITH US; LATE AND SOON. **13–14. Proteus . . . Triton**: sea-gods. According to the *Odyssey*, Proteus can take on any shape he wills; Triton, the lower part of whose body is that of a fish, is usually shown blowing a trumpet of conch shell.

METHOUGHT I SAW
THE FOOTSTEPS OF A THRONE

Methought I saw the footsteps of a throne
Which mists and vapours from mine eyes did shroud—
Nor view of who might sit thereon allowed;
But all the steps and ground about were strown
With sights the ruefullest that flesh and bone
Ever put on; a miserable crowd,
Sick, hale, old, young, who cried before that cloud,
"Thou art our king, O Death! to thee we groan."
Those steps I clomb; the mists before me gave
Smooth way; and I beheld the face of one   10
Sleeping alone within a mossy cave,
With her face up to heaven; that seemed to have
Pleasing remembrance of a thought foregone;
A lovely Beauty in a summer grave!

1802–04 (1807)

SCORN NOT THE SONNET;
CRITIC, YOU HAVE FROWNED

Scorn not the Sonnet; Critic, you have frowned,
Mindless of its just honours; with this key
Shakspeare unlocked his heart; the melody
Of this small lute gave ease to Petrarch's wound;
A thousand times this pipe did Tasso sound;
With it Camöens soothed an exile's grief;
The Sonnet glittered a gay myrtle leaf
Amid the cypress with which Dante crowned
His visionary brow: a glow-worm lamp,
It cheered mild Spenser, called from Faery-land   10
To struggle through dark ways; and when a damp
Fell round the path of Milton, in his hand
The Thing became a trumpet; whence he blew
Soul-animating strains—alas, too few!

1820–27 (1827)

NUNS FRET NOT AT THEIR
CONVENT'S NARROW ROOM

Nuns fret not at their convent's narrow room;
And hermits are contented with their cells;
And students with their pensive citadels;
Maids at the wheel, the weaver at his loom,
Sit blithe and happy; bees that soar for bloom,
High as the highest Peak of Furness-fells,
Will murmur by the hour in foxglove bells:
In truth the prison, unto which we doom
Ourselves, no prison is: and hence for me,
In sundry moods, 'twas pastime to be bound   10
Within the Sonnet's scanty plot of ground;
Pleased if some Souls (for such there needs must be)
Who have felt the weight of too much liberty,
Should find brief solace there, as I have found.

1802–04 (1807)

PERSONAL TALK

I

I am not One who much or oft delight
To season my fireside with personal talk,—
Of friends, who live within an easy walk,
Or neighbours, daily, weekly, in my sight:
And, for my chance-acquaintance, ladies bright,
Sons, mothers, maidens withering on the stalk,
These all wear out of me, like Forms with chalk
Painted on rich men's floors, for one feast-night.
Better than such discourse doth silence long,
Long, barren silence, square with my desire;   10
To sit without emotion, hope, or aim,
In the loved presence of my cottage-fire,
And listen to the flapping of the flame,
Or kettle whispering its faint under-song.

NUNS FRET NOT AT THEIR CONVENT'S NARROW ROOM. **6. fells:**
here, mountains. The word can also be used to mean a stretch
of high open land, as in *The Prelude*, VI. 235.

SCORN NOT THE SONNET; CRITIC, YOU HAVE FROWNED. **4.
Petrarch's wound:** his love for Laura. The word "wound"
is itself an allusion to Petrarchan love conventions. **5.
Tasso:** Torquato Tasso, 1544–95, Italian poet. **6. Camöens:**
Luis de Camoëns, 1524–80, Portuguese poet. He spent
much of his life as a soldier in Portugal's overseas possessions, and in his poems he speaks of himself as having been
banished from the court and from the woman he loves.

II

"Yet life," you say, "is life; we have seen and see,
And with a living pleasure we describe;
And fits of sprightly malice do but bribe
The languid mind into activity.
Sound sense, and love itself, and mirth and glee
Are fostered by the comment and the gibe."   20
Even be it so: yet still among your tribe,
Our daily world's true Worldlings, rank not me!
Children are blest, and powerful; their world lies
More justly balanced; partly at their feet,
And part far from them:—sweetest melodies
Are those that are by distance made more sweet;
Whose mind is but the mind of his own eyes,
He is a Slave; the meanest we can meet!

III

Wings have we,—and as far as we can go
We may find pleasure: wilderness and wood,   30
Blank ocean and mere sky, support that mood
Which with the lofty sanctifies the low.
Dreams, books, are each a world; and books, we know,
Are a substantial world, both pure and good:
Round these, with tendrils strong as flesh and blood,
Our pastime and our happiness will grow.
There find I personal themes, a plenteous store,
Matter wherein right voluble I am,
To which I listen with a ready ear;
Two shall be named, pre-eminently dear,—   40
The gentle Lady married to the Moor;
And heavenly Una with her milk-white Lamb.

IV

Nor can I not believe but that hereby
Great gains are mine; for thus I live remote
From evil-speaking; rancour, never sought,
Comes to me not; malignant truth, or lie.
Hence have I genial seasons, hence have I
Smooth passions, smooth discourse, and joyous
    thought:
And thus from day to day my little boat
Rocks in its harbour, lodging peaceably.   50
Blessings be with them—and eternal praise,
Who gave us nobler loves, and nobler cares—
The Poets, who on earth have made us heirs
Of truth and pure delight by heavenly lays!
Oh! might my name be numbered among theirs,
Then gladly would I end my mortal days.

1802–06 (1807)

PERSONAL TALK. **41. Lady . . . Moor:** Desdemona in
Othello. **42. Una . . . Lamb:** in the *Faerie Queene,* Book I.

YEW-TREES

Derived from the *Aeneid,* VI. 268–89. See Coleridge's
comment in *Biographia Literaria,* Chapter XXII, below,
pp. 488–89.

There is a Yew-tree, pride of Lorton Vale,
Which to this day stands single, in the midst
Of its own darkness, as it stood of yore:
Not loth to furnish weapons for the bands
Of Umfraville or Percy ere they marched
To Scotland's heaths; or those that crossed the sea
And drew their sounding bows at Azincour,
Perhaps at earlier Crecy, or Poictiers.
Of vast circumference and gloom profound
This solitary Tree! a living thing   10
Produced too slowly ever to decay;
Of form and aspect too magnificent
To be destroyed. But worthier still of note
Are those fraternal Four of Borrowdale,
Joined in one solemn and capacious grove;
Huge trunks! and each particular trunk a growth
Of intertwisted fibres serpentine
Up-coiling, and inveterately convolved;
Nor uninformed with Phantasy, and looks
That threaten the profane; a pillared shade,   20
Upon whose grassless floor of red-brown hue,
By sheddings from the pining umbrage tinged
Perennially—beneath whose sable roof
Of boughs, as if for festal purpose, decked
With unrejoicing berries—ghostly Shapes
May meet at noontide; Fear and trembling Hope,
Silence and Foresight; Death the Skeleton
And Time the Shadow;—there to celebrate,
As in a natural temple scattered o'er
With altars undisturbed of mossy stone,   30
United worship; or in mute repose
To lie, and listen to the mountain flood
Murmuring from Glaramara's inmost caves.

1803 (1815)

YEW-TREES. **4–8. weapons . . . Poictiers:** Bows were made
from yew wood. Umfraville and Percy are names of leaders
in the medieval wars with Scotland; the battles of Agincourt
(1415), Crécy (1346), and Poitiers (1356) took place during
the Hundred Years' War with France. **22. pining umbrage:**
withering foliage. Umbrage suggests foliage that casts a
deep shade. **33. Glaramara:** a mountain at the head of the
valley of Borrowdale.

THE GREEN LINNET

Beneath these fruit-tree boughs that shed
Their snow-white blossoms on my head,
With brightest sunshine round me spread
  Of spring's unclouded weather,
In this sequestered nook how sweet
To sit upon my orchard-seat!
And birds and flowers once more to greet,
  My last year's friends together.

One have I marked, the happiest guest
In all this covert of the blest: 10
Hail to Thee, far above the rest
  In joy of voice and pinion!
Thou, Linnet! in thy green array,
Presiding Spirit here to-day,
Dost lead the revels of the May;
  And this is thy dominion.

While birds, and butterflies, and flowers,
Make all one band of paramours,
Thou, ranging up and down the bowers,
  Art sole in thy employment: 20
A Life, a Presence like the Air,
Scattering thy gladness without care,
Too blest with any one to pair;
  Thyself thy own enjoyment.

Amid yon tuft of hazel trees,
That twinkle to the gusty breeze,
Behold him perched in ecstasies,
  Yet seeming still to hover;
There! where the flutter of his wings
Upon his back and body flings   30
Shadows and sunny glimmerings,
  That cover him all over.

My dazzled sight he oft deceives,
A Brother of the dancing leaves;
Then flits, and from the cottage-eaves
  Pours forth his song in gushes;
As if by that exulting strain
He mocked and treated with disdain
The voiceless Form he chose to feign,
  While fluttering in the bushes.   40

1803 (1807)

ODE TO DUTY

The motto is adapted from Seneca, *Moral Epistles*, 120. 10. It may be translated, "Now I am virtuous not by deliberation, but by habit, having been brought to such a point that I can not only act rightly, but cannot act except rightly." The poem is somewhat unexpected in Wordsworth at this period, being difficult to associate with *The Prelude*, on which he was spending his main effort. But it has affinities with earlier utterances, such as "Resolution and Independence" or the brave fortitude of "Michael." It shows the influence of Roman Stoic philosophy, which held that absolute moral law pervades the cosmos. After the fifth stanza there was originally another stanza, omitted from the poem after 1807:

> Yet not the less would I throughout
> Still act according to the voice
> Of my own wish; and feel past doubt
> That my submissiveness was choice:
> Not seeking in the school of pride
> For "precepts over dignified,"
> Denial and restraint I prize
> No farther than they breed a second Will more wise.

"Jam non consilio bonus, sed more eò perductus, ut non tantum rectè facere possim, sed nisi rectè facere non possim."

Stern Daughter of the Voice of God!
O Duty! if that name thou love
Who art a light to guide, a rod
To check the erring, and reprove;
Thou, who art victory and law
When empty terrors overawe;
From vain temptation dost set free;
And calm'st the weary strife of frail humanity!

There are who ask not if thine eye
Be on them; who, in love and truth,   10
Where no misgiving is, rely
Upon the genial sense of youth:
Glad Hearts! without reproach or blot;
Who do thy work, and know it not:
Oh! if through confidence misplaced
They fail, thy saving arms, dread Power! around them
  cast.

Serene will be our days and bright,
And happy will our nature be,
When love is an unerring light,
And joy its own security.   20
And they a blissful course may hold
Even now, who, not unwisely bold,
Live in the spirit of this creed;
Yet seek thy firm support, according to their need.

I, loving freedom, and untried;
No sport of every random gust,
Yet being to myself a guide,
Too blindly have reposed my trust:
And oft, when in my heart was heard
Thy timely mandate, I deferred   30
The task, in smoother walks to stray;
But thee I now would serve more strictly, if I may.

Through no disturbance of my soul,
Or strong compunction in me wrought,
I supplicate for thy control;
But in the quietness of thought:
Me this unchartered freedom tires;
I feel the weight of chance-desires:
My hopes no more must change their name,
I long for a repose that ever is the same.   40

Stern Lawgiver! yet thou dost wear
The Godhead's most benignant grace;
Nor know we anything so fair
As is the smile upon thy face:
Flowers laugh before thee on their beds
And fragrance in thy footing treads;
Thou dost preserve the stars from wrong;
And the most ancient heavens, through Thee, are fresh
     and strong.

To humbler functions, awful Power!
I call thee: I myself commend
Unto thy guidance from this hour;   50
Oh, let my weakness have an end!
Give unto me, made lowly wise,
The spirit of self-sacrifice;
The confidence of reason give;
And in the light of truth thy Bondman let me live!

1804 (1807)

ODE TO DUTY. **55. reason:** not merely logical inference. The
word is used in a broader sense, as in Milton and the
Christian Humanist tradition, and refers to man's intellectual
powers generally, especially his capability for immediate
ethical insight. Cf. "Character of the Happy Warrior,"
below, l. 27.

THE SMALL CELANDINE

There is a Flower, the lesser Celandine,
That shrinks, like many more, from cold and rain;
And, the first moment that the sun may shine,
Bright as the sun himself, 'tis out again!

When hailstones have been falling, swarm on swarm,
Or blasts the green field and the trees distrest,
Oft have I seen it muffled up from harm,
In close self-shelter, like a Thing at rest.

But lately, one rough day, this Flower I passed
And recognised it, though an altered form,   10
Now standing forth an offering to the blast,
And buffeted at will by rain and storm.

I stopped, and said with inly-muttered voice,
"It doth not love the shower, nor seek the cold:
This neither is its courage nor its choice,
But its necessity in being old.

"The sunshine may not cheer it, nor the dew;
It cannot help itself in its decay;
Stiff in its members, withered, changed of hue."
And, in my spleen, I smiled that it was grey.   20

To be a Prodigal's Favourite—then, worse truth,
A Miser's Pensioner—behold our lot!
O Man, that from thy fair and shining youth
Age might but take the things Youth needed not!

1804 (1807)

I WANDERED LONELY
AS A CLOUD

See Dorothy's *Grasmere Journal*, April 15, 1804, below,
p. 362, and Coleridge's discussion in *Biographia Literaria*,
Chapter XXII, below, p. 482.

I wandered lonely as a cloud
That floats on high o'er vales and hills,
When all at once I saw a crowd,
A host, of golden daffodils;
Beside the lake, beneath the trees,
Fluttering and dancing in the breeze.

Continuous as the stars that shine
And twinkle on the milky way,
They stretched in never-ending line
Along the margin of a bay: 10
Ten thousand saw I at a glance,
Tossing their heads in sprightly dance.

The waves beside them danced; but they
Out-did the sparkling waves in glee:
A poet could not but be gay,
In such a jocund company:
I gazed—and gazed—but little thought
What wealth the show to me had brought:

For oft, when on my couch I lie
In vacant or in pensive mood,   20
They flash upon that inward eye
Which is the bliss of solitude;
And then my heart with pleasure fills,
And dances with the daffodils.

                                        1804 (1807)

SHE WAS A PHANTOM
OF DELIGHT

She was a Phantom of delight
When first she gleamed upon my sight;
A lovely Apparition, sent
To be a moment's ornament;
Her eyes as stars of Twilight fair;
Like Twilight's, too, her dusky hair;
But all things else about her drawn
From May-time and the cheerful Dawn;
A dancing Shape, an Image gay,
To haunt, to startle, and way-lay.   10

I saw her upon nearer view,
A Spirit, yet a Woman too!
Her household motions light and free,
And steps of virgin-liberty;
A countenance in which did meet
Sweet records, promises as sweet;
A Creature not too bright or good
For human nature's daily food;
For transient sorrows, simple wiles,
Praise, blame, love, kisses, tears, and smiles.   20

SHE WAS A PHANTOM OF DELIGHT. **1. She:** Mary Hutchinson,
Wordsworth's wife.

And now I see with eye serene
The very pulse of the machine;
A Being breathing thoughtful breath,
A Traveller between life and death;
The reason firm, the temperate will,
Endurance, foresight, strength, and skill;
A perfect Woman, nobly planned,
To warn, to comfort, and command;
And yet a Spirit still, and bright
With something of angelic light.   30

                                        1804 (1807)

ELEGIAC STANZAS

SUGGESTED BY A PICTURE
OF PEELE CASTLE, IN A STORM,
PAINTED BY SIR GEORGE BEAUMONT

Peele Castle stands on an island about a mile from the
coast and can be seen from Rampside, where Wordsworth spent four weeks visiting a cousin in 1794. The
poem describes two paintings of the castle, one that of
Sir George Beaumont (ll. 43–52) and the other that which
Wordsworth would have painted (ll. 13–28) before the
loss of his dearly loved brother John, a ship's captain, who
was drowned at sea February 5, 1805. (See the letters of
Richard and William Wordsworth, February 7 and 11,
1805, below, pp. 354–55.) In Wordsworth's poetry the
sea is a recurrent symbolic rendering of ultimate reality.
See, for example, "Intimations of Immortality," "It is a
Beauteous Evening," and *Excursion*, IV. ll. 32–47.

Beaumont was a landscape painter and wealthy patron
of the arts. He and Lady Beaumont were close friends of
the Wordsworths.

I was thy neighbour once, thou rugged Pile!
Four summer weeks I dwelt in sight of thee:
I saw thee every day; and all the while
Thy Form was sleeping on a glassy sea.

So pure the sky, so quiet was the air!
So like, so very like, was day to day!
Whene'er I looked, thy Image still was there;
It trembled, but it never passed away.

22. machine: Cf. *Hamlet*, II. ii. 124: "Thine evermore . . .
whilst this Machine is to him."

How perfect was the calm! it seemed no sleep;
No mood, which season takes away, or brings: 10
I could have fancied that the mighty Deep
Was even the gentlest of all gentle Things.

Ah! then, if mine had been the Painter's hand,
To express what then I saw; and add the gleam,
The light that never was, on sea or land,
The consecration, and the Poet's dream;

I would have planted thee, thou hoary Pile
Amid a world how different from this!
Beside a sea that could not cease to smile;
On tranquil land, beneath a sky of bliss.   20

Thou shouldst have seemed a treasure-house divine
Of peaceful years; a chronicle of heaven;—
Of all the sunbeams that did ever shine
The very sweetest had to thee been given.

A Picture had it been of lasting ease,
Elysian quiet, without toil or strife;
No motion but the moving tide, a breeze,
Or merely silent Nature's breathing life.

Such, in the fond illusion of my heart,
Such Picture would I at that time have made: 30
And seen the soul of truth in every part,
A steadfast peace that might not be betrayed.

So once it would have been,—'tis so no more;
I have submitted to a new control:
A power is gone, which nothing can restore;
A deep distress hath humanised my Soul.

Not for a moment could I now behold
A smiling sea, and be what I have been:
The feeling of my loss will n'er be old;
This, which I know, I speak with mind serene.   40

Then, Beaumont, Friend! who would have been the
    Friend,
If he had lived, of Him whom I deplore,
This work of thine I blame not, but commend;
This sea in anger, and that dismal shore.

O 'tis a passionate Work!—yet wise and well,
Well chosen is the spirit that is here;
That Hulk which labours in the deadly swell,
This rueful sky, this pageantry of fear!

And this huge Castle, standing here sublime,
I love to see the look with which it braves,   50
Cased in the unfeeling armour of old time,
The lightning, the fierce wind, and trampling waves.

Farewell, farewell the heart that lives alone,
Housed in a dream, at distance from the Kind!
Such happiness, wherever it be known,
Is to be pitied; for 'tis surely blind.

But welcome fortitude, and patient cheer,
And frequent sights of what is to be borne!
Such sights, or worse, as are before me here.—
Not without hope we suffer and we mourn.   60

                                        1805 (1807)

STEPPING WESTWARD

❦

The poem illustrates Wordsworth's ability to evoke
something visionary and symbolic from a rather
commonplace occurrence. Wordsworth and Dorothy
were on a tour of Scotland, and Dorothy's manuscript
Recollections of a Tour Made in Scotland in A.D. 1803 reads:
"We have never had a more delightful walk than this
evening. Ben Lomond and the three pointed-topped
mountains of Loch Lomond, which we had seen from
the Garrison, were very majestic under the clear sky, the
lake perfectly calm, the air sweet and mild. . . . The sun
had been set for some time, when, being within a quarter
of a mile of the ferryman's hut, our path having led us
close to the shore of the calm lake, we met two neatly
dressed women, without hats, who had probably been
taking their Sunday evening's walk. One of them said
to us in a friendly, soft tone of voice, 'What! you are
stepping westward?' I cannot describe how affecting
this simple expression was in that remote place, with the
western sky in front, *yet glowing with the departed sun.*"
The poem was written twenty-one months later
(June, 1805).

❦

"What, you are stepping westward?"—"Yea."
—'Twould be a *wildish* destiny,
If we, who thus together roam
In a strange Land, and far from home,
Were in this place the guests of Chance:
Yet who would stop, or fear to advance,
Though home or shelter he had none,
With such a sky to lead him on?

The dewy ground was dark and cold;
Behind, all gloomy to behold;   10
And stepping westward seemed to be
A kind of *heavenly* destiny:

ELEGIAC STANZAS. **54. Kind:** humankind.

I liked the greeting; 'twas a sound
Of something without place or bound;
And seemed to give me spiritual right
To travel through that region bright.

The voice was soft, and she who spake
Was walking by her native lake:
The salutation had to me
The very sound of courtesy: 20
Its power was felt; and while my eye
Was fixed upon the glowing Sky,
The echo of the voice enwrought
A human sweetness with the thought
Of travelling through the world that lay
Before me in my endless way.

                              1805 (1807)

THE SOLITARY REAPER

Dorothy's *Recollections of a Tour Made in Scotland in
A.D. 1803* reads: "As we descended, the scene became
more fertile, our way being pleasantly varied; through
coppices or open fields, and passing farmhouses, though
always with an intermixture of uncultivated ground. It
was harvest-time, and the fields were quietly (might I
be allowed to say pensively?) enlivened by small companies of reapers. It is not uncommon in the more lonely
parts of the Highlands to see a *single* person so employed.
The following poem was suggested to Wm. by a
beautiful sentence in Thomas Wilkinson's *Tour in
Scotland*." (She then inserts the poem Wordsworth wrote
two years later.)
    The sentence in Wilkinson is, "Passed a female who
was reaping alone; she sung in Erse, as she bended over
her sickle; the sweetest human voice I ever heard: her
strains were tenderly melancholy, and felt delicious, long
after they were heard no more."

Behold her, single in the field,
Yon solitary Highland Lass!
Reaping and singing by herself;
Stop here, or gently pass!
Alone she cuts and binds the grain,
And sings a melancholy strain;
O listen! for the Vale profound
Is overflowing with the sound.

No Nightingale did ever chaunt
More welcome notes to weary bands   10

Of travellers in some shady haunt,
Among Arabian sands:
A voice so thrilling ne'er was heard
In spring-time from the Cuckoo-bird,
Breaking the silence of the seas
Among the farthest Hebrides.

Will no one tell me what she sings?—
Perhaps the plaintive numbers flow
For old, unhappy, far-off things,
And battles long ago: 20
Or is it some more humble lay,
Familiar matter of to-day?
Some natural sorrow, loss, or pain,
That has been, and may be again?

Whate'er the theme, the Maiden sang
As if her song could have no ending;
I saw her singing at her work,
And o'er the sickle bending;—
I listened, motionless and still;
And, as I mounted up the hill,   30
The music in my heart I bore,
Long after it was heard no more.

                              1805 (1807)

CHARACTER OF
THE HAPPY WARRIOR

"The above verses were written soon after tidings had
been received of the Death of Lord Nelson, which event
directed the Author's thoughts to the subject" (W. W.).
    "For the sake of such of my friends as may happen to
read this note I will add, that many elements of the
character here portrayed were found in my brother
John" (W. W.).

Who is the happy Warrior? Who is he
That every man in arms should wish to be?
—It is the generous Spirit, who, when brought
Among the tasks of real life, hath wrought
Upon the plan that pleased his boyish thought:
Whose high endeavours are an inward light
That makes the path before him always bright:
Who, with a natural instinct to discern
What knowledge can perform, is diligent to learn;

Abides by this resolve, and stops not there,   10
But makes his moral being his prime care;
Who, doomed to go in company with Pain,
And Fear, and Bloodshed, miserable train!
Turns his necessity to glorious gain;
In face of these doth exercise a power
Which is our human nature's highest dower;
Controls them and subdues, transmutes, bereaves
Of their bad influence, and their good receives:
By objects, which might force the soul to abate
Her feeling, rendered more compassionate;   20
Is placable—because occasions rise
So often that demand such sacrifice;
More skilful in self-knowledge, even more pure,
As tempted more; more able to endure,
As more exposed to suffering and distress;
Thence, also, more alive to tenderness.
—'Tis he whose law is reason; who depends
Upon that law as on the best of friends;
Whence, in a state where men are tempted still
To evil for a guard against worse ill,   30
And what in quality or act is best
Doth seldom on a right foundation rest,
He labours good on good to fix, and owes
To virtue every triumph that he knows:
—Who, if he rise to station of command,
Rises by open means; and there will stand
On honourable terms, or else retire,
And in himself possess his own desire;
Who comprehends his trust, and to the same
Keeps faithful with a singleness of aim;   40
And therefore does not stoop, nor lie in wait
For wealth, or honours, or for worldly state;
Whom they must follow; on whose head must fall,
Like showers of manna, if they come at all:
Whose powers shed round him in the common strife,
Or mild concerns of ordinary life,
A constant influence, a peculiar grace;
But who, if he be called upon to face
Some awful moment to which Heaven has joined
Great issues, good or bad for human kind,   50
Is happy as a Lover; and attired
With sudden brightness, like a Man inspired;
And, through the heat of conflict, keeps the law
In calmness made, and sees what he foresaw;
Or if an unexpected call succeed,
Come when it will, is equal to the need:
—He who, though thus endued as with a sense
And faculty for storm and turbulence,
Is yet a Soul whose master-bias leans

To homefelt pleasures and to gentle scenes;   60
Sweet images! which, wheresoe'er he be,
Are at his heart; and such fidelity
It is his darling passion to approve;
More brave for this, that he hath much to love:—
'Tis, finally, the Man, who, lifted high,
Conspicuous object in a Nation's eye,
Or left unthought-of in obscurity,—
Who, with a toward or untoward lot,
Prosperous or adverse, to his wish or not—
Plays, in the many games of life, that one   70
Where what he most doth value must be won:
Whom neither shape of danger can dismay,
Nor thought of tender happiness betray;
Who, not content that former worth stand fast,
Looks forward, persevering to the last,
From well to better, daily self-surpast:
Who, whether praise of him must walk the earth
For ever, and to noble deeds give birth,
Or he must fall, to sleep without his fame,
And leave a dead unprofitable name—   80
Finds comfort in himself and in his cause;
And, while the mortal mist is gathering, draws
His breath in confidence of Heaven's applause:
This is the happy Warrior; this is He
That every Man in arms should wish to be.

1805-06 (1807)

COMPOSED BY THE SIDE OF GRASMERE LAKE

Clouds, lingering yet, extend in solid bars
Through the grey west; and lo! these waters, steeled
By breezeless air to smoothest polish, yield
A vivid repetition of the stars;
Jove, Venus, and the ruddy crest of Mars
Amid his fellows beauteously revealed
At happy distance from earth's groaning field,
Where ruthless mortals wage incessant wars.
Is it a mirror?—or the nether Sphere
Opening to view the abyss in which she feeds   10
Her own calm fires?—But list! a voice is near;
Great Pan himself low-whispering through the reeds,
"Be thankful, thou; for, if unholy deeds
Ravage the world, tranquillity is here!"

1807 (1819)

SURPRISED BY JOY—
IMPATIENT AS THE WIND

Surprised by joy—impatient as the Wind
I turned to share the transport—Oh! with whom
But Thee, deep buried in the silent tomb,
That spot which no vicissitude can find?
Love, faithful love, recalled thee to my mind—
But how could I forget thee? Through what power,
Even for the least division of an hour,
Have I been so beguiled as to be blind
To my most grievous loss!—That thought's return
Was the worst pang that sorrow ever bore,   10
Save one, one only, when I stood forlorn,
Knowing my heart's best treasure was no more;
That neither present time, nor years unborn
Could to my sight that heavenly face restore.

1812–15 (1815)

LAODAMIA

⌘

"The incident of the trees growing and withering put the subject into my thoughts, and I wrote with the hope of giving it a loftier tone than, so far as I know, has been given to it by any of the Ancients who have treated it. It cost me more trouble than almost anything of equal length I have ever written" (W. W.). The main source of the poem is the sixth book of the *Aeneid*. Protesilaus is an example of duty. The "just gods" (l. 160) are an embodiment of Stoic law.

In his *Memoirs of William Wordsworth*, 1851, II, 67, Christopher Wordsworth comments:

"In 'LAODAMIA' the subordination of what is sensual to what is spiritual, and the subjection of the human passions to the government of reason, is taught in language of exquisite delicacy and grace, well fitted to the solemnity and sanctity of the subject, at the same time that the balance between the claims of affection and duty is preserved with a steady hand. Laodamia forfeits the favour of heaven by a passionate abuse of it. But the trees on the tomb of Protesilaus pay a natural homage to the affections, by withering at the sight of Troy."

The poem is discussed by Walter Savage Landor in "Southey and Porson," below, pp. 559–60.

⌘

"With sacrifice before the rising morn
Vows have I made by fruitless hope inspired;
And from the infernal Gods, 'mid shades forlorn
Of night, my slaughtered Lord have I required:

Celestial pity I again implore;—
Restore him to my sight—great Jove, restore!"

So speaking, and by fervent love endowed
With faith, the Suppliant heavenward lifts her hands;
While, like the sun emerging from a cloud,
Her countenance brightens—and her eye expands;   10
Her bosom heaves and spreads, her stature grows;
And she expects the issue in repose.

O terror! what hath she perceived?—O joy!
What doth she look on?—whom doth she behold?
Her Hero slain upon the beach of Troy?
His vital presence? his corporeal mould?
It is—if sense deceive her not—'tis He!
And a God leads him, wingèd Mercury!

Mild Hermes spake—and touched her with his wand
That calms all fear; "Such grace hath crowned thy prayer,   20
Laodamía! that at Jove's command
Thy Husband walks the paths of upper air:
He comes to tarry with thee three hours' space;
Accept the gift, behold him face to face!"

Forth sprang the impassioned Queen her Lord to clasp;
Again that consummation she essayed;
But unsubstantial Form eludes her grasp
As often as that eager grasp was made.
The Phantom parts—but parts to re-unite,
And re-assume his place before her sight.   30

"Protesiláus, lo! thy guide is gone!
Confirm, I pray, the vision with thy voice:
This is our palace,—yonder is thy throne;
Speak, and the floor thou tread'st on will rejoice.
Not to appal me have the gods bestowed
This precious boon; and blest a sad abode."

"Great Jove, Laodamía! doth not leave
His gifts imperfect:—Spectre though I be,
I am not sent to scare thee or deceive;
But in reward of thy fidelity.   40
And something also did my worth obtain;
For fearless virtue bringeth boundless gain.

"Thou knowest, the Delphic oracle foretold
That the first Greek who touched the Trojan strand
Should die; but me the threat could not withhold:
A generous cause a victim did demand;

LAODAMIA. **18. Mercury:** Also called Hermes, Mercury was the messenger of the gods. **46. generous:** noble.

And forth I leapt upon the sandy plain;
A self-devoted chief—by Hector slain."

"Supreme of Heroes—bravest, noblest, best!
Thy matchless courage I bewail no more,   50
Which then, when tens of thousands were deprest
By doubt, propelled thee to the fatal shore;
Thou found'st—and I forgive thee—here thou art—
A nobler counsellor than my poor heart.

"But thou, though capable of sternest deed,
Wert kind as resolute, and good as brave;
And he, whose power restores thee, hath decreed
Thou shouldst elude the malice of the grave:
Redundant are thy locks, thy lips as fair
As when their breath enriched Thessalian air.   60

"No Spectre greets me,—no vain Shadow this;
Come, blooming Hero, place thee by my side!
Give, on this well-known couch, one nuptial kiss
To me, this day, a second time thy bride!"
Jove frowned in heaven: the conscious Parcæ threw
Upon those roseate lips a Stygian hue.

"This visage tells thee that my doom is past:
Nor should the change be mourned, even if the joys
Of sense were able to return as fast
And surely as they vanish. Earth destroys   70
Those raptures duly—Erebus disdains:
Calm pleasures there abide—majestic pains.

"Be taught, O faithful Consort, to control
Rebellious passion: for the Gods approve
The depth, and not the tumult, of the soul;
A fervent, not ungovernable, love.
Thy transports moderate; and meekly mourn
When I depart, for brief is my sojourn—"

"Ah wherefore?—Did not Hercules by force
Wrest from the guardian Monster of the tomb   80
Alcestis, a reanimated corse,
Given back to dwell on earth in vernal bloom?

59. **Redundant:** abundant. 65. **Parcæ:** the Fates. 66. **Stygian:** pertaining to the river Styx that flows seven times around the lower world; hence infernal, deathly. 71. **Erebus:** in the lower world, a dark place through which the souls of the dead passed to Hades. 79–81. **Hercules . . . Alcestis:** Admetus, husband of Alcestis, was mortally ill. He was spared by the Fates on condition that someone should die in his place. Alcestis offered herself, but she was rescued by Hercules, who defeated Death in a wrestling match.

Medea's spells dispersed the weight of years,
And Æson stood a youth 'mid youthful peers.

"The Gods to us are merciful—and they
Yet further may relent: for mightier far
Than strength of nerve and sinew, or the sway
Of magic potent over sun and star,
Is love, though oft to agony distrest,   89
And though his favourite seat be feeble woman's breast.

"But if thou goest, I follow—" "Peace!" he said,—
She looked upon him and was calmed and cheered;
The ghastly colour from his lips had fled;
In his deportment, shape, and mien, appeared
Elysian beauty, melancholy grace,
Brought from a pensive though a happy place.

He spake of love, such love as Spirits feel
In worlds whose course is equable and pure;
No fears to beat away—no strife to heal—
The past unsighed for, and the future sure;   100
Spake of heroic arts in graver mood
Revived, with finer harmony pursued;

Of all that is most beauteous—imaged there
In happier beauty; more pellucid streams,
An ampler ether, a diviner air,
And fields invested with purpureal gleams;
Climes which the sun, who sheds the brightest day
Earth knows, is all unworthy to survey.

Yet there the Soul shall enter which hath earned
That privilege by virtue.—"Ill," said he,   110
"The end of man's existence I discerned,
Who from ignoble games and revelry
Could draw, when we had parted, vain delight,
While tears were thy best pastime, day and night;

"And while my youthful peers before my eyes
(Each hero following his peculiar bent)
Prepared themselves for glorious enterprise
By martial sports,—or, seated in the tent,
Chieftains and kings in council were detained;
What time the fleet at Aulis lay enchained.   120

"The wished-for wind was given:—I then revolved
The oracle, upon the silent sea;
And, if no worthier led the way, resolved
That, of a thousand vessels, mine should be

83–84. **Medea's . . . youth:** The sorceress Medea restored youth to her father-in-law, Aeson. 120. **Aulis:** where the Greek fleet, ready to sail toward Troy, was becalmed until Agamemnon offered in sacrifice his daughter Iphigenia.

The foremost prow in pressing to the strand,—
Mine the first blood that tinged the Trojan sand.

"Yet bitter, oft-times bitter, was the pang
When of thy loss I thought, belovèd Wife!
On thee too fondly did my memory hang,
And on the joys we shared in mortal life,—   130
The paths which we had trod—these fountains, flowers;
My new-planned cities, and unfinished towers.

"But should suspense permit the Foe to cry,
'Behold they tremble!—haughty their array,
Yet of their number no one dares to die?'
In soul I swept the indignity away:
Old frailties then recurred:—but lofty thought,
In act embodied, my deliverance wrought.

"And Thou, though strong in love, art all too weak
In reason, in self-government too slow;   140
I counsel thee by fortitude to seek
Our blest re-union in the shades below.
The invisible world with thee hath sympathised;
Be thy affections raised and solemnised.

"Learn, by a mortal yearning, to ascend—
Seeking a higher object. Love was given,
Encouraged, sanctioned, chiefly for that end;
For this the passion to excess was driven—
That self might be annulled: her bondage prove
The fetters of a dream opposed to love."—   150

Aloud she shrieked! for Hermes reappears!
Round the dear Shade she would have clung—'tis vain:
The hours are past—too brief had they been years;
And him no mortal effort can detain:
Swift, toward the realms that know not earthly day,
He through the portal takes his silent way,
And on the palace-floor a lifeless corse She lay.

Thus, all in vain exhorted and reproved,
She perished; and, as for a wilful crime,
By the just Gods whom no weak pity moved,   160
Was doomed to wear out her appointed time,
Apart from happy Ghosts, that gather flowers
Of blissful quiet 'mid unfading bowers.

—Yet tears to human suffering are due;
And mortal hopes defeated and o'erthrown
Are mourned by man, and not by man alone,
As fondly he believes.—Upon the side
Of Hellespont (such faith was entertained)

A knot of spiry trees for ages grew
From out the tomb of him for whom she died;   170
And ever, when such stature they had gained
That Ilium's walls were subject to their view,
The trees' tall summits withered at the sight;
A constant interchange of growth and blight!

                              1814 (1815)

FROM

THE EXCURSION

〜〜

At least since 1798, Wordsworth had, at the vehement and repeated urging of Coleridge, planned to write a long, philosophic poem to be called *The Recluse; Or, Views of Nature, Man, and Society.* Except for *The Excursion,* which was to be the second of three parts, not much of *The Recluse* was actually composed.

The "Prospectus" had been written by 1800 and was originally intended to be the conclusion of the first book of *The Recluse.* Wordsworth published it with *The Excursion* in 1814 as "a kind of Prospectus of the design and scope" of *The Recluse* as a whole. It may be noticed that it is theologically less orthodox and more vague than *The Excursion* itself. Book I was first completed in 1798 as a separate poem to be called "The Ruined Cottage." The rural setting is not that of the Lake District but the southwest of England. Most of Books II, III, and IV were originally drafted in 1806.

After Book I *The Excursion* consists mainly of philosophical and religious conversation between four persons, the Wanderer, the Poet, the Solitary, and the Parson, each of whom speaks for a different side of Wordsworth himself. The general aim of the conversation is to answer the arguments of the Solitary, who is a skeptic, and to bring him religious consolation.

For discussion of the poem, see Coleridge in *Biographia Literaria,* Chapter XXII, below, pp. 478–82, *Table Talk,* July 21, 1832, below, p. 517, letter to Wordsworth, May 30, 1815, below, p. 529; Francis Jeffrey, below, pp. 366–70; and Hazlitt, review of *The Excursion,* below, pp. 613–17.

In a Preface to the poem Wordsworth wrote:

"It may be proper to state whence the poem, of which *The Excursion* is a part, derives its Title of *The Recluse.*— Several years ago, when the Author retired to his native mountains, with the hope of being enabled to construct a literary Work that might live, it was a reasonable thing that he should take a review of his own mind, and examine how far Nature and Education had qualified him for such employment. As subsidiary to this preparation, he undertook to record, in verse, the origin and progress of his own powers, as far as he was acquainted with them. That Work, addressed to a dear Friend, most

---

168. Hellespont: the Dardanelles.

172. Ilium: the city of Troy.

distinguished for his knowledge and genius, and to whom the Author's Intellect is deeply indebted, has been long finished; and the result of the investigation which gave rise to it was a determination to compose a philosophical poem, containing views of Man, Nature, and Society; and to be Entitled, The Recluse; as having for its principal subject the sensations and opinions of a poet living in retirement.—The preparatory poem is biographical, and conducts the history of the Author's mind to the point when he was emboldened to hope that his faculties were sufficiently matured for entering upon the arduous labour which he had proposed to himself; and the two Works have the same kind of relation to each other, if he may so express himself, as the ante-chapel has to the body of a gothic church. Continuing this allusion, he may be permitted to add, that his minor Pieces, which have been long before the Public, when they shall be properly arranged, will be found by the attentive Reader to have such connection with the main Work as may give them claim to be likened to the little cells, oratories, and sepulchral recesses, ordinarily included in those edifices."

PROSPECTUS

On Man, on Nature, and on Human Life,
Musing in solitude, I oft perceive
Fair trains of imagery before me rise,
Accompanied by feelings of delight
Pure, or with no unpleasing sadness mixed;
And I am conscious of affecting thoughts
And dear remembrances, whose presence soothes
Or elevates the Mind, intent to weigh
The good and evil of our mortal state.
—To these emotions, whencesoe'er they come,   10
Whether from breath of outward circumstance,
Or from the Soul—an impulse to herself—
I would give utterance in numerous verse.
Of Truth, of Grandeur, Beauty, Love, and Hope,
And melancholy Fear subdued by Faith;
Of blessèd consolations in distress;
Of moral strength, and intellectual Power;
Of joy in widest commonalty spread;
Of the individual Mind that keeps her own
Inviolate retirement, subject there   20
To Conscience only, and the law supreme
Of that Intelligence which governs all—
I sing:—"fit audience let me find though few!"

So prayed, more gaining than he asked, the Bard—
In holiest mood. Urania, I shall need
Thy guidance, or a greater Muse, if such
Descend to earth or dwell in highest heaven!
For I must tread on shadowy ground, must sink
Deep—and, aloft ascending, breathe in worlds
To which the heaven of heavens is but a veil.   30
All strength—all terror, single or in bands,
That ever was put forth in personal form—
Jehovah—with his thunder, and the choir
Of shouting Angels, and the empyreal thrones—
I pass them unalarmed. Not Chaos, not
The darkest pit of lowest Erebus,
Nor aught of blinder vacancy, scooped out
By help of dreams—can breed such fear and awe
As fall upon us often when we look
Into our Minds, into the Mind of Man—   40
My haunt, and the main region of my song.
—Beauty—a living Presence of the earth,
Surpassing the most fair ideal Forms
Which craft of delicate Spirits hath composed
From earth's materials—waits upon my steps;
Pitches her tents before me as I move,
An hourly neighbour. Paradise, and groves
Elysian, Fortunate Fields—like those of old
Sought in the Atlantic Main—why should they be
A history only of departed things,   50
Or a mere fiction of what never was?
For the discerning intellect of Man,
When wedded to this goodly universe
In love and holy passion, shall find these
A simple produce of the common day.
—I, long before the blissful hour arrives,
Would chant, in lonely peace, the spousal verse
Of this great consummation:—and, by words
Which speak of nothing more than what we are,
Would I arouse the sensual from their sleep   60
Of Death, and win the vacant and the vain
To noble raptures; while my voice proclaims
How exquisitely the individual Mind

THE EXCURSION. PROSPECTUS. **23–30. I . . . veil:** strongly influenced by *Paradise Lost*, VII. 1–15, from which the phrase "fit audience . . . few" is adapted. Urania was the Muse

of astronomy and was associated with hymns of praise to the gods. Milton invokes her as a "sister" of "Eternal Wisdom," and thinks of Wisdom as an aspect of God's creative power. In Milton the "heaven of heavens" is the abode of God and angels. **36. Erebus:** See "Laodamia," above, n. 71. **48. Fortunate Fields:** Islands of the Blessed, supposed in ancient myth to lie beyond the Pillars of Hercules (the two promontories at the east end of the Strait of Gibraltar).

(And the progressive powers perhaps no less
Of the whole species) to the external World
Is fitted:—and how exquisitely, too—
Theme this but little heard of among men—
The external World is fitted to the Mind;
And the creation (by no lower name
Can it be called) which they with blended might   70
Accomplish:—this is our high argument.
—Such grateful haunts foregoing, if I oft
Must turn elsewhere—to travel near the tribes
And fellowships of men, and see ill sights
Of madding passions mutually inflamed;
Must hear Humanity in fields and groves
Pipe solitary anguish; or must hang
Brooding above the fierce confederate storm
Of sorrow, barricadoed evermore
Within the walls of cities—may these sounds   80
Have their authentic comment; that even these
Hearing, I be not downcast or forlorn!—
Descend, prophetic Spirit! that inspir'st
The human Soul of universal earth,
Dreaming on things to come; and dost possess
A metropolitan temple in the hearts
Of mighty Poets; upon me bestow
A gift of genuine insight; that my Song
With star-like virtue in its place may shine,
Shedding benignant influence, and secure   90
Itself from all malevolent effect
Of those mutations that extend their sway
Throughout the nether sphere!—And if with this
I mix more lowly matter; with the thing
Contemplated, describe the Mind and Man
Contemplating; and who, and what he was—
The transitory Being that beheld
This Vision; when and where, and how he lived;
Be not this labour useless. If such theme
May sort with highest objects, then—dread Power!
Whose gracious favour is the primal source   101
Of all illumination,—may my Life
Express the image of a better time,
More wise desires, and simpler manners;—nurse
My Heart in genuine freedom:—all pure thoughts
Be with me;—so shall thy unfailing love
Guide, and support, and cheer me to the end!

1798 (1814)

83-85. **Descend . . . come:** See Shakespeare's Sonnet CVII.
86. **metropolitan:** of or pertaining to a chief bishop;
hence leading, preeminent.

BOOK FIRST
THE WANDERER

Argument

A summer forenoon.—The Author reaches a ruined
Cottage upon a Common, and there meets with a
revered Friend, the Wanderer, of whose education and
course of life he gives an account.—The Wanderer,
while resting under the shade of the Trees that surround
the Cottage, relates the History of its last Inhabitant.

'Twas summer, and the sun had mounted high:
Southward the landscape indistinctly glared
Through a pale steam; but all the northern downs,
In clearest air ascending, showed far off
A surface dappled o'er with shadows flung
From brooding clouds; shadows that lay in spots
Determined and unmoved, with steady beams
Of bright and pleasant sunshine interposed;
To him most pleasant who on soft cool moss
Extends his careless limbs along the front   10
Of some huge cave, whose rocky ceiling casts
A twilight of its own, an ample shade,
Where the wren warbles, while the dreaming man,
Half conscious of the soothing melody,
With side-long eye looks out upon the scene,
By power of that impending covert, thrown,
To finer distance. Mine was at that hour
Far other lot, yet with good hope that soon
Under a shade as grateful I should find
Rest, and be welcomed there to livelier joy.   20
Across a bare wide Common I was toiling
With languid steps that by the slippery turf
Were baffled; nor could my weak arm disperse
The host of insects gathering round my face,
And ever with me as I paced along.

Upon that open moorland stood a grove,
The wished-for port to which my course was bound.
Thither I came, and there, amid the gloom
Spread by a brotherhood of lofty elms,
Appeared a roofless Hut; four naked walls   30
That stared upon each other!—I looked round,
And to my wish and to my hope espied
The Friend I sought; a Man of reverend age,
But stout and hale, for travel unimpaired.

BOOK I. 21. **Common:** In the Lake District, the common
lands of the villages are usually bare, upland pasture.

There was he seen upon the cottage-bench,
Recumbent in the shade, as if asleep;
An iron-pointed staff lay at his side.

  Him had I marked the day before—alone
And stationed in the public way, with face
Turned toward the sun then setting, while that staff 40
Afforded, to the figure of the man
Detained for contemplation or repose,
Graceful support; his countenance as he stood
Was hidden from my view, and he remained
Unrecognised; but, stricken by the sight,
With slackened footsteps I advanced, and soon
A glad congratulation we exchanged
At such unthought-of meeting.—For the night
We parted, nothing willingly; and now
He by appointment waited for me here,   50
Under the covert of these clustering elms.

  We were tried Friends: amid a pleasant vale,
In the antique market-village where was passed
My school-time, an apartment he had owned,
To which at intervals the Wanderer drew,
And found a kind of home or harbour there.
He loved me; from a swarm of rosy boys
Singled out me, as he in sport would say,
For my grave looks, too thoughtful for my years.
As I grew up, it was my best delight   60
To be his chosen comrade. Many a time,
On holidays, we rambled through the woods:
We sate—we walked; he pleased me with report
Of things which he had seen; and often touched
Abstrusest matter, reasonings of the mind
Turned inward; or at my request would sing
Old songs, the product of his native hills;
A skilful distribution of sweet sounds,
Feeding the soul, and eagerly imbibed
As cool refreshing water, by the care   70
Of the industrious husbandman, diffused
Through a parched meadow-ground, in time of drought.
Still deeper welcome found his pure discourse:
How precious when in riper days I learned
To weigh with care his words, and to rejoice
In the plain presence of his dignity!

  Oh! many are the Poets that are sown
By Nature; men endowed with highest gifts,
The vision and the faculty divine;
Yet wanting the accomplishment of verse,   80
(Which in the docile season of their youth,
It was denied them to acquire, through lack

Of culture and the inspiring aid of books,
Or haply by a temper too severe,
Or a nice backwardness afraid of shame)
Nor having e'er, as life advanced, been led
By circumstance to take unto the height
The measure of themselves, these favoured Beings,
All but a scattered few, live out their time,
Husbanding that which they possess within,   90
And go to the grave, unthought of. Strongest minds
Are often those of whom the noisy world
Hears least; else surely this Man had not left
His graces unrevealed and unproclaimed.
But, as the mind was filled with inward light,
So not without distinction had he lived,
Beloved and honoured—far as he was known.
And some small portion of his eloquent speech,
And something that may serve to set in view
The feeling pleasures of his loneliness,   100
His observations, and the thoughts his mind
Had dealt with—I will here record in verse;
Which, if with truth it correspond, and sink
Or rise as venerable Nature leads,
The high and tender Muses shall accept
With gracious smile, deliberately pleased,
And listening Time reward with sacred praise.

  Among the hills of Athol he was born;
Where, on a small hereditary farm,
An unproductive slip of rugged ground,   110
His Parents, with their numerous offspring, dwelt;
A virtuous household, though exceeding poor!
Pure livers were they all, austere and grave,
And fearing God; the very children taught
Stern self-respect, a reverence for God's word,
And an habitual piety, maintained
With strictness scarcely known on English ground.

  From his sixth year, the Boy of whom I speak,
In summer, tended cattle on the hills;
But, through the inclement and the perilous days 120
Of long-continuing winter, he repaired,
Equipped with satchel, to a school, that stood
Sole building on a mountain's dreary edge,
Remote from view of city spire, or sound
Of minster clock! From that bleak tenement
He, many an evening, to his distant home
In solitude returning, saw the hills
Grow larger in the darkness; all alone
Beheld the stars come out above his head,

108. Athol: in Scotland.

And travelled through the wood, with no one near
To whom he might confess the things he saw.   131

So the foundations of his mind were laid.
In such communion, not from terror free,
While yet a child, and long before his time,
Had he perceived the presence and the power
Of greatness; and deep feelings had impressed
So vividly great objects that they lay
Upon his mind like substances, whose presence
Perplexed the bodily sense. He had received
A precious gift; for, as he grew in years,   140
With these impressions would he still compare
All his remembrances, thoughts, shapes, and forms;
And, being still unsatisfied with aught
Of dimmer character, he thence attained
An active power to fasten images
Upon his brain; and on their pictured lines
Intensely brooded, even till they acquired
The liveliness of dreams. Nor did he fail,
While yet a child, with a child's eagerness
Incessantly to turn his ear and eye   150
On all things which the moving seasons brought
To feed such appetite—nor this alone
Appeased his yearning:—in the after-day
Of boyhood, many an hour in caves forlorn,
And 'mid the hollow depths of naked crags
He sate, and even in their fixed lineaments,
Or from the power of a peculiar eye,
Or by creative feeling overborne,
Or by predominance of thought oppressed,
Even in their fixed and steady lineaments   160
He traced an ebbing and a flowing mind,
Expression ever varying!
                              Thus informed,
He had small need of books; for many a tale
Traditionary, round the mountains hung,
And many a legend, peopling the dark woods,
Nourished Imagination in her growth,
And gave the Mind that apprehensive power
By which she is made quick to recognise
The moral properties and scope of things.
But eagerly he read, and read again,   170
Whate'er the minister's old shelf supplied;
The life and death of martyrs, who sustained,
With will inflexible, those fearful pangs
Triumphantly displayed in records left
Of persecution, and the Covenant—times

Whose echo rings through Scotland to this hour!
And there, by lucky hap, had been preserved
A straggling volume, torn and incomplete,
That left half-told the preternatural tale,
Romance of giants, chronicle of fiends,   180
Profuse in garniture of wooden cuts
Strange and uncouth; dire faces, figures dire,
Sharp-kneed, sharp-elbowed, and lean-ankled too,
With long and ghostly shanks—forms which once seen
Could never be forgotten!
                              In his heart,
Where Fear sate thus, a cherished visitant,
Was wanting yet the pure delight of love
By sound diffused, or by the breathing air,
Or by the silent looks of happy things,
Or flowing from the universal face   190
Of earth and sky. But he had felt the power
Of Nature, and already was prepared,
By his intense conceptions, to receive
Deeply the lesson deep of love which he,
Whom Nature, by whatever means, has taught
To feel intensely, cannot but receive.

Such was the Boy—but for the growing Youth
What soul was his, when, from the naked top
Of some bold headland, he beheld the sun
Rise up, and bathe the world in light! He looked—
Ocean and earth, the solid frame of earth   201
And ocean's liquid mass, in gladness lay
Beneath him:—Far and wide the clouds were touched,
And in their silent faces could he read
Unutterable love. Sound needed none,
Nor any voice of joy; his spirit drank
The spectacle: sensation, soul, and form,
All melted into him; they swallowed up
His animal being; in them did he live,
And by them did he live; they were his life.   210
In such access of mind, in such high hour
Of visitation from the living God,
Thought was not; in enjoyment it expired.
No thanks he breathed, he proffered no request;
Rapt into still communion that transcends
The imperfect offices of prayer and praise,
His mind was a thanksgiving to the power
That made him; it was blessedness and love!

A Herdsman on the lonely mountain tops,
Such intercourse was his, and in this sort   220

175. **Covenant:** in Scottish history, the National Covenant of 1638 for the defense of Presbyterianism against Episco-pacy, which had been introduced under James I and Charles I.

Was his existence oftentimes *possessed*.
O then how beautiful, how bright, appeared
The written promise! Early had he learned
To reverence the volume that displays
The mystery, the life which cannot die;
But in the mountains did he *feel* his faith.
All things, responsive to the writing, there
Breathed immortality, revolving life,
And greatness still revolving; infinite:
There littleness was not; the least of things   230
Seemed infinite; and there his spirit shaped
Her prospects, nor did he believe,—he *saw*.
What wonder if his being thus became
Sublime and comprehensive! Low desires,
Low thoughts had there no place; yet was his heart
Lowly; for he was meek in gratitude,
Oft as he called those ecstasies to mind,
And whence they flowed; and from them he
     acquired
Wisdom, which works thro' patience; thence he
     learned
In oft-recurring hours of sober thought   240
To look on Nature with a humble heart,
Self-questioned where it did not understand,
And with a superstitious eye of love.

So passed the time; yet to the nearest town
He duly went with what small overplus
His earnings might supply, and brought away
The book that most had tempted his desires
While at the stall he read. Among the hills
He gazed upon that mighty orb of song,
The divine Milton. Lore of different kind,   250
The annual savings of a toilsome life,
His School-master supplied; books that explain
The purer elements of truth involved
In lines and numbers, and, by charm severe,
(Especially perceived where nature droops
And feeling is suppressed) preserve the mind
Busy in solitude and poverty.
These occupations oftentimes deceived
The listless hours, while in the hollow vale,
Hollow and green, he lay on the green turf   260
In pensive idleness. What could he do,
Thus daily thirsting, in that lonesome life,
With blind endeavours? Yet, still uppermost,
Nature was at his heart as if he felt,
Though yet he knew not how, a wasting power
In all things that from her sweet influence
Might tend to wean him. Therefore with her hues,

Her forms, and with the spirit of her forms,
He clothed the nakedness of austere truth.
While yet he lingered in the rudiments   270
Of science, and among her simplest laws,
His triangles—they were the stars of heaven,
The silent stars! Oft did he take delight
To measure the altitude of some tall crag
That is the eagle's birth-place, or some peak
Familiar with forgotten years, that shows
Inscribed upon its visionary sides,
The history of many a winter storm,
Or obscure records of the path of fire.

     And thus before his eighteenth year was told,   280
Accumulated feelings pressed his heart
With still increasing weight; he was o'erpowered
By Nature; by the turbulence subdued
Of his own mind; by mystery and hope,
And the first virgin passion of a soul
Communing with the glorious universe.
Full often wished he that the winds might rage
When they were silent: far more fondly now
Than in his earlier season did he love
Tempestuous nights—the conflict and the sounds   290
That live in darkness. From his intellect
And from the stillness of abstracted thought
He asked repose; and, failing oft to win
The peace required, he scanned the laws of light
Amid the roar of torrents, where they send
From hollow clefts up to the clearer air
A cloud of mist, that smitten by the sun
Varies its rainbow hues. But vainly thus,
And vainly by all other means, he strove
To mitigate the fever of his heart.   300

     In dreams, in study, and in ardent thought,
Thus was he reared; much wanting to assist
The growth of intellect, yet gaining more,
And every moral feeling of his soul
Strengthened and braced, by breathing in content
The keen, the wholesome, air of poverty,
And drinking from the well of homely life.
—But, from past liberty, and tried restraints,
He now was summoned to select the course
Of humble industry that promised best   310
To yield him no unworthy maintenance.
Urged by his Mother, he essayed to teach
A village-school—but wandering thoughts were then
A misery to him; and the Youth resigned
A task he was unable to perform.

That stern yet kindly Spirit, who constrains
The Savoyard to quit his naked rocks,
The free-born Swiss to leave his narrow vales,
(Spirit attached to regions mountainous
Like their own stedfast clouds) did now impel   320
His restless mind to look abroad with hope
—An irksome drudgery seems it to plod on,
Through hot and dusty ways, or pelting storm,
A vagrant Merchant under a heavy load
Bent as he moves, and needing frequent rest;
Yet do such travellers find their own delight;
And their hard service, deemed debasing now,
Gained merited respect in simpler times;
When squire, and priest, and they who round them
    dwelt
In rustic sequestration—all dependent   330
Upon the PEDLAR's toil—supplied their wants,
Or pleased their fancies, with the wares he brought.
Not ignorant was the Youth that still no few
Of his adventurous countrymen were led
By perseverance in this track of life
To competence and ease:—to him it offered
Attractions manifold;—and this he chose.
—His Parents on the enterprise bestowed
Their farewell benediction, but with hearts
Foreboding evil. From his native hills   340
He wandered far; much did he see of men,
Their manners, their enjoyments, and pursuits,
Their passions and their feelings; chiefly those
Essential and eternal in the heart,
That, 'mid the simpler forms of rural life,
Exist more simple in their elements,
And speak a plainer language. In the woods,
A lone Enthusiast, and among the fields,
Itinerant in this labour, he had passed
The better portion of his time; and there   350
Spontaneously had his affections thriven
Amid the bounties of the year, the peace
And liberty of nature; there he kept
In solitude and solitary thought
His mind in a just equipoise of love.
Serene it was, unclouded by the cares
Of ordinary life; unvexed, unwarped
By partial bondage. In his steady course,
No piteous revolutions had he felt,
No wild varieties of joy and grief.   360
Unoccupied by sorrow of its own,
His heart lay open; and, by nature tuned
And constant disposition of his thoughts
To sympathy with man, he was alive

To all that was enjoyed where'er he went,
And all that was endured; for, in himself
Happy, and quiet in his cheerfulness,
He had no painful pressure from without
That made him turn aside from wretchedness
With coward fears. He could *afford* to suffer   370
With those whom he saw suffer. Hence it came
That in our best experience he was rich,
And in the wisdom of our daily life.
For hence, minutely, in his various rounds,
He had observed the progress and decay
Of many minds, of minds and bodies too;
The history of many families;
How they had prospered; how they were o'erthrown
By passion or mischance, or such misrule
Among the unthinking masters of the earth   380
As makes the nations groan.
                    This active course
He followed till provision for his wants
Had been obtained;—the Wanderer then resolved
To pass the remnant of his days, untasked
With needless services, from hardship free.
His calling laid aside, he lived at ease:
But still he loved to pace the public roads
And the wild paths; and, by the summer's warmth
Invited, often would he leave his home
And journey far, revisiting the scenes   390
That to his memory were most endeared.
—Vigorous in health, of hopeful spirits, undamped
By worldly-mindedness or anxious care;
Observant, studious, thoughtful, and refreshed
By knowledge gathered up from day to day;
Thus had he lived a long and innocent life.

    The Scottish Church, both on himself and those
With whom from childhood he grew up, had held
The strong hand of her purity; and still
Had watched him with an unrelenting eye.   400
This he remembered in his riper age
With gratitude, and reverential thoughts.
But by the native vigour of his mind,
By his habitual wanderings out of doors,
By loneliness, and goodness, and kind works,
Whate'er, in docile childhood or in youth,
He had imbibed of fear or darker thought
Was melted all away; so true was this,
That sometimes his religion seemed to me
Self-taught, as of a dreamer in the woods;   410
Who to the model of his own pure heart
Shaped his belief, as grace divine inspired,

And human reason dictated with awe.
—And surely never did there live on earth
A man of kindlier nature. The rough sports
And teasing ways of children vexed not him;
Indulgent listener was he to the tongue
Of garrulous age; nor did the sick man's tale,
To his fraternal sympathy addressed,
Obtain reluctant hearing.
       Plain his garb;   420
Such as might suit a rustic Sire, prepared
For sabbath duties; yet he was a man
Whom no one could have passed without remark.
Active and nervous was his gait; his limbs
And his whole figure breathed intelligence.
Time had compressed the freshness of his cheek
Into a narrower circle of deep red,
But had not tamed his eye; that, under brows
Shaggy and grey, had meanings which it brought
From years of youth; which, like a Being made 430
Of many Beings, he had wondrous skill
To blend with knowledge of the years to come,
Human, or such as lie beyond the grave.

So was He framed; and such his course of life
Who now, with no appendage but a staff,
The prized memorial of relinquished toils,
Upon that cottage-bench reposed his limbs,
Screened from the sun. Supine the Wanderer lay,
His eyes as if in drowsiness half shut,
The shadows of the breezy elms above   440
Dappling his face. He had not heard the sound
Of my approaching steps, and in the shade
Unnoticed did I stand some minutes' space.
At length I hailed him, seeing that his hat
Was moist with water-drops, as if the brim
Had newly scooped a running stream. He rose,
And ere our lively greeting into peace
Had settled, "'Tis," said I, "a burning day:
My lips are parched with thirst, but you, it seems,
Have somewhere found relief." He, at the word, 450
Pointing towards a sweet-briar, bade me climb
The fence where that aspiring shrub looked out
Upon the public way. It was a plot
Of garden ground run wild, its matted weeds
Marked with the steps of those, whom, as they passed,
The gooseberry trees that shot in long lank slips,
Or currants, hanging from their leafless stems,
In scanty strings, had tempted to o'erleap
The broken wall. I looked around, and there,
Where two tall hedge-rows of thick alder boughs 460

Joined in a cold damp nook, espied a well
Shrouded with willow-flowers and plumy fern.
My thirst I slaked, and, from the cheerless spot
Withdrawing, straightway to the shade returned
Where sate the old Man on the cottage-bench;
And, while, beside him, with uncovered head,
I yet was standing, freely to respire,
And cool my temples in the fanning air,
Thus did he speak. "I see around me here
Things which you cannot see: we die, my Friend, 470
Nor we alone, but that which each man loved
And prized in his peculiar nook of earth
Dies with him, or is changed; and very soon
Even of the good is no memorial left.
—The Poets, in their elegies and songs
Lamenting the departed, call the groves,
They call upon the hills and streams to mourn,
And senseless rocks; nor idly; for they speak,
In these their invocations, with a voice
Obedient to the strong creative power   480
Of human passion. Sympathies there are
More tranquil, yet perhaps of kindred birth,
That steal upon the meditative mind,
And grow with thought. Beside yon spring I stood,
And eyed its waters till we seemed to feel
One sadness, they and I. For them a bond
Of brotherhood is broken: time has been
When, every day, the touch of human hand
Dislodged the natural sleep that binds them up
In mortal stillness; and they ministered   490
To human comfort. Stooping down to drink,
Upon the slimy foot-stone I espied
The useless fragment of a wooden bowl,
Green with the moss of years, and subject only
To the soft handling of the elements:
There let it lie—how foolish are such thoughts!
Forgive them;—never—never did my steps
Approach this door but she who dwelt within
A daughter's welcome gave me, and I loved her
As my own child. Oh, Sir! the good die first, 500
And they whose hearts are dry as summer dust
Burn to the socket. Many a passenger
Hath blessed poor Margaret for her gentle looks,
When she upheld the cool refreshment drawn
From that forsaken spring; and no one came
But he was welcome; no one went away
But that it seemed she loved him. She is dead,
The light extinguished of her lonely hut,
The hut itself abandoned to decay,
And she forgotten in the quiet grave.   510

"I speak," continued he, "of One whose stock
Of virtues bloomed beneath this lowly roof.
She was a Woman of a steady mind,
Tender and deep in her excess of love;
Not speaking much, pleased rather with the joy
Of her own thoughts: by some especial care
Her temper had been framed, as if to make
A Being, who by adding love to peace
Might live on earth a life of happiness.
Her wedded Partner lacked not on his side   520
The humble worth that satisfied her heart:
Frugal, affectionate, sober, and withal
Keenly industrious. She with pride would tell
That he was often seated at his loom,
In summer, ere the mower was abroad
Among the dewy grass,—in early spring,
Ere the last star had vanished.—They who passed
At evening, from behind the garden fence
Might hear his busy spade, which he would ply,
After his daily work, until the light   530
Had failed, and every leaf and flower were lost
In the dark hedges. So their days were spent
In peace and comfort; and a pretty boy
Was their best hope, next to the God in heaven.

"Not twenty years ago, but you I think
Can scarcely bear it now in mind, there came
Two blighting seasons, when the fields were left
With half a harvest. It pleased Heaven to add
A worse affliction in the plague of war:
This happy Land was stricken to the heart!   540
A Wanderer then among the cottages,
I, with my freight of winter raiment, saw
The hardships of that season: many rich
Sank down, as in a dream, among the poor;
And of the poor did many cease to be,
And their place knew them not. Meanwhile, abridged
Of daily comforts, gladly reconciled
To numerous self-denials, Margaret
Went struggling on through those calamitous years
With cheerful hope, until the second autumn,   550
When her life's Helpmate on a sick-bed lay,
Smitten with perilous fever. In disease
He lingered long; and, when his strength returned,
He found the little he had stored, to meet
The hour of accident or crippling age,
Was all consumed. A second infant now
Was added to the troubles of a time
Laden, for them and all of their degree,
With care and sorrow: shoals of artisans

From ill-requited labour turned adrift   560
Sought daily bread from public charity,
They, and their wives and children—happier far
Could they have lived as do the little birds
That peck along the hedge-rows, or the kite
That makes her dwelling on the mountain rocks!

"A sad reverse it was for him who long
Had filled with plenty, and possessed in peace,
This lonely Cottage. At the door he stood,
And whistled many a snatch of merry tunes
That had no mirth in them; or with his knife   570
Carved uncouth figures on the heads of sticks—
Then, not less idly, sought, through every nook
In house or garden, any casual work
Of use or ornament; and with a strange,
Amusing, yet uneasy, novelty,
He mingled, where he might, the various tasks
Of summer, autumn, winter, and of spring.
But this endured not; his good humour soon
Became a weight in which no pleasure was:
And poverty brought on a petted mood   580
And a sore temper: day by day he drooped,
And he would leave his work—and to the town
Would turn without an errand his slack steps;
Or wander here and there among the fields.
One while he would speak lightly of his babes,
And with a cruel tongue: at other times
He tossed them with a false unnatural joy:
And 'twas a rueful thing to see the looks
Of the poor innocent children. 'Every smile,'
Said Margaret to me, here beneath these trees,   590
'Made my heart bleed.' "
                              At this the Wanderer paused;
And, looking up to those enormous elms,
He said, " 'Tis now the hour of deepest noon.
At this still season of repose and peace,
This hour when all things which are not at rest
Are cheerful; while this multitude of flies
With tuneful hum is filling all the air;
Why should a tear be on an old Man's cheek?
Why should we thus, with an untoward mind,
And in the weakness of humanity,   600
From natural wisdom turn our hearts away;
To natural comfort shut our eyes and ears;
And, feeding on disquiet, thus disturb
The calm of nature with our restless thoughts?"

580. petted: sulky, peevish.

He spake with somewhat of a solemn tone:
But, when he ended, there was in his face
Such easy cheerfulness, a look so mild,
That for a little time it stole away
All recollection; and that simple tale
Passed from my mind like a forgotten sound.   610
A while on trivial things we held discourse,
To me soon tasteless. In my own despite,
I thought of that poor Woman as of one
Whom I had known and loved. He had rehearsed
Her homely tale with such familiar power,
With such an active countenance, an eye
So busy, that the things of which he spake
Seemed present; and, attention now relaxed,
A heart-felt chillness crept along my veins.
I rose; and, having left the breezy shade,   620
Stood drinking comfort from the warmer sun,
That had not cheered me long—ere, looking round
Upon that tranquil Ruin, I returned,
And begged of the old Man that, for my sake,
He would resume his story.   He replied,
"It were a wantonness, and would demand
Severe reproof, if we were men whose hearts
Could hold vain dalliance with the misery
Even of the dead; contented thence to draw
A momentary pleasure, never marked   630
By reason, barren of all future good.
But we have known that there is often found
In mournful thoughts, and always might be found,
A power to virtue friendly; wer't not so,
I am a dreamer among men, indeed
An idle dreamer! 'Tis a common tale,
An ordinary sorrow of man's life,
A tale of silent suffering, hardly clothed
In bodily form.—But without further bidding
I will proceed.   While thus it fared with them,   640
To whom this cottage, till those hapless years,
Had been a blessed home, it was my chance
To travel in a country far remote;
And when these lofty elms once more appeared
What pleasant expectations lured me on
O'er the flat Common!—With quick step I reached
The threshold, lifted with light hand the latch;
But, when I entered, Margaret looked at me
A little while; then turned her head away
Speechless,—and, sitting down upon a chair,   650
Wept bitterly. I wist not what to do,
Nor how to speak to her. Poor Wretch! at last

She rose from off her seat, and then,—O Sir!
I cannot *tell* how she pronounced my name:—
With fervent love, and with a face of grief
Unutterably helpless, and a look
That seemed to cling upon me, she enquired
If I had seen her husband. As she spake
A strange surprise and fear came to my heart,
Nor had I power to answer ere she told   660
That he had disappeared—not two months gone.
He left his house: two wretched days had past,
And on the third, as wistfully she raised
Her head from off her pillow, to look forth,
Like one in trouble, for returning light,
Within her chamber-casement she espied
A folded paper, lying as if placed
To meet her waking eyes. This tremblingly
She opened—found no writing, but beheld
Pieces of money carefully enclosed,   670
Silver and gold. 'I shuddered at the sight,'
Said Margaret, 'for I knew it was his hand
That must have placed it there; and ere that day
Was ended, that long anxious day, I learned,
From one who by my husband had been sent
With the sad news, that he had joined a troop
Of soldiers, going to a distant land.
—He left me thus—he could not gather heart
To take a farewell of me; for he feared
That I should follow with my babes, and sink   680
Beneath the misery of that wandering life.'

"This tale did Margaret tell with many tears:
And, when she ended, I had little power
To give her comfort, and was glad to take
Such words of hope from her own mouth as served
To cheer us both. But long we had not talked
Ere we built up a pile of better thoughts,
And with a brighter eye she looked around
As if she had been shedding tears of joy.
We parted.—'Twas the time of early spring;   690
I left her busy with her garden tools;
And well remember, o'er that fence she looked,
And, while I paced along the foot-way path,
Called out, and sent a blessing after me,
With tender cheerfulness, and with a voice
That seemed the very sound of happy thoughts.

"I roved o'er many a hill and many a dale,
With my accustomed load; in heat and cold,
Through many a wood and many an open ground,
In sunshine and in shade, in wet and fair,   700

Drooping or blithe of heart, as might befal;
My best companions now the driving winds,
And now the 'trotting brooks' and whispering trees,
And now the music of my own sad steps,
With many a short-lived thought that passed between,
And disappeared.
        I journeyed back this way,
When, in the warmth of midsummer, the wheat
Was yellow; and the soft and bladed grass,
Springing afresh, had o'er the hay-field spread
Its tender verdure. At the door arrived,   710
I found that she was absent. In the shade,
Where now we sit, I waited her return.
Her cottage, then a cheerful object, wore
Its customary look,—only, it seemed,
The honeysuckle, crowding round the porch,
Hung down in heavier tufts; and that bright weed,
The yellow stone-crop, suffered to take root
Along the window's edge, profusely grew
Blinding the lower panes. I turned aside,
And strolled into her garden. It appeared   720
To lag behind the season, and had lost
Its pride of neatness. Daisy-flowers and thrift
Had broken their trim border-lines, and straggled
O'er paths they used to deck: carnations, once
Prized for surpassing beauty, and no less
For the peculiar pains they had required,
Declined their languid heads, wanting support.
The cumbrous bind-weed, with its wreaths and bells,
Had twined about her two small rows of peas,
And dragged them to the earth.
              Ere this an hour   730
Was wasted.—Back I turned my restless steps;
A stranger passed; and, guessing whom I sought,
He said that she was used to ramble far.—
The sun was sinking in the west; and now
I sate with sad impatience. From within
Her solitary infant cried aloud;
Then, like a blast that dies away self-stilled,
The voice was silent. From the bench I rose;
But neither could divert nor soothe my thoughts.
The spot, though fair, was very desolate—   740
The longer I remained, more desolate:
And, looking round me, now I first observed
The corner stones, on either side the porch,
With dull red stains discoloured, and stuck o'er
With tufts and hairs of wool, as if the sheep,

744. stains: from the dyes splashed on the sheep to indicate
to whom they belong.

That fed upon the Common, thither came
Familiarly, and found a couching-place
Even at her threshold. Deeper shadows fell
From these tall elms; the cottage-clock struck eight;—
I turned, and saw her distant a few steps.   750
Her face was pale and thin—her figure, too,
Was changed. As she unlocked the door, she said,
'It grieves me you have waited here so long,
But, in good truth, I've wandered much of late;
And, sometimes—to my shame I speak—have need
Of my best prayers to bring me back again.'
While on the board she spread our evening meal,
She told me—interrupting not the work
Which gave employment to her listless hands—
That she had parted with her elder child;   760
To a kind master on a distant farm
Now happily apprenticed.—'I perceive
You look at me, and you have cause; to-day
I have been travelling far; and many days
About the fields I wander, knowing this
Only, that what I seek I cannot find;
And so I waste my time: for I am changed;
And to myself,' said she, 'have done much wrong
And to this helpless infant. I have slept
Weeping, and weeping have I waked; my tears   770
Have flowed as if my body were not such
As others are; and I could never die.
But I am now in mind and in my heart
More easy; and I hope,' said she, 'that God
Will give me patience to endure the things
Which I behold at home.'
                It would have grieved
Your very soul to see her. Sir, I feel
The story linger in my heart; I fear
'Tis long and tedious; but my spirit clings
To that poor Woman:—so familiarly   780
Do I perceive her manner, and her look,
And presence; and so deeply do I feel
Her goodness, that, not seldom, in my walks
A momentary trance comes over me;
And to myself I seem to muse on One
By sorrow laid asleep; or borne away,
A human being destined to awake
To human life, or something very near
To human life, when he shall come again
For whom she suffered. Yes, it would have grieved
Your very soul to see her: evermore   791
Her eyelids drooped, her eyes downward were cast;
And, when she at her table gave me food,
She did not look at me. Her voice was low,

Her body was subdued. In every act
Pertaining to her house-affairs, appeared
The careless stillness of a thinking mind
Self-occupied; to which all outward things
Are like an idle matter. Still she sighed,
But yet no motion of the breast was seen, 800
No heaving of the heart. While by the fire
We sate together, sighs came on my ear,
I knew not how, and hardly whence they came.

 "Ere my departure, to her care I gave,
For her son's use, some tokens of regard,
Which with a look of welcome she received;
And I exhorted her to place her trust
In God's good love, and seek his help by prayer.
I took my staff, and, when I kissed her babe,
The tears stood in her eyes. I left her then 810
With the best hope and comfort I could give:
She thanked me for my wish;—but for my hope
It seemed she did not thank me.
       I returned,
And took my rounds along this road again
When on its sunny bank the primrose flower
Peeped forth, to give an earnest of the Spring.
I found her sad and drooping: she had learned
No tidings of her husband; if he lived,
She knew not that he lived; if he were dead,
She knew not he was dead. She seemed the same 820
In person and appearance; but her house
Bespake a sleepy hand of negligence;
The floor was neither dry nor neat, the hearth
Was comfortless, and her small lot of books,
Which, in the cottage-window, heretofore
Had been piled up against the corner panes
In seemly order, now, with straggling leaves
Lay scattered here and there, open or shut,
As they had chanced to fall. Her infant Babe
Had from its Mother caught the trick of grief, 830
And sighed among its playthings. I withdrew,
And once again entering the garden saw,
More plainly still, that poverty and grief
Were now come nearer to her: weeds defaced
The hardened soil, and knots of withered grass:
No ridges there appeared of clear black mold,
No winter greenness; of her herbs and flowers,
It seemed the better part were gnawed away
Or trampled into earth; a chain of straw,
Which had been twined about the slender stem 840
Of a young apple-tree, lay at its root;
The bark was nibbled round by truant sheep.

—Margaret stood near, her infant in her arms,
And, noting that my eye was on the tree,
She said, 'I fear it will be dead and gone
Ere Robert come again.' When to the House
We had returned together, she enquired
If I had any hope:—but for her babe
And for her little orphan boy, she said,
She had no wish to live, that she must die 850
Of sorrow. Yet I saw the idle loom
Still in its place; his sunday garments hung
Upon the self-same nail; his very staff
Stood undisturbed behind the door.
        And when,
In bleak December, I retraced this way,
She told me that her little babe was dead,
And she was left alone. She now, released
From her maternal cares, had taken up
The employment common through these wilds, and
  gained,
By spinning hemp, a pittance for herself; 860
And for this end had hired a neighbour's boy
To give her needful help. That very time
Most willingly she put her work aside,
And walked with me along the miry road,
Heedless how far; and, in such piteous sort
That any heart had ached to hear her, begged
That, wheresoe'er I went, I still would ask
For him whom she had lost. We parted then—
Our final parting; for from that time forth
Did many seasons pass ere I returned 870
Into this tract again.
      Nine tedious years;
From their first separation, nine long years,
She lingered in unquiet widowhood;
A Wife and Widow. Needs must it have been
A sore heart-wasting! I have heard, my Friend,
That in yon arbour oftentimes she sate
Alone, through half the vacant sabbath day;
And, if a dog passed by, she still would quit
The shade, and look abroad. On this old bench
For hours she sate; and evermore her eye 880
Was busy in the distance, shaping things
That made her heart beat quick. You see that path,
Now faint,—the grass has crept o'er its grey line;
There, to and fro, she paced through many a day
Of the warm summer, from a belt of hemp
That girt her waist, spinning the long-drawn thread
With backward steps. Yet ever as there passed
A man whose garments showed the soldier's red,
Or crippled mendicant in sailor's garb,

The little child who sate to turn the wheel   890
Ceased from his task; and she with faltering voice
Made many a fond enquiry; and when they,
Whose presence gave no comfort, were gone by,
Her heart was still more sad. And by yon gate,
That bars the traveller's road, she often stood,
And when a stranger horseman came, the latch
Would lift, and in his face look wistfully:
Most happy, if, from aught discovered there
Of tender feeling, she might dare repeat
The same sad question. Meanwhile her poor Hut 900
Sank to decay; for he was gone, whose hand,
At the first nipping of October frost,
Closed up each chink, and with fresh bands of straw
Chequered the green-grown thatch. And so she lived
Through the long winter, reckless and alone;
Until her house by frost, and thaw, and rain,
Was sapped; and while she slept, the nightly damps
Did chill her breast; and in the stormy day
Her tattered clothes were ruffled by the wind,
Even at the side of her own fire. Yet still   910
She loved this wretched spot, nor would for worlds
Have parted hence; and still that length of road,
And this rude bench, one torturing hope endeared,
Fast rooted at her heart: and here, my Friend,—
In sickness she remained; and here she died;
Last human tenant of these ruined walls!"

    The old Man ceased: he saw that I was moved;
From that low bench, rising instinctively
I turned aside in weakness, nor had power
To thank him for the tale which he had told.   920
I stood, and leaning o'er the garden wall
Reviewed that Woman's sufferings; and it seemed
To comfort me while with a brother's love
I blessed her in the impotence of grief.
Then towards the cottage I returned; and traced
Fondly, though with an interest more mild,
That secret spirit of humanity
Which, 'mid the calm oblivious tendencies
Of nature, 'mid her plants, and weeds, and flowers,
And silent overgrowings, still survived.   930
The old Man, noting this, resumed, and said,
"My Friend! enough to sorrow you have given,
The purposes of wisdom ask no more:
Nor more would she have craved as due to One
Who, in her worst distress, had ofttimes felt
The unbounded might of prayer; and learned, with soul
Fixed on the Cross, that consolation springs,
From sources deeper far than deepest pain,

For the meek Sufferer. Why then should we read
The forms of things with an unworthy eye?   940
She sleeps in the calm earth, and peace is here.
I well remember that those very plumes,
Those weeds, and the high spear-grass on that wall,
By mist and silent rain-drops silvered o'er,
As once I passed, into my heart conveyed
So still an image of tranquillity,
So calm and still, and looked so beautiful
Amid the uneasy thoughts which filled my mind,
That what we feel of sorrow and despair
From ruin and from change, and all the grief   950
That passing shows of Being leave behind,
Appeared an idle dream, that could maintain,
Nowhere, dominion o'er the enlightened spirit
Whose meditative sympathies repose
Upon the breast of Faith. I turned away,
And walked along my road in happiness."

    He ceased. Ere long the sun declining shot
A slant and mellow radiance, which began
To fall upon us, while, beneath the trees,
We sate on that low bench: and now we felt,   960
Admonished thus, the sweet hour coming on.
A linnet warbled from those lofty elms,
A thrush sang loud, and other melodies,
At distance heard, peopled the milder air.
The old Man rose, and, with a sprightly mien
Of hopeful preparation, grasped his staff;
Together casting then a farewell look
Upon those silent walls, we left the shade;
And, ere the stars were visible, had reached
A village-inn,—our evening resting-place.   970

from BOOK SECOND

In this extract the Solitary is speaking. He is telling of a
poor and aged man who was heartlessly treated by the
housewife paid by the parish to support him. Sent out to
dig turf upon the moor, he lost his way in a storm, and
lay exposed to the weather for twenty-four hours. As the
extract begins, he has just been found by a search party.

"So was he lifted gently from the ground,
And with their freight homeward the shepherds moved
Through the dull mist, I following—when a step,
A single step, that freed me from the skirts   830
Of the blind vapour, opened to my view

Glory beyond all glory ever seen
By waking sense or by the dreaming soul!
The appearance, instantaneously disclosed,
Was of a mighty city—boldly say
A wilderness of building, sinking far
And self-withdrawn into a boundless depth,
Far sinking into splendor—without end!
Fabric it seemed of diamond and of gold,
With alabaster domes, and silver spires,   840
And blazing terrace upon terrace, high
Uplifted; here, serene pavilions bright,
In avenues disposed; there, towers begirt
With battlements that on their restless fronts
Bore stars—illumination of all gems!
By earthly nature had the effect been wrought
Upon the dark materials of the storm
Now pacified; on them, and on the coves
And mountain-steeps and summits, whereunto
The vapours had receded, taking there   850
Their station under a cerulean sky.
Oh, 'twas an unimaginable sight!
Clouds, mists, streams, watery rocks and emerald turf,
Clouds of all tincture, rocks and sapphire sky,
Confused, commingled, mutually inflamed,
Molten together, and composing thus,
Each lost in each, that marvellous array
Of temple, palace, citadel, and huge
Fantastic pomp of structure without name,
In fleecy folds voluminous, enwrapped.   860
Right in the midst, where interspace appeared
Of open court, an object like a throne
Under a shining canopy of state
Stood fixed; and fixed resemblances were seen
To implements of ordinary use.
But vast in size, in substance glorified;
Such as by Hebrew Prophets were beheld
In vision—forms uncouth of mightiest power
For admiration and mysterious awe.
This little Vale, a dwelling-place of Man,   870
Lay low beneath my feet; 'twas visible—
I saw not, but I felt that it was there.
That which I *saw* was the revealed abode
Of Spirits in beatitude: my heart
Swelled in my breast.—'I have been dead,' I cried,
'And now I live! Oh! wherefore *do* I live?'
And with that pang I prayed to be no more!—
—But I forget our Charge, as utterly
I then forgot him:—there I stood and gazed:
The apparition faded not away,   880
And I descended.

from BOOK THIRD

The Poet has criticized the ancient Stoics who closed
the heart against "admiration, and all sense of joy." As
the Poet ceases the Solitary replies.

    His countenance gave notice that my zeal
Accorded little with his present mind;
I ceased, and he resumed.—"Ah! gentle Sir,
Slight, if you will, the *means;* but spare to slight   360
The *end* of those, who did, by system, rank,
As the prime object of a wise man's aim,
Security from shock of accident,
Release from fear; and cherished peaceful days
For their own sakes, as mortal life's chief good,
And only reasonable felicity.
What motive drew, what impulse, I would ask,
Through a long course of later ages, drove,
The hermit to his cell in forest wide;
Or what detained him, till his closing eyes   370
Took their last farewell of the sun and stars,
Fast anchored in the desert?—Not alone
Dread of the persecuting sword, remorse,
Wrongs unredressed, or insults unavenged
And unavengeable, defeated pride,
Prosperity subverted, maddening want,
Friendship betrayed, affection unreturned,
Love with despair, or grief in agony;—
Not always from intolerable pangs
He fled; but, compassed round by pleasure, sighed 380
For independent happiness; craving peace,
The central feeling of all happiness,
Not as a refuge from distress or pain,
A breathing-time, vacation, or a truce,
But for its absolute self; a life of peace,
Stability without regret or fear;
That hath been, is, and shall be evermore!—
Such the reward he sought; and wore out life,
There, where on few external things his heart
Was set, and those his own; or, if not his,   390
Subsisting under nature's stedfast law.

    "What other yearning was the master tie
Of the monastic brotherhood, upon rock
Aërial, or in green secluded vale,
One after one, collected from afar,

An undissolving fellowship?—What but this,
The universal instinct of repose,
The longing for confirmed tranquillity,
Inward and outward; humble, yet sublime:
The life where hope and memory are as one;   400
Where earth is quiet and her face unchanged
Save by the simplest toil of human hands
Or seasons' difference; the immortal Soul
Consistent in self-rule; and heaven revealed
To meditation in that quietness!—

from BOOK FOURTH

In the first extract (ll. 66–122) the Wanderer is speaking,
having just offered a prayer that God will continue to
inspire him to seek "Repose and hope among eternal
things."

In the second passage (ll. 718–62) the Wanderer is
asserting that religious belief is an instinct of the imagination, and further that the forms of nature continually
rouse in man the idea of a God. The same speech continues in lines 847–87.

Lines 941–94 present the Wanderer comparing scientific and rationalistic thinkers with the instinctive piety of
his humble forbears ("these," l. 943), and the Wanderer
continues his speech in the passage beginning at line 1058.

"And what are things eternal?—powers depart,"
The grey-haired Wanderer stedfastly replied,
Answering the question which himself had asked,
"Possessions vanish, and opinions change,
And passions hold a fluctuating seat: 70
But, by the storms of circumstance unshaken,
And subject neither to eclipse nor wane,
Duty exists;—immutably survive,
For our support, the measures and the forms,
Which an abstract intelligence supplies;
Whose kingdom is, where time and space are not.
Of other converse which mind, soul, and heart,
Do, with united urgency, require,
What more that may not perish?—Thou, dread source,
Prime, self-existing cause and end of all   80
That in the scale of being fill their place;
Above our human region, or below,
Set and sustained;—thou, who didst wrap the cloud
Of infancy around us, that thyself,
Therein, with our simplicity awhile

Might'st hold, on earth, communion undisturbed;
Who from the anarchy of dreaming sleep,
Or from its death-like void, with punctual care,
And touch as gentle as the morning light,
Restor'st us, daily, to the powers of sense   90
And reason's stedfast rule—thou, thou alone
Art everlasting, and the blessed Spirits,
Which thou includest, as the sea her waves:
For adoration thou endur'st; endure
For consciousness the motions of thy will;
For apprehension those transcendent truths
Of the pure intellect, that stand as laws
(Submission constituting strength and power)
Even to thy Being's infinite majesty!
This universe shall pass away—a work   100
Glorious! because the shadow of thy might,
A step, or link, for intercourse with thee.
Ah! if the time must come, in which my feet
No more shall stray where meditation leads,
By flowing stream, through wood, or craggy wild,
Loved haunts like these; the unimprisoned Mind
May yet have scope to range among her own,
Her thoughts, her images, her high desires.
If the dear faculty of sight should fail,
Still, it may be allowed me to remember   110
What visionary powers of eye and soul
In youth were mine; when, stationed on the top
Of some huge hill—expectant, I beheld
The sun rise up, from distant climes returned
Darkness to chase, and sleep; and bring the day
His bounteous gift! or saw him toward the deep
Sink, with a retinue of flaming clouds
Attended; then, my spirit was entranced
With joy exalted to beatitude;
The measure of my soul was filled with bliss,   120
And holiest love; as earth, sea, air, with light,
With pomp, with glory, with magnificence!

                    . . .

"The lively Grecian, in a land of hills,
Rivers and fertile plains, and sounding shores,—
Under a cope of sky more variable,   720
Could find commodious place for every God,
Promptly received, as prodigally brought,
From the surrounding countries, at the choice
Of all adventurers. With unrivalled skill,
As nicest observation furnished hints
For studious fancy, his quick hand bestowed
On fluent operations a fixed shape;
Metal or stone, idolatrously served.

And yet—triumphant o'er this pompous show
Of art, this palpable array of sense,   730
On every side encountered; in despite
Of the gross fictions chanted in the streets
By wandering Rhapsodists; and in contempt
Of doubt and bold denial hourly urged
Amid the wrangling schools—a SPIRIT hung,
Beautiful region! o'er thy towns and farms,
Statues and temples, and memorial tombs;
And emanations were perceived; and acts
Of immortality, in Nature's course,
Exemplified by mysteries, that were felt   740
As bonds, on grave philosopher imposed
And armed warrior; and in every grove
A gay or pensive tenderness prevailed,
When piety more awful had relaxed.
—'Take, running river, take these locks of mine'—
Thus would the Votary say—'this severed hair,
My vow fulfilling, do I here present,
Thankful for my beloved child's return.
Thy banks, Cephisus, he again hath trod,
Thy murmurs heard; and drunk the crystal lymph 750
With which thou dost refresh the thirsty lip,
And, all day long, moisten these flowery fields!'
And doubtless, sometimes, when the hair was shed
Upon the flowing stream, a thought arose
Of Life continuous, Being unimpaired;
That hath been, is, and where it was and is
There shall endure,—existence unexposed
To the blind walk of mortal accident;
From diminution safe and weakening age;
While man grows old, and dwindles, and decays; 760
And countless generations of mankind
Depart; and leave no vestige where they trod.

. . .

"Once more to distant ages of the world
Let us revert, and place before our thoughts
The face which rural solitude might wear
To the unenlightened swains of pagan Greece.   850
—In that fair clime, the lonely herdsman, stretched
On the soft grass through half a summer's day,
With music lulled his indolent repose:
And, in some fit of weariness, if he,
When his own breath was silent, chanced to hear
A distant strain, far sweeter than the sounds
Which his poor skill could make, his fancy fetched,
Even from the blazing chariot of the sun,

A beardless Youth, who touched a golden lute,
And filled the illumined groves with ravishment.   860
The nightly hunter, lifting a bright eye
Up towards the crescent moon, with grateful heart
Called on the lovely wanderer who bestowed
That timely light, to share his joyous sport:
And hence, a beaming Goddess with her Nymphs,
Across the lawn and through the darksome grove,
Not unaccompanied with tuneful notes
By echo multiplied from rock or cave,
Swept in the storm of chase; as moon and stars
Glance rapidly along the clouded heaven,   870
When winds are blowing strong. The traveller slaked
His thirst from rill or gushing fount, and thanked
The Naiad. Sunbeams, upon distant hills
Gliding apace, with shadows in their train,
Might, with small help from fancy, be transformed
Into fleet Oreads sporting visibly.
The Zephyrs fanning, as they passed, their wings,
Lacked not, for love, fair objects whom they wooed
With gentle whisper. Withered boughs grotesque,
Stripped of their leaves and twigs by hoary age,   880
From depth of shaggy covert peeping forth
In the low vale, or on steep mountain side;
And, sometimes, intermixed with stirring horns
Of the live deer, or goat's depending beard,—
These were the lurking Satyrs, a wild brood
Of gamesome Deities; or Pan himself,
The simple shepherd's awe-inspiring God!"

. . .

"Now, shall our great Discoverers," he exclaimed,
Raising his voice triumphantly, "obtain
From sense and reason less than these obtained,
Though far misled? Shall men for whom our age
Unbaffled powers of vision hath prepared,
To explore the world without and world within,
Be joyless as the blind? Ambitious spirits—
Whom earth, at this late season, hath produced
To regulate the moving spheres, and weigh
The planets in the hollow of their hand;   950
And they who rather dive than soar, whose pains
Have solved the elements, or analysed
The thinking principle—shall they in fact
Prove a degraded Race? and what avails
Renown, if their presumption make them such?
Oh! there is laughter at their work in heaven!

BOOK IV. **735. schools:** of philosophy.

876. **Oreads:** mountain nymphs. **884. depending:** hanging,
drooping.

Inquire of ancient Wisdom; go, demand
Of mighty Nature, if 'twas ever meant
That we should pry far off yet be unraised;
That we should pore, and dwindle as we pore,   960
Viewing all objects unremittingly
In disconnexion dead and spiritless;
And still dividing, and dividing still,
Break down all grandeur, still unsatisfied
With the perverse attempt, while littleness
May yet become more little; waging thus
An impious warfare with the very life
Of our own souls!

          And if indeed there be
An all-pervading Spirit, upon whom
Our dark foundations rest, could he design   970
That this magnificent effect of power,
The earth we tread, the sky that we behold
By day, and all the pomp which night reveals;
That these—and that superior mystery
Our vital frame, so fearfully devised,
And the dread soul within it—should exist
Only to be examined, pondered, searched,
Probed, vexed, and criticised?—Accuse me not
Of arrogance, unknown Wanderer as I am,
If, having walked with Nature threescore years,   980
And offered, far as frailty would allow,
My heart a daily sacrifice to Truth,
I now affirm of Nature and of Truth,
Whom I have served, that their DIVINITY
Revolts, offended at the ways of men
Swayed by such motives, to such ends employed;
Philosophers, who, though the human soul
Be of a thousand faculties composed,
And twice ten thousand interests, do yet prize
This soul, and the transcendent universe,   990
No more than as a mirror that reflects
To proud Self-love her own intelligence;
That one, poor, finite object, in the abyss
Of infinite Being, twinkling restlessly!

          . . .

  "Within the soul a faculty abides,
That with interpositions, which would hide
And darken, so can deal that they become   1060
Contingencies of pomp; and serve to exalt
Her native brightness. As the ample moon,
In the deep stillness of a summer even
Rising behind a thick and lofty grove,
Burns, like an unconsuming fire of light,
In the green trees; and, kindling on all sides

Their leafy umbrage, turns the dusky veil
Into a substance glorious as her own,
Yea, with her own incorporated, by power
Capacious and serene. Like power abides   1070
In man's celestial spirit; virtue thus
Sets forth and magnifies herself; thus feeds
A calm, a beautiful, and silent fire,
From the encumbrances of mortal life,
From error, disappointment—nay, from guilt;
And sometimes, so relenting justice wills,
From palpable oppressions of despair."

  The Solitary by these words was touched
With manifest emotion, and exclaimed;
"But how begin? and whence?—'The Mind is free—
Resolve,' the haughty Moralist would say,   1081
'This single act is all that we demand.'
Alas! such wisdom bids a creature fly
Whose very sorrow is, that time hath shorn
His natural wings!—To friendship let him turn
For succour; but perhaps he sits alone
On stormy waters, tossed in a little boat
That holds but him, and can contain no more!
Religion tells of amity sublime
Which no condition can preclude; of One   1090
Who sees all suffering, comprehends all wants,
All weakness fathoms, can supply all needs:
But is that bounty absolute?—His gifts,
Are they not, still, in some degree, rewards
For acts of service? Can his love extend
To hearts that own not him? Will showers of grace,
When in the sky no promise may be seen,
Fall to refresh a parched and withered land?
Or shall the groaning Spirit cast her load
At the Redeemer's feet?"

          In rueful tone,   1100
With some impatience in his mien, he spake:
Back to my mind rushed all that had been urged
To calm the Sufferer when his story closed;
I looked for counsel as unbending now;
But a discriminating sympathy
Stooped to this apt reply:—

          "As men from men
Do, in the constitution of their souls,
Differ, by mystery not to be explained;
And as we fall by various ways, and sink
One deeper than another, self-condemned,   1110
Through manifold degrees of guilt and shame;
So manifold and various are the ways

1103. **Sufferer:** the Solitary.

Of restoration, fashioned to the steps
Of all infirmity, and tending all
To the same point, attainable by all—
Peace in ourselves, and union with our God.
For you, assuredly, a hopeful road
Lies open: we have heard from you a voice
At every moment softened in its course
By tenderness of heart; have seen your eye,   1120
Even like an altar lit by fire from heaven,
Kindle before us.—Your discourse this day,
That, like the fabled Lethe, wished to flow
In creeping sadness, through oblivious shades
Of death and night, has caught at every turn
The colours of the sun. Access for you
Is yet preserved to principles of truth,
Which the imaginative Will upholds
In seats of wisdom, not to be approached
By the inferior Faculty that moulds,   1130
With her minute and speculative pains,
Opinion, ever changing!
                    I have seen
A curious child, who dwelt upon a tract
Of inland ground, applying to his ear
The convolutions of a smooth-lipped shell;
To which, in silence hushed, his very soul
Listened intensely; and his countenance soon
Brightened with joy; for from within were heard
Murmurings, whereby the monitor expressed
Mysterious union with its native sea.   1140
Even such a shell the universe itself
Is to the ear of Faith; and there are times,
I doubt not, when to you it doth impart
Authentic tidings of invisible things;
Of ebb and flow, and ever-during power;
And central peace, subsisting at the heart
Of endless agitation. Here you stand,
Adore, and worship, when you know it not;
Pious beyond the intention of your thought;
Devout above the meaning of your will.   1150
—Yes, you have felt, and may not cease to feel.
The estate of man would be indeed forlorn
If false conclusions of the reasoning power
Made the eye blind, and closed the passages
Through which the ear converses with the heart.
Has not the soul, the being of your life,
Received a shock of awful consciousness,
In some calm season, when these lofty rocks

At night's approach bring down the unclouded sky,
To rest upon their circumambient walls;   1160
A temple framing of dimensions vast,
And yet not too enormous for the sound
Of human anthems,—choral song, or burst
Sublime of instrumental harmony,
To glorify the Eternal! What if these
Did never break the stillness that prevails
Here,—if the solemn nightingale be mute,
And the soft woodlark here did never chant
Her vespers,—Nature fails not to provide
Impulse and utterance. The whispering air   1170
Sends inspiration from the shadowy heights,
And blind recesses of the caverned rocks;
The little rills, and waters numberless,
Inaudible by daylight, blend their notes
With the loud streams: and often, at the hour
When issue forth the first pale stars, is heard,
Within the circuit of this fabric huge,
One voice—the solitary raven, flying
Athwart the concave of the dark blue dome,
Unseen, perchance above all power of sight—   1180
An iron knell! with echoes from afar
Faint—and still fainter—as the cry, with which
The wanderer accompanies her flight
Through the calm region, fades upon the ear,
Diminishing by distance till it seemed
To expire; yet from the abyss is caught again,
And yet again recovered!"

AFTER-THOUGHT

⁓⁓

This is the last of a sonnet series, *The River Duddon*, in which Wordsworth follows the river from its source in the mountains to its end in the sea. Compare *The Excursion*, IV. 754–62.

⁓⁓

I thought of Thee, my partner and my guide,
As being past away.—Vain sympathies!
For, backward, Duddon! as I cast my eyes,
I see what was, and is, and will abide;
Still glides the Stream, and shall for ever glide;
The Form remains, the Function never dies;
While we, the brave, the mighty, and the wise,
We Men, who in our morn of youth defied

1123. **Lethe:** river that flows through Hades. Its waters, when drunk, cause oblivion.

AFTER-THOUGHT. **1. Thee:** the River Duddon.

The elements, must vanish;—be it so!
Enough, if something from our hands have power   10
To live, and act, and serve the future hour;
And if, as toward the silent tomb we go,
Through love, through hope, and faith's transcendent
     dower,
We feel that we are greater than we know.

                                    1818–20 (1820)

INSIDE OF
KING'S COLLEGE CHAPEL,
CAMBRIDGE

Tax not the royal Saint with vain expense,
With ill-matched aims the Architect who planned—
Albeit labouring for a scanty band
Of white-robed Scholars only—this immense
And glorious Work of fine intelligence!
Give all thou canst; high Heaven rejects the lore
Of nicely-calculated less or more;
So deemed the man who fashioned for the sense
These lofty pillars, spread that branching roof
Self-poised, and scooped into ten thousand cells,   10
Where light and shade repose, where music dwells
Lingering—and wandering on as loth to die;
Like thoughts whose very sweetness yieldeth proof
That they were born for immortality.

                                    1820 (1822)

MUTABILITY

❦

This, like the preceding poem, is from the *Ecclesiastical
Sonnets*, a sequence narrating the history of the Church in
England. It comes immediately before a sonnet entitled
"Old Abbeys," in which Wordsworth reflects upon the
ruins of monasteries destroyed or abandoned during the
Reformation. The first reference of the title, then, is to
the mutability of the Church, of which churches are the
visible signs. In ecclesiastical history the word "dissolution" (l.1) especially brings to mind Henry VIII dissolving
the monastic orders.

❦

INSIDE OF KING'S COLLEGE CHAPEL, CAMBRIDGE. **1. royal
Saint**: Henry VI, 1421–71, who founded King's College.
The beautiful chapel is much larger than was required
for the relatively small student body.

From low to high doth dissolution climb,
And sink from high to low, along a scale
Of awful notes, whose concord shall not fail;
A musical but melancholy chime,
Which they can hear who meddle not with crime,
Nor avarice, nor over-anxious care.
Truth fails not; but her outward forms that bear
The longest date do melt like frosty rime,
That in the morning whitened hill and plain
And is no more; drop like the tower sublime   10
Of yesterday, which royally did wear
His crown of weeds, but could not even sustain
Some casual shout that broke the silent air,
Or the unimaginable touch of Time.

                                    1821 (1822)

TO A SKYLARK

Ethereal minstrel! pilgrim of the sky!
Dost thou despise the earth where cares abound?
Or, while the wings aspire, are heart and eye
Both with thy nest upon the dewy ground?
Thy nest which thou canst drop into at will,
Those quivering wings composed, that music still!

Leave to the nightingale her shady wood;
A privacy of glorious light is thine;
Whence thou dost pour upon the world a flood
Of harmony, with instinct more divine;   10
Type of the wise who soar, but never roam;
True to the kindred points of Heaven and Home!

                                    1825 (1827)

EXTEMPORE EFFUSION
UPON THE DEATH
OF JAMES HOGG

❦

James Hogg, a Scottish poet known as "The Ettrick
Shepherd," died November 21, 1835. Sir Walter Scott
and George Crabbe had died in 1832, Coleridge and
Lamb in 1834, and Felicia Hemans (a minor poetess) in
1835.

❦

When first, descending from the moorlands,
I saw the Stream of Yarrow glide
Along a bare and open valley,
The Ettrick Shepherd was my guide.

When last along its banks I wandered,
Through groves that had begun to shed
Their golden leaves upon the pathways,
My steps the Border-minstrel led.

The mighty Minstrel breathes no longer,
'Mid mouldering ruins low he lies;   10
And death upon the braes of Yarrow,
Has closed the Shepherd-poet's eyes:

Nor has the rolling year twice measured,
From sign to sign, its steadfast course,
Since every mortal power of Coleridge
Was frozen at its marvellous source;

The rapt One, of the godlike forehead,
The heaven-eyed creature sleeps in earth:
And Lamb, the frolic and the gentle,
Has vanished from his lonely hearth.   20

Like clouds that rake the mountain-summits,
Or waves that own no curbing hand,
How fast has brother followed brother,
From sunshine to the sunless land!

Yet I, whose lids from infant slumber
Were earlier raised, remain to hear
A timid voice, that asks in whispers,
"Who next will drop and disappear?"

Our haughty life is crowned with darkness,
Like London with its own black wreath,   30
On which with thee, O Crabbe! forth-looking,
I gazed from Hampstead's breezy heath.

As if but yesterday departed,
Thou too art gone before; but why,
O'er ripe fruit, seasonably gathered,
Should frail survivors heave a sigh?

Mourn rather for that holy Spirit,
Sweet as the spring, as ocean deep;
For Her who, ere her summer faded,
Has sunk into a breathless sleep.   40

No more of old romantic sorrows,
For slaughtered Youth or love-lorn Maid!
With sharper grief is Yarrow smitten,
And Ettrick mourns with her their Poet dead.

1835 (1835)

SO FAIR, SO SWEET, WITHAL SO SENSITIVE

This poem, composed when Wordsworth was seventy-four, may be compared with "Lines Written in Early Spring" forty-six years earlier, and lines 13–15 with "Intimations of Immortality," lines 12–13.

So fair, so sweet, withal so sensitive,
Would that the little Flowers were born to live,
Conscious of half the pleasure which they give;

That to this mountain-daisy's self were known
The beauty of its star-shaped shadow, thrown
On the smooth surface of this naked stone!

And what if hence a bold desire should mount
High as the Sun, that he could take account
Of all that issues from his glorious fount!

So might he ken how by his sovereign aid   10
These delicate companionships are made;
And how he rules the pomp of light and shade;

And were the Sister-power that shines by night
So privileged, what a countenance of delight
Would through the clouds break forth on human sight!

Fond fancies! wheresoe'r shall turn thine eye
On earth, air, ocean, or the starry sky,
Converse with Nature in pure sympathy;

All vain desires, all lawless wishes quelled,
Be Thou to love and praise alike impelled,   20
Whatever boon is granted or withheld.

1844 (1845)

EXTEMPORE EFFUSION. **2. Yarrow:** river in southeastern Scotland. **8. Border-minstrel:** Scott. **37. Spirit:** Felicia Hemans, who died at the age of forty-two.

PREFACE
TO THE SECOND EDITION
OF THE LYRICAL BALLADS (1800)

❧

Though Wordsworth later remarked that he "never cared a straw about the theory," he spent considerable effort in writing this Preface, and he revised it especially in 1802 and 1805. The final text—the one that appeared in the *Poems* of 1849–50—is given here.

The Preface is intended to refute neoclassic theory and practice as Wordsworth understood it. In doing so he provides a basic premise for the nineteenth-century view of poetry. He argues that poetry is essentially speech, as opposed to craft or artifice, and that it differs from ordinary speech mainly by the presence of passion (with the "addition" of meter). The more natural or spontaneous it is, the better.

But in Wordsworth critical assumptions combined with a democratic and primitivistic faith. He was convinced that urban ways of life, particularly the more artificial manners of the upper classes, were unnatural and corrupting. The essential grandeur of human nature—which he wishes both to disclose and to address—may be found in humble and rustic people, whose language is also purer and more permanent. Therefore they are to be the source both of his subject matter and of his diction.

The Preface rose out of conversations with Coleridge, who urged Wordsworth to write it, and is to that extent a joint production. Coleridge, however, was never quite satisfied with Wordsworth's remarks on diction, and in *Biographia Literaria*, Chapters XVII–XIX, he discussed and criticized them at length.

❧

The first Volume of these Poems has already been submitted to general perusal. It was published, as an experiment, which, I hoped might be of some use to ascertain, how far, by fitting to metrical arrangement a selection of the real language of men in a state of vivid sensation, that sort of pleasure and that quantity of pleasure may be imparted, which a Poet may rationally endeavour to impart.

I had formed no very inaccurate estimate of the probable effect of those Poems: I flattered myself that they who should be pleased with them would read them with more than common pleasure: and, on the other hand, I was well aware, that by those who should dislike them, they would be read with more than common dislike. The result has differed from my expectation in this only, that a greater number have been pleased than I ventured to hope I should please.

Several of my Friends are anxious for the success of these Poems, from a belief, that, if the views with which they were composed were indeed realised, a class of Poetry would be produced, well adapted to interest mankind permanently, and not unimportant in the quality, and in the multiplicity of its moral relations: and on this account they have advised me to prefix a systematic defence of the theory upon which the Poems were written. But I was unwilling to undertake the task, knowing that on this occasion the Reader would look coldly upon my arguments, since I might be suspected of having been principally influenced by the selfish and foolish hope of *reasoning* him into an approbation of these particular Poems: and I was still more unwilling to undertake the task, because, adequately to display the opinions, and fully to enforce the arguments, would require a space wholly disproportionate to a preface. For, to treat the subject with the clearness and coherence of which it is susceptible, it would be necessary to give a full account of the present state of the public taste in this country, and to determine how far this taste is healthy or depraved; which, again, could not be determined, without pointing out in what manner language and the human mind act and re-act on each other, and without retracing the revolutions, not of literature alone, but likewise of society itself. I have therefore altogether declined to enter regularly upon this defence; yet I am sensible, that there would be something like impropriety in abruptly obtruding upon the Public, without a few words of introduction, Poems so materially different from those upon which general approbation is at present bestowed.

It is supposed, that by the act of writing in verse an Author makes a formal engagement that he will gratify certain known habits of association; that he not only thus apprises the Reader that certain classes of ideas and expressions will be found in his book, but that others will be carefully excluded. This exponent or symbol held forth by metrical language must in different eras of literature have excited very different expectations: for example, in the age of Catullus, Terence, and Lucretius, and that of Statius or Claudian; and in our own country, in the age of Shakspeare and Beaumount and Fletcher, and that of Donne and Cowley, or Dryden, or Pope. I will not take upon me to determine the exact import of the promise which, by the act of writing in verse, an Author in the present day makes to his reader: but it will undoubtedly appear to many persons that I have not fulfilled the terms of an engagement thus voluntarily contracted. They who have been accustomed to the gaudiness and inane phraseology of many modern writers, if they persist in reading this book to its conclusion, will, no doubt, frequently have to struggle with feelings of strangeness and awkwardness: they will look round for poetry, and will be induced to inquire by what species of courtesy these attempts can be permitted to assume that title. I hope therefore the reader will not censure me for attempting to state what I have proposed to myself to perform; and also (as far as the limits of a preface will permit) to explain some of the chief reasons which have determined me in the choice of my purpose: that at least he may be spared any unpleasant feeling of disappointment, and that I myself may be protected from one of the most dishonourable accusations which can be brought against an Author; namely, that of an indolence which prevents him from endeavouring to ascertain what is his duty, or, when his duty is ascertained, prevents him from performing it.

The principal object, then, proposed in these Poems was to choose incidents and situations from common life, and to relate or describe them, throughout, as far as was possible in a selection of language really used by men, and, at the same time, to throw over them a certain colouring of imagination, whereby ordinary things should be presented to the mind in an unusual aspect; and, further, and above all, to make these incidents and situations interesting by tracing in them, truly though not ostentatiously, the primary laws of our nature: chiefly, as far as regards the manner in which we associate ideas in a state of excitement. Humble and rustic life was generally chosen, because, in that condition, the essential passions of the heart find a better soil in which they can attain their maturity, are less under restraint, and speak a plainer and more emphatic language; because in that condition of life our elementary feelings co-exist in a state of greater simplicity, and, consequently, may be more accurately contemplated, and more forcibly communicated; because the manners of rural life germinate from those elementary feelings, and, from the necessary character of rural occupations, are more easily comprehended, and are more durable; and, lastly, because in that condition the passions of men are incorporated with the beautiful and permanent forms of Nature. The language, too, of these men has been adopted (purified indeed from what appear to be its real defects, from all lasting and rational causes of dislike or disgust)

because such men hourly communicate with the best objects from which the best part of language is originally derived; and because, from their rank in society and the sameness and narrow circle of their intercourse, being less under the influence of social vanity, they convey their feelings and notions in simple and unelaborated expressions. Accordingly, such a language, arising out of repeated experience and regular feelings, is a more permanent, and a far more philosophical language, than that which is frequently substituted for it by Poets, who think that they are conferring honour upon themselves and their art, in proportion as they separate themselves from the sympathies of men, and indulge in arbitrary and capricious habits of expression, in order to furnish food for fickle tastes, and fickle appetites, of their own creation.[1]

I cannot, however, be insensible to the present outcry against the triviality and meanness, both of thought and language, which some of my contemporaries have occasionally introduced into their metrical compositions; and I acknowledge that this defect, where it exists, is more dishonourable to the Writer's own character than false refinement or arbitrary innovation, though I should contend at the same time, that it is far less pernicious in the sum of its consequences. From such verses the Poems in these volumes will be found distinguished at least by one mark of difference, that each of them has a worthy *purpose*. Not that I always began to write with a distinct purpose formally conceived; but habits of meditation have, I trust, so prompted and regulated my feelings, that my descriptions of such objects as strongly excite those feelings, will be found to carry along with them a *purpose*. If this opinion be erroneous, I can have little right to the name of a Poet. For all good poetry is the spontaneous overflow of powerful feelings: and though this be true, Poems to which any value can be attached were never produced on any variety of subjects but by a man who, being possessed of more than usual organic sensibility, had also thought long and deeply. For our continued influxes of feeling are modified and directed by our thoughts, which are indeed the representatives of all our past feelings; and, as by contemplating the relation of these general representatives to each other, we discover what is really important to men, so, by the repetition and continuance of this act, our feelings

PREFACE, SECOND EDITION OF THE LYRICAL BALLADS. **1. creation:** "It is worth while here to observe, that the affecting parts of Chaucer are almost always expressed in language pure and universally intelligible even to this day" (W. W.).

will be connected with important subjects, till at length, if we be originally possessed of much sensibility, such habits of mind will be produced, that, by obeying blindly and mechanically the impulses of those habits, we shall describe objects, and utter sentiments, of such a nature, and in such connection with each other, that the understanding of the Reader must necessarily be in some degree enlightened, and his affections strengthened and purified.

It has been said that each of these poems has a purpose. Another circumstance must be mentioned which distinguishes these Poems from the popular Poetry of the day; it is this, that the feeling therein developed gives importance to the action and situation, and not the action and situation to the feeling.[2]

A sense of false modesty shall not prevent me from asserting, that the Reader's attention is pointed to this

mark of distinction, far less for the sake of these particular Poems than from the general importance of the subject. The subject is indeed important! For the human mind is capable of being excited without the application of gross and violent stimulants; and he must have a very faint perception of its beauty and dignity who does not know this, and who does not further know, that one being is elevated above another in proportion as he possesses this capability. It has therefore appeared to me, that to endeavour to produce or enlarge this capability is one of the best services in which, at any period, a Writer can be engaged; but this service, excellent at all times, is especially so at the present day. For a multitude of causes, unknown to former times, are now acting with a combined force to blunt the discriminating powers of the mind, and, unfitting it for all voluntary exertion, to reduce it to a state of almost savage torpor. The most effective of these causes are the great national events which are daily taking place, and the increasing accumulation of men in cities, where the uniformity of their occupations produces a craving for extraordinary incident, which the rapid communication of intelligence hourly gratifies. To this tendency of life and manners the literature and theatrical exhibitions of the country have conformed themselves. The invaluable works of our elder writers, I had almost said the works of Shakspeare and Milton, are driven into neglect by frantic novels, sickly and stupid German Tragedies, and deluges of idle and extravagant stories in verse.[3]— When I think upon this degrading thirst after outrageous stimulation, I am almost ashamed to have spoken of the feeble endeavour made in these volumes to counteract it; and, reflecting upon the magnitude of the general evil, I should be oppressed with no dishonourable melancholy, had I not a deep impression of certain inherent and indestructible qualities of the human mind, and likewise of certain powers in the great and permanent objects that act upon it, which are equally inherent and indestructible: and were there not added to this impression a belief, that the time is approaching when the evil will be systematically opposed, by men of greater powers, and with far more distinguished success.

Having dwelt thus long on the subjects and aim of

2. **feeling:** The text above is that of the last revision. The edition of 1800 was more explicit: "I have said that each of these poems has a purpose. I have also informed my Reader what this purpose will be found principally to be: namely to illustrate the manner in which our feelings and ideas are associated in a state of excitement. But speaking in less general language, it is to follow the fluxes and refluxes of the mind when agitated by the great and simple affections of our nature. This object I have endeavoured in these short essays to attain by various means; by tracing the maternal passion through many of its more subtle windings, as in the poems of the IDIOT BOY and the MAD MOTHER; by accompanying the last struggles of a human being at the approach of death, cleaving in solitude to life and society, as in the Poem of the FORSAKEN INDIAN; by shewing, as in the stanzas entitled WE ARE SEVEN, the perplexity and obscurity which in childhood attend our notion of death, or rather our utter inability to admit that notion; or by displaying the strength of fraternal, or to speak more philosophically, of moral attachment when early associated with the great and beautiful objects of nature, as in THE BROTHERS; or, as in the Incident of SIMON LEE, by placing my Reader in the way of receiving from ordinary moral sensations another and more salutary impression than we are accustomed to receive from them. It has also been part of my general purpose to attempt to sketch characters under the influence of less impassioned feelings, as in the OLD MAN TRAVELLING, THE TWO THIEVES, &c., characters of which the elements are simple, belonging rather to nature than to manners, such as exist now and will probably always exist, and which from their constitution may be distinctly and profitably contemplated. I will not abuse the indulgence of my Reader by dwelling longer upon this subject; but it is proper that I should mention another circumstance," and the text of 1800 continues as above to the word "feeling" at the close of the paragraph. Thereafter the 1800 text adds, "My meaning will be rendered perfectly intelligible by referring my Reader to the Poems entitled POOR SUSAN and the CHILDLESS FATHER, particularly the last stanza of the latter Poem."

3. **frantic . . . verse:** the sentimental and sensational literature of the time, such as the "Gothic" novels of Mrs. Ann Radcliffe and M. G. Lewis or the translation from German of melodramas by August von Kotzebue.

these Poems, I shall request the Reader's permission to apprise him of a few circumstances relating to their *style*, in order, among other reasons, that he may not censure me for not having performed what I never attempted. The Reader will find that personifications of abstract ideas rarely occur in these volumes, and are utterly rejected, as an ordinary device to elevate the style, and raise it above prose. My purpose was to imitate, and, as far as is possible, to adopt the very language of men; and assuredly such personifications do not make any natural or regular part of that language. They are, indeed, a figure of speech occasionally prompted by passion, and I have made use of them as such; but have endeavoured utterly to reject them as a mechanical device of style, or as a family language which Writers in metre seem to lay claim to by prescription. I have wished to keep the Reader in the company of flesh and blood, persuaded that by so doing I shall interest him. Others who pursue a different track will interest him likewise; I do not interfere with their claim, but wish to prefer a claim of my own. There will also be found in these volumes little of what is usually called poetic diction; as much pains has been taken to avoid it as is ordinarily taken to produce it; this has been done for the reason already alleged, to bring my language near to the language of men; and further, because the pleasure which I have proposed to myself to impart, is of a kind very different from that which is supposed by many persons to be the proper object of poetry. Without being culpably particular, I do not know how to give my Reader a more exact notion of the style in which it was my wish and intention to write, than by informing him that I have at all times endeavoured to look steadily at my subject; consequently, there is I hope in these Poems little falsehood of description, and my ideas are expressed in language fitted to their respective importance. Something must have been gained by this practice, as it is friendly to one property of all poetry, namely, good sense: but it has necessarily cut me off from a large portion of phrases and figures of speech which from father to son have long been regarded as the common inheritance of Poets. I have also thought it expedient to restrict myself still further, having abstained from the use of many expressions, in themselves proper and beautiful, but which have been foolishly repeated by bad Poets, till such feelings of disgust are connected with them as it is scarcely possible by any art of association to overpower.

If in a poem there should be found a series of lines, or even a single line, in which the language, though naturally arranged, and according to the strict laws of metre, does not differ from that of prose, there is a numerous class of critics, who, when they stumble upon these prosaisms, as they call them, imagine that they have made a notable discovery, and exult over the Poet as over a man ignorant of his own profession. Now these men would establish a canon of criticism which the Reader will conclude he must utterly reject, if he wishes to be pleased with these volumes. And it would be a most easy task to prove to him, that not only the language of a large portion of every good poem, even of the most elevated character, must necessarily, except with reference to the metre, in no respect differ from that of good prose, but likewise that some of the most interesting parts of the best poems will be found to be strictly the language of prose when prose is well written. The truth of this assertion might be demonstrated by innumerable passages from almost all the poetical writings, even of Milton himself. To illustrate the subject in a general manner, I will here adduce a short composition of Gray, who was at the head of those who, by their reasonings, have attempted to widen the space of separation betwixt Prose and Metrical composition, and was more than any other man curiously elaborate in the structure of his own poetic diction.

> In vain to me the smiling mornings shine,
> And reddening Phoebus lifts his golden fire:
> The birds in vain their amorous descant join,
> Or cheerful fields resume their green attire.
> These ears, alas! for other notes repine;
> *A different object do these eyes require;*
> *My lonely anguish melts no heart but mine;*
> *And in my breast the imperfect joys expire;*
> Yet morning smiles the busy race to cheer,
> And new-born pleasure brings to happier men;
> The fields to all their wonted tribute bear;
> To warm their little loves the birds complain.
> *I fruitless mourn to him that cannot hear,*
> *And weep the more because I weep in vain.*[4]

It will easily be perceived, that the only part of this Sonnet which is of any value is the lines printed in Italics; it is equally obvious, that, except in the rhyme, and in the use of the single word "fruitless" for fruitlessly, which is so far a defect, the language of these lines does in no respect differ from that of prose.

By the foregoing quotation it has been shown that

---

4. **In . . . vain:** "Sonnet on the Death of Richard West."

the language of Prose may yet be well adapted to Poetry; and it was previously asserted that a large portion of the language of every good poem can in no respect differ from that of good Prose. We will go further. It may be safely affirmed, that there neither is, nor can be, any *essential* difference between the language of prose and metrical composition. We are fond of tracing the resemblance between Poetry and Painting, and, accordingly, we call them Sisters: but where shall we find bonds of connection sufficiently strict to typify the affinity betwixt metrical and prose composition? They both speak by and to the same organs; the bodies in which both of them are clothed may be said to be of the same substance, their affections are kindred, and almost identical, not necessarily differing even in degree; Poetry [5] sheds no tears "such as Angels weep," but natural and human tears; she can boast of no celestial ichor that distinguishes her vital juices from those of prose; the same human blood circulates through the veins of them both.

If it be affirmed that rhyme and metrical arrangement of themselves constitute a distinction which overturns what has just been said on the strict affinity of metrical language with that of prose, and paves the way for other artificial distinctions which the mind voluntarily admits, I answer that the language of such Poetry as is here recommended is, as far as is possible, a selection of the language really spoken by men; that this selection, wherever it is made with true taste and feeling, will of itself form a distinction far greater than would at first be imagined, and will entirely separate the composition from the vulgarity and meanness of ordinary life; and, if metre be superadded thereto, I believe that a dissimilitude will be produced altogether sufficient for the gratification of a rational mind. What other distinction would we have? Whence is it to come? And where is it to exist? Not, surely, where the Poet speaks through the mouths of his characters: it cannot be necessary here, either for elevation of style, or any of its supposed ornaments: for, if the Poet's subject be judiciously chosen, it will naturally,

and upon fit occasion, lead him to passions the language of which, if selected truly and judiciously, must necessarily be dignified and variegated, and alive with metaphors and figures. I forbear to speak of an incongruity which would shock the intelligent Reader, should the Poet interweave any foreign splendour of his own with that which the passion naturally suggests: it is sufficient to say that such addition is unnecessary. And, surely, it is more probable that those passages, which with propriety abound with metaphors and figures, will have their due effect, if, upon other occasions where the passions are of a milder character, the style also be subdued and temperate.

But, as the pleasure which I hope to give by the Poems now presented to the Reader must depend entirely on just notions upon this subject, and as it is in itself of high importance to our taste and moral feelings, I cannot content myself with these detached remarks. And if, in what I am about to say, it shall appear to some that my labour is unnecessary, and that I am like a man fighting a battle without enemies, such persons may be reminded, that, whatever be the language outwardly holden by men, a practical faith in the opinions which I am wishing to establish is almost unknown. If my conclusions are admitted, and carried as far as they must be carried if admitted at all, our judgments concerning the works of the greatest Poets both ancient and modern will be far different from what they are at present, both when we praise, and when we censure: and our moral feelings influencing and influenced by these judgments will, I believe, be corrected and purified.

Taking up the subject, then, upon general grounds, let me ask, what is meant by the word Poet? What is a Poet? To whom does he address himself? And what language is to be expected from him?—He is a man speaking to men: a man, it is true, endowed with more lively sensibility, more enthusiasm and tenderness, who has a greater knowledge of human nature, and a more comprehensive soul, than are supposed to be common among mankind; a man pleased with his own passions and volitions, and who rejoices more than other men in the spirit of life that is in him; delighting to contemplate similar volitions and passions as manifested in the goings-on of the universe, and habitually impelled to create them where he does not find them. To these qualities he has added a disposition to be affected more than other men by absent things as if they were present; an ability of conjuring up in himself passions which are indeed far from being the same

5. **Poetry:** "I here use the word 'Poetry' (though against my own judgment) as opposed to the word Prose, and synonymous with metrical composition. But much confusion has been introduced into criticism by the contradistinction of Poetry and Prose, instead of the more philosophical one of Poetry and Matter of Fact, or Science. The only strict antithesis to Prose is Metre; nor is this, in truth, a *strict* antithesis, because lines and passages of metre so naturally occur in writing prose, that it would be scarcely possible to avoid them, even were it desirable" (W. W.).

as those produced by real events, yet (especially in those parts of the general sympathy which are pleasing and delightful) do more nearly resemble the passions produced by real events, than anything which, from the motions of their own minds merely, other men are accustomed to feel in themselves:—whence, and from practice, he has acquired a greater readiness and power in expressing what he thinks and feels, and especially those thoughts and feelings which, by his own choice, or from the structure of his own mind, arise in him without immediate external excitement.

But whatever portion of this faculty we may suppose even the greatest Poet to possess, there cannot be a doubt that the language which it will suggest to him, must often, in liveliness and truth, fall short of that which is uttered by men in real life, under the actual pressure of those passions, certain shadows of which the Poet thus produces, or feels to be produced, in himself.

However exalted a notion we would wish to cherish of the character of a Poet, it is obvious, that while he describes and imitates passions, his employment is in some degree mechanical, compared with the freedom and power of real and substantial action and suffering. So that it will be the wish of the Poet to bring his feelings near to those of the persons whose feelings he describes, nay, for short spaces of time, perhaps, to let himself slip into an entire delusion, and even confound and identify his own feelings with theirs; modifying only the language which is thus suggested to him by a consideration that he describes for a particular purpose, that of giving pleasure. Here, then, he will apply the principle of selection which has been already insisted upon. He will depend upon this for removing what would otherwise be painful or disgusting in the passion; he will feel that there is no necessity to trick out or to elevate nature: and, the more industriously he applies this principle, the deeper will be his faith that no words, which *his* fancy or imagination can suggest, will be to be compared with those which are the emanations of reality and truth.

But it may be said by those who do not object to the general spirit of these remarks, that, as it is impossible for the Poet to produce upon all occasions language as exquisitely fitted for the passion as that which the real passion itself suggests, it is proper that he should consider himself as in the situation of a translator, who does not scruple to substitute excellencies of another kind for those which are unattainable by him; and endeavours occasionally to surpass his original, in order to make some amends for the general inferiority

to which he feels that he must submit. But this would be to encourage idleness and unmanly despair. Further, it is the language of men who speak of what they do not understand; who talk of Poetry as of a matter of amusement and idle pleasure; who will converse with us as gravely about a *taste* for Poetry, as they express it, as if it were a thing as indifferent as a taste for rope-dancing, or Frontiniac or Sherry. Aristotle, I have been told, has said that Poetry is the most philosophic of all writing: it is so: its object is truth, not individual and local, but general, and operative; not standing upon external testimony, but carried alive into the heart by passion; truth which is its own testimony, which gives competence and confidence to the tribunal to which it appeals, and receives them from the same tribunal. Poetry is the image of man and nature. The obstacles which stand in the way of the fidelity of the Biographer and Historian, and of their consequent utility, are incalculably greater than those which are to be encountered by the Poet who comprehends the dignity of his art. The Poet writes under one restriction only, namely, the necessity of giving immediate pleasure to a human Being possessed of that information which may be expected from him, not as a lawyer, a physician, a mariner, an astronomer, or a natural philosopher, but as a Man. Except this one restriction, there is no object standing between the Poet and the image of things; between this, and the Biographer and Historian, there are a thousand.

Nor let this necessity of producing immediate pleasure be considered as a degradation of the Poet's art. It is far otherwise. It is an acknowledgment of the beauty of the universe, an acknowledgment the more sincere, because not formal, but indirect; it is a task light and easy to him who looks at the world in the spirit of love: further, it is a homage paid to the native and naked dignity of man, to the grand elementary principle of pleasure, by which he knows, and feels, and lives, and moves. We have no sympathy but what is propagated by pleasure: I would not be misunderstood; but wherever we sympathise with pain, it will be found that the sympathy is produced and carried on by subtle combinations with pleasure. We have no knowledge, that is, no general principles drawn from the contemplation of particular facts, but what has been built up by pleasure, and exists in us by pleasure alone. The Man of science, the Chemist and Mathematician, whatever difficulties and disgusts they may have had to struggle with, know and feel this. However painful may be the objects with which the Anatomist's knowledge is connected, he feels that his knowledge is pleasure; and where he has no pleasure he has no knowledge. What then does the Poet? He considers man and the objects that surround him as acting and reacting upon each other, so as to produce an infinite complexity of pain and pleasure; he considers man in his own nature and in his ordinary life as contemplating this with a certain quantity of immediate knowledge, with certain convictions, intuitions, and deductions, which from habit acquire the quality of intuitions; he considers him as looking upon this complex scene of ideas and sensations, and finding every where objects that immediately excite in him sympathies which, from the necessities of his nature, are accompanied by an overbalance of enjoyment.

To this knowledge which all men carry about with them, and to these sympathies in which, without any other discipline than that of our daily life, we are fitted to take delight, the Poet principally directs his attention. He considers man and nature as essentially adapted to each other, and the mind of man as naturally the mirror of the fairest and most interesting properties of nature. And thus the Poet, prompted by this feeling of pleasure, which accompanies him through the whole course of his studies, converses with general nature, with affections akin to those, which, through labour and length of time, the Man of science has raised up in himself, by conversing with those particular parts of nature which are the objects of his studies. The knowledge both of the Poet and the Man of science is pleasure; but the knowledge of the one cleaves to us as a necessary part of our existence, our natural and unalienable inheritance; the other is a personal and individual acquisition, slow to come to us, and by no habitual and direct sympathy connecting us with our fellow-beings. The Man of science seeks truth as a remote and unknown benefactor; he cherishes and loves it in his solitude: the Poet, singing a song in which all human beings join with him, rejoices in the presence of truth as our visible friend and hourly companion. Poetry is the breath and finer spirit of all knowledge; it is the impassioned expression which is in the countenance of all Science. Emphatically may it be said of the Poet, as Shakspeare hath said of man, that "he looks before and after." He is the rock of defence for human nature; an upholder and preserver, carrying every where with him relationship and love. In spite of difference of soil and climate, of language and manners, of laws and customs; in spite of things silently gone out of mind, and things violently destroyed; the Poet binds together by passion and knowledge the vast empire of human society, as it is spread over the whole earth, and over all time. The objects of the Poet's thoughts are every where; though the eyes and senses of man are, it is true, his favourite guides, yet he will follow wheresoever he can find an atmosphere of sensation in which to move his wings. Poetry is the first and last of all knowledge—it is as immortal as the heart of man. If the labours of Men of science should ever create any material revolution, direct or indirect, in our condition, and in the impressions which we habitually receive, the Poet will sleep then no more than at present; he will be ready to follow the steps of the Man of science, not only in those general indirect effects, but he will be at his side, carrying sensation into the midst of the objects of the science itself. The remotest discoveries of the Chemist, the Botanist, or Mineralogist, will be as proper objects of the Poet's art as any upon which it can be employed, if the time should ever come when these things shall be familiar to us, and the relations under which they are contemplated by the followers of these respective sciences shall be manifestly and palpably material to us as enjoying and suffering beings. If the time should ever come when what is now called science, thus familiarized to men, shall be ready to put on, as it were, a form of flesh and blood, the Poet will lend his divine spirit to aid the transfiguration, and will welcome the Being thus produced, as a dear and genuine inmate of the household of man.—It is not, then, to be supposed that any one, who holds that sublime notion of Poetry which I have attempted to convey, will break in upon the sanctity and truth of his pictures by transitory and accidental ornaments, and endeavour to excite admiration of himself by arts, the necessity of which must manifestly depend upon the assumed meanness of his subject.

What has been thus far said applies to Poetry in general, but especially to those parts of composition where the Poet speaks through the mouths of his characters; and upon this point it appears to authorise the conclusion that there are few persons of good sense, who would not allow that the dramatic parts of composition are defective, in proportion as they deviate from the real language of nature, and are coloured by a diction of the Poet's own, either peculiar to him as an individual Poet or belonging simply to Poets in general; to a body of men who, from the circumstance of their composition being in metre it is expected will employ a particular language.

It is not, then, in the dramatic parts of composition that we look for this distinction of language; but still it may be proper and necessary where the Poet speaks to us in his own person and character. To this I answer by referring the Reader to the description before given of a Poet. Among the qualities there enumerated as principally conducing to form a Poet, is implied nothing differing in kind from other men, but only in degree. The sum of what was said is, that the Poet is chiefly distinguished from other men by a greater promptness to think and feel without immediate external excitement, and a greater power in expressing such thoughts and feelings as are produced in him in that manner. But these passions and thoughts and feelings are the general passions and thoughts and feelings of men. And with what are they connected? Undoubtedly with our moral sentiments and animal sensations, and with the causes which excite these; with the operations of the elements, and the appearances of the visible universe; with storm and sunshine, with the revolutions of the seasons, with cold and heat, with loss of friends and kindred, with injuries and resentments, gratitude and hope, with fear and sorrow. These, and the like, are the sensations and objects which the Poet describes, as they are the sensations of other men, and the objects which interest them. The poet thinks and feels in the spirit of human passions. How, then, can his language differ in any material degree from that of all other men who feel vividly and see clearly? It might be *proved* that it is impossible. But supposing that this were not the case, the Poet might then be allowed to use a peculiar language when expressing his feelings for his own gratification, or that of men like himself. But Poets do not write for Poets alone, but for men. Unless, therefore, we are advocates for that admiration which subsists upon ignorance, and that pleasure which arises from hearing what we do not understand, the Poet must descend from this supposed height; and, in order to excite rational sympathy, he must express himself as other men express themselves. To this it may be added, that while he is only selecting from the real language of men, or, which amounts to the same thing, composing accurately in the spirit of such selection, he is treading upon safe ground, and we know what we are to expect from him. Our feelings are the same with respect to metre; for, as it may be proper to remind the Reader, the distinction of metre is regular and uniform, and not, like that which is produced by what is usually called POETIC DICTION, arbitrary, and subject to infinite caprices, upon which

no calculation whatever can be made. In the one case, the Reader is utterly at the mercy of the Poet, respecting what imagery or diction he may choose to connect with the passion; whereas, in the other, the metre obeys certain laws, to which the Poet and Reader both willingly submit because they are certain, and because no interference is made by them with the passion, but such as the concurring testimony of ages has shown to heighten and improve the pleasure which co-exists with it.

It will now be proper to answer an obvious question, namely, Why, professing these opinions, have I written in verse? To this, in addition to such answer as is included in what has been already said, I reply, in the first place, Because, however I may have restricted myself, there is still left open to me what confessedly constitutes the most valuable object of all writing, whether in prose or verse; the great and universal passions of men, the most general and interesting of their occupations, and the entire world of nature before me—to supply endless combinations of forms and imagery. Now, supposing for a moment that whatever is interesting in these objects may be as vividly described in prose, why should I be condemned for attempting to superadd to such description, the charm which, by the consent of all nations, is acknowledged to exist in metrical language? To this, by such as are yet unconvinced, it may be answered that a very small part of the pleasure given by Poetry depends upon the metre, and that it is injudicious to write in metre, unless it be accompanied with the other artificial distinctions of style with which metre is usually accompanied, and that, by such deviation, more will be lost from the shock which will thereby be given to the Reader's associations than will be counterbalanced by any pleasure which he can derive from the general power of numbers. In answer to those who still contend for the necessity of accompanying metre with certain appropriate colours of style in order to the accomplishment of its appropriate end, and who also, in my opinion, greatly underrate the power of metre in itself, it might, perhaps, as far as relates to these Volumes, have been almost sufficient to observe that poems are extant, written upon more humble subjects, and in a still more naked and simple style, which have continued to give pleasure from generation to generation. Now, if nakedness and simplicity be a defect, the fact here mentioned affords a strong presumption that poems somewhat less naked and simple are capable of affording pleasure at the present day; and, what I wished *chiefly* to attempt, at present, was to justify myself for having written under the impression of this belief.

But various causes might be pointed out why, when the style is manly, and the subject of some importance, words metrically arranged will long continue to impart such a pleasure to mankind as he who proves the extent of that pleasure will be desirous to impart. The end of Poetry is to produce excitement in co-existence with an overbalance of pleasure; but, by the supposition, excitement is an unusual and irregular state of the mind; ideas and feelings do not, in that state, succeed each other in accustomed order. If the words, however, by which this excitement is produced be in themselves powerful, or the images and feelings have an undue proportion of pain connected with them, there is some danger that the excitement may be carried beyond its proper bounds. Now the co-presence of something regular, something to which the mind has been accustomed in various moods and in a less excited state, cannot but have great efficacy in tempering and restraining the passion by an intertexture of ordinary feeling, and of feeling not strictly and necessarily connected with the passion. This is unquestionably true; and hence, though the opinion will at first appear paradoxical, from the tendency of metre to divest language, in a certain degree, of its reality, and thus to throw a sort of half-consciousness of unsubstantial existence over the whole composition, there can be little doubt but that more pathetic situations and sentiments, that is, those which have a greater proportion of pain connected with them, may be endured in metrical composition, especially in rhyme, than in prose. The metre of the old ballads is very artless, yet they contain many passages which would illustrate this opinion; and, I hope, if the following Poems be attentively perused, similar instances will be found in them. This opinion may be further illustrated by appealing to the reader's own experience of the reluctance with which he comes to the re-perusal of the distressful parts of *Clarissa Harlowe,* or *The Gamester,*[6] while Shakspeare's writings, in the most pathetic scenes, never act upon us, as pathetic, beyond the bounds of pleasure—an effect which, in a much greater degree than might at first be imagined, is to be ascribed to small, but continual and regular impulses of pleasurable surprise from the metrical arrangement.—On the other hand (what it must be allowed will much more frequently happen) if the

Poet's words should be incommensurate with the passion, and inadequate to raise the Reader to a height of desirable excitement, then, (unless the Poet's choice of his metre has been grossly injudicious) in the feelings of pleasure which the Reader has been accustomed to connect with metre in general, and in the feeling, whether cheerful or melancholy, which he has been accustomed to connect with that particular movement of metre, there will be found something which will greatly contribute to impart passion to the words, and to effect the complex end which the Poet proposes to himself.

If I had undertaken a SYSTEMATIC defence of the theory here maintained, it would have been my duty to develop the various causes upon which the pleasure received from metrical language depends. Among the chief of these causes is to be reckoned a principle which must be well known to those who have made any of the Arts the object of accurate reflection; namely, the pleasure which the mind derives from the perception of similitude in dissimilitude. This principle is the great spring of the activity of our minds, and their chief feeder. From this principle the direction of the sexual appetite, and all the passions connected with it, take their origin: it is the life of our ordinary conversation; and upon the accuracy with which similitude in dissimilitude, and dissimilitude in similitude are perceived, depend our taste and our moral feelings. It would not be a useless employment to apply this principle to the consideration of metre, and to show that metre is hence enabled to afford much pleasure, and to point out in what manner that pleasure is produced. But my limits will not permit me to enter upon this subject, and I must content myself with a general summary.

I have said that poetry is the spontaneous overflow of powerful feelings; it takes its origin from emotion recollected in tranquillity: the emotion is contemplated till, by a species of re-action, the tranquillity gradually disappears, and an emotion, kindred to that which was before the subject of contemplation, is gradually produced, and does itself actually exist in the mind. In this mood successful composition generally begins, and in a mood similar to this it is carried on; but the emotion, of whatever kind, and in whatever degree, from various causes, is qualified by various pleasures, so that in describing any passions whatsoever, which are voluntarily described, the mind will, upon the whole, be in a state of enjoyment. If Nature be thus cautious to preserve in a state of enjoyment a being so employed,

---

6. The Gamester: a domestic tragedy by Edward Moore, first performed in 1753.

the Poet ought to profit by the lesson held forth to him, and ought especially to take care, that, whatever passions he communicates to his Reader, those passions, if his Reader's mind be sound and vigorous, should always be accompanied with an overbalance of pleasure. Now the music of harmonious metrical language, the sense of difficulty overcome, and the blind association of pleasure which has been previously received from works of rhyme or metre of the same or similar construction, an indistinct perception perpetually renewed of language closely resembling that of real life, and yet, in the circumstance of metre, differing from it so widely—all these imperceptibly make up a complex feeling of delight, which is of the most important use in tempering the painful feeling always found intermingled with powerful descriptions of the deeper passions. This effect is always produced in pathetic and impassioned poetry; while, in lighter compositions, the ease and gracefulness with which the Poet manages his numbers are themselves confessedly a principal source of the gratification of the reader. All that it is *necessary* to say, however, upon this subject, may be effected by affirming, what few persons will deny, that, of two descriptions, either of passions, manners, or characters each of them equally well executed, the one in prose and the other in verse, the verse will be read a hundred times where the prose is read once.[7]

Having thus explained a few of my reasons for writing in verse, and why I have chosen subjects from common life, and endeavoured to bring my language near to the real language of men, if I have been too minute in pleading my own cause, I have at the same time been treating a subject of general interest; and for this reason a few words shall be added with reference solely to these particular poems, and to some defects which will probably be found in them. I am sensible

7. once: At this point the edition of 1800 added: "We see that Pope, by the power of verse alone, has contrived to render the plainest common sense interesting, and even frequently to invest it with the appearance of passion. In consequence of these convictions I related in metre the Tale of GOODY BLAKE and HARRY GILL which is one of the rudest of this collection. I wished to draw attention to the truth that the power of the human imagination is sufficient to produce such changes even in our physical nature as might almost appear miraculous. The truth is an important one; the fact (for it is a *fact*) is a valuable illustration of it. And I have the satisfaction of knowing that it has been communicated to many hundreds of people who would never have heard of it, had it not been narrated as a Ballad, and in a more impressive metre than is usual in Ballads."

that my associations must have sometimes been particular instead of general, and that, consequently, giving to things a false importance, I may have sometimes written upon unworthy subjects; but I am less apprehensive on this account, than that my language may frequently have suffered from those arbitrary connections of feelings and ideas with particular words and phrases, from which no man can altogether protect himself. Hence I have no doubt, that, in some instances, feelings, even of the ludicrous, may be given to my Readers by expressions which appeared to me tender and pathetic. Such faulty expressions, were I convinced they were faulty at present, and that they must necessarily continue to be so, I would willingly take all reasonable pains to correct. But it is dangerous to make these alterations on the simple authority of a few individuals, or even of certain classes of men; for where the understanding of an Author is not convinced, or his feelings altered, this cannot be done without great injury to himself: for his own feelings are his stay and support; and, if he set them aside in one instance, he may be induced to repeat this act till his mind shall lose all confidence in itself, and become utterly debilitated. To this it may be added, that the Critic ought never to forget that he is himself exposed to the same errors as the Poet, and, perhaps, in a much greater degree: for there can be no presumption in saying of most readers, that it is not probable they will be so well acquainted with the various stages of meaning through which words have passed, or with the fickleness or stability of the relations of particular ideas to each other; and, above all, since they are so much less interested in the subject, they may decide lightly and carelessly.

Long as the Reader has been detained, I hope he will permit me to caution him against a mode of false criticism which has been applied to Poetry, in which the language closely resembles that of life and nature. Such verses have been triumphed over in parodies, of which Dr. Johnson's stanza is a fair specimen:—

> I put my hat upon my head
> And walked into the Strand,
> And there I met another man
> Whose hat was in his hand.

Immediately under these lines let us place one of the most justly-admired stanzas of the "Babes in the Woods."

> These pretty Babes with hand in hand
> Went wandering up and down;
> But never more they saw the man
> Approaching from the town.

In both these stanzas the words, and the order of the words, in no respect differ from the most unimpassioned conversation. There are words in both, for example, "the Strand," and "the Town," connected with none but the most familiar ideas; yet the one stanza we admit as admirable, and the other as a fair example of the superlatively contemptible. Whence arises this difference? Not from the metre, not from the language, not from the order of the words; but the *matter* expressed in Dr. Johnson's stanza is contemptible. The proper method of treating trivial and simple verses, to which Dr. Johnson's stanza would be a fair parallelism, is not to say, that is a bad kind of poetry, or, this is not poetry; but, this wants sense; it is neither interesting in itself, nor can *lead* to anything interesting; the images neither originate in that sane state of feeling which arises out of thought, nor can excite thought or feeling in the Reader. This is the only sensible manner of dealing with such verses. Why trouble yourself about the species till you have previously decided upon the genus? Why take pains to prove that an ape is not a Newton, when it is self-evident that he is not a man?

One request I must make of my reader, which is, that in judging these Poems he would decide by his own feelings genuinely, and not by reflection upon what will probably be the judgment of others. How common is it to hear a person say, I myself do not object to this style of composition, or this or that expression, but, to such and such classes of people it will appear mean or ludicrous! This mode of criticism, so destructive of all sound unadulterated judgment, is almost universal: let the Reader then abide, independently, by his own feelings, and, if he finds himself affected, let him not suffer such conjectures to interfere with his pleasure.

If an Author, by any single composition, has impressed us with respect for his talents, it is useful to consider this as affording a presumption, that on other occasions, where we have been displeased, he, nevertheless may not have written ill or absurdly; and further, to give him so much credit for this one composition as may induce us to review what has displeased us, with more care than we should otherwise have bestowed upon it. This is not only an act of justice, but, in our decisions upon poetry especially, may conduce, in a high degree, to the improvement of our own taste; for an *accurate* taste in poetry, and in all the other arts, as Sir Joshua Reynolds has observed, is an *acquired* talent, which can only be produced by thought and a long-continued intercourse with the best models of composition. This is mentioned, not with so ridiculous a purpose as to prevent the most inexperienced Reader from judging for himself (I have already said that I wish him to judge for himself); but merely to temper the rashness of decision, and to suggest that, if Poetry be a subject on which much time has not been bestowed, the judgment may be erroneous; and that, in many cases, it necessarily will be so.

Nothing would, I know, have so effectually contributed to further the end which I have in view, as to have shown of what kind the pleasure is, and how that pleasure is produced, which is confessedly produced by metrical composition essentially different from that which I have here endeavoured to recommend: for the Reader will say that he has been pleased by such composition; and what more can be done for him? The power of any art is limited; and he will suspect, that, if it be proposed to furnish him with new friends, that can be only upon condition of his abandoning his old friends. Besides, as I have said, the Reader is himself conscious of the pleasure which he has received from such composition, composition to which he has peculiarly attached the endearing name of Poetry; and all men feel an habitual gratitude, and something of an honourable bigotry, for the objects which have long continued to please them: we not only wish to be pleased, but to be pleased in that particular way in which we have been accustomed to be pleased. There is in these feelings enough to resist a host of arguments; and I should be the less able to combat them successfully, as I am willing to allow, that, in order entirely to enjoy the Poetry which I am recommending, it would be necessary to give up much of what is ordinarily enjoyed. But, would my limits have permitted me to point out how this pleasure is produced, many obstacles might have been removed, and the Reader assisted in perceiving that the powers of language are not so limited as he may suppose; and that it is possible for Poetry to give other enjoyments, of a purer, more lasting, and more exquisite nature. This part of the subject has not been altogether neglected, but it has not been so much my present aim to prove, that the interest excited by some other kinds of poetry is less vivid, and less worthy of the nobler powers of the mind, as to offer reasons for presuming, that if my purpose were fulfilled, a species of poetry would be produced, which is genuine poetry; in its nature well adapted to interest mankind permanently, and likewise important in the multiplicity and quality of its moral relations.

From what has been said, and from a perusal of the Poems, the reader will be able clearly to perceive the object which I had in view: he will determine how far it has been attained; and, what is a much more important question, whether it be worth attaining: and upon the decision of these two questions will rest my claim to the approbation of the Public.

APPENDIX

∽◦∾

"See p. 327—'by what is usually called POETIC DICTION'" (W. W.).
This Appendix was added in 1802.

∽◦∾

Perhaps, as I have no right to expect that attentive perusal, without which, confined, as I have been, to the narrow limits of a Preface, my meaning cannot be thoroughly understood, I am anxious to give an exact notion of the sense in which the phrase poetic diction has been used; and for this purpose, a few words shall here be added, concerning the origin and characteristics of the phraseology, which I have condemned under that name.

The earliest poets of all nations generally wrote from passion excited by real events; they wrote naturally, and as men: feeling powerfully as they did, their language was daring, and figurative. In succeeding times, Poets, and Men ambitious of the fame of Poets, perceiving the influence of such language, and desirous of producing the same effect without being animated by the same passion, set themselves to a mechanical adoption of these figures of speech, and made use of them, sometimes with propriety, but much more frequently applied them to feelings and thoughts with which they had no natural connection whatsoever. A language was thus insensibly produced, differing materially from the real language of men in *any situation*. The Reader or Hearer of this distorted language found himself in a perturbed and unusual state of mind: when affected by the genuine language of passion he had been in a perturbed unusual state of mind also: in both cases he was willing that his common judgment and understanding should be laid asleep, and he had no instinctive and infallible perception of the true to make him reject the false; the one served as a passport for the other. The emotion was in both cases delightful, and no wonder if he confounded the one with the other, and believed them both to be produced by the same, or similar causes. Besides, the Poet spake to him in the character of a man to be looked up to, a man of genius and authority. Thus, and from a variety of other causes, this distorted language was received with admiration; and Poets, it is probable, who had before contented themselves for the most part with misapplying only expressions which at first had been dictated by real passion, carried the abuse still further, and introduced phrases composed apparently in the spirit of the original figurative language of passion, yet altogether of their own invention, and characterised by various degrees of wanton deviation from good sense and Nature.

It is indeed true, that the language of the earliest Poets was felt to differ materially from ordinary language, because it was the language of extraordinary occasions; but it was really spoken by men, language which the Poet himself had uttered when he had been affected by the events which he described, or which he had heard uttered by those around him. To this language it is probable that metre of some sort or other was early superadded. This separated the genuine language of Poetry still further from common life, so that whoever read or heard the poems of these earliest Poets felt himself moved in a way in which he had not been accustomed to be moved in real life, and by causes manifestly different from those which acted upon him in real life. This was the great temptation to all the corruptions which have followed: under the protection of this feeling succeeding Poets constructed a phraseology which had one thing, it is true, in common with the genuine language of poetry, namely, that it was not heard in ordinary conversation; that it was unusual. But the first Poets, as I have said, spake a language which, though unusual, was still the language of men. This circumstance, however, was disregarded by their successors; they found that they could please by easier means: they became proud of modes of expression which they themselves had invented, and which were uttered only by themselves. In process of time metre became a symbol or promise of this unusual language, and whoever took upon him to write in metre, according as he possessed more or less of true poetic genius, introduced less or more of this adulterated phraseology into his compositions, and the true and the false were inseparably interwoven until, the taste of men becoming gradually perverted, this language was received as a natural language: and at length, by the influence of books upon men, did to a certain degree really become so. Abuses of this kind were imported from one nation to another, and with the progress of refinement this diction became daily more and more corrupt, thrusting out of sight the plain humanities of Nature by a motley masquerade of tricks, quaintnesses, hieroglyphics, and enigmas.

It would not be uninteresting to point out the causes of the pleasure given by this extravagant and absurd diction. It depends upon a great variety of causes, but upon none, perhaps, more than its influence in impressing a notion of the peculiarity and exaltation of the Poet's character, and in flattering the Reader's self-love by bringing him nearer to a sympathy with that character; an effect which is accomplished by unsettling ordinary habits of thinking, and thus assisting the Reader to approach to that perturbed and dizzy state of mind in which if he does not find himself, he imagines that he is *balked* of a peculiar enjoyment which poetry can and ought to bestow.

The sonnet quoted from Gray, in the Preface, except the lines printed in Italics, consists of little else but this diction, though not of the worst kind; and indeed, if one may be permitted to say so, it is far too common in the best writers both ancient and modern. Perhaps in no way, by positive example, could more easily be given a notion of what I mean by the phrase *poetic diction* than by referring to a comparison between the metrical paraphrase which we have of passages in the Old and New Testament, and those passages as they exist in our common Translation. See Pope's "Messiah" throughout; Prior's "Did sweeter sounds adorn my flowing tongue," &c. &c. "Though I speak with the tongues of men and of angels," &c. &c. 1st Corinthians, chap. xiii. By way of immediate example take the following of Dr. Johnson :—

"Turn on the prudent Ant thy heedless eyes,
Observe her labours, Sluggard, and be wise;
No stern command, no monitory voice,
Prescribes her duties, or directs her choice;
Yet, timely provident, she hastes away
To snatch the blessings of a plenteous day;
When fruitful Summer loads the teeming plain,
She crops the harvest, and she stores the grain.
How long shall sloth usurp thy useless hours,
Unnerve thy vigour, and enchain thy powers?
While artful shades thy downy couch enclose,
And soft solicitation courts repose,
Amidst the drowsy charms of dull delight,
Year chases year with unremitted flight,
Till Want now following, fraudulent and slow,
Shall spring to seize thee, like an ambush'd foe."

From this hubbub of words pass to the original. "Go to the Ant, thou Sluggard, consider her ways, and be wise: which having no guide, overseer, or ruler, provideth her meat in the summer, and gathereth her food in the harvest. How long wilt thou sleep, O Sluggard? when wilt thou arise out of thy sleep? Yet a little sleep, a little slumber, a little folding of the hands to sleep. So shall thy poverty come as one that travelleth, and thy want as an armed man." Proverbs, chap. vi.

One more quotation, and I have done. It is from Cowper's Verses supposed to be written by Alexander Selkirk :[1]—

"Religion! what treasure untold
Resides in that heavenly word!
More precious than silver and gold,
Or all that this earth can afford.
But the sound of the church-going bell
These valleys and rocks never heard,
Ne'er sighed at the sound of a knell,
Or smiled when a sabbath appeared.

"Ye winds, that have made me your sport
Convey to this desolate shore
Some cordial endearing report
Of a land I must visit no more.
My Friends, do they now and then send
A wish or a thought after me?
O tell me I yet have a friend.
Though a friend I am never to see."

This passage is quoted as an instance of three different styles of composition. The first four lines are poorly expressed; some Critics would call the language prosaic; the fact is, it would be bad prose, so bad, that it is scarcely worse in metre. The epithet "church-going" applied to a bell, and that by so chaste a writer as Cowper, is an instance of the strange abuses which Poets have introduced into their language, till they and their Readers take them as matters of course, if they do not single them out expressly as objects of admiration. The two lines "Ne'er sighed at the sound," &c., are, in my opinion, an instance of the language of passion wrested from its proper use, and, from the mere circumstance of the composition being in metre, applied upon an occasion that does not justify such violent

APPENDIX TO PREFACE, SECOND EDITION. **1. Alexander Selkirk:** the prototype of Robinson Crusoe, marooned from 1704 to 1709 on Juan Fernandez Island in the Pacific Ocean.

expressions; and I should condemn the passage, though perhaps few Readers will agree with me, as vicious poetic diction. The last stanza is throughout admirably expressed: it would be equally good whether in prose or verse, except that the Reader has an exquisite pleasure in seeing such natural language so naturally connected with metre. The beauty of this stanza tempts me to conclude with a principle which ought never to be lost sight of, and which has been my chief guide in all I have said,—namely, that in works of *imagination and sentiment*, for of these only have I been treating, in proportion as ideas and feelings are valuable, whether the composition be in prose or in verse, they require and exact one and the same language. Metre is but adventitious to composition, and the phraseology for which that passport is necessary, even where it may be graceful at all, will be little valued by the judicious.

from PREFACE
TO THE EDITION OF 1815

This was the Preface to the *Poems* of 1815 (a collected edition of Wordsworth's poems).

. . . Imagination, in the sense of the word as giving title to a class[1] of the following Poems, has no reference to images that are merely a faithful copy, existing in the mind, of absent external objects; but is a word of higher import, denoting operations of the mind upon those objects, and processes of creation or of composition, governed by certain fixed laws. I proceed to illustrate my meaning by instances. A parrot *hangs* from the wires of his cage by his beak or by his claws; or a monkey from the bough of a tree by his paws or his tail. Each creature does so literally and actually. In the first Eclogue of Virgil, the Shepherd, thinking of the time when he is to take leave of his farm, thus addresses his goats:—

> Non ego vos posthac viridi projectus in antro
> Dumosa *pendere* procul de rupe videbo.[2]

> half way up
> *Hangs* one who gathers samphire,[3]

is the well-known expression of Shakespeare delineating an ordinary image upon the cliffs of Dover. In these two instances is a slight exertion of the faculty which I denominate imagination, in the use of one word: neither the goats nor the samphire-gatherer do literally hang, as does the parrot or the monkey; but, presenting to the senses something of such an appearance, the mind in its activity, for its own gratification, contemplates them as hanging.

> As when far off at Sea a Fleet descried
> *Hangs* in the clouds, by equinoctial winds
> Close sailing from Bengala or the Isles
> Of Ternate or Tydore, whence Merchants bring
> Their spicy drugs; they on the trading flood
> Through the wide Ethiopian to the Cape
> Ply, stemming nightly toward the Pole: so seem'd
> Far off the flying Fiend.[4]

Here is the full strength of the imagination involved in the word, *hangs*, and exerted upon the whole image: First, the fleet, an aggregate of many ships, is represented as one mighty person, whose track, we know and feel, is upon the waters; but, taking advantage of its appearance to the senses, the Poet dares to represent it as *hanging in the clouds*, both for the gratification of the mind in contemplating the image itself, and in reference to the motion and appearance of the sublime object to which it is compared.

From images of sight we will pass to those of sound:

> Over his own sweet voice the Stock-dove *broods;*[5]

of the same bird,

> His voice was *buried* among trees,
> Yet to be come at by the breeze;[6]

> O, Cuckoo! shall I call thee *Bird*,
> Or but a wandering *Voice?*[7]

PREFACE TO THE EDITION OF 1815. **1.** In this and subsequent editions, Wordsworth classified and grouped his poems according to a complex scheme. The categories referred sometimes to subject matter and sometimes to the faculty of the mind that Wordsworth thought predominant in the writing of the poem. Thus, one group is entitled "Poems of the Imagination."

2. Non . . . videbo: lines 76–77: "Never again shall I, lying in some mossy cave, see you in the far distance hanging from a bushy crag." **3. half . . . samphire:** *King Lear*, IV. vi. 16. **4. As . . . Fiend:** *Paradise Lost*, II. 636–43. **5. Over . . . broods:** "Resolution and Independence," l. 5. **6. His . . . breeze:** "O Nightingale! Thou Surely Art," ll. 13–14. **7. O . . . Voice:** "To the Cuckoo," ll. 3–4.

The stock-dove is said to *coo*, a sound well imitating the note of the bird; but, by the intervention of the metaphor *broods*, the affections are called in by the imagination to assist in marking the manner in which the bird reiterates and prolongs her soft note, as if herself delighting to listen to it, and participating of a still and quiet satisfaction, like that which may be supposed inseparable from the continuous process of incubation. "His voice was buried among trees," a metaphor expressing the love of *seclusion* by which this Bird is marked; and characterizing its note as not partaking of the shrill and the piercing, and therefore more easily deadened by the intervening shade; yet a note so peculiar, and withal so pleasing, that the breeze, gifted with that love of the sound which the Poet feels, penetrates the shade in which it is entombed, and conveys it to the ear of the listener.

> Shall I call thee Bird
> Or but a wandering Voice?

This concise interrogation characterizes the seeming ubiquity of the voice of the cuckoo, and dispossesses the creature almost of a corporeal existence; the Imagination being tempted to this exertion of her power by a consciousness in the memory that the cuckoo is almost perpetually heard throughout the season of spring, but seldom becomes an object of sight.

Thus far of images independent of each other, and immediately endowed by the mind with properties that do not inhere in them, upon an incitement from properties and qualities the existence of which is inherent and obvious. These processes of imagination are carried on either by conferring additional properties upon an object, or abstracting from it some of those which it actually possesses, and thus enabling it to react upon the mind which hath performed the process, like a new existence.

I pass from the Imagination acting upon an individual image to a consideration of the same faculty employed upon images in a conjunction by which they modify each other. The Reader has already had a fine instance before him in the passage quoted from Virgil, where the apparently perilous situation of the goat, hanging upon the shaggy precipice, is contrasted with that of the shepherd, contemplating it from the seclusion of the cavern in which he lies stretched at ease and in security. Take these images separately, and how unaffecting the picture compared with that produced by their being thus connected with, and opposed to, each other!

> As a huge Stone is sometimes seen to lie
> Couched on the bald top of an eminence,
> Wonder to all who do the same espy
> By what means it could thither come, and whence;
> So that it seems a thing endued with sense,
> Like a sea-beast crawled forth, which on a shelf
> Of rock or sand reposeth, there to sun himself.

> Such seemed this Man; not all alive or dead,
> Nor all asleep, in his extreme old age.

> . . . .

> Motionless as a cloud the old Man stood,
> That heareth not the loud winds when they call,
> And moveth altogether if it move at all.[8]

In these images, the conferring, the abstracting, and the modifying powers of the Imagination, immediately and mediately acting, are all brought into conjunction. The stone is endowed with something of the power of life to approximate it to the sea-beast; and the sea-beast stripped of some of its vital qualities to assimilate it to the stone; which intermediate image is thus treated for the purpose of bringing the original image, that of the stone, to a nearer resemblance to the figure and condition of the aged Man; who is divested of so much of the indications of life and motion as to bring him to the point where the two objects unite and coalesce in just comparison. After what has been said, the image of the cloud need not be commented upon.

Thus far of an endowing or modifying power: but the Imagination also shapes and *creates;* and how? By innumerable processes; and in none does it more delight than in that of consolidating numbers into unity, and dissolving and separating unity into number, —alternations proceeding from, and governed by, a sublime consciousness of the soul in her own mighty and almost divine powers. Recur to the passage already cited from Milton. When the compact Fleet, as one Person, has been introduced "Sailing from Bengala," "They," *i.e.*, the "merchants," representing the fleet resolved into a multitude of ships, "ply" their voyage towards the extremities of the earth: "So" (referring to the word "As" in the commencement) "seemed the flying Fiend"; the image of his person acting to recombine the multitude of ships into one body,—the point from which the comparison set out. "So seemed," and to whom seemed? To the heavenly Muse who dictates the poem, to the eye of

8. **As . . . all:** "Resolution and Independence," ll. 57–65, 75–77.

the Poet's mind, and to that of the Reader, present at one moment in the wide Ethiopian, and the next in the solitudes, then first broken in upon, of the infernal regions!

Modo me Thebis, modo ponit Athenis.[9]

Hear again this mighty poet,—speaking of the Messiah going forth to expel from Heaven the rebellious angels,

> Attended by ten thousand, thousand Saints
> He onward came: far off his coming shone,— [10]

the retinue of saints, and the person of the Messiah himself, lost almost and merged in the splendour of the indefinite abstraction, "His coming!"

As I do not mean here to treat this subject further than to throw some light upon the present volumes, and especially upon one division of them, I shall spare myself and the reader the trouble of considering the Imagination as it deals with thoughts and sentiments, as it regulates the composition of characters, and determines the course of actions: I will not consider it (more than I have already done by implication) as that power which, in the language of one of my most esteemed Friends, "draws all things to one, which makes things animate or inanimate, beings with their attributes, subjects with their accessaries, take one colour and serve to one effect." [11] The grand store-house of enthusiastic and meditative Imagination, of poetical, as contradistinguished from human and dramatic Imagination, is the prophetic and lyrical parts of the Holy Scriptures, and the works of Milton, to which I cannot forbear to add those of Spenser. I select these writers in preference to those of ancient Greece and Rome because the anthropomorphitism of the Pagan religion subjected the minds of the greatest poets in those countries too much to the bondage of definite form; from which the Hebrews were preserved by their abhorrence of idolatry. This abhorrence was almost as strong in our great epic Poet, both from circumstances of his life, and from the constitution of his mind. However imbued the surface might be with classical literature, he was a Hebrew in soul; and all things tended in him towards the sublime. Spenser, of a gentler nature, maintained his freedom by aid of his allegorical spirit, at one time inciting him to create persons out of abstractions; and, at another, by a superior effort of genius, to give the universality and permanence of abstractions to his human beings, by means of attributes and emblems that belong to the highest moral truths and the purest sensations,—of which his character of Una is a glorious example. Of the human and dramatic imagination the works of Shakespeare are an inexhaustible source.

> I tax not you, ye elements, with unkindness,
> I never gave you kingdoms, called you daughters.[12]

And if, bearing in mind the many Poets distinguished by this prime quality, whose names I omit to mention; yet justified by recollection of the insults which the ignorant, the incapable and the presumptuous have heaped upon these and my other writings, I may be permitted to anticipate the judgement of posterity upon myself, I shall declare (censurable, I grant, if the notoriety of the fact above stated does not justify me) that I have given, in these unfavourable times, evidence of exertions of this faculty upon its worthiest objects, the external universe, the moral and religious sentiments of Man, his natural affections, and his acquired passions; which have the same ennobling tendency as the productions of men in this kind, worthy to be holden in undying remembrance.

ESSAY SUPPLEMENTARY
TO THE PREFACE
OF 1815

With the young of both sexes, Poetry is, like love, a passion; but, for much the greater part of those who have been proud of its power over their minds, a necessity soon arises of breaking the pleasing bondage; or it relaxes of itself;—the thoughts being occupied in domestic cares, or the time engrossed by business. Poetry then becomes only an occasional recreation; while to those whose existence passes away in a course of fashionable pleasure, it is a species of luxurious amusement. In middle and declining age, a scattered number of serious persons resort to poetry, as to religion, for a protection against the pressure of trivial employments, and as a consolation for the afflictions of life. And, lastly, there are many, who, having been

---

9. Modo . . . Athenis: Horace, *Epistles*, II. 213: "A poet can put you in Thebes, or Athens, or wherever he wishes." 10. Attended . . . shone: *Paradise Lost*, VI. 767–68. 11. "draws . . . effect": "Charles Lamb upon the genius of Hogarth" (W. W.).

12. I . . . daughters: *King Lear*, III. ii. 16–17.

enamoured of this art in their youth, have found leisure, after youth was spent, to cultivate general literature; in which poetry has continued to be comprehended *as a study.*

Into the above classes the Readers of poetry may be divided; Critics abound in them all; but from the last only can opinions be collected of absolute value, and worthy to be depended upon, as prophetic of the destiny of a new work. The young, who in nothing can escape delusion, are especially subject to it in their intercourse with Poetry. The cause, not so obvious as the fact is unquestionable, is the same as that from which erroneous judgments in this art, in the minds of men of all ages, chiefly proceed; but upon Youth it operates with peculiar force. The appropriate business of poetry, (which, nevertheless, if genuine, is as permanent as pure science,) her appropriate employment, her privilege and her *duty*, is to treat of things not as they *are*, but as they *appear*: not as they exist in themselves, but as they *seem* to exist to the *senses*, and to the *passions*. What a world of delusion does this acknowledged obligation prepare for the inexperienced! what temptations to go astray are here held forth for them whose thoughts have been little disciplined by the understanding, and whose feelings revolt from the sway of reason!—When a juvenile Reader is in the height of his rapture with some vicious passage, should experience throw in doubts, or common-sense suggest suspicions, a lurking consciousness that the realities of the Muse are but shows, and that her liveliest excitements are raised by transient shocks of conflicting feeling and successive assemblages of contradictory thoughts—is ever at hand to justify extravagance, and to sanction absurdity. But, it may be asked, as these illusions are unavoidable, and, no doubt, eminently useful to the mind as a process, what good can be gained by making observations, the tendency of which is to diminish the confidence of youth in its feelings, and thus to abridge its innocent and even profitable pleasures? The reproach implied in the question could not be warded off, if Youth were incapable of being delighted with what is truly excellent; or, if these errors always terminated of themselves in due season. But, with the majority, though their force be abated, they continue through life. Moreover, the fire of youth is too vivacious an element to be extinguished or damped by a philosophical remark; and, while there is no danger that what has been said will be injurious or painful to the ardent and the confident, it may prove beneficial to those who, being

enthusiastic, are, at the same time, modest and ingenuous. The intimation may unite with their own misgivings to regulate their sensibility, and to bring in, sooner than it would otherwise have arrived, a more discreet and sound judgment.

If it should excite wonder that men of ability, in later life, whose understandings have been rendered acute by practice in affairs, should be so easily and so far imposed upon when they happen to take up a new work in verse, this appears to be the cause;—that, having discontinued their attention to poetry, whatever progress may have been made in other departments of knowledge, they have not, as to this art, advanced in true discernment beyond the age of youth. If, then, a new poem fall in their way, whose attractions are of that kind which would have enraptured them during the heat of youth, the judgment not being improved to a degree that they shall be disgusted, they are dazzled; and prize and cherish the faults for having had power to make the present time vanish before them, and to throw the mind back, as by enchantment, into the happiest season of life. As they read, powers seem to be revived, passions are regenerated, and pleasures restored. The Book was probably taken up after an escape from the burden of business, and with a wish to forget the world, and all its vexations and anxieties. Having obtained this wish, and so much more, it is natural that they should make report as they have felt.

If Men of mature age, through want of practice, be thus easily beguiled into admiration of absurdities, extravagances, and misplaced ornaments, thinking it proper that their understandings should enjoy a holiday, while they are unbending their minds with verse, it may be expected that such Readers will resemble their former selves also in strength of prejudice, and an inaptitude to be moved by the unostentatious beauties of a pure style. In the higher poetry, an enlightened Critic chiefly looks for a reflection of the wisdom of the heart and the grandeur of the imagination. Wherever these appear, simplicity accompanies them; Magnificence herself when legitimate, depending upon a simplicity of her own, to regulate her ornaments. But it is a well-known property of human nature, that our estimates are ever governed by comparisons, of which we are conscious with various degrees of distinctness. Is it not, then, inevitable (confining these observations to the effects of style merely) that an eye, accustomed to the glaring hues of diction by which such Readers are caught and excited, will for the most part be rather repelled than attracted by an original Work, the colouring of which is disposed according to a pure and refined scheme of harmony? It is in the fine arts as in the affairs of life, no man can *serve* (*i.e.* obey with zeal and fidelity) two Masters.

As Poetry is most just to its own divine origin when it administers the comforts and breathes the spirit of religion, they who have learned to perceive this truth, and who betake themselves to reading verse for sacred purposes, must be preserved from numerous illusions to which the two Classes of Readers, whom we have been considering, are liable. But, as the mind grows serious from the weight of life, the range of its passions is contracted accordingly; and its sympathies become so exclusive, that many species of high excellence wholly escape, or but languidly excite, its notice. Besides, men who read from religious or moral inclinations, even when the subject is of that kind which they approve, are beset with misconceptions and mistakes peculiar to themselves. Attaching so much importance to the truths which interest them, they are prone to overrate the Authors by whom those truths are expressed and enforced. They come prepared to impart so much passion to the Poet's language, that they remain unconscious how little, in fact, they receive from it. And, on the other hand, religious faith is to him who holds it so momentous a thing, and error appears to be attended with such tremendous consequences, that, if opinions touching upon religion occur which the Reader condemns, he not only cannot sympathise with them, however animated the expression, but there is, for the most part, an end put to all satisfaction and enjoyment. Love, if it before existed, is converted into dislike; and the heart of the Reader is set against the Author and his book.—To these excesses, they, who from their professions ought to be the most guarded against them, are perhaps the most liable; I mean those sects whose religion, being from the calculating understanding, is cold and formal. For when Christianity, the religion of humility, is founded upon the proudest faculty of our nature, what can be expected but contradictions? Accordingly, believers of this cast are at one time contemptuous; at another, being troubled, as they are and must be, with inward misgivings, they are jealous and suspicious;—and at all seasons, they are under temptation to supply by the heat with which they defend their tenets, the animation which is wanting to the constitution of the religion itself.

Faith was given to man that his affections, detached from the treasures of time, might be inclined to settle upon those of eternity;—the elevation of his nature, which this habit produces on earth, being to him a presumptive evidence of a future state of existence; and giving him a title to partake of its holiness. The religious man values what he sees chiefly as an "imperfect shadowing forth" of what he is incapable of seeing. The concerns of religion refer to indefinite objects, and are too weighty for the mind to support them without relieving itself by resting a great part of the burthen upon words and symbols. The commerce between Man and his Maker cannot be carried on but by a process where much is represented in little, and the Infinite Being accommodates himself to a finite capacity. In all this may be perceived the affinity between religion and poetry; between religion—making up the deficiencies of reason by faith; and poetry—passionate for the instruction of reason; between religion—whose element is infinitude, and whose ultimate trust is the supreme of things, submitting herself to circumscription, and reconciled to substitutions; and poetry—ethereal and transcendent, yet incapable to sustain her existence without sensuous incarnation. In this community of nature may be perceived also the lurking incitements of kindred error;—so that we shall find that no poetry has been more subject to distortion, than that species, the argument and scope of which is religious; and no lovers of the art have gone farther astray than the pious and the devout.

Whither then shall we turn for that union of qualifications which must necessarily exist before the decisions of a critic can be of absolute value? For a mind at once poetical and philosophical; for a critic whose affections are as free and kindly as the spirit of society, and whose understanding is severe as that of dispassionate government? Where are we to look for that initiatory composure of mind which no selfishness can disturb? For a natural sensibility that has been tutored into correctness without losing anything of its quickness; and for active faculties, capable of answering the demands which an Author of original imagination shall make upon them, associated with a judgment that cannot be duped into admiration by aught that is worthy of it?—among those and those only, who, never having suffered their youthful love of poetry to remit much of its force, have applied to the consideration of the laws of this art the best power of their understandings. At the same time it must be observed—that, as this Class comprehends the only judgments which are trust-worthy, so does it include the most erroneous and perverse. For to be mistaught is worse than to be untaught; and no perverseness equals that which is supported by system, no errors are so difficult to root out as those which the understanding has pledged its credit to uphold. In this Class are contained censors, who, if they be pleased with what is good, are pleased with it only by imperfect glimpses, and upon false principles; who, should they generalise rightly, to a certain point, are sure to suffer for it in the end; who, if they stumble upon a sound rule, are fettered by misapplying it, or by straining it too far; being incapable of perceiving when it ought to yield to one of higher order. In it are found critics too petulant to be passive to a genuine poet, and too feeble to grapple with him; men, who take upon them to report of the course which *he* holds whom they are utterly unable to accompany,—confounded if he turn quick upon the wing, dismayed if he soar steadily "into the region;"—men of palsied imaginations and indurated hearts; in whose minds all healthy action is languid, who therefore feed as the many direct them, or, with the many, are greedy after vicious provocatives;—judges, whose censure is auspicious, and whose praise ominous! In this class meet together the two extremes of best and worst.

The observations presented in the foregoing series are of too ungracious a nature to have been made without reluctance; and, were it only on this account, I would invite the reader to try them by the test of comprehensive experience. If the number of judges who can be confidently relied upon be in reality so small, it ought to follow that partial notice only, or neglect, perhaps long continued, or attention wholly inadequate to their merits—must have been the fate of most works in the higher departments of poetry; and that, on the other hand, numerous productions have blazed into popularity, and have passed away, leaving scarcely a trace behind them: it will be further found, that when Authors shall have at length raised themselves into general admiration and maintained their ground, errors and prejudices have prevailed concerning their genius and their works, which the few who are conscious of those errors and prejudices would deplore; if they were not recompensed by perceiving that there are select Spirits for whom it is ordained that their fame shall be in the world an existence like that of Virtue, which owes its being to the struggles it makes, and its vigour to the enemies whom it provokes; —a vivacious quality, ever doomed to meet with opposition, and still triumphing over it; and, from the

nature of its dominion, incapable of being brought to the sad conclusion of Alexander, when he wept that there were no more worlds for him to conquer.

Let us take a hasty retrospect of the poetical literature of this Country for the greater part of the last two centuries, and see if the facts support these inferences.

Who is there that now reads the "Creation" of Dubartas?[1] Yet all Europe once resounded with his praise; he was caressed by kings; and, when his Poem was translated into our language, the "Faery Queen" faded before it. The name of Spenser, whose genius is of a higher order than even that of Ariosto, is at this day scarcely known beyond the limits of the British Isles. And if the value of his works is to be estimated from the attention now paid to them by his countrymen, compared with that which they bestow on those of some other writers, it must be pronounced small indeed.

"The laurel, meed of mighty conquerors
And poets *sage*"—

are his own words; but his wisdom has, in this particular, been his worst enemy: while its opposite, whether in the shape of folly or madness, has been *their* best friend. But he was a great power, and bears a high name: the laurel has been awarded to him.

A dramatic Author, if he write for the stage, must adapt himself to the taste of the audience, or they will not endure him; accordingly the mighty genius of Shakspeare was listened to. The people were delighted: but I am not sufficiently versed in stage antiquities to determine whether they did not flock as eagerly to the representation of many pieces of contemporary Authors, wholly undeserving to appear upon the same boards. Had there been a formal contest for superiority among dramatic writers, that Shakspeare, like his predecessors Sophocles and Euripides, would have often been subject to the mortification of seeing the prize adjudged to sorry competitors, becomes too probable, when we reflect that the admirers of Settle and Shadwell[2]

ESSAY SUPPLEMENTARY TO THE PREFACE OF 1815. **1. Dubartas:** Guillaume de Salluste, Seigneur Du Bartas, 1544–90, French Huguenot poet. *La Semaine*, published in 1578 and afterwards translated into English, is an epic on the creation of the world. **2. Settle and Shadwell:** Elkanah Settle, 1648–1724, author of *The Empress of Morocco* and other inflated dramas in the heroic style, and Thomas Shadwell, 1642?–92, author of dramatic works on contemporary life. Both were enemies and objects of satire to Dryden.

were, in a later age, as numerous, and reckoned as respectable in point of talent, as those of Dryden. At all events, that Shakspeare stooped to accommodate himself to the People, is sufficiently apparent; and one of the most striking proofs of his almost omnipotent genius, is, that he could turn to such glorious purpose those materials which the prepossessions of the age compelled him to make use of. Yet even this marvellous skill appears not to have been enough to prevent his rivals from having some advantage over him in public estimation; else how can we account for passages and scenes that exist in his works, unless upon a supposition that some of the grossest of them, a fact which in my own mind I have no doubt of, were foisted in by the Players, for the gratification of the many?

But that his Works, whatever might be their reception upon the stage, made but little impression upon the ruling Intellects of the time, may be inferred from the fact that Lord Bacon, in his multifarious writings, nowhere either quotes or alludes to him. His dramatic excellence enabled him to resume possession of the stage after the Restoration; but Dryden tells us that in his time two of the plays of Beaumont and Fletcher were acted for one of Shakspeare's. And so faint and limited was the perception of the poetic beauties of his dramas in the time of Pope, that, in his Edition of the Plays, with a view of rendering to the general reader a necessary service, he printed between inverted commas those passages which he thought most worthy of notice.

At this day, the French Critics have abated nothing of their aversion to this darling of our Nation: "the English, with their bouffon de Shakspeare,"[3] is as familiar an expression among them as in the time of Voltaire. Baron Grimm[4] is the only French writer who seems to have perceived his infinite superiority to the first names of the French Theatre; an advantage which the Parisian Critic owed to his German blood and German education. The most enlightened Italians, though well acquainted with our language, are wholly incompetent to measure the proportions of Shakspeare. The Germans only, of foreign nations, are approaching towards a knowledge and feeling of what he is. In some respects they have acquired a superiority over the fellow-countrymen of the Poet: for among us it is a

current, I might say, an established opinion, that Shakspeare is justly praised when he is pronounced to be "a wild irregular genius, in whom great faults are compensated by great beauties." How long may it be before this misconception passes away, and it becomes universally acknowledged that the judgment of Shakspeare in the selection of his materials, and in the manner in which he has made them, heterogeneous as they often are, constitute a unity of their own, and contribute all to one great end, is not less admirable than his imagination, his invention, and his intuitive knowledge of human Nature?

There is extant a small Volume of miscellaneous poems, in which Shakspeare expresses his own feelings in his own person. It is not difficult to conceive that the Editor, George Steevens,[5] should have been insensible to the beauties of one portion of that Volume, the Sonnets; though in no part of the writings of this Poet is found, in an equal compass, a greater number of exquisite feelings felicitously expressed. But, from regard to the Critic's own credit, he would not have ventured to talk of an act of parliament not being strong enough to compel the perusal of those little pieces, if he had not known that the people of England were ignorant of the treasures contained in them: and if he had not, moreover, shared the too common propensity of human nature to exult over a supposed fall into the mire of a genius whom he had been compelled to regard with admiration, as an inmate of the celestial regions— "there sitting where he durst not soar."[6]

Nine years before the death of Shakspeare, Milton was born; and early in life he published several small poems, which, though on their first appearance they were praised by a few of the judicious, were afterwards neglected to that degree, that Pope in his youth could borrow from them without risk of its being known. Whether these poems are at this day justly appreciated, I will not undertake to decide: nor would it imply a severe reflection upon the mass of readers to suppose the contrary; seeing that a man of the acknowledged genius of Voss,[7] the German poet, could suffer their spirit to evaporate; and could change their character, as is done in the translation made by him of the most popular of those pieces. At all events, it is certain that these Poems of Milton are now much read, and loudly

3. bouffon de Shakspeare: buffoon Shakespeare. 4. Baron Grimm: Friedrich Melchior, Baron von Grimm, 1723–1807, lived in Paris and carried on a *Literary Correspondence*— comprising descriptions of new books and opinions on literature and politics—with various German rulers who subscribed to his letters. They were not published until 1812.

5. George Steevens: 1736–1800. His edition of Shakespeare's plays appeared in 1773, and an additional volume included the Poems in 1780. 6. "there . . . soar": *Paradise Lost*, IV. 839. 7. Voss: Johan Heinrich Voss, 1751–1826.

praised; yet were they little heard of till more than 150 years after their publication; and of the Sonnets, Dr. Johnson, as appears from Boswell's Life of him, was in the habit of thinking and speaking as contemptuously as Steevens wrote upon those of Shakspeare.

About the time when the Pindaric odes of Cowley and his imitators, and the productions of that class of curious thinkers whom Dr. Johnson has strangely styled metaphysical Poets, were beginning to lose something of that extravagant admiration which they had excited, the "Paradise Lost" made its appearance. "Fit audience find though few," was the petition addressed by the Poet to his inspiring Muse. I have said elsewhere that he gained more than he asked; this I believe to be true; but Dr. Johnson has fallen into a gross mistake when he attempts to prove, by the sale of the work, that Milton's Countrymen were *"just to it"* upon its first appearance. Thirteen hundred Copies were sold in two years; an uncommon example, he asserts, of the prevalence of genius in opposition to so much recent enmity as Milton's public conduct had excited. But, be it remembered that, if Milton's political and religious opinions, and the manner in which he announced them, had raised him many enemies, they had procured him numerous friends; who, as all personal danger was passed away at the time of publication, would be eager to procure the master-work of a man whom they revered, and whom they would be proud of praising. Take, from the number of purchasers, persons of this class, and also those who wished to possess the Poem as a religious work, and but few I fear would be left who sought for it on account of its poetical merits. The demand did not immediately increase; "for," says Dr. Johnson, "many more readers" (he means persons in the habit of reading poetry) "than were supplied at first the Nation did not afford." How careless must a writer be who can make this assertion in the face of so many existing title-pages to belie it! Turning to my own shelves, I find the folio of Cowley, seventh edition, 1681. A book near it is Flatman's Poems,[8] fourth edition, 1686; Waller, fifth edition, same date. The Poems of Norris[9] of Bemerton not long after went, I believe, through nine editions. What further demand there might be for these works I do not know; but I well remember, that, twenty-five

years ago, the booksellers' stalls in London swarmed with the folios of Cowley. This is not mentioned in disparagement of that able writer and amiable man; but merely to show—that, if Milton's work were not more read, it was not because readers did not exist at the time. The early editions of the "Paradise Lost" were printed in a shape which allowed them to be sold at a low price, yet only three thousand copies of the Work were sold in eleven years; and the Nation, says Dr. Johnson, had been satisfied from 1623 to 1664, that is, forty-one years, with only two editions of the Works of Shakspeare; which probably did not together make one thousand Copies; facts adduced by the critic to prove the "paucity of Readers."—There were readers in multitudes; but their money went for other purposes, as their admiration was fixed elsewhere. We are authorised, then, to affirm, that the reception of the "Paradise Lost," and the slow progress of its fame, are proofs as striking as can be desired that the positions which I am attempting to establish are not erroneous.—How amusing to shape to one's self such a critique as a Wit of Charles's days, or a Lord of the Miscellanies or trading Journalist of King William's time, would have brought forth, if he had set his faculties industriously to work upon this Poem, everywhere impregnated with *original* excellence.

So strange indeed are the obliquities of admiration, that they whose opinions are much influenced by authority will often be tempted to think that there are no fixed principles[10] in human nature for this art to rest upon. I have been honoured by being permitted to peruse in MS. a tract composed between the period of the Revolution and the close of that century. It is the Work of an English Peer of high accomplishments, its object to form the character and direct the studies of his son. Perhaps nowhere does a more beautiful treatise of the kind exist. The good sense and wisdom of the thoughts, the delicacy of the feelings, and the charm of the style, are, throughout, equally conspicuous. Yet the Author, selecting among the Poets of his own country those whom he deems most worthy of his son's perusal, particularises only Lord Rochester, Sir John Denham, and Cowley. Writing about the same time, Shaftesbury, an author at present unjustly depreciated, describes the English Muses as only yet lisping in their cradles.

---

8. **Flatman's Poems:** Thomas Flatman, 1637–88, author of *Poems and Songs* (1674). **9. Norris:** John Norris, 1657–1711, Rector of Bemerton. His *Collection of Miscellanies* (1687) went through five editions by 1710.

10. "This opinion seems actually to have been entertained by Adam Smith, the worst critic, David Hume not excepted, that Scotland, a soil to which this sort of weed seems natural, has produced" (W. W.).

The arts by which Pope, soon afterwards, contrived to procure to himself a more general and a higher reputation than perhaps any English Poet ever attained during his life-time, are known to the judicious. And as well known is it to them, that the undue exertion of those arts is the cause why Pope has for some time held a rank in literature, to which, if he had not been seduced by an over-love of immediate popularity, and had confided more in his native genius, he never could have descended. He bewitched the nation by his melody, and dazzled it by his polished style, and was himself blinded by his own success. Having wandered from humanity in his Eclogues with boyish inexperience, the praise, which these compositions obtained, tempted him into a belief that Nature was not to be trusted, at least in pastoral Poetry. To prove this by example, he put his friend Gay upon writing those Eclogues which their author intended to be burlesque. The instigator of the work, and his admirers, could perceive in them nothing but what was ridiculous. Nevertheless, though these Poems contain some detestable passages, the effect, as Dr. Johnson well observes, "of reality and truth became conspicuous even when the intention was to show them grovelling and degraded." The Pastorals, ludicrous to such as prided themselves upon their refinement, in spite of those disgusting passages, "became popular, and were read with delight, as just representations of rural manners and occupations."

Something less than sixty years after the publication of the "Paradise Lost" appeared Thomson's "Winter;" which was speedily followed by his other Seasons. It is a work of inspiration; much of it is written from himself, and nobly from himself. How was it received? "It was no sooner read," says one of his contemporary biographers,[11] "than universally admired: those only excepted who had not been used to feel, or to look for anything in poetry, beyond a *point* of satirical or epigrammatic wit, a smart *antithesis* richly trimmed with rhyme, or the softness of an *elegiac* complaint. To such his manly classical spirit could not readily commend itself; till, after a more attentive perusal, they had got the better of their prejudices, and either acquired or affected a truer taste. A few others stood aloof, merely because they had long before fixed the articles of their poetical creed, and resigned themselves to an absolute despair of ever seeing anything new and

original. These were somewhat mortified to find their notions disturbed by the appearance of a poet, who seemed to owe nothing but to nature and his own genius. But, in a short time, the applause became unanimous; every one wondering how so many pictures, and pictures so familiar, should have moved them but faintly to what they felt in his descriptions. His digressions too, the overflowings of a tender benevolent heart, charmed the reader no less; leaving him in doubt, whether he should more admire the Poet or love the Man."

This case appears to bear strongly against us:—but we must distinguish between wonder and legitimate admiration. The subject of the work is the changes produced in the appearances of nature by the revolution of the year: and, by undertaking to write in verse, Thomson pledged himself to treat his subject as became a Poet. Now, it is remarkable that, excepting the nocturnal Reverie of Lady Winchilsea,[12] and a passage or two in the "Windsor Forest" of Pope, the poetry of the period intervening between the publication of the "Paradise Lost" and the "Seasons" does not contain a single new image of external nature; and scarcely presents a familiar one from which it can be inferred that the eye of the Poet had been steadily fixed upon his object, much less that his feelings had urged him to work upon it in the spirit of genuine imagination. To what a low state knowledge of the most obvious and important phenomena had sunk, is evident from the style in which Dryden has executed a description of Night in one of his Tragedies, and Pope his translation of the celebrated moonlight scene in the "Iliad." A blind man, in the habit of attending accurately to descriptions casually dropped from the lips of those around him, might easily depict these appearances with more truth. Dryden's lines are vague, bombastic, and senseless; those of Pope, though he had Homer to guide him, are throughout false and contradictory. The verses of Dryden, once highly celebrated, are forgotten; those of Pope still retain their hold upon public estimation,—nay, there is not a passage of descriptive poetry, which at this day finds so many and such ardent admirers. Strange to think of an enthusiast, as may have been the case with thousands, reciting those verses under the cope of a moonlight sky, without having his raptures in the least disturbed by a

11. one . . . biographers: Patrick Murdoch, who died in 1774, a friend of Thomson. His brief life of the poet was included in the memorial edition of the *Works* (1762), and in most later editions of the *Seasons*.

12. nocturnal . . . Winchilsea: Anne Finch, Countess of Winchilsea, 1661–1720. Her "Nocturnal Reverie" was published in 1713.

suspicion of their absurdity!—If these two distinguished writers could habitually think that the visible universe was of so little consequence to a poet, that it was scarcely necessary for him to cast his eyes upon it, we may be assured that those passages of the elder poets which faithfully and poetically describe the phenomena of nature, were not at that time holden in much estimation, and that there was little accurate attention paid to those appearances.

Wonder is the natural product of Ignorance; and as the soil was *in such good condition* at the time of the publication of the "Seasons," the crop was doubtless abundant. Neither individuals nor nations become corrupt all at once, nor are they enlightened in a moment. Thomson was an inspired poet, but he could not work miracles; in cases where the art of seeing had in some degree been learned, the teacher would further the proficiency of his pupils, but he could do little *more:* though so far does vanity assist men in acts of self-deception, that many would often fancy they recognized a likeness when they knew nothing of the original. Having shown that much of what his biographer deemed genuine admiration must in fact have been blind wonderment—how is the rest to be accounted for?—Thomson was fortunate in the very title of his poem, which seemed to bring it home to the prepared sympathies of every one: in the next place, notwithstanding his high powers, he writes a vicious style; and his false ornaments are exactly of that kind which would be most likely to strike the undiscerning. He likewise abounds with sentimental common-places, that, from the manner in which they were brought forward, bore an imposing air of novelty. In any well-used copy of the "Seasons" the book generally opens of itself with the rhapsody on love, or with one of the stories (perhaps "Damon and Musidora"[13]); these also are prominent in our collections of Extracts, and are the parts of his Work which, after all, were probably most efficient in first recommending the author to general notice. Pope, repaying praises which he had received, and wishing to extol him to the highest, only styles him, "an elegant and philosophical Poet;" nor are we able to collect any unquestionable proofs that the true characteristics of Thomson's genius as an imaginative poet were perceived, till the elder Warton,[14] almost forty years after the publication of the "Seasons," pointed them out by a note in his Essay on

the "Life and Writings of Pope." In the "Castle of Indolence" (of which Gray speaks so coldly) these characteristics were almost as conspicuously displayed, and in verse more harmonious, and diction more pure. Yet that fine poem was neglected on its appearance, and is at this day the delight only of a few!

When Thomson died, Collins breathed forth his regrets in an Elegiac Poem, in which he pronounces a poetical curse upon *him* who should regard with insensibility the place where the Poet's remains were deposited. The Poems of the mourner himself have now passed through innumerable editions, and are universally known; but if, when Collins died, the same kind of imprecation had been pronounced by a surviving admirer, small is the number whom it would not have comprehended. The notice which his poems attained during his lifetime was so small, and of course the sale so insignificant, that not long before his death he deemed it right to repay to the bookseller the sum which he had advanced for them, and threw the edition into the fire.

Next in importance to the "Seasons" of Thomson, though at considerable distance from that work in order of time, come the "Reliques of Ancient English Poetry;" collected, new-modelled, and in many instances (if such a contradiction in terms may be used) composed by the Editor, Dr. Percy.[15] This work did not steal silently into the world, as is evident from the number of legendary tales, that appeared not long after its publication; and had been modelled, as the authors persuaded themselves, after the old Ballad. The Compilation was however ill suited to the then existing taste of city society; and Dr. Johnson, 'mid the little senate to which he gave laws, was not sparing in his exertions to make it an object of contempt. The critic triumphed, the legendary imitators were deservedly disregarded, and, as undeservedly, their ill-imitated models sank, in this country, into temporary neglect; while Bürger,[16] and other able writers of Germany, were translating or imitating these Reliques, and composing, with the aid of inspiration thence derived, poems which are the delight of the German nation. Dr. Percy was so abashed by the ridicule flung upon his labours from the ignorance and insensibility of the persons with whom he lived, that, though while he

13. **"Damon and Musidora"**: from "Summer." **14. the elder Warton:** Joseph Warton, 1722–1800, literary critic and brother of the more renowned Thomas Warton, 1728–90.

15. **Dr. Percy:** Thomas Percy, 1729–1811. His *Reliques*, published in 1765, brought to the attention of the literary world the old English ballads. Percy had rewritten or added to many of the ballads he printed. **16. Bürger:** Gottfried August Bürger, 1748–94.

was writing under a mask he had not wanted resolution to follow his genius into the regions of true simplicity and genuine pathos (as is evinced by the exquisite ballad of "Sir Cauline" and by many other pieces), yet when he appeared in his own person and character as a poetical writer, he adopted, as in the tale of the "Hermit of Warkworth," a diction scarcely in any one of its features distinguishable from the vague, the glossy, and unfeeling language of his day. I mention this remarkable fact with regret, esteeming the genius of Dr. Percy in this kind of writing superior to that of any other man by whom in modern times it has been cultivated. That even Bürger (to whom Klopstock[17] gave, in my hearing, a commendation which he denied to Goethe and Schiller, pronouncing him to be a genuine poet, and one of the few among the Germans whose works would last) had not the fine sensibility of Percy, might be shown from many passages, in which he has deserted his original only to go astray. For example,

> "Now daye was gone, and night was come,
> And all were fast asleepe,
> All save the Lady Emeline,
> Who sate in her bowre to weepe:
>
> "And soone she heard her true Love's voice
> Low whispering at the walle,
> Awake, awake, my dear Ladye,
> 'Tis I thy true-love call."[18]

Which is thus tricked out and dilated:

> "Als nun die Nacht Gebirg' und Thal
> Vermummt in Rabenschatten,
> Und Hochburgs Lampen überall
> Schon ausgeflimmert hatten,
> Und alles tief entschlafen war;
> Doch nur das Fräulein immerdar,
> Voll Fieberangst, noch wachte,
> Und seinen Ritter dachte:
> Da horch! Ein süsser Liebeston
> Kam leis' empor geflogen.
> 'Ho, Trudchen, ho! Da bin ich schon!
> Frisch auf! Dich angezogen!'"

But from humble ballads we must ascend to heroics. All hail, Macpherson![19] hail to thee, Sire, of Ossian!

17. Klopstock: Friedrich Gottlieb Klopstock, 1724–1803, a German poet of considerable reputation whom Wordsworth had met in 1798. 18. "Now . . . call": from the ballad entitled "The Child of Elle" in Percy's Reliques. 19. Macpherson: James Macpherson, 1736–96, in 1762 published Fingal, and in 1763 Temora, both purporting to be translations from the Gaelic epics of a poet named Ossian. They were spurious.

The Phantom was begotten by the snug embrace of an impudent Highlander upon a cloud of tradition—it travelled southward, where it was greeted with acclamation, and the thin Consistence took its course through Europe, upon the breath of popular applause. The Editor of the "Reliques" had indirectly preferred a claim to the praise of invention, by not concealing that his supplementary labours were considerable! how selfish his conduct, contrasted with that of the disinterested Gael, who, like Lear, gives his kingdom away, and is content to become a pensioner upon his own issue for a beggarly pittance!—Open this far-famed Book!—I have done so at random, and the beginning of the "Epic Poem Temora," in eight Books, presents itself. "The blue waves of Ullin roll in light. The green hills are covered with day. Trees shake their dusky heads in the breeze. Grey torrents pour their noisy streams. Two green hills with aged oaks surround a narrow plain. The blue course of a stream is there. On its banks stood Cairbar of Atha. His spear supports the king; the red eyes of his fear are sad. Cormac rises on his soul with all his ghastly wounds." Precious memorandums from the pocket-book of the blind Ossian!

If it be unbecoming, as I acknowledge that for the most part it is, to speak disrespectfully of Works that have enjoyed for a length of time a widely-spread reputation, without at the same time producing irrefragable proofs of their unworthiness, let me be forgiven upon this occasion.—Having had the good fortune to be born and reared in a mountainous country, from my very childhood I have felt the falsehood that pervades the volumes imposed upon the world under the name of Ossian. From what I saw with my own eyes, I knew that the imagery was spurious. In nature everything is distinct, yet nothing defined into absolute independent singleness. In Macpherson's work, it is exactly the reverse; everything (that is not stolen) is in this manner defined, insulated, dislocated, deadened,—yet nothing distinct. It will always be so when words are substituted for things. To say that the characters never could exist, that the manners are impossible, and that a dream has more substance than the whole state of society, as there depicted, is doing nothing more than pronouncing a censure which Macpherson defied; when, with the steeps of Morven[20] before his eyes, he could talk so familiarly of his Car-borne heroes;—of Morven, which, if one may judge

20. Morven: in Macpherson's writings, the kingdom of Fingal. Morven is also the name of a mountain in Scotland.

from its appearance at the distance of a few miles, contains scarcely an acre of ground sufficiently accommodating for a sledge to be trailed along its surface.—Mr. Malcolm Laing[21] has ably shown that the diction of this pretended translation is a motley assemblage from all quarters; but he is so fond of making out parallel passages as to call poor Macpherson to account for his "*ands*" and his "*buts!*" and he has weakened his argument by conducting it as if he thought that every striking resemblance was a *conscious* plagiarism. It is enough that the coincidences are too remarkable for its being probable or possible that they could arise in different minds without communication between them. Now as the Translators of the Bible, and Shakspeare, Milton, and Pope, could not be indebted to Macpherson, it follows that he must have owed his fine feathers to them; unless we are prepared gravely to assert, with Madame de Staël, that many of the characteristic beauties of our most celebrated English Poets are derived from the ancient Fingallian; in which case the modern translator would have been but giving back to Ossian his own.—It is consistent that Lucien Buonaparte, who could censure Milton for having surrounded Satan in the infernal regions with courtly and regal splendour, should pronounce the modern Ossian to be the glory of Scotland;—a country that has produced a Dunbar, a Buchanan, a Thomson, and a Burns! These opinions are of ill omen for the Epic ambition of him who has given them to the world.

Yet, much as those pretended treasures of antiquity have been admired, they have been wholly uninfluential upon the literature of the Country. No succeeding writer appears to have caught from them a ray of inspiration; no author, in the least distinguished, has ventured formally to imitate them—except the boy, Chatterton, on their first appearance. He had perceived, from the successful trials which he himself had made in literary forgery, how few critics were able to distinguish between a real ancient medal and a counterfeit of modern manufacture; and he set himself to the work of filling a magazine with *Saxon Poems*, counterparts of those of Ossian, as like his as one of his misty stars is to another. This incapability to amalgamate with the literature of the Island, is, in my estimation, a decisive proof that the book is essentially unnatural; nor should I require any other to demonstrate it to be a

21. **Mr. Malcolm Laing:** The discussion by Laing, 1762–1818, of the authenticity of the Ossian poems was attached to his *History of Scotland*, 1802.

forgery, audacious as worthless.—Contrast, in this respect, the effect of Macpherson's publication with the "Reliques" of Percy, so unassuming, so modest in their pretensions!—I have already stated how much Germany is indebted to this latter work; and for our own country, its poetry has been absolutely redeemed by it. I do not think that there is an able writer in verse of the present day who would not be proud to acknowledge his obligations to the "Reliques;" I know that it is so with my friends; and, for myself, I am happy in this occasion to make a public avowal of my own.

Dr. Johnson, more fortunate in his contempt of the labours of Macpherson than those of his modest friend, was solicited not long after to furnish Prefaces biographical and critical for the works of some of the most eminent English Poets. The booksellers took upon themselves to make the collection; they referred probably to the most popular miscellanies, and, unquestionably, to their books of accounts; and decided upon the claim of authors to be admitted into a body of the most eminent, from the familiarity of their names with the readers of that day, and by the profits, which, from the sale of his works, each had brought and was bringing to the Trade. The Editor was allowed a limited exercise of discretion, and the Authors whom he recommended are scarcely to be mentioned without a smile. We open the volume of Prefatory Lives, and to our astonishment the *first* name we find is that of Cowley!—What is become of the morning-star of English Poetry? Where is the bright Elizabethan constellation? Or, if names be more acceptable than images, where is the ever-to-be honoured Chaucer? where is Spenser? where Sidney? and, lastly, where he, whose rights as a poet, contra-distinguished from those which he is universally allowed to possess as a dramatist, we have vindicated,— where Shakspeare?—These, and a multitude of others not unworthy to be placed near them, their contemporaries and successors, we have *not*. But in their stead, we have (could better be expected when precedence was to be settled by an abstract of reputation at any given period made, as in this case before us?) Roscommon, and Stepney, and Phillips, and Walsh, and Smith, and Duke, and King, and Spratt—Halifax, Granville, Sheffield, Congreve, Broome, and other reputed Magnates—metrical writers utterly worthless and useless, except for occasions like the present, when their productions are referred to as evidence what a small quantity of brain is necessary to procure a considerable stock of admiration, provided the aspirant will accommodate himself to the likings and fashions of his day.

As I do not mean to bring down this retrospect to our own times, it may with propriety be closed at the era of this distinguished event. From the literature of other ages and countries, proofs equally cogent might have been adduced, that the opinions announced in the former part of this Essay are founded upon truth. It was not an agreeable office, nor a prudent undertaking, to declare them; but their importance seemed to render it a duty. It may still be asked, where lies the particular relation of what has been said to these Volumes?—The question will be easily answered by the discerning Reader who is old enough to remember the taste that prevailed when some of these poems were first published, seventeen years ago; who has also observed to what degree the poetry of this Island has since that period been coloured by them; and who is further aware of the unremitting hostility with which, upon some principle or other, they have each and all been opposed. A sketch of my own notion of the constitution of Fame has been given; and, as far as concerns myself, I have cause to be satisfied. The love, the admiration, the indifference, the slight, the aversion, and even the contempt, with which these Poems have been received, knowing, as I do, the source within my own mind, from which they have proceeded, and the labour and pains, which, when labour and pains appeared needful, have been bestowed upon them, must all, if I think consistently, be received as pledges and tokens, bearing the same general impression, though widely different in value;—they are all proofs that for the present time I have not laboured in vain; and afford assurances, more or less authentic, that the products of my industry will endure.

If there be one conclusion more forcibly pressed upon us than another by the review which has been given of the fortunes and fate of poetical Works, it is this,—that every author, as far as he is great and at the same time *original*, has had the task of *creating* the taste by which he is to be enjoyed: so has it been, so will it continue to be. This remark was long since made to me by the philosophical Friend for the separation of whose poems from my own I have previously expressed my regret. The predecessors of an original Genius of a high order will have smoothed the way for all that he has in common with them;—and much he will have in common; but, for what is peculiarly his own, he will be called upon to clear and often to shape his own road:—he will be in the condition of Hannibal among the Alps.

And where lies the real difficulty of creating that taste by which a truly original poet is to be relished? Is it in breaking the bonds of custom, in overcoming the prejudices of false refinement, and displacing the aversions of inexperience? Or, if he labour for an object which here and elsewhere I have proposed to myself, does it consist in divesting the reader of the pride that induces him to dwell upon those points wherein men differ from each other, to the exclusion of those in which all men are alike, or the same; and in making him ashamed of the vanity that renders him insensible of the appropriate excellence which civil arrangements, less unjust than might appear, and Nature illimitable in her bounty, have conferred on men who may stand below him in the scale of society? Finally, does it lie in establishing that dominion over the spirits of readers by which they are to be humbled and humanised, in order that they may be purified and exalted?

If these ends are to be attained by the mere communication of *knowledge*, it does *not* lie here.—TASTE, I would remind the reader, like IMAGINATION, is a word which has been forced to extend its services far beyond the point to which philosophy would have confined them. It is a metaphor, taken from a *passive* sense of the human body, and transferred to things which are in their essence *not* passive,—to intellectual *acts* and *operations*. The word, Imagination, has been overstrained, from impulses honourable to mankind, to meet the demands of the faculty which is perhaps the noblest of our nature. In the instance of Taste, the process has been reversed; and from the prevalence of dispositions at once injurious and discreditable, being no other than that selfishness which is the child of apathy, —which, as Nations decline in productive and creative power, makes them value themselves upon a presumed refinement of judging. Poverty of language is the primary cause of the use which we make of the word, Imagination; but the word, Taste, has been stretched to the sense which it bears in modern Europe by habits of self-conceit, inducing that inversion in the order of things whereby a passive faculty is made paramount among the faculties conversant with the fine arts. Proportion and congruity, the requisite knowledge being supposed, are subjects upon which taste may be trusted; it is competent to this office;—for in its intercourse with these the mind is *passive*, and is affected painfully or pleasurably as by an instinct. But the profound and the exquisite in feeling, the lofty and universal in thought and imagination; or, in ordinary language, the pathetic and the sublime;—are neither of them, accurately speaking, objects of a faculty which could ever without a sinking in the spirit of Nations have been designated by the metaphor—*Taste*. And why? Because without the exertion of a co-operating *power* in the mind of the Reader, there can be no adequate sympathy with either of these emotions: without this auxiliary impulse, elevated or profound passion cannot exist.

Passion, it must be observed, is derived from a word which signifies *suffering*: but the connection which suffering has with effort, with exertion, and *action*, is immediate and inseparable. How strikingly is this property of human nature exhibited by the fact, that, in popular language, to be in a passion, is to be angry!— But,

> "Anger in hasty *words* or *blows*
> Itself discharges on its foes."

To be moved, then, by a passion, is to be excited, often to external, and always to internal, effort; whether for the continuance and strengthening of the passion, or for its suppression, accordingly as the course which it takes may be painful or pleasurable. If the latter, the soul must contribute to its support, or it never becomes vivid,—and soon languishes, and dies. And this brings us to the point. If every great poet with whose writings men are familiar, in the highest exercise of his genius, before he can be thoroughly enjoyed, has to call forth and to communicate *power*, this service, in a still greater degree, falls upon an original writer, at his first appearance in the world.—Of genius the only proof is, the act of doing well what is worthy to be done, and what was never done before: Of genius, in the fine arts, the only infallible sign is the widening the sphere of human sensibility, for the delight, honour, and benefit of human nature. Genius is the introduction of a new element into the intellectual universe: or, if that be not allowed, it is the application of powers to objects on which they had not before been exercised, or the employment of them in such a manner as to produce effects hitherto unknown. What is all this but an advance, or a conquest, made by the soul of the poet? Is it to be supposed that the reader can make progress of this kind, like an Indian prince or general—stretched on his palanquin, and borne by his slaves? No; he is invigorated and inspirited by his leader, in order that he may exert himself; for he cannot proceed in quiescence, he cannot be carried like a dead weight. Therefore to create taste is to call forth and bestow power, of which knowledge is the effect; and *there* lies the true difficulty.

As the pathetic participates of an *animal* sensation, it might seem—that, if the springs of this emotion were genuine, all men, possessed of competent knowledge of the facts and circumstances, would be instantaneously affected. And, doubtless, in the works of every true poet will be found passages of that species of excellence, which is proved by effects immediate and universal. But there are emotions of the pathetic that are simple and direct, and others—that are complex and revolutionary; some—to which the heart yields with gentleness; others—against which it struggles with pride; these varieties are infinite as the combinations of circumstance and the constitutions of character. Remember, also, that the medium through which, in poetry, the heart is to be affected, is language; a thing subject to endless fluctuations and arbitrary associations. The genius of the poet melts these down for his purpose; but they retain their shape and quality to him who is not capable of exerting, within his own mind, a corresponding energy. There is also a meditative, as well as a human, pathos; an enthusiastic, as well as an ordinary, sorrow; a sadness that has its seat in the depths of reason, to which the mind cannot sink gently of itself— but to which it must descend by treading the steps of thought. And for the sublime,—if we consider what are the cares that occupy the passing day, and how remote is the practice and the course of life from the sources of sublimity, in the soul of Man, can it be wondered that there is little existing preparation for a poet charged with a new mission to extend its kingdom, and to augment and spread its enjoyments?

Away, then, with the senseless iteration of the word, *popular*, applied to new works in poetry, as if there were no test of excellence in this first of the fine arts but that all men should run after its productions, as if urged by an appetite, or constrained by a spell!—The qualities of writing best fitted for eager reception are either such as startle the world into attention by their audacity and extravagance; or they are chiefly of a superficial kind, lying upon the surfaces of manners; or arising out of a selection and arrangement of incidents, by which the mind is kept upon the stretch of curiosity, and the fancy amused without the trouble of thought. But in everything which is to send the soul into herself, to be admonished of her weakness, or to be made conscious of her power;—wherever life and nature are described as operated upon by the creative or abstracting virtue of the imagination; wherever the instinctive wisdom of antiquity and her heroic passions uniting, in the heart of the poet, with the meditative wisdom of later ages, have produced that accord of sublimated humanity, which is at once a history of the remote past and a prophetic enunciation of the remotest future, *there*, the poet must reconcile himself for a season to few and scattered hearers.—Grand thoughts (and Shakspeare must often have sighed over this truth), as they are most naturally and most fitly conceived in solitude, so can they not be brought forth in the midst of plaudits, without some violation of their sanctity. Go to a silent exhibition of the productions of the sister Art,[22] and be convinced that the qualities which dazzle at first sight, and kindle the admiration of the multitude, are essentially different from those by which permanent influence is secured. Let us not shrink from following up these principles as far as they will carry us, and conclude with observing —that there never has been a period, and perhaps never will be, in which vicious poetry, of some kind or other, has not excited more zealous admiration, and been far more generally read, than good; but this advantage attends the good, that the *individual*, as well as the species, survives from age to age; whereas, of the depraved, though the species be immortal, the individual quickly *perishes:* the object of present admiration vanishes, being supplanted by some other as easily produced; which, though no better, brings with it at least the irritation of novelty,—with adaptation, more or less skilful, to the changing humours of the majority of those who are most at leisure to regard poetical works when they first solicit their attention.

Is it the result of the whole, that, in the opinion of the Writer, the judgment of the People is not to be respected? The thought is most injurious; and, could the charge be brought against him, he would repel it with indignation. The People have already been justified, and their eulogium pronounced by implication, when it was said, above—that, of *good* poetry, the *individual*, as well as the species, *survives*. And how does it survive but through the People? What preserves it but their intellect and their wisdom?

—Past and future, are the wings
On whose support, harmoniously conjoined,
Moves the great Spirit of human knowledge—MS.[23]

The voice that issues from this Spirit, is that Vox Populi which the Deity inspires. Foolish must he be who can mistake for this a local acclamation, or a transitory outcry—transitory though it be for years, local

though from a Nation. Still more lamentable is his error who can believe that there is anything of divine infallibility in the clamour of that small though loud portion of the community, ever governed by factitious influence, which, under the name of the PUBLIC, passes itself, upon the unthinking, for the PEOPLE. Towards the Public, the Writer hopes that he feels as much deference as it is entitled to: but to the People, philosophically characterised, and to the embodied spirit of their knowledge, so far as it exists and moves, at the present, faithfully supported by its two wings, the past and the future, his devout respect, his reverence, is due. He offers it willingly and readily; and, this done, takes leave of his Readers, by assuring them—that, if he were not persuaded that the contents of these Volumes, and the Work to which they are subsidiary, evince something of the "Vision and the Faculty divine"; and that, both in words and things, they will operate in their degree, to extend the domain of sensibility for the delight, the honour, and the benefit of human nature, notwithstanding the many happy hours which he has employed in their composition, and the manifold comforts and enjoyments they have procured to him, he would not, if a wish could do it, save them from immediate destruction;—from becoming at this moment, to the world, as a thing that had never been.

LETTERS

OF WORDSWORTH AND HIS FAMILY

Dorothy Wordsworth to Mary Hutchinson (?)[1]
Racedown,[*June, 1797.*]

. . . You had a great loss in not seeing Coleridge. He is a wonderful man. His conversation teems with soul, mind, and spirit. Then he is so benevolent, so good tempered and cheerful, and, like William, interests himself so much about every little trifle. At first I thought him very plain, that is, for about three minutes: he is pale and thin, has a wide mouth, thick lips, and not very good teeth, longish loose-growing half-curling rough black hair. But if you hear him speak for five minutes you think no more of

---

22. sister Art: painting. **23. Past . . . knowledge:** *Prelude,* VI. 448–50.

LETTERS. **1. Mary Hutchinson:** the woman who later became Wordsworth's wife. She was probably the recipient of this letter.

them. His eye is large and full, not dark but grey; such an eye as would receive from a heavy soul the dullest expression; but it speaks every emotion of his animated mind; it has more of the "poet's eye in a fine frenzy rolling" than I ever witnessed. He has fine dark eyebrows, and an overhanging forehead.

The first thing that was read after he came was William's new poem *The Ruined Cottage*[2] with which he was much delighted; and after tea he repeated to us two acts and a half of his tragedy *Osorio*. The next morning William read his tragedy *The Borderers*. . . .

W. W. to Charles James Fox[3]
Grasmere, Westmoreland January 14th 1801

Sir,

It is not without much difficulty, that I have summoned the courage to request your acceptance of these Volumes. Should I express my real feelings, I am sure that I should seem to make a parade of diffidence and humility.

Several of the poems contained in these Volumes are written upon subjects, which are the common property of all Poets, and which, at some period of your life, must have been interesting to a man of your sensibility, and perhaps may still continue to be so. It would be highly gratifying to me to suppose that even in a single instance the manner in which I have treated these general topics should afford you any pleasure; but such a hope does not influence me upon the present occasion; in truth I do not feel it. Besides, I am convinced that there must be many things in this collection, which may impress you with an unfavorable idea of my intellectual powers. I do not say this with a wish to degrade myself; but I am sensible that this must be the case, from the different circles in which we have moved, and the different objects with which we have been conversant.

Being utterly unknown to you as I am, I am well aware, that if I am justified in writing to you at all, it is

2. **The Ruined Cottage**: the story of Margaret, later included in Book First of *The Excursion*. 3. **Charles James Fox**: 1749–1806, statesman and orator. With this letter Wordsworth sent a copy of the second edition of the *Lyrical Ballads*. Fox wrote a cordial letter in reply. He preferred "Harry Gill," "We Are Seven," "Her Eyes Are Wild," and "The Idiot Boy," and thought less highly of "Michael" because "I am no great friend to blank verse for subjects which are to be treated with simplicity."

necessary, my letter should be short; but I have feelings within me which I hope will so far shew themselves in this Letter as to excuse the trespass which I am afraid I shall make. In common with the whole of the English People I have observed in your public character a constant predominance of sensibility of heart. Necessitated as you have been from your public situation to have much to do with men in bodies, and in classes, and accordingly to contemplate them in that relation, it has been your praise that you have not thereby been prevented from looking upon them as individuals, and that you have habitually left your heart open to be influenced by them in that capacity. This habit cannot but have made you dear to Poets; and I am sure that, if since your first entrance into public life there has been a single true poet living in England, he must have loved you.

But were I assured that I myself had a just claim to the title of a Poet, all the dignity being attached to the Word which belongs to it, I do not think that I should have ventured for that reason to offer these volumes to you: at present it is solely on account of two poems in the second volume, the one entitled "*The Brothers*," and the other "*Michael*," that I have been emboldened to take this liberty.

It appears to me that the most calamitous effect, which has followed the measures which have lately been pursued in this country, is a rapid decay of the domestic affections among the lower orders of society. This effect the present Rulers of this Country are not conscious of, or they disregard it. For many years past, the tendency of society amongst almost all the nations of Europe has been to produce it. But recently by the spreading of manufactures through every part of the country, by the heavy taxes upon postage, by workhouses, Houses of Industry, and the invention of Soup-shops &c. &c. superadded to the encreasing disproportion between the price of labour and that of the necessaries of life, the bonds of domestic feeling among the poor, as far as the influence of these things has extended, have been weakened, and in innumerable instances entirely destroyed. The evil would be the less to be regretted, if these institutions were regarded only as palliatives to a disease; but the vanity and pride of their promoters are so subtly interwoven with them, that they are deemed great discoveries and blessings to humanity. In the mean time parents are separated from their children, and children from their parents; the wife no longer prepares with her own hands a meal for her husband, the produce of his labour;

there is little doing in his house in which his affections can be interested, and but little left in it which he can love. I have two neighbours, a man and his wife, both upwards of eighty years of age; they live alone; the husband has been confined to his bed many months and has never had, nor till within these few weeks has ever needed, any body to attend to him but his wife. She has recently been seized with a lameness which has often prevented her from being able to carry him his food to his bed; the neighbours fetch water for her from the well, and do other kind offices for them both, but her infirmities encrease. She told my Servant two days ago that she was afraid they must both be boarded out among some other Poor of the parish (they have long been supported by the parish) but she said, it was hard, having kept house together so long, to come to this, and she was sure that "it would burst her heart." I mention this fact to shew how deeply the spirit of independence is, even yet, rooted in some parts of the country. These people could not express themselves in this way without an almost sublime conviction of the blessings of independent domestic life. If it is true, as I believe, that this spirit is rapidly disappearing, no greater curse can befal a land.

I earnestly entreat your pardon for having detained you so long. In the two poems, "*The Brothers*" and "*Michael*" I have attempted to draw a picture of the domestic affections as I know they exist amongst a class of men who are now almost confined to the North of England. They are small independent *proprietors* of land here called statesmen, men of respectable education who daily labour on their own little properties. The domestic affections will always be strong amongst men who live in a country not crowded with population, if these men are placed above poverty. But if they are proprietors of small estates, which have descended to them from their ancestors, the power which these affections will acquire amongst such men is inconceivable by those who have only had an opportunity of observing hired labourers, farmers, and the manufacturing Poor. Their little tract of land serves as a kind of permanent rallying point for their domestic feelings, as a tablet upon which they are written which makes them objects of memory in a thousand instances when they would otherwise be forgotten. It is a fountain fitted to the nature of social man from which supplies of affection, as pure as his heart was intended for, are daily drawn. This class of men is rapidly disappearing. You, Sir, have a consciousness, upon which every good man will congratulate you,

that the whole of your public conduct has in one way or other been directed to the preservation of this class of men, and those who hold similar situations. You have felt that the most sacred of all property is the property of the Poor. The two Poems which I have mentioned were written with a view to shew that men who do not wear fine cloaths can feel deeply. "Pectus enim est quod disertos facit, et vis mentis. Ideoque imperitis quoque, si modo sint aliquo affectu concitati, verba non desunt."[4] The poems are faithful copies from nature; and I hope, whatever effect they may have upon you, you will at least be able to perceive that they may excite profitable sympathies in many kind and good hearts, and may in some small degree enlarge our feelings of reverence for our species, and our knowledge of human nature, by shewing that our best qualities are possessed by men whom we are too apt to consider, not with reference to the points in which they resemble us, but to those in which they manifestly differ from us. I thought, at a time when these feelings are sapped in so many ways that the two poems might co-operate, however feebly, with the illustrious efforts which you have made to stem this and other evils with which the country is labouring, and it is on this account alone that I have taken the liberty of thus addressing you.

Wishing earnestly that the time may come when the country may perceive what it has lost by neglecting your advice, and hoping that your latter days may be attended with health and comfort.

I remain, With the highest respect and admiration,
Your most obedient and humble Serv[t]

W WORDSWORTH

W. W. to John Wilson[5]
[*June, 1802.*]

MY DEAR SIR,

Had it not been for a very amiable modesty you would not have imagined that your letter could give me any offence. It was on many accounts highly grateful to me. I was pleased to find that I had given so much pleasure to an ingenuous and able mind, and

4. **"Pectus . . . desunt"**: from Quintilian, *Inst. Orat.*, X. vii. 15: "For eloquence comes from passion and force of imagination. The most uneducated men are never lacking in words if they are stirred by some passion." 5. **John Wilson**: Wilson, 1785–1854, was a critic. Seventeen years old at this time, he had sent Wordsworth an admiring letter about the *Lyrical Ballads*.

I further considered the enjoyment which you had had from my Poems as an earnest that others might be delighted with them in the same, or a like manner. It is plain from your letter that the pleasure which I have given you has not been blind or unthinking; you have studied the poems, and prove that you have entered into the spirit of them. They have not given you a cheap or vulgar pleasure; therefore I feel that you are entitled to my kindest thanks for having done some violence to your natural diffidence in the communication which you have made to me.

There is scarcely any part of your letter that does not deserve particular notice; but partly from some constitutional infirmities, and partly from certain habits of mind, I do not write any letters unless upon business, not even to my dearest friends. Except during absence from my own family I have not written five letters of friendship during the last five years. I have mentioned this in order that I may retain your good opinion, should my letter be less minute than you are entitled to expect. You seem to be desirous of my opinion on the influence of natural objects in forming the character of Nations. This cannot be understood without first considering their influence upon men in general, first, with reference to such subjects as are common to all countries; and, next, such as belong exclusively to any particular country, or in a greater degree to it than to another. Now it is manifest that no human being can be so besotted and debased by oppression, penury, or any other evil which unhumanizes man as to be utterly insensible to the colours, forms, or smell of flowers, the [? voices] and motions of birds and beasts, the appearances of the sky and heavenly bodies, the genial warmth of a fine day, the terror and uncomfortableness of a storm, etc. etc. How dead soever many full-grown men may outwardly seem to these things, all are more or less affected by them; and in childhood, in the first practice and exercise of their senses, they must have been not the nourishers merely, but often the fathers of their passions. There cannot be a doubt that in tracts of country where images of danger, melancholy, and grandeur, or loveliness, softness, and ease prevail, they will make themselves felt powerfully in forming the characters of the people, so as to produce a uniformity of national character, where the nation is small and is not made up of men who, inhabiting different soils, climates, etc., by their civil usages and relations, materially interfere with each other. It was so formerly, no doubt, in the Highlands of

Scotland; but we cannot perhaps observe it in our own island at the present day, because, even in the most sequestered places, by manufactures, traffic, religion, law, interchange of inhabitants, etc., distinctions are done away which would otherwise have been strong and obvious. This complex state of society does not, however, prevent the characters of individuals from frequently receiving a strong bias, not merely from the impressions of general nature, but also from local objects and images. But it seems that to produce these effects, in the degree in which we frequently find them to be produced, there must be a peculiar sensibility of original organization combining with moral accidents, as is exhibited in *The Brothers* and in *Ruth;* I mean, to produce this in a marked degree; not that I believe that any man was ever brought up in the country without loving it, especially in his better moments, or in a district of particular grandeur or beauty, without feeling some stronger attachment to it on that account than he would otherwise have felt. I include, you will observe, in these considerations, the influence of climate, changes in the atmosphere and elements, and the labours and occupations which particular districts require.

You begin what you say upon *The Idiot Boy* with this observation, that nothing is a fit subject for poetry which does not please. But here follows a question, Does not please whom? Some have little knowledge of natural imagery of any kind, and, of course, little relish for it; some are disgusted with the very mention of the words 'pastoral poetry,' 'sheep,' or 'shepherds'; some cannot tolerate a poem with a ghost or any supernatural agency in it; others would shrink from an animated description of the pleasures of love, as from a thing carnal and libidinous; some cannot bear to see delicate and refined feelings ascribed to men in low conditions of society, because their vanity and self-love tell them that these belong only to themselves and men like themselves in dress, station, and way of life; others are disgusted with the naked language of some of the most interesting passions of men, because either it is indelicate, or gross, or vulgar; as many fine ladies could not bear certain expressions in *The Mother*[6] and *The Thorn*, and as in the instance of Adam Smith, who, we are told, could not endure the ballad of *Clym of the Clough*, because the author had not written like a gentleman. Then there are professional and national prejudices forevermore. Some take no interest in the

6. **The Mother:** "Her Eyes Are Wild."

description of a particular passion or quality, as love of solitariness, we will say, genial activity of fancy, love of nature, religion, and so forth, because they have [little or] nothing of it in themselves; and so on without end. I return then to [the] question, please whom? or what? I answer, human nature, as it has been [and ever] will be. But where are we to find the best measure of this? I answer, [from with]in; by stripping our own hearts naked, and by looking out of ourselves to[wards men] who lead the simplest lives, and those most according to nature; men who have never known false refinements, wayward and artificial desires, false criticisms, effeminate habits of thinking and feeling, or who, having known these things, have outgrown them. This latter class is the most to be depended upon, but it is very small in number. People in our rank in life are perpetually falling into one sad mistake, namely, that of supposing that human nature and the persons they associate with are one and the same thing. Whom do we generally associate with? Gentlemen, persons of fortune, professional men, ladies, persons who can afford to buy, or can easily procure, books of half-a-guinea price, hot-pressed, and printed upon superfine paper, These persons are, it is true, a part of human nature, but we err lamentably if we suppose them to be fair representatives of the vast mass of human existence. And yet few ever consider books but with reference to their power of pleasing these persons and men of a higher rank; few descend lower, among cottages and fields, and among children. A man must have done this habitually before his judgment upon *The Idiot Boy* would be in any way decisive with me. I *know* I have done this myself habitually; I wrote the poem with exceeding delight and pleasure, and whenever I read it I read it with pleasure. You have given me praise for having reflected faithfully in my Poems the feelings of human nature. I would fain hope that I have done so. But a great Poet ought to do more than this: he ought, to a certain degree, to rectify men's feelings, to give them new compositions of feeling, to render their feelings more sane, pure, and permanent, in short, more consonant to nature, that is, to eternal nature, and the great moving spirit of things. He ought to travel before men occasionally as well as at their sides. I may illustrate this by a reference to natural objects. What false notions have prevailed from generation to generation as to the true character of the Nightingale. As far as my Friend's[7] Poem in the

7. **Friend**: Coleridge.

Lyrical Ballads is read, it will contribute greatly to rectify these. You will recollect a passage in Cowper, where, speaking of rural sounds, he says,

And *even* the boding owl
That hails the rising moon has charms for me.

Cowper was passionately fond of natural objects, yet you see he mentions it as a marvellous thing that he could connect pleasure with the cry of the owl. In the same poem he speaks in the same manner of that beautiful plant, the gorse; making in some degree an amiable boast of his loving it, *unsightly* and unsmooth as it is. There are many aversions of this kind, which, though they have some foundation in nature, have yet so slight a one that, though they may have prevailed hundreds of years, a philosopher will look upon them as accidents. So with respect to many moral feelings, either of love or dislike. What excessive admiration was paid in former times to personal prowess and military success; it is so with the latter even at the present day, but surely not nearly so much as heretofore. So with regard to birth, and innumerable other modes of sentiment, civil and religious. But you will be inclined to ask by this time how all this applies to *The Idiot Boy*. To this I can only say that the loathing and disgust which many people have at the sight of an idiot, is a feeling which, though having some foundation in human nature, is not necessarily attached to it in any virtuous degree, but is owing in a great measure to a false delicacy, and, if I may say it without rudeness, a certain want of comprehensiveness of thinking and feeling. Persons in the lower classes of society have little or nothing of this: if an idiot is born in a poor man's house, it must be taken care of, and cannot be boarded out, as it would be by gentlefolks, or sent to a public or private asylum for such unfortunate beings. [Poor people,] seeing frequently among their neighbours such objects, easily [forget] whatever there is of natural disgust about them, and have [therefore] a sane state, so that without pain or suffering they [perform] their duties towards them. I could with pleasure pursue this subject, but I must now strictly adopt the plan which I proposed to myself when I began to write this letter, namely, that of setting down a few hints or memorandums, which you will think of for my sake.

I have often applied to idiots, in my own mind, that sublime expression of Scripture, that *their life is hidden with God*. They are worshipped, probably from a feeling of this sort, in several parts of the East. Among the

Alps, where they are numerous, they are considered, I believe, as a blessing to the family to which they belong. I have, indeed, often looked upon the conduct of fathers and mothers of the lower classes of society towards idiots as the great triumph of the human heart. It is there that we see the strength, disinterestedness, and grandeur of love; nor have I ever been able to contemplate an object that calls out so many excellent and virtuous sentiments without finding it hallowed thereby, and having something in me which bears down before it, like a deluge, every feeble sensation of disgust and aversion.

There are, in my opinion, several important mistakes in the latter part of your letter which I could have wished to notice; but I find myself much fatigued. These refer both to the Boy and the Mother. I must content myself simply with observing that it is probable that the principle cause of your dislike to this particular poem lies in the *word* Idiot. If there had been any such word in our language *to which we had attached passion*, as lack-wit, half-wit, witless, etc., I should have certainly employed it in preference; but there is no such word. Observe (this is entirely in reference to this particular poem), my 'Idiot' is not one of those who cannot articulate, or of those that are usually disgusting in their persons:

> Whether in cunning or in joy,
> And then his words were not a few, etc.

and the last speech at the end of the poem. The boy whom I had in my mind was by no means disgusting in his appearance, quite the contrary; and I have known several with imperfect faculties who are handsome in their persons and features. There is one, at present, within a mile of my own house, remarkably so, though [he has something] of a stare and vacancy in his countenance. A friend of mine knowing that some persons had a dislike to the poem, such as you have expressed, advised me to add a stanza, describing the person of the boy [so as] entirely to separate him in the imaginations of my readers from that class of idiots who are disgusting in their persons; but the narration in the poem is so rapid and impassioned, that I could not find a place in which to insert the stanza without checking the progress of the poem and [so leaving] a deadness upon the feeling. This poem has, I know, frequently produced the same effect as it did upon you and your friends; but there are many also to whom it affords exquisite delight, and who, indeed, prefer it to any other of my poems. This proves

that the feelings there delineated are such as men *may* sympathise with. This is enough for my purpose. It is not enough for me as Poet, to delineate merely such feelings as all men *do* sympathise with; but it is also highly desirable to add to these others, such as all men *may* sympathise with, and such as there is reason to believe they would be better and more moral beings if they did sympathise with.

I conclude with regret, because I have not said one half of [what I intended] to say; but I am sure you will deem my excuse sufficient, [when I] inform you that my head aches violently, and I am in other respects unwell. I must, however, again give you my warmest thanks for your kind letter. I shall be happy to hear from you again, and do not think it unreasonable that I should request a letter from you when I feel that the answer which I may make to it will not perhaps be above three or four lines. This I mention to you with frankness, and you will not take it ill after what I have before said of my remissness in writing letters. I am, dear sir, with great respect,

> Yours sincerely,
> W. WORDSWORTH.

W. W. to Sara Hutchinson[8]
Monday 14th June [1802]

MY DEAR SARA

I am exceedingly sorry that the latter part of the Leech-gatherer[9] has displeased you, the more so because I cannot take to myself (that being the case) much pleasure or satisfaction in having pleased you in the former part. I will explain to you in prose my feeling in writing that Poem, and then you will be better able to judge whether the fault be mine or yours or partly both. I describe myself as having been exalted to the highest pitch of delight by the joyousness and beauty of Nature and then as depressed, even in the midst of those beautiful objects, to the lowest dejection and despair. A young Poet in the midst of the happiness of Nature is described as overwhelmed by the thought of the miserable reverses which have befallen the

---

8. Sara Hutchinson: shortly to become Wordsworth's sister-in-law. **9. Leech-gatherer:** "Resolution and Independence." An early draft of the poem had been sent to Mary and Sara Hutchinson. Here Wordsworth replies to their criticisms, though the passages to which they objected were later deleted.

happiest of all men, viz Poets—I think of this till I am so deeply impressed by it, that I consider the manner in which I was rescued from my dejection and despair almost as an interposition of Providence. "Now whether it was by peculiar grace A leading from above"—A person reading this Poem with feelings like mine will have been awed and controuled, expecting almost something spiritual or supernatural—What is brought forward? "A lonely place, a Pond" "by which an old man *was*, far from all house or home"—not stood, not sat, but *"was"*—the figure presented in the most naked simplicity possible. This feeling of spirituality or supernaturalness is again referred to as being strong in my mind in this passage—*"How came he here* thought I or what can he be doing?" I then describe him, whether ill or well is not for me to judge with perfect confidence, but this I can *confidently* affirm, that, though I believe God has given me a strong imagination, I cannot conceive a figure more impressive than that of an old Man like this, the survivor of a Wife and ten children, travelling alone among the mountains and all lonely places, carrying with him his own fortitude, and the necessities which an unjust state of society has entailed upon him. You say and Mary (that is you can say no more than that) the Poem is *very well* after the introduction of the old man; this is not true, if it is not more than very well it is very bad, there is no intermediate state. You speak of his speech as tedious: everything is tedious when one does not read with the feelings of the Author—*"The Thorn"* is tedious to hundreds; and so is the *Idiot Boy* to hundreds. It is in the character of the old man to tell his story in a manner which an *impatient* reader must necessarily feel as tedious. But Good God! Such a figure, in such a place, a pious self-respecting, miserably infirm, and Old Man telling such a tale!

My dear Sara, it is not a matter of indifference whether you are pleased with this figure and his employment; it may be comparatively so, whether you are pleased or not with *this Poem;* but it is of the utmost importance that you should have had pleasure from contemplating the fortitude, independence, persevering spirit, and the general moral dignity of this old man's character. Your feelings upon the Mother, and the Boys with the Butterfly, were not indifferent: it was an affair of whole continents of moral sympathy. I will talk more with you on this when we meet—at present, farewell and Heaven for ever bless you!

W. W.

W. W. to Thomas De Quincey
Grasmere, near Kendal, Westmoreland,
July 29th, 1803.

DEAR SIR,

Your Letter [10] dated May 31, I did not receive till the day before yesterday (owing I presume to the remissness of Messrs Longman and Rees, in forwarding it). I am much concerned at this; as though I am sure you would not suppose me capable of neglecting such a Letter, yet still my silence must needs have caused you some uneasiness.

It is impossible not to be pleased when one is told that one has given so much pleasure; and it is to me a still higher gratification to find that my poems have impressed a stranger with such favourable ideas of my character as a man. Having said this, which is easily said, I find some difficulty in replying more particularly to your Letter.

It is needless to say that it would be out of nature were I not to have kind feelings towards one who expresses sentiments of such profound esteem and admiration of my writings as you have done. You can have no doubt but that these sentiments however conveyed to me must have been acceptable; and I assure you that they are still more welcome coming from yourself. You will thus perceive that the main end which you proposed to yourself in writing to me is answered, viz., that I am already kindly disposed towards you. My friendship it is not in my power to give; this is a gift which no man can make, it is not in our own power: a sound and healthy friendship is the growth of time and circumstance, it will spring up and thrive like a wild-flower when these favour, and when they do not it is in vain to look for it.

I do not suppose that I am saying anything which you do not know as well as myself. I am simply reminding you of a commonplace truth which your high admiration of me may have robbed, perhaps, of that weight which it ought to have with you.

And this leads me to what gave me great concern,

10. **Letter:** forwarded to Wordsworth by Longman and Rees, publishers of the third edition (1802) of the *Lyrical Ballads*. De Quincey was seventeen and had not yet met Wordsworth. In his letter he enthusiastically praised the *Lyrical Ballads* and solicited Wordsworth's friendship: "what claim can I urge to a fellowship with such a society as yours, beaming (as it does) with genius so wild and so magnificent?"

I mean the very unreasonable value which you set upon my writings, compared with those of others. You are young and ingenuous, and I wrote with a hope of pleasing the young, the ingenuous and the unworldly, above all others; but sorry indeed should I be to stand in the way of the proper influence of other writers. You will know that I allude to the great names of past times, and above all to those of our own Country.

I have taken the liberty of saying this much to hasten on the time when you will value my poems not less, but those of others, more. That time, I know, would come of itself, and may come sooner for what I have said, which at all events I am sure you cannot take ill.

How many things there are in a man's character, of which his writings, however miscellaneous, or voluminous, will give no idea. How many thousand things which go to making up the value of a practically moral man, concerning not one of which any conclusion can be drawn from what he says of himself or of others in the World's Ear. You probably would never guess from anything you know of me, that I am the most lazy and impatient Letter-writer in the world. You will perhaps have observed that the first two or three Lines of this sheet are in a tolerably fair, legible hand, and, now every Letter, from A to Z, is a complete rout, one upon the heels of the other. Indeed so difficult do I find it to master this ill habit of idleness and impatience, that I have long since ceased to write any Letters but upon business. In justice to myself and you I have found myself obliged to mention this, lest you should think me unkind if you find me a slovenly and sluggish Correspondent.

I am going with my friend Coleridge and my Sister upon a tour into Scotland for six weeks or two months. This will prevent me hearing from you as soon as I could wish, as most likely we shall set off in a few days. If however you write immediately, I may have the pleasure of receiving your Letter before our departure; if we are gone, I shall order it to be sent after me.—I need not add that it will give me great pleasure to see you at Grasmere, if you should ever come this way.

I am, dear sir, with great sincerity and esteem,

Yours sincerely,

W. WORDSWORTH.

P.S.—I have just looked my letter over, and find that towards the conclusion I have been in a most unwarrantable hurry, especially in what I have said on seeing you here. I seem to have expressed myself absolutely with coldness. This is not in my feeling, I assure you. I shall indeed be very happy to see you at Grasmere, if you ever find it convenient to visit this delightful country. You speak of yourself as being very young; and therefore may have many engagements of great importance with respect to your worldly concerns and future happiness in life. Do not neglect these on any account; but if consistent with these and your other duties you could find time to visit this country, which is no great distance from your present residence, I should, I repeat it, be very happy to see you.

W. W.

Richard Wordsworth[11] *to W. W.*
Staple Inn, 7th Febry 1805.

MY DEAR BROTHER,

It is with the most painful concern that I inform you of the Loss of the Ship Abergavenny, off Weymouth last night.

I am acquainted with but few of the particulars of this melancholy Event. I am told that a great number of Persons have perished, and that our Brother John is amongst that number. Mr. Joseph Wordsworth[12] is amongst those who have been saved. The Ship struck against a Rock, and went to the Bottom. You will impart this to Dorothy in the best manner you can, and remember me most affectly to her, and your wife, believe me yours most sincerely,

RD WORDSWORTH.

W. W. to R. W.
Monday evening 11th Febry —05

MY DEAR BROTHER,

The lamentable news which your Letter has brought has now been known to us seven hours during which time I have done all in my power to alleviate the distress of poor Dorothy and my Wife.—Mary and I were walking out when the Letter came; it was brought by Sarah Hutchinson who had come from Kendal where she was staying, to be of use in the house and to comfort

---

11. **Richard Wordsworth:** Wordsworth's older brother, a lawyer in London. 12. **Joseph Wordsworth:** a cousin of William and Richard. He was in the service of the East India Company.

us; so that I had no power of breaking the force of the shock to Dorothy or to Mary. They are both very ill, Dorothy especially, on whom this loss of her beloved Brother will long take deep hold. I shall do my best to console her; but John was very dear to me and my heart will never forget him. God rest his soul! When you can bear to write do inform us not generally but as minutely as possible of the manner of this catastrophe. It would comfort us in this lonely place, though at present nobody in the house but myself could bear to hear a word on the subject. It is indeed a great affliction to us!

God bless you my dear Brother; Dorothy's and Mary's best love. We wish you were with us. God keep the rest of us together! the set is now broken. Farewell.

Dear Brother
Wm Wordsworth.

W. W. to Lady Beaumont
Coleorton, Tuesday May 21st 1807.

Pray excuse this villainous paper, I cannot find any other of the folio size.

My dear Lady Beaumont,

Though I am to see you soon I cannot but write a word or two, to thank you for the interest you take in my Poems [13] as evinced by your solicitude about their immediate reception. I write partly to thank you for this and to express the pleasure it has given me, and partly to remove any uneasiness from your mind which the disappointments you sometimes meet with in this labour of love may occasion. I see that you have many battles to fight for me; more than in the ardour and confidence of your pure and elevated mind you had ever thought of being summoned to; but be assured that this opposition is nothing more than what I distinctly foresaw that you and my other Friends would have to encounter. I say this, not to give myself credit for an eye of prophecy, but to allay any vexatious thoughts on my account which this opposition may have produced in you. It is impossible that any expectations can be lower than mine concerning the immediate effect of this little work upon what is called the Public. I do not here take into consideration the envy and malevolence, and all the bad passions which always stand in the way of a work of any merit from a living

13. **Poems:** *Poems in Two Volumes* (1807).

Poet; but merely think of the pure absolute honest ignorance, in which all worldlings of every rank and situation must be enveloped, with respect to the thoughts, feelings, and images, on which the life of my Poems depends. The things which I have taken, whether from within or without,—what have they to do with routs, dinners, morning calls, hurry from door to door, from street to street, on foot or in Carriage; with Mr. Pitt or Mr. Fox, Mr. Paul or Sir Francis Burdett, the Westminster Election or the Borough of Honiton; in a word, for I cannot stop to make my way through the hurry of images that present themselves to me, what have they to do with endless talking about things nobody cares anything for except as far as their own vanity is concerned, and this with persons they care nothing for but as their vanity or *selfishness* is concerned; what have they to do (to say all at once) with a life without love? in such a life there can be no thought; for we have no thought (save thoughts of pain) but as far as we have love and admiration. It is an awful truth, that there neither is, nor can be, any genuine enjoyment of Poetry among nineteen out of twenty of those persons who live, or wish to live, in the broad light of the world—among those who either are, or are striving to make themselves, people of consideration in society. This is a truth, and an awful one, because to be incapable of a feeling of Poetry in my sense of the word is to be without love of human nature and reverence for God.

Upon this I shall insist elsewhere; at present let me confine myself to my object, which is to make you, my dear Friend, as easy-hearted as myself with respect to these Poems. Trouble not yourself upon their present reception; of what moment is that compared with what I trust is their destiny, to console the afflicted, to add sunshine to daylight by making the happy happier, to teach the young and the gracious of every age, to see, to think and feel, and therefore to become more actively and securely virtuous; this is their office, which I trust they will faithfully perform long after we (that is, all that is mortal of us) are mouldered in our graves. I am well aware how far it would seem to many I overrate my own exertions when I speak in this way, in direct connection with the Volumes I have just made public.

I am not, however, afraid of such censure, insignificant as probably the majority of those poems would appear to very respectable persons; I do not mean London wits and witlings, for these have too many bad passions about them to be respectable even if they had more intellect than the benign laws of providence will allow to such a heartless existence as theirs is; but grave, kindly-natured, worthy persons, who would be pleased if they could. I hope that these Volumes are not without some recommendations, even for Readers of this class, but their imagination has slept; and the voice which is the voice of my Poetry without Imagination cannot be heard.

Leaving these, I was going to say a word to such Readers as Mr. Rogers. Such!—how would he be offended if he knew I considered him only as a representative of a class, and not an unique! 'Pity,' says Mr. R., 'that so many trifling things should be admitted to obstruct the view of those that have merit;' now, let this candid judge take, by way of example, the sonnets, which, probably, with the exception of two or three other Poems for which I will not contend appear to him the most trifling, as they are the shortest, I would say to him, omitting things of higher consideration, there is one thing which must strike you at once if you will only read these poems,—that those to Liberty, at least, have a connection with, or a bearing upon, each other, and therefore, if individually they want weight, perhaps, as a Body, they may not be so deficient, at least this ought to induce you to suspend your judgement, and qualify it so far as to allow that the writer aims at least at comprehensiveness. But dropping this, I would boldly say at once, that these Sonnets, while they each fix the attention upon some important sentiment separately considered, do at the same time collectively make a Poem on the subject of civil Liberty and national independence, which, either for simplicity of style or grandeur of moral sentiment, is, alas! likely to have few parallels in the Poetry of the present day. Again, turn to the 'Moods of my own Mind'. There is scarcely a Poem here of above thirty Lines, and very trifling these poems will appear to many; but, omitting to speak of them individually, do they not, taken collectively, fix the attention upon a subject eminently poetical, viz., the interest which objects in nature derive from the predominance of certain affections more or less permanent, more or less capable of salutary renewal in the mind of the being contemplating these objects? This is poetic, and essentially poetic, and why? because it is creative.

But I am wasting words, for it is nothing more than you know, and if said to those for whom it is intended, it would not be understood.

I see by your last Letter that Mrs. Fermor has entered into the spirit of these 'Moods of my own Mind.'

Your transcript from her Letter gave me the greatest pleasure; but I must say that even she has something yet to receive from me. I say this with confidence, from her thinking that I have fallen below myself in the Sonnet beginning—'With ships the sea was sprinkled far and nigh.' As to the other which she objects to, I will only observe that there is a misprint in the last line but two, 'And *though* this wilderness' for 'And *through* this wilderness'—that makes it unintelligible. This latter Sonnet for many reasons, though I do not abandon it, I will not now speak of; but upon the other, I could say something important in conversation, and will attempt now to illustrate it by a comment which I feel will be very inadequate to convey my meaning. There is scarcely one of my Poems which does not aim to direct the attention to some moral sentiment, or to some general principle, or law of thought, or of our intellectual constitution. For instance in the present case, who is there that has not felt that the mind can have no rest among a multitude of objects, of which it either cannot make one whole, or from which it cannot single out one individual, whereupon may be concentrated the attention divided among or distracted by a multitude? After a certain time we must either select one image or object, which must put out of view the rest wholly, or must subordinate them to itself while it stands forth as a Head:

> Now glowed the firmament
> With living sapphires! Hesperus, that *led*
> The starry host, rode brightest; till the Moon,
> Rising in clouded majesty, at length,
> Apparent *Queen*, unveiled *her peerless* light,
> And o'er the dark her silver mantle threw.[14]

Having laid this down as a general principle, take the case before us. I am represented in the Sonnet as casting my eyes over the sea, sprinkled with a multitude of Ships, like the heavens with stars, my mind may be supposed to float up and down among them in a kind of dreamy indifference with respect either to this or that one, only in a pleasurable state of feeling with respect to the whole prospect. 'Joyously it showed,' this continued till that feeling may be supposed to have passed away, and a kind of comparative listlessness or apathy to have succeeded, as at this line, 'Some veering up and down, one knew not why.' All at once, while I am in this state, comes forth an object, an individual, and my mind, sleepy and unfixed, is

---

14. Now . . . threw: from *Paradise Lost*, IV. 604–09.

awakened and fastened in a moment. 'Hesperus, that *led* The starry host,' is a poetical object, because the glory of his own Nature gives him the pre-eminence the moment he appears; he calls forth the poetic faculty, receiving its exertions as a tribute; but this Ship in the Sonnet may, in a manner still more appropriate, be said to come upon a mission of the poetic Spirit, because in its own appearance and attributes it is barely sufficiently distinguish[ed] to rouse the creative faculty of the human mind; to exertions at all times welcome, but doubly so when they come upon us when in a state of remissness. The mind being once fixed and rouzed, all the rest comes from itself; it is merely a lordly Ship, nothing more:

> This ship was nought to me, nor I to her,
> Yet I pursued her with a lover's look.

My mind wantons with grateful joy in the exercise of its own powers, and, loving its own creation,

> This ship to all the rest I did prefer,

making her a sovereign or a regent, and thus giving body and life to all the rest; mingling up this idea with fondness and praise—

> where she comes the winds must stir;

and concluding the whole with

> On went She, and due north her journey took.

Thus taking up again the Reader with whom I began, letting him know how long I must have watched this favourite Vessel, and inviting him to rest his mind as mine is resting.

Having said so much upon a mere 14 lines, which Mrs. Fermor did not approve, I cannot but add a word or two upon my satisfaction in finding that my mind has so much in common with hers, and that we participate so many of each other's pleasures. I collect this from her having singled out the two little Poems, the Daffodils, and the Rock crowned with snowdrops. I am sure that whoever is much pleased with either of these quiet and tender delineations must be fitted to walk through the recesses of my poetry with delight, and will there recognise, at every turn, something or other in which, and over which, it has that property and right which knowledge and love confer. The line, 'Come, blessed barrier, etc.,' in the sonnet upon Sleep, which Mrs. F. points out, had before been mentioned to me by Coleridge, and indeed by almost everybody who had heard it, as eminently beautiful. My letter

(as this 2nd sheet, which I am obliged to take, admonishes me) is growing to an enormous length; and yet, saving that I have expressed my calm confidence that these Poems will live, I have said nothing which has a particular application to the object of it, which was to remove all disquiet from your mind on account of the condemnation they may at present incur from that portion of my contemporaries who are called the Public. I am sure, my dear Lady Beaumont, if you attach any importance [to it] it can only be from an apprehension that it may affect me, upon which I have already set you at ease, or from a fear that this present blame is ominous of their future or final destiny. If this be the case, your tenderness for me betrays you; be assured that the decision of these persons has nothing to do with the Question; they are altogether incompetent judges. These people in the senseless hurry of their idle lives do not *read* books, they merely snatch a glance at them that they may talk about them. And even if this were not so, never forget what I believe was observed to you by Coleridge, that every great and original writer, in proportion as he is great or original, must himself create the taste by which he is to be relished; he must teach the art by which he is to be seen; this, in a certain degree, even to all persons, however wise and pure may be their lives, and however unvitiated their taste; but for those who dip into books in order to give an opinion of them, or talk about them to take up an opinion—for this multitude of unhappy, and misguided, and misguiding beings, an entire regeneration must be produced; and if this be possible, it must be a work *of time*. To conclude, my ears are stone-dead to this idle buzz, and my flesh as insensible as iron to these petty stings; and after what I have said I am sure yours will be the same. I doubt not that you will share with me an invincible confidence that my writings (and among them these little Poems) will co-operate with the benign tendencies in human nature and society, wherever found; and that they will, in their degree, be efficacious in making men wiser, better, and happier. Farewell; I will not apologise for this Letter, though its length demands an apology. Believe me, eagerly wishing for the happy day when I shall see you and Sir George here, most affectionately yours,

WM WORDSWORTH.

Do not hurry your coming hither on our account: my Sister regrets that she did not press this upon you, as you say in your Letter, 'we cannot *possibly* come before the first week in June'; from which we infer that your kindness will induce you to make sacrifices for our sakes. Whatever pleasure we may have in thinking of Grasmere, we have no impatience to be gone, and think with full as much regret at leaving Coleorton. I had, for myself, indeed, a wish to be at Grasmere with as much of the summer before me as might be, but to this I attach no importance whatever, as far as the gratification of that wish interferes with any inclination or duty of yours. I could not be satisfied without seeing you here, and shall have great pleasure in waiting.

DOROTHY WORDSWORTH

from GRASMERE JOURNAL

〜〜〜

Born on Christmas, 1771, Dorothy Wordsworth was twenty months younger than William. After the death of their father in 1783, she lived with relatives until Wordsworth was able to set up housekeeping in 1795 at Racedown, a farmhouse in Dorset. Thereafter she stayed with him for the rest of his life, surviving him by five years. Coleridge described her in a letter of about 1797: "Her manners are simple, ardent, impressive—.

> In every motion her most innocent soul
> Outbeams so brightly, that who saw would say,
> Guilt was a thing impossible in her.—
> ["The Destiny of Nations," ll. 173–75].

Her information various—her eye watchful in minutest observation of nature—and her taste a perfect electrometer—it bends, protrudes, and draws in, at subtlest beauties & most recondite faults." Her journal gives an intimate, unrivaled picture both of Wordsworth and of the milieu in which he lived and worked.

〜〜〜

New Year's Day [1802]. We walked, Wm. and I, towards Martindale.
January 2nd, Saturday. It snowed all day. We walked near to Dalemain in the snow.[1]
On Saturday, January 23rd, we left Eusemere at 10 o'clock in the morning, I behind Wm., Mr. Clarkson

DOROTHY WORDSWORTH. **1. We . . . snow:** Wordsworth and Dorothy were visiting friends, the Clarksons, at their house called Eusemere on Ullswater. They stayed there until January 23.

on his Galloway.[2] The morning not very promising, the wind cold. The mountains large and dark, but only thinly streaked with snow; a strong wind. We dined in Grisdale on ham, bread, and milk. We parted from Mr. C. at one o'clock—it rained all the way home. We struggled with the wind, and often rested as we went along. A hail shower met us before we reached the Tarn, and the way often was difficult over the snow; but at the Tarn the view closed in. We saw nothing but mists and snow: and at first the ice on the Tarn below us cracked and split, yet without water, a dull grey white. We lost our path, and could see the Tarn no longer. We made our way out with difficulty, guided by a heap of stones which we well remembered. We were afraid of being bewildered in the mists, till the darkness should overtake us. We were long before we knew that we were in the right track, but thanks to William's skill we knew it long before we could see our way before us. There was no footmark upon the snow either of man or beast. We saw 4 sheep before we had left the snow region. The Vale of Grasmere, when the mists broke away, looked soft and grave, of a yellow hue. It was dark before we reached home. We were not very much tired. My inside was sore with the cold—we had both of us been much heated upon the mountains but we caught no cold. O how comfortable and happy we felt ourselves, sitting by our own fire, when we had got off our wet clothes and had dressed ourselves fresh and clean. We found 5 £ from Montagu and 20 £ from Christopher.[3] We talked about the Lake of Como, read in the *Descriptive Sketches*, looked about us, and felt that we were happy. We indulged dear thoughts about home—poor Mary![4] we were sad to think of the contrast for her.

[*January*] 24*th, Sunday.* We went into the orchard as soon as breakfast was over. Laid out the situation for our new room, and sauntered a while. We had Mr. Clarkson's turkey for dinner; the night before we had boiled the gizzard and some mutton, and made a nice piece of cookery for Wm.'s supper.[5] Wm. walked in the morning, I wrote to Coleridge. After dinner I lay down till tea time—I rose refreshed and

2. Galloway: a pony. **3. Montagu . . . Christopher:** Basil Montagu, a friend to whom Wordsworth had loaned a large portion of his capital, and Christopher Wordsworth, the youngest brother of William and Dorothy, who had promised Dorothy a present of £20 each year "until he marries." **4. Mary:** Mary Hutchinson. **5. dinner . . . supper:** Four meals were usually served in the early nineteenth century: breakfast, dinner in the middle of the day, tea, and supper in the late evening.

better. Wm. could not beat away sleep when I was gone. We went late to bed.

January 29th, Friday. Wm. was very unwell. Worn out with his bad night's rest. He went to bed—I read to him, to endeavour to make him sleep. Then I came into the other room, and read the first book of *Paradise Lost.* After dinner we walked to Ambleside—found Lloyds at Luff's[6]—we stayed and drank tea by ourselves. A heart-rending letter from Coleridge[7]—we were sad as we could be. Wm. wrote to him. We talked about Wm.'s going to London. It was a mild afternoon—there was an unusual softness in the prospects as we went, a rich yellow upon the fields, and a soft grave purple on the waters. When we returned many stars were out, the clouds were moveless, in the sky soft purple, the Lake of Rydale calm, Jupiter behind, Jupiter at least *we* call him, but William says we always call the largest star Jupiter. When we came home we both wrote to C. I was stupefied.

January 30th, Saturday. A cold dark morning. William chopped wood—I brought it in a basket. A cold wind. Wm. slept better, but he thinks he looks ill—he is shaving now. He asks me to set down the story of Barbara Wilkinson's turtle dove. Barbara is an old maid. She had two turtle doves. One of them died, the first year I think. The other bird continued to live alone in its cage for 9 years, but for one whole year it had a companion and daily visitor—a little mouse, that used to come and feed with it; and the dove would caress it, and cower over it with its wings, and make a loving noise to it. The mouse, though it did not testify equal delight in the dove's company, yet it was at perfect ease. The poor mouse disappeared, and the dove was left solitary till its death. It died of a short sickness, and was buried under a tree with funeral ceremony by Barabara and her maidens, and one or two others.

On *Saturday, 30th,* Wm. worked at *The Pedlar*[8] all the morning. He kept the dinner waiting till four o'clock. He was much tired. We were preparing to walk when a heavy rain came on.

[*January*] 31st, *Sunday.* Wm. had slept very ill—he was tired and had a bad headache. We walked round the two lakes. Grasmere was very soft, and Rydale was extremely beautiful from the western side. Nab

Scar was just topped by a cloud which, cutting it off as high as it could be cut off, made the mountain look uncommonly lofty. We sate down a long time in different places. I always love to walk that way, because it is the way I first came to Rydale and Grasmere, and because our dear Coleridge did also. When I came with Wm., 6½ years ago, it was just at sunset. There was a rich yellow light on the waters, and the Islands were reflected there. To-day it was grave and soft, but not perfectly calm. William says it was much such a day as when Coleridge came with him. The sun shone out before we reached Grasmere. We sate by the roadside at the foot of the Lake, close to Mary's dear name, which she had cut herself upon the stone. Wm. cut at it with his knife to make it plainer. We amused ourselves for a long time in watching the breezes, some as if they came from the bottom of the lake, spread in a circle, brushing along the surface of the water, and growing more delicate, as it were thinner, and of a *paler* colour till they died away. Others spread out like a peacock's tail, and some went right forward this way and that in all directions. The lake was still where these breezes were not, but they made it all alive. I found a strawberry blossom in a rock. The little slender flower had more courage than the green leaves, for *they* were but half expanded and half grown, but the blossom was spread full out. I uprooted it rashly, and I felt as if I had been committing an outrage, so I planted it again. It will have but a stormy life of it, but let it live if it can. We found Calvert[9] here. I brought a handkerchief full of mosses, which I placed on the chimneypiece when C. was gone. He dined with us, and carried away the encyclopaedias. After they were gone, I spent some time in trying to reconcile myself to the change, and in rummaging out and arranging some other books in their places. One good thing is this—there is a nice elbow place for William, and he may sit for the picture of John Bunyan any day. Mr. Simpson[10] drank tea with us. We paid our rent to Benson. William's head was bad after Mr. S. was gone. I petted him on the carpet and began a letter to Sara.

February 2nd, Tuesday. A fine clear morning, but sharp and cold. Wm. went into the orchard after breakfast to chop wood. I walked backwards and forwards on the platform. Molly called me down to

---

6. Lloyds at Luff's: Charles and Mrs. Lloyd visiting Captain and Mrs. Luff. **7. heart-rending . . . Coleridge:** Coleridge was ill in London. **8. The Pedlar:** "The Ruined Cottage," which Wordsworth was rewriting. It was finally published as Book First of *The Excursion.*

9. Calvert: William Calvert, a friend, brother of Raisley Calvert, who had bequeathed Wordsworth a small legacy. **10. Mr. Simpson:** the Rev. Joseph Sympson, a familiar acquaintance.

Charles Lloyd—he brought me flower seeds from his Brother. William not quite well. We walked into Easedale—were turned back in the open field by the sight of a cow—every horned cow puts me in terror. We walked as far as we could, having crossed the footbridge, but it was dirty, and we turned back—walked backwards and forwards between Goody Bridge and Butterlip How. William wished to break off composition,[11] and was unable, and so did himself harm. The sun shone, but it was cold. After dinner William worked at *The Pedlar*. After tea I read aloud the eleventh book of *Paradise Lost*. We were much impressed, and also melted into tears. The papers came in soon after I had laid aside the book—a good thing for my Wm. I worked a little to-day at putting the linen into repair that came in the box. Molly washing.

[*February*] *3rd, Wednesday*. A rainy morning. We walked to Rydale for letters, found one from Mrs. Cookson and Mary H. It snowed upon the hills. We sate down on the wall at the foot of White Moss. Sate by the fire in the evening. Wm. tired, and did not compose. He went to bed soon, and could not sleep. I wrote to Mary H. Sent off the letter by Fletcher.[12] Wrote also to Coleridge. Read Wm. to sleep after dinner, and read to him in bed till ½ past one.

[*February*] *4th, Thursday*. I was very sick, bad headach and unwell—I lay in bed till 3 o'clock, that is I lay down as soon as breakfast was over. It was a terribly wet day. William sate in the house all day. Fletcher's boy did not come home. I worked at Montagu's shirts. Wm. thought a little about *The Pedlar*. I slept in the sitting room. Read Smollet's life.

[*February*] *5th, Friday*. A cold snowy morning. Snow and hail showers. We did not walk. Wm. cut wood a little. I read the story of [?] in Warly [?]. Sara's parcel came with waistcoat. The Chaucer not only misbound, but a leaf or two wanting. I wrote about it to Mary and wrote to Soulby. We received the waistcoat, shoes and gloves from Sara by the waggon. William not well. Sate up late at *The Pedlar*.

February 6th, Saturday. William had slept badly. It snowed in the night, and was on Saturday, as Molly[13]

expressed it, a cauld clash.[14] William went to Rydale for letters—he came home with two very affecting letters from Coleridge—resolved to try another climate. I was stopped in my writing, and made ill by the letters. William a bad headache; he made up a bed on the floor, but could not sleep—I went to his bed and slept not—better when I rose—wrote again after tea, and translated two or three of Lessing's[15] *Fables*.

[*February*] *7th, Sunday*. A fine clear frosty morning. The eaves drop with the heat of the sun all day long. The ground thinly covered with snow. The road black, rocks bluish. Before night the Island was quite green; the sun had melted all the snow upon it. Mr. Simpson called before Wm. had done shaving—William had had a bad night and was working at his poem. We sate by the fire, and did not walk, but read *The Pedlar*, thinking it done; but lo! though Wm. could find fault with no one part of it, it was uninteresting, and must be altered. Poor Wm.!

[*February*] *12th, Friday*. A very fine, bright, clear, hard frost. Wm. working again. I recopied *The Pedlar*, but poor Wm. all the time at work. Molly tells me "What! little Sally's gone to visit at Mr. Simpson's. They say she's very smart, she's got on a new bed-gown that her Cousin gave her, it's a very bonny one, they tell me, but I've not seen it. Sally and me's in luck." In the afternoon a poor woman came, *she said*, to beg some rags for her husband's leg, which had been wounded by a slate from the roof in the great wind—but she has been used to go a-begging, for she has often come here. Her father lived to the age of 105. She is a woman of strong bones, with a complexion that has been beautiful, and remained very fresh last year, but now she looks broken, and her little boy—a pretty little fellow, and whom I have loved for the sake of Basil[16]—looks thin and pale. I observed this to her. "Ay," says she, "we have all been ill. Our house was unroofed in the storm nearly, and so we lived in it for more than a week." The child wears a ragged drab coat and a fur cap, poor little fellow, I think he seems scarcely at all grown since the first time I saw him. William was with me; we met him

---

11. **William . . . composition**: Many of the walks and excursions mentioned by Dorothy were actually sessions in which Wordsworth was at work. He usually composed out-of-doors, half-chanting to himself as he walked along and not writing down his verses until later. Revision was usually an indoor job. 12. **Fletcher**: the mail carrier. 13. **Molly**: Molly Fisher, their part-time servant.

14. **clash**: in northern dialect, a quantity or mass of something wet, as a *clash* of mud, a *clash* of oatmeal. 15. **Lessing**: Gotthold Ephraim Lessing, 1729–81. Dorothy had learned German in 1798–99 in Germany. 16. **Basil**: Basil Montagu, Jr., son of Wordsworth's friend, had lived with William and Dorothy for approximately two years at Racedown and Alfoxden. He is the child in "Anecdote for Fathers."

in a lane going to Skelwith Bridge. He looked very pretty. He was walking lazily, in the deep narrow lane, over-shadowed with the hedgerows, his meal poke hung over his shoulder. He said he "was going a laiting." Poor creatures! He now wears the same coat he had on at that time. When the woman was gone, I could not help thinking that we are not half thankful enough that we are placed in that condition of life in which we are. We do not so often bless God for this, as we wish for this £50, that £100, etc. etc. We have not, however, to reproach ourselves with ever breathing a murmur. This woman's was but a *common* case. The snow still lies upon the ground. Just at the closing in of the day, I heard a cart pass the door, and at the same time the dismal sound of a crying infant. I went to the window, and had light enough to see that a man was driving a cart, which seemed not to be very full, and that a woman with an infant in her arms was following close behind and a dog close to her. It was a wild and melancholy sight. Wm. rubbed his table after candles were lighted, and we sate a long time with the windows unclosed; I almost finished writing *The Pedlar;* but poor Wm. wore himself and me out with labour. We had an affecting conversation. Went to bed at 12 o'clock.

[*February*] *16th, Tuesday.* A fine morning, but I had persuaded myself not to expect Wm.;[17] I believe because I was afraid of being disappointed. I ironed all day. He came in just at tea time, had only seen Mary H. for a couple of hours between Eamont Bridge and Hartshorn Tree. Mrs. C.[18] better. He had had a difficult journey over Kirkstone, and came home by Threlkeld—his mouth and breath were very cold when he kissed me. We spent a sweet evening. He was better, had altered *The Pedlar.* We went to bed pretty soon. Mr. Graham said he wished Wm. had been with him the other day—he was riding in a post-chaise and he heard a strange cry that he could not understand, the sound continued, and he called to the chaise driver to stop. It was a little girl that was crying as if her heart would burst. She had got up behind the chaise, and her cloak had been caught by the wheel, and was jammed in, and it hung there. She was crying after it. Poor thing. Mr. Graham took her into the chaise, and the cloak was released from the wheel, but the child's misery did not cease, for her cloak was torn to rags;

it had been a miserable cloak before, but she had no other, and it was the greatest sorrow that could befal her. Her name was Alice Fell.[19] She had no parents, and belonged to the next town. At the next town, Mr. G. left money with some respectable people in the town, to buy her a new cloak.

[*March 4th*], *Thursday.* Before we had quite finished breakfast Calvert's man brought the horses for Wm. We had a deal to do, to shave, pens to make, poems to put in order for writing, to settle the dress, pack up etc., and the man came before the pens were made, and he was obliged to leave me with only two. Since he has left me at half-past 11 (it is now 2) I have been putting the drawers into order, laid by his clothes which we had thrown here and there and everywhere, filed two months' newspapers and got my dinner, 2 boiled eggs and 2 apple tarts. I have set Molly on to clear the garden a little, and I myself have helped. I transplanted some snowdrops—the Bees are busy. Wm. has a nice bright day. It was hard frost in the night. The Robins are singing sweetly. Now for my walk. I *will* be busy. I *will* look well, and be well when he comes back to me.[20] O the Darling! Here is one of his bitten apples. I can hardly find in my heart to throw it into the fire. I must wash myself, then off. I walked round the two Lakes, crossed the stepping-stones at Rydale foot. Sate down where we always sit. I was full of thoughts about my darling. Blessings on him. I came home at the foot of our own hill under Loughrigg. They are making sad ravages in the woods. Benson's wood is going, and the wood above the River. The wind has blown down a small fir tree on the Rock that terminates John's path—I suppose the wind of Wednesday night. I read German after my return till tea time. I worked and read the L. B., enchanted with the *Idiot Boy.* Wrote to Wm., then went to bed. It snowed when I went to bed.

[*March 13th,*] *Saturday Morning.* It was as cold as ever it has been all winter, very hard frost. I baked pies bread and seed cake for Mr. Simpson. William finished *Alice Fell*, and then he wrote the poem of *The Beggar Woman*, taken from a woman whom I had seen in May (now nearly 2 years ago) when John and he were at Gallow Hill. I sate with him at intervals all the morning, took down his stanzas, etc. After dinner we walked to Rydale for letters—it was terribly

---

17. **to expect Wm.:** Wordsworth had been the last two days in Penrith, visiting Mary Hutchinson. 18. **Mrs. C.:** Mrs. Clarkson.

19. **Alice Fell:** See the poem "Alice Fell," above. 20. **back to me:** Wordsworth was away March 4 to March 7.

cold—we had 2 or 3 brisk hail showers—the hail stones looked clean and pretty upon the dry clean road. Little Peggy Simpson was standing at the door catching the hail stones in her hand—she grows very like her mother. When she is sixteen years old I dare say that to her Grandmother's eye she will seem as like to what her mother was, as any rose in her garden is like the rose that grew there years before. No letters at Rydale. We drank tea as soon as we reached home. After tea I read to William that account of the little boy belonging to the tall woman, and an unlucky thing it was, for he could not escape from those very words, and so he could not write the poem. He left it unfinished, and went tired to bed. In our walk from Rydale he had got warmed with the subject, and had half cast the poem.

[*March 14th,*] *Sunday Morning*. William had slept badly—he got up at nine o'clock, but before he rose he had finished *The Beggar Boys*, and while we were at breakfast that is (for I had breakfasted) he, with his basin of broth before him untouched, and a little plate of bread and butter he wrote the Poem to a Butterfly! [21] He ate not a morsel, nor put on his stockings, but sate with his shirt neck unbuttoned, and his waistcoat open while he did it. The thought first came upon him as we were talking about the pleasure we both always feel at the sight of a butterfly. I told him that I used to chase them a little, but that I was afraid of brushing the dust off their wings, and did not catch them. He told me how they used to kill all the white ones when he went to school because they were Frenchmen. Mr. Simpson came in just as he was finishing the Poem. After he was gone I wrote it down and the other poems, and I read them all over to him. We then called at Mr. Olliff's—Mr. O. walked with us to within sight of Rydale—the sun shone very pleasantly, yet it was extremely cold. We dined and then Wm. went to bed. I lay upon the fur gown before the fire, but I could not sleep—I lay there a long time. It is now halfpast 5—I am going to write letters—I began to write to Mrs. Rawson. William rose without having slept—we sate comfortably by the fire till he began to try to alter *The Butterfly*, and tired himself—he went to bed tired.

[*April*] *15th, Thursday*. It was a threatening, misty morning, but mild. We set off after dinner from Eusemere. Mrs. Clarkson went a short way with us, but turned back. The wind was furious, and we thought we must have returned. We first rested in the large

boat-house, then under a furze bush opposite Mr. Clarkson's. Saw the plough going in the field. The wind seized our breath. The Lake was rough. There was a boat by itself floating in the middle of the bay below Water Millock. We rested again in the Water Millock Lane. The hawthorns are black and green, the birches here and there greenish, but there is yet more of purple to be seen on the twigs. We got over into a field to avoid some cows—people working. A few primroses by the roadside—woodsorrel flower, the anemone, scentless violets, strawberries, and that starry, yellow flower which Mrs. C. calls pile wort. When we were in the woods beyond Gowbarrow Park we saw a few daffodils [22] close to the water-side. We fancied that the lake had floated the seeds ashore, and that the little colony had so sprung up. But as we went along there were more and yet more; and at last, under the boughs of the trees, we saw that there was a long belt of them along the shore, about the breadth of a country turnpike road. I never saw daffodils so beautiful. They grew among the mossy stones about and about them; some rested their heads upon these stones as on a pillow for weariness; and the rest tossed and reeled and danced, and seemed as if they verily laughed with the wind, that blew upon them over the lake; they looked so gay, ever glancing, ever changing. This wind blew directly over the lake to them. There was here and there a little knot, and a few stragglers a few yards higher up; but they were so few as not to disturb the simplicity, unity, and life of that one busy highway. We rested again and again. The bays were stormy, and we heard the waves at different distances, and in the middle of the water, like the sea. Rain came on—we were wet when we reached Luff's, but we called in. Luckily all was chearless and gloomy, so we faced the storm—we *must* have been wet if we had waited—put on dry clothes at Dobson's. I was very kindly treated by a young woman, the landlady looked sour, but it is her way. She gave us a goodish supper, excellent ham and potatoes. We paid 7/- when we came away. William was sitting by a bright fire when I came downstairs. He soon made his way to the library, piled up in a corner of the window. He brought out a volume of Enfield's *Speaker*, another miscellany, and an odd volume of Congreve's plays. We had a glass of warm rum and water. We enjoyed ourselves, and wished for Mary. It rained and blew, when we went to bed. N.B. Deer in Gowbarrow Park like skeletons.

---

21. **to a Butterfly:** "To a Butterfly" ("Stay near me—do not take thy flight!").

22. **daffodils:** See "I Wandered Lonely as a Cloud," above.

April 16th, Friday (Good Friday). When I undrew my curtains in the morning, I was much affected by the beauty of the prospect, and the change. The sun shone, the wind had passed away, the hills looked chearful, the river was very bright as it flowed into the lake. The church rises up behind a little knot of rocks, the steeple not so high as an ordinary three-story house. Trees in a row in the garden under the wall. After Wm. had shaved we set forward; the valley is at first broken by little rocky woody knolls that make retiring places, fairy valleys in the vale; the river winds along under these hills, travelling, not in a bustle but not slowly, to the lake. We saw a fisherman in the flat meadow on the other side of the water. He came towards us, and threw his line over the two-arched bridge. It is a bridge of a heavy construction, almost bending inwards in the middle, but it is grey, and there is a look of ancientry in the architecture of it that pleased me. As we go on the vale opens out more into one vale, with somewhat of a cradle bed. Cottages, with groups of trees, on the side of the hills. We passed a pair of twin Children, 2 years old. Sate on the next bridge which we crossed—a single arch. We rested again upon the turf, and looked at the same bridge. We observed arches in the water, occasioned by the large stones sending it down in two streams. A sheep came plunging through the river, stumbled up the bank, and passed close to us, it had been frightened by an insignificant little dog on the other side. Its fleece dropped a glittering shower under its belly. Primroses by the road-side, pile wort that shone like stars of gold in the sun, violets, strawberries, retired and half-buried among the grass. When we came to the foot of Brothers Water, I left William sitting on the bridge, and went along the path on the right side of the Lake through the wood. I was delighted with what I saw. The water under the boughs of the bare old trees, the simplicity of the mountains, and the exquisite beauty of the path. There was one grey cottage. I repeated *The Glow-worm*, as I walked along. I hung over the gate, and thought I could have stayed for ever. When I returned, I found William writing a poem descriptive of the sights and sounds we saw and heard.[23] There was the gentle flowing of the stream, the glittering, lively lake, green fields without a living creature to be seen on them, behind us, a flat pasture with 42 cattle feeding; to our left, the road leading to the hamlet. No smoke there, the sun shone on the bare roofs. The people were at work ploughing, harrowing, and sowing; lasses spreading dung, a dog's barking now and then, cocks crowing, birds twittering, the snow in patches at the top of the highest hills, yellow palms, purple and green twigs on the birches, ashes with their glittering spikes quite bare. The hawthorn a bright green, with black stems under the oak. The moss of the oak glossy. We then went on, passed two sisters at work (*they first passed us*), one with two pitchforks in her hand, the other had a spade. We had some talk with them. They laughed aloud after we were gone, perhaps half in wantonness, half boldness. William finished his poem before we got to the foot of Kirkstone. There we ate our dinner. There were hundreds of cattle in the vale. The walk up Kirkstone was very interesting. The becks among the rocks were all alive. Wm. showed me the little mossy streamlet which he had before loved when he saw its bright green track in the snow. The view above Ambleside very beautiful. There we sate and looked down on the green vale. We watched the crows at a little distance from us become white as silver as they flew in the sunshine, and when they went still further, they looked like shapes of water passing over the green fields. The whitening of Ambleside church is a great deduction from the beauty of it, seen from this point. We called at the Luff's, the Boddingtons there. Did not go in, and went round by the fields. I pulled off my stockings, intending to wade the beck, but I was obliged to put them on, and we climbed over the wall at the bridge. The post passed us. No letters! Rydale Lake was in its own evening brightness: the Islands and Points distinct. Jane Ashburner came up to us when we were sitting upon the wall. We rode in her cart to Tom Dawson's. All well. The garden looked pretty in the half-moonlight, half-daylight. As we went up the vale of Brother's Water more and more cattle feeding, 100 of them.

May 7th, Friday. William had slept uncommonly well, so, feeling himself strong, he fell to work at *The Leech Gatherer*[24]; he wrote hard at it till dinner time, then he gave over, tired to death—he had finished the poem. I was making Derwent's[25] frocks. After dinner we sate in the orchard. It was a thick, hazy,

---

23. **a poem . . . heard:** See "Written in March" (actually written in April), above.

24. **The Leech Gatherer:** later called "Resolution and Independence." 25. **Derwent:** Derwent Coleridge, Coleridge's child.

dull air. The thrush sang almost continually; the little birds were more than usually busy with their voices. The sparrows are now full fledged. The nest is so full that they lie upon one another, they sit quietly in their nest with closed mouths. I walked to Rydale after tea, which we drank by the kitchen fire. The evening very dull—a terrible kind of threatening brightness at sunset above Easedale. The sloe-thorn beautiful in the hedges, and in the wild spots higher up among the hawthorns. No letters. William met me. He had been digging in my absence, and cleaning the well. We walked up beyond Lewthwaites. A very dull sky; coolish; crescent moon now and then. I had a letter brought me from Mrs. Clarkson while we were walking in the orchard. I observed the sorrel leaves opening at about 9 o'clock. William went to bed tired with thinking about a poem.

May 8th, Saturday Morning. We sowed the scarlet beans in the orchard, and read *Henry V.* there. William lay on his back on the seat. I wept "For names, sounds, faiths, delights and duties lost"—taken from a poem upon Cowley's wish to retire to the Plantations. Read in the Review. I finished Derwent's frocks. After dinner William added a step to the orchard steps.

May 9th, Sunday Morning. The air considerably colder to-day, but the sun shone all day. William worked at *The Leech Gatherer* almost incessantly from morning till tea-time. I copied *The Leech Gatherer* and other poems for Coleridge. I was oppressed and sick at heart, for he wearied himself to death. After tea he wrote two stanzas in the manner of Thomson's *Castle of Indolence*, and was tired out. Bad news of Coleridge.

FRANCIS JEFFREY

∼⌣∼

Francis Jeffrey, 1773–1850, was editor of *The Edinburgh Review,* the most influential magazine of the period. It was devoted to the interests of the Whig party. Motivated partly by politics—Wordsworth was by this time a Tory—Jeffrey's attacks also result from an honest application of eighteenth-century premises of taste. They reveal how Wordsworth was read by a great many people in his time. His poetry had an immediate appeal for some contemporaries—for example, Coleridge, Lamb, Haydon, Crabb Robinson—and for the gifted minds of a younger generation, such as De Quincey, John Wilson, and Keats. The tone of Jeffrey's criticism is not unduly harsh in an age of savage reviewing. See, for example, Hazlitt on *The Excursion,* below, pp. 613–17, or J. G. Lockhart on Keats, below, pp.1247–49.

∼⌣∼

from REVIEW OF *POEMS*[1] BY GEORGE CRABBE

(Edinburgh Review, *April, 1808*)

It is not quite fair, perhaps, thus to draw a detailed parallel between a living poet, and one whose reputation has been sealed by death,[2] and by the immutable sentence of a surviving generation. Yet there are so few of his contemporaries to whom Mr. Crabbe bears any resemblance, that we can scarcely explain our opinion of his merit, without comparing him to some of his predecessors. There is one set of writers, indeed from whose works those of Mr. Crabbe might receive all that elucidation which results from contrast, and from an entire opposition in all points of taste and opinion. We allude now to the Wordsworths, and the Southeys, and Coleridges, and all that ambitious fraternity, that, with good intentions and extraordinary talents, are labouring to bring back our poetry to the fantastical oddity and puling childishness of Withers, Quarles, or Marvel.[3] These gentlemen write a great deal about rustic life, as well as Mr. Crabbe; and they even agree with him in dwelling much on its discomforts; but nothing can be more opposite than the views they take of the subject, or the manner in which they execute their representation of them.

Mr. Crabbe exhibits the common people of England pretty much as they are, and as they must appear to every one who will take the trouble of examining into their condition; at the same time that he renders his sketches in a very high degree interesting and beautiful —by selecting what is most fit for description—by grouping them into such forms as must catch the attention or awake the memory—and by scattering over the whole such traits of moral sensibility, of sarcasm, and of deep reflection, as every one must feel to be natural, and own to be powerful. The gentlemen of the new school, on the other hand, scarcely ever condescend to take their subjects from any description of persons at all known to the common inhabitants of the world; but invent for themselves certain whimsical and unheard-of beings, to whom they impute

FRANCIS JEFFREY. **1. Poems:** The poems describe rural life in a remarkably realistic way. **2. one . . . death:** Jeffrey has been comparing Crabbe with Oliver Goldsmith, 1728–74. **3. Withers, Quarles, or Marvel:** George Wither (or Withers), 1588–1667, Francis Quarles, 1592–1644, Andrew Marvell, 1621–78.

some fantastical combination of feelings, and then labour to excite our sympathy for them, either by placing them in incredible situations, or by some strained and exaggerated moralisation of a vague and tragical description. Mr. Crabbe, in short, shows us something which we have all seen, or may see, in real life; and draws from it such feelings and such reflections as every human being must acknowledge that it is calculated to excite. He delights us by the truth, and vivid and picturesque beauty of his representations, and by the force and pathos of the sensations with which we feel that they are connected. Mr. Wordsworth and his associates, on the other hand, introduce us to beings whose existence was not previously suspected by the acutest observers of nature; and excite an interest for them—where they do excite any interest—more by an eloquent and refined analysis of their own capricious feelings, than by any obvious or intelligible ground of sympathy in their situation.

Those who are acquainted with the Lyrical Ballads, or the more recent publications of Mr. Wordsworth, will scarcely deny the justice of this representation; but in order to vindicate it to such as do not enjoy that advantage, we must beg leave to make a few hasty references to the former, and by far the least exceptional of those productions.

A village schoolmaster, for instance, is a pretty common poetical character. Goldsmith has drawn him inimitably; so has Shenstone,[4] with the slight change of sex; and Mr. Crabbe, in two passages, has followed their footsteps. Now. Mr. Wordsworth has a village schoolmaster also—a personage who makes no small figure in three or four of his poems.[5] But by what traits is this worthy old gentleman delineated by the new poet? No pedantry—no innocent vanity of learning—no mixture of indulgence with the pride of power, and of poverty with the consciousness of rare acquirements. Every feature which belongs to the situation, or marks the character in common apprehension, is scornfully discarded by Mr. Wordsworth; who represents his grey-haired rustic pedagogue as a sort of half crazy, sentimental person, overrun with fine feelings, constitutional merriment, and a most humorous melancholy. Here are the two stanzas in which this consistent and intelligible character is portrayed. The diction is at least as new as the conception.

> "The sighs which Matthew heav'd were sighs
>   Of one tir'd out with *fun* and *madness;*
> The tears which came to Matthew's eyes
>   Were tears of light—*the oil of gladness,*
>
> "Yet sometimes, when the secret cup
>   Of still and serious thought went round
> He seem'd as if he *drank it up,*
>   He felt with spirit so profound.
> Thou *soul* of God's best *earthly mould,*" &c.[6]

A frail damsel again is a character common enough in all poems; and one upon which many fine and pathetic lines have been expended. Mr. Wordsworth has written more than three hundred on the subject: but, instead of new images of tenderness, or delicate representation of intelligible feelings, he has contrived to tell us nothing whatever of the unfortunate fair one, but that her name is Martha Ray;[7] and that she goes up to the top of a hill, in a red cloak, and cries "O misery!" All the rest of the poem is filled with a description of an old thorn and a pond, and of the silly stories which the neighbouring old women told about them.

The sports of childhood, and the untimely death of promising youth, is also a common topic for poetry. Mr. Wordsworth has made some blank verse about it; but, instead of the delightful and picturesque sketches with which so many authors of moderate talent have presented us on this inviting subject, all that he is pleased to communicate of *his* rustic child, is, that he used to amuse himself with shouting to the owls, and hearing them answer. To make amends for this brevity, the process of his mimicry is most accurately described.

> —— "With fingers interwoven, both hands
> Press'd closely palm to palm, and to his mouth
> Uplifted, he, as through an instrument,
> Blew mimic hootings to the silent owls,
> That they might answer him."—[8]

This is all we hear of him; and for the sake of this one accomplishment, we are told, that the author has frequently stood mute, and gazed on his grave for half an hour together!

Love, and the fantasies of lovers, have afforded an ample theme to poets of all ages. Mr. Wordsworth, however, has thought fit to compose a piece, illustrating

---

4. **Shenstone:** William Shenstone, 1714–63, author of *The Schoolmistress.* **5. three . . . poems:** the "Matthew" poems—"Expostulation and Reply," "Matthew," "The Two April Mornings," and "The Fountain."

6. **"The sighs . . . mould":** "Matthew," ll. 21–29. "Oil," l. 24, was changed to "dew" in the edition of 1815. **7. Martha Ray:** See "The Thorn," above. **8. "With . . . him":** "There Was a Boy," ll. 7–11.

this copious subject by one single thought. A lover trots away to see his mistress one fine evening, gazing all the way on the moon; when he comes to her door,

> "O mercy! to myself I cried,
> If Lucy should be dead."[9]

And there the poem ends!

Now, we leave it to any reader of common candour and discernment to say, whether these representations of character and sentiment are drawn from that eternal and universal standard of truth and nature, which every one is knowing enough to recognise, and no one great enough to depart from with impunity; or whether they are not formed, as we have ventured to allege, upon certain fantastic and affected peculiarities in the mind or fancy of the author, into which it is most improbable that many of his readers will enter, and which cannot, in some cases, be comprehended without much effort and explanation. Instead of multiplying instances of these wide and wilful aberrations from ordinary nature it may be more satisfactory to produce the author's own admission of the narrowness of the plan upon which he writes, and of the very extraordinary circumstances which he himself sometimes thinks it necessary for his readers to keep in view, if they would wish to understand the beauty or propriety of his delineations.

A pathetic tale of guilt or superstition may be told, we are apt to fancy, by the poet himself, in his general character of poet, with full as much effect as by any other person. An old nurse, at any rate, or a monk or parish clerk, is always at hand to give grace to such a narration. None of these, however, would satisfy Mr. Wordsworth. He has written a long poem of this sort, in which he thinks it indispensably necessary to apprise the reader, that he has endeavoured to represent the language and sentiments of a particular character—of which character he adds, "the reader will have a general notion, if he has ever known a man, *a captain of a small trading vessel*, for example, who, being *past the middle age of life*, has retired upon *an annuity, or small independent income*, to some *village* or country town, of which he was *not a native*, or in which he had not been accustomed to live!"[10]

Now, we must be permitted to doubt, whether, among all the readers of Mr. Wordsworth (few or

many), there is a single individual who has had the happiness of knowing a person of this very peculiar description; or who is capable of forming any sort of conjecture of the particular disposition and turn of thinking which such a combination of attributes would be apt to produce. To us, we will confess, the *annonce* appears as ludicrous and absurd as it would be in the author of an ode or an epic to say, "Of this piece the reader will necessarily form a very erroneous judgment, unless he is apprised, that it was written by a pale man in a green coat—sitting cross-legged on an oaken stool—which a scratch on his nose, and a spelling dictionary on the table."[11]

from REVIEW OF *THE EXCURSION*

(Edinburgh Review, *November, 1814*)

This will never do! It bears no doubt the stamp of the author's heart and fancy: But unfortunately not half so visibly as that of his peculiar system. His former poems were intended to recommend that system, and to bespeak favour for it by their individual merit;—but this, we suspect, must be recommended by the system—and can only expect to succeed where it has been previously established. It is longer, weaker, and tamer, than any of Mr. Wordsworth's other productions; with less boldness of originality, and less even of that extreme simplicity and lowliness of tone which wavered so prettily, in the Lyrical Ballads, between silliness and pathos. We have imitations of Cowper, and even of Milton here; engrafted on the natural drawl of the Lakers—and all diluted into harmony by that profuse and irrepressible wordiness which deluges all the blank verse of this school of poetry, and lubricates and weakens the whole structure of their style.

Though it fairly fills four hundred and twenty

---

9. **"O . . . dead"**: "Strange Fits of Passion Have I Known," ll. 27–28. 10. **"the reader . . . live!"**: quoted from Wordsworth's note to "The Thorn."

11. **table:** "Some of our readers may have a curiosity to know in what manner this old annuitant captain does actually express himself in the village of his adoption. For their gratification, we annex the first two stanzas of his; in which, with all the attention we have been able to bestow, we have been utterly unable to detect any traits that can be supposed to characterise either a seaman, an annuitant, or a stranger in a country town. It is a style, on the contrary, which we should ascribe, without hesitation, to a certain poetical fraternity in the West of England; and which we verily believe, never was, and never will be, used by any one out of that fraternity." (Jeffrey's note. He goes on to quote the beginning of "The Thorn.")

good quarto pages, without note, vignette, or any sort of extraneous assistance, it is stated in the title—with something of an imprudent candour—to be but "a portion" of a larger work; and in the preface, where an attempt is rather unsuccessfully made to explain the whole design, it is still more rashly disclosed, that it is but "*a part of the second part, of a long and laborious work*"—which is to consist of three parts!

What Mr. Wordsworth's ideas of length are, we have no means of accurately judging: But we cannot help suspecting that they are liberal to a degree that will alarm the weakness of most modern readers. As far as we can gather from the preface, the entire poem[12]—or one of them (for we really are not sure whether there is to be one or two), is of a biographical nature; and is to contain the history of the author's mind, and of the origin and progress of his poetical powers, up to the period when they were sufficiently matured to qualify him for the great work on which he has been so long employed. Now, the quarto before us contains an account of one of his youthful rambles in the vales of Cumberland, and occupies precisely the period of three days! So that, by the use of a very powerful *calculus*, some estimate may be formed of the probable extent of the entire biography.

This small specimen, however, and the statements with which it is prefaced, have been sufficient to set our minds at rest in one particular. The case of Mr. Wordsworth, we perceive, is now manifestly hopeless; and we give him up as altogether incurable, and beyond the power of criticism. We cannot, indeed, altogether omit taking precautions now and then against the spreading of the malady;—but for himself, though we shall watch the progress of his symptoms as a matter of professional curiosity and instruction, we really think it right not to harass him any longer with nauseous remedies,—but rather to throw in cordials and lenitives, and wait in patience for the natural termination of the disorder. In order to justify this desertion of our patient, however, it is proper to state why we despair of the success of a more active practice.

A man who has been for twenty years at work on such matter as is now before us, and who comes complacently forward with a whole quarto of it, after all the admonitions he has received, cannot reasonably be expected to "change his hand, or check his pride,"

upon the suggestion of far weightier monitors than we can pretend to be. Inveterate habit must now have given a kind of sanctity to the errors of early taste; and the very powers of which we lament the perversion, have probably become incapable of any other application. The very quantity, too, that he has written, and is at this moment working up for publication upon the old pattern, makes it almost hopeless to look for any change of it. All this is so much capital already sunk in the concern; which must be sacrificed if that be abandoned: and no man likes to give up for lost the time and talent and labour which he has embodied in any permanent production. We were not previously aware of these obstacles to Mr. Wordsworth's conversion; and, considering the peculiarities of his former writings merely as the result of certain wanton and capricious experiments on public taste and indulgence, conceived it to be our duty to discourage their repetition by all the means in our power. We now see clearly, however, how the case stands;—and, making up our minds, though with the most sincere pain and reluctance, to consider him as finally lost to the good cause of poetry, shall endeavour to be thankful for the occasional gleams of tenderness and beauty which the natural force of his imagination and affections must still shed over all his productions,—and to which we shall ever turn with delight, in spite of the affectation and mysticism and prolixity, with which they are so abundantly contrasted.

Long habits of seclusion, and an excessive ambition of originality, can alone account for the disproportion which seems to exist between this author's taste and his genius; or for the devotion with which he has sacrificed so many precious gifts at the shrine of those paltry idols which he has set up for himself among his lakes and his mountains. Solitary musings, amidst such scenes, might no doubt be expected to nurse up the mind to the majesty of poetical conception—(though it is remarkable, that all the greater poets lived, or had lived, in the full current of society):—But the collision of equal minds,—the admonition of prevailing impressions—seems necessary to reduce its redundancies, and repress that tendency to extravagance or puerility, into which the self-indulgence and self-admiration of genius is so apt to be betrayed, when it is allowed to wanton, without awe or restraint, in the triumph and delight of its own intoxication. That its flights should be graceful and glorious in the eyes of men, it seems almost to be necessary that they should be made in the consciousness that men's eyes are to

12. **entire poem:** i.e., *The Recluse*. See headnote to *The Excursion*, above.

behold them,—and that the inward transport and vigour by which they are inspired, should be tempered by an occasional reference to what will be thought of them by those ultimate dispensers of glory. An habitual and general knowledge of the few settled and permanent maxims, which form the canon of general taste in all large and polished societies—a certain tact, which informs us at once that many things, which we still love and are moved by in secret, must necessarily be despised as childish, or derided as absurd, in all such societies—though it will not stand in the place of genius, seems necessary to the success of its exertions; and though it will never enable any one to produce the higher beauties of art, can alone secure the talent which does produce them from errors that must render it useless. Those who have most of the talent, however, commonly acquire this knowledge with the greatest facility;—and if Mr. Wordsworth, instead of confining himself almost entirely to the society of the dalesmen and cottagers, and little children, who form the subjects of his book, had condescended to mingle a little more with the people that were to read and judge of it, we cannot help thinking that its texture might have been considerably improved: At least it appears to us to be absolutely impossible, that any one who had lived or mixed familiarly with men of literature and ordinary judgment in poetry (of course we exclude the coadjutors and disciples of his own school), could ever have fallen into such gross faults, or so long mistaken them for beauties. His first essays we looked upon in a good degree as poetical paradoxes, —maintained experimentally, in order to display talent, and court notoriety;—and so maintained, with no more serious belief in their truth, than is usually generated by an ingenious and animated defence of other paradoxes. But when we find that he has been for twenty years exclusively employed upon articles of this very fabric, and that he has still enough of raw material on hand to keep him so employed for twenty years to come, we cannot refuse him the justice of believing that he is a sincere convert to his own system, and must ascribe the peculiarities of his composition, not to any transient affectation, or accidental caprice of imagination, but to a settled perversity of taste or understanding, which has been fostered, if not altogether created, by the circumstances to which we have alluded.

The volume before us, if we were to describe it very shortly, we should characterise as a tissue of moral and devotional ravings, in which innumerable changes are rung upon a few simple and familiar ideas:

—But with such an accompaniment of long words, long sentences, and unwieldy phrases—and such a hubbub of strained raptures and fantastical sublimities, that it is often difficult for the most skilful and attentive student to obtain a glimpse of the author's meaning —and altogether impossible for an ordinary reader to conjecture what he is about. Moral and religious enthusiasm, though undoubtedly poetical emotions, are at the same time but dangerous inspirers of poetry; nothing being so apt to run into interminable dulness or mellifluous extravagance, without giving the unfortunate author the slightest intimation of his danger. His laudable zeal for the efficacy of his preachments, he very naturally mistakes for the ardour of poetical inspiration;—and, while dealing out the high words and glowing phrases which are so readily supplied by themes of this description, can scarcely avoid believing that he is eminently original and impressive:— All sorts of commonplace notions and expressions are sanctified in his eyes, by the sublime ends for which they are employed; and the mystical verbiage of the Methodist pulpit is repeated, till the speaker entertains no doubt that he is the chosen organ of divine truth and persuasion. But if such be the common hazards of seeking inspiration from those potent fountains, it may easily be conceived what chance Mr. Wordsworth had of escaping their enchantment,—with his natural propensities to wordiness, and his unlucky habit of debasing pathos with vulgarity. The fact accordingly is, that in this production he is more obscure than a Pindaric poet of the seventeenth century; and more verbose "than even himself of yore;" while the wilfulness with which he persists in choosing his examples of intellectual dignity and tenderness exclusively from the lowest ranks of society, will be sufficiently apparent, from the circumstance of his having thought fit to make his chief prolocutor in this poetical dialogue, and chief advocate of Providence and virtue, *an old Scotch Pedlar*—retired indeed from business—but still rambling about in his former haunts, and gossiping among his old customers, without his pack on his shoulders. The other persons of the drama are, a retired military chaplain, who has grown half an atheist and half a misanthrope—the wife of an unprosperous weaver—a servant girl with her natural child—a parish pauper, and one or two other personages of equal rank and dignity.

The character of the work is decidedly didactic; and more than nine tenths of it are occupied with a species of dialogue, or rather a series of long sermons or harangues, which pass between the pedlar, the author, the old chaplain, and a worthy vicar, who entertains the whole party at dinner on the last day of their excursion. The incidents which occur in the course of it are as few and trifling as can well be imagined;—and those which the different speakers narrate in the course of their discourses, are introduced rather to illustrate their arguments or opinions, than for any interest they are supposed to possess of their own.—The doctrine which the work is intended to enforce, we are by no means certain that we have discovered. In so far as we can collect, however, it seems to be neither more nor less than the old familiar one, that a firm belief in the providence of a wise and beneficent Being must be our great stay and support under all afflictions and perplexities upon earth—and that there are indications of his power and goodness in all the aspects of the visible universe, whether living or inanimate—every part of which should therefore be regarded with love and reverence, as exponents of those great attributes. We can testify, at least, that these salutary and important truths are inculcated at far greater length, and with more repetitions, than in any ten volumes of sermons that we ever perused. It is also maintained, with equal conciseness and originality, that there is frequently much good sense, as well as much enjoyment, in the humbler conditions of life; and that, in spite of great vices and abuses, there is a reasonable allowance both of happiness and goodness in society at large. If there be any deeper or more recondite doctrines in Mr. Wordsworth's book, we must confess that they have escaped us;—and, convinced as we are of the truth and soundness of those to which we have alluded, we cannot help thinking that they might have been better enforced with less parade and prolixity. His effusions on what may be called the physiognomy of external nature, or its moral and theological expression, are eminently fantastic, obscure, and affected.—It is quite time, however, that we should give the reader a more particular account of this singular performance.

It opens with a picture of the author toiling across a bare common in a hot summer day, and reaching at last a ruined hut surrounded with tall trees, where he meets by appointment with a hale old man, with an iron-pointed staff lying beside him. Then follows a retrospective account of their first acquaintance—formed, it seems, when the author was at a village school; and his aged friend occupied "one room,—the fifth part of a house" in the neighbourhood. After this, we have the history of this reverend person at no small

length. He was born, we are happy to find, in Scotland —among the hills of Athol; and his mother, after his father's death, married the parish schoolmaster—so that he was taught his letters betimes: But then, as it is here set forth with much solemnity,

"From his sixth year, the boy of whom I speak,
In summer, tended cattle on the hills!"

And again, a few pages after, that there may be no risk of mistake as to a point of such essential importance—

"From early childhood, even, as hath been said,
From his *sixth year*, he had been sent abroad,
In summer—to tend herds! Such was his task!"[13]

In the course of this occupation it is next recorded, that he acquired such a taste for rural scenery and open air, that when he was sent to teach a school in a neighbouring village, he found it "a misery to him;" and determined to embrace the more romantic occupation of a Pedlar—or, as Mr. Wordsworth more musically expresses it,

"A vagrant merchant, bent beneath his load;"[14]

—and in the course of his peregrinations had acquired a very large acquaintance, which, after he had given up dealing, he frequently took a summer ramble to visit.

The author, on coming up to this interesting personage, finds him sitting with his eyes half shut;—and, not being quite sure whether he is asleep or awake, stands "some minutes' space" in silence beside him. "At length," says he, with his own delightful simplicity—

"At length I hail'd him—*seeing that his hat
Was moist* with water-drops, as if the brim
Had newly scoop'd a running stream!—
———''Tis,' said I, 'a burning day!
My lips are parch'd with thirst;—but you, I guess,
Have somewhere found relief.'"

Upon this, the benevolent old man points him out, not a running stream, but a well in a corner, to which the author repairs; and, after minutely describing its situation, beyond a broken wall, and between two

13. "From early . . . task!": the 1814 text, for which I. 197–99 were substituted in 1827. 14. "A vagrant . . . load": the 1814 text of I. 324, revised (and made less musical) in 1837.

alders that "grew in a cold, damp nook," he thus faithfully chronicles the process of his return:—

"My thirst I slak'd; and from the cheerless spot
    Withdrawing, straightway to the shade return'd,
    Where sate the old man on the cottage bench."

The Pedlar then gives an account of the last inhabitants of the deserted cottage beside them. These were, a good industrious weaver and his wife and children. They are very happy for a while; till sickness and want of work came upon them; and then the father enlisted as a soldier, and the wife pined in that lonely cottage—growing every year more careless and desponding, as her anxiety and fears for her absent husband, of whom no tidings ever reached her, accumulated. Her children died, and left her cheerless and alone; and at last she died also; and the cottage fell to decay. We must say, that there is very considerable pathos in the telling of this simple story; and that they who can get over the repugnance excited by the triteness of its incidents, and the lowness of its objects, will not fail to be struck with the author's knowledge of the human heart, and the power he possesses of stirring up its deepest and gentlest sympathies. His prolixity, indeed, it is not so easy to get over. This little story fills about twenty-five quarto pages; and abounds, of course, with mawkish sentiment, and details of preposterous minuteness. When the tale is told, the travellers take their staves, and end their first day's journey, without further adventure, at a little inn.

. . . .

Nobody can be more disposed to do justice to the great powers of Mr. Wordsworth than we are; and, from the first time that he came before us, down to the present moment, we have uniformly testified in their favour, and assigned indeed our high sense of their value as the chief ground of the bitterness with which we resented their perversion. That perversion, however, is now far more visible than their original dignity; and while we collect the fragments, it is impossible not to mourn over the ruins from which we are condemned to pick them. If any one should doubt of the existence of such a perversion, or be disposed to dispute about the instances we have hastily brought forward, we would just beg leave to refer him to the general plan and character of the poem now before us. Why should Mr. Wordsworth have made his hero a superannuated Pedlar? What but the most wretched affectation, or provoking perversity of taste, could induce any one to place his chosen advocate of wisdom and virtue in so absurd and fantastic a condition? Did Mr. Wordsworth really imagine, that his favourite doctrines were likely to gain any thing in point of effect or authority, by being put into the mouth of a person accustomed to higgle about tape, or brass sleeve-buttons? Or is it not plain, that, independent of the ridicule and disgust which such a personification must excite in many of his readers, its adoption exposes his work throughout to the charge of revolting incongruity, and utter disregard of probability or nature? For, after he has thus wilfully debased his moral teacher by a low occupation, is there one word that he puts into his mouth, or one sentiment of which he makes him the organ, that has the most remote reference to that occupation? Is there any thing in his learned, abstract, and logical harangues, that savours of the calling that is ascribed to him? Are any of their materials such as a pedlar could possibly have dealt in? Are the manners, the diction, the sentiments, in any, the very smallest degree, accommodated to a person in that condition? or are they not eminently and conspicuously such as could not by possibility belong to it? A man who went about selling flannel and pocket-handkerchiefs in this lofty diction, would soon frighten away all his customers; and would infallibly pass either for a madman, or for some learned and affected gentleman, who, in a frolic, had taken up a character which he was peculiarly ill qualified for supporting.

The absurdity in this case, we think, is palpable and glaring: but it is exactly of the same nature with that which infects the whole substance of the work—a puerile ambition of singularity engrafted on an unlucky predilection for truisms; and an affected passion for simplicity and humble life, most awkwardly combined with a taste for mystical refinements, and all the gorgeousness of obscure phraseology. His taste for simplicity is evinced by sprinkling up and down his interminable declamations a few descriptions of baby-houses, and of old hats with wet brims; and his amiable partiality for humble life, by assuring us that a wordy rhetorician, who talks about Thebes, and allegorizes all the heathen mythology, was once a pedlar—and making him break in upon his magnificent orations with two or three awkward notices of something that he had seen when selling winter raiment about the country—or of the changes in the state of society, which had almost annihilated his former calling.

The following extracts are reminiscences of Wordsworth in his old age. Rawnsley's paper, a minor classic in its way, was written in 1881. The recollections of the persons interviewed go back from forty to sixty years. For other personal descriptions and accounts of Wordsworth, see Hazlitt, "My First Acquaintance with Poets," below, p. 686, and Haydon's *Autobiography*, March 7, 1821, April 13, 1821, and May 23, 1821, below, p. 773.

H. D. RAWNSLEY

from REMINISCENCES OF WORDSWORTH AMONG THE PEASANTRY OF WESTMORELAND

I asked whether Mr. Wordsworth was much thought of. He[1] replied, 'Latterly, but we thowt li'le eneuf on him. He was nowt to li'le Hartley.[2] Li'le Hartley was a philosopher, you see; Wudsworth was a poet. Ter'ble girt difference betwixt them two wayses, ye kna.' I asked whether he had ever found that poems of Mr. Wordsworth were read in the cottages, whether he had read them himself. 'Well you see, blessed barn, there's pomes and pomes, and Wudsworth's was not for sec as us. I never did see his pomes—not as I can speak to in any man's house in these parts, but,' he added, 'ye kna there's bits in t' papers fra time to time bearing his naame.'

This unpopularity of Wordsworth's poems among the peasantry was strangely corroborated that very same day by an old man whom I met on the road, who said he had often seen the poet, and had once been present and heard him make a long speech, and that was at the laying of the foundation of the Boys' Schoolroom at Bowness, which was built by one Mr. Bolton of Storrs Hall.

On that occasion Mr. 'Wudsworth talked lang and weel eneuf,' and he remembered that he 'had put a pome he had written into a bottle wi' some coins in the hollow of the foundation-stone.'

I asked him whether he had ever seen or read any of the poet's works, and he answered, 'Nay, not likely;

for Wudsworth wasn't a man as wreat on separate bits, saäme as Hartley Coleridge, and was niver a frequenter of public-houses, or owt of that sort.' But he added, 'He was a good writer, he supposed, and he was a man folks thowt a deal on i' t' dale: he was sic a weel-meäning, decent, quiet man.'

But to return to my host at the public. Wordsworth, in his opinion, was not fond of children, nor animals. He would come round the garden, but never 'say nowt.' Sometimes, but this was seldom, he would say, 'Oh! you're planting peas?' or, 'Where are you setting onions?' but only as a master would ask a question of a servant. He had, he said, never seen him out of temper once, neither in the garden, nor when he was along o' Miss Dorothy in her invalid chair.[3] But, he added, 'What went on i' t' hoose I can't speak till'; meaning that as an outdoor servant he had no sufficiently accurate knowledge of the in-door life to warrant his speaking of it. Wordsworth was not an early riser, had no particular flower he was fondest of that he could speak to; never was heard to sing or whistle a tune in his life; there 'was noa two words about that, though he bummed a deäl';—of this more presently.

'He was a plain man, plainly dressed, and so was she, ya mun kna. But eh, blessed barn! he was fond o' his own childer, and fond o' Dorothy, especially when she was faculty strucken, poor thing; and as for his wife, there was noa two words about their being truly companionable; and Wudsworth was a silent man wi'out a doubt, but he was not aboon bein' tender and quite *monstrable* [demonstrative] at times in his oan family.'

I asked about Mr. Wordsworth's powers of observation. Had he noticed in his garden walks how he stooped down and took this or that flower, or smelt this or that herb? (I have heard since that the poet's sense of smell was limited.) 'Na, he wadna speak to that, but Mr. Wudsworth was what you might call a vara practical-eyed man, a man as seemed to see aw that was stirrin'.'

Perhaps the most interesting bit of information I obtained, before our pleasant chat was at an end, was a description of the way in which the poet composed on the grass terrace at Rydal Mount. 'Eh! blessed barn, my informant continued, 'I think I can see him at it

---

H. D. RAWNSLEY. **1. He:** a former gardener's boy at Rydal Mount. **2. Hartley:** Coleridge's son.

3. invalid chair: From 1831 to her death in 1855 Dorothy was in feeble health, often bedridden, and her mental faculties were impaired.

now, He was ter'ble thrang with visitors and folks, you mun kna, at times, but if he could git awa fra them for a spell, he was out upon his gres walk; and then he would set his heäd a bit forrad, and put his hands behint his back. And then he would start a bumming, and it was bum, bum, bum, stop; then bum, bum, bum reet down till t'other end, and then he'd set down and git a bit o' paper out and write a bit; and then he git up, and bum, bum, bum, and goa on bumming for long enough right down and back agean. I suppose, ya kna, the bumming helped him out a bit. However, his lips was always goan' whoale time he was upon the gres walk. He was a kind mon, there's no two words about that: if any one was sick i' the plaace, he wad be off to see til 'em.'

. . .

'Well, George,[4] what sort o' a man in personal appearance was Mr. Wordsworth?'

'He was what you might ca' a ugly man,—mak of John Rigg much,—much about seame height, 6 feet or 6 feet 2,[5]—smaller, but deal rougher in the face.'

I knew John Rigg by sight, and can fancy from the pictures of the poet that the likeness is striking in the brow and profile.

'But he was,' continued George, 'numbledy in t' kneas, walked numbledy, ye kna, but that might o' wussened wi' age.' In George's mind age accounted for most of the peculiarities he had noticed in the poet, but George's memory could go back fifty years, and he ought to have remembered Wordsworth as hale and hearty. 'He wozn't a man as said a deal to common fwoak. But he talked a deal to hissen. I often seead his lips a gaäin', and he'd a deal o' mumblin' to hissel, and 'ud stop short and be a lookin' down upo' the ground, as if he was in a thinkin' waäy. But that might ha' growed on him wi' age, an' aw, ye kna.'

. . .

'He was yan as keppit his heäd doon and eyes upo' t' ground, and mumbling to hissel; but why, why, he 'ud never pass folks draining, or ditching, or walling a cottage, but what he'd stop and say, "Eh dear, but it's a pity to move that stean, and doant ya think ya might leave that tree?" I[6] mind there was a walling

chap just going to shoot a girt stean to bits wi' powder i' t' grounds at Rydal, and he came up and saaved it, and wreat summat on it.'

'But what was his reason,' I asked, 'for stopping the wallers or ditchers, or tree-cutters, at their work?'

'Well, well, he couldn't bide to see t' faäce o' things altered, ye kna. It was all along of him that Grasmere folks have their Common open. Ye may ga now reet up to t' sky ower Guzedale, wi'out liggin' leg to t' fence, and all through him. He said it was a pity to enclose it and run walls over it, and the quality backed him, and he won. Fwoaks was angry eneuf, and wreat rhymes aboot it; but why, why, it's a deal pleasanter for them as walks up Grisedale, ye kna, let alean reets o' foddering and goosage for freemen i' Gersmer.'

'Wudsworth,' my kindly giant[7] replied, 'for a' he hed nea pride nor nowt, was a man who was quite one to hissel, ye kna. He was not a man as fwoaks could crack wi', nor not a man as could crack wi' fwoaks. But thear was anudder thing as kep' fwoaks off, he hed a terr'ble girt deep voice, and ye med see his faace agaan for lang eneuf. I've knoan folks, village lads and lasses, coming ower by t' auld road aboon what runs fra Gersmer to Rydal, flayt a'most to death there by t' Wishing Gate to hear t' girt voice a groanin' and mutterin' and thunderin' of a still evening. And he hed a way of standin' quite still by t' rock there in t' path under Rydal, and fwoaks could hear sounds like a wild beast coming fra t' rocks, and childer were scared fit to be dëad a'most.'

SIR HENRY TAYLOR

from AUTOBIOGRAPHY (1885)

They[1] were as little alike in their aspect as in their genius. The only thing common to both countenances was that neither expressed a limitation. You might not have divined from either frontispiece the treasures of the volume—it was not likely that you should—but when you knew that there they were, there was

---

4. **George:** a laborer who saw Wordsworth frequently. 5. **6 feet 2:** Actually Wordsworth was a little over 5 feet 9 inches by Haydon's measurement. Haydon thought it "a very fine, heroic proportion." 6. **I:** The person speaking was a builder in Westmorland.

7. **giant:** a farmer and hotel keeper in the Rydal neighborhood. SIR HENRY TAYLOR. 1. **They:** Wordsworth and Sir Walter Scott. Taylor's description is of Wordsworth in 1831.

nothing but what harmonized with your knowledge. Both were the faces of considerable men. Scott's had a character of rusticity. Wordsworth's was a face which did not assign itself to any class. It was a hardy, weather-beaten old face, which might have belonged to a nobleman, a yeoman, a mariner, or a philosopher; for there was so much of a man that you lost sight of superadded distinctions. For my own part I should not, judging by his face, have guessed him to be a poet. To my eyes there was more of strength than refinement in the face. But I think he took a different view of it himself. Whatever view he took, if occasion arose, he would be sure to disclose it; for his thoughts went naked. I was once discussing with him the merits of a picture of himself, hanging on the wall in Lockhart's house in London. Some one had said it was like—

"Yes," he replied, "I cannot deny that there is a likeness; such a likeness as the artist could produce; it is like me so far as *he* could go in me; it is like if you suppose all the finer faculties of the mind to be withdrawn: that, I should say, is Wordsworth, the chancellor of the exchequer—Wordsworth, the Speaker of the House of Commons."

In this there was not more vanity than belongs to other men; the difference being that what there was, like everything else in him, was wholly indisguised. He naturally took an interest in his own looks, and wished to take the most favorable view of them; as most men do, though most men do not make mention of it. And there is something to be said for his view. Perhaps what was wanting was only *physical* refinement. It was a rough, gray face, full of rifts and clefts and fissures, out of which, some one said, you might expect lichens to grow.

THOMAS CARLYLE

from REMINISCENCES (1881)

On a summer morning (let us call it 1840 then) I was apprised by Taylor[1] that Wordsworth had come to town, and would meet a small party of us at a certain tavern in St. James's Street, at breakfast, to which I was invited for the given day and hour. We had a pretty little room, quiet though looking streetward (tavern's name is quite lost to me); the morning sun was pleasantly tinting the opposite houses, a balmy, calm, and sunlight morning. Wordsworth, I think, arrived just along with me; we had still five minutes of sauntering and miscellaneous talking before the whole were assembled. I do not positively remember any of them, except that James Spedding was there, and that the others, not above five or six in whole, were polite intelligent quiet persons, and, except Taylor and Wordsworth, not of any special distinction in the world. Breakfast was pleasant, fairly beyond the common of such things. Wordsworth seemed in good tone, and, much to Taylor's satisfaction, talked a great deal; about "poetic" correspondents of his own (i.e. correspondents for the sake of his poetry; especially one such who had sent him, from Canton, an excellent chest of tea; correspondent grinningly applauded by us all); then about ruralities and miscellanies; "Countess of Pembroke" antique she-Clifford, glory of those northern parts, who was not new to any of us, but was set forth by Wordsworth with gusto and brief emphasis. . . . These were the first topics. Then finally about literature, literary laws, practices, observances, at considerable length, and turning wholly on the mechanical part, including even a good deal of shallow enough etymology, from me and others, which was well received. On all this Wordsworth enlarged with evident satisfaction, and was joyfully reverent of the "wells of English undefiled;" though stone dumb as to the deeper rules and wells of Eternal Truth and Harmony, which you were to try and set forth by said undefiled wells of English or what other speech you had! To me a little disappointing, but not much; though it would have given me pleasure had the robust veteran man emerged a little out of vocables into things, now and then, as he never once chanced to do. For the rest, he talked well in his way; with veracity, easy brevity and force, as a wise tradesman would of his tools and workshop,—and as no unwise one could. His voice was good, frank and sonorous, though practically clear distinct and forcible rather than melodious; the tone of him businesslike, sedately confident; no discourtesy, yet no anxiety about being courteous. A fine wholesome rusticity, fresh as his mountain breezes, sat well on the stalwart veteran, and on all he said and did. You would have said he was a usually taciturn man; glad to unlock himself to audience sympathetic and intelligent, when such offered itself. His face bore marks of much,

THOMAS CARLYLE. **1. Taylor:** Sir Henry Taylor.

not always peaceful, meditation; the look of it not bland or benevolent so much as close impregnable and hard: a man *multa tacere loquive paratus*,[2] in a world where he had experienced no lack of contradictions as he strode along! The eyes were not very brilliant, but they had a quiet clearness; there was enough of brow and well shaped; rather too much of cheek ("horse face" I have heard satirists say); face of squarish shape and decidedly longish, as I think the head itself was (its "length" going horizontal); he was large-boned, lean, but still firm-knit tall and strong-looking when he stood, a right good old steel-grey figure, with rustic simplicity and dignity about him, and a vivacious strength looking through him which might have suited one of those old steel-grey markgrafs[3] whom Henry the Fowler set up to ward the "marches" and do battle with the intrusive heathen in a stalwart and judicious manner.

2. **multa . . . paratus:** ready to speak or be silent.

3. **markgrafs:** military lords, keepers of the borders or "marches" in Germany.

SIR WALTER SCOTT

1771–1832

THOUGH SCOTT is remembered primarily for his novels, he wrote prolifically in many other forms. These range from poetry—and even a few plays—to biography, folklore, and editions of past writers (Dryden and Swift). He was also an antiquarian and was versed not only in British but in European history as a whole. In addition he had a career as a lawyer, holding various legal posts throughout his life.

Born in Edinburgh at a time when that city had begun to rival London and Paris as a cultural center, Scott read widely as a boy—possibly all the more because he was somewhat lame. He learned Latin, French, Italian and, what was unusual at the time, some German. Yet he was no recluse; through a calm but constant effort of will, he lived as active an outdoor life as he could. After he became a member of the bar at the age of twenty-one, he devoted spare time to riding about the Scottish Border country, and a few years later brought out a three-volume collection of ballads, *Minstrelsy of the Scottish Border* (1802–03). This was followed by his own poetic romances, *The Lay of the Last Minstrel* (1805), *Marmion* (1808), and *The Lady of the Lake* (1810).

Meanwhile, in 1805 he had entered as a silent partner in a bookselling business, John Ballantyne and Co., and also helped promote the founding of the Tory journal *The Quarterly Review* (1809). He bought a farm on the Tweed (1812) and began to build a mansion, "Abbotsford," where he hoped to live the life of a country gentleman. He turned down the offer of the Poet Laureateship (1813), which was then awarded to Robert Southey, partly at Scott's suggestion. The immense popularity of Byron's verse romances had now somewhat dimmed the appeal of Scott's; partly for this reason, and partly from the need for more money to maintain his position and new commitments, he turned to prose fiction. The decision to do so proved fortunate, and *Waverly*

(1814) was followed by another twenty novels in the course of the next ten years, including *Guy Mannering* (1815), *Rob Roy* (1817), *The Heart of Midlothian* (1818), *Ivanhoe* (1819), *Kenilworth* (1821), and *Quentin Durward* (1823). Those dealing with the recent past of Scotland are still valued for their blend of character, folklore, scenic atmosphere, and humor, while his historical fiction virtually created—certainly established—the historical novel as an important literary form. Though all of these works were written anonymously, Scott was often guessed to be the author and in 1819 was created a baronet.

Because of unwise management, the bookselling business in which he was a partner failed in 1826, and Scott found himself liable for £130,000. He could have gone into bankruptcy and thus been able to keep whatever he earned afterwards. Instead he assumed full obligation, and he worked constantly at his writing until his death six years later (1832). When he died, the debts, through the sale of Scott's copyrights, were finally discharged.

Though his poetic romances were extremely popular when first published, contemporary reviewers were not complimentary. They found these poems rambling in construction and undistinguished in style. Later critics, so far as they deal with this poetry at all, have been more favorably disposed, but the tendency is to prefer the lyrics or songs to the narrative verse. In any case his novels have completely overshadowed his poetry ever since they first appeared.

Bibliography

J. G. Lockhart's *Life* (1837–38) is one of the great English biographies and the basis of all subsequent lives of Scott. Among the many more recent biographies, that by John Buchan, 1932, is excellent.

FROM

THE LAY OF
THE LAST MINSTREL

ᴄᴡᴄ

Scott's hope, as he said in the Preface to this poem, was to "illustrate the customs and manners which anciently prevailed on the Borders of Scotland and England. The inhabitants living in a state partly pastoral and partly warlike, and combining habits of constant depredation with the influence of a rude spirit of chivalry, were often engaged in scenes highly susceptible of poetical ornament. As the description of scenery and manners was more the object of the author than a combined and regular narrative, the plan of the ancient metrical romance was adopted, which allows greater latitude in this respect than would be consistent with the dignity of a regular poem." The narrator of the poem is "an ancient minstrel, the last of the race, who as he is supposed to have survived the Revolution, might have caught something of the refinement of modern poetry, without losing the simplicity of his original model. The date of the tale itself is about the middle of the sixteenth century, when most of the persons actually flourished." The meter is patterned after that of Coleridge's "Christabel" (see below, p. 414), which, though still unpublished, Scott had recently heard a friend of Coleridge recite. *The Lay of the Last Minstrel*, written 1802–04, was published in 1805. The poem is in six cantos. The famous lines beginning "Breathes there the man, with soul so dead," are the opening lines of Canto VI.

ᴄᴡᴄ

INTRODUCTION

The way was long, the wind was cold,
The Minstrel was infirm and old;
His withered cheek and tresses gray
Seemed to have known a better day;
The harp, his sole remaining joy,
Was carried by an orphan boy.
The last of all the bards was he,
Who sung of Border chivalry;
For, well-a-day! their date was fled,
His tuneful brethren all were dead;   10
And he, neglected and oppressed,
Wished to be with them and at rest.
No more on prancing palfrey borne,
He carolled, light as lark at morn;
No longer courted and caressed,
High placed in hall, a welcome guest,
He poured, to lord and lady gay,
The unpremeditated lay:
Old times were changed, old manners gone;
A stranger filled the Stuarts' throne;   20
The bigots of the iron time
Had called his harmless art a crime.
A wandering harper, scorned and poor,
He begged his bread from door to door,
And tuned, to please a peasant's ear,
The harp a king had loved to hear.
He passed where Newark's stately tower
Looks out from Yarrow's birchen bower:
The Minstrel gazed with wishful eye—
No humbler resting-place was nigh.   30
With hesitating step at last
The embattled portal arch he passed,
Whose ponderous grate and massy bar
Had oft rolled back the tide of war,
But never closed the iron door
Against the desolate and poor.
The Duchess marked his weary pace,
His timid mien, and reverend face,
And bade her page the menials tell
That they should tend the old man well: 40
For she had known adversity,
Though born in such a high degree;
In pride of power, in beauty's bloom,
Had wept o'er Monmouth's bloody tomb!

When kindness had his wants supplied,
And the old man was gratified,
Began to rise his minstrel pride;
And he began to talk anon
Of good Earl Francis, dead and gone,
And of Earl Walter, rest him God!   50
A braver ne'er to battle rode;
And how full many a tale he knew
Of the old warriors of Buccleuch:
And, would the noble Duchess deign
To listen to an old man's strain,
Though stiff his hand, his voice though weak,
He thought even yet, the sooth to speak,
That, if she loved the harp to hear,
He could make music to her ear.

The humble boon was soon obtained;   60
The aged Minstrel audience gained.
But when he reached the room of state
Where she with all her ladies sate,
Perchance he wished his boon denied:
For, when to tune his harp he tried,
His trembling hand had lost the ease
Which marks security to please;
And scenes, long past, of joy and pain
Came wildering o'er his aged brain—
He tried to tune his harp in vain.   70

The pitying Duchess praised its chime,
And gave him heart, and gave him time,
Till every string's according glee
Was blended into harmony.
And then, he said, he would full fain
He could recall an ancient strain
He never thought to sing again.
It was not framed for village churls,
But for high dames and mighty earls;
He had played it to King Charles the Good   80
When he kept court in Holyrood;
And much he wished, yet feared, to try
The long-forgotten melody.
Amid the strings his fingers strayed,
And an uncertain warbling made,
And oft he shook his hoary head.
But when he caught the measure wild,
The old man raised his face and smiled;

THE LAY OF THE LAST MINSTREL. INTRODUCTION. **20–21. stranger . . . bigots:** Cromwell and the Puritans. **44. Monmouth:** James Scott, Duke of Monmouth, 1649–85, illegitimate son of Charles II, executed for leading a rebellion against James II. **80. Charles the Good:** Charles I. **81. Holyrood:** royal palace in Edinburgh.

And lightened up his faded eye
With all a poet's ecstasy!   90
In varying cadence, soft or strong,
He swept the sounding chords along:
The present scene, the future lot,
His toils, his wants, were all forgot;
Cold diffidence and age's frost
In the full tide of song were lost;
Each blank, in faithless memory void,
The poet's glowing thoughts supplied;
And, while his harp responsive rung,
'Twas thus the LATEST MINSTREL sung.   100

                  1802–04 (1805)

from CANTO VI

Breathes there the man, with soul so dead,
Who never to himself hath said,
   This is my own, my native land?
Whose heart hath ne'er within him burned,
As home his footsteps he hath turned
   From wandering on a foreign strand?
If such there breathe, go, mark him well;
For him no minstrel raptures swell;
High though his titles, proud his name,
Boundless his wealth as wish can claim,—   10
Despite those titles, power, and pelf,
The wretch, concentred all in self,
Living, shall forfeit fair renown,
And, doubly dying, shall go down
To the vile dust from whence he sprung,
Unwept, unhonored, and unsung.

O Caledonia, stern and wild,
Meet nurse for a poetic child!
Land of brown heath and shaggy wood,
Land of the mountain and the flood,   20
Land of my sires! what mortal hand
Can e're untie the filial band
That knits me to thy rugged strand!
Still, as I view each well-known scene,
Think what is now and what hath been,
Seems as to me, of all bereft,
Sole friends thy woods and streams were left;
And thus I love them better still,
Even in extremity of ill.

By Yarrow's stream still let me stray,   30
Though none should guide my feeble way;
Still feel the breeze down Ettrick break,
Although it chill my withered cheek;
Still lay my head by Teviot-stone,
Though there, forgotten and alone,
The bard may draw his parting groan.

                  1802–04 (1805)

HUNTING SONG

Waken, lords and ladies gay,
On the mountain dawns the day,
All the jolly chase is here,
With hawk, and horse, and hunting-spear!
Hounds are in their couples yelling,
Hawks are whistling, horns are knelling,
Merrily, merrily, mingle they,
"Waken, lords and ladies gay."

Waken, lords and ladies gay,
The mist has left the mountain gray,   10
Springlets in the dawn are steaming,
Diamonds on the brake are gleaming:
And foresters have busy been,
To track the buck in thicket green;
Now we come to chant our lay,
"Waken, lords and ladies gay."

Waken, lords and ladies gay,
To the greenwood haste away;
We can show you where he lies,
Fleet of foot, and tall of size;   20
We can show the marks he made,
When 'gainst the oak his antlers fray'd;
You shall see him brought to bay,
"Waken, lords and ladies gay."

Louder, louder chant the lay,
Waken, lords and ladies gay!
Tell them youth, and mirth, and glee,
Run a course as well as we;
Time, stern huntsman! who can baulk,
Stanch as hound, and fleet as hawk: 30
Think of this, and rise with day,
Gentle lords and ladies gay.

                    (1808)

CANTO VI. **17. Caledonia:** Scotland.

30–34. Yarrow's . . . Teviot-stone: The Yarrow and
the Ettrick are in Selkirkshire, Scotland, and the Teviot in
Roxburghshire. HUNTING SONG. **5. couples:** leashes.

FROM

MARMION

⌒⌒⌒

Published in 1808, *Marmion* was planned as a romantic tale of chivalry centering on the Battle of Flodden in Northumberland (1513) where the English defeated the Scots. "Where Shall the Lover Rest?" is sung by a youth, Fitz-Eustace, to help while away a tedious night of waiting. "Lochinvar," sung by Lady Heron to the Scottish king, is based on the folk ballad "Katharine Janfarie."

⌒⌒⌒

WHERE SHALL THE LOVER REST

Where shall the lover rest,
　Whom the fates sever
From his true maiden's breast,
　Parted forever?
Where, through groves deep and high,
　Sounds the far billow,
Where early violets die,
　Under the willow.

Chorus

Eleu loro, etc. Soft shall be his pillow.

There, through the summer day,　　　　10
　Cool streams are laving;
There, while the tempests sway,
　Scarce are boughs waving;
There thy rest shalt thou take,
　Parted forever,
Never again to wake,
　Never, O never!

Chorus

Eleu loro, etc. Never, O never!

Where shall the traitor rest,
　He the deceiver,
Who could win maiden's breast,　　　　20
　Ruin and leave her?

In the lost battle,
　Borne down by the flying,
Where mingles war's rattle
　With groans of the dying.

Chorus

Eleu loro, etc. There shall he be lying.

Her wing shall the eagle flap
　O'er the false-hearted:
His warm blood the wolf shall lap,　　　30
　Ere life be parted.
Shame and dishonor sit
　By his grave ever;
Blessing shall hallow it,—
　Never, O never!

Chorus

Eleu loro, etc. Never, O never!

　　　　　　　　　　　　　1806 (1808)

LOCHINVAR

Oh! Young Lochinvar is come out of the west,
Through all the wide Border his steed was the best;
And save his good broadsword he weapons had none,
He rode all unarmed and he rode all alone.
So faithful in love and so dauntless in war,
There never was knight like the young Lochinvar.

He stayed not for brake and he stopped not for stone,
He swam the Eske river where ford there was none;
But ere he alighted at Netherby gate
The bride had consented, the gallant came late:　　10
For a laggard in love and a dastard in war
Was to wed the fair Ellen of brave Lochinvar.

So bodly he entered the Netherby Hall,
Among bridesmen, and kinsmen, and brothers, and all:
Then spoke the bride's father, his hand on his sword,—
For the poor craven bridegroom said never a word,—
"Oh! come ye in peace here, or come ye in war,
Or to dance at our bridal, young Lord Lochinvar?"—

"I long wooed your daughter, my suit you denied;
Love swells like the Solway, but ebbs like its tide— 20

MARMION. WHERE SHALL THE LOVER REST. **9. Eleu loro:** should read "Eleu horo," nonsense syllables conventional in Celtic verse, equivalent to "hey-diddle-diddle."

LOCHINVAR. **7. brake:** thicket. **20. Solway:** the Firth of Solway, noted for the violent swell of its tides.

And now am I come, with this lost love of mine,
To lead but one measure, drink one cup of wine.
There are maidens in Scotland more lovely by far,
That would gladly be bride to the young Lochinvar."

The bride kissed the goblet; the knight took it up,
He quaffed off the wine, and he threw down the cup.
She looked down to blush, and she looked up to sigh,
With a smile on her lips and a tear in her eye.
He took her soft hand ere her mother could bar,—
"Now tread we a measure!" said young Lochinvar. 30

So stately his form, and so lovely her face,
That never a hall such a galliard did grace;
While her mother did fret, and her father did fume,
And the bridegroom stood dangling his bonnet and
    plume;
And the bride-maidens whispered, " 'T were better
    by far
To have matched our fair cousin with young Lochin-
    var."

One touch to her hand and one word in her ear,
When they reached the hall-door, and the charger
    stood near;
So light to the croupe the fair lady he swung,
So light to the saddle before her he sprung!   40
"She is won! we are gone, over bank, bush, and scaur;
They'll have fleet steeds that follow," quoth young
    Lochinvar.

There was mounting 'mong Græmes of the Netherby
    clan;
Forsters, Fenwicks, and Musgraves, they rode and
    they ran:
There was racing and chasing on Cannobie Lee,
But the lost bride of Netherby ne'er did they see.
So daring in love and so dauntless in war,
Have ye e'er heard of gallant like young Lochinvar?

                                        1806 (1808)

from INTRODUCTION
TO CANTO SIXTH

And well our Christian sires of old
Loved when the year its course had rolled,
And brought blithe Christmas back again,
With all his hospitable train.

Domestic and religious rite
Gave honor to the holy night;
On Christmas eve the bells were rung;   30
On Christmas eve the mass was sung:
That only night in all the year
Saw the stoled priest the chalice rear.
The damsel donned her kirtle sheen;
The hall was dressed with holly green;
Forth to the wood did merrymen go,
To gather in the mistletoe.
Then opened wide the baron's hall
To vassal, tenant, serf, and all;
Power laid his rod of rule aside,   40
And Ceremony doffed his pride.
The heir, with roses in his shoes,
That night might village partner choose;
The lord, underogating, share
The vulgar game of "post and pair,"
All hailed, with uncontrolled delight
And general voice, the happy night
That to the cottage, as the crown,
Brought tidings of salvation down.

    The fire, with well-dried logs supplied,   50
Went roaring up the chimney wide;
The huge hall-table's oaken face,
Scrubbed till it shone, the day to grace,
Bore then upon its massive board
No mark to part the squire and lord.
Then was brought in the lusty brawn
By old blue-coated serving-man;
Then the grim boar's-head frowned on high,
Crested with bays and rosemary.
Well can the green-garbed ranger tell   60
How, when, and where, the monster fell,
What dogs before his death he tore,
And all the baiting of the boar.
The wassail round, in good brown bowls
Garnished with ribbons, blithely trowls.
There the huge sirloin reeked; hard by
Plum-porridge stood and Christmas pie;
Nor failed old Scotland to produce
At such high tide her savory goose.
Then came the merry maskers in,   70
And carols roared with blithesome din;
If unmelodious was the song,
It was a hearty note and strong.
Who lists may in their mumming see

---

32. **galliard:** a dance. 39. **croupe:** seat behind the rider of
the horse. 41. **scaur:** cliff.

INTRODUCTION TO CANTO SIXTH. 65. **trowls:** trolls.

Traces of ancient mystery;
White shirts supplied the masquerade,
And smutted cheeks the visors made;
But oh! what maskers, richly dight,
Can boast of bosoms half so light!
England was merry England when   80
Old Christmas brought his sports again.
'Twas Christmas broached the mightiest ale,
'Twas Christmas told the merriest tale;
A Christmas gambol oft could cheer
The poor man's heart through half the year.

    Still linger in our northern clime
Some remnants of the good old time,
And still within our valleys here
We hold the kindred title dear,
Even when, perchance, its far-fetched claim   90
To Southron ear sounds empty name;
For course of blood, our proverbs deem,
Is warmer than the mountain-stream.
And thus my Christmas still I hold
Where my great-grandsire came of old,
With amber beard and flaxen hair
And reverent apostolic air,
The feast and holy-tide to share,
And mix sobriety with wine,
And honest mirth with thoughts divine.   100

1806 (1808)

FROM

THE LADY OF THE LAKE

The narrative poem from which the following three lyrics are taken is laid in the vicinity of Loch Katrine in the western Highlands. The "Hail to the Chief" is sung by boatmen as they bring their chief to shore. Scott's lyric was intended as an imitation of boatsongs generally sung by highlanders, which, he said "are so adapted as to keep time with the sweep of the oars." The "Coronach" is sung by women over the body of their dead chief. Scott explains that "The *Coronach* of the highlanders, like the *Ululatus* of the Romans, and the *Ululoo* of the Irish, was a wild expression of lamentation, poured forth by the mourners over the body of a departed friend. When the words of it were articulate, they expressed the praises of the deceased, and the loss the clan would sustain by his death."

HAIL TO THE CHIEF

Hail to the Chief who in triumph advances!
  Honored and blessed be the ever-green Pine!
Long may the tree, in his banner that glances,
  Flourish, the shelter and grace of our line!
    Heaven send it happy dew,
    Earth lend it sap anew,
  Gayly to bourgeon and broadly to grow,
    While every Highland glen
    Sends our shout back again,
  "Roderigh Vich Alpine dhu, ho! ieroe!"   10

Ours is no sapling, chance-sown by the fountain,
  Blooming at Beltane, in winter to fade;
When the whirlwind has stripped every leaf on the
    mountain,
  The more shall Clan-Alpine exult in her shade.
    Moored in the rifted rock,
    Proof to the tempest's shock,
  Firmer he roots him the ruder it blow;
    Menteith and Breadalbane, then,
    Echo his praise again,
  "Roderigh Vich Alpine dhu, ho! ieroe!"   20

Proudly our pibroch has thrilled in Glen Fruin,
  And Bannochar's groans to our slogan replied;
Glen-Luss and Ross-dhu, they are smoking in ruin,
  And the rest of Loch Lomond lie dead on her side.
    Widow and Saxon maid
    Long shall lament our raid,
  Think of Clan-Alpine with fear and with woe;
    Lennox and Leven-glen
    Shake when they hear again,
  "Roderigh Vich Alpine dhu, ho! ieroe!"   30

THE LADY OF THE LAKE. HAIL TO THE CHIEF. **10.** "**Roderigh . . . ieroe!**": "Black Roderick, descendant of Alpine" (Scott's note). **12. Beltane:** May Day. **18. Menteith and Breadalbane:** districts in Perthshire, Scotland. **21. pibroch:** battle music played on the bagpipe. **22. Bannochar:** Bannachra castle, in Glen Fruin (near Loch Lomond) stronghold of the Colquhouns. In 1603 the Colquhouns were massacred. **23. Glen-Luss and Ross-dhu:** Glen Luss is Luss, in Dunbartonshire, on the west side of Loch Lomond. Ross-dhu is Rossdhu, also in Dunbartonshire on the west side of Loch Lomond. **28. Lennox:** "The Lennox, as the district is called which encircles the lower extremity of Loch Lomond, was peculiarly exposed to the incursions of the mountaineers, who inhabited the inaccessible fastnesses at the upper end of the lake, and the neighboring district of Loch Katrine. These were often marked by circumstances of great ferocity" (Scott's note).

Row, vassals, row, for the pride of the Highlands!
  Stretch to your oars for the ever-green Pine!
O that the rosebud that graces yon islands
  Were wreathed in a garland around him to twine!
    O that some seedling gem,
    Worthy such noble stem
Honored and blessed in their shadow might grow!
    Loud should Clan-Alpine then
    Ring from her deepmost glen,
"Roderigh Vich Alpine dhu, ho! ieroe!"   40

                              1809–10 (1810)

CORONACH

He is gone on the mountain,
  He is lost to the forest,
Like a summer-dried fountain,
  When our need was the sorest.
The font, reappearing,
  From the rain-drops shall borrow,
But to us comes no cheering,
  To Duncan no morrow!

The hand of the reaper
  Takes the ears that are hoary,   10
But the voice of the weeper
  Wails manhood in glory.
The autumn winds rushing
  Waft the leaves that are searest,
But our flower was in flushing,
  When blighting was nearest.

Fleet foot on the correi,
  Sage counsel in cumber,
Red hand in the foray,
  How sound is thy slumber!   20
Like the dew on the mountain,
  Like the foam on the river,
Like the bubble on the fountain,
  Thou art gone, and forever.

                              1809–10 (1810)

CORONACH. **17. correi:** hollow in a hillside to which the
game flee for hiding. **18. cumber:** trouble.

HYMN TO THE VIRGIN

Ave Maria! maiden mild!
  Listen to a maiden's prayer!
Thou canst hear though from the wild,
  Thou canst save amid despair.
Safe may we sleep beneath thy care,
  Though banished, outcast, and reviled—
Maiden! hear a maiden's prayer;
  Mother, hear a suppliant child!
                              Ave Maria!

Ave Maria! undefiled!
  The flinty couch we now must share   10
Shall seem with down of eider piled,
  If thy protection hover there.
The murky cavern's heavy air
  Shall breathe of balm if thou hast smiled;
Then, Maiden! hear a maiden's prayer,
  Mother, list a suppliant child!
                              Ave Maria!

Ave Maria! stainless styled!
  Foul demons of the earth and air,
From this their wonted haunt exiled,
  Shall flee before thy presence fair.   20
We bow us to our lot of care,
  Beneath thy guidance reconciled:
Hear for a maid a maiden's prayer,
  And for a father hear a child!
                              Ave Maria!
                              1809–10 (1810)

FROM

ROKEBY

The narrative poem from which these two lyrics are
taken was composed between September 15 and
December 31, 1812. The action takes place during the
English civil war (1642–49). Allen-a-Dale was the
minstrel of Robin Hood's band.

THE ROVER'S FAREWELL

"A weary lot is thine, fair maid,
  A weary lot is thine!
To pull the thorn thy brow to braid,
  And press the rue for wine!

A lightsome eye, a soldier's mien,
   A feather of the blue,
A doublet of the Lincoln green,—
   No more of me you knew,
          My love!
No more of me you knew.   10

"This morn is merrry June, I trow,
   The rose is budding fain;
But she shall bloom in winter snow
   Ere we two meet again."
He turned his charger as he spake
   Upon the river shore,
He gave his bridle-reins a shake,
   Said, "Adieu for evermore,
          My love!
And adieu for evermore."   20

           1812 (1813)

ALLEN-A-DALE

Allen-a-Dale has no fagot for burning,
Allen-a-Dale has no furrow for turning,
Allen-a-Dale has no fleece for the spinning,
Yet Allen-a-Dale has red gold for the winning.
Come, read me my riddle! come, hearken my tale!
And tell me the craft of bold Allen-a-Dale.

The Baron of Ravensworth prances in pride,
And he views his domains upon Arkindale side.
The mere for his net and the land for his game,
The chase for the wild and the park for the tame; 10
Yet the fish of the lake and the deer of the vale
Are less free to Lord Dacre than Allen-a-Dale!

Allen-a-Dale was ne'er belted a knight,
Though his spur be as sharp and his blade be as bright;
Allen-a-Dale is no baron or lord,
Yet twenty tall yeomen will draw at his word;
And the best of our nobles his bonnet will vail,
Who at Rere-cross on Stanmore meets Allen-a-Dale!

Allen-a-Dale to his wooing is come;
The mother, she asked of his household and home: 20
"Though the castle of Richmond stand fair on the hill,
My hall," quoth bold Allen, "shows gallanter still;
'T is the blue vault of heaven, with its crescent so pale
And with all its bright spangles!" said Allen-a-Dale.

ROKEBY. ALLEN-A-DALE. **17. vail:** take off.

The father was steel and the mother was stone;
They lifted the latch and they bade him be gone;
But loud on the morrow their wail and their cry:
He had laughed on the lass with his bonny black eye,
And she fled to the forest to hear a love-tale,
And the youth it was told by was Allen-a-Dale! 30

           1812 (1813)

AND WHAT THOUGH WINTER WILL PINCH SEVERE

From Scott's novel, *Old Mortality* (1816), Chapter XIX.

And what though winter will pinch severe
   Through locks of gray and a cloak that's old,
Yet keep up thy heart, bold cavalier,
   For a cup of sack shall fence the cold.

For time will rust the brightest blade,
   And years will break the strongest bow;
Was never wight so starkly made,
   But time and years would overthrow.

           1816 (1816)

PROUD MAISIE

From Scott's novel, *The Heart of Midlothian* (1818), Chapter XLV.

   Proud Maisie is in the wood
      Walking so early;
   Sweet Robin sits on the bush,
      Singing so rarely.

   "Tell me, thou bonny bird,
      When shall I marry me?"—
   "When six braw gentlemen
      Kirkward shall carry ye."

AND WHAT THOUGH WINTER WILL PINCH SEVERE. **4. sack:** dry Spanish wine. **7. starkly:** strongly. PROUD MAISIE. **7. braw:** handsome.

"Who makes the bridal bed,
   Birdie, say truly?"—   10
"The gray-headed sexton
   That delves the grave duly.

"The glow-worm o'er grave and stone
   Shall light thee steady,
The owl from the steeple sing,
   'Welcome, proud lady.' "

                    1818 (1818)

GLEE FOR KING CHARLES

From Scott's novel, *Woodstock, or the Cavalier: A Tale of the Year 1651* (1826), Chapter XX.

Bring the bowl which you boast,
   Fill it up to the brim;
'T is to him we love most,
   And to all who love him.

Brave gallants, stand up,
   And avaunt ye, base carles!
Were there death in the cup,
   Here's a health to King Charles!

Though he wanders through dangers,
   Unaided, unknown,   10
Dependent on strangers,
   Estranged from his own;
Though 't is under our breath
   Amidst forfeits and perils,
Here's to honor and faith,
   And a health to King Charles!

Let such honors abound
   As the time can afford,
The knee on the ground,
   And the hand on the sword;   20
But the time shall come round
   When, 'mid Lords, Dukes, and Earls,
The loud trumpet shall sound,
   Here's a health to King Charles!

                    1826 (1826)

GLEE FOR KING CHARLES. **6. carles:** fellows, churls.

SAMUEL TAYLOR COLERIDGE

1772–1834

COLERIDGE is a major poet and one of the foremost English critics. His writings on religion, morals, and politics exerted a potent influence throughout the nineteenth century. His note-books, marginalia, and letters, the arsenals of his more formal prose, contain further speculation and insight on every conceivable subject—philosophy, art, landscape, language, love, Shakespeare, psychology, dreams—speculation and insight that still illuminate and suggest, fostering what Coleridge would have termed the germinal or generative powers of the mind. In view of this, it seems strange indeed that he has so often been judged an example of great gifts come to little.

The impression is perhaps excusable in his contemporaries. For one thing, they had the extraordinary experience of meeting him and hearing his spontaneous, thought-packed eloquence, the effect further heightened by sonorous tones and vivid countenance. Sometimes after an evening with him they felt he had just talked a book greater than any he had written. Then, too, much of his finest work was not published. Most of the Shakespeare criticism, for example, was given only in lectures; it comes down to us in reports from others and his own notes printed after his death. There is also the simple fact that his conceptions were often beyond the mental horizon of readers in his time. Finally, he repeatedly planned a major work that would set forth the whole of his philosophy in a systematic way. As the years passed without this work appearing, he felt guilty and apologetic, and his self-reproach influenced both his contemporaries and later critics who themselves performed much less than he. Few men have accomplished more.

He was born in 1772 in the country town of Ottery St. Mary, where his father was vicar.

The youngest of fourteen children, he was his father's favorite and quickly developed a precocious confidence in talking with adults. The activity naturally fostered his intellectual powers, and being otherwise a lonely child—his older brothers resented him—he passed much of his time reading. "So I became a *dreamer*," Coleridge later explained to his friend, Thomas Poole,

> and acquired an indisposition to all bodily activity—and I was fretful, and inordinately passionate, and . . . slothful . . . with a memory & understanding forced into almost an unnatural ripeness, I was flattered & wondered at by all the old women—& so became very vain, and despised most of the boys, that were at all near my own age—and before I was eight years old, I was a *character*.
>
> [*Letters*, ed. Griggs, I, 347–48].

When his father died in 1781 he was admitted to Christ's Hospital in London—a famous school that gave free education to orphans—where he first met Charles Lamb. In after years Wordsworth (see *Prelude*, VI. 264 ff., above, p. 239) and Lamb ("Christ's Hospital Five and Thirty Years Ago," below, pp. 584–85) tended to picture Coleridge's school years as unhappy. But roaming through the city fascinated him endlessly; there were pleasant holidays and afternoons bathing in the river; he surrounded himself with friends and disciples—throughout his life he made friends with extraordinary ease; and he exploited the libraries and the opportunities for intellectual discussion. In fact, he used to stop passers-by, especially clergymen, and engage them in argument. And he fell in love with Mary Evans, the sister of a schoolmate. For a while he planned to become a cobbler, and then a doctor, but he was also writing poems. Moreover, he plunged deep into metaphysical and scientific studies, of which one result was that he briefly "sported infidel" and received a thrashing from his teacher.

He went to Jesus College, Cambridge, and continued for a while to win prizes and medals. But he was soon caught up in the free, convivial life and in the excitement of politics—for how could one pore over the classics when society was being regenerated in France and good men oppressed in England?—and he failed in competition for an important scholarship. Overwhelmed by debts and unrequited love, not knowing where to turn, he took the desperate and bizarre step of enlisting in the cavalry. He was rescued by his brothers and returned briefly to Cambridge, but left in 1794 without a degree.

In that same year he met Robert Southey, and the two young poets, with some friends, originated the scheme they called Pantisocracy. Like many radicals of that time and since, they believed that in order to eradicate from the human heart selfishness, superstition, and vice one need only establish the right sort of social milieu. They decided to found an ideal community on the banks of the Susquehanna River in Pennsylvania. Property would be held in common. The men would earn their living by agriculture, which would take only a few hours each day, and spend their remaining time in literary activities. In his Pantisocratic zeal he fell into an engagement with Sarah Fricker, the sister of Southey's betrothed. She was an unfortunate choice. Prim, conventional, intellectually limited, she could not understand him, and she was a scold; but prodded by Southey (for which Coleridge never quite forgave him) he finally married, "resolved," as he said, "but wretched." Meanwhile, Pantisocracy was crumbling. In spite of desperate efforts, including a play (*The Fall of Robespierre*, jointly written with Southey) and political lectures in Bristol, money could not be raised. Moreover, Southey began to propose modifications—he wanted, for example, to keep a servant in the midst of this utopian brotherhood—and finally renounced the scheme. Coleridge took his revenge in a very long letter that begins, "You are *lost* to *me*, because you are lost to Virtue," but that was small compensation. However impractical it was, Pantisocracy had been for a while the dearest object of his life.

He had met Wordsworth in 1795. A casual correspondence ensued, and then he paid a brief visit to Wordsworth and his sister at Racedown. Their friendship knit at once, and the same year (1797) the Wordsworths rented Alfoxden House close to the village of Nether Stowey, near Bristol, where Coleridge had a cottage next to another friend, the sturdy, sensible Thomas Poole. Hazlitt, in "My First Acquantance with Poets," evokes the way of their life together— the eager talk, the reading of poems, the myriad plans for the future. To Wordsworth Coleridge brought unwearied belief in his genius—he was already convinced that Wordsworth was the greatest poet since Milton—and a devoted criticism of even the least details of his writing. And, of course, the two poets gave each other intellectual support and excitement. During the eighteen months of their daily intimacy Coleridge produced his finest poetry.

He had no regular income and was reluctantly intending to become a Unitarian minister when in January, 1798, the wealthy Staffordshire manufacturers of china, Thomas and Josiah Wedgwood, offered him a life annuity of £150, the purpose being to free him to use his marvelous talents as he liked. The first result was a trip to Germany (1798–99) to learn the language. Returning to England, he settled in Keswick, not very far from the Wordsworths in Grasmere, and devoted himself mainly to metaphysical studies, for the first time reading Kant. Now began a period of illness and misery deepening year after year. He felt a hopeless, yearning love for Sara Hutchinson, later Wordsworth's sister-in-law. From 1800 to approximately 1816 he suffered chronic physical pain, chiefly neuralgic and rheumatic at first. To relieve the symptoms he started habitually to take opium in the liquid form of laudanum. At this time opium was commonly prescribed and freely available. (Coleridge had employed it occasionally at least as far back as college and doubtless earlier.) Its effects were not clearly understood, and he could not have known that it would bring on severe physical and psychological disturbances, necessitating more opium in a downward spiral of pain and despair. Neither was it accepted that addiction cannot be cured simply by an effort of will. Hence though Coleridge might not reproach himself for the ignorance in which he began using the drug, he felt intolerable disgust and guilt at his lifelong inability to break the habit. (See the letter to Joseph Cottle, April, 26 1814, printed below, p. 529.) In 1804 he sailed to Malta, hoping the climate might restore his health, but after a little over two years returned more wretched than ever. He separated from his wife, though he always continued to support her, and lived henceforth mostly in London.

He began in 1808 a course of lectures, the first of many, on the principles of poetry. The effort only partially succeeded—he was disorganized and often listless—but in his physical condition to have lectured at all was a triumph. He then conceived a periodical, *The Friend*, which would offer subscribers essays on morality, taste, and religion. These appeared irregularly from 1808–10, but the enterprise lost money. At this time Coleridge was staying with the Wordsworths in Grasmere and dictating his essays to Sara Hutchinson. When he left, he learned from a mutual friend that Wordsworth had spoken contemptuously of him, and, though he said nothing directly to Wordsworth, he felt their friendship sundered and his life in ruins. In 1813 Byron, who was on the committee in charge of the Drury Lane Theatre, used his influence to have Coleridge's play *Remorse* performed, and he received £400, the largest sum he earned from any single work. In 1815, at the nadir of his misfortunes, he wrote his great but uneven book, the *Biographia Literaria*.

From his distressing physical and psychological state he was delivered by the attention of a kindly physician, James Gillman, in whose care Coleridge placed himself in 1816, living thereafter in his household. While Coleridge could never quit opium altogether, he managed to reduce greatly the amount he consumed, and he recovered much of his former energy and spirits. Henceforth for a few years publications came forth regularly: *The Statesman's Manual* and another *Lay Sermon* on politics and economics (1816–17); poems collected as *Sibylline Leaves* (1817); a play, *Zapolya* (1817); *The Friend* revised and enlarged (1818) to include the important *Treatise on Method;* and two courses of lectures in 1818–19, for which he wrote *On Poesy or Art*. In 1820 he was deeply distressed when his son Hartley lost through intemperance a fellowship at Oriel College, Oxford. Soon afterwards he completed his *Aids to Reflection* (1825). His last work was *On the Constitution of Church and State* (1830).

Meanwhile, in his later years he was working more steadily than before on the *Opus Maximum* that was to contain the definitive exposition of his religious and philosophic position. This remained unfinished, but through his other writings, old and new, and through the young friends and disciples who came to hear him, his fame spread and with it his influence on the thought of the age. Gillman's house in suburban Highgate became an object of literary and philosophic pilgrimage, and if visitors could not always understand or recollect what they had heard, the difficulty stemmed perhaps, as his nephew Henry Nelson Coleridge said, from his "exhaustive, cyclical mode of discoursing" as he worked out "an almost miraculous logic." There were also less philosophic evenings. On one occasion Coleridge, after much claret, threw a glass through the window pane and then took aim with a fork at a wine glass placed on a tumbler. His roseate face was

> lit up with animation, his large grey eye beaming, his white hair floating, and his whole frame, as it were, radiating with intense interest, as he poised the fork in his hand, and launched it at the fragile object.
>
> [William Jerdan, *Autobiography* (1852–53), IV, 233].

On the way home he affirmed that Theodore Hook, a companion who had also broken a window, was as true a genius as Dante. Old friends continued loyal. Lamb came round on Sundays. Wordsworth and Coleridge (reconciled in 1812) toured Europe together in 1828, and Wordsworth visited him again in London during the winter of 1830–31, by which time Coleridge's health was visibly failing. But his power of intellect he kept to the end, which came quietly in 1834. The memory he left was unforgettable. In an age that especially celebrated the ideal of genius he had been the living example.

2

In mode and technique Coleridge's poetry exhibits an astonishing variety and inventiveness, outward manifestations of his complex temperament and restless intellect. It is typical that each of his three finest poems, "The Ancient Mariner," "Christabel," and "Kubla Khan," is a different technical experiment. In the same few years when he was composing these prototypes of romantic witchery, Coleridge also originated and perfected work of an opposite type, the personal, descriptive, quietly meditative "conversation poem," of which examples are "The Eolian Harp," "This Lime-Tree Bower My Prison," "Frost at Midnight," and "The Nightingale." And in contrast to this he also essayed the sublime in "France: An Ode." With "Dejection" he produced what is, after Wordsworth's "Intimations of Immortality," the finest irregular ode in the language; and even when he had ceased to write verse except occasionally, he continued to experiment with technique (see the metrical effects of "The Knight's Tomb," below, p. 439), while his long-matured philosophic power could result in the intense energy of abstract phrase one finds, for example, in "Human Life." Finally, in such strange utterances as "Limbo" and "Ne Plus Ultra" Coleridge combines dream with philosophic speculation and portentous emotion, briefly developing a mode of visionary poetry completely his own.

As a form, the "conversation poem" derives from the descriptive-meditative verse of the eighteenth century, from such poems as Goldsmith's "Deserted Village" or Gray's "Elegy in a Country Churchyard." These present the speaker reflecting upon a particular scene, and passages of description alternate with passages of meditation touched with feeling. They are organized as argument and hence, however pensive, still convey something of the impression of public discourse, an impression to which their relatively generalized imagery, balanced syntax, and regular patterns of versification are altogether suitable. Coleridge transformed this species of poem in ways that will be found generally typical of verse in the Romantic period. The "scene" has become a moment in domestic life and thus establishes at once an intimate relation between ourselves and the speaker. Images are more particularized and realistic—in fact, they are rendered with a spare and delicate exactitude not quite duplicated in any other poet—and at the same time are more suggestive. Sentences have the spontaneous, unpredictable structure of conversation. At the end of "Frost at Midnight," for example, Coleridge addresses his child,

> Therefore all seasons shall be sweet to thee,
> Whether the summer clothe the general earth
> With greenness,

and one anticipates that the sentence will be balanced by a reference to winter equally generalized. Instead the poem suddenly veers into a different mode of speech as it delineates a series of winter sights and sounds:

> or the redbreast sit and sing
> Betwixt the tufts of snow on the bare branch
> Of mossy apple-tree, while the nigh thatch
> Smokes in the sun-thaw; whether the eve-drops fall
> Heard only in the trances of the blast . . .

As the images succeed each other they become increasingly suggestive and might even be called symbols, provided that one uses the term in Coleridge's sense. For he defined a symbol not as a sign that stands for something other than itself but as a living part or instance in the larger reality it manifests. Thus at the end of the poem the "secret ministry of frost" is a benevolent, beauty-making power in nature, and the last line and a half, the

> silent icicles,
> Quietly shining to the quiet Moon,

more complexly reveal the intercommunion of all things in an organic universe.

Finally, the conversation poems are not organized as argument; instead they develop an inward, relaxed movement of the mind by associations that seem psychologically natural or probable. "Frost at Midnight" passes from a hearthside scene to recollections of schooldays, the associative link being a peculiar film of fire on the grate. The same thing occurs in "The Eolian Harp" where the wind blowing through the lute suggests the thought that through all natural objects

> sweeps
> Plastic and vast, one intellectual breeze,
> At once the Soul of each, and God of all.

To put it another way, in exhibiting the casual flow of consciousness Coleridge takes advantage of the basic convention of Romantic poetry, namely that poetry is spontaneous utterance. That convention exerts a far-reaching control over the way poetry will be read. It means that the poem enacts the process of its creation, and its beauty of form lies in natural movement and growth.

If "The Ancient Mariner," "Christabel," and "Kubla Khan" are achievements of a larger power and implication than the conversation poems, that is partly because in dream or narrative the imagination plays more freely and ambiguously than in discourse. But the same general qualities persist—the colloquial freedom, the exact and suggestive imagery, the philosophical and psychological depth. The new element in these poems is, of course, the strangeness of setting and supernatural action, for which Coleridge drew upon years of out-of-the-way reading. He was to write no such poems again, and by 1802, with "Dejection," he had completed about four-fifths of his total poetic work. Probably the main cause of this slackened activity was the prolonged suffering described above; as he once remarked, "When a Man is unhappy, he writes damned bad Poetry." Contributing factors were his intense admiration of Wordsworth, before whose work and character he felt himself a lesser poet, and the abstract habits of thinking engendered by critical and philosophical studies. (See his letters of March 25, 1801, and July 19, 1802, below, pp. 525 and 526.) But in these same studies he was also grappling with issues that could easily seem more important even than poetry.

In criticism his achievement had and still has fundamental importance. This is true not only because of his discussions of Wordsworth and Shakespeare, or his treatment of particular questions such as poetic diction or the psychological effect of meter, or because with the concept of organic form he supplied a rationale for the working premises of Romantic poetry. His criticism had a purpose still more general and significant, for he showed the possibility of making critical judgments that are objective—that is, universally valid—yet not necessarily based on a standard external to the work of art. Virtually all thinkers from Aristotle through the eighteenth century had followed the premise that art should be primarily an imitation of "nature," the word "nature" meaning a system of stable norms the human mind could know. With the development in the eighteenth century of empirical philosophy and psychology, however, it appeared that man cannot know things as they are in themselves; we possess only "ideas" of things, formed out of experience, linked by processes of association, and further associated with pleasures and pains. As experience varies in each individual, so necessarily do ideas and associations, and hence differing opinions will inevitably be formed as to the truth and pleasurableness of a particular work of art. Disputes cannot be settled because there is no objective criterion. This critical relativism, if argued rigidly, concludes that criticism is little more than a form of self-expression.

Coleridge countered with a genetic theory that grounds the value of a poem (or any work of art) in the processes of mind by which it was created, and which it continues to activate in the reader. He had adopted this position, he says, by the time he was twenty-five, as a result of controversies rising out of his devotion to Bowles's sonnets:

> I labored at a solid foundation, on which permanently to ground my opinions, in the component faculties of the human mind itself, and their comparative dignity and importance. According to the faculty or source, from which the pleasure given by any poem or passage was derived, I estimated the merit of such poem or passage.

The point of view was given a complex elaboration in later years, usually in connection with the theory of the "imagination." In a brilliant passage at the end of Chapter XIV of *Biographia Literaria* Coleridge begins:

> The poet, described in *ideal* perfection, brings the whole soul of man into activity, with the subordination of its faculties to each other, according to their relative worth and dignity.

Here the value of a poem varies not only with the "comparative dignity and importance" of the faculties it calls forth but also with the degree to which the activity is total and at the same time ordered, drawing upon all functions of the mind (sensation, intellection, emotion, etc.) in such a way that, instead of conflicting, they contribute to the same end. In this, however, there is still insufficient basis for ranking *King Lear* above any sonnet. Especially in "On Poesy or Art" Coleridge enriches his aesthetics by stressing (what was already explicit in previous writings) that beauty is a process of the same kind as that in nature itself, where the form subsumes increasing numbers of constituent elements as it develops. (In this sense, indeed, the artist imitates nature—not its visible things but its inner working.) Accordingly, the process of ordering achieves a heightened value as it controls and emerges through a larger number of "parts"—the term variously referring to images, characters, or actions in a play; lines, colors, or shapes in a painting; notes or melodies in music, and the like. This is the point of Coleridge's various formulations in "On the Principles of Genial Criticism" and "On Poesy or Art" that beauty is "multëity in unity" or the "unity of the manifold." In these phrases the emphasis still falls, however, on the power of the artist's mind to assimilate and unify, without sufficient consideration of the worth of the "parts" as such. To common sense it would be a defect in Coleridge's aesthetic theory to have stopped there, but he did not. He assimilated in his aesthetic theories his philosophic doctrines of "reason" and "ideas." Perhaps the most satisfactory single expression of this comes in a passage in *The Statesman's Manual* where Coleridge describes the function of the imagination in creating symbols. The imagination is

> that reconciling and mediatory power, which incorporating the reason in images of the sense, and organizing (as it were) the flux of the senses by the permanence and self-circling energies of the reason, gives birth to a system of symbols.

If one thinks of a work of art, the living educt of the imagination, as a "symbol" (and a system of lesser symbols) it thus acquires unique value not simply because it unifies and orders, but because in doing so it enacts the essential process of reality. This gives art the highest conceivable function, but for the moment we need remind ourselves only that a work of art embodies *ideas* (the "self-circling energies of the reason"). By an *idea* Coleridge meant an ultimate knowledge not derivable from the senses. Though ideas may be awakened in us by other modes of discourse, they can be contemplated only in symbols.

Coleridge's brilliant but scattered remarks on the imagination can be (and have been) interpreted in any number of ways. We can see them for what they are, and find their underlying coherence, only if we consider them in the light of his approach to the human mind as a whole (the imagination, as he said, is not a separate "faculty" of mind but rather a "completing power" that works through and by means of the entire mind). Here we should remember Coleridge's lifelong philosophical hope to overcome the sharp division between the human mind and the external world—a separation of mind and nature that had been widely taken for granted since the time of Descartes, a century and a half before, despite attempt after attempt to reason it away.

Hoping to show that the mind was capable of a direct knowledge of nature in *all* of its aspects, he asserted a twofold function of the mind and made a basic distinction between *reason* and *understanding* (in this he followed Kant but with considerable modifications). Through our senses we learn about the concrete, material world; and the organization of this empirical kind of knowledge, the generalization on the basis of our sense impressions, is the function of the "understanding." Beyond this are the ultimate forms and laws of nature that, for Plato, were the essential reality, and the existence of which is either denied by the empiricist or else regarded by him and so many others as completely unknowable by the human intellect. To Coleridge, direct insight into the universal forms (Plato's "ideas") was possible through the "reason." Here he is very much a Platonist. On the other hand, he was, even more, an "organicist"; reality for Coleridge did not consist in the universal forms alone but in their living immanence within the concrete world. That living process in nature as a whole ("the eternal act of Creation") is "repeated" or "echoed" within the "finite mind" by the activity of the imagination. It joins the insights of reason with the knowledge of the empirical world that we get from the senses and the "understanding" and conceives them as "one." (It is in trying to emphasize this all-important function of the imagination as the unifying agent of the mind that Coleridge makes his distinction between imagination and mere fancy.) Though his concept of the imagination has a broad philosophical context, it is naturally central in his literary criticism, and all the more since he conceives of art, at its best, as a "mediatress between, and reconciler of, nature and man." There especially, as it "brings the whole soul of man into activity," the imagination "reveals itself in the balance or reconciliation of opposite or discordant qualities; of sameness, with difference; of the general, with the concrete; the idea, with the image; the individual, with the representative" (*Biographia Literaria*, Chap. XIV).

After 1820 Coleridge turned almost exclusively to religious questions. Himself of devout Christian faith, he opposed both the main types of Protestant Christianity to be found in England at this time—the emotionalism and biblical fundamentalism of contemporary Methodists and also the current Anglican theology, which tended to be aridly rational and utilitarian—and he sought to provide a more adequate ground and mode of religious faith. In this he was prophetic. A double onslaught was soon to descend from Germany—philosophical idealism, which dissolved the historical and dogmatic elements of Christianity, and critical and historical study of the Bible—and, as Basil Willey says, Coleridge provided a bulwark before most Englishmen knew their faith was attacked. He emphasized, with qualifications (some of them apparent chiefly in manuscripts still unpublished), the subjective grounds of Christian belief. That is, we believe doctrines (original sin, for example) proposed by the Bible and the Church not primarily because they are confirmed by outward evidence but because they interpret and answer to direct self-awareness. The organ of faith he terms the "higher reason," which is not so much reason in the sense given in the preceding paragraph, but rather a synthesis of intellect, will, and feeling, a total self-experience taken as showing the possibility of faith and the ground for it. We know our predicament, our needs, our freedom to choose whether to believe or not, and if we "try it" faith will confirm itself in further experience. The standpoint is in some respects akin to Christian existentialism in the present day.

Coleridge's influence appears in diverse realms of thought throughout the nineteenth century. His subjective or "spiritual" religion was given mainly in *Aids to Reflection* and in two posthumous works, *Confessions of an Inquiring Spirit*, on interpreting the Bible, and *Literary Remains*. In *On the Constitution of Church and State* he developed the idea of a church fulfilling an educative function of highest consequence in the life of the state. These writings nourished the Broad Church movement throughout the century, but his thought also shows affinities with the opposite party of Tractarians led by Edward Pusey, John Keble, and John Henry Newman, from whom present-day High-Church Anglicanism descends. In discussing ethics and politics Coleridge for a while provided virtually the sole philosophical counterweight to the utilitarian teachings of Jeremy Bentham and James Mill, with their neglect of spiritual values, and if subsequently so many of the major Victorians—Carlyle, Arnold, Ruskin— also resisted utilitarianism, that is partly because Coleridge showed the way. His influence on literary criticism has been even more obvious. The public lectures on Shakespeare, given from 1808 to 1819, remain the finest example of the approach to Shakespeare by the psychological analysis of character that typified most Shakespeare criticism to at least the First World War. At the same time his description of the poetic imagination and his insistence on the organic unity of a work of art have, especially through the early writings of I. A. Richards, supplied ideas and slogans to the so-called New Criticism, which has, on the whole, been otherwise unsympathetic to Romanticism.

It is, of course, ironic that premises originally suggested by Romantic poetry should so often have been used to discredit it, but the irony is typical. Coleridge's disciples have usually been specialists in a particular line of thought or interest. They have naturally adopted not the whole context of his thinking but brilliant insights capable of further development and practical application. There is no reason to think that this activity of adoption has come to an end. The unpredictable, suggestive prose retains undiminished its fostering power in other minds. But Coleridge himself is greater than any specialized use that can be made of him. His final example and challenge lie in the extent to which at any particular moment he was able to bring to bear the whole of his concerns. Thus his formal treatise "On Method" begins with a discussion of Mistress Quickly in *Henry IV*, and his perceptions in reading Shakespeare continually rose out of speculations in religion, morals, and psychology. If his thought never settled into the systematic unity for which he struggled, it is a continual example of the spontaneous unifying of richly diverse materials that he prized in art—and above all in Shakespeare. He is the Shakespeare of ideas.

Bibliography

Though Southey called it a "Dutch attempt at German sublimity," discerning readers such as Lamb and Hazlitt quickly recognized the power of "The Ancient Mariner" and, to a lesser extent, of Coleridge's other poems. Through most of his life his prose writings had little circulation. To his contemporaries he was famous chiefly for his talk. The fame lived on (slightly damaged by Carlyle's brilliant but acerb description in *The Life of John Sterling*, 1851), but after his death intellectual developments began, as it were, to catch up with him, and through most of the nineteenth century he seemed important primarily as a political and religious thinker. John Stuart Mill, treating him mainly from the former point of view in *The Westminster Review*, 1840 (reprinted in *Dissertations and Discussions*, 1857), felt that in significance and influence he was the chief rival to Bentham, and early testimonials to the impact of his religious thought appear in such writers—prominent in the life of their time—as Thomas Arnold, F. D. Maurice, and Julius Hare. By 1885 John Tulloch, tracing the history of *Religious Thought in Britain in the Nineteenth Century*, found it necessary to reckon extensively with Coleridge. More recently he has come to the fore especially as a critic and poet.

Several attempts have been made in the twentieth century to survey his total achievement. Of these one may mention John H. Muirhead, *Coleridge as Philosopher*, 1930; and the essay by Basil Willey, "Samuel Taylor Coleridge," in *Nineteenth Century Studies*, 1949. Also of general interest are the sections on Coleridge in D. G. James, *The Romantic Comedy*, 1948.

Most writing, however, has focused mainly on one aspect. His political thought is studied by Alfred Cobban, *Edmund Burke and the Revolt Against the Eighteenth Century*, 1929; Crane Brinton, *Political Ideas of the English Romanticists*, 1926; and John Colmer, *Coleridge: Critic of Society*, 1959, which gives not so much an analysis as a running summary of his political essays. The place and impact of Coleridge in English religious thought are traced by Vernon F. Storr, *Development of English Theology in the Nineteenth Century, 1800–1860*, 1913; Herbert Stewart, "Place of Coleridge in English Theology," *Harvard Theological Review*, Vol. XI, No. 1, January, 1918; and Charles R. Sanders, *Coleridge and the Broad Church Movement*, 1942. James D. Boulger, *Coleridge as a Religious Thinker*, 1961, describes and analyzes his religious philosophy. J. Robert Barth, S.J., *Coleridge and Christian Doctrine*, 1969, synthesizes his theological ideas from the scattered, often unpublished sources. Two books dealing primarily with the sources of his philosophical principles are Claud Howard, *Coleridge's Idealism*, 1924, relating him to seventeenth-century English thinkers, and René Wellek, *Immanuel Kant in England*, 1931, emphasizing his use of German idealism.

His aesthetics have received more ample treatment. Systematic discussion was inaugurated by John Shawcross in his extended and very valuable introduction to his edition of the *Biographia Literaria*, 1907. Thereafter came I. A. Richards, *Coleridge on Imagination*, 1934; Gordon McKenzie, *Organic Unity in Coleridge*, 1939; and W. J. Bate, "Coleridge on the Function of Art," in *Perspectives of Criticism*, ed. Harry Levin, 1950, certainly the single most incisive discussion of relevant questions. M. H. Abrams, *The Mirror and the Lamp*, 1953, places him in a wider context of critics and critical assumptions in the Romantic period. James V. Baker, *The Sacred River*, 1957, traces the chronological development of Coleridge's aesthetics, with particular emphasis on the role he assigns to the unconscious. Three recent comprehensive studies are R. H. Fogle, *The Idea of Coleridge's Criticism*, 1962; Joseph A. Appleyard, *Coleridge's Philosophy of Literature, 1791–1819*, 1965; and J. R. de J. Jackson, *Method and Imagination in Coleridge's Criticism*, 1969. R. W. Babcock, *The Genesis of Shakespeare Idolatry, 1766–99*, 1931, explores the eighteenth-century background of his approach to Shakespeare, and see also Augustus Ralli, *A History of Shakespearian Criticism*, 2 vols., 1932.

For an introductory appreciation of his poetry one may turn to Humphrey House, *Coleridge*, 1953. In *The Road to Xanadu*, 1927, John Livingston Lowes exhaustively traces the literary sources of "The Ancient Mariner" and "Kubla Khan." The book is perhaps the most brilliant example in English of this kind of scholarship. N. P. Stallknecht's *Strange Seas of Thought*, 1945, includes a chapter on the philosophical implications of "The Ancient Mariner," especially in relation to the last sections of Wordsworth's *Prelude*. In an essay in *Scrutiny*, March, 1941, D. W. Harding approaches the poem from a psychoanalytic point of view. There are several studies of symbolism in the poem: Maud Bodkin, *Archetypal Patterns in Poetry*, 1934, elaborates a Jungian scheme; G. Wilson Knight, *The Starlit Dome*, 1941, offers his typically oracular, often suggestive interpretations of imagery; Robert Penn Warren's introduction to his edition of the poem, 1946, concentrates on symbolic patterns and interrelations. Of "Kubla Khan" there are interpretations in several of the books already cited, and see also Elisabeth Schneider, *Coleridge, Opium and Kubla Khan*, 1953, especially for her accounts of the effect of opium on dreams and the literary tradition that lies behind the poem. Marshall Suther, *The Dark Night of Samuel Taylor Coleridge*, 1960, is, from one point of view, a sustained explication of "Dejection," though Suther is also suggesting why Coleridge declined as a poet. J. B. Beer, *Coleridge the Visionary*, 1959, concentrates on "The Ancient Mariner," "Christabel," and "Kubla Khan." A comprehensive treatment is Max F. Schulz, *The Poetic Voices of Coleridge*, 1963.

His poetry is readily available, the standard text being *Poems*, ed. E. H. Coleridge, 1912. Prose edited in the twentieth century includes *Biographia Literaria*, ed. John Shawcross, 2 vols., 1907, printed with "On the Principles of Genial Criticism" and "On Poesy or Art;" *Treatise on Method*, ed. Alice D. Snyder, 1934, the text being the first version, not the revision that appeared in *The Friend* (1818); *Political Tracts of Wordsworth, Coleridge and Shelley*, ed. R. J. White, 1953, for the *Lay Sermons*; and *Confessions of an Inquiring Spirit*, ed. H. StJ. Hart, 1956. H. N. Coleridge's collection of marginalia and lecture notes, *Literary Remains*, 4 vols., 1836–39, has not been reprinted, but the same material is comprised in three more carefully edited works: *Shakespeare Criticism*, ed. T. M. Raysor, 2 vols., 1930; *Miscellaneous Criticism*, ed. T. M. Raysor, 1936; *Coleridge on the Seventeenth Century*, ed. Roberta Florence Brinkley, 1955. For other prose writings nineteenth-century editions must be used: *Aids to Reflection, On the Constitution of Church and State, The Friend*, and the newspaper articles collected by Sara Coleridge as *Essays on His Own Times*, 1850. There is also H. N. Coleridge's sampling of the *Table Talk*, 1836.

In the twentieth century a number of works have been published for the first time: *Coleridge on Logic and Learning*, ed. Alice D. Snyder, 1929, prints from philosophical manuscripts; and *Philosophical Lectures*, ed. Kathleen Coburn, 1949, in which Coleridge expounds not his own views but the history of philosophy. Selections from his notebooks were printed by E. H. Coleridge, *Anima Poetae*, 1895. *Notebooks*, ed. Kathleen Coburn, 2 vols., 1957–61, prints the text completely, through 1808. Coleridge's *Letters* were first edited by E. H. Coleridge, 1895. These were supplemented by *Unpublished Letters*, ed. E. L. Griggs, 1932, and more recently *Letters*, ed. E. L. Griggs, 4 vols., 1956–59, collects correspondence to 1819.

An excellent biography is Lawrence Hanson's *Life*, 1938, but it goes only to 1800. For the years thereafter see E. K. Chambers, *Coleridge*, 1938, which, however, is unsympathetic, or the older *Coleridge* by James Dykes Campbell, 1894. One should also mention H. M. Margoliouth, *Wordsworth and Coleridge*, 1953, and especially the brilliantly distilled biography, equally valuable for its critical insights, by W. J. Bate, *Coleridge*, 1968.

SONNET

TO THE RIVER OTTER

Dear native Brook! wild Streamlet of the West!
  How many various-fated years have past,
  What happy and what mournful hours, since last
I skimm'd the smooth thin stone along thy breast,
Numbering its light leaps! yet so deep imprest
Sink the sweet scenes of childhood, that mine eyes
  I never shut amid the sunny ray,
But straight with all their tints thy waters rise,
  Thy crossing plank, thy marge with willows grey,
And bedded sand that vein'd with various dyes   10
Gleam'd through thy bright transparence! On my way,
  Visions of Childhood! oft have ye beguil'd
Lone manhood's cares, yet waking fondest sighs:
  Ah! that once more I were a careless Child!

?1793 (1796)

LINES

TO A BEAUTIFUL SPRING IN A VILLAGE

Once more! sweet Stream! with slow foot wandering
    near,
I bless thy milky waters cold and clear.
Escap'd the flashing of the noontide hours,
With one fresh garland of Pierian flowers
(Ere from thy zephyr-haunted brink I turn)
My languid hand shall wreath thy mossy urn.
For not through pathless grove with murmur rude
Thou soothest the sad wood-nymph, Solitude;
Nor thine unseen in cavern depths to well,
The Hermit-fountain of some dripping cell!   10

Pride of the Vale! thy useful streams supply
The scatter'd cots and peaceful hamlet nigh.

LINES. **4. Pierian:** The Pierian spring, in Macedonia, was sacred to the Muses. Its waters gave poetic inspiration.

The elfin tribe around thy friendly banks
With infant uproar and soul-soothing pranks,
Releas'd from school, their little hearts at rest,
Launch paper navies on thy waveless breast.
The rustic here at eve with pensive look
Whistling lorn ditties leans upon his crook,
Or, starting, pauses with hope-mingled dread
To list the much-lov'd maid's accustom'd tread: 20
She, vainly mindful of her dame's command,
Loiters, the long-fill'd pitcher in her hand.

Unboastful Stream! thy fount with pebbled falls
The faded form of past delight recalls,
What time the morning sun of Hope arose,
And all was joy; save when another's woes
A transient gloom upon my soul imprest,
Like passing clouds impictur'd on thy breast.
Life's current then ran sparkling to the noon,
Or silvery stole beneath the pensive Moon: 30
Ah! now it works rude brakes and thorns among,
Or o'er the rough rock bursts and foams along!

                                        1794 (1796)

PANTISOCRACY

❧

For an account of Coleridge's scheme to found an ideal
community in America see the Introduction.

❧

No more my visionary soul shall dwell
On joys that were; no more endure to weigh
The shame and anguish of the evil day,
Wisely forgetful! O'er the ocean swell
Sublime of Hope, I seek the cottag'd dell
Where Virtue calm with careless step may stray,
And dancing to the moonlight roundelay,
The wizard Passions weave an holy spell.
Eyes that have ach'd with Sorrow! Ye shall weep
Tears of doubt-mingled joy, like theirs who start   10
From Precipices of distemper'd sleep,
On which the fierce-eyed Fiends their revels keep,
And see the rising Sun, and feel it dart
New rays of pleasance trembling to the heart.

                                        1794 (1849)

TO A YOUNG ASS

ITS MOTHER BEING TETHERED NEAR IT

Poor little Foal of an oppressèd race!
I love the languid patience of thy face:
And oft with gentle hand I give thee bread,
And clap thy ragged coat, and pat thy head.
But what thy dulled spirits hath dismay'd,
That never thou dost sport along the glade?
And (most unlike the nature of things young)
That earthward still thy moveless head is hung?
Do thy prophetic fears anticipate,
Meek Child of Misery! thy future fate?   10
The starving meal, and all the thousand aches
"Which patient Merit of the Unworthy takes"?
Or is thy sad heart thrill'd with filial pain
To see thy wretched mother's shorten'd chain?
And truly, very piteous is *her* lot—
Chain'd to a log within a narrow spot,
Where the close-eaten grass is scarcely seen,
While sweet around her waves the tempting green!

Poor Ass! thy master should have learnt to show
Pity—best taught by fellowship of Woe!   20
For much I fear me that *He* lives like thee,
Half famish'd in a land of Luxury!
How *askingly* its footsteps hither bend?
It seems to say, "And have I then *one* friend?"
Innocent foal! thou poor despis'd forlorn!
I hail thee *Brother*—spite of the fool's scorn!
And fain would take thee with me, in the Dell
Of Peace and mild Equality to dwell,
Where Toil shall call the charmer Health his bride,
And Laughter tickle Plenty's ribless side!   30
How thou wouldst toss thy heels in gamesome play,
And frisk about, as lamb or kitten gay!
Yea! and more musically sweet to me
Thy dissonant harsh bray of joy would be,
Than warbled melodies that soothe to rest
The aching of pale Fashion's vacant breast!

                                        1794 (1794)

TO A YOUNG ASS. **12.** "**Which . . . takes**": *Hamlet*, III.
i. 74. **27–28. Dell . . . Equality**: the Pantisocratic community.

TO
THE REV. W. L. BOWLES

⌒⌒⌒

William Lisle Bowles, 1762–1850, published in 1789 a
volume of sonnets noticeable in their day for gently
plaintive sentiment and natural observation. Coleridge
read them in his seventeenth year and at once became
an ardent admirer. In the *Biographia Literaria*, Chapter I,
he says, "I made, within less than a year and a half,
more than forty transcriptions, as the best presents I could offer
to those, who had in any way won my regard," and he
still believes that "of the then living poets, Bowles and
Cowper were . . . the first who combined natural
thoughts with natural diction; the first who reconciled
the heart with the head."

⌒⌒⌒

My heart has thank'd thee, BOWLES! for those soft
      strains
   Whose sadness soothes me, like the murmuring
   Of wild-bees in the sunny showers of spring!
For hence not callous to the mourner's pains

Through Youth's gay prime and thornless paths I went:
   And when the mightier Throes of mind began,
   And drove me forth, a thought-bewilder'd man,
Their mild and manliest melancholy lent

A mingled charm, such as the pang consign'd
   To slumber, though the big tear it renew'd;   10
   Bidding a strange mysterious PLEASURE brood
Over the wavy and tumultuous mind,

As the great SPIRIT erst with plastic sweep
Mov'd on the darkness of the unform'd deep.

                              1794–96 (1796)

THE EOLIAN HARP

COMPOSED AT CLEVEDON, SOMERSETSHIRE

My pensive Sara! thy soft cheek reclined
Thus on mine arm, most soothing sweet it is

To sit beside our Cot, our Cot o'ergrown
With white-flower'd Jasmin, and the broad-leav'd
      Myrtle,
(Meet emblems they of Innocence and Love!)
And watch the clouds, that late were rich with light,
Slow saddening round, and mark the star of eve
Serenely brilliant (such should Wisdom be)
Shine opposite! How exquisite the scents
Snatch'd from yon bean-field! and the world *so*
      hush'd!   10
The stilly murmur of the distant Sea
Tells us of silence.

                    And that simplest Lute,
Placed length-ways in the clasping casement, hark!
How by the desultory breeze caress'd,
Like some coy maid half yielding to her lover,
It pours such sweet upbraiding, as must needs
Tempt to repeat the wrong! And now, its strings
Boldlier swept, the long sequacious notes
Over delicious surges sink and rise,
Such a soft floating witchery of sound   20
As twilight Elfins make, when they at eve
Voyage on gentle gales from Fairy-Land,
Where Melodies round honey-dropping flowers,
Footless and wild, like birds of Paradise,
Nor pause, nor perch, hovering on untam'd wing!
O! the one Life within us and abroad,
Which meets all motion and becomes its soul,
A light in sound, a sound-like power in light,
Rhythm in all thought, and joyance every where—
Methinks, it should have been impossible   30
Not to love all things in a world so fill'd;
Where the breeze warbles, and the mute still air
Is Music slumbering on her instrument.

   And thus, my Love! as on the midway slope
Of yonder hill I stretch my limbs at noon,
Whilst through my half-clos'd eye-lids I behold
The sunbeams dance, like diamonds, on the main,
And tranquil muse upon tranquillity;
Full many a thought uncall'd and undetain'd,
And many idle flitting phantasies,   40
Traverse my indolent and passive brain,
As wild and various as the random gales
That swell and flutter on this subject Lute!

TO THE REV. W. L. BOWLES. **13–14 Spirit . . . deep:** See
Genesis 1:2. THE EOLIAN HARP. **Eolian Harp:** Named for
Eolus, god of the winds, the Eolian harp is a box with strings
across its open ends. When placed at an open window, it
makes music as the breeze passes through it.

24. birds of Paradise: They were once thought to have
no feet and to spend almost their whole lives in air, feeding
on dew and nectar. **26–33. O! . . . instrument:** not in the
text of 1796. These lines first appeared in 1817. **37. the
main:** the sea. Clevedon is on the Bristol Channel.

And what if all of animated nature
Be but organic Harps diversely fram'd,
That tremble into thought, as o'er them sweeps
Plastic and vast, one intellectual breeze,
At once the Soul of each, and God of all?

But thy more serious eye a mild reproof
Darts, O beloved Woman! nor such thoughts   50
Dim and unhallow'd dost thou not reject,
And biddest me walk humbly with my God.
Meek Daughter in the family of Christ!
Well hast thou said and holily disprais'd
These shapings of the unregenerate mind;
Bubbles that glitter as they rise and break
On vain Philosophy's aye-babbling spring.
For never guiltless may I speak of him,
The Incomprehensible! save when with awe
I praise him, and with Faith that inly *feels*;   60
Who with his saving mercies healed me,
A sinful and most miserable man,
Wilder'd and dark, and gave me to possess
Peace, and this Cot, and thee, heart-honour'd Maid!

                                        1795 (1796)

REFLECTIONS
ON HAVING LEFT
A PLACE OF RETIREMENT

Low was our pretty Cot: our tallest Rose
Peep'd at the chamber-window. We could hear
At silent noon, and eve, and early morn,
The Sea's faint murmur. In the open air
Our Myrtles blossom'd; and across the porch
Thick Jasmins twined: the little landscape round
Was green and woody, and refresh'd the eye.
It was a spot which you might aptly call
The Valley of Seclusion! Once I saw
(Hallowing his Sabbath-day by quietness)   10
A wealthy son of Commerce saunter by,
Bristowa's citizen: methought, it calm'd
His thirst of idle gold, and made him muse
With wiser feelings: for he paus'd, and look'd
With a pleas'd sadness, and gaz'd all around,
Then eyed our Cottage, and gaz'd round again,
And sigh'd, and said, it was a Blessed Place.
And we *were* bless'd. Oft with patient ear

Long-listening to the viewless sky-lark's note
(Viewless, or haply for a moment seen   20
Gleaming on sunny wings) in whisper'd tones
I've said to my Beloved, "Such, sweet Girl!
The inobtrusive song of Happiness,
Unearthly minstrelsy! then only heard
When the Soul seeks to hear; when all is hush'd,
And the Heart listens!"
                    But the time, when first
From that low Dell, steep up the stony Mount
I climb'd with perilous toil and reach'd the top,
Oh! what a goodly scene! *Here* the bleak mount,
The bare bleak mountain speckled thin with sheep;   30
Grey clouds, that shadowing spot the sunny fields;
And river, now with bushy rocks o'er-brow'd,
Now winding bright and full, with naked banks;
And seats, and lawns, the Abbey and the wood,
And cots, and hamlets, and faint city-spire;
The Channel *there*, the Islands and white sails,
Dim coasts, and cloud-like hills, and shoreless Ocean—
It seem'd like Omnipresence! God, methought,
Had built him there a Temple: the whole World
Seem'd *imag'd* in its vast circumference: 40
No *wish* profan'd my overwhelmed heart.
Blest hour! It was a luxury,—to be!

Ah! quiet Dell! dear Cot, and Mount sublime!
I was constrain'd to quit you. Was it right,
While my unnumber'd brethren toil'd and bled,
That I should dream away the entrusted hours
On rose-leaf beds, pampering the coward heart
With feelings all too delicate for use?
Sweet is the tear that from some Howard's eye
Drops on the cheek of one he lifts from earth: 50
And he that works me good with unmov'd face,
Does it but half: he chills me while he aids,
My benefactor, not my brother man!
Yet even this, this cold beneficence
Praise, praise it, O my Soul! oft as thou scann'st
The sluggard Pity's vision-weaving tribe!
Who sigh for Wretchedness, yet shun the Wretched,
Nursing in some delicious solitude
Their slothful loves and dainty sympathies!
I therefore go, and join head, heart, and hand,   60
Active and firm, to fight the bloodless fight
Of Science, Freedom, and the Truth in Christ.

ON HAVING LEFT A PLACE OF RETIREMENT. **12. Bristowa:**
Bristol.

49. **Howard:** John Howard, 1726?–90, philanthropist. Cf.
Lamb's "Christ's Hospital Five and Thirty Years Ago,"
below, n. 23.

Yet oft when after honourable toil
Rests the tir'd mind, and waking loves to dream,
My spirit shall revisit thee, dear Cot!
Thy Jasmin and thy window-peeping Rose,
And Myrtles fearless of the mild sea-air.
And I shall sigh fond wishes—sweet Abode!
Ah!—had none greater! And that all had such!
It might be so—but the time is not yet.   70
Speed it, O Father! Let thy Kingdom come!

1795 (1796)

ODE
TO THE DEPARTING YEAR

To the edition of 1797 Coleridge prefixed the following
argument: "The Ode commences with an address to the
Divine Providence that regulates into one vast harmony
all the events of time, however calamitous some of them
may appear to mortals. The second Strophe calls on
men to suspend their private joys and sorrows, and devote them for a while to the cause of human nature in
general. The first Epode speaks of the Empress of
Russia, who died of an apoplexy on the 17th of November 1796; having just concluded a subsidiary treaty with
the Kings combined against France. The first and second
Antistrophe describe the Image of the Departing Year,
etc., as in a vision. The second Epode prophesies, in
anguish of spirit, the downfall of this country."

I

Spirit who sweepest the wild Harp of Time!
    It is most hard, with an untroubled ear
    Thy dark inwoven harmonies to hear!
Yet, mine eye fix'd on Heaven's unchanging clime
Long had I listen'd, free from mortal fear,
    With inward stillness, and a bowed mind;
    When lo! its folds far waving on the wind,
I saw the train of the Departing Year!
    Starting from my silent sadness
    Then with no unholy madness,   10
Ere yet the enter'd cloud foreclos'd my sight,
I rais'd the impetuous song, and solemnis'd his flight.

II

    Hither, from the recent tomb,
    From the prison's direr gloom,

    From Distemper's midnight anguish;
And thence, where Poverty doth waste and languish;
    Or where, his two bright torches blending,
        Love illumines Manhood's maze;
    Or where o'er cradled infants bending,
        Hope has fix'd her wishful gaze;   20
        Hither, in perplexed dance,
Ye Woes! ye young-eyed Joys! advance!
By Time's wild harp, and by the hand
        Whose indefatigable sweep
        Raises its fateful strings from sleep,
I bid you haste, a mix'd tumultuous band!
        From every private bower,
        And each domestic hearth,
        Haste for one solemn hour;
And with a loud and yet a louder voice,   30
O'er Nature struggling in portentous birth,
        Weep and rejoice!
Still echoes the dread Name that o'er the earth
Let slip the storm, and woke the brood of Hell:
    And now advance in saintly Jubilee
Justice and Truth! They too have heard thy spell,
    They too obey thy name, divinest Liberty!

III

I mark'd Ambition in his war-array!
    I heard the mailed Monarch's troublous cry—
"Ah! wherefore does the Northern Conqueress stay!
Groans not her chariot on its onward way?"   41

ODE TO THE DEPARTING YEAR. **33. Name:** "The name of
Liberty, which at the commencement of the French Revolution was both the occasion and the pretext of unnumbered
crimes and horrors" (Coleridge's note). **40. Northern
Conqueress:** Catherine the Great of Russia, 1729–96. "A
subsidiary Treaty had been just concluded; and Russia was to
have furnished more effectual aid than that of pious mani-
festoes to the Powers combined against France. I rejoice—not
over the deceased Woman (I never dared figure the Russian
Sovereign to my imagination under the dear and venerable
Character of WOMAN—WOMAN, that complex term for
Mother, Sister, Wife!) I rejoice, as at the disenshrining of a
Daemon! I rejoice, as at the extinction of the evil Principle
impersonated! This very day, six years ago, the massacre of
Ismail was perpetrated. THIRTY THOUSAND HUMAN BEINGS,
MEN, WOMEN, AND CHILDREN, murdered in cold blood,
for no other crime than that their garrison had defended the
place with perseverance and bravery. Why should I recall the
poisoning of her husband, her iniquities in Poland, or her late
unmotivated attack on Persia, the desolating ambition of her
public life, or the libidinous excesses of her private hours!
I have no wish to qualify myself for the office of Historio-
grapher to the King of Hell!" (C).

Fly, mailed Monarch, fly!
   Stunn'd by Death's twice mortal mace,
   No more on Murder's lurid face
The insatiate Hag shall gloat with drunken eye!
   Manes of the unnumber'd slain!
   Ye that gasp'd on Warsaw's plain!
   Ye that erst at Ismail's tower,
When human ruin choked the streams,
   Fell in Conquest's glutted hour,   50
Mid women's shrieks and infants' screams!
   Spirits of the uncoffin'd slain,
   Sudden blasts of triumph swelling,
Oft, at night, in misty train,
   Rush around her narrow dwelling!
The exterminating Fiend is fled—
   (Foul her life, and dark her doom)
Mighty armies of the dead
   Dance, like death-fires, round her tomb!
   Then with prophetic song relate,   60
Each some Tyrant-Murderer's fate!

IV

Departing Year! 'twas on no earthly shore
   My soul beheld thy Vision! Where alone,
   Voiceless and stern, before the cloudy throne,
Aye Memory sits: thy robe inscrib'd with gore,
   With many an unimaginable groan
   Thou storiedst thy sad hours! Silence ensued,
   Deep silence o'er the ethereal multitude,
Whose locks with wreaths, whose wreaths with
     glories shone.
   Then, his eye wild ardours glancing,   70
   From the choired gods advancing,
The Spirit of the Earth made reverence meet,
And stood up, beautiful, before the cloudy seat.

V

Throughout the blissful throng,
   Hush'd were harp and song:
Till wheeling round the throne the Lampads seven,
   (The mystic Words of Heaven)
   Permissive signal make:
The fervent Spirit bow'd, then spread his wings and
   spake!

"Thou in stormy blackness throning   80
   Love and uncreated Light,
By the Earth's unsolaced groaning,
   Seize thy terrors, Arm of might!
By Peace with proffer'd insult scared,
   Masked Hate and envying Scorn!
   By years of Havoc yet unborn!
And Hunger's bosom to the frost-winds bared!
   But chief by Afric's wrongs,
   Strange, horrible, and foul!
By what deep guilt belongs   90
To the deaf Synod, 'full of gifts and lies!'
By Wealth's insensate laugh! by Torture's howl!
   Avenger, rise!
For ever shall the thankless Island scowl,
   Her quiver full, and with unbroken bow?
Speak! from thy storm-black Heaven O speak aloud!
   And on the darkling foe
Open thine eye of fire from some uncertain cloud!
   O dart the flash! O rise and deal the blow!
The Past to thee, to thee the Future cries!   100
   Hark! how wide Nature joins her groans below!
   Rise, God of Nature! rise."

VI

The voice had ceas'd, the Vision fled;
Yet still I gasp'd and reel'd with dread.
And ever, when the dream of night
Renews the phantom to my sight,
Cold sweat-drops gather on my limbs;
   My ears throb hot; my eye-balls start;
My brain with horrid tumult swims;
   Wild is the tempest of my heart;   110
And my thick and struggling breath
Imitates the toil of death!
No stranger agony confounds
   The Soldier on the war-field spread,
When all foredone with toil and wounds,
   Death-like he dozes among heaps of dead!
(The strife is o'er, the day-light fled,
   And the night-wind clamours hoarse!
See! the starting wretch's head
   Lies pillow'd on a brother's corse!)   120

---

46. Manes: spirits. **63. Vision:** "Thy image in a vision"
(C). **76. Lampads:** See Revelation 4:5. Cf. "Ne Plus
Ultra," below, ll. 18–21.

88. Afric's wrongs: slavery. **91. Synod:** In early version
Coleridge wrote "Senate." **'full . . . lies!':** "gifts used i[n]
Scripture for Corruption" (C.).

VII

Not yet enslaved, not wholly vile,
O Albion! O my mother Isle!
Thy valleys, fair as Eden's bowers
Glitter green with sunny showers;
Thy grassy upland's gentle swells
  Echo to the bleat of flocks;
(Those grassy hills, those glittering dells
  Proudly ramparted with rocks)
And Ocean mid his uproar wild
    Speaks safety to his Island-child!   130
    Hence for many a fearless age
    Has social Quiet lov'd thy shore;
    Nor ever proud Invader's rage
Or sack'ed thy towers, or stain'd thy fields with gore.

VIII

Abandon'd of Heaven! mad Avarice thy guide,
At cowardly distance, yet kindling with pride—
Mid thy herds and thy corn-fields secure thou hast
  stood,
And join'd the wild yelling of Famine and Blood!
The nations curse thee! They with eager wondering
  Shall hear Destruction like a vulture, scream!   140
  Strange-eyed Destruction! who with many a dream
Of central fires through nether seas up-thundering
  Soothes her fierce solitude; yet as she lies
    By livid fount, or red volcanic stream,
    If ever to her lidless dragon-eyes,
  O Albion! thy predestin'd ruins rise,
The fiend-hag on her perilous couch doth leap,
Muttering distemper'd triumph in her charmed sleep.

IX

    Away, my soul, away!
  In vain, in vain the Birds of warning sing—   150
And hark! I hear the famish'd brood of prey
Flap their lank pennons on the groaning wind!
    Away, my soul, away!
  I unpartaking of the evil thing,
    With daily prayer and daily toil
    Soliciting for food my scanty soil,
  Have wail'd my country with a loud Lament.
Now I recentre my immortal mind
  In the deep Sabbath of meek self-content;
Cleans'd from the vaporous passions that bedim   160
God's Image, sister of the Seraphim.

               1796 (1796)

THIS LIME-TREE BOWER MY PRISON

[ADDRESSED TO CHARLES LAMB,
OF THE INDIA HOUSE, LONDON]

In the June of 1797 some long-expected friends paid a
visit to the author's cottage; and on the morning of their
arrival, he met with an accident, which disabled him
from walking during the whole time of their stay. One
evening, when they had left him for a few hours, he composed the following lines in the garden-bower [C.].

Well, they are gone, and here must I remain,
This lime-tree bower my prison! I have lost
Beauties and feelings, such as would have been
Most sweet to my remembrance even when age
Had dimm'd mine eyes to blindness! They, meanwhile,
Friends, whom I never more may meet again,
On springy heath, along the hill-top edge,
Wander in gladness, and wind down, perchance,
To that still roaring dell, of which I told;
The roaring dell, o'erwooded, narrow, deep,   10
And only speckled by the mid-day sun;
Where its slim trunk the ash from rock to rock
Flings arching like a bridge;—that branchless ash,
Unsunn'd and damp, whose few poor yellow leaves
Ne'er tremble in the gale, yet tremble still,
Fann'd by the water-fall! and there my friends
Behold the dark green file of long lank weeds,
That all at once (a most fantastic sight!)
Still nod and drip beneath the dripping edge
Of the blue clay-stone.

            Now, my friends emerge 20
Beneath the wide wide Heaven—and view again
The many-steepled tract magnificent
Of hilly fields and meadows, and the sea,
With some fair bark, perhaps, whose sails light up
The slip of smooth clear blue betwixt two Isles
Of purple shadow! Yes! they wander on
In gladness all; but thou, methinks, most glad,
My gentle-hearted Charles! for thou hast pined
And hunger'd after Nature, many a year,
In the great City pent, winning thy way   30
With sad yet patient soul, through evil and pain
And strange calamity! Ah! slowly sink

THIS LIME-TREE BOWER MY PRISON. **32. strange calamity:**
Mary Lamb's fit of insanity in which she killed her mother
in 1796.

Behind the western ridge, thou glorious Sun!
Shine in the slant beams of the sinking orb,
Ye purple heath-flowers! richlier burn, ye clouds!
Live in the yellow light, ye distant groves!
And kindle, thou blue Ocean! So my friend
Struck with deep joy may stand, as I have stood,
Silent with swimming sense; yea, gazing round
On the wide landscape, gaze till all doth seem   40
Less gross than bodily; and of such hues
As veil the Almighty Spirit, when yet he makes
Spirits perceive his presence.

                        A delight
Comes sudden on my heart, and I am glad
As I myself were there! Nor in this bower,
This little lime-tree bower, have I not mark'd
Much that has sooth'd me. Pale beneath the blaze
Hung the transparent foliage; and I watch'd
Some broad and sunny leaf, and lov'd to see
The shadow of the leaf and stem above   50
Dappling its sunshine! And that walnut-tree
Was richly ting'd, and a deep radiance lay
Full on the ancient ivy, which usurps
Those fronting elms, and now, with blackest mass
Makes their dark branches gleam a lighter hue
Through the late twilight: and though now the bat
Wheels silent by, and not a swallow twitters,
Yet still the solitary humble-bee
Sings in the bean-flower! Henceforth I shall know
That Nature ne'er deserts the wise and pure;   60
No plot so narrow, be but Nature there,
No waste so vacant, but may well employ
Each faculty of sense, and keep the heart
Awake to Love and Beauty! and sometimes
'Tis well to be bereft of promis'd good,
That we may lift the soul, and contemplate
With lively joy the joys we cannot share.
My gentle-hearted Charles! when the last rook
Beat its straight path along the dusky air
Homewards, I blest it! deeming its black wing   70
(Now a dim speck, now vanishing in light)
Had cross'd the mighty Orb's dilated glory,
While thou stood'st gazing; or, when all was still,
Flew creeking o'er thy head, and had a charm
For thee, my gentle-hearted Charles, to whom
No sound is dissonant which tells of Life.

                              1797 (1800)

THE RIME
OF THE ANCIENT MARINER

The circumstances surrounding the origin of the poem
were set forth by Coleridge at the start of Chapter XIV
of *Biographia Literaria*, below, pp. 452–53, and by Wordsworth on two occasions. One, dating from 1843, is
given above in the headnote to Wordsworth's "We Are
Seven." The second was in approximately 1835, when
Wordsworth told the Rev. Alexander Dyce that

> *The Ancient Mariner* was founded on a strange
> dream, which a friend of Coleridge had, who fancied
> he saw a skeleton ship, with figures in it. We had both
> determined to write some poetry for a monthly
> magazine, the profits of which were to defray the
> expenses of a little excursion we were to make to-
> gether. *The Ancient Mariner* was intended for this
> periodical, but was too long. I had very little share in
> the composition of it, for I soon found that the style of
> Coleridge and myself would not assimilate. Besides
> the lines (in the fourth part):
>
> > And thou art long, and lank, and brown,
> > As is the ribbed sea-sand—
>
> I wrote the stanza (in the first part):
>
> > He holds him with his glittering eye—
> > The Wedding-Guest stood still,
> > And listens like a three-years' child:
> > The Mariner hath his will—
>
> and four or five lines more in different parts of the
> poem, which I could not now point out. The idea of
> *"shooting an albatross" was mine; for I had been reading
> Shelvocke's Voyages, which probably Coleridge never
> saw.* I also suggested the reanimation of the dead
> bodies, to work the ship.

The poem was first printed in the *Lyrical Ballads* of
1798, where it included numerous archaic words and
spellings. These overly self-conscious imitations of the old
English ballads were drastically pruned in later versions,
and the marginal glosses were added along with the
Latin motto. Because of its strangeness and obscurity
the poem was condemned by early reviewers, so much
so that Wordsworth thought it injured the sale of the
volume. For later readers it is Coleridge's supreme
poetic achievement, and compared with even the finest
poetry he had written previously it shows an unpredictable change in mode and a sudden, immense stride to a
new level of accomplishment. At least some part of this
is doubtless attributable to friendship with Wordsworth.

In the *Table Talk* (May 31, 1830) Coleridge replied
to a couple of criticisms:

> Mrs. Barbauld [1743–1825, poet and essayist] once
> told me that she admired the Ancient Mariner very

much, but that there were two faults in it,—it was improbable, and had no moral. As for the probability, I owned that that might admit some question; but as to the want of a moral, I told her that in my own judgment the poem had too much; and that the only, or chief fault, if I might say so, was the obtrusion of the moral sentiment so openly on the reader as a principle or cause of action in a work of pure imagination. It ought to have had no more moral than the *Arabian Nights'* tale of the merchant's sitting down to eat dates by the side of a well and throwing the shells aside, and lo! a genie starts up and says he *must* kill the aforesaid merchant *because* one of the date shells had, it seems, put out the eye of the genie's son.

He does not, of course, mean to deny that the poem has significance. His point is rather that a moral becomes obtrusive (he is probably thinking of the final stanzas), and this always tends to limit the possible meanings of an imaginative work. Nevertheless the poem has an inexhaustible fascination. Whatever else may be said about it, we can today see it as one of the first of many works in the nineteenth century in which the paradox is recognized that a guilty act may be the prelude to profound spiritual discovery and knowledge.

⁓

Facile credo, plures esse Naturas invisibiles quam visibiles in rerum universitate. Sed horum omnium familiam quis nobis enarrabit? et gradus et cognationes et discrimina et singulorum munera? Quid agunt? quae loca habitant? Harum rerum notitiam semper ambivit ingenium humanum, nunquam attigit. Juvat, interea, non diffiteor, quandoque in animo, tanquam in tabulà, majoris et melioris mundi imaginem contemplari: ne mens assuefacta hodiernae vitae minutiis se contrahat nimis, et tota subsidat in pusillas cogitationes. Sed veritati interea invigilandum est, modusque servandus, ut certa ab incertis, diem a nocte, distinguamus.

[Thomas Burnet,[1] *Archaeologiae Philosophicae* (1692), p. 68].

THE RIME OF THE ANCIENT MARINER. **Motto:** "I readily believe that there are more invisible than visible things in the universe. But who shall describe for us their families, their ranks, relationships, distinguishing features and functions? What do they do? Where do they live? The human mind has always circled about knowledge of these things, but never attained it. I do not doubt, however, that it is sometimes good to contemplate in the mind, as in a picture, the image of a greater and better world; otherwise the intellect, habituated to the petty things of daily life, may too much contract itself, and wholly sink down to trivial thoughts. But meanwhile we must be vigilant for truth and keep proportion, that we may distinguish the certain from the uncertain, day from night." **1. Thomas Burnet:** Anglican divine, 1635?–1715.

Argument

How a Ship having passed the Line was driven by storms to the cold Country towards the South Pole; and how from thence she made her course to the tropical Latitude of the Great Pacific Ocean; and of the strange things that befell; and in what manner the Ancyent Marinere came back to his own Country.

Part I

An ancient Mariner meeteth three Gallants bidden to a wedding-feast, and detaineth one.

It is an ancient Mariner,
And he stoppeth one of three.
"By thy long grey beard and glittering eye,
Now wherefore stopp'st thou me?

The Bridegroom's doors are opened wide,
And I am next of kin;
The guests are met, the feast is set:
May'st hear the merry din."

He holds him with his skinny hand,
"There was a ship," quoth he.   10
"Hold off! unhand me, grey-beard loon!"
Eftsoons his hand dropt he.

The Wedding-Guest is spellbound by the eye of the old seafaring man, and constrained to hear his tale.

He holds him with his glittering eye—
The Wedding-Guest stood still,
And listens like a three years' child:
The Mariner hath his will.

The Wedding-Guest sat on a stone:
He cannot choose but hear;
And thus spake on that ancient man,
The bright-eyed Mariner.   20

"The ship was cheered, the harbour cleared,
Merrily did we drop
Below the kirk, below the hill,
Below the lighthouse top.

The Mariner tells how the ship sailed southward with a good wind and fair weather, till it reached the line.

The Sun came up upon the left,
Out of the sea came he!
And he shone bright, and on the right
Went down into the sea.

12. Eftsoons: at once.

Higher and higher every day,
Till over the mast at noon—"   30
The Wedding-Guest here beat his breast,
For he heard the loud bassoon.

The bride hath paced into the hall,
Red as a rose is she;
Nodding their heads before her goes
The merry minstrelsy.

The Wedding-Guest he beat his breast,
Yet he cannot choose but hear;
And thus spake on that ancient man,
The bright-eyed Mariner.   40

"And now the STORM-BLAST came, and
    he
Was tyrannous and strong:
He struck with his o'ertaking wings,
And chased us south along.

With sloping masts and dipping prow,
As who pursued with yell and blow
Still treads the shadow of his foe,
And forward bends his head,
The ship drove fast, loud roared the
    blast,
And southward aye we fled.   50

And now there came both mist and
    snow,
And it grew wondrous cold:
And ice, mast-high, came floating by,
As green as emerald.

And through the drifts the snowy clifts
Did send a dismal sheen:
Nor shapes of men nor beasts we ken—
The ice was all between.

The ice was here, the ice was there,
The ice was all around: 60
It cracked and growled, and roared and
    howled,
Like noises in a swound!

At length did cross an Albatross,
Thorough the fog it came;
As if it had been a Christian soul,
We hailed it in God's name.

It ate the food it ne'er had eat,
And round and round it flew.
The ice did split with a thunder-fit;
The helmsman steered us through!   70

And a good south wind sprung up
    behind;
The Albatross did follow,
And every day, for food or play,
Came to the mariners' hollo!

In mist or cloud, on mast or shroud,
It perched for vespers nine;
Whiles all the night, through fog-smoke
    white,
Glimmered the white Moon-shine."

"God save thee, ancient Mariner!
From the fiends, that plague thee thus!—
Why look'st thou so?"—With my
    cross-bow   81
I shot the ALBATROSS.

Part II

The Sun now rose upon the right:
Out of the sea came he,
Still hid in mist, and on the left
Went down into the sea.

And the good south wind still blew
    behind,
But no sweet bird did follow,
Nor any day for food or play
Came to the mariners' hollo!   90

And I had done a hellish thing,
And it would work 'em woe:
For all averred, I had killed the bird
That made the breeze to blow.
Ah wretch! said they, the bird to slay,
That made the breeze to blow!

Nor dim nor red, like God's own head,
The glorious Sun uprist:
Then all averred, I had killed the bird
That brought the fog and mist.   100
'Twas right, said they, such birds to slay,
That bring the fog and mist.

---

30. over . . . noon: The ship has reached the equator. **62.
swound:** swoon.

83. upon the right: The ship has rounded the Horn and
is heading north into the Pacific.

The fair breeze blew, the white foam flew,
The furrow followed free; *alliteration*
We were the first that ever burst
Into that silent sea.

The fair breeze
continues;
the ship enters
the Pacific
Ocean, and
sails north-
ward, even
till it reaches
the Line.

Down dropt the breeze, the sails dropt down,
'Twas sad as sad could be;
And we did speak only to break
The silence of the sea!   110

*The ship hath
been suddenly
becalmed.*

All in a hot and copper sky,
The bloody Sun, at noon,
Right up above the mast did stand,
No bigger than the Moon.

Day after day, day after day,
We stuck, nor breath nor motion;
As idle as a painted ship
Upon a painted ocean.   *like a painting on canvas*

*And the Alba-
tross begins to
be avenged.*

Water, water, every where,
And all the boards did shrink;   120
Water, water, every where,
Nor any drop to drink.

The very deep did rot: O Christ!
That ever this should be!
Yea, slimy things did crawl with legs
Upon the slimy sea.

*A Spirit had
followed
them; one of
the invisible
inhabitants of
this planet,
neither de-
parted souls
nor angels;
concerning
whom the
learned Jew,
Josephus, and
the Platonic
Constantino-
politan,
Michael
Psellus, may
be consulted.
They are very
numerous,
and there is
no climate or
element with-
out one or
more.*

About, about, in reel and rout
The death-fires danced at night;
The water, like a witch's oils,
Burnt green, and blue and white.   130   *electricity + sea creatures in water*

And some in dreams assured were
Of the Spirit that plagued us so;
Nine fathom deep he had followed us
From the land of mist and snow.

And every tongue, through utter drought,
Was withered at the root;   *striking*
We could not speak, no more than if
We had been choked with soot.

128. **death-fires:** St. Elmo's fire, atmospheric electricity causing an appearance of lights in the rigging. Sailors believed they were omens of death. (See John Livingston Lowes, *The Road to Xanadu* [1927], pp. 85–86.)

Ah! well a-day! what evil looks
Had I from old and young!   140
Instead of the cross, the Albatross
About my neck was hung.

The shipmates,
in their sore
distress, would
fain throw the
whole guilt on
the ancient
Mariner: in
sign whereof
they hang the
dead sea-bird
round his
neck.

Part III

There passed a weary time. Each throat
Was parched, and glazed each eye.
A weary time! a weary time!
How glazed each weary eye,
When looking westward, I beheld
A something in the sky.

*The ancient
Mariner be-
holdeth a sign
in the element
afar off.*

At first it seemed a little speck,
And then it seemed a mist;   150
It moved and moved, and took at last
A certain shape, I wist.

A speck, a mist, a shape, I wist!
And still it neared and neared:
As if it dodged a water-sprite,
It plunged and tacked and veered.

With throats unslaked, with black lips baked,
We could nor laugh nor wail;
Through utter drought all dumb we stood!
I bit my arm, I sucked the blood,   160
And cried, A sail! a sail!

*At its nearer
approach, it
seemeth him
to be a ship;
and at a dear
ransom he
freeth his
speech from
the bonds of
thirst.*   *so chapped bec. so thirsty*

With throats unslaked, with black lips baked,
Agape they heard me call:
Gramercy! they for joy did grin,
And all at once their breath drew in,
As they were drinking all.

A flash of joy;

See! see! (I cried) she tacks no more!
Hither to work us weal;
Without a breeze, without a tide,
She steadies with upright keel!   170

*And horror
follows. For
can it be a
ship that
comes onward
without wind
or tide?*

The western wave was all a-flame.
The day was well nigh done!
Almost upon the western wave
Rested the broad bright Sun;
When that strange shape drove suddenly
Betwixt us and the Sun.   *perversion of what one expects to see in real life*

152. **wist:** knew. 168. **weal:** good.

It seemeth him but the skeleton of a ship.

And straight the Sun was flecked with bars,
(Heaven's Mother send us grace!)
As if through a dungeon-grate he peered
With broad and burning face.   180

Alas! (thought I, and my heart beat loud)

And its ribs are seen as bars on the face of the setting Sun.

How fast she nears and nears!
Are those *her* sails that glance in the Sun,
Like restless gossameres?

The Spectre-Woman and her Death-mate, and no other on board the skeleton ship.

Are those *her* ribs through which the Sun
Did peer, as through a grate?
And is that Woman all her crew?
Is that a DEATH? and are there two?
Is DEATH that woman's mate?

Her lips were red, *her* looks were free,
Her locks were yellow as gold: 191
Her skin was as white as leprosy,
The Night-mare LIFE-IN-DEATH was she,
Who thicks man's blood with cold.

Like vessel, like crew!

Death and Life-in-Death have diced for the ship's crew, and she (the latter) winneth the ancient Mariner.

The naked hulk alongside came,
And the twain were casting dice;
"The game is done! I've won! I've won!"
Quoth she, and whistles thrice.

No twilight within the courts of the Sun.

The Sun's rim dips; the stars rush out:
At one stride comes the dark;   200
With far-heard whisper, o'er the sea,
Off shot the spectre-bark.

At the rising of the Moon,

We listened and looked sideways up!
Fear at my heart, as at a cup,
My life-blood seemed to sip!
The stars were dim, and thick the night,
The steersman's face by his lamp gleamed white;
From the sails the dew did drip—
Till clomb above the eastern bar
The horned Moon, with one bright star   210
Within the nether tip.

210–11. **Moon . . . tip:** "It is a common superstition among sailors that something evil is about to happen whenever a star dogs the moon" (C.).

One after another,

One after one, by the star-dogged Moon,
Too quick for groan or sigh,
Each turned his face with a ghastly pang,
And cursed me with his eye.

His shipmates drop down dead.

Four times fifty living men,
(And I heard nor sigh nor groan)
With heavy thump, a lifeless lump,
They dropped down one by one.

But Life-in-Death begins her work on the ancient Mariner.

The souls did from their bodies fly,—
They fled to bliss or woe!   221
And every soul, it passed me by,
Like the whizz of my cross-bow!

Part IV

The Wedding-Guest feareth that a Spirit is talking to him;

"I fear thee, ancient Mariner!
I fear thy skinny hand!
And thou art long, and lank, and brown,
As is the ribbed sea-sand.

I fear thee and thy glittering eye,
And thy skinny hand, so brown."—
Fear not, fear not, thou Wedding-Guest!   230
This body dropt not down.

But the ancient Mariner assureth him of his bodily life, and proceedeth to relate his horrible penance.

Alone, alone, all, all alone,
Alone on a wide wide sea!
And never a saint took pity on
My soul in agony.

He despiseth the creatures of the calm,

The many men, so beautiful!
And they all dead did lie:
And a thousand thousand slimy things
Lived on; and so did I.

And envieth that *they* should live, and so many lie dead.

I looked upon the rotting sea,   240
And drew my eyes away;
I looked upon the rotting deck,
And there the dead men lay.

I looked to heaven, and tried to pray;
But or ever a prayer had gusht,
A wicked whisper came, and made
My heart as dry as dust.

I closed my lids, and kept them close,
And the balls like pulses beat;
For the sky and the sea, and the sea and
    the sky   250
Lay like a load on my weary eye,
And the dead were at my feet.

But the curse
liveth for him
in the eye of
the dead men.

The cold sweat melted from their
    limbs,
Nor rot nor reek did they:
The look with which they looked on me
Had never passed away.

An orphan's curse would drag to hell
A spirit from on high;

In his lone-
liness and
fixedness he
yearneth to-
wards the
journeying
Moon, and the
stars that still
sojourn, yet
still move
onward; and
every where
the blue sky
belongs to
them, and is
their ap-
pointed rest,
and their
native country
and their own
natural homes,
which they
enter unan-
nounced, as
lords that are
certainly ex-
pected and yet
there is a silent
joy at their
arrival.

But oh! more horrible than that
Is the curse in a dead man's eye!   260
Seven days, seven nights, I saw that
    curse,
And yet I could not die.

The moving Moon went up the sky,
And no where did abide:
Softly she was going up,
And a star or two beside—

Her beams bemocked the sultry main,
Like April hoar-frost spread;
But where the ship's huge shadow
    lay,
The charmed water burnt alway   270
A still and awful red.

By the light
of the Moon
he beholdeth
God's crea-
tures of the
great calm.

Beyond the shadow of the ship,
I watched the water-snakes:
They moved in tracks of shining white,
And when they reared, the elfish light
Fell off in hoary flakes.

Within the shadow of the ship
I watched their rich attire:
Blue, glossy green, and velvet black,
They coiled and swam; and every track
Was a flash of golden fire.   281

Their beauty
and their
happiness.

O happy living things! no tongue
Their beauty might declare:
A spring of love gushed from my
    heart,

He blesseth
them in his
heart.

And I blessed them unaware:
Sure my kind saint took pity on me,
And I blessed them unaware.

The spell
begins to
break.

The self-same moment I could pray;
And from my neck so free
The Albatross fell off, and sank   290
Like lead into the sea.

Part V

Oh sleep! it is a gentle thing,
Beloved from pole to pole!
To Mary Queen the praise be given!
She sent the gentle sleep from Heaven,
That slid into my soul.

By grace of
the holy
Mother, the
ancient
Mariner is
refreshed with
rain.

The silly buckets on the deck,
That had so long remained,
I dreamt that they were filled with dew;
And when I awoke, it rained.   300

My lips were wet, my throat was cold,
My garments all were dank;
Sure I had drunken in my dreams,
And still my body drank.

I moved, and could not feel my limbs:
I was so light—almost
I thought that I had died in sleep,
And was a blessed ghost.

He heareth
sounds and
seeth strange
sights and
commotions
in the sky and
the element.

And soon I heard a roaring wind:
It did not come anear;   310
But with its sound it shook the sails,
That were so thin and sere.

The upper air burst into life!
And a hundred fire-flags sheen,
To and fro they were hurried about!
And to and fro, and in and out,
The wan stars danced between.

And the coming wind did roar more
    loud,
And the sails did sigh like sedge;
And the rain poured down from one
    black cloud;   320
The Moon was at its edge.

The thick black cloud was cleft, and still
The Moon was at its side:
Like waters shot from some high crag,
The lightning fell with never a jag,
A river steep and wide.

The loud wind never reached the ship,
Yet now the ship moved on!
Beneath the lightning and the Moon
The dead men gave a groan.   330

They groaned, they stirred, they all
     uprose,
Nor spake, nor moved their eyes;
It had been strange, even in a dream,
To have seen those dead men rise.

The helmsman steered, the ship moved
     on;
Yet never a breeze up-blew;
The mariners all 'gan work the ropes,
Where they were wont to do;
They raised their limbs like lifeless tools—
We were a ghastly crew.   340

The body of my brother's son
Stood by me, knee to knee:
The body and I pulled at one rope,
But he said nought to me.

"I fear thee, ancient Mariner!"
Be calm, thou Wedding-Guest!
'Twas not those souls that fled in pain,
Which to their corses came again,
But a troop of spirits blest:

For when it dawned—they dropped
     their arms,   350
And clustered round the mast;
Sweet sounds rose slowly through their
     mouths,
And from their bodies passed.

Around, around, flew each sweet sound,
Then darted to the Sun;
Slowly the sounds came back again,
Now mixed, now one by one.

Sometimes a-dropping from the sky
I heard the sky-lark sing;
Sometimes all little birds that are,   360
How they seemed to fill the sea and air
With their sweet jargoning!

And now 'twas like all instruments,
Now like a lonely flute;
And now it is an angel's song,
That makes the heavens be mute.

The bodies of the ship's crew are inspired and the ship moves on;

But not by the souls of the men, nor by dæmons of earth or middle air, but by a blessed troop of angelic spirits, sent down by the invocation of the guardian saint.

It ceased; yet still the sails made on
A pleasant noise till noon,
A noise like of a hidden brook
In the leafy month of June,   370
That to the sleeping woods all night
Singeth a quiet tune.

Till noon we quietly sailed on,
Yet never a breeze did breathe:
Slowly and smoothly went the ship,
Moved onward from beneath.

Under the keel nine fathom deep,
From the land of mist and snow,
The spirit slid: and it was he
That made the ship to go.   380
The sails at noon left off their tune,
And the ship stood still also.

The Sun, right up above the mast,
Had fixed her to the ocean:
But in a minute she 'gan stir,
With a short uneasy motion—
Backwards and forwards half her length
With a short uneasy motion.

Then like a pawing horse let go,
She made a sudden bound: 390
It flung the blood into my head,
And I fell down in a swound.

How long in that same fit I lay,
I have not to declare;
But ere my living life returned,
I heard and in my soul discerned
Two voices in the air.

"Is it he?" quoth one, "Is this the man?
By him who died on cross,
With his cruel bow he laid full low   400
The harmless Albatross.

The spirit who bideth by himself
In the land of mist and snow,
He loved the bird that loved the man
Who shot him with his bow."

The other was a softer voice,
As soft as honey-dew:
Quoth he, "The man hath penance done,
And penance more will do."

The lonesome Spirit from the south-pole carries on the ship as far as the Line, in obedience to the angelic troop, but still requireth vengeance.

The Polar Spirit's fellow-dæmons, the invisible inhabitants of the element, take part in his wrong; and two of them relate, one to the other, that penance long and heavy for the ancient Mariner hath been accorded to the Polar Spirit, who returneth southward.

362. jargoning: warbling.

394. I . . . declare: I cannot say.

Part VI

FIRST VOICE

"But tell me, tell me! speak again,   410
Thy soft response renewing—
What makes that ship drive on so fast?
What is the ocean doing?"

SECOND VOICE

"Still as a slave before his lord,
The ocean hath no blast;
His great bright eye most silently
Up to the Moon is cast—

If he may know which way to go;
For she guides him smooth or grim.
See, brother, see! how graciously   420
She looketh down on him."

FIRST VOICE

"But why drives on that ship so fast,
Without or wave or wind?"

SECOND VOICE

"The air is cut away before,
And closes from behind.

Fly, brother, fly! more high, more
high!
Or we shall be belated:
For slow and slow that ship will go,
When the Mariner's trance is abated."

I woke, and we were sailing on   430
As in a gentle weather:
'Twas night, calm night, the moon was
high;
The dead men stood together.

All stood together on the deck,
For a charnel-dungeon fitter:
All fixed on me their stony eyes,
That in the Moon did glitter.

The pang, the curse, with which they
died,
Had never passed away:
I could not draw my eyes from theirs,
Nor turn them up to pray.   441

The Mariner hath been cast into a trance; for the angelic power causeth the vessel to drive northward faster than human life could endure.

The supernatural motion is retarded; the Mariner awakes, and his penance begins anew.

435. **charnel-dungeon:** a dungeon where the bodies of the
dead are piled as in a charnel house.

The curse is finally expiated.

And now this spell was snapt: once
more
I viewed the ocean green,
And looked far forth, yet little saw
Of what had else been seen—

Like one, that on a lonesome road
Doth walk in fear and dread,
And having once turned round walks
on,
And turns no more his head;
Because he knows, a frightful fiend   450
Doth close behind him tread.

But soon there breathed a wind on me,
Nor sound nor motion made:
Its path was not upon the sea,
In ripple or in shade.

It raised my hair, it fanned my cheek
Like a meadow-gale of spring—
It mingled strangely with my fears,
Yet it felt like a welcoming.

Swiftly, swiftly flew the ship,   460
Yet she sailed softly too:
Sweetly, sweetly blew the breeze—
On me alone it blew.

And the ancient Mariner beholdeth his native country.

Oh! dream of joy! is this indeed
The light-house top I see?
Is this the hill? is this the kirk?
Is this mine own countree?

We drifted o'er the harbour-bar,
And I with sobs did pray—
O let me be awake, my God!   470
Or let me sleep alway.

The harbour-bay was clear as glass,
So smoothly it was strewn!
And on the bay the moonlight lay,
And the shadow of the Moon.

The rock shone bright, the kirk no less,
That stands above the rock:
The moonlight steeped in silentness
The steady weathercock.

And the bay was white with silent
light,   480
Till rising from the same,
Full many shapes, that shadows were,
In crimson colours came.

The angelic spirits leave the dead bodies,

And appear in
their own
forms of light.

A little distance from the prow
Those crimson shadows were:
I turned my eyes upon the deck—
Oh, Christ! what saw I there!

Each corse lay flat, lifeless and flat,
And, by the holy rood!
A man all light, a seraph-man,   490
On every corse there stood.

This seraph-band, each waved his hand:
It was a heavenly sight!
They stood as signals to the land,
Each one a lovely light;

This seraph-band, each waved his hand,
No voice did they impart—
No voice; but oh! the silence sank
Like music on my heart.

But soon I heard the dash of oars,   500
I heard the Pilot's cheer;
My head was turned perforce away
And I saw a boat appear.

The Pilot and the Pilot's boy,
I heard them coming fast:
Dear Lord in Heaven! it was a joy
The dead men could not blast.

I saw a third—I heard his voice:
It is the Hermit good!
He singeth loud his godly hymns   510
That he makes in the wood.
He'll shrieve my soul, he'll wash away
The Albatross's blood.

Part VII

The Hermit of
the Wood,

This Hermit good lives in that wood
Which slopes down to the sea.
How loudly his sweet voice he rears!
He loves to talk with marineres
That come from a far countree.

He kneels at morn, and noon, and eve—
He hath a cushion plump: 520
It is the moss that wholly hides
The rotted old oak-stump.

The skiff-boat neared: I heard them
    talk,
"Why, this is strange, I trow!
Where are those lights so many and
    fair,
That signal made but now?"

Approacheth
the ship with
wonder.

"Strange, by my faith!" the Hermit
    said—
"And they answered not our cheer!
The planks looked warped! and see
    those sails,
How thin they are and sere!   530
I never saw aught like to them,
Unless perchance it were

Brown skeletons of leaves that lag
My forest-brook along;
When the ivy-tod is heavy with snow,
And the owlet whoops to the wolf be-
    low,
That eats the she-wolf's young."

"Dear Lord! it hath a fiendish look—
(The Pilot made reply)
I am a-feared"—"Push on, push on!"
Said the Hermit cheerily.   541

The boat came closer to the ship,
But I nor spake nor stirred;
The boat came close beneath the ship,
And straight a sound was heard.

The ship
suddenly
sinketh.

Under the water it rumbled on,
Still louder and more dread:
It reached the ship, it split the bay;
The ship went down like lead.

The ancient
Mariner is
saved in the
Pilot's boat.

Stunned by that loud and dreadful
    sound,   550
Which sky and ocean smote,
Like one that hath been seven days
    drowned
My body lay afloat;
But swift as dreams, myself I found
Within the Pilot's boat.

Upon the whirl, where sank the ship,
The boat spun round and round;
And all was still, save that the hill
Was telling of the sound.

---

489. rood: cross.

535. ivy-tod: ivy bush.

I moved my lips—the Pilot shrieked
And fell down in a fit;   561
The holy Hermit raised his eyes,
And prayed where he did sit.

I took the oars: the Pilot's boy,
Who now doth crazy go,
Laughed loud and long, and all the
          while
His eyes went to and fro.
"Ha! ha!" quoth he, "full plain I see,
The Devil knows how to row."

And now, all in my own countree,   570
I stood on the firm land!
The Hermit stepped forth from the
          boat,
And scarcely he could stand.

*The ancient
Mariner
earnestly en-
treateth the
Hermit to
shrieve him;
and the
penance of
life falls on
him.*

"O shrieve me, shrieve me, holy man!"
The Hermit crossed his brow.
"Say quick," quoth he, "I bid thee
          say—
What manner of man art thou?"

Forthwith this frame of mine was
          wrenched
With a woful agony,
Which forced me to begin my tale;
And then it left me free.   581

*And ever and
anon through
out his future
life an agony
constraineth
him to travel
from land to
land;*

Since then, at an uncertain hour,
That agony returns:
And till my ghastly tale is told,
This heart within me burns.

I pass, like night, from land to land;
I have strange power of speech;
That moment that his face I see,
I know the man that must hear me:
To him my tale I teach.   590

What loud uproar bursts from that
          door!
The wedding-guests are there:
But in the garden-bower the bride
And bride-maids singing are:
And hark the little vesper bell,
Which biddeth me to prayer!

O Wedding-Guest! this soul hath been
Alone on a wide wide sea:
So lonely 'twas, that God himself
Scarce seemed there to be.   600

O sweeter than the marriage-feast,
'Tis sweeter far to me,
To walk together to the kirk
With a goodly company!—

To walk together to the kirk,
And all together pray,
While each to his great Father bends,
Old men, and babes, and loving friends
And youths and maidens gay!

*And to teach,
by his own
example, love
and reverence
to all things
that God made
and loveth.*

Farewell, farewell! but this I tell   610
To thee, thou Wedding-Guest!
He prayeth well, who loveth well
Both man and bird and beast.

He prayeth best, who loveth best
All things both great and small;
For the dear God who loveth us,
He made and loveth all.

The Mariner, whose eye is bright,
Whose beard with age is hoar,   619
Is gone: and now the Wedding-Guest
Turned from the bridegroom's door.

He went like one that hath been stunned,
And is of sense forlorn:
A sadder and a wiser man,
He rose the morrow morn.

                    1797–98 (1798)

CHRISTABEL

That the poem is unfinished has often been explained by asserting that Coleridge did not know how it was to end. But Coleridge denied this:

The reason of my not finishing Christabel is not, that I don't know how to do it—for I have, as I always had, the whole plan entire from beginning to end in my mind; but I fear I could not carry on with equal success the execution of the idea, an extremely subtle and difficult one

[*Table Talk*, July 6, 1833].

575. **crossed his brow:** made the sign of the cross on his forehead.

Friends and acquaintances recorded various continuations and interpretations that they claimed to have heard from Coleridge. On scrutiny, however, these testimonials turn out to be suggestive rather than authoritative. (They are collected and assessed by Arthur H. Nethercott, *The Road to Tryermaine*, 1939, a book wholly devoted to this enigmatic poem.) Geraldine is clearly a witch, presumably both a type of vampire and a lamia or serpent-woman, but as in Keats's *Lamia*, below, p. 1188, the poem develops a complex attitude toward this figure. Though clearly an evil and supernatural being—for example, at line 129 she cannot cross the threshold of the castle because it has been blessed against evil spirits—there are also indications that her actions are not voluntary and that she herself suffers and pities. One of the most suggestive hints toward the total meaning of the intended poem, recorded both by Derwent Coleridge, the poet's son, and by James Gillman, in whose family Coleridge lived for many years, is that the tale was to be of vicarious suffering undergone by Christabel, "founded on the notion, that the virtuous of this world save the wicked." In other words, the tale would be based ultimately on the Christian doctrine of the Atonement.

The "celebrated poets" referred to in the Preface are Sir Walter Scott and Lord Byron. Both had heard "Christabel" recited, Scott in 1801 and Byron in 1811, and both had adopted a similar though less varied meter in works already published. (See Scott's *The Lay of the Last Minstrel* and the headnote to it, above, p. 376.) The metrical principle Coleridge describes in the Preface was not altogether new, as he claims, since it had been employed in much old English verse. However it was certainly unlike anything to which Coleridge's readers were accustomed. Compare his account of a similar meter in Wordsworth's *The White Doe of Rylstone*: it is "rather dramatic than lyric, i.e. not such an arrangement of syllables, not such a meter, as acts a priori and with complete self-subsistence . . . but depending for it's beauty always, and often even for it's metrical existence, on the *sense* and *passion*" (*Letters*, ed. E. L. Griggs, III, 112).

∽

PREFACE

The first part of the following poem was written in the year 1797, at Stowey, in the county of Somerset. The second part, after my return from Germany, in the year 1800, at Keswick, Cumberland. It is probable that if the poem had been finished at either of the former periods, or if even the first and second part had been published in the year 1800, the impression of its originality would have been much greater than I dare at present expect. But for this I have only my own indolence to blame. The dates are mentioned for the exclusive purpose of precluding charges of plagiarism or servile imitation from myself. For there is amongst us a set of critics, who seem to hold, that every possible thought and image is traditional; who have no notion that there are such things as fountains in the world, small as well as great; and who would therefore charitably derive every rill they behold flowing, from a perforation made in some other man's tank. I am confident, however, that as far as the present poem is concerned, the celebrated poets whose writings I might be suspected of having imitated, either in particular passages, or in the tone and the spirit of the whole, would be among the first to vindicate me from the charge, and who, on any striking coincidence, would permit me to address them in this doggerel version of two monkish Latin hexameters.

> 'Tis mine and it is likewise yours;
> But an if this will not do;
> Let it be mine, good friend! for I
> Am the poorer of the two.

I have only to add that the metre of Christabel is not, properly speaking, irregular, though it may seem so from its being founded on a new principle: namely, that of counting in each line the accents, not the syllables. Though the latter may vary from seven to twelve, yet in each line the accents will be found to be only four. Nevertheless, this occasional variation in number of syllables is not introduced wantonly, or for the mere ends of convenience, but in correspondence with some transition in the nature of the imagery or passion.

Part I

'Tis the middle of night by the castle clock,
And the owls have awakened the crowing cock;
Tu—whit!——Tu—whoo!
And hark, again! the crowing cock,
How drowsily it crew.

Sir Leoline, the Baron rich,
Hath a toothless mastiff bitch;
From her kennel beneath the rock
She maketh answer to the clock,
Four for the quarters, and twelve for the hour;   10
Ever and aye, by shine and shower,
Sixteen short howls, not over loud;
Some say, she sees my lady's shroud.

Is the night chilly and dark?
The night is chilly, but not dark.

The thin gray cloud is spread on high,
It covers but not hides the sky.
The moon is behind, and at the full;
And yet she looks both small and dull.
The night is chill, the cloud is gray: 20
'Tis a month before the month of May,
And the Spring comes slowly up this way.

The lovely lady, Christabel,
Whom her father loves so well,
What makes her in the wood so late,
A furlong from the castle gate?
She had dreams all yesternight
Of her own betrothed knight;
And she in the midnight wood will pray
For the weal of her lover that's far away. 30

She stole along, she nothing spoke,
The sighs she heaved were soft and low,
And naught was green upon the oak
But moss and rarest misletoe:
She kneels beneath the huge oak tree,
And in silence prayeth she.

The lady sprang up suddenly,
The lovely lady, Christabel!
It moaned as near, as near can be,
But what it is she cannot tell.— 40
On the other side it seems to be,
Of the huge, broad-breasted, old oak tree.

The night is chill; the forest bare;
Is it the wind that moaneth bleak?
There is not wind enough in the air
To move away the ringlet curl
From the lovely lady's cheek—
There is not wind enough to twirl
The one red leaf, the last of its clan,
That dances as often as dance it can, 50
Hanging so light, and hanging so high,
On the topmost twig that looks up at the sky.

Hush, beating heart of Christabel!
Jesu, Maria, shield her well!
She folded her arms beneath her cloak,
And stole to the other side of the oak.
  What sees she there?

There she sees a damsel bright,
Drest in a silken robe of white,
That shadowy in the moonlight shone: 60
The neck that made that white robe wan,
Her stately neck, and arms were bare;
Her blue-veined feet unsandal'd were,
And wildly glittered here and there
The gems entangled in her hair.
I guess, 'twas frightful there to see
A lady so richly clad as she—
Beautiful exceedingly!

Mary mother, save me now!
(Said Christabel,) And who art thou? 70

The lady strange made answer meet,
And her voice was faint and sweet:—
Have pity on my sore distress,
I scarce can speak for weariness:
Stretch forth thy hand, and have no fear!
Said Christabel, How camest thou here?
And the lady, whose voice was faint and sweet,
Did thus pursue her answer meet:—

My sire is of a noble line,
And my name is Geraldine: 80
Five warriors seized me yestermorn,
Me, even me, a maid forlorn:
They choked my cries with force and fright,
And tied me on a palfrey white.
The palfrey was as fleet as wind,
And they rode furiously behind.
They spurred amain, their steeds were white:
And once we crossed the shade of night.
As sure as Heaven shall rescue me,
I have no thought what men they be; 90
Nor do I know how long it is
(For I have lain entranced I wis)
Since one, the tallest of the five,
Took me from the palfrey's back,
A weary woman, scarce alive.
Some muttered words his comrades spoke:
He placed me underneath this oak;
He swore they would return with haste;
Whither they went I cannot tell—
I thought I heard, some minutes past, 100
Sounds as of a castle bell.
Stretch forth thy hand (thus ended she),
And help a wretched maid to flee.

Then Christabel stretched forth her hand,
And comforted fair Geraldine:
O well, bright dame! may you command
The service of Sir Leoline;
And gladly our stout chivalry

Will he send forth and friends withal
To guide and guard you safe and free   110
Home to your noble father's hall.

She rose: and forth with steps they passed
That strove to be, and were not, fast.
Her gracious stars the lady blest,
And thus spake on sweet Christabel:
All our household are at rest,
The hall as silent as the cell;
Sir Leoline is weak in health,
And may not well awakened be,
But we will move as if in stealth,   120
And I beseech your courtesy,
This night, to share your couch with me.

They crossed the moat, and Christabel
Took the key that fitted well;
A little door she opened straight,
All in the middle of the gate;
The gate that was ironed within and without,
Where an army in battle array had marched out.
The lady sank, belike through pain,
And Christabel with might and main   130
Lifted her up, a weary weight,
Over the threshold of the gate:
Then the lady rose again,
And moved, as she were not in pain.

So free from danger, free from fear,
They crossed the court: right glad they were.
And Christabel devoutly cried
To the lady by her side,
Praise we the Virgin all divine
Who hath rescued thee from thy distress!   140
Alas, alas! said Geraldine,
I cannot speak for weariness.
So free from danger, free from fear,
They crossed the court: right glad they were.

Outside her kennel, the mastiff old
Lay fast asleep, in moonshine cold.
The mastiff old did not awake,
Yet she an angry moan did make!
And what can ail the mastiff bitch?
Never till now she uttered yell   150
Beneath the eye of Christabel.
Perhaps it is the owlet's scritch:
For what can ail the mastiff bitch?

CHRISTABEL. **152. scritch:** screech.

They passed the hall, that echoes still,
Pass as lightly as you will!
The brands were flat, the brands were dying,
Amid their own white ashes lying;
But when the lady passed, there came
A tongue of light, a fit of flame;
And Christabel saw the lady's eye,   160
And nothing else saw she thereby,
Save the boss of the shield of Sir Leoline tall,
Which hung in a murky old niche in the wall.
O softly tread, said Christabel,
My father seldom sleepeth well.

Sweet Christabel her feet doth bare,
And jealous of the listening air
They steal their way from stair to stair,
Now in glimmer, and now in gloom,
And now they pass the Baron's room,   170
As still as death, with stifled breath!
And now have reached her chamber door;
And now doth Geraldine press down
The rushes of the chamber floor.

The moon shines dim in the open air,
And not a moonbeam enters here.
But they without its light can see
The chamber carved so curiously,
Carved with figures strange and sweet,
All made out of the carver's brain,   180
For a lady's chamber meet:
The lamp with twofold silver chain
Is fastened to an angel's feet.

The silver lamp burns dead and dim;
But Christabel the lamp will trim.
She trimmed the lamp, and made it bright,
And left it swinging to and fro,
While Geraldine, in wretched plight,
Sank down upon the floor below.

O weary lady, Geraldine,   190
I pray you, drink this cordial wine!
It is a wine of virtuous powers;
My mother made it of wild flowers.

And will your mother pity me,
Who am a maiden most forlorn?
Christabel answered—Woe is me!
She died the hour that I was born.
I have heard the grey-haired friar tell
How on her death-bed she did say,
That she should hear the castle-bell   200

Strike twelve upon my wedding-day.
O mother dear! that thou wert here!
I would, said Geraldine, she were!

But soon with altered voice, said she—
"Off, wandering mother! Peak and pine!
I have power to bid thee flee."
Alas! what ails poor Geraldine?
Why stares she with unsettled eye?
Can she the bodiless dead espy?
And why with hollow voice cries she,   210
"Off, woman, off! this hour is mine—
Though thou her guardian spirit be,
Off, woman, off! 'tis given to me."

Then Christabel knelt by the lady's side,
And raised to heaven her eyes so blue—
Alas! said she, this ghastly ride—
Dear lady! it hath wildered you!
The lady wiped her moist cold brow,
And faintly said, " 'tis over now!"

Again the wild-flower wine she drank: 220
Her fair large eyes 'gan glitter bright,
And from the floor whereon she sank,
The lofty lady stood upright:
She was most beautiful to see,
Like a lady of a far countree.

And thus the lofty lady spake—
"All they who live in the upper sky,
Do love you, holy Christabel!
And you love them, and for their sake
And for the good which me befel,   230
Even I in my degree will try,
Fair maiden, to requite you well.
But now unrobe yourself; for I
Must pray, ere yet in bed I lie."

Quoth Christabel, So let it be!
And as the lady bade, did she.
Her gentle limbs did she undress,
And lay down in her loveliness.

But through her brain of weal and woe
So many thoughts moved to and fro,   240
That vain it were her lids to close;
So half-way from the bed she rose,
And on her elbow did recline
To look at the lady Geraldine.

205. **Peak:** grow thin.

Beneath the lamp the lady bowed,
And slowly rolled her eyes around;
Then drawing in her breath aloud,
Like one that shuddered, she unbound
The cincture from beneath her breast:
Her silken robe, and inner vest,   250
Dropt to her feet, and full in view,
Behold! her bosom and half her side——
A sight to dream of, not to tell!
O shield her! shield sweet Christabel!

Yet Geraldine nor speaks nor stirs;
Ah! what a stricken look was hers!
Deep from within she seems half-way
To lift some weight with sick assay,
And eyes the maid and seeks delay;
Then suddenly, as one defied,   260
Collects herself in scorn and pride,
And lay down by the Maiden's side!—
And in her arms the maid she took,
            Ah wel-a-day!
And with low voice and doleful look
These words did say:
"In the touch of this bosom there worketh a spell,
Which is lord of thy utterance, Christabel!
Thou knowest to-night, and wilt know to-morrow,
This mark of my shame, this seal of my sorrow; 270
            But vainly thou warrest,
                For this is alone in
                Thy power to declare,
                That in the dim forest
                Thou heard'st a low moaning,
And found'st a bright lady, surpassingly fair;
And didst bring her home with thee in love and in
            charity,
To shield her and shelter her from the damp air."

The Conclusion to Part I

It was a lovely sight to see
The lady Christabel, when she   280
Was praying at the old oak tree.
    Amid the jagged shadows
    Of mossy leafless boughs,
    Kneeling in the moonlight,
    To make her gentle vows;
Her slender palms together prest,
Heaving sometimes on her breast;
Her face resigned to bliss or bale—
Her face, oh call it fair not pale,

And both blue eyes more bright than clear,   290
Each about to have a tear.

With open eyes (ah woe is me!)
Asleep, and dreaming fearfully,
Fearfully dreaming, yet, I wis,
Dreaming that alone, which is—
O sorrow and shame! Can this be she,
The lady, who knelt at the old oak tree?
And lo! the worker of these harms,
That holds the maiden in her arms,
Seems to slumber still and mild,   300
As a mother with her child.

A star hath set, a star hath risen,
O Geraldine! since arms of thine
Have been the lovely lady's prison.
O Geraldine! one hour was thine—
Thou'st had thy will! By tairn and rill,
The night-birds all that hour were still.
But now they are jubilant anew,
From cliff and tower, tu—whoo! tu—whoo!
Tu—whoo! tu—whoo! from wood and fell!   310

And see! the lady Christabel
Gathers herself from out of her trance;
Her limbs relax, her countenance
Grows sad and soft; the smooth thin lids
Close o'er her eyes; and tears she sheds—
Large tears that leave the lashes bright!
And oft the while she seems to smile
As infants at a sudden light!

Yea, she doth smile, and she doth weep,
Like a youthful hermitess,   320
Beauteous in a wilderness,
Who, praying always, prays in sleep.
And, if she move unquietly,
Perchance, 'tis but the blood so free
Comes back and tingles in her feet.
No doubt, she hath a vision sweet.
What if her guardian spirit 'twere,
What if she knew her mother near?
But this she knows, in joys and woes,
That saints will aid if men will call: 330
For the blue sky bends over all!

1797

Part II

Each matin bell, the Baron saith,
Knells us back to a world of death.
These words Sir Leoline first said,
When he rose and found his lady dead:
These words Sir Leoline will say
Many a morn to his dying day!

And hence the custom and law began
That still at dawn the sacristan,
Who duly pulls the heavy bell,   340
Five and forty beads must tell
Between each stroke—a warning knell,
Which not a soul can choose but hear
From Bratha Head to Wyndermere.

Saith Bracy the bard, So let it knell!
And let the drowsy sacristan
Still count as slowly as he can!
There is no lack of such, I ween,
As well fill up the space between.
In Langdale Pike and Witch's Lair,   350
And Dungeon-ghyll so foully rent,
With ropes of rock and bells of air
Three sinful sextons' ghosts are pent,
Who all give back, one after t'other,
The death-note to their living brother;
And oft too, by the knell offended,
Just as their one! two! three! is ended,
The devil mocks the doleful tale
With a merry peal from Borodale.

The air is still! through mist and cloud   360
That merry peal comes ringing loud;
And Geraldine shakes off her dread,
And rises lightly from the bed;
Puts on her silken vestments white,
And tricks her hair in lovely plight,
And nothing doubting of her spell
Awakens the lady Christabel.
"Sleep you, sweet lady Christabel?
I trust that you have rested well."

And Christabel awoke and spied   370
The same who lay down by her side—
O rather say, the same whom she
Raised up beneath the old oak tree!

344. Bratha Head to Wyndermere: These and the places
named in lines 350–59 are in the Lake District of England.
350. Pike: peak. **351. ghyll:** ravine forming the bed of
a stream. **365. plight:** fold or plait.

306. tairn: mountain pool. **310. fell:** moor or barren
upland.

Nay, fairer yet! and yet more fair!
For she belike hath drunken deep
Of all the blessedness of sleep!
And while she spake, her looks, her air
Such gentle thankfulness declare,
That (so it seemed) her girded vests
Grew tight beneath her heaving breasts.   380
"Sure I have sinn'd!" said Christabel,
"Now heaven be praised if all be well!"
And in low faltering tones, yet sweet,
Did she the lofty lady greet
With such perplexity of mind
As dreams too lively leave behind.

So quickly she rose, and quickly arrayed
Her maiden limbs, and having prayed
That He, who on the cross did groan,
Might wash away her sins unknown,   390
She forwith led fair Geraldine
To meet her sire, Sir Leoline.

The lovely maid and the lady tall
Are pacing both into the hall,
And pacing on through page and groom,
Enter the Baron's presence-room.

The Baron rose, and while he prest
His gentle daughter to his breast,
With cheerful wonder in his eyes
The lady Geraldine espies,   400
And gave such welcome to the same,
As might beseem so bright a dame!

But when he heard the lady's tale,
And when she told her father's name,
Why waxed Sir Leoline so pale,
Murmuring o'er the name again,
Lord Roland de Vaux of Tryermaine?

Alas! they had been friends in youth;
But whispering tongues can poison truth;
And constancy lives in realms above;   410
And life is thorny; and youth is vain;
And to be wroth with one we love
Doth work like madness in the brain.
And thus it chanced, as I divine,
With Roland and Sir Leoline.
Each spake words of high disdain
And insult to his heart's best brother:
They parted—ne'er to meet again!
But never either found another
To free the hollow heart from paining—   420

They stood aloof, the scars remaining,
Like cliffs which had been rent asunder;
A dreary sea now flows between;—
But neither heat, nor frost, nor thunder,
Shall wholly do away, I ween,
The marks of that which once hath been.

Sir Leoline, a moment's space,
Stood gazing on the damsel's face:
And the youthful Lord of Tryermaine
Came back upon his heart again.   430

O then the Baron forgot his age,
His noble heart swelled high with rage;
He swore by the wounds in Jesu's side
He would proclaim it far and wide,
With trump and solemn heraldry,
That they, who thus had wronged the dame,
Were base as spotted infamy!
"And if they dare deny the same,
My herald shall appoint a week,
And let the recreant traitors seek   440
My tourney court—that there and then
I may dislodge their reptile souls
From the bodies and forms of men!"
He spake: his eye in lightning rolls!
For the lady was ruthlessly seized; and he kenned
In the beautiful lady the child of his friend!

And now the tears were on his face,
And fondly in his arms he took
Fair Geraldine, who met the embrace,
Prolonging it with joyous look.   450
Which when she viewed, a vision fell
Upon the soul of Christabel,
The vision of fear, the touch and pain!
She shrunk and shuddered, and saw again—
(Ah, woe is me! Was it for thee,
Thou gentle maid! such sights to see?)

Again she saw that bosom old,
Again she felt that bosom cold,
And drew in her breath with a hissing sound:
Whereat the Knight turned wildly round,   460
And nothing saw, but his own sweet maid
With eyes upraised, as one that prayed.

The touch, the sight, had passed away,
And in its stead that vision blest,
Which comforted her after-rest
While in the lady's arms she lay,

445. kenned: knew, recognized.

Had put a rapture in her breast,
And on her lips and o'er her eyes
Spread smiles like light!
                With new surprise,
"What ails then my beloved child?"   470
The Baron said—His daughter mild
Made answer, "All will yet be well!"
I ween, she had no power to tell
Aught else: so mighty was the spell.

Yet he, who saw this Geraldine,
Had deemed her sure a thing divine:
Such sorrow with such grace she blended,
As if she feared she had offended
Sweet Christabel, that gentle maid!
And with such lowly tones she prayed   480
She might be sent without delay
Home to her father's mansion.
                  "Nay!
Nay, by my soul!" said Leoline.
"Ho! Bracy the bard, the charge be thine!
Go thou, with music sweet and loud,
And take two steeds with trappings proud,
And take the youth whom thou lov'st best
To bear thy harp, and learn thy song,
And clothe you both in solemn vest,
And over the mountains haste along,   490
Lest wandering folk, that are abroad,
Detain you on the valley road.

"And when he has crossed the Irthing flood,
My merry bard! he hastes, he hastes
Up Knorren Moor, through Halegarth Wood,
And reaches soon that castle good
Which stands and threatens Scotland's wastes.

"Bard Bracy? bard Bracy! your horses are fleet,
Ye must ride up the hall, your music so sweet,
More loud than your horses' echoing feet!   500
And loud and loud to Lord Roland call,
Thy daughter is safe in Langdale hall!
Thy beautiful daughter is safe and free—
Sir Leoline greets thee thus through me!
He bids thee come without delay
With all thy numerous array
And take thy lovely daughter home:
And he will meet thee on the way
With all his numerous array
White with their panting palfreys' foam: 510
And, by mine honour! I will say,

473. ween: suppose, imagine.

That I repent me of the day
When I spake words of fierce disdain
To Roland de Vaux of Tryermaine!—
—For since that evil hour hath flown,
Many a summer's sun hath shone;
Yet ne'er found I a friend again
Like Roland de Vaux of Tryermaine."

The lady fell, and clasped his knees,
Her face upraised, her eyes o'erflowing;   520
And Bracy replied, with faltering voice,
His gracious Hail on all bestowing!—
"Thy words, thou sire of Christabel,
Are sweeter than my harp can tell;
Yet might I gain a boon of thee,
This day my journey should not be,
So strange a dream hath come to me,
That I had vowed with music loud
To clear yon wood from thing unblest,
Warned by a vision in my rest!   530
For in my sleep I saw that dove,
That gentle bird, whom thou dost love,
And call'st by thy own daughter's name—
Sir Leoline! I saw the same
Fluttering, and uttering fearful moan,
Among the green herbs in the forest alone.
Which when I saw and when I heard,
I wonder'd what might ail the bird;
For nothing near it could I see,
Save the grass and green herbs underneath the old tree.

"And in my dream methought I went   541
To search out what might there be found;
And what the sweet bird's trouble meant,
That thus lay fluttering on the ground.
I went and peered, and could descry
No cause for her distressful cry;
But yet for her dear lady's sake
I stooped, methought, the dove to take,
When lo! I saw a bright green snake
Coiled around its wings and neck.   550
Green as the herbs on which it couched,
Close by the dove's its head it crouched;
And with the dove it heaves and stirs,
Swelling its neck as she swelled hers!
I woke; it was the midnight hour,
The clock was echoing in the tower;
But though my slumber was gone by,
This dream it would not pass away—
It seems to live upon my eye!
And thence I vowed this self-same day   560

With music strong and saintly song
To wander through the forest bare,
Lest aught unholy loiter there."

Thus Bracy said: the Baron, the while,
Half-listening heard him with a smile;
Then turned to Lady Geraldine,
His eyes made up of wonder and love;
And said in courtly accents fine,
"Sweet maid, Lord Roland's beauteous dove,
With arms more strong than harp or song,   570
Thy sire and I will crush the snake!"
He kissed her forehead as he spake,
And Geraldine in maiden wise
Casting down her large bright eyes,
With blusing cheek and courtesy fine
She turned her from Sir Leoline;
Softly gathering up her train,
That o'er her right arm fell again;
And folded her arms across her chest,
And couched her head upon her breast,   580
And looked askance at Christabel——
Jesu, Maria, shield her well!

A snake's small eye blinks dull and shy;
And the lady's eyes they shrunk in her head,
Each shrunk up to a serpent's eye,
And with somewhat of malice, and more of dread,
At Christabel she looked askance!—
One moment—and the sight was fled!
But Christabel in dizzy trance
Stumbling on the unsteady ground   590
Shuddered aloud, with a hissing sound;
And Geraldine again turned round,
And like a thing, that sought relief,
Full of wonder and full of grief,
She rolled her large bright eyes divine
Wildly on Sir Leoline.

The maid, alas! her thoughts are gone,
She nothing sees—no sight but one!
The maid, devoid of guile and sin,
I know not how, in fearful wise,   600
So deeply had she drunken in
That look, those shrunken serpent eyes,
That all her features were resigned
To this sole image in her mind:
And passively did imitate
That look of dull and treacherous hate!
And thus she stood, in dizzy trance,
Still picturing that look askance

With forced unconscious sympathy
Full before her father's view——   610
As far as such a look could be
In eyes so innocent and blue!

And when the trance was o'er, the maid
Paused awhile, and inly prayed:
Then falling at the Baron's feet,
"By my mother's soul do I entreat
That thou this woman send away!"
She said: and more she could not say:
For what she knew she could not tell,
O'er-mastered by the mighty spell.   620

Why is thy cheek so wan and wild,
Sir Leoline? Thy only child
Lies at thy feet, thy joy, thy pride,
So fair, so innocent, so mild;
The same, for whom thy lady died!
O by the pangs of her dear mother
Think thou no evil of thy child!
For her, and thee, and for no other,
She prayed the moment ere she died:
Prayed that the babe for whom she died,   630
Might prove her dear lord's joy and pride!
    That prayer her deadly pangs beguiled,
        Sir Leoline!
    And wouldst thou wrong thy only child,
        Her child and thine?

Within the Baron's heart and brain
If thoughts, like these, had any share,
They only swelled his rage and pain,
And did but work confusion there.
His heart was cleft with pain and rage,   640
His cheeks they quivered, his eyes were wild,
Dishonoured thus in his old age;
Dishonoured by his only child,
And all his hospitality
To the wronged daughter of his friend
By more than woman's jealousy
Brought thus to a disgraceful end—
He rolled his eye with stern regard
Upon the gentle minstrel bard,
And said in tones abrupt, austere—   650
"Why, Bracy! dost thou loiter here?
I bade thee hence!" The bard obeyed;
And turning from his own sweet maid,
The aged knight, Sir Leoline,
Led forth the lady Geraldine!

1800

The Conclusion to Part II

A little child, a limber elf,
Singing, dancing to itself,
A fairy thing with red round cheeks,
That always finds, and never seeks,
Makes such a vision to the sight   660
As fills a father's eyes with light;
And pleasures flow in so thick and fast
Upon his heart, that he at last
Must needs express his love's excess
With words of unmeant bitterness.
Perhaps 'tis pretty to force together
Thoughts so all unlike each other;
To mutter and mock a broken charm,
To dally with wrong that does no harm.
Perhaps 'tis tender too and pretty   670
At each wild word to feel within
A sweet recoil of love and pity.
And what, if in a world of sin
(O sorrow and shame should this be true!)
Such giddiness of heart and brain
Comes seldom save from rage and pain,
So talks as it's most used to do.

                                        1797–1801 (1816)

FROST AT MIDNIGHT

The Frost performs its secret ministry,
Unhelped by any wind. The owlet's cry
Came loud—and hark, again! loud as before.
The inmates of my cottage, all at rest,
Have left me to that solitude, which suits
Abstruser musings: save that at my side
My cradled infant slumbers peacefully.
'Tis calm indeed! so calm, that it disturbs
And vexes meditation with its strange
And extreme silentness. Sea, hill, and wood,   10
This populous village! Sea, and hill, and wood,
With all the numberless goings-on of life,
Inaudible as dreams! the thin blue flame
Lies on my low-burnt fire, and quivers not;
Only that film, which fluttered on the grate,

Still flutters there, the sole unquiet thing.
Methinks, its motion in this hush of nature
Gives it dim sympathies with me who live,
Making it a companionable form,
Whose puny flaps and freaks the idling Spirit   20
By its own moods interprets, every where
Echo or mirror seeking of itself,
And makes a toy of Thought.

                              But O! how oft,
How oft, at school, with most believing mind,
Presageful, have I gazed upon the bars,
To watch that fluttering *stranger*! and as oft
With unclosed lids, already had I dreamt
Of my sweet birth-place, and the old church-tower,
Whose bells, the poor man's only music, rang
From morn to evening, all the hot Fair-day,   30
So sweetly, that they stirred and haunted me
With a wild pleasure, falling on mine ear
Most like articulate sounds of things to come!
So gazed I, till the soothing things, I dreamt,
Lulled me to sleep, and sleep prolonged my dreams!
And so I brooded all the following morn,
Awed by the stern preceptor's face, mine eye
Fixed with mock study on my swimming book:
Save if the door half opened, and I snatched
A hasty glance, and still my heart leaped up,   40
For still I hoped to see the *stranger's* face,
Townsman, or aunt, or sister more beloved,
My play-mate when we both were clothed alike!

Dear Babe, that sleepest cradled by my side,
Whose gentle breathings, heard in this deep calm,
Fill up the interspersed vacancies
And momentary pauses of the thought!
My babe so beautiful! it thrills my heart
With tender gladness, thus to look at thee,
And think that thou shalt learn far other lore,   50
And in far other scenes! For I was reared
In the great city, pent 'mid cloisters dim,
And saw nought lovely but the sky and stars.
But *thou*, my babe! shalt wander like a breeze
By lakes and sandy shores, beneath the crags
Of ancient mountain, and beneath the clouds,
Which image in their bulk both lakes and shores
And mountain crags: so shalt thou see and hear
The lovely shapes and sounds intelligible

FROST AT MIDNIGHT. **15. film:** bit of soot fluttering on the bar of the grate. "In all parts of the kingdom these films are called *strangers* and supposed to portend the arrival of some absent friend" (C.).

24. school: Christ's Hospital, London, which Coleridge entered at the age of nine. **28. birth-place:** the country town of Ottery St. Mary, Devonshire. **52. cloisters:** of his school.

Of that eternal language, which thy God   60
Utters, who from eternity doth teach
Himself in all, and all things in himself.
Great universal Teacher! he shall mould
Thy spirit, and by giving make it ask.

Therefore all seasons shall be sweet to thee,
Whether the summer clothe the general earth
With greenness, or the redbreast sit and sing
Betwixt the tufts of snow on the bare branch
Of mossy apple-tree, while the nigh thatch
Smokes in the sun-thaw; whether the eave-drops fall
Heard only in the trances of the blast,   71
Or if the secret ministry of frost
Shall hang them up in silent icicles,
Quietly shining to the quiet Moon.

1798 (1798)

FRANCE: AN ODE

The poem was prompted by the French invasion of Switzerland in the spring of 1798. This attack on a country renowned for its republican institutions and for the sturdy independence of its people shocked many Englishmen who had previously sympathized with the Revolution.

In 1802 the poem was reprinted in *The Morning Post* with an Argument prefixed:

First Stanza. An invocation to those objects in Nature the contemplation of which had inspired the Poet with a devotional love of Liberty. *Second Stanza.* The exultation of the Poet at the commencement of the French Revolution, and his unqualified abhorrence of the Alliance against the Republic. *Third Stanza.* The blasphemies and horrors during the domination of the Terrorists regarded by the Poet as a transient storm, and as the natural consequence of the former despotism and of the foul superstition of Popery. Reason, indeed, began to suggest many apprehensions; yet still the Poet struggled to retain the hope that France would make conquests by no other means than by presenting to the observation of Europe a people more happy and better instructed than under other forms of Government. *Fourth Stanza.* Switzerland, and the Poet's recantation. *Fifth Stanza.* An address to Liberty, in which the poet expresses his conviction that those feelings and that grand *ideal* of Freedom which the mind attains by its contemplation of its individual nature, and of the sublime surrounding objects (see Stanza the First) do not belong to men, as a society, nor can possibly be either gratified or realised, under any form of human

government; but belong to the individual man, so far as he is pure, and inflamed with the love and adoration of God in Nature.

The poem may be read in connection with Wordsworth's account in *The Prelude* of his early partisanship for the Revolution and gradual change of mind. The thought underlying the first stanza, that natural objects ceaselessly inculcate the idea of liberty, recurs in Wordsworth's sonnet "To Toussaint L'Ouverture."

I

Ye Clouds! that far above me float and pause,
        Whose pathless march no mortal may controul!
    Ye Ocean-Waves! that, wheresoe'er ye roll,
Yield homage only to eternal laws!
Ye Woods! that listen to the night-birds singing,
        Midway the smooth and perilous slope reclined,
Save when your own imperious branches swinging,
        Have made a solemn music of the wind!
Where, like a man beloved of God,
Through glooms, which never woodman trod,   10
        How oft, pursuing fancies holy,
My moonlight way o'er flowering weeds I wound,
        Inspired, beyond the guess of folly,
By each rude shape and wild unconquerable sound!
O ye loud Waves! and O ye Forests high!
    And O ye Clouds that far above me soared!
Thou rising Sun! thou blue rejoicing Sky!
    Yea, every thing that is and will be free!
    Bear witness for me, wheresoe'er ye be,
    With what deep worship I have still adored   20
        The spirit of divinest Liberty.

II

When France in wrath her giant-limbs upreared,
    And with that oath, which smote air, earth, and sea,
    Stamped her strong foot and said she would be free,
Bear witness for me, how I hoped and feared!
With what a joy my lofty gratulation
    Unawed I sang, amid a slavish band:
And when to whelm the disenchanted nation,
    Like fiends embattled by a wizard's wand,
        The Monarchs marched in evil day,   30
        And Britain joined the dire array;

60. **eternal language:** the forms of nature.

FRANCE: AN ODE. **28. disenchanted nation:** the French people released at the Revolution from the evil spell that had held them in chains. **30-31. Monarchs . . . array:** Allied in the war against France were Prussia and Austria (from 1792) and England and Holland (from 1793).

Though dear her shores and circling ocean,
Though many friendships, many youthful loves
    Had swoln the patriot emotion
And flung a magic light o'er all her hills and groves;
Yet still my voice, unaltered, sang defeat
    To all that braved the tyrant-quelling lance,
And shame too long delayed and vain retreat!
For ne'er, O Liberty! with partial aim
I dimmed thy light or damped thy holy flame;   40
    But blessed the paeans of delivered France,
And hung my head and wept at Britain's name.

III

"And what," I said, "though Blasphemy's loud scream
    With that sweet music of deliverance strove!
    Though all the fierce and drunken passions wove
A dance more wild than e'er was maniac's dream!
    Ye storms, that round the dawning East assembled,
The Sun was rising, though ye hid his light!"
    And when, to soothe my soul, that hoped and
        trembled,
The dissonance ceased, and all seemed calm and bright;
    When France her front deep-scarr'd and gory   51
    Concealed with clustering wreaths of glory;
        When, insupportably advancing,
    Her arm made mockery of the warrior's ramp;
        While timid looks of fury glancing,
    Domestic treason, crushed beneath her fatal stamp,
Writhed like a wounded dragon in his gore;
    Then I reproached my fears that would not flee;
"And soon," I said, "shall Wisdom teach her lore
In the low huts of them that toil and groan!   60
And, conquering by her happiness alone,
    Shall France compel the nations to be free,
Till Love and Joy look round, and call the Earth their
    own."

IV

Forgive me, Freedom! O forgive those dreams!
    I hear thy voice, I hear thy loud lament,
    From bleak Helvetia's icy caverns sent—
I hear thy groans upon her blood-stained streams!
    Heroes, that for your peaceful country perished,
And ye that, fleeing, spot your mountain-snows
    With bleeding wounds; forgive me, that I cherished
One thought that ever blessed your cruel foes!   71

66. Helvetia: Switzerland.

To scatter rage, and traitorous guilt,
    Where Peace her jealous home had built;
        A patriot-race to disinherit
Of all that made their stormy wilds so dear;
        And with inexpiable spirit
To taint the bloodless freedom of the mountaineer—
O France, that mockest Heaven, adulterous, blind,
    And patriot only in pernicious toils!
Are these thy boasts, Champion of human kind?   80
    To mix with Kings in the low lust of sway,
Yell in the hunt, and share the murderous prey;
To insult the shrine of Liberty with spoils
    From freemen torn; to tempt and to betray?

V

    The Sensual and the Dark rebel in vain,
    Slaves by their own compulsion! In mad game
    They burst their manacles and wear the name
        Of Freedom, graven on a heavier chain!
    O Liberty! with profitless endeavour
Have I pursued thee, many a weary hour;   90
    But thou nor swell'st the victor's strain, nor ever
Didst breathe thy soul in forms of human power.
    Alike from all, howe'er they praise thee,
    (Nor prayer, nor boastful name delays thee)
        Alike from Priestcraft's harpy minions,
    And factious Blasphemy's obscener slaves,
        Thou speedest on thy subtle pinions,
The guide of homeless winds, and playmate of the
        waves!
And there I felt thee!—on that sea-cliff's verge,
    Whose pines, scarce travelled by the breeze above,
Had made one murmur with the distant surge!   101
Yes, while I stood and gazed, my temples bare,
And shot my being through earth, sea, and air,
    Possessing all things with intensest love,
        O Liberty! my spirit felt thee there.

1798 (1798)

LEWTI

OR THE CIRCASSIAN LOVE-CHAUNT

❧

A revision and enlargement of Wordsworth's youthful
poem "Beauty and Moonlight."

❧

At midnight by the stream I roved,
To forget the form I loved.
Image of Lewti! from my mind
Depart; for Lewti is not kind.
The Moon was high, the moonlight gleam
   And the shadow of a star
Heaved upon Tamaha's stream;
   But the rock shone brighter far,
The rock half sheltered from my view
By pendent boughs of tressy yew.—   10
So shines my Lewti's forehead fair,
Gleaming through her sable hair.
Image of Lewti! from my mind
Depart; for Lewti is not kind.

   I saw a cloud of palest hue,
      Onward to the moon it passed;
   Still brighter and more bright it grew,
   With floating colours not a few,
      Till it reached the moon at last:
Then the cloud was wholly bright,   20
With a rich and amber light!
And so with many a hope I seek,
     And with such joy I find my Lewti;
And even so my pale wan cheek
     Drinks in as deep a flush of beauty!
Nay, treacherous image! leave my mind,
If Lewti never will be kind.

   The little cloud—it floats away,
     Away it goes; away so soon!
Alas! it has no power to stay: 30
Its hues are dim, its hues are grey—
     Away it passes from the moon!
How mournfully it seems to fly,
     Ever fading more and more,
To joyless regions of the sky—
     And now 'tis whiter than before!
As white as my poor cheek will be,
     When, Lewti! on my couch I lie,
A dying man for love of thee.
Nay, treacherous image! leave my mind—   40
And yet, thou didst not look unkind.

   I saw a vapour in the sky,
Thin, and white, and very high;
I ne'er beheld so thin a cloud:
     Perhaps the breezes that can fly
     Now below and now above,
Have snatched aloft the lawny shroud
     Of Lady fair—that died for love.

For maids, as well as youths, have perished
From fruitless love too fondly cherished.   50
Nay, treacherous image! leave my mind—
For Lewti never will be kind.

Hush! my heedless feet from under
     Slip the crumbling banks for ever:
Like echoes to a distant thunder,
     They plunge into the gentle river.
The river-swans have heard my tread,
And startle from their reedy bed.
O beauteous birds! methinks ye measure
     Your movements to some heavenly tune!   60
O beauteous birds! 'tis such a pleasure
     To see you move beneath the moon,
I would it were your true delight
To sleep by day and wake all night.

I know the place where Lewti lies,
When silent night has closed her eyes:
     It is a breezy jasmine-bower,
The nightingale sings o'er her head:
     Voice of the Night! had I the power
That leafy labyrinth to thread,   70
And creep, like thee, with soundless tread,
I then might view her bosom white
Heaving lovely to my sight,
As these two swans together heave
On the gently-swelling wave.

Oh! that she saw me in a dream,
     And dreamt that I had died for care;
All pale and wasted I would seem,
     Yet fair withal, as spirits are!
I'd die indeed, if I might see   80
Her bosom heave, and heave for me!
Soothe, gentle image! soothe my mind!
To-morrow Lewti may be kind.

                       1798 (1798)

FEARS IN SOLITUDE

WRITTEN IN APRIL 1798,
DURING THE ALARM OF AN INVASION[1]

A green and silent spot, amid the hills,
A small and silent dell! O'er stiller place
No singing sky-lark ever poised himself.

FEARS IN SOLITUDE. **1. invasion:** the French project of invading England.

The hills are heathy, save that swelling slope,
Which hath a gay and gorgeous covering on,
All golden with the never-bloomless furze,
Which now blooms most profusely: but the dell,
Bathed by the mist, is fresh and delicate
As vernal corn-field, or the unripe flax,
When, through its half-transparent stalks, at eve,   10
The level sunshine glimmers with green light.
Oh! 'tis a quiet spirit-healing nook!
Which all, methinks, would love; but chiefly he,
The humble man, who, in his youthful years,
Knew just so much of folly, as had made
His early manhood more securely wise!
Here he might lie on fern or withered heath,
While from the singing lark (that sings unseen
The minstrelsy that solitude loves best),
And from the sun, and from the breezy air,   20
Sweet influences trembled o'er his frame;
And he, with many feelings, many thoughts,
Made up a meditative joy, and found
Religious meanings in the forms of Nature!
And so, his senses gradually wrapt
In a half sleep, he dreams of better worlds,
And dreaming hears thee still, O singing lark,
That singest like an angel in the clouds!

My God! it is a melancholy thing
For such a man, who would full fain preserve   30
His soul in calmness, yet perforce must feel
For all his human brethren—O my God!
It weighs upon the heart, that he must think
What uproar and what strife may now be stirring
This way or that way o'er these silent hills—
Invasion, and the thunder and the shout,
And all the crash of onset; fear and rage,
And undetermined conflict—even now,
Even now, perchance, and in his native isle:
Carnage and groans beneath this blessed sun!   40
We have offended, Oh! my countrymen!
We have offended very grievously,
And been most tyrannous. From east to west
A groan of accusation pierces Heaven!
The wretched plead against us; multitudes
Countless and vehement, the sons of God,
Our brethren! Like a cloud that travels on,
Steamed up from Cairo's swamps of pestilence,
Even so, my countrymen! have we gone forth
And borne to distant tribes slavery and pangs,   50
And, deadlier far, our vices, whose deep taint

6. **furze:** an evergreen shrub with yellow flowers.

With slow perdition murders the whole man,
His body and his soul! Meanwhile, at home,
All individual dignity and power
Engulfed in Courts, Committees, Institutions,
Associations and Societies,
A vain, speech-mouthing, speech-reporting Guild,
One Benefit-Club for mutual flattery,
We have drunk up, demure as at a grace,
Pollutions from the brimming cup of wealth;   60
Contemptuous of all honourable rule,
Yet bartering freedom and the poor man's life
For gold, as at a market! The sweet words
Of Christian promise, words that even yet
Might stem destruction, were they wisely preached,
Are muttered o'er by men, whose tones proclaim
How flat and wearisome they feel their trade:
Rank scoffers some, but most too indolent
To deem them falsehoods or to know their truth.
Oh! blasphemous! the Book of Life is made   70
A superstitious instrument, on which
We gabble o'er the oaths we mean to break;
For all must swear—all and in every place,
College and wharf, council and justice-court;
All, all must swear, the briber and the bribed,
Merchant and lawyer, senator and priest,
The rich, the poor, the old man and the young;
All, all make up one scheme of perjury,
That faith doth reel; the very name of God
Sounds like a juggler's charm; and, bold with joy,   80
Forth from his dark and lonely hiding-place,
(Portentous sight!) the owlet Atheism,
Sailing on obscene wings athwart the noon,
Drops his blue-fringed lids, and holds them close,
And hooting at the glorious sun in Heaven,
Cries out, "Where is it?"

Thankless too for peace,
(Peace long preserved by fleets and perilous seas)
Secure from actual warfare, we have loved
To swell the war-whoop, passionate for war!
Alas! for ages ignorant of all   90
Its ghastlier workings, (famine or blue plague,
Battle, or siege, or flight through wintry snows,)
We, this whole people, have been clamorous
For war and bloodshed; animating sports,
The which we pay for as a thing to talk of,
Spectators and not combatants! No guess
Anticipative of a wrong unfelt,
No speculation on contingency,
However dim and vague, too vague and dim

To yield a justifying cause; and forth, 100
(Stuffed out with big preamble, holy names,
And adjurations of the God in Heaven,)
We send our mandates for the certain death
Of thousands and ten thousands! Boys and girls,
And women, that would groan to see a child
Pull off an insect's leg, all read of war,
The best amusement for our morning meal!
The poor wretch, who has learnt his only prayers
From curses, who knows scarcely words enough
To ask a blessing from his Heavenly Father, 110
Becomes a fluent phraseman, absolute
And technical in victories and defeats,
And all our dainty terms for fratricide;
Terms which we trundle smoothly o'er our tongues
Like mere abstractions, empty sounds to which
We join no feeling and attach no form!
As if the soldier died without a wound;
As if the fibres of this godlike frame
Were gored without a pang; as if the wretch,
Who fell in battle, doing bloody deeds, 120
Passed off to Heaven, translated and not killed;
As though he had no wife to pine for him,
No God to judge him! Therefore, evil days
Are coming on us, O my countrymen!
And what if all-avenging Providence,
Strong and retributive, should make us know
The meaning of our words, force us to feel
The desolation and the agony
Of our fierce doings?

                Spare us yet awhile,
Father and God! O! spare us yet awhile! 130
Oh! let not English women drag their flight
Fainting beneath the burthen of their babes,
Of the sweet infants, that but yesterday
Laughed at the breast! Sons, brothers, husbands, all
Who ever gazed with fondness on the forms
Which grew up with you round the same fire-side,
And all who ever heard the sabbath-bells
Without the infidel's scorn, make yourselves pure!
Stand forth! be men! repel an impious foe,
Impious and false, a light yet cruel race, 140
Who laugh away all virtue, mingling mirth
With deeds of murder; and still promising
Freedom, themselves too sensual to be free,
Poison life's amities, and cheat the heart
Of faith and quiet hope, and all that soothes,
And all that lifts the spirit! Stand we forth;
Render them back upon the insulted ocean,

And let them toss as idly on its waves
As the vile sea-weed, which some mountain-blast
Swept from our shores! And oh! may we return 150
Not with a drunken triumph, but with fear,
Repenting of the wrongs with which we stung
So fierce a foe to frenzy!

                     I have told,
O Britons! O my brethren! I have told
Most bitter truth, but without bitterness.
Nor deem my zeal or factious or mistimed;
For never can true courage dwell with them,
Who, playing tricks with conscience, dare not look
At their own vices. We have been too long
Dupes of a deep delusion! Some, belike, 160
Groaning with restless enmity, expect
All change from change of constituted power;
As if a Government had been a robe,
On which our vice and wretchedness were tagged
Like fancy-points and fringes, with the robe
Pulled off at pleasure. Fondly these attach
A radical causation to a few
Poor drudges of chastising Providence,
Who borrow all their hues and qualities
From our own folly and rank wickedness, 170
Which gave them birth and nursed them. Others,
    meanwhile,
Dote with a mad idolatry; and all
Who will not fall before their images,
And yield them worship, they are enemies
Even of their country!

                  Such have I been deemed.—
But, O dear Britain! O my Mother Isle!
Needs must thou prove a name most dear and holy
To me, a son, a brother, and a friend,
A husband, and a father! who revere
All bonds of natural love, and find them all 180
Within the limits of thy rocky shores.
O native Britain! O my Mother Isle!
How shouldst thou prove aught else but dear and holy
To me, who from thy lakes and mountain-hills,
Thy clouds, thy quiet dales, thy rocks and seas,
Have drunk in all my intellectual life,
All sweet sensations, all ennobling thoughts,
All adoration of the God in nature,
All lovely and all honourable things,
Whatever makes this mortal spirit feel 190
The joy and greatness of its future being?
There lives nor form nor feeling in my soul
Unborrowed from my country! O divine

And beauteous island! thou hast been my sole
And most magnificent temple, in the which
I walk with awe, and sing my stately songs,
Loving the God that made me!—

                May my fears,
My filial fears, be vain! and may the vaunts
And menace of the vengeful enemy
Pass like the gust, that roared and died away   200
In the distant tree: which heard, and only heard
In this low dell, bowed not the delicate grass.

But now the gentle dew-fall sends abroad
The fruit-like perfume of the golden furze:
The light has left the summit of the hill,
Though still a sunny gleam lies beautiful,
Aslant the ivied beacon. Now farewell,
Farewell, awhile, O soft and silent spot!
On the green sheep-track, up the heathy hill,
Homeward I wind my way; and lo! recalled   210
From bodings that have well-nigh wearied me,
I find myself upon the brow, and pause
Startled! And after lonely sojourning
In such a quiet and surrounded nook,
This burst of prospect, here the shadowy main,
Dim-tinted, there the mighty majesty
Of that huge amphitheatre of rich
And elmy fields, seems like society—
Conversing with the mind, and giving it
A livelier impulse and a dance of thought!   220
And now, beloved Stowey! I behold
Thy church-tower, and, methinks, the four huge elms
Clustering, which mark the mansion of my friend;
And close behind them, hidden from my view,
Is my own lowly cottage, where my babe
And my babe's mother dwell in peace! With light
And quickened footsteps thitherward I tend,
Remembering thee, O green and silent dell!
And grateful, that by nature's quietness
And solitary musings, all my heart   230
Is softened, and made worthy to indulge
Love, and the thoughts that yearn for human kind.

                          1798 (1798)

THE NIGHTINGALE

A CONVERSATION POEM, APRIL, 1798

No cloud, no relique of the sunken day
Distinguishes the West, no long thin slip

Of sullen light, no obscure trembling hues.
Come, we will rest on this old mossy bridge!
You see the glimmer of the stream beneath,
But hear no murmuring: it flows silently,
O'er its soft bed of verdure. All is still,
A balmy night! and though the stars be dim,
Yet let us think upon the vernal showers
That gladden the green earth, and we shall find   10
A pleasure in the dimness of the stars.
And hark! the Nightingale begins its song,
"Most musical, most melancholy" bird!
A melancholy bird? Oh! idle thought!
In Nature there is nothing melancholy.
But some night-wandering man whose heart was
    pierced
With the remembrance of a grievous wrong,
Or slow distemper, or neglected love,
(And so, poor wretch! filled all things with himself,
And made all gentle sounds tell back the tale   20
Of his own sorrow) he, and such as he,
First named these notes a melancholy strain.
And many a poet echoes the conceit;
Poet who hath been building up the rhyme
When he had better far have stretched his limbs
Beside a brook in mossy forest-dell,
By sun or moon-light, to the influxes
Of shapes and sounds and shifting elements
Surrendering his whole spirit, of his song
And of his fame forgetful! so his fame   30
Should share in Nature's immortality,
A venerable thing! and so his song
Should make all Nature lovelier, and itself
Be loved like Nature! But 'twill not be so;
And youths and maidens most poetical,
Who lose the deepening twilights of the spring
In ball-rooms and hot theatres, they still
Full of meek sympathy must heave their sighs
O'er Philomela's pity-pleading strains.

My Friend, and thou, our Sister! we have learnt   40
A different lore: we may not thus profane
Nature's sweet voices, always full of love
And joyance! 'Tis the merry Nightingale

THE NIGHTINGALE. **13. "Most . . . melancholy"**: Milton's
"Il Penseroso," l. 62. **39. Philomela**: The daughter of a
king of Athens, Philomela was violated by Tereus, who cut
out her tongue. After taking a fearful revenge she fled, with
Tereus pursuing, and was changed, according to Ovid, into
a nightingale. **40. Friend . . . Sister**: William and
Dorothy Wordsworth.

That crowds, and hurries, and precipitates
With fast thick warble his delicious notes,
As he were fearful that an April night
Would be too short for him to utter forth
His love-chant, and disburthen his full soul
Of all its music!

          And I know a grove
Of large extent, hard by a castle huge,   50
Which the great lord inhabits not; and so
This grove is wild with tangling underwood,
And the trim walks are broken up, and grass,
Thin grass and king-cups grow within the paths.
But never elsewhere in one place I knew
So many nightingales; and far and near,
In wood and thicket, over the wide grove,
They answer and provoke each other's song,
With skirmish and capricious passagings,
And murmurs musical and swift jug jug,   60
And one low piping sound more sweet than all—
Stirring the air with such a harmony,
That should you close your eyes, you might almost
Forget it was not day! On moonlight bushes,
Whose dewy leaflets are but half-disclosed,
You may perchance behold them on the twigs,
Their bright, bright eyes, their eyes both bright and
     full,
Glistening, while many a glow-worm in the shade
Lights up her love-torch.

          A most gentle Maid,
Who dwelleth in her hospitable home   70
Hard by the castle, and at latest eve
(Even like a Lady vowed and dedicate
To something more than Nature in the grove)
Glides through the pathways; she knows all their notes,
That gentle Maid! and oft, a moment's space,
What time the moon was lost behind a cloud,
Hath heard a pause of silence; till the moon
Emerging, hath awakened earth and sky
With one sensation, and those wakeful birds
Have all burst forth in choral minstrelsy,   80
As if some sudden gale had swept at once
A hundred airy harps! And she hath watched
Many a nightingale perch giddily
On blossomy twig still swinging from the breeze,
And to that motion tune his wanton song
Like tipsy Joy that reels with tossing head.

59. passagings: fencing matches and also runs or flourishes
of music.

Farewell, O Warbler! till to-morrow eve,
And you, my friends! farewell, a short farewell!
We have been loitering long and pleasantly,
And now for our dear homes.—That strain again! 90
Full fain it would delay me! My dear babe,
Who, capable of no articulate sound,
Mars all things with his imitative lisp,
How he would place his hand beside his ear,
His little hand, the small forefinger up,
And bid us listen! And I deem it wise
To make him Nature's play-mate. He knows well
The evening-star; and once, when he awoke
In most distressful mood (some inward pain
Had made up that strange thing, an infant's dream—)
I hurried with him to our orchard-plot,   101
And he beheld the moon, and, hushed at once,
Suspends his sobs, and laughs most silently,
While his fair eyes, that swam with undropped tears,
Did glitter in the yellow moon-beam! Well!—
It is a father's tale: But if that Heaven
Should give me life, his childhood shall grow up
Familiar with these songs, that with the night
He may associate joy.—Once more, farewell, 109
Sweet Nightingale! once more, my friends! farewell.

               1798 (1798)

THE BALLAD OF THE DARK LADIÉ

In a manuscript list of poems Coleridge indicated that
there were (or were intended to be?) 190 lines in this
ballad. But the ballad was either never finished or never
published in full, and we have only the fragments given
below.

A FRAGMENT

    Beneath yon birch with silver bark,
    And boughs so pendulous and fair,
    The brook falls scatter'd down the rock:
        And all is mossy there!

    And there upon the moss she sits,
    The Dark Ladié in silent pain;
    The heavy tear is in her eye,
        And drops and swells again.

90. That . . . again: Cf. *Twelfth Night*, I. i. 4.

Three times she sends her little page
Up the castled mountain's breast,   10
If he might find the Knight that wears
    The Griffin for his crest.

The sun was sloping down the sky,
And she had linger'd there all day,
Counting moments, dreaming fears—
    Oh wherefore can he stay?

She hears a rustling o'er the brook,
She sees far off a swinging bough!
"'Tis He! 'Tis my betrothed Knight!
    Lord Falkland, it is Thou!"   20

She springs, she claps him round the neck,
She sobs a thousand hopes and fears,
Her kisses glowing on his cheeks
    She quenches with her tears.

    .   .   .

"My friends with rude ungentle words
They scoff and bid me fly to thee!
O give me shelter in thy breast!
    O shield and shelter me!

"My Henry, I have given thee much,
I gave what I can ne'er recall,   30
I gave my heart, I gave my peace,
    O Heaven! I gave thee all."

The Knight made answer to the Maid,
While to his heart he held her hand,
"Nine castles hath my noble sire,
    None statelier in the land.

"The fairest one shall be my love's,
The fairest castle of the nine!
Wait only till the stars peep out,
    The fairest shall be thine: 40

"Wait only till the hand of eve
Hath wholly closed yon western bars,
And through the dark we two will steal
    Beneath the twinkling stars!"—

"The dark? the dark? No! not the dark?
The twinkling stars? How, Henry? How?"
O God! 'twas in the eye of noon
    He pledged his sacred vow!

And in the eye of noon my love
Shall lead me from my mother's door,   50
Sweet boys and girls all clothed in white
    Strewing flowers before:

But first the nodding minstrels go
With music meet for lordly bowers,
The children next in snow-white vests,
    Strewing buds and flowers!

And then my love and I shall pace,
My jet black hair in pearly braids,
Between our comely bachelors
    And blushing bridal maids.   60

    .   .   .

1798 (1834)

KUBLA KHAN

OR, A VISION IN A DREAM
A FRAGMENT

This poem is now usually interpreted as a poem about the act of poetic creation in which the major images are symbolic.

The following fragment is here published at the request of a poet of great and deserved celebrity,[1] and, as far as the Author's own opinions are concerned, rather as a psychological curiosity, than on the ground of any supposed *poetic* merits.

In the summer of the year 1797, the Author, then in ill health, had retired to a lonely farm-house between Porlock and Linton, on the Exmoor confines of Somerset and Devonshire. In consequence of a slight indisposition, an anodyne had been prescribed, from the effects of which he fell asleep in his chair at the moment that he was reading the following sentence, or words of the same substance, in "Purchas's Pilgrimage": "Here the Khan Kubla commanded a palace to be built, and a stately garden thereunto. And thus ten miles of fertile ground were inclosed with a wall."[2] The Author continued for about three hours in a

KUBLA KHAN. **1. celebrity:** Lord Byron. **2. "Here . . . wall":** "In Xamdu did Cublai Can build a stately Palace, encompassing sixteene miles of plaine ground with a wall, wherein are fertile Meddowes, pleasant Springs, delightful Streames, and all sorts of beasts of chase and game, and in the middest thereof a sumptuous house of pleasure" (Samuel Purchas, *Purchas His Pilgrimage* [1613]). Kublai Khan, 1216-94, was a grandson of Genghis Khan and emperor of China

profound sleep, at least of the external senses,[3] during which time he has the most vivid confidence, that he could not have composed less than from two to three hundred lines; if that indeed can be called composition in which all the images rose up before him as *things*, with a parallel production of the correspondent expressions, without any sensation or consciousness of effort. On awaking he appeared to himself to have a distinct recollection of the whole, and taking his pen, ink, and paper, instantly and eagerly wrote down the lines that are here preserved. At this moment he was unfortunately called out by a person on business from Porlock, and detained by him above an hour, and on his return to his room, found, to his no small surprise and mortification, that though he still retained some vague and dim recollection of the general purport of the vision, yet, with the exception of some eight or ten scattered lines and images, all the rest had passed away like the images on the surface of a stream into which a stone has been cast, but, alas! without the after restoration of the latter!

> Then all the charm
> Is broken—all that phantom-world so fair
> Vanishes, and a thousand circlets spread,
> And each mis-shape[s] the other. Stay awhile,
> Poor youth! who scarcely dar'st lift up thine eyes—
> The stream will soon renew its smoothness, soon
> The visions will return! And lo, he stays,
> And soon the fragments dim of lovely forms
> Come trembling back, unite, and now once more
> The pool becomes a mirror.[4]

Yet from the still surviving recollections in his mind, the Author has frequently purposed to finish for himself what had been originally, as it were, given to him. Σαμερον αδιον ασω[5]: but the to-morrow is yet to come.

In Xanadu did Kubla Khan
A stately pleasure-dome decree:
Where Alph, the sacred river, ran
Through caverns measureless to man
   Down to a sunless sea.
So twice five miles of fertile ground
With walls and towers were girdled round:

And there were gardens bright with sinuous rills,
Where blossomed many an incense-bearing tree;
And here were forests ancient as the hills,   10
Enfolding sunny spots of greenery.

But oh! that deep romantic chasm which slanted
Down the green hill athwart a cedarn cover!
A savage place! as holy and enchanted
As e'er beneath a waning moon was haunted
By woman wailing for her demon-lover!
And from this chasm, with ceaseless turmoil seething,
As if this earth in fast thick pants were breathing,
A mighty fountain momently was forced:
Amid whose swift half-intermitted burst   20
Huge fragments vaulted like rebounding hail,
Or chaffy grain beneath the thresher's flail:
And 'mid these dancing rocks at once and ever
It flung up momently the sacred river.
Five miles meandering with a mazy motion
Through wood and dale the sacred river ran,
Then reached the caverns measureless to man,
And sank in tumult to a lifeless ocean:
And 'mid this tumult Kubla heard from far
Ancestral voices prophesying war!   30
   The shadow of the dome of pleasure
   Floated midway on the waves;
   Where was heard the mingled measure
   From the fountain and the caves.
It was a miracle of rare device,
A sunny pleasure-dome with caves of ice!

   A damsel with a dulcimer
   In a vision once I saw:
   It was an Abyssinian maid,
   And on her dulcimer she played,   40
   Singing of Mount Abora.
   Could I revive within me
   Her symphony and song,
   To such a deep delight 'twould win me,
That with music loud and long,
I would build that dome in air,
That sunny dome! those caves of ice!
And all who heard should see them there,
And all should cry, Beware! Beware!
His flashing eyes, his floating hair!   50
Weave a circle round him thrice,
And close your eyes with holy dread,
For he on honey-dew hath fed,
And drunk the milk of Paradise.

1797 or 1798 (1816)

---

3. **sleep . . . senses:** In a manuscript note Coleridge explains that the sleep was actually a reverie brought on by opium. 4. **Then . . . mirror:** from Coleridge's "The Picture; or, The Lover's Resolution," ll. 91–100. 5. Σαμερον αδιον ασω: Theocritus, I. 145: "I'll sing you a sweeter song another day."

TO ASRA

Are there two things, of all which men possess,
That are so like each other and so near,
As mutual Love seems like to Happiness?
Dear Asra, woman beyond utterance dear!
This Love, which ever welling at my heart,
Now in its living fount doth heave and fall,
Now overflowing pours thro' every part
Of all my frame, and fills and changes all,
Like vernal waters springing up through snow,
This Love that seeming great beyond the power   10
Of growth, yet seemeth ever more to grow,
Could I transmute the whole to one rich Dower
Of Happy Life, and give it all to Thee,
Thy lot, methinks, were Heaven, thy age, Eternity!

1801 (1893)

DEJECTION: AN ODE

༺᠊ᨆ᠊༻

The poem is a revision and drastic condensation of a verse-letter to Sara Hutchinson written after hearing the opening stanzas of Wordsworth's "Intimations of Immortality." "Dejection: An Ode" evidences Coleridge's uncanny power to grasp and precisely to render psychological states and also his vivid and delicate perception of natural objects. The statement that "we receive but what we give" (l. 47) indicates that his conception of the imagination was not identical with Wordsworth's (cf. "To William Wordsworth," ll. 18–19, below). The "afflictions" (l. 82) to which Coleridge refers in explaining his loss of joy and imaginative power would be chiefly illness, domestic unhappiness, and hopeless love for Sara Hutchinson. He often said that these, and especially illness, drove him to metaphysical studies ("abstruse research," l. 89), which in turn further sapped his poetic gift.

༺᠊ᨆ᠊༻

Late, late yestreen I saw the new Moon,
    With the old Moon in her arms;
And I fear, I fear, my Master dear!
We shall have a deadly storm.

Ballad of Sir Patrick Spence

TO ASRA. **Asra:** Sara Hutchinson.

I

Well! If the Bard was weather-wise, who made
    The grand old ballad of Sir Patrick Spence,
    This night, so tranquil now, will not go hence
Unroused by winds, that ply a busier trade
Than those which mould yon cloud in lazy flakes,
Or the dull sobbing draft, that moans and rakes
Upon the strings of this Æolian lute,
        Which better far were mute.
    For lo! the New-moon winter-bright!
    And overspread with phantom light,   10
    (With swimming phantom light o'erspread
    But rimmed and circled by a silver thread)
I see the old Moon in her lap, foretelling
    The coming-on of rain and squally blast.
And oh! that even now the gust were swelling,
    And the slant night-shower driving loud and fast!
Those sounds which oft have raised me, whilst they awed,
        And sent my soul abroad,
Might now perhaps their wonted impulse give,
Might startle this dull pain, and make it move and live!   20

II

A grief without a pang, void, dark, and drear,
    A stifled, drowsy, unimpassioned grief,
    Which finds no natural outlet, no relief,
        In word, or sigh, or tear—
O Lady! in this wan and heartless mood,
To other thoughts by yonder throstle woo'd,
    All this long eve, so balmy and serene,
Have I been gazing on the western sky,
    And its peculiar tint of yellow green:
And still I gaze—and with how blank an eye!   30
And those thin clouds above, in flakes and bars,
That give away their motion to the stars;
Those stars, that glide behind them or between,
Now sparkling, now bedimmed, but always seen:
Yon crescent Moon, as fixed as if it grew
In its own cloudless, starless lake of blue;
I see them all so excellently fair,
I see, not feel, how beautiful they are!

DEJECTION: AN ODE. **7. Æolian lute:** See note to "The Eolian Harp," above.

III

My genial spirits fail;
   And what can these avail   40
To lift the smothering weight from off my breast?
   It were a vain endeavour,
   Though I should gaze for ever
On that green light that lingers in the west:
I may not hope from outward forms to win
The passion and the life, whose fountains are within.

IV

O Lady! we receive but what we give,
And in our life alone does Nature live:
Ours is her wedding garment, ours her shroud!
   And would we aught behold, of higher worth,   50
Than that inanimate cold world allowed
To the poor loveless ever-anxious crowd,
   Ah! from the soul itself must issue forth
A light, a glory, a fair luminous cloud
     Enveloping the Earth—
And from the soul itself must there be sent
   A sweet and potent voice, of its own birth,
Of all sweet sounds the life and element!

V

O pure of heart! thou need'st not ask of me
What this strong music in the soul may be!   60
What, and wherein it doth exist,
This light, this glory, this fair luminous mist,
This beautiful and beauty-making power.
   Joy, virtuous Lady! Joy that ne'er was given,
Save to the pure, and in their purest hour,
Life, and Life's effluence, cloud at once and shower,
Joy, Lady! is the spirit and the power,
Which wedding Nature to us gives in dower
   A new Earth and new Heaven,
Undreamt of by the sensual and the proud—   70
Joy is the sweet voice, Joy the luminous cloud—
     We in ourselves rejoice!
And thence flows all that charms or ear or sight,
   All melodies the echoes of that voice,
All colours a suffusion from that light.

VI

There was a time when, though my path was rough,
   This joy within me dallied with distress,
And all misfortunes were but as the stuff
   Whence Fancy made me dreams of happiness:
For hope grew round me, like the twining vine,   80
And fruits, and foliage, not my own, seemed mine.
But now afflictions bow me down to earth:
Nor care I that they rob me of my mirth;
   But oh! each visitation
Suspends what nature gave me at my birth,
   My shaping spirit of Imagination.
For not to think of what I needs must feel,
   But to be still and patient, all I can;
And haply by abstruse research to steal
   From my own nature all the natural man—   90
   This was my sole resource, my only plan:
Till that which suits a part infects the whole,
And now is almost grown the habit of my soul.

VII

Hence, viper thoughts, that coil around my mind,
   Reality's dark dream!
I turn from you, and listen to the wind,
   Which long has raved unnoticed. What a scream
Of agony by torture lengthened out
That lute sent forth! Thou Wind, that rav'st without,
   Bare crag, or mountain-tairn, or blasted tree,   100
Or pine-grove whither woodman never clomb,
Or lonely house, long held the witches' home,
   Methinks were fitter instruments for thee,
Mad Lutanist! who in this month of showers,
Of dark-brown gardens, and of peeping flowers,
Mak'st Devils' yule, with worse than wintry song,
The blossoms, buds, and timorous leaves among.
   Thou Actor, perfect in all tragic sounds!
Thou mighty Poet, e'en to frenzy bold!
     What tell'st thou now about?   110
     'Tis of the rushing of an host in rout,
With groans, of trampled men, with smarting
       wounds—
At once they groan with pain, and shudder with the
       cold!
But hush! there is a pause of deepest silence!
   And all that noise, as of a rushing crowd,

---

39. **genial:** Cf. Wordsworth's "Tintern Abbey," above,
l. 113.

100. **tairn:** pool.

With groans, and tremulous shudderings—all is
    over—
  It tells another tale, with sounds less deep and loud!
    A tale of less affright,
    And tempered with delight,
As Otway's self had framed the tender lay,—   120
    'Tis of a little child
    Upon a lonesome wild,
Not far from home, but she hath lost her way:
And now moans low in bitter grief and fear,
And now screams loud, and hopes to make her mother
    hear.

VIII

'Tis midnight, but small thoughts have I of sleep:
Full seldom may my friend such vigils keep!
Visit her, gentle Sleep! with wings of healing,
    And may this storm be but a mountain-birth,
May all the stars hang bright above her dwelling, 130
    Silent as though they watched the sleeping Earth!
        With light heart may she rise,
        Gay fancy, cheerful eyes,
Joy lift her spirit, joy attune her voice;
To her may all things live, from pole to pole,
Their life the eddying of her living soul!
    O simple spirit, guided from above,
    Dear Lady! friend devoutest of my choice,
        Thus mayest thou ever, evermore rejoice.

1802 (1802)

HYMN BEFORE SUN-RISE,
IN THE VALE OF CHAMOUNI

❧

When first printed in *The Morning Post*, September 11,
1802, an introductory note was supplied:

Chamouni is one of the highest mountain valleys of
the Barony of Faucigny in the Savoy Alps; and exhibits a kind of fairy world, in which the wildest
appearances (I had almost said horrors) of Nature
alternate with the softest and most beautiful. The
chain of Mont Blanc is its boundary; and besides the
Arve it is filled with sounds from the Arveiron, which
rushes from the melted glaciers, like a giant, mad

120. **Otway:** Thomas Otway, 1652–85. The original draft
read "William's" and the reference is to Wordsworth's
"Lucy Gray," above. 129. **mountain-birth:** perhaps a local
and brief storm of the sort common in mountain regions.

with joy, from a dungeon, and forms other torrents
of snow-water, having their rise in the glaciers which
slope down into the valley. The beautiful *Gentiana
major*, or greater gentian, with blossoms of the brightest blue, grows in large companies a few steps from
the never-melted ice of the glaciers. I thought it an
affecting emblem of the boldness of human hope,
venturing near, and, as it were, leaning over the brink
of the grave. Indeed, the whole vale, its every light,
its every sound, must needs impress every mind not
utterly callous with the thought—Who *would* be,
who *could* be an Atheist in this valley of wonders! If
any of the readers of the *Morning Post* have visited
this vale in their journeys among the Alps, I am confident that they will not find the sentiments and feelings
expressed, or attempted to be expressed, in the following poem, extravagant.

Coleridge was never at Chamouni. The fact is, of
course, irrelevant to the integrity of an imaginative
work, but it has been enmeshed in a web of controversy
about the poem. In 1834 De Quincey asserted that
Coleridge had an "unacknowledged obligation" to a
poem (of some twenty lines) by Frederike Brun, 1765–
1835, a German poetess. Though charges of plagiarism
can hardly be made relevant to a much longer poem in a
different language, they have resulted in minute study of
the literary sources of the poem. For further discussion
see two letters by A. P. Rossiter in *The Times Literary
Supplement*, September 28 and October 26, 1951, and
Letters, ed. E. L. Griggs, II, 864–65 and 865 n.

❧

Hast thou a charm to stay the morning-star
In his steep course? So long he seems to pause
On thy bald awful head, O sovran BLANC,
The Arve and Arveiron at thy base
Rave ceaselessly; but thou, most awful Form!
Risest from forth thy silent sea of pines,
How silently! Around thee and above
Deep is the air and dark, substantial, black,
An ebon mass: methinks thou piercest it,
As with a wedge! But when I look again,   10
It is thine own calm home, thy crystal shrine,
Thy habitation from eternity!
O dread and silent Mount! I gazed upon thee,
Till thou, still present to the bodily sense,
Didst vanish from my thought: entranced in prayer
I worshipped the Invisible alone.

  Yet, like some sweet beguiling melody,
So sweet, we know not we are listening to it,
Thou, the meanwhile, wast blending with my
    Thought,
Yea, with my Life and Life's own secret joy: 20

Till the dilating Soul, enrapt, transfused,
Into the mighty vision passing—there
As in her natural form, swelled vast to Heaven!

Awake, my soul! not only passive praise
Thou owest! not alone these swelling tears,
Mute thanks and secret ecstasy! Awake,
Voice of sweet song! Awake, my heart, awake!
Green vales and icy cliffs, all join my Hymn.

Thou first and chief, sole sovereign of the Vale!
O struggling with the darkness all the night,   30
And visited all night by troops of stars,
Or when they climb the sky or when they sink:
Companion of the morning-star at dawn,
Thyself Earth's rosy star, and of the dawn
Co-herald: wake, O wake, and utter praise!
Who sank thy sunless pillars deep in Earth?
Who filled thy countenance with rosy light?
Who made thee parent of perpetual streams?

And you, ye five wild torrents fiercely glad!
Who called you forth from night and utter death, 40
From dark and icy caverns called you forth,
Down those precipitous, black, jagged rocks,
For ever shattered and the same for ever?
Who gave you your invulnerable life,
Your strength, your speed, your fury, and your joy,
Unceasing thunder and eternal foam?
And who commanded (and the silence came),
Here let the billows stiffen, and have rest?

Ye Ice-falls! ye that from the mountain's brow
Adown enormous ravines slope amain—   50
Torrents, methinks, that heard a mighty voice,
And stopped at once amid their maddest plunge!
Motionless torrents! silent cataracts!
Who made you glorious as the Gates of Heaven
Beneath the keen full moon? Who bade the sun
Clothe you with rainbows? Who, with living flowers
Of loveliest blue, spread garlands at your feet?—
GOD! let the torrents, like a shout of nations,
Answer! and let the ice-plains echo, GOD!
GOD! sing ye meadow-streams with gladsome voice! 60
Ye pine-groves, with your soft and soul-like sounds!
And they too have a voice, yon piles of snow,
And in their perilous fall shall thunder, GOD!

Ye living flowers that skirt the eternal frost!
Ye wild goats sporting round the eagle's nest!
Ye eagles, play-mates of the mountain-storm!

Ye lightnings, the dread arrows of the clouds!
Ye signs and wonders of the element!
Utter forth God, and fill the hills with praise!

Thou too, hoar Mount! with thy sky-pointing peaks,
Oft from whose feet the avalanche, unheard,   71
Shoots downward, glittering through the pure serene
Into the depth of clouds, that veil thy breast—
Thou too again, stupendous Mountain! thou
That as I raise my head, awhile bowed low
In adoration, upward from thy base
Slow travelling with dim eyes suffused with tears,
Solemnly seemest, like a vapoury cloud,
To rise before me—Rise, O ever rise,
Rise like a cloud of incense from the Earth!   80
Thou kingly Spirit throned among the hills,
Thou dread ambassador from Earth to Heaven,
Great Hierarch! tell thou the silent sky,
And tell the stars, and tell yon rising sun
Earth, with her thousand voices, praises GOD.

1802 (1802)

THE PAINS OF SLEEP

❧

Dreams of the kind described here were frequently
experienced by Coleridge and were so horrifying to him
that he was at times afraid to attempt sleep.

❧

Ere on my bed my limbs I lay,
It hath not been my use to pray
With moving lips or bended knees;
But silently, by slow degrees,
My spirit I to Love compose,
In humble trust mine eye-lids close,
With reverential resignation,
No wish conceived, no thought exprest,
Only a sense of supplication;
A sense o'er all my soul imprest   10
That I am weak, yet not unblest,
Since in me, round me, every where
Eternal Strength and Wisdom are.

But yester-night I prayed aloud
In anguish and in agony,
Up-starting from the fiendish crowd
Of shapes and thoughts that tortured me:
A lurid light, a trampling throng,

Sense of intolerable wrong,
And whom I scorned, those only strong!   20
Thirst of revenge, the powerless will
Still baffled, and yet burning still!
Desire with loathing strangely mixed
On wild or hateful objects fixed.
Fantastic passions! maddening brawl!
And shame and terror over all!
Deeds to be hid which were not hid,
Which all confused I could not know
Whether I suffered, or I did:
For all seemed guilt, remorse or woe,   30
My own or others still the same
Life-stifling fear, soul-stifling shame.

So two nights passed: the night's dismay
Saddened and stunned the coming day.
Sleep, the wide blessing, seemed to me
Distemper's worst calamity.
The third night, when my own loud scream
Had waked me from the fiendish dream,
O'ercome with sufferings strange and wild,
I wept as I had been a child;   40
And having thus by tears subdued
My anguish to a milder mood,
Such punishments, I said, were due
To natures deepliest stained with sin,—
For aye entempesting anew
The unfathomable hell within,
The horror of their deeds to view,
To know and loathe, yet wish and do!
Such griefs with such men well agree,
But wherefore, wherefore fall on me?   50
To be beloved is all I need,
And whom I love, I love indeed.

1803 (1816)

PHANTOM

Two passages in Coleridge's Malta notebooks make clear that the poem attempts to describe Sara Hutchinson as she appeared in a dream.

All look and likeness caught from earth,
All accident of kin and birth,
Had pass'd away. There was no trace
Of aught on that illumined face,

Uprais'd beneath the rifted stone,
But of one spirit all her own;—
She, she herself, and only she,
Shone through her body visibly.

1805 (1834)

WHAT IS LIFE?

Resembles life what once was deem'd of light,
Too ample in itself for human sight?
An absolute self—an element ungrounded—
All that we see, all colours of all shade
By encroach of darkness made?—
Is very life by consciousness unbounded?
And all the thoughts, pains, joys of mortal breath,
A war-embrace of wrestling life and death?

1805 (1829)

TO WILLIAM WORDSWORTH

COMPOSED ON THE NIGHT
AFTER HIS RECITATION OF A POEM ON
THE GROWTH OF AN INDIVIDUAL MIND

Before leaving for Malta in 1804 Coleridge had seen only the first five books of Wordsworth's *Prelude*. When he returned in the late summer of 1806, in ill health and despondent, Wordsworth read him (over a period of almost two weeks) the rest of the poem, which had been completed in 1805. The reading ended the night of January 7, 1807.

Friend of the wise! and Teacher of the Good!
Into my heart have I received that Lay
More than historic, that prophetic Lay
Wherein (high theme by thee first sung aright)
Of the foundations and the building up
Of a Human Spirit thou hast dared to tell
What may be told, to the understanding mind
Revealable; and what within the mind
By vital breathings secret as the soul
Of vernal growth, oft quickens in the heart   10
Thoughts all too deep for words!—

TO WILLIAM WORDSWORTH.   **11. Thoughts . . . words:**
Cf. the last line of Wordsworth's "Intimations of Immortality," above.

Theme hard as
high!
Of smiles spontaneous, and mysterious fears
(The first-born they of Reason and twin-birth),
Of tides obedient to external force,
And currents self-determined, as might seem,
Or by some inner Power; of moments awful,
Now in thy inner life, and now abroad,
When power streamed from thee, and thy soul re-
ceived
The light reflected, as a light bestowed—
Of fancies fair, and milder hours of youth,   20
Hyblean murmurs of poetic thought
Industrious in its joy, in vales and glens
Native or outland, lakes and famous hills!
Or on the lonely high-road, when the stars
Were rising; or by secret mountain-streams,
The guides and the companions of thy way!

Of more than Fancy, of the Social Sense
Distending wide, and man beloved as man,
Where France in all her towns lay vibrating
Like some becalmed bark beneath the burst   30
Of Heaven's immediate thunder, when no cloud
Is visible, or shadow on the main.
For thou wert there, thine own brows garlanded,
Amid the tremor of a realm aglow,
Amid a mighty nation jubilant,
When from the general heart of human kind
Hope sprang forth like a full-born Deity!
——Of that dear Hope afflicted and struck down,
So summoned homeward, thenceforth calm and sure
From the dread watch-tower of man's absolute self,
With light unwaning on her eyes, to look   41
Far on—herself a glory to behold,
The Angel of the vision! Then (last strain)
Of Duty, chosen Laws controlling choice,
Action and joy!—An Orphic song indeed,
A song divine of high and passionate thoughts
To their own music chaunted!

O great Bard!
Ere yet that last strain dying awed the air,
With stedfast eye I viewed thee in the choir
Of ever-enduring men. The truly great   50
Have all one age, and from one visible space
Shed influence! They, both in power and act,

Are permanent, and Time is not with them,
Save as it worketh for them, they in it.
Nor less a sacred Roll, than those of old,
And to be placed, as they, with gradual fame
Among the archives of mankind, thy work
Makes audible a linked lay of Truth,
Of Truth profound a sweet continuous lay,
Not learnt, but native, her own natural notes!   60
Ah! as I listened with a heart forlorn,
The pulses of my being beat anew:
And even as Life returns upon the drowned,
Life's joy rekindling roused a throng of pains—
Keen pangs of Love, awakening as a babe
Turbulent, with an outcry in the heart;
And fears self-willed, that shunned the eye of Hope;
And Hope that scarce would know itself from Fear;
Sense of past Youth, and Manhood come in vain,
And Genius given, and Knowledge won in vain;   70
And all which I had culled in wood-walks wild,
And all which patient toil had reared, and all,
Commune with thee had opened out—but flowers
Strewed on my corse, and borne upon my bier
In the same coffin, for the self-same grave!

That way no more! and ill beseems it me,
Who came a welcomer in herald's guise,
Singing of Glory, and Futurity,
To wander back on such unhealthful road,
Plucking the poisons of self-harm! And ill   80
Such intertwine beseems triumphal wreaths
Strew'd before thy advancing!

Nor do thou,
Sage Bard! impair the memory of that hour
Of thy communion with my nobler mind
By pity or grief, already felt too long!
Nor let my words import more blame than needs.
The tumult rose and ceased: for Peace is nigh
Where Wisdom's voice has found a listening heart.
Amid the howl of more than wintry storms,
The Halcyon hears the voice of vernal hours   90
Already on the wing.

Eve following eve,
Dear tranquil time, when the sweet sense of Home
Is sweetest! moments for their own sake hailed
And more desired, more precious, for thy song,
In silence listening, like a devout child,

---

13. **Reason:** Cf. "Reason," below. **21. Hyblean:** from
Hybla, a Sicilian town noted for its honey. **45. Orphic:**
enchanting and oracular, from the legendary poet Orpheus.
Cf. *The Prelude,* I. 233, above.

90. **Halcyon:** a legendary bird capable of calming the winter
sea so that it can nest on the water.

My soul lay passive, by thy various strain
Driven as in surges now beneath the stars,
With momentary stars of my own birth,
Fair constellated foam, still darting off
Into the darkness; now a tranquil sea,   100
Outspread and bright, yet swelling to the moon.

And when—O Friend! my comforter and guide!
Strong in thyself, and powerful to give strength!—
Thy long sustained Song finally closed,
And thy deep voice had ceased—yet thou thyself
Wert still before my eyes, and round us both
That happy vision of beloved faces—
Scarce conscious, and yet conscious of its close
I sate, my being blended in one thought
(Thought was it? or aspiration? or resolve?)   110
Absorbed, yet hanging still upon the sound—
And when I rose, I found myself in prayer.

1807 (*1817*)

HUMAN LIFE

ON THE DENIAL OF IMMORTALITY

If dead, we cease to be; if total gloom
    Swallow up life's brief flash for aye, we fare
As summer-gusts, of sudden birth and doom,
    Whose sound and motion not alone declare,
But are their whole of being! If the breath
    Be Life itself, and not its task and tent,
If even a soul like Milton's can know death;
    O Man! thou vessel purposeless, unmeant,
Yet drone-hive strange of phantom purposes!
    Surplus of Nature's dread activity,   10
Which, as she gazed on some nigh-finished vase,
    Retreating slow, with meditative pause,
    She formed with restless hands unconsciously.
Blank accident! nothing's anomaly!
    If rootless thus, thus substanceless thy state,
Go, weigh thy dreams, and be thy hopes, thy fears,
    The counter-weights!—Thy laughter and thy tears
    Mean but themselves, each fittest to create
And to repay the other! Why rejoices
    Thy heart with hollow joy for hollow good?   20
    Why cowl thy face beneath the mourner's hood?
Why waste thy sighs, and thy lamenting voices,
    Image of Image, Ghost of Ghostly Elf,
That such a thing as thou feel'st warm or cold?
Yet what and whence thy gain, if thou withhold

These costless shadows of thy shadowy self?
Be sad! be glad! be neither! seek, or shun!
Thou hast no reason why! Thou canst have none;
Thy being's being is contradiction.

?*1815* (*1817*)

"Limbo" and "Ne Plus Ultra" follow each other here
as they do in one of Coleridge's notebooks. The dates
are conjectural, and the whole of "Limbo" was not
published in Coleridge's lifetime. Traditionally located
on the border of Hell, Limbo was the abode of the just
pagans who had died before the coming of Christ, and
of unbaptized infants. More loosely it may suggest a
state of unreal existence, neither non-being nor being.
Ne plus ultra literally means "nothing more beyond."
The phrase is usually applied in the sense of "perfection,"
but here it has the opposite sense of "positive Negation."
With the last two lines of "Limbo" compare a passage
in the Comment to Aphorism XIX on spiritual religion in the *Aids to Reflection*:

Besides that dissolution of our earthly tabernacle
which we call death, there is another death, not the
mere negation of life, but its positive opposite. And
as there is a mystery of life, and an assimilation to the
principle of life, even to him who is the Life; so is
there a mystery of death, and an assimilation to the
principle of evil; a fructifying of the corrupt seed, of
which death is the germination. Thus the regeneration to spiritual life is at the same time a redemption from the spiritual death.

LIMBO

The sole true Something—This! In Limbo's Den
It frightens Ghosts, as here Ghosts frighten men.
Thence cross'd unseiz'd—and shall some fated hour
Be pulveris'd by Demogorgon's power,
And given as poison to annihilate souls—
Even now it shrinks them—they shrink in as Moles
(Nature's mute monks, live mandrakes of the ground)
Creep back from Light—then listen for its sound;—
See but to dread, and dread they know not why—
The natural alien of their negative eye.   10

LIMBO. 4. **Demogorgon:** a mysterious and awful being of
the nether world. See note to Shelley's *Prometheus Unbound*,
II. iv.

'Tis a strange place, this Limbo!—not a Place,
Yet name it so;—where Time and weary Space
Fettered from flight, with night-mare sense of fleeing,
Strive for their last crepuscular half-being;—
Lank Space, and scytheless Time with branny hands
Barren and soundless as the measuring sands,
Not mark'd by flit of Shades,—unmeaning they
As moonlight on the dial of the day!
But that is lovely—looks like Human Time,—
An Old Man with a steady look sublime,   20
That stops his earthly task to watch the skies;
But he is blind—a Statue hath such eyes;—
Yet having moonward turn'd his face by chance,
Gazes the orb with moon-like countenance,
With scant white hairs, with foretop bald and high,
He gazes still,—his eyeless face all eye;—
As 'twere an organ full of silent sight,
His whole face seemeth to rejoice in light!
Lip touching lip, all moveless, bust and limb—
He seems to gaze at that which seems to gaze on him!
   No such sweet sights doth Limbo den immure,   31
Wall'd round, and made a spirit-jail secure,
By the mere horror of blank Naught-at-all,
Whose circumambience doth these ghosts enthral.
A lurid thought is growthless, dull Privation,
Yet that is but a Purgatory curse;
Hell knows a fear far worse,
A fear—a future state;—'tis positive Negation!

                                          1817 (1893)

NE PLUS ULTRA

      Sole Positive of Night!
      Antipathist of Light!
Fate's only essence! primal scorpion rod—
The one permitted opposite of God!—
Condensed blackness and abysmal storm
      Compacted to one sceptre
         Arms the Grasp enorm—
         The Intercepter—
The Substance that still casts the shadow Death!—
      The Dragon foul and fell—   10
         The unrevealable,
And hidden one, whose breath
Gives wind and fuel to the fires of Hell!
      Ah! sole despair
      Of both th' eternities in Heaven!

Sole interdict of all-bedewing prayer,
      The all-compassionate!
      Save to the Lampads Seven
Reveal'd to none of all th' Angelic State,
      Save to the Lampads Seven,   20
      That watch the throne of Heaven!

                                          ?1826 (1834)

THE KNIGHT'S TOMB

⌒⌒⌒

Composed as a metrical experiment, these lines were
repeated by a mutual friend to Sir Walter Scott, who
quoted them in *Ivanhoe* and *Castle Dangerous.*

⌒⌒⌒

Where is the grave of Sir Arthur O'Kellyn?
Where may the grave of that good man be?—
By the side of a spring, on the breast of Helvellyn,
Under the twigs of a young birch tree!
The oak that in summer was sweet to hear,
And rustled its leaves in the fall of the year,
And whistled and roared in the winter alone,
Is gone,—and the birch in its stead is grown.—
The Knight's bones are dust,
And his good sword rust;—   10
His soul is with the saints, I trust.

                                          ?1817 (1834)

ON DONNE'S POETRY

With Donne, whose muse on dromedary trots,
Wreathe iron pokers into true-love knots;
Rhyme's sturdy cripple, fancy's maze and clue,
Wit's forge and fire-blast, meaning's press and screw.

                                          ?1818 (1836)

NE PLUS ULTRA. **18. Lampads Seven:** See "Ode to the
Departing Year," above, ll. 76–77, and note. THE KNIGHT'S
TOMB. **3. Helvellyn:** mountain in the Lake District.

WORK WITHOUT HOPE

LINES COMPOSED 21ST FEBRUARY 1825

All nature seems at work. Slugs leave their lair—
The bees are stirring—birds are on the wing—
And Winter slumbering in the open air,
Wears on his smiling face a dream of Spring!
And I the while, the sole unbusy thing,
Nor honey make, nor pair, nor build, nor sing.

  Yet well I ken the banks where amaranths blow,
Have traced the font whence streams of nectar flow.
Bloom, O ye amaranths! bloom for whom ye may,
For me ye bloom not! Glide, rich streams, away!　10
With lips unbrightened, wreathless brow, I stroll:
And would you learn the spells that drowse my soul?
Work without Hope draws nectar in a sieve,
And Hope without an object cannot live.

1825 (1828)

CONSTANCY
TO AN IDEAL OBJECT

Since all that beat about in Nature's range,
Or veer or vanish; why should'st thou remain
The only constant in a world of change,
O yearning Thought! that liv'st but in the brain?
Call to the Hours, that in the distance play,
The faery people of the future day——
Fond Thought! not one of all that shining swarm
Will breathe on thee with life-enkindling breath,
Till when, like strangers shelt'ring from a storm,
Hope and Despair meet in the porch of Death!　10
Yet still thou haunt'st me; and though well I see,
She is not thou, and only thou art she,
Still, still as though some dear embodied Good,
Some living Love before my eyes there stood
With answering look a ready ear to lend,
I mourn to thee and say—"Ah! loveliest friend!
That this the meed of all my toils might be,
To have a home, an English home, and thee!"
Vain repetition! Home and Thou are one.
The peacefull'st cot, the moon shall shine upon,　20
Lulled by the thrush and wakened by the lark,
Without thee were but a becalmed bark,

CONSTANCY TO AN IDEAL OBJECT. **12. She:** Sara Hutchinson.

Whose Helmsman on an ocean waste and wide
Sits mute and pale his mouldering helm beside.

And art thou nothing? Such thou art, as when
The woodman winding westward up the glen
At wintry dawn, where o'er the sheep-track's maze
The viewless snow-mist weaves a glist'ning haze,
Sees full before him, gliding without tread,
An image with a glory round its head;　30
The enamoured rustic worships its fair hues,
Nor knows he makes the shadow, he pursues!

?1826 (1828)

PHANTOM OR FACT

A DIALOGUE IN VERSE

AUTHOR

A lovely form there sate beside my bed,
And such a feeding calm its presence shed,
A tender love so pure from earthly leaven,
That I unnethe the fancy might control,
'Twas my own spirit newly come from heaven,
Wooing its gentle way into my soul!
But ah! the change—It had not stirr'd, and yet—
Alas! that change how fain would I forget!
That shrinking back, like one that had mistook!
That weary, wandering, disavowing look!　10
'Twas all another, feature, look, and frame,
And still, methought, I knew, it was the same!

FRIEND

This riddling tale, to what does it belong?
Is't history? vision? or an idle song?
Or rather say at once, within what space
Of time this wild disastrous change took place?

AUTHOR

Call it a moment's work (and such it seems)
This tale's a fragment from the life of dreams;
But say, that years matur'd the silent strife,
And 'tis a record from the dream of life.　20

?1830 (1834)

30. glory: halo. Coleridge refers to a phenomenon in which on a misty day a walker may see his own figure projected by the sun into the mist before him, with a ring of light around the head. PHANTOM OR FACT. **4. unnethe:** with difficulty.

DESIRE

Where true Love burns Desire is Love's pure flame;
It is the reflex of our earthly frame,
That takes its meaning from the nobler part,
And but translates the language of the heart.

?1830 (1834)

REASON

❧

The passage from Dante reads, "Thou thyself makest
thyself dense with false imagining, so thou seest not that
which thou wouldst see, if thou hadst cast it off."

❧

["Finally, what is Reason? You have often asked me: and
this is my answer":—]

Whene'er the mist, that stands 'twixt God and thee,
Defecates to a pure transparency,
That intercepts no light and adds no stain—
There Reason is, and then begins her reign!

But alas!
——— "tu stesso, ti fai grosso
Col falso immaginar, si che non vedi
Cio che vedresti, se l'avessi scosso."

Dante, *Paradiso*, I. 88–90
1830 (1830)

SELF-KNOWLEDGE

—E coelo descendit γνῶθι σεαυτόν—JUVENAL, xi. 27.

Γνῶθι σεαυτόν!—and is this the prime
And heaven-sprung adage of the olden time!—

REASON. **Motto: "Finally . . . answer":** This passage,
including the prose, the four-line poem, and the quotation
from Dante, was printed by Coleridge as the conclusion
to a long prose work entitled *On the Constitution of
Church and State*. It is the final paragraph of that work.
The sentences in prose are presumably intended by Coleridge as an introduction to the verse. They are not printed
in brackets by Coleridge. SELF-KNOWLEDGE. **Motto:** "From
heaven descended *know thyself*." The maxim was attributed
to various sages, including Socrates, and to the Delphic
oracle. It was inscribed in gold letters over the portico of
the temple of Delphi.

Say, canst thou make thyself?—Learn first that trade;—
Haply thou mayst know what thyself had made.
What hast thou, Man, that thou dar'st call thine
    own?—
What is there in thee, Man, that can be known?—
Dark fluxion, all unfixable by thought,
A phantom dim of past and future wrought,
Vain sister of the worm,—life, death, soul, clod—
Ignore thyself, and strive to know thy God!   10

1832 (1834)

EPITAPH

Stop, Christian passer-by!—Stop, child of God,
And read with gentle breast. Beneath this sod
A poet lies, or that which once seem'd he.
O, lift one thought in prayer for S. T. C.;
That he who many a year with toil of breath
Found death in life, may here find life in death!
Mercy for praise—to be forgiven for fame
He ask'd, and hoped, through Christ. Do thou the
    same!

1833 (1834)

FROM

ON THE PRINCIPLES
OF GENIAL CRITICISM

ESSAY THIRD

❧

Coleridge began this short work (published in a Bristol
periodical in 1814) in order to call public attention to
the paintings—then being exhibited in Bristol—by the
American artist, Washington Allston. With the second
essay he started to turn to more general problems of art,
and then did so entirely in the much longer third essay,
printed below, the first really comprehensive exposition
of his philosophy of art that survives. His discussion here
may be compared with the more developed treatment
in "On Poesy or Art," below, p. 491, written four
years later.

❧

EPITAPH. **7. for:** instead of.

PEDANTRY consists in the use of words unsuitable to the time, place, and company. The language of the market would be as pedantic in the schools as that of the schools in the market. The mere man of the world, who insists that in a philosophic investigation of principles and general laws, no other terms should be used, than occur in common conversation, and with no greater definiteness, is at least as much a *pedant* as the man of learning, who, perhaps overrating the acquirements of his auditors, or deceived by his own familiarity with technical phrases, talks at the wine-table with his eye fixed on his study or laboratory; even though, instead of desiring his wife to make the tea, he should bid her add to the usual quantum sufficit of Thea Sinensis the Oxyd of Hydrogen, saturated with Calorique.[1] If (to use the old metaphor) both smell of the shop, yet the odour from the Russia-leather bindings of the good old *authentic-looking* folios and quartos is less annoying than the steams from the tavern or tallow-vat. Nay, though the pedantry should originate in vanity, yet a good-natured man would more easily tolerate the Fox-brush of ostentatious erudition ("the fable is somewhat musty")[2] than the Sans-culotterie[3] of a contemptuous ignorance, that assumes a merit from mutilation by a self-consoling grin at the pompous incumbrance of tails.

In a philosophic disquisition, besides the necessity of confining many words of ordinary use to one definite sense, the writer has to make his choice between two difficulties, whenever his purpose requires him to wean his reader's attention from the *degrees* of things, which alone form the dictionary of common life, to the *kind*, independent of *degree*: as when, for instance, a chemist discourses on the heat in ice, or on latent or fixed light. In this case, he must either use old words with new meanings, the plan adopted by Dr. Darwin in his Zoonomia;[4] or he must borrow from the schools, or himself coin a nomenclature exclusively appropriated to his subject, after the example of the French chemists, and indeed of all eminent natural philosophers and historians in all countries. There seems to me little ground for hesitation as to which of the two shall be preferred: it being clear, that the former is a twofold exertion of mind in one and the same act. The reader is obliged, not only to recollect the new definition, but—which is incomparably more difficult and perplexing—to unlearn and keep out of view the old and habitual meaning: an evil, for which the *semblance* of eschewing pedantry is a very poor and inadequate compensation. I have, therefore, in two or three instances ventured on a disused or scholastic term, where without it I could not have avoided confusion or ambiguity. Thus, to express in one word what belongs to the senses or the recipient and more passive faculty of the soul, I have re-introduced the word *sensuous*, used, among many others of our elder writers, by Milton, in his exquisite definition of poetry, as "simple, sensuous, passionate":[5] because the term *sensual* is seldom used at present, except in a bad sense, and *sensitive* would convey a different meaning. Thus too I have restored the words, *intuition* and *intuitive*, to their original sense—"an intuition," says Hooker, "that is, a direct and immediate beholding or presentation of an object to the mind through the senses or the imagination."—Thus geometrical truths are all intuitive, or accompanied by an intuition. Nay, in order to express "*the many*," as simply contra-distinguished from "*the one*," I have hazarded the smile of the reader, by introducing to his acquaintance, from the forgotten terminology of the old schoolmen, the phrase, *multëity*, because I felt that I could not substitute *multitude*, without more or less connecting with it the notion of "a *great* many." Thus the Philosopher of the later Platonic, or Alexandrine school, named the triangle the first-born of beauty, it being the first and simplest symbol of *multëity in unity*. These are, I believe, the only liberties of this kind which I have found it necessary to attempt in the present essay: partly, because its object will be attained sufficiently for my present purpose, by attaching a clear and distinct meaning to the different terms used by us, in our appreciation of works of art, and partly because I am about to put to the press a large volume on the LOGOS,[6] or the communicative intelligence in nature and in man, together with, and as preliminary to, a Commentary on the Gospel of

PRINCIPLES OF GENIAL CRITICISM. ESSAY III. **1. quantum . . . Calorique:** a sufficient amount of Chinese tea with hot water. Calorique was the name of a supposed fluid that produced heat. **2. "the fable . . . musty":** *Hamlet*, III. ii. 359. **3. Sans-culotterie:** from the French *sans culotte*, literally, "without breeches." During the Revolution the term was applied to Paris revolutionists from the poorer classes and came to mean an extreme democrat, a radical or Jacobin. **4. Zoonomia:** a work (1794–96) by Erasmus Darwin (grandfather of Charles) offering a comprehensive survey of the senses and biological instincts of the animal world.

5. "simple . . . passionate": from Milton's treatise "Of Education" (1644). **6. volume on the logos:** Coleridge's intended *magnum opus*, never completed.

St. John; and in this work I have labored to give real and adequate definitions of all the component faculties of our moral and intellectual being, exhibiting constructively the origin, development, and destined functions of each. And now with silent wishes that these explanatory pre-notices may be attributed to their true cause, a sense of respect for the understanding of my reflecting readers, I proceed to my promised and more amusing task, that of establishing, illustrating, and exemplifying the distinct powers of the different modes of pleasure excited by the works of nature or of human genius with their exponent and appropriable terms. "Harum indagatio subtilitatum etsi non est utilis ad machinas farinarias conficiendas, exuit animum tamen inscitiæ rubigine, acuitque ad alia."[7]—*Scaliger, Exerc.* 307, §3.

AGREEABLE.—We use this word in two senses; in the first for whatever agrees with our nature, for that which is congruous with the primary constitution of our senses. Thus green is naturally agreeable to the eye. In this sense the word expresses, at least involves, a pre-established harmony between the organs and their appointed objects. In the second sense, we convey by the word *agreeable*, that the thing has by force of habit (thence called a second nature) been made to agree with us; or that it has become agreeable to us by its recalling to our minds some one or more things that were dear and pleasing to us; or lastly, on account of some after pleasure or advantage, of which it has been the constant cause or occasion. Thus by force of custom men *make* the taste of tobacco, which was at first hateful to the palate, agreeable to them; thus too, as our Shakspeare observes,

"Things base and vile, holding no quality,
Love can transpose to form and dignity—"[8]

the crutch that had supported a revered parent, after the first anguish of regret, becomes agreeable to the affectionate child; and I once knew a very sensible and accomplished Dutch gentleman, who, spite of his own sense of the ludicrous nature of the feeling, was more delighted by the first grand concert of frogs he heard in this country, than he had been by Catalina

singing in the compositions of Cimarosa.[9] The last clause needs no illustrations, as it comprises all the objects that are agreeable to us, only because they are the means by which we gratify our smell, touch, palate, and mere bodily feeling.

The BEAUTIFUL, contemplated in its essentials, that is, in *kind* and not in *degree*, is that in which the *many*, still seen as many, becomes one. Take a familiar instance, one of a thousand. The frost on a window-pane has by accident crystallized into a striking resemblance of a tree or a sea-weed. With what pleasure we trace the parts, and their relations to each other, and to the whole! Here is the stalk or trunk, and here the branches or sprays—sometimes even the buds or flowers. Nor will our pleasure be less, should the caprice of the crystallization represent some object disagreeable to us, provided only we can see or fancy the component parts each in relation to each, and all forming a whole. A lady would see an admirably painted tiger with pleasure, and at once pronounce it beautiful,—nay, an owl, a frog, or a toad, who would have shrieked or shuddered at the sight of the things themselves. So far is the Beautiful from depending wholly on association, that it is frequently produced by the mere removal of associations. Many a sincere convert to the beauty of various insects, as of the dragon-fly, the fangless snake, &c., has Natural History made, by exploding the terror or aversion that had been connected with them.

The most general definition of beauty, therefore, is—that I may fulfil my threat of plaguing my readers with hard words—Multëity in Unity. Now it will be always found, that whatever is the definition of the *kind*, independent of degree, becomes likewise the definition of the highest degree of that kind. An old coach-wheel lies in the coachmaker's yard, disfigured with tar and dirt (I purposely take the most trivial instances)—if I turn away my attention from these, and regard the *figure* abstractly, "still," I might say to my companion, "there is beauty in that wheel, and you yourself would not only admit, but would feel it, had you never seen a wheel before. See how the rays proceed from the centre to the circumferences, and how many different images are distinctly comprehended at one glance, as forming one whole, and each part in some harmonious relation to each and to all." But imagine the polished golden wheel of the chariot

7. **"Harum . . . alia":** "The investigation of this subtle point, though of no use for constructing food-making machines, removes the mold of ignorance from the mind, and sharpens it for other purposes" (J. C. Scaliger, 1484–1558). 8. **"Things . . . dignity":** *A Midsummer Night's Dream,* I. i. 232–33.

9. **Catalina . . . Cimarosa:** Angelica Catalani, 1780–1849, soprano; Domenico Cimarosa, 1749–1801, Italian composer.

of the Sun, as the poets have described it: then the figure, and the real thing so figured, exactly coincide. There is nothing heterogeneous, nothing to abstract from: by its perfect smoothness and circularity in width, each part is (if I may borrow a metaphor from a sister sense) as perfect a melody, as the whole is a complete harmony. This, we should say, is beautiful throughout. Of all "the many," which I actually see, each and all are really reconciled into unity: while the effulgence from the whole coincides with, and seems to represent, the effluence of delight from my own mind in the intuition of it.

It seems evident then, first, that beauty is harmony, and subsists only in composition,[10] and secondly, that the first species of the Agreeable can alone be a component part of the beautiful, that namely which is naturally consonant with our senses by the pre-established harmony between nature and the human mind; and thirdly, that even of this species, those objects only can be admitted (according to rule the first) which belong to the eye and ear, because they alone are susceptible of distinction of parts. Should an Englishman gazing on a mass of cloud rich with the rays of the rising sun exclaim, even without distinction of, or reference to its form, or its relation to other objects, how beautiful! I should have no quarrel with him. First, because by the law of association there is in all visual beholdings at least an indistinct subsumption of form and relation: and, secondly, because even in the coincidence between the sight and the object there is an approximation to the reduction of the many into one. But who, that heard a Frenchman call the flavor of a leg of mutton a beautiful taste, would not immediately recognize him for a Frenchman, even though there should be neither grimace or characteristic nasal twang? The result, then, of the whole is that the shapely (i.e. *formosus*) joined with the naturally agreeable, constitutes what, speaking accurately, we mean by the word beautiful (i.e. *pulcher*).

But we are conscious of faculties far superior to the highest impressions of sense; we have life and free-will.—What then will be the result, when the Beautiful, arising from regular form, is so modified by the perception of life and spontaneous action, as that the latter only shall be the object of our conscious *perception*, while the former merely acts, and yet does effectively

10. **in composition:** that is, when there are different parts being harmonized together (a single element alone is not "beautiful").

act, on our feelings? With pride and pleasure I reply by referring my reader to the group in Mr. Allston's grand picture of the "Dead Man reviving from the touch of the bones of the Prophet Elisha," beginning with the slave at the head of the reviving body, then proceeding to the daughter clasping her swooning mother; to the mother, the wife of the reviving man; then to the soldier behind who supports her; to the two figures eagerly conversing: and lastly, to the exquisitely graceful girl who is bending downward, and whose hand nearly touches the thumb of the slave! You will find, what you had not suspected, that you have here before you a circular group. But by what variety of life, motion, and passion is all the stiffness, that would result from an obvious regular figure, swallowed up, and the figure of the group as much concealed by the action and passion, as the skeleton, which gives the form of the human body, is hidden by the flesh and its endless outlines!

In Raphael's admirable Galatea (the print of which is doubtless familiar to most of my readers) the circle is perceived at first sight; but with what multiplicity of rays and chords within the area of the circular group, with what elevations and depressions of the circumference, with what an endless variety and sportive wildness in the component figure, and in the junctions of the figures, is the balance, the perfect reconciliation, effected between these two conflicting principles of the FREE LIFE, and of the confining FORM! How entirely is the stiffness that would have resulted from the obvious regularity of the latter, *fused* and (if I may hazard so bold a metaphor) almost *volatilized* by the interpenetration and electrical flashes of the former.

But I shall recur to this consummate work for more specific illustrations hereafter: and have indeed in some measure offended already against the laws of method, by anticipating materials which rather belong to a more advanced stage of the disquisition. It is time to recapitulate, as briefly as possible, the arguments already advanced, and having summed up the result, to leave behind me this, the only portion of these essays, which, as far as the subject itself is concerned, will demand any *effort* of attention from a reflecting and intelligent reader. And let me be permitted to remind him, that the distinctions, which it is my object to prove and elucidate, have not merely a foundation in nature and the noblest faculties of the human mind, but are likewise the very ground-work, nay, an indispensable condition, of all *rational* enquiry concerning the Arts. For it is self-evident, that whatever may be judged of differently by different persons, in the very same degree of moral and intellectual cultivation, extolled by one and condemned by another, without any error being assignable to either, can never be an object of general principles: and *vice versâ*, that whatever can be brought to the test of general principles presupposes a distinct origin from these pleasures and tastes, which, for the wisest purposes, are made to depend on local and transitory fashions, accidental associations, and the peculiarities of individual temperament: to all which the philosopher, equally with the well-bred man of the world, applies the old adage, *de gustibus non est disputandum*.[11] Be it, however, observed that "de gustibus" is by no means the same as "de gustu," nor will it escape the scholar's recollection, that taste, in its metaphorical use, was first adopted by the Romans, and unknown to the less luxurious Greeks, who designated this faculty, sometimes by the word αἴσθησις, and sometimes by φιλοκαλία[12]—"ἀνδρῶν τῶν καθ' ἡμᾶς φιλοκαλώτατος γεγονώς—i.e. endowed by nature with the most exquisite taste of any man of our age," says Porphyry of his friend, Castricius. Still, this metaphor, borrowed from the pregustatores of the old Roman Banquets, is singularly happy and appropriate. In the palate, the perception of the object and its qualities is involved in the *sensation*, in the mental taste it is involved in the *sense*. We have a *sensation* of sweetness, in a healthy palate, from honey; a *sense* of beauty, in an uncorrupted taste, from the view of the rising or setting sun.

RECAPITULATION. *Principle the First*. That which has become, or which has been *made* agreeable to us, from causes not contained in its own nature, or in its original conformity to the human organs and faculties; that which is not pleasing for its own sake, but by connection or association with some other thing, separate or separable from it, is neither beautiful, nor capable of being a component part of Beauty: though it may greatly increase the sum of our pleasure, when it does not interfere with the beauty of the object, nay, even when it detracts from it. A moss-rose, with a sprig of myrtle and jasmine, is not more *beautiful* from having been plucked from the garden, or presented to us by the hand of the woman we love, but is abundantly more delightful. The total pleasure received

from one of Mr. Bird's finest pictures may, without any impeachment of our taste, be the greater from his having introduced into it the portrait of one of our friends, or from our pride in him as our townsman, or from our knowledge of his personal qualities; but the amiable artist would rightly consider it a coarse compliment, were it affirmed, that the *beauty* of the piece, or its merit as a work of genius, was the more perfect on this account, I am conscious that I look with a stronger and more pleasureable emotion at Mr. Allston's large landscape, in the spirit of Swiss scenery, from its having been the occasion of my first acquaintance with him in Rome. This may or may not be a compliment to *him*; but the true compliment to the picture was made by a lady of high rank and cultivated taste, who declared, in my hearing, that she never stood before that landscape without seeming to feel the breeze blow out of it upon her. But the most striking instance is afforded by the portrait of a departed or absent friend or parent; which is endeared to us, and more delightful, from some awkward position of the limbs, which had defied the contrivances of art to render it picturesque, but which was the characteristic habit of the original.

Principle the Second.—That which is naturally agreeable and consonant to human nature, so that the exceptions may be attributed to disease or defect; that, the pleasure from which is contained in the immediate impression; cannot, indeed, with strict propriety, be called beautiful, exclusive of its relations, but one among the component parts of beauty, in whatever instance it is susceptible of existing as a part of a whole. This, of course, excludes the mere objects of the taste, smell, and feeling, though the sensation from these, especially from the latter when organized into touch, may secretly, and without our consciousness, enrich and vivify the perceptions and images of the eye and ear; which alone are true organs of sense, their sensations in a healthy or uninjured state being too faint to be noticed by the mind. We may, indeed, in common conversation, call purple a beautiful color, or the tone of a single note on an excellent piano-forte a beautiful tone; but if we were questioned, we should agree that a rich or delightful color; a rich, or sweet, or clear tone; would have been more appropriate—and this with less hesitation in the latter instance than in the former, because the single tone is more manifestly of the nature of a *sensation*, while color is the medium which seems to blend sensation and perception, so as to hide, as it were, the former in the latter; the direct opposite of which takes place in the lower senses of

---

11. de . . . disputandum: It is impossible to argue about tastes. Coleridge goes on to make a distinction between tastes and taste (*de gustu*). 12. αἴσθησις: sensation; φιλοκαλία: love for the beautiful.

feeling, smell, and taste. (In strictness, there is even in these an ascending scale. The smell is less sensual and more sentient than mere feeling, the taste than the smell, and the eye than the ear: but between the ear and the taste exists the chasm or break, which divides the beautiful and the elements of beauty from the merely agreeable.) When I reflect on the manner in which smoothness, richness of sound, &c., enter into the formation of the beautiful, I am induced to suspect that they act negatively rather than positively. Something there must be to realize the form, something in and by which the *forma informans* reveals itself: and these, less than any that could be substituted, and in the least possible degree, distract the attention, in the least possible degree obscure the idea, of which they (composed into outline and surface) are the symbol. An illustrative hint may be taken from a pure crystal, as compared with an opaque, semi-opaque or clouded mass, on the one hand, and with a perfectly transparent body, such as the air, on the other. The crystal is lost in the light, which yet it contains, embodies, and gives a shape to; but which passes shapeless through the air, and, in the ruder body, is either quenched or dissipated.

Principle the Third. The safest definition, then, of Beauty, as well as the oldest, is that of Pythagoras: THE REDUCTION OF MANY TO ONE—or, as finely expressed by the sublime disciple of Ammonius, τὸ ἄμερες ὄν, ἐν πολλοῖς φανταζόμενον,[13] of which the following may be offered as both paraphrase and corollary. *The sense of beauty subsists in simultaneous intuition of the relation of parts, each to each, and of all to a whole: exciting an immediate and absolute complacency, without intervenence, therefore, of any interest, sensual or intellectual.* The BEAUTIFUL is thus at once distinguished both from the AGREEABLE, which is beneath it, and from the GOOD, which is above it: for both these have an interest necessarily attached to them: both act on the WILL, and excite a desire for the actual existence of the image or idea contemplated: while the sense of beauty rests gratified in the mere contemplation or intuition, regardless whether it be a fictitious Apollo, or a real Antinous.[14]

The Mystics meant the same, when they define beauty as the subjection of matter to spirit so as to be transformed into a symbol, in and through which the spirit reveals itself; and declare *that* the *most* beautiful, where the most obstacles to a full manifestation have been most perfectly overcome. I would that the readers, for whom alone I write (*intelligibilia enim, non intellectum adfero*)[15] had Raphael's Galatea, or his School of Athens, before them! or that the Essay might be read by some imaginative student, warm from admiration of the King's College Chapel at Cambridge, or of the exterior and interior of York Cathedral! I deem the sneers of a host of petty critics, unalphabeted in the life and truth of things, and as devoid of sound learning as of intuitive taste, well and wisely hazarded for the prospect of communicating the pleasure, which to such minds the following passage of Plotinus will not fail to give—Plotinus, a name venerable even to religion with the great Cosmus, Lorenzo de Medici, Ficinus, Politian, Leonardo da Vinci, and Michael Angelo, but now known only as a name to the majority even of our most learned Scholars!—Plotinus, difficult indeed, but under a rough and austere rind concealing fruit worthy of Paradise; and if obscure, "*at tenet umbra Deum!*"[16] Ὅταν οὖν καὶ ἡ αἴσθησις τὸ ἐν σώμασιν εἶδος ἴδη συνδησάμενον καὶ κρατῆσαν τῆς φύσεως τῆς ἐναντίας, καὶ μορφὴν ἐπ' ἄλλαις μορφαῖς ἐκπρεπῶς ἐποχουμένην, συνελοῦσα ἀθρόον αὐτὸ τὸ πολλαχῆ ἀνήνεγκέ τε καὶ ἔδωκε τῷ ἔνδον σύμφωνον καὶ συναρμόττον καὶ φίλον.[17] A divine passage, faintly represented in the following lines, written many years ago by the writer, though without reference to, or recollection of, the above.

"O lady! we *receive* but what we *give*,
And in *our* life alone does nature live!
Ours is her wedding-garment, ours her shroud!
And would we aught behold of higher worth,
Than that inanimate cold world allow'd
To the poor, loveless, ever-anxious crowd:
Ah! from the soul itself must issue forth
A light, a glory, a fair luminous cloud,
    Enveloping the earth!

13. τὸ . . . φανταζόμενον: "The one (undivided) conceived as it is issuing through diversity (or multiplicity)." 14. Antinous: a youth beloved by the Roman emperor Hadrian and a frequent subject of sculpture.

15. intelligibilia . . . adfero: "I offer matters that are intelligible but not the intellect capable of conceiving them." 16. "at . . . Deum!": "yet the shadow holds a god." 17. Ὅταν . . . φίλον: "When the mind perceives the form intertwining itself through the concrete, and conquering the material reality opposed to it: then, after perceiving the dominating form, visibly emerging, the mind—collecting together the diverse materials into a harmonized union—bestows this form it has perceived back upon the object—a form quite in keeping with the character of the inner principle [actually there] in the object" (Plotinus, *Ennead*, I. 6; trans. W. J. Bate, ed., *Criticism: The Major Texts*, p. 374).

And from the soul itself must there be sent
A sweet and powerful voice, of its own birth,
    Of all sweet sounds the life and element!
O pure of heart! thou need'st not ask of me,
What this strong music in the soul may be;
    What and wherein it doth subsist,
This light, this glory, this fair luminous mist,
    This beautiful, and beauty-making power!
Joy, O beloved! joy, that ne'er was given,
Save to the pure and in their purest hour,
Life of our life, the parent and the birth,
Which, wedding nature to us, gives in dower
    A new heaven and new earth,
Undreamt of by the sensual and the proud—
This is the strong voice, this the luminous cloud!
    Our inmost selves rejoice:
And thence flows all that glads or ear or sight,
All melodies the echoes of that voice,
All colors a suffusion from that light,
And its celestial tint of yellow-green:
And still I gaze—and with how blank an eye!
And those thin clouds above, in flakes and bars,
That give away their motion to the stars;
Those stars, that glide behind them or between,
Now sparkling, now bedimm'd, but always seen;
Yon crescent moon, that seems as if it grew
In its own starless, cloudless lake of blue—
I see them all, so excellently fair!
I see, not feel, how beautiful they are." [18]

S. T. C. MS. Poem.

SCHOLIUM. We have sufficiently distinguished the beautiful from the agreeable, by the sure criterion, that, when we find an object agreeable, the *sensation* of pleasure always precedes the judgement, and is its determining cause. We *find* it agreeable. But when we declare an object beautiful, the contemplation or intuition of its beauty precedes the *feeling* of complacency, in order of nature at least: nay, in great depression of spirits may even exist without sensibly producing it.—

"A grief without a pang, void, dark, and drear!
A stifled, drowsy, unimpassion'd grief,
    That finds no natural outlet, no relief
        In word, or sigh, or tear!
O dearest lady! in this heartless mood,
To other thoughts by yon sweet throstle woo'd!
All this long eve, so balmy and serene,
Have I been gazing at the western sky."

18. "O lady!... are": These and the following lines are from Coleridge's "Dejection: An Ode," printed above.

Now the least reflection convinces us that our sensations, whether of pleasure or of pain, are the incommunicable parts of our nature; such as can be reduced to no universal rule; and in which therefore we have no right to expect that others should agree with us, or to blame them for disagreement. That the Greenlander prefers train oil to olive oil, and even to wine, we explain at once by our knowledge of the climate and productions to which he has been habituated. Were the man as enlightened as Plato, his palate would still find that most agreeable to which it had been most accustomed. But when the Iroquois Sachem, after having been led to the most perfect specimens of architecture in Paris, said that he saw nothing so beautiful as the cook's shops, we attribute this without hesitation to savagery of intellect, and infer with certainty that the sense of the beautiful was either altogether dormant in his mind, or at best very imperfect. The Beautiful, therefore, not originating in the sensations, must belong to the intellect: and therefore we *declare* an object beautiful, and feel an inward right to *expect* that others should coincide with us. But we feel no right to *demand* it: and this leads us to that, which hitherto we have barely touched upon, and which we shall now attempt to illustrate more fully, namely, to the distinction of the Beautiful from the Good.

Let us suppose Milton in company with some stern and prejudiced Puritan, contemplating the front of York Cathedral, and at length expressing his admiration of its beauty. We will suppose it too at that time of his life, when his religious opinions, feelings, and prejudices most nearly coincided with those of the rigid Antiprelatists.—P. Beauty; I am sure, it is not the beauty of holiness. M. True; but yet it is beautiful.—P. It delights not me. What is it good for? Is it of any use but to be stared at?—M. Perhaps not! but still it is beautiful.—P. But call to mind the pride and wanton vanity of those cruel shavelings, that wasted the labor and substance of so many thousand poor creatures in the erection of this haughty pile.—M. I do. But still it is very beautiful.—P. Think how many score of places of worship, incomparably better suited both for prayer and preaching, and how many faithful ministers might have been maintained, to the blessing of tens of thousands, to them and their children's children, with the treasures lavished on this worthless mass of stone and cement.—M. Too true! but nevertheless it is *very* beautiful.—P. And it is not merely useless; but it feeds the pride of the prelates, and keeps alive the popish and carnal spirit among the people.—M. Even so! and I presume not to question the wisdom, nor detract from the pious zeal, of the first Reformers of Scotland, who for these reasons destroyed so many fabrics, scarce inferior in beauty to this now before our eyes. But I did not call it *good*, nor have I told thee, brother! that if this were levelled with the ground, and existed only in the works of the modeller or engraver, that I should desire to reconstruct it. The GOOD consists in the congruity of a thing with the laws of the reason and the nature of the will, and in its fitness to determine the latter to actualize the former: and it is always discursive. The Beautiful arises from the perceived harmony of an object, whether sight or sound, with the inborn and constitutive rules of the judgement and imagination: and it is always intuitive. As light to the eye, even such is beauty to the mind, which cannot but have complacency in whatever is perceived as pre-configured to its living faculties. Hence the Greeks called a beautiful object καλόν quasi καλοῦν, i.e. *calling on* the soul, which receives instantly, and welcomes it as something connatural. Πάλιν οὖν ἀναλαβόντες, λέγωμεν τί δῆτα ἐστὶ τὸ ἐν τοῖς σώμασι καλόν. Πρῶτον ἔστι μὲν γάρ τι καὶ βολῇ τῇ πρώτῃ αἰσθητὸν γινόμενον, καὶ ἡ ψυχὴ ὥσπερ συνεῖσα λέγει, καὶ ἐπιγνοῦσα ἀποδέχεται, καὶ οἷον συναρμόττεται. Πρὸς δὲ τὸ αἰσχρὸν προσβαλοῦσα ἀνίλλεται, καὶ ἀρνεῖται καὶ ἀνανεύει ἐπ᾽ αὐτοῦ οὐ συμφωνοῦσα, καὶ ἀλλοτριουμένη.[19]—PLOTIN: Ennead. I. Lib. 6.

FROM

BIOGRAPHIA LITERARIA

❧

Biographia Literaria was an attempt to carry out two separate projects Coleridge had long in mind. As far back as 1803 he had decided to "write my metaphysical works, as *my life*, & *in* my Life—intermixed with all the other events or history of the mind and fortunes of S. T. Coleridge." Though the autobiographical portions are sketchy and the metaphysical disquisitions only

19. Πάλιν . . . ἀλλοτριουμένη: "To return, then: let us state what physical beauty [the beauty in objects] is. It is something perceived esthetically, in the first impression of it. And the soul, as if understanding it, responds to it, and is drawn, as it were, into harmony with it. On the other hand, in confronting the ugly, the soul shrinks back, refusing and rejecting it because of the lack of harmony with it, and a feeling of estrangement" (trans. Bate, p. 375).

slightly autobiographical, this corresponds roughly to Chapters I–XIII, and the root intention seems to have been a more philosophically oriented equivalent in prose to Wordsworth's *The Prelude*. The second project originated with Wordsworth's Preface to the *Lyrical Ballads* (1800). This had been, Coleridge said, "half a child of my own brain," but in a few years he began to suspect a "radical difference" in his and Wordsworth's opinions, particularly with regard to the proper diction of poetry. To clarify his own position he wrote the extended discussion of poetry and of Wordsworth that makes up Chapters XIV–XXII. The book was composed in 1815 but was not published until 1817. It went unnoticed by the general public and was not reprinted in Coleridge's lifetime.

❧

from CHAPTER IV

During the last year of my residence at Cambridge, I became acquainted with Mr. Wordsworth's first publication entitled "Descriptive Sketches"; and seldom, if ever, was the emergence of an original poetic genius above the literary horizon more evidently announced. In the form, style, and manner of the whole poem, and in the structure of the particular lines and periods, there is an harshness and acerbity connected and combined with words and images all a-glow, which might recall those products of the vegetable world, where gorgeous blossoms rise out of the hard and thorny rind and shell, within which the rich fruit was elaborating. The language was not only peculiar and strong, but at times knotty and contorted, as by its own impatient strength; while the novelty and struggling crowd of images, acting in conjunction with the difficulties of the style, demanded always a greater closeness of attention, than poetry, (at all events, than descriptive poetry) has a right to claim. It not seldom therefore justified the complaint of obscurity. In the following extract I have sometimes fancied, that I saw an emblem of the poem itself, and of the author's genius as it was then displayed.

"'Tis storm; and hid in mist from hour to hour,
All day the floods a deepening murmur pour;
The sky is veiled, and every cheerful sight:
Dark is the region as with coming night;
And yet what frequent bursts of overpowering light!
Triumphant on the bosom of the storm,
Glances the fire-clad eagle's wheeling form;
Eastward, in long perspective glittering, shine
The wood-crowned cliffs that o'er the lake recline;

Wide o'er the Alps a hundred streams unfold,
At once to pillars turn'd that flame with gold;
Behind his sail the peasant strives to shun
The West, that burns like one dilated sun,
Where in a mighty crucible expire
The mountains, glowing hot, like coals of fire."[1]

The poetic PSYCHE, in its process to full development, undergoes as many changes as its Greek name-sake, the butterfly.[2] And it is remarkable how soon genius clears and purifies itself from the faults and errors of its earliest products; faults which, in its earliest compositions, are the more obtrusive and confluent, because as heterogeneous elements, which had only a temporary use, they constitute the very *ferment*, by which themselves are carried off. Or we may compare them to some diseases, which must work on the humours, and be thrown out on the surface, in order to secure the patient from their future recurrence. I was in my twenty-fourth year, when I had the happiness of knowing Mr. Wordsworth personally, and while memory lasts, I shall hardly forget the sudden effect produced on my mind, by his recitation of a manuscript poem, which still remains unpublished,[3] but of which the stanza, and tone of style, were the same as those of the "Female Vagrant," as originally printed in the first volume of the "Lyrical Ballads." There was here no mark of strained thought, or forced diction, no crowd or turbulence of imagery; and, as the poet hath himself well described in his lines "on re-visiting the Wye,"[4] manly reflection, and human associations had given both variety, and an additional interest to natural objects, which in the passion and appetite of the first love they had seemed to him neither to need or permit. The occasional obscurities, which had risen from an imperfect controul over the resources of his native language, had almost wholly disappeared, together with that worse defect of arbitrary and illogical phrases, at once hackneyed, and fantastic, which hold so distinguished a place in the *technique* of ordinary poetry, and will, more or less, alloy the earlier poems of the truest genius, unless the attention has been specifically directed to their worthlessness and incongruity. I did not perceive anything particular

in the mere style of the poem alluded to during its recitation, except indeed such difference as was not separable from the thought and manner; and the Spenserian stanza, which always, more or less, recalls to the reader's mind Spenser's own style, would doubtless have authorized, in my then opinion, a more frequent descent to the phrases of ordinary life, than could without an ill effect have been hazarded in the heroic couplet. It was not however the freedom from false taste, whether as to common defects, or to those more properly his own, which made so unusual an impression on my feelings immediately, and subsequently on my judgement. It was the union of deep feeling with profound thought; the fine balance of truth in observing, with the imaginative faculty in modifying the objects observed; and above all the original gift of spreading the tone, the *atmosphere*, and with it the depth and height of the ideal world around forms, incidents, and situations, of which, for the common view, custom had bedimmed all the lustre, had dried up the sparkle and the dew drops. "To find no contradiction in the union of old and new; to contemplate the ANCIENT of days and all his works with feelings as fresh, as if all had then sprang forth at the first creative fiat; characterizes the mind that feels the riddle of the world, and may help to unravel it. To carry on the feelings of childhood into the powers of manhood; to combine the child's sense of wonder and novelty with the appearances, which every day for perhaps forty years had rendered familiar;

'With sun and moon and stars throughout the year,
And man and woman;'[5]

this is the character and privilege of genius, and one of the marks which distinguish genius from talents. And therefore is it the prime merit of genius and its most unequivocal mode of manifestation, so to represent familiar objects as to awaken in the minds of others a kindred feeling concerning them and that freshness of sensation which is the constant accompaniment of mental, no less than of bodily, convalescence. Who has not a thousand times seen snow fall on water? Who has not watched it with a new feeling, from the time that he has read Burns' comparison of sensual pleasure

'To snow that falls upon a river
A moment white—then gone for ever!'[6]

BIOGRAPHIA LITERARIA. CHAPTER IV. **1.** " **'Tis . . . fire**": from Wordsworth's *Descriptive Sketches*, ll. 332–47. **2. butterfly:** Psyche (the soul) was traditionally represented thus. Cf. Keats's "Ode to Psyche," below. **3. poem . . . unpublished:** "Guilt and Sorrow." A part of it was published as "The Female Vagrant" in the *Lyrical Ballads* of 1798. **4. lines . . . Wye:** "Tintern Abbey."

5. '**With . . . woman**': from Milton's sonnet "To Mr. Cyriack Skinner Upon His Blindness." **6.** '**To . . . ever!**': from Burns's "Tam o' Shanter," ll. 61–62.

In poems, equally as in philosophic disquisitions, genius produces the strongest impressions of novelty, while it rescues the most admitted truths from the impotence caused by the very circumstance of their universal admission. Truths of all others the most awful and mysterious, yet being at the same time of universal interest, are too often considered as *so* true, that they lose all the life and efficiency of truth, and lie bed-ridden in the dormitory of the soul, side by side with the most despised and exploded errors."—THE FRIEND, p. 76, No. 5.

This excellence, which in all Mr. Wordsworth's writings is more or less predominant, and which constitutes the character of his mind, I no sooner felt, than I sought to understand. Repeated meditations led me first to suspect, (and a more intimate analysis of the human faculties, their appropriate marks, functions, and effects matured my conjecture into full conviction,) that fancy and imagination were two distinct and widely different faculties, instead of being, according to the general belief, either two names with one meaning, or, at furthest, the lower and higher degree of one and the same power. It is not, I own, easy to conceive a more opposite translation of the Greek *Phantasia* than the Latin *Imaginatio*; but it is equally true that in all societies there exists an instinct of growth, a certain collective, unconscious good sense working progressively to desynonymize[7] those words originally of the same meaning, which the conflux of dialects had supplied to the more homogeneous languages, as the Greek and German: and which the same cause, joined with accidents of translation from original works of different countries, occasion in mixt languages like our own. The first and most important point to be proved is, that two conceptions perfectly distinct are confused under one and the same word, and (this done)

to appropriate that word exclusively to one meaning, and the synonyme (should there be one) to the other. But if (as will be often the case in the arts and sciences) no synonyme exists, we must either invent or borrow a word. In the present instance the appropriation has already begun, and been legitimated in the derivative adjective: Milton had a highly *imaginative*, Cowley a very *fanciful* mind. If therefore I should succeed in establishing the actual existences of two faculties generally different, the nomenclature would be at once determined. To the faculty by which I had characterized Milton, we should confine the term *imagination*; while the other would be contradistinguished as *fancy*. Now were it once fully ascertained, that this division is no less grounded in nature, than that of delirium from mania, or Otway's

"Lutes, lobsters, seas of milk, and ships of amber,"

from Shakespear's

"What! have his daughters brought him to this pass?"[8]

or from the preceding apostrophe to the elements; the theory of the fine arts, and of poetry in particular, could not, I thought, but derive some additional and important light. It would in its immediate effects furnish a torch of guidance to the philosophical critic; and ultimately to the poet himself. In energetic minds, truth soon changes by domestication into power; and from directing in the discrimination and appraisal of the product, becomes influencive in the production. To admire on principle, is the only way to imitate without loss of originality.

It has been already hinted, that metaphysics and psychology have long been my hobby-horse. But to have a hobby-horse, and to be vain of it, are so commonly found together, that they pass almost for the

---

7. **desynonymize:** "This is effected either by giving to the one word a general, and to the other an exclusive use; as 'to put on the back' and 'to indorse'; or by an actual distinction of meanings, as 'naturalist,' and 'physician'; or by difference of relation, as 'I' and 'Me' (each of which the rustics of our different provinces still use in all the cases singular of the first personal pronoun). Even the mere difference, or corruption, in the *pronunciation* of the same word, if it have become general, will produce a new word with a distinct signification; thus 'property' and 'propriety'; the latter of which, even to the time of Charles II, was the *written* word for all the senses of both. Thus too 'mister' and 'master,' both hasty pronunciations of the same word 'magister,' 'mistress,' and 'miss,' 'if' and 'give,' &c. &c. There is a sort of *minim immortal* among the animalcula infusoria which has not naturally either birth, or death, absolute beginning, or

absolute end: for at a certain period a small point appears on its back, which deepens and lengthens till the creature divides into two, and the same process recommences in each of the halves now become integral. This may be a fanciful, but it is by no means a bad emblem of the formation of words, and may facilitate the conception, how immense a nomenclature may be organized from a few simple sounds by rational beings in a social state. For each new application, or excitement of the same sound, will call forth a different sensation, which cannot but affect the pronunciation. The after recollection of the sound, without the same vivid sensation, will modify it still further; till at length all trace of the original likeness is worn away" (C.). 8. **"Lutes . . . pass?":** Thomas Otway's *Venice Preserved*, V. ii. 151; *King Lear*, III. iv. 63.

same. I trust therefore, that there will be more good humour than contempt, in the smile with which the reader chastises my self-complacency, if I confess myself uncertain, whether the satisfaction from the perception of a truth new to myself may not have been rendered more poignant by the conceit, that it would be equally so to the public. There was a time, certainly, in which I took some little credit to myself, in the belief that I had been the first of my countrymen, who had pointed out the diverse meaning of which the two terms were capable, and analyzed the faculties to which they should be appropriated. Mr. W. Taylor's recent volume of synonymes I have not yet seen;[9] but his specification of the terms in question has been clearly shown to be both insufficient and erroneous by Mr. Wordsworth in the Preface added to the late collection of his "Lyrical Ballads and other poems." The explanation which Mr. Wordsworth has himself given will be found to differ from mine, chiefly perhaps, as our objects are different. It could scarcely indeed happen otherwise, from the advantage I have enjoyed of frequent conversation with him on a subject to which a poem of his own first directed my attention, and my conclusions concerning which, he had made more lucid to myself by many happy instances drawn from the operation of natural objects on the mind. But it was Mr. Wordsworth's purpose to consider the influences of fancy and imagination as they are manifested in poetry, and from the different effects to conclude their diversity in kind; while it is my object to investigate the seminal principle, and then from the

kind to deduce the degree. My friend has drawn a masterly sketch of the branches with their *poetic* fruitage. I wish to add the trunk, and even the roots as far as they lift themselves above ground, and are visible to the naked eye of our common consciousness.

Yet even in this attempt I am aware, that I shall be obliged to draw more largely on the reader's attention, than so immethodical a miscellany can authorize; when in such a work (the *Ecclesiastical Polity*) of such a mind as Hooker's, the judicious author, though no less admirable for the perspicuity than for the port and dignity of his language; and though he wrote for men of learning in a learned age; saw nevertheless occasion to anticipate and guard against "complaints of obscurity," as often as he was about to trace his subject "to the highest well-spring and fountain." Which, (continues he) "because men are not accustomed to, the pains we take are more needful a great deal, than acceptable; and the matters we handle, seem by reason of newness (till the mind grow better acquainted with them) dark and intricate." I would gladly therefore spare both myself and others this labor, if I knew how without it to present an intelligible statement of my poetic creed; not as my *opinions*, which weigh for nothing, but as deductions from established premises conveyed in such a form, as is calculated either to effect a fundamental conviction, or to receive a fundamental confutation. If I may dare once more adopt the words of Hooker, "they, unto whom we shall seem tedious, are in no wise injured by us, because it is in their own hands to spare that labor, which they

9. **not yet seen:** "I ought to have added, with the exception of a single sheet which I accidentally met with at the printer's. Even from this scanty specimen, I found it impossible to doubt the talent, or not to admire the ingenuity of the author. That his distinctions were for the greater part unsatisfactory to *my* mind, proves nothing against their accuracy; but it may possibly be serviceable to him, in case of a second edition, if I take this opportunity of suggesting the query; whether he may not have been occasionally misled, by having assumed, as to me he appeared to have done, the non-existence of *any* absolute synonymes in our language? Now I cannot but think, that there are many which remain for our posterity to distinguish and appropriate, and which I regard as so much reversionary wealth in our mother-tongue. When two distinct meanings are confounded under one or more words, (and such must be the case, as sure as our knowledge is progressive and of course imperfect) erroneous consequences will be drawn, and what is true in one sense of the word will be affirmed as true in toto. Men of research, startled by the consequences, seek in the things themselves (whether in or out of the mind) for a knowledge of the fact, and having discovered the difference,

remove the equivocation either by the substitution of a new word, or by the appropriation of one of the two or more words, that had before been used promiscuously. When this distinction has been so naturalized and of such general currency that the language itself does as it were *think* for us (like the sliding rule which is the mechanic's safe substitute for arithmetical knowledge) we then say, that it is evident to *common sense*. Common sense, therefore, differs in different ages. What was born and christened in the schools passes by degrees into the world at large, and becomes the property of the market and the tea-table. At least I can discover no other meaning of the term, *common sense*, if it is to convey any specific difference from sense and judgement *in genere*, and where it is not used scholastically for the *universal reason*. Thus in the reign of Charles II, the philosophic world was called to arms by the moral sophisms of Hobbes, and the ablest writers exerted themselves in the detection of an error, which a school-boy would now be able to confute by the mere recollection, that *compulsion* and *obligation* conveyed two ideas perfectly disparate, and that what appertained to the one, had been falsely transferred to the other by a mere confusion of terms" (C.).

are not willing to endure." Those at least, let me be permitted to add, who have taken so much pains to render me ridiculous for a perversion of taste, and have supported the charge by attributing strange notions to me on no other authority than their own conjectures, owe it to themselves as well as to me not to refuse their attention to my own statement of the theory, which I *do* acknowledge; or shrink from the trouble of examining the grounds on which I rest it, or the arguments which I offer in its justification.

from CHAPTER XIII

The IMAGINATION then, I consider either as primary, or secondary. The primary IMAGINATION I hold to be the living Power and prime Agent of all human Perception, and as a repetition in the finite mind of the eternal act of creation in the infinite I AM. The secondary Imagination I consider as an echo of the former, co-existing with the conscious will, yet still as identical with the primary in the *kind* of its agency, and differing only in *degree*, and in the *mode* of its operation. It dissolves, diffuses, dissipates, in order to recreate; or where this process is rendered impossible, yet still at all events it struggles to idealize and to unify. It is essentially *vital*, even as all objects (*as* objects) are essentially fixed and dead.

FANCY, on the contrary, has no other counters to play with, but fixities and definites. The Fancy is indeed no other than a mode of Memory emancipated from the order of time and space; while it is blended with, and modified by that empirical phenomenon of the will, which we express by the word CHOICE. But equally with the ordinary memory the Fancy must receive all its materials ready made from the law of association.

CHAPTER XIV

Occasion of the Lyrical Ballads, and the objects originally proposed—Preface to the second edition—The ensuing controversy, its causes and acrimony—Philosophic definitions of a poem and poetry with scholia.

During the first year[1] that Mr. Wordsworth and I were neighbours, our conversations turned frequently on the two cardinal points of poetry, the power of

CHAPTER XIV. I. **first year:** 1797.

exciting the sympathy of the reader by a faithful adherence to the truth of nature, and the power of giving the interest of novelty by the modifying colors of imagination. The sudden charm, which accidents of light and shade, which moon-light or sun-set diffused over a known and familiar landscape, appeared to represent the practicability of combining both. These are the poetry of nature. The thought suggested itself (to which of us I do not recollect) that a series of poems might be composed of two sorts. In the one, the incidents and agents were to be, in part at least, supernatural; and the excellence aimed at was to consist in the interesting of the affections by the dramatic truth of such emotions, as would naturally accompany such situations, supposing them real. And real in *this* sense they have been to every human being who, from whatever source of delusion, has at any time believed himself under supernatural agency. For the second class, subjects were to be chosen from ordinary life; the characters and incidents were to be such, as will be found in every village and its vicinity, where there is a meditative and feeling mind to seek after them, or to notice them, when they present themselves.

In this idea originated the plan of the "Lyrical Ballads"; in which it was agreed, that my endeavours should be directed to persons and characters supernatural, or at least romantic; yet so as to transfer from our inward nature a human interest and a semblance of truth sufficient to procure for these shadows of imagination that willing suspension of disbelief for the moment, which constitutes poetic faith. Mr. Wordsworth, on the other hand, was to propose to himself as his object, to give the charm of novelty to things of every day, and to excite a feeling analogous to the supernatural, by awakening the mind's attention from the lethargy of custom, and directing it to the loveliness and the wonders of the world before us; an inexhaustible treasure, but for which, in consequence of the film of familiarity and selfish solicitude we have eyes, yet see not, ears that hear not, and hearts that neither feel nor understand.

With this view I wrote "The Ancient Mariner," and was preparing among other poems, "The Dark Ladie," and the "Christabel," in which I should have more nearly realized my ideal, than I had done in my first attempt. But Mr. Wordsworth's industry had proved so much more successful, and the number of his poems so much greater, that my compositions, instead of forming a balance, appeared rather an interpolation of heterogeneous matter. Mr. Wordsworth added two or three poems written in his own character, in the impassioned, lofty, and sustained diction, which is characteristic of his genius. In this form the "Lyrical Ballads" were published; and were presented by him, as an *experiment*, whether subjects, which from their nature rejected the usual ornaments and extra-colloquial style of poems in general, might not be so managed in the language of ordinary life as to produce the pleasurable interest, which it is the peculiar business of poetry to impart. To the second edition he added a preface of considerable length; in which, notwithstanding some passages of apparently a contrary import, he was understood to contend for the extension of this style to poetry of all kinds, and to reject as vicious and indefensible all phrases and forms of style that were not included in what he (unfortunately, I think, adopting an equivocal expression) called the language of *real* life. From this preface, prefixed to poems in which it was impossible to deny the presence of original genius, however mistaken its direction might be deemed, arose the whole long-continued controversy.[2] For from the conjunction of perceived power with supposed heresy I explain the inveteracy and in some instances, I grieve to say, the acrimonious passions, with which the controversy has been conducted by the assailants.

Had Mr. Wordsworth's poems been the silly, the childish things, which they were for a long time described as being; had they been really distinguished from the compositions of other poets merely by meanness of language and inanity of thought; had they indeed contained nothing more than what is found in the parodies and pretended imitations of them; they must have sunk at once, a dead weight, into the slough of oblivion, and have dragged the preface along with them. But year after year increased the number of Mr. Wordsworth's admirers. They were found too not in the lower classes of the reading public, but chiefly among young men of strong sensibility and meditative minds; and their admiration (inflamed perhaps in some degree by opposition) was distinguished by its intensity, I might almost say, by its *religious* fervor. These facts, and the intellectual energy of the author, which was more or less consciously felt, where it was outwardly and even boisterously denied, meeting with sentiments of aversion to his opinions, and of alarm at their consequences, produced

an eddy of criticism, which would of itself have borne up the poems by the violence, with which it whirled them round and round. With many parts of this preface, in the sense attributed to them, and which the words undoubtedly seem to authorize, I never concurred; but on the contrary objected to them as erroneous in principle, and as contradictory (in appearance at least) both to other parts of the same preface, and to the author's own practice in the greater number of the poems themselves. Mr. Wordsworth in his recent collection has, I find, degraded this prefatory disquisition to the end of his second volume, to be read or not at the reader's choice. But he has not, as far as I can discover, announced any change in his poetic creed. At all events, considering it as the source of a controversy, in which I have been honored more than I deserve by the frequent conjunction of my name with his, I think it expedient to declare once for all, in what points I coincide with his opinions, and in what points I altogether differ. But in order to render myself intelligible I must previously, in as few words as possible, explain my ideas, first, of a POEM; and secondly, of POETRY itself, in *kind*, and in *essence*.

The office of philosophical *disquisition* consists in just *distinction;* while it is the privilege of the philosopher to preserve himself constantly aware, that distinction is not division. In order to obtain adequate notions of any truth, we must intellectually separate its distinguishable parts; and this is the technical *process* of philosophy. But having so done, we must then restore them in our conceptions to the unity, in which they actually co-exist; and this is the *result* of philosophy. A poem contains the same elements as a prose composition; the difference therefore must consist in a different combination of them, in consequence of a different object being proposed. According to the difference of the object will be the difference of the combination. It is possible, that the object may be merely to facilitate the recollection of any given facts or observations by artificial arrangement; and the composition will be a poem, merely because it is distinguished from prose by metre, or by rhyme, or by both conjointly. In this, the lowest sense, a man might attribute the name of a poem to the well-known enumeration of the days in the several months;

"Thirty days hath September,
April, June, and November," &c.

and others of the same class and purpose. And as a particular pleasure is found in anticipating the

---

2. **controversy:** over the theory of poetic language developed in Wordsworth's Preface, above.

recurrence of sounds and quantities, all compositions that have this charm super-added, whatever be their contents, *may* be entitled poems.

So much for the superficial *form*. A difference of object and contents supplies an additional ground of distinction. The immediate purpose may be the communication of truths; either of truth absolute and demonstrable, as in works of science; or of facts experienced and recorded, as in history. Pleasure, and that of the highest and most permanent kind, may *result* from the *attainment* of the end; but it is not itself the immediate end. In other works the communication of pleasure may be the immediate purpose; and though truth, either moral or intellectual, ought to be the *ultimate* end, yet this will distinguish the character of the author, not the class to which the work belongs. Blest indeed is that state of society, in which the immediate purpose would be baffled by the perversion of the proper ultimate end; in which no charm of diction or imagery could exempt the Bathyllus even of an Anacreon, or the Alexis of Virgil,[3] from disgust and aversion!

But the communication of pleasure may be the immediate object of a work not metrically composed; and that object may have been in a high degree attained, as in novels and romances. Would then the mere superaddition of metre, with or without rhyme, entitle *these* to the name of poems? The answer is, that nothing can permanently please, which does not contain in itself the reason why it is so, and not otherwise. If metre be superadded, all other parts must be made consonant with it. They must be such, as to justify the perpetual and distinct attention to each part, which an exact correspondent recurrence of accent and sound are calculated to excite. The final definition then, so deduced, may be thus worded. A poem is that species of composition, which is opposed to works of science, by proposing for its *immediate* object pleasure, not truth; and from all other species (having *this* object in common with it) it is discriminated by proposing to itself such delight from the *whole*, as is compatible with a distinct gratification from each component *part*.

Controversy is not seldom excited in consequence of the disputants attaching each a different meaning to the same word; and in few instances has this been more striking, than in disputes concerning the present subject. If a man chooses to call every composition a poem, which is rhyme, or measure, or both, I must leave his opinion uncontroverted. The distinction is at least competent to characterize the writer's intention. If it were subjoined, that the whole is likewise entertaining or affecting, as a tale, or as a series of interesting reflections, I of course admit this as another fit ingredient of a poem, and an additional merit. But if the definition sought for be that of a *legitimate* poem, I answer, it must be one, the parts of which mutually support and explain each other; all in their proportion harmonizing with, and supporting the purpose and known influences of metrical arrangement. The philosophic critics of all ages coincide with the ultimate judgement of all countries, in equally denying the praises of a just poem, on the one hand, to a series of striking lines or distiches, each of which, absorbing the whole attention of the reader to itself, disjoins it from its context, and makes it a separate whole, instead of an harmonizing part; and on the other hand, to an unsustained composition, from which the reader collects rapidly the general result, unattracted by the component parts. The reader should be carried forward, not merely or chiefly by the mechanical impulse of curiosity, or by a restless desire to arrive at the final solution; but by the pleasureable activity of mind excited by the attractions of the journey itself. Like the motion of a serpent, which the Egyptians made the emblem of intellectual power; or like the path of sound through the air; at every step he pauses and half recedes, and from the retrogressive movement collects the force which again carries him onward. "Præcipitandus est *liber* spiritus,"[4] says Petronius Arbiter most happily. The epithet, *liber*, here balances the preceding verb; and it is not easy to conceive more meaning condensed in fewer words.

But if this should be admitted as a satisfactory character of a poem, we have still to seek for a definition of poetry. The writings of PLATO, and Bishop TAYLOR, and the "Theoria Sacra" of BURNET,[5] furnish undeniable proofs that poetry of the highest kind may exist without metre, and even without the contra-distinguishing objects of a poem. The first chapter of Isaiah (indeed a very large portion of the whole book) is poetry in

---

3. **Bathyllus . . . Virgil:** Bathyllus was a beautiful youth celebrated by the Greek lyricist Anacreon; Alexis, a boy loved by the shepherd Corydon in Vergil's second *Eclogue*.

4. **"Præcipitandus . . . spiritus":** "The free spirit must be hurled onward" (*Satyricon*, in the Loeb Classical Library, p. 252). Petronius was a Roman satirist of the first century A.D. 5. **Burnet:** the *Telluris Theoria Sacra* (1681–89)—"The Sacred Theory of the Earth"—by Thomas Burnet.

the most emphatic sense; yet it would be not less irrational than strange to assert, that pleasure, and not truth, was the immediate object of the prophet. In short, whatever *specific* import we attach to the word, poetry, there will be found involved in it, as a necessary consequence, that a poem of any length neither can be, or ought to be, all poetry. Yet if an harmonious whole is to be produced, the remaining parts must be preserved *in keeping* with the poetry; and this can be no otherwise effected than by such a studied selection and artificial arrangement, as will partake of *one,* though not a *peculiar* property of poetry. And this again can be no other than the property of exciting a more continuous and equal attention than the language of prose aims at, whether colloquial or written.

My own conclusions on the nature of poetry, in the strictest use of the word, have been in part anticipated in the preceding disquisition on the fancy and imagination. What is poetry? is so nearly the same question with, what is a poet? that the answer to the one is involved in the solution of the other. For it is a distinction resulting from the poetic genius itself, which sustains and modifies the images, thoughts, and emotions of the poet's own mind.

The poet, described in *ideal* perfection, brings the whole soul of man into activity, with the subordination of its faculties to each other, according to their relative worth and dignity. He diffuses a tone and spirit of unity, that blends, and (as it were) *fuses,* each into each, by that synthetic and magical power, to which we have exclusively appropriated the name of imagination. This power, first put in action by the will and understanding, and retained under their irremissive, though gentle and unnoticed, controul (*laxis effertur habenis*)[6] reveals itself in the balance or reconciliation of opposite or discordant qualities: of sameness, with difference; of the general, with the concrete; the idea, with the image; the individual, with the representative; the sense of novelty and freshness, with old and familiar objects; a more than usual state of emotion, with more than usual order; judgement ever awake and steady self-possession, with enthusiasm and feeling profound or vehement; and while it blends and harmonizes the natural and the artificial, still subordinates art to nature; the manner to the matter; and our admiration of the poet to our sympathy with the poetry. "Doubtless," as Sir John Davies observes of the soul (and his

words may with slight alteration be applied, and even more appropriately, to the poetic IMAGINATION)

Doubtless this could not be, but that she turns
   Bodies to spirit by sublimation strange,
As fire converts to fire the things it burns,
   As we our food into our nature change.

From their gross matter she abstracts their forms,
   And draws a kind of quintessence from things;
Which to her proper nature she transforms,
   To bear them light on her celestial wings.

Thus does she, when from individual states
   She doth abstract the universal kinds;
Which then re-clothed in divers names and fates
   Steal access through our senses to our minds.[7]

Finally, GOOD SENSE[8] is the BODY of poetic genius, FANCY its DRAPERY, MOTION its LIFE, and IMAGINATION the soul that is everywhere, and in each; and forms all into one graceful and intelligent whole.

CHAPTER XV

The specific symptoms of poetic power elucidated in a critical analysis of Shakespeare's Venus and Adonis, and Lucrece.

In the application of these principles to purposes of practical criticism as employed in the appraisal of works more or less imperfect, I have endeavoured to discover what the qualities in a poem are, which may be deemed promises and specific symptoms of poetic power, as distinguished from general talent determined to poetic composition by accidental motives, by an act of the will, rather than by the inspiration of a genial and productive nature. In this investigation, I could not, I thought, do better, than keep before me the earliest work of the greatest genius, that perhaps human nature has yet produced, our *myriad-minded* Shakespeare. I mean the "Venus and Adonis," and the "Lucrece"; works which give at once strong

6. **laxis . . . habenis:** "Driven on with loose reins" (Vergil, *Georgics,* II. 364).

7. **Doubtless . . . minds:** "On the Soul of Man," Sect. IV.
8. **good sense:** Cf. Coleridge's letter to Lady Beaumont (June, 1814): "The sum total of all intellectual excellence is good sense and method. When these have passed into the instinctive readiness of habit, when the wheel revolves so rapidly that we cannot see it revolve at all, then we call the combination Genius. But in all modes alike, and in all professions, the two sole component parts even of *Genius,* are GOOD SENSE and METHOD."

promises of the strength, and yet obvious proofs of the immaturity, of his genius. From these I abstracted the following marks, as characteristics of original poetic genius in general.

1. In the "Venus and Adonis," the first and most obvious excellence is the perfect sweetness of the versification; its adaptation to the subject; and the power displayed in varying the march of the words without passing into a loftier and more majestic rhythm than was demanded by the thoughts, or permitted by the propriety of preserving a sense of melody predominant. The delight in richness and sweetness of sound, even to a faulty excess, if it be evidently original, and not the result of an easily imitable mechanism, I regard as a highly favourable promise in the compositions of a young man. "The man that hath not music in his soul"[1] can indeed never be a genuine poet. Imagery (even taken from nature, much more when transplanted from books, as travels, voyages, and works of natural history); affecting incidents; just thoughts; interesting personal or domestic feelings; and with these the art of their combination or intertexture in the form of a poem; may all by incessant effort be acquired as a trade, by a man of talents and much reading, who, as I once before observed, has mistaken an intense desire of poetic reputation for a natural poetic genius; the love of the arbitrary end for a possession of the peculiar means. But the sense of musical delight, with the power of producing it, is a gift of imagination; and this together with the power of reducing multitude into unity of effect, and modifying a series of thoughts by some one predominant thought or feeling, may be cultivated and improved, but can never be learned. It is in these that "poeta nascitur non fit."[2]

2. A second promise of genius is the choice of subjects very remote from the private interests and circumstances of the writer himself. At least I have found, that where the subject is taken immediately from the author's personal sensations and experiences, the excellence of a particular poem is but an equivocal mark, and often a fallacious pledge, of genuine poetic power. We may perhaps remember the tale of the statuary, who had acquired considerable reputation for the legs of his goddesses, though the rest of the statue accorded but indifferently with ideal beauty; till his wife, elated by her husband's praises, modestly acknowledged that she herself had been his constant model. In the "Venus and Adonis" this proof of poetic power exists even to excess. It is throughout as if a superior spirit more intuitive, more intimately conscious, even than the characters themselves, not only of every outward look and act, but of the flux and reflux of the mind in all its subtlest thoughts and feelings, were placing the whole before our view; himself meanwhile unparticipating in the passions, and actuated only by that pleasureable excitement, which had resulted from the energetic fervor of his own spirit in so vividly exhibiting, what it had so accurately and profoundly contemplated. I think, I should have conjectured from these poems, that even then the great instinct, which impelled the poet to the drama, was secretly working in him, prompting him by a series and never broken chain of imagery, always vivid and, because unbroken, often minute; by the highest effort of the picturesque in words, of which words are capable, higher perhaps than was ever realized by any other poet, even Dante not excepted; to provide a substitute for that visual language, that constant intervention and running comment by tone, look and gesture, which in his dramatic works he was entitled to expect from the players. His "Venus and Adonis" seem at once the characters themselves, and the whole representation of those characters by the most consummate actors. You seem to be told nothing, but to see and hear everything. Hence it is, that from the perpetual activity of attention required on the part of the reader; from the rapid flow, the quick change, and the playful nature of the thoughts and images; and above all from the alienation, and, if I may hazard such an expression, the utter *aloofness* of the poet's own feelings, from those of which he is at once the painter and the analyst; that though the very subject cannot but detract from the pleasure of a delicate mind, yet never was poem less dangerous on a moral account. Instead of doing as Ariosto,[3] and as, still more offensively, Wieland[4] has done, instead of degrading and deforming passion into appetite, the trials of love into the struggles of concupiscence; Shakespeare has here represented the animal impulse itself, so as to preclude all sympathy with it, by dissipating the reader's notice among the thousand outward images, and now beautiful, now fanciful circumstances, which form its dresses and its scenery;

CHAPTER XV. 1. "The . . . soul": See *The Merchant of Venice*, V. i. 83. 2. "poeta . . . fit": "A poet is born, not made."

3. **Ariosto:** Ludovico Ariosto, 1474–1533, Italian poet.
4. **Wieland:** C. M. Wieland, 1733–1813, German poet.

or by diverting our attention from the main subject by those frequent witty or profound reflections, which the poet's ever active mind has deduced from, or connected with, the imagery and the incidents. The reader is forced into too much action to sympathize with the merely passive of our nature. As little can a mind thus roused and awakened be brooded on by mean and indistinct emotion, as the low, lazy mist can creep upon the surface of a lake, while a strong gale is driving it onward in waves and billows.

3. It has been before observed that images, however beautiful, though faithfully copied from nature, and as accurately represented in words, do not of themselves characterize the poet. They become proofs of original genius only as far as they are modified by a predominant passion; or by associated thoughts or images awakened by that passion; or when they have the effect of reducing multitude to unity, or succession to an instant; or lastly, when a human and intellectual life is transferred to them from the poet's own spirit,

"Which shoots its being through earth, sea, and air."[5]

In the two following lines for instance, there is nothing objectionable, nothing which would preclude them from forming, in their proper place, part of a descriptive poem:

"Behold yon row of pines, that shorn and bow'd
Bend from the sea-blast, seen at twilight eve."

But with a small alteration of rhythm, the same words would be equally in their place in a book of topography, or in a descriptive tour. The same image will rise into semblance of poetry if thus conveyed:

"Yon row of bleak and visionary pines,
By twilight glimpse discerned, mark! how they flee
From the fierce sea-blast, all their tresses wild
Streaming before them."

I have given this as an illustration, by no means as an instance, of that particular excellence which I had in view, and in which Shakespeare even in his earliest, as in his latest, works surpasses all other poets. It is by this, that he still gives a dignity and a passion to the objects which he presents. Unaided by any previous

excitement, they burst upon us at once in life and in power.

"Full many a glorious morning have I seen
Flatter the mountain tops with sovereign eye."

Shakespeare, Sonnet 33rd.

"Not mine own fears, nor the prophetic soul
Of the wide world dreaming on things to come—

. . .

The mortal moon hath her eclipse endur'd,
And the sad augurs mock their own presage;
Incertainties now crown themselves assur'd,
And Peace proclaims olives of endless age.
Now with the drops of this most balmy time
My Love looks fresh, and DEATH to me subscribes!
Since spite of him, I'll live in this poor rhyme,
While he insults o'er dull and speechless tribes.
And thou in this shalt find thy monument,
When tyrants' crests, and tombs of brass are spent."

Sonnet 107.

As of higher worth, so doubtless still more characteristic of poetic genius does the imagery become, when it moulds and colors itself to the circumstances, passion, or character, present and foremost in the mind. For unrivalled instances of this excellence, the reader's own memory will refer him to the LEAR, OTHELLO, in short to which not of the "*great, ever living, dead man's*" dramatic works? "Inopem me copia fecit."[6] How true it is to nature, he has himself finely expressed in the instance of love in Sonnet 98.

"From you have I been absent in the spring,
When proud pied April drest in all its trim
Hath put a spirit of youth in every thing,
That heavy Saturn laugh'd and leap'd with him.
Yet nor the lays of birds, nor the sweet smell
Of different flowers in odour and in hue,
Could make me any summer's story tell,
Or from their proud lap pluck them, where they grew:
Nor did I wonder at the lilies white,
Nor praise the deep vermilion in the rose;
They were, tho' sweet, but figures of delight,
Drawn after you, you pattern of all those.
Yet seem'd it winter still, and, you away,
As with your shadow I with these did play!"

5. **"Which . . . air"**: from Coleridge's "France: An Ode," l. 103.

6. **"Inopem . . . fecit"**: "Abundance has made me poor."

Scarcely less sure, or if a less valuable, not less indispensable mark

Γονίμου μὲν ποιητοῦ——
——ὅστις ῥῆμα γενναῖον λάκοι,[7]

will the imagery supply, when, with more than the power of the painter, the poet gives us the liveliest image of succession with the feeling of simultaneousness!

"With this, he breaketh from the sweet embrace
Of those fair arms, that held him to her heart,
And homeward through the dark lawns runs apace:
*Look! how a bright star shooteth from the sky,
So glides he in the night from Venus' eye.*"[8]

4. The last character I shall mention, which would prove indeed but little, except as taken conjointly with the former; yet without which the former could scarce exist in a high degree, and (even if this were possible) would give promises only of transitory flashes and a meteoric power; is DEPTH, and ENERGY of THOUGHT. No man was ever yet a great poet, without being at the same time a profound philosopher. For poetry is the blossom and the fragrancy of all human knowledge, human thoughts, human passions, emotions, language. In Shakespeare's *poems* the creative power and the intellectual energy wrestle as in a war embrace. Each in its excess of strength seems to threaten the extinction of the other. At length in the DRAMA they were reconciled, and fought each with its shield before the breast of the other. Or like two rapid streams, that, at their first meeting within narrow and rocky banks, mutually strive to repel each other and intermix reluctantly and in tumult; but soon finding a wider channel and more yielding shores blend, and dilate, and flow on in one current and with one voice. The "Venus and Adonis" did not perhaps allow the display of the deeper passions. But the story of Lucretia seems to favor and even demand their intensest workings. And yet we find in *Shakespeare's* management of the tale neither pathos, nor any other *dramatic* quality. There is the same minute and faithful imagery as in the former poem, in the same vivid colors, inspirited by the same impetuous vigor of thought, and diverging and contracting with the same activity of the assimilative and of the modifying faculties; and with a yet larger display, a yet wider

range of knowledge and reflection; and lastly, with the same perfect dominion, often *domination*, over the whole world of language. What then shall we say? even this; that Shakespeare, no mere child of nature; no automaton of genius; no passive vehicle of inspiration possessed by the spirit, not possessing it; first studied patiently, meditated deeply, understood minutely, till knowledge, become habitual and intuitive, wedded itself to his habitual feelings, and at length gave birth to that stupendous power, by which he stands alone, with no equal or second in his own class; to that power which seated him on one of the two glory-smitten summits of the poetic mountain, with Milton as his compeer, not rival. While the former darts himself forth, and passes into all the forms of human character and passion, the one Proteus[9] of the fire and the flood; the other attracts all forms and things to himself, into the unity of his own IDEAL. All things and modes of action shape themselves anew in the being of MILTON; while SHAKESPEARE becomes all things, yet for ever remaining himself. O what great men hath thou not produced, England! my country! truly indeed—

"Must *we* be free or die, who speak the tongue,
Which SHAKESPEARE spake; the faith and morals hold,
Which MILTON held. In every thing we are sprung
Of earth's first blood, have titles manifold!"[10]

Wordsworth

CHAPTER XVII

Examination of the tenets peculiar to Mr. Wordsworth —Rustic life (above all, *low* and rustic life) especially unfavorable to the formation of a human diction— The *best* parts of language the product of philosophers, not of clowns or shepherds—Poetry essentially ideal and generic—The language of Milton as much the language of *real* life, yea, incomparably more so than that of the cottager.

As far then as Mr. Wordsworth in his preface contended, and most ably contended, for a reformation in our poetic diction, as far as he has evinced the truth of passion, and the *dramatic* propriety of those figures and metaphors in the original poets, which, stripped

---

7. Γονίμου . . . λάκοι: "You'll never find a true poet who would speak a noble word" (Aristophanes, *Frogs*, ll. 96–97). 8. "With . . . eye": *Venus and Adonis*, ll. 811–15.

9. **Proteus:** Son of Poseidon, Proteus could take any shape at will. 10. "Must . . . manifold!": from Wordsworth's sonnet, "It Is Not to Be Thought Of," ll. 11–14.

of their justifying reasons, and converted into mere artifices of connection or ornament, constitute the characteristic falsity in the poetic style of the moderns; and as far as he has, with equal acuteness and clearness, pointed out the process by which this change was effected, and the resemblances between that state into which the reader's mind is thrown by the pleasureable confusion of thought from an unaccustomed train of words and images; and that state which is induced by the natural language of empassioned feeling; he undertook a useful task, and deserves all praise, both for the attempt and for the execution. The provocations to this remonstrance in behalf of truth and nature were still of perpetual recurrence before and after the publication of this preface. I cannot likewise but add, that the comparison of such poems of merit, as have been given to the public within the last ten or twelve years, with the majority of those produced previously to the appearance of that preface, leave no doubt on my mind, that Mr. Wordsworth is fully justified in believing his efforts to have been by no means ineffectual. Not only in the verses of those who have professed their admiration of his genius, but even of those who have distinguished themselves by hostility to his theory, and depreciation of his writings, are the impressions of his principles plainly visible. It is possible, that with these principles others may have been blended, which are not equally evident; and some which are unsteady and subvertible from the narrowness or imperfection of their basis. But it is more than possible, that these errors of defect or exaggeration, by kindling and feeding the controversy, may have conduced not only to the wider propagation of the accompanying truths, but that, by their frequent presentation to the mind in an excited state, they may have won for them a more permanent and practical result. A man will borrow a part from his opponent the more easily, if he feels himself justified in continuing to reject a part. While there remain important points in which he can still feel himself in the right, in which he still finds firm footing for continued resistance, he will gradually adopt those opinions, which were the least remote from his own convictions, as not less congruous with his own theory than with that which he reprobates. In like manner with a kind of instinctive prudence, he will abandon by little and little his weakest posts, till at length he seems to forget that they had ever belonged to him, or affects to consider them at most as accidental and "petty annexments," the removal of which leaves the citadel unhurt and unendangered.

My own differences from certain supposed parts of Mr. Wordsworth's theory ground themselves on the assumption, that his words had been rightly interpreted, as purporting that the proper diction for poetry in general consists altogether in a language taken, with due exceptions, from the mouths of men in real life, a language which actually constitutes the natural conversation of men under the influence of natural feelings. My objection is, first, that in *any* sense this rule is applicable only to *certain* classes of poetry; secondly, that even to these classes it is not applicable, except in such a sense, as hath never by any one (as far as I know or have read) been denied or doubted; and lastly, that as far as, and in that degree in which it is *practicable*, yet as a *rule* it is useless, if not injurious, and therefore either need not, or ought not to be practised. The poet informs his reader, that he had generally chosen *low and rustic* life; but not *as* low and rustic, or in order to repeat that pleasure of doubtful moral effect, which persons of elevated rank and of superior refinement oftentimes derive from a happy *imitation* of the rude unpolished manners and discourse of their inferiors. For the pleasure so derived may be traced to three exciting causes. The first is the naturalness, in *fact*, of the things represented. The second is the apparent naturalness of the *representation*, as raised and qualified by an imperceptible infusion of the author's own knowledge and talent, which infusion does, indeed, constitute it an *imitation* as distinguished from a mere *copy*.[1] The third cause may be found in the reader's conscious feeling of his superiority awakened by the contrast presented to him; even as for the same purpose the kings and great barons of yore retained sometimes *actual* clowns and fools, but more frequently shrewd and witty fellows in that *character*. These, however, were not Mr. Wordsworth's objects. *He* chose low and rustic life, "because in that condition the essential passions of the heart find a better soil, in which they can attain their maturity, are less under restraint, and speak a plainer and more emphatic language; because in that condition of life our elementary feelings coexist in a state of greater simplicity, and consequently may be more accurately contemplated, and more forcibly communicated; because the manners of rural life germinate from those elementary feelings;

CHAPTER XVII. **I. imitation . . . copy:** For Coleridge's distinction between "imitation" and "copy," see "On Poesy or Art," below.

and from the necessary character of rural occupations are more easily comprehended, and are more durable; and lastly, because in that condition the passions of men are incorporated with the beautiful and permanent forms of nature."

Now it is clear to me, that in the most interesting of the poems, in which the author is more or less dramatic, as "the Brothers," "Michael," "Ruth," "the Mad Mother,"[2] &c., the persons introduced are by no means taken *from low or rustic life* in the common acceptation of those words; and it is not less clear, that the sentiments and language, as far as they can be conceived to have been really transferred from the minds and conversation of such persons, are attributable to causes and circumstances not necessarily connected with "their occupations and abode." The thoughts, feelings, language, and manners of the shepherd-farmers in the vales of Cumberland and Westmoreland, as far as they are actually adopted in those poems, may be accounted for from causes, which will and do produce the same results in *every* state of life, whether in town or country. As the two principal I rank that INDEPENDENCE, which raises a man above servitude, or daily toil for the profit of others, yet not above the necessity of industry and a frugal simplicity of domestic life; and the accompanying unambitious, but solid and religious, EDUCATION, which has rendered few books familiar, but the Bible, and the liturgy or hymn book. To this latter cause, indeed, which is so far *accidental*, that it is the blessing of particular countries and a particular age, not the product of particular places or employments, the poet owes the show of probability, that his personages might really feel, think, and talk with any tolerable resemblance to his representation. It is an excellent remark of Dr. Henry More's,[3] (Enthusiasmus triumphatus, Sec. XXXV.), that "a man of confined education, but of good parts, by constant reading of the Bible will naturally form a more winning and commanding rhetoric than those that are learned; the intermixture of tongues and of artificial phrases debasing *their* style."

It is, moreover, to be considered that to the formation of healthy feelings, and a reflecting mind, *negations*

involve impediments not less formidable than sophistication and vicious intermixture. I am convinced, that for the human soul to prosper in rustic life a certain vantage-ground is pre-requisite. It is not every man that is likely to be improved by a country life or by country labors. Education, or original sensibility, or both, must pre-exist, if the changes, forms, and incidents of nature are to prove a sufficient stimulant. And where these are not sufficient, the mind contracts and hardens by want of stimulants: and the man becomes selfish, sensual, gross, and hard-hearted. Let the management of the POOR LAWS in Liverpool, Manchester, or Bristol be compared with the ordinary dispensation of the poor rates in agricultural villages, where the *farmers* are the overseers and guardians of the poor. If my own experience have not been particularly unfortunate, as well as that of the many respectable country clergymen with whom I have conversed on the subject, the result would engender more than scepticism concerning the desireable influences of low and rustic life in and for itself. Whatever may be concluded on the other side, from the stronger local attachments and enterprising spirit of the Swiss, and other mountaineers, applies to a particular mode of pastoral life, under forms of property that permit and beget manners truly republican, not to rustic life in general, or to the absence of artificial cultivation. On the contrary the mountaineers, whose manners have been so often eulogized, are in general better educated and greater readers than men of equal rank elsewhere. But where this is not the case, as among the peasantry of North Wales, the ancient mountains, with all their terrors and all their glories, are pictures to the blind, and music to the deaf.

I should not have entered so much into detail upon this passage, but here seems to be the point, to which all the lines of difference converge as to their source and centre. (I mean, as far as, and in whatever respect, my poetic creed *does* differ from the doctrines promulged in this preface.) I adopt with full faith the principle of Aristotle,[4] that poetry as poetry is

---

2. **"the Mad Mother"**: in later editions entitled, "Her Eyes Are Wild." 3. **More**: Henry More, 1614–87, was one of the Cambridge Platonists, a group of seventeenth-century English philosophers who strongly influenced Coleridge. They reacted against Puritan rigidity and dogmatism and also the materialism of Hobbes.

4. **principle of Aristotle**: in the *Poetics*, 9. 3–4. "Poetry, therefore, is a more philosophical and higher thing than history: for poetry tends to express the universal, history the particular. By the universal I mean how a person of a certain type will on occasion speak or act, according to the law of probability or necessity; and it is this universality at which poetry aims" (trans. S. H. Butcher, 1895, rev. ed. 1911).

essentially [5] *ideal*, that it avoids and excludes all *accident*; that its apparent individualities of rank, character, or occupation must be *representative* of a class; and that the *persons* of poetry must be clothed with *generic* attributes, with the *common* attributes of the class: not with such as one gifted individual might *possibly* possess, but such as from his situation it is most probable before-hand that he *would* possess. If my premises are right and my deductions legitimate, it follows that there can be no *poetic* medium between the swains of Theocritus and those of an imaginary golden age.

The characters of the vicar and the shepherd-mariner in the poem of "THE BROTHERS," that of the shepherd of Greenhead Ghyll in the "MICHAEL," have all the verisimilitude and representative quality, that the purposes of poetry can require. They are persons of a known and abiding class, and their manners and sentiments the natural product of circumstances common to the class. Take "MICHAEL" for instance:

"An old man stout of heart, and strong of limb:
His bodily frame had been from youth to age
Of an unusual strength: his mind was keen,

5. essentially: "Say not that I am recommending abstractions; for these class-characteristics which constitute the instructiveness of a character, are so modified and particularized in each person of the Shakespearean Drama, that life itself does not excite more distinctly that sense of individuality which belongs to real existence. Paradoxical as it may sound, one of the essential properties of Geometry is not less essential to dramatic excellence; and Aristotle has accordingly required of the poet an involution of the universal in the individual. The chief differences are, that in Geometry it is the universal truth, which is uppermost in the consciousness; in poetry the individual form, in which the truth is clothed. With the ancients, and not less with the elder dramatists of England and France, both comedy and tragedy were considered as kinds of poetry. They neither sought in comedy to make us laugh merely; much less to make us laugh by wry faces, accidents of jargon, *slang* phrases for the day, or the clothing of common-place morals drawn from the shops or mechanic occupations of their characters. Nor did they condescend in tragedy to wheedle away the applause of the spectators, by representing before them facsimiles of their own mean selves in all their existing meanness, or to work on the sluggish sympathies by a pathos not a whit more respectable than the maudlin tears of drunkenness. Their tragic scenes were meant to *affect* us indeed; but yet within the bounds of pleasure, and in union with the activity both of our understanding and imagination. They wished to transport the mind to a sense of its possible greatness, and to implant the germs of that greatness, during the temporary oblivion of the worthless 'thing we are,' and of the peculiar state in which each man *happens* to be, suspending our individual recollections and lulling them to sleep amid the music of nobler thoughts" (C.).

Intense, and frugal, apt for all affairs,
And in his shepherd's calling he was prompt
And watchful more than ordinary men.
Hence he had learnt the meaning of all winds,
Of blasts of every tone; and oftentimes
When others heeded not, he heard the South
Make subterraneous music, like the noise
Of bagpipers on distant Highland hills.
The shepherd, at such warning, of his flock
Bethought him, and he to himself would say,
The winds are now devising work for me!
And truly at all times the storm, that drives
The traveller to a shelter, summon'd him
Up to the mountains. He had been alone
Amid the heart of many thousand mists,
That came to him and left him on the heights.
So liv'd he, till his eightieth year was pass'd.
And grossly that man errs, who should suppose
That the green vallies, and the streams and rocks,
Were things indifferent to the shepherd's thoughts.
Fields, where with chearful spirits he had breath'd
The common air; the hills, which he so oft
Had climb'd with vigorous steps; which had impress'd
So many incidents upon his mind
Of hardship, skill or courage, joy or fear;
Which, like a book, preserved the memory
Of the dumb animals, whom he had sav'd,
Had fed or shelter'd, linking to such acts,
So grateful in themselves, the certainty
Of honorable gain; these fields, these hills
Which were his living being, even more
Than his own blood—what could they less? had laid
Strong hold on his affections, were to him
A pleasureable feeling of blind love,
The pleasure which there is in life itself."

On the other hand, in the poems which are pitched at a lower note, as the "HARRY GILL," "IDIOT BOY," the *feelings* are those of human nature in general; though the poet has judiciously laid the *scene* in the country, in order to place *himself* in the vicinity of interesting images, without the necessity of ascribing a sentimental perception of their beauty to the persons of his drama. In the "Idiot Boy," indeed, the mother's character is not so much a real and native product of a "situation where the essential passions of the heart find a better soil, in which they can attain their maturity and speak a plainer and more emphatic language," as it is an impersonation of an instinct abandoned by judgement. Hence the two following charges seem to me not wholly groundless: at least, they are the only plausible objections, which I have heard to that fine poem. The one is, that the author has not, in the poem itself, taken sufficient care to preclude from the reader's fancy the disgusting images of *ordinary morbid idiocy*, which yet it was by no means his intention to represent. He has even by the "burr, burr, burr," uncounteracted by any preceding description of the boy's beauty, assisted in recalling them. The other is, that the idiocy of the *boy* is so evenly balanced by the folly of the *mother*, as to present to the general reader rather a laughable burlesque on the blindness of anile dotage, than an analytic display of maternal affection in its ordinary workings.

In the "Thorn" the poet himself acknowledges in a note the necessity of an introductory poem, in which he should have pourtrayed the character of the person from whom the words of the poem are supposed to proceed: a superstitious man moderately imaginative, of slow faculties and deep feelings, "a captain of a small trading vessel, for example, who, being past the middle age of life, had retired upon an annuity, or small independent income, to some village or country town of which he was not a native, or in which he had not been accustomed to live. Such men having nothing to do become credulous and talkative from indolence." But in a poem, still more in a lyric poem (and the NURSE in Shakespeare's Romeo and Juliet alone prevents me from extending the remark even to dramatic *poetry*, if indeed the Nurse itself can be deemed altogether a case in point) it is not possible to imitate truly a dull and garrulous discourser, without repeating the effects of dullness and garrulity. However this may be, I dare assert, that the parts (and these form the far larger portion of the whole) which might as well or still better have proceeded from the poet's own imagination, and have been spoken in his own character, are those which have given, and which will continue to give, universal delight; and that the passages exclusively appropriate to the supposed narrator, such as the last couplet of the third stanza;[6] the seven last lines of the tenth;[7] and the five following

stanzas, with the exception of the four admirable lines at the commencement of the fourteenth, are felt by many unprejudiced and unsophisticated hearts, as sudden and unpleasant sinkings from the height to which the poet had previously lifted them, and to which he again re-elevates both himself and his reader.

'Tis now some two-and-twenty years
Since she (her name is Martha Ray)
Gave, with a maiden's true good will,
Her company to Stephen Hill;
And she was blithe and gay,
And she was happy, happy still
Whene'er she thought of Stephen Hill.

And they had fix'd the wedding-day,
The morning that must wed them both;
But Stephen to another maid
Had sworn another oath;
And, with this other maid, to church
Unthinking Stephen went—
Poor Martha! on that woeful day
A pang of pitiless dismay
Into her soul was sent;
A fire was kindled in her breast,
Which might not burn itself to rest.

They say, full six months after this,
While yet the summer leaves were green,
She to the mountain-top would go,
And there was often seen.
'Tis said a child was in her womb,
As now to any eye was plain;
She was with child, and she was mad;
Yet often she was sober sad
From her exceeding pain.
Oh me! ten thousand times I'd rather
That he had died, that cruel father!

. . .

Last Christmas when we talked of this,
Old farmer Simpson did maintain,
That in her womb the infant wrought
About its mother's heart, and brought
Her senses back again:
And, when at last her time drew near,
Her looks were calm, her senses clear.
No more I know, I wish I did,
And I would tell it all to you:
For what became of this poor child
There's none that ever knew:
And if a child was born or no,
There's no one that could ever tell;
And if 'twas born alive or dead,
There's no one knows, as I have said:
But some remember well,
That Martha Ray about this time
Would up the mountain often climb"   (C.).

---

6. **couplet . . . stanza:** "I've measured it from side to side; 'Tis three feet long, and two feet wide" (C.). **7. tenth:**

"Nay, rack your brain—'tis all in vain,
I'll tell you every thing I know;
But to the Thorn, and to the Pond
Which is a little step beyond,
I wish that you would go:
Perhaps when you are at the place,
You something of her tale may trace.
I'll give you the best help I can:
Before you up the mountain go,
Up to the dreary mountain-top,
I'll tell you all I know.

If then I am compelled to doubt the theory, by which the choice of *characters* was to be directed, not only *à priori*, from grounds of reason, but both from the few instances in which the poet himself *need* be supposed to have been governed by it, and from the comparative inferiority of those instances; still more must I hesitate in my assent to the sentence which immediately follows the former citation; and which I can neither admit as particular fact, or as general rule. "The language too of these men is adopted (purified indeed from what appear to be its real defects, from all lasting and rational causes of dislike or disgust) because such men hourly communicate with the best objects from which the best part of language is originally derived; and because, from their rank in society and the sameness and narrow circle of their intercourse, being less under the action of social vanity, they convey their feelings and notions in simple and unelaborated expressions." To this I reply; that a rustic's language, purified from all provincialism and grossness, and so far reconstructed as to be made consistent with the rules of grammar (which are in essence no other than the laws of universal logic, applied to psychological materials) will not differ from the language of any other man of common-sense, however learned or refined he may be, except as far as the notions, which the rustic has to convey, are fewer and more indiscriminate. This will become still clearer, if we add the consideration (equally important though less obvious) that the rustic, from the more imperfect developement of his faculties, and from the lower state of their cultivation, aims almost solely to convey *insulated facts*, either those of his scanty experience or his traditional belief; while the educated man chiefly seeks to discover and express those *connections* of things, or those relative *bearings* of fact to fact, from which some more or less general law is deducible. For *facts* are valuable to a wise man, chiefly as they lead to the discovery of the indwelling *law*, which is the true *being* of things, the sole solution of their modes of existence, and in the knowledge of which consists our dignity and our power.

As little can I agree with the assertion, that from the objects with which the rustic hourly communicates the best part of language is formed. For first, if to communicate with an object implies such an acquaintance with it, as renders it capable of being discriminately reflected on; the distinct knowledge of an uneducated rustic would furnish a very scanty vocabulary. The few things, and modes of action, requisite for his

bodily conveniences, would alone be individualized; while all the rest of nature would be expressed by a small number of confused general terms. Secondly, I deny that the words and combinations of words derived from the objects, with which the rustic is familiar, whether with distinct or confused knowledge, can be justly said to form the *best* part of language. It is more than probable, that many classes of the brute creation possess discriminating sounds, by which they can convey to each other notices of such objects as concern their food, shelter, or safety. Yet we hesitate to call the aggregate of such sounds a language, otherwise than metaphorically. The best part of human language, properly so called, is derived from reflection on the acts of the mind itself. It is formed by a voluntary appropriation of fixed symbols to internal acts, to processes and results of imagination, the greater part of which have no place in the consciousness of uneducated man; though in civilized society, by imitation and passive remembrance of what they hear from their religious instructors and other superiors, the most uneducated share in the harvest which they neither sowed or reaped. If the history of the phrases in hourly currency among our peasants were traced, a person not previously aware of the fact would be surprised at finding so large a number, which three or four centuries ago were the exclusive property of the universities and the schools; and, at the commencement of the Reformation, had been transferred from the school to the pulpit, and thus gradually passed into common life. The extreme difficulty, and often the impossibility, of finding words for the simplest moral and intellectual processes in the languages of uncivilized tribes has proved perhaps the weightiest obstacle to the progress of our most zealous and adroit missionaries. Yet these tribes are surrounded by the same nature as our peasants are; but in still more impressive forms; and they are, moreover, obliged to *particularize* many more of them. When, therefore, Mr. Wordsworth adds, "accordingly, such a language" (meaning, as before, the language of rustic life purified from provincialism) "arising out of repeated experience and regular feelings, is a more permanent, and a far more philosophical language, than that which is frequently substituted for it by poets, who think they are conferring honor upon themselves and their art in proportion as they indulge in arbitrary and capricious habits of expression:" it may be answered, that the language, which he has in view, can be attributed to rustics with no greater right, than the style of Hooker or

Bacon to Tom Brown or Sir Roger L'Estrange.[8] Doubtless, if what is peculiar to each were omitted in each, the result must needs be the same. Further, that the poet, who uses an illogical diction, or a style fitted to excite only the low and changeable pleasure of wonder by means of groundless novelty, substitutes a language of *folly* and *vanity*, not for that of the *rustic*, but for that of *good sense* and *natural feeling*.

Here let me be permitted to remind the reader, that the positions, which I controvert, are contained in the sentences—"*a selection of the* REAL *language of men;*"— "*the language of these men*" (i.e. men in low and rustic life) "*I propose to myself to imitate, and, as far as is possible, to adopt the very language of men.*" "*Between the language of prose and that of metrical composition, there neither is, nor can be any essential difference.*" It is against these exclusively that my opposition is directed.

I object, in the very first instance, to an equivocation in the use of the word "real." Every man's language varies, according to the extent of his knowledge, the activity of his faculties, and the depth or quickness of his feelings. Every man's language has, first, its *individualities*; secondly, the common properties of the *class* to which he belongs; and thirdly, words and phrases of *universal* use. The language of Hooker, Bacon, Bishop Taylor, and Burke differs from the common language of the learned class only by the superior number and novelty of the thoughts and relations which they had to convey. The language of Algernon Sidney differs not at all from that, which every well-educated gentleman would wish to write, and (with due allowances for the undeliberateness, and less connected train, of thinking natural and proper to conversation) such as he would wish to talk. Neither one nor the other differ half so much from the general language of cultivated society, as the language of Mr. Wordsworth's homeliest composition differs from that of a common peasant. For "real" therefore, we must substitute *ordinary*, or *lingua communis*. And this, we have proved, is no more to be found in the phraseology of low and rustic life than in that of any other class. Omit the peculiarities of each, and the result of course must be common to all. And assuredly the omissions and changes to be made in the language of rustics, before it could be transferred to any species of poem, except the drama or other professed imitation, are at least as numerous

and weighty, as would be required in adapting to the same purpose the ordinary language of tradesmen and manufacturers. Not to mention, that the language so highly extolled by Mr. Wordsworth varies in every county, nay in every village, according to the accidental character of the clergyman, the existence or non-existence of schools; or even, perhaps, as the exciseman, publican, or barber, happen to be, or not to be, zealous politicians, and readers of the weekly newspaper *pro bono publico*. Anterior to cultivation, the lingua communis of every country, as Dante has well observed, exists every where in parts, and no where as a whole.

Neither is the case rendered at all more tenable by the addition of the words, *in a state of excitement*. For the nature of a man's words, where he is strongly affected by joy, grief, or anger, must necessarily depend on the number and quality of the general truths, conceptions and images, and of the words expressing them, with which his mind had been previously stored. For the property of passion is not to *create*; but to set in increased activity. At least, whatever new connections of thoughts or images, or (which is equally, if not more than equally, the appropriate effect of strong excitement) whatever generalizations of truth or experience, the heat of passion may produce; yet the terms of their conveyance must have pre-existed in his former conversations, and are only collected and crowded together by the unusual stimulation. It is indeed very possible to adopt in a poem the unmeaning repetitions, habitual phrases, and other blank counters, which an unfurnished or confused understanding interposes at short intervals, in order to keep hold of his subject, which is still slipping from him, and to give him time for recollection; or in mere aid of vacancy, as in the scanty companies of a country stage the same player pops backwards and forwards, in order to prevent the appearance of empty spaces, in the procession of Macbeth, or Henry VIIIth. But what assistance to the poet, or ornament to the poem, these can supply, I am at a loss to conjecture. Nothing assuredly can differ either in origin or in mode more widely from the *apparent* tautologies of intense and turbulent feeling, in which the passion is greater and of longer endurance than to be exhausted or satisfied by a single representation of the image or incident exciting it. Such repetitions I admit to be a beauty of the highest kind; as illustrated by Mr. Wordsworth himself from the song of Deborah. "*At her feet he bowed, he fell, he lay down; at her feet he bowed, he fell; where he bowed, there he fell down dead.*"[9]

---

8. **Brown . . . L'Estrange:** Tom Brown, 1663–1704, remembered for his satiric sketches of London life, and Sir Roger L'Estrange, 1616–1704, a prolific writer who was also one of the first English journalists.

9. **"At . . . dead":** Judges, 5:27.

CHAPTER XVIII

Language of metrical composition, why and wherein
essentially different from that of prose—Origin and
elements of metre—Its necessary consequences, and
the conditions thereby imposed on the metrical writer
in the choice of his diction.

I conclude, therefore, that the attempt is impracticable;
and that, were it not impracticable, it would still be
useless. For the very power of making the selection
implies the previous possession of the language
selected. Or where can the poet have lived? And by
what rules could he direct his choice, which would not
have enabled him to select and arrange his words by
the light of his own judgement? We do not adopt
the language of a class by the mere adoption of such
words exclusively, as that class would use, or at least
understand; but likewise by following the *order*, in
which the words of such men are wont to succeed each
other. Now this order, in the intercourse of uneducated
men, is distinguished from the diction of their superiors
in knowledge and power, by the greater *disjunction* and
separation in the component parts of that, whatever it
be, which they wish to communicate. There is a want
of that prospectiveness of mind, that *surview*, which
enables a man to foresee the whole of what he is to
convey, appertaining to any one point; and by this
means so to subordinate and arrange the different parts
according to their relative importance, as to convey
it at once, and as an organized whole.

Now I will take the first stanza, on which I have
chanced to open, in the Lyrical Ballads. It is one the
most simple and the least peculiar in its language.

> "In distant countries have I been,
> And yet I have not often seen
> A healthy man, a man full grown,
> Weep in the public roads alone.
> But such a one, on English ground,
> And in the broad highway, I met;
> Along the broad highway he came,
> His cheeks with tears were wet:
> Sturdy he seem'd, though he was sad;
> And in his arms a lamb he had."[1]

The words here are doubtless such as are current in
all ranks of life; and of course not less so in the hamlet
and cottage than in the shop, manufactory, college, or
palace. But is this the *order*, in which the rustic would
have placed the words? I am grievously deceived, if
the following less *compact* mode of commencing the
same tale be not a far more faithful copy. "I have been
in a many parts, far and near, and I don't know that I
ever saw before a man crying by himself in the public
road; a grown man I mean, that was neither sick nor
hurt," &c., &c. But when I turn to the following stanza
in "The Thorn":

> "At all times of the day and night
> This wretched woman thither goes,
> And she is known to every star,
> And every wind that blows:
> And there, beside the thorn, she sits,
> When the blue day-light's in the skies;
> And when the whirlwind's on the hill,
> Or frosty air is keen and still;
> And to herself she cries,
> Oh misery! Oh misery!
> Oh woe is me! Oh misery!"

and compare this with the language of ordinary men;
or with that which I can conceive at all likely to proceed, in *real* life, from *such* a narrator, as is supposed in
the note to the poem; compare it either in the succession
of the images or of the sentences; I am reminded of
the sublime prayer and hymn of praise, which MILTON,[2]
in opposition to an established liturgy, presents as a
fair *specimen* of common extemporary devotion, and
such as we might expect to hear from every self-
inspired minister of a conventicle! And I reflect with
delight, how little a mere theory, though of his own
workmanship, interferes with the processes of genuine
imagination in a man of true poetic genius, who possesses, as Mr. Wordsworth, if ever man did, most
assuredly does possess,

"THE VISION AND THE FACULTY DIVINE."[3]

One point then alone remains, but that the most
important; its examination having been, indeed, my
chief inducement for the preceding inquisition. "*There
neither is or can be any essential difference between the
language of prose and metrical composition.*" Such is
Mr. Wordsworth's assertion. Now prose itself, at
least in all argumentative and consecutive works,
differs, and ought to differ, from the language of

CHAPTER XVIII. 1. "In . . . had": "The Last of the Flock,"
ll. 1–10.

2. **Milton:** possibly the prayer of Adam and Eve in *Paradise
Lost*, V. 153–208. 3. "The . . . divine": from Wordsworth's *The Excursion*, I. 79.

conversation; even as reading ought to differ from talking. Unless therefore the difference denied be that of the mere *words*, as materials common to all styles of writing, and not of the *style* itself in the universally admitted sense of the term, it might be naturally presumed that there must exist a still greater between the ordonnance of poetic composition and that of prose, than is expected to distinguish prose from ordinary conversation.

There are not, indeed, examples wanting in the history of literature, of apparent paradoxes that have summoned the public wonder as new and startling truths, but which on examination have shrunk into tame and harmless *truisms*; as the eyes of a cat, seen in the dark, have been mistaken for flames of fire. But Mr. Wordsworth is among the last men, to whom a delusion of this kind would be attributed by anyone, who had enjoyed the slightest opportunity of understanding his mind and character. Where an objection has been anticipated by such an author as natural, his answer to it must needs be interpreted in some sense which either is, or has been, or is capable of being controverted. My object then must be to discover some other meaning for the term *"essential difference"* in this place, exclusive of the indistinction and community of the words themselves. For whether there ought to exist a class of words in the English, in any degree resembling the poetic dialect of the Greek and Italian, is a question of very subordinate importance. The number of such words would be small indeed, in our language; and even in the Italian and Greek, they consist not so much of different words, as of slight differences in the *forms* of declining and conjugating the same words; forms, doubtless, which having been, at some period more or less remote, the common grammatic flexions of some tribe or province, had been accidentally appropriated to poetry by the general admiration of certain master intellects, the first established lights of inspiration, to whom that dialect happened to be native.

Essence, in its primary signification, means the principle of *individuation*, the inmost principle of the possibility of any thing, as that particular thing. It is equivalent to the *idea* of a thing, when ever we use the word, idea, with philosophic precision. Existence, on the other hand, is distinguished from essence, by the superinduction of *reality*. Thus we speak of the essence, and essential properties of a circle; but we do not therefore assert, that any thing, which really exists, is mathematically circular. Thus too, without any

tautology we contend for the *existence* of the Supreme Being; that is, for a reality correspondent to the idea. There is, next, a *secondary* use of the word essence, in which it signifies the point or ground of contra-distinction between two modifications of the same substance or subject. Thus we should be allowed to say, that the style of architecture of Westminster Abbey is *essentially* different from that of St. Paul's, even though both had been built with blocks cut into the same form, and from the same quarry. Only in this latter sense of the term must it have been *denied* by Mr. Wordsworth (for in this sense alone is it *affirmed* by the general opinion) that the language of poetry (i.e. the formal construction, or architecture, of the words and phrases) is *essentially* different from that of prose. Now the burthen of the proof lies with the oppugner, not with the supporters of the common belief. Mr. Wordsworth, in consequence, assigns as the proof of his position, "that not only the language of a large portion of every good poem, even of the most elevated character, must necessarily, except with reference to the metre, in no respect differ from that of good prose, but likewise that some of the most interesting parts of the best poems will be found to be strictly the language of prose, when prose is well written. The truth of this assertion might be demonstrated by innumerable passages from almost all the poetical writings even of Milton himself." He then quotes Gray's sonnet—

> "In vain to me the smiling mornings shine,
> And reddening Phœbus lifts his golden fire;
> The birds in vain their amorous descant join,
> Or chearful fields resume their green attire.
> These ears, alas! for other notes repine;
> *A different object do these eyes require;*
> *My lonely anguish melts no heart but mine;*
> *And in my breast the imperfect joys expire.*
> Yet morning smiles the busy race to cheer,
> And newborn pleasure brings to happier men:
> The fields to all their wonted tribute bear,
> To warm their little loves the birds complain.
> *I fruitless mourn to him that cannot hear,*
> *And weep the more because I weep in vain,*"

and adds the following remark:—"It will easily be perceived, that the only part of this Sonnet, which is of any value, is the lines printed in italics. It is equally obvious, that, except in the rhyme, and in the use of the single word 'fruitless' for 'fruitlessly,' which is so far a defect, the language of these lines does in no respect differ from that of prose."

An idealist defending his system by the fact, that

when asleep we often believe ourselves awake, was well answered by his plain neighbour, "Ah, but when awake do we ever believe ourselves asleep?"—Things identical must be convertible. The preceding passage seems to rest on a similar sophism. For the question is not, whether there may not occur in prose an order of words, which would be equally proper in a poem; nor whether there are not beautiful lines and sentences of frequent occurrence in good poems, which would be equally becoming as well as beautiful in good prose; for neither the one nor the other has ever been either denied or doubted by any one. The true question must be, whether there are not modes of expression, a *construction*, and an *order* of sentences, which are in their fit and natural place in a serious prose composition, but would be disproportionate and heterogeneous in metrical poetry; and, vice versa, whether in the language of a serious poem there may not be an arrangement both of words and sentences, and a use and selection of (what are called) *figures of speech*, both as to their kind, their frequency, and their occasions, which on a subject of equal weight would be vicious and alien in correct and manly prose. I contend that in both cases this unfitness of each for the place of the other frequently will and ought to exist.

And first from the *origin* of metre. This I would trace to the balance in the mind effected by that spontaneous effort which strives to hold in check the workings of passion. It might be easily explained likewise in what manner this salutary antagonism is assisted by the very state, which it counteracts; and how this balance of antagonists became organized into *metre* (in the usual acceptation of that term) by a supervening act of the will and judgement, consciously and for the foreseen purpose of pleasure. Assuming these principles, as the data of our argument, we deduce from them two legitimate conditions, which the critic is entitled to expect in every metrical work. First, that, as the *elements* of metre owe their existence to a state of increased excitement, so the metre itself should be accompanied by the natural language of excitement. Secondly, that as these elements are formed into metre *artificially*, by a *voluntary* act, with the design and for the purpose of blending *delight* with emotion, so the traces of present *volition* should throughout the metrical language be proportionately discernible, Now these two conditions must be reconciled and co-present. There must be not only a partnership, but a union; an interpenetration of passion and of will, of *spontaneous* impulse and of *voluntary* purpose. Again, this union

can be manifested only in a frequency of forms and figures of speech (originally the offspring of passion, but now the adopted children of power) greater than would be desired or endured, where the emotion is not voluntarily encouraged and kept up for the sake of that pleasure, which such emotion, so tempered and mastered by the will, is found capable of communicating. It not only dictates, but of itself tends to produce, a more frequent employment of picturesque and vivifying language, than would be natural in any other case, in which there did not exist, as there does in the present, a previous and well understood, though tacit, *compact* between the poet and his reader, that the latter is entitled to expect, and the former bound to supply, this species and degree of pleasureable excitement. We may in some measure apply to this union the answer of POLIXENES, in the Winter's Tale, to PERDITA's neglect of the streaked gilly-flowers, because she had heard it said,

"There is an art which, in their piedness, shares
With great creating nature.
                    Pol: Say there be;
Yet nature is made better by no mean,
But nature makes that mean; so, ev'n that art,
Which, you say, adds to nature, is an art,
That nature makes. You see, sweet maid, we marry
A gentler scyon to the wildest stock;
And make conceive a bark of ruder kind
By bud of nobler race. This is an art,
Which does mend nature—change it rather; but
The art itself is nature." [4]

Secondly, I argue from the EFFECTS of metre. As far as metre acts in and for itself, it tends to increase the vivacity and susceptibility both of the general feelings and of the attention. This effect it produces by the continued excitement of surprize, and by the quick reciprocations of curiosity still gratified and still re-excited, which are too slight indeed to be at any one moment objects of distinct consciousness, yet become considerable in their aggregate influence. As a medicated atmosphere, or as wine during animated conversation; they act powerfully, though themselves unnoticed. Where, therefore, correspondent food and appropriate matter are not provided for the attention and feelings thus roused, there must needs be a disappointment felt; like that of leaping in the dark from the last step of a stair-case, when we had prepared our muscles for a leap of three or four.

4. **"There . . . nature":** *The Winter's Tale*, IV. iv. 87–97.

The discussion on the powers of metre in the preface is highly ingenious and touches at all points on truth. But I cannot find any statement of its powers considered abstractly and separately. On the contrary Mr. Wordsworth seems always to estimate metre by the powers, which it exerts during (and, as I think, in *consequence of*) its combination with other elements of poetry. Thus the previous difficulty is left unanswered, *what* the elements are, with which it must be combined in order to produce its own effects to any pleasureable purpose. Double and tri-syllable rhymes, indeed, form a lower species of wit, and, attended to exclusively for their own sake, may become a source of momentary amusement; as in poor Smart's distich to the Welsh 'Squire who had promised him a hare:

"Tell me, thou son of great Cadwallader!
Hast sent the hare? or hast thou swallow'd her?"[5]

But for any *poetic* purposes, metre resembles (if the aptness of the simile may excuse its meanness) yeast, worthless or disagreeable by itself, but giving vivacity and spirit to the liquor with which it is proportionally combined.

The reference to the "Children in the Wood," by no means satisfies my judgement. We all willingly throw ourselves back for awhile into the feelings of our childhood. This ballad, therefore, we read under such recollections of our own childish feelings, as would equally endear to us poems, which Mr. Wordsworth himself would regard as faulty in the opposite extreme of gaudy and technical ornament. Before the invention of printing, and in a still greater degree, before the introduction of writing, metre, especially *alliterative* metre (whether alliterative at the beginning of the words, as in "Pierce Plouman,"[6] or at the end as in rhymes) possessed an independent value as assisting the recollection, and consequently the preservation, of *any* series of truths or incidents. But I am not convinced by the collation of facts, that the "Children in the Wood" owes either its preservation, or its popularity, to its metrical form. Mr. Marshal's repository affords a number of tales in prose inferior in pathos and general merit, some of as old a date, and many as widely popular. "TOM HICKATHRIFT," "JACK THE GIANT-KILLER," "GOODY TWO-SHOES," and "LITTLE RED RIDING-HOOD"

are formidable rivals. And that they have continued in prose, cannot be fairly explained by the assumption, that the comparative meanness of their thoughts and images precluded even the humblest forms of metre. The scene of GOODY TWO-SHOES in the church is perfectly susceptible of metrical narration; and, among the Θαύματα θαυμαστότατα[7] even of the present age, I do not recollect a more astonishing image than that of the *"whole rookery, that flew out of the giant's beard,"* scared by the tremendous voice, with which this monster answered the challenge of the heroic TOM HICKATHRIFT!

If from these we turn to compositions universally, and independently of all early associations, beloved and admired; would "THE MARIA," "THE MONK," or "THE POOR MAN'S ASS" of Sterne, be read with more delight, or have a better chance of immortality, had they without any change in the diction been composed in rhyme, than in their present state? If I am not grossly mistaken, the general reply would be in the negative. Nay, I will confess, that, in Mr. Wordsworth's own volumes, the "ANECDOTE FOR FATHERS," "SIMON LEE," "ALICE FELL," "THE BEGGARS," and "THE SAILOR'S MOTHER," notwithstanding the beauties which are to be found in each of them where the poet interposes the music of his own thoughts, would have been more delightful to me in prose, told and managed, as by Mr. Wordsworth they would have been, in a moral essay, or pedestrian tour.

Metre in itself is simply a stimulant of the attention, and therefore excites the question: Why is the attention to be thus stimulated? Now the question cannot be answered by the pleasure of the metre itself: for this we have shown to be *conditional*, and dependent on the appropriateness of the thoughts and expressions, to which the metrical form is superadded. Neither can I conceive any other answer that can be rationally given, short of this: I write in metre, because I am about to use a language different from that of prose. Besides, where the language is not such, how interesting soever the reflections are, that are capable of being drawn by a philosophic mind from the thoughts or incidents of the poem, the metre itself must often become feeble. Take the last three stanzas of "THE SAILOR'S MOTHER," for instance. If I could for a moment abstract from the effect produced on the author's feelings, as a man, by the incident at the time of its real occurrence, I would dare appeal to his own judgement, whether

---

5. **"Tell . . . her?"**: Christopher Smart, 1722–71, "To the Rev. Mr. Powell," ll. 13–14. 6. **"Pierce Plouman"**: *The Vision Concerning Piers Plowman*, a fourteenth-century poem attributed to William Langland.

7. Θαύματα θαυμαστότατα: "wonder of wonders."

in the *metre* itself he found a sufficient reason for *their* being written *metrically*?

> "And, thus continuing, she said,
> I had a son, who many a day
> Sailed on the seas; but he is dead;
> In Denmark he was cast away:
> And I have travelled far as Hull, to see
> What clothes he might have left, or other property.
>
> The bird and cage they both were his:
> 'Twas my son's bird; and neat and trim
> He kept it: many voyages
> This singing-bird hath gone with him;
> When last he sailed he left the bird behind;
> As it might be, perhaps, from bodings of his mind.
>
> He to a fellow-lodger's care
> Had left it, to be watched and fed,
> Till he came back again; and there
> I found it when my son was dead;
> And now, God help me for my little wit!
> I trail it with me, Sir! he took so much delight in it."

If disproportioning the emphasis we read these stanzas so as to make the rhymes perceptible, even *tri-syllable* rhymes could scarcely produce an equal sense of oddity and strangeness, as we feel here in finding *rhymes at all* in sentences so exclusively collo-quial. I would further ask whether, but for that vision-ary state, into which the figure of the woman and the susceptibility of his own genius had placed the poet's imagination, (a state, which spreads its influence and coloring over all, that co-exists with the exciting cause, and in which

> "The simplest, and the most familiar things
> Gain a strange power of spreading awe around them,") [8]

I would ask the poet whether he would not have felt an abrupt downfall in these verses from the preceding stanza?

> "The ancient spirit is not dead;
> Old times, thought I, are breathing there;
> Proud was I that my country bred
> Such strength, a dignity so fair:
> She begged an alms, like one in poor estate;
> I looked at her again, nor did my pride abate."

It must not be omitted, and is besides worthy of notice, that those stanzas furnish the only fair instance that I have been able to discover in all Mr. Wordsworth's

writings, of an *actual* adoption, or true imitation, of the *real* and *very* language of *low and rustic life*, freed from provincialisms.

Thirdly, I deduce the position from all the causes elsewhere assigned, which render metre the proper form of poetry, and poetry imperfect and defective without metre. Metre therefore having been connected with *poetry* most often and by a peculiar fitness, what-ever else is combined with *metre* must, though it be not itself *essentially* poetic, have nevertheless some property in common with poetry, as an intermedium of affinity, a sort (if I may dare borrow a well-known phrase from technical chemistry) of *mordaunt* [9] between it and the super-added metre. Now poetry, Mr. Wordsworth truly affirms, does always imply PASSION: which word must be here understood in its general sense, as an excited state of the feelings and faculties. And as every passion has its proper pulse, so will it likewise have its character-istic modes of expression. But where there exists that degree of genius and talent which entitles a writer to aim at the honors of a poet, the very *act* of poetic composi-tion *itself* is, and is *allowed* to imply and to produce, an unusual state of excitement, which of course justifies and demands a correspondent difference of language, as truly, though not perhaps in as marked a degree, as the excitement of love, fear, rage, or jealousy. The vividness of the descriptions or declamations in DONNE or DRYDEN is as much and as often derived from the force and fervor of the describer, as from the reflections, forms or incidents, which constitute their subject and materials. The wheels take fire from the mere rapidity of their motion. To what extent, and under what modifications, this may be admitted to act, I shall attempt to define in an after remark on Mr. Wordsworth's reply to this objection, or rather on his objection to this reply, as already anticipated in his preface.

Fourthly, and as intimately connected with this, if not the same argument in a more general form, I adduce the high spiritual instinct of the human being impelling us to seek unity by harmonious adjustment, and thus establishing the principle, that *all* the parts of an organized whole must be assimilated to the more *important* and *essential* parts. This and the preceding arguments may be strengthened by the reflection, that the composition of a poem is among the *imitative* arts; and that imitation, as opposed to copying, consists

---

8. "The . . . them": echoes Coleridge's play *Remorse*, IV. i. 72–73.

9. **mordaunt:** mordant. In dyeing cloths the mordant fixes coloring matters to the fabric, hence a binding link.

either in the interfusion of the SAME throughout the radically DIFFERENT, or of the different throughout a base radically the same.

Lastly, I appeal to the practice of the best poets, of all countries and in all ages, as *authorizing* the opinion (*deduced* from all the foregoing) that in every import of the word ESSENTIAL, which would not here involve a mere truism, there may be, is, and ought to be an *essential* difference between the language of prose and of metrical composition.

In Mr. Wordsworth's criticism of GRAY's Sonnet, the readers' sympathy with his praise or blame of the different parts is taken for granted rather perhaps too easily. He has not, at least, attempted to win or compel it by argumentative analysis. In *my* conception at least, the lines rejected as of no value do, with the exception of the two first, differ as much and as little from the language of common life, as those which he has printed in italics as possessing genuine excellence. Of the five lines thus honourably distinguished, two of them differ from prose, even more widely than the lines which either precede or follow, in the *position* of the words.

> "*A different object do these eyes require;*
> My lonely anguish melts no heart but mine;
> *And in my breast the imperfect joys expire.*"

But were it otherwise, what would this prove, but a truth, of which no man ever doubted? Videlicet, that there are sentences, which would be equally in their place both in verse and prose. Assuredly it does not prove the point, which alone requires proof; namely, that there are not passages, which would suit the one and not suit the other. The first line of this sonnet is distinguished from the ordinary language of men by the epithet to morning. (For we will set aside, at present, the consideration, that the particular word "*smiling*" is hackneyed and (as it involves a sort of personification) not quite congruous with the common and material attribute of *shining*.) And, doubtless, this adjunction of epithets for the purpose of additional description, where no particular attention is demanded for the quality of the thing, would be noticed as giving a poetic cast to a man's conversation. Should the sportsman exclaim, "*Come boys! the rosy morning calls you up,*" he will be supposed to have some song in his head. But no one suspects this, when he says, "A wet morning shall not confine us to our beds." This then is either a defect in poetry, or it is not. Whoever should decide in the *affirmative*, I would request him to re-peruse

any one poem of any confessedly great poet from Homer to Milton, or from Æschylus to Shakespeare; and to strike out (in thought I mean) every instance of this kind. If the number of these fancied erasures did not startle him; or if he continued to deem the work improved by their total omission; he must advance reasons of no ordinary strength and evidence, reasons grounded in the essence of human nature. Otherwise, I should not hesitate to consider him as a man not so much *proof against* all authority, as *dead to* it.

The second line,

> "And reddening Phœbus lifts his golden fire;"

has indeed almost as many faults as words. But then it is a bad line, not because the language is distinct from that of prose; but because it conveys incongruous images, because it confounds the cause and the effect, the real *thing* with the personified *representative* of the thing; in short, because it differs from the language of GOOD SENSE! That the "Phœbus" is hackneyed, and a school-boy image, is an *accidental* fault, dependent on the age in which the author wrote, and not deduced from the nature of the thing. That it is part of an exploded mythology, is an objection more deeply grounded. Yet when the torch of ancient learning was re-kindled, so cheering were its beams, that our eldest poets, cut off by Christianity from all *accredited* machinery, and deprived of all *acknowledged* guardians and symbols of the great objects of nature, were naturally induced to adopt, as a *poetic* language, those fabulous personages, those forms of the [10] supernatural in nature, which had given them such dear delight in the poems of their great masters. Nay, even at this day what scholar of genial taste will not so far sympathize with them, as to read with pleasure in PETRARCH, CHAUCER, or SPENSER, what he would perhaps condemn as puerile in a modern poet?

I remember no poet, whose writings would safelier stand the test of Mr. Wordsworth's theory, than SPENSER. Yet will Mr. Wordsworth say, that the style of the following stanza is either undistinguished from prose, and the language of ordinary life? Or that it is

---

10. **forms of the:** Here Coleridge adds in a note, "But still more by the mechanical system of philosophy which has needlessly infected our theological opinions, and teaching us to consider the world in its relation to God, as of a building to its mason, leaves the idea of omnipresence a mere abstract notion in the state-room of our reason."

vicious, and that the stanzas are *blots* in the "Faery Queen"?

> "By this the northern waggoner had set
> His sevenfold teme behind the steadfast starre,
> That was in ocean waves yet never wet,
> But firme is fixt, and sendeth light from farre
> To all that in the wild deep wandering are:
> And chearful chanticleer with his note shrill
> Had warned once that Phœbus' fiery carre
> In haste was climbing up the easterne hill,
> Full envious that night so long his roome did fill."
>
> > Book I. Can. 2. St. 2.

> "At last the golden orientall gate
> Of greatest heaven gan to open fayre,
> And Phœbus fresh, as brydegrome to his mate,
> Came dauncing forth, shaking his deawie hayre,
> And hurl'd his glist'ring beams through gloomy ayre:
> Which when the wakeful elfe perceived, streightway
> He started up, and did him selfe prepayre
> In sun-bright armes and battailous array;
> For with that pagan proud he combat will that day."
>
> > B.I. Can. 5. St. 2.

On the contrary to how many passages, both in hymn books and in blank verse poems, could I, (were it not invidious), direct the reader's attention, the style of which is most *unpoetic, because,* and only because, it is the style of *prose*? He will not suppose me capable of having in my mind such verses, as

> "I put my hat upon my head
> And walk'd into the Strand;
> And there I met another man,
> Whose hat was in his hand."[11]

To such specimens it would indeed be a fair and full reply, that these lines are not bad, because they are *unpoetic;* but because they are empty of all sense and feeling; and that it were an idle attempt to prove that an ape is not a Newton, when it is evident that he is not a man. But the sense shall be good and weighty, the language correct and dignified, the subject interesting and treated with feeling; and yet the style shall, notwithstanding all these merits, be justly blamable as *prosaic,* and solely because the words and the order of the words would find their appropriate place in prose, but are not suitable to *metrical* composition. The "Civil Wars" of Daniel is an instructive, and even interesting work; but take the following stanzas (and from the

hundred instances which abound I might probably have selected others far more striking):

> "And to the end we may with better ease
> Discern the true discourse, vouchsafe to shew
> What were the times foregoing near to these,
> That these we may with better profit know.
> Tell how the world fell into this disease;
> And how so great distemperature did grow;
> So shall we see with what degrees it came;
> How things at full do soon wax out of frame."

> "Ten kings had from the Norman conqu'ror reign'd
> With intermixt and variable fate,
> When England to her greatest height attain'd
> Of power, dominion, glory, wealth, and state;
> After it had with much ado sustain'd
> The violence of princes, with debate
> For titles and the often mutinies
> Of nobles for their ancient liberties."

> "For first, the Norman, conqu'ring all by might,
> By might was forc'd to keep what he had got;
> Mixing our customs and the form of right
> With foreign constitutions he had brought;
> Mast'ring the mighty, humbling the poorer wight,
> By all severest means that could be wrought;
> And, making the succession doubtful, rent
> His new-got state, and left it turbulent."
>
> > B.I. St. VII. VIII. & IX

Will it be contended on the one side, that these lines are mean and senseless? Or on the other, that they are not prosaic, and for *that* reason unpoetic? This poet's well-merited epithet is that of the "*well-languaged Daniel;*" but likewise, and by the consent of his contemporaries no less than of all succeeding critics, the "prosaic Daniel." Yet those, who thus designate this wise and amiable writer, from the frequent incorrespondency of his diction to his metre in the majority of his compositions, not only deem them valuable and interesting on other accounts; but willingly admit, that there are to be found throughout his poems, and especially in his *Epistles* and in his *Hymen's Triumph,* many and exquisite specimens of that style which, as the *neutral ground* of prose and verse, is common to both. A fine and almost faultless extract, eminent, as for other beauties, so for its perfection in this species of diction, may be seen in LAMB's Dramatic Specimens, &c.,[12] a work of various interest from the nature of the

---

11. "I . . . hand": Samuel Johnson's extemporaneous parody of the ballad-stanza (Boswell, *Life of Johnson,* ed. G. B. Hill and L. F. Powell, 1934, II, 136).

12. **Dramatic Specimens, &c.:** Charles Lamb's *Specimens of English Dramatic Poets Contemporary with Shakespeare* (1808).

selections themselves, (all from the plays of Shake-speare's contemporaries), and deriving a high additional value from the notes, which are full of just and original criticism, expressed with all the freshness of originality.

Among the possible effects of practical adherence to a theory, that aims to *identify* the style of prose and verse, (if it does not indeed claim for the latter a yet nearer resemblance to the average style of men in the vivâ voce intercourse of real life) we might anticipate the following as not the least likely to occur. It will happen, as I have indeed before observed, that the metre itself, the sole acknowledged difference, will occasionally become metre to the eye only. The existence of *prosaisms*, and that they detract from the merit of a poem, *must* at length be conceded, when a number of successive lines can be rendered, even to the most delicate ear, unrecognizable as verse, or as having even been intended for verse, by simply transcribing them as prose; when, if the poem be in blank verse, this can be effected without any alteration, or at most by merely restoring one or two words to their proper places, from which they have been transplanted[13] for no assignable cause or reason but that of the author's convenience; but, if it be in rhyme, by the mere exchange of the final word of each line for some other of the same meaning, equally appropriate, dignified, and euphonic.

The answer or objection in the preface to the anticipated remark "that metre paves the way to other distinctions," is contained in the following words. "The distinction of rhyme and metre is voluntary and uniform, and not, like that produced by (what is called) poetic diction, arbitrary, and subject to infinite caprices, upon which no calculation whatever can be made. In the one case the reader is utterly at the mercy of the poet respecting what imagery or diction he may choose to connect with the passion." But is this a *poet*, of whom a poet is speaking? No surely! rather of a fool or madman: or at best of a vain or ignorant phantast! And might not brains so wild and so deficient make just the same havock with rhymes and metres, as they are supposed to effect with modes and figures of speech? How is the reader at the *mercy* of such men? If he continue to read their nonsense, is it not his own

fault? The ultimate end of criticism is much more to establish the principles of writing, than to furnish *rules* how to pass judgement on what has been written by others; if indeed it were possible that the two could be separated. But if it be asked, by what principles the poet is to regulate his own style, if he do not adhere closely to the sort and order of words which he hears in the market, wake, high-road, or plough-field? I reply; by principles, the ignorance or neglect of which would convict him of being no *poet*, but a silly or presumptuous usurper of the name! By the principles of grammar, logic, psychology! In one word by such a knowledge of the facts, material and spiritual, that most appertain to his art, as, if it have been governed and applied by *good sense*, and rendered instinctive by habit, becomes the representative and reward of our past conscious reasonings, insights, and conclusions, and acquires the name of TASTE. By what *rule* that does not leave the reader at the poet's mercy, and the poet at his own, is the latter to distinguish between the language suitable to *suppressed*, and the language, which is characteristic of *indulged*, anger? Or between that of rage and that of jealousy? Is it obtained by wandering about in search of angry or jealous people in uncultivated society, in order to copy their words? Or not far rather by the power of imagination proceeding upon the *all in each* of human nature? By *meditation*, rather than by *observation?* And by the latter in consequence only of the former? As eyes, for which the former has pre-determined their field of vision, and to which, as to *its* organ, it communicates a microscopic power? There is not, I firmly believe, a man now living, who has, from his own inward experience, a clearer intuition, than Mr. Wordsworth himself, that the last mentioned are the true sources of *genial* discrimination. Through the same process and by the same creative agency will the poet distinguish the degree and kind of the excitement produced by the very act of poetic composition. As intuitively will he know, what differences of style it at once inspires and justifies; what intermixture of conscious volition is natural to that state; and in what instances such figures and colors of speech degenerate into mere creatures of an arbitrary purpose, cold technical artifices of ornament or connection. For, even as truth is its own light and evidence, discovering at once itself and falsehood, so is it the prerogative of poetic genius to distinguish by parental instinct its proper offspring from the changelings, which the gnomes of vanity or the fairies of fashion may have laid in its cradle or called by its names. Could a rule be

---

13. transplanted: "As the ingenious gentleman under the influence of the Tragic Muse contrived to dislocate, 'I wish you a good morning, Sir! Thank you, Sir, and I wish you the same,' into two blank-verse heroics:—To you a morning good, good Sir! I wish. You, Sir! I thank: to you the same wish I" (C.).

given from *without*, poetry would cease to be poetry, and sink into a mechanical art. It would be μόρφωσις, not ποίησις.[14] The *rules* of the IMAGINATION are themselves the very powers of growth and production. The *words*, to which they are reducible, present only the outlines and external appearance of the fruit. A deceptive counterfeit of the superficial form and colors may be elaborated; but the marble peach feels cold and heavy, and *children* only put it to their mouths. We find no difficulty in admitting as excellent, and the legitimate language of poetic fervor self-impassioned, DONNE's apostrophe to the Sun in the second stanza of his "Progress of the Soul:"

"Thee, eye of heaven! this great soul envies not:
By thy male force is all, we have, begot.
In the first East thou now beginn'st to shine,
Suck'st early balm and island spices there,
And wilt anon in thy loose-rein'd career
At Tagus, Po, Seine, Thames, and Danow dine,
And see at night this western world of mine:
Yet hast thou not more nations seen than she,
Who before thee one day began to be,
And, thy frail light being quench'd, shall long, long out-
        live thee!"

Or the next stanza but one:

"Great destiny, the commissary of God,
That hast mark'd out a path and period
For ev'ry thing! Who, where we offspring took,
Our ways and ends see'st at one instant: thou
Knot of all causes! Thou, whose changeless brow
Ne'er smiles or frowns! O! vouchsafe thou to look,
And shew my story in thy eternal book, &c."

As little difficulty do we find in excluding from the honors of unaffected warmth and elevation the madness prepense of pseudo-poesy, or the startling *hysteric* of weakness overexerting itself, which bursts on the unprepared reader in sundry odes and apostrophes to abstract terms. Such are the Odes to Jealousy, to Hope, to Oblivion, and the like, in Dodsley's collection[15] and the magazines of that day, which seldom fail to remind me of an Oxford copy of verses on the two SUTTONS,[16] commencing with

"INOCULATION, heavenly maid! descend!"

---

14. μόρφωσις . . . ποίησις: "fashioning, not poetry." 15. **Dodsley's collection:** *Poems by Several Hands* (1748–58). Robert Dodsley, 1703–64, was a London bookseller. 16. **Suttons:** Daniel and Robert Sutton helped popularize inoculation in the eighteenth century.

It is not to be denied that men of undoubted talents, and even poets of true, though not of first-rate, genius, have from a mistaken theory deluded both themselves and others in the opposite extreme. I once read to a company of sensible and well-educated women the introductory period of Cowley's preface to his "*Pindaric Odes, written in imitation of the style and manner of the odes of Pindar.*" "If, (says Cowley), a man should undertake to translate Pindar, word for word, it would be thought that one madman had translated another; as may appear, when he, that understands not the original, reads the verbal traduction of him into Latin prose, than which nothing seems more raving." I then proceeded with his own free version of the second Olympic, composed for the charitable purpose of *rationalizing* the Theban Eagle.

"Queen of all harmonious things,
Dancing words and speaking strings,
What God, what hero, wilt thou sing?
What happy man to equal glories bring?
Begin, begin thy noble choice,
And let the hills around reflect the image of thy voice.
Pisa does to Jove belong,
Jove and Pisa claim thy song.
The fair first-fruits of war, th' Olympic games.
Alcides offer'd up to Jove;
Alcides too thy strings may move!
But, oh! what man to join with these can worthy prove?
Join Theron boldly to their sacred names;
Theron the next honor claims;
Theron to no man gives place,
Is first in Pisa's and in Virtue's race;
Theron there, and he alone,
Ev'n his own swift forefathers has outgone."

One of the company exclaimed, with the full assent of the rest, that if the original were madder than this, it must be incurably mad. I then translated the ode from the Greek, and as nearly as possible, word for word; and the impression was, that in the general movement of the periods, in the form of the connections and transitions, and in the sober majesty of lofty sense, it appeared to them to approach more nearly, than any other poetry they had heard, to the style of our Bible in the prophetic books. The first strophe will suffice as a specimen:

"Ye harp-controuling hymns! (or) ye hymns the sove-
        reigns of harps!
What God? what Hero?
What Man shall we celebrate?
Truly Pisa indeed is of Jove,

But the Olympiad (or the Olympic games) did Hercules
   establish,
The first-fruits of the spoils of war.
But Theron for the four-horsed car,
That bore victory to him,
It behoves us now to voice aloud:
The Just, the Hospitable,
The Bulwark of Agrigentum,
Of renowned fathers
The Flower, even him
Who preserves his native city erect and safe."

But are such rhetorical caprices condemnable only
for their deviation from the language of real life?
and are they by no other means to be precluded, but
by the rejection of all distinctions between prose and
verse, save that of metre? Surely good sense, and a
moderate insight into the constitution of the human
mind, would be amply sufficient to prove, that such
language and such combinations are the native produce
neither of the fancy nor of the imagination; that their
operation consists in the excitement of surprise by the
juxta-position and *apparent* reconciliation of widely
different or incompatible things. As when, for instance,
the hills are made to reflect the image of a *voice*. Surely,
no unusual taste is requisite to see clearly, that this
compulsory juxta-position is not produced by the
presentation of impressive or delightful forms to
the inward vision, nor by any sympathy with the
modifying powers with which the genius of the
poet had united and inspirited all the objects of his
thought; that it is therefore a species of *wit*, a pure work
of the *will*, and implies a leisure and self-possession
both of thought and of feeling, incompatible with
the steady fervor of a mind possessed and filled
with the grandeur of its subject. To sum up the whole
in one sentence. When a poem, or a part of a poem,
shall be adduced, which is evidently vicious in the
figures and contexture of its style, yet for the condemnation of which no reason can be assigned, except that
it differs from the style in which men actually converse,
then, and not till then, can I hold this theory to be
either plausible, or practicable, or capable of furnishing
either rule, guidance, or precaution, that might not,
more easily and more safely, as well as more naturally,
have been deduced in the author's own mind from
considerations of grammar, logic, and the truth and
nature of things, confirmed by the authority of works,
whose fame is not of ONE country nor of ONE age.

from CHAPTER XX

. . . From Mr. Wordsworth's more elevated compositions, which already form three-fourths of his works;
and will, I trust, constitute hereafter a still larger proportion;—from these, whether in rhyme or blank-
verse, it would be difficult and almost superfluous to
select instances of a diction peculiarly his own, of a
style which cannot be imitated, without its being at
once recognised as originating in Mr. Wordsworth.
It would not be easy to open on any one of his loftier
strains, that does not contain examples of this; and
more in proportion as the lines are more excellent,
and most like the author. For those, who may happen
to have been less familiar with his writings, I will give
three specimens taken with little choice. The first
from the lines on the "BOY OF WINANDER-MERE,"—who

"Blew mimic hootings to the silent owls,
That they might answer him. And they would shout
Across the watery vale, and shout again,
With long halloos and screams, and echoes loud
Redoubled and redoubled; concourse wild
Of mirth and jocund din. And when it chanced,
That pauses of deep silence mock'd his skill,
Then sometimes in that silence, while he hung
Listening, a gentle shock of mild surprize
Has carried far into his heart the voice
Of mountain-torrents; or the visible scene
Would enter unawares into his mind
With all its solemn imagery, its rocks,
Its woods, and that uncertain heaven, received
Into the bosom of the steady lake." [1]

CHAPTER XX. **I. "Blew . . . lake":** from Wordsworth's
"There Was a Boy," ll. 10–25. To the word *scene* in the
quotation Coleridge appends the following note: "Mr.
Wordsworth's having judiciously adopted 'concourse wild'
[l. 5] in this passage for 'a wild scene' as it stood in the former
edition, encourages me to hazard a remark which I certainly
should not have made in the works of a poet less austerely
accurate in the use of words than he is, to his own great
honor. It respects the propriety of the word 'scene' even in
the sentence in which it is retained. Dryden, and he only in
his more careless verses, was the first, as far as my researches
have discovered, who for the convenience of rhyme used
this word in the vague sense which has been since too
current even in our best writers and which (unfortunately, I
think) is given as its first explanation in Dr. Johnson's
Dictionary, and therefore would be taken by an incautious
reader as its proper sense. In Shakespeare and Milton the
word is never used without some clear reference, proper or
metaphorical, to the theatre. Thus Milton:

The second shall be that noble imitation of Drayton[2] (if it was not rather a coincidence) in the "JOANNA."

"When I had gazed perhaps two minutes' space,
Joanna, looking in my eyes, beheld
That ravishment of mine, and laughed aloud.
The rock, like something starting from a sleep,
Took up the lady's voice, and laughed again!
That ancient woman seated on HELM-CRAG
Was ready with her cavern; HAMMAR-SCAR
And the tall steep of SILVER-HOW sent forth
A noise of laughter; southern LOUGHRIGG heard,
And FAIRFIELD answered with a mountain tone.
HELVELLYN far into the clear blue sky
Carried the lady's voice!—old SKIDDAW blew
His speaking trumpet!—back out of the clouds
From GLARAMARA southward came the voice:
And KIRKSTONE tossed it from his misty head!"

The third, which is in rhyme, I take from the "Song at the feast of Brougham Castle, upon the restoration of Lord Clifford the shepherd to the estates of his ancestors."

"Now another day is come,
Fitter hope, and nobler doom;
He hath thrown aside his crook,
And hath buried deep his book;
Armour rusting in the halls
On the blood of Clifford calls;

'Cedar and pine, and fir and branching palm
A sylvan *scene;* and as the ranks ascend
Shade above shade, a woody *theatre*
Of stateliest view' (*Paradise Lost*, IV. 139–42).
I object to any extension of its meaning, because the word is already more equivocal than might be wished; inasmuch as in the limited use which I recommend it may still signify two different things; namely the scenery, and the characters and actions presented on the stage during the presence of particular scenes. It can therefore be preserved from obscurity only by keeping the original signification full in the mind. Thus Milton again: 'Prepare thee for another scene' (*Paradise Lost*, XI. 633)." **2. Drayton:**

Which Copland scarce had spoke, but quickly every hill
Upon her verge that stands, the neighbouring vallies fill;
Helvillon from his height, it through the mountains threw,
From whom as soon again, the sound Dunbalrase drew,
From whose stone-trophied head, it on the Wendross went,
Which, tow'rds the sea again, resounded it to Dent.
That Brodwater, therewith within her banks astound,
In sailing to the sea told it to Egremound,
Whose buildings, walks and streets, with echoes loud and long,
Did mightily commend old Copland for her song!
(Drayton's *Polyolbion:* Song XXX.)

'Quell the Scot,' exclaims the lance!
'Bear me to the heart of France,'
Is the longing of the shield—
Tell thy name, thou trembling field!—
Field of death, where'er thou be,
Groan thou with our victory!
Happy day, and mighty hour,
When our shepherd, in his power,
Mailed and horsed, with lance and sword,
To his ancestors restored,
Like a re-appearing star,
Like a glory from afar,
First shall head the flock of war!"

"Alas! the fervent harper did not know
That for a tranquil soul the lay was framed,
Who, long compelled in humble walks to go,
Was softened into feeling, soothed, and tamed.

Love had he found in huts where poor men lie;
His daily teachers had been woods and rills;
The silence that is in the starry sky,
The sleep that is among the lonely hills."[3]

The words themselves, in the foregoing extracts, are no doubt sufficiently common for the greater part. (But in what poem are they not so, if we except a few misadventurous attempts to translate the arts and sciences into verse?) In the "Excursion" the number of polysyllabic (or what the common people call, *dictionary*) words is more than usually great. And so must it needs be, in proportion to the number and variety of an author's conceptions, and his solicitude to express them with precision. But are those words *in those places* commonly employed in real life to express the same thought or outward thing? Are they the style used in the ordinary intercourse of spoken words? No! nor are the modes of connections; and still less the breaks and transitions. Would any but a poet—at least could any one without being conscious that he had expressed himself with noticeable vivacity—have described a bird singing loud by, "The thrush is *busy in the wood?*"—or have spoken of boys with a string of club-moss round their rusty hats, as the boys "*with their green coronal?*"—or have translated a beautiful May-day into "*Both earth and sky keep jubilee?*" or have brought all the different marks and circumstances of a sea-loch before the mind, as the actions of a living and acting power? Or have represented the reflection of the sky in the water, as "*That uncertain heaven received into*

---

3. "Now . . . hills": "Song at the Feast of Brougham Castle," ll. 138–64.

the bosom of the steady lake?" Even the grammatical construction is not unfrequently peculiar; as *"The wind, the tempest roaring high, the tumult of a tropic sky, might well be dangerous food to him, a youth* to whom was given, &c." There is a peculiarity in the frequent use of the ἀσυνάρτητον [4] (i.e. the omission of the connective particle before the last of several words, or several sentences used grammatically as single words, all being in the same case and governing or governed by the same verb) and not less in the construction of words by apposition (*to him, a youth*). In short, were there excluded from Mr. Wordsworth's poetic compositions all, that a literal adherence to the theory of his preface *would* exclude, two-thirds at least of the marked beauties of his poetry must be erased. For a far greater number of lines would be sacrificed than in any other recent poet; because the pleasure received from Wordsworth's poems being less derived either from excitement of curiosity or the rapid flow of narration, the *striking* passages form a larger proportion of their value. I do not adduce it as a fair criterion of comparative excellence, nor do I even think it such; but merely as matter of fact. I affirm, that from no contemporary writer could so many lines be quoted, without reference to the poem in which they are found, for their own independent weight or beauty. From the sphere of my own experience I can bring to my recollection three persons of no every-day powers and acquirements, who had read the poems of others with more, and more unalloyed pleasure, and had thought more highly of their authors, as poets; who yet have confessed to me, that from no modern work had so many passages started up anew in their minds at different times, and as different occasions had awakened a meditative mood.

CHAPTER XXII

The characteristic defects of Wordsworth's poetry, with the principles from which the judgement, that they are defects, is deduced—Their proportion to the beauties—For the greatest part characteristic of his theory only.

If Mr. Wordsworth have set forth principles of poetry which his arguments are insufficient to support, let him and those who have adopted his sentiments be set right by the confutation of these arguments, and by the substitution of more philosophical principles. and still let the due credit be given to the portion and importance of the truths, which are blended with his theory; truths, the too exclusive attention to which had occasioned its errors, by tempting him to carry those truths beyond their proper limits. If his mistaken theory have at all influenced his poetic compositions, let the effects be pointed out, and the instances given. But let it likewise be shown, how far the influence has acted; whether diffusively, or only by starts; whether the number and importance of the poems and passages thus infected be great or trifling compared with the sound portion; and lastly, whether they are inwoven into the texture of his works, or are loose and separable. The result of such a trial would evince beyond a doubt, what it is high time to announce decisively and aloud, that the *supposed* characteristics of Mr. Wordsworth's poetry, whether admired or reprobated; whether they are simplicity or simpleness; faithful adherence to essential nature, or wilful selections from human nature of its meanest forms and under the least attractive associations; are as little the *real* characteristics of his poetry at large, as of his genius and the constitution of his mind.

In a comparatively small number of poems he chose to try an experiment; and this experiment we will suppose to have failed. Yet even in these poems it is impossible not to perceive that the natural *tendency* of the poet's mind is to great objects and elevated conceptions. The poem entitled "Fidelity" is for the greater part written in language, as unraised and naked as any perhaps in the two volumes. Yet take the following stanza and compare it with the preceding stanzas of the same poem.

> "There sometimes doth a leaping fish
> Send through the tarn a lonely cheer;
> The crags repeat the raven's croak,
> In symphony austere;
> Thither the rainbow comes—the cloud—
> And mists that spread the flying shroud;
> And sun-beams; and the sounding blast,
> That if it could would hurry past;
> But that enormous barrier binds it fast."

Or compare the four last lines of the concluding stanza with the former half.

> "Yes, proof was plain that since the day
> On which the traveller thus had died,
> The dog had watched about the spot,
> Or by his master's side

4. ἀσυνάρτητον: lack of smooth juncture.

> How nourish'd there through such long time
> He knows, who gave that love sublime,
> And gave that strength of feeling, great
> Above all human estimate!"

Can any candid and intelligent mind hesitate in determining, which of these best represents the tendency and native character of the poet's genius? Will he not decide that the one was written because the poet *would* so write, and the other because he could not so entirely repress the force and grandeur of his mind, but that he must in some part or other of *every* composition write otherwise? In short, that his only disease is the being out of his element; like the swan, that, having amused himself, for a while, with crushing the weeds on the river's bank, soon returns to his own majestic movements on its reflecting and sustaining surface. Let it be observed that I am here supposing the imagined judge, to whom I appeal, to have already decided against the poet's theory, as far as it is different from the principles of the art, generally acknowledged.

I cannot here enter into a detailed examination of Mr. Wordsworth's works; but I will attempt to give the main results of my own judgement, after an acquaintance of many years, and repeated perusals. And though, to appreciate the defects of a great mind it is necessary to understand previously its characteristic excellences, yet I have already expressed myself with sufficient fulness, to preclude most of the ill effects that might arise from my pursuing a contrary arrangement. I will therefore commence with what I deem the prominent *defects* of his poems hitherto published.

The first *characteristic, though only occasional* defect, which I appear to myself to find in these poems is the INCONSTANCY of the *style.* Under this name I refer to the sudden and unprepared transitions from lines or sentences of peculiar felicity (at all events striking and original) to a style, not only unimpassioned but undistinguished. He sinks too often and too abruptly to that style, which I should place in the second division of language, dividing it into the three species; *first,* that which is peculiar to poetry; *second,* that which is only proper in prose; and *third,* the neutral or common to both. There have been works, such as Cowley's Essay on Cromwell, in which prose and verse are intermixed (not as in the Consolation of Boetius, or the Argenis of Barclay, by the insertion of poems supposed to have been spoken or composed on occasions previously related in prose, but) the poet passing from one to the other, as the nature of the thoughts or his own

feelings dictated. Yet this mode of composition does not satisfy a cultivated taste. There is something unpleasant in the being thus obliged to alternate states of feeling so dissimilar, and this too in a species of writing, the pleasure from which is in part derived from the preparation and previous expectation of the reader. A portion of that awkwardness is felt which hangs upon the introduction of songs in our modern comic operas; and to prevent which the judicious Metastasio[1] (as to whose exquisite *taste* there can be no hesitation, whatever doubts may be entertained as to his *poetic genius*) uniformly placed the ARIA at the end of the scene, at the same time that he almost always raises and impassions the style of the recitative immediately preceding. Even in real life, the difference is great and evident between words used as the *arbitrary marks* of thought, our smooth market-coin of intercourse, with the image and superscription worn out by currency; and those which convey pictures either borrowed from *one* outward object to enliven and particularize some *other;* or used allegorically to body forth the inward state of the person speaking; or such as are at least the exponents of his peculiar turn and unusual extent of faculty. So much so indeed, that in the social circles of private life we often find a striking use of the latter put a stop to the general flow of conversation, and by the excitement arising from concentered attention produce a sort of damp and interruption for some minutes after. But in the perusal of works of literary *art,* we *prepare* ourselves for such language; and the business of the writer, like that of a painter whose subject requires unusual splendor and prominence, is so to raise the lower and neutral tints, that what in a different style would be the *commanding* colors, are here used as the means of that gentle *degradation* requisite in order to produce the effect of a *whole.* Where this is not achieved in a poem, the metre merely reminds the reader of his claims in order to disappoint them; and where this defect occurs frequently, his feelings are alternately startled by anticlimax and hyperclimax.

I refer the reader to the exquisite stanzas cited for another purpose from the blind Highland Boy;[2] and

---

CHAPTER XXII. **1. Metastasio:** Pietro Metastasio, 1698–1782, Italian dramatist and librettist. **2. stanzas . . . Boy:** "The Blind Highland Boy," ll. 46–70, quoted by Coleridge in Chapter XX.

then annex, as being in my opinion instances of this *disharmony* in style, the two following:

> "And one, the rarest, was a shell,
> Which he, poor child, had studied well:
> The shell of a green turtle, thin
> And hollow;—you might sit therein,
>   It was so wide, and deep."

> "Our Highland Boy oft visited
> The house which held this prize; and, led
> By choice or chance, did thither come
> One day, when no one was at home,
>   And found the door unbarred."

Or page 172, vol. I.

> "'Tis gone—forgotten—*let me do
> My best*. There was a smile or two—
> I can remember them, I see
> The smiles worth all the world to me.
> Dear Baby, I must lay thee down:
> Thou troublest me with strange alarms;
> Smiles hast thou, sweet ones of thine own;
> I cannot keep thee in my arms;
> For they confound me: *as it is*,
> I have forgot those smiles of his!"[3]

Or page 269, vol. I.

> "Thou hast a nest, for thy love and thy rest,
> And though little troubled with sloth
> Drunken lark! thou would'st be loth
> To be such a traveller as I.
>   Happy, happy liver!
> *With a soul as strong as a mountain river
> Pouring out praise to th'Almighty giver!*
> Joy and jollity be with us both!
> Hearing thee or else some other,
>   As merry a brother
> I on the earth will go plodding on
> By myself chearfully till the day is done."[4]

The incongruity, which I appear to find in this passage, is that of the two noble lines in italics with the preceding and following. So vol. II. page 30.

> "Close by a pond, upon the further side,
> He stood alone; a minute's space, I guess,
> I watch'd him, he continuing motionless:
> To the pool's further margin then I drew,
> He being all the while before me full in view."[5]

Compare this with the repetition of the same image, in the next stanza but two.

> "And, still as I drew near with gentle pace,
> Beside the little pond or moorish flood
> Motionless as a cloud the old man stood,
> That heareth not the loud winds as they call,
> And moveth altogether, if it move at all."

Or lastly, the second of the three following stanzas, compared both with the first and the third.

> "My former thoughts returned; the fear that kills;
> And hope that is unwilling to be fed;
> Cold, pain, and labour, and all fleshly ills;
> And mighty poets in their misery dead.
> But now, perplex'd by what the old man had said,
> My question eagerly did I renew,
> 'How is it that you live, and what is it you do?'

> He with a smile did then his words repeat;
> And said, that gathering leeches far and wide
> He travell'd; stirring thus about his feet
> The waters of the ponds where they abide.
> 'Once I could meet with them on every side,
> But they have dwindled long by slow decay;
> Yet still I persevere, and find them where I may.'

> While he was talking thus, the lonely place,
> The old man's shape, and speech, all troubled me:
> In my mind's eye I seemed to see him pace
> About the weary moors continually,
> Wandering about alone and silently."

Indeed this fine poem is *especially* characteristic of the author. There is scarce a defect or excellence in his writings of which it would not present a specimen. But it would be unjust not to repeat that this defect is only occasional. From a careful reperusal of the two volumes of poems, I doubt whether the objectionable passages would amount in the whole to one hundred lines; not the eighth part of the number of pages. In the "EXCURSION" the feeling of incongruity is seldom excited by the diction of any passage considered in itself, but by the sudden superiority of some other passage forming the context.

The second defect I can generalize with tolerable accuracy, if the reader will pardon an uncouth and new-coined word. There is, I should say, not seldom a *matter-of-factness* in certain poems. This may be divided into, *first*, a laborious minuteness and fidelity in the representation of objects, and their positions, as they appeared to the poet himself; *secondly*, the insertion of accidental circumstances, in order to the full explanation of his living characters, their dispositions and

---

3. "'Tis . . . his!": "The Emigrant Mother," ll. 55–64. 4. "Thou . . . done": "To a Skylark," ll. 18–25, 28–31. 5. "Close . . . view": "Resolution and Independence," from a deleted stanza that came at line 57.

actions; which circumstances might be necessary to establish the probability of a statement in real life, where nothing is taken for granted by the hearer; but appear superfluous in poetry, where the reader is willing to believe for his own sake. To this *accidentality* I object, as contravening the essence of poetry, which Aristotle pronounces to be σπουδαιότατον καὶ φιλοσοφώτατον γένος,[6] the most intense, weighty and philosophical product of human art; adding, as the *reason*, that it is the most catholic and abstract. The following passage from Davenent's prefatory letter[7] to Hobbs well expresses this truth. "When I considered the actions which I meant to describe, (those inferring the persons), I was again persuaded rather to choose those of a former age, than the present; and in a century so far removed, as might preserve me from their improper examinations, who know not the requisites of a poem, nor how much pleasure they lose, (and even the pleasures of heroic poesy are not unprofitable), who take away the liberty of a poet, and fetter his feet in the shackles of an historian. For why should a poet doubt in story to mend the intrigues of fortune by more delightful conveyances of probable fictions, because austere historians have entered into bond to truth? An obligation, which were in poets as foolish and unnecessary, as in the bondage of false martyrs, who lie in chains for a mistaken opinion. *But by this I would imply, that truth narrative and past is the idol of historians, (who worship a dead thing), and truth operative, and by effects continually alive, is the mistress of poets, who hath not her existence in matter, but in reason.*"

For this minute accuracy in the painting of local imagery, the lines[8] in the EXCURSION, pp. 96, 97, and 98, may be taken, if not as a striking instance, yet as an illustration of my meaning. It must be some strong motive (as, for instance, that the description was necessary to the intelligibility of the tale) which could induce me to describe in a number of verses what a draughtsman could present to the eye with incomparably greater satisfaction by half a dozen strokes of his pencil, or the painter with as many touches of his brush. Such descriptions too often occasion in the mind of a reader, who is determined to understand his author, a feeling of labor, not very dissimilar to that, with which he would construct a diagram, line by

line, for a long geometrical proposition. It seems to be like taking the pieces of a dissected map out of its box. We first look at one part, and then at another, then join and dove-tail them; and when the successive acts of attention have been completed, there is a retrogressive effort of mind to behold it as a whole. The poet should paint to the imagination, not to the fancy; and I know no happier case to exemplify the distinction between these two faculties. Master-pieces of the former mode of poetic painting abound in the writings of Milton, ex. gr.

> "The fig-tree; not that kind for fruit renown'd,
> But such as at this day, to Indians known,
> In Malabar or Decan spreads her arms
> Branching so broad and long, that in the ground
> The bended twigs take root, *and daughters grow*
> *About the mother tree, a pillar'd shade*
> *High over-arch'd, and* ECHOING WALKS BETWEEN:
> *There oft the Indian Herdsman, shunning heat,*
> *Shelters in cool, and tends his pasturing herds*
> *At loop holes cut through thickest shade.*"

> Milton, P. L. 9. 1100.

This is *creation* rather than *painting*, or if painting, yet such, and with such co-presence of the whole picture flash'd at once upon the eye, as the sun paints in a camera obscura. But the poet must likewise understand and command what Bacon calls the *vestigia communia*[9] of the senses, the latency of all in each, and more especially as by a magical *penna duplex*,[10] the excitement of vision by sound and the exponents of sound. Thus "THE ECHOING WALKS BETWEEN," may be almost said to reverse the fable in tradition of the head of Memnon,[11] in the Egyptian statue. Such may be deservedly entitled the *creative words* in the world of imagination.

The second division respects an apparent minute adherence to *matter-of-fact* in characters and incidents; *a biographical* attention to probability, and an *anxiety* of explanation and retrospect. Under this head I shall deliver, with no feigned diffidence, the results of my best reflection on the great point of controversy between Mr. Wordsworth and his objectors; namely, on THE CHOICE OF HIS CHARACTERS. I have already declared and, I trust, justified, my utter dissent from

---

6. σπουδαιότατον . . . γένος: the most serious and philosophical form. 7. **Davenent's . . . letter:** William Davenant, 1606–68, Preface to *Gondibert* (1651). 8. **lines:** *The Excursion*, III. 50ff.

9. **vestigia communia:** common traces. 10. **penna duplex:** double wing (or feather). 11. **Memnon:** The colossal statue near Thebes, Egypt, was said to have uttered a musical sound when the rays of the morning sun first struck it.

the mode of argument which his critics have hitherto employed. To *their* question, Why did you chuse such a character, or a character from such a rank of life? the poet might in my opinion fairly retort: why with the conception of my character did you make wilful choice of mean or ludicrous associations not furnished by me, but supplied from your own sickly and fastidious feelings? How was it, indeed, probable, that such arguments could have any weight with an author, whose plan, whose guiding principle, and main object it was to attack and subdue that state of association, which leads us to place the chief value on those things in which man DIFFERS from man, and to forget or disregard the high dignities, which belong to HUMAN NATURE, the sense and the feeling, which *may* be, and *ought* to be, found in *all* ranks? The feelings with which, as Christians, we contemplate a mixed congregation rising or kneeling before their common Maker: Mr. Wordsworth would have us entertain at *all* times, as men, and as readers; and by the excitement of this lofty, yet prideless impartiality in *poetry*, he might hope to have encouraged its continuance in *real life*. The praise of good men be his! In real life, and, I trust, even in my imagination, I honor a virtuous and wise man, without reference to the presence or absence of artificial advantages. Whether in the person of an armed baron, a laurel'd bard, &c., or of an old pedlar, or still older leach-gatherer, the same qualities of head and heart must claim the same reverence. And even in poetry I am not conscious, that I have ever suffered my feelings to be disturbed or offended by any thoughts or images, which the poet himself has not presented.

But yet I object nevertheless and for the following reasons. First, because the object in view, as an *immediate* object, belongs to the moral philosopher, and would be pursued, not only more appropriately, but in my opinion with far greater probability of success, in sermons or moral essays, than in an elevated poem. It seems, indeed, to destroy the main fundamental distinction, not only between a poem and prose, but even between philosophy and works of fiction, inasmuch as it proposes *truth* for its immediate object, instead of *pleasure*. Now till the blessed time shall come, when truth itself shall be pleasure, and both shall be so united, as to be distinguishable in words only, not in feeling, it will remain the poet's office to proceed upon that state of association, which actually exists as *general*; instead of attempting first to *make* it what it ought to be, and then to let the pleasure follow. But here is unfortunately a small

Hysteron-Proteron.[12] For the communication of pleasure is the introductory means by which alone the poet must expect to moralize his readers. Secondly: though I were to admit, for a moment, *this* argument to be groundless: yet how is the moral effect to be produced, by merely attaching the name of some low profession to powers which are *least* likely, and to qualities which are assuredly not *more* likely, to be found in it? The poet, speaking in his own person, may at once delight and improve us by sentiments, which teach us the independence of goodness, of wisdom, and even of genius, on the favors of fortune. And having made a due reverence before the throne of Antonine, he may bow with equal awe before Epictetus among his fellow-slaves—

> ——————— "and rejoice
> In the plain presence of his dignity."[13]

Who is not at once delighted and improved, when the POET Wordsworth himself exclaims,

> "O many are the poets that are sown
> By Nature; man endowed with highest gifts,
> The vision and the faculty divine,
> Yet wanting the accomplishment of verse,
> Nor having e'er, as life advanced, been led
> By circumstance to take unto the height
> The measure of themselves, these favor'd beings,
> All but a scatter'd few, live out their time
> Husbanding that which they possess within,
> And go to the grave unthought of. Strongest minds
> Are often those of whom the noisy world
> Hears least."[14]

To use a colloquial phrase, such sentiments, in such language, do one's heart good; though I for my part, have not the fullest faith in the *truth* of the observation. On the contrary I believe the instances to be exceedingly rare; and should feel almost as strong an objection to introduce such a character in a poetic fiction, as a pair of black swans on a lake in a fancy-landscape. When I think how many, and how much better books than Homer, or even than Herodotus, Pindar or Eschylus, could have read, are in the power of almost every man, in a country where almost every man is instructed to read and write; and how restless, how difficultly hidden, the powers of genius are; and yet find even in situations the most favorable, according to Mr. Wordsworth, for the formation

12. **Hysteron-Proteron:** inversion of the logical order. 13. **"and . . . dignity":** *The Excursion*, I. 75–76. 14. **"O . . . least":** *The Excursion*, I. 77–80, 86–93.

of a pure and poetic language; in situations which ensure familiarity with the grandest objects of the imagination; but *one* BURNS, among the shepherds of *Scotland*, and not a single poet of humble life among those of *English* lakes and mountains; I conclude, that POETIC GENIUS is not only a very delicate but a very rare plant.

But be this as it may, the feelings with which

> "I think of CHATTERTON, the marvellous boy,
> The sleepless soul, that perished in his pride;
> Of BURNS, that walk'd in glory and in joy
> Behind his plough upon the mountain-side"—[15]

are widely different from those with which I should read a *poem*, where the author, having occasion for the character of a poet and a philosopher in the fable of his narration, had chosen to make him a *chimney-sweeper*; and then, in order to remove all doubts on the subject, had *invented* an account of his birth, parentage and education, with all the strange and fortunate accidents which had concurred in making him at once poet, philosopher, and sweep! Nothing but biography can justify this. If it be admissible even in a *Novel*, it must be one in the manner of De Foe's, that were meant to pass for histories, not in the manner of Fielding's: in the life of Moll Flanders, or Colonel Jack, not in a Tom Jones, or even a Joseph Andrews. Much less then can it be legitimately introduced in a *poem*, the characters of which, amid the strongest individualization, must still remain representative. The precepts of Horace,[16] on this point, are grounded on the nature both of poetry and of the human mind. They are not more peremptory, than wise and prudent. For in the first place a deviation from them perplexes the reader's feelings, and all the circumstances, which are feigned in order to make such accidents less improbable, divide and disquiet his faith, rather than aid and support it. Spite of all attempts, the fiction *will* appear, and unfortunately not as *fictitious* but as *false*. The reader not only *knows*, that the sentiments and language are the poet's own, and his own too in his *artificial* character, *as poet*; but by the fruitless endeavours to make him think the contrary, he is not even suffered to *forget* it. The effect is similar to that produced by an epic poet, when the fable and the characters are *derived* from Scripture history, as in

the *Messiah*[17] of *Klopstock*, or in *Cumberland's Calvary*;[18] and not merely *suggested* by it, as in the Paradise Lost of Milton. That *illusion*, contra-distinguished from *delusion*, that *negative* faith, which simply permits the images presented to work by their own force, without either denial or affirmation of their real existence by the judgement, is rendered impossible by their immediate neighbourhood to words and facts of known and absolute truth. A faith, which transcends even historic belief, must absolutely *put out* this mere poetic Analogon[19] of faith, as the summer sun is said to extinguish our household fires, when it shines full upon them. What would otherwise have been yielded to as pleasing fiction, is repelled as revolting falsehood. The effect produced in this latter case by the solemn belief of the reader, is in a less degree brought about in the instances, to which I have been objecting, by the baffled attempts of the author to *make* him believe.

Add to all the foregoing the seeming uselessness both of the project and of the anecdotes from which it is to derive support. Is there one word, for instance, attributed to the pedlar in the "EXCURSION," characteristic of a *pedlar*? One sentiment, that might not more plausibly, even without the aid of any previous explanation, have proceeded from any wise and beneficent old man, of a rank or profession in which the language of learning and refinement are natural and to be expected? Need the rank have been at all particularized, where nothing follows which the knowledge of that rank is to explain or illustrate? When on the contrary this information renders the man's language, feelings, sentiments, and information a riddle, which must itself be solved by episodes of anecdote? Finally when this, and this alone, could have induced a genuine *poet* to inweave in a poem of the loftiest style, and on subjects the loftiest and of most universal interest, such minute matters of fact, (not unlike those furnished for the obituary of a magazine by the friends of some obscure *ornament of society lately deceased* in some obscure town), as

> "Among the hills of Athol he was born:
> There, on a small hereditary farm,
> An unproductive slip of rugged ground,
> His Father dwelt; and died in poverty;
> While he, whose lowly fortune I retrace,

---

15. "I . . . side": "Resolution and Independence," ll. 43–46. 16. precepts of Horace: *Ars Poetica*, 2. 114ff.

17. Messiah: the *Messias* by the German poet F. G. Klopstock, 1724–1803. 18. Calvary: a Miltonic imitation (1792) by Richard Cumberland, 1732–1811. 19. Analogon: analogue.

The youngest of three sons, was yet a babe,
A little one—unconscious of their loss.
But, ere he had outgrown his infant days,
His widowed mother, for a second mate,
Espoused the teacher of the Village School;
Who on her offspring zealously bestowed
Needful instruction."

"From his sixth year, the Boy of whom I speak,
In summer tended cattle on the hills;
But, through the inclement and the perilous days
Of long-continuing winter, he repaired
To his step-father's school,"—&c. [20]

For all the admirable passages interposed in this narration, might, with trifling alterations, have been far more appropriately, and with far greater verisimilitude, told of a poet in the character of a poet; and without incurring another defect which I shall now mention, and a sufficient illustration of which will have been here anticipated.

Third; an undue predilection for the *dramatic* form in certain poems, from which one or other of two evils result. Either the thoughts and diction are different from that of the poet, and then there arises an incongruity of style; or they are the same and indistinguishable, and then it presents a species of ventriloquism, where two are represented as talking, while in truth one man only speaks.

The fourth class of defects is closely connected with the former; but yet are such as arise likewise from an intensity of feeling disproportionate to *such* knowledge and value of the objects described, as can be fairly anticipated of men in general, even of the most cultivated classes; and with which therefore few only, and those few particularly circumstanced, can be supposed to sympathize. In this class, I comprise occasional prolixity, repetition, and an eddying, instead of progression, of thought. As instances, see pages 27, 28, and 62 of the Poems, Vol. I,[21] and the first eighty lines of the Sixth Book of the Excursion.

Fifth and last; thoughts and images too great for the subject. This is an approximation to what might be called *mental* bombast, as distinguished from verbal:

for, as in the latter there is a disproportion of the expressions to the thoughts, so in this there is a disproportion of thought to the circumstance and occasion. This, by the bye, is a fault of which none but a man of genius is capable. It is the awkwardness and strength of Hercules with the distaff of Omphale.[22]

It is a well-known fact, that bright colors in motion both make and leave the strongest impressions on the eye. Nothing is more likely too, than that a vivid image or visual spectrum, thus originated, may become the link of association in recalling the feelings and images that had accompanied the original impression. But if we describe this in such lines, as

"They flash upon that inward eye,
Which is the bliss of solitude!"[23]

in what words shall we describe the joy of retrospection, when the images and virtuous actions of a whole well-spent life, pass before that conscience which is indeed the *inward* eye: which is indeed "*the bliss of solitude?*" Assuredly we seem to sink most abruptly, not to say burlesquely, and almost as in a *medly*, from this couplet to—

"And then my heart with pleasure fills,
And dances with the *daffodils.*"

The second instance is from Vol. II. page 12, where the poet, having gone out for a day's tour of pleasure, meets early in the morning with a knot of *gypsies,* who had pitched their blanket-tents and straw-beds, together with their children and asses, in some field by the road-side. At the close of the day on his return our tourist found them in the same place. "Twelve hours," says he,

"Twelve hours, twelve bounteous hours are gone, while I
Have been a traveller under open sky,
Much witnessing of change and cheer,
Yet as I left I find them here!"[24]

Whereat the poet, without seeming to reflect that the poor tawny wanderers might probably have been tramping for weeks together through road and lane, over moor and mountain, and consequently must have been right glad to rest themselves, their children and

---

20. "Among . . . school,"—&c.: *The Excursion,* I. 108ff. Coleridge cites the edition of 1814. Wordsworth later changed many of the passages Coleridge had criticized. **21. pages . . . Vol. I.:** Pages 27 and 28 of *Poems,* 2 vols. (1815), contain "Anecdote for Fathers," ll. 13–52. Page 62 is blank, but Coleridge's daughter Sara suggested that her father was thinking of "Song at the Feast of Brougham Castle," ll. 78–101, on p. 62 of Vol. II.

22. Hercules . . . Omphale: Omphale, queen of Lydia, purchased Hercules as a slave and put him to women's work. Cf. Ovid, *Heroides,* 9. 52ff. **23.** "They . . . solitude!": "I Wandered Lonely as a Cloud," ll. 21–22. **24.** "Twelve . . . here!": "Gipsies," ll. 9–12.

cattle, for one whole day; and overlooking the obvious truth, that such repose might be quite as necessary for *them*, as a walk of the same continuance was pleasing or healthful for the more fortunate poet; expresses his indignation in a series of lines, the diction and imagery of which would have been rather above, than below the mark, had they been applied to the immense empire of China improgressive for thirty centuries:

"The weary SUN betook himself to rest:—
—Then issued VESPER from the fulgent west,
Outshining, like a visible God,
The glorious path in which he trod!
And now, ascending, after one dark hour,
And one night's diminution of her power,
Behold the mighty MOON! this way
She looks, as if at them—but they
Regard not her:—oh, better wrong and strife,
Better vain deeds or evil than such life!
The silent HEAVENS have goings on:
The STARS have tasks!—but *these* have none!"

The last instance of this defect (for I know no other than these already cited) is from the Ode,[25] page 351, Vol. II., where, speaking of a child, "a six years' darling of a pigmy size," he thus addresses him:

"Thou best philosopher, who yet dost keep
Thy heritage! Thou eye among the blind,
That, deaf and silent, read'st the eternal deep,
Haunted for ever by the Eternal Mind,—
Mighty Prophet! Seer blest!
On whom those truths do rest,
Which we are toiling all our lives to find!
Thou, over whom thy immortality
Broods like the day, a master o'er the slave,
A presence that is not to be put by!"

Now here, not to stop at the daring spirit of metaphor which connects the epithets "deaf and silent," with the apostrophized *eye*: or (if we are to refer it to the preceding word, philosopher) the faulty and equivocal syntax of the passage; and without examining the propriety of making a "master *brood* o'er a slave," or the *day* brood *at all*; we will merely ask, what does all this mean? In what sense is a child of that age a *philosopher*? In what sense does he *read* "the eternal deep"? In what sense is he declared to be "*for ever haunted*" by the Supreme Being? or so inspired as to deserve the splendid titles of a *mighty prophet*, a *blessed seer*? By reflection? by knowledge? by conscious intuition? or by *any* form or modification of consciousness?

25. Ode: "Intimations of Immortality."

These would be tidings indeed; but such as would presuppose an immediate revelation to the inspired communicator, and require miracles to authenticate his inspiration. Children at this age give us no such information of themselves; and at what time were we dipped in the Lethe, which has produced such utter oblivion of a state so godlike? There are many of us that still possess some remembrances, more or less distinct, respecting themselves at six years old; pity that the worthless straws only should float, while treasures, compared with which all the mines of Golconda and Mexico were but straws, should be absorbed by some unknown gulf into some unknown abyss.

But if this be too wild and exorbitant to be suspected as having been the poet's meaning; if these mysterious gifts, faculties, and operations, are *not* accompanied with consciousness; who *else* is conscious of them? or how can it be called the child, if it be no part of the child's conscious being? For aught I know, the thinking Spirit within me may be *substantially* one with the principle of life, and of vital operation. For aught I know, it might be employed as a secondary agent in the marvellous organization and organic movements of my body. But, surely, it would be strange language to say, that *I* construct my *heart!* or that *I* propel the finer influences through my *nerves!* or that *I* compress my brain, and draw the curtains of sleep round my own eyes! SPINOZA and BEHMEN[26] were, on different systems, both Pantheists; and among the ancients there were philosophers, teachers of the EN KAI ΠAN,[27] who not only taught that God was All, but that this All constituted God. Yet not even these would confound the *part*, as a part, with the Whole, *as* the whole. Nay, in no system is the distinction between the individual and God, between the Modification, and the one only Substance, more sharply drawn, than in that of SPINOZA. JACOBI indeed relates of LESSING, that, after a conversation with him at the house of the poet, GLEIM[28] (the Tyrtæus and Anacreon of the German Parnassus) in which conversation L. had avowed privately to Jacobi his reluctance to admit any *personal* existence of the Supreme Being, or the *possibility* of personality except in a finite Intellect, and while they were sitting at table,

26. Behmen: Jakob Böhme, 1575–1624, the German mystic, whose name in England was commonly spelled "Behmen."
27. EN KAI ΠAN: one and all. **28. Jacobi; Lessing; Gleim:** the German philosopher F. H. Jacobi, 1743–1819, and the German poets Gotthold Ephraim Lessing, 1729–81, and Johann Gleim, 1718–1803.

a shower of rain came on unexpectedly. Gleim expressed his regret at the circumstance, because they had meant to drink their wine in the garden: upon which Lessing in one of his half-earnest half-joking moods, nodded to Jacobi, and said, "It is *I*, perhaps, that am doing *that*," i.e. *raining!* and J. answered, "or perhaps I"; Gleim contented himself with staring at them both, without asking for any explanation.

So with regard to this passage. In what sense can the magnificent attributes, above quoted, be appropriated to a *child*, which would not make them equally suitable to a *bee*, or a *dog*, or a *field of corn*: or even to a ship, or to the wind and waves that propel it? The omnipresent Spirit works equally in them, as in the child; and the child is equally unconscious of it as they. It cannot surely be, that the four lines, immediately following, are to contain the explanation?

> "To whom the grave
> Is but a lonely bed without the sense or sight
> Of day or the warm light,
> A place of thought where we in waiting lie." [29]

Surely, it cannot be that this wonder-rousing apostrophe is but a comment on the little poem, "We are seven?" that the whole meaning of the passage is reducible to the assertion, that a *child*, who by the bye at six years old would have been better instructed in most Christian families, has no other notion of death than that of lying in a dark, cold place? And still, I hope, not as in a *place of thought!* not the frightful notion of lying *awake* in his grave! The analogy between death and sleep is too simple, too natural, to render so horrid a belief possible for children; even had they not been in the habit, as all Christian children are, of hearing the latter term used to express the former. But if the child's belief be only, that "he is not dead, but sleepeth:" [30] wherein does it differ from that of his father and mother, or any other adult and instructed person? To form an idea of a thing's becoming nothing; or of nothing becoming a thing; is impossible to all finite beings alike, of whatever age, and however educated or uneducated. Thus it is with splendid paradoxes in general. If the words are taken in the common sense, they convey an absurdity; and if, in contempt of dictionaries and custom, they are so interpreted as to avoid the absurdity, the meaning dwindles into some bald truism. Thus you must at once understand the words *contrary* to their common import, in order to

arrive at any *sense*; and *according* to their common import, if you are to receive from them any feeling of *sublimity* or *admiration*.

Though the instances of this defect in Mr. Wordsworth's poems are so few, that for themselves it would have been scarce just to attract the reader's attention toward them; yet I have dwelt on it, and perhaps the more for this very reason. For being so very few, they cannot sensibly detract from the reputation of an author, who is even characterized by the number of profound truths in his writings, which will stand the severest analysis; and yet few as they are, they are exactly those passages which his *blind* admirers would be most likely, and best able, to imitate. But Wordsworth, where he is indeed Wordsworth, may be mimicked by Copyists, he may be plundered by Plagiarists; but he can not be imitated, except by those who are not born to be imitators. For without his depth of feeling and his imaginative power his *sense* would want its vital warmth and peculiarity; and without his strong sense, his *mysticism* would become *sickly*—mere fog, and dimness!

To these defects which, as appears by the extracts, are only occasional, I may oppose, with far less fear of encountering the dissent of any candid and intelligent reader, the following (for the most part correspondent) excellences. First, an austere purity of language both grammatically and logically; in short a perfect appropriateness of the words to the meaning. Of how high value I deem this, and how particularly estimable I hold the example at the present day, has been already stated: and in part too the reasons on which I ground both the moral and intellectual importance of habituating ourselves to a strict accuracy of expression. It is noticeable, how limited an acquaintance with the masterpieces of art will suffice to form a correct and even a sensitive taste, where none but masterpieces have been seen and admired: while on the other hand, the most correct notions, and the widest acquaintance with the works of excellence of all ages and countries, will not perfectly secure us against the contagious familiarity with the far more numerous offspring of tastelessness or of a perverted taste. If this be the case, as it notoriously is, with the arts of music and painting, much more difficult will it be to avoid the infection of multiplied and daily examples in the practice of an art, which uses words, and words only, as its instruments. In poetry, in which every line, every phrase, may pass the ordeal of deliberation and deliberate choice, it is possible, and barely possible, to attain

29. "To . . . lie": omitted by Wordsworth in later editions.
30. "he . . . sleepeth": Matthew 9:24.

that ultimatum which I have ventured to propose as the infallible test of a blameless style; its *untranslatableness* in words of the same language without injury to the meaning. Be it observed, however, that I include in the *meaning* of a word not only its correspondent object, but likewise all the associations which it recalls. For language is framed to convey not the object alone, but likewise the character, mood and intentions of the person who is representing it. In poetry it *is* practicable to preserve the diction uncorrupted by the affectations and misappropriations, which promiscuous authorship, and reading not promiscuous only because it is disproportionally most conversant with the compositions of the day, have rendered general. Yet even to the poet, composing in his own province, it is an arduous work: and as the result and pledge of a watchful good sense, of fine and luminous distinction, and of complete self-possession, may justly claim all the honor which belongs to an attainment equally difficult and valuable, and the more valuable for being rare. It is at *all* times the proper food of the understanding; but in an age of corrupt eloquence it is both food and antidote.

In prose I doubt whether it be even possible to preserve our style wholly unalloyed by the vicious phraseology which meets us everywhere, from the sermon to the newspaper, from the harangue of the legislator to the speech from the convivial chair, announcing a *toast* or sentiment. Our chains rattle, even while we are complaining of them. The poems of Boetius rise high in our estimation when we compare them with those of his contemporaries, as Sidonius Apollinarius, &c. They might even be referred to a purer age, but that the prose, in which they are set, as jewels in a crown of lead or iron, betrays the true age of the writer. Much however may be effected by education. I believe not only from grounds of reason, but from having in great measure assured myself of the fact by actual though limited experience, that, to a youth led from his first boyhood to investigate the meaning of every word and the reason of its choice and position, Logic presents itself as an old acquaintance under new names.

On some future occasion, more especially demanding such disquisition, I shall attempt to prove the close connection between veracity and habits of mental accuracy; the beneficial after-effects of verbal precision in the preclusion of fanaticism, which masters the feelings more especially by indistinct watch-words; and to display the advantages which language alone, at least which language with incomparably greater ease and certainty than any other means, presents to the instructor of impressing modes of intellectual energy so constantly, so imperceptibly, and as it were by such elements and atoms, as to secure in due time the formation of a second nature. When we reflect, that the cultivation of the judgement is a positive command of the moral law, since the reason can give the *principle* alone, and the conscience bears witness only to the *motive*, while the application and effects must depend on the judgement: when we consider, that the greater part of our success and comfort in life depends on distinguishing the similar from the same, that which is peculiar in each thing from that which it has in common with others, so as still to select the most probable, instead of the merely possible or positively unfit, we shall learn to value earnestly and with a practical seriousness a mean, already prepared for us by nature and society, of teaching the young mind to think well and wisely by the same unremembered process and with the same never forgotten results, as those by which it is taught to speak and converse. Now how much warmer the interest is, how much more genial the feelings of reality and practicability, and thence how much stronger the impulses to imitation are, which a *contemporary* writer, and especially a contemporary *poet*, excites in youth and commencing manhood, has been treated of in the earlier pages of these sketches. I have only to add, that all the praise which is due to the exertion of such influence for a purpose so important, joined with that which must be claimed for the infrequency of the same excellence in the same perfection, belongs in full right to Mr. Wordsworth. I am far however from denying that we have poets whose *general* style possesses the same excellence, as Mr. Moore, Lord Byron, Mr. Bowles,[31] and, in all his later and more important works, our laurel-honoring Laureate.[32] But there are none, in whose works I do not appear to myself to find *more* exceptions, than in those of Wordsworth. Quotations or specimens would here be wholly out of place, and must be left for the critic who doubts and would invalidate the justice of this eulogy so applied.

The second characteristic excellence of Mr. W's work is: a correspondent weight and sanity of the Thoughts and Sentiments, won—not from books, but

---

31. Bowles: William Lisle Bowles, 1762–1850, whose sonnets had moved and influenced Coleridge when he was young. **32. Laureate:** Robert Southey.

—from the poet's own meditative observation. They are *fresh* and have the dew upon them. His muse, at least when in her strength of wing, and when she hovers aloft in her proper element,

> "Makes audible a linked lay of truth,
> Of truth profound a sweet continuous lay,
> Not learnt, but native, her own natural notes!" [33]

S. T. C.

Even throughout his smaller poems there is scarcely one, which is not rendered valuable by some just and original reflection.

See page 25, vol. 2nd.: [34] or the two following passages in one of his humblest compositions.

> "O Reader! had you in your mind
> Such stores as silent thought can bring,
> O gentle Reader! you would find
> A tale in every thing;"

and

> "I've heard of hearts unkind, kind deeds
> With coldness still returning;
> Alas! the gratitude of men
> Has oftener left me mourning" [35]

or in a still higher strain the six beautiful quatrains, page 134.

> "Thus fares it still in our decay:
> And yet the wiser mind
> Mourns less for what age takes away
> Than what it leaves behind.
>
> The Blackbird in the summer trees,
> The Lark upon the hill,
> Let loose their carols when they please,
> Are quiet when they will.
>
> With nature never do *they* wage
> A foolish strife; they see
> A happy youth, and their old age
> Is beautiful and free!
>
> But we are pressed by heavy laws;
> And often, glad no more,
> We wear a face of joy, because
> We have been glad of yore.

> If there is one, who need bemoan
> His kindred laid in earth,
> The household hearts that were his own,
> It is the man of mirth.
>
> My days, my Friend, are almost gone,
> My life has been approved,
> And many love me; but by none
> Am I enough beloved." [36]

or the sonnet on Buonaparte, page 202, vol. 2; [37] or finally (for a volume would scarce suffice to exhaust the instances) the last stanza of the poem on the withered Celandine, vol. 2, p. 212.

> "To be a prodigal's favorite—then, worse truth,
> A miser's pensioner—behold our lot!
> O man! that from thy fair and shining youth
> Age might but take the things youth needed not." [38]

Both in respect of this and of the former excellence, Mr. Wordsworth strikingly resembles Samuel Daniel, one of the golden writers of our golden Elizabethan age, now most causelessly neglected: Samuel Daniel, whose diction bears no mark of time, no distinction of age, which has been, and as long as our language shall last, will be so far the language of the to-day and for ever, as that it is more intelligible to us, than the transitory fashions of our own particular age. A similar praise is due to his sentiments. No frequency of perusal can deprive them of their freshness. For though they are brought into the full day-light of every reader's comprehension; yet are they drawn up from depths which few in any age are priviledged to visit, into which few in any age have courage or inclination to descend. If Mr. Wordsworth is not equally with Daniel alike intelligible to all readers of average understanding in all passages of his works, the comparative difficulty does not arise from the greater impurity of the ore, but from the nature and uses of the metal. A poem is not necessarily obscure, because it does not aim to be popular. It is enough, if a work be perspicuous to those for whom it is written, and

> "Fit audience find, though few." [39]

To the "Ode on the intimation of immortality from recollections of early childhood" the poet might have

---

33. "Makes . . . notes!": Coleridge's "To William Wordsworth," ll. 58–60. **34. page 25, vol. 2nd.:** "Star-Gazers," ll. 9–24. **35.** "O . . . mourning": "Simon Lee," ll. 65–68, 93–96.

36. "Thus . . . beloved": "The Fountain," ll. 33–56. **37. page 202, vol. 2:** "I grieved for Buonaparté." **38.** "To . . . not": "The Small Celandine" ("There is a flower"), ll. 21–24. **39.** "Fit . . . few": *Paradise Lost*, VII. 31.

prefixed the lines which Dante addresses to one of his own Canzoni—

"Canzon, io credo, che saranno radi
Che tua ragione intendan bene,
Tanto lor sei faticoso ed alto." [40]

"O lyric song, there will be few, think I,
Who may thy import understand aright:
Thou art for *them* so arduous and so high!"

But the ode was intended for such readers only as had been accustomed to watch the flux and reflux of their inmost nature, to venture at times into the twilight realms of consciousness, and to feel a deep interest in modes of inmost being, to which they know that the attributes of time and space are inapplicable and alien, but which yet can not be conveyed save in symbols of time and space. For such readers the sense is sufficiently plain, and they will be as little disposed to charge Mr. Wordsworth with believing the Platonic pre-existence in the ordinary interpretation of the words, as I am to believe, that Plato himself ever meant or taught it.

> Πολλά μοι ὑπ' ἀγκῶ-
> νος ὠκέα βέλη
> ἔνδον ἐντὶ φαρέτρας
> φωνᾶντα συνετοῖσιν· ἐς
> δὲ τὸ πᾶν ἑρμηνέων
> χατίζει. σοφὸς ὁ πολ-
> λὰ εἰδὼς φυᾷ.
> μαθόντες δέ, λάβροι
> παγγλωσσίᾳ, κόρακες ὥς,
> ἄκραντα γαρύετον
> Διὸς πρὸς ὄρνιχα θεῖον. [41]

Third (and wherein he soars far above Daniel) the sinewy strength and originality of single lines and paragraphs: the frequent curiosa felicitas of his diction, of which I need not here give specimens, having anticipated them in a preceding page. This beauty, and as eminently characteristic of Wordsworth's poetry, his rudest assailants have felt themselves compelled to acknowledge and admire.

Fourth; the perfect truth of nature in his images and descriptions, as taken immediately from nature, and proving a long and genial intimacy with the very spirit which gives the physiognomic expression to all the works of nature. Like a green field reflected in a calm and perfectly transparent lake, the image is distinguished from the reality only by its greater softness and lustre. Like the moisture or the polish on a pebble, genius neither distorts nor false-colours its objects; but on the contrary brings out many a vein and many a tint, which escapes the eye of common observation, thus raising to the rank of gems what had been often kicked away by the hurrying foot of the traveller on the dusty high road of custom.

Let me refer to the whole description of skating, vol. I., page 42 to 47, [42] especially to the lines

"So through the darkness and the cold we flew,
And not a voice was idle: with the din
Meanwhile the precipices rang aloud;
The leafless trees and every icy crag
Tinkled like iron; while the distant hills
Into the tumult sent an alien sound
Of melancholy, not unnoticed, while the stars
Eastward were sparkling clear, and in the west
The orange sky of evening died away."

Or to the poem on the green linnet, vol. I. page 244. What can be more accurate yet more lovely than the two concluding stanzas?

"Upon yon tuft of hazel trees,
That twinkle to the gusty breeze,
Behold him perched in ecstasies,
    Yet seeming still to hover;
There! where the flutter of his wings
Upon his back and body flings
Shadows and sunny glimmerings,
    That cover him all over.

While thus before my eyes he gleams,
A brother of the leaves he seems;
When in a moment forth he teems
    His little song in gushes:
As if it pleased him to disdain
And mock the form which he did feign,
While he was dancing with the train
    Of leaves among the bushes."

Or the description of the blue-cap, and of the noontide silence, page 284; [43] or the poem to the cuckoo,

---

40. "Canzon . . . alto": *Il Convivio*, II. i. 53–55. 41. Πολλά . . . θεῖον: Pindar, *Olympian Odes*, II. 149–59: "In the quiver beneath my arm I have many swift arrows to sing for those with understanding. But the multitude need interpreters. The true poet is he who knows much by the gift of nature, but those who have only learnt verse as a craft and are violent and immoderate in their speech waste their breath in chattering like crows against the divine bird of Zeus."

42. vol. I. . . . . 47: the poem called "Influence of Natural Objects," later published in *The Prelude*, I. 401–63. 43. page 284: "The Kitten and the Falling Leaves."

page 299;[44] or, lastly, though I might multiply the references to ten times the number, to the poem, so completely Wordsworth's, commencing

"Three years she grew in sun and shower," &c.

Fifth: a meditative pathos, a union of deep and subtle thought with sensibility; a sympathy with man as man; the sympathy indeed of a contemplator, rather than a fellow-sufferer or co-mate, (spectator, haud particeps)[45] but of a contemplator, from whose view no difference of rank conceals the sameness of the nature; no injuries of wind or weather, or toil, or even of ignorance, wholly disguise the human face divine. The superscription and the image of the Creator still remain legible to *him* under the dark lines, with which guilt or calamity had cancelled or cross-barred it. Here the man and the poet lose and find themselves in each other, the one as glorified, the latter as substantiated. In this mild and philosophic pathos, Wordsworth appears to me without a compeer. Such he *is*: so he *writes*. See vol. I. page 134 to 136,[46] or that most affecting composition, the "Affliction of Margaret——of——," page 165 to 168, which no mother, and, if I may judge by my own experience, no parent can read without a tear. Or turn to that genuine lyric, in the former edition, entitled "The Mad Mother,"[47] page 174 to 178, of which I cannot refrain from quoting two of the stanzas, both of them for their pathos, and the former for the fine transition in the two concluding lines of the stanza, so expressive of that deranged state, in which from the increased sensibility the sufferer's attention is abruptly drawn off by every trifle, and in the same instant plucked back again by the one despotic thought, bringing home with it, by the blending, *fusing* power of Imagination and Passion, the alien object to which it had been so abruptly diverted, no longer an alien but an ally and an inmate.

"Suck, little babe, oh suck again!
It cools my blood; it cools my brain:
Thy lips, I feel them, baby! they
Draw from my heart the pain away.
Oh! press me with thy little hand;
It loosens something at my chest:
About that tight and deadly band
I feel thy little fingers prest.
The breeze I see is in the tree!
It comes to cool my babe and me."

"Thy father cares not for my breast,
'Tis thine, sweet baby, there to rest,
'Tis all thine own!—and, if its hue
Be changed, that was so fair to view,
'Tis fair enough for thee, my dove!
My beauty, little child, is flown,
But thou wilt live with me in love;
And what if my poor cheek be brown?
'Tis well for me, thou canst not see
How pale and wan it else would be."

Last, and pre-eminently, I challenge for this poet the gift of IMAGINATION in the highest and strictest sense of the word. In the play of *Fancy*, Wordsworth, to my feelings, is not always graceful, and sometimes *recondite*. The *likeness* is occasionally too strange, or demands too peculiar a point of view, or is such as appears the creature of predetermined research, rather than spontaneous presentation. Indeed his fancy seldom displays itself, as mere and unmodified fancy. But in imaginative power, he stands nearest of all modern writers to Shakespeare and Milton; and yet in a kind perfectly unborrowed and his own. To employ his own words, which are at once an instance and an illustration, he does indeed to all thoughts and to all objects

"—————————— add the gleam,
The light that never was, on sea or land,
The consecration, and the poet's dream."[48]

I shall select a few examples as most obviously manifesting this faculty; but if I should ever be fortunate enough to render my analysis of imagination, its origin and characters, thoroughly intelligible to the reader, he will scarcely open on a page of this poet's works without recognising, more or less, the presence and the influences of this faculty.

From the poem on the Yew Trees, vol. I. page 303, 304.

"But worthier still of note
Are those fraternal four of Borrowdale,
Joined in one solemn and capacious grove:
Huge trunks!—and each particular trunk a growth
Of intertwisted fibres serpentine
Up-coiling, and inveterately convolved,—
Not uninformed with phantasy, and looks
That threaten the profane;—a pillared shade,
Upon whose grassless floor of red-brown hue,
By sheddings from the pinal umbrage tinged
Perennially—beneath whose sable roof

44. **page 299:** "To the Cuckoo" ("O blithe new-comer"). 45. **spectator . . . particeps:** a spectator, hardly a participant. 46. **page . . . 136:** "'Tis said that some have died for love." 47. **"The Mad Mother":** "Her Eyes Are Wild."

48. **"add . . . dream":** "Elegiac Stanzas Suggested by a Picture of Peele Castle," ll. 14–16.

Of boughs, as if for festal purpose decked
With unrejoicing berries, ghostly shapes
May meet at noontide—FEAR and trembling HOPE,
SILENCE and FORESIGHT—DEATH, the skeleton,
And TIME, the shadow—there to celebrate,
As in a natural temple scattered o'er
With altars undisturbed of mossy stone,
United worship; or in mute repose
To lie, and listen to the mountain flood
Murmuring from Glaramara's inmost caves."

The effect of the old man's figure in the poem of
Resignation and Independence, vol. II. page 33.

"While he was talking thus, the lonely place,
The old man's shape, and speech, all troubled me:
In my mind's eye I seemed to see him pace
About the weary moors continually,
Wandering about alone and silently."

Or the 8th, 9th, 19th, 26th, 31st, and 33d,[49] in the
collection of miscellaneous sonnets—the sonnet on
the subjugation of Switzerland,[50] page 210, or the last
ode,[51] from which I especially select the two following
stanzas or paragraphs, page 349 to 350.

[Coleridge here quotes ll. 58–76, and 133–171, printed
above.]

And since it would be unfair to conclude with an
extract, which, though highly characteristic, must yet,
from the nature of the thoughts and the subject, be
interesting, or perhaps intelligible, to but a limited
number of readers; I will add, from the poet's last
published work, a passage equally Wordsworthian;
of the beauty of which, and of the imaginative power
displayed therein, there can be but one opinion, and
one feeling. See "White Doe," page 5.

"Fast the church-yard fills;—anon
Look again and they are gone;
The cluster round the porch, and the folk
Who sate in the shade of the prior's oak!
And scarcely have they disappear'd,
Ere the prelusive hymn is heard;—
With one consent the people rejoice,
Filling the church with a lofty voice!

They sing a service which they feel,
For 'tis the sun-rise of their zeal;
And faith and hope are in their prime
In great Eliza's golden time.

"A moment ends the fervent din,
And all is hushed, without and within;
For though the priest, more tranquilly,
Recites the holy liturgy,
The only voice which you can hear
Is the river murmuring near.
When soft!—the dusky trees between,
And down the path through the open green,
Where is no living thing to be seen;
And through yon gateway, where is found,
Beneath the arch with ivy bound,
Free entrance to the church-yard ground;
And right across the verdant sod,
Towards the very house of God;
Comes gliding in with lovely gleam,
Comes gliding in serene and slow,
Soft and silent as a dream,
A solitary doe!
White she is as lily of June,
And beauteous as the silver moon
When out of sight the clouds are driven
And she is left alone in heaven!
Or like a ship some gentle day
In sunshine sailing far away—
A glittering ship, that hath the plain
Of ocean for her own domain.

. . .

"What harmonious pensive changes
Wait upon her as she ranges
Round and through this pile of state
Overthrown and desolate!
Now a step or two her way
Is through space of open day,
Where the enamoured sunny light
Brightens her that was so bright;
Now doth a delicate shadow fall,
Falls upon her like a breath,
From some lofty arch or wall,
As she passes underneath."[52]

The following analogy will, I am apprehensive,
appear dim and fantastic, but in reading Bartram's
Travels[53] I could not help transcribing the following
lines as a sort of allegory, or connected simile and
metaphor of Wordsworth's intellect and genius.—

49. 8th . . . 33d: "Where Lies the Land," "Even as a
Dragon's Eye," "O Mountain Stream," "Composed Upon
Westminster Bridge," "Methought I Saw the Footsteps of
a Throne," and "It is a Beauteous Evening." 50. sonnet
. . . Switzerland: "Thoughts of a Briton on the Sub-
jugation of Switzerland." 51. last ode: "Intimations of
Immortality."

52. "Fast . . . underneath": *The White Doe of Rylstone*,
I. 31–66, 79–90. 53. Bartram's Travels: William Bartram's
*Travels Through North and South Carolina and the Cherokee
Country* (1792), p. 36.

"The soil is a deep, rich, dark mould, on a deep stratum of tenacious clay; and that on a foundation of rocks, which often break through both strata, lifting their back above the surface. The trees which chiefly grow here are the gigantic black oak; magnolia magni-floria; fraxinus excelsior; platane; and a few stately tulip trees." What Mr. Wordsworth *will* produce, it is not for me to prophecy: but I could pronounce with the liveliest convictions what he is capable of producing. It is the FIRST GENUINE PHILOSOPHIC POEM.

The preceding criticism will not, I am aware, avail to overcome the prejudices of those, who have made it a business to attack and ridicule Mr. Wordsworth's compositions.

Truth and prudence might be imaged as concentric circles. The poet may perhaps have passed beyond the latter, but he has confined himself far within the bounds of the former, in designating these critics, as too petulant to be passive to a genuine poet, and too feeble to grapple with him;—"men of palsied imaginations, in whose minds all healthy action is languid;—who, therefore, feed as the many direct them, or with the many are greedy after vicious provocatives."[54]

Let not Mr. Wordsworth be charged with having expressed himself too indignantly, till the wantonness and the systematic and malignant perseverance of the aggressions have been taken into fair consideration. I myself heard the commander in chief[55] of this unmanly warfare make a boast of his private admiration of Wordsworth's genius. I have heard him declare, that whoever came into his room would probably find the Lyrical Ballads lying open on his table, and that (speaking exclusively of those written by Mr. Wordsworth himself) he could nearly repeat the whole of them by heart. *But* a Review, in order to be a saleable article, must be *personal, sharp,* and *pointed*: and, *since then*, the poet has made himself, and with himself all who were, or were supposed to be, his friends and admirers, the object of the critic's revenge—how? by having spoken of a work so conducted in the terms which it deserved! I once heard a clergyman in boots and buckskin avow, that he would cheat his own father *in a horse*. A moral system of a similar nature seems to have been adopted by too many anonymous critics. As we used to say at school, in reviewing they *make* being rogues: and he, who complains, is to be laughed at for his ignorance of *the game*. With the pen out of

their hand they are *honorable men*. They exert indeed power (which is to that of the injured party who should attempt to expose their glaring perversions and misstatements, as twenty to one) to write down, and (where the author's circumstances permit) to *impoverish* the man, whose learning and genius they themselves in private have repeatedly admitted. They knowingly strive to make it impossible for the man even to publish[56] any future work without exposing himself to all the wretchedness of debt and embarrassment. But this is all *in their vocation*: and, bating what they do in their *vocation*, "*who can say that black is the white of their eye?*"

So much for the detractors from Wordsworth's merits. On the other hand, much as I might wish for their fuller sympathy, I dare not flatter myself, that the freedom with which I have declared my opinions concerning both his theory and his defects, most of which are more or less connected with his theory, either as cause or effect, will be satisfactory or pleasing to *all* the poet's admirers and advocates. More indiscriminate than mine their admiration may be: deeper and more sincere it can not be. But I have advanced no opinion either for praise or censure, other than as texts introductory to the reasons which compel me to form it. Above all, I was fully convinced that such a criticism was not only wanted; but that, if executed with adequate ability, it must conduce, in no mean degree, to Mr. Wordsworth's *reputation*. His *fame* belongs to another age, and can neither be accelerated nor retarded. How small the proportion of the defects are to the beauties, I have repeatedly declared; and that no one of them originates in deficiency of poetic genius. Had they been more and greater, I should still, as a friend to his literary character in the present age, consider an analytic display of them as *pure gain*; if only it removed, as surely to all reflecting minds even the foregoing analysis must have removed, the strange mistake, so slightly grounded, yet so widely and industriously propagated, of Mr. Wordsworth's turn for SIMPLICITY! I am not half so much irritated by hearing his enemies abuse him for vulgarity of style, subject, and conception; as I am disgusted with the

---

54. **"men . . . provocatives"**: Wordsworth's "Essay Supplementary to the Preface." 55. **commander in chief**: Francis Jeffrey, 1773–1850, editor of *The Edinburgh Review*.

56. **publish**: "Not many months ago an eminent bookseller was asked what he thought of ———? The answer was: 'I have heard his powers very highly spoken of by some of our first-rate men; but I would not have a work of his if any one would give it me: for he is spoken but slightly of, or not at all, in the Quarterly Review: and the Edinburgh, you know, is decided to cut him up!'" (C.).

gilded side of the same meaning, as displayed by some affected admirers, with whom he is, forsooth, a *sweet, simple poet!* and *so* natural, that little master Charles and his younger sister are *so* charmed with them, that they play at "Goody Blake," or at "Johnny and Betty Foy!"

Were the collection of poems, published with these biographical sketches, important enough, (which I am not vain enough to believe), to deserve such a distinction; EVEN AS I HAVE DONE, SO WOULD I BE DONE UNTO.

For more than eighteen months have the volume of Poems, entitled SIBYLLINE LEAVES, and the present volumes, up to this page, been printed, and ready for publication. But, ere I speak of myself in the tones, which are alone natural to me under the circumstances of late years, I would fain present myself to the Reader as I was in the first dawn of my literary life:

"When Hope grew round me, like the climbing vine,
And fruits and foliage, not my own, seem'd mine!"[57]

For this purpose I have selected from the letters, which I wrote home from Germany, those which appeared likely to be most interesting, and at the same time most pertinent to the title of this work.

ON POESY OR ART

❧

The most complete single expression of Coleridge's aesthetics, "On Poesy or Art" consists of notes for a lecture of 1818. Delivering the lecture, he would naturally have expanded these elliptical, fragmentary statements, some of which the footnotes below attempt to interpret. (Many of them are from W. J. Bate, ed. *Criticism: the Major Texts*, 1952.) The lecture was strongly influenced by the German philosopher Friedrich Wilhelm von Schelling, particularly his "On the Relation of the Formative Arts to Nature," in which there are many closely parallel passages.

❧

Man communicates by articulation of sounds, and paramountly by the memory in the ear; nature by the impression of bounds and surfaces on the eye, and through the eye it gives significance and appropriation, and thus the conditions of memory, or the capability of being remembered, to sounds, smells, &c. Now Art, used collectively for painting, sculpture, architecture

57. "When . . . mine!": Coleridge's "Dejection: An Ode," ll. 80–81.

and music, is the mediatress between, and reconciler of, nature and man. It is, therefore, the power of humanizing nature, of infusing the thoughts and passions of man into every thing which is the object of his contemplation; color, form, motion, and sound, are the elements which it combines, and it stamps them into unity in the mould of a moral idea.[1]

The primary art is writing;—primary, if we regard the purpose abstracted from the different modes of realizing it, those steps of progression of which the instances are still visible in the lower degrees of civilization. First, there is mere gesticulation; then rosaries or *wampum*; then picture-language; then hieroglyphics, and finally alphabetic letters. These all consist of a translation of man into nature, of a substitution of the visible for the audible.

The so called music of savage tribes as little deserves the name of art for the understanding, as the ear warrants it for music. Its lowest state is a mere expression of passion by sounds which the passion itself necessitates; —the highest amounts to no more than a voluntary

ON POESY OR ART. 1. **idea:** "This rather abrupt opening, and the two following paragraphs, might be paraphrased as follows: Men communicate with each other by sounds, particularly words. We learn about nature, however, primarily through our eyes; and what we see gives order and point to what we pick up about nature through our other senses. Now art seeks to present nature to man in the most meaningful way possible. It draws upon both sight and sound, therefore; it puts what it portrays in such a way as to strike home to man's feelings, and also gives it meaning to the *mind* by presenting it in the light of thought and ideals. Art, by doing this, is thus a coming together of nature and man—a mediator between them. Now the kind of thing art tries to do is shown, in its most basic form, by writing (that is, *visible signs* or *symbols*—whether in the forms of gestures, picture-language, or alphabetic letters). For here especially what man is trying to do is to take his own thoughts and feelings and translate them into forms that are visible. He is *creating symbols*, in other words, visible signs, like the forms we see in nature, and through which we learn about nature, but which are now being used to represent *human* thoughts and feelings. Consequently, language—particularly written language—is a more distinctively human medium than are the sounds used by music or the forms or colors used by the visual arts (which, as contrasted with language, are more directly taken over from nature). That is, articulate speech, particularly written speech, is more *symbolic*, more creative; the human mind has entered more into the making of it as a medium of communication. Thus, in that 'union or reconciliation' of nature and man which art tries to bring about, poetry especially, of all the arts, represents more of a reaching out on the part of the human mind—more of a coming forward to meet nature, so to speak" (W. J. Bate, ed., *Criticism: The Major Texts*, p. 393n.).

reproduction of these sounds in the absence of the occasioning causes, so as to give the pleasure of contrast,—for example, by the various outcries of battle in the song of security and triumph. Poetry also is purely human; for all its materials are from the mind, and all its products are for the mind. But it is the apotheosis of the former state, in which by excitement of the associative power passion itself imitates order, and the order resulting produces a pleasureable passion, and thus it elevates the mind by making its feelings the object of its reflexion. So likewise, whilst it recalls the sights and sounds that had accompanied the occasions of the original passions, poetry impregnates them with an interest not their own by means of the passions, and yet tempers the passion by the calming power which all distinct images exert on the human soul. In this way poetry is the preparation for art, inasmuch as it avails itself of the forms of nature to recall, to express, and to modify the thoughts and feelings of the mind. Still, however, poetry can only act through the intervention of articulate speech, which is so peculiarly human, that in all languages it constitutes the ordinary phrase by which man and nature are contradistinguished. It is the original force of the word "brute," and even "mute" and "dumb" do not convey the absence of sound, but the absence of articulated sounds.

As soon as the human mind is intelligibly addressed by an outward image exclusively of articulate speech, so soon does art commence. But please to observe that I have laid particular stress on the words "human mind,"—meaning to exclude thereby all results common to man and all other sentient creatures, and consequently confining myself to the effect produced by the congruity of the animal impression with the reflective powers of the mind; so that not the thing presented, but that which is re-presented by the thing, shall be the source of the pleasure. In this sense nature itself is to a religious observer the art of God; and for the same cause art itself might be defined as of a middle quality between a thought and a thing, or, as I said before, the union and reconciliation of that which is nature with that which is exclusively human. It is the figured language of thought, and is distinguished from nature by the unity of all the parts in one thought or idea. Hence nature itself would give us the impression of a work of art, if we could see the thought which is present at once in the whole and in every part; and a work of art will be just in proportion as it adequately conveys the thought, and rich in proportion to the variety of parts which it holds in unity.

If, therefore, the term "mute" be taken as opposed not to sound but to articulate speech, the old definition of painting will in fact be the true and best definition of the Fine Arts in general, that is, *muta poesis*, mute poesy, and so of course poesy. And, as all languages perfect themselves by a gradual process of desynonymizing words originally equivalent, I have cherished the wish to use the word "poesy" as the generic or common term, and to distinguish that species of poesy which is not *muta poesis* by its usual name "poetry"; while of all the other species which collectively form the Fine Arts, there would remain this as the common definition,—that they all, like poetry, are to express intellectual purposes, thoughts, conceptions, and sentiments which have their origin in the human mind,—not, however, as poetry does, by means of articulate speech, but as nature or the divine art does, by form, color, magnitude, proportion, or by sound, that is, silently or musically.

Well! it may be said—but who has ever thought otherwise? We all know that art is the imitatress of nature. And, doubtless, the truths which I hope to convey would be barren truisms, if all men meant the same by the words "imitate" and "nature." But it would be flattering mankind at large, to presume that such is the fact. First, to imitate. The impression on the wax is not an imitation, but a copy, of the seal; the seal itself is an imitation.[2] But, further, in order to

2. **imitation:** "One of Coleridge's favorite distinctions is that between 'imitation' and 'copy.' A true imitation, as distinct from a copy, will have different materials with which to work; and the form that is drawn out from those materials will, in some ways, also be different from that of the original. Thus, if an animal is carved in profile onto a seal, not only are the materials different—metal instead of bone, flesh, and skin, and a two-dimensional plane instead of three dimensions—but the form also (except for the similarity in the outline of the profile) may in many ways be different because of circumstances determined by the nature of the *materials*. . . . Because an imitation starts with an acknowledged difference, we then—as we note similarities in it with the original—work from diversity up to similarity (or unity); this process of *unity working through diversity* is not only a characteristic of what is beautiful in nature, as well as of an 'imitation' of it, but also, on a third level, it characterizes that act of mind by which we compare the imitation and the original. But if we see a 'copy'—which is never as good as the original, and can never compete with nature herself on her own grounds—then this process of moving from diversity to similarity is turned around, and we move backwards We *start* with similarity (the unity between the original and the copy), and then begin to note the differences, as in Coleridge's example of seeing wax figures" (Bate, p. 395n.).

form a philosophic conception, we must seek for the kind, as the heat in ice, invisible light, &c., whilst, for practical purposes, we must have reference to the degree. It is sufficient that philosophically we understand that in all imitation two elements must coexist, and not only coexist, but must be perceived as coexisting. These two constituent elements are likeness and unlikeness, or sameness and difference, and in all genuine creations of art there must be a union of these disparates. The artist may take his point of view where he pleases, provided that the desired effect be perceptibly produced,—that there be likeness in the difference, difference in the likeness, and a reconcilement of both in one. If there be likeness to nature without any check of difference, the result is disgusting, and the more complete the delusion, the more loathsome the effect. Why are such simulations of nature, as waxwork figures of men and women, so disagreeable?[3] Because, not finding the motion and the life which we expected, we are shocked as by a falsehood, every circumstance of detail, which before induced us to be interested, making the distance from truth more palpable. You set out with a supposed reality and are disappointed and disgusted with the deception; whilst, in respect to a work of genuine imitation, you begin with an acknowledged total difference, and then every touch of nature gives you the pleasure of an approximation to truth. The fundamental principle of all this is undoubtedly the horror of falsehood and the love of truth inherent in the human breast. The Greek tragic dance rested on these principles, and I can deeply sympathize in imagination with the Greeks in this favorite part of their theatrical exhibitions, when I call to mind the pleasure I felt in beholding the combat of the Horatii and Curiatii most exquisitely danced in Italy to the music of Cimarosa.

Secondly, as to nature. We must imitate nature! yes, but what in nature,—all and everything? No, the beautiful in nature. And what then is the beautiful? What is beauty? It is, in the abstract, the unity of the manifold, the coalescence of the diverse; in the concrete, it is the union of the shapely (*formosum*) with the vital. In the dead organic it depends on regularity of form, the first and lowest species of which is the triangle with

all its modifications, as in crystals, architecture, &c.; in the living organic it is not mere regularity of form, which would produce a sense of formality; neither is it subservient to any thing beside itself. It may be present in a disagreeable object, in which the proportion of the parts constitutes a whole; it does not arise from association, as the agreeable does, but sometimes lies in the rupture of association; it is not different to different individuals and nations, as has been said, nor is it connected with the ideas of the good, or the fit, or the useful. The sense of beauty is intuitive, and beauty itself is all that inspires pleasure without, and aloof from, and even contrarily to, interest.

If the artist copies the mere nature, the *natura naturata*, what idle rivalry! If he proceeds only from a given form, which is supposed to answer to the notion of beauty, what an emptiness, what an unreality there always is in his productions, as in Cipriani's pictures![4] Believe me, you must master the essence, the *natura naturans*, which presupposes a bond between nature in the higher sense and the soul of man.[5]

The wisdom in nature is distinguished from that in man by the co-instantaneity of the plan and the execution; the thought and the product are one, or are given at once; but there is no reflex act, and hence there is no moral responsibility. In man there is reflexion, freedom, and choice; he is, therefore, the head of the visible creation. In the objects of nature[6] are presented, as in a mirror, all the possible elements, steps, and processes of intellect antecedent to consciousness, and therefore to the full development of the intelligential act; and man's mind is the very focus of all the rays of intellect which are scattered throughout the images of nature. Now so to place these images, totalized, and fitted to the limits of the human mind, as to elicit from, and to superinduce upon, the forms themselves

---

3. disagreeable: Cf. Coleridge's letter to C. Mathews, June, 1814: "What a marble peach on a mantle-piece, that you take up deluded, & put down with pettish disgust, is compared with a fruit-piece of Vanhuysen's, even such is a mere *Copy* of nature compared with a true histrionic *Imitation*."

4. Cipriani's pictures: Giovanni Battista Cipriani, 1727–85. **5. Believe . . . man:** "The standpoint is opposed both to complete naturalism and to abstractionism in art. If the artist simply copies the *natura naturata*, he is fighting a losing battle with nature. If, on the other hand, he takes some abstract form for beauty, and tries to *project* it upon his object, the form will still be bodiless and unreal" (Bate, p. 396n.). The artist must exhibit things in their full, individual concreteness, but he must also disclose the idea or law that works within. That he can do this presupposes that nature is essentially not matter but a dynamic system of ideas or laws. **6. objects of nature:** here considered as arranged in an ascending series from the simplest forms of matter, through the more complex structures of crystals, the lowest forms of organic life, etc.

the moral reflexions to which they approximate, to make the external internal, the internal external, to make nature thought, and thought nature,—this is the mystery of genius in the Fine Arts. Dare I add that the genius must act on the feeling, that body is but a striving to become mind,—that it is mind in its essence![7]

In every work of art there is a reconcilement of the external with the internal; the conscious is so impressed on the unconscious as to appear in it; as compare mere letters inscribed on a tomb with figures themselves constituting the tomb. He who combines the two is the man of genius; and for that reason he must partake of both. Hence there is in genius itself an unconscious activity; nay, that is the genius in the man of genius. And this is the true exposition of the rule that the artist must first eloign himself from nature in order to return to her with full effect. Why this? Because if he were to begin by mere painful copying, he would produce masks only, not forms breathing life. He must out of his own mind create forms according to the severe laws of the intellect, in order to generate in himself that co-ordination of freedom and law, that involution of obedience in the prescript, and of the prescript in the impulse to obey, which assimilates him to nature, and enables him to understand her. He merely absents himself for a season from her, that his own spirit, which has the same ground with nature, may learn her unspoken language in its main radicals, before he approaches to her endless compositions of them. Yes, not to acquire cold notions—lifeless technical rules— but living and life-producing ideas, which shall contain their own evidence, the certainty that they are essentially one with the germinal causes in nature,—his consciousness being the focus and mirror of both,—for this does the artist for a time abandon the external real in order to return to it with a complete sympathy with its internal and actual. For of all we see, hear, feel

and touch the substance is and must be in ourselves; and therefore there is no alternative in reason between the dreary (and thank heaven! almost impossible) belief that every thing around us is but a phantom, or that the life which is in us is in them likewise; and that to know is to resemble, when we speak of objects out of ourselves, even as within ourselves to learn is, according to Plato, only to recollect;—the only effective answer to which, that I have been fortunate to meet with, is that which Pope has consecrated for future use in the line—

"And coxcombs vanquish Berkeley with a grin!"[8]

The artist must imitate that which is within the thing, that which is active through form and figure, and discourses to us by symbols—the *Natur-geist*, or spirit of nature, as we unconsciously imitate those whom we love; for so only can he hope to produce any work truly natural in the object and truly human in the effect. The idea which puts the form together cannot itself be the form. It is above form, and is its essence, the universal in the individual, or the individuality itself,—the glance and the exponent of the indwelling power.

Each thing that lives has its moment of self-exposition, and so has each period of each thing, if we remove the disturbing forces of accident. To do this is the business of ideal art, whether in images of childhood, youth, or age, in man or in woman. Hence a good portrait is the abstract of the personal; it is not the likeness for actual comparison, but for recollection. This explains why the likeness of a very good portrait is not always recognized; because some persons never abstract, and amongst these are especially to be numbered the near relations and friends of the subject, in consequence of the constant pressure and check exercised on their minds by the actual presence of the original. And each thing that only appears to live has also its possible position of relation to life, as nature herself testifies, who where she cannot be, prophesies her being in the crystallized metal, or the inhaling plant.

The charm, the indispensable requisite, of sculpture is unity of effect. But painting rests in a material remoter from nature, and its compass is therefore greater. Light and shade give external, as well as internal, being even with all its accidents, whilst sculpture is confined to the latter. And here I may observe that the subjects chosen for works of art, whether in sculpture or

---

7. **essence:** "By having consciousness and the ability to reflect and choose, man's mind serves as a focal point at which the unfolding processes of nature can become awake, so to speak, and achieve, through awareness, the ability to see, distinguish, and know themselves. In this sense, whatever awareness is attained by the human mind may be regarded as a *completing* of nature, as a coming of nature to a head and a *flowering out* of it into a *self-consciousness that is able to evaluate and select* and is therefore morally responsible. Genius in art is found in the ability to weld the processes of nature ('the external') to the completing thought and moral evaluating of the mind ('the internal') in such a way that they become entirely one" (Bate, p. 396n.).

8. **"And . . . grin!":** not by Pope but John Brown, 1715– 66, "Essay on Satire," II. 224.

painting, should be such as really are capable of being expressed and conveyed within the limits of those arts. Moreover they ought to be such as will affect the spectator by their truth, their beauty, or their sublimity, and therefore they may be addressed to the judgement, the senses, or the reason. The peculiarity of the impression which they may make, may be derived either from color and form, or from proportion and fitness, or from the excitement of the moral feelings; or all these may be combined. Such works as do combine these sources of effect must have the preference in dignity.

Imitation of the antique may be too exclusive, and may produce an injurious effect on modern sculpture; —1st, generally, because such an imitation cannot fail to have a tendency to keep the attention fixed on externals rather than on the thought within;—2ndly, because, accordingly, it leads the artist to rest satisfied with that which is always imperfect, namely, bodily form, and circumscribes his views of mental expression to the ideas of power and grandeur only;—3rdly, because it induces an effort to combine together two incongruous things, that is to say, modern feelings in antique forms;—4thly, because it speaks in a language, as it were, learned and dead, the tones of which, being unfamiliar, leave the common spectator cold and unimpressed;—and lastly, because it necessarily causes a neglect of thoughts, emotions and images of profounder interest and more exalted dignity, as motherly, sisterly, and brotherly love, piety, devotion, the divine become human,—the Virgin, the Apostle, the Christ. The artist's principle in the statue of a great man should be the illustration of departed merit; and I cannot but think that a skilful adoption of modern habiliments would, in many instances, give a variety and force of effect which a bigoted adherence to Greek or Roman costume precludes. It is, I believe, from artists finding Greek models unfit for several important modern purposes, that we see so many allegorical figures on monuments and elsewhere. Painting was, as it were, a new art, and being unshackled by old models it chose its own subjects, and took an eagle's flight. And a new field seems opened for modern sculpture in the symbolical expression of the ends of life, as in Guy's monument, Chantrey's children in Worcester Cathedral,[9] &c.

9. **Guy's . . . Cathedral:** the statue of Thomas Guy, 1645?–1742, founder of Guy's Hospital, London; and the statuary group, "The Sleeping Children," by Sir Francis Chantrey in the Lichfield (not Worcester) Cathedral.

Architecture exhibits the greatest extent of the difference from nature which may exist in works of art. It involves all the powers of design, and is sculpture and painting inclusively. It shews the greatness of man, and should at the same time teach him humility.

Music is the most entirely human of the fine arts, and has the fewest *analoga* in nature. Its first delightfulness is simple accordance with the ear; but it is an associated thing, and recalls the deep emotions of the past with an intellectual sense of proportion. Every human feeling is greater and larger than the exciting cause,—a proof, I think, that man is designed for a higher state of existence; and this is deeply implied in music, in which there is always something more and beyond the immediate expression.

With regard to works in all the branches of the fine arts, I may remark that the pleasure arising from novelty must of course be allowed its due place and weight. This pleasure consists in the identity of two opposite elements, that is to say—sameness and variety. If in the midst of the variety there be not some fixed object for the attention, the unceasing succession of the variety will prevent the mind from observing the difference of the individual objects; and the only thing remaining will be the succession, which will then produce precisely the same effect as sameness. This we experience when we let the trees or hedges pass before the fixed eye during a rapid movement in a carriage, or, on the other hand, when we suffer a file of soldiers or ranks of men in procession to go on before us without resting the eye on any one in particular. In order to derive pleasure from the occupation of the mind, the principle of unity must always be present, so that in the midst of the multeity the centripetal force be never suspended, nor the sense be fatigued by the predominance of the centrifugal force. This unity in multeity I have elsewhere stated as the principle of beauty. It is equally the source of pleasure in variety, and in fact a higher term including both. What is the seclusive or distinguishing term between them?

Remember that there is a difference between form as proceeding, and shape as superinduced;—the latter is either the death or the imprisonment of the thing;—the former is its self-witnessing and self-effected sphere of agency. Art would or should be the abridgment of nature. Now the fulness of nature is without character, as water is purest when without taste, smell, or color; but this is the highest, the apex only,—it is not the whole.[10] The object of art is to give the whole *ad hominem*; hence each step of nature hath its ideal, and hence the possibility of a climax up to the perfect form of a harmonized chaos.

To the idea of life victory or strife is necessary; as virtue consists not simply in the absence of vices, but in the overcoming of them. So it is in beauty. The sight of what is subordinated and conquered heightens the strength and the pleasure; and this should be exhibited by the artist either inclusively in his figure, or else out of it, and beside it to act by way of supplement and contrast. And with a view to this, remark the seeming identity of body and mind in infants, and thence the loveliness of the former; the commencing separation in boyhood, and the struggle of equilibrium in youth: thence onward the body is first simply indifferent; then demanding the translucency of the mind not to be worse than indifferent; and finally all that presents the body as body becoming almost of an excremental nature.

FROM

SHAKESPEAREAN CRITICISM

꙰

The following extracts come from Coleridge's marginalia and lecture notes, and from shorthand reports of his public lectures delivered between 1808 and 1819.

꙰

THE CHARACTER OF HAMLET[1]

1. Shakespeare's mode of conceiving characters out of his own intellectual and moral faculties, by conceiving any one intellectual or moral faculty in morbid excess and then placing himself, thus mutilated and

---

10. **Now . . . whole:** "The ultimate forms of nature, when they are abstracted from their function as controlling agents, lack the character, the definite outlines, that arise from limitation—from being brought to focus, in other words, into specific individuality. But this ultimate 'apex' is not the 'whole' of nature; the 'whole' includes the process Coleridge elsewhere described as 'totality dawning into individuation.' The aim of art is to render the *whole*, this *active process*, meaningful *ad hominem*—that is, meaningful to human nature, to human feelings" (Bate, p. 399n.). SHAKESPEAREAN CRITICISM. 1. This extract is from notes for a lecture of 1813.

diseased, under given circumstances. This we shall have repeated occasion to re-state and enforce. In Hamlet I conceive him to have wished to exemplify the moral necessity of a due balance between our attention to outward objects and our meditation on inward thoughts—a due balance between the real and the imaginary world. In Hamlet this balance does not exist—his thoughts, images, and fancy [being] far more vivid than his perceptions, and his very perceptions instantly passing thro' the medium of his contemplations, and acquiring as they pass a form and color not naturally their own. Hence great, enormous, intellectual activity, and a consequent proportionate aversion to real action, with all its symptoms and accompanying qualities.

> Action is transitory, a step, a blow, etc.[2]

Then as in the first instance proceed with a cursory survey thro' the play, with comments, etc.

(1) The easy language of ordinary life, contrasted with the direful music and wild rhythm of the opening of *Macbeth*. Yet the armour, the cold, the dead silence, all placing the mind in the state congruous with tragedy.

(2) The admirable judgement and yet confidence in his own marvellous powers in introducing the ghost twice, each rising in solemnity and awfulness before its third appearance to Hamlet himself.

(3) Shakespeare's tenderness with regard to all innocent superstition: no Tom Paine declarations and pompous philosophy.

(4) The first words that Hamlet speaks—

> A little more than kin, and less than kind.

He begins with that play of words, the complete absence of which characterizes *Macbeth* . . . [?]. No one can have heard quarrels among the vulgar but must have noticed the close connection of punning with angry contempt. Add too what is highly characteristic of superfluous activity of mind, a sort of playing with a thread or watch chain or snuff box.

(5) And [note] how the character develops itself in the next speech—the aversion to externals, the betrayed habit of brooding over the world within him, and the prodigality of beautiful words, which are, as it were, the half embodyings of thoughts, that make them more than thoughts, give them an outness, a reality *sui generis*, and yet retain their correspondence and shadowy approach to the images and movements within.

---

2. **Action . . . etc.:** Wordsworth's *The Borderers*, III. 1539.

(6) The first soliloquy [I. ii.

O, that this too too solid flesh would melt.]

[The] reasons why *taedium vitae* oppresses minds like Hamlet's: the exhaustion of bodily feeling from perpetual exertion of mind; that all mental form being indefinite and ideal, realities must needs become cold, and hence it is the indefinite that combines with passion.

(7) And in this mood the relation is made [by Horatio, who tells Hamlet of his father's ghost], of which no more than [that] it is a perfect model of dramatic narration and dramatic style, the purest poetry and yet the most natural language, equally distant from the inkhorn and the provincial plough.

(8) Hamlet's running into long reasonings [while waiting for the ghost], carrying off the impatience and uneasy feelings of expectation by running away from the *particular* in[to] the *general*. This aversion to personal, individual concerns, and escape to generalizations and general reasonings a most important characteristic.

Besides that, it does away with surprizing all the ill effects that the two former appearances of the ghost would have produced by rendering the ghost an expected phenomenon, and restores to it all the suddenness essential to the effect.

(9) The ghost [is] a superstition connected with the most [sacred?] truths of revealed religion and, therefore, O how contrasted from the withering and wild language of the [witches in] *Macbeth*.

(10) The instant and over violent resolve of Hamlet —how he wastes in the efforts of resolving the energies of action. Compare this with the . . . [?] of Medea; and [note] his quick relapse into the satirical and ironical vein [after the ghost disappears].

(11) Now comes the difficult task, [interpreting the jests of Hamlet when his companions overtake him].

The familiarity, comparative at least, of a brooding mind with shadows is something. Still more the necessary alternation when one muscle long strained is relaxed; the antagonist comes into action of itself. Terror [is] closely connected with the ludicrous; the latter [is] the common mode by which the mind tries to emancipate itself from terror. The laugh is rendered by nature itself the language of extremes, even as tears are. Add too, Hamlet's wildness is but *half-false*. O that subtle trick to pretend the *acting* only when we are very near *being* what we act. And this explanation

of the same with Ophelia's vivid images [describing Hamlet's desperation when he visits her]; nigh akin to, and productive of, temporary mania. [See II. i. 75–100, the speeches of] Ophelia, [which were just mentioned,] proved by [Hamlet's wildness at Ophelia's grave, v. i. 248–78].

(12) Hamlet's character, as I have conceived [it, is] described by himself [in the soliloquy after the players leave him—

O, what a rogue and peasant slave am I, etc.]

. . .

The[3] seeming inconsistencies in the conduct and character of Hamlet have long exercised the conjectural ingenuity of critics; and as we are always loth to suppose that the cause of defective apprehension is in ourselves, the mystery has been too commonly explained by the very easy process of supposing that it is, in fact, inexplicable, and by resolving the difficulty into the capricious and irregular genius of Shakespeare.

Mr. Coleridge, in his *third* lecture, has effectually exposed the shallow and stupid arrogance of this vulgar and indolent decision. He has shewn that the intricacies of Hamlet's character may be traced to Shakespeare's deep and accurate science in mental philosophy. That this character must have some common connection with the laws of our nature, was assumed by the lecturer from the fact that Hamlet was the darling of every country where literature was fostered. He thought it essential to the understanding of Hamlet's character that we should reflect on the constitution of our own minds. Man was distinguished from the animal in proportion as thought prevailed over sense; but in healthy processes of the mind, a balance was maintained between the impressions of outward objects and the inward operations of the intellect: if there be an overbalance in the contemplative faculty, man becomes the creature of meditation, and loses the power of action. Shakespeare seems to have conceived a mind in the highest degree of excitement, with this over-powering activity of intellect, and to have placed him in circumstances where he was obliged to act on the spur of the moment. Hamlet, though brave and careless of death, had contracted a morbid sensibility from this overbalance in the mind, producing the lingering and vacillating delays of

3. The following extract is from a newspaper report of Coleridge's lecture; the lecture was presumably based on the notes above.

procrastination, and wasting in the energy of resolving the energy of acting. Thus the play of *Hamlet* offers a direct contrast to that of *Macbeth*: the one proceeds with the utmost slowness, the other with breathless and crowded rapidity.

The effect of this overbalance of imagination is beautifully illustrated in the inward brooding of Hamlet—the effect of a superfluous activity of thought. His mind, unseated from its healthy balance, is for ever occupied with the world within him, and abstracted from external things; his words give a substance to shadows, and he is dissatisfied with commonplace realities. It is the nature of thought to be indefinite, while definiteness belongs to reality. The sense of sublimity arises, not from the sight of an outward object, but from the reflection upon it; not from the impression, but from the idea. Few have seen a celebrated waterfall without feeling something of disappointment: it is only subsequently, by reflection, that the idea of the waterfall comes full into the mind, and brings with it a train of sublime associations. Hamlet felt this: in him we see a mind that keeps itself in a state of abstraction, and beholds external objects as hieroglyphics. His soliloquy, "Oh that this too, too solid flesh would melt," arises from a craving after the indefinite: a disposition or temper which most easily besets men of genius; a morbid craving for that which is not. The self-delusion common to this temper of mind was finely exemplified in the character which Hamlet gives of himself: "It cannot be, but I am pigeon-liver'd, and lack gall, to make oppression bitter." He mistakes the seeing his chains for the breaking of them; and delays action, till action is of no use; and he becomes the victim of circumstances and accident.

The lecturer, in descending to particulars, took occasion to defend from the common charge of improbable eccentricity, the scene which follows Hamlet's interview with the Ghost. He showed that after the mind has been stretched beyond its usual pitch and tone, it must either sink into exhaustion and inanity, or seek relief by change. Persons conversant with deeds of cruelty contrive to escape from their conscience by connecting something of the ludicrous with them, and by inventing grotesque terms, and a certain technical phraseology, to disguise the horror of their practices.

The terrible, however paradoxical it may appear, will be found to touch on the verge of the ludicrous. Both arise from the perception of something out of the common nature of things,—something out of place: if from this we can abstract danger, the uncommonness alone remains, and the sense of the ridiculous is excited. The close alliance of these opposites appears from the circumstance that laughter is equally the expression of extreme anguish and horror as of joy: in the same manner that there are tears of joy as well as tears of sorrow, so there is a laugh of terror as well as a laugh of merriment. These complex causes will naturally have produced in Hamlet the disposition to escape from his own feelings of the overwhelming and supernatural by a wild transition to the ludicrous,—a sort of cunning bravado, bordering on the flights of delirium.

STAGE ILLUSION[4]

We commence with *The Tempest* as a specimen of the romantic drama. But whatever play of Shakespeare's we had selected there is one preliminary point to be first settled, as the indispensable condition not only of just and genial criticism, but of all consistency in our opinions. This point is contained in the words, probable, natural. We are all in the habit of praising Shakespeare or of hearing him extolled for his fidelity to nature. Now what are we to understand by these words in their application to the drama? Assuredly not the ordinary meaning of them. Farquhar,[5] the most ably, and if we except a few sentences in one of Dryden's prefaces (written for a particular purpose and in contradiction to the opinions elsewhere supported by him) first exposed the ludicrous absurdities involved in the supposition, and demolished as with the single sweep of a careless hand the whole edifice of French criticism respecting the so-called unities of time and place. But a moment's reflection suffices to make every man conscious of what every man must have before felt, that the drama is an *imitation* of reality, not a *copy*—and that imitation is contradistinguished from copy by this: that a certain quantum of difference is essential to the former, and an indispensable condition and cause of the pleasure we derive from it; while in a copy it is a defect, contravening its name and purpose. If illustration were needed, it should be sufficient to ask why we prefer a fruit view of Van Huysum's to a

4. from lecture notes of 1817. Cf. the letter to Daniel Stuart, below, May 13, 1816. **5. Farquhar:** George Farquhar, 1678–1707, in *A Discourse upon Comedy* (1702).

marble peach on a mantel-piece, or why we prefer an historical picture of West to Mrs. Salmon's wax-figure gallery. Not only that we ought, but that we actually do, all of us judge of the drama under this impression, we need no other proof than the impassive slumber of our sense of probability when we hear an actor announce himself as a Greek, Roman, Venetian, or Persian in good mother English. And how little our great dramatist feared awakening in it we have a lively instance in proof in Portia's answer to Nerissa's question, "What say you then to Falconbridge, the young baron of England?"—to which she replies, "You know I say nothing to him; for he understands not me, nor I him. He hath neither Latin, French, or Italian, and you will come into the court and swear that I have a poor pennyworth in the English."[6]

Still, however, there is a sort of improbability with which we are shocked in dramatic representation no less than in the narrative of real life. Consequently, there must be rules respecting it; and as rules are nothing but means to an end previously ascertained (the inattention to which simple truth has been the occasion of all the pedantry of the French school), we must first ascertain what the immediate end or object of the drama is. Here I find two extremes in critical decision: the French, which evidently presupposes that a perfect delusion is to be aimed at—an opinion which now needs no fresh confutation; the opposite, supported by Dr. Johnson,[7] supposes the auditors throughout as in the full and positive reflective knowledge of the contrary. In evincing the impossibility of delusion, he makes no sufficient allowance for an intermediate state, which we distinguish by the term illusion.

In what this consists I cannot better explain than by referring you to the highest degree of it; namely, dreaming. It is laxly said that during sleep we take our dreams for realities, but this is irreconcilable with the nature of sleep, which consists in a suspension of the voluntary and, therefore, of the comparative power. The fact is that we pass no judgement either way: we simply do not judge them to be unreal, in consequence of which the images act on our minds, as far as they act at all, by their own force as images. Our state while we are dreaming differs from that in which we are in the perusal of a deeply interesting novel in the degree

rather than in the kind, and from three causes: First, from the exclusion of all outward impressions on our senses the images in sleep become proportionally more vivid than they can be when the organs of sense are in their active state. Secondly, in sleep the sensations, and with these the emotions and passions which they counterfeit, are the causes of our dream-images, while in our waking hours our emotions are the effects of the images presented to us. (Apparitions [are] *so detectible*.) Lastly, in sleep we pass at once by a sudden collapse into this supension of will and the comparative power: whereas in an interesting play, read or represented, we are brought up to this point, as far as it is requisite or desirable, gradually, by the art of the poet and the actors; and with the consent and positive aidance of our own will. We *choose* to be deceived. The rule, therefore, may be easily inferred. Whatever tends to prevent the mind from placing it[self] or from being gradually placed in this state in which the images have a negative reality must be a defect, and consequently anything that must force itself on the auditors' mind as improbable, not because it *is* improbable (for that the whole play is foreknown to be) but because it cannot but *appear* as such.

But this again depends on the degree of excitement in which the mind is supposed to be. Many things would be intolerable in the first scene of a play that would not at all interrupt our enjoyment in the height of the interest. The narrow cockpit may hold

> The vasty fields of France, or we may cram
> Within its wooden O the very casques
> That did affright the air at Agincourt?[8]

And again, on the other hand, many obvious improbabilities will be endured as belonging to the groundwork of the story rather than to the drama, in the first scenes, which would disturb or disentrance us from all illusion in the acme of our excitement, as, for instance, Lear's division of his realm and banishment of Cordelia. But besides this dramatic probability, all the other excellencies of the drama, as unity of interest, with distinctness and subordination of the characters, appropriateness of style, nay, and the charm of language and sentiment for their own sakes, yet still as far as they tend to increase the inward excitement, are all means to this chief end, that of producing and supporting this willing illusion.

---

6. **"What . . . English"**: *The Merchant of Venice*, I. ii. 59–64. **7. supported by Dr. Johnson**: Coleridge refers to Johnson's discussion of the unities of time and place in his "Preface to Shakespeare."

8. **The . . . Agincourt**: *Henry V*, Prologue, ll. 12–14.

ANCIENT AND MODERN ART

Ancients, statuesque; moderns, picturesque. Ancients, rhythm and melody; moderns, harmony. Ancients, the finite, and, therefore, grace, elegance, proportion, fancy, dignity, majesty,—whatever is capable of being definitely conveyed by defined forms or thoughts. The moderns, the infinite and [the] indefinite as the vehicle of the infinite; hence more [devoted] to the passions, the obscure hopes and fears—the wandering thro' [the] infinite, grander moral feelings, more august conceptions of man as man, the future rather than the present,—sublimity.

MECHANIC AND ORGANIC FORM

The subject of the present lecture is no less than a question submitted to your understandings, emancipated from national prejudice: Are the plays of *Shakespeare* works of rude uncultivated genius, in which the splendor of the parts compensates, if aught can compensate, for the barbarous shapelessness and irregularity of the whole? To which not only the French critics, but even his own English admirers, say [yes]. Or is the form equally admirable with the matter, the judgement of the great poet not less deserving of our wonder than his genius? Or to repeat the question in other words, is Shakespeare a great dramatic poet on account only of those beauties and excellencies which he possesses in common with the ancients, but with diminished claims to our love and honor to the full extent of his difference from them? Or are these very differences additional proofs of poetic wisdom, at once results and symbols of living power as contrasted with lifeless mechanism, of free and rival originality as contradistinguished from servile imitation, or more accurately, [from] a blind copying of effects instead of a true imitation of the essential principles? Imagine not I am about to oppose genius to rules. No! the comparative value of these rules is the very cause to be tried. The spirit of poetry, like all other living powers, must of necessity circumscribe itself by rules, were it only to unite power with beauty. It must embody in order to reveal itself; but a living body is of necessity an organized one,—and what is

organization, but the connection of parts to a whole, so that each part is at once end and means! This is no discovery of criticism; it is a necessity of the human mind—and all nations have felt and obeyed it, in the invention of metre and measured sounds as the vehicle and involucrum of poetry, itself a fellow-growth from the same life, even as the bark is to the tree.

No work of true genius dare want its appropriate form; neither indeed is there any danger of this. As it must not, so neither can it, be lawless! For it is even this that constitutes it genius—the power of acting creatively under laws of its own origination. How then comes it that not only single Zoili,[9] but whole nations have combined in unhesitating condemnation of our great dramatist, as a sort of African nature, fertile in beautiful monsters, as a wild heath where islands of fertility look greener from the surrounding waste, where the loveliest plants now shine out among unsightly weeds and now are choked by their parasitic growth, so intertwined that we cannot disentangle the weed without snapping the flower. In this statement I have had no reference to the vulgar abuse of Voltaire, save as far as his charges are coincident with the decisions of his commentators and (so they tell you) his almost idolatrous admirers. The true ground of the mistake, as has been well remarked by a continental critic,[10] lies in the confounding mechanical regularity with organic form. The form is mechanic when on any given material we impress a pre-determined form, not necessarily arising out of the properties of the material, as when to a mass of wet clay we give whatever shape we wish it to retain when hardened. The organic form, on the other hand, is innate; it shapes as it develops itself from within, and the fullness of its development is one and the same with the perfection of its outward form. Such is the life, such the form. Nature, the prime genial artist, inexhaustible in diverse powers, is equally inexhaustible in forms. Each exterior is the physiognomy of the being within, its true image reflected and thrown out from the concave mirror. And even such is the appropriate excellence of her chosen poet, of our own Shakespeare, himself a nature humanized, a genial understanding directing self-consciously a power and an implicit wisdom deeper than consciousness.

---

9. Zoili: from Zoilus, classical example of a bad critic. **10. critic:** A. W. Schlegel, 1767–1845.

POETRY IS IDEAL[11]

It would be necessary for him, in the first place, to enquire whether poetry ought to be a *copy*, or only an *imitation* of what is true nature? According to every effect he had been able to trace, he was of opinion that the pleasure we receive arose, not from its being a *copy*, but from its being an imitation; and the word imitation itself means always a combination of a certain degree of dissimilitude with a certain degree of similitude. If it were merely the same as looking at a glass reflection, we should receive no pleasure. A waxen image after once it had been seen pleased no longer, or very little, but when the resemblance of a thing was given upon canvas or a flat surface, then we were delighted.

In poetry it is still more so; the difference there is of a higher character. We take the purest parts and combine them with our own minds, with our own hopes, with our own inward yearnings after perfection, and, being frail and imperfect, we wish to have a shadow, a sort of prophetic existence present to us, which tells us what we are not, but yet, blending in us much that we are, promises great things of what we may be. It is the truth (and poetry results from that instinct—the effort of perfecting ourselves), the conceiving that which is imperfect to be perfect and blending the nobler mind with the meaner object.

Thus, of Shakespeare he had often heard it said that he was a close copier of nature, that he was a child of nature, like a Dutch painter copying exactly the object before him. He was a child of nature, but it was of human nature and of the most important of human nature. In the meanest characters, it was still Shakespeare; it was not the mere Nurse in *Romeo and Juliet*, or the Dogberry in *Much Ado about Nothing*, or the blundering Constable in *Measure for Measure*, but it was this great and mighty being changing himself into the Nurse or the blundering Constable, that gave delight. We know that no Nurse talked exactly in that way, tho' particular sentences might be to that purpose. He might compare it to Proteus,[12] who now flowed, a river; now raged, a fire; now roared, a lion—he assumed all changes, but still in the stream, in the fire, in the beast, it was not only the resemblance, but

it was the divinity that appeared in it, and assumed the character.

Coleridge included music and painting under the great genus of poetry, and we could not understand those unless we first impressed upon the mind that they are ideal, and not the mere copy of things, but the contemplation of mind upon things. When you look upon a portrait, you must not compare it with the face when present, but with the recollection of the face. It refers not so much to the senses, as to the ideal sense of the friend not present.

THE GRANDEST EFFORTS OF POETRY[13]

In my mind, what have often been censured as Shakespeare's conceits are completely justifiable, as belonging to the state, age, or feeling of the individual. Sometimes, when they cannot be vindicated on these grounds, they may well be excused by the taste of his own and of the preceding age; as for instance, in Romeo's speech,

> "Here's much to do with hate, but more with love:—
> Why then, O brawling love! O loving hate!
> O anything, of nothing first created!
> O heavy lightness! serious vanity!
> Misshapen chaos of well-seeming forms!
> Feather of lead, bright smoke, cold fire, sick health!
> Still-waking sleep, that is not what it is!"
>
> Act I., Scene I [ll. 182–88].

I dare not pronounce such passages as these to be absolutely unnatural, not merely because I consider the author a much better judge than I can be, but because I can understand and allow for an effort of the mind, when it would describe what it cannot satisfy itself with the description of, to reconcile opposites and qualify contradictions, leaving a middle state of mind more strictly appropriate to the imagination than any other, when it is, as it were, hovering between images. As soon as it is fixed on one image, it becomes understanding; but while it is unfixed and wavering between them, attaching itself permanently to none, it is imagination. Such is the fine description of Death in Milton:—

> "The other shape,
> If shape it might be call'd, that shape had none
> Distinguishable in member, joint, or limb,

---

11. from a shorthand report of a lecture of 1811. **12. Proteus:** in Greek mythology, a sea-god who could assume many different identities.

13. from a shorthand report of a lecture of 1811.

Or substance might be call'd, that shadow seem'd,
For each seem'd either: black it stood as night;
Fierce as ten furies, terrible as hell,
And shook a dreadful dart: what seem'd his head
The likeness of a kingly crown had on.''

Paradise Lost, Book II [ll. 666–73]

The grandest efforts of poetry are where the imagination is called forth, not to produce a distinct form, but a strong working of the mind, still offering what is still repelled, and again creating what is again rejected; the result being what the poet wishes to impress, namely, the substitution of a sublime feeling of the unimaginable for a mere image. I have sometimes thought that the passage just read might be quoted as exhibiting the narrow limit of painting, as compared with the boundless power of poetry: painting cannot go beyond a certain point; poetry rejects all control, all confinement. Yet we know that sundry painters have attempted pictures of the meeting between Satan and Death at the gates of Hell; and how was Death represented? Not as Milton has described him, but by the most defined thing that can be imagined—a skeleton, the dryest and hardest image that it is possible to discover; which, instead of keeping the mind in a state of activity, reduces it to the merest passivity,—an image, compared with which a square, a triangle, or any other mathematical figure, is a luxuriant fancy.

FROM

THE STATESMAN'S MANUAL

❧

The Statesman's Manual (1816) was the first of two "Lay Sermons" addressed to the governing classes and devoted to political and economic issues. The work is undeniably eccentric; it attempts to show that in such matters the Bible is the best guide. But in doing so it also reminds that these issues must be decided by ultimate moral and religious ends. Moreover, the work contains brilliant passages of philosophic and literary insight. The penetrating characterization of Satan in *Paradise Lost* may be compared with the admiration expressed by Blake and Shelley (pp. 70, 981,) and with the Romantic idealization of the "Byronic hero" (see Introduction to Byron, below, p. 782).

❧

IDEAS

Notions,[1] the depthless abstractions of fleeting *phœnomena*, the shadows of sailing vapors, the colorless repetitions of rainbows, have effected their utmost when they have added to the distinctness of our knowledge. For this very cause they are of themselves adverse to lofty emotion, and it requires the influence of a light and warmth, not their own, to make them crystallize into a semblance of growth. But every principle is actualized by an idea; and every idea is living, productive, partaketh of infinity, and (as Bacon has sublimely observed) containeth an endless power of semination. Hence it is, that science, which consists wholly in ideas and principles, is power. *Scientia et potentia* (saith the same philosopher) *in idem coincidunt*.[2] Hence too it is, that notions, linked arguments, reference to particular facts and calculations of prudence, influence only the comparatively few, the men of leisurely minds who have been trained up to them: and even these few they influence but faintly. But for the reverse, I appeal to the general character of the doctrines which have collected the most numerous sects, and acted upon the moral being of the converts with a force that might well seem supernatural. The great principles of our religion, the sublime ideas spoken out everywhere in the Old and New Testament, resemble the fixed stars, which appear of the same size to the naked as to the armed eye; the magnitude of which the telescope may rather seem to diminish than to increase. At the annunciation of principles, of ideas, the soul of man awakes and starts up, as an exile in a far distant land at the unexpected sounds of his native language, when after long years of absence, and almost of oblivion, he is suddenly addressed in his own mother-tongue. He weeps for joy, and embraces the speaker as his brother. How else can we explain the fact so honorable to Great Britain, that the poorest amongst us will contend with as much enthusiasm as the richest for the rights of property? These rights are the spheres and necessary conditions of free agency. But free agency contains the idea of the free will; and in this he intuitively knows the sublimity, and the infinite hopes, fears, and capabilities of his

THE STATESMAN'S MANUAL. **1. Notions:** products of the understanding as opposed to reason. See Introduction, p. 393. **2. Scientia . . . coincidunt:** *Novum Organum*, I. 3: "Knowledge and power meet in one."

own nature. On what other ground but the cognateness of ideas and principles to man as man does the nameless soldier rush to the combat in defence of the liberties or the honor of his country?—Even men woefully neglectful of the precepts of religion will shed their blood for its truth.

SYMBOL AND ALLEGORY

. . . And in nothing is Scriptural history more strongly contrasted with the histories of highest note in the present age, than in its freedom from the hollowness of abstractions. While the latter present a shadow-fight of things and quantities, the former gives us the history of men, and balances the important influence of individual minds with the previous state of the national morals and manners, in which, as constituting a specific susceptibility, it presents to us the true cause both of the influence itself, and of the weal or woe that were its consequents. How should it be otherwise? The histories and political economy of the present and preceding century partake in the general contagion of its mechanic philosophy, and are the product of an unenlivened generalizing understanding. In the Scriptures they are the living educts of the imagination; of that reconciling and mediatory power, which incorporating the reason in images of the sense, and organizing (as it were) the flux of the senses by the permanence and self-circling energies of the reason, gives birth to a system of symbols, harmonious in themselves, and consubstantial with the truths of which they are the conductors. These are the *wheels* which Ezekiel beheld, when the hand of the Lord was upon him, and he saw visions of God as he sate among the captives by the river of Chebar. *Whithersoever the Spirit was to go, the* wheels *went, and thither was their spirit to go:—for the spirit of the living creature was in the* wheels *also.*[3] . . . A hunger-bitten and idea-less philosophy naturally produces a starveling and comfortless religion. It is among the miseries of the present age that it recognizes no *medium* between literal and metaphorical. Faith is either to be buried in the dead letter, or its name and honors usurped by a counterfeit product of the mechanical understanding, which in the blindness of self-complacency confounds symbols

with allegories.[4] Now an allegory is but a translation of abstract notions into a picture-language, which is itself nothing but an abstraction from objects of the senses; the principal being more worthless even than its phantom proxy, both alike unsubstantial, and the former shapeless to boot. On the other hand a symbol (ὅ ἔστιν ἀεὶ ταυτηγόρικον)[5] is characterized by a translucence of the special in the individual, or of the general in the special, or of the universal in the general; above all by the translucence of the eternal through and in the temporal. It always partakes of the reality which it renders intelligible; and while it enunciates the whole, abides itself as a living part in that unity of which it is the representative. The other are but empty echoes which the fancy arbitrarily associates with apparitions of matter, less beautiful but not less shadowy than the sloping orchard or hillside pasture-field seen in the transparent lake below. Alas, for the flocks that are to be led forth to such pastures!

SATANIC SELF-IDOLATRY

. . . But in its utmost abstraction and consequent state of reprobation, the will becomes Satanic pride and rebellious self-idolatry in the relations of the spirit to itself, and remorseless despotism relatively to others; the more hopeless as the more obdurate by its subjugation of sensual impulses, by its superiority to toil and pain and pleasure; in short, by the fearful resolve to find in itself alone the one absolute motive of action, under which all other motives from within and from without must be either subordinated or crushed.

This is the character which Milton has so philosophically as well as sublimely embodied in the Satan of his Paradise Lost. Alas! too often has it been embodied in real life. Too often has it given a dark and savage grandeur to the historic page. And wherever it has appeared, under whatever circumstances of time and country, the same ingredients have gone to its composition; and it has been identified by the same attributes. Hope in which there is no cheerfulness; stedfastness within and immovable resolve, with outward restlessness and whirling activity; violence with guile;

---

3. **Whithersoever . . . also:** Ezekiel 1:20.

4. **Faith . . . allegories:** biblical fundamentalism or a narrowly rational approach to the Bible that, treating it as allegory, restates its teachings in abstract terms and in doing so interprets them according to its own wishes. **5.** ὅ . . . ταυτηγόρικον: "that which is significant of the same thing always."

temerity with cunning; and, as the result of all, interminableness of object with perfect indifference of means; these are the qualities that have constituted the commanding genius; these are the marks, that have characterized the masters of mischief, the liberticides, and mighty hunters of mankind, from Nimrod[6] to Buonaparte. And from inattention to the possibility of such a character as well as from ignorance of its elements, even men of honest intentions too frequently become fascinated. Nay, whole nations have been so far duped by this want of insight and reflection as to regard with palliative admiration, instead of wonder and abhorrence, the Molochs[7] of human nature, who are indebted for the larger portion of their meteoric success to their total want of principle, and who surpass the generality of their fellow creatures in one act of courage only, that of daring to say with their whole heart, "Evil, be thou my good!"—[8]All system so far is power; and a systematic criminal, self-consistent and entire in wickedness, who entrenches villany within villany, and barricadoes crime by crime, has removed a world of obstacles by the mere decision, that he will have no obstacles, but those of force and brute matter.

FROM

THE FRIEND

∽∾∾

These extracts come from the revised edition of Coleridge's *The Friend* (1818), first published in periodical form in 1809–10 and first collected in 1812. But the essay here entitled "On Radicals and Republicans" was actually given as a speech at Bristol in 1795 and published as *Conciones ad Populum* in the same year. Coleridge republished it to show that he was never at any period of his life "a convert to the Jacobinical system."

∽∾∾

HIS PROSE STYLE

An author's pen, like children's legs, improves by exercise. That part of the blame which rests in myself, I am exerting my best faculties to remove. A man long accustomed to silent and solitary meditation, in proportion as he increases the power of thinking in long and connected trains, is apt to lose or lessen the talent of communicating his thoughts with grace and perspicuity. Doubtless too, I have in some measure injured my style, in respect to it's facility and popularity, from having almost confined my reading, of late years, to the works of the ancients and those of the elder writers in the modern languages. We insensibly imitate what we habitually admire; and an aversion to the epigrammatic unconnected periods of the fashionable Anglo-gallican taste has too often made me willing to forget, that the stately march and difficult evolutions, which characterize the eloquence of Hooker, Bacon, Milton, and Jeremy Taylor, are, notwithstanding their intrinsic excellence, still less suited to a periodical essay. This fault I am now endeavouring to correct; though I can never so far sacrifice my judgement to the desire of being immediately popular, as to cast my sentences in the French moulds, or affect a style which an ancient critic would have deemed purposely invented for persons troubled with the asthma to read, and for those to comprehend who labour under the more pitiable asthma of a short-witted intellect. It cannot but be injurious to the human mind never to be called into effort: the habit of receiving pleasure without any exertion of thought, by the mere excitement of curiosity and sensibility, may be justly ranked among the worst effects of habitual novel reading. It is true that these short and unconnected sentences are easily and instantly understood: but it is equally true, that wanting all the cement of thought as well as of style, all the connections and (if you will forgive so trivial a metaphor) all the hooks-and-eyes of the memory, they are as easily forgotten: or rather, it is scarcely possible that they should be remembered.—Nor is it less true, that those who confine their reading to such books dwarf their own faculties, and finally reduce their understandings to a deplorable imbecility. . . . Like idle morning visitors, the brisk and breathless periods hurry in and hurry off in quick and profitless succession; each indeed for the moments of it's stay prevents the pain of vacancy, while it indulges the love of sloth; but all together they leave the mistress of the house (the soul, I mean) flat and exhausted, incapable of attending to her own concerns, and unfitted for the conversation of more rational guests.

6. **Nimrod**: Genesis 10:9. 7. **Molochs**: In the Old Testament Moloch was an idol to which children were sacrificed. Cf. *Paradise Lost*, I. 392–96. 8. **"Evil . . . good!"**: *Paradise Lost*, IV. 110.

ON RADICALS
AND REPUBLICANS

Companies resembling the present will, from a variety of circumstances, consist chiefly of the zealous advocates for freedom. It will therefore be our endeavour, not so much to excite the torpid, as to regulate the feelings of the ardent: and above all, to evince the necessity of bottoming on fixed principles, that so we may not be the unstable patriots of passion or accident, nor hurried away by names of which we have not sifted the meaning, and by tenets of which we have not examined the consequences. The times are trying; and in order to be prepared against their difficulties, we should have acquired a prompt facility of adverting in all our doubts to some grand and comprehensive truth. In a deep and strong soil must that tree fix its roots, the height of which is to *reach to heaven, and the sight of it to the ends of all the earth.*

The example of France is indeed a warning to Britain. A nation wading to its rights through blood, and marking the track of freedom by devastation! Yet let us not embattle our feelings against our reason. Let us not indulge our malignant passions under the mask of humanity. Instead of railing with infuriate declamation against these excesses, we shall be more profitably employed in tracing them to their sources. French freedom is the beacon which if it guides to equality should shew us likewise the dangers that throng the road.

The annals of the French revolution have recorded in letters of blood, that the knowledge of the few cannot counteract the ignorance of the many; that the light of philosophy, when it is confined to a small minority, points out the possessors as the victims, rather than the illuminators, of the multitude. The patriots of France either hastened into the dangerous and gigantic error of making certain evil the means of contingent good, or were sacrificed by the mob, with whose prejudices and ferocity their unbending virtue forbade them to assimilate. Like Samson, the people were strong—like Samson the people were blind. "Those two massy pillars" of the temple of oppression, their monarchy and aristocracy,

With horrible convulsion to and fro
They tugg'd, *they* shook—till down they came and drew
The whole roof after them with burst of thunder
Upon the heads of all who sat beneath,

Lords, ladies, captains, counsellors, *and* priests,
Their choice nobility![1]

The Girondists,[2] who were the first republicans in power, were men of enlarged views and great literary attainments; but they seem to have been deficient in that vigour and daring activity, which circumstances made necessary. Men of genius are rarely either prompt in action or consistent in general conduct. Their early habits have been those of contemplative indolence; and the day-dreams, with which they have been accustomed to amuse their solitude, adapt them for splendid speculation, not temperate and practicable counsels. Brissot,[3] the leader of the Gironde party, is entitled to the character of a virtuous man, and an eloquent speaker; but he was rather a sublime visionary, than a quick-eyed politician; and his excellences equally with his faults rendered him unfit for the helm in the stormy hour of revolution. Robespierre,[4] who displaced him, possessed a glowing ardour that still remembered the end, and a cool ferocity that never either overlooked or scrupled the means. What that end was, is not known: that it was a wicked one, has by no means been proved. I rather think, that the distant prospect, to which he was travelling, appeared to him grand and beautiful; but that he fixed his eye on it with such intense eagerness as to neglect the foulness of the road. If, however, his first intentions were pure, his subsequent enormities yield us a melancholy proof, that it is not the character of the possessor which directs the power, but the power which shapes and depraves the character of the possessor. In Robespierre, its influence was assisted by the properties of his disposition.—Enthusiasm, even in the gentlest temper, will frequently generate sensations of an unkindly order. If we clearly perceive any one thing to be of vast and infinite importance to ourselves and all mankind, our first feelings impel us to turn with angry contempt from those, who doubt and oppose it. The ardour of undisciplined benevolence seduces us into malignity: and whenever our hearts are warm, and our objects great and excellent, intolerance is the sin that does most easily beset us. But this enthusiasm in Robespierre was blended with gloom, and suspiciousness, and inordinate vanity. His dark imagination

THE FRIEND. **1. With . . . nobility:** Milton's *Samson Agonistes*, ll. 1649–54. **2. Girondists:** the moderate party suppressed by the Jacobins in 1793. **3. Brissot:** Jacques Pierre Brissot, 1754–93. **4. Robespierre:** Maximilien François Robespierre, 1758–94, leader of the Jacobins. He was thought to be the main instigator of the Reign of Terror.

was still brooding over supposed plots against freedom; —to prevent tyranny he became a tyrant,—and having realised the evils which he suspected, a wild and dreadful tyrant.—And thus, his ear deafened to the whispers of conscience by the clamorous plaudits of the mob, he despotized in all the pomp of patrotism, and masqueraded on the bloody stage of revolution, a Caligula[5] with the cap of liberty on his head.

It has been affirmed, and I believe with truth, that the system of terrorism by suspending the struggles of contrarient factions communicated an energy to the operations of the republic, which had been hitherto unknown, and without which it could not have been preserved. The system depended for its existence on the general sense of its necessity, and when it had answered its end, it was soon destroyed by the same power that had given it birth—popular opinion. It must not however be disguised, that at all times, but more especially when the public feelings are wavy and tumultuous, artful demagogues may create this opinion: and they, who are inclined to tolerate evil as the means of contingent good, should reflect, that if the excesses of terrorism gave to the republic that efficiency and repulsive force which its circumstances made necessary, they likewise afforded to the hostile courts the most powerful support, and excited that indignation and horror which every where precipitated the subject into the designs of the ruler. Nor let it be forgotten that these excesses perpetuated the war in La Vendée[6] and made it more terrible, both by the accession of numerous partizans, who had fled from the persecution of Robespierre, and by inspiring the Chouans with fresh fury, and an unsubmitting spirit of revenge and desperation.

Revolutions are sudden to the unthinking only. Strange rumblings and confused noises still precede these earthquakes and hurricanes of the moral world. The process of revolution in France has been dreadful, and should incite us to examine with an anxious eye the motives and manners of those, whose conduct and opinions seem calculated to forward a similar event in our own country. The oppositionists to "things as they are," are divided into many and different classes. To delineate them with an unflattering accuracy may be a delicate, but it is a necessary, task, in order that we may enlighten, or at least be aware of, the misguided men who have enlisted under the banners of liberty, from no principles or with bad ones: whether they be those, who

> admire they know not what,
> And know not whom, but as one leads to the other;—[7]

or whether those,

> Whose end is private hate, not help to freedom,
> Adverse and turbulent when she would lead
> To virtue.

The majority of democrats appear to me to have attained that portion of knowledge in politics, which infidels possess in religion. I would by no means be supposed to imply that the objections of both are equally unfounded, but that they both attribute to the system which they reject, all the evils existing under it; and that both contemplating truth and justice in the nakedness of abstraction, condemn constitutions and dispensations without having sufficiently examined the natures, circumstances, and capacities of their recipients.

The first class among the professed friends of liberty is composed of men, who unaccustomed to the labour of thorough investigation, and not particularly oppressed by the burthens of state, are yet impelled by their feelings to disapprove of its grosser depravities, and prepared to give an indolent vote in favour of reform. Their sensibilities not braced by the co-operation of fixed principles, they offer no sacrifices to the divinity of active virtue. Their political opinions depend with weather-cock uncertainty on the winds of rumour, that blow from France. On the report of French victories they blaze into republicanism, at a tale of French excesses they darken into aristocrats. These dough-baked patriots are not however useless. This oscillation of political opinion will retard the day of revolution, and it will operate as a preventive to its excesses. Indecisiveness of character, though the effect of timidity, is almost always associated with benevolence.

Wilder features characterize the second class. Sufficiently possessed of natural sense to despise the priest, and of natural feeling to hate the oppressor, they listen only to the inflammatory harangues of

---

5. **Caligula:** Gaius Caesar, tyrannous and insane emperor of Rome, 37–41 A.D. 6. **war . . . Vendée:** a royalist insurrection against the revolutionary government in Paris. The Chouans were guerrilla fighters on the royalist side.

7. **admire . . . other:** *Paradise Regained*, III. 52–53. The next quotation has not been traced. Coleridge may be quoting from a poet who echoes Milton, since the phrase "adverse and turbulent" occurs in *Samson Agonistes*, l. 1040.

some mad-headed enthusiast, and imbibe from them poison, not food; rage, not liberty. Unillumined by philosophy, and stimulated to a lust of revenge by aggravated wrongs, they would make the altar of freedom stream with blood, while the grass grew in the desolated halls of justice.

We contemplate those principles with horror. Yet they possess a kind of wild justice well calculated to spread them among the grossly ignorant. To unenlightened minds, there are terrible charms in the idea of retribution, however savagely it be inculcated. The groans of the oppressors make fearful yet pleasant music to the ear of him, whose mind is darkness, and into whose soul the iron has entered.

This class, at present, is comparatively small—yet soon to form an overwhelming majority, unless great and immediate efforts are used to lessen the intolerable grievances of our poor brethren, and infuse into their sorely wounded hearts the healing qualities of knowledge. For can we wonder that men should want humanity, who want all the circumstances of life that humanize? Can we wonder that with the ignorance of brutes they should unite their ferocity? Peace and comfort be with these! But let us shudder to hear from men of dissimilar opportunities sentiments of similar revengefulness. The purifying alchemy of education may transmute the fierceness of an ignorant man into virtuous energy; but what remedy shall we apply to him whom plenty has not softened, whom knowledge has not taught benevolence? This is one among the many fatal effects which result from the want of fixed principles.

There is a third class among the friends of freedom, who possess not the wavering character of the first description, nor the ferocity last delineated. They pursue the interests of freedom steadily, but with narrow and self-centering views: they anticipate with exultation the abolition of privileged orders, and of acts that persecute by exclusion from the right of citizenship. Whatever is above them they are most willing to drag down; but every proposed alteration that would elevate their poorer brethren, they rank among the dreams of visionaries: as if there were any thing in the superiority of lord to gentleman so mortifying in the barrier, so fatal to happiness in the consequences, as the more real distinction of master and servant, of rich man and of poor. Wherein am I made worse by my ennobled neighbour? Do the childish titles of aristocracy detract from my domestic comforts, or prevent my intellectual acquisitions? But

those institutions of society which should condemn me to the necessity of twelve hours' daily toil, would make my soul a slave, and sink the rational being in the mere animal. It is a mockery of our fellow-creatures' wrongs to call them equal in rights, when by the bitter compulsion of their wants we make them inferior to us in all that can soften the heart, or dignify the understanding. Let us not say that this is the work of time—that it is impracticable at present, unless we each in our individual capacities do strenuously and perseveringly endeavour to diffuse among our domestics those comforts and that illumination which far beyond all political ordinances are the true equalizers of men.

We turn with pleasure to the contemplation of that small but glorious band, whom we may truly distinguish by the name of thinking and disinterested patriots. These are the men who have encouraged the sympathetic passions till they have become irresistible habits, and made their duty a necessary part of their self-interest, by the long-continued cultivation of that moral taste which derives our most exquisite pleasures from the contemplation of possible perfection, and proportionate pain from the perception of existing depravity. Accustomed to regard all the affairs of man as a process, they never hurry and they never pause. Theirs is not that twilight of political knowledge which gives us just light enough to place one foot before the other; as they advance the scene still opens upon them, and they press right onward with a vast and various landscape of existence around them. Calmness and energy mark all their actions. Convinced that vice originates not in the man, but in the surrounding circumstances; not in the heart, but in the understanding; the Christian patriot is hopeless concerning no one;—to correct a vice or generate a virtuous conduct he pollutes not his hands with the scourge of coercion; but by endeavouring to alter circumstances would remove, or by strengthening the intellect disarm, the temptation. The unhappy children of vice and folly, whose tempers are adverse to their own happiness as well as to the happiness of others, will at times awaken a natural pang; but he looks forward with gladdened heart to that glorious period when justice shall have established the universal fraternity of love. These soul-ennobling views bestow the virtues which they anticipate. He whose mind is habitually impressed with them soars above the present state of humanity, and may be justly said to dwell in the presence of the Most High.

> Would the forms
> Of servile custom cramp *the patriot's* power?
> Would sordid policies, the barbarous growth
> Of ignorance and rapine, bow *him* down
> To tame pursuits, to indolence and fear?
> Lo!—*he* appeals to nature, to the winds
> And rolling waves, the sun's unwearied course,
> The elements and seasons: all declare
> For what the Eternal Maker has ordain'd
> The powers of man: we feel within ourselves
> His energy divine: he tells the heart
> He meant, he made, us to behold and love
> What he beholds and loves, the general orb
> Of life and being—to be great like him,
> Beneficent and active.[8]

That general illumination should precede revolution, is a truth as obvious, as that the vessel should be cleansed before we fill it with a pure liquor. But the mode of diffusing it is not discoverable with equal facility. We certainly should never attempt to make proselytes by appeals to the selfish feelings, and consequently, should plead for the oppressed, not to them. The author of an essay on political justice considers private societies as the sphere of real utility;—that (each one illuminating those immediately beneath him,) truth by a gradual descent may at last reach the lowest order. But this is rather plausible than just or practicable. Society as at present constituted does not resemble a chain that ascends in a continuity of links. Alas! between the parlour and the kitchen, the coffee-room and the tap, there is a gulf that may not be passed. He would appear to me to have adopted the best as well as the most benevolent mode of diffusing truth, who, uniting the zeal of the Methodist with the views of the philosopher, should be personally among the poor, and teach them their duties in order that he may render them susceptible of their rights.

Yet by what means can the lower classes be made to learn their duties, and urged to practise them? The human race may perhaps possess the capability of all excellence; and truth, I doubt not, is omnipotent to a mind already disciplined for its reception; but assuredly the over-worked labourer, skulking into an ale-house, is not likely to exemplify the one, or prove the other. In that barbarous tumult of inimical interests, which the present state of society exhibits, religion appears to offer the only means universally efficient. The perfectness of future men is indeed a benevolent tenet, and

may operate on a few visionaries, whose studious habits supply them with employment, and seclude them from temptation. But a distant prospect, which we are never to reach, will seldom quicken our footsteps, however lovely it may appear; and a blessing, which not ourselves but posterity are destined to enjoy, will scarcely influence the actions of any—still less of the ignorant, the prejudiced, and the selfish.

Preach the Gospel to the poor. By its simplicity it will meet their comprehension, by its benevolence soften their affections, by its precepts it will direct their conduct, by the vastness of its motives insure their obedience. The situation of the poor is perilous: they are indeed both

> from within and from without
> Unarmed to all temptations.[9]

Prudential reasonings will in general be powerless with them. For the incitements of this world are weak in proportion as we are wretched:—

> The world is not my friend, nor the world's law.
> The world has got no law to make me rich.[10]

They too, who live from hand to mouth, will most frequently become improvident. Possessing no stock of happiness they eagerly seize the gratifications of the moment, and snatch the froth from the wave as it passes by them. Nor is the desolate state of their families a restraining motive, unsoftened as they are by education, and benumbed into selfishness by the torpedo touch of extreme want. Domestic affections depend on association. We love an object if, as often as we see or recollect it, an agreeable sensation arises in our minds. But alas! how should he glow with the charities of father and husband, who gaining scarcely more than his own necessities demand, must have been accustomed to regard his wife and children, not as the soothers of finished labour, but as rivals for the insufficient meal? In a man so circumstanced the tyranny of the present can be overpowered only by the tenfold mightiness of the future. Religion will cheer his gloom with her promises, and by habituating his mind to anticipate an infinitely great revolution hereafter, may prepare it even for the sudden reception of a less degree of melioration in this world.

But if we hope to instruct others, we should familiarize our own minds to some fixed and determinate

---

8. **Would . . . active:** Mark Akenside, 1721–70, *The Pleasures of Imagination*, III. 615–29.

9. **from . . . temptations:** an adaptation of *Paradise Lost*, IV. 64–65. 10. **The world is . . . rich:** *Romeo and Juliet*, V. i. 72–73.

principles of action. The world is a vast labyrinth, in which almost every one is running a different way, and almost every one manifesting hatred to those who do not run the same way. A few indeed stand motionless, and not seeking to lead themselves or others out of the maze, laugh at the failures of their brethren. Yet with little reason: for more grossly than the most bewildered wanderer does he err, who never aims to go right. It is more honourable to the head, as well as to the heart, to be misled by our eagerness in the pursuit of truth, than to be safe from blundering by contempt of it. The happiness of mankind is the end of virtue, and truth is the knowledge of the means; which he will never seriously attempt to discover, who has not habitually interested himself in the welfare of others. The searcher after truth must love and be beloved; for general benevolence is a necessary motive to constancy of pursuit; and this general benevolence is begotten and rendered permanent by social and domestic affections. Let us beware of that proud philosophy, which affects to inculcate philanthropy while it denounces every home-born feeling by which it is produced and nutured. The paternal and filial duties discipline the heart and prepare it for the love of all mankind. The intensity of private attachments encourages, not prevents, universal benevolence. The nearer we approach to the sun, the more intense his heat: yet what corner of the system does he not cheer and vivify?

The man who would find truth, must likewise seek it with a humble and simple heart, otherwise he will be precipitate and overlook it; or he will be prejudiced, and refuse to see it. To emancipate itself from the tyranny of association, is the most arduous effort of the mind, particularly in religious and political disquisitions. The assertors of the system have associated with it the preservation of order and public virtue; the oppugners, imposture and wars and rapine. Hence, when they dispute, each trembles at the consequences of the other's opinions instead of attending to his train of arguments. Of this however we may be certain, whether we be Christians or infidels, aristocrats or republicans, that our minds are in a state insusceptible of knowledge, when we feel an eagerness to detect the falsehood of an adversary's reasonings, not a sincere wish to discover if there be truth in them;—when we examine an argument in order that we may answer it, instead of answering because we have examined it.

Our opponents are chiefly successful in confuting the theory of freedom by the practices of its advocates:

from our lives they draw the most forcible arguments against our doctrines. Nor have they adopted an unfair mode of reasoning. In a science the evidence suffers neither diminution nor increase from the actions of its professors; but the comparative wisdom of political systems depends necessarily on the manners and capacities of the recipients. Why should all things be thrown into confusion to acquire that liberty which a faction of sensualists and gamblers will neither be able nor willing to preserve?

A system of fundamental reform will scarcely be effected by massacres mechanized into revolution. We cannot therefore inculcate on the minds of each other too often or with too great earnestness the necessity of cultivating benevolent affections. We should be cautious how we indulge the feelings even of virtuous indignation. Indignation is the handsome brother of anger and hatred. The temple of despotism, like that of Tescalipoca, the Mexican deity, is built of human skulls, and cemented with human blood;—let us beware that we be not transported into revenge while we are levelling the loathsome pile; lest when we erect the edifice of freedom we but vary the style of architecture, not change the materials. Let us not wantonly offend even the prejudices of our weaker brethren, nor by ill-timed and vehement declarations of opinion excite in them malignant feelings towards us. The energies of the mind are wasted in these intemperate effusions. Those materials of projectile force, which now carelessly scattered explode with an offensive and useless noise, directed by wisdom and union might heave rocks from their base,—or perhaps (apart from the metaphor) might produce the desired effect without the convulsion.

For this subdued sobriety of temper a practical faith in the doctrine of philosophical necessity[11] seems the only preparative. That vice is the effect of error and the offspring of surrounding circumstances, the object therefore of condolence not of anger, is a proposition easily understood, and as easily demonstrated. But to make it spread from the understanding to the affections, to call it into action, not only in the great exertions of patriotism, but in the daily and hourly occurrences of

---

11. **doctrine . . . necessity:** the doctrine that everything, including human action and feeling, was determined in a rigid sequence of cause and effect was current among radicals of the time. The argument showed that to reform social institutions would change the nature of man. Coleridge soon ceased to be a necessitarian. (See his letter to Thomas Poole, below, March 16, 1801.)

social life, requires the most watchful attentions of the most energetic mind. It is not enough that we have once swallowed these truths;—we must feed on them, as insects on a leaf, till the whole heart be coloured by their qualities, and shew its food in every the minutest fibre.

Finally, in the spirit of the Apostle,

Watch ye! Stand fast in the principles of which ye have been convinced! Quit yourselves like men! Be strong! Yet let all things be done in the spirit of love![12]

THE SPEECH
OF EDUCATED MEN

What is that which first strikes us, and strikes us at once, in a man of education, and which, among educated men, so instantly distinguishes the man of superior mind, that (as was observed with eminent propriety of the late Edmund Burke) "we cannot stand under the same archway during a shower of rain, without finding him out"?[13] Not the weight or novelty of his remarks; not any unusual interest of facts communicated by him; for we may suppose both the one and the other precluded by the shortness of our intercourse, and the triviality of the subjects. The difference will be impressed and felt, though the conversation should be confined to the state of the weather or the pavement. Still less will it arise from any peculiarity in his words and phrases. For if he be, as we now assume, a well-educated man as well as a man of superior powers, he will not fail to follow the golden rule of Julius Cæsar, *insolens verbum, tanquam scopulum, evitare*.[14] Unless where new things necessitate new terms, he will avoid an unusual word as a rock. It must have been among the earliest lessons of his youth, that the breach of this precept, at all times hazardous, becomes ridiculous in the topics of ordinary conversation. There remains but one other point of distinction possible; and this must be, and in fact is, the true cause of the impression made on us. It is the unpremeditated and evidently habitual arrangement of his words, grounded on the habit of foreseeing, in each integral

part, or (more plainly) in every sentence, the whole that he then intends to communicate. However irregular and desultory his talk, there is method in the fragments.

Listen, on the other hand, to an ignorant man, though perhaps shrewd and able in his particular calling, whether he be describing or relating. We immediately perceive, that his memory alone is called into action; and that the objects and events recur in the narration in the same order, and with the same accompaniments, however accidental or impertinent, in which they had first occurred to the narrator. The necessity of taking breath, the efforts of recollection, and the abrupt rectification of its failures, produce all his pauses; and with exception of the "and then," the "and there," and the still less significant, "and so," they constitute likewise all his connections.

FROM

ESSAYS ON HIS OWN TIMES

WILLIAM PITT THE YOUNGER

This essay appeared in *The Morning Post*, March 19, 1800. It is printed after *The Friend* in order to keep Coleridge's political writings together. The explanation of Pitt's character was echoed by Macaulay in his more famous essay on Pitt for the *Encyclopaedia Britannica* (1859).

Plutarch, in his comparative biography of Rome and Greece, has generally chosen for each pair of lives the two contemporaries who most nearly resemble each other. His work would perhaps have been more interesting, if he had adopted the contrary arrangement and selected those rather, who had attained to the possession of similar influence or similar fame, by means, actions, and talents, the most dissimilar. For power is the sole object of philosophical attention in man, as in inanimate nature: and in the one equally as in the other, we understand it more intimately, the more diverse the circumstances are with which we have observed it co-exist. In our days the two persons, who appear to have influenced the interests and actions of

---

12. **Watch . . . love:** 1 Corinthians 16:13–14. **13. "we . . . out":** The remark is by Samuel Johnson in Boswell's *Life*, ed. G. B. Hill and L. F. Powell, IV. 275. **14. insolens . . . evitare:** "to avoid an unusual word like a rock." The phrase is adapted by Coleridge from a quotation in Aulus Gellius, *Noctes Atticae*, I. x. 4, and Macrobius, *Saturnalia*, I. 5.

men the most deeply and the most diffusively are beyond doubt the Chief Consul of France,[1] and the Prime Minister of Great Britain; and in these two are presented to us similar situations with the greatest dissimilitude of characters.

William Pitt was the younger son of Lord Chatham; a fact of no ordinary importance in the solution of his character, of no mean significance in the heraldry of morals and intellect. His father's rank, fame, political connections, and parental ambition were his mould;— he was cast, rather than grew. A palpable election, a conscious predestination controlled the free agency, and transfigured the individuality of his mind; and that, which he *might have been*, was compelled into that, which he *was to be*. From his early childhood it was his father's custom to make him stand up on a chair, and declaim before a large company; by which exercise, practised so frequently, and continued for so many years, he acquired a premature and unnatural dexterity in the combination of words, which must of necessity have diverted his attention from present objects, obscured his impressions, and deadened his genuine feelings. Not the *thing* on which he was speaking, but the praises to be gained by the speech, were present to his intuition; hence he associated all the operations of his faculties with words, and his pleasures with the surprise excited by them.

But an inconceivably large portion of human knowledge and human power is involved in the science and management of *words*; and an education of words, though it destroys genius, will often create, and always foster, talent. The young Pitt was conspicuous far beyond his fellows, both at school and at college. He was always full grown: he had neither the promise nor the awkwardness of a growing intellect. Vanity, early satiated, formed and elevated itself into a love of power; and in losing this colloquial vanity he lost one of the prime links that connect the individual with the species, too early for the affections, though not too early for the understanding. At college he was a severe student; his mind was founded and elemented in words and generalities, and these too formed all the super-structure. That revelry and that debauchery, which are so often fatal to the powers of intellect, would probably have been serviceable to him; they would have given him a closer communion with realities, they would have induced a greater presentness to present objects. But Mr. Pitt's conduct was correct, unimpressibly correct.

ESSAYS ON HIS OWN TIMES. **1. Chief . . . France:** Napoleon.

His after-discipline in the special pleader's office, and at the bar, carried on the scheme of his education with unbroken uniformity. His first political connections were with the Reformers, but those who accuse him of sympathising or coalescing with their intemperate or visionary plans, misunderstand his character, and are ignorant of the historical facts. Imaginary situations in an imaginary state of things rise up in minds that possess a power and facility in combining images.— Mr. Pitt's ambition was conversant with old situations in the old state of things, which furnish nothing to the imagination, though much to the wishes. In his endeavours to realise his father's plan of reform, he was probably as sincere as a being, who had derived so little knowledge from actual impressions, could be. But his sincerity had no living root of affection; while it was propped up by his love of praise and immediate power, so long it stood erect and no longer. He became a member of the Parliament—supported the popular opinions, and in a few years, by the influence of the popular party, was placed in that high and awful rank in which he now is. The fortunes of his country, we had almost said, the fates of the world, were placed in his wardship—we sink in prostration before the inscrutable dispensations of Providence, when we reflect in whose wardship the fates of the world were placed!

The influencer of his country and of his species was a young man, the creature of another's pre-determination, sheltered and weather-fended from all the elements of experience; a young man, whose feet had never wandered; whose very eye had never turned to the right or to the left; whose whole track had been as curveless as the motion of a fascinated reptile! It was a young man, whose heart was solitary, because he had existed always amid objects of futurity, and whose imagination too was unpopulous, because those objects of hope, to which his habitual wishes had transferred, and as it were *projected*, his existence, were all familiar and long established objects!—A plant sown and reared in a hot-house, for whom the very air that surrounded him, had been regulated by the thermometer of previous purpose; to whom the light of nature had penetrated only through glasses and covers; who had had the sun without the breeze; whom no storm had shaken; on whom no rain had pattered; on whom the dews of heaven had not fallen!—A being, who had had no feelings connected with man or nature, no spontaneous impulses, no unbiassed and desultory studies, no genuine science, nothing that constitutes individuality in intellect, nothing that teaches brother-hood in affection! Such was the man—such, and so denaturalised the spirit, on whose wisdom and philanthropy the lives and living enjoyments of so many millions of human beings were made unavoidably dependent. From this time a real enlargement of mind became almost impossible. Pre-occupations, intrigue, the undue passion and anxiety, with which all facts must be surveyed; the crowd and confusion of those facts, none of them seen, but all communicated, and by that very circumstance, and by the necessity of perpetually classifying them, transmuted into words and generalities; pride, flattery, irritation, artificial power; these, and circumstances resembling these, necessarily render the heights of office barren heights, which command indeed a vast and extensive prospect, but attract so many clouds and vapours, that most often all prospect is precluded. Still, however, Mr. Pitt's situation, however inauspicious for his real being, was favourable to his fame. He heaped period on period; persuaded himself and the nation, that extemporaneous arrangement of sentences was eloquence; and that eloquence implied wisdom. His father's struggles for freedom, and his own attempts, gave him an almost unexampled popularity; and his office necessarily associated with his name all the great events, that happened during his Administration. There were not however wanting men, who saw through this delusion; and refusing to attribute the industry, integrity, and enterprising spirit of our merchants, the agricultural improvements of our land-holders, the great inventions of our manufacturers, or the valour and skilfulness of our sailors to the merits of a minister, they have continued to decide on his character from those acts and those merits, which belong to him and to him alone. Judging him by this standard, they have been able to discover in him no one proof or symptom of a commanding genius. They have discovered him never controlling, never creating, events, but always yielding to them with rapid change, and sheltering himself from inconsistency by perpetual indefiniteness. In the Russian war, they saw him abandoning meanly what he had planned weakly, and threatened insolently. In the debates on the Regency, they detected the laxity of his constitutional principles, and received proofs that his eloquence consisted not in the ready application of a general system to particular questions, but in the facility of arguing for or against any question by specious generalities, without reference to any system. In these debates, he combined what is most dangerous in

democracy, with all that is most degrading in the old superstitions of monarchy; and taught an inherency of the office in the person, in order to make the office itself a nullity, and the Premiership, with its accompanying majority, the sole and permanent power of the State. And now came the French Revolution. This was a new event; the old routine of reasoning, the common trade of politics were to become obsolete. He appeared wholly unprepared for it: half favouring, half condemning, ignorant of what he favoured, and why he condemned, he neither displayed the honest enthusiasm and fixed principle of Mr. Fox, nor the intimate acquaintance with the general nature of man, and the consequent *prescience* of Mr. Burke.

After the declaration of war, long did he continue in the common cant of office, in declamation about the Scheldt and Holland, and all the vulgar causes of common contests! and when at last the immense genius of his new supporter had beat him out of these *words* (words signifying *places* and *dead objects*, and signifying nothing more), he adopted other words in their places, other generalities—Atheism and Jacobinism—phrases, which he learnt from Mr. Burke, but without learning the philosophical definitions and involved consequences, with which that great man accompanied those words. Since the death of Mr. Burke, the forms and the sentiments, and the tone of the French have undergone many and important changes: how, indeed, is it possible that it should be otherwise, while man is the creature of experience! But still Mr. Pitt proceeds in an endless repetition of the same *general phrases*. This is his element; deprive him of general and abstract phrases, and you reduce him to silence. But you cannot deprive him of them. Press him to specify an *individual* fact of advantage to be derived from a war, and he answers, Security! Call upon him to particularize a crime, and he exclaims —Jacobinism! Abstractions defined by abstractions! Generalities defined by generalities! As a minister of finance, he is still, as ever, the man of words and abstractions! Figures, custom-house reports, imports and exports, commerce and revenue—all flourishing, all splendid! Never was such a prosperous country, as England, under his administration! Let it be objected, that the agriculture of the country is, by the over-balance of commerce, and by various and complex causes, in such a state, that the country hangs as a pensioner for bread on its neighbours, and a bad season uniformly threatens us with famine—This (it is replied) is owing to our PROSPERITY—all *prosperous* nations are in great distress for food!—still PROSPERITY, still GENERAL PHRASES, unenforced by one *single image*, one *single fact* of real national amelioration; of any one comfort enjoyed, where it was not before enjoyed; of any one class of society becoming healthier, wiser, or happier. These are *things*, these are realities; and these Mr. Pitt has neither the imagination to body forth, nor the sensibility to feel for. Once indeed, in an evil hour, intriguing for popularity, he suffered himself to be persuaded to evince a talent for the Real, the Individual; and he brought in his POOR BILL!! When we hear the minister's talent for finance so loudly trumpeted, we turn involuntarily to his POOR BILL —to that acknowledged abortion—that unanswerable evidence of his ignorance respecting all the fundamental relations and actions of property, and of the social union!

As his reasonings, even so is his eloquence. One character pervades his whole being. Words on words, finely arranged, and so dexterously consequent, that the whole bears the semblance of argument, and still keeps awake a sense of surprise; but when all is done, nothing rememberable has been said; no one philosophical remark, no one image, not even a pointed aphorism. Not a sentence of Mr. Pitt's has ever been quoted, or formed the favourite phrase of the day—a thing unexampled in any man of equal reputation. But while he speaks, the effect varies according to the character of his auditor. The man of no talent is swallowed up in surprise; and when the speech is ended, he remembers his feelings, but nothing distinct of that which produced them—(how opposite an effect to that of nature and genius, from whose works the idea still remains, when the feeling is passed away— remains to connect itself with the other feelings, and combine with new impressions!) The mere man of talent hears him with admiration—the mere man of genius with contempt—the philosopher neither admires nor contemns, but listens to him with a deep and solemn interest, tracing in the effects of his eloquence the power of words and phrases, and that peculiar constitution of human affairs in their present state, which so eminently favours this power.

Such appears to us to be the prime minister of Great Britain, whether we consider him as a statesman or as an orator. The same character betrays itself in his private life; the same coldness to realities, and to all whose excellence relates to reality. He has patronised no science, he has raised no man of genius from obscurity; he counts no one prime work of God among

his friends. From the same source he has no attachment to female society, no fondness for children, no perceptions of beauty in natural scenery; but he is fond of convivial indulgences, of that stimulation, which, keeping up the glow of self-importance and the sense of internal power, gives feelings without the mediation of ideas.

These are the elements of his mind; the accidents of his fortune, the circumstances that enabled such a mind to acquire and retain such a power, would form a subject of a philosophical history, and that too of no scanty size. We can scarcely furnish the chapter of contents to a work, which would comprise subjects so important and delicate, as the causes of the diffusion and intensity of secret influence; the machinery and state intrigue of marriages; the overbalance of the commercial interest; the panic of property struck by the late revolution; the short-sightedness of the careful; the carelessness of the far-sighted; and all those many and various events which have given to a decorous profession of religion, and a seemliness of private morals, such an unwonted weight in the attainment and preservation of public power. We are unable to determine whether it be more consolatory or humiliating to human nature, that so many complexities of event, situation, character, age, and country, should be necessary in order to the production of a Mr. Pitt.

FROM

AIDS TO REFLECTION

◦✺◦

Aids to Reflection (1825) is the most important of Coleridge's writings on religion. It is composed as a series of aphorisms (many of them taken from the Anglican divine Robert Leighton, 1611–84) with commentary. The fragmentary method is typical of Coleridge, but the separate items, some of which are extended essays, are placed to create a progressive argument. Moreover, the aphoristic style itself tends to stimulate reflection, which was one of Coleridge's chief objects. By "reflection" he means a "looking down into" one's own mind, feeling, and experience, the "ultimate object" being "to discover the living fountain and spring-head of the Christian faith in the believer himself" (see Introduction, pp. 393–94). The extracts given here come from the first section, "Introductory Aphorisms," except for the passage on "Mystics and Mysticism," which is from the Conclusion.

◦✺◦

Aphorism I

In philosophy equally as in poetry, it is the highest and most useful prerogative of genius to produce the strongest impressions of novelty, while it rescues admitted truths from the neglect caused by the very circumstance of their universal admission. Extremes meet. Truths, of all others the most awful and interesting, are too often considered as so true, that they lose all the power of truth, and lie bed-ridden in the dormitory of the soul, side by side with the most despised and exploded errors.

Aphorism XVIII

Examine the journals of our zealous missionaries, I will not say among the Hottentots or Esquimaux, but in the highly civilized, though fearfully uncultivated, inhabitants of ancient India. How often, and how feelingly, do they describe the difficulty of rendering the simplest chain of thought intelligible to the ordinary natives, the rapid exhaustion of their whole power of attention, and with what distressful effort it is exerted while it lasts! Yet it is among these that the hideous practices of self-torture chiefly prevail. O if folly were no easier than wisdom, it being often so very much more grievous, how certainly might these unhappy slaves of superstition be converted to Christianity! But, alas! to swing by hooks passed through the back, or to walk in shoes with nails of iron pointed upwards through the soles—all this is so much less difficult, demands so much less exertion of the will than to reflect, and by reflection to gain knowledge and tranquillity!

COMMENT   It is not true, that ignorant persons have no notion of the advantages of truth and knowledge. They confess, they see and bear witness to, these advantages in the conduct, the immunities, and the superior powers of the possessors. Were they attainable by pilgrimages the most toilsome, or penances the most painful, we should assuredly have as many pilgrims and self-tormentors in the service of true religion, as now exist under the tyranny of papal or Brahman superstition.

Aphorism XIX

In countries enlightened by the gospel, however, the most formidable and (it is to be feared) the most frequent impediment to men's turning the mind inwards upon themselves, is that they are afraid of what they shall find there. There is an aching hollowness in the bosom, a dark cold speck at the heart, an obscure and boding sense of a somewhat, that must be kept out of sight of the conscience; some secret lodger, whom they can neither resolve to eject or retain.

COMMENT   Few are so obdurate, few have sufficient strength of character, to be able to draw forth an evil tendency or immoral practice into distinct consciousness, without bringing it in the same moment before an awaking conscience. But for this very reason it becomes a duty of conscience to form the mind to a habit of distinct consciousness. An unreflecting Christian walks in twilight among snares and pitfalls! He entreats the heavenly Father not to lead him into temptation, and yet places himself on the very edge of it, because he will not kindle the torch which his Father had given into his hands, as a mean of prevention, and lest he should pray too late.

ON SENSIBILITY

If prudence, though practically inseparable from morality, is not to be confounded with the moral principle; still less may sensibility, that is, a constitutional quickness of sympathy with pain and pleasure, and a keen sense of the gratifications that accompany social intercourse, mutual endearments, and reciprocal preferences, be mistaken, or deemed a substitute, for either. Sensibility is not even a sure pledge of a good heart, though among the most common meanings of that many-meaning and too commonly misapplied expression.

So far from being either morality, or one with the moral principle, it ought not even to be placed in the same rank with prudence. For prudence is at least an offspring of the understanding; but sensibility (the sensibility, I mean, here spoken of) is for the greater part a quality of the nerves, and a result of individual bodily temperament.

Prudence is an active principle, and implies a sacrifice of self, though only to the same self projected, as it were, to a distance. But the very term sensibility, marks its passive nature; and in its mere self, apart from choice and reflection, it proves little more than the coincidence or contagion of pleasurable or painful sensations in different persons.

Alas! how many are there in this over-stimulated

age, in which the occurrence of excessive and unhealthy sensitiveness is so frequent, as even to have reversed the current meaning of the word, nervous. How many are there whose sensibility prompts them to remove those evils alone, which by hideous spectacle or clamorous outcry are present to their senses and disturb their selfish enjoyments. Provided the dunghill is not before their parlour window, they are well contented to know that it exists, and perhaps as the hotbed on which their own luxuries are reared. Sensibility is not necessarily benevolence. Nay, by rendering us tremblingly alive to trifling misfortunes, it frequently prevents it, and induces an effeminate selfishness instead,

— pampering the coward heart
With feelings all too delicate for use.

Sweet are the tears, that from a Howard's eye
Drop on the cheek of one, he lifts from earth:
And he, who works me good with unmoved face,
Does it but half. He chills me, while he aids,
My benefactor, not my brother man.
But even this, this cold benevolence,
Seems worth, seems manhood, when there rise before me
The sluggard pity's vision-weaving tribe,
Who sigh for wretchedness yet shun the wretched,
Nursing in some delicious solitude
Their slothful loves and dainty sympathies.[1]

Lastly, where virtue is, sensibility is the ornament and becoming attire of virtue. On certain occasions it may almost be said to become virtue. But sensibility and all the amiable qualities may likewise become, and too often have become, the panders of vice, and the instruments of seduction.

So must it needs be with all qualities that have their rise only in parts and fragments of our nature. A man of warm passions may sacrifice half his estate to rescue a friend from prison: for he is naturally sympathetic, and the more social part of his nature happened to be uppermost. The same man shall afterwards exhibit the same disregard of money in an attempt to seduce that friend's wife or daughter.

All the evil achieved by Hobbes and the whole school of materialists will appear inconsiderable if it be compared with the mischief effected and occasioned by the sentimental philosophy of Sterne,[2] and his numerous imitators. The vilest appetites and the most remorseless inconstancy towards their objects, acquired the titles of the *heart, the irresistible feelings, the too tender sensibility:* and if the frosts of prudence, the icy chains of human law thawed and vanished at the genial warmth of human nature, who could help it? It was an amiable weakness!

About this time, too, the profanation of the word, Love, rose to its height. The French naturalists, Buffon[3] and others, borrowed it from the sentimental novelists: the Swedish and English philosophers took the contagion; and the muse of science condescended to seek admission into the saloons of fashion and frivolity, rouged like a harlot, and with the harlot's wanton leer. I know not how the annals of guilt could be better forced into the service of virtue, than by such a comment on the present paragraph, as would be afforded by a selection from the sentimental correspondence produced in courts of justice within the last thirty years, fairly translated into the true meaning of the words, and the actual object and purpose of the infamous writers.

Do you in good earnest aim at dignity of character? By all the treasures of a peaceful mind, by all the charms of an open countenance, I conjure you, O youth! turn away from those who live in the twilight between vice and virtue. Are not reason, discrimination, law, and deliberate choice, the distinguishing characters of humanity? Can aught then worthy of a human being proceed from a habit of soul, which would exclude all these and (to borrow a metaphor from paganism) prefer the den of Trophonius[4] to the temple and oracles of the God of light? Can any thing manly, I say, proceed from those, who for law and light would substitute shapeless feelings, sentiments, impulses, which as far as they differ from the vital workings in the brute animals owe the difference to their former connexion with the proper virtues of humanity; as dendrites derive the outlines, that constitute their value above other clay-stones, from the casual neighbourhood and pressure of the plants, the names of which they assume! Remember, that love itself in its highest earthly bearing, as the ground of the marriage union, becomes love by an inward fiat of the will, by a completing and sealing act of moral election, and lays claim to permanence only under the form of duty.

AIDS TO REFLECTION. **1. pampering . . . sympathies:** Coleridge's "Reflections on Having Left a Place of Retirement," ll. 47–59. **2. Sterne:** Laurence Sterne's *Tristram Shandy* (1760–67) and *A Sentimental Journey* (1768).

3. Buffon: Georges Louis Leclerc de Buffon, 1707–88, author of the *Histoire Naturelle* (1749–88). **4. Trophonius:** Trophonius, who built the first temple of Apollo at Delphi, was worshipped after death in a cave of Boeotia.

Aphorism XXVII

Exclusively of the abstract sciences, the largest and worthiest portion of our knowledge consists of aphorisms: and the greatest and best of men is but an aphorism.

MYSTICS AND MYSTICISM

Antinous.—"What do you call mysticism? And do you use the word in a good or in a bad sense?"

Nous.—"In the latter only; as far, at least, as we are concerned with it. When a man refers to inward feelings and experiences, of which mankind at large are not conscious, as evidences of the truth of any opinion—such a man I call a Mystic: and the grounding of any theory or belief on accidents and anomalies of individual sensations or fancies, and the use of peculiar terms invented, or perverted from their ordinary significations, for the purpose of expressing these idiosyncrasies and pretended facts of interior consciousness, I name Mysticism. Where the error consists simply in the Mystic's attaching to these anomalies of his individual temperament the character of reality, and in receiving them as permanent truths, having a subsistence in the divine mind, though revealed to himself alone; but entertains this persuasion without demanding or expecting the same faith in his neighbours—I should regard it as a species of enthusiasm, always indeed to be deprecated, but yet capable of co-existing with many excellent qualities both of head and heart. But when the Mystic, by ambition or still meaner passions, or (as sometimes is the case) by an uneasy and self-doubting state of mind which seeks confirmation in outward sympathy, is led to impose his faith, as a duty, on mankind generally: and when with such views he asserts that the same experiences would be vouchsafed, the same truths revealed, to every man but for his secret wickedness and unholy will;—such a Mystic is a fanatic, and in certain states of the public mind a dangerous member of society. And most so in those ages and countries in which fanatics of elder standing are allowed to persecute the fresh competitor. For under these predicaments, Mysticism, though originating in the singularities of an individual nature, and therefore essentially anomalous, is nevertheless highly contagious. It is apt to collect a swarm and cluster *circum fana*, around the new fane; and therefore merits the name of *fanaticism*, or as the Germans say, *Schwärmerey*, that is swarm-making."

FROM

SPECIMENS OF THE TABLE TALK OF SAMUEL TAYLOR COLERIDGE

Jotted down after evenings with Coleridge, these records of his conversation were published one year after his death by his nephew, Henry Nelson Coleridge.

July 2. 1830.

Every man is born an Aristotelian, or a Platonist. I do not think it possible that any one born an Aristotelian can become a Platonist; and I am sure no born Platonist can ever change into an Aristotelian. They are the two classes of men, beside which it is next to impossible to conceive a third. The one considers reason a quality, or attribute; the other considers it a power. I believe that Aristotle never could get to understand what Plato meant by an idea. There is a passage, indeed, in the Eudemian Ethics which looks like an exception; but I doubt not of its being spurious, as that whole work is supposed by some to be. With Plato ideas are constitutive in themselves.

Aristotle was, and still is, the sovereign lord of the understanding; the faculty judging by the senses. He was a conceptualist, and never could raise himself into that higher state, which was natural to Plato, and has been so to others, in which the understanding is distinctly contemplated, and, as it were, looked down upon from the throne of actual ideas, or living, inborn, essential truths.

Yet what a mind was Aristotle's—only not the greatest that ever animated the human form!—the parent of science, properly so called, the master of criticism, and the founder or editor of logic! But he confounded science with philosophy, which is an error. Philosophy is the middle state between science, or knowledge, and sophia, or wisdom.

July 27. 1830.

John Thelwall[1] had something very good about him. We were once sitting in a beautiful recess in the Quantocks,[2] when I said to him, "Citizen John, this is a fine place to talk treason in!"—"Nay! Citizen Samuel," replied he, "it is rather a place to make a man forget that there is any necessity for treason!"

Thelwall thought it very unfair to influence a child's mind by inculcating any opinions before it should have come to years of discretion, and be able to choose for itself. I showed him my garden, and told him it was my botanical garden. "How so?" said he, "it is covered with weeds."—"Oh," I replied, "*that* is only because it has not yet come to its age of discretion and choice. The weeds, you see, have taken the liberty to grow, and I thought it unfair in me to prejudice the soil towards roses and strawberries."

September 21. 1830.

I do not know whether I deceive myself, but it seems to me that the young men, who were my contemporaries, fixed certain principles in their minds, and followed them out to their legitimate consequences, in a way which I rarely witness now. No one seems to have any distinct convictions, right or wrong; the mind is completely at sea, rolling and pitching on the waves of facts and personal experiences. Mr.——is, I suppose, one of the rising young men of the day; yet he went on talking, the other evening, and making remarks with great earnestness, some of which were palpably irreconcilable with each other. He told me that facts gave birth to, and were the absolute ground of, principles; to which I said, that unless he had a principle of selection, he would not have taken notice of those facts upon which he grounded his principle. You must have a lantern in your hand to give light, otherwise all the materials in the world are useless, for you cannot find them; and if you could, you could not arrange them. "But then," said Mr.——, "*that* principle of selection came from facts!"—"To be sure!" I replied; "but there must have been again an antecedent light to see those antecedent facts. The relapse may be carried in imagination backwards for ever,—but go back as you may, you cannot come to a

man without a previous aim or principle." He then asked me what I had to say to Bacon's induction: I told him I had a good deal to say, if need were; but that it was perhaps enough for the occasion to remark, that what he was evidently taking for the Baconian *in*duction was mere *de*duction—a very different thing.

September 28. 1830.

Why need we talk of a fiery hell? If the will, which is the law of our nature, were withdrawn from our memory, fancy, understanding, and reason, no other hell could equal, for a spiritual being, what we should then feel, from the anarchy of our powers. It would be conscious madness—a horrid thought!

July 21. 1832.

I cannot help regretting that Wordsworth did not first publish his thirteen books on the growth of an individual mind[3]—superior, as I used to think, upon the whole, to the Excursion. You may judge how I felt about them by my own poem upon the occasion.[4] Then the plan laid out, and, I believe, partly suggested by me, was, that Wordsworth should assume the station of a man in mental repose, one whose principles were made up, and so prepared to deliver upon authority a system of philosophy. He was to treat man as man,— a subject of eye, ear, touch, and taste, in contact with external nature, and informing the senses from the mind, and not compounding a mind out of the senses; then he was to describe the pastoral and other states of society, assuming something of the Juvenalian spirit as he approached the high civilization of cities and towns, and opening a melancholy picture of the present state of degeneracy and vice; thence he was to infer and reveal the proof of, and necessity for, the whole state of man and society being subject to, and illustrative of, a redemptive process in operation, showing how this idea reconciled all the anomalies, and promised future glory and restoration. Something of this sort was, I think, agreed on. It is, in substance, what I have been all my life doing in my system of philosophy.[5]

SPECIMENS OF THE TABLE TALK. 1. **John Thelwall:** English radical, 1764–1834. 2. **Quantocks:** the Quantock Hills, near Nether Stowey.

3. **thirteen . . . mind:** *The Prelude.* 4. **poem . . . occasion:** "To William Wordsworth." 5. **plan . . . philosophy:** Cf. Coleridge's letter to Wordsworth, below, May 30, 1815.

I think Wordsworth possessed more of the genius of a great philosophic poet than any man I ever knew, or, as I believe, has existed in England since Milton; but it seems to me that he ought never to have abandoned the contemplative position, which is peculiarly—perhaps I might say exclusively—fitted for him. His proper title is *Spectator ab extra.*[6]

August 6. 1832.

You will find this a good gage or criterion of genius,—whether it progresses and evolves, or only spins upon itself. Take Dryden's Achitophel and Zimri,[7]—Shaftesbury and Buckingham; every line adds to or modifies the character, which is, as it were, a-building up to the very last verse; whereas, in Pope's Timon,[8] &c. the first two or three couplets contain all the pith of the character, and the twenty or thirty lines that follow are so much evidence or proof of overt acts of jealousy, or pride, or whatever it may be that is satirized. In like manner compare Charles Lamb's exquisite criticisms on Shakespeare with Hazlitt's round and round imitations of them.

April 7. 1833.

In Shakespeare one sentence begets the next naturally; the meaning is all inwoven. He goes on kindling like a meteor through the dark atmosphere; yet, when the creation in its outline is once perfect, then he seems to rest from his labour, and to smile upon his work, and tell himself that it is very good. You see many scenes and parts of scenes which are simply Shakespeare's, disporting himself in joyous triumph and vigorous fun after a great achievement of his highest genius.

July 3. 1833.

The definition of good prose is—proper words in their proper places;—of good verse—the most proper words in their proper places. The propriety is in either case relative. The words in prose ought to express the intended meaning, and no more; if they attract attention to themselves, it is, in general, a fault. In the very best styles, as Southey's, you read page after page, understanding the author perfectly, without once taking notice of the medium of communication;—it is as if he had been speaking to you all the while. But in verse you must do more; there the words, the *media*, must be beautiful, and ought to attract your notice—yet not so much and so perpetually as to destroy the unity which ought to result from the whole poem. This is the general rule, but, of course, subject to some modifications, according to the different kinds of prose or verse. Some prose may approach towards verse, as oratory, and therefore a more studied exhibition of the *media* may be proper; and some verse may border more on mere narrative, and there the style should be simpler. But the great thing in poetry is, *quocumque modo,*[9] to effect a unity of impression upon the whole; and a too great fulness and profusion of point in the parts will prevent this. Who can read with pleasure more than a hundred lines or so of Hudibras[10] at one time? Each couplet or quatrain is so whole in itself, that you can't connect them. There is no fusion,—just as it is in Seneca.

June 28. 1834.

You may not understand my system, or any given part of it,—or by a determined act of wilfulness, you may, even though perceiving a ray of light, reject it in anger and disgust:—but this I will say,—that if you once master it, or any part of it, you cannot hesitate to acknowledge it as the truth. You cannot be sceptical about it.

The metaphysical disquisition at the end of the first volume of the "Biographia Literaria" is unformed and immature;—it contains the fragments of the truth, but it is not fully thought out. It is wonderful to myself to think how infinitely more profound my views now are, and yet how much clearer they are withal. The circle is completing; the idea is coming round to, and to be, the common sense.

---

6. **Spectator . . . extra:** "watcher from without." 7. **Achitophel and Zimri:** satiric portraits of Shaftesbury and Buckingham in *Absalom and Achitophel*, ll. 150–74, 544–68. 8. **Timon:** *Moral Essays*, IV. 99–168.

9. **quocumque modo:** "in whatever way." 10. **Hudibras:** comic narrative poem by Samuel Butler, 1612–80.

FROM

ANIMA POETAE

❧

Anima Poetae, a selection from Coleridge's notebooks, was published by Ernest Hartley Coleridge in 1895.

❧

The elder languages were fitter for poetry because they expressed only prominent ideas with clearness, the others but darkly. . . . Poetry gives most pleasure when only generally and not perfectly understood. It was so by me with Gray's "Bard" and Collins' Odes. The "Bard" once intoxicated me, and now I read it without pleasure. From this cause it is that what I call metaphysical poetry gives me so much delight [1799].

Poetry which excites us to artificial feelings makes us callous to real ones [1796].

Materialists unwilling to admit the mysterious element of our nature make it all mysterious—nothing mysterious in nerves, eyes, etc., but that nerves think, etc.! Stir up the sediment into the transparent water, and so make all opaque [1801].

Quære, whether or no too great definiteness of terms in any language may not consume too much of the vital and idea-creating force in distinct, clear, full-made images, and so prevent originality. For original might be distinguished from positive thought [?1801–02].

The unspeakable comfort to a good man's mind, nay, even to a criminal, to be *understood*,—to have some one that understands one,—and who does not feel that, on earth, no one does? The hope of this, always more or less disappointed, gives the passion to friendship [1801–02].

Take away from sounds the sense of outness, and what a horrible disease would every minute become! A drive over a pavement would be exquisite torture. What, then, is sympathy if the feelings be not disclosed? An inward reverberation of the stifled cry of distress [1802–03].

Socinianism,[1] moonlight; methodism, a stove. O for some sun to unite heat and light! [1802–03].

Nothing affects me much at the moment it happens. It either stupefies me, and I, perhaps, look at a merry-make and dance-the-hay[2] of flies, or listen entirely to the loud click of the great clock, or I am simply indifferent, not without some sense of philosophical self-complacency. For a thing at the moment is but a thing of the moment; it must be taken up into the mind, diffuse itself through the whole multitude of shapes and thoughts, not one of which it leaves un-tinged, between not one of which and it some new thought is not engendered. Now this is a work of time, but the body feels it quicker with me [1803].

One excellent use of communication of sorrow to a friend is this, that in relating what ails us, we ourselves first know exactly what the real grief is, and see it for itself in its own form and limits. Unspoken grief is a misty medley of which the real affliction only plays the first fiddle, blows the horn to a scattered mob of obscure feelings. Perhaps, at certain moments, a single, almost insignificant sorrow may, by association, bring together all the little relics of pain and dis-comfort, bodily and mental, that we have endured even from infancy [1803].

A most unpleasant dispute with Wordsworth and Hazlitt. I spoke, I fear, too contemptuously; but they spoke so irreverently, so malignantly of the Divine Wisdom that it overset me. Hazlitt, how easily raised to rage and hatred self-projected! but who shall find the force that can drag him out of the depths into one expression of kindness, into the showing of one gleam of the light of love on his countenance. Peace be with him! But *thou*, dearest Wordsworth—and what if Ray, Durham, Paley[3] have carried the observation of the aptitude of things too far, too habitually into pedantry? O how many worse pedantries! how few so harmless, with so much efficient good! Dear William, pardon pedantry in others, and avoid it in yourself, instead of scoffing and reviling at pedantry

ANIMA POETAE. **1. Socinianism:** narrowly rationalizing theology; see Introduction, p. 393. **2. dance-the-hay:** a country dance with couples interweaving. **3. Ray, Durham, Paley:** John Ray, 1627–1705, William Derham, 1657–1735, and William Paley, 1743–1805, naturalist divines who attempted to prove the wisdom and benevolence of God by the evidence of design in nature.

in good men and a good cause, and *becoming* a pedant yourself in a bad cause—even by that very act becoming one. But, surely, always to look at the superficies of objects for the purpose of taking delight in their beauty, and sympathy with their real or imagined life, is as deleterious to the health and manhood of intellect as always to be peering and unravelling contrivances may be to the simplicity of the affections and the grandeur and unity of the imagination. O dearest William! would Ray or Durham have spoken of God as you spoke of Nature? [1803].

This evening, and indeed all this day, I ought to have been reading and filling the margins of Malthus.[4]

I had begun and found it pleasant. Why did I neglect it? Because I ought not to have done this. The same applies to the reading and writing of letters, essays, etc. Surely this is well worth a serious analysis, that, by understanding, I may attempt to heal it. For it is a deep and wide disease in my moral nature, at once elm-and-oak-rooted. Is it love of liberty, of spontaneity, or what? These all express, but do not explain, the fact.

After I had got into bed last night I said to myself that I had been pompously enunciating as a difficulty a problem of easy and common solution,—viz., that it was the effect of association. From infancy up to manhood, under parents, schoolmasters, inspectors, etc., our pleasures and pleasant self-chosen pursuits (self-chosen because pleasant, and not originally pleasant because self-chosen) have been forcibly inter-rupted, and dull, unintelligible rudiments or painful tasks imposed upon us instead. Now all duty is felt as a *command*, and every command is of the nature of an offence. Duty, therefore, by the law of association being felt as a command from without, would naturally call up the sensation of the pain roused from the commands of parents and schoolmasters. But I awoke this morning at half past one, and as soon as disease permitted me to think at all, the shallowness and sophistry of this solution flashed upon me at once. I saw that the phenomenon occurred far, far too early; I have observed it in infants of two or three months old, and in Hartley[5] I have seen it turned up and laid bare to the unarmed eye of the merest common sense.

The fact is, that interruption of itself is painful, because, and as far as, it acts as *disruption*. And thus without any reference to, or distinct recollection of, my former theory I saw great reason to attribute the effect, wholly, to the streamy nature of the associative faculty, and the more, as it is evident that they labor under this defect who are most reverie-ish and streamy—Hartley, for instance, and myself. This seems to me no common corroboration of my former thought on the origin of moral evil in general [1804].

Days and weeks and months pass on, and now a year—and the sea, the sea, and the breeze have their influence on me, and [so, too, has the association with] good and sensible men. I feel a pleasure upon me, and I am, to the outward view, cheerful, and have myself no distinct consciousness of the contrary, for I use my faculties, not, indeed, at once, but freely. But, oh! I am never happy, never deeply gladdened. I know not —I have forgotten—what the *joy* is of which the heart is full, as of a deep and quiet fountain overflowing insensibly, or the gladness of joy when the fountain overflows ebullient [1804].

It is often said that books are companions. They are so, dear, very dear companions. But I often, when I read a book that delights me on the whole, feel a pang that the author is not present, that I cannot *object* to him this and that, express my sympathy and gratitude for this part, and mention some facts that self-evidently overset a second, start a doubt about a third, or confirm and carry [on] a fourth thought. At times I become restless, for my nature is very social [1804].

There are two sorts of talkative fellows whom it would be injurious to confound, and I, S. T. Coleridge, am the latter, The first sort is of those who use five hundred words more than needs to express an idea— that is not my case. Few men, I will be bold to say, put more meaning into their words than I, or choose them more deliberately and discriminately. The second sort is of those who use five hundred more ideas, images, reasons, etc., than there is any need of to arrive at their object, till the only object arrived at is that the mind's eye of the bystander is dazzled with colors succeeding so rapidly as to leave one vague impression that there has been a great blaze of colors all about something. Now this is my case, and a grievous fault it is. My illustrations swallow up my thesis. I feel too intensely the omnipresence of all in each, platonically

---

4. Malthus: Thomas Robert Malthus, 1766–1834, best known for his *Essay on the Principle of Population* (1798) in which he prophesied that the world's population, increasing in geometrical progression, would outstrip the means of supporting it. **5. Hartley:** Coleridge's son.

speaking; or, psychologically, my brain-fibres, or the spiritual light which abides in the brain-marrow, as visible light appears to do in sundry rotten mackerel and other *smashy* matters, is of too general an affinity with all things, and though it perceives the *difference* of things, yet is eternally pursuing the likenesses, or, rather, that which is common [between them]. Bring me two things that seem the very same, and then I am quick enough [not only] to show the difference, even to hair-splitting, but to go on from circle to circle till I break against the shore of my hearers' patience, or have my concentricals dashed to nothing by a snore. That is my ordinary mishap. At Malta, however, no one can charge me with one or the other. I have earned the general character of being a quiet well-meaning man, rather dull indeed! and who would have thought that he had been a *poet!* "O, a very wretched poetaster, ma'am! As to the reviews, 'tis well known he half ruined himself in paying cleverer fellows than himself to write them," etc. [1804].

How far might one imagine all the theory of association out of a system of growth, by applying to the brain and soul what we know of an embryo? One tiny particle combines with another its like, and, so, lengthens and thickens, and this is, at once, memory and increasing vividness of impression. One might make a very amusing allegory of an embryo soul up to birth! Try! it is promising! You have not above three hundred volumes to write before you come to it, and as you write, perhaps, a volume once in ten years, you have ample time.

My dear fellow! never be ashamed of scheming— you can't think of living less than 4000 years, and that would nearly suffice for your present schemes. To be sure, if they go on in the same ratio to the performance, then a small difficulty arises; but never mind! look at the bright side always and die in a dream! Oh! [1804].

Modern poetry is characterized by the poets' *anxiety* to be always striking. There is the same march in the Greek and Latin poets. Claudian,[6] who had powers to have been anything—observe in him this anxious, craving vanity! Every line, nay, every word, stops, looks full in your face, and asks and *begs* for praise! As in a Chinese painting, there are no distances, no perspective, but all is in the foreground; and this is nothing but vanity. I am pleased to think that, when

a mere stripling, I had formed the opinion that true taste was virtue, and that bad writing was bad feeling [1805].

In looking at objects of Nature while I am thinking, as at yonder moon dim-gleaming through the dewy window-pane, I seem rather to be seeking, as it were *asking* for, a symbolical language for something within me that already and for ever exists, than observing anything new. Even when that latter is the case, yet still I have always an obscure feeling as if that new phenomena were the dim awaking of a forgotten or hidden truth of my inner nature. It is still interesting as a word—a symbol [1805].

One of the strangest and most painful peculiarities of my nature (unless others have the same, and, like me, hide it, from the same inexplicable feeling of causeless shame and sense of a sort of guilt, joined with the apprehension of being feared and shrunk from as a something transnatural) I will here record—and my motive, or, rather, impulse, to do this seems an effort to eloign and abalienate[7] it from the dark adyt[8] of my own being by a visual outness, and not the wish for others to see it. It consists in a sudden second sight of some hidden vice, past, present, or to come, of the person or persons with whom I am about to form a close intimacy—which never deters me, but rather (as all these transnaturals) urges me on, just like the feeling of an eddy-torrent to a swimmer. I see it as a vision, feel it as a prophecy, not as one *given* me by any other being, but as an act of my own spirit, of the absolute *noumenon*,[9] which, in so doing, seems to have offended against some law of its being, and to have acted the traitor by a commune with full consciousness independent of the tenure or inflected state of association, cause, and effect, etc. [1811–12].

LETTERS

⤮

Coleridge was a voluminous correspondent and, unlike Wordsworth, was completely himself in his letters. He commonly wrote at length, and since much of it naturally concerns business and personal matters of interest only to biographers, his letters are given here mainly

---

6. Claudian: Latin poet, 365?–408?

7. abalienate: remove. **8. adyt:** sanctuary, secret place. **9. noumenon:** thing-in-itself.

in extracts rather than wholes. Three letters (to Humphry Davy, Feb. 3, 1801; to Joseph Cottle, April 26, 1814; to Wordsworth, May 30, 1815) are given entirely, for the interest of the subject matter and also for illustration of his extraordinary capacity for friendship. Of his correspondents John Thelwall was a noted radical with whom he carried on a mainly epistolary friendship for a few years (see *Table Talk*, July 27, 1830); Joseph Cottle, a Bristol poet and printer, gave him financial aid in his early years (including publishing the *Lyrical Ballads*); Humphry Davy, later a famous chemist, was a friend for several years; Thomas Poole, one of his closest friends, was a businessman in Nether Stowey; William Godwin, the philosopher and novelist, was famous in his time chiefly for *Political Justice* (1793); William Sotheby, a minor writer, was a friend from their first meeting in 1802; Thomas Clarkson, noted for his effort to have the slave trade abolished, was, with his wife, a friend of both Wordsworth and Coleridge; and Daniel Stuart was owner and editor of *The Morning Post*, for which Coleridge wrote a large number of articles.

～～

To John Thelwall
Saturday Nov. 19th [1796]

Your portrait of yourself interested me—As to me, my face, unless when animated by immediate eloquence, expresses great Sloth, & great, indeed almost ideotic, good nature. 'Tis a mere carcase of a face: fat, flabby, & expressive chiefly of inexpression.—Yet, I am told, that my eyes, eyebrows, & forehead are physiognomically good—; but of this the Deponent knoweth not. As to my shape, 'tis a good shape enough, if measured—but my gait is awkward, & the walk, & the *Whole man* indicates *indolence capable of energies.*—I am, & ever have been, a great reader—& have read almost every thing—a library-cormorant—I am *deep* in all out of the way books, whether of the monkish times, or of the puritanical aera—I have read & digested most of the Historical Writers—; but I do not *like* History. Metaphysics, & Poetry, & "Facts of mind"—(i.e. Accounts of all the strange phantasms that ever possessed your philosophy-dreamers from Tauth [Thoth],[1] the Egyptian to Taylor,[2] the English Pagan,) are my darling Studies.—In short, I seldom read except to amuse myself—& I am almost always reading.——Of useful knowledge, I am a so-so chemist, & I love

chemistry——all else is *blank*,—but I *will* be (please God) an Horticulturist & a Farmer. I compose very little—& I absolutely hate composition. Such is my dislike, that even a sense of Duty is sometimes too weak to overpower it.

I cannot breathe thro' my nose—so my mouth, with sensual thick lips, is almost always open. In conversation I am impassioned, and oppose what I deem [error] with an eagerness, which is often mistaken for personal asperity——but I am ever so swallowed up in the *thing*, that I perfectly forget my *opponent*. Such am I.

To John Thelwall
December 17th. Saturday Night. [1796]

. . . I feel strongly, and I think strongly; but I seldom feel without thinking, or think without feeling. Hence tho' my poetry has in general a *hue* of tenderness, or Passion over it, yet it seldom exhibits unmixed & simple tenderness or Passion. My philosophical opinions are blended with, or deduced from, my feelings: & this, I think, peculiarizes my style of Writing. And like every thing else, it is sometimes a beauty, and sometimes a fault. But do not let us introduce an act of Uniformity against Poets—I have room enough in *my* brain to admire, aye & almost equally, the *head* and fancy of Akenside, and the *heart* and fancy of Bowles,[3] the solemn Lordliness of Milton, & the divine Chit chat of Cowper: and whatever a man's excellence is, that will be likewise his fault.

To Joseph Cottle
[c. 3 July 1797]

Wordsworth & his exquisite Sister are with me—She is a woman indeed!—in mind, I mean, & heart—for her person is such, that if you expected to see a pretty woman, you would think her ordinary—if you expected to find an ordinary woman, you would think her pretty!—But her manners are simple, ardent, impressive—.

> In every motion her most innocent soul
> Outbeams so brightly, that who saw would say,
> Guilt was a thing impossible in her.—[4]

LETTERS. **1. Thoth:** in Egyptian mythology, the scribe of the gods and inventor of numbers (hence a god of magic and wisdom). **2. Taylor:** Thomas Taylor, 1758–1835, known especially for his translations from the Neoplatonists.

3. Akenside . . . Bowles: See *The Friend*, above, n. 8, and *Biographia Literaria*, above, Chap. XXII, n. 30. **4. In . . . her:** Coleridge's "Destiny of Nations," ll. 173–75.

Her information various—her eye watchful in minutest observation of nature—and her taste a perfect electrometer—it bends, protrudes, and draws in, at subtlest beauties & most recondite faults.

To John Thelwall
Saturday Morning [*14 October 1797*]

—I can *at times* feel strongly the beauties, you describe, in themselves, & for themselves—but more frequently *all things* appear little—all the knowledge, that can be acquired, child's play——the universe itself—what but an immense heap of *little* things?—I can contemplate nothing but parts, & parts are all *little*—!—My mind feels as if it ached to behold & know something *great*—something *one & indivisible*—and it is only in the faith of this that rocks or waterfalls, mountains or caverns give me the sense of sublimity or majesty!—But in this faith *all things* counterfeit infinity!—

To Humphry Davy
Tuesday, Feb. 3. 1801

MY DEAR DAVY

I can scarcely reconcile it to my Conscience to make you pay postage for another Letter. O what a fine Unveiling of modern Politics it would be, if there were published a minute Detail of all the sums received by Government from the Post Establishment, and of all the outlets, in which the sums so received, flowed out again—and on the other hand all the domestic affections that had been stifled, all the intellectual progress that would have been, but is not, on account of this heavy Tax, &c &c——The *Letters* of a nation ought to be payed for, as an article of national expence.——Well—but I did not take up this paper to flourish away in splenetic Politics.——

A Gentleman resident here, his name Calvert,[5] an idle, good-hearted, and ingenious man, has a great desire to commence fellow-student with me & Wordsworth in Chemistry.—He is an intimate friend of Wordsworth's—& he has proposed to Wordsworth to take a house which he (Calvert) has nearly built, called Windy Brow, in a delicious situation, scarce half a mile from Grieta Hall, the residence of S. T.

---

5. **Calvert:** William Calvert, brother of Raisley Calvert, who had bequeathed Wordsworth a legacy.

Coleridge Esq./ and so for him (Calvert) to live with them, i.e. Wordsworth & his Sister.—In this case he means to build a little Laboratory &c.—Wordsworth has not quite decided, but is strongly inclined to adopt the scheme, because he and his Sister have before lived with Calvert on the same footing, and are much attached to him; because my Health is so precarious, and so much injured by Wet, and his health too is, like little potatoes, no great things, and therefore Grasmere (13 miles from Keswick) is too great a distance for us to enjoy each other's Society without inconvenience as much as it would be profitable for us both; & likewise because he feels it more & more necessary for him to have some intellectual pursuit less closely connected with deep passion, than Poetry, & is of course desirous too not to be so wholly ignorant of knowleges so exceedingly important—. However whether Wordsworth come or no, Calvert & I have determined to begin & go on. Calvert is a man of sense, and some originality/ and is besides what is well called a *handy* man. He is a good practical mechanic &c—and is desirous to lay out any sum of money that may be necessary. You know how long, how ardently I have wished to initiate myself in Chemical science—both for it's own sake, and in no small degree likewise, my beloved friend!—that I may be able to sympathize with *all*, that you do and think.—Sympathize blindly with it all I do even *now*, God knows! from the very middle of my heart's heart—; but I would fain sympathize with you in the Light of Knowlege.—This opportunity therefore is exceedingly precious to me—as on my own account I could not afford any the least additional expence, having been already by long & successive Illnesses thrown behind hand so much, that for the next 4 or five months, I fear, let me work as hard as I can, I shall not be able to do what my heart within me *burns* to do—that is, *concenter* my free mind to the affinities of the Feelings with Words & Ideas under the title of "Concerning Poetry & the nature of the Pleasures derived from it."——I have faith, that I do understand this subject / and I am sure, that if I write what I ought to do on it, the Work would supersede all the Books of Metaphysics hitherto written / and all the Books of Morals too.—To whom shall a young man utter *his* Pride, if not to a young man whom he loves?——

I beg you therefore, my dear Davy! to write to me a long Letter when you are at leisure, informing me 1. What Books it will be well for Mr Calvert to purchase. 2. Directions for a convenient little Laboratory—

and 3rdly—to what amount the apparatus would run in expence, and whether or no you would be so good as to superintend it's making at Bristol.—Fourthly, give me your advice how to *begin*——and fifthly & lastly & mostly do send a *drop* of hope to my parched Tongue, that you will, if you can, come & visit me in the Spring.—Indeed, indeed, you ought to see this Country, this divine Country—and then the Joy you would send into *me*!

The Shape of this paper will convince you with what eagerness I began this Letter—I really did not see that it was not a Sheet.

I have been *thinking* vigorously during my Illness—so that I cannot say, that my long long wakeful nights have been all lost to me. The subject of my meditations ha[s] been the Relations of Thoughts to Things, in the language of Hume, of Ideas to Impressions: I may be truly described in the words of Descartes. I have been "res cogitans, id est, dubitans, affirmans, negans, p[auca] intelligens, multa ignorans, volens, nolens, imaginans etia[m,] et sentiens—"[6] & I please myself with believing, that [you] will receive no small pleasure from the result of [my] broodings, altho' I expect in you (in some points) [a] determined opponent —but I say of my mind, in this respect,

> "Manet imperterritus ille
> Hostem magnanimum opperiens, et mole suâ stat."[7]

Every poor fellow has his proud hour sometimes—& this, I suppose, is mine.—

I am better in every respect than I was; but am still *very feeble*. The Weather has been woefully against me for the last fortnight, it having rained here almost incessantly—I take large quantities of Bark, but the effect is (to express myself with the dignity of Science) $X = 0000000$: and I shall not gather strength or t[hat] suffusion of bloom which belongs to my healthy state, till I can walk out.

God bless you, me dear Davy! & your ever affectionate Friend, S. T. Coleridge.

P.S. An electrical machine & a number of little nick nacks connected with it Mr Calvert has.——*Write*.

To Thomas Poole
Monday Night. [*16 March 1801*]
[*Endorsed March 16, 1801*]

MY DEAR FRIEND

The interval since my last Letter has been filled up by me in the most intense Study. If I do not greatly delude myself, I have not only completely extricated the notions of Time, and Space; but have overthrown the doctrine of Association, as taught by Hartley,[8] and with it all the irreligious metaphysics of modern Infidels—especially, the doctrine of Necessity.[9]—This I have *done*; but I trust, that I am about to do more—namely, that I shall be able to evolve all the five senses, that is, to deduce them from *one sense*, & to state their growth, & the causes of their difference——& in this evolvement to solve the process of Life & Consciousness.——I write this to you only; & I pray you, mention what I have written to no one.—At Wordsworth's advice or rather fervent intreaty I have intermitted the pursuit—the intensity of thought, & the multitude of minute experiments with Light & Figure, have made me so nervous & feverish, that I cannot sleep as long as I ought & have been used to do; & the Sleep, which I have, is made up of Ideas so connected, & so little different from the operations of Reason, that it does not afford me the due Refreshment. I shall therefore take a Week's respite; & make Christabel ready for the Press—which I shall publish by itself—in order to get rid of all my engagements with Longman[10] —My German Book I have suffered to remain suspended, chiefly because the thoughts which had employed my sleepless nights during my illness were *imperious* over me, & tho' Poverty was staring me in the face, yet I dared behold my Image miniatured in the pupil of her hollow eye, so steadily did I look her in the Face!——for it seemed to me a Suicide of my very soul to divert my attention from Truths so important, which came to me almost as a Revelation / Likewise, I cannot express to you, dear Friend of my heart!—the loathing, which I once or twice felt, when I attempted to write, merely for the Bookseller, without any sense of the moral utility of what I was writing.—I shall therefore, as I said, immediately publish my CHRISTABEL,

---

6. "res . . . sentiens": Descartes' *Meditatio Tertia*: "a thing that thinks, that is, doubts, affirms, denies, understands a few matters, is ignorant of many, wills, refuses, imagines, and perceives." 7. "Manet . . . stat": "he remains unterrified, awaiting his noble foe, and stands fast in his bulk."

8. Hartley: David Hartley's *Observations on Man* (1749), argued that the mind and emotions are formed wholly by processes of association. 9. Necessity: See *The Friend*, above, n. 11. 10. Longman: the publisher.

with two Essays annexed to it, on the Praeternatural—and on Metre. This done I shall propose to Longman instead of my Travels (which tho' nearly done I am exceedingly anxious not to publish, because it brings me forward in a *personal* way, as a man who relates little adventures of himself to *amuse* people—& thereby exposes me to sarcasm & the malignity of anonymous Critics, & is besides *beneath me*—I say, *beneath me* / for to whom should a young man utter the pride of his Heart if not to the man whom he loves more than all others?) I shall propose to Longman to accept instead of these Travels a work on the originality & merits of Locke, Hobbes, & Hume / which work I mean as a *Pioneer* to my greater work, and as exhibiting a proof that I have not formed opinions without an attentive Perusal of the works of my Predecessors from Aristotle to Kant.—I am *confident*, that I can prove that the Reputation of these three men has been wholly un-merited, & I have in what I have already written traced the whole history of the causes that effected this reputation entirely to Wordsworth's satisfaction.

To Thomas Poole
Monday Night [23 March 1801]

My opinion is this—that deep Thinking is attainable only by a man of deep Feeling, and that all Truth is a species of Revelation. The more I understand of Sir Isaac Newton's works, the more boldly I dare utter to my own mind & therefore to *you*, that I believe the Souls of 500 Sir Isaac Newtons would go to the making up of a Shakspere or a Milton. But if it please the Almighty to grant me health, hope, and a steady mind, (always the 3 clauses of my hourly prayers) before my 30th year I will thoroughly under-stand the whole of Newton's Works—At present, I must content myself with endeavouring to make myself entire master of his easier work, that on Optics.[11] I am exceedingly delighted with the beauty & neatness of his experiments, & with the accuracy of his *immediate* Deductions from them—but the opinions founded on these Deductions, and indeed his whole Theory is, I am persuaded, so exceedingly superficial as without impropriety to be deemed false. Newton was a mere materialist—*Mind* in his system is always passive—a lazy Looker-on on an external World. If the mind be

not *passive*, if it be indeed made in God's Image, & that too in the sublimest sense—the Image of the *Creator*—there is ground for suspicion, that any system built on the passiveness of the mind must be false, as a system. I need not observe, My dear Friend, how unutterably silly & contemptible these Opinions would be, if written to any but to another Self. I assure you, solemnly assure you, that you & Words-worth are the only men on Earth to whom I would have uttered a word on this subject—. It is a rule, by which I hope to direct all my literary efforts, to let my Opinions & my Proofs go together. It is *insolent* to *differ* from the public *opinion* in *opinion*, if it be only *opinion*. It is sticking up little *i by itself i* against the whole alphabet, But one *word* with *meaning* in it is worth the whole alphabet together—such is a sound Argument, an incontrovertible Fact.—

To William Godwin
Greta Hall, Keswick,
Wednesday, March 25, 1801

. . . The Poet is dead in me—my imagination (or rather the Somewhat that had been imaginative) lies, like a Cold Snuff on the circular Rim of a Brass Candle-stick, without even a stink of Tallow to remind you that it was once cloathed & mitred with Flame. That is past by!—I was once a Volume of Gold Leaf, rising & riding on every breath of Fancy—but I have beaten myself back into weight & density, & now I sink in quick-silver, yea, remain squat and square on the earth amid the hurricane, that makes Oaks and Straws join in one Dance, fifty yards high in the Element.

Have you seen the second Volume of the Lyrical Ballads, & the Preface prefixed to the First?——I should judge of a man's Heart, and Intellect precisely according to the degree & intensity of the admiration, with which he read those poems——Perhaps, instead of Heart I should have said Taste, but when I think of The Brothers, of Ruth, and of Michael, I recur to the expres-sion, & am enforced to say *Heart*. If I die, and the Book-sellers will give you any thing for my Life, be sure to say—"Wordsworth descended on him, like the Γνῶθι σεαυτόν[12] from Heaven; by shewing to him what true Poetry was, he made him know, that he himself was no Poet."

11. **Optics:** *Optics* (1704).

12. **Γνῶθι σεαυτόν:** "Know thyself."

To William Godwin

Friday Morning, Jan. 22. 1802
King's Street, Covent Garden—

. . . Partly from ill-health, & partly from an unhealthy & reverie-like vividness of *Thoughts,* & (pardon the pedantry of the phrase) a diminished Impressibility from *Things,* my ideas, wishes, & feelings are to a diseased degree disconnected from *motion & action.* In plain & natural English, I am a dreaming & therefore an indolent man—. I am a Starling self-incaged, & always in the Moult, & my whole Note is, Tomorrow, & tomorrow, & tomorrow. The same causes, that have robbed me to so great a degree of the self-impelling self-directing Principle, have deprived me too of the due powers of Resistances to Impulses from without. If I might so say, I am, as an *acting* man, a creature of mere Impact.

To William Sotheby

Tuesday, July 13, 1802.
Greta Hall, Keswick

. . . It is easy to cloathe Imaginary Beings with our own Thoughts & Feelings; but to send ourselves out of ourselves, to *think* ourselves in to the Thoughts and Feelings of Beings in circumstances wholly & strangely different from our own / hoc labor, hoc opus / and who has atchieved it? Perhaps only Shakespere. Metaphisics is a word, that you, my dear Sir! are no great Friend to / but yet you will agree, that a great Poet must be, implicitè if not explicitè, a profound Metaphysician. He may not have it in logical coherence, in his Brain & Tongue; but he must have it by *Tact /* for all sounds, & forms of human nature he must have the *ear* of a wild Arab listening in the silent Desert, the eye of a North American Indian tracing the footsteps of an Enemy upon the Leaves that strew the Forest—; the *Touch* of a Blind Man feeling the face of a darling Child—

To William Sotheby

July 19, [1802.] Keswick

. . . I wished to force myself out of metaphysical trains of Thought—which, when I trusted myself to my own Ideas, came upon me uncalled—& when I wished to write a poem, beat up Game of far other kind—instead of a Covey of poetic Partridges with whirring wings of music, or wild Ducks *shaping* their rapid flight in forms always regular (a still better image of Verse) up came a metaphysical Bustard, urging it's slow, heavy, laborious, earth-skimming Flight, over dreary & level Wastes.

To William Sotheby

Friday, Sept. 10, 1802.
Greta Hall, Keswick

. . . Bowles's Stanzas on Navigation[13] are among the best in that second Volume / but the whole volume is woefully inferior to it's Predecessor. There reigns thro' all the blank verse poems such a perpetual trick of *moralizing* every thing—which is very well, occasionally—but never to see or describe any interesting appearance in nature, without connecting it by dim analogies with the moral world, proves faintness of Impression. Nature has her proper interest; & he will know what it is, who believes & feels, that every Thing has a Life of it's own, & that we are all *one Life.* A Poet's *Heart & Intellect* should be *combined, intimately* combined & *unified,* with the great appearances in Nature—& not merely held in solution & loose mixture with them, in the shape of formal Similies. I do not mean to *exclude* these formal Similies—there are moods of mind, in which they are natural—pleasing moods of mind, & such as a Poet will often have, & sometimes express; but they are not his highest, & most appropriate moods. They are "Sermoni propiora"[14] which I once translated—"*Properer for a Sermon.*" The truth is—Bowles has indeed the *sensibility* of a poet; but he has not the *Passion* of a great Poet. His latter Writings all want *native* Passion—Milton here & there supplies him with an appearance of it—but he has no native Passion, because he is not a Thinker—& has probably weakened his Intellect by the haunting Fear of becoming extravagant. . . . It has struck [me] with great force lately, that the Psalms afford a most compleat answer to those, who state the Jehovah of the Jews, as a personal & national

13. **Bowles's . . . Navigation:** W. L. Bowles's, "The Spirit of Navigation Discovery," *Sonnets and Other Poems,* 2 vols. (1802). The "predecessor" was *Sonnets* (1789), which Coleridge had once admired enormously. 14. **"Sermoni propiora":** Horace, *Satires,* I. iv. 42: "more suited to conversation."

God—& the Jews, as differing from the Greeks, only in calling the minor Gods, Cherubim & Seraphim—& confining the word God to their Jupiter. It must occur to every Reader that the Greeks in their religious poems address always the Numina Loci, the Genii, the Dryads, the Naiads, &c &c—All natural Objects were *dead*—mere hollow Statues—but there was a Godkin or Goddessling *included* in each—In the Hebrew Poetry you find nothing of this poor Stuff —as poor in genuine Imagination, as it is mean in Intellect— / At best, it is but Fancy, or the aggregating Faculty of the mind—not *Imagination*, or the *modifying*, and *co-adunating* Faculty.[15] This the Hebrew Poets appear to me to have possessed beyond all others—& next to them the English. In the Hebrew Poets each Thing has a Life of it's own, & yet they are all one Life. In God they move & live, & *have* their Being—not *had*, as the cold System of Newtonian Theology represents but *have*.

To Robert Southey
Sunday, Aug. 14. 1803

. . . There is a state of mind, wholly unnoticed, as far as I know, by any Physical or Metaphysical Writer hitherto, & which yet is necessary to the explanation of some of the most important phaenomena of Sleep & Disease / it is a transmutation of the *succession* of *Time* into the *juxtaposition* of *Space*, by which the smallest Impulses, if quickly & regularly recurrent, *aggregate* themselves—& attain a kind of visual magnitude with a correspondent Intensity of general Feeling. —The simplest Illustration would be the *circle* of Fire made by whirling round a live Coal—only here the mind is passive. Suppose the same effect produced ab intra—& you have a clue to the whole mystery of frightful Dreams, & Hypochondriacal Delusions.—I merely *hint* this; but I could detail the whole process, complex as it is.—Instead of "an imaginary aggravation &c" it would be better to say—"an *aggregation* of slight Feelings by the force of a diseasedly retentive Imagination."*As to the apprehension of Danger*—it would belong to my Disease, if it could belong to me. But Sloth, Carelessness, Resignation—in all things that have reference to mortal Life—is not merely *in* me; it is *me*.

15. **Imagination . . . Faculty:** This was the first occasion on which Coleridge distinguished imagination from fancy.

To Thomas Wedgwood
Greta Hall, Keswick.
Sept. 16. [1803.] Friday

MY DEAR WEDGWOOD

I reached home on yesterday noon; & it was not a Post Day.—William Hazlitt is a thinking, observant, original man, of great power as a Painter of Character Portraits, & far more in the manner of the old Painters, than any living Artist, but the Object must be *before* him / he has no imaginative memory. So much for his Intellectuals.—His manners are to 99 in 100 singularly repulsive—: brow-hanging, shoe-contemplative, *strange* / Sharp seemed to like him / but Sharp saw him only for half an hour, & that walking—he is, I verily believe, kindly-natured—is very fond of, attentive to, & patient with, children / but he is jealous, gloomy, & of an irritable Pride—& addicted to women, as objects of sexual Indulgence. With all this, there is much good in him—he is disinterested, an enthusiastic Lover of the great men, who have been before us—he says things that are his own in a way of his own—& tho' from habitual Shyness & the Outside & bearskin at least of misanthropy, he is strangely confused & dark in his conversation & delivers himself of almost all his conceptions with a Forceps, yet he says more than any man, I ever knew, yourself only excepted, that is his own in a way of his own—& oftentimes when he has warmed his mind, & the synovial juice has come out & spread over his joints, he will gallop for half an hour together with real Eloquence. He sends wellheaded & well-feathered Thoughts straight forwards to the mark with a Twang of the Bow-string.

To Thomas Poole
Friday, Oct. 14. 1803.
Greta Hall, Keswick

. . . I now see very little of Wordsworth: my own Health makes it inconvenient & unfit for me to go thither one third as often, as I used to do—and Wordsworth's Indolence, &c keeps him at home. Indeed, were I an irritable man, and an unthinking one, I should probably have considered myself as having been very unkindly used by him in this respect—for I was at one time confined for two months, & he never came in to see me / me, who had ever payed such unremitting attentions to him. But we must take the good & the ill together; & by seriously & habitually reflecting on our own faults & endeavouring to amend them we shall then find little difficulty in confining our attention as far as it acts on our Friends' characters, to their good Qualities.—Indeed, I owe it to Truth & Justice as well as to myself to say, that the concern, which I have felt in this instance, and one or two other more *crying* instances, of Self-involution in Wordsworth, has been almost wholly a Feeling of friendly Regret, & disinterested Apprehension—I saw him more & more benetted in hypochondriacal Fancies, living wholly among *Devotees*—having every the minutest Thing, almost his very Eating & Drinking, done for him by his Sister, or Wife—& I trembled, lest a Film should rise, and thicken on his moral Eye. —The habit too of writing such a multitude of small Poems was in this instance hurtful to him—such Things as that Sonnet of his in Monday's Morning Post, about Simonides & the Ghost [16]— / I rejoice therefore with a deep & true Joy, that he has at length yielded to my urgent & repeated—almost unremitting—requests & remonstrances—& will go on with the Recluse [17] exclusively.—A Great Work, in which he will sail; on an open Ocean, & a steady wind; unfretted by short tacks, reefing, & hawling & disentangling the ropes—— great work necessarily comprehending his attention & Feelings within the circle of great objects & elevated Conceptions—this is his natural Element—the having been out of it has been his Disease—to return into it is the specific Remedy, both Remedy & Health. It is what Food is to Famine. I have seen enough, positively to give me feelings of hostility towards the plan of several of the Poems in the L. Ballads: & I really consider it as a misfortune, that Wordsworth ever deserted his former mountain Track to wander in Lanes & allies; tho'in the event it may prove to have been a great Benefit to him. He will steer, I trust, the middle course.—But he found himself to be, or rather to be called, the Head & founder of a *Sect* in Poetry: & assuredly he has written—& published in the M. Post, as W. L. D.[18] & sometimes with no signature—poems written with a *sectarian* spirit, & in a sort of Bravado.—

16. **Simonides . . . Ghost**: in *Poetical Works*, ed. De Selincourt and Darbishire, III. 408. Wordsworth never reprinted the sonnet. **17. Recluse**: that is, *The Prelude*, on which Wordsworth was working and which was to be a part of the never finished *Recluse*. **18. W. L. D.**: The initials with which Wordsworth signed the sonnets in *The Morning Post* have been thought to stand for "Wordsworthius Libertati dedicavit."

To Thomas Clarkson
13 Oct. 1806. Bury St Edmonds

Your third Question admits—in consequence of the preceding—of a briefer and more immediate Answer. What is the difference between the Reason, and the Understanding?—I would reply, that that Faculty of the Soul which apprehends and retains the mere notices of Experience, as for instance that such an object has a triangular figure, that it is of such or such a magnitude, and of such and such a color, and consistency, with the anticipation of meeting the same under the same circumstances, in other words, all the mere φαινόμενα [19] of our nature, we may call the Understanding. But all such notices, as are characterized by UNIVERSALITY and NECESSITY, as that every Triangle *must* in all places and at all times have it's two sides greater than it's third—and which are evidently not the effect of any Experience, but the condition of all Experience, & that indeed without which Experience itself would be inconceivable, we may call Reason— and this class of knowlege was called by the Ancients Νοούμενα [20] in distinction from the former, or φαινόμενα. Reason is therefore most eminently the Revelation of an immortal soul, and it's best Synonime —it is the forma formans, which contains in itself the law of it's own conceptions. Nay, it is highly probable, that the contemplation of essential Form as remaining the same thro' all varieties of color and magnitude and developement, as in the acorn even as in the Oak, first gave to the Mind the ideas, by which it explained to itself those notices of it's Immortality revealed to it by it's conscience.

To Thomas Poole
28 Jan. 1810. Grasmere, Kendal.

. . . Of Parentheses I may be too fond—and will be on my guard in this respect—. But I am certain that no work of empassioned & eloquent reasoning ever did or could subsist without them—They are the *drama* of Reason—& present the thought growing, instead

19. **φαινόμενα**: "appearances." 20. **Νοούμενα**: "conceptions."

of a mere Hortus siccus. The aversion to them is one of the numberless symptoms of a feeble Frenchified Public.

To Joseph Cottle
April 26, 1814

You have poured oil in the raw and festering Wound of an old friend's Conscience, Cottle! but it is oil of Vitriol! I but barely glanced at the middle of the first page of your Letter, & have seen no more of it—not from resentment (God forbid!) but from the state of my bodily & mental sufferings, that scarcely permitted human fortitude to let in a new visitor of affliction. The object of my present reply is to state the case just as it is—first, that for years the anguish of my spirit has been indescribable, the sense of my danger *staring*, but the conscience of my GUILT worse, far far worse than all!—I have prayed with drops of agony on my Brow, trembling not only before the Justice of my Maker, but even before the Mercy of my Redeemer. "I gave thee so many Talents. What hast thou done with them"?—Secondly—that it is false & cruel to say, (overwhelmed as I am with the sense of my direful Infirmity) that I attempt or ever have attempted to *disguise* or conceal the cause. On the contrary, not only to friends have I stated the whole Case with tears & the very bitterness of shame; but in two instances I have warned young men, mere acquaintances who had spoken of having taken Laudanum, of the direful Consequences, by an ample exposition of it's tremendous effects on myself—Thirdly, tho' before God I dare not lift up my eyelids, & only do not despair of his Mercy because to despair would be adding crime to crime; yet to my fellowmen I may say, that I was seduced into the ACCURSED Habit ignorantly.—I had been almost bed-ridden for many months with swellings in my knees—in a medical Journal I unhappily met with an account of a cure performed in a similar case (or what to me appeared so) by rubbing in of Laudanum, at the same time taking a given dose internally—It acted like a charm, like a miracle! I recovered the use of my Limbs, of my appetite, of my Spirits—& this continued for near a fortnight—At length, the unusual Stimulus subsided—the complaint returned—the supposed remedy was recurred to——but I can not go thro' the dreary history—suffice it to say, that effects were produced, which acted on me by *Terror & Cowardice*

of PAIN & sudden Death, not (so help me God!) by any temptation of Pleasure, or expectation or desire of exciting pleasurable Sensations. On the very contrary, Mrs Morgan & her Sister[21] will bear witness so far, as to say that the longer I abstained, the higher my spirits were, the keener my enjoyments—till the moment, the direful moment, arrived, when my pulse began to fluctuate, my Heart to palpitate, & such a dreadful *falling-abroad*, as it were, of my whole frame, such intolerable Restlessness & incipient Bewilderment, that in the last of my several attempts to abandon the dire poison, I exclaimed in agony, what I now repeat in seriousness & solemnity—"I am too poor to hazard this! Had I but a few hundred Pounds, but 200 £, half to send to Mrs Coleridge, & half to place myself in a private madhouse, where I could procure nothing but what a Physician thought proper, & where a medical attendant could be constantly with me for two or three months (in less than that time Life or Death would be determined) then there might be Hope. Now there is none!"—O God! how willingly would I place myself under Dr Fox in his Establishment—for my Case is a species of madness, only that it is a derangement, an utter impotence of the *Volition*, & not of the intellectual Faculties—You bid me rouse myself—go, bid a man paralytic in both arms rub them briskly together, & that will cure him. Alas! (he would reply) that I cannot move my arms is my Complaint & my misery.—

My friend, Wade,[22] is not at home—& I sent off all the little money, I had—or I would with this have inclosed the 10 £ received from you.—

May God bless you | & | Your affectionate & | most afflicted

S. T. COLERIDGE.—

To William Wordsworth
30 May 1815

MY HONORED FRIEND

On my return from Devizes, whither I had gone to procure some Vaccine matter (the Small Pox having appeared in Calne, and Mrs M's[23] Sister believing

---

21. **Mrs Morgan . . . Sister:** Charlotte Brent and Mary, wife of John Morgan, friends of Coleridge from 1795. 22. **Wade:** Josiah Wade, a tradesman and a friend of Coleridge from 1795. 23. **Mrs M's:** Morgan's (see n. 21 above).

herself never to have had it) I found your Letter:[24] and I will answer it immediately, tho' to answer it as I could wish to do would require more recollection and arrangement of Thought than is always to be commanded on the Instant. But I dare not trust my own habit of procrastination—and do what I would, it would be impossible in a single Letter to give more than *general* convictions. But even after a tenth or twentieth Letter I should still be disquieted as knowing how poor a substitute must Letters be for a vivâ voce examination of a Work with it's Author, Line by Line. It is most uncomfortable from many, many Causes, to express anything but sympathy, and gratulation to an absent friend, to whom for the more substantial Third of a Life we have been habituated to look up: especially, where our Love, tho' increased by many and different influences, yet begun and throve and knit it's Joints in the perception of his Superiority. It is not in *written words*, but by the hundred modifications that Looks make, and Tone, and denial of the FULL sense of the very words used, that one can reconcile the struggle between sincerity and diffidence, between the Persuasion, that I am in the Right, and that as deep tho' not so vivid conviction, that it may be the positiveness of Ignorance rather than the Certainty of Insight. Then come the Human Frailties—the dread of giving pain, or exciting suspicions of alteration and Dyspathy[25]—in short, the almost inevitable Insincerities between imperfect Beings, however sincerely attached to each other. It is hard (and I am *Protestant* enough to doubt whether it is right) to confess the whole Truth even of one's Self—Human Nature scarce endures it even *to* one's Self!—but to me it is still harder to do this of and to a revered Friend.—

But to your Letter. First, I had never determined to print the Lines addressed to you—I lent them to L. Beaumont[26] on her promise that they should be copied and returned—& not knowing any copy in my own possession I sent for them, because I was making a *Mss* Collection of *all* my poems, publishable or unpublishable—& still more perhaps, for the Handwriting[27] of the only perfect Copy, that entrusted to her Ladyship.—Most assuredly, I never once thought

of printing them without having consulted you—and since I lit on the first rude draught, and corrected it as well as I could, I wanted no additional reason for it's not being published in my Life Time, than it's *personality* respecting myself—After the opinions, I had given publicly, for the preference of the Lycidas (moral no less than poetical) to Cowley's Monody, I could not have printed it consistently—. It is for the Biographer, not the Poet, to give the *accidents* of *individual* Life. Whatever is not representative, generic, may be indeed most poetically exprest, but is not Poetry. Otherwise, I confess, your prudential Reasons would not have weighed with me, except as far as my name might haply injure your reputation—for there is nothing in the Lines as far [as] your Powers are concerned, which I have not as fully expressed elsewhere— and I hold it a miserable cowardice to withhold a deliberate opinion only because the Man is alive.

2ndly. for the EXCURSION. I feared that had I been silent concerning the Excursion, Lady B. would have drawn some strange inference—& yet I had scarcely sent off the Letter before I repented that I had not run that risk rather than have approach to Dispraise communicated to you by a third person—. But what did my criticism amount to, reduced to it's full and naked Sense?—This: that *comparatively* with the *former* Poem[28] the excursion, as far as it was new to me, had disappointed my expectations—that the Excellences were so many and of so high a class, that it was impossible to attribute the inferiority, if any such really existed, to any flagging of the Writer's own genius— and that I conjectured that it might have been occasioned by the influence of self-established Convictions having given to certain Thoughts and Expressions a depth & force which they had not for readers in general.—In order therefore to explain the *disappointment* I must recall to your mind what my *expectations* were: and as these again were founded on the supposition, that (in whatever order it might be published) the Poem on the growth of your own mind was as the ground-plat[29] and the Roots, out of which the Recluse was to have sprung up as the Tree—as far as the same Sap in both, I expected them doubtless to have formed one compleat Whole, but in matter, form, and product to be different, each not only a distinct but a different Work.—In the first I had found "themes by thee first sung aright"—

---

24. **Letter:** in which Wordsworth asked Coleridge not to print his poem "To William Wordsworth." He also asked for criticism of the recently published *Excursion*. 25. **Dyspathy:** opposite of sympathy. 26. **L. Beaumont:** Sir George and Lady Beaumont were mutual friends of Wordsworth and Coleridge. 27. **Handwriting:** of Sara Hutchinson; see Introduction, p. 387.

28. **former Poem:** *The Prelude*. 29. **ground-plat:** ground-plan.

Of Smiles spontaneous and mysterious fears
(The first-born they of Reason and Twin-birth),
Of Tides obedient to external force,
And Currents self-determin'd, as might seem,
Or by some central Breath; of moments aweful,
Now in thy inner Life, and now abroad,
When Power stream'd from thee, and thy Soul received
The Light reflected as a Light bestow'd!—
Of Fancies fair, and milder Hours of Youth,
Hyblaean murmurs of poetic Thought
Industrious in it's Joy, in vales and Glens
Native or outland, Lakes and famous Hills!
Or on the lonely High-road, when the stars
Were rising; or by secret mountain streams,
The Guides and the Companions of thy Way.

Of more than *Fancy*—of the *social sense*
Distending wide and man beloved as man:
Where France in all her towns lay vibrating
Ev'n as a Bark becalm'd beneath the Burst
Of Heaven's immediate Thunder, when no Cloud
Is visible, or Shadow on the Main!—
For Thou wert there, thy own Brows garlanded,
Amid the tremor of a realm aglow,
Amid a mighty Nation jubilant,
When from the general Heart of Human Kind
HOPE sprang forth, like a full-born Deity!
Of that dear Hope afflicted, and amaz'd,
So homeward summon'd! thenceforth calm & sure
From the dread Watch tower of man's absolute Self
With Light unwaning on her Eyes, to look
Far on!—herself a Glory to behold,
The Angel of the Vision!—Then (last strain!)
Of Duty! Chosen Laws controlling choice!
Action and Joy!—AN ORPHIC SONG INDEED,
A SONG DIVINE OF HIGH AND PASSIONATE TRUTHS
TO THEIR OWN MUSIC CHAUNTED![30]

Indeed thro' the whole of that Poem με Αὔρα τις
εἰσέπνευσε μυστικωτάτη.[31] This I considered as
"the EXCURSION"; and the second as "THE RECLUSE" I
had (from what I had at different times gathered from
your conversation on the Plan) anticipated as commencing with you set down and settled in an abiding
Home, and that with the Description of that Home you
were to begin a *Philosophical Poem*, the result and fruits
of a Spirit so fram'd & so disciplin'd, as had been told
in the former. Whatever in Lucretius is Poetry is not
philosophical, whatever is philosophical is not Poetry:
and in the very Pride of confident Hope I looked forward to the Recluse, as the *first* and *only* true Phil.

Poem in existence. Of course, I expected the Colors,
Music, imaginative Life, and Passion of *Poetry*; but the
matter and arrangement of *Philosophy*—not doubting
from the advantages of the Subject that the Totality
of a System was not only capable of being harmonized
with, but even calculated to aid, the unity (Beginning,
Middle, and End) of a *Poem*. Thus, whatever the Length
of the Work might be, still it was a *determinate* Length:
of the subjects announced each would have it's own
appointed place, and excluding repetitions each would
relieve & rise in interest above the other—. I supposed
you first to have meditated the faculties of Man in the
abstract, in their correspondence with his Sphere of
action, and first, in the Feeling, Touch, and Taste,
then in the Eye, & last in the Ear, to have laid a solid
and immoveable foundation for the Edifice by removing the sandy Sophisms of Locke, and the Mechanic
Dogmatists, and demonstrating that the Senses were
living growths and developements of the Mind &
Spirit in a much juster as well as higher sense, than the
mind can be said to be formed by the Senses—. Next,
I understand that you would take the Human Race
in the concrete, have exploded the absurd notion of
Pope's Essay on Man, Darwin,[32] and all the countless
Believers—even (strange to say) among Xtians of
Man's having progressed from an Ouran Outang state—
so contrary to all History, to all Religion, nay, to all
Possibility—to have affirmed a Fall in some sense, as a
fact, the possibility of which cannot be understood
from the nature of the Will, but the reality of which is
attested by Experience & Conscience—Fallen men
contemplated in the different ages of the World, and
in the different states—Savage—Barbarous—Civilized
—the lonely Cot, or Borderer's Wigwam—the
Village—the Manufacturing Town—Sea-port—City
—Universities—and not disguising the sore evils,
under which the whole Creation groans, to point out
however a manifest Scheme of Redemption from this
Slavery, of Reconciliation from this Enmity with
Nature—what are the Obstacles, the *Antichrist* that
must be & already is—and to conclude by a grand
didactic swell on the necessary identity of a true
Philosophy with true Religion, agreeing in the results
and differing only as the analytic and synthetic process,
as discursive from intuitive, the former chiefly useful
as perfecting the latter—in short, the necessity of a
general revolution in the modes of developing &
disciplining the human mind by the substitution of

30. Of smiles . . . chaunted: "To William Wordsworth,"
ll. 12–47. 31. με . . . μυστικωτάτη: Aristophanes, *Frogs*,
l. 314: "Breathed over me in a mystic way."

32. Darwin: Erasmus Darwin, 1731–1802.

Life, and Intelligence (considered in it's different powers from the Plant up to that state in which the difference of Degree becomes a new kind (man, self-consciousness) but yet not by essential opposition) for the philosophy of mechanism which in every thing that is most worthy of the human Intellect strikes *Death*, and cheats itself by mistaking clear Images for distinct conceptions, and which idly demands Conceptions where Intuitions alone are possible or adequate to the majesty of the Truth.—In short, Facts elevated into Theory—Theory into Laws—& Laws into living & intelligent Powers— true Idealism necessarily perfecting itself in Realism, & Realism refining itself into Idealism.—

Such or something like this was the Plan,[33] I had supposed that you were engaged on—. Your own words will therefore explain my feelings—viz—that your object "was not to convey recondite or refined truths but to place commonplace Truths in an interesting point of View."[34] Now this I supposed to have been in your two Volumes of Poems,[35] as far as was desirable, or p[ossible,] without an insight into the whole Truth—. How can common [trut]hs be made permanently interesting but by being *bottomed* in our common nature—it is only by the profoundest Insight into Numbers and Quantity that a sublimity & even religious Wonder become attached to the simplest operations of Arithmetic, the most evident properties of the Circle or Triangle—.

I have only to finish a Preface which I shall have done in two or at farthest three days—and I will then, dismissing all comparison either with the Poem on the Growth of your own Support, or with the imagined Plan of the Recluse, state fairly my main Objections to the Excursion as it is—But it would have been alike unjust both to you and to myself, if I had led you to suppose that any disappointment, I may have felt, arose wholly or chiefly from the Passages, I do not like—or from the Poem considered irrelatively.—

Allston[36] lives at 8, Buckingham Place, Fitzroy Square—He has lost his wife—& been most unkindly treated—& most unfortunate—I hope, you will call on him.—Good God! to think of such a Grub as

Daw[e][37] with more than he can do—and such a Genius as Allston, without a Single Patron!

God bless you!—I am & never have been other than your | most affectionate

S. T. COLERIDGE.—

To Daniel Stuart

J. Gillman's, Esqre, Surgeon
High Gate, Monday, 13 May 1816.

DEAR STUART

It is among the feeblenesses of our Nature, that we are often to a certain degree acted on by stories gravely asserted, of which we yet do most religiously disbelieve every Syllable—nay, which perhaps, we happen to know to be false. The truth is, that Images and Thoughts possess a power in and of themselves, independent of that act of the Judgement or Understanding by which we affirm or deny the existence of a reality correspondent to them. Such is the ordinary state of the mind in Dreams. It is not strictly accurate to say, that we believe our dreams to be actual while we are dreaming. We neither believe it or disbelieve it —with the will the comparing power is suspended, and without the comparing power any act of Judgement, whether affirmation or denial, is impossible. The Forms and Thoughts act merely by their own inherent power; and the strong feelings at times apparently connected with them are in point of fact bodily sensations, which are the causes or occasions of the Images, not (as when we are awake) the effects of them. Add to this a voluntary Lending of the Will to this suspension of one of it's own operations (i.e. that of comparison & consequent decision concerning the reality of any sensuous Impression) and you have the true Theory of Stage Illusion—equally distant from the absurd notion of the French Critics, who ground their principles on the presumption of an absolute *Delusion*, and of Dr Johnson[38] who would persuade us that our Judgements are as broad awake during the most masterly representation of the deepest scenes of Othello, as a philosopher would be during the exhibition of a Magic Lantern with Punch & Joan, & Pull Devil Pull Baker, &c on it's painted Slides. —Now as Extremes always meet, this Dogma of our

---

33. **Plan:** Cf. *Table Talk*, above, July 21, 1832. This, of course, is not a plan Wordsworth ever had in mind, but an outline of Coleridge's own philosophic system. 34. **"was . . . View":** quoting from Wordsworth's letter. 35. **Poems:** *Poems*, 2 vols. (1815). 36. **Allston:** Washington Allston, 1779–1843, American painter. Cf. the headnote to "On the Principles of Genial Criticism."

37. **Daw[e]:** George Dawe, 1781–1829, portrait and historical painter. 38. **Dr Johnson:** See *Shakespearean Criticism*, above, n. 7.

dogmatic Critic and soporific Irenist[39] would lead by inevitable consequence to that very doctrine of the Unities maintained by the French Belle Lettrists, which it was the object of his strangely over-rated contradictory & most illogical Preface (to Shakespear) to overthrow.

To William Sotheby

9 Waterloo Plains, Ramsgate,
November 9, 1828.

MY DEAR SIR

It is a not unfrequent tragico-whimsical fancy with me to imagine myself as the survivor of

> "This breathing House not built with hands,
> This body that does me grievous wrong"— [40]

and an Assessor at it's dissection—infusing, as spirits may be supposed to have the power of doing, this and that thought into the mind of the Anatomist. Ex. gr. Be so good as to give a cut just *there*, right across the umbilical region—there lurks the fellow that for so many years tormented me on my first waking! or—a stab *there*, I beseech you, it was the seat and source of that dreaded subsultus[41] which so often threw my Book out of my hand, or drove my pen in a blur over the paper on which I was writing! But above all and over all has risen and hovered the strong half-wish, half-belief, that there would be found if not the justifying reason yet the more than the palliation and excuse—if

not the necessitating *cause*, yet the originating occasion, of my heaviest—and in truth they, are so bad that without vanity or self-delusion I might be allowed to call them my *only* offences against others, viz. Sins of Omission. O if in addition to the disturbing accidents and Taxes on my Time resulting from my almost constitutional pain and difficulty in uttering and in persisting to utter, NO! if in addition to the distractions of narrow and embarrassed Circumstances, and of a poor man constrained to be under obligation to generous and affectionate Friends only one degree richer than himself, the calls of the day forcing me away in my most genial hours from a work in which my very heart and soul were buried, to a five guinea task, which fifty persons might have done better, at least, more effectually for the purpose; if in addition to these, and half a score other intrusive Draw-backs, it were possible to convey without inflicting the sensations, which (suspended by the stimulus of earnest conversation or of rapid motion) annoy and at times overwhelm me as soon as I sit down alone, with my pen in my hand, and my head bending and body compressed, over my table (I cannot say, desk)—I dare believe that in the mind of a competent Judge what I have performed will excite more surprize than what I have omitted to do, or failed in doing. . . . Because you possess my inward respect, I would not stand in a worse light, than the knowledge of the whole truth would place me, or forfeit more of your esteem than my conscience assents to. I need not tell you, that pecuniary motives either do not act at all—or are of that class of stimulants which act only as Narcotics: and as to what *people in general* think about me, my mind and spirit are too awfully occupied with the concerns of another Tribunal, before which I stand momently, to be much affected by it one way or other.

39. Irenist: alluding to Johnson's play *Irene* (1749). **40. "This . . . wrong":** Coleridge's "Youth and Age," ll. 8–9. **41. subsultus:** a convulsive or twitching motion.

ROBERT SOUTHEY

1774–1843

IF SO FEW readers today have even a tolerably clear idea of Southey, it is by no means because he was a non-entity. He was not; and in any case writers less gifted than he are recalled more distinctly and justly. The explanation is that he is remembered for widely different reasons, and when they are added together they create not only a mixed but a rather bizarre impression.

Most of us first hear of Southey as one of the "Lake Poets," a friend of Wordsworth and Coleridge. This naturally sets up expectations. But the poetry of his that we read seems by comparison thin and pedestrian, and we are not tempted to go any further—even though we learn that this industrious man wrote an enormous amount, including several now forgotten epics. (His poetry alone fills ten volumes.) We next encounter him as the butt of some of Byron's choicest ridicule, particularly in the brilliant Dedication of *Don Juan* and the parody Byron wrote of Southey's *Vision of Judgement*. Byron, in fact, cannot seem to leave poor Southey alone, and the hapless victim begins to remind us of Colley Cibber (Southey's early eighteenth-century predecessor as poet laureate) whom Alexander Pope ceaselessly ridiculed. Southey is indeed Byron's Colley Cibber; even the name (pronounced as it is spelled, and not "Suthey") takes on a comic overtone from Byron's reference to his verbosity (*Don Juan*, I, 1633–36):

> Thou shalt believe in Milton, Dryden, Pope;
> Thou shalt not set up Wordsworth, Coleridge, Southey;
> Because the first is craz'd beyond all hope,
> The second drunk, the third so quaint and *mouthey*.

On the other hand, we find that Southey was respected by writers of the time that we do admire—even Shelley, whose poetry was so different. He was a learned man, well read in history as well as literature. Moreover, whatever we think of his poetry, his prose is very readable. In fact some of it (for example, his *Life of Nelson* and his *Life of John Wesley*) equals any written during the period, and it has qualities that are refreshing and unexpected—a direct simplicity, a manly and honest frankness ("His prose," Byron admitted, "is perfect"). To this series of conflicting images of Southey should be added another: his interest in childhood, which he shared with Blake, Wordsworth, and Lamb, and which shows itself in two perennial favorites of children, "The Cataract of Lodore" and his nursery tale "The Three Bears," from his miscellany called *The Doctor*.

Southey's younger years, as Byron often reminds us, were in strange contrast to his later ones. Born in Bristol in 1774, the son of an unsuccessful linen draper, he was sent to Westminster School, from which he was later expelled when a school paper published an essay in which he attacked flogging. At Oxford, where he was supposed to be studying for the clergy, he became broadly Unitarian in his religious views. An ardent sympathizer with the French Revolution, he also wrote at Oxford a sort of epic, *Joan of Arc*, celebrating the heroine as a champion of liberty, and a drama, *Wat Tyler*, eulogizing the English Peasant's Revolt of 1381. Coleridge, on a visit to Oxford, became a friend, and the two young men evolved their scheme of "pantisocracy"—an ideal community to be founded on the banks of the Susquehanna, in America, in which property would be shared equally. It was not long before the more practical Southey began to cool toward the scheme. Meanwhile, however, the two men married the Fricker sisters—Southey marrying Edith and Coleridge Sara. After a visit to Portugal and Spain, Southey, at Coleridge's suggestion, settled (in 1803) in the Lake Country, in a house (Greta Hall, which still stands) on the outskirts of Keswick.

Southey's epic poems now began to appear steadily. (He also studied law in a desultory way and then abandoned it.) These longer poems, typical of the Romantic predilection for the remote, included *Thalaba the Destroyer* (1801), *Madoc* (1805), *The Curse of Kehama* (1810), and *Roderick the Last of the Goths* (1814). During this period Southey, like Wordsworth and Coleridge, gradually shed the radical politics of his youth, and in 1813 he was appointed poet laureate. The position was not much coveted, and Southey's predecessor (Henry James Pye, known as "Poetical Pye"; cf. Byron's Dedication to *Don Juan*, l. 8, and note) had long been a subject of laughter because of his laborious dullness. Of Southey's official poems as laureate, the *Vision of Judgment* (1821), on the death of George III, is best known because of Byron's parody of it. Prose writing was increasingly occupying his attention. The best of it, aside from the lives of Nelson (1813) and John Wesley (1820), is to be found in his *History of Brazil* (1810–19), the *History of the Peninsular Wars* (1823–32), the *Lives of the British Admirals* (1833–40), and his miscellany *The Doctor* (1837–47), some of which was published after his death. Though he wrote so incessantly as to give the impression of a mechanical temperament, Southey was high-strung, and he was able to keep writing partly by having several projects going at the same time and turning rapidly from one to another when his energy flagged. Overwork, anxiety, and domestic unhappiness (two of his children and then his wife died) helped to produce a breakdown in his mid-sixties, and his memory left him four years before his death in 1843.

Bibliography

Editions include the *Poetical Works*, 10 vols., 1837–38, and selections by Edward Dowden, 1895, 1930, and by M. H. Fitzgerald, 1909. The prose works are not collected. Editions of selections and single works include *Select Prose*, ed. Jacob Zeitlin, 1916; the *Life of Nelson*, ed. H. B. Butler, 1911, and ed. Geoffrey Callender, 1922; the *Life of Wesley*, ed. M. H. Fitzgerald, 2 vols., 1925; *Letters*, ed. Fitzgerald, 1912; and an abridgement of *The Doctor* also by Fitzgerald, 1930.

For biography and commentary see Edward Dowden's *Southey*, 1879, in the English Men of Letters Series; William Haller, *Early Life of Robert Southey*, 1917; Jack Simmons, *Southey*, 1945; and Geoffrey Carnall, *Robert Southey and His Age*, 1960.

THE BATTLE OF BLENHEIM

Blenheim, in Bavaria, was the scene of the battle (1704) where the French were defeated by the allied army of the English, Dutch, and Austrians under the command of the Duke of Marlborough and Prince Eugene of Austria.

It was a summer evening,
  Old Kaspar's work was done,
And he before his cottage door
  Was sitting in the sun,
And by him sported on the green
His little grandchild Wilhelmine.

She saw her brother Peterkin
  Roll something large and round,
Which he beside the rivulet
  In playing there had found;   10
He came to ask what he had found,
That was so large, and smooth, and round.

Old Kaspar took it from the boy,
  Who stood expectant by;
And then the old man shook his head,
  And, with a natural sigh,
"'Tis some poor fellow's skull," said he,
  "Who fell in the great victory.

"I find them in the garden,
  For there's many here about;   20
And often when I go to plough,
  The ploughshare turns them out!
For many a thousand men," said he,
  "Were slain in that great victory."

"Now tell us what 'twas all about,"
  Young Peterkin, he cries;
And little Wilhelmine looks up
  With wonder-waiting eyes;
"Now tell us all about the war,
And what they fought each other for."   30

"It was the English," Kaspar cried,
  "Who put the French to rout;
But what they fought each other for,
  I could not well make out;
But everybody said," quoth he,
  "That 'twas a famous victory.

"My father lived at Blenheim then,
    Yon little stream hard by;
They burnt his dwelling to the ground,
    And he was forced to fly;   40
So with his wife and child he fled,
Nor had he where to rest his head.

"With fire and sword the country round
    Was wasted far and wide,
And many a childing mother then,
    And new-born baby died;
But things like that, you know, must be
At every famous victory.

"They say it was a shocking sight
    After the field was won;   50
For many thousand bodies here
    Lay rotting in the sun;
But things like that, you know, must be
After a famous victory.

"Great praise the Duke of Marlbro' won,
    And our good Prince Eugene."
"Why 'twas a very wicked thing!"
    Said little Wilhelmine.
"Nay, nay, my little girl," quoth he,
"It was a famous victory.   60

"And everybody praised the Duke
    Who this great fight did win."
"But what good came of it at last?"
    Quoth little Peterkin.
"Why that I cannot tell," said he,
"But 'twas a famous victory."

                              1798 (1798)

MY DAYS AMONG THE DEAD
ARE PAST

~~~

Wordsworth thought these lines a touching expression
of the character of Southey, who spent most of his time
in his library of fourteen thousand books. "His dearly
prized books," said his son, "remained a pleasure to
him" even after he had lost the ability to read with
comprehension; and before his death "he would walk
slowly round his library, looking at them and taking
them down mechanically."

~~~

My days among the dead are past;
    Around me I behold,
Where'er these casual eyes are cast,
    The mighty minds of old;
My never-failing friends are they,
With whom I converse day by day.

With them I take delight in weal,
    And seek relief in woe;
And while I understand and feel
    How much to them I owe,   10
My cheeks have often been bedew'd
With tears of thoughtful gratitude.

My thoughts are with the dead, with them
    I live in long-past years,
Their virtues love, their faults condemn,
    Partake their hopes and fears,
And from their lessons seek and find
Instruction with an humble mind.

My hopes are with the dead, anon
    My place with them will be,   20
And I with them shall travel on
    Through all futurity;
Yet leaving here a name, I trust,
That will not perish in the dust.

                              1818 (1823)

THE CATARACT OF LODORE

~~~

These lines, written by Southey for his children, have
continued to charm readers long after most of what he
wrote has been forgotten. The famous waterfall of
Lodore (so disappointing to tourists who visit the Lake
Country in the summer; for by then it has almost dried
up) is near the head of Derwentwater, not far from
where Southey lived in Keswick.

~~~

"How does the water
    Come down at Lodore?"
My little boy ask'd me
    Thus, once on a time;
And moreover he task'd me
    To tell him in rhyme.
Anon at the word,
There first came one daughter

And then came another,
    To second and third   10
The request of their brother,
And to hear how the water
Comes down at Lodore,
With its rush and its roar,
    As many a time
They had seen it before.
So I told them in rhyme,
For of rhymes I had store:
And 'twas in my vocation
    For their recreation   20
That so I should sing;
Because I was Laureate
To them and the King.

From its sources which well
    In the tarn on the fell;
    From its fountains
    In the mountains,
Its rills and its gills;
Through moss and through brake,
    It runs and it creeps   30
For awhile, till it sleeps
    In its own little lake.
And thence at departing,
Awakening and starting,
It runs through the reeds
And away it proceeds,
Through meadow and glade,
    In sun and in shade,
And through the wood-shelter,
Among crags in its flurry,   40
    Helter-skelter,
    Hurry-scurry.
Here it comes sparkling,
And there it lies darkling;
Now smoking and frothing
Its tumult and wrath in,
    Till in this rapid race
    On which it is bent,
    It reaches the place
    Of its steep descent.   50

    The cataract strong
    Then plunges along,
    Striking and raging
    As if a war waging

Its caverns and rocks among:
    Rising and leaping,
    Sinking and creeping,
    Swelling and sweeping,
    Showering and springing,
    Flying and flinging,   60
    Writhing and ringing,
    Eddying and whisking,
    Spouting and frisking,
    Turning and twisting,
    Around and around
    With endless rebound!
    Smiting and fighting,
    A sight to delight in;
    Confounding, astounding,
Dizzying and deafening the ear with its sound. 70

    Collecting, projecting,
    Receding and speeding,
    And shocking and rocking,
    And darting and parting,
    And threading and spreading,
    And whizzing and hissing,
    And dripping and skipping,
    And hitting and splitting,
    And shining and twining,
    And rattling and battling,   80
    And shaking and quaking,
    And pouring and roaring,
    And waving and raving,
    And tossing and crossing,
    And flowing and going,
    And running and stunning,
    And foaming and roaming,
    And dinning and spinning,
    And dropping and hopping,
    And working and jerking,   90
    And guggling and struggling,
    And heaving and cleaving,
    And moaning and groaning;
    And glittering and frittering,
    And gathering and feathering,
    And whitening and brightening,
    And quivering and shivering,
    And hurrying and skurrying,
    And thundering and floundering;

Dividing and gliding and sliding,   100
And falling and brawling and sprawling,
And driving and riving and striving,
And sprinkling and twinkling and wrinkling,

And sounding and bounding and rounding,
And bubbling and troubling and doubling,
And grumbling and rumbling and tumbling,
And clattering and battering and shattering;

Retreating and beating and meeting and sheeting,
Delaying and straying and playing and spraying,
Advancing and prancing and glancing and dancing,
Recoiling, turmoiling and toiling and boiling,   111
And gleaming and streaming and steaming and
        beaming,
And rushing and flushing and brushing and gushing,
And flapping and rapping and clapping and slapping,
And curling and whirling and purling and twirling,
And thumping and plumping and bumping and
        jumping,
And dashing and flashing and splashing and clashing;
And so never ending, but always descending,
Sounds and motions forever and ever are blending,
All at once and all o'er, with a mighty uproar;   120
And this way the water comes down at Lodore.

                                    1820 (1823)

FROM

THE LIFE OF HORATIO, LORD NELSON

∽⚬⚬∽

This direct, simple biography has been widely regarded
as the finest piece of biographical writing of the period.
It was published in 1813.

∽⚬⚬∽

from

THE BATTLE OF TRAFALGAR

Unremitting exertions were made to equip the
ships which Nelson had chosen, and especially to refit
the *Victory*, which was once more to bear his flag.
Before he left London he called at his upholsterer's,
where the coffin which Captain Hallowell had given
him was deposited, and desired that its history might
be engraven upon the lid, saying it was highly probable
he might want it on his return. He seemed, indeed,
to have been impressed with an expectation that he
should fall in the battle. In a letter to his brother,
written immediately after his return, he had said:

"We must not talk of Sir Robert Calder's battle.
I might not have done so much with my small force.
If I had fallen in with them, you might probably have
been a lord before I wished, for I know they meant
to make a dead set at the *Victory*." Nelson had once
regarded the prospect of death with gloomy satisfaction; it was when he anticipated the upbraidings of his
wife and the displeasure of his venerable father. The
state of his feelings now was expressed in his private
journal in these words: "Friday night (Sept. 13th),
at half-past ten, I drove from dear, dear Merton, where
I left all which I hold dear in this world, to go to serve
my king and country. May the great God whom I
adore enable me to fulfill the expectations of my
country! And, if it is His good pleasure that I should
return, my thanks will never cease being offered up
to the throne of His mercy. If it is His good providence
to cut short my days upon earth, I bow with the greatest
submission; relying that He will protect those so dear
to me, whom I may leave behind! His will be done.
Amen! Amen! Amen!"

Early on the following morning he reached Portsmouth; and, having despatched his business on shore,
endeavored to elude the populace by taking a byway to
the beach; but a crowd collected in his train, pressing
forward to obtain a sight of his face;—many were in
tears, and many knelt down before him, and blessed
him as he passed. England has had many heroes, but
never one who so entirely possessed the love of his
fellow-countrymen as Nelson. All men knew that
his heart was as humane as it was fearless; that there
was not in his nature the slightest alloy of selfishness
or cupidity; but that, with perfect and entire devotion,
he served his country with all his heart, and with all
his soul, and with all his strength; and, therefore, they
loved him as truly and as fervently as he loved England.

.   .   .

About half-past nine in the morning of the 19th,
the *Mars*, being the nearest to the fleet of the ships
which formed the line of communication with the
frigates in shore, repeated the signal that the enemy
were coming out of port. The wind was at this time
very light, with partial breezes, mostly from the
S.S.W. Nelson ordered the signal to be made for a
chase in the south-east quarter. About two, the repeating ships announced that the enemy were at sea.
All night the British fleet continued under all sail,
steering to the south-east. At daybreak they were in
the entrance of the Straits, but the enemy were not in

sight. About seven, one of the frigates made signal that the enemy were bearing north. Upon this the *Victory* hove-to, and shortly afterwards Nelson made sail again to the northward. In the afternoon the wind blew fresh from the south-west, and the English began to fear that the foe might be forced to return to port.

A little before sunset, however, Blackwood, in the *Euryalus*, telegraphed that they appeared determined to go to the westward. "And that," said the Admiral in his diary, "they shall not do, if it is in the power of Nelson and Bronte to prevent them." Nelson had signified to Blackwood that he depended upon him to keep sight of the enemy. They were observed so well that all their motions were made known to him, and, as they wore twice, he inferred that they were aiming to keep the port of Cadiz open, and would retreat there as soon as they saw the British fleet; for this reason he was very careful not to approach near enough to be seen by them during the night. At daybreak the combined fleets were distinctly seen from the *Victory's* deck, formed in a close line of battle ahead, on the starboard tack, about twelve miles to leeward, and standing to the south. Our fleet consisted of twenty-seven sail of the line and four frigates; theirs of thirty-three and seven large frigates. Their superiority was greater in size and weight of metal than in numbers. They had four thousand troops on board; and the best riflemen who could be procured, many of them Tyrolese, were dispersed through the ships. Little did the Tyrolese, and little did the Spaniards at that day, imagine what horrors the wicked tyrant whom they served was preparing for their country.

Soon after daylight Nelson came upon deck. The 21st of October was a festival in his family, because on that day his uncle, Captain Suckling, in the *Dreadnought*, with two other line-of-battle ships, had beaten off a French squadron of four sail of the line and three frigates. Nelson, with that sort of superstition from which few persons are entirely exempt, had more than once expressed his persuasion that this was to be the day of his battle also; and he was well pleased at seeing his prediction about to be verified. The wind was now from the west,—light breezes, with a long heavy swell. Signal was made to bear down upon the enemy in two lines; and the fleet set all sail. Collingwood, in the *Royal Sovereign*, led the lee-line of thirteen ships; the *Victory* led the weather-line of fourteen. Having seen that all was as it should be, Nelson retired to his cabin, and wrote this prayer:—

"May the Great God, whom I worship, grant to my country, and for the benefit of Europe in general, a great and glorious victory; and may no misconduct in any one tarnish it; and may humanity after victory be the predominant feature in the British fleet. For myself individually, I commit my life to Him that made me, and may His blessing alight on my endeavors for serving my country faithfully! To Him I resign myself, and the just cause which is intrusted to me to defend. Amen, Amen, Amen."

．　．　．

Blackwood went on board the *Victory* about six. He found him in good spirits, but very calm; not in that exhilaration which he had felt upon entering into battle at Aboukir and Copenhagen; he knew that his own life would be particularly aimed at, and seems to have looked for death with almost as sure an expectation as for victory. His whole attention was fixed upon the enemy. They tacked to the northward, and formed their line on the larboard tack; thus bringing the shoals of Trafalgar and St. Pedro under the lee of the British, and keeping the port of Cadiz open for themselves. This was judiciously done: and Nelson, aware of all the advantages which it gave them, made signal to prepare to anchor.

Villeneuve[1] was a skilful seaman, worthy of serving a better master and a better cause. His plan of defence was as well conceived, and as original, as the plan of attack. He formed the fleet in a double line, every alternate ship being about a cable's length[2] to windward of her second ahead and astern. Nelson, certain of a triumphant issue to the day, asked Blackwood what he should consider as a victory. That officer answered that, considering the handsome way in which battle was offered by the enemy, their apparent determination for a fair trial of strength, and the situation of the land, he thought it would be a glorious result if fourteen were captured. He replied: "I shall not be satisfied with less than twenty." Soon afterwards he asked him if he did not think there was a signal wanting. Captain Blackwood made answer that he thought the whole fleet seemed very clearly to understand what they were about. These words were scarcely spoken before that signal was made which will be remembered as long as the language or even the memory of England shall endure—Nelson's last signal: "ENGLAND EXPECTS EVERY

THE LIFE OF NELSON. **1. Villeneuve:** the French commander. **2. cable's length:** 100 fathoms (600 feet).

MAN TO DO HIS DUTY!" It was received throughout the fleet with a shout of answering acclamation, made sublime by the spirit which it breathed and the feeling which it expressed. "Now," said Lord Nelson, "I can do no more. We must trust to the great Disposer of all events and the justice of our cause. I thank God for this great opportunity of doing my duty."

He wore that day, as usual, his Admiral's frock-coat, bearing on the left breast four stars of the different orders with which he was invested. Ornaments which rendered him so conspicuous a mark for the enemy were beheld with ominous apprehensions by his officers. It was known that there were riflemen on board the French ships, and it could not be doubted but that his life would be particularly aimed at. They communicated their fears to each other, and the surgeon, Mr. Beatty, spoke to the chaplain, Dr. Scott, and to Mr. Scott, the public secretary, desiring that some person would entreat him to change his dress or cover the stars; but they knew that such a request would highly displease him. "In honor I gained them," he had said when such a thing had been hinted to him formerly, "and in honor I will die with them." Mr. Beatty, however, would not have been deterred by any fear of exciting his displeasure from speaking to him himself upon a subject in which the weal of England, as well as the life of Nelson, was concerned; but he was ordered from the deck before he could find an opportunity. This was a point upon which Nelson's officers knew that it was hopeless to remonstrate or reason with him; but both Blackwood and his own captain, Hardy, represented to him how advantageous to the fleet it would be for him to keep out of action as long as possible; and he consented at last to let the *Leviathan* and the *Téméraire*, which were sailing abreast of the *Victory*, be ordered to pass ahead. Yet even here the last infirmity of this noble mind[3] was indulged; for these ships could not pass ahead if the *Victory* continued to carry all her sail; and so far was Nelson from shortening sail, that it was evident he took pleasure in pressing on, and rendering it impossible for them to obey his own orders. A long swell was setting into the Bay of Cadiz: our ships, crowding all sail, moved majestically before it, with light winds from the south-west. The sun shone on the sails of the enemy; and their well-formed line, with their numerous three-deckers, made an appearance which any other assailants would

have thought formidable; but the British sailors only admired the beauty and the splendor of the spectacle; and, in full confidence of winning what they saw, remarked to each other, what a fine sight yonder ships would make at Spithead![4]

The French admiral, from the *Bucentaure*, beheld the new manner in which his enemy was advancing, Nelson and Collingwood each leading his line; and, pointing them out to his officers, he is said to have exclaimed that such conduct could not fail to be successful. Yet Villeneuve had made his own dispositions with the utmost skill, and the fleets under his command waited for the attack with perfect coolness. Ten minutes before twelve they opened their fire. Eight or nine of the ships immediately ahead of the *Victory*, and across her bows, fired single guns at her, to ascertain whether she was yet within their range. As soon as Nelson perceived that their shot passed over him, he desired Blackwood and Captain Prowse, of the *Sirius*, to repair to their respective frigates, and on their way to tell all the captains of the line-of-battle ships that he depended on their exertions, and that, if by the prescribed mode of attack they found it impracticable to get into action immediately, they might adopt whatever they thought best, provided it led them quickly and closely alongside an enemy. As they were standing on the front of the poop, Blackwood took him by the hand, saying he hoped soon to return and find him in possession of twenty prizes. He replied, "God bless you, Blackwood; I shall never see you again."

Nelson's column was steered about two points more to the north than Collingwood's, in order to cut off the enemy's escape into Cadiz. The lee line, therefore, was first engaged. "See," cried Nelson, pointing to the *Royal Sovereign*, as she steered right for the centre of the enemy's line, cut through it astern of the *Santa Anna*, three-decker, and engaged her at the muzzle of her guns on the starboard side; "see how that noble fellow Collingwood carries his ship into action!" Collingwood, delighted at being first in the heat of the fire, and knowing the feelings of his Commander and old friend, turned to his captain and exclaimed, "Rotherham, what would Nelson give to be here!" Both these brave officers perhaps at this moment thought of Nelson with gratitude for a circumstance which had occurred on the preceding day. Admiral Collingwood, with some of the captains, having gone

---

3. last . . . mind: love of fame (paraphrased from Milton's "Lycidas," l. 71).

4. Spithead: an anchorage near the naval base at Portsmouth.

on board the *Victory* to receive instructions, Nelson inquired of him where his captain was, and was told in reply that they were not upon good terms with each other. "Terms!" said Nelson, "good terms with each other!" Immediately he sent a boat for Captain Rotherham, led him, as soon as he arrived, to Collingwood, and saying, "Look, yonder are the enemy!" bade them shake hands like Englishmen.

The enemy continued to fire a gun at a time at the *Victory* till they saw that a shot had passed through her main-topgallant sail; then they opened their broadsides, aiming chiefly at her rigging, in the hope of disabling her before she could close with them. Nelson, as usual, had hoisted several flags, lest one should be shot away. The enemy showed no colors till late in the action, when they began to feel the necessity of having them to strike.[5] For this reason the *Santissima Trinidad*—Nelson's old acquaintance,[6] as he used to call her—was distinguishable only by her four decks; and to the bow of this opponent he ordered the *Victory* to be steered. Meantime an incessant raking fire was kept up upon the *Victory*. The admiral's secretary was one of the first who fell: he was killed by a cannon-shot, while conversing with Hardy. Captain Adair, of the marines, with the help of a sailor, endeavored to remove the body from Nelson's sight, who had a great regard for Mr. Scott; but he anxiously asked, "Is that poor Scott that's gone?" and being informed that it was indeed so, exclaimed, "Poor fellow!" Presently a double-headed shot struck a party of marines, who were drawn up on the poop, and killed eight of them: upon which Nelson immediately desired Captain Adair to disperse his men round the ship, that they might not suffer so much from being together. A few minutes afterwards a shot struck the forebrace bits on the quarterdeck, and passed between Nelson and Hardy, a splinter from the bit tearing off Hardy's buckle and bruising his foot. Both stopped, and looked anxiously at each other, each supposing the other to be wounded. Nelson then smiled, and said, "This is too warm work, Hardy, to last long."

The *Victory* had not yet returned a single gun: fifty of her men had been by this time killed or wounded, and her main-topmast, with all her studding sails and their booms, shot away. Nelson declared that, in all his battles, he had seen nothing which surpassed the cool courage of his crew on this occasion. At four

minutes after twelve she opened her fire from both sides of her deck. It was not possible to break the enemy's line without running on board one of their ships: Hardy informed him of this, and asked which he would prefer. Nelson replied: "Take your choice, Hardy, it does not signify much." The master was then ordered to put the helm to port, and the *Victory* ran on board the *Redoubtable*, just as her tiller ropes were shot away. The French ship received her with a broadside, then instantly let down her lower-deck ports for fear of being boarded through them, and never afterwards fired a great gun during the action. Her tops, like those of all the enemy's ships, were filled with riflemen. Nelson never placed musketry in his tops; he had a strong dislike to the practice, not merely because it endangers setting fire to the sails, but also because it is a murderous sort of warfare, by which individuals may suffer, and a commander now and then be picked off, but which never can decide the fate of a general engagement.

Captain Harvey, in the *Téméraire*, fell on board the *Redoubtable* on the other side; another enemy was in like manner on board the *Téméraire*; so that these four ships formed as compact a tier as if they had been moored together, their heads all lying the same way. The lieutenants of the *Victory*, seeing this, depressed their guns of the middle and lower decks, and fired with a diminished charge, lest the shot should pass through and injure the *Téméraire*; and because there was danger that the *Redoubtable* might take fire from the lower-deck guns, the muzzles of which touched her side when they were run out, the fireman of each gun stood ready with a bucket of water, which, as soon as the gun was discharged, he dashed into the hole made by the shot. An incessant fire was kept up from the *Victory* from both sides; her larboard guns playing upon the *Bucentaure* and the huge *Santissima Trinidad*.

It had been part of Nelson's prayer that the British fleet might be distinguished by humanity in the victory which he expected. Setting an example himself, he twice gave orders to cease firing upon the *Redoubtable*, supposing that she had struck, because her great guns were silent; for, as she carried no flag, there was no means of instantly ascertaining the fact. From this ship, which he had thus twice spared, he received his death. A ball fired from her mizzen-top, which, in the then situation of the two vessels, was not more than fifteen yards from that part of the deck where he was standing, struck the epaulette on his left shoulder,—about a

<hr>

5. strike: lower the colors in token of surrender. **6. acquaintance:** He had fought her before, at the Battle of the Nile.

quarter after one, just in the heat of the action. He fell upon his face, on the spot which was covered with his poor secretary's blood. Hardy, who was a few steps from him, turning round, saw three men raising him up. "They have done for me at last, Hardy," said he. "I hope not!" cried Hardy. "Yes," he replied; "my back bone is shot through." Yet even now, not for a moment losing his presence of mind, he observed, as they were carrying him down the ladder, that the tiller ropes, which had been shot away, were not yet replaced, and ordered that new ones should be rove immediately:—then, that he might not be seen by the crew, he took out his handkerchief, and covered his face and his stars.—Had he but concealed these badges of honor from the enemy, England, perhaps, would not have had cause to receive with sorrow the news of the battle of Trafalgar.

The cockpit was crowded with wounded and dying men, over whose bodies he was with some difficulty conveyed, and laid upon a pallet in the midshipmen's berth. It was soon perceived, upon examination, that the wound was mortal. This, however, was concealed from all, except Captain Hardy, the chaplain, and the medical attendants. He himself being certain, from the sensation in his back, and the gush of blood which he felt momently within his breast, that no human care could avail him, insisted that the surgeon should leave him, and attend to those to whom he might be useful; "for," said he, "you can do nothing for me." All that could be done was to fan him with paper, and frequently to give him lemonade to alleviate his intense thirst. He was in great pain, and expressed much anxiety for the event of the action, which now began to declare itself. As often as a ship struck, the crew of the *Victory* hurraed, and at every hurra a visible expression of joy gleamed in the eyes and marked the countenance of the dying hero. But he became impatient to see Captain Hardy; and as that officer, though often sent for, could not leave the deck, Nelson feared that some fatal cause prevented him, and repeatedly cried: "Will no one bring Hardy to me? He must be killed! He is surely dead!" An hour and ten minutes elapsed from the time when Nelson received his wound before Hardy could come to him. They shook hands in silence; Hardy in vain struggling to suppress the feelings of that most painful and yet sublimest moment. "Well, Hardy," said Nelson, "how goes the day with us?"—"Very well," replied Hardy, "ten ships have struck, but five of the van have tacked, and show an intention to bear down upon the *Victory*. I have

called two or three of our fresh ships round, and have no doubt of giving them a drubbing."—"I hope," said Nelson, "none of our ships have struck?" Hardy answered, "There was no fear of that." Then, and not till then, Nelson spoke of himself. "I am a dead man, Hardy," said he; "I am going fast; it will be all over with me soon. Come nearer to me. Let my dear Lady Hamilton have my hair and all other things belonging to me." Hardy observed that he hoped Mr. Beatty could yet hold out some prospect of life. "Oh no!" he replied, "it is impossible; my back is shot through. Beatty will tell you so." Captain Hardy then once more shook hands with him, and with a heart almost bursting hastened upon deck.

By this time all feeling below the breast was gone; and Nelson, having made the surgeon ascertain this, said to him: "You know I am gone. I know it. I feel something rising in my breast"—putting his hand on his left side—"which tells me so." And upon Beatty's inquiring whether his pain was very great, he replied, "So great that he wished he was dead. Yet," said he in a lower voice, "one would like to live a little longer too!" And after a few minutes, in the same under-tone, he added: "What would become of poor Lady Hamilton if she knew my situation?" Next to his country she occupied his thoughts. Captain Hardy, some fifty minutes after he had left the cockpit, returned, and again taking the hand of his dying friend and Commander, congratulated him on having gained a complete victory. How many of the enemy were taken he did not know, as it was impossible to perceive them distinctly, but fourteen or fifteen at least. "That's well!" cried Nelson, "but I bargained for twenty." And then in a stronger voice he said: "Anchor, Hardy, anchor." Hardy upon this hinted that Admiral Collingwood would take upon himself the direction of affairs. "Not while I live, Hardy," said the dying Nelson, ineffectually endeavoring to raise himself from the bed. "Do you anchor." His previous order for preparing to anchor had shown how clearly he foresaw the necessity of this. Presently, calling Hardy back, he said to him, in a low voice, "Don't throw me overboard"; and he desired that he might be buried by his parents, unless it should please the king to order otherwise. Then, reverting to private feelings: "Take care of my dear Lady Hamilton, Hardy; take care of poor Lady Hamilton.—Kiss me, Hardy," said he. Hardy knelt down, and kissed his cheek: and Nelson said, "Now I am satisfied. Thank God, I have done my duty." Hardy stood over him in silence for a moment or two, then knelt again, and kissed his forehead. "Who is that?" said Nelson; and being informed, he replied, "God bless you, Hardy." And Hardy then left him—forever.

Nelson now desired to be turned upon his right side, and said: "I wish I had not left the deck; for I shall soon be gone." Death was, indeed, rapidly approaching. He said to the chaplain: "Doctor, I have *not* been a *great* sinner"; and, after a short pause, "Remember that I leave Lady Hamilton, and my daughter, Horatia, as a legacy to my country." His articulation now became difficult; but he was distinctly heard to say, "Thank God, I have done my duty!" These words he had repeatedly pronounced; and they were the last words he uttered. He expired at thirty minutes after four,—three hours and a quarter after he had received his wound.

. . .

The death of Nelson was felt in England as something more than a public calamity: men started at the intelligence, and turned pale, as if they had heard of the loss of a dear friend. An object of our admiration and affection, of our pride and our hopes, was suddenly taken from us; and it seemed as if we had never, till then, known how deeply we loved and reverenced him. What the country had lost in its great naval hero—the greatest of our own, and of all former times—was scarcely taken into the account of grief. So perfectly, indeed, had he performed his part, that the maritime war, after the battle of Trafalgar, was considered at an end; the fleets of the enemy were not merely defeated, but destroyed; new navies must be built, and a new race of seamen reared for them, before the possibility of their invading our shores could again be contemplated. It was not, therefore, from any selfish reflection upon the magnitude of our loss that we mourned for him: the general sorrow was of a higher character. The people of England grieved that funeral ceremonies, public monuments, and posthumous rewards, were all which they could now bestow upon him, whom the king, the legislature, and the nation, would alike have delighted to honor; whom every tongue would have blessed; whose presence in every village through which he might have passed would have wakened the church bells, have given school-boys a holiday, have drawn children from their sports to gaze upon him, and "old men from the chimney corner," to look upon Nelson ere they died. The victory of Trafalgar was celebrated, indeed, with the usual forms of rejoicing, but they were without joy; for such already was the glory of the British navy, through Nelson's surpassing genius, that it scarcely seemed to receive any addition from the most signal victory that ever was achieved upon the seas; and the destruction of this mighty fleet, by which all the maritime schemes of France were totally frustrated, hardly appeared to add to our security or strength; for while Nelson was living, to watch the combined squadrons of the enemy, we felt ourselves as secure as now, when they were no longer in existence.

There was reason to suppose, from the appearances upon opening the body, that in the course of nature he might have attained, like his father, to a good old age. Yet he cannot be said to have fallen prematurely whose work was done, nor ought he to be lamented who died so full of honors and at the height of human fame. The most triumphant death is that of the martyr; the most awful that of the martyred patriot; the most splendid that of the hero in the hour of victory; and if the chariot and the horses of fire had been vouchsafed for Nelson's translation, he could scarcely have departed in a brighter blaze of glory. He has left us, not indeed his mantle of inspiration, but a name and an example which are at this hour inspiring thousands of the youth of England —a name which is our pride, and an example which will continue to be our shield and our strength. Thus it is that the spirits of the great and the wise continue to live and to act after them, verifying in this sense the language of the old mythologist:

*Τοί μεν δαίμονες εἰσί, Διός μεγάλου διὰ βουλὰς
Ἐσθλοί, ἐπιχθόνιοι, φύλακες θνητῶν ἀνθρώπων.*[7]

FROM

THE LIFE OF WESLEY

WESLEY'S DEATH

Upon his eighty-sixth birthday, he says, "I now find I grow old. My sight is decayed, so that I cannot read a small print, unless in a strong light. My strength is decayed; so that I walk much slower than I did some

7. Τοί . . . ἀνθρώπων: "Some there are that we may call pure spirits living on earth—kind guardians of mortal men, preserving them from harm" (Hesiod, *Works and Days*, ll. 122-23).

years since. My memory of names, whether of persons or places, is decayed, till I stop a little to recollect them. What I should be afraid of is, if I took thought for the morrow, that my body should weigh down my mind, and create either stubbornness, by the decrease of my understanding, or peevishness, by the increase of bodily infirmities. But thou shalt answer for me, O Lord, my God!" His strength now diminished so much that he found it difficult to preach more than twice a day; and for many weeks he abstained from his five o'clock morning sermons, because a slow and settled fever parched his mouth. Finding himself a little better, he resumed the practice, and hoped to hold on a little longer; but, at the beginning of the year 1790, he writes, "I am now an old man, decayed from head to foot. My eyes are dim; my right hand shakes much; my mouth is hot and dry every morning; I have a lingering fever almost every day; my motion is weak and slow. However, blessed be God! I do not slack my labors; I can preach and write still." In the middle of the same year he closed his cash account-book with the following words, written with a tremulous hand, so as to be scarcely legible: "For upwards of eighty-six years I have kept my accounts exactly: I will not attempt it any longer, being satisfied with the continued conviction, that I save all I can, and give all I can; that is, all I have." His strength was now quite gone, and no glasses would help his sight. "But I feel no pain," he says, "from head to foot; only, it seems, nature is exhausted, and, humanly speaking, will sink more and more, till

The weary springs of life stand still at last."

On the 1st of February, 1791, he wrote his last letter to America. It shows how anxious he was that his followers should consider themselves as one united body. "See," said he, "that you never give place to one thought of separating from your brethren in Europe. Lose no opportunity of declaring to all men, that the Methodists are one people in all the world, and that it is their full determination so to continue." He expressed also a sense that his hour was almost come. "Those that desire to write," said he, "or say anything to me, have no time to lose; for time has shaken me by the hand, and death is not far behind":—

words which his father had used in one of the last letters that he addressed to his sons at Oxford. On the 17th of that month he took cold after preaching at Lambeth. For some days he struggled against an increasing fever, and continued to preach till the Wednesday following when he delivered his last sermon. From that time he became daily weaker and more lethargic, and on the 2nd of March he died in peace, being in the eighty-eighth year of his age, and the sixty-fifth of his ministry.

During his illness he said, "Let me be buried in nothing but what is woolen; and let my corpse be carried in my coffin into the chapel." Some years before, he had prepared a vault for himself, and for those itinerant preachers who might die in London. In his will he directed that six poor men should have twenty shillings each for carrying his body to the grave; "for I particularly desire," said he, "that there may be no hearse, no coach, no escutcheon, no pomp except the tears of them that loved me, and are following me to Abraham's bosom. I solemnly adjure my executors, in the name of God, punctually to observe this." At the desire of many of his friends, his body was carried into the chapel the day preceding the interment, and there lay in a kind of state becoming the person, dressed in his clerical habit, with gown, cassock, and band; the old clerical cap on his head; a Bible in one hand, and a white handkerchief in the other. The face was placid, and the expression which death had fixed upon his venerable features was that of a serene and heavenly smile. The crowds who flocked to see him were so great, that it was thought prudent, for fear of accidents, to accelerate the funeral, and perform it between five and six in the morning. The intelligence, however, could not be kept entirely secret, and several hundred persons attended at that unusual hour. Mr. Richardson, who performed the service, had been one of his preachers almost thirty years. When he came to that part of the service, "Forasmuch as it hath pleased Almighty God to take unto himself the soul of our dear *brother*," his voice changed, and he substituted the word *father;* and the feeling with which he did this was such, that the congregation, who were shedding silent tears, burst at once into loud weeping.

1819–20 (1820)

WALTER SAVAGE LANDOR

1775–1864

FROM his own time to the present day, Landor, whose long life spanned three generations, has always fascinated other writers. They have found a vicarious relief in the leonine independence, courage, and integrity of the man. "That deep-mouthed Boeotian Savage Landor," as Byron called him, was completely indifferent to fashions of theme and style, and in this—and in other ways (including his incongruities and unpredictabilities)—he remained always and uniquely the same. His political liberalism, unlike that of so many of his contemporaries, never wavered. It was fiery when he was a schoolboy at Rugby and just as much so when he died at the age of eighty-nine. In his youth his ideals of style, in prose as well as verse, were a classical purity of diction and a stately but condensed simplicity of idiom that were somewhat apart from the mainstream of Romanticism. He adhered to these ideals, unconcerned with changes of taste, while two generations of Romantic writers and thirty years of the Victorian era passed by. Meanwhile his personal life, colorfully erratic, was also of a piece; hotheaded and yet stiffly haughty, sensitive and extremely generous (Dickens presents an affectionate caricature of him as Boythorn in *Bleak House*), he quarreled violently with almost everyone he knew, from boyhood to old age, while still remaining a loyal friend.

Born in Warwick, the son of a physician and a wealthy heiress, Landor was sent to school at Rugby. When he was about to be expelled for rebellion, his family quietly withdrew him. Privately tutored, he became an accomplished scholar in Greek and Latin, went on to Oxford and, an ardent republican, was sent away the next year because he fired a gun at the window of a Tory whom he detested. He was told he could return after a year's absence, but disdainfully refused. This led to a quarrel with his father, who nevertheless gave him a small pension (he was now nineteen), and Landor then devoted himself to poetry. His first volume appeared the next year (1795). It was followed three years later by his long poem, *Gebir* (which he also wrote and published in a Latin version). Shortly afterward, on the death of his father (1805), he inherited an estate that allowed him to live with complete independence. He then went to Spain in order to fight in the national army there against Napoleon, and supported a regiment for a while at his own expense. Returning to England, he brought out his tragedy *Count Julian* (1812).

His marriage, in 1811, was followed by an attempt at gentleman farming and, more import-ant, efforts to alleviate the situation of the peasantry. A few years later he left England for an extended tour of the Continent and eventually settled in Florence. There he began work on his *Imaginary Conversations*, the first collection of which was published from 1824 to 1829. Others were added as time passed. Together they comprise one of the monuments of English prose, and range, in the characters presented, from the ancient Greeks to men of Landor's own day, and from kings and generals to poets, theologians, and philosophers. Shrewd psychological insight combined with an informed historical sense are distilled into a purity and calm of language equalled by only three other writers in the last hundred years of English prose: Arnold, Newman, and Santayana. After a quarrel with his wife, Landor separated from her in 1835 and returned to England, living at Bath for several years. Perhaps his most ambitious work of this period was *Pericles and Aspasia* (1836), which consists of imaginary letters from the time of Pericles concerned with art, philosophy, and politics. Others works included a trilogy of three historical plays, *Andrea of Hungary*, *Giovanna of Naples*, and *Fra Rupert* (1839–40), and his verse tales and dialogues on Greek subjects, *The Hellenics of Walter Savage Landor* (1846–47).

At length he returned to Florence, almost as full of vigor in his eighties as when he was middle-aged. But he was in precarious financial straits. With impetuous, Lear-like generosity, he had signed over to his children most of his property, and it was only through the efforts of Robert Browning that he was able to secure from them an allowance. He died at the age of eighty-nine, six years after he had returned to Italy.

Bibliography

The standard edition is the *Complete Works*, ed. T. E. Welby and Stephen Wheeler, 16 vols., 1927–36. The *Poetical Works* have also been edited separately by Stephen Wheeler, 3 vols., 1937. Selections include those by Sir Sidney Colvin (Golden Treasury Series), 1882; Havelock Ellis, 1933; E. K. Chambers, 1946; and *Imaginary Conversations*, ed. Ernest de Selincourt, 1933.

For biography and general criticism, see John Forster, *Walter Savage Landor*, 2 vols., 1869; Sir Sidney Colvin, *Landor* (English Men of Letters Series), 1881; and especially R. H. Super, *Walter Savage Landor*, 1954, and Malcolm Elwin, *Landor: A Replevin*, 1958 (superseding Elwin's earlier *Savage Landor*, 1941). For a fine condensed criticism, see the section on Landor (pp. 229–45) in Douglas Bush, *Mythology and the Romantic Tradition*, 1937.

ROSE AYLMER

⟡

These lines, admired and often recited by Charles Lamb, were written when Landor heard of the death of Rose Aylmer (only daughter of the fourth Baron Aylmer; hence "sceptred race") in Calcutta in 1800. They were later carved on her tomb.

⟡

Ah what avails the sceptred race,
    Ah what the form divine!
What every virtue, every grace!
    Rose Aylmer, all were thine.
Rose Aylmer, whom these wakeful eyes
    May weep, but never see,
A night of memories and of sighs
    I consecrate to thee.

                1800 (1806)

MOTHER, I CANNOT MIND MY WHEEL

Mother, I cannot mind my wheel;
    My fingers ache, my lips are dry:
Oh! if you felt the pain I feel!
    But oh, who ever felt as I?

No longer could I doubt him true;
    All other men may use deceit:
He always said my eyes were blue,
    And often swore my lips were sweet.

                (1806)

A FIESOLAN IDYL

Here, where precipitate Spring with one light bound
Into hot Summer's lusty arms expires;
And where go forth at morn, at eve, at night,
Soft airs, that want the lute to play with them,
And softer sighs, that know not what they want;

A FIESOLAN IDYL. I. **Here:** The town of Fiesole, where Landor had a villa, is just outside Florence.

Under a wall, beneath an orange-tree
Whose tallest flowers could tell the lowlier ones
Of sights in Fiesole right up above,
While I was gazing a few paces off
At what they seem'd to show me with their nods,   10
Their frequent whispers and their pointing shoots,
A gentle maid came down the garden-steps
And gathered the pure treasure in her lap.
I heard the branches rustle, and stept forth
To drive the ox away, or mule, or goat,
(Such I believed it must be); for sweet scents
Are the swift vehicles of stil sweeter thoughts,
And nurse and pillow the dull memory
That would let drop without them her best stores.
They bring me tales of youth and tones of love,   20
And 'tis and ever was my wish and way
To let all flowers live freely, and all die,
Whene'er their Genius bids their souls depart,
Among their kindred in their native place.
I never pluck the rose; the violet's head
Hath shaken with my breath upon its bank
And not reproacht me; the ever-sacred cup
Of the pure lily hath between my hands
Felt safe, unsoil'd, nor lost one grain of gold.
I saw the light that made the glossy leaves   30
More glossy; the fair arm, the fairer cheek
Warmed by the eye intent on its pursuit;
I saw the foot, that, altho half-erect
From its gray slipper, could not lift her up
To what she wanted: I held down a branch
And gather'd her some blossoms; since their hour
Was come, and bees had wounded them, and flies
Of harder wing were working their way thro
And scattering them in fragments under foot.
So crisp were some, they rattled unevolved,   40
Others, ere broken off, fell into shells,
For such appear the petals when detacht,
Unbending, brittle, lucid, white like snow,
And like snow not seen thro, by eye or sun:
Yet every one her gown received from me
Was fairer than the first. I thought not so,
But so she praised them to reward my care.
I said, "You find the largest."
                       "This indeed,"
Cried she, "is large and sweet." She held one forth,
Whether for me to look at or to take   50
She knew not, nor did I; but taking it
Would best have solved (and this she felt) her doubts.
I dared not touch it; for it seemed a part
Of her own self; fresh, full, the most mature

Of blossoms, yet a blossom; with a touch
To fall, and yet unfallen. She drew back
The boon she tender'd, and then, finding not
The ribbon at her waist to fix it in,
Dropt it, as loth to drop it, on the rest.

(*1831*)

PLEASURE! WHY THUS DESERT THE HEART

Pleasure! why thus desert the heart
    In its spring-tide!
I could have seen her, I could part,
    And but have sigh'd!

O'er every youthful charm to stray,
    To gaze, to touch—
Pleasure! why take so much away,
    Or give so much!

(*1831*)

ABSENCE

Ianthe! you resolve to cross the sea!
    A path forbidden *me!*
Remember, while the Sun his blessing sheds
    Upon the mountain-heads,
How often we have watcht him laying down
    His brow, and dropt our own
Against each other's, and how faint and short
    And sliding the support!
What will succeed it now? Mine is unblest,
    Ianthe! nor will rest   10
But on the very thought that swells with pain.
    O bid me hope again!
O give me back what Earth, what (without you)
    Not Heaven itself can do—
One of the golden days that we have past,
    And let it be my last!
Or else the gift would be, however sweet,
    Fragile and incomplete.

(*1831*)

DIRCE

Stand close around, ye Stygian set,
    With Dirce in one boat conveyed!
Or Charon, seeing, may forget
    That he is old and she a shade.

(*1831*)

HOMAGE

Away, my verse; and never fear,
    As men before such beauty do;
On you she will not look severe,
    She will not turn her eyes from you.
Some happier graces could I lend
    That in her memory you should live,
Some little blemishes might blend,
    For it would please her to forgive.

(*1831*)

SO LATE REMOVED

So late removed from him she swore,
    With clasping arms and vows and tears,
In life and death she would adore,
    While memory, fondness, bliss, endears—

Can she forswear! can she forget?
    Strike, mighty Love! strike Vengeance!—soft!
Conscience must come, and bring Regret—
    These let her feel! nor these too oft!

(*1831*)

PAST RUINED ILION HELEN LIVES

Past ruined Ilion Helen lives,
    Alcestis rises from the shades;
Verse calls them forth; 'tis verse that gives
    Immortal youth to mortal maids.

PAST RUINED ILION HELEN LIVES. **1. Ilion:** Troy. **2. Alcestis
. . . shades:** Alcestis died to save her husband, Admetus,
king of Thessaly, and was then allowed to return from
Hades.

Soon shall Oblivion's deepening veil
  Hide all the peopled hills you see,
The gay, the proud, while lovers hail
  In distant ages you and me.

The tear for fading beauty check,
  For passing glory cease to sigh;   10
One form shall rise above the wreck,
  One name, Ianthe, shall not die.

                     (*1831*)

MILD IS
THE PARTING YEAR

Mild is the parting year, and sweet
  The odour of the falling spray;
Life passes on more rudely fleet,
  And balmless is its closing day.

I wait its close, I court its gloom,
  But mourn that never must there fall
Or on my breast or on my tomb
  The tear that would have soothed it all.

                     (*1831*)

EPITAPH AT FIESOLÈ

Lo! where the four mimosas blend their shade,
In calm repose at last is Landor laid;
For ere he slept he saw them planted here
By her his soul had ever held most dear,
And he had lived enough when he had dried her tear.

                     (*1831*)

THE MAID'S LAMENT

I loved him not; and yet, now he is gone,
  I feel I am alone.
I checked him while he spoke; yet could he speak,
  Alas! I would not check.
For reasons not to love him once I sought,
  And wearied all my thought
To vex myself and him: I now would give
  My lóve, could he but live
Who lately lived for me, and, when he found
  'Twas vain, in holy ground   10

He hid his face amid the shades of death!
  I waste for him my breath
Who wasted his for me! but mine returns,
  And this lorn bosom burns
With stifling heat, heaving it up in sleep,
  And waking me to weep
Tears that had melted his soft heart: for years
  Wept he as bitter tears!
Merciful God! such was his latest prayer,
  These may she never share.   20
Quieter is his breath, his breast more cold,
  Than daisies in the mould,
Where children spell, athwart the churchyard gate,
  His name and life's brief date.
Pray for him, gentle souls, whoe'er you be,
  And, oh! pray too for me!

                     (*1834*)

THE HAMADRYAD

Rhaicos was born amid the hills wherefrom
Gnidos the light of Caria is discern'd,
And small are the white-crested that play near,
And smaller onward are the purple waves.
Thence festal choirs were visible, all crown'd
With rose and myrtle if they were inborn;
If from Pandion sprang they, on the coast
Where stern Athenè raised her citadel,
Then olive was intwined with violets
Cluster'd in bosses, regular and large.   10
For various men wore various coronals;
But one was their devotion; 'twas to her
Whose laws all follow, her whose smile withdraws
The sword from Ares, thunderbolt from Zeus,
And whom in his chill cave the mutable
Of mind, Poseidon, the sea-king, reveres,
And whom his brother, stubborn Dis, hath pray'd
To turn in pity the averted check
Of her he bore away, with promises,
Nay, with loud oath before dread Styx itself,   20
To give her daily more and sweeter flowers
Than he made drop from her on Enna's dell.
  Rhaicos was looking from his father's door
At the long trains that hastened to the town

THE HAMADRYAD. **2. Gnidos . . . Caria:** in southwestern Asia Minor, next to the Aegean Sea. **10. bosses:** projecting or raised ornaments. **14. Ares:** Mars. **19. her:** Proserpine.

From all the valleys, like bright rivulets
Gurgling with gladness, wave outrunning wave,
And thought it hard he might not also go
And offer up one prayer, and press one hand,
He knew not whose. The father call'd him in,
And said, "Son Rhaicos! those are idle games;   30
Long enough I have lived to find them so."
And ere he ended sighed, as old men do
Always, to think how idle such games are.
"I have not yet," thought Rhaicos in his heart,
And wanted proof.
        "Suppose thou go and help
Echeion at the hill, to bark yon oak
And lop its branches off, before we delve
About the trunk and ply the root with axe:
This we may do in winter."
        Rhaicos went;
For thence he could see farther, and see more   40
Of those who hurried to the city-gate.
Echeion he found there with naked arm
Swart-hair'd, strong-sinew'd, and his eyes intent
Upon the place where first the axe should fall:
He held it upright. "There are bees about,
Or wasps, or hornets," said the cautious eld,
"Look sharp, O son of Thallinos!" The youth
Inclined his ear, afar, and warily,
And cavern'd in his hand. He heard a buzz
At first, and then the sound grew soft and clear,   50
And then divided into what seem'd tune,
And there were words upon it, plaintive words.
He turn'd, and said, "Echeion! do not strike
That tree: it must be hollow; for some god
Speaks from within. Come thyself near." Again
Both turn'd toward it: and behold! there sat
Upon the moss below, with her two palms
Pressing it, on each side, a maid in form.
Downcast were her long eyelashes, and pale
Her cheek, but never mountain-ash display'd   60
Berries of color like her lip so pure,
Nor were the anemones about her hair
Soft, smooth, and wavering like the face beneath.
"What dost thou here?" Echeion, half-afraid,
Half-angry cried. She lifted up her eyes,
But nothing spake she. Rhaicos drew one step
Backward, for fear came likewise over him,
But not such fear: he panted, gasp'd, drew in
His breath, and would have turn'd it into words,
But could not into one.
        "O send away   70
That sad old man!" said she. The old man went

Without a warning from his master's son,
Glad to escape, for sorely he now fear'd.
And the axe shone behind him in their eyes.
    Hamad. And wouldst thou too shed the most
        innocent
Of blood? No vow demands it; no god wills
The oak to bleed.
    Rhaicos.   Who art thou? whence? why here?
And whither wouldst thou go? Among the robed
In white or saffron, or the hue that most
Resembles dawn or the clear sky, is none   80
Array'd as thou art. What so beautiful
As that gray robe which clings about thee close,
Like moss to stones adhering, leaves to trees,
Yet lets thy bosom rise and fall in turn,
As, touch'd by zephyrs, fall and rise the boughs
Of graceful platan by the river-side?
    Hamad. Lovest thou well thy father's house?
    Rhaicos.   Indeed
I love it, well I love it, yet would leave
For thine, where'er it be, my father's house,
With all the marks upon the door, that show   90
My growth at every birthday since the third,
And all the charms, o'erpowering evil eyes,
My mother nail'd for me against my bed,
And the Cydonian bow (which thou shalt see)
Won in my race last spring from Eutychos.
    Hamad. Bethink thee what it is to leave a home
Thou never yet hast left, one night, one day.
    Rhaicos. No, 'tis not hard to leave it; 'tis not hard
To leave, O maiden, that paternal home,
If there be one on earth whom we may love   100
First, last, forever; one who says that she
Will love forever too. To say which word,
Only to say it, surely is enough—
It shows such kindness—if 'twere possible
We at the moment think she would indeed.
    Hamad. Who taught thee all this folly at thy age?
    Rhaicos. I have seen lovers and have learn'd to love.
    Hamad. But wilt thou spare the tree?
    Rhaicos. My father wants
The bark; the tree may hold its place awhile.   109
    Hamad. Awhile! thy father numbers then my days?
    Rhaicos. Are there no others where the moss beneath
Is quite as tufty? Who would send thee forth
Or ask thee why thou tarriest? Is thy flock
Anywhere near?

86. platan: plane tree. **94. Cydonian:** Cydonia, in Crete,
was noted for its skilled archers.

Hamad.   I have no flock: I kill
Nothing that breathes, that stirs, that feels the air,
The sun, the dew. Why should the beautiful
(And thou art beautiful) disturb the source
Whence springs all beauty? Hast thou never heard
Of Hamadryads?
     Rhaicos.   Heard of them I have:
Tell me some tale about them. May I sit   120
Beside thy feet? Art thou not tired? The herbs
Are very soft; I will not come too nigh;
Do but sit there, nor tremble so, nor doubt.
Stay, stay an instant: let me first explore
If any acorn of last year be left
Within it; thy thin robe too ill protects
Thy dainty limbs against the harm one small
Acorn may do. Here's none. Another day
Trust me; till then let me sit opposite.
     Hamad. I seat me; be thou seated, and content.   130
     Rhaicos. O sight for gods! ye men below! adore
The Aphroditè. *Is* she there below?
Or sits she here before me, as she sate
Before the shepherd on those heights that shade
The Hellespont, and brought his kindred woe?
     Hamad. Reverence the higher Powers; nor deem
          amiss
Of her who pleads to thee, and would repay—
Ask not how much—but very much. Rise not;
No, Rhaicos, no! Without the nuptial vow
Love is unholy. Swear to me that none   140
Of mortal maids shall ever taste thy kiss,
Then take thou mine; then take it, not before.
     Rhaicos. Hearken, all gods above! O Aphroditè!
O Herè! Let my vow be ratified!
But wilt thou come into my father's house?
     Hamad. Nay; and of mine I cannot give thee part.
     Rhaicos. Where is it?
     Hamad.   In this oak.
     Rhaicos.   Ay; now begins
The tale of Hamadryad; tell it through.
     Hamad. Pray of thy father never to cut down
My tree; and promise him, as well thou mayst,   150
That every year he shall receive from me
More honey than will buy him nine fat sheep,
More wax than he will burn to all the gods.
Why fallest thou upon thy face? Some thorn
May scratch it, rash young man! Rise up; for shame!
     Rhaicos. For shame I can not rise. O pity me!
I dare not sue for love—but do not hate!
Let me once more behold thee—not once more,
But many days: let me love on—unloved!

I aimed too high: on my head the bolt   160
Falls back, and pierces to the very brain.
     Hamad. Go—rather go, than make me say I love.
     Rhaicos. If happiness is immortality,
(And whence enjoy it else the gods above?)
I am immortal too: my vow is heard:
Hark! on the left—Nay, turn not from me now,
I claim my kiss.
     Hamad. Do men take first, then claim?
Do thus the seasons run their course with them?

Her lips were seal'd, her head sank on his breast.
'Tis said that laughs were heard within the wood: 170
But who should hear them?—and whose laughs? and
          why?

Savory was the smell, and long past noon,
Thallinos! in thy house; for marjoram,
Basil and mint, and thyme and rosemary,
Were sprinkled on the kid's well-roasted length,
Awaiting Rhaicos. Home he came at last,
Not hungry, but pretending hunger keen,
With head and eyes just o'er the maple plate.
"Thou seest but badly, coming from the sun,
Boy Rhaicos!" said the father. "That oak's bark   180
Must have been tough, with little sap between;
It ought to run; but it and I are old."
Rhaicos, although each morsel of the bread
Increased by chewing, and the meat grew cold
And tasteless to his palate, took a draught
Of gold-bright wine, which, thirsty as he was,
He thought not of until his father filled
The cup, averring water was amiss,
But wine had been at all times poured on kid—
It was religion.
          He, thus fortified,   190
Said, not quite boldly, and not quite abashed,
"Father, that oak is Jove's own tree: that oak
Year after year will bring thee wealth from wax
And honey. There is one who fears the gods
And the gods love;—that one" (he blushed, nor
          said
What one) "hast promised this, and may do more.
Thou hast not many moons to wait until
The bees have done their best: if then there come
Nor wax nor honey, let the tree be hewn."
     "Zeus hath bestowed on thee a prudent mind," 200
Said the glad sire; "but look thou often there,
And gather all the honey thou canst find
In every crevice, over and above
What has been promised; would they reckon that?"

Rhaicos went daily; but the nymph as oft,
Invisible. To play at love, she knew,
Stopping its breathings when it breathes most soft,
Is sweeter than to play on any pipe.
She played on his: she fed upon his sighs;
They pleased her when they gently waved her hair,
Cooling the pulses of her purple veins,   211
And when her absence brought them out, they pleased.
Even among the fondest of them all,
What mortal or immortal maid is more
Content with giving happiness than pain?
One day he was returning from the wood
Despondently. She pitied him, and said,
"Come back!" and twined her fingers in the hem
Above his shoulder. Then she led his steps
To a cool rill that ran o'er level sand   220
Through lentisk and through oleander; there
Bathed she his feet, lifting them on her lap
When bathed, and drying them in both her hands.
He dared complain; for those who most are loved
Most dare it; but not harsh was his complaint.
"O thou inconstant!" said he, "if stern law
Bind thee, or will, stronger than sternest law,
O, let me know henceforward when to hope
The fruit of love that grows for me but here."
He spake; and plucked it from its pliant stem.   230
"Impatient Rhaicos! Why thus intercept
The answer I would give? There is a bee
Whom I have fed, a bee who knows my thoughts
And executes my wishes: I will send
That messenger. If ever thou art false,
Drawn by another, own it not, but drive
My bee away; then shall I know my fate,
And—for thou must be wretched—weep at thine.
But often as my heart persuades to lay
Its cares on thine and throb itself to rest,   240
Expect her with thee, whether it be morn
Or eve, at any time when woods are safe."
    Day after day the Hours beheld them blest,
And season after season: years had past,
Blest were they still. He who asserts that Love
Ever is sated of sweet things, the same
Sweet things he fretted for in earlier days,
Never, by Zeus! loved he a hamadryad.

    The night had now grown longer, and perhaps
The Hamadryads find them lone and dull   250

221. lentisk: the mastic tree of southern Europe; **oleander**:
a poisonous evergreen shrub.

Among their woods; one did, alas! She called
Her faithful bee: 'twas when all bees should sleep,
And all did sleep but hers. She was sent forth
To bring that light which never wintry blast
Blows out, nor rain nor snow extinguishes,
The light that shines from loving eyes upon
Eyes that love back, till they can see no more.

    Rhaicos was sitting at his father's hearth:
Between them stood the table, not o'erspread
With fruits which autumn now profusely bore,   260
Nor anise cakes, not odorous wine; but there
The draft-board was expanded; at which game
Triumphant sat old Thallinos; the son
Was puzzled, vex'd, discomfited, distraught.
A buzz was at his ear: up went his hand,
And it was heard no longer. The poor bee
Return'd, (but not until the morn shone bright)
And found the Hamadryad with her head
Upon her aching wrist, and showed one wing
Half-broken off, the other's meshes marr'd,   270
And there were bruises which no eye could see
Saving a Hamadryad's.
                      At this sight
Down fell the languid brow, both hands fell down,
A shriek was carried to the ancient hall
Of Thallinos: he heard it not: his son
Heard it, and ran forthwith into the wood.
No bark was on the tree, no leaf was green,
The trunk was riven through. From that day forth
Nor word nor whisper sooth'd his ear, nor sound
Even of insect wing; but loud laments   280
The woodmen and the shepherds one long year
Heard day and night; for Rhaicos would not quit
The solitary place, but moan'd and died.
Hence milk and honey wonder not, O guest,
To find set duly on the hollow stone.

                                    (*1846*)

TWENTY YEARS HENCE
MY EYES MAY GROW

Twenty years hence my eyes may grow
If not quite dim, yet rather so,
Still yours from others they shall know
        Twenty years hence.

Twenty years hence tho' it may hap
That I be call'd to take a nap
In a cool cell where thunder-clap
        Was never heard,

There breathe but o'er my arch of grass
A not too sadly sigh'd *Alas*,   10
And I shall catch, ere you can pass,
        That wingèd word.

                (*1846*)

DEATH STANDS ABOVE ME

Death stands above me, whispering low
    I know not what into my ear:
Of his strange language all I know
    Is, there is not a word of fear.

                (*1853*)

DYING SPEECH
OF AN OLD PHILOSOPHER

I strove with none; for none was worth my strife:
    Nature I loved, and, next to Nature, Art:
I warmed both hands before the fire of life;
    It sinks; and I am ready to depart.

                (*1853*)

WELL I REMEMBER
HOW YOU SMILED

Well I remember how you smiled
    To see me write your name upon
The soft sea-sand. "*O! what a child!
You think you're writing upon stone!*"
I have since written what no tide
    Shall ever wash away, what men
Unborn shall read o'er ocean wide
    And find Ianthe's name again.

                (*1863*)

FROM

IMAGINARY
CONVERSATIONS

from SOUTHEY AND PORSON

One of the longest of Landor's conversations, this centers on the poetry of Wordsworth. The participants are Landor's (and of course Wordsworth's) friend Robert Southey and the great classical scholar Richard Porson (1759–1808), Regius Professor of Greek at Cambridge. From this dialogue a few parts have been omitted that are concerned mainly with revealing Porson's own character—his pride and impatience. Inclined to be censorious about Wordsworth, Porson is far more testy, as might be expected from such a scholar, about the reviews in the popular quarterlies. With all his impatience, he is at least a man of mind; and Landor, while giving rein to Porson's irritabilities about other matters, shows his fairness and his strictness of standard leading him to reconsider as the dialogue with Southey progresses.

PORSON. I suspect, Mr. Southey, you are angry with me for the freedom with which I have spoken of your poetry and Wordsworth's.

SOUTHEY. What could have induced you to imagine it, Mr. Professor? You have indeed bent your eyes upon me, since we have been together, with somewhat of fierceness and defiance; I presume that you fancied me to be a commentator; and I am not irritated at a mistake. You wrong me, in your belief that an opinion on my poetical works hath molested me; but you afford me more than compensation in supposing me acutely sensible of any injustice done to Wordsworth. If we must converse at all upon these topics, we will converse on him. What man ever existed, who spent a more retired, a more inoffensive, a more virtuous life, or who adorned it with nobler studies?

PORSON. I believe, none; I have always heard it; and those who attack him with virulence or with levity, are men of as little morality as reflexion. I have demonstrated that one of them, he who wrote the *Pursuits of Literature*,[1] could not construe a Greek sentence or scan a verse; and I have fallen on the very *Index* from which he drew out his forlorn hope on the parade. This is incomparably the most impudent fellow I have met with in the course of my reading, which has lain, you know, in a province where impudence is no rarity. I am sorry to say that we critics who write for the learned, have sometimes set a bad example to our younger brothers, the critics who write for the public: but if they were considerate and prudent, they would find out that a deficiency in weight and authority might in some measure be compensated by deference and decorum. Not to mention the refuse of the literary world, the sweeping of booksellers' shops, the dust thrown up by them in a corner to blow by pinches on new publications; not to tread upon or disturb this filth, the greatest of our critics now living are only great men comparatively; which they betray when they look disdainfully on the humbler in judgement and intellect: for if these were not humbler, what would they themselves be? A little wit, or as that is not always at hand, a little impudence instead of it, throws its briar over dry and deep lacunes: a little grease covers a great quantity of poor broth. Instead of any thing in this way I would seriously recommend to the employer of our critics, young and old, that he oblige them to pursue a course of study such as the following; namely, that under the superintendence of some respectable student from the university, they first read and examine the contents of the book; a thing greatly more useful in criticism than is generally thought; secondly, that they carefully write them down, number them, and range them under their several heads; thirdly, that they mark every beautiful, every faulty, every ambiguous, every uncommon, expression. This being completed, that they inquire what author, ancient or modern, has treated the same subject; that they compare them, first in smaller, afterwards in larger, portions, noting every defect in precision, and its causes, every excellence, and its nature. . . .

. . . Adverse as I have shewn myself to the style and manner of Wordsworth, I never thought that all his reviewers put together could compose any thing equal to the worst paragraph in his volumes. I have spoken vehemently against him, and mildly against

them; because he could do better, they never could. If he thinks me his enemy it is through modesty: if they think me their friend it is through impudence. The same people would treat me with as little reverence as they treat him with, if any thing I write were popular, or could become so. It is by fixing on such works that they are carried with them into the doorway. The porter of Cleopatra would not have admitted the asps if they had not been under the figs.

Shew me, if you can, Mr. Southey, a temperate, accurate, solid exposition, of any English work whatever, in any English review.

SOUTHEY. Not having at hand so many numbers as it would be requisite to turn over, I must decline the challenge.

PORSON. I have observed the same man extoll in private, the very book on whose ruin he dined the day before.

SOUTHEY. His judgement then may be ambiguous, but you must not deny him the merit of gratitude. If you blame the poor and vicious for abusing the solaces of poverty and vice, how much more should you censure those who administer to them the means of such indulgence.

PORSON. The publications which excite the most bustle and biting from these fellows, are always the best, as the fruit on which the flies gather is the ripest. Periodical critics were never so plentiful as they now are. There is hardly a young author who does not make his first attempt in some review; shewing his teeth, hanging by his tail, pleased and pleasing by the volubility of his chatter, and doing his best to get a penny for his exhibitor and a nut for his own pouch, by the facetiousness of the tricks he performs upon our heads and shoulders. From all I can recollect of what I noticed when I turned over such matters, a wellsized and useful volume might be compiled and published annually, containing the incorrect expressions, and omitting the opinions, of our booksellers' boys, the reviewers.

.   .   .

Returning, Mr. Southey, to the difficulty, or rather to the rarity, of an accurate and just survey of poetical and other literary works, I do not see why we should not borrow an idea from geometricians and astronomers, why we also should not have our triangles and quadrants, why in short we should not measure out writings by small portions at a time, and compare the brighter parts of two authors, page for instance by

IMAGINARY CONVERSATIONS. SOUTHEY AND PORSON. **1. he . . . Literature:** Thomas James Mathias; the *Pursuits of Literature* (1794) is a satire on writers of the time.

page. The minor beauties, the complexion and con-texture, may be considered at last, and more at large. Daring geniusses, ensigns and undergraduates, members of Anacreontic and Pindaric[2] clubs, will scoff at me. Painters who can draw nothing correctly, hold Raffael in contempt, and appeal to the sublimity of Michael-Angelo and the splendour of Titian, ignorant that these great men were great by science first, and employed in painting, at all times, the very means I propose for criticism. Venus and the damned submitted to the same squaring.

Such a method would be very useful to critics in general, and even the wisest and most impartial would be much improved by it; although few, either by these means or any, are likely to be quite correct, or quite unanimous, on the merits of any two authors whatsoever.

SOUTHEY. Those who are learners would be teachers; while those who have learned much would procure them at any price. It is only when we have mounted high, that we are sensible of wanting a hand.

PORSON. On the subject of poetry in particular, there are some questions not yet sufficiently discussed: I will propose two. First, admitting that in all the tragedies of Sophocles there was (which I believe) twice or thrice as much good poetry as in the Iliad, does it follow that he was as admirable a poet as Homer?

SOUTHEY. I doubt it; so much do I attribute to the conception and formation of a novel and vast design, and so wide is the difference I see between the completion of one very great, and the perfection of many smaller. Would even these have existed without Homer? I think not.

PORSON. My next question is, whether a poet is to be judged from the quantity of his bad poetry or from the quality of his best?

SOUTHEY. I should certainly say from the latter: because it must be in poetry as in sculpture and painting; he who arrives at a high degree of excellence in these arts, will have made more models, more sketches and designs, than he who has reached but a lower; and the

2. **Anacreontic and Pindaric:** Anacreon, sixth century B.C., was a Greek poet whose lyrics, surviving only in a few fragments, celebrate love and wine. Pindar, *c.* 522–442 B.C., was a Greek lyric poet. A misunderstanding of his system of versification gave rise to the English irregular ode—also called the "false Pindaric"—in the eighteenth century. Wordsworth's "Ode: Intimations of Immortality" is the finest poem in the genre.

conservation of them, whether by accident or by choice, can injure and affect in no manner his more perfect and elaborate works. A drop of sealing-wax, falling by chance or negligence, may efface a fine impression: but what is well done in poetry is never to be effaced by what is ill done afterwards. Even the bad poetry of a good poet hath something in it which renders it more valuable, to a judge of these matters, than what passes for much better, and what in many essential points is truly so. I will however keep to the argument, not having lost sight of my illustration, in alluding to designs and sketches. Many men would leave themselves pennyless, to purchase an early and rude drawing by Raffael; some arabesque, some nose upon a gryphen, or gryphen upon a nose, and never would inquire whether the painter had kept it in his portfolio or had cast it away. The same persons, and others whom we call much wiser, exclame loudly against any literary sketch unworthy of a leaf among the productions of its author. No ideas are so trivial, so incorrect, so incoherent, but they may have entered the idle fancy, and have taken a higher place than they ought in the warm imagination, of the best poets. We find in Dante, as you just now remarked, a prodigious quantity of them; and indeed not a few in Virgil, grave as he is, and stately. Infantine and petty there is hardly any thing in the Iliad, but the dull and drowthy stop us unexpectedly now and then. The boundaries of mind lie beyond these writers, altho their splendour lets us see nothing on the further side. In so wide and untrodden a creation as that of Shakespear, can we wonder or complain that sometimes we are bewildered and entangled in the exuberance of fertility? Drybrained men upon the Continent, the trifling wits of the theatre, accurate however and expert calculators, tell us that his beauties are balanced by his faults. The poetical opposition, the liberal whig wiseacres, puffing for popularity, cry cheerily against them, *his faults are balanced by his beauties.* In reality, all the faults that ever were committed in poetry would be but as air to earth, if we could weigh them against one single thought or image, such as almost every scene exhibits, in every drama of this unrivalled genius.

PORSON. A third question . . What is the reason why, when not only the glory of great kings and statesmen, but even of great philosophers, is much enhanced by two or three good apophthegms, that of a great poet is lowered by them, even if he should invest them with good verse? for certainly the dignity of a great poet is thought to be lowered by the writing of epigrams.

SOUTHEY. The great poet could do better things; the others could not. People in this apparent act of injustice do real justice, and conferr high honour where it is due, without intending or knowing it.

All writers have afforded some information, or have excited some sentiment or idea, somewhere. This alone should exempt the humblest of them from revilings, unless it appear that he hath misapplied his powers, from insolence or from malice. In that case, whatever sentence may be passed upon him, I consider it no honour to be the executioner. What must we think of those who travel far and wide, that, before they go to rest, they may burst into the arbour of a recluse, whose weakest thoughts are benevolence, whose worst are purity? On his poetry I shall say nothing, unless you lead me to it, wishing you however to examine it analytically and severely.

PORSON. I will not dissemble or deny, that to compositions of a new kind, like Wordsworth's, we come without scales and weights, and without the means of making an assay.

SOUTHEY. Mr. Porson, it does not appear to me, that anything more is necessary in the first instance, than to interrogate our hearts in what manner they have been affected. If the ear is satisfied; if at one moment a tumult is aroused in the breast, and tranquillized at another, with a perfect consciousness of equal power exerted in both cases; if we rise up from the perusal of the work with a strong excitement to thought, to imagination, to sensibility; above all, if we sat down with some propensities toward evil, and walk away with much stronger toward good, in the midst of a world, which we never had entered, and of which we never had dreamed before, shall we perversely put on again the *old man* of criticism, and deny that we have been conducted by a most beneficent and most potent genius? Nothing proves to me so manifestly in what a pestiferous condition are its lazarettos, as when I observe how little hath been objected against those who have substituted words for things, and how much against those who have reinstated things for words.

Let Wordsworth prove to the world, that there may be animation without blood and broken bones, and tenderness remote from the stews. Some may doubt it; for even things the most evident are often but little perceived and strangely estimated. Swift ridiculed the music of Handel and the generalship of Marlborough,

Pope the style of Middleton and the scholarship of Bentley, Gray the abilities of Shaftesbury and the eloquence of Rousseau. Shakespear hardly found those who would collect his tragedies; Milton was read from godliness; Virgil was antiquated and rustic, Cicero Asiatic.[3] What a rabble has persecuted my friend, in these latter times the glory of our country! An elephant is born to be consumed by ants in the midst of his unapproachable solitudes. Wordsworth is the prey of Jeffrey.[4] Why repine? and not rather amuse ourselves with allegories, and recollect that God in the creation left his noblest creature at the mercy of a serpent.

PORSON. In my opinion your friend is verbose; not indeed without something for his words to rest upon, but from a resolution to gratify and indulge his capacity. He pursues his thoughts too far; and considers more how he may shew them entirely, than how he may shew them advantageously. Good men may utter whatever comes uppermost, good poets may not. It is better, but it is also more difficult, to make a selection of thoughts, than to accumulate them. He who has a splendid sideboard, should likewise have an iron chest with a double lock upon it, and should hold in reserve a greater part than he displays.

Wordsworth goes out of his way to be attacked. He picks up a piece of dirt, throws it on the carpet in the midst of the company, and cries *This is a better man than any of you.* He does indeed mould the base material into what form he chooses; but why not rather invite us to contemplate it, than challenge us to condemn it? This surely is false taste.

SOUTHEY. The principal and the most general accusation against Wordsworth is, that the vehicle of his thoughts is unequal to them. Now did ever the judges at the Olympic games say, *We would have awarded to you the meed of victory, if your chariot had been equal to your horses: it is true they have won; but the people is displeased at a car neither new nor richly gilt, and without a gryphen or sphynx engraven on the axle?*

You admire simplicity in Euripides; you censure it in Wordsworth; believe me, sir, it arises in neither from penury of thought, which seldom has produced it, but from the strength of temperance, and at the suggestion of principle. Some of his critics are sincere in their censure, and are neither invidious nor unlearned;

3. **Asiatic:** ornate in style. 4. **Jeffrey:** See selections, above, from Francis Jeffrey, 1773–1840, editor of *The Edinburgh Review.*

but their optics have been formed upon other objects, altogether dissimilar, and they are (permitt me an expression not the worse for daily use) entirely out of their element. His very clearness puzzles and perplexes them, and they imagine that straitness is distortion, as children on seeing a wand dipt in limpid and still water.

PORSON. Fleas know not whether they are on the body of a giant, or upon one of ordinary size. Clear writers, like clear fountains, do not seem so deep as they are: the turbid look most profound.

SOUTHEY. Ignorance however has not been single-handed the enemy of Wordsworth, but Petulance and Malignity have accompanied her, and have been unremittent in their attacks. Small poets, small critics, lawyers, who have much time upon their hands, and hanging heavily, come forward unfeed against him; such is the spirit of patriotism, rushing everywhere for the public good. Most of these have tried their fortune at some little lottery-office of literature, and, receiving a blank, have chewed upon it harshly and wryly. We, like jackdaws, are very amicable creatures while we all are together in the dust; but let any one gain a battlement or steeple, and behold! the rest fly about him at once, and beat him down.

Take up a poem of Wordsworth's and read it; I would rather say, read them all; and, knowing that a mind like yours must grasp closely what comes within it, I will then appeal to you whether any poet of our country, since Shakespear, has exerted a greater variety of powers with less strain and less ostentation. I would however, with his permission, lay before you for this purpose a poem which is yet unpublished and incomplete.

PORSON. Pity, with his abilities, he does not imitate the ancients somewhat more.

SOUTHEY. Whom did they imitate? If his genius is equal to theirs he has no need of a guide. He also will be an ancient; and the very counterparts of those, who now decry him, will extoll him a thousand years hence in malignity to the moderns. The ancients have always been opposed to them; just as, at routs and dances, elderly beauties to younger. It would be wise to contract the scene of action, and to decide the business in both cases by couples.

. . .

PORSON. I ought not to have interrupted you so long, in your attempt to prove Wordsworth shall I say the rival or the resembler of the ancients?

SOUTHEY. Such excursions are not unseasonable in such discussions, and lay in a store of good humour for them. Your narrative has amused me exceedingly. As you call upon me to return with you to the point we set out from, I hope I may assert without a charge of paradox, that whatever is good in poetry is common to all good poets, however wide may be the diversity of manner. Nothing can be more dissimilar than the three Greek tragedians: but would you prefer the closest and best copier of Homer to the worst (whichever he be) amongst them? Let us avoid what is indifferent or doubtful, and embrace what is good, whether we see it in another or not; and if we have contracted any peculiarity while our muscles and bones were softer, let us hope finally to outgrow it. Our feelings and modes of thinking forbid and exclude a very frequent imitation of the old classics, not to mention our manners, which have a nearer connexion than is generally known to exist with the higher poetry. When the occasion permitted it, Wordsworth has not declined to treat a subject as an ancient poet of equal vigour would have treated it. Let me repeat to you his Laodamia.[5]

PORSON. After your animated recital of this classic poem, I begin to think more highly of you both. It is pleasant to find two poets living as brothers, and particularly when the palm lies between them, with hardly a third in sight. Those who have ascended to the summit of the mountain, sit quietly and familiarly side by side; it is only those who are climbing with briars about their legs, that kick and scramble. Yours is a temper found less frequently in our country than in others. The French poets indeed must stick together to keep themselves warm. By employing courteous expressions mutually, they indulge their vanity rather than their benevolence, and bring the spirit of contest into action gaily and safely. Among the Romans we find Virgil, Horace, and several of their contemporaries, intimately united and profuse of reciprocal praise. Ovid, Cicero, and Pliny are authors the least addicted to censure, and the most ready to offer their testimony in favour of abilities in Greek or countryman. These are the three Romans, the least amiable of nations, and (one excepted) the least sincere, with whom I should have liked best to spend an evening.

SOUTHEY. Ennius and old Cato, I am afraid, would have run away with your first affections.

PORSON. Old Cato! he, like a wafer, must have

5. **Laodamia:** See under Wordsworth, above.

been well wetted to be good for any thing. Such gentlemen as old Cato we meet every day in St. Mary Axe, and wholesomer wine than his wherever there are sloes and turnips. Ennius could converse without ignorance about Scipio, and without jealousy about Homer.

SOUTHEY. And I think he would not have disdained to nod his head on reading the *Laodamia*.

PORSON. You have recited a most spirited thing indeed. I never had read it. Now to give you a proof that I have been attentive, I will remark two passages that offend me. In the first stanza,

> With sacrifice before the rising morn
> Performed, my slaughtered lord have I required;
> And in thick darkness, amid shades forlorn,
> Him of the infernal Gods have I desired.[6]

The second line and the fourth terminate too much alike: *have I required* and *have I desired* are worse than prosaic; besides which there are four words together of equal length in each.

> He spake of love, such love as spirits feel
> In worlds whose course is equable and pure;
> No fears to beat away, no strife to heal,
> The past unsighed for, and the future sure;
> Spake, as a witness, of a second birth
> For all that is most perfect upon earth.

In a composition such as Sophocles might have exulted to own, and in a stanza the former part of which might have been heard with shouts of rapture in the regions it describes, how unseasonable is the allusion to *witness* and *second birth!* which things, however holy and venerable in themselves, come stinking and reeking to us from the conventicle. I desire to see Laodamia in the silent and gloomy mansion of her beloved Protesilaus; not elbowed by the godly butchers in Tottenham-court-road, nor smelling devoutly of ratafia among the sugar-bakers' wives at Blackfriars.

Mythologies should be kept distinct: the fireplace of one should never be subject to the smoke of another. The Gods of different countries, when they come together unexpectedly, are jealous Gods, and, as our old women say, *turn the house out of windows*.

A current of rich and bright thoughts runs throughout the poem. Pindar himself would not, on that

subject, have braced one to more vigour and freshness, nor Euripides have inspired into it more tenderness and passion. I am not insensible to that warmly chaste morality which is the soul of it, nor indifferent to the benefits that literature on many occasions has derived from Christianity. But poetry is a luxury to which, if she tolerates and permits it, she accepts no invitation: she beats down your gates and citadels, levels your high places, and eradicates your groves. For which reason I dwell more willingly with those authors, who cannot mix and confound the manners they represent. The hope that we may rescue at Herculaneum a great number of them, hath, I firmly believe, kept me alive. Reasonably may the best be imagined to exist in a library of some thousands. It will be recorded to the infamy of the kings and princes now reigning, or rather of those whose feet put into motion their rocking-horses, that they never have made a common cause in behalf of learning, but on the contrary have made a common cause against it. The earth opened her entrails before them, conjured them to receive again, while it was possible, the glories of their species . . . and they turned their backs. They pretend that it is not their business or their duty to interfere in the internal affairs of other nations. This is not an internal affair of any: it interests all; it belongs to all; and these scrupulous men have no scruple to interfere in giving their countenance and assistence, when a province is to be torn away or a people to be invaded.

SOUTHEY. To neglect what is recoverable in the authors of antiquity, is like rowing away from a crew that is making its escape from shipwreck.

PORSON. The most contemptible of the Medicean family did more for the advancement of letters than the whole body of existing potentates. If their delicacy is shocked or alarmed at the idea of making a proposal to send scientific and learned men thither, let them send a brace of pointers as internuncios, and the property is their own. Twenty men in seven years might retrieve the worst losses we experience from the bigotry of popes and califs. I do not intend to assert, that every Herculanean manuscript might within that period be unfolded; but the three first legible sentences might be; which is quite sufficient to inform the scholar, whether a further attempt on the scroll would repay his trouble. There are fewer than thirty Greek authors worth inquiring for; they exist beyond doubt, and beyond doubt they may, by attention, patience, and skill, be brought to light.

SOUTHEY. You and I, Mr. Porson, are truly and

---

6. **With . . . desired:** "The memory of Porson was extraordinary and quite capable of this repetition" (Landor's note).

sincerely concerned in the loss of such treasures: but how often have we heard much louder lamentations than ours, from gentlemen who, if they were brought again to light, would never cast their eyes over them, even in the bookseller's window. I have been present at homilies on the corruption and incredulity of the age, and principally on the violation of the sabbath, from sleek clergymen, canons of cathedrals, who were at the gaming-table the two first hours of it, on that very day; and I have listened to others on the loss of the classics, from men who never read one half of what is remaining to us of Cicero and Livius. The Greek language is almost unknown out of England and northern Germany: in the remainder of the world, exclusive of Greece, I doubt whether fifty scholars ever read one page of it without a version. Give fifteen to Italy, twelve to the Netherlands, five to France; the remainder will hardly be collected in Sweden, Russia, Austria: as for Spain and Portugal, we might as well look for them among the Moors and Negroes. The knowledge of books written in our own language is extending daily in our country, which, whatever dissatisfaction or disgust its rulers may occasion in you, contains four-fifths of the learned and scientific men now on earth.

PORSON. This position is, I think, incontrovertible: but although the knowledge too of Greek is extending in England, I doubt whether it is to be found in such large masses as formerly. Schools and universities, like rills and torrents, roll down some grains of it every season; but the lumps have been long stored up in cabinets. I delight in the diffusion of learning; yet, I must confess it, I am most gratified and transported at finding a large portion of it in one place: just as I would rather have a solid pat of butter at breakfast, than a splash of grease upon the table-cloth that covers half of it. Do not attempt to defend the idle and inconsiderate knaves who manage our affairs for us; or defend them on some other ground: prove, if you please, that they have, one after another, been incessantly occupied in rendering us more moral, more prosperous, more free; but abstain, sir, from any allusion to their solicitude on the improvement of our literary condition. With a smaller sum than is annually expended on the appointment of some silly and impertinent young envoy, we might restore all, or nearly all, those writers of immortal name, whose disappearance has been the regret of Genius for four entire centuries. In my opinion a few thousand pounds, laid out on such an undertaking, would be laid out

as creditably as on a Persian carpet or a Turkish tent; as creditably as on a collar of rubies and a ball-dress of Brussels-lace for our Lady in the manger, or as on gilding, for the adoration of princesses and their capuchins,[7] the posteriors and anteriors of saint Januarius.[8]

(*1823*)

EPICTETUS AND SENECA

ᴄᴡᴄ

Landor here imagines a dialogue between the two most famous Stoics of ancient Rome. Epictetus, a former slave, lived as he taught, following Stoic principles with heroic simplicity. The philosopher Seneca served as tutor to Nero, later became one of the emperor's chief advisers, and, after taking part in a conspiracy against Nero, was ordered by the emperor to take his own life.

ᴄᴡᴄ

SENECA. Epictetus! I desired your master Epaphroditus to send you hither, having been much pleased with his report of your conduct, and much surprised at the ingenuity of your writings.

EPICTETUS. Then I am afraid, my friend! . . .

SENECA. *My friend!* are these the expressions . . . Well, let it pass . . philosophers must bear bravely . . the people expect it.

EPICTETUS. Are philosophers then only philosophers for the people? and, instead of instructing them, must they play tricks before them? Give me rather the gravity of dancing dogs: their motions are for the rabble; their reverential eyes and pendent paws are under the pressure of awe at a master; but they are dogs, and not below their destinies.

SENECA. Epictetus! I will give you three talents to let me take that sentiment for my own.

EPICTETUS. I would give thee twenty, if I had them, to make it thine.

---

7. **capuchins:** friars of the order of St. Francis. 8. **Januarius:** referring to the head and dried blood in Naples of St. Januarius (martyred under Emperor Diocletian). The blood was said to become liquid on certain days of the year. Porson's heavy jest also involves a learned pun typical of him: a pun on the Roman deity Janus, who had two heads facing opposite directions (the implication probably being that a second head of St. Januarius is kept for the purpose of exhibiting the miracle).

SENECA. You mean, by lending to it the graces of my language.

EPICTETUS. I mean, by lending it to thy conduct. And now let me console and comfort thee, under the calamity I brought on thee by calling thee *my friend.* If thou art not my friend, why send for me? Enemy I can have none: being a slave, Fortune has now done with me.

SENECA. Continue then your former observations. What were you saying?

EPICTETUS. That which thou interruptedst.

SENECA. What was it?

EPICTETUS. I should have remarked that, if thou foundest ingenuity in my writings, thou must have discovered in them some deviation from the plain homely truths of Zeno and Cleanthes.[1]

SENECA. We all swerve a little from them.

EPICTETUS. In practise too?

SENECA. Yes, even in practise, I am afraid.

EPICTETUS. Often?

SENECA. Too often.

EPICTETUS. Strange! I have been attentive, and yet have remarked but one difference among you great personages at Rome.

SENECA. What difference fell under your observation?

EPICTETUS. Crates and Zeno and Cleanthes taught us, that our desires were to be subdued by philosophy alone. In this city, their acute and inventive scholars take us aside, and shew us that there is not only one way, but two.

SENECA. Two ways?

EPICTETUS. They whisper in our ear, *These two ways are philosophy and enjoyment: the wiser man will take the readier, or, not finding it, the alternative.* Thou reddenest.

SENECA. Monsterous degeneracy!

EPICTETUS. What magnificent rings! Pardon me! I did not notice them until thou liftedst up thy hands to heaven, in detestation of such effeminacy and impudence.

SENECA. The rings are not amiss: my rank rivets them upon my fingers: I am forced to wear them. Our emperor[2] gave me one, Epaphroditus another, Tigellinus the third. I cannot lay them aside a single

day, for fear of offending the Gods, and those whom they love the most worthily.

EPICTETUS. Altho they make thee stretch out thy fingers, like the arms and legs of one of us slaves upon a cross.

SENECA. O horrible! find some other resemblance.

EPICTETUS. The extremities of a figleaf.

SENECA. Ignoble!

EPICTETUS. The claws of a toad, trodden on or stoned.

SENECA. You have great need, Epictetus, of an instructor in eloquence and rhetoric: you want topics and tropes and figures.

EPICTETUS. I have no room for 'em. They make such a buz in the house, a man's own wife cannot understand what he says to her.

SENECA. Let us reason a little upon style: I would set you right, and remove from before you the prejudices of a somewhat rustic education. We may adorn the simplicity of the wisest.

EPICTETUS. Thou canst not adorn simplicity. What is naked or defective is susceptible of decoration: what is decorated is simplicity no longer. Thou mayest give another thing in exchange for it; but if thou wert master of it, thou wouldest preserve it inviolate. It is no wonder that we mortals, so little able as we are to see truth, should be less able to express it.

SENECA. You have formed at present no idea of style.

EPICTETUS. I never think about it. First I consider whether what I am about to say is true; then whether I can say it with brevity, in such a manner as that others shall see it as clearly as I do in the light of truth: for if they survey it as an ingenuity, my desire is ungratified, my duty unfulfilled.

SENECA. We must attract the attention of readers, by novelty and force and grandeur of expression.

EPICTETUS. We must so. Nothing is so grand as truth, nothing so forcible, nothing so novel.

SENECA. Sonorous sentences are wanted, to awaken the lethargy of indolence.

EPICTETUS. Awaken it to what? here lies the question; and a weighty one it is. If thou awakenest men where they can see nothing, and do no work, it is better to let them rest: but will not they, thinkest thou, look up at a rainbow, unless they are called to it by a clap of thunder?

SENECA. Your early youth, Epictetus, has been I will not say neglected, but cultivated with rude instruments and unskilful hands.

EPICTETUS AND SENECA. **1. Zeno and Cleanthes:** Zeno, fourth century B.C., was the founder of the Stoic philosophy, and Cleanthes was a follower of the next century. **2. emperor:** Nero.

EPICTETUS. I thank God for it. Those rude instruments have left the turf lying yet toward the sun; and those unskilful hands have plucked out only the docks.

SENECA. We hope and believe that we have attained a vein of eloquence, brighter and more varied than has been hitherto laid open to the world.

EPICTETUS. Than any in the Greek?

SENECA. We trust so.

EPICTETUS. Than your Cicero's?

SENECA. If the declaration may be made without an offence to modesty. Surely you cannot estimate or value the eloquence of that noble pleader.

EPICTETUS. Imperfectly; not being born in Italy: and the noble pleader is a much less man with me than the noble philosopher. I regret that having farms and villas, he would not keep his distance from the pumping up of foul words, against thieves, cutthroats, and other rogues; and that he lied, sweated, and thumped his head and thighs, in favour of those who were no better.

SENECA. Senators must have clients,[3] and must protect them.

EPICTETUS. Innocent or guilty.

SENECA. Doubtless.

EPICTETUS. If I regret what is, and may not be, I regret much more what both is and must be. However it is an amiable thing, and no small merit in the wealthy, even to trifle and play at their leisure-hours with Philosophy. It cannot be expected that any such a personage should espouse her, or should recommend her as an inseparable mate to his heir.

SENECA. I would.

EPICTETUS. Yes, Seneca, but thou hast no son to make the match for; and thy recommendation, I suspect, would be given him before he could consummate the marriage. Every man wishes his sons to be philosophers while they are very young; but takes especial care, as they grow older, to teach them its insufficiency and unfitness for their intercourse with mankind. The paternal voice says, *You must not be particular: you are about to have a profession to live by: follow those who have thriven the best in it.* Now among these, whatever be the profession, canst thou point out to me one single philosopher?

SENECA. Not just now . . nor, upon reflexion, do I think it feasible.

EPICTETUS. Thou indeed mayest live much to

thy ease and satisfaction with philosophy, having (they say) two thousand talents.[4]

SENECA. And a trifle to spare . . pressed upon me by that godlike youth, my pupil Nero.

EPICTETUS. Seneca! where God hath placed a mine, he hath placed the materials of an earthquake.

SENECA. A true philosopher is beyond the reach of Fortune.

EPICTETUS. The false one thinks himself so. Fortune troubles her head very little about philosophers; but she remembers where she hath set a rich man, and she laughs to see the Destinies at his door.

(1828)

PETER THE GREAT AND ALEXIS

❧

The speakers are Peter I (1672–1725), tsar of Russia, and his son Alexius (1690–1718), the heir to the crown. Relations between them were poisoned by personal antipathies and also by political hostility. Peter was attempting to modernize Russia and Alexius allowed himself to become a tool of Peter's opponents. In order to escape the control of his father, Alexius eventually took refuge with the emperor Charles VI of Austria. He was lured back to Russia with a promise that he would not be punished, but in fact he was compelled to make a confession that he had imagined rebellion and wished for his father's death. (Even with the resources of despotism it was impossible to show that Alexius had taken steps to bring about these wishes.) The investigation under torture of his conduct was deliberately continued until he died of successive beatings.

❧

PETER. And so, after flying from thy father's house, thou art returned again from Vienna . . after this affront in the face of Europe, thou darest to appear before me.

ALEXIS. My emperor and father! I am brought before your Majesty, not at my own desire.

PETER. I believe it well.

ALEXIS. I would not anger you.

PETER. What hope hadst thou, rebel, in thy flight to Vienna?

---

3. **clients:** followers, supporters.

4. **talents:** The value of a talent, which was a large unit of money, was equivalent to about £200 in 1828 and about £1,600 in 1964. The total sum of 2000 talents would thus be equivalent to about $4½ million now.

ALEXIS. The hope of peace and privacy; the hope of security; and above all things, of never more offending you.

PETER. That hope thou hast accomplished.

Thou imaginedst then that my brother of Austria would maintain thee at his court . . speak!

ALEXIS. No, sir! I imagined that he would have afforded me a place of refuge.

PETER. Didst thou then take money with thee?

ALEXIS. A few gold pieces.

PETER. How many?

ALEXIS. About sixty.

PETER. He would have given thee promises for half the money; but the double of it does not purchase a house: ignorant wretch!

ALEXIS. I knew as much as that; altho my birth did not appear to destine me to purchase a house anywhere; and hitherto your liberality, my father, hath supplied my wants of every kind.

PETER. Not of wisdom, not of duty, not of spirit, not of courage, not of ambition. I have educated thee among my guards and horses, among my drums and trumpets, among my flags and masts. When thou wert a child, and couldst hardly walk, I have taken thee into the arsenal, tho children should not enter, according to regulations; I have there rolled cannon-balls before thee over iron plates; and I have shewn thee bright new arms, bayonets and sabres; and I have pricked the back of my hand until the blood came out in many places; and I have made thee lick it; and I have then done the same to thine. Afterwards, from thy tenth year, I have mixt gunpowder in thy grog; I have peppered thy peaches; I have poured bilgewater (with a little good wholesome tar in it) upon thy melons; I have brought out girls to mock thee and cocker thee, and talk like mariners, to make thee braver. Nothing would do. Nay, recollect thee! I have myself led thee forth to the window when fellows were hanged and shot; and I have shewn thee every day the halves and quarters of bodies; and I have sent an orderly or chamberlain for the heads; and I have pulled the cap up from over the eyes; and I have made thee, in spite of thee, look stedfastly upon them . . incorrigible coward!

And now another word with thee about thy scandalous flight from the palace; in time of quiet too! To the point! did he, or did he not? did my brother of Austria invite thee?

ALEXIS. May I answer without doing an injury or disservice to his Imperial Majesty?

PETER. Thou mayest. What injury canst thou or any one do, by the tongue, to such as he is?

ALEXIS. At the moment, no; he did not: nor indeed can I assert that he at any time invited me: but he had said that he pitied me.

PETER. About what? hold thy tongue . . let that pass. Princes never pity but when they would make traitors: then their hearts grow tenderer than tripe. He pitied thee, kind soul, when he would throw thee at thy father's head; but finding thy father too strong for him, he now commiserates the parent, laments the son's rashness and disobedience, and would not make God angry for the world. At first however there must have been some overture on his part; otherwise thou art too shame-faced for intrusion. Come . . thou hast never had wit enough to lie . . tell me the truth, the whole truth.

ALEXIS. He said that, if ever I wanted an asylum, his court was open to me.

PETER. Open! so is the tavern; but folks pay for what they get there. Open truly! and didst thou find it so?

ALEXIS. He received me kindly.

PETER. I see he did.

ALEXIS. Derision, O my father, is not the fate I merit.

PETER. True, true! it was not intended.

ALEXIS. Kind father! punish me then as you will.

PETER. Villain! wouldst thou kiss my hand too? Art thou ignorant that the Austrian threw thee away from him, with the same indifference as he would the outermost leaf of a sandy sunburnt lettuce?

ALEXIS. Alas! I am not ignorant of this.

PETER. He dismissed thee at my order. If I had demanded from him his daughter, to be the bed-fellow of a Kalmuc,[1] he would have given her, and praised God.

ALEXIS. O father! is his baseness my crime?

PETER. No; thine is greater. Thy intention, I know, is to subvert the institutions it has been the labour of my lifetime to establish. Thou hast never rejoiced at my victories.

ALEXIS. I have rejoiced at your happiness and your safety.

PETER. Liar! coward! traitor! when the Polanders and Swedes fell before me, didst thou from thy soul congratulate me? didst thou get drunk, at home or

PETER THE GREAT AND ALEXIS. I. **Kalmuc:** a member of a Mongol tribe of Sungaria.

abroad, or praise the Lord of Hosts and saint Nicolas? Wert thou not silent and civil and low-spirited?

ALEXIS. I lamented the irretrievable loss of human life; I lamented that the bravest and noblest were swept away the first; that the gentlest and most domestic were the earliest mourners; that frugality was supplanted by intemperance; that order was succeded by confusion; and that your Majesty was destroying the glorious plans you alone were capable of devising.

PETER. I destroy them! how? of what plans art thou speaking?

ALEXIS. Of civilizing the Muscovites. The Polanders in part were civilized: the Swedes were more so than any other nation on the continent; and so excellently versed were they in military science, and so personally courageous, that every man you killed cost you seven or eight.

PETER. Thou liest; nor six. And civilized forsooth! why, the robes of the metropolitan, him at Upsal, are not worth three ducats, between Jew and Livornese.[2] I have no notion that Poland and Sweden shall be the only countries that produce great princes. What right have they to such monarchs as Gustavus and Sobieski? All Europe ought to look to this, before discontent becomes general, and the people does to us what we have the privilege of doing to the people. I am wasting my words: there is no arguing with positive fools like thee. So thou wouldst have desired me to let the Polanders and Swedes lie still and quiet! two such powerful nations!

ALEXIS. For that reason and others I would have gladly seen them rest, until our own people had increased in numbers and prosperity.

PETER. And thus thou disputest my right before my face, to the exercise of the supreme power.

ALEXIS. Sir! God forbid!

PETER. God forbid indeed! what care such villains as thou art what God forbids! He forbids the son to be disobedient to the father: he forbids . . he forbids . . twenty things. I do not wish, and will not have, a successor who dreams of dead people.

ALEXIS. My father! I have dreamt of none such.

PETER. Thou hast; and hast talked about them . . Scythians I think they call 'em. Now who told thee, Mr. Professor, that the Scythians were a happier people than we are; that they were inoffensive; that they were free; that they wandered with their carts from pasture

to pasture, from river to river; that they traded with good faith; that they fought with good courage; that they injured none, invaded none, and feared none? At this rate I have effected nothing. The great founder of Rome, I heard in Holland, slew his brother for despising the weakness of his walls: and shall the founder of this better place spare a degenerate son, who prefers a vagabond life to a civilized one, a cart to a city, a Scythian to a Muscovite? Have I not shaved my people, and breeched them? have I not formed them into regular armies, with bands of music and havresacs? Are bows better than cannon? shepherds than dragoons, mare's milk than brandy, raw steaks than broiled? Wouldst thou have ever eaten stockfish but for me, or have jumped with joy at the roe of a red herring? Thine are doctrines that strike at the root of politeness and sound government. Every prince in Europe is interested in rooting them out by fire and sword. There is no other way with false doctrines: breath against breath does little.

ALEXIS. Sire, I never have attempted to disseminate my opinions.

PETER. How couldst thou? the seed would fall only upon granite. Those however who caught it brought it to me.

ALEXIS. Never have I undervalued Civilization: on the contrary, I regretted whatever impeded it. In my opinion, the evils that have been attributed to it, sprang from its imperfections and voids; and that no nation has yet acquired it more than very scantily.

PETER. How so? give me thy reasons; thy fancies rather; for reason thou hast none.

ALEXIS. When I find the first of men, in rank and genius, hating one another, and becoming slanderers and liars in order to lower and vilify an oppenent; when I hear the God of mercy invoked to massacres, and thanked for furthering what he reprobates and condemns, I look back in vain on any barbarous people for worse barbarism. Soldiers, it is said in ancient mythology, sprang from dragon's teeth, sown by Cadmus, who introduced letters. It would appear that these also came from the same sack as the soldiers, and were only the rottenest of the fangs, kept till the last. I have expressed my admiration of our forefathers, who, not being Christians, were yet more virtuous than those who are so; more temperate, more just, more sincere, more chaste, more peaceable.

PETER. Malignant atheist!

ALEXIS. Indeed, my father, were I malignant I must also be an atheist; for malignity is contrary to

---

2. Livornese: of Livorno (Leghorn), Italy, formerly a great banking center.

the command, and incompatible with the belief, of God.

PETER. Am I Czar of Muscovy, and hear discourses on reason and religion! from my own son too! No, by the Holy Trinity! thou art no son of mine. . . . If thou touchest my knee again, I crack thy knuckles with this tobacco-stopper: I wish it were a sledge-hammer for thy sake. Off, sycophant! Off, run-away slave!

ALEXIS. Father! father! my heart is broken! if I have offended, forgive me!

PETER. The state requires thy signal punishment.

ALEXIS. If the state requires it, be it so: but let my father's anger cease!

PETER. The world shall judge between us. I will brand thee with infamy.

ALEXIS. Until now, O father! I never had a proper sense of glory. Hear me, O Czar! let not a thing so vile as I am stand between you and the world! Let none accuse you!

PETER. Accuse me! rebel! accuse me! traitor!

ALEXIS. Let none speak ill of you, O my father! The public voice shakes the palace; the public voice penetrates the grave; the public voice precedes the chariot of Almighty God, and is heard at the judgment-seat.

PETER. Let it go to the devil! I will have none of it here in Petersburgh. Our church says nothing about it; our laws forbid it. As for thee, unnatural brute, I have no more to do with thee neither.

Ho there! Chancellor! What! come at last! wert napping, or counting thy ducats?

CHANCELLOR. Your Majesty's will and pleasure!

PETER. Is the senate assembled in that room?

CHANCELLOR. Every member, sire.

PETER. Conduct this youth with thee, and let them judge him: thou understandest me.

CHANCELLOR. Your Majesty's commands are the breath of our nostrils.

PETER. If these rascals are remiss, I will try my new cargo of Livonian[3] hemp upon them.

CHANCELLOR (returning.) Sire! sire!

PETER. Speak, fellow! Surely they have not condemned him to death, without giving themselves time to read the accusation, that thou comest back so quickly.

CHANCELLOR. No, sire! nor has either been done.

PETER. Then thy head quits thy shoulders.

CHANCELLOR. O sire!

PETER. Curse thy silly sires! what art thou about?

CHANCELLOR. Alas! he fell.

PETER. Tie him up to thy chair then. Cowardly beast! what made him fall?

CHANCELLOR. The hand of Death . . the name of father.

PETER. Thou puzzlest me; prythee speak plainlier.

CHANCELLOR. We told him that his crime was proven and manifest . . that his life was forfeited.

PETER. So far, well enough.

CHANCELLOR. He smiled . .

PETER. He did! did he! Impudence shall do him little good. Who could have expected it from that smock-face! Go on . . what then?

CHANCELLOR. He said calmly, but not without sighing twice or thrice, *Lead me to the scaffold: I am weary of life: nobody loves me.*

I condoled with him, and wept upon his hand, holding the paper against my bosom. He took the corner of it between his fingers, and said, *Read me this paper: read my death-warrant. Your silence and tears have signified it; yet the law has its forms. Do not keep me in suspense . . my father says, too truly, I am not courageous . . but the death that leads me to my God shall never terrify me.*

PETER. I have seen these white-livered knaves die resolutely: I have seen them quietly fierce like white ferrets, with their watery eyes and tiny teeth. You read it.

CHANCELLOR. In part, sire! When he heard your Majesty's name, accusing him of treason, and attempts at rebellion and parricide, he fell speechless. We raised him up: he was motionless; he was dead!

PETER. Inconsiderate and barbarous varlet as thou art, dost thou recite this ill accident to a father! and to one who has not dined! Bring me a glass of brandy.

CHANCELLOR. And it please your Majesty, might I call a . . a . .

PETER. Away, and bring it: scamper! All equally and alike shall obey and serve me.

Hearkye! bring the bottle with it: I must cool myself . . and . . hearkye! a rasher of bacon on thy life! and some pickled sturgeon, and good strong cheese.

---

3. **Livonian:** of Livonia, a Russian district near the Baltic.

(*1828*)

CHARLES LAMB

1775–1834

PERHAPS no writer of the time had a wider circle of literary friends. At his famous Wednesday (sometimes Thursday) evenings one might meet Hazlitt, Coleridge, Hunt, Haydon, as well as others whose names are now largely forgotten. In the midst Lamb himself, slightly tipsy, stuttered puns and opinions, "fine, piquant, deep, eloquent things in half a dozen half sentences," said Hazlitt. "His jests scald like tears: and he probes a question with a play upon words." But not until the age of forty-five, when he began to write his Elia essays for *The London Magazine*, did he adopt a form that permitted full literary use of the gifts displayed in conversation and marvelous letters to his friends. The personal essay had been created as a genre by Montaigne, who, in his last writings, offers an intimate, freely digressive inventory of his tastes, opinions, and traits. Something of the same familiar tone was caught by Lamb's seventeenth-century favorites, Robert Burton and Sir Thomas Browne. But in Lamb or Hazlitt the genre becomes characteristically Romantic. Where Montaigne studied himself to know man, the Romantic essayists evoke a feeling, a particular character, or a moment of experience, not so much for any intellectual purpose as for endowing their own lives with heightened vividness. That is, their encounters, pleasures, moods, and friendships are rescued from time and rendered with a depth and distinctness not usually present or realized in experience as it comes to us. Accordingly these essays are much more concrete and circumstantial than familiar essays had been in the past, and they are saturated with personal emotion, especially nostalgia. Indeed, the dominant theme is the comparison of past with present, the backward look to the self in youth or early manhood, often with regret or wonder at the strange changes that have come upon it. In "New Year's Eve," for example, the sentences that end, "From what have I not fallen, if the child I

567

remember was indeed myself," create this complex nostalgia. Here, as often, it comes perilously near sentimentality, though laced, as usual in Lamb, with unpredictable sidelights of humor. And Lamb's humor is of many kinds, from verbal wit to fantastic elaboration and sheer caprice.

The son of a lawyer's clerk, Lamb was the youngest of seven children, of whom only three lived past childhood. At the age of seven he entered Christ's Hospital, a free school founded in 1552 for children of the poor. He left at the age of fourteen and went to work, eventually finding a place in the accounting department of the East India House, a trading company, where he stayed until retirement in 1825. Most of his writing was accomplished during evenings and holidays; his works, as he liked to say, were some hundred folio ledgers of accounts in the India House. Except for a few tours and visits, he passed his life in London with his sister, to whose happiness he had dedicated himself with silent heroism. In 1796, during a fit of insanity, Mary Lamb had stabbed her mother to death, and rather than leave her in an asylum, Lamb had persuaded the court to release her to his care. In after years her attacks returned occasionally (though most of the time she was well), and she and her brother always lived under the constant threat of her affliction. The strain easily accounts for much in Lamb's behavior—the wild humor (see the account under Haydon, below, of the "immortal dinner"), the excessive drinking, the frequent depression.

His first literary productions were poems in a volume with Coleridge's. These were followed by more poems (in a joint publication with a friend Charles Lloyd) and a novel, *Rosamund Gray* (1798), in which Shelley found "much knowledge of the sweetest and deepest parts of our nature." Four years later a blank verse tragedy, *John Woodvil*, was published. It has fine speeches, but the Drury Lane Theater declined to produce it. Like much drama in the period, it imitates the Elizabethans but lacks their understanding of the theater. He next attempted a farce, *Mr. H.*, which, when produced (1806), was hissed off the stage. (Lamb joined in the hisses in order, he said, to conceal that he was the author.) Meanwhile, he was a constant reader of Shakespeare and other Elizabethan and Jacobean playwrights, and the fruits of this appeared in the *Tales from Shakespeare* (1807), written with his sister, and *Specimens of the English Dramatic Poets* (1808), an anthology of scenes from Shakespeare's contemporaries with brief criticisms. Lamb took particular pride in this work, which was, he said, "the first to draw the public attention to the old English Dramatists."

The Elia essays were inaugurated in *The London Magazine* in August, 1820. Lamb had, of course, already published many miscellaneous essays, such as "On the Tragedies of Shakespeare" (1811). For the most part these had exhibited an admirable discursive prose, but without the range and complexity of tones and styles that now emerged. Lamb's Elia style is often described as easy, intimate, slightly quaint or archaic in diction, relatively loose or spontaneous in sentence structure (a favorite punctuation is the dash). But usually the greater a literary figure, the less it is possible to generalize about his style. In the range of tones open to essayists in his time—for every age necessarily has limitations, which are also its opportunities—there was very little Lamb could not reach.

Bibliography

The standard edition of the collected *Works* is that by E. V. Lucas, 1903–05. One should also mention the edition of Alfred Ainger, 1883–88. For the most complete gathering of his *Letters* one again turns to E. V. Lucas, 3 vols., 1935. For biography and criticism see *The Life of Charles Lamb*, by E. V. Lucas, 1905, revised 1921. Edmund Blunden, in *Charles Lamb: His Life Recorded by His Contemporaries*, 1934, collects notices and memorials of Lamb and his works that appeared in his own lifetime or shortly after. Denys Thompson, "Our Debt to Lamb," in *Determinations*, ed. F. R. Leavis, 1934 (or in "The Essayist at Large," *Reading and Discrimination*, 1934), delivers a stimulating attack on Lamb. More sympathetic criticisms include those of Walter Pater, *Appreciations*, 1889, and Edmund Blunden, *Charles Lamb and His Contemporaries*, 1933. For study of his prose style see Jules Derocquigny, *Charles Lamb*, 1904.

THE OLD FAMILIAR FACES

Where are they gone, the old familiar faces?

I had a mother, but she died, and left me,
Died prematurely in a day of horrors—
All, all are gone, the old familiar faces.

I have had playmates, I have had companions,
In my days of childhood, in my joyful school-days—
All, all are gone, the old familiar faces.

I have been laughing, I have been carousing,
Drinking late, sitting late, with my bosom cronies—
All, all are gone, the old familiar faces.   10

I loved a love once, fairest among women.
Closed are her doors on me, I must not see her—
All, all are gone, the old familiar faces.

I have a friend, a kinder friend has no man.
Like an ingrate, I left my friend abruptly;
Left him, to muse on the old familiar faces.

Ghost-like, I paced round the haunts of my childhood.
Earth seemed a desert I was bound to traverse,
Seeking to find the old familiar faces.

Friend of my bosom, thou more than a brother!   20
Why wert not thou born in my father's dwelling?
So might we talk of the old familiar faces.

For some they have died, and some they have left me,
And some are taken from me; all are departed;
All, all are gone, the old familiar faces.

1798 (1798)

PARENTAL RECOLLECTIONS

A child's a plaything for an hour;
    Its pretty tricks we try
For that or for a longer space;
    Then tire, and lay it by.

But I know one, that to itself
    All seasons could control;
That would have mocked the sense of pain
    Out of a grieved soul.

Thou straggler into loving arms,
    Young climber up of knees,   10
When I forget thy thousand ways,
    Then life and all shall cease.

(1809)

THE OLD FAMILIAR FACES. **3. Died . . . horrors:** Lamb's sister, Mary, killed their mother in a fit of insanity. See Introduction, above. **11. love . . . women:** Ann Simmons. Cf. the "Alice" of Lamb's "Dream-Children," below. **14. friend:** possibly Charles Lloyd, a minor poet.

20. Friend . . . bosom: Coleridge. **24. taken from me:** Mary Lamb in the asylum.

WRITTEN AT CAMBRIDGE

I was not trained in Academic bowers,
And to those learned streams I nothing owe
Which copious from those twin fair founts do flow;
Mine have been anything but studious hours.
Yet I can fancy, wandering 'mid thy towers,
Myself a nursling, Granta, of thy lap;
My brow seems tightening with the Doctor's cap,
And I walk *gowned;* feel unusual powers.
Strange forms of logic clothe my admiring speech,
Old Ramus' ghost is busy at my brain;   10
And my skull teems with notions infinite.
Be still, ye reeds of Camus, while I teach
Truths, which transcend the searching Schoolmen's
    vein,
And half had staggered that stout Stagirite!

1819 (1819)

ON THE
TRAGEDIES OF SHAKSPEARE

CONSIDERED WITH REFERENCE TO THEIR
FITNESS FOR STAGE REPRESENTATION

Taking a turn the other day in the Abbey,[1] I was struck with the affected attitude of a figure, which I do not remember to have seen before, and which upon examination proved to be a whole-length of the celebrated Mr. Garrick.[2] Though I would not go so far with some good catholics abroad as to shut players altogether out of consecrated ground, yet I own I was not a little scandalized at the introduction of theatrical airs and gestures into a place set apart to remind us of the saddest realities. Going nearer, I found inscribed under this harlequin figure the following lines:—

To paint fair Nature, by divine command,
Her magic pencil in his glowing hand,
A Shakspeare rose; then, to expand his fame
Wide o'er this breathing world, a Garrick came.

Though sunk in death the forms the Poet drew,
The Actor's genius bade them breathe anew;
Though, like the bard himself, in night they lay,
Immortal Garrick call'd them back to day:
And till Eternity with pow'r sublime
Shall mark the mortal hour of hoary Time,
Shakspeare and Garrick like twin-stars shall shine,
And earth irradiate with a beam divine.

It would be an insult to my readers' understandings to attempt any thing like a criticism on this farrago of false thoughts and nonsense. But the reflection it led me into was a kind of wonder, how, from the days of the actor here celebrated to our own, it should have been the fashion to compliment every performer in his turn, that has had the luck to please the town in any of the great characters of Shakspeare, with the notion of possessing a *mind congenial with the poet's:* how people should come thus unaccountably to confound the power of originating poetical images and conceptions with the faculty of being able to read or recite the same when put into words;[3] or what connection that absolute mastery over the heart and soul of man, which a great dramatic poet possesses, has with those low tricks upon the eye and ear, which a player by observing a few general effects, which some common passion, as grief, anger, &c. usually has upon the gestures and exterior, can so easily compass. To know the internal workings and movements of a great mind, of an Othello or a Hamlet for instance, the *when* and the *why* and the *how far* they should be moved; to what pitch a passion is becoming; to give the reins and to pull in the curb exactly at the moment when the drawing in or the slackening is most graceful; seems to demand a reach of intellect of a vastly different extent from that which is employed upon the bare imitation of the signs of these passions in the countenance or gesture, which signs are usually observed to be most lively and emphatic in the weaker sort of minds, and which signs can after all but indicate some passion, as I said before, anger, or grief, generally;

WRITTEN AT CAMBRIDGE. **6. Granta:** Cambridge. **10. Ramus:** Petrus Ramus, 1515–72, was thought to have developed a non-Aristotelian form of logic. ON THE TRAGEDIES OF SHAKSPEARE. **1. Abbey:** Westminster Abbey, where many celebrated people are buried. **2. Mr. Garrick:** David Garrick, 1717–79, was the most famous actor of the eighteenth century.

3. words: "It is observable that we fall into this confusion only in *dramatic* recitations. We never dream that the gentleman who reads Lucretius in public with great applause, is therefore a great poet and philosopher; nor do we find that Tom Davies, the bookseller, who is recorded to have recited the Paradise Lost better than any man in England in his day (though I cannot help thinking there must be some mistake in this tradition) was therefore, by his intimate friends, set upon a level with Milton" (Lamb's note).

but of the motives and grounds of the passion, wherein it differs from the same passion in low and vulgar natures, of these the actor can give no more idea by his face or gesture than the eye (without a metaphor) can speak, or the muscles utter intelligible sounds. But such is the instantaneous nature of the impressions which we take in at the eye and ear at a playhouse, compared with the slow apprehension oftentimes of the understanding in reading, that we are apt not only to sink the play-writer in the consideration which we pay to the actor, but even to identify in our minds in a perverse manner, the actor with the character which he represents. It is difficult for a frequent playgoer to disembarrass the idea of Hamlet from the person and voice of Mr. K.[4] We speak of Lady Macbeth, while we are in reality thinking of Mrs. S.[5] Nor is this confusion incidental alone to unlettered persons, who, not possessing the advantage of reading, are necessarily dependent upon the stage-player for all the pleasure which they can receive from the drama, and to whom the very idea of *what an author is* cannot be made comprehensible without some pain and perplexity of mind: the error is one from which persons otherwise not meanly lettered, find it almost impossible to extricate themselves.

Never let me be so ungrateful as to forget the very high degree of satisfaction which I received some years back from seeing for the first time a tragedy of Shakspeare performed, in which those two great performers sustained the principal parts. It seemed to embody and realize conceptions which had hitherto assumed no distinct shape. But dearly do we pay all our life after for this juvenile pleasure, this sense of distinctness. When the novelty is past, we find to our cost that instead of realizing an idea, we have only materialized and brought down a fine vision to the standard of flesh and blood. We have let go a dream, in quest of an unattainable substance.

How cruelly this operates upon the mind, to have its free conceptions thus crampt and pressed down to the measure of a strait-lacing actuality, may be judged from that delightful sensation of freshness, with which we turn to those plays of Shakspeare which have escaped being performed, and to those passages in the acting plays of the same writer which have happily been left out in the performance. How far the very custom of hearing any thing *spouted*, withers and blows

upon a fine passage, may be seen in those speeches from Henry the Fifth, &c. which are current in the mouths of school-boys from their being to be found in *Enfield Speakers*, and such kind of books. I confess myself utterly unable to appreciate that celebrated soliloquy in Hamlet, beginning "To be or not to be," or to tell whether it be good, bad, or indifferent, it has been so handled and pawed about by declamatory boys and men, and torn so inhumanly from its living place and principle of continuity in the play, till it is become to me a perfect dead member.

It may seem a paradox, but I cannot help being of opinion that the plays of Shakspeare are less calculated for performance on a stage, than those of almost any other dramatist whatever. Their distinguishing excellence is a reason that they should be so. There is so much in them, which comes not under the province of acting, with which eye, and tone, and gesture, have nothing to do.

The glory of the scenic art is to personate passion, and the turns of passion; and the more coarse and palpable the passion is, the more hold upon the eyes and ears of the spectators the performer obviously possesses. For this reason, scolding scenes, scenes where two persons talk themselves into a fit of fury, and then in a surprising manner talk themselves out of it again, have always been the most popular upon our stage. And the reason is plain, because the spectators are here most palpably appealed to, they are the proper judges in this war of words, they are the legitimate ring that should be formed round such "intellectual prize-fighters." Talking is the direct object of the imitation here. But in all the best dramas, and in Shakspeare above all, how obvious it is, that the form of *speaking*, whether it be in soliloquy or dialogue, is only a medium, and often a highly artificial one, for putting the reader or spectator into possession of that knowledge of the inner structure and workings of mind in a character, which he could otherwise never have arrived at *in that form of composition* by any gift short of intuition. We do here as we do with novels written in the *epistolary form*. How many improprieties, perfect solecisms in letter-writing, do we put up with in Clarissa[6] and other books, for the sake of the delight which that form upon the whole gives us.

But the practice of stage representation reduces every thing to a controversy of elocution. Every

---

4. Mr. K.: John Philip Kemble, 1757-1823. **5. Mrs. S.:** Sarah Siddons, 1755-1831.

6. Clarissa: Samuel Richardson's epistolary novel, *Clarissa Harlowe*.

character, from the boisterous blasphemings of Bajazet[7] to the shrinking timidity of womanhood, must play the orator. The love-dialogues of Romeo and Juliet, those silver-sweet sounds of lovers' tongues by night; the more intimate and sacred sweetness of nuptial colloquy between an Othello or a Posthumus with their married wives, all those delicacies which are so delightful in the reading, as when we read of those youthful dalliances in Paradise—

> ———————— As beseem'd
> Fair couple link'd in happy nuptial league,
> Alone:[8]

by the inherent fault of stage representation, how are these things sullied and turned from their very nature by being exposed to a large assembly; when such speeches as Imogen addresses to her lord,[9] come drawling out of the mouth of a hired actress, whose courtship, though nominally addressed to the personated Posthumus, is manifestly aimed at the spectators, who are to judge of her endearments and her returns of love.

The character of Hamlet is perhaps that by which, since the days of Betterton,[10] a succession of popular performers have had the greatest ambition to distinguish themselves. The length of the part may be one of their reasons. But for the character itself, we find it in a play, and therefore we judge it a fit subject of dramatic representation. The play itself abounds in maxims and reflexions beyond any other, and therefore we consider it as a proper vehicle for conveying moral instruction. But Hamlet himself—what does he suffer meanwhile by being dragged forth as the public schoolmaster, to give lectures to the crowd! Why, nine parts in ten of what Hamlet does, are transactions between himself and his moral sense, they are the effusions of his solitary musings, which he retires to holes and corners and the most sequestered parts of the palace to pour forth; or rather, they are the silent meditations with which his bosom is bursting, reduced to *words* for the sake of the reader, who must else remain ignorant of what is passing there. These profound sorrows, these light-and-noise-abhorring ruminations, which the tongue scarce dares utter to deaf walls and chambers, how can they be represented by a gesticulating actor, who comes and mouths them out before an audience, making four hundred people

his confidants at once. I say not that it is the fault of the actor so to do; he must pronounce them *ore rotundo*,[11] he must accompany them with his eye, he must insinuate them into his auditory by some trick of eye, tone, or gesture, or he fails. *He must be thinking all the while of his appearance, because he knows that all the while the spectators are judging of it.* And this is the way to represent the shy, negligent, retiring Hamlet.

It is true that there is no other mode of conveying a vast quantity of thought and feeling to a great portion of the audience, who otherwise would never earn it for themselves by reading, and the intellectual acquisition gained this way may, for aught I know, be inestimable; but I am not arguing that Hamlet should not be acted, but how much Hamlet is made another thing by being acted. I have heard much of the wonders which Garrick performed in this part; but as I never saw him, I must have leave to doubt whether the representation of such a character came within the province of his art. Those who tell me of him, speak of his eye, of the magic of his eye, and of his commanding voice: physical properties, vastly desirable in an actor, and without which he can never insinuate meaning into an auditory,—but what have they to do with Hamlet? what have they to do with intellect? In fact, the things aimed at in theatrical representation, are to arrest the spectator's eye upon the form and the gesture, and so to gain a more favourable hearing to what is spoken: it is not what the character is, but how he looks; not what he says, but how he speaks it. I see no reason to think that if the play of Hamlet were written over again by some such writer as Banks or Lillo,[12] retaining the process of the story, but totally omitting all the poetry of it, all the divine features of Shakspeare, his stupendous intellect; and only taking care to give us enough of passionate dialogue, which Banks or Lillo were never at a loss to furnish; I see not how the effect could be much different upon an audience, nor how the actor has it in his power to represent Shakspeare to us differently from his representation of Banks or Lillo. Hamlet would still be a youthful accomplished prince, and must be gracefully personated; he might be puzzled in his mind, wavering in his conduct, seemingly-cruel to Ophelia, he might see a ghost, and start at it, and address it kindly when he found it to be his father; all this in the poorest and most homely

---

7. **Bajazet:** in Marlowe's *Tamburlaine.* 8. **As . . . Alone:** *Paradise Lost,* IV. 338–40. 9. **Imogen . . . lord:** *Cymbeline,* I. i. 10. **Betterton:** Thomas Betterton, 1635?–1710.

11. **ore rotundo:** with round mouth, loud and clear. 12. **Banks or Lillo:** John Banks, fl. 1696, or George Lillo, 1693?–1739, hack dramatists.

language of the servilest creeper after nature that ever consulted the palate of an audience; without troubling Shakspeare for the matter: and I see not but there would be room for all the power which an actor has, to display itself. All the passions and changes of passion might remain: for those are much less difficult to write or act than is thought, it is a trick easy to be attained, it is but rising or falling a note or two in the voice, a whisper with a significant forboding look to announce its approach, and so contagious the counterfeit appearance of any emotion is, that let the words be what they will, the look and tone shall carry it off and make it pass for deep skill in the passions.

It is common for people to talk of Shakspeare's plays being *so natural;* that every body can understand him. They are natural indeed, they are grounded deep in nature, so deep that the depth of them lies out of the reach of most of us. You shall hear the same persons say that George Barnwell[13] is very natural, and Othello is very natural, that they are both very deep; and to them they are the same kind of thing. At the one they sit and shed tears, because a good sort of young man is tempted by a naughty woman to commit *a trifling peccadillo,* the murder of an uncle or so,[14] that is all, and so comes to an untimely end, which is *so moving;* and at the other, because a blackamoor in a fit of jealousy kills his innocent white wife: and the odds are that ninety-nine out of a hundred would willingly behold the same catastrophe happen to both the heroes, and have thought the rope more due to Othello than to Barnwell. For of the texture of Othello's mind, the inward construction marvellously laid open with all its

strengths and weaknesses, its heroic confidences and its human misgivings, its agonies of hate springing from the depths of love, they see no more than the spectators at a cheaper rate, who pay their pennies a-piece to look through the man's telescope in Leicester-fields, see into the inward plot and topography of the moon. Some dim thing or other they see, they see an actor personating a passion, of grief, or anger, for instance, and they recognize it as a copy of the usual external effects of such passions; or at least as being true to *that symbol of the emotion which passes current at the theatre for it,* for it is often no more than that: but of the grounds of the passion, its correspondence to a great or heroic nature, which is the only worthy object of tragedy,—that common auditors know any thing of this, or can have any such notions dinned into them by the mere strength of an actor's lungs,—that apprehensions foreign to them should be thus infused into them by storm, I can neither believe, nor understand how it can be possible.

We talk of Shakspeare's admirable observation of life, when we should feel, that not from a petty inquisition into those cheap and every-day characters which surrounded him, as they surround us, but from his own mind, which was, to borrow a phrase of Ben Jonson's, the very "sphere of humanity,"[15] he fetched those images of virtue and of knowledge, of which every one of us recognizing a part, think we comprehend in our natures the whole; and oftentimes mistake the powers which he positively creates in us, for nothing more than indigenous faculties of our own minds, which only waited the application of corresponding virtues in him to return a full and clear echo of the same.

To return to Hamlet.—Among the distinguishing features of that wonderful character, one of the most interesting (yet painful) is that soreness of mind which makes him treat the intrusions of Polonius with harshness, and that asperity which he puts on in his interviews with Ophelia. These tokens of an unhinged mind (if they be not mixed in the latter case with a profound artifice of love, to alienate Ophelia by affected discourtesies, so to prepare her mind for the breaking off of that loving intercourse, which can no longer find a place amidst business so serious as that which he has to do) are parts of his character, which to reconcile

13. **George Barnwell:** in Lillo's *The London Merchant* (1731). In the play a courtesan, Millwood, seduces a young apprentice, Barnwell, and leads him to rob his master and kill his uncle, for which he and Millwood are executed.

14. **uncle or so:** "If this note could hope to meet the eye of any of the Managers, I would intreat and beg of them, in the name of both the Galleries, that this insult upon the morality of the common people of London should cease to be eternally repeated in the holiday weeks. Why are the 'Prentices of this famous and well-governed city, instead of an amusement, to be treated over and over again with a nauseous sermon of George Barnwell? Why *at the end of their vistoes* are we to place the *gallows?* Were I an uncle, I should not much like a nephew of mine to have such an example placed before his eyes. It is really making uncle-murder too trivial to exhibit it as done upon such slight motives;—it is attributing too much to such characters as Millwood;—it is putting things into the heads of good young men, which they would never otherwise have dreamed of. Uncles that think any thing of their lives, should fairly petition the Chamberlain against it" (L.).

15. **"sphere of humanity":** See Jonson's "To the Immortal Memory, and Friendship of that Noble Pair, Sir Lucius Cary and Sir H. Morison," l. 53.

with our admiration of Hamlet, the most patient consideration of his situation is no more than necessary; they are what we *forgive afterwards*, and explain by the whole of his character, but *at the time* they are harsh and unpleasant. Yet such is the actor's necessity of giving strong blows to the audience, that I have never seen a player in this character, who did not exaggerate and strain to the utmost these ambiguous features,— these temporary deformities in the character. They make him express a vulgar scorn at Polonius which utterly degrades his gentility, and which no explanation can render palateable; they make him shew contempt, and curl up the nose at Ophelia's father,— contempt in its very grossest and most hateful form; but they get applause by it: it is natural, people say; that is, the words are scornful, and the actor expresses scorn, and that they can judge of: but why so much scorn, and of that sort, they never think of asking.

So to Ophelia.—All the Hamlets that I have ever seen, rant and rave at her as if she had committed some great crime, and the audience are highly pleased, because the words of the part are satirical, and they are enforced by the strongest expression of satirical indignation of which the face and voice are capable. But then, whether Hamlet is likely to have put on such brutal appearances to a lady whom he loved so dearly, is never thought on. The truth is, that in all such deep affections as had subsisted between Hamlet and Ophelia, there is a stock of *supererogatory love*, (if I may venture to use the expression) which in any great grief of heart, especially where that which preys upon the mind cannot be communicated, confers a kind of indulgence upon the grieved party to express itself, even to its heart's dearest object, in the language of a temporary alienation; but it is not alienation, it is a distraction purely, and so it always makes itself to be felt by that object: it is not anger, but grief assuming the appearance of anger,—love awkwardly counterfeiting hate, as sweet countenances when they try to frown: but such sternness and fierce disgust as Hamlet is made to shew, is no counterfeit, but the real face of absolute aversion,— of irreconcileable alienation. It may be said he puts on the madman; but then he should only so far put on this counterfeit lunacy as his own real distraction will give him leave; that is, incompletely, imperfectly; not in that confirmed, practised way, like a master of his art, or as Dame Quickly would say, "like one of those harlotry players."[16]

16. Dame . . . players: 1 *Henry IV*, II. iv. 437.

I mean no disrespect to any actor, but the sort of pleasure which Shakspeare's plays give in the acting seems to me not at all to differ from that which the audience receive from those of other writers; and, *they being in themselves essentially so different from all others*, I must conclude that there is something in the nature of acting which levels all distinctions. And in fact, who does not speak indifferently of the Gamester[17] and of Macbeth as fine stage performances, and praise the Mrs. Beverley in the same way as the Lady Macbeth of Mrs. S.? Belvidera, and Calista, and Isabella, and Euphrasia,[18] are they less liked than Imogen, or than Juliet, or than Desdemona? Are they not spoken of and remembered in the same way? Is not the female performer as great (as they call it) in one as in the other? Did not Garrick shine, and was he not ambitious of shining in every drawling tragedy that his wretched day produced,—the productions of the Hills and the Murphys and the Browns,[19]—and shall he have that honour to dwell in our minds for ever as an inseparable concomitant with Shakspeare? A kindred mind! O who can read that affecting sonnet of Shakspeare which alludes to his profession as a player:—

Oh for my sake do you with Fortune chide,
The guilty goddess of my harmless deeds,
That did not better for my life provide
Than public means which public custom breeds—
Thence comes it that my name receives a brand;
And almost thence my nature is subdued
To what it works in, like the dyer's hand——[20]

Or that other confession:—

Alas! 'tis true, I have gone here and there,
And made myself a motly to thy view,
Gor'd mine own thoughts, sold cheap what is most dear—[21]

Who can read these instances of jealous self-watchfulness in our sweet Shakspeare, and dream of any congeniality between him and one that, by every tradition of him, appears to have been as mere a player as ever existed; to have had his mind tainted with the lowest players'

17. **the Gamester:** by Edward Moore, 1712–57. **18. Belvidera . . . Euphrasia:** Belvidera, in *Venice Preserved* (1682), by Thomas Otway; Calista, in *The Fair Penitent* (1703), by Nicholas Rowe; Isabella, in *The Fatal Marriage* (1694), by Thomas Southerne; Euphrasia, in *The Grecian Daughter* (1772), by Arthur Murphy. **19. Hills . . . Browns:** Dr. John Hill, 1716?–1775; Arthur Murphy, 1727–1805; Rev. John Brown, 1715–66. **20. Oh . . . hand:** Sonnet CXI. **21. Alas! . . . dear:** Sonnet CX.

vices,—envy and jealousy, and miserable cravings after applause; one who in the exercise of his profession was jealous even of the women-performers that stood in his way; a manager full of managerial tricks and stratagems and finesse: that any resemblance should be dreamed of between him and Shakspeare,—Shakspeare who, in the plenitude and consciousness of his own powers, could with that noble modesty, which we can neither imitate nor appreciate, express himself thus of his own sense of his own defects:—

> Wishing me like to one more rich in hope,
> Featur'd like him, like him with friends possest;
> Desiring *this man's art, and that man's scope.*[22]

I am almost disposed to deny to Garrick the merit of being an admirer of Shakspeare. A true lover of his excellencies he certainly was not; for would any true lover of them have admitted into his matchless scenes such ribald trash as Tate and Cibber,[23] and the rest of them, that

> With their darkness durst affront his light,[24]

have foisted into the acting plays of Shakspeare? I believe it impossible that he could have had a proper reverence for Shakspeare, and have condescended to go through that interpolated scene[25] in Richard the Third, in which Richard tries to break his wife's heart by telling her he loves another woman, and says, "if she survives this she is immortal." Yet I doubt not he delivered this vulgar stuff with as much anxiety of emphasis as any of the genuine parts: and for acting, it is as well calculated as any. But we have seen the part of Richard lately produce great fame to an actor by his manner of playing it, and it lets us into the secret of acting, and of popular judgments of Shakspeare derived from acting. Not one of the spectators who have witnessed Mr. C.'s[26] exertions in that part, but has come away with a proper conviction that Richard is a very wicked man, and kills little children in their beds, with something like the pleasure which the giants and ogres in children's books are represented to have taken in that practice; moreover, that he is very

close and shrewd and devilish cunning, for you could see that by his eye.

But is in fact this the impression we have in reading the Richard of Shakspeare? Do we feel any thing like disgust, as we do at that butcher-like representation of him that passes for him on the stage? A horror at his crimes blends with the effect which we feel, but how is it qualified, how is it carried off, by the rich intellect which he displays, his resources, his wit, his buoyant spirits, his vast knowledge and insight into characters, the poetry of his part,—not an atom of all which is made perceivable in Mr. C.'s way of acting it. Nothing but his crimes, his actions, is visible; they are prominent and staring; the murderer stands out, where is the lofty genius, the man of vast capacity,—the profound, the witty, accomplished Richard?

The truth is, the Characters of Shakspeare are so much the objects of meditation rather than of interest or curiosity as to their actions, that while we are reading any of his great criminal characters,—Macbeth, Richard, even Iago,—we think not so much of the crimes which they commit, as of the ambition, the aspiring spirit, the intellectual activity, which prompts them to over-leap those moral fences. Barnwell is a wretched murderer; there is a certain fitness between his neck and the rope; he is the legitimate heir to the gallows; nobody who thinks at all can think of any alleviating circumstances in his case to make him a fit object of mercy. Or to take an instance from the higher tragedy, what else but a mere assassin is Glenalvon![27] Do we think of any thing but of the crime which he commits, and the rack which he deserves? That is all which we really think about him. Whereas in corresponding characters in Shakspeare so little do the actions comparatively affect us, that while the impulses, the inner mind in all its perverted greatness, solely seems real and is exclusively attended to, the crime is comparatively nothing. But when we see these things represented, the acts which they do are comparatively every thing, their impulses nothing. The state of sublime emotion into which we are elevated by those images of night and horror which Macbeth is made to utter, that solemn prelude with which he entertains the time till the bell shall strike which is to call him to murder Duncan,—when we no longer read it in a book, when we have given up that vantage-ground of abstraction which reading possesses over seeing, and come to see a man in his bodily shape before our eyes

---

22. **Wishing . . . scope:** Sonnet XXIX. **23. Tate and Cibber:** Nahum Tate, 1652–1715; Colley Cibber, 1671–1757. Both men revised Shakespeare's plays, Tate going so far as to give *Lear* a happy ending. **24. With . . . light:** *Paradise Lost*, I. 391. **25. interpolated scene:** by Cibber. **26. Mr. C.:** George Frederick Cooke, 1756–1811.

27. **Glenalvon:** in *Douglas* by John Home, 1722–1808.

actually preparing to commit a murder, if the acting be true and impressive, as I have witnessed it in Mr. K.'s performance of that part, the painful anxiety about the act, the natural longing to prevent it while it yet seems unperpetrated, the too close pressing semblance of reality, give a pain and an uneasiness which totally destroy all the delight which the words in the book convey, where the deed doing never presses upon us with the painful sense of presence: it rather seems to belong to history,—to something past and inevitable, if it has any thing to do with time at all. The sublime images, the poetry alone, is that which is present to our minds in the reading.

So to see Lear acted,—to see an old man tottering about the stage with a walking-stick, turned out of doors by his daughters in a rainy night, has nothing in it but what is painful and disgusting. We want to take him into shelter and relieve him. That is all the feeling which the acting of Lear ever produced in me. But the Lear of Shakspeare cannot be acted. The contemptible machinery by which they mimic the storm which he goes out in, is not more inadequate to represent the horrors of the real elements, than any actor can be to represent Lear: they might more easily propose to personate the Satan of Milton upon a stage, or one of Michael Angelo's terrible figures. The greatness of Lear is not in corporal dimension, but in intellectual: the explosions of his passion are terrible as a volcano: they are storms turning up and disclosing to the bottom that sea, his mind, with all its vast riches. It is his mind which is laid bare. This case of flesh and blood seems too insignificant to be thought on; even as he himself neglects it. On the stage we see nothing but corporal infirmities and weakness, the impotence of rage; while we read it, we see not Lear, but we are Lear,—we are in his mind, we are sustained by a grandeur which baffles the malice of daughters and storms; in the aberrations of his reason, we discover a mighty irregular power of reasoning, immethodized from the ordinary purposes of life, but exerting its powers, as the wind blows where it listeth, at will upon the corruptions and abuses of mankind. What have looks, or tones, to do with that sublime identification of his age with that of the *heavens themselves*, when in his reproaches to them for conniving at the injustice of his children, he reminds them that "they themselves are old." What gesture shall we appropriate to this? What has the voice or the eye to do with such things? But the play is beyond all art, as the tamperings with it shew: it is too hard and stony; it must have love-scenes,

and a happy ending. It is not enough that Cordelia is a daughter, she must shine as a lover too. Tate has put his hook in the nostrils of this Leviathan, for Garrick and his followers, the showmen of the scene, to draw the mighty beast about more easily. A happy ending!—as if the living martyrdom that Lear had gone through,—the flaying of his feelings alive, did not make a fair dismissal from the stage of life the only decorous thing for him. If he is to live and be happy after, if he could sustain this world's burden after, why all this pudder and preparation,—why torment us with all this unnecessary sympathy? As if the childish pleasure of getting his gilt robes and sceptre again could tempt him to act over again his misused station,—as if at his years, and with his experience, any thing was left but to die.

Lear is essentially impossible to be represented on a stage. But how many dramatic personages are there in Shakspeare, which though more tractable and feasible (if I may so speak) than Lear, yet from some circumstance, some adjunct to their character, are improper to be shewn to our bodily eye. Othello for instance. Nothing can be more soothing, more flattering to the nobler parts of our natures, than to read of a young Venetian lady of highest extraction, through the force of love and from a sense of merit in him whom she loved, laying aside every consideration of kindred, and country, and colour, and wedding with a *coal-black Moor*—(for such he is represented, in the imperfect state of knowledge respecting foreign countries in those days, compared with our own, or in compliance with popular notions, though the Moors are now well enough known to be by many shades less unworthy of a white woman's fancy)—it is the perfect triumph of virtue over accidents, of the imagination over the senses. She sees Othello's colour in his mind. But upon the stage, when the imagination is no longer the ruling faculty, but we are left to our poor unassisted senses, I appeal to every one that has seen Othello played, whether he did not, on the contrary, sink Othello's mind in his colour; whether he did not find something extremely revolting in the courtship and wedded caresses of Othello and Desdemona; and whether the actual sight of the thing did not over-weigh all that beautiful compromise which we make in reading;—and the reason it should do so is obvious, because there is just so much reality presented to our senses as to give a perception of disagreement, with not enough of belief in the internal motives,—all that which is unseen,—to overpower and reconcile the first and obvious prejudices.[28] What we see upon a stage is body and bodily action; what we are conscious of in reading is almost exclusively the mind, and its movements: and this I think may sufficiently account for the very different sort of delight with which the same play so often affects us in the reading and the seeing.

It requires little reflection to perceive, that if those characters in Shakspeare which are within the precincts of nature, have yet something in them which appeals too exclusively to the imagination, to admit of their being made objects to the senses without suffering a change and a diminution,—that still stronger the objection must lie against representing another line of characters, which Shakspeare has introduced to give a wildness and a supernatural elevation to his scenes, as if to remove them still farther from that assimilation to common life in which their excellence is vulgarly supposed to consist. When we read the incantations of those terrible beings the Witches in Macbeth, though some of the ingredients of their hellish composition savour of the grotesque, yet is the effect upon us other than the most serious and appalling that can be imagined? Do we not feel spell-bound as Macbeth was? Can any mirth accompany a sense of their presence? We might as well laugh under a consciousness of the principle of Evil himself being truly and really present with us. But attempt to bring these beings on to a stage, and you turn them instantly into so many old women, that men and children are to laugh at. Contrary to the old saying, that "seeing is believing," the sight actually destroys the faith: and the mirth in which we indulge at their expense, when we see these creatures upon a stage, seems to be a sort of indemnification which we make to ourselves for the terror which they put us in when reading made them an object of belief,—when we surrendered up our reason to the poet, as children to their nurses and their elders; and we laugh at our fears, as children who thought they saw something in the dark, triumph when the bringing in of a candle discovers the vanity of their fears. For this exposure of supernatural agents upon a stage is truly bringing in a candle to expose their own delusiveness. It is the solitary taper and the book that generates a faith in these terrors: a ghost by chandelier light, and in good company, deceives no spectators,—a ghost that can be measured by the eye, and his human dimensions made out at leisure. The sight of a well-lighted house, and a well-dressed audience, shall arm the most nervous child against any apprehensions: as Tom Brown[29] says of the impenetrable skin of Achilles with his impenetrable armour over it, "Bully Dawson would have fought the devil with such advantages."

Much has been said, and deservedly, in reprobation of the vile mixture which Dryden has thrown into the Tempest:[30] doubtless without some such vicious alloy, the impure ears of that age would never have sate out to hear so much innocence of love as is contained in the sweet courtship of Ferdinand and Miranda. But is the Tempest of Shakspeare at all a subject for stage representation? It is one thing to read of an enchanter, and to believe the wondrous tale while we are reading it; but to have a conjuror brought before us in his conjuring-gown, with his spirits about him, which none but himself and some hundred of favoured spectators before the curtain are supposed to see, involves such a quantity of the *hateful incredible*, that all our reverence for the author cannot hinder us from perceiving such gross attempts upon the senses to be in the highest degree childish and inefficient. Spirits and fairies cannot be represented, they cannot even be painted,—they can only be believed. But the elaborate and anxious provision of scenery, which the luxury of the age demands, in these cases works a quite contrary effect to what is intended. That which in comedy, or plays of familiar life, adds so much to the life of the imitation, in plays which appeal to the higher faculties, positively destroys the illusion which it is introduced to aid. A parlour or a drawing-room,—a library opening into a garden,—a garden with an alcove in it,—a street, or the piazza of Covent-garden, does well enough in a scene; we are content to give as much credit to it as it demands; or rather, we think little about it,—it is little more than reading at the top

28. **prejudices:** "The error of supposing that because Othello's colour does not offend us in the reading, it should also not offend us in the seeing, is just such a fallacy as supposing that an Adam and Eve in a picture shall affect us just as they do in the poem. But in the poem we for a while have Paradisaical senses given us, which vanish when we see a man and his wife without clothes in the picture. The painters themselves feel this, as is apparent by the aukward shifts they have recourse to, to make them look not quite naked; by a sort of prophetic anachronism, antedating the invention of fig-leaves. So in the reading of the play, we see with Desdemona's eyes; in the seeing of it, we are forced to look with our own" (L.).

29. **Tom Brown:** Thomas Brown, 1663–1704. Lamb cites his "Observations on Virgil, Ovid and Homer." **30. Dryden . . . Tempest:** Lamb refers to the revision of Shakespeare's *Tempest* by Dryden and Sir William Davenant in 1667.

of a page, "Scene, a Garden;" we do not imagine ourselves there, but we readily admit the imitation of familiar objects. But to think by the help of painted trees and caverns, which we know to be painted, to transport our minds to Prospero, and his island and his lonely cell;[31] or by the aid of a fiddle dexterously thrown in, in an interval of speaking, to make us believe that we hear those supernatural noises of which the isle was full:—the Orrery Lecturer at the Haymarket[32] might as well hope, by his musical glasses cleverly stationed out of sight behind his apparatus, to make us believe that we do indeed hear the chrystal spheres ring out that chime, which if it were to inwrap our fancy long, Milton thinks,

Time would run back and fetch the age of gold,
And speckled vanity
Would sicken soon and die,
And leprous Sin would melt from earthly mould;
Yea Hell itself would pass away,
And leave its dolorous mansions to the peering day.[33]

The Garden of Eden, with our first parents in it, is not more impossible to be shewn on a stage, than the Enchanted Isle, with its no less interesting and innocent first settlers.

The subject of Scenery is closely connected with that of the Dresses, which are so anxiously attended to on our stage. I remember the last time I saw Macbeth played, the discrepancy I felt at the changes of garment which he varied,—the shiftings and re-shiftings, like a Romish priest at mass. The luxury of stage-improvements, and the importunity of the public eye, require this. The coronation robe of the Scottish monarch was fairly a counterpart to that which our King wears when he goes to the Parliament-house,—just so full and cumbersome, and set out with ermine and pearls. And if things must be represented, I see not what to find fault with in this. But in reading, what robe are we conscious of? Some dim images of royalty—a crown and sceptre, may float before our eyes, but who shall describe the fashion of it? Do we see in our mind's eye what Webb or any other robe-maker could

pattern? This is the inevitable consequence of imitating every thing, to make all things natural. Whereas the reading of a tragedy is a fine abstraction. It presents to the fancy just so much of external appearances as to make us feel that we are among flesh and blood, while by far the greater and better part of our imagination is employed upon the thoughts and internal machinery of the character. But in acting, scenery, dress, the most contemptible things, call upon us to judge of their naturalness.

Perhaps it would be no bad similitude, to liken the pleasure which we take in seeing one of these fine plays acted, compared with that quiet delight which we find in the reading of it, to the different feelings with which a reviewer, and a man that is not a reviewer, reads a fine poem. The accursed critical habit,—the being called upon to judge and pronounce, must make it quite a different thing to the former. In seeing these plays acted, we are affected just as judges. When Hamlet compares the two pictures of Gertrude's first and second husband, who wants to see the pictures? But in the acting, a miniature must be lugged out; which we know not to be the picture, but only to shew how finely a miniature may be represented. This shewing of every thing, levels all things: it makes tricks, bows, and curtesies, of importance. Mrs. S. never got more fame by any thing than by the manner in which she dismisses the guests in the banquet-scene in Macbeth: it is as much remembered as any of her thrilling tones or impressive looks. But does such a trifle as this enter into the imaginations of the readers of that wild and wonderful scene? Does not the mind dismiss the feasters as rapidly as it can? Does it care about the gracefulness of the doing it? But by acting, and judging of acting, all these non-essentials are raised into an importance, injurious to the main interest of the play.

I have confined my observations to the tragic parts of Shakspeare. It would be no very difficult task to extend the enquiry to his comedies; and to shew why Falstaff, Shallow, Sir Hugh Evans, and the rest, are equally incompatible with stage representation. The length to which this Essay has run, will make it, I am afraid, sufficiently distasteful to the Amateurs of the Theatre, without going any deeper into the subject at present.

(1811)

31. cell: "It will be said these things are done in pictures. But pictures and scenes are very different things. Painting is a world of itself, but in scene-painting there is the attempt to deceive; and there is the discordancy, never to be got over, between painted scenes and real people" (L.). 32. Orrery . . . Haymarket: Lectures on astronomy were delivered at the Theatre Royal, Haymarket. 33. Time . . . day: from "On the Morning of Christ's Nativity," ll. 135–40.

CHRIST'S HOSPITAL FIVE AND THIRTY YEARS AGO

In Mr. Lamb's "Works," published a year or two since, I find a magnificent eulogy on my old school,[1] such as it was, or now appears to him to have been, between the years 1782 and 1789. It happens, very oddly, that my own standing at Christ's was nearly corresponding with his; and, with all gratitude to him for his enthusiasm for the cloisters, I think he has contrived to bring together whatever can be said in praise of them, dropping all the other side of the argument most ingeniously.

I remember L. at school; and can well recollect that he had some peculiar advantages, which I and others of his school-fellows had not. His friends lived in town, and were near at hand; and he had the privilege of going to see them, almost as often as he wished, through some invidious distinction, which was denied to us. The present worthy sub-treasurer[2] to the Inner Temple can explain how that happened. He had his tea and hot rolls in a morning, while we were battening upon our quarter of a penny loaf—our *crug*[3]—moistened with attenuated small beer, in wooden piggins,[4] smacking of the pitched leathern jack it was poured from. Our Monday's milk porritch, blue and tasteless, and the pease soup of Saturday, coarse and choking, were enriched for him with a slice of "extraordinary bread and butter," from the hot-loaf of the Temple. The Wednesday's mess of millet, somewhat less repugnant (we had three banyan[5] to four meat days in the week), was endeared to his palate with a lump of double-refined,[6] and a smack of ginger (to make it go down the more glibly) or the fragrant cinnamon. In lieu of our *half-pickled* Sundays, or *quite fresh* boiled beef on Thursdays (strong as *caro equina*[7]), with detestable marigolds floating in the pail to poison the broth—our scanty mutton crags[8] on Fridays—and rather more savory, but grudging, portions of the same flesh, rotten-roasted[9] or rare, on the Tuesdays (the only dish which excited our appetites, and disappointed our stomachs, in almost equal proportion)—he had his hot plate of roast veal, or the more tempting griskin[10]

CHRIST'S HOSPITAL FIVE AND THIRTY YEARS AGO. **1. eulogy . . . school:** "Recollections of Christ's Hospital." **2. sub-treasurer:** Randal Norris, a lifelong friend. **3. crug:** slang for bread. **4. piggins:** small pails. **5. banyan:** meatless. **6. double-refined:** sugar. **7. caro equina:** horse flesh. **8. crags:** necks. **9. rotten-roasted:** over-roasted. **10. griskin:** pork chop.

(exotics unknown to our palates), cooked in the paternal kitchen (a great thing), and brought him daily by his maid or aunt! I remember the good old relative (in whom love forbade pride) squatting down upon some odd stone in a by-nook of the cloisters, disclosing the viands (of higher regale than those cates which the ravens ministered to the Tishbite[11]), and the contending passions of L. at the unfolding. There was love for the bringer; shame for the thing brought, and the manner of its bringing; sympathy for those who were too many to share in it; and, at top of all, hunger (eldest, strongest of the passions!) predominant, breaking down the stony fences of shame, and awkwardness, and a troubling over-consciousness.

I was a poor friendless boy. My parents and those who should care for me, were far away. Those few acquaintances of theirs, which they could reckon upon being kind to me in the great city, after a little forced notice, which they had the grace to take of me on my first arrival in town, soon grew tired of my holiday visits. They seemed to them to recur too often, though I thought them few enough; and one after another, they all failed me, and I felt myself alone among six hundred playmates.

O the cruelty of separating a poor lad from his early homestead! The yearnings which I used to have towards it in those unfledged years! How, in my dreams, would my native town (far in the west) come back, with its church, and trees, and faces! How I would wake weeping, and in the anguish of my heart exclaim upon sweet Calne in Wiltshire!

To this late hour of my life, I trace impressions left by the recollection of those friendless holidays. The long warm days of summer never return but they bring with them a gloom from the haunting memory of those *whole-day-leaves*, when, by some strange arrangement, we were turned out, for the live-long day, upon our own hands, whether we had friends to go to, or none. I remember those bathing-excursions to the New-River, which L. recalls with such relish, better, I think, than he can—for he was a home-seeking lad, and did not much care for such water-pastimes. How merrily we would sally forth into the fields; and strip under the first warmth of the sun; and wanton like young dace in the streams; getting us appetites for noon, which those of us that were pennyless (our scanty morning crust long since exhausted) had not the means

---

11. Tishbite: See I Kings 17:1–6; *Paradise Regained*, II. 266–70.

of allaying—while the cattle, and the birds, and the fishes, were at feed about us, and we had nothing to satisfy our cravings—the very beauty of the day, and the exercise of the pastime, and the sense of liberty, setting a keener edge upon them!—How faint and languid, finally, we would return, towards nightfall, to our desired morsel, half-rejoicing, half-reluctant, that the hours of our uneasy liberty had expired!

It was worse in the days of winter, to go prowling about the streets objectless—shivering at cold windows of print-shops, to extract a little amusement; or haply, as a last resort, in the hope of a little novelty, to pay a fifty-times repeated visit (where our individual faces should be as well known to the warden as those of his own charges) to the Lions in the Tower—to whose levée by courtesy immemorial, we had a prescriptive title to admission.

L.'s governor (so we called the patron who presented us to the foundation) lived in a manner under his paternal roof. Any complaint which he had to make was sure of being attended to. This was understood at Christ's, and was an effectual screen to him against the severity of masters, or worse tyranny of the monitors. The oppressions of these young brutes are heart-sickening to call to recollection. I have been called out of my bed, and *waked for the purpose*, in the coldest winter nights—and this not once, but night after night—in my shirt, to receive the discipline of a leathern thong, with eleven other sufferers, because it pleased my callow overseer, when there has been any talking heard after we were gone to bed, to make the six last beds in the dormitory, where the youngest children of us slept, answerable for an offense they neither dared to commit, nor had the power to hinder. The same execrable tyranny drove the younger part of us from the fires, when our feet were perishing with snow; and, under the cruelest penalties, forbad the indulgence of a drink of water, when we lay in sleepless summer nights, fevered with the season, and the day's sports.

There was one H——, who, I learned, in after days was seen expiating some maturer offense in the hulks.[12] (Do I flatter myself in fancying that this might be the planter of that name, who suffered—at Nevis, I think, or St. Kits,—some few years since? My friend Tobin was the benevolent instrument of bringing him to the gallows.) This petty Nero actually branded a boy, who had offended him, with a red hot iron; and nearly

starved forty of us, with exacting contributions, to the one-half of our bread, to pamper a young ass, which, incredible as it may seem, with the connivance of the nurse's daughter (a young flame of his) he had contrived to smuggle in, and keep upon the leads of the *ward*, as they called our dormitories. This game went on for better than a week, till the foolish beast, not able to fare well but he must cry roast meat[13]—happier than Caligula's minion,[14] could he have kept his own counsel—but, foolisher, alas! than any of his species in the fables—waxing fat, and kicking, in the fulness of bread,[15] one unlucky minute would needs proclaim his good fortune to the world below; and, laying out his simple throat, blew such a ram's-horn blast, as (toppling down the walls of his own Jericho[16]) set concealment any longer at defiance. The client was dismissed, with certain attentions, to Smithfield; but I never understood that the patron underwent any censure on the occasion. This was in the stewardship of L.'s admired Perry.

Under the same *facile* administration, can L. have forgotten the cool impunity with which the nurses used to carry away openly, in open platters, for their own tables, one out of two of every hot joint, which the careful matron had been seeing scrupulously weighed out for our dinners? These things were daily practiced in that magnificent apartment, which L. (grown connoisseur since, we presume) praises so highly for the grand paintings "by Verrio, and others," with which it is "hung round and adorned."[17] But the sight of sleek well-fed blue-coat boys[18] in pictures was, at that time, I believe, little consolatory to him, or us, the living ones, who saw the better part of our provisions carried away before our faces by harpies;[19] and ourselves reduced (with the Trojan in the hall of Dido)

To feed our mind with idle portraiture.[20]

L. has recorded the repugnance of the school to *gags*, or the fat of fresh beef boiled; and sets it down to some superstition. But these unctuous morsels are

---

12. **hulks:** ships used as prisons.

13. **not . . . meat:** He couldn't keep quiet about his luck. 14. **Caligula's minion:** his horse, Incitatus, which he raised to the rank of consul. 15. **waxing . . . bread:** a web of allusions. See Deuteronomy 32:15; Ezekiel 16:49; *Hamlet*, III. iii. 80. 16. **toppling . . . Jericho:** See Joshua 6. 17. **"by . . . adorned":** quoted from Lamb's earlier essay. See n. 1, above. 18. **blue-coat boys:** Students at Christ's Hospital wore blue coats as a school uniform. 19. **harpies:** See *Aeneid*, III. 247–57. 20. **Trojan . . . portraiture:** *Aeneid*, I. 464.

never grateful to young palates (children are universally fat-haters) and in strong, coarse, boiled meats, *unsalted*, are detestable. A *gag-eater* in our time was equivalent to a goul and held in equal detestation. ——suffered under the imputation:

—— 'Twas said
He ate strange flesh.[21]

He was observed, after dinner, carefully to gather up the remnants left at his table (not many, nor very choice fragments, you may credit me)—and, in an especial manner, these disreputable morsels, which he would convey away and secretly stow in the settle that stood at his bed-side. None saw when he ate them. It was rumored that he privately devoured them in the night. He was watched, but no traces of such midnight practices were discoverable. Some reported that, on leave-days, he had been seen to carry out of the bounds a large blue check handkerchief, full of something. This then must be the accursed thing.[22] Conjecture next was at work to imagine how he could dispose of it. Some said he sold it to the beggars. This belief generally prevailed. He went about moping. None spake to him. No one would play with him. He was excommunicated; put out of the pale of the school. He was too powerful a boy to be beaten, but he underwent every mode of that negative punishment, which is more grievous than many stripes. Still he persevered. At length he was observed by two of his schoolfellows, who were determined to get at the secret, and had traced him one leave-day for that purpose, to enter a large worn-out building, such as there exist specimens of in Chancery-lane, which are let out to various scales of pauperism with open door, and a common staircase. After him they silently slunk in, and followed by stealth up four flights, and saw him tap at a poor wicket, which was opened by an aged woman, meanly clad. Suspicion was now ripened into certainty. The informers had secured their victim. They had him in their toils. Accusation was formally preferred, and retribution most signal was looked for. Mr. Hathaway, the then steward (for this happened a little after my time), with that patient sagacity which tempered all his conduct, determined to investigate the matter, before he proceeded to sentence. The result was that the supposed mendicants, the receivers or purchasers of the mysterious scraps, turned out to be

the parents of——, an honest couple come to decay,—whom this seasonable supply had, in all probability, saved from mendicancy; and that this young stork, at the expense of his own good name, had all this while been only feeding the old birds!—The governors on this occasion, much to their honor, voted a present relief to the family of——, and presented him with a silver medal. The lesson which the steward read upon RASH JUDGMENT, on the occasion of publicly delivering the medal to——, I believe, would not be lost upon his auditory. I had left school then, but I remember——. He was a tall, shambling youth, with a cast in his eye, not at all calculated to conciliate hostile prejudices. I have since seen him carrying a baker's basket. I think I heard he did not do quite so well by himself, as he had done by the old folks.

I was a hypochondriac lad; and the sight of a boy in fetters, upon the day of my first putting on the blue clothes, was not exactly fitted to assuage the natural terrors of initiation. I was of tender years, barely turned of seven; and had only read of such things in books, or seen them but in dreams. I was told he had *run away*. This was the punishment for the first offence. As a novice I was soon after taken to see the dungeons. These were little, square, Bedlam cells, where a boy could just lie at his length upon straw and a blanket—a mattress, I think, was afterwards substituted—with a peep of light, let in askance, from a prison-orifice at top, barely enough to read by. Here the poor boy was locked in by himself all day, without sight of any but the porter who brought him his bread and water—who *might not speak to him*;—or of the beadle, who came twice a week to call him out to receive his periodical chastisement, which was almost welcome, because it separated him for a brief interval from solitude:—and here he was shut by himself *of nights*, out of the reach of any sound, to suffer whatever horrors the weak nerves, and superstition incident to his time of life, might subject him to.[23] This was the penalty

23. **subject him to:** "One or two instances of lunacy, or attempted suicide, accordingly, at length convinced the governors of the impolicy of this part of the sentence, and the midnight torture of the spirits was dispensed with. This fancy of dungeons for children was a sprout of Howard's brain, for which (saving the reverence due to Holy Paul) methinks I could willingly spit upon his statue" (L.). Lamb's reference is to John Howard, 1726?–90, philanthropist and prison reformer, and probably to a story that Howard had once locked his own child in an outhouse while he went to see a visitor. There is a statue of Howard in St. Paul's Cathedral.

21. **'Twas . . . flesh:** *Antony and Cleopatra*, I. iv. 67–68.
22. **accursed thing:** See Joshua 7:13.

for the second offence.—Wouldst thou like, reader, to see what became of him in the next degree?

The culprit, who had been a third time an offender, and whose expulsion was at this time deemed irreversible, was brought forth, as at some solemn *auto da fe*,[24] arrayed in uncouth and most appalling attire—all trace of his late "watchet weeds"[25] carefully effaced, he was exposed in a jacket, resembling those which London lamplighters formerly delighted in, with a cap of the same. The effect of this divestiture was such as the ingenious devisers of it could have anticipated. With his pale and frighted features, it was as if some of those disfigurements in Dante[26] had seized upon him. In this disguisement he was brought into the hall (*L.'s favorite state-room*), where awaited him the whole number of his school-fellows, whose joint lessons and sports he was thenceforth to share no more; the awful presence of the steward, to be seen for the last time; of the executioner beadle, clad in his state robe for the occasion; and of two faces more, of direr import, because never but in these extremities visible. These were governors; two of whom, by choice, or charter, were always accustomed to officiate at these *Ultima Supplicia;*[27] not to mitigate (so at least we understood it), but to enforce the uttermost stripe. Old Bamber Gascoigne, and Peter Aubert, I remember, were colleagues on one occasion, when the beadle turning rather pale, a glass of brandy was ordered to prepare him for the mysteries. The scourging was, after the old Roman fashion, long and stately. The lictor[28] accompanied the criminal quite round the hall. We were generally too faint with attending to the previous disgusting circumstances, to make accurate report with our eyes of the degree of corporal punishment inflicted. Report, of course, gave out the back knotty and livid. After scourging, he was made over, in his *San Benito*,[29] to his friends, if he had any (but commonly such poor runagates were friendless), or to his parish officer, who, to enhance the effect of the scene, had his station allotted to him on the outside of the hall gate.

These solemn pageantries were not played off so often as to spoil the general mirth of the community. We had plenty of exercise and recreation *after* school hours; and, for myself, I must confess, that I was never happier, than *in* them. The Upper and Lower Grammar Schools were held in the same room; and an imaginary line only divided their bounds. Their character was as different as that of the inhabitants of the two sides of the Pyrenees. The Rev. James Boyer[30] was the Upper Master; but the Rev. Matthew Field presided over that portion of the apartment of which I had the good fortune to be a member. We lived a life as careless as birds. We talked and did just what we pleased, and nobody molested us. We carried an accidence, or a grammar, for form; but for any trouble it gave us, we might take two years in getting through the verbs deponent, and another two in forgetting all that we had learned about them. There was now and then the formality of saying a lesson, but if you had not learned it, a brush across the shoulders (just enough to disturb a fly) was the sole remonstrance. Field never used the rod; and in truth he wielded the cane with no great good will—holding it "like a dancer."[31] It looked in his hands rather like an emblem than an instrument of authority; and an emblem, too, he was ashamed of. He was a good easy man, that did not care to ruffle his own peace, nor perhaps set any great consideration upon the value of juvenile time. He came among us, now and then, but often staid away whole days from us; and when he came, it made no difference to us—he had his private room to retire to, the short time he staid, to be out of the sound of our noise. Our mirth and uproar went on. We had classics of our own, without being beholden to "insolent Greece or haughty Rome,"[32] that passed current among us—*Peter Wilkins—The Adventures of the Hon. Capt. Robert Boyle—The Fortunate Blue Coat Boy*—and the like. Or we cultivated a turn for mechanic or scientific operations; making little sun-dials of paper; or weaving those ingenious parentheses, called *cat-cradles;* or making dry peas to dance upon the end of a tin pipe; or studying the art military over that laudable game "French and English,"[33] and a hundred other such devices to pass

---

24. auto da fe: act of the faith, the name given to the execution (by strangling or burning) of heretics by the Spanish Inquisition. **25. "watchet weeds":** blue clothes. Lamb may be quoting William Collins, "Ode to the Manners," l. 68. **26. disfigurements in Dante:** See *Inferno*, Cantos XXVIII and XXX. **27. Ultima Supplicia:** extreme punishments. **28. lictor:** Roman official who executed the punishments decreed by the magistrates. **29. San Benito:** garment worn by the victims at an auto-da-fé.

30. Rev. James Boyer: Cf. Coleridge's account of him in Chapter I of *Biographia Literaria*. **31. "like a dancer":** *Antony and Cleopatra*, III. ii. 36. **32. "insolent . . . Rome":** Ben Jonson, "To the Memory of . . . William Shakespeare," l. 39. **33. "French and English":** a game for two persons. A sheet of paper is covered with dots, and the players take turns drawing a line across the paper with their eyes closed. The one who obliterates the most dots wins.

away the time—mixing the useful with the agreeable —as would have made the souls of Rousseau and John Locke[34] chuckle to have seen us.

Matthew Field belonged to that class of modest divines who affect to mix in equal proportion the *gentleman*, the *scholar*, and the *Christian;* but, I know not how, the first ingredient is generally found to be the predominating dose in the composition. He was engaged in gay parties, or with his courtly bow at some episcopal levée, when he should have been attending upon us. He had for many years the classical charge of a hundred children, during the four or five first years of their education; and his very highest form seldom proceeded further than two or three of the introductory fables of Phædrus. How things were suffered to go on thus, I cannot guess. Boyer, who was the proper person to have remedied these abuses, always affected, perhaps felt, a delicacy in interfering in a province not strictly his own. I have not been without my suspicions that he was not altogether displeased at the contrast we presented to his end of the school. We were a sort of Helots to his young Spartans.[35] He would sometimes, with ironic deference, send to borrow a rod of the Under Master, and then, with sardonic grin, observe to one of his upper boys, "how neat and fresh the twigs looked." While his pale students were battering their brains over Xenophon and Plato, with a silence as deep as that enjoined by the Samite,[36] we were enjoying ourselves at our ease in our little Goshen.[37] We saw a little into the secrets of his discipline, and the prospect did but the more reconcile us to our lot. His thunders rolled innocuous for us; his storms came near, but never touched us; contrary to Gideon's miracle,[38] while all around were drenched, our fleece was dry. His boys turned out the better scholars; we, I suspect, have the advantage in temper. His pupils cannot speak of him without something of terror allaying their gratitude; the remembrance of Field comes back with all the soothing images of indolence, and summer slumbers, and work like play, and innocent idleness, and Elysian exemptions, and life itself a "playing holiday."[39]

Though sufficiently removed from the jurisdiction of Boyer, we were near enough (as I have said) to understand a little of his system. We occasionally heard sounds of the *Ululantes*,[40] and caught glances of Tartarus. B. was a rabid pedant. His English style was crampt to barbarism. His Easter anthems (for his duty obliged him to those periodical flights) were grating as *scrannel pipes*.[41] He would laugh, ay, and heartily, but then it must be at Flaccus's quibble about *Rex*[42]—— or at the *tristis severitas in vultu*,[43] or *inspicere in patinas*,[44] of Terence—thin jests, which at their first broaching hardly have had *vis*[45] enough to move Roman muscle. —He had two wigs, both pedantic, but of differing omen. The one serene, smiling, fresh powdered, betokening a mild day. The other, an old discolored, unkempt, angry caxon,[46] denoting frequent and bloody execution. Woe to the school, when he made his morning appearance in his *passy*, or *passionate wig*. No comet expounded surer.—J. B. had a heavy hand. I have known him double his knotty fist at a poor trembling child (the maternal milk hardly dry upon its lips) with a "Sirrah, do you presume to set your wits at me?"—Nothing was more common than to see him make a headlong entry into the schoolroom, from his inner recess, or library, and, with turbulent eye, singling out a lad, roar out, "Od's my life, Sirrah" (his favorite adjuration), "I have a great mind to whip you,"—then, with as sudden a retracting impulse, fling back into his lair—and, after a cooling lapse of some minutes (during which all but the culprit had totally forgotten the context) drive headlong out

34. Rousseau . . . Locke: cited here for their views on education. **35. Helots . . . Spartans:** Spartan parents would exhibit a drunken Helot or slave to their sons as a warning. **36. Samite:** Pythagoras of Samos forbade his students to speak until they had listened to his lectures for five years. **37. Goshen:** See Exodus 8:22. **38. Gideon's miracle:** Judges 6:37–38. **39. "playing holiday":** 1 *Henry IV*, I. ii. 227.

40. Ululantes: wailers. **41. scrannel pipes:** thin, harsh pipes (from Milton's "Lycidas," l. 124). "In this and every thing B. was the antipodes of his co-adjutor. While the former was digging his brains for crude anthems, worth a pig-nut, F. would be recreating his gentlemanly fancy in the more flowery walks of the Muses. A little dramatic effusion of his, under the name of Vertumnus and Pomona, is not yet forgotten by the chroniclers of that sort of literature. It was accepted by Garrick, but the town did not give it their sanction.—B. used to say of it, in a way of half-compliment, half-irony, that it was *too classical for representation*" (L.). **42. Flaccus's . . . Rex:** In Horace (Flaccus), *Satires*, I. vii. 34–35, there is a pun (quibble) on *rex* as meaning "king" and as a man's name. **43. tristis . . . vultu:** "sad severity in his countenance," a phrase used of an outright liar by one of the characters in Terence, *Andria*, V. ii. 16. **44. inspicere in patinas:** "look into the stewpans," a phrase from Terence, *Adelphi*, III. iii. 74, where a slave parodies a father's advice to his son to look into the lives of men as into a mirror. **45. vis:** strength, force. **46. caxon:** a kind of wig.

again, piecing out his imperfect sense, as if it had been some Devil's Litany, with the expletory yell—*"and I WILL, too."*—In his gentler moods, when the *rabidus furor* [47] was assuaged, he had resort to an ingenious method, peculiar, for what I have heard, to himself, of whipping the boy, and reading the Debates, at the same time; a paragraph, and a lash between; which in those times, when parliamentary oratory was most at a height and flourishing in these realms, was not calculated to impress the patient with a veneration for the diffuser graces of rhetoric.

Once, and but once, the uplifted rod was known to fall ineffectual from his hand—when droll squinting W—— having been caught putting the inside of the master's desk to a use for which the architect had clearly not designed it, to justify himself, with great simplicity averred, that *he did not know that the thing had been forewarned.* This exquisite irrecognition of any law antecedent to the *oral* or *declaratory*, struck so irresistibly upon the fancy of all who heard it (the pedagogue himself not excepted) that remission was unavoidable.

L. has given credit to B's great merits as an instructor. Coleridge, in his *Literary Life*, has pronounced a more intelligible and ample encomium on them. The author of *The Country Spectator* [48] doubts not to compare him with the ablest teachers of antiquity. Perhaps we cannot dismiss him better than with the pious ejaculation of C—when he heard that his old master was on his death-bed—"Poor J. B.!—may all his faults be forgiven; and may he be wafted to bliss by little cherub boys, all head and wings, with no *bottoms* to reproach his sublunary infirmities."

Under him were many good and sound scholars bred.—First Grecian [49] of my time was Lancelot Pepys Stevens, kindest of boys and men, since Co-grammar-master (and inseparable companion) with Dr. T——e. [50] What an edifying spectacle did this brace of friends present to those who remembered the anti-socialities of their predecessors!—You never met the one by chance in the street without a wonder, which was quickly dissipated by the almost immediate

sub-appearance of the other. Generally arm in arm, these kindly coadjutors lightened for each other the toilsome duties of their profession, and when, in advanced age, one found it convenient to retire, the other was not long in discovering that it suited him to lay down the fasces [51] also. Oh, it is pleasant, as it is rare, to find the same arm linked in yours at forty, which at thirteen helped it to turn over the *Cicero De Amicitia*, [52] or some tale of Antique Friendship, which the young heart even then was burning to anticipate!—Co-Grecian with S. was Th—, who has since executed with ability various diplomatic functions at the Northern courts. Th— was a tall, dark, saturnine youth, sparing of speech, with raven locks. Thomas Fanshaw Middleton followed him (now Bishop of Calcutta) a scholar and a gentleman in his teens. He has the reputation of an excellent critic; and is author (besides *The Country Spectator*) of *A Treatise on the Greek Article, against Sharpe.*—M. is said to bear his mitre in India, where the *regni novitas* [53] (I dare say) sufficiently justifies the bearing. A humility quite as primitive as that of Jewel or Hooker [54] might not be exactly fitted to impress the minds of those Anglo-Asiatic diocesans with a reverence for home institutions, and the church which those fathers watered. The manners of M. at school, though firm, were mild, and unassuming.—Next to M. (if not senior to him) was Richards, author of *The Aboriginal Britons,* the most spirited of the Oxford Prize Poems; a pale, studious Grecian.—Then followed poor S——, ill-fated M——! of these the Muse is silent.

> Finding some of Edward's race
> Unhappy, pass their annals by. [55]

Come back into memory, like as thou wert in the day-spring of thy fancies, with hope like a fiery column before thee [56]—the dark pillar not yet turned—Samuel Taylor Coleridge—Logician, Metaphysician, Bard!—How have I seen the casual passer through the Cloisters

---

47. **rabidus furor:** raging fury (Catullus, *Carmina,* LXIII. 38). 48. **author . . . Spectator:** Thomas Fanshaw Middleton, 1769–1822, conducted the magazine *The Country Spectator* in 1792–93 and later became Bishop of Calcutta. 49. **Grecian:** one of the better students, who would be sent to the University with a scholarship. 50. **Dr. T——e:** Arthur William Trollope, 1768–1827, succeeded Boyer as Upper Grammar Master.

51. **fasces:** the bundle of rods enclosing an axe with the blade projecting that was carried before the Roman magistrate as a symbol of authority; here, the birch rod. 52. **De Amicitia:** Cicero's essay *On Friendship.* 53. **regni novitas:** newness of reign. See the *Aeneid,* I. 562. 54. **Jewel or Hooker:** John Jewel, 1522–71, Bishop of Salisbury, and Richard Hooker, 1534?–1600, early apologists for the Anglican Church. 55. **Finding . . . by:** slightly altered from Matthew Prior's *Carmen Seculare for the Year 1700,* ll. 104–05. "Edward's race" would be the students at Christ's Hospital, which was founded by Edward VI. 56. **fiery . . . thee:** See Exodus 13:21.

stand still, intranced with admiration (while he weighed the disproportion between the *speech* and the *garb* of the young Mirandula),[57] to hear thee unfold, in thy deep and sweet intonations, the mysteries of Jamblichus, or Plotinus[58] (for even in those years thou waxedst not pale at such philosophic draughts), or reciting Homer in his Greek, or Pindar——while the walls of the old Grey Friars re-echoed to the accents of the *inspired charity-boy!*—Many were the "wit-combats" (to dally awhile with the words of old Fuller), between him and C. V. LeG——, "which two I behold like a Spanish great galleon, and an English man of war; Master Coleridge, like the former, was built far higher in learning, solid, but slow in his performances. C. V. L., with the English man of war, lesser in bulk, but lighter in sailing, could turn with all tides, tack about, and take advantage of all winds, by the quickness of his wit and invention."[59]

Nor shalt thou, their compeer, be quickly forgotten, Allen, with the cordial smile, and still more cordial laugh, with which thou wert wont to make the old Cloisters shake, in thy cognition of some poignant jest of theirs; or the anticipation of some more material, and, peradventure, practical one, of thine own. Extinct are those smiles, with that beautiful countenance, with which (for thou wert the *Nireus formosus*[60] of the school), in the days of thy maturer waggery, thou didst disarm the wrath of infuriated town-damsel, who, incensed by provoking pinch, turning tigress-like round, suddenly converted by thy angel-look, exchanged the half-formed terrible "*bl*——," for a gentler greeting—"*bless thy handsome face!*"

Next follow two, who ought to be now alive, and the friends of Elia—the junior Le G—— and F——; who impelled, the former by a roving temper, the latter by too quick a sense of neglect—ill capable of enduring the slights poor sizars are sometimes subject to in our seats of learning—exchanged their Alma Mater for the camp; perishing, one by climate, and one on the plains of Salamanca:—Le G——, sanguine, volatile, sweet-natured; F——, dogged, faithful,

anticipative of insult, warm-hearted, with something of the old Roman height about him.

Fine, frank-hearted Fr——, the present master of Hertford, with Marmaduke T——, mildest of missionaries—and both my good friends still—close the catalogue of Grecians in my time.

(1820)

NEW YEAR'S EVE

Every man hath two birth-days: two days, at least, in every year, which set him upon revolving the lapse of time, as it affects his mortal duration. The one is that which in an especial manner he termeth *his*. In the gradual desuetude of old observances, this custom of solemnizing our proper birth-day hath nearly passed away, or is left to children, who reflect nothing at all about the matter, nor understand any thing in it beyond cake and orange. But the birth of a New Year is of an interest too wide to be pretermitted by king or cobbler. No one ever regarded the First of January with indifference. It is that from which all date their time, and count upon what is left. It is the nativity of our common Adam.

Of all sound of all bells—(bells, the music nighest bordering upon heaven)—most solemn and touching is the peal which rings out the Old Year. I never hear it without a gathering-up of my mind to a concentration of all the images that have been diffused over the past twelvemonth; all I have done or suffered, performed or neglected—in that regretted time. I begin to know its worth, as when a person dies. It takes a personal colour; nor was it a poetical flight in a contemporary, when he exclaimed

I saw the skirts of the departing Year.[1]

It is no more than what in sober sadness every one of us seems to be conscious of, in that awful leave-taking. I am sure I felt it, and all felt it with me, last night; though some of my companions affected rather to manifest an exhilaration at the birth of the coming year, than any very tender regrets for the decease of its predecessor. But I am none of those who

Welcome the coming, speed the parting guest.[2]

---

57. **Mirandula:** Picco della Mirandola, 1463–94, Italian philosopher. 58. **Jamblichus . . . Plotinus:** Neoplatonic philosophers of the third and fourth centuries A.D. 59. "**which . . . invention**": Lamb adapts a famous account in Thomas Fuller, *The Worthies in England* (1662), under "Warwickshire," of the arguments between Shakespeare and Ben Jonson. 60. **Nireus formosus:** beautiful Nireus. He was the most beautiful of the Greeks at Troy. See the *Iliad*, II. 673.

NEW YEAR'S EVE. **I. I . . . Year:** Coleridge's "Ode to the Departing Year," l. 8, as printed in 1797. **2. Welcome . . . guest:** Pope's *Odyssey*, XV. 84, and also his imitation of Horace, *Satires*, II. ii. 160.

I am naturally, beforehand, shy of novelties; new books, new faces, new years,—from some mental twist which makes it difficult in me to face the prospective. I have almost ceased to hope; and am sanguine only in the prospects of other (former) years. I plunge into foregone visions and conclusions. I encounter pell-mell with past disappointments. I am armour-proof against old discouragements. I forgive, or overcome in fancy, old adversaries. I play over again *for love*, as the gamesters phrase it, games, for which I once paid so dear. I would scarce now have any of those untoward accidents and events of my life reversed. I would no more alter them than the incidents of some well-contrived novel. Methinks, it is better that I should have pined away seven of my goldenest years, when I was thrall to the fair hair, and fairer eyes, of Alice W——n, than that so passionate a love-adventure should be lost. It was better that our family should have missed that legacy, which old Dorrell cheated us of, than that I should have at this moment two thousand pounds *in banco*, and be without the idea of that specious old rogue.

In a degree beneath manhood, it is my infirmity to look back upon those early days. Do I advance a paradox, when I say, that, skipping over the intervention of forty years, a man may have leave to love *himself*, without the imputation of self-love?

If I know aught of myself, no one whose mind is introspective—and mine is painfully so—can have a less respect for his present identity, than I have for the man Elia. I know him to be light, and vain, and humorsome; a notorious * * *; addicted to * * * * *: averse from counsel, neither taking it nor offering it;— * * * besides; a stammering buffoon; what you will; lay it on, and spare not; I subscribe to it all, and much more, than thou canst be willing to lay at his door ——but for the child Elia—that "other me," there, in the back-ground—I must take leave to cherish the remembrance of that young master—with as little reference, I protest, to this stupid changeling of five-and-forty, as if it had been a child of some other house, and not of my parents. I can cry over its patient small-pox at five, and rougher medicaments. I can lay its poor fevered head upon the sick pillow at Christ's,[3] and wake with it in surprise at the gentle posture of maternal tenderness hanging over it, that unknown had watched its sleep. I know how it shrank from any the least colour of falsehood.—God help thee, Elia,

how art thou changed! Thou art sophisticated.[4]—I know how honest, how courageous (for a weakling) it was—how religious, how imaginative, how hopeful! From what have I not fallen, if the child I remember was indeed myself,—and not some dissembling guardian, presenting a false identity, to give the rule to my unpractised steps, and regulate the tone of my moral being!

That I am fond of indulging, beyond a hope of sympathy, in such retrospection, may be the symptom of some sickly idiosyncrasy. Or is it owing to another cause; simply, that being without wife or family, I have not learned to project myself enough out of myself; and having no offspring of my own to dally with, I turn back upon memory, and adopt my own early idea, as my heir and favourite? If these speculations seem fantastical to thee, reader—(a busy man, perchance), if I tread out of the way of thy sympathy, and am singularly-conceited only, I retire, impenetrable to ridicule, under the phantom cloud of Elia.

The elders, with whom I was brought up, were of a character not likely to let slip the sacred observance of any old institution; and the ringing out of the Old Year was kept by them with circumstances of peculiar ceremony.—In those days the sound of those midnight chimes, though it seemed to raise hilarity in all around me, never failed to bring a train of pensive imagery into my fancy. Yet I then scarce conceived what it meant, or thought of it as a reckoning that concerned me. Not childhood alone, but the young man till thirty, never feels practically that he is mortal. He knows it indeed, and, if need were, he could preach a homily on the fragility of life; but he brings it not home to himself, any more than in a hot June we can appropriate to our imagination the freezing days of December. But now, shall I confess a truth?—I feel these audits but too powerfully. I begin to count the probabilities of my duration, and to grudge at the expenditure of moments and shortest periods, like miser's farthings. In proportion as the years both lessen and shorten, I set more count upon their periods, and would fain lay my ineffectual finger upon the spoke of the great wheel. I am not content to pass away "like a weaver's shuttle."[5] Those metaphors solace me not, nor sweeten the unpalatable draught of mortality. I care not to be carried with the tide, that

---

3. **Christ's:** Christ's Hospital. See the foregoing essay.

4. **God . . . sophisticated:** Cf. Shakespeare's *A Midsummer Night's Dream*, III. i. 121–22. 5. **"like . . . shuttle":** Job 8:6.

smoothly bears human life to eternity; and reluct at the inevitable course of destiny. I am in love with this green earth; the face of town and country; the unspeakable rural solitudes, and the sweet security of streets. I would set up my tabernacle here. I am content to stand still at the age to which I am arrived; I, and my friends: to be no younger, no richer, no handsomer. I do not want to be weaned by age; or drop, like mellow fruit, as they say, into the grave.—Any alteration, on this earth of mine, in diet or in lodging, puzzles and discomposes me. My household-gods plant a terrible fixed foot, and are not rooted up without blood. They do not willingly seek Lavinian shores.[6] A new state of being staggers me.

Sun, and sky, and breeze, and solitary walks, and summer holidays, and the greenness of fields, and the delicious juices of meats and fishes, and society, and the cheerful glass, and candle-light, and fire-side conversations, and innocent vanities, and jests, and *irony itself*—do these things go out with life?

Can a ghost laugh, or shake his gaunt sides, when you are pleasant with him?

And you, my midnight darlings, my Folios! must I part with the intense delight of having you (huge armfuls) in my embraces? Must knowledge come to me, if it come at all, by some awkward experiment of intuition, and no longer by this familiar process of reading?

Shall I enjoy friendships there, wanting the smiling indications which point me to them here,—the recognisable face—the "sweet assurance of a look"—?[7]

In winter this intolerable disinclination to dying—to give it its mildest name—does more especially haunt and beset me. In a genial August noon, beneath a sweltering sky, death is almost problematic. At those times do such poor snakes as myself enjoy an immortality. Then we expand and burgeon. Then are we as strong again, as valiant again, as wise again, and a great deal taller. The blast that nips and shrinks me, puts me in thoughts of death. All things allied to the insubstantial, wait upon that master feeling; cold, numbness, dreams, perplexity; moonlight itself, with its shadowy and spectral appearances,—that cold ghost of the sun, or Phœbus' sickly sister, like that

innutritious one denounced in the Canticles:[8]—I am none of her minions—I hold with the Persian.[9]

Whatsoever thwarts, or puts me out of my way, brings death into my mind. All partial evils, like humours, run into that capital plague-sore.—I have heard some profess an indifference to life. Such hail the end of their existence as a port of refuge; and speak of the grave as of some soft arms, in which they may slumber as on a pillow. Some have wooed death — — — but out upon thee, I say, thou foul, ugly phantom! I detest, abhor, execrate, and (with Friar John[10]) give thee to six-score thousand devils, as in no instance to be excused or tolerated, but shunned as a universal viper; to be branded, proscribed, and spoken evil of! In no way can I be brought to digest thee, thou thin melancholy *Privation*, or more frightful and confounding *Positive!*

Those antidotes, prescribed against the fear of thee, are altogether frigid and insulting, like thyself. For what satisfaction hath a man, that he shall "lie down with kings and emperors in death,"[11] who in his lifetime never greatly coveted the society of such bedfellows?—or, forsooth, that, "so shall the fairest face appear?"[12]—why, to comfort me, must Alice W——n be a goblin? More than all, I conceive disgust at those impertinent and misbecoming familiarities, inscribed upon your ordinary tombstones. Every dead man must take upon himself to be lecturing me with his odious truism, that "such as he now is, I must shortly be." Not so shortly, friend, perhaps, as thou imaginest. In the meantime I am alive. I move about. I am worth twenty of thee. Know thy betters! Thy New Years' Days are past. I survive, a jolly candidate for 1821. Another cup of wine—and while that turncoat bell, that just now mournfully chanted the obsequies of 1820 departed, with changed notes lustily rings in a successor, let us attune to its peal the song made on a like occasion, by hearty, cheerful Mr Cotton.—[13]

The New Year

Hark, the cock crows, and yon bright star
Tells us, the day himself's not far;
And see where, breaking from the night,
He gilds the western hills with light.

6. Lavinian shores: Italy (see the *Aeneid*, I. 2). After the fall of Troy, Aeneas with his companions brought his household gods to Italy and founded Rome. **7. "sweet . . . look":** adapted from Matthew Roydon's elegy on Sir Philip Sidney.

8. Canticles: Song of Solomon 8:8: "We have a little sister, and she hath no breasts." **9. Persian:** sun-worshipper. **10. Friar John:** in Rabelais' *Gargantua*. **11. "lie . . . death":** See Job 3:13–14. **12. "so . . . appear?":** David Mallett's "William and Margaret," l. 9. **13. Mr Cotton:** Charles Cotton, 1630–87.

With him old Janus doth appear,
Peeping into the future year,
With such a look as seems to say,
The prospect is not good that way.
Thus do we rise ill sights to see,
And 'gainst ourselves to prophesy;
When the prophetic fear of things
A more tormenting mischief brings,
More full of soul-tormenting gall,
Than direst mischiefs can befall.
But stay! but stay! methinks my sight,
Better inform'd by clearer light,
Discerns sereneness in that brow,
That all contracted seem'd but now.
His revers'd face may show distaste,
And frown upon the ills are past;
But that which this way looks is clear,
And smiles upon the New-born Year.
He looks too from a place so high,
The Year lies open to his eye;
And all the moments open are
To the exact discoverer.
Yet more and more he smiles upon
The happy revolution.
Why should we then suspect or fear
The influences of a year,
So smiles upon us the first morn,
And speaks us good so soon as born?
Plague on't! the last was ill enough,
This cannot but make better proof;
Or, at the worst, as we brush'd through
The last, why so we may this too;
And then the next in reason shou'd
Be superexcellently good:
For the worst ills (we daily see)
Have no more perpetuity,
Than the best fortunes that do fall;
Which also bring us wherewithal
Longer their being to support,
Than those do of the other sort:
And who has one good year in three,
And yet repines at destiny,
Appears ungrateful in the case,
And merits not the good he has.
Then let us welcome the New Guest
With lusty brimmers of the best;
Mirth always should Good Fortune meet,
And renders e'en Disaster sweet:
And though the Princess turn her back,
Let us but line ourselves with sack,
We better shall by far hold out,
Till the next Year she face about.

How say you, reader—do not these verses smack of the rough magnanimity of the old English vein? Do they not fortify like a cordial; enlarging the heart, and productive of sweet blood, and generous spirits, in the concoction? Where be those puling fears of death, just now expressed or affected?—Passed like a cloud—absorbed in the purging sunlight of clear poetry—clean washed away by a wave of genuine Helicon,[14] your only Spa for these hypochondries—And now another cup of the generous! and a merry New Year, and many of them, to you all, my masters!

(*1821*)

DREAM-CHILDREN

A REVERIE

Children love to listen to stories about their elders when *they* were children; to stretch their imagination to the conception of a traditionary great-uncle, or grandame whom they never saw. It was in this spirit that my little ones crept about me the other evening to hear about their great-grandmother Field who lived in a great house in Norfolk (a hundred times bigger than that in which they and papa lived) which had been the scene—so at least it was generally believed in that part of the country—of the tragic incidents which they had lately become familiar with from the ballad of *The Children in the Wood*. Certain it is that the whole story of the children and their cruel uncle was to be seen fairly carved out in wood upon the chimney-piece of the great hall, the whole story down to the Robin Redbreasts, till a foolish rich person pulled it down to set up a marble one of modern invention in its stead, with no story upon it. Here Alice put out one of her dear mother's looks, too tender to be called upbraiding. Then I went on to say how religious and how good their great-grandmother Field was, how beloved and respected by every body, though she was not indeed the mistress of this great house, but had only the charge of it (and yet in some respects she might be said to be the mistress of it too) committed to her by the owner, who preferred living in a newer and more

14. **Helicon:** a mountain, home of Apollo and the Muses. Lamb apparently confuses Mt. Helicon with the fountain that rises on it, Hippocrene, the waters of which give poetic inspiration.

fashionable mansion which he had purchased somewhere in the adjoining county; but still she lived in it in a manner as if it had been her own, and kept up the dignity of the great house in a sort while she lived, which afterwards came to decay, and was nearly pulled down, and all its old ornaments stripped and carried away to the owner's other house, where they were set up, and looked as awkward as if some one were to carry away the old tombs they had seen lately at the Abbey, and stick them up in Lady C.'s tawdry gilt drawing-room. Here John smiled, as much as to say, "that would be foolish indeed." And then I told how, when she came to die, her funeral was attended by a concourse of all the poor, and some of the gentry too, of the neighborhood for many miles round, to show their respect for her memory, because she had been such a good and religious woman; so good indeed that she knew all the Psaltery, by heart, ay, and a great part of the Testament besides. Here little Alice spread her hands. Then I told what a tall, upright, graceful person their great-grandmother Field once was; and how in her youth she was esteemed the best dancer—here Alice's little right foot played an involuntary movement, till, upon my looking grave, it desisted—the best dancer, I was saying, in the county, till a cruel disease, called a cancer, came, and bowed her down with pain; but it could never bend her good spirits, or make them stoop, but they were still upright, because she was so good and religious. Then I told how she was used to sleep by herself in a lone chamber of the great lone house; and how she believed that an apparition of two infants was to be seen at midnight gliding up and down the great staircase near where she slept, but she said "those innocents would do her no harm"; and how frightened I used to be, though in those days I had my maid to sleep with me, because I was never half so good or religious as she—and yet I never saw the infants. Here John expanded all his eye-brows and tried to look courageous. Then I told how good she was to all her grand-children, having us to the great-house in the holydays, where I in particular used to spend many hours by myself, in gazing upon the old busts of the Twelve Cæsars, that had been Emperors of Rome, till the old marble heads would seem to live again, or I to be turned into marble with them; how I never could be tired with roaming about that huge mansion, with its vast empty rooms, with their worn-out hangings, fluttering tapestry, and carved oaken panels, with the gilding almost rubbed out—sometimes in the spacious old-fashioned gardens, which I had almost to myself, unless when now and then a solitary gardening man would cross me—and how the nectarines and peaches hung upon the walls, without my ever offering to pluck them, because they were forbidden fruit, unless now and then,—and because I had more pleasure in strolling about among the old melancholy-looking yew trees, or the firs, and picking up the red berries, and the fir apples, which were good for nothing but to look at—or in lying about upon the fresh grass, with all the fine garden smells around me—or basking in the orangery, till I could almost fancy myself ripening too along with the oranges and the limes in that grateful warmth—or in watching the dace that darted to and fro in the fishpond, at the bottom of the garden, with here and there a great sulky pike hanging midway down the water in silent state, as if it mocked at their impertinent friskings,—I had more pleasure in these busy-idle diversions than in all the sweet flavors of peaches, nectarines, oranges, and such like common baits of children. Here John slyly deposited back upon the plate a bunch of grapes, which, not unobserved by Alice, he had meditated dividing with her, and both seemed willing to relinquish them for the present as irrelevant. Then in somewhat a more heightened tone, I told how, though their great-grandmother Field loved all her grand-children, yet in an especial manner she might be said to love their uncle, John L——,[1] because he was so handsome and spirited a youth, and a king to the rest of us; and, instead of moping about in solitary corners, like some of us, he would mount the most mettlesome horse he could get, when but an imp no bigger than themselves, and make it carry him half over the county in a morning, and join the hunters when there were any out—and yet he loved the old great house and gardens too, but had too much spirit to be always pent up within their boundaries—and how their uncle grew up to man's estate as brave as he was handsome, to the admiration of every body, but of their great-grandmother Field especially; and how he used to carry me upon his back when I was a lame-footed boy—for he was a good bit older than me—many a mile when I could not walk for pain;—and how in after life he became lame-footed too, and I did not always (I fear) make allowances enough for him when he was impatient, and in pain, nor remember sufficiently how considerate he had been to me when

DREAM-CHILDREN: **1. John L——**: Lamb's brother John had died on October 26, 1821.

I was lame-footed; and how when he died, though he had not been dead an hour, it seemed as if he had died a great while ago, such a distance there is betwixt life and death; and how I bore his death as I thought pretty well at first, but afterwards it haunted and haunted me; and though I did not cry or take it to heart as some do, and as I think he would have done if I had died, yet I missed him all day long, and knew not till then how much I had loved him. I missed his kindness, and I missed his crossness, and wished him to be alive again, to be quarrelling with him (for we quarrelled sometimes) rather than not have him again, and was as uneasy without him, as he their poor uncle must have been when the doctor took off his limb. Here the children fell a-crying, and asked if their little mourning which they had on was not for uncle John, and they looked up, and prayed me not to go on about their uncle, but to tell them some stories about their pretty dead mother. Then I told how for seven long years, in hope sometimes, sometimes in despair, yet persisting ever, I courted the fair Alice W——n; and, as much as children could understand, I explained to them what coyness, and difficulty, and denial meant in maidens—when suddenly, turning to Alice, the soul of the first Alice looked out at her eyes with such a reality of re-presentment, that I became in doubt which of them stood there before me, or whose that bright hair was; and while I stood gazing, both the children gradually grew fainter to my view, receding, and still receding till nothing at last but two mournful features were seen in the uttermost distance, which, without speech, strangely impressed upon me the effects of speech: "We are not of Alice, nor of thee, nor are we children at all. The children of Alice called Bartrum father. We are nothing; less than nothing, and dreams. We are only what might have been, and must wait upon the tedious shores of Lethe[2] millions of ages before we have existence, and a name"——and immediately awaking, I found myself quietly seated in my bachelor arm-chair, where I had fallen asleep, with the faithful Bridget[3] unchanged by my side—but John L. (or James Elia) was gone forever.

(1822)

---

2. **tedious . . . Lethe:** See the *Aeneid*, VI. 748–51. They are waiting at the river of Lethe to return to mortal life.
3. **Bridget:** name given Lamb's sister Mary in the Elia essays.

A DISSERTATION UPON ROAST PIG

Mankind, says a Chinese manuscript, which my friend M. was obliging enough to read and explain to me, for the first seventy thousand ages ate their meat raw, clawing or biting it from the living animal, just as they do in Abyssinia to this day. This period is not obscurely hinted at by their great Confucius in the second chapter of his *Mundane Mutations*, where he designates a kind of golden age by the term Cho-fang, literally the Cooks' Holiday. The manuscript goes on to say, that the art of roasting, or rather broiling (which I take to be the elder brother) was accidentally discovered in the manner following. The swineherd, Ho-ti, having gone out into the woods one morning, as his manner was, to collect mast[1] for his hogs, left his cottage in the care of his eldest son Bo-bo, a great lubberly boy, who being fond of playing with fire, as younkers of his age commonly are, let some sparks escape into a bundle of straw, which kindling quickly, spread the conflagration over every part of their poor mansion till it was reduced to ashes. Together with the cottage (a sorry antediluvian make-shift of a building, you may think it), what was of much more importance, a fine litter of new-farrowed pigs, no less than nine in number, perished. China pigs have been esteemed a luxury all over the East from the remotest periods that we read of. Bo-bo was in the utmost consternation, as you may think, not so much for the sake of the tenement, which his father and he could easily build up again with a few dry branches, and the labor of an hour or two, at any time, as for the loss of the pigs. While he was thinking what he should say to his father, and wringing his hands over the smoking remnants of one of those untimely sufferers, an odor assailed his nostrils, unlike any scent which he had before experienced. What could it proceed from? —not from the burnt cottage—he had smelt that smell before—indeed this was by no means the first accident of the kind which had occurred through the negligence of this unlucky young fire-brand. Much less did it resemble that of any known herb, weed, or flower. A premonitory moistening at the same time overflowed his nether lip. He knew not what to think. He next stooped down to feel the pig, if there were any signs

---

A DISSERTATION UPON ROAST PIG. **1. mast:** nuts, especially beechnuts.

of life in it. He burnt his fingers, and to cool them he applied them in his booby fashion to his mouth. Some of the crumbs of the scorched skin had come away with his fingers, and for the first time in his life (in the world's life indeed, for before him no man had known it) he tasted—*crackling!*[2] Again he felt and fumbled at the pig. It did not burn him so much now, still he licked his fingers from a sort of habit. The truth at length broke into his slow understanding, that it was the pig that smelt so, and the pig that tasted so delicious; and, surrendering himself up to the new-born pleasure, he fell to tearing up whole handfuls of the scorched skin with the flesh next it, and was cramming it down his throat in his beastly fashion, when his sire entered amid the smoking rafters, armed with retributory cudgel, and finding how affairs stood, began to rain blows upon the young rogue's shoulders, as thick as hail-stones, which Bo-bo heeded not any more than if they had been flies. The tickling pleasure, which he experienced in his lower regions, had rendered him quite callous to any inconveniences he might feel in those remote quarters. His father might lay on, but he could not beat him from his pig, till he had fairly made an end of it, when, becoming a little more sensible of his situation, something like the following dialogue ensued.

"You graceless whelp, what have you got there devouring? Is it not enough that you have burnt me down three houses with your dog's tricks, and be hanged to you, but you must be eating fire, and I know not what—what have you got there, I say?"

"O father, the pig, the pig, do come and taste how nice the burnt pig eats."

The ears of Ho-ti tingled with terror. He cursed his son, and he cursed himself that ever he should beget a son that should eat burnt pig.

Bo-bo, whose scent was wonderfully sharpened since morning, soon raked out another pig, and fairly rending it asunder, thrust the lesser half by main force into the fists of Ho-ti, still shouting out "Eat, eat, eat the burning pig, father, only taste—O Lord,"— with such-like barbarous ejaculations, cramming all the while as if he would choke.

Ho-ti trembled in every joint while he grasped the abominable thing, wavering whether he should not put his son to death for an unnatural young monster, when the crackling scorching his fingers, as it had done his son's, and applying the same remedy to them, he

2. **crackling:** the crisp skin of roasted pork.

in his turn tasted some of its flavor, which, make what sour mouths he would for a pretence, proved not altogether displeasing to him. In conclusion (for the manuscript here is a little tedious) both father and son fairly sat down to the mess, and never left off till they had despatched all that remained of the litter.

Bo-bo was strictly enjoined not to let the secret escape, for the neighbors would certainly have stoned them for a couple of abominable wretches, who could think of improving upon the good meat which God had sent them. Nevertheless, strange stories got about. It was observed that Ho-ti's cottage was burnt down now more frequently than ever. Nothing but fires from this time forward. Some would break out in broad day, others in the night-time. As often as the sow farrowed, so sure was the house of Ho-ti to be in a blaze; and Ho-ti himself, which was the more remarkable, instead of chastising his son, seemed to grow more indulgent to him than ever. At length they were watched, the terrible mystery discovered, and father and son summoned to take their trial at Pekin, then an inconsiderable assize town. Evidence was given, the obnoxious food itself produced in court, and verdict about to be pronounced, when the foreman of the jury begged that some of the burnt pig, of which the culprits stood accused, might be handed into the box. He handled it, and they all handled it, and burning their fingers, as Bo-bo and his father had done before them, and nature prompting to each of them the same remedy, against the face of all the facts, and the clearest charge which judge had ever given—to the surprise of the whole court, townsfolk, strangers, reporters, and all present—without leaving the box, or any manner of consultation whatever, they brought in a simultaneous verdict of Not Guilty.

The judge, who was a shrewd fellow, winked at the manifest iniquity of the decision: and, when the court was dismissed, went privily, and bought up all the pigs that could be had for love or money. In a few days his Lordship's town house was observed to be on fire. The thing took wing, and now there was nothing to be seen but fires in every direction. Fuel and pigs grew enormously dear all over the district. The insurance offices one and all shut up shop. People built slighter and slighter every day, until it was feared that the very science of architecture would in no long time be lost to the world. Thus this custom of firing houses continued, till in process of time, says my manuscript, a sage arose, like our Locke, who made a discovery, that the flesh of swine, or indeed of any other animal, might be cooked (*burnt*, as they called it) without the necessity of consuming a whole house to dress it. Then first began the rude form of a gridiron. Roasting by the string, or spit, came in a century or two later; I forget in whose dynasty. By such slow degrees, concludes the manuscript, do the most useful and seemingly the most obvious arts, make their way among mankind.

Without placing too implicit faith in the account above given, it must be agreed, that if a worthy pretext for so dangerous an experiment as setting houses on fire (especially in these days) could be assigned in favor of any culinary object, that pretext and excuse might be found in ROAST PIG.

Of all the delicacies in the whole *mundus edibilis*,[3] I will maintain it to be the most delicate—*princeps obsoniorum*.[4]

I speak not of your grown porkers—things between pig and pork—those hobbydehoys—but a young and tender suckling—under a moon old—guiltless as yet of the sty—with no original speck of the *amor immunditiæ*,[5] the hereditary failing of the first parent, yet manifest—his voice as yet not broken, but something between a childish treble, and a grumble—the mild forerunner, or *præludium*,[6] of a grunt.

He must be roasted. I am not ignorant that our ancestors ate them seethed, or boiled—but what a sacrifice of the exterior tegument!

There is no flavor comparable, I will contend, to that of the crisp, tawny, well-watched, not over-roasted, *crackling*, as it is well called—the very teeth are invited to their share of the pleasure at this banquet in overcoming the coy, brittle resistance—with the adhesive oleaginous—O call it not fat—but an indefinable sweetness growing up to it—the tender blossoming of fat—fat cropped in the bud—taken in the shoot—in the first innocence—the cream and quintessence of the child-pig's yet pure food——the lean, no lean, but a kind of animal manna—or, rather, fat and lean (if it must be so) blended and running into each other, that both together make but one ambrosian result, or common substance.

Behold him, while he is doing—it seemeth rather a refreshing warmth, than a scorching heat, that he is so passive to. How equably he twirleth round the string!—Now he is just done. To see the extreme

sensibility of that tender age, he hath wept out his pretty eyes—radiant jellies—shooting stars[7]—

See him in the dish, his second cradle, how meek he lieth!—wouldst thou have had this innocent grow up to the grossness and indocility which too often accompany maturer swinehood? Ten to one he would have proved a glutton, a sloven, an obstinate, disagreeable animal—wallowing in all manner of filthy conversation—from these sins he is happily snatched away—

Ere sin could blight, or sorrow fade,
Death came with timely care—— [8]

his memory is odoriferous—no clown curseth, while his stomach half rejecteth, the rank bacon—no coalheaver bolteth him in reeking sausages—he hath a fair sepulchre in the grateful stomach of the judicious epicure—and for such a tomb might be content to die.

He is the best of sapors.[9] Pine-apple is great. She is indeed almost too transcendent—a delight, if not sinful, yet so like to sinning, that really a tender-conscienced person would do well to pause—too ravishing for mortal taste, she woundeth and excoriateth the lips that approach her—like lovers' kisses, she biteth—she is a pleasure bordering on pain from the fierceness and insanity of her relish—but she stoppeth at the palate —she meddleth not with the appetite—and the coarsest hunger might barter her consistently for a mutton chop.

Pig—let me speak his praise—is no less provocative of the appetite, than he is satisfactory to the criticalness of the censorious palate. The strong man may batten on him, and the weakling refuseth not his mild juices.

Unlike to mankind's mixed characters, a bundle of virtues and vices, inexplicably intertwisted, and not to be unravelled without hazard, he is—good throughout. No part of him is better or worse than another. He helpeth, as far as his little means extend, all around. He is the least envious of banquets. He is all neighbors' fare.

I am one of those who freely and ungrudgingly impart a share of the good things of this life which fall to their lot (few as mine are in this kind), to a friend. I protest I take as great an interest in my friend's pleasures, his relishes, and proper satisfactions, as in mine own. "Presents," I often say, "endear Absents."

---

3. **mundus edibilis:** world of edibles. 4. **princeps obsoniorum:** prince of delicacies. 5. **amor immunditiæ:** love of filth. 6. **præludium:** prelude.

7. **jellies . . . stars:** an allusion to the superstition that shooting stars leave jellies where they fall. 8. **Ere . . . care:** Coleridge's "Epitaph on an Infant," ll. 1–2. 9. **sapors:** savors.

Hares, pheasants, partridges, snipes, barn-door chicken (those "tame villatic fowl"[10]), capons, plovers, brawn, barrels of oysters, I dispense as freely as I receive them. I love to taste them, as it were, upon the tongue of my friend. But a stop must be put somewhere. One would not, like Lear, "give everything." I make my stand upon pig. Methinks it is an ingratitude to the Giver of all good flavors, to extra-domiciliate, or send out of the house, slightingly, (under pretext of friendship, or I know not what) a blessing so particularly adapted, predestined, I may say, to my individual taste.—It argues an insensibility.

I remember a touch of conscience in this kind at school. My good old aunt, who never parted from me at the end of a holiday without stuffing a sweet-meat, or some nice thing, into my pocket, had dismissed me one evening with a smoking plum-cake, fresh from the oven. In my way to school (it was over London bridge) a gray-headed old beggar saluted me (I have no doubt at this time of day that he was a counterfeit). I had no pence to console him with, and in the vanity of self-denial, and the very coxcombry of charity, school-boy-like, I made him a present of—the whole cake! I walked on a little, buoyed up, as one is on such occasions, with a sweet soothing of self-satisfaction; but before I had got to the end of the bridge, my better feelings returned, and I burst into tears, thinking how ungrateful I had been to my good aunt, to go and give her good gift away to a stranger, that I had never seen before, and who might be a bad man for aught I knew; and then I thought of the pleasure my aunt would be taking in thinking that I—I myself, and not another—would eat her nice cake—and what should I say to her the next time I saw her—how naughty I was to part with her pretty present—and the odor of that spicy cake came back upon my recollection, and the pleasure and the curiosity I had taken in seeing her make it, and her joy when she sent it to the oven, and how disappointed she would feel that I had never had a bit of it in my mouth at last—and I blamed my impertinent spirit of alms-giving, and out-of-place hypocrisy of goodness, and above all I wished never to see the face again of that insidious, good-for-nothing, old gray impostor.

Our ancestors were nice in their methods of sacrificing these tender victims. We read of pigs whipt to death with something of a shock, as we hear of any other obsolete custom. The age of discipline is gone by,

or it would be curious to inquire (in a philosophical light merely) what effect this process might have towards intenerating and dulcifying a substance, naturally so mild and dulcet as the flesh of young pigs. It looks like refining a violet. Yet we should be cautious, while we condemn the inhumanity, how we censure the wisdom of the practice. It might impart a gusto—

I remember an hypothesis, argued upon by the young students, when I was at St. Omer's,[11] and maintained with much learning and pleasantry on both sides, "Whether, supposing that the flavor of a pig who obtained his death by whipping (*per flagellationem extremam*[12]) superadded a pleasure upon the palate of a man more intense than any possible suffering we can conceive in the animal, is man justified in using that method of putting the animal to death?" I forget the decision.

His sauce should be considered. Decidedly, a few bread crumbs, done up with his liver and brains, and a dash of mild sage. But, banish, dear Mrs. Cook, I beseech you, the whole onion tribe. Barbecue[13] your whole hogs to your palate, steep them in shalots, stuff them out with plantations of the rank and guilty garlic; you cannot poison them, or make them stronger than they are—but consider, he is a weakling—a flower.

(*1822*)

OLD CHINA

I have an almost feminine partiality for old china. When I go to see any great house, I inquire for the china-closet, and next for the picture gallery. I cannot defend the order of preference, but by saying that we have all some taste or other, of too ancient a date to admit of our remembering distinctly that it was an acquired one. I can call to mind the first play, and the first exhibition, that I was taken to; but I am not conscious of a time when china jars and saucers were introduced into my imagination.

I had no repugnance then—why should I now have? —to those little, lawless, azure-tinctured grotesques, that under the notion of men and women, float about, uncircumscribed by any element, in that world before perspective—a china tea-cup.

10. **"tame . . . fowl":** Milton's *Samson Agonistes*, l. 1695.

11. **St. Omer's:** a Jesuit college in France. Lamb himself was never there. 12. **per . . . extremam:** by extreme flagellation. 13. **Barbecue:** roast whole.

I like to see my old friends—whom distance cannot diminish—figuring up in the air (so they appear to our optics), yet on *terra firma* still—for so we must in courtesy interpret that speck of deeper blue, which the decorous artist, to prevent absurdity, has made to spring up beneath their sandals.

I love the men with women's faces, and the women, if possible, with still more womanish expressions.

Here is a young and courtly Mandarin, handing tea to a lady from a salver—two miles off. See how distance seems to set off respect! And here the same lady, or another—for likeness is identity on tea-cups—is stepping into a little fairy boat, moored on the hither side of this calm garden river, with a dainty mincing foot, which in a right angle of incidence (as angles go in our world) must infallibly land her in the midst of a flowery mead—a furlong off on the other side of the same strange stream!

Farther on—if far or near can be predicated of their world—see horses, trees, pagodas, dancing the hays.[1]

Here—a cow and rabbit couchant, and co-extensive —so objects show, seen through the lucid atmosphere of fine Cathay.

I was pointing out to my cousin last evening, over our Hyson[2] (which we are old fashioned enough to drink unmixed still of an afternoon), some of these *speciosa miracula*[3] upon a set of extraordinary old blue china (a recent purchase) which we were now for the first time using; and could not help remarking, how favorable circumstances had been to us of late years, that we could afford to please the eye sometimes with trifles of this sort—when a passing sentiment seemed to over-shade the brows of my companion. I am quick at detecting these summer clouds in Bridget.[4]

"I wish the good old times would come again," she said, "when we were not quite so rich. I do not mean that I want to be poor; but there was a middle state;"—so she was pleased to ramble on,—"in which I am sure we were a great deal happier. A purchase is but a purchase, now that you have money enough and to spare. Formerly it used to be a triumph. When we coveted a cheap luxury (and, O! how much ado I had to get you to consent in those times!) we were used to have a debate two or three days before, and to weigh the *for* and *against*, and think what we might spare it out of, and what saving we could hit upon,

that should be an equivalent. A thing was worth buying then, when we felt the money that we paid for it.

"Do you remember the brown suit, which you made to hang upon you, till all your friends cried shame upon you, it grew so thread-bare—and all because of that folio Beaumont and Fletcher, which you dragged home late at night from Barker's in Covent-garden? Do you remember how we eyed it for weeks before we could make up our minds to the purchase, and had not come to a determination till it was near ten o'clock of the Saturday night, when you set off from Islington, fearing you should be too late—and when the old bookseller with some grumbling opened his shop, and by the twinkling taper (for he was setting bed-wards) lighted out the relic from his dusty treasures— and when you lugged it home, wishing it were twice as cumbersome—and when you presented it to me— and when we were exploring the perfectness of it (*collating* you called it)—and while I was repairing some of the loose leaves with paste, which your impatience would not suffer to be left till day-break— was there no pleasure in being a poor man? or can those neat black clothes which you wear now, and are so careful to keep brushed, since we have become rich and finical, give you half the honest vanity with which you flaunted it about in that over-worn suit—your old corbeau[5]—for four or five weeks longer than you should have done, to pacify your conscience for the mighty sum of fifteen—or sixteen shillings was it?—a great affair we thought it then—which you had lavished on the old folio. Now you can afford to buy any book that pleases you, but I do not see that you ever bring me home any nice old purchases now.

"When you came home with twenty apologies for laying out a less number of shillings upon that print after Lionardo, which we christened the *Lady Blanch;* when you looked at the purchase, and thought of the money—and thought of the money, and looked again at the picture—was there no pleasure in being a poor man? Now, you have nothing to do but to walk into Colnaghi's, and buy a wilderness of Lionardos. Yet do you?

"Then, do you remember our pleasant walks to Enfield, and Potter's Bar, and Waltham,[6] when we had a holyday—holydays, and all other fun, are gone, now we are rich—and the little hand-basket in which I used to deposit our day's fare of savory cold lamb

---

OLD CHINA. **1. hays:** a rustic dance with couples interweaving. **2. Hyson:** a variety of green tea. **3. speciosa miracula:** shining wonders. **4. Bridget:** Mary Lamb.

5. corbeau: a dark green cloth, almost black. **6. Enfield . . . Waltham:** suburbs on the north of London.

and salad—and how you would pry about at noon-tide for some decent house, where we might go in, and produce our store—only paying for the ale that you must call for—and speculate upon the looks of the landlady, and whether she was likely to allow us a table-cloth—and wish for such another honest hostess, as Izaak Walton has described[7] many a one on the pleasant banks of the Lea, when he went a-fishing—and sometimes they would prove obliging enough, and sometimes they would look grudgingly upon us—but we had cheerful looks still for one another, and would eat our plain food savorily, scarcely grudging Piscator his Trout Hall? Now, when we go out a day's pleasuring, which is seldom moreover, we *ride* part of the way—and go into a fine inn, and order the best of dinners, never debating the expense—which, after all, never has half the relish of those chance country snaps,[8] when we were at the mercy of uncertain usage, and a precarious welcome.

"You are too proud to see a play anywhere now but in the pit. Do you remember where it was we used to sit, when we saw *The Battle of Hexham*, and *The Surrender of Calais*, and Bannister and Mrs. Bland in *The Children in the Wood*—when we squeezed out our shillings a-piece to sit three or four times in a season in the one-shilling gallery—where you felt all the time that you ought not to have brought me—and more strongly I felt obligation to you for having brought me—and the pleasure was the better for a little shame—and when the curtain drew up, what cared we for our place in the house, or what mattered it where we were sitting, when our thoughts were with Rosalind in Arden, or with Viola at the Court of Illyria? You used to say that the gallery was the best place of all for enjoying a play socially—that the relish of such exhibitions must be in proportion to the infrequency of going—that the company we met there, not being in general readers of plays, were obliged to attend the more, and did attend, to what was going on, on the stage—because a word lost would have been a chasm, which it was impossible for them to fill up. With such reflections we consoled our pride then—and I appeal to you, whether, as a woman, I met generally with less attention and accommodation than I have done since in more expensive situations in the house? The getting in indeed, and the crowding up those inconvenient staircases, was bad enough,—but there was still a law of civility to women recognized to quite as great an extent as we ever found in the other passages—and how a little difficulty overcome heightened the snug seat, and the play, afterwards! Now we can only pay our money, and walk in. You cannot see, you say, in the galleries now. I am sure we saw, and heard too, well enough then—but sight, and all, I think, is gone with our poverty.

"There was pleasure in eating strawberries, before they became quite common—in the first dish of peas, while they were yet dear—to have them for a nice supper, a treat. What treat can we have now? If we were to treat ourselves now—that is, to have dainties a little above our means, it would be selfish and wicked. It is the very little more that we allow ourselves beyond what the actual poor can get at, that makes what I call a treat—when two people living together, as we have done, now and then indulge themselves in a cheap luxury, which both like; while each apologizes, and is willing to take both halves of the blame to his single share. I see no harm in people making much of themselves in that sense of the word. It may give them a hint how to make much of others. But now—what I mean by the word—we never do make much of ourselves. None but the poor can do it. I do not mean the veriest poor of all, but persons as we were, just above poverty.

"I know what you were going to say, that it is mighty pleasant at the end of the year to make all meet—and much ado we used to have every Thirty-first Night of December to account for our exceedings—many a long face did you make over your puzzled accounts, and in contriving to make it out how we had spent so much—or that we had not spent so much—or that it was impossible we should spend so much next year—and still we found our slender capital decreasing—but then, betwixt ways, and projects, and compromises of one sort or another, and talk of curtailing this charge, and doing without that for the future—and the hope that youth brings, and laughing spirits (in which you were never poor till now), we pocketed up our loss, and in conclusion, with 'lusty brimmers' (as you used to quote it out of *hearty cheerful Mr. Cotton*, as you called him), we used to welcome in the 'coming guest.'[9] Now we have no reckoning at all at the end of the old year—no flattering promises about the new year doing better for us."

7. Izaak . . . described: in *The Complete Angler*, 1653. Piscator, mentioned below, is the fisherman in this book. 8. snaps: snacks.

9. 'lusty . . . guest': the phrases were quoted in Lamb's essay, "New Year's Eve," above. See Charles Cotton's poem there and also n. 2.

Bridget is so sparing of her speech, on most occasions, that when she gets into a rhetorical vein, I am careful how I interrupt it. I could not help, however, smiling at the phantom of wealth which her dear imagination had conjured up out of a clear income of poor—— hundred pounds a year. "It is true we were happier when we were poorer, but we were also younger, my cousin. I am afraid we must put up with the excess, for if we were to shake the superflux into the sea, we should not much mend ourselves. That we had much to struggle with, as we grew up together, we have reason to be most thankful. It strengthened, and knit our compact closer. We could never have been what we have been to each other, if we had always had the sufficiency which you now complain of. The resisting power—those natural dilations of the youthful spirit, which circumstances cannot straiten—with us are long since passed away. Competence to age is supplementary youth; a sorry supplement indeed, but I fear the best that is to be had. We must ride, where we formerly walked: live better, and lie softer—and shall be wise to do so—than we had means to do in those good old days you speak of. Yet could those days return—could you and I once more walk our thirty miles a day—could Bannister and Mrs. Bland again be young, and you and I be young to see them—could the good old one-shilling gallery days return—they are dreams, my cousin, now—but could you and I at this moment, instead of this quiet argument, by our well-carpeted fire-side, sitting on this luxurious sofa—be once more struggling up those inconvenient stair-cases, pushed about, and squeezed, and elbowed by the poorest rabble of poor gallery scramblers—could I once more hear those anxious shrieks of yours —and the delicious *Thank God, we are safe*, which always followed when the topmost stair, conquered, let in the first light of the whole cheerful theatre down beneath us—I know not the fathom line that ever touched a descent so deep as I would be willing to bury more wealth in than Crœsus had, or the great Jew R——[10] is supposed to have, to purchase it. And now do just look at that merry little Chinese waiter holding an umbrella, big enough for a bed-tester, over the head of that pretty insipid half-Madonna-ish chit of a lady in that very blue summer-house."

(*1823*)

10. **R——**: Nathan Meyer Rothschild, 1777–1836, founder of the English branch of the Rothschild banking firm.

LETTERS

To William Wordsworth
[*P.M. 30th January 1801.*]

Thanks for your Letter and Present. I had already borrowed your second volume.[1] What most please me are, the Song of Lucy.[2] . . . *Simon's sickly daughter in the Sexton made me cry.*[3] Next to these are the description of the continuous Echoes in the story of Joanna's laugh,[4] where the mountains and all the scenery absolutely seem alive—and that fine Shakesperian character of the Happy Man, in the Brothers,

> —— that creeps about the fields,
> Following his fancies by the hour, to bring
> Tears down his cheek, or solitary smiles
> Into his face, *until the Setting Sun*
> *Write Fool upon his forehead.*[5]

I will mention one more: the delicate and curious feeling in the wish for the Cumberland Beggar, that he may have about him the melody of Birds, altho' he hear them not. Here the mind knowingly passes a fiction upon herself, first substituting her own feelings for the Beggar's, and, in the same breath detecting the fallacy, will not part with the wish.[6]— The Poet's Epitaph is disfigured, to my taste by the vulgar satire upon parsons and lawyers in the beginning, and the coarse epithet of pin point in the 6th stanza.[7] All the rest is eminently good, and your own. I will just add that it appears to me a fault in the Beggar, that the instructions conveyed in it are too direct and like a lecture: they don't slide into the mind of the reader, while he is imagining no such matter. An intelligent reader finds a sort of insult in being told, I will teach you how to think upon this subject. This fault, if I am right, is in a ten-thousandth worse degree to be found in Sterne and many many novelists & modern poets, who continually put a sign post up to

LETTERS. **1. second volume:** the second edition of Wordsworth's *Lyrical Ballads*, published in 1800. **2. Song of Lucy:** See "Lucy Gray," above, under Wordsworth. **3. Simon's . . . cry:** in Wordsworth's "To a Sexton," l. 14. **4. Echoes . . . laugh:** In "To Joanna," ll. 51–65. **5. that . . . forehead:** "The Brothers," ll. 108–12. **6. Cumberland . . . wish:** See "The Old Cumberland Beggar," above, ll. 184–85. **7. The . . . stanza:** "A Poet's Epitaph," above. Lamb refers to stanzas ii and iii and line 24, which originally read, "Thy pinpoint of a soul."

shew where you are to feel. They set out with assuming their readers to be stupid. Very different from Robinson Crusoe, the Vicar of Wakefield, Roderick Random, and other beautiful bare narratives. There is implied an unwritten compact between Author and reader; I will tell you a story, and I suppose you will understand it. Modern novels 'St. Leons'[8] and the like are full of such flowers as these 'Let not my reader suppose,' 'Imagine, *if you can*'—modest!—&c.—I will here have done with praise and blame. I have written so much, only that you may not think I have passed over your book without observation.—I am sorry that Coleridge has christened his Ancient Marinere 'a poet's Reverie'— it is as bad as Bottom the Weaver's declaration that he is not a Lion but only the scenical representation of a Lion.[9] What new idea is gained by this Title, but one subversive of all credit, which the tale should force upon us, of its truth? For me, I was never so affected with any human Tale. After first reading it, I was totally possessed with it for many days—I dislike all the miraculous part of it, but the feelings of the man under the operation of such scenery dragged me along like Tom Piper's magic whistle.[10] I totally differ from your idea that the Marinere should have had a character and profession. This is a Beauty in Gulliver's Travels, where the mind is kept in a placid state of little wonderments; but the Ancient Marinere undergoes such Trials, as overwhelm and bury all individuality or memory of what he was, like the state of a man in a Bad dream, one terrible peculiarity of which is: that all consciousness of personality is gone. Your other observation[11] is I think as well a little unfounded: the Marinere from being conversant in supernatural events *has* acquired a supernatural and strange cast of *phrase*, eye, appearance, &c. which frighten the wedding guest. You will excuse my remarks, because I am hurt and vexed that you should think it necessary, with a prose apology, to open the eyes of dead men that cannot see. To sum up a general opinion of the second vol.—I do not feel any one poem in it so forcibly as the Ancient

Marinere, the Mad Mother,[12] and the Lines at Tintern Abbey in the first.—I could, too, have wished the Critical preface had appeared in a separate treatise. All its dogmas are true and just, and most of them new, *as* criticism. But they associate a *diminishing* idea with the Poems which follow, as having been written for *Experiment* on the public taste, more than having sprung (as they must have done) from living and daily circumstances.—I am prolix, because I am gratifyed in the opportunity of writing to you, and I don't well know when to leave off. I ought before this to have reply'd to your very kind invitation into Cumberland. With you and your Sister I could gang any where. But I am afraid whether I shall ever be able to afford so desperate a Journey. Separate from the pleasure of your company, I don't much care if I never see a mountain in my life. I have passed all my days in London, until I have formed as many and intense local attachments, as any of you mountaineers can have done with dead nature. The Lighted shops of the Strand and Fleet Street, the innumerable trades, tradesmen and customers, coaches, waggons, playhouses, all the bustle and wickedness round about Covent Garden, the very women of the Town, the Watchmen, drunken scenes, rattles,—life awake, if you awake, at all hours of the night, the impossibility of being dull in Fleet Street, the crowds, the very dirt & mud, the Sun shining upon houses and pavements, the print shops, the old book stalls, parsons cheap'ning books, coffee houses, steams of soups from kitchens, the pantomimes, London itself a pantomime and a masquerade,—all these things work themselves into my mind and feed me, without a power of satiating me. The wonder of these sights impells me into nightwalks about her crowded streets, and I often shed tears in the motley Strand from fulness of joy at so much Life.—All these emotions must be strange to you. So are your rural emotions to me. But consider, what must I have been doing all my life, not to have lent great portions of my heart with usury to such scenes?——

My attachments are all local, purely local. I have no passion (or have had none since I was in love, and then it was the spurious engendering of poetry & books) to groves and vallies. The rooms where I was born, the furniture which has been before my eyes all my life, a book case which has followed me about (like a faithful dog, only exceeding him in knowledge)

---

8. **'St. Leons':** William Godwin's novel, *St. Leon* (1799). **9. Bottom . . . Lion:** *A Midsummer Night's Dream*, III. i. 40–45; V. i. 222–23. **10. Tom . . . whistle:** the piper who freed the town of Hamelin from rats. When the townsmen would not pay, he drew their children away by his piping. **11. Your . . . observation:** In a note to the *Lyrical Ballads* Wordsworth had said that the mariner "has no distinct character, either in his profession of Mariner, or as a human being who having been long under the controul of supernatural impressions might be supposed himself to partake of something supernatural."

12. **Mad Mother:** "Her Eyes Are Wild."

wherever I have moved—old chairs, old tables, streets, squares, where I have sunned myself, my old school,—these are my mistresses. Have I not enough, without your mountains? I do not envy you. I should pity you, did I not know, that the Mind will make friends of any thing. Your sun & moon and skys and hills & lakes affect me no more, or scarcely come to me in more venerable characters, than as a gilded room with tapestry and tapers, where I might live with handsome visible objects. I consider the clouds above me but as a roof, beautifully painted but unable to satisfy the mind, and at last, like the pictures of the apartment of a connoisseur, unable to afford him any longer a pleasure. So fading upon me, from disuse, have been the Beauties of Nature, as they have been confinedly called; so ever fresh & green and warm are all the inventions of men and assemblies of men in this great city. I should certainly have laughed with dear Joanna.

Give my kindest love, *and my sister's*, to D.[13] & your*self* and a kiss from me to little Barbara Lewthwaite.[14]

C. LAMB.

Thank you for Liking my Play!![15]

To Thomas Manning[16]
Feb. 15, 1801.

I had need be cautious henceforward what opinion I give of the 'Lyrical Ballads.' All the North of England are in a turmoil. Cumberland and Westmoreland[17] have already declared a state of war. I lately received from Wordsworth a copy of the second volume, accompanied by an acknowledgement of having received from me many months since a copy of a certain Tragedy, with excuses for not having made any acknowledgement sooner, it being owing to an 'almost insurmountable aversion from Letter-writing.' This letter I answered in due form and time, and enumerated several of the passages which had most affected me, adding, unfortunately, that no single piece had moved me so forcibly as the *Ancient Mariner, The*

Mad Mother, or the *Lines at Tintern Abbey*. The Post did not sleep a moment. I received almost instantaneously a long letter of four sweating pages from my Reluctant Letter-Writer, the purport of which was, that he was sorry his 2d vol. had not given me more pleasure (Devil a hint did I give that it had *not pleased me*), and 'was compelled to wish that my range of sensibility was more extended, being obliged to believe that I should receive large influxes of happiness and happy Thoughts' (I suppose from the L. B.)—With a deal of stuff about a certain Union of Tenderness and Imagination, which in the sense he used Imagination was not the characteristic of Shakspeare, but which Milton possessed in a degree far exceeding other Poets: which Union, as the highest species of Poetry, and chiefly deserving that name, 'He was most proud to aspire to'; then illustrating the said Union by two quotations from his own 2d vol. (which I had been so unfortunate as to miss). 1st Specimen—a father addresses his son:

> When thou
> First camest into the World, as it befalls
> To new-born Infants, thou didst sleep away
> Two days: and *Blessings from Thy father's Tongue*
> *Then fell upon thee.*[18]

The lines were thus undermarked, and then followed 'This Passage, as combining in an extraordinary degree that Union of Imagination and Tenderness which I am speaking of, I consider as one of the Best I ever wrote!'

2d Specimen.—A youth, after years of absence, revisits his native place, and thinks (as most people do) that there has been strange alteration in his absence:—

> And that the rocks
> And everlasting Hills themselves were changed.[19]

You see both these are good Poetry: but after one has been reading Shakspeare twenty of the best years of one's life, to have a fellow start up, and prate about some unknown quality, which Shakspeare possessed in a degree inferior to Milton and *somebody else*!! This was not to be *all* my castigation. Coleridge, who had not written to me some months before, starts up from his bed of sickness to reprove me for my hardy presumption: four long pages, equally sweaty and more tedious, came from him; assuring me that, when

---

13. **D.:** Wordsworth's sister, Dorothy. 14. **Barbara Lewthwaite:** in Wordsworth's poem "The Pet Lamb." 15. **my Play:** *John Woodvil*. 16. **Thomas Manning:** One of Lamb's closest friends, Manning was a mathematician. 17. **Cumberland and Westmoreland:** Coleridge lived at Keswick, in Cumberland County; Wordsworth, at Grasmere in Westmorland.

18. **When . . . thee:** "Michael," ll. 339–43. 19. **And . . . changed:** "The Brothers," ll. 88–89.

the works of a man of true genius such as W. undoubtedly was, do not please me at first sight, I should
suspect the fault to lie 'in me and not in them,' etc.
etc. etc. etc. etc. What am I to do with such people?
I certainly shall write them a very merry Letter.
Writing to *you*, I may say that the 2d vol. has no such
pieces as the three I enumerated. It is full of original
thinking and an observing mind, but it does not often
make you laugh or cry.—It too artfully aims at simplicity of expression. And you sometimes doubt if
Simplicity be not a cover for Poverty. The best Piece
in it I will send you, being *short*. I have grievously
offended my friends in the North by declaring my
undue preference; but I need not fear you:—

> She dwelt among the untrodden ways
> Beside the Springs of Dove,
> A maid whom there were few [none] to praise
> And very few to love.
>
> A violet, by a mossy stone,
> Half hidden from the eye.
> Fair as a star when only one
> Is shining in the sky.
>
> She lived unknown; and few could know,
> When Lucy ceased to be.
> But she is in the [her] grave, and oh!
> The difference to me.

This is choice and genuine, and so are many, many
more. But one does not like to have 'em rammed
down one's throat. 'Pray, take it—it's very good—let
me help you—eat faster.' . . .

To Thomas Manning
24th Sept., 1802, London.

MY DEAR MANNING,

Since the date of my last letter, I have been a traveller.
A strong desire seized me of visiting remote regions.
My first impulse was to go and see Paris. It was a
trivial objection to my aspiring mind, that I did not
understand a word of the language, since I certainly
intend some time in my life to see Paris, and equally
certainly never intend to learn the language; therefore
that could be no objection. However, I am very glad
I did not go, because you had left Paris (I see) before I
could have set out. I believe, Stoddart[20] promising to
go with me another year prevented that plan. My next
scheme, (for to my restless, ambitious mind London
was become a bed of thorns) was to visit the far-
famed Peak in Derbyshire, where the Devil sits, they
say, without breeches. *This* my purer mind rejected as
indelicate. And my final resolve was a tour to the
Lakes. I set out with Mary to Keswick, without giving
Coleridge any notice; for my time being precious did
not admit of it. He received us with all the hospitality in
the world, and gave up his time to show us all the
wonders of the country. He dwells upon a small hill by
the side of Keswick, in a comfortable house, quite enveloped on all sides by a net of mountains: great
floundering bears and monsters they seemed, all couchant and asleep. We got in in the evening, travelling in a
post-chaise from Penrith, in the midst of a gorgeous
sunshine, which transmuted all the mountains into colours, purple, &c. &c. We thought we had got into Fairy
Land. But that went off (as it never came again—while
we stayed we had no more fine sunsets); and we
entered Coleridge's comfortable study just in the dusk,
when the mountains were all dark with clouds upon
their heads. Such an impression I never received from
objects of sight before, nor do I suppose I can ever again.
Glorious creatures, fine old fellows, Skiddaw, &c.
I never shall forget ye, how ye lay about that night,
like an intrenchment; gone to bed, as it seemed for
the night, but promising that ye were to be seen in
the morning. Coleridge had got a blazing fire in his
study; which is a large, antique, ill-shaped room, with
an old-fashioned organ, never played upon, big enough
for a church, shelves of scattered folios, an Æolian
harp, and an old sofa, half-bed, &c. And all looking
out upon the last fading view of Skiddaw and his
broad-breasted brethren: what a night! Here we stayed
three full weeks, in which time I visited Wordsworth's
cottage, where we stayed a day or two with the Clarksons[21] (good people and most hospitable, at whose
house we tarried one day and night), and saw Lloyd.[22]
The Wordsworths were gone to Calais. They have
since been in London and past much time with us:
he is now gone into Yorkshire to be married to a girl of
small fortune,[23] but he is in expectation of augmenting
his own in consequence of the death of Lord Lonsdale,
who kept him out of his own in conformity with a

---

20. **Stoddart:** Sir John Stoddart, 1773–1856, a writer and
lawyer. His sister Sarah married William Hazlitt.

21. **Clarksons:** Thomas Clarkson, 1760–1846, a leader in the
agitation against slavery. He and his wife were friends of the
Wordsworths. 22. **Lloyd:** Charles Lloyd, a minor poet
and a friend. 23. **girl . . . fortune:** Mary Hutchinson.

plan my lord had taken up in early life of making everybody unhappy.[24] So we have seen Keswick, Grasmere, Ambleside, Ulswater (where the Clarksons live), and a place at the other end of Ulswater—I forget the name—to which we travelled on a very sultry day, over the middle of Helvellyn. We have clambered up to the top of Skiddaw, and I have waded up the bed of Lodore.[25] In fine, I have satisfied myself, that there is such a thing as that which tourists call *romantic*, which I very much suspected before: they make such a spluttering about it, and toss their splendid epithets around them, till they give as dim a light as at four o'clock next morning the lamps do after an illumination. Mary was excessively tired, when she got about half-way up Skiddaw, but we came to a cold rill (than which nothing can be imagined more cold, running over cold stones), and with the reinforcement of a draught of cold water she surmounted it most manfully. Oh, its fine black head, and the bleak air atop of it, with a prospect of mountains all about, and about, making you giddy; and then Scotland afar off, and the border countries so famous in song and ballad! It was a day that will stand out, like a mountain, I am sure, in my life. But I am returned (I have now

24. **Lord Lonsdale . . . unhappy:** At the death of Words-worth's father, Lord Lonsdale owed him a large sum which he refused to pay. His heir paid the debt. 25. **Lodore:** the famous waterfall. See Southey's "The Cataract of Lodore," above.

been come home near three weeks—I was a month out), and you cannot conceive the degradation I felt at first, from being accustomed to wander free as air among mountains, and bathe in rivers without being controlled by any one, to come home and *work*. I felt very *little*. I had been dreaming I was a very great man. But that is going off, and I find I shall conform in time to that state of life to which it has pleased God to call me. Besides, after all, Fleet-Street and the Strand are better places to live in for good and all than among Skiddaw. Still, I turn back to those great places where I wandered about, participating in their greatness. After all, I could not *live* in Skiddaw. I could spend a year—two, three years—among them, but I must have a prospect of seeing Fleet-Street at the end of that time, or I should mope and pine away, I know. Still, Skiddaw is a fine creature. My habits are changing, I think: *i.e.* from drunk to sober. Whether I shall be happier or not remains to be proved. I shall certainly be more happy in a morning; but whether I shall not sacrifice the fat, and the marrow, and the kidneys, *i.e.* the night, the glorious care-drowning night, that heals all our wrongs, pours wine into our mortifications, changes the scene from indifferent and flat to bright and brilliant!—O Manning, if I should have formed a diabolical resolution, by the time you come to England, of not admitting any spirituous liquors into my house, will you be my guest on such shameworthy terms? Is life, with such limitations, worth trying? . . .

THOMAS CAMPBELL

1777–1844

CAMPBELL was a lifelong liberal in his political sentiments, but gently conservative—even timid—in his sense of literary propriety. Born in Glasgow, he was the son of a merchant whose firm had engaged in trade with the colony of Virginia and had now lost most of its business as a result of the American Revolution. While at school and at the University of Glasgow he won prizes for verse-writing, spent his vacations as a tutor in the Highlands, and shared the general interest of his time in folklore and balladry. He then went to Edinburgh to study law, meanwhile writing his poem *The Pleasures of Hope* (published in 1799). It at once became popular. The topics were pleasing and relevant to liberal and humanitarian readers (one of the themes is the effect on people deprived of hope, and Campbell focuses on such examples as Negro slavery and the partition of Poland), while, at the same time, its conventionality of mode and idiom disturbed no one. The speculative, musing approach that the title suggests had been familiar for half a century—since, in fact, Mark Akenside's *Pleasures of the Imagination* (1744) and Thomas Warton's *Pleasures of Melancholy* (1745). In versification, moreover, Campbell was even less bold than his eighteenth-century predecessors, most of whom, including Akenside and Warton, had written in blank verse. Campbell adhered to the closed heroic couplet, and through practice and constant polish he attained a reasonable condensation in using it (a few lines have become proverbial—"Like angel-visits, few and far between," or "'Tis distance lends enchantment to the view"). The poem raised expectations that Campbell lacked the self-confidence to try to satisfy. He traveled abroad, writing a few lyrics ("Ye Mariners of England" and "Hohenlinden" date from this time). These were followed by a narrative poem about the Indians in the New World, *Gertrude of Wyoming* (1809), written in Spenserian stanzas. From now on his productiveness waned. He occupied himself with work on his *Specimens of the British Poets*, including with these selections some perceptive short biographies and a very respectable prefatory essay. He became editor in 1820 of *The New Monthly Magazine* and took part in founding the University of London. Personal troubles mounted. His remaining son (the other had died in childhood) became insane; his wife died in 1828; and Campbell, himself not well, retired from public life. He lived on to 1844, writing occasionally, but the only work that survives from these years is his *Letters from the South* (1837), written during a trip to France and Algeria.

Bibliography

The standard edition is the *Complete Poetical Works*, ed. J. L. Robertson, 1907. For selections, see *Poems*, ed. Lewis Campbell, 1904 (Golden Treasury Series). A fine, brief contemporary discussion is Hazlitt's "Mr. Campbell, Mr. Crabbe" in *The Spirit of the Age*, 1825. For biography and more recent criticism, see J. C. Hadden, *Thomas Campbell*, 1899, and W. M. Dixon, "Thomas Campbell," in *An Apology for the Arts*, 1944.

FROM

THE PLEASURES
OF HOPE

HOPE ABIDETH

At summer eve, when Heaven's ethereal bow
Spans with bright arch the glittering hills below,
Why to yon mountain turns the musing eye,
Whose sunbright summit mingles with the sky?
Why do those cliffs of shadowy tint appear
More sweet than all the landscape smiling near?
'Tis distance lends enchantment to the view,
And robes the mountains in its azure hue.
Thus, with delight, we linger to survey
The promised joys of life's unmeasured way;   10
Thus, from afar, each dim-discovered scene
More pleasing seems than all the past hath been;
And every form, that Fancy can repair
From dark oblivion, glows divinely there.
    What potent spirit guides the raptured eye
To pierce the shades of dim futurity?

Can Wisdom lend, with all her heavenly power,
The pledge of Joy's anticipated hour?
Ah, no! she darkly sees the fate of man—
Her dim horizon bounded to a span;   20
Or, if she hold an image to the view,
'Tis Nature pictured too severely true.
With thee, sweet Hope! resides the heavenly light
That pours remotest rapture on the sight:
Thine is the charm of life's bewildered way,
That calls each slumbering passion into play.
Waked by thy touch, I see the sister band,
On tiptoe watching, start at thy command,
And fly where'er thy mandate bids them steer,
To Pleasure's path or Glory's bright career.   30
    Primeval Hope, the Aönian Muses say,
When Man and Nature mourned their first decay;
When every form of death, and every woe,
Shot from malignant stars to earth below;
When Murder bared his arm, and rampant War
Yoked the red dragons of her iron car;
When Peace and Mercy, banished from the plain,
Sprung on the viewless winds to heaven again;
All, all forsook the friendless, guilty mind,
But Hope, the charmer, lingered still behind.   40

POLAND

Where barbarous hordes of Scythian mountains roam,
Truth, Mercy, Freedom, yet shall find a home;   340
Where'er degraded Nature bleeds and pines,
From Guinea's coast to Sibir's dreary mines,
Truth shall pervade the unfathomed darkness there,
And light the dreadful features of despair.—
Hark! the stern captive spurns his heavy load,
And asks the image back that Heaven bestowed!
Fierce in his eye the fire of valor burns,
And, as the slave departs, the man returns.
Oh! sacred Truth! thy triumph ceased awhile,
And Hope, thy sister, ceased with thee to smile,   350
When leagued Oppression poured to Northern wars
Her whiskered pandoors and her fierce hussars,
Waved her dread standard to the breeze of morn,
Pealed her loud drum, and twanged her trumpet horn;
Tumultuous horror brooded o'er her van,
Presaging wrath to Poland—and to man!
Warsaw's last champion from her height surveyed,
Wide o'er the fields, a waste of ruin laid;
"Oh! Heaven!" he cried, "my bleeding country save!
Is there no hand on high to shield the brave?   360
Yet, though destruction sweep those lovely plains,
Rise, fellow-men! our country yet remains!
By that dread name, we wave the sword on high!
And swear for her to live!—with her to die!"
He said, and on the rampart-heights arrayed
His trusty warriors, few, but undismayed;
Firm-paced and slow, a horrid front they form,
Still as the breeze, but dreadful as the storm;
Low murmuring sounds along their banners fly,
Revenge, or death,—the watchword and reply;   370
Then pealed the notes, omnipotent to charm,
And the loud tocsin tolled their last alarm!
In vain, alas! in vain, ye gallant few!
From rank to rank your volleyed thunder flew:
Oh, bloodiest picture in the book of Time,
Sarmatia fell, unwept, without a crime;
Found not a generous friend, a pitying foe,
Strength in her arms, nor mercy in her woe!

Dropped from her nerveless grasp the shattered spear,
Closed her bright eye, and curbed her high career;—
Hope, for a season, bade the world farewell,   381
And Freedom shrieked as Kosciusko fell!
The sun went down, nor ceased the carnage there,
Tumultuous murder shook the midnight air;
On Prague's proud arch the fires of ruin glow,
His blood-dyed waters murmuring far below;
The storm prevails, the rampart yields a way;
Bursts the wild cry of horror and dismay!
Hark! as the smoldering piles with thunder fall,
A thousand shrieks for hopeless mercy call!   390
Earth shook; red meteors flashed along the sky,
And conscious Nature shuddered at the cry!
Oh! righteous Heaven; ere Freedom found a grave,
Why slept the sword omnipotent to save?
Where was thine arm, O Vengeance! where thy rod,
That smote the foes of Zion and of God;
That crushed proud Ammon, when his iron car
Was yoked in wrath, and thundered from afar?
Where was the storm that slumbered till the host
Of blood-stained Pharaoh left their trembling coast;
Then bade the deep in wild commotion flow,   401
And heaved an ocean on their march below?
Departed spirits of the mighty dead!
Ye that at Marathon and Leuctra bled!
Friends of the world! restore your swords to man,
Fight in his sacred cause, and lead the van!
Yet for Sarmatia's tears of blood atone,
And make her arm puissant as your own!
Oh! once again to Freedom's cause return
The patriot Tell—the Bruce of Bannockburn!   410
Yes! thy proud lords, unpitied land, shall see
That man hath yet a soul—and dare be free!
A little while, along thy saddening plains,
The starless night of Desolation reigns;
Truth shall restore the light by Nature given.
And, like Prometheus, bring the fire of Heaven!
Prone to the dust Oppression shall be hurled,
Her name, her nature, withered from the world!

1796–99 (1799)

THE PLEASURES OF HOPE. POLAND. **342. Guinea's . . . Sibir's:**
Guinea, Africa, from which slaves were commonly exported,
and the mines of Siberia. **351. leagued Oppression:**
the division of Poland in the 1790's by Russia, Prussia, and
Austria. **352. pandoors:** an Austrian regiment noted for its
fierceness. **357. champion:** Thaddeus Kosciusko, taken
prisoner in 1794. **376. Sarmatia:** Poland.

400–02. Pharoah . . . below: See Exodus 14.

YE MARINERS OF ENGLAND

A NAVAL ODE

I

Ye Mariners of England,
That guard our native seas,
Whose flag has braved, a thousand years,
The battle and the breeze—
Your glorious standard launch again
To match another foe!
And sweep through the deep,
While the stormy winds do blow,—
While the battle rages loud and long,
And the stormy winds do blow.   10

II

The spirit of your fathers
Shall start from every wave!
For the deck it was their field of fame,
And Ocean was their grave.
Where Blake and mighty Nelson fell,
Your manly hearts shall glow,
As ye sweep through the deep,
While the stormy winds do blow,—
While the battle rages loud and long,
And the stormy winds do blow.   20

III

Britannia needs no bulwarks,
No towers along the steep;
Her march is o'er the mountain-waves,
Her home is on the deep.
With thunders from her native oak,
She quells the floods below,
As they roar on the shore,
When the stormy winds do blow,—
When the battle rages loud and long,
And the stormy winds do blow.   30

IV

The meteor flag of England
Shall yet terrific burn,
Till danger's troubled night depart,
And the star of peace return.

Then, then, ye ocean-warriors!
Our song and feast shall flow
To the fame of your name,
When the storm has ceased to blow,—
When the fiery fight is heard no more,
And the storm has ceased to blow.   40

1799–1800 (*1801*)

HOHENLINDEN

Hohenlinden, Bavaria, was the scene of a fierce battle
(December, 1800) where the French defeated the
Austrians.

On Linden, when the sun was low,
All bloodless lay th' untrodden snow,
And dark as winter was the flow
Of Iser, rolling rapidly.

But Linden saw another sight,
When the drum beat at dead of night,
Commanding fires of death to light
The darkness of her scenery.

By torch and trumpet fast array'd,
Each horseman drew his battle-blade,   10
And furious every charger neigh'd,
To join the dreadful revelry.

Then shook the hills with thunder riven,
Then rush'd the steed to battle driven,
And louder than the bolts of heaven,
Far flash'd the red artillery.

But redder yet that light shall glow
On Linden's hills of stainèd snow,
And bloodier yet the torrent flow
Of Iser, rolling rapidly.   20

'Tis morn; but scarce yon level sun
Can pierce the war-clouds, rolling dun,
Where furious Frank and fiery Hun,
Shout in their sulphurous canopy.

The combat deepens. On, ye brave,
Who rush to glory, or the grave!
Wave, Munich! all thy banners wave,
And charge with all thy chivalry!

Few, few, shall part where many meet!
The snow shall be their winding-sheet, 30
And every turf beneath their feet
Shall be a soldier's sepulchre.

1802 (1802)

THE BATTLE OF THE BALTIC

⌒⋇⌒

In this battle of April, 1801, Nelson and the British fleet defeated the Danes. The poem is an attempt, said Campbell, "to write an English ballad on the battle of Copenhagen, as much as possible in that plain, strong style peculiar to our old ballads, which tell us the when, where, and how the event happened—without gaud or ornament but what the subject essentially and easily affords."

⌒⋇⌒

Of Nelson and the North,
Sing the glorious day's renown,
When to battle fierce came forth
All the might of Denmark's crown,
And her arms along the deep proudly shone;
By each gun and lighted brand,
In a bold determined hand,
And the Prince of all the land
Led them on.

Like leviathans afloat, 10
Lay their bulwarks on the brine,
While the sign of battle flew
On the lofty British line;
It was ten of April morn by the chime.
As they drifted on their path,
There was silence deep as death;
And the boldest held his breath,
For a time.

But the might of England flush'd
To anticipate the scene; 20
And her van the fleeter rush'd
O'er the deadly space between.
"Hearts of oak!" our captain cried, when each gun
From its adamantine lips
Spread a death-shade round the ships,
Like the hurricane eclipse
Of the sun.

Again! again! again!
And the havoc did not slack,
Till a feeble cheer the Dane 30
To our cheering sent us back;
Their shots along the deep slowly boom—
Then ceased—and all is wail,
As they strike the shatter'd sail,
Of in conflagration pale
Light the gloom.

Out spoke the victor then,
As he hail'd them o'er the wave;
"Ye are brothers! ye are men!
And we conquer but to save; 40
So peace instead of death let us bring.
But yield, proud foe, thy fleet,
With the crews, at England's feet,
And make submission meet
To our King."

Then Denmark bless'd our chief,
That he gave her wounds repose;
And the sounds of joy and grief
From her people wildly rose,
As Death withdrew his shades from the day. 50
While the sun look'd smiling bright
O'er a wide and woful sight,
Where the fires of funeral light
Died away.

Now joy, old England, raise!
For the tidings of thy might,
By the festal cities' blaze,
Whilst the wine-cup shines in light;
And yet amidst that joy and uproar,
Let us think of them that sleep, 60
Full many a fathom deep,
By thy wild and stormy steep,
Elsinore!

Brave hearts! to Britain's pride
Once so faithful and so true,
On the deck of fame that died,
With the gallant good Riou;
Soft sigh the winds of Heaven o'er their grave!
While the billow mournful rolls,
And the mermaid's song condoles, 70
Singing glory to the souls
Of the brave!

1804–05 (1809)

THE LAST MAN

⤬

Compare Byron's poem, "Darkness," below, the
subject of which Campbell suggested to him.

⤬

All worldly shapes shall melt in gloom,
   The Sun himself must die,
Before this mortal shall assume
   Its Immortality!
I saw a vision in my sleep
That gave my spirit strength to sweep
   Adown the gulf of Time!
I saw the last of human mold
That shall Creation's death behold,
   As Adam saw her prime!   10

The Sun's eye had a sickly glare,
   The Earth with age was wan,
The skeletons of nations were
   Around that lonely man!
Some had expired in fight,—the brands
Still rusted in their bony hands;
   In plague and famine some!
Earth's cities had no sound nor tread;
And ships were drifting with the dead
   To shores where all was dumb!   20

Yet, prophet-like, that lone one stood
   With dauntless words and high,
That shook the sere leaves from the wood
   As if a storm passed by,
Saying, "We are twins in death, proud Sun!
Thy face is cold, thy race is run,
   'Tis Mercy bids thee go;
For thou ten thousand thousand years
Hast seen the tide of human tears,
   That shall no longer flow.   30

"What though beneath thee man put forth
   His pomp, his pride, his skill,
And arts that made fire, flood, and earth,
   The vassals of his will?
Yet mourn I not thy parted sway,
Thou dim discrownèd king of day,

For all those trophied arts
And triumphs that beneath thee sprang
Healed not a passion or a pang
   Entailed on human hearts.   40

"Go, let oblivion's curtain fall
   Upon the stage of men,
Nor with thy rising beams recall
   Life's tragedy again.
Its piteous pageants bring not back,
Nor waken flesh upon the rack
   Of pain anew to writhe—
Stretched in disease's shapes abhorred,
Or mown in battle by the sword
   Like grass beneath the scythe.   50

"E'en I am weary in yon skies
   To watch thy fading fire;
Test of all sumless agonies,
   Behold not me expire!
My lips that speak thy dirge of death—
Their rounded gasp and gurgling breath
   To see thou shalt not boast;
The eclipse of Nature spreads my pall,—
The majesty of Darkness shall
   Receive my parting ghost!   60

"This spirit shall return to Him
   Who gave its heavenly spark;
Yet think not, Sun, it shall be dim
   When thou thyself art dark!
No! it shall live again, and shine
In bliss unknown to beams of thine.
   By him recalled to breath
Who captive led captivity,
Who robbed the grave of Victory,
   And took the sting from Death!   70

"Go, Sun, while Mercy holds me up
   On Nature's awful waste
To drink this last and bitter cup
   Of grief that man shall taste—
Go, tell the night that hides thy face
Thou saw'st the last of Adam's race
   On Earth's sepulchral clod
The darkening universe defy
To quench his immortality
   Or shake his trust in God!"   80

1823 (1823)

WILLIAM HAZLITT

1778–1830

HAZLITT, as we have only recently begun to appreciate, was the most representative critic of English Romanticism and (together with Coleridge, or not far below him) one of the two most distinguished. Throughout the Victorian period and the first third of the twentieth century, the works of his that were most generally read were his familiar essays. It is through our present-day interest in the history of ideas, and above all our new attempt to understand Romanticism more comprehensively, that we find ourselves turning more to Hazlitt as a critic while still valuing him as a general essayist.

A political liberal all his life (and here he was typical of the central impulse of Romanticism), he was born in 1778, the son of a Unitarian minister who supported American independence and the principles of the French Revolution. When Hazlitt was five the family went for a while to America and then returned to England, settling at Wem, Shropshire, where his father preached. At the age of fifteen he was sent to the Unitarian College in Hackney, a suburb of London, where his tutor was Joseph Priestley (remembered for his discovery of oxygen but also well known at the time both as a literary critic and as a political liberal). Hazlitt hesitated between philosophy and painting. He ended by trying both, and his first publication (1805) was his *Essay on the Principles of Human Action*. Meanwhile it was necessary to earn a living, and he turned naturally and easily to journalism. He wrote on almost every topic—literature, the visual arts, economics, politics, and casual subjects. As he neared the age of forty he began to concentrate more on literature. Particularly notable was his book *Characters of Shakespeare's Plays* (1817), bought and eagerly read by his admiring young disciple, Keats. Then Hazlitt began his series of lectures (1818) at the Surrey Institution (which Keats also attended when he could)—the *Lectures on the English Poets* (two of which, "On Shakspeare and Milton" and "On the Living Poets," are printed below). These were followed the next year by the *Lectures on the English Comic Writers* (1819), the opening lecture of which ("On Wit and Humour") is presented here.

Hazlitt's manner in company was alternately shy and aggressive. Coleridge once spoke of him as "brow-hanging, shoe-contemplative, *strange*." There is something altogether typical in the way in which, as he entered a house, he would try to toss his hat cavalierly onto a table, only for it to slip over the table and fall on the floor. Something of this awkward aggressiveness has been attributed to his personal life during his thirties—his stormy marriage to Sarah Stoddart, and his hope, after divorce, to marry the flighty daughter of his landlord, Sarah Walker, which was disappointed when she proved faithless. On the other hand, much that took place in his personal life was itself a by-product of his constant self-dissatisfaction.

He had cultivated the knack of writing with extraordinary speed (his collected works fill twenty-one volumes). If we are bothered at times by his comparative diffuseness and his constant tendency to quote, we should remember that he had to earn his living through journalism and was very often paid by the page. As he neared the age of fifty he became gentler and more nostalgic (see, for example, the essay "On Reading Old Books," below), though, if attacked in the press (and much journalistic writing of the time turned into political and critical warfare), he could prove himself a master of satiric invective—almost invariably in a good cause. In 1824 Hazlitt married again, but his wife reluctantly left him because of constant conflict with his son by his first marriage. He continued to write with independence and energy, admired by many of the younger writers. His last words, when he died at the age of fifty-two (1830), are among the famous "last words" of history: "Well, I've had a happy life."

His life was one of vigor and of unswerving fidelity (whatever the occasional harshnesses) to principles and ideals. "I know," said Keats back in 1817, when he was twenty-two and Hazlitt thirty-nine, "he thinks himself not estimated by ten people in the world—I wish he knew he *is*"; and the more he thought of it in the months that followed the more he felt that among the "three things to rejoice at in this age" was "Hazlitt's depth of taste."

Hazlitt is representative of so much in English Romanticism because of his range: he speaks for the visual arts as well as literature; he is concerned with economics and political theory. He went to plays, reviewed them, and knew the practical side of the theater, as Coleridge, for example, did not; and unlike Wordsworth, Coleridge, and Shelley, but like Byron, Lamb, and especially Keats, he had a rich sense of the comic in literature. Moreover, in expressing Romantic ideals, he went back to eighteenth-century English philosophy, assimilating it as English Romanticism in general assimilated it. He is an empiricist in the tradition of John Locke, constantly reverting to concrete experience. As he approaches literature and the arts from this fertile tradition, he gives a coherent expression to almost every serious critical value held in his own period except for the more transcendental, mystical, or abstractly theoretical ones. The Romantic exploration of the psychology of the imagination is an example. More than any other writer in the period, Hazlitt offered a penetrating development of this psychology in the line of the British empirical philosophers of the eighteenth century, who had, after all, virtually invented it. This was no mystical concept of the imagination but an informed appreciation of the fluid process by which the mind fuses direct concrete experiences (a good summary of the position is found in the two essays "On Genius and Common Sense," printed below).

He also developed, further than any other critic, the concept of the sympathetic nature of the imagination for which eighteenth-century British psychology had laid the foundations. The standpoint is described in his first work, the *Essay on the Principles of Human Action*. Once we depart from our direct experience of the immediate present, or our memory of the past, we have no way of conceiving our "identity" as individuals—above all of conceiving our future—except through an act of the imagination by which we "sympathize" or become "identified" with a mere idea or concept (the idea of our "future" being—something that does not yet exist in fact). The same act of imagination by which we jump forward to our still nonexistent future is equally capable of identifying itself with others. "I could not love myself, if I were not capable of loving others." It is this ability of the poet to get outside himself, and to present life and nature with vivid concreteness, that Hazlitt most values in art. Shakespeare especially is held up as an example of this capacity of mind (see in particular the lecture "On Shakspeare and Milton," below), and it is through Hazlitt more than anyone else that Keats was able to evolve his own suggestive concept of the poet of "Negative Capability."

In Hazlitt's approach to the literature of his own day, his gusto and relish were combined with psychological shrewdness and a firm, but by no means narrow, standard based on the greatest writings of the English past. He saw at once the self-defeating factors in the sentimentalism and emotionalism of the extreme (and lesser) Romantic writers: self-defeating because ultimately egocentric and therefore working against both objectivity and range or diversity of experience. We see this also in his writings on some of the major figures, especially Wordsworth and Byron.

If at times he becomes too impatient, we should remember the reason—his idolization of the objective, dramatic imagination as typified by the Shakespearean drama, in contrast to the more introspective and lyrical character of later poetry. Here, too, he was typical of his age. English Romanticism had a divided soul. Despite its achievements, it also looked back nostalgically on the "larger forms" (epic and dramatic tragedy) of earlier periods—as indeed has every generation since the Romantics.

Bibliography

The standard edition of Hazlitt is the *Collected Works*, ed. P. P. Howe, 21 vols., 1930–34. Of numerous selected editions, the best for general purposes is *Selected Essays*, ed. Geoffrey Keynes, 1930, and for the criticism, *Hazlitt on English Literature*, ed. Jacob Zeitlin, 1926. An edition of the letters is now being prepared by H. M. Sikes. For a straight chronological account of his life, see *The Life of William Hazlitt* by P. P. Howe, 1922; rev. 1947. However, much of the material in earlier biographies, which were relatively unconcerned with Hazlitt's ideas, has been superseded by Herschel Baker's magisterial *William Hazlitt*, 1962, which is concerned with every facet of Hazlitt's life and work, including the political writing. Our appreciation of Hazlitt as a critic is, as stated above, relatively recent. Aside from Baker's comprehensive work, the best short discussion of this aspect of Hazlitt is W. J. Bate's introduction to him in *Criticism: The Major Texts*, 1952, pp. 281–92, which, together with Bate's earlier studies of the Romantic concept of the "sympathetic imagination" and John Bullitt's "Hazlitt and the Romantic Conception of the Imagination" (*Philological Quarterly*, XXIV [1945], 343–61), was influential in calling attention to Hazlitt's premises as a psychological critic. Other studies of Hazlitt's criticism include H. W. Garrod, "The Place of Hazlitt in English Criticism," in *The Profession of Poetry*, 1929; Elisabeth Schneider, *Aesthetics of William Hazlitt*, 1933; the section on Hazlitt in René Wellek, *History of Modern Criticism*, 1955; and W. P. Albrecht, "Hazlitt's Preference for Tragedy," *PMLA*, LXXI (1956), 1042–51. On other aspects of Hazlitt, see Stewart C. Wilcox, *Hazlitt in the Workshop*, 1943, and W. P. Albrecht, *William Hazlitt and the Malthusian Controversy*, 1950.

from ESSAY ON THE PRINCIPLES OF HUMAN ACTION

Hazlitt's first work, the basis of which had been thought out before he was twenty, is on a subject that had fascinated British philosophers for over a century: the question whether human nature is automatically and essentially selfish (a standpoint assumed by Thomas Hobbes and his followers) or whether it is capable of genuine "disinterestedness" (freedom from self-interest). Hazlitt's work is an argument, as its subtitle states, "In Defense of the Natural Disinterestedness of the Human Mind," and his procedure is to analyze the imagination and the nature of what comprises "identity."

It is the design of the following Essay to shew that the human mind is naturally disinterested, or that it is naturally interested in the welfare of others in the same way, and from the same direct motives, by which we are impelled to the pursuit of our own interest.

The objects in which the mind is interested may be either past or present, or future. These last alone can be the objects of rational or voluntary pursuit; for neither the past, nor present can be altered for the better, or worse by any efforts of the will. It is only from the interest excited in him by future objects that man becomes a moral agent, or is denominated selfish, or the contrary, according to the manner in which he is affected by what relates to his own *future* interest, or that of others. I propose then to shew that the mind is naturally interested in it's own welfare in a peculiar mechanical manner, only as far as relates to it's past, or present impressions.

·   ·   ·

Those who have maintained the doctrine of the natural selfishness of the human mind have always taken it for granted as a self-evident principle that *a man must love himself*, or that it is not less absurd to ask why a man should be interested in his own personal welfare, than it would be to ask why a man in a state of actual enjoyment, or suffering likes what gives him pleasure, and dislikes what gives him pain. They say, that no such necessity, nor any positive reason whatever can be conceived to exist for my promoting the welfare of another, since I cannot possibly feel the pleasures, or pains which another feels without first becoming that other, that our interests must be as necessarily distinct as we ourselves are, that the good which I do to another, in itself and for it's own sake can be nothing to me. *Good* is a term relative only to the being who enjoys it. The good which he does not feel must be matter of perfect indifference to him. How can I be required to make a painful exertion, or sacrifice a present convenience to serve another, if I am to be nothing the better for it? I waste my powers out of myself without sharing in the effects which they produce. Whereas when I sacrifice my present ease or convenience, for the sake of a greater good to myself at a future period, the same being who suffers afterwards enjoys, both the loss and the gain are mine, I am upon the whole a gainer in real enjoyment, and am therefore justified to myself: I act with a view to an end in which I have a real, substantial interest. The human soul, continue some of the writers, naturally thirsts after happiness; it either enjoys, or seeks to enjoy. It constantly reaches forward towards the possession of happiness, it strives to draw it to itself, and to be absorbed in it. But as the mind cannot enjoy any good but what it possesses within itself, neither can it seek to produce any good but what it can enjoy: it is just as idle to suppose that the love of happiness or good should prompt any being to give up his own interest for the sake of another, as it would be to attempt to allay violent thirst by giving water to another to drink.

Now I can conceive that a man must be necessarily interested in his own actual feelings, whatever these may be, merely because he feels them. He cannot help receiving pain from what gives him pain, or pleasure from what gives him pleasure. But I cannot conceive how he can have the same necessary, absolute interest in whatever relates to himself, or in his own pleasures and pains, generally speaking, whether he feels them, or not. . . .

All voluntary action, that is all action proceeding from a will, or effort of the mind to produce a certain event must relate to the future, or to those things, the existence of which is problematical, undetermined, and therefore capable of being affected by the means made use of with a view to their production, or the contrary. But that which is future, which does not yet exist can excite no interest in itself, nor act upon the mind in any way but by means of the imagination. The direct primary motive, or impulse which determines the mind to the volition of any thing must therefore in all cases depend on the *idea* of that thing as conceived of by the imagination, and on the idea solely. For the thing itself is a non-entity. By the very act of it's being *willed*, it is supposed not to exist. It neither is any thing, nor can be the cause of any thing. We are never interested in the things themselves which are the real, ultimate, practical objects of volition: the feelings of desire, aversion, &c. connected with voluntary action are always excited by the ideas of those things before they exist. The true impulse to voluntary action can only exist in the mind of a being capable of foreseeing the consequences of things, of being interested in them from the imaginary impression thus made upon his mind, and of making choice of the means necessary to produce, or prevent what he desires or dreads. This distinction must be absolute and universally applicable, if it is so at all. The motives by which I am impelled to the pursuit of my own welfare can no more be the result of a direct impression of the thing which is the object of desire, or aversion, of any positive communication between my present, and future feelings, or of a sort of hypostatical[1] union between the interests of the being acting, and the being acted upon, than the motives by which I am interested in the welfare of others can be so. It is true I have a real, positive interest in my actual feelings which I have not in those of others. But actual pleasure, and pain are not the objects of voluntary action. It can be to no purpose, it is downright nonsense to will that which actually exists, which is impressed on my senses to exist, or not to exist, since it will exist neither more nor less for my willing it, or not willing it. Our shrinking from that which gives us pain could not in any respect be considered as an act of volition, or reason, if we did not know that the same object which gives us pain will continue to give us pain while we remain in contact with it. The mere mechanical movement which generally accompanies much pain does not appear to me to have any thing more to do with self-love properly so called than the convulsive motions or distortions of the muscles caused by bodily disease.—In other words the object of volition is never the cause of volition.

ESSAY ON THE PRINCIPLES OF HUMAN ACTION. **1. hypostatical:** a purely conceptual union that is being regarded as concretely real and personal.

The motive, or internal impression impelling me to the pursuit of any object is by the supposition incompatible with any such interest as belongs to the actual enjoyment of any good, or to the idea of *possession*. The real object of any particular volition is always a mere physical consequence of that volition, since it is willed for that very reason that otherwise it would not exist at all, and since the effect which the mind desires to produce by any voluntary action must be subsequent to that action. It cannot therefore exert any power over my present volitions, and actions, unless we suppose it to act before it exists, which is absurd. For there is no faculty in the mind by which future impressions can excite in it a presentiment of themselves in the same way that past impressions act upon it by means of memory. When we say that future objects act upon the mind by means of the imagination, it is not meant that such objects exercise a real power over the imagination, but merely that it is by means of this faculty that we can foresee the probable or necessary consequences of things, and are interested in them.

. . .

Considering mankind in this two-fold relation, as they are to themselves, or as they appear to one another, as the subjects of their own thoughts, or the thoughts of others, we shall find the origin of that wide and absolute distinction which the mind feels in comparing itself with others to be confined to two faculties, viz. *sensation*, or rather consciousness, and *memory*. The operation of both these faculties is of a perfectly exclusive and individual nature; and so far as their operation extends (but no farther) is man a personal, or if you will, a selfish being. The sensation excited in me by a piece of red-hot iron striking against any part of my body is simple, absolute, terminating in itself, not representing any thing beyond itself, nor capable of being represented by any other sensation or communicated to any other being. The same sensation may indeed be excited in another by the same means, but this sensation does not imply any reference to, or consciousness of mine: there is no communication between my nerves, and another's brain, by means of which he can be affected with my sensations as I am myself. The only notice or perception which another can have of this sensation in me or which I can have of a similar sensation in another is by means of the imagination. I can form an imaginary idea of that pain as existing out of myself: but I can only feel it as a

sensation when it is actually impressed on myself. Any impression made on another can neither be the cause nor object of sensation to me. The impression or idea left in my mind by this sensation, and afterwards excited either by seeing iron in the same state, or by any other means is properly an idea of memory. This idea necessarily refers to some previous impression in my own mind, and can only exist in consequence of that impression: it cannot be derived from any impression made on another. I do not *remember* the feelings of any one but myself. I may remember the objects which must have caused such or such feelings in others, or the outward signs of passion which accompanied them: these however are but the recollection of my own immediate impressions, of what I saw or heard; and I can only form an idea of the feelings themselves after they have ceased, as I must do at the time by means of the imagination. But though we should take away all power of imagination from the human mind, my own feelings must leave behind them certain traces, or representations of themselves retaining the same properties, and having the same immediate connection with the conscious principle. On the other hand if I wish to anticipate my own future feelings, whatever these may be, I must do so by means of the same faculty, by which I conceive of those of others whether past or future. I have no distinct or separate faculty on which the events and feelings of my future being are impressed beforehand, and which shews as in an inchanted mirror to me and me alone the reversed picture of my future life. It is absurd to suppose that the feelings which I am to have hereafter should excite certain correspondent impressions, or presentiments of themselves before they exist, or act mechanically upon my mind by a secret sympathy. I can only abstract myself from my present being and take an interest in my future being in the same sense and manner, in which I can go out of myself entirely and enter into the minds and feelings of others. In short there neither is nor can be any principle belonging to the individual which antecedently gives him the same sort of connection with his future being that he has with his past, or that reflects the impressions of his future feelings backwards with the same kind of consciousness that his past feelings are transmitted forwards through the channels of memory. The size of the river as well as it's taste depends on the water that has already fallen into it. It cannot roll back it's course, nor can the stream next the source be affected by the water that falls into it afterwards. Yet we call both the same river. Such is the nature of personal identity.[2] If this account be true (and for my own part the only perplexity that crosses my mind in thinking of it arises from the utter impossibility of conceiving of the contrary supposition) it will follow that those faculties which may be said to constitute self, and the operations of which convey that idea to the mind draw all their materials from the past and present. But all voluntary action must relate solely and exclusively to the future. That is, all those impressions or ideas with which selfish, or more properly speaking, personal feelings must be naturally connected are just those which have nothing at all to do with the motives of action.

If indeed it were possible for the human mind to alter the present or the past, so as either to recal what was done, or, to give it a still greater reality, to make it exist over again and in some more emphatical sense, then man might with some pretence of reason be supposed naturally incapable of being impelled to the pursuit of any *past* or *present* object but from the mechanical excitement of personal motives. It might in this case be pretended that the impulses of imagination and sympathy are of too light, unsubstantial, and remote a nature to influence our real conduct, and that nothing is worthy of the concern of a wise man in which he has not this direct, unavoidable, and homefelt interest. This is however too absurd a supposition to be dwelt on for a moment. I do not *will* that to be which already exists as an object of sense, nor that to have

been which has already existed, and is become an object of memory. Neither can I will a thing not to be which actually exists, or that which has really existed not to have been. The only proper objects of voluntary action are (by necessity) future events: these can excite no possible interest in the mind but by means of the imagination; and these make the same direct appeal to that faculty whether they relate to ourselves, or others, as the eye receives with equal directness the impression of our own external form, or that of others.

1798–1805 (1805)

from OBSERVATIONS ON MR. WORDSWORTH'S POEM *THE EXCURSION*

～∽✺∽～

This review was published in two parts in *The Round Table* (August and October, 1814). Reviewers in this period quoted long extracts from the work being considered. The extracts and some of Hazlitt's comments on them have been omitted. Compare Hazlitt's discussion of Wordsworth below in the lecture "On the Living Poets."

～∽✺∽～

The poem of The *Excursion* resembles that part of the country in which the scene is laid. It has the same vastness and magnificence, with the same nakedness and confusion. It has the same overwhelming, oppressive power. It excites or recalls the same sensations which those who have traversed that wonderful scenery must have felt. We are surrounded with the constant sense and superstitious awe of the collective power of matter, of the gigantic and eternal forms of nature, on which, from the beginning of time, the hand of man has made no impression. Here are no dotted lines, no hedge-row beauties, no box-tree borders, no gravel walks, no square mechanic inclosures; all is left loose and irregular in the rude chaos of aboriginal nature. The boundaries of hill and valley are the poet's only geography, where we wander with him incessantly over deep beds of moss and waving fern, amidst the troops of red-deer and wild animals. Such is the severe simplicity of Mr. Wordsworth's taste, that we doubt whether he would not reject a druidical temple, or time-hallowed ruin as too modern

---

2. identity: Hazlitt adds the following note: "Suppose a number of men employed to cast a mound into the sea. As far as it has gone, the workmen pass backwards and forwards on it, it stands firm in it's place, and though it recedes farther and farther from the shore, it is still joined to it. A man's personal identity and self-interest have just the same principle and extent, and can reach no farther than his actual existence. But if a man of a metaphysical turn, seeing that the pier was not yet finished, but was to be continued to a certain point and in a certain direction, should take it into his head to insist that what was already built and what was to be built were the same pier, that the one must afford as good footing as the other, and should accordingly walk over the pier-head on the solid foundation of his metaphysical hypothesis—he would argue a great deal more ridiculously, but not a whit more absurdly than those who found a principle of absolute self-interest on a man's future identity with his present being. But say you, the comparison does not hold in this, that the man *can* extend his thoughts (and that very wisely too) beyond the present moment, whereas in the other case he cannot move a single step forwards. Grant it. This will only shew that the mind has wings as well as feet, which of itself is a sufficient answer to the selfish hypothesis."

and artificial for his purpose. He only familiarises himself or his readers with a stone, covered with lichens, which has slept in the same spot of ground from the creation of the world, or with the rocky fissure between two mountains caused by thunder, or with a cavern scooped out by the sea. His mind is, as it were, coëval with the primary forms of things; his imagination holds immediately from nature, and "owes no allegiance" but "to the elements."

The *Excursion* may be considered as a philosophical pastoral poem,—as a scholastic romance. It is less a poem on the country, than on the love of the country. It is not so much a description of natural objects, as of the feelings associated with them; not an account of the manners of rural life, but the result of the poet's reflections on it. He does not present the reader with a lively succession of images or incidents, but paints the outgoings of his own heart, the shapings, of his own fancy. He may be said to create his own materials; his thoughts are his real subject. His understanding broods over that which is "without form and void," and "makes it pregnant." [1] He sees all things in himself. He hardly ever avails himself of remarkable objects or situations, but, in general, rejects them as interfering with the workings of his own mind, as disturbing the smooth, deep, majestic current of his own feelings. Thus his descriptions of natural scenery are not brought home distinctly to the naked eye by forms and circumstances, but every object is seen through the medium of innumerable recollections, is clothed with the haze of imagination like a glittering vapour, is obscured with the excess of glory, has the shadowy brightness of a waking dream. The image is lost in the sentiment, as sound in the multiplication of echoes.

> "And visions, as prophetic eyes avow,
> Hang on each leaf, and cling to every bough."

In describing human nature, Mr. Wordsworth equally shuns the common 'vantage-grounds of popular story, of striking incident, or fatal catastrophe, as cheap and vulgar modes of producing an effect. He scans the human race as the naturalist measures the earth's zone, without attending to the picturesque points of view, the abrupt inequalities of surface. He contemplates the passions and habits of men, not in their extremes, but in their first elements; their follies and vices, not at their height, with all their

embossed evils upon their heads, but as lurking in embryo,—the seeds of the disorder inwoven with our very constitution. He only sympathises with those simple forms of feeling, which mingle at once with his own identity, or with the stream of general humanity. To him the great and the small are the same; the near and the remote; what appears, and what only is. The general and the permanent, like the Platonic ideas, are his only realities. All accidental varieties and individual contrasts are lost in an endless continuity of feeling, like drops of water in the ocean-stream! An intense intellectual egotism swallows up every thing. Even the dialogues introduced in the present volume are soliloquies of the same character, taking different views of the subject. The recluse, the pastor, and the pedlar, are three persons in one poet. We ourselves disapprove of these "interlocutions between Lucius and Caius" as impertinent babbling, where there is no dramatic distinction of character. But the evident scope and tendency of Mr. Wordsworth's mind is the reverse of dramatic. It resists all change of character, all variety of scenery, all the bustle, machinery, and pantomime of the stage, or of real life,—whatever might relieve, or relax, or change the direction of its own activity, jealous of all competition. The power of his mind preys upon itself. It is as if there were nothing but himself and the universe. He lives in the busy solitude of his own heart; in the deep silence of thought. His imagination lends life and feeling only to "the bare trees and mountains bare"; peoples the viewless tracts of air, and converses with the silent clouds!

. . .

There is, in fact, in Mr. Wordsworth's mind an evident repugnance to admit anything that tells for itself, without the interpretation of the poet,—a fastidious antipathy to immediate effect—a systematic unwillingness to share the palm with his subject. Where, however, he has a subject presented to him, "such as the meeting soul may pierce," and to which he does not grudge to lend the aid of his fine genius, his powers of description and fancy seem to be little inferior to those of his classical predecessor, Akenside. Among several others which we might select we give the following passage, describing the religion of ancient Greece:

[Hazlitt here quotes *Excursion*, IV. 851–87 and 941–92]

. . .

Mr. Wordsworth's writings exhibit all the internal

OBSERVATIONS ON MR. WORDSWORTH'S *Excursion*. **1. "without . . . pregnant"**: Genesis 1:2; *Paradise Lost*, I. 22.

power, without the external form of poetry. He has scarcely any of the pomp and decoration and scenic effect of poetry: no gorgeous palaces nor solemn temples awe the imagination; no cities rise "with glistering spires and pinnacles adorned";[2] we meet with no knights pricked forth on airy steeds; no hair-breadth 'scapes and perilous accidents by flood or field. Either from the predominant habit of his mind not requiring the stimulus of outward impressions, or from the want of an imagination teeming with various forms, he takes the common every-day events and objects of nature, or rather seeks those that are the most simple and barren of effect; but he adds to them a weight of interest from the resources of his own mind, which makes the most insignificant things serious and even formidable. All other interests are absorbed in the deeper interest of his own thoughts, and find the same level. His mind magnifies the littleness of his subject, and raises its meanness; lends it his strength, and clothes it with borrowed grandeur. With him, a molehill, covered with wild thyme, assumes the importance of "the great vision of the guarded mount";[3] a puddle is filled with preternatural faces, and agitated with the fiercest storms of passion.

The extreme simplicity which some persons have objected to in Mr. Wordsworth's poetry, is to be found only in the subject and the style: the sentiments are subtle and profound. In the latter respect, his poetry is as much above the common standard or capacity, as in the other it is below it. His poems bear a distant resemblance to some of Rembrandt's landscapes, who, more than any other painter, created the medium through which he saw nature, and out of the stump of an old tree, a break in the sky, and a bit of water, could produce an effect almost miraculous.

Mr. Wordsworth's poems in general are the history of a refined and contemplative mind, conversant only with itself and nature. An intense feeling of the associations of this kind is the peculiar and characteristic feature of all his productions. He has described the love of nature better than any other poet. This sentiment, inly felt in all its force, and sometimes carried to an excess, is the source both of his strength and of his weakness. However we may sympathise with Mr. Wordsworth in his attachment to groves and fields, we cannot extend the same admiration to their inhabitants, or to the manners of country life in general. . . .

All country people hate each other. They have so little comfort, that they envy their neighbours the smallest pleasure or advantage, and nearly grudge themselves the necessaries of life. From not being accustomed to enjoyment, they become hardened and adverse to it—stupid, for want of thought—selfish, for want of society. There is nothing good to be had in the country, or, if there is, they will not let you have it. They had rather injure themselves than oblige any one else. Their common mode of life is a system of wretchedness and self-denial, like what we read of among barbarous tribes. You live out of the world. You cannot get your tea and sugar without sending to the next town for it: you pay double, and have it of the worst quality. The small-beer is sure to be sour—the milk skimmed—the meat bad, or spoiled in the cooking. You cannot do a single thing you like; you cannot walk out or sit at home, or write or read, or think or look as if you did, without being subject to impertinent curiosity. The apothecary annoys you with his complaisance; the parson with his superciliousness. If you are poor, you are despised; if you are rich, you are feared and hated. If you do any one a favour, the whole neighbourhood is up in arms; the clamour is like that of a rookery; and the person himself, it is ten to one, laughs at you for your pains, and takes the first opportunity of shewing you that he labours under no uneasy sense of obligation. There is a perpetual round of mischief-making and backbiting for want of any better amusement. There are no shops, no taverns, no theatres, no opera, no concerts, no pictures, no public-buildings, no crowded streets, no noise of coaches, or of courts of law,—neither courtiers nor courtesans, no literary parties, no fashionable routs, no society, no books, or knowledge of books. Vanity and luxury are the civilisers of the world, and sweeteners of human life. Without objects either of pleasure or action, it grows harsh and crabbed: the mind becomes stagnant, the affections callous, and the eye dull. Man left to himself soon degenerates into a very disagreeable person. Ignorance is always bad enough; but rustic ignorance is intolerable. Aristotle has observed, that tragedy purifies the affections by terror and pity. If so, a company of tragedians should be established at the public expence, in every village or hundred, as a better mode of education than either Bell's or Lancaster's.[4] The benefits of knowledge are never so

---

2. with . . . adorned: *Paradise Lost*, III. 550. 3. "the . . . mount": "Lycidas," l. 161.

4. **Bell's or Lancaster's:** Andrew Bell, 1753–1832, and Joseph Lancaster, 1770–1838, education reformers.

well understood as from seeing the effects of ignorance, in their naked, undisguised state, upon the common country people. Their selfishness and insensibility are perhaps less owing to the hardships and privations, which make them, like people out at sea in a boat, ready to devour one another, than to their having no idea of anything beyond themselves and their immediate sphere of action. They have no knowledge of, and consequently can take no interest in, anything which is not an object of their senses, and of their daily pursuits. They hate all strangers, and have generally a nickname for the inhabitants of the next village. The two young noblemen in Guzman d'Alfarache,[5] who went to visit their mistresses only a league out of Madrid, were set upon by the peasants, who came round them calling out, "*A wolf.*" Those who have no enlarged or liberal ideas, can have no disinterested or generous sentiments. Persons who are in the habit of reading novels and romances, are compelled to take a deep interest in, and to have their affections strongly excited by, fictitious characters and imaginary situations; their thoughts and feelings are constantly carried out of themselves, to persons they never saw, and things that never existed: history enlarges the mind, by familiarising us with the great vicissitudes of human affairs, and the catastrophes of states and kingdoms; the study of morals accustoms us to refer our actions to a general standard of right and wrong; and abstract reasoning, in general, strengthens the love of truth, and produces an inflexibility of principle which cannot stoop to low trick and cunning. Books, in Lord Bacon's phrase, are "a discipline of humanity."[6] Country people have none of these advantages, nor any others to supply the place of them. Having no circulating libraries to exhaust their love of the marvellous, they amuse themselves with fancying the disasters and disgraces of their particular acquaintance. Having no hump-backed *Richard* to excite their wonder and abhorrence, they make themselves a bug-bear of their own, out of the first obnoxious person they can lay their hands on. Not having the fictitious distresses and gigantic crimes of poetry to stimulate their imagination and their passions, they vent their whole stock of spleen, malice, and invention, on their friends and next-door neighbours. They get up a little pastoral drama at home, with fancied events, but real characters. All

their spare time is spent in manufacturing and propagating the lie for the day, which does its office, and expires. The next day is spent in the same manner. It is thus that they embellish the simplicity of rural life! The common people in civilised countries are a kind of domesticated savages. They have not the wild imagination, the passions, the fierce energies, or dreadful vicissitudes of the savage tribes, nor have they the leisure, the indolent enjoyments and romantic superstitions, which belonged to the pastoral life in milder climates, and more remote periods of society. They are taken out of a state of nature, without being put in possession of the refinements of art. The customs and institutions of society cramp their imaginations without giving them knowledge. If the inhabitants of the mountainous districts described by Mr. Wordsworth are less gross and sensual than others, they are more selfish. Their egotism becomes more concentrated, as they are more insulated, and their purposes more inveterate, as they have less competition to struggle with. The weight of matter which surrounds them, crushes the finer sympathies. Their minds become hard and cold, like the rocks which they cultivate. The immensity of their mountains makes the human form appear little and insignificant. Men are seen crawling between Heaven and earth, like insects to their graves. Nor do they regard one another more than flies on a wall. Their physiognomy expresses the materialism of their character, which has only one principle—rigid self-will. They move on with their eyes and foreheads fixed, looking neither to the right nor to the left, with a heavy slouch in their gait, and seeming as if nothing would divert them from their path. We do not admire this plodding pertinacity, always directed to the main chance. There is nothing which excites so little sympathy in our minds, as exclusive selfishness. If our theory is wrong, at least it is taken from pretty close observation, and is, we think, confirmed by Mr. Wordsworth's own account.

Of the stories contained in the latter part of the volume, we like that of the Whig and Jacobite friends, and of the good knight, Sir Alfred Irthing, the best. The last reminded us of a fine sketch of a similar character in the beautiful poem of *Hart Leap Well.* To conclude,—if the skill with which the poet had chosen his materials had been equal to the power which he has undeniably exerted over them, if the objects (whether persons or things) which he makes use of as the vehicle of his sentiments, had been such as to convey them in all their depth and force, then

5. **Guzman d'Alfarache**: a novel by Mateo Aleman (1599).
6. "a . . . humanity": "Of Marriage and Single Life."

the production before us might indeed "have proved a monument," as he himself wishes it, worthy of the author, and of his country. Whether, as it is, this very original and powerful performance may not rather remain like one of those stupendous but half-finished structures, which have been suffered to moulder into decay, because the cost and labour attending them exceeded their use or beauty, we feel that it would be presumptuous in us to determine.

1814 (1814)

from ON THE CHARACTER
OF ROUSSEAU

∾᙭᙭↝

Published in *The Examiner*, April 14, 1816.

↜᙭᙭∾

Madame de Stael, in her Letters on the Writings and Character of Rousseau, gives it as her opinion, "that the imagination was the first faculty of his mind, and that this faculty even absorbed all the others." And she farther adds, "Rousseau had great strength of reason on abstract questions, or with respect to objects, which have no reality but in the mind." Both these opinions are radically wrong. Neither imagination nor reason can properly be said to have been the original predominant faculties of his mind. The strength both of imagination and reason, which he possessed, was borrowed from the excess of another faculty; and the weakness and poverty of reason and imagination, which are to be found in his works, may be traced to the same source, namely, that these faculties in him were artificial, secondary, and dependant, operating by a power not theirs, but lent to them. The only quality which he possessed in an eminent degree, which alone raised him above ordinary men, and which gave to his writings and opinions an influence greater, perhaps, than has been exerted by any individual in modern times, was extreme sensibility, or an acute and even morbid feeling of all that related to his own impressions, to the objects and events of his life. He had the most intense consciousness of his own existence. No object that had once made an impression on him was ever after effaced. Every feeling in his mind became a passion. His craving after excitement was an appetite and a disease. His interest in his own thoughts

and feelings was always wound up to the highest pitch; and hence the enthusiasm which he excited in others. He owed the power which he exercised over the opinions of all Europe, by which he created numberless disciples, and overturned established systems, to the tyranny which his feelings, in the first instance, exercised over himself. The dazzling blaze of his reputation was kindled by the same fire that fed upon his vitals.[1] His ideas differed from those of other men only in their force and intensity. His genius was the effect of his temperament. He created nothing, he demonstrated nothing, by a pure effort of the understanding. His fictitious characters are modifications of his own being, reflections and shadows of himself. His speculations are the obvious exaggerations of a mind, giving a loose to its habitual impulses, and moulding all nature to its own purposes. Hence his enthusiasm and his eloquence, bearing down all opposition. Hence the warmth and the luxuriance, as well as the sameness of his descriptions. Hence the frequent verboseness of his style; for passion lends force and reality to language, and makes words supply the place of imagination. Hence the tenaciousness of his logic, the acuteness of his observations, the refinement and the inconsistency of his reasoning. Hence his keen penetration, and his strange want of comprehension of mind: for the same intense feeling which enabled him to discern the first principles of things, and seize some one view of a subject in all its ramifications, prevented him from admitting the operation of other causes which interfered with his favourite purpose, and involved him in endless wilful contradictions. Hence his excessive egotism, which filled all objects with himself, and would have occupied the universe with his smallest interest. Hence his jealousy and suspicion of others; for no attention, no respect or sympathy, could come up to the extravagant claims of his self-love. Hence his dissatisfaction with himself and with all around him; for nothing could satisfy his ardent longings after good, his restless appetite of being. Hence his feelings, overstrained and exhausted, recoiled upon themselves, and produced his love of

ON THE CHARACTER OF ROUSSEAU. **1. fire . . . vitals:** "He did more towards the French Revolution than any other man. Voltaire, by his wit and penetration, had rendered superstition contemptible, and tyranny odious: but it was Rousseau who brought the feeling of irreconcilable enmity to rank and privileges, *above humanity*, home to the bosom of every man, —identified it with all the pride of intellect, and with the deepest yearnings of the human heart" (H.).

silence and repose, his feverish aspirations after the quiet and solitude of nature. Hence in part also his quarrel with the artificial institutions and distinctions of society, which opposed so many barriers to the unrestrained indulgence of his will, and allured his imagination to scenes of pastoral simplicity or of savage life, where the passions were either not excited or left to follow their own impulse,—where the petty vexations and irritating disappointments of common life had no place,—and where the tormenting pursuits of arts and sciences were lost in pure animal enjoyment, or indolent repose. Thus he describes the first savage wandering for ever under the shade of magnificent forests, or by the side of mighty rivers, smit with the unquenchable love of nature!

. . .

Rousseau, in all his writings, never once lost sight of himself. He was the same individual from first to last. The spring that moved his passions never went down, the pulse that agitated his heart never ceased to beat. It was this strong feeling of interest, accumulating in his mind, which overpowers and absorbs the feelings of his readers. He owed all his power to sentiment. The writer who most nearly resembles him in our own times is the author of the *Lyrical Ballads*. We see no other difference between them, than that the one wrote in prose and the other in poetry; and that prose is perhaps better adapted to express those local and personal feelings, which are inveterate habits in the mind, than poetry, which embodies its imaginary creations. We conceive that Rousseau's exclamation, "*Ah, voila de la pervenche*,"[2] comes more home to the mind than Mr. Wordsworth's discovery of the linnet's nest "with five blue eggs," or than his address to the cuckoo,[3] beautiful as we think it is; and we will confidently match the Citizen of Geneva's adventures on the Lake of Bienne against the Cumberland Poet's floating dreams on the Lake of Grasmere. Both create an interest out of nothing, or rather out of their own feelings; both weave numberless recollections into one sentiment; both wind their own being round whatever object occurs to them. . . .

1816 (1816)

---

2. **"Ah, . . . pervenche"**: "Ah! there are periwinkles" (*Confessions*, Pt. I, Bk. VI). 3. **with . . . cuckoo**: "The Sparrow's Nest" and "To the Cuckoo."

ON GUSTO

Published in *The Examiner*, May 26, 1816.

Gusto in art is power or passion defining any object. —It is not so difficult to explain this term in what relates to expression (of which it may be said to be the highest degree) as in what relates to things without expression, to the natural appearances of objects, as mere colour or form. In one sense, however, there is hardly any object entirely devoid of expression, without some character of power belonging to it, some precise association with pleasure or pain: and it is in giving this truth of character from the truth of feeling, whether in the highest or lowest degree, but always in the highest degree of which the subject is capable, that gusto consists.

There is a gusto in the colouring of Titian. Not only do his heads seem to think—his bodies seem to feel. This is what the Italians mean by the *morbidezza* of his flesh-colour. It seems sensitive and alive all over; not merely to have the look and texture of flesh, but the feeling in itself. For example, the limbs of his female figures have a luxurious softness and delicacy, which appears conscious of the pleasure of the beholder. As the objects themselves in nature would produce an impression on the sense, distinct from every other object, and having something divine in it, which the heart owns and the imagination consecrates, the objects in the picture preserve the same impression, absolute, unimpaired, stamped with all the truth of passion, the pride of the eye, and the charm of beauty. Rubens makes his flesh-colour like flowers; Albano's[1] is like ivory; Titian's is like flesh, and like nothing else. It is as different from that of other painters, as the skin is from a piece of white or red drapery thrown over it. The blood circulates here and there, the blue veins just appear, the rest is distinguished throughout only by that sort of tingling sensation to the eye, which the body feels within itself. This is gusto.—Vandyke's flesh-colour, though it has great truth and purity, wants gusto. It has not the internal character, the living principle in it. It is a smooth surface, not a warm,

---

ON GUSTO. 1. **Albano**: Francesco Albani, 1578–1660.

moving mass. It is painted without passion, with indifference. The hand only has been concerned. The impression slides off from the eye, and does not, like the tones of Titian's pencil, leave a sting behind it in the mind of the spectator. The eye does not acquire a taste or appetite for what it sees. In a word, gusto in painting is where the impression made on one sense excites by affinity those of another.

Michael Angelo's forms are full of gusto. They every where obtrude the sense of power upon the eye. His limbs convey an idea of muscular strength, of moral grandeur, and even of intellectual dignity: they are firm, commanding, broad, and massy, capable of executing with ease the determined purposes of the will. His faces have no other expression than his figures, conscious power and capacity. They appear only to think what they shall do, and to know that they can do it. This is what is meant by saying that his style is hard and masculine. It is the reverse of Correggio's, which is effeminate. That is, the gusto of Michael Angelo consists in expressing energy of will without proportionable sensibility, Correggio's in expressing exquisite sensibility without energy of will. In Correggio's faces as well as figures we see neither bones nor muscles, but then what a soul is there, full of sweetness and of grace—pure, playful, soft, angelical! There is sentiment enough in a hand painted by Correggio to set up a school of history painters. Whenever we look at the hands of Correggio's women or of Raphael's, we always wish to touch them.

Again, Titian's landscapes have a prodigious gusto, both in the colouring and forms. We shall never forget one that we saw many years ago in the Orleans Gallery of Acteon hunting. It had a brown, mellow, autumnal look. The sky was of the colour of stone. The winds seemed to sing through the rustling branches of the trees, and already you might hear the twanging of bows resound through the tangled mazes of the wood. Mr. West,[2] we understand, has this landscape. He will know if this description of it is just. The landscape background of the St. Peter Martyr is another well-known instance of the power of this great painter to give a romantic interest and an appropriate character to the objects of his pencil, where every circumstance adds to the effect of the scene,—the bold trunks of the tall forest trees, the trailing ground plants, with that cold convent spire rising in the distance, amidst the blue sapphire mountains and the golden sky.

Rubens has a great deal of gusto in his Fauns and Satyrs, and in all that expresses motion, but in nothing else. Rembrandt has it in everything; everything in his pictures has a tangible character. If he puts a diamond in the ear of a Burgomaster's wife, it is of the first water; and his furs and stuffs are proof against a Russian winter. Raphael's gusto was only in expression; he had no idea of the character of anything but the human form. The dryness and poverty of his style in other respects is a phenomenon in the art. His trees are like sprigs of grass stuck in a book of botanical specimens. Was it that Raphael never had time to go beyond the walls of Rome? That he was always in the streets, at church, or in the bath? He was not one of the Society of Arcadians.[3]

Claude's landscapes, perfect as they are, want gusto. This is not easy to explain. They are perfect abstractions of the visible images of things; they speak the visible language of nature truly. They resemble a mirror or miscroscope. To the eye only they are more perfect than any other landscapes that ever were or will be painted; they give more of nature, as cognizable by one sense alone; but they lay an equal stress on all visible impressions; they do not interpret one sense by another; they do not distinguish the character of different objects as we are taught, and can only be taught, to distinguish them by their effect on the different senses. That is, his eye wanted imagination: it did not strongly sympathize with his other faculties. He saw the atmosphere, but he did not feel it. He painted the trunk of a tree or a rock in the foreground as smooth—with as complete an abstraction of the gross, tangible impression, as any other part of the picture; his trees are perfectly beautiful, but quite immoveable; they have a look of enchantment. In short, his landscapes are unequalled imitations of nature, released from its

---

2. **Mr. West:** Benjamin West, 1738–1820, president of the Royal Academy. Cf. Keats's remarks (in the "Negative Capability" letter to his brothers, below, December 21–27, 1817) on West's picture, "Death on the Pale Horse," where in effect he criticizes West, as Hazlitt does Vandyke above, for lack of gusto or "intensity."

3. **not . . . Arcadians:** "Raphael not only could not paint a landscape; he could not paint people in a landscape. He could not have painted the heads or the figures, or even the dresses of the St. Peter Martyr. His figures have always an *in-door* look, that is, a set, determined, voluntary, dramatic character, arising from their own passions, or a watchfulness of those of others, and want that wild uncertainty of expression, which is connected with the accidents of nature and the changes of the elements. He has nothing *romantic* about him" (H.).

subjection to the elements,—as if all objects were become a delightful fairy vision, and the eye had rarefied and refined away the other senses.

The gusto in the Greek statues is of a very singular kind. The sense of perfect form nearly occupies the whole mind, and hardly suffers it to dwell on any other feeling. It seems enough for them *to be*, without acting or suffering. Their forms are ideal, spiritual. Their beauty is power. By their beauty they are raised above the frailties of pain or passion; by their beauty they are deified.

The infinite quantity of dramatic invention in Shakspeare takes from his gusto. The power he delights to shew is not intense, but discursive. He never insists on any thing as much as he might, except a quibble. Milton has great gusto. He repeats his blow twice; grapples with and exhausts his subject. His imagination has a double relish of its objects, an inveterate attachment to the things he describes, and to the words describing them.

—— "Or where Chineses drive
With sails and wind their *cany* waggons *light.*"

.   .   .

"Wild above rule or art, *enormous* bliss." [4]

There is a gusto in Pope's compliments, in Dryden's satires, and Prior's tales; and among prose-writers, Boccaccio and Rabelais had the most of it. We will only mention one other work which appears to us to be full of gusto, and that is the *Beggar's Opera*. If it is not, we are altogether mistaken in our notions on this delicate subject.

1816 (1816)

FROM

LECTURES ON THE ENGLISH POETS

These lectures were delivered at the Surrey Institution early in 1818. "On Shakspeare and Milton" was the third lecture and "On the Living Poets" was the last.

4. **"Or . . . bliss"**: *Paradise Lost*, III. 438–39; V. 297.

ON SHAKSPEARE AND MILTON

In looking back to the great works of genius in former times, we are sometimes disposed to wonder at the little progress which has since been made in poetry, and in the arts of imitation in general. But this is perhaps a foolish wonder. Nothing can be more contrary to the fact, than the supposition that in what we understand by the *fine arts*, as painting, and poetry, relative perfection is only the result of repeated efforts in successive periods, and that what has been once well done, constantly leads to something better. What is mechanical, reducible to rule, or capable of demonstration, is progressive, and admits of gradual improvement: what is not mechanical, or definite, but depends on feeling, taste, and genius, very soon becomes stationary, or retrograde, and loses more than it gains by transfusion. The contrary opinion is a vulgar error, which has grown up, like many others, from transferring an analogy of one kind to something quite distinct, without taking into the account the difference in the nature of the things, or attending to the difference of the results. For most persons, finding what wonderful advances have been made in biblical criticism, in chemistry, in mechanics, in geometry, astronomy, &c. *i.e.* in things depending on mere inquiry and experiment, or on absolute demonstration, have been led hastily to conclude, that there was a general tendency in the efforts of the human intellect to improve by repetition, and, in all other arts and institutions, to grow perfect and mature by time. We look back upon the theological creed of our ancestors, and their discoveries in natural philosophy, with a smile of pity: science, and the arts connected with it, have all had their infancy, their youth, and manhood, and seem to contain in them no principle of limitation or decay: and, inquiring no farther about the matter, we infer, in the intoxication of our pride, and the height of our self-congratulation, that the same progress has been made, and will continue to be made, in all other things which are the work of man. The fact, however, stares us so plainly in the face, that one would think the smallest reflection must suggest the truth, and overturn our sanguine theories. The greatest poets, the ablest orators, the best painters, and the finest sculptors that the world ever saw, appeared soon after the birth of these arts, and lived in a state of society which was, in other respects, comparatively barbarous. Those arts, which depend on individual genius and incommunicable power, have always leaped at once from infancy to manhood, from the first rude dawn of invention to their meridian height and dazzling lustre, and have in general declined ever after. This is the peculiar distinction and privilege of each, of science and of art:—of the one, never to attain its utmost limit of perfection; and of the other, to arrive at it almost at once. Homer, Chaucer, Spenser, Shakspeare, Dante, and Ariosto, (Milton alone was of a later age, and not the worse for it)—Raphael, Titian, Michael Angelo, Correggio, Cervantes, and Boccaccio, the Greek sculptors and tragedians,—all lived near the beginning of their arts—perfected, and all but created them. These giant-sons of genius stand indeed upon the earth, but they tower above their fellows; and the long line of their successors, in different ages, does not interpose any object to obstruct their view, or lessen their brightness. In strength and stature they are unrivalled; in grace and beauty they have not been surpassed. In after-ages, and more refined periods, (as they are called) great men have arisen, one by one, as it were by throes and at intervals; though in general the best of these cultivated and artificial minds were of an inferior order; as Tasso and Pope, among poets; Guido and Vandyke, among painters. But in the earlier stages of the arts, as soon as the first mechanical difficulties had been got over, and the language was sufficiently acquired, they rose by clusters, and in constellations, never so to rise again!

The arts of painting and poetry are conversant with the world of thought within us, and with the world of sense around us—with what we know, and see, and feel intimately. They flow from the sacred shrine of our own breasts, and are kindled at the living lamp of nature. But the pulse of the passions assuredly beat as high, the depths and soundings of the human heart were as well understood three thousand, or three hundred years ago, as they are at present: the face of nature, and "the human face divine" shone as bright then as they have ever done. But it is *their* light, reflected by true genius on art, that marks out its path before it, and sheds a glory round the Muses' feet, like that which

"Circled Una's angel face,
And made a sunshine in the shady place."[1]

The four greatest names in English poetry, are almost the four first we come to—Chaucer, Spenser,

Shakspeare, and Milton. There are no others that can really be put in competition with these. The two last have had justice done them by the voice of common fame. Their names are blazoned in the very firmament of reputation; while the two first (though "the fault has been more in their stars than in themselves that they are underlings") either never emerged far above the horizon, or were too soon involved in the obscurity of time. The three first of these are excluded from Dr. Johnson's Lives of the Poets (Shakspeare indeed is so from the dramatic form of his compositions): and the fourth, Milton, is admitted with a reluctant and churlish welcome.

In comparing these four writers together, it might be said that Chaucer excels as the poet of manners, or of real life; Spenser, as the poet of romance; Shakspeare as the poet of nature (in the largest use of the term); and Milton, as the poet of morality. Chaucer most frequently describes things as they are; Spenser, as we wish them to be; Shakspeare, as they would be; and Milton as they ought to be. As poets, and as great poets, imagination, that is, the power of feigning things according to nature, was common to them all: but the principle or moving power, to which this faculty was most subservient in Chaucer, was habit, or inveterate prejudice; in Spenser, novelty, and the love of the marvellous; in Shakspeare, it was the force of passion, combined with every variety of possible circumstances; and in Milton, only with the highest. The characteristic of Chaucer is intensity; of Spenser, remoteness; of Milton, elevation; of Shakspeare, every thing.—It has been said by some critic, that Shakspeare was distinguished from the other dramatic writers of his day only by his wit; that they had all his other qualities but that; that one writer had as much sense, another as much fancy, another as much knowledge of character, another the same depth of passion, and another as great a power of language. This statement is not true; nor is the inference from it well-founded, even if it were. This person does not seem to have been aware that, upon his own shewing, the great distinction of Shakspeare's genius was its virtually including the genius of all the great men of his age, and not his differing from them in one accidental particular. But to have done with such minute and literal trifling.

The striking peculiarity of Shakspeare's mind was its generic quality, its power of communication with all other minds—so that it contained a universe of thought and feeling within itself, and had no one peculiar bias, or exclusive excellence more than another.

He was just like any other man, but that he was like all other men. He was the least of an egotist that it was possible to be. He was nothing in himself; but he was all that others were, or that they could become.[2] He not only had in himself the germs of every faculty and feeling, but he could follow them by anticipation, intuitively, into all their conceivable ramifications, through every change of fortune or conflict of passion, or turn of thought. He had "a mind reflecting ages past," and present:—all the people that ever lived are there. There was no respect of persons with him. His genius shone equally on the evil and on the good, on the wise and the foolish, the monarch and the beggar: "All corners of the earth, kings, queens, and states, maids, matrons, nay, the secrets of the grave," are hardly hid from his searching glance. He was like the genius of humanity, changing places with all of us at pleasure, and playing with our purposes as with his own. He turned the globe round for his amusement, and surveyed the generations of men, and the individuals as they passed, with their different concerns, passions, follies, vices, virtues, actions, and motives—as well those that they knew, as those which they did not know, or acknowledge to themselves. The dreams of childhood, the ravings of despair, were the toys of his fancy. Airy beings waited at his call, and came at his bidding. Harmless fairies "nodded to him, and did him curtesies": and the night-hag bestrode the blast at the command of "his so potent art." The world of spirits lay open to him, like the world of real men and women: and there is the same truth in his delineations of the one as of the other; for if the preternatural characters he describes could be supposed to exist, they would speak, and feel, and act, as he makes them. He had only to think of any thing in order to become that thing, with all the circumstances belonging to it. When he conceived of a character, whether real or imaginary, he not only entered into all its thoughts and feelings, but seemed instantly, and as if by touching a secret spring, to be surrounded with all the same objects, "subject to the same skyey influences," the same local, outward, and unforeseen accidents which would occur in reality. Thus the character of Caliban not only stands before us with a language and manners of its own, but the scenery and situation of the enchanted

island he inhabits, the traditions of the place, its strange noises, its hidden recesses, "his frequent haunts and ancient neighbourhood," are given with a miraculous truth of nature, and with all the familiarity of an old recollection. The whole "coheres semblably together" in time, place, and circumstance. In reading this author, you do not merely learn what his characters say,—you see their persons. By something expressed or understood, you are at no loss to decypher their peculiar physiognomy, the meaning of a look, the grouping, the bye-play, as we might see it on the stage. A word, an epithet paints a whole scene, or throws us back whole years in the history of the person represented. So (as it has been ingeniously remarked) when Prospero describes himself as left alone in the boat with his daughter, the epithet which he applies to her, "Me and thy *crying* self," flings the imagination instantly back from the grown woman to the helpless condition of infancy, and places the first and most trying scene of his misfortunes before us, with all that he must have suffered in the interval. How well the silent anguish of Macduff is conveyed to the reader, by the friendly expostulation of Malcolm—"What! man, ne'er pull your hat upon your brows!" Again, Hamlet, in the scene with Rosencrans and Guildenstern, somewhat abruptly concludes his fine soliloquy on life by saying, "Man delights not me, nor woman neither, though by your smiling you seem to say so." Which is explained by their answer—"My lord, we had no such stuff in our thoughts. But we smiled to think, if you delight not in man, what lenten entertainment the players shall receive from you, whom we met on the way":—as if while Hamlet was making this speech, his two old schoolfellows from Wittenberg had been really standing by, and he had seen them smiling by stealth, at the idea of the players crossing their minds. It is not "a combination and a form" of words, a set speech or two, a preconcerted theory of a character, that will do this: but all the persons concerned must have been present in the poet's imagination, as at a kind of rehearsal; and whatever would have passed through their minds on the occasion, and have been observed by others, passed through his, and is made known to the reader.—I may add in passing, that Shakspeare always gives the best directions for the costume and carriage of his heroes. Thus to take one example, Ophelia gives the following account of Hamlet; and as Ophelia had seen Hamlet, I should think her word ought to be taken against that of any modern authority.

---

2. **He . . . become:** Cf. Keats's letters below on "Negative Capability" and on his Shakespearean ideal of the "characterless" poet (Dec. 21–27, 1817, and Oct. 27, 1818).

"*Ophelia.*   My lord, as I was reading in my closet,
Prince Hamlet, with his doublet all unbrac'd,
No hat upon his head, his stockings loose,
Ungartred, and down-gyved to his ancle,
Pale as his shirt, his knees knocking each other,
And with a look so piteous,
As if he had been sent from hell
To speak of horrors, thus he comes before me.
     Polonius.   Mad for thy love!
     Oph.   My lord, I do not know,
But truly I do fear it.
     Pol.   What said he?
     Oph.   He took me by the wrist, and held me hard
Then goes he to the length of all his arm;
And with his other hand thus o'er his brow,
He falls to such perusal of my face,
As he would draw it: long staid he so;
At last, a little shaking of my arm,
And thrice his head thus waving up and down,
He rais'd a sigh so piteous and profound,
As it did seem to shatter all his bulk,
And end his being. That done, he lets me go,
And with his head over his shoulder turn'd,
He seem'd to find his way without his eyes;
For out of doors he went without their help,
And to the last bended their light on me."

                                   Act. II. Scene I.

How after this airy, fantastic idea of irregular grace
and bewildered melancholy any one can play Hamlet,
as we have seen it played, with strut, and stare, and
antic right-angled sharp-pointed gestures, it is difficult
to say, unless it be that Hamlet is not bound, by the
prompter's cue, to study the part of Ophelia. The
account of Ophelia's death begins thus:

"There is a willow hanging o'er a brook,
That shows its hoary leaves in the glassy stream."—

Now this is an instance of the same unconscious power
of mind which is as true to nature as itself. The leaves
of the willow are, in fact, white underneath, and it is
this part of them which would appear "hoary" in
the reflection in the brook. The same sort of intuitive
power, the same faculty of bringing every object in
nature, whether present or absent, before the mind's
eye, is observable in the speech of Cleopatra, when
conjecturing what were the employment of Antony
in his absence:—"He's speaking now, or murmuring,
where's my serpent of old Nile?" How fine to make
Cleopatra have this consciousness of her own character,
and to make her feel that it is this for which Antony
is in love with her! She says, after the battle of Actium,

when Antony has resolved to risk another fight,
"It is my birth-day; I had thought to have held it
poor: but since my lord is Antony again, I will be
Cleopatra." What other poet would have thought of
such a casual resource of the imagination, or would
have dared to avail himself of it? The thing happens
in the play as it might have happened in fact.—That
which, perhaps, more than any thing else distinguishes
the dramatic productions of Shakspeare from all
others, is this wonderful truth and individuality of
conception. Each of his characters is as much itself,
and as absolutely independent of the rest, as well as
of the author, as if they were living persons, not fictions
of the mind. The poet may be said, for the time, to
identify himself with the character he wishes to represent, and to pass from one to another, like the same
soul successively animating different bodies. By an
art like that of the ventriloquist, he throws his imagination out of himself, and makes every word appear to
proceed from the mouth of the person in whose name
it is given. His plays alone are properly expressions of
the passions, not descriptions of them. His characters
are real beings of flesh and blood; they speak like men,
not like authors. One might suppose that he had stood
by at the time, and overheard what passed. As in our
dreams we hold conversations with ourselves, make remarks, or communicate intelligence, and have no idea
of the answer which we shall receive, and which we ourselves make, till we hear it: so the dialogues in Shakspeare
are carried on without any consciousness of what is
to follow, without any appearance of preparation or
premeditation. The gusts of passion come and go
like sounds of music borne on the wind. Nothing is
made out by formal inference and analogy, by climax
and antithesis: all comes, or seems to come, immediately
from nature. Each object and circumstance exists in
his mind, as it would have existed in reality: each
several train of thought and feeling goes on of itself,
without confusion or effort. In the world of his imagination, every thing has a life, a place, and being of its
own!

Chaucer's characters are sufficiently distinct from
one another, but they are too little varied in themselves,
too much like identical propositions. They are consistent, but uniform; we get no new idea of them from
first to last; they are not placed in different lights, nor
are their subordinate *traits* brought out in new situations; they are like portraits or physiognomical studies,
with the distinguishing features marked with inconceivable truth and precision, but that preserve the same

unaltered air and attitude. Shakspeare's are historical figures, equally true and correct, but put into action, where every nerve and muscle is displayed in the struggle with others, with all the effect of collision and contrast, with every variety of light and shade. Chaucer's characters are narrative, Shakspeare's dramatic, Milton's epic. That is, Chaucer told only as much of his story as he pleased, as was required for a particular purpose. He answered for his characters himself. In Shakspeare they are introduced upon the stage, are liable to be asked all sorts of questions, and are forced to answer for themselves. In Chaucer we perceive a fixed essence of character. In Shakspeare there is a continual composition and decomposition of its elements, a fermentation of every particle in the whole mass, by its alternate affinity or antipathy to other principles which are brought in contact with it. Till the experiment is tried, we do not know the result, the turn which the character will take in its new circumstances. Milton took only a few simple principles of character, and raised them to the utmost conceivable grandeur, and refined them from every base alloy. His imagination, "nigh sphered in Heaven," claimed kindred only with what he saw from that height, and could raise to the same elevation with itself. He sat retired and kept his state alone, "playing with wisdom"; while Shakspeare mingled with the crowd, and played the host, "to make society the sweeter welcome."

The passion in Shakspeare is of the same nature as his delineation of character. It is not some one habitual feeling or sentiment preying upon itself, growing out of itself, and moulding every thing to itself; it is passion modified by passion, by all the other feelings to which the individual is liable, and to which others are liable with him; subject to all the fluctuations of caprice and accident; calling into play all the resources of the understanding and all the energies of the will; irritated by obstacles or yielding to them; rising from small beginnings to its utmost height; now drunk with hope, now stung to madness, now sunk in despair, now blown to air with a breath, now raging like a torrent. The human soul is made the sport of fortune, the prey of adversity: it is stretched on the wheel of destiny, in restless ecstacy. The passions are in a state of projection. Years are melted down to moments, and every instant teems with fate. We know the results, we see the process. Thus after Iago has been boasting to himself of the effect of his poisonous suggestions on the mind of Othello, "which, with a little act upon the blood, will work like mines of sulphur," he adds—

"Look where he comes! not poppy, nor mandragora,
Nor all the drowsy syrups of the East,
Shall ever medicine thee to that sweet sleep
Which thou ow'dst yesterday."—

And he enters at this moment, like the crested serpent, crowned with his wrongs and raging for revenge! The whole depends upon the turn of a thought. A word, a look, blows the spark of jealousy into a flame; and the explosion is immediate and terrible as a volcano. The dialogues in Lear, in Macbeth, that between Brutus and Cassius, and nearly all those in Shakspeare, where the interest is wrought up to its highest pitch, afford examples of this dramatic fluctuation of passion. The interest in Chaucer is quite different; it is like the course of a river, strong, and full, and increasing. In Shakspeare, on the contrary, it is like the sea, agitated this way and that, and loud-lashed by furious storms; while in the still pauses of the blast, we distinguish only the cries of despair, or the silence of death! Milton, on the other hand, takes the imaginative part of passion—that which remains after the event, which the mind reposes on when all is over, which looks upon circumstances from the remotest elevation of thought and fancy, and abstracts them from the world of action to that of contemplation. The objects of dramatic poetry affect us by sympathy, by their nearness to ourselves, as they take us by surprise, or force us upon action, "while rage with rage doth sympathise"; the objects of epic poetry affect us through the medium of the imagination, by magnitude and distance, by their permanence and universality. The one fill us with terror and pity, the other with admiration and delight. There are certain objects that strike the imagination, and inspire awe in the very idea of them, independently of any dramatic interest, that is, of any connection with the vicissitudes of human life. For instance, we cannot think of the pyramids of Egypt, of a Gothic ruin, or an old Roman encampment, without a certain emotion, a sense of power and sublimity coming over the mind. The heavenly bodies that hung over our heads wherever we go, and "in their untroubled element shall shine when we are laid in dust, and all our cares forgotten," affect us in the same way. Thus Satan's address to the Sun[3] has an epic, not a dramatic interest; for though the second person in the dialogue makes no answer and feels no concern, yet the eye of that vast luminary is upon him, like the eye of heaven, and seems conscious of what he says, like an universal

---

3. **Satan's . . . Sun:** *Paradise Lost*, IV. 31–32.

presence. Dramatic poetry and epic, in their perfection, indeed, approximate to and strengthen one another. Dramatic poetry borrows aid from the dignity of persons and things, as the heroic does from human passion, but in theory they are distinct.—When Richard II. calls for the looking-glass to contemplate his faded majesty in it, and bursts into that affecting exclamation: "Oh, that I were a mockery-king of snow, to melt away before the sun of Bolingbroke," we have here the utmost force of human passion, combined with the ideas of regal splendour and fallen power. When Milton says of Satan:

> "—— His form had not yet lost
> All her original brightness, nor appear'd
> Less than archangel ruin'd, and th' excess
> Of glory obscur'd;"— 4

the mixture of beauty, of grandeur, and pathos, from the sense of irreparable loss, of never-ending, unavailing regret, is perfect.

The great fault of a modern school of poetry is, that it is an experiment to reduce poetry to a mere effusion of natural sensibility; or what is worse, to divest it both of imaginary splendour and human passion, to surround the meanest objects with the morbid feelings and devouring egotism of the writers' own minds.5 Milton and Shakspeare did not so understand poetry. They gave a more liberal interpretation both to nature and art. They did not do all they could to get rid of the one and the other, to fill up the dreary void with the Moods of their own Minds.6 They owe their power over the human mind to their having had a deeper sense than others of what was grand in the objects of nature, or affecting in the events of human life. But to the men I speak of there is nothing interesting, nothing heroical, but themselves. To them the fall of gods or of great men is the same. They do not enter into the feeling. They cannot understand the terms. They are even debarred from the last poor, paltry consolation of an unmanly triumph over fallen greatness; for their minds reject, with a convulsive effort and intolerable loathing, the very idea that there ever was, or was thought to be, any thing superior to themselves. All that has ever excited the attention or admiration of the world, they look upon with the most perfect indifference; and they are surprised to find that the world repays their indifference with scorn. "With what measure they mete, it has been meted to them again."— 7

Shakspeare's imagination is of the same plastic kind as his conception of character or passion. "It glances from heaven to earth, from earth to heaven." Its movement is rapid and devious. It unites the most opposite extremes; or, as Puck says, in boasting of his own feats, "puts a girdle round about the earth in forty minutes."8 He seems always hurrying from his subject, even while describing it; but the stroke, like the lightning's, is sure as it is sudden. He takes the widest possible range, but from that very range he has his choice of the greatest variety and aptitude of materials. He brings together images the most alike, but placed at the greatest distance from each other; that is, found in circumstances of the greatest dissimilitude. From the remoteness of his combinations, and the celerity with which they are effected, they coalesce the more indissolubly together. The more the thoughts are strangers to each other, and the longer they have been kept asunder, the more intimate does their union seem to become. Their felicity is equal to their force. Their likeness is made more dazzling by their novelty. They startle, and take the fancy prisoner in the same instant. I will mention one or two which are very striking, and not much known, out of Troilus and Cressida.9 Æneas says to Agamemnon,

> "I ask that I may waken reverence,
> And on the cheek be ready with a blush
> Modest as morning, when she coldly eyes
> The youthful Phœbus."

Ulysses urging Achilles to shew himself in the field, says—

> "No man is the lord of anything,
> Till he communicate his parts to others:
> Nor doth he of himself know them for aught,
> Till he behold them formed in the applause,
> Where they're extended! which like an arch reverberates
> The voice again, or like a gate of steel,
> Fronting the sun, receives and renders back
> Its figure and its heat."

---

4. **"His . . . obscur'd"**: *Paradise Lost*, I. 591–94. 5. **egotism . . . minds:** Cf. the conclusion to the selection above from Hazlitt's "On the Character of Rousseau." 6. **Moods . . . Minds:** Hazlitt is mocking the phrase Wordsworth used as the title of one of the sections in his *Poems* (1807): "Moods of My Own Mind."

7. **"With . . . again"**: Mark 4:24; Luke 6:38. 8. **"It glances . . . minutes"**: *A Midsummer Night's Dream*, V. i. 13 and II. i. 175. 9. **mention . . . Cressida:** I. iii. 227–30; III. iii. 115–23; III. iii. 222.

Patroclus gives the indolent warrior the same advice.

> "Rouse yourself; and the weak wanton Cupid
> Shall from your neck unloose his amorous fold,
> And like a dew-drop from the lion's mane
> Be shook to air."

Shakspeare's language and versification are like the rest of him. He has a magic power over words: they come winged at his bidding; and seem to know their places. They are struck out at a heat, on the spur of the occasion, and have all the truth and vividness which arise from an actual impression of the objects. His epithets and single phrases are like sparkles, thrown off from an imagination, fired by the whirling rapidity of its own motion. His language is hieroglyphical.[10] It translates thoughts into visible images. It abounds in sudden transitions and elliptical expressions. This is the source of his mixed metaphors, which are only abbreviated forms of speech. These, however, give no pain from long custom. They have, in fact, become idioms in the language. They are the building, and not the scaffolding to thought. We take the meaning and effect of a well-known passage entire, and no more stop to scan and spell out the particular words and phrases, than the syllables of which they are composed. In trying to recollect any other author, one sometimes stumbles, in case of failure, on a word as good. In Shakspeare, any other word but the true one, is sure to be wrong. If any body, for instance, could not recollect the words of the following description,

> "—— Light thickens,
> And the crow makes wing to the rooky wood,"[11]

he would be greatly at a loss to substitute others for them equally expressive of the feeling. These remarks, however, are strictly applicable only to the impassioned parts of Shakspeare's language, which flowed from the warmth and originality of his imagination, and were his own. The language used for prose conversation and ordinary business is sometimes technical, and involved in the affectation of the time. Compare, for example, Othello's apology to the senate, relating "his whole course of love," with some of the preceding parts relating to his appointment, and the official dispatches from Cyprus. In this respect, "the business of the state does him offence." His versification is no less powerful, sweet, and varied. It has every occasional excellence, of sullen intricacy, crabbed and perplexed, or of the smoothest and loftiest expansion—from the ease and familiarity of measured conversation to the lyrical sounds

> "—— Of ditties highly penned,
> Sung by a fair queen in a summer's bower,
> With ravishing division to her lute."[12]

It is the only blank verse in the language, except Milton's, that for itself is readable. It is not stately and uniformly swelling like his, but varied and broken by the inequalities of the ground it has to pass over in its uncertain course,

> "And so by many winding nooks it strays,
> With willing sport to the wild ocean."[13]

It remains to speak of the faults of Shakspeare. They are not so many or so great as they have been represented; what there are, are chiefly owing to the following causes:—The universality of his genius was, perhaps, a disadvantage to his single works; the variety of his resources, sometimes diverting him from applying them to the most effectual purposes. He might be said to combine the powers of Æschylus and Aristophanes, of Dante and Rabelais, in his own mind. If he had been only half what he was, he would perhaps have appeared greater. The natural ease and indifference of his temper made him sometimes less scrupulous than he might have been. He is relaxed and careless in critical places; he is in earnest throughout only in Timon, Macbeth, and Lear. Again, he had no models of acknowledged excellence constantly in view to stimulate his efforts, and by all that appears, no love of fame. He wrote for the "great vulgar and the small," in his time, not for posterity. If Queen Elizabeth and the maids of honour laughed heartily at his worst jokes, and the catcalls in the gallery were silent at his best passages, he went home satisfied, and slept the next night well. He did not trouble himself about Voltaire's criticisms. He was willing to take advantage of the ignorance of the age in many things; and if his plays pleased others, not to quarrel with them himself. His very facility

10. **language . . . hieroglyphical:** Cf. Keats's remark in a review of "Edmund Kean as a Shakespearean Actor" (*The Champion*, December, 1817): "The spiritual is felt when the very letters and points of charactered language show like the hieroglyphics of beauty;—the mysterious signs of an immortal free-masonry!" 11. **"Light . . . wood":** *Macbeth*, III. ii. 50.

12. **"Of . . . lute":** 1 *Henry IV*, III. i. 208–10. 13. **"And . . . ocean":** *The Two Gentlemen of Verona*, II. vii. 31.

of production would make him set less value on his own excellences, and not care to distinguish nicely between what he did well or ill. His blunders in chronology and geography do not amount to above half a dozen, and they are offences against chronology and geography, not against poetry. As to the unities, he was right in setting them at defiance. He was fonder of puns than became so great a man. His barbarisms were those of his age. His genius was his own. He had no objection to float down with the stream of common taste and opinion: he rose above it by his own buoyancy, and an impulse which he could not keep under, in spite of himself or others, and "his delights did shew most dolphin-like."

He had an equal genius for comedy and tragedy; and his tragedies are better than his comedies, because tragedy is better than comedy. His female characters, which have been found fault with as insipid, are the finest in the world. Lastly, Shakspeare was the least of a coxcomb of any one that ever lived, and much of a gentleman.

Shakspeare discovers in his writings little religious enthusiasm, and an indifference to personal reputation; he had none of the bigotry of his age, and his political prejudices were not very strong. In these respects, as well as in every other, he formed a direct contrast to Milton. Milton's works are a perpetual invocation to the Muses; a hymn to Fame. He had his thoughts constantly fixed on the contemplation of the Hebrew theocracy, and of a perfect commonwealth; and he seized the pen with a hand just warm from the touch of the ark of faith. His religious zeal infused its character into his imagination; so that he devotes himself with the same sense of duty to the cultivation of his genius, as he did to the exercise of virtue, or the good of his country. The spirit of the poet, the patriot, and the prophet, vied with each other in his breast. His mind appears to have held equal communion with the inspired writers, and with the bards and sages of ancient Greece and Rome;—

> "Blind Thamyris, and blind Mæonides,
> And Tiresias, and Phineus, prophets old."[14]

He had a high standard, with which he was always comparing himself, nothing short of which could satisfy his jealous ambition. He thought of nobler forms and nobler things than those he found about him. He lived apart, in the solitude of his own thoughts,

14. "Blind . . . old": *Paradise Lost*, III. 35-36.

carefully excluding from his mind whatever might distract its purposes or alloy its purity, or damp its zeal. "With darkness and with dangers compassed round," he had the mighty models of antiquity always present to his thoughts, and determined to raise a monument of equal height and glory, "piling up every stone of lustre from the brook," for the delight and wonder of posterity. He had girded himself up, and as it were, sanctified his genius to this service from his youth. "For after," he says, "I had from my first years, by the ceaseless diligence and care of my father, been exercised to the tongues, and some sciences as my age could suffer, by sundry masters and teachers, it was found that whether aught was imposed upon me by them, or betaken to of my own choice, the style by certain vital signs it had, was likely to live; but much latelier, in the private academies of Italy, perceiving that some trifles which I had in memory, composed at under twenty or thereabout, met with acceptance above what was looked for; I began thus far to assent both to them and divers of my friends here at home, and not less to an inward prompting which now grew daily upon me, that by labour and intense study (which I take to be my portion in this life), joined with the strong propensity of nature, I might perhaps leave something so written to after-times as they should not willingly let it die. The accomplishment of these intentions, which have lived within me ever since I could conceive myself anything worth to my country, lies not but in a power above man's to promise; but that none hath by more studious ways endeavoured, and with more unwearied spirit that none shall, that I dare almost aver of myself, as far as life and free leisure will extend. Neither do I think it shame to covenant with any knowing reader, that for some few years yet, I may go on trust with him toward the payment of what I am now indebted, as being a work not to be raised from the heat of youth or the vapours of wine; like that which flows at waste from the pen of some vulgar amourist, or the trencher fury of a rhyming parasite, nor to be obtained by the invocation of Dame Memory and her Siren daughters, but by devout prayer to that eternal spirit who can enrich with all utterance and knowledge, and sends out his Seraphim with the hallowed fire of his altar, to touch and purify the lips of whom he pleases: to this must be added industrious and select reading, steady observation, and insight into all seemly and generous arts and affairs. Although it nothing content me to have disclosed thus much beforehand; but that I trust hereby to make it manifest with what small willingness I endure to interrupt the pursuit of no less hopes than these, and leave a calm and pleasing solitariness, fed with cheerful and confident thoughts, to embark in a troubled sea of noises and hoarse disputes, from beholding the bright countenance of truth in the quiet and still air of delightful studies."[15]

So that of Spenser:

"The noble heart that harbours virtuous thought,
    And is with child of glorious great intent,
Can never rest until it forth have brought
    The eternal brood of glory excellent."[16]

Milton, therefore, did not write from casual impulse, but after a severe examination of his own strength, and with a resolution to leave nothing undone which it was in his power to do. He always labours, and almost always succeeds. He strives hard to say the finest things in the world, and he does say them. He adorns and dignifies his subject to the utmost: he surrounds it with every possible association of beauty or grandeur, whether moral, intellectual, or physical. He refines on his descriptions of beauty; loading sweets on sweets, till the sense aches at them; and raises his images of terror to a gigantic elevation, that "makes Ossa like a wart." In Milton, there is always an appearance of effort: in Shakspeare, scarcely any.

Milton has borrowed more than any other writer, and exhausted every source of imitation, sacred or profane; yet he is perfectly distinct from every other writer. He is a writer of centos, and yet in originality scarcely inferior to Homer. The power of his mind is stamped on every line. The fervour of his imagination melts down and renders malleable, as in a furnace, the most contradictory materials. In reading his works, we feel ourselves under the influence of a mighty intellect, that the nearer it approaches to others, becomes more distinct from them. The quantity of art in him shews the strength of his genius: the weight of his intellectual obligations would have oppressed any other writer. Milton's learning has the effect of intuition. He describes objects, of which he could only have read in books, with the vividness of actual observation. His imagination has the force of nature. He makes words tell as pictures.

"Him followed Rimmon, whose delightful seat
Was fair Damascus, on the fertile banks
Of Abbana and Pharphar, lucid streams."[17]

The word *lucid* here gives to the idea all the sparkling effect of the most perfect landscape.

And again:

"As when a vulture on Imaus bred,
Whose snowy ridge the roving Tartar bounds,
Dislodging from a region scarce of prey,
To gorge the flesh of lambs and yeanling kids
On hills where flocks are fed, flies towards the springs
Of Ganges or Hydaspes, Indian streams;
But in his way lights on the barren plains
Of Sericana, where Chineses drive
With sails and wind their cany waggons light."[18]

If Milton had taken a journey for the express purpose, he could not have described this scenery and mode of life better. Such passages are like demonstrations of natural history. Instances might be multiplied without end.

We might be tempted to suppose that the vividness with which he describes visible objects, was owing to their having acquired an unusual degree of strength in his mind, after the privation of his sight; but we find the same palpableness and truth in the descriptions which occur in his early poems. In Lycidas he speaks of "the great vision of the guarded mount," with that preternatural weight of impression with which it would present itself suddenly to "the pilot of some small night-foundered skiff": and the lines in the Penseroso, describing "the wandering moon,"

"Riding near her highest noon,
Like one that had been led astray
Through the heaven's wide pathless way,"[19]

are as if he had gazed himself blind in looking at her. There is also the same depth of impression in his descriptions of the objects of all the different senses, whether colours, or sounds, or smells—the same absorption of his mind in whatever engaged his attention at the time. It has been indeed objected to Milton, by a common perversity of criticism, that his ideas were musical rather than picturesque, as if because they were in the highest degree musical, they must be (to keep the sage critical balance even, and to allow no one man to possess two qualities at the same time)

---

15. **"For . . . studies"**: Preface to Book II, *Reason of Church Government*. **16. "The . . . excellent"**: *Faerie Queene*, I. v. i.

17. **"Him . . . streams"**: *Paradise Lost*, I. 467–69. **18. "As . . . light"**: *Paradise Lost*, III. 431–39. **19. "Riding . . . way"**: *"Il Penseroso,"* ll. 67–70.

proportionably deficient in other respects. But Milton's poetry is not cast in any such narrow, common-place mould; it is not so barren of resources. His worship of the Muse was not so simple or confined. A sound arises "like a steam of rich distilled perfumes"; we hear the pealing organ, but the incense on the altars is also there, and the statues of the gods are ranged around! The ear indeed predominates over the eye, because it is more immediately affected, and because the language of music blends more immediately with, and forms a more natural accompaniment to, the variable and indefinite associations of ideas conveyed by words. But where the associations of the imagination are not the principal thing, the individual object is given by Milton with equal force and beauty. The strongest and best proof of this, as a characteristic power of his mind, is, that the persons of Adam and Eve, of Satan, &c. are always accompanied, in our imagination, with the grandeur of the naked figure; they convey to us the ideas of sculpture. As an instance, take the following:

> "———— He soon
> Saw within ken a glorious Angel stand,
> The same whom John saw also in the sun:
> His back was turned, but not his brightness hid;
> Of beaming sunny rays a golden tiar
> Circled his head, nor less his locks behind
> Illustrious on his shoulders fledge with wings
> Lay waving round; on some great charge employ'd
> He seem'd, or fix'd in cogitation deep.
> Glad was the spirit impure, as now in hope
> To find who might direct his wand'ring flight
> To Paradise, the happy seat of man,
> His journey's end, and our beginning woe.
> But first he casts to change his proper shape,
> Which else might work him danger or delay
> And now a stripling cherub he appears,
> Not of the prime, yet such as in his face
> Youth smiled celestial, and to every limb
> Suitable grace diffus'd, so well he feign'd:
> Under a coronet his flowing hair
> In curls on either cheek play'd; wings he wore
> Of many a colour'd plume sprinkled with gold,
> His habit fit for speed succinct, and held
> Before his decent steps a silver wand." [20]

The figures introduced here have all the elegance and precision of a Greek statue; glossy and impurpled, tinged with golden light, and musical as the strings of Memnon's harp!

Again, nothing can be more magnificent than the portrait of Beelzebub:

> "With Atlantean shoulders fit to bear
> The weight of mightiest monarchies:" [21]

Or the comparison of Satan, as he "lay floating many a rood," to "that sea beast,"

> "Leviathan, which God of all his works
> Created hugest that swim the ocean-stream!" [22]

What a force of imagination is there in this last expression! What an idea it conveys of the size of that hugest of created beings, as if it shrunk up the ocean to a stream, and took up the sea in its nostrils as a very little thing! Force of style is one of Milton's greatest excellences. Hence, perhaps, he stimulates us more in the reading, and less afterwards. The way to defend Milton against all impugners, is to take down the book and read it.

Milton's blank verse is the only blank verse in the language (except Shakspeare's) that deserves the name of verse. Dr. Johnson, who had modelled his ideas of versification on the regular sing-song of Pope, condemns the Paradise Lost as harsh and unequal. I shall not pretend to say that this is not sometimes the case; for where a degree of excellence beyond the mechanical rules of art is attempted, the poet must sometimes fail. But I imagine that there are more perfect examples in Milton of musical expression, or of an adaptation of the sound and movement of the verse to the meaning of the passage, than in all our other writers, whether of rhyme or blank verse, put together, (with the exception already mentioned). Spenser is the most harmonious of our stanza writers, as Dryden is the most sounding and varied of our rhymists. But in neither is there any thing like the same ear for music, the same power of approximating the varieties of poetical to those of musical rhythm, as there is in our great epic poet. The sound of his lines is moulded into the expression of the sentiment, almost of the very image. They rise or fall, pause or hurry rapidly on, with exquisite art, but without the least trick or affectation, as the occasion seems to require.

The following are some of the finest instances:

> "———— His hand was known
> In Heaven by many a tower'd structure high;——
> Nor was his name unheard or unador'd
> In ancient Greece: and in the Ausonian land

20. "He . . . wand": *Paradise Lost*, III. 621–44.

21. "With . . . monarchies": *Paradise Lost*, II. 306–07. 22. "Leviathan . . . ocean-stream!": *Paradise Lost*, I. 200–02.

Men called him Mulciber: and how he fell
From Heaven, they fabled, thrown by angry Jove
Sheer o'er the chrystal battlements; from morn
To noon he fell, from noon to dewy eve,
A summer's day; and with the setting sun
Dropt from the zenith like a falling star
On Lemnos, the Ægean isle: thus they relate,
Erring."—

"———— But chief the spacious hall
Thick swarm'd, both on the ground and in the air,
Brush'd with the hiss of rustling wings. As bees
In spring time, when the sun with Taurus rides,
Pour forth their populous youth about the hive
In clusters; they among fresh dews and flow'rs
Fly to and fro: or on the smoothed plank,
The suburb of their straw-built citadel,
New rubb'd with balm, expatiate and confer
Their state affairs. So thick the airy crowd
Swarm'd and were straiten'd; till the signal giv'n,
Behold a wonder! They but now who seem'd
In bigness to surpass earth's giant sons,
Now less than smallest dwarfs, in narrow room
Throng numberless, like that Pygmean race
Beyond the Indian mount, or fairy elves,
Whose midnight revels by a forest side
Or fountain, some belated peasant sees,
Or dreams he sees, while over-head the moon
Sits arbitress, and nearer to the earth
Wheels her pale course: they on their mirth and dance
Intent, with jocund music charm his ear;
At once with joy and fear his heart rebounds."[23]

I can only give another instance, though I have some difficulty in leaving off.

"Round he surveys (and well might, where he stood
So high above the circling canopy
Of night's extended shade) from th' eastern point
Of Libra to the fleecy star that bears
Andromeda far off Atlantic seas
Beyond the horizon: then from pole to pole
He views in breadth, and without longer pause
Down right into the world's first region throws
His flight precipitant, and winds with ease
Through the pure marble air his oblique way
Amongst innumerable stars that shone
Stars distant, but nigh hand seem'd other worlds;
Or other worlds they seem'd or happy isles,"[24] &c.

The verse, in this exquisitely modulated passage, floats up and down as if it had itself wings. Milton has himself given us the theory of his versification—

"Such as the meeting soul may pierce
In notes with many a winding bout
Of linked sweetness long drawn out."[25]

Dr. Johnson and Pope would have converted his vaulting Pegasus into a rocking-horse.[26] Read any other blank verse but Milton's,—Thomson's, Young's, Cowper's, Wordsworth's,—and it will be found, from the want of the same insight into "the hidden soul of harmony," to be mere lumbering prose.

To proceed to a consideration of the merits of Paradise Lost, in the most essential point of view, I mean as to the poetry of character and passion. I shall say nothing of the fable, or of other technical objections or excellences; but I shall try to explain at once the foundation of the interest belonging to the poem. I am ready to give up the dialogues in Heaven, where, as Pope justly observes, "God the Father turns a school-divine"; nor do I consider the battle of the angels as the climax of sublimity, or the most successful effort of Milton's pen. In a word, the interest of the poem arises from the daring ambition and fierce passions of Satan, and from the account of the paradisaical happiness, and the loss of it by our first parents. Three-fourths of the work are taken up with these characters, and nearly all that relates to them is unmixed sublimity and beauty. The two first books alone are like two massy pillars of solid gold.

Satan is the most heroic subject that ever was chosen for a poem; and the execution is as perfect as the design is lofty. He was the first of created beings, who, for endeavouring to be equal with the highest, and to divide the empire of heaven with the Almighty, was hurled down to hell. His aim was no less than the throne of the universe; his means, myriads of angelic armies bright, the third part of the heavens, whom he lured after him with his countenance, and who durst defy the Omnipotent in arms. His ambition was the greatest, and his punishment was the greatest; but not so his despair, for his fortitude was as great as his sufferings. His strength of mind was matchless as his strength of body; the vastness of his designs did not surpass the firm, inflexible determination with which he submitted to his irreversible doom, and final loss of all good. His power of action and of suffering was equal. He was the greatest power that was ever overthrown, with the

23. "His . . . rebounds": *Paradise Lost,* I. 732–47 and 762–88. 24. "Round . . . isles": *Paradise Lost,* III. 555–67.

25. "Such . . . out": "L'Allegro," ll. 138–40. 26. Pegasus . . . rocking-horse: Cf. Keats's "Sleep and Poetry," below, ll. 186–87: "They sway'd about upon a rocking horse, / And thought it Pegasus."

strongest will left to resist or to endure. He was baffled, not confounded. He stood like a tower; or

> "————— As when Heaven's fire
> Hath scathed the forest oaks or mountain pines." [27]

He was still surrounded with hosts of rebel angels, armed warriors, who own him as their sovereign leader, and with whose fate he sympathises as he views them round, far as the eye can reach; though he keeps aloof from them in his own mind, and holds supreme counsel only with his own breast. An outcast from Heaven, Hell trembles beneath his feet, Sin and Death are at his heels, and mankind are his easy prey.

> "All is not lost; th' unconquerable will,
> And study of revenge, immortal hate,
> And courage never to submit or yield,
> And what else is not to be overcome," [28]

are still his. The sense of his punishment seems lost in the magnitude of it; the fierceness of tormenting flames is qualified and made innoxious by the greater fierceness of his pride; the loss of infinite happiness to himself is compensated in thought, by the power of inflicting infinite misery on others. Yet Satan is not the principle of malignity, or of the abstract love of evil—but of the abstract love of power, of pride, of self-will personified, to which last principle all other good and evil, and even his own, are subordinate. From this principle he never once flinches. His love of power and contempt for suffering are never once relaxed from the highest pitch of intensity. His thoughts burn like a hell within him; but the power of thought holds dominion in his mind over every other consideration. The consciousness of a determined purpose, of "that intellectual being, those thoughts that wander through eternity," though accompanied with endless pain, he prefers to nonentity, to "being swallowed up and lost in the wide womb of uncreated night." He expresses the sum and substance of all ambition in one line. "Fallen cherub, to be weak is miserable, doing or suffering!" After such a conflict as his, and such a defeat, to retreat in order, to rally, to make terms, to exist at all, is something; but he does more than this—he founds a new empire in hell, and from it conquers this new world, whither he bends his undaunted flight, forcing his way through nether and surrounding fires. The poet has not in all this given us a mere shadowy

outline; the strength is equal to the magnitude of the conception. The Achilles of Homer is not more distinct; the Titans were not more vast; Prometheus chained to his rock was not a more terrific example of suffering and of crime. Wherever the figure of Satan is introduced, whether he walks or flies, "rising aloft incumbent on the dusky air," it is illustrated with the most striking and appropriate images: so that we see it always before us, gigantic, irregular, portentous, uneasy, and disturbed—but dazzling in its faded splendour, the clouded ruins of a god. The deformity of Satan is only in the depravity of his will; he has no bodily deformity to excite our loathing or disgust. The horns and tail are not there, poor emblems of the unbending, unconquered spirit, of the writhing agonies within. Milton was too magnanimous and open an antagonist to support his argument by the bye-tricks of a hump and cloven foot; to bring into the fair field of controversy the good old catholic prejudices of which Tasso and Dante have availed themselves, and which the mystic German critics would restore. He relied on the justice of his cause, and did not scruple to give the devil his due. Some persons may think that he has carried his liberality too far, and injured the cause he professed to espouse by making him the chief person in his poem. Considering the nature of his subject, he would be equally in danger of running into this fault, from his faith in religion, and his love of rebellion; and perhaps each of these motives had its full share in determining the choice of his subject.

Not only the figure of Satan, but his speeches in council, his soliloquies, his address to Eve, his share in the war in heaven, or in the fall of man, shew the same decided superiority of character. To give only one instance, almost the first speech he makes:

"Is this the region, this the soil, the clime,
Said then the lost archangel, this the seat
That we must change for Heaven; this mournful gloom
For that celestial light? Be it so, since he
Who now is sov'rain can dispose and bid
What shall be right: farthest from him is best,
Whom reason hath equal'd, force hath made supreme
Above his equals. Farewel happy fields,
Where joy for ever dwells: Hail horrors, hail
Infernal world, and thou profoundest Hell,
Receive thy new possessor: one who brings
A mind not to be chang'd by place or time.
The mind is its own place, and in itself
Can make a Heav'n of Hell, a Hell of Heav'n.
What matter where, if I be still the same,
And what I should be, all but less than he

---

27. "As . . . pines": *Paradise Lost*, I. 612–13. **28. "All . . . overcome"**: *Paradise Lost*, I. 106–09.

Whom thunder hath made greater? Here at least
We shall be free; th' Almighty hath not built
Here for his envy, will not drive us hence:
Here we may reign secure, and in my choice
To reign is worth ambition, though in Hell:
Better to reign in Hell, than serve in Heaven." [29]

The whole of the speeches and debates in Pandemonium are well worthy of the place and the occasion —with Gods for speakers, and angels and archangels for hearers. There is a decided manly tone in the arguments and sentiments, an eloquent dogmatism, as if each person spoke from thorough conviction; an excellence which Milton probably borrowed from his spirit of partisanship, or else his spirit of partisanship from the natural firmness and vigour of his mind. In this respect Milton resembles Dante, (the only modern writer with whom he has any thing in common) and it is remarkable that Dante, as well as Milton, was a political partisan. That approximation to the severity of impassioned prose which has been made an objection to Milton's poetry, and which is chiefly to be met with in these bitter invectives, is one of its great excellences. The author might here turn his philippics against Salmasius to good account. The rout in Heaven is like the fall of some mighty structure, nodding to its base, "with hideous ruin and combustion down." But, perhaps, of all the passages in Paradise Lost, the description of the employments of the angels during the absence of Satan, some of whom "retreated in a silent valley, sing with notes angelical to many a harp their own heroic deeds and hapless fall by doom of battle," is the most perfect example of mingled pathos and sublimity.—What proves the truth of this noble picture in every part, and that the frequent complaint of want of interest in it is the fault of the reader, not of the poet, is that when any interest of a practical kind takes a shape that can be at all turned into this, (and there is little doubt that Milton had some such in his eye in writing it,) each party converts it to its own purposes, feels the absolute identity of these abstracted and high speculations; and that, in fact, a noted political writer of the present day has exhausted nearly the whole account of Satan in the Paradise Lost, by applying it to a character whom he considered as after the devil, (though I do not know whether he would make even that exception) the greatest enemy of the human race. This may serve to shew that Milton's Satan is not a very insipid personage.

Of Adam and Eve it has been said, that the ordinary reader can feel little interest in them, because they have none of the passions, pursuits, or even relations of human life, except that of man and wife, the least interesting of all others, if not to the parties concerned, at least to the by-standers. The preference has on this account been given to Homer, who, it is said, has left very vivid and infinitely diversified pictures of all the passions and affections, public and private, incident to human nature—the relations of son, of brother, parent, friend, citizen, and many others. Longinus preferred the Iliad to the Odyssey, on account of the greater number of battles it contains; but I can neither agree to his criticism, nor assent to the present objection. It is true, there is little action in this part of Milton's poem; but there is much repose, and more enjoyment. There are none of the every-day occurrences, contentions, disputes, wars, fightings, feuds, jealousies, trades, professions, liveries, and common handicrafts of life; "no kind of traffic; letters are not known; no use of service, of riches, poverty, contract, succession, bourne, bound of land, tilth, vineyard none; no occupation, no treason, felony, sword, pike, knife, gun, nor need of any engine." So much the better; thank Heaven, all these were yet to come. But still the die was cast, and in them our doom was sealed. In them

> "The generations were prepared; the pangs,
> The internal pangs, were ready, the dread strife
> Of poor humanity's afflicted will,
> Struggling in vain with ruthless destiny." [30]

In their first false step we trace all our future woe, with loss of Eden. But there was a short and precious interval between, like the first blush of morning before the day is overcast with tempest, the dawn of the world, the birth of nature from "the unapparent deep," with its first dews and freshness on its cheek, breathing odours. Theirs was the first delicious taste of life, and on them depended all that was to come of it. In them hung trembling all our hopes and fears. They were as yet alone in the world, in the eye of nature, wondering at their new being, full of enjoyment and enraptured with one another, with the voice of their Maker walking in the garden, and ministering angels attendant on their steps, winged messengers from heaven like rosy clouds descending in their sight. Nature played around them her virgin fancies wild; and spread for

---

29. **"Is . . . Heaven":** *Paradise Lost*, I. 242–63.

30. **"The . . . destiny":** Wordsworth's *Excursion*, VI. 554–57.

them a repast where no crude surfeit reigned. Was there nothing in this scene, which God and nature alone witnessed, to interest a modern critic? What need was there of action, where the heart was full of bliss and innocence without it! They had nothing to do but feel their own happiness, and "know to know no more." "They toiled not, neither did they spin; yet Solomon in all his glory was not arrayed like one of these." All things seem to acquire fresh sweetness, and to be clothed with fresh beauty in their sight. They tasted as it were for themselves and us, of all that there ever was pure in human bliss. "In them the burthen of the mystery, the heavy and the weary weight of all this unintelligible world, is lightened." They stood awhile perfect, but they afterwards fell, and were driven out of Paradise, tasting the first fruits of bitterness as they had done of bliss. But their pangs were such as a pure spirit might feel at the sight—their tears "such as angels weep." The pathos is of that mild contemplative kind which arises from regret for the loss of unspeakable happiness, and resignation to inevitable fate. There is none of the fierceness of intemperate passion, none of the agony of mind and turbulence of action, which is the result of the habitual struggles of the will with circumstances, irritated by repeated disappointment, and constantly setting its desires most eagerly on that which there is an impossibility of attaining. This would have destroyed the beauty of the whole picture. They had received their unlooked-for happiness as a free gift from their Creator's hands, and they submitted to its loss, not without sorrow, but without impious and stubborn repining.

"In either hand the hast'ning angel caught
Our ling'ring parents, and to th' eastern gate
Led them direct, and down the cliff as fast
To the subjected plain; then disappear'd.
They looking back, all th' eastern side beheld
Of Paradise, so late their happy seat,
Wav'd over by that flaming brand, the gate
With dreadful faces throng'd, and fiery arms:
Some natural tears they dropt, but wip'd them soon;
The world was all before them, where to choose
Their place of rest, and Providence their guide." [31]

1818 (1818)

31. **"In . . . guide":** *Paradise Lost,* XII. 637–47.

from ON THE LIVING POETS

Genius is the heir of fame; but the hard condition on which the bright reversion must be earned is the loss of life. Fame is the recompense not of the living, but of the dead. The temple of fame stands upon the grave: the flame that burns upon its altars is kindled from the ashes of great men. Fame itself is immortal, but it is not begot till the breath of genius is extinguished. For fame is not popularity, the shout of the multitude, the idle buzz of fashion, the venal puff, the soothing flattery of favour or of friendship; but it is the spirit of a man surviving himself in the minds and thoughts of other men, undying and imperishable. It is the power which the intellect exercises over the intellect, and the lasting homage which is paid to it, as such, independently of time and circumstances, purified from partiality and evil-speaking. Fame is the sound which the stream of high thoughts, carried down to future ages, makes as it flows—deep, distant, murmuring evermore like the waters of the mighty ocean. He who has ears truly touched to this music, is in a manner deaf to the voice of popularity.—The love of fame differs from mere vanity in this, that the one is immediate and personal, the other ideal and abstracted. It is not the direct and gross homage paid to himself, that the lover of true fame seeks or is proud of; but the indirect and pure homage paid to the eternal forms of truth and beauty as they are reflected in his mind, that gives him confidence and hope. The love of nature is the first thing in the mind of the true poet: the admiration of himself the last. A man of genius cannot well be a coxcomb; for his mind is too full of other things to be much occupied with his own person. He who is conscious of great powers in himself, has also a high standard of excellence with which to compare his efforts: he appeals also to a test and judge of merit, which is the highest, but which is too remote, grave, and impartial, to flatter his self-love extravagantly, or puff him up with intolerable and vain conceit. This, indeed, is one test of genius and of real greatness of mind, whether a man can wait patiently and calmly for the award of posterity, satisfied with the unwearied exercise of his faculties, retired within the sanctuary of his own thoughts; or whether he is eager to forestal his own immortality, and mortgage it for a newspaper puff. He who thinks much of himself, will be in danger of being forgotten by the rest of the world: he who is always trying to lay violent hands on reputation, will not secure the best and most lasting. If the restless candidate for praise takes no pleasure, no sincere and heartfelt delight in his works, but as they are admired and applauded by others, what should others see in them to admire or applaud? They cannot be expected to admire them because they are *his*; but for the truth and nature contained in them, which must first be inly felt and copied with severe delight, from the love of truth and nature, before it can ever appear there. Was Raphael, think you, when he painted his pictures of the Virgin and Child in all their inconceivable truth and beauty of expression, thinking most of his subject or of himself? Do you suppose that Titian, when he painted a landscape, was pluming himself on being thought the finest colourist in the world, or making himself so by looking at nature? Do you imagine that Shakspeare, when he wrote Lear or Othello, was thinking of any thing but Lear and Othello? Or that Mr. Kean, when he plays these characters, is thinking of the audience?—No: he who would be great in the eyes of others, must first learn to be nothing in his own. The love of fame, as it enters at times into his mind, is only another name for the love of excellence; or it is the ambition to attain the highest excellence, sanctioned by the highest authority—that of time.

Those minds, then, which are the most entitled to expect it, can best put up with the postponement of their claims to lasting fame. They can afford to wait. They are not afraid that truth and nature will ever wear out; will lose their gloss with novelty, or their effect with fashion. If their works have the seeds of immortality in them, they will live; if they have not, they care little about them as theirs. They do not complain of the start which others have got of them in the race of everlasting renown, or of the impossibility of attaining the honours which time alone can give, during the term of their natural lives. They know that no applause, however loud and violent, can anticipate or over-rule the judgment of posterity; that the opinion of no one individual, nor of any one generation, can have the weight, the authority (to say nothing of the force of sympathy and prejudice), which must belong to that of successive generations. The brightest living reputation cannot be equally imposing to the imagination, with that which is covered and rendered venerable with the hoar of innumerable ages. No modern production can have the same atmosphere of sentiment around it, as the remains of classical antiquity. But then our moderns may console themselves with the reflection, that they will be old in their turn, and will either be remembered with still increasing honours, or quite forgotten!

I would speak of the living poets as I have spoken of the dead (for I think highly of many of them); but I cannot speak of them with the same reverence, because I do not feel it; with the same confidence, because I cannot have the same authority to sanction my opinion. I cannot be absolutely certain that any body, twenty years hence, will think any thing about any of them; but we may be pretty sure that Milton and Shakspeare will be remembered twenty years hence. We are, therefore, not without excuse if we husband our enthusiasm a little, and do not prematurely lay out our whole stock in untried ventures, and what may turn out to be false bottoms. I have myself out-lived one generation of favourite poets, the Darwins, the Hayleys, the Sewards.[1] Who reads them now?—If, however, I have not the verdict of posterity to bear me out in bestowing the most unqualified praises on their immediate successors, it is also to be remembered, that neither does it warrant me in condemning them. Indeed, it was not my wish to go into this ungrateful part of the subject; but something of the sort is expected from me, and I must run the gauntlet as well as I can. Another circumstance that adds to the difficulty of doing justice to all parties is, that I happen to have had a personal acquaintance with some of these jealous votaries of the Muses; and that is not the likeliest way to imbibe a high opinion of the rest. Poets do not praise one another in the language of hyperbole. I am afraid, therefore, that I labour under a degree of prejudice against some of the most popular poets of the day, from an early habit of deference to the critical opinions of some of the least popular. I cannot say that I ever learnt much about Shakspeare or Milton, Spenser or Chaucer, from these professed guides; for I never heard them say much about them. They were always talking of themselves and one another. Nor am I certain that this sort of personal intercourse with living authors, while it takes away all real relish or freedom of opinion with regard to their contemporaries, greatly enhances our respect for themselves. Poets are not ideal beings; but have their prose-sides, like the commonest of the people. We often

ON THE LIVING POETS. **1. Darwins . . . Sewards:** Erasmus Darwin, 1731–1802, grandfather of Charles and author of poetry with scientific themes; William Hayley, 1745–1820, a minor poet remembered mainly for his autobiography; and Anna Seward, 1747–1809, the "Swan of Lichfield," a friend of Dr. Johnson.

hear persons say, What they would have given to have seen Shakspeare! For my part, I would give a great deal not to have seen him; at least, if he was at all like any body else that I have ever seen. But why should he; for his works are not! This is, doubtless, one great advantage which the dead have over the living. It is always fortunate for ourselves and others, when we are prevented from exchanging admiration for knowledge. The splendid vision that in youth haunts our idea of the poetical character, fades, upon acquaintance, into the light of common day; as the azure tints that deck the mountain's brow are lost on a nearer approach to them.

. . .

The first poetess I can recollect is Mrs. Barbauld,[2] with whose works I became acquainted before those of any other author, male or female, when I was learning to spell words of one syllable in her story-books for children. I became acquainted with her poetical works long after in Enfield's Speaker; and remember being much divided in my opinion at that time, between her Ode to Spring and Collins's Ode to Evening. I wish I could repay my childish debt of gratitude in terms of appropriate praise. She is a very pretty poetess; and, to my fancy, strews the flowers of poetry most agreeably round the borders of religious controversy. She is a neat and pointed prose-writer. Her "Thoughts on the Inconsistency of Human Expectations," is one of the most ingenious and sensible essays in the language. There is the same idea in one of Barrow's Sermons.[3]

Mrs. Hannah More[4] is another celebrated modern poetess, and I believe still living. She has written a great deal which I have never read.

Miss Baillie[5] must make up this trio of female poets. Her tragedies and comedies, one of each to illustrate each of the passions, separately from the rest, are heresies in the dramatic art. She is a Unitarian in poetry. With her the passions are, like the French republic, one and indivisible: they are not so in nature, or in Shakspeare. Mr. Southey has, I believe, somewhere expressed an opinion, that the Basil of Miss Baillie is superior to Romeo and Juliet. I shall not stay to contradict him. On the other hand, I prefer her De Montfort, which was condemned on the stage, to some later tragedies, which have been more fortunate—to the Remorse, Bertram, and lastly, Fazio. There is in the chief character of that play a nerve, a continued unity of interest, a setness of purpose and precision of outline which John Kemble alone was capable of giving; and there is all the grace which women have in writing. In saying that De Montfort was a character which just suited Mr. Kemble, I mean to pay a compliment to both. He was not "a man of no mark or likelihood": and what he could be supposed to do particularly well, must have a meaning in it. As to the other tragedies just mentioned, there is no reason why any common actor should not "make mouths in them at the invisible event,"—one as well as another. Having thus expressed my sense of the merits of the authoress, I must add, that her comedy of the Election, performed last summer at the Lyceum with indifferent success, appears to me the perfection of baby-house theatricals. Every thing in it has such a *do-me-good* air, is so insipid and amiable. Virtue seems such a pretty playing at make-believe, and vice is such a naughty word. It is a theory of some French author, that little girls ought not to be suffered to have dolls to play with, to call them *pretty dears*, to admire their black eyes and cherry cheeks, to lament and bewail over them if they fall down and hurt their faces, to praise them when they are good, and scold them when they are naughty. It is a school of affectation: Miss Baillie has profited of it. She treats her grown men and women as little girls treat their dolls—makes moral puppets of them, pulls the wires, and they talk virtue and act vice, according to their cue and the title prefixed to each comedy or tragedy, not from any real passions of their own, or love either of virtue or vice.

The transition from these to Mr. Rogers's Pleasures of Memory,[6] is not far: he is a very lady-like poet. He is an elegant, but feeble writer. He wraps up obvious thoughts in a glittering cover of fine words; is full of enigmas with no meaning to them; is studiously inverted, and scrupulously far-fetched; and his verses are poetry, chiefly because no particle, line, or syllable of them reads like prose. He differs from Milton in this respect, who is accused of having inserted a number

---

2. **Mrs. Barbauld:** Anna Letitia Barbauld, 1743–1825, remembered because of her remark to Coleridge about the "Rime of the Ancient Mariner"—that it was "improbable" and that it "lacked a moral." 3. **Barrow's Sermons:** Isaac Barrow, 1630–77, whose style Coleridge as well as Hazlitt admired. 4. **Mrs. Hannah More:** A friend of Johnson, Hannah More, 1745–1833, wrote prolifically in all forms— poetry, drama, the novel, letters, and religious tracts. 5. **Miss Baillie:** Joanna Baillie, 1762–1851, a Scottish poetess and playwright. Her play *De Monfort* was produced in 1800 by the Shakespearean actor, John Philip Kemble, mentioned below.

6. **Mr. Rogers's . . . Memory:** Samuel Rogers, 1763–1855. His *Pleasures of Memory* was published in 1792.

of prosaic lines in Paradise Lost. This kind of poetry, which is a more minute and inoffensive species of the Della Cruscan,[7] is like the game of asking what one's thoughts are like. It is a tortuous, tottering, wriggling, fidgetty translation of every thing from the vulgar tongue, into all the tantalizing, teasing, tripping, lisping *mimminee-pimminee* of the highest brilliancy and fashion of poetical diction. You have nothing like truth of nature or simplicity of expression. The fastidious and languid reader is never shocked by meeting, from the rarest chance in the world, with a single homely phrase or intelligible idea. You cannot see the thought for the ambiguity of the language, the figure for the finery, the picture for the varnish. . . .

Campbell's Pleasures of Hope[8] is of the same school, in which a painful attention is paid to the expression in proportion as there is little to express, and the decomposition of prose is substituted for the composition of poetry. How much the sense and keeping in the ideas are sacrificed to a jingle of words and epigrammatic turn of expression, may be seen in such lines as the following:—one of the characters, an old invalid, wishes to end his days under

> "Some hamlet shade, to yield his sickly form
> Health in the breeze, and shelter in the storm."

Now the antithesis here totally fails: for it is the breeze, and not the tree, or as it is quaintly expressed, *hamlet shade*, that affords health, though it is the tree that affords shelter in or from the storm. Instances of the same sort of *curiosa infelicitas* are not rare in this author. His verses on the Battle of Hohenlinden have considerable spirit and animation. His Gertrude of Wyoming is his principal performance. It is a kind of historical paraphrase of Mr. Wordsworth's poem of Ruth. It shews little power, or power enervated by extreme fastidiousness. . . .

Tom Moore is a poet of a quite different stamp. He is as heedless, gay, and prodigal of his poetical wealth, as the other is careful, reserved, and parsimonious. The genius of both is national. Mr. Moore's Muse is another Ariel, as light, as tricksy, as indefatigable, and as humane a spirit. His fancy is for ever on the wing, flutters in the gale, glitters in the sun. Every thing lives, moves, and sparkles in his poetry, while over all love waves his purple light. His thoughts are as restless, as many, and as bright as the insects that people the sun's beam. "So work the honey-bees," extracting liquid sweets from opening buds; so the butterfly expands its wings to the idle air; so the thistle's silver down is wafted over summer seas. An airy voyager on life's stream, his mind inhales the fragrance of a thousand shores, and drinks of endless pleasures under halcyon skies. Wherever his footsteps tend over the enamelled ground of fairy fiction—

> "Around him the bees in play flutter and cluster,
> And gaudy butterflies frolic around."

The fault of Mr. Moore is an exuberance of involuntary power. His facility of production lessens the effect of, and hangs as a dead weight upon, what he produces. His levity at last oppresses. The infinite delight he takes in such an infinite number of things, creates indifference in minds less susceptible of pleasure than his own. He exhausts attention by being inexhaustible. His variety cloys; his rapidity dazzles and distracts the sight. The graceful ease with which he lends himself to every subject, the genial spirit with which he indulges in every sentiment, prevents him from giving their full force to the masses of things, from connecting them into a whole. He wants intensity, strength, and grandeur. His mind does not brood over the great and permanent; it glances over the surfaces, the first impressions of things, instead of grappling with the deep-rooted prejudices of the mind, its inveterate habits, and that "perilous stuff that weighs upon the heart." His pen, as it is rapid and fanciful, wants momentum and passion. It requires the same principle to make us thoroughly like poetry, that makes us like ourselves so well, the feeling of continued identity. The impressions of Mr. Moore's poetry are detached, desultory, and physical. Its gorgeous colours brighten and fade like the rainbow's. Its sweetness evaporates like the effluvia exhaled from beds of flowers! His gay laughing style, which relates to the immediate pleasures of love or wine, is better than his sentimental and romantic vein. His Irish melodies are not free from affectation and a certain sickliness of pretension. His serious descriptions are apt to run into flowery tenderness. His pathos sometimes melts into a mawkish sensibility, or crystallizes into all the prettinesses of allegorical language, and glittering hardness of external imagery. But he has wit at will, and of the first quality.

---

7. **Della Cruscan**: the name assumed by a minor school of poets in the late eighteenth century. It was taken from a famous academy founded in Florence in 1582, the purpose of which had been to refine the Italian language. 8. **Campbell's . . . Hope**: See selections from Thomas Campbell, above.

His satirical and burlesque poetry is his best: it is first-rate. His Twopenny Post-Bag is a perfect "nest of spicery"; where the Cayenne is not spared. The politician there sharpens the poet's pen. In this too, our bard resembles the bee—he has its honey and its sting.

Mr. Moore ought not to have written Lalla Rookh, even for three thousand guineas. His fame is worth more than that. He should have minded the advice of Fadladeen. It is not, however, a failure, so much as an evasion and a consequent disappointment of public expectation. He should have left it to others to break conventions with nations, and faith with the world. He should, at any rate, have kept his with the public. Lalla Rookh is not what people wanted to see whether Mr. Moore could do; namely, whether he could write a long epic poem. It is four short tales. The interest, however, is often high-wrought and tragic, but the execution still turns to the effeminate and voluptuous side. . . .

If Mr. Moore has not suffered enough personally, Lord Byron (judging from the tone of his writings) might be thought to have suffered too much to be a truly great poet. If Mr. Moore lays himself too open to all the various impulses of things, the outward shews of earth and sky, to every breath that blows, to every stray sentiment that crosses his fancy; Lord Byron shuts himself up too much in the impenetrable gloom of his own thoughts, and buries the natural light of things in "nook monastic." The Giaour, the Corsair, Childe Harold, are all the same person, and they are apparently all himself. The everlasting repetition of one subject, the same dark ground of fiction, with the darker colours of the poet's mind spread over it, the unceasing accumulation of horrors on horror's head, steels the mind against the sense of pain, as inevitably as the unwearied Siren sounds and luxurious monotony of Mr. Moore's poetry make it inaccessible to pleasure. Lord Byron's poetry is as morbid as Mr. Moore's is careless and dissipated. He has more depth of passion, more force and impetuosity, but the passion is always of the same unaccountable character, at once violent and sullen, fierce and gloomy. It is not the passion of a mind struggling with misfortune, or the hopelessness of its desires, but of a mind preying upon itself, and disgusted with, or indifferent to all other things. There is nothing less poetical than this sort of unaccommodating selfishness. There is nothing more repulsive than this sort of ideal absorption of all the interests of others, of the good and ills of life, in

the ruling passion and moody abstraction of a single mind, as if it would make itself the centre of the universe, and there was nothing worth cherishing but its intellectual diseases. It is like a cancer, eating into the heart of poetry. But still there is power; and power rivets attention and forces admiration. "He hath a demon": and that is the next thing to being full of the God. His brow collects the scattered gloom: his eye flashes livid fire that withers and consumes. But still we watch the progress of the scathing bolt with interest, and mark the ruin it leaves behind with awe. Within the contracted range of his imagination, he has great unity and truth of keeping. He chooses elements and agents congenial to his mind, the dark and glittering ocean, the frail bark hurrying before the storm, pirates and men that "house on the wild sea with wild usages." He gives the tumultuous eagerness of action, and the fixed despair of thought. In vigour of style and force of conception, he in one sense surpasses every writer of the present day. His indignant apothegms are like oracles of misanthropy. He who wishes for "a curse to kill with," may find it in Lord Byron's writings. Yet he has beauty lurking underneath his strength, tenderness sometimes joined with the phrenzy of despair. A flash of golden light sometimes follows from a stroke of his pencil, like a falling meteor. The flowers that adorn his poetry bloom over charnel-houses and the grave!

There is one subject on which Lord Byron is fond of writing, on which I wish he would not write—Buonaparte. Not that I quarrel with his writing for him, or against him, but with his writing both for him and against him. What right has he to do this? Buonaparte's character, be it what else it may, does not change every hour according to his Lordship's varying humour. He is not a pipe for Fortune's finger, or for his Lordship's Muse, to play what stop she pleases on. Why should Lord Byron now laud him to the skies in the hour of his success, and then peevishly wreak his disappointment on the God of his idolatry? The man he writes of does not rise or fall with circumstances: but "looks on tempests and is never shaken." Besides, he is a subject for history, and not for poetry.

"Great princes' favourites their fair leaves spread,
    But as the marigold at the sun's eye,
And in themselves their pride lies buried;
    For at a frown they in their glory die.
The painful warrior, famoused for fight,
    After a thousand victories once foil'd,

Is from the book of honour razed quite,
    And all the rest forgot for which he toil'd." [9]

If Lord Byron will write any thing more on this hazardous theme, let him take these lines of Shakspeare for his guide, and finish them in the spirit of the original—they will then be worthy of the subject.

Walter Scott is the most popular of all the poets of the present day, and deservedly so. He describes that which is most easily and generally understood with more vivacity and effect than any body else. He has no excellences, either of a lofty or recondite kind, which lie beyond the reach of the most ordinary capacity to find out; but he has all the good qualities which all the world agree to understand. His style is clear, flowing, and transparent: his sentiments, of which his style is an easy and natural medium, are common to him with his readers. He has none of Mr. Wordsworth's *idiosyncracy*. He differs from his readers only in a greater range of knowledge and facility of expression. His poetry belongs to the class of *improvisatori* poetry. It has neither depth, height, nor breadth in it; neither uncommon strength, nor uncommon refinement of thought, sentiment, or language. It has no originality. But if this author has no research, no moving power in his own breast, he relies with the greater safety and success on the force of his subject. He selects a story such as is sure to please, full of incidents, characters, peculiar manners, costume, and scenery; and he tells it in a way that can offend no one. He never wearies or disappoints you. He is communicative and garrulous; but he is not his own hero. He never obtrudes himself on your notice to prevent your seeing the subject. What passes in the poem, passes much as it would have done in reality. The author has little or nothing to do with it. Mr. Scott has great intuitive power of fancy, great vividness of pencil in placing external objects and events before the eye. The force of his mind is picturesque, rather than *moral*. He gives more of the features of nature than the soul of passion. He conveys the distinct outlines and visible changes in outward objects, rather than "their mortal consequences." He is very inferior to Lord Byron in intense passion, to Moore in delightful fancy, to Mr. Wordsworth in profound sentiment: but he has more picturesque power than any of them; that is, he places the objects themselves, about which *they* might feel and think, in a much more striking point of view, with greater

variety of dress and attitude, and with more local truth and colouring. His imagery is Gothic and grotesque. The manners and actions have the interest and curiosity belonging to a wild country and a distant period of time. Few descriptions have a more complete reality, a more striking appearance of life and motion, than that of the warriors in the Lady of the Lake, who start up at the command of Rhoderic Dhu, from their concealment under the fern, and disappear again in an instant. The Lay of the Last Minstrel and Marmion are the first, and perhaps the best of his works. The Goblin Page, in the first of these, is a very interesting and inscrutable little personage. In reading these poems, I confess I am a little disconcerted, in turning over the page, to find Mr. Westall's pictures, which always seem *fac-similies* of the persons represented, with ancient costume and a theatrical air. This may be a compliment to Mr. Westall, but it is not one to Walter Scott. The truth is, there is a modern air in the midst of the antiquarian research of Mr. Scott's poetry. It is history or tradition in masquerade. Not only the crust of old words and images is worn off with time,—the substance is grown comparatively light and worthless. The forms are old and uncouth; but the spirit is effeminate and frivolous. This is a deduction from the praise I have given to his pencil for extreme fidelity, though it has been no obstacle to its drawing-room success. He has just hit the town between the romantic and the fashionable; and between the two, secured all classes of readers on his side. In a word, I conceive that he is to the great poet, what an excellent mimic is to a great actor. There is no determinate impression left on the mind by reading his poetry. It has no results. The reader rises up from the perusal with new images and associations, but he remains the same man that he was before. A great mind is one that moulds the minds of others. Mr. Scott has put the Border Minstrelsy and scattered traditions of the country into easy, animated verse. But the Notes to his poems are just as entertaining as the poems themselves, and his poems are only entertaining.

Mr. Wordsworth is the most original poet now living. He is the reverse of Walter Scott in his defects and excellences. He has nearly all that the other wants, and wants all that the other possesses. His poetry is not external, but internal; it does not depend upon tradition, or story, or old song; he furnishes it from his own mind, and is his own subject. He is the poet of mere sentiment. Of many of the Lyrical Ballads, it is not possible to speak in terms of too high praise, such

9. "Great . . . toil'd": Shakespeare's Sonnet XXV, ll. 5–12.

as Hart-leap Well, the Banks of the Wye, Poor Susan, parts of the Leech-gatherer, the lines to a Cuckoo, to a Daisy, the Complaint, several of the Sonnets, and a hundred others of inconceivable beauty, of perfect originality and pathos. They open a finer and deeper vein of thought and feeling than any poet in modern times has done, or attempted. He has produced a deeper impression, and on a smaller circle, than any other of his contemporaries. His powers have been mistaken by the age, nor does he exactly understand them himself. He cannot form a whole. He has not the constructive faculty. He can give only the fine tones of thought, drawn from his mind by accident or nature, like the sounds drawn from the Æolian harp by the wandering gale.—He is totally deficient in all the machinery of poetry. His *Excursion*, taken as a whole, notwithstanding the noble materials thrown away in it, is a proof of this. The line labours, the sentiment moves slow, but the poem stands stock-still. The reader makes no way from the first line to the last. It is more than any thing in the world like Robinson Crusoe's boat, which would have been an excellent good boat, and would have carried him to the other side of the globe, but that he could not get it out of the sand where it stuck fast. I did what little I could to help to launch it at the time, but it would not do. I am not, however, one of those who laugh at the attempts or failures of men of genius. It is not my way to cry "Long life to the conqueror." Success and desert are not with me synonymous terms; and the less Mr. Wordsworth's general merits have been understood, the more necessary is it to insist upon them. This is not the place to repeat what I have already said on the subject. The reader may turn to it in the Round Table. I do not think, however, there is any thing in the larger poem equal to many of the detached pieces in the Lyrical Ballads. As Mr. Wordsworth's poems have been little known to the public, or chiefly through garbled extracts from them, I will here given an entire poem (one that has always been a favourite with me), that the reader may know what it is that the admirers of this author find to be delighted with in his poetry. Those who do not feel the beauty and the force of it, may save themselves the trouble of inquiring farther.

[Hazlitt here quotes the whole of Wordsworth's "Hart-Leap Well," printed under Wordsworth, above.]

Mr. Wordsworth is at the head of that which has been denominated the Lake school of poetry; a school which, with all my respect for it, I do not think sacred from criticism or exempt from faults, of some of which faults I shall speak with becoming frankness; for I do not see that the liberty of the press ought to be shackled, or freedom of speech curtailed, to screen either its revolutionary or renegado extravagances. This school of poetry had its origin in the French revolution, or rather in those sentiments and opinions which produced that revolution; and which sentiments and opinions were indirectly imported into this country in translations from the German about that period. Our poetical literature had, towards the close of the last century, degenerated into the most trite, insipid, and mechanical of all things, in the hands of the followers of Pope and the old French school of poetry. It wanted something to stir it up, and it found that something in the principles and events of the French revolution. From the impulse it thus received, it rose at once from the most servile imitation and tamest common-place, to the utmost pitch of singularity and paradox. The change in the belles-lettres was as complete, and to many persons as startling, as the change in politics, with which it went hand in hand. There was a mighty ferment in the heads of statesmen and poets, kings and people. According to the prevailing notions, all was to be natural and new. Nothing that was established was to be tolerated. All the common-place figures of poetry, tropes, allegories, personifications, with the whole heathen mythology, were instantly discarded; a classical allusion was considered as a piece of antiquated foppery; capital letters were no more allowed in print, than letters-patent of nobility were permitted in real life; kings and queens were dethroned from their rank and station in legitimate tragedy or epic poetry, as they were decapitated elsewhere; rhyme was looked upon as a relic of the feudal system, and regular metre was abolished along with regular government. Authority and fashion, elegance or arrangement, were hooted out of countenance, as pedantry and prejudice. Every one did that which was good in his own eyes. The object was to reduce all things to an absolute level; and a singularly affected and outrageous simplicity prevailed in dress and manners, in style and sentiment. A striking effect produced where it was least expected, something new and original, no matter whether good, bad, or indifferent, whether mean or lofty, extravagant or childish, was all that was aimed at, or considered as compatible with sound philosophy and an age of reason. The licentiousness grew extreme: Coryate's

Crudities[10] were nothing to it. The world was to be turned topsy-turvy; and poetry, by the good will of our Adamwits, was to share its fate and begin *de novo*. It was a time of promise, a renewal of the world and of letters; and the Deucalions, who were to perform this feat of regeneration, were the present poet-laureat[11] and the two authors of the Lyrical Ballads. The Germans, who made heroes of robbers, and honest women of cast-off mistresses, had already exhausted the extravagant and marvellous in sentiment and situation: our native writers adopted a wonderful simplicity of style and matter. The paradox they set out with was, that all things are by nature equally fit subjects for poetry; or that if there is any preference to be given, those that are the meanest and most unpromising are the best, as they leave the greatest scope for the unbounded stores of thought and fancy in the writer's own mind. Poetry had with them "neither buttress nor coigne of vantage to make its pendant bed and procreant cradle." It was not "born so high: its aiery buildeth in the cedar's top, and dallies with the wind, and scorns the sun." It grew like a mushroom out of the ground; or was hidden in it like a truffle, which it required a particular sagacity and industry to find out and dig up. They founded the new school on a principle of sheer humanity, on pure nature void of art. It could not be said of these sweeping reformers and dictators in the republic of letters, that "in their train walked crowns and crownets; that realms and islands, like plates, dropt from their pockets": but they were surrounded, in company with the Muses, by a mixed rabble of idle apprentices and Botany Bay[12] convicts, female vagrants, gipsies, meek daughters in the family of Christ, of ideot boys and mad mothers, and after them "owls and night-ravens flew." They scorned "degrees, priority, and place, insisture, course, proportion, season, form, office, and custom in all line of order":—the distinctions of birth, the vicissitudes of fortune, did not enter into their abstracted, lofty, and levelling calculation of human nature. He who was more than man, with them was none. They claimed kindred only with the commonest of the people: peasants, pedlars, and village-barbers were their oracles and bosom friends. Their poetry, in the

10. **Coryate's Crudities:** Thomas Coryate, 1577?–1617, *Coryats Crudities Hastily Gobbled up in Five Months Travells in France, Savoy, Italian, Rhetia, etc.* (1611). Coryate's style reaches constantly for unusual effects. 11. **poet-laureat:** Southey. 12. **Botany Bay:** penal colony in Australia. Southey wrote four "Botany-Bay Eclogues."

extreme to which it professedly tended, and was in effect carried, levels all distinctions of nature and society; has "no figures nor no fantasies," which the prejudices of superstition or the customs of the world draw in the brains of men; "no trivial fond records" of all that has existed in the history of past ages; it has no adventitious pride, pomp, or circumstance, to set it off; "the marshal's truncheon, nor the judge's robe"; neither tradition, reverence, nor ceremony, "that to great one 'longs": it breaks in pieces the golden images of poetry, and defaces its armorial bearings, to melt them down in the mould of common humanity or of its own upstart self-sufficiency. They took the same method in their new-fangled "metre ballad-mongering" scheme, which Rousseau did in his prose paradoxes— of exciting attention by reversing the established standards of opinion and estimation in the world. They were for bringing poetry back to its primitive simplicity and state of nature, as he was for bringing society back to the savage state: so that the only thing remarkable left in the world by this change, would be the persons who had produced it. A thorough adept in this school of poetry and philanthropy is jealous of all excellence but his own. He does not even like to share his reputation with his subject; for he would have it all proceed from his own power and originality of mind. Such a one is slow to admire any thing that is admirable; feels no interest in what is most interesting to others, no grandeur in any thing grand, no beauty in anything beautiful. He tolerates only what he himself creates; he sympathizes only with what can enter into no competition with him, with "the bare trees and mountains bare, and grass in the green field." He sees nothing but himself and the universe. He hates all greatness and all pretensions to it, whether well or ill-founded. His egotism is in some respects a madness; for he scorns even the admiration of himself, thinking it a presumption in any one to suppose that he has taste or sense enough to understand him. He hates all science and all art; he hates chemistry, he hates conchology; he hates Voltaire; he hates Sir Isaac Newton; he hates wisdom; he hates wit; he hates metaphysics, which he says are unintelligible, and yet he would be thought to understand them; he hates prose; he hates all poetry but his own; he hates the dialogues in Shakspeare; he hates music, dancing, and painting; he hates Rubens, he hates Rembrandt; he hates Raphael, he hates Titian; he hates Vandyke; he hates the antique; he hates the Apollo Belvidere; he hates the Venus of Medicis. This is the reason that so few people take an interest in his writings, because he takes an interest in nothing that others do!—The effect has been perceived as something odd; but the cause or principle has never been distinctly traced to its source before, as far as I know. The proofs are to be found every where—in Mr. Southey's Botany Bay Eclogues, in his book of Songs and Sonnets, his Odes and Inscriptions, so well parodied in the Anti-Jacobin Review, in his Joan of Arc, and last, though not least, in his Wat Tyler:

> "When Adam delved, and Eve span,
> Where was then the gentleman?"[13]

(—or the poet laureat either, we may ask?)—In Mr. Coleridge's Ode to an Ass's Foal, in his Lines to Sarah, his Religious Musings; and in his and Mr. Wordsworth's Lyrical Ballads, *passim.*

Of Mr. Southey's larger epics, I have but a faint recollection at this distance of time, but all that I remember of them is mechanical and extravagant, heavy and superficial. His affected, disjointed style is well imitated in the Rejected Addresses. The difference between him and Sir Richard Blackmore[14] seems to be, that the one is heavy and the other light, the one solemn and the other pragmatical, the one phlegmatic and the other flippant; and that there is no Gay in the present time to give a Catalogue Raisonné of the performances of the living undertaker of epics. Kehama is a loose sprawling figure, such as we see cut out of wood or paper, and pulled or jerked with wire or thread, to make sudden and surprising motions, without meaning, grace, or nature in them. By far the best of his works are some of his shorter personal compositions, in which there is an ironical mixture of the quaint and serious, such as his lines on a picture of Gaspar Poussin, the fine tale of Gualberto, his Description of a Pig, and the Holly-tree, which is an affecting, beautiful, and modest retrospect on his own character. May the aspiration with which it concludes be fulfilled!—But the little he has done of true and sterling excellence, is overloaded by the quantity of indifferent matter which he turns out every year, "prosing or versing,"

with equally mechanical and irresistible facility. His Essays, or political and moral disquisitions, are not so full of original matter as Montaigne's. They are second or third rate compositions in that class.

It remains that I should say a few words of Mr. Coleridge; and there is no one who has a better right to say what he thinks of him than I have.[15] "Is there here any dear friend of Cæsar? To him I say, that Brutus's love to Cæsar was no less than his." But no matter.—His Ancient Mariner is his most remarkable performance, and the only one that I could point out to any one as giving an adequate idea of his great natural powers. It is high German, however, and in it he seems to "conceive of poetry but as a drunken dream, reckless, careless, and heedless, of past, present, and to come." His tragedies (for he has written two) are not answerable to it; they are, except a few poetical passages, drawling sentiment and metaphysical jargon. He has no genuine dramatic talent. There is one fine passage in his Christabel, that which contains the description of the quarrel between Sir Leoline and Sir Roland de Vaux of Tryermaine, who had been friends in youth.

> "Alas! they had been friends in youth,
> But whispering tongues can poison truth;
> And constancy lives in realms above;
> And life is thorny; and youth is vain;
> And to be wroth with one we love,
> Doth work like madness in the brain:
> And thus it chanc'd as I divine,
> With Roland and Sir Leoline.
> Each spake words of high disdain
> And insult to his heart's best brother,
> And parted ne'er to meet again!
> But neither ever found another
> To free the hollow heart from paining—
>
> They stood aloof, the scars remaining,
> Like cliffs which had been rent asunder:
> A dreary sea now flows between,
> But neither heat, nor frost, nor thunder,
> Shall wholly do away I ween
> The marks of that which once hath been.

13. **"When . . . gentleman?":** a popular rhyme of the middle ages. Southey had quoted it in *Wat Tyler* (II. i. 1–2), a youthful drama expressing radical political sentiments. Southey's politics having changed completely, his enemies maliciously published *Wat Tyler* in 1817, and liberal writers liked to remind him of it. **14. Sir Richard Blackmore:** An early eighteenth-century poet, satirized by Pope for his long, stiffly written epics.

15. **than I have:** Coleridge had been one of Hazlitt's first heroes (cf. the description of their first meeting in "My First Acquaintance with Poets," below). This in itself helps to explain his feeling of anticlimax as Coleridge grew older. But it is partly because Hazlitt felt slighted by Coleridge that, over the years, his remarks also began to take on such a personal, almost aggrieved tone. His tart review of Coleridge's *Biographia Literaria* in 1817 further estranged them. With this brief discussion of Coleridge, cf. Hazlitt's essay on him in *The Spirit of the Age,* below.

Sir Leoline a moment's space
Stood gazing on the damsel's face;
And the youthful lord of Tryermaine
Came back upon his heart again."

[ll. 408–30]

It might seem insidious if I were to praise his ode entitled Fire, Famine, and Slaughter, as an effusion of high poetical enthusiasm, and strong political feeling. His Sonnet to Schiller conveys a fine compliment to the author of the Robbers, and an equally fine idea of the state of youthful enthusiasm in which he composed it.

"Schiller! that hour I would have wish'd to die,
  If through the shudd'ring midnight I had sent
  From the dark dungeon of the tower time-rent,
That fearful voice, a famish'd father's cry—

That in no after moment aught less vast
  Might stamp me mortal! A triumphant shout
  Black Horror scream'd, and all her goblin rout
From the more with'ring scene diminish'd pass'd.

Ah! Bard tremendous in sublimity!
  Could I behold thee in thy loftier mood,
Wand'ring at eve, with finely frenzied eye,
  Beneath some vast old tempest-swinging wood!
  Awhile, with mute awe gazing, I would brood,
Then weep aloud in a wild ecstacy!"—

His *Conciones ad Populum*, Watchman, &c. are dreary trash. Of his Friend, I have spoken the truth elsewhere. But I may say of him here, that he is the only person I ever knew who answered to the idea of a man of genius. He is the only person from whom I ever learnt any thing. There is only one thing he could learn from me in return, but *that* he has not. He was the first poet I ever knew. His genius at that time had angelic wings, and fed on manna. He talked on for ever; and you wished him to talk on for ever. His thoughts did not seem to come with labour and effort; but as if borne on the gusts of genius, and as if the wings of his imagination lifted him from off his feet. His voice rolled on the ear like the pealing organ, and its sound alone was the music of thought. His mind was clothed with wings; and raised on them, he lifted philosophy to heaven. In his descriptions, you then saw the progress of human happiness and liberty in bright and never-ending succession, like the steps of Jacob's ladder, with airy shapes ascending and descending, and with the voice of God at the top of the ladder. And shall I, who heard

him then, listen to him now? Not I! . . . .[16] That spell is broke; that time is gone for ever; that voice is heard no more: but still the recollection comes rushing by with thoughts of long-past years, and rings in my ears with never-dying sound.

"What though the radiance which was once so bright,
Be now for ever taken from my sight,
Though nothing can bring back the hour
Of glory in the grass, of splendour in the flow'r;
  I do not grieve, but rather find
    Strength in what remains behind;
    In the primal sympathy,
    Which having been, must ever be;
    In the soothing thoughts that spring
    Out of human suffering;
In years that bring the philosophic mind!"—

I have thus gone through the task I intended, and have come at last to the level ground. I have felt my subject gradually sinking from under me as I advanced, and have been afraid of ending in nothing. The interest has unavoidably decreased at almost every successive step of the progress, like a play that has its catastrophe in the first or second act. This, however, I could not help. I have done as well as I could.

1818 (1818)

ON WIT AND HUMOUR

This is the opening lecture in Hazlitt's *Lectures on the English Comic Writers* (1819).

Man is the only animal that laughs and weeps; for he is the only animal that is struck with the difference between what things are, and what they ought to be. We weep at what thwarts or exceeds our desires in serious matters: we laugh at what only disappoints our expectations in trifles. We shed tears from sympathy with real and necessary distress; as we burst into laughter from want of sympathy with that which is unreasonable and unnecessary, the absurdity of which provokes our spleen or mirth, rather than any serious reflections on it.

16. **Not I!** . . . . : The dots are Hazlitt's and do not mark an omission.

To explain the nature of laughter and tears, is to account for the condition of human life; for it is in a manner compounded of these two! It is a tragedy or a comedy—sad or merry, as it happens. The crimes and misfortunes that are inseparable from it, shock and wound the mind when they once seize upon it, and when the pressure can no longer be borne, seek relief in tears: the follies and absurdities that men commit, or the odd accidents that befal them, afford us amusement from the very rejection of these false claims upon our sympathy, and end in laughter. If every thing that went wrong, if every vanity or weakness in another gave us a sensible pang, it would be hard indeed: but as long as the disagreeableness of the consequences of a sudden disaster is kept out of sight by the immediate oddity of the circumstances, and the absurdity or unaccountableness of a foolish action is the most striking thing in it, the ludicrous prevails over the pathetic, and we receive pleasure instead of pain from the farce of life which is played before us, and which discomposes our gravity as often as it fails to move our anger or our pity!

Tears may be considered as the natural and involuntary resource of the mind overcome by some sudden and violent emotion, before it has had time to reconcile its feelings to the change of circumstances: while laughter may be defined to be the same sort of convulsive and involuntary movement, occasioned by mere surprise or contrast (in the absence of any more serious emotion), before it has time to reconcile its belief to contradictory appearances. If we hold a mask before our face, and approach a child with this disguise on, it will at first, from oddity and incongruity of the appearance, be inclined to laugh; if we go nearer to it, steadily, and without saying a word, it will begin to be alarmed, and be half inclined to cry: if we suddenly take off the mask, it will recover from its fears, and burst out a-laughing; but if, instead of presenting the old well-known countenance, we have concealed a satyr's head or some frightful caricature behind the first mask, the suddenness of the change will not in this case be a source of merriment to it, but will convert its surprise into an agony of consternation, and will make it scream out for help, even though it may be convinced that the whole is a trick at bottom.

The alternation of tears and laughter, in this little episode in common life, depends almost entirely on the greater or less degree of interest attached to the different changes of appearance. The mere suddenness of the transition, the mere baulking our expectations, and turning them abruptly into another channel, seems to give additional liveliness and gaiety to the animal spirits; but the instant the change is not only sudden, but threatens serious consequences, or calls up the shape of danger, terror supersedes our disposition to mirth, and laughter gives place to tears. It is usual to play with infants, and make them laugh by clapping your hands suddenly before them; but if you clap your hands too loud, or too near their sight, their countenances immediately change, and they hide them in the nurse's arms. Or suppose the same child, grown up a little older, comes to a place, expecting to meet a person it is particularly fond of, and does not find that person there, its countenance suddenly falls, its lips begin to quiver, its cheek turns pale, its eye glistens, and it vents its little sorrow (grown too big to be concealed) in a flood of tears. Again, if the child meets the same person unexpectedly after long absence, the same effect will be produced by an excess of joy, with different accompaniments; that is, the surprise and the emotion excited will make the blood come into his face, his eyes sparkle, his tongue falter or be mute, but in either case the tears will gush to his relief, and lighten the pressure about his heart. On the other hand, if a child is playing at hide-and-seek, or blindman's-buff, with persons it is ever so fond of, and either misses them where it had made sure of finding them, or suddenly runs up against them where it had least expected it, the shock or additional impetus given to the imagination by the disappointment or the discovery, in a matter of this indifference, will only vent itself in a fit of laughter. The transition here is not from one thing of importance to another, or from a state of indifference to a state of strong excitement; but merely from one impression to another that we did not at all expect, and when we had expected just the contrary. The mind having been led to form a certain conclusion, and the result producing an immediate solution of continuity in the chain of our ideas, this alternate excitement and relaxation of the imagination, the object also striking upon the mind more vividly in its loose unsettled state, and before it has had time to recover and collect itself, causes that alternate excitement and relaxation, or irregular convulsive movement of the muscular and nervous system, which constitutes physical laughter. The *discontinuous* in our sensations produces a correspondent jar and discord in the frame. The steadiness of our faith and of our features begins to give way at the same time. We turn with an incredulous smile from a story that staggers our belief: and we are ready to split our sides with laughing at an extravagance that sets all common sense and serious concern at defiance.

To understand or define the ludicrous, we must first know what the serious is. Now the serious is the habitual stress which the mind lays upon the expectation of a given order of events, following one another with a certain regularity and weight of interest attached to them. When this stress is increased beyond its usual pitch of intensity, so as to overstrain the feelings by the violent opposition of good to bad, or of objects to our desires, it becomes the pathetic or tragical. The ludicrous, or comic, is the unexpected loosening or relaxing this stress below its usual pitch of intensity, by such an abrupt transposition of the order of our ideas, as taking the mind unawares, throws it off its guard, startles it into a lively sense of pleasure, and leaves no time nor inclination for painful reflections.

The essence of the laughable then is the incongruous, the disconnecting one idea from another, or the jostling of one feeling against another. The first and most obvious cause of laughter is to be found in the simple succession of events, as in the sudden shifting of a disguise, or some unlooked-for accident, without any absurdity of character or situation. The accidental contradiction between our expectations and the event can hardly be said, however, to amount to the ludicrous: it is merely laughable. The ludicrous is where there is the same contradiction between the object and our expectations, heightened by some deformity of inconvenience, that is, by its being contrary to what is customary or desirable; as the ridiculous, which is the highest degree of the laughable, is that which is contrary not only to custom but to sense and reason, or is a voluntary departure from what we have a right to expect from those who are conscious of absurdity and propriety in words, looks, and actions.

Of these different kinds or degrees of the laughable, the first is the most shallow and short-lived; for the instant the immediate surprise of a thing's merely happening one way or another is over, there is nothing to throw us back upon our former expectation, and renew our wonder at the event a second time. The second sort, that is, the ludicrous arising out of the improbable or distressing, is more deep and lasting, either because the painful catastrophe excites a greater curiosity, or because the old impression, from its habitual hold on the imagination, still recurs mechanically, so that it is longer before we can seriously make up our minds to the unaccountable deviation from it.

The third sort, or the ridiculous arising out of absurdity as well as improbability, that is, where the defect or weakness is of a man's own seeking, is the most refined of all, but not always so pleasant as the last, because the same contempt and disapprobation which sharpens and subtilises our sense of the impropriety, adds a severity to it inconsistent with perfect ease and enjoyment. This last species is properly the province of satire. The principle of contrast is, however, the same in all the stages, in the simply laughable, the ludicrous, the ridiculous; and the effect is only the more complete, the more durably and pointedly this principle operates.

To give some examples in these different kinds. We laugh, when children, at the sudden removing of a pasteboard mask: we laugh, when grown up, more gravely at the tearing off the mask of deceit. We laugh at absurdity; we laugh at deformity. We laugh at a bottle-nose in a caricature; at a stuffed figure of an alderman in a pantomine, and at the tale of Slaukenbergius.[1] A giant standing by a dwarf makes a contemptible figure enough. Rosinante and Dapple[2] are laughable from contrast, as their masters from the same principle make two for a pair. We laugh at the dress of foreigners, and they at ours. Three chimney-sweepers meeting three Chinese in Lincoln's-inn Fields, they laughed at one another till they were ready to drop down. Country people laugh at a person because they never saw him before. Any one dressed in the height of the fashion, or quite out of it, is equally an object of ridicule. One rich source of the ludicrous is distress with which we cannot sympathise from its absurdity or insignificance. Women laugh at their lovers. We laugh at a damned author, in spite of our teeth, and though he may be our friend. "There is something in the misfortunes of our best friends that pleases us." We laugh at people on the top of a stage-coach, or in it, if they seem in great extremity. It is hard to hinder children from laughing at a stammerer, at a negro, at a drunken man, or even at a madman. We laugh at mischief. We laugh at what we do not believe. We say that an argument or an assertion that is very absurd, is quite ludicrous. We laugh to shew our satisfaction with ourselves, or our contempt for those about us, or to conceal our envy or our ignorance. We laugh at fools, and at those who pretend to be wise—at extreme simplicity, awkwardness, hypocrisy, and affectation.

ON WIT AND HUMOUR. 1. **Slaukenbergius:** in Sterne's *Tristram Shandy*, Bk. IV. 2. **Rosinante and Dapple:** Don Quixote's horse and Sancho Panza's donkey.

"They were talking of me," says Scrub, "for they laughed *consumedly*." Lord Foppington's insensibility to ridicule, and airs of ineffable self-conceit, are no less admirable; and Joseph Surface's[3] cant maxims of morality, when once disarmed of their power to do hurt, become sufficiently ludicrous.—We laugh at that in others which is a serious matter to ourselves; because our self-love is stronger than our sympathy, sooner takes the alarm, and instantly turns our heedless mirth into gravity, which only enhances the jest to others. Some one is generally sure to be the sufferer by a joke. What is sport to one, is death to another. It is only very sensible or very honest people, who laugh as freely at their own absurdities as at those of their neighbours. In general the contrary rule holds, and we only laugh at those misfortunes in which we are spectators, not sharers. The injury, the disappointment, shame, and vexation that we feel, put a stop to our mirth; while the disasters that come home to us, and excite our repugnance and dismay, are an amusing spectacle to others. The greater resistance we make, and the greater the perplexity into which we are thrown, the more lively and *piquant* is the intellectual display of cross-purposes to the by-standers. Our humiliation is their triumph. We are occupied with the disagreeable-ness of the result instead of its oddity or unexpectedness. Others see only the conflict of motives, and the sudden alternation of events; we feel the pain as well, which more than counterbalances the speculative entertainment we might receive from the contemplation of our abstract situation.

You cannot force people to laugh: you cannot give a reason why they should laugh: they must laugh of themselves, or not at all. As we laugh from a spontaneous impulse, we laugh the more at any restraint upon this impulse. We laugh at a thing merely because we ought not. If we think we must not laugh, this perverse impediment makes our temptation to laugh the greater; for by endeavouring to keep the obnoxious image out of sight, it comes upon us more irresistibly and repeatedly; and the inclination to indulge our mirth, the longer it is held back, collects its force, and breaks out the more violently in peals of laughter. In like manner, any thing we must not think of makes us laugh, by its coming upon us by stealth and unawares, and from the very efforts we make to exclude it. A secret, a loose word, a wanton jest, make people laugh. Aretine laughed himself to death at hearing a lascivious story. Wickedness is often made a substitute for wit; and in most of our good old comedies, the intrigue of the plot and the double meaning of the dialogue go hand-in-hand, and keep up the ball with wonderful spirit between them. The consciousness, however it may arise, that there is something that we ought to look grave at, is almost always a signal for laughing outright: we can hardly keep our countenance at a sermon, a funeral, or a wedding. What an excellent old custom was that of throwing the stocking! What a deal of innocent mirth has been spoiled by the disuse of it!—It is not an easy matter to preserve decorum in courts of justice. The smallest circumstance that interferes with the solemnity of the proceedings, throws the whole place into an uproar of laughter. People at the point of death often say smart things. Sir Thomas More jested with his executioner. Rabelais and Wycherley both died with a *bon-mot* in their mouths.

Misunderstandings, (*malentendus*) where one person means one thing, and another is aiming at something else, are another great source of comic humour, on the same principle of ambiguity and contrast. There is a high-wrought instance of this in the dialogue between Aimwell and Gibbet, in the Beaux' Stratagem, where Aimwell mistakes his companion for an officer in a marching regiment, and Gibbet takes it for granted that the gentleman is a highwayman. The alarm and consternation occasioned by some one saying to him, in the course of common conversation, "I apprehend you," is the most ludicrous thing in that admirably natural and powerful performance, Mr. Emery's Robert Tyke.[4] Again, unconsciousness in the person himself of what he is about, or of what others think of him, is also a great heightener of the sense of absurdity. It makes it come the fuller home upon us from his insensibility to it. His simplicity sets off the satire, and gives it a finer edge. It is a more extreme case still where the person is aware of being the object of ridicule, and yet seems perfectly reconciled to it as a matter of course. So wit is often the more forcible and pointed for being dry and serious, for it then seems as if the speaker himself had no intention in it, and we were the first to find it out. Irony, as a species of wit, owes its

---

3. **Scrub . . . Foppington . . . Surface:** characters in George Farquhar's *Beaux Stratagem* (1707), Sir John Vanbrugh's *Relapse* (1697), and Richard B. Sheridan's *School for Scandal* (1777).

4. **Robert Tyke:** in Thomas Morton's *School of Reform* (1815).

force to the same principle. In such cases it is the contrast between the appearance and the reality, the suspense of belief, and the seeming incongruity, that gives point to the ridicule, and makes it enter the deeper when the first impression is overcome. Excessive impudence, as in the Liar; or excessive modesty, as in the hero of She Stoops to Conquer; or a mixture of the two, as in the Busy Body,[5] are equally amusing. Lying is a species of wit and humour. To lay any thing to a person's charge from which he is perfectly free, shews spirit and invention; and the more incredible the effrontery, the greater is the joke.

There is nothing more powerfully humorous than what is called *keeping* in comic character, as we see it very finely exemplified in Sancho Panza and Don Quixote. The proverbial phlegm and the romantic gravity of these two celebrated persons may be regarded as the height of this kind of excellence. The deep feeling of character strengthens the sense of the ludicrous. Keeping in comic character is consistency in absurdity; a determined and laudable attachment to the incongruous and singular. The regularity completes the contradiction; for the number of instances of deviation from the right line, branching out in all directions, shews the inveteracy of the original bias to any extravagance or folly, the natural improbability, as it were, increasing every time with the multiplication of chances for a return to common sense, and in the end mounting up to an incredible and unaccountably ridiculous height, when we find our expectations as invariably baffled. The most curious problem of all, is this truth of absurdity to itself. That reason and good sense should be consistent, is not wonderful: but that caprice, and whim, and fantastical prejudice, should be uniform and infallible in their results, is the surprising thing. But while this characteristic clue to absurdity helps on the ridicule, it also softens and harmonises its excesses; and the ludicrous is here blended with a certain beauty and decorum, from this very truth of habit and sentiment, or from the principle of similitude in dissimilitude. The devotion to nonsense, and enthusiasm about trifles, is highly affecting as a moral lesson: it is one of the striking weaknesses and greatest happinesses of our nature. That which excites so lively and lasting an interest in itself, even though it should not be wisdom, is not despicable in the sight of reason and

humanity. We cannot suppress the smile on the lip; but the tear should also stand ready to start from the eye. The history of hobby-horses is equally instructive and delightful; and after the pair I have just alluded to, My Uncle Toby's is one of the best and gentlest that "ever lifted leg!" The inconveniences, odd accidents, falls, and bruises, to which they expose their riders, contribute their share to the amusement of the spectators; and the blows and wounds that the Knight of the Sorrowful Countenance received in his many perilous adventures, have applied their healing influence to many a hurt mind.—In what relates to the laughable, as it arises from unforeseen accidents or self-willed scrapes, the pain, the shame, the mortification, and utter helplessness of situation, add to the joke, provided they are momentary, or overwhelming only to the imagination of the sufferer. Malvolio's punishment and apprehensions are as comic, from our knowing that they are not real, as Christopher Sly's drunken transformation and short-lived dream of happiness are for the like reason. Parson Adams's fall into the tub at the 'Squire's, or his being discovered in bed with Mrs. Slipslop, though pitiable, are laughable accidents:[6] nor do we read with much gravity of the loss of his Æschylus, serious as it was to him at the time.—A Scotch clergyman, as he was going to church, seeing a spruce conceited mechanic who was walking before him, suddenly covered all over with dirt, either by falling into the kennel, or by some other calamity befalling him, smiled and passed on: but afterwards seeing the same person, who had stopped to refit, seated directly facing him in the gallery, with a look of perfect satisfaction and composure, as if nothing of the sort had happened to him, the idea of his late disaster and present self-complacency struck him so powerfully, that, unable to resist the impulse, he flung himself back in the pulpit, and laughed till he could laugh no longer. I remember reading a story in an odd number of the European Magazine, of an old gentleman who used to walk out every afternoon, with a gold-headed cane, in the fields opposite Baltimore House, which were then open, only with foot-paths crossing them. He was frequently accosted by a beggar with a wooden leg, to whom he gave money, which only made him more importunate. One day, when he was more troublesome than usual, a well-dressed person

---

5. **Liar . . . Busy Body:** The references are to Samuel Foote's *Liar* (1762), Goldsmith's *She Stoops to Conquer* (1773), and Susannah Centilivre's *Busybody* (1709).

6. **Malvolio's . . . accidents:** Malvolio in *Twelfth Night*, Sly in *The Taming of the Shrew*, and Adams and Mrs. Slipslop in Fielding's *Joseph Andrews* (1742).

happening to come up, and observing how saucy the fellow was, said to the gentleman, "Sir, if you will lend me your cane for a moment, I'll give him a good threshing for his impertinence." The old gentleman, smiling at the proposal, handed him his cane, which the other no sooner was going to apply to the shoulders of the culprit, than he immediately whipped off his wooden leg, and scampered off with great alacrity, and his chastiser after him as hard as he could go. The faster the one ran, the faster the other followed him, brandishing the cane, to the great astonishment of the gentleman who owned it, till having fairly crossed the fields, they suddenly turned a corner, and nothing more was seen of either of them.

In the way of mischievous adventure, and a wanton exhibition of ludicrous weakness in character, nothing is superior to the comic parts of the Arabian Nights' Entertainments. To take only the set of stories of the Little Hunchback, who was choked with a bone, and the Barber of Bagdad and his seven brothers,—there is that of the tailor who was persecuted by the miller's wife, and who, after toiling all night in the mill, got nothing for his pains:—of another who fell in love with a fine lady who pretended to return his passion, and inviting him to her house, as the preliminary condition of her favour, had his eyebrows shaved, his clothes stripped off, and being turned loose into a winding gallery, he was to follow her, and by over-taking obtain all his wishes, but, after a turn or two, stumbled on a trap-door, and fell plump into the street, to the great astonishment of the spectators and his own, shorn of his eyebrows, naked, and without a ray of hope left:—that of the castle-building pedlar, who, in kicking his wife, the supposed daughter of an emperor, kicks down his basket of glass, the brittle foundation of his ideal wealth, his good fortune, and his arrogance:—that, again, of the beggar who dined with the Barmecide, and feasted with him on the names of wines and dishes: and, last and best of all, the inimitable story of the Impertinent Barber himself, one of the seven, and worthy to be so; his pertinacious, incredible, teasing, deliberate, yet unmeaning folly, his wearing out the patience of the young gentleman whom he is sent for to shave, his preparations and his professions of speed, his taking out an astrolabe to measure the height of the sun while his razors are getting ready, his dancing the dance of Zimri and singing the song of Zamtout, his disappointing the young man of an assignation, following him to the place of rendezvous, and alarming the master of

the house in his anxiety for his safety, by which his unfortunate patron loses his hand in the affray, and this is felt as an awkward accident. The danger which the same loquacious person is afterwards in, of losing his head for want of saying who he was, because he would not forfeit his character of being "justly called the Silent," is a consummation of the jest, though, if it had really taken place, it would have been carrying the joke too far. There are a thousand instances of the same sort in the Thousand and One Nights, which are an inexhaustible mine of comic humour and invention, and which, from the manners of the East which they describe, carry the principle of callous indifference in a jest as far as it can go. The serious and marvellous stories in that work, which have been so much admired and so greedily read, appear to me monstrous and abortive fictions, like disjointed dreams, dictated by a preternatural dread of arbitrary and despotic power, as the comic and familiar stories are rendered proportionably amusing and interesting from the same principle operating in a different direction, and producing endless uncertainty and vicissitude, and an heroic contempt for the untoward accidents and petty vexations of human life. It is the gaiety of despair, the mirth and laughter of a respite during pleasure from death. The strongest instances of effectual and harrowing imagination, are in the story of Amine and her three sisters, whom she led by her side as a leash of hounds, and of the *goul* who nibbled grains of rice for her dinner, and preyed on human carcasses. In this condemnation of the serious parts of the Arabian Nights, I have nearly all the world, and in particular the author of the Ancient Mariner, against me, who must be allowed to be a judge of such matters, and who said, with a subtlety of philosophical conjecture which he alone possesses, "That if I did not like them, it was because I did not dream." On the other hand, I have Bishop Atterbury on my side, who, in a letter to Pope, fairly confesses that "he could not read them in his old age."

There is another source of comic humour which has been but little touched on or attended to by the critics —not the infliction of casual pain, but the pursuit of uncertain pleasure and idle gallantry. Half the business and gaiety of comedy turns upon this. Most of the adventures, difficulties, demurs, hair-breadth 'scapes, disguises, deceptions, blunders, disappointments, successes, excuses, all the dextrous manœuvres, artful inuendos, assignations, billets-doux, *double entendres*, sly allusions, and elegant flattery, have an eye to this—

to the obtaining of those "favours secret, sweet, and precious," in which love and pleasure consist, and which when attained, and the *equivoque* is at an end, the curtain drops, and the play is over. All the attractions of a subject that can only be glanced at indirectly, that is a sort of forbidden ground to the imagination, except under severe restrictions, which are constantly broken through; all the resources it supplies for intrigue and invention; the bashfulness of the clownish lover, his looks of alarm and petrified astonishment; the foppish affectation and easy confidence of the happy man; the dress, the airs, the languor, the scorn, and indifference of the fine lady; the bustle, pertness, loquaciousness, and tricks of the chambermaid; the impudence, lies, and roguery of the valet; the matchmaking and unmaking; the wisdom of the wise; the sayings of the witty, the folly of the fool; "the soldier's, scholar's, courtier's eye, tongue, sword, the glass of fashion and the mould of form," have all a view to this. It is the closet in Blue-Beard. It is the life and soul of Wycherley, Congreve, Vanbrugh, and Farquhar's plays. It is the salt of comedy, without which it would be worthless and insipid. It makes Horner decent, and Millamant divine. It is the jest between Tattle and Miss Prue. It is the bait with which Olivia, in the Plain Dealer,[7] plays with honest Manly. It lurks at the bottom of the catechism which Archer teaches Cherry, and which she learns by heart. It gives the finishing grace to Mrs. Amlet's confession—"Though I'm old, I'm chaste." Valentine and his Angelica would be nothing without it; Miss Peggy would not be worth a gallant; and Slender's "sweet Anne Page" would be no more! "The age of comedy would be gone, and the glory of our playhouses extinguished for ever." Our old comedies would be invaluable, were it only for this, that they keep alive this sentiment, which still survives in all its fluttering grace and breathless palpitations on the stage.

Humour is the describing the ludicrous as it is in itself; wit is the exposing it, by comparing or contrasting it with something else. Humour is, as it were, the growth of nature and accident; wit is the product of art and fancy. Humour, as it is shewn in books, is an imitation of the natural or acquired absurdities of mankind, or of the ludicrous in accident, situation, and character: wit is the illustrating and heightening the sense of that absurdity by some sudden and unexpected likeness or opposition of one thing to another, which sets off the quality we laugh at or despise in a still more contemptible or striking point of view. Wit, as distinguished from poetry, is the imagination or fancy inverted, and so applied to given objects, as to make the little look less, the mean more light and worthless; or to divert our admiration or wean our affections from that which is lofty and impressive, instead of producing a more intense admiration and exalted passion, as poetry does. Wit may sometimes, indeed, be shewn in compliments as well as satire; as in the common epigram—

"Accept a miracle, instead of wit:
See two dull lines with Stanhope's pencil writ."[8]

But then the mode of paying it is playful and ironical, and contradicts itself in the very act of making its own performance an humble foil to another's. Wit hovers round the borders of the light and trifling, whether in matters of pleasure or pain; for as soon as it describes the serious seriously, it ceases to be wit, and passes into a different form. Wit is, in fact, the eloquence of indifference, or an ingenious and striking exposition of those evanescent and glancing impressions of objects which affect us more from surprise or contrast to the train of our ordinary and literal preconceptions, than from any thing in the objects themselves exciting our necessary sympathy or lasting hatred. The favourite employment of wit is to add littleness to littleness, and heap contempt on insignificance by all the arts of petty and incessant warfare; or if it ever affects to aggrandise, and use the language of hyperbole, it is only to betray into derision by a fatal comparison, as in the mockheroic; or if it treats of serious passion, it must do it so as to lower the tone of intense and high-wrought sentiment, by the introduction of burlesque and familiar circumstances. To give an instance or two. Butler, in his Hudibras, compares the change of night into day, to the change of colour in a boiled lobster.

"The sun had long since, in the lap
Of Thetis, taken out his nap;
And, like a lobster boil'd, the morn
From black to red, began to turn:
When Hudibras, whom thoughts and aching
'Twixt sleeping kept all night, and waking,
Began to rub his drowsy eyes,
And from his couch prepared to rise,
Resolving to dispatch the deed
He vow'd to do with trusty speed."[9]

---

7. **Plain Dealer**: by William Wycherly (1677).

8. "**Accept . . . writ**": Joseph Spence's *Anecdotes* (1820), p. 378. 9. "**The . . . speed**": *Hudibras*, II. ii. 29–38.

Compare this with the following stanzas in Spenser, treating of the same subject:—

"By this the Northern Waggoner had set
His seven-fold team behind the stedfast star,
That was in Ocean waves yet never wet,
But firm is fix'd and sendeth light from far
To all that in the wide deep wand'ring are:
And cheerful chanticleer with his note shrill,
Had warned once that Phœbus' fiery car
In haste was climbing up the eastern hill,
Full envious that night so long his room did fill.

At last the golden oriental gate
Of greatest heaven 'gan to open fair,
And Phœbus, fresh as bridegroom to his mate,
Came dancing forth, shaking his dewy hair,
And hurl'd his glist'ring beams through gloomy air:
Which when the wakeful elf perceiv'd, straitway
He started up and did himself prepare
In sun-bright arms and battailous array,
For with that pagan proud he combat will that day."[10]

In this last passage, every image is brought forward that can give effect to our natural impression of the beauty, the splendour, and solemn grandeur of the rising sun; pleasure and power wait on every line and word: whereas, in the other, the only memorable thing is a grotesque and ludicrous illustration of the alteration which takes place from darkness to gorgeous light, and that brought from the lowest instance, and with associations that can only disturb and perplex the imagination in its conception of the real object it describes. There cannot be a more witty, and at the same time degrading comparison, than that in the same author, of the Bear turning round the pole-star to a bear tied to a stake:—

"But now a sport more formidable
Had raked together village rabble;
'Twas an old way of recreating
Which learned butchers call bear-baiting,
A bold adventrous exercise
With ancient heroes in high prize,
For authors do affirm it came
From Isthmian or Nemæan game;
Others derive it from the Bear
That's fixed in Northern hemisphere,
And round about his pole does make
A circle like a bear at stake,
That at the chain's end wheels about
And overturns the rabble rout."[11]

I need not multiply examples of this sort.—Wit or ludicrous invention produces its effect oftenest by comparison, but not always. It frequently effects its purposes by unexpected and subtle distinctions. For instance, in the first kind, Mr. Sheridan's description of Mr. Addington's administration as the fag-end of Mr. Pitt's, who had remained so long on the treasury bench that, like Nicias in the fable, "he left the sitting part of the man behind him,"[12] is as fine an example of metaphorical wit as any on record. The same idea seems, however, to have been included in the old well-known nickname of the *Rump* Parliament. Almost as happy an instance of the other kind of wit, which consists in sudden retorts, in turns upon an idea, and diverting the train of your adversary's argument abruptly and adroitly into another channel, may be seen in the sarcastic reply of Porson, who hearing some one observe, that "certain modern poets would be read and admired when Homer and Virgil were forgotten," made answer—"And not till then!" Sir Robert Walpole's definition of the gratitude of place-expectants, "That it is a lively sense of *future* favours," is no doubt wit, but it does not consist in the finding out any coincidence or likeness, but in suddenly transposing the order of time in the common account of this feeling, so as to make the professions of those who pretend to it correspond more with their practice. It is filling up a blank in the human heart with a word that explains its hollowness at once. Voltaire's saying, in answer to a stranger who was observing how tall his trees grew—"That they had nothing else to do"— was a quaint mixture of wit and humour, making it out as if they really led a lazy, laborious life; but there was here neither allusion or metaphor. Again, that master-stroke in Hudibras is sterling wit and profound satire, where speaking of certain religious hypocrites he says, that they

"Compound for sins they are inclin'd to,
By damning those they have no mind to,"[13]

but the wit consists in the truth of the character, and in the happy exposure of the ludicrous contradiction between the pretext and the practice; between their lenity towards their own vices, and their severity to those of others. The same principle of nice distinction must be allowed to prevail in those lines of the same

10. "By . . . day": *Faerie Queene,* I. ii. 1 and I. v. 2.
11. "But . . . rout": *Hudibras,* I. i. 675–88.

12. "he . . . him": R. B. Sheridan's "Speech on the Definitive Treaty of Peace, May 14, 1794." 13. "Compound . . . to": *Hudibras,* II. i. 215–16.

author, where he is professing to expound the dreams of judicial astrology.

> "There's but the twinkling of a star
> Betwixt a man of peace and war,
> A thief and justice, fool and knave,
> A huffing officer and a slave;
> A crafty lawyer and pickpocket;
> A great philosopher and a blockhead;
> A formal preacher and a player;
> A learn'd physician and man slayer." [14]

The finest piece of wit I know of, is in the lines of Pope on the Lord Mayor's show—

> "Now night descending, the proud scene is o'er,
> But lives in Settle's numbers one day more." [15]

This is certainly as mortifying an inversion of the idea of poetical immortality as could be thought of; it fixes the *maximum* of littleness and insignificance: but it is not by likeness to any thing else that it does this, but by literally taking the lowest possible duration of ephemeral reputation, marking it (as with a slider) on the scale of endless renown, and giving a rival credit for it as his loftiest praise. In a word, the shrewd separation or disentangling of ideas that seem the same, or where the secret contradiction is not sufficiently suspected, and is of a ludicrous and whimsical nature, is wit just as much as the bringing together those that appear at first sight totally different. There is then no sufficient ground for admitting Mr. Locke's celebrated definition of wit, which he makes to consist in the finding out striking and unexpected resemblances in things so as to make pleasant pictures in the fancy, while judgment and reason, according to him, lie the clean contrary way, in separating and nicely distinguishing those wherein the smallest difference is to be found. [16]

On this definition Harris, [17] the author of Hermes, has very well observed that the demonstrating the equality of the three angles of a right-angled triangle to two right ones, would, upon the principle here stated, be a piece of wit instead of an act of the judgment or understanding, and Euclid's Elements a collection of epigrams. On the contrary it has appeared, that the detection and exposure of difference, particularly where this implies nice and subtle observation, as in discriminating between pretence and practice, between appearance and reality, is common to wit and satire with judgment and reasoning, and certainly the comparing and connecting our ideas together is an essential part of reason and judgment, as well as of wit and fancy. —Mere wit, as opposed to reason or argument, consists in striking out some casual and partial coincidence which has nothing to do, or at least implies no necessary connection with the nature of the things, which are forced into a seeming analogy by a play upon words, or some irrelevant conceit, as in puns, riddles, alliteration, &c. The jest, in all such cases, lies in the sort of mock-identity, or nominal resemblance, established by the intervention of the same words

---

14. "There's . . . slayer": *Hudibras*, II. iii. 957–64. 15. "Now . . . more": *Dunciad*, I. 89–90. 16. Mr. Locke's . . . found: "His words are—'If in having our ideas in the memory ready at hand consists quickness of parts, in this of having them unconfused, and being able nicely to distinguish one thing from another, where there is but the least difference, consists in a great measure the exactness of judgment and clearness of reason, which is to be observed in one man above another. And hence, perhaps, may be given some reason of that common observation, that men who have a great deal of wit and prompt memories, have not always the clearest judgment or deepest reason. For wit lying mostly in the assemblage of ideas, and putting them together with quickness and variety, wherein can be found any resemblance or congruity, thereby to

make up pleasant pictures and agreeable visions in the fancy; judgment, on the contrary, lies quite on the other side, in separating carefully one from another, ideas wherein can be found the least difference, thereby to avoid being misled by similitude, and by affinity to take one thing for another.' (Essay [Concerning Human Understanding], I, 143.) This definition, such as it is, Mr. Locke took without acknowledgment from Hobbes, who says in his Leviathan, 'This difference of quickness in imagining is caused by the difference of men's passions, that love and dislike some one thing, some another, and therefore some men's thoughts run one way, some another, and are held to and observed differently the things that pass through their imagination. And whereas in this succession of thoughts there is nothing to observe in the things they think on, but either in what they be like one another, or in what they be unlike, those that observe their similitudes, in case they be such as are but rarely observed by others, are said to have a good wit, by which is meant on this occasion a good fancy. But they that observe their differences and dissimilitudes, which is called distinguishing and discerning and judging between thing and thing; in case such discerning be not easy, are said to have a good judgment; and particularly in matter of conversation and business, wherein times, places, and persons are to be discerned, this virtue is called discretion. The former, that is, fancy, without the help of judgment, is not commended for a virtue; but the latter, which is judgment or discretion, is commended for itself, without the help of fancy.' *Leviathan*, p. 32" (H.). 17. Harris: James Harris, author of *Hermes; or, A Philosophical Inquiry Concerning Universal Grammar* (1751).

expressing different ideas, and countenancing as it were, by a fatality of language, the mischievous insinuation which the person who has the wit to take advantage of it wishes to convey. So when the disaffected French wits applied to the new order of the *Fleur du lys* the *double entendre* of *Compagnons d' Ulysse*, or companions of Ulysses, meaning the animal into which the fellow-travellers of the hero of the Odyssey were transformed, this was a shrewd and biting intimation of a galling truth (if truth it were) by a fortuitous concourse of letters of the alphabet, jumping in "a foregone conclusion," but there was no proof of the thing, unless it was self-evident. And, indeed, this may be considered as the best defence of the contested maxim—That *ridicule is the test of truth;*[18] viz. that it does not contain or attempt a formal proof of it, but owes its power of conviction to the bare suggestion of it, so that if the thing when once hinted is not clear in itself, the satire fails of its effect and falls to the ground. The sarcasm here glanced at the character of the new or old French noblesse may not be well founded; but it is so like truth, and "comes in such a questionable shape," backed with the appearance of an identical proposition, that it would require a long train of facts and laboured arguments to do away the impression, even if we were sure of the honesty and wisdom of the person who undertook to refute it. A flippant jest is as good a test of truth as a solid bribe; and there are serious sophistries,

"Soul-killing lies, and truths that work small good,"[19]

as well as idle pleasantries. Of this we may be sure, that ridicule fastens on the vulnerable points of a cause, and finds out the weak sides of an argument; if those who resort to it sometimes rely too much on its success, those who are chiefly annoyed by it almost always are so with reason, and cannot be too much on their guard against deserving it. Before we can laugh at a thing, its absurdity must at least be open and palpable to common apprehension. Ridicule is necessarily built on certain supposed facts, whether true or false, and on their inconsistency with certain acknowledged maxims, whether right or wrong. It is, therefore, a fair test, if not of philosophical or abstract truth, at least of what is truth according to public opinion and common sense; for it can only expose to instantaneous contempt that which is condemned by public opinion, and is hostile to the common sense of mankind. Or to put it differently, it is the test of the quantity of truth that there is in our favourite prejudices.—To shew how nearly allied wit is thought to be to truth, it is not unusual to say of any person—"Such a one is a man of sense, for though he said nothing, he laughed in the right place."—Alliteration comes in here under the head of a certain sort of verbal wit; or, by pointing the expression, sometimes points the sense. Mr. Grattan's[20] wit or eloquence (I don't know by what name to call it) would be nothing without this accompaniment. Speaking of some ministers whom he did not like, he said, "Their only means of government are the guinea and the gallows." There can scarcely, it must be confessed, be a more effectual mode of political conversion than one of these applied to a man's friends, and the other to himself. The fine sarcasm of Junius[21] on the effect of the supposed ingratitude of the Duke of Grafton at court—"The instance might be painful, but the principle would please"—notwithstanding the profound insight into human nature it implies, would hardly pass for wit without the alliteration, as some poetry would hardly be acknowledged as such without the rhyme to clench it. A quotation or a hackneyed phrase dextrously turned or wrested to another purpose, has often the effect of the liveliest wit. An idle fellow who had only fourpence left in the world, which had been put by to pay for the baking some meat for his dinner, went and laid it out to buy a new string for a guitar. An old acquaintance on hearing this story, repeated those lines out of the Allegro—

"And ever against *eating* cares
Lap me in soft Lydian airs."

The reply of the author of the periodical paper called the World to a lady at church, who seeing him look thoughtful, asked what he was thinking of—"The next World,"—is a perversion of an established formula of language, something of the same kind.— Rhymes are sometimes a species of wit, where there is an alternate combination and resolution or decomposition of the elements of sound, contrary to

18. **ridicule . . . truth:** a maxim popularized by the Earl of Shaftesbury, 1671–1713, and often cited in the early eighteenth century as a moral justification of satire. 19. **"Soul-killing . . . good":** Lamb's *John Woodvil*, II. ii.

20. **Mr. Grattan:** Henry Grattan, 1746–1820, an Irish statesman and orator. 21. **fine . . . Junius:** in the *Letters of Junius*, No. 49. "Junius" was the pseudonym of a writer of political satires in the form of letters. They appeared in *The Public Advertiser*, a periodical, from 1760 to 1771. His identity remains conjectural.

our usual division and classification of them in ordinary speech, not unlike the sudden separation and re-union of the component parts of the machinery in a pantomime. The author who excels infinitely the most in this way is the writer of Hudibras. He also excels in the invention of single words and names which have the effect of wit by sounding big, and meaning nothing: —"full of sound and fury, signifying nothing." But of the artifices of this author's burlesque style I shall have occasion to speak hereafter.—It is not always easy to distinguish between the wit of words and that of things. "For thin partitions do their bounds divide." Some of the late Mr. Curran's[22] bon mots or jeux d'esprit, might be said to owe their birth to this sort of equivocal generation; or were a happy mixture of verbal wit and a lively and picturesque fancy, of legal acuteness in detecting the variable applications of words, and of a mind apt at perceiving the ludicrous in external objects. "Do you see any thing ridiculous in this wig?" said one of his brother judges to him. "Nothing but the head," was the answer. Now here instantaneous advantage was taken of the slight technical ambiguity in the construction of language, and the matter-of-fact flung into the scale as a thumping makeweight. After all, verbal and accidental strokes of wit, though the most surprising and laughable, are not the best and most lasting. That wit is the most refined and effectual, which is founded on the detection of unexpected likeness or distinction in things, rather than in words. It is more severe and galling, that is, it is more unpardonable though less surprising, in proportion as the thought suggested is more complete and satisfactory, from its being inherent in the nature of the things themselves. Hæret lateri lethalis arundo.[23] Truth makes the greatest libel; and it is that which barbs the darts of wit. The Duke of Buckingham's saying, "Laws are not, like women, the worse for being old,"[24] is an instance of a harmless truism and the utmost malice of wit united. This is, perhaps, what has been meant by the distinction between true and false wit. Mr. Addison, indeed, goes so far as to make it the exclusive test of true wit that it will bear translation into another language, that is to say, that it does not depend at all on the form of expression.[25] But this is by no means the case. Swift would hardly have allowed of such a strait-laced theory, to make havoc with his darling conundrums; though there is no one whose serious wit is more that of things, as opposed to a mere play either of words or fancy. I ought, I believe, to have noticed before, in speaking of the difference between wit and humour, that wit is often pretended absurdity, where the person overacts or exaggerates a certain part with a conscious design to expose it as if it were another person, as when Mandrake in the Twin Rivals says, "This glass is too big, carry it away, I'll drink out of the bottle."[26] On the contrary, when Sir Hugh Evans says very innocently, "'Od's plessed will, I will not be absence at the grace,"[27] though there is here a great deal of humour, there is no wit. This kind of wit of the humorist, where the person makes a butt of himself, and exhibits his own absurdities or foibles purposely in the most pointed and glaring lights, runs through the whole of the character of Falstaff, and is, in truth, the principle on which it is founded. It is an irony directed against one's-self. Wit is, in fact, a voluntary act of the mind, or exercise of the invention, shewing the absurd and ludicrous consciously, whether in ourselves or another. Cross-readings, where the blunders are designed, are wit: but if any one were to light upon them through ignorance or accident, they would be merely ludicrous.

It might be made an argument of the intrinsic superiority of poetry or imagination to wit, that the former does not admit of mere verbal combinations. Whenever they do occur, they are uniformly blemishes. It requires something more solid and substantial to raise admiration or passion. The general forms and aggregate masses of our ideas must be brought more into play, to give weight and magnitude. Imagination may be said to be the finding out something similar in things generally alike, or with like feelings attached to them; while wit principally aims at finding out something that seems the same, or amounts to a momentary deception where you least expected it, viz. in things totally opposite. The reason why more slight and partial, or merely accidental and nominal resemblances serve the purposes of wit, and indeed characterise its essence as a distinct operation and faculty of the mind, is, that the object of ludicrous poetry is naturally to let down and lessen; and it is easier to let down than to raise up, to weaken than to

22. **Mr. Curran:** John Philpot Curran, 1750–1817, Irish orator and statesman. **23. Hæret . . . arundo:** "The deadly reed clings closely to her side" (Aeneid, IV. 73). **24. "Laws . . . old":** "Speech on the Dissolution of Parliament" (1670). **25. Mr. Addison . . . expression:** Spectator, No. 61.

26. **"This . . . bottle":** George Farquhar's Twin Rivals (1702), II. ii. **27. "'Od's . . . grace":** The Merry Wives of Windsor, I. i. 276.

strengthen, to disconnect our sympathy from passion and power, than to attach and rivet it to any object of grandeur or interest, to startle and shock our pre-conceptions by incongruous and equivocal combinations, than to confirm, enforce, and expand them by powerful and lasting associations of ideas, or striking and true analogies. A slight cause is sufficient to produce a slight effect. To be indifferent or sceptical, requires no effort; to be enthusiastic and in earnest, requires a strong impulse, and collective power. Wit and humour (comparatively speaking, or taking the extremes to judge of the gradations by) appeal to our indolence, or vanity, our weakness, and insensibility; serious and impassioned poetry appeals to our strength, our magnanimity, our virtue, and humanity. Any thing is sufficient to heap contempt upon an object; even the bare suggestion of a mischievous allusion to what is improper, dissolves the whole charm, and puts an end to our admiration of the sublime or beautiful. Reading the finest passage in Milton's Paradise Lost in a false tone, will make it seem insipid and absurd. The cavilling at, or invidiously pointing out, a few slips of the pen, will embitter the pleasure, or alter our opinion of a whole work, and make us throw it down in disgust. The critics are aware of this vice and infirmity in our nature, and play upon it with periodical success. The meanest weapons are strong enough for this kind of warfare, and the meanest hands can wield them. Spleen can subsist on any kind of food. The shadow of a doubt, the hint of an inconsistency, a word, a look, a syllable, will destroy our best-formed convictions. What puts this argument in as striking a point of view as any thing, is the nature of parody or burlesque, the secret of which lies merely in transposing or applying at a venture to any thing, or to the lowest objects, that which is applicable only to certain given things, or to the highest matters. "From the sublime to the ridiculous, there is but one step." The slightest want of unity of impression destroys the sublime; the detection of the smallest incongruity is an infallible ground to rest the ludicrous upon. But in serious poetry, which aims at rivetting our affections, every blow must tell home. The missing a single time is fatal, and undoes the spell. We see how difficult it is to sustain a continued flight of impressive sentiment: how easy it must be then to travestie or burlesque it, to flounder into nonsense, and be witty by playing the fool. It is a common mistake, however, to suppose that parodies degrade, or imply a stigma on the subject: on the contrary, they in general imply something serious or sacred in

the originals. Without this, they would be good for nothing; for the immediate contrast would be wanting, and with this they are sure to tell. The best parodies are, accordingly, the best and most striking things reversed. Witness the common travesties of Homer and Virgil. Mr. Canning's court parodies on Mr. Southey's popular odes, are also an instance in point (I do not know which were the cleverest); and the best of the Rejected Addresses is the parody on Crabbe, though I do not certainly think that Crabbe is the most ridiculous poet now living.

Lear and the Fool are the sublimest instance I know of passion and wit united, or of imagination unfolding the most tremendous sufferings, and of burlesque on passion playing with it, aiding and relieving its intensity by the most pointed, but familiar and indifferent illustrations of the same thing in different objects, and on a meaner scale. The Fool's reproaching Lear with "making his daughters his mothers," his snatches of proverbs and old ballads, "The hedge-sparrow fed the cuckoo so long, that it had its head bit off by its young," and "Whoop jug, I know when the horse follows the cart," are a running commentary of trite truisms, pointing out the extreme folly of the infatuated old monarch, and in a manner reconciling us to its inevitable consequences.

Lastly, there is a wit of sense and observation, which consists in the acute illustration of good sense and practical wisdom, by means of some far-fetched conceit or quaint imagery. The matter is sense, but the form is wit. Thus the lines in Pope—

"'Tis with our judgments as our watches, none
    Go just alike; yet each believes his own—"[28]

are witty, rather than poetical; because the truth they convey is a mere dry observation on human life, without elevation or enthusiasm, and the illustration of it is of that quaint and familiar kind that is merely curious and fanciful. Cowley is an instance of the same kind in almost all his writings. Many of the jests and witticisms in the best comedies are moral aphorisms and rules for the conduct of life, sparkling with wit and fancy in the mode of expression. The ancient philosophers also abounded in the same kind of wit, in telling home truths in the most unexpected manner.— In this sense Æsop was the greatest wit and moralist that ever lived. Ape and slave, he looked askance at human nature, and beheld its weaknesses and errors

28. "'Tis . . . own": *Essay on Criticism*, ll. 9-10.

transferred to another species. Vice and virtue were to him as plain as any objects of sense. He saw in man a talking, absurd, obstinate, proud, angry animal; and clothed these abstractions with wings, or a beak, or tail, or claws, or long ears, as they appeared embodied in these hieroglyphics in the brute creation. His moral philosophy is natural history. He makes an ass bray wisdom, and a frog croak humanity. The store of moral truth, and the fund of invention in exhibiting it in eternal forms, palpable and intelligible, and delightful to children and grown persons, and to all ages and nations, are almost miraculous. The invention of a fable is to me the most enviable exertion of human genius: it is the discovering a truth to which there is no clue, and which, when once found out, can never be forgotten. I would rather have been the author of Æsop's Fables, than of Euclid's Elements!—That popular entertainment, Punch and the Puppet-show, owes part of its irresistible and universal attraction to nearly the same principle of inspiring inanimate and mechanical agents with sense and consciousness. The drollery and wit of a piece of wood is doubly droll and farcical. Punch is not merry in himself, but "he is the cause of heartfelt mirth in other men."[29] The wires and pulleys that govern his motions are conductors to carry off the spleen, and all "that perilous stuff that weighs upon the heart." If we see a number of people turning the corner of a street, ready to burst with secret satisfaction, and with their faces bathed in laughter, we know what is the matter—that they are just come from a puppet-show. Who can see three little painted, patched-up figures, no bigger than one's thumb, strut, squeak and gibber, sing, dance, chatter, scold, knock one another about the head, give themselves airs of importance, and "imitate humanity most abominably," without laughing immoderately? We overlook the farce and mummery of human life in little, and for nothing; and what is still better, it costs them who have to play in it nothing. We place the mirth, and glee, and triumph, to our own account; and we know that the bangs and blows they have received go for nothing, as soon as the showman puts them up in his box and marches off quietly with them, as jugglers of a less amusing description sometimes march off with the wrongs and rights of mankind in their pockets!—I have heard no bad judge of such matters say, that "he liked a comedy better than a tragedy, a farce better

than a comedy, a pantomime better than a farce, but a puppet-show best of all." I look upon it, that he who invented puppet-shows was a greater benefactor to his species, than he who invented Operas!

I shall conclude this imperfect and desultory sketch of wit and humour with Barrow's[30] celebrated description of the same subject. He says, "—But first it may be demanded, what the thing we speak of is, or what this facetiousness doth import; to which question I might reply, as Democritus did to him that asked the definition of a man—'tis that which we all see and know; and one better apprehends what it is by acquaintance, than I can inform him by description. It is, indeed, a thing so versatile and multiform, appearing in so many shapes, so many postures, so many garbs, so variously apprehended by several eyes and judgments, that it seemeth no less hard to settle a clear and certain notice thereof, than to make a portrait of Proteus, or to define the figure of fleeting air. Sometimes it lieth in pat allusion to a known story, or in seasonable application of a trivial saying, or in forging an apposite tale: sometimes it playeth in words and phrases, taking advantage from the ambiguity of their sense, or the affinity of their sound: sometimes it is wrapped in a dress of luminous expression; sometimes it lurketh under an odd similitude. Sometimes it is lodged in a sly question, in a smart answer; in a quirkish reason; in a shrewd intimation; in cunningly diverting or cleverly restoring an objection: sometimes it is couched in a bold scheme of speech; in a tart irony; in a lusty hyperbole; in a startling metaphor; in a plausible reconciling of contradictions, or in acute nonsense: sometimes a scenical representation of persons or things, a counterfeit speech, a mimical look or gesture passeth for it; sometimes an affected simplicity, sometimes a presumptuous bluntness giveth it being; sometimes it riseth only from a lucky hitting upon what is strange: sometimes from a crafty wresting obvious matter to the purpose: often it consisteth in one knows not what, and springeth up one can hardly tell how. Its ways are unaccountable and inexplicable, being answerable to the numberless rovings of fancy and windings of language. It is, in short, a manner of speaking out of the simple and plain way (such as reason teacheth and knoweth things by), which by a pretty surprising uncouthness in conceit or expression doth affect and amuse the fancy, shewing in it some wonder, and

---

29. "he . . . men": a misquotation of II *Henry IV*, I. ii. 11–12.

30. **Barrow:** Isaac Barrow (see "On the Living Poets," above, n. 3).

breathing some delight thereto. It raiseth admiration, as signifying a nimble sagacity of apprehension, a special felicity of invention, a vivacity of spirit, and reach of wit more than vulgar: it seeming to argue a rare quickness of parts, that one can fetch in remote conceits applicable; a notable skill that he can dextrously accommodate them to a purpose before him, together with a lively briskness of humour, not apt to damp those sportful flashes of imagination. (Whence in Aristotle such persons are termed ἐπιδέξιοι, dexterous men and εὐτρόποι, men of facile or versatile manners, who can easily turn themselves to all things, or turn all things to themselves.) It also procureth delight by gratifying curiosity with its rareness or semblance of difficulty (as monsters, not for their beauty but their rarity; as juggling tricks, not for their use but their abstruseness, are beheld with pleasure;) by diverting the mind from its road of serious thoughts; by instilling gaiety and airiness of spirit; by provoking to such dispositions of spirit, in way of emulation or complaisance, and by seasoning matter, otherwise distasteful or insipid, with an unusual and thence grateful tang."— *Barrow's Works, Serm.* 14.

I will only add by way of general caution, that there is nothing more ridiculous than laughter without a cause, nor any thing more troublesome than what are called laughing people. A professed laugher is as contemptible and tiresome a character as a professed wit: the one is always contriving something to laugh at, the other is always laughing at nothing. An excess of levity is as impertinent as an excess of gravity. A character of this sort is well personified by Spenser, in the Damsel of the Idle Lake—

> "——— Who did assay
> To laugh at shaking of the leavés light." [31]

Any one must be mainly ignorant or thoughtless, who is surprised at every thing he sees; or wonderfully conceited, who expects every thing to conform to his standard of propriety. Clowns and idiots laugh on all occasions; and the common failing of wishing to be thought satirical often runs through whole families in country places, to the great annoyance of their neighbours. To be struck with incongruity in whatever comes before us, does not argue great comprehension or refinement of perception, but rather a looseness and flippancy of mind and temper, which prevents the individual from connecting any two ideas steadily

or consistently together. It is owing to a natural crudity and precipitateness of the imagination, which assimilates nothing properly to itself. People who are always laughing, at length laugh on the wrong side of their faces; for they cannot get others to laugh with them. In like manner, an affectation of wit by degrees hardens the heart, and spoils good company and good manners. A perpetual succession of good things puts an end to common conversation. There is no answer to a jest, but another; and even where the ball can be kept up in this way without ceasing, it tires the patience of the by-standers, and runs the speakers out of breath. Wit is the salt of conversation, not the food.

The four chief names for comic humour out of our own language are Aristophanes and Lucian among the ancients, Moliere and Rabelais among the moderns. Of the two first I shall say, for I know but little. I should have liked Aristophanes better, if he had treated Socrates less scurvily, for he has treated him most scurvily both as to wit and argument. His Plutus and his Birds are striking instances, the one of dry humour, the other of airy fancy.—Lucian is a writer who appears to deserve his full fame: he has the licentious and extravagant wit of Rabelais, but directed more uniformly to a purpose; and his comic productions are interspersed with beautiful and eloquent descriptions, full of sentiment, such as the exquisite account of the fable of the halcyon put into the mouth of Socrates, and the heroic eulogy on Bacchus, which is conceived in the highest strain of glowing panegyric.

The two other authors I proposed to mention are modern, and French. Moliere, however, in the spirit of his writings, is almost as much an English as a French author—quite a *barbare* in all in which he really excelled. He was unquestionably one of the greatest comic geniuses that ever lived; a man of infinite wit, gaiety, and invention—full of life, laughter, and whim. But it cannot be denied, that his plays are in general mere farces, without scrupulous adherence to nature, refinement of character, or common probability. The plots of several of them could not be carried on for a moment without a perfect collusion between the parties to wink at contradictions, and act in defiance of the evidence of their senses. For instance, take the *Medecin malgrè lui* (the Mock Doctor), in which a common wood-cutter takes upon himself, and is made successfully to support through a whole play, the character of a learned physician, without exciting the least suspicion; and yet, notwithstanding the absurdity of the plot, it is one of the most laughable and truly comic productions

31. **"Who . . . light":** *Faerie Queene*, II. vi. 7.

that can well be imagined. The rest of his lighter pieces, the *Bourgeois Gentilhomme*, *Monsieur Pourceaugnac*, *George Dandin*, (or Barnaby Brittle,) &c. are of the same description—gratuitous assumptions of character, and fanciful and outrageous caricatures of nature. He indulges at his peril in the utmost license of burlesque exaggeration; and gives a loose to the intoxication of his animal spirits. With respect to his two most laboured comedies, the Tartuffe and Misanthrope, I confess that I find them rather hard to get through: they have much of the improbability and extravagance of the others, united with the endless common-place prosing of French declamation. What can exceed, for example, the absurdity of the Misanthrope, who leaves his mistress, after every proof of her attachment and constancy, for no other reason than that she will not submit to the *technical formality* of going to live with him in a wilderness? The characters, again, which Celimene gives of her female friends, near the opening of the play, are admirable satires, (as good as Pope's characters of women,) but not exactly in the spirit of comic dialogue. The strictures of Rousseau on this play, in his Letter to D'Alembert, are a fine specimen of the best philosophical criticism.—The same remarks apply in a greater degree to the Tartuffe. The long speeches and reasonings in this play tire one almost to death: they may be very good logic, or rhetoric, or philosophy, or any thing but comedy. If each of the parties had retained a special pleader to speak his sentiments, they could not have appeared more verbose or intricate. The improbability of the character of Orgon is wonderful. This play is in one point of view invaluable, as a lasting monument of the credulity of the French to all verbal professions of wisdom or virtue; and its existence can only be accounted for from that astonishing and tyrannical predominance which words exercise over things in the mind of every Frenchman. The *Ecole des Femmes*, from which Wycherley has borrowed his Country Wife, with the true spirit of original genius, is, in my judgment, the masterpiece of Moliere. The set speeches in the original play, it is true, would not be borne on the English stage, nor indeed on the French, but that they are carried off by the verse. The *Critique de l'Ecole des Femmes*, the dialogue of which is prose, is written in a very different style. Among other things, this little piece contains an exquisite, and almost unanswerable defence of the superiority of comedy over tragedy. Moliere was to be excused for taking this side of the question.

A writer of some pretensions among ourselves has reproached the French with "an equal want of books and men." There is a common French print, in which Moliere is represented reading one of his plays in the presence of the celebrated Ninon de l'Enclos, to a circle of the wits and first men of his own time. Among these are the great Corneille; the tender, faultless Racine; Fontaine, the artless old man, unconscious of immortality; the accomplished St. Evremond; the Duke de la Rochefoucault, the severe anatomiser of the human breast; Boileau, the flatterer of courts and judge of men! Were these men nothing? They have passed for men (and great ones) hitherto, and though the prejudice is an old one, I should hope it may still last our time.

Rabelais is another name that might have saved this unjust censure. The wise sayings and heroic deeds of Gargantua and Pantagruel ought not to be set down as nothing. I have already spoken my mind at large of this author; but I cannot help thinking of him here, sitting in his easy chair, with an eye languid with excess of mirth, his lip quivering with a new-born conceit, and wiping his beard after a well-seasoned jest, with his pen held carelessly in his hand, his wine-flagons, and his books of law, of school divinity, and physic before him, which were his jest-books, whence he drew endless stores of absurdity; laughing at the world and enjoying it by turns, and making the world laugh with him again, for the last three hundred years, at his teeming wit and its own prolific follies. Even to those who have never read his works, the name of Rabelais is a cordial to the spirits, and the mention of it cannot consist with gravity or spleen!

1819 (1819)

ON GENIUS AND COMMON SENSE

From *Table Talk, or Original Essays on Men and Manners* (1821–22).

We hear it maintained by people of more gravity than understanding, that genius and taste are strictly reducible to rules, and that there is a rule for every thing. So far is it from being true that the finest breath of fancy is a definable thing, that the plainest common sense is only what Mr. Locke would have called a *mixed mode*, subject to a particular sort of acquired and undefinable tact. It is asked, "If you do not know the rule by which a thing is done, how can you be sure of doing it a second time?" And the answer is, "If you do not know the muscles by the help of which you walk, how is it you do not fall down at every step you take?" In art, in taste, in life, in speech, you decide from feeling, and not from reason; that is, from the impression of a number of things on the mind, which impression is true and well-founded, though you may not be able to analyse or account for it in the several particulars. In a gesture you use, in a look you see, in a tone you hear, you judge of the expression, propriety, and meaning from habit, not from reason or rules; that is to say, from innumerable instances of like gestures, looks, and tones, in innumerable other circumstances, variously modified, which are too many and too refined to be all distinctly recollected, but which do not therefore operate the less powerfully upon the mind and eye of taste. Shall we say that these impressions (the immediate stamp of nature) do not operate in a given manner till they are classified and reduced to rules, or is not the rule itself grounded upon the truth and certainty of that natural operation? How then can the distinction of the understanding as to the manner in which they operate be necessary to their producing their due and uniform effect upon the mind? If certain effects did not regularly arise out of certain causes in mind as well as matter, there could be no rule given for them: nature does not follow the rule, but suggests it. Reason is the interpreter and critic of nature and genius, not their lawgiver and judge. He must be a poor creature indeed whose practical convictions do not in almost all cases outrun his deliberate understanding, or who does not feel and know much more than he can give a reason for.—Hence the distinction between eloquence and wisdom, between ingenuity and common sense. A man may be dextrous and able in explaining the grounds of his opinions, and yet may be a mere sophist, because he only sees one half of a subject. Another may feel the whole weight of a question, nothing relating to it may be lost upon him, and yet he may be able to give no account of the manner in which it affects him, or to drag his reasons from their silent lurking-places. This last will be a wise man, though neither a logician nor rhetorician. Goldsmith was a fool to Dr. Johnson in argument; that is, in assigning the specific grounds of his opinions: Dr. Johnson was a fool to Goldsmith in the fine tact, the airy, intuitive faculty with which he skimmed the surfaces of things, and unconsciously formed his opinions.

Common sense is the just result of the sum-total of such unconscious impressions in the ordinary occurrences of life, as they are treasured up in the memory, and called out by the occasion. Genius and taste depend much upon the same principle exercised on loftier ground and in more unusual combinations.

I am glad to shelter myself from the charge of affectation or singularity in this view of an often debated but ill-understood point, by quoting a passage from Sir Joshua Reynolds's Discourses,[1] which is full, and, I think, conclusive to the purpose. He says,

"I observe, as a fundamental ground common to all the Arts with which we have any concern in this Discourse, that they address themselves only to two faculties of the mind, its imagination and its sensibility.

"All theories which attempt to direct or to control the Arts, upon any principles falsely called rational, which we form to ourselves upon a supposition of what ought in reason to be the end or means of Art, independent of the known first effect produced by objects on the imagination, must be false and delusive. For though it may appear bold to say it, the imagination is here the residence of truth. If the imagination be affected, the conclusion is fairly drawn; if it be not affected, the reasoning is erroneous, because the end is not obtained; the effect itself being the test, and the only test, of the truth and efficacy of the means.

"There is in the commerce of life, as in Art, a sagacity which is far from being contradictory to right reason, and is superior to any occasional exercise of that faculty; which supersedes it; and does not wait for the slow progress of deduction, but goes at once, by what appears a kind of intuition, to the conclusion. A man endowed with this faculty feels and acknowledges the truth, though it is not always in his power, perhaps, to give a reason for it; because he cannot recollect and bring before him all the materials that gave birth to his opinions; for very many and very intricate considerations may unite to form the principle, even of small and minute parts, involved in, or dependent on, a great system of things:—though these in process of time are forgotten, the right impression still remains fixed in his mind.

"This impression is the result of the accumulated experience of our whole life, and has been collected, we do not always know how, or when. But this mass of collective observation, however acquired, ought

ON GENIUS AND COMMON SENSE. **1. passage . . . Discourses:** Discourse No. 13, delivered in 1786.

to prevail over that reason, which, however powerfully exerted on any particular occasion, will probably comprehend but a partial view of the subject; and our conduct in life, as well as in the arts, is or ought to be generally governed by this habitual reason: it is our happiness that we are enabled to draw on such funds. If we were obliged to enter into a theoretical deliberation on every occasion before we act, life would be at a stand, and Art would be impracticable.

"It appears to me therefore" (continues Sir Joshua) "that our first thoughts, that is, the effect which any thing produces on our minds, on its first appearance, is never to be forgotten; and it demands for that reason, because it is the first, to be laid up with care. If this be not done, the artist may happen to impose on himself by partial reasoning; by a cold consideration of those animated thoughts which proceed, not perhaps from caprice or rashness (as he may afterwards conceit), but from the fulness of his mind, enriched with the copious stores of all the various inventions which he had ever seen, or had ever passed in his mind. These ideas are infused into his design, without any conscious effort; but if he be not on his guard, he may reconsider and correct them, till the whole matter is reduced to a common-place invention.

"This is sometimes the effect of what I mean to caution you against; that is to say, an unfounded distrust of the imagination and feeling, in favour of narrow, partial, confined, argumentative theories, and of principles that seem to apply to the design in hand; without considering those general impressions on the fancy in which real principles of *sound reason*, and of much more weight and importance, are involved, and, as it were, lie hid under the appearance of a sort of vulgar sentiment. Reason, without doubt, must ultimately determine every thing; at this minute it is required to inform us when that very reason is to give way to feeling."

Mr. Burke, by whom the foregoing train of thinking was probably suggested, has insisted on the same thing, and made rather a perverse use of it in several parts of his Reflections on the French Revolution; and Windham in one of his Speeches has clenched it into an aphorism—"There is nothing so true as habit." Once more I would say, common sense is tacit reason. Conscience is the same tacit sense of right and wrong, or the impression of our moral experience and moral apprehensions on the mind, which, because it works unseen, yet certainly, we suppose to be an instinct, implanted in the mind; as we sometimes attribute the violent operations of our passions, of which we can neither trace the source nor assign the reason, to the instigation of the Devil!

I shall here try to go more at large into this subject, and to give such instances and illustrations of it as occur to me.

One of the persons who had rendered themselves obnoxious to Government, and been included in a charge for high treason in the year 1794, had retired soon after into Wales to write an epic poem and enjoy the luxuries of a rural life. In his peregrinations through that beautiful scenery, he had arrived one fine morning at the inn at Llangollen, in the romantic valley of that name. He had ordered his breakfast, and was sitting at the window in all the dalliance of expectation, when a face passed of which he took no notice at the instant— but when his breakfast was brought in presently after, he found his appetite for it gone, the day had lost its freshness in his eye, he was uneasy and spiritless; and without any cause that he could discover, a total change had taken place in his feelings. While he was trying to account for this odd circumstance, the same face passed again—it was the face of Taylor the spy; and he was no longer at a loss to explain the difficulty. He had before caught only a transient glimpse, a passing side-view of the face; but though this was not sufficient to awaken a distinct idea in his memory, his feelings, quicker and surer, had taken the alarm; a string had been touched that gave a jar to his whole frame, and would not let him rest, though he could not at all tell what was the matter with him. To the flitting, shadowy, half-distinguished profile that had glided by his window was linked unconsciously and mysteriously, but inseparably, the impression of the trains that had been laid for him by this person;—in this brief moment, in this dim, illegible shorthand of the mind he had just escaped the speeches of the Attorney and Solicitor-General over again; the gaunt figure of Mr. Pitt glared by him; the walls of a prison enclosed him; and he felt the hands of the executioner near him, without knowing it till the tremor and disorder of his nerves gave information to his reasoning faculties that all was not well within. That is, the same state of mind was recalled by one circumstance in the series of association that had been produced by the whole set of circumstances at the time, though the manner in which this was done was not immediately perceptible. In other words, the feeling of pleasure or pain, of good or evil, is revived, and acts instantaneously upon the mind, before we have time to recollect the precise objects which have originally given birth to it. The incident here mentioned was merely, then, one case of what the learned understand by the *association of ideas:* but all that is meant by feeling or common sense is nothing but the different cases of the association of ideas, more or less true to the impression of the original circumstances, as reason begins with the more formal developement of those circumstances, or pretends to account for the different cases of the association of ideas. But it does not follow that the dumb and silent pleading of the former (though sometimes, nay often mistaken) is less true than that of its babbling interpreter, or that we are never to trust its dictates without consulting the express authority of reason. Both are imperfect, both are useful in their way, and therefore both are best together, to correct or to confirm one another. It does not appear that in the singular instance above mentioned, the sudden impression on the mind was superstition or fancy, though it might have been thought so, had it not been proved by the event to have a real physical and moral cause. Had not the same face returned again, the doubt would never have been properly cleared up, but would have remained a puzzle ever after, or perhaps have been soon forgot.—By the law of association, as laid down by physiologists, any impression in a series can recall any other impression in that series without going through the whole in order: so that the mind drops the intermediate links, and passes on rapidly and by stealth to the more striking effects of pleasure or pain which have naturally taken the strongest hold of it. By doing this habitually and skilfully with respect to the various impressions and circumstances with which our experience makes us acquainted, it forms a series of unpremeditated conclusions on almost all subjects that can be brought before it, as just as they are of ready application to human life; and common sense is the name of this body of unassuming but practical wisdom. Common sense, however, is an impartial, instinctive result of truth and nature, and will therefore bear the test and abide the scrutiny of the most severe and patient reasoning. It is indeed incomplete without it. By ingrafting reason on feeling, we "make assurance double sure."

"'Tis the last key-stone that makes up the arch—
Then stands it a triumphal mark! Then men
Observe the strength, the height, the why and when
It was erected: and still walking under,
Meet some new matter to look up, and wonder." [2]

But reason, not employed to interpret nature, and to improve and perfect common sense and experience, is, for the most part, a building without a foundation. —The criticism exercised by reason then on common sense may be as severe as it pleases, but it must be as patient as it is severe. Hasty, dogmatical, self-satisfied reason is worse than idle fancy, or bigotted prejudice. It is systematic, ostentatious in error, closes up the avenues of knowledge, and "shuts the gates of wisdom on mankind." It is not enough to shew that there is no reason for a thing, that we do not see the reason of it: if the common feeling, if the involuntary prejudice sets in strong in favour of it, if, in spite of all we can do, there is a lurking suspicion on the side of our first impressions, we must try again, and believe that truth is mightier than we. So, in offering a definition of any subject, if we feel a misgiving that there is any fact or circumstance omitted, but of which we have only a vague apprehension, like a name we cannot recollect, we must ask for more time, and not cut the matter short by an arrogant assumption of the point in dispute. Common sense thus acts as a check-weight on sophistry, and suspends our rash and superficial judgments. On the other hand, if not only no reason can be given for a thing, but every reason is clear against it, and we can account from ignorance, from authority, from interest, from different causes, for the prevalence of an opinion or sentiment, then we have a right to conclude that we have mistaken a prejudice for an instinct, or have confounded a false and partial impression with the fair and unavoidable inference from general observation. Mr. Burke said that we ought not to reject every prejudice, but should separate the husk of prejudice from the truth it encloses, and so try to get at the kernel within; and thus far he was right.[3] But he was wrong in insisting that we are to cherish our prejudices, "because they are prejudices:" for if they are all well-founded, there is no occasion to inquire into their origin or use; and he who sets out to philosophise upon them, or make the separation Mr. Burke talks of in this spirit and with this previous determination, will be very likely to mistake a maggot or a rotten canker for the precious kernel of truth, as was indeed the case with our political sophist.

There is nothing more distinct than common sense and vulgar opinion. Common sense is only a judge of things that fall under common observation, or

2. "'Tis . . . wonder": paraphrased from Ben Jonson's "Epistle to Sir Edward Sackville," ll. 136–42.

3. **Mr. Burke . . . right:** *Reflections on the Revolution in France, Select Works,* ed. E. J. Payne (1892), II, 102.

immediately come home to the business and bosoms of men. This is of the very essence of its principle, the basis of its pretensions. It rests upon the simple process of feeling, it anchors in experience. It is not, nor it cannot be, the test of abstract, speculative opinions. But half the opinions and prejudices of mankind, those which they hold in the most unqualified approbation and which have been instilled into them under the strongest sanctions, are of this latter kind, that is, opinions, not which they have ever thought, known, or felt one tittle about, but which they have taken up on trust from others, which have been palmed on their understandings by fraud or force, and which they continue to hold at the peril of life, limb, property, and character, with as little warrant from common sense in the first instance as appeal to reason in the last. The *ultima ratio regum* proceeds upon a very different plea. Common sense is neither priestcraft nor state-policy. Yet "there's the rub that makes absurdity of so long life;" and, at the same time, gives the sceptical philosophers the advantage over us. Till nature has fair play allowed it, and is not adulterated by political and polemical quacks (as it so often has been), it is impossible to appeal to it as a defence against the errors and extravagances of mere reason. If we talk of common sense, we are twitted with vulgar prejudice, and asked how we distinguish the one from the other: but common and received opinion is indeed "a compost heap" of crude notions, got together by the pride and passions of individuals, and reason is itself the thrall or manumitted slave of the same lordly and besotted masters, dragging its servile chain, or committing all sorts of Saturnalian licences, the moment it feels itself freed from it.—If ten millions of Englishmen are furious in thinking themselves right in making war upon thirty millions of Frenchmen, and if the last are equally bent upon thinking the others always in the wrong, though it is a common and national prejudice, both opinions cannot be the dictate of good sense: but it may be the infatuated policy of one or both governments to keep their subjects always at variance. If a few centuries ago all Europe believed in the infallibility of the Pope, this was not an opinion derived from the proper exercise or erroneous direction of the common sense of the people: common sense had nothing to do with it—they believed whatever their priests told them. England at present is divided into Whigs and Tories, Churchmen and Dissenters: both parties have numbers on their side; but common sense and party-spirit are two different things. Sects and heresies are upheld partly by sympathy, and partly by the love of contradiction: if there was nobody of a different way of thinking, they would fall to pieces of themselves. If a whole court say the same thing, this is no proof that they think it, but that the individual at the head of the court has said it: if a mob agree for a while in shouting the same watch-word, this is not to me an example of the *sensus communis*; they only repeat what they have heard repeated by others. If indeed a large proportion of the people are in want of food, of clothing, of shelter, if they are sick, miserable, scorned, oppressed, and if each feeling it in himself, they all say so with one voice and one heart, and lift up their hands to second their appeal, this I should say was but the dictate of common sense, the cry of nature. But to wave this part of the argument, which it is needless to push farther, I believe that the best way to instruct mankind is not by pointing out to them their mutual errors, but by teaching them to think rightly on indifferent matters, where they will listen with patience in order to be amused, and where they do not consider a definition or a syllogism as the greatest injury you can offer them.

There is no rule for expression. It is got at solely by *feeling*, that is, on the principle of the association of ideas, and by transferring what has been found to hold good in one case (with the necessary modifications) to others. A certain look has been remarked strongly indicative of a certain passion or trait of character, and we attach the same meaning to it or are affected in the same pleasurable or painful manner by it, where it exists in a less degree, though we can define neither the look itself nor the modification of it. Having got the general clue, the exact result may be left to the imagination to vary, to extenuate or aggravate it according to circumstances. In the admirable profile of Oliver Cromwell after——, the drooping eyelids, as if drawing a veil over the fixed, penetrating glance, the nostrils somewhat distended, and lips compressed so as hardly to let the breath escape him, denote the character of the man for high-reaching policy and deep designs as plainly as they can be written. How is it that we decipher this expression in the face? First, by feeling it: and how is it that we feel it? Not by pre-established rules, but by the instinct of analogy, by the principle of association, which is subtle and sure in proportion as it is variable and indefinite. A circumstance, apparently of no value, shall alter the whole interpretation to be put upon an expression or action; and it shall alter it thus powerfully because in proportion to its very insignificance it shews a strong general principle at work that extends in its ramifications to the smallest things. This in fact will make all the difference between minuteness and subtlety or refinement; for a small or trivial effect may in given circumstances imply the operation of a great power. Stillness may be the result of a blow too powerful to be resisted; silence may be imposed by feelings too agonising for utterance. The minute, the trifling and insipid, is that which is little in itself, in its causes and its consequences: the subtle and refined is that which is slight and evanescent at first sight, but which mounts up to a mighty sum in the end, which is an essential part of an important whole, which has consequences greater than itself, and where more is meant than meets the eye or ear. We complain sometimes of littleness in a Dutch picture, where there are a vast number of distinct parts and objects, each small in itself, and leading to nothing else. A sky of Claude's cannot fall under this censure, where one imperceptible gradation is as it were the scale to another, where the broad arch of heaven is piled up of endlessly intermediate gold and azure tints, and where an infinite number of minute, scarce noticed particulars blend and melt into universal harmony. The subtlety in Shakespear, of which there is an immense deal every where scattered up and down, is always the instrument of passion, the vehicle of character. The action of a man pulling his hat over his forehead is indifferent enough in itself, and, generally speaking, may mean any thing or nothing: but in the circumstances in which Macduff is placed, it is neither insignificant nor equivocal.

"What! man, ne'er pull your hat upon your brows," &c.

It admits but of one interpretation or inference, that which follows it:—

"Give sorrow words: the grief that does not speak,
Whispers the o'er-fraught heart, and bids it break."[4]

The passage in the same play, in which Duncan and his attendants are introduced commenting on the beauty and situation of Macbeth's castle, though familiar in itself, has been often praised for the striking contrast it presents to the scenes which follow.—The same look in different circumstances may convey a totally different expression. Thus the eye turned round to look at you without turning the head indicates generally slyness or suspicion: but if this is combined with large expanded eye-lids or fixed eye-brows, as we see it in Titian's pictures, it will denote calm contemplation or piercing sagacity, without any thing of meanness or fear of being observed. In other cases, it may imply merely indolent enticing voluptuousness, as in Lely's[5] portraits of women. The languor and weakness of the eye-lids gives the amorous turn to the expression. How should there be a rule for all this beforehand, seeing it depends on circumstances ever varying, and scarce discernible but by their effect on the mind? Rules are applicable to abstractions, but expression is concrete and individual. We know the meaning of certain looks, and we feel how they modify one another in conjunction. But we cannot have a separate rule to judge of all their combinations in different degrees and circumstances, without foreseeing all those combinations, which is impossible: or, if we did foresee them, we should only be where we are, that is, we could only make the rule as we now judge without it, from imagination and the feeling of the moment. The absurdity of reducing expression to a preconcerted system was perhaps never more evidently shewn than in a picture of the Judgment of Solomon by so great a man as N. Poussin, which I once heard admired for the skill and discrimination of the artist in making all the women, who are ranged on one side, in the greatest alarm at the sentence of the judge, while all the men on the opposite side see through the design of it. Nature does not go to work or cast things in a regular mould in this sort of way. I once heard a person remark of another—"He has an eye like a vicious horse." This was a fair analogy. We all, I believe, have noticed the look of an horse's eye, just before he is going to bite or kick. But will any one, therefore, describe to me exactly what that look is? It was the same acute observer that said of a self-sufficient prating music-master—"He talks on all subjects *at sight*"— which expressed the man at once by an allusion to his profession. The coincidence was indeed perfect. Nothing else could compare to the easy assurance with which this gentleman would volunteer an explanation of things of which he was most ignorant; but the *nonchalance* with which a musician sits down to a harpsicord to play a piece he has never seen before. My physiognomical friend would not have hit on this mode of illustration without knowing the profession of the subject of his criticism; but having this hint given him,

---

4. **"What! . . . break"**: *Macbeth*, IV. iii. 209–11.

5. **Lely:** Sir Peter Lely, 1618–80, a Dutch painter who lived in England.

it instantly suggested itself to his "sure trailing." The manner of the speaker was evident; and the association of the music-master sitting down to play at sight, lurking in his mind, was immediately called out by the strength of his impression of the character. The feeling of character, and the felicity of invention in explaining it, were nearly allied to each other. The first was so wrought up and running over, that the transition to the last was easy and unavoidable. When Mr. Kean[6] was so much praised for the action of Richard in his last struggle with his triumphant antagonist, where he stands, after his sword is wrested from him, with his hands stretched out, "as if his will could not be disarmed, and the very phantoms of his despair had a withering power," he said that he borrowed it from seeing the last efforts of Painter in his fight with Oliver. This assuredly did not lessen the merit of it. Thus it ever is with the man of real genius. He has the feeling of truth already shrined in his own breast, and his eye is still bent on nature to see how she expresses herself. When we thoroughly understand the subject, it is easy to translate from one language into another. Raphael, in muffling up the figure of Elymas the Sorcerer in his garments, appears to have extended the idea of blindness even to his clothes. Was this design? Probably not; but merely the feeling of analogy thoughtlessly suggesting this device, which being so suggested was retained and carried on, because it flattered or fell in with the original feeling. The tide of passion, when strong, overflows and gradually insinuates itself into all nooks and corners of the mind. Invention (of the best kind) I therefore do not think so distinct a thing from feeling, as some are apt to imagine. The springs of pure feeling will rise and fill the moulds of fancy that are fit to receive it. There are some striking coincidences of colour in well-composed pictures, as in a straggling weed in the foreground streaked with blue or red to answer to a blue or red drapery, to the tone of the flesh or an opening in the sky:—not that this was intended, or done by rule (for then it would presently become affected and ridiculous), but the eye being imbued with a certain colour, repeats and varies it from a natural sense of harmony, a secret craving and appetite for beauty, which in the same manner soothes and gratifies the eye of taste, though the cause is not understood. *Tact, finesse*, is nothing but the being completely aware

of the feeling belonging to certain situations, passions, &c. and the being consequently sensible to their slightest indications or movements in others. One of the most remarkable instances of this sort of faculty is the following story, told of Lord Shaftesbury, the grandfather of the author of the Characteristics. He had been to dine with Lady Clarendon and her daughter, who was at that time privately married to the Duke of York (afterwards James II.) and as he returned home with another nobleman who had accompanied him, he suddenly turned to him, and said, "Depend upon it, the Duke has married Hyde's daughter." His companion could not comprehend what he meant; but on explaining himself, he said, "Her mother behaved to her with an attention and a marked respect that it is impossible to account for in any other way; and I am sure of it." His conjecture shortly afterwards proved to be the truth. This was carrying the prophetic spirit of common sense as far as it could go.—

THE SAME
SUBJECT CONTINUED

Genius or originality is, for the most part, *some strong quality in the mind, answering to and bringing out some new and striking quality in nature.*

Imagination is, more properly, the power of carrying on a given feeling into other situations, which must be done best according to the hold which the feeling itself has taken of the mind.[1] In new and unknown combinations, the impression must act by sympathy, and not by rule; but there can be no sympathy, where there is no passion, no original interest. The personal interest may in some cases oppress and circumscribe the imaginative faculty, as in the instance of Rousseau: but in general the strength and consistency of the imagination will be in proportion to the strength and depth of feeling; and it is rarely that a man even of lofty genius will be able to do more than carry on his own feelings and character, or some prominent and ruling passion, into fictitious and uncommon situations. Milton has by allusion embodied a great part of his political and personal history in the chief characters

---

6. **Mr. Kean:** Edmund Kean, 1787–1833, one of the foremost Shakespearean actors of the time.

THE SAME SUBJECT CONTINUED. 1. **Imagination . . . mind:** "I do not here speak of the figurative or fanciful exercise of the imagination which consists in finding out some striking object or image to illustrate another" (H.).

and incidents of Paradise Lost. He has, no doubt, wonderfully adapted and heightened them, but the elements are the same; you trace the bias and opinions of the man in the creations of the poet. Shakespear (almost alone) seems to have been a man of genius, raised above the definition of genius. "Born universal heir to all humanity," he was "as one, in suffering all who suffered nothing;" with a perfect sympathy with all things, yet alike indifferent to all: who did not tamper with nature or warp her to his own purposes; who "knew all qualities with a learned spirit," instead of judging of them by his own predilections; and was rather "a pipe for the Muse's finger to play what stop she pleased," than anxious to set up any character or pretensions of his own. His genius consisted in the faculty of transforming himself at will into whatever he chose: his originality was the power of seeing every object from the exact point of view in which others would see it. He was the Proteus[2] of human intellect. Genius in ordinary is a more obstinate and less versatile thing. It is sufficiently exclusive and self-willed, quaint and peculiar. It does some one thing by virtue of doing nothing else: it excels in some one pursuit by being blind to all excellence but its own. It is just the reverse of the cameleon; for it does not borrow, but lend its colour to all about it: or like the glow-worm, discloses a little circle of gorgeous light in the twilight of obscurity, in the night of intellect, that surrounds it. So did Rembrandt. If ever there was a man of genius, he was one, in the proper sense of the term. He lived in and revealed to others a world of his own, and might be said to have invented a new view of nature. He did not discover things *out of* nature, in fiction or fairy land, or make a voyage to the moon "to descry new lands, rivers, or mountains in her spotty globe," but saw things *in* nature that every one had missed before him, and gave others eyes to see them with. This is the test and triumph of originality, not to shew us what has never been, and what we may therefore very easily never have dreamt of, but to point out to us what is before our eyes and under our feet, though we have had no suspicion of its existence, for want of sufficient strength of intuition, of determined grasp of mind to seize and retain it. Rembrandt's conquests were not over the *ideal*, but the real. He did not contrive

a new story or character, but we nearly owe to him a fifth part of painting, the knowledge of *chiaroscuro*—a distinct power and element in art and nature. He had a steadiness, a firm keeping of mind and eye, that first stood the shock of "fierce extremes" in light and shade, or reconciled the greatest obscurity and the greatest brilliancy into perfect harmony; and he therefore was the first to hazard this appearance upon canvas, and give full effect to what he saw and delighted in. He was led to adopt this style of broad and startling contrast from its congeniality to his own feelings: his mind grappled with that which afforded the best exercise to its master-powers: he was bold in act, because he was urged on by a strong native impulse. Originality is then nothing but nature and feeling working in the mind. A man does not affect to be original: he is so, because he cannot help it, and often without knowing it. This extraordinary artist indeed might be said to have had a particular organ for colour. His eye seemed to come in contact with it as a feeling, to lay hold of it as a substance, rather than to contemplate it as a visual object. The texture of his landscapes is "of the earth, earthy"—his clouds are humid, heavy, slow; his shadows are "darkness that may be felt," a "palpable obscure;" his lights are lumps of liquid splendour! There is something more in this than can be accounted for from design or accident: Rembrandt was not a man made up of two or three rules and directions for acquiring genius.

I am afraid I shall hardly write so satisfactory a character of Mr. Wordsworth, though he, too, like Rembrandt, has a faculty of making something out of nothing, that is, out of himself, by the medium through which he sees and with which he clothes the barrenest subject. Mr. Wordsworth is the last man to "look abroad into universality," if that alone constituted genius: he looks at home into himself, and is "content with riches fineless." He would in the other case be "poor as winter," if he had nothing but general capacity to trust to. He is the greatest, that is, the most original poet of the present day, only because he is the greatest egotist. He is "self-involved, not dark." He sits in the centre of his own being, and there "enjoys bright day." He does not waste a thought on others. Whatever does not relate exclusively and wholly to himself, is foreign to his views. He contemplates a whole-length figure of himself, he looks along the unbroken line of his personal identity. He thrusts aside all other objects, all other interests with scorn and impatience, that he may repose on his own being,

2. **Proteus:** a sea-god in Greek mythology who was able to assume different shapes at will. Cf. Hazlitt's remarks on Shakespeare's genius in the lecture "On Shakespeare and Milton," above, and Keats's letter to Woodhouse, below, October 27, 1818.

that he may dig out the treasures of thought contained in it, that he may unfold the precious stores of a mind for ever brooding over itself. His genius is the effect of his individual character. He stamps that character, that deep individual interest, on whatever he meets. The object is nothing but as it furnishes food for internal meditation, for old associations. If there had been no other being in the universe, Mr. Wordsworth's poetry would have been just what it is. If there had been neither love nor friendship, neither ambition nor pleasure nor business in the world, the author of the Lyrical Ballads need not have been greatly changed from what he is— might still have "kept the noiseless tenour of his way," retired in the sanctuary of his own heart, hallowing the Sabbath of his own thoughts. With the passions, the pursuits, and imaginations of other men, he does not profess to sympathise, but "finds tongues in the trees, books in the running brooks, sermons in stones, and good in every thing." With a mind averse from outward objects, but ever intent upon its own workings, he hangs a weight of thought and feeling upon every trifling circumstance connected with his past history. The note of the cuckoo sounds in his ear like the voice of other years; the daisy spreads its leaves in the rays of boyish delight, that stream from his thoughtful eyes; the rainbow lifts its proud arch in heaven but to mark his progress from infancy to manhood; an old thorn is buried, bowed down under the mass of associations he has wound about it; and to him, as he himself beautifully says,

—— "The meanest flow'r that blows can give
Thoughts that do often lie too deep for tears."[3]

It is this power of habitual sentiment, or of transferring the interest of our conscious existence to whatever gently solicits attention, and is a link in the chain of association, without rousing our passions or hurting our pride, that is the striking feature in Mr. Wordsworth's mind and poetry. Others have felt and shown this power before, as Withers,[4] Burns, &c. but none have felt it so intensely and absolutely as to lend to it the voice of inspiration, as to make it the foundation of a new style and school in poetry. His strength, as it so often happens, arises from the excess of his weakness. But he has opened a new avenue to the human heart, has explored another secret haunt and nook of nature,

"sacred to verse, and sure of everlasting fame." Compared with his lines, Lord Byron's stanzas are but exaggerated commonplace, and Walter Scott's poetry (not his prose) old wives' fables. There is no one in whom I have been more disappointed than in the writer here spoken of, nor with whom I am more disposed on certain points to quarrel: but the love of truth and justice which obliges me to do this, will not suffer me to blench his merits. Do what he can, he cannot help being an original-minded man. His poetry is not servile. While the cuckoo returns in the spring, while the daisy looks bright in the sun, while the rainbow lifts its head above the storm—

"Yet I'll remember thee, Glencairn,
And all that thou hast done for me!"[5]

Sir Joshua Reynolds, in endeavouring to show that there is no such thing as proper originality, a spirit emanating from the mind of the artist and shining through his works, has traced Raphael through a number of figures which he has borrowed from Masaccio and others. This is a bad calculation. If Raphael had only borrowed those figures from others, would he, even in Sir Joshua's sense, have been entitled to the praise of originality? Plagiarism, I presume, in so far as it is plagiarism, is not originality. Salvator is considered by many as a great genius. He was what they call an irregular genius. My notion of genius is not exactly the same as theirs. It has also been made a question whether there is not more genius in Rembrandt's *Three Trees* than in all Claude Lorraine's landscapes? I do not know how that may be: but it was enough for Claude to have been a perfect landscape-painter.

Capacity is not the same thing as genius. Capacity may be described to relate to the quantity of knowledge, however acquired; genius to its quality and the mode of acquiring it. Capacity is a power over given ideas or combinations of ideas; genius is the power over those which are not given, and for which no obvious or precise rule can be laid down. Or capacity is power of any sort: genius is power of a different sort from what has yet been shown. A retentive memory, a clear understanding is capacity, but it is not genius. The admirable Crichton was a person of prodigious capacity; but there is no proof (that I know) that he had an atom of genius. His verses that remain are dull

3. "The . . . tears": "Intimations of Immortality," ll. 201–02. 4. Withers: a slip for George Wither, the seventeenth-century poet.

5. "Yet . . . me!": Burns's "Lament for James, Earl of Glencairn," ll. 79–80.

and sterile. He could learn all that was known of any subject: he could do any thing if others could show him the way to do it. This was very wonderful: but that is all you can say of it. It requires a good capacity to play well at chess: but, after all, it is a game of skill, and not of genius. Know what you will of it, the understanding still moves in certain tracks in which others have trod before it, quicker or slower, with more or less comprehension and presence of mind. The greatest skill strikes out nothing for itself, from its own peculiar resources; the nature of the game is a thing determinate and fixed: there is no royal or poetical road to check-mate your adversary. There is no place for genius but in the indefinite and unknown. The discovery of the binomial theorem was an effort of genius; but there was none shown in Jedediah Buxton's being able to multiply 9 figures by 9 in his head. If he could have multiplied 90 figures by 90 instead of 9, it would have been equally useless toil and trouble.[6] He is a man of capacity who possesses considerable intellectual riches: he is a man of genius who finds out a vein of new ore. Originality is the seeing nature differently from others, and yet as it is in itself. It is not singularity or affectation, but the discovery of new and valuable truth. All the world do not see the whole meaning of any object they have been looking at. Habit blinds them to some things: short-sightedness to others. Every mind is not a gauge and measure of truth. Nature has her surface and her dark recesses. She is deep, obscure, and infinite. It is only minds on whom she makes her fullest impressions that can penetrate her shrine or unveil her *Holy of Holies*. It is only those whom she has filled with her spirit that have the boldness or the power to reveal her mysteries to others. But nature has a thousand aspects, and one man can only draw out one of them.

Whoever does this, is a man of genius. One displays her force, another her refinement, one her power of harmony, another her suddenness of contrast, one her beauty of form, another her splendour of colour. Each does that for which he is best fitted by his particular genius, that is to say, by some quality of mind in which the quality of the object sinks deepest, where it finds the most cordial welcome, is perceived to its utmost extent, and where again it forces its way out from the fulness with which it has taken possession of the mind of the student. The imagination gives out what it has first absorbed by congeniality of temperament, what it has attracted and moulded into itself by elective affinity, as the loadstone draws and impregnates iron. A little originality is more esteemed and sought for than the greatest acquired talent, because it throws a new light upon things, and is peculiar to the individual. The other is common; and may be had for the asking, to any amount.

1821 (1821)

ON READING OLD BOOKS

꙰

This and "The Fight" below represent two aspects of a form of writing in which Hazlitt excelled—the familiar essay: the first essay consisting of general literary reminiscence and the latter of direct vivid reporting. "On Reading Old Books" was published in *The London Magazine*, February, 1821, and republished the same year in Hazlitt's *Plain Speaker*.

꙰

I hate to read new books. There are twenty or thirty volumes that I have read over and over again, and these are the only ones that I have any desire ever to read at all. It was a long time before I could bring myself to sit down to the Tales of My Landlord, but now that author's[1] works have made a considerable addition to my scanty library. I am told that some of Lady Morgan's are good, and have been recommended to look into Anastasius;[2] but I have not yet ventured

6. **toil and trouble:** "The only good thing I ever heard come of this man's singular faculty of memory was the following. A gentleman was mentioning his having been sent up to London from the place where he lived to see Garrick act. When he went back into the country, he was asked what he thought of the player and the play. 'Oh!' he said, 'he did not know: he had only seen a little man strut about the stage, and repeat 7956 words.' We all laughed at this, but a person in one corner of the room, holding one hand to his forehead, and seeming mightily delighted, called out, 'Ay, indeed! And pray, was he found to be correct?' This was the super-erogation of literal matter-of-fact curiosity. Jedediah Buxton's counting the number of words was idle enough; but here was a fellow who wanted some one to count them over again to see if he was correct" (H.).

ON READING OLD BOOKS. **1. that author:** Sir Walter Scott. **2. Lady Morgan's . . . Anastasius:** Lady Morgan, 1783–1859, was known for her Irish stories; *Anastasius* (1819) was a picaresque novel about the modern Near East by Thomas Hope.

upon that task. A lady, the other day, could not refrain from expressing her surprise to a friend, who said he had been reading Delphine:[3]—she asked,—If it had not been published some time back? Women judge of books as they do of fashions or complexions, which are admired only "in their newest gloss." That is not my way. I am not one of those who trouble the circulating libraries much, or pester the booksellers for mail-coach copies of standard periodical publications. I cannot say that I am greatly addicted to black-letter, but I profess myself well versed in the marble bindings of Andrew Millar, in the middle of the last century; nor does my taste revolt at Thurloe's State Papers, in Russia leather; or an ample impression of Sir William Temple's Essays, with a portrait after Sir Godfrey Kneller in front. I do not think altogether the worse of a book for having survived the author a generation or two. I have more confidence in the dead than the living. Contemporary writers may generally be divided into two classes—one's friends or one's foes. Of the first we are compelled to think too well, and of the last we are disposed to think too ill, to receive much genuine pleasure from the perusal, or to judge fairly of the merits of either. One candidate for literary fame, who happens to be of our acquaintance, writes finely, and like a man of genius; but unfortunately has a foolish face, which spoils a delicate passage:—another inspires us with the highest respect for his personal talents and character, but does not quite come up to our expectations in print. All these contradictions and petty details interrupt the calm current of our reflections. If you want to know what any of the authors were who lived before our time, and are still objects of anxious inquiry, you have only to look into their works. But the dust and smoke and noise of modern literature have nothing in common with the pure, silent air of immortality.

When I take up a work that I have read before (the oftener the better) I know what I have to expect. The satisfaction is not lessened by being anticipated. When the entertainment is altogether new, I sit down to it as I should to a strange dish,—turn and pick out a bit here and there, and am in doubt what to think of the composition. There is a want of confidence and security to second appetite. New-fangled books are also like made-dishes in this respect, that they are generally little else than hashes and *rifaccimentos* of what has been served up entire and in a more natural state

at other times. Besides, in thus turning to a well-known author, there is not only an assurance that my time will not be thrown away, or my palate nauseated with the most insipid or vilest trash,—but I shake hands with, and look an old, tried, and valued friend in the face,—compare notes, and chat the hours away. It is true, we form dear friendships with such ideal guests—dearer, alas! and more lasting, than those with our most intimate acquaintance. In reading a book which is an old favourite with me (say the first novel I ever read) I not only have the pleasure of imagination and of a critical relish of the work, but the pleasures of memory added to it. It recals the same feelings and associations which I had in first reading it, and which I can never have again in any other way. Standard productions of this kind are links in the chain of our conscious being. They bind together the different scattered divisions of our personal identity. They are landmarks and guides in our journey through life. They are pegs and loops on which we can hang up, or from which we can take down, at pleasure, the wardrobe of a moral imagination, the relics of our best affections, the tokens and records of our happiest hours. They are "for thoughts and for remembrance!" They are like Fortunatus's Wishing-Cap—they give us the best riches—those of Fancy; and transport us, not over half the globe, but (which is better) over half our lives, at a word's notice!

My father Shandy solaced himself with Bruscambille.[4] Give me for this purpose a volume of Peregrine Pickle or Tom Jones. Open either of them any where —at the Memoirs of Lady Vane, or the adventures at the masquerade with Lady Bellaston, or the disputes between Thwackum and Square, or the escape of Molly Seagrim, or the incident of Sophia and her muff, or the edifying prolixity of her aunt's lecture— and there I find the same delightful, busy, bustling scene as ever, and feel myself the same as when I was first introduced into the midst of it. Nay, sometimes the sight of an odd volume of these good old English authors on a stall, or the name lettered on the back among others on the shelves of a library, answers the purpose, revives the whole train of ideas, and sets "the puppets dallying." Twenty years are struck off the list, and I am a child again. A sage philosopher, who was not a very wise man, said, that he should like very

---

3. **Delphine:** a novel (1802) by Mme de Staël.

4. **father . . . Bruscambille:** imaginary author of a "pro-logue" on long noses in Sterne's mockery of pedantic learning in *Tristram Shandy*, Bk. III, Chap. XXXV.

well to be young again, if he could take his experience along with him. This ingenious person did not seem to be aware, by the gravity of his remark, that the great advantage of being young is to be without this weight of experience, which he would fain place upon the shoulders of youth, and which never comes too late with years. Oh! what a privilege to be able to let this hump, like Christian's burthen, drop from off one's back, and transport one's self, by the help of a little musty duodecimo, to the time when "ignorance was bliss," and when we first got a peep at the rarée-show of the world, through the glass of fiction—gazing at mankind, as we do at wild beasts in a menagerie, through the bars of their cages,—or at curiosities in a museum, that we must not touch! For myself, not only are the old ideas of the contents of the work brought back to my mind in all their vividness, but the old associations of the faces and persons of those I then knew, as they were in their life-time—the place where I sat to read the volume, the day when I got it, the feeling of the air, the fields, the sky—return, and all my early impressions with them. This is better to me—those places, those times, those persons, and those feelings that come across me as I retrace the story and devour the page, are to me better far than the wet sheets of the last new novel from the Ballantyne press, to say nothing of the Minerva press in Leadenhall-street. It is like visiting the scenes of early youth. I think of the time "when I was in my father's house, and my path ran down with butter and honey,"—when I was a little, thoughtless child, and had no other wish or care but to con my daily task, and be happy!—Tom Jones, I remember, was the first work that broke the spell. It came down in numbers once a fortnight, in Cooke's pocket-edition, embellished with cuts. I had hitherto read only in school-books, and a tiresome ecclesiastical history (with the exception of Mrs. Radcliffe's Romance of the Forest): but this had a different relish with it,—"sweet in the mouth," though not "bitter in the belly." It smacked of the world I lived in, and in which I was to live—and shewed me groups, "gay creatures" not "of the element," but of the earth; not "living in the clouds," but travelling the same road that I did;—some that had passed on before me, and others that might soon overtake me. My heart had palpitated at the thoughts of a boarding-school ball, or gala-day at Midsummer or Christmas: but the world I had found out in Cooke's edition of the British Novelists was to me a dance through life, a perpetual gala-day. The six-penny numbers of this

work regularly contrived to leave off just in the middle of a sentence, and in the nick of a story, where Tom Jones discovers Square behind the blanket; or where Parson Adams, in the inextricable confusion of events, very undesignedly gets to bed to Mrs. Slip-slop. Let me caution the reader against this impression of Joseph Andrews; for there is a picture of Fanny in it which he should not set his heart on, lest he should never meet with any thing like it; or if he should, it would, perhaps, be better for him that he had not. It was just like —— ——! With what eagerness I used to look forward to the next number, and open the prints! Ah! never again shall I feel the enthusiastic delight with which I gazed at the figures, and anticipated the story and adventures of Major Bath and Commodore Trunnion, of Trim and my Uncle Toby, of Don Quixote and Sancho and Dapple, of Gil Blas and Dame Lorenza Sephora, of Laura and the fair Lucretia, whose lips open and shut like buds of roses. To what nameless ideas did they give rise,—with what airy delights I filled up the outlines, as I hung in silence over the page! —Let me still recal them, that they may breathe fresh life into me, and that I may live that birthday of thought and romantic pleasure over again! Talk of the *ideal*! This is the only true ideal—the heavenly tints of Fancy reflected in the bubbles that float upon the spring-tide of human life.

Oh! Memory! shield me from the world's poor strife,
And give those scenes thine everlasting life!

The paradox with which I set out is, I hope, less startling than it was; the reader will, by this time, have been let into my secret. Much about the same time, or I believe rather earlier, I took a particular satisfaction in reading Chubb's Tracts,[5] and I often think I will get them again to wade through. There is a high gusto of polemical divinity in them; and you fancy that you hear a club of shoemakers at Salisbury, debating a disputable text from one of St. Paul's Epistles in a workmanlike style, with equal shrewdness and per-tinacity. I cannot say much for my metaphysical studies, into which I launched shortly after with great ardour, so as to make a toil of a pleasure. I was presently entangled in the briars and thorns of subtle distinc-tions,—of "fate, free-will, fore-knowledge absolute," though I cannot add that "in their wandering mazes I found no end;"[6] for I did arrive at some very satis-factory and potent conclusions; nor will I go so far,

5. **Chubb's Tracts:** deistic writings (1754) by Thomas Chubb. **6.** "**fate . . . end":** *Paradise Lost,* II. 560–61.

however ungrateful the subject might seem, as to exclaim with Marlowe's Faustus—"Would I had never seen Wittenberg, never read book"—that is, never studied such authors as Hartley, Hume, Berkeley, &c. Locke's Essay on the Human Understanding is, however, a work from which I never derived either pleasure or profit; and Hobbes, dry and powerful as he is, I did not read till long afterwards. I read a few poets, which did not much hit my taste,—for I would have the reader understand, I am deficient in the faculty of imagination; but I fell early upon French romances and philosophy, and devoured them tooth-and-nail. Many a dainty repast have I made of the New Eloise;[7] —the description of the kiss; the excursion on the water; the letter of St. Preux, recalling the time of their first loves; and the account of Julia's death; these I read over and over again with unspeakable delight and wonder. Some years after, when I met with this work again, I found I had lost nearly my whole relish for it (except some few parts) and was, I remember, very much mortified with the change in my taste, which I sought to attribute to the smallness and gilt edges of the edition I had bought, and its being perfumed with rose-leaves. Nothing could exceed the gravity, the solemnity with which I carried home and read the Dedication to the Social Contract, with some other pieces of the same author, which I had picked up at a stall in a coarse leathern cover. Of the Confessions I have spoken elsewhere, and may repeat what I have said—"Sweet is the dew of their memory, and pleasant the balm of their recollection!" Their beauties are not "scattered like stray-gifts o'er the earth," but sown thick on the page, rich and rare. I wish I had never read the Emilius, or read it with less implicit faith. I had no occasion to pamper my natural aversion to affectation or pretence, by romantic and artificial means. I had better have formed myself on the model of Sir Fopling Flutter.[8] There is a class of persons whose virtues and most shining qualities sink in, and are concealed by, an absorbent ground of modesty and reserve; and such a one I do, without vanity, profess myself. Now these are the very persons who are likely to attach themselves to the character of Emilius, and of whom it is sure to be the bane. This dull, phlegmatic, retiring humour is not in a fair way to be corrected, but confirmed and rendered desperate,

by being in that work held up as an object of imitation, as an example of simplicity and magnanimity—by coming upon us with all the recommendations of novelty, surprise, and superiority to the prejudices of the world—by being stuck upon a pedestal, made amiable, dazzling, a *leurre de dupe*! The reliance on solid worth which it inculcates, the preference of sober truth to gaudy tinsel, hangs like a mill-stone round the neck of the imagination—"a load to sink a navy"—impedes our progress, and blocks up every prospect in life. A man, to get on, to be successful, conspicuous, applauded, should not retire upon the centre of his conscious resources, but be always at the circumference of appearances. He must envelop himself in a halo of mystery—he must ride in an equipage of opinion—he must walk with a train of self-conceit following him—he must not strip himself to a buff-jerkin, to the doublet and hose of his real merits, but must surround himself with a *cortege* of prejudices, like the signs of the Zodiac—he must seem any thing but what he is, and then he may pass for any thing he pleases. The world love to be amused by hollow professions, to be deceived by flattering appearances, to live in a state of hallucination; and can forgive every thing but the plain, downright, simple honest truth—such as we see it chalked out in the character of Emilius. —To return from this digression, which is a little out of place here.

Books have in a great measure lost their power over me; nor can I revive the same interest in them as formerly. I perceive when a thing is good, rather than feel it. It is true,

"Marcian Colonna is a dainty book;"[9]

and the reading of Mr. Keats's Eve of Saint Agnes lately made me regret that I was not young again. The beautiful and tender images there conjured up, "come like shadows—so depart." The "tiger-moth's wings," which he has spread over his rich poetic blazonry, just flit across my fancy; the gorgeous twilight window which he has painted over again in his verse, to me "blushes" almost in vain "with blood of queens and kings." I know how I should have felt at one time in reading such passages; and that is all. The sharp luscious flavour, the fine *aroma* is fled, and nothing but the stalk, the bran, the husk of literature is left. If any one were to ask me what I read now, I

7. **New Eloise:** Rousseau's *La Nouvelle Héloïse* (1761). 8. **Sir Fopling Flutter:** character in George Etherege's play, *The Man of Mode* (1676).

9. **"Marcian . . . book":** Charles Lamb's sonnet "To Barry Cornwall," l. 5.

might answer with my Lord Hamlet in the play—
"Words, words, words."—"What is the matter?"—
—"*Nothing*!"—They have scarce a meaning. But it was
not always so. There was a time when to my thinking,
every word was a flower or a pearl, like those which
dropped from the mouth of the little peasant-girl
in the Fairy tale, or like those that fall from the great
preacher in the Caledonian Chapel! I drank of the
stream of knowledge that tempted, but did not mock
my lips, as of the river of life, freely. How eagerly I
slaked my thirst of German sentiment, "as the hart
that panteth for the water-springs;" how I bathed and
revelled, and added my floods of tears to Goethe's
Sorrows of Werter, and to Schiller's Robbers—

"Giving my stock of more to that which had too
    much!" [10]

I read, and assented with all my soul to Coleridge's
fine Sonnet,[11] begining—

"Schiller! that hour I would have wish'd to die,
If through the shuddering midnight I had sent,
From the dark dungeon of the tow'r time-rent,
That fearful voice, a famish'd father's cry!"

I believe I may date my insight into the mysteries of
poetry from the commencement of my acquaintance
with the authors of the Lyrical Ballads; at least, my
discrimination of the higher sorts—not my predilection
for such writers as Goldsmith or Pope: nor do I
imagine they will say I got my liking for the Novelists,
or the comic writers,—for the characters of Valentine,
Tattle, or Miss Prue, from them. If so, I must have
got from them what they never had themselves. In
points where poetic diction and conception are concerned, I may be at a loss, and liable to be imposed
upon: but in forming an estimate of passages relating
to common life and manners, I cannot think I am a
plagiarist from any man. I there "know my cue without a prompter." I may say of such studies—*Intus et
in cute.*[12] I am just able to admire those literal touches
of observation and description, which persons of
loftier pretensions over-look and despise. I think I
comprehend something of the characteristic part of
Shakspeare; and in him indeed, all is characteristic,
even the nonsense and poetry. I believe it was the
celebrated Sir Humphrey Davy who used to say, that

Shakspeare was rather a metaphysician than a poet.
At any rate, it was not ill said. I wish that I had sooner
known the dramatic writers contemporary with
Shakspeare; for in looking them over about a year
ago, I almost revived my old passion for reading, and
my old delight in books, though they were very nearly
new to me. The Periodical Essayists I read long ago.
The Spectator I like extremely: but the Tatler took
my fancy most. I read the others soon after, the
Rambler, the Adventurer, the World, the Connoisseur:
I was not sorry to get to the end of them, and have no
desire to go regularly through them again. I consider
myself a thorough adept in Richardson. I like the
longest of his novels best, and think no part of them
tedious; nor should I ask to have any thing better to do
than to read them from beginning to end, to take them
up when I chose, and lay them down when I was tired,
in some old family mansion in the country, till every
word and syllable relating to the bright Clarissa, the
divine Clementina, the beautiful Pamela, "with every
trick and line of their sweet favour," were once more
"graven in my heart's tables."[13] I have a sneaking
kindness for Mackenzie's Julia de Roubigné—for the
deserted mansion, and straggling gilliflowers on the
mouldering garden-wall; and still more for his Man
of Feeling; not that it is better, nor so good; but at the
time I read it, I sometimes thought of the heroine,
Miss Walton, and of Miss —— together, and "that
ligament, fine as it was, was never broken!"—One
of the poets that I have always read with most pleasure,
and can wander about in for ever with a sort of voluptuous indolence, is Spenser; and I like Chaucer even
better. The only writer among the Italians I can pretend
to any knowledge of, is Boccacio, and of him I cannot
express half my admiration. His story of the Hawk
I could read and think of from day to day, just as I
would look at a picture of Titian's!—

I remember, as long ago as the year 1798, going to
a neighbouring town (Shrewsbury, where Farquhar
has laid the plot of his Recruiting Officer) and bringing

---

10. "Giving . . . much!": *As You Like It*, II. i. 48. 11.
Sonnet: Hazlitt quotes the sonnet entire near the close of
the lecture "On the Living Poets," above. 12. Intus . . .
cute: "I know you inside and out" (Persius, *Satires*, 3. 30).

13. heart's tables: "During the peace of Amiens, a young
English officer, of the name of Lovelace, was presented at
Buonaparte's levee. Instead of the usual question, 'Where
have you served, Sir?' the First Consul immediately addressed
him, 'I perceive your name, Sir, is the same as that of
the hero of Richardson's Romance!' Here was a Consul.
The young man's uncle, who was called Lovelace, told me
this anecdote while we were stopping together at Calais.
I had also been thinking that his was the same name as that
of the hero of Richardson's Romance. This is one of my
reasons for liking Buonaparte" (H.).

home with me, "at one proud swoop," a copy of Milton's Paradise Lost, and another of Burke's Reflections on the French Revolution—both which I have still; and I still recollect, when I see the covers, the pleasure with which I dipped into them as I returned with my double prize. I was set up for one while. That time is past "with all its giddy raptures:" but I am still anxious to preserve its memory, "embalmed with odours."—With respect to the first of these works, I would be permitted to remark here in passing, that it is a sufficient answer to the German criticism which has since been started against the character of Satan (*viz*. that it is not one of disgusting deformity, or pure, defecated malice) to say that Milton has there drawn, not the abstract principle of evil, not a devil incarnate, but a fallen angel. This is the scriptural account, and the poet has followed it. We may safely retain such passages as that well-known one—

> —— His form had not yet lost
> All her original brightness; nor appear'd
> Less than archangel ruin'd; and the excess
> Of glory obscur'd—— [14]

for the theory, which is opposed to them, "falls flat upon the grunsel edge, and shames its worshippers." Let us hear no more then of this monkish cant, and bigotted outcry for the restoration of the horns and tail of the devil!—Again, as to the other work, Burke's Reflections, I took a particular pride and pleasure in it, and read it to myself and others for months afterwards. I had reason for my prejudice in favour of this author. To understand an adversary is some praise: to admire him is more. I thought I did both: I knew I did one. From the first time I ever cast my eyes on any thing of Burke's (which was an extract from his Letter to a Noble Lord in a three-times a week paper, The St. James's Chronicle, in 1796), I said to myself, "This is true eloquence: this is a man pouring out his mind on paper." All other style seemed to me pedantic and impertinent. Dr. Johnson's was walking on stilts; and even Junius's (who was at that time a favourite with me) with all his terseness, shrunk up into little antithetic points and well-trimmed sentences. But Burke's style was forked and playful as the lightning, crested like the serpent. He delivered plain things on a plain ground; but when he rose, there was no end of his flights and circumgyrations—and in this very Letter, "he, like an eagle in a dove-cot, fluttered *his*

Volscians" (the Duke of Bedford and the Earl of Lauderdale) "in Corioli." I did not care for his doctrines. I was then, and am still, proof against their contagion; but I admired the author, and was considered as not a very staunch partisan of the opposite side, though I thought myself that an abstract proposition was one thing—a masterly transition, a brilliant metaphor, another. I conceived too that he might be wrong in his main argument, and yet deliver fifty truths in arriving at a false conclusion. I remember Coleridge assuring me, as a poetical and political set-off to my sceptical admiration, that Wordsworth had written an Essay on Marriage, which, for manly thought and nervous expression, he deemed incomparably superior. As I had not, at that time, seen any specimens of Mr. Wordsworth's prose style, I could not venture my doubts on the subject. If there are greater prose-writers than Burke, they either lie out of my course of study, or are beyond my sphere of comprehension. I am too old to be a convert to a new mythology of genius. The niches are occupied, the tables are full. If such is still my admiration of this man's misapplied powers, what must it have been at a time when I myself was in vain trying, year after year, to write a single Essay, nay, a single page or sentence; when I regarded the wonders of his pen with the longing eyes of one who was dumb and a changeling; and when, to be able to convey the slightest conception of my meaning to others in words, was the height of an almost hopeless ambition! But I never measured others' excellences by my own defects: though a sense of my own incapacity, and of the steep, impassable ascent from me to them, made me regard them with greater awe and fondness. I have thus run through most of my early studies and favourite authors, some of whom I have since criticised more at large. Whether those observations will survive me, I neither know nor do I much care: but to the works themselves, "worthy of all acceptation," and to the feelings they have always excited in me since I could distinguish a meaning in language, nothing shall ever prevent me from looking back with gratitude and triumph. To have lived in the cultivation of an intimacy with such works, and to have familiarly relished such names, is not to have lived quite in vain.

There are other authors whom I have never read, and yet whom I have frequently had a great desire to read, from some circumstance relating to them. Among these is Lord Clarendon's History of the Grand Rebellion, after which I have a hankering, from hearing it spoken of by good judges—from my interest in

---

14. His . . . obscur'd: *Paradise Lost*, I. 591.

the events, and knowledge of the characters from other sources, and from having seen fine portraits of most of them. I like to read a well-penned character, and Clarendon is said to have been a master in this way. I should like to read Froissart's Chronicles, Hollingshed and Stowe, and Fuller's Worthies. I intend, whenever I can, to read Beaumont and Fletcher all through. There are fifty-two of their plays, and I have only read a dozen or fourteen of them. A Wife for a Month, and Thierry and Theodoret, are, I am told, delicious, and I can believe it. I should like to read the speeches in Thucydides, and Guicciardini's History of Florence, and Don Quixote in the original. I have often thought of reading the Loves of Persiles and Sigismunda, and the Galatea of the same author. But I somehow reserve them like "another Yarrow."[15] I should also like to read the last new novel (if I could be sure it was so) of the author of Waverley:—no one would be more glad than I to find it the best!

1820–21 (1821)

THE FIGHT

◇∿◇

Published in *The New Monthly Magazine*, February, 1822. The subject is one of the famous prizefights in the period, that which took place between Tom Hickman (nicknamed the "Gas Man") and Bill Neate on December 11, 1821. As was the custom at that time, the fighting was done with bare fists and kept up until one of the boxers was knocked out or acknowledged defeat. This essay is famous as the ancestor of present-day sports reporting.

∿∿◇

Where there's a will, there's a way,—I said to myself, as I walked down Chancery-lane, about half-past six o'clock on Monday the 10th of December, to inquire at Jack Randall's where the fight the next day was to be; and I found "the proverb" nothing "musty"[1] in the present instance. I was determined to see this fight, come what would, and see it I did, in great style. It was my *first fight*, yet it more than answered my expectations. Ladies! it is to you I dedicate this description;

nor let it seem out of character for the fair to notice the exploits of the brave. Courage and modesty are the old English virtues; and may they never look cold and askance on one another! Think, ye fairest of the fair, loveliest of the lovely kind, ye practicers of soft enchantment, how many more ye kill with poisoned baits than ever fell in the ring; and listen with subdued air and without shuddering, to a tale only tragic in appearance, and sacred to the FANCY.[2]

I was going down Chancery-lane, thinking to ask at Jack Randall's where the fight was to be, when looking through the glass-door of the *Hole in the Wall*, I heard a gentleman asking the same question *at* Mrs. Randall, as the author of Waverley would express it. Now Mrs. Randall stood answering the gentleman's question, with the authenticity of the lady of the Champion of the Light Weights. Thinks I, I'll wait till this person comes out, and learn from him how it is. For to say a truth, I was not fond of going into this house of call for heroes and philosophers, ever since the owner of it (for Jack is no gentleman) threatened once upon a time to kick me out of doors for wanting a mutton-chop at his hospitable board, when the conqueror in thirteen battles was more full of *blue ruin* than of good manners. I was the more mortified at this repulse, inasmuch as I had heard Mr. James Simpkins, hosier in the Strand, one day when the character of the *Hole in the Wall* was brought in question, observe—"The house is a very good house, and the company quite genteel: I have been there myself!" Remembering this unkind treatment of mine host, to which mine hostess was also a party, and not wishing to put her in unquiet thoughts at a time jubilant like the present, I waited at the door, when, who should issue forth but my friend Jo. Toms, and turning suddenly up Chancery-lane with that quick jerk and impatient stride which distinguishes a lover of the FANCY, I said, "I'll be hanged if that fellow is not going to the fight, and is on his way to get me to go with him." So it proved in effect, and we agreed to adjourn to my lodgings to discuss measures with that cordiality which makes old friends like new, and new friends like old, on great occasions. We are cold to others only when we are dull in ourselves, and have neither thoughts nor feelings to impart to them. Give man a topic in his head, a throb of pleasure in his heart, and he will be glad to share it with the first person he meets. Toms and I, though we seldom meet, were an

---

15. **"another Yarrow"**: referring to Wordsworth's "Yarrow Unvisited." THE FIGHT. 1. **"the proverb"** . . . **"musty"**: *Hamlet*, III. ii. 359.

2. **the Fancy**: Boxing was popularly spoken of thus.

alter idem[3] on this memorable occasion, and had not an idea that we did not candidly impart; and "so carelessly did we fleet the time,"[4] that I wish no better, when there is another fight, than to have him for a companion on my journey down, and to return with my friend Jack Pigott, talking of what was to happen or of what did happen, with a noble subject always at hand, and liberty to digress to others whenever they offered. Indeed, on my repeating the lines from Spenser in an involuntary fit of enthusiasm,

> What more felicity can fall to creature,
> Than to enjoy delight with liberty?[5]

my last-named ingenious friend stopped me by saying that this, translated into the vulgate, meant "*Going to see a fight.*"

Jo. Toms and I could not settle about the method of going down. He said there was a caravan, he understood, to start from Tom Belcher's at two, which would go there *right out* and back again the next day. Now I never travel at night, and said I should get a cast to Newbury by one of the mails. Jo. swore the thing was impossible, and I could only answer that I had made up my mind to it. In short, he seemed to me to waver, said he only came to see if I was going, had letters to write, a cause coming on the day after, and faintly said at parting (for I was bent on setting out that moment)—"Well, we meet at Philippi!"[6] I made the best of my way to Piccadilly. The mail coach stand was bare. "They are all gone," said I—"this is always the way with me—in the instant I lose the future—if I had not stayed to pour that last cup of tea, I should have been just in time"—and cursing my folly and ill-luck together, without inquiring at the coach-office whether the mails were gone or not, I walked on in despite, and to punish my own dilatoriness and want of determination. At any rate I would not turn back: I might get to Hounslow, or perhaps farther, to be on my road the next morning. I passed Hyde Park Corner (my Rubicon), and trusted to fortune. Suddenly I heard the clattering of a Brentford stage, and the fight rushed full upon my fancy. I argued (not unwisely) that even a Brentford coachman was better company than my own thoughts (such as they were just then), and at his invitation mounted the box with him. I immediately stated my case to him—namely, my quarrel with myself for missing the Bath or Bristol mail, and my determination to get on in consequence as well as I could, without any disparagement or insulting comparison between longer or shorter stages. It is a maxim with me that stage-coaches, and consequently stage-coachmen, are respectable in proportion to the distance they have to travel: so I said nothing on that subject to my Brentford friend. Any incipient tendency to an abstract proposition, or (as he might have construed it) to a personal reflection of this kind, was however nipped in the bud; for I had no sooner declared indignantly that I had missed the mails, than he flatly denied that they were gone along, and lo! at the instant three of them drove by in rapid, provoking, orderly succession, as if they would devour the ground before them. Here again I seemed in the contradictory situation of the man in Dryden who exclaims,

> I follow Fate, which does too hard pursue![7]

If I had stopped to inquire at the White Horse Cellar, which would not have taken me a minute, I should now have been driving down the road in all the dignified unconcern and *ideal* perfection of mechanical conveyance. The Bath mail I had set my mind upon, and I had missed it, as I missed every thing else, by my own absurdity, in putting the will for the deed, and aiming at ends without employing means. "Sir," said he of the Brentford, "the Bath mail will be up presently, my brother-in-law drives it, and I will engage to stop him if there is a place empty." I almost doubted my good genius; but, sure enough, up it drove like lightning, and stopped directly at the call of the Brentford Jehu.[8] I would not have believed this possible, but the brother-in-law of a mail-coach driver is himself no mean man. I was transferred without loss of time from the top of one coach to that of the other, desired the guard to pay my fare to the Brentford coachman for me as I had no change, was accommodated with a great coat, put up my umbrella to keep off a drizzling mist, and we began to cut through the air like an arrow. The milestones disappeared one after another, the rain kept off; Tom Turtle, the trainer, sat before me on the coach-box, with whom I exchanged civilities as a gentleman going to the fight; the passion that had transported me an hour before was subdued to pensive regret and conjectural musing on the next

---

3. **alter idem:** another and the same. 4. "**so . . . time**": *As You Like It*, I. i. 124. 5. **What . . . liberty:** Spenser's "Muiopotmos," ll. 209–10. 6. "**Well . . . Philippi!**": *Julius Caesar*, IV. iii. 287.

7. **I . . . pursue:** *The Indian Emperor*, IV. iii. 8. **Jehu:** coachman.

day's battle; I was promised a place inside at Reading, and upon the whole, I thought myself a lucky fellow. Such is the force of imagination! On the outside of any other coach on the 10th of December, with a Scotch mist drizzling through the cloudy moonlight air, I should have been cold, comfortless, impatient, and, no doubt, wet through; but seated on the Royal mail, I felt warm and comfortable, the air did me good, the ride did me good, I was pleased with the progress we had made, and confident that all would go well through the journey. When I got inside at Reading, I found Turtle and a stout valetudinarian, whose costume bespoke him of one of the FANCY, and who had risen from a three months' sick bed to get into the mail to see the fight. They were intimate, and we fell into a lively discourse. My friend the trainer was confined in his topics to fighting dogs and men, to bears and badgers; beyond this he was "quite chap-fallen," had not a word to throw at a dog, or indeed very wisely fell asleep, when any other game was started. The whole art of training (I, however, learnt from him) consists in two things—exercise and abstinence, abstinence and exercise, repeated alternately and without end. A yolk of an egg with a spoonful of rum in it is the first thing in the morning, and then a walk of six miles till breakfast. This meal consists of a plentiful supply of tea and toast and beefsteaks. Then another six or seven miles till dinner-time and another supply of solid beef or mutton with a pint of porter, and perhaps, at the utmost, a couple of glasses of sherry. Martin trains on water, but this increases his infirmity on another very dangerous side. The Gas-man takes now and then a chirping glass (under the rose[9]) to console him, during a six weeks' probation, for the absence of Mrs. Hickman—an agreeable woman, with (I understand) a pretty fortune of two hundred pounds. How matter presses on me! What stubborn things are facts! How inexhaustible is nature and art! "It is well," as I once heard Mr. Richmond observe, "to see a variety." He was speaking of cock-fighting as an edifying spectacle. I cannot deny but that one learns more of what *is* (I do not say of what *ought to be*) in this desultory mode of practical study, than from reading the same book twice over, even though it should be a moral treatise. Where was I? I was sitting at dinner with the candidate for the honors of the ring, "where good digestion waits on appetite, and health on both." Then follows an hour of social chat and native glee;

and afterwards, to another breathing over heathy hill or dale. Back to supper, and then to bed, and up by six again—Our hero

> Follows so the ever-running sun
> With profitable ardor.[10]

to the day that brings him victory or defeat in the green fairy circle. Is not this life more sweet than mine? I was going to say; but I will not libel any life by comparing it to mine, which is (at the date of these presents) bitter as coloquintida and the dregs of aconitum!

The invalid in the Bath mail soared a pitch above the trainer, and did not sleep so sound, because he had "more figures and more fantasies." We talked the hours away merrily. He had faith in surgery, for he had had three ribs set right, that had been broken in a *turn-up* at Belcher's, but thought physicians old women, for they had no antidote in their catalogue for brandy. An indigestion is an excellent common-place for two people that never met before. By way of ingratiating myself, I told him the story of my doctor, who, on my earnestly representing to him that I thought his regimen had done me harm, assured me that the whole pharmacopeia contained nothing comparable to the prescription he had given me; and, as a proof of its undoubted efficacy, said that "he had had one gentleman with my complaint under his hands for the last fifteen years." This anecdote made my companion shake the rough sides of his three great coats with boisterous laughter; and Turtle, starting out of his sleep, swore he knew how the fight would go, for he had had a dream about it. Sure enough the rascal told us how the three first rounds went off, but "his dream," like others, "denoted a foregone conclusion."[11] He knew his men. The moon now rose in silver state, and I ventured, with some hesitation, to point out this object of placid beauty, with the blue serene beyond, to the man of science, to which his ear he "seriously inclined," the more as it gave promise *d'un beau jour* for the morrow, and showed the ring undrenched by envious showers, arrayed in sunny smiles. Just then, all going on well, I thought of my friend Toms, whom I had left behind, and said innocently, "There was a blockhead of a fellow I left in town, who said there was no possibility of getting down by the mail, and talked of going by a caravan from Belcher's at two in the morning, after he had

9. under the rose: secretly.

10. Follows . . . ardor: *Henry V*, IV. i. 293. 11. "denoted . . . conclusion": *Othello*, III. iii. 428.

written some letters."—"Why," said he of the lapels, "I should not wonder if that was the very person we saw running about like mad from one coach-door to another, and asking if anyone had seen a friend of his, a gentleman going to the fight, whom he had missed stupidly enough by staying to write a note."—"Pray, Sir," said my fellow-traveller, "had he a plaid-cloak on?"—"Why, no," said I, "not at the time I left him, but he very well might afterwards, for he offered to lend me one." The plaid-cloak and the letter decided the thing. Joe, sure enough, was in the Bristol mail, which preceded us by about fifty yards. This was droll enough. We had now but a few miles to our place of destination, and the first thing I did on alighting at Newbury, both coaches stopping at the same time, was to call out, "Pray, is there a gentleman in that mail of the name of Toms?"—"No," said Joe borrowing something of the vein of Gilpin,[12] "for I have just got out."—"Well!" says he, "this is lucky; but you don't know how vexed I was to miss you; for," added he, lowering his voice, "do you know when I left you I went to Belcher's to ask about the caravan, and Mrs. Belcher said very obligingly, she couldn't tell about that, but there were two gentlemen who had taken places by the mail and were gone on in a landau, and she could frank us.[13] It's a pity I didn't meet with you; we could then have got down for nothing. But *mum's the word.*" It's the devil for anyone to tell me a secret, for it's sure to come out in print. I do not care so much to gratify a friend, but the public ear is too great a temptation for me.

Our present business was to get beds and supper at an inn; but this was no easy task. The public-houses were full, and where you saw a light at a private house, and people poking their heads out of the casement to see what was going on, they instantly put them in and shut the window, the moment you seemed advancing with a suspicious overture for accommodation. Our guard and coachman thundered away at the outer gate of the Crown for some time without effect—such was the greater noise within;—and when the doors were unbarred, and we got admittance, we found a party assembled in the kitchen round a good hospitable fire, some sleeping, others drinking, others talking on politics and on the fight. A tall English yeoman

(something like Matthews in the face, and quite as great a wag)—

A lusty man to ben an abbot able,[14]—

was making such a prodigious noise about rents and taxes, and the price of corn now and formerly, that he had prevented us from being heard at the gate. The first thing I heard him say was to a shuffling fellow who wanted to be off a bet for a shilling glass of brandy and water—"Confound it, man, don't be *insipid!*" Thinks I, that is a good phrase. It was a good omen. He kept it up so all night, nor flinched with the approach of morning. He was a fine fellow, with sense, wit, and spirit, a hearty body and a joyous mind, freespoken, frank, convivial—one of that true English breed that went with Harry the Fifth to the siege of Harfleur—"standing like greyhounds in the slips,"[15] etc. We ordered tea and eggs (beds were soon found to be out of the question) and this fellow's conversation was *sauce piquante.* It did one's heart good to see him brandish his oaken towel[16] and to hear him talk. He made mince-meat of a drunken, stupid, red-faced, quarrelsome, *frowsy* farmer, whose nose "he moralized into a thousand similes," making it out a firebrand like Bardolph's.[17] "I'll tell you what my friend," says he, "the landlady has only to keep you here to save fire and candle. If one was to touch your nose, it would go off like a piece of charcoal." At this the other only grinned like an idiot, the sole variety in his purple face being his little peering gray eyes and yellow teeth, called for another glass, swore he would not stand it, and after many attempts to provoke his humorous antagonist to single combat, which the other turned off (after working him up to a ludicrous pitch of choler) with great adroitness, he fell quietly asleep with a glass of liquor in his hand, which he could not lift to his head. His laughing persecutor made a speech over him, and turning to the opposite side of the room, while they were all sleeping in the midst of this "loud and furious fun," said, "There's a scene, by G–d, for Hogarth to paint. I think he and Shakspeare were our two best men at copying life." This confirmed me in my good opinion of him. Hogarth, Shakspeare, and Nature, were just enough for him (indeed for any man) to know. I said, "You read Cobbett, don't you? At

---

12. **vein of Gilpin:** in a jesting vein (from William Cowper's comic ballad "John Gilpin"). 13. **frank us:** give us free passage.

14. **A . . . able:** Prologue, *Canterbury Tales,* l. 167. 15. **"standing . . . slips":** *Henry V,* III. i. 31. 16. **towel:** club. 17. **firebrand . . . Bardolph's:** like the fiery red nose of Bardolph in *Henry IV.*

least," says I, "you talk just as well as he writes." He seemed to doubt this. But I said, "We have an hour to spare: if you'll get pen, ink, and paper, and keep on talking, I'll write down what you say; and if it doesn't make a capital Political Register, I'll forfeit my head. You have kept me alive tonight, however. I don't know what I should have done without you." He did not dislike this view of the thing, nor my asking if he was not about the size of Jem Belcher; and told me soon afterwards, in the confidence of friendship, that "the circumstance which had given him nearly the greatest concern in his life, was Cribb's beating Jem after he had lost his eye by racket-playing."—The morning dawns; that dim but yet clear light appears, which weighs like solid bars of metal on the sleepless eyelids; the guests drop down from their chambers one by one—but it was too late to think of going to bed now (the clock was on the stroke of seven), we had nothing for it but to find a barber's (the pole that glittered in the morning sun lighted us to his shop), and then a nine miles' march to Hungerford. The day was fine, the sky was blue, the mists were retiring from the marshy ground, the path was tolerably dry, the sitting-up all night had not done us much harm—at least the cause was good; we talked of this and that with amicable difference, roving and sipping of many subjects, but still invariably we returned to the fight. At length, a mile to the left of Hungerford, on a gentle eminence, we saw the ring surrounded by covered carts, gigs, and carriages, of which hundreds had passed us on the road; Toms gave a youthful shout, and we hastened down a narrow lane to the scene of action.

Reader, have you ever seen a fight? If not, you have a pleasure to come, at least if it is a fight like that between the Gas-man and Bill Neate. The crowd was very great when we arrived on the spot; open carriages were coming up, with streamers flying and music playing, and the country-people were pouring in over hedge and ditch in all directions, to see their hero beat or be beaten. The odds were still on Gas, but only about five to four. Gully[18] had been down to try Neate, and had backed him considerably, which was a damper to the sanguine confidence of the adverse party. About two hundred thousand pounds were pending. The Gas says he has 3000*l.* which were promised him by different gentlemen if he had won. He had presumed too much on himself, which had made others presume on him. This spirited and formidable

young fellow seems to have taken for his motto the old maxim, that "there are three things necessary to success in life—*Impudence! Impudence! Impudence!*" It is so in matters of opinion, but not in the FANCY, which is the most practical of all things, though even here confidence is half the battle, but only half. Our friend had vapored and swaggered too much, as if he wanted to grin and bully his adversary out of the fight. "Alas! the Bristol man was not so tamed!"[19] —"This is *the grave-digger*" (would Tom Hickman exclaim in the moments of intoxication from gin and success, showing his tremendous right hand), "this will send many of them to their long homes; I haven't done with them yet!" Why should he—though he had licked four of the best men within the hour, yet why should he threaten to inflict dishonorable chastisement on my old master Richmond, a veteran going off the stage, and who has borne his sable honors meekly? Magnanimity, my dear Tom, and bravery, should be inseparable. Or why should he go up to his antagonist, the first time he ever saw him at the Fives Court, and measuring him from head to foot with a glance of contempt, as Achilles surveyed Hector, say to him, "What, are you Bill Neate? I'll knock more blood out of that great carcase of thine, this day fortnight, than you ever knock'd out of a bullock's!" It was not manly, 'twas not fighter-like. If he was sure of the victory (as he was not), the less said about it the better. Modesty should accompany the FANCY as its shadow. The best men were always the best behaved. Jem Belcher, the Game Chicken (before whom the Gas-man could not have lived) were civil, silent men. So is Cribb, so is Tom Belcher, the most elegant of sparrers, and not a man for everyone to take by the nose. I enlarged on this topic in the mail (while Turtle was asleep), and said very wisely (as I thought) that impertinence was a part of no profession. A boxer was bound to beat his man, but not to thrust his fist, either actually or by implication, in everyone's face. Even a highwayman, in the way of trade, may blow out your brains, but if he uses foul language at the same time, I should say he was no gentleman. A boxer, I would infer, need not be a blackguard or a coxcomb, more than another. Perhaps I press this point too much on a fallen man—Mr. Thomas Hickman has by this time learnt that first of all lessons, "That man was made to mourn."[20] He has lost nothing by the late fight but

18. **Gully:** a retired prizefighter, John Gully.

19. **"Alas! . . . tamed!":** paraphrased from William Cowper's *The Task*, II. 322. **20. "That . . . mourn":** title of a poem by Robert Burns.

his presumption; and that every man may do as well without! By an over-display of this quality, however, the public had been prejudiced against him, and the *knowing-ones* were taken in. Few but those who had bet on him wished Gas to win. With my own pre-possessions on the subject, the result of the 11th of December appeared to me as fine a piece of poetical justice as I had ever witnessed. The difference of weight between the two combatants (14 stone[21] to 12) was nothing to the sporting men. Great, heavy, clumsy, long-armed Bill Neate kicked the beam in the scale of the Gas-man's vanity. The amateurs were frightened at his big words, and thought that they would make up for the difference of six feet and five feet nine. Truly, the FANCY are not men of imagination. They judge of what has been, and cannot conceive of any-thing that is to be. The Gas-man had won hitherto; therefore he must beat a man half as big again as him-self—and that to a certainty. Besides, there are as many feuds, factions, prejudices, pedantic notions in the FANCY as in the state or in the schools. Mr. Gully is almost the only cool, sensible man among them, who exercises an unbiased discretion, and is not a slave to his passions in these matters. But enough of reflections, and to our tale. The day, as I have said, was fine for a December morning. The grass was wet, and the ground miry, and ploughed up with multitudinous feet, except that, within the ring itself, there was a spot of virgin-green closed in and unprofaned by vulgar tread, that shone with dazzling brightness in the mid-day sun. For it was now noon, and we had an hour to wait. This is the trying time. It is then the heart sickens, as you think what the two champions are about, and how a time will determine their fate. After the first blow is struck, there is no opportunity for nervous apprehen-sions; you are swallowed up in the immediate interest of the scene—but

> Between the acting of a dreadful thing
> And the first motion, all the interim is
> Like a phantasma, or a hideous dream.[22]

I found it so as I felt the sun's rays clinging to my back, and saw the white wintry clouds sink below the verge of the horizon. "So, I thought, my fairest hopes have faded from my sight!—so will the Gas-man's glory, or that of his adversary, vanish in an hour." The *swells* were parading in their white box-coats, the outer

ring was cleared with some bruises on the heads and shins of the rustic assembly (for the *cockneys* had been distanced by the sixty-six miles); the time drew near, I had got a good stand; a bustle, a buzz, ran through the crowd, and from the opposite side entered Neate, between his second and bottle-holder. He rolled along, swathed in his loose great coat, his knock-knees bending under his huge bulk; and, with a modest cheerful air, threw his hat into the ring. He then just looked around, and began quietly to undress; when from the other side there was a similar rush and an opening made, and the Gas-man came forward with a conscious air of anticipated triumph, too much like the cock-of-the-walk. He strutted about more than became a hero, sucked oranges with a supercilious air, and threw away the skin with a toss of his head, and went up and looked at Neate, which was an act of supererogation. The only sensible thing he did was, as he strode away from the modern Ajax, to fling out his arms, as if he wanted to try whether they would do their work that day. By this time they had stripped, and presented a strong contrast in appearance. If Neate was like Ajax, "with Atlantean shoulders, fit to bear"[23] the pugilistic reputation of all Bristol, Hickman might be compared to Diomed, light, vigorous, elastic, and his back glistened in the sun, as he moved about, like a panther's hide. There was now a dead pause—attention was awe-struck. Who at that moment, big with a great event, did not draw his breath short—did not feel his heart throb? All was ready. They tossed up for the sun,[24] and the Gas-man won. They were led up to the *scratch*[25]—shook hands, and went at it.

In the first round everyone thought it was all over. After making play a short time, the Gas-man flew at his adversary like a tiger, struck five blows in as many seconds, three first, and then following him as he staggered back, two more, right and left, and down he fell, a mighty ruin. There was a shout, and I said, "There is no standing this." Neate seemed like a lifeless lump of flesh and bone, round which the Gas-man's blows played with the rapidity of electricity or light-ning, and you imagined he would only be lifted up to be knocked down again. It was as if Hickman held a sword or a fire in that right hand of his, and directed it against an unarmed body. They met again, and Neate

21. **stone:** A stone is 14 pounds. 22. **Between . . . dream:** *Julius Caesar,* II. i. 63–65.

23. **"with . . . bear":** *Paradise Lost,* II. 306. 24. **sun:** to see which had to face it in the fight. 25. **scratch:** the line where the fight began.

seemed, not cowed, but particularly cautious. I saw his teeth clenched together and his brows knit close against the sun. He held both his arms at full length straight before him, like two sledge-hammers, and raised his left an inch or two higher. The Gas-man could not get over this guard—they struck mutually and fell, but without advantage on either side. It was the same in the next round; but the balance of power was thus restored—the fate of the battle was suspended. No one could tell how it would end. This was the only moment in which opinion was divided; for in the next, the Gas-man aiming a mortal blow at his adversary's neck, with his right hand, and failing from the length he had to reach, the other returned it with his left at full swing, planted a tremendous blow on his cheek-bone and eyebrow, and made a red ruin of that side of his face. The Gas-man went down, and there was another shout—a roar of triumph as the waves of fortune rolled tumultuously from side to side. This was a settler. Hickman got up, and "grinned horrible a ghastly smile,"[26] yet he was evidently dashed in his opinion of himself; it was the first time he had been so punished; all one side of his face was perfect scarlet, and his right eye was closed in dingy blackness, as he advanced to the fight, less confident, but still determined. After one or two rounds, not receiving another such remembrancer, he rallied and went at it with his former impetuosity. But in vain. His strength had been weakened,—his blows could not tell at such a distance,—he was obliged to fling himself at his adversary, and could not strike from his feet; and almost as regularly as he flew at him with his right hand, Neate warded the blow, or drew back out of its reach, and felled him with the return of his left. There was little cautious sparring—no half-hits—no tapping and trifling, none of the *petit-maîtreship*[27] of the art—they were almost all knock-down blows:—the fight was a good stand-up fight. The wonder was the half-minute time. If there had been a minute or more allowed between each round, it would have been intelligible how they should by degrees recover strength and resolution; but to see two men smashed to the ground, smeared with gore, stunned, senseless, the breath beaten out of their bodies, and then, before you recover from the shock, to see them rise up with new strength and courage, stand steady to inflict or receive mortal offence, and rush upon each other "like two clouds over

the Caspian"[28]—this is the most astonishing thing of all.—This is the high and heroic state of man! From this time forward the event became more certain every round; and about the twelfth it seemed as if it must have been over. Hickman generally stood with his back to me; but in the scuffle, he had changed positions, and Neate just then made a tremendous lunge at him, and hit him full in the face. It was doubtful whether he would fall backwards or forwards; he hung suspended for a second or two, and then fell back, throwing his hands in the air, and with his face lifted up to the sky. I never saw anything more terrific than his aspect just before he fell. All traces of life, of natural expression, were gone from him. His face was like a human skull, a death's head, spouting blood. The eyes were filled with blood, the nose streamed with blood, the mouth gaped blood. He was not like an actual man, but like a preternatural, spectral appearance, or like one of the figures in Dante's *Inferno*. Yet he fought on after this for several rounds, still striking the first desperate blow, and Neate standing on the defensive, and using the same cautious guard to the last, as if he had still all his work to do; and it was not till the Gas-man was so stunned in the seventeenth or eighteenth round, that his senses forsook him, and he could not come to time, that the battle was declared over. Ye who despise the FANCY, do something to show as much *pluck*, or as much self-possession as this, before you assume a superiority which you have never given a single proof of by any one action in the whole course of your lives! —When the Gas-man came to himself, the first words he uttered were, "Where am I? What is the matter?" —"Nothing is the matter, Tom—you have lost the battle, but you are the bravest man alive." And Jackson whispered to him, "I am collecting a purse for you, Tom."—Vain sounds, and unheard at that moment! Neate instantly went up and shook him cordially by the hand, and seeing some old acquaintance, began to flourish with his fists, calling out, "Ah, you always said I couldn't fight—What do you think now?" But all in good humor, and without any appearance of arrogance; only it was evident Bill Neate was pleased that he had won the fight. When it was over I asked Cribb if he did not think it was a good one. He said, "*Pretty well!*" The carrier-pigeons now mounted into the air, and one of them flew with the news of her husband's victory to the bosom of Mrs. Neate. Alas, for Mrs. Hickman!

---

26. "grinned . . . smile": Death in *Paradise Lost*, II. 846. 27. petit-maîtreship: foppery or affectation.

28. "like . . . Caspian": *Paradise Lost*, II. 714–16.

Mais au revoir, as Sir Fopling Flutter[29] says. I went down with Toms; I returned with Jack Pigott, whom I met on the ground. Toms is a rattlebrain; Pigott is a sentimentalist. Now, under favor, I am a sentimentalist too—therefore I say nothing, but that the interest of the excursion did not flag as I came back. Pigott and I marched along the causeway leading from Hungerford to Newbury, now observing the effect of a brilliant sun on the tawny meads or moss-colored cottages, now exulting in the fight, now digressing to some topic of general and elegant literature. My friend was dressed in character for the occasion, or like one of the FANCY; that is, with a double portion of greatcoats, clogs,[30] and overalls: and just as we had agreed with a couple of country-lads to carry his superfluous wearing-apparel to the next town, we were overtaken by a return post-chaise, into which I got, Pigott preferring a seat on the bar.[31] There were two strangers already in the chaise, and on their observing they supposed I had been to the fight, I said I had, and concluded they had done the same. They appeared, however, a little shy and sore on the subject; and it was not till after several hints dropped, and questions put, that it turned out that they had missed it. One of these friends had undertaken to drive the other there in his gig: they had set out, to make sure work, the day before at three in the afternoon. The owner of the one-horse vehicle scorned to ask his way, and drove right on to Bagshot, instead of turning off at Hounslow: there they stopped all night, and set off the next day across the country to Reading, from whence they took coach, and got down within a mile or two of Hungerford, just half an hour after the fight was over. This might be safely set down as one of the miseries of human life. We parted with these two gentlemen who had been to see the fight, but had returned as they went, at Wolhampton, where we were promised beds (an irresistible temptation, for Pigott had passed the preceding night at Hungerford as we had done at Newbury), and we turned into an old bow-windowed parlor with carpet and a snug-fire; and after devouring a quantity of tea, toast, eggs, sat down to consider, during an hour of philosophic leisure, what we should have for supper. In the midst of an Epicurean deliberation between a roasted fowl and mutton chops with mashed potatoes, we were interrupted by an inroad of Goths and Vandals

—*O procul este profani*[32]—not real flash-men,[33] but interlopers, noisy pretenders, butchers from Tothill-fields, brokers from Whitechapel, who called immediately for pipes and tobacco, hoping it would not be disagreeable to the gentlemen, and began to insist that it was *a cross*.[34] Pigott withdrew from the smoke and noise into another room, and left me to dispute the point with them for a couple of hours *sans intermission* by the dial. The next morning we rose refreshed; and on observing that Jack had a pocket volume in his hand, in which he read in the intervals of our discourse, I inquired what it was, and learned to my particular satisfaction that it was a volume of the New Eloise.[35] Ladies, after this, will you contend that a love for the FANCY is incompatible with the cultivation of sentiment?—We jogged on as before, my friend setting me up in a genteel drab greatcoat and green silk handkerchief (which I must say became me exceedingly), and after stretching our legs for a few miles, and seeing Jack Randall, Ned Turner, and Scroggins pass on the top of one of the Bath coaches, we engaged with the driver of the second to take us to London for the usual fee. I got inside, and found three other passengers. One of them was an old gentleman with an aquiline nose, powdered hair, and a pigtail, and who looked as if he had played many a rubber at the Bath rooms.[36] I said to myself, he is very like Mr. Windham; I wish he would enter into conversation, that I might hear what fine observations would come from those finely-turned features. However, nothing passed, till, stopping to dine at Reading, some inquiry was made by the company about the fight, and I gave (as the reader may believe) an eloquent and animated description of it. When we got into the coach again, the old gentleman, after a graceful exordium, said he had, when a boy, been to a fight between the famous Broughton and George Stevenson, who was called the *Fighting Coachman*, in the year 1770, with the late Mr. Windham. This beginning flattered the spirit of prophecy within me and rivetted my attention. He went on—"George Stevenson was coachman to a friend of my father's. He was an old man when I saw him some years afterwards. He took hold of his own arm and said, 'there was muscle here once, but now it is no more than this

---

29. **Sir Fopling Flutter:** in George Etherege's play, *The Man of Mode* (1676), III. ii. 30. **clogs:** shoes with wooden soles. 31. **bar:** driver's seat.

32. **O . . . profani:** "Away, O profane ones" (*Aeneid*, VI. 258). 33. **flash-men:** patrons of sport, especially boxing. 34. **a cross:** prearranged. 35. **New Eloise:** a novel by Rousseau. 36. **rubber . . . rooms:** games of cards at the resort town of Bath.

young gentleman's.' He added, 'Well, no matter; I have been here long, I am willing to go hence, and I hope I have done no more harm than another man.' Once," said my unknown companion, "I asked him if he had ever beat Broughton. He said Yes; that he had fought with him three times, and the last time he fairly beat him, though the world did not allow it. 'I'll tell you how it was, master. When the seconds lifted us up in the last round, we were so exhausted that neither of us could stand, and we fell upon one another, and as Master Broughton fell uppermost, the mob gave it in his favor, and he was said to have won the battle. But,' says he, 'the fact was, that as his second (John Cuthbert) lifted him up, he said to him, "I'll fight no more, I've had enough;" which,' says Stevenson, 'you know gave me the victory. And to prove to you that this was the case, when John Cuthbert was on his death-bed, and they asked him if there was anything on his mind which he wished to confess, he answered, "Yes, that there was one thing he wished to set right, for that certainly Master Stevenson won that last fight with Master Broughton; for he whispered him as he lifted him up in the last round of all, that he had had enough." ' This," said the Bath gentleman, "was a bit of human nature;" and I have written this account of the fight on purpose that it might not be lost to the world. He also stated as a proof of the candor of mind in this class of men, that Stevenson acknowledged that Broughton could have beat him in his best day; but that he (Broughton) was getting old in their last rencounter. When we stopped in Piccadilly, I wanted to ask the gentleman some questions about the late Mr. Windham, but had not courage. I got out, resigned my coat and green silk handkerchief to Pigott (loth to part with these ornaments of life), and walked home in high spirits.

P.S. Toms called upon me the next day, to ask me if I did not think the fight was a complete thing. I said I thought it was. I hope he will relish my account of it.

1821–22 (1822)

MY FIRST ACQUAINTANCE WITH POETS

This was first published in *The Liberal*, April, 1823. With the portrait of Coleridge here, compare Hazlitt's remarks on him above in the lecture "On the Living Poets" and in *The Spirit of the Age* below.

My father was a Dissenting Minister at W——m[1] in Shropshire; and in the year 1798 (the figures that compose that date are to me like the "dreaded name of Demogorgon"[2]) Mr. Coleridge came to Shrewsbury, to succeed Mr. Rowe in the spiritual charge of a Unitarian congregation there. He did not come till late on the Saturday afternoon before he was to preach; and Mr. Rowe, who himself went down to the coach in a state of anxiety and expectation, to look for the arrival of his successor, could find no one at all answering the description but a round-faced man in a short black coat (like a shooting-jacket) which hardly seemed to have been made for him, but who seemed to be talking at a great rate to his fellow-passengers. Mr. Rowe had scarce returned to give an account of his disappointment, when the round-faced man in black entered, and dissipated all doubts on the subject by beginning to talk. He did not cease while he staid; nor has he since, that I know of. He held the good town of Shrewsbury in delightful suspense for three weeks that he remained there, "fluttering the *proud Salopians* like an eagle in a dove-cote;"[3] and the Welsh mountains that skirt the horizon with their tempestuous confusion, agree to have heard no such mystic sounds since the days of

High-born Hoel's harp or soft Llewellyn's lay![4]

As we passed along between W——m and Shrewsbury, and I eyed their blue tops seen through the wintry branches, or the red rustling leaves of the sturdy oak-trees by the roadside, a sound was in my ears as of a Siren's song; I was stunned, startled with it, as from deep sleep; but I had no notion then that I should ever be able to express my admiration to others in motley imagery or quaint allusion, till the light of his genius shone into my soul, like the sun's rays glittering in the puddles of the road. I was at that time dumb, inarticulate, helpless, like a worm by the way-side, crushed, bleeding, lifeless; but now, bursting from the deadly bands that "bound them,

> With Styx nine times round them,"[5]

my ideas float on winged words, and as they expand their plumes, catch the golden light of other years. My soul has indeed remained in its original bondage, dark, obscure, with longings infinite and unsatisfied; my heart, shut up in the prison-house of this rude clay, has never found, nor will it ever find, a heart to speak to; but that my understanding also did not remain dumb and brutish, or at length found a language to express itself, I owe to Coleridge. But this is not to my purpose.

My father lived ten miles from Shrewsbury, and was in the habit of exchanging visits with Mr. Rowe, and with Mr. Jenkins of Whitchurch (nine miles farther on) according to the custom of Dissenting Ministers in each other's neighborhood. A line of communication is thus established, by which the flame of civil and religious liberty is kept alive, and nourishes its smouldering fire unquenchable, like the fires in the *Agamemnon* of Æschylus, placed at different stations, that waited for ten long years to announce with their blazing pyramids the destruction of Troy. Coleridge had agreed to come over and see my father, according to the courtesy of the country, as Mr. Rowe's probable successor; but in the meantime I had gone to hear him preach the Sunday after his arrival. A poet and a philosopher getting up into a Unitarian pulpit to preach the Gospel, was a romance in these degenerate days, a sort of revival of the primitive spirit of Christianity, which was not to be resisted.

It was in January, 1798, that I rose one morning before daylight, to walk ten miles in the mud, and went to hear this celebrated person preach. Never, the longest day I have to live, shall I have such another walk as this cold, raw, comfortless one, in the winter of the year 1798. *Il y a des impressions que ni le tems ni les circonstances peuvent effacer. Dusse-je vivre des siècles entiers, le doux tems de ma jeunesse ne peut renaître pour moi, ni s'effacer jamais dans ma mémoire.*[6] When I got there, the organ was playing the 100th Psalm, and, when it was done, Mr. Coleridge rose and gave out his text, "And he went up into the mountain to pray, HIMSELF, ALONE."[7] As he gave out this text, his voice "rose like a steam of rich distilled perfumes,"[8] and when he came to the two last words, which he pronounced loud, deep, and distinct, it seemed to me, who was then young, as if the sounds had echoed from the bottom of the human heart, and as if that prayer might have floated in solemn silence through the universe. The idea of St. John came into mind, "of one crying in the wilderness, who had his loins girt about, and whose food was locusts and wild honey."[9] The preacher then launched into his subject, like an eagle dallying with the wind. The sermon was upon peace and war; upon church and state—not their alliance, but their separation—on the spirit of the world and the spirit of Christianity, not as the same, but as opposed to one another. He talked of those who had "inscribed the cross of Christ on banners dripping with human gore." He made a poetical and pastoral excursion,—and to show the fatal effects of war, drew a striking contrast between the simple shepherd boy, driving his team afield, or sitting under the hawthorn, piping to his flock, "as though he should never be old," and the same poor country-lad, crimped, kidnapped, brought into town, made drunk at an ale-house, turned into a wretched drummer-boy, with his hair sticking on end with powder and pomatum, a long cue at his back, and tricked out in the loathsome finery of the profession of blood.

> Such were the notes our once-lov'd poet sung.[10]

And for myself, I could not have been more delighted if I had heard the music of the spheres. Poetry and Philosophy had met together, Truth and Genius had embraced, under the eye and with the sanction of Religion. This was even beyond my hopes. I returned home well satisfied. The sun that was still laboring pale and wan through the sky, obscured by thick mists, seemed an emblem of the *good cause;* and the

---

5. **With . . . them:** Pope's "Ode on St. Cecilia's Day," l. 90.

6. **Il y a . . . mémoire:** from Rousseau's *Confessions,* Pt. II, Bk. VII: "There are impressions that neither time nor circumstances can efface. Were I to live entire centuries, the sweet times of my youth could not return for me, nor ever be effaced from my memory." 7. **"And . . . Alone":** John 6:15. 8. **"rose . . . perfumes":** Milton's "Comus," l. 556. 9. **"of . . . honey":** Matthew 3:3-4. 10. **Such . . . sung:** Pope's "Epistle to Robert, Earl of Oxford," l. 1.

cold dank drops of dew that hung half melted on the beard of the thistle, had something genial and refreshing in them; for there was a spirit of hope and youth in all nature, that turned everything into good. The face of nature had not then the brand of Jus Divinum[11] on it:

Like to that sanguine flower inscrib'd with woe.[12]

On the Tuesday following, the half-inspired speaker came. I was called down into the room where he was, and went half-hoping, half-afraid. He received me very graciously, and I listened for a long time without uttering a word. I did not suffer in his opinion by my silence. "For those two hours," he afterwards was pleased to say, "he was conversing with W. H.'s forehead!" His appearance was different from what I had anticipated from seeing him before. At a distance, and in the dim light of the chapel, there was to me a strange wildness in his aspect, a dusky obscurity, and I thought him pitted with the small-pox. His complexion was at that time clear, and even bright—

As are the children of yon azure sheen.[13]

His forehead was broad and high, light as if built of ivory, with large projecting eyebrows, and his eyes rolling beneath them like a sea with darkened lustre. "A certain tender bloom his face o'erspread," a purple tinge as we see it in the pale thoughtful complexions of the Spanish portrait-painters, Murillo and Velasquez. His mouth was gross, voluptuous, open, eloquent; his chin good-humored and round; but his nose, the rudder of the face, the index of the will, was small, feeble, nothing—like what he has done. It might seem that the genius of his face as from a height surveyed and projected him (with sufficient capacity and huge aspiration) into the world unknown of thought and imagination, with nothing to support or guide his veering purpose, as if Columbus had launched his adventurous course for the New World in a scallop, without oars or compass. So at least I comment on it after the event. Coleridge in his person was rather above the common size, inclining to the corpulent, or like Lord Hamlet, "somewhat fat and pursy."[14] His hair (now, alas! gray) was then black and glossy

as the raven's, and fell in smooth masses over his forehead. This long pendulous hair is peculiar to enthusiasts, to those whose minds tend heavenward; and is traditionally inseparable (though of a different color) from the pictures of Christ. It ought to belong, as a character, to all who preach *Christ crucified*, and Coleridge was at that time one of those!

It was curious to observe the contrast between him and my father, who was a veteran in the cause, and then declining into the vale of years. He had been a poor Irish lad, carefully brought up by his parents, and sent to the University of Glasgow (where he studied under Adam Smith) to prepare him for his future destination. It was his mother's proudest wish to see her son a Dissenting Minister. So if we look back to past generations (as far as eye can reach) we see the same hopes, fears, wishes, followed by the same disappointments, throbbing in the human heart; and so we may see them (if we look forward) rising up forever, and disappearing, like vaporish bubbles, in the human breast! After being tossed about from congregation to congregation in the heats of the Unitarian controversy, and squabbles about the American war, he had been relegated to an obscure village, where he was to spend the last thirty years of his life, far from the only converse that he loved, the talk about disputed texts of Scripture and the cause of civil and religious liberty. Here he passed his days, repining but resigned, in the study of the Bible, and the perusal of the Commentators,—huge folios, not easily got through, one of which would outlast a winter! Why did he pore on these from morn to night (with the exception of a walk in the fields or a turn in the garden to gather brocoli-plants or kidney-beans of his own rearing, with no small degree of pride and pleasure)? Here was "no figures nor no fantasies,"—neither poetry nor philosophy—nothing to dazzle, nothing to excite modern curiosity; but to his lacklustre eyes there appeared, within the pages of the ponderous, unwieldy, neglected tomes, the sacred name of JEHOVAH in Hebrew capitals: pressed down by the weight of the style, worn to the last fading thinness of the understanding, there were glimpses, glimmering notions of the patriarchal wanderings, with palm-trees hovering on the horizon, and processions of camels at the distance of three thousand years; there was Moses with the Burning Bush, the number of the Twelve Tribes, types, shadows, glosses on the law and the prophets; there were discussions (dull enough) on the age of Methuselah, a mighty speculation! there were outlines,

---

11. **Jus Divinum:** divine law. **12. Like . . . woe:** "Lycidas," l. 106. The flower is the Greek hyacinth with markings like the Greek word for woe (AI). **13. As . . . sheen:** This and the next quote are from James Thomson's "Castle of Indolence," Pt. II, l. 295, and Pt. I. l. 507. **14. "somewhat . . . pursy":** *Hamlet*, V. ii. 298.

rude guesses at the shape of Noah's Ark and of the riches of Solomon's Temple; questions as to the date of the creation, predictions of the end of all things; the great lapses of time, the strange mutations of the globe were unfolded with the voluminous leaf, as it turned over; and though the soul might slumber with an hieroglyphic veil of inscrutable mysteries drawn over it, yet it was in a slumber ill-exchanged for all the sharpened realities of sense, wit, fancy, or reason. My father's life was comparatively a dream; but it was a dream of infinity and eternity, of death, the resurrection, and a judgment to come!

No two individuals were ever more unlike than were the host and his guest. A poet was to my father a sort of nondescript: yet whatever added grace to the Unitarian cause was to him welcome. He could hardly have been more surprised and pleased, if our visitor had worn wings. Indeed, his thoughts had wings; and as the silken sounds rustled round our little wainscoted parlor, my father threw back his spectacles over his forehead, his white hairs mixing with its sanguine hue; and a smile of delight beamed across his rugged cordial face, to think that Truth had found a new ally in Fancy![15] Besides, Coleridge seemed to take considerable notice of me, and that of itself was enough. He talked very familiarly, but agreeably, and glanced over a variety of subjects. At dinner-time he grew more animated, and dilated in a very edifying manner on Mary Wolstonecraft and Mackintosh. The last, he said, he considered (on my father's speaking of his *Vindiciæ Gallicæ* as a capital performance) as a clever scholastic man—a master of the topics,—or as the ready warehouseman of letters, who knew exactly where to lay his hand on what he wanted, though the goods were not his own. He thought him no match for Burke, either in style or matter. Burke was a metaphysician, Mackintosh a mere logician. Burke was an orator (almost a poet) who reasoned in figures, because he had an eye for nature: Mackintosh, on the other hand, was a rhetorician, who had only an eye to commonplaces. On this I ventured to say that I had always entertained a great opinion of Burke, and that (as far as I could find) the speaking of him with contempt might be made the test of a vulgar democratical mind.

This was the first observation I ever made to Coleridge, and he said it was a very just and striking one. I remember the leg of Welsh mutton and the turnips on the table that day had the finest flavor imaginable. Coleridge added that Mackintosh and Tom Wedgwood (of whom, however, he spoke highly) had expressed a very indifferent opinion of his friend Mr. Wordsworth, on which he remarked to them—"He strides on so far before you, that he dwindles in the distance!" Godwin had once boasted to him of having carried on an argument with Mackintosh for three hours with dubious success; Coleridge told him—"If there had been a man of genius in the room, he would have settled the question in five minutes." He asked me if I had ever seen Mary Wolstonecraft, and I said, I had once for a few moments, and that she seemed to me to turn off Godwin's objections to something she advanced with quite a playful, easy air. He replied, that "this was only one instance of the ascendancy which people of imagination exercised over those of mere intellect." He did not rate Godwin very high[16] (this was caprice or prejudice, real or affected) but he had a great idea of Mrs. Wolstonecraft's powers of conversation, none at all of her talent for book-making. We talked a little about Holcroft.[17] He had been asked if he was not much struck *with* him, and he said, he thought himself in more danger of being struck *by* him. I complained that he would not let me get on at all, for he required a definition of even the commonest word, exclaiming, "What do you mean by a *sensation*, Sir? What do you mean by an *idea?*" This, Coleridge said, was barricadoing the road to truth:—it was setting up a turnpike-gate at every step we took. I forget a great number of things, many more than I remember; but the day passed off pleasantly, and the next morning Mr. Coleridge was to return to Shrewsbury. When I came down to breakfast, I found that he had just received a letter from his friend, T. Wedgwood, making him an offer of £150 a year if he chose to waive his present pursuit, and devote himself entirely to the study of poetry and philosophy. Coleridge seemed to make up his mind to close with this proposal

15. **Fancy:** "My father was one of those who mistook his talent after all. He used to be very much dissatisfied that I preferred his Letters to his Sermons. The last were forced and dry; the first came naturally from him. For ease, half-plays on words, and a supine, monkish, indolent pleasantry, I have never seen them equalled" (H.)

16. **rate . . . high:** "He complained in particular of the presumption of his attempting to establish the future immortality of man, 'without' (as he said) 'knowing what Death was, or what Life was'—and the tone in which he pronounced these two words seemed to convey a complete image of both" (H.). 17. **Holcroft:** Thomas Holcroft, 1745–1809, friend of Godwin. Hazlitt edited and completed the autobiography (1816) left by him.

in the act of tying on one of his shoes. It threw an additional damp on his departure. It took the wayward enthusiast quite from us to cast him into Deva's winding vales, or by the shores of old romance. Instead of living at ten miles distance, and being the pastor of a Dissenting congregation at Shrewsbury, he was henceforth to inhabit the Hill of Parnassus, to be a Shepherd on the Delectable Mountains. Alas! I knew not the way thither, and felt very little gratitude for Mr. Wedgwood's bounty. I was pleasantly relieved from this dilemma; for Mr. Coleridge asking for a pen and ink, and going to a table to write something on a bit of card, advanced towards me with undulating step, and giving me the precious document, said that that was his address, *Mr. Coleridge, Nether-Stowey, Somersetshire;* and that he should be glad to see me there in a few weeks' time, and, if I chose, would come half-way to meet me. I was not less surprised than the shepherd-boy (this simile is to be found in *Cassandra*) when he sees a thunder-bolt fall close at his feet. I stammered out my acknowledgments and acceptance of this offer (I thought Mr. Wedgwood's annuity a trifle to it) as well as I could; and this mighty business being settled, the poet-preacher took leave, and I accompanied him six miles on the road. It was a fine morning in the middle of winter, and he talked the whole way. The scholar in Chaucer is described as going

Sounding on his way.[18]

So Coleridge went on his. In digressing, in dilating, in passing from subject to subject, he appeared to me to float in air, to slide on ice. He told me in confidence (going along) that he should have preached two sermons before he accepted the situation at Shrewsbury, one on Infant Baptism, the other on the Lord's Supper, showing that he could not administer either, which would have effectually disqualified him for the object in view. I observed that he continually crossed me on the way by shifting from one side of the foot-path to the other. This struck me as an odd movement; but I did not at that time connect it with any instability of purpose or involuntary change of principle, as I have done since. He seemed unable to keep on in a straight line. He spoke slightingly of Hume (whose Essay on Miracles he said was stolen from an objection started in one of South's Sermons—*Credat Judæus*

Apella! [19]). I was not very much pleased at this account of Hume, for I had just been reading, with infinite relish, that completest of all metaphysical *choke-pears,* his *Treatise on Human Nature,* to which the *Essays,* in point of scholastic subtlety and close reasoning, are mere elegant trifling, light summer-reading. Coleridge even denied the excellence of Hume's general style, which I think betrayed a want of taste or candor. He however made me amends by the manner in which he spoke of Berkeley. He dwelt particularly on his *Essay on Vision* as a masterpiece of analytical reasoning. So it undoubtedly is. He was exceedingly angry with Dr. Johnson for striking the stone with his foot, in allusion to this author's Theory of Matter and Spirit, and saying, "Thus I confute him, Sir." Coleridge drew a parallel (I don't know how he brought about the connection) between Bishop Berkeley and Tom Paine. He said the one was an instance of a subtle, the other of an acute, mind, than which no two things could be more distinct. The one was a shop-boy's quality, the other the characteristic of a philosopher. He considered Bishop Butler[20] as a true philosopher, a profound and conscientious thinker, a genuine reader of nature and his own mind. He did not speak of his *Analogy,* but of his *Sermons at the Rolls' Chapel,* of which I had never heard. Coleridge somehow always contrived to prefer the *unknown* to the *known.* In this instance he was right. The *Analogy* is a tissue of sophistry, of wire-drawn, theological special-pleading; the *Sermons* (with the Preface to them) are in a fine vein of deep, matured reflection, a candid appeal to our observation of human nature, without pedantry and without bias. I told Coleridge I had written a few remarks, and was sometimes foolish enough to believe that I had made a discovery on the same subject (the *Natural Disinterestedness of the Human Mind*[21])—and I tried to explain my view of it to Coleridge, who listened with great willingness, but I did not succeed in making myself understood. I sat down to the task shortly afterwards for the twentieth time, got new

---

19. South's . . . Apella: Robert South, 1634–1716, famous for his direct and homely style. The Latin tag is from Horace's *Satires,* I. v. 101: "Let the Jew Apella believe it; I shall not"—the implication being that a Jew, as contrasted with a Roman, was credulous. **20. Bishop Butler:** Bishop Joseph Butler, 1692–1752, who was to influence Hazlitt's philosophical thinking (especially the conception of "identity" developed in Hazlitt's *Principles of Human Action*). **21. Natural . . . Mind:** the premise of his *Principles of Human Action,* a selection from which is printed above.

18. Sounding . . . way: Prologue to the *Canterbury Tales,* l. 307.

pens and paper, determined to make clear work of it, wrote a few meagre sentences in the skeleton-style of a mathematical demonstration, stopped half-way down the second page; and, after trying in vain to pump up any words, images, notions, apprehensions, facts, or observations, from that gulf of abstraction in which I had plunged myself for four or five years preceding, gave up the attempt as labor in vain, and shed tears of helpless despondency on the blank unfinished paper. I can write fast enough now. Am I better than I was then? Oh no! One truth discovered, one pang of regret at not being able to express it, is better than all the fluency and flippancy in the world. Would that I could go back to what I then was! Why can we not revive past times as we can revisit old places? If I had the quaint Muse of Sir Philip Sidney to assist me, I would write a *Sonnet to the Road between W——m and Shrewsbury,* and immortalize very step of it by some fond enigmatical conceit. I would swear that the very milestones had ears, and that Harmer-hill stooped with all its pines, to listen to a poet, as he passed! I remember but one other topic of discourse in this walk. He mentioned Paley,[22] praised the naturalness and clearness of his style, but condemned his sentiments, thought him a mere time-serving casuist, and said that "the fact of his work on Moral and Political Philosophy being made a text-book in our Universities was a disgrace to the national character." We parted at the six-mile stone; and I returned homeward, pensive but much pleased. I had met with unexpected notice from a person, whom I believed to have been prejudiced against me. "Kind and affable to me had been his condescension, and should be honored ever with suitable regard."[23] He was the first poet I had known, and he certainly answered to that inspired name. I had heard a great deal of his powers of conversation, and was not disappointed. In fact, I never met with anything at all like them, either before or since. I could easily credit the accounts which were circulated of his holding forth to a large party of ladies and gentlemen, an evening or two before, on the Berkeleian Theory, when he made the whole material universe look like a transparency of fine words; and another story (which I believe he has somewhere told himself) of his being asked to a party at Birmingham, of his smoking tobacco and going to sleep after dinner on

a sofa, where the company found him, to their no small surprise, which was increased to wonder when he started up of a sudden, and rubbing his eyes, looked about him, and launched into a three-hours' description of the third heaven, of which he had had a dream, very different from Mr. Southey's Vision of Judgment and also from that other Vision of Judgment,[24] which Mr. Murray, the Secretary of the Bridge-street Junto,[25] has taken into his especial keeping!

On my way back, I had a sound in my ears—it was the voice of Fancy: I had a light before me—it was the face of Poetry. The one still lingers there, the other has not quitted my side! Coleridge in truth met me half-way on the ground of philosophy, or I should not have been won over to his imaginative creed. I had an uneasy, pleasurable sensation all the time, till I was to visit him. During those months the chill breath of winter gave me a welcoming; the vernal air was balm and inspiration to me. The golden sunsets, the silver star of evening, lighted me on my way to new hopes and prospects. *I was to visit Coleridge in the spring.* This circumstance was never absent from my thoughts, and mingled with all my feelings. I wrote to him at the time proposed, and received an answer postponing my intended visit for a week or two, but very cordially urging me to complete my promise then. This delay did not damp, but rather increased, my ardor. In the meantime I went to Llangollen Vale, by way of initiating myself in the mysteries of natural scenery; and I must say I was enchanted with it. I had been reading Coleridge's description of England, in his fine *Ode on the Departing Year,* and I applied it, *con amore,*[26] to the objects before me. That valley was to me (in a manner) the cradle of a new existence: in the river that winds through it, my spirit was baptized in the waters of Helicon!

I returned home, and soon after set out on my journey with unworn heart and untried feet. My way lay through Worcester and Gloucester, and by Upton, where I thought of Tom Jones and the adventure of the muff.[27] I remember getting completely wet through one day, and stopping at an inn (I think it was at Tewkesbury) where I sat up all night to read Paul

---

22. **Paley:** William Paley, 1743–1805. Coleridge frequently assailed his writings. 23. **"Kind . . . regard":** paraphrased from *Paradise Lost,* VIII. 648–50.

24. **other . . . Judgment:** that of Byron, below. 25. **Mr. Murray . . . Junto:** Charles Murray, lawyer for the Constitutional Association for Opposing Disloyal and Seditious Pamphlets. His office was at 6 New Bridge Street, Blackfriars. 26. **con amore:** with love. 27. **adventure . . . muff:** Bk. X, Chap. V.

and Virginia.[28] Sweet were the showers in early youth that drenched my body, and sweet the drops of pity that fell upon the books I read! I recollect a remark of Coleridge's upon this very book, that nothing could show the gross indelicacy of French manners and the entire corruption of their imagination more strongly than the behavior of the heroine in the last fatal scene, who turns away from a person on board the sinking vessel, that offers to save her life, because he has thrown off his clothes to assist him in swimming. Was this a time to think of such a circumstance? I once hinted to Wordsworth, as we were sailing in his boat on Grasmere lake, that I thought he had borrowed the idea of his *Poems on the Naming of Places* from the local inscriptions of the same kind in Paul and Virginia. He did not own the obligation, and stated some distinction without a difference, in defence of his claim to originality. Any the slightest variation would be sufficient for this purpose in his mind; for whatever *he* added or omitted would inevitably be worth all that any one else had done, and contain the marrow of the sentiment. I was still two days before the time fixed for my arrival, for I had taken care to set out early enough. I stopped these two days at Bridgewater, and when I was tired of sauntering on the banks of its muddy river, returned to the inn, and read Camilla.[29] So I have loitered my life away, reading books, looking at pictures, going to plays, hearing, thinking, writing on what pleased me best. I have wanted only one thing to make me happy; but wanting that, have wanted everything!

I arrived, and was well received. The country about Nether Stowey is beautiful, green and hilly, and near the sea-shore. I saw it but the other day, after an interval of twenty years, from a hill near Taunton. How was the map of my life spread out before me, as the map of the country lay at my feet! In the afternoon Coleridge took me over to All-Foxden, a romantic old family-mansion of the St. Aubins, where Wordsworth lived. It was then in the possession of a friend of the poet's, who gave him the free use of it.[30] Somehow that period (the time just after the French Revolution) was not a time when *nothing was given for nothing*. The mind opened, and a softness might be perceived coming over the heart of individuals, beneath "the scales that fence" our self-interest. Wordsworth himself was from home, but his sister kept house, and set before us a frugal repast; and we had free access to her brother's poems, the *Lyrical Ballads*, which were still in manuscript, or in the form of *Sybilline Leaves*. I dipped into a few of these with great satisfaction, and with the faith of a novice. I slept that night in an old room with blue hangings, and covered with the round-faced family-portraits of the age of George I and II and from the wooded declivity of the adjoining park that overlooked my window, at the dawn of day, could

Hear the loud stag speak.[31]

In the outset of life (and particularly at this time I felt it so) our imagination has a body to it. We are in a state between sleeping and waking, and have indistinct but glorious glimpses of strange shapes, and there is always something to come better than what we see. As in our dreams the fulness of the blood gives warmth and reality to the coinage of the brain, so in youth our ideas are clothed, and fed, and pampered with our good spirits; we breathe thick with thoughtless happiness, the weight of future years presses on the strong pulses of the heart, and we repose with undisturbed faith in truth and good. As we advance, we exhaust our fund of enjoyment and of hope. We are no longer wrapped in *lamb's-wool*, lulled in Elysium. As we taste the pleasures of life, their spirit evaporates, the sense palls; and nothing is left but the phantoms, the lifeless shadows of what *has been!*

That morning, as soon as breakfast was over, we strolled out into the park, and seating ourselves on the trunk of an old ash-tree that stretched along the ground, Coleridge read aloud with a sonorous and musical voice *The Ballad of Betty Foy*. I was not critically or skeptically inclined. I saw touches of truth and nature, and took the rest for granted. But in *The Thorn*, the *Mad Mother*,[32] and the *Complaint of a Poor Indian Woman*, I felt that deeper power and pathos which have been since acknowledged,

In spite of pride, in erring reason's spite,[33]

as the characteristics of this author; and the sense of a new style and a new spirit in poetry came over me. It had to me something of the effect that arises from the

---

28. **Paul and Virginia:** a novel by Jacques Henri Bernardin de Saint-Pierre, 1737–1814. 29. **Camilla:** *or, A Picture of Youth*, a novel by Frances (Burney) S'Arblay, 1752–1840. 30. **free . . . it:** Not at all. He rented the place for £23 a year, which was only slightly under the rent usually paid for a house of this sort in the country.

31. **Hear . . . speak:** Ben Jonson's "To Sir Robert Wroth," l. 22. 32. **Mad Mother:** later called "Her Eyes Are Wild." 33. **In . . . spite:** Pope's *Essay on Man*, I. 293.

turning up of the fresh soil, or of the first welcome breath of spring:

While yet the trembling year is unconfirmed.[34]

Coleridge and myself walked back to Stowey that evening, and his voice sounded high

Of Providence, foreknowledge, will, and fate,
Fix'd fate, free-will, foreknowledge absolute,[35]

as we passed through echoing grove, by fairy stream or waterfall, gleaming in the summer moonlight! He lamented that Wordsworth was not prone enough to believe in the traditional superstitions of the place, and that there was a something corporeal, a *matter-of-fact-ness*, a clinging to the palpable, or often to the petty, in his poetry, in consequence. His genius was not a spirit that descended to him through the air; it sprung out of the ground like a flower, or unfolded itself from a green spray, on which the gold-finch sang. He said, however (if I remember right) that this objection must be confined to his descriptive pieces, that his philosophic poetry had a grand and comprehensive spirit in it, so that his soul seemed to inhabit the universe like a palace, and to discover truth by intuition, rather than by deduction. The next day Wordsworth arrived from Bristol at Coleridge's cottage. I think I see him now. He answered in some degree to his friend's description of him, but was more gaunt and Don Quixote-like. He was quaintly dressed (according to the *costume* of that unconstrained period) in a brown fustian jacket and striped pantaloons. There was something of a roll, a lounge, in his gait, not unlike his own Peter Bell. There was a severe, worn pressure of thought about his temples, a fire in his eye (as if he saw something in objects more than the outward appearance), an intense high narrow forehead, a Roman nose, cheeks furrowed by strong purpose and feeling, and a convulsive inclination to laughter about the mouth, a good deal at variance with the solemn, stately expression of the rest of his face. Chantry's bust wants the marking traits; but he was teased into making it regular and heavy: Haydon's head of him, introduced into *The Entrance of Christ into Jerusalem*, is the most like his drooping weight of thought and expression. He sat down and talked very naturally and freely, with a mixture of clear, gushing accents in his voice, a deep

gutteral intonation, and a strong tincture of the northern *burr*, like the crust on wine. He instantly began to make havoc of the half of a Cheshire cheese on the table, and said triumphantly that "his marriage with experience had not been so productive as Mr. Southey's in teaching him a knowledge of the good things of life." He had been to see the *Castle Spectre*, by Monk Lewis, while at Bristol, and described it very well. He said "it fitted the taste of the audience like a glove." This *ad captandum*[36] merit was however by no means a recommendation of it, according to the severe principles of the new school, which reject rather than court popular effect. Wordsworth, looking out of the low, latticed window, said, "How beautifully the sun sets on that yellow bank!" I thought within myself, "With what eyes these poets see nature!" and ever after, when I saw the sunset stream upon the objects facing it, conceived I had made a discovery, or thanked Mr. Wordsworth for having made one for me! We went over to All-Foxden again the day following, and Wordsworth read us the story of Peter Bell in the open air; and the comment made upon it by his face and voice was very different from that of some later critics! Whatever might be thought of the poem, "his face was as a book where men might read strange matters,"[37] and he announced the fate of his hero in prophetic tones. There is a *chaunt* in the recitation both of Coleridge and Wordsworth, which acts as a spell upon the hearer, and disarms the judgment. Perhaps they have deceived themselves by making habitual use of this ambiguous accompaniment. Coleridge's manner is more full, animated, and varied; Wordsworth's more equable, sustained, and internal. The one might be termed more *dramatic*, the other more *lyrical*. Coleridge has told me that he himself liked to compose in walking over uneven ground, or breaking through the straggling branches of a copsewood; whereas Wordsworth always wrote (if he could) walking up and down a straight gravel-walk, or in some spot where the continuity of his verse met with no collateral interruption. Returning that same evening, I got into a metaphysical argument with Wordsworth, while Coleridge was explaining the different notes of the nightingale to his sister, in which we neither of us succeeded in making ourselves perfectly clear and intelligible. Thus I passed three weeks at Nether Stowey and in the neighborhood, generally devoting the afternoons to

---

34. **While . . . unconfirmed:** James Thomson's *The Seasons*, "Spring," l. 293. 35. **Of . . . absolute:** *Paradise Lost*, II. 559–60.

36. **ad captandum:** for capturing applause. 37. **"his . . . matters":** *Macbeth*, I. v. 63.

a delightful chat in an arbor made of bark by the poet's friend Tom Poole, sitting under two fine elm-trees, and listening to the bees humming round us, while we quaffed our *flip*. It was agreed, among other things, that we should make a jaunt down the Bristol-Channel, as far as Linton. We set off together on foot, Coleridge, John Chester, and I. This Chester was a native of Nether Stowey, one of those who were attracted to Coleridge's discourse as flies are to honey, or bees in swarming-time to the sound of a brass pan. He "followed in the chase like a dog who hunts, not like one that made up the cry."[38] He had on a brown cloth coat, boots, and corduroy breeches, was low in stature, bow-legged, had a drag in his walk like a drover, which he assisted by a hazel switch, and kept on a sort of trot by the side of Coleridge, like a running footman by a state coach, that he might not lose a syllable or sound, that fell from Coleridge's lips. He told me his private opinion, that Coleridge was a wonderful man. He scarcely opened his lips, much less offered an opinion the whole way: yet of the three, had I to choose during that journey, I would be John Chester. He afterwards followed Coleridge into Germany, where the Kantean philosophers were puzzled how to bring him under any of their categories. When he sat down at table with his idol, John's felicity was complete; Sir Walter Scott's or Mr. Blackwood's, when they sat down at the same table with the King, was not more so. We passed Dunster on our right, a small town between the brow of a hill and the sea. I remember eying it wistfully as it lay below us: contrasted with the woody scene around, it looked as clear, as pure, as *embrowned* and ideal as any landscape I have seen since, of Gasper Poussin's or Domenichino's. We had a long day's march—(our feet kept time to the echoes of Coleridge's tongue)—through Minehead and by the Blue Anchor, and on to Linton, which we did not reach till near midnight, and where we had some difficulty in making a lodgment. We however knocked the people of the house up at last, and we were repaid for our apprehensions and fatigue by some excellent rashers of fried bacon and eggs. The view in coming along had been splendid.We walked for miles and miles on dark brown heaths overlooking the channel, with the Welsh hills beyond, and at times descended into little sheltered valleys close by the sea-side, with a smuggler's face scowling by us, and then had to ascend conical hills with a path winding up through a coppice to a barren

top, like a monk's shaven crown, from one of which I pointed out to Coleridge's notice the bare masts of a vessel on the very edge of the horizon and within the red-orbed disk of the setting sun, like his own spectre-ship in the *Ancient Mariner*. At Linton the character of the sea-coast becomes more marked and rugged. There is a place called the *Valley of Rocks* (I suspect this was only the poetical name for it) bedded among precipices overhanging the sea, with rocky caverns beneath, into which the waves dash, and where the sea-gull forever wheels its screaming flight. On the tops of these are huge stones thrown transverse, as if an earthquake had tossed them there, and behind these is a fretwork of perpendicular rocks, something like the *Giant's Causeway*. A thunder-storm came on while we were at the inn, and Coleridge was running out bare-headed to enjoy the commotion of the elements in the *Valley of Rocks*, but as if in spite, the clouds only muttered a few angry sounds, and let fall a few refreshing drops. Coleridge told me that he and Wordsworth were to have made this place the scene of a prose-tale, which was to have been in the manner of, but far superior to, the *Death of Abel*, but they had relinquished the design. In the morning of the second day, we breakfasted luxuriously in an old-fashioned parlor, on tea, toast, eggs, and honey, in the very sight of the bee-hives from which it had been taken, and a garden full of thyme and wild flowers that had produced it. On this occasion Coleridge spoke of Virgil's Georgics, but not well. I do not think he had much feeling for the classical or elegant. It was in this room that we found a little worn-out copy of the *Seasons*, lying in a window-seat, on which Coleridge exclaimed, "*That* is true fame!" He said Thomson was a great poet, rather than a good one; his style was as meretricious as his thoughts were natural. He spoke of Cowper as the best modern poet. He said the *Lyrical Ballads* were an experiment about to be tried by him and Wordsworth, to see how far the public taste would endure poetry written in a more natural and simple style than had hitherto been attempted; totally discarding the artifices of poetical diction, and making use only of such words as had probably been common, in the most ordinary language since the days of Henry II. Some comparison was introduced between Shakespear and Milton. He said "he hardly knew which to prefer. Shakespear appeared to him a mere stripling in the art; he was as tall and as strong, with infinitely more activity than Milton, but he never appeared to have come to man's estate; or if he had, he would not have been a man,

---

38. "followed . . . cry": *Othello*, II. iii. 370.

but a monster." He spoke with contempt of Gray, and with intolerance of Pope. He did not like the versification of the latter. He observed that "the ears of these couplet-writers might be charged with having short memories, that could not retain the harmony of whole passages." He thought little of Junius as a writer; he had a dislike of Dr. Johnson; and a much higher opinion of Burke as an orator and politician, than of Fox or Pitt. He however thought him very inferior in richness of style and imagery to some of our elder prose-writers, particularly Jeremy Taylor. He liked Richardson, but not Fielding; nor could I get him to enter into the merits of *Caleb Williams*.[39] In short, he was profound and discriminating with respect to those authors whom he liked, and where he gave his judgment fair play; capricious, perverse, and prejudiced in his antipathies and distastes. We loitered on the "ribbed sea-sands,"[40] in such talk as this, a whole morning, and I recollect met with a curious sea-weed, of which John Chester told us the country name! A fisherman gave Coleridge an account of a boy that had been drowned the day before, and that they had tried to save him at the risk of their own lives. He said "he did not know how it was that they ventured, but, sir, we have a *nature* towards one another." This expression, Coleridge remarked to me, was a fine illustration of that theory of disinterestedness which I (in common with Butler[41]) had adopted. I broached to him an argument of mine to prove that *likeness* was not mere association of ideas. I said that the mark in the sand put one in mind of a man's foot, not because it was part of a former impression of a man's foot (for it was quite new) but because it was like the shape of a man's foot. He assented to the justness of this distinction (which I have explained at length elsewhere, for the benefit of the curious), and John Chester listened; not from any interest in the subject, but because he was astonished that I should be able to suggest anything to Coleridge that he did not already know. We returned

on the third morning, and Coleridge remarked the silent cottage-smoke curling up the valleys where, a few evenings before, we had seen the lights gleaming through the dark.

In a day or two after we arrived at Stowey, we set out, I on my return home, and he for Germany. It was a Sunday morning, and he was to preach that day for Dr. Toulmin of Taunton. I asked him if he had prepared anything for the occasion? He said he had not even thought of the text, but should as soon as we parted. I did not go to hear him,—this was a fault,—but we met in the evening at Bridgewater. The next day we had a long day's walk to Bristol, and sat down, I recollect, by a well-side on the road, to cool ourselves and satisfy our thirst, when Coleridge repeated to me some descriptive lines of his tragedy of *Remorse*, which I must say became his mouth and that occasion better than they, some years after, did Mr. Elliston's and the Drury-lane boards,—

Oh memory! shield me from the world's poor strife,
And give those scenes thine everlasting life.

I saw no more of him for a year or two, during which period he had been wandering in the Hartz Forest in Germany; and his return was cometary, meteorous, unlike his setting out. It was not till some time after that I knew his friends Lamb and Southey. The last always appears to me (as I first saw him) with a common-place-book under his arm, and the first with a *bon-mot* in his mouth. It was at Godwin's that I met him with Holcroft and Coleridge, where they were disputing fiercely which was the best—*Man as he was, or man as he is to be.* "Give me," says Lamb, "man as he is *not* to be." This saying was the beginning of a friendship between us, which I believe still continues. —Enough of this for the present.

But there is matter for another rhyme,
And I to this may add a second tale.[42]

1823 (1823)

---

39. **Caleb Williams:** At this point Hazlitt's note adds, "He had no idea of pictures, of Claude or Raphael, and at this time I had as little as he. He sometimes gives a striking account at present of the Cartoons at Pisa, by Buffamalco and others; of one in particular where Death is seen in the air brandishing his scythe, and the great and mighty of the earth shudder at his approach, while the beggars and the wretched kneel to him as their deliverer. He would of course understand so broad and fine a moral as this at any time." 40. **"ribbed sea-sands":** "Rime of the Ancient Mariner," l. 227. 41. **Butler:** See n. 20, above.

42. **But . . . tale:** Wordsworth's "Hart-Leap Well," ll. 95-96.

FROM

THE SPIRIT OF THE AGE

⤳⤳⤳

Written in 1824 and published in 1825, *The Spirit of the Age* is a collection of literary portraits of contemporary figures. The essay on Byron was written just before Lord Byron's death.

⤳⤳⤳

MR. COLERIDGE

The present is an age of talkers, and not of doers; and the reason is, that the world is growing old. We are so far advanced in the Arts and Sciences, that we live in retrospect, and doat on past achievements. The accumulation of knowledge has been so great, that we are lost in wonder at the height it has reached, instead of attempting to climb or add to it; while the variety of objects distracts and dazzles the looker-on. What *niche* remains unoccupied? What path untried? What is the use of doing anything, unless we could do better than all those who have gone before us? What hope is there of this? We are like those who have been to see some noble monument of art, who are content to admire without thinking of rivalling it; or like guests after a feast, who praise the hospitality of the donor "and thank the bounteous Pan"[1]— perhaps carrying away some trifling fragments; or like the spectators of a mighty battle, who still hear its sound afar off, and the clashing of armour and the neighing of the war-horse and the shout of victory is in their ears, like the rushing of innumerable waters!

Mr. Coleridge has "a mind reflecting ages past"[2]; his voice is like the echo of the congregated roar of the "dark rearward and abyss" of thought. He who has seen a mouldering tower by the side of a crystal lake, hid by the mist, but glittering in the wave below, may conceive the dim, gleaming, uncertain intelligence of his eye: he who has marked the evening clouds uprolled (a world of vapours), has seen the picture of his mind, unearthly, unsubstantial, with gorgeous tints and ever-varying forms—

> That which was now a horse, even with a thought
> The rack dislimns, and makes it indistinct
> As water is in water.[3]

Our author's mind is (as he himself might express it) *tangential*. There is no subject on which he has not touched, none on which he has rested. With an understanding fertile, subtle, expansive, "quick, forgetive, apprehensive," beyond all living precedent, few traces of it will perhaps remain. He lends himself to all impressions alike; he gives up his mind and liberty of thought to none. He is a general lover of art and science, and wedded to no one in particular. He pursues knowledge as a mistress, with outstretched hands and winged speed; but as he is about to embrace her, his Daphne turns—alas! not to a laurel! Hardly a speculation has been left on record from the earliest time, but it is loosely folded up in Mr. Coleridge's memory, like a rich, but somewhat tattered piece of tapestry: we might add (with more seeming than real extravagance), that scarce a thought can pass through the mind of man, but its sound has at some time or other passed over his head with rustling pinions. On whatever question or author you speak, he is prepared to take up the theme with advantage—from Peter Abelard down to Thomas Moore, from the subtlest metaphysics to the politics of the *Courier*. There is no man of genius, in whose praise he descants, but the critic seems to stand above the author, and "what in him is weak, to strengthen, what is low, to raise and support": nor is there any work of genius that does not come out of his hands like an illuminated Missal, sparkling even in its defects. If Mr. Coleridge had not been the most impressive talker of his age, he would probably have been the finest writer; but he lays down his pen to make sure of an auditor, and mortgages the admiration of posterity for the stare of an idler. If he had not been a poet, he would have been a powerful logician; if he had not dipped his wing in the Unitarian controversy, he might have soared to the very summit of fancy. But in writing verse, he is trying to subject the Muse to *transcendental* theories: in his abstract reasoning, he misses his way by strewing it with flowers. All that he has done of moment, he had done twenty years ago: since then, he may be said to have lived on the sound of his own voice. Mr. Coleridge is too rich in

---

THE SPIRIT OF THE AGE. MR. COLERIDGE. **1. "and . . . Pan"**: paraphrased from Milton's "Comus," l. 176. **2. "a . . . past"**: from the poem on Shakespeare (signed "J. M. S.") in the second folio (1632).

3. That . . . water: *Antony and Cleopatra*, IV. xiv. 9–11.

intellectual wealth, to need to task himself to any drudgery: he has only to draw the sliders of his imagination, and a thousand subjects expand before him, startling him with their brilliancy, or losing themselves in endless obscurity—

"And by the force of blear illusion,
They draw him on to his confusion." [4]

What is the little he could add to the stock, compared with the countless stores that lie about him, that he should stoop to pick up a name, or to polish an idle fancy? He walks abroad in the majesty of an universal understanding, eyeing the "rich strond," or golden sky above him, and "goes sounding on his way," in eloquent accents, uncompelled and free!

Persons of the greatest capacity are often those, who for this reason do the least; for surveying themselves from the highest point of view, amidst the infinite variety of the universe, their own share in it seems trifling, and scarce worth a thought, and they prefer the contemplation of all that is, or has been, or can be, to the making a coil about doing what, when done, is no better than vanity. It is hard to concentrate all our attention and efforts on one pursuit, except from ignorance of others; and without this concentration of our faculties, no great progress can be made in any one thing. It is not merely that the mind is not capable of the effort; it does not think the effort worth making. Action is one; but thought is manifold. He whose restless eye glances through the wide compass of nature and art, will not consent to have "his own nothings monstered" [5]: but he must do this, before he can give his whole soul to them. The mind, after "letting contemplation have its fill," or

"Sailing with supreme dominion
Through the azure deep of air," [6]

sinks down on the ground, breathless, exhausted, powerless, inactive; or if it must have some vent to its feelings, seeks the most easy and obvious; is soothed by friendly flattery, lulled by the murmur of immediate applause, thinks as it were aloud, and babbles in its dreams! A scholar (so to speak) is a more disinterested and abstracted character than a mere author. The first looks at the numberless volumes of a library, and says, "All these are mine": the other points to a single

volume (perhaps it may be an immortal one) and says, "My name is written on the back of it." This is a puny and groveling ambition, beneath the lofty amplitude of Mr. Coleridge's mind. No, he revolves in his wayward soul, or utters to the passing wind, or discourses to his own shadow, things mightier and more various! —Let us draw the curtain, and unlock the shrine.

Learning rocked him in his cradle, and while yet a child,

"He lisped in numbers, for the numbers came." [7]

At sixteen [8] he wrote his *Ode on Chatterton*, and he still reverts to that period with delight, not so much as it relates to himself (for that string of his own early promise of fame rather jars than otherwise) but as exemplifying the youth of a poet. Mr. Coleridge talks of himself, without being an egotist, for in him the individual is always merged in the abstract and general. He distinguished himself at school and at the University by his knowledge of the classics, and gained several prizes for Greek epigrams. How many men are there (great scholars, celebrated names in literature) who having done the same thing in their youth, have no other idea all the rest of their lives but of this achievement, of a fellowship and dinner, and who, installed in academic honours, would look down on our author as a mere strolling bard! At Christ's Hospital, where he was brought up, he was the idol of those among his school-fellows, who mingled with their bookish studies the music of thought and of humanity; and he was usually attended round the cloisters by a group of these (inspiring and inspired) whose hearts, even then, burnt within them as he talked, and where the sounds yet linger to mock ELIA [9] on his way, still turning pensive to the past! One of the finest and rarest parts of Mr. Coleridge's conversation, is when he expatiates on the Greek tragedians (not that he is not well acquainted, when he pleases, with the epic poets, or the philosophers, or orators, or historians of antiquity) —on the subtle reasonings and melting pathos of Euripides, on the harmonious gracefulness of Sophocles, tuning his love-laboured song, like sweetest warblings from a sacred grove; on the high-wrought trumpet-tongued eloquence of Æschylus, whose Prometheus, above all, is like an Ode to Fate, and a pleading with

---

4. "And . . . confusion": *Macbeth*, III. v. 28–29. 5. "his . . . monstered": *Coriolanus*, III. ii. 81. 6. "Sailing . . . air": Gray's "Progress of Poesy," ll. 116–17.

7. "He . . . came": Pope's "Epistle to Dr. Arbuthnot," l. 128. 8. sixteen: actually at the age of seventeen ("Monody on the Death of Chatterton"). 9. Elia: a reference to Charles Lamb's "Christ's Hospital Five and Thirty Years Ago."

Providence, his thoughts being let loose as his body is chained on his solitary rock, and his afflicted will (the emblem of mortality)

"Struggling in vain with ruthless destiny."[10]

As the impassioned critic speaks and rises in his theme, you would think you heard the voice of the Man hated by the Gods, contending with the wild winds as they roar, and his eye glitters with the spirit of Antiquity!

Next, he was engaged with Hartley's tribes of mind,[11] "etherial braid, thought-woven,"—and he busied himself for a year or two with vibrations and vibratiuncles and the great law of association that binds all things in its mystic chain, and the doctrine of Necessity (the mild teacher of Charity) and the Millennium, anticipative of a life to come—and he plunged deep into the controversy on Matter and Spirit, and, as an escape from Dr. Priestley's[12] Materialism, where he felt himself imprisoned by the logician's spell, like Ariel in the cloven pine-tree, he became suddenly enamoured of Bishop Berkeley's[13] fairy-world,[14] and used in all companies to build the universe, like a brave poetical fiction, of fine words—and he was deep-read in Malebranche, and in Cudworth's Intellectual System (a huge pile of learning, unwieldy, enormous) and in Lord Brook's hieroglyphic theories, and in Bishop Butler's Sermons, and in the Duchess of Newcastle's fantastic folios, and in Clarke and South and Tillotson,[15]

and all the fine thinkers and masculine reasoners of that age—and Leibnitz's[16] *Pre-Established Harmony* reared its arch above his head, like the rainbow in the cloud, covenanting with the hopes of man—and then he fell plump, ten thousand fathoms down (but his wings saved him harmless) into the *hortus siccus*[17] of Dissent, where he pared religion down to the standard of reason and stripped faith of mystery, and preached Christ crucified and the Unity of the Godhead, and so dwelt for a while in the spirit with John Huss and Jerome of Prague and Socinus and old John Zisca, and ran through Neal's History of the Puritans, and Calamy's Non-Conformists' Memorial, having like thoughts and passions with them—but then Spinoza[18] became his God, and he took up the vast chain of being in his hand, and the round world became the centre and the soul of all things in some shadowy sense, forlorn of meaning, and around him he beheld the living traces and the sky-pointing proportions of the mighty Pan—but poetry redeemed him from this spectral philosophy, and he bathed his heart in beauty, and gazed at the golden light of heaven, and drank of the spirit of the universe, and wandered at eve by fairy-stream or fountain,

"—— When he saw nought but beauty,
When he heard the voice of that Almighty One
In every breeze that blew, or wave that murmured"—[19]

and wedded with truth in Plato's shade, and in the writings of Proclus and Plotinus saw the ideas of things

10. **"Struggling . . . destiny":** Wordsworth's *Excursion*, VI. 557. 11. **Hartley's . . . mind:** David Hartley, whose *Observations on Man* (1749) derived all mind from the mechanical association of ideas, altruism being the final stage of the process. 12. **Dr. Priestley:** Joseph Priestley, 1733–1804. A follower of Hartley and a noted scientist (he was the discoverer of oxygen), Priestley was Hazlitt's tutor at the Unitarian College in Hackney. 13. **Bishop Berkeley:** the philosopher George Berkeley, 1685–1753. 14. **fairy-world:** Hazlitt adds the following note: "Mr. Coleridge named his eldest son (the writer of some beautiful Sonnets) after Hartley, and the second after Berkeley. The third was called Derwent, after the river of that name. Nothing can be more characteristic of his mind than this circumstance. All his ideas are indeed like a river, flowing on for ever, and still murmuring as it flows, discharging its waters and still replenished—'And so by many winding nooks it strays, / With willing sport to the wild ocean!'" The quotation is from *The Two Gentlemen of Verona*, II. vii. 31–32. 15. **Malebranche . . . Tillotson:** seventeenth-century philosophical and religious writers: Nicholas Malebranche, who wrote *De la Recherche de la Vérité* (1674); the Cambridge Platonist, Ralph Cudworth, of whose *True*

Intellectual System of the Universe (1678) Hazlitt is thinking; Fulke Greville, Lord Brooke, 1554–1628; Bishop Joseph Butler (see n. 20 to "My First Acquaintance with Poets," above); Margaret Cavendish, Duchess of Newcastle, 1624–74; and three noted Restoration writers of sermons, Samuel Clarke, Robert South, and John Tillotson. 16. **Leibnitz:** the German philosopher Gottfried Wilhelm Leibnitz, 1646–1716, who conceived the universe as consisting of independent "monads" kept in relation by a pre-established harmony. 17. **hortus siccus:** the "dry garden" of religious dissent from the Church of England. 18. **Huss . . . Spinoza:** John Huss, 1373–1415, the noted religious reformer of Bohemia, and his follower, Jerome of Prague; Socinus, the name taken by Lelio and Fausto Sozzini, sixteenth-century Italian theologians who viewed Christ not as a part of the Godhead but as an inspired prophet of God's word; John Zisca, 1360–1424, leader of the Hussites (followers of John Huss); Daniel Neal, author of the *History of the Puritans* (1732–38); Edmund Calamy, 1671–1732, Nonconformist writer; and the great philosopher of pantheism, Baruch Spinoza, 1632–77, who conceived all things as manifestations of one infinite substance, God. 19. **"When . . . murmured":** Coleridge's *Remorse*, IV. ii. 100–02.

in the eternal mind, and unfolded all mysteries with the Schoolmen and fathomed the depths of Duns Scotus and Thomas Aquinas, and entered the third heaven with Jacob Behmen, and walked hand in hand with Swedenborg[20] through the pavilions of the New Jerusalem, and sung his faith in the promise and in the word in his *Religious Musings*—and lowering himself from that dizzy height,[21] poised himself on Milton's wings, and spread out his thoughts in charity with the glad prose of Jeremy Taylor, and wept over Bowles's[22] Sonnets, and studied Cowper's blank verse, and betook himself to Thomson's Castle of Indolence, and sported with the wits of Charles the Second's days and of Queen Anne, and relished Swift's style and that of the John Bull (Arbuthnot's we mean, not Mr. Croker's), and dallied with the British Essayists and Novelists, and knew all qualities of more modern writers with a learned spirit, Johnson, and Goldsmith, and Junius, and Burke, and Godwin, and the Sorrows of Werther, and Jean Jacques Rousseau, and Voltaire, and Marivaux, and Crebillon,[23] and thousands more—now "laughed with Rabelais in his easy chair" or pointed to Hogarth, or afterwards dwelt on Claude's classic scenes, or spoke with rapture of Raphael, and compared the women at Rome to figures that had walked out of his pictures, or visited the Oratory of Pisa, and described the works of Giotto and Ghirlandaio and Massaccio, and gave the moral of the picture of the Triumph of Death, where the beggars and the wretched invoke his dreadful dart, but the rich and mighty of the earth quail and shrink before it; and in that land of siren sights and sounds, saw a dance of peasant girls, and was charmed with lutes and gondolas,—or wandered into Germany and lost himself in the labyrinths of the Hartz Forest and of the Kantean philosophy, and amongst the cabalistic names of Fichte and Schelling and Lessing,[24] and God knows who—this was long after, but all the former while, he had nerved his heart and filled his eyes with tears, as he hailed the rising orb of liberty, since quenched in darkness and in blood, and had kindled his affections at the blaze of the French Revolution, and sang for joy when the towers of the Bastile and the proud places of the insolent and the oppressor fell, and would have floated his bark, freighted with fondest fancies, across the Atlantic wave with Southey[25] and others to seek for peace and freedom—

"In Philarmonia's undivided dale!"[26]

Alas! "Frailty, thy name is *Genius!*"—What is become of all this mighty heap of hope, of thought, of learning, and humanity? It has ended in swallowing doses of oblivion and in writing paragraphs in the *Courier.*—Such, and so little is the mind of man!

It was not to be supposed that Mr. Coleridge could keep on at the rate he set off; he could not realize all he knew or thought, and less could not fix his desultory ambition; other stimulants supplied the place, and kept up the intoxicating dream, the fever and the madness of his early impressions. Liberty (the philosopher's and the poet's bride) had fallen a victim, meanwhile, to the murderous practices of the hag, Legitimacy. Proscribed by court-hirelings, too romantic for the herd of vulgar politicians, our enthusiast stood at bay, and at last turned on the pivot of a subtle casuistry to the *unclean side:* but his discursive reason would not let

20. **Proclus . . . Swedenborg:** Proclus, 410–85, and Plotinus, 204–70, were Neoplatonists who influenced Coleridge profoundly—especially Plotinus; Scotus, 1265?–1308, and Aquinas, 1227–74, were two great medieval theologians; Behmen is the common English spelling of Jakob Böhme, 1575–1624, the German Neoplatonist; and Emmanuel Swedenborg, 1688–1772, was a Swedish scientist, philosopher, and founder of theosophy (God is infinite love and wisdom, and the end of creation is the approximation of man to God). 21. **lowering . . . height:** With this portion of Hazlitt's magnificent sentence, compare his remark eight years earlier (1817), in his review of Coleridge's *Biographia Literaria:* "Mr. Coleridge has ever since, from the combined forces of poetic levity and metaphysic bathos, been trying to fly, not in the air, but under ground—playing at hawk and buzzard between sense and nonsense,—floating or sinking in fine Kantean categories, in a state of suspended animation 'twixt dreaming and awake,—quitting the plain ground of 'history and particular facts' for the first butterfly theory, fancy-bred from the maggots of his brain,—going up in an air-balloon filled with fetid gas from the writings of Jacob Behmen and the mystics, and coming down in a parachute made of the soiled and fashionable leaves of the Morning Post,—promising us an account of the Intellectual System of the Universe, and putting us off with a reference to a promised dissertation on the Logos, introductory to an intended commentary on the entire Gospel of St. John." 22. **Bowles:** William Lisle Bowles, 1762–1850. 23. **Marivaux . . . Crebillon:** Pierre Carlet de Marivaux, 1688–1763, and P. J. de Crébillon, 1674–1762, French dramatists.

24. **Fichte . . . Lessing:** J. G. Fichte, 1762–1814, and Friedrich W. J. Schelling, 1775–1854, German idealistic philosophers who followed Kant; and Gotthold Ephraim Lessing, 1729–81, the German critic and playwright. 25. **across . . . Southey:** their scheme of "pantisocracy" (see Introduction to Coleridge, above). 26. **"In . . . dale!":** paraphrased from Coleridge's "Monody on the Death of Chatterton," l. 140.

him trammel himself into a poet-laureate or stamp-distributor, and he stopped, ere he had quite passed that well-known "bourne from whence no traveller returns"—and so has sunk into torpid, uneasy repose, tantalized by useless resources, haunted by vain imaginings, his lips idly moving, but his heart for ever still, or, as the shattered chords vibrate of themselves, making melancholy music to the ear of memory! Such is the fate of genius in an age, when in the unequal contest with sovereign wrong, every man is ground to powder who is not either a born slave, or who does not willingly and at once offer up the yearnings of humanity and the dictates of reason as a welcome sacrifice to besotted prejudice and loathsome power.

Of all Mr. Coleridge's productions, the *Ancient Mariner* is the only one that we could with confidence put into any person's hands, on whom we wished to impress a favourable idea of his extraordinary powers. Let whatever other objections be made to it, it is unquestionably a work of genius—of wild, irregular, overwhelming imagination, and has that rich, varied movement in the verse, which gives a distant idea of the lofty or changeful tones of Mr. Coleridge's voice. In the *Christabel*, there is one splendid passage on divided friendship.[27] The *Translation of Schiller's Wallenstein* is also a masterly production in its kind, faithful and spirited. Among his smaller pieces there are occasional bursts of pathos and fancy, equal to what we might expect from him; but these form the exception, and not the rule. Such, for instance, is his affecting Sonnet to the author of the Robbers. . . .[28]

His Tragedy, entitled *Remorse*, is full of beautiful and striking passages, but it does not place the author in the first rank of dramatic writers. But if Mr. Coleridge's works do not place him in that rank, they injure instead of conveying a just idea of the man, for he himself is certainly in the first class of general intellect.

If our author's poetry is inferior to his conversation, his prose is utterly abortive. Hardly a gleam is to be found in it of the brilliancy and richness of those stores of thought and language that he pours out incessantly, when they are lost like drops of water in the ground. The principal work, in which he has attempted to embody his general views of things, is the FRIEND, of which, though it contains some noble passages and fine trains of thought, prolixity and obscurity are the most frequent characteristics.

No two persons can be conceived more opposite in character or genius than the subject of the present and of the preceding sketch. Mr. Godwin, with less natural capacity, and with fewer acquired advantages, by concentrating his mind on some given object, and doing what he had to do with all his might, has accomplished much, and will leave more than one monument of a powerful intellect behind him; Mr. Coleridge, by dissipating his, and dallying with every subject by turns, has done little or nothing to justify to the world or to posterity, the high opinion which all who have ever heard him converse, or known him intimately, with one accord entertain of him. Mr. Godwin's faculties have kept at home, and plied their task in the workshop of the brain, diligently and effectually: Mr. Coleridge's have gossiped away their time, and gadded about from house to house, as if life's business were to melt the hours in listless talk. Mr. Godwin is intent on a subject, only as it concerns himself and his reputation; he works it out as a matter of duty, and discards from his mind whatever does not forward his main object as impertinent and vain. Mr. Coleridge, on the other hand, delights in nothing but episodes and digressions, neglects whatever he undertakes to perform, and can act only on spontaneous impulses, without object or method. "He cannot be constrained by mastery." While he should be occupied with a given pursuit, he is thinking of a thousand other things; a thousand tastes, a thousand objects tempt him, and distract his mind, which keeps open house, and entertains all comers; and after being fatigued and amused with morning calls from idle visitors, finds the day consumed and its business unconcluded. Mr. Godwin, on the contrary, is somewhat exclusive and unsocial in his habits of mind, entertains no company but what he gives his whole time and attention to, and wisely writes over the doors of his understanding, his fancy, and his senses—"No admittance except on business." He has none of that fastidious refinement and false delicacy, which might lead him to balance between the endless variety of modern attainments. He does not throw away his life (nor a single half-hour of it) in adjusting the claims of different accomplishments, and in choosing between them or making himself master of them all. He sets about his task, (whatever it may be) and goes through it with spirit and fortitude. He has the happiness to think an author the greatest character in the world, and himself the greatest author in it.

27. **passage . . . friendship:** that quoted by Hazlitt near the end of his lecture "On the Living Poets." 28. **ellipsis:** Hazlitt here quotes the sonnet to Schiller ("To the Author of *The Robbers*").

Mr. Coleridge, in writing an harmonious stanza, would stop to consider whether there was not more grace and beauty in a *Pas de trois*, and would not proceed till he had resolved this question by a chain of metaphysical reasoning without end. Not so Mr. Godwin. That is best to him, which he can do best. He does not waste himself in vain aspirations and effeminate sympathies. He is blind, deaf, insensible to all but the trump of Fame. Plays, operas, painting, music, ball-rooms, wealth, fashion, titles, lords, ladies, touch him not—all these are no more to him than to the magician in his cell, and he writes on to the end of the chapter, through good report and evil report. *Pingo in eternitatem*—is his motto. He neither envies nor admires what others are, but is contented to be what he is, and strives to do the utmost he can. Mr. Coleridge has flirted with the Muses as with a set of mistresses: Mr. Godwin has been married twice, to Reason and to Fancy, and has to boast no short-lived progeny by each. So to speak, he has *valves* belonging to his mind, to regulate the quantity of gas admitted into it, so that like the bare, unsightly, but well-compacted steam-vessel, it cuts its liquid way, and arrives at its promised end: while Mr. Coleridge's bark, "taught with the little nautilus to sail," the sport of every breath, dancing to every wave,

> "Youth at its prow, and Pleasure at its helm,"[29]

flutters its gaudy pennons in the air, glitters in the sun, but we wait in vain to hear of its arrival in the destined harbour. Mr. Godwin, with less variety and vividness, with less subtlety and susceptibility both of thought and feeling, has had firmer nerves, a more determined purpose, a more comprehensive grasp of his subject, and the results are as we find them. Each has met with his reward: for justice has, after all, been done to the pretensions of each; and we must, in all cases, use means to ends!

It was a misfortune to any man of talent to be born in the latter end of the last century. Genius stopped the way of Legitimacy, and therefore it was to be abated, crushed, or set aside as a nuisance. The spirit of the monarchy was at variance with the spirit of the age. The flame of liberty, the light of intellect was to be extinguished with the sword—or with slander, whose edge is sharper than the sword. The war between power and reason was carried on by the first of these abroad —by the last at home. No quarter was given (then or now) by the Government-critics, the authorised censors

of the press, to those who followed the dictates of independence, who listened to the voice of the tempter, Fancy. Instead of gathering fruits and flowers, immortal fruits and amaranthine flowers, they soon found themselves beset not only by a host of prejudices, but assailed with all the engines of power, by nicknames, by lies, by all the arts of malice, interest and hypocrisy, without the possibility of their defending themselves "from the pelting of the pitiless storm," that poured down upon them from the strong-holds of corruption and authority. The philosophers, the dry abstract reasoners, submitted to this reverse pretty well, and armed themselves with patience "as with triple steel" to bear discomfiture, persecution, and disgrace. But the poets, the creatures of sympathy, could not stand the frowns both of king and people. They did not like to be shut out when places and pensions, when the critic's praises, and the laurel-wreath were about to be distributed. They did not stomach being *sent to Coventry*, and Mr. Coleridge sounded a retreat for them by the help of casuistry, and a musical voice.— "His words were hollow, but they pleased the ear" of his friends of the Lake School, who turned back disgusted and panic-struck from the dry desert of unpopularity, like Hassan the camel driver,

> "And curs'd the hour, and curs'd the luckless day,
> When first from Shiraz' walls they bent their way."[30]

They are safely inclosed there, but Mr. Coleridge did not enter with them; pitching his tent upon the barren waste without, and having no abiding place nor city of refuge.

1824 (1825)

LORD BYRON

Lord Byron and Sir Walter Scott are among writers now living the two, who would carry away a majority of suffrages as the greatest geniuses of the age. The former would, perhaps, obtain the preference with the fine gentlemen and ladies (squeamishness apart)—the latter with the critics and the vulgar. We shall treat of them in the same connection, partly on account of their distinguished pre-eminence, and partly because

---

29. **"Youth . . . helm"**: Gray's "The Bard," II. 2.

30. **"And . . . way"**: William Collins' "Persian Eclogues," II. 3–4.

they afford a complete contrast to each other. In their poetry, in their prose, in their politics, and in their tempers, no two men can be more unlike.

If Sir Walter Scott may be thought by some to have been

"Born universal heir to all humanity,"

it is plain Lord Byron can set up no such pretension. He is, in a striking degree, the creature of his own will. He holds no communion with his kind; but stands alone, without mate or fellow—

"As if a man were author of himself,
And owned no other kin." [1]

He is like a solitary peak, all access to which is cut off not more by elevation than distance. He is seated on a lofty eminence, "cloud-capt," or reflecting the last rays of setting suns; and in his poetical moods, reminds us of the fabled Titans, retired to a ridgy steep, playing on their Pan's-pipes, and taking up ordinary men and things in their hands with haughty indifference. He raises his subject to himself, or tramples on it; he neither stoops to, nor loses himself in it. He exists not by sympathy, but by antipathy. He scorns all things, even himself. Nature must come to him to sit for her picture —he does not go to her. She must consult his time, his convenience, and his humour; and wear a *sombre* or a fantastic garb, or his Lordship turns his back upon her. There is no ease, no unaffected simplicity of manner, no "golden mean." All is strained, or petulant in the extreme. His thoughts are sphered and crystalline; his style "prouder than when blue Iris bends"; his spirit fiery, impatient, wayward, indefatigable. Instead of taking his impressions from without, in entire and almost unimpaired masses, he moulds them according to his own temperament, and heats the materials of his imagination in the furnace of his passions.—Lord Byron's verse glows like a flame, consuming every thing in its way; Sir Walter Scott's glides like a river, clear, gentle, harmless. The poetry of the first scorches, that of the last scarcely warms. The light of the one proceeds from an internal source, ensanguined, sullen, fixed; the other reflects the hues of Heaven, or the face of nature, glancing vivid and various. The productions of the Northern Bard have the rust and the freshness of antiquity about them; those of the Noble Poet cease to startle from their extreme ambition of novelty,

both in style and matter. Sir Walter's rhymes are "silly sooth"—

"And dally with the innocence of thought,
Like the old age"— [2]

his Lordship's Muse spurns *the olden time*, and affects all the supercilious airs of a modern fine lady and an upstart. The object of the one writer is to restore us to truth and nature: the other chiefly thinks how he shall display his own power, or vent his spleen, or astonish the reader either by starting new subjects and trains of speculation, or by expressing old ones in a more striking and emphatic manner than they have been expressed before. He cares little what it is he says, so that he can say it differently from others. This may account for the charges of plagiarism which have been repeatedly brought against the Noble Poet—if he can borrow an image or sentiment from another, and heighten it by an epithet or an allusion of greater force and beauty than is to be found in the original passage, he thinks he shows his superiority of execution in this in a more marked manner than if the first suggestion had been his own. It is not the value of the observation itself he is solicitous about; but he wishes to shine by contrast—even nature only serves as a foil to set off his style. He therefore takes the thoughts of others (whether contemporaries or not) out of their mouths, and is content to make them his own, to set his stamp upon them, by imparting to them a more meretricious gloss, a higher relief, a greater loftiness of tone, and a characteristic inveteracy of purpose. Even in those collateral ornaments of modern style, slovenliness, abruptness, and eccentricity (as well as in terseness and significance), Lord Byron, when he pleases, defies competition and surpasses all his contemporaries. Whatever he does, he must do in a more decided and daring manner than any one else—he lounges with extravagance, and yawns so as to alarm the reader! Self-will, passion, the love of singularity, a disdain of himself and of others (with a conscious sense that this is among the ways and means of procuring admiration) are the proper categories of his mind: he is a lordly writer, is above his own reputation, and condescends to the Muses with a scornful grace!

Lord Byron, who in his politics is a *liberal*, in his genius is haughty and aristocratic: Walter Scott, who is an aristocrat in principle, is popular in his writings, and is (as it were) equally *servile* to nature and to opinion.

LORD BYRON. **1. "As . . . kin"**: *Coriolanus*, V. iii. 36–37.   **2. "And . . . age"**: *Twelfth Night*, II. iv. 47–49.

The genius of Sir Walter is essentially imitative, or "denotes a foregone conclusion"[3]: that of Lord Byron is self-dependent; or at least requires no aid, is governed by no law, but the impulses of its own will. We confess, however much we may admire independence of feeling and erectness of spirit in general or practical questions, yet in works of genius we prefer him who bows to the authority of nature, who appeals to actual objects, to mouldering superstitions, to history, observation, and tradition, before him who only consults the pragmatical and restless workings of his own breast, and gives them out as oracles to the world. We like a writer (whether poet or prose-writer) who takes in (or is willing to take in) the range of half the universe in feeling, character, description, much better than we do one who obstinately and invariably shuts himself up in the Bastile of his own ruling passions. In short, we had rather be Sir Walter Scott (meaning thereby the Author of Waverley) than Lord Byron, a hundred times over. And for the reason just given, namely, that he casts his descriptions in the mould of nature, ever-varying, never tiresome, always interesting and always instructive, instead of casting them constantly in the mould of his own individual impressions. He gives us man as he is, or as he was, in almost every variety of situation, action, and feeling. Lord Bryon makes man after his own image, woman after his own heart; the one is a capricious tyrant, the other a yielding slave; he gives us the misanthrope and the voluptuary by turns; and with these two characters, burning or melting in their own fires, he makes out everlasting centos of himself. He hangs the cloud, the film of his existence over all outward things—sits in the centre of his thoughts, and enjoys dark night, bright day, the glitter and the gloom "in cell monastic"—we see the mournful pall, the crucifix, the death's heads, the faded chaplet of flowers, the gleaming tapers, the agonized growl of genius, the wasted form of beauty—but we are still imprisoned in a dungeon, a curtain intercepts our view, we do not breathe freely the air of nature or of our own thoughts—the other admired author draws aside the curtain, and the veil of egotism is rent, and he shows us the crowd of living men and women, the endless groups, the landscape back-ground, the cloud and the rainbow, and enriches our imaginations and relieves one passion by another, and expands and lightens reflection, and takes away that tightness at the breast which arises from thinking or wishing to think that there is nothing in the world out of a man's self!—In this point of view, the Author of Waverley is one of the greatest teachers of morality that ever lived, by emancipating the mind from petty, narrow, and bigotted prejudices: Lord Byron is the greatest pamperer of those prejudices, by seeming to think there is nothing else worth encouraging but the seeds or the full luxuriant growth of dogmatism and self-conceit. In reading the *Scotch Novels*, we never think about the author, except from a feeling of curiosity respecting our unknown benefactor: in reading Lord Byron's works, he himself is never absent from our minds. The colouring of Lord Byron's style, however rich and dipped in Tyrian dyes, is nevertheless opaque, is in itself an object of delight and wonder: Sir Walter Scott's is perfectly transparent. In studying the one, you seem to gaze at the figures cut in stained glass, which exclude the view beyond, and where the pure light of Heaven is only a means of setting off the gorgeousness of art: in reading the other, you look through a noble window at the clear and varied landscape without. Or to sum up the distinction in one word, Sir Walter Scott is the most *dramatic* writer now living; and Lord Byron is the least so. It would be difficult to imagine that the Author of Waverley is in the smallest degree a pedant; as it would be hard to persuade ourselves that the author of Childe Harold and Don Juan is not a coxcomb, though a provoking and sublime one. In this decided preference given to Sir Walter Scott over Lord Byron, we distinctly include the prose-works of the former; for we do not think his poetry alone by any means entitles him to that precedence. Sir Walter in his poetry, though pleasing and natural, is a comparative trifler: it is in his anonymous productions that he has shown himself for what he is!—

Intensity is the great and prominent distinction of Lord Byron's writings. He seldom gets beyond force of style, nor has he produced any regular work or masterly whole. He does not prepare any plan beforehand, nor revise and retouch what he has written with polished accuracy. His only object seems to be to stimulate himself and his readers for the moment—to keep both alive, to drive away *ennui*, to substitute a feverish and irritable state of excitement for listless indolence or even calm enjoyment. For this purpose he pitches on any subject at random without much thought or delicacy—he is only impatient to begin—and takes care to adorn and enrich it as he proceeds

3. "denotes . . . conclusion": *Othello*, III. iii. 428.

with "thoughts that breathe and words that burn."[4] He composes (as he himself has said) whether he is in the bath, in his study, or on horseback—he writes as habitually as others talk or think—and whether we have the inspiration of the Muse or not, we always find the spirit of the man of genius breathing from his verse. He grapples with his subject, and moves, penetrates, and animates it by the electric force of his own feelings. He is often monotonous, extravagant, offensive; but he is never dull, or tedious, but when he writes prose. Lord Byron does not exhibit a new view of nature, or raise insignificant objects into importance by the romantic associations with which he surrounds them; but generally (at least) takes common-place thoughts and events, and endeavours to express them in stronger and statelier language than others. His poetry stands like a Martello tower by the side of his subject. He does not, like Mr. Wordsworth, lift poetry from the ground, or create a sentiment out of nothing. He does not describe a daisy or a periwinkle, but the cedar or the cypress: not "poor men's cottages, but princes' palaces."[5] His *Childe Harold* contains a lofty and impassioned review of the great events of history, of the mighty objects left as wrecks of time, but he dwells chiefly on what is familiar to the mind of every schoolboy; has brought out few new traits of feeling or thought; and has done no more than justice to the reader's preconceptions by the sustained force and brilliancy of his style and imagery.

Lord Byron's earlier productions, *Lara*, the *Corsair*, &c. were wild and gloomy romances, put into rapid and shining verse. They discover the madness of poetry, together with the inspiration: sullen, moody, capricious, fierce, inexorable, gloating on beauty, thirsting for revenge, hurrying from the extremes of pleasure to pain, but with nothing permanent, nothing healthy or natural. The gaudy decorations and the morbid sentiments remind one of flowers strewed over the face of death! In his *Childe Harold* (as has been just observed) he assumes a lofty and philosophic tone, and "reasons high of providence, fore-knowledge, will, and fate."[6] He takes the highest points in the history of the world, and comments on them from a more commanding eminence: he shows us the crumbling monuments of time, he invokes the great names, the mighty spirit of antiquity. The universe is changed into

a stately mausoleum:—in solemn measures he chaunts a hymn to fame. Lord Byron has strength and elevation enough to fill up the moulds of our classical and time-hallowed recollections, and to rekindle the earliest aspirations of the mind after greatness and true glory with a pen of fire. The names of Tasso, of Ariosto, of Dante, of Cincinnatus, of Cæsar, of Scipio, lose nothing of their pomp or their lustre in his hands, and when he begins and continues a strain of panegyric on such subjects, we indeed sit down with him to a banquet of rich praise, brooding over imperishable glories,

"Till Contemplation has her fill."[7]

Lord Byron seems to cast himself indignantly from "this bank and shoal of time," or the frail tottering bark that bears up modern reputation, into the huge sea of ancient renown, and to revel there with untired, outspread plume. Even this in him is spleen—his contempt of his contemporaries makes him turn back to the lustrous past, or project himself forward to the dim future!—Lord Byron's tragedies, Faliero, Sardanapalus, &c. are not equal to his other works. They want the essence of the drama. They abound in speeches and descriptions, such as he himself might make either to himself or others, lolling on his couch of a morning, but do not carry the reader out of the poet's mind to the scenes and events recorded. They have neither action, character, nor interest, but are a sort of *gossamer* tragedies, spun out, and glittering, and spreading a flimsy veil over the face of nature. Yet he spins them on. Of all that he has done in this way the *Heaven and Earth* (the same subject as Mr. Moore's *Loves of the Angels*) is the best. We prefer it even to *Manfred*. *Manfred* is merely himself, with a fancy-drapery on: but in the dramatic fragment published in the *Liberal*,[8] the space between Heaven and Earth, the stage on which his characters have to pass to and fro, seems to fill his Lordship's imagination; and the Deluge, which he has so finely described, may be said to have drowned all his own idle humours.

We must say we think little of our author's turn for satire. His "English Bards and Scotch Reviewers" is dogmatical and insolent, but without refinement or point. He calls people names, and tries to transfix a character with an epithet, which does not stick, because

---

4. **"thoughts . . . burn"**: Gray's "Progress of Poesy," l. 110. 5. **"poor . . . palaces"**: *The Merchant of Venice*, I. ii. 14. 6. **"reasons . . . fate"**: *Paradise Lost*, II. 558.

7. **"Till . . . fill"**: John Dyer's "Grongar Hill," l. 26. **8. Liberal:** a periodical published briefly in Italy (1822–23) by Byron and Leigh Hunt.

it has no other foundation than his own petulance and spite; or he endeavours to degrade by alluding to some circumstance of external situation. He says of Mr. Wordsworth's poetry, that "it is his aversion."[9] That may be: but whose fault is it? This is the satire of a lord, who is accustomed to have all his whims or dislikes taken for gospel, and who cannot be at the pains to do more than signify his contempt or displeasure. If a great man meets with a rebuff which he does not like, he turns on his heel, and this passes for a repartee. The Noble Author says of a celebrated barrister and critic, that he was "born in a garret sixteen stories high." The insinuation is not true; or if it were, it is low. The allusion degrades the person who makes, not him to whom it is applied. This is also the satire of a person of birth and quality, who measures all merit by external rank, that is, by his own standard. So his Lordship, in a "Letter to the Editor of My Grandmother's Review," addresses him fifty times as "*my dear Robarts*"; nor is there any other wit in the article. This is surely a mere assumption of superiority from his Lordship's rank, and is the sort of *quizzing* he might use to a person who came to hire himself as a valet to him at *Long's*[10]—the waiters might laugh, the public will not. In like manner, in the controversy about Pope, he claps Mr. Bowles on the back with a coarse facetious familiarity, as if he were his chaplain whom he had invited to dine with him, or was about to present to a benefice. The reverend divine might submit to the obligation, but he has no occasion to subscribe to the jest. If it is a jest that Mr. Bowles should be a parson, and Lord Byron a peer, the world knew this before; there was no need to write a pamphlet to prove it.

The *Don Juan* indeed has great power; but its power is owing to the force of the serious writing, and to the oddity of the contrast between that and the flashy passages with which it is interlarded. From the sublime to the ridiculous there is but one step. You laugh and are surprised that any one should turn round and *travestie* himself: the drollery is in the utter discontinuity of ideas and feelings. He makes virtue serve as a foil to vice; *dandyism* is (for want of any other) a variety of genius. A classical intoxication is followed by the splashing of soda-water, by frothy effusions of ordinary bile. After the lightning and the hurricane, we are introduced to the interior of the cabin and the contents of wash-hand basins. The solemn hero of tragedy plays *Scrub* in the farce. This is "very tolerable and not to be endured."[11] The Noble Lord is almost the only writer who has prostituted his talents in this way. He hallows in order to desecrate; takes a pleasure in defacing the images of beauty his hands have wrought; and raises our hopes and our belief in goodness to Heaven only to dash them to the earth again, and break them in pieces the more effectually from the very height they have fallen. Our enthusiasm for genius or virtue is thus turned into a jest by the very person who has kindled it, and who thus fatally quenches the sparks of both. It is not that Lord Byron is sometimes serious and sometimes trifling, sometimes profligate, and sometimes moral—but when he is most serious and most moral, he is only preparing to mortify the unsuspecting reader by putting a pitiful *hoax* upon him. This is a most unaccountable anomaly. It is as if the eagle were to build its eyry in a common sewer, or the owl were seen soaring to the mid-day sun. Such a sight might make one laugh, but one would not wish or expect it to occur more than once.[12]

In fact, Lord Byron is the spoiled child of fame as well as fortune. He has taken a surfeit of popularity, and is not contented to delight, unless he can shock the public. He would force them to admire in spite of decency and common sense—he would have them read what they would read in no one but himself, or he would not give a rush for their applause. He is to be "a chartered libertine,"[13] from whom insults are favours, whose contempt is to be a new incentive to admiration. His Lordship is hard to please: he is equally averse to notice or neglect, enraged at censure and scorning praise. He tries the patience of the town to the very utmost, and when they show signs of weariness or disgust, threatens to *discard* them. He says he will write on, whether he is read or not. He would never write another page, if it were not to court popular applause, or to affect a superiority over it. In this respect also, Lord Byron presents a striking contrast to Sir Walter Scott. The latter takes what part of the public favour falls to his share, without grumbling (to be sure he has no reason to complain); the former is always quarrelling with the world about his *modicum*

9. "it . . . aversion": See *Don Juan*, below, III. 845–48. 10. Long's: restaurant in Bond Street.

11. "very . . . endured": *Much Ado About Nothing*, III. iii. 37. 12. more than once: "This censure applies to the first Cantos of Don Juan much more than to the last. It has been called a Tristram Shandy in rhyme: it is rather a poem written about itself" (H.). 13. "a . . . libertine": *Henry V*, I. i. 48.

of applause, the *spolia opima* of vanity, and ungraciously throwing the offerings of incense heaped on his shrine back in the faces of his admirers. Again, there is no taint in the writings of the Author of Waverley, all is fair and natural and *above-board:* he never outrages the public mind. He introduces no anomalous character: broaches no staggering opinion. If he goes back to old prejudices and superstitions as a relief to the modern reader, while Lord Byron floats on swelling paradoxes—

"Like proud seas under him"; [14]

if the one defers too much to the spirit of antiquity, the other panders to the spirit of the age, goes to the very edge of extreme and licentious speculation, and breaks his neck over it. Grossness and levity are the playthings of his pen. It is a ludicrous circumstance that he should have dedicated his *Cain* to the worthy Baronet! Did the latter ever acknowledge the obligation? We are not nice, not very nice; but we do not particularly approve those subjects that shine chiefly from their rottenness: nor do we wish to see the Muses drest out in the flounces of a false or questionable philosophy, like *Portia* and *Nerissa* in the garb of Doctors of Law. We like metaphysics as well as Lord Byron; but not to see them making flowery speeches, nor dancing a measure in the fetters of verse. We have as good as hinted, that his Lordship's poetry consists mostly of a tissue of superb common-places; even his paradoxes are *common-place*. They are familiar in the schools: they are only new and striking in his dramas and stanzas, by being out of place. In a word, we think that poetry moves best within the circle of nature and received opinion: speculative theory and subtle casuistry are forbidden ground to it. But Lord Byron often wanders into this ground wantonly, wilfully, and unwarrantably. The only apology we can conceive for the spirit of some of Lord Byron's writings, is the spirit of some of those opposed to him. They would provoke a man to write anything. "Farthest from them is best." [15] The extravagance and license of the one seems a proper antidote to the bigotry and narrowness of the other. The first *Vision of Judgment* was a set-off to the second, though

"None but itself could be its parallel." [16]

14. **"Like . . . him":** John Fletcher's *Two Noble Kinsmen,* II. ii. 23. 15. **"Farthest . . . best":** *Paradise Lost*, I. 247. 16. **"None . . . parallel":** Burke (from whose *Letters on a Regicide Peace* Hazlitt apparently takes the line) quotes it from Lewis Theobald's play, *The Double Falsehood* (1727).

Perhaps the chief cause of most of Lord Byron's errors is, that he is that anomaly in letters and in society, a Noble Poet. It is a double privilege, almost too much for humanity. He has all the pride of birth and genius. The strength of his imagination leads him to indulge in fantastic opinions; the elevation of his rank sets censure at defiance. He becomes a pampered egotist. He has a seat in the House of Lords, a niche in the Temple of Fame. Every-day mortals, opinions, things are not good enough for him to touch or think of. A mere nobleman is, in his estimation, but "the tenth transmitter of a foolish face" [17]: a mere man of genius is no better than a worm. His Muse is also a lady of quality. The people are not polite enough for him: the Court not sufficiently intellectual. He hates the one and despises the other. By hating and despising others, he does not learn to be satisfied with himself. A fastidious man soon grows querulous and splenetic. If there is nobody but ourselves to come up to our idea of fancied perfection, we easily get tired of our idol. When a man is tired of what he is, by a natural perversity he sets up for what he is not. If he is a poet, he pretends to be a metaphysician: if he is a patrician in rank and feeling, he would fain be one of the people. His ruling motive is not the love of the people, but of distinction; not of truth, but of singularity. He patronizes men of letters out of vanity, and deserts them from caprice, or from the advice of friends. He embarks in an obnoxious publication to provoke censure, and leaves it to shift for itself for fear of scandal. We do not like Sir Walter's gratuitous servility: we like Lord Byron's preposterous *liberalism* little better. He may affect the principles of equality, but he resumes his privilege of peerage, upon occasion. His Lordship has made great offers of service to the Greeks—money and horses. He is at present in Cephalonia, waiting the event!

We had written thus far when news came of the death of Lord Byron, and put an end at once to a strain of somewhat peevish invective, which was intended to meet his eye, not to insult his memory. Had we known that we were writing his epitaph, we must have done it with a different feeling. As it is, we think it better and more like himself, to let what we had written stand, than to take up our leaden shafts, and try to melt them into "tears of sensibility," or mould them into dull praise, and an affected show of candour.

17. **"the . . . face":** from Richard Savage's *The Bastard* (1728).

We were not silent during the author's life-time, either for his reproof or encouragement (such as we could give, and *he* did not disdain to accept) nor can we now turn undertakers' men to fix the glittering plate upon his coffin, or fall into the procession of popular woe.—Death cancels every thing but truth; and strips a man of every thing but genius and virtue. It is a sort of natural canonization. It makes the meanest of us sacred—it installs the poet in his immortality, and lifts him to the skies. Death is the great assayer of the sterling ore of talent. At his touch the drossy particles fall off, the irritable, the personal, the gross, and mingle with the dust—the finer and more ethereal part mounts with the winged spirit to watch over our latest memory, and protect our bones from insult. We consign the least worthy qualities to oblivion, and cherish the nobler and imperishable nature with double pride and fondness. Nothing could show the real superiority of genius in a more striking point of view than the idle contests and the public indifference about the place of Lord Byron's interment, whether in Westminster Abbey or his own family-vault. A king must have a coronation—a nobleman a funeral-procession.—The man is nothing without the pageant. The poet's cemetery is the human mind, in which he sows the seeds of never-ending thought—his monument is to be found in his works:

> "Nothing can cover his high fame but Heaven;
> No pyramids set off his memory,
> But the eternal substance of his greatness." [18]

Lord Byron is dead: he also died a martyr to his zeal in the cause of freedom, for the last, best hopes of man. Let that be his excuse and his epitaph!

1824 (1825)

18. **"Nothing . . . greatness"**: Beaumont and Fletcher's *The False One* (1647), II. i.

THOMAS MOORE

1779–1852

REMEMBERED NOW primarily for a half-dozen lyrics, Moore had an immense reputation in his own lifetime, though the later years of this generous, convivial writer were not very happy. Born in Dublin, the son of a grocer and wine merchant, he attended Trinity College there and then went to London to study law at the Middle Temple. He had already begun to write poetry, and had brought with him from Ireland a translation of the odes of the Greek poet Anacreon which was published a year later (1800). He quickly became a social success—he was also an accomplished singer.

Through the influence of his friend and patron, Lord Moira, in 1803 he was appointed Admiralty Registrar at Bermuda, where he found life dull, especially after what he had seen of London society. As was possible to do at the time, he appointed a deputy to handle the work, returning to England by way of the United States and Canada (see the letter to his mother, below, written from Niagara Falls). Another volume of poems (1806) was followed by some satires of modest quality, and, far more important, by the happy project of writing lyrics for a series of Irish airs—his *Irish Melodies* that continued to appear, in installments, throughout the next twenty-five years (1808–34). Moore himself composed some of the music for them. He married in 1811 and at about that time became a friend of Byron, who later sent him his memoirs to be published after his death. (These were destroyed, however, after heated arguments with Byron's executors, and Moore confined himself to publishing his valuable *Letters and Journals of Lord Byron* in 1830.) At the height of his fame Moore brought out *Lalla Rookh* (1817), a series of four Oriental tales recited to an Indian princess, Lalla Rookh ("tulip cheek"), on her journey from Delhi to her marriage in Cashmere. For the poem the publisher, Longman, paid

Moore the highest sum ever before paid for a single poem (£3,000, or approximately £15,000 to £17,000 in present-day currency). Unfortunately the next year his deputy in Bermuda disappeared after having embezzled twice that amount, and Moore was legally responsible for the theft. In order to avoid arrest for debt, Moore spent some time on the continent, writing prolifically while he was there and returning in 1822 when the matter was adjusted. For several years he worked on a long *History of Ireland* (four volumes of which appeared from 1835 to 1846), hoping to call English attention to the evils of the British government there and also to give the Irish more pride in their past. Though he lacked the qualifications of an historian and found the writing slow and burdensome, he struggled on with it. Meanwhile he had never quite recovered financially. His only remaining child died in 1846. After this final blow Moore failed steadily and he was mentally helpless in his last years.

Bibliography

Editions of the poetry include *Poetical Works*, ed. A. D. Godley, 1910; *Selected Poems*, ed. C. L. Falkiner (Golden Treasury), 1903; *Irish Melodies and Songs*, ed. S. L. Gwynn (Muses' Library), 1908; and *Lyrics and Satires*, ed. Sean O'Faoláin, 1929. Selections from the *Diary* have been edited by J. B. Priestley, 1925. For biography and criticism, see especially the fine study by Howard Mumford Jones, *The Harp That Once: A Chronicle of the Life of Thomas Moore*, 1937. Earlier biographies include A. J. Symington, *Thomas Moore*, 1880, and S. L. Gwynn, *Thomas Moore* (English Men of Letters Series), 2nd ed., 1924.

FROM

NATIONAL AIRS

OFT, IN THE STILLY NIGHT

SCOTCH AIR

Oft, in the stilly night,
  Ere Slumber's chain has bound me,
Fond Memory brings the light
  Of other days around me;
    The smiles, the tears,
    Of boyhood's years,
  The words of love then spoken;
    The eyes that shone,
    Now dimmed and gone,
  The cheerful hearts now broken!   10

Thus, in the stilly night,
  Ere Slumber's chain has bound me,
Sad Memory brings the light
  Of other days around me.

When I remember all
  The friends, so linked together,
I've seen around me fall,
  Like leaves in wintry weather;
    I feel like one
    Who treads alone   20
  Some banquet hall deserted,
    Whose lights are fled,
    Whose garlands dead,
  And all but he departed!
Thus, in the stilly night,
  Ere Slumber's chain has bound me,
Sad Memory brings the light
  Of other days around me.

(1815)

HARK! THE VESPER HYMN IS STEALING

RUSSIAN AIR

Hark! the vesper hymn is stealing
  O'er the waters soft and clear;
Nearer yet and nearer pealing,
  And now bursts upon the ear:
    Jubilate, Amen.
Farther now, now farther stealing,
  Soft it fades upon the ear:
    Jubilate, Amen.

Now, like moonlight waves retreating
  To the shore, it dies along;   10
Now, like angry surges meeting,
  Breaks the mingled tide of song:
    Jubilate, Amen.
Hush! again, like waves retreating
  To the shore, it dies along:
    Jubilate, Amen.

           (*1815*)

FROM

LALLA ROOKH

THE LIGHT OF THE HARAM

Who has not heard of the Vale of Cashmere,
  With its roses the brightest that earth ever gave,
Its temples, and grottos, and fountains as clear
  As the love-lighted eyes that hang over their wave?

Oh! to see it at sunset, when warm o'er the lake
  Its splendor at parting a summer eve throws,
Like a bride, full of blushes, when lingering to take
  A last look of her mirror at night ere she goes!
When the shrines through the foliage are gleaming
    half shown,
And each hallows the hour by some rites of its own. 10
Here the music of prayer from a minaret swells,
  Here the Magian his urn, full of perfume, is swinging,
And here, at the altar, a zone of sweet bells

Round the waist of some fair Indian dancer is
  ringing.
Or to see it by moonlight, when mellowly shines
The light o'er its palaces, gardens, and shrines;
When the water-falls gleam, like a quick fall of
  stars,
And the nightingale's hymn from the Isle of Chenars
Is broken by laughs and light echoes of feet
From the cool, shining walks where the young people
  meet: 20
Or at morn, when the magic of daylight awakes
A new wonder each minute, as slowly it breaks,—
Hills, cupolas, fountains, called forth every one
Out of darkness, as if but just born of the Sun.
When the Spirit of Fragrance is up with the day,
From his Haram of night-flowers stealing away;
And the wind, full of wantonness, woos like a lover
The young aspen-trees, till they tremble all over.
When the East is as warm as the light of first hopes,
  And Day, with his banner of radiance unfurled, 30
Shines in through the mountainous portal that opes,
  Sublime, from that valley of bliss to the world!

But never yet, by night or day,
  In dew of spring or summer's ray,
  Did the sweet valley shine so gay
As now it shines—all love and light,
Visions by day and feasts by night!
A happier smile illumes each brow,
  With quicker spread each heart uncloses,
And all its ecstasy, for now   40
  The valley holds its Feast of Roses;
The joyous time, when pleasures pour
Profusely round and, in their shower,
Hearts open, like the season's rose,—
  The floweret of a hundred leaves,
Expanding while the dew-fall flows,
  And every leaf its balm receives.

'Twas when the hour of evening came
  Upon the lake, serene and cool,
When Day had hid his sultry flame   50
  Behind the palms of Baramoule,
When maids began to lift their heads,
Refreshed from their embroidered beds,
Where they had slept the sun away,
And waked to moonlight and to play.
All were abroad—the busiest hive
On Bela's hills is less alive,
When saffron beds are full in flower,
Than looked the valley in that hour.

LALLA ROOKH. THE LIGHT OF THE HARAM. **12. Magian:**
priest. **13. zone:** girdle.

A thousand restless torches played   60
Through every grove and island shade;
A thousand sparkling lamps were set
On every dome and minaret;
And fields and pathways, far and near,
Were lighted by a blaze so clear,
That you could see, in wandering round,
The smallest rose-leaf on the ground.
Yet did the maids and matrons leave
Their veils at home, that brilliant eve;
And there were glancing eyes about,   70
And cheeks, that would not dare shine out
    In open day, but thought they might
Look lovely then, because 'twas night.
And all were free, and wandering,
    And all exclaimed to all they met,
That never did the summer bring
    So gay a Feast of Roses yet;
The moon had never shed a light
    So clear as that which blessed them there;
The roses ne'er shone half so bright,   80
    Nor they themselves looked half so fair.

And what a wilderness of flowers!
It seemed as though from all the bowers
And fairest fields of all the year,
The mingled spoil were scattered here.
The lake, too, like a garden breathes,
    With the rich buds that o'er it lie,—
As if a shower of fairy wreaths
    Had fallen upon it from the sky!
And then the sound of joy:—the beat   90
Of tabors and of dancing feet;
The minaret-crier's chant of glee
Sung from his lighted gallery,
And answered by a ziraleet
From neighboring Haram, wild and sweet;
The merry laughter, echoing
From gardens, where the silken swing
Wafts some delighted girl above
The top leaves of the orange-grove;
Or, from those infant groups at play   100
Among the tents that line the way,
Flinging, unawed by slave or mother,
Handfuls of roses at each other.
Then the sounds from the lake:—the low whispering
        in boats,
    As they shoot through the moonlight; the dipping
        of oars;

94. ziraleet: chorus of women.

And the wild, airy warbling that everywhere floats,
Through the groves, round the islands, as if all the
        shores,
Like those of Kathay, uttered music, and gave
An answer in song to the kiss of each wave.
But the gentlest of all are those sounds, full of feeling,
That soft from the lute of some lover are stealing, 111
Some lover, who knows all the heart-touching power
Of a lute and a sigh in this magical hour.
Oh! best of delights as it everywhere is
To be near the loved *One*,—what a rapture is his
Who in moonlight and music thus sweetly may glide
O'er the Lake of Cashmere, with that *One* by his side!
If woman can make the worst wilderness dear,
Think, think what a heaven she must make of Cashmere!

                                        1814–17 (1817)

FROM

IRISH MELODIES

THE HARP THAT ONCE THROUGH TARA'S HALLS

Tara, northwest of Dublin, was the seat of many of the
ancient Irish kings.

The harp that once through Tara's halls
    The soul of music shed,
Now hangs as mute on Tara's walls
    As if that soul were fled.—
So sleeps the pride of former days,
    So glory's thrill is o'er,
And hearts that once beat high for praise
    Now feel that pulse no more!

No more to chiefs and ladies bright
    The harp of Tara swells;   10
The chord alone that breaks at night
    Its tale of ruin tells.
Thus freedom now so seldom wakes,
    The only throb she gives
Is when some heart indignant breaks,
    To show that still she lives.

                                        (1834)

LET ERIN REMEMBER
THE DAYS OF OLD

Let Erin remember the days of old,
    Ere her faithless sons betrayed her;
When Malachi wore the collar of gold,
    Which he won from her proud invader,
When her kings, with standard of green unfurled
    Led the Red-Branch Knights to danger;—
Ere the emerald gem of the western world
    Was set in the crown of a stranger.
On Lough Neagh's bank as the fisherman strays,
    When the clear cold eve's declining,   10
He sees the round towers of other days
    In the wave beneath him shining;
Thus shall memory often, in dreams sublime,
    Catch a glimpse of the days that are over;
Thus, sighing, look thro' the waves of time
    For the long-faded glories they cover.

                                        (*1834*)

SHE IS FAR FROM THE LAND

She is far from the land where her young hero sleeps,
    And lovers are round her, sighing:
But coldly she turns from their gaze, and weeps,
    For her heart in his grave is lying.

She sings the wild song of her dear native plains,
    Every one which he loved awaking;—
Ah! little they think who delight in her strains
    How the heart of the Minstrel is breaking.

He had lived for his love, for his country he died,
    They were all that to life had intwined him;   10
Nor soon shall the tears of his country be dried,
    Nor long will his love stay behind him.

O, make her a grave where the sunbeams rest,
    When they promise a glorious morrow;
They'll shine o'er her sleep, like a smile from the West,
    From her own loved island of sorrow.

                                        (*1834*)

BELIEVE ME, IF ALL THOSE
ENDEARING YOUNG CHARMS

Believe me, if all those endearing young charms,
    Which I gaze on so fondly today,
Were to change by tomorrow, and fleet in my arms,
    Like fairy gifts fading away,
Thou wouldst still be adored, as this moment thou art,
    Let thy loveliness fade as it will,
And around the dear ruin each wish of my heart
    Would intwine itself verdantly still.

It is not while beauty and youth are thine own,
    And my cheeks unprofaned by a tear,   10
That the fervor and faith of a soul can be known,
    To which time will but make thee more dear;
No, the heart that has truly loved never forgets,
    But as truly loves on to the close,
As the sunflower turns on her god, when he sets,
The same look which she turned when he rose.

                                        (*1834*)

DEAR HARP OF MY COUNTRY

Dear Harp of my Country! in darkness I found thee,
    The cold chain of silence had hung o'er thee long,
When proudly, my own Island Harp, I unbound thee,
    And gave all thy chords to light, freedom, and song!
The warm lay of love and the light note of gladness
    Have wakened thy fondest, thy liveliest thrill;
But, so oft hast thou echoed the deep sigh of sadness,
    That even in thy mirth it will steal from thee still.

Dear Harp of my Country! farewell to thy numbers,
    This sweet wreath of song is the last we shall
        twine!   10
Go, sleep with the sunshine of Fame on thy slumbers,
    Till touched by some hand less unworthy than mine;
If the pulse of the patriot, soldier, or lover,
    Have throbbed at our lay, 'tis thy glory alone;
I was but as the wind, passing heedlessly over,
    And all the wild sweetness I waked was thy own.

                                        (*1834*)

IRISH MELODIES. LET ERIN REMEMBER. **3. Malachi:** tenth-century Irish king who conquered a Danish invader and took from him a gold collar. **6. Red-Branch:** a legendary group of Irish knights. **9. Lough Neagh:** lake in Antrim County, Ireland.

Letter To His Mother
Niagara Falls,
July 24, 1804

MY DEAREST MOTHER,—

I have seen the Falls, and am all rapture and amazement. I cannot give you a better idea of what I felt than by transcribing what I wrote off hastily in my journal on returning. "Arrived at Chippewa, within three miles of the Falls, on Saturday, July 21, to dinner. That evening walked towards the Falls, but got no farther than the rapids, which gave us a prelibation of the grandeur we had to expect. Next day, Sunday, July 22, went to visit the Falls. Never shall I forget the impression I felt at the first glimpse of them which we got as the carriage passed over the hill that overlooks them. We were not near enough to be agitated by the terrific effects of the scene; but saw through the trees this mighty flow of waters descending with calm magnificence, and received enough of its grandeur to set imagination on the wing; imagination which, even at Niagara, can outrun reality. I felt as if approaching the very residence of the Deity; the tears started into my eyes; and I remained, for moments after we had lost sight of the scene, in that delicious absorption which pious enthusiasm alone can produce. We arrived at the New Ladder and descended to the bottom. Here all its awful sublimities rushed full upon me. But the former exquisite sensation was gone. I now saw all. The string that had been touched by the first impulse, and which *fancy* would have kept for ever in vibration, now rested at *reality*. Yet, though there was no more to imagine, there was much to feel. My whole heart and soul ascended towards the Divinity in a swell of devout admiration, which I never before experienced. Oh! bring the atheist here, and he cannot return an atheist! I pity the man who can coldly sit down to write a description of these ineffable wonders: much more do I pity him who can submit them to the admeasurement of gallons and yards. It is impossible by pen or pencil to convey even a faint idea of their magnificence. Painting is lifeless; and the most burning words of poetry have all been lavished upon inferior and ordinary subjects. We must have new combinations of language to describe the Falls of Niagara. . . ."

Ever, ever your

TOM

LEIGH HUNT

1784–1859

HUNT was a picturesque figure. To begin with, he reminds us of Micawber in *David Copperfield*. There are the same carefree optimism, generosity to the young and less fortunate; the chaotic household and noisy children; the same hand-to-mouth way of life, innocent affectations, and impracticality. At the same time, though only a middling poet himself, he was a perceptive critic, had read very widely, did much to arouse the interest of younger writers in Italian poetry, and in his lifelong political liberalism showed a genuine courage.

His parents came from America. His father had been a clergyman in the West Indies and his mother was a Quakeress from Philadelphia. Reared in a humanitarian atmosphere, Hunt and his older brother John plunged into journalism, arguing for the abolition of slavery and child labor, for prison reform, religious toleration, and greater political equality. An attack on the dissolute Prince Regent in their magazine *The Examiner* (1811), by which time Hunt was twenty-seven, resulted in a fine and imprisonment for two years. Allowed to have his family and books with him in prison, he received visitors, translated Italian poetry, and, in a typical gesture of defiance, had his room covered with wallpaper to make it look like a rose-bower.

He had been writing verse, off and on, since boyhood, and now made a more serious effort in his *Story of Rimini* (1816), based on the tale of the lovers Paolo and Francesca (which Dante relates in the fifth canto of his *Inferno*). Here especially Hunt used the open couplet and the mixture of sentiment and jaunty colloquialism with which his style is so often associated and which Keats, in his early verse, caught from Hunt. Meanwhile Hunt was befriending Keats and Shelley, and also serving as the main object of attack in a series of articles, the "Cockney School of Poetry," published by the Tory *Blackwood's Magazine*. Other poems followed, especially

Foliage (1818), *Hero and Leander; and Bacchus and Ariadne* (1819), as well as critical and familiar essays. Shelley and Byron asked him to Italy to help with a new magazine, the *Liberal*, but Shelley, Hunt's principal patron, was drowned almost immediately after Hunt arrived (1822). The project was soon abandoned, and Hunt and his family at length returned to England (1825). Though he continued to write poetry, the work of his later years for which he is principally remembered is his prose, especially his *Lord Byron and Some of His Contemporaries* (1828), *Imagination and Fancy* (1844), and his eminently readable *Autobiography* (1850).

Bibliography

In addition to the relatively full edition of 1870–72, in 7 vols., there are the almost complete collection of the poems, ed. H. S. Milford, 1922; the *Essays of Leigh Hunt*, ed. A. Symons, 1887, revised 1903; and the three carefully edited volumes by L. H. and C. W. Houtchens, the *Dramatic Criticism*, 1949, *Literary Criticism*, 1956, and *Political and Occasional Essays*, 1962. The letters were edited, though incompletely, by Thornton Hunt (2 vols., 1862). Hunt's *Autobiography* can be consulted in the editions by Roger Ingpen, 1903, Edmund Blunden, 1928, and especially J. E. Morpurgo, 1948.

The best biography is Edmund Blunden, *Leigh Hunt and His Circle*, 1930. (For those who read French, Louis Landré's detailed *Leigh Hunt*, 2 vols., 1935–36, may also be recommended.) A brief general discussion is to be found in C. D. Thorpe, "Leigh Hunt as Man of Letters: An Essay in Revaluation," prefixed to the Houtchens edition (above) of the *Literary Criticism*.

And Honor, and the Muse with growing wings,
And Love Domestic, smiling equably.

1813 (1813)

TO HAMPSTEAD

WRITTEN DURING THE AUTHOR'S
IMPRISONMENT, AUGUST, 1813

Sweet upland, to whose walks, with fond repair,
Out of thy western slope I took my rise
Day after day, and on these feverish eyes
Met the moist fingers of the bathing air;—
If health, unearned of thee, I may not share,
Keep it, I pray thee, where my memory lies,
In thy green lanes, brown dells, and breezy skies,
Till I return, and find thee doubly fair.

Wait then my coming, on that lighthouse land,
Health, and the joy that out of nature springs,   10
And Freedom's air-blown locks;—but stay with me,
Friendship, frank entering with the cordial hand,

FROM

THE STORY OF RIMINI

from CANTO III

The poem is founded on the episode of Paolo and Francesca in Dante's *Inferno*, Canto V. The following passage is given as a sample of Hunt's idiom and his loose flowing couplet, both of which strongly influenced Keats's *Endymion*.

A noble range it was, of many a rood,
Walled round with trees, and ending in a wood:
Indeed the whole was leafy; and it had
A winding stream about it, clear and glad,
That danced from shade to shade, and on its way
Seemed smiling with delight to feel the day.
There was the pouting rose, both red and white,
The flamy heart's-ease, flushed with purple light,
Blush-hiding strawberry, sunny-coloured box,   390
Hyacinth, handsome with his clustering locks,
The lady lily, looking gently down,
Pure lavender, to lay in bridal gown,
The daisy, lovely on both sides,—in short,
All the sweet cups to which the bees resort,
With plots of grass, and perfumed walks between
Of citron, honeysuckle and jessamine,
With orange, whose warm leaves so finely suit,
And look as if they'd shade a golden fruit;
And midst the flowers, turfed round beneath a shade
Of circling pines, a babbling fountain played,   401
And 'twixt their shafts you saw the water bright,
Which through the darksome tops glimmered with
     showering light.
So now you walked beside an odorous bed
Of gorgeous hues, white, azure, golden, red;
And now turned off into a leafy walk,
Close and continuous, fit for lovers' talk;
And now pursued the stream, and as you trod
Onward and onward o'er the velvet sod,
Felt on your face an air, watery and sweet,   410
And a new sense in your soft-lighting feet;
And then perhaps you entered upon shades,
Pillowed with dells and uplands 'twixt the glades,
Through which the distant palace, now and then,
Looked lordly forth with many-windowed ken;
A land of trees, which reaching round about,
In shady blessing stretched their old arms out,
With spots of sunny opening, and with nooks,
To lie and read in, sloping into brooks,
Where at her drink you started the slim deer,   420
Retreating lightly with a lovely fear.
And all about, the birds kept leafy house,
And sung and sparkled in and out the boughs;
And all about, a lovely sky of blue
Clearly was felt, or down the leaves laughed through.
And here and there, in every part, were seats,
Some in the open walks, some in retreats;
With bowering leaves o'erhead, to which the eye
Looked up half sweetly and half awfully,—

Places of nestling green, for poets made,   430
Where when the sunshine struck a yellow shade,
The slender trunks, to inward peeping sight
Thronged in dark pillars up the gold green light.

But 'twixt the wood and flowery walks, halfway,
And formed of both, the loveliest portion lay,
A spot, that struck you like enchanted ground:—
It was a shallow dell, set in a mound
Of sloping shrubs, that mounted by degrees,
The birch and poplar mixed with heavier trees;
From under which, sent through a marble spout,   440
Betwixt the dark wet green, a rill gushed out,
Whose low sweet talking seemed as if it said
Something eternal to that happy shade:
The ground within was lawn, with plots of flowers
Heaped towards the centre, and with citron bowers;
And in the midst of all, clustered about
With bay and myrtle, and just gleaming out,
Lurked a pavilion,—a delicious sight,
Small, marble, well-proportioned, mellowy white
With yellow vine-leaves sprinkled,—but no more,—
And a young orange either side the door.   451
The door was to the wood, forward, and square,
The rest was domed at top, and circular;
And through the dome the only light came in,
Tinged, as it entered, with the vine-leaves thin.

It was a beauteous piece of ancient skill,
Spared from the rage of war, and perfect still;
By most supposed the work of fairy hands,
Famed for luxurious taste, and choice of lands,—
Alcina, or Morgana,—who from fights   460
And errant fame inveigled amorous knights,
And lived with them in a long round of blisses,
Feasts, concerts, baths, and bower-enshaded kisses.
But 'twas a temple, as its sculpture told,
Built to the Nymphs that haunted there of old;
For o'er the door was carved a sacrifice
By girls and shepherds brought, with reverent eyes,
Of sylvan drinks and foods, simple and sweet,
And goats with struggling horns and planted feet:
And on a line with this ran round about   470
A like relief, touched exquisitely out,
That shewed, in various scenes, the nymphs themselves;
Some by the water side on bowery shelves
Leaning at will,—some in the water sporting
With sides half swelling forth, and looks of courting,—
Some in a flowery dell, hearing a swain
Play on his pipe, till the hills ring again,—

Some tying up their long moist hair,—some sleeping
Under the trees, with fauns and satyrs peeping,—
Or, sidelong-eyed, pretending not to see   480
The latter in the brakes come creepingly,
While their forgotten urns, lying about
In the green herbage, let the water out.
Never, be sure, before or since was seen
A summer-house so fine in such a nest of green.

1812–16 (1816)

TO THE GRASSHOPPER
AND THE CRICKET

❧

This poem was composed in a sonnet-writing contest
with Keats, for whose poem see below, p. 1127.

❧

Green little vaulter in the sunny grass,
Catching your heart up at the feel of June,
Sole voice that's heard amidst the lazy noon,
When even the bees lag at the summoning brass,
And you, warm little housekeeper, who class
With those who think the candles come too soon,
Loving the fire, and with your tricksome tune
Nick the glad silent moments as they pass;

Oh sweet and tiny cousins, that belong,
One to the fields, the other to the hearth,   10
Both have your sunshine; both, though small, are
     strong
At your clear hearts; and both seem given to earth
To ring in thoughtful ears this natural song—
In doors and out, summer and winter, mirth.

1816 (1817)

THE NILE

❧

Written in another sonnet contest with both Shelley and
Keats. Their efforts (the time-limit was fifteen minutes)
were far inferior to that of Hunt, who had proposed the
subject and continued working on the sonnet for several
hours.

❧

It flows through old hushed Egypt and its sands,
Like some grave mighty thought threading a dream,
And times and things, as in that vision, seem
Keeping along it their eternal stands,—
Caves, pillars, pyramids, the shepherd bands
That roamed through the young world, the glory
     extreme
Of high Sesostris, and that southern beam,
The laughing queen that caught the world's great hands.

Then comes a mightier silence, stern and strong,
As of a world left empty of its throng,   10
And the void weighs on us; and then we wake,
And hear the fruitful stream lapsing along
'Twixt villages, and think how we shall take
Our own calm journey on for human sake.

1818 (1818)

ON A LOCK OF MILTON'S HAIR

❧

Hunt had just acquired this lock for his collection. Compare Keats's lines on the same subject written extemporaneously at the suggestion of Hunt, below, p. 1149.

❧

It lies before me there, and my own breath
     Stirs its thin outer threads, as though beside
     The living head I stood in honored pride,
Talking of lovely things that conquer death.
Perhaps he pressed it once, or underneath
     Ran his fine fingers, when he leant, blank-eyed,
     And saw, in fancy, Adam and his bride
With their heaped locks, or his own Delphic wreath.

There seems a love in hair, though it be dead.
It is the gentlest, yet the strongest thread   10
     Of our frail plant,—a blossom from the tree
Surviving the proud trunk;—as if it said,
     Patience and Gentleness is Power. In me
     Behold affectionate eternity.

1818 (1818)

ABOU BEN ADHEM

Abou Ben Adhem, may his tribe increase!
Awoke one night from a deep dream of peace,
And saw, within the moonlight in his room,
Making it rich, and like a lily in bloom,
An angel writing in a book of gold:—

Exceeding peace had made Ben Adhem bold,
And to the presence in the room he said,
"What writest thou?"—The vision raised its head,
And with a look made of all sweet accord,
Answered, "The names of those who love the Lord."
"And is mine one?" said Abou. "Nay, not so,"   11
Replied the angel. Abou spoke more low,
But cheerly still; and said, "I pray thee then,
"Write me as one that loves his fellow men."

The angel wrote and vanished. The next night
It came again with a great wakening light,
And showed the names whom love of God had blessed,
And lo! Ben Adhem's name led all the rest.

1834 (1844)

RONDEAU

ᗛᨆᗝ

"Jenny" was Jane Welsh Carlyle, wife of Thomas
Carlyle. Hunt, on a visit, had announced that one of
Carlyle's works had just been accepted for publication.

ᗛᨆᗝ

Jenny kissed me when we met,
   Jumping from the chair she sat in;
Time, you thief, who love to get
   Sweets into your list, put that in:
Say I'm weary, say I'm sad,
   Say that health and wealth have missed me,
Say I'm growing old, but add,
   Jenny kissed me.

1838 (1838)

from WHAT IS POETRY?

Poetry is a passion, because it seeks the deepest impressions; and because it must undergo, in order to convey them.

It is a passion for truth, because without truth the impression would be false or defective.

It is a passion for beauty, because its office is to exalt and refine by means of pleasure, and because beauty is nothing but the loveliest form of pleasure.

It is a passion for power, because power is impression triumphant, whether over the poet as desired by himself, or over the reader as affected by the poet.

It embodies and illustrates its impressions by imagination, or images of the objects of which it treats, and other images brought in to throw light on those objects, in order that it may enjoy and impart the feeling of their truth in its utmost conviction and affluence.

It illustrates them by fancy, which is a lighter play of imagination or the feeling of analogy coming short of seriousness, in order that it may laugh with what it loves, and show how it can decorate it with fairy ornament.

It modulates what it utters because in running the whole round of beauty it must needs include beauty of sound, and because, in the height of its enjoyment, it must show the perfection of its triumph, and make difficulty itself become part of its facility and joy.

And lastly, Poetry shapes this modulation into uniformity for its outline and variety for its parts, because it thus realizes the last idea of beauty itself, which includes the charm of diversity within the flowing round of habit and ease.

Poetry is imaginative passion. The quickest and subtlest test of the possession of its essence is in expression; the variety of things to be expressed shows the amount of its resources; and the continuity of the song completes the evidence of its strength and greatness. He who has thought, feeling, expression, imagination, action, character, and continuity, all in the largest amount and highest degree, is the greatest poet.

Poetry includes whatsoever of painting can be made visible to the mind's eye, and whatsoever of music can be conveyed by sound and proportion without singing or instrumentation. But it far surpasses those divine arts in suggestiveness, range, and intellectual wealth;—the first, in expression of thought, combination of images, and the triumph over space and time; the second, in all that can be done by speech apart from the tones and modulations of pure sound. Painting and music, however, include all those portions of the gift of poetry that can be expressed and heightened by the visible and melodious. Painting, in a certain apparent manner, is things themselves; music, in a certain audible manner, is their very emotion and grace. Music and painting are proud to be related to poetry, and poetry loves and is proud of them.

Poetry begins where matter of fact or of science ceases to be merely such and to exhibit a further truth, that is to say, the connection it has with the world of emotion, and its power to produce imaginative pleasure. Inquiring of a gardener, for instance, what flower it is we see yonder, he answers, "A lily." This is matter of fact. The botanist pronounces it to be of the order of "Hexandria monogynia." This is matter of science. It is the "lady" of the garden, says Spenser; and here we begin to have a poetical sense of its fairness and grace. It is

> The plant and flower of *light*,[1]

says Ben Jonson; and poetry then shows us the beauty of the flower in all its mystery and splendor.

If it be asked how we know perceptions like these to be true, the answer is, by the fact of their existence—by the consent and delight of poetic readers. And as feeling is the earliest teacher, and perception the only final proof of things the most demonstrable by science, so the remotest imaginations of the poets may often be found to have the closest connection with matter of fact; perhaps might always be so, if the subtlety of our perceptions were a match for the causes of them. Consider this image of Ben Jonson's—of a lily being the flower of light. Light, undecomposed, is white; and as the lily is white, and light is white, and whiteness itself is nothing *but* light, the two things, so far, are not merely similar, but identical. A poet might add, by an analogy drawn from the connection of light and color, that there is a "golden dawn" issuing out of the white lily in the rich yellow of the stamens. I have no desire to push this similarity farther than it may be worth. Enough has been stated to show that, in poetical as well as in other analogies, "the same feet of Nature", as Bacon says, may be seen "treading in different paths";[2] and that the most scornful, that is

to say, dullest disciple of fact, should be cautious how he betrays the shallowness of his philosophy by discerning no poetry in its depths.

But the poet is far from dealing only with these subtle and analogical truths. Truth of every kind belongs to him, provided it can bud into any kind of beauty, or is capable of being illustrated and impressed by the poetic faculty. Nay, the simplest truth is often so beautiful and impressive of itself that one of the greatest proofs of his genius consists in his leaving it to stand alone, illustrated by nothing but the light of its own tears or smiles, its own wonder, might, or playfulness. Hence the complete effect of many a simple passage in our old English ballads and romances, and of the passionate sincerity in general of the greatest early poets, such as Homer and Chaucer, who flourished before the existence of a "literary world," and were not perplexed by a heap of notions and opinions, or by doubts how emotion ought to be expressed. The greatest of their successors never write equally to the purpose, except when they can dismiss everything from their minds but the like simple truth. In the beautiful poem of *Sir Eger, Sir Graham, and Sir Gray-Steel* (see it in Ellis's *Specimens* or Laing's *Early Metrical Tales*) a knight thinks himself disgraced in the eyes of his mistress:—

> Sir Eger said, "If it be so,
> Then wot I well I must forgo
> Love-liking, and manhood, all clean!"
> *The water rushed out of his een!*[3]

. . .

There are different kinds and degrees of imagination, some of them necessary to the formation of every true poet, and all of them possessed by the greatest. Perhaps they may be enumerated as follows: first, that which presents to the mind any object or circumstance in everyday life, as when we imagine a man holding a sword, or looking out of a window; second, that which presents real, but not everyday circumstances, as King Alfred tending the loaves, or Sir Philip Sidney giving up the water to the dying soldier; third, that which combines character and events directly imitated from real life, with imitative realities of its own invention, as the probable parts of the histories of Priam and Macbeth, or what may be called natural fiction as distinguished from supernatural; fourth, that which conjures up things and events not to be found in nature, as Homer's gods and Shakespeare's witches, enchanted

WHAT IS POETRY? **1. The . . . light:** "To the Immortall Memorie, and Friendship of that Noble Pair, Sir Lucius Cary, and Sir H. Morison," l. 72. **2. "the same . . . paths":** *Advancement of Learning*, II. v. 3.

3. Sir . . . een: ll. 773–76.

horses and spears, Ariosto's hippogriff, &c.; fifth, that which, in order to illustrate or aggravate one image, introduces another: sometimes in simile, as when Homer compares Apollo descending in his wrath at noon-day to the coming of night-time; sometimes in metaphor, or simile comprised in a word, as in Milton's "motes that *people* the sunbeams"; sometimes in concentrating into a word the main history of any person or thing, past or even future, as in the "starry Galileo" of Byron, and that ghastly foregone conclusion of the epithet "murdered" applied to the yet living victim in Keats's story from Boccaccio,—

> So the two brothers and their *murdered* man
> Rode towards fair Florence;—[4]

sometimes in the attribution of a certain representative quality which makes one circumstance stand for others, as in Milton's gray-fly winding its *"sultry horn,"*[5] which epithet contains the heat of a summer's day; sixth, that which reverses this process, and makes a variety of circumstances take color from one, like nature seen with jaundiced or glad eyes, or under the influence of storm or sunshine; as when in *Lycidas,* or the Greek pastoral poets, the flowers and the flocks are made to sympathize with a man's death; or, in the Italian poet, the river flowing by the sleeping Angelica seems talking of love—

> Parea che l'erba le fiorisse intorno,
> E d'amor ragionasse quella riva![6]
> *Orlando Innamorato,* Canto iii.

or in the voluptuous homage paid to the sleeping Imogen by the very light in the chamber and the reaction of her own beauty upon itself; or in the "witch element" of the tragedy of *Macbeth* and the May-day night of *Faust;* seventh, and last, that which by a single expression, apparently of the vaguest kind, not only meets but surpasses in its effect the extremest force of the most particular description; as in that exquisite passage of Coleridge's *Christabel* where the unsuspecting object of the witch's malignity is bidden to go to bed:—

> Quoth Christabel, So let it be!
> And as the lady bade, did she.
> Her gentle limbs did she undress,
> *And lay down in her loveliness;*—[7]

a perfect verse surely, both for feeling and music. The very smoothness and gentleness of the limbs is in the series of the letter *l*'s.

I am aware of nothing of the kind surpassing that most lovely inclusion of physical beauty in moral, neither can I call to mind any instances of the imagination that turns accompaniments into accessories, superior to those I have alluded to. Of the class of comparison, one of the most touching (many a tear must it have drawn from parents and lovers) is in a stanza which has been copied into the *Friar of Orders Gray* out of Beaumont and Fletcher:—

> Weep no more, lady, weep no more,
>   Thy sorrow is in vain;
> *For violets plucked the sweetest showers*
>   *Will ne'er make grow again.*[8]

And Shakespeare and Milton abound in the very grandest; such as Antony's likening his changing fortunes to the cloudrack; Lear's appeal to the old age of the heavens; Satan's appearance in the horizon, like a fleet "hanging in the clouds"; and the comparisons of him with the comet and the eclipse. Nor unworthy of this glorious company for its extraordinary combination of delicacy and vastness is that enchanting one of Shelley's in the Adonais:—

> Life, like a dome of many-colored glass,
> Stains the white radiance of eternity.[9]

I multiply these particulars in order to impress upon the reader's mind the great importance of imagination in all its phases as a constituent part of the highest poetic faculty.

1844 (1844)

PROEM TO SELECTIONS FROM KEATS

Keats was born a poet of the most poetical kind. All his feelings came to him through a poetical medium, or were speedily colored by it. He enjoyed a jest as heartily as any one, and sympathized with the lowliest commonplace; but the next minute his thoughts were

---

4. So . . . Florence: *Isabella,* st. xxvii. **5. "sultry horn":** "Lycidas," l. 28. **6. Parea . . . riva:** "It appeared that the grass flourished about them, and that that bank discoursed of love." **7. Quoth . . . loveliness:** *Christabel,* ll. 235-38.

8. Weep . . . again: *The Queen of Corinth,* III. ii. 1-4.
9. Life . . . eternity: st. lii, printed below under Shelley.

in a garden of enchantment, with nymphs, and fauns, and shapes of exalted humanity:

Elysian beauty, melancholy grace.[1]

It might be said of him that he never beheld an oak tree without seeing the Dryad. His fame may now forgive the critics who disliked his politics, and did not understand his poetry. Repeated editions of him in England, France, and America attest its triumphant survival of all obloquy; and there can be no doubt that he has taken a permanent station among the British poets, of a very high, if not thoroughly mature, description.

Keats's early poetry, indeed, partook plentifully of the exuberance of youth; and even in most of his later, his sensibility, sharpened by mortal illness, tended to a morbid excess. His region is "a wilderness of sweets," —flowers of all hue, and "weeds of glorious feature,"— where, as he says, the luxuriant soil brings

The pipy hemlock to strange undergrowth.[2]

But there also is the "rain-scented eglantine," and bushes of May-flowers, with bees, and myrtle, and bay,—and endless paths into forests haunted with the loveliest as well as gentlest beings; and the gods live in the distance, amid notes of majestic thunder. I do not say that no "surfeit" is ever there; but I do, that there is no end of the "nectared sweets." In what other English poet (however superior to him in other respects) are you so *certain* of never opening a page without lighting upon the loveliest imagery and the most eloquent expressions? Name one. Compare any succession of their pages at random, and see if the young poet is not sure to present his stock of beauty; crude it may be, in many instances; too indiscriminate in general; never, perhaps, thoroughly perfect in cultivation; but there it is, exquisite of its kind, and filling envy with despair. He died at five-and-twenty; he had not revised his earlier works, nor given his genius its last pruning. His *Endymion*, in resolving to be free from all critical trammels, had no versification; and his last noble fragment, *Hyperion*, is not faultless,— but it is nearly so. *The Eve of St. Agnes* betrays morbidity only in one instance (noticed in the comment). Even in his earliest productions, which are to be considered as those of youth just emerging from boyhood, are to be found passages of as masculine a beauty as ever

were written. Witness the *Sonnet on Reading Chapman's Homer*,—epical in the splendor and dignity of its images, and terminating with the noblest Greek simplicity. Among his finished productions, however, of any length, *The Eve of St. Agnes* still appears to me the most delightful and complete specimen of his genius. It stands midway between his most sensitive ones (which, though of rare beauty, occasionally sink into feebleness) and the less generally characteristic majesty of the fragment of *Hyperion*. Doubtless his greatest poetry is to be found in *Hyperion;* and had he lived, there is as little doubt he would have written chiefly in that strain; rising superior to those languishments of love which made the critics so angry, and which they might so easily have pardoned at his time of life. But *The Eve of St. Agnes* had already bid most of them adieu,—exquisitely loving as it is. It is young, but full-grown poetry of the rarest description; graceful as the beardless Apollo; glowing and gorgeous with the colors of romance. I have therefore reprinted the whole of it in the present volume, together with the comment alluded to in the Preface; especially as, in addition to felicity of treatment, its subject is in every respect a happy one, and helps to "paint" this our bower of "poetry with delight." Melancholy, it is true, will "break in" when the reader thinks of the early death of such a writer; but it is one of the benevolent provisions of nature that all good things tend to pleasure in the recollection, when the bitterness of their loss is past, their own sweetness embalms them.

A thing of beauty is a joy forever.[3]

While writing this paragraph, a hand-organ out-of-doors has been playing one of the mournfullest and loveliest of the airs of Bellini—another genius who died young. The sound of music always gives a feeling either of triumph or tenderness to the state of mind in which it is heard: in this instance it seemed like one departed spirit come to bear testimony to another, and to say how true indeed may be the union of sorrowful and sweet recollections.

Keats knew the youthful faults of his poetry as well as any man, as the reader may see by the preface to *Endymion*, and its touching though manly acknowledgment of them to critical candor. I have this moment read it again, after a lapse of years, and have been astonished to think how anybody could answer such an appeal to the mercy of strength, with the cruelty

PROEM TO SELECTIONS FROM KEATS. **1. Elysian . . . grace:** Wordsworth's "Laodamia," l. 95. **2. The . . . undergrowth:** *Endymion*, I. 241.

3. A . . . forever: the opening line of *Endymion*.

of weakness. All the good for which Mr. Gifford[4] pretended to be zealous, he might have effected with pain to no one, and glory to himself; and therefore all the evil he mixed with it was of his own making. But the secret at the bottom of such unprovoked censure is exasperated inferiority. Young poets, upon the whole, —at least very young poets,—had better not publish at all. They are pretty sure to have faults; and jealousy and envy are as sure to find them out, and wreak upon them their own disappointments. The critic is often an unsuccessful author, almost always an inferior one to a man of genius, and possesses his sensibility neither to beauty nor to pain. If he does,—if by any chance he is a man of genius himself (and such things have been), sure and certain will be his regret, some day, for having given pains which he might have turned into noble pleasures; and nothing will console him but that very charity towards himself, the grace of which can only be secured to us by our having denied it to no one.

Let the student of poetry observe that in all the luxury of *The Eve of St. Agnes* there is nothing of the conventional craft of artificial writers; no heaping up of words or similes for their own sakes or the rhyme's sake; no gaudy commonplaces; no borrowed airs of earnestness; no tricks of inversion; no substitution of reading or of ingenious thoughts for feeling or spontaneity; no irrelevancy or unfitness of any sort. All flows out of sincerity and passion. The writer is as much in love with the heroine as his hero is; his description of the painted window, however gorgeous, has not an untrue or superfluous word; and the only speck of a fault in the whole poem arises from an excess of emotion.

1844 (1844)

FROM

LORD BYRON AND SOME OF HIS CONTEMPORARIES

. . . Lord Byron had naturally as much regard for his title as any other nobleman; perhaps more, because he had professed not to care about it. Besides, he had a poetical imagination. Mr. Shelley, who, though he

4. **Mr. Gifford:** William Gifford, editor of *The Quarterly Review*, which had savagely attacked Keats's *Endymion* in 1818.

had not known him longer, had known him more intimately, was punctilious in giving him his title, and told me very plainly that he thought it best for all parties. His oldest acquaintances, it is true, behaved in this respect, as it is the custom to behave in great familiarity of intercourse. Mr. Shelley did not choose to be so familiar; and he thought, that although I had acted differently in former times, a long suspension of intercourse would give farther warrant to a change, desirable on many accounts, quite unaffected, and intended to be acceptable. I took care, accordingly, not to accompany my new punctilio with any air of study or gravity. In every other respect things appeared the same as before. We laughed, and chatted, and rode out, and were as familiar as need be; and I thought he regarded the matter just as I wished. However, he did not like it.

This may require some explanation. Lord Byron was very proud of his rank. M. Beyle ("Count Stendhal"), when he saw him at the opera in Venice, made this discovery at a glance; and it was a discovery no less subtle than true. He would appear sometimes as jealous of his title, as if he had usurped it. A friend told me, that an Italian apothecary having sent him one day a packet of medicines, addressed to "Mons. Byron," this mock-heroic mistake aroused his indignation, and he sent back the physic to learn better manners. His coat of arms was fixed up in front of his bed. I have heard that it was a joke with him to mystify the sense of the motto to his fair friend, who wished particularly to know what "Crede Byron" meant. The motto, it must be acknowledged, was awkward. The version, to which her Italian helped her, was too provocative of comment to be allowed. There are mottoes, as well as scutcheons, of pretence, which must often occasion the bearers much taunt and sarcasm, especially from indignant ladies. Custom, indeed, and the interested acquiescence of society, enable us to be proud of imputed merits, though we contradict them every day of our life: otherwise it would be wonderful how people could adorn their equipages, and be continually sealing their letters with maxims and stately moralities, ludicrously inapplicable. It would be like wearing ironical papers in their hats.

But Lord Byron, besides being a lord, was a man of letters, and he was extremely desirous of the approbation of men of letters. He loved to enjoy the privileges of his rank, and at the same time to be thought above them. It is true, if he thought you not above them yourself, he was the better pleased. On this account among others, no man was calculated to delight him in a higher degree than Thomas Moore; who with every charm he wished for in a companion, and a reputation for independence and liberal opinion, admired both genius and title for their own sakes. But his Lordship did not always feel quite secure of the bon-mots of his brother wit. His conscience had taught him suspicion; and it was a fault with him and his *côterie*, as it is with most, that they all talked too much of one another behind their backs. But "admiration at all events" was his real motto. If he thought you an admirer of titles, he was well pleased that you should add that homage to the other, without investigating it too nicely. If not, he was anxious that you should not suppose him anxious about the matter. When he beheld me, therefore, in the first instance, taking such pains to show my philosophy, he knew very well that he was secure, address him as I might; but now that he found me grown older, and suspected from my general opinions and way of life, that my experience, though it adopted the style of the world when mixing with it, partook less of it than ever in some respects, he was chagrined at this change in my appellatives. He did not feel so at once; but the more we associated, and the greater insight he obtained into the tranquil and unaffected conclusions I had come to on a great many points, upon which he was desirous of being thought as indifferent as myself, the less satisfied he became with it. At last, thinking I had ceased to esteem him, he petulantly bantered me on the subject. I knew, in fact, that under all the circumstances, neither of us could afford a change back again to the old entire familiarity; he, because he would have regarded it as a triumph warranting very peculiar consequences, and such as would by no means have saved me from the penalties of the previous offence; and I, because I was under certain disadvantages, that would not allow me to indulge him. With any other man, I would not have stood it out. It would have ill become the very sincerity of my feelings. But even the genius of Lord Byron did not enable him to afford being conceded to. He was so annoyed one day at Genoa at not succeeding in bantering me out of my epistolary proprieties, that he addressed me a letter beginning, "Dear Lord Hunt." This sally made me laugh heartily. I told him so; and my unequivocal relish of the joke pacified him; so that I heard no more on the subject.

The familiarities of my noble acquaintance, which I had taken at first for a compliment and a cordiality, were dealt out in equal portions to all who came near him. They proceeded upon that royal instinct of an immeasurable distance between the parties, the safety of which, it is thought, can be compromised by no appearance of encouragement. The farther you are off, the more securely the personage may indulge your good opinion of him. The greater his merits, and the more transporting his condescension, the less can you be so immodest as to have pretensions of your own. You may be intoxicated into familiarity. That is excusable, though not desirable. But not to be intoxicated any how,—not to show any levity, and yet not to be possessed with a seriousness of the pleasure, is an offence. When I agreed to go to Italy and join in setting up the proposed work, Shelley, who was fond of giving his friends appellations, happened to be talking one day with Lord Byron of the mystification which the name of "Leigh Hunt" would cause the Italians; and passing from one fancy to another, he proposed that they should translate it into Leontius. Lord Byron approved of this conceit, and at Pisa was in the habit of calling me so. I liked it; especially as it seemed a kind of new link with my beloved friend, then, alas! no more. I was pleased to be called in Italy, what he would have called me there had he been alive: and the familiarity was welcome to me from Lord Byron's mouth, partly because it pleased himself, partly because it was not of a wordly fashion, and the link with my friend was thus rendered compatible. In fact, had Lord Byron been what I used to think him, he might have called me what he chose; and I should have been as proud to be at his call, as I endeavoured to be pleased. As it was, there was something not unsocial nor even unenjoying in our intercourse, nor was there any appearance of constraint; but, upon the whole, it was not pleasant: it was not cordial. There was a sense of mistake on both sides. However, this came by degrees. At first there was hope, which I tried hard to indulge; and there was always some joking going forward; some melancholy mirth, which a spectator might have taken for pleasure.

Our manner of life was this. Lord Byron, who used to sit up at night, writing Don Juan (which he did under the influence of gin and water), rose late in the morning. He breakfasted; read; lounged about, singing an air, generally out of Rossini, and in a swaggering style, though in a voice at once small and veiled; then took a bath, and was dressed; and coming down-stairs, was heard, still singing, in the court-yard, out of which the garden ascended at the back of the house. The servants at the same time brought out two or three chairs. My study, a little room in a corner, with an orange-tree peeping in at the window, looked upon this court-yard. I was generally at my writing when he came down, and either acknowledged his presence by getting up and saying something from the window, or he called out "Leontius!" and came halting up to the window with some joke, or other challenge to conversation. (Readers of good sense will do me the justice of discerning where any thing is spoken of in a tone of objection, and where it is only brought in as requisite to the truth of the picture.) His dress, as at Monte-Nero, was a nankin[1] jacket, with white waist-coat and trowsers, and a cap, either velvet or linen, with a shade to it. In his hand was a tobacco-box, from which he helped himself like unto a shipman, but for a different purpose; his object being to restrain the pinguifying impulses of hunger. Perhaps also he thought it good for the teeth. We then lounged about, or sat and talked, Madame Guiccioli with her sleek tresses descending after her toilet to join us. The garden was small and square, but plentifully stocked with oranges and other shrubs; and, being well watered, looked very green and refreshing under the Italian sky. The lady generally attracted us up into it, if we had not been there before. . . .

He had a delicate white hand, of which he was proud; and he attracted attention to it by rings. He thought a hand of this description almost the only mark remaining now-a-days of a gentleman; of which it certainly is not, nor of a lady either; though a coarse one implies handiwork. He often appeared holding a handker-chief, upon which his jewelled fingers lay imbedded, as in a picture. He was as fond of fine linen as a quaker; and had the remnant of his hair oiled and trimmed with all the anxiety of a Sardanapalus.

The visible character to which this effeminacy gave rise appears to have indicated itself as early as his travels in the Levant, where the Grand Signior is said to have taken him for a woman in disguise. But he had tastes of a more masculine description. He was fond of swimming to the last, and used to push out to a good distance in the Gulf of Genoa. He was also, as I have before mentioned, a good horseman; and he liked to have a great dog or two about him, which is not a habit observable in timid men. . . .

Mr. Shelley, when he died, was in his thirtieth year. His figure was tall and slight, and his constitution consumptive. He was subject to violent spasmodic pains, which would sometimes force him to lie on the ground till they were over; but he had always a kind word to give to those about him, when his pangs allowed him to speak. In his organization, as well as in some other respects, he resembled the German poet, Schiller. Though well-turned, his shoulders were bent a little, owing to premature thought and trouble. The same causes had touched his hair with grey; and though his habits of temperance and exercise gave him a remarkable degree of strength, it is not supposed that he could have lived many years. He used to say, that he had lived three times as long as the calendar gave out; which he would prove, between jest and earnest, by some remarks on Time,

"That would have puzzled that stout Stagyrite."[2]

Like the Stagyrite's, his voice was high and weak. His eyes were large and animated, with a dash of wild-ness in them; his face small, but well-shaped, par-ticularly the mouth and chin, the turn of which was very sensitive and graceful. His complexion was naturally fair and delicate, with a colour in the cheeks. He had brown hair, which, though tinged with grey, surmounted his face well, being in considerable quan-tity, and tending to a curl. His side-face upon the whole was deficient in strength, and his features would not have told well in a bust; but when fronting and looking at you attentively, his aspect had a certain seraphical character that would have suited a portrait of John the Baptist, or the angel whom Milton describes as holding a reed "tipt with fire." Nor would the most religious mind, had it known him, have objected to the comparison; for, with all his scepticism, Mr. Shelley's disposition may be truly said to have been any thing but irreligious. A person of much eminence for piety in our times has well observed, that the greatest want of religious feeling is not to be found among the greatest infidels, but among those who never think of religion but as a matter of course. The leading feature of Mr. Shelley's character, may be said to have been a natural piety. He was pious towards nature, towards his friends, towards the whole human race, towards the meanest insect of the forest. He did himself an injustice with the public, in using the popular name of the Supreme Being inconsiderately. He identified it

---

LORD BYRON AND SOME OF HIS CONTEMPORARIES. **1. nankin:** a buff or yellow cotton fabric.

2. "That . . . Stagyrite": from Charles Lamb's "Written at Cambridge," l. 14. The Stagirite is Aristotle, born at Stagira.

solely with the most vulgar and tyrannical notions of a God made after the worst human fashion; and did not sufficiently reflect, that it was often used by a juster devotion to express a sense of the great Mover of the universe. An impatience in contradicting wordly and pernicious notions of a supernatural power, led his own aspirations to be misconstrued; for though, in the severity of his dialectics, and particularly in moments of despondency, he sometimes appeared to be hopeless of what he most desired,—and though he justly thought, that a Divine Being would prefer the increase of benevolence and good before any praise, or even recognition of himself, (a reflection worth thinking of by the intolerant,) yet there was in reality no belief to which he clung with more fondness than that of some great pervading "Spirit of Intellectual Beauty;" as may be seen in his aspirations on that subject. He said to me in the cathedral at Pisa, while the organ was playing, "What a divine religion might be found out, if charity were really made the principle of it, instead of faith!"

Music affected him deeply. He had also a delicate perception of the beauties of sculpture. It is not one of the least evidences of his conscientious turn of mind, that with the inclination, and the power, to surround himself in Italy with all the graces of life, he made no sort of attempt that way; finding other use for his money, and not always satisfied with himself for indulging even in the luxury of a boat. When he bought elegancies of any kind, it was to give away. Boating was his great amusement. He loved the mixture of action and repose which he found in it; and delighted to fancy himself gliding away to Utopian isles, and bowers of enchantment. But he would give up any pleasure to do a deed of kindness. . . .

[Shelley] passed his days like a hermit. He rose early in the morning, walked and read before breakfast, took that meal sparingly, wrote and studied the greater part of the morning, walked and read again, dined on vegetables (for he took neither meat nor wine,) conversed with his friends, (to whom his house was ever open,) again walked out, and usually finished with reading to his wife till ten o'clock, when he went to bed. This was his daily existence. His book was generally Plato, or Homer, or one of the Greek tragedians, or the Bible, in which last he took a great, though peculiar, and often admiring interest. One of his favourite parts was the book of Job. The writings attributed to Solomon he thought too Epicurean, in the modern sense of the word; and in his notions of St. Paul, he agreed with the writer of the work entitled "Not Paul but Jesus." For his Christianity, in the proper sense of the word, he went to the gospel of St. James, and to the sermon on the Mount by Christ himself, for whose truly divine spirit he entertained the greatest reverence. There was nothing which embittered his reviewers against him more than the knowledge of this fact, and his refusal to identify their superstitions and worldly use of the Christian doctrines with the just idea of a great Reformer and advocate of the many; one, whom they would have been the first to cry out against, had he appeared now. His want of faith, indeed, in one sense of the word, and his exceeding faith in the existence of goodness and the great doctrine of charity, formed a comment, the one on the other, very formidable to the less troublesome constructions of the orthodox.

Some alarmists at Marlow said, that if he went on at this rate, he would make all the poor people infidels. He went on, till ill health and calumny, and the love of his children, forced him abroad. During his residence at Marlow, Mr. Shelley published a "Proposal for putting Reform to the Vote" throughout England; for which purpose, as an earnest of his sincerity, he offered to contribute a hundred pounds. This hundred pounds (which owing to his liberal habits he could very ill spare at the time) he would have done his best to supply, by saving and economizing. It was not uncommon with him to give away all his ready money, and be compelled to take a journey on foot, or on the top of a stage, no matter during what weather. His constitution, though naturally consumptive, had attained, by temperance and exercise, to a surprising power of resisting fatigue. As an instance of his extraordinary generosity, an acquaintance of his, a man of letters, enjoyed from him at that period a pension of a hundred a-year; and he continued to enjoy it, till fortune rendered it superfluous. But the princeliness of his disposition was seen most in his behaviour to his friend, the writer of this memoir, who is proud to relate, that Mr. Shelley once made him a present of fourteen hundred pounds, to extricate him from debt.

Mr. Keats, when he died, had just completed his four-and-twentieth year. He was under the middle height; and his lower limbs were small in comparison with his upper, but neat and well-turned. His shoulders were very broad for his size: he had a face, in which energy and sensibility were remarkably mixed up, an eager power checked and made patient by ill health.

Every feature was at once strongly cut, and delicately alive. If there was any faulty expression, it was in the mouth, which was not without something of a character of pugnacity. The face was rather long than otherwise; the upper lip projected a little over the under; the chin was bold, the cheeks sunken; the eyes mellow and glowing; large, dark and sensitive. At the recital of a noble action, or a beautiful thought, they would suffuse with tears, and his mouth trembled. In this, there was ill health as well as imagination, for he did not like these betrayals of emotion; and he had great personal as well as moral courage. His hair, of a brown colour, was fine, and hung in natural ringlets. The head was a puzzle for the phrenologists, being remarkably small in the skull; a singularity which he had in common with Lord Byron and Mr. Shelley, none of whose hats I could get on. Mr. Keats was sensible of the disproportion above noticed, between his upper and lower extremities; and he would look at his hand, which was faded, and swollen in the veins, and say it was the hand of a man of fifty. He was a seven months' child: his mother, who was a lively woman, passionately fond of amusement, is supposed to have hastened her death by too great an inattention to hours and seasons. Perhaps she hastened that of her son.

Mr. Keats's origin was of the humblest description; he was born October 29, 1796, at a livery-stables in Moorfields, of which his grandfather was the proprietor. I am very incurious, and did not know this till the other day. He never spoke of it, perhaps out of a personal soreness which the world had exasperated. After receiving the rudiments of a classical education at Mr. Clarke's school at Enfield, he was bound apprentice to Mr. Hammond, a surgeon, in Church-street, Edmonton; and his enemies having made a jest even of this, he did not like to be reminded of it; at once disdaining them for their meanness, and himself for being sick enough to be moved by them. Mr. Clarke, junior, his schoolmaster's son, a reader of genuine discernment, had encouraged with great warmth the genius that he saw in the young poet; and it was to Mr. Clarke I was indebted for my acquaintance with him. I shall never forget the impression made upon me by the exuberant specimens of genuine though young poetry that were laid before me, and the promise of which was seconded by the fine fervid countenance of the writer. We became intimate on the spot, and I found the young poet's heart as warm as his imagination. We read and walked together, and used to write verses of an evening upon a given subject. No imaginative pleasure was left unnoticed by us, or unenjoyed; from the recollection of the bards and patriots of old, to the luxury of a summer rain at our window, or the clicking of the coal in winter-time. Not long afterwards, having the pleasure of entertaining at dinner Mr. Godwin, Mr. Hazlitt, and Mr. Basil Montague, I showed them the verses of my young friend, and they were pronounced to be as extraordinary as I thought them. One of them was that noble sonnet on first reading Chapman's Homer, which terminates with so energetic a calmness, and which completely announced the new poet taking possession.

1823–28 (1828)

THOMAS DE QUINCEY

1785–1859

ESTIMATES of De Quincey's stature as a writer tend to vary more than those of the other principal men of letters of the time, and not merely from period to period but also within each generation, from his own to the present. There are readers who place him among the half-dozen greatest writers of the age; others find him self-involved, sprawling, and even tedious; while a third group grants him an undeniable interest but almost entirely for psychological or historical reasons.

This wide difference of reaction can be readily explained by two characteristics of his work. One of them is the rather specialized interest of the subject matter in at least three-quarters of his vast output. Even though he skillfully trod the brink between information or analysis and journalism, this necessarily lessened his audience in his own day, and with the passing of time much of this writing has inevitably been supplanted—a hazard foreseen by De Quincey himself. A second explanation is the ornately erudite self-consciousness of his approach, which, whatever its merits, tends to specialize his appeal even more. For the burden of the reader's attention becomes shifted to the style and, through the style, to the man. De Quincey's virtuosity, moreover, is one that often seems to request such attention. Even the elaborately worked-up simplicity of Landor is more willing—to use Hazlitt's phrase—to "share the palm with the subject." There are any number of possible explanations or justifications, including the fundamental one that the result is worth it. But to solicit the reader by ostentation can prove a mixed blessing to a writer. Moreover, since demand has a way of provoking counter-demand, the reader may impose standards or hang back with reservations that would not otherwise come to mind. Finally, the differences with which De Quincey is viewed are probably deepened by the quasi-moral distaste that many people have always felt before any highly mannered form of expression.

Yet admirers continue to appear (Edmund Wilson is an example in our own day); and they tend as a rule to be widely informed readers—often men active themselves in the profession of letters—who prize variety in literature and justly regard De Quincey as a fascinating example of one form of it. They delight in the very combination of qualities that seems to unsettle other readers—a character at once shy and persistent, urbane and childlike; and, in his writing, the brilliant interplay of fact and fancy, of learning and impressionism, of precision and sheer bulk of output, and an almost businesslike professional competence unpredictably graced by a rakish and Bohemian cast. Again, at least for our century, there is the historical fascination of his position as a writer both of and yet apart from the English Romantic period: so much of it, and in so many ways, that we should naturally expect to find other writers of the period like him, and yet apart from it for the blunt, challenging reason that we cannot. Finally—to mention a category that cuts across mere differences of period—he is in the line of Defoe, Johnson, Goldsmith, and Hazlitt, as one of the great literary journalists of the English language.

De Quincey was born August 15, 1785, the fifth of eight children of a well-to-do Manchester merchant, and quickly proved precocious. At the age of thirteen, for example, he could write Greek fluently. In another two years he could talk in it with equal ease. "From my birth," he later said, "I was made an intellectual creature, and intellectual in the highest sense my pursuits and pleasures have been." His father died when he was seven. His mother, who had been left a very substantial income, was keenly interested in the education of the children, especially of her gifted son. The family soon moved to Bath; and De Quincey attended a school there, another in Wiltshire, and still another in Manchester. By this time he was fifteen, chafing against the leisureliness of his preparation and, perhaps hoping his actions would prompt permission to enter Oxford earlier than had been planned, he ran away. His mother and uncle then granted him a generous allowance with which he hoped to spend time in Wales studying independently. But he soon left his retreat in Wales, disappeared into the vagrant world of London, felt the delights of independence, and haunted Soho and the moneylenders. He was finally allowed to apply for Worcester College, Oxford, which he entered shortly after he became eighteen (1803).

His health by this time had become rather shaky. He suffered painful attacks of neuralgia and "gnawing pains in the stomach," which may have been a symptom of ulcers. It was now (1804) that he began to take opium in the liquid form of laudanum. At Oxford his work was brilliant and spasmodic. He appeared reluctant to hurry to a degree, left the university at times, and, during this period, became acquainted with Coleridge and visited Wordsworth in the Lake Country. Leaving Oxford permanently in 1808, without a degree, he began the study of law at the Middle Temple, but again departed for the Lake Country (1809). There he leased Dove Cottage (the house in which Wordsworth had lived), arriving in that small place with twenty-nine chests full of books. It was "a great pleasure to us all," said Dorothy Wordsworth, "to have access to such a library," and was especially a "solid advantage to my brother." Here in Grasmere he was to stay until 1820. Within a few years his friendship with Wordsworth was to begin a slow decline. It later faded completely when De Quincey brought out his vivid, tactless reminiscences of Wordsworth and the Grasmere circle (1834). Meanwhile, De Quincey's eleven years at Grasmere (1809–20) were occupied with reading, study, financial problems, and repeated struggles against the growing influence of opium. In 1816 he married the daughter of a nearby farmer, the patient and loyal Margaret Simpson, with whom he was to have eight children. Some settled work was obviously necessary. For a short while he served as editor of the *Westmorland Gazette*. Finally, determined to make a fresh start, he left the Lake Country for London in 1820.

It was in London, at the age of thirty-five, that De Quincey's literary career began. The lateness of the start in one so precocious is another of the many incongruities of his life and work. Charles Lamb, whom he had met years before, introduced him to the proprietors of *The London Magazine*. There and in other periodicals, particularly *Blackwood's* and *Tait's Magazine*, his writing appeared in growing volume. The bulk is astonishing (by no means all of it was included in the fourteen volumes he began to collect near the end of his life). Equally astonishing is the range—biography and history (here one should cite especially his noble biographical series on "The Caesars"); criticism and rhetorical analysis; political economy, metaphysics, translation, and imaginative prose. We have mentioned that his subjects were relatively specialized in comparison with those of most of the major writers of the day. But as the lives of some of the most learned scholars illustrate—men whose work is distributed throughout scores of reviews and articles—specialization is quite compatible with range. We can regret that so much of De Quincey's writing falls into the category of what he later called the "literature of knowledge" —writing that he distinguished from "literature of power," or imaginative writing. For literature of knowledge inevitably dates. It is always possible for later industry or reflection to pick holes or to add something further. As De Quincey himself says (in the essay on the subject printed below), let the material of the literature of knowledge be "even partially revised, let it be but expanded,—nay, even let its teaching be but placed in a better order,—and instantly it is superseded. Whereas the feeblest works in the Literature of Power, surviving at all, survive as finished and unalterable amongst men." Moreover, De Quincey had few of the advantages of later, more happily placed scholars. He was forced more to rely on his wits; he was also, since he made his living from his work, encouraged to write rapidly; spreading himself widely, he naturally left himself open. Thus, for example, he plunged into German literature and philosophy, and wrote of it—as a rule—with sharpness and dispatch. But there were inevitably some snap judgments, and they have been treated mercilessly. (There is the at least partial excuse of strong temptation since De Quincey's manner sometimes displayed a certain parade.) We should remember, however, how few English-speaking people of his day knew anything at all of German literature or German philosophy.

Of the vast number of works that continued to appear after he left for London, only three were ever published in book form. By far the most famous was his *Confessions of an English Opium-Eater* (1822; first printed in *The London Magazine* in 1821, considerably enlarged in

1856). The two other works published separately were never much regarded: a novel of the Thirty Years' War, *Klosterheim* (1832), which De Quincey viewed as hack writing and left out of his collected works, and, when he was almost sixty, his *Logic of Political Economy* (1844). Though the *Confessions* had been published anonymously, the identity of the author became quickly known, for the book at once attracted wide attention. It was indeed unique. To its personal and confessional interest it added a sustained power of musical effect and of descriptive dream-painting. The *Confessions*, together with the later *Suspiria de Profundis*, are De Quincey's principal efforts in impassioned prose, though literally hundreds of short passages in his other writings could be cited as equally good examples. More than any other writer of the time he established the ideal—and the example—of the "prose poem"; and it is mainly for this reason that, during the era of Ruskin and Pater in the late nineteenth century, he was praised with more consensus than he has been at any other time. Still, it should be remembered that the *Confessions*, the *Suspiria*, and a few shorter works (or passages) that can be bracketed with them, make up only a fraction of the work of one of the four or five greatest of English journalists. Of the many accounts of him during these years two may be cited. One is by Thomas Hood, who was working as an editor of *The London Magazine*, and who was accustomed to finding De Quincey, when he called on him in the small room where he wrote, "in the midst of a German Ocean of literature in a storm, flooding all the floor, the table, billows of books tossing, tumbling, surging open"; and, rising, the small chinalike figure would then talk for a full hour, looking at the opposite side of the room as though he "seemed to be less speaking than reading from 'a hand-writing on the wall'." Another glimpse, often quoted, is that of Carlyle, who thought that De Quincey, sitting by the candlelight, could be easily mistaken "for the beautifullest little child; blue-eyed, sparkling face, had there not been a something, too, which said, '*Eccovi*—this child has been in hell'."

After he had established himself as a periodical writer, De Quincey began to drift northward again, alternating between the Lake Country and Edinburgh. Throughout all these years his struggles with opium continued. Meanwhile two of his children died and then his wife (1837). Now fifty-two, he settled in a cottage outside Edinburgh and continued to write prolifically. And, as ever, this short, gentle man, so youthful in appearance, was admired for his unfailing, elaborately old-fashioned courtliness. A writer of the time described him as "marvellously fine in point of conversation, looking like an old beggar, but with the manners of a prince." He paid little attention to domestic details, and his daughters came to his rescue. In 1853, when he was sixty-eight, he started to collect his works. Before the edition was completed, he died on December 8, 1859.

Bibliography

When asked to collect his works, De Quincey had said, "The thing is absolutely, insuperably, and forever impossible." The first comprehensive edition was published in America by a

Boston firm, 20 vols., 1851–59. While it was appearing, De Quincey himself began to assemble his collected works, omitting many articles and revising others. This edition, completed after his death, he entitled *Selections Grave and Gay; Writings Published and Unpublished*, 14 vols., 1853–60. Expanded editions, in 1862–63, 1871, and 1878, were followed by what remains, though incomplete, the standard edition: *Collected Writings*, ed. David Masson, 14 vols., 1889–90. Mention should also be made of the *Uncollected Writings*, with preface and notes by James Hogg, 2 vols., 1890, and *Posthumous Writings*, ed. A. H. Japp, 2 vols., 1891–93. Many of De Quincey's more ephemeral essays are still to be found only in the periodicals where they originally appeared. Among the numerous volumes of selections are the *Selected Writings*, ed. Philip Van Doren Stern, 1937, and, on specific subjects, *Essays on Style, Rhetoric and Language*, ed. F. N. Scott, 1893, and *De Quincey's Literary Criticism*, ed. Helen Darbishire, 1909. Both the original and revised versions of *The Confessions of an English Opium-Eater* are readily obtained. The edition by Malcolm Elwin, 1956, deserves special mention. De Quincey's letters, of which no complete edition exists, appear in *Life and Writings*, by H. A. Page (A. H. Japp), cited immediately below among biographies; W. H. Bonner, *De Quincey at Work*, 1936, and especially J. E. Jordan, *De Quincey to Wordsworth*, 1963.

The first real biography was that by H. A. Page (A. H. Japp), *Life and Writings, with Unpublished Correspondence*, 2 vols., 1877, rev. one vol., 1890. The short biography by David Masson, in the English Men of Letters Series, 1881, was essentially an abridgment of Japp's work. Recent biographies include the admirably distilled *De Quincey*, 1935, by Malcolm Elwin in the Great Lives Series; the detailed biography by Horace A. Eaton, 1936, and the biographical-critical study by Edward Sackville-West, 1936. An excellent brief discussion is J. C. Metcalf, *De Quincey: A Portrait*, 1940. General studies of his style include Lane Cooper, *The Prose Poetry of Thomas De Quincey*, 1902, and, of his thought and criticism, S. K. Proctor, *Thomas De Quincey's Theory of Literature*, 1943, J. E. Jordan, *Thomas De Quincey, Literary Critic*, 1952, and J. Hillis Miller, *The Disappearance of God: Five Nineteenth-Century Writers*, 1963, pp. 17–80.

FROM

CONFESSIONS OF AN ENGLISH OPIUM-EATER

Confessions of an English Opium-Eater was first published anonymously in *The London Magazine* (September and October, 1821), then as a separate book (1822), and considerably enlarged over thirty years later (1856). "The entire *Confessions*," said De Quincey, "were designed to convey a narrative of my own experience as an opium-eater, drawn up with entire simplicity and fidelity to the facts; from which they can in no respect have deviated, except by such trifling inaccuracies of date, etc., as the memoranda I have with me in London would not, in all cases, enable me to reduce to certainty." The early part of the book consists primarily of autobiographical narrative. The selections below are from the two concluding parts of the work.

from THE PLEASURES OF OPIUM

It is so long since I first took opium that if it had been a trifling incident in my life I might have forgotten its date: but cardinal events are not to be forgotten; and from circumstances connected with it I remember that it must be referred to the autumn of 1804. During that season I was in London, having come thither for the first time since my entrance at college. And my introduction to opium arose in the following way. From an early age I had been accustomed to wash my head in cold water at least once a day: being suddenly seized with toothache, I attributed it to some relaxation caused by an accidental intermission of that practice; jumped out of bed; plunged my head into a basin of cold water; and with hair thus wetted went to sleep. The next morning, as I need hardly say, I awoke with excruciating rheumatic pains of the head and face, from which I had hardly any respite for about twenty days. On the twenty-first day, I think it was, and on a Sunday, that I went out into the streets, rather to run away, if possible, from my torments, than with any distinct purpose. By accident I met a college acquaintance who recommended opium. Opium! dread agent of unimaginable pleasure and pain! I had heard of it as I had of manna or of ambrosia, but no further: how unmeaning a sound was it at that time! what solemn chords does it now strike upon my heart! what heart-quaking vibrations of sad and happy remembrances! Reverting for a moment to these, I feel a mystic importance attached to the minutest circumstances connected with the place and the time, and the man, if man he was, that first laid open to me the Paradise of Opium-eaters. It was a Sunday afternoon, wet and cheerless: and a duller spectacle this earth of ours has not to show than a rainy Sunday in London. My road homewards lay through Oxford Street; and near "the *stately* Pantheon," [1] as Mr. Wordsworth has obligingly called it, I saw a druggist's shop. The druggist, unconscious minister of celestial pleasures!—as if in sympathy with the rainy Sunday, looked dull and stupid, just as any mortal druggist might be expected to look on a Sunday: and, when I asked for the tincture of opium, he gave it to me as any other man might do: and furthermore, out of my shilling, returned me what seemed to be real copper halfpence, taken out of a real wooden drawer. Nevertheless, in spite of such indications of humanity, he has ever since existed in my mind as the beatific vision of an immortal druggist, sent down to earth on a special mission to myself. And it confirms me in this way of considering him, that, when I next came up to London, I sought him near the stately Pantheon, and found him not: and thus to me, who knew not his name (if indeed he had one), he seemed

rather to have vanished from Oxford Street than to have removed in any bodily fashion. The reader may choose to think of him as, possibly, no more than a sublunary druggist: it may be so: but my faith is better: I believe him to have evanesced, or evaporated. So unwillingly would I connect any mortal remembrances with that hour, and place, and creature, that first brought me acquainted with the celestial drug.

Arrived at my lodgings, it may be supposed that I lost not a moment in taking the quantity prescribed. I was necessarily ignorant of the whole art and mystery of opium-taking: and, what I took, I took under every disadvantage. But I took it:—and in an hour, oh! heavens! what a revulsion! what an upheaving, from its lowest depths, of the inner spirit! what an apocalypse of the world within me! That my pains had vanished, was now a trifle in my eyes:—this negative effect was swallowed up in the immensity of those positive effects which had opened before me—in the abyss of divine enjoyment thus suddenly revealed. Here was a panacea —a $\phi\acute{a}\rho\mu\alpha\kappa\sigma\nu$ $\nu\eta\pi\epsilon\nu\theta\acute{e}s$ [2] for all human woes: here was the secret of happiness, about which philosophers had disputed for so many ages, at once discovered: happiness might now be bought for a penny, and carried in the waistcoat pocket: portable ecstasies might be had corked up in a pint bottle: and peace of mind could be sent down in gallons by the mail-coach. But, if I talk in this way, the reader will think I am laughing: and I can assure him, that nobody will laugh long who deals much with opium: its pleasures even are of a grave and solemn complexion; and in his happiest state, the opium-eater cannot present himself in the character of "L'Allegro": even then, he speaks and thinks as becomes "Il Penseroso." [3] Nevertheless, I have a very reprehensible way of jesting at times in the midst of my own misery: and, unless when I am checked by some more powerful feelings, I am afraid I shall be guilty of this indecent practice even in these annals of suffering or enjoyment. The reader must allow a little to my infirm nature in this respect: and with a few indulgences of that sort, I shall endeavor to be as grave, if not drowsy, as fits a theme like opium, so antimercurial as it really is, and so drowsy as it is falsely reputed.

And, first, one word with respect to its bodily effects: for upon all that has been hitherto written on

CONFESSIONS OF AN ENGLISH OPIUM-EATER. **1.** "**the . . Pantheon**": Wordsworth's "Power of Music," l. 3.

2. Φάρμακον νηπενθές: sorrow-banishing drug (*Odyssey*, IV. 220– 21). **3.** "**L'Allegro**" . . . "**Il Penseroso**": a reference to Milton's famous pair of poems, "L'Allegro" and "Il Penseroso," that contrast the cheerful and the reflective type of man.

the subject of opium, whether by travellers in Turkey, who may plead their privilege of lying as an old immemorial right, or by professors of medicine, writing *ex cathedra*,[4]—I have but one emphatic criticism to pronounce—Lies! lies! lies! I remember once, in passing a bookstall, to have caught these words from a page of some satiric author:—"By this time I became convinced that the London newspapers spoke truth at least twice a week, *viz.*, on Tuesday and Saturday, and might safely be depended upon for——the list of bankrupts." In like manner, I do by no means deny that some truths have been delivered to the world in regard to opium: thus it has been repeatedly affirmed by the learned that opium is a dusky brown in color; and this, take notice, I grant: secondly, that it is rather dear; which I also grant, for in my time, East-India opium has been three guineas a pound, and Turkey eight: and, thirdly, that if you eat a good deal of it, most probably you must do what is particularly disagreeable to any man of regular habits, *viz.*, die. These weighty propositions are, all and singular, true: I cannot gainsay them: and truth ever was, and will be, commendable. But in these three theorems, I believe we have exhausted the stock of knowledge as yet accumulated by man on the subject of opium. And therefore, worthy doctors, as there seems to be room for further discoveries, stand aside, and allow me to come forward and lecture on this matter.

First, then, it is not so much affirmed as taken for granted by all who ever mention opium, formally or incidentally, that it does, or can, produce intoxication. Now, reader, assure yourself, *meo periculo*,[5] that no quantity of opium ever did, or could intoxicate. As to the tincture of opium (commonly called laudanum) *that* might certainly intoxicate if a man could bear to take enough of it; but why? because it contains so much proof spirit, and not because it contains so much opium. But crude opium, I affirm peremptorily, is incapable of producing any state of body at all resembling that which is produced by alcohol: and not in *degree* only incapable, but even in *kind*: it is not in the quantity of its effects merely, but in the quality, that it differs altogether. The pleasure given by wine is always mounting, and tending to a crisis, after which it declines: that from opium, when once generated, is stationary for eight or ten hours: the first, to borrow a technical distinction from medicine, is a case of acute

—the second, of chronic pleasure: the one is a flame, the other a steady and equable glow. But the main distinction lies in this, and whereas wine disorders the mental faculties, opium, on the contrary, if taken in a proper manner, introduces amongst them the most exquisite order, legislation, and harmony. Wine robs a man of his self-possession: opium greatly invigorates it. Wine unsettles and clouds the judgment, and gives a preternatural brightness and a vivid exaltation to the contempts and the admirations, the loves and the hatreds, of the drinker: opium on the contrary communicates serenity and equipoise to all the faculties, active or passive: and with respect to the temper and moral feelings in general, it gives simply that sort of vital warmth which is approved by the judgment, and which would probably always accompany a bodily constitution of primeval or antediluvian health.

. . .

With respect to the torpor supposed to follow, or rather, if we were to credit the numerous pictures of Turkish opium-eaters, to accompany the practice of opium-eating, I deny that also. Certainly opium is classed under the head of narcotics; and some such effect it may produce in the end: but the primary effects of opium are always, and in the highest degree, to excite and stimulate the system: this first stage of its action always lasted with me, during my noviciate, for upwards of eight hours; so that it must be the fault of the opium eater himself if he does not so time his exhibition of the dose, to speak medically, as that the whole weight of its narcotic influence may descend upon his sleep. Turkish opium-eaters, it seems, are absurd enough to sit, like so many equestrian statues, on logs of wood as stupid as themselves. But that the reader may judge of the degree in which opium is likely to stupify the faculties of an Englishman, I shall, by way of treating the question illustratively, rather than argumentatively, describe the way in which I myself often passed an opium evening in London, during the period between 1804 and 1812. It will be seen that at least opium did not move me to seek solitude, and much less to seek inactivity, or the torpid state of self-involution ascribed to the Turks. I give this account at the risk of being pronounced a crazy enthusiast or visionary: but I regard *that* little: I must desire my reader to bear in mind that I was a hard student, and at severe studies for all the rest of my time: and certainly I had a right occasionally to relaxations as well as other people: these, however, I allowed myself but seldom.

4. ex cathedra: from a position of authority. **5. meo periculo:** at my own risk.

The late Duke of [Norfolk] used to say, "Next Friday, by the blessing of Heaven, I purpose to be drunk": and in like manner I used to fix beforehand how often, within a given time, and when, I would commit a debauch of opium. This was seldom more than once in three weeks: for at that time I could not have ventured to call every day (as I did afterwards) for *"a glass of laudanum negus, warm, and without sugar."* No: as I have said, I seldom drank laudanum, at that time, more than once in three weeks: this was usually on a Tuesday or a Saturday night . . .

I shall be charged with mysticism, Behmenism, quietism,[6] etc., but *that* shall not alarm me. Sir H. Vane, the Younger, was one of our wisest men: and let my readers see if he, in his philosophical works, be half as unmystical as I am.—I say, then, that it has often struck me that the scene itself was somewhat typical of what took place in such a reverie. The town of L[iverpool] represented the earth, with its sorrows and its graves left behind, yet not out of sight, nor wholly forgotten. The ocean, in everlasting but gentle agitation, and brooded over by a dove-like calm, might not unfitly typify the mind and the mood which then swayed it. For it seemed to me as if then first I stood at a distance, and aloof from the uproar of life, as if the tumult, the fever, and the strife, were suspended; a respite granted from the secret burthens of the heart; a sabbath of repose; a resting from human labors. Here were the hopes which blossom in the paths of life, reconciled with the peace which is in the grave; motions of the intellect as unwearied as the heavens, yet for all anxieties a halcyon calm: a tranquillity that seemed no product of inertia, but as if resulting from mighty and equal antagonisms; infinite activities, infinite repose.

Oh! just, subtle, and mighty opium! that to the hearts of poor and rich alike, for the wounds that will never heal, and for "the pangs that tempt the spirit to rebel,"[7] bringest an assuaging balm; eloquent opium! that with thy potent rhetoric stealest away the purposes of wrath; and to the guilty man for one night givest back the hopes of his youth, and hands washed pure from blood; and to the proud man a brief oblivion for

Wrongs unredress'd and insults unavenged;[8]

that summonest to the chancery of dreams, for the triumphs of suffering innocence, false witnesses; and confoundest perjury; and dost reverse the sentences of unrighteous judges:—thou buildest upon the bosom of darkness, out of the fantastic imagery of the brain, cities and temples beyond the art of Phidias and Praxiteles—beyond the splendor of Babylon and Hekatompylos:[9] and "from the anarchy of dreaming sleep,"[10] callest into sunny light the faces of long-buried beauties, and the blessed household countenances, cleansed from the "dishonors of the grave."[11] Thou only givest these gifts to man; and thou hast the keys of Paradise, oh, just, subtle, and mighty opium!

from THE PAINS OF OPIUM

Many years ago, when I was looking over Piranesi's Antiquities of Rome, Mr. Coleridge, who was standing by, described to me a set of plates by that artist, called his *Dreams*, and which record the scenery of his own visions during the delirium of a fever: some of them (I describe only from memory of Mr. Coleridge's account) representing vast Gothic halls, on the floor of which stood all sorts of engines and machinery, wheels, cables, pulleys, levers, catapults, etc., etc., expressive of enormous power put forth, and resistance overcome. Creeping along the sides of the walls, you perceived a staircase; and upon it, groping his way upwards, was Piranesi himself: follow the stairs a little further, and you perceive it come to a sudden abrupt termination, without any balustrade, and allowing no step onwards to him who had reached the extremity, except into the depths below. Whatever is to become of poor Piranesi, you suppose, at least, that his labors must in some way terminate here. But raise your eyes, and behold a second flight of stairs still higher: on which again Piranesi is perceived, but this time standing on the very brink of the abyss. Again elevate your eye, and a still more aerial flight of stairs is beheld: and again is poor Piranesi busy on his aspiring labors: and so on, until the unfinished stairs and Piranesi both are lost in the upper gloom of the hall.—With the same power of endless growth and self-reproduction did my architecture proceed in dreams. In the early stage of my malady, the splendors of my dreams were indeed chiefly

---

6. **Behmenism, quietism:** "Behmenism," from the German mystic Jakob Behmen (or Böhme); "quietism" refers to a mystical indifference to the material world. 7. **"the . . . rebel":** Wordsworth's "Dedication" to the *White Doe of Rylstone*, l. 36. 8. **Wrongs . . . unavenged:** Wordsworth's *Excursion*, III. 374.

9. **Hekatompylos** Hundred-gated Thebes. 10. **"from . . . sleep":** *Excursion*, IV. 87. 11. **"dishonors . . . grave":** I Corinthians 15:43.

architectural: and I beheld such pomp of cities and palaces as was never yet beheld by the waking eye, unless in the clouds. From a great modern poet I cite part of a passage which describes, as an appearance actually beheld in the clouds, what in many of its circumstances I saw frequently in sleep:

> The appearance, instantaneously disclosed,
> Was of a mighty city—boldly say
> A wilderness of building, sinking far
> And self-withdrawn into a wondrous depth,
> Far sinking into splendor—without end!
> Fabric it seem'd of diamond, and of gold,
> With alabaster domes, and silver spires,
> And glazing terrace upon terrace, high
> Uplifted; here, serene pavilions bright
> In avenues disposed; there, towers begirt
> With battlements that on their restless fronts
> Bore stars—illumination of all gems!
> By earthly nature had the effect been wrought
> Upon the dark materials of the storm
> Now pacified: on them, and on the coves,
> And mountain-steeps and summits, whereunto
> The vapors had receded,—taking there
> Their station under a cerulean sky,[12] etc., etc.

The sublime circumstance—"battlements that on their *restless* fronts bore stars,"—might have been copied from my architectural dreams, for it often occurred.—We hear it reported of Dryden, and of Fuseli in modern times, that they thought proper to eat raw meat for the sake of obtaining splendid dreams: how much better for such a purpose to have eaten opium, which yet I do not remember that any poet is recorded to have done, except the dramatist Shadwell: and in ancient days, Homer is, I think, rightly reputed to have known the virtues of opium.[13]

To my architecture succeeded dreams of lakes and silvery expanses of water:—these haunted me so much that I feared, though possibly it will appear ludicrous to a medical man, that some dropsical state or tendency of the brain might thus be making itself, to use a metaphysical word, *objective;* and the sentient organ *project* itself as its own object.—For two months I suffered greatly in my head—a part of my bodily structure which had hitherto been so clear from all touch or taint of weakness, physically, I mean, that I used to say of it, as the last Lord Orford said of his stomach, that it seemed likely to survive the rest of my person.—Till now I had never felt headache even, or

12. The . . . sky: *Excursion*, II. 834–51. 13. Homer . . . opium: *Odyssey*, IV. 220–21.

any the slightest pain, except rheumatic pains caused by my own folly. However, I got over this attack, though it must have been verging on something very dangerous.

The waters now changed their character,—from translucent lakes, shining like mirrors, they now became seas and oceans. And now came a tremendous change, which, unfolding itself slowly like a scroll, through many months, promised an abiding torment; and, in fact, never left me until the winding up of my case. Hitherto the human face had mixed often in my dreams, but not despotically, nor with any special power of tormenting. But now that which I have called the tyranny of the human face began to unfold itself. Perhaps some part of my London life might be answerable for this. Be that as it may, now it was that upon the rocking waters of the ocean the human face began to appear: the sea appeared paved with innumerable faces, upturned to the heavens: faces, imploring, wrathful, despairing, surged upwards by thousands, by myriads, by generations, by centuries:—my agitation was infinite,—my mind tossed—and surged with the ocean.

May, 1818.

The Malay has been a fearful enemy for months. I have been every night, through his means, transported into Asiatic scenes. I know not whether others share in my feelings on this point; but I have often thought that if I were compelled to forego England, and to live in China, and among Chinese manners and modes of life and scenery, I should go mad. The causes of my horror lie deep and some of them must be common to others. Southern Asia, in general, is the seat of awful images and associations. As the cradle of the human race, it would alone have a dim and reverential feeling connected with it. But there are other reasons. No man can pretend that the wild, barbarous, and capricious superstitions of Africa, or of savage tribes elsewhere, affect him in the way that he is affected by the ancient, monumental, cruel, and elaborate religions of Indostan, etc. The mere antiquity of Asiatic things, of their institutions, histories, modes of faith, etc., is so impressive, that to me the vast age of the race and name overpowers the sense of youth in the individual. A young Chinese seems to me an antediluvian man renewed. Even Englishmen, though not bred in any knowledge of such institutions, cannot but shudder at the mystic sublimity of *castes* that have flowed apart, and refused to mix, through such immemorial tracts of time; nor can any man fail to be awed by the names of the Ganges or the Euphrates. It contributes much to these feelings that southern Asia is, and has been for thousands of years, the part of the earth most swarming with human life; the great *officina gentium*.[14] Man is a weed in those regions. The vast empires also, into which the enormous population of Asia has always been cast, give a further sublimity to the feelings associated with all Oriental names or images. In China, over and above what it has in common with the rest of southern Asia, I am terrified by the modes of life, by the manners, and the barrier of utter abhorrence, and want of sympathy, placed between us by feelings deeper than I can analyze. I could sooner live with lunatics, or brute animals. All this, and much more than I can say, or have time to say, the reader must enter into before he can comprehend the un-imaginable horror which these dreams of Oriental imagery, and mythological tortures, impressed upon me. Under the connecting feeling of tropical heat and vertical sunlights, I brought together all creatures, birds, beasts, reptiles, all trees and plants, usages and appearances, that are found in all tropical regions, and assembled them together in China or Indostan. From kindred feelings, I soon brought Egypt and all her gods under the same law. I was stared at, hooted at, grinned at, chattered at, by monkeys, by paroquets, by cocka-toos. I ran into pagodas: and was fixed for centuries at the summit, or in secret rooms; I was the idol; I was the priest; I was worshipped; I was sacrificed. I fled from the wrath of Brama through all the forests of Asia: Vishnu hated me: Seeva laid wait for me. I came suddenly upon Isis and Osiris: I had done a deed, they said, which the ibis and the crocodile trembled at. I was buried for a thousand years in stone coffins, with mummies and sphinxes, in narrow chambers at the heart of eternal pyramids. I was kissed, with cancerous kisses, by crocodiles; and laid, confounded with all un-utterable slimy things, amongst reeds and Nilotic mud.

Some slight abstraction I thus attempt of my oriental dreams, which filled me always with such amazement at the monstrous scenery, that horror seemed absorbed for a while in sheer astonishment. Sooner or later came a reflux of feeling that swallowed up the astonishment, and left me, not so much in terror, as in hatred and abomination of what I saw. Over every form, and threat, and punishment, and dim sightless incarceration, brooded a sense of eternity and infinity that drove me into an oppression as of madness. Into these dreams

14. **officina gentium:** swarm of nations.

only, it was, with one or two slight exceptions, that any circumstances of physical horror entered. All before had been moral and spiritual terrors. But here the main agents were ugly birds, or snakes, or crocodiles; especially the last. The cursed crocodile became to me the object of more horror than almost all the rest. I was compelled to live with him; and (as was always the case almost in my dreams) for centuries. I escaped sometimes, and found myself in Chinese houses, with cane tables, etc. All the feet of the tables, sofas, etc., soon became instinct with life: the abominable head of the crocodile, and his leering eyes, looked out at me, multiplied into a thousand repetitions: and I stood loathing and fascinated. And so often did this hideous reptile haunt my dreams, that many times the very same dream was broken up in the very same way: I heard gentle voices speaking to me (I hear everything when I am sleeping); and instantly I awoke: it was broad noon; and my children were standing, hand in hand, at my bedside; come to show me their colored shoes, or new frocks, or to let me see them dressed for going out. I protest that so awful was the transition from the damned crocodile, and the other unutterable monsters and abortions of my dreams, to the sight of innocent *human* natures and of infancy, that, in the mighty and sudden revulsion of mind, I wept, and could not forbear it, as I kissed their faces.

Then suddenly would come a dream of far different character,—a tumultuous dream,—commencing with a music such as now I often heard in sleep,—music of preparation and of awakening suspense. The undulations of fast-gathering tumults were like the opening of the Coronation Anthem; and, like *that*, gave the feeling of a multitudinous movement, of infinite cavalcades filing off, and the tread of innumerable armies. The morning was come of a mighty day—a day of crisis and of ultimate hope for human nature, then suffering mysterious eclipse, and laboring in some dread extremity. Somewhere, but I knew not where—somehow, but I knew not how—by some beings, but I knew not by whom—a battle, a strife, an agony, was traveling through all its stages—was evolving itself, like the catastrophe of some mighty drama, with which my sympathy was the more insupportable, from deepening confusion as to its local scene, its cause, its nature, and its undecipherable issue. I (as is usual in dreams where, of necessity, we make ourselves central to every movement) had the power, and yet had not the power, to decide it. I had the power, if I could raise myself to will it; and yet again had not the power, for the weight of twenty Atlantics was upon me, or the oppression of inexpiable guilt. "Deeper than ever plummet sounded,"[15] I lay inactive. Then, like a chorus, the passion deepened. Some greater interest was at stake, some mightier cause, than ever yet the sword had pleaded, or trumpet had proclaimed. Then came sudden alarms; hurrying to and fro; trepidations of innumerable fugitives, I knew not whether from the good cause or the bad; darkness and lights; tempest and human faces; and at last, with the sense that all was lost, female forms, and the features that were worth all the world to me; and but a moment allowed—and clasped hands, with heart-breaking partings, and then—everlasting farewells! and, with a sigh such as the caves of hell sighed when the incestuous mother uttered the abhorred name of Death, the sound was reverberated—everlasting farewells! and again, and yet again reverberated—everlasting farewells!

And I awoke in struggles, and cried aloud, "I will sleep no more!"

1820–22 (1821–22)

ON THE KNOCKING
AT THE GATE IN *MACBETH*

❧

This famous interpretation of *Macbeth*, II. ii–iii, was published originally in *The London Magazine* (Oct. 1823).

❧

From my boyish days I had always felt a great perplexity on one point in *Macbeth*. It was this:—the knocking at the gate which succeeds to the murder of Duncan produced to my feelings an effect for which I never could account. The effect was that it reflected back upon the murderer a peculiar awfulness and a depth of solemnity; yet, however obstinately I endeavored with my understanding to comprehend this, for many years I never could see *why* it should produce such an effect.

Here I pause for one moment to exhort the reader never to pay any attention to his understanding when it stands in opposition to any other faculty of his mind. The mere understanding, however useful and indispensable, is the meanest faculty in the human mind, and the most to be distrusted; and yet the great majority of

people trust to nothing else,—which may do for ordinary life, but not for philosophical purposes. Of this, out of ten thousand instances that I might produce, I will cite one. Ask of any person whatsoever who is not previously prepared for the demand by a knowledge of the perspective, to draw in the rudest way the commonest appearance which depends upon the laws of that science,—as, for instance, to represent the effect of two walls standing at right angles to each other, or the appearance of the houses on each side of a street as seen by a person looking down the street from one extremity. Now, in all cases, unless the person has happened to observe in pictures how it is that artists produce these effects, he will be utterly unable to make the smallest approximation to it. Yet why? For he has actually seen the effect every day of his life. The reason is that he allows his understanding to overrule his eyes. His understanding, which includes no intuitive knowledge of the laws of vision, can furnish him with no reason why a line which is known and can be proved to be a horizontal line should not *appear* a horizontal line: a line that made any angle with the perpendicular less than a right angle would seem to him to indicate that his houses were all tumbling down together. Accordingly, he makes the line of his houses a horizontal line, and fails, of course, to produce the effect demanded. Here, then, is one instance out of many in which not only the understanding is allowed to overrule the eyes, but where the understanding is positively allowed to obliterate the eyes, as it were; for not only does the man believe the evidence of his understanding in opposition to that of his eyes, but (what is monstrous) the idiot is not aware that his eyes ever gave such evidence. He does not know that he has seen (and therefore, *quoad*,[1] his consciousness has *not* seen) that which he *has* seen every day of his life.

But to return from this digression. My understanding could furnish no reason why the knocking at the gate in *Macbeth* should produce any effect, direct or reflected. In fact, my understanding said positively that it could *not* produce any effect. But I knew better; I felt that it did; and I waited and clung to the problem until further knowledge should enable me to solve it. At length, in 1812, Mr. Williams made his *début* on the stage of Ratcliffe Highway, and executed those unparalleled murders which have procured for him such a brilliant and undying reputation. On which murders,

---

15. "Deeper . . . sounded": *The Tempest*, V. i. 56.

ON THE KNOCKING AT THE GATE IN *Macbeth*. **1. quoad:** on this account, in this respect.

by the way, I must observe, that in one respect they have had an ill effect, by making the connoisseur in murder very fastidious in his taste, and dissatisfied by anything that has been since done in that line. All other murders look pale by the deep crimson of his; and, as an amateur once said to me in a querulous tone, "There has been absolutely nothing *doing* since his time, or nothing that's worth speaking of." But this is wrong; for it is unreasonable to expect all men to be great artists, and born with the genius of Mr. Williams. Now it will be remembered that in the first of these murders (that of the Marrs) the same incident (of a knocking at the door soon after the work of extermination was complete) did actually occur which the genius of Shakespeare has invented; and all good judges, and the most eminent dilettanti, acknowledged the felicity of Shakespeare's suggestion as soon as it was actually realized. Here, then, was a fresh proof that I was right in relying on my own feeling, in opposition to my understanding; and I again set myself to study the problem. At length I solved it to my own satisfaction; and my solution is this:—Murder, in ordinary cases, where the sympathy is wholly directed to the case of the murdered person, is an incident of coarse and vulgar horror; and for this reason,—that it flings the interest exclusively upon the natural but ignoble instinct by which we cleave to life: an instinct which, as being indispensable to the primal law of self-preservation, is the same in kind (though different in degree) amongst all living creatures. This instinct, therefore, because it annihilates all distinctions, and degrades the greatest of men to the level of "the poor beetle that we tread on," exhibits human nature in its most abject and humiliating attitude. Such an attitude would little suit the purposes of the poet. What then must he do? He must throw the interest on the murderer. Our sympathy must be with *him* (of course I mean a sympathy of comprehension, a sympathy by which we enter into his feelings, and are made to understand them,—not a sympathy of pity or approbation). In the murdered person, all strife of thought, all flux and reflux of passion and of purpose, are crushed by one overwhelming panic; the fear of instant death smites him "with its petrific mace." But in the murderer, such a murderer as a poet will condescend to, there must be raging some great storm of passion,—jealousy, ambition, vengeance, hatred,—which will create a hell within him; and into this hell we are to look.

In *Macbeth*, for the sake of gratifying his own enormous and teeming faculty of creation, Shake-speare has introduced two murderers: and, as usual in his hands, they are remarkably discriminated: but,—though in Macbeth the strife of mind is greater than in his wife, the tiger spirit not so awake, and his feelings caught chiefly by contagion from her,—yet, as both were finally involved in the guilt of murder, the murderous mind of necessity is finally to be presumed in both. This was to be expressed; and, on its own account, as well as to make it a more proportionable antagonist to the unoffending nature of their victim, "the gracious Duncan," and adequately to expound "the deep damnation of his taking off," this was to be expressed with peculiar energy. We were to be made to feel that the human nature,—*i.e.*, the divine nature of love and mercy, spread through the hearts of all creatures, and seldom utterly withdrawn from man,—was gone, vanished, extinct, and that the fiendish nature had taken its place. And, as this effect is marvellously accomplished in the *dialogues* and *soliloquies* themselves, so it is finally consummated by the expedient under consideration; and it is to this that I now solicit the reader's attention. If the reader has ever witnessed a wife, daughter, or sister in a fainting fit, he may chance to have observed that the most affecting moment in such a spectacle is *that* in which a sigh and a stirring announce the recommencement of suspended life. Or, if the reader has ever been present in a vast metropolis on the day when some great national idol was carried in funeral pomp to his grave, and, chancing to walk near the course through which it passed, has felt powerfully, in the silence and desertion of the streets, and in the stagnation of ordinary business, the deep interest which at that moment was possessing the heart of man,—if all at once he should hear the death-like stillness broken up by the sound of wheels rattling away from the scene, and making known that the transitory vision was dissolved, he will be aware that at no moment was his sense of the complete suspension and pause in ordinary human concerns so full and affecting as at that moment when the suspension ceases, and the goings-on of human life are suddenly resumed. All action in any direction is best expounded, measured, and made apprehensible, by reaction. Now, apply this to the case in *Macbeth*. Here, as I have said, the retiring of the human heart and the entrance of the fiendish heart was to be expressed and made sensible. Another world has stepped in; and the murderers are taken out of the region of human things, human purposes, human desires. They are transfigured: Lady Macbeth is "unsexed"; Macbeth has forgot that he was born of woman; both are conformed to the image of devils; and the world of devils is suddenly revealed. But how shall this be conveyed and made palpable? In order that a new world may step in, this world must for a time disappear. The murderers and the murder must be insulated—cut off by an immeasurable gulf from the ordinary tide and succession of human affairs —locked up and sequestered in some deep recess; we must be made sensible that the world of ordinary life is suddenly arrested, laid asleep, tranced, racked into a dread armistice; time must be annihilated, relation to things without abolished; and all must pass self-withdrawn into a deep syncope and suspension of earthly passion. Hence it is that, when the deed is done, when the work of darkness is perfect, then the world of darkness passes away like a pageantry in the clouds: the knocking at the gate is heard, and it makes known audibly that the reaction has commenced; the human has made its reflux upon the fiendish; the pulses of life are beginning to beat again, and the re-establishment of the goings-on of the world in which we live first makes us profoundly sensible of the awful parenthesis that had suspended them.

O mighty poet! Thy works are not as those of other men, simply and merely great works of art, but are also like the phenomena of nature, like the sun and the sea, the stars and the flowers, like frost and snow, rain and dew, hailstorm and thunder, which are to be studied with entire submission of our own faculties, and in the perfect faith that in them there can be no too much or too little, nothing useless or inert, but that, the farther we press in our discoveries, the more we shall see proofs of design and self-supporting arrangement where the careless eye had seen nothing but accident!

1823 (1823)

from RECOLLECTIONS OF CHARLES LAMB

This essay was first printed in *Tait's Magazine* (April and June, 1838).

Amongst the earliest literary acquaintances I made was that with the inimitable Charles Lamb: inimitable, I say, but the word is too limited in its meaning; for, as is said of Milton in that well-known life of him attached

to all common editions of the *Paradise Lost* (Fenton's, I think), "in both senses he was above imitation." Yes; it was as impossible to the moral nature of Charles Lamb that he should imitate another as, in an intellectual sense, it was impossible that any other should successfully imitate him. To write with patience even, not to say genially, for Charles Lamb it was a very necessity of his constitution that he should write from his own wayward nature; and that nature was so peculiar that no other man, the ablest at mimicry, could counterfeit its voice. But let me not anticipate; for these were opinions about Lamb which I had not when I first knew him, nor could have had by any reasonable title. "Elia," be it observed, the exquisite "Elia," was then unborn; Lamb had as yet published nothing to the world which proclaimed him in his proper character of a most original man of genius: at best, he could have been thought no more than a man of talent—and of talent moving in a narrow path, with a power rather of mimicking the quaint and the fantastic than any large grasp over catholic beauty. And, therefore, it need not offend the most doting admirer of Lamb as he is *now* known to us, a brilliant star forever fixed in the firmament of English Literature, that I acknowledge myself to have sought his acquaintance rather under the reflex honor he had enjoyed of being known as Coleridge's friend than for any which he yet held directly and separately in his own person. My earliest advances towards this acquaintance had an inauspicious aspect; and it may be worth while reporting the circumstances, for they were characteristic of Charles Lamb; and the immediate result was—that we parted, not perhaps (as Lamb says of his philosophic friend R. and the Parisians) "with mutual contempt," but at least with coolness; and, on my part, with something that might have even turned to disgust—founded, however, entirely on my utter misapprehension of Lamb's character and his manners—had it not been for the winning goodness of Miss Lamb,[1] before which all resentment must have melted in a moment.

It was either late in 1804 or early in 1805, according to my present computations, that I had obtained from a literary friend a letter of introduction to Mr. Lamb. All that I knew of his works was his play of *John Woodvil*, which I had bought in Oxford, and perhaps *I* only had bought throughout that great University, at the time of my matriculation there, about the Christmas of 1803. Another book fell into my hands on that

RECOLLECTIONS OF CHARLES LAMB. 1. **Miss Lamb:** Lamb's sister, Mary.

same morning, I recollect—the *Gebir* of Mr. Walter Savage Landor, which astonished me by the splendor of its descriptions (for I had opened accidentally upon the sea-nymph's marriage with Tamor, the youthful brother of Gebir)—and I bought this also. Afterwards, when placing these two most unpopular of books on the same shelf with the other far holier idols of my heart, the joint poems of Wordsworth and Coleridge as then associated in the *Lyrical Ballads*—poems not equally unknown, perhaps a *little* better known, but only with the result of being more openly scorned, rejected—I could not but smile internally at the fair prospect I had of congregating a library which no man had read but myself. *John Woodvil* I had almost studied, and Miss Lamb's pretty *High-Born Helen*, and the ingenious imitations of Burton; these I had read, and, to a certain degree, must have admired, for some parts of them had settled without effort in my memory. I had read also the *Edinburgh*[2] notice of them; and with what contempt may be supposed from the fact that my veneration for Wordsworth transcended all that I felt for any created being, past or present; insomuch that, in the summer, or spring rather, of that same year, and full eight months before I first went to Oxford, I had ventured to address a letter to him, through his publishers, the Messrs. Longman (which letter, Miss Wordsworth in after years assured me they believed to be the production of some person much older than I represented myself), and that in due time I had been honored by a long answer from Wordsworth; an honor which, I well remember, kept me awake, from mere excess of pleasure, through a long night in June, 1803. It was not to be supposed that the very feeblest of admirations could be shaken by mere scorn and contumely, unsupported by any shadow of a reason. Wordsworth, therefore, could not have suffered in any man's opinion from the puny efforts of this new autocrat amongst reviewers; but what was said of Lamb, though not containing one iota of criticism, either good or bad, had certainly more point and cleverness. The supposition that *John Woodvil* might be a lost drama, recovered from the age of Thespis,[3] and entitled to the hircus,[4] etc., must, I should think, have won a smile from Lamb himself; or why say "Lamb himself," which means "*even* Lamb," when he would have been the *very* first to laugh (as he was afterwards among the first to hoot at his own farce), provided only he could detach his mind from the ill-nature and hard contempt which accompanied the wit. This wit had certainly not dazzled my eyes in the slightest degree. So far as I was left at leisure by a more potent order of poetry to think of the *John Woodvil* at all, I had felt and acknowledged a delicacy and tenderness in the situations as well as the sentiments, but disfigured, as I thought, by quaint, grotesque, and *mimetic* phraseology. The main defect, however, of which I complained, was defect of power. I thought Lamb had no right to take his station amongst the inspired writers who had just then risen to throw new blood into our literature, and to breathe a breath of life through the worn-out, or, at least, torpid organization of the national mind. He belonged, I thought, to the old literature; and, as a poet, he certainly does. There were in his verses minute scintillations of genius—now and then, even a subtle sense of beauty; and there were shy graces, lurking half-unseen, like violets in the shade. But there was no power on a colossal scale; no breadth; no choice of great subjects; no wrestling with difficulty; no creative energy. So I thought then; and so I should think now, if Lamb were viewed chiefly as a poet. Since those days he has established his right to a seat in any company. But why? and in what character? As "Elia":—the essays of "Elia" are as exquisite a gem amongst the jewelry of literature as any nation can show. . . .

For the reasons, therefore, I have given, never thinking of Charles Lamb as a poet, and, at that time, having no means for judging of him in any other character, I had requested the letter of introduction to him rather with a view to some further knowledge of Coleridge (who was then absent from England) than from any special interest about Lamb himself. However, I felt the extreme discourtesy of approaching a man and asking for his time and civility under such an avowal: and the letter, therefore, as I believe, or as I requested, represented me in the light of an admirer. I hope it did; for that character might have some excuse for what followed, and heal the unpleasant impression likely to be left by a sort of *fracas* which occurred at my first meeting with Lamb. This was so characteristic of Lamb that I have often laughed at it since I came to know what *was* characteristic of Lamb.

But first let me describe my brief introductory call upon him at the India House. I had been told that he was never to be found at home except in the evenings; and to have called then would have been, in a manner, forcing myself upon his hospitalities, and at a moment

2. **Edinburgh:** *The Edinburgh Review.* 3. **Thespis:** Thespis, (sixth century B.C., was supposedly the founder of the Greek drama. 4. **hircus:** goat.

when he might have confidential friends about him; besides that, he was sometimes tempted away to the theatres. I went, therefore, to the India House; made inquiries amongst the servants; and, after some trouble (for *that* was early in his Leadenhall Street career, and possibly he was not much known), I was shown into a small room, or else a small section of a large one (thirty-four years affects one's remembrance of some circumstances), in which was a very lofty writing desk, separated by a still higher railing from that part of the floor on which the profane—the laity, like myself—were allowed to approach the *clerus*, or clerkly rulers of the room. Within the railing sat, to the best of my remembrance, six quill-driving gentlemen; not gentlemen whose duty or profession it was merely to drive the quill, but who were then driving it—*gens de plume*,[5] such *in esse*, as well as *in posse*—in act as well as habit; for, as if they supposed me a spy sent by some superior power to report upon the situation of affairs as surprised by me, they were all too profoundly immersed in their oriental studies to have any sense of my presence. Consequently, I was reduced to a necessity of announcing myself and my errand. I walked, therefore, into one of the two open doorways of the railing, and stood closely by the high stool of him who occupied the first place within the little aisle. I touched his arm, by way of recalling him from his lofty Leadenhall speculation to this sublunary world; and, presenting my letter, asked if that gentleman (pointing to the address) were really a citizen of the present room; for I had been repeatedly misled, by the directions given me, into wrong rooms. The gentleman smiled; it was a smile not to be forgotten. This was Lamb. And here occurred a *very*, *very* little incident—one of those which pass so fugitively that they are gone and hurrying away into Lethe almost before your attention can have arrested them; but it was an incident which, to me, who happened to notice it, served to express the courtesy and delicate consideration of Lamb's manner. The seat upon which he sat was a very high one; so absurdly high, by the way, that I can imagine no possible use or sense in such an altitude, unless it were to restrain the occupant from playing truant at the fire by opposing Alpine difficulties to his descent.

Whatever might be the original purpose of this aspiring seat, one serious dilemma arose from it, and this it was which gave the occasion to Lamb's act of courtesy. Somewhere there is an anecdote, meant to

illustrate the ultra-obsequiousness of the man,—either I have heard of it in connection with some actual man known to myself, or it is told in a book of some historical coxcomb,—that, being on horseback, and meeting some person or other whom it seemed advisable to flatter, he actually dismounted, in order to pay his court by a more ceremonious bow. In Russia, as we all know, this was, at one time, upon meeting any of the Imperial family, an act of legal necessity: and there, accordingly, but there only, it would have worn no ludicrous aspect. Now, in this situation of Lamb's, the act of descending from his throne, a very elaborate process, with steps and stages analogous to those on horseback—of slipping your right foot out of the stirrup, throwing your leg over the crupper, etc.—was, to all intents and purposes, the same thing as dismounting from a great elephant of a horse. Therefore it both was, and was felt to be by Lamb, supremely ludicrous. On the other hand, to have sate still and stately upon this aerial station, to have bowed condescendingly from this altitude, would have been—not ludicrous indeed; performed by a very superb person and supported by a superb bow, it might have been vastly fine, and even terrifying to many young gentlemen under sixteen; but it would have had an air of ungentlemanly assumption. Between these extremes, therefore, Lamb had to choose;—between appearing ridiculous himself for a moment, by going through a ridiculous evolution which no man could execute with grace; or, on the other hand, appearing lofty and assuming, in a degree which his truly humble nature (for he was the humblest of men in the pretensions which he put forward for himself) must have shrunk from with horror. Nobody who knew Lamb can doubt how the problem was solved: he began to dismount instantly; and, as it happened that the very first *round* of his descent obliged him to turn his back upon me as if for a sudden purpose of flight, he had an excuse for laughing; which he did heartily—saying, at the same time, something to this effect: that I must not judge from first appearances; that he should revolve upon me; that he was not going to fly; and other facetiæ, which challenged a general laugh from the clerical brotherhood.

When he had reached the basis of terra firma on which I was standing, naturally, as a mode of thanking him for his courtesy, I presented my hand; which, in a general case, I should certainly not have done; for I cherished, in an ultra-English degree, the English custom (a wise custom) of bowing in frigid silence on a first introduction to a stranger; but, to a man of literary

5. **gens de plume:** people of the pen.

talent, and one who had just practiced so much kindness in my favor at so probable a hazard to himself of being laughed at for his pains, I could not maintain that frosty reserve. Lamb took my hand; did not absolutely reject it: but rather repelled my advance by his manner. This, however, long afterwards I found, was only a habit derived from his too great sensitiveness to the variety of people's feelings, which run through a gamut so infinite of degrees and modes as to make it unsafe for any man who respects himself to be too hasty in his allowances of familiarity. Lamb had, as he was entitled to have, a high self-respect; and me he probably suspected (as a young Oxonian) of some aristocratic tendencies. The letter of introduction, containing (I imagine) no matters of business, was speedily run through; and I instantly received an invitation to spend the evening with him. Lamb was not one of those who catch at the chance of escaping from a bore by fixing some distant day, when accidents (in duplicate proportion, perhaps, to the number of intervening days) may have carried you away from the place: he sought to benefit by no luck of that kind; for he was, with his limited income—and I say it deliberately—positively the most hospitable man I have known in this world. That night, the same night, I was to come and spend the evening with him. I had gone to the India House with the express purpose of accepting whatever invitation he should give me; and, therefore, I accepted this, took my leave, and left Lamb in the act of resuming his aerial position.

I was to come so early as to drink tea with Lamb; and the hour was seven. He lived in the Temple; and I, who was not then, as afterwards I became, a student and member of "the Honorable Society of the Middle Temple," did not know much of the localities. However, I found out his abode, not greatly beyond my time: nobody had been asked to meet me,—which a little surprised me, but I was glad of it; for, besides Lamb, there was present his sister, Miss Lamb, of whom, and whose talents and sweetness of disposition, I had heard. I turned the conversation, upon the first opening which offered, to the subject of Coleridge; and many of my questions were answered satisfactorily, because seriously, by Miss Lamb. But Lamb took a pleasure in baffling me, or in throwing ridicule upon the subject. Out of this grew the matter of our affray. We were speaking of *The Ancient Mariner*. Now, to explain what followed, and a little to excuse myself, I must beg the reader to understand that I was under twenty years of age, and that my admiration for Coleridge (as,

in perhaps a still greater degree, for Wordsworth) was literally in no respect short of a religious feeling: it had, indeed, all the sanctity of religion, and all the tenderness of a human veneration. Then, also, to imagine the strength which it would derive from circumstances that do not exist now, but did then, let the reader further suppose a case—not such as he may have known since that era about Sir Walter Scotts and Lord Byrons, where every man you could possibly fall foul of, early or late, night or day, summer of winter, was in perfect readiness to feel and express his sympathy with the admirer—but when no man, beyond one or two in each ten thousand, had so much as heard of either Coleridge or Wordsworth, and that one, or those two, knew them only to scorn them, trample on them, spit upon them. Men so abject in public estimation, I maintain, as that Coleridge and that Wordsworth, had not existed before, have not existed since, will not exist again. We have heard in old times of donkeys insulting effete or dying lions by kicking them; but in the case of Coleridge and Wordsworth it was effete donkeys that kicked living lions. They, Coleridge and Wordsworth, were the Pariahs [6] of literature in those days: as much scorned wherever they were known; but escaping that scorn only because they were as little known as Pariahs, and even more obscure.

Well, after this bravura, by way of conveying my sense of the real position then occupied by these two authors—a position which thirty and odd years have altered, by a revolution more astonishing and total than ever before happened in literature or in life—let the reader figure to himself the sensitive horror with which a young person, carrying his devotion about with him, of necessity, as the profoundest of secrets, like a primitive Christian amongst a nation of Pagans, or a Roman Catholic convert amongst the bloody idolators of Japan—in Oxford, above all places, hoping for no sympathy, and feeling a daily grief, almost a shame, in harboring this devotion to that which, nevertheless, had done more for the expansion and sustenance of his own inner mind than all literature besides—let the reader figure, I say, to himself, the shock with which such a person must recoil from hearing the very friend and associate of these authors utter what seemed at that time a burning ridicule of all which belonged to them—their books, their thoughts, their places, their persons. This had gone on for some time before we came upon the ground of *The Ancient Mariner;* I had

6. **Pariahs:** outcasts.

been grieved, perplexed, astonished; and how else could I have felt reasonably, knowing nothing of Lamb's propensity to mystify a stranger; he, on the other hand, knowing nothing of the depth of my feelings on these subjects, and that they were not so much mere literary preferences as something that went deeper than life or household affections? At length, when he had given utterance to some ferocious canon of judgment, which seemed to question the entire value of the poem, I said, perspiring (I dare say) in this detestable crisis—"But, Mr. Lamb, good heavens! how is it possible you can allow yourself in such opinions? What instance could you bring from the poem that would bear you out in these insinuations?"—"Instances?" said Lamb: "oh, I'll instance you, if you come to that. Instance, indeed! Pray, what do you say to this—

> The many men so beautiful,
> And they all dead did lie?[7]

So beautiful, indeed! Beautiful! Just think of such a gang of Wapping vagabonds,[8] all covered with pitch, and chewing tobacco; and the old gentleman himself—what do you call him?—the bright-eyed fellow?"[9] What more might follow I never heard; for, at this point, in a perfect rapture of horror, I raised my hands —both hands—to both ears; and, without stopping to think or to apologize, I endeavored to restore equanimity to my disturbed sensibilities by shutting out all further knowledge of Lamb's impieties. At length he seemed to have finished; so I, on my part, thought I might venture to take off the embargo: and in fact he *had* ceased; but no sooner did he find me restored to my hearing than he said with a most sarcastic smile—which he could assume upon occasion—"If you please, sir, we'll say grace before we begin." I know not whether Lamb were really piqued or not at the mode by which I had expressed my disturbance: Miss Lamb certainly was not; her goodness led her to pardon me, and to treat me—in whatever light she might really view my almost involuntary rudeness—as the party who had suffered wrong; and, for the rest of the evening, she was so pointedly kind and conciliatory in her manner that I felt greatly ashamed of my boyish failure in self-command. . . .

1838 (1838)

7. **The . . . lie:** "Rime of the Ancient Mariner," ll. 236–37. 8. **Wapping vagabonds:** Wapping was a rough area of London on the bank of the Thames. 9. **bright-eyed fellow:** "Rime of the Ancient Mariner," l. 20.

FROM

SUSPIRIA DE PROFUNDIS

Suspiria de Profundis (or "sighs from the depths" of the soul) was conceived, says De Quincey, as a sequel to the *Confessions of an English Opium-Eater*. The *Suspiria* and *The English Mail-Coach* (see below) are culminating examples of the prose of dream-painting that began with the *Confessions*. The central theme of the work is the necessity of pain and sorrow for the growth of the human soul.

DREAMING

In 1821, as a contribution to a periodical work,— in 1822, as a separate volume,—appeared the "Confessions of an English Opium-Eater." The object of that work was to reveal something of the grandeur which belongs *potentially* to human dreams. Whatever may be the number of those in whom this faculty of dreaming splendidly can be supposed to lurk, there are not, perhaps, very many in whom it is developed. He whose talk is of oxen will probably dream of oxen; and the condition of human life which yokes so vast a majority to a daily experience incompatible with much elevation of thought oftentimes neutralizes the tone of grandeur in the reproductive faculty of dreaming, even for those whose minds are populous with solemn imagery. Habitually to dream magnificently, a man must have a constitutional determination to reverie. This in the first place; and even this, where it exists strongly, is too much liable to disturbance from the gathering agitation of our present English life. Already, what by the procession through fifty years of mighty revolutions amongst the kingdoms of the earth, what by the continual development of vast physical agencies,— steam in all its applications, light getting under harness as a slave for man, powers from heaven descending upon education and accelerations of the press, powers from hell (as it might seem, but these also celestial) coming round upon artillery and the forces of destruction,—the eye of the calmest observer is troubled; the brain is haunted as if by some jealousy of ghostly beings moving amongst us; and it becomes too evident that, unless this colossal pace of advance can be retarded (a thing not to be expected), or, which is happily more probable, can be met by counter-forces of corresponding magnitude,—forces in the direction of religion or profound philosophy that shall radiate centrifugally against this storm of life so perilously centripetal towards the vortex of the merely human,—left to itself, the natural tendency of so chaotic a tumult must be to evil; for some minds to lunacy, for others a reagency of fleshly torpor. How much this fierce condition of eternal hurry upon an arena too exclusively human in its interests is likely to defeat the grandeur which is latent in all men, may be seen in the ordinary effect from living too constantly in varied company. The word *dissipation*, in one of its uses, expresses that effect; the action of thought and feeling is consciously dissipated and squandered. To reconcentrate them into meditative habits, a necessity is felt by all observing persons for sometimes retiring from crowds. No man ever will unfold the capacities of his own intellect who does not at least checker his life with solitude. How much solitude, so much power. Or, if not true in that rigor of expression, to this formula undoubtedly it is that the wise rule of life must approximate.

LEVANA AND OUR LADIES OF SORROW

Oftentimes at Oxford I saw Levana in my dreams. I knew her by her Roman symbols. Who is Levana? Reader, that do not pretend to have leisure for very much scholarship, you will not be angry with me for telling you. Levana was the Roman goddess that performed for the new-born infant the earliest office of ennobling kindness,—typical, by its mode, of that grandeur which belongs to man everywhere, and of that benignity in powers invisible which even in Pagan worlds sometimes descends to sustain it. At the very moment of birth, just as the infant tasted for the first time the atmosphere of our troubled planet, it was laid on the ground. *That* might bear different interpretations. But immediately, lest so grand a creature should grovel there for more than one instant, either the paternal hand, as proxy for the goddess Levana, or some near kinsman, as proxy for the father, raised it upright, bade it look erect as the king of all this world, and presented its forehead to the stars, saying, perhaps, in his heart, "Behold what is greater than yourselves!" This symbolic act represented the function of Levana. And that mysterious lady, who never revealed her face (except to me in dreams), but always acted by delegation, had her name from the Latin verb (as still it is the Italian verb) *levare*, to raise aloft.

This is the explanation of Levana. And hence it has arisen that some people have understood by Levana the tutelary power that controls the education of the nursery. She, that would not suffer at his birth even a prefigurative or mimic degradation for her awful ward, far less could be supposed to suffer the real degradation attaching to the non-development of his powers. She therefore watches over human education. Now, the word *edŭco*, with the penultimate short, was derived (by a process often exemplified in the crystallization of languages) from the word *edūco*, with the penultimate long. Whatsoever *educes*, or develops, *educates*. By the education of Levana, therefore, is meant,—not the poor machinery that moves by spelling books and grammars, but by that mighty system of central forces hidden in the deep bosom of human life, which by passion, by strife, by temptation, by the energies of resistance, works forever upon children,—resting not day or night, any more than the mighty wheel of day and night themselves, whose moments, like restless spokes, are glimmering forever as they revolve.

If, then, *these* are the ministries by which Levana works, how profoundly must she reverence the agencies of grief! But you, reader, think that children generally are not liable to grief such as mine. There are two senses in the word *generally*,—the sense of Euclid, where it means *universally* (or in the whole extent of the *genus*), and a foolish sense of this word, where it means *usually*. Now, I am far from saying that children universally are capable of grief like mine. But there are more than you ever heard of who die of grief in this island of ours. I will tell you a common case. The rules of Eton require that a boy on the *foundation* should be there twelve years: he is superannuated at eighteen; consequently he must come at six. Children torn away from mothers and sisters at that age not unfrequently die. I speak of what I know. The complaint is not entered by the registrar as grief; but *that* it is. Grief of that sort, and at that age, has killed more than ever have been counted amongst its martyrs.

Therefore it is that Levana often communes with the powers that shake man's heart; therefore it is that she dotes upon grief. "These ladies," said I softly to myself, on seeing the ministers with whom Levana was conversing, "these are the Sorrows; and they are three in number: as the *Graces* are three, who dress man's life with beauty; the *Parcæ* are three, who weave the dark arras of man's life in their mysterious loom always with colors sad in part, sometimes angry with tragic crimson and black; the *Furies* are three, who visit with retributions called from the other side of the grave offences that walk upon this; and once even the *Muses* were but three, who fit the harp, the trumpet, or the lute, to the great burdens of man's impassioned creations. These are the Sorrows; all three of whom I know." The last words I say *now;* but in Oxford I said, "one of whom I know, and the others too surely I *shall* know." For already, in my fervent youth, I saw (dimly relieved upon the dark background of my dreams) the imperfect lineaments of the awful Sisters.

These Sisters—by what name shall we call them? If I say simply "The Sorrows," there will be a chance of mistaking the term; it might be understood of individual sorrow,—separate cases of sorrow,—whereas I want a term expressing the mighty abstractions that incarnate themselves in all individual sufferings of man's heart, and I wish to have these abstractions presented as impersonations,—that is, as clothed with human attributes of life, and with functions pointing to flesh. Let us call them, therefore, *Our Ladies of Sorrow.*

I know them thoroughly, and have walked in all their kingdoms. Three sisters they are, of one mysterious household; and their paths are wide apart; but of their dominion there is no end. Them I saw often conversing with Levana, and sometimes about myself. Do they talk, then? O no! Mighty phantoms like these disdain the infirmities of language. They may utter voices through the organs of man when they dwell in human hearts, but amongst themselves is no voice nor sound; eternal silence reigns in *their* kingdoms. They spoke not as they talked with Levana; they whispered not; they sang not; though oftentimes methought they *might* have sung: for I upon earth had heard their mysteries oftentimes deciphered by harp and timbrel, by dulcimer and organ. Like God, whose servants they are, they utter their pleasure not by sounds that perish, or by words that go astray, but by signs in heaven, by changes on earth, by pulses in secret rivers, heraldries painted on darkness, and hieroglyphics written on the tablets of the brain. *They* wheeled in mazes; *I* spelled the steps. *They* telegraphed[1] from afar; *I* read the signals. *They* conspired together; and on the mirrors of darkness *my* eye traced the plots. *Theirs* were the symbols; *mine* are the words.

What is it the Sisters are? What is it that they do? Let me describe their form and their presence, if form it were that still fluctuated in its outline, or presence it were that forever advanced to the front or forever receded amongst shades.

The eldest of the three is named *Mater Lachrymarum,* Our Lady of Tears. She it is that night and day raves and moans, calling for vanished faces. She stood in Rama, where a voice was heard of lamentation,— Rachel weeping for her children, and refusing to be comforted.[2] She it was that stood in Bethlehem on the night when Herod's sword[3] swept its nurseries of Innocents, and the little feet were stiffened forever which, heard at times as they trotted along floors overhead, woke pulses of love in household hearts that were not unmarked in heaven. Her eyes are sweet and subtle, wild and sleepy, by turns; oftentimes rising to the clouds, oftentimes challenging the heavens. She wears a diadem round her head. And I knew by childish memories that she could go abroad upon the winds, when she heard the sobbing of litanies, or the thundering of organs, and when she beheld the mustering of summer clouds. This Sister, the elder, it is that carries keys more than papal at her girdle, which open every cottage and every palace. She, to my knowledge, sat all last summer by the bedside of the blind beggar, him that so often and so gladly I talked with, whose pious daughter, eight years old, with the sunny countenance, resisted the temptations of play and village mirth, to travel all day long on dusty roads with her afflicted father. For this did God send her a great reward. In the springtime of the year, and whilst yet her own spring was budding, He recalled her to himself. But her blind father mourns forever over *her:* still he dreams at midnight that the little guiding hand is locked within his own; and still he wakens to a darkness that is *now* within a second and a deeper darkness. This *Mater Lachrymarum* also has been sitting all this winter of 1844–5 within the bedchamber of the Czar, bringing before his eyes a daughter (not less pious) that vanished to God not less suddenly, and left behind her a darkness not less profound.[4] By the power of the keys it is that Our Lady of Tears glides, a ghostly intruder, into the chambers of sleepless men, sleepless women, sleepless children, from Ganges to the Nile, from Nile to Mississippi. And her, because she is the first-born of her house, and has the widest empire, let us honor with the title of "Madonna."

2. **Rachel . . . comforted:** Jeremiah 31:15. 3. **Herod's sword:** Matthew 2:16. 4. **daughter . . . profound:** Alexandra, daughter of Czar Nicholas I, had died in August, 1844.

SUSPIRIA DE PROFUNDIS. LEVANA AND OUR LADIES. I. **telegraphed:** signaled.

The second Sister is called *Mater Suspiriorum*, Our Lady of Sighs. She never scales the clouds, nor walks abroad upon the winds. She wears no diadem. And her eyes, if they were ever seen, would be neither sweet nor subtle; no man could read their story; they would be found filled with perishing dreams, and with wrecks of forgotten delirium. But she raises not her eyes; her head, on which sits a dilapidated turban, droops forever, forever fastens on the dust. She weeps not. She groans not. But she sighs inaudibly at intervals. Her sister, Madonna, is oftentimes stormy and frantic, raging in the highest against heaven, and demanding back her darlings. But Our Lady of Sighs never clamors, never defies, dreams not of rebellious aspirations. She is humble to abjectness. Hers is the meekness that belongs to the hopeless. Murmur she may, but it is in her sleep. Whisper she may, but it is to herself in the twilight. Mutter she does at times, but it is in solitary places that are desolate as she is desolate, in ruined cities, and when the sun has gone down to his rest. This Sister is the visitor of the Pariah, of the Jew, of the bondsman to the oar in the Mediterranean galleys; of the English criminal in Norfolk Island, blotted out from the books of remembrance in sweet far-off England; of the baffled penitent reverting his eyes forever upon a solitary grave, which to him seems the altar overthrown of some past and bloody sacrifice, on which altar no oblations can now be availing, whether towards pardon that he might implore, or towards reparation that he might attempt. Every slave that at noonday looks up to the tropical sun with timid reproach, as he points with one hand to the earth, our general mother, but for *him* a stepmother, as he points with the other hand to the Bible, our general teacher, but against *him* sealed and sequestered; every woman sitting in darkness, without love to shelter her head, or hope to illumine her solitude, because the heaven-born instincts kindling in her nature germs of holy affections, which God implanted in her womanly bosom, having been stifled by social necessities, now burn sullenly to waste, like sepulchral lamps amongst the ancients; every nun defrauded of her unreturning May-time by wicked kinsman, whom God will judge; every captive in every dungeon; all that are betrayed, and all that are rejected; outcasts by traditionary law, and children of *hereditary* disgrace: all these walk with Our Lady of Sighs. She also carries a key; but she needs it little. For her kingdom is chiefly amongst the tents of Shem, and the houseless vagrant of every clime. Yet in the very highest ranks of man she finds chapels of her own; and

even in glorious England there are some that, to the world, carry their heads as proudly as the reindeer, who yet secretly have received her mark upon their foreheads.

But the third Sister, who is also the youngest ——! Hush! whisper whilst we talk of *her*. Her kingdom is not large, or else no flesh should live: but within that kingdom all power is hers. Her head, turreted like that of Cybele, rises almost beyond the reach of sight. She droops not; and her eyes, rising so high, *might* be hidden by distance. But, being what they are, they cannot be hidden: through the treble veil of crape which she wears the fierce light of a blazing misery, that rests not for matins or for vespers, for noon of day or noon of night, for ebbing or for flowing tide, may be read from the very ground. She is the defier of God. She also is the mother of lunacies, and the suggestress of suicides. Deep lie the roots of her power; but narrow is the nation that she rules. For she can approach only those in whom a profound nature has been upheaved by central convulsions; in whom the heart trembles and the brain rocks under conspiracies of tempest from without and tempest from within. Madonna moves with uncertain steps, fast or slow, but still with tragic grace. Our Lady of Sighs creeps timidly and stealthily. But this youngest Sister moves with incalculable motions, bounding, and with tiger's leaps. She carries no key; for, though coming rarely amongst men, she storms all doors at which she is permitted to enter at all. And *her* name is *Mater Tenebrarum*,—Our Lady of Darkness.

1845–49 (1849)

from JOAN OF ARC

〜⌒〜

First printed in *Tait's Magazine* for March and April, 1847.

〜⌒〜

What is to be thought of *her*? What is to be thought of the poor shepherd girl from the hills and forests of Lorraine, that—like the Hebrew shepherd boy[1] from the hills and forests of Judea—rose suddenly out of the quiet, out of the safety, out of the religious inspiration, rooted in deep pastoral solitudes, to a station in the van of armies, and to the more perilous station at the right

JOAN OF ARC. **1. shepherd boy:** David.

hand of kings? The Hebrew boy inaugurated his patriotic mission by an *act*, by a victorious *act*, such as no man could deny. But so did the girl of Lorraine, if we read her story as it was read by those who saw her nearest. Adverse armies bore witness to the boy as no pretender; but so they did to the gentle girl. Judged by the voices of all who saw them *from a station of good-will*, both were found true and loyal to any promises involved in their first acts. Enemies it was that made the difference between their subsequent fortunes. The boy rose to a splendor and a noonday prosperity, both personal and public, that rang through the records of his people, and became a byword among his posterity for a thousand years, until the sceptre was departing from Judah. The poor, forsaken girl, on the contrary, drank not herself from that cup of rest which she had secured for France. She never sang together with the songs that rose in her native Domrémy as echoes to the departing steps of invaders. She mingled not in the festal dances at Vaucouleurs which celebrated in rapture the redemption of France. No! for her voice was then silent; no! for her feet were dust. Pure, innocent, noble-hearted girl! whom, from earliest youth, ever I believed in as full of truth and self-sacrifice, this was amongst the strongest pledges for *thy* truth, that never once—no, not for a moment of weakness—didst thou revel in the vision of coronets and honor from man. Coronets for thee! Oh no! Honors, if they come when all is over, are for those that share thy blood. Daughter of Domrémy, when the gratitude of thy king shall awaken, thou wilt be sleeping the sleep of the dead. Call her, King of France, but she will not hear thee. Cite her by the apparitors to come and receive a robe of honor, but she will be found *en contumace*.[2] When the thunders of universal France, as even yet may happen, shall proclaim the grandeur of the poor shepherd girl that gave up all for her country, thy ear, young shepherd girl, will have been deaf for five centuries. To suffer and to do, that was thy portion in this life; that was thy destiny; and not for a moment was it hidden from thyself. Life, thou saidst, is short; and the sleep which is in the grave is long; let me use that life, so transitory, for the glory of those heavenly dreams destined to comfort the sleep which is so long! This pure creature—pure from every suspicion of even a visionary self-interest, even as she was pure in senses more obvious—never once did this holy child, as regarded herself, relax from her belief in the darkness

that was travelling to meet her. She might not prefigure the very manner of her death; she saw not in vision, perhaps, the aerial altitude of the fiery scaffold, the spectators without end on every road pouring into Rouen as to a coronation, the surging smoke, the volleying flames, the hostile faces all around, the pitying eye that lurked but here and there, until nature and imperishable truth broke loose from artificial restraints; —these might not be apparent through the mists of the hurrying future.

. . .

The shepherd girl that had delivered France—she, from her dungeon, she, from her baiting at the stake, she, from her duel with fire, as she entered her last dream—saw Domrémy, saw the fountain of Domrémy, saw the pomp of forests in which her childhood had wandered. That Easter festival which man had denied to her languishing heart—that resurrection of springtime, which the darkness of dungeons had intercepted from *her*, hungering after the glorious liberty of forests—were by God given back into her hands, as jewels that had been stolen from her by robbers. With those, perhaps (for the minutes of dreams can stretch into ages), was given back to her by God the bliss of childhood. By special privilege for *her* might be created, in this farewell dream, a second childhood, innocent as the first; but not, like *that*, sad with the gloom of a fearful mission in the rear. This mission had now been fulfilled. The storm was weathered; the skirts even of that mighty storm were drawing off. The blood that she was to reckon for had been exacted; the tears that she was to shed in secret had been paid to the last. The hatred to herself in all eyes had been faced steadily, had been suffered, had been survived. And in her last fight upon the scaffold she had triumphed gloriously; victoriously she had tasted the stings of death. For all, except this comfort from her farewell dream, she had died—died, amidst the tears of ten thousand enemies— died, amidst the drums and trumpets of armies—died, amidst peals redoubling upon peals, volleys upon volleys, from the saluting clarions of martyrs.

Bishop of Beauvais! because the guilt-burdened man is in dreams haunted and waylaid by the most frightful of his crimes, and because upon that fluctuating mirror —rising (like the mocking mirrors of *mirage* in Arabian deserts) from the fens of death—most of all are reflected the sweet countenances which the man has laid in ruins; therefore I know, bishop, that you also, entering your final dream, saw Domrémy. That fountain, of which the witnesses spoke so much, showed itself to your eyes

2. **en contumace:** in default (by not appearing).

in pure morning dews: but neither dews, nor the holy dawn, could cleanse away the bright spots of innocent blood upon its surface. By the fountain, bishop, you saw a woman seated, that hid her face. But, as *you* draw near, the woman raises her wasted features. Would Domrémy know them again for the features of her child? Ah, but *you* know them, bishop, well! Oh, mercy! what a groan was *that* which the servants, waiting outside the bishop's dream at his bedside, heard from his laboring heart, as at this moment he turned away from the fountain and the woman, seeking rest in the forests afar off. Yet not *so* to escape the woman, whom once again he must behold before he dies. In the forests to which he prays for pity, will he find a respite? What a tumult, what a gathering of feet is there! In glades where only wild deer should run, armies and nations are assembling; towering in the fluctuating crowd are phantoms that belong to departed hours. There is the great English Prince, Regent of France. There is my Lord of Winchester, the princely cardinal, that died and made no sign. There is the Bishop of Beauvais, clinging to the shelter of thickets. What building is that which hands so rapid are raising? Is it a martyr's scaffold? Will they burn the child of Domrémy a second time? No: it is a tribunal that rises to the clouds; and two nations stand around it, waiting for a trial. Shall my Lord of Beauvais sit again upon the judgment seat, and again number the hours for the innocent? Ah no! he is the prisoner at the bar. Already all is waiting: the mighty audience is gathered, the Court is hurrying to their seats, the witnesses are arrayed, the trumpets are sounding, the judge is taking his place. Oh! but this is sudden. My lord, have you no counsel? "Counsel I have none: in heaven above, or on earth beneath, counsellor there is none now that would take a brief from *me*: all are silent." Is it, indeed, come to this? Alas! the time is short, the tumult is wondrous, the crowd stretches away into infinity; but yet I will search in it for somebody to take your brief: I know of somebody that will be your counsel. Who is this that cometh from Domrémy? Who is she in bloody coronation robes from Rheims? Who is she that cometh with blackened flesh from walking the furnaces of Rouen? This is she, the shepherd girl, counsellor that had none for herself, whom I choose, bishop, for yours. She it is, I engage, that shall take my lord's brief. She it is, bishop, that would plead for you: yes, bishop, SHE— when heaven and earth are silent.

1847 (1847)

LITERATURE OF KNOWLEDGE AND LITERATURE OF POWER

First printed in *The North Briton Review*, August, 1848, as part of a review of *The Works of Alexander Pope*, ed. W. Roscoe, 1847.

What is it that we mean by *literature?* Popularly, and amongst the thoughtless, it is held to include everything that is printed in a book. Little logic is required to disturb *that* definition. The most thoughtless person is easily made aware that in the idea of *literature* one essential element is,—some relation to a general and common interest of man, so that what applies only to a local or professional or merely personal interest, even though presenting itself in the shape of a book, will not belong to literature. So far the definition is easily narrowed; and it is as easily expanded. For not only is much that takes a station in books not literature, but, inversely, much that really *is* literature never reaches a station in books. The weekly sermons of Christendom, that vast pulpit literature which acts so extensively upon the popular mind—to warn, to uphold, to renew, to comfort, to alarm—does not attain the sanctuary of libraries in the ten-thousandth part of its extent. The drama, again, as for instance the finest of Shakespeare's plays in England and all leading Athenian plays in the noontide of the Attic stage,[1] operated as a literature on the public mind, and were (according to the strictest letter of that term) *published* through the audiences that witnessed[2] their representation, some time before they were published as things to be read; and they were published in this scenical mode of publication with much more effect than they could have had as books during ages of costly copying or of costly printing.

Books, therefore, do not suggest an idea co-extensive and interchangeable with the idea of literature, since much literature, scenic, forensic, or didactic (as from

LITERATURE OF KNOWLEDGE AND LITERATURE OF POWER. **1. Attic stage:** the Athenian stage. **2. published . . . witnessed:** "Charles I, for example, when Prince of Wales, and many others in his father's court, gained their known familiarity with Shakespeare—not through the original quartos, so slenderly diffused, nor through the first folio of 1623, but through the court representations of his chief dramas at Whitehall" (De Quincey's note).

lectures and public orators), may never come into books, and much that *does* come into books may connect itself with no literary interest. But a far more important correction, applicable to the common vague idea of literature, is to be sought, not so much in a better definition of literature, as in a sharper distinction of the two functions which it fulfils. In that great social organ which, collectively, we call literature, there may be distinguished two separate offices, that may blend and often *do* so, but capable, severally, of a severe insulation, and naturally fitted for reciprocal repulsion. There is, first, the literature of *knowledge*, and, secondly, the literature of *power*. The function of the first is to *teach*; the function of the second is to *move*: the first is a rudder; the second an oar or a sail. The first speaks to the *mere* discursive understanding; the second speaks ultimately, it may happen, to the higher understanding, or reason, but always *through* affections of pleasure and sympathy. Remotely it may travel towards an object seated in what Lord Bacon calls *dry* light;[3] but proximately it does and must operate—else it ceases to be a literature of *power*—on and through that *humid* light which clothes itself in the mists and glittering *iris*[4] of human passions, desires, and genial emotions. Men have so little reflected on the higher functions of literature as to find it a paradox if one should describe it as a mean or subordinate purpose of books to give information. But this is a paradox only in the sense which makes it honorable to be paradoxical. Whenever we talk in ordinary language of seeking information or gaining knowledge, we understand the words as connected with something of absolute novelty. But it is the grandeur of all truth which *can* occupy a very high place in human interests that it is never absolutely novel to the meanest of minds: it exists eternally, by way of germ or latent principle, in the lowest as in the highest, needing to be developed but never to be planted. To be capable of transplantation is the immediate criterion of a truth that ranges on a lower scale. Besides which, there is a rarer thing than truth, namely, *power*, or deep sympathy with truth. What is the effect, for instance, upon society, of children? By the pity, by the tenderness, and by the peculiar modes of admiration, which connect themselves with the helplessness, with the innocence, and with the simplicity of children, not only are the primal affections strengthened and continually renewed, but the qualities which are dearest in the sight of heaven— the frailty, for instance, which appeals to forbearance,

the innocence which symbolizes the heavenly, and the simplicity which is most alien from the worldly— are kept up in perpetual remembrance, and their ideals are continually refreshed. A purpose of the same nature is answered by the higher literature, *viz.*, the literature of power. What do you learn from *Paradise Lost?* Nothing at all. What do you learn from a cookery-book? Something new, something that you did not know before, in every paragraph. But would you therefore put the wretched cookery-book on a higher level of estimation than the divine poem? What you owe to Milton is not any knowledge, of which a million separate items are still but a million of advancing steps on the same earthly level; what you owe is *power*, that is, exercise and expansion to your own latent capacity of sympathy with the infinite, where every pulse and each separate influx is a step upwards, a step ascending as upon a Jacob's ladder[5] from earth to mysterious altitudes above the earth. *All* the steps of knowledge, from first to last, carry you further on the same plane, but could never raise you one foot above your ancient level of earth; whereas the very *first* step in power is a flight, is an ascending movement into another element where earth is forgotten.

Were it not that human sensibilities are ventilated and continually called out into exercise by the great phenomena of infancy, or of real life as it moves through chance and change, or of literature as it recombines these elements in the mimicries of poetry, romance, etc., it is certain that, like any animal power or muscular energy falling into disuse, all such sensibilities would gradually droop and dwindle. It is in relation to these great *moral* capacities of man that the literature of power, as contradistinguished from that of knowledge, lives and has its field of action. It is concerned with what is highest in man; for the Scriptures themselves never condescended to deal by suggestion or cooperation with the mere discursive understanding: when speaking of man in his intellectual capacity, the Scriptures speak not of the understanding, but of *"the understanding heart,"*[6]—making the heart, *i.e.*, the great *intuitive* (or non-discursive) organ, to be the interchangeable formula for man in his highest state of capacity for the infinite. Tragedy, romance, fairy tale, or epopee,[7] all alike restore to man's mind the ideals of justice, of hope, of truth, of mercy, of retribution, which else (left to the support of daily life in its realities)

---

3. **Lord Bacon . . . light:** *Apothegms*, 268. 4. **iris:** rainbow.

5. **Jacob's ladder:** Genesis 28:12. 6. **"the understanding heart":** 1 Kings 3:9–12. 7. **epopee:** epic.

would languish for want of sufficient illustration. What is meant, for instance, by *poetic justice?*[8]—It does not mean a justice that differs by its object from the ordinary justice of human jurisprudence; for then it must be confessedly a very bad kind of justice; but it means a justice that differs from common forensic justice by the degree in which it *attains* its object, a justice that is more omnipotent over its own ends, as dealing—not with the refractory elements of earthly life, but with the elements of its own creation, and with materials flexible to its own purest preconceptions. It is certain that, were it not for the Literature of Power, these ideals would often remain amongst us as mere arid notional forms; whereas, by the creative forces of man put forth in literature, they gain a vernal life of restoration, and germinate into vital activities. The commonest novel, by moving in alliance with human fears and hopes, with human instincts of wrong and right, sustains and quickens those affections. Calling them into action, it rescues them from torpor. And hence the pre-eminency, over all authors that merely *teach* of the meanest that moves, or that teaches, if at all, indirectly *by* moving. The very highest work that has ever existed in the literature of Knowledge is but a *provisional* work: a book upon trial and sufferance, and *quamdiu bene se gesserit.*[9] Let its teaching be even partially revised, let it be but expanded,—nay, even let its teaching be but placed in a better order,—and instantly it is superseded. Whereas the feeblest works in the Literature of Power, surviving at all, survive as finished and unalterable amongst men. For instance, the *Principia* of Sir Isaac Newton was a book *militant* on earth from the first. In all stages of its progress it would have to fight for its existence: 1st, as regards absolute truth; 2dly, when that combat was over, as regards its form or mode of presenting the truth. And as soon as a La Place, or anybody else, builds higher upon the foundations laid by this book, effectually he throws it out of the sunshine into decay and darkness; by weapons won from this book he superannuates and destroys this book, so that soon the name of Newton remains as a mere *nominis umbra*,[10] but his book, as a living power, has transmigrated into other forms. Now, on the contrary, the *Iliad*, the *Prometheus* of Æschylus, the *Othello* or *King Lear*, the *Hamlet* or *Macbeth*, and the *Paradise Lost* are not militant but

triumphant forever as long as the languages exist in which they speak or can be taught to speak. They never *can* transmigrate into new incarnations. To reproduce *these* in new forms, or variations, even if in some things they should be improved, would be to plagiarize. A good steam engine is properly superseded by a better. But one lovely pastoral valley is not superseded by another, nor a statue of Praxiteles by a statue of Michael Angelo. These things are separated not by imparity, but by disparity. They are not thought of as unequal under the same standard, but as different in *kind*, and, if otherwise equal, as equal under a different standard. Human works of immortal beauty and works of nature in one respect stand on the same footing: they never absolutely repeat each other, never approach so near as not to differ; and they differ not as better and worse, or simply by more and less: they differ by undecipherable and incommunicable differences, that cannot be caught by mimicries, that cannot be reflected in the mirror of copies, that cannot become ponderable in the scales of vulgar comparison.

1848 (1848)

FROM

THE ENGLISH MAIL-COACH

First printed in parts in *Blackwood's Magazine* (Oct. and Dec., 1849) and then thoroughly revised (1854) and unified under the present title when De Quincey was preparing the collected edition of his works.

SECTION II—THE VISION OF SUDDEN DEATH

What is to be taken as the predominant opinion of man, reflective and philosophic, upon SUDDEN DEATH? It is remarkable that, in different conditions of society, sudden death has been variously regarded as the consummation of an earthly career most fervently to be desired, or, again, as that consummation which is with most horror to be deprecated. Cæsar the Dictator, at his last dinner-party (*cœna*), on the very evening before his assassination, when the minutes of his earthly career were numbered, being asked what death, in *his* judgment, might be pronounced the most eligible, replied "That which should be most sudden." On the

---

8. poetic justice: The neoclassic critical doctrine that at the end of the play the good should be rewarded and the bad punished. **9. quamdiu . . . gesserit:** as long as it conducted itself well. **10. nominis umbra:** shadow of a name.

other hand, the divine Litany of our English Church, when breathing forth supplications, as if in some representative character, for the whole human race prostrate before God, places such a death in the very van of horrors: "From lightning and tempest; from plague, pestilence, and famine; from battle and murder, and from SUDDEN DEATH—*Good Lord, deliver us.*" Sudden death is here made to crown the climax in a grand ascent of calamities; it is ranked among the last of curses; and yet by the noblest of Romans it was ranked as the first of blessings. In that difference most readers will see little more than the essential difference between Christianity and Paganism. But this, on consideration, I doubt The Christian Church may be right in its estimate of sudden death; and it is a natural feeling, though after all it may also be an infirm one, to wish for a quiet dismissal from life, as that which *seems* most reconcilable with meditation, with penitential retrospects, and with the humilities of farewell prayer. There does not, however, occur to me any direct scriptural warrant for this earnest petition of the English Litany, unless under a special construction of the word *sudden*. It seems a petition indulged rather and conceded to human infirmity than exacted from human piety. It is not so much a doctrine built upon the eternities of the Christian system as a plausible opinion built upon special varieties of physical temperament. Let that, however, be as it may, two remarks suggest themselves as prudent restraints upon a doctrine which else *may* wander, and *has* wandered, into an uncharitable superstition. The first is this: that many people are likely to exaggerate the horror of a sudden death from the disposition to lay a false stress upon words or acts simply because by an accident they have become *final* words or acts. If a man dies, for instance, by some sudden death when he happens to be intoxicated, such a death is falsely regarded with peculiar horror, as though the intoxication were suddenly exalted into a blasphemy. But *that* is unphilosophic. The man was or he was not, *habitually* a drunkard. If not, if his intoxication were a solitary accident, there can be no reason for allowing special emphasis to this act simply because through misfortune it became his final act. Nor, on the other hand, if it were no accident, but one of his *habitual* transgressions, will it be the more habitual or the more a transgression because some sudden calamity, surprising him, has caused this habitual transgression to be also a final one. Could the man have had any reason even dimly to foresee his own sudden death, there would have been a new feature in his act

of intemperance—a feature of presumption and irreverence, as in one that, having known himself drawing near to the presence of God, should have suited his demeanor to an expectation so awful. But this is no part of the case supposed. And the only new element in the man's act is not any element of special immorality, but simply of special misfortune.

The other remark has reference to the meaning of the word *sudden*. Very possibly Cæsar and the Christian Church do not differ in the way supposed,—that is, do not differ by any difference of doctrine as between Pagan and Christian views of the moral temper appropriate to death; but perhaps they are contemplating different cases. Both contemplate a violent death, a βιαθάνατος—death that is βίαιος, or, in other words, death that is brought about, not by internal and spontaneous change, but by active force having its origin from without. In this meaning the two authorities agree. Thus far they are in harmony. But the difference is that the Roman by the word *sudden* means *unlingering*, whereas the Christian Litany by *sudden death* means a *death without warning*, consequently without any available summons to religious preparation. The poor mutineer who kneels down to gather into his heart the bullets from twelve fire-locks of his pitying comrades dies by a most sudden death in Cæsar's sense; one shock, one mighty spasm, one (possibly *not* one) groan, and all is over. But, in the sense of the Litany, the mutineer's death is far from sudden: his offence originally, his imprisonment, his trial, the interval between his sentence and its execution, having all furnished him with separate warnings of his fate—having all summoned him to meet it with solemn preparation.

Here at once, in this sharp verbal distinction, we comprehend the faithful earnestness with which a holy Christian Church pleads on behalf of her poor departing children that God would vouchsafe to them the last great privilege and distinction possible on a death-bed— *viz.*, the opportunity of untroubled preparation for facing this mighty trial. Sudden death, as a mere variety in the modes of dying where death in some shape is inevitable, proposes a question of choice which, equally in the Roman and the Christian sense, will be variously answered according to each man's variety of temperament. Meantime, one aspect of sudden death there is, one modification, upon which no doubt can arise, that of all martyrdoms it is the most agitating—*viz.*, where it surprises a man under circumstances which offer (or which seem to offer) some hurrying, flying, inappreciably minute chance of evading it. Sudden as the danger which it affronts must be any effort by which such an evasion can be accomplished. Even *that*, even the sickening necessity for hurrying in extremity where all hurry seems destined to be vain,—even that anguish is liable to a hideous exasperation in one particular case: *viz.*, where the appeal is made not exclusively to the instinct of self-preservation, but to the conscience, on behalf of some other life besides your own, accidentally thrown upon *your* protection. To fail, to collapse in a service merely your own, might seem comparatively venial; though, in fact, it is far from venial. But to fail in a case where Providence has suddenly thrown into your hands the final interests of another,—a fellow-creature shuddering between the gates of life and death: this, to a man of apprehensive conscience, would mingle the misery of an atrocious criminality with the misery of a bloody calamity. You are called upon, by the case supposed, possibly to die, but to die at the very moment when, by any even partial failure or effeminate collapse of your energies, you will be self-denounced as a murderer. You had but the twinkling of an eye for your effort, and that effort might have been unavailing; but to have risen to the level of such an effort would have rescued you, though not from dying, yet from dying as a traitor to your final and farewell duty.

The situation here contemplated exposes a dreadful ulcer, lurking far down in the depths of human nature. It is not that men generally are summoned to face such awful trials. But potentially, and in shadowy outline, such a trial is moving subterraneously in perhaps all men's natures. Upon the secret mirror of our dreams such a trial is darkly projected, perhaps, to every one of us. That dream, so familiar to childhood, of meeting a lion, and, through languishing prostration in hope and the energies of hope, that constant sequel of lying down before the lion publishes the secret frailty of human nature—reveals its deep-seated falsehood to itself—records its abysmal treachery. Perhaps not one of us escapes that dream; perhaps, as by some sorrowful doom of man, that dream repeats for every one of us, through every generation, the original temptation in Eden. Every one of us, in this dream, has a bait offered to the infirm places of his own individual will; once again a snare is presented for tempting him into captivity to a luxury of ruin; once again, as in aboriginal Paradise, the man falls by his own choice; again, by infinite iteration, the ancient earth groans to Heaven, through her secret caves, over the weakness of her child. "Nature, from her seat, sighing through all her works,"

again "gives signs of woe that all is lost";[1] and again the counter-sigh is repeated to the sorrowing heavens for the endless rebellion against God. It is not without probability that in the world of dreams every one of us ratifies for himself the original transgression. In dreams, perhaps under some secret conflict of the midnight sleeper, lighted up to the consciousness at the time, but darkened to the memory as soon as all is finished, each several child of our mysterious race completes for himself the treason of the aboriginal fall.

The incident, so memorable in itself by its features of horror, and so scenical by its grouping for the eye, which furnished the text for this reverie upon *Sudden Death* occurred to myself in the dead of night, as a solitary spectator, when seated on the box of the Manchester and Glasgow mail, in the second or third summer after Waterloo. I find it necessary to relate the circumstances, because they are such as could not have occurred unless under a singular combination of accidents. In those days, the oblique and lateral communications with many rural post-offices were so arranged, either through necessity or through defect of system, as to make it requisite for the main north-western mail (*i.e.*, the *down* mail) on reaching Manchester to halt for a number of hours; how many, I do not remember; six or seven, I think; but the result was that, in the ordinary course, the mail recommenced its journey northwards about midnight. Wearied with the long detention at a gloomy hotel, I walked out about eleven o'clock at night for the sake of fresh air; meaning to fall in with the mail and resume my seat at the post-office. The night, however, being yet dark, as the moon had scarcely risen, and the streets being at that hour empty, so as to offer no opportunities for asking the road, I lost my way, and did not reach the post-office until it was considerably past midnight; but, to my great relief (as it was important for me to be in Westmoreland by the morning), I saw in the huge saucer eyes of the mail, blazing through the gloom, an evidence that my chance was not yet lost. Past the time it was; but, by some rare accident, the mail was not even yet ready to start. I ascended to my seat on the box, where my cloak was still lying as it had lain at the Bridgewater Arms. I had left it there in imitation of a nautical discoverer, who leaves a bit of bunting on the shore of his discovery, by way of warning off the ground the whole human race, and notifying to the

THE ENGLISH MAIL-COACH. SECTION II. **I. "Nature . . . lost"**: *Paradise Lost*, IX. 782–84.

Christian and the heathen worlds, with his best compliments, that he has hoisted his pocket-handkerchief once and forever upon that virgin soil: thenceforward claiming the *jus dominii*[2] to the top of the atmosphere above it, and also the right of driving shafts to the centre of the earth below it; so that all people found after this warning either aloft in upper chambers of the atmosphere, or groping in subterraneous shafts, or squatting audaciously on the surface of the soil, will be treated as trespassers—kicked, that is to say, or decapitated, as circumstances may suggest, by their very faithful servant, the owner of the said pocket-handkerchief. In the present case, it is probable that my cloak might not have been respected, and the *jus gentium*[3] might have been cruelly violated in my person—for, in the dark, people commit deeds of darkness, gas being a great ally of morality; but it so happened that on this night there was no other outside passenger; and thus the crime, which else was but too probable, missed fire for want of a criminal.

Having mounted the box, I took a small quantity of laudanum, having already travelled two hundred and fifty miles—*viz.*, from a point seventy miles beyond London. In the taking of laudanum there is nothing extraordinary. But by accident it drew upon me the special attention of my assessor on the box, the coachman. And in *that* also there was nothing extraordinary. But by accident, and with great delight, it drew my own attention to the fact that this coachman was a monster in point of bulk, and that he had but one eye. In fact, he had been foretold by Virgil as

Monstrum horrendum, informe, ingens, cui lumen
    ademptum.[4]

He answered to the conditions in every one of the items:—1, a monster he was; 2, dreadful; 3, shapeless; 4, huge; 5, who had lost an eye. But why should *that* delight *me*? Had he been one of the Calendars[5] in *The Arabian Nights*, and had paid down his eye as the price of his criminal curiosity, what right had *I* to exult in his misfortune? I did *not* exult; I delighted in no man's punishment, though it were even merited. But these personal distinctions (Nos. 1, 2, 3, 4, 5) identified in an instant an old friend of mine whom I had known in the south for some years as the most masterly of mail-coachmen. He was the man in all Europe that could

(if *any* could) have driven six-in-hand full gallop over *Al Sirat*[6]—that dreadful bridge of Mahomet, with no side battlements, and of *extra* room not enough for a razor's edge—leading right across the bottomless gulf. Under this eminent man, whom in Greek I cognominated Cyclops *Diphrélates* (Cyclops the Charioteer), I, and others known to me, studied the diphrelatic art. Excuse, reader, a word too elegant to be pedantic. As a pupil, though I paid extra fees, it is to be lamented that I did not stand high in his esteem. It showed his dogged honesty (though, observe, not his discernment) that he could not see my merits. Let us excuse his absurdity in this particular by remembering his want of an eye. Doubtless *that* made him blind to my merits. In the art of conversation, however, he admitted that I had the whip-hand of him. On the present occasion great joy was at our meeting. But what was Cyclops doing here? Had the medical men recommended northern air, or how? I collected, from such explanations as he volunteered, that he had an interest at stake in some suit-at-law now pending at Lancaster; so that probably he had got himself transferred to this station for the purpose of connecting with his professional pursuits an instant readiness for the calls of his lawsuit.

Meantime, what are we stopping for? Surely we have now waited long enough. Oh, this procrastinating mail, and this procrastinating post-office! Can't they take a lesson upon that subject from *me*? Some people have called *me* procrastinating. Yet you are witness, reader, that I was here kept waiting for the post-office. Will the post-office lay its hand on its heart, in its moments of sobriety, and assert that ever it waited for me? What are they about? The guard tells me that there is a large extra accumulation of foreign mails this night, owing to irregularities caused by war, by wind, by weather, in the packet service, which as yet does not benefit at all by steam. For an *extra* hour, it seems, the post-office has been engaged in threshing out the pure wheaten correspondence of Glasgow, and winnowing it from the chaff of all baser intermediate towns. But at last all is finished. Sound your horn, guard! Manchester, good-bye! we've lost an hour by your criminal conduct at the post-office: which, however, though I do not mean to part with a serviceable ground of complaint, and one which really *is* such for the horses, to me secretly is an advantage, since it compels us to look sharply for this lost hour amongst the next eight or nine, and to recover it (if we can) at the rate of one mile extra per hour. Off

---

2. **jus dominii:** law of ownership. 3. **jus gentium:** law of nations. 4. **Monstrum . . . ademptum:** *Aeneid*, III. 658: "A horrid monster, deformed, huge, whose eye has been taken out" (Polyphemus). 5. **Calendars:** a Mohammedan order of begging friars.

6. **Al Sirat:** bridge from Hades to Paradise.

we are at last, and at eleven miles an hour; and for the moment I detect no changes in the energy or in the skill of Cyclops.

From Manchester to Kendal, which virtually (though not in law) is the capital of Westmoreland, there were at this time seven stages of eleven miles each. The first five of these, counting from Manchester, terminate in Lancaster; which is therefore fifty-five miles north of Manchester, and the same distance exactly from Liverpool. The first three stages terminate in Preston (called, by way of distinction from other towns of that name, *Proud* Preston), at which place it is that the separate roads from Liverpool and from Manchester to the north become confluent. Within these first three stages lay the foundation, the progress, and termination of our night's adventure. During the first stage, I found out that Cyclops was mortal: he was liable to the shocking affection of sleep—a thing which previously I had never suspected. If a man indulges in the vicious habit of sleeping, all the skill in aurigation[7] of Apollo himself, with the horses of Aurora to execute his notions, avails him nothing. "Oh, Cyclops!" I exclaimed, "thou art mortal. My friend, thou snorest." Through the first eleven miles, however, this infirmity—which I grieve to say that he shared with the whole Pagan Pantheon—betrayed itself only by brief snatches. On waking up, he made an apology for himself which, instead of mending matters, laid open a gloomy vista of coming disasters. The summer assizes, he reminded me, were now going on at Lancaster: in consequence of which for three nights and three days he had not lain down on a bed. During the day he was waiting for his own summons as a witness on the trial in which he was interested, or else, lest he should be missing at the critical moment, was drinking with the other witnesses under the pastoral surveillance of the attorneys. During the night, or that part of it which at sea would form the middle watch, he was driving. This explanation certainly accounted for his drowsiness, but in a way which made it much more alarming; since now, after several days' resistance to this infirmity, at length he was steadily giving way. Throughout the second stage he grew more and more drowsy. In the second mile of the third stage he surrendered himself finally and without a struggle to his perilous temptation. All his past resistance had but deepened the weight of this final oppression. Seven atmospheres of sleep rested upon him; and, to

7. **aurigation:** chariot driving.

consummate the case, our worthy guard, after singing *Love Amongst the Roses* for perhaps thirty times, without invitation and without applause, had in revenge moodily resigned himself to slumber—not so deep, doubtless, as the coachman's, but deep enough for mischief. And thus at last, about ten miles from Preston, it came about that I found myself left in charge of his Majesty's London and Glasgow mail, then running at the least twelve miles an hour.

What made this negligence less criminal than else it must have been thought was the condition of the roads at night during the assizes. At that time, all the law business of populous Liverpool, and also of populous Manchester, with its vast cincture of populous rural districts, was called up by ancient usage to the tribunal of Lilliputian Lancaster. To break up this old traditional usage required, 1, a conflict with powerful established interests; 2, a large system of new arrangements, and 3, a new parliamentary statute. But as yet this change was merely in contemplation. As things were at present, twice in the year[8] so vast a body of business rolled northwards from the southern quarter of the county that for a fortnight at least it occupied the severe exertions of two judges in its despatch. The consequence of this was that every horse available for such a service, along the whole line of road, was exhausted in carrying down the multitudes of people who were parties to the different suits. By sunset, therefore, it usually happened that, through utter exhaustion amongst men and horses, the road sank into profound silence. Except the exhaustion in the vast adjacent county of York from a contested election, no such silence succeeding to no such fiery uproar was ever witnessed in England.

On this occasion the usual silence and solitude prevailed along the road. Not a hoof nor a wheel was to be heard. And, to strengthen this false luxurious confidence in the noiseless roads, it happened also that the night was one of peculiar solemnity and peace. For my own part, though slightly alive to the possibilities of peril, I had so far yielded to the influence of the mighty calm as to sink into a profound reverie. The month was August; in the middle of which lay my own birthday—a festival to every thoughtful man suggesting solemn and often sigh-born thoughts. The county was my own native county—upon which, in its southern section, more than upon any equal area known to man past or

8. **twice . . . year:** "There were at that time only two assizes even in the most populous counties—viz., the Lent Assizes and the Summer Assizes" (De Q.).

present, had descended the original curse of labor in its heaviest form, not mastering the bodies only of men, as of slaves, or criminals in mines, but working through the fiery will. Upon no equal space of earth was, or ever had been, the same energy of human power put forth daily. At this particular season also of the assizes, that dreadful hurricane of flight and pursuit, as it might have seemed to a stranger, which swept to and from Lancaster all day long, hunting the county up and down, and regularly subsiding back into silence about sunset, could not fail (when united with this permanent distinction of Lancashire as the very metropolis and citadel of labor) to point the thoughts pathetically upon that countervision of rest, of saintly repose from strife and sorrow, towards which, as to their secret haven, the profounder aspirations of man's heart are in solitude continually travelling. Obliquely upon our left we were nearing the sea; which also must, under the present circumstances, be repeating the general state of halcyon repose. The sea, the atmosphere, the light, bore each an orchestral part in this universal lull. Moonlight and the first timid tremblings of the dawn were by this time blending; and the blendings were brought into a still more exquisite state of unity by a slight silvery mist, motionless and dreamy, that covered the woods and fields, but with a veil of equable transparency. Except the feet of our own horses,—which, running on a sandy margin of the road, made but little disturbance,—there was no sound abroad. In the clouds and on the earth prevailed the same majestic peace; and, in spite of all that the villain of a schoolmaster has done for the ruin of our sublimer thoughts, which are the thoughts of our infancy, we still believe in no such nonsense as a limited atmosphere. Whatever we may swear with our false feigning lips, in our faithful hearts we still believe, and must forever believe, in fields of air traversing the total gulf between earth and the central heavens. Still, in the confidence of children that tread without fear *every* chamber in their father's house, and to whom no door is closed, we, in that Sabbatic vision which sometimes is revealed for an hour upon nights like this, ascend with easy steps from the sorrow-stricken fields of earth upwards to the sandals of God.

Suddenly, from thoughts like these I was awakened to a sullen sound, as of some motion on the distant road. It stole upon the air for a moment; I listened in awe; but then it died away. Once roused, however, I could not but observe with alarm the quickened motion of our horses. Ten years' experience had made my eye learned in the valuing of motion; and I saw

that we were now running thirteen miles an hour. I pretend to no presence of mind. On the contrary, my fear is that I am miserably and shamefully deficient in that quality as regards action. The palsy of doubt and distraction hangs like some guilty weight of dark unfathomed remembrances upon my energies when the signal is flying for *action*. But, on the other hand, this accursed gift I have, as regards *thought*, that in the first step towards the possibility of a misfortune I see its total evolution; in the radix of the series I see too certainly and too instantly its entire expansion; in the first syllable of the dreadful sentence I read already the last. It was not that I feared for ourselves. *Us* our bulk and impetus charmed against peril in any collision. And I had ridden through too many hundreds of perils that were frightful to approach, that were matter of laughter to look back upon, the first face of which was horror, the parting face a jest—for any anxiety to rest upon *our* interests. The mail was not built, I felt assured, nor bespoke, that could betray *me* who trusted to its protection. But any carriage that we could meet would be frail and light in comparison of ourselves. And I remarked this ominous accident of our situation, —we were on the wrong side of the road. But then, it may be said, the other party, if other there was, might also be on the wrong side; and two wrongs might make a right. *That* was not likely. The same motive which had drawn *us* to the right-hand side of the road—*viz.*, the luxury of the soft beaten sand as contrasted with the paved centre—would prove attractive to others. The two adverse carriages would therefore, to a certainty, be travelling on the same side; and from this side, as not being ours in law, the crossing over to the other would, of course, be looked for from *us*.[9] Our lamps, still lighted, would give the impression of vigilance on our part. And every creature that met us would rely upon *us* for quartering.[10] All this, and if the separate links of the anticipation had been a thousand times more, I saw, not discursively, or by effort, or by succession, but by one flash of horrid simultaneous intuition.

Under this steady though rapid anticipation of the

9. **from us:** "It is true that, according to the law of the case as established by legal precedents, all carriages were required to give way before royal equipages, and therefore before the mail as one of them. But this only increased the danger, as being a regulation very imperfectly made known, very unequally enforced, and therefore often embarrassing the movements on both sides" (De Q.). 10. **quartering:** "This is the technical word, and, I presume, derived from the French *cartayer*, to evade a rut or any obstacle" (De Q.).

evil which *might* be gathering ahead, ah! what a sullen mystery of fear, what a sigh of woe, was that which stole upon the air, as again the far-off sound of a wheel was heard! A whisper it was—a whisper from, perhaps, four miles off—secretly announcing a ruin that, being foreseen, was not the less inevitable; that, being known, was not therefore healed. What could be done—who was it that could do it—to check the storm-flight of these maniacal horses? Could I not seize the reins from the grasp of the slumbering coachman? You, reader, think that it would have been in *your* power to do so. And I quarrel not with your estimate of yourself. But, from the way in which the coachman's hand was viced between his upper and lower thigh, this was impossible. Easy was it? See, then, that bronze equestrian statue. The cruel rider has kept the bit in his horse's mouth for two centuries. Unbridle him for a minute, if you please, and wash his mouth with water. Easy was it? Unhorse me, then, that imperial rider; knock me those marble feet from those marble stirrups of Charlemagne.

The sounds ahead strengthened, and were now too clearly the sounds of wheels. Who and what could it be? Was it industry in a taxed cart? Was it youthful gaiety in a gig? Was it sorrow that loitered, or joy that raced? For as yet the snatches of sound were too intermitting, from distance, to decipher the character of the motion. Whoever were the travellers, something must be done to warn them. Upon the other party rests the active responsibility, but upon *us*—and, woe is me! that *us* was reduced to my frail opium-shattered self—rests the responsibility of warning. Yet, how should this be accomplished? Might I not sound the guard's horn? Already, on the first thought, I was making my way over the roof of the guard's seat. But this, from the accident which I have mentioned, of the foreign mails being piled upon the roof, was a difficult and even dangerous attempt to one cramped by nearly three hundred miles of outside travelling. And, fortunately, before I had lost much time in the attempt, our frantic horses swept round an angle of the road which opened upon us that final stage where the collision must be accomplished and the catastrophe sealed. All was apparently finished. The court was sitting; the case was heard; the judge had finished; and only the verdict was yet in arrear.

Before us lay an avenue straight as an arrow, six hundred yards, perhaps, in length; and the umbrageous trees, which rose in a regular line from either side, meeting high overhead, gave to it the character of a cathedral aisle. These trees lent a deeper solemnity to the early light; but there was still light enough to perceive, at the further end of this Gothic aisle, a frail reedy gig, in which were seated a young man, and by his side a young lady. Ah, young sir! what are you about? If it is requisite that you should whisper your communications to this young lady—though really I see nobody, at an hour and on a road so solitary, likely to overhear you—is it therefore requisite that you should carry your lips forward to hers? The little carriage is creeping on at one mile an hour; and the parties within it, being thus tenderly engaged, are naturally bending down their heads. Between them and eternity, to all human calculation, there is but a minute and a half. Oh heavens! what is it that I shall do? Speaking or acting, what help can I offer? Strange it is, and to a mere auditor of the tale might seem laughable, that I should need a suggestion from the *Iliad* to prompt the sole resource that remained. Yet so it was. Suddenly I remembered the shout of Achilles, and its effect.[11] But could I pretend to shout like the son of Peleus, aided by Pallas? No: but then I needed not the shout that should alarm all Asia militant; such a shout would suffice as might carry terror into the hearts of two thoughtless young people and one gig-horse. I shouted—and the young man heard me not. A second time I shouted—and now he heard me, for now he raised his head.

Here, then, all had been done that, by me, *could* be done; more on *my* part was not possible. Mine had been the first step; the second was for the young man; the third was for God. If, said I, this stranger is a brave man, and if indeed he loves the young girl at his side—or, loving her not, if he feels the obligation, pressing upon every man worthy to be called a man, of doing his utmost for a woman confided to his protection—he will at least make some effort to save her. If *that* fails, he will not perish the more, or by a death more cruel, for having made it; and he will die as a brave man should, with his face to the danger, and with his arm about the woman that he sought in vain to save. But, if he makes no effort,—shrinking without a struggle from his duty,—he himself will not the less certainly perish for this business of poltroonery. He will die no less: and why not? Wherefore should we grieve that there is one craven less in the world? No; *let* him perish, without a pitying thought of ours wasted upon him; and, in that case, all our grief will be reserved for

11. **shout . . . effect:** *Iliad*, XVIII. 217–31. The shout terrified the Trojans, allowing the Greeks the opportunity to gather strength.

the fate of the helpless girl who now, upon the least shadow of failure in *him*, must by the fiercest of translations—must without time for a prayer—must within seventy seconds—stand before the judgment-seat of God.

But craven he was not: sudden had been the call upon him, and sudden was his answer to the call. He saw, he heard, he comprehended, the ruin that was coming down: already its gloomy shadow darkened above him; and already he was measuring his strength to deal with it. Ah! what a vulgar thing does courage seem when we see nations buying it and selling it for a shilling a-day:[12] ah! what a sublime thing does courage seem when some fearful summons on the great deeps of life carries a man, as if running before a hurricane, up to the giddy crest of some tumultuous crisis from which lie two courses, and a voice says to him audibly, "One way lies hope; take the other, and mourn forever!" How grand a triumph if, even then, amidst the raving of all around him, and the frenzy of the danger, the man is able to confront his situation—is able to retire for a moment into solitude with God, and to seek his counsel from *Him*!

For seven seconds, it might be, of his seventy, the stranger settled his countenance steadfastly upon us, as if to search and value every element in the conflict before him. For five seconds more of his seventy he sat immovably, like one that mused on some great purpose. For five more, perhaps, he sat with eyes upraised, like one that prayed in sorrow, under some extremity of doubt, for light that should guide him to the better choice. Then suddenly he rose; stood upright; and, by a powerful strain upon the reins, raising his horse's fore-feet from the ground, he slewed[13] him round on the pivot of his hind-legs, so as to plant the little equipage in a position nearly at right angles to ours. Thus far his condition was not improved; except as a first step had been taken towards the possibility of a second. If no more were done, nothing was done; for the little carriage still occupied the very centre of our path, though in an altered direction. Yet even now it may not be too late: fifteen of the seventy seconds may still be unexhausted; and one almighty bound may avail to clear the ground. Hurry, then, hurry! for the flying moments—*they* hurry. Oh, hurry, hurry, my brave young man! for the cruel hoofs of our horses— *they* also hurry! Fast are the flying moments, faster are

the hoofs of our horses. But fear not for *him*, if human energy can suffice; faithful was he that drove to his terrific duty; faithful was the horse to *his* command. One blow, one impulse given with voice and hand, by the stranger, one rush from the horse, one bound as if in the act of rising to a fence, landed the docile creature's fore-feet upon the crown or arching centre of the road. The larger half of the little equipage had then cleared our overtowering shadow: *that* was evident even to my own agitated sight. But it mattered little that one wreck should float off in safety if upon the wreck that perished were embarked the human freightage. The rear part of the carriage—was *that* certainly beyond the line of absolute ruin? What power could answer the question? Glance of eye, thought of man, wing of angel, which of these had speed enough to sweep between the question and the answer, and divide the one from the other? Light does not tread upon the steps of light more indivisibly than did our all-conquering arrival upon the escaping efforts of the gig. *That* must the young man have felt too plainly. His back was now turned to us; not by sight could he any longer communicate with the peril; but, by the dreadful rattle of our harness, too truly had his ear been instructed that all was finished as regarded any effort of *his*. Already in resignation he had rested from his struggle; and perhaps in his heart he was whispering, "Father, which art in heaven, do Thou finish above what I on earth have attempted." Faster than ever mill-race we ran past them in our inexorable flight. Oh, raving of hurricanes that must have sounded in their young ears at the moment of our transit! Even in that moment the thunder of collision spoke aloud. Either with the swingle-bar,[14] or with the haunch of our near leader, we had struck the off-wheel of the little gig, which stood rather obliquely, and not quite so far advanced as to be accurately parallel with the near-wheel. The blow, from the fury of our passage, resounded terrifically. I rose in horror, to gaze upon the ruins we might have caused. From my elevated station I looked down, and looked back upon the scene; which in a moment told its own tale, and wrote all its records on my heart forever.

Here was the map of the passion[15] that now had finished. The horse was planted immovably, with his fore-feet upon the paved crest of the central road. He

12. **shilling a-day:** the English soldier's daily wage. 13. **slewed:** turned.

14. **swingle-bar:** a whiffletree—a horizontal crossbar to which the ends of the traces of a harness are attached. 15. **the passion:** terror.

of the whole party might be supposed untouched by the passion of death. The little cany[16] carriage—partly, perhaps, from the violent torsion of the wheels in its recent movement, partly from the thundering blow we had given to it—as if it sympathized with human horror, was all alive with tremblings and shiverings. The young man trembled not, nor shivered. He sat like a rock. But *his* was the steadiness of agitation frozen into rest by horror. As yet he dared not to look round; for he knew that, if anything remained to do, by him it could no longer bedone. And as yet he knew not for certain if their safety were accomplished. But the lady——

But the lady——! Oh, heavens! will that spectacle ever depart from my dreams, as she rose and sank upon her seat, sank and rose, threw up her arms wildly to heaven, clutched at some visionary object in the air, fainting, praying, raving, despairing? Figure to yourself, reader, the elements of the case; suffer me to recall before your mind the circumstances of that unparalleled situation. From the silence and deep peace of this saintly summer night—from the pathetic blending of this sweet moonlight, dawnlight, dreamlight—from the manly tenderness of this flattering, whispering, murmuring love—suddenly as from the woods and fields—suddenly as from the chambers of the air opening in revelation—suddenly as from the ground yawning at her feet, leaped upon her, with the flashing of cataracts, Death the crowned phantom, with all the equipage of his terrors, and the tiger roar of his voice.

The moments were numbered; the strife was finished; the vision was closed. In the twinkling of an eye, our flying horses had carried us to the termination of the umbrageous aisle; at the right angles we wheeled into our former direction; the turn of the road carried the scene out of my eyes in an instant, and swept it into my dreams forever.

SECTION III—DREAM-FUGUE:

FOUNDED ON THE PRECEDING
THEME OF SUDDEN DEATH

Whence the sound
Of instruments, that made melodious chime,
Was heard, of harp and organ; and who moved
Their stops and chords was seen; his volant touch
Instinct through all proportions, low and high,
Fled and pursued transverse the resonant fugue.
—*Paradise Lost* [XI. 558–63].

16. **cany:** as though made of cane.

Tumultuosissimamente

Passion of sudden death! that once in youth I read and interpreted by the shadows of thy averted signs![1] —rapture of panic taking the shape (which amongst tombs in churches I have seen) of woman bursting her sepulchral bonds—of woman's Ionic form bending forward from the ruins of her grave with arching foot, with eyes upraised, with clasped adoring hands—waiting, watching, trembling, praying for the trumpet's call to rise from dust forever! Ah, vision too fearful of shuddering humanity on the brink of almighty abysses!—vision that didst start back, that didst reel away, like a shrivelling scroll from before the wrath of fire racing on the wings of the wind! Epilepsy so brief of horror, wherefore is it that thou canst not die? Passing so suddenly into darkness, wherefore is it that still thou sheddest thy sad funeral blights upon the gorgeous mosaics of dreams? Fragment of music too passionate, heard once, and heard no more, what aileth thee, that thy deep rolling chords come up at intervals through all the worlds of sleep, and after forty years have lost no element of horror?

I

Lo, it is summer—almighty summer! The everlasting gates of life and summer are thrown open wide; and on the ocean, tranquil and verdant as a savannah, the unknown lady from the dreadful vision and I myself are floating—she upon a fairy pinnace, and I upon an English three-decker. Both of us are wooing gales of festal happiness within the domain of our common country, within that ancient watery park, within the pathless chase of ocean, where England takes her pleasure as a huntress through winter and summer, from the rising to the setting sun. Ah, what a wilderness of floral beauty was hidden, or was suddenly revealed, upon the tropic islands through which the pinnace moved! And upon her deck what a bevy of human flowers: young women how lovely, young men how noble, that were dancing together, and slowly drifting towards *us* amidst music and incense, amidst blossoms from forests and gorgeous corymbi[2] from vintages,

SECTION III. **1. averted signs:** "I read the course and changes of the lady's agony in the succession of her involuntary gestures; but it must be remembered that I read all this from the rear, never once catching the lady's full face, and even her profile imperfectly" (De Q.). **2. corymbi:** bunches of flowers or fruit.

amidst natural carolling, and the echoes of sweet girlish laughter. Slowly the pinnace nears us, gaily she hails us, and silently she disappears beneath the shadow of our mighty bows. But then, as at some signal from heaven, the music, and the carols, and the sweet echoing of girlish laughter—all are hushed. What evil has smitten the pinnace, meeting or overtaking her? Did ruin to our friends couch within our own dreadful shadow? Was our shadow the shadow of death? I looked over the bow for an answer, and behold! the pinnace was dismantled; the revel and the revellers were found no more; the glory of the vintage was dust; and the forests with their beauty were left without a witness upon the seas. "But where," and I turned to our crew—"where are the lovely women that danced beneath the awning of flowers and clustering corymbi? Whither have fled the noble young men that danced with *them?*" Answer there was none. But suddenly the man at the mast-head, whose countenance darkened with alarm, cried out, "Sail on the weather beam! Down she comes upon us: in seventy seconds she also will founder."

II

I looked to the weather side, and the summer had departed. The sea was rocking, and shaken with gathering wrath. Upon its surface sat mighty mists, which grouped themselves into arches and long cathedral aisles. Down one of these, with the fiery pace of a quarrel[3] from a cross-bow, ran a frigate right athwart our course. "Are they mad?" some voice exclaimed from our deck. "Do they woo their ruin?" But in a moment, as she was close upon us, some impulse of a heady current or local vortex gave a wheeling bias to her course, and off she forged without a shock. As she ran past us, high aloft amongst the shrouds stood the lady of the pinnace. The deeps opened ahead in malice to receive her, towering surges of foam ran after her, the billows were fierce to catch her. But far away she was borne into desert spaces of the sea: whilst still by sight I followed her, as she ran before the howling gale, chased by angry sea-birds and by maddening billows; still I saw her, as at the moment when she ran past us, standing amongst the shrouds, with her white draperies streaming before the wind. There she stood, with hair dishevelled, one hand clutched amongst the tackling—rising, sinking, fluttering, trembling, praying; there

3. **quarrel:** bolt used as an arrow in a crossbow.

for leagues I saw her as she stood, raising at intervals one hand to heaven, amidst the fiery crests of the pursuing waves and the raving of the storm; until at last, upon a sound from afar of malicious laughter and mockery, all was hidden forever in driving showers; and afterwards, but when I knew not, nor how.

III

Sweet funeral bells from some incalculable distance, wailing over the dead that die before the dawn, awakened me as I slept in a boat moored to some familiar shore. The morning twilight even then was breaking; and, by the dusky revelations which it spread, I saw a girl, adorned with a garland of white roses about her head for some great festival, running along the solitary strand in extremity of haste. Her running was the running of panic; and often she looked back as to some dreadful enemy in the rear. But, when I leaped ashore, and followed on her steps to warn her of a peril in front, alas! from me she fled as from another peril, and vainly I shouted to her of quicksands that lay ahead. Faster and faster she ran; round a promontory of rocks she wheeled out of sight; in an instant I also wheeled round it, but only to see the treacherous sands gathering above her head. Already her person was buried; only the fair young head and the diadem of white roses around it were still visible to the pitying heavens; and, last of all, was visible one white marble arm. I saw by the early twilight this fair young head, as it was sinking down to darkness—saw this marble arm, as it rose above her head and her treacherous grave, tossing, faltering, rising, clutching, as at some false deceiving hand stretched out from the clouds—saw this marble arm uttering her dying hope, and then uttering her dying despair. The head, the diadem, the arm—these all had sunk; at last over these also the cruel quicksand had closed; and no memorial of the fair young girl remained on earth, except my own solitary tears, and the funeral bells from the desert seas, that, rising again more softly, sang a requiem over the grave of the buried child, and over her blighted dawn.

I sat, and wept in secret the tears that men have ever given to the memory of those that died before the dawn, and by the treachery of earth, our mother. But suddenly the tears and funeral bells were hushed by a shout as of many nations, and by a roar as from some great king's artillery, advancing rapidly along the valleys, and heard afar by echoes from the mountains. "Hush!" I said, as I bent my ear earthwards to listen—

"hush!—this either is the very anarchy of strife, or else"—and then I listened more profoundly, and whispered as I raised my head—"or else, oh heavens! it is *victory* that is final, victory that swallows up all strife."

IV

Immediately, in trance, I was carried over land and sea to some distant kingdom, and placed upon a triumphal car, amongst companions crowned with laurel. The darkness of gathering midnight, brooding over all the land, hid from us the mighty crowds that were weaving restlessly about ourselves as a centre: we heard them, but saw them not. Tidings had arrived, within an hour, of a grandeur that measured itself against centuries; too full of pathos they were, too full of joy, to utter themselves by other language than by tears, by restless anthems, and *Te Deums*[4] reverberated from the choirs and orchestras of earth. These tidings we that sat upon the laurelled car had it for our privilege to publish amongst all nations. And already, by signs audible through the darkness, by snortings and tramplings, our angry horses, that knew no fear or fleshly weariness, upbraided us with delay. Wherefore *was* it that we delayed? We waited for a secret word, that should bear witness to the hope of nations as now accomplished forever. At midnight the secret word arrived; which word was—*Waterloo and Recovered Christendom!* The dreadful word shone by its own light; before us it went; high above our leaders' heads it rode, and spread a golden light over the paths which we traversed. Every city, at the presence of the secret word, threw open its gates. The rivers were conscious as we crossed. All the forests, as we ran along their margins, shivered in homage to the secret word. And the darkness comprehended it.[5]

Two hours after midnight we approached a mighty Minster. Its gates, which rose to the clouds, were closed. But, when the dreadful word that rode before us reached them with its golden light, silently they moved back upon their hinges; and at a flying gallop our equipage entered the grand aisle of the cathedral. Headlong was our pace; and at every altar, in the little chapels and oratories to the right hand and left of our course, the lamps, dying or sickening, kindled anew in sympathy with the secret word that was flying past.

Forty leagues we might have run in the cathedral, and as yet no strength of morning light had reached us, when before us we saw the aerial galleries of organ and choir. Every pinnacle of fretwork, every station of advantage amongst the traceries, was crested by white-robed choristers that sang deliverance; that wept no more tears, as once their fathers had wept; but at intervals that sang together to the generations, saying,

"Chant the deliverer's praise in every tongue,"

and receiving answers from afar,

"Such as once in heaven and earth were sung."

And of their chanting was no end; of our headlong pace was neither pause nor slackening.

Thus as we ran like torrents—thus as we swept with bridal rapture over the Campo Santo[6] of the cathedral graves—suddenly we became aware of a vast necropolis rising upon the far-off horizon—a city of sepulchres, built within the saintly cathedral for the warrior dead that rested from their feuds on earth. Of purple granite was the necropolis; yet, in the first minute, it lay like a purple stain upon the horizon, so mighty was the distance. In the second minute it trembled through many changes, growing into terraces and towers of wondrous altitude, so mighty was the pace. In the third minute already, with our dreadful gallop, we were entering its suburbs. Vast sarcophagi rose on every side, having towers and turrets that, upon the limits of the central aisle, strode forward with haughty intrusion, that ran back with mighty shadows into answering recesses. Every sarcophagus showed many bas-reliefs—bas-reliefs of battles and of battle-fields; battles from forgotten ages, battles from yesterday; battle-fields that, long since, nature had healed and reconciled to herself with the sweet oblivion of flowers; battle-fields that were yet angry and crimson with carnage. Where the terraces ran, there did *we* run; where the towers curved, there did *we* curve. With the flight of swallows our horses swept round every angle. Like rivers in flood wheeling round headlands, like hurricanes that ride into the secrets of forests, faster than ever light unwove the mazes of darkness, our flying equipage carried earthly passions, kindled warrior instincts, amongst the dust that lay around us—dust oftentimes of our noble fathers that had slept in God from Crécy to Trafalgar.[7] And

---

4. **Te Deums:** from the Christian hymn, *Te Deum laudamus*, "we praise Thee, O God." 5. **darkness . . . it:** John 1:5.

6. **Campo Santo:** cemetery at Pisa. 7. **Crécy . . . Trafalgar:** the battles of Crécy (1346) and Trafalgar (1805).

now had we reached the last sarcophagus, now were we abreast of the last bas-relief, already had we recovered the arrow-like flight of the illimitable central aisle, when coming up this aisle to meet us we beheld afar off a female child, that rode in a carriage as frail as flowers. The mists which went before her hid the fawns that drew her, but could not hide the shells and tropic flowers with which she played—but could not hide the lovely smiles by which she uttered her trust in the mighty cathedral, and in the cherubim that looked down upon her from the mighty shafts of its pillars. Face to face she was meeting us; face to face she rode, as if danger there were none. "Oh, baby!" I exclaimed, "shalt thou be the ransom for Waterloo? Must we, that carry tidings of great joy to every people, be messengers of ruin to thee!" In horror I rose at the thought; but then also, in horror at the thought, rose one that was sculptured on a bas-relief—a Dying Trumpeter. Solemnly from the field of battle he rose to his feet; and, unslinging his stony trumpet, carried it, in his dying anguish, to his stony lips—sounding once, and yet once again; proclamation that, in *thy* ears, oh baby! spoke from the battlements of death. Immediately deep shadows fell between us, and aboriginal silence. The choir had ceased to sing. The hoofs of our horses, the dreadful rattle of our harness, the groaning of our wheels, alarmed the graves no more. By horror the bas-relief had been unlocked unto life. By horror we, that were so full of life, we men and our horses, with their fiery fore-legs rising in mid air to their everlasting gallop, were frozen to a bas-relief. Then a third time the trumpet sounded; the seals were taken off all pulses; life, and the frenzy of life, tore into their channels again; again the choir burst forth in sunny grandeur, as from the muffling of storms and darkness; again the thunderings of our horses carried temptation into the graves. One cry burst from our lips, as the clouds, drawing off from the aisle, showed it empty before us. —"Whither has the infant fled?—is the young child caught up to God?" Lo! afar off, in a vast recess, rose three mighty windows to the clouds; and on a level with their summits, at height insuperable to man, rose an altar of purest alabaster. On its eastern face was trembling a crimson glory. A glory was it from the reddening dawn that now streamed *through* the windows? Was it from the crimson robes of the martyrs painted *on* the windows? Was it from the bloody bas-reliefs of earth? There suddenly, within that crimson radiance, rose the apparition of a woman's head, and then of a woman's figure. The child it was—grown up

to woman's height. Clinging to the horns of the altar, voiceless she stood—sinking, rising, raving, despairing; and behind the volume of incense that, night and day, streamed upwards from the altar, dimly was seen the fiery font, and the shadow of that dreadful being who should have baptized her with the baptism of death. But by her side was kneeling her better angel, that hid his face with wings; that wept and pleaded for *her*; that prayed when *she* could *not*; that fought with Heavens by tears for *her* deliverance; which also, as he raised his immortal countenance from his wings, I saw, by the glory in his eye, that from Heaven he had won at last.

V

Then was completed the passion of the mighty fugue. The golden tubes of the organ, which as yet had but muttered at intervals—gleaming amongst clouds and surges of incense—threw up, as from fountains unfathomable, columns of heart-shattering music. Choir and anti-choir were filling fast with unknown voices. Thou also, Dying Trumpeter, with thy love that was victorious, and thy anguish that was finishing, didst enter the tumult; trumpet and echo—farewell love, and farewell anguish—rang through the dreadful *sanctus*. Oh, darkness of the grave! that from the crimson altar and from the fiery font wert visited and searched by the effulgence in the angel's eye—were these indeed thy children? Pomps of life, that, from the burials of centuries, rose again to the voice of perfect joy, did ye indeed mingle with the festivals of Death? Lo! as I looked back for seventy leagues through the mighty cathedral, I saw the quick and the dead that sang together to God, together that sang to the generations of man. All the hosts of jubilation, like armies that ride in pursuit, moved with one step. Us, that, with laurelled heads, were passing from the cathedral, they overtook, and, as with a garment, they wrapped us round with thunders greater than our own. As brothers we moved together; to the dawn that advanced, to the stars that fled; rendering thanks to God in the highest—that, having hid His face through one generation behind thick clouds of War, once again was ascending, from the Campo Santo of Waterloo was ascending, in the visions of Peace; rendering thanks for thee, young girl! whom having overshadowed with His ineffable passion of death, suddenly did God relent, suffered thy angel to turn aside His arm, and even in thee, sister unknown! shown to me for a moment only to be hidden for ever, found an occasion to glorify His goodness. A thousand times, amongst the phantoms of sleep, have I seen thee entering the gates of the golden dawn, with the secret word riding before thee, with the armies of the grave behind thee,—seen thee sinking, rising, raving, despairing; a thousand times in the worlds of sleep have seen thee followed by God's angel through storms, through desert seas, through the darkness of quick-sands, through dreams and the dreadful revelations that are in dreams; only that at the last, with one sling of His victorious arm, He might snatch thee back from ruin, and might emblazon in thy deliverance the endless resurrections of His love!

1849 (1849)

THOMAS LOVE PEACOCK

1785–1866

PEACOCK, though he began his literary career as a poet, is remembered principally for the genial satiric humor of his novels, *Headlong Hall* (1816), *Melincourt* (1817), *Nightmare Abbey* (1818), *Maid Marian* (1822), *The Misfortunes of Elphin* (1829), and *Crotchet Castle* (1831). It is typical of this shrewd, realistic man that he should have been successful in business and yet widely liked and respected by writers of the time.

Peacock's famous essay "The Four Ages of Poetry" may not have been completely serious, but he probably took for granted the general assumption, which had been widely held for seventy years, that poetry suffers a decline in both vigor and range as a civilization matures. Discussion of the subject had become very active in the middle eighteenth century, especially in England, and continued through the Romantic period. Where are the "greater genres," such as the epic and the tragic drama, and why do poets, as the generations pass, turn to smaller subjects and forms and tend to look inward and write about their personal feelings? Why do poets like Homer and Shakespeare appear in the earlier stages of a culture? Several eighteenth-century writers, such as David Hume and Edward Young, offered the explanation that the mere existence of great poets or artists in the past intimidates later writers. Others, like Oliver Goldsmith, follow the cyclical interpretation of history popularized for the eighteenth century by the Italian philosopher Vico. Goldsmith, anticipating Peacock, suggested that, as knowledge increases and becomes more precise, writers turn to prose and much of the talent previously devoted to the arts generally becomes diverted to philosophy, criticism, history, and science. The Romantic movement is what it is partly because of the shadow of that general apprehension. For it is often nostalgic—even primitivistically so—as it looks back to the past. In its search for a "tradition" on which to reground itself, it often leaps over the intervening century and a half to the art and literature from the late Middle Ages through the Renaissance. Its ideals include the spontaneity, intensity, and scope of earlier art. Meanwhile it is confident that it can also open new areas to art.

In rephrasing the concept of the decline of poetry, Peacock goes back to the classical myth of the "Four Ages." Where he becomes less serious is in his discussion of the poetry of his own day, and in the puckish exaggeration of his statement that poetry served merely as the "mental rattle" to awaken the intellect in the infancy of a culture and that we need not expect "maturity of mind to make a serious business of the playthings of its childhood." The latter remark in particular was almost deliberately calculated to excite Peacock's good friend Shelley, whose "Defence of Poetry" (see below, p. 1072) is an eloquent answer.

Bibliography

Peacock's *Works* have been edited in 10 vols. by Herbert F. B. Brett-Smith and C. E. Jones, 1924–34. Separate editions include *Poems*, ed. Brimley Johnson, 1906, *Novels*, ed. David Garnett, 1963, *Four Ages of Poetry*, ed. Herbert F. B. Brett-Smith, 1921, and, with introduction and notes, Shelley and Peacock, *Defence of Poetry and Four Ages of Poetry*, ed. J. E. Jordan, 1965. *The Pleasures of Peacock*, ed. Ben Ray Redman, 1947, is a selection from Peacock's writings. Peacock's *Letters to Edward Hookham and Percy B. Shelley* have been edited by Richard Garnett, 1910. Still the best biography is Carl Van Doren, *The Life of Thomas Love Peacock*, 1911. The most extended critical discussion is Jean-Jacques Mayoux, *Un épicurien anglais: Thomas Love Peacock*, 1932, which has not been translated.

THE FOUR AGES
OF POETRY

Poetry, like the world, may be said to have four ages, but in a different order: the first age of poetry being the age of iron; the second, of gold; the third, of silver; and the fourth, of brass.

The first, or iron age of poetry, is that in which rude bards celebrate in rough numbers the exploits of ruder chiefs, in days when every man is a warrior, and when the great practical maxim of every form of society, "to keep what we have and to catch what we can," is not yet disguised under names of justice and forms of law, but is the naked motto of the naked sword, which is the only judge and jury in every question of *meum* and

tuum. In these days, the only three trades flourishing (besides that of priest which flourishes always) are those of king, thief, and beggar: the beggar being for the most part a king deject, and the thief a king expectant. The first question asked of a stranger is, whether he is a beggar or a thief: the stranger, in reply, usually assumes the first, and awaits a convenient opportunity to prove his claim to the second appellation.

The natural desire of every man to engross to himself as much power and property as he can acquire by any of the means which might makes right, is accompanied by the no less natural desire of making known to as many people as possible the extent to which he has been a winner in this universal game. The successful warrior becomes a chief; the successful chief becomes a king: his next want is an organ to disseminate the fame of his achievements and the extent of his possessions; and this

organ he finds in a bard, who is always ready to celebrate the strength of his arm, being first duly inspired
by that of his liquor. This is the origin of poetry, which,
like all other trades, takes its rise in the demand for the
commodity, and flourishes in proportion to the extent
of the market.

Poetry is thus in its origin panegyrical. The first rude
songs of all nations appear to be a sort of brief historical
notices, in a strain of tumid hyperbole, of the exploits
and possessions of a few pre-eminent individuals. They
tell us how many battles such an one has fought, how
many helmets he has cleft, how many breastplates he
has pierced, how many widows he has made, how
much land he has appropriated, how many houses he
has demolished for other people, what a large one he
has built for himself, how much gold he has stowed
away in it, and how liberally and plentifully he pays,
feeds, and intoxicates the divine and immortal bards,
the sons of Jupiter, but for whose everlasting songs the
names of heroes would perish.

This is the first stage of poetry before the invention of
written letters. The numerical modulation is at once
useful as a help to memory, and pleasant to the ears of
uncultured men, who are easily caught by sound: and
from the exceeding flexibility of the yet unformed language, the poet does no violence to his ideas in sub-
jecting them to the fetters of number. The savage indeed
lisps in numbers,[1] and all rude and uncivilized people
express themselves in the manner which we call poetical.

The scenery by which he is surrounded, and the
superstitions which are the creed of his age, form the
poet's mind. Rocks, mountains, seas, unsubdued
forests, unnavigable rivers, surround him with forms
of power and mystery, which ignorance and fear have
peopled with spirits, under multifarious names of gods,
goddesses, nymphs, genii, and dæmons. Of all these
personages marvellous tales are in existence: the
nymphs are not indifferent to handsome young men,
and the gentlemen genii are much troubled and very
troublesome with a propensity to be rude to pretty
maidens: the bard therefore finds no difficulty in
tracing the genealogy of his chief to any of the deities in
his neighbourhood with whom the said chief may be
most desirous of claiming relationship.

In this pursuit, as in all others, some of course will
attain a very marked pre-eminence; and these will be
held in high honour, like Demodocus in the Odyssey,

and will be consequently inflated with boundless
vanity, like Thamyris[2] in the Iliad. Poets are as yet the
only historians and chroniclers of their time, and the
sole depositories of all the knowledge of their age; and
though this knowledge is rather a crude congeries of
traditional phantasies than a collection of useful truths,
yet, such as it is, they have it to themselves. They are
observing and thinking, while others are robbing and
fighting: and though their object be nothing more
than to secure a share of the spoil, yet they accomplish
this end by intellectual, not by physical, power: their
success excites emulation to the attainment of intellectual eminence: thus they sharpen their own wits and
awaken those of others, at the same time that they
gratify vanity and amuse curiosity. A skilful display
of the little knowledge they have gains them credit for
the possession of much more which they have not.
Their familiarity with the secret history of gods and
genii obtains for them, without much difficulty, the
reputation of inspiration; thus they are not only historians but theologians, moralists, and legislators:
delivering their oracles *ex cathedrâ*, and being indeed
often themselves (as Orpheus and Amphion[3]) regarded
as portions and emanations of divinity: building cities
with a song, and leading brutes with a symphony;
which are only metaphors for the faculty of leading
multitudes by the nose.

The golden age of poetry finds its materials in the age
of iron. This age begins when poetry begins to be
retrospective; when something like a more extended
system of civil polity is established; when personal
strength and courage avail less to the aggrandizing of
their possessor and to the making and marring of kings
and kingdoms, and are checked by organized bodies,
social institutions, and hereditary successions. Men also
live more in the light of truth and within the interchange of observation; and thus perceive that the
agency of gods and genii is not so frequent among themselves as, to judge from the songs and legends of the
past time, it was among their ancestors. From these
two circumstances, really diminished personal power,
and apparently diminished familiarity with gods and
genii, they very easily and naturally deduce two conclusions: 1st, That men are degenerated, and 2nd, That
they are less in favour with the gods. The people of the

THE FOUR AGES OF POETRY. **1. lisps . . . numbers:** Peacock
echoes Pope's "Epistle to Dr. Arbuthnot," l. 128.

2. Demodocus . . . Thamyris: minstrels in the Homeric
poems. **3. Orpheus and Amphion:** Orpheus was able to
charm the beasts through his music; Amphion, by playing
the harp, drew into place the stones that made the wall of
Thebes.

petty states and colonies, which have now acquired stability and form, which owed their origin and first prosperity to the talents and courage of a single chief, magnify their founder through the mists of distance and tradition, and perceive him achieving wonders with a god or goddess always at his elbow. They find his name and his exploits thus magnified and accompanied in their traditionary songs, which are their only memorials. All that is said of him is in this character. There is nothing to contradict it. The man and his exploits and his tutelary deities are mixed and blended in one invariable association. The marvellous too is very much like a snowball: it grows as it rolls downward, till the little nucleus of truth which began its descent from the summit is hidden in the accumulation of superinduced hyperbole.

When tradition, thus adorned and exaggerated, has surrounded the founders of families and states with so much adventitious power and magnificence, there is no praise which a living poet can, without fear of being kicked for clumsy flattery, address to a living chief, that will not still leave the impression that the latter is not so great a man as his ancestors. The man must in this case be praised through his ancestors. Their greatness must be established, and he must be shown to be their worthy descendant. All the people of a state are interested in the founder of their state. All states that have harmonized into a common form of society, are interested in their respective founders. All men are interested in their ancestors. All men love to look back into the days that are past. In these circumstances traditional national poetry is reconstructed and brought like chaos into order and form. The interest is more universal: understanding is enlarged: passion still has scope and play: character is still various and strong: nature is still unsubdued and existing in all her beauty and magnificence, and men are not yet excluded from her observation by the magnitude of cities or the daily confinement of civic life: poetry is more an art: it requires greater skill in numbers, greater command of language, more extensive and various knowledge, and greater comprehensiveness of mind. It still exists without rivals in any other department of literature; and even the arts, painting and sculpture certainly, and music probably, are comparatively rude and imperfect. The whole field of intellect is its own. It has no rivals in history, nor in philosophy, nor in science. It is cultivated by the greatest intellects of the age, and listened to by all the rest. This is the age of Homer, the golden age of poetry. Poetry has now attained its perfection:

it has attained the point which it cannot pass: genius therefore seeks new forms for the treatment of the same subjects: hence the lyric poetry of Pindar and Alcæus, and the tragic poetry of Æschylus and Sophocles. The favour of kings, the honour of the Olympic crown, the applause of present multitudes, all that can feed vanity and stimulate rivalry, await the successful cultivator of this art, till its forms become exhausted, and new rivals arise around it in new fields of literature, which gradually acquire more influence as, with the progress of reason and civilization, facts become more interesting than fiction: indeed the maturity of poetry may be considered the infancy of history. The transition from Homer to Herodotus is scarcely more remarkable than that from Herodotus to Thucydides: in the gradual dereliction of fabulous incident and ornamented language, Herodotus is as much a poet in relation to Thucydides as Homer is in relation to Herodotus. The history of Herodotus is half a poem: it was written while the whole field of literature yet belonged to the Muses, and the nine books of which it was composed were therefore of right, as well as of courtesy, superinscribed with their nine names.

Speculations, too, and disputes, on the nature of man and of mind; on moral duties and on good and evil; on the animate and inanimate components of the visible world; begin to share attention with the eggs of Leda and the horns of Io,[4] and to draw off from poetry a portion of its once undivided audience.

Then comes the silver age, or the poetry of civilized life. This poetry is of two kinds, imitative and original. The imitative consists in recasting, and giving an exquisite polish to, the poetry of the age of gold: of this Virgil is the most obvious and striking example. The original is chiefly comic, didactic, or satiric: as in Menander, Aristophanes, Horace, and Juvenal. The poetry of this age is characterized by an exquisite and fastidious selection of words, and a laboured and somewhat monotonous harmony of expression: but its monotony consists in this, that experience having exhausted all the varieties of modulation, the civilized poetry selects the most beautiful, and prefers the repetition of these to ranging through the variety of all. But the best expression being that into which the idea naturally falls, it requires the utmost labour and care

---

4. **Leda . . . Io:** referring to the Greek myths in which Zeus, in the form of a swan, wooed Leda and fathered Castor and Pollux, and, in order to conceal Io from his wife Hera, transformed her into a heifer.

so to reconcile the inflexibility of civilized language and the laboured polish of versification with the idea intended to be expressed, that sense may not appear to be sacrificed to sound. Hence numerous efforts and rare success.

This state of poetry is however a step towards its extinction. Feeling and passion are best painted in, and roused by, ornamental and figurative language; but the reason and the understanding are best addressed in the simplest and most unvarnished phrase. Pure reason and dispassionate truth would be perfectly ridiculous in verse, as we may judge by versifying one of Euclid's demonstrations. This will be found true of all dispassionate reasoning whatever, and all reasoning that requires comprehensive views and enlarged combinations. It is only the more tangible points of morality, those which command assent at once, those which have a mirror in every mind, and in which the severity of reason is warmed and rendered palatable by being mixed up with feeling and imagination, that are applicable even to what is called moral poetry: and as the sciences of morals and of mind advance towards perfection, as they become more enlarged and comprehensive in their views, as reason gains the ascendancy in them over imagination and feeling, poetry can no longer accompany them in their progress, but drops into the back ground, and leaves them to advance alone.

Thus the empire of thought is withdrawn from poetry, as the empire of facts had been before. In respect of the latter, the poet of the age of iron celebrates the achievements of his contemporaries; the poet of the age of gold celebrates the heroes of the age of iron; the poet of the age of silver re-casts the poems of the age of gold: we may here see how very slight a ray of historical truth is sufficient to dissipate all the illusions of poetry. We know no more of the men than of the gods of the Iliad; no more of Achilles than we do of Thetis; no more of Hector and Andromache than we do of Vulcan and Venus: these belong altogether to poetry; history has no share in them: but Virgil knew better than to write an epic about Cæsar; he left him to Livy; and travelled out of the confines of truth and history into the old regions of poetry and fiction.

Good sense and elegant learning, conveyed in polished and somewhat monotonous verse, are the perfection of the original and imitative poetry of civilized life. Its range is limited, and when exhausted, nothing remains but the *crambe repetita*[5] of common-

place, which at length becomes thoroughly wearisome, even to the most indefatigable readers of the newest new nothings.

It is now evident that poetry must either cease to be cultivated, or strike into a new path. The poets of the age of gold have been imitated and repeated till no new imitation will attract notice: the limited range of ethical and didactic poetry is exhausted: the associations of daily life in an advanced state of society are of very dry, methodical, unpoetical matters-of-fact: but there is always a multitude of listless idlers, yawning for amusement, and gaping for novelty: and the poet makes it his glory to be foremost among their purveyors.

Then comes the age of brass, which, by rejecting the polish and the learning of the age of silver, and taking a retrograde stride to the barbarisms and crude traditions of the age of iron, professes to return to nature and revive the age of gold. This is the second childhood of poetry. To the comprehensive energy of the Homeric Muse, which, by giving at once the grand outline of things, presented to the mind a vivid picture in one or two verses, inimitable alike in simplicity and magnificence, is substituted a verbose and minutely-detailed description of thoughts, passions, actions, persons, and things, in that loose rambling style of verse, which any one may write, *stans pede in uno*,[6] at the rate of two hundred lines in an hour. To this age may be referred all the poets who flourished in the decline of the Roman Empire. The best specimen of it, though not the most generally known, is the Dionysiaca of Nonnus, which contains many passages of exceeding beauty in the midst of masses of amplification and repetition.

The iron age of classical poetry may be called the bardic; the golden, the Homeric; the silver, the Virgilian; and the brass, the Nonnic.

Modern poetry has also its four ages: but "it wears its rue with a difference."[7]

To the age of brass in the ancient world succeeded the dark ages, in which the light of the Gospel began to spread over Europe, and in which, by a mysterious and inscrutable dispensation, the darkness thickened with the progress of the light. The tribes that overran the Roman Empire brought back the days of barbarism, but with this difference, that there were many books in the world, many places in which they were preserved, and occasionally some one by whom they were read,

---

5. **crambe repetita:** cabbage served up once again.

6. **stans . . . uno:** standing on one foot. 7. "**it . . . difference**": *Hamlet*, IV. v. 181.

who indeed (if he escaped being burned *pour l'amour de Dieu*,[8]) generally lived an object of mysterious fear, with the reputation of magician, alchymist, and astrologer. The emerging of the nations of Europe from this superinduced barbarism, and their settling into new forms of polity, was accompanied, as the first ages of Greece had been, with a wild spirit of adventure, which, co-operating with new manners and new superstitions, raised up a fresh crop of chimæras, not less fruitful, though far less beautiful, than those of Greece. The semi-deification of women by the maxims of the age of chivalry, combining with these new fables, produced the romance of the middle ages. The founders of the new line of heroes took the place of the demi-gods of Grecian poetry. Charlemagne and his Paladins, Arthur and his knights of the round table, the heroes of the iron age of chivalrous poetry, were seen through the same magnifying mist of distance, and their exploits were celebrated with even more extravagant hyperbole. These legends, combined with the exaggerated love that pervades the songs of the troubadours, the reputation of magic that attached to learned men, the infant wonders of natural philosophy, the crazy fanaticism of the crusades, the power and privileges of the great feudal chiefs, and the holy mysteries of monks and nuns, formed a state of society in which no two laymen could meet without fighting, and in which the three staple ingredients of lover, prize-fighter, and fanatic, that composed the basis of the character of every true man, were mixed up and diversified, in different individuals and classes, with so many distinctive excellencies, and under such an infinite motley variety of costume, as gave the range of a most extensive and picturesque field to the two great constituents of poetry, love and battle.

From these ingredients of the iron age of modern poetry, dispersed in the rhymes of minstrels and the songs of the troubadours, arose the golden age, in which the scattered materials were harmonized and blended about the time of the revival of learning; but with this peculiar difference, that Greek and Roman literature pervaded all the poetry of the golden age of modern poetry, and hence resulted a heterogeneous compound of all ages and nations in one picture; an infinite licence, which gave to the poet the free range of the whole field of imagination and memory. This was carried very far by Ariosto, but farthest of all by Shakespeare and his contemporaries, who used time and

locality merely because they could not do without them, because every action must have its when and where: but they made no scruple of deposing a Roman Emperor by an Italian Count, and sending him off in the disguise of a French pilgrim to be shot with a blunderbuss by an English archer. This makes the old English drama very picturesque, at any rate, in the variety of costume, and very diversified in action and character; though it is a picture of nothing that ever was seen on earth except a Venetian carnival.

The greatest of English poets, Milton, may be said to stand alone between the ages of gold and silver, combining the excellencies of both; for with all the energy, and power, and freshness of the first, he united all the studied and elaborate magnificence of the second.

The silver age succeeded; beginning with Dryden, coming to perfection with Pope, and ending with Goldsmith, Collins, and Gray.

Cowper divested verse of its exquisite polish; he thought in metre, but paid more attention to his thoughts than his verse. It would be difficult to draw the boundary of prose and blank verse between his letters and his poetry.

The silver age was the reign of authority; but authority now began to be shaken, not only in poetry but in the whole sphere of its dominion. The contemporaries of Gray and Cowper were deep and elaborate thinkers. The subtle scepticism of Hume, the solemn irony of Gibbon, the daring paradoxes of Rousseau, and the biting ridicule of Voltaire, directed the energies of four extraordinary minds to shake every portion of the reign of authority. Enquiry was roused, the activity of intellect was excited, and poetry came in for its share of the general result. The changes had been rung on lovely maid and sylvan shade, summer heat and green retreat, waving trees and sighing breeze, gentle swains and amorous pains, by versifiers who took them on trust, as meaning something very soft and tender, without much caring what: but with this general activity of intellect came a necessity for even poets to appear to know something of what they professed to talk of. Thomson and Cowper looked at the trees and hills which so many ingenious gentlemen had rhymed about so long without looking at them at all, and the effect of the operation on poetry was like the discovery of a new world. Painting shared the influence, and the principles of picturesque beauty were explored by adventurous essayists with indefatigable pertinacity. The success which attended these experiments, and the

---

8. pour . . . Dieu: for the love of God.

pleasure which resulted from them, had the usual effect of all new enthusiasms, that of turning the heads of a few unfortunate persons, the patriarchs of the age of brass, who, mistaking the prominent novelty for the all-important totality, seem to have ratiocinated much in the following manner: "Poetical genius is the finest of all things, and we feel that we have more of it than any one ever had. The way to bring it to perfection is to cultivate poetical impressions exclusively. Poetical impressions can be received only among natural scenes: for all that is artificial is anti-poetical. Society is artificial, therefore we will live out of society. The mountains are natural, therefore we will live in the mountains. There we shall be shining models of purity and virtue, passing the whole day in the innocent and amiable occupation of going up and down hill, receiving poetical impressions, and communicating them in immortal verse to admiring generations." To some such perversion of intellect we owe that egregious confraternity of rhymesters, known by the name of the Lake Poets; who certainly did receive and communicate to the world some of the most extraordinary poetical impressions that ever were heard of, and ripened into models of public virtue, too splendid to need illustration. They wrote verses on a new principle; saw rocks and rivers in a new light; and remaining studiously ignorant of history, society, and human nature, cultivated the phantasy only at the expence of the memory and the reason; and contrived, though they had retreated from the world for the express purpose of seeing nature as she was, to see her only as she was not, converting the land they lived in into a sort of fairy-land, which they peopled with mysticisms and chimæras. This gave what is called a new tone to poetry, and conjured up a herd of desperate imitators, who have brought the age of brass prematurely to its dotage.

The descriptive poetry of the present day has been called by its cultivators a return to nature. Nothing is more impertinent than this pretension. Poetry cannot travel out of the regions of its birth, the uncultivated lands of semi-civilized men. Mr. Wordsworth, the great leader of the returners to nature, cannot describe a scene under his own eyes without putting into it the shadow of a Danish boy or the living ghost of Lucy Gray, or some similar phantastical parturition of the moods of his own mind.

In the origin and perfection of poetry, all the associations of life were composed of poetical materials. With us it is decidedly the reverse. We know too that there are no Dryads in Hyde-park nor Naiads in the Regent's-canal. But barbaric manners and supernatural interventions are essential to poetry. Either in the scene, or in the time, or in both, it must be remote from our ordinary perceptions. While the historian and the philosopher are advancing in, and accelerating, the progress of knowledge, the poet is wallowing in the rubbish of departed ignorance, and raking up the ashes of dead savages to find gewgaws and rattles for the grown babies of the age. Mr. Scott digs up the poachers and cattle-stealers of the ancient border. Lord Byron cruizes for thieves and pirates on the shores of the Morea and among the Greek Islands. Mr. Southey wades through ponderous volumes of travels and old chronicles, from which he carefully selects all that is false, useless, and absurd, as being essentially poetical; and when he has a commonplace book full of monstrosities, strings them into an epic. Mr. Wordsworth picks up village legends from old women and sextons; and Mr. Coleridge, to the valuable information acquired from similar sources, superadds the dreams of crazy theologians and the mysticisms of German metaphysics, and favours the world with visions in verse, in which the quadruple elements of sexton, old woman, Jeremy Taylor, and Emanuel Kant, are harmonized into a delicious poetical compound. Mr. Moore presents us with a Persian, and Mr. Campbell[9] with a Pennsylvanian tale, both formed on the same principle as Mr. Southey's epics, by extracting from a perfunctory and desultory perusal of a collection of voyages and travels, all that useful investigation would not seek for and that common sense would reject.

These disjointed relics of tradition and fragments of second-hand observation, being woven into a tissue of verse, constructed on what Mr. Coleridge calls a new principle (that is, no principle at all), compose a modern-antique compound of frippery and barbarism, in which the puling sentimentality of the present time is grafted on the misrepresented ruggedness of the past into a heterogeneous congeries of unamalgamating manners, sufficient to impose on the common readers of poetry, over whose understandings the poet of this class possesses that commanding advantage, which, in all circumstances and conditions of life, a man who knows something, however little, always possesses over one who knows nothing.

A poet in our times is a semi-barbarian in a civilized

9. **Mr. Moore . . . Mr. Campbell:** Thomas Moore's *Lalla Rookh* (1817) and Thomas Campbell's *Gertrude of Wyoming* (1809).

community. He lives in the days that are past. His ideas, thoughts, feelings, associations, are all with barbarous manners, obsolete customs, and exploded superstitions. The march of his intellect is like that of a crab, backward. The brighter the light diffused around him by the progress of reason, the thicker is the darkness of antiquated barbarism, in which he buries himself like a mole, to throw up the barren hillocks of his Cimmerian labours. The philosophic mental tranquillity which looks round with an equal eye on all external things, collects a store of ideas, discriminates their relative value, assigns to all their proper place, and from the materials of useful knowledge thus collected, appreciated, and arranged, forms new combinations that impress the stamp of their power and utility on the real business of life, is diametrically the reverse of that frame of mind which poetry inspires, or from which poetry can emanate. The highest inspirations of poetry are resolvable into three ingredients: the rant of unregulated passion, the whining of exaggerated feeling, and the cant of factitious sentiment: and can therefore serve only to ripen a splendid lunatic like Alexander, a puling driveller like Werter,[10] or a morbid dreamer like Wordsworth. It can never make a philosopher, nor a statesman, nor in any class of life an useful or rational man. It cannot claim the slightest share in any one of the comforts and utilities of life of which we have witnessed so many and so rapid advances. But though not useful, it may be said it is highly ornamental, and deserves to be cultivated for the pleasure it yields. Even if this be granted, it does not follow that a writer of poetry in the present state of society is not a waster of his own time, and a robber of that of others. Poetry is not one of those arts which, like painting, require repetition and multiplication, in order to be diffused among society. There are more good poems already existing than are sufficient to employ that portion of life which any mere reader and recipient of poetical impressions should devote to them, and these having been produced in poetical times, are far superior in all the characteristics of poetry to the artificial reconstructions of a few morbid ascetics in unpoetical times. To read the promiscuous rubbish of

the present time to the exclusion of the select treasures of the past, is to substitute the worse for the better variety of the same mode of enjoyment.

But in whatever degree poetry is cultivated, it must necessarily be to the neglect of some branch of useful study: and it is a lamentable spectacle to see minds, capable of better things, running to seed in the specious indolence of these empty aimless mockeries of intellectual exertion. Poetry was the mental rattle that awakened the attention of intellect in the infancy of civil society: but for the maturity of mind to make a serious business of the playthings of its childhood, is as absurd as for a full-grown man to rub his gums with coral, and cry to be charmed to sleep by the jingle of silver bells.

As to that small portion of our contemporary poetry, which is neither descriptive, nor narrative, nor dramatic, and which, for want of a better name, may be called ethical, the most distinguished portion of it, consisting merely of querulous, egotistical rhapsodies, to express the writer's high dissatisfaction with the world and every thing in it, serves only to confirm what has been said of the semi-barbarous character of poets, who from singing dithyrambics and "Io Triumphe," while society was savage, grow rabid, and out of their element, as it becomes polished and enlightened.

Now when we consider that it is not the thinking and studious, and scientific and philosophical part of the community, not to those whose minds are bent on the pursuit and promotion of permanently useful ends and aims, that poets must address their minstrelsy, but to that much larger portion of the reading public, whose minds are not awakened to the desire of valuable knowledge, and who are indifferent to any thing beyond being charmed, moved, excited, affected, and exalted: charmed by harmony, moved by sentiment, excited by passion, affected by pathos, and exalted by sublimity: harmony, which is language on the rack of Procrustes;[11] sentiment, which is canting egotism in the mask of refined feeling; passion, which is the commotion of a weak and selfish mind; pathos, which is the whining of an unmanly spirit; and sublimity, which is the inflation of an empty head: when we consider that the great and permanent interests of human society become more and more the main spring of intellectual pursuit; that in proportion as they become so, the subordinacy of the

---

10. **Alexander . . . Werter:** The reference to Alexander is unclear. It has been suggested that Peacock refers to Alexander Cruden, an eighteenth-century scholar who made a concordance to the Bible, became obsessed with correcting printer's proofs and other errors, and was occasionally confined for insanity. By Werter is intended Werther, the subject of the famous novel by Goethe.

11. **Procrustes:** who placed his victims on a bed and cut them to fit it if they were too tall or stretched them if they were too short.

ornamental to the useful will be more and more seen and acknowledged; and that therefore the progress of useful art and science, and of moral and political knowledge, will continue more and more to withdraw attention from frivolous and unconducive, to solid and conducive studies: that therefore the poetical audience will not only continually diminish in the proportion of its number to that of the rest of the reading public, but will also sink lower and lower in the comparison of intellectual acquirement: when we consider that the poet must still please his audience, and must therefore continue to sink to their level, while the rest of the community is rising above it: we may easily conceive that the day is not distant, when the degraded state of every species of poetry will be as generally recognized as that of dramatic poetry has long been: and this not from any decrease either of intellectual power, or intellectual acquisition, but because intellectual power and intellectual acquisition have turned themselves into other and better channels, and have abandoned the cultivation and the fate of poetry to the degenerate fry of modern rhymesters, and their olympic judges, the magazine critics, who continue to debate and promulgate oracles about poetry, as if it were still what it was in the Homeric age, the all-in-all of intellectual progression, and as if there were no such things in existence as mathematicians, astronomers, chemists, moralists, metaphysicians, historians, politicians, and political economists, who have built into the upper air of intelligence a pyramid, from the summit of which they see the modern Parnassus far beneath them, and, knowing how small a place it occupies in the comprehensiveness of their prospect, smile at the little ambition and the circumscribed perceptions with which the drivellers and mountebanks upon it are contending for the poetical palm and the critical chair.

(1820)

BENJAMIN ROBERT HAYDON

1786–1846

THE DELIGHTFUL AUTOBIOGRAPHY of the painter Benjamin Robert Haydon, and the enormous *Diary* on which his autobiography draws, contain anecdotes of many of his contemporaries, including Wordsworth, Keats, Shelley, and Lamb, the most famous of which is his account of the "immortal dinner" (printed below), one of the most picturesque of its kind since Boswell's *Life of Johnson*.

A man of titanic energy, Haydon dreamed of revolutionizing painting and took as his subjects famous historical episodes that he portrayed on huge canvases—"The Judgement of Solomon," "Xenophon on His Retreat with the Ten Thousand," "The Banishment of Aristides," and others. Of these the best known is his "Christ's Entry into Jerusalem," in which he inserted, as spectators, the faces of Wordsworth, Lamb, Hazlitt, and Keats. Meanwhile he had become known as a champion of the "Elgin Marbles," the great collection of sculpture from the Parthenon that Lord Elgin had brought back from Greece and that is now in the British Museum. Elgin had been willing to sell the collection to the nation. But the famous art critic Richard Payne Knight claimed that many of the statues were Roman replacements dating from the time of the Emperor Hadrian. A committee of the House of Commons was set up in order to decide the matter; and though out of courtesy to Payne Knight, Haydon was not asked to testify, an article that Haydon then wrote was influential in the decision to buy the collection.

Haydon's brave and confused life (he suffered from poor eyes, constant debts, and personal frustrations) finally ended in 1846, when he shot himself before the unfinished canvas of another of his large paintings, a scene from the life of Alfred the Great.

Bibliography

Several editions are available of the *Autobiography* (edited after his death by Tom Taylor, 1853), especially the 2 volume reprint, 1926, with an introduction by Aldous Huxley; the edition of A. P. D. Penrose, 1927; and the convenient Oxford edition by Edmund Blunden, 1927. The *Diary* has been recently edited by W. B. Pope, 5 vols., 1960–63. Writings on Haydon include George Paston (pseud.), *B. R. Haydon and His Friends*, 1905; Eric George, *The Life and Death of Benjamin Robert Haydon, 1786–1846*, 1948; Clarke Olney, *Benjamin Robert Haydon*, 1952; and the introduction written by Huxley mentioned above.

FROM

AUTOBIOGRAPHY

HAYDON'S FIRST SIGHT
OF THE ELGIN MARBLES

[1808] Wilkie[1] proposed that we should go and see the Elgin Marbles . . . I agreed, dressed, and away we went to Park Lane.[2] I had no more notion of what I was to see than of anything I had never heard of, and walked in with the utmost nonchalance.

This period of our lives was one of great happiness. Painting all day; then dining at the Old Slaughter Chop House; then going to the Academy until eight to fill up the evening; then going home to tea—that blessing of a studious man—talking over our respective exploits, what he had been doing, and what I had done, and then, frequently, to relieve our minds fatigued by their eight and twelve hours' work, giving vent to the most extraordinary absurdities. Often have we made rhymes on odd names, and shouted with laughter at each new line that was added. Sometimes, lazily inclined after a good dinner, we have lounged about near Drury Lane or Covent Garden, hesitating whether to go in, and often have I (knowing first that there was nothing I wished to see) assumed a virtue I did not possess, and pretending moral superiority, preached to Wilkie on the weakness of not resisting such temptations for the sake of our art and our duty, and marched him

off to his studies when he was longing to see *Mother Goose*.

One night when *I* was dying to go in, he dragged *me* away to the Academy and insisted on my working, to which I agreed on the promise of a stroll afterwards. As soon as we had finished, out we went, and in passing a penny show in the piazza, we fired up and determined to go in. We entered and slunk away in a corner; while waiting for the commencement of the show, in came all our student friends, one after the other. We shouted out at each one as he arrived, and then popped our heads down in our corner again, much to the indignation of the chimney-sweeps and vegetable boys who composed the audience, but at last we were discovered, and then we all joined in applauding the entertainment of *Pull Devil, Pull Baker*, and at the end raised such a storm of applause, clapping our hands, stamping our feet, and shouting with all the power of a dozen pair of lungs, that to save our heads from the fury of the sweeps we had to run downstairs as if the devil indeed was trying to catch us. After this boisterous amusement, we retired to my rooms and drank tea, talking away on art, starting principles, arguing long and fiercely, and at midnight separating, to rest, rise and work again until the hour of dinner brought us once more together, again to draw, argue or laugh.

Young, strong and enthusiastic, with no sickness, no debilities, full of hope, believing all the world as honourable as ourselves, wishing harm to no one, and incredulous of any wishing harm to us, we streamed on in a perpetual round of innocent enjoyment, and I look back on these hours as the most uninterrupted by envy, the least harassed by anxiety, and the fullest of unalloyed pleasure, of all that have crossed the path of my life.

AUTOBIOGRAPHY. **1. Wilkie:** David (later Sir David) Wilkie, 1785–1841, Scottish painter. **2. Park Lane:** the home on Park Lane of Thomas Bruce, Earl of Elgin, 1766–1841.

Such being the condition of our minds, no opportunity for improvement was ever granted to the one which he did not directly share with the other; and naturally when Wilkie got this order for the marbles his first thought was that I would like to go.

To Park Lane then we went, and after passing through the hall and thence into an open yard, entered a damp, dirty pent-house where lay the marbles ranged within sight and reach. The first thing I fixed my eyes on was the wrist of a figure in one of the female groups, in which were visible, though in a feminine form, the radius and ulna. I was astonished, for I had never seen them hinted at in any female wrist in the antique. I darted my eye to the elbow, and saw the outer condyle visibly affecting the shape as in nature. I saw that the arm was in repose and the soft parts in relaxation. That combination of nature and idea which I had felt was so much wanting for high art was here displayed to mid-day conviction. My heart beat! If I had seen nothing else I had beheld sufficient to keep me to nature for the rest of my life. But when I turned to the Theseus and saw that every form was altered by action or repose,— when I saw that the two sides of his back varied, one side stretched from the shoulder-blade being pulled forward, and the other side compressed from the shoulder-blade being pushed close to the spine as he rested on his elbow, with the belly flat because the bowels fell into the pelvis as he sat,—and when, turning to the Ilissus, I saw the belly protruded, from the figure lying on its side,—and again, when in the figure of the fighting metope[3] I saw the muscle shown under the one arm-pit in that instantaneous action of darting out, and left out in the other arm-pits because not wanted— when I saw, in fact, the most heroic style of art combined with all the essential detail of actual life, the thing was done at once and for ever. . . . I felt the future, I foretold that they would prove themselves the finest things on earth, that they would overturn the false beau-ideal, where nature was nothing, and would establish the true beau-ideal, of which nature alone is the basis.

I shall never forget the horses' heads—the feet in the metopes! I felt as if a divine truth had blazed inwardly upon my mind and I knew that they would at last rouse the art of Europe from its slumber in the darkness.

I do not say this *now*, when all the world acknowledges it, but I said it then, *when no one would believe me.*

I went home in perfect excitement, Wilkie trying to moderate my enthusiasm with his national caution.

Utterly disgusted at my wretched attempt at the heroic in the form and action of my Dentatus,[4] I dashed out the abominable mass and breathed as if relieved of a nuisance. I passed the evening in a mixture of torture and hope; all night I dozed and dreamed of the marbles. I rose at five in a fever of excitement, tried to sketch the Theseus from memory, did so, and saw that I comprehended it. I worked that day and another and another, fearing that I was deluded. At last I got an order for myself; I rushed away to Park Lane; the impression was more vivid than before. I drove off to Fuseli,[5] and fired him to such a degree that he ran upstairs, put on his coat and away we sallied. I remember that first a coal-cart with eight horses stopped us as it struggled up one of the lanes of the Strand; then a flock of sheep blocked us up; Fuseli, in a fury of haste and rage, burst into the middle of them, and they got between his little legs and jostled him so much that I screamed with laughter in spite of my excitement. He swore all along the Strand like a little fury. At last we came to Park Lane. Never shall I forget his uncompromising enthusiasm. He strode about saying, "De Greeks were godes! de Greeks were godes!" We went back to his house, where I dined with him, and we passed the evening in looking over Quintilian and Pliny. Immortal period of my sanguine life! To look back on those hours has been my solace in the bitterest afflictions. Had Fuseli always acted about the marbles as honestly as he did then, it would have been well for his reputation; but when he was left to his own reflections he remembered what he had always said of things on very different principles, and when I called again he began to back out, so I left him after recalling what he had felt before he had time to be cautious. He did not behave with the same grandeur of soul that West[6] did. He, too, was in the decline of life; he, too, used to talk of art above nature, and of the beau-ideal; but he nobly acknowledged that he knew nothing until he saw the marbles, and bowed his venerable head before them as if in reverence of their majesty.

Peace and honour to his memory! There was more

---

3. **metope:** a square space in the frieze of the Parthenon, above the columns but below the main figures.

4. **Dentatus:** Haydon's painting of the assassination of the Roman statesman L. S. Dentatus, which Lord Mulgrave had commissioned. 5. **Fuseli:** Henry Fuseli, 1741–1825, celebrated Swiss painter who lived in England. 6. **West:** Benjamin West, 1738–1820, American painter living in England and president of the Royal Academy.

true feeling in his submission to their principles than in all Fuseli's boastful sneers.

It is curious that the godlike length of limb in the Greek productions put me in mind of Fuseli's general notions of the heroic, and there is justice in the idea. But as he had not nature for his guide his indefinite impressions ended in manner and bombast. The finest ideas of form in imitative art must be based on a knowledge of the component parts of that form, or an artist is, as Petrarch says: "In alto mar senza governo."[7]

I expressed myself warmly to Lord Mulgrave and asked him if he thought he could get me leave to draw from the marbles. He spoke to Lord Elgin, and on the condition that my drawings were not to be engraved permission was granted to me. Conscious I had the power, like a puppy I did not go for some days, and when I went was told that Lord Elgin had changed his mind. The pain I felt at the loss of such an opportunity taught me a lesson for life; for never again did I lose one moment in seeking the attainment of an object when an opportunity offered. However, I applied again to Lord Mulgrave and he in time induced Lord Elgin to admit me. For three months I drew until I had mastered the forms of these divine works and brought my hand and mind into subjection.

I saw that the essential was selected in them and the superfluous rejected; that first, all the causes of action were known and then all of those causes wanted for any particular action were selected; that then skin covered the whole and the effect of the action, relaxation, purpose or gravitation was shown on the skin. This appeared, as far as I could see *then*, to be the principle. For Dentatus I selected all the muscles requisite for human action, no more nor less, and then the members wanted for *his* action, and no more nor less.

I put a figure in the corner of a lower character, that is, more complicated in its forms, having parts not essential, and this showed the difference between the form of a hero and common man. The wiseacres of the time quizzed me, of course, for placing a naked soldier in a Roman army, a thing never done by any artist. Raffaele did so in Constantine's battle, but they had nothing to do with Raffaele and perhaps never heard of Raffaele's battle.

I drew at the marbles ten, fourteen, and fifteen hours at a time; staying often till twelve at night, holding a candle and my board in one hand and drawing with the other; and so I should have stayed till morning had not the sleepy porter come yawning in to tell me it was twelve o'clock, and then often have I gone home, cold, benumbed and damp, my clothes steaming up as I dried them; and so, spreading my drawings on the floor and putting a candle on the ground, I have drank my tea at one in the morning with ecstacy as its warmth trickled through my frame, and looked at my picture and dwelt on my drawings, and pondered on the change of empires and thought that I had been contemplating what Socrates looked at and Plato saw,—and then, lifted up with my own high urgings of soul, I have prayed God to enlighten my mind to discover the principles of those divine things,—and then I have had inward assurances of future glory, and almost fancying divine influence in my room have lingered to my mattress bed and soon dozed into a rich, balmy slumber. Oh, those were days of luxury and rapture and uncontaminated purity of mind! No sickness, no debility, no fatal, fatal weakness of sight. I arose with the sun and opened my eyes to its light only to be conscious of my high pursuit; I sprang from my bed, dressed as if possessed, and passed the day, the noon, and the night in the same dream of abstracted enthusiasm; secluded from the world, regardless of its feelings, unimpregnable to disease, insensible to contempt, a being of elevated passions, a spirit that

"Fretted the pigmy body to decay,
And o'erinformed its tenement of clay."[8]

THE CONTROVERSY
OVER THE ELGIN MARBLES

April 29th [*1815*].—This week has really been a week of great delight. Never have I had such irresistible and perpetual urgings of future greatness. I have been like a man with air-balloons under his armpits, and ether in his soul. While I was painting, walking, or thinking, beaming flashes of energy followed and impressed me. O God! grant they may not be presumptuous feelings. Grant they may be the fiery anticipations of a great soul born to realise them. They came over me, and shot across me, and shook me, till I lifted up my heart and thanked God.

May 2nd.—Went to the Institution last night to see the Vandykes and the Rembrandts lighted by lamps.

---

7. **"In . . . governo":** "On the high sea without a tiller" (Petrarch, *Rime*, CXXXII. 11).

8. **"Fretted . . . clay":** Dryden's "Absalom and Achitophel," ll. 157–58.

Was amazingly impressed with the care, diligence, and complete finish of the works of these great men. Came home, and looked at my own picture. It must be done so, and there is an end. The beauty of the women, the exquisite fresh, nosegay sweetness of their looks, their rich crimson velvet, and white satin, and lace, and muslin, and diamonds, with their black eyes and peachy complexions, and snowy necks, and delicious forms, and graceful motions, and sweet nothingness of conversation, bewildered and distracted me. What the nobility have to enjoy in this world! What has not the prince? But they do not seem happy; they want the stimulus of action: their minds preying on themselves seek refuge in novelty, and often sacrifice principle to procure it.

It was about this time that that admirable system of exhibiting the great works of the old masters was begun at the British Institution, and nothing showed so much the want of noble feeling on the part of the Academy as the way in which its members received the resolution. Lawrence[9] was looking at the Gevartius when I was there, and as he turned round, to my wonder, his face was boiling with rage as he grated out between his teeth, "I suppose they think we want teaching." I met Stothard in my rounds, who said, "this will destroy us." "No," I replied, "it will certainly rouse me." "Why," said he, "perhaps that is the right way to take it." On the minds of the people the effect was prodigious. All classes were benefited, and so was the fame of the old masters themselves, for now their finest works were brought forth to the world from odd corners and rooms where they had never perfectly been seen.

It was also about this time that whispers and rumours began to spread in the art against the Elgin Marbles, and very quickly reached my ears. I was up in a moment, and ready to fight to the last gasp in their defence, for having studied them night and day it was natural that I should feel astonished at hearing from various quarters that their beauty, truth, and originality were questioned by a great authority in matters of art. As this difference of opinion eventually led to a great battle, I may as well in this place give a slight memoir of these divine fragments.

Lord Elgin, who was a man of fine taste, on receiving his appointment as Ambassador to the Porte, in 1800,

9. **Lawrence:** Sir Thomas Lawrence, 1769–1830, English painter, who later succeeded Benjamin West as president of the Royal Academy (1820).

consulted with Harrison of Chester how he could render his influence at Constantinople available for the improvement of art, with reference to the glorious remains of Athens.

Harrison told me (in 1821) that he immediately advised his Lordship to procure, if possible, casts from the Ionic columns at the angle of the pediment, to show how the Greeks turned the volute round at that point, and also suggested that sculpture would be greatly benefited by casts from any fine works remaining. Lord Elgin, thus advised, having first failed in obtaining the support of the Government (who with all their love for the arts did not feel themselves at all justified in advancing the public money for such objects), and being unable to meet the enormous demands of English artists and moulders, proceeded to obtain on his road the assistance of foreign artists, who were more moderate in their terms. After much trouble he at last established at Athens six moulders and artists to draw, cast, and mould everything valuable in art, whether sculpture, architecture, or inscription.

So far Lord Elgin entertained no further notions, but when his artists informed him of the daily ravages of the Turks, and added that, during their stay, several works of sculpture had been injured, fired at, and even pounded into lime to build houses with;—when he found that of a whole temple existing in Stuart's day, near the Ilissus, not a stone was then to be seen;—when he learnt that all the English travellers who came to Athens, with their natural love of little bits, broke off arms or noses to bring home as relics;—he naturally concluded that in fifty years' time, at such a rate of devastation, scarce a fragment of architecture or sculpture would remain.

His position was a delicate one. Suspended between the desire of saving from ruin and enriching his country with works which he felt were unequalled in beauty, and his dread of that which he knew would be immediately imputed to him, viz. having taken advantage of his public power and position to further private and pecuniary objects, he was tormented, as all men are tormented who, contemplating a service to their fellow-creatures, feel the sad certainty that, for a time, they will stir up their hatred and provoke persecution instead of receiving legitimate gratitude and reward.

With the energy of a daring will he resolved that the bold step was the only rational one, and having made up his mind he directly applied to the Porte for leave to mould and remove, and for a special licence to dig and excavate. Who will censure his resolution and decision?

No one will now; but everyone did then. A hue and cry was raised. It was swelled by Byron.[10] Lord Elgin was lampooned, abused, and every motive imputed to him but the one by which alone he was impelled.

But Lord Elgin was a man not easily daunted; he put up his scaffoldings in spite of epigrams, and commenced removing what remained of the sculptures and architecture. After nearly five years of constant anxieties and disappointment those remains of matchless beauty, the glorious Elgin Marbles, were at last got down to the *Piræus*,—at last they were embarked,—at last the ship set sail, and while, with a fair wind and shining sun, she was scudding away for old England, the pilot ran her on a rock, and down went marbles and ship in many fathoms water! Here was a misery; but Hamilton, Lord Elgin's secretary, who was with them, did not despair. He hired a set of divers from the opposite coast of Asia Minor, and after immense perseverance recovered every case. Not a fragment was missing; again they started; again the winds blew, and the sun shone, and after many weeks they were at last safely landed and lodged in Richmond Gardens, to set the whole art in an uproar.

Lord Elgin, who little knew the political state of art, was not prepared for any opposition. Innocent noble! he believed that the marbles had only to be seen to be appreciated! He little knew that there was a Royal Academy which never risked injury to its preponderance for the sake of art. He little knew that there were societies of dilettanti, who frowned at any man who presumed to form a collection unless under their sanction, so that they should share any repute which might accrue. He little knew that an eminent scholar, who was forming a collection of bronzes, which he meant to leave to the nation, and who having, like most eminent scholars, an intense admiration of what was ancient, believed that nobody but himself knew anything of art or nature, would become jealous at this sudden irruption into what he considered his exclusive domain.

However little poor Lord Elgin knew of these matters, he soon discovered that we *had* a Royal Academy, that we *had* societies of dilettanti, and that we *had* an eminent scholar collecting bronzes, whose *ipse dixit* no one dared dispute, be he what he might in rank, station or talent; and Lord Elgin soon discovered also that this eminent scholar, with the natural jealousy of a collector, meant to take the field against the originality, beauty, nature and skill of his Lordships' marbles. At the first dinner-party at which Lord Elgin met him, he cried out in a loud voice, "You have lost your labour, my Lord Elgin; your marbles are over-rated; they are not Greek, they are Roman of the time of Hadrian." Lord Elgin, totally unprepared for such an assault, did not reply, for he did not know what to say.

If Payne Knight[11] had no foundation but historical evidence for such an opinion, his evidence was shallow indeed, and if it proceeded from his knowledge as a connoisseur, the perfection of the works he wished to traduce at once proved that his judgment, taste, and feeling, were utterly beneath notice. But such was the effect of Payne Knight's opinion, that the marbles went down in fashionable estimation from that hour. Government cooled, and artists became frightened because an eminent scholar, jealous of their possessor, denied the superiority of these glorious remains. Lord Elgin, feeling this, in utter despair removed them to Park Lane, built a shed over them and left them, as he feared, to an unmerited fate. Many melancholy, many poetical moments did I enjoy there, musing on these mighty fragments piled on each other, covered with dirt, dripping with damp, and utterly neglected for seasons together. But I gained from these sublime relics the leading principles of my practice, and I saw that the union of nature and idea was here so perfect, that the great artist, in his works, seemed more like an agent of the Creator to express vitality by marble than a mere human genius.

Yet notwithstanding the excellence of these divine works, notwithstanding that their faithfulness to nature was distinctly proved by comparison with the forms of the finest boxers of the day, notwithstanding that their beauty was proclaimed by the mighty voice of public approbation, the learned despot of dinner-parties would not be beaten, and eight years passed over in apathy on the part of the British Government.

WORDSWORTH AND SIR WALTER SCOTT

March 7th [*1821*].—Sir Walter Scott, Lamb, Wilkie, and Proctor have been with me all the morning, and a most delightful morning have we had. Scott operated on us like champagne and whisky mixed. In the course of conversation he alluded to *Waverley;* there was a dead silence. Wilkie, who was talking to him, stopped,

---

10. **swelled by Byron:** In *Childe Harold's Pilgrimage,* Canto II, Byron attacks Lord Elgin for having stripped the Acropolis in Athens of its famous classical sculptures.

11. **Payne Knight:** See Introduction, above.

and looked so agitated you would have thought that he was the author. I was bursting to have a good round at him, but as this was his first visit I did not venture. It is singular how success and the want of it operate on two extraordinary men—Walter Scott and Wordsworth. Scott enters a room and sits at table with the coolness and self-possession of conscious fame; Wordsworth with a mortified elevation of head, as if fearful he was not estimated as he deserved.

Scott is always cool and very amusing; Wordsworth often egotistical and overwhelming. Scott can afford to talk of trifles, because he knows the world will think him a great man who condescends to trifle; Wordsworth must always be eloquent and profound, because he knows that he is considered childish and puerile. Scott seems to wish to appear less than he really is, while Wordsworth struggles to be thought, at the moment, greater than he is suspected to be.

This is natural. Scott's disposition is the effect of success operating on a genial temperament, while Wordsworth's evidently arises from the effect of unjust ridicule wounding an intense self-esteem.

I think that Scott's success would have made Wordsworth insufferable, while Wordsworth's failures would not have rendered Scott a whit less delightful.

Scott is the companion of Nature in all her feelings and freaks, while Wordsworth follows her like an apostle, sharing her solemn moods and impressions.

WORDSWORTH

April 13th [1821].—I had a cast made yesterday of Wordsworth's face.[12] He bore it like a philosopher. John Scott[13] was to meet him at breakfast, and just as he came in the plaster was put on. Wordsworth was sitting in the other room in my dressing-gown, with his hands folded, sedate, solemn and still. I stepped in to Scott and told him as a curiosity to take a peep, that he might say the first sight he ever had of so great a poet was in this stage towards immortality.

I opened the door slowly, and there he sat innocent and unconscious of our plot, in mysterious stillness and silence.

When he was relieved he came in to breakfast with his usual cheerfulness and delighted us by his bursts of inspiration. At one time he shook us both in explaining the principles of his system, his views of man, and his object in writing.

Wordsworth's faculty is in describing those far-reaching and intense feelings and glimmerings and doubts and fears and hopes of man, as referring to what he might be before he was born or what he may be hereafter.

He is a great being and will hereafter be ranked as one who had a portion of the spirit of the mighty ones, especially Milton, but who did not possess the power of using that spirit otherwise than with reference to himself and so as to excite a reflex action only: this is, in my opinion, his great characteristic.

We afterwards called on Hunt, and as Hunt had previously attacked him and had now re-formed his opinions the meeting was interesting.

Hunt paid him the highest compliments, and told him that as he grew wiser and got older he found his respect for his powers and enthusiasm for his genius increase.

Hunt was very ill or it would have been his place to have called on Wordsworth. Here, again, he really burst forth with burning feelings; I never heard him so eloquent before.

I afterwards sauntered along with him to West-end Lane and so on to Hampstead, with great delight. Never did any man so beguile the time as Wordsworth. His purity of heart, his kindness, his soundness of principle, his information, his knowledge, and the intense and eager feelings with which he pours forth all he knows affect, interest and enchant one. I do not know anyone I would be so inclined to worship as a purified being.

Last night I had been at an insipid rout. The contrast was vivid. There the beauty of the women was the only attraction.

In speaking of Lucien Buonaparte[14] as we sauntered along, I said of his poem, that the materials were ill-arranged as referring to an end. "I don't care for that," said Wordsworth, "if there are good things in a poem." Here he was decidedly wrong; but he did not say this with reference to this particular poem, because he thought little of it.

May 23rd.—Breakfasted with Wordsworth, and

12. **face:** For a life-mask to be used as a model in painting Wordsworth's face into "Christ's Entry into Jerusalem" (see Introduction, above). For the same purpose Haydon also took the more famous life-mask of Keats, reproduced in most of the biographies of Keats. 13. **John Scott:** editor of *The London Magazine.*

14. **Lucien Buonaparte:** younger brother of Napoleon. The poem mentioned was his two-volume epic, *Charlemagne, ou l'Église délivrée* (1814).

spent a delightful two hours. Speaking of Burke, Fox, and Pitt, he said: "You always went from Burke with your mind filled; from Fox with your feelings excited; and from Pitt with wonder at his having had the power to make the worse appear the better reason." "Pitt," he said, "preferred power to principle."

I say it is not so. Pitt at a crisis of danger sacrificed his consistency for the sake of his sovereign and country. Which is more just?

Wordsworth has one and perhaps the greatest part of the great genius; but he has not the *lucidus ordo*,[15] and he undervalues it, which is wrong. In phrenological development he is without constructiveness while imagination is as big as an egg.

KEATS

About this time [*October, 1816*] I met John Keats at Leigh Hunt's, and was amazingly interested by his prematurity of intellectual and poetical power.

I read one or two of his sonnets and formed a very high idea of his genius. After a short time I liked him so much that a general invitation on my part followed, and we became extremely intimate. He visited my painting-room at all times, and at all times was welcome.

He was below the middle size, with a low forehead, and an eye that had an inward look, perfectly divine, like a Delphian priestess who saw visions. The greatest calamity for Keats was his being brought before the world by a set who had so much the habit of puffing each other that every one connected with it suffered in public estimation. Hence every one was inclined to disbelieve his genius. After the first criticism in the *Quarterly* somebody from Dartmouth sent him £25. I told Mrs Hoppner this, and begged her to go to Gifford[16] and endeavour to prevent his assault on *Endymion*. She told me she found him writing with his green shade before his eyes, totally insensible to all reproach or entreaty. "How can you, Gifford, dish up in this dreadful manner a youth who has never offended you?" "It has done him good," replied Gifford; "he has had £25 from Devonshire." Mrs Hoppner was extremely intimate with Gifford, and she told me she had a great mind to snatch the manuscript from the table and throw it in the fire. She left Gifford in a great passion, but without producing the least effect.

One evening (19th November, 1816) after a most eager interchange of thoughts I received from Keats his sonnet, beginning "Great spirits now on earth are sojourning." I thanked him, and he wrote, "Your letter has filled me with a proud pleasure, and shall be kept by me as a stimulus to exertion. I begin to fix my eye on one horizon. The idea of your sending it to Wordsworth puts me out of breath. You know with what reverence I would send my well wishes to him."

As I was walking one day with him in the Kilburn meadows, he said: "Haydon, what a pity it is there is not a human dusthole."

His brother (who died in a consumption, and to whom Keats alludes in the lines in his beautiful "Ode to the Nightingale," "and youth grows pale and spectre thin, and dies") told me some interesting things about his infancy, which they got from a servant whom they were obliged to find out to ascertain his brother's age before he could come to his property.

He was when an infant a most violent and ungovernable child. At five years of age or thereabouts, he once got hold of a naked sword and shutting the door swore nobody should go out. His mother wanted to do so, but he threatened her so furiously she began to cry, and was obliged to wait till somebody through the window saw her position and came to her rescue.

An old lady (Mrs Grafty, of Craven Street, Finsbury) told his brother George—when in reply to her question, "what John was doing," he told her he had determined to become a poet—that this was very odd, because when he could just speak, instead of answering questions put to him he would always make a rhyme to the last word people said, and then laugh. As he grew up he was apprenticed to an apothecary, in which position he led a wretched life, translated Ovid without having ever been properly taught Latin, and read Shakespeare, Spenser and Chaucer. He used sometimes to say to his brother he feared he should never be a poet, and if he was not he would destroy himself. He used to suffer such agonies at this apprehension that his brother said they really feared he would execute his threat. At last his master, weary of his disgust, gave him up his time. During his mother's last illness his devoted attachment interested all. He sat up whole nights with her in a great chair, would suffer nobody to give her medicine or even cook her food but himself, and read novels to her in her intervals of ease.

Keats was the only man I ever met with who seemed and looked conscious of a high calling, except Words-worth. Byron and Shelley were always sophisticating about their verses: Keats sophisticated about nothing.

---

15. lucidus ordo: the clear order, or capacity of methodical arrangement. **16. Gifford:** William Gifford, 1756–1826, editor of the Tory review *The Quarterly*.

MEETING WITH SHELLEY
AT LEIGH HUNT'S

[*January, 1817.*] . . . It was now, too, I was first invited to meet Shelley, and readily accepted the invitation.[17] I went a little after the time, and seated myself in the place kept for me at the table, right opposite Shelley himself, as I was told after, for I did not then know what hectic, spare, weakly yet intellectual-looking creature it was, carving a bit of brocoli or cabbage on his plate, as if it had been the substantial wing of a chicken. —— and his wife and her sister, Keats, Horace Smith, and myself made up the party.

In a few minutes Shelley opened the conversation by saying in the most feminine and gentle voice, "As to that detestable religion, the Christian——" I looked astounded, but casting a glance round the table easily saw by ——'s expression of ecstasy and the women's simper, I was to be set at that evening *vi et armis*.[18] No reply, however, was made to this sally during dinner, but when the dessert came and the servant was gone, to it we went like fiends. —— and —— were deists. I felt exactly like a stag at bay and resolved to gore without mercy. Shelley said the Mosaic and Christian dispensations were inconsistent. I swore they were not, and that the Ten Commandments had been the foundation of all the codes of law in the earth. Shelley denied it. —— backed him. I affirmed they were— neither of us using an atom of logic. Shelley said Shakespeare could not have been a Christian because of the dialogue in *Cymbeline*.[19]

> "*Gaoler*. For look you, sir, you know not the way you should go.
> "*Posthumus*. Yes indeed I do, fellow.
> "*Gaol*. Your death has eyes in his head then, and I have never seen him so pictured: you must either be directed by some who take upon themselves to know, or take upon yourself that I am sure you do not know, or jump the after inquiry on your own peril, and how you shall speed on your journey's end, I think you will never return to tell me.
> "*Post*. I tell ye, fellow, there are none want eyes to direct them the way I am going, but such as wink, and will not use them.

> "*Gaol*. What an infinite mock is this, that a man should have the best use of eyes to see the way of blindness."

I replied, that proved nothing; you might as well argue Shakespeare was in favour of murder because, when he makes a murderer, he is ready to murder, as infer he did not believe in another world, or in Christianity, because he has put sophistry about men's state after death in the mouth of his gaoler.

I argued that his own will might be inferred to contain his own belief, and there he says, "In Jesus Christ hoping and assuredly believing, I, W. Shakespeare, etc. . . ." Shelley said, "That was a mere matter of form." I said, "That opinion was mere matter of inference, and, if quotation were argument, I would give two passages to one in my favour." They sneered, and I at once quoted:

> "Though justice be thy plea, consider this,
> That in the course of justice none of us
> Should see salvation."[20]

And again:

> "Why all the souls that are were forfeit once,
> And he that might th' advantage best have took,
> Found out the remedy."[21]

Neither ——, Keats nor —— said a word to this; but still Shelley, —— and —— kept at it till, finding I was a match for them in argument, they became personal, and so did I. We said unpleasant things to each other, and when I retired to the other room for a moment I overheard them say, "Haydon is fierce." "Yes," said ——; "the question always irritates him." As the women were dressing to go, —— said to me with a look of nervous fear, "Are these creatures to be d——d, Haydon?" "Good Heaven," said I, "what a morbid view of Christianity."

The assertion of —— that these sort of discussions irritated me is perfectly true; but it was not so much the question as their manner of treating it. I never heard any sceptic but Hazlitt discuss the matter with the gravity such a question demanded. The eternity of the human soul is not a joke, as —— was always inclined to make of it, not in reality—for the thought wrenched his being to the very midriff—but apparently that he might conceal his frightful apprehensions. For he was by nature

---

17. invitation: Haydon was a fiercely orthodox man, and he and Leigh Hunt argued constantly about religion. There is little doubt that Hunt asked Haydon to this dinner in order to goad him. His portrait of Shelley was naturally colored by this unfortunate experience. **18. vi et armis:** with force and arms. **19. Cymbeline:** V. iv. 180–96.

20. "Though . . . salvation": *The Merchant of Venice*, IV. i. 198–200. **21. "Why . . . remedy":** *Measure for Measure*, II. ii. 73–75.

gloomy, and all his wit and jokes, and flowers and green fields were only so many desperate efforts to break through the web which hung round and impeded him. Luckily for me I was deeply impressed with the denunciations, the promises, the hopes, the beauty of Christianity. I received an impression at an early age which has never been effaced, and never will, and which neither the insidious efforts of —— nor the sophistry of Shelley ever for a moment shook. My irritation proceeded not from my fear of them, but from my being unable to command my feelings when I heard Voltaire almost worshipped in the very same breath that had called St Paul, Mister Paul, and when, with a smile of ineffable superiority, it was intimated that he was a cunning fellow. I used to say, "Let us go on without appellations of that kind. I detest them." "Oh, the question irritates you," was the reply. "And always will when so conducted," was my answer. "I am like Johnson;[22] I will not suffer so awful a question as the truth or falsehood of Christianity to be treated like a new farce, and if you persist I will go." . . .

After this dinner I made up my mind to subject myself no more to the chance of these discussions, but gradually to withdraw from the whole party.

THE IMMORTAL DINNER

On December 28th [1817] the immortal dinner came off in my painting-room, with Jerusalem[23] towering up behind us as a background. Wordsworth was in fine cue, and we had a glorious set-to—on Homer, Shakespeare, Milton and Virgil. Lamb got exceedingly merry and exquisitely witty; and his fun in the midst of Wordsworth's solemn intonations of oratory was like the sarcasm and wit of the fool in the intervals of Lear's passion. He made a speech and voted me absent, and made them drink my health. "Now," said Lamb, "you old lake poet, you rascally poet, why do you call Voltaire dull?" We all defended Wordsworth, and affirmed there was a state of mind when Voltaire would be dull. "Well," said Lamb, "here's Voltaire—the Messiah of the French nation, and a very proper one too."

He then, in a strain of humour beyond description, abused me for putting Newton's head into my picture;

"a fellow," said he, "who believed nothing unless it was as clear as the three sides of a triangle." And then he and Keats agreed he had destroyed all the poetry of the rainbow by reducing it to the prismatic colours. It was impossible to resist him, and we all drank "Newton's health, and confusion to mathematics." It was delightful to see the good humour of Wordsworth in giving in to all our frolics without affectation and laughing as heartily as the best of us.

By this time other friends joined, amongst them poor Ritchie[24] who was going to penetrate by Fezzan to Timbuctoo. I introduced him to all as "a gentleman going to Africa." Lamb seemed to take no notice; but all of a sudden he roared out: "Which is the gentleman we are going to lose?" We then drank the victim's health, in which Ritchie joined.

In the morning of this delightful day, a gentleman, a perfect stranger, had called on me. He said he knew my friends, had an enthusiasm for Wordsworth, and begged I would procure him the happiness of an introduction. He told me he was a comptroller of stamps, and often had correspondence with the poet. I thought it a liberty; but still, as he seemed a gentleman, I told him he might come.

When we retired to tea we found the comptroller. In introducing him to Wordsworth I forgot to say who he was. After a little time the comptroller looked down, looked up and said to Wordsworth: "Don't you think, sir, Milton was a great genius?" Keats looked at me, Wordsworth looked at the comptroller. Lamb who was dozing by the fire turned round and said: "Pray, sir, did you say Milton was a great genius?" "No, sir; I asked Mr Wordsworth if he were not." "Oh," said Lamb, "then you are a silly fellow." "Charles! my dear Charles!" said Wordsworth; but Lamb, perfectly innocent of the confusion he had created, was off again by the fire.

After an awful pause the comptroller said: "Don't you think Newton a great genius?" I could not stand it any longer. Keats put his head into my books. Ritchie squeezed in a laugh. Wordsworth seemed asking himself: "Who is this?" Lamb got up, and taking a candle, said: "Sir, will you allow me to look at your phrenological development?" He then turned

22. Johnson: Samuel Johnson. 23. Jerusalem: Haydon's vast canvas, "Christ's Entry into Jerusalem," on which he was working.

24. Ritchie: a young surgeon, Joseph Ritchie, who was planning to set off on an expedition to discover the source of the Niger River in Africa and died on the journey in 1819.

his back on the poor man, and at every question of the comptroller he chaunted:

> "Diddle diddle dumpling, my son John
> Went to bed with his breeches on."

The man in office, finding Wordsworth did not know who he was, said in a spasmodic and half-chuckling anticipation of assured victory: "I have had the honour of some correspondence with you, Mr Wordsworth." "With me, sir?" said Wordsworth, "not that I remember." "Don't you, sir? I am a comptroller of stamps." There was a dead silence, the comptroller evidently thinking that was enough. While we were waiting for Wordsworth's reply, Lamb sung out:

> "Hey diddle diddle,
> The cat and the fiddle."

"My dear Charles!" said Wordsworth.

> "Diddle diddle dumpling, my son John,"

chaunted Lamb, and then rising, exclaimed: "Do let me have another look at that gentleman's organs." Keats and I hurried Lamb into the painting-room, shut the door and gave way to inextinguishable laughter. Monkhouse[25] followed and tried to get Lamb away. We went back, but the comptroller was irreconcilable. We soothed and smiled and asked him to supper. He stayed though his dignity was sorely affected. However, being a good-natured man, we parted all in good humour, and no ill effects followed.

All the while, until Monkhouse succeeded, we could hear Lamb struggling in the painting-room and calling at intervals: "Who is that fellow? Allow me to see his organs once more."

It was indeed an immortal evening. Wordsworth's fine intonation as he quoted Milton and Virgil, Keats' eager inspired look, Lamb's quaint sparkle of lambent humour, so speeded the stream of conversation, that in my life I never passed a more delightful time. All our fun was within bounds. Not a word passed that an apostle might not have listened to. It was a night worthy of the Elizabethan age, and my solemn Jerusalem flashing up by the flame of the fire, with Christ hanging over us like a vision, all made up a picture which will long glow upon

> "that inward eye
> Which is the bliss of solitude."[26]

25. **Monkhouse:** Thomas Monkhouse, 1783–1825, a cousin of Mrs. Wordsworth. 26. **"that . . . solitude":** Wordsworth's "I Wandered Lonely as a Cloud," above, ll. 21–22.

Keats made Ritchie promise he would carry his *Endymion* to the great desert of Sahara and fling it in the midst.

Poor Ritchie went to Africa, and died, as Lamb foresaw, in 1819. Keats died in 1821, at Rome. C. Lamb is gone, joking to the last. Monkhouse is dead, and Wordsworth and I are the only two now living (1841) of that glorious party.

MEETING WITH WORDSWORTH IN 1842

22nd [May].—Wordsworth called to-day, and we went to church together. There was no seat to be got at the chapel near us, belonging to the rectory of Paddington, and we sat among publicans and sinners. I determined to try him, so advised our staying, as we could hear more easily. He agreed like a Christian; and I was much interested in seeing his venerable white head close to a servant in livery, and on the same level. The servant in livery fell asleep, and so did Wordsworth. I jogged him at the Gospel, and he opened his eyes and read well. A preacher preached when we expected another, so it was a disappointment. We afterwards walked to Rogers's across the park. He had a party to lunch, so I went into the pictures, and sucked Rembrandt, Reynolds, Veronese, Raffaele, Bassan and Tintoretto. Wordsworth said, "Haydon is downstairs." "Ah," said Rogers, "he is better employed than chattering nonsense upstairs." As Wordsworth and I crossed the park, we said, "Scott, Wilkie, Keats, Hazlitt, Beaumont, Jackson, Charles Lamb are all gone; we only are left." He said, "How old are you?" "Fifty-six," I replied. "How old are you?" "Seventy-three," he said; "in my seventy-third year. I was born in 1770." "And I in 1786." "You have many years before you." "I trust I have; and you, too, I hope. Let us cut out Titian, who was ninety-nine." "Was he ninety-nine?" said Wordsworth. "Yes," said I, "and his death was a moral; for as he lay dying of the plague, he was plundered, and could not help himself." We got on Wakley's abuse. We laughed at him. I quoted his own beautiful address to the stock dove. He said, once in a wood, Mrs Wordsworth and a lady were walking, when the stock dove was cooing. A farmer's wife coming by said to herself, "Oh, I do like stock doves!" Mrs Wordsworth, in all her enthusiasm for Wordsworth's poetry, took the old woman to her heart; "but," continued the old woman, "Some like them in a pie; for my part there's nothing like 'em stewed in onions."

GEORGE GORDON, LORD BYRON

1788–1824

NO POET fascinated his contemporaries more than Byron. Everything conspired to rivet attention—his genius, wealth, and personal beauty, his sudden fame and fall, the breath of scandal, the heroic death, and something incalculable, defiant, and reckless in his character and style of life. Recollections of him abound, yet his nature was so complex and various that few students are sure they understand him; his contemporaries often painted utterly diverse portraits. John Galt remembered Byron at the age of twenty-one, sailing through the Mediterranean on his first eastern travels.

> Amidst the shrouds and rattlings, in the tranquillity of the moonlight . . . he seemed almost apparitional, suggesting dim reminiscences of him who shot the albatross. He was as a mystery in a winding-sheet, crowned with a halo.

This may be partly satire—Galt was usually disposed more to irony than enthusiasm—but it evokes the legendary Byron of Byronism, a figure of weird communings, shadowy guilt, the allure of a fallen Lucifer. "Mad, bad and dangerous to know," Lady Caroline Lamb wrote in her diary after meeting him. The words apply much more to herself, yet there was some truth in them. But it is also true that the sensibility of the age hungered for such a figure. The novelist Lady Blessington approached with a thrill of dread when in 1823 she met him for the first time, but was much disappointed. He was not sublimely glowering but instead lonely, gossipy, very eager to please, and not even a good horseman.

> I had expected to find him a dignified, cold, reserved and haughty person, but nothing can be more different; for were I to point out the prominant defect of Lord Byron, I should say it was flippancy, and a total want of that natural self-possession and dignity which ought to characterize a man of birth and education.

But he was a brilliant talker. Of these two Byrons, the first emerges primarily in *Childe Harold* and the second in his masterpiece, *Don Juan*.

Celebrity had been granted all at once with the publication of the first two cantos of *Childe Harold* in 1812, when, as he said, "I awoke one morning and found myself famous." The poem owed its extraordinary success partly to its vivid narrative of travel. But there was also the character of the hero, whom readers naturally identified with Byron himself. Sated with pleasures and proudly concealing a hopeless sorrow, Harold had exiled himself from his native land and carried his "marble heart" through Portugal, Spain, Albania, and Greece. The mixture of crime, pride, apathy, exotic experience, and secret woe proved irresistible. Byron, said the Duchess of Devonshire, was "really the only topic of almost every conversation—the men jealous of him, the women of each other." For three years he was lionized in high society, the sensation he caused being constantly rekindled by further poems with fatal heroes and by scandalous amours with Lady Caroline Lamb and Lady Oxford, supplemented by darker rumours of misconduct with his half-sister, and by alarming opinions in politics and religion.

At times Byron gloomily believed that blood inheritance is fate. His immediate ancestors were not reassuring. His grandfather had been an admiral, known in the fleet as "Foul-weather Jack" because storms blew wherever he sailed. His great-uncle, the "Wicked Lord," had killed a friend in a drunken duel. His father was a notorious libertine who had married twice and squandered the fortunes of both wives. He died when Byron was three, and the boy lived with his mother in straitened circumstances until, at the age of ten, he came into the family estate, Newstead Abbey. These years were embittered by the unpredictable rages of his mother and by his own acute self-consciousness arising from a slight, possibly congenital lameness. At the age of thirteen he went to Harrow, "a fat, bashful boy" with a temper. By graduation he had formed close friendships, suffered from unrequited love for Mary Chaworth, and been judged a bad influence and lazy, though with oratorical promise. He attended Trinity College, Cambridge, and in vacations composed the lyrics published in *Hours of Idleness* (1807). These inoffensive pieces by a college boy and a lord were indulgently received, except by *The Edinburgh Review*. Incensed by this one gap in the garland of praise, Byron added some lines on the editor, Francis Jeffrey, to a literary satire he had in hand. This was published as *English Bards and Scotch Reviewers* (1809), a brash and lively performance, though Byron certainly overestimated its effect. In 1809–11 he went on the tour narrated in Cantos I and II of *Childe Harold*, which were written en route.

In the years of social luster that followed something self-destructive was at work in him. Regency society would tolerate a great deal, but Byron recklessly, half-deliberately courted the opprobrium that finally burst upon him. His downfall came with the scandal attending legal separation from his wife. He had married the virtuous and wealthy Annabella Milbanke in 1815 without feeling any strong attraction. Possibly he wished a more regular life—he always liked domesticity—but he may also have sought the aid of marriage as he struggled to control his attraction to his half-sister, Augusta Leigh. Miss Milbanke probably imagined she would reform him. Biographers do not precisely agree as to the cause of the separation, but there was a rich multiplicity of sufficient reasons. For one thing, Byron's habits were always eccentric: he went on prolonged diets, taking nothing but biscuits and soda water; he stayed up all night; he kept loaded pistols by his bedside; he would not endure the sight of a woman eating. Now more than ever he was given to moods, hysteria, and strange rages. He could be brutal. And then Lady Byron was prim and naive; Byron enjoyed shocking her, and he had a fatal need of confession. He threw out hints—almost certainly well-founded—of his relations with his half-sister. Lady Byron naturally did not specify this charge in the articles of separation, but, whether with calculation or not, the story was spread, and "the excommunicating voice of society," as Tom Moore put it, became imperious. If Byron stepped into a room, "Countesses and Ladies of fashion" left "in crowds."

Quitting England in 1816, Byron made his way through the Low Countries to Switzerland. Here he met Shelley, whose influence shows itself in Canto III of *Childe Harold*. He went on to Venice, whence he shocked his London friends by vivid letters describing profligate adventures. The scene of riot was further heightened by a menagerie that wandered freely in his palace; dogs, monkeys, peacocks, guinea hens, a crane, and other beasts shrieked and disputed the grand stairway. But at this time he composed some of his finest work, *Manfred*, *Childe Harold* IV, *Beppo*, and the start of *Don Juan*.

In 1819 he more or less settled down, living with the Countess Guiccioli. Through her family he became involved in Italian politics—minor conspiracies against Austrian rule—and though these intrigues came to nothing, Byron was watched and, through his Italian friends, harrassed by the police. At the urging of Shelley, he consented to participate in a radical journal to be called *The Liberal* and edited by Leigh Hunt, who came to Italy in 1822 for the purpose. But after Shelley drowned, Byron and Hunt learned to detest each other and *The Liberal* soon ceased. Meanwhile Byron was still working on *Don Juan* and sending his brilliant, racy letters to friends in England. For throughout his career personal friends—such as John Cam Hobhouse, Douglas Kinnaird, and Tom Moore—remained loyal. Better than anyone else they knew his generosity, courage, honesty, ebullient vitality, and wit.

Though he had long been famous throughout Europe, the last scene of Byron's life brought new renown. Revolution against Turkish rule had broken out in Greece. In England a committee to aid the rebellion was formed. Byron decided to support its work and sailed for Greece in 1823, carrying money and medical supplies. Settling down in the marshy, fever-ridden town of Missolonghi, he attempted to get together an artillery brigade. He bravely persevered against dishonesty, incompetence, and mutiny, but not much could be accomplished. On the ninth of

April, 1824, he came down with fever. The illness persisted and by the eighteenth he was delirious. Coming to consciousness for a moment, he saw his friends gathered around weeping. "*O questa e una bella scena*," he said half-smiling ("Oh, what a beautiful scene!"). When the news reached England, a fifteen-year-old boy, Alfred Tennyson, went into the woods near his home and wrote on a rock, "Byron is dead." On that day, he later said, "the whole world seemed to be darkened for me."

2

Byron had a profound impact on European and especially French Romanticism. The influence derived mainly from a character type (the "Byronic hero") first portrayed in Cantos I and II of *Childe Harold* and thereafter developed variously in *The Corsair, Lara, Childe Harold* III and IV, *Manfred*, and *Cain*. This figure had prototypes in the Gothic novels of the eighteenth century, in Chateaubriand's *René*, and in the characters of Milton's Satan and of Napoleon as seen through Romantic eyes. The Byronic hero continued to haunt nineteenth-century literature and philosophy. He is a man greater than others in emotion, capability, and suffering. Only among wild and vast forms of nature—the ocean, the precipices and glaciers of the Alps—can he find a counterpart to his own titanic passions. Driven by a demon within, he is fatal to himself and others; for no one can resist his hypnotic fascination and authority. He has committed a sin that itself expresses his superiority: lesser men could not even conceive a like transgression. Against his own suffering he brings to bear a superhuman pride and fortitude. Indeed, without the horror of his fate there could not be the splendor of self-assertion and self-mastery in which he experiences a strange joy and triumph.

Here Byron sounds a note that reverberates throughout the nineteenth century, running below or outside the surface optimism. For the Byronic hero has a metaphysical significance. He defies the Power that made and doomed him. Or rather he must submit, but will not acquiesce in the vast wrong. He embodies an ultimate refusal, a rebellion against the injustice, as it is felt, of limitation, suffering, and death. But he also embodies a form of nihilism; he can find no ground of action or value outside his own will. (See Coleridge's analysis of this type of character in *The Statesman's Manual*, above, pp. 503–04.) To put it another way, from the sin he has committed he derives freedom. Even though he is racked by guilt, he knows the mere conventionality and groundlessness of moral codes. It is to this complex figure that the French *Byronisme* alludes, and because of its vogue Goethe could take Byron as the symbol of his revolutionary age. But this interpretation of Byron now appears incomplete, and we are likely to prefer the comedy of *The Vision of Judgment* and *Don Juan*, a side of his work that his contemporaries underrated or ignored.

The characteristic notes of Byron's style are speed, vehemence, and, in the nonsatiric poems, melodrama. The speed results partly from his clarity of conception and phrase, which themselves express the vigorous positivism of his mind. It also depends on the rather abstract and general vocabulary he employs.

> Roll on, thou deep and dark blue Ocean—roll!
> Ten thousand fleets sweep over thee in vain;
> Man marks the earth with ruin—his control
> Stops with the shore.
>
> [*Childe Harold's Pilgrimage*, IV. 1603–06].

The succession of general terms—Ocean, fleets, Man, earth, ruin, control—together with the relative scarcity of concrete, particular detail, allow a boldness and freedom of motion quite unlike, for example, the Keatsian massing of imagery. Furthermore, Byron often adopts phrases that are ready-to-hand, stock counters that the mind registers without effort. There is, for example, the famous stanza on Napoleon after his downfall.

> Yet well thy soul hath brooked the turning tide
> With that untaught innate philosophy,
> Which, be it wisdom, coldness, or deep pride,
> Is gall and wormwood to an enemy.
> When the whole host of hatred stood hard by,
> To watch and mock thee shrinking, thou hast smiled
> With a sedate and all-enduring eye;—
> When fortune fled her spoiled and favorite child,
> He stood unbowed beneath the ills upon him piled.
>
> [*Childe Harold's Pilgrimage*, III. 343–51].

That it is effective cannot be questioned, but the effect includes ideas of a "turning tide," smiling on enemies, being the spoiled child of fortune, and standing "unbowed" beneath accumulated ills. The stanza is more obviously clichéd than most, but this nonchalant or offhand way of writing is typical. Neither can we say that Byron succeeds in spite of it. He succeeds because of it. The pace it allows is indispensable to our sense of sweeping panoramas, stark confrontations, and explosive turns of feeling. But it may also remind us that all poetry cannot receive the same kind of attention, and that Byron's must usually be read rapidly. He did not labor to be condensed in implication—the suggestive richness of phrase we find in the best work of his contemporaries and of poets today. If this is a limitation, it is shared by Chaucer and Homer, and cannot by itself remove him from the first rank.

By vehemence is meant something that includes and goes beyond vigor of phrase. His conceptions are unbounded. His imagination multiplies, heightens, and leaves reason (or reasonableness) gasping behind. "Ten thousand fleets" sweep over the ocean; Napoleon's enemies are "the whole host of hatred." In other stanzas of *Childe Harold* a quiet state of mind "is a hell"; aspiration "once kindled" is "quenchless evermore"; as for Byron himself, "What am I?

Nothing"; his feelings are "crushed," but in creating the figure of Harold he can still "glow." One of his most characteristic effects is the awe-inspiring gesture or utmost assertion grandly posed:

> Fare thee well! and if forever,
> Still forever, fare *thee well*
>
> ["Fare Thee Well," ll. 1–2].

> Where rose the mountains, there to him were friends;
> Where rolled the ocean, thereon was his home
>
> [*Childe Harold's Pilgrimage*, III. 109–10].

We are dealing, then, with a poetry that carries all its moods to an extreme pitch. The reader is constantly assailed by prodigious, theatrical emotions.

We find it altogether enjoyable. Our critical, qualifying intelligence takes a holiday, having no anxiety to expose the splendid scene. For in his heavy roles we do not think Byron wholly serious. (It is a significant paradox that he can be deeply serious when he is satiric or ironic.) *Childe Harold, Manfred,* and most of the lyrics stand to a more earnest poetry as melodrama does to tragedy. That is, they are single-minded, sensational, extreme, and agreeably harrowing. There is plainly the gusto of emotion relished for its own sake. Melodrama need not, however, be cheap or crude. It is an inferior art only when measured against the very highest. And it gives a distinct pleasure of its own, a kind of entertainment we cannot have in whatever art engages us more profoundly. The pleasure is of release and expansion, a romp and revel of emotions that are ordinarily inhibited or qualified. In saying this, however, we are suggesting the probable response of readers today. Byron's contemporaries saw these poems in an altogether different light.

No more than with *Childe Harold* can the brilliance of style in *Don Juan* and the other satiric poems be exhibited in extracts. One must listen to the sound of the voice. In some ways it is the same voice, swift, varying, unpredictable. But it is persistently ironic. The vocabulary has been expanded to include more colloquial and even slangy ranges. The stanza is more flexible; there is a much wider arc of tones. In short, Byron finally created a style that allowed him to say anything he liked, which in poetry is an ultimate and rare achievement. As Virginia Woolf phrases it, this style is

> an elastic shape which will hold whatever you choose to put into it. Thus he could write out his mood as it came to him; he could say whatever came into his head. . . . Still, it doesn't seem an easy example to follow.

Indeed it was not easy to follow; *Don Juan* had many imitators, now forgotten. "Like all free and easy things," Mrs. Woolf adds, "only the skilled and mature really bring them off successfully."

Bibliography

The powerful impact of Byron in roughly the years 1812–42 has an important place in the history of nineteenth-century Europe. Since then his reputation has gone through marked vicissitudes, and the curve of assessment in England and America has more or less reduplicated itself on the Continent, except that Europeans have always given him the higher rating. In summary, his reputation reached its zenith in the years closely following his death and its nadir in the 1850's. The years of glory can be explained partly by the romantic interest of his poetry for that generation, and partly by his death in the Greek struggle for independence. For at his death he became a hero of human freedom. To understand the decline, we need only keep in mind whatever is immoral or irreligious in his life and poetry while also meditating on the cluster of attitudes we call Victorianism. The subsequent rebirth of esteem, culminating about 1880, was connected with the critical reaction against middle-class smugness and with reviving sympathy for revolutionary politics. Also at this time began the swing of taste that finally brought *Don Juan* to the fore. Early admiration had been directed wholly to the "romantic" Byron of *Lara*, *The Corsair*, and *Childe Harold*. Nowadays we too often go to the opposite extreme, relishing the satire of *Beppo*, *The Vision of Judgment*, and *Don Juan* while ignoring the boldness of declamation and stance in *Childe Harold*.

Frantic enthusiasm is seldom felt by reviewers, and throughout his career Byron's kept cooler heads than the general public. By the time of his death they had more or less agreed on his distinctive qualities. Their portrait was of great abilities perverted. Conceding originality, daring, vividness, and strength, they were put off by his immorality, skepticism, and what Scott called "misanthropical ennui." As Shelley (whose reactions to Byron alternated violently) said in a letter of 1818: "Contemplating in the distorted mirror of his own thoughts the nature and the destiny of man, what can he behold but objects of contempt and despair?" Moreover, almost all critics pointed to the restricted range of his sensibility; the same characters were duplicated from poem to poem. In a letter Lamb remarked, "To be a Poet is to be . . . the whole Man—not a petty portion of occasional low passion worked up into a permanent form of Humanity. Shakespeare has thrust such rubbishly feelings in a corner, the dark dusty heart of Don John in the Much Ado." To represent the estimates of these critics one might pick out Francis Jeffrey in *The Edinburgh Review*, XXXVI, February, 1822, or Hazlitt (above in "On the Living Poets" and "Lord Byron"). One may also mention the numerous remarks of Goethe. They are important in themselves and also because they were caught up by later readers, notably Matthew Arnold. Goethe sees Byron as an embodiment of the *Zeitgeist* (the spirit of the age). He is a demonic and revolutionary force, but he lacks a complete culture and depth of reflection.

In 1831 Thomas Babington Macaulay (in *The Edinburgh Review*, LIII, June; reprinted in *Critical and Miscellaneous Essays*, 1841) stressed the split in Byron between Romantic and neo-classical attitudes. Macaulay was the first important critic to make this observation, which has been assumed and elaborated ever since. (An extended discussion from this point of view is that of William Calvert, *Byron: Romantic Paradox*, 1935.) Macaulay is not, however, especially sympathetic, and with the rise of Tennyson in 1842 Byron's verse began to seem slovenly. Moreover, many readers, such as Thackeray, thought him insincere: "That man *never* wrote from his heart. He got up rapture and enthusiasm with an eye on the public." That Byron was not a finished craftsman was admitted throughout the later nineteenth century ("craftsmanship" referred at this time mainly to phrasing and versification, not to narrative skills), but Byron returned to fame as a great battler in the liberal cause, "a passionate and dauntless soldier," Arnold said, who had "shattered himself to pieces" on the "impregnable Philistinism" of the "purblind and hideous" middle class. Moreover, what John Morley called Byron's "poetical *worldliness*," his "energetic interest in real transactions," could seem a reviving salt air in the Victorian palace of art. Four essays may be taken to represent the favorable opinion of the later nineteenth century: A. C. Swinburne, Preface to *Selection from the Works of Lord Byron*, 1866 (reprinted in *Essays and Studies*, 1875); John Morley, "Byron," in *The Fortnightly Review*, XIV, December, 1870 (reprinted in *Critical Miscellanies*, 1886); John Ruskin, "Fiction, Fair and Foul," Part III, in *The Nineteenth Century*, VIII, September, 1880 (reprinted in *Works*, Library Edition, XXXIV); Matthew Arnold, Preface to *Poems of Byron*, 1881 (reprinted in *Essays in Criticism, Second Series*, 1888).

Writing on Byron in the twentieth century has been largely given over to clarifying particular questions. E. W. Marjarum, *Byron as Skeptic and Believer*, 1938, concentrates on his religious opinions. E. J. Lovell, Jr., *Byron: The Record of a Quest*, 1949, takes up Byron's attitude to nature in studies of particular topics. Two books devoted to *Don Juan* are P. G. Trueblood, *The Flowering of Byron's Genius*, 1945, and E. F. Boyd, *Byron's Don Juan: A Critical Study*, 1945, and there is also the Variorum Edition, edited by T. G. Steffan and W. W. Pratt, 4 vols., 1957. The first volume, by Steffan, gives a detailed history of the writing of the poem, analyzes Byron's method of composition, and concludes with a series of excellent critical essays. The next two volumes comprise the text, and the final volume, by Pratt, contains notes and a history of critical estimates. There is also G. M. Ridenour, *The Style of Don Juan*, 1960. Two books survey the vicissitudes of Byron's reputation and of Byronism: S. C. Chew, *Byron in England*, 1924, and Edmond Estève, *Byron et le romantisme français*, 1907. A general critical assessment is attempted by Andrew Rutherford, *Byron*, 1961, and see also T. S. Eliot, "Byron," in *On Poetry and Poets*, 1957. L. A. Marchand expertly surveys the entire canon in *Byron's Poetry: A Critical Introduction*, 1965.

Biographical studies have proliferated remarkably. Leslie Marchand, *Byron*, 1957, is the standard work. A readable short life is that by Peter Quennell, *Byron*, 1934, in the Great Lives Series. The most noteworthy mines of new information in the past twenty years have been the Italian sources for his life in Italy. These were utilized by Iris Origo, *The Last Attachment*, 1949, on his affair with the Countess Guiccioli, and by C. L. Cline, *Byron, Shelley, and Their Pisan Circle*, 1952. Controversy still rages over the cause of the separation from Lady Byron. Harriet Beecher Stowe began the discussion as long ago as 1869 when she charged that he had committed incest with his half-sister. Her purpose in this revelation was "to weaken the evil influence of his writings, and shorten his expiation in another world." The effect of the dispute she initiated has naturally been to attract new readers in shoals. E. J. Lovell, Jr., *His Very Self and Voice*, 1954, is not a biography but gathers together a great many recollections of his conversation. G. Wilson Knight, *Lord Byron: Christian Virtues*, 1952, is a suggestive and original portrait of Byron's character.

With the exception of *Don Juan*, for which see the Variorum Edition, the standard edition includes both the *Poetry*, ed. E. H. Coleridge, 7 vols., and *The Letters and Journals*, ed. R. E. Prothero, 6 vols., 1898–1904. Other collections of letters are *Lord Byron in His Letters*, ed. V. H. Collins, 1927, and, in paperback, *Selected Letters*, ed. Jacques Barzun, 1953. *Byron: A Self-Portrait*, ed. Peter Quennell, 2 vols., 1950, has a selection of letters and the complete text of the diaries and "Detached Thoughts."

LACHIN Y GAIR

"Lachin y Gair, or, as it is pronounced in the Erse, Loch na Garr, towers proudly preeminent in the Northern Highlands. One of our modern tourists mentions it as the highest mountain, perhaps, in Great Britain. Be that as it may, it is certainly one of the most sublime and picturesque amongst our 'Caledonian Alps.' Its appearance is of a dusky hue, but the summit is the seat of eternal snows. Near Lachin y Gair I spent some of the early part of my life, the recollection of which has given birth to the following stanzas" (from Byron's Preface to *Hours of Idleness*).

Away, ye gay landscapes, ye gardens of roses!
  In you let the minions of luxury rove;
Restore me the rocks, where the snow-flake reposes,
  Though still they are sacred to freedom and love:
Yet, Caledonia, beloved are thy mountains,
  Round their white summits though elements war;
Though cataracts foam 'stead of smooth-flowing
    fountains,
  I sigh for the valley of dark Loch na Garr.

Ah! there my young footsteps in infancy wander'd;
  My cap was the bonnet, my cloak was the plaid; 10
On chieftains long perish'd my memory ponder'd,
  As daily I strode through the pine-cover'd glade;

I sought not my home till the day's dying glory
  Gave place to the rays of the bright polar star;
For fancy was cheer'd by traditional story,
  Disclosed by the natives of dark Loch na Garr.

"Shades of the dead! have I not heard your voices
  Rise on the night-rolling breath of the gale?"
Surely the soul of the hero rejoices,
  And rides on the wind, o'er his own Highland vale.
Round Loch na Garr while the stormy mist gathers, 21
  Winter presides in his cold icy car:
Clouds there encircle the forms of my fathers;
  They dwell in the tempests of dark Loch na Garr.

"Ill-starr'd, though brave, did no visions foreboding
  Tell you that fate had forsaken your cause?"
Ah! were you destined to die at Culloden,
  Victory crown'd not your fall with applause:
Still were you happy in death's earthly slumber,
  You rest with your clan in the caves of Braemar; 30
The pibroch resounds, to the piper's loud number,
  Your deeds on the echoes of dark Loch na Garr.

Years have roll'd on, Loch na Garr, since I left you,
  Years must elapse ere I tread you again:
Nature of verdure and flow'rs has bereft you,
  Yet still are you dearer than Albion's plain.

LACHIN Y GAIR. **25–26. "Ill-starr'd . . . cause?":** an allusion, said Byron, to his maternal ancestors, descended from the Scottish royalty. **31. pibroch:** bagpipe music.

England! thy beauties are tame and domestic
  To one who has roved o'er the mountains afar:
Oh for the crags that are wild and majestic!
  The steep frowning glories of dark Loch na Garr. 40

1807 (1807)

WHEN WE TWO PARTED

When we two parted
  In silence and tears,
Half broken-hearted
  To sever for years,
Pale grew the cheek and cold,
  Colder thy kiss;
Truly that hour foretold
  Sorrow to this.

The dew of the morning
  Sunk chill on my brow—   10
It felt like the warning
  Of what I feel now.
Thy vows are all broken,
  And light is thy fame:
I hear thy name spoken,
  And share in its shame.

They name thee before me,
  A knell to mine ear;
A shudder comes o'er me—
  Why wert thou so dear?   20
They know not I knew thee,
  Who knew thee too well:—
Long, long shall I rue thee,
  Too deeply to tell.

In secret we met—
  In silence I grieve,
That thy heart could forget,
  Thy spirit deceive.
If I should meet thee
  After long years,   30
How should I greet thee?—
  With silence and tears.

1808 (1816)

FROM

ENGLISH BARDS
AND SCOTCH REVIEWERS

An attack by *The Edinburgh Review* on Byron's early book of poems, *Hours of Idleness*, prompted Byron to complete and publish this satire, parts of which had already been composed. He uses the heroic couplet of Dryden and Pope (though in a far less condensed manner), and has as his ultimate model Pope's great satire on writers and critics, the *Dunciad*. Within a few years Byron wrote in the margin that he regretted the publication of the satire, "not only on account of the injustice of much of the critical and some of the personal part of it, but the tone and temper are such as I cannot approve." Hazlitt comments on the poem in "Lord Byron," above, pp. 697–98.

Oh! Southey! Southey! cease thy varied song!
A bard may chant too often and too long:
As thou art strong in verse, in mercy, spare!
A fourth, alas! were more than we could bear.
But if, in spite of all the world can say,
Thou still wilt verseward plod thy weary way; 230
If still in Berkley ballads most uncivil,
Thou wilt devote old women to the devil,
The babe unborn thy dread intent may rue:
"God help thee," Southey, and thy readers too.

  Next comes the dull disciple of thy school,
That mild apostate from poetic rule,
The simple Wordsworth, framer of a lay
As soft as evening in his favorite May,
Who warns his friend "to shake off toil and trouble,
And quit his books for fear of growing double;"
Who, both by precept and example, shows   241
That prose is verse, and verse is merely prose;
Convincing all, by demonstration plain,

Poetic souls delight in prose insane;
And Christmas stories tortured into rhyme
Contain the essence of the true sublime.

ENGLISH BARDS AND SCOTCH REVIEWERS. **230. plod . . . way:** an echo of Gray's *Elegy*, l. 3: "The ploughman homeward plods his weary way." **239–40. "to . . . double":** Wordsworth's "The Tables Turned," ll. 1–4.

Thus, when he tells the tale of Betty Foy,
The idiot mother of "an idiot boy;"
A moon-struck, silly lad, who lost his way,
And, like his bard, confounded night with day; 250
So close on each pathetic part he dwells,
And each adventure so sublimely tells,
That all who view the "idiot in his glory"
Conceive the bard the hero of the story.

Shall gentle Coleridge pass unnoticed here,
To turgid ode and tumid stanza dear?
Though themes of innocence amuse him best,
Yet still obscurity's a welcome guest.
If Inspiration should her aid refuse
To him who takes a pixy for a muse,   260
Yet none in lofty numbers can surpass
The bard who soars to elegize an ass.
So well the subject suits his noble mind,
He brays the laureat of the long-ear'd kind.

Oh! wonder-working Lewis! monk, or bard,
Who fain wouldst make Parnassus a churchyard!
Lo! wreaths of yew, not laurel, bind thy brow,
Thy muse a sprite, Apollo's sexton thou!
Whether on ancient tombs thou tak'st thy stand,
By gibb'ring spectres hail'd, thy kindred band;
Or tracest chaste descriptions on thy page,   271
To please the females of our modest age;
All hail, M. P.! from whose infernal brain
Thin-sheeted phantoms glide, a grisly train;
At whose command "grim women" throng in
     crowds,
And kings of fire, of water, and of clouds,
With "small gray men," "wild yagers," and what
     not,
To crown with honor thee and Walter Scott;
Again all hail! if tales like thine may please,
St. Luke alone can vanquish the disease;   280
Even Satan's self with thee might dread to dwell,
And in thy skull discern a deeper hell.

. . .

Health to immortal Jeffrey! once, in name,
England could boast a judge almost the same;

In soul so like, so merciful, yet just,   440
Some think that Satan has resign'd his trust,
And given the spirit to the world again,
To sentence letters, as he sentenced men.
With hand less mighty, but with heart as black,
With voice as willing to decree the rack;
Bred in the courts betimes, though all that law
As yet hath taught him is to find a flaw;
Since well instructed in the patriot school
To rail at party, though a party tool,
Who knows, if chance his patrons should restore
Back to the sway they forfeited before,   451
His scribbling toils some recompense may meet,
And raise this Daniel to the judgment-seat?
Let Jeffreys' shade indulge the pious hope,
And greeting thus, present him with a rope:
"Heir to my virtues! man of equal mind!
Skill'd to condemn as to traduce mankind,
This cord receive, for thee reserved with care,
To wield in judgment, and at length to wear."

. . .

To the famed throng now paid the tribute due,
Neglected genius! let me turn to you.   800
Come forth, oh Campbell! give thy talents scope;
Who dares aspire if thou must cease to hope?
And thou, melodious Rogers! rise at last,
Recall the pleasing memory of the past;
Arise! let blest remembrance still inspire,
And strike to wonted tones thy hallow'd lyre;
Restore Apollo to his vacant throne,
Assert thy country's honor and thine own.
What! must deserted Poesy still weep
Where her last hopes with pious Cowper sleep? 810
Unless, perchance, from his cold bier she turns,
To deck the turf that wraps her minstrel, Burns!

. . .

There be who say, in these enlighten'd days,
That splendid lies are all the poet's praise;   850
That strain'd invention, ever on the wing,
Alone impels the modern bard to sing:
'Tis true, that all who rhyme—nay, all who write,
Shrink from that fatal word to genius—trite;

---

248. **"an idiot boy":** See Wordsworth's poem, above.
260. **pixy:** referring to Coleridge's "Songs of the Pixies."
262. **ass:** See Coleridge's "To a Young Ass," above. 265.
Lewis: M. G. Lewis, author of the Gothic novel *The Monk*
(1796). 273. **M. P.:** Lewis was a member of Parliament.
277. **yagers:** huntsmen. 438. **Jeffrey:** Francis Jeffrey, selections from whom are printed above. 439. **same:** George

Jeffreys, the "hanging judge" in the "Bloody Assizes"
(1685). 801–02. **Campbell . . . hope:** referring to Thomas
Campbell's *Pleasures of Hope*, excerpts from which are
printed above. 803. **Rogers:** Samuel Rogers, 1763–1855.
810. **Cowper:** William Cowper, 1731–1800.

Yet Truth sometimes will lend her noblest fires,
And decorate the verse herself inspires:
This fact in Virtue's name let Crabbe attest;
Though nature's sternest painter, yet the best.

                                1807–1809 (1809)

WRITTEN AFTER SWIMMING
FROM SESTOS TO ABYDOS

Caught by the legend of Leander, who nightly swam
across the Hellespont from Abydos to visit Hero (a
priestess of Aphrodite in Sestos), Byron and a friend
decided to duplicate the feat at least once. The direct
distance is only a mile, but the current is strong and
hazardous; the entire distance covered therefore came to
four miles.

I

If, in the month of dark December,
    Leander, who was nightly wont
(What maid will not the tale remember?)
    To cross thy stream, broad Hellespont!

II

If, when the wintry tempest roared,
    He sped to Hero, nothing loth,
And thus of old thy current poured,
    Fair Venus! how I pity both!

III

For *me*, degenerate modern wretch,
    Though in the genial month of May   10
My dripping limbs I faintly stretch,
    And think I've done a feat to-day.

IV

But since he crossed the rapid tide,
    According to the doubtful story,
To woo,—and—Lord knows what beside,
    And swam for Love, as I for Glory;

V

'Twere hard to say who fared the best:
    Sad mortals! thus the Gods still plague you!
He lost his labour, I my jest:
    For he was drowned, and I've the ague.   20

                                1810 (1812)

MAID OF ATHENS, ERE WE PART

The Greek refrain means: "My life, I love you."

Ζώη μοῦ, σᾶς ἀγαπῶ.

Maid of Athens, ere we part,
Give, oh give me back my heart!
Or, since that has left my breast,
Keep it now, and take the rest!
Hear my vow before I go,
Ζώη μοῦ, σᾶς ἀγαπῶ.

By those tresses unconfined,
Woo'd by each Ægean wind;
By those lids whose jetty fringe
Kiss thy soft cheeks' blooming tinge;   10
By those wild eyes like the roe,
Ζώη μοῦ, σᾶς ἀγαπῶ.

By that lip I long to taste;
By that zone-encircled waist;
By all the token-flowers that tell
What words can never speak so well;
By love's alternate joy and woe,
Ζώη μοῦ, σᾶς ἀγαπῶ.

Maid of Athens, I am gone:
Think of me, sweet! when alone.   20
Though I fly to Istambol,
Athens holds my heart and soul:
Can I cease to love thee? No!
Ζώη μοῦ, σᾶς ἀγαπῶ.

                                1810 (1812)

857. Crabbe: George Crabbe, one of whose verse tales is
printed above.

SHE WALKS IN BEAUTY

This and the next four lyrics are from Byron's volume of *Hebrew Melodies.*

She walks in beauty, like the night
  Of cloudless climes and starry skies;
And all that's best of dark and bright
  Meet in her aspect and her eyes:
Thus mellow'd to that tender light
  Which heaven to gaudy day denies.

One shade the more, one ray the less,
  Had half impaired the nameless grace
Which waves in every raven tress,
  Or softly lightens o'er her face;   10
Where thoughts serenely sweet express
  How pure, how dear their dwelling-place.

And on that cheek, and o'er that brow,
  So soft, so calm, yet eloquent,
The smiles that win, the tints that glow,
  But tell of days in goodness spent,
A mind at peace with all below,
  A heart whose love is innocent!

                      1814 (1815)

OH! SNATCH'D AWAY
IN BEAUTY'S BLOOM

Oh! snatch'd away in beauty's bloom,
On thee shall press no ponderous tomb;
  But on thy turf shall roses rear
  Their leaves, the earliest of the year;
And the wild cypress wave in tender gloom:

And oft by yon blue gushing stream
  Shall Sorrow lean her drooping head,
And feed deep thought with many a dream,
  And lingering pause and lightly tread;
  Fond wretch! as if her step disturb'd the dead!

SHE WALKS IN BEAUTY. 1. **She:** Lady Wilmot Horton, a cousin of Byron by marriage, whom he met at a ball the night before he wrote these lines.

Away! we know that tears are vain,   11
  That death nor heeds nor hears distress:
Will this unteach us to complain?
  Or make one mourner weep the less?
And thou—who tell'st me to forget,
Thy looks are wan, thine eyes are wet.

                      1814 (1815)

MY SOUL IS DARK

My soul is dark—Oh! quickly string
  The harp I yet can brook to hear;
And let thy gentle fingers fling
  Its melting murmurs o'er mine ear.
If in this heart a hope be dear,
  That sound shall charm it forth again:
If in these eyes there lurk a tear,
  'Twill flow, and cease to burn my brain.

But bid the strain be wild and deep,
  Nor let thy notes of joy be first: 10
I tell thee, minstrel, I must weep,
  Or else this heavy heart will burst;
For it hath been by sorrow nursed,
  And ach'd in sleepless silence long;
And now 'tis doom'd to know the worst,
  And break at once—or yield to song.

                      1814 (1815)

SONG OF SAUL
BEFORE HIS LAST BATTLE

For the story see I Samuel 31.

Warriors and chiefs! should the shaft or the sword
Pierce me in leading the host of the Lord,
Heed not the corse, though a king's, in your path:
Bury your steel in the bosoms of Gath!

Thou who art bearing my buckler and bow,
Should the soldiers of Saul look away from the foe,
Stretch me that moment in blood at thy feet!
Mine be the doom which they dared not to meet.

Farewell to others, but never we part,
Heir to my royalty, son of my heart!   10
Bright is the diadem, boundless the sway,
Or kingly the death, which awaits us today!

                                        1815 (1815)

THE DESTRUCTION
OF SENNACHERIB

For the story see II Kings 18–19.

The Assyrian came down like the wolf on the fold,
And his cohorts were gleaming in purple and gold;
And the sheen of their spears was like stars on the sea,
When the blue wave rolls nightly on deep Galilee.

Like the leaves of the forest when summer is green,
That host with their banners at sunset were seen:
Like the leaves of the forest when autumn hath blown,
That host on the morrow lay wither'd and strown.

For the Angel of Death spread his wings on the blast,
And breathed in the face of the foe as he pass'd;   10
And the eyes of the sleepers wax'd deadly and chill,
And their hearts but once heaved, and forever grew still!

And there lay the steed with his nostril all wide,
But through it there roll'd not the breath of his pride;
And the foam of his gasping lay white on the turf,
And cold as the spray of the rock-beating surf.

And there lay the rider distorted and pale,
With the dew on his brow, and the rust on his mail:
And the tents were all silent, the banners alone,
The lances unlifted, the trumpet unblown.   20

And the widows of Ashur are loud in their wail,
And the idols are broke in the temple of Baal;
And the might of the Gentile, unsmote by the sword,
Hath melted like snow in the glance of the Lord!

                                        1815 (1815)

STANZAS FOR MUSIC

This poem was written on the death of Byron's school friend, the Duke of Dorset. In a letter of the next year, Byron speaks of these lines as "the truest, though the most melancholy, I ever wrote."

There's not a joy the world can give like that it takes
      away,
When the glow of early thought declines in feeling's
      dull decay;
'Tis not on youth's smooth cheek the blush alone, which
      fades so fast,
But the tender bloom of heart is gone, ere youth itself
      be past.

Then the few whose spirits float above the wreck of
      happiness
Are driven o'er the shoals of guilt or ocean of excess:
The magnet of their course is gone, or only points in
      vain
The shore to which their shiver'd sail shall never
      stretch again.

Then the mortal coldness of the soul like death itself
      comes down;
It cannot feel for others' woes, it dare not dream its
      own;   10
That heavy chill has frozen o'er the fountain of our tears,
And though the eye may sparkle still, 'tis where the ice
      appears.

Though wit may flash from fluent lips, and mirth
      distract the breast,
Through midnight hours that yield no more their
      former hope of rest;
'Tis but as ivy-leaves around the ruin'd turret wreath,
All green and wildly fresh without, but worn and gray
      beneath.

Oh could I feel as I have felt,—or be what I have been,
Or weep as I could once have wept o'er many a vanish'd
      scene;
As springs in deserts found seem sweet, all brackish
      though they be,
So, midst the wither'd waste of life, those tears would
      flow to me.   20

                                        1815 (1816)

SONNET ON CHILLON

Published with Byron's verse tale, *The Prisoner of Chillon*, the scene of which is a dungeon in a castle on Lake Geneva.

Eternal Spirit of the chainless Mind!
Brightest in dungeons, Liberty! thou art,
For there thy habitation is the heart—
The heart which love of thee alone can bind;
And when thy sons to fetters are consign'd—
To fetters, and the damp vault's dayless gloom,
Their country conquers with their martyrdom,
And Freedom's fame finds wings on every wind.
Chillon! thy prison is a holy place,
And thy sad floor an altar—for 'twas trod,   10
Until his very steps have left a trace
Worn, as if thy cold pavement were a sod,
By Bonnivard! May none those marks efface!
For they appeal from tyranny to God.

1816 (1816)

FARE THEE WELL

Addressed to Lady Byron (March, 1816), with whom a deed of separation was being signed.

"Alas! they had been friends in youth;
But whispering tongues can poison truth;
And constancy lives in realms above:
And life is thorny; and youth is vain;
And to be wroth with one we love,
Doth work like madness in the brain;
But never either found another
To free the hollow heart from paining—
They stood aloof, the scars remaining,
Like cliffs which had been rent asunder;
A dreary sea now flows between,
But neither heat, nor frost, nor thunder,
Shall wholly do away, I ween,
The marks of that which once hath been."

—Coleridge's *Christabel.*

Fare thee well! and if forever,
Still forever, fare *thee well:*
Even though unforgiving, never
'Gainst thee shall my heart rebel.

Would that breast were bared before thee
Where thy head so oft hath lain,
While that placid sleep came o'er thee
Which thou ne'er canst know again:

Would that breast, by thee glanced over,
Every inmost thought could show!   10
Then thou wouldst at last discover
'Twas not well to spurn it so.

Though the world for this commend thee—
Though it smile upon the blow,
Even its praises must offend thee,
Founded on another's woe:

Though my many faults defaced me,
Could no other arm be found,
Than the one which once embraced me,
To inflict a cureless wound?   20

Yet, oh yet, thyself deceive not;
Love may sink by slow decay,
But by sudden wrench, believe not
Hearts can thus be torn away:

Still thine own its life retaineth,
Still must mine, though bleeding, beat;
And the undying thought which paineth
Is—that we no more may meet.

These are words of deeper sorrow
Than the wail above the dead;   30
Both shall live, but every morrow
Wakes us from a widow'd bed.

And when thou wouldst solace gather,
When our child's first accents flow,
Wilt thou teach her to say "Father!"
Though his care she must forego?

When her little hands shall press thee,
When her lip to thine is press'd,
Think of him whose prayer shall bless thee,
Think of him thy love had bless'd!   40

Should her lineaments resemble
Those thou never more may'st see,
Then thy heart will softly tremble
With a pulse yet true to me.

All my faults perchance thou knowest,
    All my madness none can know;
All my hopes, where'er thou goest,
    Wither, yet with *thee* they go.

Every feeling hath been shaken;
    Pride, which not a world could bow,   50
Bows to thee—by thee forsaken,
    Even my soul forsakes me now:

But 'tis done—all words are idle—
    Words from me are vainer still;
But the thoughts we cannot bridle
    Force their way without the will.

Fare thee well! thus disunited,
    Torn from every nearer tie,
Sear'd in heart, and lone, and blighted,
    More than this I scarce can die.   60

                  1816 (1816)

STANZAS TO AUGUSTA

⟳⟲

To Augusta Leigh, Byron's half-sister.

⟳⟲

Though the day of my destiny's over,
    And the star of my fate hath declined,
Thy soft heart refused to discover
    The faults which so many could find;
Though thy soul with my grief was acquainted,
    It shrunk not to share it with me,
And the love which my spirit hath painted
    It never hath found but in *thee*.

Then when nature around me is smiling,
    The last smile which answers to mine,   10
I do not believe it beguiling,
    Because it reminds me of thine;
And when winds are at war with the ocean,
    As the breasts I believed in with me,
If their billows excite an emotion,
    It is that they bear me from *thee*.

Though the rock of my last hope is shiver'd,
    And its fragments are sunk in the wave,

Though I feel that my soul is deliver'd
    To pain—it shall not be its slave.   20
There is many a pang to pursue me:
    They may crush, but they shall not contemn;
They may torture, but shall not subdue me;
    'Tis of *thee* that I think—not of them.

Though human, thou didst not deceive me,
    Though woman, thou didst not forsake,
Though loved, thou forborest to grieve me,
    Though slander'd, thou never couldst shake;
Though trusted, thou didst not disclaim me,
    Though parted, it was not to fly,   30
Though watchful, 'twas not to defame me,
    Nor, mute, that the world might belie.

Yet I blame not the world, nor despise it,
    Nor the war of the many with one;
If my soul was not fitted to prize it,
    'Twas folly not sooner to shun:
And if dearly that error hath cost me,
    And more than I once could foresee,
I have found that, whatever it lost me,
    It could not deprive me of *thee*.   40

From the wreck of the past, which hath perish'd,
    Thus much I at least may recall,
It hath taught me that what I most cherish'd
    Deserved to be dearest of all:
In the desert a fountain is springing,
    In the wide waste there still is a tree,
And a bird in the solitude singing,
    Which speaks to my spirit of *thee*.

                  1816 (1816)

STANZAS FOR MUSIC

I

There be none of Beauty's daughters
    With a magic like thee;
And like music on the waters
    Is thy sweet voice to me:
When, as if its sound were causing
The charmed Ocean's pausing,
The waves lie still and gleaming,
And the lulled winds seem dreaming:

II

And the Midnight Moon is weaving
    Her bright chain o'er the deep;   10
Whose breast is gently heaving,
    As an infant's asleep:
So the spirit bows before thee,
To listen and adore thee;
With a full but soft emotion,
Like the swell of Summer's ocean.

                    1816 (1816)

DARKNESS

∽∾∾

The theme of *The Last Man* (a popular novel that appeared in 1806) fascinated several writers at the time, including Thomas Campbell, Thomas Hood, and Mrs. Shelley. Byron's poem is probably the best-known example.

∾∾∾

I had a dream, which was not all a dream.
The bright sun was extinguished, and the stars
Did wander darkling in the eternal space,
Rayless, and pathless, and the icy Earth
Swung blind and blackening in the moonless air;
Morn came and went—and came, and brought no day,
And men forgot their passions in the dread
Of this their desolation; and all hearts
Were chill'd into a selfish prayer for light:
And they did live by watchfires—and the thrones,   10
The palaces of crowned kings—the huts,
The habitations of all things which dwell,
Were burnt for beacons; cities were consumed,
And men were gather'd round their blazing homes
To look once more into each other's face;
Happy were those who dwelt within the eye
Of the volcanos, and their mountain-torch:
A fearful hope was all the World contain'd;
Forests were set on fire—but hour by hour
They fell and faded—and the crackling trunks   20
Extinguished with a crash—and all was black.
The brows of men by the despairing light
Wore an unearthly aspect, as by fits
The flashes fell upon them; some lay down
And hid their eyes and wept; and some did rest

Their chins upon their clenched hands, and smiled;
And others hurried to and fro, and fed
Their funeral piles with fuel, and looked up
With mad disquietude on the dull sky,
The pall of a past World; and then again   30
With curses cast them down upon the dust,
And gnash'd their teeth and howled: the wild birds shrieked
And, terrified, did flutter on the ground,
And flap their useless wings; the wildest brutes
Came tame and tremulous; and vipers crawled
And twined themselves among the multitude,
Hissing, but stingless—they were slain for food:
And War, which for a moment was no more,
Did glut himself again:—a meal was bought
With blood, and each sate sullenly apart   40
Gorging himself in gloom: no Love was left;
All earth was but one thought—and that was Death
Immediate and inglorious; and the pang
Of famine fed upon all entrails—men
Died, and their bones were tombless as their flesh;
The meagre by the meagre were devoured,
Even dogs assailed their masters, all save one,
And he was faithful to a corse, and kept
The birds and beasts and famished men at bay,
Till hunger clung them, or the dropping dead   50
Lured their lank jaws; himself sought out no food,
But with a piteous and perpetual moan,
And a quick desolate cry, licking the hand
Which answered not with a caress—he died.
The crowd was famished by degrees; but two
Of an enormous city did survive,
And they were enemies: they met beside
The dying embers of an altar-place
Where had been heaped a mass of holy things
For an unholy usage; they raked up,   60
And shivering scraped with their cold skeleton hands
The feeble ashes, and their feeble breath
Blew for a little life, and made a flame
Which was a mockery; then they lifted up
Their eyes as it grew lighter, and beheld
Each other's aspects—saw, and shrieked, and died—
Even of their mutual hideousness they died,
Unknowing who he was upon whose brow
Famine had written Fiend. The World was void,
The populous and the powerful was a lump,   70
Seasonless, herbless, treeless, manless, lifeless—
A lump of death—a chaos of hard clay.
The rivers, lakes, and ocean all stood still,
And nothing stirred within their silent depths;

Ships sailorless lay rotting on the sea,
And their masts fell down piecemeal: as they dropped
They slept on the abyss without a surge—
The waves were dead; the tides were in their grave,
The Moon, their mistress, had expired before;
The winds were withered in the stagnant air, 80
And the clouds perished; Darkness had no need
Of aid from them—She was the Universe.

1816 (1816)

PROMETHEUS

Titan! to whose immortal eyes
The sufferings of mortality,
Seen in their sad reality,
Were not as things that gods despise;
What was thy pity's recompense?
A silent suffering, and intense;
The rock, the vulture, and the chain,
All that the proud can feel of pain,
The agony they do not show,
The suffocating sense of woe, 10
Which speaks but in its loneliness,
And then is jealous lest the sky
Should have a listener, nor will sigh
Until its voice is echoless.

Titan! to thee the strife was given
Between the suffering and the will,
Which torture where they cannot kill;
And the inexorable Heaven,
And the deaf tyranny of Fate,
The ruling principle of Hate, 20
Which for its pleasure doth create
The things it may annihilate,
Refused thee even the boon to die:
The wretched gift eternity
Was thine—and thou hast borne it well.
All that the Thunderer wrung from thee
Was but the menace which flung back
On him the torments of thy rack;
The fate thou didst so well foresee,
But would not to appease him tell; 30
And in thy silence was his sentence,
And in his soul a vain repentance,

PROMETHEUS. **26. Thunderer:** Jupiter. **29. fate:** Prometheus foresaw the downfall of Jupiter. Cf. Shelley's *Prometheus Unbound*, below, I. 371–74.

And evil dread so ill dissembled,
That in his hand the lightnings trembled.

Thy Godlike crime was to be kind,
To render with thy precepts less
The sum of human wretchedness,
And strengthen man with his own mind;
But baffled as thou wert from high,
Still in thy patient energy, 40
In the endurance, and repulse
Of thine impenetrable spirit,
Which Earth and Heaven could not convulse,
A mighty lesson we inherit:
Thou art a symbol and a sign
To mortals of their fate and force;
Like thee, man is in part divine,
A troubled stream from a pure source;
And man in portions can foresee
His own funereal destiny; 50
His wretchedness, and his resistance,
And his sad unallied existence:
To which his spirit may oppose
Itself—and equal to all woes,
And a firm will, and a deep sense,
Which even in torture can descry
Its own concenter'd recompense,
Triumphant where it dares defy,
And making death a victory.

1816 (1816)

FROM

CHILDE HAROLD'S PILGRIMAGE

୶ଈଚ

Written intermittently throughout nine years (1809–17), from the age of twenty-one to twenty-nine, the inspired travelogue of *Childe Harold* established Byron's fame not only in the British Isles but throughout Europe. At the center of its enormous appeal was both the description of things seen and the character of Harold himself, who seemed to many an embodiment of the Romantic ideal. He was a nobleman ("Childe" was a medieval title given the eldest son of a noble family until he received his actual title) and yet at the same time a "self-exile"—a man who has had all the advantages, and has experienced life widely and deeply, but is now weary of empty pleasures and seeks solace and self-forgetfulness in travel. Thus as Harold (or Byron) stands before breathtaking scenes of nature or monuments of the past, such as Rome and Venice, he touches electrically the feelings of his readers. It is not simply that the scenes grip attention and offer an escape, as any vividly rendered poem of travel might do. Harold also voices powerfully the sense of life's staleness and needless complication that underlies the wish to escape. To this should be added the immense sweep and verve of Byron's language. Its boldness (typical is the line about the Battle of Waterloo: "How that red rain hath made the harvest grow!") could and still can appeal to every level of reader, from those who know little poetry to the most sophisticated. Moreover its rapid clarity and directness are not necessarily without depths. Passages that hold us as description may also function as symbols expressing the character of Harold himself—for example, the wild turbulence of natural scenes or the sense of greatness in ruin conveyed by the images of Rome and Venice. And there is more irony than one might at first expect. Reading the line cited above, one may keep in mind that to Byron the "harvest" of Waterloo was the reactionary settlement of 1815. Or there are the lines on Venice that begin Canto IV:

> I stood in Venice, on the "Bridge of Sighs";
> A palace and a prison on each hand.

At one level, palace and prison suggest romantically interesting extremes of exaltation and misery. But the yoking of the two terms reflects Byron's hatred of monarchy, and this is accomplished while the lines still remain literal description.

Cantos I and II of *Childe Harold* had focused on Spain, Portugal, the Greek Isles, and Albania. In Canto III Byron starts in Belgium and moves down the Rhine to Switzerland and the Alps. In his Preface to the first two cantos he had maintained that Harold was completely "the child of imagination." Now, however, after the scandal and separation from his wife that forced Byron to leave England, Harold becomes more openly Byron's mouthpiece, and the conception of Harold as a "self-exile" is emphasized. With Canto IV, the final part of the poem, Byron moves south to Italy. His use of the Spenserian stanza in the last two cantos is altogether original. In the first two he had followed the eighteenth-century imitators of Spenser, even to using archaic words as they did; and the stanzas have a corresponding stiffness. But with Canto III he employs it with a flexibility and a dashing energy that this normally slow-paced and majestic stanza has been given by no other writer.

❧

from CANTO III

I

Is thy face like thy mother's, my fair child!
Ada! sole daughter of my house and heart?

When last I saw thy young blue eyes they smiled,
And then we parted,—not as now we part,
But with a hope.—
　　　　　　Awaking with a start,
The waters heave around me; and on high
The winds lift up their voices: I depart,
Whither I know not; but the hour's gone by,
When Albion's lessening shores could grieve or glad
　　mine eye.

II

Once more upon the waters! yet once more!　　10
And the waves bound beneath me as a steed
That knows his rider. Welcome to their roar!
Swift be their guidance, wheresoe'er it lead!
Though the strained mast should quiver as a reed,
And the rent canvas fluttering strew the gale,
Still must I on; for I am as a weed,
Flung from the rock, on Ocean's foam to sail
Where'er the surge may sweep, the tempest's breath
　　prevail.

III

In my youth's summer I did sing of one,
The wandering outlaw of his own dark mind;　　20
Again I seize the theme, then but begun,
And bear it with me, as the rushing wind
Bears the cloud onwards: in that tale I find
The furrows of long thought, and dried-up tears,
Which, ebbing, leave a sterile track behind,
O'er which all heavily the journeying years
Plod the last sands of life,—where not a flower appears.

IV

Since my young days of passion—joy, or pain,
Perchance my heart and harp have lost a string,
And both may jar: it may be, that in vain　　30
I would essay as I have sung to sing.
Yet, though a dreary strain, to this I cling;
So that it wean me from the weary dream
Of selfish grief or gladness—so it fling
Forgetfulness around me—it shall seem
To me, though to none else, a not ungrateful theme.

CHILDE HAROLD'S PILGRIMAGE. CANTO III. **3. last I saw:** Augusta Ada was only a month old when Byron's wife left him (Jan. 1816), and Byron never saw her afterwards. **10. Once more . . . once more:** an echo of Henry's address to his troops in *Henry V*, III. i. 1. **19. youth's summer:** referring to Canto I, written seven years before when Byron was twenty-one.

V

He, who grown aged in this world of woe,
In deeds, not years, piercing the depths of life,
So that no wonder waits him; nor below
Can love or sorrow, fame, ambition, strife,   40
Cut to his heart again with the keen knife
Of silent, sharp endurance: he can tell
Why thought seeks refuge in lone caves, yet rife
With airy images, and shapes which dwell
Still unimpaired, though old, in the soul's haunted cell.

VI

'Tis to create, and in creating live
A being more intense that we endow
With form our fancy, gaining as we give
The life we image, even as I do now.
What am I? Nothing: but not so art thou,   50
Soul of my thought! with whom I traverse earth,
Invisible, but gazing, as I glow
Mixed with thy spirit, blended with thy birth,
And feeling still with thee in my crushed feelings'
    dearth.

VII

Yet must I think less wildly:—I *have* thought
Too long and darkly, till my brain became,
In its own eddy boiling and o'erwrought,
A whirling gulf of phantasy and flame:
And thus, untaught in youth my heart to tame,
My springs of life were poisoned. 'Tis too late!   60
Yet am I changed; though still enough the same
In strength to bear what time cannot abate,
And feed on bitter fruits without accusing Fate.

VIII

Something too much of this:—but now 'tis past,
And the spell closes with its silent seal.
Long absent Harold reappears at last;
He of the breast which fain no more would feel,
Wrung with the wounds which kill not, but ne'er
    heal;

Yet Time, who changes all, had altered him
In soul and aspect as in age: years steal   70
Fire from the mind as vigor from the limb;
And life's enchanted cup but sparkles near the brim.

IX

His had been quaffed too quickly, and he found
The dregs were wormwood; but he filled again,
And from a purer fount, on holier ground,
And deemed its spring perpetual; but in vain!
Still round him clung invisible a chain
Which galled forever, fettering though unseen,
And heavy though it clanked not; worn with pain,
Which pined although it spoke not, and grew keen,
Entering with every step he took through many a
    scene.   81

X

Secure in guarded coldness, he had mixed
Again in fancied safety with his kind,
And deemed his spirit now so firmly fixed
And sheathed with an invulnerable mind,
That, if no joy, no sorrow lurked behind;
And he, as one, might 'midst the many stand
Unheeded, searching through the crowd to find
Fit speculation; such as in strange land
He found in wonder-works of God and Nature's
    hand.   90

XI

But who can view the ripened rose, nor seek
To wear it? who can curiously behold
The smoothness and the sheen of beauty's cheek,
Nor feel the heart can never all grow old?
Who can contemplate Fame through clouds unfold
The star which rises o'er her steep, nor climb?
Harold, once more within the vortex, rolled
On with the giddy circle, chasing Time,
Yet with a nobler aim than in his youth's fond prime.

XII

But soon he knew himself the most unfit   100
Of men to herd with man; with whom he held
Little in common; untaught to submit

---

51. **Soul . . . thought:** product of my imagination
(Harold). 64. **Something . . . this:** a quote from *Hamlet*,
III. ii. 79. 65. **silent seal:** seal of silence.

99. **fond:** foolish.

His thoughts to others, though his soul was quelled
In youth by his own thoughts; still uncompelled,
He would not yield dominion of his mind
To spirits against whom his own rebelled;
Proud though in desolation; which could find
A life within itself, to breathe without mankind.

XIII

Where rose the mountains, there to him were friends;
Where rolled the ocean, thereon was his home; 110
Where a blue sky, and glowing clime, extends,
He had the passion and the power to roam;
The desert, forest, cavern, breaker's foam,
Were unto him companionship; they spake
A mutual language, clearer than the tome
Of his land's tongue, which he would oft forsake
For Nature's pages glassed by sunbeams on the lake.

XIV

Like the Chaldean, he could watch the stars,
Till he had peopled them with beings bright
As their own beams; and earth, and earth-born jars,
And human frailties, were forgotten quite: 121
Could he have kept his spirit to that flight
He had been happy; but this clay will sink
Its spark immortal, envying it the light
To which it mounts, as if to break the link
That keeps us from yon heaven which woos us to its
    brink.

XV

But in man's dwellings he became a thing
Restless and worn, and stern and wearisome,
Drooped as a wild-born falcon with clipt wing,
To whom the boundless air alone were home: 130
Then came his fit again, which to o'ercome,
As eagerly the barred-up bird will beat
His breast and beak against his wiry dome
Till the blood tinge his plumage, so the heat
Of his impeded soul would through his bosom eat.

XVI

Self-exiled Harold wanders forth again,
With nought of hope left, but with less of gloom;
The very knowledge that he lived in vain,
That all was over on this side the tomb,
Had made Despair a smilingness assume,   140
Which, though 'twere wild,—as on the plundered
    wreck
When mariners would madly meet their doom
With draughts intemperate on the sinking deck,—
Did yet inspire a cheer, which he forbore to check.

XVII

Stop!—for thy tread is on an empire's dust!
An earthquake's spoil is sepulchred below!
Is the spot marked with no colossal bust?
Nor column trophied for triumphal show?
None; but *the moral's truth* tells simpler so,
As the ground was before, thus let it be;—   150
How that red rain hath made the harvest grow!
And is this all the world has gained by thee,
Thou first and last of fields! king-making Victory?

XVIII

And Harold stands upon this place of skulls,
The grave of France, the deadly Waterloo!
How in an hour the power which gave annuls
Its gifts, transferring fame as fleeting too!
In "pride of place" here last the eagle flew,
Then tore with bloody talon the rent plain,
Pierced by the shaft of banded nations through; 160
Ambition's life and labors all were vain;
He wears the shattered links of the world's broken
    chain.

XIX

Fit retribution! Gaul may champ the bit
And foam in fetters;—but is Earth more free?
Did nations combat to make *one* submit;
Or league to teach all kings true sovereignty?

---

117. **glassed:** made glass-like. 118. **Chaldean:** Babylonians of the Second Empire, noted for their skill in astronomy and astrology. 131. **fit:** mood. Cf. *Macbeth*, III. iv. 21; "Then comes my fit again."

145. **Stop:** Harold is now at the field of Waterloo; the battle had been fought the year before. 153. **king-making:** establishing the kings of Europe more firmly. 158. **"pride of place":** term in falconry meaning the highest point of flight; the "eagle" refers to Napoleon. 162. **chain:** Napoleon was now imprisoned at St. Helena. 163. **Gaul:** France.

What! shall reviving Thraldom again be
The patched-up idol of enlightened days?
Shall we, who struck the Lion down, shall we
Pay the Wolf homage? proffering lowly gaze   170
And servile knees to thrones? No; *prove* before ye
    praise!

XX

If not, o'er one fallen despot boast no more!
In vain fair cheeks were furrowed with hot tears
For Europe's flowers long rooted up before
The trampler of her vineyards; in vain, years
Of death, depopulation, bondage, fears,
Have all been borne, and broken by the accord
Of roused-up millions; all that most endears
Glory, is when the myrtle wreathes a sword
Such as Harmodius drew on Athens' tyrant lord.   180

XXI

There was a sound of revelry by night,
And Belgium's capital had gathered then
Her Beauty and her Chivalry, and bright
The lamps shone o'er fair women and brave men;
A thousand hearts beat happily; and when
Music arose with its voluptuous swell,
Soft eyes looked love to eyes which spake again,
And all went merry as a marriage bell;
But hush! hark! a deep sound strikes like a rising knell!

XXII

Did ye not hear it?—No; 'twas but the wind,   190
Or the car rattling o'er the stony street;
On with the dance! let joy be unconfined;
No sleep till morn, when Youth and Pleasure meet
To chase the glowing Hours with flying feet—

168. enlightened days: those of the eighteenth century.
169. Lion: Napoleon. **170 Wolf:** any of the various victors.
171. prove: expect empirical proof. **180. Harmodius:** whose
dagger, when he slew the tyrant Hipparchus, was hid in
myrtle, the symbol of love. **181. revelry:** A famous ball
was given in Brussels by the Duchess of Richmond on the
eve (June 15, 1815) of the Battle of Quatre-Bras, two days
before the Battle of Waterloo. (It is also one of the famous
episodes in Thackeray's *Vanity Fair*, Chap. XXX.) Throughout the nineteenth century these stanzas on Waterloo were
among the favorite "elocution pieces" recited at school
commencements.

But hark!—that heavy sound breaks in once more,
As if the clouds its echo would repeat;
And nearer, clearer, deadlier than before!
Arm! Arm! it is—it is—the cannon's opening roar!

XXIII

Within a windowed niche of that high hall
Sate Brunswick's fated chieftain; he did hear   200
That sound the first amidst the festival,
And caught its tone with Death's prophetic ear;
And when they smiled because he deemed it near,
His heart more truly knew that peal too well
Which stretched his father on a bloody bier,
And roused the vengeance blood alone could
    quell;
He rushed into the field, and, foremost fighting, fell.

XXIV

Ah! then and there was hurrying to and fro,
And gathering tears, and tremblings of distress,
And cheeks all pale, which but an hour ago   210
Blushed at the praise of their own loveliness;
And there were sudden partings, such as press
The life from out young hearts, and choking sighs
Which ne'er might be repeated; who could guess
If ever more should meet those mutual eyes,
Since upon night so sweet such awful morn could
    rise!

XXV

And there was mounting in hot haste: the steed,
The mustering squadron, and the clattering car,
Went pouring forward with impetuous speed,
And swiftly forming in the ranks of war;   220
And the deep thunder peal on peal afar;
And near, the beat of the alarming drum
Roused up the soldier ere the morning star;
While thronged the citizens with terror dumb,
Or whispering, with white lips—"The foe! They come!
    they come!"

200. fated chieftain: Frederick William, Duke of Brunswick, destined to be killed in the battle of Quatre-Bras; his
father (l. 205) had died a few years before in the Battle of
Auerstädt (1806).

XXVI

And wild and high the *Cameron's Gathering* rose!
The war-note of Lochiel, which Albyn's hills
Have heard, and heard, too, have her Saxon foes:—
How in the noon of night that pibroch thrills,
Savage and shrill! But with the breath which fills
Their mountain-pipe, so fill the mountaineers   231
With the fierce native daring which instils
The stirring memory of a thousand years,
And Evan's, Donald's fame rings in each clansman's
    ears!

XXVII

And Ardennes waves above them her green leaves,
Dewy with Nature's tear-drops, as they pass,
Grieving, if aught inanimate e'er grieves,
Over the unreturning brave,—alas!
Ere evening to be trodden like the grass
Which now beneath them, but above shall grow
In its next verdure, when this fiery mass   241
Of living valor, rolling on the foe
And burning with high hope, shall moulder cold and
    low.

.   .   .

XXXIV

There is a very life in our despair,
Vitality of poison,—a quick root
Which feeds these deadly branches; for it were   300
As nothing did we die; but life will suit
Itself to sorrow's most detested fruit,
Like to the apples on the Dead Sea's shore,
All ashes to the taste. Did man compute
Existence by enjoyment, and count o'er
Such hours 'gainst years of life,—say, would he name
    threescore?

226. **Cameron's Gathering:** the war song of the Cameron
Clan. 227. **Lochiel:** title of the chief of the Camerons.
Albyn: Gaelic name for Scotland. 228. **Saxon:** English.
229. **pibroch:** bagpipe music, especially the battle call of the
pipes (cf. "Lachin y Gair," above, l. 31). 234. **Evan's,
Donald's:** Sir Evan Cameron, who fought for James II,
and his descendant, Donald, who fought for the Young
Pretender. 235. **Ardennes:** the same forest area that served
as a battlefield in the two world wars.

XXXV

The Psalmist numbered out the years of man:
They are enough; and if thy tale be *true*,
Thou, who didst grudge him even that fleeting span,
More than enough, thou fatal Waterloo!   310
Millions of tongues record thee, and anew
Their children's lips shall echo them, and say—
"Here, where the sword united nations drew,
Our countrymen were warring on that day!"
And this is much, and all which will not pass away.

XXXVI

There sunk the greatest, nor the worst of men,
Whose spirit, antithetically mixt,
One moment of the mightiest, and again
On little objects with like firmness fixt;
Extreme in all things! hadst thou been betwixt,   320
Thy throne had still been thine, or never been;
For daring made thy rise as fall: thou seek'st
Even now to re-assume the imperial mien,
And shake again the world, the thunderer of the scene!

XXXVII

Conqueror and captive of the earth art thou!
She trembles at thee still, and thy wild name
Was ne'er more bruited in men's minds than now
That thou art nothing, save the jest of fame,
Who wooed thee once, thy vassal, and became
The flatterer of thy fierceness, till thou wert   330
A god unto thyself! nor less the same
To the astounded kingdoms all inert,
Who deemed thee for a time whate'er thou didst assert.

XXXVIII

Oh, more or less than man—in high or low,
Battling with nations, flying from the field;
Now making monarchs' necks thy footstool, now
More than thy meanest soldier taught to yield;
An empire thou couldst crush, command, rebuild,
But govern not thy pettiest passion, nor,
However deeply in men's spirits skilled,   340

307. **Psalmist:** Psalms 90:10. 316. **greatest . . . men:**
Napoleon. 317. **antithetically mixt:** composed of opposite
extremes. 327. **bruited:** sounded abroad.

Look through thine own, nor curb the lust of war,
Nor learn that tempted fate will leave the loftiest star.

XXXIX

Yet well thy soul hath brooked the turning tide
With that untaught innate philosophy,
Which, be it wisdom, coldness, or deep pride,
Is gall and wormwood to an enemy.
When the whole host of hatred stood hard by,
To watch and mock thee shrinking, thou hast smiled
With a sedate and all-enduring eye;—
When fortune fled her spoiled and favorite child,
He stood unbowed beneath the ills upon him piled. 351

XL

Sager than in thy fortunes; for in them
Ambition steeled thee on too far to show
That just habitual scorn, which could contemn
Men and their thoughts; 'twas wise to feel, not so
To wear it ever on thy lip and brow,
And spurn the instruments thou wert to use
Till they were turned unto thine overthrow:
'Tis but a worthless world to win or lose;
So hath it proved to thee, and all such lot who choose.

XLI

If, like a tower upon a headland rock,   361
Thou hadst been made to stand or fall alone,
Such scorn of man had helped to brave the shock;
But men's thoughts were the steps which paved thy
          throne,
Their admiration thy best weapon shone;
The part of Philip's son was thine, not then
(Unless aside thy purple had been thrown)
Like stern Diogenes to mock at men;
For sceptred cynics earth were far too wide a den.

XLII

But quiet to quick bosoms is a hell,   370
And *there* hath been thy bane; there is a fire
And motion of the soul which will not dwell
In its own narrow being, but aspire

366. **Philip's son:** Alexander the Great. 368. **Diogenes:**
the famous Greek Cynic.

Beyond the fitting medium of desire;
And, but once kindled, quenchless evermore,
Preys upon high adventure, nor can tire
Of aught but rest; a fever at the core,
Fatal to him who bears, to all who ever bore.

. . .

LXXII

I live not in myself, but I become   680
Portion of that around me; and to me
High mountains are a feeling, but the hum
Of human cities torture: I can see
Nothing to loathe in nature, save to be
A link reluctant in a fleshly chain,
Classed among creatures, when the soul can flee,
And with the sky, the peak, the heaving plain
Of ocean, or the stars, mingle, and not in vain.

LXXIII

And thus I am absorbed, and this is life:
I look upon the peopled desert past,   690
As on a place of agony and strife,
Where, for some sin, to sorrow I was cast,
To act and suffer, but remount at last
With a fresh pinion; which I feel to spring,
Though young, yet waxing vigorous as the blast
Which it would cope with, on delighted wing,
Spurning the clay-cold bonds which round our being
          cling.

LXXIV

And when, at length, the mind shall be all free
From what it hates in this degraded form,
Reft of its carnal life, save what shall be   700
Existent happier in the fly and worm,—
When elements to elements conform,
And dust is as it should be, shall I not
Feel all I see, less dazzling, but more warm?
The bodiless thought? the spirit of each spot?
Of which, even now, I share at times the immortal lot?

682. **mountains . . . feeling:** Harold is now in Switzerland. Cf. Wordsworth's "Tintern Abbey," above, ll. 77–79:
"the tall rock, / The mountain . . . were then to me /
An appetite; a feeling and a love."

LXXV

Are not the mountains, waves, and skies, a part
Of me and of my soul, as I of them?
Is not the love of these deep in my heart
With a pure passion? should I not contemn   710
All objects, if compared with these? and stem
A tide of suffering, rather than forego
Such feelings for the hard and worldly phlegm
Of those whose eyes are only turned below,
Gazing upon the ground, with thoughts which dare not
    glow?

LXXVI

But this is not my theme; and I return
To that which is immediate, and require
Those who find contemplation in the urn,
To look on one, whose dust was once all fire,
A native of the land where I respire   720
The clear air for a while—a passing guest,
Where he became a being,—whose desire
Was to be glorious; 'twas a foolish quest,
The which to gain and keep, he sacrificed all rest.

LXXVII

Here the self-torturing sophist, wild Rousseau,
The apostle of affliction, he who threw
Enchantment over passion, and from woe
Wrung overwhelming eloquence, first drew
The breath which made him wretched; yet he knew
How to make madness beautiful, and cast   730
O'er erring deeds and thoughts a heavenly hue
Of words, like sunbeams, dazzling as they past
The eyes, which o'er them shed tears feelingly and fast.

LXXVIII

His love was passion's essence:—as a tree
On fire by lightning, with ethereal flame
Kindled he was, and blasted; for to be
Thus, and enamored, were to him the same.
But his was not the love of living dame,
Nor of the dead who rise upon our dreams,
But of ideal beauty, which became   740
In him existence, and o'erflowing teems
Along his burning page, distempered though it seems.

725. **Rousseau:** Jean Jacques Rousseau, 1712–78, who was born in Geneva.

LXXIX

This breathed itself to life in Julie, *this*
Invested her with all that's wild and sweet;
This hallowed, too, the memorable kiss
Which every morn his fevered lip would greet,
From hers, who but with friendship his would meet;
But to that gentle touch through brain and breast
Flashed the thrilled spirit's love-devouring heat;
In that absorbing sigh perchance more blest   750
Than vulgar minds may be with all they seek possest.

LXXX

His life was one long war with self-sought foes,
Or friends by him self-banished; for his mind
Had grown suspicion's sanctuary, and chose,
For its own cruel sacrifice, the kind,
'Gainst whom he raged with fury strange and blind.
But he was phrensied,—wherefore, who may know?
Since cause might be which skill could never find;
But he was phrensied by disease or woe,
To that worst pitch of all, which wears a reasoning
    show.   760

LXXXI

For then he was inspired, and from him came,
As from the Pythian's mystic cave of yore,
Those oracles which set the world in flame,
Nor ceased to burn till kingdoms were no more:
Did he not this for France? which lay before
Bowed to the inborn tyranny of years?
Broken and trembling to the yoke she bore,
Till by the voice of him and his compeers
Roused up to too much wrath which follows o'er-
    grown fears?

LXXXII

They made themselves a fearful monument!   770
The wreck of old opinions—things which grew,
Breathed from the birth of time: the veil they rent,
And what behind it lay, all earth shall view.

743. **Julie:** heroine of Rousseau's *La Nouvelle Héloïse*. 745. **memorable kiss:** Rousseau, for a while, would take a daily walk in order to encounter the Comtesse d'Houdetot "for the sake of the single kiss that was the common salutation of French acquaintance." 762. **Pythian's . . . cave:** the oracle of Apollo at Delphi.

But good with ill they also overthrew,
Leaving but ruins, wherewith to rebuild
Upon the same foundations, and renew
Dungeons and thrones, which the same hour refilled,
As heretofore, because ambition was self-willed.

LXXXIII

But this will not endure, nor be endured!
Mankind have felt their strength, and made it felt.
They might have used it better, but allured   781
By their new vigor, sternly have they dealt
On one another; pity ceased to melt
With her once natural charities. But they,
Who in oppression's darkness caved had dwelt,
They were not eagles, nourished with the day;
What marvel then, at times, if they mistook their
    prey?

LXXXIV

What deep wounds ever closed without a scar?
The heart's bleed longest, and but heal to wear
That which disfigures it; and they who war   790
With their own hopes, and have been vanquished,
    bear
Silence, but not submission: in his lair
Fixed Passion holds his breath, until the hour
Which shall atone for years; none need despair:
It came, it cometh, and will come,—the power
To punish or forgive—in *one* we shall be slower.

. . . .

CXIII

I have not loved the world, nor the world me;
I have not flattered its rank breath, nor bowed 1050
To its idolatries a patient knee,
Nor coined my cheek to smiles, nor cried aloud
In worship of an echo; in the crowd
They could not deem me one of such; I stood
Amongst them, but not of them; in a shroud
Of thoughts which were not their thoughts, and still
    could,
Had I not filed my mind, which thus itself subdued.

1057. filed: defiled. Byron gives a note to *Macbeth*, III. i.
65: "For Banquo's issue have I filed my mind."

CXIV

I have not loved the world, nor the world me,—
But let us part fair foes; I do believe,   1059
Though I have found them not, that there may be
Words which are things, hopes which will not
    deceive,
And virtues which are merciful, nor weave
Snares for the failing; I would also deem
O'er others' griefs that some sincerely grieve;
That two, or one, are almost what they seem,
That goodness is no name, and happiness no dream.

CXV

My daughter! with thy name this song begun;
My daughter! with thy name thus much shall end;
I see thee not, I hear thee not, but none
Can be so wrapt in thee; thou art the friend   1070
To whom the shadows of far years extend:
Albeit my brow thou never shouldst behold,
My voice shall with thy future visions blend,
And reach into thy heart, when mine is cold,
A token and a tone, even from thy father's mould.

CXVI

To aid thy mind's development, to watch
Thy dawn of little joys, to sit and see
Almost thy very growth, to view thee catch
Knowledge of objects,—wonders yet to thee!
To hold thee lightly on a gentle knee,   1080
And print on thy soft cheek a parent's kiss,—
This, it should seem, was not reserved for me;
Yet this was in my nature: as it is,
I know not what is there, yet something like to this.

CXVII

Yet, though dull hate as duty should be taught,
I know that thou wilt love me; though my name
Should be shut from thee, as a spell still fraught
With desolation, and a broken claim:
Though the grave closed between us,—'twere the
    same,
I know that thou wilt love me; though to drain
My blood from out thy being were an aim,   1091
And an attainment,—all would be in vain,—
Still thou wouldst love me, still that more than life
    retain.

CXVIII

The child of love, though born in bitterness,
And nurtured in convulsion,—of thy sire
These were the elements, and thine no less.
As yet such are around thee, but thy fire
Shall be more tempered, and thy hope far higher.
Sweet be thy cradled slumbers! O'er the sea
And from the mountains where I now respire,
Fain would I waft such blessing upon thee,   1101
As, with a sign, I deem thou might'st have been to me.

from CANTO IV

I

I stood in Venice, on the "Bridge of Sighs";
A palace and a prison on each hand:
I saw from out the wave her structures rise
As from the stroke of the enchanter's wand:
A thousand years their cloudy wings expand
Around me, and a dying Glory smiles
O'er the far times, when many a subject land
Looked to the winged Lion's marble piles,
Where Venice sate in state, throned on her hundred
    isles!

II

She looks a sea Cybele, fresh from ocean,   10
Rising with her tiara of proud towers
At airy distance, with majestic motion,
A ruler of the waters and their powers:
And such she was;—her daughters had their dowers
From spoils of nations, and the exhaustless East
Poured in her lap all gems in sparkling showers.
In purple was she robed, and of her feast
Monarchs partook, and deemed their dignity increased.

III

In Venice Tasso's echoes are no more,
And silent rows the songless gondolier;   20
Her palaces are crumbling to the shore,
And music meets not always now the ear:
Those days are gone—but Beauty still is here.
States fall, arts fade—but Nature doth not die,
Nor yet forget how Venice once was dear,
The pleasant place of all festivity,
The revel of the earth, the masque of Italy!

IV

But unto us she hath a spell beyond
Her name in story, and her long array
Of mighty shadows, whose dim forms despond   30
Above the dogeless city's vanished sway;
Ours is a trophy which will not decay
With the Rialto; Shylock and the Moor,
And Pierre, cannot be swept or worn away—
The keystones of the arch! though all were o'er,
For us repeopled were the solitary shore.

V

The beings of the mind are not of clay;
Essentially immortal, they create
And multiply in us a brighter ray
And more beloved existence: that which Fate   40
Prohibits to dull life, in this our state
Of mortal bondage, by these spirits supplied,
First exiles, then replaces what we hate;
Watering the heart whose early flowers have died,
And with a fresher growth replenishing the void.

. . .

CANTO IV. **1.** **"Bridge of Sighs"**: the famous covered bridge over which the prisoners walked from the Doge's Palace to the San Marco Prison. **8. winged Lion:** emblem of St. Mark, patron saint of Venice. **10. Cybele:** mother of the gods, often pictured as wearing a crown (or "tiara," l. 11) of towers. Speaking of the image, Byron writes: "Sabellicus [a Renaissance historian of Venice] describing the appearance of Venice, has made use of the above image, which would not be poetical were it not true."

19-20. Tasso's . . . gondolier: The gondoliers had once the custom of reciting lines from Tasso's *Jerusalem Delivered*. **27. masque of Italy:** the lavish entertainment center of Italy. **31. dogeless:** The last doge had been deposed by Napoleon. **33. Rialto:** site of the Venetian exchange and center of commercial activity, and associated with the characters mentioned above—Shylock, Othello ("the Moor"), and Pierre (in Thomas Otway's play *Venice Preserved*). **36. were:** would be.

XIII

Before St. Mark still glow his steeds of brass,
Their gilded collars glittering in the sun;   110
But is not Doria's menace come to pass?
Are they not *bridled?*—Venice, lost and won,
Here thirteen hundred years of freedom done,
Sinks, like a seaweed, into whence she rose!
Better be whelmed beneath the waves, and shun,
Even in destruction's depth, her foreign foes,
From whom submission wrings an infamous repose.

XIV

In youth she was all glory,—a new Tyre;
Her very by-word sprung from victory,
The "Planter of the Lion," which through fire   120
And blood she bore o'er subject earth and sea;
Though making many slaves, herself still free,
And Europe's bulwark 'gainst the Ottomite;
Witness Troy's rival, Candia! Vouch it, ye
Immortal waves that saw Lepanto's fight!
For ye are names no time nor tyranny can blight.

XV

Statues of glass—all shivered—the long file
Of her dead Doges are declined to dust;
But where they dwelt, the vast and sumptuous pile
Bespeaks the pageant of their splendid trust;   130
Their sceptre broken, and their sword in rust,
Have yielded to the stranger: empty halls,
Thin streets, and foreign aspects, such as must
Too oft remind her who and what enthrals,
Have flung a desolate cloud o'er Venice' lovely walls.

109–11. steeds . . . menace: the bronze horses at the front of St. Mark's Church; Doria, the Genoese commander, had vowed in 1379 to "bridle" them. 120. "Planter . . . Lion": Byron assumes that the nickname of the Venetians (Pantaloni) was given them because of their enterprise as colonists: wherever they arrived they would "plant" their flag, bearing the emblem of the lion (see l. 8, above). The word, however, probably comes from the name of St. Pantaleon. 124. Candia: in Crete. Candia was defended by Venice for twenty-four years against the Turks ("the Ottomite"); "Troy's rival" since Troy held out against the Greeks for only ten years. 125. Lepanto's fight: naval victory against the Turks (1571) in which the losses of the Turks were more than three times those of the Venetians and the other Europeans.

XVI

When Athens' armies fell at Syracuse,
And fettered thousands bore the yoke of war,
Redemption rose up in the Attic muse,
Her voice their only ransom from afar:
See! as they chant the tragic hymn, the car   140
Of the o'ermastered victor stops, the reins
Fall from his hands, his idle scimitar
Starts from its belt—he rends his captive's chains,
And bids him thank the bard for freedom and his strains.

XVII

Thus, Venice, if no stronger claim were thine,
Were all thy proud historic deeds forgot,
Thy choral memory of the bard divine,
Thy love of Tasso, should have cut the knot
Which ties thee to thy tyrants; and thy lot
Is shameful to the nations,—most of all,   150
Albion! to thee: the ocean queen should not
Abandon ocean's children; in the fall
Of Venice, think of thine, despite thy watery wall.

XVIII

I loved her from my boyhood; she to me
Was as a fairy city of the heart,
Rising like water-columns from the sea,
Of joy the sojourn, and of wealth the mart;
And Otway, Radcliffe, Schiller, Shakspeare's art,
Had stamped her image in me, and even so,
Although I found her thus, we did not part;   160
Perchance even dearer in her day of woe,
Than when she was a boast, a marvel, and a show.

XIX

I can repeople with the past—and of
The present there is still for eye and thought,
And meditation chastened down, enough;
And more, it may be, than I hoped or sought;
And of the happiest moments which were wrought
Within the web of my existence, some
From thee, fair Venice! have their colors caught:

136–38. Syracuse . . . muse: When the Athenians were defeated at Syracuse, Sicily, those who were able to recite passages from Euripides were allowed to go free. 158. Otway . . . Shakspeare's: See n. 33, above. Byron also refers to Schiller's *Ghost-Seer* and to Ann Radcliffe, the scene of whose novels was often laid in Italy.

There are some feelings time cannot benumb   170
Nor torture shake, or mine would now be cold and
    dumb.

. . .

LXXVIII

Oh Rome! my country! city of the soul!
The orphans of the heart must turn to thee,
Lone mother of dead empires! and control
In their shut breasts their petty misery.
What are our woes and sufferance? Come and see
The cypress, hear the owl, and plod your way
O'er steps of broken thrones and temples, Ye!   700
Whose agonies are evils of a day—
A world is at our feet as fragile as our clay.

LXXIX

The Niobe of nations! there she stands,
Childless and crownless, in her voiceless woe;
An empty urn within her withered hands,
Whose holy dust was scattered long ago;
The Scipios' tomb contains no ashes now;
The very sepulchres lie tenantless
Of thy heroic dwellers: dost thou flow,
Old Tiber! through a marble wilderness?   710
Rise, with thy yellow waves, and mantle her distress.

LXXX

The Goth, the Christian, Time, War, Flood, and
    Fire,
Have dealt upon the seven-hilled city's pride;
She saw her glories star by star expire,
And up the steep barbarian monarchs ride,
Where the car climbed the Capitol; far and wide
Temple and tower went down, nor left a site:
Chaos of ruins! who shall trace the void,

694. Rome: Throughout the stanzas on Rome, Byron is
also thinking of the theme and sweep of Edward Gibbon's
great history *The Decline and Fall of the Roman Empire.*
703. Niobe: typifying perpetual grief for what is lost.
Because she boasted of her superiority, as a mother of twelve,
to Latona, mother of Apollo and Artemis, her children were
slain. Changed into a stone, she continued to weep in streams
flowing down the rock. **707. Scipio's tomb:** discovered
and despoiled in an excavation of 1780. **716. car . . . Capitol:** where the triumphal chariot (in the celebrations of the
Roman Triumphs) climbed up from the Sacred Way in the
Forum to the Capitoline Hill.

O'er the dim fragments cast a lunar light,   719
And say, "here was, or is," where all is doubly night?

. . . .

XCIII

What from this barren being do we reap?
Our senses narrow, and our reason frail,   830
Life short, and truth a gem which loves the deep,
And all things weighed in custom's falsest scale;
Opinion an omnipotence,—whose veil
Mantles the earth with darkness, until right
And wrong are accidents, and men grow pale
Lest their own judgments should become too bright,
And their free thoughts be crimes, and earth have too
    much light.

XCIV

And thus they plod in sluggish misery,
Rotting from sire to son, and age to age,
Proud of their trampled nature, and so die,   840
Bequeathing their hereditary rage
To the new race of inborn slaves, who wage
War for their chains, and rather than be free,
Bleed gladiator-like, and still engage
Within the same arena where they see
Their fellows fall before, like leaves of the same tree.

XCV

I speak not of men's creeds—they rest between
Man and his Maker—but of things allowed,
Averred, and known, and daily, hourly seen:—
The yoke that is upon us doubly bowed,   850
And the intent of tyranny avowed,
The edict of Earth's rulers, who are grown
The apes of him who humbled once the proud,
And shook them from their slumbers on the throne—
Too glorious, were this all his mighty arm had done.

XCVI

Can tyrants but by tyrants conquered be,
And Freedom find no champion and no child
Such as Columbia saw arise when she
Sprung forth a Pallas, armed and undefiled?

853. him . . . proud: Napoleon. **859. Sprung . . . Pallas:**
Pallas Athena, in Greek mythology, sprang full grown from
the forehead of Zeus.

Or must such minds be nourished in the wild,   860
Deep in the unpruned forest, 'midst the roar
Of cataracts, where nursing Nature smiled
On infant Washington? Has Earth no more
Such seeds within her breast, or Europe no such shore?

XCVII

But France got drunk with blood to vomit crime,
And fatal have her Saturnalia been
To Freedom's cause, in every age and clime;
Because the deadly days which we have seen,
And vile Ambition, that built up between
Man and his hopes an adamantine wall,   870
And the base pageant last upon the scene,
Are grown the pretext for the eternal thrall
Which nips life's tree, and dooms man's worst—his
    second fall.

XCVIII

Yet, Freedom! yet thy banner, torn but flying,
Streams like the thunder-storm *against* the wind;
Thy trumpet voice, though broken now and dying,
The loudest still the tempest leaves behind;
Thy tree hath lost its blossoms, and the rind,
Chopped by the axe, looks rough and little worth,
But the sap lasts, and still the seed we find   880
Sown deep, even in the bosom of the North:
So shall a better Spring less bitter fruit bring forth.

.   .   .

CXLIII

A ruin—yet what ruin! from its mass
Walls, palaces, half-cities, have been reared;   1280
Yet oft the enormous skeleton ye pass,
And marvel where the spoil could have appeared.
Hath it indeed been plundered, or but cleared?
Alas! developed, opens the decay,
When the colossal fabric's form is neared:

It will not bear the brightness of the day,
Which streams too much on all, years, man, have reft
    away.

CXLIV

But when the rising moon begins to climb
Its topmost arch, and gently pauses there;
When the stars twinkle through the loops of time,
And the low night-breeze waves along the air   1291
The garland-forest, which the gray walls wear,
Like laurels on the bald first Cæsar's head;
When the light shines serene but doth not glare,
Then in this magic circle raise the dead:
Heroes have trod this spot—'tis on their dust ye tread.

CXLV

"While stands the Coliseum, Rome shall stand;
When falls the Coliseum, Rome shall fall;
And when Rome falls—the World." From our own
    land
Thus spake the pilgrims o'er this mighty wall   1300
In Saxon times, which we are wont to call
Ancient; and these three mortal things are still
On their foundations, and unaltered all;
Rome and her Ruin past Redemption's skill,
The World, the same wide den—of thieves, or what
    ye will.

.   .   .

CLXXVII

Oh! that the desert were my dwelling-place,
With one fair Spirit for my minister,
That I might all forget the human race,
And, hating no one, love but only her!

1293. laurels . . . head: Byron added the following note:
"Suetonius informs us that Julius Caesar was particularly
gratified by that decree of the senate which enabled him
to wear a wreath of laurel on all occasions. He was anxious,
not to show that he was the conqueror of the world, but
to hide that he was bald," (Suetonius, *Lives of the Caesars*,
I. 45). **1297–99. "While . . . World":** Byron states: "This
is quoted in *The Decline and Fall of the Roman Empire* as a
proof that the Coliseum was entire when seen by the Anglo-
Saxon pilgrims at the end of the seventh or the beginning of
the eighth century." (See Gibbon, Chap. LXXI.) Gibbon
cites the passage from Bede's *Glossarium*, who heard it
from Anglo-Saxon pilgrims to Rome. **1586. one fair Spirit:**
possibly a reference to his sister Augusta (see "Stanzas to
Augusta," above).

866. her Saturnalia: the Reign of Terror. **871. base pageant:**
the pageant (1815) of the Congress of Vienna, the Holy
Alliance, and the Second Treaty of Paris, all of which secured
the power of the sovereigns of Europe still more firmly (cf.
the phrase on Waterloo, "king-making Victory," above, III.
153). **881. North:** England.

Ye elements!—in whose ennobling stir
I feel myself exalted—Can ye not   1590
Accord me such a being? Do I err
In deeming such inhabit many a spot?
Though with them to converse can rarely be our lot.

CLXXVIII

There is a pleasure in the pathless woods,
There is a rapture on the lonely shore,
There is society, where none intrudes,
By the deep sea, and music in its roar:
I love not man the less, but Nature more,
From these our interviews, in which I steal
From all I may be, or have been before,   1600
To mingle with the Universe, and feel
What I can ne'er express, yet cannot all conceal.

CLXXIX

Roll on, thou deep and dark blue Ocean—roll!
Ten thousand fleets sweep over thee in vain;
Man marks the earth with ruin—his control
Stops with the shore; upon the watery plain
The wrecks are all thy deed, nor doth remain
A shadow of man's ravage, save his own,
When, for a moment, like a drop of rain,
He sinks into thy depths with bubbling groan,   1610
Without a grave, unknelled, uncoffined, and unknown.

CLXXX

His steps are not upon thy paths,—thy fields
Are not a spoil for him,—thou dost arise
And shake him from thee; the vile strength he wields
For earth's destruction thou dost all despise,
Spurning him from thy bosom to the skies,
And send'st him, shivering in thy playful spray
And howling, to his gods, where haply lies
His petty hope in some near port or bay,   1619
And dashest him again to earth:—there let him lay.

CLXXXI

The armaments which thunderstrike the walls
Of rock-built cities, bidding nations quake,
And monarchs tremble in their capitals,
The oak leviathans, whose huge ribs make
Their clay creator the vain title take
Of lord of thee, and arbiter of war—
These are thy toys, and, as the snowy flake,
They melt into thy yeast of waves, which mar
Alike the Armada's pride or spoils of Trafalgar.   1629

CLXXXII

Thy shores are empires, changed in all save thee—
Assyria, Greece, Rome, Carthage, what are they?
Thy waters washed them power while they were
          free,
And many a tyrant since; their shores obey
The stranger, slave, or savage; their decay
Has dried up realms to deserts:—not so thou;—
Unchangeable, save to thy wild waves' play,
Time writes no wrinkle on thine azure brow:
Such as creation's dawn beheld, thou rollest now.

CLXXXIII

Thou glorious mirror, where the Almighty's form
Glasses itself in tempests; in all time,—   1640
Calm or convulsed, in breeze, or gale, or storm,
Icing the pole, or in the torrid clime
Dark-heaving—boundless, endless, and sublime,
The image of Eternity, the throne
Of the Invisible; even from out thy slime
The monsters of the deep are made; each zone
Obeys thee; thou goest forth, dread, fathomless, alone.

CLXXXIV

And I have loved thee, Ocean! and my joy
Of youthful sports was on thy breast to be
Borne, like thy bubbles, onward: from a boy   1650
I wantoned with thy breakers—they to me

1620. **lay:** lie. This famous solecism was quite conscious on the part of Byron, who enjoyed affecting an aristocratic negligence that would cause surprise in readers less confident of their social position.

1624. **oak leviathans:** warships. 1629. **Armada's . . . Trafalgar:** Half of the Spanish Armada (1588) was lost in storm; so were many of the French ships taken by Nelson at Trafalgar (1805). 1640. **Glasses:** mirrors.

Were a delight; and if the freshening sea
Made them a terror—'twas a pleasing fear,
For I was as it were a child of thee,
And trusted to thy billows far and near,
And laid my hand upon thy mane—as I do here.

CLXXXV

My task is done, my song hath ceased, my theme
Has died into an echo; it is fit
The spell should break of this protracted dream.
The torch shall be extinguished which hath lit   1660
My midnight lamp—and what is writ, is writ;
Would it were worthier! but I am not now
That which I have been—and my visions flit
Less palpably before me—and the glow
Which in my spirit dwelt is fluttering, faint, and low.

CLXXXVI

Farewell! a word that must be, and hath been—
A sound which makes us linger;—yet—farewell!
Ye! who have traced the Pilgrim to the scene
Which is his last, if in your memories dwell
A thought which once was his, if on ye swell   1670
A single recollection, not in vain
He wore his sandal-shoon and scallop-shell;
Farewell! with *him* alone may rest the pain,
If such there were—with *you*, the moral of his strain.

                                        1817 (1818)

MANFRED

A DRAMATIC POEM

ᵔᵛᵔ

Written in 1816 and early 1817, *Manfred* follows
Prometheus and *Childe Harold* in again presenting the
"Byronic hero," now more frankly an outcast tortured
by guilt and remorse for some mysterious and unfor-
givable crime. *Manfred* was conceived, said Byron to his
publisher John Murray, simply as "a poem, for it is *no
drama*." In fact Byron said that he purposely rendered it
"quite impossible for the stage, for which my intercourse
with Drury Lane has given me the greatest contempt."

1672. sandal-shoon . . . scallop-shell: traditional emblems
of the pilgrim—sandals for travel on land, and a scallop-shell
(worn on the hat) symbolizing travel by sea.

If the characters are not very distinct, even for a "closet
drama" or "reading play," Byron's justification is that
this is a special form of "dramatic poem"—an extended
soliloquy, for the purpose of psychological revelation, in
which the "characters" are openly conceived as expressions of inner debate and varying moods within Manfred
himself. From the beginning critics have noted resemblances between *Manfred* and Marlowe's *The Tragical
History of Doctor Faustus* (which Byron said he had not
read; see his letter to Murray, below, Oct. 12, 1817),
Goethe's *Faust*, and the *Prometheus* of Aeschylus. But
Goethe, in his review of *Manfred*, paid special tribute to
the originality of the work:

> Byron's tragedy, *Manfred*, was to me a wonderful
> phenomenon, and one that closely touched me. This
> singular intellectual poet has taken my *Faustus* to him-
> self, and extracted from it the strangest nourishment
> for his hypochondriac humor. He has made use of the
> impelling principles in his own way, for his own pur-
> poses, so that no one of them remains the same; and it
> is particularly on this account that I cannot enough
> admire his genius. The whole is in this way so com-
> pletely formed anew that it would be an interesting
> task for the critic to point out, not only the alterations
> he has made, but their degree of resemblance with,
> or dissimilarity to, the original; in the course of which
> I cannot deny that the gloomy heat of an unbounded
> and exuberant despair becomes at last oppressive to us.
> Yet is the dissatisfaction we feel always connected with
> esteem and admiration.

ᵔᵛᵔ

"There are more things in heaven and earth, Horatio,
Than are dreamt of in your philosophy."

DRAMATIS PERSONÆ

MANFRED
CHAMOIS HUNTER
ABBOT OF ST. MAURICE
MANUEL
HERMAN
WITCH OF THE ALPS
ARIMANES
NEMESIS
THE DESTINIES
SPIRITS, &C.

The SCENE *of the Drama is amongst the Higher Alps
—partly in the Castle of Manfred, and partly in the
Mountains.*

MANFRED. **Motto:** *Hamlet,* I. v. 166–67.

ACT I

SCENE I

MANFRED *alone.—Scene, a Gothic Gallery.*
Time, Midnight.

MANFRED.   The lamp must be replenished, but even
    then
It will not burn so long as I must watch:
My slumbers—if I slumber—are not sleep,
But a continuance of enduring thought,
Which then I can resist not: in my heart
There is a vigil, and these eyes but close
To look within; and yet I live, and bear
The aspect and the form of breathing men.
But grief should be the instructor of the wise;
Sorrow is knowledge: they who know the most   10
Must mourn the deepest o'er the fatal truth,
The tree of knowledge is not that of life.
Philosophy and science, and the springs
Of wonder, and the wisdom of the world,
I have essayed, and in my mind there is
A power to make these subject to itself—
But they avail not: I have done men good,
And I have met with good even among men—
But this availed not: I have had my foes,
And none have baffled, many fallen before me—   20
But this availed not:—Good, or evil, life,
Powers, passions, all I see in other beings,
Have been to me as rain unto the sands,
Since that all-nameless hour. I have no dread,
And feel the curse to have no natural fear,
Nor fluttering throb, that beats with hopes or wishes,
Or lurking love of something on the earth.
Now to my task.—
            Mysterious agency!
Ye spirits of the unbounded universe!
Whom I have sought in darkness and in light—   30
Ye, who do compass earth about, and dwell
In subtler essence—ye, to whom the tops
Of mountains inaccessible are haunts,
And earth's and ocean's caves familiar things—
I call upon ye by the written charm
Which gives me power upon you—Rise! Appear!
        [*A pause.*]
They come not yet.—Now by the voice of him
Who is the first among you—by this sign,
Which makes you tremble—by the claims of him

Who is undying,—Rise! Appear——Appear!   40
        [*A pause.*]
If it be so—Spirits of earth and air,
Ye shall not thus elude me: by a power,
Deeper than all yet urged, a tyrant-spell,
Which had its birthplace in a star condemned,
The burning wreck of a demolished world,
A wandering hell in the eternal space
By the strong curse which is upon my soul,
The thought which is within me and around me,
I do compel ye to my will—Appear!

    [*A star is seen at the darker end of the gallery:*
    it is stationary; and a voice is heard singing.]

FIRST SPIRIT

Mortal! to thy bidding bowed,   50
From my mansion in the cloud,
Which the breath of twilight builds
And the summer's sunset gilds
With the azure and vermilion,
Which is mixed for my pavilion;
Though thy quest may be forbidden,
On a star-beam I have ridden:
To thine adjuration bowed,
Mortal—be thy wish avowed!

VOICE OF THE SECOND SPIRIT

Mont Blanc is the monarch of mountains;   60
    They crowned him long ago
On a throne of rocks, in a robe of clouds,
    With a diadem of snow.
Around his waist are forests braced,
    The avalanche in his hand;
But ere it fall, that thundering ball
    Must pause for my command.
The glacier's cold and restless mass
    Moves onward day by day;
But I am he who bids it pass,   70
    Or with its ice delay.
I am the spirit of the place,
    Could make the mountain bow
And quiver to his caverned base—
    And what with me would *thou?*

VOICE OF THE THIRD SPIRIT

In the blue depth of the waters,
    Where the wave hath no strife,

Where the wind is a stranger,
   And the sea-snake hath life,
Where the mermaid is decking   80
   Her green hair with shells,
Like the storm on the surface
   Came the sound of thy spells;
O'er my calm Hall of Coral
   The deep echo rolled—
To the Spirit of Ocean
   Thy wishes unfold!

FOURTH SPIRIT

Where the slumbering earthquake
   Lies pillowed on fire,
And the lakes of bitumen   90
   Rise boilingly higher;
Where the roots of the Andes
   Strike deep in the earth,
As their summits to heaven
   Shoot soaringly forth;
I have quitted my birthplace
   Thy bidding to bide—
Thy spell hath subdued me,
   Thy will be my guide!

FIFTH SPIRIT

I am the rider of the wind,   100
   The stirrer of the storm;
The hurricane I left behind
   Is yet with lightning warm;
To speed to thee, o'er shore and sea
   I swept upon the blast:
The fleet I met sailed well, and yet
   'Twill sink ere night be past.

SIXTH SPIRIT

My dwelling is the shadow of the night,
Why doth thy magic torture me with light?

SEVENTH SPIRIT

The star which rules thy destiny   110
Was ruled, ere earth began, by me:
It was a world as fresh and fair
As e'er revolved round sun in air;
Its course was free and regular,
Space bosomed not a lovelier star.

The hour arrived—and it became
A wandering mass of shapeless flame,
A pathless comet, and a curse,
The menace of the universe;
Still rolling on with innate force,   120
Without a sphere, without a course,
A bright deformity on high,
The monster of the upper sky!
And thou! beneath its influence born—
Thou worm! whom I obey and scorn—
Forced by a power (which is not thine,
And lent thee but to make thee mine)
For this brief moment to descend,
Where these weak spirits round thee bend
And parley with a thing like thee——   130
What wouldst thou, child of clay! with me?

THE SEVEN SPIRITS

Earth, ocean, air, night, mountains, winds, thy star,
   Are at thy beck and bidding, child of clay!
Before thee at thy quest their spirits are—
   What wouldst thou with us, son of mortals—say?
   MAN.   Forgetfulness—
   FIRST SPIRIT.   Of what—of whom—and why?
   MAN.   Of that which is within me; read it there—
Ye know it, and I cannot utter it.
   SPIRIT.   We can but give thee that which we possess:
Ask of us subjects, sovereignty, the power   141
O'er earth—the whole, or portion—or a sign
Which shall control the elements, whereof
We are the dominators,—each and all,
These shall be thine.
   MAN.   Oblivion, self-oblivion.
Can ye not wring from out the hidden realms
Ye offer so profusely what I ask?
   SPIRIT.   It is not in our essence, in our skill;
But—thou may'st die.
   MAN.   Will death bestow it on me?
   SPIRIT.   We are immortal, and do not forget;   150
We are eternal; and to us the past
Is, as the future, present. Art thou answered?
   MAN.   Ye mock me—but the power which brought
      ye here
Hath made you mine. Slaves, scoff not at my will!
The mind, the spirit, the Promethean spark,
The lightning of my being, is as bright,
Pervading, and far darting as your own,
And shall not yield to yours, though cooped in clay!
Answer, or I will teach you what I am.

SPIRIT.   We answer as we answered; our reply   160
Is even in thine own words.
    MAN.   Why say ye so?
    SPIRIT.   If, as thou say'st, thine essence be as ours,
We have replied in telling thee, the thing
Mortals call death hath naught to do with us.
    MAN.   I then have called ye from your realms in
      vain;
Ye cannot, or ye will not, aid me.
    SPIRIT.   Say,
What we possess we offer; it is thine:
Bethink ere thou dismiss us; ask again—
Kingdom, and sway, and strength, and length of days—
    MAN.   Accursed! what have I to do with days?   170
They are too long already.—Hence—begone!
    SPIRIT.   Yet pause: being here, our will would do
      thee service;
Bethink thee, is there then no other gift
Which we can make not worthless in thine eyes?
    MAN.   No, none: yet stay—one moment, ere we
      part,
I would behold ye face to face. I hear
Your voices, sweet and melancholy sounds,
As music on the waters; and I see
The steady aspect of a clear large star;
But nothing more. Approach me as ye are,   180
Or one, or all, in your accustomed forms.
    SPIRIT.   We have no forms, beyond the elements
Of which we are the mind and principle:
But choose a form—in that we will appear.
    MAN.   I have no choice; there is no form on earth
Hideous or beautiful to me. Let him,
Who is most powerful of ye, take such aspect
As unto him may seem most fitting—Come!
    SEVENTH SPIRIT (*appearing in the shape of a beautiful
      female figure*).   Behold!
    MAN.   Oh God! if it be thus, and *thou*
Art not a madness and a mockery,   190
I yet might be most happy, I will clasp thee,
And we again will be—
           [*The figure vanishes.*]
              My heart is crushed!
    [MANFRED *falls senseless.*]

(*A voice is heard in the Incantation which follows.*)

When the moon is on the wave,
  And the glow-worm in the grass,
And the meteor on the grave,
  And the wisp on the morass;

When the falling stars are shooting,
And the answered owls are hooting,
And the silent leaves are still
In the shadow of the hill,   200
Shall my soul be upon thine,
With a power and with a sign.

Though thy slumber may be deep,
Yet thy spirit shall not sleep;
There are shades that will not vanish,
There are thoughts thou canst not banish;
By a power to thee unknown,
Thou canst never be alone;
Thou art wrapt as with a shroud,
Thou art gathered in a cloud;   210
And forever shalt thou dwell
In the spirit of this spell.

Though thou seest me not pass by,
Thou shalt feel me with thine eye
As a thing that, though unseen,
Must be near thee, and hath been;
And when in that secret dread
Thou hast turned around thy head,
Thou shalt marvel I am not
As thy shadow on the spot,   220
And the power which thou dost feel
Shall be what thou must conceal.

And a magic voice and verse
Hath baptized thee with a curse;
And a spirit of the air
Hath begirt thee with a snare;
In the wind there is a voice
Shall forbid thee to rejoice;
And to thee shall night deny
All the quiet of her sky;   230
And the day shall have a sun,
Which shall make thee wish it done.

From thy false tears I did distil
An essence which hath strength to kill;
From thy own heart I then did wring
The black blood in its blackest spring;
From thy own smile I snatched the snake,
For there it coiled as in a brake;
From thy own lip I drew the charm
Which gave all these their chiefest harm;   240
In proving every poison known,
I found the strongest was thine own.

ACT I. **Sc. i. 238. brake:** thicket.

By thy cold breast and serpent smile,
By thy unfathomed gulfs of guile,
By that most seeming virtuous eye,
By thy shut soul's hypocrisy;
By the perfection of thine art
Which passed for human thine own heart;
By thy delight in others' pain,
And by thy brotherhood of Cain,   250
I call upon thee! and compel
Thyself to be thy proper hell!

And on thy head I pour the vial
Which doth devote thee to this trial;
Nor to slumber, nor to die,
Shall be in thy destiny;
Though thy death shall still seem near
To thy wish, but as a fear;
Lo! the spell now works around thee,
And the clankless chain hath bound thee;   260
O'er thy heart and brain together
Hath the word been passed—now wither!

SCENE II

The Mountain of the Jungfrau.—Time, Morning.—

MANFRED *alone upon the Cliffs.*

MAN.   The spirits I have raised abandon me,
The spells which I have studied baffle me,
The remedy I recked of tortured me;
I lean no more on superhuman aid;
It hath no power upon the past, and for
The future, till the past be gulfed in darkness,
It is not of my search.—My mother earth!
And thou fresh breaking day, and you, ye mountains,
Why are ye beautiful? I cannot love ye.
And thou, the bright eye of the universe,   10
That openest over all, and unto all
Art a delight—thou shin'st not on my heart.
And you, ye crags, upon whose extreme edge
I stand, and on the torrent's brink beneath
Behold the tall pines dwindled as to shrubs
In dizziness of distance; when a leap,
A stir, a motion, even a breath, would bring
My breast upon its rocky bosom's bed

To rest forever—wherefore do I pause?
I feel the impulse—yet I do not plunge;   20
I see the peril—yet do not recede;
And my brain reels—and yet my foot is firm:
There is a power upon me which withholds,
And makes it my fatality to live;
If it be life to wear within myself
This barrenness of spirit, and to be
My own soul's sepulchre, for I have ceased
To justify my deeds unto myself—
The last infirmity of evil. Ay,
Thou winged and cloud-cleaving minister,   30
    [*An eagle passes.*]
Whose happy flight is highest into heaven,
Well may'st thou swoop so near me—I should be
Thy prey, and gorge thine eaglets; thou art gone
Where the eye cannot follow thee; but thine
Yet pierces downward, onward, or above,
With a pervading vision.—Beautiful!
How beautiful is all this visible world!
How glorious in its action and itself!
But we, who name ourselves its sovereigns, we,
Half dust, half deity, alike unfit   40
To sink or soar, with our mixed essence make
A conflict of its elements, and breathe
The breath of degradation and of pride,
Contending with low wants and lofty will,
Till our mortality predominates,
And men are—what they name not to themselves,
And trust not to each other. Hark! the note,
    [*The Shepherd's pipe in the distance is heard.*]
The natural music of the mountain reed—
For here the patriarchal days are not
A pastoral fable—pipes in the liberal air,   50
Mixed with the sweet bells of the sauntering herd;
My soul would drink these echoes. Oh, that I were
The viewless spirit of a lovely sound,
A living voice, a breathing harmony,
A bodiless enjoyment—born and dying
With the blest tone which made me!

Enter from below a CHAMOIS HUNTER.
CHAMOIS HUNTER.   Even so
This way the chamois leapt: her nimble feet
Have baffled me; my gains today will scarce
Repay my break-neck travail.—What is here?
Who seems not of my trade, and yet hath reached   60

---

252. Thyself . . . hell: Cf. Satan in *Paradise Lost*, I.
254–55: "The mind is its own place, and in itself / Can make
a Heaven of Hell, a Hell of Heaven."

Sc. ii. 37–38. beautiful . . . glorious: an echo of *Hamlet*,
II. ii. 286–87.

A height which none even of our mountaineers,
Save our best hunters, may attain: his garb
Is goodly, his mien manly, and his air
Proud as a free-born peasant's, at this distance:
I will approach him nearer.

    MAN. (*not perceiving the other*).   To be thus—
Gray-haired with anguish, like these blasted pines,
Wrecks of a single winter, barkless, branchless,
A blighted trunk upon a cursed root,
Which but supplies a feeling to decay—
And to be thus, eternally but thus,   70
Having been otherwise! Now furrowed o'er
With wrinkles, ploughed by moments,—not by
    years,—
And hours, all tortured into ages—hours
Which I outlive!—Ye toppling crags of ice!
Ye avalanches, whom a breath draws down
In mountainous o'erwhelming, come and crush me!
I hear ye momently above, beneath,
Crash with a frequent conflict; but ye pass,
And only fall on things that still would live;
On the young flourishing forest, or the hut   80
And hamlet of the harmless villager.

    C. HUN.   The mists begin to rise from up the valley;
I'll warn him to descend, or he may chance
To lose at once his way and life together.

    MAN.   The mists boil up around the glaciers; clouds
Rise curling fast beneath me, white and sulphury,
Like foam from the roused ocean of deep hell,
Whose every wave breaks on a living shore,
Heaped with the damned like pebbles.—I am giddy.

    C. HUN.   I must approach him cautiously; if near,
A sudden step will startle him, and he   91
Seems tottering already.

    MAN.   Mountains have fallen,
Leaving a gap in the clouds, and with the shock
Rocking their Alpine brethren; filling up
The ripe green valleys with destruction's splinters;
Damming the rivers with a sudden dash,
Which crushed the waters into mist and made
Their fountains find another channel—thus,
Thus, in its old age, did Mount Rosenberg—
Why stood I not beneath it?   100

    C. HUN.   Friend! have a care,
Your next step may be fatal!—for the love
Of him who made you, stand not on that brink!

    MAN. (*not hearing him*).   Such would have been for
    me a fitting tomb;
My bones had then been quiet in their depth;
They had not then been strewn upon the rocks

For the wind's pastime—as thus—thus they shall be—
In this one plunge.—Farewell, ye opening heavens!
Look not upon me thus reproachfully—
Ye were not meant for me—Earth! take these atoms!

    [*As* MANFRED *is in act to spring from the cliff, the*
    CHAMOIS HUNTER *seizes and retains him with a*
    sudden grasp.]

    C. HUN.   Hold, madman!—though aweary of thy
    life,   110
Stain not our pure vales with thy guilty blood:
Away with me—I will not quit my hold.

    MAN.   I am most sick at heart—nay, grasp me not—
I am all feebleness—the mountains whirl
Spinning around me—I grow blind—
        What art thou?

    C. HUN.   I'll answer that anon. Away with me!
The clouds grow thicker—there—now lean on me—
Place your foot here—here, take this staff, and cling
A moment to that shrub—now give me your hand,
And hold fast by my girdle—softly—well—   120
The Chalet will be gained within an hour:
Come on, we'll quickly find a surer footing,
And something like a pathway, which the torrent
Hath washed since winter.—Come, 'tis bravely done—
You should have been a hunter.—Follow me.
[*As they descend the rocks with difficulty, the scene closes.*]

ACT II

SCENE I

A Cottage amongst the Bernese Alps.

MANFRED *and the* CHAMOIS HUNTER.

    C. HUN.   No, no—yet pause—thou must not yet go
    forth:
Thy mind and body are alike unfit
To trust each other, for some hours, at least;
When thou art better, I will be thy guide—
But whither?

    MAN.   It imports not: I do know
My route full well, and need no further guidance.

115. blind: In the remaining lines of the scene, and in those
that open the next act, Byron is recalling the scene in *King
Lear* (IV. vi) in which Gloucester hopes to plunge from
the Cliffs of Dover.

C. HUN.   Thy garb and gait bespeak thee of high
 lineage—
One of the many chiefs, whose castled crags
Look o'er the lower valleys—which of these
May call thee lord? I only know their portals; 10
My way of life leads me but rarely down
To bask by the huge hearths of those old halls,
Carousing with the vassals; but the paths,
Which step from out our mountains to their doors,
I know from childhood—which of these is thine?
 MAN.   No matter.
 C. HUN.   Well, sir, pardon me the ques-
 tion,
And be of better cheer. Come, taste my wine;
'Tis of an ancient vintage; many a day
'T has thawed my veins among our glaciers, now
Let it do thus for thine. Come, pledge me fairly. 20
 MAN.   Away, away! there's blood upon the brim!
Will it then never—never sink in the earth?
 C. HUN.   What dost thou mean? thy senses wander
 from thee.
 MAN.   I say 'tis blood—my blood! the pure warm
 stream
Which ran in the veins of my fathers, and in ours
When we were in our youth, and had one heart,
And loved each other as we should not love,
And this was shed: but still it rises up,
Coloring the clouds, that shut me out from heaven,
Where thou art not—and I shall never be. 30
 C. HUN.   Man of strange words, and some half-
 maddening sin,
Which makes thee people vacancy, whate'er
Thy dread and sufferance be, there's comfort yet—
The aid of holy men, and heavenly patience—
 MAN.   Patience and patience! Hence—that word
 was made
For brutes of burthen, not for birds of prey;
Preach it to mortals of a dust like thine,—
I am not of thine order.
 C. HUN.   Thanks to heaven!
I would not be of thine for the free fame
Of William Tell; but whatsoe'er thine ill, 40
It must be borne, and these wild starts are useless.
 MAN.   Do I not bear it?—Look on me—I live.
 C. HUN.   This is convulsion, and no healthful life.
 MAN.   I tell thee, man! I have lived many years,
Many long years, but they are nothing now
To those which I must number: ages—ages—
Space and eternity—and consciousness,
With the fierce thirst of death—and still unslaked!

C. HUN.   Why, on thy brow the seal of middle age
Hath scarce been set; I am thine elder far. 50
 MAN.   Think'st thou existence doth depend on time?
It doth; but actions are our epochs: mine
Have made my days and nights imperishable,
Endless, and all alike, as sands on the shore,
Innumerable atoms; and one desert,
Barren and cold, on which the wild waves break,
But nothing rests, save carcasses and wrecks,
Rocks, and the salt-surf weeds of bitterness.
 C. HUN.   Alas! he's mad—but yet I must not leave
 him.
 MAN.   I would I were—for then the things I see 60
Would be but a distempered dream.
 C. HUN.   What is it
That thou dost see, or think thou look'st upon?
 MAN.   Myself, and thee—a peasant of the Alps—
Thy humble virtues, hospitable home,
And spirit patient, pious, proud, and free;
Thy self-respect, grafted on innocent thoughts;
Thy days of health, and nights of sleep; thy toils,
By danger dignified, yet guiltless; hopes
Of cheerful old age and a quiet grave,
With cross and garland over its green turf, 70
And thy grandchildren's love for epitaph;
This do I see—and then I look within—
It matters not—my soul was scorched already!
 C. HUN.   And wouldst thou then exchange thy lot
 for mine?
 MAN.   No, friend! I would not wrong thee, nor
 exchange
My lot with living being: I can bear—
However wretchedly, 'tis still to bear—
In life what others could not brook to dream,
But perish in their slumber.
 C. HUN.   And with this—
This cautious feeling for another's pain, 80
Canst thou be black with evil?—say not so.
Can one of gentle thoughts have wreaked revenge
Upon his enemies?
 MAN.   Oh! no, no, no!
My injuries came down on those who loved me—
On those whom I best loved: I never quelled
An enemy, save in my just defence—
But my embrace was fatal.
 C. HUN.   Heaven give thee rest!
And penitence restore thee to thyself;
My prayers shall be for thee.

ACT II. **Sc. i. 85. quelled:** killed.

MAN.   I need them not—
But can endure thy pity. I depart—   90
'Tis time—farewell!—Here's gold, and thanks for thee;
No words—it is thy due. Follow me not—
I know my path—the mountain peril's past:
And once again I charge thee, follow not!

[*Exit* MANFRED.]

SCENE II

A lower Valley in the Alps.—A Cataract.

Enter MANFRED.

MAN.   It is not noon—the sunbow's rays still arch
The torrent with the many hues of heaven,
And roll the sheeted silver's waving column
O'er the crag's headlong perpendicular,
And fling its lines of foaming light along
And to and fro, like the pale courser's tail,
The giant steed, to be bestrode by death,
As told in the Apocalypse. No eyes
But mine now drink this sight of loveliness;
I should be sole in this sweet solitude,   10
And with the Spirit of the place divide
The homage of these waters.—I will call her.
   [MANFRED *takes some of the water into the palm of his*
   hand, and flings it into the air muttering the adjuration.
   After a pause, the WITCH OF THE ALPS *rises beneath*
         the arch of the sunbow of the torrent.]
Beautiful spirit! with thy hair of light,
And dazzling eyes of glory, in whose form
The charms of earth's least mortal daughters grow
To an unearthly stature, in an essence
Of purer elements; while the hue of youth,—
Carnationed like a sleeping infant's cheek,
Rocked by the beating of her mother's heart,
Or the rose tints, which summer's twilight leaves   20
Upon the lofty glacier's virgin snow,
The blush of earth embracing with her heaven,—
Tinge thy celestial aspect, and make tame
The beauties of the sunbow which bends o'er thee.
Beautiful Spirit! in thy calm clear brow,
Wherein is glassed serenity of soul,
Which of itself shows immortality,
I read that thou wilt pardon to a son
Of Earth, whom the abstruser powers permit
At times to commune with them—if that he   30

Sc. ii. 8. told . . . Apocalypse: Revelation 6:8.

Avail him of his spells—to call thee thus,
And gaze on thee a moment.
   WITCH OF THE ALPS.   Son of Earth!
I know thee, and the powers which give thee power;
I know thee for a man of many thoughts,
And deeds of good and ill, extreme in both,
Fatal and fated in thy sufferings.
I have expected this—what wouldst thou with me?
   MAN.   To look upon thy beauty—nothing further.
The face of the earth hath maddened me, and I
Take refuge in her mysteries, and pierce   40
To the abodes of those who govern her—
But they can nothing aid me. I have sought
From them what they could not bestow, and now
I search no further.
   WITCH.   What could be the quest
Which is not in the power of the most powerful,
The rulers of the invisible?
   MAN.   A boon;
But why should I repeat it? 'twere in vain.
   WITCH.   I know not that; let thy lips utter it.
   MAN.   Well, though it torture me, 'tis but the same;
My pangs shall find a voice. From my youth upwards
My spirit walked not with the souls of men,   51
Nor looked upon the earth with human eyes;
The thirst of their ambition was not mine,
The aim of their existence was not mine;
My joys, my griefs, my passions, and my powers,
Made me a stranger; though I wore the form
I had no sympathy with breathing flesh,
Nor midst the creatures of clay that girded me
Was there but one who—but of her anon.
I said with men, and with the thoughts of men,   60
I held but slight communion; but instead,
My joy was in the wilderness,—to breathe
The difficult air of the iced mountain's top,
Where the birds dare not build, nor insect's wing
Flit o'er the herbless granite; or to plunge
Into the torrent, and to roll along
On the swift whirl of the new breaking wave
Of river-stream, or ocean, in their flow.
In these my early strength exulted; or
To follow through the night the moving moon,   70
The stars and their development; or catch
The dazzling lightnings till my eyes grew dim;
Or to look, list'ning, on the scattered leaves,
While autumn winds were at their evening song.
These were my pastimes, and to be alone;
For if the beings, of whom I was one,—
Hating to be so,—crossed me in my path,

I felt myself degraded back to them,
And was all clay again. And then I dived,
In my lone wanderings, to the caves of death,   80
Searching its cause in its effect; and drew
From withered bones, and skulls, and heaped up dust,
Conclusions most forbidden. Then I passed
The nights of years in sciences untaught,
Save in the old time; and with time and toil,
And terrible ordeal, and such penance
As in itself hath power upon the air,
And spirits that do compass air and earth,
Space, and the people infinite, I made
Mine eyes familiar with Eternity,   90
Such as, before me, did the Magi, and
He who from out their fountain dwellings raised
Eros and Anteros, at Gadara,
As I do thee;—and with my knowledge grew
The thirst of knowledge, and the power and joy
Of this most bright intelligence, until—
   WITCH.   Proceed.
   MAN.   Oh! I but thus prolonged my words,
Boasting these idle attributes, because
As I approach the core of my heart's grief—   100
But to my task. I have not named to thee
Father, or mother, mistress, friend, or being,
With whom I wore the chain of human ties;
If I had such, they seemed not such to me;
Yet there was one—
   WITCH.   Spare not thyself—proceed.
   MAN.   She was like me in lineaments; her eyes,
Her hair, her features, all, to the very tone
Even of her voice, they said were like to mine;
But softened all, and tempered into beauty:
She had the same lone thoughts and wanderings,   110
The quest of hidden knowledge, and a mind
To comprehend the universe: nor these
Alone, but with them gentler powers than mine,
Pity, and smiles, and tears—which I had not;
And tenderness—but that I had for her;
Humility—and that I never had.
Her faults were mine—her virtues were her own—
I loved her, and destroyed her!
   WITCH.   With thy hand?
   MAN.   Not with my hand, but heart—which broke
    her heart;

92–93. He . . . Gadara: Iamblichus, the Neoplatonic
philosopher (fourth century). At Gadara, in Syria, he summoned up Eros, god of love, and Anteros, the god who
avenged unrequited love, to explain why the two springs
there were named after them.

It gazed on mine, and withered. I have shed   120
Blood, but not hers—and yet her blood was shed;
I saw—and could not stanch it.
   WITCH.   And for this—
A being of the race thou dost despise,
The order, which thine own would rise above,
Mingling with us and ours,—thou dost forego
The gifts of our great knowledge, and shrink'st
   back
In recreant mortality—Away!
   MAN.   Daughter of air! I tell thee, since that hour—
But words are breath—look on me in my sleep,
Or watch my watchings—Come and sit by me!   130
My solitude is solitude no more,
But peopled with the furies;—I have gnashed
My teeth in darkness till returning morn,
Then cursed myself till sunset;—I have prayed
For madness as a blessing—'tis denied me.
I have affronted death—but in the war
Of elements the waters shrunk from me,
And fatal things passed harmless; the cold hand
Of an all-pitiless demon held me back,
Back by a single hair, which would not break.   140
In fantasy, imagination, all
The affluence of my soul—which one day was
A Crœsus in creation—I plunged deep,
But, like an ebbing wave, it dashed me back
Into the gulf of my unfathomed thought.
I plunged amidst mankind—Forgetfulness
I sought in all, save where 'tis to be found,
And that I have to learn; my sciences,
My long-pursued and superhuman art,
Is mortal here: I dwell in my despair—   150
And live—and live forever.
   WITCH.   It may be
That I can aid thee.
   MAN.   To do this thy power
Must wake the dead, or lay me low with them.
Do so—in any shape—in any hour—
With any torture—so it be the last.
   WITCH.   That is not in my province; but if thou
Wilt swear obedience to my will, and do
My bidding, it may help thee to thy wishes.
   MAN.   I will not swear—Obey! and whom? the
   spirits
Whose presence I command, and be the slave   160
Of those who served me—Never!
   WITCH.   Is this all?
Hast thou no gentler answer?—Yet bethink thee,
And pause ere thou rejectest.

MAN.   I have said it.
WITCH.   Enough! I may retire then—say!
MAN.   Retire!

[*The* WITCH *disappears.*]

MAN. (*alone*).   We are the fools of time and terror.
   Days
Steal on us, and steal from us; yet we live,
Loathing our life, and dreading still to die.
In all the days of this detested yoke—
This vital weight upon the struggling heart,
Which sinks with sorrow, or beats quick with pain,
Or joy that ends in agony or faintness—   171
In all the days of past and future, for
In life there is no present, we can number
How few—how less than few—wherein the soul
Forbears to pant for death, and yet draws back
As from a stream in winter, though the chill
Be but a moment's. I have one resource
Still in my science—I can call the dead,
And ask them what it is we dread to be:
The sternest answer can but be the grave,   180
And that is nothing. If they answer not—
The buried prophet answered to the Hag
Of Endor; and the Spartan Monarch drew
From the Byzantine maid's unsleeping spirit
An answer and his destiny—he slew
That which he loved, unknowing what he slew,
And died unpardoned—though he called in aid
The Phyxian Jove, and in Phigalia roused
The Arcadian Evocators to compel
The indignant shadow to depose her wrath,   190
Or fix her term of vengeance—she replied
In words of dubious import, but fulfilled.
If I had never lived, that which I love
Had still been living; had I never loved,
That which I love would still be beautiful,
Happy and giving happiness. What is she?
What is she now?—a sufferer for my sins—
A thing I dare not think upon—or nothing.
Within few hours I shall not call in vain—
Yet in this hour I dread the thing I dare: 200

182–83: **buried . . . Endor:** For the story see I Samuel
28. **183–85. Spartan . . . destiny:** Pausanius, who by mistake killed his mistress, Cleonice. The priests summoned up
her spirit so that he could ask pardon, and the spirit told
Pausanius he would soon be freed from his troubles. Soon
afterwards he died.

Until this hour I never shrunk to gaze
On spirit, good or evil—now I tremble,
And feel a strange cold thaw upon my heart.
But I can act even what I most abhor,
And champion human fears.—The night approaches.
                            [*Exit.*]

SCENE III

The Summit of the Jungfrau Mountain.

Enter FIRST DESTINY.

FIRST DESTINY.   The moon is rising broad, and round,
   and bright;
And here on snows, where never human foot
Of common mortal trod, we nightly tread,
And leave no traces: o'er the savage sea,
The glassy ocean of the mountain ice,
We skim its rugged breakers, which put on
The aspect of a tumbling tempest's foam,
Frozen in a moment—a dead whirlpool's image:
And this most steep fantastic pinnacle,
The fretwork of some earthquake—where the clouds
Pause to repose themselves in passing by—   11
Is sacred to our revels, or our vigils;
Here do I wait my sisters, on our way
To the Hall of Arimanes, for tonight
Is our great festival—'tis strange they come not.

A VOICE *without, singing*

   The captive usurper,
      Hurled down from the throne,
   Lay buried in torpor,
      Forgotten and lone;
   I broke through his slumbers,   20
      I shivered his chain,
   I leagued him with numbers—
      He's tyrant again!
With the blood of a million he'll answer my care,
With a nation's destruction—his flight and despair.

SECOND VOICE, *without*

The ship sailed on, the ship sailed fast,
But I left not a sail, and I left not a mast;
There is not a plank of the hull or the deck,
And there is not a wretch to lament o'er his wreck;

Sc. iii. **16. captive usurper:** Napoleon.

Save one, whom I held, as he swam, by the hair,   30
And he was a subject, well worthy my care;
A traitor on land, and a pirate at sea—
But I saved him to wreak further havoc for me!

FIRST DESTINY, *answering*

The city lies sleeping;
    The morn, to deplore it,
May dawn on it weeping:
    Sullenly, slowly,
The black plague flew o'er it—
    Thousands lie lowly;
Tens of thousands shall perish;   40
    The living shall fly from
The sick they should cherish;
    But nothing can vanquish
The touch that they die from.
    Sorrow and anguish,
And evil and dread,
    Envelop a nation;
The blest are the dead,
    Who see not the sight
Of their own desolation;   50
    This work of a night—
This wreck of a realm—this deed of my doing—
For ages I've done, and shall still be renewing!

Enter the SECOND *and* THIRD DESTINIES.

THE THREE

Our hands contain the hearts of men,
    Our footsteps are their graves;
We only give to take again
    The spirits of our slaves!

FIRST DESTINY.   Welcome!—Where's Nemesis?
SECOND DESTINY.   At
    some great work;
But what I know not, for my hands were full.
THIRD DESTINY.   Behold she cometh.   60

Enter NEMESIS.

FIRST DES.   Say, where hast thou been?
My sisters and thyself are slow tonight.
    NEMESIS.   I was detained repairing shattered thrones,
Marrying fools, restoring dynasties,
Avenging men upon their enemies,
And making them repent their own revenge;

Goading the wise to madness; from the dull
Shaping our oracles to rule the world
Afresh, for they were waxing out of date,
And mortals dared to ponder for themselves,   70
To weigh kings in the balance, and to speak
Of freedom, the forbidden fruit.—Away!
We have outstayed the hour—mount we our clouds.

[*Exeunt.*]

SCENE IV

The Hall of ARIMANES. ARIMANES *on his Throne, a
Globe of Fire, surrounded by the* SPIRITS.

Hymn of the SPIRITS

Hail to our Master!—Prince of earth and air!
    Who walks the clouds and waters—in his hand
The sceptre of the elements, which tear
    Themselves to chaos at his high command!
He breatheth—and a tempest shakes the sea;
    He speaketh—and the clouds reply in thunder;
He gazeth—from his glance the sunbeams flee;
    He moveth—earthquakes rend the world asunder.
Beneath his footsteps the volcanoes rise;
    His shadow is the pestilence; his path   10
The comets herald through the crackling skies;
    And planets turn to ashes at his wrath.
To him war offers daily sacrifice;
    To him death pays his tribute; life is his,
With all its infinite of agonies—
    And his the spirit of whatever is!

Enter the DESTINIES *and* NEMESIS.

FIRST DES.   Glory to Arimanes! on the earth
His power increaseth—both my sisters did
His bidding, nor did I neglect my duty!
    SECOND DES.   Glory to Arimanes! we who bow   20
The necks of men, bow down before his throne!
    THIRD DES.   Glory to Arimanes! we await His nod!
    NEM.   Sovereign of sovereigns! we are thine,
And all that liveth, more or less, is ours,
And most things wholly so; still to increase
Our power, increasing thine, demands our care,

Sc. iv. Setting of scene. Arimanes: Ahriman, the principle
of evil in the ancient Persian religion.

And we are vigilant. Thy late commands
Have been fulfilled to the utmost.

          Enter MANFRED.

A SPIRIT.   What is here?
A mortal!—Thou most rash and fatal wretch,
Bow down and worship!   30
   SECOND SPIRIT.   I do know the man—
A magian of great power, and fearful skill!
   THIRD SPIRIT.   Bow down and worship, slave!—
     What, know'st thou not
Thine and our sovereign?—Tremble, and obey!
  ALL THE SPIRITS.   Prostrate thyself, and thy con-
    demned clay,
Child of the earth! or dread the worst.
   MAN.   I know it;
And yet ye see I kneel not.
   FOURTH SPIRIT.   'Twill be taught thee.
   MAN.   'Tis taught already;—many a night on the
    earth,
On the bare ground, have I bowed down my face,
And strewed my head with ashes; I have known   40
The fullness of humiliation, for
I sunk before my vain despair, and knelt
To my own desolation.
   FIFTH SPIRIT.   Dost thou dare
Refuse to Arimanes on his throne
What the whole earth accords, beholding not
The terror of his glory?—Crouch, I say.
   MAN.   Bid *him* bow down to that which is above
    him,
The overruling Infinite—the Maker
Who made him not for worship—let him kneel,
And we will kneel together.
   THE SPIRITS.   Crush the worm!   50
Tear him to pieces!—
   FIRST DES.   Hence! avaunt!—he's mine.
Prince of the powers invisible! This man
Is of no common order, as his port
And presence here denote; his sufferings
Have been of an immortal nature, like
Our own; his knowledge, and his powers and will,
As far as is compatible with clay,
Which clogs the ethereal essence, have been such
As clay hath seldom borne; his aspirations
Have been beyond the dwellers of the earth,   60
And they have only taught him what we know—
That knowledge is not happiness, and science
But an exchange of ignorance for that

Which is another kind of ignorance.
This is not all—the passions, attributes
Of earth and heaven, from which no power, nor being,
Nor breath from the worm upwards is exempt,
Have pierced his heart, and in their consequence
Made him a thing which I, who pity not,
Yet pardon those who pity. He is mine,   70
And thine, it may be; be it so, or not,
No other spirit in this region hath
A soul like his—or power upon his soul.
   NEM.   What doth he here then?
   FIRST DES.   Let him answer
    that.
   MAN.   Ye know what I have known; and without
    power
I could not be amongst ye: but there are
Powers deeper still beyond—I come in quest
Of such, to answer unto what I seek.
   NEM.   What wouldst thou?
   MAN.   Thou canst not reply
    to me.
Call up the dead—my question is for them.   80
   NEM.   Great Arimanes, doth thy will avouch
The wishes of this mortal?
   ARIMANES.   Yea.
   NEM.   Whom wouldst
    thou
Uncharnel?
   MAN.   One without a tomb—call up
Astarte.

                      NEMESIS

      Shadow! or spirit!
        Whatever thou art,
      Which still doth inherit
        The whole or a part
      Of the form of thy birth,
      Of the mould of thy clay,   90
      Which returned to the earth,—
        Reappear to the day!
      Bear what thou borest,
        The heart and the form,
      And the aspect thou worest
        Redeem from the worm.
Appear!—Appear!—Appear!
Who sent thee there requires thee here!

   [*The* PHANTOM OF ASTARTE *rises and stands in the midst.*]

MAN. Can this be death? there's bloom upon her
  cheek;
But now I see it is no living hue,   100
But a strange hectic—like the unnatural red
Which autumn plants upon the perished leaf.
It is the same! Oh, God! that I should dread
To look upon the same—Astarte!—No,
I cannot speak to her—but bid her speak—
Forgive me or condemn me.

NEMESIS

By the power which hath broken
  The grave which enthralled thee,
Speak to him who hath spoken,
  Or those who have called thee!
    MAN.   She is silent, 110
And in that silence I am more than answered.
    NEM. My power extends no further. Prince of Air!
It rests with thee alone—command her voice.
    ARI. Spirit—obey this sceptre!
    NEM.   Silent still!
She is not of our order, but belongs
To the other powers. Mortal! thy quest is vain,
And we are baffled also.
    MAN.   Hear me, hear me—
Astarte! my beloved! speak to me:
I have so much endured—so much endure—
Look on me! the grave hath not changed thee more 120
Than I am changed for thee. Thou lovedst me
Too much, as I loved thee: we were not made
To torture thus each other, though it were
The deadliest sin to love as we have loved.
Say that thou loath'st me not—that I do bear
This punishment for both—that thou wilt be
One of the blessed—and that I shall die;
For hitherto all hateful things conspire
To bind me in existence—in a life
Which makes me shrink from immortality—   130
A future like the past. I cannot rest.
I know not what I ask, nor what I seek:
I feel but what thou art, and what I am;
And I would hear yet once before I perish
The voice which was my music—Speak to me!
For I have called on thee in the still night,
Startled the slumbering birds from the hushed boughs,
And woke the mountain wolves, and made the caves
Acquainted with thy vainly echoed name,
Which answered me—many things answered me—

Spirits and men—but thou wert silent all.   141
Yet speak to me! I have outwatched the stars,
And gazed o'er heaven in vain in search of thee.
Speak to me! I have wandered o'er the earth,
And never found thy likeness—Speak to me!
Look on the fiends around—they feel for me:
I fear them not, and feel for thee alone—
Speak to me! though it be in wrath;—but say—
I reck not what—but let me hear thee once—
This once—once more!
    PHANTOM OF ASTARTE. Manfred!
    MAN.   Say on, say on—
I live but in the sound—it is thy voice!   151
    PHAN. Manfred! Tomorrow ends thine earthly ills.
Farewell!
    MAN. Yet one word more—am I forgiven?
    PHAN. Farewell!
    MAN.   Say, shall we meet again?
    PHAN. Farewell!
    MAN.   One word for mercy!
    Say, thou lovest me.
    PHAN. Manfred!

[*The* SPIRIT OF ASTARTE *disappears.*]

    NEM. She's gone, and will not be recalled;
Her words will be fulfilled. Return to the earth.
    A SPIRIT. He is convulsed.—This is to be a mortal
And seek the things beyond mortality.   160
    ANOTHER SPIRIT. Yet, see, he mastereth himself, and
  makes
His torture tributary to his will.
Had he been one of us, he would have made
An awful spirit.
    NEM.   Hast thou further question
Of our great sovereign, or his worshippers?
    MAN. None.
    NEM.   Then for a time farewell.
    MAN. We meet then! Where? On the earth?—
Even as thou wilt: and for the grace accorded
I now depart a debtor. Fare ye well!
    [*Exit* MANFRED.]

(*Scene closes.*)

ACT III

SCENE I

A Hall in the Castle of MANFRED.

MANFRED *and* HERMAN.

MAN.   What is the hour?
HERMAN.   It wants but one till sun-
set,
And promises a lovely twilight.
MAN.   Say,
Are all things so disposed of in the tower
As I directed?
HER.   All, my lord, are ready:
Here is the key and casket.
MAN.   It is well:
Thou may'st retire.

                    [*Exit* HERMAN.]

MAN. (*alone*). There is a calm upon me—
Inexplicable stillness! which till now,
Did not belong to what I knew of life.
If that I did not know philosophy
To be of all our vanities the motliest,   10
The merest word that ever fooled the ear
From out the schoolman's jargon, I should deem
The golden secret, the sought "Kalon," found,
And seated in my soul. It will not last,
But it is well to have known it, though but once:
It hath enlarged my thoughts with a new sense,
And I within my tablets would note down
That there is such a feeling. Who is there?

              Re-enter HERMAN.
HER.   My lord, the abbot of St. Maurice craves
To greet your presence.

          Enter the ABBOT OF ST. MAURICE.
ABBOT.   Peace be with Count Man-
fred!   20
MAN.   Thanks, holy father! welcome to these walls;
Thy presence honors them, and blesseth those
Who dwell within them.

ACT III. **Sc. i. 13.** "**Kalon**": the beautiful, or the supreme
good.

ABBOT.   Would it were so, Count!—
But I would fain confer with thee alone.
MAN.   Herman, retire—What would my reverend
guest?
ABBOT.   Thus, without prelude:—Age and zeal, my
office,
And good intent, must plead my privilege;
Our near, though not acquainted neighborhood,
May also be my herald. Rumors strange,
And of unholy nature, are abroad,   30
And busy with thy name; a noble name
For centuries: may he who bears it now
Transmit it unimpaired!
MAN.   Proceed,—I listen.
ABBOT.   'Tis said thou holdest converse with the
things
Which are forbidden to the search of man;
That with the dwellers of the dark abodes,
The many evil and unheavenly spirits
Which walk the valley of the shade of death,
Thou communest. I know that with mankind,
Thy fellows in creation, thou dost rarely   40
Exchange thy thoughts, and that thy solitude
Is as an anchorite's, were it but holy.
MAN.   And what are they who do avouch these
things?
ABBOT.   My pious brethren—the sacred peasantry—
Even thy own vassals—who do look on thee
With most unquiet eyes. Thy life's in peril.
MAN.   Take it.
ABBOT.   I come to save, and not destroy:
I would not pry into thy secret soul;
But if these things be sooth, there still is time
For penitence and pity: reconcile thee   50
With the true church, and through the church to
heaven.
MAN.   I hear thee. This is my reply: whate'er
I may have been, or am, doth rest between
Heaven and myself. I shall not choose a mortal
To be my mediator. Have I sinned
Against your ordinances? prove and punish!
ABBOT.   My son! I did not speak of punishment,
But penitence and pardon;—with thyself
The choice of such remains—and for the last,
Our institutions and our strong belief   60
Have given me power to smooth the path from sin
To higher hope and better thoughts; the first
I leave to heaven,—"Vengeance is mine alone!"

63. "Vengeance . . . alone!": Romans 12:19.

So saith the Lord, and with all humbleness
His servant echoes back the awful word.

   MAN.   Old man! there is no power in holy men,
Nor charm in prayer, nor purifying form
Of penitence, nor outward look, nor fast,
Nor agony—nor, greater than all these,
The innate tortures of that deep despair,   70
Which is remorse without the fear of hell,
But all in all sufficient to itself
Would make a hell of heaven—can exorcise
From out the unbounded spirit the quick sense
Of its own sins, wrongs, sufferance, and revenge
Upon itself; there is no future pang
Can deal that justice on the self-condemned
He deals on his own soul.

   ABBOT.   All this is well;
For this will pass away, and be succeeded
By an auspicious hope, which shall look up   80
With calm assurance to that blessed place,
Which all who seek may win, whatever be
Their earthly errors, so they be atoned:
And the commencement of atonement is
The sense of its necessity. Say on—
And all our church can teach thee shall be taught;
And all we can absolve thee shall be pardoned.

   MAN.   When Rome's sixth emperor was near his
      last,
The victim of a self-inflicted wound,
To shun the torments of a public death   90
From senates once his slaves, a certain soldier,
With show of loyal pity, would have stanched
The gushing throat with his officious robe;
The dying Roman thrust him back, and said—
Some empire still in his expiring glance—
"It is too late—is this fidelity?"

   ABBOT.   And what of this?

   MAN.   I answer with the
      Roman—
"It is too late!"

   ABBOT.   It never can be so,
To reconcile thyself with thy own soul,
And thy own soul with heaven. Hast thou no hope?
'Tis strange—even those who do despair above,   101
Yet shape themselves some fantasy on earth,
To which frail twig they cling, like drowning men.

   MAN.   Ay—father! I have had those earthly visions,
And noble aspirations in my youth,

---

73. **hell of heaven:** See I. i. above, n. 252. **88. Rome's . . .
emperor:** Nero.

To make my own the mind of other men,
The enlightener of nations; and to rise
I knew not whither—it might be to fall;
But fall, even as the mountain-cataract,
Which, having leapt from its more dazzling height,
Even in the foaming strength of its abyss   111
(Which casts up misty columns that become
Clouds raining from the re-ascended skies),
Lies low but mighty still.—But this is past,
My thoughts mistook themselves.

   ABBOT.   And wherefore so?

   MAN.   I could not tame my nature down; for he
Must serve who fain would sway; and soothe, and
      sue,
And watch all time, and pry into all place,
And be a living lie, who would become
A mighty thing amongst the mean, and such   120
The mass are; I disdained to mingle with
A herd, though to be leader—and of wolves.
The lion is alone, and so am I.

   ABBOT.   And why not live and act with other men?

   MAN.   Because my nature was averse from life;
And yet not cruel; for I would not make,
But find a desolation. Like the wind,
The red-hot breath of the most lone simoom,
Which dwells but in the desert, and sweeps o'er
The barren sands which bear no shrubs to blast,   130
And revels o'er their wild and arid waves,
And seeketh not, so that it is not sought,
But being met is deadly,—such hath been
The course of my existence; but there came
Things in my path which are no more.

   ABBOT.   Alas!
I 'gin to fear that thou art past all aid
From me and from my calling; yet so young,
I still would—

   MAN.   Look on me! there is an order
Of mortals on the earth, who do become
Old in their youth, and die ere middle age,   140
Without the violence of warlike death;
Some perishing of pleasure, some of study,
Some worn with toil, some of mere weariness,
Some of disease, and some insanity,
And some of withered or of broken hearts;
For this last is a malady which slays
More than are numbered in the lists of fate,
Taking all shapes, and bearing many names.
Look upon me! for even of all these things
Have I partaken; and of all these things,   150
One were enough; then wonder not that I

Am what I am, but that I ever was,
Or having been, that I am still on earth.
    ABBOT.   Yet, hear me still—
    MAN.   Old man! I do re-
spect
Thine order, and revere thine years; I deem
Thy purpose pious, but it is in vain:
Think me not churlish; I would spare thyself,
Far more than me, in shunning at this time
All further colloquy; and so—farewell.
             [*Exit* MANFRED.]

    ABBOT.   This should have been a noble creature: he
Hath all the energy which would have made   161
A goodly frame of glorious elements,
Had they been wisely mingled; as it is,
It is an awful chaos—light and darkness,
And mind and dust, and passions and pure thoughts
Mixed, and contending without end or order,—
All dormant or destructive; he will perish,
And yet he must not; I will try once more.
For such are worth redemption; and my duty
Is to dare all things for a righteous end.   170
I'll follow him—but cautiously, though surely.
            [*Exit* ABBOT.]

SCENE II

Another Chamber.

MANFRED *and* HERMAN.

    HER.   My lord, you bade me wait on you at sunset:
He sinks behind the mountain.
    MAN.   Doth he so?
I will look on him.
    [MANFRED *advances to the Window of the Hall.*]
              Glorious orb! the idol
Of early nature, and the vigorous race
Of undiseased mankind, the giant sons
Of the embrace of angels, with a sex
More beautiful than they, which did draw down
The erring spirits who can ne'er return.—
Most glorious orb! that wert a worship, ere
The mystery of thy making was revealed!   10
Thou earliest minister of the Almighty,
Which gladdened, on their mountain tops, the hearts

Sc. ii. 6. embrace of angels: Genesis 6:2–4. 12–13. gladdened . . . shepherds: The Chaldeans generally were
interested in astronomy (see *Childe Harold*, III. n. 118).

Of the Chaldean shepherds, till they poured
Themselves in orisons! Thou material God!
And representative of the unknown—
Who chose thee for his shadow! Thou chief star!
Center of many stars! which mak'st our earth
Endurable, and temperest the hues
And hearts of all who walk within thy rays!
Sire of the seasons! Monarch of the climes,   20
And those who dwell in them! for near or far,
Our inborn spirits have a tint of thee
Even as our outward aspects;—thou dost rise,
And shine, and set in glory. Fare thee well!
I ne'er shall see thee more. As my first glance
Of love and wonder was for thee, then take
My latest look; thou wilt not beam on one
To whom the gifts of life and warmth have been
Of a more fatal nature. He is gone:
I follow.   30
            [*Exit* MANFRED.]

SCENE III

The Mountains—The Castle of MANFRED *at some
distance—A Terrace before a Tower.—Time, Twilight.*

HERMAN, MANUEL, *and other Dependants of* MANFRED.

    HER.   'Tis strange enough; night after night, for
    years,
He hath pursued long vigils in this tower,
Without a witness. I have been within it,—
So have we all been ofttimes; but from it,
Or its contents, it were impossible
To draw conclusions absolute, of aught
His studies tend to. To be sure, there is
One chamber where none enter: I would give
The fee of what I have to come these three years,
To pore upon its mysteries.
    MANUEL.   'Twere dangerous: 10
Content thyself with what thou know'st already.
    HER.   Ah! Manuel! thou art elderly and wise,
And couldst say much; thou hast dwelt within the
    castle—
How many years is 't?
    MANUEL.   Ere Count Manfred's birth,
I served his father, whom he nought resembles.
    HER.   There be more sons in like predicament.
But wherein do they differ?

Sc. iii. 9. fee of: ownership of.

MANUEL.   I speak not
Of features or of form, but mind and habits;
Count Sigismund was proud, but gay and free,—
A warrior and a reveller; he dwelt not   20
With books and solitude, nor made the night
A gloomy vigil, but a festal time,
Merrier than day; he did not walk the rocks
And forests like a wolf, nor turn aside
From men and their delights.
    HER.   Beshrew the hour,
But those were jocund times! I would that such
Would visit the old walls again; they look
As if they had forgotten them.
    MANUEL.   These walls
Must change their chieftain first. Oh! I have seen
Some strange things in them, Herman.
    HER.   Come,   be
        friendly,   30
Relate me some to while away our watch:
I've heard thee darkly speak of an event
Which happened hereabouts, by this same tower.
    MANUEL.   That was a night indeed! I do remember
'Twas twilight, as it may be now, and such
Another evening;—yon red cloud, which rests
On Eigher's pinnacle, so rested then,—
So like that it might be the same; the wind
Was faint and gusty, and the mountain snows
Began to glitter with the climbing moon;   40
Count Manfred was, as now, within his tower,—
How occupied, we knew not, but with him
The sole companion of his wanderings
And watchings—her, whom of all earthly things
That lived, the only thing he seemed to love,—
As he, indeed, by blood was bound to do,
The lady Astarte, his—
                    Hush! who comes here?

Enter the ABBOT.

    ABBOT.   Where is your master?
    HER.   Yonder   in   the
        tower.
    ABBOT.   I must speak with him.
    MANUEL.   'T is impossible;
He is most private, and must not be thus   50
Intruded on.
    ABBOT.   Upon myself I take
The forfeit of my fault, if fault there be—
But I must see him.

37. **Eigher's pinnacle:** a mountain east of the Jungfrau.

HER.   Thou hast seen him once
This eve already.
    ABBOT.   Herman! I command thee,
Knock, and apprize the Count of my approach.
    HER.   We dare not.
    ABBOT.   Then it seems I must be herald
Of my own purpose.
    MANUEL.   Reverend father stop—
I pray you pause.
    ABBOT.   Why so?
    MANUEL.   But step this way,
And I will tell you further.

[*Exeunt.*]

SCENE IV

Interior of the Tower.

MANFRED *alone.*

    MAN.   The stars are forth, the moon above the tops
Of the snow-shining mountains.—Beautiful!
I linger yet with nature, for the night
Hath been to me a more familiar face
Than that of man; and in her starry shade
Of dim and solitary loveliness,
I learned the language of another world.
I do remember me, that in my youth,
When I was wandering,—upon such a night
I stood within the Coliseum's wall,   10
'Midst the chief relics of almighty Rome;
The trees which grew along the broken arches
Waved dark in the blue midnight, and the stars
Shone through the rents of ruin; from afar
The watch-dog bayed beyond the Tiber; and
More near from out the Cæsar's palace came
The owl's long cry, and, interruptedly,
Of distant sentinels the fitful song
Begun and died upon the gentle wind.
Some cypresses beyond the time-worn breach   20
Appeared to skirt the horizon, yet they stood
Within a bowshot. Where the Cæsars dwelt,
And dwell the tuneless birds of night, amidst
A grove which springs through levelled battlements,
And twines its roots with the imperial hearths,
Ivy usurps the laurel's place of growth;
But the gladiator's bloody Circus stands,
A noble wreck in ruinous perfection,

While Cæsar's chambers, and the Augustan halls,
Grovel on earth in indistinct decay.   30
And thou didst shine, thou rolling moon, upon
All this, and cast a wide and tender light,
Which softened down the hoar austerity
Of rugged desolation, and filled up,
As 't were anew, the gaps of centuries;
Leaving that beautiful which still was so,
And making that which was not, till the place
Became religion, and the heart ran o'er
With silent worship of the great of old,—
The dead but sceptred sovereigns, who still rule   40
Our spirits from their urns.
                                        'Twas such a night!
'Tis strange that I recall it at this time;
But I have found our thoughts take wildest flight
Even at the moment when they should array
Themselves in pensive order.

                    Enter the ABBOT
ABBOT.   My good lord!
I crave a second grace for this approach;
But yet let not my humble zeal offend
By its abruptness—all it hath of ill
Recoils on me; its good in the effect
May light upon your head—could I say *heart*—   50
Could I touch *that*, with words or prayers, I should
Recall a noble spirit which hath wandered;
But is not yet all lost.
    MAN.   Thou know'st me not;
My days are numbered, and my deeds recorded;
Retire, or 't will be dangerous—Away!
    ABBOT.   Thou dost not mean to menace me?
    MAN.   Not I;
I simply tell thee peril is at hand,
And would preserve thee.
    ABBOT.   What dost thou mean?
    MAN.   Look there!
What dost thou see?
    ABBOT.   Nothing.
    MAN.   Look there, I say,
And steadfastly;—now tell me what thou seest.   60
    ABBOT.   That which should shake me, but I fear it
        not:
I see a dusk and awful figure rise,
Like an infernal god, from out the earth;
His face wrapt in a mantle, and his form
Robed as with angry clouds: he stands between
Thyself and me—but I do fear him not.
    MAN. Thou hast no cause; he shall not harm thee, but

His sight may shock thine old limbs into palsy.
I say to thee—Retire!
    ABBOT.   And I reply—
Never—till I have battled with this fiend:—   70
What doth he here?
    MAN.   Why—ay—what doth he here?
I did not send for him,—he is unbidden.
    ABBOT.   Alas! lost mortal! what with guests like
        these
Hast thou to do? I tremble for thy sake:
Why doth he gaze on thee, and thou on him?
Ah! he unveils his aspect: on his brow
The thunder-scars are graven: from his eye
Glares forth the immortality of hell—
Avaunt!—
    MAN.   Pronounce—what is thy mission?
    SPIRIT.   Come!
    ABBOT.   What art thou, unknown being? answer!—
        speak!   80
    SPIRIT.   The genius of this mortal.—Come! 'tis time.
    MAN.   I am prepared for all things, but deny
The power which summons me. Who sent thee here?
    SPIRIT.   Thou 'lt know anon—Come! come!
    MAN.   I have
        commanded
Things of an essence greater far than thine,
And striven with thy masters. Get thee hence!
    SPIRIT.   Mortal! thine hour is come—Away! I say.
    MAN.   I knew, and know my hour is come, but not
To render up my soul to such as thee:
Away! I'll die as I have lived—alone.   90
    SPIRIT.   Then I must summon up my brethren.—
        Rise!

            [*Other* SPIRITS *rise up.*]

    ABBOT.   Avaunt! ye evil ones!—Avaunt! I say;
Ye have no power where piety hath power,
And I do charge ye in the name—
    SPIRIT.   Old man!
We know ourselves, our mission, and thine order;
Waste not thy holy words on idle uses,
It were in vain: this man is forfeited.
Once more I summon him—Away! Away!
    MAN.   I do defy ye,—though I feel my soul
Is ebbing from me, yet I do defy ye;   100
Nor will I hence, while I have earthly breath
To breathe my scorn upon ye—earthly strength
To wrestle, though with spirits; what ye take
Shall be ta'en limb by limb.

SPIRIT.   Reluctant mortal!
Is this the Magian who would so pervade
The world invisible, and make himself
Almost our equal? Can it be that thou
Art thus in love with life? the very life
Which made thee wretched!
    MAN.   Thou false fiend, thou
      liest!
My life is in its last hour,—*that* I know,   110
Nor would redeem a moment of that hour;
I do not combat against death, but thee
And thy surrounding angels; my past power,
Was purchased by no compact with thy crew,
But by superior science—penance, daring,
And length of watching, strength of mind, and skill
In knowledge of our fathers—when the earth
Saw men and spirits walking side by side,
And gave ye no supremacy: I stand
Upon my strength—I do defy—deny—   120
Spurn back, and scorn ye!—
    SPIRIT.   But thy many crimes
Have made thee—
    MAN.   What are they to such as thee?
Must crimes be punished but by other crimes,
And greater criminals?—Back to thy hell!
Thou hast no power upon me, *that* I feel;
Thou never shalt possess me, *that* I know:
What I have done is done; I bear within
A torture which could nothing gain from thine:
The mind which is immortal makes itself
Requital for its good or evil thoughts,—   130
Is its own origin of ill and end
And its own place and time; its innate sense,
When stripped of this mortality, derives
No color from the fleeting things without,
But is absorbed in sufferance or in joy,
Born from the knowledge of its own desert.
Thou didst not tempt me, and thou couldst not tempt
    me;
I have not been thy dupe, nor am thy prey—
But was my own destroyer, and will be
My own hereafter.—Back, ye baffled fiends!—   140
The hand of death is on me—but not yours.

[*The* DEMONS *disappear.*]

Sc. iv. 129-32. mind . . . time: Byron for the third time
in *Manfred* recurs to the lines in *Paradise Lost* cited above
(I. i. n. 252; cf. III. i. 73).

ABBOT.   Alas! how pale thou art—thy lips are
    white—
And thy breast heaves—and in thy gasping throat
The accents rattle: Give thy prayers to heaven—
Pray—albeit but in thought,—but die not thus.
    MAN.   'Tis over—my dull eyes can fix thee not;
But all things swim around me, and the earth
Heaves as it were beneath me. Fare thee well!
Give me thy hand.
    ABBOT.   Cold—cold—even to the heart—
But yet one prayer—Alas! how fares it with thee? 150
    MAN.   Old man! 'tis not so difficult to die.
      [MANFRED *expires.*]

ABBOT.   He's gone—his soul hath ta'en its earthless
    flight;
Whither? I dread to think—but he is gone.

1816–17 (1817)

SO, WE'LL GO
NO MORE A-ROVING

So, we'll go no more a-roving
    So late into the night,
Though the heart be still as loving,
    And the moon be still as bright.

For the sword outwears its sheath,
    And the soul outwears the breast,
And the heart must pause to breathe,
    And love itself have rest.

Though the night was made for loving,
    And the day returns too soon,   10
Yet we'll go no more a-roving
    By the light of the moon.

1817 (1830)

MY BOAT IS ON THE SHORE

My boat is on the shore,
    And my bark is on the sea;
But, before I go, Tom Moore,
    Here's a double health to thee!

Here's a sigh to those who love me,
　　And a smile to those who hate;
And, whatever sky's above me,
　　Here's a heart for every fate.

Though the ocean roar around me,
　　Yet it still shall bear me on;　　　　10
Though a desert should surround me,
　　It hath springs that may be won.

Were 't the last drop in the well,
　　As I gasp'd upon the brink,
Ere my fainting spirit fell,
　　'T is to thee that I would drink.

With that water, as this wine,
　　The libation I would pour
Should be—peace with thine and mine,
　　And a health to thee, Tom Moore.　　20

　　　　　　　　　　　　1817 (1821)

FROM

DON JUAN

Begun in July, 1818, Byron's garrulous masterpiece broke off only with his death. In *Beppo* he had already explored the attractions of the style he uses in *Don Juan*. It is a style derived from Italian poets of the Renaissance and from the contemporary imitation by John Hookham Frere, who wrote under the pseudonym of "Whistle-craft" (1817). The Italians—Pulci, Boiardo, Berni, Ariosto—provided a model for the colloquial idiom, the volubility, the brisk and easy handling of ottava rima (a stanza of eight pentameter lines rhyming *abababcc*), the profuse incident, the wandering plot, and, above all, the medley of realism and romance, of sentiment and buffoonery. "Whistlecraft" contributed especially by suggesting the use of "Hudibrastic" rhymes (see note to Canto I, ll. 175–76), the most obvious of Byron's verbal high jinks, though he also amuses himself with puns, slang, deliberate solecisms, cacaphony, and the like. That Byron happened upon these writers is one of the most significant accidents in literary history; only with their help did he achieve full expression of his complex personality.

Even his friends were troubled by what they deemed the immorality of the poem, and the reviewers could hardly find words for their disgust. A few defenders, Byron most of all, praised his candid honesty, his knowledge of human nature, and his vigor in exposing cant and hypocrisy. But discussion at this level is rarely useful. Episodes that are shocking are not necessarily immoral and, on the other hand, honesty need not imply sensitivity and depth. More thoughtful readers saw that the poem might be described not as immoral but as nihilistic. They lacked experience with such concepts and temperaments, and naturally did not use phrases that come to mind nowadays. But they recognized that Byron seemed unable to believe in anything, not even in his own emotions. He was strangely driven to undercut his strongest passages. This is the point of Hazlitt's comment on the poem in "Lord Byron," above. When Byron is most serious and most moral he at once begins to doubt the sincerity of his emotion and the grounds for it, and so "You laugh and are surprised that anyone should turn round and *travestie* himself. . . . A classical intoxication is followed by the splashing of soda water, by frothy effusions of ordinary bile." From this point of view, the rapid fluctuation of feeling and attitude is not merely wit; it expresses Byron's tendency to nihilism. But to overemphasize this aspect would be to be owl-solemn and blind. The poem has marvelous verve, color, and movement. It does not create character in much depth, but the marionettes it animates have a fierce energy of passion. It is fascinated by the world. It is sunny, exuberant, and very amusing. The best comment on it is Byron's. "Confess, confess—you dog," he writes to a friend, "it may be bawdy but is it not good English? it may be profligate but is it not *life*, is it not *the thing*?"

The two main figures in the poem are Juan and the narrator. The Don Juan of legend had been a heartless seducer, but in Byron women are the aggressors and Juan simply responds—we are made to feel—as any man would in the circumstances. Indeed, his character is remarkably generalized. We can list traits—courage, kindliness, generosity, idealism—but the upshot seems to be that he is an average sort of fellow subject to average instincts and illusions. Moreover, though he learns to cope with the increasingly complicated society into which he is introduced, he does not change in any fundamental respect. That is, his experience takes no hold upon him. He does not reflect; he is not troubled by guilt; he scarcely remembers his past; he meets each new vicissitude freshly, directly, and fearlessly. He has an invulnerable youth, and for this reason we find him immensely likable, though we cannot be profoundly engaged in his adventures.

The results of a dissolute and various course of life such as his are summed up in another personage, equally attractive and much more distinct. He is the narrator. We may, if we like, call him Byron; for Byron deliberately alludes to himself. But we should keep in mind that just as much as Juan he is a character in a literary work. The narrator is older than Juan, more experienced, worldly-wise, disillusioned, garrulous and gossipy, a middle-aged society rake with a liking for drink and metaphysical speculation. He editorializes, digresses, and runs unstoppably on in vigorous, amusing talk. The episode with Haidée, for example, is rather as though the love of Romeo and Juliet were told by Pandarus.

When Juan woke he found some good things ready,
　　A bath, a breakfast, and the finest eyes
That ever made a youthful heart less steady,
　　Besides her maid's, as pretty for their size;

But I have spoken of all this already—
　　And repetition's tiresome and unwise—
Well—Juan, after bathing in the sea,
　　Came always back to coffee and Haidée

　　　　　　　　　　　　　　　　[II. 1361–68].

The connoisseurship of maiden's eyes is not Juan's but
the narrator's. He is not only unforgettable in his own
right but essential in the total expression. He makes
possible the strange union of romantic incident and disillusioned commentary, of sympathy, shrewdness, and
laughter.

　　The selection below comprises all of Canto I, the shipwreck from Canto II, the love of Juan and Haidée in
Cantos II–IV, and Juan's adventures in English high
society in Cantos XI–XVI (including all of Canto XVI).

FRAGMENT

On the back of the Poet's MS. of Canto I.

I would to heaven that I were so much clay,
　　As I am blood, bone, marrow, passion, feeling—
Because at least the past were pass'd away—
　　And for the future—(but I write this reeling,
Having got drunk exceedingly to-day,
　　So that I seem to stand upon the ceiling)
I say—the future is a serious matter—
And so—for God's sake—hock and soda-water!

DEDICATION

I

Bob Southey! You 're a poet—Poet-laureate,
　　And representative of all the race;
Although 't is true that you turned out a Tory at
　　Last,—yours has lately been a common case;
And now, my Epic Renegade! what are ye at?
　　With all the Lakers, in and out of place?
A nest of tuneful persons, to my eye
Like "four and twenty Blackbirds in a pye;

DON JUAN. **Fragment. 8. hock:** Rhine wine (from the
German *Hochheimer*). Hock and soda water were frequently
drunk to relieve a hangover. DEDICATION. **6. Lakers:** the
"Lake Poets." **8. pye:** Byron puns on the name of the
pedestrian Henry James Pye, 1745–1813, the poet laureate
just before Robert Southey. He was derisively known as
"Poetical Pye." The "pye being opened" (his position as poet
laureate available) the blackbirds all began to sing "before
the King."

II

"Which pye being opened they began to sing"　　　　10
　　(This old song and new simile holds good),
"A dainty dish to set before the King,"
　　Or Regent, who admires such kind of food;—
And Coleridge, too, has lately taken wing,
　　But like a hawk encumbered with his hood,—
Explaining metaphysics to the nation—
I wish he would explain his Explanation.

III

You, Bob! are rather insolent, you know,
　　At being disappointed in your wish
To supersede all warblers here below,
　　And be the only Blackbird in the dish;　　　　20
And then you overstrain yourself, or so,
　　And tumble downward like the flying fish
Gasping on deck, because you soar too high, Bob,
And fall, for lack of moisture quite a-dry, Bob!

IV

And Wordsworth, in a rather long "Excursion"
　　(I think the quarto holds five hundred pages),
Has given a sample from the vasty version
　　Of his new system to perplex the sages;
'T is poetry—at least by his assertion,
　　And may appear so when the dog-star rages—　　30
And he who understands it would be able
To add a story to the Tower of Babel.

V

You—Gentlemen! by dint of long seclusion
　　From better company, have kept your own
At Keswick, and, through still continued fusion
　　Of one another's minds, at last have grown
To deem as a most logical conclusion,
　　That Poesy has wreaths for you alone:
There is a narrowness in such a notion,
Which makes me wish you'd change your lakes for
　　ocean.　　　　　　　　　　　　　　　　40

12. Regent: the Prince Regent, later George IV. **24. a-dry,
Bob:** In the slang of Byron's time, "dry Bob" indicated
incomplete intercourse. **35. Keswick:** in the Lake Country.

VI

I would not imitate the petty thought,
  Nor coin my self-love to so base a vice,
For all the glory your conversion brought,
  Since gold alone should not have been its price.
You have your salary: was 't for that you wrought?
  And Wordsworth has his place in the Excise.
You're shabby fellows—true—but poets still,
And duly seated on the immortal hill.

VII

Your bays may hide the baldness of your brows—
  Perhaps some virtuous blushes;—let them go—   50
To you I envy neither fruit nor boughs—
  And for the fame you would engross below,
The field is universal, and allows
  Scope to all such as feel the inherent glow;
Scott, Rogers, Campbell, Moore, and Crabbe, will try
'Gainst you the question with posterity.

VIII

For me, who, wandering with pedestrian Muses,
  Contend not with you on the winged steed,
I wish your fate may yield ye, when she chooses,
  The fame you envy, and the skill you need;   60
And recollect a poet nothing loses
  In giving to his brethren their full meed
Of merit, and complaint of present days
Is not the certain path to future praise.

IX

He that reserves his laurels for posterity
  (Who does not often claim the bright reversion)
Has generally no great crop to spare it, he
  Being only injured by his own assertion;
And although here and there some glorious rarity
  Arise like Titan from the sea's immersion,   70
The major part of such appellants go
To—God knows where—for no one else can know.

X

If, fallen in evil days on evil tongues,
  Milton appealed to the Avenger, Time,

If Time, the Avenger, execrates his wrongs,
  And makes the word "Miltonic" mean "*sublime*,"
He deigned not to belie his soul in songs,
  Nor turn his very talent to a crime;
He did not loathe the Sire to laud the Son,
But closed the tyrant-hater he begun.   80

XI

Think'st thou, could he—the blind Old Man—arise,
  Like Samuel from the grave, to freeze once more
The blood of monarchs with his prophecies,
  Or be alive again—again all hoar
With time and trials, and those helpless eyes,
  And heartless daughters—worn—and pale—and
    poor;
Would *he* adore a sultan? *he* obey
The intellectual eunuch Castlereagh?

XII

Cold-blooded, smooth-faced, placid miscreant!
  Dabbling its sleek young hands in Erin's gore,   90
And thus for wider carnage taught to pant,
  Transferred to gorge upon a sister shore,
The vulgarest tool that Tyranny could want,
  With just enough of talent, and no more,
To lengthen letters by another fixed,
And offer poison long already mixed.

XIII

An orator of such set trash of phrase
  Ineffably—legitimately vile,
That even its grossest flatterers dare not praise,
  Nor foes—all nations—condescend to smile;   100
Not even a sprightly blunder's spark can blaze
  From that Ixion grindstone's ceaseless toil,
That turns and turns to give the world a notion
Of endless torments and perpetual motion.

79. Sire . . . Son: Milton, opposed to Charles I, did not change his politics after the Restoration, as did so many, and praise Charles II. **82. Like Samuel:** I Samuel 27:13-14. **88. Castlereagh:** Robert Stewart, Viscount Castlereagh, 1769-1822, Foreign Minister in the Tory government, had repressed the United Irish rebellion. Hence the reference to "Erin's gore" in line 90. **102. Ixion:** chained to an eternally revolving wheel for ingratitude to Zeus.

46. Wordsworth . . . Excise: Wordsworth had the position of Distributor of Stamps for Westmorland. **73. fallen . . . tongues:** from *Paradise Lost*, VII. 25-26.

XIV

A bungler even in its disgusting trade,
  And botching, patching, leaving still behind
Something of which its masters are afraid,
  States to be curbed, and thoughts to be confined,
Conspiracy or Congress to be made—
  Cobbling at manacles for all mankind—   110
A tinkering slave-maker, who mends old chains,
With God and man's abhorrence for its gains.

XV

If we may judge of matter by the mind,
  Emasculated to the marrow *It*
Hath but two objects, how to serve, and bind,
  Deeming the chain it wears even men may fit,
Eutropius of its many masters,—blind
  To worth as freedom, wisdom as to wit,
Fearless—because *no* feeling dwells in ice,
Its very courage stagnates to a vice.   120

XVI

Where shall I turn me not to *view* its bonds,
  For I will never *feel* them;—Italy!
Thy late reviving Roman soul desponds
  Beneath the lie this State-thing breathed o'er thee—
Thy clanking chain, and Erin's yet green wounds,
  Have voices—tongues to cry aloud for me.
Europe has slaves, allies, kings, armies still,
And Southey lives to sing them very ill.

XVII

Meantime, Sir Laureate, I proceed to dedicate,
  In honest simple verse, this song to you.   130
And, if in flattering strains I do not predicate,
  'T is that I will retain my "buff and blue";
My politics as yet are all to educate:
  Apostasy 's so fashionable, too,
To keep *one* creed's a task grown quite Herculean:
Is it not so, my Tory, Ultra-Julian?

Venice, September 16, 1818

117. **Eutropius:** a eunuch who became minister in the court of Arcadius in the Byzantine Empire (see Gibbon's *Decline and Fall of the Roman Empire*, Chap. XXXII). 124. **lie:** After supporting Genoa, Castlereagh favored its annexation by Piedmont. 132. **"buff and blue":** colors of Charles James Fox and the Whig Club. 136. **Ultra-Julian:** referring to Julian the Apostate.

CANTO THE FIRST

I

I want a hero: an uncommon want,
  When every year and month sends forth a new one,
Till, after cloying the gazettes with cant,
  The age discovers he is not the true one:
Of such as these I should not care to vaunt,
  I'll therefore take our ancient friend Don Juan—
We all have seen him, in the pantomime,
Sent to the devil somewhat ere his time.

II

Vernon, the butcher Cumberland, Wolfe, Hawke,
  Prince Ferdinand, Granby, Burgoyne, Keppel, Howe,   10
Evil and good, have had their tithe of talk,
  And filled their sign-posts then, like Wellesley now;
Each in their turn like Banquo's monarchs stalk,
  Followers of fame, "nine farrow" of that sow:
France, too, had Buonaparté and Dumourier
Recorded in the Moniteur and Courier.

III

Barnave, Brissot, Condorcet, Mirabeau,
  Pétion, Clootz, Danton, Marat, La Fayette,
Were French, and famous people, as we know;
  And there were others, scarce forgotten yet,   20
Joubert, Hoche, Marceau, Lannes, Desaix, Moreau,
  With many of the military set,
Exceedingly remarkable at times,
But not at all adapted to my rhymes.

IV

Nelson was once Britannia's god of war,
  And still should be so, but the tide is turned;

CANTO I. **7. pantomine:** The Don Juan legend had been a popular subject in pantomime for a century and a half. **9–10. Vernon . . . Howe:** military and naval heroes of the eighteenth century. **12. Wellesley:** the Duke of Wellington. **13. Banquo's monarchs:** *Macbeth*, IV. i. 112–13. **14. "nine farrow":** "Pour in sow's blood, that hath eaten / Her nine farrow" (*Macbeth*, IV. i. 65). **15. Dumourier:** Charles Dumourier was a French general. **16. Moniteur . . . Courier:** French newspapers. **17–21. Barnave . . . Moreau:** The first list is of men active in the French Revolution; the second (l. 21), military heroes in the Napoleonic wars.

There's no more to be said of Trafalgar,
  'T is with our hero quietly inurned;
Because the army 's grown more popular,
  At which the naval people are concerned,   30
Besides, the prince is all for the land-service,
Forgetting Duncan, Nelson, Howe, and Jervis.

V

Brave men were living before Agamemnon
  And since, exceeding valorous and sage,
A good deal like him too, though quite the same none;
  But then they shone not on the poet's page,
And so have been forgotten:—I condemn none,
  But can't find any in the present age
Fit for my poem (that is, for my new one);
So, as I said, I'll take my friend Don Juan.   40

VI

Most epic poets plunge "in medias res"
  (Horace makes this the heroic turnpike road),
And then your hero tells, whene'er you please,
  What went before—by way of episode,
While seated after dinner at his ease,
  Beside his mistress in some soft abode,
Palace, or garden, paradise, or cavern,
Which serves the happy couple for a tavern.

VII

That is the usual method, but not mine—
  My way is to begin with the beginning;   50
The regularity of my design
  Forbids all wandering as the worst of sinning,
And therefore I shall open with a line
  (Although it cost me half an hour in spinning)
Narrating somewhat of Don Juan's father,
And also of his mother, if you'd rather.

VIII

In Seville was he born, a pleasant city,
  Famous for oranges and women—he

Who has not seen it will be much to pity,
  So says the proverb—and I quite agree;   60
Of all the Spanish towns is none more pretty,
  Cadiz, perhaps—but that you soon may see:—
Don Juan's parents lived beside the river,
A noble stream, and called the Guadalquivir.

IX

His father's name was Jóse—*Don*, of course,
  A true Hidalgo, free from every stain
Of Moor or Hebrew blood, he traced his source
  Through the most Gothic gentlemen of Spain;
A better cavalier ne'er mounted horse,
  Or, being mounted, e'er got down again,   70
Than Jóse, who begot our hero, who
Begot—but that's to come—Well, to renew:

X

His mother was a learned lady, famed
  For every branch of every science known—
In every Christian language ever named,
  With virtues equalled by her wit alone:
She made the cleverest people quite ashamed,
  And even the good with inward envy groan,
Finding themselves so very much exceeded
In their own way by all the things that she did.   80

XI

Her memory was a mine: she knew by heart
  All Calderon and greater part of Lopé,
So that if any actor missed his part
  She could have served him for the prompter's copy;
For her Feinagle's were an useless art,
  And he himself obliged to shut up shop—he
Could never make a memory so fine as
That which adorned the brain of Donna Inez.

73. **learned lady:** In addition to satirizing the bluestocking generally (a common object of attack by writers of the time, especially Hazlitt), Byron is also ridiculing the intellectual pretensions of his wife. 82. **Calderon . . . Lopé:** Calderon de la Barca, 1600–81, and Lopé de Vega, 1562–1635, the great dramatists of Spain. 85. **Feinagle:** Gregor von Feinagle, who had recently lectured in London (1811) on a system of memory.

32. **Duncan . . . Jervis:** British admirals in the war against Napoleon. 33. **Brave . . . Agamemnon:** a quote from Horace (*Odes*, IV. ix. 25–26). 41. **"in medias res":** The poet, says Horace (*Art of Poetry*, l. 148), should, in beginning his story, plunge "into the midst of things."

XII

Her favorite science was the mathematical,
 Her noblest virtue was her magnanimity; 90
Her wit (she sometimes tried at wit) was Attic all,
 Her serious sayings darkened to sublimity;
In short, in all things she was fairly what I call
 A prodigy—her morning dress was dimity,
Her evening silk, or, in the summer, muslin,
And other stuffs, with which I won't stay puzzling.

XIII

She knew the Latin—that is, "the Lord's prayer,"
 And Greek—the alphabet—I'm nearly sure;
She read some French romances here and there,
 Although her mode of speaking was not pure; 100
For native Spanish she had no great care,
 At least her conversation was obscure;
Her thoughts were theorems, her words a problem,
As if she deemed that mystery would ennoble 'em.

XIV

She liked the English and the Hebrew tongue,
 And said there was analogy between 'em;
She proved it somehow out of sacred song,
 But I must leave the proofs to those who've seen 'em,
But this I heard her say, and can't be wrong,
 And all may think which way their judgments lean
  'em, 110
"'T is strange—the Hebrew noun which means 'I am,'
The English always use to govern d—n."

XV

Some women use their tongues—she *looked* a lecture,
 Each eye a sermon, and her brow a homily,
An all-in-all sufficient self-director,
 Like the lamented late Sir Samuel Romilly,
The Law's expounder, and the State's corrector,
 Whose suicide was almost an anomaly—
One sad example more, that "All is vanity,"—
(The jury brought their verdict in "Insanity.") 120

116. Sir Samuel Romilly: who had served as lawyer to
Lady Byron. He had committed suicide in 1818 after the
death of his wife (hence Byron's gibe that the verdict of the
jury, after his death, was "insanity").

XVI

In short, she was a walking calculation,
 Miss Edgeworth's novels stepping from their covers,
Or Mrs. Trimmer's books on education,
 Or "Cœlebs' Wife" set out in quest of lovers,
Morality's prim personification,
 In which not Envy's self a flaw discovers;
To others' share let "female errors fall,"
For she had not even one—the worst of all.

XVII

Oh! she was perfect past all parallel—
 Of any modern female saint's comparison; 130
So far above the cunning powers of hell,
 Her guardian angel had given up his garrison;
Even her minutest motions went as well
 As those of the best time-piece made by Harrison:
In virtues nothing earthly could surpass her,
Save thine "incomparable oil," Macassar!

XVIII

Perfect she was, but as perfection is
 Insipid in this naughty world of ours,
Where our first parents never learned to kiss
 Till they were exiled from their earlier bowers, 140
Where all was peace, and innocence, and bliss
 (I wonder how they got through the twelve hours),
Don Jóse, like a lineal son of Eve,
Went plucking various fruit without her leave.

XIX

He was a mortal of the careless kind,
 With no great love for learning, or the learned,
Who chose to go where'er he had a mind,
 And never dreamed his lady was concerned;

122–24. Miss Edgeworth's . . . "Cœlebs' Wife": another
reflection on women of literary pretension (cf. l. 73, above).
Maria Edgeworth, 1767–1849, was a popular novelist; Sarah
Trimmer, 1741–1810, wrote flaccid books on the education
of children; Hannah More, 1745–1833, was the writer of the
sermonizing novel to which Byron refers, *Cœlebs in Search of
a Wife* (1800). **127. "female errors fall':** a quote from
Pope's *Rape of the Lock*, II. 17. **134. Harrison:** John Harrison,
who designed a chronometer for determining the longitude at
sea. **136. Macassar:** hair oil from the island of Macassar, from
which the lace coverings on top of the backs of plush chairs
were called "anti-macassars" throughout the Victorian era.

The world, as usual, wickedly inclined
  To see a kingdom or a house o'erturned,　　150
Whispered he had a mistress, some said *two*,
But for domestic quarrels *one* will do.

XX

Now Donna Inez had, with all her merit,
  A great opinion of her own good qualities;
Neglect, indeed, requires a saint to bear it,
  And such, indeed, she was in her moralities;
But then she had a devil of a spirit,
  And sometimes mixed up fancies with realities,
And let few opportunities escape
Of getting her liege lord into a scrape.　　160

XXI

This was an easy matter with a man
  Oft in the wrong, and never on his guard;
And even the wisest, do the best they can,
  Have moments, hours, and days, so unprepared,
That you might "brain them with their lady's fan";
  And sometimes ladies hit exceeding hard,
And fans turn into falchions in fair hands,
And why and wherefore no one understands.

XXII

'T is pity learned virgins ever wed
  With persons of no sort of education,　　170
Or gentlemen, who, though well born and bred,
  Grow tired of scientific conversation;
I don't choose to say much upon this head,
  I'm a plain man, and in a single station,
But—Oh! ye lords of ladies intellectual,
Inform us truly, have they not hen-pecked you all?

XXIII

Don Jóse and his lady quarrelled—*why*,
  Not any of the many could divine,
Though several thousand people chose to try,
  'T was surely no concern of theirs nor mine;　　180

I loathe that low vice—curiosity;
  But if there's anything in which I shine,
'T is in arranging all my friends' affairs,
Not having, of my own, domestic cares.

XXIV

And so I interfered, and with the best
  Intentions, but their treatment was not kind;
I think the foolish people were possessed,
  For neither of them could I ever find,
Although their porter afterwards confessed—
  But that's no matter, and the worst 's behind,　　190
For little Juan o'er me threw, down stairs,
A pail of housemaid's water unawares.

XXV

A little curly-headed, good-for-nothing,
  And mischief-making monkey from his birth;
His parents ne'er agreed except in doting
  Upon the most unquiet imp on earth;
Instead of quarrelling, had they been but both in
  Their senses, they'd have sent young master forth
To school, or had him soundly whipped at home,
To teach him manners for the time to come.　　200

XXVI

Don Jóse and the Donna Inez led
  For some time an unhappy sort of life,
Wishing each other, not divorced, but dead;
  They lived respectably as man and wife,
Their conduct was exceedingly well-bred,
  And gave no outward signs of inward strife,
Until at length the smothered fire broke out,
And put the business past all kind of doubt.

XXVII

For Inez called some druggists and physicians,
  And tried to prove her loving lord was *mad*,　　210
But as he had some lucid intermissions,
  She next decided he was only *bad*;
Yet when they asked her for her depositions,
  No sort of explanation could be had,

---

165. "brain . . . fan": a quote from *Henry V*, II. iii. 25–26.
175–76. intellectual . . . all: one of the most brilliant examples of "Hudibrastic" rhyme, in which a polysyllabic word is rhymed by two or three monosyllabic words altogether unexpected.

210. prove . . . mad: a hope of Lady Byron before the separation.

Save that her duty both to man and God
Required this conduct—which seemed very odd.

XXVIII

She kept a journal, where his faults were noted,
   And opened certain trunks of books and letters,
All which might, if occasion served, be quoted;
   And then she had all Seville for abettors,   220
Besides her good old grandmother (who doted);
   The hearers of her case became repeaters,
Then advocates, inquisitors, and judges,
Some for amusement, others for old grudges.

XXIX

And then this best and meekest woman bore
   With such serenity her husband's woes,
Just as the Spartan ladies did of yore,
   Who saw their spouses killed, and nobly chose
Never to say a word about them more—
   Calmly she heard each calumny that rose,   230
And saw *his* agonies with such sublimity,
That all the world exclaimed, "What magnanimity!"

XXX

No doubt this patience, when the world is damning us,
   Is philosophic in our former friends;
'T is also pleasant to be deemed magnanimous,
   The more so in obtaining our own ends;
And what the lawyers call a "*malus animus*"
   Conduct like this by no means comprehends:
Revenge in person 's certainly no virtue,
But then 't is not *my* fault, if *others* hurt you.   240

XXXI

And if our quarrels should rip up old stories,
   And help them with a lie or two additional,
I'm not to blame, as you well know—no more is
   Any one else—they were become traditional;
Besides, their resurrection aids our glories
   By contrast, which is what we just were wishing all:
And science profits by this resurrection—
Dead scandals form good subjects for dissection.

XXXII

Their friends had tried at reconciliation,
   Then their relations, who made matters worse,   250
('T were hard to tell upon a like occasion
   To whom it may be best to have recourse—
I can't say much for friend or yet relation):
   The lawyers did their utmost for divorce,
But scarce a fee was paid on either side
Before, unluckily, Don Jóse died.

XXXIII

He died: and most unluckily, because,
   According to all hints I could collect
From counsel learned in those kinds of laws
   (Although their talk 's obscure and circumspect),
His death contrived to spoil a charming cause;   261
   A thousand pities also with respect
To public feeling, which on this occasion
Was manifested in a great sensation.

XXXIV

But ah! he died; and buried with him lay
   The public feeling and the lawyers' fees:
His house was sold, his servants sent away,
   A Jew took one of his two mistresses,
A priest the other—at least so they say:
   I asked the doctors after his disease—   270
He died of the slow fever called the tertian,
And left his widow to her own aversion.

XXXV

Yet Jóse was an honorable man,
   That I must say, who knew him very well;
Therefore his frailties I 'll no further scan,
   Indeed there were not many more to tell:
And if his passions now and then outran
   Discretion, and were not so peaceable
As Numa's (who was also named Pompilius),
He had been ill brought up, and was born bilious.   280

XXXVI

Whate'er might be his worthlessness or worth,
   Poor fellow! he had many things to wound him,

---

215. duty . . . God: Byron echoes a letter from his wife as
she was seeking separation from him ("I deem it *my duty to
God* to act as I am acting"). **237. "malus animus":**
malice aforethought.

279. Numa: Numa Pompilius, second king of Rome, whose
peaceful reign lasted forty-three years.

Let's own—since it can do no good on earth—
　It was a trying moment that which found him
Standing alone beside his desolate hearth,
　Where all his household gods lay shivered round him:
No choice was left his feelings or his pride,
Save death or Doctors' Commons—so he died.

XXXVII

Dying intestate, Juan was sole heir
　To a chancery suit, and messuages and lands,　290
Which, with a long minority and care,
　Promised to turn out well in proper hands:
Inez became sole guardian, which was fair,
　And answered but to nature's just demands;
An only son left with an only mother
Is brought up much more wisely than another.

XXXVIII

Sagest of women, even of widows, she
　Resolved that Juan should be quite a paragon,
And worthy of the noblest pedigree:
　(His sire was of Castile, his dam from Aragon).　300
Then for accomplishments of chivalry,
　In case our lord the king should go to war again,
He learned the arts of riding, fencing, gunnery,
And how to scale a fortress—or a nunnery.

XXXIX

But that which Donna Inez most desired,
　And saw into herself each day before all
The learned tutors whom for him she hired,
　Was, that his breeding should be strictly moral:
Much into all his studies she inquired,
　And so they were submitted first to her, all,　310
Arts, sciences, no branch was made a mystery
To Juan's eyes, excepting natural history.

XL

The languages, especially the dead,
　The sciences, and most of all the abstruse,
The arts, at least all such as could be said
　To be the most remote from common use,
In all these he was much and deeply read:
　But not a page of anything that 's loose,

Or hints continuation of the species,
Was ever suffered, lest he should grow vicious.　320

XLI

His classic studies made a little puzzle,
　Because of filthy loves of gods and goddesses,
Who in the earlier ages raised a bustle,
　But never put on pantaloons or bodices;
His reverend tutors had at times a tussle,
　And for their Æneids, Iliads, and Odysseys,
Were forced to make an odd sort of apology,
For Donna Inez dreaded the Mythology.

XLII

Ovid's a rake, as half his verses show him,
　Anacreon's morals are a still worse sample,　330
Catullus scarcely has a decent poem,
　I don't think Sappho's Ode a good example,
Although Longinus tells us there is no hymn
　Where the sublime soars forth on wings more ample;
But Virgil's songs are pure, except that horrid one
Beginning with "Formosum Pastor Corydon."

XLIII

Lucretius' irreligion is too strong
　For early stomachs, to prove wholesome food;
I can't help thinking Juvenal was wrong,
　Although no doubt his real intent was good,　340
For speaking out so plainly in his song,
　So much indeed as to be downright rude;
And then what proper person can be partial
To all those nauseous epigrams of Martial?

XLIV

Juan was taught from out the best edition,
　Expurgated by learned men, who place,

336. **"Formosum . . . Corydon":** concerned with the love of the shepherd Corydon for "the fair Alexis" (Vergil, *Buccolics*, Eclogue II). 337–39: **Lucretius' . . . Juvenal:** referring to the *De Rerum Natura* of Lucretius, which advances the theory that the universe was created by the chance combination of atoms, and the satires of Juvenal that denounced the vices of Roman society. 346–51. **Expurgated . . . appendix:** a practice actually followed by some eighteenth- and many nineteenth-century editors and scholars.

288. **Doctors' Commons:** divorce courts. 290. **messuages:** legal term for the dwelling house and adjacent buildings.

Judiciously, from out the schoolboy's vision,
    The grosser parts; but, fearful to deface
Too much their modest bard by this omission,
    And pitying sore this mutilated case,   350
They only add them all in an appendix,
Which saves, in fact, the trouble of an index;

XLV

For there we have them all "at one fell swoop,"
    Instead of being scattered through the pages;
They stand forth marshalled in a handsome troop,
    To meet the ingenuous youth of future ages,
Till some less rigid editor shall stoop
    To call them back into their separate cages,
Instead of standing staring all together,
Like garden gods—and not so decent either.   360

XLVI

The Missal too (it was the family Missal)
    Was ornamented in a sort of way
Which ancient mass-books often are, and this all
    Kinds of grotesques illumined; and how they,
Who saw those figures on the margin kiss all,
    Could turn their optics to the text and pray,
Is more than I know—But Don Juan's mother
Kept this herself, and gave her son another.

XLVII

Sermons he read, and lectures he endured,
    And homilies, and lives of all the saints;   370
To Jerome and to Chrysostom inured,
    He did not take such studies for restraints;
But how faith is acquired, and then insured,
    So well not one of the aforesaid paints
As Saint Augustine in his fine Confessions,
Which make the reader envy his transgressions.

361. **Missal:** the book containing the service of the Mass
for the entire year, often beautifully illuminated in the
Middle Ages. 371. **Jerome . . . inured:** the ascetic life
of St. Jerome (340?–420) in the desert, and the moral writings
of St. John Chrysostom (347?–407). 375–76. **Saint Augustine . . . transgressions:** "See his *Confessions*, I. ix. By
the representation which Saint Augustine gives of himself
in his youth, it is easy to see that he was what we should call
a rake. He avoided the school as the plague; he loved nothing
but gaming and public shows; he robbed his father of everything he could find; he invented a thousand lies to escape
the rod, which they were obliged to make use of to punish
his irregularities" (B.).

XLVIII

This, too, was a sealed book to little Juan—
    I can't but say that his mamma was right,
If such an education was the true one.
    She scarcely trusted him from out her sight;   380
Her maids were old, and if she took a new one,
    You might be sure she was a perfect fright,
She did this during even her husband's life—
I recommend as much to every wife.

XLIX

Young Juan waxed in godliness and grace;
    At six a charming child, and at eleven
With all the promise of as fine a face
    As e'er to man's maturer growth was given.
He studied steadily and grew apace,
    And seemed, at least, in the right road to heaven,
For half his days were passed at church, the other   391
Between his tutors, confessor, and mother.

L

At six, I said, he was a charming child,
    At twelve he was a fine, but quiet boy;
Although in infancy a little wild,
    They tamed him down amongst them: to destroy
His natural spirit not in vain they toiled,
    At least it seemed so; and his mother's joy
Was to declare how sage, and still, and steady,
Her young philosopher was grown already.   400

LI

I had my doubts, perhaps I have them still,
    But what I say is neither here nor there:
I knew his father well, and have some skill
    In character—but it would not be fair
From sire to son to augur good or ill:
    He and his wife were an ill sorted pair—
But scandal 's my aversion—I protest
Against all evil speaking, even in jest.

LII

For my part I say nothing—nothing—but
    This I will say—my reasons are my own—   401
That if I had an only son to put
    To school (as God be praised that I have none),

'T is not with Donna Inez I would shut
   Him up to learn his catechism alone,
No—no—I'd send him out betimes to college,
For there it was I picked up my own knowledge.

LIII

For there one learns—'t is not for me to boast,
   Though I acquired—but I pass over *that*,
As well as all the Greek I since have lost:
   I say that there's the place—but *"Verbum sat,"*   420
I think I picked up too, as well as most,
   Knowledge of matters—but no matter *what*—
I never married—but, I think, I know
That sons should not be educated so.

LIV

Young Juan now was sixteen years of age,
   Tall, handsome, slender, but well knit: he seemed
Active, though not so sprightly, as a page;
   And everybody but his mother deemed
Him almost man; but she flew in a rage
   And bit her lips (for else she might have screamed)
If any said so, for to be precocious   431
Was in her eyes a thing the most atrocious.

LV

Amongst her numerous acquaintance, all
   Selected for discretion and devotion,
There was the Donna Julia, whom to call
   Pretty were but to give a feeble notion
Of many charms in her as natural
   As sweetness to the flower, or salt to ocean,
Her zone to Venus, or his bow to Cupid,
(But this last simile is trite and stupid).   440

LVI

The darkness of her Oriental eye
   Accorded with her Moorish origin;
(Her blood was not all Spanish, by the by;
   In Spain, you know, this is a sort of sin).
When proud Granada fell, and, forced to fly,
   Boabdil wept, of Donna Julia's kin

420. **"Verbum sat"**: "A word [to the wise] is sufficient."
429. **flew in a rage**: Here Donna Inez is given a trait of Byron's mother rather than his wife. 446. **Boabdil**: last Moorish king of Grenada, defeated by Ferdinand in 1492.

Some went to Africa, some stayed in Spain,
Her great great grandmamma chose to remain.

LVII

She married (I forget the pedigree)
   With an Hidalgo, who transmitted down   450
His blood less noble than such blood should be;
   At such alliances his sires would frown,
In that point so precise in each degree
   That they bred *in and in*, as might be shown,
Marrying their cousins—nay, their aunts, and nieces,
Which always spoils the breed, if it increases.

LVIII

This heathenish cross restored the breed again,
   Ruined its blood, but much improved its flesh;
For from a root the ugliest in old Spain
   Sprung up a branch as beautiful as fresh;   460
The sons no more were short, the daughters plain:
   But there's a rumor which I fain would hush,
'T is said that Donna Julia's grandmamma
Produced her Don more heirs at love than law.

LIX

However this might be, the race went on
   Improving still through every generation,
Until it centered in an only son,
   Who left an only daughter: my narration
May have suggested that this single one
   Could be but Julia (whom on this occasion   470
I shall have much to speak about), and she
Was married, charming, chaste, and twenty-three.

LX

Her eye (I'm very fond of handsome eyes)
   Was large and dark, suppressing half its fire
Until she spoke, then through its soft disguise
   Flashed an expression more of pride than ire,
And love than either; and there would arise
   A something in them which was not desire,
But would have been, perhaps, but for the soul
Which struggled through and chastened down the
   whole.   480

450. **Hidalgo**: a lesser Spanish nobleman.

LXI

Her glossy hair was clustered o'er a brow
   Bright with intelligence, and fair, and smooth;
Her eyebrow's shape was like the aërial bow,
   Her cheek all purple with the beam of youth,
Mounting, at times, to a transparent glow,
   As if her veins ran lightning; she, in sooth,
Possessed an air and grace by no means common:
Her stature tall—I hate a dumpy woman.

LXII

Wedded she was some years, and to a man
   Of fifty, and such husbands are in plenty;   490
And yet, I think, instead of such a ONE
   'T were better to have TWO of five-and-twenty,
Especially in countries near the sun:
   And now I think on 't, "mi vien in mente,"
Ladies even of the most uneasy virtue
Prefer a spouse whose age is short of thirty.

LXIII

'T is a sad thing, I cannot choose but say,
   And all the fault of that indecent sun,
Who cannot leave alone our helpless clay,
   But will keep baking, broiling, burning on,   500
That howsoever people fast and pray,
   The flesh is frail, and so the soul undone:
What men call gallantry, and gods adultery,
Is much more common where the climate's sultry.

LXIV

Happy the nations of the moral North!
   Where all is virtue, and the winter season
Sends sin, without a rag on, shivering forth
   ('T was snow that brought St. Anthony to reason);
Where juries cast up what a wife is worth,
   By laying whate'er sum, in mulct, they please on
The lover, who must pay a handsome price,   511
Because it is a marketable vice.

494. "mi . . . mente": It comes into my mind. 508. St.
Anthony: a slip for St. Francis, as Byron suspected when he
wrote on the proof-sheets: "I am not sure it was not St.
Francis who had the wife of snow—in that case the line must
run, 'St. Francis back to reason.'" The publisher, however,
failed to make the change.

LXV

Alfonso was the name of Julia's lord,
   A man well looking for his years, and who
Was neither much beloved nor yet abhorred:
   They lived together as most people do,
Suffering each others' foibles by accord,
   And not exactly either *one* or *two;*
Yet he was jealous, though he did not show it,
For jealousy dislikes the world to know it.   520

LXVI

Julia was—yet I never could see why—
   With Donna Inez quite a favorite friend;
Between their tastes there was small sympathy,
   For not a line had Julia ever penned:
Some people whisper (but, no doubt, they lie,
   For malice still imputes some private end)
That Inez had, ere Don Alfonso's marriage,
Forgot with him her very prudent carriage;

LXVII

And that still keeping up the old connexion,
   Which time had lately rendered much more chaste,
She took his lady also in affection,   531
   And certainly this course was much the best:
She flattered Julia with her sage protection,
   And complimented Don Alfonso's taste;
And if she could not (who can?) silence scandal,
At least she left it a more slender handle.

LXVIII

I can't tell whether Julia saw the affair
   With other people's eyes, or if her own
Discoveries made, but none could be aware
   Of this, at least no symptom e'er was shown;   540
Perhaps she did not know, or did not care,
   Indifferent from the first, or callous grown:
I'm really puzzled what to think or say,
She kept her counsel in so close a way.

LXIX

Juan she saw, and, as a pretty child,
   Caressed him often—such a thing might be
Quite innocently done, and harmless styled,
   When she had twenty years, and thirteen he;

But I am not so sure I should have smiled
　　When he was sixteen, Julia twenty-three;　　550
These few short years make wondrous alterations,
Particularly amongst sun-burnt nations.

LXX

Whate'er the cause might be, they had become
　　Changed; for the dame grew distant, the youth shy,
Their looks cast down, their greetings almost dumb,
　　And much embarrassment in either eye;
There surely will be little doubt with some
　　That Donna Julia knew the reason why,
But as for Juan, he had no more notion
Than he who never saw the sea of ocean.　　560

LXXI

Yet Julia's very coldness still was kind,
　　And tremulously gentle her small hand
Withdrew itself from his, but left behind
　　A little pressure, thrilling, and so bland
And slight, so very slight, that to the mind
　　'T was but a doubt; but ne'er magician's wand
Wrought change with all Armida's fairy art
Like what this light touch left on Juan's heart.

LXXII

And if she met him, though she smiled no more,
　　She looked a sadness sweeter than her smile,　　570
As if her heart had deeper thoughts in store
　　She must not own, but cherished more the while
For that compression in its burning core;
　　Even innocence itself has many a wile,
And will not dare to trust itself with truth,
And love is taught hypocrisy from youth.

LXXIII

But passion most dissembles, yet betrays
　　Even by its darkness; as the blackest sky
Foretells the heaviest tempest, it displays
　　Its workings through the vainly guarded eye,　　580
And in whatever aspect it arrays
　　Itself, 't is still the same hypocrisy:
Coldness or anger, even disdain or hate,
Are masks it often wears, and still too late.

567. Armida's fairy art: Armida is an enchantress in Tasso's
Jerusalem Delivered.

LXXIV

Then there were sighs, the deeper for suppression,
　　And stolen glances, sweeter for the theft,
And burning blushes, though for no transgression,
　　Tremblings when met, and restlessness when left;
All these are little preludes to possession,
　　Of which young passion cannot be bereft,　　590
And merely tend to show how greatly love is
Embarrassed at first starting with a novice.

LXXV

Poor Julia's heart was in an awkward state;
　　She felt it going, and resolved to make
The noblest efforts for herself and mate,
　　For honor's, pride's, religion's, virtue's sake.
Her resolutions were most truly great,
　　And almost might have made a Tarquin quake:
She prayed the Virgin Mary for her grace,
As being the best judge of a lady's case.　　600

LXXVI

She vowed she never would see Juan more,
　　And next day paid a visit to his mother,
And looked extremely at the opening door,
　　Which, by the Virgin's grace, let in another;
Grateful she was, and yet a little sore—
　　Again it opens, it can be no other,
'T is surely Juan now—No! I'm afraid
That night the Virgin was no further prayed.

LXXVII

She now determined that a virtuous woman
　　Should rather face and overcome temptation,　　610
That flight was base and dastardly, and no man
　　Should ever give her heart the least sensation;
That is to say, a thought beyond the common
　　Preference, that we must feel upon occasion,
For people who are pleasanter than others,
But then they only seem so many brothers.

598. Tarquin: one of a Roman family, noted for its cruel
arrogance, to which belonged the fifth and seventh kings
of Rome.

LXXVIII

And even if by chance—and who can tell?
   The devil's so very sly—she should discover
That all within was not so very well,
   And, if still free, that such or such a lover   620
Might please perhaps, a virtuous wife can quell
   Such thoughts, and be the better when they're over;
And if the man should ask, 't is but denial:
I recommend young ladies to make trial.

LXXIX

And then there are such things as love divine,
   Bright and immaculate, unmixed and pure,
Such as the angels think so very fine,
   And matrons, who would be no less secure,
Platonic, perfect, "just such love as mine":
   Thus Julia said—and thought so, to be sure;   630
And so I'd have her think, were I the man
On whom her reveries celestial ran.

LXXX

Such love is innocent, and may exist
   Between young persons without any danger:
A hand may first, and then a lip be kist;
   For my part, to such doings I'm a stranger,
But *hear* these freedoms form the utmost list
   Of all o'er which such love may be a ranger:
If people go beyond, 't is quite a crime,
But not my fault—I tell them all in time.   640

LXXXI

Love, then, but love within its proper limits
   Was Julia's innocent determination
In young Don Juan's favor, and to him its
   Exertion might be useful on occasion;
And, lighted at too pure a shrine to dim its
   Ethereal lustre, with what sweet persuasion
He might be taught, by love and her together—
I really don't know what, nor Julia either.

LXXXII

Fraught with this fine intention, and well fenced
   In mail of proof—her purity of soul,   650
She, for the future of her strength convinced,

And that her honor was a rock, or mole,
Exceeding sagely from that hour dispensed
   With any kind of troublesome control;
But whether Julia to the task was equal
Is that which must be mentioned in the sequel.

LXXXIII

Her plan she deemed both innocent and feasible,
   And, surely, with a stripling of sixteen
Not scandal's fangs could fix on much that's seizable,
   Or if they did so, satisfied to mean   660
Nothing but what was good, her breast was peaceable:
   A quiet conscience makes one so serene!
Christians have burnt each other, quite persuaded
That all the Apostles would have done as they did.

LXXXIV

And if in the mean time her husband died,
   But Heaven forbid that such a thought should cross
Her brain, though in a dream! (and then she sighed)
   Never could she survive that common loss;
But just suppose that moment should betide,
   I only say suppose it—*inter nos.*   670
(This should be *entre nous,* for Julia thought
In French, but then the rhyme would go for nought.)

LXXXV

I only say, suppose this supposition:
   Juan being then grown up to man's estate
Would fully suit a widow of condition,
   Even seven years hence it would not be too late;
And in the interim (to pursue this vision)
   The mischief, after all, could not be great,
For he would learn the rudiments of love,
I mean the seraph way of those above.   680

LXXXVI

So much for Julia. Now we'll turn to Juan.
   Poor little fellow! he had no idea
Of his own case, and never hit the true one;
   In feelings quick as Ovid's Miss Medea,
He puzzled over what he found a new one,

684. Ovid's . . . Medea: the sudden fondness of Medea
for Jason in the *Metamorphoses,* VII. 10–12.

But not as yet imagined it could be a
Thing quite in course, and not at all alarming,
Which, with a little patience, might grow charming.

LXXXVII

Silent and pensive, idle, restless, slow,
His home deserted for the lonely wood,   690
Tormented with a wound he could not know,
His, like all deep grief, plunged in solitude:
I'm fond myself of solitude or so,
But then, I beg it may be understood,
By solitude I mean a Sultan's, not
A hermit's, with a haram for a grot.

LXXXVIII

"Oh Love! in such a wilderness as this,
Where transport and security entwine,
Here is the empire of thy perfect bliss,
And here thou art a god indeed divine."   700
The bard I quote from does not sing amiss,
With the exception of the second line,
For that same twining "transport and security"
Are twisted to a phrase of some obscurity.

LXXXIX

The poet meant, no doubt, and thus appeals
To the good sense and senses of mankind,
The very thing which everybody feels,
As all have found on trial, or may find,
That no one likes to be disturbed at meals
Or love.—I won't say more about "entwined"   710
Or "transport," as we knew all that before,
But beg "Security" will bolt the door.

XC

Young Juan wandered by the glassy brooks,
Thinking unutterable things; he threw
Himself at length within the leafy nooks
Where the wild branch of the cork forest grew;
There poets find materials for their books,
And every now and then we read them through,
So that their plan and prosody are eligible,
Unless, like Wordsworth, they prove unintelligible.

697–700. "Oh . . . divine": Thomas Campbell's *Gertrude of Wyoming* (1809), III. i. 1–4.

XCI

He, Juan (and not Wordsworth), so pursued   721
His self-communion with his own high soul,
Until his mighty heart, in its great mood,
Had mitigated part, though not the whole
Of its disease; he did the best he could
With things not very subject to control,
And turned, without perceiving his condition,
Like Coleridge, into a metaphysician.

XCII

He thought about himself, and the whole earth,
Of man the wonderful, and of the stars,   730
And how the deuce they ever could have birth;
And then he thought of earthquakes, and of wars,
How many miles the moon might have in girth,
Of air-balloons, and of the many bars
To perfect knowledge of the boundless skies;—
And then he thought of Donna Julia's eyes.

XCIII

In thoughts like these true wisdom may discern
Longings sublime, and aspirations high,
Which some are born with, but the most part learn
To plague themselves withal, they know not why:
'T was strange that one so young should thus concern
His brain about the action of the sky;   742
If *you* think 't was philosophy that this did,
I can't help thinking puberty assisted.

XCIV

He pored upon the leaves, and on the flowers,
And heard a voice in all the winds; and then
He thought of wood-nymphs and immortal bowers,
And how the goddesses came down to men:
He missed the pathway, he forgot the hours,
And when he looked upon his watch again,   750
He found how much old Time had been a winner—
He also found that he had lost his dinner.

XCV

Sometimes he turned to gaze upon his book,
Boscan, or Garcilasso;—by the wind

754. Boscan, or Garcilasso: Juan Boscan and Garcilaso de la Vega, Spanish poets of the early sixteenth century.

Even as the page is rustled while we look,
　　So by the poesy of his own mind
Over the mystic leaf his soul was shook,
　　As if 't were one whereon magicians bind
Their spells, and give them to the passing gale
According to some good old woman's tale.　　760

XCVI

Thus would he while his lonely hours away
　　Dissatisfied, nor knowing what he wanted;
Nor glowing reverie, nor poet's lay,
　　Could yield his spirit that for which it panted,
A bosom whereon he his head might lay,
　　And hear the heart beat with the love it granted,
With—several other things which I forget,
Or which, at least, I need not mention yet.

XCVII

Those lonely walks, and lengthening reveries,
　　Could not escape the gentle Julia's eyes;　　770
She saw that Juan was not at his ease;
　　But that which chiefly may, and must surprise,
Is, that the Donna Inez did not tease
　　Her only son with question or surmise;
Whether it was she did not see, or would not,
Or, like all very clever people, could not.

XCVIII

This may seem strange, but yet 't is very common;
　　For instance—gentlemen, whose ladies take
Leave to o'erstep the written rights of woman,
　　And break the—Which commandment is 't they
　　　　break?　　780
(I have forgot the number, and think no man
　　Should rashly quote, for fear of a mistake.)
I say, when these same gentlemen are jealous,
They make some blunder, which their ladies tell us.

XCIX

A real husband always is suspicious,
　　But still no less suspects in the wrong place,
Jealous of some one who had no such wishes,
　　Or pandering blindly to his own disgrace,
By harboring some dear friend extremely vicious;
　　The last indeed's infallibly the case:　　790
And when the spouse and friend are gone off wholly,
He wonders at their vice, and not his folly.

C

Thus parents also are at times short-sighted;
　　Though watchful as the lynx, they ne'er discover,
The while the wicked world beholds delighted,
　　Young Hopeful's mistress, or Miss Fanny's lover,
Till some confounded escapade has blighted
　　The plan of twenty years, and all is over;
And then the mother cries, the father swears,
And wonders why the devil he got heirs.　　800

CI

But Inez was so anxious, and so clear
　　Of sight, that I must think, on this occasion,
She had some other motive much more near
　　For leaving Juan to this new temptation,
But what that motive was, I shan't say here;
　　Perhaps to finish Juan's education,
Perhaps to open Don Alfonso's eyes,
In case he thought his wife too great a prize.

CII

It was upon a day, a summer's day;—
　　Summer 's indeed a very dangerous season,　　810
And so is spring about the end of May;
　　The sun, no doubt, is the prevailing reason;
But whatsoe'er the cause is, one may say,
　　And stand convicted of more truth than treason,
That there are months which nature grows more merry
　　in,—
March has its hares, and May must have its heroine.

CIII

'T was on a summer's day—the sixth of June:—
　　I like to be particular in dates,
Not only of the age, and year, but moon;
　　They are a sort of post-house, where the Fates　820
Change horses, making history change its tune,
　　Then spur away o'er empires and o'er states,
Leaving at last not much besides chronology,
Excepting the post-obits of theology.

CIV

'T was on the sixth of June, about the hour
　　Of half-past six—perhaps still nearer seven—

When Julia sate within as pretty a bower
   As e'er held houri in that heathenish heaven
Described by Mahomet, and Anacreon Moore,
   To whom the lyre and laurels have been given, 830
With all the trophies of triumphant song—
He won them well, and may he wear them long!

CV

She sate, but not alone; I know not well
   How this same interview had taken place,
And even if I knew, I should not tell—
   People should hold their tongues in any case;
No matter how or why the thing befell,
   But there were she and Juan, face to face—
When two such faces are so, 't would be wise,
But very difficult, to shut their eyes.   540

CVI

How beautiful she looked! her conscious heart
   Glowed in her cheek, and yet she felt no wrong.
Oh Love! how perfect is thy mystic art,
   Strengthening the weak, and trampling on the
     strong!
How self-deceitful is the sagest part
   Of mortals whom thy lure hath led along!—
The precipice she stood on was immense,
So was her creed in her own innocence.

CVII

She thought of her own strength, and Juan's youth,
   And of the folly of all prudish fears,   850
Victorious virtue, and domestic truth,
   And then of Don Alfonso's fifty years:
I wish these last had not occurred, in sooth,
   Because that number rarely much endears,
And through all climes, the snowy and the sunny,
Sounds ill in love, whate'er it may in money.

CVIII

When people say, "I 've told you *fifty* times,"
   They mean to scold, and very often do;

829. **Anacreon Moore:** Thomas Moore (see selections from him above) was sometimes called thus because his first publication was a translation of Anacreon's odes. The reference in line 828 is apparently to the tale "Paradise and the Peri," in Moore's *Lalla Rookh* (1817).

When poets say, "I 've written *fifty* rhymes,"
   They make you dread that they'll recite them too;
In gangs of *fifty*, thieves commit their crimes;   861
   At *fifty* love for love is rare, 't is true,
But then, no doubt, it equally as true is,
A good deal may be bought for *fifty* Louis.

CIX

Julia had honor, virtue, truth, and love
   For Don Alfonso; and she inly swore,
By all the vows below to powers above,
   She never would disgrace the ring she wore,
Nor leave a wish which wisdom might reprove;
   And while she pondered this, besides much more,
One hand on Juan's carelessly was thrown,   871
Quite by mistake—she thought it was her own;

CX

Unconsciously she leaned upon the other,
   Which played within the tangles of her hair;
And to contend with thoughts she could not smother
   She seemed, by the distraction of her air,
'T was surely very wrong in Juan's mother
   To leave together this imprudent pair,
She who for many years had watched her son so—
I'm very certain *mine* would not have done so.   880

CXI

The hand which still held Juan's, by degrees
   Gently, but palpably confirmed its grasp,
As if it said, "Detain me, if you please";
   Yet there 's no doubt she only meant to clasp
His fingers with a pure Platonic squeeze;
   She would have shrunk as from a toad or asp,
Had she imagined such a thing could rouse
A feeling dangerous to a prudent spouse.

CXII

I cannot know what Juan thought of this,
   But what he did, is much what you would do; 890
His young lip thanked it with a grateful kiss,
   And then, abashed at its own joy, withdrew
In deep despair, lest he had done amiss,—
   Love is so very timid when 't is new:
She blushed, and frowned not, but she strove to speak,
And held her tongue, her voice was grown so weak.

CXIII

The sun set, and up rose the yellow moon;
   The devil 's in the moon for mischief; they
Who called her CHASTE, methinks, began too soon
   Their nomenclature; there is not a day,   900
The longest, not the twenty-first of June,
   Sees half the business in a wicked way,
On which three single hours of moonshine smile—
And then she looks so modest all the while.

CXIV

There is a dangerous silence in that hour,
   A stillness, which leaves room for the full soul
To open all itself, without the power
   Of calling wholly back its self-control;
The silver light which, hallowing tree and tower,
   Sheds beauty and deep softness o'er the whole, 910
Breathes also to the heart, and o'er it throws
A loving languor, which is not repose.

CXV

And Julia sate with Juan, half embraced
   And half retiring from the glowing arm,
Which trembled like the bosom where 't was placed;
   Yet still she must have thought there was no harm,
Or else 't were easy to withdraw her waist;
   But then the situation had its charm,
And then—— God knows what next—I can't go on;
I 'm almost sorry that I e'er begun.   920

CXVI

Oh Plato! Plato! you have paved the way,
   With your confounded phantasies, to more
Immoral conduct by the fancied sway
   Your system feigns o'er the controlless core
Of human hearts, than all the long array
   Of poets and romancers:—You're a bore,
A charlatan, a coxcomb—and have been,
At best, no better than a go-between.

CXVII

And Julia's voice was lost, except in sighs,
   Until too late for useful conversation;   930
The tears were gushing from her gentle eyes,
   I wish, indeed, they had not had occasion;

But who, alas! can love, and then be wise?
   Not that remorse did not oppose temptation;
A little still she strove, and much repented,
And whispering "I will ne'er consent"—consented.

CXVIII

'T is said that Xerxes offered a reward
   To those who could invent him a new pleasure:
Methinks the requisition 's rather hard,
   And must have cost his majesty a treasure;   940
For my part, I 'm a moderate-minded bard,
   Fond of a little love (which I call leisure);
I care not for new pleasures, as the old
Are quite enough for me, so they but hold.

CXIX

Oh Pleasure! you 're indeed a pleasant thing,
   Although one must be damned for you, no doubt:
I make a resolution every spring
   Of reformation, ere the year run out,
But somehow, this my vestal vow takes wing,
   Yet still, I trust, it may be kept throughout;   950
I 'm very sorry, very much ashamed,
And mean, next winter, to be quite reclaimed.

CXX

Here my chaste Muse a liberty must take—
   Start not; still chaster reader—she 'll be nice hence-
Forward, and there is no great cause to quake;
   This liberty is a poetic licence,
Which some irregularity may make
   In the design, and as I have a high sense
Of Aristotle and the Rules, 't is fit
To beg his pardon when I err a bit.   960

CXXI

This licence is to hope the reader will
   Suppose from June the sixth (the fatal day
Without whose epoch my poetic skill
   For want of facts would all be thrown away),
But keeping Julia and Don Juan still
   In sight, that several months have passed; we 'll say

---

937. Xerxes: King of Persia from 486 to 465 B.C., who invaded Greece and was defeated at the Battle of Salamis.

'T was in November, but I'm not so sure
About the day—the era 's more obscure.

CXXII

We 'll talk of that anon.—'T is sweet to hear
  At midnight on the blue and moonlit deep   970
The song and oar of Adria's gondolier,
  By distance mellowed, o'er the waters sweet;
'T is sweet to see the evening star appear;
  'T is sweet to listen as the night-winds creep
From leaf to leaf; 't is sweet to view on high
The rainbow, based on ocean, span the sky.

CXXIII

'T is sweet to hear the watch-dog's honest bark
  Bay deep-mouthed welcome as we draw near home;
'T is sweet to know there is an eye will mark
  Our coming, and look brighter when we come;
'T is sweet to be awakened by the lark,   981
  Or lulled by falling waters; sweet the hum
Of bees, the voice of girls, the song of birds,
The lisp of children, and their earliest words.

CXXIV

Sweet is the vintage, when the showering grapes
  In Bacchanal profusion reel to earth,
Purple and gushing; sweet are our escapes
  From civic revelry to rural mirth;
Sweet to the miser are his glittering heaps,
  Sweet to the father is his first-born's birth,   990
Sweet is revenge—especially to women,
Pillage to soldiers, prize-money to seamen.

CXXV

Sweet is a legacy, and passing sweet
  The unexpected death of some old lady
Or gentleman of seventy years complete,
  Who 've made "us youth" wait too—too long
    already

For an estate, or cash, or country seat,
  Still breaking, but with stamina so steady
That all the Israelites are fit to mob its
Next owner for their double-damned post-obits. 1000

CXXVI

'T is sweet to win, no matter how, one's laurels,
  By blood or ink; 't is sweet to put an end
To strife; 't is sometimes sweet to have our quarrels,
  Particularly with a tiresome friend:
Sweet is old wine in bottles, ale in barrels;
  Dear is the helpless creature we defend
Against the world; and dear the schoolboy spot
We ne'er forget, though there we are forgot.

CXXVII

But sweeter still than this, than these, than all,
  Is first and passionate love—it stands alone,   1010
Like Adam's recollection of his fall;
  The tree of knowledge has been plucked—all 's
    known—
And life yields nothing further to recall
  Worthy of this ambrosial sin, so shown,
No doubt in fable, as the unforgiven
Fire which Prometheus filched for us from heaven.

CXXVIII

Man 's a strange animal, and makes strange use
  Of his own nature, and the various arts,
And likes particularly to produce
  Some new experiment to show his parts;   1020
This is the age of oddities let loose,
  Where different talents find their different marts;
You'd best begin with truth, and when you've lost your
Labor, there 's a sure market for imposture.

CXXIX

What opposite discoveries we have seen!
  (Signs of true genius, and of empty pockets.)
One makes new noses, one a guillotine,
  One breaks your bones, one sets them in their sockets;

---

971. Adria's gondolier: the gondoliers of Venice (on the Adriatic), who used to sing as they rowed. Cf. *Childe Harold*, IV. 19–20. **994. some old lady:** Byron is also taking a dig at his mother-in-law, who, as he told John Murray, was "not of those who die—the amiable only do; and those whose death would *do good* live." **996. "us youth":** a quote from the aging Falstaff ("They hate us youth," 1 *Henry IV*, II. ii. 93).

1000. post-obits: debts contracted by the heir to an estate which then become due as soon as he receives the estate.

But vaccination certainly has been
    A kind antithesis to Congreve's rockets,   1030
With which the Doctor paid off an old pox,
By borrowing a new one from an ox.

CXXX

Bread has been made (indifferent) from potatoes;
    And galvanism has set some corpses grinning,
But has not answered like the apparatus
    Of the Humane Society's beginning,
By which men are unsuffocated gratis:
    What wondrous new machines have late been spin-
        ning!
I said the small pox has gone out of late;
Perhaps it may be followed by the great.   1040

CXXXI

'T is said the great came from America;
    Perhaps it may set out on its return,—
The population there so spreads, they say
    'T is grown high time to thin it in its turn,
With war, or plague, or famine, any way,
    So that civilisation they may learn;
And which in ravage the more loathsome evil is—
Their real lues, or our pseudo-syphilis?

CXXXII

This is the patent age of new inventions
    For killing bodies, and for saving souls,   1050
All propagated with the best intentions;
    Sir Humphrey Davy's lantern, by which coals
Are safely mined for in the mode he mentions,
    Tombuctoo travels, voyages to the Poles,
Are ways to benefit mankind, as true
Perhaps, as shooting them at Waterloo.

1030. **Congreve's rockets:** an artillery shell invented by Sir
Charles Congreve (1818) and used at the Battle of Leipzig.
1031. **Doctor:** Edward Jenner, 1749–1823, who established
the use of cowpox injections as vaccination for smallpox.
1034. **galvanism . . . corpses:** Luigi Galvani, 1737–98,
was the first to observe that muscular movements can be
produced by an electric current (the phenomenon known
as galvanism). His nephew had recently applied it to the body
of a murderer. 1036–37. **Humane Society's . . . gratis:** a
society organized to rescue and revive victims of drowning.
1040. **followed . . . great:** followed by syphilis (the "great
pox"). 1052. **Sir Humphrey Davy's lantern:** safety lamp
for coal miners. 1054. **Tombuctoo . . . Poles:** books of
travel that had recently appeared.

CXXXIII

Man 's a phenomenon, one knows not what,
    And wonderful beyond all wondrous measure;
'T is pity though, in this sublime world, that
    Pleasure 's a sin, and sometimes sin 's a pleasure;
Few mortals know what end they would be at,   1061
    But whether glory, power, or love, or treasure,
The path is through perplexing ways, and when
The goal is gained, we die, you know—and then——

CXXXIV

What then?—I do not know, no more do you—
    And so good night.—Return we to our story:
'T was in November, when fine days are few,
    And the far mountains wax a little hoary,
And clap a white cape on their mantles blue;
    And the sea dashes round the promontory,   1070
And the loud breaker boils against the rock,
And sober suns must set at five o'clock.

CXXXV

'T was, as the watchmen say, a cloudy night;
    No moon, no stars, the wind was low or loud
By gusts, and many a sparkling hearth was bright
    With the piled wood, round which the family crowd;
There 's something cheerful in that sort of light,
    Even as a summer sky 's without a cloud;
I 'm fond of fire, and crickets, and all that,
A lobster salad, and champagne, and chat.   1080

CXXXVI

'T was midnight—Donna Julia was in bed,
    Sleeping, most probably,—when at her door
Arose a clatter might awake the dead,
    If they had never been awoke before,
And that they have been so we all have read,
    And are to be so, at the least, once more;—
The door was fastened, but with voice and fist
First knocks were heard, then "Madam—Madam—
    hist!

CXXXVII

"For God's sake, Madam—Madam—here's my master,
    With more than half the city at his back—   1090

Was ever heard of such a curst disaster!
   'T is not my fault—I kept good watch—Alack!
Do pray undo the bolt a little faster—
   They 're on the stair just now, and in a crack
Will all be here; perhaps he yet may fly—
   Surely the window 's not so *very* high!"

CXXXVIII

By this time Don Alfonso was arrived,
   With torches, friends, and servants in great number;
The major part of them had long been wived,   1099
   And therefore paused not to disturb the slumber
Of any wicked woman, who contrived
   By stealth her husband's temples to encumber:
Examples of this kind are so contagious,
Were *one* not punished, *all* would be outrageous.

CXXXIX

I can't tell how, or why, or what suspicion
   Could enter into Don Alfonso's head;
But for a cavalier of his condition
   It surely was exceedingly ill-bred,
Without a word of previous admonition,
   To hold a levee round his lady's bed,   1110
And summon lackeys, armed with fire and sword,
To prove himself the thing he most abhorred.

CXL

Poor Donna Julia! starting as from sleep
   (Mind—that I do not say—she had not slept),
Began at once to scream, and yawn, and weep;
   Her maid, Antonia, who was an adept,
Contrived to fling the bed-clothes in a heap,
   As if she had just now from out them crept:
I can't tell why she should take all this trouble
To prove her mistress had been sleeping double.   1120

CXLI

But Julia mistress, and Antonia maid,
   Appeared like two poor harmless women, who
Of goblins, but still more of men afraid,
   Had thought one man might be deterred by two,
And therefore side by side were gently laid,
   Until the hours of absence should run through,
And truant husband should return, and say,
"My dear, I was the first who came away."

CXLII

Now Julia found at length a voice, and cried,
   "In heaven's name, Don Alfonso, what d' ye mean?
Has madness seized you? would that I had died   1131
   Ere such a monster's victim I had been!
What may this midnight violence betide,
   A sudden fit of drunkenness or spleen?
Dare you suspect me, whom the thought would kill?
Search, then, the room!"—Alfonso said, "I will."

CXLIII

He searched, *they* searched, and rummaged everywhere,
   Closet and clothes-press, chest and window-seat.
And found much linen, lace, and several pair
   Of stockings, slippers, brushes, combs, complete,
With other articles of ladies fair,   1141
   To keep them beautiful, or leave them neat:
Arras they pricked and curtains with their swords,
And wounded several shutters, and some boards.

CXLIV

Under the bed they searched, and there they found—
   No matter what—it was not that they sought;
They opened windows, gazing if the ground
   Had signs of footmarks, but the earth said nought;
And then they stared each other's faces round:
   'T is odd, not one of all these seekers thought,   1150
And seems to me almost a sort of blunder,
Of looking *in* the bed as well as under.

CXLV

During this inquisition Julia's tongue
   Was not asleep—"Yes, search and search," she cried,
"Insult on insult heap, and wrong on wrong!
   It was for this that I became a bride!
For this in silence I have suffered long
   A husband like Alfonso at my side;
But now I 'll bear no more, nor here remain,
If there be law or lawyers in all Spain.   1160

CXLVI

"Yes, Don Alfonso! husband now no more.
   If ever you indeed deserved the name,
Is 't worthy of your years?—you have three-score—
   Fifty, or sixty, it is all the same—

Is 't wise or fitting, causeless to explore
  For facts against a virtuous woman's fame?
Ungrateful, perjured, barbarous Don Alfonso,
How dare you think your lady would go on so?

CXLVII

"Is it for this I have disdained to hold
  The common privileges of my sex?   1170
That I have chosen a confessor so old
  And deaf, that any other it would vex,
And never once he has had cause to scold,
  But found my very innocence perplex
So much, he always doubted I was married—
How sorry you will be when I 've miscarried!

CXLVIII

"Was it for this that no Cortejo e'er
  I yet have chosen from out the youth of Seville?
Is it for this I scarce went anywhere,
  Except to bull-fights, mass, play, rout, and revel?
Is it for this, whate'er my suitors were,   1181
  I favored none—nay, was almost uncivil?
Is it for this that General Count O'Reilly,
Who took Algiers, declares I used him vilely?

CXLIX

"Did not the Italian Musico Cazzani
  Sing at my heart six months at least in vain?
Did not his countryman, Count Corniani,
  Call me the only virtuous wife in Spain?
Were there not also Russians, English, many?
  The Count Strongstroganoff I put in pain,   1190
And Lord Mount Coffeehouse, the Irish peer,
Who killed himself for love (with wine) last year.

CL

"Have I not had two bishops at my feet?
  The Duke of Ichar, and Don Fernan Nunez?
And is it thus a faithful wife you treat?
  I wonder in what quarter now the moon is:

1177. **Cortejo:** the acknowledged lover of a married woman.
1183. **General Count O'Reilly:** Byron's note: "Donna
Julia here made a mistake. Count O'Reilly did not take
Algiers—but Algiers very nearly took him: he and his
army and fleet retreated with great loss, and not much
credit, from before that city, in the year 1775." 1187.
Corniani: from Italian *cornuto* (cuckold).

I praise your vast forbearance not to beat
  Me also, since the time so opportune is—
Oh, valiant man! with sword drawn and cocked
    trigger,
Now, tell me, don't you cut a pretty figure?   1200

CLI

"Was it for this you took your sudden journey,
  Under pretence of business indispensable,
With that sublime of rascals your attorney,
  Whom I see standing there, and looking sensible
Of having played the fool? though both I spurn, he
  Deserves the worst, his conduct 's less defensible,
Because, no doubt, 't was for his dirty fee,
And not from any love to you nor me.

CLII

"If he comes here to take a deposition,
  By all means let the gentleman proceed;   1210
You 've made the apartment in a fit condition:—
  There 's pen and ink for you, sir, when you need—
Let everything be noted with precision,
  I would not you for nothing should be fee'd—
But as my maid 's undrest, pray turn your spies out."
"Oh!" sobbed Antonia, "I could tear their eyes out."

CLIII

"There is the closet, there the toilet, there
  The antechamber—search them under, over;
There is the sofa, there the great arm-chair,
  The chimney—which would really hold a lover.
I wish to sleep, and beg you will take care   1221
  And make no further noise, till you discover
The secret cavern of this lurking treasure—
And when 't is found, let me, too, have that pleasure.

CLIV

"And now, Hidalgo! now that you have thrown
  Doubt upon me, confusion over all,
Pray have the courtesy to make it known
  Who is the man you search for? how d' ye call
Him? what 's his lineage? let him but be shown—
  I hope he 's young and handsome—is he tall?   1230
Tell me—and be assured, that since you stain
Mine honor thus, it shall not be in vain.

CLV

"At least, perhaps, he has not sixty years,
  At that age he would be too old for slaughter,
Or for so young a husband's jealous fears—
  (Antonia! let me have a glass of water.)
I am ashamed of having shed these tears,
  They are unworthy of my father's daughter;
My mother dreamed not in my natal hour,
That I should fall into a monster's power.   1240

CLVI

"Perhaps 't is of Antonia you are jealous,
  You saw that she was sleeping by my side,
When you broke in upon us with your fellows;
  Look where you please—we 've nothing, sir, to hide;
Only another time, I trust, you 'll tell us,
  Or for the sake of decency abide
A moment at the door, that we may be
Dressed to receive so much good company.

CLVII

"And now, sir, I have done, and say no more;
  The little I have said may serve to show   1250
The guileless heart in silence may grieve o'er
  The wrongs to whose exposure it is slow:—
I leave you to your conscience as before,
  'T will one day ask you, *why* you used me so?
God grant you feel not then the bitterest grief!
Antonia! where 's my pocket-handkerchief?"

CLVIII

She ceased, and turned upon her pillow; pale
  She lay, her dark eyes flashing through their tears,
Like skies that rain and lighten; as a veil,   1259
  Waved and o'ershading her wan cheek, appears
Her streaming hair; the black curls strive, but fail,
  To hide the glossy shoulder, which uprears
Its snow through all;—her soft lips lie apart,
And louder than her breathing beats her heart.

CLIX

The Senhor Don Alfonso stood confused;
  Antonia bustled round the ransacked room,
And, turning up her nose, with looks abused
  Her master, and his myrmidons, of whom

Not one, except the attorney, was amused;
  He, like Achates, faithful to the tomb,   1270
So there were quarrels, cared not for the cause,
Knowing they must be settled by the laws.

CLX

With prying snub-nose, and small eyes, he stood,
  Following Antonia's motions here and there,
With much suspicion in his attitude;
  For reputations he had little care;
So that a suit or action were made good,
  Small pity had he for the young and fair,
And ne'er believed in negatives, till these
Were proved by competent false witnesses.   1280

CLXI

But Don Alfonso stood with downcast looks,
  And, truth to say, he made a foolish figure;
When, after searching in five hundred nooks,
  And treating a young wife with so much rigor,
He gained no point, except some self-rebukes,
  Added to those his lady with such vigor
Had poured upon him for the last half hour,
Quick, thick, and heavy—as a thunder-shower.

CLXII

At first he tried to hammer an excuse,
  To which the sole reply was tears and sobs,   1290
And indications of hysterics, whose
  Prologue is always certain throes, and throbs,
Gasps, and whatever else the owners choose:
  Alfonso saw his wife, and thought of Job's;
He saw too, in perspective, her relations,
And then he tried to muster all his patience.

CLXIII

He stood in act to speak, or rather stammer,
  But sage Antonia cut him short before
The anvil of his speech received the hammer,   1299
  With "Pray, sir, leave the room and say no more,
Or madam dies."—Alfonso muttered, "D—n her."
  But nothing else, the time of words was o'er;

1270. **Achates:** the faithful friend of Aeneas who accompanied him from Troy to Rome. **1294. Job's:** "Thou speakest as one of the foolish women speaketh" (Job 2:10).

He cast a rueful look or two, and did,
He knew not wherefore, that which he was bid.

CLXIV

With him retired his *"posse comitatus,"*
    The attorney last, who lingered near the door
Reluctantly, still tarrying there as late as
    Antonia let him—not a little sore
At this most strange and unexplained *"hiatus"*
    In Don Alfonso's facts, which just now wore   1310
An awkward look; as he revolved the case,
The door was fastened in his legal face.

CLXV

No sooner was it bolted, than—Oh shame!
    Oh sin! Oh sorrow! and Oh womankind!
How can you do such things and keep your fame,
    Unless this world, and t' other too, be blind?
Nothing so dear as an unfilched good name!
    But to proceed—for there is more behind:
With much heartfelt reluctance be it said,
Young Juan slipped, half-smothered, from the bed.

CLXVI

He had been hid—I don't pretend to say   1321
    How, nor can I indeed describe the where—
Young, slender, and packed easily, he lay,
    No doubt, in little compass, round or square;
But pity him I neither must nor may
    His suffocation by that pretty pair;
'T were better, sure, to die so, than be shut
With maudlin Clarence in his Malmsey butt.

CLXVII

And, secondly, I pity not, because
    He had no business to commit a sin,   1330
Forbid by heavenly, fined by human laws,
    At least 't was rather early to begin;
But at sixteen the conscience rarely gnaws
    So much as when we call our old debts in
At sixty years, and draw the accompts of evil,
And find a deuced balance with the devil.

1305. **"posse comitatus"**: company. 1328. **maudlin . . . butt:** referring to the story that the Duke of Clarence was drowned in a butt of Malmsey wine (*Richard III*, I. iv).

CLXVIII

Of his position I can give no notion:
    'T is written in the Hebrew Chronicle,
How the physicians, leaving pill and potion,
    Prescribed, by way of blister, a young belle,   1340
When old King David's blood grew dull in motion,
    And that the medicine answered very well;
Perhaps 't was in a different way applied,
For David lived, but Juan nearly died.

CLXIX

What 's to be done? Alfonso will be back
    The moment he has sent his fools away.
Antonia's skill was put upon the rack,
    But no device could be brought into play—
And how to parry the renewed attack?
    Besides, it wanted but few hours of day: 1350
Antonia puzzled; Julia did not speak,
But pressed her bloodless lip to Juan's cheek.

CLXX

He turned his lip to hers, and with his hand
    Called back the tangles of her wandering hair;
Even then their love they could not all command,
    And half forgot their danger and despair:
Antonia's patience now was at a stand—
    "Come, come, 't is no time now for fooling there,"
She whispered in great wrath—"I must deposit
This pretty gentleman within the closet: 1360

CLXXI

"Pray, keep your nonsense for some luckier night—
    Who can have put my master in this mood?
What will become on 't—I 'm in such a fright,
    The devil's in the urchin, and no good—
Is this a time for giggling? this a plight?
    Why, don't you know that it may end in blood?
You 'll lose your life, and I shall lose my place,
My mistress all, for that half-girlish face.

CLXXII

"Had it but been for a stout cavalier
    Of twenty-five or thirty—(come, make haste)   1370

1338. **Hebrew Chronicle:** in I Kings 1: 1–3.

But for a child, what piece of work is here!
  I really, madam, wonder at your taste—
(Come, sir, get in)—my master must be near:
  There, for the present, at the least, he's fast,
And if we can but till the morning keep
  Our counsel—(Juan, mind, you must not sleep)."

CLXXIII

Now, Don Alfonso entering, but alone,
  Closed the oration of the trusty maid:
She loitered, and he told her to be gone,
  An order somewhat sullenly obeyed;   1380
However, present remedy was none,
  And no great good seemed answered if she staid;
Regarding both with slow and sidelong view,
She snuffed the candle, curtsied, and withdrew.

CLXXIV

Alfonso paused a minute—then begun
  Some strange excuses for his late proceeding:
He would not justify what he had done,
  To say the best, it was extreme ill-breeding;
But there were ample reasons for it, none
  Of which he specified in this his pleading: 1390
His speech was a fine sample, on the whole,
Of rhetoric, which the learned call "*rigmarole.*"

CLXXV

Julia said nought; though all the while there rose
  A ready answer, which at once enables
A matron, who her husband's foible knows,
  By a few timely words to turn the tables,
Which, if it does not silence, still must pose,—
  Even if it should comprise a pack of fables;
'T is to retort with firmness, and when he
Suspects with *one*, do you reproach with *three*.   1400

CLXXVI

Julia, in fact, had tolerable grounds,—
  Alfonso's loves with Inez were well known;
But whether 't was that one's own guilt confounds—
  But that can't be, as has been often shown,
A lady with apologies abounds;—
  It might be that her silence sprang alone
From delicacy to Don Juan's ear,
To whom she knew his mother's fame was dear.

CLXXVII

There might be one more motive, which makes two,
  Alfonso ne'er to Juan had alluded,—   1410
Mentioned his jealousy, but never who
  Had been the happy lover, he concluded,
Concealed amongst his premises; 't is true,
  His mind the more o'er this its mystery brooded;
To speak of Inez now were, one may say,
Like throwing Juan in Alfonso's way.

CLXXVIII

A hint, in tender cases, is enough;
  Silence is best: besides there is a *tact*—
(That modern phrase appears to me sad stuff,
  But it will serve to keep my verse compact)—   1420
Which keeps, when pushed by questions rather rough,
  A lady always distant from the fact:
The charming creatures lie with such a grace,
There's nothing so becoming to the face.

CLXXIX

They blush, and we believe them, at least I
  Have always done so; 't is of no great use,
In any case, attempting a reply,
  For then their eloquence grows quite profuse;
And when at length they're out of breath, they sigh,
  And cast their languid eyes down, and let loose
A tear or two, and then we make it up;   1431
And then—and then—and then—sit down and sup.

CLXXX

Alfonso closed his speech, and begged her pardon,
  Which Julia half withheld, and then half granted,
And laid conditions, he thought very hard, on,
  Denying several little things he wanted:
He stood like Adam lingering near his garden,
  With useless penitence perplexed and haunted,
Beseeching she no further would refuse,
When, lo! he stumbled o'er a pair of shoes.   1440

CLXXXI

A pair of shoes!—what then? not much, if they
  Are such as fit with ladies' feet, but these

(No one can tell how much I grieve to say)
   Were masculine; to see them, and to seize,
Was but a moment's act.—Ah! well-a-day!
   My teeth begin to chatter, my veins freeze—
Alfonso first examined well their fashion,
And then flew out into another passion.

CLXXXII

He left the room for his relinquished sword,
   And Julia instant to the closet flew.   1450
"Fly, Juan, fly; for heaven's sake—not a word—
   The door is open—you may yet slip through
The passage you so often have explored—
   Here is the garden-key—Fly—fly—Adieu!
Haste—haste! I hear Alfonso's hurrying feet—
Day has not broke—there's no one in the street."

CLXXXIII

None can say that this was not good advice,
   The only mischief was, it came too late;
Of all experience 't is the usual price,
   A sort of income-tax laid on by fate: 1460
Juan had reached the room-door in a trice,
   And might have done so by the garden-gate,
But met Alfonso in his dressing-gown,
Who threatened death—so Juan knocked him down.

CLXXXIV

Dire was the scuffle, and out went the light;
   Antonia cried out "Rape!" and Julia "Fire!"
But not a servant stirred to aid the fight.
   Alfonso, pommelled to his heart's desire,
Swore lustily he 'd be revenged this night;
   And Juan, too, blasphemed an octave higher; 1470
His blood was up: though young, he was a Tartar,
And not at all disposed to prove a martyr.

CLXXXV

Alfonso's sword had dropped ere he could draw it,
   And they continued battling hand to hand,
For Juan very luckily ne'er saw it;
   His temper not being under great command,
If at that moment he had chanced to claw it,
   Alfonso's days had not been in the land
Much longer.—Think of husbands', lovers' lives!
And how ye may be doubly widows—wives!   1480

CLXXXVI

Alfonso grappled to detain the foe,
   And Juan throttled him to get away,
And blood ('t was from the nose) began to flow;
   At last, as they more faintly wrestling lay,
Juan contrived to give an awkward blow,
   And then his only garment quite gave way;
He fled, like Joseph, leaving it; but there,
I doubt, all likeness ends between the pair.

CLXXXVII

Lights came at length, and men, and maids, who found
   An awkward spectacle their eyes before;   1490
Antonia in hysterics, Julia swooned,
   Alfonso leaning, breathless, by the door;
Some half-torn drapery scattered on the ground,
   Some blood, and several footsteps, but no more:
Juan the gate gained, turned the key about,
And liking not the inside, locked the out.

CLXXXVIII

Here ends this canto.—Need I sing, or say,
   How Juan, naked, favored by the night,
Who favors what she should not, found his way,
   And reached his home in an unseemly plight? 1500
The pleasant scandal which arose next day,
   The nine days' wonder which was brought to light,
And how Alfonso sued for a divorce,
Were in the English newspapers, of course.

CLXXXIX

If you would like to see the whole proceedings,
   The depositions and the cause at full,
The names of all the witnesses, the pleadings
   Of counsel to nonsuit, or to annul,
There 's more than one edition, and the readings
   Are various, but they none of them are dull; 1510
The best is that in short-hand ta'en by Gurney,
Who to Madrid on purpose made a journey.

1487. Joseph: who fled from the advances of Potiphar's wife (Genesis 39:12). **1511. short-hand . . . Gurney:** William Brodie Gurney, the shorthand writer for Parliament, reported many of the more important trials.

CXC

But Donna Inez, to divert the train
  Of one of the most circulating scandals
That had for centuries been known in Spain,
  At least since the retirement of the Vandals,
First vowed (and never had she vowed in vain)
  To Virgin Mary several pounds of candles;
And then, by the advice of some old ladies,
She sent her son to be shipped off from Cadiz.   1520

CXCI

She had resolved that he should travel through
  All European climes, by land or sea,
To mend his former morals, and get new,
  Especially in France and Italy
(At least this is the thing most people do).
  Julia was sent into a convent: she
Grieved, but, perhaps, her feelings may be better
Shown in the following copy of her Letter:—

CXCII

"They tell me 't is decided you depart:
  'T is wise—'t is well, but not the less a pain;   1530
I have no further claim on your young heart,
  Mine is the victim, and would be again:
To love too much has been the only art
  I used;—I write in haste, and if a stain
Be on this sheet, 't is not what it appears;
My eyeballs burn and throb, but have no tears.

CXCIII

"I loved, I love you, for this love have lost
  State, station, heaven, mankind's, my own esteem,
And yet cannot regret what it hath cost,
  So dear is still the memory of that dream;   1540
Yet, if I name my guilt, 't is not to boast,
  None can deem harshlier of me than I deem:
I trace this scrawl because I cannot rest—
I 've nothing to reproach or to request.

CXCIV

"Man's love is of man's life a thing apart,
  'T is woman's whole existence; man may range

1545–46. **Man's . . . existence:** This famous remark is
paraphrased by Byron from Mme de Staël's *De l'influence des
passions* (1796).

The court, camp, church, the vessel, and the mart;
  Sword, gown, gain, glory, offer in exchange
Pride, fame, ambition, to fill up his heart,
  And few there are whom these cannot estrange;
Men have all these resources, we but one,   1551
To love again, and be again undone.

CXCV

"You will proceed in pleasure, and in pride,
  Beloved and loving many; all is o'er
For me on earth, except some years to hide
  My shame and sorrow deep in my heart's core:
These I could bear, but cannot cast aside
  The passion which still rages as before,—
And so farewell—forgive me, love me—No,
That word is idle now—but let it go.   1560

CXCVI

"My breast has been all weakness, is so yet;
  But still I think I can collect my mind;
My blood still rushes where my spirit 's set,
  As roll the waves before the settled wind;
My heart is feminine, nor can forget—
  To all, except one image, madly blind;
So shakes the needle, and so stands the pole,
As vibrates my fond heart to my fixed soul.

CXCVII

"I have no more to say, but linger still,
  And dare not set my seal upon this sheet,   1570
And yet I may as well the task fulfil,
  My misery can scarce be more complete:
I had not lived till now, could sorrow kill;
  Death shuns the wretch who fain the blow would
    meet,
And I must even survive this last adieu,
And bear with life to love and pray for you!"

CXCVIII

This note was written upon gilt-edged paper
  With a neat little crow-quill, slight and new;
Her small white hand could hardly reach the taper,
  It trembled as magnetic needles do,   1580
And yet she did not let one tear escape her:
  The seal a sun-flower; "*Elle vous suit partout*,"

1582. **"Elle . . . partout":** She follows you everywhere.

The motto, cut upon a white cornelian;
The wax was superfine, its hue vermilion.

CXCIX

This was Don Juan's earliest scrape; but whether
  I shall proceed with his adventures is
Dependent on the public altogether;
  We 'll see, however, what they say to this,
Their favor in an author's cap 's a feather,
  And no great mischief 's done by their caprice;
And if their approbation we experience,   1591
Perhaps they 'll have some more about a year hence.

CC

My poem 's epic, and is meant to be
  Divided in twelve books; each book containing,
With love, and war, a heavy gale at sea,
  A list of ships, and captains, and kings reigning,
New characters; the episodes are three:
  A panoramic view of hell 's in training
After the style of Virgil and of Homer,
So that my name of Epic 's no misnomer.   1600

CCI

All these things will be specified in time,
  With strict regard to Aristotle's rules,
The *Vade Mecum* of the true sublime,
  Which makes so many poets, and some fools:
Prose poets like blank-verse, I 'm fond of rhyme,
  Good workmen never quarrel with their tools;
I 've got new mythological machinery,
And very handsome supernatural scenery.

CCII

There 's only one slight difference between
  Me and my epic brethren gone before,   1610
And here the advantage is my own, I ween
  (Not that I have not several merits more,
But this will more peculiarly be seen);
  They so embellish, that 't is quite a bore
Their labyrinth of fables to thread through,
Whereas this story 's actually true.

1603. Vade Mecum: a handbook or guide (literally "go with me").

CCIII

If any person doubt it, I appeal
  To history, tradition, and to facts,
To newspapers, whose truth all know and feel,
  To plays in five, and operas in three acts: 1620
All these confirm my statement a good deal,
  But that which more completely faith exacts
Is, that myself, and several now in Seville,
Saw Juan's last elopement with the devil.

CCIV

If ever I should condescend to prose,
  I 'll write poetical commandments, which
Shall supersede beyond all doubt all those
  That went before; in these I shall enrich
My text with many things that no one knows,
  And carry precept to the highest pitch: 1630
I 'll call the work "Longinus o'er a Bottle,
Or, Every Poet his *own* Aristotle."

CCV

Thou shalt believe in Milton, Dryden, Pope;
  Thou shalt not set up Wordsworth, Coleridge,
    Southey;
Because the first is crazed beyond all hope,
  The second drunk, the third so quaint and mouthy:
With Crabbe it may be difficult to cope,
  And Campbell's Hippocrene is somewhat drouthy:
Thou shalt not steal from Samuel Rogers, nor
Commit—flirtation with the muse of Moore.   1640

CCVI

Thou shalt not covet Mr. Sotheby's Muse,
  His Pegasus, nor anything that 's his;

1631–32. "Longinus . . . Aristotle": On the basis of Aristotle's *Poetics*, with its analysis of structure and of component parts, classical and Renaissance critics constructed laws or "rules" for poetry. Longinus' *On the Sublime* is the great classical prototype of criticism that stresses inspiration and emotional power. **1638. Hippocrene:** fountain on Mt. Helicon that inspired the poet who drank from it. **1641. Mr. Sotheby:** William Sotheby, 1757–1833, was known for his translations.

Thou shalt not bear false witness like "the Blues"—
  (There's one, at least, is very fond of this);
Thou shalt not write, in short, but what I choose;
  This is true criticism, and you may kiss—
Exactly as you please, or not,—the rod;
And if you don't, I'll lay it on, by G—d!

CCVII

If any person should presume to assert
  This story is not moral, first, I pray,   1650
That they will not cry out before they're hurt,
  Then that they'll read it o'er again, and say
(But, doubtless, nobody will be so pert),
  That this is not a moral tale, though gay;
Besides, in Canto Twelfth, I mean to show,
The very place where wicked people go.

CCVIII

If, after all, there should be some so blind
  To their own good this warning to despise,
Led by some tortuosity of mind,
  Not to believe my verse and their own eyes,   1660
And cry that they "the moral cannot find,"
  I tell him, if a clergyman, he lies;
Should captains the remark, or critics, make,
They also lie too—under a mistake.

CCIX

The public approbation I expect,
  And beg they'll take my word about the moral,
Which I with their amusement will connect
  (So children cutting teeth receive a coral);
Meantime they'll doubtless please to recollect
  My epical pretensions to the laurel: 1670
For fear some prudish readers should grow skittish,
I've bribed my grandmother's review—the British.

CCX

I sent it in a letter to the Editor,
  Who thanked me duly by return of post—
I'm for a handsome article his creditor;
  Yet, if my gentle Muse he please to roast,

And break a promise after having made it her,
  Denying the receipt of what it cost,
And smear his page with gall instead of honey,
  All I can say is—that he had the money.   1680

CCXI

I think that with this holy new alliance
  I may ensure the public, and defy
All other magazines of art or science,
  Daily, or monthly, or three monthly; I
Have not essayed to multiply their clients,
  Because they tell me 't were in vain to try,
And that the Edinburgh Review and Quarterly
Treat a dissenting author very martyrly.

CCXII

"*Non ego hoc ferrem calida juventâ*
  Consule Planco," Horace said, and so   1690
Say I; by which quotation there is meant a
  Hint that some six or seven good years ago
(Long ere I dreamt of dating from the Brenta)
  I was most ready to return a blow,
And would not brook at all this sort of thing
In my hot youth—when George the Third was King.

CCXIII

But now at thirty years my hair is gray—
  (I wonder what it will be like at forty?
I thought of a peruke the other day—)
  My heart is not much greener; and, in short, I   1700
Have squandered my whole summer while 't was May,
  And feel no more the spirit to retort; I
Have spent my life, both interest and principal,
And deem not, what I deemed, my soul invincible.

CCXIV

No more—no more—Oh! never more on me
  The freshness of the heart can fall like dew,
Which out of all the lovely things we see
  Extracts emotions beautiful and new,
Hived in our bosoms like the bag o' the bee.
  Think'st thou the honey with those objects grew?

1643. "the Blues": blue stockings (women with literary interests; see above, I. 73). One of Byron's later works is a short satirical play in verse called *The Blues: A Literary Eclogue* (1821). **1644. There's one:** Byron's wife.

1689–90. "Non . . . Planco": "I should not have borne this in the heat of youth when Plancus was consul" (Horace, *Odes*, III. xiv. 27–28). **1693. Brenta:** river near Venice.

Alas! 't was not in them, but in thy power   1711
To double even the sweetness of a flower.

CCXV

No more—no more—Oh! never more, my heart,
   Canst thou be my sole world, my universe!
Once all in all, but now a thing apart,
   Thou canst not be my blessing or my curse:
The illusion 's gone for ever, and thou art
   Insensible, I trust, but none the worse,
And in thy stead I 've got a deal of judgment
Though heaven knows how it ever found a lodgment.

CCXVI

My days of love are over; me no more   1721
   The charms of maid, wife, and still less of widow,
Can make the fool of which they made before,—
   In short, I must not lead the life I did do;
The credulous hope of mutual minds is o'er,
   The copious use of claret is forbid too,
So for a good old-gentlemanly vice,
I think I must take up with avarice.

CCXVII

Ambition was my idol, which was broken
   Before the shrines of Sorrow, and of Pleasure;   1730
And the two last have left me many a token
   O'er which reflection may be made at leisure;
Now, like Friar Bacon's brazen head, I 've spoken
   "Time is, Time was, Time 's past":—a chymic
      treasure
Is glittering youth, which I have spent betimes—
My heart in passion, and my head on rhymes.

CCXVIII

What is the end of fame? 't is but to fill
   A certain portion of uncertain paper:
Some liken it to climbing up a hill,
   Whose summit, like all hills, is lost in vapor;   1740
For this men write, speak, preach, and heroes kill,
   And bards burn what they call their "midnight
      taper,"

To have, when the original is dust,
A name, a wretched picture, and worse bust.

CCXIX

What are the hopes of man? Old Egypt's King
   Cheops erected the first pyramid
And largest, thinking it was just the thing
   To keep his memory whole, and mummy hid:
But somebody or other rummaging,
   Burglariously broke his coffin's lid: 1750
Let not a monument give you or me hopes,
Since not a pinch of dust remains of Cheops.

CCXX

But I, being fond of true philosophy,
   Say very often to myself, "Alas!
All things that have been born were born to die,
   And flesh (which Death mows down to hay) is grass;
You 've passed your youth not so unpleasantly,
   And if you had it o'er again—'t would pass—
So thank your stars that matters are no worse,
And read your Bible, sir, and mind your purse."   1760

CCXXI

But for the present, gentle reader! and
   Still gentler purchaser! the bard—that 's I—
Must, with permission, shake you by the hand,
   And so your humble servant, and good-bye!
We meet again, if we should understand
   Each other; and if not, I shall not try
Your patience further than by this short sample—
'T were well if others followed my example.

CCXXII

"Go, little book, from this my solitude!
   I cast thee on the waters—go thy ways!   1770
And if, as I believe, thy vein be good,
   The world will find thee after many days."
When Southey's read, and Wordsworth understood,
   I can't help putting in my claim to praise—
The four first rhymes are Southey's, every line:
For God's sake, reader! take them not for mine!

1818 (1819)

1721. **days . . . over:** Byron paraphrases Horace (*Odes*, IV. i. 29–32). 1733. **Bacon's . . . head:** a brass head that could speak, made by Roger Bacon (cf. Robert Greene, *Friar Bacon and Friar Bungay*, IV. i).

1752. **Cheops:** the builder of the Great Pyramid. 1769–72. **"Go . . . days":** quoted from the last stanza of Southey's *Epilogue to the Lay of the Laureate*.

They won't lay out their money on the dead—
It costs three francs for every mass that's said.   440

from CANTO THE SECOND

Juan sails from Cadiz. The boat is caught in a storm and
goes down.

LII

Then rose from sea to sky the wild farewell—
　　Then shriek'd the timid, and stood still the brave—
Then some leap'd overboard with dreadful yell,   411
　　As eager to anticipate their grave;
And the sea yawn'd around her like a hell,
　　And down she suck'd with her the whirling wave,
Like one who grapples with his enemy,
And strives to strangle him before he die.

LIII

And first one universal shriek there rush'd,
　　Louder than the loud ocean, like a crash
Of echoing thunder; and then all was hush'd,
　　Save the wild wind and the remorseless dash   420
Of billows; but at intervals there gush'd,
　　Accompanied with a convulsive splash,
A solitary shriek, the bubbling cry
Of some strong swimmer in his agony.

LIV

The boats, as stated, had got off before,
　　And in them crowded several of the crew;
And yet their present hope was hardly more
　　Than what it had been, for so strong it blew
There was slight chance of reaching any shore;
　　And then they were too many, though so few—
Nine in the cutter, thirty in the boat,   431
Were counted in them when they got afloat.

LV

All the rest perish'd; near two hundred souls
　　Had left their bodies; and what's worse, alas!
When over Catholics the ocean rolls,
　　They must wait several weeks before a mass
Takes off one peck of purgatorial coals,
　　Because, till people know what's come to pass,

LVI

Juan got into the long-boat, and there
　　Contrived to help Pedrillo to a place;
It seem'd as if they had exchanged their care,
　　For Juan wore the magisterial face
Which courage gives, while poor Pedrillo's pair
　　Of eyes were crying for their owner's case:
Battista, though (a name call'd shortly Tita),
Was lost by getting at some aqua-vita.

LVII

Pedro, his valet, too, he tried to save,
　　But the same cause, conducive to his loss,   450
Left him so drunk, he jump'd into the wave
　　As o'er the cutter's edge he tried to cross,
And so he found a wine-and-watery grave;
　　They could not rescue him although so close,
Because the sea ran higher every minute,
And for the boat—the crew kept crowding in it.

LVIII

A small old spaniel—which had been Don Jóse's,
　　His father's, whom he loved, as ye may think,
For on such things the memory reposes
　　With tenderness—stood howling on the brink,   460
Knowing, (dogs have such intellectual noses!)
　　No doubt, the vessel was about to sink;
And Juan caught him up, and ere he stepp'd
Off, threw him in, then after him he leap'd.

LIX

He also stuff'd his money where he could
　　About his person, and Pedrillo's too,
Who let him do, in fact, whate'er he would,
　　Not knowing what himself to say, or do,
As every rising wave his dread renew'd;
　　But Juan, trusting they might still get through,   470
And deeming there were remedies for any ill,
Thus re-embark'd his tutor and his spaniel.

CANTO II. **442. Pedrillo:** Juan's tutor.

LX

'Twas a rough night, and blew so stiffly yet,
  That the sail was becalm'd between the seas,
Though on the wave's high top too much to set,
  They dared not take it in for all the breeze:
Each sea curl'd o'er the stern, and kept them wet,
  And made them bale without a moment's ease,
So that themselves as well as hopes were damp'd,
And the poor little cutter quickly swamp'd.   480

LXI

Nine souls more went in her: the long-boat still
  Kept above water, with an oar for mast,
Two blankets stitch'd together, answering ill
  Instead of sail, were to the oar made fast:
Though every wave roll'd menacing to fill,
  And present peril all before surpass'd,
They grieved for those who perish'd with the cutter,
And also for the biscuit-casks and butter.

LXII

The sun rose red and fiery, a sure sign
  Of the continuance of the gale: to run   490
Before the sea until it should grow fine,
  Was all that for the present could be done:
A few tea-spoonfuls of their rum and wine
  Were served out to the people, who begun
To faint, and damaged bread wet through the bags,
And most of them had little clothes but rags.

LXIII

They counted thirty, crowded in a space
  Which left scarce room for motion or exertion;
They did their best to modify their case,
  One half sate up, though numb'd with the immer-
    sion,
While t'other half were laid down in their place,   501
  At watch and watch; thus, shivering like the tertian
Ague in its cold fit, they fill'd their boat,
With nothing but the sky for a great coat.

LXIV

'Tis very certain the desire of life
  Prolongs it: this is obvious to physicians,

When patients, neither plagued with friends nor wife,
  Survive through very desperate conditions,
Because they still can hope, nor shines the knife
  Nor shears of Atropos before their visions: 510
Despair of all recovery spoils longevity,
And makes men's miseries of alarming brevity.

LXV

'Tis said that persons living on annuities
  Are longer lived than others,—God knows why,
Unless to plague the grantors,—yet so true it is,
  That some, I really think, *do* never die;
Of any creditors the worst a Jew it is,
  And *that*'s their mode of furnishing supply:
In my young days they lent me cash that way,
Which I found very troublesome to pay.   520

LXVI

'Tis thus with people in an open boat,
  They live upon the love of life, and bear
More than can be believed, or even thought,
  And stand like rocks the tempest's wear and tear;
And hardship still has been the sailor's lot,
  Since Noah's ark went cruising here and there;
She had a curious crew as well as cargo,
Like the first old Greek privateer, the Argo.

LXVII

But man is a carnivorous production,
  And must have meals, at least one meal a day;   530
He cannot live, like woodcocks, upon suction,
  But, like the shark and tiger, must have prey;
Although his anatomical construction
  Bears vegetables, in a grumbling way,
Your labouring people think beyond all question
Beef, veal, and mutton, better for digestion.

LXVIII

And thus it was with this our hapless crew;
  For on the third day there came on a calm,
And though at first their strength it might renew,
  And lying on their weariness like balm,   540

---

510. Atropos: in Greek mythology one of the three Fates
that cut the thread of life.

Lull'd them like turtles sleeping on the blue
  Of ocean, when they woke they felt a qualm,
And fell all ravenously on their provision,
  Instead of hoarding it with due precision.

LXIX

The consequence was easily foreseen—
  They ate up all they had, and drank their wine,
In spite of all remonstrances, and then
  On what, in fact, next day were they to dine?
They hoped the wind would rise, these foolish men!
  And carry them to shore; these hopes were fine,
But as they had but one oar, and that brittle,   551
It would have been more wise to save their victual.

LXX

The fourth day came, but not a breath of air,
  And Ocean slumber'd like an unwean'd child:
The fifth day, and their boat lay floating there,
  The sea and sky were blue, and clear, and mild—
With their one oar (I wish they had had a pair)
  What could they do? and hunger's rage grew wild:
So Juan's spaniel, spite of his entreating,
Was kill'd, and portion'd out for present eating.   560

LXXI

On the sixth day they fed upon his hide,
  And Juan, who had still refused, because
The creature was his father's dog that died,
  Now feeling all the vulture in his jaws,
With some remorse received (though first denied)
  As a great favour one of the fore-paws,
Which he divided with Pedrillo, who
Devour'd it, longing for the other too.

LXXII

The seventh day, and no wind—the burning sun
  Blister'd and scorch'd, and, stagnant on the sea,   570
They lay like carcasses; and hope was none,
  Save in the breeze that came not; savagely
They glared upon each other—all was done,
  Water, and wine, and food,—and you might see
The longings of the cannibal arise
(Although they spoke not) in their wolfish eyes.

LXXIII

At length one whisper'd his companion, who
  Whisper'd another, and thus it went round,
And then into a hoarser murmur grew,
  An ominous, and wild, and desperate sound;   580
And when his comrade's thought each sufferer knew,
  'Twas but his own, suppress'd till now, he found:
And out they spoke of lots of flesh and blood,
And who should die to be his fellow's food.

LXXIV

But ere they came to this, they that day shared
  Some leathern caps, and what remain'd of shoes;
And then they look'd around them, and despair'd,
  And none to be the sacrifice would choose;
At length the lots were torn up, and prepared,
  But of materials that must shock the Muse—   590
Having no paper, for the want of better,
They took by force from Juan Julia's letter.

LXXV

Then lots were made, and mark'd, and mix'd, and handed
  In silent horror, and their distribution
Lull'd even the savage hunger which demanded,
  Like the Promethean vulture, this pollution;
None in particular had sought or plann'd it,
  'Twas nature gnaw'd them to this resolution,
By which none were permitted to be neuter—
And the lot fell on Juan's luckless tutor.   600

LXXVI

He but requested to be bled to death:
  The surgeon had his instruments, and bled
Pedrillo, and so gently ebb'd his breath,
  You hardly could perceive when he was dead.
He died as born, a Catholic in faith,
  Like most in the belief in which they're bred,
And first a little crucifix he kiss'd,
And then held out his jugular and wrist.

LXXVII

The surgeon, as there was no other fee,
  Had his first choice of morsels for his pains;   610

But being thirstiest at the moment, he
   Preferr'd a draught from the fast-flowing veins:
Part was divided, part thrown in the sea,
   And such things as the entrails and the brains
Regaled two sharks, who follow'd o'er the billow—
The sailors ate the rest of poor Pedrillo.

LXXVIII

The sailors ate him, all save three or four,
   Who were not quite so fond of animal food;
To these was added Juan, who, before
   Refusing his own spaniel, hardly could   620
Feel now his appetite increased much more;
   'Twas not to be expected that he should,
Even in extremity of their disaster,
Dine with them on his pastor and his master.

LXXIX

'Twas better that he did not; for, in fact,
   The consequence was awful in the extreme;
For they, who were most ravenous in the act,
   Went raging mad—Lord! how they did blaspheme!
And foam and roll, with strange convulsions rack'd,
   Drinking salt-water like a mountain-stream,   630
Tearing, and grinning, howling, screeching, swearing,
And, with hyæna-laughter, died despairing.

LXXX

Their numbers were much thinn'd by this infliction,
   And all the rest were thin enough, Heaven knows;
And some of them had lost their recollection,
   Happier than they who still perceived their woes;
But others ponder'd on a new dissection,
   As if not warn'd sufficiently by those
Who had already perish'd, suffering madly,
For having used their appetites so sadly.   640

LXXXI

And next they thought upon the master's mate,
   As fattest; but he saved himself, because,
Besides being much averse from such a fate,
   There were some other reasons: the first was,
He had been rather indisposed of late;
   And that which chiefly proved his saving clause,
Was a small present made to him at Cadiz,
By general subscription of the ladies.

LXXXII

Of poor Pedrillo something still remain'd,
   But was used sparingly,—some were afraid,   650
And others still their appetites constrain'd,
   Or but at times a little supper made;
All except Juan, who throughout abstain'd,
   Chewing a piece of bamboo, and some lead:
At length they caught two boobies, and a noddy,
And then they left off eating the dead body.

LXXXIII

And if Pedrillo's fate should shocking be,
   Remember Ugolino condescends
To eat the head of his arch-enemy
   The moment after he politely ends   660
His tale: if foes be food in hell, at sea
   'Tis surely fair to dine upon our friends,
When shipwreck's short allowance grows too scanty,
Without being much more horrible than Dante.

LXXXIV

And the same night there fell a shower of rain,
   For which their mouths gaped, like the cracks of
     earth
When dried to summer dust; till taught by pain,
   Men really know not what good water's worth;
If you had been in Turkey or in Spain,
   Or with a famish'd boat's-crew had your berth, 670
Or in the desert heard the camel's bell,
You'd wish yourself where Truth is—in a well.

LXXXV

It pour'd down torrents, but they were no richer
   Until they found a ragged piece of sheet,
Which served them as a sort of spongy pitcher,
   And when they deem'd its moisture was complete,
They wrung it out, and though a thirsty ditcher
   Might not have thought the scanty draught so sweet
As a full pot of porter, to their thinking
They ne'er till now had known the joys of drinking.

655. boobies . . . noddy: sea birds. 658–61. Ugolino . . .
tale: After relating his story, the hungry Ugolino returns
to gnawing the skull of his enemy (Dante, *Inferno*, XXXIII.
76–78).

LXXXVI

And their baked lips, with many a bloody crack,   681
   Suck'd in the moisture, which like nectar stream'd;
Their throats were ovens, their swoln tongues were
   black
   As the rich man's in hell, who vainly scream'd
To beg the beggar, who could not rain back
   A drop of dew, when every drop had seem'd
To taste of heaven—If this be true, indeed,
Some Christians have a comfortable creed.

LXXXVII

There were two fathers in this ghastly crew,
   And with them their two sons, of whom the one
Was more robust and hardy to the view,   691
   But he died early; and when he was gone,
His nearest messmate told his sire, who threw
   One glance at him, and said, "Heaven's will be done!
I can do nothing," and he saw him thrown
Into the deep without a tear or groan.

LXXXVIII

The other father had a weaklier child,
   Of a soft cheek, and aspect delicate;
But the boy bore up long, and with a mild
   And patient spirit held aloof his fate;   700
Little he said, and now and then he smiled,
   As if to win a part from off the weight
He saw increasing on his father's heart,
With the deep deadly thought, that they must part.

LXXXIX

And o'er him bent his sire, and never raised
   His eyes from off his face, but wiped the foam
From his pale lips, and ever on him gazed,
   And when the wish'd-for shower at length was come,
And the boy's eyes, which the dull film half glazed,
   Brighten'd, and for a moment seem'd to roam,   710
He squeezed from out a rag some drops of rain
Into his dying child's mouth—but in vain.

XC

The boy expired—the father held the clay,
   And look'd upon it long, and when at last

Death left no doubt, and the dead burthen lay
   Stiff on his heart, and pulse and hope were past,
He watch'd it wistfully, until away
   'Twas borne by the rude wave wherein 'twas cast;
Then he himself sunk down all dumb and shivering,
And gave no sign of life, save his limbs quivering. 720

XCI

Now overhead a rainbow, bursting through
   The scattering clouds, shone, spanning the dark sea,
Resting its bright base on the quivering blue;
   And all within its arch appear'd to be
Clearer than that without, and its wide hue
   Wax'd broad and waving, like a banner free,
Then changed like to a bow that's bent, and then
Forsook the dim eyes of these shipwreck'd men.

XCII

It changed, of course; a heavenly chameleon,
   The airy child of vapour and the sun,   730
Brought forth in purple, cradled in vermilion,
   Baptized in molten gold, and swathed in dun,
Glittering like crescents o'er a Turk's pavilion,
   And blending every colour into one,
Just like a black eye in a recent scuffle
(For sometimes we must box without the muffle).

XCIII

Our shipwreck'd seamen thought it a good omen—
   It is as well to think so, now and then;
'Twas an old custom of the Greek and Roman,
   And may become of great advantage when   740
Folks are discouraged; and most surely no men
   Had greater need to nerve themselves again
Than these, and so this rainbow look'd like hope—
Quite a celestial kaleidoscope.

. . .

[When only four remain alive, they see land.]

CIV

The shore looked wild, without a trace of man,
   And girt by formidable waves; but they
Were mad for land, and thus their course they ran,
   Though right ahead the roaring breakers lay:

A reef between them also now began
  To show its boiling surf and bounding spray,   830
But finding no place for their landing better,
They ran the boat for shore,—and overset her.

CV

But in his native stream, the Guadalquivir,
  Juan to lave his youthful limbs was wont;
And having learnt to swim in that sweet river,
  Had often turned the art to some account:
A better swimmer you could scarce see ever,
  He could, perhaps, have passed the Hellespont,
As once (a feat on which ourselves we prided)
Leander, Mr. Ekenhead, and I did.   840

CVI

So here, though faint, emaciated, and stark,
  He buoyed his boyish limbs, and strove to ply
With the quick wave, and gain, ere it was dark,
  The beach which lay before him, high and dry:
The greatest danger here was from a shark,
  That carried off his neighbor by the thigh;
As for the other two, they could not swim,
So nobody arrived on shore but him.

CVII

Nor yet had he arrived but for the oar,
  Which, providentially for him, was washed   850
Just as his feeble arms could strike no more,
  And the hard wave o'erwhelmed him as 'twas dashed
Within his grasp; he clung to it, and sore
  The waters beat while he thereto was lashed;
At last, with swimming, wading, scrambling, he
Rolled on the beach, half senseless, from the sea:

CVIII

There, breathless, with his digging nails he clung
  Fast to the sand, lest the returning wave,
From whose reluctant roar his life he wrung,
  Should suck him back to her insatiate grave: 860
And there he lay, full length, where he was flung,
  Before the entrance of a cliff-worn cave,
With just enough of life to feel its pain,
And deem that it was saved, perhaps in vain.

840. Mr. Ekenhead: Lt. Ekenhead of the marines, who
swam across the Hellespont with Byron (see the poem above,
"Written After Swimming from Sestos to Abydos").

CIX

With slow and staggering effort he arose,
  But sunk again upon his bleeding knee
And quivering hand; and then he looked for those
  Who long had been his mates upon the sea;
But none of them appeared to share his woes,
  Save one, a corpse, from out the famished three,
Who died two days before, and now had found   871
An unknown barren beach for burial-ground.

CX

And as he gazed, his dizzy brain spun fast,
  And down he sunk; and as he sunk, the sand
Swam round and round, and all his senses passed:
  He fell upon his side, and his stretched hand
Drooped dripping on the oar (their jury-mast),
  And, like a withered lily, on the land
His slender frame and pallid aspect lay,
As fair a thing as e'er was formed of clay.   880

CXI

How long in his damp trance young Juan lay
  He knew not, for the earth was gone for him,
And time had nothing more of night nor day
  For his congealing blood, and senses dim;
And how this heavy faintness passed away
  He knew not, till each painful pulse and limb,
And tingling vein, seemed throbbing back to life,
For Death, though vanquished, still retired with strife.

CXII

His eyes he opened, shut, again unclosed,
  For all was doubt and dizziness; he thought   890
He still was in the boat, and had but dozed,
  And felt again with his despair o'erwrought,
And wished it death in which he had reposed,
  And then once more his feelings back were brought,
And slowly by his swimming eyes was seen
A lovely female face of seventeen.

CXIII

'Twas bending close o'er his, and the small mouth
  Seemed almost prying into his for breath;
And chafing him, the soft warm hand of youth
  Recalled his answering spirits back from death; 900

And, bathing his chill temples, tried to soothe
    Each pulse to animation, till beneath
Its gentle touch and trembling care, a sigh
To these kind efforts made a low reply.

CXIV

Then was the cordial poured, and mantle flung
    Around his scarce-clad limbs; and the fair arm
Raised higher the faint head which o'er it hung;
    And her transparent cheek, all pure and warm,
Pillowed his death-like forehead; then she wrung
    His dewy curls, long drenched by every storm; 910
And watched with eagerness each throb that drew
A sigh from his heaved bosom—and hers, too.

CXV

And lifting him with care into the cave,
    The gentle girl, and her attendant,—one
Young, yet her elder, and of brow less grave,
    And more robust of figure—then begun
To kindle fire, and as the new flames gave
    Light to the rocks that roofed them, which the sun
Had never seen, the maid, or whatsoe'er
She was, appeared distinct, and tall, and fair.   920

CXVI

Her brow was overhung with coins of gold,
    That sparkled o'er the auburn of her hair,
Her clustering hair, whose longer locks were rolled
    In braids behind; and though her stature were
Even of the highest for a female mold,
    They nearly reached her heel; and in her air
There was a something which bespoke command,
As one who was a lady in the land.

CXVII

Her hair, I said, was auburn; but her eyes
    Were black as death, their lashes the same hue, 930
Of downcast length, in whose silk shadow lies
    Deepest attraction; for when to the view
Forth from its raven fringe the full glance flies,
    Ne'er with such force the swiftest arrow flew;
'Tis as the snake late coiled, who pours his length,
And hurls at once his venom and his strength.

CXVIII

Her brow was white and low, her cheek's pure dye
    Like twilight rosy still with the set sun;
Short upper lip—sweet lips! that make us sigh
    Ever to have seen such; for she was one   940
Fit for the model of a statuary
    (A race of mere impostors, when all's done—
I've seen much finer women, ripe and real,
Than all the nonsense of their stone ideal).

.   .   .

CXXIII

And these two tended him, and cheer'd him both
    With food and raiment, and those soft attentions,
Which are—(as I must own)—of female growth,
    And have ten thousand delicate inventions: 980
They made a most superior mess of broth,
    A thing which poesy but seldom mentions,
But the best dish that e'er was cook'd since Homer's
Achilles order'd dinner for new comers.

CXXIV

I'll tell you who they were, this female pair,
    Lest they should seem princesses in disguise;
Besides, I hate all mystery, and that air
    Of clap-trap, which your recent poets prize;
And so, in short, the girls they really were
    They shall appear before your curious eyes,   990
Mistress and maid; the first was only daughter
Of an old man, who lived upon the water.

CXXV

A fisherman he had been in his youth,
    And still a sort of fisherman was he;
But other speculations were, in sooth,
    Added to his connexion with the sea,
Perhaps not so respectable, in truth:
    A little smuggling, and some piracy,
Left him, at last, the sole of many masters
Of an ill-gotten million of piastres.   1000

984. Achilles . . . dinner: the feast Achilles gave to the
Greek ambassadors (*Iliad*, IX. 193–94).

CXXVI

A fisher, therefore, was he,—though of men,
　　Like Peter the Apostle,—and he fish'd
For wandering merchant vessels, now and then,
　　And sometimes caught as many as he wish'd;
The cargoes he confiscated, and gain
　　He sought in the slave-market too, and dish'd
Full many a morsel for that Turkish trade,
By which, no doubt, a good deal may be made.

CXXVII

He was a Greek, and on his isle had built
　　(One of the wild and smaller Cyclades)　　　1010
A very handsome house from out his guilt,
　　And there he lived exceedingly at ease;
Heavens knows what cash he got or blood he spilt,
　　A sad old fellow was he, if you please;
But this I know, it was a spacious building,
Full of barbaric carving, paint, and gilding.

CXXVIII

He had an only daughter, call'd Haidée,
　　The greatest heiress of the Eastern Isles;
Besides, so very beautiful was she,
　　Her dowry was as nothing to her smiles:　　　1020
Still in her teens, and like a lovely tree
　　She grew to womanhood, and between whiles
Rejected several suitors, just to learn
How to accept a better in his turn.

CXXIX

And walking out upon the beach, below
　　The cliff, towards sunset, on that day she found,
Insensible,—not dead, but nearly so,—
　　Don Juan, almost famish'd, and half drown'd;
But being naked, she was shock'd, you know,
　　Yet deem'd herself in common pity bound,　　　1030
As far as in her lay, "to take him in,
A stranger" dying, with so white a skin.

CXXX

But taking him into her father's house
　　Was not exactly the best way to save,
But like conveying to the cat the mouse,
　　Or people in a trance into their grave;

Because the good old man had so much "νοῦς,"
　　Unlike the honest Arab thieves so brave,
He would have hospitably cured the stranger,
And sold him instantly when out of danger.　　　1040

CXXXI

And therefore, with her maid, she thought it best
　　(A virgin always on her maid relies)
To place him in the cave for present rest:
　　And when, at last, he open'd his black eyes,
Their charity increased about their guest;
　　And their compassion grew to such a size,
It open'd half the turnpike gates to heaven—
(St. Paul says, 'tis the toll which must be given).

CXXXII

They made a fire,—but such a fire as they
　　Upon the moment could contrive with such　　　1050
Materials as were cast up round the bay,—
　　Some broken planks, and oars, that to the touch
Were nearly tinder, since, so long they lay
　　A mast was almost crumbled to a crutch;
But, by God's grace, here wrecks were in such plenty,
That there was fuel to have furnish'd twenty.

CXXXIII

He had a bed of furs, and a pelisse,
　　For Haidée stripp'd her sables off to make
His couch; and, that he might be more at ease,
　　And warm, in case by chance he should awake,
They also gave a petticoat apiece,　　　1061
　　She and her maid,—and promised by daybreak
To pay him a fresh visit, with a dish
For breakfast, of eggs, coffee, bread, and fish.

CXXXIV

And thus they left him to his lone repose:
　　Juan slept like a top, or like the dead,
Who sleep at last, perhaps (God only knows),
　　Just for the present; and in his lull'd head
Not even a vision of his former woes
　　Throbb'd in accursed dreams, which sometimes
　　　　spread　　　1070
Unwelcome visions of our former years,
Till the eye, cheated, opens thick with tears.

1037. "νοῦς": Greek for mind or spirit.

CXXXV

Young Juan slept all dreamless:—but the maid,
    Who smooth'd his pillow, as she left the den
Look'd back upon him, and a moment staid,
    And turn'd, believing that he call'd again.
He slumber'd; yet she thought, at least she said
    (The heart will slip, even as the tongue and pen),
He had pronounced her name—but she forgot
That at this moment Juan knew it not.   1080

CXXXVI

And pensive to her father's house she went,
    Enjoining silence strict to Zoe, who
Better than her knew what, in fact, she meant,
    She being wiser by a year or two:
A year or two's an age when rightly spent,
    And Zoe spent hers, as most women do,
In gaining all that useful sort of knowledge
Which is acquired in Nature's good old college.

    .   .   .

[The next morning Haidée and her maid, Zoe, return. They provide Juan with some clothes and breakfast.]

CLXI

And then fair Haidée tried her tongue at speaking,
    But not a word could Juan comprehend,
Although he listen'd so that the young Greek in
    Her earnestness would ne'er have made an end;
And, as he interrupted not, went eking
    Her speech out to her protégé and friend,
Till pausing at the last her breath to take,
She saw he did not understand Romaic.

CLXII

And then she had recourse to nods, and signs,
    And smiles, and sparkles of the speaking eye,   1290
And read (the only book she could) the lines
    Of his fair face, and found, by sympathy,
The answer eloquent, where the soul shines
    And darts in one quick glance a long reply;
And thus in every look she saw exprest
A world of words, and things at which she guess'd.

1288. Romaic: the modern Greek language.

CLXIII

And now, by dint of fingers and of eyes,
    And words repeated after her, he took
A lesson in her tongue; but by surmise,
    No doubt, less of her language than her look: 1300
As he who studies fervently the skies
    Turns oftener to the stars than to his book,
Thus Juan learn'd his alpha beta better
From Haidée's glance than any graven letter.

CLXIV

'Tis pleasing to be school'd in a strange tongue
    By female lips and eyes—that is, I mean,
When both the teacher and the taught are young,
    As was the case, at least, where I have been;
They smile so when one's right, and when one's wrong
    They smile still more, and then there intervene
Pressure of hands, perhaps even a chaste kiss;— 1311
I learn'd the little that I know by this:

CLXV

That is, some words of Spanish, Turk, and Greek,
    Italian not at all, having no teachers;
Much English I cannot pretend to speak,
    Learning that language chiefly from its preachers,
Barrow, South, Tillotson, whom every week
    I study, also Blair, the highest reachers
Of eloquence in piety and prose—
I hate your poets, so read none of those.   1320

CLXVI

As for the ladies, I have nought to say,
    A wanderer from the British world of fashion,
Where I, like other "dogs, have had my day,"
    Like other men, too, may have had my passion—
But that, like other things, has pass'd away,
    And all her fools whom I *could* lay the lash on:
Foes, friends, men, women, now are nought to me
But dreams of what has been, no more to be.

CLXVII

Return we to Don Juan. He begun
    To hear new words, and to repeat them; but 1330

Some feelings, universal as the sun,
　　Were such as could not in his breast be shut
More than within the bosom of a nun:
　　He was in love,—as you would be, no doubt,
With a young benefactress,—so was she,
Just in the way we very often see.

CLXVIII

And every day by daybreak—rather early
　　For Juan, who was somewhat fond of rest—
She came into the cave, but it was merely
　　To see her bird reposing in his nest;　　　1340
And she would softly stir his locks so curly,
　　Without disturbing her yet slumbering guest,
Breathing all gently o'er his cheek and mouth,
As o'er a bed of roses the sweet south.

CLXIX

And every morn his colour freshlier came,
　　And every day help'd on his convalescence;
'Twas well, because health in the human frame
　　Is pleasant, besides being true love's essence,
For health and idleness to passion's flame
　　Are oil and gunpowder; and some good lessons
Are also learnt from Ceres and from Bacchus,　1351
Without whom Venus will not long attack us.

CLXX

While Venus fills the heart (without heart really
　　Love, though good always, is not quite so good),
Ceres presents a plate of vermicelli,—
　　For love must be sustain'd like flesh and blood,—
While Bacchus pours out wine, or hands a jelly:
　　Eggs, oysters, too, are amatory food;
But who is their purveyor from above　　　1359
Heaven knows,—it may be Neptune, Pan, or Jove.

CLXXI

When Juan woke he found some good things ready,
　　A bath, a breakfast, and the finest eyes
That ever made a youthful heart less steady,
　　Besides her maid's, as pretty for their size;

1351–52. **Ceres . . . us:** Without Ceres (goddess of food) and Bacchus (god of wine) we are in little danger from Venus (love).

But I have spoken of all this already—
　　And repetition's tiresome and unwise,—
Well—Juan, after bathing in the sea,
Came always back to coffee and Haidée.

CLXXII

Both were so young, and one so innocent,
　　That bathing pass'd for nothing; Juan seem'd　1370
To her, as 'twere, the kind of being sent,
　　Of whom these two years she had nightly dream'd,
A something to be loved, a creature meant
　　To be her happiness, and whom she deem'd
To render happy; all who joy would win
Must share it,—Happiness was born a twin.

CLXXIII

It was such pleasure to behold him, such
　　Enlargement of existence to partake
Nature with him, to thrill beneath his touch,
　　To watch him slumbering, and to see him wake:
To live with him for ever were too much;　　1381
　　But then the thought of parting made her quake:
He was her own, her ocean-treasure, cast
Like a rich wreck—her first love, and her last.

CLXXIV

And thus a moon roll'd on, and fair Haidée
　　Paid daily visits to her boy, and took
Such plentiful precautions, that still he
　　Remain'd unknown within his craggy nook;
At last her father's prows put out to sea,
　　For certain merchantmen upon the look,　　1390
Not as of yore to carry off an Io,
But three Ragusan vessels bound for Scio.

CLXXV

Then came her freedom, for she had no mother,
　　So that, her father being at sea, she was
Free as a married woman, or such other
　　Female, as where she likes may freely pass,
Without even the encumbrance of a brother,
　　The freest she that ever gazed on glass:
I speak of Christian lands in this comparison,
Where wives, at least, are seldom kept in garrison.

1391–92: **Io . . . Scio:** Io was carried off by Phoenician merchants; Ragusa is on the Adriatic shore of Yugoslavia; Scio is an Aegean isle.

CLXXVI

Now she prolong'd her visits and her talk   1401
  (For they must talk), and he had learnt to say
So much as to propose to take a walk,—
  For little had he wander'd since the day
On which, like a young flower snapp'd from the stalk,
  Drooping and dewy on the beach he lay,—
And thus they walk'd out in the afternoon,
And saw the sun set opposite the moon.

CLXXVII

It was a wild and breaker-beaten coast,
  With cliffs above, and a broad sandy shore,   1410
Guarded by shoals and rocks as by an host,
  With here and there a creek, whose aspect wore
A better welcome to the tempest-tost;
  And rarely ceased the haughty billow's roar,
Save on the dead long summer days, which make
The outstretch'd ocean glitter like a lake.

CLXXVIII

And the small ripple spilt upon the beach
  Scarcely o'erpass'd the cream of your champagne,
When o'er the brim the sparkling bumpers reach,
  That spring-dew of the spirit! the heart's rain! 1420
Few things surpass old wine; and they may preach
  Who please,—the more because they preach in
      vain,—
Let us have wine and woman, mirth and laughter,
Sermons and soda-water the day after.

CLXXIX

Man, being reasonable, must get drunk;
  The best of life is but intoxication:
Glory, the grape, love, gold, in these are sunk
  The hopes of all men, and of every nation;
Without their sap, how branchless were the trunk
  Of life's strange tree, so fruitful on occasion: 1430
But to return,—Get very drunk; and when
You wake with headache, you shall see what then.

CLXXX

Ring for your valet—bid him quickly bring
  Some hock and soda-water, then you'll know

A pleasure worthy Xerxes the great king;
  For not the blest sherbet, sublimed with snow,
Nor the first sparkle of the desert-spring,
  Nor Burgundy in all its sunset glow,
After long travel, ennui, love, or slaughter,
Vie with that draught of hock and soda-water.   1440

CLXXXI

The coast—I think it was the coast that I
  Was just describing—Yes, it *was* the coast—
Lay at this period quiet as the sky,
  The sands untumbled, the blue waves untost,
And all was stillness, save the sea-bird's cry,
  And dolphin's leap, and little billow crost
By some low rock or shelve, that made it fret
Against the boundary it scarcely wet.

CLXXXII

And forth they wander'd, her sire being gone,
  As I have said, upon an expedition;   1450
And mother, brother, guardian, she had none,
  Save Zoe, who, although with due precision
She waited on her lady with the sun,
  Thought daily service was her only mission,
Bringing warm water, wreathing her long tresses,
And asking now and then for cast-off dresses.

CLXXXIII

It was the cooling hour, just when the rounded
  Red sun sinks down behind the azure hill,
Which then seems as if the whole earth it bounded,
  Circling all nature, hush'd, and dim, and still,   1460
With the far mountain-crescent half surrounded
  On one side, and the deep sea calm and chill
Upon the other, and the rosy sky,
With one star sparkling through it like an eye.

CLXXXIV

And thus they wander'd forth, and hand in hand,
  Over the shining pebbles and the shells,
Glided along the smooth and harden'd sand,
  And in the worn and wild receptacles
Work'd by the storms, yet work'd as it were plann'd,
  In hollow halls, with sparry roofs and cells,   1470
They turn'd to rest; and, each clasp'd by an arm,
Yielded to the deep twilight's purple charm.

CLXXXV

They look'd up to the sky, whose floating glow
    Spread like a rosy ocean, vast and bright;
They gazed upon the glittering sea below,
    Whence the broad moon rose circling into sight;
They heard the waves splash, and the wind so low,
    And saw each other's dark eyes darting light
Into each other—and, beholding this,
Their lips drew near, and clung into a kiss;   1480

CLXXXVI

A long, long kiss, a kiss of youth, and love,
    And beauty, all concentrating like rays
Into one focus, kindled from above;
    Such kisses as belong to early days,
Where heart, and soul, and sense, in concert move,
    And the blood's lava, and the pulse a blaze,
Each kiss a heart-quake,—for a kiss's strength,
I think, it must be reckon'd by its length.

CLXXXVII

By length I mean duration; theirs endured
    Heaven knows how long—no doubt they never
        reckon'd;   1490
And if they had, they could not have secured
    The sum of their sensations to a second:
They had not spoken; but they felt allured,
    As if their souls and lips each other beckon'd,
Which, being join'd, like swarming bees they clung—
Their hearts the flowers from whence the honey sprung.

CLXXXVIII

They were alone, but not alone as they
    Who shut in chambers think it loneliness;
The silent ocean, and the starlight bay,
    The twilight glow, which momently grew less,
The voiceless sands, and dropping caves, that lay 1501
    Around them, made them to each other press,
As if there were no life beneath the sky
Save theirs, and that their life could never die.

CLXXXIX

They fear'd no eyes nor ears on that lone beach,
    They felt no terrors from the night, they were
All in all to each other: though their speech
    Was broken words, they *thought* a language there,—

And all the burning tongues the passions teach
    Found in one sigh the best interpreter   1510
Of nature's oracle—first love,—that all
Which Eve has left her daughters since her fall.

CXC

Haidée spoke not of scruples, ask'd no vows,
    Nor offer'd any; she had never heard
Of plight and promises to be a spouse,
    Or perils by a loving maid incurr'd;
She was all which pure ignorance allows,
    And flew to her young mate like a young bird;
And never having dreamt of falsehood, she
Had not one word to say of constancy.   1520

CXCI

She loved, and was beloved—she adored,
    And she was worshipp'd; after nature's fashion,
Their intense souls, into each other pour'd,
    If souls could die, had perish'd in that passion,—
But by degreees their senses were restored,
    Again to be o'ercome, again to dash on;
And, beating 'gainst *his* bosom, Haidée's heart
Felt as if never more to beat apart.

CXCII

Alas! they were so young, so beautiful,
    So lonely, loving, helpless, and the hour   1530
Was that in which the heart is always full,
    And, having o'er itself no further power,
Prompts deeds eternity cannot annul,
    But pays off moments in an endless shower
Of hell-fire—all prepared for people giving
Pleasure or pain to one another living.

CXCIII

Alas! for Juan and Haidée! they were
    So loving and so lovely—till then never,
Excepting our first parents, such a pair
    Had run the risk of being damn'd for ever;   1540
And Haidée, being devout as well as fair,
    Had, doubtless, heard about the Stygian river,
And hell and purgatory—but forgot
Just in the very crisis she should not.

1542. Stygian: the river Styx in Hades.

CXCIV

They look upon each other, and their eyes
  Gleam in the moonlight; and her white arm clasps
Round Juan's head, and his around her lies
  Half buried in the tresses which it grasps;
She sits upon his knee, and drinks his sighs,
  He hers, until they end in broken gasps;   1550
And thus they form a group that's quite antique,
Half naked, loving, natural, and Greek.

CXCV

And when those deep and burning moments pass'd,
  And Juan sunk to sleep within her arms,
She slept not, but all tenderly, though fast,
  Sustain'd his head upon her bosom's charms;
And now and then her eye to heaven is cast,
  And then on the pale cheek her breast now warms,
Pillow'd on her o'erflowing heart, which pants
With all it granted, and with all it grants.   1560

CXCVI

An infant when it gazes on a light,
  A child the moment when it drains the breast,
A devotee when soars the Host in sight,
  An Arab with a stranger for a guest,
A sailor when the prize has struck in fight,
  A miser filling his most hoarded chest,
Feel rapture; but not such true joy are reaping
As they who watch o'er what they love while sleeping.

CXCVII

For there it lies so tranquil, so beloved,
  All that it hath of life with us is living;   1570
So gentle, stirless, helpless, and unmoved,
  And all unconscious of the joy 'tis giving;
All it hath felt, inflicted, pass'd, and proved,
  Hush'd into depths beyond the watcher's diving;
There lies the thing we love with all its errors
And all its charms, like death without its terrors.

CXCVIII

The lady watch'd her lover—and that hour
  Of Love's, and Night's, and Ocean's solitude,

O'erflow'd her soul with their united power;
  Amidst the barren sand and rocks so rude   1580
She and her wave-worn love had made their bower,
  Where nought upon their passion could intrude,
And all the stars that crowded the blue space
Saw nothing happier than her glowing face.

CXCIX

Alas! the love of women! it is known
  To be a lovely and a fearful thing;
For all of theirs upon that die is thrown,
  And if 'tis lost, life hath no more to bring
To them but mockeries of the past alone,
  And their revenge is as the tiger's spring,   1590
Deadly, and quick, and crushing; yet, as real
Torture is theirs, what they inflict they feel.

CC

They are right; for man, to man so oft unjust,
  Is always so to women; one sole bond
Awaits them, treachery is all their trust;
  Taught to conceal, their bursting hearts despond
Over their idol, till some wealthier lust
  Buys them in marriage—and what rests beyond?
A thankless husband, next a faithless lover,
Then dressing, nursing, praying, and all's over.   1600

CCI

Some take a lover, some take drams or prayers,
  Some mind their household, others dissipation,
Some run away, and but exchange their cares,
  Losing the advantage of a virtuous station;
Few changes e'er can better their affairs,
  Theirs being an unnatural situation,
From the dull palace to the dirty hovel:
Some play the devil, and then write a novel.

CCII

Haidée was Nature's bride, and knew not this;
  Haidée was Passion's child, born where the sun
Showers triple light, and scorches even the kiss   1611
  Of his gazelle-eyed daughters; she was one

1608. write a novel: After her affair with Byron, Caroline
Lamb wrote a novel, *Glenarvon* (1816), in which she included
the farewell letter Byron had written her.

Made but to love, to feel that she was his
    Who was her chosen: what was said or done
Elsewhere was nothing. She had nought to fear,
Hope, care, nor love beyond, her heart beat *here*.

.   .   .

CCVIII

But Juan! had he quite forgotten Julia?
    And should he have forgotten her so soon?
I can't but say it seems to me mostly truly a
    Perplexing question; but, no doubt, the moon 1660
Does these things for us, and whenever newly a
    Strong palpitation rises, 'tis her boon,
Else how the devil is it that fresh features
Have such a charm for us poor human creatures?

CCIX

I hate inconstancy—I loathe, detest,
    Abhor, condemn, abjure the mortal made
Of such quicksilver clay that in his breast
    No permanent foundation can be laid;
Love, constant love, has been my constant guest,
    And yet last night, being at a masquerade,   1670
I saw the prettiest creature, fresh from Milan,
Which gave me some sensations like a villain.

CCX

But soon Philosophy came to my aid,
    And whisper'd, "Think of every sacred tie!"
"I will, my dear Philosophy!" I said,
    "But then her teeth, and then, oh, Heaven! her eye!
I'll just inquire if she be wife or maid,
    Or neither—out of curiosity."
"Stop!" cried Philosophy, with air so Grecian
(Though she was masqued then as a fair Venetian); 1680

CCXI

"Stop!" so I stopp'd.—But to return: that which
    Men call inconstancy is nothing more
Than admiration due where nature's rich
    Profusion with young beauty covers o'er
Some favour'd object; and as in the niche
    A lovely statue we almost adore,
This sort of adoration of the real
Is but a heightening of the "beau ideal."

CCXII

'Tis the perception of the beautiful,
    A fine extension of the faculties,   1690
Platonic, universal, wonderful,
    Drawn from the stars, and filter'd through the skies,
Without which life would be extremely dull;
    In short, it is the use of our own eyes,
With one or two small senses added, just
To hint that flesh is form'd of fiery dust.

CCXIII

Yet 'tis a painful feeling, and unwilling,
    For surely if we always could perceive
In the same object graces quite as killing
    As when she rose upon us like an Eve,   1700
'Twould save us many a heart-ache, many a shilling
    (For we must get them any how, or grieve),
Whereas, if one sole lady pleased for ever,
How pleasant for the heart, as well as liver!

CCXIV

The heart is like the sky, a part of heaven,
    But changes night and day, too, like the sky;
Now o'er it clouds and thunder must be driven,
    And darkness and destruction as on high:
But when it hath been scorch'd, and pierced, and riven,
    Its storms expire in water-drops; the eye   1710
Pours forth at last the heart's blood turn'd to tears,
Which make the English climate of our years.

CCXV

The liver is the lazaret of bile,
    But very rarely executes its function,
For the first passion stays there such a while,
    That all the rest creep in and form a junction,
Like knots of vipers on a dunghill's soil,
    Rage, fear, hate, jealousy, revenge, compunction,
So that all mischiefs spring up from this entrail,
Like earthquakes from the hidden fire call'd "central."

CCXVI

In the mean time, without proceeding more   1721
    In this anatomy, I've finish'd now
Two hundred and odd stanzas as before,
    That being about the number I'll allow

Each canto of the twelve, or twenty-four;
   And, laying down my pen, I make my bow,
Leaving Don Juan and Haidée to plead
For them and theirs with all who deign to read.

                   1818–19 (1819)

from CANTO THE THIRD

I

Hail, Muse! *et cætera.*—We left Juan sleeping,
   Pillow'd upon a fair and happy breast,
And watch'd by eyes that never yet knew weeping,
   And loved by a young heart, too deeply blest
To feel the poison through her spirit creeping,
   Or know who rested there, a foe to rest,
Had soil'd the current of her sinless years,
And turn'd her pure heart's purest blood to tears!

II

Oh, Love! what is it in this world of ours
   Which makes it fatal to be loved? Ah why   10
With cypress branches hast thou wreathed thy bowers,
   And made thy best interpreter a sigh?
As those who dote on odours pluck the flowers,
   And place them on their breast—but place to die—
Thus the frail beings we would fondly cherish
Are laid within our bosoms but to perish.

III

In her first passion woman loves her lover,
   In all the others all she loves is love,
Which grows a habit she can ne'er get over,
   And fits her loosely—like an easy glove,   20
As you may find, whene'er you like to prove her:
   One man alone at first her heart can move;
She then prefers him in the plural number,
Not finding that the additions much encumber.

IV

I know not if the fault be men's or theirs;
   But one thing's pretty sure; a woman planted

CANTO III. **26. planted:** abandoned (from the French *planter là*).

(Unless at once she plunge for life in prayers)
   After a decent time must be gallanted;
Although, no doubt, her first of love affairs
   Is that to which her heart is wholly granted;   30
Yet there are some, they say, who have had *none*,
But those who have ne'er end with only *one*.

V

'Tis melancholy, and a fearful sign
   Of human frailty, folly, also crime,
That love and marriage rarely can combine,
   Although they both are born in the same clime;
Marriage from love, like vinegar from wine—
   A sad, sour, sober beverage—by time
Is sharpen'd from its high celestial flavour,
Down to a very homely household savour.   40

VI

There's something of antipathy, as 'twere,
   Between their present and their future state;
A kind of flattery that's hardly fair
   Is used until the truth arrives too late—
Yet what can people do, except despair?
   The same things change their names at such a rate;
For instance—passion in a lover's glorious,
But in a husband is pronounced uxorious.

VII

Men grow ashamed of being so very fond;
   They sometimes also get a little tired   50
(But that, of course, is rare), and then despond:
   The same things cannot always be admired,
Yet 'tis "so nominated in the bond,"
   That both are tied till one shall have expired.
Sad thought! to lose the spouse that was adorning
Our days, and put one's servants into mourning.

VIII

There's doubtless something in domestic doings
   Which forms, in fact, true love's antithesis;
Romances paint at full length people's wooings,
   But only give a bust of marriages;   60
For no one cares for matrimonial cooings,
   There's nothing wrong in a connubial kiss:
Think you, if Laura had been Petrarch's wife,
He would have written sonnets all his life?

IX

All tragedies are finish'd by a death,
    All comedies are ended by a marriage;
The future states of both are left to faith,
    For authors fear description might disparage
The worlds to come of both, or fall beneath,
    And then both worlds would punish their miscar-
        riage;   70
So leaving each their priest and prayer-book ready,
They say no more of Death or of the Lady.

X

The only two that in my recollection
    Have sung of heaven and hell, or marriage, are
Dante and Milton, and of both the affection
    Was hapless in their nuptials, for some bar
Of fault or temper ruin'd the connexion
    (Such things, in fact, it don't ask much to mar);
But Dante's Beatrice and Milton's Eve
Were not drawn from their spouses, you conceive. 80

XI

Some persons say that Dante meant theology
    By Beatrice, and not a mistress—I,
Although my opinion may require apology,
    Deem this a commentator's phantasy,
Unless indeed it was from his own knowledge he
    Decided thus, and show'd good reason why;
I think that Dante's more abstruse ecstatics
Meant to personify the mathematics.

XII

Haidée and Juan were not married, but
    The fault was theirs, not mine: it is not fair,   90
Chaste reader, then, in any way to put
    The blame on me, unless you wish they were;
Then if you'd have them wedded, please to shut
    The book which treats of this erroneous pair,
Before the consequences grow too awful;
'Tis dangerous to read of loves unlawful.

72. **Death . . . Lady:** "Death and the Lady" was a popular eighteenth-century ballad. 75–76. **Dante . . . nuptials:** Byron notes Dante's disparaging remark about his wife (*Inferno*, XVI. 45) and adds: "Milton's first wife ran away from him within the first month. If she had not, what would John Milton have done?" 88. **mathematics:** another allusion to Lady Byron's interests (cf. above, I. 89).

XIII

Yet they were happy,—happy in the illicit
    Indulgence of their innocent desires;
But more imprudent grown with every visit,
    Haidée forgot the island was her sire's;   100
When we have what we like, 'tis hard to miss it,
    At least in the beginning, ere one tires;
Thus she came often, not a moment losing,
Whilst her piratical papa was cruising.

XIV

Let not his mode of raising cash seem strange,
    Although he fleeced the flags of every nation,
For into a prime minister but change
    His title, and 'tis nothing but taxation;
But he, more modest, took an humbler range
    Of life, and in an honester vocation   110
Pursued o'er the high seas his watery journey,
And merely practised as a sea-attorney.

XV

The good old gentleman had been detain'd
    By winds and waves, and some important captures;
And, in the hope of more, at sea remain'd,
    Although a squall or two had damp'd his raptures,
By swamping one of the prizes; he had chain'd
    His prisoners, dividing them like chapters
In number'd lots; they all had cuffs and collars,
And averaged each from ten to a hundred dollars. 120

.   .   .

XIX

Then having settled his marine affairs,
    Despatching single cruisers here and there,
His vessel having need of some repairs,
    He shaped his course to where his daughter fair
Continued still her hospitable cares;
    But that part of the coast being shoal and bare,   150
And rough with reefs which ran out many a mile,
His port lay on the other side o' the isle.

XX

And there he went ashore without delay,
    Having no custom-house nor quarantine

To ask him awkward questions on the way
   About the time and place where he had been:
He left his ship to be hove down next day,
   With orders to the people to careen;
So that all hands were busy beyond measure,
In getting out goods, ballast, guns, and treasure.   160

. . .

XXVII

He saw his white walls shining in the sun,
   His garden trees all shadowy and green;   210
He heard his rivulet's light bubbling run,
   The distant dog-bark; and perceived between
The umbrage of the wood so cool and dun,
   The moving figures, and the sparkling sheen
Of arms (in the East all arm)—and various dyes
Of colour'd garbs, as bright as butterflies.

XXVIII

And as the spot where they appear he nears,
   Surprised at these unwonted signs of idling,
He hears—alas! no music of the spheres,
   But an unhallow'd, earthly sound of fiddling!   220
A melody which made him doubt his ears,
   The cause being past his guessing or unriddling;
A pipe, too, and a drum, and shortly after,
A most unoriental roar of laughter.

. . .

XXXVIII

He did not know (alas! how men will lie)
   That a report (especially the Greeks)
Avouch'd his death (such people never die),
   And put his house in mourning several weeks,—
But now their eyes and also lips were dry;   301
   The bloom, too, had return'd to Haidée's cheeks.
Her tears, too, being return'd into their fount,
She now kept house upon her own account.

XXXIX

Hence all this rice, meat, dancing, wine, and fiddling,
   Which turn'd the isle into a place of pleasure;
The servants all were getting drunk or idling,
   A life which made them happy beyond measure.

Her father's hospitality seem'd middling,
   Compared with what Haidée did with his treasure;
'Twas wonderful how things went on improving,   311
While she had not one hour to spare from loving.

XL

Perhaps you think in stumbling on this feast,
   He flew into a passion, and in fact
There was no mighty reason to be pleased;
   Perhaps you prophesy some sudden act,
The whip, the rack, or dungeon at the least,
   To teach his people to be more exact,
And that, proceeding at a very high rate,
He show'd the royal *penchants* of a pirate.   320

XLI

You're wrong.—He was the mildest manner'd man
   That ever scuttled ship or cut a throat;
With such true breeding of a gentleman,
   You never could divine his real thought,
No courtier could, and scarcely woman can
   Gird more deceit within a petticoat;
Pity he loved adventurous life's variety,
He was so great a loss to good society.

. . .

LIII

He was a man of a strange temperament,
   Of mild demeanour though of savage mood,
Moderate in all his habits, and content
   With temperance in pleasure, as in food,   420
Quick to perceive, and strong to bear, and meant
   For something better, if not wholly good;
His country's wrongs and his despair to save her
Had stung him from a slave to an enslaver.

LIV

The love of power, and rapid gain of gold,
   The hardness by long habitude produced,
The dangerous life in which he had grown old,
   The mercy he had granted oft abused,
The sights he was accustom'd to behold,
   The wild seas, and wild men with whom he cruised,
Had cost his enemies a long repentance,   431
And made him a good friend, but bad acquaintance.

LV

But something of the spirit of old Greece
    Flash'd o'er his soul a few heroic rays,
Such as lit onward to the Golden Fleece
    His predecessors in the Colchian days;
'Tis true he had no ardent love for peace—
    Alas! his country show'd no path to praise:
Hate to the world and war with every nation
He waged, in vengeance of her degradation.   440

LVI

Still o'er his mind the influence of the clime
    Shed its Ionian elegance, which show'd
Its power unconsciously full many a time,—
    A taste seen in the choice of his abode,
A love of music and of scenes sublime,
    A pleasure in the gentle stream that flow'd
Past him in crystal, and a joy in flowers,
Bedew'd his spirit in his calmer hours.

LVII

But whatsoe'er he had of love reposed
    On that beloved daughter; she had been   450
The only thing which kept his heart unclosed
    Amidst the savage deeds he had done and seen,
A lonely pure affection unopposed:
    There wanted but the loss of this to wean
His feelings from all milk of human kindness,
And turn him like the Cyclops mad with blindness.

LVIII

The cubless tigress in her jungle raging
    Is dreadful to the shepherd and the flock;
The ocean when its yeasty war is waging
    Is awful to the vessel near the rock;   460
But violent things will sooner bear assuaging,
    Their fury being spent by its own shock,
Than the stern, single, deep, and wordless ire
Of a strong human heart, and in a sire.

436. Colchian days: Colchis, on the Black Sea, to which
the Argonauts sailed. **456. Cyclops . . . blindness:**
Ulysses put out the eye of Polyphemus, king of the Cyclops,
with a red-hot brand.

LIX

It is a hard although a common case
    To find our children running restive—they
In whom our brightest days we would retrace,
    Our little selves re-formed in finer clay,
Just as old age is creeping on apace,
    And clouds come o'er the sunset of our day,   470
They kindly leave us, though not quite alone,
But in good company—the gout or stone.

LX

Yet a fine family is a fine thing
    (Provided they don't come in after dinner);
'Tis beautiful to see a matron bring
    Her children up (if nursing them don't thin her);
Like cherubs round an altar-piece they cling
    To the fire-side (a sight to touch a sinner).
A lady with her daughters or her nieces
Shine like a guinea and seven-shilling pieces.   480

LXI

Old Lambro pass'd unseen a private gate,
    And stood within his hall at eventide;
Meantime the lady and her lover sate
    At wassail in their beauty and their pride:
An ivory inlaid table spread with state
    Before them, and fair slaves on every side;
Gems, gold, and silver, form'd the service mostly,
Mother of pearl and coral the less costly.

.   .   .

LXVII

Haidée and Juan carpeted their feet
    On crimson satin, border'd with pale blue;   530
Their sofa occupied three parts complete
    Of the apartment—and appear'd quite new;
The velvet cushions (for a throne more meet)
    Were scarlet, from whose glowing centre grew
A sun emboss'd in gold, whose rays of tissue,
Meridian-like, were seen all light to issue.

LXVIII

Crystal and marble, plate and porcelain,
    Had done their work of splendour; Indian mats

And Persian carpets, which the heart bled to stain,
  Over the floors were spread; gazelles and cats,   540
And dwarfs and blacks, and such like things that gain
  Their bread as ministers and favourites—(that's
To say, by degradation)—mingled there
  As plentiful as in a court, or fair.

. . .

LXXVIII

And now they were diverted by their suite,
  Dwarfs, dancing-girls, black eunuchs, and a poet,
Which made their new establishment complete;
  The last was of great fame, and liked to show it:
His verses rarely wanted their due feet—   621
  And for his theme—he seldom sung below it,
He being paid to satirise or flatter,
As the psalm says, "inditing a good matter."

LXXIX

He praised the present, and abused the past,
  Reversing the good custom of old days,
An Eastern anti-jacobin at last
  He turn'd, preferring pudding to *no* praise—
For some few years his lot had been o'ercast
  By his seeming independent in his lays,   630
But now he sung the Sultan and the Pacha
With truth like Southey, and with verse like Crashaw.

LXXX

He was a man who had seen many changes,
  And always changed as true as any needle;
His polar star being one which rather ranges,
  And not the fix'd—he knew the way to wheedle:
So vile he 'scaped the doom which oft avenges;
  And being fluent (save indeed when fee'd ill),
He lied with such a fervour of intention—
There was no doubt he earn'd his laureate pension. 640

. . .

LXXXIII

But now being lifted into high society,
  And having pick'd up several odds and ends
Of free thoughts in his travels, for variety,
  He deem'd, being in a lone isle, among friends, 660
That without any danger of a riot, he
  Might for long lying make himself amends;
And singing as he sung in his warm youth,
Agree to a short armistice with truth.

LXXXIV

He had travell'd 'mongst the Arabs, Turks, and Franks,
  And knew the self-loves of the different nations;
And having lived with people of all ranks,
  Had something ready upon most occasions—
Which got him a few presents and some thanks.
  He varied with some skill his adulations;   670
To "do at Rome as Romans do," a piece
Of conduct was which he observed in Greece.

LXXXV

Thus, usually, when he was asked to sing,
  He gave the different nations something national;
'Twas all the same to him—"God save the king,"
  Or "*Ça ira*," according to the fashion all:
His muse made increment of anything,
  From the high lyric down to the low rational:
If Pindar sang horse-races, what should hinder
Himself from being as pliable as Pindar?   680

LXXXVI

In France, for instance, he would write a chanson;
  In England a six canto quarto tale;
In Spain he'd make a ballad or romance on
  The last war—much the same in Portugal;
In Germany, the Pegasus he'd prance on
  Would be old Goethe's—(see what says De Staël);
In Italy he'd ape the "Trecentisti;"
In Greece, he'd sing some sort of hymn like this t' ye:

---

632. Crashaw: The style of Richard Crashaw and of the metaphysical poets generally was disliked at this time because of its obscurity.

676. "Ça ira": song of the French revolutionists ("It will go on"). **686. see . . . De Staël:** referring to Mme de Staël's remark (*De l'Allemagne* [1818], I. 227) that Goethe "represented the entire literature of Germany." **687. "Trecentisti":** writers of the 1300's (i.e., the fourteenth century).

1

The isles of Greece, the isles of Greece!
　　Where burning Sappho loved and sung, 690
Where grew the arts of war and peace,
　　Where Delos rose, and Phœbus sprung!
Eternal summer gilds them yet,
But all, except their sun, is set.

2

The Scian and the Teian muse,
　　The hero's harp, the lover's lute,
Have found the fame your shores refuse;
　　Their place of birth alone is mute
To sounds which echo further west
Than your sires' "Islands of the Blest."　700

3

The mountains look on Marathon—
　　And Marathon looks on the sea;
And musing there an hour alone,
　　I dream'd that Greece might still be free;
For standing on the Persians' grave,
I could not deem myself a slave.

4

A king sate on the rocky brow
　　Which looks o'er sea-born Salamis;
And ships, by thousands, lay below,
　　And men in nations;—all were his!　710
He counted them at break of day—
And when the sun set where were they?

5

And where are they? and where art thou,
　　My country? On thy voiceless shore
The heroic lay is tuneless now—
　　The heroic bosom beats no more!
And must thy lyre, so long divine,
Degenerate into hands like mine?

6

'Tis something, in the dearth of fame,
　　Though link'd among a fetter'd race,　720
To feel at least a patriot's shame,
　　Even as I sing, suffuse my face;
For what is left the poet here?
For Greeks a blush—for Greece a tear.

7

Must *we* but weep o'er days more blest?
　　Must *we* but blush?—Our fathers bled.
Earth! render back from out thy breast
　　A remnant of our Spartan dead!
Of the three hundred grant but three,
To make a new Thermopylæ!　730

8

What, silent still? and silent all?
　　Ah! no;—the voices of the dead
Sound like a distant torrent's fall,
　　And answer, "Let one living head,
But one arise,—we come, we come!"
'Tis but the living who are dumb.

9

In vain—in vain: strike other chords;
　　Fill high the cup with Samian wine!
Leave battles to the Turkish hordes,
　　And shed the blood of Scio's vine!　740
Hark! rising to the ignoble call—
How answer each bold Bacchanal!

10

You have the Pyrrhic dance as yet,
　　Where is the Pyrrhic phalanx gone?
Of two such lessons, why forget
　　The nobler and the manlier one?
You have the letters Cadmus gave—
Think ye he meant them for a slave?

692. **Delos . . . Phœbus:** Phœbus Apollo was born on the isle of Delos. 695. **Scian . . . Teian:** Homer was the poet of Scio's isle, Anacreon that of Teos. 700. **"Islands . . . Blest":** the Cape Verde Islands or the Canaries. 707. **king:** Xerxes, who watched the Battle of Salamis from Mt. Aegaleos.

730. **Thermopylæ:** where Leonidas and the Spartans held the pass against the Persians. 743. **Pyrrhic dance:** a military dance invented by Pyrrhus, king of Epirus (after whom the "pyrrhic victory" is named—a victory where the losses are so great as to amount to a defeat). 747. **Cadmus:** Phoenician prince said to have introduced the alphabet into Greece.

II

Fill high the bowl with Samian wine!
    We will not think of themes like these!   750
It made Anacreon's song divine:
    He served—but served Polycrates—
A tyrant; but our masters then
Were still, at least, our countrymen.

12

The tyrant of the Chersonese
    Was freedom's best and bravest friend;
That tyrant was Miltiades!
    Oh! that the present hour would lend
Another despot of the kind!
Such claims as his were sure to bind.   760

13

Fill high the bowl with Samian wine!
    On Suli's rock, and Parga's shore,
Exists the remnant of a line
    Such as the Doric mothers bore;
And there, perhaps, some seed is sown,
The Heracleidan blood might own.

14

Trust not for freedom to the Franks—
    They have a king who buys and sells;
In native swords, and native ranks,
    The only hope of courage dwells: 770
But Turkish force, and Latin fraud,
Would break your shield, however broad.

15

Fill high the bowl with Samian wine!
    Our virgins dance beneath the shade—
I see their glorious black eyes shine;
    But gazing on each glowing maid,
My own the burning tear-drop laves,
To think such breasts must suckle slaves.

16

Place me on Sunium's marbled steep,
    Where nothing, save the waves and I,   780
May hear our mutual murmurs sweep;
    There, swan-like, let me sing and die:
A land of slaves shall ne'er be mine—
Dash down yon cup of Samian wine!

LXXXVII

Thus sung, or would, or could, or should have sung,
    The modern Greek, in tolerable verse;
If not like Orpheus quite, when Greece was young,
    Yet in these times he might have done much worse:
His strain display'd some feeling—right or wrong;
    And feeling, in a poet, is the source   790
Of others' feelings; but they are such liars,
And take all colours—like the hands of dyers.

LXXXVIII

But words are things, and a small drop of ink,
    Falling like dew, upon a thought, produces
That which makes thousands, perhaps millions, think;
    'Tis strange, the shortest letter which man uses
Instead of speech, may form a lasting link
    Of ages; to what straits old Time reduces
Frail man when paper—even a rag like this,
Survives himself, his tomb, and all that's his!   800

.   .   .

XCIII

All are not moralists, like Southey, when
    He prated to the world of "Pantisocrasy;"
Or Wordsworth unexcised, unhired, who then
    Season'd his pedlar poems with democracy;

752. Polycrates: tyrant of Samos, whom Anacreon served as court poet. **755–57. Chersonese . . . Miltiades:** Miltiades, tyrant of Chersonesus, led the Greeks at Marathon. **762. Suli's . . . Parga's:** Suli and Parga are towns in Albania. **766. Heracleidan:** descendants of Hercules. **767. Franks:** West Europeans generally.

779. Sunium: hill looking over the sea on the southeastern edge of Attica. On the crest stand the ruins of the ancient temple of Poseidon. **790–92. poet . . . dyers:** Byron echoes Shakespeare's remark in the sonnets (CXI, ll. 6–7) that his "nature is subdued / To what it works in, like the dyer's hand." **834. "Pantisocrasy":** Coleridge's and Southey's scheme to settle in America on the banks of the Susquehanna and try to allow "human perfectibility" to develop (see Introduction to Coleridge).

Or Coleridge, long before his flighty pen
　Let to the Morning Post its aristocracy;
When he and Southey, following the same path,
Espoused two partners (milliners of Bath).　　　840

XCIV

Such names at present cut a convict figure,
　The very Botany Bay in moral geography;
Their loyal treason, renegado rigour,
　Are good manure for their more bare biography,
Wordsworth's last quarto, by the way, is bigger
　Than any since the birthday of typography;
A drowsy frowzy poem, call'd the "Excursion,"
Writ in a manner which is my aversion.

XCV

He there builds up a formidable dyke
　Between his own and others' intellect;　　　850
But Wordsworth's poem, and his followers, like
　Joanna Southcote's Shiloh, and her sect,
Are things which in this century don't strike
　The public mind,—so few are the elect;
And the new births of both their stale virginities
Have proved but dropsies, taken for divinities.

XCVI

But let me to my story: I must own,
　If I have any fault, it is digression—
Leaving my people to proceed alone,
　While I soliloquize beyond expression;　　　860
But these are my addresses from the throne,
　Which put off business to the ensuing session:
Forgetting each omission is a loss to
The world, not quite so great as Ariosto.

XCVII

I know that what our neighbours call "longueurs,"
　(We've not so good a word, but have the thing,

In that complete perfection which insures
　An epic from Bob Southey every Spring—)
Form not the true temptation which allures
　The reader; but 'twould not be hard to bring　　　870
Some fine examples of the epopée,
To prove its grand ingredient is ennui.

XCVIII

We learn from Horace, "Homer sometimes sleeps;"
　We feel without him, Wordsworth sometimes
　　wakes,—
To show with what complacency he creeps,
　With his dear "Waggoners," around his lakes.
He wishes for "a boat" to sail the deeps—
　Of ocean?—No, of air; and then he makes
Another outcry for "a little boat,"
And drivels seas to set it well afloat.　　　880

XCIX

If he must fain sweep o'er the ethereal plain,
　And Pegasus runs restive in his "Waggon,"
Could he not beg the loan of Charles's Wain?
　Or pray Medea for a single dragon?
Or if, too classic for his vulgar brain,
　He fear'd his neck to venture such a nag on,
And he must needs mount nearer to the moon,
Could not the blockhead ask for a balloon?

C

"Pedlars," and "Boats," and "Waggons!" Oh! ye
　　shades
　Of Pope and Dryden, are we come to this?　　　890
That trash of such sort not alone evades
　Contempt, but from the bathos' vast abyss
Floats scumlike uppermost, and these Jack Cades
　Of sense and song above your graves may hiss—

---

840. Espoused two partners: Southey married Edith Fricker and Coleridge her sister, Sara (they were not milliners however). **842. Botany Bay:** penal colony in New South Wales, Australia. **852. Joanna Southcote:** started a religious sect in 1801. She announced she was to become the "mother of Shiloh," and there was wide interest. She was suffering from dropsy, however, and died of it the next year. **864. Ariosto:** Lodovico Ariosto, 1474–1533, Italian poet. **865. "longueurs":** tediousness.

871. epopée: epic. **873. "Homer . . . sleeps":** Horace, Art of Poetry, l. 359. **876. "Waggoners":** referring to Wordsworth's poem The Waggoner (1819). **877–78: wishes . . . ocean:** in the opening of Wordsworth's "Peter Bell." **883. Charles's Wain:** the constellation Charles's (i.e. Charlemagne's) wagon, now generally known as the Big Dipper. **884. Medea . . . dragon:** Medea flees in a chariot drawn by dragons at the end of Euripides' Medea. **893. Jack Cades:** Cade led a rebellious mob of commoners (1450) against Henry VI.

The "little boatman" and his "Peter Bell"
Can sneer at him who drew "Achitophel!"

CI

T' our tale.—The feast was over, the slaves gone,
  The dwarfs and dancing girls had all retired;
The Arab lore and poet's song were done,
  And every sound of revelry expired;   900
The lady and her lover, left alone,
  The rosy flood of twilight's sky admired;—
Ave Maria! o'er the earth and sea,
That heavenliest hour of Heaven is worthiest thee!

CII

Ave Maria! blessed be the hour!
  The time, the clime, the spot, where I so oft
Have felt that moment in its fullest power
  Sink o'er the earth so beautiful and soft,
While sung the deep bell in the distant tower,
  Or the faint dying day-hymn stole aloft,   910
And not a breath crept through the rosy air,
And yet the forest leaves seem'd stirr'd with prayer.

CIII

Ave Maria! 'tis the hour of prayer!
  Ave Maria! 'tis the hour of love!
Ave Maria! may our spirits dare
  Look up to thine and to thy Son's above!
Ave Maria! oh that face so fair!
  Those downcast eyes beneath the Almighty dove—
What though 'tis but a pictured image?—strike—
That painting is no idol,—'tis too like.   920

CIV

Some kinder casuists are pleased to say,
  In nameless print—that I have no devotion;
But set those persons down with me to pray,
  And you shall see who has the properest notion
Of getting into heaven the shortest way;
  My altars are the mountains and the ocean,
Earth, air, stars,—all that springs from the great Whole,
Who hath produced, and will receive the soul.

CV

Sweet hour of twilight!—in the solitude
  Of the pine forest, and the silent shore   930
Which bounds Ravenna's immemorial wood,
  Rooted where once the Adrian wave flow'd o'er,
To where the last Cæsarean fortress stood,
  Evergreen forest! which Boccaccio's lore
And Dryden's lay made haunted ground to me,
How have I loved the twilight hour and thee!

CVI

The shrill cicalas, people of the pine,
  Making their summer lives one ceaseless song,
Were the sole echoes, save my steed's and mine,
  And vesper bell's that rose the boughs along;   940
The spectre huntsman of Onesti's line,
  His hell-dogs, and their chase, and the fair throng
Which learn'd from this example not to fly
From a true lover,—shadow'd my mind's eye.

CVII

Oh, Hesperus! thou bringest all good things—
  Home to the weary, to the hungry cheer,
To the young bird the parent's brooding wings,
  The welcome stall to the o'erlabour'd steer;
Whate'er of peace about our hearthstone clings,
  Whate'er our household gods protect of dear,   950
Are gather'd round us by thy look of rest;
Thou bring'st the child, too, to the mother's breast.

CVIII

Soft hour! which wakes the wish and melts the heart
  Of those who sail the seas, on the first day
When they from their sweet friends are torn apart;
  Or fills with love the pilgrim on his way
As the far bell of vesper makes him start,
  Seeming to weep the dying day's decay;
Is this a fancy which our reason scorns?
Ah! surely nothing dies but something mourns!   960

1819–20 (1821)

896. **"Achitophel!"**: referring to Wordsworth's slighting remarks on Dryden in the "Essay, Supplementary to the Preface" (1815), above. One of Dryden's finest satires was "Absalom and Achitophel" (1681).

934–35. **Boccaccio's . . . lay:** Dryden's *Theodore and Honoria* is a verse adaptation of a tale in Boccaccio's *Decameron* (Day V, Table 8). **937. cicalas:** cicadas. **941. Onesti's line:** in Boccaccio's tale mentioned above (l. 934). **945. Hesperus:** the evening star. Byron here echoes a famous fragment by Sappho (No. 104 in *Pretarum Lesbiorum Fragmenta*, ed. Edgar Lobel and Denys Page, 1955).

from CANTO THE FOURTH

I

Nothing so difficult as a beginning
  In poesy, unless perhaps the end;
For oftentimes when Pegasus seems winning
  The race, he sprains a wing, and down we tend,
Like Lucifer when hurl'd from heaven for sinning;
  Our sin the same, and hard as his to mend,
Being pride, which leads the mind to soar too far,
Till our own weakness shows us what we are.

II

But time, which brings all beings to their level,
  And sharp Adversity, will teach at last   10
Man,—and, as we would hope,—perhaps the devil,
  That neither of their intellects are vast:
While youth's hot wishes in our red veins revel,
  We know not this—the blood flows on too fast:
But as the torrent widens towards the ocean,
We ponder deeply on each past emotion.

III

As boy, I thought myself a clever fellow,
  And wish'd that others held the same opinion;
They took it up when my days grew more mellow,
  And other minds acknowledge my dominion: 20
Now my sere fancy "falls into the yellow
  Leaf," and Imagination droops her pinion,
And the sad truth which hovers o'er my desk
Turns what was once romantic to burlesque.

IV

And if I laugh at any mortal thing,
  'Tis that I may not weep; and if I weep,
'Tis that our nature cannot always bring
  Itself to apathy, for we must steep
Our hearts first in the depths of Lethe's spring,
  Ere what we least wish to behold will sleep: 30
Thetis baptized her mortal son in Styx;
A mortal mother would on Lethe fix.

CANTO IV. **21–22. "falls . . . Leaf":** *Macbeth*, V. iii. 22–23.
29. Lethe: the river of forgetfulness in Hades. **31. Thetis . . .
Styx:** Thetis, mother of Achilles, dipped him in the river
Styx in order to make him invulnerable.

V

Some have accused me of a strange design
  Against the creed and morals of the land,
And trace it in this poem every line:
  I don't pretend that I quite understand
My own meaning when I would be *very* fine;
  But the fact is that I have nothing plann'd,
Unless it were to be a moment merry,
A novel word in my vocabulary.   40

VI

To the kind reader of our sober clime
  This way of writing will appear exotic;
Pulci was sire of the half-serious rhyme,
  Who sang when chivalry was more Quixotic,
And revell'd in the fancies of the time,
  True knights, chaste dames, huge giants, kings
    despotic;
But all these, save the last, being obsolete,
I chose a modern subject as more meet.

VII

How I have treated it, I do not know;
  Perhaps no better than they have treated me   50
Who have imputed such designs as show
  Not what they saw, but what they wish'd to see:
But if it gives them pleasure, be it so;
  This is a liberal age, and thoughts are free:
Meantime Apollo plucks me by the ear,
And tells me to resume my story here.

VIII

Young Juan and his lady-love were left
  To their own hearts' most sweet society;
Even Time the pitiless in sorrow cleft
  With his rude scythe such gentle bosoms; he   60
Sigh'd to behold them of their hours bereft,
  Though foe to love; and yet they could not be
Meant to grow old, but die in happy spring,
Before one charm or hope had taken wing.

. . .

43. Pulci: Luigi Pulci, 1432–84, to whose burlesque poem,
Morgante Maggiore, Byron refers. Byron translated a canto
from it in the winter of 1819–20.

XXVI

Juan and Haidée gazed upon each other
  With swimming looks of speechless tenderness,
Which mix'd all feelings, friend, child, lover, brother,
  All that the best can mingle and express
When two pure hearts are pour'd in one another,
  And love too much, and yet cannot love less;
But almost sanctify the sweet excess
By the immortal wish and power to bless.

XXVII

Mix'd in each other's arms, and heart in heart,
  Why did they not then die?—they had lived too long
Should an hour come to bid them breathe apart;   211
  Years could but bring them cruel things or wrong;
The world was not for them, nor the world's art
  For beings passionate as Sappho's song;
Love was born *with* them, *in* them, so intense,
It was their very spirit—not a sense.

XXVIII

They should have lived together deep in woods,
  Unseen as sings the nightingale; they were
Unfit to mix in these thick solitudes
  Call'd social, haunts of Hate, and Vice, and Care:
How lonely every freeborn creature broods!   221
  The sweetest song-birds nestle in a pair;
The eagle soars alone; the gull and crow
Flock o'er their carrion, just like men below.

XXIX

Now pillow'd cheek to cheek, in loving sleep,
  Haidée and Juan their siesta took,
A gentle slumber, but it was not deep,
  For ever and anon a something shook
Juan, and shuddering o'er his frame would creep;
  And Haidée's sweet lips murmur'd like a brook   230
A wordless music, and her face so fair
Stirr'd with her dream, as rose-leaves with the air;

XXX

Or as the stirring of a deep clear stream
  Within an Alpine hollow, when the wind
Walks o'er it, was she shaken by the dream,
  The mystical usurper of the mind—

O'erpowering us to be whate'er may seem
  Good to the soul which we no more can bind;
Strange state of being! (for 'tis still to be),
Senseless to feel, and with seal'd eyes to see.   240

XXXI

She dream'd of being alone on the sea-shore,
  Chain'd to a rock; she knew not how, but stir
She could not from the spot, and the loud roar
  Grew, and each wave rose roughly, threatening her;
And o'er her upper lip they seem'd to pour,
  Until she sobb'd for breath, and soon they were
Foaming o'er her lone head, so fierce and high—
Each broke to drown her, yet she could not die.

XXXII

Anon—she was released, and then she stray'd
  O'er the sharp shingles with her bleeding feet,   250
And stumbled almost every step she made;
  And something roll'd before her in a sheet,
Which she must still pursue howe'er afraid:
  'Twas white and indistinct, nor stopp'd to meet
Her glance nor grasp, for still she gazed and grasp'd,
And ran, but it escaped her as she clasp'd.

XXXIII

The dream changed:—in a cave she stood, its walls
  Were hung with marble icicles; the work
Of ages on its water-fretted halls,
  Where waves might wash, and seals might breed and
      lurk;   260
Her hair was dripping, and the very balls
  Of her black eyes seem'd turn'd to tears, and mirk
The sharp rocks look'd below each drop they caught,
Which froze to marble as it fell,—she thought.

XXXIV

And wet, and cold, and lifeless at her feet,
  Pale as the foam that froth'd on his dead brow,
Which she essay'd in vain to clear, (how sweet
  Were once her cares, how idle seem'd they now!)
Lay Juan, nor could aught renew the beat
  Of his quench'd heart; and the sea dirges low   270
Rang in her sad ears like a mermaid's song,
And that brief dream appear'd a life too long.

XXXV

And gazing on the dead, she thought his face
    Faded, or alter'd into something new—
Like to her father's features, till each trace
    More like and like to Lambro's aspect grew—
With all his keen worn look and Grecian grace;
    And starting, she awoke, and what to view?
Oh! Powers of Heaven! what dark eye meets she there?
'Tis—'tis her father's—fix'd upon the pair!   280

XXXVI

Then shrieking, she arose, and shrieking fell,
    With joy and sorrow, hope and fear, to see
Him whom she deem'd a habitant where dwell
    The ocean-buried, risen from death, to be
Perchance the death of one she loved too well:
    Dear as her father had been to Haidée,
It was a moment of that awful kind——
I have seen such—but must not call to mind.

XXXVII

Up Juan sprung to Haidée's bitter shriek,
    And caught her falling, and from off the wall   290
Snatch'd down his sabre, in hot haste to wreak
    Vengeance on him who was the cause of all:
Then Lambro, who till now forebore to speak,
    Smiled scornfully, and said, "Within my call,
A thousand scimitars await the word;
Put up, young man, put up your silly sword."

XXXVIII

And Haidée clung around him; "Juan, 'tis—
    'Tis Lambro—'tis my father! Kneel with me—
He will forgive us—yes—it must be—yes.
    Oh! dearest father, in this agony   300
Of pleasure and of pain—even while I kiss
    Thy garment's hem with transport, can it be
That doubt should mingle with my filial joy?
Deal with me as thou wilt, but spare this boy."

XXXIX

High and inscrutable the old man stood,
    Calm in his voice, and calm within his eye—
Not always signs with him of calmest mood:
    He look'd upon her, but gave no reply;

Then turn'd to Juan, in whose cheek the blood
    Oft came and went, as there resolved to die;   310
In arms, at least, he stood, in act to spring
On the first foe whom Lambro's call might bring.

XL

"Young man, your sword;" so Lambro once more said:
    Juan replied, "Not while this arm is free."
The old man's cheek grew pale, but not with dread,
    And drawing from his belt a pistol, he
Replied, "Your blood be then on your own head."
    Then look'd close at the flint, as if to see
'Twas fresh—for he had lately used the lock—
And next proceeded quietly to cock.   320

XLI

It has a strange quick jar upon the ear,
    That cocking of a pistol, when you know
A moment more will bring the sight to bear
    Upon your person, twelve yards off, or so;
A gentlemanly distance, not too near,
    If you have got a former friend for foe;
But after being fired at once or twice,
The ear becomes more Irish, and less nice.

XLII

Lambro presented, and one instant more
    Had stopp'd this Canto, and Don Juan's breath,   330
When Haidée threw herself her boy before;
    Stern as her sire: "On me," she cried, "let death
Descend—the fault is mine; this fatal shore
    He found—but sought not. I have pledged my faith;
I love him—I will die with him: I knew
Your nature's firmness—know your daughter's too."

XLIII

A minute past, and she had been all tears,
    And tenderness, and infancy; but now
She stood as one who champion'd human fears—
    Pale, statue-like, and stern, she woo'd the blow;
And tall beyond her sex, and their compeers,   341
    She drew up to her height, as if to show
A fairer mark; and with a fix'd eye scann'd
Her father's face—but never stopp'd his hand.

---

328. Irish . . . nice: bolder and less exacting. **341. their compeers:** equal in height to Juan and Lambro.

XLIV

He gazed on her, and she on him; 'twas strange
    How like they look'd! the expression was the same;
Serenely savage, with a little change
    In the large dark eye's mutual-darted flame;
For she, too, was as one who could avenge,
    If cause should be—a lioness, though tame;   350
Her father's blood before her father's face
Boil'd up, and proved her truly of his race.

XLV

I said they were alike, their features and
    Their stature, differing but in sex and years;
Even to the delicacy of their hand
    There was resemblance, such as true blood wears;
And now to see them, thus divided, stand
    In fix'd ferocity, when joyous tears,
And sweet sensations, should have welcomed both,
Show what the passions are in their full growth.   360

XLVI

The father paused a moment, then withdrew
    His weapon, and replaced it; but stood still,
And looking on her, as to look her through,
    "Not *I*," he said, "have sought this stranger's ill;
Not *I* have made this desolation: few
    Would bear such outrage, and forbear to kill;
But I must do my duty—how thou hast
Done thine, the present vouches for the past.

XLVII

"Let him disarm; or, by my father's head,
    His own shall roll before you like a ball!"   370
He raised his whistle as the word he said,
    And blew; another answer'd to the call,
And rushing in disorderly, though led,
    And arm'd from boot to turban, one and all,
Some twenty of his train came, rank on rank;
He gave the word, "Arrest or slay the Frank."

XLVIII

Then, with a sudden movement, he withdrew
    His daughter; while compress'd within his clasp,
Twixt her and Juan interposed the crew;
    In vain she struggled in her father's grasp—   380

His arms were like a serpent's coil: then flew
    Upon their prey, as darts an angry asp,
The file of pirates; save the foremost, who
Had fallen, with his right shoulder half cut through.

XLIX

The second had his cheek laid open; but
    The third, a wary, cool old sworder, took
The blows upon his cutlass, and then put
    His own well in; so well, ere you could look,
His man was floor'd, and helpless at his foot,
    With the blood running like a little brook   390
From two smart sabre gashes, deep and red—
One on the arm, the other on the head.

L

And then they bound him where he fell, and bore
    Juan from the apartment: with a sign
Old Lambro bade them take him to the shore,
    Where lay some ships which were to sail at nine.
They laid him in a boat, and plied the oar
    Until they reach'd some galliots, placed in line;
On board of one of these, and under hatches,
They stow'd him, with strict orders to the watches.

LI

The world is full of strange vicissitudes,   401
    And here was one exceedingly unpleasant;
A gentleman so rich in the world's goods,
    Handsome and young, enjoying all the present,
Just at the very time when he least broods
    On such a thing is suddenly to sea sent,
Wounded and chain'd, so that he cannot move,
And all because a lady fell in love.

LII

Here I must leave him, for I grow pathetic,
    Moved by the Chinese nymph of tears, green tea!
Than whom Cassandra was not more prophetic;   411
    For if my pure libations exceed three,
I feel my heart become so sympathetic,
    That I must have recourse to black Bohea:
'Tis pity wine should be so deleterious,
For tea and coffee leave us much more serious,

398. galliots: small galleys.

LIII

Unless when qualified with thee, Cogniac!
　　Sweet Naïad of the Phlegethontic rill!
Ah! why the liver wilt thou thus attack,
　　And make, like other nymphs, thy lovers ill?　　420
I would take refuge in weak punch, but *rack*
　　(In each sense of the word), whene'er I fill
My mild and midnight beakers to the brim,
Wakes me next morning with its synonym.

LIV

I leave Don Juan for the present, safe—
　　Not sound, poor fellow, but severely wounded;
Yet could his corporal pangs amount to half
　　Of those with which his Haidée's bosom bounded!
She was not one to weep, and rave, and chafe,
　　And then give way, subdued because surrounded;
Her mother was a Moorish maid from Fez,　　431
Where all is Eden, or a wilderness.

LV

There the large olive rains its amber store
　　In marble fonts; there grain, and flour, and fruit,
Gush from the earth until the land runs o'er;
　　But there, too, many a poison-tree has root,
And midnight listens to the lion's roar,
　　And long, long deserts scorch the camel's foot,
Or heaving whelm the helpless caravan;
And as the soil is, so the heart of man.　　440

LVI

Afric is all the sun's, and as her earth
　　Her human clay is kindled; full of power
For good or evil, burning from its birth,
　　The Moorish blood partakes the planet's hour,
And like the soil beneath it will bring forth:
　　Beauty and love were Haidée's mother's dower;
But her large dark eye show'd deep Passion's force,
Though sleeping like a lion near a source.

LVII

Her daughter, temper'd with a milder ray,
　　Like summer clouds all silvery, smooth, and fair,

Till slowly charged with thunder they display　　451
　　Terror to earth, and tempest to the air,
Had held till now her soft and milky way;
　　But overwrought with passion and despair,
The fire burst forth from her Numidian veins,
Even as the Simoom sweeps the blasted plains.

LVIII

The last sight which she saw was Juan's gore,
　　And he himself o'ermaster'd and cut down;
His blood was running on the very floor
　　Where late he trod, her beautiful, her own;　　460
Thus much she view'd an instant and no more,—
　　Her struggles ceased with one convulsive groan;
On her sire's arm, which until now scarce held
Her writhing, fell she like a cedar fell'd.

LIX

A vein had burst, and her sweet lips' pure dyes
　　Were dabbled with the deep blood which ran o'er;
And her head droop'd, as when the lily lies
　　O'ercharged with rain: her summon'd handmaids
　　　bore
Their lady to her couch with gushing eyes;
　　Of herbs and cordials they produced their store,
But she defied all means they could employ,　　471
Like one life could not hold, nor death destroy.

LX

Days lay she in that state unchanged, though chill—
　　With nothing livid, still her lips were red;
She had no pulse, but death seem'd absent still;
　　No hideous sign proclaim'd her surely dead;
Corruption came not in each mind to kill
　　All hope; to look upon her sweet face bred
New thoughts of life, for it seem'd full of soul—
She had so much, earth could not claim the whole.　480

．　．　．

LXIX

Twelve days and nights she wither'd thus; at last,
　　Without a groan, or sigh, or glance, to show
A parting pang, the spirit from her passed:
　　And they who watch'd her nearest could not know

---

418. Phlegethontic rill: Phlegethon was a river of fire in Hades. **421. rack:** arrack (distilled spirit, generally rum).

456. Simoom: a strong, hot desert wind (cf. *Manfred*, III. i 128).

The very instant, till the change that cast
  Her sweet face into shadow, dull and slow,   550
Glazed o'er her eyes—the beautiful, the black—
  Oh! to possess such lustre—and then lack!

LXX

She died, but not alone; she held within
  A second principle of life, which might
Have dawn'd a fair and sinless child of sin;
  But closed its little being without light,
And went down to the grave unborn, wherein
  Blossom and bough lie wither'd with one blight;
In vain the dews of Heaven descend above
The bleeding flower and blasted fruit of love.   560

LXXI

Thus lived—thus died she; never more on her
  Shall sorrow light, or shame. She was not made
Through years or moons the inner weight to bear,
  Which colder hearts endure till they are laid
By age in earth: her days and pleasures were
  Brief, but delightful—such as had not staid
Long with her destiny; but she sleeps well
By the sea-shore, whereon she loved to dwell.

LXXII

That isle is now all desolate and bare,
  Its dwellings down, its tenants pass'd away;   570
None but her own and father's grave is there,
  And nothing outward tells of human clay;
Ye could not know where lies a thing so fair,
  No stone is there to show, no tongue to say,
What was; no dirge, except the hollow sea's,
Mourns o'er the beauty of the Cyclades.

                                    1819–20 (*1821*)

from CANTO THE ELEVENTH

〜✺〜

After many vicissitudes Juan arrives in London as an
emissary of Queen Catherine the Great of Russia. He
enters London society and meets the bluestockings
(ladies with literary interests).

〜✺〜

L

The Blues, that tender tribe, who sigh o'er sonnets,
  And with the pages of the last Review
Line the interior of their heads or bonnets,
  Advanced in all their azure's highest hue:
They talk'd bad French or Spanish, and upon its
  Late authors ask'd him for a hint or two;
And which was softest, Russian or Castilian?
And whether in his travels he saw Ilion?   400

LI

Juan, who was a little superficial,
  And not in literature a great Drawcansir,
Examined by this learned and especial
  Jury of matrons, scarce knew what to answer:
His duties warlike, loving, or official,
  His steady application as a dancer,
Had kept him from the brink of Hippocrene,
Which now he found was blue instead of green.

LII

However, he replied at hazard, with
  A modest confidence and calm assurance,   410
Which lent his learned lucubrations pith,
  And pass'd for arguments of good endurance.
That prodigy, Miss Araminta Smith
  (Who at sixteen translated "Hercules Furens"
Into as furious English), with her best look,
Set down his sayings in her common-place book.

LIII

Juan knew several languages—as well
  He might—and brought them up with skill, in time
To save his fame with each accomplish'd belle,
  Who still regretted that he did not rhyme.   420
There wanted but this requisite to swell
  His qualities (with them) into sublime:
Lady Fitz-Frisky, and Miss Mævia Mannish,
Both long'd extremely to be sung in Spanish.

CANTO XI. **402. Drawcansir:** loud braggart in George
Villiers' satirical play, *The Rehearsal* (1671). **414. "Hercules
Furens":** a play by Seneca.

LIV

However, he did pretty well, and was
  Admitted as an aspirant to all
The coteries, and, as in Banquo's glass,
  At great assemblies or in parties small,
He saw ten thousand living authors pass,
  That being about their average numeral;   430
Also the eighty "greatest living poets,"
  As every paltry magazine can show it's.

LV

In twice five years the "greatest living poet,"
  Like to the champion in the fisty ring,
Is call'd on to support his claim, or show it,
  Although 'tis an imaginary thing.
Even I—albeit I'm sure I did not know it,
  Nor sought of foolscap subjects to be king,—
Was reckon'd, a considerable time,
  The grand Napoleon of the realms of rhyme.   440

LVI

But Juan was my Moscow, and Faliero
  My Leipsic, and my Mont Saint Jean seems Cain:
"La Belle Alliance" of dunces down at zero,
  Now that the Lion's fall'n, may rise again:
But I will fall at least as fell my hero;
  Nor reign at all, or as a *monarch* reign;
Or to some lonely isle of gaolers go,
  With turncoat Southey for my turnkey Lowe.

LVII

Sir Walter reign'd before me; Moore and Campbell
  Before and after: but now grown more holy,   450
The Muses upon Sion's hill must ramble
  With poets almost clergymen, or wholly;
And Pegasus has a psalmodic amble
  Beneath the very Reverend Rowley Powley,

Who shoes the glorious animal with stilts,
A modern Ancient Pistol—by the hilts!

LVIII

Still he excels that artificial hard
  Labourer in the same vineyard, though the vine
Yields him but vinegar for his reward,—
  That neutralised dull Dorus of the Nine;   460
That swarthy Sporus, neither man nor bard;
  That ox of verse, who *ploughs* for every line:—
Cambyses' roaring Romans beat at least
The howling Hebrews of Cybele's priest.—

LIX

Then there's my gentle Euphues; who, they say,
  Sets up for being a sort of *moral me;*
He'll find it rather difficult some day
  To turn out both, or either, it may be.
Some persons think that Coleridge hath the sway;
  And Wordsworth has supporters, two or three;   470
And that deep-mouth'd Bœotian "Savage Landor"
Has taken for a swan rogue Southey's gander.

LX

John Keats, who was kill'd off by one critique,
  Just as he really promised something great,
If not intelligible, without Greek
  Contrived to talk about the Gods of late,
Much as they might have been supposed to speak.
  Poor fellow! His was an untoward fate;
'Tis strange the mind, that fiery particle,
Should let itself be snuff'd out by an article.   480

.   .   .

---

427. Banquo's glass: *Macbeth*, IV. i. 112–13. **441–43.
Moscow . . . Alliance:** Byron is comparing his works to
the defeats of Napoleon (Moscow, Leipzig, and Mont Saint
Jean); "La Belle Alliance" was the name of the farmhouse
where the victors met after Napoleon's final defeat at Waterloo. **448. Lowe:** Sir Hudson Lowe, governor of St. Helena
during Napoleon's exile. **449. Sir Walter:** Scott. **454.
Powley:** the Rev. George Croly, 1780–1860, sometimes
called, because of his stilted style, "Cambyses Croly."

456. Ancient Pistol: 1 *Henry IV*, II. iv. 97. **458–61.
Labourer . . . Sporus:** Henry Hart Milman, author of
the poem *Fall of Jerusalem* (1820); the allusion to Dorus is
unclear; and Sporus was a hermaphrodite and a favorite of
Nero. **465. Euphues:** Bryan Waller Procter (he wrote
under the pen name of "Barry Cornwall"), whom Byron
calls "my . . . Euphues" (after John Lyly's famous *Euphues*,
with its excessively mannered style) because Procter had
been praised in a review as a Byron free from "profligacy."
471. Bœotian . . . Landor: Walter Savage Landor (see
selections above); the people of Bœotia, in central Greece,
were considered dull and boorish by the Athenians. **473.
Keats . . . critique:** alluding to the attack on *Endymion* in
The Quarterly Review by J. W. Croker. **476–77. Gods . . .
speak:** Byron refers to *Hyperion* (printed below), which,
he said, "seems actually inspired by the Titans, and is as
sublime as Aeschylus."

LXXXVII

But how shall I relate in other cantos
    Of what befell our hero in the land,   690
Which 'tis the common cry and lie to vaunt as
    A moral country? But I hold my hand—
For I disdain to write an Atalantis;
    But 'tis as well at once to understand
You are *not* a moral people, and you know it
Without the aid of too sincere a poet.

LXXXVIII

What Juan saw and underwent shall be
    My topic, with of course the due restriction
Which is required by proper courtesy;
    And recollect the work is only fiction,   700
And that I sing of neither mine nor me,
    Though every scribe, in some slight turn of diction,
Will hint allusions never *meant*. Ne'er doubt
This—when I speak, I *don't hint*, but *speak out*.

LXXXIX

Whether he married with the third or fourth
    Offspring of some sage husband-hunting countess,
Or whether with some virgin of more worth
    (I mean in Fortune's matrimonial bounties)
He took to regularly peopling Earth,
    Of which your lawful, awful wedlock fount is,—
Or whether he was taken in for damages,   711
For being too excursive in his homages,—

XC

Is yet within the unread events of time.
    Thus far, go forth, thou lay, which I will back
Against the same given quantity of rhyme,
    For being as much the subject of attack
As ever yet was any work sublime,
    By those who love to say that white is black.
So much the better!—I may stand alone,   719
But would not change my free thoughts for a throne.

1822–23 (1823)

693. Atalantis: referring to an early eighteenth-century book of scandal in the form of a novel, *The New Atalantis* (1709) by Mrs. Manley.

from CANTO THE TWELFTH

I

Of all the barbarous middle ages, that
    Which is most barbarous is the middle age
Of man! it is—I really scarce know what;
    But when we hover between fool and sage,
And don't know justly what we would be at—
    A period something like a printed page,
Black letter upon foolscap, while our hair
Grows grizzled, and we are not what we were;—

II

Too old for youth,—too young, at thirty-five,
    To herd with boys, or hoard with good threescore,—
I wonder people should be left alive;   11
    But since they are, that epoch is a bore:
Love lingers still, although 'twere late to wive:
    And as for other love, the illusion's o'er;
And money, that most pure imagination,
Gleams only through the dawn of its creation.

III

O Gold! Why call we misers miserable?
    Theirs is the pleasure that can never pall;
Theirs is the best bower anchor, the chain cable
    Which holds fast other pleasures great and small. 20
Ye who but see the saving man at table,
    And scorn his temperate board, as none at all,
And wonder how the wealthy can be sparing,
Know not what visions spring from each cheese-paring.

IV

Love or lust makes man sick, and wine much sicker;
    Ambition rends, and gaming gains a loss;
But making money, slowly first, then quicker,
    And adding still a little through each cross
(Which *will* come over things), beats love or liquor,
    The gamester's counter, or the statesman's *dross*. 30
O Gold! I still prefer thee unto paper,
Which makes bank credit like a bark of vapour.

V

Who hold the balance of the world? Who reign
    O'er congress, whether royalist or liberal?

Who rouse the shirtless patrios of Spain?
  (That make old Europe's journals squeak and gibber
      all).
Who keep the world, both old and new, in pain
  Or pleasure? Who make politics run glibber all?
The shade of Buonaparte's noble daring?—
Jew Rothschild, and his fellow-Christian, Baring.   40

VI

Those, and the truly liberal Lafitte,
  Are the true lords of Europe. Every loan
Is not a merely speculative hit,
  But seats a nation or upsets a throne.
Republics also get involved a bit;
  Columbia's stock hath holders not unknown
On 'Change, and even thy silver soil, Peru,
Must get itself discounted by a Jew.

VII

Why call the miser miserable? as
  I said before: the frugal life is his,   50
Which in a saint or cynic ever was
  The theme of praise: a hermit would not miss
Canonization for the self-same cause,
  And wherefore blame gaunt wealth's austerities?
Because, you'll say, nought calls for such a trial;—
Then there's more merit in his self-denial.

VIII

He is your only poet;—passion, pure
  And sparkling on from heap to heap, displays,
Possess'd, the ore, of which mere hopes allure
  Nations athwart the deep: the golden rays   60
Flash up in ingots from the mine obscure;
  On him the diamond pours its brilliant blaze;
While the mild emerald's beam shades down the dies
Of other stones, to soothe the miser's eyes.

IX

The lands on either side are his; the ship
  From Ceylon, Inde, or far Cathay, unloads
For him the fragrant produce of each trip;
  Beneath his cars of Ceres groan the roads,
And the vine blushes like Aurora's lip;
  His very cellars might be kings' abodes;   70
While he, despising every sensual call,
Commands—the intellectual lord of all.

X

Perhaps he hath great projects in his mind,
  To build a college, or to found a race,
A hospital, a church,—and leave behind
  Some dome surmounted by his meagre face:
Perhaps he fain would liberate mankind
  Even with the very ore which makes them base;
Perhaps he would be wealthiest of his nation,
Or revel in the joys of calculation.   80

XI

But whether all, or each, or none of these
  May be the hoarder's principle of action,
The fool will call such mania a disease:—
  What is his *own*? Go—look at each transaction,
Wars, revels, loves—do these bring men more ease
  Than the mere plodding through each "vulgar
      fraction"?
Or do they benefit mankind? Lean miser!
Let spendthrifts' heirs inquire of yours—who's wiser?

                                    1822–23 (1823)

from CANTO THE FOURTEENTH

Still in England, Juan moves into high society. He stays
at the country mansion of Lord Henry Amundeville and
his wife, Lady Adeline.

CANTO XII. **36. squeak and gibber:** quoted from *Hamlet*,
I. i. 115–16. **40. Rothschild . . . Baring:** Nathan Rothschild,
1777–1836, in charge of the London house of the famous
banking firm, and Alexander Baring, Baron Ashburton,
1774–1848, head of the banking firm of Baring Brothers.
41. Lafitte: Jacques Lafitte, 1767–1844, governor of the
Bank of France and a liberal member of the Chamber of
Deputies.

74. build . . . race: the antithesis echoes a line of Pope,
"Die, and endow a college, or a cat" (*Moral Essays*, III.
96). **86. "vulgar fraction":** Byron puns on the term used
for common fractions (as distinct from decimal fractions).

XLI

No marvel then he was a favourite;
   A full-grown Cupid, very much admired;
A little spoilt, but by no means so quite;
   At least he kept his vanity retired.
Such was his tact, he could alike delight
   The chaste, and those who are not so much inspired.
The Duchess of Fitz-Fulke, who loved "*tracasserie*,"
Began to treat him with some small "*agacerie*."

XLII

She was a fine and somewhat full-blown blonde,
   Desirable, distinguish'd, celebrated   330
For several winters in the grand, *grande monde*.
   I'd rather not say what might be related
Of her exploits, for this were ticklish ground;
   Besides there might be falsehood in what's stated;
Her late performance had been a dead set
At Lord Augustus Fitz-Plantagenet.

XLIII

This noble personage began to look
   A little black upon this new flirtation;
But such small licences must lovers brook,
   Mere freedoms of the female corporation.   340
Woe to the man who ventures a rebuke!
   'Twill but precipitate a situation
Extremely disagreeable, but common
To calculators when they count on woman.

XLIV

The circle smiled, then whisper'd, and then sneer'd;
   The Misses bridled, and the matrons frown'd;
Some hoped things might not turn out as they fear'd;
   Some would not deem such women could be found;
Some ne'er believed one half of what they heard;
   Some look'd perplex'd, and others look'd profound;
And several pitied with sincere regret   351
Poor Lord Augustus Fitz-Plantagenet.

XLV

But what is odd, none ever named the duke,
   Who, one might think, was something in the affair:
True, he was absent, and, 'twas rumour'd, took
   But small concern about the when, or where,
Or what his consort did: if he could brook
   Her gaieties, none had a right to stare:
Theirs was that best of unions, past all doubt,
Which never meets, and therefore can't fall out.   360

XLVI

But, oh! that I should ever pen so sad a line!
   Fired with an abstract love of virtue, she,
My Dian of the Ephesians, Lady Adeline,
   Began to think the duchess' conduct free;
Regretting much that she had chosen so bad a line,
   And waxing chiller in her courtesy.
Look'd grave and pale to see her friend's fragility,
For which most friends reserve their sensibility.

XLVII

There's nought in this bad world like sympathy;
   'Tis so becoming to the soul and face,   370
Sets to soft music the harmonious sigh,
   And robes sweet friendship in a Brussels lace.
Without a friend, what were humanity,
   To hunt our errors up with a good grace?
Consoling us with—"Would you had thought twice!
Ah! if you had but follow'd my advice!"

. . .

LI

The Lady Adeline's serene severity
   Was not confined to feeling for her friend,
Whose fame she rather doubted with posterity,
   Unless her habits should begin to mend:
But Juan also shared in her austerity,
   But mix'd with pity, pure as e'er was penn'd:
His inexperience moved her gentle ruth,
And (as her junior by six weeks) his youth.

. . .

CANTO XIV. **327. "tracasserie"**: mischief-making. **328.**
"agacerie": flirtatiousness.

363. Dian of the Ephesians: the tutelary goddess of
ancient Ephesus.

LXXXV

Our gentle Adeline had one defect—
    Her heart was vacant, though a splendid mansion;
Her conduct had been perfectly correct,
    As she had seen nought claiming its expansion.
A wavering spirit may be easier wreck'd,
    Because 'tis frailer, doubtless, than a stanch one;
But when the latter works its own undoing,
Its inner crash is like an earthquake's ruin.   680

LXXXVI

She loved her lord, or thought so; but *that* love
    Cost her an effort, which is a sad toil,
The stone of Sysiphus, if once we move
    Our feelings 'gainst the nature of the soil.
She had nothing to complain of, or reprove,
    No bickerings, no connubial turmoil:
Their union was a model to behold,
Serene and noble,—conjugal, but cold.

LXXXVII

There was no great disparity of years,
    Though much in temper; but they never clash'd:
They moved like stars united in their spheres,   691
    Or like the Rhone by Leman's waters wash'd,
Where mingled and yet separate appears
    The river from the lake, all bluely dash'd
Through the serene and placid glassy deep,
Which fain would lull its river-child to sleep.

LXXXVIII

Now when she once had ta'en an interest
    In anything, however she might flatter
Herself that her intentions were the best,
    Intense intentions are a dangerous matter: 700
Impressions were much stronger than she guess'd,
    And gather'd as they run like growing water
Upon her mind: the more so, as her breast
Was not at first too readily impress'd.

             1822–23 (1823)

---

683. Sysiphus: condemned in Hades to roll forever a great stone up a hill from which, after he reached the top, it continually rolled back.

from CANTO THE FIFTEENTH

XXVIII

When Adeline, in all her growing sense
    Of Juan's merits and his situation,
Felt on the whole an interest intense,—
    Partly perhaps because a fresh sensation,   220
Or that he had an air of innocence,
    Which is for innocence a sad temptation—
As women hate half measures, on the whole,
She 'gan to ponder how to save his soul.

XXIX

She had a good opinion of advice,
    Like all who give and eke receive it gratis,
For which small thanks are still the market price,
    Even where the article at highest rate is:
She thought upon the subject twice or thrice,
    And morally decided, the best state is   230
For morals, marriage; and this question carried,
She seriously advised him to get married.

XXX

Juan replied, with all becoming deference,
    He had a predilection for that tie;
But that, at present, with immediate reference
    To his own circumstances, there might lie
Some difficulties, as in his own preference,
    Or that of her to whom he might apply:
That still he'd wed with such or such a lady,
If that they were not married all already.   240

XXXI

Next to the making matches for herself,
    And daughters, brothers, sisters, kith or kin,
Arranging them like books on the same shelf,
    There's nothing women love to dabble in
More (like a stock-holder in growing pelf)
    Than match-making in general: 'tis no sin
Certes, but a preventative, and therefore
That is, no doubt, the only reason wherefore.

XXXII

But never yet (except of course a miss
    Unwed, or mistress never to be wed,   250
Or wed already, who object to this)
    Was there chaste dame who had not in her head
Some drama of the marriage unities,
    Observed as strictly both at board and bed
As those of Aristotle, though sometimes
They turn out melodramas or pantomimes.

XXXIII

They generally have some only son,
    Some heir to a large property, some friend
Of an old family, some gay Sir John,
    Or grave Lord George, with whom perhaps might
        end   260
A line, and leave posterity undone,
    Unless a marriage was applied to mend
The prospect and their morals: and besides,
They have at hand a blooming glut of brides.

XXXIV

From these they will be careful to select,
    For this an heiress, and for that a beauty;
For one a songstress who hath no defect,
    For 'tother one who promises much duty;
For this a lady no one can reject,
    Whose sole accomplishments were quite a booty;
A second for her excellent connexions;   271
A third, because there can be no objections.

. . .

XL

But Adeline determined Juan's wedding
    In her own mind, and that's enough for woman:
But then, with whom? There was the sage Miss Read-
        ing,
    Miss Raw, Miss Flaw, Miss Showman, and Miss
        Knowman,
And the two fair co-heiresses Giltbedding.
    She deem'd his merits something more than com-
        mon:
All these were unobjectionable matches,   319
And might go on, if well wound up, like watches.

XLI

There was Miss Millpond, smooth as summer's sea,
    That usual paragon, an only daughter,
Who seem'd the cream of equanimity,
    Till skimm'd—and then there was some milk and
        water,
With a slight shade of blue too, it might be,
    Beneath the surface; but what did it matter?
Love's riotous, but marriage should have quiet,
And being consumptive, live on a milk diet.

XLII

And then there was the Miss Audacia Shoestring,
    A dashing demoiselle of good estate,   330
Whose heart was fix'd upon a star or blue string;
    But whether English dukes grew rare of late,
Or that she had not harp'd upon the true string,
    By which such sirens can attract our great,
She took up with some foreign younger brother,
A Russ or Turk—the one's as good as t'other.

XLIII

And then there was—but why should I go on,
    Unless the ladies should go off?—there was
Indeed a certain fair and fairy one,
    Of the best class, and better than her class,—   340
Aurora Raby, a young star who shone
    O'er life, too sweet an image for such glass,
A lovely being, scarcely form'd or moulded,
A rose with all its sweetest leaves yet folded;

XLIV

Rich, noble, but an orphan; left an only
    Child to the care of guardians good and kind;
But still her aspect had an air so lonely!
    Blood is not water; and where shall we find

CANTO XV. **321–25: Miss Millpond . . . blue:** another reference to the humorless Annabella Milbanke whom Byron had married (1815); the "shade of blue" is another allusion to one of Byron's favorite objects of attack, the characteristics of the bluestockings. (Cf. I. 73, 1642; XI. 393; and XVI. 417.)

Feelings of youth like those which overthrown lie
    By death, when we are left, alas! behind,   350
To feel, in friendless palaces, a home
Is wanting, and our best ties in the tomb?

XLV

Early in years, and yet more infantine
    In figure, she had something of sublime
In eyes which sadly shone, as seraphs' shine.
    All youth—but with an aspect beyond time;
Radiant and grave—as pitying man's decline;
    Mournful—but mournful of another's crime,
She look'd as if she sat by Eden's door,
And grieved for those who could return no more. 360

XLVI

She was a Catholic, too, sincere, austere,
    As far as her own gentle heart allow'd,
And deem'd that fallen worship far more dear
    Perhaps because 'twas fallen: her sires were proud
Of deeds and days when they had fill'd the ear
    Of nations, and had never bent or bow'd
To novel power; and as she was the last,
She held their old faith and old feelings fast.

XLVII

She gazed upon a world she scarcely knew,
    As seeking not to know it; silent, lone,   370
As grows a flower, thus quietly she grew,
    And kept her heart serene within its zone.
There was awe in the homage which she drew;
    Her spirit seem'd as seated on a throne
Apart from the surrounding world, and strong
In its own strength—most strange in one so young!

XLVIII

Now it so happen'd, in the catalogue
    Of Adeline, Aurora was omitted,
Although her birth and wealth had given her vogue
    Beyond the charmers we have already cited;   380
Her beauty also seem'd to form no clog
    Against her being mention'd as well fitted,
By many virtues, to be worth the trouble
Of single gentlemen who would be double.

XLIX

And this omission, like that of the bust
    Of Brutus at the pageant of Tiberius,
Made Juan wonder, as no doubt he must.
    This he express'd half smiling and half serious;
When Adeline replied with some disgust,
    And with an air, to say the least, imperious,   390
She marvell'd "what he saw in such a baby
As that prim, silent, cold Aurora Raby?"

L

Juan rejoin'd—"She was a Catholic,
    And therefore fittest, as of his persuasion;
Since he was sure his mother would fall sick,
    And the Pope thunder excommunication,
If—" But here Adeline, who seem'd to pique
    Herself extremely on the inoculation
Of others with her own opinions, stated—
As usual—the same reason which she late did.   400

LI

And wherefore not? A reasonable reason,
    If good, is none the worse for repetition;
If bad, the best way's certainly to tease on,
    And amplify: you lose much by concision,
Whereas insisting in or out of season
    Convinces all men, even a politician;
Or—what is just the same—it wearies out.
So the end's gain'd, what signifies the route?

LII

Why Adeline had this slight prejudice—
    For prejudice it was—against a creature   410
As pure as sanctity itself from vice,
    With all the added charm of form and feature,
For me appears a question far too nice,
    Since Adeline was liberal by nature;
But nature's nature, and has more caprices
Than I have time, or will, to take to pieces.

385–86. bust / Of Brutus: classic example of something conspicuous by its absence. Tiberius Caesar refused to allow the bust of Brutus to be carried in the funeral procession of Junia, Brutus' sister; and the result, said Tacitus, was to make the people more conscious of Brutus than if the bust had been there.

LIII

Perhaps she did not like the quiet way
  With which Aurora on those baubles look'd,
Which charm most people in their earlier day:
  For there are few things by mankind less brook'd, 420
And womankind too, if we so may say,
  Than finding thus their genius stand rebuked,
Like "Anthony's by Cæsar," by the few
Who look upon them as they ought to do.

LIV

It was not envy—Adeline had none;
  Her place was far beyond it, and her mind.
It was not scorn—which could not light on one
  Whose greatest *fault* was leaving few to find.
It was not jealousy, I think: but shun
  Following the "ignes fatui" of mankind.   430
It was not——but 'tis easier far, alas!
To say what it was not than what it was.

                      1822–23 (1824)

CANTO THE SIXTEENTH

I

The antique Persians taught three useful things,
  To draw the bow, to ride, and speak the truth.
This was the mode of Cyrus, best of kings—
  A mode adopted since by modern youth.
Bows have they, generally with two strings;
  Horses they ride without remorse or ruth;
At speaking truth perhaps they are less clever,
But draw the long bow better now than ever.

II

The cause of this effect, or this defect,—
  "For this effect defective comes by cause,"—   10
Is what I have not leisure to inspect;
  But this I must say in my own applause,
Of all the Muses that I recollect,
  Whate'er may be her follies or her flaws,
In some things, mine's beyond all contradiction
The most sincere that ever dealt in fiction.

III

And as she treats all things, and ne'er retreats
  From anything, this epic will contain
A wilderness of the most rare conceits,
  Which you might elsewhere hope to find in vain. 20
'Tis true there be some bitters with the sweets,
  Yet mix'd so slightly, that you can't complain,
But wonder they so few are, since my tale is
"De rebus cunctis et quibusdam aliis."

IV

But of all truths which she has told, the most
  True is that which she is about to tell.
I said it was a story of a ghost—
  What then? I only know it so befell.
Have you explored the limits of the coast,
  Where all the dwellers of the earth must dwell? 30
'Tis time to strike such puny doubters dumb as
The sceptics who would not believe Columbus.

V

Some people would impose now with authority,
  Turpin's or Monmouth Geoffry's Chronicle;
Men whose historical superiority
  Is always greatest at a miracle.
But Saint Augustine has the great priority,
  Who bids all men believe the impossible,
Because 'tis so. Who nibble, scribble, quibble, he
Quiets at once with "*quia* impossibile."   40

VI

And therefore, mortals, cavil not at all;
  Believe:—if 'tis improbable, you *must*,

---

423. "Anthony's by Cæsar": "under him / My genius is rebuk'd, as it is said / Mark Antony's was by Caesar" (*Macbeth*, III. i. 55–57). CANTO XVI. **1–3. Persians . . . Cyrus:** told by Herodotus (I. 136) and especially in Xenophon's account of Cyrus, frequently regarded in the ancient world as a model king because of his virtues (*Cyropaedia*, VIII. 7). **8. draw . . . bow:** slang for exaggerating the truth. **10. "For . . . cause":** quoted from Polonius in *Hamlet*, II. ii. 103.

24. "De . . . aliis": Byron combines the titles of two works attributed to St. Thomas Aquinas, *De Omnibus Rebus* ("Concerning All Things") and *De Quibusdam Aliis* ("Concerning Some Others"). **34. Turpin's . . . Chronicle:** referring to a chronicle mistakenly attributed to Turpin, Archbishop of Rheims (*c.* 770), and the *Historia Regum Britanniae* by Geoffrey of Monmouth (1100–54). **40. "quia impossibile"**: Tertullian, not St. Augustine, argued that Christianity is certain "*because* it is impossible."

And if it is impossible, you *shall:*
   'Tis always best to take things upon trust.
I do not speak profanely, to recall
   Those holier mysteries which the wise and just
Receive as gospel, and which grow more rooted,
As all truths must, the more they are disputed:

VII

I merely mean to say what Johnson said,
   That in the course of some six thousand years,   50
All nations have believed that from the dead
   A visitant at intervals appears;
And what is strangest upon this strange head,
   Is, that whatever bar the reason rears
'Gainst such belief, there's something stronger still
In its behalf, let those deny who will.

VIII

The dinner and the soirée too were done,
   The supper too discuss'd, the dames admired,
The banqueteers had dropp'd off one by one—
   The song was silent, and the dance expired: 60
The last thin petticoats were vanish'd, gone
   Like fleecy clouds into the sky retired,
And nothing brighter gleam'd through the saloon
Than dying tapers—and the peeping moon.

IX

The evaporation of a joyous day
   Is like the last glass of champagne, without
The foam which made its virgin bumper gay;
   Or like a system coupled with a doubt;
Or like a soda bottle when its spray
   Has sparkled and let half its spirit out;   70
Or like a billow left by storms behind,
Without the animation of the wind;

X

Or like an opiate, which brings troubled rest,
   Or none; or like—like nothing that I know
Except itself;—such is the human breast;
   A thing, of which similitudes can show

No real likeness,—like the old Tyrian vest
   Dyed purple, none at present can tell how,
If from a shell-fish or from cochineal.
So perish every tyrant's robe piece-meal!   80

XI

But next to dressing for a rout or ball,
   Undressing is a woe; our robe de chambre
May sit like that of Nessus, and recall
   Thoughts quite as yellow, but less clear than amber.
Titus exclaim'd, "I've lost a day!" Of all
   The nights and days most people can remember,
(I have had of both, some not to be disdain'd,)
I wish they'd state how many they have gain'd.

XII

And Juan, on retiring for the night,
   Felt restless, and perplex'd, and compromised: 90
He thought Aurora Raby's eyes more bright
   Than Adeline (such is advice) advised;
If he had known exactly his own plight,
   He probably would have philosophised;
A great resource to all, and ne'er denied
Till wanted; therefore Juan only sigh'd.

XIII

He sigh'd;—the next resource is the full moon,
   Where all sighs are deposited; and now
It happen'd luckily, the chaste orb shone
   As clear as such a climate will allow;   100
And Juan's mind was in the proper tone
   To hail her with the apostrophe—"O thou!"
Of amatory egotism the *Tuism,*
Which further to explain would be a truism.

XIV

But lover, poet, or astronomer,
   Shepherd, or swain, whoever may behold,
Feel some abstraction when they gaze on her:
   Great thoughts we catch from thence (besides a cold

---

49. **what Johnson said:** "There is no people, rude or learned, among whom apparitions of the dead are not related and believed . . . some who deny it with their tongues, confess it with their fears" (*Rasselas*, Chap. XXX). 79. **shell-fish . . . cochineal:** the dye known as Tyrian purple made from shellfish or from the kermes berry (cochineal). 83. **Nessus:** Hercules was killed by wearing the poisoned tunic of the centaur Nessus. 85. **"I've . . . day!":** said by the Roman Emperor Titus if a day passed without his granting a favor. 103. **Tuism:** from the Latin "tu" (you), as opposed to egoism.

Sometimes, unless my feelings rather err);
  Deep secrets to her rolling light are told;   110
The ocean's tides and mortals' brains she sways,
And also hearts, if there be truth in lays.

XV

Juan felt somewhat pensive, and disposed
  For contemplation rather than his pillow:
The Gothic chamber, where he was enclosed,
  Let in the rippling sound of the lake's billow,
With all the mystery by midnight caused:
  Below his window waved (of course) a willow;
And he stood gazing out on the cascade
That flash'd and after darken'd in the shade.   120

XVI

Upon his table or his toilet,—*which*
  Of these is not exactly ascertain'd,—
(I state this, for I am cautious to a pitch
  Of nicety, where a fact is to be gain'd,)
A lamp burn'd high, while he leant from a niche,
  Where many a Gothic ornament remain'd,
In chisell'd stone and painted glass, and all
That time has left our fathers of their hall.

XVII

Then, as the night was clear though cold, he threw
  His chamber door wide open—and went forth   130
Into a gallery, of a sombre hue,
  Long, furnish'd with old pictures of great worth,
Of knights and dames heroic and chaste too,
  As doubtless should be people of high birth.
But by dim lights the portraits of the dead
Have something ghastly, desolate, and dread.

XVIII

The forms of the grim knight and pictured saint
  Look living in the moon; and as you turn
Backward and forward to the echoes faint
  Of your own footsteps—voices from the urn   140
Appear to wake, and shadows wild and quaint
  Start from the frames which fence their aspects stern,
As if to ask how you can dare to keep
A vigil there, where all but death should sleep.

XIX

And the pale smile of beauties in the grave,
  The charms of other days, in starlight gleams,
Glimmer on high; their buried locks still wave
  Along the canvas; their eyes glance like dreams
On ours, or spars within some dusky cave,
  But death is imaged in their shadowy beams.   150
A picture is the past; even ere its frame
Be gilt, who sate hath ceased to be the same.

XX

As Juan mused on mutability,
  Or on his mistress—terms synonymous—
No sound except the echo of his sigh
  Or step ran sadly through that antique house;
When suddenly he heard, or thought so, nigh,
  A supernatural agent—or a mouse,
Whose little nibbling rustle will embarrass
Most people as it plays along the arras.   160

XXI

It was no mouse, but lo! a monk, array'd
  In cowl and beads, and dusky garb, appear'd,
Now in the moonlight, and now lapsed in shade,
  With steps that trod as heavy, yet unheard;
His garments only a slight murmur made;
  He moved as shadowy as the Sisters weird,
But slowly; and as he passed Juan by,
Glanced, without pausing, on him a bright eye.

XXII

Juan was petrified; he had heard a hint
  Of such a spirit in these halls of old,   170
But thought, like most men, there was nothing in 't
  Beyond the rumour which such spots unfold,
Coin'd from surviving superstition's mint,
  Which passes ghosts in currency like gold,
But rarely seen, like gold compared with paper.
And did he see this? or was it a vapour?

XXIII

Once, twice, thrice pass'd, repass'd—the thing of air,
  Or earth beneath, or heaven, or t'other place;

166. Sisters weird: the witches in *Macbeth*.

And Juan gazed upon it with a stare,
  Yet could not speak or move; but, on its base   180
As stands a statue, stood: he felt his hair
  Twine like a knot of snakes around his face;
He tax'd his tongue for words, which were not granted,
To ask the reverend person what he wanted.

XXIV

The third time, after a still longer pause,
  The shadow pass'd away—but where? the hall
Was long, and thus far there was no great cause
  To think his vanishing unnatural:
Doors there were many, through which, by the laws
  Of physics, bodies whether short or tall   190
Might come or go; but Juan could not state
Through which the spectre seem'd to evaporate.

XXV

He stood—how long he knew not, but it seem'd
  An age—expectant, powerless, with his eyes
Strain'd on the spot where first the figure gleam'd;
  Then by degrees recall'd his energies,
And would have pass'd the whole off as a dream,
  But could not wake; he was, he did surmise,
Waking already, and return'd at length
Back to his chamber, shorn of half his strength.   200

XXVI

All there was as he left it: still his taper
  Burnt, and not *blue*, as modest tapers use,
Receiving sprites with sympathetic vapour;
  He rubb'd his eyes, and they did not refuse
Their office; he took up an old newspaper;
  The paper was right easy to peruse;
He read an article the king attacking,
And a long eulogy of "patent blacking."

XXVII

This savour'd of this world; but his hand shook—
  He shut his door, and after having read   210

A paragraph, I think about Horne Tooke,
  Undrest, and rather slowly went to bed.
There, couch'd all snugly on his pillow's nook,
  With what he had seen his phantasy he fed;
And though it was no opiate, slumber crept
Upon him by degrees, and so he slept.

XXVIII

He woke betimes; and, as may be supposed,
  Ponder'd upon his visitant or vision,
And whether it ought not to be disclosed,
  At risk of being quizz'd for superstition.   220
The more he thought, the more his mind was posed;
  In the mean time, his valet, whose precision
Was great, because his master brook'd no less,
Knock'd to inform him it was time to dress.

XXIX

He dress'd; and like young people he was wont
  To take some trouble with his toilet, but
This morning rather spent less time upon 't;
  Aside his very mirror soon was put;
His curls fell negligently o'er his front,
  His clothes were not curb'd to their usual cut,   230
His very neckcloth's Gordian knot was tied
Almost an hair's breadth too much on one side.

XXX

And when he walk'd down into the saloon,
  He sate him pensive o'er a dish of tea,
Which he perhaps had not discover'd soon,
  Had it not happen'd scalding hot to be,
Which made him have recourse unto his spoon;
  So much distrait he was, that all could see
That something was the matter—Adeline
The first—but *what* she could not well divine.   240

XXXI

She look'd, and saw him pale, and turn'd as pale
  Herself; then hastily look'd down, and mutter'd
Something, but what's not stated in my tale.
  Lord Henry said, his muffin was ill-butter'd;

---

202. **not blue:** According to legend, a ghost was present if a candle burned blue. 208. **"patent blacking":** advertisements for shoe or boot blacking, often written in doggerel rhyme, the ancestor of present-day singing commercials.

211. **Horne Tooke:** a philologist, 1736–1812, more widely known as a supporter of political reform.

The Duchess of Fitz-Fulke play'd with her veil,
    And look'd at Juan hard, but nothing utter'd.
Aurora Raby with her large dark eyes
Survey'd him with a kind of calm surprise.

XXXII

But seeing him all cold and silent still,
    And everybody wondering more or less,   250
Fair Adeline enquired "If he were ill?"
    He started, and said, "Yes—no—rather—yes."
The family physician had great skill,
    And being present, now began to express
His readiness to feel his pulse and tell
The cause, but Juan said, "He was quite well."

XXXIII

"Quite well; yes,—no."—These answers were mysteri-
        ous,
    And yet his looks appear'd to sanction both,
However they might savour of delirious;
    Something like illness of a sudden growth   260
Weigh'd on his spirit, though by no means serious:
    But for the rest, as he himself seem'd loth
To state the case, it might be ta'en for granted
It was not the physician that he wanted.

XXXIV

Lord Henry, who had now discuss'd his chocolate,
    Also the muffin whereof he complain'd,
Said, Juan had not got his usual look elate,
    At which he marvell'd, since it had not rain'd;
Then ask'd her Grace what news were of the duke of
        late?
    Her Grace replied, *his* Grace was rather pain'd   270
With some slight, light, hereditary twinges
Of gout, which rusts aristocratic hinges.

XXXV

Then Henry turn'd to Juan, and address'd
    A few words of condolence on his state:
"You look," quoth he, "as if you had had your rest
    Broke in upon by the Black Friar of late."

276. Black Friar: the ghost said to haunt Byron's home, Newstead Abbey.

"What Friar?" said Juan; and he did his best
    To put the question with an air sedate,
Or careless; but the effort was not valid
To hinder him from growing still more pallid.   280

XXXVI

"Oh! have you never heard of the Black Friar?
    The spirit of these walls?"—"In truth not I."
"Why Fame—but Fame you know's sometimes a liar—
    Tells an odd story, of which by and by:
Whether with time the spectre has grown shyer,
    Or that our sires had a more gifted eye
For such sights, though the tale is half believed,
The Friar of late has not been oft perceived.

XXXVII

"The last time was——"—"I pray," said Adeline—
    (Who watch'd the changes of Don Juan's brow,   290
And from its context thought she could divine
    Connexions stronger than he chose to avow
With this same legend)—"if you but design
    To jest, you'll choose some other theme just now,
Because the present tale has oft been told,
And is not much improved by growing old."

XXXVIII

"Jest!" quoth Milor; "why, Adeline, you know
    That we ourselves—'twas in the honeymoon—
Saw——"—"Well, no matter, 'twas so long ago;
    But, come, I'll set your story to a tune."   300
Graceful as Dian, when she draws her bow,
    She seiz'd her harp, whose strings were kindled soon
As touch'd, and plaintively began to play
The air of "'Twas a Friar of Orders Gray."

XXXIX

"But add the words," cried Henry, "which you made;
    For Adeline is half a poetess,"
Turning round to the rest, he smiling said.
    Of course the others could not but express
In courtesy their wish to see display'd
    By one *three* talents, for there were no less— 310
The voice, the words, the harper's skill, at once
Could hardly be united by a dunce.

XL

After some fascinating hesitation,—
   The charming of these charmers, who seem bound,
I can't tell why, to this dissimulation,—
   Fair Adeline, with eyes fix'd on the ground
At first, then kindling into animation,
   Added her sweet voice to the lyric sound,
And sang with much simplicity,—a merit
Not the less precious, that we seldom hear it.   320

1

Beware! beware! of the Black Friar,
   Who sitteth by Norman stone,
For he mutters his prayer in the midnight air,
   And his mass of the days that are gone.
When the Lord of the Hill, Amundeville,
   Made Norman Church his prey,
And expell'd the friars, one friar still
   Would not be driven away.

2

Though he came in his might, with King
      Henry's right,
   To turn church lands to lay,   330
With sword in hand, and torch to light
   Their walls, if they said nay;
A monk remain'd, unchased, unchain'd,
   And he did not seem form'd of clay,
For he's seen in the porch, and he's seen in the
      church,
   Though he is not seen by day.

3

And whether for good, or whether for ill,
   It is not mine to say;
But still with the house of Amundeville
   He abideth night and day.   340
By the marriage-bed of their lords, 'tis said,
   He flits on the bridal eve;
And 'tis held as faith, to their bed of death
   He comes—but not to grieve.

4

When an heir is born, he's heard to mourn,
   And when aught is to befall
That ancient line, in the pale moonshine
   He walks from hall to hall.

His form you may trace, but not his face,
   'Tis shadow'd by his cowl;   350
But his eyes may be seen from the folds
      between,
   And they seem of a parted soul.

5

But beware! beware! of the Black Friar,
   He still retains his sway,
For he is yet the church's heir
   Whoever may be the lay.
Amundeville is lord by day,
   But the monk is lord by night;
Nor wine nor wassail could raise a vassal
   To question that friar's right.   360

6

Say nought to him as he walks the hall,
   And he'll say nought to you;
He sweeps along in his dusky pall,
   As o'er the grass the dew.
Then grammercy! for the Black Friar;
   Heaven sain him! fair or foul,
And whatsoe'er may be his prayer,
   Let ours be for his soul.

XLI

The lady's voice ceased, and the thrilling wires
   Died from the touch that kindled them to sound;
And the pause follow'd, which when song expires 371
   Pervades a moment those who listen round;
And then of course the circle much admires,
   Nor less applauds, as in politeness bound,
The tones, the feeling, and the execution,
To the performer's diffident confusion.

XLII

Fair Adeline, though in a careless way,
   As if she rated such accomplishment
As the mere pastime of an idle day,
   Pursued an instant for her own content,   380
Would now and then, as 'twere *without* display,
   Yet *with* display in fact, at times relent
To such performances with haughty smile,
To show she *could*, if it were worth her while.

XLIII

Now this (but we will whisper it aside)
    Was—pardon the pedantic illustration—
Trampling on Plato's pride with greater pride,
    As did the Cynic on some like occasion:
Deeming the sage would be much mortified,
    Or thrown into a philosophic passion,   390
For a spoilt carpet—but the "Attic bee"
Was much consoled by his own repartee.

XLIV

Thus Adeline would throw into the shade
    (By doing easily, whene'er she chose,
What dilettanti do with vast parade)
    Their sort of *half profession;* for it grows
To something like this when too oft display'd;
    And that it is so, everybody knows,
Who have heard Miss That or This, or Lady T'other,
Show off—to please their company or mother.   400

XLV

Oh! the long evenings of duets and trios!
    The admirations and the speculations;
The "Mamma Mia's!" and the "Amor Mio's!"
    The "Tanti palpiti's" on such occasions:
The "Lasciami's," and quavering "Addio's!"
    Amongst our own most musical of nations;
With "Tu mi chamas's" from Portingale,
To soothe our ears, lest Italy should fail.

XLVI

In Babylon's bravuras—as the Home-
    Heart-ballads of Green Erin or Gray Highlands,   410
That bring Lochaber back to eyes that roam
    O'er far Atlantic continents or islands,

The calentures of music which o'ercome
    All mountaineers with dreams that they are nigh
        lands,
No more to be beheld but in such visions—
Was Adeline well versed, as compositions.

XLVII

She also had a twilight tinge of *"Blue,"*
    Could write rhymes, and compose more than she
        wrote,
Made epigrams occasionally too
    Upon her friends, as everybody ought.   420
But still from that sublimer azure hue,
    So much the present dye, she was remote;
Was weak enough to deem Pope a great poet,
And what was worse, was not ashamed to show it.

XLVIII

Aurora—since we are touching upon taste,
    Which now-a-days is the thermometer
By whose degrees all characters are class'd—
    Was more Shakespearian, if I do not err.
The worlds beyond this world's perplexing waste
    Had more of her existence, for in her   430
There was a depth of feeling to embrace
Thoughts, boundless, deep, but silent too as Space.

XLIX

Not so her gracious, graceful, graceless Grace,
    The full-grown Hebe of Fitz-Fulke, whose mind,
If she had any, was upon her face,
    And that was of a fascinating kind.
A little turn for mischief you might trace
    Also thereon,—but that's not much; we find
Few females without some such gentle leaven,
For fear we should suppose us quite in heaven.   440

L

I have not heard she was at all poetic,
    Though once she was seen reading the "Bath Guide,"
And "Hayley's Triumphs," which she deem'd pathetic,
    Because she said *her temper* had been tried

387–91. Trampling . . . "Attic bee": Diogenes the Cynic, invited to Plato's house, said as he trod on the carpets, "I trample on the pride of Plato," to which Plato responded, "Yes, with pride of another sort." "Attic bee" was a nickname given to Plato because of a story that bees had settled for a moment on his lips when he was a baby, thus signifying that his language would have the sweetness of honey. **404–05. "Tanti palpiti's" . . . "Addio's!":** the "heart throbs," "permit me's," and "farewells." **410. Erin . . . Highlands:** the Irish ballads of Thomas Moore and the Scottish ones of Sir Walter Scott.

413. calentures: fevers. **417. tinge of "Blue":** intellectual pretensions (see note to XV. 321, above). **434. Hebe:** goddess of youth. **442. "Bath Guide":** a satiric poem, *The New Bath Guide,* by Christopher Anstey. **443. "Hayley's Triumphs":** William Hayley's poems, *The Triumphs of Temper* and the *Triumph of Music.*

So much, the bard had really been prophetic
  Of what she had gone through with—since a bride.
But of all verse, what most insured her praise
  Were sonnets to herself, or "bouts rimés."

LI

'Twere difficult to say what was the object
  Of Adeline, in bringing this same lay   450
To bear on what appear'd to her the subject
  Of Juan's nervous feelings on that day.
Perhaps she merely had the simple project
  To laugh him out of his supposed dismay;
Perhaps she might wish to confirm him in it,
  Though why I cannot say—at least this minute.

LII

But so far the immediate effect
  Was to restore him to his self-propriety,
A thing quite necessary to the elect,
  Who wish to take the tone of their society: 460
In which you cannot be too circumspect,
  Whether the mode be persiflage or piety,
But wear the newest mantle of hypocrisy,
  Of pain of much displeasing the gynocracy.

LIII

And therefore Juan now began to rally
  His spirits, and without more explanation
To jest upon such themes in many a sally.
  Her Grace, too, also seized the same occasion,
With various similar remarks to tally,
  But wish'd for a still more detail'd narration   470
Of this same mystic friar's curious doings,
About the present family's deaths and wooings.

LIV

Of these few could say more than has been said;
  They pass'd as such things do, for superstition
With some, while others, who had more in dread
  The theme, half credited the strange tradition;
And much was talk'd on all sides on that head:
  But Juan, when cross-question'd on the vision,
Which some supposed (though he had not avow'd it)
Had stirr'd him, answer'd in a way to cloud it.   480

---

448. "bouts rimés": poetry written with predetermined
rhymes. **464. gynocracy:** government by women.

LV

And then, the mid-day having worn to one,
  The company prepared to separate;
Some to their several pastimes, or to none,
  Some wondering 'twas so early, some so late.
There was a goodly match too, to be run
  Between some greyhounds on my lord's estate,
And a young race-horse of old pedigree,
Match'd for the spring, whom several went to see.

LVI

There was a picture-dealer who had brought
  A special Titian, warranted original,   490
So precious that it was not to be bought,
  Though princes the possessor were besieging all.
The king himself had cheapen'd it, but thought
  The civil list he deigns to accept (obliging all
His subjects by his gracious acceptation)—
Too scanty, in these times of low taxation.

LVII

But as Lord Henry was a connoisseur,—
  The friend of artists, if not arts,—the owner,
With motives the most classical and pure,
  So that he would have been the very donor,   500
Rather than seller, had his wants been fewer,
  So much he deem'd his patronage an honour,
Had brought the capo d'opera, not for sale,
But for his judgment—never known to fail.

LVIII

There was a modern Goth, I mean a Gothic
  Bricklayer of Babel, call'd an architect,
Brought to survey these grey walls, which though so
    thick,
  Might have from time acquired some slight defect;
Who, after rummaging the Abbey through thick
  And thin, produced a plan whereby to erect   510
New buildings of correctest conformation,
And throw down old, which he call'd *restoration*.

---

488. Match'd . . . spring: entered for the spring races
in a "Match Book." **494. civil list:** the names of those receiving grants from Parliament. **503. capo d'opera:** chief work
or masterpiece.

LIX

The cost would be a trifle—an "old song,"
  Set to some thousands ('tis the usual burden
Of that same tune, when people hum it long)—
  The price would speedily repay its worth in
An edifice no less sublime than strong,
  By which Lord Henry's good taste would go forth in
Its glory, through all ages shining sunny,
For Gothic daring shown in English money.   520

LX

There were two lawyers busy on a mortgage
  Lord Henry wish'd to raise for a new purchase;
Also a lawsuit upon tenures burgage,
  And one on tithes, which sure are Discord's torches,
Kindling Religion till she throws down *her* gage,
  "Untying" squires "to fight against the churches;"
There was a prize ox, a prize pig, and ploughman,
For Henry was a sort of Sabine showman.

LXI

There were two poachers caught in a steel trap,
  Ready for gaol, their place of convalescence;   530
There was a country girl in a close cap
  And scarlet cloak (I hate the sight to see, since—
Since—since—in youth, I had the sad mishap—
  But luckily I have paid few parish fees since):
That scarlet cloak, alas! unclosed with rigour,
Presents the problem of a double figure.

LXII

A reel within a bottle is a mystery,
  One can't tell how it e'er got in or out;
Therefore the present piece of natural history
  I leave to those who are fond of solving doubt;   540
And merely state, though not for the consistory,
  Lord Henry was a justice, and that Scout
The constable, beneath a warrant's banner,
Had bagg'd this poacher upon Nature's manor.

523. **tenures burgage:** property for which yearly rent was paid the lord of the borough. 526. **"Untying . . . churches":** quoted from *Macbeth*, IV. i. 50–53. 528. **Sabine showman:** gentleman farmer (from Horace's "Sabine Farm").

LXIII

Now justices of peace must judge all pieces
  Of mischief of all kinds, and keep the game
And morals of the country from caprices
  Of those who have not a licence for the same;
And of all things, excepting tithes and leases,
  Perhaps these are most difficult to tame: 550
Preserving partridges and pretty wenches
Are puzzles to the most precautious benches.

LXIV

The present culprit was extremely pale,
  Pale as if painted so; her cheek being red
By nature, as in higher dames less hale
  'Tis white, at least when they just rise from bed.
Perhaps she was ashamed of seeming frail,
  Poor soul! for she was country born and bred,
And knew no better in her immorality
Than to wax white—for blushes are for quality.   560

LXV

Her black, bright, downcast, yet espiègle eye,
  Had gather'd a large tear into its corner,
Which the poor thing at times essay'd to dry,
  For she was not a sentimental mourner
Parading all her sensibility,
  Nor insolent enough to scorn the scorner,
But stood in trembling, patient tribulation,
To be call'd up for her examination.

LXVI

Of course these groups were scatter'd here and there,
  Not nigh the gay saloon of ladies gent.   570
The lawyers in the study; and in air
  The prize pig, ploughman, poachers; the men sent
From town, viz. architect and dealer, were
  Both busy (as a general in his tent
Writing despatches) in their several stations,
Exulting in their brilliant lucubrations.

LXVII

But this poor girl was left in the great hall,
  While Scout, the parish guardian of the frail,

561. **espiègle:** roguish. 570. **gent:** gentle.

Discuss'd (he hated beer yclept the "small")
   A mighty mug of *moral* double ale.   580
She waited until Justice could recall
   Its kind attentions to their proper pale,
To name a thing in nomenclature rather
Perplexing for most virgins—a child's father.

LXVIII

You see here was enough of occupation
   For the Lord Henry, link'd with dogs and horses.
There was much bustle too, and preparation
   Below stairs on the score of second courses;
Because, as suits their rank and situation,
   Those who in counties have great land resources 590
Have "public days," when all men may carouse,
Though not exactly what's call'd "open house."

LXIX

But once a week or fortnight, *un*invited
   (Thus we translate a *general invitation*),
All country gentlemen, esquired or knighted,
   May drop in without cards, and take their station
At the full board, and sit alike delighted
   With fashionable wines and conversation;
And, as the isthmus of the grand connexion,
Talk o'er themselves the past and next election.   600

LXX

Lord Henry was a great electioneerer,
   Burrowing for boroughs like a rat or rabbit.
But county contests cost him rather dearer,
   Because the neighbouring Scotch Earl of Giftgabbit
Had English influence, in the self-same sphere here;
   His son, the Honourable Dick Dicedrabbit,
Was member for the "other interest" (meaning
The same self-interest, with a different leaning).

LXXI

Courteous and cautious therefore in his county,
   He was all things to all men, and dispensed   610
To some civility, to others bounty,
   And promises to all—which last commenced
To gather to a somewhat large amount, he
   Not calculating how much they condensed;
But what with keeping some, and breaking others,
His word had the same value as another's.

LXXII

A friend to freedom and freeholders—yet
   No less a friend to government—he held,
That he exactly the just medium hit
   'Twixt place and patriotism—albeit compell'd, 620
Such was his sovereign's pleasure, (though unfit,
   He added modestly, when rebels rail'd,)
To hold some sinecures he wish'd abolish'd,
But that with them all law would be demolish'd.

LXXIII

He was "free to confess"—(whence comes this phrase?
   Is 't English? No—'tis only parliamentary)
That innovation's spirit now-a-days
   Had made more progress than for the last century.
He would not tread a factious path to praise,
   Though for the public weal disposed to venture high;
As for his place, he could but say this of it,   631
That the fatigue was greater than the profit.

LXXIV

Heaven, and his friends, knew that a private life
   Had ever been his sole and whole ambition;
But could he quit his king in times of strife,
   Which threaten'd the whole country with perdition?
When demagogues would with a butcher's knife
   Cut through and through (oh! damnable incision!)
The Gordian or the Geordi-an knot, whose strings
Have tied together commons, lords, and kings.   640

LXXV

Sooner "come place into the civil list
   And champion him to the utmost—"he would keep
     it,
Till duly disappointed or dismiss'd:
   Profit he cared not for, let others reap it;
But should the day come when place ceased to exist,
   The country would have far more cause to weep it:
For how could it go on? Explain who can!
He gloried in the name of Englishman.

LXXVI

He was as independent—ay, much more—
   Than those who were not paid for independence, 650

As common soldiers, or a common —— shore,
   Have in their several arts or parts ascendance
O'er the irregulars in lust or gore,
   Who do not give professional attendance.
Thus on the mob all statesmen are as eager
To prove their pride, as footmen to a beggar.

LXXVII

All this (save the last stanza) Henry said,
   And thought. I say no more—I've said too much;
For all of us have either heard or read—
   Off—or *upon* the hustings—some slight such   660
Hints from the independent heart or head
   Of the official candidate. I'll touch
No more on this—the dinner-bell hath rung,
And grace is said; the grace I *should* have *sung*—

LXXVIII

But I'm too late, and therefore must make play.
   'Twas a great banquet, such as Albion old
Was wont to boast—as if a glutton's tray
   Were something very glorious to behold.
But 'twas a public feast and public day,—
   Quite full, right dull, guests hot, and dishes cold,
Great plenty, much formality, small cheer,   671
And everybody out of their own sphere.

LXXIX

The squires familiarly formal, and
   My lords and ladies proudly condescending;
The very servants puzzling how to hand
   Their plates—without it might be too much bending
From their high places by the sideboard's stand—
   Yet, like their master's, fearful of offending.
For any deviation from the graces
Might cost both man and master too—their *places*. 680

LXXX

There were some hunters bold, and coursers keen,
   Whose hounds ne'er err'd, nor greyhounds deign'd
    to lurch;
Some deadly shots too, Septembrizers, seen
   Earliest to rise, and last to quit the search

Of the poor partridge through this stubble screen.
   There were some massy members of the church,
Takers of tithes, and makers of good matches,
And several who sung fewer psalms than catches.

LXXXI

There were some country wags too—and, alas!
   Some exiles from the town, who had been driven
To gaze, instead of pavement, upon grass,   691
   And rise at nine in lieu of long eleven.
And lo! upon that day it came to pass,
   I sate next that o'erwhelming son of heaven,
The very powerful parson, Peter Pith,
The loudest wit I e'er was deafen'd with.

LXXXII

I knew him in his livelier London days,
   A brilliant diner out, though but a curate;
And not a joke he cut but earn'd its praise,
   Until preferment, coming at a sure rate,   700
(O Providence! how wondrous are thy ways!
   Who would suppose thy gifts sometimes obdurate?)
Gave him, to lay the devil who looks o'er Lincoln,
A fat fen vicarage, and nought to think on.

LXXXIII

His jokes were sermons, and his sermons jokes;
   But both were thrown away amongst the fens;
For wit hath no great friend in aguish folks.
   No longer ready ears and short-hand pens
Imbibed the gay bon-mot, or happy hoax:
   The poor priest was reduced to common sense, 710
Or to coarse efforts very loud and long,
To hammer a hoarse laugh from the thick throng.

LXXXIV

There *is* a difference, says the song, "between
   A beggar and a queen," or *was* (of late
The latter worse used of the two we've seen—
   But we'll say nothing of affairs of state);

683. Septembrizers: those who massacred the political
prisoners in Paris in September, 1792.

695. Peter Pith: Rev. Sidney Smith, who wrote the *Peter
Plymley Letters* (1807). **703. devil . . . Lincoln:** the gargoyle
on St. Hugh's Chapel of Lincoln Cathedral.

A difference "'twixt a bishop and a dean,"
　　A difference between crockery ware and plate,
As between English beef and Spartan broth—
And yet great heroes have been bred by both.　　720

LXXXV

But of all nature's discrepancies, none
　　Upon the whole is greater than the difference
Beheld between the country and the town,
　　Of which the latter merits every preference
From those who have few resources of their own,
　　And only think, or act, or feel, with reference
To some small plan of interest or ambition—
Both which are limited to no condition.

LXXXVI

But "en avant!" The light loves languish o'er
　　Long banquets and too many guests, although　　730
A slight repast makes people love much more,
　　Bacchus and Ceres being, as we know,
Even from our grammar upwards, friends of yore
　　With vivifying Venus, who doth owe
To these the invention of champagne and truffles:
Temperance delights her, but long fasting ruffles.

LXXXVII

Dully pass'd o'er the dinner of the day;
　　And Juan took his place, he knew not where,
Confused, in the confusion, and distrait,
　　And sitting as if nail'd upon his chair:　　740
Though knives and forks clang'd round as in a fray,
　　He seem'd unconscious of all passing there,
Till some one, with a groan, exprest a wish
(Unheeded twice) to have a fin or fish.

LXXXVIII

On which, at the *third* asking of the bans,
　　He started; and perceiving smiles around
Broadening to grins, he colour'd more than once,
　　And hastily—as nothing can confound
A wise man more than laughter from a dunce—
　　Inflicted on the dish a deadly wound,　　750
And with such hurry, that, ere he could curb it,
He had paid his neighbour's prayer with half a turbot.

734. vivifying Venus: Cf. Terence, *Eunuchus*, IV. v. 6:
"Without food and drink, love perishes."

LXXXIX

This was no bad mistake, as it occurr'd,
　　The supplicator being an amateur;
But others, who were left with scarce a third,
　　Were angry—as they well might, to be sure.
They wonder'd how a young man so absurd
　　Lord Henry at his table should endure;
And this, and his not knowing how much oats
Had fall'n last market, cost his host three votes.　　760

XC

They little knew, or might have sympathised,
　　That he the night before had seen a ghost,
A prologue which but slightly harmonised
　　With the substantial company engross'd
By matter, and so much materialised,
　　That one scarce knew at what to marvel most
Of two things—how (the question rather odd is)
Such bodies could have souls, or souls such bodies.

XCI

But what confused him more than smile or stare,
　　From all the 'squires and 'squiresses around,　　770
Who wonder'd at the abstraction of his air,
　　Especially as he had been renown'd
For some vivacity among the fair,
　　Even in the country circle's narrow bound—
(For little things upon my lord's estate
Were good small talk for others still less great)—

XCII

Was, that he caught Aurora's eye on his,
　　And something like a smile upon her cheek.
Now this he really rather took amiss:
　　In those who rarely smile, their smile bespeaks　　780
A strong external motive; and in this
　　Smile of Aurora's there was nought to pique
Or hope, or love, with any of the wiles
Which some pretend to trace in ladies' smiles.

XCIII

'Twas a mere quiet smile of contemplation,
　　Indicative of some surprise and pity;
And Juan grew carnation with vexation,
　　Which was not very wise, and still less witty,

Since he had gain'd at least her observation,
　　A most important outwork of the city—　　790
As Juan should have known, had not his senses
By last night's ghost been driven from their defences.

XCIV

But what was bad, she did not blush in turn,
　　Nor seem embarrass'd—quite the contrary;
Her aspect was as usual, still—*not* stern—
　　And she withdrew, but cast not down, her eye,
Yet grew a little pale—with what? concern?
　　I know not; but her colour ne'er was high—
Though sometimes faintly flush'd—and always clear,
As deep seas in a sunny atmosphere.　　800

XCV

But Adeline was occupied by fame
　　This day; and watching, witching, condescending
To the consumers of fish, fowl, and game,
　　And dignity with courtesy so blending,
As all must blend whose part it is to aim
　　(Especially as the sixth year is ending)
At their lord's, son's, or similar connexion's
Safe conduct through the rocks of reëlections.

XCVI

Though this was most expedient on the whole,
　　And usual—Juan, when he cast a glance　　810
On Adeline while playing her grand rôle,
　　Which she went through as though it were a dance,
Betraying only now and then her soul
　　By a look scarce perceptibly askance
(Of weariness or scorn), began to feel
Some doubt how much of Adeline was *real;*

XCVII

So well she acted all and every part
　　By turns—with that vivacious versatility,
Which many people take for want of heart.
　　They err—'tis merely what is call'd mobility,　　820

820. mobility: "I am not sure that mobility is English; but it is expressive of a quality which rather belongs to other climates, though it is sometimes seen to a great extent in our own. It may be defined as an excessive susceptibility of immediate impressions—at the same time without *losing* the past; and is, though sometimes apparently useful to the possessor, a most painful and unhappy attribute" (B.).

A thing of temperament and not of art,
　　Though seeming so, from its supposed facility;
And false—though true; for surely they're sincerest
Who are strongly acted on by what is nearest.

XCVIII

This makes your actors, artists, and romancers,
　　Heroes sometimes, though seldom—sages never;
But speakers, bards, diplomatists, and dancers,
　　Little that's great, but much of what is clever;
Most orators, but very few financiers,
　　Though all Exchequer chancellors endeavour,　　830
Of late years, to dispense with Cocker's rigours,
And grow quite figurative with their figures.

XCIX

The poets of arithmetic are they
　　Who, though they prove not two and two to be
Five, as they might do in a modest way,
　　Have plainly made it out that four are three,
Judging by what they take, and what they pay.
　　The Sinking Fund's unfathomable sea,
That most unliquidating liquid, leaves
The debt unsunk, yet sinks all it receives.　　840

C

While Adeline dispensed her airs and graces,
　　The fair Fitz-Fulke seem'd very much at ease;
Though too well bred to quiz men to their faces,
　　Her laughing blue eyes with a glance could seize
The ridicules of people in all places—
　　That honey of your fashionable bees—
And store it up for mischievous enjoyment;
And this at present was her kind employment.

CI

However, the day closed, as days must close;
　　The evening also waned—and coffee came.　　850
Each carriage was announced, and ladies rose,
　　And curtsying off, as curtsies country dame,
Retired: with most unfashionable bows
　　Their docile esquires also did the same,
Delighted with their dinner and their host,
But with the Lady Adeline the most.

831. Cocker's rigours: Cocker's *Arithmetic*, which, though first published in 1677, was still being used.

CII

Some praised her beauty: others her great grace;
  The warmth of her politeness, whose sincerity
Was obvious in each feature of her face,
  Whose traits were radiant with the rays of verity.
Yes: *she* was truly worthy *her* high place!   861
  No one could envy her deserved prosperity.
And then her dress—what beautiful simplicity
Draperied her form with curious felicity!

CIII

Meanwhile sweet Adeline deserved their praises,
  By an impartial indemnification
For all her past exertion and soft phrases,
  In a most edifying conversation,
Which turn'd upon their late guests' miens and faces,
  And families, even to the last relation;   870
Their hideous wives, their horrid selves and dresses,
And truculent distortion of their tresses.

CIV

True, *she* said little—'twas the rest that broke
  Forth into universal epigram;
But then 'twas to the purpose what she spoke:
  Like Addison's "faint praise," so wont to damn,
Her own but served to set off every joke,
  As music chimes in with a melodrame.
How sweet the task to shield an absent friend!
I ask but this of mine, to—*not* defend.   880

CV

There were but two exceptions to this keen
  Skirmish of wits o'er the departed; one
Aurora, with her pure and placid mien;
  And Juan, too, in general behind none
In gay remark on what he had heard or seen,
  Sate silent now, his usual spirits gone:
In vain he heard the others rail or rally,
He would not join them in a single sally.

CVI

'Tis true he saw Aurora look as though
  She approved his silence; she perhaps mistook   890
Its motive for that charity we owe
  But seldom pay the absent, nor would look
Farther; it might or it might not be so.
  But Juan, sitting silent in his nook,
Observing little in his reverie,
Yet saw this much, which he was glad to see.

CVII

The ghost at least had done him this much good,
  In making him as silent as a ghost,
If in the circumstances which ensued
  He gain'd esteem where it was worth the most.   900
And certainly Aurora had renew'd
  In him some feelings he had lately lost
Or harden'd; feelings which, perhaps ideal,
Are so divine, that I must deem them real:—

CVIII

The love of higher things and better days;
  The unbounded hope, and heavenly ignorance
Of what is call'd the world, and the world's ways;
  The moments when we gather from a glance
More joy than from all future pride or praise,
  Which kindle manhood, but can ne'er entrance 910
The heart in an existence of its own,
Of which another's bosom is the zone.

CIX

Who would not sigh *Aἰaî τὰν Κυθέρειαν*
  That *hath* a memory, or that *had* a heart?
Alas! *her* star must fade like that of Dian:
  Ray fades on ray, as years on years depart.
Anacreon only had the soul to tie an
  Unwithering myrtle round the unblunted dart
Of Eros: but though thou hast play'd us many tricks,
Still we respect thee, "Alma Venus Genetrix!"   920

---

876. Addison's "faint praise": from Pope's portrait of
Addison in the *Epistle to Dr. Arbuthnot:* "Damn with faint
praise, assent with civil leer" (l. 201).

913. Aἰaî τὰν Κυθέρειαν: from Bion, "woe, woe for
Cytherea" (*Epitaphios Adonidos*, l. 28). **920. "Alma . . .
Genetrix!"**: from the opening of Lucretius, *De Rerum
Natura:* "Mother of Aeneas and his people, joy of men and
gods, life-giving Venus."

CX

And full of sentiments, sublime as billows
  Heaving between this world and worlds beyond,
Don Juan, when the midnight hour of pillows
  Arrived, retired to his; but to despond
Rather than rest. Instead of poppies, willows
  Waved o'er his couch; he meditated, fond
Of those sweet bitter thoughts which banish sleep,
And make the worldling sneer, the youngling weep.

CXI

The night was as before: he was undrest,
  Saving his night-gown, which is an undress;   930
Completely "sans culotte," and without vest;
  In short, he hardly could be clothed with less:
But apprehensive of his spectral guest,
  He sate with feelings awkward to express
(By those who have not had such visitations),
Expectant of the ghost's fresh operations.

CXII

And not in vain he listen'd;—Hush! what's that?
  I see—I see—Ah, no!—'tis not—yet 'tis—
Ye powers! it is the—the—the—Pooh! the cat!
  The devil may take that stealthy pace of his!   940
So like a spiritual pit-a-pat,
  Or tiptoe of an amatory Miss,
Gliding the first time to a rendezvous,
And dreading the chaste echoes of her shoe.

CXIII

Again—what is 't? The wind? No, no,—this time
  It is the sable Friar as before,
With awful footsteps regular as rhyme,
  Or (as rhymes may be in these days) much more.
Again through shadows of the night sublime,
  When deep sleep fell on men, and the world wore
The starry darkness round her like a girdle   951
Spangled with gems—the monk made his blood curdle.

CXIV

A noise like to wet fingers drawn on glass,
  Which sets the teeth on edge; and a slight clatter
Like showers which on the midnight gusts will pass,
  Sounding like very supernatural water,

Came over Juan's ear, which throbb'd, alas!
  For immaterialism's a serious matter;
So that even those whose faith is the most great
In souls immortal, shun them tête-à-tête.   960

CXV

Were his eyes open?—Yes! and his mouth too.
  Surprise has this effect—to make one dumb,
Yet leave the gate which eloquence slips through
  As wide, as if a long speech were to come.
Nigh and more nigh the awful echoes drew,
  Tremendous to a mortal tympanum:
His eyes were open, and (as was before
Stated) his mouth. What open'd next?—the door.

CXVI

It open'd with a most infernal creak,
  Like that of hell. "Lasciate ogni speranza   970
Voi ch' entrate!" The hinge seem'd to speak,
  Dreadful as Dante's *rima*, or this stanza;
Or—but all words upon such themes are weak:
  A single shade's sufficient to entrance a
Hero—for what is substance to a spirit?
Or how is 't *matter* trembles to come near it?

CXVII

The door flew wide, not swiftly,—but, as fly
  The sea-gulls, with a steady, sober flight—
And then swung back; nor close—but stood awry,
  Half letting in long shadows on the light,   980
Which still in Juan's candlesticks burn'd high,
  For he had two, both tolerably bright,
And in the doorway, darkening darkness, stood
The sable Friar in his solemn hood.

CXVIII

Don Juan shook, as erst he had been shaken
  The night before; but being sick of shaking,
He first inclined to think he had been mistaken;
  And then to be ashamed of such mistaking;
His own internal ghost began to awaken
  Within him, and to quell his corporal quaking—
Hinting that soul and body on the whole   991
Were odds against a disembodied soul.

970–71. "Lasciate . . . entrate!": "Abandon hope, all ye
who enter here" (*Inferno*, III. 9).

CXIX

And then his dread grew wrath, and his wrath fierce,
   And he arose, advanced—the shade retreated;
But Juan, eager now the truth to pierce,
   Followed, his veins no longer cold, but heated,
Resolved to thrust the mystery carte and tierce,
   At whatsoever risk of being defeated:
The ghost stopp'd, menaced, then retired, until
He reach'd the ancient wall, then stood stone still.

CXX

Juan put forth one arm—Eternal powers!   1001
   It touch'd no soul, nor body, but the wall,
On which the moonbeams fell in silvery showers,
   Chequer'd with all the tracery of the hall;
He shudder'd, as no doubt the bravest cowers
   When he can't tell what 'tis that doth appal.
How odd, a single hobgoblin's nonentity
Should cause more fear than a whole host's identity!

CXXI

But still the shade remain'd: the blue eyes glared,
   And rather variably for stony death;   1010
Yet one thing rather good the grave had spared,
   The ghost had a remarkably sweet breath.
A straggling curl show'd he had been fair-hair'd;
   A red lip, with two rows of pearls beneath,
Gleam'd forth, as through the casement's ivy shroud
The moon peep'd, just escaped from a grey cloud.

CXXII

And Juan, puzzled, but still curious, thrust
   His other arm forth—Wonder upon wonder!
It press'd upon a hard but glowing bust,
   Which beat as if there was a warm heart under.   1020
He found, as people on most trials must,
   That he had made at first a silly blunder,
And that in his confusion he had caught
Only the wall, instead of what he sought.

CXXIII

The ghost, if ghost it were, seem'd a sweet soul
   As ever lurk'd beneath a holy hood:

997. carte and tierce: positions of thrust in fencing.

A dimpled chin, a neck of ivory, stole
   Forth into something much like flesh and blood;
Back fell the sable frock and dreary cowl,
   And they reveal'd—alas! that e'er they should!   1030
In full, voluptuous, but *not o'er*grown bulk,
The phantom of her frolic Grace—Fitz-Fulke!

               1822–23 (1824)

THE VISION
OF JUDGMENT

⟨✺⟩

This was written at Ravenna, Italy, in 1821, as a satirical counterpart to a poem of the same name by Robert Southey which had appeared a few months before. The occasion of Southey's fulsome *Vision of Judgment* was the death of George III. In it the poet, in a trance, sees George III proceed to Heaven from the tomb and there be judged. Before such virtue the various accusers (such as John Wilkes and Junius, who had attacked him on earth) are helpless. The "absolvers" that come to praise him include the spirit of Washington, who testifies that George III was one who ever acted "with upright heart, as befitted a sovereign / True to his sacred trust." The king is then beatified "and to bliss everlasting appointed." Southey's preface included a heavy attack on Byron and on what he called "the Satanic School." He then sat back with happy confidence that he had sent a "stone from my sling which has smitten their Goliath in the forehead. I have fastened his name upon the gibbet for reproach and ignominy as long as it shall endure." Byron's publisher, John Murray, refused to take Byron's parody; but after a year the poem was printed by John Hunt (brother of Leigh Hunt) in his journal *The Liberal* (Oct., 1822), for which, shortly after Byron's death, he was fined £100.

⟨✺⟩

I

Saint Peter sat by the celestial gate:
   His keys were rusty, and the lock was dull,
So little trouble had been given of late;
   Not that the place by any means was full,
But since the Gallic era "eighty-eight"
   The devils had ta'en a longer, stronger pull,
And "a pull altogether," as they say
At sea—which drew most souls another way.

THE VISION OF JUDGMENT. **5. "eighty-eight":** taken as the opening year of the French Revolution, which is more frequently dated from the fall of the Bastille in 1789.

II

The angels all were singing out of tune,
   And hoarse with having little else to do,   10
Excepting to wind up the sun and moon,
   Or curb a runaway young star or two,
Or wild colt of a comet, which too soon
   Broke out of bounds o'er th' ethereal blue,
Splitting some planet with its playful tail,
As boats are sometimes by a wanton whale.

III

The guardian seraphs had retired on high,
   Finding their charges past all care below;
Terrestrial business fill'd nought in the sky
   Save the recording angel's black bureau;   20
Who found, indeed, the facts to multiply
   With such rapidity of vice and woe,
That he had stripp'd off both his wings in quills,
And yet was in arrear of human ills.

IV

His business so augmented of late years,
   That he was forced, against his will no doubt,
(Just like those cherubs, earthly ministers,)
   For some resource to turn himself about,
And claim the help of his celestial peers,
   To aid him ere he should be quite worn out   30
By the increased demand for his remarks;
Six angels and twelve saints were named his clerks.

V

This was a handsome board—at least for heaven;
   And yet they had even then enough to do,
So many conquerors' cars were daily driven,
   So many kingdoms fitted up anew;
Each day too slew its thousands six or seven,
   Till at the crowning carnage, Waterloo,
They threw their pens down in divine disgust—
The page was so besmear'd with blood and dust.   40

VI

This by the way; 'tis not mine to record
   What angels shrink from: even the very devil
On this occasion his own work abhorr'd,
   So surfeited with the infernal revel:

Though he himself had sharpen'd every sword,
   It almost quench'd his innate thirst of evil.
(Here Satan's sole good work deserves insertion—
'Tis, that he has both generals in reversion.)

VII

Let's skip a few short years of hollow peace,
   Which peopled earth no better, hell as wont,   50
And heaven none—they form the tyrant's lease,
   With nothing but new names subscribed upon 't;
'Twill one day finish: meantime they increase,
   "With seven heads and ten horns," and all in front,
Like Saint John's foretold beast; but ours are born
Less formidable in the head than horn.

VIII

In the first year of freedom's second dawn
   Died George the Third; although no tyrant, one
Who shielded tyrants, till each sense withdrawn
   Left him nor mental nor external sun: 60
A better farmer ne'er brush'd dew from lawn,
   A worse king never left a realm undone!
He died—but left his subjects still behind,
One half as mad—and t'other no less blind.

IX

He died! his death made no great stir on earth;
   His burial made some pomp; there was profusion
Of velvet, gilding, brass, and no great dearth
   Of aught but tears—save those shed by collusion.
For these things may be bought at their true worth;
   Of elegy there was the due infusion—   70
Bought also; and the torches, cloaks, and banners,
Heralds, and relics of old Gothic manners,

X

Form'd a sepulchral melodrame. Of all
   The fools who flock'd to swell or see the show,
Who cared about the corpse? The funeral
   Made the attraction, and the black the woe.

48. generals in reversion: Wellington and Napoleon are
his (Satan's) for future possession. **55. Saint John's . . .
beast:** Revelation 13. **57. second dawn:** 1820, marked
by the new rise of revolutionary spirit in southern
Europe.

There throbb'd not there a thought which pierced the
    pall;
  And when the gorgeous coffin was laid low,
It seem'd the mockery of hell to fold
The rottenness of eighty years in gold.   80

XI

So mix his body with the dust! It might
  Return to what it *must* far sooner, were
The natural compound left alone to fight
  Its way back into earth, and fire, and air;
But the unnatural balsams merely blight
  What nature made him at his birth, as bare
As the mere million's base unmummied clay—
Yet all his spices but prolong decay.

XII

He's dead—and upper earth with him has done;
  He's buried; save the undertaker's bill,   90
Or lapidary scrawl, the world is gone
  For him, unless he left a German will;
But where's the proctor who will ask his son?
  In whom his qualities are reigning still,
Except that household virtue, most uncommon,
Of constancy to a bad, ugly woman.

XIII

"God save the king!" It is a large economy
  In God to save the like; but if he will
Be saving, all the better; for not one am I
  Of those who think damnation better still: 100
I hardly know too if not quite alone am I
  In this small hope of bettering future ill
By circumscribing, with some slight restriction,
The eternity of hell's hot jurisdiction.

XIV

I know this is unpopular; I know
  'Tis blasphemous; I know one may be damn'd
For hoping no one else may e'er be so;
  I know my catechism; I know we're cramm'd

With the best doctrines till we quite o'erflow;
  I know that all save England's church have shamm'd,
And that the other twice two hundred churches   111
And synagogues have made a *damn'd* bad purchase.

XV

God help us all! God help me too! I am,
  God knows, as helpless as the devil can wish,
And not a whit more difficult to damn,
  Than is to bring to land a late-hook'd fish,
Or to the butcher to purvey the lamb;
  Not that I'm fit for such a noble dish,
As one day will be that immortal fry
Of almost everybody born to die.   120

XVI

Saint Peter sat by the celestial gate,
  And nodded o'er his keys; when, lo! there came
A wondrous noise he had not heard of late—
  A rushing sound of wind, and stream, and flame;
In short, a roar of things extremely great,
  Which would have made aught save a saint exclaim;
But he, with first a start and then a wink,
Said, "There's another star gone out, I think!"

XVII

But ere he could return to his repose,
  A cherub flapp'd his right wing o'er his eyes— 130
At which St. Peter yawn'd, and rubb'd his nose:
  "Saint porter," said the angel, "prithee rise!"
Waving a goodly wing, which glow'd, as glows
  An earthly peacock's tail, with heavenly dyes:
To which the saint replied, "Well, what's the matter?
"Is Lucifer come back with all this clatter?"

XVIII

"No," quoth the cherub; "George the Third is dead."
  "And who *is* George the Third?" replied the apostle:
"*What George? what Third?*" "The king of England,"
    said
  The angel. "Well! he won't find kings to jostle 140
Him on his way; but does he wear his head;
  Because the last we saw here had a tustle,
And ne'er would have got into heaven's good graces,
Had he not flung his head in all our faces.

92. German will: The will of George I (great-grandfather
of George III) had been disregarded by his son; "German"
because George I was of the German House of Hanover.

141–42. wear . . . last: Louis XVI had lost his head at the
guillotine (1793).

XIX

"He was, if I remember, king of France;
　That head of his, which could not keep a crown
On earth, yet ventured in my face to advance
　A claim to those of martyrs—like my own:
If I had had my sword, as I had once
　When I cut ears off, I had cut him down;　　150
But having but my *keys*, and not my brand,
I only knock'd his head from out his hand.

XX

"And then he set up such a headless howl,
　That all the saints came out and took him in;
And there he sits by St. Paul, cheek by jowl;
　That fellow Paul—the parvenù! The skin
Of St. Bartholomew, which makes his cowl
　In heaven, and upon earth redeem'd his sin
So as to make a martyr, never sped
Better than did this weak and wooden head.　　160

XXI

"But had it come up here upon its shoulders,
　There would have been a different tale to tell:
The fellow-feeling in the saint's beholders
　Seems to have acted on them like a spell;
And so this very foolish head heaven solders
　Back on its trunk: it may be very well,
And seems the custom here to overthrow
Whatever has been wisely done below."

XXII

The angel answer'd, "Peter! do not pout:
　The king who comes has head and all entire,　　170
And never knew much what it was about—
　He did as doth the puppet—by its wire,
And will be judged like all the rest, no doubt:
　My business and your own is not to inquire
Into such matters, but to mind our cue—
Which is to act as we are bid to do."

XXIII

While thus they spake, the angelic caravan,
　Arriving like a rush of mighty wind,

Cleaving the fields of space, as doth the swan
　Some silver stream (say Ganges, Nile, or Inde,　　180
Or Thames, or Tweed), and 'midst them an old man
　With an old soul, and both extremely blind,
Halted before the gate, and in his shroud
Seated their fellow-traveller on a cloud.

XXIV

But bringing up the rear of this bright host
　A Spirit of a different aspect waved
His wings, like thunder-clouds above some coast
　Whose barren beach with frequent wrecks is paved;
His brow was like the deep when tempest-toss'd;
　Fierce and unfathomable thoughts engraved　　190
Eternal wrath on his immortal face,
And *where* he gazed a gloom pervaded space.

XXV

As he drew near, he gazed upon the gate
　Ne'er to be enter'd more by him or Sin,
With such a glance of supernatural hate,
　As made Saint Peter wish himself within;
He patter'd with his keys at a great rate,
　And sweated through his apostolic skin:
Of course his perspiration was but ichor,
Or some such other spiritual liquor.　　200

XXVI

The very cherubs huddled all together,
　Like birds when soars the falcon; and they felt
A tingling to the tip of every feather,
　And form'd a circle like Orion's belt
Around their poor old charge; who scarce knew
　　whither
　His guards had led him, though they gently dealt
With royal manes (for by many stories,
And true, we learn the angels all are Tories).

XXVII

As things were in this posture, the gate flew
　Asunder, and the flashing of its hinges　　210
Flung over space an universal hue
　Of many-colour'd flame, until its tinges

156–57. skin . . . Bartholomew: St. Bartholomew had
been skinned alive before his death.

199. ichor: the ethereal liquid that flowed in the veins of
the gods.

Reach'd even our speck of earth, and made a new
   Aurora borealis spread its fringes
O'er the North Pole; the same seen, when ice-bound,
By Captain Parry's crew, in "Melville's Sound."

XXVIII

And from the gate thrown open issued beaming
   A beautiful and mighty Thing of Light,
Radiant with glory, like a banner streaming
   Victorious from some world-o'erthrowing fight:
My poor comparisons must needs be teeming   221
   With earthly likenesses, for here the night
Of clay obscures our best conceptions, saving
Johanna Southcote, or Bob Southey raving.

XXIX

'Twas the archangel Michael: all men know
   The make of angels and archangels, since
There's scarce a scribbler has not one to show,
   From the fiends' leader to the angels' prince.
There also are some altar-pieces, though
   I really can't say that they much evince   230
One's inner notions of immortal spirits;
But let the connoisseurs explain *their* merits.

XXX

Michael flew forth in glory and in good;
   A goodly work of him from whom all glory
And good arise; the portal past—he stood;
   Before him the young cherubs and saints hoary—
(I say *young*, begging to be understood
   By looks, not years; and should be very sorry
To state, they were not older than St. Peter,
But merely that they seem'd a little sweeter).   240

XXXI

The cherubs and the saints bow'd down before
   That arch-angelic hierarch, the first
Of essences angelical, who wore
   The aspect of a god; but this ne'er nursed
Pride in his heavenly bosom, in whose core
   No thought, save for his Master's service, durst

Intrude, however glorified and high;
He knew him but the viceroy of the sky.

XXXII

He and the sombre silent Spirit met—
   They knew each other both for good and ill;   250
Such was their power, that neither could forget
   His former friend and future foe; but still
There was a high, immortal, proud regret
   In either's eye, as if 'twere less their will
Than destiny to make the eternal years
Their date of war, and their "champ clos" the spheres.

XXXIII

But here they were in neutral space: we know
   From Job, that Satan hath the power to pay
A heavenly visit thrice a year or so;
   And that the "sons of God," like those of clay,   260
Must keep him company; and we might show
   From the same book, in how polite a way
The dialogue is held between the Powers
Of Good and Evil—but 'twould take up hours.

XXXIV

And this is not a theologic tract,
   To prove with Hebrew and with Arabic
If Job be allegory or a fact,
   But a true narrative; and thus I pick
From out the whole but such and such an act
   As sets aside the slightest thought of trick.   270
'Tis every tittle true, beyond suspicion,
And accurate as any other vision.

XXXV

The spirits were in neutral space, before
   The gate of heaven; like eastern thresholds is
The place where Death's grand cause is argued o'er,
   And souls despatch'd to that world or to this;
And therefore Michael and the other wore
   A civil aspect: though they did not kiss,
Yet still between his Darkness and his Brightness
There pass'd a mutual glance of great politeness.   280

---

216. Parry: Captain Edward Parry, who wrote an account of his voyage of 1819-20 to discover a northwest passage. **224. Southcote:** See *Don Juan*, above, III. n. 852.

256. "champ clos": closed field at a tournament. **258. Job:** Job 1:2.

XXXVI

The Archangel bow'd, not like a modern beau,
   But with a graceful oriental bend,
Pressing one radiant arm just where below
   The heart in good men is supposed to tend.
He turn'd as to an equal, not too low,
   But kindly; Satan met his ancient friend
With more hauteur, as might an old Castilian
Poor noble meet a mushroom rich civilian.

XXXVII

He merely bent his diabolic brow
   An instant; and then raising it, he stood   290
In act to assert his right or wrong, and show
   Cause why King George by no means could or
     should
Make out a case to be exempt from woe
   Eternal, more than other kings, endued
With better sense and hearts, whom history mentions,
Who long have "paved hell with their good intentions."

XXXVIII

Michael began: "What wouldst thou with this man,
   Now dead, and brought before the Lord? What ill
Hath he wrought since his mortal race began,
   That thou canst claim him? Speak! and do thy will,
If it be just: if in this earthly span   301
   He hath been greatly failing to fulfil
His duties as a king and mortal, say,
And he is thine; if not, let him have way."

XXXIX

"Michael!" replied the Prince of Air, "even here,
   Before the Gate of him thou servest, must
I claim my subject: and will make appear
   That as he was my worshipper in dust,
So shall he be in spirit, although dear
   To thee and thine, because nor wine nor lust   310
Were of his weaknesses; yet on the throne
He reign'd o'er millions to serve me alone.

XL

"Look to our earth, or rather mine; it was,
   Once, more thy master's: but I triumph not

In this poor planet's conquest; nor, alas!
   Need he thou servest envy me my lot:
With all the myriads of bright worlds which pass
   In worship round him, he may have forgot
Yon weak creation of such paltry things:
I think few worth damnation save their kings,—   320

XLI

"And these but as a kind of quit-rent, to
   Assert my right as lord: and even had
I such an inclination, 'twere (as you
   Well know) superfluous; they are grown so bad,
That hell has nothing better left to do
   Than leave them to themselves: so much more mad
And evil by their own internal curse,
Heaven cannot make them better, nor I worse.

XLII

"Look to the earth, I said, and say again:
   When this old, blind, mad, helpless, weak, poor
     worm   330
Began in youth's first bloom and flush to reign,
   The world and he both wore a different form,
And much of earth and all the watery plain
   Of ocean call'd him king: through many a storm
His isles had floated on the abyss of time;
For the rough virtues chose them for their clime.

XLIII

"He came to his sceptre young; he leaves it old:
   Look to the state in which he found his realm,
And left it; and his annals too behold,
   How to a minion first he gave the helm;   340
How grew upon his heart a thirst for gold,
   The beggar's vice, which can but overwhelm
The meanest hearts; and for the rest, but glance
Thine eye along America and France.

XLIV

"'Tis true, he was a tool from first to last
   (I have the workmen safe); but as a tool

321. quit-rent: a small fixed rent paid by a freeholder to his feudal superior as a substitute for services he would otherwise be expected to render. **340. minion . . . helm:** the Earl of Bute, George's former tutor, whom he had made Prime Minister in 1762.

So let him be consumed. From out the past
  Of ages, since mankind have known the rule
Of monarchs—from the bloody rolls amass'd
  Of sin and slaughter—from the Cæsar's school,   350
Take the worst pupil; and produce a reign
More drench'd with gore, more cumber'd with the
    slain.

XLV

"He ever warr'd with freedom and the free:
  Nations as men, home subjects, foreign foes,
So that they utter'd the word 'Liberty!'
  Found George the Third their first opponent. Whose
History was ever stain'd as his will be
  With national and individual woes?
I grant his household abstinence; I grant
His neutral virtues, which most monarchs want;   360

XLVI

"I know he was a constant consort; own
  He was a decent sire, and middling lord.
All this is much, and most upon a throne;
  As temperance, if at Apicius' board,
Is more than at an anchorite's supper shown.
  I grant him all the kindest can accord;
And this was well for him, but not for those
Millions who found him what oppression chose.

XLVII

"The New World shook him off; the Old yet groans
  Beneath what he and his prepared, if not   370
Completed: he leaves heirs on many thrones
  To all his vices, without what begot
Compassion for him—his tame virtues; drones
  Who sleep, or despots who have now forgot
A lesson which shall be re-taught them, wake
Upon the thrones of earth; but let them quake!

XLVIII

"Five millions of the primitive, who hold
  The faith which makes ye great on earth, implored

A *part* of that vast *all* they held of old,—
  Freedom to worship—not alone your Lord,   380
Michael, but you, and you, Saint Peter! Cold
  Must be your souls, if you have not abhorr'd
The foe to Catholic participation
In all the license of a Christian nation.

XLIX

"True! he allow'd them to pray God; but as
  A consequence of prayer, refused the law
Which would have placed them upon the same base
  With those who did not hold the saints in awe."
But here Saint Peter started from his place,
  And cried, "You may the prisoner withdraw: 390
Ere heaven shall ope her portals to this Guelph,
While I am guard, may I be damn'd myself!

L

"Sooner will I with Cerberus exchange
  My office (and *his* is no sinecure)
Than see this royal Bedlam bigot range
  The azure fields of heaven, of that be sure!"
"Saint!" replied Satan, "you do well to avenge
  The wrongs he made your satellites endure;
And if to this exchange you should be given,
I'll try to coax *our* Cerberus up to heaven."   400

LI

Here Michael interposed: "Good saint! and devil!
  Pray, not so fast; you both outrun discretion.
Saint Peter! you were wont to be more civil!
  Satan! excuse this warmth of his expression,
And condescension to the vulgar's level:
  Even saints sometimes forget themselves in session.
Have you got more to say?"—"No."—"If you please,
I'll trouble you to call your witnesses."

LII

Then Satan turn'd and waved his swarthy hand,
  Which stirr'd with its electric qualities   410
Clouds farther off than we can understand,
  Although we find him sometimes in our skies;

---

364. Apicius' board: Apicius was a famous Roman epicure of the time of Augustus and Tiberius. **377–83. Five . . . participation:** the five million Irish; George III had opposed the bill (1795) to allow Catholics to hold office.

391. Guelph: family name of the House of Hanover. **393. Cerberus:** the three-headed dog at the entrance to Hades.

Infernal thunder shook both sea and land
  In all the planets, and hell's batteries
Let off the artillery, which Milton mentions
  As one of Satan's most sublime inventions.

LIII

This was a signal unto such damn'd souls
  As have the privilege of their damnation
Extended far beyond the mere controls
  Of worlds past, present, or to come; no station   420
Is theirs particularly in the rolls
  Of hell assign'd; but where their inclination
Or business carries them in search of game,
They may range freely—being damn'd the same.

LIV

They're proud of this—as very well they may,
  It being a sort of knighthood, or gilt key
Stuck in their loins; or like to an "entré"
  Up the back stairs, or such free-masonry.
I borrow my comparisons from clay,
  Being clay myself. Let not those spirits be   430
Offended with such base low likenesses;
We know their posts are nobler far than these.

LV

When the great signal ran from heaven to hell—
  About ten million times the distance reckon'd
From our sun to its earth, as we can tell
  How much time it takes up, even to a second,
For every ray that travels to dispel
  The fogs of London, through which, dimly
    beacon'd,
The weathercocks are gilt some thrice a year,
If that the *summer* is not too severe:—   440

LVI

I say that I can tell—'twas half a minute:
  I know the solar beams take up more time
Ere, packed up for their journey, they begin it;
  But then their telegraph is less sublime,
And if they ran a race, they would not win it
  'Gainst Satan's couriers bound for their own clime.

The sun takes up some years for every ray
To reach its goal—the devil not half a day.

LVII

Upon the verge of space, about the size
  Of half-a-crown, a little speck appear'd   450
(I've seen a something like it in the skies
  In the Ægean, ere a squall); it near'd,
And, growing bigger, took another guise;
  Like an aërial ship it tack'd, and steer'd,
Or *was* steer'd (I am doubtful of the grammar
Of the last phrase, which makes the stanza stammer;—

LVIII

But take your choice); and then it grew a cloud;
  And so it was—a cloud of witnesses.
But such a cloud! No land e'er saw a crowd
  Of locusts numerous as the heavens saw these;   460
They shadow'd with their myriads space; their loud
  And varied cries were like those of wild geese
(If nations may be liken'd to a goose),
And realised the phrase of "hell broke loose."

LIX

Here crash'd a sturdy oath of stout John Bull,
  Who damn'd away his eyes as heretofore:
There Paddy brogued "By Jasus!"—"What's your
    wull?"
    The temperate Scot exclaim'd: the French ghost
    swore
In certain terms I shan't translate in full,
  As the first coachman will; and 'midst the war,   470
The voice of Jonathan was heard to express,
"*Our* president is going to war, I guess."

LX

Besides there were the Spaniard, Dutch, and Dane;
  In short, an universal shoal of shades,
From Otaheite's isle to Salisbury Plain,
  Of all climes and professions, years and trades,
Ready to swear against the good king's reign,
  Bitter as clubs in cards are against spades:
All summon'd by this grand "subpœna," to
Try if kings mayn't be damn'd like me or you.   480

---

415. Milton mentions: *Paradise Lost*, VI. 484–85. **426. gilt
key:** insignia, worn at the belt, of the Lord Chamberlain.

464. "hell . . . loose": *Paradise Lost*, IV. 918. **475.
Otaheite's isle:** the earlier name for Tahiti.

LXI

When Michael saw this host, he first grew pale,
   As angels can; next, like Italian twilight,
He turn'd all colours—as a peacock's tail,
   Or sunset streaming through a Gothic skylight
In some old abbey, or a trout not stale,
   Or distant lightning on the horizon *by* night,
Or a fresh rainbow, or a grand review
Of thirty regiments in red, green, and blue.

LXII

Then he address'd himself to Satan: "Why—
   My good old friend, for such I deem you, though 490
Our different parties make us fight so shy,
   I ne'er mistake you for a *personal* foe;
Our difference is *political*, and I
   Trust that, whatever may occur below,
You know my great respect for you: and this
Makes me regret whate'er you do amiss—

LXIII

"Why, my dear Lucifer, would you abuse
   My call for witnesses? I did not mean
That you should half of earth and hell produce;
   'Tis even superfluous, since two honest, clean, 500
True testimonies are enough: we lose
   Our time, nay, our eternity, between
The accusation and defence: if we
Hear both, 'twill stretch our immortality."

LXIV

Satan replied, "To me the matter is
   Indifferent, in a personal point of view:
I can have fifty better souls than this
   With far less trouble than we have gone through
Already; and I merely argued his
   Late majesty of Britain's case with you   510
Upon a point of form: you may dispose
Of him; I've kings enough below, God knows!"

LXV

Thus spoke the Demon (late call'd "multifaced"
   By multo-scribbling Southey). "Then we'll call
One or two persons of the myriads placed
   Around our congress, and dispense with all

The rest," quoth Michael: "Who may be so graced
   As to speak first? there's choice enough—who shall
It be?" Then Satan answer'd, "There are many;
But you may choose Jack Wilkes as well as any." 520

LXVI

A merry, cock-eyed, curious-looking sprite
   Upon the instant started from the throng,
Dress'd in a fashion now forgotten quite;
   For all the fashions of the flesh stick long
By people in the next world; where unite
   All the costumes since Adam's, right or wrong,
From Eve's fig-leaf down to the petticoat,
Almost as scanty, of days less remote.

LXVII

The spirit look'd around upon the crowds
   Assembled, and exclaim'd, "My friends of all   530
The spheres, we shall catch cold amongst these clouds;
   So let's to business: why this general call?
If those are freeholders I see in shrouds,
   And 'tis for an election that they bawl,
Behold a candidate with unturn'd coat!
Saint Peter, may I count upon your vote?"

LXVIII

"Sir," replied Michael, "you mistake; these things
   Are of a former life, and what we do
Above is more august; to judge of kings
   Is the tribunal met: so now you know."   540
"Then I presume those gentlemen with wings,"
   Said Wilkes, "are cherubs; and that soul below
Looks much like George the Third, but to my mind
A good deal older—Bless me! is he blind?"

LXIX

"He is what you behold him, and his doom
   Depends upon his deeds," the Angel said.
"If you have aught to arraign in him, the tomb
   Gives license to the humblest beggar's head
To lift itself against the loftiest."—"Some,"
   Said Wilkes, "don't wait to see them laid in lead, 550
For such a liberty—and I, for one,
Have told them what I thought beneath the sun."

520. **Wilkes:** John Wilkes, 1727–97, popular liberal who had
attacked George III and was elected Lord Mayor of London;
also known for his licentiousness.

LXX

"*Above* the sun repeat, then, what thou hast
   To urge against him," said the Archangel. "Why,"
Replied the spirit, "since old scores are past,
   Must I turn evidence? In faith, not I.
Besides, I beat him hollow at the last,
   With all his Lords and Commons: in the sky
I don't like ripping up old stories, since
His conduct was but natural in a prince.   560

LXXI

"Foolish, no doubt, and wicked, to oppress
   A poor unlucky devil without a shilling;
But then I blame the man himself much less
   Than Bute and Grafton, and shall be unwilling
To see him punish'd here for their excess,
   Since they were both damn'd long ago, and still in
Their place below: for me, I have forgiven,
And vote his 'habeas corpus' into heaven."

LXXII

"Wilkes," said the Devil, "I understand all this;
   You turn'd to half a courtier ere you died,   570
And seem to think it would not be amiss
   To grow a whole one on the other side
Of Charon's ferry; you forget that *his*
   Reign is concluded; whatsoe'er betide,
He won't be sovereign more: you've lost your labour
For at the best he will but be your neighbour.

LXXIII

"However, I knew what to think of it,
   When I beheld you in your jesting way
Flitting and whispering round about the spit
   Where Belial, upon duty for the day,   580
With Fox's lard was basting William Pitt,
   His pupil; I knew what to think, I say:
That fellow even in hell breeds farther ills;
I'll have him *gagg'd*—'twas one of his own bills.

LXXIV

"Call Junius!" From the crowd a shadow stalk'd,
   And at the name there was a general squeeze,
So that the very ghosts no longer walk'd
   In comfort, at their own aërial ease,
But were all ramm'd, and jamm'd (but to be balk'd,
   As we shall see), and jostled hands and knees,   590
Like wind compress'd and pent within a bladder,
Or like a human colic, which is sadder.

LXXV

The shadow came—a tall, thin, grey-hair'd figure,
   That look'd as it had been a shade on earth;
Quick in its motions, with an air of vigour,
   But nought to mark its breeding or its birth:
Now it wax'd little, then again grew bigger,
   With now an air of gloom, or savage mirth;
But as you gazed upon its features, they
Changed every instant—to *what*, none could say.   600

LXXVI

The more intently the ghosts gazed, the less
   Could they distinguish whose the features were;
The Devil himself seem'd puzzled even to guess;
   They varied like a dream—now here, now there;
And several people swore from out the press,
   They knew him perfectly; and one could swear
He was his father: upon which another
Was sure he was his mother's cousin's brother:

LXXVII

Another, that he was a duke, or knight,
   An orator, a lawyer, or a priest,   610
A nabob, a man-midwife; but the wight
   Mysterious changed his countenance at least
As oft as they their minds: though in full sight
   He stood, the puzzle only was increased;
The man was a phantasmagoria in
Himself—he was so volatile and thin.

---

585. Junius: pseudonym of a writer (probably Sir Philip Francis, mentioned below in l. 632) who had attacked George III and his supporters. The title page carried the words, *Stat Nominis Umbra* ("A shadow stands for the name"); hence Byron's reference to the "shadow" in line 593.

564. Grafton: The Duke of Grafton was an early minister of George III. **581. Fox's lard:** Charles James Fox was quite corpulent.

LXXVIII

The moment that you had pronounced him *one*,
  Presto! his face changed, and he was another;
And when that change was hardly well put on,
  It varied, till I don't think his own mother   620
(If that he had a mother) would her son
  Have known, he shifted so from one to t'other;
Till guessing from a pleasure grew a task,
At this epistolary "Iron Mask."

LXXIX

For sometimes he like Cerberus would seem—
  "Three gentlemen at once" (as sagely says
Good Mrs. Malaprop); then you might deem
  That he was not even *one; now* many rays
Were flashing round him; and now a thick steam
  Hid him from sight—like fogs on London days: 630
Now Burke, now Tooke, he grew to people's fancies,
And certes often like Sir Philip Francis.

LXXX

I've an hypothesis—'tis quite my own;
  I never let it out till now, for fear
Of doing people harm about the throne,
  And injuring some minister or peer,
On whom the stigma might perhaps be blown;
  It is—my gentle public, lend thine ear!
'Tis, that what Junius we are wont to call
Was *really, truly,* nobody at all.   640

LXXXI

I don't see wherefore letters should not be
  Written without hands, since we daily view
Them written without heads; and books, we see,
  Are fill'd as well without the latter too:
And really till we fix on somebody
  For certain sure to claim them as his due,
Their author, like the Niger's mouth, will bother
The world to say if *there* be mouth or author.

LXXXII

"And who and what art thou?" the Archangel said.
  "For *that* you may consult my title-page,"   650
Replied this mighty shadow of a shade:
  "If I have kept my secret half an age,
I scarce shall tell it now."—"Canst thou upbraid,"
  Continued Michael, "George Rex, or allege
Aught further?" Junius answer'd, "You had better
First ask him for *his* answer to my letter:

LXXXIII

"My charges upon record will outlast
  The brass of both his epitaph and tomb."
"Repent'st thou not," said Michael, "of some past
  Exaggeration? something which may doom   660
Thyself if false, as him if true? Thou wast
  Too bitter—is it not so?—in thy gloom
Of passion?"—"Passion!" cried the phantom dim,
"I loved my country, and I hated him.

LXXXIV

"What I have written, I have written: let
  The rest be on his head or mine!" So spoke
Old "Nominis Umbra;" and while speaking yet,
  Away he melted in celestial smoke.
Then Satan said to Michael, "Don't forget
  To call George Washington, and John Horne Tooke,
And Franklin;"—but at this time there was heard 671
A cry for room, though not a phantom stirr'd.

LXXXV

At length with jostling, elbowing, and the aid
  Of cherubim appointed to that post,
The devil Asmodeus to the circle made
  His way, and look'd as if his journey cost
Some trouble. When his burden down he laid,
  "What's this?" cried Michael; "why, 'tis not a
    ghost?"
"I know it," quoth the incubus; "but he
Shall be one, if you leave the affair to me.   680

---

624. "Iron Mask": that is, his identity is still unknown (as that of the "Man in the Iron Mask," imprisoned in the Bastille by Louis XIV, was unknown). **627. Mrs. Malaprop:** who absurdly misused her words (in R. B. Sheridan's *School for Scandal*); hence the term "malapropism." **647. Niger's mouth:** allusion to explorations in Nigeria.

665. What . . . written: John 19:22. **670. John Horne Tooke:** See *Don Juan*, above, XVI. n. 211; Tooke had opposed the Tory policy toward the American colonies. **675. Asmodeus:** devil.

LXXXVI

"Confound the renegado! I have sprain'd
  My left wing, he's so heavy; one would think
Some of his works about his neck were chain'd.
  But to the point; while hovering o'er the brink
Of Skiddaw (where as usual it still rain'd),
  I saw a taper, far below me, wink,
And stooping, caught this fellow at a libel—
No less on history than the Holy Bible.

LXXXVII

"The former is the devil's scripture, and
  The latter yours, good Michael: so the affair   690
Belongs to all of us, you understand.
  I snatch'd him up just as you see him there,
And brought him off for sentence out of hand:
  I've scarcely been ten minutes in the air—
At least a quarter it can hardly be:
I dare say that his wife is still at tea."

LXXXVIII

Here Satan said, "I know this man of old,
  And have expected him for some time here;
A sillier fellow you will scarce behold,
  Or more conceited in his petty sphere: 700
But surely it was not worth while to fold
  Such trash below your wing, Asmodeus dear:
We had the poor wretch safe (without being bored
With carriage) coming of his own accord.

LXXXIX

"But since he's here, let's see what he has done."
  "Done!" cried Asmodeus, "he anticipates
The very business you are now upon,
  And scribbles as if head clerk to the Fates.
Who knows to what his ribaldry may run,   709
  When such an ass as this, like Balaam's, prates?"
"Let's hear," quoth Michael, "what he has to say:
You know we're bound to that in every way."

XC

Now the bard, glad to get an audience, which
  By no means often was his case below,

Began to cough, and hawk, and hem, and pitch
  His voice into that awful note of woe
To all unhappy hearers within reach
  Of poets when the tide of rhyme's in flow;
But stuck fast with his first hexameter,
Not one of all whose gouty feet would stir.   720

XCI

But ere the spavin'd dactyls could be spurr'd
  Into recitative, in great dismay
Both cherubim and seraphim were heard
  To murmur loudly through their long array;
And Michael rose ere he could get a word
  Of all his founder'd verses under way,
And cried, "For God's sake stop, my friend! 'twere best—
Non Di, non homines—you know the rest."

XCII

A general bustle spread throughout the throng,
  Which seem'd to hold all verse in detestation;   730
The angels had of course enough of song
  When upon service; and the generation
Of ghosts had heard too much in life, not long
  Before, to profit by a new occasion:
The monarch, mute till then, exclaim'd, "What! what!
Pye come again? No more—no more of that!"

XCIII

The tumult grew; an universal cough
  Convulsed the skies, as during a debate,
When Castlereagh has been up long enough
  (Before he was first minister of state,   740
I mean—the *slaves hear now*); some cried "Off, off!"
  As at a farce; till, grown quite desperate,
The bard Saint Peter pray'd to interpose
(Himself an author) only for his prose.

XCIV

The varlet was not an ill-favour'd knave;
  A good deal like a vulture in the face,

685. Skiddaw: mountain in the Lake Country; hence near Southey's home. 710. Balaam: Numbers 22:28.

728. Non . . . homines: allusion to Horace's *Art of Poetry*, ll. 372–73 ("Neither gods nor men can tolerate a mediocre poet"). 735. What! What!: Byron imitates the repetitively ejaculatory speech of George III. 736. Pye: Henry James Pye, poet laureate before Southey (see Dedication to *Don Juan*, l. 8).

With a hook nose and a hawk's eye, which gave
   A smart and sharper-looking sort of grace
To his whole aspect, which, though rather grave,
   Was by no means so ugly as his case;   750
But that, indeed, was hopeless as can be,
Quite a poetic felony "*de se.*"

XCV

Then Michael blew his trump, and still'd the noise
   With one still greater, as is yet the mode
On earth besides; except some grumbling voice,
   Which now and then will make a slight inroad
Upon decorous silence, few will twice
   Lift up their lungs when fairly overcrow'd;
And now the bard could plead his own bad cause,
With all the attitudes of self-applause.   760

XCVI

He said—(I only give the heads)—he said,
   He meant no harm in scribbling; 'twas his way
Upon all topics; 'twas, besides, his bread,
   Of which he butter'd both sides; 'twould delay
Too long the assembly (he was pleased to dread),
   And take up rather more time than a day,
To name his works—he would but cite a few—
"Wat Tyler"—"Rhymes on Blenheim"—"Water-
   loo."

XCVII

He had written praises of a regicide;
   He had written praises of all kings what ever;   770
He had written for republics far and wide,
   And then against them bitterer than ever:
For pantisocracy he once had cried
   Aloud, a scheme less moral than 'twas clever;
Then grew a hearty anti-jacobin—
Had turn'd his coat—and would have turn'd his skin.

XCVIII

He had sung against all battles, and again
   In their high praise and glory; he had call'd
Reviewing "the ungentle craft," and then
   Become as base a critic as e'er crawl'd—   780

Fed, paid, and pamper'd by the very men
   By whom his muse and morals had been maul'd:
He had written much blank verse, and blanker prose,
   And more of both than anybody knows.

XCIX

He had written Wesley's life:—here turning round
   To Satan, "Sir, I'm ready to write yours,
In two octavo volumes, nicely bound,
   With notes and preface, all that most allures
The pious purchaser; and there's no ground
   For fear, for I can choose my own reviewers: 790
So let me have the proper documents,
That I may add you to my other saints."

C

Satan bow'd, and was silent. "Well, if you,
   With amiable modesty, decline
My offer, what says Michael? There are few
   Whose memoirs could be render'd more divine.
Mine is a pen of all work; not so new
   As it was once, but I would make you shine
Like your own trumpet. By the way, my own
Has more of brass in it, and is as well blown.   800

CI

"But talking about trumpets, here's my Vision!
   Now you shall judge, all people; yes, you shall
Judge with my judgment, and by my decision
   Be guided who shall enter heaven or fall.
I settle all these things by intuition,
   Times present, past, to come, heaven, hell, and all,
Like King Alfonso. When I thus see double,
I save the Deity some worlds of trouble."

CII

He ceased, and drew forth an MS.; and no
   Persuasion on the part of devils, saints,   810
Or angels, now could stop the torrent; so
   He read the first three lines of the contents;

---

752. felony "de se": suicide. **769. regicide:** a poem about
Henry Marten, one of the judges who had condemned
Charles I to death. **779. "the ungentle craft":** in his *Re-
mains of Henry Kirke White* (1808).

807. Alfonso: "King Alfonso [of Castile], speaking of the
Ptolomean system, said that 'had he been consulted at the
creation of the world, he would have spared the Maker
some absurdities' " (B.).

But at the fourth, the whole spiritual show
   Had vanish'd, with variety of scents,
Ambrosial and sulphureous, as they sprang,
Like lightning, off from his "melodious twang."

CIII

Those grand heroics acted as a spell:
   The angels stopp'd their ears and plied their pinions;
The devils ran howling, deafen'd, down to hell;
   The ghosts fled, gibbering, for their own domin-
      ions—   820
(For 'tis not yet decided where they dwell,
   And I leave every man to his opinions);
Michael took refuge in his trump—but, lo!
His teeth were set on edge, he could not blow!

CIV

Saint Peter, who has hitherto been known
   For an impetuous saint, upraised his keys,
And at the fifth line knock'd the poet down;
   Who fell like Phaëton, but more at ease,
Into his lake, for there he did not drown;
   A different web being by the Destinies   830
Woven for the Laureate's final wreath, whene'er
Reform shall happen either here or there.

CV

He first sank to the bottom—like his works,
   But soon rose to the surface—like himself;
For all corrupted things are buoy'd like corks,
   By their own rottenness, light as an elf,
Or wisp that flits o'er a morass: he lurks,
   It may be, still, like dull books on a shelf,
In his own den, to scrawl some "Life" or "Vision,"
As Welborn says—"the devil turn'd precisian."   840

CVI

As for the rest, to come to the conclusion
   Of this true dream, the telescope is gone

---

816. **"melodious twang"**: Byron cites John Aubrey, who in his *Miscellanies* (1696) spoke of a ghost that disappeared with "a curious perfume, and most melodious twang." **828. Phaëton**: Apollo's son, who had tried to drive the chariot of the sun. **840. Welborn**: Welborn figures in Massinger's play, *A New Way To Pay Old Debts* (1633). **precisian**: Puritan.

Which kept my optics free from all delusion,
   And show'd me what I in my turn have shown;
All I saw farther, in the last confusion,
   Was, that King George slipp'd into heaven for one;
And when the tumult dwindled to a calm,
I left him practising the hundredth psalm.

                          1821 (1822)

ON THIS DAY I COMPLETE MY THIRTY-SIXTH YEAR

Byron wrote this on January 22, 1824, three months before his death. His friend Count Pietro Gamba relates the following account:

> This morning Lord Byron came from his bedroom into the apartment where Colonel Stanhope and some friends were assembled, and said with a smile—"You were complaining, the other day, that I never write any poetry now:—this is my birthday, and I have just finished something, which I think, is better than what I usually write.' He then produced these noble and affecting verses, which were afterwards found written in his journals, with only the following introduction: 'Jan. 22; on this day I complete my 36th year."

'Tis time this heart should be unmoved,
   Since others it hath ceased to move:
Yet, though I cannot be beloved,
     Still let me love!

My days are in the yellow leaf;
   The flowers and fruits of love are gone;
The worm, the canker, and the grief
     Are mine alone!

The fire that on my bosom preys
   Is lone as some volcanic isle;
No torch is kindled at its blaze—   10
     A funeral pile.

The hope, the fear, the jealous care,
   The exalted portion of the pain
And power of love, I cannot share,
     But wear the chain.

ON THIS DAY I COMPLETE MY THIRTY-SIXTH YEAR. **5. yellow leaf**: *Macbeth*, V. ii. 22.

But 'tis not *thus*—and 'tis not *here*—
  Such thoughts should shake my soul, nor *now*,
Where glory decks the hero's bier,
    Or binds his brow.   20

The sword, the banner, and the field,
  Glory and Greece, around me see!
The Spartan, borne upon his shield,
    Was not more free.

Awake! (not Greece—she *is* awake!)
  Awake, my spirit! Think through *whom*
Thy life-blood tracks its parent lake,
    And then strike home!

Tread those reviving passions down,
  Unworthy manhood!—unto thee   30
Indifferent should the smile or frown
    Of beauty be.

If thou regrett'st thy youth, *why live?*
  The land of honorable death
Is here:—up to the field, and give
    Away thy breath!

Seek out—less often sought than found—
  A soldier's grave, for thee the best;
Then look around, and choose thy ground,
    And take thy rest.   40

1824 (1824)

LETTERS

❧

Byron was one of the finest talkers of his age. His letters enable us still to hear the free monologues, the vigorous, downright, tumbling words and sentences that fascinated his contemporaries. Not all his moods got into his letters. One does not find, for example, the consciously sublime poses and gestures of *Childe Harold* or *Manfred*, or the nostalgic idealism that partly emerges in the handling of the love of Haidée and Juan. Perhaps there are some things one is more willing to express in poetry than to private friends. But with such exceptions the letters are the intimate and continuing revelation of a man fascinated by the world, by people—their habits, vagaries, and hypocrisies—and, most of all, by himself. Yet the letters are not just conversational and self-

23. **Spartan . . . shield:** The Spartan was carried thus from the battlefield if killed or wounded. 26. **Think . . . whom:** A reference to Byron's ancestry, which, on his mother's side, went back to the early Scottish kings.

revealing. In the best of them Byron was engaged in vivifying his life by writing it down, reliving his experience under the conditions of art and style—the special art and style he created for his letters. This meant that experience took on incomparable clearness—character, energy, and idea caught in a bright foreground, the shadowy depths eliminated—accompanied by unpredictable shifts of feeling and ripples of irony. Like the letters of Keats, those of Byron have a literary interest in their own right.

Of Byron's correspondents Francis Hodgson (1781–1852) was a friend and fellow poet; John Murray (1778–1843) was Byron's publisher; John Cam Hobhouse (1786–1869), an intimate friend, traveled with Byron in 1809–10 and was best man at his wedding; Douglas Kinnaird (1788–1830) was a friend and also Byron's banker; and Richard Belgrave Hoppner (1786–1872) was English Consul at Venice. Asterisks in the texts indicate deletions no longer recoverable made by the original editors or by the custodians of the manuscripts. Ellipses indicate that matter has been omitted by the present editor. Where no salutations or closings appear, the letters are printed as they are given in Thomas Moore's *The Life, Letters and Journals of Lord Byron* (1830), the original texts of these letters having been lost.

❧

To Francis Hodgson
Lisbon, July 16, 1809.

Thus far have we pursued our route, and seen all sorts of marvellous sights, palaces, convents, etc.;—which, being to be heard in my friend Hobhouse's forthcoming Book of Travels, I shall not anticipate by smuggling any account whatsoever to you in a private and clandestine manner. I must just observe, that the village of Cintra in Estremadura is the most beautiful, perhaps, in the world.

I am very happy here, because I loves oranges, and talks bad Latin to the monks, who understand it, as it is like their own,—and I goes into society (with my pocket-pistols), and I swims in the Tagus all across at once, and I rides on an ass or a mule, and swears Portuguese, and have got a diarrhoea and bites from the mosquitoes. But what of that? Comfort must not be expected by folks that go a pleasuring.

When the Portuguese are pertinacious, I say *Carracho!* —the great oath of the grandees, that very well supplies the place of "Damme,"—and, when dissatisfied with my neighbour, I pronounce him *Ambra di merdo*. With these two phrases, and a third, *Avra bouro*, which signifieth "Get an ass," I am universally understood to be a person of degree and a master of languages. How merrily we lives that travellers be!—if we had food and raiment. But, in sober sadness, any thing is better than England, and I am infinitely amused with my pilgrimage as far as it has gone.

To-morrow we start to ride post near 400 miles as far as Gibraltar, where we embark for Melita and Byzantium. A letter to Malta will find me, or to be forwarded, if I am absent. Pray embrace the Drury and Dwyer, and all the Ephesians you encounter. I am writing with Butler's donative pencil, which makes my bad hand worse. Excuse illegibility.

Hodgson! send me the news, and the deaths and defeats, and capital crimes and the misfortunes of one's friends; and let us hear of literary matters, and the controversies and the criticisms. All this will be pleasant— *Suave mari magno*,[1] etc. Talking of that, I have been sea-sick, and sick of the sea. Adieu.

Yours faithfully

To His Mother
Prevesa, November 12, 1809.

MY DEAR MOTHER,

I have now been some time in Turkey: this place is on the coast, but I have traversed the interior of the province of Albania on a visit to the Pacha. I left Malta in the *Spider*, a brig of war, on the 21st of September, and arrived in eight days at Prevesa. I thence have been about 150 miles, as far as Tepaleen, his Highness's country palace, where I stayed three days. The name of the Pacha is *Ali*, and he is considered a man of the first abilities: he governs the whole of Albania (the ancient Illyricum), Epirus, and part of Macedonia. His son, Vely Pacha, to whom he has given me letters, governs the Morea, and has great influence in Egypt; in short, he is one of the most powerful men in the Ottoman empire. When I reached Yanina, the capital, after a journey of three days over the mountains, through a country of the most picturesque beauty, I found that Ali Pacha was with his army in Illyricum, besieging Ibrahim Pacha in the castle of Berat. He had heard that an Englishman of rank was in his dominions, and had left orders in Yanina with the commandant to provide a house, and supply me with every kind of necessary

gratis; and, though I have been allowed to make presents to the slaves, etc., I have not been permitted to pay for a single article of houshold consumption.

I rode out on the vizier's horses, and saw the palaces of himself and grandsons: they are splendid, but too much ornamented with silk and gold. I then went over the mountains through Zitza, a village with a Greek monastery (where I slept on my return), in the most beautiful situation (always excepting Cintra, in Portugal) I ever beheld. In nine days I reached Tepaleen. Our journey was much prolonged by the torrents that had fallen from the mountains, and intersected the roads. I shall never forget the singular scene on entering Tepaleen at five in the afternoon, as the sun was going down. It brought to my mind (with some change of *dress*, however) Scott's description of Branksome Castle in his *Lay*,[2] and the feudal system. The Albanians, in their dresses, (the most magnificent in the world, consisting of a long *white kilt*, gold-worked cloak, crimson velvet gold-laced jacket and waistcoat, silver-mounted pistols and daggers), the Tartars with their high caps, the Turks in their vast pelisses and turbans, the soldiers and black slaves with the horses, the former in groups in an immense large open gallery in front of the palace, the latter placed in a kind of cloister below it, two hundred steeds ready caparisoned to move in a moment, couriers entering or passing out with the despatches, the kettle-drums beating, boys calling the hour from the minaret of the mosque, altogether, with the singular appearance of the building itself, formed a new and delightful spectacle to a stranger. I was conducted to a very handsome apartment, and my health inquired after by the vizier's secretary, *à-la-mode Turque!*

The next day I was introduced to Ali Pacha. I was dressed in a full suit of staff uniform, with a very magnificent sabre, etc. The vizier received me in a large room paved with marble; a fountain was playing in the centre; the apartment was surrounded by scarlet ottomans. He received me standing, a wonderful compliment from a Mussulman, and made me sit down on his right hand. I have a Greek interpreter for general use, but a physician of Ali's named Femlario, who understands Latin, acted for me on this occasion. His first question was, why, at so early an age, I left my country?—(the Turks have no idea of travelling for amusement). He then said, the English minister, Captain Leake, had told him I was of a great family, and desired his respects

LETTERS. **1. Suave . . . magno**: from Lucretius, II. 1. The entire citation would be: "It is sweet when on a great sea the winds trouble the waters to gaze from land on the great toils of others."

2. lay: *Lay of the Last Minstrel*, Canto I.

to my mother; which I now, in the name of Ali Pacha, present to you. He said he was certain I was a man of birth, because I had small ears, curling hair, and little white hands, and expressed himself pleased with my appearance and garb. He told me to consider him as a father whilst I was in Turkey, and said he looked on me as his son. Indeed, he treated me like a child, sending me almonds and sugared sherbert, fruit and sweetmeats, twenty times a day. He begged me to visit him often, and at night, when he was at leisure. I then, after coffee and pipes, retired for the first time. I saw him thrice afterwards. It is singular that the Turks, who have no hereditary dignities, and few great families, except the Sultans, pay so much respect to birth; for I found my pedigree more regarded than my title.

To-day I saw the remains of the town of Actium, near which Antony lost the world, in a small bay, where two frigates could hardly manoeuvre: a broken wall is the sole remnant. On another part of the gulf stand the ruins of Nicopolis, built by Augustus in honour of his victory. Last night I was at a Greek marriage; but this and a thousand things more I have neither time nor *space* to describe.

His highness is sixty years old, very fat, and not tall, but with a fine face, light blue eyes, and a white beard; his manner is very kind, and at the same time he possesses that dignity which I find universal amongst the Turks. He has the appearance of anything but his real character, for he is a remorseless tyrant, guilty of the most horrible cruelties, very brave, and so good a general that they call him the Mahometan Buonaparte. Napoleon has twice offered to make him King of Epirus, but he prefers the English interest, and abhors the French, as he himself told me. He is of so much consequence, that he is much courted by both, the Albanians being the most warlike subjects of the Sultan, though Ali is only nominally dependent on the Porte; he has been a mighty warrior, but is as barbarous as he is successful, roasting rebels, etc., etc. Buonaparte sent him a snuff-box with his picture. He said the snuff-box was very well, but the picture he could excuse, as he neither liked it nor the original. His ideas of judging of a man's birth from ears, hands, etc., were curious enough. To me he was, indeed, a father, giving me letters, guards, and every possible accommodation. Our next conversations were of war and travelling, politics and England. He called my Albanian soldier, who attends me, and told him to protect me at all hazard; his name is Viscillie, and, like all the Albanians, he is brave, rigidly honest, and faithful; but they are cruel, though not treacherous, and have several vices but no meannesses. They are, perhaps, the most beautiful race, in point of countenance, in the world; their women are sometimes handsome also, but they are treated like slaves, *beaten*, and, in short, complete beasts of burden; they plough, dig, and sow. I found them carrying wood, and actually repairing the highways. The men are all soldiers, and war and the chase their sole occupations. The women are the labourers, which after all is no great hardship in so delightful a climate. Yesterday, the 11th of November, I bathed in the sea; to-day is so hot that I am writing in a shady room of the English consul's, with three doors wide open, no fire, or even *fireplace*, in the house, except for culinary purposes.

I am going to-morrow, with a guard of fifty men, to Patras in the Morea, and thence to Athens, where I shall winter. Two days ago I was nearly lost in a Turkish ship of war, owing to the ignorance of the captain and crew, though the storm was not violent. Fletcher[3] yelled after his wife, the Greeks called on all the saints, the Mussulmans on Alla; the captain burst into tears and ran below deck, telling us to call on God; the sails were split, the main-yard shivered, the wind blowing fresh, the night setting in, and all our chance was to make Corfu, which is in possession of the French, or (as Fletcher pathetically termed it) "a watery grave." I did what I could to console Fletcher, but finding him incorrigible, wrapped myself up in my Albanian capote (an immense cloak), and lay down on deck to wait the worst. I have learnt to philosophise in my travels; and if I had not, complaint was useless. Luckily the wind abated, and only drove us on the coast of Suli, on the main land, where we landed, and proceeded, by the help of the natives, to Prevesa again; but I shall not trust Turkish sailors in future, though the Pacha had ordered one of his own galliots to take me to Patras. I am therefore going as far as Missolonghi by land, and there have only to cross a small gulf to get to Patras.

Fletcher's next epistle will be full of marvels. We were one night lost for nine hours in the mountains in a thunderstorm, and since nearly wrecked. In both cases Fletcher was sorely bewildered, from apprehensions of famine and banditti in the first, and drowning in the second instance. His eyes were a little hurt by the lightning, or crying (I don't know which), but are now recovered. When you write, address to me at Mr. Strane's, English consul, Patras, Morea.

I could tell you I know not how many incidents that

---

3. **Fletcher:** Byron's servant.

I think would amuse you, but they crowd on my mind as much as they would swell my paper, and I can neither arrange them in the one, nor put them down on the other, except in the greatest confusion. I like the Albanians much; they are not all Turks; some tribes are Christians. But their religion makes little difference in their manner or conduct. They are esteemed the best troops in the Turkish service. I lived on my route, two days at once, and three days again, in a barrack at Salora, and never found soldiers so tolerable, though I have been in the garrisons of Gibraltar and Malta, and seen Spanish, French, Sicilian, and British troops in abundance. I have had nothing stolen, and was always welcome to their provision and milk. Not a week ago an Albanian chief, (every village has its chief, who is called Primate,) after helping us out of the Turkish galley in her distress, feeding us, and lodging my suite, consisting of Fletcher, a Greek, two Athenians, a Greek priest, and my companion, Mr. Hobhouse, refused any compensation but a written paper stating that I was well received; and when I pressed him to accept a few sequins, "No," he replied; "I wish you to love me, not to pay me." These are his words.

It is astonishing how far money goes in this country. While I was in the capital I had nothing to pay by the vizier's order; but since, though I have generally had sixteen horses, and generally six or seven men, the expense has not been *half* as much as staying only three weeks in Malta, though Sir A. Ball, the governor, gave me a house for nothing, and I had only *one servant*. By the by, I expect Hanson to remit regularly; for I am not about to stay in this province for ever. Let him write to me at Mr. Strane's, English consul, Patras. The fact is, the fertility of the plains is wonderful, and specie is scarce, which makes this remarkable cheapness. I am going to Athens, to study modern Greek, which differs much from the ancient, though radically similar. I have no desire to return to England, nor shall I, unless compelled by absolute want, and Hanson's neglect; but I shall not enter into Asia for a year or two, as I have much to see in Greece, and I may perhaps cross into Africa, at least the Egyptian part. Fletcher, like all Englishmen, is very much dissatisfied, though a little reconciled to the Turks by a present of eighty piastres from the vizier, which, if you consider every thing, and the value of specie here, is nearly worth ten guineas English. He has suffered nothing but from cold, heat, and vermin, which those who lie in cottages and cross mountains in a cold country must undergo, and of which I have equally partaken with himself; but he is

not valiant, and is afraid of robbers and tempests. I have no one to be remembered to in England, and wish to hear nothing from it, but that you are well, and a letter or two on business from Hanson, whom you may tell to write. I will write when I can, and beg you to believe me,

> Your affectionate son,
> BYRON.

P.S.—I have some very "magnifiques" Albanian dresses, the only expensive articles in this country. They cost fifty guineas each, and have so much gold, they would cost in England two hundred.

I have been introduced to Hussein Bey, and Mahmout Pacha, both little boys, grandchildren of Ali, at Yanina; they are totally unlike our lads, have painted complexions like rouged dowagers, large black eyes, and features perfectly regular. They are the prettiest little animals I ever saw, and are broken into the court ceremonies already. The Turkish salute is a slight inclination of the head, with the hand on the heart; intimates always kiss. Mahmout is ten years old, and hopes to see me again; we are friends without understanding each other, like many other folks, though from a different cause. He has given me a letter to his father in the Morea, to whom I have also letters from Ali Pacha.

To Francis Hodgson
Newstead Abbey, Sept. 3, 1811.

MY DEAR HODGSON,

I will have nothing to do with your immortality; we are miserable enough in this life, without the absurdity of speculating upon another. If men are to live, why die at all? and if they die, why disturb the sweet and sound sleep that "knows no waking"? "Post Mortem nihil est, ipsaque Mors nihil . . . quaeris quo jaceas post obitum loco? Quo *non* Nata jacent."[4]

As to revealed religion, Christ came to save men; but a good Pagan will go to heaven, and a bad Nazarene to hell; "Argal"[5] (I argue like the gravedigger) why are not all men Christians? or why are any? If mankind

---

4. **"Post . . . jacent"**: from Seneca's *Troades*, ll. 397–402: "After death is nothing, and death itself is nothing . . . You ask, where shall you lie after death? Where the unborn lie." 5. **"Argal"**: perversion of Latin *ergo*, "therefore." Byron is thinking of *Hamlet*, V. i. 21.

may be saved who never heard or dreamt, at Timbuctoo, Otaheite, Terra Incognita, etc., of Galilee and its Prophet, Christianity is of no avail: if they cannot be saved without, why are not all orthodox? It is a little hard to send a man preaching to Judaea, and leave the rest of the world—Negers and what not—*dark* as their complexions, without a ray of light for so many years to lead them on high; and who will believe that God will damn men for not knowing what they were never taught? I hope I am sincere; I was so at least on a bed of sickness in a far-distant country, when I had neither friend, nor comforter, nor hope, to sustain me. I looked to death as a relief from pain, without a wish for an after-life, but a confidence that the God who punishes in this existence had left that last asylum for the weary.

'Ον ὁ θεὸς ἀγαπάει ἀποθνήσκει νέος⁶

I am no Platonist, I am nothing at all; but I would sooner be a Paulician, Manichean, Spinozist, Gentile, Pyrrhonian, Zoroastrian, than one of the seventy-two villainous sects who are tearing each other to pieces for the love of the Lord and hatred of each other. Talk of Galileeism? Show me the effects—are you better, wiser, kinder by your precepts? I will bring you ten Mussulmans shall shame you in all goodwill towards men, prayer to God, and duty to their neighbours. And is there a Talapoin, or a Bonze,⁷ who is not superior to a fox-hunting curate? But I will say no more on this endless theme; let me live, well if possible, and die without pain. The rest is with God, who assuredly, had He *come* or *sent*, would have made Himself manifest to nations, and intelligible to all.

I shall rejoice to see you. My present intention is to accept Scrope Davies's invitation; and then, if you accept mine, we shall meet *here* and *there*. Did you know poor Matthews? I shall miss him much at Cambridge.

To Thomas Moore
Newstead Abbey, Sept. 20, 1814.

> Here's to her who long
>   Hath waked the poet's sigh!
> The girl who gave to song
>   What gold could never buy.

My dear Moore,

I am going to be married—that is, I am accepted, and one usually hopes the rest will follow. My mother of the Gracchi⁸ (that *are* to be), *you* think too strait-laced for me, although the paragon of only children, and invested with "golden opinions of all sorts of men,"⁹ and full of "most blest conditions"¹⁰ as Desdemona herself. Miss Milbanke is the lady, and I have her father's invitation to proceed there in my elect capacity,—which, however, I cannot do till I have settled some business in London, and got a blue coat.

She is said to be an heiress, but of that I really know nothing certainly, and shall not enquire. But I do know, that she has talents and excellent qualities; and you will not deny her judgment, after having refused six suitors and taken me.

Now, if you have anything to say against this, pray do; my mind's made up, positively fixed, determined, and therefore I will listen to reason, because now it can do no harm. Things may occur to break it off, but I will hope not. In the meantime, I tell you (a *secret*, by the by,—at least, till I know she wishes it to be public) that I have proposed and am accepted. You need not be in a hurry to wish me joy, for one mayn't be married for months. I am going to town to-morrow: but expect to be here, on my way there, within a fortnight.

If this had not happened, I should have gone to Italy. In my way down, perhaps, you will meet me at Nottingham, and come over with me here. I need not say that nothing will give me greater pleasure. I must, of course, reform thoroughly; and, seriously, if I can contribute to her happiness, I shall secure my own. She is so good a person, that—that—in short, I wish I was a better.

Ever,

To The Countess of ——
Albany, October 5, 1814.

*Dear Lady * *,*

Your recollection and invitation do me great honour; but I am going to be "married, and can't come." My intended is two hundred miles off, and the moment my business here is arranged, I must set out in a great hurry to be happy. Miss Milbanke is the good-natured person

---

6. Ὂν . . . νέος: from Menander: "Whom the gods love die young." 7. Talapoin . . . Bonze: Buddhist monk of the Far East.

8. Gracchi: the brothers Tiberius and Gaius, 163–21 B.C., champions of the plebians against the Roman senate. 9. "golden . . . men": *Macbeth*, I. vii. 33. 10. "most . . . conditions": *Othello*, II. ii. 253.

who has undertaken me, and, of course, I am very much in love, and as silly as all single gentlemen must be in that sentimental situation. I have been accepted these three weeks; but when the event will take place, I don't exactly know. It depends partly upon lawyers, who are never in a hurry. One can be sure of nothing; but, at present, there appears no other interruption to this intention, which seems as mutual as possible, and now no secret, though I did not tell first,—and all our relatives are congratulating away to right and left in the most fatiguing manner.

You perhaps know the lady. She is niece to Lady Melbourne, and cousin to Lady Cowper and others of your acquaintance, and has no fault, except being a great deal too good for me, and that I must pardon, if nobody else should. It might have been *two* years ago, and, if it had, would have saved me a world of trouble. She has employed the interval in refusing about half a dozen of my particular friends, (as she did me once, by the way,) and has taken me at last, for which I am very much obliged to her. I wish it was well over, for I do hate bustle, and there is no marrying without some;— and then, I must not marry in a black coat, they tell me, and I can't bear a blue one.

Pray forgive me for scribbling all this nonsense. You know I must be serious all the rest of my life, and this is a parting piece of buffoonery, which I write with tears in my eyes, expecting to be agitated. Believe me, most seriously and sincerely your obliged servant,

BYRON.

P.S.—My best rems. to Lord * * on his return.

To S. T. Coleridge
13 Terrace, Piccadilly,
October 18th, 1815.

DEAR SIR,

Your letter I have just received. I will willingly do whatever you direct about the volumes in question [11]— the sooner the better: it shall not be for want of endeavour on my part, as a negotiator with the 'Trade' (to talk technically) that you are not enabled to do yourself justice. Last spring I saw Wr. Scott. He repeated to me a considerable portion of an unpublished poem of yours—the wildest and finest I ever heard in that kind of composition. The title [12] he did not mention, but I think the heroine's name was Geraldine. At all events, the 'toothless mastiff bitch' and the 'witch Lady,' the description of the hall, the lamp suspended from the image, and more particularly of the girl herself as she went forth in the evening—all took a hold on my imagination which I never shall wish to shake off. I mention this, not for the sake of boring you with compliments, but as a prelude to the hope that this poem is or is to be in the volumes you are now about to publish. I do not know that even "Love" or the "Antient Mariner" are so impressive—and to me there are few things in our tongue beyond these two productions.

Wr. Scott is a staunch and sturdy admirer of yours, and with a just appreciation of your capacity deplored to me the want of inclination and exertion which prevented you from giving full scope to your mind. I will answer your question as to the "Beggar's Bush" [13] tomorrow or next day. I shall see Rae and Dibdin (the acting Mrs.) tonight for that purpose.

Oh—your tragedy [14]—I do not wish to hurry you, but I am indeed very anxious to have it under consideration. It is a field in which there are none living to contend against you and in which I should take a pride and pleasure in seeing you compared with the dead. I say this *not* disinterestedly, but as a *Committee*man. We have nothing even tolerable, except a tragedy of Sotheby's, which shall not interfere with yours when ready. You can have no idea what trash there is in the four hundred *fallow* dramas now lying on the shelves of D[rury] L[ane]. I never thought so highly of good writers as lately, since I have had an opportunity of comparing them with the bad.

Ever yours truly,

BYRON.

To Leigh Hunt
13, Terrace, Piccadilly,
September–October 30, 1815.

MY DEAR HUNT,

Many thanks for your books, of which you already know my opinion. Their external splendour should

11. **volumes in question:** *Biographia Literaria* and *Sibylline Leaves.* Coleridge was about to offer the copyright to the London booksellers and hoped Byron would use his influence to secure a good price.

12. **title:** "Christabel." 13. **"Beggar's Bush":** Coleridge hoped to obtain employment revising for the contemporary stage this comedy by Beaumont and Fletcher. At this time Byron was a member of the Sub-Committee of Management of the Drury Lane Theater. 14. **tragedy:** Zapolya.

not disturb you as inappropriate—they have still more within than without. I take leave to differ with you on Wordsworth, as freely as I once agreed with you; at that time I gave him credit for a promise, which is unfulfilled. I still think his capacity warrants all you say of *it* only, but that his performances since *Lyrical Ballads* are miserably inadequate to the ability which lurks within him: there is undoubtedly much natural talent spilt over the *Excursion;* but it is rain upon rocks—where it stands and stagnates, or rain upon sands—where it falls without fertilizing. Who can understand him? Let those who do, make him intelligible. Jacob Behmen, Swedenborg, and Joanna Southcote, are mere types of this arch-apostle of mystery and mysticism. But I have done,—no, I have not done, for I have two petty, and perhaps unworthy objections in small matters to make to him, which, with his pretensions to accurate observation, and fury against Pope's false translation of "the Moonlight scene in Homer," I wonder he should have fallen into;—these be they:—He says of Greece in the body of his book—that it is a land of

> "*Rivers, fertile plains,* and *sounding* shores,
>   Under a cope of *variegated* sky."

The rivers are dry half the year, the plains are barren, and the shores *still* and *tideless* as the Mediterranean can make them; the sky is any thing but variegated, being for months and months but "darkly, deeply, beautifully blue."—The next is in his notes, where he talks of our "Monuments crowded together in the busy, etc., of a large town," as compared with the "still seclusion of a Turkish cemetery in some *remote* place." This is pure stuff; for *one* monument in our churchyards there are *ten* in the Turkish, and so crowded, that you cannot walk between them; that is, divided merely by a path or road; and as to "*remote* places," men never take the trouble in a barbarous country, to carry their dead very far; they must have lived near to where they were buried. There are no cemeteries in "remote places," except such as have the cypress and the tombstone still left, where the olive and the habitation of the living have perished.

These things I was struck with, as coming peculiarly in my own way; and in both of these he is wrong; yet I should have noticed neither, but for his attack on Pope for a like blunder, and a peevish affectation about him of despising a popularity which he will never obtain. I write in great haste, and, I doubt, *not* much to the purpose; but you have it hot and hot, just as it comes, and so let it go. By-the-way, both he and you go too far against Pope's "So when the moon," etc.; it is no translation, I know; but it is not such false description as asserted. I have read it on the spot; there is a burst, and a lightness, and a glow about the night in the Troad, which makes the "planets vivid," and the "pole glowing." The moon is—at least the sky is, clearness itself; and I know no more appropriate expression for the expansion of such a heaven—o'er the scene —the plain—the sky—Ida—the Hellespont—Simois—Scamander—and the Isles—than that of a "flood of glory." I am getting horribly lengthy, and must stop: to the whole of your letter "I say ditto to Mr. Burke," as the Bristol candidate cried by way of electioneering harangue.[15] You need not speak of morbid feelings and vexations to me; I have plenty; but I must blame partly the times, and chiefly myself: but let us forget them. *I* shall be very apt to do so when I see you next. Will you come to the theatre and see our new management? You shall cut it up to your heart's content, root and branch, afterwards, if you like; but come and see it! If not, I must come and see you.

Ever yours, very truly and affectionately,

BYRON.

P.S.—Not a word from Moore for these two months. Pray let me have the rest of *Rimini*. You have two excellent points in that poem—originality and Italianism. I will back you as a bard against half the fellows on whom you have thrown away much good criticism and eulogy; but don't let your bookseller publish in *quarto*; it is the worst size possible for circulation. I say this on bibliopolical authority.

Again, yours ever,

B.

To John Murray
Venice, November 25, 1816.

DEAR SIR,

It is some months since I have heard from or of you —I think not since I left Diodati. From Milan I wrote once or twice; but have been here some little time, and intend to pass the winter without removing. I was much

---

15. ditto . . . harangue: In 1774 Burke was campaigning for election to Parliament from Bristol. His running mate, Henry Cruger, simply cried "ditto to Mr. Burke" when it was his turn to speak.

pleased with the Lago di Garda, and with Verona, particularly the amphitheatre, and a sarcophagus in a Convent garden, which they show as Juliet's; they insist on the *truth* of her history. Since my arrival at Venice, the lady of the Austrian governor told me that between Verona and Vicenza there are still ruins of the castle of the *Montecchi*,[16] and a chapel once appertaining to the Capulets. Romeo seems to have been of *Vicenza* by the tradition; but I was a good deal surprised to find so firm a faith in Bandello's novel, which seems really to have been founded on a fact.

Venice pleases me as much as I expected, and I expected much. It is one of those places which I know before I see them, and has always haunted me the most after the East. I like the gloomy gaiety of their gondolas, and the silence of their canals. I do not even dislike the evident decay of the city, though I regret the singularity of its vanished costume; however, there is much left still; the Carnival, too, is coming.

St. Mark's, and indeed Venice, is most alive at night. The theatres are not open till *nine*, and the society is proportionably late. All this is to my taste; but most of your countrymen miss and regret the rattle of hackney coaches, without which they can't sleep.

I have got remarkably good apartments in a private house; I see something of the inhabitants (having had a good many letters to some of them); I have got my gondola; I read a little, and luckily could speak Italian (more fluently though than accurately) long ago. I am studying, out of curiosity, the *Venetian* dialect, which is very naïve, and soft, and peculiar, though not at all classical; I go out frequently, and am in very good contentment.

The *Helen* of Canova (a bust which is in the house of Madame the Countess d'Albrizzi, whom I know) is, without exception, to my mind, the most perfectly beautiful of human conceptions, and far beyond my ideas of human execution.

> In this beloved marble view
>   Above the works and thoughts of Man,
> What Nature *could*, but *would not*, do,
>   And Beauty and Canova *can*!
> Beyond Imagination's power,
>   Beyond the Bard's defeated art,
> With Immortality her dower,
>   Behold the *Helen* of the *heart*!

Talking of the "heart" reminds me that I have fallen in love, which, except falling into the Canal (and that would be useless as I can swim), is the best (or worst) thing I could do. I am therefore in love—fathomless love; but lest you should make some splendid mistake, and envy me the possession of some of those princesses or countesses with whose affections your English voyagers are apt to invest themselves, I beg leave to tell you, that my goddess is only the wife of a "Merchant of Venice"; but then she is pretty as an Antelope, is but two-and-twenty years old, has the large, black, Oriental eyes, with the Italian countenance, and dark glossy hair, of the curl and colour of Lady Jersey's. Then she has the voice of a lute, and the song of a Seraph (though not quite so sacred), besides a long postscript of graces, virtues, and accomplishments, enough to furnish out a new chapter for Solomon's Song. But her great merit is finding out mine—there is nothing so amiable as discernment. Our little arrangement is completed; the usual oaths having been taken, and everything fulfilled according to the "understood relations" of such liaisons.

The general race of women appear to be handsome; but in Italy, as on almost all the Continent, the highest orders are by no means a well-looking generation, and indeed reckoned by their countrymen very much otherwise. Some are exceptions, but most of them as ugly as Virtue herself.

If you write, address to me here, *poste restante*, as I shall probably stay the winter over. I never see a newspaper, and know nothing of England, except in a letter now and then from my Sister. Of the M.S. sent you I know nothing, except that you have received it, and are to publish it, etc., etc.: but when, where, and how, you leave me to guess. But it don't much matter.

I suppose you have a world of works passing through your process for next year? When does Moore's poem[17] appear? I sent a letter for him, addressed to your care, the other day.

So Mr. Frere[18] is married; and you tell me in a former letter that he had "nearly forgotten that he was so." He is fortunate.

> Yours ever, and very truly,
> B.

---

16. **Montecchi:** the family of the Montagues in *Romeo and Juliet.*

17. **poem:** *Lalla Rookh.* 18. **Mr. Frere:** John Hookham Frere, 1769–1846, diplomat and author. He was married on September 12, 1816.

To John Murray
Rome, May 9, 1817.

MY DEAR SIR,

. . . Southey's *Wat Tyler*[19] is rather awkward, but the Goddess Nemesis has done well. He is—I will not say what, but I wish he was something else. I hate all intolerance, but most the intolerance of Apostacy, and the wretched vehemence with which a miserable creature, who has contradicted himself, lies to his own heart, and endeavours to establish his sincerity by proving himself a rascal—*not* for changing his opinions, but for persecuting those who are of less malleable matter. It is no disgrace to Mr. Southey to have written *Wat Tyler*, and afterwards to have written his birthday or Victory odes (I speak only of their *politics*), but it is something, for which I have no words, for this man to have endeavoured to bring to the stake (for such would he do) men who think as he thought, and for no reason but because they think so still, when he has found it convenient to think otherwise. Opinions are made to be changed, or how is truth to be got at? We don't arrive at it by standing on one leg, or on the first day of our setting out, but, though we may jostle one another on the way, that is no reason why we should strike or trample. *Elbowing's* enough. I am all for moderation, which profession of faith I beg leave to conclude by wishing Mr. Southey damned—not as a poet but as a politician. There is a place in Michael Angelo's last judgment in the Sistine Chapel which would just suit him, and may the like await him in that of our Lord and (*not his*) Saviour Jesus Christ— Amen! . . . .

Yours ever truly,

B. . . .

To John Murray
September 15, 1817.

DEAR SIR,

I enclose a sheet for correction, if ever you get to another edition. You will observe that the blunder in

printing makes it appear as if the Chateau[20] was *over* St. Gingo, instead of being on the opposite shore of the Lake, over Clarens. So, separate the paragraphs, otherwise my *to*pography will seem as inaccurate as your *ty*pography on this occasion.

The other day I wrote to convey my proposition with regard to the 4th and concluding canto.[21] I have gone over and extended it to one hundred and fifty stanzas, which is almost as long as the two first were originally, and longer by itself than any of the smaller poems except *The Corsair*. Mr. Hobhouse has made some very valuable and accurate notes of considerable length, and you may be sure I will do for the text all that I can to finish with decency. I look upon *Childe Harold* as my best; and as I begun, I think of concluding with it. But I make no resolution on that head, as I broke my former intention with regard to *The Corsair*. However, I fear that I shall never do better; and yet, not being thirty years of age, for some moons to come, one ought to be progressive as far as Intellect goes for many a good year. But I have had a devilish deal of wear and tear of mind and body in my time, besides having published too often and much already. God grant me some judgement! to do what may be most fitting in that and every thing else, for I doubt my own exceedingly.

I have read *Lallah Rookh*, but not with sufficient attention yet, for I ride about, and lounge, and ponder, and—two or three other things; so that my reading is very desultory, and not so attentive as it used to be. I am very glad to hear of its popularity, for Moore is a very noble fellow in all respects, and will enjoy it without any of the bad feeling which success—good or evil —sometimes engenders in the men of rhyme. Of the poem itself, I will tell you my opinion when I have mastered it: I say of the *poem*, for I don't like the *prose* at all—at all; and in the meantime, the "Fire worshippers" is the best, and the "Veiled Prophet" the worst, of the volume.

With regard to poetry in general, I am convinced, the more I think of it, that he and *all* of us—Scott, Southey, Wordsworth, Moore, Campbell, I,—are all in the wrong, one as much as another; that we are upon a wrong revolutionary poetical system, or systems, not worth a damn in itself, and from which none but Rogers and Crabbe are free; and that the present and next generations will finally be of this

---

19. Wat Tyler: a drama filled with revolutionary ardor. After composing it in 1794 Southey laid it aside and forgot about it. The manuscript was found and surreptitiously published in 1817, by which time Southey was poet laureate and a prominent writer for the Tory press. The embarrassing relic of his youthful sentiments gave malicious and unconcealed joy to the radicals.

20. Chateau: of Chillon, in Byron's "The Prisoner of Chillon." **21. canto:** of *Childe Harold*.

opinion. I am the more confirmed in this by having lately gone over some of our classics, particularly *Pope*, whom I tried in this way.—I took Moore's poems and my own and some others, and went over them side by side with Pope's, and I was really astonished (I ought not to have been so) and mortified at the ineffable distance in point of sense, harmony, effect, and even *Imagination*, passion and *Invention*, between the little Queen Anne's man, and us of the Lower Empire. Depend upon it, it is all Horace then, and Claudian now, among us; and if I had to begin again, I would model myself accordingly. Crabbe's the man, but he has got a coarse and impracticable subject, and Rogers, the Grandfather of living Poetry, is retired upon half-pay, (I don't mean as a Banker),—

> Since pretty Miss Jaqueline,
> With her nose aquiline,

and has done enough, unless he were to do as he did formerly.

To John Murray
October 12, 1817.

DEAR SIR,

. . . Many thanks for the *Edinburgh Review* which is very kind about *Manfred*, and defends its originality,[22] which I did not know that any body had attacked. I *never read*, and do not know that I ever saw, the *Faustus* of Marlow, and had, and have, no Dramatic works by me in English, except the recent things you sent me; but I heard Mr. Lewis translate verbally some scenes of *Goethe's Faust* (which were some good, and some bad) last summer;—which is all I know of the history of that magical personage; and as to the germs of *Manfred*, they may be found in the Journal which I sent to Mrs. Leigh (part of which you saw) when I went over first the Dent de Jamont [*sic*], and then the Wengeren [*sic*] or Wengeberg Alp and Sheideck and made the giro of the Jungfrau, Shreckhorn, etc., etc., shortly before I left Switzerland. I have the whole scene of *Manfred* before me, as if it was but yesterday, and could point it out, spot by spot, torrent and all.

22. **originality:** *Blackwood's Edinburgh Magazine* for July, 1817, had suggested that the general conception of *Manfred* and many particular features were borrowed from Marlowe's *Dr. Faustus*. In the issue of August, 1817, *The Edinburgh Review* came to Byron's defense: in *Dr. Faustus* there is to be found nothing "of the pride, the abstraction, and the heart-rooted misery in which [Byron's] originality consists."

Of the *Prometheus* of Aeschylus I was passionately fond as a boy (it was one of the Greek plays we read thrice a year at Harrow);—indeed that and the *Medea* were the only ones, except the *Seven before Thebes*, which ever much pleased me. As to the *Faustus* of Marlow, I never read, never saw, nor heard of it—at least, thought of it, except that I think Mr. Gifford mentioned, in a note of his which you sent me, something about the catastrophe; but not as having any thing to do with mine, which may or may not resemble it, for any thing I know.

The *Prometheus*, if not exactly in my plan, has always been so much in my head, that I can easily conceive its influence over all or any thing that I have written;—but I deny Marlow and his progeny, and beg that you will do the same. . . .

> Yours ever truly,
> B. . . .

To Hobhouse
Venice, November 11th, 1818.

DEAR HOBHOUSE,

By the favour of Lord Lauderdale (who tells me, by the way, that you have made some very good speeches, and are to turn out an orator—*seriously*), I have sent an "Oeuvre" of "Poeshie," which will not arrive probably till some time after this letter—though they start together—as the letter is rather the youngest of the two. It is addressed to you at Mr. Murray's. I request you to read—and having read—and if possible approved, to obtain the largest, or (if large be undeserved) the fairest price from him, or anyone else.

There are, firstly, the first canto of Don Juan (in the style of Beppo and Pulci—forgive me for putting Pulci *second*, it is a slip—"Ego et Rex meus") containing two hundred *Octaves*, and a dedication in verse of a dozen to Bob Southey, *bitter* as necessary—I mean the dedication; I will tell you why.

On his return from Switzerland, two years ago, he said that Shelley and I "had formed a League of Incest, and practised our precepts with, &c." He lied like a rascal, for they[23] *were not sisters, one* being Godwin's daughter, by Mary Wollstonecraft, and the other the daughter of the present Mrs. G. by a *former* husband. The attack contains no allusion to the cause; but some good verses, and all political and poetical.

Besides this "*pome*," there is "Mazeppa," and an

23. **they:** Mary Shelley and Claire Clairmont.

Ode on Venice. The last not very intelligible, and you may omit it if you like. Don Juan and Mazeppa are perhaps better; you will see.

The whole consists of between two and three thousand lines. You can consult Douglas K. about the price thereof, and your own judgement, and whose else you like, about their merits.

As one of the poems is as free as La Fontaine; and bitter in politics, too; the damned cant and Toryism of the day may make Murray pause; in that case you will take any bookseller who *bids best*. When I say *free*, I mean that freedom which Ariosto, Boiardo, and Voltaire—Pulci, Berni, all the best Italian and French—as well as Pope and Prior amongst the English—permitted themselves; but no improper words, nor *phrases;* merely some situations which are taken from life. However, you will see to all this when the MSS. arrive. . . .

B.

To John Murray
Venice, January 25, 1819.

DEAR SIR,

You will do me the favour to print privately (for private distribution) fifty copies of *Don Juan*. The list of the men to whom I wish it to be presented, I will send hereafter. The other two poems had best be added to the collective edition: I do not approve of *their* being published separately. *Print Don Juan entire*, omitting, of course, the lines on Castlereagh, as I am not on the spot to meet him. I have a second Canto ready, which will be sent by and bye. By this post, I have written to Mr. Hobhouse, addressed to your care.

Yours very truly,

B.

P.S.—I have acquiesced in the request and representation; and having done so, it is idle to detail my arguments in favour of my own Self-love and "Poeshie;" but I *protest*. If the poem has poetry, it would stand; if not, fall: the rest is "leather and prunella," and has never yet affected any human production "pro or con". Dullness is the only annihilator in such cases. As to the Cant of the day, I despise it, as I have ever done all its other finical fashions, which become you as paint became the Antient Britons. If you admit this prudery, you must omit half Ariosto, La Fontaine, Shakespeare, Beaumont, Fletcher, Massinger, Ford, all the Charles Second writers; in short,

something of most who have written before Pope and are worth reading, and much of Pope himself. *Read him*—most of you *don't*—but *do*—and I will forgive you; though the inevitable consequence would be that you would burn all I have ever written, and all your other wretched Claudians of the day (except Scott and Crabbe) into the bargain. I wrong Claudian, who *was* a *poet*, by naming him with such fellows; but he was the *ultimus Romanorum*, the tail of the Comet, and these persons are the tail of an old Gown cut into a waistcoat for Jackey; but being both *tails*, I have compared one with the other, though very unlike, like all Similies. I write in a passion and a Sirocco, and I was up till six this morning at the Carnival; but I *protest*, as I did in my former letter.

To John Murray
Venice, April 6, 1819.

DEAR SIR,

The Second Canto of *Don Juan* was sent, on Saturday last, by post, in 4 packets, two of 4, and two of three sheets each, containing in all two hundred and seventeen stanzas, octave measure. But I will permit no curtailments, except those mentioned about Castlereagh and the two *Bobs* in the Introduction. You sha'n't make *Canticles* of my Cantos. The poem will please, if it is lively; if it is stupid, it will fail; but I will have none of your damned cutting and slashing. If you please, you may publish *anonymously*; it will perhaps be better, but I will battle my way against them all, like a Porcupine.

So you and Mr. Foscolo,[24] etc., want me to undertake what you call a "great work?" an Epic poem, I suppose, or some such pyramid. I'll try no such thing; I hate tasks. And then "seven or eight years!" God send us all well this day three months, let alone years. If one's years can't be better employed than in sweating poesy, a man had better be a ditcher. And works, too!—is *Childe Harold* nothing? You have so many "*divine*" poems, is it nothing to have written a *Human* one? without any of your worn-out machinery. Why, man, I could have spun the thoughts of the four cantos of that poem into twenty, had I wanted to book-make, and its passion into as many modern tragedies. Since you want *length*, you shall have enough of *Juan*, for I'll make 50 cantos.

24. **Mr. Foscolo:** Ugo Foscolo, 1778–1827, Italian patriot and writer, who lived in England from 1816 until his death.

And Foscolo, too! Why does *he* not do something more than the *Letters of Ortis*, and a tragedy, and pamphlets? He has good fifteen years more at his command than I have: what has he done all that time:— proved his Genius, doubtless, but not fixed its fame, nor done his utmost.

Besides, I mean to write my best work in *Italian*, and it will take me nine years more thoroughly to master the language; and then if my fancy exist, and I exist too, I will try what I *can* do *really*. As to the Estimation of the English which you talk of, let them calculate what it is worth, before they insult me with their insolent condescension.

I have not written for their pleasure. If they are pleased, it is that they chose to be so; I have never flattered their opinions, nor their pride; nor will I. Neither will I make "Ladies books" *al dilettar le femine e la plebe.*[25] I have written from the fullness of my mind, from passion, from impulse, from many motives, but not for their "sweet voices."

I know the precise worth of popular applause, for few Scribblers have had more of it; and if I chose to swerve into their paths, I could retain it, or resume it, or increase it. But I neither love ye, nor fear ye; and though I buy with ye and sell with ye, and talk with ye, I will neither eat with ye, drink with ye, nor pray with ye. They made me, without my search, a species of popular Idol; they, without reason or judgement, beyond the caprice of their good pleasure, threw down the Image from its pedestal; it was not broken with the fall, and they would, it seems, again replace it—but they shall not.

You ask about my health: about the beginning of the year I was in a state of great exhaustion, attended by such debility of Stomach that nothing remained upon it; and I was obliged to reform my "way of life," which was conducting me from the "yellow leaf" to the Ground, with all deliberate speed. I am better in health and morals, and very much yours ever,

Bⁿ

P.S.—Tell Mrs. Leigh I have never had "my Sashes," and I want some tooth-powder, the red, by all or any means.

To John Murray
Venice, May 15, 1819.

DEAR SIR,

. . . The story of Shelley's agitation[26] is true. I can't tell what seized him, for he don't want courage. He was once with me in a gale of Wind, in a small boat, right under the rocks between Meillerie and St. Gingo. We were five in the boat—a servant, two boatmen, and ourselves. The sail was mismanaged, and the boat was filling fast. He can't swim. I stripped off my coat— made him strip off his and take hold of an oar, telling him that I thought (being myself an expert swimmer) I could save him, if he would not struggle when I took hold of him—unless we got smashed against the rocks, which were high and sharp, with an awkward surf on them at that minute. We were then about a hundred yards from shore, and the boat in peril. He answered me with the greatest coolness, that "he had no notion of being saved, and that I would have enough to do to save myself, and begged not to trouble me." Luckily, the boat righted, and, baling, we got round a point into St. Gingo, where the inhabitants came down and embraced the boatmen on their escape, the Wind having been high enough to tear up some huge trees from the Alps above us, as we saw next day.

And yet the same Shelley, who was as cool as it was possible to be in such circumstances, (of which I am no judge myself, as the chance of swimming naturally gives self possession when near shore), certainly had the fit of phantasy which Polidori describes, though *not exactly* as he describes it. . . .

I am, yours very truly,

B.

---

26. **agitation:** The story is told in the Preface to Dr. John Polidori's ghost story, "The Vampire": "It appears that one evening Lord B., Mr. P. B. Shelley, two ladies, and the gentleman before alluded to . . . began relating ghost stories, when his Lordship having recited the beginning of *Christabel*, then unpublished, the whole took so strong a hold of Mr. Shelley's mind, that he suddenly started up, and ran out of the room. The physician and Lord Byron followed, and discovered him leaning against a mantel-piece, with cold drops of perspiration trickling down his face. After having given him something to refresh him, upon enquiring the cause of his alarm, they found that his wild imagination having pictured to him the bosom of one of the ladies with eyes (which was reported of a lady in the neighbourhood where he lived), he was obliged to leave the room in order to destroy the impression." Polidori was for a while Byron's traveling companion and personal physician.

---

25. **al . . . plebe:** to please women and the mob.

To John Murray
Bologna, June 7, 1819.

DEAR SIR

. . . I have been picture-gazing this morning at the famous Domenichino and Guido, both of which are superlative. I afterwards went to the beautiful Cimetery of Bologna, beyond the walls, and found, besides the superb burial-ground, an original of a *Custode*, who reminded me of the grave-digger in Hamlet. He has a collection of Capuchins' skulls, labelled on the forehead, and taking down one of them, said, "This was Brother Desiderio Berro, who died at forty—one of my best friends. I begged his head of his brethren after his decease, and they gave it me. I put it in lime and then boiled it. Here it is, teeth and all, in excellent preservation. He was the merriest, cleverest fellow I ever knew. Wherever he went, he brought joy; and when any one was melancholy, the sight of him was enough to make him cheerful again. He walked so actively, you might have taken him for a dancer—he joked—he laughed—oh! he was such a Frate as I never saw before, nor ever shall again!"

He told me that he had himself planted all the cypresses in the Cimetery; that he had the greatest attachment to them and to his dead people; that since 1801 they had buried fifty three thousand persons. In showing some older monuments, there was that of a Roman girl of twenty, with a bust by Bernini. She was a princess Barberini, dead two centuries ago: he said that, on opening her grave, they had found her hair complete, and "as yellow as gold." Some of the epitaphs at Ferrara pleased me more than the more splendid monuments at Bologna; for instance—

> "Martini Luigi
> Implora pace."

> "Lucrezia Picini
> Implora eterna quiete."

Can any thing be more full of pathos? Those few words say all that can be said or sought: the dead had had enough of life; all they wanted was rest, and this they "*implore.*" There is all the helplessness, and humble hope, and deathlike prayer, that can arise from the grave—"*implora pace.*" I hope, whoever may survive me, and shall see me put in the foreigners' burying-ground at the Lido, within the fortress by the Adriatic, will see those two words, and no more, put over me. I

trust they won't think of "pickling, and bringing me home to Clod or Blunderbuss Hall."[27] I am sure my bones would not rest in an English grave, or my clay mix with the earth of that country. I believe the thought would drive me mad on my deathbed, could I suppose that any of my friends would be base enough to convey my carcass back to your soil. I would not even feed your worms, if I could help it. . . .

> Yours truly,
> B. . . .

To The Hon. Augusta Leigh
Ravenna, July 26th, 1819.

MY DEAREST AUGUSTA,

I am at too great a distance to scold you, but I *will* ask you whether *your* letter of the 1st July *is an answer* to the letter I wrote you before I quitted Venice? What? is it come to *this*? Have you no memory? or no heart? You *had* both—and I *have* both—at least for *you*.

I write this presuming that you received *that* letter. Is it that you fear? Do not be afraid of the past; the world has its own affairs without thinking of *ours* and you may write safely. If you do, address as usual to *Venice*. My house is not in St. Marc's but on the Grand Canal, within sight of the Rialto Bridge.

I do not like at all this pain in your side and always think of your mother's constitution. You must always be to me the first consideration in the world. Shall I come to *you*? or would a warm climate do you good? If so say the word, and I will provide you and your whole family (including that precious luggage your husband) with the means of making an agreeable journey. You need not fear about *me*. I am much altered and should be little trouble to you, nor would I give you more of my company than you like. I confess after three and a half—and *such years*! and *such a year* as preceded those three years!—it would be a relief to me to see you again, and if it would be so to you I will come to you. Pray answer me, and recollect that I will do as you like in everything, even to returning to England, which is not the pleasantest of residences were *you* out of it.

I write from Ravenna. I came here on account of a Countess Guiccioli, a girl of twenty married to a very rich old man of sixty about a year ago. With her last winter I had a *liaison* according to the good old Italian

27. "**pickling . . . Hall**": Sheridan's *The Rivals*, V. iii.

custom. She miscarried in May and sent for me here, and here I have been these two months. She is pretty, a great coquette, extremely vain, excessively affected, clever enough, without the smallest principle, with a good deal of imagination and some passion. She had set her heart on carrying me off from Venice out of vanity, and succeeded, and having made herself the subject of general conversation has greatly contributed to her recovery. Her husband is one of the richest nobles of Ravenna, three-score years of age. This is his third wife. You may suppose what *esteem* I entertain for *her*. Perhaps it is about equal on both sides. I have my saddle-horses here and there is good riding in the forest. With these, and my carriage which is here also, and the sea, and my books, and the lady, the time passes. I am very fond of riding and always *was out* of England. But I hate your Hyde Park, and your turnpike roads, and must have forests, downs, or desarts to expatiate in. I detest *knowing* the road one is to go, and being interrupted by your damned finger-posts, or a blackguard roaring for twopence at a turnpike.

I send you a sonnet which this faithful lady had made for the nuptials of one of her relations in which she swears the most *alarming constancy* to her husband. Is not this good? You may suppose my *face* when she shewed it to me. I could not help laughing—one of *our* laughs. All this is very absurd, but you see that I have good morals at bottom.

She is an equestrian too, but a bore in her rides, for she can't guide her horse and he runs after mine, and tries to bite him, and then she begins screaming in a high hat and sky-blue riding habit, making a most absurd figure, and embarrassing me and both our grooms, who have the devil's own work to keep her from tumbling, or having her clothes torn off by the trees and thickets of the pine forest. I fell a little in love with her intimate friend, a certain Geltruda (that is *Gertrude*) who is very young and seems very well disposed to be perfidious; but alas! *her* husband is jealous, and the G. also detected me in an illicit squeezing of hands, the consequence of which was that the friend was whisked off to Bologna for a few days, and since her return I have never been able to see her but twice, with a dragon of a mother in law and a barbarian husband by her side, besides my own dear precious *Amica*, who hates all flirting but her own. But I have a priest who befriends me and the Gertrude says a good deal with her great black eyes, so that perhaps . . . but alas! I mean to give up these things altogether. I have now given you some account of my present state. The guide-book will tell you about Ravenna. I can't tell how long or short may be my stay. Write to me—love me—as ever

Yours most affect^ly, B.

P.S.—*This* affair is *not* in the least expensive, being all in the wealthy line, but troublesome, for the lady is imperious, and exigeante. However there are hopes that we may quarrel. When we do you shall hear.

To John Murray
Bologna, August 29, 1819.

DEAR SIR,

I have been in a rage these two days, and am still bilious therefrom. You shall hear. A Captain of Dragoons, Ostheid, Hanoverian by birth, in the Papal troops at present, whom I had obliged by a loan when nobody would lend him a Paul, recommended a horse to me, on sale by a Lieutenant Rossi, an officer who unites the sale of cattle to the purchase of men. I bought it. The next day, on shoeing the horse, we discovered the *thrush*,[28]—the animal being warranted sound. I sent to reclaim the contract and the money. The Lieutenant desired to speak with me in person. I consented. He came. It was his own particular request. He began a story. I asked him if he would return the money. He said no—but he would exchange. He asked an exorbitant price for his other horses. I told him that he was a thief. He said he was an *officer* and a man of honour, and pulled out a Parmesan passport signed by General Count Neipperg. I answered, that as he was an officer, I would treat him as such; and that as to his being a Gentleman, he might prove it by returning the money: as for his Parmesan passport, I should have valued it more if it had been a Parmesan Cheese. He answered in high terms, and said that if it were in the *morning* (it was about eight o'clock in the evening) he would have *satisfaction*. I then lost my temper: "As for THAT," I replied, "you shall have it directly,—it will be *mutual* satisfaction, I can assure you. You are a thief, and, as you say, an officer; my pistols are in the next room loaded; take one of the candles, examine, and make your choice of weapons." He replied, that *pistols* were *English weapons*; *he* always fought with the *Sword*. I told him that I was able to accommodate him, having three regimental swords in a drawer near us; and he might take the longest and put himself on guard.

28. **thrush**: an infection of the hoof.

All this passed in presence of a third person. He then said *No*; but tomorrow morning he would give me the meeting at any time or place. I answered that it was not usual to appoint meetings in the presence of witnesses, and that we had best speak man to man, and fix time and instruments. But as the man present was leaving the room, the Lieutenant Rossi, before he could shut the door after him, ran out roaring "help and murder" most lustily, and fell into a sort of hysteric in the arms of about fifty people, who all saw that I had no weapon of any sort or kind about me, and followed him, asking him what the devil was the matter with him. Nothing would do: he ran away without his hat, and went to bed, ill of the fright. He then tried his complaint at the police, which dismissed it as frivolous. He is, I believe, gone away, or going.

The horse was warranted, but, I believe, so worded that the villain will not be obliged to refund, according to law. He endeavoured to raise up an indictment of assault and battery, but as it was in a public inn, in a frequented street, there were too many witnesses to the contrary; and, as a military man, he has not cut a martial figure, even in the opinion of the Priests. He ran off in such a hurry that he left his hat, and never missed it till he got to his hostel or inn. The facts are as I tell you: I can assure you, he began by "coming Captain Grand over me," or I should never have thought of trying his "cunning in fence;" but what could I do? He talked of "honour, and satisfaction, and his commission"—he produced a military passport: there are severe punishments for *regular duels* on the continent, and trifling ones for *rencontres*, so that it is best to fight it out directly; he had robbed, and then wanted to insult me;—what could I do? My patience was gone, and the weapons at hand, fair and equal: besides, it was just after dinner, when my digestion is bad, and I don't like to be disturbed. His friend Ostheid is at Forli; we shall meet on my way back to Ravenna. The Hanoverian seems the greater rogue of the two; and if my valour does not ooze away like Acres's— "Odds flints and triggers!"[29] if it should be a rainy morning, and my stomach in disorder, there may be something for the obituary.

Now pray, "Sir Lucius, do not you look upon me as a very ill used gentleman?"[30] I send my Lieutenant to match Hobhouse's *Major* Cartwright: "and so good

morrow to you, good Master Lieutenant." With regard to other things I will write soon, but I have ✱✱✱ incessantly for these last three months, and quarrelling and fooling till I can scribble no more.

Yours,

B.

To Kinnaird
Venice, Octr 26, 1819.

MY DEAR DOUGLAS,

My late expenditure has arisen from living at a distance from Venice and being obliged to keep up two establishments, from frequent journeys and buying some furniture and books as well as a horse or two— and not from any renewal of the *Epicurean* system as you suspect.

I have been faithful to my honest liaison with Countess Guiccioli and I can assure you that she has never cost me directly or indirectly a sixpence, indeed the circumstances of herself and family render this no merit. I never offered her but one present—a broach of brilliants and she sent it back to me with her *own hair* in it (✱✱✱✱ but *that* is an Italian custom) and a note to say that she was not in the habit of receiving presents of that value, but hoped that I would not consider her sending it back as an affront, nor the value diminished by the enclosure.

· · ·

As to subscribing to Manchester—if I do that I will write a letter to Burdett for publication to accompany the Subscription[31] which shall be more radical than anything yet rooted—but I feel lazy. I have thought of this for some time but, alas, the air of this cursed Italy enervates and disfranchises the thoughts of a man after nearly four years of respiration to say nothing of emission.

As to "Don Juan", confess, confess—you dog and be candid—that it is the sublime of *that there* sort of writing—it may be bawdy but is it not good English? It may be profligate but is it not *life*, is it not *the thing*? Could any man have written it who has not lived in the world?—and [f]ooled in a post-chaise?—in a hackney coach?—in a gondola?—against a wall?—in a court

---

29. "Odds . . . triggers!": *The Rivals*, III. iv: "Odds flints, pans, and triggers! I'll challenge him directly." **30. "Sir Lucius . . . gentleman?"**: *The Rivals*, III. iv.

31. Burdett . . . Subscription: A subscription was being collected to aid the victims of the Peterloo Massacre (see headnote to Shelley's "Mask of Anarchy"). Sir Francis Burdett took a leading part in the subsequent political attacks on the government.

carriage?—in a vis à vis?—on a table?—and under it? I have written about a hundred stanzas of a third Canto, but it is a damned modest—the outcry has frighted me. I have such projects for the Don but the Cant is so much stronger than the *, nowadays, that the benefit of experience in a man who had well weighed the worth of both monosyllables must be lost to despairing posterity. After all, what stuff this outcry is. Lalla Rookh and Little are more dangerous than my burlesque poem can be. Moore has been here; we got tipsy together, and were very amicable. He is gone on to Rome. I put my life (in MS.) into his hands (not for publication), you, or anybody else may see it at his return. It only comes up to 1816. He is a noble fellow and looks quite fresh and poetical, nine years (the age of a poem's education) my senior. He looks younger—this comes from marriage, and being settled in the country. I want to go to South America—I have written to Hobhouse all about it. I wrote to my wife three months ago, under cover to Murray. Has she got the letter—or is the letter got into Blackwood's Magazine?

You ask after my christmas pye. Remit it anyhow —circulars is the best. You are right about income—I must have it all. How the devil do I know that I may live a year or a month? I wish I knew, that I might regulate my spending in more ways than one. As it is one always thinks that there is but a span. A man may as well break or be damned for a large sum as a small one. I should be loth to pay the devil or any other creditor more than six pence in the pound.

              B.

To John Murray
Ravenna, February 21, 1820.

DEAR MURRAY,

   . . . By Saturday's post I sent you four packets, containing Cantos third and fourth of D[on] J[uan]; recollect that these two cantos reckon only as *one* with you and me, being, in fact, the third Canto cut into two, because I found it too long. Remember this, and don't imagine that there could be any other motive.[32] The whole is about 225 Stanzas, more or less, and a lyric of 96 lines, so that they are no longer than the first *single* cantos: but the truth is, that I made the first too long, and should have cut those down also had I thought better. Instead of saying in future for so many cantos, say so many *Stanzas* or pages: it was Jacob Tonson's way, and certainly the best: it prevents

32. motive: Byron was paid by the canto.

mistakes. I might have sent you a dozen cantos of 40 Stanzas each,—those of the *Minstrel* (Beattie's) are no longer,—and ruined you at once, if you don't suffer as it is; but recollect you are not *pinned down* to anything you say in a letter, and that, calculating even these two cantos as *one* only (which they were and are to be reckoned), you are not bound by your offer: act as may seem fair to all parties.

I have finished my translation of the first Canto of the "*Morgante Maggiore*" of Pulci, which I will transcribe and send: it is the parent, not only of *Whistlecraft*,[33] but of all jocose Italian poetry. You must print it side by side with the original Italian, because I wish the reader to judge of the fidelity: it is stanza for stanza, and often line for line, if not word for word.

You ask me for a volume of manners, etc., on Italy: perhaps I am in the case to know more of them than most Englishmen, because I have lived among the natives, and in parts of the country where Englishmen never resided before (I speak of Romagna and this place particularly); but there are many reasons why I do not choose to touch in print on such a subject. I have lived in their houses and in the heart of their families, sometimes merely as "*amico di casa,*" and sometimes as "*Amico di cuore*" of the *Dama*, and in neither case do I feel myself authorized in making a book of them. Their moral is not your moral; their life is not your life; you would not understand it; it is not English, nor French, nor German, which you would all understand. The Conventual education, the Cavalier Servitude, the habits of thought and living are so entirely different, and the difference becomes so much more striking the more you live intimately with them, that I know not how to make you comprehend a people, who are at once temperate and profligate, serious in their character and buffoons in their amusements, capable of impressions and passions, which are at once *sudden* and *durable* (what you find in no other nation), and who actually have *no society* (what we would call so), as you may see by their Comedies: they have no real comedy, not even in Goldoni; and that is because they have no Society to draw it from.

Their Conversazioni are not Society at all. They go to the theatre to talk, and into company to hold their tongues. The *women* sit in a circle, and the men gather into groupes, or they play at dreary *Faro* or "*Lotto*

33. Whistlecraft: Imitating the style of Pulci, John Hookham Frere wrote *Prospectus and Specimen of an Intended National Work, by William and Robert Whistlecraft.* Byron said that this was his immediate model in writing *Beppo.*

reale," for small sums. Their Académie are Concerts like our own, with better music and more form. Their best things are the Carnival balls and masquerades, when every body runs mad for six weeks. After their dinners and suppers, they make extempore verses and buffoon one another; but it is in a humour which you would not enter into, ye of the North.

In their houses it is better. I should know something of the matter, having had a pretty general experience among their women, from the fisherman's wife up to the *Nobil' Donna,* whom I serve. Their system has its rules, and its fitnesses, and decorums, so as to be reduced to a kind of discipline or game at hearts, which admits few deviations, unless you wish to lose it. They are extremely tenacious, and jealous as furies; not permitting their lovers even to marry if they can help it, and keeping them always close to them in public as in private whenever they can. In short, they transfer marriage to adultery, and strike the *not* out of that commandment. The reason is, that they marry for their parents, and love for themselves. They exact fidelity from a lover as a debt of honour, while they pay the husband as a tradesman, that is, not at all. You hear a person's character, male or female, canvassed, not as depending on their conduct to their husbands or wives, but to their mistress or lover. And—and—that's all. If I wrote a quarto, I don't know that I could do more than amplify what I have here noted. It is to be observed that while they do all this, the greatest outward respect is to be paid to the husbands, not only by the ladies, but by their *Serventi—* particularly if the husband serves no one himself (which is not often the case, however): so that you would often suppose them relations—the *Servente* making the figure of one adopted into the family. Sometimes the ladies run a little restive and elope, or divide, or make a scene; but this is at starting, generally, when they know no better, or when they fall in love with a foreigner, or some such anomaly,—and is always reckoned unnecessary and extravagant. . . .

Yours ever,

B.

To Richard Belgrave Hoppner
Ravenna, April 22.ᵈ 1820.

MY DEAR HOPPNER,

. . . About Allegra,[34] I can only say to Claire—that I so totally disapprove of the mode of Children's

34. **Allegra:** daughter of Byron and Claire Clairmont.

treatment in their family,[35] that I should look upon the Child as going into a hospital. Is it not so? Have they *reared* one?[36] Her health here has hitherto been *excellent,* and her temper not bad; she is sometimes vain and obstinate, but always clean and cheerful, and as, in a year or two, I shall either send her to England, or put her in a Convent for education, these defects will be remedied as far as they can in human nature. But the Child shall not quit me again to perish of Starvation, and green fruit, or be taught to believe that there is no Deity. Whenever there is convenience of vicinity and access, her Mother can always have her with her; otherwise no. It was so stipulated from the beginning.

The Girl is not so well off as with you, but far better than with them; the fact is she is spoilt, being a great favourite with every body on account of the fairness of her Skin, which shines among their dusky children like the milky way, but there is no comparison of her situation now, and that under Elise, or with them. She has grown considerably, is very clean, and lively. She has plenty of air and exercise at home, and she goes out daily with Mᶜ Guiccioli in her carriage to the Corso.

The paper is finished and so must the letter be.

Yours ever,

B.

My best respects to Mrs. H. and the little boy—and Dorville.

To John Murray
Ravenna, February 16, 1821.

DEAR MORAY,

. . . The 5ᵗʰ is so far from being the last of *D. J.,* that it is hardly the beginning. I meant to take him the tour of Europe, with a proper mixture of siege, battle, and adventure, and to make him finish as *Anacharsis Cloots* in the French revolution. To how many cantos this may extend, I know not, nor whether (even if I live) I shall complete it; but this was my notion: I meant to have made him a *Cavalier Servente* in Italy, and a cause for a divorce in England, and a Sentimental "Werther-faced man" in Germany, so as to show the different ridicules of the society in each of those countries, and to have displayed him gradually *gâté*

35. **family:** the Shelleys. 36. **reared one:** Shelley and his wife had lost three children.

and *blasé* as he grew older, as is natural. But I had not quite fixed whether to make him end in Hell, or in an unhappy marriage, not knowing which would be the severest. The Spanish tradition says Hell: but it is probably only an Allegory of the other state. You are now in possession of my notions on the subject. . . .

Yours, ever and truly,

B.

To Percy Bysshe Shelley
Ravenna, April 26, 1821.

The child[37] continues doing well, and the accounts are regular and favourable. It is gratifying to me that you and Mrs. Shelley do not disapprove of the step which I have taken, which is merely temporary.

I am very sorry to hear what you say of Keats—is it *actually* true? I did not think criticism had been so killing. Though I differ from you essentially in your estimate of his performances, I so much abhor all unnecessary pain, that I would rather he had been seated on the highest peak of Parnassus than have perished in such a manner. Poor fellow! though with such inordinate self-love he would probably have not been very happy. I read the review of *Endymion* in the *Quarterly*. It was severe,—but surely not so severe as many reviews in that and other journals upon others.

I recollect the effect on me of the *Edinburgh* on my first poem;[38] it was rage, and resistance, and redress—but not despondency nor despair. I grant that those are not amiable feelings; but, in this world of bustle and broil, and especially in the career of writing, a man should calculate upon his powers of *resistance* before he goes into the arena.

"Expect not life from pain nor danger free,
    Nor deem the doom of man reversed for thee."[39]

You know my opinion of *that second-hand* school of poetry. You also know my high opinion of your own poetry,—because it is of *no school.* I read *Cenci*—but, besides that I think the *subject* essentially *un*dramatic, I am not an admirer of our old dramatists, *as models.* I deny that the English have hitherto had a drama at all. Your *Cenci,* however, was a work of power, and

poetry. As to *my* drama, pray revenge yourself upon it, by being as free as I have been with yours.

I have not yet got your *Prometheus,* which I long to see. I have heard nothing of mine, and do not know that it is yet published. I have published a pamphlet on the Pope controversy, which you will not like. Had I known Keats was dead—or that he was alive and so sensitive—I should have omitted some remarks upon his poetry, to which I was provoked by his *attack* upon *Pope,* and my disapprobation of *his own* style of writing.

You want me to undertake a great poem—I have not the inclination nor the power. As I grow older, the indifference—*not* to life, for we love it by instinct—but to the stimuli of life, increases. Besides, this late failure of the Italians has latterly disappointed me for many reasons,—some public, some personal. My respects to Mrs. S.

Yours ever,

B.

P.S.—Could not you and I contrive to meet this summer? Could not you take a run here *alone*?

To John Murray
Ravenna, September 24th 1821.

DEAR MURRAY,

I have been thinking over our late correspondence, and wish to propose to you the following articles for our future:—

1stly That you shall write to me of yourself, of the health, wealth, and welfare of all friends; but of *me* (*quoad me*) little or nothing.

2dly That you shall send me Soda powders, tooth-powder, tooth-brushes, or any such anti-odontalgic or chemical articles, as heretofore, *ad libitum,* upon being re-imbursed for the same.

3dly That you shall *not* send me any modern, or (as they are called) *new,* publications in *English* whatsoever, save and excepting any writing, prose or verse, of (or reasonably presumed to be of) Walter Scott, Crabbe, Moore, Campbell, Rogers, Gifford, Joanna Baillie, *Irving* (the American), Hogg, Wilson (*Isle of Palms* Man), or *any* especial *single* work of fancy which is thought to be of considerable merit; *Voyages* and *travels,* provided that they are *neither in Greece, Spain, Asia Minor, Albania, nor Italy,* will be welcome: having

37. **child:** Allegra had been placed in a convent to be educated. 38. **Edinburgh . . . poem:** an unfavorable review of *Hours of Idleness* (1809) in *The Edinburgh Review.* 39. **"Expect . . . thee":** from Samuel Johnson's *Vanity of Human Wishes,* ll. 155-56.

travelled the countries mentioned, I know that what is said of them can convey nothing further which I desire to know about them.—No other English works whatsoever.

4^{thly} That you send me *no periodical works* whatsoever—*no Edinburgh, Quarterly, Monthly,* nor any Review, Magazine, Newspaper, English or foreign, of any description.

5th,ly That you send me *no* opinions whatsoever, either *good, bad,* or *indifferent,* of yourself, or your friends, or others, concerning any work, or works, of mine, past, present, or to come.

6^{thly} That all negotiations in matters of business between you and me pass through the medium of the Honble Douglas Kinnaird, my friend and trustee, or Mr. Hobhouse, as *Alter Ego,* and tantamount to myself during my absence, or presence.

Some of these propositions may at first seem strange, but they are founded. The quantity of trash I have received as books is incalculable, and neither amused nor instructed. Reviews and Magazines are at the best but ephemeral and superficial reading: *who thinks* of the *grand article* of *last year* in any *given review*? in the next place, if they regard *myself,* they tend to increase *Egotism*; if favourable, I do not deny that the praise *elates,* and if unfavourable, that the abuse *irritates*—the latter may conduct me to inflict a species of Satire, which would neither do good to you nor to your friends: *they* may smile *now,* and so may *you*; but if I took you all in hand, it would not be difficult to cut you up like gourds. I did as much by as powerful people at nineteen years old,[40] and I know little as yet, in three and thirty, which should prevent me from making all your ribs Gridirons for your hearts, if such were my propensity. But it is *not*: Therefore let me hear none of your provocations. If any thing occurs so very *gross* as to require my notice, I shall hear of it from my personal friends. For the rest, I merely request to be left in ignorance.

The same applies to opinions, *good, bad,* or *indifferent,* of persons in conversation or correspondence: these do not *interrupt,* but they *soil* the *current* of my *Mind.* I am sensitive enough, but *not* till I am *touched*; and *here* I am beyond the touch of the short arms of literary England, except the few feelers of the Polypus that crawl over the Channel in the way of Extract.

All these precautions *in* England would be useless: the libeller or the flatterer would there reach me in spite of all; but in Italy we know little of literary England, and think less, except what reaches us through some garbled and brief extract in some miserable Gazette. For *two years* (excepting two or three articles cut out and sent to *you,* by the post) I never read a newspaper which was not forced upon me by some accident, and know, upon the whole, as little of England as you all do of Italy, and God knows *that* is little enough, with all your travels, etc., etc., etc. The English travellers *know Italy* as *you* know Guernsey: how much is *that*?

If any thing occurs so violently gross or personal as to require notice, Mr. D^s Kinnaird will let me *know;* but of *praise* I desire to hear *nothing.*

You will say, "to what tends all this?" I will answer THAT;—to keep my mind *free and unbiassed* by all paltry and personal irritabilities of praise or censure;—to let my Genius take its natural direction, while my feelings are like the dead, who know nothing and feel nothing of all or aught that is said or done in their regard.

If you can observe these conditions, you will spare yourself and others some pain: let me not be worked upon to rise up; for if I do, it will not be for a little: if you can *not* observe these conditions, we shall cease to be correspondents, but *not friends*; for I shall always be

Yours ever and truly,

BYRON.

P.S.—I have taken these resolutions not from any irritation against *you* or *yours,* but simply upon reflection that all reading, either praise or censure, of myself has done me harm. When I was in Switzerland and Greece, I was out of the way of hearing either, and *how I wrote there*! In Italy I am out of the way of it too; but latterly, partly through my fault, and partly through your kindness in wishing to send me the *newest* and most periodical publications, I have had a crowd of reviews, etc., thrust upon me, which have bored me with their jargon, of one kind or another, and taken off my attention from greater objects. You have also sent me a parcel of trash of poetry, for no reason that I can conceive, unless to provoke me to write a new *English Bards.* Now *this* I wish to avoid; for if ever I *do,* it will be a strong production; and I desire peace, as long as the fools will keep their nonsense out of my way.

---

40. I . . . **old:** in *English Bards and Scotch Reviewers.*

To Thomas Moore
Pisa, March 4, 1822.

. . . I am sorry you think *Werner* even *approaching* to any fitness for the stage, which, with my notions upon it, is very far from my present object. With regard to the publication, I have already explained that I have no exorbitant expectations of either fame or profit in the present instances; but wish them published because they are written, which is the common feeling of all scribblers.

With respect to "Religion," can I never convince you that *I* have no such opinions as the characters in that drama, which seems to have frightened every body? Yet *they* are nothing to the expressions in Goethe's *Faust* (which are ten times hardier), and not a whit more bold than those of Milton's Satan. My ideas of a character may run away with me: like all imaginative men, I, of course, embody myself with the character while I *draw* it, but not a moment after the pen is from off the paper.

I am no enemy to religion, but the contrary. As a proof, I am educating my natural daughter a strict Catholic in a convent of Romagna; for I think people can never have *enough* of religion, if they are to have any. I incline, myself, very much to the Catholic doctrines; but if I am to write a drama, I must make my characters speak as I conceive them likely to argue.

As to poor Shelley, who is another bugbear to you and the world, he is, to my knowledge, the *least* selfish and the mildest of men—a man who has made more sacrifices of his fortune and feelings for others than any I ever heard of. With his speculative opinions I have nothing in common, nor desire to have.

The truth is, my dear Moore, you live near the *stove* of society, where you are unavoidably influenced by its heat and its vapours. I did so once—and too much —and enough to give a colour to my whole future existence. As my success in society was *not* inconsiderable, I am surely not a prejudiced judge upon the subject, unless in its favour; but I think it, as now constituted, *fatal* to all great original undertakings of every kind. I never courted it *then*, when I was young and high in blood, and one of its "curled darlings;" and do you think I would do so *now*, when I am living in a clearer atmosphere? One thing *only* might lead me back to it, and that is, to try once more if I could do any good in *politics*; but *not* in the petty politics I see now preying upon our miserable country.

Do not let me be misunderstood, however. If you speak your *own* opinions, they ever had, and will have, the greatest weight with *me*. But if you merely *echo* the *monde*, (and it is difficult not to do so, being in its favour and its ferment,) I can only regret that you should ever repeat any thing to which I cannot pay attention.

But I am prosing. The gods go with you, and as much immortality of all kinds as may suit your present and all other existence.

Yours,

To Lady——
Atbaro, November 10, 1822.

The Chevalier persisted in declaring himself an ill-used gentleman, and describing you as a kind of cold Calypso, who lead astray people of an amatory disposition without giving them any sort of compensation, contenting yourself, it seems, with only making *one* fool instead of two, which is the more approved method of proceeding on such occasions. For my part, I think you are quite right; and be assured from me that a woman (as society is constituted in England) who gives any advantage to a man may expect a lover, but will sooner or later find a tyrant; and this is not the man's fault either, perhaps, but is the necessary and natural result of the circumstances of society, which, in fact, tyrannise over the man equally with the woman; that is to say, if either of them have any feeling or honour.

You can write to me at your leisure and inclination. I have always laid it down as a maxim, and found it justified by experience, that a man and a woman make far better friendships than can exist between two of the same sex; but *these* with this condition, that they never have made, or are to make, love with each other. Lovers may, and, indeed, generally *are* enemies, but they can be friends; because there must always be a spice of jealousy and a something of self in all their speculations.

Indeed, I rather look upon love altogether as a sort of hostile transaction, very necessary to make or to break matches, and keep the world going, but by no means a sinecure to the parties concerned.

Now, as my love perils are, I believe, pretty well over, and yours, by all accounts, are never to begin, we shall be the best friends imaginable, as far as both are concerned; and with this advantage, that we may both fall to loving right and left through all our acquaintance, without either sullenness or sorrow from that amiable passion, which are its inseparable attendants.

Believe me, etc.,

N. B.[41]

To Henri Beyle
Genoa, May 29, 1823.

SIR,

At present, that I know to whom I am indebted for a very flattering mention in the *Rome, Naples, and Florence*, in 1817, by Mons. Stendhal, it is fit that I should return my thanks (however undesired or undesirable) to Mons. Beyle, with whom I had the honour of being acquainted at Milan, in 1816. You only did me too much honour in what you were pleased to say in that work; but it has hardly given me less pleasure than the praise itself, to become at length aware (which I have done by mere accident) that I am indebted for it to one of whose good opinion I was really ambitious. So many changes have taken place since that period in the Milan circle, that I hardly dare recur to it;—some dead, some banished, and some in the Austrian dungeons.—Poor Pellico![42] I trust that, in his iron solitude, his Muse is consoling him in part—one day to delight us again, when both she and her Poet are restored to freedom.

Of your works I have only seen *Rome*, etc., the Lives of Haydn and Mozart, and the *brochure* on Racine and Shakespeare. The *Histoire de la Peinture* I have not yet the good fortune to possess.

There is one part of your observations in the pamphlet which I shall venture to remark upon;—it regards Walter Scott. You say that "his character is little worthy of enthusiasm," at the same time that you mention his productions in the manner they deserve. I have known Walter Scott long and well, and in occasional situations which call forth the *real* character —and I can assure you that his character *is* worthy of admiration—that of all men he is the most *open*, the most *honourable*, the most *amiable*. With his politics I have nothing to do: they differ from mine, which renders it difficult for me to speak of them. But he is

perfectly sincere in them: and Sincerity may be humble, but she cannot be servile. I pray you, therefore, to correct or soften that passage. You may, perhaps, attribute this officiousness of mine to a false affectation of *candour*, as I happen to be a writer also. Attribute it to what motive you please, but *believe the truth*. I say that Walter Scott is as nearly a thorough good man as man can be, because I *know* it by experience to be the case.

If you do me the honour of an answer, may I request a speedy one?—because it is possible (though not yet decided) that circumstances may conduct me once more to Greece. My present address is Genoa, where an answer will reach me in a short time, or be forwarded to me wherever I may be.

I beg you to believe me, with a lively recollection of our brief acquaintance, and the hope of one day renewing it,

Your ever obliged

and obedient humble servant,

NOEL BYRON.

To Johann Wolfgang Von Goethe
Leghorn, July 24, 1823.

ILLUSTRIOUS SIR,

I cannot thank you as you ought to be thanked for the lines which my young friend, Mr. Sterling, sent me of yours; and it would but ill become me to pretend to exchange verses with him who, for fifty years, has been the undisputed sovereign of European literature. You must therefore accept my most sincere acknowledgements in prose—and in hasty prose too; for I am at present on my voyage to Greece once more, and surrounded by hurry and bustle, which hardly allow a moment even to gratitude and admiration to express themselves.

I sailed from Genoa some days ago, was driven back by a gale of wind, and have since sailed again and arrived here, "Leghorn," this morning, to receive on board some Greek passengers for their struggling country.

Here also I found your lines and Mr. Sterling's letter; and I could not have had a more favourable omen, a more agreeable surprise, than a word of Goethe, written by his own hand.

I am returning to Greece, to see if I can be of any little use there: if ever I come back, I will pay a visit to Weimar, to offer the sincere homage of one of the

---

41. **N. B.:** When Lady Noel, Byron's mother-in-law, died, he inherited the Noel coat of arms and henceforth often signed his name Noel Byron. 42. **Pellico:** Silvio Pellico, 1788–1854, a writer and political liberal, was imprisoned at the outbreak of the Neapolitan Revolution in 1820.

many millions of your admirers. I have the honour to be, ever and most respectfully, y[our]

> Obliged adm[irer] and se[rvant],
> NOEL BYRON.

To The Countess Guiccioli
October 7 [1823].

Pietro[43] has told you all the gossip of the island,—our earthquakes, our politics, and present abode in a pretty village. As his opinions and mine on the Greeks are nearly similar, I need say little on that subject. I was a fool to come here; but, being here, I must see what is to be done.

October —.

We are still in Cephalonia, waiting for news of a more accurate description; for all is contradiction and division in the reports of the state of the Greeks. I shall fulfil the object of my mission from the Committee, and then return into Italy; for it does not seem likely that, as an individual, I can be of use to them;—at least no other foreigner has yet appeared to be so, nor does it seem likely that any will be at present.

Pray be as cheerful and tranquil as you can; and be assured that there is nothing here that can excite any thing but a wish to be with you again,—though we are very kindly treated by the English here of all descriptions. Of the Greeks, I can't say much good hitherto, and I do not like to speak ill of them, though they do of one another.

October 29.

You may be sure that the moment I can join you again, will be as welcome to me as at any period of our recollection. There is nothing very attractive here to divide my attention; but I must attend to the Greek cause, both from honour and inclination. Messrs. B[rowne] and T[relawny] are both in the Morea, where they have been very well received, and both of them write in good spirits and hopes. I am anxious to hear how the Spanish cause will be arranged, as I think it may have an influence on the Greek contest. I wish that both were fairly and favourably settled, that I might return to Italy, and talk over with you *our*, or rather Pietro's adventures, some of which are rather amusing, as also some of the incidents of our

voyages and travels. But I reserve them, in the hope that we may laugh over them together at no very distant period.

To Hobhouse
10bre 27th, 1823.

DEAR HOBHOUSE,

I embark for Messalonghi. Douglas K[innair]d and Bowring can tell you the rest. I particularly require and entreat you to desire Douglas K[innair]d to send me credits to the uttermost, that I may get the Greeks to keep the field. Never mind *me*, so that the *cause goes on*; if that is well, all is well. Douglas must send me *my money* (Rochdale Manor included, if the sale is completed, and the purchase money paid); the Committee must furnish *their* money, and the monied people *theirs*; with these we will soon have men enough, and all that.

> Yrs. ever, N. B.

P.S. Mavrocordato's[44] letter says, that my presence will "*electrify* the troops," so I am going over to "electrify" the Suliotes, as George Primrose went to Holland "to teach the Dutch English, who were fond of it to distraction."[45]

To Thomas Moore
Cephalonia, December 27, 1823.

I received a letter from you some time ago. I have been too much employed latterly to write as I could wish, and even now must write in haste.

I embark for Missolonghi to join Mavrocordato in four-and-twenty hours. The state of parties (but it were a long story) has kept me here till *now*; but now that Mavrocordato (their Washington, or their Kosciusko) is employed again, I can act with a *safe conscience*. I carry money to pay the squadron, etc., and I have influence with the Suliotes, *supposed* sufficient to keep them in harmony with some of the dissentients; —for there are plenty of differences, but trifling.

---

43. Pietro: Count Gamba, brother of the Countess Guiccioli.

44. Mavrocordato: Prince Alexander Mavrocordato, 1791–1865, a Greek revolutionary leader. **45. George . . . distraction:** In Goldsmith's novel *The Vicar of Wakefield*, George Primrose, son of the Vicar, is persuaded to sail to Holland to teach the Dutch English. He forgets that he himself will have to know Dutch.

It is imagined that we shall attempt either Patras, or the castles on the Straits; and it seems, by most accounts, that the Greeks, at any rate, the Suliotes, who are in affinity with me of "bread and salt,"—expect that I should march with them, and—be it even so! If any thing in the way of fever, fatigue, famine, or otherwise, should cut short the middle age of a brother warbler,—like Garcilasso de la Vega, Kleist, Korner, Joukoffsky (a Russian nightingale—see Bowring's Anthology), or Thersander, or,—or somebody else—but never mind—I pray you to remember me in your "smiles and wine."[46]

I have hopes that the cause will triumph; but, whether it does or no, still "Honor must be minded as strictly as milk diet." I trust to observe both.

Ever, etc.
BYRON.

To His Highness Yusuff Pasha
Missolonghi, January 23, 1824.

HIGHNESS!

A vessel, in which a friend and some domestics of mine were embarked, was detained a few days ago, and released by order of your Highness. I have now to thank you; not for liberating the vessel, which, as carrying a neutral flag, and being under British protection, no one had a right to detain; but for having treated my friends with so much kindness while they were in your hands.

In the hope, therefore, that it may not be altogether displeasing to your Highness, I have requested the governor of this place to release four Turkish prisoners, and he has humanely consented to do so. I lose no time, therefore, in sending them back, in order to make as early a return as I could for your courtesy on the late occasion. These prisoners are liberated without any conditions: but should the circumstance find a place in your recollection, I venture to beg, that your Highness will treat such Greeks as may henceforth fall into your hands with humanity; more especially since the horrors of war are sufficiently great in themselves, without being aggravated by wanton cruelties on either side.

NOEL BYRON.

---

46. "smiles and wine": Tom Moore's "The Legacy."

To The Hon. Augusta Leigh
Messolonghi, [Monday] Feb 23ᵈ *1824.*

MY DEAREST AUGUSTA,

I received a few days ago yours and Lady B's report of Ada's health, with other letters from England for which I ought to be and am (I hope) sufficiently thankful, as they were of great comfort and I wanted some, having been recently unwell, but am now much better. So that you need not be alarmed.

You will have heard of our journeys and escapes, and so forth, perhaps with some exaggeration; but it is all very well now, and I have been for some time in Greece, which is in as good a state as could be expected considering circumstances. But I will not plague you with politics, wars, or earthquakes, though we had another very smart one three nights ago, which produced a scene ridiculous enough, as no damage was done except to those who stuck fast in the scuffle to get first out of the doors or windows, amongst whom some recent importations, fresh from England, who had been used to quieter elements, were rather squeezed in the press for precedence.

I have been obtaining the release of about nine and twenty Turkish prisoners—men, women, and children—and have sent them at my own expense home to their friends, but one, a pretty little girl of nine years of age named Hato or Hatagée, has expressed a strong wish to remain with me, or under my care, and I have nearly determined to adopt her. If I thought that Lady B. would let her come to England as a Companion to Ada—(they are about the same age), and we could easily provide for her; if not, I can send her to Italy for education. She is very lively and quick, and with great black oriental eyes, and Asiatic features. All her brothers were killed in the Revolution; her mother wishes to return to her husband who is at Prevesa, but says that she would rather entrust the child to me in the present state of the Country. Her extreme youth and sex have hitherto saved her life, but there is no saying what might occur in the course of the war (and of such a war), and I shall probably commit her to the charge of some English lady in the islands for the present. The Child herself has the same wish, and seems to have a decided character for her age. You can mention this matter if you think it worth while. I merely wish her to be respectably educated and treated, and, if my years and all things be considered, I presume it would be difficult to conceive me to have any other views.

With regard to Ada's health, I am glad to hear that it is so much better. But I think it right that Lady B. should be informed, and guard against it accordingly, that her description of much of her indisposition and tendencies very nearly resemble my own at a similar age, except that I was much more impetuous. Her preference of prose (strange as it may seem) was and indeed is mine (for I hate reading verse, and always did), and I never invented anything but "boats—ships" and generally relating to the Ocean. I shewed the report to Col. Stanhope, who was struck with the resemblance of parts of it to the paternal line even now. But it is also fit, though unpleasant, that I should mention that my recent attack, and a very severe one, had a strong appearance of epilepsy. Why—I know not, for it is late in life—its first appearance at thirty-six— and, as far as I know, it is not hereditary, and it is that it may not become so, that you should tell Lady B. to take some precautions in the case of Ada. My attack has not yet returned, and I am fighting it off with abstinence and exercise, and thus far with success; if merely casual, it is all very well.[47]

EDWARD JOHN TRELAWNY

from RECOLLECTIONS

༺◆༻

Edward John Trelawny (1792–1881) was himself an adventurer who might have stepped out of a Byronic tale (see *Adventures of a Younger Son*, 1831, his freely inventive autobiography). He first met Shelley and Byron in January, 1822, when both were living in Pisa. His *Recollections of the Last Days of Shelley and Byron* (1858) is easily the most vivid of the many descriptions of the two poets in Italy. It should be kept in mind, however, that Trelawny was writing many years after the events he describes and also that he disliked Byron. As a result, his recollections can be both biased and inaccurate. Nevertheless, the overall portrait of Byron in Italy closely resembles that given by other acquaintances. For further extracts from Trelawny see under Shelley, below, p. 1096.

༺◆༻

47. **well:** This letter was found unfinished on Byron's writing table after his death.

In external appearance Byron realised that ideal standard with which imagination adorns genius. He was in the prime of life, thirty-five; of middle height, five feet eight and a half inches; regular features, without a stain or furrow on his pallid skin, his shoulders broad, chest open, body and limbs finely proportioned. His small, highly-finished head and curly hair, had an airy and graceful appearance from the massiveness and length of his throat: you saw his genius in his eyes and lips. In short, Nature could do little more than she had done for him, both in outward form and in the inward spirit she had given to animate it. But all these rare gifts to his jaundiced imagination only served to make his one personal defect (lameness) the more apparent, as a flaw is magnified in a diamond when polished; and he brooded over that blemish as sensitive minds will brood until they magnify a wart into a wen.

His lameness certainly helped to make him sceptical, cynical, and savage. There was no peculiarity in his dress, it was adapted to the climate; a tartan jacket braided,—he said it was the Gordon pattern, and that his mother was of that ilk. A blue velvet cap with a gold band, and very loose nankeen trousers, strapped down so as to cover his feet: his throat was not bare, as represented in drawings. At three o'clock, one of his servants announced that his horses were at the door, which broke off his discussion with Shelley, and we all followed him to the hall. At the outer door, we found three or four very ordinary-looking horses; they had holsters on the saddles, and many other superfluous trappings, such as the Italians delight in, and Englishmen eschew. Shelley, and an Irish visitor just announced, mounted two of these sorry jades. I luckily had my own cattle. Byron got into a calèche, and did not mount his horse until we had cleared the gates of the town, to avoid, as he said, being stared at by the "d—d Englishers", who generally congregated before his house on the Arno. After an hour or two of slow riding and lively talk,—for he was generally in good spirits when on horseback,—we stopped at a small *podere* on the roadside, and dismounting went into the house, in which we found a table with wine and cakes. From thence we proceeded into the vineyard at the back; the servant brought two brace of pistols, a cane was stuck in the ground and a five-paul piece, the size of half-a-crown, placed in a slit at the top of the cane. Byron, Shelley, and I, fired at fifteen paces, and one of us generally hit the cane or the coin: our firing was pretty equal; after five or six shots each, Byron pocketed the battered money and sauntered about the grounds. We then remounted. On our return homewards, Shelley urged Byron to complete something he had begun. Byron smiled and replied:

"John Murray, my patron and paymaster, says my plays won't act. I don't mind that, for I told him they were not written for the stage——but he adds, my poesy won't sell: that I do mind, for I have an 'itching palm'. He urges me to resume my old 'Corsair style, to please the ladies'."

Shelley indignantly answered:

"That is very good logic for a bookseller, but not for an author: the shop interest is to supply the ephemeral demand of the day. It is not for him but for you 'to put a ring in the monster's nose' to keep him from mischief."

Byron smiling at Shelley's warmth, said:

"John Murray is right, if not righteous: all I have yet written has been for women-kind; you must wait until I am forty, their influence will then die a natural death, and I will show the men what I can do."

Shelley replied:

"Do it now—write nothing but what your conviction of its truth inspires you to write; you should give counsel to the wise, and not take it from the foolish. Time will reverse the judgment of the vulgar. Cotemporary criticism only represents the amount of ignorance genius has to contend with."

I was then and afterwards pleased and surprised at Byron's passiveness and docility in listening to Shelley —but all who heard him felt the charm of his simple, earnest manner; while Byron knew him to be exempt from the egotism, pedantry, coxcombry, and, more than all, the rivalry of authorship, and that he was the truest and most discriminating of his admirers.

Byron looking at the western sky, exlaimed:

"Where is the green your friend the Laker talks such fustian about," meaning Coleridge:

"Gazing on the western sky,
And its peculiar tint of yellow green."
                    "Dejection: an Ode" [ll. 28–29].

"Who ever," asked Byron, "saw a green sky?"

Shelley was silent, knowing that if he replied, Byron would give vent to his spleen. So I said, "The sky in England is oftener green than blue."

"Black, you mean," rejoined Byron; and this discussion brought us to his door.

As he was dismounting he mentioned two odd words that would rhyme. I observed on the felicity he had shown in this art, repeating a couplet out of *Don Juan;* he was both pacified and pleased at this, and putting his hand on my horse's crest, observed:

"If you are curious in these matters, look in Swift. I will send you a volume; he beats us all hollow, his rhymes are wonderful."

And then we parted for that day, which I have been thus particular in recording, not only as it was the first of our acquaintance, but as containing as fair a sample as I can give of his appearance, ordinary habits, and conversation.

.   .   .

In 1809, he first left England, rode on horseback through Spain and Portugal, 400 miles, crossed the Mediterranean on board a frigate, and landed in Greece; where he passed two years in sauntering through a portion of that small country: this, with a trip to Smyrna, Constantinople, Malta, and Gibraltar, generally on board our men-of-war, where you have all the ease, comfort, and most of the luxuries of your own homes;—this is the extent of the voyages and travels he was so proud of. Anything more luxurious than sailing on those seas, and riding through those lands, and in such a blessed climate, I know from experience, is not to be found in this world. Taking into account the result of these travels as shown in his works, he might well boast; he often said, if he had ever written a line worth preserving, it was Greece that inspired it. After this trip he returned to England, and remained there some years, four or five; then abandoned it for ever, passed through the Netherlands, went up the Rhine, paused for some months in Switzerland, crossed the Alps into Italy, and never left that peninsula until the last year of his life. He was never in France, for when he left England, Paris was in the hands of the Allies, and he said he could not endure to witness a country associated in his mind with so many glorious deeds of arts and arms, bullied by "certain rascal officers, slaves in authority, the knaves of justice!"

To return, however, to his travels. If you look at a map you will see what a narrow circle comprises his wanderings. Any man might go, and many have gone without the aid of steam, over the same ground in a few months—even if he had to walk with a knapsack, where Byron rode. The Pilgrim moved about like a Pasha, with a host of attendants, and all that he and they required on the journey. So far as I could learn from Fletcher, his yeoman bold—and he had been with him from the time of his first leaving England,—

Byron wherever he was, so far as it was practicable, pursued the same lazy, dawdling habits he continued during the time I knew him. He was seldom out of his bed before noon, when he drank a cup of very strong green tea, without sugar or milk. At two he ate a biscuit and drank soda-water. At three he mounted his horse and sauntered along the road—and generally the same road,—if alone, racking his brains for fitting matter and rhymes for the coming poem, he dined at seven, as frugally as anchorites are said in storybooks to have done, at nine he visited the family of Count Gamba, on his return home he sat reading or composing until two or three o'clock in the morning, and then to bed, often feverish, restless and exhausted—to dream, as he said, more than to sleep.

Something very urgent, backed by the importunity of those who had influence over him, could alone induce him to break through the routine I have described, for a day, and it was certain to be resumed on the next,—he was constant in this alone.

His conversation was anything but literary, except when Shelley was near him. The character he most commonly appeared in was of the free and easy sort, such as had been in vogue when he was in London, and George IV was Regent; and his talk was seasoned with anecdotes of the great actors on and off the stage, boxers, gamblers, duellists, drunkards, etc., etc., appropriately garnished with the slang and scandal of that day. Such things had all been in fashion, and were at that time considered accomplishments by gentlemen; and of this tribe of Mohawks the Prince Regent was the chief, and allowed to be the most perfect specimen. Byron, not knowing the tribe was extinct, still prided himself on having belonged to it; of nothing was he more indignant, than of being treated as a man of letters, instead of as a Lord and a man of fashion: this prevented foreigners and literary people from getting on with him, for they invariably so offended. His long absence had not effaced the mark John Bull brands his children with; the instant he loomed above the horizon, on foot or horseback, you saw at a glance he was a Britisher. He did not understand foreigners, nor they him; and, during the time I knew him, he associated with no Italians except the family of Count Gamba.

. . .

He seemed to take an especial pleasure in making a clean breast to every newcomer, as if to mock their previous conceptions of him, and to give the lie to the portraits published of him. He said to me, as we were riding together alone, shortly after I knew him:

"Now, confess, you expected to find me a 'Timon of Athens', or a 'Timur the Tartar'; or did you think I was a mere singsong driveller of poesy, full of what I heard Braham at a rehearsal call '*Entusamusy*'; and are you not mystified at finding me what I am,—a man of the world—never in earnest—laughing at all things mundane."

. . . In his perverse and moody humours, Byron would give vent to his Satanic vein. After a long silence, one day on horseback, he began:

"I have a conscience, although the world gives me no credit for it; I am now repenting, not of the few sins I have committed, but of the many I have not committed. There are things, too, we should not do, if they were not forbidden. My *Don Juan* was cast aside and almost forgotten, until I heard that the pharisaic synod in John Murray's back parlour had pronounced it as highly immoral, and unfit for publication. 'Because thou art virtuous thinkest thou there shall be no more cakes and ale?' Now my brain is throbbing and must have vent. I opined gin was inspiration, but cant is stronger. To-day I had another letter warning me against the Snake (Shelley). He, alone, in this age of humbug, dares stem the current, as he did to-day the flooded Arno in his skiff, although I could not observe he made any progress. The attempt is better than being swept along as all the rest are, with the filthy garbage scoured from its banks."

. . .

On coming near Lonza, a small islet, converted into one of their many dungeons by the Neapolitan government, I said to Byron:

"There is a sight that would curdle the milky blood of a poet-laureate."

"If Southey was here," he answered, "he would sing hosannas to the Bourbons. Here kings and governors are only the jailors and hangmen of the detestable Austrian barbarians. What dolts and drivellers the people are to submit to such universal despotism. I should like to see, from this our ark, the world submerged, and all the rascals on it drowning like rats."

I put a pencil and paper in his hand, saying:

"Perpetuate your curses on tyranny, for poets like ladies generally side with the despots."

He readily took the paper and set to work. I walked the deck and prevented his being disturbed. He looked as crest-fallen as a riotous boy, suddenly pounced upon by a master and given an impossible task, scrawling and scratching out, sadly perplexed. After a long spell, he said:

"You think it is as easy to write poetry as smoke a segar,—look, it's only doggerel. Extemporising verses is nonsense; poetry is a distinct faculty,—it won't come when called,—you may as well whistle for a wind; a Pythoness was primed when put upon her tripod. I must chew the cud before I write. I have thought over most of my subjects for years before writing a line."

He did not, however, give up the task, and sat pondering over the paper for nearly an hour; then gnashing his teeth, he tore up what he had written, and threw the fragments overboard.

Seeing I looked disappointed:

"You might as well ask me to describe an earthquake, whilst the ground was trembling under my feet. Give me time,—I can't forget the theme: but for this Greek business I should have been at Naples writing a Fifth canto of *Childe Harold*, expressly to give vent to my detestation of the Austrian tyranny in Italy."

Knowing and sympathising with Byron's sensitiveness, his associates avoided prying into the cause of his lameness; so did strangers, from good breeding or common humanity. It was generally thought his halting gait originated in some defect of the right foot or ankle—the right foot was the most distorted, and it had been made worse in his boyhood by vain efforts to set it right. He told me that for several years he wore steel splints, which so wrenched the sinews and tendons of his leg, that they increased his lameness; the foot was twisted inwards, only the edge touched the ground, and that leg was shorter than the other. His shoes were peculiar—very high heeled, with the soles uncommonly thick on the inside and pared thin on the outside—the toes were stuffed with cotton-wool, and his trousers were very large below the knee and strapped down so as to cover his feet. The peculiarity of his gait was now accounted for; he entered a room with a sort of run, as if he could not stop, then planted his best leg well forward, throwing back his body to keep his balance. In early life whilst his frame was light and elastic, with the aid of a stick he might have tottered along for a mile or two; but after he had waxed heavier, he seldom attempted to walk more than a few hundred yards, without squatting down or leaning against the first wall, bank, rock, or tree at hand, never sitting on the ground, as it would have been difficult for him to get up again. In the company of strangers, occasionally, he would make desperate efforts to conceal his infirmity, but the hectic flush on his face, his swelling veins, and quivering nerves betrayed him, and he suffered for many days after such exertions. Disposed to fatten, incapable of taking exercise to check the tendency, what could he do? If he added to his weight, his feet would not have supported him; in this dilemma he was compelled to exist in a state of semi-starvation; he was less than eleven stone when at Genoa, and said he had been fourteen at Venice. The pangs of hunger which travellers and shipwrecked mariners have described were nothing to what he suffered; their privations were temporary, his were for life, and more unendurable, as he was in the midst of abundance.

PERCY BYSSHE SHELLEY

1792–1822

IN EVERY GENERATION a few persons nobly dream of transforming the world. Through persuasion, example, and leadership, they aspire to bring in at last the age of justice and truth. But often they do not know how to begin; they would not like to be ridiculous; and the impulse ravels out. Shelley acted. If his efforts, when he was not yet twenty-one, were not always very practical, that was partly because he had boundless confidence in the reasonableness of mankind. For example, when he was eighteen years old and at Oxford, he sent copies of his pamphlet *The Necessity of Atheism* to all officials and professors at Oxford, perhaps to many at Cambridge, and to every bishop in the kingdom. He thought that a university is dedicated to open discussion: if his argument was logical, readers would be converted; if it was not, then the error would be exposed in argument. The Oxford authorities took a different view, and Shelley was expelled, having attended college only six months. Not long afterwards he went to Ireland, intending a modest beginning to the great task of world reformation by organizing the Irish into "the society of peace and love." When this had no success, he settled in Wales, where, now nineteen, he distributed another pamphlet (a "Declaration of Rights") partly by enclosing it in dark green bottles which he cast into the ocean and partly by balloons. During this time he was writing *Queen Mab*. In this long work the Queen of Faery delivers to the disembodied soul of the beautiful Ianthe a series of lectures on social, political, and religious themes. She declares that Necessity rules the world and will gradually usher in a happier time. Kings, priests, religion, and commerce are pernicious and must perish. Women will be emancipated; love will be tender and true; man will become vegetarian and hence healthy and gentle; and the earth will shift its position. The poem is the first significant revelation of a strange and extraordinary genius. It is erudite, philosophical, fantastic, lyrical, and filled with visionary splendors and didacticism. As for Shelley's wild hope and purpose that the poem should renovate society, it had, like other of Shelley's poems, a wide circulation in radical movements throughout the nineteenth century, and its total effect would naturally be impossible to estimate.

Shelley descended from a line of country gentlemen. He was born on August 4, 1792, the oldest of seven children, of whom the next five were girls. In after years his sisters remembered how in childhood he would hold them spellbound with stories of an old alchemist with a long beard who lived in the attic, or a great old snake in the garden. He wrote poems, some of which were printed locally, and he gave proof of an unusually intellectual, generous, and courageous temperament. At the age of eleven he was sent to Eton. In later years he could identify the bullying and fagging he experienced with other forms of hatred and tyranny, and see himself even in school as its victim and dedicated foe, living by a higher ideal than the conforming minions of a cruel system. From the point of view of the other boys, however, Shelley was an odd fish. Having grown up with his sisters, he had no taste for sports and did not know how to fight. Moreover, his appearance seemed girlish and his restlessly inquiring intelligence led him to acts that were judged eccentric. One Etonian remembered him as "notorious for setting fire to old trees with burning glasses." But he had good friends. The persecution relaxed as he grew older. He pursued literary avocations, churning out a romance and a small volume of poems. (Both productions show a taste for gothic horrors—ghosts, demons, gloom, black whirlwinds —which thereafter persisted as elements of Shelley's imagination.) And during the few months in college he eagerly continued what his father called his "printing freaks," turning out poems, a romance, a novel, and an "Essay on Love."

The history of Shelley's marriages illustrates the platitude that actions have a different meaning as they are seen from inside or out. To outsiders it appears that Shelley married a girl of sixteen and then, three years later, abandoned her and his children to run off with a girl of seventeen. (And this second love, for which so much had been sacrificed, also turned out to be mutable.) To Shelley himself it appeared that he was inspired by the highest ideals throughout. Not long after leaving Oxford in 1811 he had married Harriet Westbrook, knowing that he was not in love and disbelieving in marriage as an institution. The main reason for his marriage was that he had unsettled her religious opinions and thus exposed her to the persecuting tyranny of schoolmates and teachers. "Gratitude and admiration all demand that I should love her *for ever*." It was, in short, an incredible act of gallantry. Besides, there would be the pleasure of "moulding a really noble soul into all that can make its nobleness useful and lovely." But this proved impossible. Harriet was pretty, loving, gentle, and faithful, but not a soul-mate. And in 1814 Shelley met the attractive and intellectual Mary, daughter of the philosopher William Godwin. It seemed wrong that the spirit of love, filling the heart with happiness and virtue, should be denied, and it seemed still more repulsive that one should be condemned to live a lie in a married parody of love. If Harriet truly loved Shelley, she would be glad of his renewed happiness and love Mary for his sake. He proposed that all three should live together, Harriet as his sister and Mary as his wife. When Harriet declined, he eloped with Mary.

The next year was a happy one for Shelley. *Alastor* was testimony to his rapidly developing creative power. Financial difficulties were relieved early in 1815 when he inherited a large income. In 1816, during a tour of Switzerland, he met Byron and also wrote the "Hymn to Intellectual Beauty" and "Mont Blanc." But when he returned to England life began to assume a different tone. Mary's half-sister committed suicide, and the situation was such that Shelley felt partly responsible. And then Harriet was found dead in circumstances that also suggested suicide. Meanwhile Mary's step-sister, Claire Clairmont, was pregnant by Byron. Since she was living with Shelley and his wife—she had nowhere else to go—scandal naturally pointed to Shelley as the father. Then the courts denied Shelley the custody of his children by Harriet, and, in 1818, in poor health and fearful that he might also lose custody of his children by Mary, he left England to live henceforth in Italy.

Prometheus Unbound, which he began within a few months, reflects the new interests and impulses that flowed in upon him—friendship with Byron, the vivid color and light of Italy, the luxuriant beauty of nature, and the painting, sculpture, and architecture of the classical world and the Renaissance. As Shelley says, his personal joy in the Italian spring was caught up and transfigured in this drama of cosmic optimism. Yet shorter lyrics composed at the same time reflect feelings of loneliness, dejection, and fatigue, feelings that persisted and deepened in the few remaining years of his life. Something of this must be attributed to chronic ill-health, but more fundamental causes were the death of two of his children, the prolonged depression and partial estrangement this occasioned in Mary, the failure of his political hopes, the public indifference to his poetry—Shelley had no audience—and, in short, the gradual disappointment of virtually every eager aspiration with which his career had begun. Yet in these years he poured forth a series of major works—*The Cenci, Epipsychidion, Adonais, Hellas*, the unfinished *Triumph of Life*, the *Defence of Poetry*—and a profusion of lyrics, some of them among the finest in English. And this was despite an astonishing commitment of energy to projects, causes, and generous efforts of every kind. He died suddenly, by drowning, July 8, 1822, when an open boat in which he and a friend were sailing was caught in a storm. He was twenty-nine years old. For his gravestone Leigh Hunt supplied a final summing-up, *Cor Cordium*, "heart of hearts."

2

To most readers Shelley's poetry makes its first appeal through qualities of style, that is, imagery, versification, tone, and rhetorical pattern. But to enjoy this style freely one must slip past the screen of generalizations that a hundred years of criticism have erected. Readers here will be offered a partial sampling of Shelley's remarkably various resources. In a passage from the "Lines Written Among the Euganean Hills" Shelley comes, except for the rhythm, near to the poetry of the twentieth century:

> On the beach of a northern sea
> Which tempests shake eternally,
> As once the wretch there lay to sleep,
> Lies a solitary heap,
> One white skull and seven dry bones,
> On the margin of the stones,
> Where a few gray rushes stand,
> Boundaries of the sea and land
>
> [ll. 45–52].

The short clauses and phrases flow easily; the language is direct and simple; the imagery concrete and spare. Yet the scene as a whole has the haunting strangeness one associates with Romantic verse. At another extreme, in *Adonais*, we encounter abstraction and driving rhetoric—

> That Light whose smile kindles the Universe,
> That Beauty in which all things work and move,
> That Benediction . . .
>
> [ll. 478–80]

and the transmuting of philosophy into symbolism:

> the pure spirit shall flow
> Back to the burning fountain whence it came
>
> [ll. 338–39].

If Shelley seems difficult in such moments, it is partly because we are not accustomed—except in Yeats—to quite this union of passion, abstraction, symbolism, and headlong speed. *Prometheus Unbound* orchestrates diverse styles by playing one speaker against the next. At Act I, ll. 316ff., Ione describes the coming of the god Mercury. The passage more or less corresponds to the stock idea of Shelley's verse. It has the singing movement, the assonance (*i* and long *o* sounds), the vivid and unusual contrasts of colors—azure, snow, gold, purple, rose, blood, ivory—the smoothly running and proliferating clauses, the oddness of detail and metaphor—feet that glow like ivory, ivory that is ensanguined—and the rapidity of imagination that, as often in Shakespeare, accumulates a rich cluster of incompletely activated associations.

> But see, where through the azure chasm
>   Of yon forked and snowy hill,
> Trampling the slant winds on high
>   With golden-sandalled feet, that glow
> Under plumes of purple dye,
> Like rose-ensanguined ivory,
>   A Shape comes now.

And this lyric is almost immediately followed by the harsh, chuckling, grating speech of the furies:

> Ha! I scent life!
>   Let me but look into his eyes!
>   The hope of torturing him smells like a heap
>   Of corpses, to a death-bird after battle.

In contrast to the melody that went before, the diction has become predominantly Anglo-Saxon (the phrase "death-bird after battle" might come from an Anglo-Saxon poem) and explosive (the labials, dentals, and gutterals in such words as hope, torturing, heap, corpses, death-bird, and battle). Then one hears the stately speech of Mercury.

>   Awful Sufferer!
>   To thee unwilling, most unwillingly
>   I come, by the great Father's will driven down,
>   To execute a doom of new revenge.

And no inventory of Shelley's resources can omit his facility in creation of myth (*Prometheus Unbound*, I. 62–64; II. ii. 40–49), architectural solemnities (III. iii. 159–66), condensed aphorism (I. 489; III. iv. 43), and direct, hard-hitting speech (I. 623–28).

What then of the usual charges that Shelley is vague, profuse, and merely emotive? Certainly they are not true without strong qualification, and they have little relevance for anyone who shares Shelley's philosophy, or rather what one might call his sense of the cosmos. It is plainly impossible to sum up complex, subtle, and changing attitudes in a few sentences, but the point is that through experiences such as those described in the "Hymn to Intellectual Beauty" Shelley intuited an eternal reality beyond and sundered from the mortal world in which we live, where all we know is fleeting, unsubstantial, and illusory. (This intuition was naturally assimilated with Platonic philosophy, but it remained deeply personal.) Poetry or art is created in or immediately after moments of visionary ascent to the eternal, and is an attempt to render such moments in words and images. That is, an artist has only the data of this world as expressive means, and using them he must attempt to convey something utterly different and ultimately quite ineffable. To the extent that he succeeds, he creates "Forms more real"—closer to ultimate reality—than the things we perceive in the world about us. But the forms art presents are only *more* real. Success is never complete. Words and images are only symbolic gestures pointing to something beyond. In fact, like all the symbols we necessarily use in thinking and speaking, they may themselves be only symbolic of deeper symbols in a series that endlessly recedes towards the "deep truth" that remains "imageless"—that is, beyond any real apprehension— and Shelley's poetry is shot through with despair not only of language but of the human mind.

Yet if poetry can go even a little way toward the "deep truth," it will not be through a few choice expressions. The only way to surmount the inherent limitations of language is to use a lot of it all at once. Hence in poems such as "To a Skylark" or *Epipsychidion* Shelley speaks in rapid series of images that, blending together, may suggest a whole of which each separately suggests only an element. To complain of profuseness and vagueness seems beside the point if they were the sole means available for meeting an overwhelming dilemma.

The dilemma had somehow to be met, if only because what poetry tries to convey has immense importance for human life. Evil may be interpreted as moral or metaphysical, or both, that is, as the selfishness, aggressiveness, and fear embedded in the human heart, or as the limitation, pain, and death to which man is subject in a cosmos indifferent to his happiness. In general, evil means the latter to Keats, the former to Blake, Coleridge, and Shelley. It is no good saying, as the radicals said (and Shelley in *Queen Mab*), that man is corrupted by society and its institutions, which, if transformed, will transform human nature; for society—Shelley gradually came to believe—is the spontaneous creation and outward mirror of human nature. A fundamental bettering of social institutions will come about only after the regeneration of man. In other words, the heart in each individual must be stirred to love what is good, and to this end art is the great instrument. It could not, of course, be effective were not a yearning for goodness and beauty already present in mankind, but at the same time it enlightens and gives force to this yearning. In rapt vision of the transcendent world, where alone man's aspiration can be finally satisfied, and sensitively responsive to all that is lovely in this world, an artist "redeems from decay the visitations of the divinity in man." By thus calling the heart to an ideal, and by bringing about what Shelley in a great phrase describes as "Transforming enlargements of the imagination," artists and poets are "the unacknowledged legislators of the world." The phrase, dulled by familiarity and read out of context, can easily be set aside as gushing and hyperbolic sentiment. But it issues from a complex argument. Shelley was deeply convinced of what T. S. Eliot calls "The Social Function of Poetry," and in the Preface to *Prometheus Unbound* and in *A Defence of Poetry* he gives the most comprehensive explanation of it that has yet been attempted.

Bibliography

Criticism of Shelley's poetry during his lifetime was almost entirely a reflection of politics. Liberals and their journals, such as Leigh Hunt's *Examiner*, praised him as a high-minded prophet of reform. Tory journals, together with most readers, were frightened or disgusted by his revolutionary ardor and by his boldness of speculation in religion and morality. One magazine recorded a common opinion when it listed him in its index as "Shelley, Percy Bysshe—an unmentionable subject." His poems had a limited sale, and for a few years after his death his name was kept alive mainly by the sympathies of radicals and his connection with Byron. But even then future fame might have been prophesied from the impression made on the best minds of the younger generation. In his teens Browning responded so strongly that he became for a while an atheist and vegetarian. Cambridge undergraduate partisans included Tennyson, Arthur Hallam, and Richard Monckton Milnes, who was later to write the first biography of Keats. When in 1839 Mrs. Shelley published the *Poetical Works* with notes, the numerous reviews were a testimony to reconsideration and wider interest. After 1850 Shelley was more popular than Byron, and since then he has held his place among the English poets—a writer often attacked, often extolled, but never forgotten.

In the later nineteenth century a stock characterization prevailed. To John Stuart Mill in "Thoughts on Poetry and Its Varieties," 1859 (reprinted in *Dissertations and Discussions*, 3 vols., 1859–67, I, 89–120), he was the purest example of a true poetic temperament, in which feeling "seizes the helm of . . . thoughts, and the succession of ideas and images becomes the mere utterance of an emotion." This is the view of Shelley as skylark or ethereal spirit pouring forth the native instincts of the heart. Its brilliantly written culmination appears in George Santayana's "Shelley," *Winds of Doctrine*, 1913. Dispraisers simply attached a different meaning to the same qualities. Hazlitt had already specified the charge in 1821: Shelley "does not grapple with the world around him . . . Bubbles are to him the only realities." Tennyson, as he grew older, found a "sort of tenuity" in his poetry. To Matthew Arnold ("Shelley," *Essays in Criticism, Second Series*, 1888), he was a "beautiful and ineffectual angel, beating in the void his luminous wings in vain." In short, the song of a skylark naturally lacks concreteness, structure, self-discipline, and common sense.

When based on this image, either praise or dispraise seems equally odious and unreal to admirers now. A general result of present-day scholarship and criticism has been to highlight complexities, that is, to illustrate that a man cannot be summed up in a metaphor. Kenneth Neill Cameron, *The Young Shelley: Genesis of a Radical*, 1950, takes up Shelley's political and religious thinking. The book supplants many previous works on this subject, all of which have given a better understanding of his thought and also made it impossible to forget how intensely he cared that society should be transformed. His philosophical interests and beliefs are explored by Floyd Stovall, *Desire and Restraint in Shelley*, 1931, and A. M. D. Hughes, *The Nascent Mind of Shelley*, 1947; J. A. Notopoulos treats a central subject with exhaustive detail in *The Platonism of Shelley*, 1949. Since Whitehead's *Science and the Modern World*, 1925, Shelley's poetic use of the scientific lore of his time has been closely studied, first by C. H. Grabo, *A Newton Among Poets*, 1930, and later by Desmond King-Hele, *Shelley: His Thought and Work*, 1960, and both writers clarify many passages that would otherwise be obscure. Grabo's *The Magic Plant*, 1936, is an extended introduction to Shelley that allows him largely to speak for himself, and in *Shelley's Major Poetry: The Fabric of a Vision*, 1948, Carlos Baker conveys a rounded portrait through analysis of major poems. Another general study is R. G. Woodman, *The Apocalyptic Vision in the Poetry of Shelley*, 1964. In forty pages of *Mythology and the Romantic Tradition*, 1937, Douglas Bush focuses on Shelley's use of classical mythology, but actually provides a distilled discussion of all major questions. *Shelley's Later Poetry* by Milton Wilson, 1959, is primarily about *Prometheus Unbound*, but it touches on other poems and topics. Analyses of particular works include especially Bennett Weaver, *Prometheus Unbound*, 1957, Earl Wasserman, *Shelley's Prometheus Unbound: A Critical Reading*, 1965, and D. H. Reiman, *Shelley's "The Triumph of Life,"* 1965. James Rieger, *The Mutiny Within*, 1967, suggestively reinterprets Shelley's character and poetry.

General qualities of style and imagination are noted in most of the books mentioned above. R. H. Fogle pursues more specialized aspects in *The Imagery of Keats and Shelley*, 1949, and Glenn O'Malley, *Shelley and Synesthesia*, 1964, carries this study further along one line. For Shelley's symbolism—its way and meaning—one should begin with W. B. Yeats's essay, "The Philosophy of Shelley's Poetry." G. W. Knight, *The Starlit Dome*, 1941, interprets the symbolism by his peculiar method of massed references, and David Perkins, *The Quest for Permanence*, 1959, relates it to other habits of style and mind in the context of individual poems. Harold Bloom, *Shelley's Mythmaking*, 1959, concentrates on a particular mode of Shelley's imagination, the mythopoeic power that inspires both local images and entire poems—the "Ode to the West Wind," for example. In *The Subtler Language*, 1959, Earl Wasserman offers a brilliant series of close explications while also investigating the structure of the major poems.

The definitive biography is that of N. I. White, *Shelley*, 2 vols., 1940, abridged as *Portrait of Shelley*, 1945. Edmund Blunden's excellent *Shelley*, 1946, should also be mentioned. The standard edition is the *Complete Works*, ed. Roger Ingpen and W. E. Peck, 10 vols., 1926–30. His *Letters* have been edited by F. L. Jones, 2 vols., 1964.

STANZAS

APRIL, 1814

❧

Shelley's first successful lyric, these stanzas refer to his unhappiness in leaving Mrs. de Boinville and her daughter, whom he had been visiting, and returning to his own home. He was already somewhat estranged from his first wife, Harriet Westbrook. "Thy friend" (l. 6) is Mrs. de Boinville; "Thy lover" (l. 7) is her daughter. In the last line the "two voices" would be those of the mother and daughter, the "one sweet smile" that of the daughter.

❧

Away! the moor is dark beneath the moon,
　Rapid clouds have drank the last pale beam of even:
Away! the gathering winds will call the darkness soon,
　And profoundest midnight shroud the serene lights
　　of heaven.

Pause not! The time is past! Every voice cries, Away!
　Tempt not with one last tear thy friend's ungentle
　　mood:
Thy lover's eye, so glazed and cold, dares not entreat
　thy stay:
Duty and dereliction guide thee back to solitude.

Away, away! to thy sad and silent home;
　Pour bitter tears on its desolated hearth;　　　　10
Watch the dim shades as like ghosts they go and come,
　And complicate strange webs of melancholy mirth.

The leaves of wasted autumn woods shall float around
　thine head:
　The blooms of dewy spring shall gleam beneath
　　thy feet:
But thy soul or this world must fade in the frost that
　binds the dead,
　Ere midnight's frown and morning's smile, ere thou
　　and peace may meet.

The cloud shadows of midnight possess their own
　repose,
　For the weary winds are silent, or the moon is in
　　the deep:
Some respite to its turbulence unresting ocean knows;
　Whatever moves, or toils, or grieves, hath its
　　appointed sleep.　　　　20

Thou in the grave shalt rest—yet till the phantoms flee
　Which that house and heath and garden made dear
　　to thee erewhile,
Thy remembrance, and repentance, and deep musings
　are not free
From the music of two voices and the light of one
　sweet smile.

1814 (1816)

TO WORDSWORTH

After the radicalism of his youth, Wordsworth's politics became increasingly conservative. In imagery, diction, and movement this sonnet shows the influence of Wordsworth's use of the form. See, for example, "O Friend! I Know Not Which Way I Must Look" and "Milton! Thou Shouldst Be Living at This Hour."

Poet of Nature, thou hast wept to know
That things depart which never may return:
Childhood and youth, friendship and love's first glow,
Have fled like sweet dreams, leaving thee to mourn.
These common woes I feel. One loss is mine
Which thou too feel'st, yet I alone deplore.
Thou wert as a lone star, whose light did shine
On some frail bark in winter's midnight roar:
Thou hast like to a rock-built refuge stood
Above the blind and battling multitude: 10
In honoured poverty thy voice did weave
Songs consecrate to truth and liberty,—
Deserting these, thou leavest me to grieve,
Thus having been, that thou shouldst cease to be.

(*1816*)

ALASTOR

OR
THE SPIRIT OF SOLITUDE

The title was suggested by Peacock after the poem was finished. He says it means "an evil genius," and that the poem treats "the spirit of solitude as a spirit of evil." But the poem does not wholly fit either this interpretation or Shelley's own in the Preface. What seems clear is that the poet, intensely committed to thought, study, and the quest of ultimate truth, ignores the human love of the Arab maiden and therefore comes to love, or is made to love (see ll. 203–05), an ideal vision. The vision is of a maid who embodies the highest excellence of his own soul, one who is an outward counterpart to all that is best within him (ll. 153–61). Thus both the poem and the Preface are important as early expressions of Shelley's doctrine of love. That the desire to possess this vision leads to death was an insight charged with metaphysical and psychological meanings elaborated by Shelley in many later poems. (See especially *Epipsychidion*, ll. 72–73.) The poem also anticipates one of Shelley's later conceptions of the poet as a visionary and isolated figure loyal to a transcendent realm of being and therefore doomed to sorrow in this mortal world of time.

PREFACE

The poem entitled *Alastor* may be considered as allegorical of one of the most interesting situations of the human mind. It represents a youth of uncorrupted feelings and adventurous genius led forth by an imagination inflamed and purified through familiarity with all that is excellent and majestic, to the contemplation of the universe. He drinks deep of the fountains of knowledge, and is still insatiate. The magnificence and beauty of the external world sinks profoundly into the frame of his conceptions, and affords to their modifications a variety not to be exhausted. So long as it is possible for his desires to point towards objects thus infinite and unmeasured, he is joyous, and tranquil, and self-possessed. But the period arrives when these objects cease to suffice. His mind is at length suddenly awakened and thirsts for intercourse with an intelligence similar to itself. He images to himself the Being whom he loves. Conversant with speculations of the sublimest and most perfect natures, the vision in which he embodies his own imaginations unites all of wonderful, or wise, or beautiful, which the poet, the philosopher, or the lover could depicture. The intellectual faculties, the imagination, the functions of sense, have their respective requisitions on the sympathy of corresponding powers in other human beings. The Poet is represented as uniting these requisitions, and attaching them to a single image. He seeks in vain for a prototype of his conception. Blasted by his disappointment, he descends to an untimely grave.

The picture is not barren of instruction to actual men. The Poet's self-centred seclusion was avenged by the furies of an irresistible passion pursuing him to speedy ruin. But that Power[1] which strikes the luminaries of the world with sudden darkness and extinction, by awakening them to too exquisite a perception of its influences, dooms to a slow and poisonous decay those meaner spirits that dare to

ALASTOR. **Preface. 1. Power:** love.

abjure its dominion. Their destiny is more abject and inglorious as their delinquency is more contemptible and pernicious. They who, deluded by no generous error, instigated by no sacred thirst of doubtful knowledge, duped by no illustrious superstition, loving nothing on this earth, and cherishing no hopes beyond, yet keep aloof from sympathies with their kind, rejoicing neither in human joy nor mourning with human grief; these, and such as they, have their apportioned curse. They languish, because none feel with them their common nature. They are morally dead. They are neither friends, nor lovers, nor fathers, nor citizens of the world, nor benefactors of their country. Among those who attempt to exist without human sympathy, the pure and tender-hearted perish through the intensity and passion of their search after its communities, when the vacancy of their spirit suddenly makes itself felt. All else, selfish, blind, and torpid, are those unforeseeing multitudes who constitute, together with their own, the lasting misery and loneliness of the world. Those who love not their fellow-beings live unfruitful lives, and prepare for their old age a miserable grave.

> "The good die first,
> And those whose hearts are dry as summer dust,
> Burn to the socket!"[2]

December 14, 1815

Nondum amabam, et amare amabam, quaerebam
quid amarem, amans amare.

Confess. St. August.

Earth, ocean, air, beloved brotherhood!
If our great Mother has imbued my soul
With aught of natural piety to feel
Your love, and recompense the boon with mine;
If dewy morn, and odorous noon, and even,
With sunset and its gorgeous ministers,
And solemn midnight's tingling silentness;
If autumn's hollow sighs in the sere wood,
And winter robing with pure snow and crowns
Of starry ice the grey grass and bare boughs;   10
If spring's voluptuous pantings when she breathes
Her first sweet kisses, have been dear to me;

If no bright bird, insect, or gentle beast
I consciously have injured, but still loved
And cherished these my kindred; then forgive
This boast, beloved brethren, and withdraw
No portion of your wonted favour now!

Mother of this unfathomable world!
Favour my solemn song, for I have loved
Thee ever, and thee only; I have watched   20
Thy shadow, and the darkness of thy steps,
And my heart ever gazes on the depth
Of thy deep mysteries. I have made my bed
In charnels and on coffins, where black death
Keeps record of the trophies won from thee,
Hoping to still these obstinate questionings
Of thee and thine, by forcing some lone ghost
Thy messenger, to render up the tale
Of what we are. In lone and silent hours,
When night makes a weird sound of its own stillness,
Like an inspired and desperate alchymist   31
Staking his very life on some dark hope,
Have I mixed awful talk and asking looks
With my most innocent love, until strange tears
Uniting with those breathless kisses, made
Such magic as compels the charmed night
To render up thy charge: . . . and, though ne'er yet
Thou hast unveiled thy inmost sanctuary,
Enough from incommunicable dream,
And twilight phantasms, and deep noon-day thought,
Has shone within me, that serenely now   41
And moveless, as a long-forgotten lyre
Suspended in the solitary dome
Of some mysterious and deserted fane,
I wait thy breath, Great Parent, that my strain
May modulate with murmurs of the air,
And motions of the forests and the sea,
And voice of living beings, and woven hymns
Of night and day, and the deep heart of man.

There was a Poet whose untimely tomb   50
No human hands with pious reverence reared,
But the charmed eddies of autumnal winds
Built o'er his mouldering bones a pyramid
Of mouldering leaves in the waste wilderness:—
A lovely youth,—no mourning maiden decked
With weeping flowers, or votive cypress wreath,
The lone couch of his everlasting sleep:—
Gentle, and brave, and generous,—no lorn bard

---

2. **"The . . . socket!"**: Wordsworth's *Excursion*, I. 519–21. **Motto. Nondum . . . amare:** Not yet did I love, and I loved to love. I sought what I should love, loving to love. **2. Mother:** Nature. **3. natural piety:** Shelley takes the phrase from Wordsworth's "My Heart Leaps Up," l. 9.

26. **obstinate questionings:** from Wordsworth's ode, "Intimations of Immortality," l. 145.

Breathed o'er his dark fate one melodious sigh:
He lived, he died, he sung, in solitude.   60
Strangers have wept to hear his passionate notes,
And virgins, as unknown he passed, have pined
And wasted for fond love of his wild eyes.
The fire of those soft orbs has ceased to burn,
And Silence, too enamoured of that voice,
Locks its mute music in her rugged cell.

   By solemn vision, and bright silver dream,
His infancy was nurtured. Every sight
And sound from the vast earth and ambient air,
Sent to his heart its choicest impulses.   70
The fountains of divine philosophy
Fled not his thirsting lips, and all of great,
Or good, or lovely, which the sacred past
In truth or fable consecrates, he felt
And knew. When early youth had passed, he left
His cold fireside and alienated home
To seek strange truths in undiscovered lands.
Many a wide waste and tangled wilderness
Has lured his fearless steps; and he has bought
With his sweet voice and eyes, from savage men,   80
His rest and food. Nature's most secret steps
He like her shadow has pursued, where'er
The red volcano overcanopies
Its fields of snow and pinnacles of ice
With burning smoke, or where bitumen lakes
On black bare pointed islets ever beat
With sluggish surge, or where the secret caves
Rugged and dark, winding among the springs
Of fire and poison, inaccessible
To avarice or pride, their starry domes   90
Of diamond and of gold expand above
Numberless and immeasurable halls,
Frequent with crystal column, and clear shrines
Of pearl, and thrones radiant with chrysolite.
Nor had that scene of ampler majesty
Than gems or gold, the varying roof of heaven
And the green earth, lost in his heart its claims
To love and wonder; he would linger long
In lonesome vales, making the wild his home,
Until the doves and squirrels would partake   100
From his innocuous hand his bloodless food,
Lured by the gentle meaning of his looks,
And the wild antelope, that starts whene'er

The dry leaf rustles in the brake, suspend
Her timid steps to gaze upon a form
More graceful than her own.
                            His wandering step,
Obedient to high thoughts, has visited
The awful ruins of the days of old:
Athens, and Tyre, and Balbec, and the waste
Where stood Jerusalem, the fallen towers   110
Of Babylon, the eternal pyramids,
Memphis and Thebes, and whatsoe'er of strange
Sculptured on alabaster obelisk,
Or jasper tomb, or mutilated sphynx,
Dark Æthiopia in her desert hills
Conceals. Among the ruined temples there,
Stupendous columns, and wild images
Of more than man, where marble daemons watch
The Zodiac's brazen mystery, and dead men
Hang their mute thoughts on the mute walls around,
He lingered, poring on memorials   121
Of the world's youth, through the long burning day
Gazed on those speechless shapes, nor, when the moon
Filled the mysterious halls with floating shades
Suspended he that task, but ever gazed
And gazed, till meaning on his vacant mind
Flashed like strong inspiration, and he saw
The thrilling secrets of the birth of time.

   Meanwhile an Arab maiden brought his food,
Her daily portion, from her father's tent,   130
And spread her matting for his couch, and stole
From duties and repose to tend his steps:—
Enamoured, yet not daring for deep awe
To speak her love:—and watched his nightly sleep,
Sleepless herself, to gaze upon his lips
Parted in slumber, whence the regular breath
Of innocent dreams arose: then, when red morn
Made paler the pale moon, to her cold home
Wildered, and wan, and panting, she returned.

---

71. **divine philosophy:** from Milton's *Comus*, l. 476.
93. **Frequent with:** crowded with. 101. **bloodless food:**
At this time Shelley was still a vegetarian.

104. **brake:** thicket. 109. **Balbec:** an ancient city (in Greek,
Heliopolis) northwest of Damascus, Syria. 112. **Memphis
and Thebes:** Memphis was the capital of the Old Kingdom
in Egypt near modern Cairo (the great pyramids were part
of the Necropolis, or "city of the dead," of Memphis);
Thebes, some hundreds of miles south in "Upper Egypt,"
was the capital of the later Egyptian Empire, or New Kingdom. 118. **daemons:** spirits (from Greek mythology) midway between gods and men. 119. **Zodiac's . . . mystery:**
Shelley refers to the picturing of the Zodiac (the imaginary
belt in the sky containing the paths of the sun, moon, and
principal planets) in a temple at Denderah, Upper Egypt.

The Poet wandering on, through Arabie 140
And Persia, and the wild Carmanian waste,
And o'er the aërial mountains which pour down
Indus and Oxus from their icy caves,
In joy and exultation held his way;
Till in the vale of Cashmire, far within
Its loneliest dell, where odorous plants entwine
Beneath the hollow rocks a natural bower,
Beside a sparkling rivulet he stretched
His languid limbs. A vision on his sleep
There came, a dream of hopes that never yet 150
Had flushed his cheek. He dreamed a veiled maid
Sate near him, talking in low solemn tones.
Her voice was like the voice of his own soul
Heard in the calm of thought; its music long,
Like woven sounds of streams and breezes, held
His inmost sense suspended in its web
Of many-coloured woof and shifting hues.
Knowledge and truth and virtue were her theme,
And lofty hopes of divine liberty,
Thoughts the most dear to him, and poesy, 160
Herself a poet. Soon the solemn mood
Of her pure mind kindled through all her frame
A permeating fire: wild numbers then
She raised, with voice stifled in tremulous sobs,
Subdued by its own pathos: her fair hands
Were bare alone, sweeping from some strange harp
Strange symphony, and in their branching veins
The eloquent blood told an ineffable tale.
The beating of her heart was heard to fill
The pauses of her music, and her breath 170
Tumultuously accorded with those fits
Of intermitted song. Sudden she rose,
As if her heart impatiently endured
Its bursting burthen: at the sound he turned,
And saw by the warm light of their own life
Her glowing limbs beneath the sinuous veil
Of woven wind, her outspread arms now bare,
Her dark locks floating in the breath of night,
Her beamy bending eyes, her parted lips
Outstretched, and pale, and quivering eagerly. 180
His strong heart sunk and sickened with excess
Of love. He reared his shuddering limbs and quelled
His gasping breath, and spread his arms to meet
Her panting bosom: . . . she drew back a while,
Then, yielding to the irresistible joy,
With frantic gesture and short breathless cry

Folded his frame in her dissolving arms.
Now blackness veiled his dizzy eyes, and night
Involved and swallowed up the vision; sleep,
Like a dark flood suspended in its course, 190
Rolled back its impulse on his vacant brain.

Roused by the shock he started from his trance—
The cold white light of morning, the blue moon
Low in the west, the clear and garish hills,
The distinct valley and the vacant woods,
Spread round him where he stood. Whither have fled
The hues of heaven that canopied his bower
Of yesternight? The sounds that soothed his sleep,
The mystery and the majesty of Earth,
The joy, the exultation? His wan eyes 200
Gaze on the empty scene as vacantly
As ocean's moon looks on the moon in heaven.
The spirit of sweet human love has sent
A vision to the sleep of him who spurned
Her choicest gifts. He eagerly pursues
Beyond the realms of dream that fleeting shade;
He overleaps the bounds. Alas! Alas!
Were limbs, and breath, and being intertwined
Thus treacherously? Lost, lost, for ever lost,
In the wide pathless desert of dim sleep, 210
That beautiful shape! Does the dark gate of death
Conduct to thy mysterious paradise,
O Sleep? Does the bright arch of rainbow clouds,
And pendent mountains seen in the calm lake,
Lead only to a black and watery depth,
While death's blue vault, with loathliest vapours hung,
Where every shade which the foul grave exhales
Hides its dead eye from the detested day,
Conducts, O Sleep, to thy delightful realms?
This doubt with sudden tide flowed on his heart, 220
The insatiate hope which it awakened, stung
His brain even like despair.

                    While daylight held
The sky, the Poet kept mute conference
With his still soul. At night the passion came,
Like the fierce fiend of a distempered dream,
And shook him from his rest, and led him forth
Into the darkness.—As an eagle grasped
In folds of the green serpent, feels her breast
Burn with the poison, and precipitates
Through night and day, tempest, and calm, and cloud,
Frantic with dizzying anguish, her blind flight 231
O'er the wide aëry wilderness: thus driven
By the bright shadow of that lovely dream,
Beneath the cold glare of the desolate night,

---

141. Carmanian waste: the Kerman desert in southern
Persia. **151. veiled maid:** Alastor, the Spirit of Solitude.

Through tangled swamps and deep precipitous dells,
Startling with careless step the moonlight snake,
He fled. Red morning dawned upon his flight,
Shedding the mockery of its vital hues
Upon his cheek of death. He wandered on
Till vast Aornos seen from Petra's steep   240
Hung o'er the low horizon like a cloud;
Through Balk, and where the desolated tombs
Of Parthian kings scatter to every wind
Their wasting dust, wildly he wandered on,
Day after day a weary waste of hours,
Bearing within his life the brooding care
That ever fed on its decaying flame.
And now his limbs were lean; his scattered hair
Sered by the autumn of strange suffering
Sung dirges in the wind; his listless hand   250
Hung like dead bone within its withered skin;
Life, and the lustre that consumed it, shone
As in a furnace burning secretly
From his dark eyes alone. The cottagers,
Who ministered with human charity
His human wants, beheld with wondering awe
Their fleeting visitant. The mountaineer,
Encountering on some dizzy precipice
That spectral form, deemed that the Spirit of wind
With lightning eyes, and eager breath, and feet   260
Disturbing not the drifted snow, had paused
In its career: the infant would conceal
His troubled visage in his mother's robe
In terror at the glare of those wild eyes,
To remember their strange light in many a dream
Of after-times; but youthful maidens, taught
By nature, would interpret half the woe
That wasted him, would call him with false names
Brother, and friend, would press his pallid hand
At parting, and watch, dim through tears, the path
Of his departure from their father's door.   271

   At length upon the lone Chorasmian shore
He paused, a wide and melancholy waste
Of putrid marshes. A strong impulse urged
His steps to the sea-shore. A swan was there,
Beside a sluggish stream among the reeds.
It rose as he approached, and with strong wings
Scaling the upward sky, bent its bright course
High over the immeasurable main.

240. **Aornos . . . Petra:** Mt. Mahabunn and the Rock of
Sogdia in Afghanistan. 242. **Balk:** province of Afghanistan
(in ancient times, Bactria). 272. **Chorasmian shore:** the
Aral Sea in central Asia.

His eyes pursued its flight.—"Thou hast a home,   280
Beautiful bird; thou voyagest to thine home,
Where thy sweet mate will twine her downy neck
With thine, and welcome thy return with eyes
Bright in the lustre of their own fond joy.
And what am I that I should linger here,
With voice far sweeter than thy dying notes,
Spirit more vast than thine, frame more attuned
To beauty, wasting these surpassing powers
In the deaf air, to the blind earth, and heaven
That echoes not my thoughts?" A gloomy smile   290
Of desperate hope wrinkled his quivering lips.
For sleep, he knew, kept most relentlessly
Its precious charge, and silent death exposed,
Faithless perhaps as sleep, a shadowy lure,
With doubtful smile mocking its own strange charms.

   Startled by his own thoughts he looked around.
There was no fair fiend near him, not a sight
Or sound of awe but in his own deep mind.
A little shallop floating near the shore
Caught the impatient wandering of his gaze.   300
It had been long abandoned, for its sides
Gaped wide with many a rift, and its frail joints
Swayed with the undulations of the tide.
A restless impulse urged him to embark
And meet lone Death on the drear ocean's waste;
For well he knew that mighty Shadow loves
The slimy caverns of the populous deep.

   The day was fair and sunny, sea and sky
Drank its inspiring radiance, and the wind
Swept strongly from the shore, blackening the waves.
Following his eager soul, the wanderer   311
Leaped in the boat, he spread his cloak aloft
On the bare mast, and took his lonely seat,
And felt the boat speed o'er the tranquil sea
Like a torn cloud before the hurricane.

   As one that in a silver vision floats
Obedient to the sweep of odorous winds
Upon resplendent clouds, so rapidly
Along the dark and ruffled waters fled
The straining boat.—A whirlwind swept it on,   320
With fierce gusts and precipitating force,
Through the white ridges of the chafed sea.
The waves arose. Higher and higher still
Their fierce necks writhed beneath the tempest's
    scourge
Like serpents struggling in a vulture's grasp.
Calm and rejoicing in the fearful war

Of wave ruining on wave, and blast on blast
Descending, and black flood on whirlpool driven
With dark obliterating course, he sate:
As if their genii were the ministers   330
Appointed to conduct him to the light
Of those beloved eyes, the Poet sate
Holding the steady helm. Evening came on,
The beams of sunset hung their rainbow hues
High 'mid the shifting domes of sheeted spray
That canopied his path o'er the waste deep;
Twilight, ascending slowly from the east,
Entwined in duskier wreaths her braided locks
O'er the fair front and radiant eyes of day;
Night followed, clad with stars. On every side   340
More horribly the multitudinous streams
Of ocean's mountainous waste to mutual war
Rushed in dark tumult thundering, as to mock
The calm and spangled sky. The little boat
Still fled before the storm; still fled, like foam
Down the steep cataract of a wintry river;
Now pausing on the edge of the riven wave;
Now leaving far behind the bursting mass
That fell, convulsing ocean: safely fled—
As if that frail and wasted human form,   350
Had been an elemental god.

                        At midnight
The moon arose: and lo! the ethereal cliffs
Of Caucasus, whose icy summits shone
Among the stars like sunlight, and around
Whose caverned base the whirlpools and the waves
Bursting and eddying irresistibly
Rage and resound for ever.—Who shall save?—
The boat fled on,—the boiling torrent drove,—
The crags closed round with black and jagged arms,
The shattered mountain overhung the sea,   360
And faster still, beyond all human speed,
Suspended on the sweep of the smooth wave,
The little boat was driven. A cavern there
Yawned, and amid its slant and winding depths
Ingulfed the rushing sea. The boat fled on
With unrelaxing speed.—"Vision and Love!"
The Poet cried aloud, "I have beheld
The path of thy departure. Sleep and death
Shall not divide us long!"

                        The boat pursued
The windings of the cavern. Daylight shone   370
At length upon that gloomy river's flow;
Now, where the fiercest war among the waves
Is calm, on the unfathomable stream

The boat moved slowly. Where the mountain, riven,
Exposed those black depths to the azure sky,
Ere yet the flood's enormous volume fell
Even to the base of Caucasus, with sound
That shook the everlasting rocks, the mass
Filled with one whirlpool all that ample chasm;
Stair above stair the eddying waters rose,   380
Circling immeasurably fast, and laved
With alternating dash the gnarled roots
Of mighty trees, that stretched their giant arms
In darkness over it. I' the midst was left,
Reflecting, yet distorting every cloud,
A pool of treacherous and tremendous calm.
Seized by the sway of the ascending stream,
With dizzy swiftness, round, and round, and round,
Ridge after ridge the straining boat arose,
Till on the verge of the extremest curve,   390
Where, through an opening of the rocky bank,
The waters overflow, and a smooth spot
Of glassy quiet mid those battling tides
Is left, the boat paused shuddering.—Shall it sink
Down the abyss? Shall the reverting stress
Of that resistless gulf embosom it?
Now shall it fall?—A wandering stream of wind,
Breathed from the west, has caught the expanded sail,
And, lo! with gentle motion, between banks
Of mossy slope, and on a placid stream,   400
Beneath a woven grove it sails, and, hark!
The ghastly torrent mingles its far roar,
With the breeze murmuring in the musical woods.
Where the embowering trees recede, and leave
A little space of green expanse, the cove
Is closed by meeting banks, whose yellow flowers
For ever gaze on their own drooping eyes,
Reflected in the crystal calm. The wave
Of the boat's motion marred their pensive task,
Which nought but vagrant bird, or wanton wind,
Or falling spear-grass, or their own decay   411
Had e'er disturbed before. The Poet longed
To deck with their bright hues his withered hair,
But on his heart its solitude returned,
And he forbore. Not the strong impulse hid
In those flushed cheeks, bent eyes, and shadowy frame
Had yet performed its ministry: it hung
Upon his life, as lightning in a cloud
Gleams, hovering ere it vanish, ere the floods
Of night close over it.

                        The noonday sun   420
Now shone upon the forest, one vast mass
Of mingling shade, whose brown magnificence

A narrow vale embosoms. There, huge caves,
Scooped in the dark base of their aëry rocks
Mocking its moans, respond and roar for ever.
The meeting boughs and implicated leaves
Wove twilight o'er the Poet's path, as led
By love, or dream, or god, or mightier Death,
He sought in Nature's dearest haunt, some bank,
Her cradle, and his sepulchre. More dark   430
And dark the shades accumulate. The oak,
Expanding its immense and knotty arms,
Embraces the light beech. The pyramids
Of the tall cedar overarching, frame
Most solemn domes within, and far below,
Like clouds suspended in an emerald sky,
The ash and the acacia floating hang
Tremulous and pale. Like restless serpents, clothed
In rainbow and in fire, the parasites,
Starred with ten thousand blossoms, flow around   440
The grey trunks, and, as gamesome infants' eyes,
With gentle meanings, and most innocent wiles,
Fold their beams round the hearts of those that love,
These twine their tendrils with the wedded boughs
Uniting their close union; the woven leaves
Make net-work of the dark blue light of day,
And the night's noontide clearness, mutable
As shapes in the weird clouds. Soft mossy lawns
Beneath these canopies extend their swells,
Fragrant with perfumed herbs, and eyed with blooms
Minute yet beautiful. One darkest glen   451
Sends from its woods of musk-rose, twined with
     jasmine,
A soul-dissolving odour, to invite
To some more lovely mystery. Through the dell,
Silence and Twilight here, twin-sisters, keep
Their noonday watch, and sail among the shades,
Like vaporous shapes half seen; beyond, a well,
Dark, gleaming, and of most translucent wave,
Images all the woven boughs above,
And each depending leaf, and every speck   460
Of azure sky, darting between their chasms;
Nor aught else in the liquid mirror laves
Its portraiture, but some inconstant star
Between one foliaged lattice twinkling fair,
Or, painted bird, sleeping beneath the moon,
Or gorgeous insect floating motionless,
Unconscious of the day, ere yet his wings
Have spread their glories to the gaze of noon.

Hither the Poet came. His eyes beheld
Their own wan light through the reflected lines   470
Of his thin hair, distinct in the dark depth
Of that still fountain; as the human heart,
Gazing in dreams over the gloomy grave,
Sees its own treacherous likeness there. He heard
The motion of the leaves, the grass that sprung
Startled and glanced and trembled even to feel
An unaccustomed presence, and the sound
Of the sweet brook that from the secret springs
Of that dark fountain rose. A Spirit seemed
To stand beside him—clothed in no bright robes   480
Of shadowy silver or enshrining light
Borrowed from aught the visible world affords
Of grace, or majesty, or mystery;—
But, undulating woods, and silent well,
And leaping rivulet, and evening gloom
Now deepening the dark shades, for speech assuming,
Held commune with him, as if he and it
Were all that was,—only . . . when his regard
Was raised by intense pensiveness, . . . two eyes,
Two starry eyes, hung in the gloom of thought,   490
And seemed with their serene and azure smiles
To beckon him.   Obedient to the light
That shone within his soul, he went, pursuing
The windings of the dell.—The rivulet
Wanton and wild, through many a green ravine
Beneath the forest flowed. Sometimes it fell
Among the moss with hollow harmony
Dark and profound. Now on the polished stones
It danced; like childhood laughing as it went:
Then, through the plain in tranquil wanderings crept,
Reflecting every herb and drooping bud   501
That overhung its quietness.—"O stream!
Whose source is inaccessibly profound,
Whither do thy mysterious waters tend?
Thou imagest my life. Thy darksome stillness,
Thy dazzling waves, thy loud and hollow gulfs,
Thy searchless fountain, and invisible course
Have each their type in me: and the wide sky,
And measureless ocean may declare as soon
What oozy cavern or what wandering cloud   510
Contains thy waters, as the universe
Tell where these living thoughts reside, when stretched
Upon thy flowers my bloodless limbs shall waste
I' the passing wind!"

426. implicated: interfolded.

484–86. But . . . assuming: The Spirit assumes the forms
of nature in order to speak with the poet.

Beside the grassy shore
Of the small stream he went; he did impress
On the green moss his tremulous step, that caught
Strong shuddering from his burning limbs. As one
Roused by some joyous madness from the couch
Of fever, he did move; yet, not like him,
Forgetful of the grave, where, when the flame   520
Of his frail exultation shall be spent,
He must descend. With rapid steps he went
Beneath the shade of trees, beside the flow
Of the wild babbling rivulet; and now
The forest's solemn canopies were changed
For the uniform and lightsome evening sky.
Grey rocks did peep from the spare moss, and stemmed
The struggling brook: tall spires of windlestrae
Threw their thin shadows down the rugged slope,
And nought but gnarled trunks of ancient pines   530
Branchless and blasted, clenched with grasping roots
The unwilling soil. A gradual change was here,
Yet ghastly. For, as fast years flow away,
The smooth brow gathers, and the hair grows thin
And white, and where irradiate dewy eyes
Had shone, gleam stony orbs:—so from his steps
Bright flowers departed, and the beautiful shade
Of the green groves, with all their odorous winds
And musical motions. Calm, he still pursued
The stream, that with a larger volume now   540
Rolled through the labyrinthine dell; and there
Fretted a path through its descending curves
With its wintry speed. On every side now rose
Rocks, which, in unimaginable forms,
Lifted their black and barren pinnacles
In the light of evening, and, its precipice
Obscuring the ravine, disclosed above,
Mid toppling stones, black gulfs and yawning caves,
Whose windings gave ten thousand various tongues
To the loud stream. Lo! where the pass expands   550
Its stony jaws, the abrupt mountain breaks,
And seems, with its accumulated crags,
To overhang the world: for wide expand
Beneath the wan stars and descending moon
Islanded seas, blue mountains, mighty streams,
Dim tracts and vast, robed in the lustrous gloom
Of leaden-coloured even, and fiery hills
Mingling their flames with twilight, on the verge

Of the remote horizon. The near scene,
In naked and severe simplicity,   560
Made contrast with the universe. A pine,
Rock-rooted, stretched athwart the vacancy
Its swinging boughs, to each inconstant blast
Yielding one only response, at each pause
In most familiar cadence, with the howl
The thunder and the hiss of homeless streams
Mingling its solemn song, whilst the broad river,
Foaming and hurrying o'er its rugged path,
Fell into that immeasurable void
Scattering its waters to the passing winds.   570

Yet the grey precipice and solemn pine
And torrent, were not all;—one silent nook
Was there. Even on the edge of that vast mountain,
Upheld by knotty roots and fallen rocks,
It overlooked in its serenity
The dark earth, and the bending vault of stars.
It was a tranquil spot, that seemed to smile
Even in the lap of horror. Ivy clasped
The fissured stones with its entwining arms,
And did embower with leaves for ever green,   580
And berries dark, the smooth and even space
Of its inviolated floor, and here
The children of the autumnal whirlwind bore,
In wanton sport, those bright leaves, whose decay,
Red, yellow, or ethereally pale,
Rivals the pride of summer. 'Tis the haunt
Of every gentle wind, whose breath can teach
The wilds to love tranquillity. One step,
One human step alone, has ever broken
The stillness of its solitude:—one voice   590
Alone inspired its echoes;—even that voice
Which hither came, floating among the winds,
And led the loveliest among human forms
To make their wild haunts the depository
Of all the grace and beauty that endued
Its motions, render up its majesty,
Scatter its music on the unfeeling storm,
And to the damp leaves and blue cavern mould,
Nurses of rainbow flowers and branching moss,
Commit the colours of that varying cheek,   600
That snowy breast, those dark and drooping eyes.

The dim and horned moon hung low, and poured
A sea of lustre on the horizon's verge
That overflowed its mountains. Yellow mist
Filled the unbounded atmosphere, and drank

---

528. windlestrae: a coarse, strong grass used for making baskets and ropes. **530. trunks:** the original editions of the poem have "roots," which is plainly a misprint. The emendation to "trunks" was suggested by W. M. Rossetti in 1870.

564. response: Shelley accented this word on the first syllable.

Wan moonlight even to fulness: not a star
Shone, not a sound was heard; the very winds,
Danger's grim playmates, on that precipice
Slept, clasped in his embrace.—O, storm of death!
Whose sightless speed divides this sullen night: 610
And thou, colossal Skeleton, that, still
Guiding its irresistible career
In thy devastating omnipotence,
Art king of this frail world, from the red field
Of slaughter, from the reeking hospital,
The patriot's sacred couch, the snowy bed
Of innocence, the scaffold and the throne,
A mighty voice invokes thee. Ruin calls
His brother Death. A rare and regal prey
He hath prepared, prowling around the world;   620
Glutted with which thou mayst repose, and men
Go to their graves like flowers or creeping worms,
Nor ever more offer at thy dark shrine
The unheeded tribute of a broken heart.

When on the threshold of the green recess
The wanderer's footsteps fell, he knew that death
Was on him. Yet a little, ere it fled,
Did he resign his high and holy soul
To images of the majestic past,
That paused within his passive being now,   630
Like winds that bear sweet music, when they breathe
Through some dim latticed chamber. He did place
His pale lean hand upon the rugged trunk
Of the old pine. Upon an ivied stone
Reclined his languid head, his limbs did rest,
Diffused and motionless, on the smooth brink
Of that obscurest chasm;—and thus he lay,
Surrendering to their final impulses
The hovering powers of life. Hope and despair,
The torturers, slept; no mortal pain or fear   640
Marred his repose, the influxes of sense,
And his own being unalloyed by pain,
Yet feebler and more feeble, calmly fed
The stream of thought, till he lay breathing there
At peace, and faintly smiling:—his last sight
Was the great moon, which o'er the western line
Of the wide world her mighty horn suspended,
With whose dun beams inwoven darkness seemed
To mingle. Now upon the jagged hills
It rests, and still as the divided frame   650
Of the vast meteor sunk, the Poet's blood,
That ever beat in mystic sympathy
With nature's ebb and flow, grew feebler still:

610. sightless: invisible.

And when two lessening points of light alone
Gleamed through the darkness, the alternate gasp
Of his faint respiration scarce did stir
The stagnate night:—till the minutest ray
Was quenched, the pulse yet lingered in his heart.
It paused—it fluttered. But when heaven remained
Utterly black, the murky shades involved   660
An image, silent, cold, and motionless,
As their own voiceless earth and vacant air.
Even as a vapour fed with golden beams
That ministered on sunlight, ere the west
Eclipses it, was now that wondrous frame—
No sense, no motion, no divinity—
A fragile lute, on whose harmonious strings
The breath of heaven did wander—a bright stream
Once fed with many-voiced waves—a dream
Of youth, which night and time have quenched for
      ever,   670
Still, dark, and dry, and unremembered now.

O, for Medea's wondrous alchemy,
Which wheresoe'er it fell made the earth gleam
With bright flowers, and the wintry boughs exhale
From vernal blooms fresh fragrance! O, that God,
Profuse of poisons, would concede the chalice
Which but one living man has drained, who now,
Vessel of deathless wrath, a slave that feels
No proud exemption in the blighting curse
He bears, over the world wanders for ever,   680
Lone as incarnate death! O, that the dream
Of dark magician in his visioned cave,
Raking the cinders of a crucible
For life and power, even when his feeble hand
Shakes in its last decay, were the true law
Of this so lovely world! But thou art fled
Like some frail exhalation; which the dawn
Robes in its golden beams,—ah! thou hast fled!
The brave, the gentle, and the beautiful,
The child of grace and genius. Heartless things   690
Are done and said i' the world, and many worms
And beasts and men live on, and mighty Earth
From sea and mountain, city and wilderness,
In vesper low or joyous orison,
Lifts still its solemn voice:—but thou art fled—

672. Medea's . . . alchemy: She restored her father-in-law to youth through a magical drink. **677. one living man:** the "Wandering Jew," who was condemned to wander the earth until the Second Coming because he had refused to let Christ rest at his house on the way to Calvary. **681. dream:** of immortality. **682. visioned cave:** cave filled with visions.

Thou canst no longer know or love the shapes
Of this phantasmal scene, who have to thee
Been purest ministers, who are, alas!
Now thou art not. Upon those pallid lips
So sweet even in their silence, on those eyes   700
That image sleep in death, upon that form
Yet safe from the worm's outrage, let no tear
Be shed—not even in thought. Nor, when those hues
Are gone, and those divinest lineaments,
Worn by the senseless wind, shall live alone
In the frail pauses of this simple strain,
Let not high verse, mourning the memory
Of that which is no more, or painting's woe
Or sculpture, speak in feeble imagery
Their own cold powers. Art and eloquence,   710
And all the shows o' the world are frail and vain
To weep a loss that turns their lights to shade.
It is a woe too "deep for tears," when all
Is reft at once, when some surpassing Spirit,
Whose light adorned the world around it, leaves
Those who remain behind, not sobs or groans,
The passionate tumult of a clinging hope;
But pale despair and cold tranquillity,
Nature's vast frame, the web of human things,
Birth and the grave, that are not as they were.   720

                              1815 (1816)

MONT BLANC

LINES WRITTEN IN THE VALE
OF CHAMOUNI

A companion poem to the "Hymn to Intellectual
Beauty" (below), "Mont Blanc" was composed at about
the same time (July, 1816) in Switzerland. Both poems
attempt to convey and elucidate an intuition of the
ultimate Being or Power of the cosmos. This Power
both transcends and descends into our world and consciousness, and to it alone can complete reality be
ascribed. The two poems obviously follow different
methods. The "Hymn to Intellectual Beauty" (intellectual in the sense of spiritual) presents its theme through
autobiography and abstract discourse, "Mont Blanc"
through description converted into symbolism (though
not allegory, in which *each* descriptive detail would
suggest an abstract meaning).
    While the "Hymn to Intellectual Beauty" is more or
less Platonic, "Mont Blanc" draws (or can be interpreted

713. **"deep for tears"**: quoted from Wordsworth's ode,
"Intimations of Immortality," l. 203.

as drawing) on several philosophical systems. Like
Coleridge, Shelley was fascinated by technical philosophy. In his poetry, however, he naturally does not
expound his ideas with systematic precision. On the
other hand, he occasionally speculates on questions caught
from philosophers, and he frequently uses philosophic
conceptions as a language in which to articulate feelings
and attitudes. Needless to say, quasi-philosophic poetry
such as this requires a special tact from the reader. He must
sense how far to press for abstract consistency and
precision. If he expects too much, the poetry will fall
apart; but if he ignores the philosophic background, he
may not understand the poetry at all. One should also
keep in mind that "Mont Blanc" voices possible speculations, not final convictions. It ends with a question, and so
does Shelley's career as a whole.
    Shelley's own note to the poem is deprecatory: "It was
composed under the immediate impression of the deep
and powerful feelings excited by the objects which it
attempts to describe; and, as an undisciplined overflowing of the soul, rests its claim to approbation on an
attempt to imitate the untamable wildness and inaccessible solemnity from which those feelings sprang."
    The river Arve runs through the Alpine valley of
Chamonix in southeast France, after rising in the Mer de
Glace glacier on the side of the Mont Blanc chain. Mont
Blanc is the highest mountain in the Alps.

I

The everlasting universe of things
Flows through the mind, and rolls its rapid waves,
Now dark—now glittering—now reflecting gloom—
Now lending splendour, where from secret springs
The source of human thought its tribute brings
Of waters,—with a sound but half its own,
Such as a feeble brook will oft assume
In the wild woods, among the mountains lone,
Where waterfalls around it leap for ever,
Where woods and winds contend, and a vast river   10
Over its rocks ceaselessly bursts and raves.

II

Thus thou, Ravine of Arve—dark, deep Ravine—
Thou many-coloured, many-voiced vale,
Over whose pines, and crags, and caverns sail
Fast cloud-shadows and sunbeams: awful scene,
Where Power in likeness of the Arve comes down
From the ice-gulfs that gird his secret throne,
Bursting through these dark mountains like the flame
Of lightning through the tempest;—thou dost lie,
Thy giant brood of pines around thee clinging,   20

Children of elder time, in whose devotion
The chainless winds still come and ever came
To drink their odours, and their mighty swinging
To hear—an old and solemn harmony;
Thine earthly rainbows stretched across the sweep
Of the aethereal waterfall, whose veil
Robes some unsculptured image; the strange sleep
Which when the voices of the desert fail
Wraps all in its own deep eternity;—
Thy caverns echoing to the Arve's commotion,   30
A loud, lone sound no other sound can tame;
Thou art pervaded with that ceaseless motion,
Thou art the path of that unresting sound—
Dizzy Ravine! and when I gaze on thee
I seem as in a trance sublime and strange
To muse on my own separate fantasy,
My own, my human mind, which passively
Now renders and receives fast influencings,
Holding an unremitting interchange
With the clear universe of things around;   40
One legion of wild thoughts, whose wandering wings
Now float above thy darkness, and now rest
Where that or thou art no unbidden guest,
In the still cave of the witch Poesy,
Seeking among the shadows that pass by,
Ghosts of all things that are, some shade of thee,
Some phantom, some faint image; till the breast
From which they fled recalls them, thou art there!

III

Some say that gleams of a remoter world
Visit the soul in sleep,—that death is slumber,   50
And that its shapes the busy thoughts outnumber
Of those who wake and live.—I look on high;—
Has some unknown omnipotence unfurled
The veil of life and death? or do I lie
In dream, and does the mightier world of sleep
Spread far around and inaccessibly
Its circles? For the very spirit fails,
Driven like a homeless cloud from steep to steep
That vanishes among the viewless gales!
Far, far above, piercing the infinite sky,   60
Mont Blanc appears,—still, snowy, and serene—
Its subject mountains their unearthly forms
Pile around it, ice and rock; broad vales between
Of frozen floods, unfathomable deeps,
Blue as the overhanging heaven, that spread
And wind among the accumulated steeps;
A desert peopled by the storms alone,

Save when the eagle brings some hunter's bone,
And the wolf tracks her there—how hideously
Its shapes are heaped around! rude, bare, and high,   70
Ghastly, and scarred, and riven.—Is this the scene
Where the old Earthquake-daemon taught her young
Ruin? Were these their toys? or did a sea
Of fire envelop once this silent snow?
None can reply—all seems eternal now.
The wilderness has a mysterious tongue
Which teaches awful doubt, or faith so mild,
So solemn, so serene, that man may be,
But for such faith, with nature reconciled;
Thou hast a voice, great Mountain, to repeal   80
Large codes of fraud and woe; not understood
By all, but which the wise, and great, and good
Interpret, or make felt, or deeply feel.

IV

The fields, the lakes, the forests, and the streams,
Ocean, and all the living things that dwell
Within the daedal earth; lightning, and rain,
Earthquake, and fiery flood, and hurricane,
The torpor of the year when feeble dreams
Visit the hidden buds, or dreamless sleep
Holds every future leaf and flower;—the bound   90
With which from that detested trance they leap;
The works and ways of man, their death and birth,
And that of him and all that his may be;
All things that move and breathe with toil and sound
Are born and die; revolve, subside, and swell.
Power dwells apart in its tranquillity,
Remote, serene, and inaccessible:
And *this*, the naked countenance of earth,
On which I gaze, even these primaeval mountains
Teach the adverting mind. The glaciers creep   100
Like snakes that watch their prey, from their far
          fountains,
Slow rolling on; there, many a precipice,
Frost and the Sun in scorn of mortal power
Have piled: dome, pyramid, and pinnacle,
A city of death, distinct with many a tower
And wall impregnable of beaming ice.
Yet not a city, but a flood of ruin
Is there, that from the boundaries of the sky

MONT BLANC **79. But . . . faith:** simply by means of such a
faith. One manuscript reads "In such a faith." **86. daedal:** a
favorite adjective meaning wonderfully contrived or formed
(from Daedalus, the artist who built the Cretan Labyrinth).

Rolls its perpetual stream; vast pines are strewing
Its destined path, or in the mangled soil   110
Branchless and shattered stand; the rocks, drawn down
From yon remotest waste, have overthrown
The limits of the dead and living world,
Never to be reclaimed. The dwelling-place
Of insects, beasts, and birds, becomes its spoil;
Their food and their retreat for ever gone,
So much of life and joy is lost. The race
Of man flies far in dread; his work and dwelling
Vanish, like smoke before the tempest's stream,
And their place is not known. Below, vast caves   120
Shine in the rushing torrents' restless gleam,
Which from those secret chasms in tumult welling
Meet in the vale, and one majestic River,
The breath and blood of distant lands, for ever
Rolls its loud waters to the ocean-waves,
Breathes its swift vapours to the circling air.

V

Mont Blanc yet gleams on high:—the power is there,
The still and solemn power of many sights,
And many sounds, and much of life and death.
In the calm darkness of the moonless nights,   130
In the lone glare of day, the snows descend
Upon that Mountain; none beholds them there,
Nor when the flakes burn in the sinking sun,
Or the star-beams dart through them:—Winds contend
Silently there, and heap the snow with breath
Rapid and strong, but silently! Its home
The voiceless lightning in these solitudes
Keeps innocently, and like vapour broods
Over the snow. The secret Strength of things
Which governs thought, and to the infinite dome
Of Heaven is as a law, inhabits thee!   141
And what were thou, and earth, and stars, and sea,
If to the human mind's imaginings
Silence and solitude were vacancy?

                                    1816 (1817)

HYMN
TO INTELLECTUAL BEAUTY

I

The awful shadow of some unseen Power
   Floats though unseen among us,—visiting

123. one . . . River: The Arve joins the Rhone.

This various world with as inconstant wing
As summer winds that creep from flower to flower,—
Like moonbeams that behind some piny mountain
      shower,
   It visits with inconstant glance
   Each human heart and countenance;
Like hues and harmonies of evening,—
   Like clouds in starlight widely spread,—
   Like memory of music fled,—   10
   Like aught that for its grace may be
Dear, and yet dearer for its mystery.

II

Spirit of BEAUTY, that dost consecrate
   With thine own hues all thou dost shine upon
   Of human thought or form,—where art thou gone?
Why dost thou pass away and leave our state,
This dim vast vale of tears, vacant and desolate?
   Ask why the sunlight not for ever
   Weaves rainbows o'er yon mountain-river,
Why aught should fail and fade that once is shown,
   Why fear and dream and death and birth   21
   Cast on the daylight of this earth
   Such gloom,—why man has such a scope
For love and hate, despondency and hope?

III

No voice from some sublimer world hath ever
   To sage or poet these responses given—
   Therefore the names of Demon, Ghost, and Heaven,
Remain the records of their vain endeavour,
Frail spells—whose uttered charm might not avail to
      sever,
   From all we hear and all we see,   30
   Doubt, chance, and mutability.
Thy light alone—like mist o'er mountains driven,
   Or music by the night-wind sent
   Through strings of some still instrument,
   Or moonlight on a midnight stream,
Gives grace and truth to life's unquiet dream.

HYMN TO INTELLECTUAL BEAUTY. **25–26. No . . . responses:**
that is, revealed religion does not give answers to the
questions in lines 13–17. **27. Demon:** See *Alastor*, above,
n. 118.

IV

Love, Hope, and Self-esteem, like clouds depart
  And come, for some uncertain moments lent.
  Man were immortal, and omnipotent,
Didst thou, unknown and awful as thou art,   40
Keep with thy glorious train firm state within his heart.
    Thou messenger of sympathies,
    That wax and wane in lovers' eyes—
Thou—that to human thought art nourishment,
    Like darkness to a dying flame!
    Depart not as thy shadow came,
    Depart not—lest the grave should be,
Like life and fear, a dark reality.

V

While yet a boy I sought for ghosts, and sped
  Through many a listening chamber, cave and ruin,
  And starlight wood, with fearful steps pursuing   51
Hopes of high talk with the departed dead.
I called on poisonous names with which our youth is
    fed;
    I was not heard—I saw them not—
    When musing deeply on the lot
Of life, at that sweet time when winds are wooing
    All vital things that wake to bring
    News of birds and blossoming,—
    Sudden, thy shadow fell on me;
I shrieked, and clasped my hands in ecstasy!   60

VI

I vowed that I would dedicate my powers
  To thee and thine—have I not kept the vow?
  With beating heart and streaming eyes, even now
I call the phantoms of a thousand hours
Each from his voiceless grave: they have in visioned
    bowers
    Of studious zeal or love's delight
    Outwatched with me the envious night—
They know that never joy illumed my brow
    Unlinked with hope that thou wouldst free
    This world from its dark slavery,
    That thou—O awful LOVELINESS,
Wouldst give whate'er these words cannot express.

49–52. While . . . dead: Cf. *Alastor*, above, ll. 18–29.

VII

The day becomes more solemn and serene
  When noon is past—there is a harmony
  In autumn, and a lustre in its sky,
Which through the summer is not heard or seen,
As if it could not be, as if it had not been!
    Thus let thy power, which like the truth
    Of nature on my passive youth
Descended, to my onward life supply   80
    Its calm—to one who worships thee,
    And every form containing thee,
    Whom, SPIRIT fair, thy spells did bind
To fear himself, and love all human kind.

                         1816 (1817)

OZYMANDIAS

I met a traveller from an antique land
Who said: Two vast and trunkless legs of stone
Stand in the desert . . . Near them, on the sand,
Half sunk, a shattered visage lies, whose frown,
And wrinkled lip, and sneer of cold command,
Tell that its sculptor well those passions read
Which yet survive, stamped on these lifeless things,
The hand that mocked them, and the heart that fed:
And on the pedestal these words appear:
"My name is Ozymandias, king of kings: 10
Look on my works, ye Mighty, and despair!"
Nothing beside remains. Round the decay
Of that colossal wreck, boundless and bare
The lone and level sands stretch far away.

                         1817 (1818)

OZYMANDIAS. **6–8. those . . . fed:** Passions survive the hand
of the artist that mocked (both imitated and derided) them
and the heart of the king that fed them.

LINES WRITTEN AMONG
THE EUGANEAN HILLS

OCTOBER, 1818

❦

One of Shelley's finest poems, these "Lines" present him looking down from the Euganean Hills to the plain of Lombardy with Venice in the distance. In this poem the scenery and color of the Italian landscape first appear in his imagery, which henceforth continues to reflect what Shelley actually saw in Italy at this time. The point is worth stressing, for readers are likely to interpret these images—hues of amethyst, azure, and gold, architectural structures glowing like fire in the sun, embowered islands, Roman ruins still standing and overgrown with flowers—as highly imaginative or idealized, when in fact they are often fairly close description. The island at the end of the poem is, of course, imaginative (and may be compared with similar retreats in *Prometheus Unbound*, III. iii, and at the end of *Epipsychidion*). In its influence on the "polluting multitude," it may be a symbol of poetry itself, which, Shelley believed, both serves as an escape from the world and at the same time educates or transforms the soul.

❦

Many a green isle needs must be
In the deep wide sea of Misery,
Or the mariner, worn and wan,
Never thus could voyage on—
Day and night, and night and day,
Drifting on his dreary way,
With the solid darkness black
Closing round his vessel's track;
Whilst above the sunless sky,
Big with clouds, hangs heavily,   10
And behind the tempest fleet
Hurries on with lightning feet,
Riving sail, and cord, and plank,
Till the ship has almost drank
Death from the o'er-brimming deep;
And sinks down, down, like that sleep
When the dreamer seems to be
Weltering through eternity;
And the dim low line before
Of a dark and distant shore   20
Still recedes, as ever still
Longing with divided will,
But no power to seek or shun,
He is ever drifted on

O'er the unreposing wave
To the haven of the grave.
What, if there no friends will greet;
What, if there no heart will meet
His with love's impatient beat;
Wander wheresoe'er he may,   30
Can he dream before that day
To find refuge from distress
In friendship's smile, in love's caress?
Then 'twill wreak him little woe
Whether such there be or no:
Senseless is the breast, and cold,
Which relenting love would fold;
Bloodless are the veins and chill
Which the pulse of pain did fill;
Every little living nerve   40
That from bitter words did swerve
Round the tortured lips and brow,
Are like sapless leaflets now
Frozen upon December's bough.

On the beach of a northern sea
Which tempests shake eternally,
As once the wretch there lay to sleep,
Lies a solitary heap,
One white skull and seven dry bones,
On the margin of the stones,   50
Where a few gray rushes stand,
Boundaries of the sea and land:
Nor is heard one voice of wail
But the sea-mews', as they sail
O'er the billows of the gale;
Or the whirlwind up and down
Howling, like a slaughtered town,
When a king in glory rides
Through the pomp of fratricides:
Those unburied bones around   60
There is many a mournful sound;
There is no lament for him,
Like a sunless vapour, dim,
Who once clothed with life and thought
What now moves nor murmurs not.

Ay, many flowering islands lie
In the waters of wide Agony:
To such a one this morn was led,
My bark by soft winds piloted:

LINES WRITTEN AMONG THE EUGANEAN HILLS. **45–65. On . . . not:** These strange and beautiful lines are probably a fantasy of Shelley's in which he imagines he has died.

'Mid the mountains Euganean 70
I stood listening to the paean
With which the legioned rooks did hail
The sun's uprise majestical;
Gathering round with wings all hoar,
Through the dewy mist they soar
Like gray shades, till the eastern heaven
Bursts, and then, as clouds of even,
Flecked with fire and azure, lie
In the unfathomable sky,
So their plumes of purple grain, 80
Starred with drops of golden rain,
Gleam above the sunlight woods,
As in silent multitudes
On the morning's fitful gale
Through the broken mist they sail,
And the vapours cloven and gleaming
Follow, down the dark steep streaming,
Till all is bright, and clear, and still,
Round the solitary hill.

Beneath is spread like a green sea 90
The waveless plain of Lombardy,
Bounded by the vaporous air,
Islanded by cities fair;
Underneath Day's azure eyes
Ocean's nursling, Venice lies,
A peopled labyrinth of walls,
Amphitrite's destined halls,
Which her hoary sire now paves
With his blue and beaming waves.
Lo! the sun upsprings behind, 100
Broad, red, radiant, half-reclined
On the level quivering line
Of the waters crystalline;
And before that chasm of light,
As within a furnace bright,
Column, tower, and dome, and spire,
Shine like obelisks of fire,
Pointing with inconstant motion
From the altar of dark ocean
To the sapphire-tinted skies; 110
As the flames of sacrifice
From the marble shrines did rise,
As to pierce the dome of gold
Where Apollo spoke of old.

Sun-girt City, thou hast been
Ocean's child, and then his queen;
Now is come a darker day,
And thou soon must be his prey,
If the power that raised thee here
Hallow so thy watery bier. 120
A less drear ruin then than now,
With thy conquest-branded brow
Stooping to the slave of slaves
From thy throne, among the waves
Wilt thou be, when the sea-mew
Flies, as once before it flew,
O'er thine isles depopulate,
And all is in its ancient state,
Save where many a palace gate
With green sea-flowers overgrown 130
Like a rock of Ocean's own,
Topples o'er the abandoned sea
As the tides change sullenly.
The fisher on his watery way,
Wandering at the close of day,
Will spread his sail and seize his oar
Till he pass the gloomy shore,
Lest thy dead should, from their sleep
Bursting o'er the starlight deep,
Lead a rapid masque of death 140
O'er the waters of his path.

Those who alone thy towers behold
Quivering through aëreal gold,
As I now behold them here,
Would imagine not they were
Sepulchres, where human forms,
Like pollution-nourished worms,
To the corpse of greatness cling,
Murdered, and now mouldering:
But if Freedom should awake 150
In her omnipotence, and shake
From the Celtic Anarch's hold
All the keys of dungeons cold,

97–98. **Amphitrite's . . . sire:** Amphitrite, wife of Poseidon (or Neptune) was the daughter of the sea-god Nereus. 111–14. **As . . . old:** at the oracle of Delphi.

116. **queen:** referring to the annual ceremony celebrating the marriage of Venice and the sea, during which a ring was cast into the water (cf. Wordsworth's "On the Extinction of the Venetian Republic," above, ll. 7–8). 117. **darker day:** Venice was now under the domination of Austria. 120. **watery bier:** Shelley echoes the phrase in Milton's "Lycidas," l. 12. The allusion is based on the belief that Venice is slowly sinking. 123. **slave of slaves:** the Austrian Emperor, Francis I. 132. **Topples:** overhangs, as if about to fall. 152. **Celtic Anarch:** the Austrian Emperor ("Celtic" having been broadly used by the Romans to apply to northern barbarians generally). The Austrians ruled much of Italy at this time.

Where a hundred cities lie
Chained like thee, ingloriously,
Thou and all thy sister band
Might adorn this sunny land,
Twining memories of old time
With new virtues more sublime;
If not, perish thou and they!—   160
Clouds which stain truth's rising day
By her sun consumed away—
Earth can spare ye: while like flowers,
In the waste of years and hours,
From your dust new nations spring
With more kindly blossoming.

Perish—let there only be
Floating o'er thy hearthless sea
As the garment of thy sky
Clothes the world immortally,   170
One remembrance, more sublime
Than the tattered pall of time,
Which scarce hides thy visage wan;—
That a tempest-cleaving Swan
Of the songs of Albion,
Driven from his ancestral streams
By the might of evil dreams,
Found a nest in thee; and Ocean
Welcomed him with such emotion
That its joy grew his, and sprung   180
From his lips like music flung
O'er a mighty thunder-fit,
Chastening terror:—what though yet
Poesy's unfailing River,
Which through Albion winds forever
Lashing with melodious wave
Many a sacred Poet's grave,
Mourn its latest nursling fled?
What though thou with all thy dead
Scarce can for this fame repay   190
Aught thine own? oh, rather say
Though thy sins and slaveries foul
Overcloud a sunlike soul?
As the ghost of Homer clings
Round Scamander's wasting springs;
As divinest Shakespeare's might
Fills Avon and the world with light
Like omniscient power which he
Imaged 'mid mortality;

As the love from Petrach's urn,   200
Yet amid yon hills doth burn,
A quenchless lamp by which the heart
Sees things unearthly;—so thou art,
Mighty spirit—so shall be
The City that did refuge thee.

Lo, the sun floats up the sky
Like thought-winged Liberty,
Till the universal light
Seems to level plain and height;
From the sea a mist has spread,   210
And the beams of morn lie dead
On the towers of Venice now,
Like its glory long ago.
By the skirts of that gray cloud
Many-domed Padua proud
Stands, a peopled solitude,
'Mid the harvest-shining plain,
Where the peasant heaps his grain
In the garner of his foe,
And the milk-white oxen slow   220
With the purple vintage strain,
Heaped upon the creaking wain,
That the brutal Celt may swill
Drunken sleep with savage will;
And the sickle to the sword
Lies unchanged, though many a lord,
Like a weed whose shade is poison,
Overgrows this region's foison,
Sheaves of whom are ripe to come
To destruction's harvest-home: 230
Men must reap the things they sow,
Force from force must ever flow,
Or worse; but 'tis a bitter woe
That love or reason cannot change
The despot's rage, the slave's revenge.

Padua, thou within whose walls
Those mute guests at festivals,
Son and Mother, Death and Sin,
Played at dice for Ezzelin,
Till Death cried, "I win, I win!"   240
And Sin cursed to lose the wager,

---

222. wain: waggon. 228. foison: harvest. 238–39. Son . . .
Ezzelin: In the play at dice Shelley echoes Coleridge's
"Rime of the Ancient Mariner," ll. 195–98; in *Paradise Lost*,
II. 648–809, Death figures as the son of Sin and commits
rape upon his mother; and the reference is to Ezzelin III,
a thirteenth-century ruler of Padua who was known for
his cruelty and who committed suicide when taken prisoner.

171. One . . . sublime: that of Byron. 175. Albion:
England.

But Death promised, to assuage her,
That he would petition for
Her to be made Vice-Emperor,
When the destined years were o'er,
Over all between the Po
And the eastern Alpine snow,
Under the mighty Austrian.
Sin smiled so as Sin only can,
And since that time, ay, long before,   250
Both have ruled from shore to shore,—
That incestuous pair, who follow
Tyrants as the sun the swallow,
As Repentance follows Crime,
And as changes follow Time.

In thine halls the lamp of learning,
Padua, now no more is burning;
Like a meteor, whose wild way
Is lost over the grave of day,
It gleams betrayed and to betray: 260
Once remotest nations came
To adore that sacred flame,
When it lit not many a hearth
On this cold and gloomy earth:
Now new fires from antique light
Spring beneath the wide world's might;
But their spark lies dead in thee,
Trampled out by Tyranny.
As the Norway woodman quells,
In the depth of piny dells,   270
One light flame among the brakes,
While the boundless forest shakes,
And its mighty trunks are torn
By the fire thus lowly born:
The spark beneath his feet is dead,
He starts to see the flames it fed
Howling through the darkened sky
With a myriad tongues victoriously,
And sinks down in fear: so thou,
O Tyranny, beholdest now   280
Light around thee, and thou hearest
The loud flames ascend, and fearest:
Grovel on the earth; ay, hide
In the dust thy purple pride!

Noon descends around me now:
'Tis the noon of autumn's glow,
When a soft and purple mist
Like a vaporous amethyst,

Or an air-dissolved star
Mingling light and fragrance, far   290
From the curved horizon's bound
To the point of Heaven's profound,
Fills the overflowing sky;
And the plains that silent lie
Underneath, the leaves unsodden
Where the infant Frost has trodden
With his morning-winged feet,
Whose bright print is gleaming yet;
And the red and golden vines,
Piercing with their trellised lines   300
The rough, dark-skirted wilderness;
The dun and bladed grass no less,
Pointing from this hoary tower
In the windless air; the flower
Glimmering at my feet; the line
Of the olive-sandalled Apennine
In the south dimly islanded;
And the Alps, whose snows are spread
High between the clouds and sun;
And of living things each one;   310
And my spirit which so long
Darkened this swift stream of song,—
Interpenetrated lie
By the glory of the sky:
Be it love, light, harmony,
Odour, or the soul of all
Which from Heaven like dew doth fall,
Or the mind which feeds this verse
Peopling the lone universe.

Noon descends, and after noon   320
Autumn's evening meets me soon,

292. **Heaven's profound:** the zenith. 306. **olive-sandalled:** olive trees at the foot of the mountains. 312. **Darkened:** with melancholy. 315-19. **Be . . . universe:** The passage is remarkably condensed: that sense of the consoling "glory" of the sky, with which the poet feels himself and all things "Interpenetrated," is itself an expression or result of his being momentarily in contact with the ultimate principle or reality of things. It may be described as love (cf. "Adonais," above, ll. 481–82: "that sustaining Love . . . through the web of being blindly wove"), light (both a physical and spiritual quality), harmony (that binds all things together in a living and changing pattern of beauty), or odor (for Shelley, the refined essence of physical beauty—body turning into spirit). Or more philosophically considered, it may be the soul of all these qualities, the Platonic One (Shelley's Intellectual Beauty) intuited through and in the physical scene. Or the ultimate reality may be conceived as thought or "mind" creating all things just as it inspires the poet's utterance.

256. **lamp of learning:** the University of Padua.

Leading the infantine moon,
And that one star, which to her
Almost seems to minister
Half the crimson light she brings
From the sunset's radiant springs:
And the soft dreams of the morn
(Which like winged winds had borne
To that silent isle, which lies
Mid remembered agonies,   330
The frail bark of this lone being)
Pass, to other sufferers fleeing,
And its ancient pilot, Pain,
Sits beside the helm again.

Other flowering isles must be
In the sea of Life and Agony:
Other spirits float and flee
O'er that gulf: even now, perhaps,
On some rock the wild wave wraps,
With folded wings they waiting sit   340
For my bark, to pilot it
To some calm and blooming cove,
Where for me, and those I love,
May a windless bower be built,
Far from passion, pain, and guilt,
In a dell mid lawny hills,
Which the wild sea-murmur fills,
And soft sunshine, and the sound
Of old forests echoing round,
And the light and smell divine   350
Of all flowers that breathe and shine:
We may live so happy there,
That the Spirits of the Air,
Envying us, may even entice
To our healing Paradise
The polluting multitude;
But their rage would be subdued
By that clime divine and calm,
And the winds whose wings rain balm
On the uplifted soul, and leaves   360
Under which the bright sea heaves;
While each breathless interval
In their whisperings musical
The inspired soul supplies
With its own deep melodies,
And the love which heals all strife
Circling, like the breath of life,
All things in that sweet abode
With its own mild brotherhood:

They, not it, would change; and soon   370
Every sprite beneath the moon
Would repent its envy vain,
And the earth grow young again.

1818 (1819)

FROM

JULIAN AND MADDALO

A CONVERSATION

PREFACE

Count Maddalo[1] is a Venetian nobleman of ancient family and of great fortune, who, without mixing much in the society of his countrymen, resides chiefly at his magnificent palace in that city. He is a person of the most consummate genius, and capable, if he would direct his energies to such an end, of becoming the redeemer of his degraded country. But it is his weakness to be proud: he derives, from a comparison of his own extraordinary mind with the dwarfish intellects that surround him, an intense apprehension of the nothingness of human life. His passions and his powers are incomparably greater than those of other men; and, instead of the latter having been employed in curbing the former, they have mutually lent each other strength. His ambition preys upon itself, for want of objects which it can consider worthy of exertion. I say that Maddalo is proud, because I can find no other word to express the concentered and impatient feelings which consume him; but it is on his own hopes and affections only that he seems to trample, for in social life no human being can be more gentle, patient, and unassuming than Maddalo. He is cheerful, frank, and witty. His more serious conversation is a sort of intoxication; men are held by it as by a spell. He has travelled much; and there is an inexpressible charm in his relation of his adventures in different countries.

Julian is an Englishman of good family, passionately attached to those philosophical notions which assert the power of man over his own mind, and the immense improvements of which, by the extinction of certain moral superstitions, human society may be yet

331. **this lone being:** the poet.

371. **Every sprite:** every human spirit. JULIAN AND MADDALO. **Preface: 1. Count Maddalo:** Maddalo is Byron and Julian is Shelley.

susceptible. Without concealing the evil in the world, he is for ever speculating how good may be made superior. He is a complete infidel, and a scoffer at all things reputed holy; and Maddalo takes a wicked pleasure in drawing out his taunts against religion. What Maddalo thinks on these matters is not exactly known. Julian, in spite of his heterodox opinions, is conjectured by his friends to possess some good qualities. How far this is possible the pious reader will determine. Julian is rather serious.

I rode one evening with Count Maddalo
Upon the bank of land which breaks the flow
Of Adria towards Venice: a bare strand
Of hillocks, heaped from ever-shifting sand,
Matted with thistles and amphibious weeds,
Such as from earth's embrace the salt ooze breeds,
Is this; an uninhabited sea-side,
Which the lone fisher, when his nets are dried,
Abandons; and no other object breaks
The waste, but one dwarf tree and some few stakes 10
Broken and unrepaired, and the tide makes
A narrow space of level sand thereon,
Where 'twas our wont to ride while day went down.
This ride was my delight. I love all waste
And solitary places; where we taste
The pleasure of believing what we see
Is boundless, as we wish our souls to be:
And such was this wide ocean, and this shore
More barren than its billows; and yet more
Than all, with a remembered friend I love   20
To ride as then I rode;—for the winds drove
The living spray along the sunny air
Into our faces; the blue heavens were bare,
Stripped to their depths by the awakening north;
And, from the waves, sound like delight broke forth
Harmonising with solitude, and sent
Into our hearts aëreal merriment.
So, as we rode, we talked; and the swift thought,
Winging itself with laughter, lingered not,
But flew from brain to brain,—such glee was ours, 30
Charged with light memories of remembered hours,
None slow enough for sadness: till we came
Homeward, which always makes the spirit tame.
This day had been cheerful but cold, and now
The sun was sinking, and the wind also.
Our talk grew somewhat serious, as may be
Talk interrupted with such raillery

As mocks itself, because it cannot scorn
The thoughts it would extinguish:—'twas forlorn,
Yet pleasing, such as once, so poets tell,   40
The devils held within the dales of Hell
Concerning God, freewill and destiny:
Of all that earth has been or yet may be,
All that vain men imagine or believe,
Or hope can paint or suffering may achieve,
We descanted, and I (for ever still
Is it not wise to make the best of ill?)
Argued against despondency, but pride
Made my companion take the darker side.
The sense that he was greater than his kind   50
Had struck, methinks, his eagle spirit blind
By gazing on its own exceeding light.
Meanwhile the sun paused ere it should alight,
Over the horizon of the mountains;—Oh,
How beautiful is sunset, when the glow
Of Heaven descends upon a land like thee,
Thou Paradise of exiles, Italy!
Thy mountains, seas, and vineyards, and the towers
Of cities they encircle!—it was ours
To stand on thee, beholding it: and then,   60
Just where we had dismounted, the Count's men
Were waiting for us with the gondola.—
As those who pause on some delightful way
Though bent on pleasant pilgrimage, we stood
Looking upon the evening, and the flood
Which lay between the city and the shore,
Paved with the image of the sky . . . the hoar
And aëry Alps towards the North appeared,
Through mist, an heaven-sustaining bulwark reared
Between the East and West; and half the sky   70
Was roofed with clouds of rich emblazonry
Dark purple at the zenith, which still grew
Down the steep West into a wondrous hue
Brighter than burning gold, even to the rent
Where the swift sun yet paused in his descent
Among the many-folded hills: they were
Those famous Euganean hills, which bear,
As seen from Lido thro' the harbour piles,
The likeness of a clump of peaked isles—
And then—as if the Earth and Sea had been   80
Dissolved into one lake of fire, were seen
Those mountains towering as from waves of flame
Around the vaporous sun, from which there came
The inmost purple spirit of light, and made
Their very peaks transparent. "Ere it fade,"

---

2. **bank of land:** the Lido. 3. **Adria:** the Adriatic Sea.

39–42. **forlorn . . . destiny:** See *Paradise Lost*, II. 555–69.

Said my companion, "I will show you soon
A better station"—so, o'er the lagune
We glided; and from that funereal bark
I leaned, and saw the city, and could mark
How from their many isles, in evening's gleam,   90
Its temples and its palaces did seem
Like fabrics of enchantment piled to Heaven.
I was about to speak, when—"We are even
Now at the point I meant," said Maddalo,
And bade the gondolieri cease to row.
"Look, Julian, on the west, and listen well
If you hear not a deep and heavy bell."
I looked, and saw between us and the sun
A building on an island; such a one
As age to age might add, for uses vile,   100
A windowless, deformed and dreary pile;
And on the top an open tower, where hung
A bell, which in the radiance swayed and swung;
We could just hear its hoarse and iron tongue:
The broad sun sunk behind it, and it tolled
In strong and black relief.—"What we behold
Shall be the madhouse and its belfry tower,"
Said Maddalo, "and ever at this hour
Those who may cross the water, hear that bell
Which calls the maniacs, each one from his cell,   110
To vespers."—"As much skill as need to pray
In thanks or hope for their dark lot have they
To their stern maker," I replied. "O ho!
You talk as in years past," said Maddalo.
"'Tis strange men change not. You were ever still
Among Christ's flock a perilous infidel,
A wolf for the meek lambs—if you can't swim
Beware of Providence." I looked on him,
But the gay smile had faded in his eye.
"And such,"—he cried, "is our mortality,   120
And this must be the emblem and the sign
Of what should be eternal and divine!—
And like that black and dreary bell, the soul,
Hung in a heaven-illumined tower, must toll
Our thoughts and our desires to meet below
Round the rent heart and pray—as madmen do
For what? they know not,—till the night of death
As sunset that strange vision, severeth
Our memory from itself, and us from all
We sought and yet were baffled." I recall   130
The sense of what he said, although I mar
The force of his expressions. The broad star
Of day meanwhile had sunk behind the hill,
And the black bell became invisible,
And the red tower looked gray, and all between

The churches, ships and palaces were seen
Huddled in gloom;—into the purple sea
The orange hues of heaven sunk silently.
We hardly spoke, and soon the gondola
Conveyed me to my lodging by the way.   140

    The following morn was rainy, cold and dim:
Ere Maddalo arose, I called on him,
And whilst I waited with his child I played;
A lovelier toy sweet Nature never made,
A serious, subtle, wild, yet gentle being,
Graceful without design and unforeseeing,
With eyes—Oh speak not of her eyes!—which seem
Twin mirrors of Italian Heaven, yet gleam
With such deep meaning, as we never see
But in the human countenance: with me   150
She was a special favourite: I had nursed
Her fine and feeble limbs when she came first
To this bleak world; and she yet seemed to know
On second sight her ancient playfellow,
Less changed than she was by six months or so;
For after her first shyness was worn out
We sate there, rolling billiard balls about,
When the Count entered. Salutations past—
"The word you spoke last night might well have cast
A darkness on my spirit—if man be   160
The passive thing you say, I should not see
Much harm in the religions and old saws
(Tho' I may never own such leaden laws)
Which break a teachless nature to the yoke:
Mine is another faith"—thus much I spoke
And noting he replied not, added: "See
This lovely child, blithe, innocent and free;
She spends a happy time with little care,
While we to such sick thoughts subjected are
As came on you last night—it is our will   170
That thus enchains us to permitted ill—
We might be otherwise—we might be all
We dream of happy, high, majestical.
Where is the love, beauty, and truth we seek
But in our mind? and if we were not weak
Should we be less in deed than in desire?"
"Ay, if we were not weak—and we aspire
How vainly to be strong!" said Maddalo:
"You talk Utopia." "It remains to know,"
I then rejoined, "and those who try may find   180
How strong the chains are which our spirit bind;
Brittle perchance as straw . . . We are assured
Much may be conquered, much may be endured,
Of what degrades and crushes us. We know

That we have power over ourselves to do
And suffer—what, we know not till we try;
But something nobler than to live and die—
So taught those kings of old philosophy
Who reigned, before Religion made men blind;
And those who suffer with their suffering kind   190
Yet feel their faith, religion." "My dear friend,"
Said Maddalo, "my judgement will not bend
To your opinion, though I think you might
Make such a system refutation-tight
As far as words go. I knew one like you
Who to this city came some months ago,
With whom I argued in this sort, and he
Is now gone mad,—and so he answered me,—
Poor fellow! but if you would like to go
We'll visit him, and his wild talk will show   200
How vain are such aspiring theories."
"I hope to prove the induction otherwise,
And that a want of that true theory, still,
Which seeks a 'soul of goodness' in things ill
Or in himself or others, has thus bowed
His being—there are some by nature proud,
Who patient in all else demand but this—
To love and be beloved with gentleness;
And being scorned, what wonder if they die
Some living death? this is not destiny   210
But man's own wilful ill."

                                      1818 (1824)

STANZAS

WRITTEN IN DEJECTION, NEAR NAPLES

I

The sun is warm, the sky is clear,
    The waves are dancing fast and bright,
Blue isles and snowy mountains wear
    The purple noon's transparent might,
    The breath of the moist earth is light,
Around its unexpanded buds;
    Like many a voice of one delight,
    The winds, the birds, the ocean floods,
The City's voice itself, is soft like Solitude's.

II

I see the Deep's untrampled floor   10
    With green and purple seaweeds strown;

I see the waves upon the shore,
    Like light dissolved in star-showers, thrown:
    I sit upon the sands alone,—
The lightning of the noontide ocean
    Is flashing round me, and a tone
    Arises from its measured motion,
How sweet! did any heart now share in my emotion.

III

Alas! I have nor hope nor health,
    Nor peace within nor calm around,   20
Nor that content surpassing wealth
    The sage in meditation found,
    And walked with inward glory crowned—
Nor fame, nor power, nor love, nor leisure.
    Others I see whom these surround—
    Smiling they live, and call life pleasure;—
To me that cup has been dealt in another measure.

IV

Yet now despair itself is mild,
    Even as the winds and waters are;
I could lie down like a tired child,   30
    And weep away the life of care
    Which I have borne and yet must bear,
Till death like sleep might steal on me,
    And I might feel in the warm air
My cheek grow cold, and hear the sea
Breathe o'er my dying brain its last monotony.

V

Some might lament that I were cold,
    As I, when this sweet day is gone,
Which my lost heart, too soon grown old,
    Insults with this untimely moan;   40
    They might lament—for I am one

STANZAS. **41–45. They . . . yet:** The lines can be variously interpreted. The sense is perhaps that some might lament his death because, though they do not love him, they regret the sorrow and failure of his life (see st. iii.). In contrast the day is enjoyed (he is not loved); it dies in "stainless glory" (he has not won "glory," l. 23); and while he would be forgotten or remembered with regret, the day lingers in the afterglow like joy in memory (or, more loosely, though the day was enjoyed, the memory of it is also joyful).

Whom men love not,—and yet regret;
　Unlike this day, which, when the sun
　Shall on its stainless glory set,
Will linger, though enjoyed, like joy in memory yet.

1818 (1824)

SONNET

LIFT NOT THE PAINTED VEIL

Lift not the painted veil which those who live
Call Life: though unreal shapes be pictured there,
And it but mimic all we would believe
With colours idly spread,—behind, lurk Fear
And Hope, twin Destinies; who ever weave
Their shadows, o'er the chasm, sightless and drear.
I knew one who had lifted it—he sought,
For his lost heart was tender, things to love,
But found them not, alas! nor was there aught
The world contains, the which he could approve.
Through the unheeding many he did move,   11
A splendour among shadows, a bright blot
Upon this gloomy scene, a Spirit that strove
For truth, and like the Preacher found it not.

1818 (1824)

PROMETHEUS UNBOUND

A LYRICAL DRAMA

IN FOUR ACTS

The myth that suggested Shelley's lyrical drama is told by Asia in II. iv. 32–100 and more sketchily in other places. It is summarized by Mrs. Shelley in a note:

Shelley followed certain classical authorities in figuring Saturn as the good principle, Jupiter the usurping evil one, and Prometheus as the regenerator, who, unable to bring mankind back to primitive innocence, used knowledge as a weapon to defeat evil, by leading mankind, beyond the state wherein they are sinless through ignorance, to that in which they are virtuous through wisdom. Jupiter punished the temerity of the Titan by chaining him to a rock of Caucasus, and causing a vulture to devour his still-renewed heart.

SONNET. **14. the Preacher:** Ecclesiastes.

There was a prophecy afloat in heaven portending the fall of Jove, the secret of averting which was known only to Prometheus; and the god offered freedom from torture on condition of its being communicated to him. According to the mythological story, this referred to the offspring of Thetis, who was destined to be greater than his father. Prometheus at last bought pardon for his crime of enriching mankind with his gifts, by revealing the prophecy. Hercules killed the vulture, and set him free; and Thetis was married to Peleus, the father of Achilles.

In using the myth Shelley altered it drastically, and he made it the vehicle of his own insights and convictions. As the outward representative of evil, Jupiter, the type of all tyrants, exhibits hate and cruelty, and also the unappeasable insecurity, distrust, and fear that make for cruelty (hence his wild glee at the start of Act III, when at last he thinks himself safe). He is also associated with such dispositions as hypocrisy, self-mistrust, and self-contempt. To overcome all this there is only one way, love; hence the crucial process that leads to Jupiter's downfall takes place in the mind of Prometheus. For Jupiter is dependent on Prometheus; he may even be conceived as a personification of the hatred Prometheus feels (thus I. 258–61 describes either Prometheus uttering the curse or Jupiter). The change of heart that dooms this god of hate has occurred by line 53 of the first act, where Prometheus feels pity for him, though the effects of this great event are not felt immediately. Once mankind, represented in Prometheus, learns through suffering to love, life takes on the utopian character celebrated in Acts III and IV. These scenes must be interpreted both as an ardent image of what man might actually become on earth and as a heavenly vision. For there is much sinewy reasoning in Shelley's drama, but there is also a directly visionary imagination of the highest intensity. That is why W. B. Yeats placed *Prometheus Unbound* "among the sacred books of the world."

The poem has traditionally been interpreted as an allegory, though without agreement in details. More recently there has been a tendency to concentrate on the imagery, explaining its sources and symbolic meaning. Shelley's own Preface emphasizes that his purpose is not so much to communicate truths—esoteric or not—but rather to present "Idealisms" (ideal examples) of "Moral excellence" and thereby to excite the emotions of admiration and love that are the springs of moral regeneration. The statement may well guide readers. Remoter meanings should be pondered—the more attentive we are the better—but what is most obvious remains most important.

For example, the encounter with Demogorgon remains profound and awakening even if we do not assign a specific allegorical or symbolic significance to this figure. Whatever he is, we understand that the human quest for answers—to such questions as the origin of the cosmos and of evil—has come to an ultimate confrontation. The effect lies partly in the tone of mind with which the confrontation is approached—the sense that such questions must be asked with a deep awareness of their awful import—and in the final refusal to give an answer. Man's striving has gone as far as it can, but the truth lies beyond human conception.

Shelley began the poem in the autumn of 1818 and finished Act I by October 8. Acts II and III were composed mainly in the early spring of 1819 in Rome, and Act IV several months later. The drama may be compared with Byron's "Prometheus," above.

PREFACE

The Greek tragic writers, in selecting as their subject any portion of their national history or mythology, employed in their treatment of it a certain arbitrary discretion. They by no means conceived themselves bound to adhere to the common interpretation or to imitate in story as in title their rivals and predecessors. Such a system would have amounted to a resignation of those claims to preference over their competitors which incited the composition. The Agamemnonian story was exhibited on the Athenian theatre with as many variations as dramas.

I have presumed to employ a similar licence. The *Prometheus Unbound* of Æschylus supposed the reconciliation of Jupiter with his victim as the price of the disclosure of the danger threatened to his empire by the consummation of his marriage with Thetis. Thetis, according to this view of the subject, was given in marriage to Peleus, and Prometheus, by the permission of Jupiter, delivered from his captivity by Hercules. Had I framed my story on this model, I should have done no more than have attempted to restore the lost drama of Æschylus; an ambition which, if my preference to this mode of treating the subject had incited me to cherish, the recollection of the high comparison such an attempt would challenge might well abate. But, in truth, I was averse from a catastrophe so feeble as that of reconciling the Champion with the Oppressor of mankind. The moral interest of the fable, which is so powerfully sustained by the sufferings and endurance of Prometheus, would be annihilated if we could conceive of him as unsaying his high language and quailing before his successful and perfidious adversary. The only imaginary being resembling in any degree Prometheus, is Satan; and Prometheus is, in my judgement, a more poetical character than Satan, because, in addition to courage, and majesty, and firm and patient opposition to omnipotent force, he is susceptible of being described as exempt from the taints of ambition, envy, revenge, and a desire for personal aggrandisement, which, in the Hero of *Paradise Lost*, interfere with the interest. The character of Satan engenders in the mind a pernicious casuistry which leads us to weigh his faults with his wrongs, and to excuse the former because the latter exceed all measure. In the minds of those who consider that magnificent fiction with a religious feeling it engenders something worse. But Prometheus is, as it were, the type of the highest perfection of moral and intellectual nature, impelled by the purest and the truest motives to the best and noblest ends.

This Poem was chiefly written upon the mountainous ruins of the Baths of Caracalla, among the flowery glades, and thickets of ordoriferous blossoming trees, which are extended in ever winding labyrinths upon its immense platforms and dizzy arches suspended in the air. The bright blue sky of Rome, and the effect of the vigorous awakening spring in that divinest climate, and the new life with which it drenches the spirits even to intoxication, were the inspiration of this drama.

The imagery which I have employed will be found, in many instances, to have been drawn from the operations of the human mind, or from those external actions by which they are expressed. This is unusual in modern poetry, although Dante and Shakespeare are full of instances of the same kind: Dante indeed more than any other poet, and with greater success. But the Greek poets, as writers to whom no resource of awakening the sympathy of their contemporaries was unknown, were in the habitual use of this power; and it is the study of their works (since a higher merit would probably be denied me) to which I am willing that my readers should impute this singularity.

One word is due in candour to the degree in which the study of contemporary writings may have tinged my composition, for such has been a topic of censure with regard to poems far more popular, and indeed more deservedly popular, than mine. It is impossible that any one who inhabits the same age with such writers as those who stand in the foremost ranks of our own, can conscientiously assure himself that his language and tone of thought may not have been modified by the study of the productions of those extraordinary intellects. It is true, that, not the spirit of their genius, but the forms in which it has manifested itself, are due less to the peculiarities of their own minds than to the peculiarity of the moral and intellectual condition of the minds among which they have been produced.

Thus a number of writers possess the form, whilst they want the spirit of those whom, it is alleged, they imitate; because the former is the endowment of the age in which they live, and the latter must be the uncommunicated lightning of their own mind.

The peculiar style of intense and comprehensive imagery which distinguishes the modern literature of England, has not been, as a general power, the product of the imitation of any particular writer. The mass of capabilities remains at every period materially the same; the circumstances which awaken it to action perpetually change. If England were divided into forty republics, each equal in population and extent to Athens, there is no reason to suppose but that, under institutions not more perfect than those of Athens, each would produce philosophers and poets equal to those who (if we except Shakespeare) have never been surpassed. We owe the great writers of the golden age of our literature to that fervid awakening of the public mind which shook to dust the oldest and most oppressive form of the Christian religion. We owe Milton to the progress and development of the same spirit: the sacred Milton was, let it ever be remembered, a republican, and a bold inquirer into morals and religion. The great writers of our own age are, we have reason to suppose, the companions and forerunners of some unimagined change in our social condition or the opinions which cement it. The cloud of mind is discharging its collected lightning, and the equilibrium between institutions and opinions is now restoring, or is about to be restored.

As to imitation, poetry is a mimetic art. It creates, but it creates by combination and representation. Poetical abstractions are beautiful and new, not because the portions of which they are composed had no previous existence in the mind of man or in nature, but because the whole produced by their combination has some intelligible and beautiful analogy with those sources of emotion and thought, and with the contemporary condition of them: one great poet is a masterpiece of nature which another not only ought to study but must study. He might as wisely and as easily determine that his mind should no longer be the mirror of all that is lovely in the visible universe, as exclude from his contemplation the beautiful which exists in the writings of a great contemporary. The pretence of doing it would be a presumption in any but the greatest; the effect, even in him, would be strained, unnatural, and ineffectual. A poet is the combined product of such internal powers as modify the nature of others; and of such external influences as excite and sustain these powers; he is not one, but both. Every man's mind is, in this respect, modified by all the objects of nature and art; by every word and every suggestion which he ever admitted to act upon his consciousness; it is the mirror upon which all forms are reflected, and in which they compose one form. Poets, not otherwise than philosophers, painters, sculptors, and musicians, are, in one sense, the creators, and, in another, the creations, of their age. From this subjection the loftiest do not escape. There is a similarity between Homer and Hesiod, between Æschylus and Euripides, between Virgil and Horace, between Dante and Petrarch, between Shakespeare and Fletcher, between Dryden and Pope; each has a generic resemblance under which their specific distinctions are arranged. If this similarity be the result of imitation, I am willing to confess that I have imitated.

Let this opportunity be conceded to me of acknowledging that I have, what a Scotch philosopher characteristically terms, "a passion for reforming the world:" what passion incited him to write and publish his book, he omits to explain. For my part I had rather be damned with Plato and Lord Bacon, than go to Heaven with Paley and Malthus.[1] But it is a mistake to suppose that I dedicate my poetical compositions solely to the direct enforcement of reform, or that I consider them in any degree as containing a reasoned system on the theory of human life. Didactic poetry is my abhorrence; nothing can be equally well expressed in prose that is not tedious and supererogatory in verse. My purpose has hitherto been simply to familiarise the highly refined imagination of the more select classes of poetical readers with beautiful idealisms of moral excellence; aware that until the mind can love, and admire, and trust, and hope, and endure, reasoned principles of moral conduct are seeds cast upon the highway of life which the unconscious passenger tramples into dust, although they would bear the harvest of his happiness. Should I live to accomplish what I purpose, that is, produce a systematical history

PROMETHEUS UNBOUND. PREFACE. **1. Paley and Malthus:** William Paley, 1743–1805, a popular apologist for Christianity, argued that the natural world gives evidence in its design of the existence and attributes of God. He also identified virtue with prudence—i.e., one obeys the Christian law from fear of hell. Thomas Robert Malthus, 1766–1834, argued that population always tends to exceed food supply and hence famine and war are inevitable. The argument was a stumbling block to reformers, since it seemed to show that man's lot could not be much improved.

of what appear to me to be the genuine elements of human society, let not the advocates of injustice and superstition flatter themselves that I should take Æschylus rather than Plato as my model.

The having spoken of myself with unaffected freedom will need little apology with the candid; and let the uncandid consider that they injure me less than their own hearts and minds by misrepresentation. Whatever talents a person may possess to amuse and instruct others, be they ever so inconsiderable, he is yet bound to exert them: if his attempt be ineffectual, let the punishment of an unaccomplished purpose have been sufficient; let none trouble themselves to heap the dust of oblivion upon his efforts; the pile they raise will betray his grave which might otherwise have been unknown.

DRAMATIS PERSONÆ

PROMETHEUS
DEMOGORGON
JUPITER
THE EARTH
OCEAN
APOLLO
MERCURY
ASIA ⎫
PANTHEA ⎬ *Oceanides*
IONE ⎭
HERCULES
THE PHANTASM OF JUPITER
THE SPIRIT OF THE EARTH
THE SPIRIT OF THE MOON
SPIRITS OF THE HOURS
SPIRITS, ECHOES, FAUNS, FURIES

ACT I

SCENE

A Ravine of Icy Rocks in the Indian Caucasus. PROMETHEUS *is discovered bound to the Precipice.* PANTHEA *and* IONE *are seated at his feet. Time, night. During the Scene, morning slowly breaks.*

PROMETHEUS. Monarch of Gods and Dæmons, and all Spirits
But One, who throng those bright and rolling worlds

ACT I. **1. Monarch:** Jupiter. **2. But One:** Prometheus, who has defied Jupiter.

Which Thou and I alone of living things
Behold with sleepless eyes! regard this Earth
Made multitudinous with thy slaves, whom thou
Requitest for knee-worship, prayer, and praise,
And toil, and hecatombs of broken hearts,
With fear and self-contempt and barren hope.
Whilst me, who am thy foe, eyeless in hate,
Hast thou made reign and triumph, to thy scorn,   10
O'er mine own misery and thy vain revenge.
Three thousand years of sleep-unsheltered hours,
And moments aye divided by keen pangs
Till they seemed years, torture and solitude,
Scorn and despair,—these are mine empire:—
More glorious far than that which thou surveyest
From thine unenvied throne, O Mighty God!
Almighty, had I deigned to share the shame
Of thine ill tyranny, and hung not here
Nailed to this wall of eagle-baffling mountain,   20
Black, wintry, dead, unmeasured; without herb,
Insect, or beast, or shape or sound of life.
Ah me! alas, pain, pain ever, for ever!

No change, no pause, no hope! Yet I endure.
I ask the Earth, have not the mountains felt?
I ask yon Heaven, the all-beholding Sun,
Has it not seen? The Sea, in storm or calm,
Heaven's ever-changing Shadow, spread below,
Have its deaf waves not heard my agony?
Ah me! alas, pain, pain ever, for ever!   30

The crawling glaciers pierce me with the spears
Of their moon-freezing crystals, the bright chains
Eat with their burning cold into my bones.
Heaven's winged hound, polluting from thy lips
His beak in poison not his own, tears up
My heart; and shapeless sights come wandering by,
The ghastly people of the realm of dream,
Mocking me: and the Earthquake-fiends are charged
To wrench the rivets from my quivering wounds
When the rocks split and close again behind: 40
While from their loud abysses howling throng
The genii of the storm, urging the rage
Of whirlwind, and afflict me with keen hail.
And yet to me welcome is day and night,
Whether one breaks the hoar frost of the morn,
Or starry, dim, and slow, the other climbs
The leaden-coloured east; for then they lead

7. **hecatombs:** huge sacrifices (in Greek "a hundred oxen").
34–36. **hound . . . heart:** The vulture, which daily tore at the entrails of Prometheus, was kissed by Jupiter on its return.

The wingless, crawling hours, one among whom
—As some dark Priest hales the reluctant victim—
Shall drag thee, cruel King, to kiss the blood   50
From these pale feet, which then might trample thee
If they disdained not such a prostrate slave.
Disdain! Ah no! I pity thee. What ruin
Will hunt thee undefended through the wide Heaven!
How will thy soul, cloven to its depth with terror,
Gape like a hell within! I speak in grief,
Not exultation, for I hate no more,
As then ere misery made me wise. The curse
Once breathed on thee I would recall. Ye Mountains,
Whose many-voiced Echoes, through the mist   60
Of cataracts, flung the thunder of that spell!
Ye icy Springs, stagnant with wrinkling frost,
Which vibrated to hear me, and then crept
Shuddering through India! Thou serenest Air,
Through which the Sun walks burning without beams!
And ye swift Whirlwinds, who on poised wings
Hung mute and moveless o'er yon hushed abyss,
As thunder, louder than your own, made rock
The orbed world! If then my words had power,
Though I am changed so that aught evil wish   70
Is dead within; although no memory be
Of what is hate, let them not lose it now!
What was that curse? for ye all heard me speak.

FIRST VOICE (*from the Mountains*)

Thrice three hundred thousand years
    O'er the Earthquake's couch we stood:
Oft, as men convulsed with fears,
    We trembled in our multitude.

SECOND VOICE (*from the Springs*)

Thunderbolts had parched our water,
    We had been stained with bitter blood,
And had run mute, 'mid shrieks of slaughter,   80
    Thro' a city and a solitude.

THIRD VOICE (*from the Air*)

I had clothed, since Earth uprose,
    Its wastes in colours not their own,

53. **pity thee:** The regeneration of Prometheus, replacing hatred by compassion, is the turning point of the action. **62–64. Ye . . . India:** a striking example of Shelley's power to create myth. **64–65. Air . . . beams:** Shelley believed that the burning rays of the sun pass through the thin air of the upper atmosphere without appearing visible as beams.

And oft had my serene repose
Been cloven by many a rending groan.

FOURTH VOICE (*from the Whirlwinds*)

We had soared beneath these mountains
    Unresting ages; nor had thunder,
Nor yon volcano's flaming fountains,
    Nor any power above or under
Ever made us mute with wonder.   90

FIRST VOICE

But never bowed our snowy crest
As at the voice of thine unrest.

SECOND VOICE

Never such a sound before
To the Indian waves we bore.
A pilot asleep on the howling sea
Leaped up from the deck in agony,
And heard, and cried, "Ah, woe is me!"
And died as mad as the wild waves be.

THIRD VOICE

By such dread words from Earth to Heaven
My still realm was never riven: 100
When its wound was closed, there stood
Darkness o'er the day like blood.

FOURTH VOICE

And we shrank back: for dreams of ruin
To frozen caves our flight pursuing
Made us keep silence—thus—and thus—
Though silence is as hell to us.

THE EARTH.   The tongueless Caverns of the craggy
    hills
Cried, "Misery!" then; the hollow Heaven replied,
"Misery!" And the Ocean's purple waves,
Climbing the land, howled to the lashing winds,   110
And the pale nations heard it, "Misery!"
PRO.   I heard a sound of voices: not the voice
Which I gave forth. Mother, thy sons and thou
Scorn him, without whose all-enduring will
Beneath the fierce omnipotence of Jove,
Both they and thou had vanished, like thin mist
Unrolled on the morning wind. Know ye not me,
The Titan? He who made his agony

The barrier to your else all-conquering foe?
Oh, rock-embosomed lawns, and snow-fed streams,
Now seen athwart frore vapours, deep below,   121
Through whose o'ershadowing woods I wandered once
With Asia, drinking life from her loved eyes;
Why scorns the spirit which informs ye, now
To commune with me? me alone, who checked,
As one who checks a fiend-drawn charioteer,
The falsehood and the force of him who reigns
Supreme, and with the groans of pining slaves
Fills your dim glens and liquid wildernesses:
Why answer ye not, still? Brethren!
     THE EARTH.   They dare not.
     PRO.   Who dares? for I would hear that curse
       again.   131
Ha, what an awful whisper rises up!
'Tis scarce like sound: it tingles through the frame
As lightning tingles, hovering ere it strike.
Speak, Spirit! from thine inorganic voice
I only know that thou art moving near
And love. How cursed I him?
     THE EARTH.   How canst thou hear
Who knowest not the language of the dead?
     PRO.   Thou art a living spirit; speak as they.
     THE EARTH.   I dare not speak like life, lest Heaven's
       fell King   140
Should hear, and link me to some wheel of pain
More torturing than the one whereon I roll.
Subtle thou art and good, and though the Gods
Hear not this voice, yet thou art more than God,
Being wise and kind: earnestly hearken now.
     PRO.   Obscurely through my brain, like shadows dim,
Sweep awful thoughts, rapid and thick. I feel
Faint, like one mingled in entwining love;
Yet 'tis not pleasure.
     THE EARTH.   No, thou canst not hear:
Thou art immortal, and this tongue is known   150
Only to those who die.
     PRO.   And what art thou,
O, melancholy Voice?

THE EARTH.   I am the Earth,
Thy mother; she within whose stony veins,
To the last fibre of the loftiest tree
Whose thin leaves trembled in the frozen air,
Joy ran, as blood within a living frame,
When thou didst from her bosom, like a cloud
Of glory, arise, a spirit of keen joy!
And at thy voice her pining sons uplifted
Their prostrate brows from the polluting dust,   160
And our almighty Tyrant with fierce dread
Grew pale, until his thunder chained thee here.
Then, see those million worlds which burn and roll
Around us: their inhabitants beheld
My sphered light wane in wide Heaven; the sea
Was lifted by strange tempest, and new fire
From earthquake-rifted mountains of bright snow
Shook its portentous hair beneath Heaven's frown;
Lightning and Inundation vexed the plains;
Blue thistles bloomed in cities; foodless toads   170
Within voluptuous chambers panting crawled:
When Plague had fallen on man, and beast, and worm,
And Famine; and black blight on herb and tree;
And in the corn, and vines, and meadow-grass,
Teemed ineradicable poisonous weeds
Draining their growth, for my wan breast was dry
With grief; and the thin air, my breath, was stained
With the contagion of a mother's hate
Breathed on her child's destroyer; ay, I heard
Thy curse, the which, if thou rememberest not,   180
Yet my innumerable seas and streams,
Mountains, and caves, and winds, and yon wide air,
And the inarticulate people of the dead,
Preserve, a treasured spell. We meditate
In secret joy and hope those dreadful words,
But dare not speak them.
     PRO.   Venerable mother!
All else who live and suffer take from thee
Some comfort; flowers, and fruits, and happy sounds,
And love, though fleeting; these may not be mine.
But mine own words, I pray, deny me not.   190
     THE EARTH.   They shall be told. Ere Babylon was dust,
The Magus Zoroaster, my dead child,
Met his own image walking in the garden.
That apparition, sole of men, he saw.

---

121. frore: frosty. 123. Asia: Mary Shelley's note: "Asia, one of the Oceanides, is the wife of Prometheus—she was, according to other mythological interpretations, the same as Venus and Nature. When the benefactor of mankind is liberated, Nature resumes the beauty of her prime, and is united to her husband, the emblem of the human race, in perfect and happy union." 129. liquid wildernesses: the oceans. 135. inorganic: bodiless. 137. love: a slip for "lovest."

159. pining sons: mankind. 192. Zoroaster: founder (fl. 1000 B.C.) of the ancient Persian religion which taught that two powers, Ormuzd (light and good) and Ahriman (darkness and evil), are locked in a struggle which will end with the final overthrow of Ahriman.

For know there are two worlds of life and death:
One that which thou beholdest; but the other
Is underneath the grave, where do inhabit
The shadows of all forms that think and live
Till death unite them and they part no more;
Dreams and the light imaginings of men,   200
And all that faith creates or love desires,
Terrible, strange, sublime and beauteous shapes.
There thou art, and dost hang, a writhing shade,
'Mid whirlwind-peopled mountains; all the gods
Are there, and all the powers of nameless worlds,
Vast, sceptred phantoms; heroes, men, and beasts;
And Demogorgon, a tremendous gloom;
And he, the supreme Tyrant, on his throne
Of burning gold. Son, one of these shall utter
The curse which all remember. Call at will   210
Thine own ghost, or the ghost of Jupiter,
Hades or Typhon, or what mightier Gods
From all-prolific Evil, since thy ruin
Have sprung, and trampled on my prostrate sons.
Ask, and they must reply: so the revenge
Of the Supreme may sweep through vacant shades,
As rainy wind through the abandoned gate
Of a fallen palace.
    PRO.   Mother, let not aught
Of that which may be evil, pass again
My lips, or those of aught resembling me.   220
Phantasm of Jupiter, arise, appear!

IONE

My wings are folded o'er mine ears:
    My wings are crossed o'er mine eyes:
Yet through their silver shade appears,
    And through their lulling plumes arise,
A Shape, a throng of sounds;
    May it be no ill to thee
O thou of many wounds!
Near whom, for our sweet sister's sake,
Ever thus we watch and wake.   230

PANTHEA

The sound is of whirlwind underground,
    Earthquake, and fire, and mountains cloven;
The shape is awful like the sound,
    Clothed in dark purple, star-inwoven.

212. **Hades or Typhon:** Pluto (god of the underworld)
and the many-headed monster, Typhon, son of the Earth,
associated with earthquakes and volcanoes.

A sceptre of pale gold
    To stay steps proud, o'er the slow cloud
His veined hand doth hold.
Cruel he looks, but calm and strong,
Like one who does, not suffers wrong.

PHANTASM OF JUPITER.   Why have the secret powers
    of this strange world   240
Driven me, a frail and empty phantom, hither
On direst storms? What unaccustomed sounds
Are hovering on my lips, unlike the voice
With which our pallid race hold ghastly talk
In darkness? And, proud sufferer, who art thou?
    PRO.   Tremendous Image, as thou art must be
He whom thou shadowest forth. I am his foe,
The Titan. Speak the words which I would hear,
Although no thought inform thine empty voice.
    THE EARTH.   Listen! And though your echoes must
        be mute,   250
Gray mountains, and old woods, and haunted springs,
Prophetic caves, and isle-surrounding streams,
Rejoice to hear what yet ye cannot speak.
    PHAN.   A spirit seizes me and speaks within:
It tears me as fire a thunder-cloud.
    PANTHEA.   See, how he lifts his mighty looks, the
        Heaven
Darkens above.
    IONE.   He speaks! O shelter me!
    PRO.   I see the curse on gestures proud and cold,
And looks of firm defiance, and calm hate,
And such despair as mocks itself with smiles,   260
Written as on a scroll: yet speak: Oh, speak!

PHANTASM

Fiend, I defy thee! with a calm, fixed mind,
    All that thou canst inflict I bid thee do;
Foul Tyrant both of Gods and Human-kind,
    One only being shalt thou not subdue.
Rain then thy plagues upon me here,
Ghastly disease, and frenzying fear;
And let alternate frost and fire
Eat into me, and be thine ire
Lightning, and cutting hail, and legioned forms   270
Of furies, driving by upon the wounding storms.

Ay, do thy worst. Thou art omnipotent.
    O'er all things but thyself I gave thee power,
And my own will. Be thy swift mischiefs sent
    To blast mankind, from yon ethereal tower.

Let thy malignant spirit move
In darkness over those I love:
On me and mine I imprecate
The utmost torture of thy hate;
And thus devote to sleepless agony,   280
This undeclining head while thou must reign on high.

But thou, who art the God and Lord: O, thou,
    Who fillest with thy soul this world of woe,
To whom all things of Earth and Heaven do bow
    In fear and worship: all-prevailing foe!
I curse thee! let a sufferer's curse
Clasp thee, his torturer, like remorse;
Till thine Infinity shall be
A robe of envenomed agony;
And thine Omnipotence a crown of pain,   290
To cling like burning gold round thy dissolving brain.

Heap on thy soul, by virtue of this Curse,
    Ill deeds, then be thou damned, beholding good;
Both infinite as is the universe,
    And thou, and thy self-torturing solitude.
An awful image of calm power
Though now thou sittest, let the hour
Come, when thou must appear to be
That which thou art internally;
And after many a false and fruitless crime   300
Scorn track thy lagging fall through boundless space
    and time.

PRO.   Were these my words, O Parent?
THE EARTH.   They were
    thine.
PRO.   It doth repent me: words are quick and vain;
Grief for awhile is blind, and so was mine.
I wish no living thing to suffer pain.

THE EARTH

Misery, Oh misery to me,
That Jove at length should vanquish thee.
Wail, howl aloud, Land and Sea,
The Earth's rent heart shall answer ye.
Howl, Spirits of the living and the dead,   310
Your refuge, your defence lies fallen and vanquished.

FIRST ECHO

Lies fallen and vanquished!

SECOND ECHO

Fallen and vanquished!

IONE

Fear not: 'tis but some passing spasm,
    The Titan is unvanquished still.
But see, where through the azure chasm
    Of yon forked and snowy hill
Trampling the slant winds on high
    With golden-sandalled feet, that glow
Under plumes of purple dye,   320
Like rose-ensanguined ivory,
    A Shape comes now,
Stretching on high from his right hand
A serpent-cinctured wand.

PANTHEA. 'Tis Jove's world-wandering herald,
Mercury.

IONE

And who are those with hydra tresses
    And iron wings that climb the wind,
Whom the frowning God represses
    Like vapours steaming up behind,
Clanging loud, an endless crowd—   330

PANTHEA

These are Jove's tempest-walking hounds,
Whom he gluts with groans and blood,
When charioted on sulphurous cloud
He bursts Heaven's bounds.

IONE

Are they now led, from the thin dead
On new pangs to be fed?

PANTHEA

The Titan looks as ever, firm, not proud.

FIRST FURY. Ha! I scent life!
SECOND FURY.   Let me but look into
    his eyes!
THIRD FURY. The hope of torturing him smells like
    a heap
Of corpses, to a death-bird after battle.   340

324. serpent-cinctured wand: the caduceus or staff of
Mercury. **326. hydra:** snakelike.

FIRST FURY. Darest thou delay, O Herald! take
    cheer, Hounds
Of Hell: what if the Son of Maia soon
Should make us food and sport—who can please long
The Omnipotent?
    MERCURY.   Back to your towers of iron,
And gnash, beside the streams of fire and wail,
Your foodless teeth. Geryon, arise! and Gorgon,
Chimæra, and thou Sphinx, subtlest of fiends
Who ministered to Thebes Heaven's poisoned wine,
Unnatural love, and more unnatural hate:
These shall perform your task.
    FIRST FURY.   Oh, mercy! mercy!
We die with our desire: drive us not back!   351
    MER. Crouch then in silence.
                      Awful Sufferer!
To thee unwilling, most unwillingly
I come, by the great Father's will driven down,
To execute a doom of new revenge.
Alas! I pity thee, and hate myself
That I can do no more: aye from thy sight
Returning, for a season, Heaven seems Hell,
So thy worn form pursues me night and day,
Smiling reproach. Wise art thou, firm and good, 360
But vainly wouldst stand forth alone in strife
Against the Omnipotent; as yon clear lamps
That measure and divide the weary years
From which there is no refuge, long have taught
And long must teach. Even now thy Torturer arms
With the strange might of unimagined pains
The powers who scheme slow agonies in Hell,
And my commission is to lead them here,
Or what more subtle, foul, or savage fiends
People the abyss, and leave them to their task. 370
Be it not so! there is a secret known
To thee, and to none else of living things,
Which may transfer the sceptre of wide Heaven,
The fear of which perplexes the Supreme:
Clothe it in words, and bid it clasp his throne
In intercession; bend thy soul in prayer,
And like a suppliant in some gorgeous fane,

Let the will kneel within thy haughty heart:
For benefits and meek submission tame
The fiercest and the mightiest.
    PRO.   Evil minds   380
Change good to their own nature. I gave all
He has; and in return he chains me here
Years, ages, night and day: whether the Sun
Split my parched skin, or in the moony night
The crystal-winged snow cling round my hair:
Whilst my beloved race is trampled down
By his thought-executing ministers.
Such is the tyrant's recompense: 'tis just:
He who is evil can receive no good;
And for a world bestowed, or a friend lost, 390
He can feel hate, fear, shame; not gratitude:
He but requites me for his own misdeed.
Kindness to such is keen reproach, which breaks
With bitter stings the light sleep of Revenge.
Submission, thou dost know I cannot try:
For what submission but that fatal word,
The death-seal of mankind's captivity,
Like the Sicilian's hair-suspended sword,
Which trembles o'er his crown, would he accept,
Or could I yield? Which yet I will not yield. 400
Let others flatter Crime, where it sits throned
In brief Omnipotence: secure are they:
For Justice, when triumphant, will weep down
Pity, not punishment, on her own wrongs,
Too much avenged by those who err. I wait,
Enduring thus, the retributive hour
Which since we spake is even nearer now.
But hark, the hell-hounds clamour: fear delay:
Behold! Heaven lowers under thy Father's frown.
    MER. Oh, that we might be spared: I to inflict 410
And thou to suffer! Once more answer me:
Thou knowest not the period of Jove's power?
    PRO. I know but this, that it must come.
    MER.   Alas!
Thou canst not count thy years to come of pain?
    PRO. They last while Jove must reign: nor more,
      nor less
Do I desire or fear.
    MER.   Yet pause, and plunge

---

342. Son of Maia: Mercury. **346–49. Geryon . . . hate:**
Geryon was a three-bodied monster killed by Hercules; the
Gorgon head turned the beholder to stone; and Chimaera
was a monster that breathed fire. The Sphinx, half lion and
half woman, killed all the passers-by who could not guess
her riddle ("What walks on four legs in the morning, two
at noon, and three at night?"); and when Oedipus guessed
the answer ("man") he lived to receive the kingdom and
to take his mother to wife (the "unnatural love" referred
to in l. 349).

387. thought-executing: thought-destroying (echoes *King
Lear*, III. iii. 4). **398. Sicilian's . . . sword:** the "sword of
Damocles," a courtier who had praised the happiness of kings.
Dionysius of Sicily suspended a sword over him by a hair at a
banquet in order to teach him how precarious that happiness
was.

Into Eternity, where recorded time,
Even all that we imagine, age on age,
Seems but a point, and the reluctant mind
Flags wearily in its unending flight,   420
Till it sink, dizzy, blind, lost, shelterless;
Perchance it has not numbered the slow years
Which thou must spend in torture, unreprieved?

PRO.   Perchance no thought can count them, yet
     they pass.

MER.   If thou might'st dwell among the Gods the
     while
Lapped in voluptuous joy?

PRO.   I would not quit
This bleak ravine, these unrepentant pains.

MER.   Alas! I wonder at, yet pity thee.

PRO.   Pity the self-despising slaves of Heaven,
Not me, within whose mind sits peace serene,   430
As light in the sun, throned: how vain is talk!
Call up the fiends.

IONE.   O, sister, look! White fire
Has cloven to the roots yon huge snow-loaded cedar;
How fearfully God's thunder howls behind!

MER.   I must obey his words and thine: alas!
Most heavily remorse hangs at my heart!

PANTHEA.   See where the child of Heaven, with
     winged feet,
Runs down the slanted sunlight of the dawn.

IONE.   Dear sister, close thy plumes over thine eyes
Lest thou behold and die: they come: they come 440
Blackening the birth of day with countless wings,
And hollow underneath, like death.

FIRST FURY.   Prometheus!

SECOND FURY.   Immortal Titan!

THIRD FURY.   Champion of
     Heaven's slaves!

PRO.   He whom some dreadful voice invokes is here,
Prometheus, the chained Titan. Horrible forms,
What and who are ye? Never yet there came
Phantasms so foul through monster-teeming Hell
From the all-miscreative brain of Jove;
Whilst I behold such execrable shapes,
Methinks I grow like what I contemplate,   450
And laugh and stare in loathsome sympathy.

FIRST FURY.   We are the ministers of pain, and fear,
And disappointment, and mistrust, and hate,
And clinging crime; and as lean dogs pursue
Through wood and lake some struck and sobbing
     fawn,

We track all things that weep, and bleed, and live,
When the great King betrays them to our will.

PRO.   Oh! many fearful natures in one name,
I know ye; and these lakes and echoes know
The darkness and the clangour of your wings.   460
But why more hideous than your loathed selves
Gather ye up in legions from the deep?

SECOND FURY.   We knew not that: Sisters, rejoice,
     rejoice!

PRO.   Can aught exult in its deformity?

SECOND FURY.   The beauty of delight makes lovers
     glad,
Gazing on one another: so are we.
As from the rose which the pale priestess kneels
To gather for her festal crown of flowers
The aëreal crimson falls, flushing her cheek,
So from our victim's destined agony   470
The shade which is our form invests us round,
Else we are shapeless as our mother Night.

PRO.   I laugh your power, and his who sent you
     here,
To lowest scorn. Pour forth the cup of pain.

FIRST FURY.   Thou thinkest we will rend thee bone
     from bone,
And nerve from nerve, working like fire within?

PRO.   Pain is my element, as hate is thine;
Ye rend me now: I care not.

SECOND FURY.   Dost imagine
We will but laugh into thy lidless eyes?

PRO.   I weigh not what ye do, but what ye suffer,
Being evil. Cruel was the power which called   481
You, or aught else so wretched, into light.

THIRD FURY.   Thou think'st we will live through
     thee, one by one,
Like animal life, and though we can obscure not
The soul which burns within, that we will dwell
Beside it, like a vain loud multitude
Vexing the self-content of wisest men:
That we will be dread thought beneath thy brain,
And foul desire round thine astonished heart,
And blood within thy labyrinthine veins   490
Crawling like agony?

PRO.   Why, ye are thus now;
Yet am I king over myself, and rule
The torturing and conflicting throngs within,
As Jove rules you when Hell grows mutinous.

417. recorded time: Macbeth, V. v. 21.

461. why . . . selves: Why more hideous than ever
before?

CHORUS OF FURIES

From the ends of the earth, from the ends of the earth,
Where the night has its grave and the morning its
    birth,
        Come, come, come!
Oh, ye who shake hills with the scream of your mirth,
When cities sink howling in ruin; and ye
Who with wingless footsteps trample the sea,   500
And close upon Shipwreck and Famine's track,
Sit chattering with joy on the foodless wreck;
        Come, come, come!
    Leave the bed, low, cold, and red,
    Strewed beneath a nation dead;
    Leave the hatred, as in ashes
      Fire is left for future burning:
    It will burst in bloodier flashes
      When ye stir it, soon returning:
    Leave the self-contempt implanted   510
In young spirits, sense-enchanted,
    Misery's yet unkindled fuel:
Leave Hell's secrets half unchanted
    To the maniac dreamer; cruel
More than ye can be with hate
    Is he with fear.
        Come, come, come!
We are steaming up from Hell's wide gate
And we burthen the blast of the atmosphere,
But vainly we toil till ye come here.   520

IONE.   Sister, I hear the thunder of new wings.
PANTHEA.   These solid mountains quiver with the
    sound
Even as the tremulous air: their shadows make
The space within my plumes more black than night.

FIRST FURY

Your call was as a winged car
Driven on whirlwinds fast and far;
It rapt us from red gulfs of war.

SECOND FURY

From wide cities, famine-wasted;

513–16. Leave . . . fear: The Furies are the source of
the idea of hell and the resulting terror that makes men
cruel. Hence Shelley's opinion that Christianity had perverted
the teaching of Christ and become a degrading superstition.
527. rapt: carried away by force.

THIRD FURY

Groans half heard, and blood untasted;

FOURTH FURY

Kingly conclaves stern and cold,   530
Where blood with gold is bought and sold;

FIFTH FURY

From the furnace, white and hot,
In which—

A FURY

      Speak not: whisper not:
I know all that ye would tell,
But to speak might break the spell
Which must bend the Invincible,
    The stern of thought;
He yet defies the deepest power of Hell.

A FURY

Tear the veil!

ANOTHER FURY

    It is torn.

CHORUS

          The pale stars of the
    morn
Shine on a misery, dire to be borne.   540
Dost thou faint, mighty Titan? We laugh thee to scorn.
Dost thou boast the clear knowledge thou waken'dst
    for man?
Then was kindled within him a thirst which outran
Those perishing waters; a thirst of fierce fever,
Hope, love, doubt, desire, which consume him for ever.

---

541. laugh . . . scorn: The Furies mock him with
his own words (above, l. 473). **544. perishing waters:**
waters of knowledge that quicken thirst but are insufficient to quench it? Perhaps they are "perishing" in the
sense that the "clear knowledge" Prometheus gave was
gradually obscured, or perhaps Shelley simply invokes a
metaphor of water vanishing as one strives to drink. The
general intent of lines 542–44 is to exhibit Prometheus
himself as a first example of a tragic irony recurrent through-

One came forth of gentle worth
Smiling on the sanguine earth;
His words outlived him, like swift poison
    Withering up truth, peace, and pity.
Look! where round the wide horizon   550
    Many a million-peopled city
Vomits smoke in the bright air.
Hark that outcry of despair!
'Tis his mild and gentle ghost
    Wailing for the faith he kindled:
Look again, the flames almost
    To a glow-worm's lamp have dwindled:
The survivors round the embers
    Gather in dread.
            Joy, joy, joy!   560
Past ages crowd on thee, but each one remembers,
And the future is dark, and the present is spread
Like a pillow of thorns for thy slumberless head.

SEMICHORUS I

Drops of bloody agony flow
From his white and quivering brow.
Grant a little respite now:
See a disenchanted nation
Springs like day from desolation;
To Truth its state is dedicate,
And Freedom leads it forth, her mate;   570
A legioned band of linked brothers
Whom Love calls children—

SEMICHORUS II

                              'Tis another's:
See how kindred murder kin:
'Tis the vintage-time for death and sin:
Blood, like new wine, bubbles within:
    Till Despair smothers
The struggling world, which slaves and tyrants win.
[*All the* FURIES *vanish, except one.*]
IONE.   Hark, sister! what a low yet dreadful groan
Quite unsuppressed is tearing up the heart
Of the good Titan, as storms tear the deep, 580

out history, that the noblest acts of man bring forth evil and
misery greater than that which existed before. **546. One:**
Christ. Shelley's vision of the compassionate teaching of
Christ leading to war and devastation was picked up by
W. B. Yeats. Cf. his "Two Songs from a Play." **547. sang-
uine:** both meanings are relevant: hopeful and blood-
stained. **567. disenchanted nation:** France, freed at the
Revolution from the evil enchantment of its previous regime.

And beasts hear the sea moan in inland caves.
Darest thou observe how the fiends torture him?
    PANTHEA.   Alas! I looked forth twice, but will no
        more.
    IONE.   What didst thou see?
    PANTHEA.   A woful sight: a youth
With patient looks nailed to a crucifix.
    IONE.   What next?
    PANTHEA.   The heaven around, the earth below
Was peopled with thick shapes of human death,
All horrible, and wrought by human hands,
And some appeared the work of human hearts,
For men were slowly killed by frowns and smiles: 590
And other sights too foul to speak and live
Were wandering by. Let us not tempt worse fear
By looking forth: those groans are grief enough.
    FURY.   Behold an emblem: those who do endure
Deep wrongs for man, and scorn, and chains, but heap
Thousandfold torment on themselves and him.
    PRO.   Remit the anguish of that lighted stare;
Close those wan lips; let that thorn-wounded brow
Stream not with blood; it mingles with thy tears!
Fix, fix those tortured orbs in peace and death,   600
So thy sick throes shake not that crucifix,
So those pale fingers play not with thy gore.
O, horrible! Thy name I will not speak,
It hath become a curse. I see, I see
The wise, the mild, the lofty, and the just,
Whom thy slaves hate for being like to thee,
Some hunted by foul lies from their heart's home,
An early-chosen, late-lamented home;
As hooded ounces cling to the driven hind;
Some linked to corpses in unwholesome cells: 610
Some—Hear I not the multitude laugh loud?—
Impaled in lingering fire: and mighty realms
Float by my feet, like sea-uprooted isles,
Whose sons are kneaded down in common blood
By the red light of their own burning homes.
    FURY.   Blood thou canst see, and fire; and canst
        hear groans;
Worse things, unheard, unseen, remain behind.
    PRO.   Worse?
    FURY.   In each human heart terror survives
The ravin it has gorged: the loftiest fear
All that they would disdain to think were true: 620
Hypocrisy and custom make their minds

584. youth: Jesus. **609. hooded ounces:** cheetahs or
leopards, used for hunting in parts of Asia, and kept hooded
until the chase begins. **619. ravin:** prey.

The fanes of many a worship, now outworn.
They dare not devise good for man's estate,
And yet they know not that they do not dare.
The good want power, but to weep barren tears.
The powerful goodness want: worse need for them.
The wise want love; and those who love want wisdom;
And all best things are thus confused to ill.
Many are strong and rich, and would be just,
But live among their suffering fellow-men   630
As if none felt: they know not what they do.
    PRO.   Thy words are like a cloud of winged snakes;
And yet I pity those they torture not.
    FURY.   Thou pitiest them? I speak no more!
                [*Vanishes.*]

    PRO.   Ah woe!
Ah woe! Alas! pain, pain ever, for ever!
I close my tearless eyes, but see more clear
Thy works within my woe-illumed mind,
Thou subtle tyrant! Peace is in the grave.
The grave hides all things beautiful and good:
I am a God and cannot find it there,   640
Nor would I seek it: for, though dread revenge,
This is defeat, fierce king, not victory.
The sights with which thou torturest gird my soul
With new endurance, till the hour arrives
When they shall be no types of things which are.
    PANTHEA.   Alas! what sawest thou?
    PRO.   There are two woes:
To speak, and to behold; thou spare me one.
Names are there, Nature's sacred watchwords, they
Were borne aloft in bright emblazonry;
The nations thronged around, and cried aloud,   650
As with one voice, Truth, liberty, and love!
Suddenly fierce confusion fell from heaven
Among them: there was strife, deceit, and fear:
Tyrants rushed in, and did divide the spoil.
This was the shadow of the truth I saw.
    THE EARTH.   I felt thy torture, son; with such mixed
       joy
As pain and virtue give. To cheer thy state
I bid ascend those subtle and fair spirits,
Whose homes are the dim caves of human thought,
And who inhabit, as birds wing the wind,   660
Its world-surrounding aether: they behold
Beyond that twilight realm, as in a glass,
The future: may they speak comfort to thee!

    PANTHEA.   Look, sister, where a troop of spirits
       gather,
Like flocks of clouds in spring's delightful weather,
Thronging in the blue air!
    IONE.   And see! more come,
Like fountain-vapours when the winds are dumb,
That climb up the ravine in scattered lines.
And, hark! is it the music of the pines?
Is it the lake? Is it the waterfall?   670
    PANTHEA.   'Tis something sadder, sweeter far than
    all.

CHORUS OF SPIRITS

From unremembered ages we
Gentle guides and guardians be
Of heaven-oppressed mortality;
And we breathe, and sicken not,
The atmosphere of human thought:
Be it dim, and dank, and gray,
Like a storm-extinguished day,
Travelled o'er by dying gleams;
   Be it bright as all between   680
Cloudless skies and windless streams,
   Silent, liquid, and serene;
As the birds within the wind,
   As the fish within the wave,
As the thoughts of man's own mind
   Float through all above the grave;
We make there our liquid lair,
Voyaging cloudlike and unpent
Through the boundless element:
Thence we bear the prophecy   690
Which begins and ends in thee!

    IONE.   More yet come, one by one: the air around
    them
Looks radiant as the air around a star.

FIRST SPIRIT

On a battle-trumpet's blast
I fled hither, fast, fast, fast,
'Mid the darkness upward cast.
From the dust of creeds outworn,
From the tyrant's banner torn,
Gathering 'round me, onward borne,

631. **they . . . do:** Luke 23:34.

688. **unpent:** unimprisoned. **690. the prophecy:** The four spirits prophesy (ll. 694–751) the ultimate triumph of love through revolution, self-transcendence, wisdom, and poetry (or the use of the creative imagination).

There was mingled many a cry—   700
Freedom! Hope! Death! Victory!
Till they faded through the sky;
And one sound, above, around,
One sound beneath, around, above,
Was moving; 'twas the soul of Love;
'Twas the hope, the prophecy,
Which begins and ends in thee.

SECOND SPIRIT

A rainbow's arch stood on the sea,
Which rocked beneath, immovably;
And the triumphant storm did flee,   710
Like a conqueror, swift and proud,
Between, with many a captive cloud,
A shapeless, dark and rapid crowd,
Each by lightning riven in half:
I heard the thunder hoarsely laugh:
Mighty fleets were strewn like chaff
And spread beneath, a hell of death,
O'er the white waters. I alit
On a great ship lightning-split,
And speeded hither on the sigh   720
Of one who gave an enemy
His plank, then plunged aside to die.

THIRD SPIRIT

I sate beside a sage's bed,
And the lamp was burning red
Near the book where he had fed,
When a Dream with plumes of flame,
To his pillow hovering came,
And I knew it was the same
Which had kindled long ago
Pity, eloquence, and woe;   730
And the world awhile below
Wore the shade its lustre made.
It has borne me here as fleet
As Desire's lightning feet:
I must ride it back ere morrow,
Or the sage will wake in sorrow.

FOURTH SPIRIT

On a poet's lips I slept
Dreaming like a love-adept
In the sound his breathing kept;

Nor seeks nor finds he mortal blisses, 740
But feeds on the aëreal kisses
Of shapes that haunt thought's wildernesses.
He will watch from dawn to gloom
The lake-reflected sun illume
The yellow bees in the ivy-bloom,
Nor heed nor see, what things they be;
But from these create he can
Forms more real than living man,
Nurslings of immortality!
One of these awakened me,   750
And I sped to succour thee.

IONE

Behold'st thou not two shapes from the east and west
Come, as two doves to one beloved nest,
Twin nurslings of the all-sustaining air
On swift still wings glide down the atmosphere?
And, hark! their sweet, sad voices! 'tis despair
Mingled with love and then dissolved in sound.

PANTHEA.   Canst thou speak, sister? all my words are
     drowned.
   IONE.   Their beauty gives me voice. See how they
     float
On their sustaining wings of skiey grain,   760
Orange and azure deepening into gold:
Their soft smiles light the air like a star's fire.

CHORUS OF SPIRITS

Hast thou beheld the form of Love?

FIFTH SPIRIT

                    As over wide dominions
I sped, like some swift cloud that wings the wide air's
     wildernesses,
That planet-crested shape swept by on lightning-
     braided pinions,
Scattering the liquid joy of life from his ambrosial
     tresses:
His footsteps paved the world with light; but as I
     passed 'twas fading,
And hollow Ruin yawned behind: great sages bound
     in madness,
And headless patriots, and pale youths who perished,
     unupbraiding,

712. **between:** that is, between the rainbow and the sea.

760. **grain:** color (cf. *Paradise Lost*, V. 285: "Sky-tinctured grain").

Gleamed in the night. I wandered o'er, till thou, O
  King of sadness,   770
Turned by thy smile the worst I saw to recollected
  gladness.

SIXTH SPIRIT

Ah, sister! Desolation is a delicate thing:
  It walks not on the earth, it floats not on the air,
But treads with lulling footstep, and fans with silent
  wing
The tender hopes which in their hearts the best and
  gentlest bear;
Who, soothed to false repose by the fanning plumes
  above
And the music-stirring motion of its soft and busy
  feet,
Dream visions of aëreal joy, and call the monster, Love,
  And wake, and find the shadow Pain, as he whom
    now we greet.

CHORUS

Though Ruin now Love's shadow be,   780
Following him, destroyingly,
  On Death's white and winged steed,
Which the fleetest cannot flee,
  Trampling down both flower and weed,
Man and beast, and foul and fair,
Like a tempest through the air;
Thou shalt quell this horseman grim,
Woundless though in heart or limb.

PRO.   Spirits! how know ye this shall be?

CHORUS

In the atmosphere we breathe,   790
As buds grow red when the snow-storms flee,
  From Spring gathering up beneath,
Whose mild winds shake the elder brake,
And the wandering herdsmen know
That the white-thorn soon will blow:

Wisdom, Justice, Love, and Peace,
When they struggle to increase,
  Are to us as soft winds be
  To shepherd boys, the prophecy
  Which begins and ends in thee.   800
IONE.   Where are the Spirits fled?
PANTHEA.   Only a sense
Remains of them, like the omnipotence
Of music, when the inspired voice and lute
Languish, ere yet the responses are mute,
Which through the deep and labyrinthine soul,
Like echoes through long caverns, wind and roll.
  PRO.   How fair these airborn shapes! and yet I feel
Most vain all hope but love; and thou art far,
Asia! who, when my being overflowed,
Wert like a golden chalice to bright wine   810
Which else had sunk into the thirsty dust.
All things are still: alas! how heavily
This quiet morning weighs upon my heart;
Though I should dream I could even sleep with grief
If slumber were denied not. I would fain
Be what it is my destiny to be,
The saviour and the strength of suffering man,
Or sink into the original gulf of things:
There is no agony, and no solace left;
Earth can console, Heaven can torment no more. 820
  PANTHEA.   Hast thou forgotten one who watches
    thee
The cold dark night, and never sleeps but when
The shadow of thy spirit falls on her?
  PRO.   I said all hope was vain but love: thou lovest.
  PANTHEA.   Deeply in truth; but the eastern star
    looks white,
And Asia waits in that far Indian vale,
The scene of her sad exile; rugged once
And desolate and frozen, like this ravine;
But now invested with fair flowers and herbs,
And haunted by sweet airs and sounds, which flow
Among the woods and waters, from the aether   831
Of her transforming presence, which would fade
If it were mingled not with thine. Farewell!

END OF THE FIRST ACT

772–75. **Desolation . . . bear:** Cf. Shelley's translation of Plato's *Symposium*, 195d: "For Homer says, that
the goddess Calamity is delicate, and that her feet are
tender. 'Her feet are soft,' he says, 'for she treads not upon
the ground, but makes her path upon the heads of men!'"
778. **monster:** because unreal. 779. **as he:** Prometheus.
780–82. **Though . . . steed:** See ll. 542–77. For Death's
steed, see Revelation 6:8. 787. **quell:** kill.

ACT II

SCENE I

Morning. A lovely Vale in the Indian Caucasus.
ASIA alone.

ASIA. From all the blasts of heaven thou hast
    descended:
Yes, like a spirit, like a thought, which makes
Unwonted tears throng to the horny eyes,
And beatings haunt the desolated heart,
Which should have learnt repose: thou hast descended
Cradled in tempests; thou dost wake, O Spring!
O child of many winds! As suddenly
Thou comest as the memory of a dream,
Which now is sad because it hath been sweet;
Like genius, or like joy which riseth up   10
As from the earth, clothing with golden clouds
The desert of our life.
This is the season, this the day, the hour;
At sunrise thou shouldst come, sweet sister mine,
Too long desired, too long delaying, come!
How like death-worms the wingless moments crawl!
The point of one white star is quivering still
Deep in the orange light of widening morn
Beyond the purple mountains: through a chasm
Of wind-divided mist the darker lake   20
Reflects it: now it wanes: it gleams again
As the waves fade, and as the burning threads
Of woven cloud unravel in pale air:
'Tis lost! and through yon peaks of cloud-like snow
The roseate sunlight quivers: hear I not
The Æolian music of her sea-green plumes
Winnowing the crimson dawn?
                [PANTHEA *enters.*]
                                I feel, I see
Those eyes which burn through smiles that fade in
    tears,
Like stars half quenched in mists of silver dew.
Beloved and most beautiful, who wearest   30
The shadow of that soul by which I live,
How late thou art! the sphered sun had climbed
The sea; my heart was sick with hope, before
The printless air felt thy belated plumes.

PANTHEA. Pardon, great Sister! but my wings
    were faint
With the delight of a remembered dream,
As are the noontide plumes of summer winds
Satiate with sweet flowers. I was wont to sleep
Peacefully, and awake refreshed and calm
Before the sacred Titan's fall, and thy   40
Unhappy love, had made, through use and pity,
Both love and woe familiar to my heart
As they had grown to thine: erewhile I slept
Under the glaucous caverns of old Ocean
Within dim bowers of green and purple moss,
Our young Ione's soft and milky arms
Locked then, as now, behind my dark, moist hair,
While my shut eyes and cheek were pressed within
The folded depth of her life-breathing bosom:
But not as now, since I am made the wind   50
Which fails beneath the music that I bear
Of thy most wordless converse; since dissolved
Into the sense with which love talks, my rest
Was troubled and yet sweet; my waking hours
Too full of care and pain.

ASIA.   Lift up thine eyes,
And let me read thy dream.

PANTHEA.   As I have said
With our sea-sister at his feet I slept.
The mountain mists, condensing at our voice
Under the moon, had spread their snowy flakes,
From the keen ice shielding our linked sleep.   60
Then two dreams came. One, I remember not.
But in the other his pale wound-worn limbs
Fell from Prometheus, and the azure night
Grew radiant with the glory of that form
Which lives unchanged within, and his voice fell
Like music which makes giddy the dim brain,
Faint with intoxication of keen joy:
"Sister of her whose footsteps pave the world
With loveliness—more fair than aught but her,
Whose shadow thou art—lift thine eyes on me."   70
I lifted them: the overpowering light
Of that immortal shape was shadowed o'er
By love; which, from his soft and flowing limbs,
And passion-parted lips, and keen, faint eyes,
Steamed forth like vaporous fire; an atmosphere
Which wrapped me in its all-dissolving power,
As the warm aether of the morning sun

ACT II. Sc. i. **26. Æolian:** like music from an Aeolian harp
on which the winds play (from Aeolus, god of the winds).
31. that soul: Prometheus.

44. glaucous: greenish blue (from Glaucus, a sea-god). **50–
52. not . . . dissolved:** not as I am since I have been
made into the wind and been dissolved, etc.

Wraps ere it drinks some cloud of wandering dew.
I saw not, heard not, moved not, only felt
His presence flow and mingle through my blood   80
Till it became his life, and his grew mine,
And I was thus absorbed, until it passed,
And like the vapours when the sun sinks down,
Gathering again in drops upon the pines,
And tremulous as they, in the deep night
My being was condensed; and as the rays
Of thought were slowly gathered, I could hear
His voice, whose accents lingered ere they died
Like footsteps of far melody: thy name
Among the many sounds alone I heard   90
Of what might be articulate; though still
I listened through the night when sound was none.
Ione wakened then, and said to me:
"Canst thou divine what troubles me to-night?
I always knew what I desired before,
Nor ever found delight to wish in vain.
But now I cannot tell thee what I seek;
I know not; something sweet, since it is sweet
Even to desire; it is thy sport, false sister;
Thou hast discovered some enchantment old,   100
Whose spells have stolen my spirit as I slept
And mingled it with thine: for when just now
We kissed, I felt within thy parted lips
The sweet air that sustained me, and the warmth
Of the life-blood, for loss of which I faint,
Quivered between our intertwining arms."
I answered not, for the Eastern star grew pale,
But fled to thee.
    ASIA.   Thou speakest, but thy words
Are as the air: I feel them not: Oh, lift
Thine eyes, that I may read his written soul!   110
    PANTHEA.   I lift them though they droop beneath
      the load
Of that they would express: what canst thou see
But thine own fairest shadow imaged there?
    ASIA.   Thine eyes are like the deep, blue, boundless
      heaven
Contracted to two circles underneath
Their long, fine lashes; dark, far, measureless,
Orb within orb, and line through line inwoven.
    PANTHEA.   Why lookest thou as if a spirit passed?
    ASIA.   There is a change: beyond their inmost
      depth
I see a shade, a shape: 'tis He, arrayed   120
In the soft light of his own smiles, which spread

120. **He:** Prometheus.

Like radiance from the cloud-surrounded moon.
Prometheus, it is thine! depart not yet!
Say not those smiles that we shall meet again
Within that bright pavilion which their beams
Shall build o'er the waste world? The dream is told.
What shape is that between us? Its rude hair
Roughens the wind that lifts it, its regard
Is wild and quick, yet 'tis a thing of air,
For through its gray robe gleams the golden dew   130
Whose stars the noon has quenched not.
    DREAM.   Follow!
      Follow!
    PANTHEA.   It is mine other dream.
    ASIA.   It disappears.
    PANTHEA.   It passes now into my mind. Methought
As we sate here, the flower-infolding buds
Burst on yon lightning-blasted almond-tree,
When swift from the white Scythian wilderness
A wind swept forth wrinkling the Earth with frost:
I looked, and all the blossoms were blown down;
But on each leaf was stamped, as the blue bells
Of Hyacinth tell Apollo's written grief,   140
O, FOLLOW, FOLLOW!
    ASIA.   As you speak, your words
Fill, pause by pause, my own forgotten sleep
With shapes. Methought among these lawns together
We wandered, underneath the young gray dawn,
And multitudes of dense white fleecy clouds
Were wandering in thick flocks along the mountains
Shepherded by the slow, unwilling wind;
And the white dew on the new-bladed grass,
Just piercing the dark earth, hung silently;
And there was more which I remember not: 150
But on the shadows of the morning clouds,
Athwart the purple mountain slope, was written
FOLLOW, O, FOLLOW! as they vanished by;
And on each herb, from which Heaven's dew had
    fallen,
The like was stamped, as with a withering fire;
A wind arose among the pines; it shook
The clinging music from their boughs, and then
Low, sweet, faint sounds, like the farewell of ghosts,

128. **regard:** glance, aspect. 132. **mine other dream:** the second of those mentioned in line 61. 136. **Scythian:** from the ancient name (Scythia) for the area north of the Black Sea. 140. **Hyacinth . . . grief:** When Hyacinthus was accidentally killed, Apollo turned him into a flower on whose petals are marked the Greek word (AI) expressing woe (this is not the flower to which the name hyacinth is now given).

Were heard: O, FOLLOW, FOLLOW, FOLLOW ME!
And then I said: "Panthea, look on me."   160
But in the depth of those beloved eyes
Still I saw, FOLLOW, FOLLOW!
    ECHO.   Follow, follow!
    PANTHEA.   The crags, this clear spring morning,
        mock our voices
As they were spirit-tongued.
    ASIA.   It is some being
Around the crags. What fine clear sounds! O, list!

ECHOES (*unseen*)

Echoes we: listen!
    We cannot stay:
As dew-stars glisten
    Then fade away—
        Child of Ocean!   170

    ASIA.   Hark! Spirits speak. The liquid responses
Of their aëreal tongues yet sound.
    PANTHEA.   I hear.

ECHOES

O, follow, follow,
    As our voice recedeth
Through the caverns hollow,
    Where the forest spreadeth;
        (*More distant.*)
    O, follow, follow!
    Through the caverns hollow,
As the song floats thou pursue,
Where the wild bee never flew,   180
Through the noontide darkness deep,
By the odour-breathing sleep
Of faint night flowers, and the waves
At the fountain-lighted caves,
While our music, wild and sweet,
Mocks thy gently falling feet,
        Child of Ocean!
    ASIA.   Shall we pursue the sound? It grows more
        faint
And distant.
    PANTHEA.   List! the strain floats nearer now.

ECHOES

In the world unknown   190
    Sleeps a voice unspoken;

By thy step alone
    Can its rest be broken;
        Child of Ocean!

    ASIA.   How the notes sink upon the ebbing wind!

ECHOES

    O, follow, follow!
    Through the caverns hollow,
As the song floats thou pursue,
By the woodland noontide dew;
By the forest, lakes, and fountains,   200
Through the many-folded mountains;
To the rents, and gulfs, and chasms,
Where the Earth reposed from spasms,
On the day when He and thou
Parted, to commingle now;
        Child of Ocean!

    ASIA.   Come, sweet Panthea, link thy hand in mine,
And follow, ere the voices fade away.

SCENE II

A Forest, intermingled with Rocks and Caverns.
ASIA *and* PANTHEA *pass into it.* Two young FAUNS
are sitting on a Rock listening.

SEMICHORUS I OF SPIRITS

The path through which that lovely twain
    Have passed, by cedar, pine, and yew,
    And each dark tree that ever grew,
    Is curtained out from Heaven's wide blue;
Nor sun, nor moon, nor wind, nor rain,
    Can pierce its interwoven bowers,
    Nor aught, save where some cloud of dew,
Drifted along the earth-creeping breeze,
Between the trunks of the hoar trees,
    Hangs each a pearl in the pale flowers   10
    Of the green laurel, blown anew;
And bends, and then fades silently,
One frail and fair anemone:
Or when some star of many a one
That climbs and wanders through steep night,
Has found the cleft through which alone
Beams fall from high those depths upon

Sc. ii. **Setting. fauns:** deities of woods and herds. They
have a human form with pointed ears, small horns, and
goats' feet.

Ere it is borne away, away,
By the swift Heavens that cannot stay,
It scatters drops of golden light,   20
Like lines of rain that ne'er unite:
And the gloom divine is all around,
And underneath is the mossy ground.

SEMICHORUS II

There the voluptuous nightingales,
    Are awake through all the broad noonday.
When one with bliss or sadness fails,
    And through the windless ivy-boughs,
Sick with sweet love, droops dying away
On its mate's music-panting bosom;
Another from the swinging blossom,   30
    Watching to catch the languid close
    Of the last strain, then lifts on high
    The wings of the weak melody,
'Till some new strain of feeling bear
    The song, and all the woods are mute;
When there is heard through the dim air
The rush of wings, and rising there
    Like many a lake-surrounded flute,
Sounds overflow the listener's brain
So sweet, that joy is almost pain.   40

SEMICHORUS I

There those enchanted eddies play
    Of echoes, music-tongued, which draw,
    By Demogorgon's mighty law,
    With melting rapture, or sweet awe,
All spirits on that secret way;
    As inland boats are driven to Ocean
Down streams made strong with mountain-thaw:
    And first there comes a gentle sound
    To those in talk or slumber bound,
    And wakes the destined soft emotion,—   50
Attracts, impels them; those who saw
    Say from the breathing earth behind
    There steams a plume-uplifting wind
Which drives them on their path, while they
    Believe their own swift wings and feet
The sweet desires within obey:
And so they float upon their way,

Until, still sweet, but loud and strong,
The storm of sound is driven along,
    Sucked up and hurrying: as they fleet   60
    Behind, its gathering billows meet
And to the fatal mountain bear
Like clouds amid the yielding air.

    FIRST FAUN.   Canst thou imagine where those
        spirits live
Which make such delicate music in the woods?
We haunt within the least frequented caves
And closest coverts, and we know these wilds,
Yet never meet them, though we hear them oft:
Where may they hide themselves?

    SECOND FAUN.   'Tis hard to tell:
I have heard those more skilled in spirits say,   70
The bubbles, which the enchantment of the sun
Sucks from the pale faint water-flowers that pave
The oozy bottom of clear lakes and pools,
Are the pavilions where such dwell and float
Under the green and golden atmosphere
Which noontide kindles through the woven leaves;
And when these burst, and the thin fiery air,
The which they breathed within those lucent domes,
Ascends to flow like meteors through the night,
They ride on them, and rein their headlong speed,   80
And bow their burning crests, and glide in fire
Under the waters of the earth again.

    FIRST FAUN.   If such live thus, have others other lives,
Under pink blossoms or within the bells
Of meadow flowers, or folded violets deep,
Or on their dying odours, when they die,
Or in the sunlight of the sphered dew?

    SECOND FAUN.   Ay, many more which we may well
        divine.
But, should we stay to speak, noontide would come,
And thwart Silenus find his goats undrawn,   90
And grudge to sing those wise and lovely songs
Of Fate, and Chance, and God, and Chaos old,
And Love, and the chained Titan's woful doom,
And how he shall be loosed, and make the earth
One brotherhood: delightful strains which cheer
Our solitary twilights, and which charm
To silence the unenvying nightingales.

51–56. those . . . obey: Prophets and seers know that all things are impelled, while those who are being impelled imagine that they are moving through their own free will.

62. fatal mountain: See Sc. iii. 90–91. thwart . . . sing: Silenus ("thwart," i.e., stubborn) was the tutor of Bacchus and was gifted with the art of prophecy. Should he find his goats "undrawn" (unmilked) by the fauns, he would begrudge singing the prophecy mentioned in lines 91–95.

SCENE III

A Pinnacle of Rock among Mountains.
ASIA *and* PANTHEA.

PANTHEA.   Hither the sound has borne us—to the
     realm
Of Demogorgon, and the mighty portal,
Like a volcano's meteor-breathing chasm,
Whence the oracular vapour is hurled up
Which lonely men drink wandering in their youth,
And call truth, virtue, love, genius, or joy,
That maddening wine of life, whose dregs they drain
To deep intoxication; and uplift,
Like Mænads who cry loud, Evoe! Evoe!
The voice which is contagion to the world.   10
ASIA.   Fit throne for such a Power! Magnificent!
How glorious art thou, Earth! And if thou be
The shadow of some spirit lovelier still,
Though evil stain its work, and it should be
Like its creation, weak yet beautiful,
I could fall down and worship that and thee.
Even now my heart adoreth: Wonderful!
Look, sister, ere the vapour dim thy brain:
Beneath is a wide plain of billowy mist,
As a lake, paving in the morning sky,   20
With azure waves which burst in silver light,
Some Indian vale. Behold it, rolling on
Under the curdling winds, and islanding
The peak whereon we stand—midway, around,
Encinctured by the dark and blooming forests,
Dim twilight lawns, and stream-illumed caves,
And wind-enchanted shapes of wandering mist;
And far on high the keen sky-cleaving mountains
From icy spires of sun-like radiance fling
The dawn, as lifted Ocean's dazzling spray,   30
From some Atlantic islet scattered up,
Spangles the wind with lamp-like water-drops.
The vale is girdled with their walls, a howl
Of cataracts from their thaw-cloven ravines,
Satiates the listening wind, continuous, vast,
Awful as silence. Hark! the rushing snow!
The sun-awakened avalanche! whose mass,
Thrice sifted by the storm, had gathered there
Flake after flake, in heaven-defying minds
As thought by thought is piled, till some great truth 40

Is loosened, and the nations echo round,
Shaken to their roots, as do the mountains now.
     PANTHEA.   Look how the gusty sea of mist is
     breaking
In crimson foam, even at our feet! it rises
As Ocean at the enchantment of the moon
Round foodless men wrecked on some oozy isle.
     ASIA.   The fragments of the cloud are scattered up;
The wind that lifts them disentwines my hair;
Its billows now sweep o'er mine eyes; my brain
Grows dizzy; see'st thou shapes within the mist?   50
     PANTHEA.   A countenance with beckoning smiles:
     there burns
An azure fire within its golden locks!
Another and another: hark! they speak!

SONG OF SPIRITS

To the deep, to the deep,
             Down, down!
Through the shade of sleep,
Through the cloudy strife
Of Death and of Life;
Through the veil and the bar
Of things which seem and are   60
Even to the steps of the remotest throne,
             Down, down!

While the sound whirls around,
             Down, down!
As the fawn draws the hound,
As the lightning the vapour,
As a weak moth the taper;
Death, despair; love, sorrow;
Time both; to-day, to-morrow;
As steel obeys the spirit of the stone,   70
             Down, down!

Through the gray, void abysm,
             Down, down!
Where the air is no prism,
And the moon and stars are not,
And the cavern-crags wear not
The radiance of Heaven,
Nor the gloom to Earth given,
Where there is One pervading, One alone,
             Down, down!   80

Sc. iii. 9. **Mænads:** wild female worshipers of Bacchus (cf.
"Ode to the West Wind," below, l. 23).

66–67. lightning . . . taper: The usual order is reversed
(read "as the vapour draws lightning and the taper draws the
moth"). **74. prism:** Light passing through a prism is resolved
into the colors of the spectrum.

In the depth of the deep,
   Down, down!
Like veiled lightning asleep,
Like the spark nursed in embers,
The last look Love remembers,
Like a diamond, which shines
On the dark wealth of mines,
A spell is treasured but for thee alone.
   Down, down!

We have bound thee, we guide thee;   90
   Down, down!
With the bright form beside thee;
Resist not the weakness,
Such strength is in meekness
That the Eternal, the Immortal,
Must unloose through life's portal
The snake-like Doom coiled underneath his throne
   By that alone.

SCENE IV

The Cave of DEMOGORGON. ASIA *and* PANTHEA.

PANTHEA. What veiled form sits on that ebon
 throne?
ASIA. The veil has fallen.
PANTHEA.   I see a mighty darkness
Filling the seat of power, and rays of gloom
Dart round, as light from the meridian sun,
Ungazed upon and shapeless; neither limb,
Nor form, nor outline; yet we feel it is
A living Spirit.
DEMOGORGON. Ask what thou wouldst know.
ASIA. What canst thou tell?
DEM.   All things thou dar'st
 demand.
ASIA. Who made the living world?
DEM.   God.
ASIA.   Who
 made all

That it contains? thought, passion, reason, will, 10
Imagination?
DEM.   God: Almighty God.
ASIA. Who made that sense which, when the winds
 of Spring
In rarest visitation, or the voice
Of one beloved heard in youth alone,
Fills the faint eyes with falling tears which dim
The radiant looks of unbewailing flowers,
And leaves this peopled earth a solitude
When it returns no more?
DEM.   Merciful God.
ASIA. And who made terror, madness, crime,
 remorse,
Which from the links of the great chain of things, 20
To every thought within the mind of man
Sway and drag heavily, and each one reels
Under the load towards the pit of death;
Abandoned hope, and love that turns to hate;
And self-contempt, bitterer to drink than blood;
Pain, whose unheeded and familiar speech
Is howling, and keen shrieks, day after day;
And Hell, or the sharp fear of Hell?
DEM.   He reigns.
ASIA. Utter his name: a world pining in pain
Asks but his name: curses shall drag him down. 30
DEM. He reigns.
ASIA.   I feel, I know it: who?
DEM.   He
 reigns.
ASIA. Who reigns? There was the Heaven and
 Earth at first,
And Light and Love; then Saturn, from whose throne
Time fell, an envious shadow: such the state
Of the earth's primal spirits beneath his sway,
As the calm joy of flowers and living leaves
Before the wind or sun has withered them
And semivital worms; but he refused
The birthright of their being, knowledge, power,
The skill which wields the elements, the thought 40
Which pierces this dim universe like light,
Self-empire, and the majesty of love;
For thirst of which they fainted. Then Prometheus
Gave wisdom, which is strength, to Jupiter,

Sc. iv. Setting. **Demogorgon:** Shelley's possible sources for Demogorgon are little help in interpretation. He could have found the name mentioned in Spenser, Milton, and Boccaccio, and he had reviewed Peacock's *Rhododaphne* in which Demogorgon is discussed at some length in a note (to VI. 159): " 'The dreaded name of Demogorgon' is familiar to every reader, in Milton's enumeration of the Powers of Chaos . . . He was the Genius of the Earth, and the Sovereign Power of the Terrestrial Dæmons. He dwelt originally with

Eternity and Chaos, till . . . he organised the chaotic elements, and surrounded the earth with the heavens . . . it was held impious to pronounce his name." **12. that sense:** by which we catch, for a moment, a glimpse of ideal beauty. **33. Saturn:** king of the Titans (the predecessors of the Olympian gods).

And with this law alone, "Let man be free,"
Clothed him with the dominion of wide Heaven.
To know nor faith, nor love, nor law; to be
Omnipotent but friendless is to reign;
And Jove now reigned; for on the race of man
First famine, and then toil, and then disease,   50
Strife, wounds, and ghastly death unseen before,
Fell; and the unseasonable seasons drove
With alternating shafts of frost and fire,
Their shelterless, pale tribes to mountain caves:
And in their desert hearts fierce wants he sent,
And mad disquietudes, and shadows idle
Of unreal good, which levied mutual war,
So ruining the lair wherein they raged.
Prometheus saw, and waked the legioned hopes
Which sleep within folded Elysian flowers,   60
Nepenthe, Moly, Amaranth, fadeless blooms,
That they might hide with thin and rainbow wings
The shape of Death; and Love he sent to bind
The disunited tendrils of that vine
Which bears the wine of life, the human heart;
And he tamed fire which, like some beast of prey,
Most terrible, but lovely, played beneath
The frown of man; and tortured to his will
Iron and gold, the slaves and signs of power,
And gems and poisons, and all subtlest forms   70
Hidden beneath the mountains and the waves.
He gave man speech, and speech created thought,
Which is the measure of the universe;
And Science struck the thrones of earth and heaven,
Which shook, but fell not; and the harmonious mind
Poured itself forth in all-prophetic song;
And music lifted up the listening spirit
Until it walked, exempt from mortal care,
Godlike, o'er the clear billows of sweet sound;
And human hands first mimicked and then mocked,
With moulded limbs more lovely than its own,   81
The human form, till marble grew divine;
And mothers, gazing, drank the love men see
Reflected in their race, behold, and perish.

61. **Nepenthe . . . Amaranth:** Nepenthe was a drug
that brought forgetfulness of sorrow, Moly the herb
given Odysseus to protect him from the spells of Circe,
and Amaranth a magic flower that never faded. **72–73.
speech . . . universe:** Cf. IV. 415–17. **80. mimicked . . .
mocked:** imitated in sculpture and then mocked by creating
statues more beautiful than the actual human form. **83–84.
mothers . . . perish:** Expectant mothers would gaze at
beautiful statues in the hope that their children would acquire
the beauty as they were being formed before birth, a beauty so
great that those who behold it perish for love. Cf. IV.412–14.

He told the hidden power of herbs and springs,
And Disease drank and slept. Death grew like sleep.
He taught the implicated orbits woven
Of the wide-wandering stars; and how the sun
Changes his lair, and by what secret spell
The pale moon is transformed, when her broad eye
Gazes not on the interlunar sea: 91
He taught to rule, as life directs the limbs,
The tempest-winged chariots of the Ocean,
And the Celt knew the Indian. Cities then
Were built, and through their snow-like columns
     flowed
The warm winds, and the azure aether shone,
And the blue sea and shadowy hills were seen.
Such, the alleviations of his state,
Prometheus gave to man, for which he hangs
Withering in destined pain: but who rains down   100
Evil, the immedicable plague, which, while
Man looks on his creation like a God
And sees that it is glorious, drives him on,
The wreck of his own will, the scorn of earth,
The outcast, the abandoned, the alone?
Not Jove: while yet his frown shook Heaven, ay, when
His adversary from adamantine chains
Cursed him, he trembled like a slave. Declare
Who is his master? Is he too a slave?
     DEM.   All spirits are enslaved which serve things
          evil: 110
Thou knowest if Jupiter be such or no.
     ASIA.   Whom calledst thou God?
     DEM.   I spoke but as
          ye speak,
For Jove is the supreme of living things.
     ASIA.   Who is the master of the slave?
     DEM.   If the
          abysm
Could vomit forth its secrets.—But a voice
Is wanting, the deep truth is imageless;
For what would it avail to bid thee gaze
On the revolving world? What to bid speak
Fate, Time, Occasion, Chance, and Change? To these
All things are subject but eternal Love.   120
     ASIA.   So much I asked before, and my heart gave
The response thou hast given; and of such truths
Each to itself must be the oracle.
One more demand; and do thou answer me

91. **interlunar:** the interval between the old and the new
moon. **94. Celt . . . Indian:** the northwest European,
crossing the sea, came to know the peoples of India.

As mine own soul would answer, did it know
That which I ask. Prometheus shall arise
Henceforth the sun of this rejoicing world:
When shall the destined hour arrive?
    DEM.   Behold!
    ASIA.   The rocks are cloven, and through the purple
      night
I see cars drawn by rainbow-winged steeds   130
Which trample the dim winds: in each there stands
A wild-eyed charioteer urging their flight.
Some look behind, as fiends pursued them there,
And yet I see no shapes but the keen stars:
Others, with burning eyes, lean forth, and drink
With eager lips the wind of their own speed,
As if the thing they loved fled on before,
And now, even now, they clasped it. Their bright locks
Stream like a comet's flashing hair: they all
Sweep onward.
    DEM.   These are the immortal Hours,   140
Of whom thou didst demand. One waits for thee.
    ASIA.   A spirit with a dreadful countenance
Checks its dark chariot by the craggy gulf.
Unlike thy brethren, ghastly charioteer,
Who art thou? Whither wouldst thou bear me? Speak!
    SPIRIT.   I am the shadow of a destiny
More dread than is my aspect: ere yon planet
Has set, the darkness which ascends with me
Shall wrap in lasting night heaven's kingless throne.
    ASIA.   What meanest thou?
    PANTHEA.   That terrible shadow
    floats   150
Up from its throne, as may the lurid smoke
Of earthquake-ruined cities o'er the sea.
Lo! it ascends the car; the coursers fly
Terrified: watch its path among the stars
Blackening the night!
    ASIA.   Thus I am answered: strange!
    PANTHEA.   See, near the verge, another chariot stays;
An ivory shell inlaid with crimson fire,
Which comes and goes within its sculptured rim
Of delicate strange tracery; the young spirit
That guides it has the dove-like eyes of hope;   160
How its soft smiles attract the soul! as light
Lures winged insects through the lampless air.

---

128. **Behold:** He answers only by pointing to the approaching chariots (l. 130). 148. **darkness:** Demogorgon. 161–62. **soul . . . air:** The classical symbol of Psyche (the soul) was a moth or butterfly.

SPIRIT

My coursers are fed with the lightning,
    They drink of the whirlwind's stream,
And when the red morning is bright'ning
    They bathe in the fresh sunbeam;
They have strength for their swiftness I deem,
Then ascend with me, daughter of Ocean.

I desire: and their speed makes night kindle;
    I fear: they outstrip the Typhoon;   170
Ere the cloud piled on Atlas can dwindle
    We encircle the earth and the moon:
    We shall rest from long labours at noon:
Then ascend with me, daughter of Ocean.

SCENE V

The Car pauses within a Cloud on the top of a snowy Mountain. ASIA, PANTHEA, *and the* SPIRIT OF THE HOUR.

SPIRIT

On the brink of the night and the morning
    My coursers are wont to respire;
But the Earth has just whispered a warning
    That their flight must be swifter than fire:
They shall drink the hot speed of desire!
    ASIA.   Thou breathest on their nostrils, but my breath
Would give them swifter speed.
    SPIRIT.   Alas! it could not.
    PANTHEA.   Oh Spirit! pause, and tell whence is the light
Which fills this cloud? the sun is yet unrisen.
    SPIRIT.   The sun will rise not until noon. Apollo 10
Is held in heaven by wonder; and the light
Which fills this vapour, as the aëreal hue
Of fountain-gazing roses fills the water,
Flows from thy mighty sister.
    PANTHEA.   Yes, I feel—
    ASIA.   What is it with thee, sister? Thou art pale.
    PANTHEA.   How thou art changed! I dare not look on thee;
I feel but see thee not. I scarce endure
The radiance of thy beauty. Some good change
Is working in the elements, which suffer

Thy presence thus unveiled. The Nereids tell   20
That on the day when the clear hyaline
Was cloven at thine uprise, and thou didst stand
Within a veined shell, which floated on
Over the calm floor of the crystal sea,
Among the Ægean isles, and by the shores
Which bear thy name; love, like the atmosphere
Of the sun's fire filling the living world,
Burst from thee, and illumined earth and heaven
And the deep ocean and the sunless caves
And all that dwells within them; till grief cast   30
Eclipse upon the soul from which it came:
Such art thou now; nor is it I alone,
Thy sister, thy companion, thine own chosen one,
But the whole world which seeks thy sympathy.
Hearest thou not sounds i' the air which speak the
    love
Of all articulate beings? Feelest thou not
The inanimate winds enamoured of thee? List!

[*Music.*]

ASIA.   Thy words are sweeter than aught else but
    his
Whose echoes they are: yet all love is sweet,
Given or returned. Common as light is love,   40
And its familiar voice wearies not ever.
Like the wide heaven, the all-sustaining air,
It·makes the reptile equal to the God:
They who inspire it most are fortunate,
As I am now; but those who feel it most
Are happier still, after long sufferings,
As I shall soon become.
PANTHEA.   List! Spirits speak.

VOICE IN THE AIR, *singing*

Life of Life! thy lips enkindle
  With their love the breath between them;
And thy smiles before they dwindle   50
  Make the cold air fire; then screen them
In those looks, where whoso gazes
Faints, entangled in their mazes.

Child of Light! thy limbs are burning
  Through the vest which seems to hide them;
As the radiant lines of morning
  Through the clouds ere they divide them;
And this atmosphere divinest
Shrouds thee wheresoe'er thou shinest.

Fair are others; none beholds thee,   60
  But thy voice sounds low and tender
Like the fairest, for it folds thee
  From the sight, that liquid splendour,
And all feel, yet see thee never,
As I feel now, lost for ever!

Lamp of Earth! where'er thou movest
  Its dim shapes are clad with brightness,
And the souls of whom thou lovest
  Walk upon the winds with lightness,
Till they fail, as I am failing,   70
Dizzy, lost, yet unbewailing!

ASIA

My soul is an enchanted boat,
  Which, like a sleeping swan, doth float
Upon the silver waves of thy sweet singing;
  And thine doth like an angel sit
  Beside a helm conducting it,
Whilst all the winds with melody are ringing.
  It seems to float ever, for ever,
  Upon that many-winding river,
  Between mountains, woods, abysses,   80
  A paradise of wildernesses!
Till, like one in slumber bound,
Borne to the ocean, I float down, around,
Into a sea profound, of ever-spreading sound:

  Meanwhile thy spirit lifts its pinions
  In music's most serene dominions;
Catching the winds that fan that happy heaven.
  And we sail on, away, afar,
  Without a course, without a star,
But, by the instinct of sweet music driven;   90
  Till through Elysian garden islets
  By thee, most beautiful of pilots,
  Where never mortal pinnace glided,
  The boat of my desire is guided:
Realms where the air we breathe is love,
Which in the winds and on the waves doth move,
Harmonizing this earth with what we feel above.

Sc. v. 20–26. The . . . name: In this passage Asia is
identified with Venus, who was born of the sea-foam.
Nereids are sea nymphs. 21. hyaline: the sea (literally something that is like glass). 42–43. Like . . . God: Cf. *Epipsychidion*, below, ll. 128–29. 51. them: themselves.

62. it: "that liquid splendor."

We have passed Age's icy caves,
   And Manhood's dark and tossing waves,
And Youth's smooth ocean, smiling to betray: 100
   Beyond the glassy gulfs we flee
   Of shadow-peopled Infancy,
Through Death and Birth, to a diviner day;
   A paradise of vaulted bowers,
   Lit by downward-gazing flowers,
   And watery paths that wind between
   Wildernesses calm and green,
Peopled by shapes too bright to see,
And rest, having beheld; somewhat like thee;
Which walk upon the sea, and chant melodiously!   110

END OF THE SECOND ACT

ACT III

SCENE I

Heaven. JUPITER *on his Throne;* THETIS *and
the other* DEITIES *assembled.*

JUPITER.   Ye congregated powers of heaven, who
      share
The glory and the strength of him ye serve,
Rejoice, henceforth I am omnipotent.
All else had been subdued to me; alone
The soul of man, like unextinguished fire,
Yet burns towards heaven with fierce reproach, and
      doubt,
And lamentation, and reluctant prayer,
Hurling up insurrection, which might make
Our antique empire insecure, though built
On eldest faith, and hell's coeval, fear;   10
And though my curses through the pendulous air,
Like snow on herbless peaks, fall flake by flake,
And cling to it; though under my wrath's night
It climbs the crags of life, step after step,
Which wound it, as ice wounds unsandalled feet,
It yet remains supreme o'er misery,
Aspiring, unrepressed, yet soon to fall:
Even now have I begotten a strange wonder,

That fatal child, the terror of the earth,
Who waits but till the destined hour arrive,   20
Bearing from Demogorgon's vacant throne
The dreadful might of ever-living limbs
Which clothed that awful spirit unbeheld,
To redescend, and trample out the spark.
Pour forth heaven's wine, Idæan Ganymede,
And let it fill the Dædal cups like fire,
And from the flower-inwoven soil divine
Ye all-triumphant harmonies arise,
As dew from earth under the twilight stars:
Drink! be the nectar circling through your veins   30
The soul of joy, ye ever-living Gods,
Till exultation burst in one wide voice
Like music from Elysian winds.
                              And thou
Ascend beside me, veiled in the light
Of the desire which makes thee one with me,
Thetis, bright image of eternity!
When thou didst cry, "Insufferable might!
God! Spare me! I sustain not the quick flames,
The penetrating presence; all my being,
Like him whom the Numidian seps did thaw   40
Into a dew with poison, is dissolved,
Sinking through its foundations:" even then
Two mighty spirits, mingling, made a third
Mightier than either, which, unbodied now,
Between us floats, felt, although unbeheld,
Waiting the incarnation, which ascends,
(Hear ye the thunder of the fiery wheels
Griding the winds?) from Demogorgon's throne.
Victory! victory! Feel'st thou not, O world,
The earthquake of his chariot thundering up   50
Olympus?

[*The Car of the* HOUR *arrives.* DEMOGORGON *descends,
and moves towards the Throne of* JUPITER.]

            Awful shape, what art thou? Speak!
DEM.   Eternity. Demand no direr name.
Descend, and follow me down the abyss.

98–103. passed . . . day: The music enables one to reverse
the sequence of time back to the "diviner day" that we left
before birth. Cf. Wordsworth's "Intimations of Immortality," above, ll. 58–65.

ACT III. Sc. i. 19. fatal child: his son by Thetis, who will
take over the throne and power of Demogorgon, thus
making the universe subject to Jupiter's will, and trample
the "spark" (I. 24) of Prometheus' rebellion. 25. Idæan
Ganymede: the boy carried off from Mt. Ida to be cupbearer
to Jupiter and the gods. 26. Dædal: cunningly contrived
(see "Mt. Blanc," above, n. 86). 40. seps: a serpent whose
bite caused putrefaction. 48. Griding: cutting through the
winds with a grating sound.

I am thy child, as thou wert Saturn's child;
Mightier than thee: and we must dwell together
Henceforth in darkness. Lift thy lightnings not.
The tyranny of heaven none may retain,
Or reassume, or hold, succeeding thee:
Yet if thou wilt, as 'tis the destiny
Of trodden worms to writhe till they are dead,   60
Put forth thy might.
    JUPITER.   Detested prodigy!
Even thus beneath the deep Titanian prisons
I trample thee! thou lingerest?
                         Mercy! mercy!
No pity, no release, no respite! Oh,
That thou wouldst make mine enemy my judge,
Even where he hangs, seared by my long revenge,
On Caucasus! he would not doom me thus.
Gentle, and just, and dreadless, is he not
The monarch of the world? What then art thou?
No refuge! no appeal!
                   Sink with me then,   70
We two will sink on the wide waves of ruin,
Even as a vulture and a snake outspent
Drop, twisted in inextricable fight,
Into a shoreless sea. Let hell unlock
Its moulded oceans of tempestuous fire,
And whelm on them into the bottomless void
This desolated world, and thee, and me,
The conqueror and the conquered, and the wreck
Of that for which they combated.
                       Ai! Ai!
The elements obey me not. I sink   80
Dizzily down, ever, for ever, down.
And, like a cloud, mine enemy above
Darkens my fall with victory! Ai, Ai!

SCENE II

The Mouth of a great River in the Island Atlantis.
OCEAN *is discovered reclining near the Shore;* APOLLO
stands beside him.

OCEAN.   He fell, thou sayest, beneath his conqueror's
    frown?
APOLLO.   Ay, when the strife was ended which
    made dim
The orb I rule, and shook the solid stars,
The terrors of his eye illumined heaven

With sanguine light, through the thick ragged skirts
Of the victorious darkness, as he fell:
Like the last glare of day's red agony,
Which, from a rent among the fiery clouds,
Burns far along the tempest-wrinkled deep.
    OCEAN.   He sunk to the abyss? To the dark void? 10
    APOLLO.   An eagle so caught in some bursting cloud
On Caucasus, his thunder-baffled wings
Entangled in the whirlwind, and his eyes
Which gazed on the undazzling sun, now blinded
By the white lightning, while the ponderous hail
Beats on his struggling form, which sinks at length
Prone, and the aëreal ice clings over it.
    OCEAN.   Henceforth the fields of heaven-reflecting
    sea
Which are my realm, will heave, unstained with
    blood,
Beneath the uplifting winds, like plains of corn   20
Swayed by the summer air; my streams will flow
Round many-peopled continents, and round
Fortunate isles; and from their glassy thrones
Blue Proteus and his humid nymphs shall mark
The shadow of fair ships, as mortals see
The floating bark of the light-laden moon
With that white star, its sightless pilot's crest,
Borne down the rapid sunset's ebbing sea;
Tracking their path no more by blood and groans,
And desolation, and the mingled voice   30
Of slavery and command; but by the light
Of wave-reflected flowers, and floating odours,
And music soft, and mild, free, gentle voices,
And sweetest music, such as spirits love.
    APOLLO.   And I shall gaze not on the deeds which
    make
My mind obscure with sorrow, as eclipse
Darkens the sphere I guide; but list, I hear
The small, clear, silver lute of the young Spirit
That sits i' the morning star.
    OCEAN.   Thou must away;
Thy steeds will pause at even, till when farewell: 40
The loud deep calls me home even now to feed it
With azure calm out of the emerald urns
Which stand for ever full beside my throne.
Behold the Nereids under the green sea,
Their wavering limbs borne on the wind-like stream,
Their white arms lifted o'er their streaming hair

Sc. ii. 24. Proteus: a sea-god under Neptune (Ocean).
27. sightless: invisible. Pilot refers to the evening star
preceding the moon.

79. Ai: Greek for "woe."

With garlands pied and starry sea-flower crowns,
Hastening to grace their mighty sister's joy.
   [*A sound of waves is heard.*]
It is the unpastured sea hungering for calm.
Peace, monster; I come now. Farewell.

APOLLO.   Farewell. 50

SCENE III

Caucasus. PROMETHEUS, HERCULES, IONE, *the* EARTH,
SPIRITS, ASIA, *and* PANTHEA, *borne in the Car with
the* SPIRIT OF THE HOUR. HERCULES *unbinds*
PROMETHEUS, *who descends.*

HERCULES. Most glorious among Spirits, thus doth
  strength
To wisdom, courage, and long-suffering love,
And thee, who art the form they animate,
Minister like a slave.
  PRO.   Thy gentle words
Are sweeter even than freedom long desired
And long delayed.
       Asia, thou light of life,
Shadow of beauty unbeheld: and ye,
Fair sister nymphs, who made long years of pain
Sweet to remember, through your love and care:
Henceforth we will not part. There is a cave, 10
All overgrown with trailing odorous plants,
Which curtain out the day with leaves and flowers,
And paved with veined emerald, and a fountain
Leaps in the midst with an awakening sound.
From its curved roof the mountain's frozen tears
Like snow, or silver, or long diamond spires,
Hang downward, raining forth a doubtful light:
And there is heard the ever-moving air,
Whispering without from tree to tree, and birds,
And bees; and all around are mossy seats, 20
And the rough walls are clothed with long soft grass;
A simple dwelling, which shall be our own;
Where we will sit and talk of time and change,
As the world ebbs and flows, ourselves unchanged.
What can hide man from mutability?
And if ye sigh, then I will smile; and thou,
Ione, shalt chant fragments of sea-music,
Until I weep, when ye shall smile away
The tears she brought, which yet were sweet to
  shed.

We will entangle buds and flowers and beams 30
Which twinkle on the fountain's brim, and make
Strange combinations out of common things,
Like human babes in their brief innocence;
And we will search, with looks and words of love,
For hidden thoughts, each lovelier than the last,
Our unexhausted spirits; and like lutes
Touched by the skill of the enamoured wind,
Weave harmonies divine, yet ever new,
From difference sweet where discord cannot be;
And hither come, sped on the charmed winds, 40
Which meet from all the points of heaven, as bees
From every flower aëreal Enna feeds,
At their known island-homes in Himera,
The echoes of the human world, which tell
Of the low voice of love, almost unheard,
And dove-eyed pity's murmured pain, and music,
Itself the echo of the heart, and all
That tempers or improves man's life, now free;
And lovely apparitions,—dim at first,
Then radiant, as the mind, arising bright 50
From the embrace of beauty (whence the forms
Of which these are the phantoms), casts on them
The gathered rays which are reality—
Shall visit us, the progeny immortal
Of Painting, Sculpture, and rapt Poesy,
And arts, though unimagined, yet to be.
The wandering voices and the shadows these
Of all that man becomes, the mediators
Of that best worship, love, by him and us
Given and returned; swift shapes and sounds, which
  grow 60
More fair and soft as man grows wise and kind,
And, veil by veil, evil and error fall:
Such virtue has the cave and place around.
   [*Turning to the* SPIRIT OF THE HOUR.]
For thee, fair Spirit, one toil remains. Ione,
Give her that curved shell, which Proteus old
Made Asia's nuptial boon, breathing within it
A voice to be accomplished, and which thou
Didst hide in grass under the hollow rock.
  IONE. Thou most desired Hour, more loved and
  lovely
Than all thy sisters, this is the mystic shell; 70
See the pale azure fading into silver

Sc. iii. 23–24. sit . . . flows: Shelley echoes *King Lear*, V.
iii. 17–19.

42. Enna: the vale in Sicily from which Proserpine was
carried off by Pluto. **43. Himera:** ancient town in northern
Sicily. **49–56. And . . . be:** Cf. I. 740–49. **51. forms:**
the Platonic forms (or "Ideas"). **65. shell:** conch-shell
trumpet.

Lining it with a soft yet glowing light:
Looks it not like lulled music sleeping there?
   SPIRIT.   It seems in truth the fairest shell of Ocean:
Its sound must be at once both sweet and strange.
   PRO.   Go, borne over the cities of mankind
On whirlwind-footed coursers: once again
Outspeed the sun around the orbed world;
And as thy chariot cleaves the kindling air,
Thou breathe into the many-folded shell,   80
Loosening its mighty music; it shall be
As thunder mingled with clear echoes: then
Return; and thou shalt dwell beside our cave.
And thou, O, Mother Earth!—
   THE EARTH.   I hear, I feel;
Thy lips are on me, and their touch runs down
Even to the adamantine central gloom
Along these marble nerves; 'tis life, 'tis joy,
And through my withered, old, and icy frame
The warmth of an immortal youth shoots down
Circling. Henceforth the many children fair   90
Folded in my sustaining arms; all plants,
And creeping forms, and insects rainbow-winged,
And birds, and beasts, and fish, and human shapes,
Which drew disease and pain from my wan bosom,
Draining the poison of despair, shall take
And interchange sweet nutriment; to me
Shall they become like sister-antelopes
By one fair dam, snow-white and swift as wind,
Nursed among lilies near a brimming stream.
The dew-mists of my sunless sleep shall float   100
Under the stars like balm: night-folded flowers
Shall suck unwithering hues in their respose:
And men and beasts in happy dreams shall gather
Strength for the coming day, and all its joy:
And death shall be the last embrace of her
Who takes the life she gave, even as a mother
Folding her child, says, "Leave me not again."
   ASIA.   Oh, mother! wherefore speak the name of
     death?
Cease they to love, and move, and breathe, and speak,
Who die?
   THE EARTH.   It would avail not to reply: 110
Thou art immortal, and this tongue is known
But to the uncommunicating dead.
Death is the veil which those who live call life:

---

105. **her:** the Earth. 113. **Death . . . life:** an echo of
Plato's *Gorgias*, 492e: "and indeed I think that Euripides may
have been right in saying: 'Who knows if life be not death
and death life?'" (trans. Jowett).

They sleep, and it is lifted: and meanwhile
In mild variety the seasons mild
With rainbow-skirted showers, and odorous winds,
And long blue meteors cleansing the dull night,
And the life-kindling shafts of the keen sun's
All-piercing bow, and the dew-mingled rain
Of the calm moonbeams, a soft influence mild,   120
Shall clothe the forests and the fields, ay, even
The crag-built deserts of the barren deep,
With ever-living leaves, and fruits, and flowers.
And thou! There is a cavern where my spirit
Was panted forth in anguish whilst thy pain
Made my heart mad, and those who did inhale it
Became mad too, and built a temple there,
And spoke, and were oracular, and lured
The erring nations round to mutual war,
And faithless faith, such as Jove kept with thee; 130
Which breath now rises, as amongst tall weeds
A violet's exhalation, and it fills
With a serener light and crimson air
Intense, yet soft, the rocks and woods around;
It feeds the quick growth of the serpent vine,
And the dark linked ivy tangling wild,
And budding, blown, or odour-faded blooms
Which star the winds with points of coloured light,
As they rain through them, and bright golden globes
Of fruit, suspended in their own green heaven,   140
And through their veined leaves and amber stems
The flowers whose purple and translucid bowls
Stand ever mantling with aëreal dew,
The drink of spirits: and it circles round,
Like the soft waving wings of noonday dreams,
Inspiring calm and happy thoughts, like mine,
Now thou art thus restored. This cave is thine.
Arise! Appear!
    [*A* SPIRIT *rises in the likeness of a winged child.*]
        This is my torch-bearer;
Who let his lamp out in old time with gazing
On eyes from which he kindled it anew   150
With love, which is as fire, sweet daughter mine,
For such is that within thine own. Run, wayward,
And guide this company beyond the peak
Of Bacchic Nysa, Mænad-haunted mountain,
And beyond Indus and its tribute rivers,
Trampling the torrent streams and glassy lakes
With feet unwet, unwearied, undelaying,
And up the green ravine, across the vale,
Beside the windless and crystalline pool,

---

154. **Nysa:** mountain where the Mænads nursed Bacchus.

Where ever lies, on unerasing waves,   160
The image of a temple, built above,
Distinct with column, arch, and architrave,
And palm-like capital, and over-wrought,
And populous with most living imagery,
Praxitelean shapes, whose marble smiles
Fill the hushed air with everlasting love.
It is deserted now, but once it bore
Thy name, Prometheus; there the emulous youths
Bore to thy honour through the divine gloom
The lamp which was thine emblem: even as those 170
Who bear the untransmitted torch of hope
Into the grave, across the night of life,
As thou hast borne it most triumphantly
To this far goal of Time. Depart, farewell.
Beside that temple is the destined cave.

SCENE IV

A Forest. In the Background a Cave. PROMETHEUS,
ASIA, PANTHEA, IONE, *and the* SPIRIT OF THE EARTH.

IONE.   Sister, it is not earthly: how it glides
Under the leaves! how on its head there burns
A light, like a green star, whose emerald beams
Are twined with its fair hair! how, as it moves,
The splendour drops in flakes upon the grass!
Knowest thou it?
    PANTHEA.   It is the delicate spirit
That guides the earth through heaven. From afar
The populous constellations call that light
The loveliest of the planets; and sometimes
It floats along the spray of the salt sea,   10
Or makes its chariot of a foggy cloud,
Or walks through fields or cities while men sleep,
Or o'er the mountain tops, or down the rivers,
Or through the green waste wilderness, as now,
Wondering at all it sees. Before Jove reigned
It loved our sister Asia, and it came
Each leisure hour to drink the liquid light
Out of her eyes, for which it said it thirsted
As one bit by a dipsas, and with her
It made its childish confidence, and told her   20
All it had known or seen, for it saw much,
Yet idly reasoned what it saw; and called her—
For whence it sprung it knew not, nor do I—

Mother, dear mother.
    THE SPIRIT OF THE EARTH (*running to* ASIA).
                    Mother, dearest mother;
May I then talk with thee as I was wont?
May I then hide my eyes in thy soft arms,
After thy looks have made them tired of joy?
May I then play beside thee the long noons,
When work is none in the bright silent air?
    ASIA.   I love thee, gentlest being, and henceforth 30
Can cherish thee unenvied: speak, I pray:
Thy simple talk once solaced, now delights.
    SPIRIT OF THE EARTH.   Mother, I am grown wiser,
        though a child
Cannot be wise like thee, within this day;
And happier too; happier and wiser both.
Thou knowest that toads, and snakes, and loathly
        worms,
And venomous and malicious beasts, and boughs
That bore ill berries in the woods, were ever
An hindrance to my walks o'er the green world:
And that, among the haunts of humankind,   40
Hard-featured men, or with proud, angry looks,
Or cold, staid gait, or false and hollow smiles,
Or the dull sneer of self-loved ignorance,
Or other such foul masks, with which ill thoughts
Hide that fair being whom we spirits call man;
And women too, ugliest of all things evil,
(Though fair, even in a world where thou art fair,
When good and kind, free and sincere like thee),
When false or frowning, made me sick at heart
To pass them, though they slept, and I unseen.   50
Well, my path lately lay through a great city
Into the woody hills surrounding it:
A sentinel was sleeping at the gate:
When there was heard a sound, so loud, it shook
The towers amid the moonlight, yet more sweet
Than any voice but thine, sweetest of all;
A long, long sound, as it would never end:
And all the inhabitants leaped suddenly
Out of their rest, and gathered in the streets,
Looking in wonder up to Heaven, while yet   60
The music pealed along. I hid myself
Within a fountain in the public square,
Where I lay like the reflex of the moon
Seen in a wave under green leaves; and soon
Those ugly human shapes and visages
Of which I spoke as having wrought me pain,
Passed floating through the air, and fading still

---

165. Praxitelean shapes: as beautiful as the statues made by
the Greek sculptor Praxiteles (fourth century B.C.). **Sc. iv.**
19. dipsas: poisonous serpent.

63. reflex: reflection.

Into the winds that scattered them; and those
From whom they passed seemed mild and lovely forms
After some foul disguise had fallen, and all   70
Were somewhat changed, and after brief surprise
And greetings of delighted wonder, all
Went to their sleep again: and when the dawn
Came, wouldst thou think that toads, and snakes, and
    efts,
Could e'er be beautiful? yet so they were,
And that with little change of shape or hue:
All things had put their evil nature off:
I cannot tell my joy, when o'er a lake
Upon a drooping bough with nightshade twined,
I saw two azure halcyons clinging downward   80
And thinning one bright bunch of amber berries,
With quick long beaks, and in the deep there lay
Those lovely forms imaged as in a sky;
So, with my thoughts full of these happy changes,
We meet again, the happiest change of all.
    ASIA. And never will we part, till thy chaste sister
Who guides the frozen and inconstant moon
Will look on thy more warm and equal light
Till her heart thaw like flakes of April snow
And love thee.
    SPIRIT OF THE EARTH.   What; as Asia loves Prome-
      theus?   90
    ASIA.   Peace, wanton, thou art yet not old enough.
Think ye by gazing on each other's eyes
To multiply your lovely selves, and fill
With sphered fires the interlunar air?
    SPIRIT OF THE EARTH.   Nay, mother, while my sister
      trims her lamp
'Tis hard I should go darkling.
    ASIA.   Listen; look!

[*The* SPIRIT OF THE HOUR *enters.*]

    PRO. We feel what thou hast heard and seen: yet
      speak.
    SPIRIT OF THE HOUR.   Soon as the sound had ceased
      whose thunder filled
The abysses of the sky and the wide earth,
There was a change: the impalpable thin air   100
And the all-circling sunlight were transformed,
As if the sense of love dissolved in them
Had folded itself round the sphered world.

---

80. **halcyons:** kingfishers, now become vegetarians. 86.
chaste sister: Diana, goddess of the moon. 96. **darkling:** in
the dark.

My vision then grew clear, and I could see
Into the mysteries of the universe:
Dizzy as with delight I floated down,
Winnowing the lightsome air with languid plumes,
My coursers sought their birthplace in the sun,
Where they henceforth will live exempt from toil,
Pasturing flowers of vegetable fire;   110
And where my moonlike car will stand within
A temple, gazed upon by Phidian forms
Of thee, and Asia, and the Earth, and me,
And you fair nymphs looking the love we feel,—
In memory of the tidings it has borne,—
Beneath a dome fretted with graven flowers,
Poised on twelve columns of resplendent stone,
And open to the bright and liquid sky.
Yoked to it by an amphisbaenic snake
The likeness of those winged steeds will mock   120
The flight from which they find repose. Alas,
Whither has wandered now my partial tongue
When all remains untold which ye would hear?
As I have said, I floated to the earth:
It was, as it is still, the pain of bliss
To move, to breathe, to be; I wandering went
Among the haunts and dwellings of mankind,
And first was disappointed not to see
Such mighty change as I had felt within
Expressed in outward things; but soon I looked,   130
And behold, thrones were kingless, and men walked
One with the other even as spirits do,
None fawned, none trampled; hate, disdain, or fear,
Self-love or self-contempt, on human brows
No more inscribed, as o'er the gate of hell,
"All hope abandon ye who enter here;"
None frowned, none trembled, none with eager fear
Gazed on another's eye of cold command,
Until the subject of a tyrant's will
Became, worse fate, the abject of his own,   140
Which spurred him, like an outspent horse, to death.
None wrought his lips in truth-entangling lines
Which smiled the lie his tongue disdained to speak;
None, with firm sneer, trod out in his own heart
The sparks of love and hope till there remained
Those bitter ashes, a soul self-consumed,
And the wretch crept a vampire among men,

---

112. **Phidian forms:** like those made by the Greek sculptor
Phidias (fifth century B.C.). 119. **amphisbaenic snake:**
possessing a head at each end and capable of moving in either
direction. 136. **"All . . . here":** the inscription over the
entrance to hell in Dante's *Inferno*, III. 9. 140. **abject . . .
own:** the abject slave of his own will.

Infecting all with his own hideous ill;
None talked that common, false, cold, hollow talk
Which makes the heart deny the *yes* it breathes,   150
Yet question that unmeant hypocrisy
With such a self-mistrust as has no name.
And women, too, frank, beautiful, and kind
As the free heaven which rains fresh light and dew
On the wide earth, past; gentle radiant forms,
From custom's evil taint exempt and pure;
Speaking the wisdom once they could not think,
Looking emotions once they feared to feel,
And changed to all which once they dared not be,
Yet being now, made earth like heaven; nor pride,
Nor jealousy, nor envy, nor ill shame,   161
The bitterest of those drops of treasured gall,
Spoilt the sweet taste of the nepenthe, love.

Thrones, altars, judgement-seats, and prisons—where-
    in,
And beside which, by wretched men were borne
Sceptres, tiaras, swords, and chains, and tomes
Of reasoned wrong, glozed on by ignorance,
Were like those monstrous and barbaric shapes,
The ghosts of a no-more-remembered fame,
Which, from their unworn obelisks, look forth   170
In triumph o'er the palaces and tombs
Of those who were their conquerors, mouldering
    round.
These imaged, to the pride of kings and priests,
A dark yet mighty faith, a power as wide
As is the world it wasted, and are now
But an astonishment; even so the tools
And emblems of its last captivity,
Amid the dwellings of the peopled earth,
Stand, not o'erthrown, but unregarded now.
And those foul shapes, abhorred by god and man,—
Which, under many a name and many a form   181
Strange, savage, ghastly, dark and execrable,
Were Jupiter, the tyrant of the world;
And which the nations, panic-stricken, served
With blood, and hearts broken by long hope, and love
Dragged to his altars soiled and garlandless,
And slain amid men's unreclaiming tears,
Flattering the thing they feared, which fear was hate,—
Frown, mouldering fast, o'er their abandoned shrines:
The painted veil, by those who were, called life,   190
Which mimicked, as with colours idly spread,
All men believed or hoped, is torn aside;
The loathsome mask has fallen, the man remains

167. **glozed on:** glossed or commented on; annotated.

Sceptreless, free, uncircumscribed, but man
Equal, unclassed, tribeless, and nationless,
Exempt from awe, worship, degree, the king
Over himself; just, gentle, wise: but man
Passionless?—no, yet free from guilt or pain,
Which were, for his will made or suffered them,
Nor yet exempt, though ruling them like slaves,   200
From chance, and death, and mutability,
The clogs of that which else might oversoar
The loftiest star of unascended heaven,
Pinnacled dim in the intense inane.

END OF THE THIRD ACT

ACT IV

SCENE

A Part of the Forest near the Cave of PROMETHEUS.
PANTHEA *and* IONE *are sleeping: they awaken gradually*
during the first Song.

VOICE OF UNSEEN SPIRITS

The pale stars are gone!
For the sun, their swift shepherd,
To their folds them compelling,
In the depths of the dawn,
Hastes, in the meteor-eclipsing array, and they flee
    Beyond his blue dwelling,
    As fawns flee the leopard.
        But where are ye?

A Train of dark FORMS *and* SHADOWS *passes*
by confusedly, singing.

Here, oh, here:
We bear the bier   10
Of the Father of many a cancelled year.
    Spectres we
    Of the dead Hours be,
We bear Time to his tomb in eternity.

    Strew, oh, strew
    Hair, not yew!
Wet the dusty pall with tears, not dew!

202. **that which else:** the human spirit. 204. **inane:** the
void of infinite space.

Be the faded flowers
Of Death's bare bowers
Spread on the corpse of the King of Hours!   20

Haste, oh, haste!
As shades are chased,
Trembling, by day, from heaven's blue waste.
We melt away,
Like dissolving spray,
From the children of a diviner day,
With the lullaby
Of winds that die
On the bosom of their own harmony!

IONE

What dark forms were they?   30

PANTHEA

The past Hours weak and gray,
With the spoil which their toil
Raked together
From the conquest but One could foil.

IONE

Have they passed?

PANTHEA

                    They have passed;
They outspeeded the blast,
While 'tis said, they are fled—

IONE

Whither, oh, whither?

PANTHEA

To the dark, to the past, to the dead.

VOICE OF UNSEEN SPIRITS

Bright clouds float in heaven,   40
Dew-stars gleam on earth,
Waves assemble on ocean,
They are gathered and driven
By the storm of delight, by the panic of glee!
They shake with emotion,
They dance in their mirth.
But where are ye?

ACT IV. **34. One:** Prometheus.

The pine boughs are singing
Old songs with new gladness,
The billows and fountains   50
Fresh music are flinging,
Like the notes of a spirit from land and from sea;
The storms mock the mountains
With the thunder of gladness.
But where are ye?

IONE.   What charioteers are these?
PANTHEA.   Where are
    their chariots?

SEMICHORUS OF HOURS

The voice of the Spirits of Air and of Earth
Have drawn back the figured curtain of sleep
Which covered our being and darkened our birth
In the deep.

A VOICE

In the deep?

SEMICHORUS II

                    Oh, below the deep. 60

SEMICHORUS I

An hundred ages we had been kept
Cradled in visions of hate and care,
And each one who waked as his brother slept,
Found the truth—

SEMICHORUS II

                    Worse than his visions were!

SEMICHORUS I

We have heard the lute of Hope in sleep;
We have known the voice of Love in dreams;
We have felt the wand of Power, and leap—

SEMICHORUS II

As the billows leap in the morning beams!

CHORUS

Weave the dance on the floor of the breeze,
Pierce with song heaven's silent light,   70
Enchant the day that too swiftly flees,
To check its flight ere the cave of Night.

Once the hungry Hours were hounds
   Which chased the day like a bleeding deer,
And it limped and stumbled with many wounds
   Through the nightly dells of the desert year.

But now, oh weave the mystic measure
   Of music, and dance, and shapes of light,
Let the Hours, and the spirits of might and pleasure,
   Like the clouds and sunbeams, unite.

A VOICE

Unite! 80

PANTHEA.   See, where the Spirits of the human mind
Wrapped in sweet sounds, as in bright veils, approach.

CHORUS OF SPIRITS

We join the throng
   Of the dance and the song,
By the whirlwind of gladness borne along;
   As the flying-fish leap
   From the Indian deep,
And mix with the sea-birds, half asleep.

CHORUS OF HOURS

Whence come ye, so wild and so fleet,
For sandals of lightning are on your feet,   90
And your wings are soft and swift as thought,
And your eyes are as love which is veiled not?

CHORUS OF SPIRITS

We come from the mind
   Of human kind
Which was late so dusk, and obscene, and blind;
   Now 'tis an ocean
   Of clear emotion,
A heaven of serene and mighty motion.

   From that deep abyss
   Of wonder and bliss,   100
Whose caverns are crystal palaces;
   From those skiey towers
   Where Thought's crowned powers
Sit watching your dance, ye happy Hours!

   From the dim recesses
   Of woven caresses,
Where lovers catch ye by your loose tresses;
   From the azure isles,
   Where sweet Wisdom smiles,
Delaying your ships with her siren wiles.   110

   From the temples high
   Of Man's ear and eye,
Roofed over Sculpture and Poesy;
   From the murmurings
   Of the unsealed springs
Where Science bedews her Dædal wings.

   Years after years,
   Through blood, and tears,
And a thick hell of hatreds, and hopes, and fears,
   We waded and flew,   120
   And the islets were few
Where the bud-blighted flowers of happiness grew.

   Our feet now, every palm,
   Are sandalled with calm,
And the dew of our wings is a rain of balm;
   And beyond our eyes
   The human love lies
Which makes all it gazes on Paradise.

CHORUS OF SPIRITS AND HOURS

Then weave the web of the mystic measure;
From the depths of the sky and the ends of the earth,
   Come, swift Spirits of might and of pleasure,   131
Fill the dance and the music of mirth,
   As the waves of a thousand streams rush by
   To an ocean of splendour and harmony!

CHORUS OF SPIRITS

Our spoil is won,
   Our task is done,
We are free to dive, or soar, or run;
   Beyond and around,
   Or within the bound
Which clips the world with darkness round.   140

   We'll pass the eyes
   Of the starry skies
Into the hoar deep to colonize:
   Death, Chaos, and Night,
   From the sound of our flight,
Shall flee, like mist from a tempest's might.

   And Earth, Air, and Light,
   And the Spirit of Might,
Which drives round the stars in their fiery flight;
   And Love, Thought, and Breath,   150
   The powers that quell Death,
Wherever we soar shall assemble beneath.

140. clips: clasps.

And our singing shall build
In the void's loose field
A world for the Spirit of Wisdom to wield;
We will take our plan
From the new world of man,
And our work shall be called the Promethean.

CHORUS OF HOURS

Break the dance, and scatter the song:
   Let some depart, and some remain.   160

SEMICHORUS I

We, beyond heaven, are driven along:

SEMICHORUS II

Us the enchantments of earth retain:

SEMICHORUS I

Ceaseless, and rapid, and fierce, and free,
With the Spirits which build a new earth and sea,
And a heaven where yet heaven could never be.

SEMICHORUS II

Solemn, and slow, and serene, and bright,
Leading the Day and outspeeding the Night,
With the powers of a world of perfect light.

SEMICHORUS I

We whirl, singing loud, round the gathering sphere,
Till the trees, and the beasts, and the clouds appear
From its chaos made calm by love, not fear.   171

SEMICHORUS II

We encircle the ocean and mountains of earth,
And the happy forms of its death and birth
Change to the music of our sweet mirth.

CHORUS OF HOURS AND SPIRITS

Break the dance, and scatter the song,
   Let some depart, and some remain,
Wherever we fly we lead along
In leashes, like starbeams, soft yet strong,
   The clouds that are heavy with love's sweet rain.

PANTHEA. Ha! they are gone!

IONE.   Yet feel you no
    delight   180
From the past sweetness?

PANTHEA.   As the bare green hill
When some soft cloud vanishes into rain,
Laughs with a thousand drops of sunny water
To the unpavilioned sky!

IONE.   Even whilst we speak
New notes arise. What is that awful sound?

PANTHEA. 'Tis the deep music of the rolling world
Kindling within the strings of the waved air
Æolian modulations.

IONE.   Listen too,
How every pause is filled with under-notes,
Clear, silver, icy, keen, awakening tones,   190
Which pierce the sense, and live within the soul,
As the sharp stars pierce winter's crystal air
And gaze upon themselves within the sea.

PANTHEA. But see where through two openings in
    the forest
Which hanging branches overcanopy,
And where two runnels of a rivulet,
Between the close moss violet-inwoven,
Have made their path of melody, like sisters
Who part with sighs that they may meet in smiles
Turning their dear disunion to an isle   200
Of lovely grief, a wood of sweet sad thoughts;
Two visions of strange radiance float upon
The ocean-like enchantment of strong sound,
Which flows intenser, keener, deeper yet
Under the ground and through the windless air.

IONE. I see a chariot like that thinnest boat,
In which the Mother of the Months is borne
By ebbing light into her western cave,
When she upsprings from interlunar dreams;
O'er which is curved an orblike canopy   210
Of gentle darkness, and the hills and woods,
Distinctly seen through that dusk aery veil,
Regard like shapes in an enchanter's glass;
Its wheels are solid clouds, azure and gold,
Such as the genii of the thunderstorm
Pile on the floor of the illumined sea
When the sun rushes under it; they roll
And move and grow as with an inward wind;
Within it sits a winged infant, white
Its countenance, like the whiteness of bright snow,
Its plumes are as feathers of sunny frost,   221
Its limbs gleam white, through the wind-flowing folds

207. Mother . . . Months: Diana, or the Moon. **213.**
Regard: appear.

Of its white robe, woof of ethereal pearl.
Its hair is white, the brightness of white light
Scattered in strings; yet its two eyes are heavens
Of liquid darkness, which the Deity
Within seems pouring, as a storm is poured
From jagged clouds, out of their arrowy lashes,
Tempering the cold and radiant air around,
With fire that is not brightness; in its hand   230
It sways a quivering moonbeam, from whose point
A guiding power directs the chariot's prow
Over its wheeled clouds, which as they roll
Over the grass, and flowers, and waves, wake sounds,
Sweet as a singing rain of silver dew.
    PANTHEA.   And from the other opening in the wood
Rushes, with loud and whirlwind harmony,
A sphere, which is as many thousand spheres,
Solid as crystal, yet through all its mass
Flow, as through empty space, music and light: 240
Ten thousand orbs involving and involved,
Purple and azure, white, and green, and golden,
Sphere within sphere; and every space between
Peopled with unimaginable shapes,
Such as ghosts dream dwell in the lampless deep,
Yet each inter-transpicuous, and they whirl
Over each other with a thousand motions,
Upon a thousand sightless axles spinning,
And with the force of self-destroying swiftness,
Intensely, slowly, solemnly roll on,   250
Kindling with mingled sounds, and many tones,
Intelligible words and music wild.
With mighty whirl the multitudinous orb
Grinds the bright brook into an azure mist
Of elemental subtlety, like light;
And the wild odour of the forest flowers,
The music of the living grass and air,
The emerald light of leaf-entangled beams
Round its intense yet self-conflicting speed,
Seem kneaded into one aëreal mass   260
Which drowns the sense. Within the orb itself,
Pillowed upon its alabaster arms,
Like to a child o'erwearied with sweet toil,
On its own folded wings, and wavy hair,
The Spirit of the Earth is laid asleep,
And you can see its little lips are moving,
Amid the changing light of their own smiles,

Like one who talks of what he loves in dream.
    IONE.   'Tis only mocking the orb's harmony.
    PANTHEA.   And from a star upon its forehead,
      shoot,   270
Like swords of azure fire, or golden spears
With tyrant-quelling myrtle overtwined,
Embleming heaven and earth united now,
Vast beams like spokes of some invisible wheel
Which whirl as the orb whirls, swifter than thought,
Filling the abyss with sun-like lightenings,
And perpendicular now, and now transverse,
Pierce the dark soil, and as they pierce and pass,
Make bare the secrets of the earth's deep heart;
Infinite mines of adamant and gold,   280
Valueless stones, and unimagined gems,
And caverns on crystalline columns poised
With vegetable silver overspread;
Wells of unfathomed fire, and water springs
Whence the great sea, even as a child, is fed,
Whose vapours clothe earth's monarch mountain-tops
With kingly, ermine snow. The beams flash on
And make appear the melancholy ruins
Of cancelled cycles: anchors, beaks of ships;
Planks turned to marble; quivers, helms, and spears,
And gorgon-headed targes, and the wheels   291
Of scythed chariots, and the emblazonry
Of trophies, standards, and armorial beasts,
Round which death laughed, sepulchred emblems
Of dead destruction, ruin within ruin!
The wrecks beside of many a city vast,
Whose population which the earth grew over
Was mortal, but not human; see, they lie,
Their monstrous works, and uncouth skeletons,
Their statues, homes and fanes; prodigious shapes 300
Huddled in gray annihilation, split,
Jammed in the hard, black deep; and over these,
The anatomies of unknown winged things,
And fishes which were isles of living scale,
And serpents, bony chains, twisted around
The iron crags, or within heaps of dust
To which the tortuous strength of their last pangs
Had crushed the iron crags; and over these

---

238. **sphere:** the Earth. 246. **inter-transpicuous:** inter-transparent. 249. **self-destroying swiftness:** Each orb whirls rapidly but on different axles which are complexly balanced so that the sphere as a whole moves forward slowly.

272. **myrtle:** symbol of love. 281. **Valueless:** priceless. 289. **cancelled cycles:** cycles of time and history. Cf. "The World's Great Age Begins Anew," in *Hellas.* 291. **gorgon-headed targes:** shields ornamented with the head of Medusa, the Gorgon, the sight of whose face turned the beholder to stone. 292. **scythed chariots:** war chariots to the wheels of which were fixed sharp revolving blades. 293. **armorial beasts:** in heraldry.

The jagged alligator, and the might
Of earth-convulsing behemoth, which once   310
Were monarch beasts, and on the slimy shores,
And weed-overgrown continents of earth,
Increased and multiplied like summer worms
On an abandoned corpse, till the blue globe
Wrapped deluge round it like a cloak, and they
Yelled, gasped, and were abolished; or some God
Whose throne was in a comet, passed, and cried,
"Be not!" And like my words they were no more.

THE EARTH

The joy, the triumph, the delight, the madness!
The boundless, overflowing, bursting gladness,   320
The vaporous exultation not to be confined!
Ha! ha! the animation of delight
Which wraps me, like an atmosphere of light,
And bears me as a cloud is borne by its own wind.

THE MOON

Brother mine, calm wanderer,
Happy globe of land and air,
Some Spirit is darted like a beam from thee,
Which penetrates my frozen frame,
And passes with the warmth of flame,
With love, and odour, and deep melody   330
Through me, through me!

THE EARTH

Ha! ha! the caverns of my hollow mountains,
My cloven fire-crags, sound-exulting fountains
Laugh with a vast and inextinguishable laughter.
The oceans, and the deserts, and the abysses,
And the deep air's unmeasured wildernesses,
Answer from all their clouds and billows, echoing after.

They cry aloud as I do. Sceptred curse,
Who all our green and azure universe
Threatenedst to muffle round with black destruction,
sending   340
A solid cloud to rain hot thunderstones,
And splinter and knead down my children's bones,
All I bring forth, to one void mass battering and
blending,—

310. **behemoth:** a large animal (probably a hippopotamus) in
Job 40:15–24. 314. **blue globe:** so called because then
the sea covered it. 325. **wanderer:** The Earth is a planet and
"wanders" through the sky as contrasted with the fixed stars.
327–28. **Spirit . . . frame:** gravitation. 333. **cloven fire-
crags:** volcanoes. 338. **Sceptred curse:** Jupiter.

Until each crag-like tower, and storied column,
Palace, and obelisk, and temple solemn,
My imperial mountains crowned with cloud, and snow,
and fire;
My sea-like forests, every blade and blossom
Which finds a grave or cradle in my bosom,
Were stamped by thy strong hate into a lifeless mire:

How art thou sunk, withdrawn, covered, drunk up
By thirsty nothing, as the brackish cup   351
Drained by a desert-troop, a little drop for all;
And from beneath, around, within, above,
Filling thy void annihilation, love
Bursts in like light on caves cloven by the thunder-ball.

THE MOON

The snow upon my lifeless mountains
Is loosened into living fountains,
My solid oceans flow, and sing, and shine:
A spirit from my heart bursts forth,
It clothes with unexpected birth   360
My cold bare bosom: Oh! it must be thine
On mine, on mine!

Gazing on thee I feel, I know
Green stalks burst forth, and bright flowers grow,
And living shapes upon my bosom move:
Music is in the sea and air,
Winged clouds soar here and there,
Dark with the rain new buds are dreaming of:
'Tis love, all love!

THE EARTH

It interpenetrates my granite mass,   370
Through tangled roots and trodden clay doth pass
Into the utmost leaves and delicatest flowers;
Upon the winds, among the clouds 'tis spread,
It wakes a life in the forgotten dead,
They breathe a spirit up from their obscurest bowers.

And like a storm bursting its cloudy prison
With thunder, and with whirlwind, has arisen
Out of the lampless caves of unimagined being:
With earthquake shock and swiftness making shiver
Thought's stagnant chaos, unremoved for ever, 380
Till hate, and fear, and pain, light-vanquished shadows,
fleeing,

380. **unremoved:** unremovable.

Leave Man, who was a many-sided mirror
Which could distort to many a shape of error
This true fair world of things, a sea reflecting love;
    Which over all his kind, as the sun's heaven
    Gliding o'er ocean, smooth, serene, and even,
Darting from starry depths radiance and life, doth
    move:

Leave Man, even as a leprous child is left,
    Who follows a sick beast to some warm cleft
Of rocks, through which the might of healing springs
    is poured;   390
    Then when it wanders home with rosy smile,
    Unconscious, and its mother fears awhile
It is a spirit, then, weeps on her child restored.

Man, oh, not men! a chain of linked thought,
    Of love and might to be divided not,
Compelling the elements with adamantine stress;
    As the sun rules, even with a tyrant's gaze,
    The unquiet republic of the maze
Of planets, struggling fierce towards heaven's free
    wilderness.

Man, one harmonious soul of many a soul,   400
    Whose nature is its own divine control,
Where all things flow to all, as rivers to the sea;
    Familiar acts are beautiful through love;
    Labour, and pain, and grief, in life's green grove
Sport like tame beasts, none knew how gentle they
    could be!

His will, with all mean passions, bad delights,
    And selfish cares, its trembling satellites,
A spirit ill to guide, but mighty to obey,
    Is as a tempest-winged ship, whose helm
    Love rules, through waves which dare not over-
    whelm,   410
Forcing life's wildest shores to own its sovereign sway.

All things confess his strength. Through the cold mass
    Of marble and of colour his dreams pass;
Bright threads whence mothers weave the robes their
    children wear;
    Language is a perpetual Orphic song,
    Which rules with Dædal harmony a throng
Of thoughts and forms, which else senseless and
    shapeless were.

394. Man . . . men: mankind, united together, not divided
into separate, self-contained individuals. 415. Orphic: from
Orpheus, whose music could entrance even trees and stones.
416. Daedal: Cf. "Mont Blanc," above, n. 86.

The lightning is his slave; heaven's utmost deep
    Gives up her stars, and like a flock of sheep
They pass before his eye, are numbered, and roll on!
    The tempest is his steed, he strides the air;   421
    And the abyss shouts from her depth laid bare,
Heaven, hast thou secrets? Man unveils me; I have none.

THE MOON

The shadow of white death has passed
From my path in heaven at last,
A clinging shroud of solid frost and sleep;
    And through my newly-woven bowers,
    Wander happy paramours,
Less mighty, but as mild as those who keep
        Thy vales more deep.   430

THE EARTH

As the dissolving warmth of dawn may fold
A half unfrozen dew-globe, green, and gold,
And crystalline, till it becomes a winged mist,
    And wanders up the vault of the blue day,
    Outlives the noon, and on the sun's last ray
Hangs o'er the sea, a fleece of fire and amethyst.

THE MOON

Thou art folded, thou art lying
In the light which is undying
Of thine own joy, and heaven's smile divine;
    All suns and constellations shower   440
    On thee a light, a life, a power
Which doth array thy sphere; thou pourest thine
        On mine, on mine!

THE EARTH

I spin beneath my pyramid of night,
Which points into the heavens dreaming delight,

418–23. lightning . . . none: perhaps the most famous of
the Romantic prophecies of man's scientific achievement.
444. pyramid of night: the shadow cast by the earth on
the side not facing the sun. One of the great modern philosophers of science, Alfred North Whitehead, comments on
this phrase and the following lines in praising Shelley's
poetic grasp of science: "This stanza could only have been
written by someone with a definite geometrical diagram
before his inward eye—a diagram which it has often been
my business to demonstrate to mathematical classes. As
evidence note especially the last line which gives poetical
imagery to the light surrounding night's pyramid. This
idea could not occur to anyone without the diagram. But
the whole poem and other poems are permeated with
touches of this kind" (Science and the Modern World, Chap. V).

Murmuring victorious joy in my enchanted sleep;
  As a youth lulled in love-dreams faintly sighing,
  Under the shadow of his beauty lying,
Which round his rest a watch of light and warmth doth
    keep.

THE MOON

As in the soft and sweet eclipse,   450
  When soul meets soul on lovers' lips,
High hearts are calm, and brightest eyes are dull;
  So when thy shadow falls on me,
  Then am I mute and still, by thee
Covered; of thy love, Orb most beautiful,
        Full, oh, too full!

Thou art speeding round the sun
Brightest world of many a one;
  Green and azure sphere which shinest
  With a light which is divinest   460
Among all the lamps of Heaven
To whom life and light is given;
  I, thy crystal paramour
  Borne beside thee by a power
Like the polar Paradise
Magnet-like of lovers' eyes;
  I, a most enamoured maiden
  Whose weak brain is overladen
With the pleasure of her love,
Maniac-like around thee move   470
  Gazing, an insatiate bride,
  On thy form from every side
Like a Mænad, round the cup
Which Agave lifted up
  In the weird Cadmæan forest.
  Brother, wheresoe'er thou soarest
I must hurry, whirl and follow
Through the heavens wide and hollow,
  Sheltered by the warm embrace
  Of thy soul from hungry space,   480
Drinking from thy sense and sight
Beauty, majesty, and might,
  As a lover or a chameleon

Grows like what it looks upon,
  As a violet's gentle eye
  Gazes on the azure sky
Until its hue grows like what it beholds,
  As a gray and watery mist
  Glows like solid amethyst
Athwart the western mountain it enfolds,   490
  When the sunset sleeps
      Upon its snow—

THE EARTH

And the weak day weeps
    That it should be so.
Oh, gentle Moon, the voice of thy delight
Falls on me like thy clear and tender light
Soothing the seaman, borne the summer night,
    Through isles for ever calm;
Oh, gentle Moon, thy crystal accents pierce
The caverns of my pride's deep universe,   500
Charming the tiger joy, whose tramplings fierce
    Made wounds which need thy balm.
  PANTHEA.   I rise as from a bath of sparkling water,
A bath of azure light, among dark rocks,
Out of the stream of sound.
  IONE.   Ah me! sweet sister,
The stream of sound has ebbed away from us,
And you pretend to rise out of its wave,
Because your words fall like the clear, soft dew
Shaken from a bathing wood-nymph's limbs and hair.
  PANTHEA.   Peace! peace! A mighty Power, which
    is as darkness,   510
Is rising out of Earth, and from the sky
Is showered like night, and from within the air
Bursts, like eclipse which had been gathered up
Into the pores of sunlight: the bright visions,
Wherein the singing spirits rode and shone,
Gleam like pale meteors through a watery night.
  IONE.   There is a sense of words upon mine ear.
  PANTHEA.   An universal sound like words: Oh, list!

DEMOGORGON

Thou, Earth, calm empire of a happy soul,
    Sphere of divinest shapes and harmonies,   520
Beautiful orb! gathering as thou dost roll
    The love which paves thy path along the skies:

---

453. thy shadow: an eclipse of the moon. **464–66: power . . . eyes:** The moon's face is turned permanently to the earth like that of a lover whose eyes are magnetized by the beloved. **474. Agave:** princess of Thebes (daughter of King Cadmus). Her son Pentheus resisted the introduction of the female Dionysiac rites in Thebes; when she found him watching the rites, she and her daughters, in a frenzy, tore him to pieces.

493. weak day weeps: the fall of dew.

THE EARTH

I hear: I am as a drop of dew that dies.

DEMOGORGON

Thou, Moon, which gazest on the nightly Earth
    With wonder, as it gazes upon thee;
Whilst each to men, and beasts, and the swift birth
    Of birds, is beauty, love, calm, harmony:

THE MOON

I hear: I am a leaf shaken by thee!

DEMOGORGON

Ye Kings of suns and stars, Dæmons and Gods,
    Aetherial Dominations, who possess   530
Elysian, windless, fortunate abodes
    Beyond Heaven's constellated wilderness:

A VOICE *from above*

Our great Republic hears, we are blest, and bless.

DEMOGORGON

Ye happy Dead, whom beams of brightest verse
    Are clouds to hide, not colours to portray,
Whether your nature is that universe
    Which once ye saw and suffered—

A VOICE *from beneath*

                                    Or as they
Whom we have left, we change and pass away.

DEMOGORGON

Ye elemental Genii, who have homes
    From man's high mind even to the central stone
Of sullen lead; from heaven's star-fretted domes   541
    To the dull weed some sea-worm battens on:

A CONFUSED VOICE

We hear: thy words waken Oblivion.

DEMOGORGON

Spirits, whose homes are flesh: ye beasts and birds,
    Ye worms, and fish; ye living leaves and buds;
Lightning and wind; and ye untameable herds,
    Meteors and mists, which throng air's solitudes:—

A VOICE

Thy voice to us is wind among still woods.

DEMOGORGON

Man, who wert once a despot and a slave;
    A dupe and a deceiver; a decay;   550
A traveller from the cradle to the grave
    Through the dim night of this immortal day:

ALL

Speak: thy strong words may never pass away.

DEMOGORGON

This is the day, which down the void abysm
At the Earth-born's spell yawns for Heaven's despotism,
    And Conquest is dragged captive through the deep:
Love, from its awful throne of patient power
In the wise heart, from the last giddy hour
    Of dread endurance, from the slippery, steep,
And narrow verge of crag-like agony, springs   560
And folds over the world its healing wings.

Gentleness, Virtue, Wisdom, and Endurance,
These are the seals of that most firm assurance
    Which bars the pit over Destruction's strength;
And if, with infirm hand, Eternity,
Mother of many acts and hours, should free
    The serpent that would clasp her with his length;
These are the spells by which to reassume
An empire o'er the disentangled doom.

To suffer woes which Hope thinks infinite;   570
To forgive wrongs darker than death or night;
    To defy Power, which seems omnipotent;
To love, and bear; to hope till Hope creates
From its own wreck the thing it contemplates;

526. swift birth: Possibly "birth" is here used in the sense of "race" or "genus." An alternative is that the phrase "swift birth" is used simply to suggest the quick birth and death, the brief puff of life, of the birds.

555. Earth-born's spell: spell of Prometheus. **565–67. And . . . length:** Infinite time (Eternity) may be unable to keep imprisoned the principle of evil. The serpent coiled around Eternity may be a symbol of cyclic recurrence that will eventually bring back the reign of Jupiter.

Neither to change, nor falter, nor repent;
This, like thy glory, Titan, is to be
Good, great and joyous, beautiful and free;
This is alone Life, Joy, Empire, and Victory.

1818–19 (1820)

SONNET: ENGLAND IN 1819

An old, mad, blind, despised, and dying king,—
Princes, the dregs of their dull race, who flow
Through public scorn,—mud from a muddy spring,—
Rulers who neither see, nor feel, nor know,
But leech-like to their fainting country cling,
Till they drop, blind in blood, without a blow,—
A people starved and stabbed in the untilled field,—
An army, which liberticide and prey
Makes as a two-edged sword to all who wield,—
Golden and sanguine laws which tempt and slay;   10
Religion Christless, Godless—a book sealed;
A Senate,—Time's worst statute unrepealed,—
Are graves, from which a glorious Phantom may
Burst, to illumine our tempestuous day.

1819 (1839)

SONG TO THE MEN
OF ENGLAND

I

Men of England, wherefore plough
For the lords who lay ye low?
Wherefore weave with toil and care
The rich robes your tyrants wear?

II

Wherefore feed, and clothe, and save,
From the cradle to the grave,
Those ungrateful drones who would
Drain your sweat—nay, drink your blood?

III

Wherefore, Bees of England, forge
Many a weapon, chain, and scourge,   10
That these stingless drones may spoil
The forced produce of your toil?

IV

Have ye leisure, comfort, calm,
Shelter, food, love's gentle balm?
Or what is it ye buy so dear
With your pain and with your fear?

V

The seed ye sow, another reaps;
The wealth ye find, another keeps;
The robes ye weave, another wears;
The arms ye forge, another bears.   20

VI

Sow seed,—but let no tyrant reap;
Find wealth,—let no impostor heap;
Weave robes,—let not the idle wear;
Forge arms,—in your defence to bear.

VII

Shrink to your cellars, holes, and cells;
In halls ye deck another dwells.
Why shake the chains ye wrought? Ye see
The steel ye tempered glance on ye.

VIII

With plough and spade, and hoe and loom,
Trace your grave, and build your tomb,   30
And weave your winding-sheet, till fair
England be your sepulchre.

1819 (1839)

SONNET: ENGLAND IN 1819. **1. king:** George III, who was to die the following year (1820). **7. stabbed . . . field:** the Manchester Massacre (see headnote to "The Mask of Anarchy," below). **10. Golden and sanguine:** paid for by gold and executed with blood. **12. statute unrepealed:** the law forbidding the Catholics to hold office.

THE MASK OF ANARCHY

WRITTEN ON THE OCCASION OF THE MASSACRE AT MANCHESTER

❦

The poem is Shelley's response to the Manchester Massacre. On August 16, 1819, a mass meeting of workers in St. Peter's Field, Manchester, was dispersed by a charge of drunken cavalry. Several persons were killed and hundreds injured. The affair caused intense indignation throughout England.

❦

I

As I lay asleep in Italy
There came a voice from over the Sea,
And with great power it forth led me
To walk in the visions of Poesy.

II

I met Murder on the way—
He had a mask like Castlereagh—
Very smooth he looked, yet grim;
Seven blood-hounds followed him:

III

All were fat; and well they might
Be in admirable plight,   10
For one by one, and two by two,
He tossed them human hearts to chew
Which from his wide cloak he drew.

IV

Next came Fraud, and he had on,
Like Eldon, an ermined gown;
His big tears, for he wept well,
Turned to mill-stones as they fell.

THE MASK OF ANARCHY. **6. Castlreagh:** Robert Stewart, Viscount Castlereagh, 1769–1822, was Foreign Secretary from 1812 to 1822. Cf. Byron's violent attack on him, above, in the Dedication to *Don Juan* (ll. 88–120). **15–16. Eldon . . . tears:** John Scott, Earl of Eldon, 1751–1838, was Lord Chancellor and was famous both for the delays in his court and for his tendency to weep easily. It was Eldon who deprived Shelley of the custody of his children by his first wife.

V

And the little children, who
Round his feet played to and fro,
Thinking every tear a gem,   20
Had their brains knocked out by them.

VI

Clothed with the Bible, as with light,
And the shadows of the night,
Like Sidmouth, next, Hypocrisy
On a crocodile rode by.

VII

And many more Destructions played
In this ghastly masquerade,
All disguised, even to the eyes,
Like Bishops, lawyers, peers, or spies.

VIII

Last came Anarchy: he rode   30
On a white horse, splashed with blood;
He was pale even to the lips,
Like Death in the Apocalypse.

IX

And he wore a kingly crown;
And in his grasp a sceptre shone;
On his brow this mark I saw—
"I AM GOD, AND KING, AND LAW!"

X

With a pace stately and fast,
Over English land he passed,
Trampling to a mire of blood   40
The adoring multitude.

XI

And a mighty troop around,
With their trampling shook the ground,
Waving each a bloody sword,
For the service of their Lord.

24. Sidmouth: Home Secretary. **33. Death . . . Apocalypse:** Revelation 6:8.

XII

And with glorious triumph, they
Rode through England proud and gay,
Drunk as with intoxication
Of the wine of desolation.

XIII

O'er fields and towns, from sea to sea,   50
Passed the Pageant swift and free,
Tearing up, and trampling down;
Till they came to London town.

XIV

And each dweller, panic-stricken,
Felt his heart with terror sicken
Hearing the tempestuous cry
Of the triumph of Anarchy.

XV

For with pomp to meet him came,
Clothed in arms like blood and flame,
The hired murderers, who did sing   60
"Thou art God, and Law, and King.

XVI

"We have waited, weak and lone
For thy coming, Mighty One!
Our purses are empty, our swords are cold,
Give us glory, and blood, and gold."

XVII

Lawyers and priests, a motley crowd,
To the earth their pale brows bowed;
Like a bad prayer not over loud,
Whispering—"Thou art Law and God."—

XVIII

Then all cried with one accord,   70
"Thou art King, and God, and Lord;
Anarchy, to thee we bow,
Be thy name made holy now!"

XIX

And Anarchy, the Skeleton,
Bowed and grinned to every one,

As well as if his education
Had cost ten millions to the nation.

XX

For he knew the Palaces
Of our Kings were rightly his;
His the sceptre, crown, and globe,   80
And the gold-inwoven robe.

XXI

So he sent his slaves before
To seize upon the Bank and Tower,
And was proceeding with intent
To meet his pensioned Parliament

XXII

When one fled past, a maniac maid,
And her name was Hope, she said:
But she looked more like Despair,
And she cried out in the air:

XXIII

"My father Time is weak and gray   90
With waiting for a better day;
See how idiot-like he stands,
Fumbling with his palsied hands!

XXIV

"He has had child after child,
And the dust of death is piled
Over every one but me—
Misery, oh, Misery!"

XXV

Then she lay down in the street,
Right before the horses' feet,
Expecting, with a patient eye,   100
Murder, Fraud, and Anarchy.

XXVI

When between her and her foes
A mist, a light, an image rose,
Small at first, and weak, and frail
Like the vapour of a vale:

XXVII

Till as clouds grow on the blast,
Like tower-crowned giants striding fast,
And glare with lightnings as they fly,
And speak in thunder to the sky,

XXVIII

It grew—a Shape arrayed in mail   110
Brighter than the viper's scale,
And upborne on wings whose grain
Was as the light of sunny rain.

XXIX

On its helm, seen far away,
A planet, like the Morning's, lay;
And those plumes its light rained through
Like a shower of crimson dew.

XXX

With step as soft as wind it passed
O'er the heads of men—so fast
That they knew the presence there,   120
And looked,—but all was empty air.

XXXI

As flowers beneath May's footstep waken,
As stars from Night's loose hair are shaken,
As waves arise when loud winds call,
Thoughts sprung where'er that step did fall.

XXXII

And the prostrate multitude
Looked—and ankle-deep in blood,
Hope, that maiden most serene,
Was walking with a quiet mien:

XXXIII

And Anarchy, the ghastly birth,   130
Lay dead earth upon the earth;
The Horse of Death tameless as wind
Fled, and with his hoofs did grind
To dust the murderers thronged behind.

110. **Shape:** probably man's dream of liberty. **112. grain:**
color.

XXXIV

A rushing light of clouds and splendour,
A sense awakening and yet tender
Was heard and felt—and at its close
These words of joy and fear arose

XXXV

As if their own indignant Earth
Which gave the sons of England birth   140
Had felt their blood upon her brow,
And shuddering with a mother's throe

XXXVI

Had turned every drop of blood
By which her face had been bedewed
To an accent unwithstood,—
As if her heart had cried aloud:

XXXVII

"Men of England, heirs of Glory,
Heroes of unwritten story,
Nurslings of one mighty Mother,
Hopes of her, and one another;   150

XXXVIII

"Rise like Lions after slumber
In unvanquishable number,
Shake your chains to earth like dew
Which in sleep had fallen on you—
Ye are many—they are few.

XXXIX

"What is Freedom?—ye can tell
That which slavery is, too well—
For its very name has grown
To an echo of your own.

XL

"'Tis to work and have such pay   160
As just keeps life from day to day
In your limbs, as in a cell
For the tyrants' use to dwell,

XLI

"So that ye for them are made
Loom, and plough, and sword, and spade,
With or without your own will bent
To their defence and nourishment.

XLII

"'Tis to see your children weak
With their mothers pine and peak,
When the winter winds are bleak,—   170
They are dying whilst I speak.

XLIII

"'Tis to hunger for such diet
As the rich man in his riot
Casts to the fat dogs that lie
Surfeiting beneath his eye;

XLIV

"'Tis to let the Ghost of Gold
Take from Toil a thousandfold
More than e'er its substance could
In the tyrannies of old.

XLV

"Paper coin—that forgery   180
Of the title-deeds, which ye
Hold to something of the worth
Of the inheritance of Earth.

XLVI

"'Tis to be a slave in soul
And to hold no strong control
Over your own wills, but be
All that others make of ye.

XLVII

"And at length when ye complain
With a murmur weak and vain
'Tis to see the Tyrant's crew   190
Ride over your wives and you—
Blood is on the grass like dew.

176. Ghost of Gold: debased paper money, which the workers were forced to take as pay.

XLVIII

"Then it is to feel revenge
Fiercely thirsting to exchange
Blood for blood—and wrong for wrong—
Do not thus when ye are strong.

XLIX

"Birds find rest, in narrow nest
When weary of their winged quest;
Beasts find fare, in woody lair
When storm and snow are in the air.   200

L

"Asses, swine, have litter spread
And with fitting food are fed;
All things have a home but one—
Thou, Oh, Englishman, hast none!

LI

"This is Slavery—savage men,
Or wild beasts within a den
Would endure not as ye do—
But such ills they never knew.

LII

"What art thou, Freedom? O! could slaves
Answer from their living graves   210
This demand—tyrants would flee
Like a dream's dim imagery:

LIII

"Thou art not, as impostors say,
A shadow soon to pass away,
A superstition, and a name
Echoing from the cave of Fame.

LIV

"For the labourer thou art bread,
And a comely table spread
From his daily labour come
In a neat and happy home.   220

LV

"Thou art clothes, and fire, and food
For the trampled multitude—

No—in countries that are free
Such starvation cannot be
As in England now we see.

LVI

"To the rich thou art a check,
When his foot is on the neck
Of his victim, thou dost make
That he treads upon a snake.

LVII

"Thou art Justice—ne'er for gold   230
May thy righteous laws be sold
As laws are in England—thou
Shield'st alike the high and low.

LVIII

"Thou art Wisdom—Freemen never
Dream that God will damn for ever
All who think those things untrue
Of which Priests make such ado.

LIX

"Thou art Peace—never by thee
Would blood and treasure wasted be
As tyrants wasted them, when all   240
Leagued to quench thy flame in Gaul.

LX

"What if English toil and blood
Was poured forth, even as a flood?
It availed, Oh, Liberty,
To dim, but not extinguish thee.

LXI

"Thou art Love—the rich have kissed
Thy feet, and like him following Christ,
Give their substance to the free
And through the rough world follow thee,

LXII

"Or turn their wealth to arms, and make   250
War for thy beloved sake

241. **Leagued . . . Gaul:** the union of European powers
against France after the Revolution. 247. **him . . . Christ:**
Matthew 19:21.

On wealth, and war, and fraud—whence they
Drew the power which is their prey.

LXIII

"Science, Poetry, and Thought
Are thy lamps; they make the lot
Of the dwellers in a cot
So serene, they curse it not.

LXIV

"Spirit, Patience, Gentleness,
All that can adorn and bless
Art thou—let deeds, not words, express   260
Thine exceeding loveliness.

LXV

"Let a great Assembly be
Of the fearless and the free
On some spot of English ground
Where the plains stretch wide around.

LXVI

"Let the blue sky overhead,
The green earth on which ye tread,
All that must eternal be
Witness the solemnity.

LXVII

"From the corners uttermost   270
Of the bounds of English coast;
From every hut, village, and town
Where those who live and suffer moan
For others' misery or their own,

LXVIII

"From the workhouse and the prison
Where pale as corpses newly risen,
Women, children, young and old
Groan for pain, and weep for cold—

LXIX

"From the haunts of daily life
Where is waged the daily strife   280
With common wants and common cares
Which sows the human heart with tares—

LXX

"Lastly from the palaces
Where the murmur of distress
Echoes, like the distant sound
Of a wind alive around

LXXI

"Those prison halls of wealth and fashion
Where some few feel such compassion
For those who groan, and toil, and wail
As must make their brethren pale—   290

LXXII

"Ye who suffer woes untold,
Or to feel, or to behold
Your lost country bought and sold
With a price of blood and gold—

LXXIII

"Let a vast assembly be,
And with great solemnity
Declare with measured words that ye
Are, as God has made ye, free—

LXXIV

"Be your strong and simple words
Keen to wound as sharpened swords,   300
And wide as targes let them be,
With their shade to cover ye.

LXXV

"Let the tyrants pour around
With a quick and startling sound,
Like the loosening of a sea,
Troops of armed emblazonry.

LXXVI

"Let the charged artillery drive
Till the dead air seems alive
With the clash of clanging wheels,
And the tramp of horses' heels.   310

LXXVII

"Let the fixed bayonet
Gleam with sharp desire to wet
Its bright point in English blood
Looking keen as one for food.

LXXVIII

"Let the horsemen's scimitars
Wheel and flash, like sphereless stars
Thirsting to eclipse their burning
In a sea of death and mourning.

LXXIX

"Stand ye calm and resolute,
Like a forest close and mute,   320
With folded arms and looks which are
Weapons of unvanquished war,

LXXX

"And let Panic, who outspeeds
The career of armed steeds
Pass, a disregarded shade
Through your phalanx undismayed.

LXXXI

"Let the laws of your own land,
Good or ill, between ye stand
Hand to hand, and foot to foot,
Arbiters of the dispute,   330

LXXXII

"The old laws of England—they
Whose reverend heads with age are gray,
Children of a wiser day;
And whose solemn voice must be
Thine own echo—Liberty!

LXXXIII

"On those who first should violate
Such sacred heralds in their state
Rest the blood that must ensue,
And it will not rest on you.

316. sphereless: stars lacking a "sphere" in which to move.
(Shelley's image derives from the Ptolemaic astronomy
according to which the heavenly bodies moved in a series of
hollow spheres that surround the earth.)

301. targes: shields.

LXXXIV

"And if then the tyrants dare   340
Let them ride among you there,
Slash, and stab, and maim, and hew,—
What they like, that let them do.

LXXXV

"With folded arms and steady eyes,
And little fear, and less surprise,
Look upon them as they slay
Till their rage has died away.

LXXXVI

"Then they will return with shame
To the place from which they came,
And the blood thus shed will speak   350
In hot blushes on their cheek.

LXXXVII

"Every woman in the land
Will point at them as they stand—
They will hardly dare to greet
Their acquaintance in the street.

LXXXVIII

"And the bold, true warriors
Who have hugged Danger in wars
Will turn to those who would be free,
Ashamed of such base company.

LXXXIX

"And that slaughter to the Nation   360
Shall steam up like inspiration,
Eloquent, oracular;
A volcano heard afar.

XC

"And these words shall then become
Like Oppression's thundered doom
Ringing through each heart and brain,
Heard again—again—again—

XCI

"Rise like Lions after slumber
In unvanquishable number—

Shake your chains to earth like dew.   370
Which in sleep had fallen on you—
Ye are many—they are few."

1819 (1832)

ODE TO THE WEST WIND

෴

This poem was conceived and chiefly written in a wood that skirts the Arno, near Florence, and on a day when that tempestuous wind, whose temperature is at once mild and animating, was collecting the vapours which pour down the autumnal rains. They began, as I foresaw, at sunset with a violent tempest of hail and rain, attended by that magnificent thunder and lightning peculiar to the Cisalpine regions.

The phenomenon alluded to at the conclusion of the third stanza is well known to naturalists. The vegetation at the bottom of the sea, of rivers, and of lakes, sympathizes with that of the land in the change of seasons, and is consequently influenced by the winds which announce it [Shelley's note].

The stanza is a modification of Dante's terza rima (a continuous succession of tercets rhyming *aba, bcb, cdc,* etc.). Few poems in English are comparable either in clarity of formal pattern or in sweep, directness and intensity of emotion, and the union of these qualities is rarer still. One may recall Coleridge's remark that the poetic imagination reconciles "a more than usual state of emotion, with more than usual order; judgement ever awake and steady self-possession, with enthusiasm and feeling profound or vehement" (*Biographia Literaria*, close of Chap. XIV).

෴

I

O wild West Wind, thou breath of Autumn's being,
Thou, from whose unseen presence the leaves dead
Are driven, like ghosts from an enchanter fleeing,

Yellow, and black, and pale, and hectic red,
Pestilence-stricken multitudes: O thou,
Who chariotest to their dark wintry bed

The winged seeds, where they lie cold and low,
Each like a corpse within its grave, until
Thine azure sister of the Spring shall blow

ODE TO THE WEST WIND. **6. chariotest:** In Shelley's verse the chariot is a recurrent image of a power that is irresistible. Cf. *Prometheus Unbound,* II. iv. 129–40, "The Sensitive Plant," below, III. 92–93; and "The Triumph of Life," below, especially ll. 47–51 and 86–169.

Her clarion o'er the dreaming earth, and fill *e*   10
(Driving sweet buds like flocks to feed in air) *f*
With living hues and odours plain and hill: *e*

Wild Spirit, which art moving everywhere;
Destroyer and preserver; hear, oh, hear!

II

Thou on whose stream, mid the steep sky's com-
        motion,
Loose clouds like earth's decaying leaves are shed,
Shook from the tangled boughs of Heaven and Ocean,

Angels of rain and lightning: there are spread
On the blue surface of thine aëry surge,
Like the bright hair uplifted from the head   20

Of some fierce Maenad, even from the dim verge
Of the horizon to the zenith's height,
The locks of the approaching storm. Thou dirge

Of the dying year, to which this closing night
Will be the dome of a vast sepulchre,
Vaulted with all thy congregated might

Of vapours, from whose solid atmosphere
Black rain, and fire, and hail will burst: oh, hear!

III

Thou who didst waken from his summer dreams
The blue Mediterranean, where he lay,   30
Lulled by the coil of his crystalline streams,

Beside a pumice isle in Baiae's bay,
And saw in sleep old palaces and towers
Quivering within the wave's intenser day,

17–21. tangled . . . Maenad: The web of associations is partly explained by a comment in Shelley's *Critical Notices* on some carved Mænad figures: "The tremendous spirit of superstition seems to have caught them in its whirlwinds and to bear them over the earth as the rapid volutions of a tempest have the ever-changing trunk of a waterspout, or as the torrent of a mountain river whirls the autumnal leaves resistlessly along in its full eddies. The hair, loose and flowing, seems caught in the tempest of their own tumultuous motion." It is probable, then, that Shelley had the image of a waterspout in mind when he wrote "tangled boughs." Mænads (in Greek, "frenzied women") were votaries of Dionysus (cf. *Prometheus Unbound*, II. iii. 9). **32. pumice:** a light, porous, gray-brown stone produced by volcanic action. **Baiae:** Baiae is a district west of Naples. Cf. Shelley's letter to Peacock, below, December 22, 1818. In ancient times Baiae was a favorite resort of the Roman emperors.

All overgrown with azure moss and flowers
So sweet, the sense faints picturing them! Thou
For whose path the Atlantic's level powers

Cleave themselves into chasms, while far below
The sea-blooms and the oozy woods which wear
The sapless foliage of the ocean, know   40

Thy voice, and suddenly grow gray with fear,
And tremble and despoil themselves: oh, hear!

IV

If I were a dead leaf thou mightest bear;
If I were a swift cloud to fly with thee;
A wave to pant beneath thy power, and share

The impulse of thy strength, only less free
Than thou, O uncontrollable! If even
I were as in my boyhood, and could be

The comrade of thy wanderings over Heaven,
As then, when to outstrip thy skiey speed   50
Scarce seemed a vision; I would ne'er have striven

As thus with thee in prayer in my sore need.
Oh, lift me as a wave, a leaf, a cloud!   *sounds like Byron melodramatic*
I fall upon the thorns of life! I bleed!

A heavy weight of hours has chained and bowed
One too like thee: tameless, and swift, and proud.

V

Make me thy lyre, even as the forest is:
What if my leaves are falling like its own!
The tumult of thy mighty harmonies

Will take from both a deep, autumnal tone,   60
Sweet though in sadness. Be thou, Spirit fierce,
My spirit! Be thou me, impetuous one!

57. thy lyre: an Aeolian lyre or harp on which the wind can play. In his "Essay on Christianity" Shelley wrote: "We are not the arbiters of every motion of our own complicated nature; we are not the masters of our own imaginations and moods of mental being. There is a Power by which we are surrounded, like the atmosphere in which some motionless lyre is suspended, which visits with its breath our silent chords at will." Cf. "A Defence of Poetry," below, para. 2, and Coleridge's "The Eolian Harp," above, ll. 44–48.

Drive my dead thoughts over the universe
Like withered leaves to quicken a new birth!
And, by the incantation of this verse,

Scatter, as from an unextinguished hearth
Ashes and sparks, my words among mankind!
Be through my lips to unawakened earth

The trumpet of a prophecy! O, Wind,
If Winter comes, can Spring be far behind?   70

                    1819 (1820)

THE INDIAN SERENADE

I

I arise from dreams of thee
In the first sweet sleep of night.
When the winds are breathing low,
And the stars are shining bright:
I arise from dreams of thee,
And a spirit in my feet
Hath led me—who knows how?
To thy chamber window, Sweet!

II

The wandering airs they faint
On the dark, the silent stream—   10
The Champak odours fail
Like sweet thoughts in a dream;
The nightingale's complaint,
It dies upon her heart;—
As I must on thine,
Oh, beloved as thou art!

III

Oh lift me from the grass!
I die! I faint! I fail!
Let thy love in kisses rain
On my lips and eyelids pale.   20
My cheek is cold and white, alas!
My heart beats loud and fast;—
Oh! press it to thine own again,
Where it will break at last.

                    1819 (1822)

THE INDIAN SERENADE. **11. Champak:** East Indian magnolia
tree with yellow flowers.

LOVE'S PHILOSOPHY

I

The fountains mingle with the river
    And the rivers with the Ocean,
The winds of Heaven mix for ever
    With a sweet emotion;
Nothing in the world is single;
    All things by a law divine
In one spirit meet and mingle.
    Why not I with thine?—

II

See the mountains kiss high Heaven,
    And the waves clasp one another;   10
No sister-flower would be forgiven
    If it disdained its brother;
And the sunlight clasps the earth,
    And the moonbeams kiss the sea:
What is all this sweet work worth
    If thou kiss not me?

                    1819 (1819)

THE SENSITIVE PLANT

The sensitive plant is the name of a variety of mimosa,
the leaves of which fan out widely from the end of each
stalk or petiole. At dark or when touched they move
upward and come together. The poem is an impromptu
fantasy on habitual themes. It may be interpreted generally in connection with the "Hymn to Intellectual
Beauty." The garden resembles similar bowers throughout Shelley's poetry, images of loving harmony whether
in the individual soul or in society. (Cf. "Lines Written
Among the Euganean Hills," ll. 343–73, *Epipsychidion*,
ll. 435–82, "The Triumph of Life," ll. 308–55, and many
passages in *Prometheus Unbound*.) This harmony is
created by the vision or presence of Intellectual Beauty
(often figured as a lady—Asia in *Prometheus Unbound*,
Emily in *Epipsychidion*, the female shape in "The Triumph of Life"), but the vision is inconstant, and when
it vanishes life becomes vacant and desolate. Nevertheless, it is a vision of what really and eternally is beyond
this world of illusion.

PART FIRST

A Sensitive Plant in a garden grew,
And the young winds fed it with silver dew,
And it opened its fan-like leaves to the light,
And closed them beneath the kisses of Night.

And the Spring arose on the garden fair,
Like the Spirit of Love felt everywhere;
And each flower and herb on Earth's dark breast
Rose from the dreams of its wintry rest.

But none ever trembled and panted with bliss
In the garden, the field, or the wilderness,   10
Like a doe in the noontide with love's sweet want,
As the companionless Sensitive Plant.

The snowdrop, and then the violet,
Arose from the ground with warm rain wet,
And their breath was mixed with fresh odour, sent
From the turf, like the voice and the instrument.

Then the pied wind-flowers and the tulip tall,
And narcissi, the fairest among them all,
Who gaze on their eyes in the stream's recess,
Till they die of their own dear loveliness;   20

And the Naiad-like lily of the vale,
Whom youth makes so fair and passion so pale
That the light of its tremulous bells is seen
Through their pavilions of tender green;

And the hyacinth purple, and white, and blue,
Which flung from its bells a sweet peal anew
Of music so delicate, soft, and intense,
It was felt like an odour within the sense;

And the rose like a nymph to the bath addressed,
Which unveiled the depth of her glowing breast,   30
Till, fold after fold, to the fainting air
The soul of her beauty and love lay bare:

And the wand-like lily, which lifted up,
As a Maenad, its moonlight-coloured cup,
Till the fiery star, which is its eye,
Gazed through clear dew on the tender sky;

And the jessamine faint, and the sweet tuberose,
The sweetest flower for scent that blows;
And all rare blossoms from every clime
Grew in that garden in perfect prime.   40

And on the stream whose inconstant bosom
Was pranked, under boughs of embowering blossom,
With golden and green light, slanting through
Their heaven of many a tangled hue,

Broad water-lilies lay tremulously,
And starry river-buds glimmered by,
And around them the soft stream did glide and dance
With a motion of sweet sound and radiance.

And the sinuous paths of lawn and of moss,
Which led through the garden along and across,   50
Some open at once to the sun and the breeze,
Some lost among bowers of blossoming trees,

Were all paved with daisies and delicate bells
As fair as the fabulous asphodels,
And flow'rets which, drooping as day drooped too,
Fell into pavilions, white, purple, and blue,
To roof the glow-worm from the evening dew.

And from this undefiled Paradise
The flowers (as an infant's awakening eyes
Smile on its mother, whose singing sweet   60
Can first lull, and at last must awaken it),

When Heaven's blithe winds had unfolded them,
As mine-lamps enkindle a hidden gem,
Shone smiling to Heaven, and every one
Shared joy in the light of the gentle sun;

For each one was interpenetrated
With the light and the odour its neighbour shed,
Like young lovers whom youth and love make dear
Wrapped and filled by their mutual atmosphere.

But the Sensitive Plant which could give small fruit
Of the love which it felt from the leaf to the root,   71
Received more than all, it loved more than ever,
Where none wanted but it, could belong to the giver,—

For the Sensitive Plant has no bright flower;
Radiance and odour are not its dower;
It loves, even like Love, its deep heart is full,
It desires what it has not, the Beautiful!

The light winds which from unsustaining wings
Shed the music of many murmurings;

THE SENSITIVE PLANT. PART I. **21. Naiad:** water nymph.

42. pranked: adorned. **54. fabulous asphodels:** The asphodel, in Greek fable, bloomed forever in the Elysian Fields.
76–77. loves . . . Beautiful: The premise is one of the
themes of Plato's *Symposium* (201–04): that love is not itself
beautiful but is the desire for the beautiful.

The beams which dart from many a star   80
Of the flowers whose hues they bear afar;

The plumed insects swift and free,
Like golden boats on a sunny sea,
Laden with light and odour, which pass
Over the gleam of the living grass;

The unseen clouds of the dew, which lie
Like fire in the flowers till the sun rides high,
Then wander like spirits among the spheres,
Each cloud faint with the fragrance it bears;

The quivering vapours of dim noontide,   90
Which like a sea o'er the warm earth glide,
In which every sound, and odour, and beam,
Move, as reeds in a single stream;

Each and all like ministering angels were
For the Sensitive Plant sweet joy to bear,
Whilst the lagging hours of the day went by
Like windless clouds o'er a tender sky.

And when evening descended from Heaven above,
And the Earth was all rest, and the air was all love,
And delight, though less bright, was far more deep,
And the day's veil fell from the world of sleep,   101

And the beasts, and the birds, and the insects were
    drowned
In an ocean of dreams without a sound;
Whose waves never mark, though they ever impress
The light sand which paves it, consciousness;

(Only overhead the sweet nightingale
Ever sang more sweet as the day might fail,
And snatches of its Elysian chant
Were mixed with the dreams of the Sensitive Plant);—

The Sensitive Plant was the earliest   110
Upgathered into the bosom of rest;
A sweet child weary of its delight,
The feeblest and yet the favourite,
Cradled within the embrace of Night.

PART SECOND

There was a Power in this sweet place,
An Eve in this Eden; a ruling Grace
Which to the flowers, did they waken or dream,
Was as God is to the starry scheme.

A Lady, the wonder of her kind,
Whose form was upborne by a lovely mind

Which, dilating, had moulded her mien and motion
Like a sea-flower unfolded beneath the ocean,

Tended the garden from morn to even:
And the meteors of that sublunar Heaven,   10
Like the lamps of the air when Night walks forth,
Laughed round her footsteps up from the Earth!

She had no companion of mortal race,
But her tremulous breath and her flushing face
Told, whilst the morn kissed the sleep from her eyes,
That her dreams were less slumber than Paradise:

As if some bright Spirit for her sweet sake
Had deserted Heaven while the stars were awake,
As if yet around her he lingering were,
Though the veil of daylight concealed him from her.

Her step seemed to pity the grass it pressed;   21
You might hear by the heaving of her breast,
That the coming and going of the wind
Brought pleasure there and left passion behind.

And wherever her aëry footstep trod,
Her trailing hair from the grassy sod
Erased its light vestige, with shadowy sweep,
Like a sunny storm o'er the dark green deep.

I doubt not the flowers of that garden sweet
Rejoiced in the sound of her gentle feet;   30
I doubt not they felt the spirit that came
From her glowing fingers through all their frame.

She sprinkled bright water from the stream
On those that were faint with the sunny beam;
And out of the cups of the heavy flowers
She emptied the rain of the thunder-showers.

She lifted their heads with her tender hands,
And sustained them with rods and osier-bands;
If the flowers had been her own infants, she
Could never have nursed them more tenderly.   40

And all killing insects and gnawing worms,
And things of obscene and unlovely forms,
She bore, in a basket of Indian woof,
Into the rough woods far aloof,—

In a basket, of grasses and wild-flowers full,
The freshest her gentle hands could pull
For the poor banished insects, whose intent,
Although they did ill, was innocent.

PART II. **38. osier-bands:** willow bands.

But the bee and the beamlike ephemeris   49
Whose path is the lightning's, and soft moths that kiss
The sweet lips of the flowers, and harm not, did she
Make her attendant angels be.

And many an antenatal tomb,
Where butterflies dream of the life to come,
She left clinging round the smooth and dark
Edge of the odorous cedar bark.

This fairest creature from earliest Spring
Thus moved through the garden ministering
All the sweet season of Summertide,
And ere the first leaf looked brown—she died!   60

PART THIRD

Three days the flowers of the garden fair,
Like stars when the moon is awakened, were,
Or the waves of Baiae, ere luminous
She floats up through the smoke of Vesuvius.

And on the fourth, the Sensitive Plant
Felt the sound of the funeral chant,
And the steps of the bearers, heavy and slow,
And the sobs of the mourners, deep and low;

The weary sound and the heavy breath,
And the silent motions of passing death,   10
And the smell, cold, oppressive, and dank,
Sent through the pores of the coffin-plank;

The dark grass, and the flowers among the grass,
Were bright with tears as the crowd did pass;
From their sighs the wind caught a mournful tone,
And sate in the pines, and gave groan for groan.

The garden, once fair, became cold and foul,
Like the corpse of her who had been its soul,
Which at first was lovely as if in sleep,
Then slowly changed, till it grew a heap   20
To make men tremble who never weep.

Swift Summer into the Autumn flowed,
And frost in the mist of the morning rode,
Though the noonday sun looked clear and bright,
Mocking the spoil of the secret night.

The rose-leaves, like flakes of crimson snow,
Paved the turf and the moss below.
The lilies were drooping, and white, and wan,
Like the head and the skin of a dying man.

And Indian plants, of scent and hue   30
The sweetest that ever were fed on dew,
Leaf by leaf, day after day,
Were massed into the common clay.

And the leaves, brown, yellow, and gray, and red,
And white with the whiteness of what is dead,
Like troops of ghosts on the dry wind passed;
Their whistling noise made the birds aghast.

And the gusty winds waked the winged seeds,
Out of their birthplace of ugly weeds,
Till they clung round many a sweet flower's stem,   40
Which rotted into the earth with them.

The water-blooms under the rivulet
Fell from the stalks on which they were set;
And the eddies drove them here and there,
As the winds did those of the upper air.

Then the rain came down, and the broken stalks
Were bent and tangled across the walks;
And the leafless network of parasite bowers
Massed into ruin; and all sweet flowers.

Between the time of the wind and the snow   50
All loathliest weeds began to grow,
Whose coarse leaves were splashed with many a speck,
Like the water-snake's belly and the toad's back.

And thistles, and nettles, and darnels rank,
And the dock, and henbane, and hemlock dank,
Stretched out its long and hollow shank,
And stifled the air till the dead wind stank.

And plants, at whose names the verse feels loath,
Filled the place with a monstrous undergrowth,
Prickly, and pulpous, and blistering, and blue,   60
Livid, and starred with a lurid dew.

And agarics, and fungi, with mildew and mould
Started like mist from the wet ground cold;
Pale, fleshy, as if the decaying dead
With a spirit of growth had been animated!

Spawn, weeds, and filth, a leprous scum,
Made the running rivulet thick and dumb,
And at its outlet flags huge as stakes
Dammed it up with roots knotted like water-snakes.

And hour by hour, when the air was still,   70
The vapours arose which have strength to kill;

PART III. **34–36. leaves . . . passed:** Cf. "Ode to the West Wind," above, ll. 2–4. **62. agarics:** a fungus.

---

49. ephemeris: a delicate insect that lives only a few hours.

At morn they were seen, at noon they were felt,
At night they were darkness no star could melt.

And unctuous meteors from spray to spray
Crept and flitted in broad noonday
Unseen; every branch on which they alit
By a venomous blight was burned and bit.

The Sensitive Plant, like one forbid,
Wept, and the tears within each lid
Of its folded leaves, which together grew,   80
Were changed to a blight of frozen glue.

For the leaves soon fell, and the branches soon
By the heavy axe of the blast were hewn;
The sap shrank to the root through every pore
As blood to a heart that will beat no more.

For Winter came: the wind was his whip:
One choppy finger was on his lip:
He had torn the cataracts from the hills
And they clanked at his girdle like manacles;

His breath was a chain which without a sound   90
The earth, and the air, and the water bound;
He came, fiercely driven, in his chariot-throne
By the tenfold blasts of the Arctic zone.

Then the weeds which were forms of living death
Fled from the frost to the earth beneath.
Their decay and sudden flight from frost
Was but like the vanishing of a ghost!

And under the roots of the Sensitive Plant
The moles and the dormice died for want:
The birds dropped stiff from the frozen air   100
And were caught in the branches naked and bare.

First there came down a thawing rain
And its dull drops froze on the boughs again;
Then there steamed up a freezing dew
Which to the drops of the thaw-rain grew;

And a northern whirlwind, wandering about
Like a wolf that had smelt a dead child out,
Shook the boughs thus laden, and heavy, and stiff,
And snapped them off with his rigid griff.

When Winter had gone and Spring came back   110
The Sensitive Plant was a leafless wreck;
But the mandrakes, and toadstools, and docks, and
    darnels,
Rose like the dead from their ruined charnels.

109. **griff:** claw.

CONCLUSION

Whether the Sensitive Plant, or that
Which within its boughs like a Spirit sat,
Ere its outward form had known decay,
Now felt this change, I cannot say.

Whether that Lady's gentle mind,
No longer with the form combined
Which scattered love, as stars do light,   120
Found sadness, where it left delight,

I dare not guess; but in this life
Of error, ignorance, and strife,
Where nothing is, but all things seem,
And we the shadows of the dream,

It is a modest creed, and yet
Pleasant if one considers it,
To own that death itself must be,
Like all the rest, a mockery.

That garden sweet, that lady fair,   130
And all sweet shapes and odours there,
In truth have never passed away:
'Tis we, 'tis ours, are changed; not they.

For love, and beauty, and delight,
There is no death nor change: their might
Exceeds our organs, which endure
No light, being themselves obscure.

                              1820 (1820)

THE CLOUD

I bring fresh showers for the thirsting flowers,
    From the seas and the streams;
I bear light shade for the leaves when laid
    In their noonday dreams.
From my wings are shaken the dews that waken
    The sweet buds every one,
When rocked to rest on their mother's breast,
    As she dances about the sun.
I wield the flail of the lashing hail,
    And whiten the green plains under,   10
And then again I dissolve it in rain,
    And laugh as I pass in thunder.

I sift the snow on the mountains below,
    And their great pines groan aghast;

And all the night 'tis my pillow white,
   While I sleep in the arms of the blast.
Sublime on the towers of my skiey bowers,
   Lightning my pilot sits;
In a cavern under is fettered the thunder,
   It struggles and howls at fits;   20
Over earth and ocean, with gentle motion,
   This pilot is guiding me,
Lured by the love of the genii that move
   In the depths of the purple sea;
Over the rills, and the crags, and the hills,
   Over the lakes and the plains,
Wherever he dream, under mountain or stream,
   The Spirit he loves remains;
And I all the while bask in Heaven's blue smile,
   Whilst he is dissolving in rains.   30

The sanguine Sunrise, with his meteor eyes,
   And his burning plumes outspread,
Leaps on the back of my sailing rack,
   When the morning star shines dead;
As on the jag of a mountain crag,
   Which an earthquake rocks and swings,
An eagle alit one moment may sit
   In the light of its golden wings.
And when Sunset may breathe, from the lit sea
     beneath,
   Its ardours of rest and of love,   40
And the crimson pall of eve may fall
   From the depth of Heaven above,
With wings folded I rest, on mine aëry nest,
   As still as a brooding dove.

That orbed maiden with white fire laden,
   Whom mortals call the Moon,
Glides glimmering o'er my fleece-like floor,
   By the midnight breezes strewn;
And wherever the beat of her unseen feet,
   Which only the angels hear,   50
May have broken the woof of my tent's thin roof,
   The stars peep behind her and peer;
And I laugh to see them whirl and flee,
   Like a swarm of golden bees,
When I widen the rent in my wind-built tent,
   Till the calm rivers, lakes, and seas,
Like strips of the sky fallen through me on high,
   Are each paved with the moon and these.

I bind the Sun's throne with a burning zone,
   And the Moon's with a girdle of pearl;   60
The volcanoes are dim, and the stars reel and swim,
   When the whirlwinds my banner unfurl.
From cape to cape, with a bridge-like shape,
   Over a torrent sea,
Sunbeam-proof, I hang like a roof,—
   The mountains its columns be.
The triumphal arch through which I march
   With hurricane, fire, and snow,
When the Powers of the air are chained to my chair,
   Is the million-coloured bow;   70
The sphere-fire above its soft colours wove,
   While the moist Earth was laughing below.

I am the daughter of Earth and Water,
   And the nursling of the Sky;
I pass through the pores of the ocean and shores;
   I change, but I cannot die.
For after the rain when with never a stain
   The pavilion of Heaven is bare,
And the winds and sunbeams with their convex
    gleams
   Build up the blue dome of air,   80
I silently laugh at my own cenotaph,
   And out of the caverns of rain,
Like a child from the womb, like a ghost from the
    tomb,
   I arise and unbuild it again.

                         1820 (1820)

TO A SKYLARK

Hail to thee, blithe Spirit!
   Bird thou never wert,
That from Heaven, or near it,
   Pourest thy full heart
In profuse strains of unpremeditated art.

   Higher still and higher
     From the earth thou springest
Like a cloud of fire;
     The blue deep thou wingest,
And singing still dost soar, and soaring ever singest.

THE CLOUD. **20. at fits:** in fits; fitfully. **33. rack:** thin flying pieces of cloud; cf. *The Tempest*, IV. i. 156: "leave
not a rack behind." **58. these:** the stars.

79. convex: The rays of light are curved by the earth's
atmosphere. **81. cenotaph:** an empty tomb or monument
honoring someone buried elsewhere; here the "blue dome
of air."

In the golden lightning   11
  Of the sunken sun,
O'er which clouds are bright'ning,
  Thou dost float and run;
Like an unbodied joy whose race is just begun.

The pale purple even
  Melts around thy flight;
Like a star of Heaven,
  In the broad daylight
Thou art unseen, but yet I hear thy shrill delight,   20

Keen as are the arrows
  Of that silver sphere,
Whose intense lamp narrows
  In the white dawn clear
Until we hardly see—we feel that it is there.

All the earth and air
  With thy voice is loud,
As, when night is bare,
  From one lonely cloud
The moon rains out her beams, and Heaven is over-
    flowed.   30

What thou art we know not;
  What is most like thee?
From rainbow clouds there flow not
  Drops so bright to see
As from thy presence showers a rain of melody.

Like a Poet hidden
  In the light of thought,
Singing hymns unbidden,
  Till the world is wrought
To sympathy with hopes and fears it heeded not: 40

Like a high-born maiden
  In a palace-tower,
Soothing her love-laden
  Soul in secret hour
With music sweet as love, which overflows her
    bower:

Like a glow-worm golden
  In a dell of dew,
Scattering unbeholden
  Its aëreal hue
Among the flowers and grass, which screen it from
    the view!   50

Like a rose embowered
  In its own green leaves,
By warm winds deflowered,
  Till the scent it gives
Makes faint with too much sweet those heavy-winged
    thieves:

Sound of vernal showers
  On the twinkling grass,
Rain-awakened flowers,
  All that ever was
Joyous, and clear, and fresh, thy music doth surpass: 60

Teach us, Sprite or Bird,
  What sweet thoughts are thine:
I have never heard
  Praise of love or wine
That panted forth a flood of rapture so divine.

Chorus Hymeneal,
  Or triumphal chant,
Matched with thine would be all
  But an empty vaunt,
A thing wherein we feel there is some hidden want.   70

What objects are the fountains
  Of thy happy strain?
What fields, or waves, or mountains?
  What shapes of sky or plain?
What love of thine own kind? what ignorance of
    pain?

With thy clear keen joyance
  Langour cannot be:
Shadow of annoyance
  Never came near thee:
Thou lovest—but ne'er knew love's sad satiety.   80

Waking or asleep,
  Thou of death must deem
Things more true and deep
  Than we mortals dream,
Or how could thy notes flow in such a crystal stream?

We look before and after,
  And pine for what is not:
Our sincerest laughter
  With some pain is fraught;
Our sweetest songs are those that tell of saddest
    thought.   90

TO A SKYLARK. **22. silver sphere:** the morning star.

55. thieves: the "warm winds" in line 53. **61. Sprite:** spirit. **66. Hymeneal:** in celebration of marriage.

Yet if we could scorn
Hate, and pride, and fear;
If we were things born
Not to shed a tear,
I know not how thy joy we ever should come near.

Better than all measures
Of delightful sound,
Better than all treasures
That in books are found,
Thy skill to poet were, thou scorner of the ground!

Teach me half the gladness 101
That thy brain must know,
Such harmonious madness
From my lips would flow
The world should listen then—as I am listening now.

1820 (1820)

ARETHUSA

This and the two following poems were written as songs
to be included in short dramas by Mary Shelley. Arethusa
was a nymph pursued even under the ocean by the river-
god Alpheus, until Artemis changed her into a fountain
rising on the island of Ortygia, near Sicily. The meter
and rhyme scheme are the same as in "The Cloud."

I

Arethusa arose
From her couch of snows
In the Acroceraunian mountains,—
From cloud and from crag,
With many a jag,
Shepherding her bright fountains.
She leapt down the rocks,
With her rainbow locks
Streaming among the streams;—
Her steps paved with green 10
The downward ravine
Which slopes to the western gleams;
And gliding and springing
She went, ever singing,
In murmurs as soft as sleep;
The Earth seemed to love her,
And Heaven smiled above her,
As she lingered towards the deep.

II

Then Alpheus bold,
On his glacier cold, 20
With his trident the mountains strook;
And opened a chasm
In the rocks—with the spasm
All Erymanthus shook.
And the black south wind
It unsealed behind
The urns of the silent snow,
And earthquake and thunder
Did rend in sunder
The bars of the springs below. 30
And the beard and the hair
Of the River-god were
Seen through the torrent's sweep,
As he followed the light
Of the fleet nymph's flight
To the brink of the Dorian deep.

III

"Oh, save me! Oh, guide me!
And bid the deep hide me,
For he grasps me now by the hair!"
The loud Ocean heard, 40
To its blue depth stirred,
And divided at her prayer;
And under the water
The Earth's white daughter
Fled like a sunny beam;
Behind her descended
Her billows, unblended
With the brackish Dorian stream:—
Like a gloomy stain
On the emerald main 50
Alpheus rushed behind,—
As an eagle pursuing
A dove to its ruin
Down the streams of the cloudy wind.

IV

Under the bowers
Where the Ocean Powers
Sit on their pearled thrones;

ARETHUSA. **24. Erymanthus:** mountain in Arcadia. **36.
Dorian deep:** the Mediterranean at the coast of Greece.

Through the coral woods
Of the weltering floods,
Over heaps of unvalued stones;   60
Through the dim beams
Which amid the streams
Weave a network of coloured light;
And under the caves,
Where the shadowy waves
Are as green as the forest's night:—
Outspeeding the shark,
And the sword-fish dark,
Under the Ocean's foam,
And up through the rifts   70
Of the mountain clifts
They passed to their Dorian home.

V

And now from their fountains
In Enna's mountains,
Down one vale where the morning basks,
Like friends once parted
Grown single-hearted,
They ply their watery tasks.
At sunrise they leap
From their cradles steep   80
In the cave of the shelving hill;
At noontide they flow
Through the woods below
And the meadows of asphodel;
And at night they sleep
In the rocking deep
Beneath the Ortygian shore;—
Like spirits that lie
In the azure sky
When they love but live no more.   90

1820 (1824)

HYMN OF APOLLO

Apollo was god of the sun, of poetry, of medicine, and
of the arts. As in *Prometheus Unbound*, the "Ode to the
West Wind," "The Cloud," and many other poems,
Shelley's imagination fuses myth with science.

60. unvalued: Cf. *Richard III*, I. iv. 27. **84. asphodel:** See
"The Sensitive Plant," above, I. n, 54.

I

The sleepless Hours who watch me as I lie,
    Curtained with star-inwoven tapestries
From the broad moonlight of the sky,
    Fanning the busy dreams from my dim eyes,—
Waken me when their Mother, the gray Dawn,
Tells them that dreams and that the moon is gone.

II

Then I arise, and climbing Heaven's blue dome,
    I walk over the mountains and the waves,
Leaving my robe upon the ocean foam;
    My footsteps pave the clouds with fire; the caves
Are filled with my bright presence, and the air   11
Leaves the green Earth to my embraces bare.

III

The sunbeams are my shafts, with which I kill
    Deceit, that loves the night and fears the day;
All men who do or even imagine ill
    Fly me, and from the glory of my ray
Good minds and open actions take new might,
Until diminished by the reign of Night.

IV

I feed the clouds, the rainbows and the flowers
    With their aethereal colours; the moon's globe
And the pure stars in their eternal bowers   21
    Are cinctured with my power as with a robe;
Whatever lamps on Earth or Heaven may shine
Are portions of one power, which is mine.

V

I stand at noon upon the peak of Heaven,
    Then with unwilling steps I wander down
Into the clouds of the Atlantic even;
    For grief that I depart they weep and frown:
What look is more delightful than the smile
With which I soothe them from the western isle?   30

VI

I am the eye with which the Universe
    Beholds itself and knows itself divine;
All harmony of instrument or verse,
    All prophecy, all medicine is mine,

All light of art or nature;—to my song
Victory and praise in its own right belong.

1820 (1824)

HYMN OF PAN

❧

Pan, the god of Nature, challenged Apollo to a musical competition.

❧

I

From the forests and highlands
    We come, we come;
From the river-girt islands,
    Where loud waves are dumb
        Listening to my sweet pipings.
The wind in the reeds and the rushes,
    The bees on the bells of thyme,
The birds on the myrtle bushes,
    The cicale above in the lime,
And the lizards below in the grass,   10
Were as silent as ever old Tmolus was,
        Listening to my sweet pipings.

II

Liquid Peneus was flowing,
    And all dark Tempe lay
In Pelion's shadow, outgrowing
    The light of the dying day,
        Speeded by my sweet pipings.
The Sileni, and Sylvans, and Fauns,
    And the Nymphs of the woods and the waves,
To the edge of the moist river-lawns,   20
    And the brink of the dewy caves,
And all that did then attend and follow,
Were silent with love, as you now, Apollo,
    With envy of my sweet pipings.

III

I sang of the dancing stars,
    I sang of the daedal Earth,

And of Heaven—and the giant wars,
    And Love, and Death, and Birth,—
        And then I changed my pipings,—
Singing how down the vale of Maenalus   30
    I pursued a maiden and clasped a reed.
Gods and men, we are all deluded thus!
    It breaks in our bosom and then we bleed:
All wept, as I think both ye now would,
If envy or age had not frozen your blood,
    At the sorrow of my sweet pipings.

1820 (1824)

TO ——

I

I fear thy kisses, gentle maiden,
    Thou needest not fear mine;
My spirit is too deeply laden
    Ever to burthen thine.

II

I fear thy mien, thy tones, thy motion,
    Thou needest not fear mine;
Innocent is the heart's devotion
    With which I worship thine.

1820 (1824)

THE TWO SPIRITS:
AN ALLEGORY

FIRST SPIRIT

O thou, who plumed with strong desire
    Wouldst float above the earth, beware!
A Shadow tracks thy flight of fire—
        Night is coming!
Bright are the regions of the air,
And among the winds and beams
    It were delight to wander there—
        Night is coming!

HYMN OF PAN. **9. cicale:** cicada or locust. **11. Tmolus:** mountain range in Asia Minor. **13–15. Peneus . . . Pelion's:** the river Peneus, in the vale of Tempe, Greece, next to Mt. Pelion. **26. daedal:** cunningly wrought (cf. "Mont Blanc," above, l. 86, and *Prometheus Unbound,* IV. 416).

30–31. Maenalus . . . reed: Pan, pursuing the nymph Syrinx, found her suddenly transformed into a reed, just as he was about to embrace her. From the reed Pan made his flute.

SECOND SPIRIT

The deathless stars are bright above;
　If I would cross the shade of night,   10
Within my heart is the lamp of love,
　　And that is day!
And the moon will smile with gentle light
On my golden plumes where'er they move;
　The meteors will linger round my flight,
　　And make night day.

FIRST SPIRIT

But if the whirlwinds of darkness waken
　Hail, and lightning, and stormy rain;
See, the bounds of the air are shaken—
　　Night is coming!   20
The red swift clouds of the hurricane
Yon declining sun have overtaken,
　The clash of the hail sweeps over the plain—
　　Night is coming!

SECOND SPIRIT

I see the light, and I hear the sound;
　I'll sail on the flood of the tempest dark,
With the calm within and the light around
　　Which makes night day:
And thou, when the gloom is deep and stark,
Look from thy dull earth, slumber-bound,   30
　My moon-like flight thou then mayst mark
　　On high, far away.

Some say there is a precipice
　Where one vast pine is frozen to ruin
O'er piles of snow and chasms of ice
　　Mid Alpine mountains;
And that the languid storm pursuing
That winged shape, for ever flies
　Round those hoar branches, aye renewing
　　Its aëry fountains.   40

Some say when nights are dry and clear,
　And the death-dews sleep on the morass,
Sweet whispers are heard by the traveller,
　　Which make night day:
And a silver shape like his early love doth pass
Upborne by her wild and glittering hair,
　And when he awakes on the fragrant grass,
　　He finds night day.

1820 (1824)

EPIPSYCHIDION

VERSES ADDRESSED TO THE NOBLE
AND UNFORTUNATE LADY, EMILIA V——,

NOW IMPRISONED IN THE CONVENT OF——

L'anima amante si slancia fuori del creato, e si crea
nell' infinito un Mondo tutto per essa, diverso assai da
questo oscuro e pauroso baratro.   Her Own Words.

As Shelley recognized in the Advertisement and
warning verses from Dante, *Epipsychidion* is one of his
most difficult poems. It conveys his doctrine of love, and,
since this conception is platonic, the best preparation for
the poem is a reading of the *Symposium*.

The poem has often been interpreted as an allegory of
Shelley's personal life. In December, 1820, he met the
beautiful Emilia Viviani, who had been placed in a convent by her father. As the allegory is worked out, the
moon (l. 277) becomes Mary Shelley; Emily, the
"Vision" (l. 322), coalesces with Emilia Viviani, who is
also the sun; and various identifications have been proposed for the other influences and astronomical bodies
to which the poem alludes. Shelley gave ample warrant
for biographical interpretation, both within the poem
and without. On the other hand, he strongly protested
against it. Perhaps the astronomical symbolism may be
approached with something of the imaginative openness
one must bring to the similar symbolism in the last act
of *Prometheus Unbound*. To the extent that one refers to
biographical information, one emphasizes the self-pity
(Shelley's literary vice) that can be found in some
passages, and, moreover, brings the poem perilously
nearer to the boundary that separates vision from daydream. Shelley wrote to Gisborne (Oct. 22, 1821) that
"The *Epipsychidion* is a mystery; as to real flesh and
blood, you know that I do not deal in those articles," and
to the same effect he told his publisher that the poem
"should not be considered as my own."

The title may perhaps be translated "this soul out of
my soul" (l. 238). The poem was composed in January
and February, 1821.

EPIPSYCHIDION. **Motto:** "The soul that loves projects itself
beyond the created world and creates in the infinite a
world all its own, very different from this obscure and
fearful abyss."

ADVERTISEMENT

The Writer of the following lines died at Florence, as he was preparing for a voyage to one of the wildest of the Sporades,[1] which he had bought, and where he had fitted up the ruins of an old building, and where it was his hope to have realised a scheme of life, suited perhaps to that happier and better world of which he is now an inhabitant, but hardly practicable in this. His life was singular; less on account of the romantic vicissitudes which diversified it, than the ideal tinge which it received from his own character and feelings. The present Poem, like the *Vita Nuova* of Dante, is sufficiently intelligible to a certain class of readers without a matter-of-fact history of the circumstances to which it relates; and to a certain other class it must ever remain incomprehensible, from a defect of a common organ of perception for the ideas of which it treats. Not but that *gran vergogna sarebbe a colui, che rimasse cosa sotto veste di figura, o di colore rettorico: e domandato non sapesse denudare le sue parole da cotal veste, in guisa che avessero verace intendimento.*[2]

The present poem appears to have been intended by the Writer as the dedication to some longer one. The stanza on the opposite page [below] is almost a literal translation from Dante's famous Canzone

Voi, ch' intendendo, il terzo ciel movete,[3] *etc.*

The presumptuous application of the concluding lines to his own composition will raise a smile at the expense of my unfortunate friend: be it a smile not of contempt, but pity.   S.

My Song, I fear that thou wilt find but few
Who fitly shall conceive thy reasoning,
Of such hard matter dost thou entertain;
Whence, if by misadventure, chance should bring
Thee to base company (as chance may do),

Quite unaware of what thou dost contain,
I prithee, comfort thy sweet self again,
My last delight! tell them that they are dull,
And bid them own that thou art beautiful.

EPIPSYCHIDION

Sweet Spirit! Sister of that orphan one,
Whose empire is the name thou weepest on,
In my heart's temple I suspend to thee
These votive wreaths of withered memory.

Poor captive bird! who, from thy narrow cage,
Pourest such music, that I might assuage
The rugged hearts of those who prisoned thee,
Were they not deaf to all sweet melody;
This song shall be thy rose: its petals pale
Are dead, indeed, my adored Nightingale!   10
But soft and fragrant is the faded blossom,
And it has no thorn left to wound thy bosom.

High, spirit-winged Heart! who dost for ever
Beat thine unfeeling bars with vain endeavour,
Till those bright plumes of thought, in which arrayed
It over-soared this low and worldly shade,
Lie shattered; and thy panting, wounded breast
Stains with dear blood its unmaternal nest!
I weep vain tears: blood would less bitter be,
Yet poured forth gladlier, could it profit thee.   20

Seraph of Heaven! too gentle to be human,
Veiling beneath that radiant form of Woman
All that is insupportable in thee
Of light, and love, and immortality!
Sweet Benediction in the eternal Curse!
Veiled Glory of this lampless Universe!
Thou Moon beyond the clouds! Thou living Form
Among the Dead! Thou Star above the Storm!
Thou Wonder, and thou Beauty, and thou Terror!
Thou Harmony of Nature's art! Thou Mirror   30
In whom, as in the splendour of the Sun,
All shapes look glorious which thou gazest on!
Ay, even the dim words which obscure thee now
Flash, lightning-like, with unaccustomed glow;
I pray thee that thou blot from this sad song
All of its much mortality and wrong,
With those clear drops, which start like sacred dew

ADVERTISEMENT. **1. Sporades:** islands in the Aegean Sea. See l. 422, below. **2. gran . . . intendimento:** "It were a shameful thing if one should rhyme under the semblance of metaphor or rhetorical similitude, and afterwards being questioned thereof, should be unable to rid his words of such semblance into such guise as could be truly understood" (trans. D. G. Rossetti). **3. Voi . . . movete:** Shelley quotes the first line of the first Canzone of Dante's *Convito*. He elsewhere translated it: "Ye who intelligent the Third Heaven move."

EPIPSYCHIDION. **1. orphan one:** probably Shelley's wife, Mary. **2. name:** Shelley himself.

From the twin lights thy sweet soul darkens through,
Weeping, till sorrow becomes ecstasy:
Then smile on it, so that it may not die.   40

    I never thought before my death to see
Youth's vision thus made perfect. Emily,
I love thee; though the world by no thin name
Will hide that love from its unvalued shame.
Would we two had been twins of the same mother!
Or, that the name my heart lent to another
Could be a sister's bond for her and thee,
Blending two beams of one eternity!
Yet were one lawful and the other true,
These names, though dear, could paint not, as is due,
How beyond refuge I am thine. Ah me!   51
I am not thine: I am a part of *thee*.

    Sweet Lamp! my moth-like Muse has burned its
        wings
Or, like a dying swan who soars and sings,
Young Love should teach Time, in his own gray style,
All that thou art. Art thou not void of guile,
A lovely soul formed to be blessed and bless?
A well of sealed and secret happiness,
Whose waters like blithe light and music are,
Vanquishing dissonance and gloom? A Star   60
Which moves not in the moving heavens, alone?
A Smile amid dark frowns? a gentle tone
Amid rude voices? a beloved light?
A Solitude, a Refuge, a Delight?
A Lute, which those whom Love has taught to play
Make music on, to soothe the roughest day
And lull fond Grief asleep? a buried treasure?
A cradle of young thoughts of wingless pleasure?
A violet-shrouded grave of Woe?—I measure
The world of fancies, seeking one like thee,   70
And find—alas! mine own infirmity.

    She met me, Stranger, upon life's rough way,
And lured me towards sweet Death; as Night by Day,
Winter by Spring, or Sorrow by swift Hope,
Led into light, life, peace. An antelope,
In the suspended impulse of its lightness,
Were less aethereally light: the brightness

Of her divinest presence trembles through
Her limbs, as underneath a cloud of dew
Embodied in the windless heaven of June   80
Amid the splendour-winged stars, the Moon
Burns, inextinguishably beautiful:
And from her lips, as from a hyacinth full
Of honey-dew, a liquid murmur drops,
Killing the sense with passion; sweet as stops
Of planetary music heard in trance.
In her mild lights the starry spirits dance,
The sunbeams of those wells which ever leap
Under the lightnings of the soul—too deep
For the brief fathom-line of thought or sense.   90
The glory of her being, issuing thence,
Stains the dead, blank, cold air with a warm shade
Of unentangled intermixture, made
By Love, of light and motion: one intense
Diffusion, one serene Omnipresence,
Whose flowing outlines mingle in their flowing,
Around her cheeks and utmost fingers glowing
With the unintermitted blood, which there
Quivers, (as in a fleece of snow-like air
The crimson pulse of living morning quiver,)   100
Continuously prolonged, and ending never,
Till they are lost, and in that Beauty furled
Which penetrates and clasps and fills the world;
Scarce visible from extreme loveliness.
Warm fragrance seems to fall from her light dress
And her loose hair; and where some heavy tress
The air of her own speed has disentwined,
The sweetness seems to satiate the faint wind;
And in the soul a wild odour is felt,
Beyond the sense, like fiery dews that melt   110
Into the bosom of a frozen bud.—
See where she stands! a mortal shape indued
With love and life and light and deity,
And motion which may change but cannot die;
An image of some bright Eternity;
A shadow of some golden dream; a Splendour
Leaving the third sphere pilotless; a tender
Reflection of the eternal Moon of Love
Under whose motions life's dull billows move;
A Metaphor of Spring and Youth and Morning;   120
A Vision like incarnate April, warning,

---

42. Youth's vision: Cf. "Hymn to Intellectual Beauty," above. **44. Will . . . shame:** will protect that love from contempt—a contempt, however, to which Shelley is indifferent (hence "unvalued"). **49. lawful . . . true:** if the latter were lawful (i.e., could I be married to both you and Mary) and the other alternative true (that we were brother and sister). **50. names:** those of "wife" or "sister." **72. She:** apparently the ideal of Beauty rather than Emilia Viviani.

86. planetary music: music of the spheres. **93. unentangled intermixture:** a union of light and motion so complete that they are not simply "entangled" but coalesced into "one intense / Diffusion." **117. third sphere:** that of Venus (love).

With smiles and tears, Frost the Anatomy
Into his summer grave.
                              Ah, woe is me!
What have I dared? where am I lifted? how
Shall I descend, and perish not? I know
That Love makes all things equal: I have heard
By mine own heart this joyous truth averred:
The spirit of the worm beneath the sod
In love and worship, blends itself with God.

Spouse! Sister! Angel! Pilot of the Fate   130
Whose course has been so starless! O too late
Beloved! O too soon adored, by me!
For in the fields of Immortality
My spirit should at first have worshipped thine,
A divine presence in a place divine;
Or should have moved beside it on this earth,
A shadow of that substance, from its birth;
But not as now:—I love thee; yes, I feel
That on the fountain of my heart a seal
Is set, to keep its waters pure and bright   140
For thee, since in those *tears* thou hast delight.
We—are we not formed, as notes of music are,
For one another, though dissimilar;
Such difference without discord, as can make
Those sweetest sounds, in which all spirits shake
As trembling leaves in a continuous air?

Thy wisdom speaks in me, and bids me dare
Beacon the rocks on which high hearts are wrecked.
I never was attached to that great sect,
Whose doctrine is, that each one should select   150
Out of the crowd a mistress or a friend,
And all the rest, though fair and wise, commend
To cold oblivion, though it is in the code
Of modern morals, and the beaten road
Which those poor slaves with weary footsteps tread,
Who travel to their home among the dead
By the broad highway of the world, and so
With one chained friend, perhaps a jealous foe,
The dreariest and the longest journey go.

True Love in this differs from gold and clay,   160
That to divide is not to take away.
Love is like understanding, that grows bright,
Gazing on many truths; 'tis like thy light,
Imagination! which from earth and sky,

And from the depths of human fantasy,
As from a thousand prisms and mirrors, fills
The Universe with glorious beams, and kills
Error, the worm, with many a sun-like arrow
Of its reverberated lightning. Narrow
The heart that loves, the brain that contemplates,   170
The life that wears, the spirit that creates
One object, and one form, and builds thereby
A sepulchre for its eternity.

Mind from its object differs most in this:
Evil from good; misery from happiness;
The baser from the nobler; the impure
And frail, from what is clear and must endure.
If you divide suffering and dross, you may
Diminish till it is consumed away;
If you divide pleasure and love and thought,   180
Each part exceeds the whole; and we know not
How much, while any yet remains unshared,
Of pleasure may be gained, of sorrow spared:
This truth is that deep well, whence sages draw
The unenvied light of hope; the eternal law
By which those live, to whom this world of life
Is as a garden ravaged, and whose strife
Tills for the promise of a later birth
The wilderness of this Elysian earth.

There was a Being whom my spirit oft   190
Met on its visioned wanderings, far aloft,
In the clear golden prime of my youth's dawn,
Upon the fairy isles of sunny lawn,
Amid the enchanted mountains, and the caves
Of divine sleep, and on the air-like waves
Of wonder-level dream, whose tremulous floor
Paved her light steps;—on an imagined shore,
Under the gray beak of some promontory
She met me, robed in such exceeding glory,
That I beheld her not. In solitudes   200
Her voice came to me through the whispering woods,
And from the fountains, and the odours deep
Of flowers, which, like lips murmuring in their sleep
Of the sweet kisses which had lulled them there,
Breathed but of *her* to the enamoured air;
And from the breezes whether low or loud,
And from the rain of every passing cloud,
And from the singing of the summer-birds,
And from all sounds, all silence. In the words
Of antique verse and high romance,—in form,   210
Sound, colour—in whatever checks that Storm

---

122. Frost the Anatomy: the skeleton of things after winter
has stripped them. **128–29. The . . . God:** Cf. *Prometheus
Unbound*, II. v. 40–43.

190. Being: Ideal Beauty.

Which with the shattered present chokes the past;
And in that best philosophy, whose taste
Makes this cold common hell, our life, a doom
As glorious as a fiery martyrdom;
Her Spirit was the harmony of truth.—

Then, from the caverns of my dreamy youth
I sprang, as one sandalled with plumes of fire,
And towards the lodestar of my one desire,
I flitted, like a dizzy moth, whose flight   220
Is as a dead leaf's in the owlet light,
When it would seek in Hesper's setting sphere
A radiant death, a fiery sepulchre,
As if it were a lamp of earthly flame.—
But She, whom prayers or tears then could not tame,
Passed, like a God throned on a winged planet,
Whose burning plumes to tenfold swiftness fan it,
Into the dreary cone of our life's shade;
And as a man with mighty loss dismayed,
I would have followed, though the grave between
Yawned like a gulf whose spectres are unseen: 231
When a voice said:—"O thou of hearts the weakest,
The phantom is beside thee whom thou seekest."
Then I—"Where?"—the world's echo answered
    "where?"
And in that silence, and in my despair,
I questioned every tongueless wind that flew
Over my tower of mourning, if it knew
Whither 'twas fled, this soul out of my soul;
And murmured names and spells which have control
Over the sightless tyrants of our fate;   240
But neither prayer nor verse could dissipate
The night which closed on her; nor uncreate
That world within this Chaos, mine and me,
Of which she was the veiled Divinity,
The world I say of thoughts that worshipped her:
And therefore I went forth, with hope and fear
And every gentle passion sick to death,
Feeding my course with expectation's breath,
Into the wintry forest of our life;
And struggling through its error with vain strife, 250
And stumbling in my weakness and my haste,
And half bewildered by new forms, I passed,
Seeking among those untaught foresters
If I could find one form resembling hers,

In which she might have masked herself from me.
There,—One, whose voice was venomed melody
Sate by a well, under blue nightshade bowers;
The breath of her false mouth was like faint flowers,
Her touch was as electric poison,—flame
Out of her looks into my vitals came,   260
And from her living cheeks and bosom flew
A killing air, which pierced like honey-dew
Into the core of my green heart, and lay
Upon its leaves; until, as hair grown gray
O'er a young brow, they hid its unblown prime
With ruins of unseasonable time.

In many mortal forms I rashly sought
The shadow of that idol of my thought.
And some were fair—but beauty dies away:
Others were wise—but honeyed words betray: 270
And One was true—oh! why not true to me?
Then, as a hunted deer that could not flee,
I turned upon my thoughts, and stood at bay,
Wounded and weak and panting; the cold day
Trembled, for pity of my strife and pain.
When, like a noonday dawn, there shone again
Deliverance. One stood on my path who seemed
As like the glorious shape which I had dreamed
As is the Moon, whose changes ever run
Into themselves, to the eternal Sun;   280
The cold chaste Moon, the Queen of Heaven's bright
    isles,
Who makes all beautiful on which she smiles,
That wandering shrine of soft yet icy flame
Which ever is transformed, yet still the same,
And warms not but illumines. Young and fair
As the descended Spirit of that sphere,
She hid me, as the Moon may hide the night
From its own darkness, until all was bright
Between the Heaven and Earth of my calm mind,
And, as a cloud charioted by the wind,   290
She led me to a cave in that wild place,
And sate beside me, with her downward face
Illumining my slumbers, like the Moon
Waxing and waning o'er Endymion.
And I was laid asleep, spirit and limb,
And all my being became bright or dim
As the Moon's image in a summer sea,
According as she smiled or frowned on me;
And there I lay, within a chaste cold bed:
Alas, I then was nor alive nor dead:—   300
For at her silver voice came Death and Life,
Unmindful each of their accustomed strife,

Masked like twin babes, a sister and a brother,
The wandering hopes of one abandoned mother,
And through the cavern without wings they flew,
And cried "Away, he is not of our crew."
I wept, and though it be a dream, I weep.

What storms then shook the ocean of my sleep,
Blotting that Moon, whose pale and waning lips
Then shrank as in the sickness of eclipse;—   310
And how my soul was as a lampless sea,
And who was then its Tempest; and when She,
The Planet of that hour, was quenched, what frost
Crept o'er those waters, till from coast to coast
The moving billows of my being fell
Into a death of ice, immovable;—
And then—what earthquakes made it gape and split,
The white Moon smiling all the while on it,
These words conceal:—If not, each word would be
The key of staunchless tears. Weep not for me!   320

At length, into the obscure Forest came
The Vision I had sought through grief and shame.
Athwart that wintry wilderness of thorns
Flashed from her motion splendour like the Morn's,
And from her presence life was radiated
Through the gray earth and branches bare and dead;
So that her way was paved, and roofed above
With flowers as soft as thoughts of budding love;
And music from her respiration spread
Like light,—all other sounds were penetrated   330
By the small, still, sweet spirit of that sound,
So that the savage winds hung mute around;
And odours warm and fresh fell from her hair
Dissolving the dull cold in the frore air:
Soft as an Incarnation of the Sun,
When light is changed to love, this glorious One
Floated into the cavern where I lay,
And called my Spirit, and the dreaming clay
Was lifted by the thing that dreamed below
As smoke by fire, and in her beauty's glow   340
I stood, and felt the dawn of my long night
Was penetrating me with living light:
I knew it was the Vision veiled from me
So many years—that it was Emily.

Twin Spheres of light who rule this passive Earth,
This world of love, this *me*; and into birth
Awaken all its fruits and flowers, and dart
Magnetic might into its central heart;

334. frore: frosty, frozen. **345. Twin Spheres:** the sun
and the moon.

And lift its billows and its mists, and guide
By everlasting laws, each wind and tide   350
To its fit cloud, and its appointed cave;
And lull its storms, each in the craggy grave
Which was its cradle, luring to faint bowers
The armies of the rainbow-winged showers;
And, as those married lights, which from the towers
Of Heaven look forth and fold the wandering globe
In liquid sleep and splendour, as a robe;
And all their many-mingled influence blend,
If equal, yet unlike, to one sweet end;—
So ye, bright regents, with alternate sway   360
Govern my sphere of being, night and day!
Thou, not disdaining even a borrowed might;
Thou, not eclipsing a remoter light;
And, through the shadow of the seasons three,
From Spring to Autumn's sere maturity,
Light it into the Winter of the tomb,
Where it may ripen to a brighter bloom.
Thou too, O Comet beautiful and fierce,
Who drew the heart of this frail Universe
Towards thine own; till, wrecked in that convulsion,
Alternating attraction and repulsion,   371
Thine went astray and that was rent in twain;
Oh, float into our azure heaven again!
Be there Love's folding-star at thy return;
The living Sun will feed thee from its urn
Of golden fire; the Moon will veil her horn
In thy last smiles; adoring Even and Morn
Will worship thee with incense of calm breath
And lights and shadows; as the star of Death
And Birth is worshipped by those sisters wild   380
Called Hope and Fear—upon the heart are piled
Their offerings,—of this sacrifice divine
A World shall be the altar.
                              Lady mine,
Scorn not these flowers of thought, the fading birth
Which from its heart of hearts that plant puts forth
Whose fruit, made perfect by the sunny eyes,
Will be as of the trees of Paradise.

The day is come, and thou wilt fly with me.
To whatsoe'er of dull mortality
Is mine, remain a vestal sister still;   390
To the intense, the deep, the imperishable,

362. Thou: the sun, not disdaining the moon though its
"might" is only "borrowed," since the moon's light is a
reflection of the sun's. **363. Thou:** the moon. **374. folding-
star:** the evening star, so called because it rises at the time
when flocks are driven into the fold.

Not mine but me, henceforth be thou united
Even as a bride, delighting and delighted.
The hour is come:—the destined Star has risen
Which shall descend upon a vacant prison.
The walls are high, the gates are strong, thick set
The sentinels—but true Love never yet
Was thus constrained: it overleaps all fence:
Like lightning, with invisible violence
Piercing its continents; like Heaven's free breath, 400
Which he who grasps can hold not; liker Death,
Who rides upon a thought, and makes his way
Through temple, tower, and palace, and the array
Of arms: more strength has Love than he or they;
For it can burst his charnel, and make free
The limbs in chains, the heart in agony,
The soul in dust and chaos.
      Emily,
A ship is floating in the harbour now,
A wind is hovering o'er the mountain's brow;
There is a path on the sea's azure floor, 410
No keel has ever ploughed that path before;
The halcyons brood around the foamless isles;
The treacherous Ocean has forsworn its wiles;
The merry mariners are bold and free:
Say, my heart's sister, wilt thou sail with me?
Our bark is as an albatross, whose nest
Is a far Eden of the purple East;
And we between her wings will sit, while Night,
And Day, and Storm, and Calm, pursue their flight,
Our ministers, along the boundless Sea, 420
Treading each other's heels, unheededly.
It is an isle under Ionian skies,
Beautiful as a wreck of Paradise,
And, for the harbours are not safe and good,
This land would have remained a solitude
But for some pastoral people native there,
Who from the Elysian, clear, and golden air
Draw the last spirit of the age of gold,
Simple and spirited; innocent and bold.
The blue Aegean girds this chosen home, 430
With ever-changing sound and light and foam,
Kissing the sifted sands, and caverns hoar;
And all the winds wandering along the shore
Undulate with the undulating tide:
There are thick woods where sylvan forms abide;
And many a fountain, rivulet, and pond,
As clear as elemental diamond,

Or serene morning air; and far beyond,
The mossy tracks made by the goats and deer
(Which the rough shepherd treads but once a year)
Pierce into glades, caverns, and bowers, and halls 441
Built round with ivy, which the waterfalls
Illumining, with sound that never fails
Accompany the noonday nightingales;
And all the place is peopled with sweet airs;
The light clear element which the isle wears
Is heavy with the scent of lemon-flowers,
Which floats like mist laden with unseen showers,
And falls upon the eyelids like faint sleep;
And from the moss violets and jonquils peep, 450
And dart their arrowy odour through the brain
Till you might faint with that delicious pain.
And every motion, odour, beam, and tone,
With that deep music is in unison:
Which is a soul within the soul—they seem
Like echoes of an antenatal dream.—
It is an isle 'twixt Heaven, Air, Earth, and Sea,
Cradled, and hung in clear tranquillity;
Bright as that wandering Eden Lucifer,
Washed by the soft blue Oceans of young air. 460
It is a favoured place. Famine or Blight,
Pestilence, War and Earthquake, never light
Upon its mountain-peaks; blind vultures, they
Sail onward far upon their fatal way:
The winged storms, chanting their thunder-psalm
To other lands, leave azure chasms of calm
Over this isle, or weep themselves in dew,
From which its fields and woods ever renew
Their green and golden immortality.
And from the sea there rise, and from the sky 470
There fall, clear exhalations, soft and bright,
Veil after veil, each hiding some delight,
Which Sun or Moon or zephyr draw aside,
Till the isle's beauty, like a naked bride
Glowing at once with love and loveliness,
Blushes and trembles at its own excess:
Yet, like a buried lamp, a Soul no less
Burns in the heart of this delicious isle,
An atom of th' Eternal, whose own smile
Unfolds itself, and may be felt, not seen 480
O'er the gray rocks, blue waves, and forests green,
Filling their bare and void interstices.—
But the chief marvel of the wilderness
Is a lone dwelling, built by whom or how

---

400. continents: that which contains it. **412. halcyons:** kingfishers.

445. airs: probably in the sense of "melodies." **459. Lucifer:** the morning star.

None of the rustic island-people know:
'Tis not a tower of strength, though with its height
It overtops the woods; but, for delight,
Some wise and tender Ocean-King, ere crime
Had been invented, in the world's young prime,
Reared it, a wonder of that simple time,   490
An envy of the isles, a pleasure-house
Made sacred to his sister and his spouse.
It scarce seems now a wreck of human art,
But, as it were Titanic; in the heart
Of Earth having assumed its form, then grown
Out of the mountains, from the living stone,
Lifting itself in caverns light and high:
For all the antique and learned imagery
Has been erased, and in the place of it
The ivy and the wild-vine interknit   500
The volumes of their many-twining stems;
Parasite flowers illume with dewy gems
The lampless halls, and when they fade, the sky
Peeps through their winter-woof of tracery
With moonlight patches, or star atoms keen,
Or fragments of the day's intense serene;—
Working mosaic on their Parian floors.
And, day and night, aloof, from the high towers
And terraces, the Earth and Ocean seem
To sleep in one another's arms, and dream   510
Of waves, flowers, clouds, woods, rocks, and all that
    we
Read in their smiles, and call reality.

    This isle and house are mine, and I have vowed
Thee to be lady of the solitude.—
And I have fitted up some chambers there
Looking towards the golden Eastern air,
And level with the living winds, which flow
Like waves above the living waves below.—
I have sent books and music there, and all
Those instruments with which high Spirits call   520
The future from its cradle, and the past
Out of its grave, and make the present last
In thoughts and joys which sleep, but cannot die,
Folded within their own eternity.
Our simple life wants little, and true taste
Hires not the pale drudge Luxury, to waste
The scene it would adorn, and therefore still,
Nature with all her children haunts the hill.
The ring-dove, in the embowering ivy, yet
Keeps up her love-lament, and the owls flit   530

494. **Titanic:** made by the Titans. **507. Parian:** marble
from the isle of Paros in the Cyclades.

Round the evening tower, and the young stars glance
Between the quick bats in their twilight dance;
The spotted deer bask in the fresh moonlight
Before our gate, and the slow, silent night
Is measured by the pants of their calm sleep.
Be this our home in life, and when years heap
Their withered hours, like leaves, on our decay,
Let us become the overhanging day,
The living soul of this Elysian isle,
Conscious, inseparable, one. Meanwhile   540
We two will rise, and sit, and walk together
Under the roof of blue Ionian weather
And wander in the meadows, or ascend
The mossy mountains, where the blue heavens bend
With lightest winds, to touch their paramour;
Or linger, where the pebble-paven shore,
Under the quick, faint kisses of the sea
Trembles and sparkles as with ecstasy,—
Possessing and possessed by all that is
Within that calm circumference of bliss,   550
And by each other, till to love and live
Be one:—or, at the noontide hour, arrive
Where some old cavern hoar seems yet to keep
The moonlight of the expired night asleep,
Through which the awakened day can never peep;
A veil for our seclusion, close as night's,
Where secure sleep may kill thine innocent lights;
Sleep, the fresh dew of languid love, the rain
Whose drops quench kisses till they burn again.
And we will talk, until thought's melody   560
Become too sweet for utterance, and it die
In words, to live again in looks, which dart
With thrilling tone into the voiceless heart,
Harmonizing silence without a sound.
Our breath shall intermix, our bosoms bound,
And our veins beat together; and our lips
With other eloquence than words, eclipse
The soul that burns between them, and the wells
Which boil under our being's inmost cells,
The fountains of our deepest life, shall be   570
Confused in Passion's golden purity,
As mountain-springs under the morning sun.
We shall become the same, we shall be one
Spirit within two frames, oh! wherefore two?
One passion in twin-hearts, which grows and grew,
Till like two meteors of expanding flame,
Those spheres instinct with it become the same,
Touch, mingle, are transfigured; ever still
Burning, yet ever inconsumable:
In one another's substance finding food,   580

Like flames too pure and light and unimbued
To nourish their bright lives with baser prey,
Which point to Heaven and cannot pass away:
One hope within two wills, one will beneath
Two overshadowing minds, one life, one death,
One Heaven, one Hell, one immortality,
And one annihilation. Woe is me!
The winged words on which my soul would pierce
Into the height of Love's rare Universe,
Are chains of lead around its flight of fire—   590
I pant, I sink, I tremble, I expire!

Weak Verses, go, kneel at your Sovereign's feet,
And say:—"We are the masters of thy slave;
What wouldest thou with us and ours and thine?"
Then call your sisters from Oblivion's cave,
All singing loud: "Love's very pain is sweet,
But its reward is in the world divine
Which, if not here, it builds beyond the grave."
So shall ye live when I am here. Then haste
Over the hearts of men, until ye meet   600
Marina, Vanna, Primus, and the rest,
And bid them love each other and be blessed:
And leave the troop which errs, and which reproves,
And come and be my guest,—for I am Love's.

1821 (1821)

ADONAIS

AN ELEGY ON THE DEATH
OF JOHN KEATS

❧

Shelley had met Keats occasionally in London and had
wished to befriend him, but he did not know him well.
Like Milton's Lycidas, the figure of Adonais is a symbol
rather than a portrait. Shelley saw in Keats a potentially
great poet whose death he attributed (incorrectly) to the
shock of savage reviews.

In form the poem is a pastoral elegy. Among the conventions of the form are a lament of nature, a procession
of mourners, a contrast between the rebirth of spring and
the fixity of grief and death, and a change of tone at the
end with the thought that the dead poet is immortal.
Even this summary recital may indicate that although
Shelley freely departs from convention he also exploits
it, and the poem gives an added pleasure to readers who
can appreciate how he does so. Without this preparation,
much that is conscious artistry may seem coldly artificial.
It is especially helpful to read Bion's *Lament for Adonis,*

which Shelley sometimes paraphrases, and, to a lesser
extent, Moschus' *Lament for Bion.* Nevertheless, the poem
remains uneven, the more so because it rises to rare
heights in stanzas of inspired Platonism.

The name Adonais immediately suggests Adonis (of
which it is the Doric form), the beautiful youth beloved
by Venus and slain by a wild boar. But it may also suggest
Adonai, a name of multiple significance in cabalistic
literature (see Earl Wasserman, *The Subtler Language,*
1959). The poem was composed in June, 1821, four
months after the death of Keats.

❧

Ἀστὴρ πρὶν μὲν ἔλαμπες ἐνὶ ζωοῖσιν Ἑῷος.
  νῦν δὲ θανὼν λάμπεις Ἕσπερος ἐν φθιμένοις.—

Plato

I

I weep for Adonais—he is dead!
O, weep for Adonais! though our tears
Thaw not the frost which binds so dear a head!
And thou, sad Hour, selected from all years
To mourn our loss, rouse thy obscure compeers,
And teach them thine own sorrow, say: "With me
Died Adonais; till the Future dares
Forget the Past, his fate and fame shall be
An echo and a light unto eternity!"

II

Where wert thou, mighty Mother, when he lay,   10
When thy Son lay, pierced by the shaft which flies
In darkness? where was lorn Urania
When Adonais died? With veiled eyes,
'Mid listening Echoes, in her Paradise
She sate, while one, with soft enamoured breath,
Rekindled all the fading melodies,
With which, like flowers that mock the corse
    beneath,
He had adorned and hid the coming bulk of Death.

ADONAIS. **Motto:**

Thou wert the morning star among the living,
  Ere thy fair light had fled;—
Now, having died, thou art, as Hesperus, giving
  New splendour to the dead.

(Shelley's translation)

5. obscure compeers: hours less remarkable and now forgotten. **10. Mother:** Urania (l. 12), the muse of astronomy
(often associated with the higher, or more sublime, forms of
poetry); the name is also applied to Aphrodite (Venus or
Love) when she is associated with spiritual love and beauty.

III

Oh, weep for Adonais—he is dead!
Wake, melancholy Mother, wake and weep!   20
Yet wherefore? Quench within their burning bed
Thy fiery tears, and let thy loud heart keep
Like his, a mute and uncomplaining sleep;
For he is gone, where all things wise and fair
Descend;—oh, dream not that the amorous Deep
Will yet restore him to the vital air;
Death feeds on his mute voice, and laughs at our despair.

IV

Most musical of mourners, weep again!
Lament anew, Urania!—He died,
Who was the Sire of an immortal strain,   30
Blind, old, and lonely, when his country's pride,
The priest, the slave, and the liberticide,
Trampled and mocked with many a loathed rite
Of lust and blood; he went, unterrified,
Into the gulf of death; but his clear Sprite
Yet reigns o'er earth; the third among the sons of
    light.

V

Most musical of mourners, weep anew!
Not all to that bright station dared to climb;
And happier they their happiness who knew,
Whose tapers yet burn through that night of time
In which suns perished; others more sublime,   41
Struck by the envious wrath of man or god,
Have sunk, extinct in their refulgent prime;
And some yet live, treading the thorny road,
Which leads, through toil and hate, to Fame's serene
    abode.

VI

But now, thy youngest, dearest one, has perished—
The nursling of thy widowhood, who grew,
Like a pale flower by some sad maiden cherished,
And fed with true-love tears, instead of dew;
Most musical of mourners, weep anew!   50
Thy extreme hope, the loveliest and the last,
The bloom, whose petals, nipped before they blew,

Died on the promise of the fruit, is waste;
The broken lily lies—the storm is overpast.

VII

To that high Capital, where kingly Death
Keeps his pale court in beauty and decay,
He came; and bought, with price of purest breath,
A grave among the eternal.—Come away!
Haste, while the vault of blue Italian day
Is yet his fitting charnel-roof! while still   60
He lies, as if in dewy sleep he lay;
Awake him not! surely he takes his fill
Of deep and liquid rest, forgetful of all ill.

VIII

He will awake no more, oh, never more!—
Within the twilight chamber spreads apace
The shadow of white Death, and at the door
Invisible Corruption waits to trace
His extreme way to her dim dwelling-place;
The eternal Hunger sits, but pity and awe
Soothe her pale rage, nor dares she to deface   70
So fair a prey, till darkness, and the law
Of change, shall o'er his sleep the mortal curtain draw.

IX

Oh, weep for Adonais!—The quick Dreams,
The passion-winged Ministers of thought,
Who were his flocks, whom near the living streams
Of his young spirit he fed, and whom he taught
The love which was its music, wander not,—
Wander no more, from kindling brain to brain,
But droop there, whence they sprung; and mourn
    their lot
Round the cold heart, where, after their sweet pain,
They ne'er will gather strength, or find a home again.   81

X

And one with trembling hands clasps his cold head,
And fans him with her moonlight wings, and cries;
"Our love, our hope, our sorrow, is not dead;
See, on the silken fringe of his faint eyes,
Like dew upon a sleeping flower, there lies

29–36. He . . . light: Milton, "third" among the great epic poets (after Homer and Dante). **51. extreme:** final.

55. Capital: Rome. **68. extreme way:** Keats's last journey. **69. eternal Hunger:** "Corruption" (l. 67).

A tear some Dream has loosened from his brain."
Lost Angel of a ruined Paradise!
She knew not 'twas her own; as with no stain
She faded, like a cloud which had outwept its rain. 90

XI

One from a lucid urn of starry dew
Washed his light limbs as if embalming them;
Another clipped her profuse locks, and threw
The wreath upon him, like an anadem,
Which frozen tears instead of pearls begem;
Another in her wilful grief would break
Her bow and winged reeds, as if to stem
A greater loss with one which was more weak;
And dull the barbed fire against his frozen cheek.

XII

Another Splendour on his mouth alit,   100
That mouth, whence it was wont to draw the
    breath
Which gave it strength to pierce the guarded wit,
And pass into the panting heart beneath
With lightning and with music: the damp death
Quenched its caress upon his icy lips;
And, as a dying meteor stains a wreath
Of moonlight vapour, which the cold night clips,
It flushed through his pale limbs, and passed to its
    eclipse.

XIII

And others came . . . Desires and Adorations,
Winged Persuasions and veiled Destinies,   110
Splendours, and Glooms, and glimmering Incarna-
    tions
Of hopes and fears, and twilight Phantasies;
And Sorrow, with her family of Sighs,
And Pleasure, blind with tears, led by the gleam
Of her own dying smile instead of eyes,
Came in slow pomp;—the moving pomp might
    seem
Like pageantry of mist on an autumnal stream.

XIV

All he had loved, and moulded into thought,
From shape, and hue, and odour, and sweet sound,
Lamented Adonais. Morning sought   120
Her eastern watch-tower, and her hair unbound,
Wet with the tears which should adorn the ground,
Dimmed the aëreal eyes that kindle day;
Afar the melancholy thunder moaned,
Pale Ocean in unquiet slumber lay,
And the wild Winds flew round, sobbing in their
    dismay.

XV

Lost Echo sits amid the voiceless mountains,
And feeds her grief with his remembered lay,
And will no more reply to winds or fountains,
Or amorous birds perched on the young green spray,
Or herdsman's horn, or bell at closing day;   131
Since she can mimic not his lips, more dear
Than those for whose disdain she pined away
Into a shadow of all sounds:—a drear
Murmur, between their songs, is all the woodmen hear.

XVI

Grief made the young Spring wild, and she threw
    down
Her kindling buds, as if she Autumn were,
Or they dead leaves; since her delight is flown,
For whom should she have waked the sullen year?
To Phoebus was not Hyacinth so dear   140
Nor to himself Narcissus, as to both
Thou, Adonais: wan they stand and sere
Amid the faint companions of their youth,
With dew all turned to tears; odour, to sighing ruth.

XVII

Thy spirit's sister, the lorn nightingale,
Mourns not her mate with such melodious pain;
Not so the eagle, who like thee could scale

---

88. Lost . . . Paradise: lost messenger (angel) of a creative
mind now gone ("a ruined Paradise"). **94. anadem:** garland.
102. guarded wit: the cautious mind. **107. clips:** embraces.

133–34. those . . . sounds: For love of Narcissus, Echo
pined away until she was only a voice. **140–41. Phoebus . . .
Narcissus:** Hyacinth was beloved and accidentally slain by
Apollo (Phoebus); Narcissus fell in love with his own
reflection in a pool. Both were changed into flowers.

Heaven, and could nourish in the sun's domain
Her mighty youth with morning, doth complain,
Soaring and screaming round her empty nest,   150
As Albion wails for thee: the curse of Cain
Light on his head who pierced thy innocent breast,
And scared the angel soul that was its earthly guest!

XVIII

Ah, woe is me! Winter is come and gone,
But grief returns with the revolving year;
The airs and streams renew their joyous tone;
The ants, the bees, the swallows reappear;
Fresh leaves and flowers deck the dead Seasons' bier;
The amorous birds now pair in every brake,
And build their mossy homes in field and brere;
And the green lizard, and the golden snake,   161
Like unimprisoned flames, out of their trance awake.

XIX

Through wood and stream and field and hill and
    Ocean
A quickening life from the Earth's heart has burst
As it has ever done, with change and motion,
From the great morning of the world when first
God dawned on Chaos; in its stream immersed,
The lamps of Heaven flash with a softer light;
All baser things pant with life's sacred thirst;
Diffuse themselves; and spend in love's delight,   170
The beauty and the joy of their renewed might.

XX

The leprous corpse, touched by this spirit tender,
Exhales itself in flowers of gentle breath;
Like incarnations of the stars, when splendour
Is changed to fragrance, they illumine death
And mock the merry worm that wakes beneath;
Nought we know, dies. Shall that alone which knows
Be as a sword consumed before the sheath
By sightless lightning?—the intense atom glows
A moment, then is quenched in a most cold respose.

148–49. **nourish . . . morning:** referring to the legend that
the eagle, in its old age, could regain its youth if it flew high
enough; the old plumage would burn away and the film
drop from its eyes. **151. Albion:** England. **159. brake:**
thicket. **160. brere:** briar.

XXI

Alas! that all we loved of him should be,   181
But for our grief, as if it had not been,
And grief itself be mortal! Woe is me!
Whence are we, and why are we? of what scene
The actors or spectators? Great and mean
Meet massed in death, who lends what life must
    borrow.
As long as skies are blue, and fields are green,
Evening must usher night, night urge the morrow,
Month follow month with woe, and year wake year
    to sorrow.

XXII

He will awake no more, oh, never more!   190
"Wake thou," cried Misery, "childless Mother, rise
Out of thy sleep, and slake, in thy heart's core,
A wound more fierce than his, with tears and sighs."
And all the Dreams that watched Urania's eyes,
And all the Echoes whom their sister's song
Had held in holy silence, cried: "Arise!"
Swift as a Thought by the snake Memory stung,
From her ambrosial rest the fading Splendour sprung.

XXIII

She rose like an autumnal Night, that springs
Out of the East, and follows wild and drear   200
The golden Day, which, on eternal wings,
Even as a ghost abandoning a bier,
Had left the Earth a corpse. Sorrow and fear
So struck, so roused, so rapt Urania;
So saddened round her like an atmosphere
Of stormy mist; so swept her on her way
Even to the mournful place where Adonais lay.

XXIV

Out of her secret Paradise she sped,
Through camps and cities rough with stone, and steel,
And human hearts, which to her aery tread   210
Yielding not, wounded the invisible
Palms of her tender feet where'er they fell:
And barbed tongues, and thoughts more sharp than
    they,
Rent the soft Form they never could repel,
Whose sacred blood, like the young tears of May,
Paved with eternal flowers that undeserving way.

XXV

In the death-chamber for a moment Death,
Shamed by the presence of that living Might,
Blushed to annihilation, and the breath
Revisited those lips, and Life's pale light   220
Flashed through those limbs, so late her dear delight.
"Leave me not wild and drear and comfortless,
As silent lightning leaves the starless night!
Leave me not!" cried Urania: her distress
Roused Death: Death rose and smiled, and met her
    vain caress.

XXVI

"Stay yet awhile! speak to me once again;
Kiss me, so long but as a kiss may live;
And in my heartless breast and burning brain
That word, that kiss, shall all thoughts else survive,
With food of saddest memory kept alive,   230
Now thou art dead, as if it were a part
Of thee, my Adonais! I would give
All that I am to be as thou now art!
But I am chained to Time, and cannot thence depart!

XXVII

"O gentle child, beautiful as thou wert,
Why didst thou leave the trodden paths of men
Too soon, and with weak hands though mighty
    heart
Dare the unpastured dragon in his den?
Defenceless as thou wert, oh, where was then
Wisdom the mirrored shield, or scorn the spear?
Or hadst thou waited the full cycle, when   241
Thy spirit should have filled its crescent sphere,
The monsters of life's waste had fled from thee like
    deer.

XXVIII

"The herded wolves, bold only to pursue;
The obscene ravens, clamorous o'er the dead;
The vultures to the conqueror's banner true
Who feed where Desolation first has fed,

And whose wings rain contagion;—how they fled,
When, like Apollo, from his golden bow
The Pythian of the age one arrow sped   250
And smiled!—The spoilers tempt no second blow,
They fawn on the proud feet that spurn them lying
    low.

XXIX

"The sun comes forth, and many reptiles spawn;
He sets, and each ephemeral insect then
Is gathered into death without a dawn,
And the immortal stars awake again;
So is it in the world of living men:
A godlike mind soars forth, in its delight
Making earth bare and veiling heaven, and when
It sinks, the swarms that dimmed or shared its light
Leave to its kindred lamps the spirit's awful night." 261

XXX

Thus ceased she: and the mountain shepherds came,
Their garlands sere, their magic mantles rent;
The Pilgrim of Eternity, whose fame
Over his living head like Heaven is bent,
An early but enduring monument,
Came, veiling all the lightnings of his song
In sorrow; from her wilds Ierne sent
The sweetest lyrist of her saddest wrong,
And Love taught Grief to fall like music from his
    tongue.   270

XXXI

Midst others of less note, came one frail Form,
A phantom among men; companionless
As the last cloud of an expiring storm
Whose thunder is its knell; he, as I guess,
Had gazed on Nature's naked loveliness,
Actaeon-like, and now he fled astray
With feeble steps o'er the world's wilderness,

238–40. dragon . . . spear: Shelley draws on the myth of Perseus. He could slay Medusa (the direct sight of whom turned the beholder to stone) only if he looked at her indirectly, through the reflection on his shield. **242. crescent sphere:** maturity. **244–46. wolves . . . vultures:** the critics.

250. Pythian: Apollo (here applied to Byron, whose satiric verses effectively scourged reviewers). **259. earth . . . heaven:** The sun lights up the earth while "veiling" the stars. **264. Pilgrim:** Byron (referring to his own wandering life and also to *Childe Harold's Pilgrimage*). **268. Ierne:** Ireland. **269. lyrist:** Thomas Moore, 1779–1852. **271. frail Form:** Shelley. **276. Actaeon-like:** Actaeon was turned into a stag and torn to pieces by his own dogs because, while hunting, he had seen Diana bathing.

And his own thoughts, along that rugged way,
Pursued, like raging hounds, their father and their
  prey.

XXXII

A pardlike Spirit beautiful and swift—   280
A Love in desolation masked;—a Power
Girt round with weakness;—it can scarce uplift
The weight of the superincumbent hour;
It is a dying lamp, a falling shower,
A breaking billow;—even whilst we speak
Is it not broken? On the withering flower
The killing sun smiles brightly: on a cheek
The life can burn in blood, even while the heart may
  break.

XXXIII

His head was bound with pansies overblown,
And faded violets, white, and pied, and blue;   290
And a light spear topped with a cypress cone,
Round whose rude shaft dark ivy-tresses grew
Yet dripping with the forest's noonday dew,
Vibrated, as the ever-beating heart
Shook the weak hand that grasped it; of that crew
He came the last, neglected and apart;
A herd-abandoned deer struck by the hunter's dart.

XXXIV

All stood aloof, and at his partial moan
Smiled through their tears; well knew that gentle
  band
Who in another's fate now wept his own,   300
As in the accents of an unknown land
He sung new sorrow; sad Urania scanned
The Stranger's mien, and murmured: "Who art
  thou?"
He answered not, but with a sudden hand
Made bare his branded and ensanguined brow,
Which was like Cain's or Christ's—oh! that it should
  be so!

280. pardlike: leopard-like. **289–92. pansies . . . ivy-tresses:** Pansies were a traditional symbol of thought, the violet of modesty, the cypress of grief, and ivy of constancy. **306. Cain's or Christ's:** wearing a mark like that given both an enemy of man (God branded Cain for killing his brother Abel) or a savior (the blood from the crown of thorns).

XXXV

What softer voice is hushed over the dead?
Athwart what brow is that dark mantle thrown?
What form leans sadly o'er the white death-bed,
In mockery of monumental stone,   310
The heavy heart heaving without a moan?
If it be He, who, gentlest of the wise,
Taught, soothed, loved, honoured the departed one,
Let me not vex, with inharmonious sighs,
The silence of that heart's accepted sacrifice.

XXXVI

Our Adonais has drunk poison—oh!
What deaf and viperous murderer could crown
Life's early cup with such a draught of woe?
The nameless worm would now itself disown:
It felt, yet could escape, the magic tone   320
Whose prelude held all envy, hate, and wrong,
But what was howling in one breast alone,
Silent with expectation of the song,
Whose master's hand is cold, whose silver lyre unstrung.

XXXVII

Live thou, whose infamy is not thy fame!
Live! fear no heavier chastisement from me,
Thou noteless blot on a remembered name!
But be thyself, and know thyself to be!
And ever at thy season be thou free
To spill the venom when thy fangs o'erflow;   330
Remorse and Self-contempt shall cling to thee;
Hot Shame shall burn upon thy secret brow,
And like a beaten hound tremble thou shalt—as now.

XXXVIII

Nor let us weep that our delight is fled
Far from these carrion kites that scream below;
He wakes or sleeps with the enduring dead;
Thou canst not soar where he is sitting now.—
Dust to the dust! but the pure spirit shall flow

310. mockery: imitation. **312. gentlest . . . wise:** Leigh Hunt, who had befriended Keats. **319. nameless worm:** The reviews that attacked Keats in *The Quarterly Review* and *Blackwood's*—in fact reviews generally at this time—were anonymous.

Back to the burning fountain whence it came,
A portion of the Eternal, which must glow   340
Through time and change, unquenchably the same,
Whilst thy cold embers choke the sordid hearth of
    shame.

XXXIX

Peace, peace! he is not dead, he doth not sleep—
He hath awakened from the dream of life—
'Tis we, who lost in stormy visions, keep
With phantoms an unprofitable strife,
And in mad trance, strike with our spirit's knife
Invulnerable nothings.—*We* decay
Like corpses in a charnel; fear and grief
Convulse us and consume us day by day,   350
And cold hopes swarm like worms within our living
    clay.

XL

He has outsoared the shadow of our night;
Envy and calumny and hate and pain,
And that unrest which men miscall delight,
Can touch him not and torture not again;
From the contagion of the world's slow stain
He is secure, and now can never mourn
A heart grown cold, a head grown gray in vain;
Nor, when the spirit's self has ceased to burn,
With sparkless ashes load an unlamented urn.   360

XLI

He lives, he wakes—'tis Death is dead, not he;
Mourn not for Adonais.—Thou young Dawn,
Turn all thy dew to splendour, for from thee
The spirit thou lamentest is not gone;
Ye caverns and ye forests, cease to moan!
Cease, ye faint flowers and fountains, and thou Air,
Which like a mourning veil thy scarf hadst thrown
O'er the abandoned Earth, now leave it bare
Even to the joyous stars which smile on its despair!

XLII

He is made one with Nature: there is heard   370
His voice in all her music, from the moan

352. **shadow . . . night:** Cf. *Epipsychidion*, above, l. 228, and *Prometheus Unbound*, IV. 444.

Of thunder, to the song of night's sweet bird;
He is a presence to be felt and known
In darkness and in light, from herb and stone,
Spreading itself where'er that Power may move
Which has withdrawn his being to its own;
Which wields the world with never-wearied love,
Sustains it from beneath, and kindles it above.

XLIII

He is a portion of the loveliness
Which once he made more lovely: he doth bear
His part, while the one Spirit's plastic stress   381
Sweeps through the dull dense world, compelling
    there,
All new successions to the forms they wear;
Torturing th' unwilling dross that checks its flight
To its own likeness, as each mass may bear;
And bursting in its beauty and its might
From trees and beasts and men into the Heaven's light.

XLIV

The splendours of the firmament of time
May be eclipsed, but are extinguished not;
Like stars to their appointed height they climb,   390
And death is a low mist which cannot blot
The brightness it may veil. When lofty thought
Lifts a young heart above its mortal lair,
And love and life contend in it, for what
Shall be its earthly doom, the dead live there
And move like winds of light on dark and stormy air.

XLV

The inheritors of unfulfilled renown
Rose from their thrones, built beyond mortal
    thought,
Far in the Unapparent. Chatterton
Rose pale,—his solemn agony had not   400
Yet faded from him; Sidney, as he fought

381. **plastic:** shaping. **384–85. Torturing . . . bear:** shaping the reluctant, material "dross" (that, by definition, resists or "checks" the spiritual) as much into the likeness of the spirit as the nature of "each mass" (each particular material) is capable of attaining. **399–404. Chatterton . . . approved:** Thomas Chatterton died of suicide at the age of seventeen; Sir Philip Sidney was killed on the battlefield at thirty-two; and Lucan, when a plot in which he had taken part against Nero was discovered, committed suicide at the age of twenty-six.

And as he fell and as he lived and loved
Sublimely mild, a Spirit without spot,
Arose; and Lucan, by his death approved:
Oblivion as they rose shrank like a thing reproved.

XLVI

And many more, whose names on Earth are dark,
But whose transmitted effluence cannot die
So long as fire outlives the parent spark,
Rose, robed in dazzling immortality.
"Thou art become as one of us," they cry,   410
"It was for thee yon kingless sphere has long
Swung blind in unascended majesty,
Silent alone amid an Heaven of Song.
Assume thy winged throne, thou Vesper of our
    throng!"

XLVII

Who mourns for Adonais? Oh, come forth,
Fond wretch! and know thyself and him aright.
Clasp with thy panting soul the pendulous Earth;
As from a centre, dart thy spirit's light
Beyond all worlds, until its spacious might
Satiate the void circumference: then shrink   420
Even to a point within our day and night;
And keep thy heart light lest it make thee sink
When hope has kindled hope, and lured thee to the
    brink.

XLVIII

Or go to Rome, which is the sepulchre,
Oh, not of him, but of our joy: 'tis nought
That ages, empires, and religions there
Lie buried in the ravage they have wrought;
For such as he can lend,—they borrow not
Glory from those who made the world their prey;
And he is gathered to the kings of thought   430
Who waged contention with their time's decay,
And of the past are all that cannot pass away.

XLIX

Go thou to Rome,—at once the Paradise,
The grave, the city, and the wilderness;

And where its wrecks like shattered mountains rise,
And flowering weeds, and fragrant copses dress
The bones of Desolation's nakedness
Pass, till the spirit of the spot shall lead
Thy footsteps to a slope of green access
Where, like an infant's smile, over the dead   440
A light of laughing flowers along the grass is spread;

L

And gray walls moulder round, on which dull Time
Feeds, like slow fire upon a hoary brand;
And one keen pyramid with wedge sublime,
Pavilioning the dust of him who planned
This refuge for his memory, doth stand
Like flame transformed to marble; and beneath,
A field is spread, on which a newer band
Have pitched in Heaven's smile their camp of death,
Welcoming him we lose with scarce extinguished
    breath.   450

LI

Here pause: these graves are all too young as yet
To have outgrown the sorrow which consigned
Its charge to each; and if the seal is set,
Here, on one fountain of a mourning mind,
Break it not thou! too surely shalt thou find
Thine own well full, if thou returnest home,
Of tears and gall. From the world's bitter wind
Seek shelter in the shadow of the tomb.
What Adonais is, why fear we to become?

LII

The One remains, the many change and pass;   460
Heaven's light forever shines, Earth's shadows fly;
Life, like a dome of many-coloured glass,
Stains the white radiance of Eternity,
Until Death tramples it to fragments.—Die,
If thou wouldst be with that which thou dost seek!
Follow where all is fled!—Rome's azure sky,
Flowers, ruins, statues, music, words, are weak
The glory they transfuse with fitting truth to speak.

---

416. **Fond:** foolish. 417. **pendulous:** hanging or floating.
424. **Rome:** where Keats was buried in the Protestant
cemetery.

444. **pyramid:** the tomb of the Roman tribune Gaius Cestus,
near the grave of Keats. 454. **mourning mind:** Shelley's
three-year-old son, William, had been buried there a year
and a half before Keats.

LIII

Why linger, why turn back, why shrink, my
    Heart?
Thy hopes are gone before: from all things here   470
They have departed; thou shouldst now depart!
A light is passed from the revolving year,
And man, and woman; and what still is dear
Attracts to crush, repels to make thee wither.
The soft sky smiles,—the low wind whispers near:
'Tis Adonais calls! oh, hasten thither,
No more let Life divide what Death can join together.

LIV

That Light whose smile kindles the Universe,
That Beauty in which all things work and move,
That Benediction which the eclipsing Curse   480
Of birth can quench not, that sustaining Love
Which through the web of being blindly wove
By man and beast and earth and air and sea,
Burns bright or dim, as each are mirrors of
The fire for which all thirst; now beams on me,
Consuming the last clouds of cold mortality.

LV

The breath whose might I have invoked in song
Descends on me; my spirit's bark is driven,
Far from the shore, far from the trembling throng
Whose sails were never to the tempest given;   490
The massy earth and sphered skies are riven!
I am borne darkly, fearfully, afar;
Whilst, burning through the inmost veil of Heaven,
The soul of Adonais, like a star,
Beacons from the abode where the Eternal are.

1821 (1821)

TO NIGHT

I

Swiftly walk o'er the western wave,
    Spirit of Night!
Out of the misty eastern cave,
Where, all the long and lone daylight,
Thou wovest dreams of joy and fear,
Which make thee terrible and dear,—
    Swift be thy flight!

484. as each: to the extent that each.

II

Wrap thy form in a mantle gray,
    Star-inwrought!
Blind with thine hair the eyes of Day;   10
Kiss her until she be wearied out,
Then wander o'er city, and sea, and land,
Touching all with thine opiate wand—
    Come, long-sought!

III

When I arose and saw the dawn,
    I sighed for thee;
When light rode high, and the dew was gone,
And noon lay heavy on flower and tree,
And the weary Day turned to his rest,
Lingering like an unloved guest,   20
    I sighed for thee.

IV

Thy brother Death came, and cried,
    Wouldst thou me?
Thy sweet child Sleep, the filmy-eyed,
Murmured like a noontide bee,
Shall I nestle near thy side?
Wouldst thou me?—And I replied,
    No, not thee!

V

Death will come when thou art dead,
    Soon, too soon—   30
Sleep will come when thou art fled;
Of neither would I ask the boon
I ask of thee, beloved Night—
Swift be thine approaching flight,
    Come soon, soon!

1821 (1824)

TIME

Unfathomable Sea! whose waves are years,
    Ocean of Time, whose waters of deep woe
Are brackish with the salt of human tears!
    Thou shoreless flood, which in thy ebb and flow

Claspest the limits of mortality,
  And sick of prey, yet howling on for more,
Vomitest thy wrecks on its inhospitable shore;
  Treacherous in calm, and terrible in storm,
    Who shall put forth on thee,
    Unfathomable Sea?   10

                                        1821 (1824)

TO ———

Music, when soft voices die,
Vibrates in the memory—
Odours, when sweet violets sicken,
Live within the sense they quicken.

Rose leaves, when the rose is dead,
Are heaped for the beloved's bed;
And so thy thoughts, when thou art gone,
Love itself shall slumber on.

                                        1821 (1824)

SONG

I

Rarely, rarely, comest thou,
  Spirit of Delight!
Wherefore hast thou left me now
  Many a day and night?
Many a weary night and day
'Tis since thou art fled away.

II

How shall ever one like me
  Win thee back again?
With the joyous and the free
  Thou wilt scoff at pain.
Spirit false! thou hast forgot
  All but those who need thee not.   10

III

As a lizard with the shade
  Of a trembling leaf,
Thou with sorrow art dismayed;
  Even the sighs of grief

Reproach thee, that thou art not near,
And reproach thou wilt not hear.

IV

Let me set my mournful ditty
  To a merry measure;   20
Thou wilt never come for pity,
  Thou wilt come for pleasure;
Pity then will cut away
Those cruel wings, and thou wilt stay.

V

I love all that thou lovest,
  Spirit of Delight!
The fresh Earth in new leaves dressed,
  And the starry night;
Autumn evening, and the morn
When the golden mists are born.   30

VI

I love snow, and all the forms
  Of the radiant frost;
I love waves, and winds, and storms,
  Everything almost
Which is Nature's, and may be
Untainted by man's misery.

VII

I love tranquil solitude,
  And such society
As is quiet, wise, and good;
  Between thee and me   40
What difference? but thou dost possess
The things I seek, not love them less.

VIII

I love Love—though he has wings,
  And like light can flee,
But above all other things,
  Spirit, I love thee—
Thou art love and life! Oh, come,
Make once more my heart thy home.

                                        1821 (1824)

TO ———. **7. thy thoughts:** presumably the "thoughts" or
remembrances of "thee" on the part of the poet.

MUTABILITY

I

The flower that smiles to-day
        To-morrow dies;
All that we wish to stay
        Tempts and then flies.
What is this world's delight?
Lightning that mocks the night,
        Brief even as bright.

II

Virtue, how frail it is!
        Friendship how rare!
Love, how it sells poor bliss   10
        For proud despair!
But we, though soon they fall,
Survive their joy, and all
        Which ours we call.

III

Whilst skies are blue and bright,
        Whilst flowers are gay,
Whilst eyes that change ere night
        Make glad the day;
Whilst yet the calm hours creep,
Dream thou—and from thy sleep   20
        Then wake to weep.

                            1821 (1824)

A LAMENT

I

O world! O life! O time!
On whose last steps I climb,
    Trembling at that where I had stood before;
When will return the glory of your prime?
    No more—Oh, never more!

II

Out of the day and night
A joy has taken flight;
    Fresh spring, and summer, and winter hoar,
Move my faint heart with grief, but with delight
    No more—Oh, never more!   10

                            1821 (1824)

SONNET:
POLITICAL GREATNESS

Nor happiness, nor majesty, nor fame,
Nor peace, nor strength, nor skill in arms or arts,
Shepherd those herds whom tyranny makes tame;
Verse echoes not one beating of their hearts,
History is but the shadow of their shame,
Art veils her glass, or from the pageant starts
As to oblivion their blind millions fleet,
Staining that Heaven with obscene imagery
Of their own likeness. What are numbers knit
By force or custom? Man who man would be,   10
Must rule the empire of himself; in it
Must be supreme, establishing his throne
On vanquished will, quelling the anarchy
Of hopes and fears, being himself alone.

                            1821 (1824)

FROM

HELLAS

The three lyrics below are sung at appropriate points in
the action by the chorus in *Hellas*, a closet drama written
to celebrate the Greek struggle for independence from
Turkey in 1821. The third lyric concludes the play.

LIFE MAY CHANGE,
BUT IT MAY FLY NOT

Life may change, but it may fly not;
Hope may vanish, but can die not;
Truth be veiled, but still it burneth;
Love repulsed,—but it returneth!

Yet were life a charnel where
Hope lay coffined with Despair;
Yet were truth a sacred lie,
Love were lust—

SONNET: POLITICAL GREATNESS. **8–9. Staining . . . likeness:**
blaspheming heaven by believing in a God who possesses
their human attributes (tyranny, jealousy, etc.).

If Liberty
Lent not life its soul of light,
Hope its iris of delight,   10
Truth its prophet's robe to wear,
Love its power to give and bear.

Fled from the folding-star of Bethlehem:
   Apollo, Pan, and Love,
   And even Olympian Jove
Grew weak, for killing Truth had glared on them;
   Our hills and seas and streams,
   Dispeopled of their dreams,   40
Their waters turned to blood, their dew to tears,
   Wailed for the golden years.

WORLDS ON WORLDS ARE ROLLING EVER

Worlds on worlds are rolling ever
   From creation to decay,
Like the bubbles on a river
   Sparkling, bursting, borne away.
But they are still immortal
Who, through birth's orient portal
And death's dark chasm hurrying to and fro,
   Clothe their unceasing flight
   In the brief dust and light
Gathered around their chariots as they go;   10
   New shapes they still may weave,
   New gods, new laws receive,
Bright or dim are they as the robes they last
   On Death's bare ribs had cast.

A power from the unknown God,
   A Promethean conqueror, came;
Like a triumphal path he trod
   The thorns of death and shame.
   A mortal shape to him
   Was like the vapour dim   20
Which the orient planet animates with light;
   Hell, Sin, and Slavery came,
   Like bloodhounds mild and tame,
Nor preyed, until their Lord had taken flight;
   The moon of Mahomet
   Arose, and it shall set:
While blazoned as on Heaven's immortal noon
   The cross leads generations on.

Swift as the radiant shapes of sleep
   From one whose dreams are Paradise   30
Fly, when the fond wretch wakes to weep,
   And Day peers forth with her blank eyes;
   So fleet, so faint, so fair,
   The Powers of earth and air

THE WORLD'S GREAT AGE BEGINS ANEW

The world's great age begins anew,
   The golden years return,
The earth doth like a snake renew
   Her winter weeds outworn:
Heaven smiles, and faiths and empires gleam,
Like wrecks of a dissolving dream.

A brighter Hellas rears its mountains
   From waves serener far;
A new Peneus rolls his fountains
   Against the morning star.   10
Where fairer Tempes bloom, there sleep
Young Cyclads on a sunnier deep.

A loftier Argo cleaves the main,
   Fraught with a later prize;
Another Orpheus sings again,
   And loves, and weeps, and dies.
A new Ulysses leaves once more
Calypso for his native shore.

Oh, write no more the tale of Troy,
   If earth Death's scroll must be!   20
Nor mix with Laian rage the joy
   Which dawns upon the free:
Although a subtler Sphinx renew
Riddles of death Thebes never knew.

---

35. folding-star: See *Epipsychidion*, above, n. 374. THE
WORLD'S GREAT AGE BEGINS ANEW. **4. weeds:** garments.
9–11. Peneus . . . Tempes: See note to "Hymn of
Pan," above, ll. 13–15. **12. Cyclads:** the Cyclades, islands
in the Aegean Sea. **13. Argo:** the ship in which Jason sailed
in quest of the Golden Fleece. **21–24. Laian . . . knew:**
Laius was the father of Oedipus. On the Sphinx and
Oedipus, see *Prometheus Unbound*, I. n. 346–49.

HELLAS. WORLDS ON WORLDS ARE ROLLING EVER. **15. A power:**
Christ. **21. orient planet:** the rising sun.

Another Athens shall arise,
  And to remoter time
Bequeath, like sunset to the skies,
  The splendour of its prime;
And leave, if nought so bright may live,
All earth can take or Heaven can give.   30

Saturn and Love their long repose
  Shall burst, more bright and good
Than all who fell, than One who rose,
  Than many unsubdued:
Not gold, not blood, their altar dowers,
But votive tears and symbol flowers.

Oh, cease! must hate and death return?
  Cease! must men kill and die?
Cease! drain not to its dregs the urn
  Of bitter prophecy.   40
The world is weary of the past,
Oh, might it die or rest at last!

                              1821 (1822)

LINES: "WHEN THE LAMP IS SHATTERED"

I

When the lamp is shattered
The light in the dust lies dead—
  When the cloud is scattered
The rainbow's glory is shed.
  When the lute is broken,
Sweet tones are remembered not;
  When the lips have spoken,
Loved accents are soon forgot.

II

  As music and splendour
Survive not the lamp and the lute,   10
  The heart's echoes render
No song when the spirit is mute:—
  No song but sad dirges,
Like the wind through a ruined cell,
  Or the mournful surges
That ring the dead seaman's knell.

33. all . . . One: the gods of the ancient world who fell, and Christ who rose.

III

When hearts have once mingled
Love first leaves the well-built nest;
  The weak one is singled
To endure what it once possessed.   20
  O Love! who bewailest
The frailty of all things here,
  Why choose you the frailest
For your cradle, your home, and your bier?

IV

Its passions will rock thee
As the storms rock the ravens on high;
  Bright reason will mock thee,
Like the sun from a wintry sky.
  From thy nest every rafter
Will rot, and thine eagle home   30
  Leave thee naked to laughter,
When leaves fall and cold winds come.

                              1822 (1824)

TO JANE: THE INVITATION

Best and brightest, come away!
Fairer far than this fair Day,
Which, like thee to those in sorrow,
Comes to bid a sweet good-morrow
To the rough Year just awake
In its cradle on the brake.
The brightest hour of unborn Spring,
Through the winter wandering,
Found, it seems, the halcyon Morn
To hoar February born.   10
Bending from Heaven, in azure mirth,
It kissed the forehead of the Earth,
And smiled upon the silent sea,
And bade the frozen streams be free,
And waked to music all their fountains,
And breathed upon the frozen mountains,
And like a prophetess of May
Strewed flowers upon the barren way,
Making the wintry world appear
Like one on whom thou smilest, dear.   20
Away, away, from men and towns,
To the wild wood and the downs—

To the silent wilderness
Where the soul need not repress
Its music lest it should not find
An echo in another's mind,
While the touch of Nature's art
Harmonizes heart to heart.
I leave this notice on my door
For each accustomed visitor:—   30
"I am gone into the fields
To take what this sweet hour yields;—
Reflection, you may come to-morrow,
Sit by the fireside with Sorrow.—
You with the unpaid bill, Despair,—
You, tiresome verse-reciter, Care,—
I will pay you in the grave,—
Death will listen to your stave.
Expectation too, be off!
To-day is for itself enough;   40
Hope, in pity mock not Woe
With smiles, nor follow where I go;
Long having lived on thy sweet food,
At length I find one moment's good
After long pain—with all your love,
This you never told me of."

Radiant Sister of the Day,
Awake! arise! and come away!
To the wild woods and the plains,
And the pools where winter rains   50
Image all their roof of leaves,
Where the pine its garland weaves
Of sapless green and ivy dun
Round stems that never kiss the sun;
Where the lawns and pastures be,
And the sandhills of the sea;—
Where the melting hoar-frost wets
The daisy-star that never sets,
And wind-flowers, and violets,
Which yet join not scent to hue,   60
Crown the pale year weak and new;
When the night is left behind
In the deep east, dun and blind,
And the blue noon is over us,
And the multitudinous
Billows murmur at our feet,
Where the earth and ocean meet,
And all things seem only one
In the universal sun.

                      1822 (1824)

TO JANE: "THE KEEN STARS WERE TWINKLING"

I

The keen stars were twinkling,
And the fair moon was rising among them,
    Dear Jane!
  The guitar was tinkling,
But the notes were not sweet till you sung them
    Again.

II

  As the moon's soft splendour
O'er the faint cold starlight of Heaven
    Is thrown,
  So your voice most tender   10
To the strings without soul had then given
    Its own.

III

  The stars will awaken,
Though the moon sleep a full hour later,
    To-night;
  No leaf will be shaken
Whilst the dews of your melody scatter
    Delight.

IV

  Though the sound overpowers,
Sing again, with your dear voice revealing   20
    A tone
  Of some world far from ours,
Where music and moonlight and feeling
    Are one.

                      1822 (1839)

WITH A GUITAR, TO JANE

❧

These verses were sent with a guitar to Jane Williams;
she and her husband Edward were close friends of the
Shelleys. The poem alludes, of course, to Shakespeare's
The Tempest.

❧

Ariel to Miranda:—Take
This slave of Music, for the sake

Of him who is the slave of thee,
And teach it all the harmony
In which thou canst, and only thou,
Make the delighted spirit glow,
Till joy denies itself again,
And, too intense, is turned to pain;
For by permission and command
Of thine own Prince Ferdinand,   10
Poor Ariel sends this silent token
Of more than ever can be spoken;
Your guardian spirit, Ariel, who,
From life to life, must still pursue
Your happiness;—for thus alone
Can Ariel ever find his own.
From Prospero's enchanted cell,
As the mighty verses tell,
To the throne of Naples, he
Lit you o'er the trackless sea,   20
Flitting on, your prow before,
Like a living meteor.
When you die, the silent Moon,
In her interlunar swoon,
Is not sadder in her cell
Than deserted Ariel.
When you live again on earth,
Like an unseen star of birth,
Ariel guides you o'er the sea
Of life from your nativity.   30
Many changes have been run
Since Ferdinand and you begun
Your course of love, and Ariel still
Has tracked your steps, and served your will;
Now, in humbler, happier lot,
This is all remembered not;
And now, alas! the poor sprite is
Imprisoned, for some fault of his,
In a body like a grave;—
From you he only dares to crave,   40
For his service and his sorrow,
A smile to-day, a song to-morrow.
The artist who this idol wrought,
To echo all harmonious thought,
Felled a tree, while on the steep
The woods were in their winter sleep,
Rocked in that repose divine
On the wind-swept Apennine;

And dreaming, some of Autumn past,
And some of Spring approaching fast,   50
And some of April buds and showers,
And some of songs in July bowers,
And all of love; and so this tree,—
O that such our death may be!—
Died in sleep, and felt no pain,
To live in happier form again:
From which, beneath Heaven's fairest star,
The artist wrought this loved Guitar,
And taught it justly to reply,
To all who question skilfully,   60
In language gentle as thine own;
Whispering in enamoured tone
Sweet oracles of woods and dells,
And summer winds in sylvan cells;
For it had learned all harmonies
Of the plains and of the skies,
Of the forests and the mountains,
And the many-voiced fountains;
The clearest echoes of the hills,
The softest notes of falling rills,   70
The melodies of birds and bees,
The murmuring of summer seas,
And pattering rain, and breathing dew,
And airs of evening; and it knew
That seldom-heard mysterious sound,
Which, driven on its diurnal round,
As it floats through boundless day,
Our world enkindles on its way.—
All this it knows, but will not tell
To those who cannot question well   80
The Spirit that inhabits it;
It talks according to the wit
Of its companions; and no more
Is heard than has been felt before,
By those who tempt it to betray
These secrets of an elder day:
But, sweetly as its answers will
Flatter hands of perfect skill,
It keeps its highest, holiest tone
For our beloved Jane alone.   90

                              1822 (1832)

WITH A GUITAR, TO JANE. **10. Prince Ferdinand:** that is,
Edward Williams. **24. interlunar:** the period between the
old and new moon.

A DIRGE

Rough wind, that moanest loud
   Grief too sad for song;
Wild wind, when sullen cloud
   Knells all the night long;
Sad storm, whose tears are vain,
Bare woods, whose branches strain,
Deep caves and dreary main,—
   Wail, for the world's wrong!

1822 (1824)

THE TRIUMPH OF LIFE

Shelley composed the poem in the spring and early summer of 1822; he died before it was completed. Mrs. Shelley said that the manuscript was in "so unfinished a state that I arranged it in its present form only with difficulty"; spaces and ellipses appear below where the text is defective. The basic metaphor of a triumph was suggested by Petrarch's allegorical *Trionfi* (see especially the first of the six poems). It refers to the processions through ancient Rome by which the senate and people honored a victorious general. At the head of the procession came the senators and magistrates, followed by trumpeters, trophies of war, and prisoners. The general was carried in a chariot drawn by four horses. Shelley also alludes to the orgiastic dances and revelry in celebration of the god Dionysus. The poem is influenced by Dante's *Divine Comedy*, not only in the use of *terza rima*, but in the handling of such episodes as the vision of Life's car or the encounter with Rousseau and, more generally, in a tendency to a more incisive and marmoreal utterance.

Swift as a spirit hastening to his task
Of glory and of good, the Sun sprang forth
Rejoicing in his splendour, and the mask

Of darkness fell from the awakened Earth—
The smokeless altars of the mountain snows
Flamed above crimson clouds, and at the birth

Of light, the Ocean's orison arose,
To which the birds tempered their matin lay.
All flowers in field or forest which unclose

THE TRIUMPH OF LIFE. **7. orison:** prayer.

Their trembling eyelids to the kiss of day,   10
Swinging their censers in the element,
With orient incense lit by the new ray

Burned slow and inconsumably, and sent
Their odorous sighs up to the smiling air;
And, in succession due, did continent,

Isle, ocean, and all things that in them wear
The form and character of mortal mould,
Rise as the Sun their father rose, to bear

Their portion of the toil, which he of old
Took as his own, and then imposed on them: 20
But I, whom thoughts which must remain untold

Had kept as wakeful as the stars that gem
The cone of night, now they were laid asleep
Stretched my faint limbs beneath the hoary stem

Which an old chestnut flung athwart the steep
Of a green Apennine: before me fled
The night; behind me rose the day; the deep

Was at my feet, and Heaven above my head,—
When a strange trance over my fancy grew
Which was not slumber, for the shade it spread   30

Was so transparent, that the scene came through
As clear as when a veil of light is drawn
O'er evening hills, they glimmer; and I knew

That I had felt the freshness of that dawn
Bathe in the same cold dew my brow and hair,
And sate as thus upon that slope of lawn

Under the self-same bough, and heard as there
The birds, the fountains and the ocean hold
Sweet talk in music through the enamoured air,
And then a vision on my brain was rolled.   40

As in that trance of wondrous thought I lay,
This was the tenour of my waking dream:—
Methought I sate beside a public way

Thick strewn with summer dust, and a great stream
Of people there was hurrying to and fro,
Numerous as gnats upon the evening gleam,

All hastening onward, yet none seemed to know
Whither he went, or whence he came, or why
He made one of the multitude, and so

Was borne amid the crowd, as through the sky   50
One of the million leaves of summer's bier;
Old age and youth, manhood and infancy,

Mixed in one mighty torrent did appear,
Some flying from the thing they feared, and some
Seeking the object of another's fear;

And others, as with steps towards the tomb,
Pored on the trodden worms that crawled beneath,
And others mournfully within the gloom

Of their own shadow walked, and called it death;
And some fled from it as it were a ghost,   60
Half fainting in the affliction of vain breath:

But more, with motions which each other crossed,
Pursued or shunned the shadows the clouds threw,
Or birds within the noonday aether lost,

Upon that path where flowers never grew,—
And, weary with vain toil and faint for thirst,
Heard not the fountains, whose melodious dew

Out of their mossy cells forever burst;
Nor felt the breeze which from the forest told
Of grassy paths and wood-lawns interspersed   70

With overarching elms and caverns cold,
And violet banks where sweet dreams brood, but they
Pursued their serious folly as of old.

And as I gazed, methought that in the way
The throng grew wilder, as the woods of June
When the south wind shakes the extinguished day,

And a cold glare, intenser than the noon,
But icy cold, obscured with blinding light
The sun, as he the stars. Like the young moon—

When on the sunlit limits of the night   80
Her white shell trembles amid crimson air,
And whilst the sleeping tempest gathers might—

Doth, as the herald of its coming, bear
The ghost of its dead mother, whose dim form
Bends in dark aether from her infant's chair,—

So came a chariot on the silent storm
Of its own rushing splendour, and a Shape
So sate within, as one whom years deform,

83–85. bear . . . chair: Within the bright crescent of the new
moon, the full moon sometimes appears faintly (the cause
being the reflection, on that part of it, of light reflected from
the earth). This was often interpreted as a sign of a coming
storm. Cf. Coleridge's "Dejection: An Ode," above, ll. 13–
14, and the motto prefixed to it from the "Ballad of Sir
Patrick Spence."

Beneath a dusky hood and double cape,
Crouching within the shadow of a tomb;   90
And o'er what seemed the head a cloud-like crape

Was bent, a dun and faint aethereal gloom
Tempering the light. Upon the chariot-beam
A Janus-visaged Shadow did assume

The guidance of that wonder-winged team;
The shapes which drew it in thick lightenings
Were lost:—I heard alone on the air's soft stream

The music of their ever-moving wings.
All the four faces of that Charioteer
Had their eyes banded; little profit brings   100

Speed in the van and blindness in the rear,
Nor then avail the beams that quench the sun,—
Or that with banded eyes could pierce the sphere

Of all that is, has been or will be done;
So ill was the car guided—but it passed
With solemn speed majestically on.

The crowd gave way, and I arose aghast,
Or seemed to rise, so mighty was the trance,
And saw, like clouds upon the thunder-blast,

The million with fierce song and maniac dance   110
Raging around—such seemed the jubilee
As when to greet some conqueror's advance

Imperial Rome poured forth her living sea
From senate-house, and forum, and theatre,
When   upon the free

Had bound a yoke, which soon they stooped to bear.
Nor wanted here the just similitude
Of a triumphal pageant, for where'er

The chariot rolled, a captive multitude
Was driven;—all those who had grown old in power
Or misery,—all who had their age subdued   121

By action or by suffering, and whose hour
Was drained to its last sand in weal or woe,
So that the trunk survived both fruit and flower;—

All those whose fame or infamy must grow
Till the great winter lay the form and name
Of this green earth with them for ever low;—

All but the sacred few who could not tame
Their spirits to the conqueror—but as soon
As they had touched the world with living flame,   130

Fled back like eagles to their native noon,
Or those who put aside the diadem
Of earthly thrones or gems . . .

Were there, of Athens or Jerusalem,
Were neither mid the mighty captives seen,
Nor mid the ribald crowd that followed them,

Nor those who went before fierce and obscene.
The wild dance maddens in the van, and those
Who lead it—fleet as shadows on the green,

Outspeed the chariot, and without repose   140
Mix with each other in tempestuous measure
To savage music, wilder as it grows,

They, tortured by their agonizing pleasure,
Convulsed and on the rapid whirlwinds spun
Of that fierce Spirit, whose unholy leisure

Was soothed by mischief since the world begun,
Throw back their heads and loose their streaming hair;
And in their dance round her who dims the sun,

Maidens and youths fling their wild arms in air
As their feet twinkle; they recede, and now   150
Bending within each other's atmosphere,

Kindle invisibly—and as they glow,
Like moths by light attracted and repelled,
Oft to their bright destruction come and go,

Till like two clouds into one vale impelled,
That shake the mountains when their lightnings mingle
And die in rain—the fiery band which held

Their natures, snaps—while the shock still may tingle;
One falls and then another in the path
Senseless—nor is the desolation single,   160

Yet ere I can say *where*—the chariot hath
Passed over them—nor other trace I find
But as of foam after the ocean's wrath

Is spent upon the desert shore;—behind,
Old men and women foully disarrayed,
Shake their gray hairs in the insulting wind,

And follow in the dance, with limbs decayed,
Seeking to reach the light which leaves them still
Farther behind and deeper in the shade.

But not the less with impotence of will   170
They wheel, though ghastly shadows interpose
Round them and round each other, and fulfil

Their work, and in the dust from whence they rose
Sink, and corruption veils them as they lie,
And past in these performs what   in those.

Struck to the heart by this sad pageantry,
Half to myself I said—"And what is this?
Whose shape is that within the car? And why—"

I would have added—"is all here amiss?—"
But a voice answered—"Life!"—I turned, and knew
(O Heaven, have mercy on such wretchedness!)   181

That what I thought was an old root which grew
To strange distortion out of the hill side,
Was indeed one of those deluded crew,

And that the grass, which methought hung so wide
And white, was but his thin discoloured hair,
And that the holes it vainly sought to hide,

Were or had been eyes:—"If thou canst, forbear
To join the dance, which I had well forborne!"
Said the grim Feature (of my thought aware).   190

"I will unfold that which to this deep scorn
Led me and my companions, and relate
The progress of the pageant since the morn;

"If thirst of knowledge shall not then abate,
Follow it thou even to the night, but I
Am weary."—Then like one who with the weight

Of his own words is staggered, wearily
He paused; and ere he could resume, I cried:
"First, who art thou?"—"Before thy memory,

"I feared, loved, hated, suffered, did and died,   200
And if the spark with which Heaven lit my spirit
Had been with purer nutriment supplied,

"Corruption would not now thus much inherit
Of what was once Rousseau,—nor this disguise
Stain that which ought to have disdained to wear it;

"If I have been extinguished, yet there rise
A thousand beacons from the spark I bore"—
"And who are those chained to the car?"—"The wise,

---

134. Athens or Jerusalem: The text is defective at this point. The allusion is probably to Socrates and Christ having not been taken captive by Life.

207. beacons: beacon fires. Cf. "Ode to the West Wind," above, ll. 66–67. The writings of Rousseau helped kindle the French Revolution.

"The great, the unforgotten,—they who wore
Mitres and helms and crowns, or wreaths of light, 210
Signs of thought's empire over thought—their lore

"Taught them not this, to know themselves; their might
Could not repress the mystery within,
And for the morn of truth they feigned, deep night

"Caught them ere evening."—"Who is he with chin
Upon his breast, and hands crossed on his chain?"—
"The child of a fierce hour; he sought to win

"The world, and lost all that it did contain
Of greatness, in its hope destroyed; and more
Of fame and peace than virtue's self can gain   220

"Without the opportunity which bore
Him on its eagle pinions to the peak
From which a thousand climbers have before

"Fallen, as Napoleon fell."—I felt my cheek
Alter, to see the shadow pass away,
Whose grasp had left the giant world so weak

That every pigmy kicked it as it lay;
And much I grieved to think how power and will
In opposition rule our mortal day,

And why God made irreconcilable   230
Good and the means of good; and for despair
I half disdained mine eyes' desire to fill

With the spent vision of the times that were
And scarce have ceased to be.—"Dost thou behold,"
Said my guide, "those spoilers spoiled, Voltaire,

"Frederick, and Paul, Catherine, and Leopold,
And hoary anarchs, demagogues, and sage
    names which the world thinks always old,

"For in the battle Life and they did wage,
She remained conqueror. I was overcome   240
By my own heart alone, which neither age,

"Nor tears, nor infamy, nor now the tomb
Could temper to its object."—"Let them pass,"
I cried, "the world and its mysterious doom

"Is not so much more glorious than it was,
That I desire to worship those who drew
New figures on its false and fragile glass

"As the old faded."—"Figures ever new
Rise on the bubble, paint them as you may;
We have but thrown, as those before us threw,   250

"Our shadows on it as it passed away.
But mark now chained to the triumphal chair
The mighty phantoms of an elder day;

"All that is mortal of great Plato there
Expiates the joy and woe his master knew not;
The star that ruled his doom was far too fair,

"And life, where long that flower of Heaven grew not,
Conquered that heart by love, which gold, or pain,
Or age, or sloth, or slavery could subdue not.

"And near him walk the   twain,   260
The tutor and his pupil, whom Dominion
Followed as tame as vulture in a chain.

"The world was darkened beneath either pinion
Of him whom from the flock of conquerors
Fame singled out for her thunder-bearing minion;

"The other long outlived both woes and wars,
Throned in the thoughts of men, and still had kept
The jealous key of Truth's eternal doors,

"If Bacon's eagle spirit had not lept
Like lightning out of darkness—he compelled   270
The Proteus shape of Nature, as it slept

"To wake, and lead him to the caves that held
The treasure of the secrets of its reign.
See the great bards of elder time, who quelled

"The passions which they sung, as by their strain
May well be known: their living melody
Tempers its own contagion to the vein

"Of these who are infected with it—I
Have suffered what I wrote, or viler pain!
And so my words have seeds of misery—   280

"Even as the deeds of others, not as theirs."
And then he pointed to a company,

236. Frederick . . . Leopold: Frederick the Great of Prussia,
Tsar Paul of Russia, Catherine the Great (Paul's predecessor),
and Leopold II of the Holy Roman Empire. **237. anarchs:**
despots.

255. joy . . . not: the "joy and woe" of love, which
Plato's "master," Socrates, "knew not" because he lived
a life of abstinence. **256–59. star . . . not:** Because of
the beauty of Aster (which in Greek means "star"), Life.
which could not conquer Plato's heart by "gold," "pain,"
etc., was able to do so through love. **261. tutor . . . pupil:**
Aristotle and Alexander the Great. **267–70. still . . . darkness:** Aristotle would still have reigned supreme over science
had it not been for the inductive method advocated by
Francis Bacon.

'Midst whom I quickly recognized the heirs
Of Caesar's crime, from him to Constantine;
The anarch chiefs, whose force and murderous snares

Had founded many a sceptre-bearing line,
And spread the plague of gold and blood abroad:
And Gregory and John, and men divine,

Who rose like shadows between man and God;
Till that eclipse, still hanging over heaven,   290
Was worshipped by the world o'er which they strode,

For the true sun it quenched—"The power was given
But to destroy," replied the leader:—"I
Am one of those who have created, even

"If it be but a world of agony."—
"Whence camest thou? and whither goest thou?
How did thy course begin?" I said, "and why?

"Mine eyes are sick of this perpetual flow
Of people, and my heart sick of one sad thought—
Speak!"—"Whence I am, I partly seem to know,   300

"And how and by what paths I have been brought
To this dread pass, methinks even thou mayest guess;—
Why this should be, my mind can compass not;

"Whither the conqueror hurries me, still less;—
But follow thou, and from spectator turn
Actor or victim in this wretchedness,

"And what thou wouldst be taught I then may learn
From thee. Now listen:—In the April prime,
When all the forest-tips began to burn

"With kindling green, touched by the azure clime
Of the young season, I was laid asleep   311
Under a mountain, which from unknown time

"Had yawned into a cavern, high and deep;
And from it came a gentle rivulet,
Whose water, like clear air, in its calm sweep

"Bent the soft grass, and kept for ever wet
The stems of the sweet flowers, and filled the grove
With sounds, which whoso hears must needs forget

"All pleasure and all pain, all hate and love,
Which they had known before that hour of rest;   320
A sleeping mother then would dream not of

"Her only child who died upon the breast
At eventide—a king would mourn no more
The crown of which his brows were dispossessed

"When the sun lingered o'er his ocean floor
To gild his rival's new prosperity.
Thou wouldst forget thus vainly to deplore

"Ills, which if ills can find no cure from thee,
The thought of which no other sleep will quell,
Nor other music blot from memory,   330

"So sweet and deep is the oblivious spell;
And whether life had been before that sleep
The Heaven which I imagine, or a Hell

"Like this harsh world in which I wake to weep,
I know not. I arose, and for a space
The scene of woods and waters seemed to keep,

"Though it was now broad day, a gentle trace
Of light diviner than the common sun
Sheds on the common earth, and all the place

"Was filled with magic sounds woven into one   340
Oblivious melody, confusing sense
Amid the gliding waves and shadows dun;

"And, as I looked, the bright omnipresence
Of morning through the orient cavern flowed,
And the sun's image radiantly intense

"Burned on the waters of the well that glowed
Like gold, and threaded all the forest's maze
With winding paths of emerald fire; there stood

"Amid the sun, as he amid the blaze
Of his own glory, on the vibrating   350
Floor of the fountain, paved with flashing rays,

"A Shape all light, which with one hand did fling
Dew on the earth, as if she were the dawn,
And the invisible rain did ever sing

"A silver music on the mossy lawn;
And still before me on the dusky grass,
Iris her many-coloured scarf had drawn:

---

284. **Caesar's crime:** the destruction of the Roman Republic and the introduction of dictatorship. The Roman emperors are the "heirs" of this crime. 288. **Gregory . . . John:** Pope Gregory the Great (c. 540–604) and apparently St. John (in which case the reference is to the Fourth Gospel, with the implication that it is departing from Christ's original teachings).

331. **oblivious:** causing oblivion. 352. **Shape:** possibly that of Ideal or Intellectual Beauty. In part, lines 308–468 put as allegory what Wordsworth expresses through metaphor in "Intimations of Immortality." 357. **Iris:** the rainbow.

"In her right hand she bore a crystal glass,
Mantling with bright Nepenthe; the fierce splendour
Fell from her as she moved under the mass   360

"Of the deep cavern, and with palms so tender,
Their tread broke not the mirror of its billow,
Glided along the river, and did bend her

"Head under the dark boughs, till like a willow
Her fair hair swept the bosom of the stream
That whispered with delight to be its pillow.

"As one enamoured is upborne in dream
O'er lily-paven lakes, mid silver mist,
To wondrous music, so this shape might seem

"Partly to tread the waves with feet which kissed   370
The dancing foam; partly to glide along
The air which roughened the moist amethyst,

"Or the faint morning beams that fell among
The trees, or the soft shadows of the trees;
And her feet, ever to the ceaseless song

"Of leaves, and winds, and waves, and birds, and bees,
And falling drops, moved in a measure new
Yet sweet, as on the summer evening breeze,

"Up from the lake a shape of golden dew
Between two rocks, athwart the rising moon,   380
Dances i' the wind, where never eagle flew;

"And still her feet, no less than the sweet tune
To which they moved, seemed as they moved to blot
The thoughts of him who gazed on them; and soon

"All that was, seemed as if it had been not;
And all the gazer's mind was strewn beneath
Her feet like embers; and she, thought by thought,

"Trampled its sparks into the dust of death;
As day upon the threshold of the east
Treads out the lamps of night, until the breath   390

"Of darkness re-illumine even the least
Of heaven's living eyes—like day she came,
Making the night a dream; and ere she ceased

"To move, as one between desire and shame
Suspended, I said—'If, as it doth seem,
Thou comest from the realm without a name

"Into this valley of perpetual dream,
Show whence I came, and where I am, and why—
Pass not away upon the passing stream.'

"'Arise and quench thy thirst,' was her reply.   400
And as a shut lily stricken by the wand
Of dewy morning's vital alchemy,

"I rose; and, bending at her sweet command,
Touched with faint lips the   cup she raised,
And suddenly my brain became as sand

"Where the first wave had more than half erased
The track of deer on desert Labrador;
Whilst the   wolf, from which they fled amazed,

"Leaves his stamp visibly upon the shore,
Until the second bursts;—so on my sight   410
Burst a new vision, never seen before,

"And the fair shape waned in the coming light,
As veil by veil the silent splendour drops
From Lucifer, amid the chrysolite

"Of sunrise, ere it tinge the mountain-tops;
And as the presence of that fairest planet,
Although unseen, is felt by one who hopes

"That his day's path may end as he began it,
In that star's smile, whose light is like the scent
Of a jonquil when evening breezes fan it,   420

"Or the soft note in which his dear lament
The Brescian shepherd breathes, or the caress
That turned his weary slumber to content;

"So knew I in that light's severe excess
The presence of that Shape which on the stream
Moved, as I moved along the wilderness,

"More dimly than a day-appearing dream,
The ghost of a forgotten form of sleep;
A light of heaven, whose half-extinguished beam

"Through the sick day in which we wake to weep
Glimmers, for ever sought, for ever lost;   431
So did that shape its obscure tenour keep

---

411. **vision:** that of Life and the car. 414. **Lucifer:** the
morning star. **chrysolite:** a gem of yellow or greenish
color. 421–22. **dear . . . shepherd:** "The favorite song,
Stanco di pascolar le pecorelle, is a Brescian national air" (Mrs.
Shelley's note).

359. **Nepenthe:** a drug giving forgetfulness of sorrow.

"Beside my path, as silent as a ghost;
But the new Vision, and the cold bright car,
With solemn speed and stunning music, crossed

"The forest, and as if from some dread war
Triumphantly returning, the loud million
Fiercely extolled the fortune of her star.

"A moving arch of victory, the vermilion
And green and azure plumes of Iris had   440
Built high over her wind-winged pavilion,

"And underneath aethereal glory clad
The wilderness, and far before her flew
The tempest of the splendour, which forbade

"Shadow to fall from leaf and stone; the crew
Seemed in that light, like atomies to dance
Within a sunbeam;—some upon the new

"Embroidery of flowers, that did enhance
The grassy vesture of the desert, played,
Forgetful of the chariot's swift advance;   450

"Others stood gazing, till within the shade
Of the great mountain its light left them dim;
Others outspeeded it; and others made

"Circles around it, like the clouds that swim
Round the high moon in a bright sea of air;
And more did follow, with exulting hymn,

"The chariot and the captives fettered there:—
But all like bubbles on an eddying flood
Fell into the same track at last, and were

"Borne onward.—I among the multitude   460
Was swept—me, sweetest flowers delayed not long;
Me, not the shadow nor the solitude;

"Me, not that falling stream's Lethean song;
Me, not the phantom of that early Form
Which moved upon its motion—but among

"The thickest billows of that living storm
I plunged, and bared my bosom to the clime
Of that cold light, whose airs too soon deform.

"Before the chariot had begun to climb
The opposing steep of that mysterious dell,   470
Behold a wonder worthy of the rhyme

464. **early Form:** the first "Shape."

"Of him who from the lowest depths of hell,
Through every paradise and through all glory,
Love led serene, and who returned to tell

"The words of hate and awe; the wondrous story
How all things are transfigured except Love;
For deaf as is a sea, which makes hoary,

"The world can hear not the sweet notes that move
The sphere whose light is melody to lovers—
A wonder worthy of his rhyme.—The grove   480

"Grew dense with shadows to its inmost covers,
The earth was gray with phantoms, and the air
Was peopled with dim forms, as when there hovers

"A flock of vampire-bats before the glare
Of the tropic sun, bringing, ere evening,
Strange night upon some Indian isle;—thus were

"Phantoms diffused around; and some did fling
Shadows of shadows, yet unlike themselves,
Behind them; some like eaglets on the wing

"Were lost in the white day; others like elves 490
Danced in a thousand unimagined shapes
Upon the sunny streams and grassy shelves;

"And others sate chattering like restless apes
On vulgar hands, . . .
Some made a cradle of the ermined capes

"Of kingly mantles; some across the tiar
Of pontiffs sate like vultures; others played
Under the crown which girt with empire

"A baby's or an idiot's brow, and made
Their nests in it. The old anatomies   500
Sate hatching their bare broods under the shade

"Of daemon wings, and laughed from their dead eyes
To reassume the delegated power,
Arrayed in which those worms did monarchize,

"Who made this earth their charnel. Others more
Humble, like falcons, sate upon the fist
Of common men, and round their heads did soar;

472. **him . . . hell:** Dante. 482. **gray with phantoms:** with the innumerable illusions, envies, and selfish ambitions that animate mankind. 504–05. **monarchize . . . charnel:** the kings and tyrants who were responsible for the death of so many, often leaving the earth a vast charnel house.

"Or like small gnats and flies, as thick as mist
On evening marshes, thronged about the brow
Of lawyers, statesmen, priest and theorist;—   510

"And others, like discoloured flakes of snow
On fairest bosoms and the sunniest hair,
Fell, and were melted by the youthful glow

'Which they extinguished; and, like tears, they were
A veil to those from whose faint lids they rained
In drops of sorrow. I became aware

"Of whence those forms proceeded which thus
        stained
The track in which we moved. After brief space,
From every form the beauty slowly waned;

"From every firmest limb and fairest face   520
The strength and freshness fell like dust, and left
The action and the shape without the grace

"Of life. The marble brow of youth was cleft
With care; and in those eyes where once hope shone,
Desire, like a lioness bereft

"Of her last cub, glared ere it died; each one
Of that great crowd sent forth incessantly
These shadows, numerous as the dead leaves blown

"In autumn evening from a poplar tree.
Each like himself and like each other were   530
At first; but some distorted seemed to be

"Obscure clouds, moulded by the casual air;
And of this stuff the car's creative ray
Wrought all the busy phantoms that were there,

"As the sun shapes the clouds; thus on the way
Mask after mask fell from the countenance
And form of all; and long before the day

"Was old, the joy which waked like heaven's glance
The sleepers in the oblivious valley, died;
And some grew weary of the ghastly dance,   540

"And fell, as I have fallen, by the wayside;—
Those soonest from whose forms most shadows passed,
And least of strength and beauty did abide."

"Then, what is life?" I cried—The cripple cast
His eye upon that car, which now had rolled
Onward, as if that look must be the last,

And answered, "Happy those for whom the gold
Of . . ."

                                          1822 (1824)

ON LIFE

Conjecturally dated 1812–14, the essay was first published in 1832.

Life and the world, or whatever we call that which
we are and feel, is an astonishing thing. The mist of
familiarity obscures from us the wonder of our being.
We are struck with admiration at some of its transient
modifications, but it is itself the great miracle. What
are changes of empires, the wreck of dynasties, with
the opinions which supported them; what is the birth
and the extinction of religious and of political systems,
to life? What are the revolutions of the globe which we
inhabit, and the operations of the elements of which
it is composed, compared with life? What is the universe
of stars, and suns, of which this inhabited earth is one,
and their motions, and their destiny, compared with
life? Life, the great miracle, we admire not, because it
is so miraculous. It is well that we are thus shielded
by the familiarity of what is at once so certain and so
unfathomable, from an astonishment which would
otherwise absorb and overawe the functions of that
which is its object.

If any artist, I do not say had executed, but had merely
conceived in his mind the system of the sun, and the
stars, and planets, they not existing, and had painted
to us in words, or upon canvas, the spectacle now
afforded by the nightly cope of heaven, and illustrated
it by the wisdom of astronomy, great would be our
admiration. Or had he imagined the scenery of this
earth, the mountains, the seas, and the rivers; the
grass, and the flowers, and the variety of the forms
and masses of the leaves of the woods, and the
colours which attend the setting and the rising sun,
and the hues of the atmosphere, turbid or serene, these
things not before existing, truly we should have been
astonished, and it would not have been a vain boast
to have said of such a man, "Non merita nome di
creatore, sennon Iddio ed il Poeta."[1] But now these
things are looked on with little wonder, and to be

ON LIFE. **1. "Non . . . Poeta"**: "None deserves the name of
creator except God and the Poet." The quotation is from
Tasso. Cf. *A Defence of Poetry*, below, n. 34.

conscious of them with intense delight is esteemed to be the distinguishing mark of a refined and extra-ordinary person. The multitude of men care not for them. It is thus with Life—that which includes all.

What is life? Thoughts and feelings arise, with or without our will, and we employ words to express them. We are born, and our birth is unremembered, and our infancy remembered but in fragments; we live on, and in living we lose the apprehension of life. How vain is it to think that words can penetrate the mystery of our being! Rightly used they may make evident our ignorance to ourselves, and this is much. For what are we? Whence do we come? and whither do we go? Is birth the commencement, is death the conclusion of our being? What is birth and death?

The most refined abstractions of logic conduct to a view of life, which, though startling to the apprehension, is, in fact, that which the habitual sense of its repeated combinations has extinguished in us. It strips, as it were, the painted curtain from this scene of things. I confess that I am one of those who am unable to refuse my assent to the conclusions of those philosophers[2] who assert that nothing exists but as it is perceived.

It is a decision against which all our persuasions struggle, and we must be long convicted before we can be convinced that the solid universe of external things is "such stuff as dreams are made of."[3] The shocking absurdities of the popular philosophy of mind and matter, its fatal consequences in morals, and their violent dogmatism concerning the source of all things, had early conducted me to materialism. This materialism is a seducing system to young and superficial minds. It allows its disciples to talk, and dispenses them from thinking. But I was discontented with such a view of things as it afforded; man is a being of high aspirations, "looking both before and after," whose "thoughts wander through eternity,"[4] disclaiming alliance with transience and decay; incapable of imagining to himself annihilation; existing but in the future and the past; being, not what he is, but what he has been and shall be. Whatever may be his true and final destination, there is a spirit within him at enmity with nothingness and dissolution. This is the character of all life and being. Each is at once the centre and the circumference; the point to which all things are referred, and the line in which all things are contained. Such contemplations as these, materialism and the popular philosophy of mind and matter alike forbid; they are only consistent with the intellectual system.

It is absurd to enter into a long recapitulation of arguments sufficiently familiar to those inquiring minds, whom alone a writer on abstruse subjects can be conceived to address. Perhaps the most clear and vigorous statement of the intellectual system is to be found in Sir William Drummond's *Academical Questions*. After such an exposition, it would be idle to translate into other words what could only lose energy and fitness by the change. Examined point by point, and word by word, the most discriminating intellects have been able to discern no train of thoughts in the process of reasoning, which does not conduct inevitably to the conclusion which has been stated.

What follows from the admission? It establishes no new truth, it gives us no additional insight into our hidden nature, neither its action nor itself. Philosophy, impatient as it may be to build, has much work yet remaining, as pioneer for the overgrowth of ages. It makes one step towards this object; it destroys error, and the roots of error. It leaves, what is too often the duty of the reformer in political and ethical questions to leave, a vacancy. It reduces the mind to that freedom in which it would have acted, but for the misuse of words and signs, the instruments of its own creation. By signs, I would be understood in a wide sense, including what is properly meant by that term, and what I peculiarly mean. In this latter sense, almost all familiar objects are signs, standing, not for themselves, but for others, in their capacity of suggesting one thought which shall lead to a train of thoughts. Our whole life is thus an education of error.

Let us recollect our sensations as children. What a distinct and intense apprehension had we of the world and of ourselves! Many of the circumstances of social life were then important to us which are now no longer so. But that is not the point of comparison on which I mean to insist. We less habitually distinguished all that we saw and felt, from ourselves. They seemed as it were to constitute one mass. There are some persons who, in this respect, are always children. Those who are subject to the state called reverie, feel as if their nature were dissolved into the surrounding universe, or as if the surrounding universe were absorbed into their being. They are conscious of no distinction. And these are states which precede, or accompany, or follow an unusually intense and vivid apprehension

2. **philosophers:** that is, Berkeley. 3. **"such . . . of":** *The Tempest*, IV. i. 156–57. 4. **"looking . . . eternity":** *Hamlet*, IV. iv. 37; *Paradise Lost*, II. 148.

of life. As men grow up this power commonly decays and they become mechanical and habitual agents. Thus feelings and then reasonings are the combined result of a multitude of entangled thoughts, and of a series of what are called impressions, planted by reiteration.

The view of life presented by the most refined deductions of the intellectual philosophy, is that of unity. Nothing exists but as it is perceived. The difference is merely nominal between those two classes of thought, which are vulgarly distinguished by the names of ideas and of external objects. Pursuing the same thread of reasoning, the existence of distinct individual minds, similar to that which is employed in now questioning its own nature, is likewise found to be a delusion. The words *I*, *you*, *they*, are not signs of any actual difference subsisting between the assemblage of thoughts this indicated, but are merely marks employed to denote the different modifications of the one mind.

Let it not be supposed that this doctrine conducts to the monstrous presumption that I, the person who now write and think, am that one mind. I am but a portion of it. The words *I* and *you*, and *they* are grammatical devices invented simply for arrangement, and totally devoid of the intense and exclusive sense usually attached to them. It is difficult to find terms adequate to express so subtle a conception as that to which the Intellectual Philosophy has conducted us. We are on that verge where words abandon us, and what wonder if we grow dizzy to look down the dark abyss of how little we know.

The relations of *things*, remain unchanged, by whatever system. By the word *things* is to be understood any object of thought, that is, any thought upon which any other thought is employed, with an apprehension of distinction. The relations of these remain unchanged; and such is the material of our knowledge.

What is the cause of life? that is, how was it produced, or what agencies distinct from life have acted or act upon life? All recorded generations of mankind have wearily busied themselves in inventing answers to this question; and the result has been,—Religion. Yet, that the basis of all things cannot be, as the popular philosophy alleges, mind, is sufficiently evident. Mind, as far as we have any experience of its properties, and beyond that experience how vain is argument! cannot create, it can only perceive. It is said also to be the cause. But cause is only a word expressing a certain state of the human mind with regard to the manner in which

two thoughts are apprehended to be related to each other. If any one desires to know how unsatisfactorily the popular philosophy employs itself upon this great question, they need only impartially reflect upon the manner in which thoughts develop themselves in their minds. It is infinitely improbable that the cause of mind, that is, of existence, is similar to mind.

ON LOVE

This fragment was written sometime between 1814 and 1819, and first published in 1829.

What is love? Ask him who lives, what is life? ask him who adores, what is God?

I know not the internal constitution of other men, nor even thine, whom I now address. I see that in some external attributes they resemble me, but when, misled by that appearance, I have thought to appeal to something in common, and unburthen my inmost soul to them, I have found my language misunderstood, like one in a distant and savage land. The more opportunities they have afforded me for experience, the wider has appeared the interval between us, and to a greater distance have the points of sympathy been withdrawn. With a spirit ill fitted to sustain such proof, trembling and feeble through its tenderness, I have everywhere sought sympathy, and have found only repulse and disappointment.

Thou demandest what is love? It is that powerful attraction towards all that we conceive, or fear, or hope beyond ourselves, when we find within our own thoughts the chasm of an insufficient void, and seek to awaken in all things that are, a community with what we experience within ourselves. If we reason, we would be understood; if we imagine, we would that the airy children of our brain were born anew within another's; if we feel, we would that another's nerves should vibrate to our own, that the beams of their eyes should kindle at once and mix and melt into our own, that lips of motionless ice should not reply to lips quivering and burning with the heart's best blood. This is Love. This is the bond and the sanction which connects not only man with man, but with every thing which exists. We are born into the world, and there is something within us which, from the instant that we live, more and more thirsts after its likeness. It is probably in correspondence with this law that the infant drains milk from the bosom of its mother; this propensity develops itself with the development of our nature. We dimly see within our intellectual nature a miniature as it were of our entire self, yet deprived of all that we condemn or depise, the ideal prototype of every thing excellent or lovely that we are capable of conceiving as belonging to the nature of man. Not only the portrait of our external being, but an assemblage of the minutest particles of which our nature is composed[1]; a mirror whose surface reflects only the forms of purity and brightness; a soul within our soul that describes a circle around its proper paradise, which pain, and sorrow, and evil dare not overleap. To this we eagerly refer all sensations, thirsting that they should resemble or correspond with it. The discovery of its antitype; the meeting with an understanding capable of clearly estimating our own; an imagination which should enter into and seize upon the subtle and delicate peculiarities which we have delighted to cherish and unfold in secret; with a frame whose nerves, like the chords of two exquisite lyres, strung to the accompaniment of one delightful voice, vibrate with the vibrations of our own; and of a combination of all these in such proportion as the type within demands; this is the invisible and unattainable point to which Love tends; and to attain which, it urges forth the powers of man to arrest the faintest shadow of that, without the possession of which there is no rest nor respite to the heart over which it rules. Hence in solitude, or in that deserted state when we are surrounded by human beings, and yet they sympathise not with us, we love the flowers, the grass, and the waters, and the sky. In the motion of the very leaves of spring, in the blue air, there is then found a secret correspondence with our heart. There is eloquence in the tongueless wind, and a melody in the flowing brooks and the rustling of the reeds beside them, which by their inconceivable relation to something within the soul, awaken the spirits to a dance of breathless rapture, and bring tears of mysterious tenderness to the eyes, like the enthusiasm of patriotic success, or the voice of one beloved singing to you alone. Sterne says that, if he were in a desert,

he would love some cypress. So soon as this want or power is dead, man becomes the living sepulchre of himself, and what yet survives is the mere husk of what once he was.

from ESSAY ON CHRISTIANITY

Composed sometime between 1813–19, the essay was first published in 1859.

. . . Whosoever is free from the contamination of luxury and licence may go forth to the fields and to the woods, inhaling joyous renovation from the breath of Spring, or catching from the odours and the sounds of Autumn some diviner mood of sweetest sadness, which improves the [solitary] heart. Whosoever is no deceiver or destroyer of his fellowmen, no liar, no flatterer, no murderer, may walk among his species, deriving from the communion with all which they contain of beautiful or of majestic, some intercourse with the Universal God. Whoever has maintained with his own heart the strictest correspondence of confidence, who dares to examine and to estimate every imagination which suggests itself to his mind, who is that which he designs to become, and only aspires to that which the divinity of his own nature shall consider and approve—he, has already seen God. We live and move and think, but we are not the creators of our own origin and existence, we are not the arbiters of every motion of our own complicated nature; we are not the masters of our own imaginations and moods of mental being. There is a Power by which we are surrounded, like the atmosphere in which some motionless lyre is suspended, which visits with its breath our silent chords, at will. Our most imperial and stupendous qualities—those on which the majesty and the power of humanity is erected—are, relatively to the inferior portion of its mechanism, indeed active and imperial; but they are the passive slaves of some higher and more omnipresent Power. This Power is God. And those who have seen God, have, in the period of their purer and more perfect nature, been harmonized by their own will to so exquisite [a] consentaneity of powers as to give forth divinest melody when the breath of universal being sweeps over their frame.

ON LOVE. **1. composed:** At this point Shelley noted, "These words are ineffectual and metaphorical. Most words are so— No help!"

A DEFENCE OF POETRY

୰

Composed in 1821 in reply to Thomas Love Peacock's *The Four Ages of Poetry* (see above, p. 758), Shelley's *Defence* was not published until 1840. Peacock had argued—only half-seriously—that poetry is the natural expression of men in primitive ages, and that it loses relevance or value in more advanced societies. Hence Shelley was mainly concerned to explain the moral (and thus the social) function of poetry. In doing so, he produced one of the most penetrating general discussions of poetry that we have.

୰

According to one mode of regarding those two classes of mental action, which are called reason and imagination, the former may be considered as mind contemplating the relations borne by one thought to another, however produced; and the latter, as mind acting upon those thoughts so as to colour them with its own light, and composing from them, as from elements, other thoughts, each containing within itself the principle of its own integrity. The one is the τὸ ποιεῖν, or the principle of synthesis, and has for its object those forms which are common to universal nature and existence itself; the other is the τὸ λογίζειν, or principle of analysis, and its action regards the relations of things, simply as relations; considering thoughts, not in their integral unity, but as the algebraical representations which conduct to certain general results. Reason is the enumeration of quantities already known; imagination is the perception of the value of those quantities, both separately and as a whole. Reason respects the differences, and imagination the similitudes of things. Reason is to imagination as the instrument to the agent, as the body to the spirit, as the shadow to the substance.

Poetry, in a general sense, may be defined to be "the expression of the imagination": and poetry is connate with the origin of man. Man is an instrument over which a series of external and internal impressions are driven, like the alternations of an ever-changing wind over an Æolian lyre, which move it by their motion to ever-changing melody. But there is a principle within the human being, and perhaps within all sentient beings, which acts otherwise than in the lyre, and produces not melody, alone, but harmony, by an internal adjustment of the sounds or motions thus excited to the impressions which excite them. It is as if the lyre could accommodate its chords to the motions of that which strikes them, in a determined proportion of sound; even as the musician can accommodate his voice to the sound of the lyre. A child at play by itself will express its delight by its voice and motions; and every inflection of tone and every gesture will bear exact relation to a corresponding antitype in the pleasurable impressions which awakened it; it will be the reflected image of that impression; and as the lyre trembles and sounds after the wind has died away, so the child seeks, by prolonging in its voice and motions the duration of the effect, to prolong also a consciousness of the cause. In relation to the objects which delight a child, these expressions are what poetry is to higher objects. The savage (for the savage is to ages what the child is to years) expresses the emotions produced in him by surrounding objects in a similar manner; and language and gesture, together with plastic or pictorial imitation, become the image of the combined effect of those objects and of his apprehension of them. Man in society, with all his passions and his pleasures, next becomes the object of the passions and pleasures of man; an additional class of emotions produces an augmented treasure of expressions; and language, gesture, and the imitative arts become at once the representation and the medium, the pencil and the picture, the chisel and the statue, the chord and the harmony. The social sympathies, or those laws from which, as from its elements, society results, begin to develop themselves from the moment that two human beings co-exist; the future is contained within the present as the plant within the seed; and equality, diversity, unity, contrast, mutual dependence, become the principles alone capable of affording the motives according to which the will of a social being is determined to action, inasmuch as he is social; and constitute pleasure in sensation, virtue in sentiment, beauty in art, truth in reasoning, and love in the intercourse of kind. Hence men, even in the infancy of society, observe a certain order in their words and actions, distinct from that of the objects and the impressions represented by them, all expression being subject to the laws of that from which it proceeds. But let us dismiss those more general considerations which might involve an inquiry into the principles of society itself, and restrict our view to the manner in which the imagination is expressed upon its forms.

In the youth of the world, men dance and sing and

imitate natural objects, observing in these actions, as in all others, a certain rhythm or order. And, although all men observe a similar, they observe not the same order in the motions of the dance, in the melody of the song, in the combinations of language, in the series of their imitations of natural objects. For there is a certain order or rhythm belonging to each of these classes of mimetic representation, from which the hearer and the spectator receive an intenser and purer pleasure than from any other: the sense of an approximation to this order has been called taste by modern writers. Every man in the infancy of art, observes an order which approximates more or less closely to that from which this highest delight results: but the diversity is not sufficiently marked, as that its gradations should be sensible, except in those instances where the predominance of this faculty of approximation to the beautiful (for so we may be permitted to name the relation between this highest pleasure and its cause) is very great. Those in whom it exists in excess are poets, in the most universal sense of the word; and the pleasure resulting from the manner in which they express the influence of society or nature upon their own minds, communicates itself to others, and gathers a sort of reduplication from that community. Their language is vitally metaphorical; that is, it marks the before unapprehended relations of things and perpetuates their apprehension, until the words which represent them become, through time, signs for portions or classes of thought instead of pictures of integral thoughts; and then, if no new poets should arise to create afresh the associations which have been thus disorganized, language will be dead to all the nobler purposes of human intercourse. These similitudes or relations are finely said by Lord Bacon to be "the same footsteps of nature impressed upon the various subjects of the world"[1]—and he considers the faculty which perceives them as the storehouse of axioms common to all knowledge. In the infancy of society every author is necessarily a poet, because language itself is poetry; and to be a poet is to apprehend the true and the beautiful, in a word, the good which exists in the relation subsisting, first between existence and perception, and secondly between perception and expression. Every original language near to its source is in itself the chaos of a cyclic poem: the copiousness of lexicography and the distinctions of grammar are

the works of a later age, and are merely the catalogue and the form of the creations of poetry.

But poets, or those who imagine and express this indestructible order, are not only the authors of language and of music, of the dance, and architecture, and statuary, and painting; they are the institutors of laws, and the founders of civil society, and the inventors of the arts of life, and the teachers, who draw into a certain propinquity with the beautiful and the true, that partial apprehension of the agencies of the invisible world which is called religion. Hence all original religions are allegorical, or susceptible of allegory, and, like Janus, have a double face of false and true. Poets, according to the circumstances of the age and nation in which they appeared, were called, in the earlier epochs of the world, legislators or prophets; a poet essentially comprises and unites both these characters. For he not only beholds intensely the present as it is, and discovers those laws according to which present things ought to be ordered, but he beholds the future in the present, and his thoughts are the germs of the flower and the fruit of latest time. Not that I assert poets to be prophets in the gross sense of the word, or that they can foretell the form as surely as they foreknow the spirit of events: such is the pretence of superstition, which would make poetry an attribute of prophecy, rather than prophecy an attribute of poetry. A poet participates in the eternal, the infinite, and the one; as far as relates to his conceptions, time and place and number are not. The grammatical forms which express the moods of time, and the difference of persons, and the distinction of place, are convertible with respect to the highest poetry without injuring it as poetry; and the choruses of Æschylus, and the Book of Job, and Dante's Paradise, would afford, more than any other writings, examples of this fact, if the limits of this essay did not forbid citation. The creations of sculpture, painting, and music are illustrations still more decisive.

Language, colour, form, and religious and civil habits of action, are all the instruments and materials of poetry;[2] they may be called poetry by that figure of speech which considers the effect as a synonym of the cause. But poetry in a more restricted sense expresses those arrangements of language, and especially metrical

A DEFENCE OF POETRY. I. "the . . . world": "De Augment. Scient., cap. i, lib. iii" (S.).

2. poetry: poetry in the more general and traditional sense ("Poesy"; cf. Coleridge's essay "On Poesy or Art," above): that is, the creative arts, and the use of the mind in a way ("making" or "shaping") analogous to that of the creative artist.

language, which are created by that imperial faculty, whose throne is curtained within the invisible nature of man. And this springs from the nature itself of language, which is a more direct representation of the actions and passions of our internal being, and is susceptible of more various and delicate combinations, than colour, form, or motion, and is more plastic and obedient to the control of that faculty of which it is a creation. For language is arbitrarily produced by the imagination, and has relation to thoughts alone; but all other materials, instruments, and conditions of art, have relations among each other, which limit and interpose between conception and expression. The former is as a mirror which reflects, the latter as a cloud which enfeebles, the light of which both are mediums of communication. Hence the fame of sculptors, painters, and musicians, although the intrinsic powers of the great masters of these arts may yield in no degree to that of those who have employed language as the hieroglyphic of their thoughts, has never equalled that of poets in the restricted sense of the term; as two performers of equal skill will produce unequal effects from a guitar and a harp. The fame of legislators and founders of religions, so long as their institutions last, alone seems to exceed that of poets in the restricted sense; but it can scarcely be a question, whether, if we deduct the celebrity which their flattery of the gross opinions of the vulgar usually conciliates, together with[3] that which belonged to them in their higher character of poets, any excess will remain.

We have thus circumscribed the word poetry within the limits of that art which is the most familiar and the most perfect expression of the faculty itself. It is necessary, however, to make the circle still narrower, and to determine the distinction between measured and unmeasured language; for the popular division into prose and verse is inadmissible in accurate philosophy.

Sounds as well as thoughts have relation both between each other and towards that which they represent, and a perception of the order of those relations has always been found connected with the perception of the order of the relations of thoughts. Hence the language of poets has ever affected a certain uniform and harmonious recurrence of sound, without which it were not poetry, and which is scarcely less indispensable to the communication of its influence, than the words themselves without reference to that peculiar

order. Hence the vanity of translation; it were as wise to cast a violet into a crucible that you might discover the formal principle of its colour and odour, as to seek to transfuse from one language into another the creations of a poet. The plant must spring again from its seed, or it will bear no flower—and this is the burthen of the curse of Babel.[4]

An observation of the regular mode of the recurrence of harmony in the language of poetical minds, together with its relation to music, produced metre, or a certain system of traditional forms of harmony of language. Yet it is by no means essential that a poet should accommodate his language to this traditional form, so that the harmony, which is its spirit, be observed. The practice is indeed convenient and popular, and to be preferred, especially in such composition as includes much action: but every great poet must inevitably innovate upon the example of his predecessors in the exact structure of his peculiar versification. The distinction between poets and prose writers is a vulgar error.[5] The distinction between philosophers and poets has been anticipated. Plato was essentially a poet—the truth and splendour of his imagery, and the melody of his language, are the most intense that it is possible to conceive. He rejected the measure of the epic, dramatic, and lyrical forms, because he sought to kindle a harmony in thoughts divested of shape and action, and he forebore to invent any regular plan of rhythm which would include, under determinate forms, the varied pauses of his style. Cicero sought to imitate the cadence of his periods, but with little success. Lord Bacon was a poet. His language has a sweet and majestic rhythm, which satisfies the sense, no less than the almost superhuman wisdom of his philosophy satisfies the intellect; it is a strain which distends, and then bursts the circumference of the reader's mind, and pours itself forth together with it into the universal element with which it has perpetual sympathy. All the authors of revolutions in opinion are not only necessarily poets as they are inventors, nor even as their words unveil the permanent analogy of things by images which participate in the life of truth; but as their periods are harmonious and rhythmical, and contain in themselves the elements of verse; being the echo of the eternal music. Nor are those supreme poets, who have employed traditional forms of rhythm

---

3. **together with:** that is, from.

4. **curse of Babel:** See Genesis 11 : 6–9. 5. **The . . . error:** Cf. Wordsworth's similar protest in his Preface to the *Lyrical Ballads*, above.

on account of the form and action of their subjects, less capable of perceiving and teaching the truth of things, than those who have omitted that form. Shakespeare, Dante, and Milton (to confine ourselves to modern writers) are philosophers of the very loftiest power.

A poem is the very image of life expressed in its eternal truth. There is this difference between a story and a poem, that a story is a catalogue of detached facts, which have no other connection than time, place, circumstance, cause and effect; the other is the creation of actions according to the unchangeable forms of human nature, as existing in the mind of the creator, which is itself the image of all other minds.[6] The one is partial, and applies only to a definite period of time, and a certain combination of events which can never again recur; the other is universal, and contains within itself the germ of a relation to whatever motives or actions have place in the possible varieties of human nature. Time, which destroys the beauty and the use of the story of particular facts, stripped of the poetry which should invest them, augments that of poetry, and for ever develops new and wonderful applications of the eternal truth which it contains. Hence epitomes have been called the moths of just history; they eat out the poetry of it. A story of particular facts is as a mirror which obscures and distorts that which should be beautiful: poetry is a mirror which makes beautiful that which is distorted.

The parts of a composition may be poetical, without the composition as a whole being a poem. A single sentence may be considered as a whole, though it may be found in the midst of a series of unassimilated portions; a single word even may be a spark of inextinguishable thought. And thus all the great historians, Herodotus, Plutarch, Livy, were poets; and although the plan of these writers, especially that of Livy, restrained them from developing this faculty in its highest degree, they made copious and ample amends for their subjection, by filling all the interstices of their subjects with living images.

Having determined what is poetry, and who are poets, let us proceed to estimate its effects upon society.

Poetry is ever accompanied with pleasure: all spirits on which it falls open themselves to receive the wisdom which is mingled with its delight. In the infancy of the world, neither poets themselves nor their auditors are fully aware of the excellence of poetry, for it acts in a divine and unapprehended manner, beyond and above consciousness; and it is reserved for future generations to contemplate and measure the mighty cause and effect in all the strength and splendour of their union. Even in modern times, no living poet ever arrived at the fulness of his fame; the jury which sits in judgment upon a poet, belonging as he does to all time, must be composed of his peers; it must be impanelled by Time from the selectest of the wise of many generations. A poet is a nightingale, who sits in darkness and sings to cheer its own solitude with sweet sounds; his auditors are as men entranced by the melody of an unseen musician, who feel that they are moved and softened, yet know not whence or why. The poems of Homer and his contemporaries were the delight of infant Greece; they were the elements of that social system which is the column upon which all succeeding civilization has reposed. Homer embodied the ideal perfection of his age in human character; nor can we doubt that those who read his verses were awakened to an ambition of becoming like to Achilles, Hector, and Ulysses; the truth and beauty of friendship, patriotism, and persevering devotion to an object, were unveiled to their depths in these immortal creations; the sentiments of the auditors must have been refined and enlarged by a sympathy with such great and lovely impersonations, until from admiring they imitated, and from imitation they identified themselves with the objects of their admiration.[7] Nor let it be objected that these characters are remote from moral perfection, and that they are by no means to be considered as edifying patterns for general imitation. Every epoch, under names more or less specious, has deified its peculiar errors; Revenge is the naked idol of the worship of a semi-barbarous age; and Self-deceit is the veiled image of unknown evil, before which luxury and satiety lie prostrate. But a

---

6. **difference . . . minds:** Shelley echoes Aristotle's premise (*Poetics*, 9. 3) that poetry "is a more philosophical and a higher thing than history. For poetry tends to express the universal, and history the particular." A literal, routine account of what actually happens will simply give the facts as they occur, without selecting and rearranging them so as to disclose the universal form or meaning. A creative art will seek the underlying pattern, choosing only those details that are directly relevant to its insight and freely arranging them so as to usher in the form or meaning it has descried.

7. **imitation . . . admiration:** Cf. Coleridge's remark ("On Poesy or Art," below) that "we unconsciously imitate those we love."

poet considers the vices of his contemporaries as a temporary dress in which his creations must be arrayed, and which cover without concealing the eternal proportions of their beauty. An epic or dramatic personage is understood to wear them around his soul, as he may the ancient armour or the modern uniform around his body; whilst it is easy to conceive a dress more graceful than either. The beauty of the internal nature cannot be so far concealed by its accidental vesture, but that the spirit of its form shall communicate itself to the very disguise, and indicate the shape it hides from the manner in which it is worn. A majestic form and graceful motions will express themselves through the most barbarous and tasteless costume. Few poets of the highest class have chosen to exhibit the beauty of their conceptions in its naked truth and splendour; and it is doubtful whether the alloy of costume, habit, etc., be not necessary to temper this planetary music for mortal ears.

The whole objection, however, of the immorality of poetry rests upon a misconception of the manner in which poetry acts to produce the moral improvement of man. Ethical science arranges the elements which poetry has created, and propounds schemes and proposes examples of civil and domestic life: nor is it for want of admirable doctrines that men hate, and despise, and censure, and deceive, and subjugate one another. But poetry acts in another and diviner manner. It awakens and enlarges the mind itself by rendering it the receptacle of a thousand unapprehended combinations of thought. Poetry lifts the veil from the hidden beauty of the world, and makes familiar objects be as if they were not familiar; it reproduces all that it represents, and the impersonations clothed in its Elysian light stand thenceforward in the minds of those who have once contemplated them, as memorials of that gentle and exalted content which extends itself over all thoughts and actions with which it co-exists. The great secret of morals is love; or a going out of our own nature, and an identification of ourselves with the beautiful which exists in thought, action, or person, not our own. A man, to be greatly good, must imagine intensely and comprehensively; he must put himself in the place of another and of many others; the pains and pleasures of his species must become his own. The great instrument of moral good is the imagination; and poetry administers to the effect by acting upon the cause. Poetry enlarges the circumference of the imagination by replenishing it with thoughts of ever new delight, which have the power of attracting

and assimilating to their own nature all other thoughts, and which form new intervals and interstices whose void forever craves fresh food. Poetry strengthens the faculty which is the organ of the moral nature of man, in the same manner as exercise strengthens a limb.[8] A poet therefore would do ill to embody his own conceptions of right and wrong, which are usually those of his place and time, in his poetical creations, which participate in neither. By this assumption of the inferior office of interpreting the effect, in which perhaps after all he might acquit himself but imperfectly, he would resign a glory in the participation of the cause. There was little danger that Homer, or any of the eternal poets, should have so far misunderstood themselves as to have abdicated this throne of their widest dominion. Those in whom the poetical faculty, though great, is less intense, as Euripides, Lucan, Tasso, Spenser, have frequently affected a moral aim, and the effect of their poetry is diminished in exact proportion to the degree in which they compel us to advert to this purpose.

Homer and the cyclic poets were followed at a certain interval by the dramatic and lyrical poets of Athens, who flourished contemporaneously with all that is most perfect in the kindred expressions of the poetical faculty; architecture, painting, music, the dance, sculpture, philosophy, and, we may add, the forms of civil life. For although the scheme of Athenian society was deformed by many imperfections which the poetry existing in chivalry and Christianity has erased from the habits and institutions of modern Europe; yet never at any other period has so much energy, beauty, and virtue, been developed; never was blind strength and stubborn form so disciplined and rendered subject to the will of man, or that will less repugnant to the dictates of the beautiful and the true, as during the century which preceded the death of Socrates. Of no other epoch in the history of our species have we records and fragments stamped so visibly with the image of the divinity in man. But it is poetry alone, in form, in action, or in language, which has rendered this epoch memorable above all others, and the storehouse of examples to everlasting time. For written poetry existed at that epoch simultaneously with the other arts, and it is an idle inquiry to demand

8. **Poetry . . . limb:** Cf. the selections, above, from Hazlitt's *Essay on the Principles of Human Action*, on the sympathetic nature of the imagination, and his remark in an essay on *Lear* that "Whoever, therefore, has a contempt for poetry, has a contempt for himself and for humanity."

which gave and which received the light, which all, as from a common focus, have scattered over the darkest periods of succeeding time. We know no more of cause and effect than a constant conjunction of events: poetry is ever found to co-exist with whatever other arts contribute to the happiness and perfection of man. I appeal to what has already been established to distinguish between the cause and the effect.

It was at the period here adverted to, that the drama had its birth; and however a succeeding writer may have equalled or surpassed those few great specimens of the Athenian drama which have been preserved to us, it is indisputable that the art itself never was understood or practised according to the true philosophy of it, as at Athens. For the Athenians employed language, action, music, painting, the dance, and religious institutions, to produce a common effect in the representation of the highest idealisms of passion and of power; each division in the art was made perfect in its kind by artists of the most consummate skill, and was disciplined into a beautiful proportion and unity one towards the other. On the modern stage a few only of the elements capable of expressing the image of the poet's conception are employed at once. We have tragedy without music and dancing; and music and dancing without the highest impersonations of which they are the fit accompaniment, and both without religion and solemnity. Religious institution has indeed been usually banished from the stage. Our system of divesting the actor's face of a mask, on which the many expressions appropriated to his dramatic character might be molded into one permanent and unchanging expression, is favorable only to a partial and inharmonious effect; it is fit for nothing but a monologue, where all the attention may be directed to some great master of ideal mimicry. The modern practice of blending comedy with tragedy, though liable to great abuse in point of practice, is undoubtedly an extension of the dramatic circle; but the comedy should be as in *King Lear*, universal, ideal, and sublime. It is perhaps the intervention of this principle which determines the balance in favor of *King Lear* against the *Oedipus Tyrannus* or the *Agamemnon*, or, if you will, the trilogies with which they are connected; unless the intense power of the choral poetry, especially that of the latter, should be considered as restoring the equilibrium. *King Lear*, if it can sustain this comparison, may be judged to be the most perfect specimen of the dramatic art existing in the world; in spite of the narrow conditions to which the poet was subjected by the ignorance

of the philosophy of the drama which has prevailed in modern Europe. Calderon, in his religious *Autos*, has attempted to fulfil some of the high conditions of dramatic representation neglected by Shakespeare; such as the establishing a relation between the drama and religion, and the accommodating them to music and dancing; but he omits the observation of conditions still more important, and more is lost than gained by a substitution of the rigidly defined and ever-repeated idealisms of a distorted superstition for the living impersonations of the truth of human passion.

But I digress.—The connexion of scenic exhibitions with the improvement or corruption of the manners of men, has been universally recognized: in other words, the presence or absence of poetry in its most perfect and universal form, has been found to be connected with good and evil in conduct or habit. The corruption which has been imputed to the drama as an effect, begins, when the poetry employed in its constitution ends: I appeal to the history of manners whether the periods of the growth of the one and the decline of the other have not corresponded with an exactness equal to any example of moral cause and effect.

The drama at Athens, or wheresoever else it may have approached to its perfection, ever co-existed with the moral and intellectual greatness of the age. The tragedies of the Athenian poets are as mirrors in which the spectator beholds himself, under a thin disguise of circumstance, stripped of all but that ideal perfection and energy which every one feels to be the internal type of all that he loves, admires, and would become. The imagination is enlarged by a sympathy with pains and passions so mighty, that they distend in their conception the capacity of that by which they are conceived; the good affections are strengthened by pity, indignation, terror, and sorrow; and an exalted calm is prolonged from the satiety of this high exercise of them into the tumult of familiar life:[9] even crime is disarmed of half its horror and all its contagion by being represented as the fatal consequence of the unfathomable agencies of nature; error is thus divested of its wilfulness; men can no longer cherish it as the creation of their choice. In a drama of the highest order there is little food for

9. The . . . life: Shelley here repeats the Aristotelian idea of *katharsis* (*Poetics*, 9. 2). The tragic drama, says Aristotle, effects "through pity and fear . . . a proper purgation of these emotions." Though sympathetic identification our emotions can be carried outside ourselves, given perspective, and "purged" of the subjective and self-centered elements.

censure or hatred; it teaches rather self-knowledge and self-respect. Neither the eye nor the mind can see itself, unless reflected upon that which it resembles. The drama, so long as it continues to express poetry, is a prismatic and many-sided mirror, which collects the brightest rays of human nature and divides and reproduces them from the simplicity of these elementary forms, and touches them with majesty and beauty, and multiplies all that it reflects, and endows it with the power of propagating its like wherever it may fall.

But in periods of the decay of social life, the drama sympathizes with that decay. Tragedy becomes a cold imitation of the form of the great masterpieces of antiquity, divested of all harmonious accompaniment of the kindred arts; and often the very form misunderstood, or a weak attempt to teach certain doctrines, which the writer considers as moral truths; and which are usually no more than specious flatteries of some gross vice or weakness, with which the author, in common with his auditors, are infected. Hence what has been called the classical and domestic drama. Addison's *Cato* is a specimen of the one; and would it were not superfluous to cite examples of the other! To such purposes poetry cannot be made subservient. Poetry is a sword of lightning, ever unsheathed, which consumes the scabbard that would contain it.[10] And thus we observe that all dramatic writings of this nature are unimaginative in a singular degree; they affect sentiment and passion, which, divested of imagination, are other names for caprice and appetite. The period in our own history of the grossest degradation of the drama is the reign of Charles II, when all forms in which poetry had been accustomed to be expressed became hymns to the triumph of kingly power over liberty and virtue. Milton stood alone illuminating an age unworthy of him. At such periods the calculating principle pervades all the forms of dramatic exhibition, and poetry ceases to be expressed upon them. Comedy loses its ideal universality: wit succeeds to humor; we laugh from self-complacency and triumph, instead of pleasure; malignity, sarcasm, and contempt, succeed to sympathetic merriment; we hardly laugh, but we smile. Obscenity, which is ever blasphemy against the divine beauty in life, becomes, from the very veil which it assumes, more active if less disgusting: it is a monster for which the corruption of society for ever brings forth new food, which it devours in secret.

The drama being that form under which a greater number of modes of expression of poetry are susceptible of being combined than any other, the connexion of poetry and social good is more observable in the drama than in whatever other form. And it is indisputable that the highest perfection of human society has ever corresponded with the highest dramatic excellence; and that the corruption or the extinction of the drama in a nation where it has once flourished, is a mark of a corruption of manners, and an extinction of the energies which sustain the soul of social life. But, as Machiavelli says of political institutions, that life may be preserved and renewed, if men should arise capable of bringing back the drama to its principles. And this is true with respect to poetry in its most extended sense: all language, institution and form, require not only to be produced but to be sustained: the office and character of a poet participates in the divine nature as regards providence, no less than as regards creation.

Civil war, the spoils of Asia, and the fatal predominance first of the Macedonian, and then of the Roman arms, were so many symbols of the extinction or suspension of the creative faculty in Greece. The bucolic writers,[11] who found patronage under the lettered tyrants of Sicily and Egypt, were the latest representatives of its most glorious reign. Their poetry is intensely melodious; like the odor of the tuberose, it overcomes and sickens the spirit with excess of sweetness; whilst the poetry of the preceding age was as a meadow-gale of June, which mingles the fragrance of all the flowers of the field, and adds a quickening and harmonizing spirit of its own, which endows the sense with a power of sustaining its extreme delight. The bucolic and erotic delicacy in written poetry is correlative with that softness in statuary, music, and the kindred arts, and even in manners and institutions, which distinguished the epoch to which I now refer. Nor is it the poetical faculty itself, or any misapplication of it, to which this want of harmony is to be imputed. An equal sensibility to the influence of the senses and the affections is to be found in the writings of Homer and Sophocles: the former, especially, has clothed sensual and pathetic images with irresistible attraction. Their superiority over these succeeding writers consists in the presence of those thoughts which belong to the inner faculties of our nature, not in the absence of those which are connected with the external: their incomparable perfection consists in a harmony of the union of all. It is

---

10. Poetry . . . it. Cf. *Adonais*, above, l. 178.

11. bucolic writers: The Greek bucolic or pastoral poets, especially Theocritus and Bion.

not what the erotic poets have, but what they have not, in which their imperfection consists. It is not inasmuch as they were poets, but inasmuch as they were not poets, that they can be considered with any plausibility as connected with the corruption of their age. Had that corruption availed so as to extinguish in them the sensibility to pleasure, passion, and natural scenery, which is imputed to them as an imperfection, the last triumph of evil would have been achieved. For the end of social corruption is to destroy all sensibility to pleasure; and, therefore, it is corruption. It begins at the imagination and the intellect as at the core, and distributes itself thence as a paralysing venom, through the affections into the very appetites, until all become a torpid mass in which hardly sense survives. At the approach of such a period, poetry ever addresses itself to those faculties which are the last to be destroyed, and its voice is heard, like the footsteps of Astræa,[12] departing from the world. Poetry ever communicates all the pleasure which men are capable of receiving: it is ever still the light of life; the source of whatever of beautiful or generous or true can have place in an evil time. It will readily be confessed that those among the luxurious citizens of Syracuse and Alexandria, who were delighted with the poems of Theocritus, were less cold, cruel, and sensual than the remnant of their tribe. But corruption must utterly have destroyed the fabric of human society before poetry can ever cease. The sacred links of that chain have never been entirely disjoined, which descending through the minds of many men is attached to those great minds, whence as from a magnet the invisible effluence is sent forth, which at once connects, animates, and sustains the life of all.[13] It is the faculty which contains within itself the seeds at once of its own and of social renovation. And let us not circumscribe the effects of the bucolic and erotic poetry within the limits of the sensibility of those to whom it was addressed. They may have perceived the beauty of those immortal compositions, simply as fragments and isolated portions: those who are more finely organized, or born in a happier age, may recognize them as episodes to that great poem, which all poets, like the co-operating thoughts of one great mind, have built up since the beginning of the world.

The same revolutions within a narrower sphere had place in ancient Rome; but the actions and forms of its social life never seem to have been perfectly saturated with the poetical element. The Romans appear to have considered the Greeks as the selectest treasuries of the selectest forms of manners and of nature, and to have abstained from creating in measured language, sculpture, music, or architecture, anything which might bear a particular relation to their own condition, whilst it should bear a general one to the universal constitution of the world. But we judge from partial evidence, and we judge perhaps partially. Ennius, Varro, Pacuvius, and Accius, all great poets, have been lost. Lucretius is in the highest, and Virgil in a very high sense, a creator. The chosen delicacy of the expressions of the latter, are as a mist of light which conceal from us the intense and exceeding truth of his conceptions of nature. Livy is instinct with poetry. Yet Horace, Catullus, Ovid, and generally the other great writers of the Virgilian age, saw man and nature in the mirror of Greece. The institutions also, and the religion of Rome were less poetical than those of Greece, as the shadow is less vivid than the substance. Hence poetry in Rome, seemed to follow, rather than accompany, the perfection of political and domestic society. The true poetry of Rome lived in its institutions; for whatever of beautiful, true, and majestic, they contained, could have sprung only from the faculty which creates the order in which they consist. The life of Camillus, the death of Regulus; the expectation of the senators, in their godlike state, of the victorious Gauls: the refusal of the republic to make peace with Hannibal, after the battle of Cannæ, were not the consequences of a refined calculation of the probable personal advantage to result from such a rhythm and order in the shows of life, to those who were at once the poets and the actors of these immortal dramas. The imagination beholding the beauty of this order, created it out of itself according to its own idea; the consequence was empire, and the reward everliving fame. These things are not the less poetry *quia carent vate sacro*.[14] They are the episodes of that cyclic poem written by Time upon the memories of men. The Past, like an inspired rhapsodist, fills the theatre of everlasting generations with their harmony.

At length the ancient system of religion and manners had fulfilled the circle of its revolutions. And the world would have fallen into utter anarchy and darkness, but that there were found poets among the authors

---

12. **Astræa:** goddess of Justice. 13. **whence . . . all:** Cf. Plato, *Ion*, 533–36.

14. **quia . . . sacro:** "because they lack a divinely inspired poet" (Horace, *Odes*, IV. ix. 28).

of the Christian and chivalric systems of manners and religion, who created forms of opinion and action never before conceived; which, copied into the imaginations of men, became as generals to the bewildered armies of their thoughts. It is foreign to the present purpose to touch upon the evil produced by these systems: except that we protest, on the ground of the principles already established, that no portion of it can be attributed to the poetry they contain.

It is probable that the poetry of Moses, Job, David, Solomon, and Isaiah, had produced a great effect upon the mind of Jesus and his disciples. The scattered fragments preserved to us by the biographers of this extraordinary person, are all instinct with the most vivid poetry. But his doctrines seem to have been quickly distorted. At a certain period after the prevalence of a system of opinions founded upon those promulgated by him, the three forms into which Plato had distributed the faculties of mind[15] underwent a sort of apotheosis, and became the object of the worship of the civilized world. Here it is to be confessed that "Light seems to thicken", and

> The crow makes wing to the rooky wood,
> Good things of day begin to droop and drowse,
> And night's black agents to their preys do rouse.[16]

But mark how beautiful an order has sprung from the dust and blood of this fierce chaos! how the world, as from a resurrection, balancing itself on the golden wings of knowledge and of hope, has reassumed its yet unwearied flight into the heaven of time. Listen to the music, unheard by outward ears, which is as a ceaseless and invisible wind, nourishing its everlasting course with strength and swiftness.

The poetry in the doctrines of Jesus Christ, and the mythology and institutions of the Celtic conquerors[17] of the Roman empire, outlived the darkness and the convulsions connected with their growth and victory, and blended themselves into a new fabric of manners and opinion. It is an error to impute the ignorance of the dark ages to the Christian doctrines or the predominance of the Celtic nations. Whatever of evil their agencies may have contained sprang from the extinction of the poetical principle, connected with the progress of despotism and superstition. Men, from causes too intricate to be here discussed, had become insensible and selfish: their own will had become feeble, and yet they were its slaves, and thence the slaves of the will of others: lust, fear, avarice, cruelty, and fraud, characterized a race amongst whom no one was to be found capable of *creating* in form, language, or institution. The moral anomalies of such a state of society are not justly to be charged upon any class of events immediately connected with them, and those events are most entitled to our approbation which could dissolve it most expeditiously. It is unfortunate for those who cannot distinguish words from thoughts, that many of these anomalies have been incorporated into our popular religion.

It was not until the eleventh century that the effects of the poetry of the Christian and chivalric systems began to manifest themselves. The principle of equality had been discovered and applied by Plato in his *Republic*,[18] as the theoretical rule of the mode in which the materials of pleasure and of power, produced by the common skill and labour of human beings, ought to be distributed among them. The limitations of this rule were asserted by him to be determined only by the sensibility of each, or the utility to result to all. Plato, following the doctrines of Timaeus and Pythagoras, taught also a moral and intellectual system of doctrine, comprehending at once the past, present, and the future condition of man. Jesus Christ divulged the sacred and eternal truths contained in these views to mankind, and Christianity, in its abstract purity, became the exoteric expression of the esoteric doctrines of the poetry and wisdom of antiquity. The incorporation of the Celtic nations with the exhausted population of the south, impressed upon it the figure of the poetry existing in their mythology and institutions. The result was a sum of the action and reaction of all the causes included in it; for it may be assumed as a maxim that no nation or religion can supersede any other without incorporating into itself a portion of that which it supersedes. The abolition of personal and

---

15. faculties of mind: Reason, the direct insight into the universal forms or "Ideas"; "passion or spirit" that, though itself distinct from reason, may aid it; and, thirdly, desire or appetite (*Republic*, IV. 435–44; cf. *Timaeus*, 69–71). Shelley's remark that this tripartite division has since become an object of worship suggests that he is thinking of the Christian Trinity. If so, the analogy is unapparent. **16. The . . . rouse:** *Macbeth*, III. ii. 50–53. **17. Celtic conquerors:** used in the Roman sense of northern Europeans generally.

18. The . . . Republic: The philosophical "guardians" or leaders of Plato's ideal commonwealth were to have no property of their own but to share in everything equally (*Republic*, Bk. III).

domestic slavery, and the emancipation of women from a great part of the degrading restraints of antiquity, were among the consequences of these events.

The abolition of personal slavery is the basis of the highest political hope that it can enter into the mind of man to conceive. The freedom of women produced the poetry of sexual love. Love became a religion, the idols of whose worship were ever present. It was as if the statues of Apollo and the Muses had been endowed with life and motion, and had walked forth among their worshippers; so that earth became peopled by the inhabitants of a diviner world. The familiar appearance and proceedings of life became wonderful and heavenly, and a paradise was created as out of the wrecks of Eden. And as this creation itself is poetry, so its creators were poets; and language was the instrument of their art: "Galeotto fù il libro, e chi lo scrisse."[19] The Provençal Trouveurs,[20] or inventors, preceded Petrarch, whose verses are as spells, which unseal the inmost enchanted fountains of the delight which is in the grief of love. It is impossible to feel them without becoming a portion of that beauty which we contemplate: it were superfluous to explain how the gentleness and the elevation of mind connected with these sacred emotions can render men more amiable, more generous and wise, and lift them out of the dull vapours of the little world of self. Dante understood the secret things of love even more than Petrarch. His *Vita Nuova* is an inexhaustible fountain of purity of sentiment and language: it is the idealized history of that period, and those intervals of his life which were dedicated to love. His apotheosis of Beatrice in Paradise, and the gradations of his own love and her loveliness, by which as by steps he feigns himself to have ascended to the throne of the Supreme Cause, is the most glorious imagination of modern poetry. The acutest critics have justly reversed the judgment of the vulgar, and the order of the great acts of the "Divine Drama," in the measure of the admiration which they accord to the Hell, Purgatory, and Paradise. The latter is a perpetual hymn of everlasting love. Love, which found a worthy poet in Plato alone of all the ancients, has been celebrated by a chorus of the greatest writers of the renovated world; and the music has penetrated the caverns of society, and its echoes still drown the dissonance of arms and superstition. At successive intervals, Ariosto, Tasso, Shakespeare, Spenser, Calderon, Rousseau, and the great writers of our own age, have celebrated the dominion of love, planting as it were trophies in the human mind of that sublimest victory over sensuality and force. The true relation borne to each other by the sexes into which human kind is distributed, has become less misunderstood; and if the error which confounded diversity with inequality of the powers of the two sexes has been partially recognized in the opinions and institutions of modern Europe, we owe this great benefit to the worship of which chivalry was the law, and poets the prophets.

The poetry of Dante may be considered as the bridge thrown over the stream of time, which unites the modern and ancient world. The distorted notions of invisible things which Dante and his rival Milton have idealized, are merely the mask and the mantle in which these great poets walk through eternity enveloped and disguised. It is a difficult question to determine how far they were conscious of the distinction which must have subsisted in their minds between their own creeds and that of the people. Dante at least appears to wish to mark the full extent of it by placing Riphaeus, whom Virgil calls *justissimus unus*,[21] in Paradise, and observing a most heretical caprice in his distribution of rewards and punishments. And Milton's poem contains within itself a philosophical refutation of that system, of which, by a strange and natural antithesis, it has been a chief popular support. Nothing can exceed the energy and magnificence of the character of Satan as expressed in *Paradise Lost*. It is a mistake to suppose that he could ever have been intended for the popular personification of evil. Implacable hate, patient cunning, and a sleepless refinement of device to inflict the extremest anguish on an enemy, these things are evil; and, although venial in a slave, are not to be forgiven in a tyrant; although redeemed by much that ennobles his defeat in one subdued, are marked by all that dishonours his conquest in the victor. Milton's Devil as a moral being is as far superior to his God, as one who perseveres in some purpose which he has conceived to be excellent in spite of adversity and torture, is to one who in the cold security of undoubted triumph inflicts

19. "Galeotto . . . scrisse": "Galeotto was the book and he who wrote it" (Dante, *Inferno*, V. 137). The line occurs near the end of the episode of Paolo and Francesca, who had fallen in love while reading a romance (*Lancelot du Lac*) in which Gallehaut acts as a go-between for the lovers. **20. Trouveurs:** troubadours.

21. Riphaeus . . . unus: Riphaeus was a Trojan and therefore a pagan (Dante, *Paradiso*, XX. 67). Vergil's phrase is from his tribute to Riphaeus, "the most just of all the Trojans" (*Aeneid*, II. 426).

the most horrible revenge upon his enemy, not from any mistaken notion of inducing him to repent of a perseverance in enmity, but with the alleged design of exasperating him to deserve new torments. Milton has so far violated the popular creed (if this shall be judged to be a violation) as to have alleged no superiority of moral virtue to his God over his Devil. And this bold neglect of a direct moral purpose is the most decisive proof of the supremacy of Milton's genius. He mingled as it were the elements of human nature as colours upon a single pallet, and arranged them in the composition of his great picture according to the laws of epic truth; that is, according to the laws of that principle by which a series of actions of the external universe and of intelligent and ethical beings is calculated to excite the sympathy of succeeding generations of mankind. The *Divina Commedia* and *Paradise Lost* have conferred upon modern mythology a systematic form; and when change and time shall have added one more superstition to the mass of those which have arisen and decayed upon the earth, commentators will be learnedly employed in elucidating the religion of ancestral Europe, only not utterly forgotten because it will have been stamped with the eternity of genius.

Homer was the first and Dante the second epic poet: that is, the second poet, the series of whose creations bore a defined and intelligible relation to the knowledge and sentiment and religion and political conditions of the age in which he lived, and of the ages which followed it: developing itself in correspondence with their development. For Lucretius had limed the wings of his swift spirit in the dregs of the sensible world;[22] and Virgil, with a modesty that ill became his genius, had affected the fame of an imitator, even whilst he created anew all that he copied; and none among the flock of mock-birds, though their notes were sweet, Apollonius Rhodius, Quintus Calaber, Nonnus, Lucan, Statius, or Claudian,[23] have sought even to fulfil a single condition of epic truth. Milton was the third epic poet.[24] For if the title of epic in its highest sense be refused to the *Aeneid*, still less can it be conceded to the *Orlando Furioso*, the *Gerusalemme Liberata*, the *Lusiad*,[25] or the *Fairy Queen*.

22. **Lucretius . . . world:** referring to the complete materialism of Lucretius' *De Rerum Natura*. 23. **Apollonius . . . Claudian:** The first three are Greek and the last three Roman poets. 24. **Milton . . . poet:** Cf. *Adonais*, l. 36. 25. **Orlando . . . Lusiad:** the principal works of the sixteenth-century poets Ariosto and Tasso, both Italian, and Camoëns (Portuguese).

Dante and Milton were both deeply penetrated with the ancient religion of the civilized world; and its spirit exists in their poetry probably in the same proportion as its forms survived in the unreformed worship of modern Europe. The one preceded and the other followed the Reformation at almost equal intervals. Dante was the first religious reformer, and Luther surpassed him rather in the rudeness and acrimony, than in the boldness of his censures of papal usurpation. Dante was the first awakener of entranced Europe; he created a language, in itself music and persuasion, out of a chaos of inharmonious barbarisms. He was the congregator of those great spirits who presided over the resurrection of learning; the Lucifer of that starry flock which in the thirteenth century shone forth from republican Italy, as from a heaven, into the darkness of the benighted world. His very words are instinct with spirit; each is as a spark, a burning atom of inextinguishable thought; and many yet lie covered in the ashes of their birth, and pregnant with a lightning which has yet found no conductor. All high poetry is infinite; it is as the first acorn, which contained all oaks potentially. Veil after veil may be undrawn, and the inmost naked beauty of the meaning never exposed. A great poem is a fountain for ever overflowing with the waters of wisdom and delight; and after one person and one age has exhausted all its divine effluence which their peculiar relations enable them to share, another and yet another succeeds, and new relations are ever developed, the source of an unforeseen and an unconceived delight.

The age immediately succeeding to that of Dante, Petrarch, and Boccaccio, was characterized by a revival of painting, sculpture, music, and architecture. Chaucer caught the sacred inspiration, and the superstructure of English literature is based upon the materials of Italian invention.

But let us not be betrayed from a defence into a critical history of poetry and its influence on society. Be it enough to have pointed out the effects of poets, in the large and true sense of the word, upon their own and all succeeding times.

But poets have been challenged to resign the civic crown to reasoners and mechanists on another plea. It is admitted that the exercise of the imagination is most delightful, but it is alleged, that that of reason is more useful. Let us examine as the grounds of this distinction, what is here meant by utility. Pleasure or good, in a general sense, is that which the consciousness of a sensitive and intelligent being seeks, and in which, when found, it acquiesces. There are two kinds of pleasure, one durable, universal and permanent; the other transitory and particular. Utility may either express the means of producing the former or the latter. In the former sense, whatever strengthens and purifies the affections, enlarges the imagination, and adds spirit to sense, is useful. But a narrower meaning may be assigned to the word utility, confining it to express that which banishes the importunity of the wants of our animal nature, the surrounding men with security of life, the dispersing the grosser delusions of superstition, and the conciliating such a degree of mutual forbearance among men as may consist with the motives of personal advantage.

Undoubtedly the promoters of utility, in this limited sense, have their appointed office in society. They follow the footsteps of poets, and copy the sketches of their creations into the book of common life. They make space, and give time. Their exertions are of the highest value, so long as they confine their administration of the concerns of the inferior powers of our nature within the limits due to the superior ones. But whilst the sceptic destroys gross superstitions, let him spare to deface, as some of the French writers have defaced, the eternal truths charactered upon the imaginations of men. Whilst the mechanist abridges, and the political economist combines labour, let them beware that their speculations, for want of correspondence with those first principles which belong to the imagination, do not tend, as they have in modern England, to exasperate at once the extremes of luxury and want. They have exemplified the saying, "To him that hath, more shall be given; and from him that hath not, the little that he hath shall be taken away."[26] The rich have become richer, and the poor have become poorer; and the vessel of the state is driven between the Scylla and Charybdis[27] of anarchy and despotism. Such are the effects which must ever flow from an unmitigated exercise of the calculating faculty.

It is difficult to define pleasure in its highest sense; the definition involving a number of apparent paradoxes. For, from an inexplicable defect of harmony in the constitution of human nature, the pain of the inferior is frequently connected with the pleasures of the superior portions of our being. Sorrow, terror, anguish, despair itself, are often the chosen expressions of an approximation to the highest good. Our sympathy in tragic fiction depends on this principle; tragedy delights by affording a shadow of the pleasure which exists in pain. This is the source also of the melancholy which is inseparable from the sweetest melody. The pleasure that is in sorrow is sweeter than the pleasure of pleasure itself. And hence the saying, "It is better to go to the house of mourning, than to the house of mirth."[28] Not that this highest species of pleasure is necessarily linked with pain. The delight of love and friendship, the ecstasy of the admiration of nature, the joy of the perception and still more of the creation of poetry, is often wholly unalloyed.

The production and assurance of pleasure in this highest sense is true utility. Those who produce and preserve this pleasure are poets or poetical philosophers.

The exertions of Locke, Hume, Gibbon, Voltaire, Rousseau,[29] and their disciples, in favour of oppressed and deluded humanity, are entitled to the gratitude of mankind. Yet it is easy to calculate the degree of moral and intellectual improvement which the world would have exhibited, had they never lived. A little more nonsense would have been talked for a century or two; and perhaps a few more men, women, and children, burnt as heretics. We might not at this moment have been congratulating each other on the abolition of the Inquisition in Spain.[30] But it exceeds all imagination to conceive what would have been the moral condition of the world if neither Dante, Petrarch, Boccaccio, Chaucer, Shakespeare, Calderón, Lord Bacon, nor Milton, had ever existed; if Raphael and Michael Angelo had never been born; if the Hebrew poetry had never been translated; if a revival of the study of Greek literature had never taken place; if no monuments of ancient sculpture had been handed down to us; and if the poetry of the religion of the ancient world had been extinguished together with its belief. The human mind could never, except by the intervention of these excitements, have been awakened to the invention of the grosser sciences, and that application of analytical reasoning to the aberrations of society, which it is now attempted to exalt over the direct expression of the inventive and creative faculty itself.

We have more moral, political and historical

26. "To . . . away": Matthew 25:29. 27. Scylla and Charybdis: a monster and a dangerous whirlpool on opposite sides of the strait between Italy and Sicily.

28. "It . . . mirth": Ecclesiastes 7:2. 29. Rousseau: "Although Rousseau has been thus classed, he was essentially a poet. The others, even Voltaire, were mere reasoners" (S.). 30. abolition . . . Spain: in 1820.

wisdom, than we know how to reduce into practice; we have more scientific and economical knowledge than can be accommodated to the just distribution of the produce which it multiplies. The poetry in these systems of thought, is concealed by the accumulation of facts and calculating processes. There is no want of knowledge respecting what is wisest and best in morals, government, and political economy, or at least, what is wiser and better than what men now practise and endure. But we let "*I dare not* wait upon *I would*, like the poor cat in the adage."[31] We want the creative faculty to imagine that which we know; we want the generous impulse to act that which we imagine; we want the poetry of life: our calculations have outrun conception; we have eaten more than we can digest. The cultivation of those sciences which have enlarged the limits of the empire of man over the external world, has, for want of the poetical faculty, proportionally circumscribed those of the internal world; and man, having enslaved the elements, remains himself a slave. To what but a cultivation of the mechanical arts in a degree disproportioned to the presence of the creative faculty, which is the basis of all knowledge, is to be attributed the abuse of all invention for abridging and combining labour, to the exasperation of the inequality of mankind? From what other cause has it arisen that the discoveries which should have lightened, have added a weight to the curse imposed on Adam? Poetry, and the principle of Self, of which money is the visible incarnation, are the God and Mammon of the world.

The functions of the poetical faculty are twofold; by one it creates new materials of knowledge and power and pleasure; by the other it engenders in the mind a desire to reproduce and arrange them according to a certain rhythm and order which may be called the beautiful and the good. The cultivation of poetry is never more to be desired than at periods when, from an excess of the selfish and calculating principle, the accumulation of the materials of external life exceed the quantity of the power of assimilating them to the internal laws of human nature. The body has then become too unwieldy for that which animates it.

Poetry is indeed something divine. It is at once the center and circumference of knowledge; it is that which comprehends all science, and that to which all science must be referred. It is at the same time the root and blossom of all other systems of thought; it is that from

which all spring, and that which adorns all; and that which, if blighted, denies the fruit and the seed, and withholds from the barren world the nourishment and the succession of the scions of the tree of life. It is the perfect and consummate surface and bloom of things; it is as the odour and the colour of the rose to the texture of the elements which compose it, as the form and splendour of unfaded beauty to the secrets of anatomy and corruption. What were virtue, love, patriotism, friendship—what were the scenery of this beautiful universe which we inhabit; what were our consolations on this side of the grave—and what were our aspirations beyond it, if poetry did not ascend to bring light and fire from those eternal regions where the owl-winged faculty of calculation dare not ever soar? Poetry is not like reasoning, a power to be exerted according to the determination of the will. A man cannot say, "I will compose poetry." The greatest poet even cannot say it; for the mind in creation is as a fading coal, which some invisible influence, like an inconstant wind, awakens to transitory brightness; this power arises from within, like the colour of a flower which fades and changes as it is developed, and the conscious portions of our natures are unprophetic either of its approach or its departure. Could this influence be durable in its original purity and force, it is impossible to predict the greatness of the results; but when composition begins, inspiration is already on the decline, and the most glorious poetry that has ever been communicated to the world is probably a feeble shadow of the original conceptions of the poet. I appeal to the greatest poets of the present day, whether it be not an error to assert that the finest passages of poetry are produced by labour and study. The toil and the delay recommended by critics can be justly interpreted to mean no more than a careful observation of the inspired moments, and an artificial connexion of the spaces between their suggestions by the intertexture of conventional expressions; a necessity only imposed by the limitedness of the poetical faculty itself; for Milton conceived the *Paradise Lost* as a whole before he executed it in portions. We have his own authority also for the muse having "dictated" to him the "unpremeditated song."[32] And let this be an answer to those who would allege the fifty-six various readings of the first line of the *Orlando Furioso*. Compositions so produced are to poetry what mosaic is to painting. The instinct and intuition of the poetical faculty is still

---

31. "I . . . adage": *Macbeth*, I. vii. 44–45.

32. muse . . . song: *Paradise Lost*, IX. 23–24.

more observable in the plastic and pictorial arts; a great statue or picture grows under the power of the artist as a child in the mother's womb; and the very mind which directs the hands in formation is incapable of accounting to itself for the origin, the gradations, or the media of the process.

Poetry is the record of the best and happiest moments of the happiest and best minds. We are aware of evanescent visitations of thought and feeling sometimes associated with place or person, sometimes regarding our own mind alone, and always arising unforeseen and departing unbidden, but elevating and delightful beyond all expression: so that even in the desire and the regret they leave, there cannot but be pleasure, participating as it does in the nature of its object. It is as it were the interpenetration of a diviner nature through our own; but its footsteps are like those of a wind over the sea, which the morning calm erases, and whose traces remain only, as on the wrinkled sand which paves it. These and corresponding conditions of being are experienced principally by those of the most delicate sensibility and the most enlarged imagination; and the state of mind produced by them is at war with every base desire. The enthusiasm of virtue, love, patriotism, and friendship, is essentially linked with such emotions; and whilst they last, self appears as what it is, an atom to a universe. Poets are not only subject to these experiences as spirits of the most refined organization, but they can colour all that they combine with the evanescent hues of this ethereal world; a word, a trait in the representation of a scene or a passion, will touch the enchanted chord, and re-animate, in those who have ever experienced these emotions, the sleeping, the cold, the buried image of the past. Poetry thus makes immortal all that is best and most beautiful in the world; it arrests the vanishing apparitions which haunt the interlunations of life, and veiling them, or in language or in form, sends them forth among mankind, bearing sweet news of kindred joy to those with whom their sisters abide—abide, because there is no portal of expression from the caverns of the spirit which they inhabit into the universe of things. Poetry redeems from decay the visitations of the divinity in man.

Poetry turns all things to loveliness; it exalts the beauty of that which is most beautiful, and it adds beauty to that which is most deformed; it marries exultation and horror, grief and pleasure, eternity and change; it subdues to union under its light yoke, all irreconcilable things. It transmutes all that it touches, and every form moving within the radiance of its presence is changed by wondrous sympathy to an incarnation of the spirit which it breathes; its secret alchemy turns to potable gold the poisonous waters which flow from death through life; it strips the veil of familiarity from the world, and lays bare the naked and sleeping beauty, which is the spirit of its forms.

All things exist as they are perceived; at least in relation to the percipient. "The mind is its own place, and of itself can make a heaven of hell, a hell of heaven." [33] But poetry defeats the curse which binds us to be subjected to the accident of surrounding impressions. And whether it spreads its own figured curtain, or withdraws life's dark veil from before the scene of things, it equally creates for us a being within our being. It makes us the inhabitants of a world to which the familiar world is a chaos. It reproduces the common universe of which we are portions and percipients, and it purges from our inward sight the film of familiarity which obscures from us the wonder of our being. It compels us to feel that which we perceive, and to imagine that which we know. It creates anew the universe, after it has been annihilated in our minds by the recurrence of impressions blunted by reiteration. It justifies the bold and true words of Tasso: *Non merita nome di creatore, se non Iddio ed il Poeta.*[34]

A poet, as he is the author to others of the highest wisdom, pleasure, virtue and glory, so he ought personally to be the happiest, the best, the wisest, and the most illustrious of men. As to his glory, let time be challenged to declare whether the fame of any other institutor of human life be comparable to that of a poet. That he is the wisest, the happiest, and the best, inasmuch as he is a poet, is equally incontrovertible; the greatest poets have been men of the most spotless virtue, of the most consummate prudence, and, if we would look into the interior of their lives, the most fortunate of men: and the exceptions, as they regard those who possessed the poetic faculty in a high yet inferior degree, will be found on consideration to confirm rather than destroy the rule. Let us for a moment stoop to the arbitration of popular breath, and usurping and uniting in our own persons the incompatible characters of accuser, witness, judge, and executioner, let us decide without trial, testimony, or form, that certain motives of those who are "there sitting where we dare not soar," [35] are reprehensible. Let us assume that

33. "The . . . heaven": *Paradise Lost*, I. 254–55. 34. **Non . . . Poeta**: See "On Life," above, n. 1. 35. "there . . . soar": *Paradise Lost*, IV. 829. Cf. *Adonais*, l. 337.

Homer was a drunkard, that Virgil was a flatterer, that Horace was a coward, that Tasso was a madman, that Lord Bacon was a peculator, that Raphael was a libertine, that Spenser was a poet laureate. It is inconsistent with this division of our subject to cite living poets, but posterity has done ample justice to the great names now referred to. Their errors have been weighed and found to have been dust in the balance; if their sins were as scarlet, they are now white as snow:[36] they have been washed in the blood of the mediator and redeemer, Time. Observe in what a ludicrous chaos the imputations of real or fictitious crime have been confused in the contemporary calumnies against poetry and poets; consider how little is as it appears—or appears as it is; look to your own motives, and judge not, lest ye be judged.[37]

Poetry, as has been said, differs in this respect from logic, that it is not subject to the control of the active powers of the mind, and that its birth and recurrence have no necessary connexion with consciousness or will. It is presumptuous to determine that these are the necessary conditions of all mental causation, when mental effects are experienced unsusceptible of being referred to them. The frequent recurrence of the poetical power, it is obvious to suppose, may produce in the mind a habit of order and harmony correlative with its own nature and with its effects upon other minds. But in the intervals of inspiration, and they may be frequent without being durable, a poet becomes a man, and is abandoned to the sudden reflux of the influences under which others habitually live. But as he is more delicately organized than other men, and sensible to pain and pleasure, both his own and that of others, in a degree unknown to them, he will avoid the one and pursue the other with an ardour proportioned to this difference. And he renders himself obnoxious to calumny, when he neglects to observe the circumstances under which these objects of universal pursuit and flight have disguised themselves in one another's garments.

But there is nothing necessarily evil in this error, and thus cruelty, envy, revenge, avarice, and the passions purely evil, have never formed any portion of the popular imputations on the lives of poets.

I have thought it most favorable to the cause of truth to set down these remarks according to the order in which they were suggested to my mind, by a con-

sideration of the subject itself, instead of observing the formality of a polemical reply; but if the view which they contain be just, they will be found to involve a refutation of the arguers against poetry, so far at least as regards the first division of the subject. I can readily conjecture what should have moved the gall of some learned and intelligent writers who quarrel with certain versifiers; I confess myself like them, unwilling to be stunned by the Theseids of the hoarse Codri[38] of the day. Bavius and Mævius undoubtedly are, as they ever were, insufferable persons. But it belongs to a philosophical critic to distinguish rather than confound.

The first part of these remarks has related to poetry in its elements and principles; and it has been shown, as well as the narrow limits assigned them would permit, that what is called poetry, in a restricted sense, has a common source with all other forms of order and of beauty, according to which the materials of human life are susceptible of being arranged, and which is poetry in a universal sense.

The second part[39] will have for its object an application of these principles to the present state of the cultivation of poetry, and a defence of the attempt to idealize the modern forms of manners and opinions, and compel them into a subordination to the imaginative and creative faculty. For the literature of England, an energetic development of which has ever preceded or accompanied a great and free development of the national will, has arisen as it were from a new birth. In spite of the low-thoughted envy which would undervalue contemporary merit, our own will be a memorable age in intellectual achievements, and we live among such philosophers and poets as surpass beyond comparison any who have appeared since the last national struggle for civil and religious liberty. The most unfailing herald, companion, and follower of the awakening of a great people to work a beneficial change in opinion or institution, is poetry. At such periods there is an accumulation of the power of communicating and receiving intense and impassioned conceptions respecting man and nature. The persons in whom this power resides may often, as far as regards many portions of their nature, have little apparent

---

36. sins . . . snow: Isaiah 1:18. 37. judge . . . judged: Matthew 7:1.

38. Codri: bad and dull poets (from Codrus, a poet attacked by Juvenal, and said to have written an atrocious tragedy about Theseus). Bavius and Maevius were mediocre poets mentioned in Vergil's *Eclogues*, III. 39. second part: never written.

correspondence with that spirit of good of which they are the ministers. But even whilst they deny and abjure, they are yet compelled to serve the power which is seated on the throne of their own soul. It is impossible to read the compositions of the most celebrated writers of the present day without being startled with the electric life which burns within their words. They measure the circumference and sound the depths of human nature with a comprehensive and all-penetrating spirit, and they are themselves perhaps the most sincerely astonished at its manifestations; for it is less their spirit than the spirit of the age. Poets are the hierophants[40] of an unapprehended inspiration; the mirrors of the gigantic shadows which futurity casts upon the present; the words which express what they understand not; the trumpets which sing to battle, and feel not what they inspire; the influence which is moved not, but moves. Poets are the unacknowledged legislators of the world.

LETTERS

To William Godwin
Keswick, January 10, 1812.

SIR

It is not otherwise to be supposed than that I should appreciate your avocations far beyond the pleasure or benefit which can accrue to me from their sacrifice. The time, however, will be small which may be mis-spent in reading this letter; and much individual pleasure as an answer might give me, I have not the vanity to imagine that it will be greater than the happiness elsewhere diffused during the time which its creation will occupy.

You complain that the generalizing character of my letter[1] renders it deficient in interest; that I am not an individual to you. Yet, intimate as I am with your character and your writings, intimacy with *yourself* must in some degree precede this exposure of my peculiarities. It is scarcely possible, however pure be the morality which he has endeavoured to diffuse, but that generalization must characterize the uninvited address of a stranger to a stranger.

I proceed to remedy the fault. I am the son of a man of fortune in Sussex. The habits of thinking of my father

and myself never coincided. Passive obedience was inculcated and enforced in my childhood. I was required to love, because it was *my duty* to love: it is scarcely necessary to remark, that coercion obviated its own intention. I was haunted with a passion for the wildest and most extravagant romances. Ancient books of Chemistry and Magic were perused with an enthusiasm of wonder, almost amounting to belief. My sentiments were unrestrained by anything within me; external impediments were numerous, and strongly applied; their effect was merely temporary.

From a reader, I became a writer of romances; before the age of seventeen I had published two, "St. Irvyne" and "Zastrozzi," each of which, though quite uncharacteristic of me as now I am, yet serves to mark the state of my mind at the period of their composition. I shall desire them to be sent to you: do not, however, consider this as any obligation to yourself to misapply your valuable time.

It is now a period of more than two years since first I saw your inestimable book on "Political Justice;" it opened to my mind fresh and more extensive views; it materially influenced my character, and I rose from its perusal a wiser and a better man. I was no longer the votary of romance; till then I had existed in an ideal world—now I found that in this universe of ours was enough to excite the interest of the heart, enough to employ the discussions of reason; I beheld, in short, that I had duties to perform. Conceive the effect which the "Political Justice" would have upon a mind before jealous of its independence and participating somewhat singularly in a peculiar susceptibility.

My age is now *nineteen*; at the period to which I allude I was at Eton. No sooner had I formed the principles which I now profess, than I was anxious to disseminate their benefits. This was done without the slightest caution. I was twice expelled, but recalled by the interference of my father. I went to Oxford. Oxonian society was insipid to me, uncongenial with my habits of thinking. I could not descend to common life: the sublime interest of poetry, lofty and exalted achievements, the proselytism of the world, the equalization of its inhabitants, were to me the soul of my soul. You can probably form some idea of the contrast exhibited to my character by those with whom I was surrounded. Classical reading and poetical writing employed me during my residence at Oxford.

In the meantime I became, in the popular sense of the word "God," an Atheist. I printed a pamphlet, avowing my opinion, and its occasion. I distributed

---

40. hierophants: priests. LETTERS. **I. letter:** Shelley had written a letter of self-introduction to Godwin on January 3.

this anonymously to men of thought and learning, wishing that Reason should decide on the case at issue; it was never my intention to deny it. Mr. Coplestone, at Oxford, among others, had the pamphlet; he showed it to the Master and the Fellows of University College, and I was sent for. I was informed, that in case I denied the publication no more would be said. I refused, and was expelled.

It will be necessary, in order to elucidate this part of my history, to inform you that I am heir by entail to an estate of £6,000 per annum. My principles have induced me to regard the law of primogeniture an evil of primary magnitude. My father's notions of family honour are incoincident with my knowledge of public good. I will never sacrifice the latter to any consideration. My father has ever regarded me as a blot, a defilement of his honour. He wished to induce me by poverty to accept of some commission in a distant regiment, and in the interim of my absence to prosecute the pamphlet, that a process of outlawry might make the estate, on his death, devolve to my younger brother. These are the leading points of the history of the man before you. Others exist, but I have thought proper to make some selection, not that it is my design to conceal or extenuate any part, but that I should by their enumeration quite outstep the bounds of modesty. Now, it is for you to judge whether, by permitting me to cultivate your friendship, you are exhibiting yourself more really useful than by the pursuance of those avocations, of which the time spent in allowing this cultivation would deprive you. I am now earnestly pursuing studious habits. I am writing "An inquiry into the causes of the failure of the French Revolution to benefit mankind." My plan is that of resolving to lose no opportunity to disseminate truth and happiness.

I am married to a woman whose views are similar to my own. To you, as the regulator and former of my mind, I must ever look with real respect and veneration.

Yours sincerely,

P. B. SHELLEY.

To Thomas Love Peacock

Naples, December [17 or 18], 1818.

MY DEAR PEACOCK,

I have received a letter from you here, dated November 1st; you see the reciprocation of letters from the

term of our travels is more slow. I entirely agree with what you say about Childe Harold. The spirit in which it is written is, if insane, the most wicked and mischievous insanity that ever was given forth. It is a kind of obstinate and self-willed folly, in which he hardens himself. I remonstrated with him in vain on the tone of mind from which such a view of things alone arises. For its real root is very different from its apparent one. Nothing can be less sublime than the true source of these expressions of contempt and desperation. The fact is, that first, the Italian Women are perhaps the most contemptible of all who exist under the moon—the most ignorant, the most disgusting, the most bigoted, the most filthy; Countesses smell so of garlick that an ordinary Englishman cannot approach them. Well, L. B. is familiar with the lowest sort of these women, the people his gondolieri pick up in the streets. He allows fathers and mothers to bargain with him for their daughters, and though this is common enough in Italy, yet for an Englishman to encourage such sickening vice is a melancholy thing. He associates with wretches who seem almost to have lost the gait and physiognomy of man, and who do not scruple to avow practices which are not only not named, but I believe seldom even conceived in England. He says he disapproves, but he endures. He is not yet an Italian and is heartily and deeply discontented with himself; and contemplating in the distorted mirror of his own thoughts the nature and the destiny of man, what can he behold but objects of contempt and despair? But that he is a great poet, I think the address to ocean proves. And he has a certain degree of candour while you talk to him, but unfortunately it does not outlast your departure. You may think how unwillingly I have left my little favourite Alba[2] in a situation where she might fall again under his authority. But I have employed arguments entreaties every thing in vain & when these fail you know I have no longer any right. No, I do not doubt, and, for his sake, I ought to hope, that his present career must end soon in some violent circumstance which must reduce our situation with respect to Alba into its antient tie.

Since I last wrote to you I have seen the ruins of Rome, the Vatican, St. Peter's, and all the miracles of ancient and modern art contained in that majestic city. The impression of it exceeds anything I have ever experienced in my travels. We staied there only a

2. **Alba:** Allegra, daughter of Claire Clairmont by Byron.

week, intending to return at the end of February, and devote two or three months to its mines of inexhaustible contemplation, to which period I refer you for a minute account of it. We visited the Forum and the ruins of the Coliseum every day. The Coliseum is unlike any work of human hands I ever saw before. It is of enormous height and circuit, and the arches built of massy stones are piled on one another, and jut into the blue air, shattered into the forms of overhanging rocks. It has been changed by time into the image of an amphitheatre of rocky hills overgrown by the wild olive, the myrtle, and the fig tree, and threaded by little paths, which wind among its ruined stairs and immeasurable galleries: the copsewood overshadows you as you wander through its labyrinths, and the wild weeds of this climate of flowers bloom under your feet. The arena is covered with grass, and pierces like the skirts of a natural plain, the chasms of the broken arches around. But a small part of the exterior circumference remains—it is exquisitely light and beautiful; and the effect of the perfection of its architecture, adorned with ranges of Corinthian pilasters, supporting a bold cornice, is such as to diminish the effect of its greatness. The interior is all ruin. I can scarcely believe that when encrusted with Dorian marble and ornamented by columns of Egyptian granite, its effect could have been so sublime and so impressive as in its present state. It is open to the sky, and it was the clear and sunny weather of the end of November in this climate when we visited it, day after day.

Near it is the arch of Constantine, or rather the arch of Trajan; for the servile and avaricious senate of degraded Rome ordered that the monument of his predecessor should be demolished in order to dedicate one to the Christian reptile, who had crept among the blood of his murdered family to the supreme power. It is exquisitely beautiful and perfect. The Forum is a plain in the middle of Rome, a kind of desert full of heaps of stones and pits; and though so near the habitations of men, is the most desolate place you can conceive. The ruins of temples stand in and around it, shattered columns and ranges of others complete, supporting cornices of exquisite workmanship, and vast vaults of shattered domes (laquearis) distinct with the regular compartments, once filled with sculptures of ivory or brass. The temples of Jupiter, and Concord, and Peace, and the Sun, and the Moon, and Vesta, are all within a short distance of this spot. Behold the wrecks of what a great nation once dedicated to the abstractions of the mind! Rome is a city, as it were,

of the dead, or rather of those who cannot die, and who survive the puny generations which inhabit and pass over the spot which they have made sacred to eternity. In Rome, at least in the first enthusiasm of your recognitions of ancient time, you see nothing of the Italians. The nature of the city assists the delusion, for its vast and antique walls describe a circumference of sixteen miles, and thus the population is thinly scattered over this space, nearly as great as London. Wide wild fields are enclosed within it, and there are grassy lanes and copses winding among the ruins, and a great green hill, lonely and bare, which overhangs the Tiber. The gardens of the modern palaces are like wild woods of cedar, and cypress, and pine, and the neglected walks are overgrown with weeds. The English burying place is a green slope near the walls, under the pyramidal tomb of Cestius, and is, I think, the most beautiful and solemn cemetery I ever beheld. To see the sun shining on its bright grass, fresh, when we first visited it, with the autumnal dews, and hear the whispering of the wind among the leaves of the trees which have overgrown the tomb of Cestius, and the soil which is stirring in the sun warm earth, and to mark the tombs, mostly of women and young people who were buried there, one might, if one were to die, desire the sleep they seem to sleep. Such is the human mind, and so it peoples with its wishes vacancy and oblivion.

I have told you little about Rome, but I reserve the Pantheon, and St. Peter's, and the Vatican, and Raffael, for my return. About a fortnight ago I left Rome, and Mary and Clare followed in three days; for it was necessary to procure lodgings here without alighting at an Inn. From my peculiar mode of travelling I saw little of the country, but could just observe that the wild beauty of the scenery and the barbarous ferocity of the inhabitants progressively increased. On entering Naples, the first circumstance that engaged my attention was an assassination. A youth ran out of a shop pursued by a woman with a bludgeon, and a man armed with a knife. The man overtook him, and with one blow in the neck laid him dead in the road. On my expressing the emotions of horror and indignation which I felt, a Calabrian priest who travelled with me laughed heartily and attempted to quiz me as what the English call a flat.[3] I never felt such an inclination to beat any one. Heaven knows I have little power, but he saw that I looked extremely displeased, and was

3. **flat**: a gullible person, one easily duped.

silent. This same man, a fellow of gigantic strength and stature, had expressed the most frantic terror of robbers on the road; he cried at the sight of my pistol, and it had been with great difficulty that the joint exertions of myself and the vetturino[4] had quieted his hysterics.

But external nature in these delightful regions contrasts with and compensates for the deformity and degradation of humanity. We have a lodging divided from the sea by the royal gardens, and from our windows we see perpetually the blue waters of the bay, forever changing, yet forever the same, and encompassed by the mountainous island of Capreæ, the lofty peaks which overhang Salerno, and the woody hill of Posilypo, whose promontories hide from us Misenum and the lofty isle Inarime, which, with its divided summit, forms the opposite horn of the bay. From the pleasant walks of the garden we see *Vesuvius;* a smoke by day and a fire by night is seen upon its summit, and the glassy sea often reflects its light or shadow. The climate is delicious. We sit without a fire, with the windows open, and have almost all the productions of an English summer. The weather is usually like what Wordsworth calls "the first fine day of March;" sometimes very much warmer, though perhaps it wants that "each minute sweeter than before,"[5] which gives an intoxicating sweetness to the awakening of the earth from its winter's sleep in England—We have made two excursions, one to Baiæ and one to Vesuvius, and we propose to visit, successively, the islands, Pæstum, Pompei, and Beneventum.

We set off an hour after sunrise one radiant morning in a little boat, there was not a cloud in the sky nor a wave upon the sea, which was so translucent that you could see the hollow caverns clothed with the glaucous sea-moss, and the leaves and branches of those delicate weeds that pave the unequal bottom of the water. As noon approached, the heat, and especially the light, became intense. We passed Posilipo, and came first to the eastern point of the bay of Puzzoli, which is within the great bay of Naples, and which again incloses that of Baiæ. Here are lofty rocks and craggy islets, with arches and portals of precipice standing in the sea, and enormous caverns, which echoed faintly with the murmur of the languid tide. This is called La Scuola di Virgilio. We then went directly across to the promontory of Misenum, leaving the precipitous islet of Nesida on the right. Here we were conducted to see the Mare Morto, and the Elysian fields; the spot on which Virgil places the scenery of the 6th Æneid. Tho extremely beautiful, as a lake, and woody hills, and this divine sky must make it, I confess my disappointment. The guide showed us an antique cemetery, where the niches used for placing the cinerary urns of the dead yet remain. We then coasted the bay of Baiæ to the left, in which we saw many picturesque and interesting ruins; but I have to remark that we never disembarked but we were disappointed—while from the boat the effect of the scenery was inexpressibly delightful. The colours of the water and the air breathe over all things here the radiance of their own beauty. After passing the Bay of Baiæ, and observing the ruins of its antique grandeur standing like rocks in the transparent sea under our boat.[6] We landed to visit lake Avernus. We passed thro the cavern of the Sybyl (not Virgil's Sybil) which pierces one of the hills which circumscribe the lake, and came to a calm and lovely basin of water, surrounded by dark woody hills, and profoundly solitary. Some vast ruins of the temple of Pluto stand on a lawny hill on one side of it, and are reflected in its windless mirror. It is far more beautiful than the Elysian fields—but there are all the materials for beauty at the latter, and the Avernus was once a chasm of deadly and pestilential vapours. About $\frac{1}{2}$ a mile from Avernus, a high hill, called Monte N[u]ovo was thrown up by Volcanic fire.

Passing onward we came to Pozzoli, the ancient Dicæarchea, where there are the columns remaining of a temple to Serapis, and the wreck of an enormous amphitheatre, changed, like the Coliseum, into a natural hill by the overteeming vegetation. Here also is the Solfatara, of which there is a poetical description in the Civil War of Petronius, beginning—Est locus, and in which the verses of the poet are infinitely finer than what he describes, for it is not a very curious place. After seeing these things we returned by moonlight to Naples in our boat. What colours there were in the sky, what radiance in the evening star, and how the moon was encompassed by a light unknown to our regions!

Our next excursion was to Vesuvius. We went to Resina in a carriage, where Mary and I mounted mules and Clare was carried in a chair on the shoulders of

---

4. **vetturino:** coachman. 5. **"the . . . before":** "To My Sister," ll. 1–2.

6. **observing . . . boat:** Cf. "Ode to the West Wind," above, ll. 32–35.

four men, much like a member of parliament after he has gained his election, and looking, with less reason, quite as frightened. So we arrived at the hermitage of St Salvador, where an old hermit, belted with rope, set forth the plates for our refreshment.

Vesuvius is, after the glaciers, the most impressive exhibition of the energies of nature I ever saw. It has not the immeasurable greatness, the overpowering magnificence, nor above all, the radiant beauty of the glaciers; but it has all their character of tremendous and irresistible strength. From Resina to the hermitage you wind up the mountain, and cross a vast stream of hardened lava, which is an actual image of the waves of the sea, changed into hard black stone by enchantment. The lines of the boiling fluid seem to hang in the air, and it is difficult to believe that the billows which seem hurrying down upon you are not actually in motion. This plain was once a sea of liquid fire. From the hermitage we crossed another stream of lava, and then went on foot up the cone—this is the only part of the ascent in which there is any difficulty, and that difficulty has been much exaggerated. It is composed of rocks of lava, and declivities of ashes; by ascending the former and descending the latter, there is very little fatigue. On the summit is a kind of irregular plain, the most horrible chaos that can be imagined; riven into ghastly chasms, and heaped up with tumuli of great stones and cinders, and enormous rocks blackened and calcined, which had been thrown from the Volcano upon one another in terrible confusion. In the midst stands the conical hill from which the volumes of smoke, and the fountains of liquid fire, are rolled forth forever. The mountain is at present in a slight state of eruption, and a thick heavy white smoke is perpetually rolled out, interrupted by enormous columns of an impenetrable black bituminous vapour, which is hurled up, fold after fold, into the sky with a deep hollow sound, and fiery stones are rained down from its darkness, and a black shower of ashes fall even on where we sate. The lava like the glacier creeps on perpetually, with a crackling sound as of suppressed fire. There are several springs of lava, and in one place it gushes precipitously over a high crag, rolling down the half melted rocks and its own over hanging waves; a cataract of quivering fire. We approached the extremity of one of the rivers of lava; it is about 20 feet in breadth and ten in height; and as the inclined plane was not rapid, its motion was very slow. We saw the masses of its dark exterior surface detach themselves as it moved, and betray the depth of the liquid flame.

In the day the fire is but slightly seen; you only observe a tremulous motion in the air, and streams and fountains of white sulphurous smoke. At length we saw the sun sink between Capreæ and Inarime, and, as the darkness increased, the effect of the fire became more beautiful. We were, as it were, surrounded by streams and cataracts of the red and radiant fire, and in the midst from the column of bituminous smoke shot up into the sky, fell the vast masses of rock white with the light of their intense heat, leaving behind them thro the dark vapour trains of splendour. We descended by torch light, and I should have enjoyed the scenery on my return, but that they conducted me, I know not how, to the hermitage in a state of intense bodily suffering, the worst effect of which was spoiling the pleasure of Mary and Clare—Our Guides on the occasion, were complete Savages. You have no idea of the horrible cries which they suddenly utter, no one knows why; the clamour, the vociferation, the tumult. Clare in her palanquin suffered most from it; and when I had gone on before, they threatened to leave her in the middle of the road, which they would have done had not my Italian servant promised them a beating, after which they became very quiet. Nothing, however, can be more picturesque than the gestures and the physiognomies of these savage people. And when, in the darkness of night, they unexpectedly begin to sing in chorus some fragments of their wild but sweet national music, the effect is exceedingly fine.

Since I wrote this, I have seen the museum of this city. Such statues! There is the Venus, an ideal shape of the most winning loveliness. A Bacchus, more sublime than any living being. A Satyr making love to a Youth in which the expressed life of the sculpture and the inconceivable beauty of the form of the youth, overcome one's repugnance to the subject. There are multitudes of wonderfully fine statues found in Herculaneum and Pompeii. We are going to see Pompeii the 1st day that the sea is waveless. Herculaneum is almost all filled up; no more excavations are made; the King bought the ground and built a palace upon it.

You don't see much of Hunt. I wish you could contrive to see him when you go to town, and ask him what he means to answer to Lord Byron's invitation.[7] He has now an opportunity, if he likes, of seeing

7. **invitation:** Hunt had been invited to Italy in order to participate in a projected magazine, *The Liberal*, in which he, Byron, and Shelley would publish their compositions.

Italy. What do you think of joining his party, and paying us a visit next year; I mean as soon as the reign of winter is dissolved? Write me your thoughts upon this. I cannot express to you the pleasure it would give me to welcome such a party.

I have depression enough of spirits and not good health, though I believe the warm air of Naples does me good. We see absolutely no one here—Adieu.

My dear Peacock,
Affectionately your friend,
P.B.S.

To Mary Shelley
Ravenna, August 10, 1821.

MY DEAREST MARY,

...We[8] ride out in the evening through the pine forests which divide this city from the sea. Our way of life is this, and I have accommodated myself to it without much difficulty. L. B. gets up at two, breakfasts—we talk read etc., until six; then we ride, and dine at eight, and after dinner sit talking till four or five in the morning. I get up at 12, and am now devoting the interval between my rising and his, to you.

L. B. is greatly improved in every respect—in genius in temper in moral views, in health in happiness. The connexion with la Guiccioli[9] has been an inestimable benefit to him. He lives in considerable splendour, but within his income, which is now about 4000 a year:—1000 of which he devotes to purposes of charity. He has had mischievous passions, but these he seems to have subdued, and he is becoming what he should be, a virtuous man. The interest which he took in the politics of Italy, and the actions he performed in consequence of it, are subjects not fit to be *written*, but are such as will delight and surprise you. He is not yet decided to go to Switzerland: a place indeed little fitted for him: the gossip and the cabals of those anglicised coteries would torment him as they did before, and might exasperate him into a relapse of libertinism, which he says he plunged into not from taste but despair. La Guiccioli and her brother (who is Lord B.'s friend and confidant, and acquiesces perfectly in her connexion with him), wish to go to Switzerland; as L. B. says merely from the novelty and pleasure of travelling. L. B. prefers Tuscany or

Lucca, and is trying to persuade them to adopt his views. He has made *me* write a long letter to her to engage her to remain—an odd thing enough for an utter stranger to write on subjects of the utmost delicacy to his friend's mistress. But it seems destined that I am always to have some active part in everybody's affairs whom I approach—I have set down in lame Italian the strongest reasons I can think of against the Swiss emigration—to tell you truth I should be very glad to accept as my fee, his establishment in Tuscany. Ravenna is a miserable place; the people are barbarous and wild, and their language the most infernal patois that you can imagine. He would be in every respect better among the Tuscans. I am afraid he would not like Florence on account of the English. What think you of Lucca for him—he would like Pisa better, if it were not for Clare,[10] but I really can hardly recommend him either for his own sake or for hers to come into such close contact with her. Gunpowder and fire ought to be kept at a respectable distance from each other. There is Lucca, Florence, Pisa, Sienna, and I think nothing more. What think you of Prato, or Pistoia, for him—no Englishman approaches those towns; but I fear that no house could be found good enough for him in that region.—I have not yet seen Allegra, but shall tomorrow or next day: as I shall ride over to Bagnacavallo for that purpose.

He has read to me one of the unpublished cantos of Don Juan, which is astonishingly fine. It sets him not above but far above all the poets of the day: every word is stamped with immortality. I despair of rivalling Lord Byron, as well I may, and there is no other with whom it is worth contending. This canto is in style, but totally, and sustained with incredible ease and power, like the end of the second canto. There is not a word which the most rigid asserter of the dignity of human nature could desire to be cancelled: it fulfills in a certain degree what I have long preached of producing something wholly new and relative to the age, and yet surpassingly beautiful. It may be vanity, but I think I see the trace of my earnest exhortation to him to create something wholly new. He has finished his *life* up to the present time and given it to Moore[11] with liberty for Moore to sell it for the best price he can get, with condition that the bookseller should publish it after his death. Moore has sold it to Murray

---

8. **We:** Shelley was visiting Byron at Ravenna. 9. **la Guiccioli:** the Countess Teresa Guiccioli, with whom Byron had formed a settled liaison.

10. **Clare:** At this time Claire Clairmont was living with the Shelleys in Pisa. 11. **life . . . Moore:** The autobiography sent to Tom Moore was burnt after Byron's death.

for *two thousand pounds*. I wish I had been in time to have interceded for a part of it for poor Hunt.—I have spoken to him of Hunt, but not with a direct view of demanding a contribution; and though I am sure that if asked it would not be refused—yet there is something in me that makes it impossible. Lord Byron and I are excellent friends, and were I reduced to poverty, or were I a writer who had no claims to a higher station than I possess—or did I possess a higher than I deserve, we should appear in all things as such, and I would freely ask him any favour. Such is not now the case. The demon of mistrust and of pride lurks between two persons in our situation poisoning the freedom of their intercourse. This is a tax, and a heavy one which we must pay for being human. I think the fault is not on my side nor is it likely, I being the weaker. I hope that in the next world these things will be better managed. What is passing in the heart of another rarely escapes the observation of one who is a strict anatomist of his own.

Write to me at Florence, where I shall remain a day at least and send me letters or news of letters. How is my little darling?[12] And how are you, and how do you get on with your book?[13] Be severe in your corrections, and expect severity from me, your sincere admirer. I flatter myself you have composed something unequalled in its kind, and that, not content with the honours of your birth and your hereditary aristocracy, you will add still higher renown to your name. Expect me at the end of my appointed time. I do not think I shall be detained. Is Clare with you, or is she coming? Have you heard anything of my poor Emilia,[14] from whom I got a letter the day of my departure, saying that her marriage was deferred for a *very short* time, on account of the illness of her sposo? How are the Williams's,[15] and Williams especially? Give my very kindest love to them, and pray take care that they do not want money.

Lord B. has here splendid apartments in the house of his mistress's husband: who is one of the richest men in Italy. *She* is divorced, with an allowance of 1200 crowns a year, a miserable pittance from a man who has 120,000 a year.—Here are two monkies, five cats, eight dogs, and ten horses, all of whom (except the horses), walk about [the] house like the masters of it. *Tita* [the] Venetian is here, and operates as my

valet; a fine fellow, with a prodigious black beard, who has stabbed two or three people, and is the most good-natured looking fellow I ever saw.

We have good rumours of the Greeks here, and a Russian war. I hardly wish the Russians to take any part in it. My maxim is with Æschylus:—τὸ δυσσεβὲς —μετὰ μὲν πλείονα τίκτει, σφετέρᾳ δ' εἰκότα γεννᾷ.[16] There is a Greek exercise for you. How should slaves produce any thing but tyranny—even as the seed produces the plant?

Adieu, dear Mary.

Yours affectionately,

S.

To Lord Byron
Pisa, Oct. 21, 1821.

My dear Lord Byron,

I should have written to you long since but that I have been led to expect you almost daily in Pisa, and that I imagined you would cross my letter on your road. Many thanks for Don Juan.—It is a poem totally of its own species, and my wonder and delight at the grace of the composition no less than the free and grand vigour of the conception of it perpetually increase. The few passages which any one might desire to be cancelled in the first and second Cantos are here reduced almost to nothing. This poem carries with it at once the stamp of originality and a defiance of imitation. Nothing has ever been written like it in English, nor, if I may venture to prophesy, will there be; without carrying upon it the mark of a secondary and borrowed light.—You unveil and present in its true deformity what is worst in human nature, and this is what the witlings of the age murmur at, conscious of their want of power to endure the scrutiny of such a light.—We are damned to the knowledge of good and evil, and it is well for us to know what we should avoid no less than what we should seek. The character of Lambro—his return—the merriment of his daughter's guests, made, as it were, in celebration of his funeral— the meeting with the lovers—and the death of Haidée, —are circumstances combined and developed in a manner that I seek elsewhere in vain. The fifth Canto, which some of your pet Zoili[17] in Albemarle St.[18]

---

12. darling: his one surviving child, Percy Florence. **13. book:** *Valperga*, a novel. **14. Emilia:** Emilia Viviani. See headnote to *Epipsychidion*, above. **15. Williams's:** Jane and Edward Williams, friends of the Shelleys.

16. τὸ . . . γεννᾷ: *Agamemnon*, ll. 758–60: "The evil deed breeds others after it like to its own race." **17. Zoili:** carping and ill-willed critics, from Zoilus, a Greek critic of the fourth century B.C. who attacked Homer. **18. Albemarle St.:** the address of John Murray, Byron's publisher.

said was *dull*, gathers instead of loses, splendour and energy—the language in which the whole is clothed—a sort of chameleon under the changing sky of the spirit that kindles it—is such as these lisping days could not have expected,—and are, believe me, in spite of the approbation which you wrest from them, little pleased to hear.

One can hardly judge from recitation, and it was not until I read it in print that I have been able to do it justice. This sort of writing only on a great plan, and perhaps in a more compact form, is what I wished you do do when I made my vows for an epic.—But I am content. You are building up a drama, such as England has not yet seen, and the task is sufficiently noble and worthy of you.

When may we expect you? The Countess G. is very patient, though sometimes she seems apprehensive that you will *never* leave Ravenna. I have suffered from my habitual disorder and from a tertian fever since I have returned, and my ill health has prevented me from showing her the attentions I could have desired in Pisa. I have heard from Hunt, who tells me that he is coming out in November, by sea I believe.—Your house is ready and all the furniture arranged. Lega, they say, is to have set off yesterday. The Countess tells me that you think of leaving Allegra for the present at the convent. Do as you think best—but I can pledge myself to find a situation for her here such as you would approve in case you change your mind.

I hear no political news but such as announces the slow victory of the spirit of the past over that of the present. The other day, a number of Heteristi,[19] escaped from the defeat in Wallachia, past through Pisa, to embark at Leghorn and join Ipsilanti[20] in Livadia. It is highly to the credit of the actual government of Tuscany, that it allowed these poor fugitives 3 livres a day each, and free quarters during their passage through these states.

Mrs. S. desires her best regards.

My dear Lord Byron,

Yours most faithfully,

P. B. SHELLEY.

To John Gisborne[21]

Lerici, June 18, 1822.

MY DEAR GISBORNE,

In my doubt as to which of your most interesting letters I shall answer, I quash the business one for the present, as the only part of it that requires an answer, requires also maturer consideration. In the first place I send you money for postage, as I intend to indulge myself in plenty of paper and no crossings. Mary will write soon; at present she suffers greatly from excess of weakness, produced by a severe miscarriage, from which she is now slowly recovering. Her situation for some hours was alarming, and as she was totally destitute of medical assistance, I took the most decisive resolutions, by dint of making her sit in ice, I succeeded in checking the hemorrhage and the fainting fits, so that when the physician arrived all danger was over, and he had nothing to do but to applaud me for my boldness. She is now doing well, and the sea-baths will soon restore her.

I have written to Ollier to send his account to you. The "Adonais" I wished to have had a fair chance, both because it is a favourite with me and on account of the memory of Keats, who was a poet of great genius, let the classic party say what it will. "Hellas" too I liked on account of the subject—one always finds some reason or other for liking one's own composition. The "Epipsychidion" I cannot look at; the person whom it celebrates was a cloud instead of a Juno; and poor Ixion[22] starts from the centaur that was the offspring of his own embrace. If you are anxious, however, to hear what I am and have been, it will tell you something thereof. It is an idealized history of my life and feelings. I think one is always in love with something or other; the error, and I confess it is not easy for spirits cased in flesh and blood to avoid it, consists in seeking in a mortal image the likeness of what is perhaps eternal. Hogg[23] is very droll and very wicked about this poem, which he says, he likes—he praises it and says:—

Tantum de medio sumptis accedit honoris.[24]

---

19. Heteristi: members of the Hetaerea, a Greek revolutionary organization prominent in the struggle for independence from Turkey. **20. Ipsilanti:** Prince Alexander Ypsilanti, 1792–1828, was president of the Hetaerea.

21. Gisborne: a retired merchant. He and his wife Maria were friends of the Shelleys. **22. cloud . . . Ixion:** A king in Thessaly, Ixion made love to Juno (Hera), wife of Zeus, on whom (in the form of a cloud) he begot the race of Centaurs. **23. Hogg:** Thomas Jefferson Hogg, Shelley's friend and subsequently his biographer. **24. Tantum . . . honoris:** "So much beauty accrues to words taken from the ordinary" (Horace, *Ars Poetica*, l. 243).

Now that, I contend, even in Latin, is not to be permitted.

Hunt is not yet arrived, but I expected him every day. I shall see little of Lord Byron, nor shall I permit Hunt to form the intermediate link between him and me. I detest all society—almost all, at least—and Lord Byron is the nucleus of all that is hateful and tiresome in it. He will be half mad to hear of these Memoirs.[25] As to me, you know my supreme indifference to such affairs, except that I must confess I am sometimes amused by the ridiculous mistakes of these writers. Tell me a little of what they say of me besides my being an Atheist. One thing I regret in it, I dread lest it should injure Hunt's prospects in the establishment of the Journal, for Lord Byron is so mentally capricious that the least impulse drives him from his anchorage. I hardly know what to think of your scheme of settling at the Land's End. Physical food is much cheaper, but you can have no intellectual food, except what is already dried and salted in folios, &c., and an unmixed diet of this sort, without any supply of fresh provisions, is bad for the spiritual digestion. The absence of care about money is certainly a great benefit, and whatever else you do, the vesting of your property in land, immediately under your own inspection, is certainly prudent. But why the Land's End? Why not choose the immediate neighbourhood of London, where the pulsations of that heart of activity and thought would reach you easily? This would be better for Henry too. I wish you would return to Italy for my own sake; but for yours, I think you are better where you are. Mrs. Gisborne's lessons would be an immense resource, if she obtained more pupils. You do not tell me how her health is. I am rejoiced to find that yours is improved, and this is a strong motive for remaining in England. As to me, Italy is more and more delightful to me, and yours and Mrs. Gisborne's presence here is almost the only accessory I could desire, though, if my wishes were not limited by my hopes, Hogg would be included. I only feel the want of those who can feel, and understand me. Whether from proximity and the continuity of domestic intercourse, Mary does not. The necessity of concealing from her thoughts that would pain her, necessitates this, perhaps. It is the curse of Tantalus, that a person possessing such excellent powers and so

pure a mind as hers, should not excite the sympathy indispensable to their application to domestic life. The Williams's are now on a visit to us, and they are people who are very pleasing to me. But words are not the instruments of our intercourse. I like Jane more and more, and I find Williams the most amiable of companions. She has a taste for music, and an elegance of form and motions that compensate in some degree for the lack of literary refinements. Mrs. Gisborne knows my gross ideas of music, and will forgive me when I say that I listen the whole evening on our terrace to the simple melodies with excessive delight. I have a boat here which was originally intended to belong equally to Williams, Trelawny, and myself, but the wish to escape from the third person induced me to become the sole proprietor. It cost me £80, and reduced me to some difficulty in point of money. However, it is swift and beautiful, and appears quite a vessel. Williams is captain, and we drive along this delightful bay in the evening wind, under the summer moon, until earth appears another world. Jane brings her guitar, and if the past and the future could be obliterated, the present would content me so well that I could say with Faust to the passing moment, "Remain, thou, thou art so beautiful."[26]

Clare is with us, and the death of her child seems to have restored her to tranquillity. Her character is somewhat altered. She is vivacious and talkative, and though she teases me sometimes, I like her. Mary is not, for the present, much discontented with her visit, which is merely temporary, and which the circumstances of the case rendered indispensable.——

Lord Byron is at Leghorn. He has fitted up a splendid vessel; a small schooner on the American model, and Trelawny is to be captain. How long the fiery spirit of our Pirate will accommodate itself to the caprice of the Poet remains to be seen.

As to Hunt, he can neither see nor feel any ill qualities from which there is a chance of his personally suffering. I write little now. It is impossible to compose except under the strong excitement of an assurance of finding sympathy in what you write. Imagine Demosthenes reciting a Philippic to the waves of the Atlantic! Lord Byron is in this respect fortunate. He touched a chord to which a million hearts responded, and the coarse music which he produced to please them disciplined him to the perfection to which he now approaches.

25. Memoirs: *Memoirs of the Life and Writings of Lord Byron, with Anecdotes of Some of His Contemporaries*, by John Watkins. The book castigates Byron (and also Shelley and Hunt) for immorality and blasphemy.

26. "Remain . . . beautiful": *Faust*, I. 1700.

I do not go on with "Charles the First."[27] I feel too little certainty of the future, and too little satisfaction with regard to the past, to undertake any subject seriously and deeply. I stand, as it were, upon a precipice, which I have ascended with great, and cannot descend without *greater*, peril, and I am content if the heaven above me is calm for the passing moment.

You don't tell me what you think of "Cain."[28] You send me the opinion of the populace, which you know I do not esteem. I have read several more of the plays of Calderon. "Los Dos Amantes del Cielo," is the finest, if I except one scene in the "Devocion de la Cruz." I read Greek and think about writing.

Remember me affectionately to Mrs. Gisborne and Henry. I do not think much of her pupils for *not* admiring Metastasio;[29] the *nil admirari*, however justly applied, seems to me a bad sign in a young person. I had rather a pupil of mine had conceived a frantic passion for Marini[30] himself, than that she had found out the critical defects of the most deficient author. When she becomes of her own accord full of genuine admiration of the finest scene in the "Purgatorio," or the opening of the "Paradiso," or some other neglected piece of excellence, hope great things.

Adieu, I must not exceed the limits of my paper however little scrupulous I seem about those of your patience.

> Ever yours most affectionately,
>
> P. B. S.

I waited three days to get this pen mended, and at last was obliged to write.

EDWARD JOHN TRELAWNY

from RECOLLECTIONS

～✦～

For further extracts from Trelawny see under Byron, above, p. 947.

～✦～

27. **"Charles the First"**: a drama. Shelley left it unfinished. 28. **"Cain"**: Byron's play. It was generally deemed blasphemous. 29. **Metastasio**: pseudonym of Pietro Trapassi, 1698–1782, a writer of opera librettos. 30. **Marini**: Giambattista Marini, 1569–1625, Italian poet who exploited a highly artificial style.

In the annals of authors I cannot find one who wrote under so many discouragements as Shelley; for even Bunyan's dungeon walls echoed the cheers of hosts of zealous disciples on the outside, whereas Shelley could number his readers on his fingers. He said, "I can only print my writings by stinting myself in food!" Published, or sold openly, they were not.

The utter loneliness in which he was condemned to pass the largest portion of his life would have paralysed any brains less subtilised by genius than his were. Yet he was social and cheerful, and, although frugal himself, most liberal to others, while to serve a friend he was ever ready to make any sacrifice. It was, perhaps, fortunate he was known to so few, for those few kept him close shorn. He went to Ravenna in 1821 on Byron's business, and, writing to his wife, makes this comment on the Pilgrim's asking him to execute a delicate commission: "But it seems destined that I am always to have some active part in the affairs of everybody whom I approach." And so he had.

. . .

Shelley's mental activity was infectious; he kept your brain in constant action. Its effect on his comrade was very striking. Williams gave up all his accustomed sports for books, and the bettering of his mind; he had excellent natural ability; and the Poet delighted to see the seeds he had sown, germinating. Shelley said he was the sparrow educating the young of the cuckoo. After a protracted labour, Ned was delivered of a five-act play. Shelley was sanguine that his pupil would succeed as a dramatic writer. One morning I was in Mrs. Williams's drawing-room, by appointment, to hear Ned read an act of his drama. I sat with an aspect as caustic as a critic who was to decide his fate. Whilst thus intent Shelley stood before us with a most woeful expression.

Mrs. Williams started up, exclaiming, "What's the matter, Percy?"

"Mary has threatened me."

"Threatened you with what?"

He looked mysterious and too agitated to reply.

Mrs. Williams repeated, "With what? to box your ears?"

"Oh, much worse than that; Mary says she will have a party; there are English singers here, the Sinclairs, and she will ask them, and everyone she or you know —oh, the horror!"

We all burst into a laugh except his friend Ned.

"It will kill me."

"Music, kill you!" said Mrs. Williams. "Why, you have told me, you flatterer, that you loved music."

"So I do. It's the company terrifies me. For pity go to Mary and intercede for me; I will submit to any other species of torture than that of being bored to death by idle ladies and gentlemen."

After various devices it was resolved that Ned Williams should wait upon the lady—he being gifted with a silvery tongue, and sympathising with the Poet in his dislike of fine ladies—and see what he could do to avert the threatened invasion of the Poet's solitude. Meanwhile, Shelley remained in a state of restless ecstasy; he could not even read or sit. Ned returned with a grave face; the Poet stood as a criminal stands at the bar, whilst the solemn arbitrator of his fate decides it. "The lady," commenced Ned, "has set her heart on having a party, and will not be baulked"; but, seeing the Poet's despair, he added, "It is to be limited to those here assembled, and some of Count Gamba's family; and instead of a musical feast—as we have no souls—we are to have a dinner." The Poet hopped off, rejoicing, making a noise I should have thought whistling, but that he was ignorant of that accomplishment.

I have seen Shelley and Byron in society, and the contrast was as marked as their characters. The former, not thinking of himself, was as much at ease as in his own home, omitting no occasion of obliging those whom he came in contact with, readily conversing with all or any who addressed him, irrespective of age or rank, dress or address. To the first party I went with Byron, as we were on our road, he said:

"It's so long since I have been in English society, you must tell me what are their present customs. Does rank lead the way, or does the ambassadress pair us off into the dining-room? Do they ask people to wine? Do we exit with the women, or stick to our claret?"

On arriving, he was flushed, fussy, embarrassed, over ceremonious, and ill at ease, evidently thinking a great deal of himself and very little of others. He had learnt his manners, as I have said, during the Regency, when society was more exclusive than even now, and consequently more vulgar.

To know an author, personally, is too often but to destroy the illusion created by his works; if you withdraw the veil of your idol's sanctuary, and see him in his night-cap, you discover a querulous old crone, a sour pedant, a supercilious coxcomb, a servile tuft-hunter, a saucy snob, or, at best, an ordinary mortal. Instead of the high-minded seeker after truth and

abstract knowledge, with a nature too refined to bear the vulgarities of life, as we had imagined, we find him full of egotism and vanity, and eternally fretting and fuming about trifles. As a general rule, therefore, it is wise to avoid writers whose works amuse or delight you, for when you see them they will delight you no more. Shelley was a grand exception to this rule. To form a just idea of his poetry, you should have witnessed his daily life; his words and actions best illustrated his writings. If his glorious conception of Gods and men constituted an atheist, I am afraid all that listened were little better. Sometimes he would run through a great work on science, condense the author's laboured exposition, and by substituting simple words for the jargon of the schools, make the most abstruse subject transparent. The cynic Byron acknowledged him to be the best and ablest man he had ever known. The truth was, Shelley loved everything better than himself. Self-preservation is, they say, the first law of nature, with him it was the last; and the only pain he ever gave his friends arose from the utter indifference with which he treated everything concerning himself. I was bathing one day in a deep pool in the Arno, and astonished the Poet by performing a series of aquatic gymnastics, which I had learnt from the natives of the South Seas. On my coming out, whilst dressing, Shelley said, mournfully:

"Why can't I swim, it seems so very easy?"

I answered, "Because you think you can't. If you determine, you will; take a header off this bank, and when you rise turn on your back, you will float like a duck; but you must reverse the arch in your spine, for it's now bent the wrong way."

He doffed his jacket and trowsers, kicked off his shoes and socks, and plunged in, and there he lay stretched out on the bottom like a conger eel, not making the least effort or struggle to save himself. He would have been drowned if I had not instantly fished him out. When he recovered his breath, he said:

"I always find the bottom of the well, and they say Truth lies there. In another minute I should have found it, and you would have found an empty shell. It is an easy way of getting rid of the body."

"What would Mrs. Shelley have said to me if I had gone back with your empty cage?"

"Don't tell Mary—not a word!" he rejoined, and then continued, "It's a great temptation; in another minute I might have been in another planet."

"But as you always find the bottom," I observed,

"you might have sunk 'deeper than did ever plummet sound'."

"I am quite easy on that subject," said the Bard. "Death is the veil, which those who live call life: they sleep, and it is lifted. Intelligence should be imperishable; the art of printing has made it so in this planet."

"Do you believe in the immortality of the spirit?"

He continued, "Certainly not; how can I? We know nothing; we have no evidence; we cannot express our inmost thoughts. They are incomprehensible even to ourselves."

"Why," I asked, "do you call yourself an atheist? it annihilates you in this world."

"It is a word of abuse to stop discussion, a painted devil to frighten the foolish, a threat to intimidate the wise and good. I used it to express my abhorrence of superstition; I took up the word, as a knight took up a gauntlet, in defiance of injustice. The delusions of Christianity are fatal to genius and originality: they limit thought."

Shelley's thirst for knowledge was unquenchable. He set to work on a book, or a pyramid of books; his eyes glistening with an energy as fierce as that of the most sordid gold-digger who works at a rock of quartz, crushing his way through all impediments, no grain of the pure ore escaping his eager scrutiny. I called on him one morning at ten, he was in his study with a German folio open, resting on the broad marble mantel-piece, over an old-fashioned fire-place, and with a dictionary in his hand. He always read standing if possible. He had promised over night to go with me, but now begged me to let him off. I then rode to Leghorn, eleven or twelve miles distant, and passed the day there; on returning at six in the evening to dine with Mrs. Shelley and the Williams's, as I had engaged to do, I went into the Poet's room and found him exactly in the position in which I had left him in the morning, but looking pale and exhausted.

"Well," I said, "have you found it?"

Shutting the book and going to the window, he replied, "No, I have lost it": with a deep sigh: "I have lost a day."

"Cheer up, my lad, and come to dinner."

Putting his long fingers through his masses of wild tangled hair, he answered faintly, "You go, I have dined—late eating don't do for me."

"What is this?" I asked as I was going out of the room, pointing to one of his bookshelves with a plate containing bread and cold meat on it.

"That,"—colouring,—"why that must be my dinner. It's very foolish; I thought I had eaten it."

Saying I was determined that he should for once have a regular meal, I lugged him into the dining-room, but he brought a book with him and read more than he ate. He seldom ate at stated periods, but only when hungry—and then like the birds, if he saw something edible lying about,—but the cupboards of literary ladies are like Mother Hubbard's, bare. His drink was water, or tea if he could get it, bread was literally his staff of life; other things he thought superfluous. An Italian who knew his way of life, not believing it possible that any human being would live as Shelley did, unless compelled by poverty, was astonished when he was told the amount of his income, and thought he was defrauded or grossly ignorant of the value of money. He, therefore, made a proposition which much amused the Poet, that he, the friendly Italian, would undertake for ten thousand crowns a-year to keep Shelley like a grand Seigneur, to provide his table with luxuries, his house with attendants, a carriage and opera box for my lady, besides adorning his person after the most approved Parisian style. Mrs. Shelley's toilette was not included in the wily Italian's estimates. The fact was, Shelley stinted himself to bare necessaries, and then often lavished the money, saved by unprecedented self-denial, on selfish fellows who denied themselves nothing; such as the great philosopher had in his eye, when he said, "It is the nature of extreme self-lovers, as they will set a house on fire, an' it were only to roast their own eggs."

Byron on our voyage to Greece, talking of England, after commenting on his own wrongs, said, "And Shelley, too, the best and most benevolent of men; they hooted him out of his country like a mad dog, for questioning a dogma. Man is the same rancorous beast now that he was from the beginning, and if the Christ they profess to worship re-appeared, they would again crucify him."

On Monday, July 8, 1822, I went with Shelley to his bankers, and then to a store. It was past one P.M. when we went on board our respective boats,—Shelley and Williams to return to their home in the Gulf of Spezzia; I in the "Bolivar" to accompany them into the offing. When we were under weigh, the guard-boat boarded us to overhaul our papers. I had not got my port clearance, the captain of the port having refused to give it to the mate, as I had often gone out without.

The officer of the Health Office consequently threatened me with forty days' quarantine. It was hopeless to think of detaining my friends. Williams had been for days fretting and fuming to be off; they had no time to spare, it was past two o'clock, and there was very little wind.

Sullenly and reluctantly I re-anchored, furled my sails, and with a ship's glass watched the progress of my friends' boat. My Genoese mate observed,—"They should have sailed this morning at three or four A.M., instead of three P.M. They are standing too much in shore; the current will set them there."

I said, "They will soon have the land-breeze."

"May-be," continued the mate, "she will soon have too much breeze; that gaff top-sail is foolish in a boat with no deck and no sailor on board." Then pointing to the S.W., "Look at those black lines and the dirty rags hanging on them out of the sky—they are a warning; look at the smoke on the water; the devil is brewing mischief."

There was a sea-fog, in which Shelley's boat was soon after enveloped, and we saw nothing more of her.

Although the sun was obscured by mists, it was oppressively sultry. There was not a breath of air in the harbour. The heaviness of the atmosphere and an unwonted stillness benumbed my senses. I went down into the cabin and sank into a slumber. I was roused up by a noise overhead and went on deck. The men were getting up a chain cable to let go another anchor. There was a general stir amongst the shipping; shifting berths, getting down yards and masts, veering out cables, hauling in of hawsers, letting go anchors, hailing from the ships and quays, boats sculling rapidly to and fro, It was almost dark, although only half-past six o'clock. The sea was of the colour, and looked as solid and smooth as a sheet of lead, and covered with an oily scum. Gusts of wind swept over without ruffling it, and big drops of rain fell on its surface, rebounding, as if they could not penetrate it. There was a commotion in the air, made up of many threatening sounds, coming upon us from the sea. Fishing-craft and coasting-vessels under bare poles rushed by us in shoals, running foul of the ships in the harbour. As yet the din and hubbub was that made by men, but their shrill pipings were suddenly silenced by the crashing voice of a thunder squall that burst right over our heads. For some time no other sounds were to be heard than the thunder, wind, and rain. When the fury of the storm, which did not last for more than twenty minutes,

had abated, and the horizon was in some degree cleared, I looked to seaward anxiously, in the hope of descrying Shelley's boat, amongst the many small craft scattered about. I watched every speck that loomed on the horizon, thinking that they would have borne up on their return to the port, as all the other boats that had gone out in the same direction had done.

I sent our Genoese mate on board some of the returning craft to make inquiries, but they all professed not to have seen the English boat. So remorselessly are the quarantine laws enforced in Italy, that, when at sea, if you render assistance to a vessel in distress, or rescue a drowning stranger, on returning to port you are condemned to a long and rigorous quarantine of fourteen or more days. The consequence is, should one vessel see another in peril, or even run it down by accident, she hastens on her course, and by general accord, not a word is said or reported on the subject. But to resume my tale. I did not leave the "Bolivar" until dark. During the night it was gusty and showery, and the lightning flashed along the coast: at daylight I returned on board, and resumed my examinations of the crews of the various boats which had returned to the port during the night. They either knew nothing, or would say nothing. My Genoese, with the quick eye of a sailor, pointed out, on board a fishing-boat, an English-made oar, that he thought he had seen in Shelley's boat, but the entire crew swore by all the saints in the calendar that this was not so. Another day was passed in horrid suspense. On the morning of the third day I rode to Pisa. Byron had returned to the Lanfranchi Palace. I hoped to find a letter from the Villa Magni: there was none. I told my fears to Hunt, and then went upstairs to Byron. When I told him, his lip quivered, and his voice faltered as he questioned me. I sent a courier to Leghorn to despatch the "Bolivar", to cruise along the coast, whilst I mounted my horse and rode in the same direction. I also despatched a courier along the coast to go as far as Nice. On my arrival at Via Reggio I heard that a punt, a water-keg, and some bottles had been found on the beach. These things I recognised as having been in Shelley's boat when he left Leghorn. Nothing more was found for seven or eight days, during which time of painful suspense I patrolled the coast with the coast-guard, stimulating them to keep a good look-out by the promise of a reward. It was not until many days after this that my worst fears were confirmed. Two bodies were found on the shore,—one near Via Reggio, which I went and examined. The face and hands, and parts of the body not protected by the dress, were fleshless. The tall slight figure, the jacket, the volume of Sophocles in one pocket, and Keats's poems in the other, doubled back, as if the reader, in the act of reading, had hastily thrust it away, were all too familiar to me to leave a doubt on my mind that this mutilated corpse was any other than Shelley's. The other body was washed on shore three miles distant from Shelley's, near the tower of Migliarino, at the Bocca Lericcio. I went there at once. This corpse was much more mutilated; it had no other covering than,—the shreds of a shirt, and that partly drawn over the head, as if the wearer had been in the act of taking it off,—a black silk handkerchief, tied sailor-fashion round the neck,— socks,—and one boot, indicating also that he had attempted to strip. The flesh, sinews, and muscles hung about in rags, like the shirt, exposing the ribs and bones. I had brought with me from Shelley's house a boot of Williams's, and this exactly matched the one the corpse had on. That, and the handkerchief, satisfied me that it was the body of Shelley's comrade. Williams was the only one of the three who could swim, and it is probable he was the last survivor. It is likewise possible, as he had a watch and money, and was better dressed than the others, that his body might have been plundered when found.

Shelley always declared that in case of wreck he would vanish instantly, and not imperil valuable lives by permitting others to aid in saving his, which he looked upon as valueless. It was not until three weeks after the wreck of the boat that a third body was found —four miles from the other two. This I concluded to be that of the sailor boy, Charles Vivian, although it was a mere skeleton, and impossible to be identified. It was buried in the sand, above the reach of the waves. I mounted my horse, and rode to the Gulf of Spezzia, put up my horse, and walked until I caught sight of the lone house on the sea-shore in which Shelley and Williams had dwelt, and where their widows still lived. Hitherto in my frequent visits—in the absence of direct evidence to the contrary—I had buoyed up their spirits by maintaining that it was not impossible but that the friends still lived; now I had to extinguish the last hope of these forlorn women. I had ridden fast, to prevent any ruder messenger from bursting in upon them. As I stood on the threshold of their house, the bearer, or rather confirmer, of news which would rack every fibre of their quivering frames to the utmost, I paused, and, looking at the sea, my memory reverted to our joyous parting only a few days before.

SHELLEY'S FUNERAL

I got a furnace made at Leghorn, of iron-bars and strong sheet-iron, supported on a stand, and laid in a stock of fuel, and such things as were said to be used by Shelley's much loved Hellenes on their funeral pyres.

On August 13, 1822, I went on board the "Bolivar", with an English acquaintance, having written to Byron and Hunt to say I would send them word when everything was ready, as they wished to be present. I had previously engaged two large feluccas, with drags and tackling, to go before, and endeavour to find the place where Shelley's boat had foundered; the captain of one of the feluccas having asserted that he was out in the fatal squall, and had seen Shelley's boat go down off Via Reggio, with all sail set. With light and fitful breezes we were eleven hours reaching our destination —the tower of Migliarino, at the Bocca Lericcio, in the Tuscan States. There was a village there, and about two miles from that place Williams was buried. So I anchored, landed, called on the officer in command, a major, and told him my object in coming, of which he was already apprised by his own government. He assured me I should have every aid from him. As it was too late in the day to commence operations, we went to the only inn in the place, and I wrote to Byron to be with us next day at noon. The major sent my letter to Pisa by a dragoon, and made arrangements for the next day. In the morning he was with us early, and gave me a note from Byron, to say he would join us as near noon as he could. At ten we went on board the commandant's boat, with a squad of soldiers in working dresses, armed with mattocks and spades, an officer of the quarantine service, and some of his crew. They had their peculiar tools, so fashioned as to do their work without coming into personal contact with things that might be infectious—long handled tongs, nippers, poles with iron hooks and spikes, and divers others that gave one a lively idea of the implements of torture devised by the holy inquisitors. Thus freighted, we started, my own boat following with the furnace, and the things I had brought from Leghorn. We pulled along the shore for some distance, and landed at a line of strong posts and railings which projected into the sea—forming the boundary dividing the Tuscan and Lucchese States. We walked along the shore to the grave, where Byron and Hunt soon joined us: they, too, had an officer and soldiers from the tower of Migliarino, an officer of the Health Office, and some dismounted dragoons, so we were surrounded by soldiers, but they kept the ground clear, and readily lent their aid. There was a considerable gathering of spectators from the neighbourhood, and many ladies richly dressed were amongst them. The spot where the body lay was marked by the gnarled root of a pine tree.

A rude hut, built of young pine-tree stems, and wattled with their branches, to keep the sun and rain out, and thatched with reeds, stood on the beach to shelter the look-out man on duty. A few yards from this was the grave, which we commenced opening—the Gulf of Spezzia and Leghorn at equal distances of twenty-two miles from us. As to fuel I might have saved myself the trouble of bringing any, for there was an ample supply of broken spars and planks cast on the shore from wrecks, besides the fallen and decaying timber in a stunted pine forest close at hand. The soldiers collected fuel whilst I erected the furnace, and then the men of the Health Office set to work, shovelling away the sand which covered the body, while we gathered round, watching anxiously. The first indication of their having found the body, was the appearance of the end of a black silk handkerchief—I grubbed this out with a stick, for we were not allowed to touch anything with our hands—then some shreds of linen were met with, and a boot with the bone of the leg and the foot in it. On the removal of a layer of brush-wood, all that now remained of my lost friend was exposed—a shapeless mass of bones and flesh. The limbs separated from the trunk on being touched.

"Is that a human body?" exclaimed Byron; "why it's more like the carcase of a sheep, or any other animal, than a man: this is a satire on our pride and folly."

I pointed to the letters E. E. W. on the black silk handkerchief.

Byron looking on, muttered, "The entrails of a worm hold together longer than the potter's clay, of which man is made. Hold! let me see the jaw," he added, as they were removing the skull, "I can recognise any one by the teeth, with whom I have talked. I always watch the lips and mouth: they tell what the tongue and eyes try to conceal."

I had a boot of Williams's with me; it exactly corresponded with the one found in the grave. The remains were removed piecemeal into the furnace.

"Don't repeat this with me," said Byron; "let my carcase rot where it falls."

The funeral pyre was now ready; I applied the fire, and the materials being dry and resinous the pine-wood burnt furiously, and drove us back. It was hot enough before, there was no breath of air, and the loose sand scorched our feet. As soon as the flames became clear, and allowed us to approach, we threw frankincense and salt into the furnace, and poured a flask of wine and oil over the body. The Greek oration was omitted, for we had lost our Hellenic bard. It was now so insufferably hot that the officers and soldiers were all seeking shade.

"Let us try the strength of these waters that drowned our friends," said Byron, with his usual audacity. "How far out do you think they were when their boat sank?"

"If you don't wish to be put into the furnace, you had better not try; you are not in condition."

He stripped, and went into the water, and so did I and my companion. Before we got a mile out, Byron was sick, and persuaded to return to the shore. My companion, too, was seized with cramp, and reached the land by my aid. At four o'clock the funeral pyre burnt low, and when we uncovered the furnace, nothing remained in it but dark-coloured ashes, with fragments of the larger bones. Poles were now put under the red-hot furnace, and it was gradually cooled in the sea. I gathered together the human ashes, and placed them in a small oak-box, bearing an inscription on a brass plate, screwed it down, and placed it in Byron's carriage. He returned with Hunt to Pisa, promising to be with us on the following day at Via Reggio. I returned with my party in the same way we came, and supped and slept at the inn. On the following morning we went on board the same boats, with the same things and party, and rowed down the little river near Via Reggio to the sea, pulled along the coast towards Massa, then landed, and began our preparations as before.

Three white wands had been stuck in the sand to mark the Poet's grave, but as they were at some distance from each other, we had to cut a trench thirty yards in length, in the line of the sticks, to ascertain the exact spot, and it was nearly an hour before we came upon the grave.

In the mean time Byron and Leigh Hunt arrived in the carriage, attended by soldiers, and the Health Officer, as before. The lonely and grand scenery that surrounded us so exactly harmonised with Shelley's genius, that I could imagine his spirit soaring over us. The sea, with the islands of Gorgona, Capraji, and Elba, was before us; old battlemented watch-towers stretched along the coast, backed by the marble-crested Apennines glistening in the sun, picturesque from their diversified outlines, and not a human dwelling was in sight. As I thought of the delight Shelley felt in such scenes of loneliness and grandeur whilst living, I felt we were no better than a herd of wolves or a pack of wild dogs, in tearing out his battered and naked body from the pure yellow sand that lay so lightly over it, to drag him back to the light of day; but the dead have no voice, nor had I power to check the sacrilege—the work went on silently in the deep and unresisting sand, not a word was spoken, for the Italians have a touch of sentiment, and their feelings are easily excited into sympathy. Even Byron was silent and thoughtful. We were startled and drawn together by a dull hollow sound that followed the blow of a mattock; the iron had struck a skull, and the body was soon uncovered. Lime had been strewn on it; this, or decomposition, had the effect of staining it of a dark and ghastly indigo colour. Byron asked me to preserve the skull for him; but remembering that he had formerly used one as a drinking-cup, I was determined Shelley's should not be so profaned. The limbs did not separate from the trunk, as in the case of Williams's body, so that the corpse was removed entire into the furnace. I had taken the precaution of having more and larger pieces of timber, in consequence of my experience of the day before of the difficulty of consuming a corpse in the open air with our appara-tus. After the fire was well kindled we repeated the ceremony of the previous day; and more wine was poured over Shelley's dead body than he had consumed during his life. This with the oil and salt made the yellow flames glisten and quiver. The heat from the sun and fire was so intense that the atmosphere was tremulous and wavy. The corpse fell open and the heart was laid bare. The frontal bone of the skull, where it had been struck with the mattock, fell off; and, as the back of the head rested on the red-hot bottom bars of the furnace, the brains literally seethed, bubbled, and boiled as in a cauldron, for a very long time.

Byron could not face this scene, he withdrew to the beach and swam off to the "Bolivar". Leigh Hunt remained in the carriage. The fire was so fierce as to produce a white heat on the iron, and to reduce its contents to grey ashes. The only portions that were not consumed were some fragments of bones, the jaw, and the skull, but what surprised us all, was that the heart remained entire. In snatching this relic from the fiery furnace, my hand was severely burnt; and had any one seen me do the act I should have been put into quarantine.

After cooling the iron machine in the sea, I collected the human ashes and placed them in a box, which I took on board the "Bolivar". Byron and Hunt retraced their steps to their home, and the officers and soldiers returned to their quarters. I liberally rewarded the men for the admirable manner in which they behaved during the two days they had been with us.

As I undertook and executed this novel ceremony, I have been thus tediously minute in describing it.

Byron's idle talk during the exhumation of Williams's remains, did not proceed from want of feeling, but from his anxiety to conceal what he felt from others. When confined to his bed and racked by spasms, which threatened his life, I have heard him talk in a much more unorthodox fashion, the instant he could muster breath to banter. He had been taught during his town-life, that any exhibition of sympathy or feeling was maudlin and unmanly, and that the appearance of daring and indifference, denoted blood and high breeding.

Shelley came of a long-lived race, and, barring accidents, there was no reason why he should not have emulated his forefathers in attaining a ripe age. He had no other complaint than occasional spasms, and these were probably caused by the excessive and almost unremitting strain on his mental powers, the solitude of his life, and his long fasts, which were not intentional, but proceeded from the abstraction and forgetfulness of himself and his wife. If food was near him, he ate it,—if not, he fasted, and it was after long fasts that he suffered from spasms. He was tall, slim, and bent from eternally poring over books; this habit had contracted his chest. His limbs were well proportioned, strong and bony—his head was very small—and his features were expressive of great sensibility, and decidedly feminine. There was nothing about him outwardly to attract notice, except his extraordinarily juvenile appearance. At twenty-nine, he still retained on his tanned and freckled cheeks, the fresh look of a boy—although his long wild locks were coming into blossom, as a polite hairdresser once said to me, whilst cutting mine.

It was not until he spoke that you could discern any-thing uncommon in him—but the first sentence he uttered, when excited by his subject, riveted your attention. The light from his very soul streamed from his eyes, and every mental emotion of which the human mind is susceptible, was expressed in his pliant and ever-changing features. He left the conviction on the minds of his audience, that however great he was as a Poet, he was greater as an orator. There was another and most rare peculiarity in Shelley,—his intellectual faculties completely mastered his material nature, and hence he unhesitatingly acted up to his own theories, if they only demanded sacrifices on his part,—it was where they implicated others that he forbore. Mrs. Shelley has observed, "Many have suggested and advocated far greater innovations in our political and social system than Shelley; but he alone practised those he approved of as just."

JOHN CLARE

1793–1864

CLARE is chronologically the last—and next to Robert Burns he is perhaps now the most famous —of the line of poets whom the eighteenth and early nineteenth centuries praised as examples of "original genius." He was born in the village of Helpston, Northamptonshire, the son of a farm laborer who could barely read (Clare's mother was completely illiterate). A sensitive child, and extremely delicate in health, he quickly acquired the knack of writing verse. When he was twenty-four he tried to get some of his poems published. At last a copy of his prospectus came to the attention of the London publisher John Taylor; and the firm of Taylor and Hessey —the same firm that was now publishing Keats—soon brought out two volumes, *Poems Descriptive of Rural Life and Scenery* (1820) and *The Village Minstrel* (1821). The first of these two volumes became immediately popular, going through four editions in a year, and during a brief visit to London Clare met Coleridge, Lamb, De Quincey, and other writers. But the second volume sold very little. Meanwhile Clare returned to Northamptonshire, married, and found himself struggling to provide for his growing family. In 1836 he had a mental breakdown and spent the following four years in a private asylum in Epping. He was then permitted to return home, but in a few months his condition became worse and he spent the remaining twenty-three years of his life in St. Andrew's Asylum in Northampton.

Clare wrote poetry almost as readily as most people talk and he seldom revised. His illness in no way lessened his creative energy or skill. The result was that he produced an enormous amount of verse, of which only four volumes appeared during his lifetime. Much of it has never been published. His earlier writings were admired especially for their exact and loving observation, their fresh, delighted way of watching the natural world. He never lost this lyrical realism, but as time went on he also developed other themes and manners. For example, "Pastoral Poesy," though entirely characteristic, shows him assimilating ideas derived mainly from Wordsworth and Coleridge. In later years he fantasied he had known and lost in his youth an ideal love. Many of his poems are to or about this, and they have a strange poignance. In his illness he is creating the memory of "Mary Joyce" because he needs an object and explanation of his deep sense of bereavement, yet the memory can only be insubstantial and elusive. In other poems such as "Invitation to Eternity" or "I Am" he explores, like his great contemporaries, the "dark passages," as Keats put it, that open on all sides as we begin to feel the mystery of man's nature and place in the cosmos. And in such a late poem as "I Lost the Love of Heaven" Clare's most obvious affinity is with the visionary lyrics of Blake.

Bibliography

The largest collected edition is *The Poems of John Clare*, ed. J. W. Tibble, 2 vols., 1935. Additional poems may be found in *Poems of John Clare's Madness*, ed. Geoffrey Grigson, 1949. Other selections are by Grigson for the Muses' Library, 1950, and by James Reeves, 1954. There are also *Sketches in the Life of John Clare, Written by Himself*, ed. Edmund Blunden, 1931; and *The Prose of John Clare*, 1951, and *Letters*, 1951, both edited by J. W. and Anne Tibble. The standard biography is *John Clare: His Life and Poetry*, 1956, also by J. W. and Anne Tibble. Critical studies include J. M. Murry, "The Poetry of John Clare," in *Countries of the Mind*, 1922, rev. ed., 1931, and J. Heath-Stubbs, "Clare and the Peasant Tradition," in *The Darkling Plain*, 1950.

IMPROMPTU ON WINTER

O winter, what a deadly foe
Art thou unto the mean and low!
What thousands now half pin'd and bare
Are forced to stand thy piercing air
All day, near numbed to death wi' cold
Some petty gentry to uphold,
Paltry proudlings hard as thee,
Dead to all humanity.
Oh, the weather's cold and snow,
Cutting winds that round me blow,   10
But much more the killing scorn!
Oh, the day that I was born
Friendless—poor as I can be,
Struck wi' death o' poverty!
But why need I the winter blame?
To me all seasons come the same:
Now winter bares each field and tree
She finds that trouble sav'd in me
Stript already, penniless,
Nothing boasting but distress;   20
And when spring chill'd nature cheers,
Still my old complaint she hears;
Summer too, in plenty blest,
Finds me poor and still distrest;
Kind autumn too, so liberal and so free,
Brings my old well-known present, Poverty.

1809–10 (1935)

SONG

One gloomy eve I roam'd about
'Neath Oxey's hazel bowers,
While timid hares were daring out,
To crop the dewy flowers;

SONG. **2. Oxey:** wooded area near Helpston.

And soothing was the scene to me,
　Right placid was my soul,
My breast was calm as summer's sea
　When waves forget to roll.

But short was even's placid smile,
　My startled soul to charm,   10
When Nelly lightly skipt the stile,
　With milk-pail on her arm:
One careless look on me she flung,
　As bright as parting day;
And like a hawk from covert sprung,
　It pounc'd my peace away.

1819–21 (1821)

PASTORAL POESY

True poesy is not in words,
　But images that thoughts express,
By which the simplest hearts are stirred
　To elevated happiness.

Mere books would be but useless things
　Where none had taste or mind to read,
Like unknown lands where beauty springs
　And none are there to heed.

But poesy is a language meet,
　And fields are every one's employ;   10
The wild flower 'neath the shepherd's feet
　Looks up and gives him joy;

A language that is ever green,
　That feelings unto all impart,
As hawthorn blossoms, soon as seen,
　Give May to every heart.

An image to the mind is brought,
　Where happiness enjoys
An easy thoughtlessness of thought
　And meets excess of joys.   20

And such is poesy; its power
　May varied lights employ,
Yet to all minds it gives the dower
　Of self-creating joy.

And whether it be hill or moor,
　I feel where'er I go
A silence that discourses more
　That any tongue can do.

Unruffled quietness hath made
　A peace in every place,   30
And woods are resting in their shade
　Of social loneliness.

The storm, from which the shepherd turns
　To pull his beaver down,
While he upon the heath sojourns,
　Which autumn pleaches brown,

Is music, ay, and more indeed
　To those of musing mind
Who through the yellow woods proceed
　And listen to the wind.   40

The poet in his fitful glee
　And fancy's many moods
Meets it as some strange melody,
　A poem of the woods,

And now a harp that flings around
　The music of the wind;
The poet often hears the sound
　When beauty fills the mind.

So would I my own mind employ,
　And my own heart impress,   50
That poesy's self's a dwelling joy
　Of humble quietness.

1824–32 (1935)

WINTER WALK

The holly bush, a sober lump of green,
Shines through the leafless shrubs all brown and grey,
And smiles at winter, be it e'er so keen,
With all the leafy luxury of May.
And oh, it is delicious, when the day
In winter's loaded garment keenly blows
And turns her back on sudden falling snows,
To go where gravel pathways creep between
Arches of evergreen that scarce let through
A single feather of the driving storm;   10
And in the bitterest day that ever blew
The walk will find some places still and warm
Where dead leaves rustle sweet and give alarm
To little birds that flirt and start away.

1832–35 (1920)

PASTORAL POESY. **36. pleaches:** bleaches. WINTER WALK. **14. flirt:** flutter.

THE VIXEN

Among the taller wood with ivy hung,
The old fox plays and dances round her young.
She snuffs and barks if any passes by
And swings her tail and turns prepared to fly.
The horseman hurries by, she bolts to see,
And turns agen, from danger never free.
If any stands she runs among the poles
And barks and snaps and drives them in the holes.
The shepherd sees them and the boy goes by
And gets a stick and progs the hole to try.   10
They get all still and lie in safety sure,
And out again when everything's secure,
And start and snap at blackbirds bouncing by
To fight and catch the great white butterfly.

1835–37 (1920)

BADGER

When midnight comes a host of dogs and men
Go out and track the badger to his den,
And put the sack within the hole, and lie
Till the old grunting badger passes by.
He comes and hears—they let the strongest loose.
The old fox hears the noise and drops the goose.
The poacher shoots and hurries from the cry,
And the old hare half wounded buzzes by.
They get a forked stick to bear him down
And clap the dogs and take him to the town,   10
And bait him all the day with many dogs,
And laugh and shout and fright the scampering hogs.
He runs along and bites at all he meets:
They shout and hollo down the noisy streets.

He turns about to face the loud uproar
And drives the rebels to their very door.
The frequent stone is hurled where'er they go;
When badgers fight, then every one's a foe.
The dogs are clapt and urged to join the fray;
The badger turns and drives them all away.   20
Though scarcely half as big, demure and small,
He fights with dogs for hours and beats them all.
The heavy mastiff, savage in the fray,
Lies down and licks his feet and turns away.

THE VIXEN. **7. poles:** weasels. **10. progs:** prods, pokes.

The bulldog knows his match and waxes cold,
The badger grins and never leaves his hold.
He drives the crowd and follows at their heels
And bites them through—the drunkard swears and
    reels.

The frighted women take the boys away,
The blackguard laughs and hurries on the fray.   30
He tries to reach the woods, an awkward race,
But sticks and cudgels quickly stop the chase.
He turns agen and drives the noisy crowd
And beats the many dogs in noises loud.
He drives away and beats them every one,
And then they loose them all and set them on.
He falls as dead and kicked by boys and men,
Then starts and grins and drives the crowd agen;
Till kicked and torn and beaten out he lies
And leaves his hold and cackles, groans, and dies.   40

1835–37 (1920)

THE PEASANT POET

He loved the brook's soft sound,
    The swallow swimming by
He loved the daisy-covered ground,
    The cloud-bedappled sky.
To him the dismal storm appeared
    The very voice of God;
And when the evening rock was reared
    Stood Moses with his rod.
And everything his eyes surveyed,
    The insects i' the brake,   10
Were creatures God Almighty made,
    He loved them for his sake—
A silent man in life's affairs,
    A thinker from a boy,
A peasant in his daily cares,
    A poet in his joy.

after 1842 (1920)

ETERNITY OF NATURE

All nature has a feeling: woods, fields, brooks
    Are life eternal; and in silence they
Speak happiness beyond the reach of books;
    There's nothing mortal in them; their decay
    Is the green life of change; to pass away

And come again in blooms revivified.
Its birth was heaven, eternal is its stay,
And with the sun and moon shall still abide
Beneath their day and night and heaven wide.

after 1842 (1935)

POETS LOVE NATURE

Poets love nature, and themselves are love,
The scorn of fools, and mock of idle pride.
The vile in nature worthless deeds approve,
They court the vile and spurn all good beside.
Poets love nature; like the calm of heaven,
Her gifts like heaven's love spread far and wide:
In all her works there are no signs of leaven,
Sorrow abashes from her simple pride.
Her flowers, like pleasures, have their season's birth,
And bloom through regions here below;   10
They are her very scriptures upon earth,
And teach us simple mirth where'er we go.
Even in prison they can solace me,
For where they bloom God is, and I am free.

after 1842 (1873)

LITTLE TROTTY WAGTAIL

Little trotty wagtail, he went in the rain,
And tittering, tottering sideways he ne'er got straight
    again,
He stooped to get a worm, and looked up to catch a
    fly,
And then he flew away ere his feathers they were dry.

Little trotty wagtail, he waddled in the mud,
And left his little footmarks, trample where he would.
He waddled in the water-pudge, and waggle went his
    tail,
And chirrupt up his wings to dry upon the garden rail.

Little trotty wagtail, you nimble all about,
And in the dimpling water-pudge you waddle in
    and out;   10
Your home is nigh at hand, and in the warm pigsty,
So, little Master Wagtail, I'll bid you a good-bye.

after 1842 (1873)

LITTLE TROTTY WAGTAIL. **Title:** A wagtail is a small bird having a long tail that bobs up and down. **7. water-pudge:** puddle.

LOVE OF NATURE

I love thee, nature, with a boundless love,
    The calm of earth, the storm of roaring woods;
The winds breathe happiness where'er I rove,
    There's life's own music in the swelling floods.
My heart is in the thunder-melting clouds,
    The snow-capt mountain, and the rolling sea;
And hear ye not the voice where darkness shrouds
    The heavens? There lives happiness for me.

Death breathes its pleasures when it speaks of him;
    My pulse beats calmer while his lightnings play. 10
My eye, with earth's delusions waxing dim,
    Clears with the brightness of eternal day.
The elements crash round me: it is he!
    Calmly I hear his voice and never start.
From Eve's posterity I stand quite free,
    Nor feel her curses rankle round my heart.

Love is not here. Hope is, and at his voice—
    The rolling thunder and the roaring sea—
My pulses leap, and with the hills rejoice;
    Then strife and turmoil are at end for me,   20
No matter where life's ocean leads me on;
    For nature is my mother, and I rest,
When tempests trouble and the sun is gone,
    Like to a weary child upon her breast.

after 1842 (1873)

CLOCK-A-CLAY

In the cowslip pips I lie,
Hidden from the buzzing fly,
While green grass beneath me lies,
Pearled with dew like fishes' eyes,
Here I lie, a clock-a-clay,
Waiting for the time of day.

While grassy forest quakes surprise,
And the wild wind sobs and sighs,
My gold home rocks as like to fall,
On its pillar green and tall;   10
When the pattering rain drives by
Clock-a-clay keeps warm and dry.

CLOCK-A-CLAY. **Title:** ladybug or ladybird. **1. pips:** blossoms. Cf. *The Tempest*, V. i. 89.

Day by day and night by night,
All the week I hide from sight;
In the cowslip pips I lie,
In rain and dew still warm and dry;
Day and night, and night and day,
Red, black-spotted clock-a-clay.

My home shakes in wind and showers,
Pale green pillar topped with flowers,   20
Bending at the wild wind's breath,
Till I touch the grass beneath;
Here I live, lone clock-a-clay,
Watching for the time of day.

after 1842 (1873)

SECRET LOVE

I hid my love when young till I
Couldn't bear the buzzing of a fly;
I hid my love to my despite
Till I could not bear to look at light:
I dare not gaze upon her face
But left her memory in each place;
Where eer I saw a wild flower lie
I kissed and bade my love good bye.

I met her in the greenest dells
Where dewdrops pearl the wood bluebells;   10
The lost breeze kissed her bright blue eye,
The bee kissed and went singing by,
A sunbeam found a passage there,
A gold chain round her neck so fair;
As secret as the wild bee's song
She lay there all the summer long.

I hid my love in field and town
Till een the breeze would knock me down,
The bees seemed singing ballads oer,
The fly's bass turned a lion's roar;   20
And even silence found a tongue,
To haunt me all the summer long;
The riddle nature could not prove
Was nothing else but secret love.

after 1842 (1920)

STANZAS

Black absence hides upon the past,
    I quite forget thy face;
And memory like the angry blast
    Will love's last smile erase.
I try to think of what has been,
    But all is blank to me;
And other faces pass between
    My early love and thee.

I try to trace thy memory now,
    And only find thy name;   10
Those inky lashes on thy brow,
    Black hair and eyes the same;
Thy round pale face of snowy dyes,
    There's nothing paints thee there.
A darkness comes before my eyes
    For nothing seems so fair.

I knew thy name so sweet and young;
    'Twas music to my ears,
A silent word upon my tongue,
    A hidden thought for years.   20
Dark hair and lashes swarthy too,
    Arched on thy forehead pale:
All else is vanished from my view
    Like voices on the gale.

after 1842 (1935)

I LOST THE LOVE
OF HEAVEN

I lost the love of heaven above,
    I spurned the lust of earth below,
I felt the sweets of fancied love,
    And hell itself my only foe.

I lost earth's joys, but felt the glow
    Of heaven's flame abound in me,
Till loveliness and I did grow
    The bard of immortality.

I loved, but woman fell away;
    I hid me from her faded flame.   10
I snatched the sun's eternal ray
    And wrote till earth was but a name.

In every language upon earth,
   On every shore, o'er every sea,
I gave my name immortal birth
   And kept my spirit with the free.

after 1842 (1924)

INVITATION TO ETERNITY

Say, wilt thou go with me, sweet maid,
Say, maiden, wilt thou go with me
Through the valley-depths of shade,
Of bright and dark obscurity;
Where the path has lost its way,
Where the sun forgets the day,
Where there's nor light nor life to see,
Sweet maiden, wilt thou go with me?

Where stones will turn to flooding streams,
Where plains will rise like ocean's waves,   10
Where life will fade like visioned dreams
And darkness darken into caves,
Say, maiden, wilt thou go with me
Through this sad non-identity
Where parents live and are forgot,
And sisters live and know us not?

Say, maiden, wilt thou go with me
In this strange death of life to be,
To live in death and be the same,
Without this life or home or name,   20
At once to be and not to be—
That was and is not—yet to see
Things pass like shadows, and the sky
Above, below, around us lie?

The land of shadows wilt thou trace,
Nor look nor know each other's face;
The present marred with reason gone,
And past and present both as one?
Say, maiden, can thy life be led
To join the living and the dead?   30
Then trace thy footsteps on with me:
We are wed to one eternity.

after 1842 (1920)

I AM

I am: yet what I am none cares or knows,
   My friends forsake me like a memory lost;
I am the self-consumer of my woes,
   They rise and vanish in oblivious host,
Like shades in love and death's oblivion lost;
And yet I am, and live with shadows tost

Into the nothingness of scorn and noise,
   Into the living sea of waking dreams,
Where there is neither sense of life nor joys,
   But the vast shipwreck of my life's esteems;   10
And e'en the dearest—that I loved the best—
Are strange—nay, rather stranger than the rest.

I long for scenes where man has never trod,
   A place where woman never smiled or wept;
There to abide with my Creator, God,
   And sleep as I in childhood sweetly slept:
Untroubling and untroubled where I lie,
The grass below—above the vaulted sky.

after 1842 (1865)

JOHN CLARE

I feel I am, I only know I am,
And plod upon the earth as dull and void;
Earth's prison chilled my body with its dram
Of dullness, and my soaring thoughts destroyed.
I fled to solitude from passion's dream,
But strife pursued: I only know I am.
I was a being created in the race
Of men, disdaining bounds of place and time;
A spirit that could travel o'er the space
Of earth and heaven like a thought sublime;   10
Tracing creation, like my Maker free,
A soul unshackled like eternity:
Spurning earth's vain and soul-debasing thrall—
But now I only know I am, that's all.

after 1842 (1935)

JOHN KEATS

1795–1821

THE YOUNGEST of the major Romantic writers, Keats died at the age of twenty-five. But he gave the surest promise of a form of genius (and a breadth of humanity) found only in the very greatest poets. It has been seriously argued that no English or American poet of the last two centuries excels him in natural endowment; and when writers think of the richly varied nature of that endowment (from effortless magic of phrase, through fertility in the use of different styles, to clairvoyance, humor, and range of sympathetic understanding) the comparison most frequently made is with Shakespeare.

The life of Keats has been told and retold even for readers who have little interest in literature. It is a moving, dramatic story. He was born on October 31, 1795, the first child of Thomas and Frances (Jennings) Keats, in what was then the northern part of London (the district of Finsbury). Thomas Keats, a young man who had come from the western part of England, probably Cornwall, worked in the stables of the prosperous John Jennings, Keats's grandfather, who provided horses for coaches that ran from London to the northern parts of England. The couple had other children—George, who later, at the age of twenty-one, emigrated to America and settled in Kentucky; Thomas (or Tom as he was always called); and then the younger sister, Fanny Keats, the only child to live to an old age. (She married a Spaniard, went to live in Spain, and died in 1889, sixty-eight years after the death of her brother. The date reminds us that if Keats had lived even into middle age he would have been a Victorian, with incalculable consequences for the literature of that period.)

At the age of eight Keats was sent to a small school in the village of Enfield, not far away. Shortly afterward his father was killed in a riding accident. Keats's mother died of tuberculosis when he was fourteen, leaving the four Keats children with only their elderly grandmother, Mrs. Jennings, to look after them.

At the school in Enfield, Keats showed little interest in books at first. Schoolmates afterward recalled mainly his terrier courage in fighting: he "was a boy whom any one . . . might easily have fancied would become great—but rather in some military capacity than in literature." But before his mother's death he had already begun to dip, on his own, into the small school library. Now he turned to books avidly. He also started to write a prose translation of Vergil's *Aeneid*. By this time he had caught the attention of Charles Cowden Clarke, an assistant at the school, which was conducted by his father; and Clarke, who was seven years older than Keats, did everything he could to encourage him. But Keats was to remain at school for only another year. His grandmother, uncertain how much longer she herself might live, appointed as guardian to her grandchildren a tea-merchant named Richard Abbey, who was later—after the grandmother's death—to use for his own purposes money they should have inherited. Abbey thought little of schools—he himself was poorly educated—and he removed Keats from Enfield and apprenticed him to a surgeon in the nearby village of Edmonton. But Keats could walk over to his old school in the evening, and he and Clarke continued to read out loud and to talk of books. Clarke tells of the time they began reading Spenser's *Faerie Queene*; Keats went through it, he says, "as a young horse would through a spring meadow—ramping." In this eagerness of response Keats already showed his remarkable gift of empathy (or sympathetic imagination). Coming across a phrase in the *Faerie Queene*, "sea-shouldering whales," he felt empathically the parting waves, and "hoisted himself up," said Clarke, "and looked burly and dominant, as he said, 'What an image that is—*sea-shouldering whales*.'"

After four years as an apprentice, Keats went in 1815 as a beginning medical student to Guy's Hospital. Here he attended lectures and helped in the hospital wards. Meanwhile he had started to write poetry. The thought of the great poets of the past became a dominant, challenging ideal, and this bold response to the "vision of greatness," to use Whitehead's phrase, never left him. At this time (Oct., 1816) he began to meet writers, most notably Leigh Hunt, whose liberal political journalism Keats had already seen at Enfield and who was also fairly well known as an "advanced" poet—a poet who was trying to do something very different from the neo-classic tradition of Dryden and Pope. Keats's early poetry, down through "Sleep and Poetry" and even *Endymion*, shows the influence of Hunt. His first book of poems (March, 1817) was published when he was twenty-two.

But he wanted to write a "long poem," one that, he said, would be a genuine "trial of Invention." He plunged into *Endymion*—an enormous effort of about four thousand lines—and finished it by the end of the year. Throughout this time he was also reading and rereading Shakespeare (one result was the famous "Negative Capability" letter, Dec. 21-27, 1817, below, p. 1209). Little satisfied with *Endymion*, he now, after this apprentice effort, hoped to begin a different sort of poetry. He was haunted by the great English poetic forms—the Miltonic epic and, above all, the Shakespearean drama. (Compare, below, "Lines on Seeing a Lock of Milton's

Hair" and "On Sitting Down to Read *King Lear* Once Again.") But also, throughout the spring of 1818, he was reflecting upon the direction that modern poetry seemed to be taking, and especially upon Wordsworth, whose poetry typified to Keats a new "thinking into the heart." (See especially the famous letter of May 3, 1818, on the "Chamber of Maiden-Thought.") Keats's ultimate hope was to combine the power and scope of the earlier poetry with the inward searching of the new. While trying to find ways to do this—which he took for granted would involve a "long preparation"—he tossed off a short romance, "Isabella," and a number of Shakespearean sonnets. His brother George married at about this time and emigrated to Kentucky; it was necessary for Keats to take George's place in looking after their brother Tom, who had tuberculosis. But Tom seemed well enough to allow Keats to go that summer (1818) on a walking tour of northern England and Scotland with his friend Charles Brown. He returned with a sore throat that was to bother him for another year, and, on returning, found that Tom was dying. Throughout the fall he nursed Tom, and, though he felt it was premature to do so, he began his most ambitious poem thus far, *Hyperion*. He was, he said, "obliged to write and plunge into abstract images" as some release from the consciousness of Tom's illness.

That same winter he met Fanny Brawne, and the relationship, after following a somewhat uneven course, resulted in an engagement a year later, in November or December of 1819. Meanwhile *Hyperion* had scarcely advanced after December, when Tom died; Keats returned briefly to romances with *The Eve of St. Agnes* and the fragment of *The Eve of Saint Mark*. Then, beginning in April, 1819, after a few weeks of despair at being unable to write, came his astonishing last five months of creative energy—the great odes of May, *Lamia* and the revised *Fall of Hyperion* in the summer of 1819, and finally the ode "To Autumn" in September.

At this point, needing money badly, he started to look for a position on a magazine. But he was feeling unwell; he had caught tuberculosis while nursing Tom, if not before. He returned to Hampstead, where he shared a house with Charles Brown (Fanny Brawne and her family lived next door), and tried fitfully to write, but his lungs soon began to hemorrhage. For eight months he put aside all writing on the advice of physicians. Finally, in August, 1820, he was told he could not survive another English winter and must go to Italy. With no money except what his publishers kindly advanced, and accompanied by a friend, Joseph Severn, a young painter, he arrived in Rome after a painful voyage of six weeks on a small ship. He suffered constantly during the last three months, not only from illness but from knowledge that everything was lost—Fanny Brawne, friends, life itself, and the hope that had inspired his career. He felt that he had scarcely crossed the threshold to the sort of poetry he sought ultimately to write; and in his despair he asked that his grave bear no name but only the words, "Here lies one whose name was writ in water." Yet in these last months (which he called his "posthumous existence") he continued to show his courage, his constant thought of others, and even flashes of his old humor. When he died in Rome on February 23, 1821, his age was twenty-five years and four months.

2

Keats's mind was the least dogmatic it is possible to be. It was dialectical and speculative. As he thought about any large question, whatever answer suggested itself was soon challenged by an opposite answer that might be equally true and that, at least, claimed attention and honest reflection. And so he found himself between antithetical beliefs, or rather possibilities of belief, and he did not try to reconcile them immediately. He distrusted the coaxing and squeezing and tenuous defining of ideas by which we seek to hold a logical position amid the weltering contrarieties of fact; he preferred to wait hopefully for further experience. A mind of this character cannot help but be speculative—that is, richly productive of points of view and generalizations but unwilling to attach "truth" to any of them. Yet at the same time Keats was not a skeptic in any thoroughgoing way. Ultimate answers may be impossible, but a genuine, altogether valuable increase in awareness, understanding, and wisdom remains open to mankind. And the adventure of experience and growth is itself deeply satisfying, even while it leads to the knowledge of possible tragedy.

A capacity for imaginative sympathy can be directed to points of view or beliefs as well as to objects or persons. So with Keats. We never find him cautiously criticizing and analyzing a proposition. Instead, he is astonishingly quick to enter with gusto into ideas, to explore them by adopting and believing them, however transiently. In other words, the same mind we described as dialectical and speculative is quick to commit itself—it is dialectical only in that commitments can be opposed, speculative in that they are not final. Keats's thinking about poetry is an example. Should poetry offer a dreamy escape, a luxurious massing of delightful sensations? For a while Keats not only thought so but welcomed the possibility. Or does the poetic imagination, as it pictures a happiness the world cannot give, intuit a truer world beyond or behind? Perhaps "The Imagination may be compared to Adam's dream—he awoke and found it truth." Here Keats shares a hope common in the Romantic era. Or should poetry dwell on the truth of man's suffering here and now, and, if so, how? Should it follow Shakespeare and portray human beings in all their complexity, diversity, and clashing purposes? Or should it seek rather to imitate the epic sweep and peerless artistry of Milton? And again, there is Wordsworth, who also knows and reveals the truth of man's experience, but who comes to it by concentrating on his own individual mind and feelings. He has not Milton's comprehensive range, but he may be deeper. And in any case, is it enough merely to reflect the truth of experience? Is poetry, after all, man's highest achievement?

For Keats's attitudes to poetry can never be isolated from his more generally moral and religious searching. One ideal is what he called "Negative Capability"—"that is when man is capable of being in uncertainties, Mysteries, doubts without any irritable reaching after fact and reason." It is especially necessary for achievement in literature, but it is also a possibly attractive human stance. It implies skepticism—for, after all, "nothing in this world is provable"—and a morally disinterested contemplation of the energies of life. But there is another sort of disinterestedness: "Very few men have ever arrived at a complete disinterestedness of Mind: very few have been influenced by a pure desire of the benefit of others · · · I can remember but two–Socrates and Jesus." Are not such "Benefactors of Humanity"—

> Who love their fellows even to the death,
> Who feel the giant agony of the world,
> And more, like slaves to poor humanity,
> Labour for mortal good—
>
> [*The Fall of Hyperion*, I. 156–59].

far nobler than poets of any kind? Yet perhaps poetry, too, can be a "friend to man." Perhaps it can fortify and console. If so, does it do this solely by incorporating with beauty and formal order its vision of things? Or must it present a doctrine that will sustain? In this case, the "human friend philosopher" may give us more than any poet. Yet "Can it be that even the greatest Philosopher ever arrived at his goal without putting aside numerous objections?" Are we really to believe in *Hyperion* that the beguiling Oceanus has fathomed the mystery of the cosmos? And the other philosopher in Keats's poetry, Apollonius in *Lamia*, has an impressive dignity and may be intellectually right, but certainly shows himself remarkably insensitive to other human beings. Yet if philosophy is not by itself the answer, reading and reflection are indispensable in forming a "Soul or Intelligence destined to possess the sense of Identity." And in partial contrast to the ideal of Negative Capability, the forming of an individual Identity or Soul in and by means of "a World of Pains and troubles" may be suggested as "a grander system of salvation than the Christian religion."

The point in this quick review is not to leave everything in the air, but to recall the complexity, range, and self-critical alertness of Keats's speculation. There was, of course, a development in his thinking. It was not that he achieved a synthesis, but he sifted to fundamentals, seeing ever more clearly what must be included in any synthesis. And meanwhile, there was a constantly increasing depth and breadth of awareness. The great image of the countenance of Moneta, in *The Fall of Hyperion*, is not an answer to anything but a powerful posing of the mystery of suffering that would have been altogether beyond the early Keats.

It was, therefore, inevitable that Keats's greatest work would be dramatic in character if not in form. That is, the poetry gives full play to diverse elements and points of view and follows the method of experience itself, from which no single or simple answer can be obtained. Hence interpretation has had to follow varying lines of emphasis, and no brief comment can be more than illustrative. Many of the later poems—the odes, "La Belle Dame," *Lamia*—may, for example, be viewed as exploring the Romantic hope to intuit and imaginatively participate in a happier realm of being. The strength of these poems is that they give complete and powerful expression to the natural human longing for a better world, a more perfect love, a lasting intensity of happiness, and yet they also remain faithful to the critical intelligence that forces us to acknowledge that dreams are only dreams and that in the sole world we know values are tragically in conflict. Life is process and death the condition of all natural fulfillment and beauty. (But natural fulfillment and beauty are as much a part of reality as death itself. Death does not make them absurd; Keats is no existentialist.) The course of Keats's greater poems reflects this habitual insight. The odes "On a Grecian Urn" and "To a Nightingale" are the living process of an experience. They begin in a happy state of mind, sweep into a culminating intensity of joy in a trance of self-forgetfulness, and return a little sorrowfully to ordinary reality. They must be read as we might read a soliloquy in a drama where the effect depends very much on the interplay and quick shift of feeling and realization. Moreover, except at the end of the "Ode on a Grecian Urn," they remain radically concrete, working with massed blocks of imagery. They imply uncanny depths of meditation and self-awareness, but they also refuse to abstract or generalize. It is largely this concrete and dramatic character that has rendered them inexhaustible to contemplation. This together with the unrivaled power of phrase makes them the greatest English lyrics of the nineteenth century.

The strength and diversity of Keats's stylistic resources have never been denied since the middle of the nineteenth century. So far as is possible, he was constantly imitated, particularly in two features—his typical line of verse (typical, that is, of *Hyperion*, *The Eve of St. Agnes*, and the odes) and his suggestive and concrete imagery. The Keatsian line is a regular, end-stopped, and slowly moving unit, bound together by complex patterns of assonance and alliteration:

> A casement high and triple-arched there was,
> All garlanded with carven imag'ries
> Of fruits, and flowers, and bunches of knot-grass,
> And diamonded with panes of quaint device
>
> [*The Eve of St. Agnes*, ll. 208–11].

Such a passage also illustrates the wealth of concrete detail Keats lovingly renders (in sharp contrast to the Wordsworthian vocabulary of general terms).

But more generally one sees in such work the mingled intensity and control Keats prized. The control is obvious in the patterning; the intensity results partly from the thick massing of imagery and partly from the degree to which Keats's imagination reinforces and extends its presentations. Assonance and alliteration are themselves a type of reinforcement, but the effect we have in mind is suggested by Keats when he remarks that "poetry should surprize by a fine excess," or by Hazlitt when he praises Milton's "gusto": Milton "repeats his blow twice; grapples with and exhausts his subject." So in the famous lines from the "Ode to a Nightingale":
the song of the nightingale

> oft-times hath
> Charm'd *magic* casements, opening *on the foam*
> Of *perilous seas*, in faery lands *forlorn*.

Another poet might have used the image in a more stripped way, referring perhaps to open casements in faery land. The Keatsian gusto or intensity in working out the image lies in the incredible enrichment conveyed by the words in italics. And the passage also reminds us that as he matured Keats learned to replace adjectives by participles, thus giving a constant suggestion of active energy.[1] Even in the abundant images of repose or slowed process objects are characteristically portrayed as doing something, as caught up in a dynamic relation with other objects. Thus the opening stanza of "To Autumn" conveys a feeling of stasis or pause while using verbs of strong activity:

> to load and bless
> With fruit the vines that round the thatch-eves run;
> To bend with apples the moss'd cottage-trees,
> And fill all fruit with ripeness to the core.

Often this sense of dynamic relationship and motion-in-stillness is given in pictorial or sculpturesque effects. The "vines that round the thatch-eves run" are an example, or the emblematic figure of joy in the "Ode on Melancholy"—

> Joy, whose hand is ever at his lips
> Bidding adieu.

One thinks also of Porphyro "Buttress'd from moonlight" or following the old Angela

> through a lowly arched way,
> Brushing the cobwebs with his lofty plume.

Often the vividness of conception merges into a direct empathy or sympathetic participation. Keats seems to sense the pressure of the song against the throat as he describes the nightingale singing "of summer in full-throated ease." A quality that fascinated poets in the 1890's is synaesthesia, the substitution of one sense for another, as in the phrase "pale and silver silence." But much more characteristic of Keats, W. J. Bate points out, is the "*substantiation* of one sense by another in order to give . . . additional dimension and depth" as in "the *moist scent* of flowers" or "*embalmed* darkness." Keats, as Douglas Bush says, "makes us simultaneously see, touch, smell, and almost hear 'hush'd, cool-rooted flowers, fragrant eyed.'" In particular, Keats appeals in an unusual degree to the senses of touch and taste, thus giving a peculiarly voluptuous and intimate sense of the object, as in the phrase "globèd peonies," where the hand seems almost to cup the flowers. To this should be added the remarkable condensation of meaning in his phrasing:

> Thou wast not born for death, immortal Bird!
> No *hungry generations* tread thee down.

And immediately after comes the great image of Ruth "in tears amid the *alien* corn."

---

[1] No poet of the period, indeed no poet in English, has so richly employed as epithets the past participle, with the intention of concentrating energy. In many of Keats's finest lines, the rhythm is lost unless the reader remembers that, at this time, the final -*ed* might be pronounced as a separate syllable—"Unclasps her warmèd jewels"; "He follow'd through a lowly archèd way"; "And purple-stainèd mouth"; or, as in the great simile in *Hyperion*, I. 72–74:

> As when, upon a trancèd summer-night,
> Those green-rob'd senators of mighty woods,
> Tall oaks, branch-charmèd by the earnest stars.

In general, where the concluding syllable (-*ed*) is not pronounced, Keats uses an apostrophe ("follow'd").

Of course, this style was not attained at once. Keats was not precocious, and his early verse shows, along with much promise, perhaps more diverse weaknesses than the apprentice work of any other poet. The self-corrective growth is astonishing. What is even more remarkable—characteristic of only the greatest artists—is that having created this style Keats went on to something else. *Lamia* is a totally new attempt. Though brilliant in its way, it would probably not have been pursued further. The direction Keats might have taken shows rather in *The Fall of Hyperion*. Here his style has become less crowded with imagery, less patterned, more discursive and colloquially relaxed. While keeping in mind that the odes are the high point of his achievement, it is possible to admire this style even more, especially for a long poem.

Bibliography

Keats profoundly influenced poets throughout the nineteenth century. The effect is already marked in Tennyson's *Poems* of 1832. At midcentury and later the Pre-Raphaelites were devoted readers and imitators. Though deliberately reacting against both Keats and Tennyson, Arnold shows their blended influence in some of his best work, notably "The Scholar Gypsy" and "Thyrsis." Walter Pater claimed Keats as a forerunner of the art for art's sake movement. But critical and popular recognition advanced much more slowly, and his reputation did not reach its present height until the 1890's and the early years of the twentieth century. For one thing, the admiration that seems inevitable to us was checked by the strangeness, as it seemed to contemporaries, of his mode or style. Readers in the early nineteenth century were naturally attuned to the poetry of the past age, and they valued the refined purity of diction and denotative clarity of which Gray's "Elegy" is a supreme example. They could think Keats's bold coinages ("*Lethe-wards* had sunk") were barbarisms, and his densely imaginative associations and persistently concrete imagery seemed obscure. Byron, according to Leigh Hunt, could not understand "the meaning of a beaker 'full of the warm south.'"

Moreover, for at least twenty-six years after his death Keats's reputation was damaged by two myths, one propagated by deliberate malice and the other by the limited taste of defenders. The first was the invention of J. G. Lockhart in a notorious series of articles "On the Cockney School of Poetry" (see extract below, p. 1247), published anonymously in *Blackwood's Magazine* in 1817 and 1818. Leigh Hunt was the object of these attacks; Keats drew himself into the field of fire by his grateful and generous poetic tributes to his friend. Lockhart portrayed an ignorant young surgeon's apprentice, Johnny Keats, a poor simpleton led into affectation and puffed with vanity by the praise of Leigh Hunt and his "'Ampstead" circle, which also included "pimpled Hazlitt" ("the Cockney Aristotle"), and B. R. Haydon ("the Cockney Raphael"). Other Tory journals joyfully seized this vivid and vicious image and it was widely repeated and seriously credited. Meanwhile Shelley did greater havoc, in the long run, with his portrait in "Adonais" of Keats as a spirit too sensitive for the shocks of life, "a pale flower by some sad maiden cherished," who had been killed by a harsh review (Shelley was thinking of the review of *Endymion* by J. W. Croker in *The Quarterly*, April, 1818). This bizarre misrepresentation lingered on, fostering a sentimental pity that blinded readers to the tough intelligence of Keats's best verse.

These myths were largely dispelled by the first biography, *Life, Letters, and Literary Remains*, by Richard Monckton Milnes, 1848, which also made clear how extraordinarily rapid Keats's development had been and hence what immense natural endowment he must have brought to bear. From this time on critics were likely to concede his greatness and to argue whether it was a greatness in promise only or actual achievement as well. But the impression of Keats was still remarkably one-sided. Even his most enthusiastic admirers could see only a poet of extraordinary sensuous richness and exotic imagination. To the Pre-Raphaelites he was a poet of the enchantment of medievalism and of vivid colors and pictorial effects. To Arnold he was an inspired interpreter of the beauty of the natural world; in "word magic" and "natural poetic felicity" he could be compared only with Shakespeare. But at the the same time, he lacked a "matured power of moral interpretation"; his poetry was not animating and fortifying.

A more rounded and just estimate of Keats gradually formed as more letters and poems were printed (for example, *The Fall of Hyperion* in 1856), but it was not generally shared until the beginning of the twentieth century. (Keats's impact on the Victorians is closely studied by G. H. Ford, *Keats and the Victorians*, 1944.) Needless to say, we do not claim that Keats was a systematic philosopher, any more than Shakespeare. But we have come to see in him a powerful intellect— quick, sensitive, open, deep—unceasingly confronting the large, unanswerable questions of human nature and destiny; he had a cast of mind that might be described as sanity raised to the highest degree. Douglas Bush remarks that if we "imagine ourselves contemporaries and in urgent need of wise advice," of the great English Romantics we would turn first to Keats. The statement itself indicates how radically our view of him has changed.

Among the many signs of a new understanding in the early twentieth century, one may pick out the significant essay, "Keats and 'Philosophy' " by A. C. Bradley, *Oxford Lectures on Poetry*, 1909. C. D. Thorpe, *The Mind of John Keats*, 1926, explored the range of ideas to which Keats was sensitively open, concentrating particularly on his self-questioning about the mode and function of his art. In less sober vein, J. M. Murry, *Keats and Shakespeare*, 1925, also tried to convey the character and profundity of Keats's mind. (Murry's *Keats*, 1955, comprises a collection of excellent essays on miscellaneous topics.) The subject is covered with magisterial sweep by Douglas Bush in "Keats and His Ideas," *The Major English Romantic Poets*, ed. C. D. Thorpe and others, 1957 (also cited in the general bibliography of Romanticism, above). Most general studies of Keats have continued the Murry and Thorpe tradition. That is, they see him as speculative and undogmatic, and explore habitual preoccupations and themes, showing polar tensions and gradual evolution in his point of view. Several of these works also relate him to other Romantic poets and to their common search for an eternal and transcendent reality in or behind the finite and changing: D. G. James, *Scepticism and Poetry*, 1937, and *The Romantic Comedy*, 1948; G. W. Knight, *The Starlit Dome*, 1941; and David Perkins, *The Quest for Permanence*,

1959. The commentary of E. C. Pettet, *On the Poetry of Keats*, 1957, covers all the poems and is particularly valuable for *Endymion*. One trait of Keats's mind is studied by W. J. Bate, *Negative Capability: The Intuitive Approach in Keats*, 1939, a pioneer essay that opened up a topic of fundamental importance. Other studies of special topics are Bernice Slote, *Keats and the Dramatic Principle*, 1958, and Ian Jack, *Keats and the Mirror of Art*, 1967, on the impact of the fine arts on Keats's poetry. The strongest effort to see a relatively definite, metaphysical system in Keats is by Earl Wasserman, *The Finer Tone*, 1953, which is a close and challenging analysis of five major poems.

Analysis of Keats's poetic style was brilliantly inaugurated by Robert Bridges in "John Keats, A Critical Essay," 1895 (reprinted in *Collected Essays*, 1929), and in the notes and appendices of Ernest de Selincourt's still valuable edition of the *Poems* (1905 and subsequent editions). W. J. Bate, *The Stylistic Development of Keats*, 1945, is a penetrating discussion of the versification. (A brief discussion of the style generally is also provided by him in "Keats's Style: Evolution Toward Qualities of Permanent Value," *The Major English Romantic Poets*, ed. Thorpe and others, 1957.) An illuminating work on the subject suggested by its title is R. H. Fogle, *The Imagery of Keats and Shelley*, 1949, and M. R. Ridley, *Keats's Craftsmanship*, 1933, is a valuable study of Keats's manuscript revisions.

The first two full-scale lives were those of Sir Sidney Colvin, 1917, and Amy Lowell, 2 vols., 1925, both still of value in their discussion of Keats's work. They have now been supplanted, however, by W. J. Bate, *John Keats*, 1963. Bate's full and moving narrative provides the best study yet available of the growth of poetic genius and of what facilitates it. In the process it offers a comprehensive critical discussion of the poems and the thought of the letters, and relates the mind and art of Keats both to the English poetic tradition as a whole and also to the intellectual background of the age. C. L. Finney, *The Evolution of Keats's Poetry*, 2 vols., 1936, provides a wealth of detail on sources and influences, particularly for the early poems. Three excellent biographies, less concerned with critical study of Keats's writing and focused more on the events of his life, are those of Dorothy Hewlett, rev. ed., 1950; Aileen Ward, 1963; and Robert Gittings, 1968. By far the best short critical biography, distilling a wealth of knowledge, is Douglas Bush, *John Keats*, 1966.

An admirable annotated edition is that of Douglas Bush, *Selected Poems and Letters*, 1959, which, though in paperback, excels more expensive editions in authenticity of text and quality of notes. Since it is not complete, it may be supplemented either by the edition of De Selincourt (mentioned above) or by the *Poetical Works*, ed. H. W. Garrod, rev. ed., 1958. Garrod offers a definitive text with full textual notes. Its defect is the complete lack of general annotation. The letters have been edited many times, most recently and accurately by H. E. Rollins, *The Letters of John Keats, 1814–1821*, 2 vols., 1958. A good selection of the letters is that edited by Lionel Trilling, 1951. *The Keats Circle: Letters and Papers and More Letters and Poems of the Keats Circle*, ed. H. E. Rollins, 2 vols., 1965, contains letters and recollections by Keats's family and friends.

IMITATION OF SPENSER

❧

Keats wrote this, his first poem, at the age of seventeen or eighteen while working as an apprentice to a surgeon in Edmonton. These descriptive stanzas are a careful imitation not of Spenser himself but of the eighteenth-century imitators of Spenser.

❧

Now Morning from her orient chamber came,
And her first footsteps touch'd a verdant hill;
Crowning its lawny crest with amber flame,
Silv'ring the untainted gushes of its rill;
Which, pure from mossy beds, did down distill,
And after parting beds of simple flowers,
By many streams a little lake did fill,
Which round its marge reflected woven bowers,
And, in its middle space, a sky that never lowers.

There the king-fisher saw his plumage bright   10
Vieing with fish of brilliant dye below;
Whose silken fins, and golden scales' light
Cast upward, through the waves, a ruby glow:
There saw the swan his neck of arched snow,
And oar'd himself along with majesty;
Sparkled his jetty eyes; his feet did show
Beneath the waves like Afric's ebony,
And on his back a fay reclined voluptuously.

Ah! could I tell the wonders of an isle
That in that fairest lake had placed been,   20
I could e'en Dido of her grief beguile;
Or rob from aged Lear his bitter teen:
For sure so fair a place was never seen,
Of all that ever charm'd romantic eye:
It seem'd an emerald in the silver sheen
Of the bright waters; or as when on high,
Through clouds of fleecy white, laughs the cœrulean sky.

And all around it dipp'd luxuriously
Slopings of verdure through the glossy tide,
Which, as it were in gentle amity,   30
Rippled delighted up the flowery side;

IMITATION OF SPENSER. **14–15. swan . . . oar'd:** a deliberate echo of *Paradise Lost*, VII. 438–40. **21. Dido:** queen of Carthage, abandoned by Aeneas. Keats had recently completed a prose translation of Vergil's *Aeneid*. **22. teen:** grief.

As if to glean the ruddy tears, it tried,
Which fell profusely from the rose-tree stem!
Haply it was the workings of its pride,
In strife to throw upon the shore a gem
Outvieing all the buds in Flora's diadem.

1814 (1817)

TO BYRON

❧

Written at the age of nineteen, and probably Keats's first sonnet, "To Byron" (typical of the so-called plaintive sonnet of the time) proves that he was by no means precocious. The sonnets printed below, written throughout the following year and a half, show his steady improvement.

❧

Byron! how sweetly sad thy melody!
   Attuning still the soul to tenderness,
   As if soft Pity, with unusual stress,
Had touch'd her plaintive lute, and thou, being by,
Hadst caught the tones, nor suffer'd them to die.
   O'ershading sorrow doth not make thee less
   Delightful: thou thy griefs dost dress
With a bright halo, shining beamily,
As when a cloud the golden moon doth veil,
   Its sides are ting'd with a resplendent glow,   10
Through the dark robe oft amber rays prevail,
   And like fair veins in sable marble flow;
Still warble, dying swan! still tell the tale,
   The enchanting tale, the tale of pleasing woe.

1814 (1848)

TO SOLITUDE

❧

Keats's first published poem, "To Solitude," was written while he was a student at Guy's Hospital. He had recently left the country village of Edmonton for the crowded tenements of the "Borough" (Southwark) where the London hospitals were situated.

❧

O Solitude! if I must with thee dwell,
   Let it not be among the jumbled heap

36. Flora: Roman goddess of flowers.

Of murky buildings; climb with me the steep,—
Nature's observatory—whence the dell,
Its flowery slopes, its river's crystal swell,
    May seem a span; let me thy vigils keep
    'Mongst boughs pavillion'd, where the deer's swift
        leap
Startles the wild bee from the fox-glove bell.
But though I'll gladly trace these scenes with thee,
    Yet the sweet converse of an innocent mind,   10
    Whose words are images of thoughts refin'd,
Is my soul's pleasure; and it sure must be
    Almost the highest bliss of human-kind,
When to thy haunts two kindred spirits flee.

                        1815 (1816)

HOW MANY BARDS

How many bards gild the lapses of time!
    A few of them have ever been the food
    Of my delighted fancy,—I could brood
Over their beauties, earthly, or sublime:
And often, when I sit me down to rhyme,
    These will in throngs before my mind intrude:
    But no confusion, no disturbance rude
Do they occasion; 'tis a pleasing chime.
So the unnumber'd sounds that evening store;
    The songs of birds—the whisp'ring of the leaves—   10
    The voice of waters—the great bell that heaves
With solemn sound,—and thousand others more,
    That distance of recognizance bereaves,
Make pleasing music, and not wild uproar.

                        1816 (1817)

TO ONE WHO HAS
BEEN LONG IN CITY PENT

To one who has been long in city pent,
    'Tis very sweet to look into the fair
    And open face of heaven,—to breathe a prayer
Full in the smile of the blue firmament.

Who is more happy, when, with heart's content,
    Fatigued he sinks into some pleasant lair
    Of wavy grass, and reads a debonair
And gentle tale of love and languishment?
Returning home at evening, with an ear
    Catching the notes of Philomel,—an eye   10
Watching the sailing cloudlet's bright career,
    He mourns that day so soon has glided by:
E'en like the passage of an angel's tear
    That falls through the clear ether silently.

                        1816 (1817)

TO

CHARLES COWDEN CLARKE

"To Charles Cowden Clarke" was written in early
September, 1816. Keats, after passing his examinations,
left Guy's Hospital for the seaside village of Margate,
hoping to use this free summer in order to see whether he
could really write poetry. He had a difficult time, was
unable to find subjects, and, rather than do nothing,
started to write poems about poetry. The best of these is
the following verse epistle addressed to Cowden Clarke.
The son of Keats's schoolmaster at Enfield, he was the
first person to introduce Keats to poetry (see Introduction above). Lines 60–67 are often cited as a tribute from
a student to a teacher.

Oft have you seen a swan superbly frowning,
And with proud breast his own white shadow
        crowning;
He slants his neck beneath the waters bright
So silently, it seems a beam of light
Come from the galaxy: anon he sports,—
With outspread wings the Naiad Zephyr courts,
Or ruffles all the surface of the lake
In striving from its crystal face to take
Some diamond water drops, and them to treasure
In milky nest, and sip them off at leisure.   10
But not a moment can he there insure them,
Nor to such downy rest can he allure them;
For down they rush as though they would be free,
And drop like hours into eternity.
Just like that bird am I in loss of time,
Whene'er I venture on the stream of rhyme;

TO SOLITUDE. **8. bee . . . bell:** See Wordsworth's "Nuns
Fret Not," above, ll. 5–7, which Keats echoes. TO ONE WHO
HAS BEEN LONG IN CITY PENT. **1. To . . . pent:** Cf. "As one
who long in populous city pent," *Paradise Lost,* IX. 445.

8. gentle tale: probably Leigh Hunt's poem, *The Story of
Rimini.* **10. Philomel:** the nightingale. TO CHARLES COWDEN
CLARKE. **6. Zephyr:** west wind.

With shatter'd boat, oar snapt, and canvas rent
I slowly sail, scarce knowing my intent;
Still scooping up the water with my fingers,
In which a trembling diamond never lingers.   20
By this, friend Charles, you may full plainly see
Why I have never penn'd a line to thee:
Because my thoughts were never free, and clear,
And little fit to please a classic ear;
Because my wine was of too poor a savour
For one whose palate gladdens in the flavour
Of sparkling Helicon:—small good it were
To take him to a desert rude, and bare,
Who had on Baiæ's shore reclin'd at ease,
While Tasso's page was floating in a breeze   30
That gave soft music from Armida's bowers,
Mingled with fragrance from her rarest flowers:
Small good to one who had by Mulla's stream
Fondled the maidens with the breasts of cream;
Who had beheld Belphœbe in a brook,
And lovely Una in a leafy nook,
And Archimago leaning o'er his book:
Who had of all that's sweet tasted, and seen,
From silv'ry ripple, up to beauty's queen;
From the sequester'd haunts of gay Titania,   40
To the blue dwelling of divine Urania:
One, who, of late, had ta'en sweet forest walks
With him who elegantly chats, and talks—
The wrong'd Libertas,—who has told you stories

Of laurel chaplets, and Apollo's glories;
Of troops chivalrous prancing through a city,
And tearful ladies made for love, and pity:
With many else which I have never known.
Thus have I thought; and days on days have flown
Slowly, or rapidly—unwilling still   50
For you to try my dull, unlearned quill.
Nor should I now, but that I've known you long;
That you first taught me all the sweets of song:
The grand, the sweet, the terse, the free, the fine;
What swell'd with pathos, and what right divine:
Spenserian vowels that elope with ease,
And float along like birds o'er summer seas;
Miltonian storms, and more, Miltonian tenderness;
Michael in arms, and more, meek Eve's fair slenderness,
Who read for me the sonnet swelling loudly   60
Up to its climax and then dying proudly?
Who found for me the grandeur of the ode,
Growing, like Atlas, stronger from its load?
Who let me taste that more than cordial dram,
The sharp, the rapier-pointed epigram?
Show'd me that epic was of all the king,
Round, vast, and spanning all like Saturn's ring?
You too upheld the veil from Clio's beauty,
And pointed out the patriot's stern duty;
The might of Alfred, and the shaft of Tell;   70
The hand of Brutus, that so grandly fell
Upon a tyrant's head. Ah! had I never seen,
Or known your kindness, what might I have been?
What my enjoyments in my youthful years,
Bereft of all that now my life endears?
And can I e'er these benefits forget?
And can I e'er repay the friendly debt?
No, doubly no;—yet should these rhymings please,
I shall roll on the grass with two-fold ease:
For I have long time been my fancy feeding   80
With hopes that you would one day think the reading
Of my rough verses not an hour misspent;
Should it e'er be so, what a rich content!
Some weeks have pass'd since last I saw the spires
In lucent Thames reflected:—warm desires
To see the sun o'erpeep the eastern dimness,
And morning shadows, streaking into slimness
Across the lawny fields, and pebbly water;
To mark the time as they grow broad, and shorter;
To feel the air that plays about the hills,   90
And sips its freshness from the little rills;

27. **Helicon:** the fountain of the muses on Mt. Helicon.
29. **Baiæ's shore:** site of an ancient resort near Naples.
Cf. "Fragment of an Ode to Maia," below, l. 3. **31.
Armida:** enchantress in Tasso's *Jerusalem Delivered*. **33.
Mulla's stream:** Mulla was an Irish river near which Spenser
lived. The three names in lines 35–37 are of characters
in the *Faerie Queene*. **40. Titania:** queen of the fairies.
41. Urania: muse of astronomy and the more sublime
kinds of poetry. **42–45. One . . . glories:** Leigh Hunt.
He is called "Libertas" because he was a political liberal
and "wrong'd" because he was jailed for an attack by his
magazine, *The Examiner*, on the dissolute Prince Regent.
"Elegantly chats, and talks" echoes the simpering colloquialism of Hunt's own poetry, and it is typical of the sort of
phrasing from which Keats later reacted. "Laurel chaplets"
refers to a practice of Hunt (and occasionally other poets) of
crowning himself with laurel: hearing such stories from Clarke,
the young Keats was enormously impressed, though later,
when he had met Hunt, and Mrs. Hunt wove them both a
crown, he became excruciatingly embarrassed. (Cf. Keats's
remark to Haydon in the letter of May 10–11, 1817, below:
"There is no greater Sin after the 7 deadly than to flatter
oneself into an idea of being a great Poet"; cf. also in Bate,
John Keats, the amusing Chapter VII, "The Laurel Crown
and the Vision of Greatness.")

63. **Atlas:** who supported the sky on his shoulders. 68. **Clio:**
the muse of history.

To see high, golden corn wave in the light
When Cynthia smiles upon a summer's night,
And peers among the cloudlets jet and white,
As though she were reclining in a bed
Of bean blossoms, in heaven freshly shed.
No sooner had I stepp'd into these pleasures
Than I began to think of rhymes and measures:
The air that floated by me seem'd to say
"Write! thou wilt never have a better day."   100
And so I did. When many lines I'd written,
Though with their grace I was not oversmitten,
Yet, as my hand was warm, I thought I'd better
Trust to my feelings, and write you a letter.
Such an attempt required an inspiration
Of a peculiar sort,—a consummation;—
Which, had I felt, these scribblings might have been
Verses from which the soul would never wean:
But many days have passed since last my heart
Was warm'd luxuriously by divine Mozart;   110
By Arne delighted, or by Handel madden'd;
Or by the song of Erin pierc'd and sadden'd:
What time you were before the music sitting,
And the rich notes to each sensation fitting.
Since I have walk'd with you through shady lanes
That freshly terminate in open plains,
And revel'd in a chat that ceased not
When at night-fall among your books we got:
No, nor when supper came, nor after that,—
Nor when reluctantly I took my hat;   120
No, nor till cordially you shook my hand
Mid-way between our homes:—your accents bland
Still sounded in my ears, when I no more
Could hear your footsteps touch the grav'ly floor.
Sometimes I lost them, and then found again;
You chang'd the footpath for the grassy plain.
In those still moments I have wish'd you joys
That well you know to honour:—"Life's very toys
With him," said I, "will take a pleasant charm;
It cannot be that aught will work him harm."   130
These thoughts now come o'er me with all their
        might:—
Again I shake your hand,—friend Charles, good night.

                                        1816 (1817)

93. **Cynthia:** the moon. 111. **Arne:** Thomas Arne, 1710–
78, noted for his operas. 112. **song of Erin:** Thomas Moore's
Irish Melodies (1807–35). 120. **took my hat:** referring to the
period when he was apprenticed to a surgeon. In the evenings
Keats would walk over from the village of Edmonton to
the school at Enfield, and Clarke and he would then read
aloud or talk about literature.

ON FIRST LOOKING
INTO CHAPMAN'S HOMER

Keats's first great sonnet was written (Oct., 1816) within
a few hours after reading, for the first time, the translation
of Homer by the Elizabethan poet George Chapman, to
which Keats's friend, Charles Cowden Clarke, introduced him. Walking back at dawn from Clarke's to
Guy's Hospital, where Keats was still working, he
apparently composed some of the sonnet on the way,
finished it as soon as he arrived at his own room, and,
before leaving for work, sent it off to Clarke. It coalesces
much of Keats's early reading at school, including an
account of William Herschel's recent discovery of the
planet Uranus (1781) and the story of the Spanish
explorers in William Robertson's *History of America*.
(For Clarke's account of the composition of the poem,
see below, pp. 1239–40.)

Much have I travell'd in the realms of gold,
    And many goodly states and kingdoms seen;
    Round many western islands have I been
Which bards in fealty to Apollo hold.
Oft of one wide expanse had I been told
    That deep-brow'd Homer ruled as his demesne;
    Yet did I never breathe its pure serene
Till I heard Chapman speak out loud and bold:
Then felt I like some watcher of the skies
    When a new planet swims into his ken;   10
Or like stout Cortez when with eagle eyes
    He star'd at the Pacific—and all his men
Look'd at each other with a wild surmise—
    Silent, upon a peak in Darien.

                                        1816 (1817)

ADDRESSED TO HAYDON

Great spirits now on earth are sojourning;
    He of the cloud, the cataract, the lake,
    Who on Helvellyn's summit, wide awake,
Catches his freshness from Archangel's wing:

ON FIRST LOOKING INTO CHAPMAN'S HOMER. 11. **Cortez:** a
mistake for Balboa. ADDRESSED TO HAYDON. **Title:** Benjamin Robert Haydon, 1786–1846, the historical painter,
had recently become a friend of Keats. 2–4. **He . . . wing:**
Wordsworth. 3. **Helvellyn:** the highest mountain in the
Lake District of England.

He of the rose, the violet, the spring,
    The social smile, the chain for Freedom's sake:
    And lo!—whose stedfastness would never take
A meaner sound than Raphael's whispering.
And other spirits there are standing apart
    Upon the forehead of the age to come;   10
These, these will give the world another heart,
    And other pulses. Hear ye not the hum
Of mighty workings?——
    Listen awhile ye nations, and be dumb.

1816 (1817)

ON THE GRASSHOPPER
AND THE CRICKET

꒰ꔛ꒱

This sonnet was written (Dec., 1816) in a sonnet-writing
contest with Hunt, who suggested the topic. See Hunt's
sonnet with the same title, above, p. 710.

꒰ꔛ꒱

The poetry of earth is never dead:
    When all the birds are faint with the hot sun,
    And hide in cooling trees, a voice will run
From hedge to hedge about the new-mown mead;
That is the Grasshopper's—he takes the lead
    In summer luxury,—he has never done
    With his delights; for when tired out with fun
He rests at ease beneath some pleasant weed.
The poetry of earth is ceasing never:
    On a lone winter evening, when the frost   10
    Has wrought a silence, from the stove there shrills
The Cricket's song, in warmth increasing ever,
    And seems to one in drowsiness half lost,
    The Grasshopper's among some grassy hills.

KEEN, FITFUL GUSTS

Keen, fitful gusts are whisp'ring here and there
    Among the bushes half leafless, and dry;
    The stars look very cold about the sky,
And I have many miles on foot to fare.
Yet feel I little of the cool bleak air,
    Or of the dead leaves rustling drearily,

5–6. He . . . sake: Hunt. 7. And lo!: Haydon.

Or of those silver lamps that burn on high,
Or of the distance from home's pleasant lair:
For I am brimfull of the friendliness
    That in a little cottage I have found;   10
Of fair-hair'd Milton's eloquent distress,
    And all his love for gentle Lycid drown'd;
Of lovely Laura in her light green dress,
    And faithful Petrarch gloriously crown'd.

1816 (1817)

I STOOD TIP-TOE

꒰ꔛ꒱

The idea for the poem came to Keats during a summer's
walk in 1816 on Hampstead Heath. He at first probably
hoped to write a narrative poem (the title was originally
Endymion) but, not finding a subject, finished it as a descriptive piece while trying in the winter of 1816 to get
together enough poems to fill out his first volume, which
was to be published shortly (March, 1817).

꒰ꔛ꒱

"Places of nestling green for Poets made."
                    Story of Rimini III. 431]

I stood tip-toe upon a little hill,
The air was cooling, and so very still,
That the sweet buds which with a modest pride
Pull droopingly, in slanting curve aside,
Their scantly leav'd, and finely tapering stems,
Had not yet lost those starry diadems
Caught from the early sobbing of the morn.
The clouds were pure and white as flocks new shorn,
And fresh from the clear brook; sweetly they slept
On the blue fields of heaven, and then there crept   10
A little noiseless noise among the leaves,
Born of the very sigh that silence heaves:
For not the faintest motion could be seen
Of all the shades that slanted o'er the green.
There was wide wand'ring for the greediest eye,
To peer about upon variety;
Far round the horizon's crystal air to skim,
And trace the dwindled edgings of its brim;

KEEN, FITFUL GUSTS. 10. cottage: Leigh Hunt's, in Hampstead. 11. Milton's . . . distress: a reference to Milton's
elegy, "Lycidas." 13. Laura: to whom Petrarch's sonnets
were written.

To picture out the quaint, and curious bending
Of a fresh woodland alley, never ending;   20
Or by the bowery clefts, and leafy shelves,
Guess where the jaunty streams refresh themselves.
I gazed awhile, and felt as light, and free
As though the fanning wings of Mercury
Had play'd upon my heels: I was light-hearted,
And many pleasures to my vision started;
So I straightway began to pluck a posey
Of luxuries bright, milky, soft and rosy.

A bush of May flowers with the bees about them;
Ah, sure no tasteful nook would be without them; 30
And let a lush laburnum oversweep them,
And let long grass grow round the roots to keep them
Moist, cool and green; and shade the violets,
That they may bind the moss in leafy nets.
A filbert hedge with wild briar overtwined,
And clumps of woodbine taking the soft wind
Upon their summer thrones; there too should be
The frequent chequer of a youngling tree,
That with a score of light green brethren shoots
From the quaint mossiness of aged roots: 40
Round which is heard a spring-head of clear waters
Babbling so wildly of its lovely daughters
The spreading blue-bells: it may haply mourn
That such fair clusters should be rudely torn
From their fresh beds, and scattered thoughtlessly
By infant hands, left on the path to die.

Open afresh your round of starry folds,
Ye ardent marigolds!
Dry up the moisture from your golden lids,
For great Apollo bids   50
That in these days your praises should be sung
On many harps, which he has lately strung;
And when again your dewiness he kisses,
Tell him, I have you in my world of blisses:
So haply when I rove in some far vale,
His mighty voice may come upon the gale.

Here are sweet peas, on tip-toe for a flight:
With wings of gentle flush o'er delicate white,
And taper fingers catching at all things,
To bind them all about with tiny rings.   60

Linger awhile upon some bending planks
That lean against a streamlet's rushy banks,
And watch intently Nature's gentle doings:
They will be found softer than ring-dove's cooings.
How silent comes the water round that bend;
Not the minutest whisper does it send

To the o'erhanging sallows: blades of grass
Slowly across the chequer'd shadows pass.
Why, you might read two sonnets, ere they reach
To where the hurrying freshnesses aye preach   70
A natural sermon o'er their pebbly beds;
Where swarms of minnows show their little heads,
Staying their wavy bodies 'gainst the streams,
To taste the luxury of sunny beams
Temper'd with coolness. How they ever wrestle
With their own sweet delight, and ever nestle
Their silver bellies on the pebbly sand.
If you but scantily hold out the hand,
That very instant not one will remain;
But turn your eye, and they are there again.   80
The ripples seem right glad to reach those cresses,
And cool themselves among the em'rald tresses;
The while they cool themselves, they freshness give,
And moisture, that the bowery green may live:
So keeping up an interchange of favours,
Like good men in the truth of their behaviours.
Sometimes goldfinches one by one will drop
From low hung branches; little space they stop;
But sip, and twitter, and their feathers sleek;
Then off at once, as in a wanton freak: 90
Or perhaps, to show their black, and golden wings,
Pausing upon their yellow flutterings.
Were I in such a place, I sure should pray
That naught less sweet, might call my thoughts away,
Than the soft rustle of a maiden's gown
Fanning away the dandelion's down;
Than the light music of her nimble toes
Patting against the sorrel as she goes.
How she would start, and blush, thus to be caught
Playing in all her innocence of thought.   100
O let me lead her gently o'er the brook,
Watch her half-smiling lips, and downward look;
O let me for one moment touch her wrist;
Let me one moment to her breathing list;
And as she leaves me may she often turn
Her fair eyes looking through her locks auburne.
What next? A tuft of evening primroses,
O'er which the mind may hover till it dozes;
O'er which it well might take a pleasant sleep,
But that 'tis ever startled by the leap   110
Of buds into ripe flowers; or by the flitting
Of diverse moths, that aye their rest are quitting;

I STOOD TIP-TOE. **67.** sallows: willows. **68.** chequer'd
shadows: Cf. Milton's "L'Allegro," l. 96, "chequer'd
shade," a phrase commonly echoed in descriptive poetry
throughout the eighteenth century.

Or by the moon lifting her silver rim
Above a cloud, and with a gradual swim
Coming into the blue with all her light.
O Maker of sweet poets, dear delight
Of this fair world, and all its gentle livers;
Spangler of clouds, halo of crystal rivers,
Mingler with leaves, and dew and tumbling streams,
Closer of lovely eyes to lovely dreams,   120
Lover of loneliness, and wandering,
Of upcast eye, and tender pondering!
Thee must I praise above all other glories
That smile us on to tell delightful stories.
For what has made the sage or poet write
But the fair paradise of Nature's light?
In the calm grandeur of a sober line,
We see the waving of the mountain pine;
And when a tale is beautifully staid,
We feel the safety of a hawthorn glade: 130
When it is moving on luxurious wings,
The soul is lost in pleasant smotherings:
Fair dewy roses brush against our faces,
And flowering laurels spring from diamond vases;
O'er head we see the jasmine and sweet briar,
And bloomy grapes laughing from green attire;
While at our feet, the voice of crystal bubbles
Charms us at once away from all our troubles:
So that we feel uplifted from the world,
Walking upon the white clouds wreath'd and curl'd.
So felt he, who first told, how Psyche went   141
On the smooth wind to realms of wonderment;
What Psyche felt, and Love, when their full lips
First touch'd; what amorous, and fondling nips
They gave each other's cheeks; with all their sighs,
And how they kist each other's tremulous eyes:
The silver lamp,—the ravishment,—the wonder—
The darkness,—loneliness,—the fearful thunder;
Their woes gone by, and both to heaven upflown,
To bow for gratitude before Jove's throne.   150
So did he feel, who pull'd the boughs aside,
That we might look into a forest wide,
To catch a glimpse of Fauns, and Dryades
Coming with softest rustle through the trees;
And garlands woven of flowers wild, and sweet,
Upheld on ivory wrists, or sporting feet:

129. **staid:** calmly composed, its energy held in balance (as distinguished from "moving" in l. 131); Keats later uses the word "stationing" to describe this effect. 141–43. **Psyche . . . Love:** For the story, see "Ode to Psyche," below. 153. **Dryades:** wood nymphs.

Telling us how fair, trembling Syrinx fled
Arcadian Pan, with such a fearful dread.
Poor nymph,—poor Pan,—how he did weep to find,
Nought but a lovely sighing of the wind   160
Along the reedy stream; a half-heard strain,
Full of sweet desolation—balmy pain.

What first inspired a bard of old to sing
Narcissus pining o'er the untainted spring?
In some delicious ramble, he had found
A little space, with boughs all woven round;
And in the midst of all, a clearer pool
Than e'er reflected in its pleasant cool,
The blue sky here, and there, serenely peeping
Through tendril wreaths fantastically creeping. 170
And on the bank a lonely flower he spied,
A meek and forlorn flower, with naught of pride,
Drooping its beauty o'er the watery clearness,
To woo its own sad image into nearness:
Deaf to light Zephyrus it would not move;
But still would seem to droop, to pine, to love.
So while the poet stood in this sweet spot,
Some fainter gleamings o'er his fancy shot;
Nor was it long ere he had told the tale
Of young Narcissus, and sad Echo's bale.   180

Where had he been, from whose warm head out-flew
That sweetest of all songs, that ever new,
That aye refreshing, pure deliciousness,
Coming ever to bless
The wanderer by moonlight? to him bringing
Shapes from the invisible world, unearthly singing
From out the middle air, from flowery nests,
And from the pillowy silkiness that rests
Full in the speculation of the stars.
Ah! surely he had burst our mortal bars;   190
Into some wond'rous region he had gone,
To search for thee, divine Endymion!
He was a Poet, sure a lover too,
Who stood on Latmus' top, what time there blew

157–62. **Syrinx . . . pain:** A nymph fleeing to escape Pan, Syrinx was changed by the water nymphs into a reed, from which Pan made a flute. 163ff. **What first inspired . . . :** Keats's conception of the origin of myth is derived primarily from Wordsworth's *Excursion*, IV. 718–62, 847–87. 190. **burst . . . bars:** an early expression of an ideal that becomes a central theme from *Endymion* through the major odes and *Lamia* to *The Fall of Hyperion*. 194. **Latmus:** is a mountain in Asia Minor. According to legend Endymion slept forever on Latmus, and the moon-goddess, Cynthia (identified with Diana, l. 197), gazed on his beauty.

Soft breezes from the myrtle vale below;
And brought in faintness solemn, sweet, and slow
A hymn from Dian's temple; while upswelling,
The incense went to her own starry dwelling.
But though her face was clear as infant's eyes,
Though she stood smiling o'er the sacrifice,   200
The Poet wept at her so piteous fate,
Wept that such beauty should be desolate:
So in fine wrath some golden sounds he won,
And gave meek Cynthia her Endymion.

Queen of the wide air; thou most lovely queen
Of all the brightness that mine eyes have seen!
As thou exceedest all things in thy shine,
So every tale, does this sweet tale of thine.
O for three words of honey, that I might
Tell but one wonder of thy bridal night!   210

Where distant ships do seem to show their keels,
Phoebus awhile delay'd his mighty wheels,
And turn'd to smile upon thy bashful eyes,
Ere he his unseen pomp would solemnize.
The evening weather was so bright, and clear,
That men of health were of unusual cheer;
Steeping like Homer at the trumpet's call,
Or young Apollo on the pedestal:
And lovely women were as fair and warm,
As Venus looking sideways in alarm.   220
The breezes were ethereal, and pure,
And crept through half-closed lattices to cure
The languid sick; it cool'd their fever'd sleep,
And soothed them into slumbers full and deep.
Soon they awoke clear eyed: nor burnt with thirsting,
Nor with hot fingers, nor with temples bursting:
And springing up, they met the wond'ring sight
Of their dear friends, nigh foolish with delight;
Who feel their arms, and breasts, and kiss and stare,
And on their placid foreheads part the hair.   230
Young men, and maidens at each other gaz'd
With hands held back, and motionless, amaz'd
To see the brightness in each other's eyes;
And so they stood, fill'd with a sweet surprise,
Until their tongues were loos'd in poesy.
Therefore no lover did of anguish die:
But the soft numbers, in that moment spoken,
Made silken ties, that never may be broken.
Cynthia! I cannot tell the greater blisses,

That follow'd thine, and thy dear shepherd's kisses:
Was there a poet born?—but now no more,   241
My wand'ring spirit must no further soar.—

1816 (1817)

SLEEP AND POETRY

Keats wrote this in November and December of 1816. It was his most ambitious poem before *Endymion*. The subject is the twofold use of the imagination for sleep (reverie or dream), on the one hand, and, on the other, for the direct apprehension of the concrete world. Implicit is the theme (especially in ll. 90–125) that fulfillment can come only through comprehensive awareness of the actual and through sympathy for it. (The interplay of dream, or reverie, and reality persists and deepens throughout all of Keats's serious poetry, especially in the great odes, *Lamia*, and *The Fall of Hyperion*.) Much of the poem, especially lines 270–312, also consists of a credo, as do so many of Keats's poems (e.g., "Lines on Seeing a Lock of Milton's Hair," "On Sitting Down To Read *King Lear* Once Again," or the "Ode to Psyche"): a statement of ultimate aim in poetry and of his own awareness of the preparation needed for it. Like "I Stood Tip-Toe" and *Endymion*, the poem is in the loose heroic couplet Keats adopted from Leigh Hunt; and like them it abounds with the lush diction that Keats also caught from Hunt but that he quickly (within another year) began to shed in favor of a firmer style. The epigraph is from "The Flower and the Leaf," ll. 17–21, which is no longer attributed to Chaucer.

As I lay in my bed slepe full unmete
Was unto me, but why that I ne might
Rest I ne wist, for there n'as erthly wight
[As I suppose] had more of hertis ese
Than I, for I n'ad sicknesse nor disese.

Chaucer

What is more gentle than a wind in summer?
What is more soothing than the pretty hummer
That stays one moment in an open flower,
And buzzes cheerily from bower to bower?
What is more tranquil than a musk-rose blowing
In a green island, far from all men's knowing?
More healthful than the leafiness of dales?
More secret than a nest of nightingales?
More serene than Cordelia's countenance?

---

218. Apollo: the statue of Apollo Belvedere (in the Vatican Museum, Rome), of which Keats had seen pictures.

SLEEP AND POETRY. **2. hummer:** hummingbird. **9. Cordelia:** youngest daughter of King Lear.

More full of visions than a high romance?   10
What but thee, Sleep? Soft closer of our eyes!
Low murmurer of tender lullabies!
Light hoverer around our happy pillows!
Wreather of poppy buds, and weeping willows!
Silent entangler of a beauty's tresses!
Most happy listener! when the morning blesses
Thee for enlivening all the cheerful eyes
That glance so brightly at the new sun-rise.

But what is higher beyond thought than thee?
Fresher than berries of a mountain tree?   20
More strange, more beautiful, more smooth, more
    regal,
Than wings of swans, than doves, than dim-seen eagle?
What is it? And to what shall I compare it?
It has a glory, and naught else can share it:
The thought thereof is awful, sweet, and holy,
Chasing away all worldliness and folly;
Coming sometimes like fearful claps of thunder,
Or the low rumblings earth's regions under;
And sometimes like a gentle whispering
Of all the secrets of some wond'rous thing   30
That breathes about us in the vacant air;
So that we look around with prying stare,
Perhaps to see shapes of light, aerial limning,
And catch soft floatings from a faint-heard hymning;
To see the laurel wreath, on high suspended,
That is to crown our name when life is ended.
Sometimes it gives a glory to the voice,
And from the heart up-strings, rejoice! rejoice!
Sounds which will reach the Framer of all things,
And die away in ardent mutterings.   40

No one who once the glorious sun has seen,
And all the clouds, and felt his bosom clean
For his great Maker's presence, but must know
What 'tis I mean, and feel his being glow:
Therefore no insult will I give his spirit,
By telling what he sees from native merit.

O Poesy! for thee I hold my pen
That am not yet a glorious denizen
Of thy wide heaven—Should I rather kneel
Upon some mountain-top until I feel   50
A glowing splendour round about me hung,
And echo back the voice of thine own tongue?
O Poesy! for thee I grasp my pen
That am not yet a glorious denizen
Of thy wide heaven; yet, to my ardent prayer,
Yield from thy sanctuary some clear air,

Smooth'd for intoxication by the breath
Of flowering bays, that I may die a death
Of luxury, and my young spirit follow
The morning sun-beams to the great Apollo   60
Like a fresh sacrifice; or, if I can bear
The o'erwhelming sweets, 'twill bring to me the fair
Visions of all places: a bowery nook
Will be elysium—an eternal book
Whence I may copy many a lovely saying
About the leaves, and flowers—about the playing
Of nymphs in woods, and fountains; and the shade
Keeping a silence round a sleeping maid;
And many a verse from so strange influence
That we must ever wonder how, and whence   70
It came. Also imaginings will hover
Round my fire-side, and haply there discover
Vistas of solemn beauty, where I'd wander
In happy silence, like the clear Meander
Through its lone vales; and where I found a spot
Of awfuller shade, or an enchanted grot,
Or a green hill o'erspread with chequer'd dress
Of flowers, and fearful from its loveliness,
Write on my tablets all that was permitted,
All that was for our human senses fitted.   80
Then the events of this wide world I'd seize
Like a strong giant, and my spirit teaze
Till at its shoulders it should proudly see
Wings to find out an immortality.

Stop and consider! life is but a day;
A fragile dew-drop on its perilous way
From a tree's summit; a poor Indian's sleep
While his boat hastens to the monstrous steep
Of Montmorenci. Why so sad a moan?
Life is the rose's hope while yet unblown;   90
The reading of an ever-changing tale;
The light uplifting of a maiden's veil;
A pigeon tumbling in clear summer air;
A laughing school-boy, without grief or care,
Riding the springy branches of an elm.

O for ten years, that I may overwhelm
Myself in poesy; so I may do the deed

---

74. Meander: a river in Asia Minor famous for its winding
course. **89. Montmorenci:**. river and waterfall in Quebec.
90–128. Life . . . fear: The conception of stages of development, derived primarily from Wordsworth (especially
"Tintern Abbey"), becomes a basic theme of Keats's poetry,
particularly in his last long poem, *The Fall of Hyperion*. See
also Keats's letter to Reynolds, below, May 3, 1818, on the
"Chamber of Maiden-Thought."

That my own soul has to itself decreed.
Then will I pass the countries that I see
In long perspective, and continually   100
Taste their pure fountains. First the realm I'll pass
Of Flora, and old Pan; sleep in the grass,
Feed upon apples red, and strawberries,
And choose each pleasure that my fancy sees;
Catch the white-handed nymphs in shady places,
To woo sweet kisses from averted faces,—
Play with their fingers, touch their shoulders white
Into a pretty shrinking with a bite
As hard as lips can make it: till agreed,
A lovely tale of human life we'll read.   110
And one will teach a tame dove how it best
May fan the cool air gently o'er my rest;
Another, bending o'er her nimble tread,
Will set a green robe floating round her head,
And still will dance with ever varied ease,
Smiling upon the flowers and the trees:
Another will entice me on, and on
Through almond blossoms and rich cinnamon;
Till in the bosom of a leafy world
We rest in silence, like two gems upcurl'd   120
In the recesses of a pearly shell.

And can I ever bid these joys farewell?
Yes, I must pass them for a nobler life,
Where I may find the agonies, the strife
Of human hearts: for lo! I see afar,
O'ersailing the blue cragginess, a car
And steeds with streamy manes—the charioteer
Looks out upon the winds with glorious fear:
And now the numerous tramplings quiver lightly
Along a huge cloud's ridge; and now with sprightly
Wheel downward come they into fresher skies,   131
Tipt round with silver from the sun's bright eyes.
Still downward with capacious whirl they glide;
And now I see them on the green-hill's side
In breezy rest among the nodding stalks.
The charioteer with wond'rous gesture talks
To the trees and mountains; and there soon appear
Shapes of delight, of mystery, and fear,
Passing along before a dusky space
Made by some mighty oaks: as they would chase   140
Some ever-fleeting music, on they sweep.
Lo! how they murmur, laugh, and smile, and weep:
Some with upholden hand and mouth severe;
Some with their faces muffled to the ear
Between their arms; some, clear in youthful bloom,
Go glad and smilingly athwart the gloom;

Some looking back, and some with upward gaze;
Yes, thousands in a thousand different ways
Flit onward—now a lovely wreath of girls
Dancing their sleek hair into tangled curls;   150
And now broad wings. Most awfully intent
The driver of those steeds is forward bent,
And seems to listen: O that I might know
All that he writes with such a hurrying glow.

The visions all are fled—the car is fled
Into the light of heaven, and in their stead
A sense of real things comes doubly strong,
And, like a muddy stream, would bear along
My soul to nothingness: but I will strive
Against all doubtings, and will keep alive   160
The thought of that same chariot, and the strange
Journey it went.
                    Is there so small a range
In the present strength of manhood, that the high
Imagination cannot freely fly
As she was wont of old? prepare her steeds,
Paw up against the light, and do strange deeds
Upon the clouds? Has she not shown us all?
From the clear space of ether, to the small
Breath of new buds unfolding? From the meaning
Of Jove's large eye-brow, to the tender greening   170
Of April meadows? Here her altar shone,
E'en in this isle; and who could paragon
The fervid choir that lifted up a noise
Of harmony, to where it aye will poise
Its mighty self of convoluting sound,
Huge as a planet, and like that roll round,
Eternally around a dizzy void?
Ay, in those days the Muses were nigh cloy'd
With honors; nor had any other care
Than to sing out and sooth their wavy hair.   180

Could all this be forgotten? Yes, a schism
Nurtured by foppery and barbarism,
Made great Apollo blush for this his land.
Men were thought wise who could not understand
His glories: with a puling infant's force
They sway'd about upon a rocking horse,
And thought it Pegasus. Ah dismal soul'd!
The winds of heaven blew, the ocean roll'd

162–67. **Is . . . clouds:** The ideal of following great
earlier poets and attaining their vigor and range is central
to Keats; cf. especially "Ode to Maia," below. **186–87:
sway'd . . . Pegasus:** Keats echoes his admired Hazlitt's
remark on eighteenth-century couplets.

Its gathering waves—ye felt it not. The blue
Bared its eternal bosom, and the dew   190
Of summer nights collected still to make
The morning precious: beauty was awake!
Why were ye not awake? But ye were dead
To things ye knew not of,—were closely wed
To musty laws lined out with wretched rule
And compass vile: so that ye taught a school
Of dolts to smooth, inlay, and clip, and fit,
Till, like the certain wands of Jacob's wit,
Their verses tallied. Easy was the task:
A thousand handicraftsmen wore the mask   200
Of Poesy. Ill-fated, impious race!
That blasphemed the bright Lyrist to his face,
And did not know it,—no, they went about,
Holding a poor, decrepid standard out
Mark'd with most flimsy mottos, and in large
The name of one Boileau!

                                    O ye whose charge
It is to hover round our pleasant hills!
Whose congregated majesty so fills
My boundly reverence, that I cannot trace
Your hallowed names, in this unholy place,   210
So near those common folk; did not their shames
Affright you? Did our old lamenting Thames
Delight you? Did ye never cluster round
Delicious Avon, with a mournful sound,
And weep? Or did ye wholly bid adieu
To regions where no more the laurel grew?
Or did ye stay to give a welcoming
To some lone spirits who could proudly sing
Their youth away, and die? 'Twas even so:
But let me think away those times of woe: 220
Now 'tis a fairer season: ye have breathed
Rich benedictions o'er us; ye have wreathed
Fresh garlands: for sweet music has been heard
In many places;—some has been upstirr'd
From out its crystal dwelling in a lake,

By a swan's ebon bill; from a thick brake,
Nested and quiet in a valley mild,
Bubbles a pipe; fine sounds are floating wild
About the earth: happy are ye and glad.
These things are doubtless: yet in truth we've had 230
Strange thunders from the potency of song;
Mingled indeed with what is sweet and strong,
From majesty: but in clear truth the themes
Are ugly clubs, the Poets' Polyphemes
Disturbing the grand sea. A drainless shower
Of light is poesy; 'tis the supreme of power;
'Tis might half slumb'ring on its own right arm.
The very archings of her eye-lids charm
A thousand willing agents to obey,
And still she governs with the mildest sway: 240
But strength alone though of the Muses born
Is like a fallen angel: trees uptorn,
Darkness, and worms, and shrouds, and sepulchres
Delight it; for it feeds upon the burrs,
And thorns of life; forgetting the great end
Of poesy, that it should be a friend
To sooth the cares, and lift the thoughts of man.

    Yet I rejoice: a myrtle fairer than
E'er grew in Paphos, from the bitter weeds
Lifts its sweet head into the air, and feeds   250
A silent space with ever sprouting green.
All tenderest birds there find a pleasant screen,
Creep through the shade with jaunty fluttering,
Nibble the little cupped flowers and sing.
Then let us clear away the choking thorns
From round its gentle stem; let the young fawns,
Yeaned in after times, when we are flown,
Find a fresh sward beneath it, overgrown
With simple flowers: let there nothing be
More boisterous than a lover's bended knee;   260
Nought more ungentle than the placid look
Of one who leans upon a closed book;
Nought more untranquil than the grassy slopes
Between two hills. All hail delightful hopes!
As she was wont, th' imagination
Into most lovely labyrinths will be gone,
And they shall be accounted poet kings
Who simply tell the most heart-easing things.

189–90. blue . . . bosom: Cf. Wordsworth's "The World
Is Too Much with Us," above, ll. 5–8. 198. Jacob's wit:
Jacob's ruse to deceive his sheep (Genesis 30:37–43). 206.
Boileau: Nicolas Boileau-Despréaux, 1636–1711, whose
verse-essay, "Art of Poetry," was a finished expression of
French neoclassic ideals. Keats had probably not yet read
it and was simply repeating a fashionable attitude toward
the poetry of the previous century. 218–19. lone . . .
die: Thomas Chatterton, 1752–70, who committed suicide
at the age of seventeen. Keats dedicated Endymion to his
memory. 224–25. some . . . lake: the "Lake Poets": Coleridge, Southey, and especially Wordsworth.

226–28. from . . . pipe: Leigh Hunt. 231–35. Strange
. . . sea: a censure of Byron's misdirected energy, though
Hunt considered it a reflection on "the morbidity that taints
. . . the Lake Poets." 246–47. a . . . man: Cf. "Ode on a
Grecian Urn," below, l. 48: "a friend to man." 257. Yeaned:
given birth.

O may these joys be ripe before I die.
Will not some say that I presumptuously   270
Have spoken? that from hastening disgrace
'Twere better far to hide my foolish face?
That whining boyhood should with reverence bow
Ere the dread thunderbolt could reach? How!
If I do hide myself, it sure shall be
In the very fane, the light of Poesy:
If I do fall, at least I will be laid
Beneath the silence of a poplar shade;
And over me the grass shall be smooth shaven;
And there shall be a kind memorial graven.   280
But off Despondence! miserable bane!
They should not know thee, who athirst to gain
A noble end, are thirsty every hour.
What though I am not wealthy in the dower
Of spanning wisdom; though I do not know
The shiftings of the mighty winds that blow
Hither and thither all the changing thoughts
Of man: though no great minist'ring reason sorts
Out the dark mysteries of human souls
To clear conceiving: yet there ever rolls   290
A vast idea before me, and I glean
Therefrom my liberty; thence too I've seen
The end and aim of Poesy. 'Tis clear
As anything most true; as that the year
Is made of the four seasons—manifest
As a large cross, some old cathedral's crest,
Lifted to the white clouds. Therefore should I
Be but the essence of deformity,
A coward, did my very eye-lids wink
At speaking out what I have dared to think.   300
Ah! rather let me like a madman run
Over some precipice; let the hot sun
Melt my Dedalian wings, and drive me down
Convuls'd and headlong! Stay! an inward frown
Of conscience bids me be more calm awhile.
An ocean dim, sprinkled with many an isle,
Spreads awfully before me. How much toil!
How many days! what desperate turmoil!
Ere I can have explored its widenesses.
Ah, what a task! upon my bended knees,   310
I could unsay those—no, impossible!
Impossible!

For sweet relief I'll dwell
On humbler thoughts, and let this strange assay
Begun in gentleness die so away.
E'en now all tumult from my bosom fades:
I turn full hearted to the friendly aids
That smooth the path of honour; brotherhood,
And friendliness the nurse of mutual good.
The hearty grasp that sends a pleasant sonnet
Into the brain ere one can think upon it;   320
The silence when some rhymes are coming out;
And when they're come, the very pleasant rout:
The message certain to be done to-morrow.
'Tis perhaps as well that it should be to borrow
Some precious book from out its snug retreat,
To cluster round it when we next shall meet.
Scarce can I scribble on; for lovely airs
Are fluttering round the room like doves in pairs;
Many delights of that glad day recalling,
When first my senses caught their tender falling.   330
And with these airs come forms of elegance
Stooping their shoulders o'er a horse's prance,
Careless, and grand—fingers soft and round
Parting luxuriant curls;—and the swift bound
Of Bacchus from his chariot, when his eye
Made Ariadne's cheek look blushingly.
Thus I remember all the pleasant flow
Of words at opening a portfolio.

Things such as these are ever harbingers
To trains of peaceful images: the stirs   340
Of a swan's neck unseen among the rushes:
A linnet starting all about the bushes:
A butterfly, with golden wings broad parted,
Nestling a rose, convuls'd as though it smarted
With over pleasure—many, many more,
Might I indulge at large in all my store
Of luxuries: yet I must not forget
Sleep, quiet with his poppy coronet:
For what there may be worthy in these rhymes
I partly owe to him: and thus, the chimes   350
Of friendly voices had just given place
To as sweet a silence, when I 'gan retrace
The pleasant day, upon a couch at ease.
It was a poet's house who keeps the keys

---

303. Dedalian: presumptuously ambitious. Dedalus invented
wax wings that melted when his son Icarus flew too close
to the sun.

312ff. For sweet relief . . . : Keats turns from his ambitions
and ideals as a poet to the living room of Hunt's cottage
("a poet's house," l. 354), with its busts of former poets
(ll. 355–58) and pictures of Sappho, Alfred the Great, Petrarch,
and others.

Of pleasure's temple. Round about were hung
The glorious features of the bards who sung

In other ages—cold and sacred busts
Smiled at each other. Happy he who trusts
To clear Futurity his darling fame!
Then there were fauns and satyrs taking aim   360
At swelling apples with a frisky leap
And reaching fingers, 'mid a luscious heap
Of vine-leaves. Then there rose to view a fane
Of liny marble, and thereto a train
Of nymphs approaching fairly o'er the sward:
One, loveliest, holding her white hand toward
The dazzling sun-rise: two sisters sweet
Bending their graceful figures till they meet
Over the trippings of a little child:
And some are hearing, eagerly, the wild   370
Thrilling liquidity of dewy piping.
See, in another picture, nymphs are wiping
Cherishingly Diana's timorous limbs;—
A fold of lawny mantle dabbling swims
At the bath's edge, and keeps a gentle motion
With the subsiding crystal: as when ocean
Heaves calmly its broad swelling smoothness o'er
Its rocky marge, and balances once more
The patient weeds; that now unshent by foam
Feel all about their undulating home.   380

Sappho's meek head was there half smiling down
At nothing; just as though the earnest frown
Of over thinking had that moment gone
From off her brow, and left her all alone.

Great Alfred's too, with anxious, pitying eyes,
As if he always listened to the sighs
Of the goaded world; and Kosciusko's worn
By horrid suffrance—mightily forlorn.
Petrarch, outstepping from the shady green,
Starts at the sight of Laura; nor can wean   390
His eyes from her sweet face. Most happy they!
For over them was seen a free display
Of out-spread wings, and from between them shone
The face of Poesy: from off her throne
She overlook'd things that I scarce could tell.
The very sense of where I was might well
Keep Sleep aloof: but more than that there came

Thought after thought to nourish up the flame
Within my breast; so that the morning light
Surprised me even from a sleepless night;   400
And up I rose refresh'd, and glad, and gay,
Resolving to begin that very day
These lines; and howsoever they be done,
I leave them as a father does his son.

1816 (1817)

AFTER DARK VAPOURS

After dark vapours have oppress'd our plains
  For a long dreary season, comes a day
  Born of the gentle South, and clears away
From the sick heavens all unseemly stains.
The anxious month, relieved of its pains,
  Takes as a long-lost right the feel of May;
  The eyelids with the passing coolness play
Like rose leaves with the drip of Summer rains.
The calmest thoughts come round us; as of leaves
  Budding—fruit ripening in stillness—Autumn suns
Smiling at eve upon the quiet sheaves—   11
Sweet Sappho's cheek—a smiling infant's breath—
  The gradual sand that through an hour-glass runs—
A woodland rivulet—a Poet's death.

1817 (1848)

ON SEEING
THE ELGIN MARBLES

⌒⌒⌒

These lines were written in early March, 1817, after
Haydon took Keats to see the celebrated figures from
the Parthenon at the British Museum. They had been
acquired from the Turks by Lord Elgin, and were later
sold by him to the British government. Their authenticity
was doubted by some; Haydon was their most fervent
champion. The sonnet suggests Keats's almost stunned
reaction to a form of visual art beyond anything he had
previously experienced.

⌒⌒⌒

---

379. unshent: unharmed. **387. Kosciusko:** Thaddeus Kosciusko, 1746–1817, Polish patriot who fought in the American Revolution and in the Polish army against the Russians;
Keats addressed a sonnet to him about the same time (Dec.,
1816) that he wrote "Sleep and Poetry."

AFTER DARK VAPOURS. **3–4. South . . . heavens:** Cf. *The
Fall of Hyperion,* I. 97–98. **10–11. fruit . . . sheaves:**
Cf. "To Autumn," below, especially ll. 2–6.

Sonnet

My spirit is too weak—mortality
    Weighs heavily on me like unwilling sleep,
    And each imagin'd pinnacle and steep
Of godlike hardship, tells me I must die
Like a sick Eagle looking at the sky.
    Yet 'tis a gentle luxury to weep
    That I have not the cloudy winds to keep,
Fresh for the opening of the morning's eye.
Such dim-conceived glories of the brain
    Bring round the heart an undescribable feud;   10
So do these wonders a most dizzy pain,
    That mingles Grecian grandeur with the rude
Wasting of old Time—with a billowy main—
    A sun—a shadow of a magnitude.

                                        1817 (1817)

ON THE SEA

It keeps eternal whisperings around
    Desolate shores, and with its mighty swell
    Gluts twice ten thousand Caverns, till the spell
Of Hecate leaves them their old shadowy sound.
Often 'tis in such gentle temper found,
    That scarcely will the very smallest shell
    Be mov'd for days from where it sometime fell,
When last the winds of Heaven were unbound.
Oh ye! who have your eye-balls vex'd and tir'd,
    Feast them upon the wideness of the Sea;   10
    Oh ye! whose ears are dinn'd with uproar rude,
Or fed too much with cloying melody—
    Sit ye near some old Cavern's Mouth and brood,
Until ye start, as if the sea-nymphs quir'd!

                                        1817 (1817)

FROM

ENDYMION:
A POETIC ROMANCE

❧

This very long poem (over four thousand lines) was
begun—and significantly finished—as what Keats himself called a deliberate "trial of Invention." He started it
in April, 1817, and completed it seven months later,
writing slowly at first but in the last half of the poem sometimes maintaining an average of fifty lines a day.

ON THE SEA. **4. Hecate:** goddess of witchcraft, often associated
with the tides.

From the time that he first began to write, he could
never forget that most of the great poets of earlier
periods wrote in what the eighteenth century called "the
greater genres" or types (such as the epic and the tragic
drama). They were like "Emperors of vast provinces," as
Keats said, while the modern poets were like the "Elector
of Hanover," each of them ruling a "petty state." Keats
instinctively felt that the ability to complete a long poem
successfully was the first important test that he had to
pass. Yet he had never been able to complete a story in
poetry; he had been writing only what he dismissed as
"short pieces" or, in longer forms, "descriptive" or
meditative poems such as "I Stood Tip-Toe" and "Sleep
and Poetry." (The ideal of a "long poem" remained
with him to the end—an ideal that he constantly tried to
attain while, as a by-product, he was able to write his
great odes and other lyrics without a sense of pressure
and self-consciousness).

The theme of *Endymion* is the search for fulfillment
and the need of the human imagination to accept the
actual world with love before the "ideal" can be known
and shared. The mortal Endymion cannot immediately
or directly attain union with the moon-goddess (any
more than the human mind can enter completely the
immortal world in the "Ode on a Grecian Urn"). Human
sympathies are necessary first; and the development to
the ideal is slow and by means of the concrete.

The first book of *Endymion* is printed below. In style
the entire poem continues the versification and idiom of
the longer poems Keats has been writing; and despite fine
lines and passages, it is, partly because of the speed with
which Keats wrote it, even more diffuse, especially in the
later books. After finishing the poem, Keats swung
violently against diffuseness. Most of his poems henceforth, certainly the poems by which he is remembered,
are distinguished by a concentrated richness of style
unsurpassed in any poet of the past two hundred years.

Printed first is Keats's short Preface, written with a full
consciousness of the poem's lacks and while he was
determining to learn to write a very different, more
demanding kind of poetry.

❧

PREFACE

Knowing within myself the manner in which this
Poem has been produced, it is not without a feeling of
regret that I make it public.

What manner I mean, will be quite clear to the
reader, who must soon perceive great inexperience, immaturity, and every error denoting a feverish attempt,
rather than a deed accomplished. The two first books,
and indeed the two last, I feel sensible are not of such
completion as to warrant their passing the press; nor
should they if I thought a year's castigation would do
them any good;—it will not: the foundations are too

sandy. It is just that this youngster should die away: a sad thought for me, if I had not some hope that while it is dwindling I may be plotting, and fitting myself for verses fit to live.

This may be speaking too presumptuously, and may deserve a punishment: but no feeling man will be forward to inflict it: he will leave me alone, with the conviction that there is not a fiercer hell than the failure in a great object. This is not written with the least atom of purpose to forestall criticisms of course, but from the desire I have to conciliate men who are competent to look, and who do look with a zealous eye, to the honour of English literature.

The imagination of a boy is healthy, and the mature imagination of a man is healthy; but there is a space of life between, in which the soul is in a ferment, the character undecided, the way of life uncertain, the ambition thick-sighted: thence proceeds mawkishness, and all the thousand bitters which those men I speak of must necessarily taste in going over the following pages.

I hope I have not in too late a day touched the beautiful mythology of Greece, and dulled its brightness: for I wish to try once more, before I bid it farewell.

BOOK I

A thing of beauty is a joy for ever:
Its loveliness increases; it will never
Pass into nothingness; but still will keep
A bower quiet for us, and a sleep
Full of sweet dreams, and health, and quiet breathing.
Therefore, on every morrow, are we wreathing
A flowery band to bind us to the earth,
Spite of despondence, of the inhuman dearth
Of noble natures, of the gloomy days,
Of all the unhealthy and o'er-darkened ways   10
Made for our searching: yes, in spite of all,
Some shape of beauty moves away the pall
From our dark spirits. Such the sun, the moon,
Trees old, and young, sprouting a shady boon
For simple sheep; and such are daffodils
With the green world they live in; and clear rills
That for themselves a cooling covert make
 Gainst the hot season; the mid forest brake,
Rich with a sprinkling of fair musk-rose blooms:

ENDYMION. **18. brake:** thicket.

And such too is the grandeur of the dooms   20
We have imagined for the mighty dead;
All lovely tales that we have heard or read:
An endless fountain of immortal drink,
Pouring unto us from the heaven's brink.

Nor do we merely feel these essences
For one short hour; no, even as the trees
That whisper round a temple become soon
Dear as the temple's self, so does the moon,
The passion poesy, glories infinite,
Haunt us till they become a cheering light   30
Unto our souls, and bound to us so fast,
That, whether there be shine, or gloom o'ercast,
They alway must be with us, or we die.

Therefore, 'tis with full happiness that I
Will trace the story of Endymion.
The very music of the name has gone
Into my being, and each pleasant scene
Is growing fresh before me as the green
Of our own vallies: so I will begin
Now while I cannot hear the city's din;   40
Now while the early budders are just new,
And run in mazes of the youngest hue
About old forests; while the willow trails
Its delicate amber; and the dairy pails
Bring home increase of milk. And, as the year
Grows lush in juicy stalks, I'll smoothly steer
My little boat, for many quiet hours,
With streams that deepen freshly into bowers.
Many and many a verse I hope to write,
Before the daisies, vermeil rimm'd and white,   50
Hide in deep herbage; and ere yet the bees
Hum about globes of clover and sweet peas,
I must be near the middle of my story.
O may no wintry season, bare and hoary,
See it half finish'd: but let Autumn bold,
With universal tinge of sober gold,
Be all about me when I make an end.
And now at once, adventuresome, I send
My herald thought into a wilderness:
There let its trumpet blow, and quickly dress   60
My uncertain path with green, that I may speed
Easily onward, thorough flowers and weed.
Upon the sides of Latmos was outspread
A mighty forest; for the moist earth fed

20. dooms: rewards, judgments. **25. essences:** for Keats the imaginative distillation of concrete experience. **63. Latmos:** Mt. Latmus, in Asia Minor.

So plenteously all weed-hidden roots
Into o'er-hanging boughs, and precious fruits.
And it had gloomy shades, sequestered deep,
Where no man went; and if from shepherd's keep
A lamb stray'd far a-down those inmost glens,
Never again saw he the happy pens   70
Whither his brethren, bleating with content,
Over the hills at every nightfall went.
Among the shepherds, 'twas believed ever,
That not one fleecy lamb which thus did sever
From the white flock, but pass'd unworried
By angry wolf, or pard with prying head,
Until it came to some unfooted plains
Where fed the herds of Pan: aye great his gains
Who thus one lamb did lose. Paths there were many,
Winding through palmy fern, and rushes fenny,   80
And ivy banks; all leading pleasantly
To a wide lawn, whence one could only see
Stems thronging all around between the swell
Of turf and slanting branches: who could tell
The freshness of the space of heaven above,
Edg'd round with dark tree tops? through which a
        dove
Would often beat its wings, and often too
A little cloud would move across the blue.

   Full in the middle of this pleasantness
There stood a marble altar, with a tress   90
Of flowers budded newly; and the dew
Had taken fairy phantasies to strew
Daisies upon the sacred sward last eve,
And so the dawned light in pomp receive.
For 'twas the morn: Apollo's upward fire
Made every eastern cloud a silvery pyre
Of brightness so unsullied, that therein
A melancholy spirit well might win
Oblivion, and melt out his essence fine
Into the winds: rain-scented eglantine   100
Gave temperate sweets to that well-wooing sun;
The lark was lost in him; cold springs had run
To warm their chilliest bubbles in the grass;
Man's voice was on the mountains; and the mass
Of nature's lives and wonders puls'd tenfold,
To feel this sun-rise and its glories old.

   Now while the silent workings of the dawn
Were busiest, into that self-same lawn
All suddenly, with joyful cries, there sped
A troop of little children garlanded;   110

Who gathering round the altar, seem'd to pry
Earnestly round as wishing to espy
Some folk of holiday: nor had they waited
For many moments, ere their ears were sated
With a faint breath of music, which ev'n then
Fill'd out its voice, and died away again.
Within a little space again it gave
Its airy swellings, with a gentle wave,
To light-hung leaves, in smoothest echoes breaking
Through copse-clad vallies,—ere their death, o'er-
        taking   120
The surgy murmurs of the lonely sea.

   And now, as deep into the wood as we
Might mark a lynx's eye, there glimmered light
Fair faces and a rush of garments white,
Plainer and plainer showing, till at last
Into the widest alley they all past,
Making directly for the woodland altar.
O kindly muse! let not my weak tongue faulter
In telling of this goodly company,
Of their old piety, and of their glee: 130
But let a portion of ethereal dew
Fall on my head, and presently unmew
My soul; that I may dare, in wayfaring,
To stammer where old Chaucer us'd to sing.

   Leading the way, young damsels danced along,
Bearing the burden of a shepherd song;
Each having a white wicker over brimm'd
With April's tender younglings: next, well trimm'd,
A crowd of shepherds with as sunburnt looks
As may be read of in Arcadian books;   140
Such as sat listening round Apollo's pipe,
When the great deity, for earth too ripe,
Let his divinity o'erflowing die
In music, through the vales of Thessaly:
Some idly trail'd their sheep-hooks on the ground,
And some kept up a shrilly mellow sound
With ebon-tipped flutes: close after these,
Now coming from beneath the forest trees,
A venerable priest full soberly,
Begirt with ministring looks: always his eye   150
Stedfast upon the matted turf he kept,
And after him his sacred vestments swept.
From his right hand there swung a vase, milk-white,
Of mingled wine, out-sparkling generous light;
And in his left he held a basket full
Of all sweet herbs that searching eye could cull:

76. **pard:** leopard.

144. **Thessaly:** a pastoral area in Greece.

Wild thyme, and valley-lillies whiter still
Than Leda's love, and cresses from the rill.
His aged head, crowned with beechen wreath,
Seem'd like a poll of ivy in the teeth   160
Of winter hoar. Then came another crowd
Of shepherds, lifting in due time aloud
Their share of the ditty. After them appear'd,
Up-followed by a multitude that rear'd
Their voices to the clouds, a fair wrought car,
Easily rolling so as scarce to mar
The freedom of three steeds of dapple brown:
Who stood therein did seem of great renown
Among the throng. His youth was fully blown,
Showing like Ganymede to manhood grown;   170
And, for those simple times, his garments were
A chieftain king's: beneath his breast, half bare,
Was hung a silver bugle, and between
His nervy knees there lay a boar-spear keen.
A smile was on his countenance; he seem'd,
To common lookers on, like one who dream'd
Of idleness in groves Elysian:
But there were some who feelingly could scan
A lurking trouble in his nether lip,
And see that oftentimes the reins would slip   180
Through his forgotten hands: then would they sigh,
And think of yellow leaves, of owlets' cry,
Of logs piled solemnly.—Ah, well-a-day,
Why should our young Endymion pine away!

Soon the assembly, in a circle rang'd,
Stood silent round the shrine: each look was chang'd
To sudden veneration: women meek
Beckon'd their sons to silence; while each cheek
Of virgin bloom paled gently for slight fear.
Endymion too, without a forest peer,   190
Stood, wan, and pale, and with an awed face,
Among his brothers of the mountain chace.
In midst of all, the venerable priest
Eyed them with joy from greatest to the least,
And, after lifting up his aged hands,
Thus spake he: "Men of Latmos! shepherd bands!
Whose care it is to guard a thousand flocks:
Whether descended from beneath the rocks
That overtop your mountains; whether come
From vallies where the pipe is never dumb;   200
Or from your swelling downs, where sweet air stirs
Blue hare-bells lightly, and where prickly furze

Buds lavish gold; or ye, whose precious charge
Nibble their fill at ocean's very marge,
Whose mellow reeds are touch'd with sounds forlorn
By the dim echoes of old Triton's horn:
Mothers and wives! who day by day prepare
The scrip, with needments, for the mountain air;
And all ye gentle girls who foster up
Udderless lambs, and in a little cup   210
Will put choice honey for a favoured youth:
Yea, every one attend! for in good truth
Our vows are wanting to our great god Pan.
Are not our lowing heifers sleeker than
Night-swollen mushrooms? Are not our wide plains
Speckled with countless fleeces? Have not rains
Green'd over April's lap? No howling sad
Sickens our fearful ewes; and we have had
Great bounty from Endymion our lord.
The earth is glad: the merry lark has pour'd   220
His early song against yon breezy sky,
That spreads so clear o'er our solemnity."

Thus ending, on the shrine he heap'd a spire
Of teeming sweets, enkindling sacred fire;
Anon he stain'd the thick and spongy sod
With wine, in honour of the shepherd-god.
Now while the earth was drinking it, and while
Bay leaves were crackling in the fragrant pile,
And gummy frankincense was sparkling bright
'Neath smothering parsley, and a hazy light   230
Spread greyly eastward, thus a chorus sang:

"O thou, whose mighty palace roof doth hang
From jagged trunks, and overshadoweth
Eternal whispers, glooms, the birth, life, death
Of unseen flowers in heavy peacefulness;
Who lov'st to see the hamadryads dress
Their ruffled locks where meeting hazels darken;
And through whole solemn hours dost sit, and
     hearken

232–306. "O . . . Lycean!": To some extent this is the first
of Keats's major odes, an example of the "odal hymn" that
becomes a central element in the poems that begin with
"Ode to Psyche" and end with "To Autumn" (see below).
Pan, often treated allegorically by earlier poets, here becomes
a symbol, as Douglas Bush says, "of the romantic imagination, of supra-mortal knowledge" (*John Keats: Selected
Poems and Letters*, ed. Douglas Bush, p. 317). Haydon re-
counts the story of Keats reciting it, when he first met
Wordsworth, and of Wordsworth's condescending reply:
"A very pretty piece of paganism." 236. hamadryads:
tree nymphs.

---

158. Leda: seduced by Zeus who came to her in the shape
of a swan; she then became mother of Helen of Troy. 170.
Ganymede: cupbearer to the gods.

The dreary melody of bedded reeds—
In desolate places, where dank moisture breeds   240
The pipy hemlock to strange overgrowth;
Bethinking thee, how melancholy loth
Thou wast to lose fair Syrinx—do thou now,
By thy love's milky brow!
By all the trembling mazes that she ran,
Hear us, great Pan!

　　"O thou, for whose soul-soothing quiet, turtles
Passion their voices cooingly 'mong myrtles,
What time thou wanderest at eventide
Through sunny meadows, that outskirt the side   250
Of thine enmossed realms: O thou, to whom
Broad leaved fig trees even now foredoom
Their ripen'd fruitage; yellow girted bees
Their golden honeycombs; our village leas
Their fairest blossom'd beans and poppied corn;
The chuckling linnet its five young unborn,
To sing for thee; low creeping strawberries
Their summer coolness; pent up butterflies
Their freckled wings; yea, the fresh budding year
All its completions—be quickly near,   260
By every wind that nods the mountain pine,
O forester divine!

　　"Thou, to whom every faun and satyr flies
For willing service; whether to surprise
The squatted hare while in half sleeping fit;
Or upward ragged precipices flit
To save poor lambkins from the eagle's maw;
Or by mysterious enticement draw
Bewildered shepherds to their path again;
Or to tread breathless round the frothy main,   270
And gather up all fancifullest shells
For thee to tumble into Naiads' cells,
And, being hidden, laugh at their out-peeping;
Or to delight thee with fantastic leaping,
The while they pelt each other on the crown
With silvery oak apples, and fir cones brown—
By all the echoes that about thee ring,
Hear us, O satyr king!

　　"O Hearkener to the loud clapping shears
While ever and anon to his shorn peers   280

A ram goes bleating: Winder of the horn,
When snouted wild-boars routing tender corn
Anger our hunstmen: Breather round our farms,
To keep off mildews, and all weather harms:
Strange ministrant of undescribed sounds,
That come a swooning over hollow grounds,
And wither drearily on barren moors:
Dread opener of the mysterious doors
Leading to universal knowledge—see,
Great son of Dryope,   290
The many that are come to pay their vows
With leaves about their brows!

　　"Be still the unimaginable lodge
For solitary thinkings; such as dodge
Conception to the very bourne of heaven,
Then leave the naked brain: be still the leaven,
That spreading in this dull and clodded earth
Gives it a touch ethereal—a new birth:
Be still a symbol of immensity;
A firmament reflected in a sea;   300
An element filling the space between;
An unknown—but no more: we humbly screen
With uplift hands our foreheads, lowly bending,
And giving out a shout most heaven rending,
Conjure thee to receive our humble Pæan,
Upon thy Mount Lycean!"

　　Even while they brought the burden to a close,
A shout from the whole multitude arose,
That lingered in the air like dying rolls
Of abrupt thunder, when Ionian shoals   310
Of dolphins bob their noses through the brine.
Meantime, on shady levels, mossy fine,
Young companies numbly began dancing
To the swift treble pipe, and humming string.
Aye, those fair living forms swam heavenly
To tunes forgotten—out of memory:
Fair creatures! whose young children's children bred
Thermopylæ its heroes—not yet dead,
But in old marbles ever beautiful.
High genitors, unconscious did they cull   320
Time's sweet first-fruits—they danc'd to weariness,
And then in quiet circles did they press
The hillock turf, and caught the latter end
Of some strange history, potent to send

---

241. pipy: tubular. **243. Syrinx:** See "I Stood Tip-Toe, n. 157–62. **247. turtles:** turtledoves. **255. poppied corn:** fields of corn interspersed with poppies. **267. maw:** crop. **272. Naiad:** water nymph.

318. Thermopylæ: the pass that Leonidas and his small army of Spartans died defending against the Persians. **319. old marbles:** Keats is thinking in particular of the Elgin Marbles.

A young mind from its bodily tenement.
Or they might watch the quoit-pitchers, intent
On either side; pitying the sad death
Of Hyacinthus, when the cruel breath
Of Zephyr slew him,—Zephyr penitent,
Who now, ere Phœbus mounts the firmament,   330
Fondles the flower amid the sobbing rain.
The archers too, upon a wider plain,
Beside the feathery whizzing of the shaft,
And the dull twanging bowstring, and the raft
Branch down sweeping from a tall ash top,
Call'd up a thousand thoughts to envelope
Those who would watch. Perhaps, the trembling knee
And frantic gape of lonely Niobe,
Poor, lonely Niobe! when her lovely young
Were dead and gone, and her caressing tongue   340
Lay a lost thing upon her paly lip,
And very, very deadliness did nip
Her motherly cheeks. Arous'd from this sad mood
By one, who at a distance loud halloo'd,
Uplifting his strong bow into the air,
Many might after brighter visions stare:
After the Argonauts, in blind amaze
Tossing about on Neptune's restless ways,
Until, from the horizon's vaulted side,
There shot a golden splendour far and wide,   350
Spangling those million poutings of the brine
With quivering ore: 'twas even an awful shine
From the exaltation of Apollo's bow;
A heavenly beacon in their dreary woe.
Who thus were ripe for high contemplating,
Might turn their steps towards the sober ring
Where sat Endymion and the aged priest
'Mong shepherds gone in eld, whose looks increas'd
The silvery setting of their mortal star.
There they discours'd upon the fragile bar   360
That keeps us from our homes ethereal;
And what our duties there: to nightly call
Vesper, the beauty-crest of summer weather;
To summon all the downiest clouds together
For the sun's purple couch; to emulate
In ministring the potent rule of fate

With speed of fire-tail'd exhalations;
To tint her pallid cheek with bloom, who cons
Sweet poesy by moonlight: besides these,
A world of other unguess'd offices.   370
Anon they wander'd, by divine converse,
Into Elysium; vieing to rehearse
Each one his own anticipated bliss.
One felt heart-certain that he could not miss
His quick gone love, among fair blossom'd boughs,
Where every zephyr-sigh pouts, and endows
Her lips with music for the welcoming.
Another wish'd, mid that eternal spring,
To meet his rosy child, with feathery sails,
Sweeping, eye-earnestly, through almond vales: 380
Who, suddenly, should stoop through the smooth
      wind,
And with the balmiest leaves his temples bind;
And, ever after, through those regions be
His messenger, his little Mercury.
Some were athirst in soul to see again
Their fellow huntsmen o'er the wide champaign
In times long past; to sit with them, and talk
Of all the chances in their earthly walk;
Comparing, joyfully, their plenteous stores
Of happiness, to when upon the moors,   390
Benighted, close they huddled from the cold,
And shar'd their famish'd scrips. Thus all out-told
Their fond imaginations,—saving him
Whose eyelids curtain'd up their jewels dim,
Endymion: yet hourly had he striven
To hide the cankering venom, that had riven
His fainting recollections. Now indeed
His senses had swoon'd off: he did not heed
The sudden silence, or the whispers low,
Or the old eyes dissolving at his woe,   400
Or anxious calls, or close of trembling palms,
Or maiden's sigh, that grief itself embalms:
But in the self-same fixed trance he kept,
Like one who on the earth had never stept.
Aye, even as dead still as a marble man,
Frozen in that old tale Arabian.

  Who whispers him so pantingly and close?
Peona, his sweet sister: of all those,
His friends, the dearest. Hushing signs she made,
And breath'd a sister's sorrow to persuade   410

---

328. Hyacinthus: a youth beloved of Apollo; while they
were playing at quoits, the jealous Zephyr blew a quoit against
the head of Hyacinthus, who was then turned by Apollo
into a flower. **334. raft:** torn. **338. Niobe:** Her twelve children were slain when she boasted that because of their number
she was superior to the wife of Zeus. **347. Argonauts:**
those who sailed with Jason in quest of the Golden Fleece.
358. eld: old age.

405–06. marble . . . Arabian: The story, from the *Arabian
Nights*, is that told by the Eldest Lady in "The Porter
and the Three Ladies of Bagdad."

A yielding up, a cradling on her care.
Her eloquence did breathe away the curse:
She led him, like some midnight spirit nurse
Of happy changes in emphatic dreams,
Along a path between two little streams,—
Guarding his forehead, with her round elbow,
From low-grown branches, and his footsteps slow
From stumbling over stumps and hillocks small;
Until they came to where these streamlets fall,
With mingled bubblings and a gentle rush,   420
Into a river, clear, brimful, and flush
With crystal mocking of the trees and sky.
A little shallop, floating there hard by,
Pointed its beak over the fringèd bank;
And soon it lightly dipt, and rose, and sank,
And dipt again, with the young couple's weight,—
Peona guiding, through the water straight,
Towards a bowery island opposite;
Which gaining presently, she steered light
Into a shady, fresh, and ripply cove,   430
Where nested was an arbour, overwove
By many a summer's silent fingering;
To whose cool bosom she was used to bring
Her playmates, with their needle broidery,
And minstrel memories of times gone by.

  So she was gently glad to see him laid
Under her favourite bower's quiet shade,
On her own couch, new made of flower leaves,
Dried carefully on the cooler side of sheaves
When last the sun his autumn tresses shook,   440
And the tann'd harvesters rich armfuls took.
Soon was he quieted to slumbrous rest:
But, ere it crept upon him, he had prest
Peona's busy hand against his lips,
And still, a sleeping, held her finger-tips
In tender pressure. And as a willow keeps
A patient watch over the stream that creeps
Windingly by it, so the quiet maid
Held her in peace: so that a whispering blade
Of grass, a wailful gnat, a bee bustling   450
Down in the blue-bells, or a wren light rustling
Among sere leaves and twigs, might all be heard.

  O magic sleep! O comfortable bird,
That broodest o'er the troubled sea of the mind
Till it is hush'd and smooth! O unconfin'd
Restraint! imprisoned liberty! great key
To golden palaces, strange minstrelsy,
Fountains grotesque, new trees, bespangled caves,
Echoing grottos, full of tumbling waves

And moonlight; aye, to all the mazy world   460
Of silvery enchantment!—who, upfurl'd
Beneath thy drowsy wing a triple hour,
But renovates and lives?—Thus, in the bower,
Endymion was calm'd to life again.
Opening his eyelids with a healthier brain,
He said: "I feel this thine endearing love
All through my bosom: thou art as a dove
Trembling its closed eyes and sleekèd wings
About me; and the pearliest dew not brings
Such morning incense from the fields of May,   470
As do those brighter drops that twinkling stray
From those kind eyes,—the very home and haunt
Of sisterly affection. Can I want
Aught else, aught nearer heaven, than such tears?
Yet dry them up, in bidding hence all fears
That, any longer, I will pass my days
Alone and sad. No, I will once more raise
My voice upon the mountain-heights; once more
Make my horn parley from their foreheads hoar:
Again my trooping hounds their tongues shall
    loll   480
Around the breathèd boar: again I'll poll
The fair-grown yew tree, for a chosen bow:
And, when the pleasant sun is setting low,
Again I'll linger in a sloping mead
To hear the speckled thrushes, and see feed
Our idle sheep. So be thou cheered, sweet,
And, if thy lute is here, softly intreat
My soul to keep in its resolved course."

  Hereat Peona, in their silver source,
Shut her pure sorrow drops with glad exclaim,   490
And took a lute, from which there pulsing came
A lively prelude, fashioning the way
In which her voice should wander. 'Twas a lay
More subtle cadenced, more forest wild
Than Dryope's lone lulling of her child;
And nothing since has floated in the air
So mournful strange. Surely some influence rare
Went, spiritual, through the damsel's hand;
For still, with Delphic emphasis, she spann'd
The quick invisible strings, even though she saw   500
Endymion's spirit melt away and thaw
Before the deep intoxication.
But soon she came, with sudden burst, upon
Her self-possession—swung the lute aside,
And earnestly said: "Brother, 'tis vain to hide

495. **Dryope:** a shepherdess who bore a child to Apollo.

That thou dost know of things mysterious,
Immortal, starry; such alone could thus
Weigh down thy nature. Hast thou sinn'd in aught
Offensive to the heavenly power? Caught
A Paphian dove upon a message sent?   510
Thy deathful bow against some deer-herd bent
Sacred to Dian? Haply, thou hast seen
Her naked limbs among the alders green;
And that, alas! is death. No, I can trace
Something more high perplexing in thy face!"

    Endymion look'd at her, and press'd her hand,
And said, "Art thou so pale, who wast so bland
And merry in our meadows? How is this?
Tell me thine ailment: tell me all amiss!—
Ah! thou hast been unhappy at the change   520
Wrought suddenly in me. What indeed more strange?
Or more complete to overwhelm surmise?
Ambition is so sluggard: 'tis no prize,
That toiling years would put within my grasp,
That I have sighed for: with so deadly gasp
No man e'er panted for a mortal love.
So all have set my heavier grief above
These things which happen. Rightly have they done:
I, who still saw the horizontal sun
Heave his broad shoulder o'er the edge of the world,
Out-facing Lucifer, and then had hurl'd   531
My spear aloft, as signal for the chace—
I, who, for very sport of heart, would race
With my own steed from Araby; pluck down
A vulture from his towery perching; frown
A lion into growling, loth retire—
To lose, at once, all my toil-breeding fire,
And sink thus low! but I will ease my breast
Of secret grief, here in this bowery nest.

    "This river does not see the naked sky,   540
Till it begins to progress silverly
Around the western border of the wood,
Whence, from a certain spot, its winding flood
Seems at the distance like a crescent moon:
And in that nook, the very pride of June,
Had I been used to pass my weary eves;
The rather for the sun unwilling leaves

So dear a picture of his sovereign power,
And I could witness his most kingly hour,
When he doth tighten up the golden reins,   550
And paces leisurely down amber plains
His snorting four. Now when his chariot last
Its beams against the zodiac-lion cast,
There blossom'd suddenly a magic bed
Of sacred ditamy, and poppies red:
At which I wondered greatly, knowing well
That but one night had wrought this flowery spell;
And, sitting down close by, began to muse
What it might mean. Perhaps, thought I, Morpheus,
In passing here, his owlet pinions shook;   560
Or, it may be, ere matron Night uptook
Her ebon urn, young Mercury, by stealth,
Had dipt his rod in it: such garland wealth
Came not by common growth. Thus on I thought,
Until my head was dizzy and distraught.
Moreover, through the dancing poppies stole
A breeze, most softly lulling to my soul;
And shaping visions all about my sight
Of colours, wings, and bursts of spangly light;
The which became more strange, and strange, and
    dim,   570
And then were gulph'd in a tumultuous swim:
And then I fell asleep. Ah, can I tell
The enchantment that afterwards befel?
Yet it was but a dream: yet such a dream
That never tongue, although it overteem
With mellow utterance, like a cavern spring,
Could figure out and to conception bring
All I beheld and felt. Methought I lay
Watching the zenith, where the milky way
Among the stars in virgin splendour pours;   580
And travelling my eye, until the doors
Of heaven appear'd to open for my flight,
I became loth and fearful to alight
From such high soaring by a downward glance:
So kept me stedfast in that airy trance,
Spreading imaginary pinions wide.
When, presently, the stars began to glide,
And faint away, before my eager view:
At which I sigh'd that I could not pursue,
And dropt my vision to the horizon's verge;   590
And lo! from opening clouds, I saw emerge
The loveliest moon, that ever silver'd o'er
A shell for Neptune's goblet: she did soar

---

510. **Paphian dove:** one sent by Aphrodite from her temple in Paphos. **512–14. Haply . . . death:** Actaeon saw Diana bathing and was transformed into a stag and killed by his own hounds. **531. Lucifer:** the morning star.

555. **ditamy:** dittany, a medicinal plant.

So passionately bright, my dazzled soul
Commingling with her argent spheres did roll
Through clear and cloudy, even when she went
At last into a dark and vapoury tent—
Whereat, methought, the lidless-eyed train
Of planets all were in the blue again.
To commune with those orbs, once more I rais'd   600
My sight right upward: but it was quite dazed
By a bright something, sailing down apace,
Making me quickly veil my eyes and face:
Again I look'd, and, O ye deities,
Who from Olympus watch our destinies!
Whence that completed form of all completeness?
Whence came that high perfection of all sweetness?
Speak, stubborn earth, and tell me where, O where
Hast thou a symbol of her golden hair?
Not oat-sheaves drooping in the western sun;   610
Not—thy soft hand, fair sister! let me shun
Such follying before thee—yet she had,
Indeed, locks bright enough to make me mad;
And they were simply gordian'd up and braided,
Leaving, in naked comeliness, unshaded,
Her pearl round ears, white neck, and orbed brow;
The which were blended in, I know not how,
With such a paradise of lips and eyes,
Blush-tinted cheeks, half smiles, and faintest sighs,
That, when I think thereon, my spirit clings   620
And plays about its fancy, till the stings
Of human neighbourhood envenom all.
Unto what awful power shall I call?
To what high fane?—Ah! see her hovering feet,
More bluely vein'd, more soft, more whitely sweet
Than those of sea-born Venus, when she rose
From out her cradle shell. The wind out-blows
Her scarf into a fluttering pavillion;
'Tis blue, and over-spangled with a million
Of little eyes, as though thou wert to shed,   630
Over the darkest, lushest blue-bell bed,
Handfuls of daisies.'—'Endymion, how strange!
Dream within dream!'—'She took an airy range,
And then, towards me, like a very maid,
Came blushing, waning, willing, and afraid,
And press'd me by the hand: Ah! 'twas too much;
Methought I fainted at the charmed touch,
Yet held my recollections, even as one
Who dives three fathoms where the waters run

Gurgling in beds of coral: for anon,   640
I felt upmounted in that region
Where falling stars dart their artillery forth,
And eagles struggle with the buffeting north
That balances the heavy meteor-stone;—
Felt too, I was not fearful, nor alone,
But lapp'd and lull'd along the dangerous sky.
Soon, as it seem'd, we left our journeying high,
And straightway into frightful eddies swoop'd;
Such as aye muster where grey time has scoop'd
Huge dens and caverns in a mountain's side;   650
There hollow sounds arous'd me, and I sigh'd
To faint once more by looking on my bliss—
I was distracted; madly did I kiss
The wooing arms which held me, and did give
My eyes at once to death: but 'twas to live,
To take in draughts of life from the gold fount
Of kind and passionate looks; to count, and count
The moments, by some greedy help that seem'd
A second self, that each might be redeem'd
And plunder'd of its load of blessedness.   660
Ah, desperate mortal! I e'en dar'd to press
Her very cheek against my crowned lip,
And, at that moment, felt my body dip
Into a warmer air: a moment more,
Our feet were soft in flowers. There was store
Of newest joys upon that alp. Sometimes
A scent of violets, and blossoming limes,
Loiter'd around us; then of honey cells,
Made delicate from all white-flower bells;
And once, above the edges of our nest,   670
An arch face peep'd,—an Oread as I guess'd.

"Why did I dream that sleep o'er-power'd me
In midst of all this heaven? Why not see,
Far off, the shadows of his pinions dark,
And stare them from me? But no, like a spark
That needs must die, although its little beam
Reflects upon a diamond, my sweet dream
Fell into nothing—into stupid sleep.
And so it was, until a gentle creep,
A careful moving caught my waking ears,   680
And up I started: Ah! my sighs, my tears,
My clenched hands:—for lo! the poppies hung
Dew-dabbled on their stalks, the ouzel sung

---

595. argent: silver. 614. gordian'd: intertwined like the
Gordian knot. 626. sea-born: so called because she rose from
the foam of the sea.

644. balances: checks, because of its strength, the flying
meteor. 654-55. give . . . death: a passion so ecstatic that it
could lead to death and thence to another kind of life. 667.
limes: linden trees. 671. Oread: mountain nymph. 683.
ouzel: a European blackbird.

A heavy ditty, and the sullen day
Had chidden herald Hesperus away,
With leaden looks: the solitary breeze
Bluster'd, and slept, and its wild self did teaze
With wayward melancholy; and I thought,
Mark me, Peona! that sometimes it brought
Faint fare-thee-wells, and sigh-shrilled adieus!—   690
Away I wander'd—all the pleasant hues
Of heaven and earth had faded: deepest shades
Were deepest dungeons; heaths and sunny glades
Were full of pestilent light; our taintless rills
Seem'd sooty, and o'er-spread with upturn'd gills
Of dying fish; the vermeil rose had blown
In frightful scarlet, and its thorns out-grown
Like spiked aloe. If an innocent bird
Before my heedless footsteps stirr'd, and stirr'd
In little journeys, I beheld in it   700
A disguis'd demon, missioned to knit
My soul with under darkness; to entice
My stumblings down some monstrous precipice:
Therefore I eager followed, and did curse
The disappointment. Time, that aged nurse,
Rock'd me to patience. Now, thank gentle heaven!
These things, with all their comfortings, are given
To my down-sunken hours, and with thee,
Sweet sister, help to stem the ebbing sea
Of weary life."
        Thus ended he, and both   710
Sat silent: for the maid was very loth
To answer; feeling well that breathed words
Would all be lost, unheard, and vain as swords
Against the enchased crocodile, or leaps
Of grasshoppers against the sun. She weeps
And wonders; struggles to devise some blame;
To put on such a look as would say, *Shame*
On this poor weakness! but, for all her strife,
She could as soon have crush'd away the life
From a sick dove. At length, to break the pause,   720
She said with trembling chance: "Is this the cause?
This all? Yet it is strange, and sad, alas!
That one who through this middle earth should pass
Most like a sojourning demi-god, and leave
His name upon the harp-string, should achieve
No higher bard than simple maidenhood,
Singing alone, and fearfully,—how the blood

Left his young cheek; and how he used to stray
He knew not where; and how he would say, *nay*,
If any said 'twas love: and yet 'twas love;   730
What could it be but love? How a ring-dove
Let fall a sprig of yew tree in his path;
And how he died: and then, that love doth scathe
The gentle heart, as northern blasts do roses;
And then the ballad of his sad life closes
With sighs, and an alas!—Endymion!
Be rather in the trumpet's mouth,—anon
Among the winds at large—that all my hearken!
Although, before the crystal heavens darken,
I watch and dote upon the silver lakes   740
Pictur'd in western cloudiness, that takes
The semblance of gold rocks and bright gold sands,
Islands, and creeks, and amber-fretted strands
With horses prancing o'er them, palaces
And towers of amethyst,—would I so teaze
My pleasant days, because I could not mount
Into those regions? The Morphean fount
Of that fine element that visions, dreams,
And fitful whims of sleep are made of, streams
Into its airy channels with so subtle,   750
So thin a breathing, not the spider's shuttle,
Circled a million times within the space
Of a swallow's nest-door, could delay a trace,
A tinting of its quality: how light
Must dreams themselves be; seeing they're more slight
Than the mere nothing that engenders them!
Then wherefore sully the entrusted gem
Of high and noble life with thoughts so sick?
Why pierce high-fronted honour to the quick
For nothing but a dream?" Hereat the youth   760
Look'd up: a conflicting of shame and ruth
Was in his plaited brow: yet, his eyelids
Widened a little, as when Zephyr bids
A little breeze to creep between the fans
Of careless butterflies: amid his pains
He seem'd to taste a drop of manna-dew,
Full palatable; and a colour grew
Upon his cheek, while thus he lifeful spake.

"Peona! ever have I long'd to slake
My thirst for the world's praises: nothing base,   770
No merely slumberous phantasm, could unlace
The stubborn canvas for my voyage prepar'd—
Though now 'tis tatter'd; leaving my bark bar'd
And sullenly drifting: yet my higher hope
Is of too wide, too rainbow-large a scope,

---

714. **enchased:** cased with armor that is embossed or engraved. **725–26. achieve . . . maidenhood:** rise no
further as a poet than a maiden singing songs.

To fret at myriads of earthly wrecks.
Wherein lies happiness? In that which becks
Our ready minds to fellowship divine,
A fellowship with essence; till we shine,
Full alchemiz'd, and free of space. Behold   780
The clear religion of heaven! Fold
A rose leaf round thy finger's taperness,
And soothe thy lips: hist, when the airy stress
Of music's kiss impregnates the free winds,
And with a sympathetic touch unbinds
Æolian magic from their lucid wombs:
Then old songs waken from enclouded tombs;
Old ditties sigh above their father's grave;
Ghosts of melodious prophecyings rave
Round every spot where trod Apollo's foot;   790
Bronze clarions awake, and faintly bruit,
Where long ago a giant battle was;
And, from the turf, a lullaby doth pass
In every place where infant Orpheus slept.
Feel we these things?—that moment have we stept
Into a sort of oneness, and our state
Is like a floating spirit's. But there are
Richer entanglements, enthralments far
More self-destroying, leading, by degrees,
To the chief intensity: the crown of these   800
Is made of love and friendship, and sits high
Upon the forehead of humanity.
All its more ponderous and bulky worth
Is friendship, whence there ever issues forth

A steady splendour; but at the tip-top,
There hangs by unseen film, an orbed drop
Of light, and that is love: its influence,
Thrown in our eyes, genders a novel sense,
At which we start and fret; till in the end,
Melting into its radiance, we blend,   810
Mingle, and so become a part of it,—
Nor with aught else can our souls interknit
So wingedly: when we combine therewith,
Life's self is nourish'd by its proper pith,
And we are nurtured like a pelican brood.
Aye, so delicious is the unsating food,
That men, who might have tower'd in the van
Of all the congregated world, to fan
And winnow from the coming step of time
All chaff of custom, wipe away all slime   820
Left by men-slugs and human serpentry,
Have been content to let occasion die,
Whilst they did sleep in love's elysium.
And, truly, I would rather be struck dumb,
Than speak against this ardent listlessness:
For I have ever thought that it might bless
The world with benefits unknowingly;
As does the nightingale, upperched high,
And cloister'd among cool and bunched leaves—
She sings but to her love, nor e'er conceives   830
How tiptoe Night holds back her dark-grey hood.
Just so may love, although 'tis understood
The mere commingling of passionate breath,
Produce more than our searching witnesseth:
What I know not: but who, of men, can tell
That flowers would bloom, or that green fruit would
      swell
To melting pulp, that fish would have bright mail,
The earth its dower of river, wood, and vale,
The meadows runnels, runnels pebble-stones,
The seed its harvest, or the lute its tones,   840
Tones ravishment, or ravishment its sweet
If human souls did never kiss and greet?

    "Now, if this earthly love has power to make
Men's being mortal, immortal; to shake
Ambition from their memories, and brim
Their measure of content: what merest whim,
Seems all this poor endeavour after fame,
To one, who keeps within his stedfast aim
A love immortal, an immortal too.
Look not so wilder'd; for these things are true,   850

777. Wherein lies happiness?: introduces a central passage
(ll. 777–802) that Keats himself thought potentially important.
He changed the lines to their present form while he was copying Book I for the printer and told his publisher, John Taylor,
that he felt, as he wrote it, "a regular stepping of the Imagination towards a Truth." Writing it thus "set before me at once
the gradations of Happiness even like a kind of Pleasure
Thermometer—and is my first Step towards the chief Attempt
in the Drama—the playing of different Natures with Joy and
Sorrow." The passage has been endlessly debated. Is this a
new phrasing of the theme of stages of development that he
describes in "Sleep and Poetry" (ll. 96–125) and is to continue
to modify to the end? Or does Keats now see this expression,
as W. J. Bate says (in *John Keats*, pp. 181–84, 192), not as a sole
theme but as a central "aspiration of the heart" destined to
interplay with other aspirations and constantly changing
circumstances and thus constituting one element in the
"playing of different Natures" toward which the drama
itself aspires? Whatever the answer, or combination of answers, the passage was important to Keats and helped a little
to redeem *Endymion*, in his own opinion, as he tried energeti-
cally to turn to other writing. **779. fellowship with essence:**
Happiness is to be found in the active sympathetic participation in a reality richly experienced and distilled into "essence"
by the imagination. **791. bruit:** make a sound.

815. pelican: fabled to nourish its young by its own blood.
831. tiptoe Night: an echo of *Romeo and Juliet*, III. v. 10.

And never can be born of atomies
That buzz about our slumbers, like brain-flies,
Leaving us fancy-sick. No, no, I'm sure,
My restless spirit never could endure
To brood so long upon one luxury,
Unless it did, though fearfully, espy
A hope beyond the shadow of a dream.
My sayings will the less obscured seem,
When I have told thee how my waking sight
Has made me scruple whether that same night   860
Was pass'd in dreaming. Hearken, sweet Peona!
Beyond the matron-temple of Latona,
Which we should see but for these darkening boughs,
Lies a deep hollow, from whose ragged brows
Bushes and trees do lean all round athwart
And meet so nearly, that with wings outraught,
And spreaded tail, a vulture could not glide
Past them, but he must brush on every side.
Some moulder'd steps lead into this cool cell,
Far as the slabbed margin of a well,   870
Whose patient level peeps its crystal eye
Right upward, through the bushes, to the sky.
Oft have I brought thee flowers, on their stalks set
Like vestal primroses, but dark velvet
Edges them round, and they have golden pits:
'Twas there I got them, from the gaps and slits
In a mossy stone, that sometimes was my seat,
When all above was faint with mid-day heat.
And there in strife no burning thoughts to heed,
I'd bubble up the water through a reed;   880
So reaching back to boy-hood: make me ships
Of moulted feathers, touchwood, alder chips,
With leaves stuck in them; and the Neptune be
Of their petty ocean. Oftener, heavily,
When love-lorn hours had left me less a child,
I sat contemplating the figures wild
Of o'er-head clouds melting the mirror through.
Upon a day, while thus I watch'd, by flew
A cloudy Cupid, with his bow and quiver;
So plainly character'd, no breeze would shiver   890
The happy chance: so happy, I was fain
To follow it upon the open plain,
And, therefore, was just going; when, behold!
A wonder, fair as any I have told—
The same bright face I tasted in my sleep,
Smiling in the clear well. My heart did leap

862. Latona: also named Leto, mother of Apollo and
Cynthia.

Through the cool depth.—It moved as if to flee—
I started up, when lo! refreshfully
There came upon my face in plenteous showers
Dew-drops, and dewy buds, and leaves, and flowers,
Wrapping all objects from my smothered sight,   901
Bathing my spirit in a new delight.
Aye, such a breathless honey-feel of bliss
Alone preserved me from the drear abyss
Of death, for the fair form had gone again.
Pleasure is oft a visitant; but pain
Clings cruelly to us, like the gnawing sloth
On the deer's tender haunches: late, and loth,
'Tis scar'd away by slow returning pleasure.
How sickening, how dark the dreadful leisure   910
Of weary days, made deeper exquisite,
By a fore-knowledge of unslumbrous night!
Like sorrow came upon me, heavier still,
Than when I wander'd from the poppy hill:
And a whole age of lingering moments crept
Sluggishly by, ere more contentment swept
Away at once the deadly yellow spleen.
Yes, thrice have I this fair enchantment seen;
Once more been tortured with renewed life.
When last the wintry gusts gave over strife   920
With the conquering sun of spring, and left the skies
Warm and serene, but yet with moistened eyes
In pity of the shatter'd infant buds,—
That time thou didst adorn, with amber studs,
My hunting cap, because I laugh'd and smil'd,
Chatted with thee, and many days exil'd
All torment from my breast;—'twas even then,
Straying about, yet, coop'd up in the den
Of helpless discontent,—hurling my lance
From place to place, and following at chance,   930
At last, by hap, through some young trees it struck,
And, plashing among bedded pebbles, stuck
In the middle of a brook,—whose silver ramble
Down twenty little falls, through reeds and bramble,
Tracing along, it brought me to a cave,
Whence it ran brightly forth, and white did lave
The nether sides of mossy stones and rock,—
'Mong which it gurgled blythe adieus, to mock
Its own sweet grief at parting. Overhead,
Hung a lush screen of drooping weeds, and spread   940
Thick, as to curtain up some wood-nymph's home.
'Ah! impious mortal, whither do I roam?'
Said I, low voic'd: 'Ah, whither! 'Tis the grot
Of Proserpine, when Hell, obscure and hot,
Doth her resign; and where her tender hands
She dabbles, on the cool and sluicy sands:

Or 'tis the cell of Echo, where she sits,
And babbles thorough silence, till her wits
Are gone in tender madness, and anon,
Faints into sleep, with many a dying tone   950
Of sadness. O that she would take my vows,
And breathe them sighingly among the boughs,
To sue her gentle ears for whose fair head,
Daily, I pluck sweet flowerets from their bed,
And weave them dyingly—send honey-whispers
Round every leaf, that all those gentle lispers
May sigh my love unto her pitying!
O charitable Echo! hear, and sing
This ditty to her!—tell her'—so I stay'd
My foolish tongue, and listening, half afraid,   960
Stood stupefied with my own empty folly,
And blushing for the freaks of melancholy.
Salt tears were coming, when I heard my name
Most fondly lipp'd, and then these accents came:
'Endymion! the cave is secreter
Than the isle of Delos. Echo hence shall stir
No sighs but sigh-warm kisses, or light noise
Of thy combing hand, the while it travelling cloys
And trembles through my labyrinthine hair.'
At that oppress'd I hurried in.—Ah! where   970
Are those swift moments? Whither are they fled?
I'll smile no more, Peona; nor will wed
Sorrow the way to death; but patiently
Bear up against it: so farewell, sad sigh;
And come instead demurest meditation,
To occupy me wholly, and to fashion
My pilgrimage for the world's dusky brink.
No more will I count over, link by link,
My chain of grief: no longer strive to find
A half-forgetfulness in mountain wind   980
Blustering about my ears: aye, thou shalt see,
Dearest of sisters, what my life shall be;
What a calm round of hours shall make my days.
There is a paly flame of hope that plays
Where'er I look: but yet, I'll say 'tis naught—
And here I bid it die. Have not I caught,
Already, a more healthy countenance?
By this the sun is setting; we may chance
Meet some of our near-dwellers with my car."

This said, he rose, faint-smiling like a star   990
Through autumn mists, and took Peona's hand:
They stept into the boat, and launch'd from land.

                                        1817 (1818)

948. **thorough:** through.

IN DREAR-NIGHTED DECEMBER

I

In drear-nighted December,
      Too happy, happy tree,
Thy branches ne'er remember
      Their green felicity:
The north cannot undo them
With a sleety whistle through them;
Nor frozen thawings glue them
      From budding at the prime.

II

In drear-nighted December,
      Too happy, happy brook,   10
Thy bubblings ne'er remember
      Apollo's summer look;
But with a sweet forgetting,
They stay their crystal fretting,
Never, never petting
      About the frozen time.

III

Ah! would 'twere so with many
      A gentle girl and boy!
But were there ever any
      Writh'd not of passed joy?   20
The feel of not to feel it,
When there is none to heal it,
Nor numbed sense to steel it,
      Was never said in rhyme.

                                        1817 (1829)

IN DREAR-NIGHTED DECEMBER. **3–4. branches . . . felicity:**
Cf. "Ode on a Grecian Urn," below, ll. 21–22. **15. petting:**
fretting.

ON SEEING A LOCK
OF MILTON'S HAIR

⌇

This and the following three poems, all written within the same week or so (late Jan., 1818), are important as expressions of Keats's new self-dedication to a conception of poetry far beyond that represented by *Endymion*—a conception exemplified by the great figures of Shakespeare and Milton. Leigh Hunt, who had just acquired the lock, suggested to Keats that he write an extemporaneous poem on it. Keats started lamely and then, after the first twenty-two lines, began to write of what was most in his mind: his need for more philosophic depth and for long preparation.

⌇

Chief of organic numbers!
  Old Scholar of the Spheres!
Thy spirit never slumbers,
  But rolls about our ears,
For ever, and for ever!
  O what a mad endeavour
    Worketh he,
Who to thy sacred and ennobled hearse
Would offer a burnt sacrifice of verse
    And melody.   10

How heavenward thou soundest,
  Live Temple of sweet noise,
And Discord unconfoundest,
  Giving Delight new joys,
And Pleasure nobler pinions!
  O, where are thy dominions?
    Lend thine ear
To a young Delian oath,—aye, by thy soul,
By all that from thy mortal lips did roll,
And by the kernel of thine earthly love,   20
Beauty, in things on earth, and things above
    I swear!

When every childish fashion
  Has vanish'd from my rhyme,
Will I, grey-gone in passion,
  Leave to an after-time
    Hymning and harmony

Of thee, and of thy works, and of thy life;
But vain is now the burning and the strife,
Pangs are in vain, until I grow high-rife   30
    With old Philosophy,
And mad with glimpses of futurity!

For many years my offerings must be hush'd;
  When I do speak, I'll think upon this hour,
Because I feel my forehead hot and flush'd,
  Even at the simplest vassal of thy power,—
    A lock of thy bright hair,—
    Sudden it came,
And I was startled, when I caught thy name
    Coupled so unaware;   40
Yet, at the moment, temperate was my blood.
I thought I had beheld it from the flood.

                                    1818 (1838)

ON SITTING DOWN TO READ
KING LEAR ONCE AGAIN

⌇

The Shakespearean drama was increasingly becoming Keats's ultimate ideal (see the "Negative Capability" letter to his brothers, below, Dec. 21–27, 1817, p. 1209), and he had been studying and underscoring Hazlitt's essay on Lear in *Characters of Shakspeare's Plays*. Sending the sonnet to his brothers, Keats wrote, "You see I am getting at it, with a sort of determination & strength."

⌇

O golden tongued Romance, with serene lute!
  Fair plumed Syren, Queen of far-away!
  Leave melodizing on this wintry day,
Shut up thine olden pages, and be mute:
Adieu! for, once again, the fierce dispute
  Betwixt damnation and impassion'd clay
  Must I burn through; once more humbly assay
The bitter-sweet of this Shakespearian fruit:
Chief Poet! and ye clouds of Albion,
  Begetters of our deep eternal theme!   10

ON SITTING DOWN TO READ KING LEAR ONCE AGAIN. **1.**
Romance: Though Keats is referring generally to romances as contrasted with the great Shakespearean tragedy, he may also be thinking specifically of his own *Endymion: A Poetic Romance*, which he was now copying and correcting for the press.

ON SEEING A LOCK OF MILTON'S HAIR. **18. Delian:** referring to the island of Delos, where Apollo was born.

When through the old oak Forest I am gone,
  Let me not wander in a barren dream,
But, when I am consumed in the fire,
  Give me new Phœnix wings to fly at my desire.

                              1818 (1838)

WHEN I HAVE FEARS

When I have fears that I may cease to be
Before my pen has glean'd my teeming brain,
Before high-piled books, in charactery,
  Hold like rich garners the full ripen'd grain;
When I behold, upon the night's starr'd face,
  Huge cloudy symbols of a high romance,
And think that I may never live to trace
  Their shadows, with the magic hand of chance;
And when I feel, fair creature of an hour,
  That I shall never look upon thee more,   10
Never have relish in the faery power
  Of unreflecting love;—then on the shore
Of the wide world I stand alone, and think
Till love and fame to nothingness do sink.

                              1818 (1848)

GOD OF THE MERIDIAN

God of the Meridian,
  And of the East and West,
To thee my soul is flown,
  And my body is earthward press'd.—
It is an awful mission,
A terrible division:
And leaves a gulph austere
To be fill'd with wordly fear.
Aye, when the soul is fled
To high above our head,   10
Affrighted do we gaze
After its airy maze,

As doth a mother wild,
When her young infant child
Is in an eagle's claws—
And is not this the cause
Of madness?—God of Song,
Thou bearest me along
Through sights I scarce can bear:
O let me, let me share   20
With the hot lyre and thee,
The staid Philosophy.
Temper my lonely hours,
And let me see thy bowers
More unalarm'd!

                              1818 (1848)

LINES ON THE
MERMAID TAVERN

Souls of Poets dead and gone,
What Elysium have ye known,
Happy field or mossy cavern,
Choicer than the Mermaid Tavern?
Have ye tippled drink more fine
Than mine host's Canary wine?
Or are fruits of Paradise
Sweeter than those dainty pies
Of venison? O generous food!
Drest as though bold Robin Hood   10
Would, with his maid Marian,
Sup and bowse from horn and can.

  I have heard that on a day
Mine host's sign-board flew away,
Nobody knew whither, till
An astrologer's old quill
To a sheepskin gave the story,
Said he saw you in your glory,
Underneath a new old sign
Sipping beverage divine,   20
And pledging with contented smack
The Mermaid in the Zodiac.

  Souls of Poets dead and gone,
What Elysium have ye known,
Happy field or mossy cavern,
Choicer than the Mermaid Tavern?

                              1818 (1820)

11–14. When . . . desire: After I have passed through the forest of romance (and am finished with *Endymion*), give me Phoenix-like wings to rise from that death or conclusion in order to fly to a new goal (the Shakespearean drama). **14. Phœnix:** the mythical bird able, after being burned, to rise with new youth from its ashes. WHEN I HAVE FEARS. **3. charactery:** writing (characters, letters).

LINES ON THE MERMAID TAVERN. **12. bowse:** drink (booze).

WHAT THE THRUSH SAID

O thou whose face hath felt the Winter's wind,
    Whose eye has seen the snow-clouds hung in mist,
    And the black elm-tops 'mong the freezing stars,
    To thee the Spring will be a harvest-time.
O thou, whose only book has been the light
    Of supreme darkness, which thou feddest on
    Night after night, when Phœbus was away,
    To thee the Spring shall be a triple morn.
O fret not after knowledge—I have none,
    And yet my song comes native with the warmth.
O fret not after knowledge—I have none,   11
    And yet the evening listens. He who saddens
At thought of idleness cannot be idle,
And he's awake who thinks himself asleep.

                    1818 (1848)

THE HUMAN SEASONS

Four seasons fill the measure of the year;
    There are four seasons in the mind of man:
He has his lusty Spring, when fancy clear
    Takes in all beauty with an easy span:
He has his Summer, when luxuriously
    Spring's honied cud of youthful thought he loves
To ruminate, and by such dreaming nigh
    His nearest unto heaven: quiet coves
His soul has in its Autumn, when his wings
    He furleth close; contented so to look   10
On mists in idleness—to let fair things
    Pass by unheeded as a threshold brook.
He has his Winter too of pale misfeature,
Or else he would forgo his mortal nature.

                    1818 (1819)

TO HOMER

Standing aloof in giant ignorance,
    Of thee I hear and of the Cyclades,

As one who sits ashore and longs perchance
    To visit dolphin-coral in deep seas.
So thou wast blind;—but then the veil was rent,
    For Jove uncurtain'd Heaven to let thee live,
And Neptune made for thee a spumy tent,
    And Pan made sing for thee his forest-hive;
Aye on the shores of darkness there is light,
    And precipices show untrodden green,   10
There is a budding morrow in midnight,
    There is a triple sight in blindness keen;
Such seeing hadst thou, as it once befel
To Dian, Queen of Earth, and Heaven, and Hell.

                    1818 (1848)

from EPISTLE TO JOHN HAMILTON REYNOLDS

☙

The poem was written extemporaneously in March, 1818, as part of a personal letter, with no thought of publication. The poem begins with casual matters and suddenly becomes more reflective.

☙

O that our dreamings all, of sleep or wake,
Would all their colours from the sunset take:
From something of material sublime,
Rather than shadow our own soul's day-time   70
In the dark void of night. For in the world
We jostle,—but my flag is not unfurl'd
On the Admiral-staff,—and to philosophize
I dare not yet! Oh, never will the prize,
High reason, and the lore of good and ill,
Be my award! Things cannot to the will
Be settled, but they tease us out of thought;
Or is it that imagination brought
Beyond its proper bound, yet still confin'd,
Lost in a sort of Purgatory blind,   80

---

5. veil . . . rent: Matthew 27:51. **14. Dian . . . Hell:** Diana was conceived in later classical mythology as taking three forms: Artemis on Earth, Selene in Heaven, and Hecate in Hell. EPISTLE TO JOHN HAMILTON REYNOLDS. **67–71: O . . . night:** O that all our dreams (whether dreams while actually asleep or daydreams and musings while awake) were a repetition (in what Keats elsewhere calls "a finer tone") of the external world rather than a projection or "shadow" of our own subjective natures. **77. tease . . . thought:** used later in the "Ode on a Grecian Urn," l. 44.

WHAT THE THRUSH SAID. **9. O . . . knowledge:** Cf. Keats's letter to Reynolds, below, February 19, 1818, of which this poem was a part. The sonnet, it should be noted, is unrhymed. TO HOMER. **2. Cyclades:** islands clustered around Delos.

Cannot refer to any standard law
Of either earth or heaven? It is a flaw
In happiness, to see beyond our bourn,-
It forces us in summer skies to mourn,
It spoils the singing of the nightingale.

Dear Reynolds! I have a mysterious tale,
And cannot speak it: the first page I read
Upon a lampit rock of green sea-weed
Among the breakers; 'twas a quiet eve,
The rocks were silent, the wide sea did weave   90
An untumultuous fringe of silver foam
Along the flat brown sand; I was at home
And should have been most happy,—but I saw
Too far into the sea, where every maw
The greater on the less feeds evermore.—
But I saw too distinct into the core
Of an eternal fierce destruction,
And so from happiness I far was gone.
Still am I sick of it, and tho', to-day,
I've gather'd young spring-leaves, and flowers gay
Of periwinkle and wild strawberry,   101
Still do I that most fierce destruction see,—
The shark at savage prey,—the hawk at pounce,—
The gentle robin, like a pard or ounce,
Ravening a worm,—Away, ye horrid moods!
Moods of one's mind! You know I hate them well.
You know I'd sooner be a clapping bell
To some Kamschatkan missionary church,
Than with these horrid moods be left i' the lurch.—
Do you get health—and Tom the same—I'll dance,
And from detested moods in new romance   111
Take refuge—Of bad lines a centaine dose
Is sure enough—and so "here follows prose."—

1818 (1848)

---

88. **lampit:** limpet, a sea mollusk. 94–105. **Too . . . worm:** The conception of what Tennyson calls "Nature red in tooth and claw" is one that Keats is still unable to reconcile even partly with man's ideals, which are also something that takes place in nature. Hence to "philosophize / I dare not yet." Compare the more mature rephrasing of the problem in the "Vale of Soul-Making" letter to George and Georgiana Keats, March 19 and April 21, 1819, below. 104. **ounce:** lynx. 108. **Kamschatkan:** Kamchatka in eastern Siberia, facing the Bering Sea. Keats had been reading about it in Buffon's *Natural History*. 111. **new romance:** probably refers to *Isabella*. 113. **"here . . . prose":** a quote from *Twelfth Night*, II. v. 154. Keats's letter at this point turns into prose.

ISABELLA

OR, THE POT OF BASIL

A STORY FROM BOCCACCIO

Begun in February but written mainly in April, 1818, *Isabella* was planned as one of a series of verse adaptations, by Keats and his friend Reynolds, of tales from Boccaccio's *Decameron*. This is the only one that Keats wrote— he was interested in other forms of writing. He described it a few months later as "too smokeable" (i.e., too green —too much in need, because of its "simplicity of knowledge," of being "smoked at the Carpenter's shaving chimney . . . Isabella is what I should call were I a reviewer 'A weak-sided Poem' with an amusing sobersadness about it"). But Keats's self-expectation was unduly high. Despite its sentimental tone, it shows a marked advance over *Endymion*: it is more condensed and proceeds with a certain vigor. As a result it was liked by many readers soon after it was published (1820), became immensely popular in the later nineteenth century, and still has admirers. The plot closely follows that in the *Decameron* (Day IV, Tale 5). The stanza form is ottava rima, perhaps better adapted to satire (see Byron's *Don Juan*) than a story of pathos.

I

Fair Isabel, poor simple Isabel!
    Lorenzo, a young palmer in Love's eye!
They could not in the self-same mansion dwell
    Without some stir of heart, some malady;
They could not sit at meals but feel how well
    It soothed each to be the other by;
They could not, sure, beneath the same roof sleep
But to each other dream, and nightly weep.

II

With every morn their love grew tenderer,
    With every eve deeper and tenderer still;   10
He might not in house, field, or garden stir,
    But her full shape would all his seeing fill;
And his continual voice was pleasanter
    To her, than noise of trees or hidden rill;
Her lute-string gave an echo of his name,
She spoilt her half-done broidery with the same.

ISABELLA. 2. **palmer . . . eye:** devotee of love.

III

He knew whose gentle hand was at the latch
  Before the door had given her to his eyes;
And from her chamber-window he would catch
  Her beauty farther than the falcon spies;   20
And constant as her vespers would he watch,
  Because her face was turn'd to the same skies;
And with sick longing all the night outwear,
To hear her morning-step upon the stair.

IV

A whole long month of May in this sad plight
  Made their cheeks paler by the break of June:
"To-morrow will I bow to my delight,
  To-morrow will I ask my lady's boon."—
"O may I never see another night,
  Lorenzo, if thy lips breathe not love's tune."—   30
So spake they to their pillows; but, alas,
Honeyless days and days did he let pass;

V

Until sweet Isabella's untouch'd cheek
  Fell sick within the rose's just domain,
Fell thin as a young mother's, who doth seek
  By every lull to cool her infant's pain:
"How ill she is," said he, "I may not speak,
  And yet I will, and tell my love all plain:
If looks speak love-laws, I will drink her tears,
And at the least 'twill startle off her cares."   40

VI

So said he one fair morning, and all day
  His heart beat awfully against his side;
And to his heart he inwardly did pray
  For power to speak; but still the ruddy tide
Stifled his voice, and puls'd resolve away—
  Fever'd his high conceit of such a bride,
Yet brought him to the meekness of a child:
Alas! when passion is both meek and wild!

VII

So once more he had wak'd and anguished
  A dreary night of love and misery,   50
If Isabel's quick eye had not been wed
  To every symbol on his forehead high;
She saw it waxing very pale and dead,
  And straight all flush'd; so, lisped tenderly,
"Lorenzo!"—here she ceas'd her timid quest,
But in her tone and look he read the rest.

VIII

"O Isabella, I can half perceive
  That I may speak my grief into thine ear;
If thou didst ever anything believe,
  Believe how I love thee, believe how near   60
My soul is to its doom: I would not grieve
  Thy hand by unwelcome pressing, would not fear
Thine eyes by gazing; but I cannot live
Another night, and not my passion shrive.

IX

"Love! thou art leading me from wintry cold,
  Lady! thou leadest me to summer clime,
And I must taste the blossoms that unfold
  In its ripe warmth this gracious morning time."
So said, his erewhile timid lips grew bold,
  And poesied with hers in dewy rhyme: 70
Great bliss was with them, and great happiness
Grew, like a lusty flower in June's caress.

X

Parting they seem'd to tread upon the air,
  Twin roses by the zephyr blown apart
Only to meet again more close, and share
  The inward fragrance of each other's heart.
She, to her chamber gone, a ditty fair
  Sang, of delicious love and honey'd dart;
He with light steps went up a western hill,
And bade the sun farewell, and joy'd his fill.   80

XI

All close they met again, before the dusk
  Had taken from the stars its pleasant veil,
All close they met, all eves, before the dusk
  Had taken from the stars its pleasant veil,
Close in a bower of hyacinth and musk,
  Unknown of any, free from whispering tale.
Ah! better had it been for ever so,
Than idle ears should pleasure in their woe.

62. fear: frighten.

XII

Were they unhappy then?—It cannot be—
  Too many tears for lovers have been shed,   90
Too many sighs give we to them in fee,
  Too much of pity after they are dead,
Too many doleful stories do we see,
  Whose matter in bright gold were best be read;
Except in such a page where Theseus' spouse
Over the pathless waves towards him bows.

XIII

But, for the general award of love,
  The little sweet doth kill much bitterness;
Though Dido silent is in under-grove,
  And Isabella's was a great distress,   100
Though young Lorenzo in warm Indian clove
  Was not embalm'd, this truth is not the less—
Even bees, the little almsmen of spring-bowers,
Know there is richest juice in poison-flowers.

XIV

With her two brothers this fair lady dwelt,
  Enriched from ancestral merchandize,
And for them many a weary hand did swelt
  In torched mines and noisy factories,
And many once proud-quiver'd loins did melt
  In blood from stinging whip;—with hollow eyes
Many all day in dazzling river stood,   111
To take the rich-ored driftings of the flood.

XV

For them the Ceylon diver held his breath,
  And went all naked to the hungry shark;
For them his ears gush'd blood; for them in death
  The seal on the cold ice with piteous bark
Lay full of darts; for them alone did seethe
  A thousand men in troubles wide and dark:
Half-ignorant, they turn'd an easy wheel,
That set sharp racks at work, to pinch and peel.   120

XVI

Why were they proud? Because their marble founts
  Gush'd with more pride than do a wretch's tears?—
Why were they proud? Because fair orange-mounts
  Were of more soft ascent than lazar stairs?—
Why were they proud? Because red-lin'd accounts
  Were richer than the songs of Grecian years?—
Why were they proud? again we ask aloud,
Why in the name of Glory were they proud?

XVII

Yet were these Florentines as self-retired
  In hungry pride and gainful cowardice,   130
As two close Hebrews in that land inspired,
  Paled in and vineyarded from beggar-spies;
The hawks of ship-mast forests—the untired
  And pannier'd mules for ducats and old lies—
Quick cat's-paws on the generous stray-away,—
Great wits in Spanish, Tuscan, and Malay.

XVIII

How was it these same ledger-men could spy
  Fair Isabella in her downy nest?
How could they find out in Lorenzo's eye
  A straying from his toil? Hot Egypt's pest   140
Into their vision covetous and sly!
  How could these money-bags see east and west?—
Yet so they did—and every dealer fair
Must see behind, as doth the hunted hare.

XIX

O eloquent and famed Boccaccio!
  Of thee we now should ask forgiving boon,
And of thy spicy myrtles as they blow,
  And of thy roses amorous of the moon,
And of thy lillies, that do paler grow
  Now they can no more hear thy ghittern's tune,
For venturing syallables that ill beseem   151
The quiet glooms of such a piteous theme.

91. **in fee:** as payment. **95. Theseus' spouse:** Ariadne, deserted by Theseus after she had helped him defeat the Minotaur (*Odyssey*, XI. 321f.). **99. under-grove:** grove in the underworld (*Aeneid*, VI. 450f.). **105–28. With . . . proud:** These stanzas were praised by G. B. Shaw as an early indictment of capitalistic "profiteers and exploiters." **109. proud-quiver'd:** bearing quivers proudly.

123–24. **orange-mounts . . . stairs:** stairs made of orange wood, pleasanter to ascend than those of a hospital (lazar). **132. Paled in:** fenced in. **140. Egypt's pest:** flies in great swarms (Exodus 8 : 21). **145. O eloquent:** Keats consciously follows Chaucer's manner in pausing to make this address to Boccaccio. **150. ghittern:** guitar.

XX

Grant thou a pardon here, and then the tale
　　Shall move on soberly, as it is meet;
There is no other crime, no mad assail
　　To make old prose in modern rhyme more sweet:
But it is done—succeed the verse or fail—
　　To honour thee, and thy gone spirit greet;
To stead thee as a verse in English tongue,
An echo of thee in the north-wind sung.　　　160

XXI

These brethren having found by many signs
　　What love Lorenzo for their sister had,
And how she lov'd him too, each unconfines
　　His bitter thoughts to other, well nigh mad
That he, the servant of their trade designs,
　　Should in their sister's love be blithe and glad,
When 'twas their plan to coax her by degrees
To some high noble and his olive-trees.

XXII

And many a jealous conference had they,
　　And many times they bit their lips alone,　　170
Before they fix'd upon a surest way
　　To make the youngster for his crime atone;
And at the last, these men of cruel clay
　　Cut Mercy with a sharp knife to the bone;
For they resolved in some forest dim
To kill Lorenzo, and there bury him.

XXIII

So on a pleasant morning, as he leant
　　Into the sun-rise, o'er the balustrade
Of the garden-terrace, towards him they bent
　　Their footing through the dews; and to him said,
"You seem there in the quiet of content,　　181
　　Lorenzo, and we are most loth to invade
Calm speculation; but if you are wise,
　　Bestride your steed while cold is in the skies.

XXIV

"To-day we purpose, aye, this hour we mount
　　To spur three leagues towards the Apennine;

159. **stead thee:** preserve in memory.

Come down, we pray thee, ere the hot sun count
　　His dewy rosary on the eglantine."
Lorenzo, courteously as he was wont,
　　Bow'd a fair greeting to these serpents' whine;　190
And went in haste, to get in readiness,
With belt, and spur, and bracing huntsman's dress.

XXV

And as he to the court-yard pass'd along,
　　Each third step did he pause, and listen'd oft
If he could hear his lady's matin-song,
　　Or the light whisper of her footstep soft;
And as he thus over his passion hung,
　　He heard a laugh full musical aloft;
When, looking up, he saw her features bright
Smile through an in-door lattice, all delight.　　200

XXVI

"Love, Isabel!" said he, "I was in pain
　　Lest I should miss to bid thee a good morrow:
Ah! what if I should lose thee, when so fain
　　I am to stifle all the heavy sorrow
Of a poor three hours' absence? but we'll gain
　　Out of the amorous dark what day doth borrow.
Good bye! I'll soon be back."—"Good bye!" said
　　she:—
And as he went she chanted merrily.

XXVII

So the two brothers and their murder'd man
　　Rode past fair Florence, to where Arno's stream 210
Gurgles through straiten'd banks, and still doth fan
　　Itself with dancing bulrush, and the bream
Keeps head against the freshets. Sick and wan
　　The brothers' faces in the ford did seem,
Lorenzo's flush with love.—They pass'd the water
Into a forest quiet for the slaughter.

XXVIII

There was Lorenzo slain and buried in,
　　There in that forest did his great love cease;
Ah! when a soul doth thus its freedom win,
　　It aches in loneliness—is ill at peace　　220
As the break-covert blood-hounds of such sin:
　　They dipp'd their swords in the water, and did tease
Their horses homeward, with convulsed spur,
Each richer by his being a murderer.

XXIX

They told their sister how, with sudden speed,
　　Lorenzo had ta'en ship for foreign lands,
Because of some great urgency and need
　　In their affairs, requiring trusty hands.
Poor Girl! put on thy stifling widow's weed,
　　And 'scape at once from Hope's accursed bands;
To-day thou wilt not see him, nor to-morrow,　　231
And the next day will be a day of sorrow.

XXX

She weeps alone for pleasures not to be;
　　Sorely she wept until the night came on,
And then, instead of love, O misery!
　′ She brooded o'er the luxury alone:
His image in the dusk she seem'd to see,
　　And to the silence made a gentle moan,
Spreading her perfect arms upon the air,
And on her couch low murmuring "Where? O where?"

XXXI

But Selfishness, Love's cousin, held not long　　241
　　Its fiery vigil in her single breast;
She fretted for the golden hour, and hung
　　Upon the time with feverish unrest—
Not long—for soon into her heart a throng
　　Of higher occupants, a richer zest,
Came tragic; passion not to be subdued,
And sorrow for her love in travels rude.

XXXII

In the mid days of autumn, on their eves
　　The breath of Winter comes from far away,　　250
And the sick west continually bereaves
　　Of some gold tinge, and plays a roundelay
Of death among the bushes and the leaves,
　　To make all bare before he dares to stray
From his north cavern. So sweet Isabel
By gradual decay from beauty fell,

XXXIII

Because Lorenzo came not. Oftentimes
　　She ask'd her brothers, with an eye all pale,
Striving to be itself, what dungeon climes
　　Could keep him off so long? They spake a tale　260

Time after time, to quiet her. Their crimes
　　Came on them, like a smoke from Hinnom's vale;
And every night in dreams they groan'd aloud,
To see their sister in her snowy shroud.

XXXIV

And she had died in drowsy ignorance,
　　But for a thing more deadly dark than all;
It came like a fierce potion, drunk by chance,
　　Which saves a sick man from the feather'd pall
For some few gasping moments; like a lance,
　　Waking an Indian from his cloudy hall　　270
With cruel pierce, and bringing him again
Sense of the gnawing fire at heart and brain.

XXXV

It was a vision.—In the drowsy gloom,
　　The dull of midnight, at her couch's foot
Lorenzo stood, and wept: the forest tomb
　　Had marr'd his glossy hair which once could shoot
Lustre into the sun, and put cold doom
　　Upon his lips, and taken the soft lute
From his lorn voice, and past his loamed ears
Had made a miry channel for his tears.　　280

XXXVI

Strange sound it was, when the pale shadow spake;
　　For there was striving, in its piteous tongue,
To speak as when on earth it was awake,
　　And Isabella on its music hung:
Languor there was in it, and tremulous shake,
　　As in a palsied Druid's harp unstrung;
And through it moan'd a ghostly under-song,
Like hoarse night-gusts sepulchral briars among.

XXXVII

Its eyes, though wild, were still all dewy bright
　　With love, and kept all phantom fear aloof　　290
From the poor girl by magic of their light,
　　The while it did unthread the horrid woof
Of the late darken'd time,—the murderous spite
　　Of pride and avarice,—the dark pine roof
In the forest,—and the sodden turfed dell,
Where, without any word, from stabs he fell.

262. Hinnom's vale: where Ahaz "burnt his children in the fire" (II Chronicles 28:3 and II Kings 23:10).

XXXVIII

Saying moreover, "Isabel, my sweet!
  Red whortle-berries droop above my head,
And a large flint-stone weighs upon my feet;
  Around me beeches and high chestnuts shed   300
Their leaves and prickly nuts; a sheep-fold bleat
  Comes from beyond the river to my bed:
Go, shed one tear upon my heather-bloom,
And it shall comfort me within the tomb.

XXXIX

"I am a shadow now, alas! alas!
  Upon the skirts of human-nature dwelling
Alone: I chant alone the holy mass,
  While little sounds of life are round me knelling,
And glossy bees at noon do fieldward pass,
  And many a chapel bell the hour is telling,   310
Paining me through: those sounds grow strange to me,
And thou art distant in Humanity.

XL

"I know what was, I feel full well what is,
  And I should rage, if spirits could go mad;
Though I forget the taste of earthly bliss,
  That paleness warms my grave, as though I had
A Seraph chosen from the bright abyss
  To be my spouse: thy paleness makes me glad;
Thy beauty grows upon me, and I feel
A greater love through all my essence steal."   320

XLI

The Spirit mourn'd "Adieu!"—dissolv'd and left
  The atom darkness in a slow turmoil;
As when of healthful midnight sleep bereft,
  Thinking on rugged hours and fruitless toil,
We put our eyes into a pillowy cleft,
  And see the spangly gloom froth up and boil:
It made sad Isabella's eyelids ache,
And in the dawn she started up awake;

XLII

"Ha! ha!" said she, "I knew not this hard life,
  I thought the worst was simple misery;   330
I thought some Fate with pleasure or with strife
  Portion'd us—happy days, or else to die;

But there is crime—a brother's bloody knife!
  Sweet Spirit, thou hast school'd my infancy:
I'll visit thee for this, and kiss thine eyes,
And greet thee morn and even in the skies."

XLIII

When the full morning came, she had devised
  How she might secret to the forest hie;
How she might find the clay, so dearly prized,
  And sing to it one latest lullaby;   340
How her short absence might be unsurmised,
  While she the inmost of the dream would try.
Resolv'd, she took with her an aged nurse,
And went into that dismal forest-hearse.

XLIV

See, as they creep along the river side,
  How she doth whisper to that aged Dame,
And, after looking round the champaign wide,
  Shows her a knife.—"What feverous hectic flame
Burns in thee, child?—What good can thee betide,
  That thou should'st smile again?"—The evening
      came,   350
And they had found Lorenzo's earthy bed;
The flint was there, the berries at his head.

XLV

Who hath not loiter'd in a green church-yard,
  And let his spirit, like a demon-mole,
Work through the clayey soil and gravel hard,
  To see scull, coffin'd bones, and funeral stole;
Pitying each form that hungry Death hath marr'd
  And filling it once more with human soul?
Ah! this is holiday to what was felt
When Isabella by Lorenzo knelt.   360

XLVI

She gaz'd into the fresh-thrown mould, as though
  One glance did fully all its secrets tell;
Clearly she saw, as other eyes would know
  Pale limbs at bottom of a crystal well;

353–400. Who . . . dethroned: These stanzas were praised
by Charles Lamb: "there is nothing more awfully simple in
diction, more nakedly grand and moving in sentiment, in
Dante, in Chaucer, or in Spenser."

Upon the murderous spot she seem'd to grow,
    Like to a native lilly of the dell:
Then with her knife, all sudden, she began
To dig more fervently than misers can.

XLVII

Soon she turn'd up a soiled glove, whereon
    Her silk had play'd in purple phantasies,   370
She kiss'd it with a lip more chill than stone,
    And put it in her bosom, where it dries
And freezes utterly unto the bone
    Those dainties made to still an infant's cries:
Then 'gan she work again; nor stay'd her care,
But to throw back at times her veiling hair.

XLVIII

That old nurse stood beside her wondering,
    Until her heart felt pity to the core
At sight of such a dismal labouring,
    And so she kneeled, with her locks all hoar,   380
And put her lean hands to the horrid thing:
    Three hours they labour'd at this travail sore;
At last they felt the kernel of the grave,
And Isabella did not stamp and rave.

XLIX

Ah! wherefore all this wormy circumstance?
    Why linger at the yawning tomb so long?
O for the gentleness of old Romance,
    The simple plaining of a minstrel's song!
Fair reader, at the old tale take a glance,
    For here, in truth, it doth not well belong   390
To speak:—O turn thee to the very tale,
And taste the music of that vision pale.

L

With duller steel than the Perséan sword
    They cut away no formless monster's head,
But one, whose gentleness did well accord
    With death, as life. The ancient harps have said,
Love never dies, but lives, immortal Lord:
    If Love impersonate was ever dead,
Pale Isabella kiss'd it, and low moan'd.
'Twas love; cold,—dead indeed, but not dethroned.

388. plaining: singing. **393. Perséan sword:** The sword of
Perseus cut off the head of the Gorgon Medusa.

LI

In anxious secrecy they took it home,   401
    And then the prize was all for Isabel:
She calm'd its wild hair with a golden comb,
    And all around each eye's sepulchral cell
Pointed each fringed lash; the smeared loam
    With tears, as chilly as a dripping well,
She drench'd away:—and still she comb'd, and kept
Sighing all day—and still she kiss'd, and wept.

LII

Then in a silken scarf,—sweet with the dews
    Of precious flowers pluck'd in Araby,   410
And divine liquids come with odorous ooze
    Through the cold serpent-pipe refreshfully,—
She wrapp'd it up; and for its tomb did choose
    A garden-pot, wherein she laid it by,
And cover'd it with mould, and o'er it set
Sweet Basil, which her tears kept ever wet.

LIII

And she forgot the stars, the moon, and sun,
    And she forgot the blue above the trees,
And she forgot the dells where waters run,
    And she forgot the chilly autumn breeze;   420
She had no knowledge when the day was done,
    And the new morn she saw not: but in peace
Hung over her sweet Basil evermore,
And moisten'd it with tears unto the core.

LIV

And so she ever fed it with thin tears,
    Whence thick, and green, and beautiful it grew,
So that it smelt more balmy than its peers
    Of Basil-tufts in Florence; for it drew
Nurture besides, and life, from human fears,
    From the fast mouldering head there shut from view:
So that the jewel, safely casketed,   431
Came forth, and in perfumed leafits spread.

LV

O Melancholy, linger here awhile!
    O Music, Music, breathe despondingly!

412. serpent-pipe: pipe used for distilling.

O Echo, Echo, from some sombre isle,
  Unknown, Lethean, sigh to us—O sigh!
Spirits in grief, lift up your heads, and smile;
  Lift up your heads, sweet Spirits, heavily,
And make a pale light in your cypress glooms,
Tinting with silver wan your marble tombs.   440

LVI

Moan hither, all ye syllables of woe,
  From the deep throat of sad Melpomene!
Through bronzed lyre in tragic order go,
  And touch the strings into a mystery;
Sound mournfully upon the winds and low;
  For simple Isabel is soon to be
Among the dead: She withers, like a palm
Cut by an Indian for its juicy balm.

LVII

O leave the palm to wither by itself;
  Let not quick Winter chill its dying hour!—   450
It may not be—those Baälites of pelf,
  Her brethren, noted the continual shower
From her dead eyes; and many a curious elf,
  Among her kindred, wonder'd that such dower
Of youth and beauty should be thrown aside
By one mark'd out to be a Noble's bride.

LVIII

And, furthermore, her brethren wonder'd much
  Why she sat drooping by the Basil green,
And why it flourish'd, as by magic touch;
  Greatly they wonder'd what the thing might mean:
They could not surely give belief, that such   461
  A very nothing would have power to wean
Her from her own fair youth, and pleasures gay,
And even remembrance of her love's delay.

LIX

Therefore they watch'd a time when they might sift
  This hidden whim; and long they watch'd in vain;
For seldom did she go to chapel-shrift,
  And seldom felt she any hunger-pain;

436. **Lethean:** from Lethe, a river in Hades whose water
produces oblivion when drunk. 442. **Melpomene:** muse
of tragedy. 451. **Baälites of pelf:** idolators worshipping
money.

And when she left, she hurried back, as swift
  As bird on wing to breast its eggs again;   470
And, patient as a hen-bird, sat her there
Beside her Basil, weeping through her hair.

LX

Yet they contriv'd to steal the Basil-pot,
  And to examine it in secret place;
The thing was vile with green and livid spot,
  And yet they knew it was Lorenzo's face:
The guerdon of their murder they had got,
  And so left Florence in a moment's space,
Never to turn again.—Away they went,
With blood upon their heads, to banishment.   480

LXI

O Melancholy, turn thine eyes away!
  O Music, Music, breathe despondingly!
O Echo, Echo, on some other day,
  From isles Lethean, sigh to us—O sigh!
Spirits of grief, sing not your "Well-a-way!"
  For Isabel, sweet Isabel, will die;
Will die a death too lone and incomplete,
Now they have ta'en away her Basil sweet.

LXII

Piteous she look'd on dead and senseless things,
  Asking for her lost Basil amorously;   490
And with melodious chuckle in the strings
  Of her lorn voice, she oftentimes would cry
After the Pilgrim in his wanderings,
  To ask him where her Basil was; and why
'Twas hid from her: "For cruel 'tis," said she,
"To steal my Basil-pot away from me."

LXIII

And so she pined, and so she died forlorn,
  Imploring for her Basil to the last.
No heart was there in Florence but did mourn
  In pity of her love, so overcast.   500
And a sad ditty of this story born
  From mouth to mouth through all the country
      pass'd:
Still is the burthen sung—"O cruelty,
To steal my Basil-pot away from me!"

1818 (1820)

FRAGMENT
OF AN ODE TO MAIA

In this unfinished poem Keats's thought of the great
poetry of early eras is put with a new richness and calm
(May 1, 1818). A year later he returned to the theme in the
first of his great odes ("Ode to Psyche").

Mother of Hermes! and still youthful Maia!
    May I sing to thee
As thou wast hymned on the shores of Baiæ?
    Or may I woo thee
In earlier Sicilian? or thy smiles
Seek as they once were sought, in Grecian isles,
    By bards who died content on pleasant sward,
    Leaving great verse unto a little clan?
O, give me their old vigour, and unheard
    Save of the quiet primrose, and the span   10
    Of heaven and few ears,
Rounded by thee, my song should die away
    Content as theirs,
Rich in the simple worship of a day.

1818 (1848)

ON VISITING
THE TOMB OF BURNS

This and the next two poems were written during the
walking tour that Keats and Charles Brown made
through Scotland and northern England during the
summer of 1818.

The town, the churchyard, and the setting sun,
    The clouds, the trees, the rounded hills all seem,
    Though beautiful, cold—strange—as in a dream,
I dreamed long ago, now new begun.

The short-liv'd, paly Summer is but won
    From Winter's ague, for one hour's gleam;
    Though sapphire-warm, their stars do never beam:
All is cold Beauty; pain is never done:
For who has mind to relish, Minos-wise,
    The Real of Beauty, free from that dead hue   10
    Sickly imagination and sick pride
Cast wan upon it? Burns! with honour due
    I oft have honour'd thee. Great shadow, hide
Thy face; I sin against thy native skies.

1818 (1848)

OLD MEG

"Old Meg" is a ballad written on the Scottish tour for
Keats's brother and sister, Tom and Fanny. Charles
Brown had been telling Keats, as they walked, of Sir
Walter Scott's account of the gypsy, Meg Merrilies, in
Guy Mannering.

Old Meg she was a Gipsy,
    And liv'd upon the Moors:
Her bed it was the brown heath turf,
    And her house was out of doors.

Her apples were swart blackberries,
    Her currants pods o' broom;
Her wine was dew of the wild white rose,
    Her book a churchyard tomb.

Her Brothers were the craggy hills,
    Her Sisters larchen trees—   10
Alone with her great family
    She liv'd as she did please.

No breakfast had she many a morn,
    No dinner many a noon,
And 'stead of supper she would stare
    Full hard against the Moon.

But every morn of woodbine fresh
    She made her garlanding,
And every night the dark glen Yew
    She wove, and she would sing.   20

FRAGMENT OF AN ODE TO MAIA. **3. Baiæ:** a Roman colony
in the Naples area. **5. earlier Sicilian:** in the pastoral vein
of Theocritus.

ON VISITING THE TOMB OF BURNS. **9. Minos-wise:** like Minos,
judge of the underworld; making an objective and complete
judgment.

And with her fingers old and brown
  She plaited Mats o' Rushes,
And gave them to the Cottagers
  She met among the Bushes.

Old Meg was brave as Margaret Queen
  And tall as Amazon:
An old red blanket cloak she wore;
  A chip hat had she on.
God rest her aged bones somewhere—
  She died full long agone!   30

1818 (1838)

THE POET

A FRAGMENT

❧

Compare Keats's observations on the sympathetic nature of the poet's imagination in his letter to Woodhouse, below, October 27, 1818, p. 1219. With lines 8–10 compare Keats's remark in a letter to Benjamin Bailey, below (Nov. 22, 1817), p. 1208, "if a Sparrow come before my Window I take part in its existence and pick about the Gravel."

❧

Where's the Poet? show him! show him,
Muses nine! that I may know him!
'Tis the man who with a man
  Is an equal, be he King,
Or poorest of the beggar-clan,
  Or any other wondrous thing
A man may be 'twixt ape and Plato;
  'Tis the man who with a bird,
Wren or Eagle, finds his way to
  All its instincts; he hath heard   10
The Lion's roaring, and can tell
  What his horny throat expresseth,
And to him the Tiger's yell
  Comes articulate and presseth
On his ear like mother-tongue.

1818 (1848)

HYPERION

❧

Most of the poem was written while Keats was nursing his dying brother Tom (Sept. to Dec. 1, 1818). Though he plunged into this work in order to distract himself, he felt that it was premature. One of his long-range ambitions was to combine the "grandeur" and "old vigour" of the great epic poems of the past with the modern "thinking into the human heart" typified by Wordsworth. (See the letter below to J. H. Reynolds, May 3, 1818, p. 1214.) Using, with some modifications, the powerful epic idiom and to some extent the versification of Milton's *Paradise Lost*, Keats apparently took as his theme (for the poem was left unfinished) the growth of consciousness. The day of the majestic Titans is coming to an end. Destined to replace them are the Olympian gods. Still unfallen is the Titan, Hyperion, god of the sun. Presumably the action of the poem was to center on his fall and final replacement by the young Apollo (to whom the fragmentary third book is devoted).

The magnificent Miltonic style, though Keats was quickly to react from it, represents an astonishing advance when compared with that of *Endymion* a year before. With it he achieves a mastery of expression and, during his remaining year of writing, turns to style after style with a creative variety unequalled by any poet within the same length of time.

❧

BOOK I

Deep in the shady sadness of a vale
Far sunken from the healthy breath of morn,
Far from the fiery noon, and eve's one star,
Sat gray-hair'd Saturn, quiet as a stone,
Still as the silence round about his lair;
Forest on forest hung about his head
Like cloud on cloud. No stir of air was there,
Not so much life as on a summer's day
Robs not one light seed from the feather'd grass,
But where the dead leaf fell, there did it rest.   10
A stream went voiceless by, still deadened more
By reason of his fallen divinity
Spreading a shade: the Naiad 'mid her reeds
Press'd her cold finger closer to her lips.

OLD MEG. **28. chip:** strip of wood fiber.

HYPERION. BOOK I. **11. voiceless:** typical of the Miltonic use of the adjective for the adverb.

Along the margin-sand large foot-marks went,
No further than to where his feet had stray'd,
And slept there since. Upon the sodden ground
His old right hand lay nerveless, listless, dead,
Unsceptred; and his realmless eyes were closed;
While his bow'd head seem'd list'ning to the Earth, 20
His ancient mother, for some comfort yet.

It seem'd no force could wake him from his place;
But there came one, who with a kindred hand
Touch'd his wide shoulders, after bending low
With reverence, though to one who knew it not.
She was a Goddess of the infant world;
By her in stature the tall Amazon
Had stood a pigmy's height: she would have ta'en
Achilles by the hair and bent his neck;
Or with a finger stay'd Ixion's wheel.   30
Her face was large as that of Memphian sphinx,
Pedestal'd haply in a palace court,
When sages look'd to Egypt for their lore.
But oh! how unlike marble was that face:
How beautiful, if sorrow had not made
Sorrow more beautiful than Beauty's self.
There was a listening fear in her regard,
As if calamity had but begun;
As if the vanward clouds of evil days
Had spent their malice, and the sullen rear   40
Was with its stored thunder labouring up.
One hand she press'd upon that aching spot
Where beats the human heart, as if just there,
Though an immortal, she felt cruel pain:
The other upon Saturn's bended neck
She laid, and to the level of his ear
Leaning with parted lips, some words she spake
In solemn tenour and deep organ tone:
Some mourning words, which in our feeble tongue
Would come in these like accents; O how frail   50
To that large utterance of the early Gods!
"Saturn, look up!—though wherefore, poor old King?
I have no comfort for thee, no not one:
I cannot say, 'O wherefore sleepest thou?'
For heaven is parted from thee, and the earth
Knows thee not, thus afflicted, for a God;
And ocean too, with all its solemn noise,

Has from thy sceptre pass'd; and all the air
Is emptied of thine hoary majesty.
Thy thunder, conscious of the new command,   60
Rumbles reluctant o'er our fallen house;
And thy sharp lightning in unpractis'd hands
Scorches and burns our once serene domain.
O aching time! O moments big as years!
All as ye pass swell out the monstrous truth,
And press it so upon our weary griefs
That unbelief has not a space to breathe.
Saturn, sleep on:—O thoughtless, why did I
Thus violate thy slumbrous solitude?
Why should I ope thy melancholy eyes?   70
Saturn, sleep on! while at thy feet I weep."
    As when, upon a tranced summer-night,
Those green-rob'd senators of mighty woods,
Tall oaks, branch-charmed by the earnest stars,
Dream, and so dream all night without a stir,
Save from one gradual solitary gust
Which comes upon the silence, and dies off,
As if the ebbing air had but one wave;
So came these words and went; the while in tears
She touch'd her fair large forehead to the ground, 80
Just where her falling hair might be outspread
A soft and silken mat for Saturn's feet.
One moon, with alteration slow, had shed
Her silver seasons four upon the night,
And still these two were postured motionless,
Like natural sculpture in cathedral cavern;
The frozen God still couchant on the earth,
And the sad Goddess weeping at his feet:
Until at length old Saturn lifted up
His faded eyes, and saw his kingdom gone,   90
And all the gloom and sorrow of the place,
And that fair kneeling Goddess; and then spake,
As with a palsied tongue, and while his beard
Shook horrid with such aspen-malady:
"O tender spouse of gold Hyperion,
Thea, I feel thee ere I see thy face;
Look up, and let me see our doom in it;
Look up, and tell me if this feeble shape
Is Saturn's; tell me, if thou hear'st the voice
Of Saturn; tell me, if this wrinkling brow,   100
Naked and bare of its great diadem,
Peers like the front of Saturn. Who had power
To make me desolate? whence came the strength?

---

23. came one: Thea, Hyperion's sister and wife. **30. Ixion:** bound to a continually turning wheel in Hades for daring to love Juno. **31. Memphian:** Memphis, the ancient capital of Lower Egypt, near which are the sphinx and pyramids. **37. regard:** look, expression. **52. poor old King:** Keats is partly recalling King Lear.

83–90. One . . . gone: typical of the sculpturesque imagery of much of the poem. **83. alteration slow:** the first of what Keats called his "Miltonic inversions," in which the adjective is placed after the noun. **94. horrid:** rough.

How was it nurtur'd to such bursting forth,
While Fate seem'd strangled in my nervous grasp?
But it is so; and I am smother'd up,
And buried from all godlike exercise
Of influence benign on planets pale,
Of admonitions to the winds and seas,
Of peaceful sway above man's harvesting,   110
And all those acts which Deity supreme
Doth ease its heart of love in.—I am gone
Away from my own bosom: I have left
My strong identity, my real self,
Somewhere between the throne, and where I sit
Here on this spot of earth. Search, Thea, search!
Open thine eyes eterne, and sphere them round
Upon all space: space starr'd, and lorn of light;
Space region'd with life-air; and barren void;
Spaces of fire, and all the yawn of hell.—   120
Search, Thea, search! and tell me, if thou seest
A certain shape or shadow, making way
With wings or chariot fierce to repossess
A heaven he lost erewhile: it must—it must
Be of ripe progress—Saturn must be King.
Yes, there must be a golden victory;
There must be Gods thrown down, and trumpets
   blown
Of triumph calm, and hymns of festival
Upon the gold clouds metropolitan,
Voices of soft proclaim, and silver stir   130
Of strings in hollow shells; and there shall be
Beautiful things made new, for the surprise
Of the sky-children; I will give command:
Thea! Thea! Thea! where is Saturn?"

This passion lifted him upon his feet,
And made his hands to struggle in the air,
His Druid locks to shake and ooze with sweat,
His eyes to fever out, his voice to cease.
He stood, and heard not Thea's sobbing deep;
A little time, and then again he snatch'd   140
Utterance thus.—"But cannot I create?
Cannot I form? Cannot I fashion forth
Another world, another universe,
To overbear and crumble this to naught?
Where is another chaos? Where?"—That word
Found way unto Olympus, and made quake
The rebel three.—Thea was startled up,
And in her bearing was a sort of hope,

As thus she quick-voic'd spake, yet full of awe.
"This cheers our fallen house: come to our friends,
O Saturn! come away, and give them heart;   151
I know the covert, for thence came I hither."
Thus brief; then with beseeching eyes she went
With backward footing through the shade a space:
He follow'd, and she turn'd to lead the way
Through aged boughs, that yielded like the mist
Which eagles cleave upmounting from their nest.

Meanwhile in other realms big tears were shed,
More sorrow like to this, and such like woe,
Too huge for mortal tongue or pen of scribe: 160
The Titans fierce, self-hid, or prison-bound,
Groan'd for the old allegiance once more,
And listen'd in sharp pain for Saturn's voice.
But one of the whole mammoth-brood still kept
His sov'reignty, and rule, and majesty;—
Blazing Hyperion on his orbed fire
Still sat, still snuff'd the incense, teeming up
From man to the sun's God; yet unsecure:
For as among us mortals omens drear
Fright and perplex, so also shuddered he—   170
Not at dog's howl, or gloom-bird's hated screech,
Or the familiar visiting of one
Upon the first toll of his passing-bell,
Or prophesyings of the midnight lamp;
But horrors, portion'd to a giant nerve,
Oft made Hyperion ache. His palace bright
Bastion'd with pyramids of glowing gold,
And touch'd with shade of bronzed obelisks,
Glar'd a blood-red through all its thousand courts,
Arches, and domes, and fiery galleries;   180
And all its curtains of Aurorian clouds
Flush'd angerly: while sometimes eagle's wings,
Unseen before by Gods or wondering men,
Darken'd the place; and neighing steeds were heard,
Not heard before by Gods or wondering men.
Also, when he would taste the spicy wreaths
Of incense, breath'd aloft from sacred hills,
Instead of sweets, his ample palate took
Savour of poisonous brass and metal sick:
And so, when harbour'd in the sleepy west,   190
After the full completion of fair day,—
For rest divine upon exalted couch
And slumber in the arms of melody,
He pac'd away the pleasant hours of ease

---

129. gold clouds metropolitan: a Miltonic device in which the noun is flanked by two adjectives. **147. rebel three:** Jupiter, Pluto, and Neptune.

171. gloom-bird's: owl's. **181. Aurorian:** from Aurora, Roman goddess of dawn. **184. neighing steeds:** Cf. *Julius Caesar*, II. ii. 23.

With stride colossal, on from hall to hall;
While far within each aisle and deep recess,
His winged minions in close clusters stood,
Amaz'd and full of fear; like anxious men
Who on wide plains gather in panting troops,
When earthquakes jar their battlements and towers.
Even now, while Saturn, rous'd from icy trance,   201
Went step for step with Thea through the woods,
Hyperion, leaving twilight in the rear,
Came slope upon the threshold of the west;
Then, as was wont, his palace-door flew ope
In smoothest silence, save what solemn tubes,
Blown by the serious Zephyrs, gave of sweet
And wandering sounds, slow-breathed melodies;
And like a rose in vermeil tint and shape,
In fragrance soft, and coolness to the eye,   210
That inlet to severe magnificence
Stood full blown, for the God to enter in.

He enter'd, but he enter'd full of wrath;
His flaming robes stream'd out beyond his heels,
And gave a roar, as if of earthly fire,
That scar'd away the meek ethereal Hours
And made their dove-wings tremble. On he flared,
From stately nave to nave, from vault to vault,
Through bowers of fragrant and enwreathed light,
And diamond-paved lustrous long arcades,   220
Until he reach'd the great main cupola;
There standing fierce beneath, he stamped his foot,
And from the basement deep to the high towers
Jarr'd his own golden region; and before
The quavering thunder thereupon had ceas'd,
His voice leapt out, despite of godlike curb,
To this result: "O dreams of day and night!
O monstrous forms! O effigies of pain!
O spectres busy in a cold, cold gloom!
O lank-ear'd Phantoms of black-weeded pools!   230
Why do I know ye? why have I seen ye? why
Is my eternal essence thus distraught
To see and to behold these horrors new?
Saturn is fallen, am I too to fall?
Am I to leave this haven of my rest,
This cradle of my glory, this soft clime,
This calm luxuriance of blissful light,
These crystalline pavilions, and pure fanes,
Of all my lucent empire? It is left
Deserted, void, nor any haunt of mine.   240
The blaze, the splendour, and the symmetry,

207. **Zephyrs:** western breezes.

I cannot see—but darkness, death and darkness.
Even here, into my centre of repose,
The shady visions come to domineer,
Insult, and blind, and stifle up my pomp.—
Fall!—No, by Tellus and her briny robes!
Over the fiery frontier of my realms
I will advance a terrible right arm
Shall scare that infant thunderer, rebel Jove,
And bid old Saturn take his throne again."—   250
He spake, and ceas'd, the while a heavier threat
Held struggle with his throat but came not forth;
For as in theatres of crowded men
Hubbub increases more they call out "Hush!"
So at Hyperion's words the Phantoms pale
Bestirr'd themselves, thrice horrible and cold;
And from the mirror'd level where he stood
A mist arose, as from a scummy marsh.
At this, through all his bulk an agony
Crept gradual, from the feet unto the crown,   260
Like a lithe serpent vast and muscular
Making slow way, with head and neck convuls'd
From over-strained might. Releas'd, he fled
To the eastern gates, and full six dewy hours
Before the dawn in season due should blush,
He breath'd fierce breath against the sleepy portals,
Clear'd them of heavy vapours, burst them wide
Suddenly on the ocean's chilly streams.
The planet orb of fire, whereon he rode
Each day from east to west the heavens through,   270
Spun round in sable curtaining of clouds;
Not therefore veiled quite, blindfold, and hid,
But ever and anon the glancing spheres,
Circles, and arcs, and broad-belting colure,
Glow'd through, and wrought upon the muffling dark
Sweet-shaped lightnings from the nadir deep
Up to the zenith,—hieroglyphics old
Which sages and keen-eyed astrologers
Then living on the earth, with labouring thought
Won from the gaze of many centuries: 280
Now lost, save what we find on remnants huge
Of stone, or marble swart; their import gone,
Their wisdom long since fled.—Two wings this orb
Possess'd for glory, two fair argent wings,
Ever exalted at the God's approach:
And now, from forth the gloom their plumes immense

246. **Tellus:** goddess of the Earth. 274. **colure:** one of two circles in the celestial sphere that intersect with each other. 276. **nadir:** lowest point of the celestial sphere, opposite the zenith. 284. **argent:** silver.

Rose, one by one, till all outspreaded were;
While still the dazzling globe maintain'd eclipse,
Awaiting for Hyperion's command.
Fain would he have commanded, fain took throne 290
And bid the day begin, if but for change.
He might not:—No, though a primeval God:
The sacred seasons might not be disturb'd.
Therefore the operations of the dawn
Stay'd in their birth, even as here 'tis told.
Those silver wings expanded sisterly,
Eager to sail their orb; the porches wide
Open'd upon the dusk demesnes of night;
And the bright Titan, phrenzied with new woes,
Unus'd to bend, by hard compulsion bent   300
His spirit to the sorrow of the time;
And all along a dismal rack of clouds,
Upon the boundaries of day and night,
He stretch'd himself in grief and radiance faint.
There as he lay, the Heaven with its stars
Look'd down on him with pity, and the voice
Of Cœlus, from the universal space,
Thus whisper'd low and solemn in his ear.
"O brightest of my children dear, earth-born
And sky-engendered, Son of Mysteries   310
All unrevealed even to the powers
Which met at thy creating; at whose joys
And palpitations sweet, and pleasures soft,
I, Cœlus, wonder, how they came and whence;
And at the fruits thereof what shapes they be,
Distinct, and visible; symbols divine,
Manifestations of that beauteous life
Diffus'd unseen throughout eternal space:
Of these new-form'd art thou, oh brightest child!
Of these, thy brethren and the Goddesses!   320
There is sad feud among ye, and rebellion
Of son against his sire. I saw him fall,
I saw my first-born tumbled from his throne!
To me his arms were spread, to me his voice
Found way from forth the thunders round his head!
Pale wox I, and in vapours hid my face.
Art thou, too, near such doom? vague fear there is:
For I have seen my sons most unlike Gods.
Divine ye were created, and divine
In sad demeanour, solemn, undisturb'd,   330
Unruffled, like high Gods, ye liv'd and ruled:
Now I behold in you fear, hope, and wrath;
Actions of rage and passion; even as
I see them, on the mortal world beneath,

307. **Cœlus:** Uranus, the sky. 323. **first-born:** Saturn.

In men who die.—This is the grief, O Son!
Sad sign of ruin, sudden dismay, and fall!
Yet do thou strive; as thou art capable,
As thou canst move about, an evident God;
And canst oppose to each malignant hour
Ethereal presence:—I am but a voice;   340
My life is but the life of winds and tides,
No more than winds and tides can I avail:—
But thou canst.—Be thou therefore in the van
Of circumstance; yea, seize the arrow's barb
Before the tense string murmur.—To the earth!
For there thou wilt find Saturn, and his woes.
Meantime I will keep watch on thy bright sun,
And of thy seasons be a careful nurse."—
Ere half this region-whisper had come down,
Hyperion arose, and on the stars   350
Lifted his curved lids, and kept them wide
Until it ceas'd; and still he kept them wide:
And still they were the same bright, patient stars.
Then with a slow incline of his broad breast,
Like to a diver in the pearly seas,
Forward he stoop'd over the airy shore,
And plung'd all noiseless into the deep night.

BOOK II

Just at the self-same beat of Time's wide wings
Hyperion slid into the rustled air,
And Saturn gain'd with Thea that sad place
Where Cybele and the bruised Titans mourn'd.
It was a den where no insulting light
Could glimmer on their tears; where their own groans
They felt, but heard not, for the solid roar
Of thunderous waterfalls and torrents hoarse,
Pouring a constant bulk, uncertain where.
Crag jutting forth to crag, and rocks that seem'd   10
Ever as if just rising from a sleep,
Forehead to forehead held their monstrous horns;
And thus in thousand hugest phantasies
Made a fit roofing to this nest of woe.
Instead of thrones, hard flint they sat upon,
Couches of rugged stone, and slaty ridge
Stubborn'd with iron. All were not assembled:
Some chain'd in torture, and some wandering.

BOOK II. **4. Cybele:** wife of Saturn and mother of the gods.

Cœus, and Gyges, and Briareüs,
Typhon, and Dolor, and Porphyrion,   20
With many more, the brawniest in assault,
Were pent in regions of laborious breath;
Dungeon'd in opaque element, to keep
Their clenched teeth still clench'd, and all their limbs
Lock'd up like veins of metal, crampt and screw'd;
Without a motion, save of their big hearts
Heaving in pain, and horribly convuls'd
With sanguine feverous boiling gurge of pulse.
Mnemosyne was straying in the world;
Far from her moon had Phœbe wandered;   30
And many else were free to roam abroad,
But for the main, here found they covert drear.
Scarce images of life, one here, one there,
Lay vast and edgeways; like a dismal cirque
Of Druid stones, upon a forlorn moor,
When the chill rain begins at shut of eve,
In dull November, and their chancel vault,
The Heaven itself, is blinded throughout night.
Each one kept shroud, nor to his neighbour gave
Or word, or look, or action of despair.   40
Creüs was one; his ponderous iron mace
Lay by him, and a shatter'd rib of rock
Told of his rage, ere he thus sank and pined.
Iäpetus another; in his grasp,
A serpent's plashy neck; its barbed tongue
Squeez'd from the gorge, and all its uncurl'd length
Dead; and because the creature could not spit
Its poison in the eyes of conquering Jove.
Next Cottus: prone he lay, chin uppermost,
As though in pain; for still upon the flint   50
He ground severe his skull, with open mouth
And eyes at horrid working. Nearest him
Asia, born of most enormous Caf,
Who cost her mother Tellus keener pangs,
Though feminine, than any of her sons:
More thought than woe was in her dusky face,

For she was prophesying of her glory;
And in her wide imagination stood
Palm-shaded temples, and high rival fanes,
By Oxus or in Ganges' sacred isles.   60
Even as Hope upon her anchor leans,
So leant she, not so fair, upon a tusk
Shed from the broadest of her elephants.
Above her, on a crag's uneasy shelve,
Upon his elbow rais'd, all prostrate else,
Shadow'd Enceladus; once tame and mild
As grazing ox unworried in the meads;
Now tiger-passion'd, lion-thoughted, wroth,
He meditated, plotted, and even now
Was hurling mountains in that second war,   70
Not long delay'd, that scar'd the younger Gods
To hide themselves in forms of beast and bird.
Not far hence Atlas; and beside him prone
Phorcus, the sire of Gorgons. Neighbour'd close
Oceanus, and Tethys, in whose lap
Sobb'd Clymene among her tangled hair.
In midst of all lay Themis, at the feet
Of Ops the queen all clouded round from sight;
No shape distinguishable, more than when
Thick night confounds the pine-tops with the clouds:
And many else whose names may not be told.   81
For when the Muse's wings are air-ward spread,
Who shall delay her flight? And she must chaunt
Of Saturn, and his guide, who now had climb'd
With damp and slippery footing from a depth
More horrid still. Above a sombre cliff
Their heads appear'd, and up their stature grew
Till on the level height their steps found ease:
Then Thea spread abroad her trembling arms
Upon the precincts of this nest of pain,   90
And sidelong fix'd her eye on Saturn's face:
There saw she direst strife; the supreme God
At war with all the frailty of grief,
Of rage, of fear, anxiety, revenge,
Remorse, spleen, hope, but most of all despair.
Against these plagues he strove in vain; for Fate
Had pour'd a mortal oil upon his head,
A disanointing poison: so that Thea,
Affrighted, kept her still, and let him pass
First onwards in, among the fallen tribe.   100

  As with us mortal men, the laden heart
Is persecuted more, and fever'd more,

19–20. Cœus . . . Porphyrion: Keats found the names of
the Titans in various sources such as Edward Baldwin (i.e.,
William Godwin), *The Pantheon*, or, *Ancient History of the
Gods of Greece and Rome* (1806), pp. 45–46. **28. gurge:**
whirlpool. **29. Mnemosyne:** memory, mother of the muses.
30. Phœbe: goddess of the moon, though Keats means to
refer to her grandmother, the Titaness, wife of Cœus.
34–35. cirque . . . stones: a recollection of the "Druid
Circle" Keats saw near Keswick on his northern walking
tour. **44. Iäpetus:** See II. n. 19–20. **45. plashy:** speckled
or splashed with color. **53. Asia:** daughter of Oceanus and
Tethys. **Caf:** Keats takes the name "Caf" from William
Beckford's novel *Vathek* (1784).

76. Clymene: daughter of Oceanus and wife of Iäpetus.
77. Themis: daughter of Cœlus. **78. Ops:** Cybele. **97.
mortal oil:** thus making him prey to human troubles.

When it is nighing to the mournful house
Where other hearts are sick of the same bruise;
So Saturn, as he walk'd into the midst,
Felt faint, and would have sunk among the rest,
But that he met Enceladus's eye,
Whose mightiness, and awe of him, at once
Came like an inspiration; and he shouted,
"Titans, behold your God!" at which some groan'd;
Some started on their feet; some also shouted;   111
Some wept, some wail'd, all bow'd with reverence;
And Ops, uplifting her black folded veil,
Show'd her pale cheeks, and all her forehead wan,
Her eye-brows thin and jet, and hollow eyes.
There is a roaring in the bleak-grown pines
When Winter lifts his voice; there is a noise
Among immortals when a God gives sign,
With hushing finger, how he means to load
His tongue with the full weight of utterless thought,
With thunder, and with music, and with pomp: 121
Such noise is like the roar of bleak-grown pines:
Which, when it ceases in this mountain'd world,
No other sound succeeds; but ceasing here,
Among these fallen, Saturn's voice therefrom
Grew up like organ, that begins anew
Its strain, when other harmonies, stopt short,
Leave the dinn'd air vibrating silverly.
Thus grew it up—"Not in my own sad breast,
Which is its own great judge and searcher out,   130
Can I find reason why ye should be thus:
Not in the legends of the first of days,
Studied from that old spirit-leaved book
Which starry Uranus with finger bright
Sav'd from the shores of darkness, when the waves
Low-ebb'd still hid it up in shallow gloom;—
And the which book ye know I ever kept
For my firm-based footstool:—Ah, infirm!
Not there, nor in sign, symbol, or portent
Of element, earth, water, air, and fire,—   140
At war, at peace, or inter-quarreling
One against one, or two, or three, or all
Each several one against the other three,
As fire with air loud warring when rain-floods
Drown both, and press them both against earth's face,
Where, finding sulphur, a quadruple wrath
Unhinges the poor world;—not in that strife,
Wherefrom I take strange lore, and read it deep,
Can I find reason why ye should be thus:
No, no-where can unriddle, though I search,   150
And pore on Nature's universal scroll
Even to swooning, why ye, Divinities,

The first-born of all shap'd and palpable Gods,
Should cower beneath what, in comparison,
Is untremendous might. Yet ye are here,
O'erwhelm'd, and spurn'd, and batter'd, ye are here!
O Titans, shall I say, 'Arise!'—Ye groan:
Shall I say 'Crouch!'—Ye groan. What can I then?
O Heaven wide! O unseen parent dear!
What can I? Tell me, all ye brethren Gods,   160
How we can war, how engine our great wrath!
O speak your counsel now, for Saturn's ear
Is all a-hunger'd. Thou, Oceanus,
Ponderest high and deep; and in thy face
I see, astonied, that severe content
Which comes of thought and musing: give us help!"

    So ended Saturn; and the God of the Sea,
Sophist and sage, from no Athenian grove,
But cogitation in his watery shades,
Arose, with locks not oozy, and began,   170
In murmurs, which his first-endeavouring tongue
Caught infant-like from the far-foamed sands,
"O ye, whom wrath consumes! who, passion-stung,
Writhe at defeat, and nurse your agonies!
Shut up your senses, stifle up your ears,
My voice is not a bellows unto ire.
Yet listen, ye who will, whilst I bring proof
How ye, perforce, must be content to stoop:
And in the proof much comfort will I give,
If ye will take that comfort in its truth.   180
We fall by course of Nature's law, not force
Of thunder, or of Jove. Great Saturn, thou
Hast sifted well the atom-universe;
But for this reason, that thou art the King,
And only blind from sheer supremacy,
One avenue was shaded from thine eyes,
Through which I wandered to eternal truth.
And first, as thou wast not the first of powers,
So art thou not the last; it cannot be:
Thou art not the beginning nor the end.   190
From chaos and parental darkness came
Light, the first fruits of that intestine broil,
That sullen ferment, which for wondrous ends

161. engine: execute. **165. astonied:** astonished. **173-243.**
"O . . . balm": This great speech of Oceanus, especially
lines 181-217 and 229-30, is often regarded as the central
theme of the poem. It can also be argued that the speech
represents a premise rather than a theme—an expression of
philosophic "disinterestedness" (an ideal Keats prized) about
what is to happen, but an expression without concrete
sympathy for the anguish of the other Titans and for the
tragedy yet to come (Hyperion's fall).

Was ripening in itself. The ripe hour came,
And with it light, and light, engendering
Upon its own producer, forthwith touch'd
The whole enormous matter into life.
Upon that very hour, our parentage,
The Heavens, and the Earth, were manifest:
Then thou first born, and we the giant race,   200
Found ourselves ruling new and beauteous realms.
Now comes the pain of truth, to whom 'tis pain;
O folly! for to bear all naked truths,
And to envisage circumstance, all calm,
That is the top of sovereignty. Mark well!
As Heaven and Earth are fairer, fairer far
Than Chaos and blank Darkness, though once chiefs;
And as we show beyond that Heaven and Earth
In form and shape compact and beautiful,
In will, in action free, companionship,   210
And thousand other signs of purer life;
So on our heels a fresh perfection treads,
A power more strong in beauty, born of us
And fated to excel us, as we pass
In glory that old Darkness: nor are we
Thereby more conquer'd, than by us the rule
Of shapeless Chaos. Say, doth the dull soil
Quarrel with the proud forests it hath fed,
And feedeth still, more comely than itself?
Can it deny the chiefdom of green groves?   220
Or shall the tree be envious of the dove
Because it cooeth, and hath snowy wings
To wander wherewithal and find its joys?
We are such forest-trees, and our fair boughs
Have bred forth, not pale solitary doves,
But eagles golden-feather'd, who do tower
Above us in their beauty, and must reign
In right thereof; for 'tis the eternal law
That first in beauty should be first in might:
Yea, by that law, another race may drive   230
Our conquerors to mourn as we do now.
Have ye beheld the young God of the Seas,
My dispossessor? Have ye seen his face?
Have ye beheld his chariot, foam'd along
By noble winged creatures he hath made?
I saw him on the calmed waters scud,
With such a glow of beauty in his eyes,
That it enforc'd me to bid sad farewell
To all my empire: farewell sad I took,
And hither came, to see how dolorous fate   240
Had wrought upon ye; and how I might best

232. young . . . Seas: Neptune.

Give consolation in this woe extreme.
Receive the truth, and let it be your balm."

Whether through poz'd conviction, or disdain,
They guarded silence, when Oceanus
Left murmuring, what deepest thought can tell?
But so it was, none answer'd for a space,
Save one whom none regarded, Clymene;
And yet she answer'd not, only complain'd,
With hectic lips, and eyes up-looking mild,   250
Thus wording timidly among the fierce:
"O Father, I am here the simplest voice,
And all my knowledge is that joy is gone,
And this thing woe crept in among our hearts,
There to remain for ever, as I fear:
I would not bode of evil, if I thought
So weak a creature could turn off the help
Which by just right should come of mighty Gods;
Yet let me tell my sorrow, let me tell
Of what I heard, and how it made me weep,   260
And know that we had parted from all hope.
I stood upon a shore, a pleasant shore,
Where a sweet clime was breathed from a land
Of fragrance, quietness, and trees, and flowers.
Full of calm joy it was, as I of grief;
Too full of joy and soft delicious warmth;
So that I felt a movement in my heart
To chide, and to reproach that solitude
With songs of misery, music of our woes;
And sat me down, and took a mouthed shell   270
And murmur'd into it, and made melody—
O melody no more! for while I sang,
And with poor skill let pass into the breeze
The dull shell's echo, from a bowery strand
Just opposite, an island of the sea,
There came enchantment with the shifting wind,
That did both drown and keep alive my ears.
I threw my shell away upon the sand,
And a wave fill'd it, as my sense was fill'd
With that new blissful golden melody.   280
A living death was in each gush of sounds,
Each family of rapturous hurried notes,
That fell, one after one, yet all at once,
Like pearl beads dropping sudden from their string:
And then another, then another strain,
Each like a dove leaving its olive perch,
With music wing'd instead of silent plumes,
To hover round my head, and make me sick
Of joy and grief at once. Grief overcame,

244. poz'd: puzzled.

And I was stopping up my frantic ears,   290
When, past all hindrance of my trembling hands,
A voice came sweeter, sweeter than all tune,
And still it cried, 'Apollo! young Apollo!

The morning-bright Apollo! young Apollo!'
I fled, it follow'd me, and cried 'Apollo!'
O Father, and O Brethren, had ye felt
Those pains of mine; O Saturn, hadst thou felt,
Ye would not call this too indulged tongue
Presumptuous, in thus venturing to be heard."

So far her voice flow'd on, like timorous brook 300
That, lingering along a pebbled coast,
Doth fear to meet the sea: but sea it met,
And shudder'd; for the overwhelming voice
Of huge Enceladus swallow'd it in wrath:
The ponderous syllables, like sullen waves
In the half-glutted hollows of reef-rocks,
Came booming thus, while still upon his arm
He lean'd; not rising, from supreme contempt.
"Or shall we listen to the over-wise,
Or to the over-foolish, Giant-Gods?   310
Not thunderbolt on thunderbolt, till all
That rebel Jove's whole armoury were spent,
Not world on world upon these shoulders piled,
Could agonize me more than baby-words
In midst of this dethronement horrible.
Speak! roar! shout! yell! ye sleepy Titans all.
Do ye forget the blows, the buffets vile?
Are ye not smitten by a youngling arm?
Dost thou forget, sham Monarch of the Waves,
Thy scalding in the seas? What, have I rous'd   320
Your spleens with so few simple words as these?
O joy! for now I see ye are not lost:
O joy! for now I see a thousand eyes
Wide-glaring for revenge!"—As this he said,
He lifted up his stature vast, and stood,
Still without intermission speaking thus:
"Now ye are flames, I'll tell you how to burn,
And purge the ether of our enemies;
How to feed fierce the crooked stings of fire,
And singe away the swollen clouds of Jove,   330
Stifling that puny essence in its tent.
O let him feel the evil he hath done;
For though I scorn Oceanus's lore,
Much pain have I for more than loss of realms:
The days of peace and slumberous calm are fled;
Those days, all innocent of scathing war,
When all the fair Existences of heaven

304. **Enceladus:** See II. n. 19–20.

Came open-eyed to guess what we would speak:—
That was before our brows were taught to frown,
Before our lips knew else but solemn sounds;   340
That was before we knew the winged thing,
Victory, might be lost, or might be won.
And be ye mindful that Hyperion,
Our brightest brother, still is undisgraced—
Hyperion, lo! his radiance is here!"

All eyes were on Enceladus's face,
And they beheld, while still Hyperion's name
Flew from his lips up to the vaulted rocks,
A pallid gleam across his features stern:
Not savage, for he saw full many a God   350
Wroth as himself. He look'd upon them all,
And in each face he saw a gleam of light,
But splendider in Saturn's, whose hoar locks
Shone like the bubbling foam about a keel
When the prow sweeps into a midnight cove.
In pale and silver silence they remain'd,
Till suddenly a splendour, like the morn,
Pervaded all the beetling gloomy steeps,
All the sad spaces of oblivion,
And every gulf, and every chasm old,   360
And every height, and every sullen depth,
Voiceless, or hoarse with loud tormented streams:
And all the everlasting cataracts,
And all the headlong torrents far and near,
Mantled before in darkness and huge shade,
Now saw the light and made it terrible.
It was Hyperion:—a granite peak
His bright feet touch'd, and there he stay'd to view
The misery his brilliance had betray'd
To the most hateful seeing of itself.   370
Golden his hair of short Numidian curl,
Regal his shape majestic, a vast shade
In midst of his own brightness, like the bulk
Of Memnon's image at the set of sun
To one who travels from the dusking East:
Sighs, too, as mournful as that Memnon's harp
He utter'd, while his hands contemplative
He press'd together, and in silence stood.
Despondence seiz'd again the fallen Gods
At sight of the dejected King of Day,   380
And many hid their faces from the light:
But fierce Enceladus sent forth his eyes

374–76. **Memnon's . . . harp:** The colossal statue of
Memnon, on the edge of the desert near the Valley of the
Kings in Egypt, for several centuries uttered a mournful
sound when the sun's rays struck it from a certain angle.

Among the brotherhood; and, at their glare,
Uprose Iäpetus, and Creüs too,
And Phorcus, sea-born, and together strode
To where he towered on his eminence.
There those four shouted forth old Saturn's name;
Hyperion from the peak loud answered, "Saturn!"
Saturn sat near the Mother of the Gods,
In whose face was no joy, though all the Gods   390
Gave from their hollow throats the name of "Saturn!"

BOOK III

Thus in alternate uproar and sad peace,
Amazed were those Titans utterly.
O leave them, Muse! O leave them to their woes;
For thou art weak to sing such tumults dire:
A solitary sorrow best befits
Thy lips, and antheming a lonely grief.
Leave them, O Muse! for thou anon wilt find
Many a fallen old Divinity
Wandering in vain about bewildered shores.
Meantime touch piously the Delphic harp,   10
And not a wind of heaven but will breathe
In aid soft warble from the Dorian flute;
For lo! 'tis for the Father of all verse.
Flush every thing that hath a vermeil hue,
Let the rose glow intense and warm the air,
And let the clouds of even and of morn
Float in voluptuous fleeces o'er the hills;
Let the red wine within the goblet boil,
Cold as a bubbling well; let faint-lipp'd shells,
On sands, or in great deeps, vermilion turn   20
Through all their labyrinths; and let the maid
Blush keenly, as with some warm kiss surpris'd.
Chief isle of the embowered Cyclades,
Rejoice, O Delos, with thine olives green,
And poplars, and lawn-shading palms, and beech,
In which the Zephyr breathes the loudest song,
And hazels thick, dark-stemm'd beneath the shade:
Apollo is once more the golden theme!
Where was he, when the Giant of the Sun
Stood bright, amid the sorrow of his peers?   30
Together had he left his mother fair
And his twin-sister sleeping in their bower,
And in the morning twilight wandered forth
Beside the osiers of a rivulet,

Full ankle-deep in lillies of the vale.
The nightingale had ceas'd, and a few stars
Were lingering in the heavens, while the thrush
Began calm-throated. Throughout all the isle
There was no covert, no retired cave
Unhaunted by the murmurous noise of waves,   40
Though scarcely heard in many a green recess.
He listen'd, and he wept, and his bright tears
Went trickling down the golden bow he held.
Thus with half-shut suffused eyes he stood,
While from beneath some cumbrous boughs hard by
With solemn step an awful Goddess came,
And there was purport in her looks for him,
Which he with eager guess began to read
Perplex'd, the while melodiously he said:
"How cam'st thou over the unfooted sea?   50
Or hath that antique mien and robed form
Mov'd in these vales invisible till now?
Sure I have heard those vestments sweeping o'er
The fallen leaves, when I have sat alone
In cool mid-forest. Surely I have traced
The rustle of those ample skirts about
These grassy solitudes, and seen the flowers
Lift up their heads, as still the whisper pass'd.
Goddess! I have beheld those eyes before,
And their eternal calm, and all that face,   60
Or I have dream'd."—"Yes," said the supreme shape,
"Thou hast dream'd of me; and awaking up
Didst find a lyre all golden by thy side,
Whose strings touch'd by thy fingers, all the vast
Unwearied ear of the whole universe
Listen'd in pain and pleasure at the birth
Of such new tuneful wonder. Is't not strange
That thou shouldst weep, so gifted? Tell me, youth,
What sorrow thou canst feel; for I am sad
When thou dost shed a tear: explain thy griefs   70
To one who in this lonely isle hath been
The watcher of thy sleep and hours of life,
From the young day when first thy infant hand
Pluck'd witless the weak flowers, till thine arm
Could bend that bow heroic to all times.
Show thy heart's secret to an ancient Power
Who hath forsaken old and sacred thrones
For prophecies of thee, and for the sake
Of loveliness new born."—Apollo then,
With sudden scrutiny and gloomless eyes,   80
Thus answer'd, while his white melodious throat
Throbb'd with the syllables.—"Mnemosyne!

---

389. Mother . . . Gods: Cybele. BOOK III. 13. Father . . .
verse: Apollo. 14. vermeil: vermillion. 29. Giant . . .
Sun: Hyperion. 34. osiers: willows.

46. Goddess: Mnemosyne (II. 29).

Thy name is on my tongue, I know not how;
Why should I tell thee what thou so well seest?
Why should I strive to show what from thy lips
Would come no mystery? For me, dark, dark,
And painful vile oblivion seals my eyes:
I strive to search wherefore I am so sad,
Until a melancholy numbs my limbs;
And then upon the grass I sit, and moan,   90
Like one who once had wings.—O why should I
Feel curs'd and thwarted, when the liegeless air
Yields to my step aspirant? why should I
Spurn the green turf as hateful to my feet?
Goddess benign, point forth some unknown thing:
Are there not other regions than this isle?
What are the stars? There is the sun, the sun!
And the most patient brilliance of the moon!
And stars by thousands! Point me out the way
To any one particular beauteous star,   100
And I will flit into it with my lyre
And make its silvery splendour pant with bliss.
I have heard the cloudy thunder: Where is power?
Whose hand, whose essence, what divinity
Makes this alarum in the elements,
While I here idle listen on the shores
In fearless yet in aching ignorance?
O tell me, lonely Goddess, by thy harp,
That waileth every morn and eventide,
Tell me why thus I rave, about these groves!   110
Mute thou remainest—mute! yet I can read
A wondrous lesson in thy silent face:
Knowledge enormous makes a God of me.
Names, deeds, grey legends, dire events, rebellions,
Majesties, sovran voices, agonies,
Creations and destroyings, all at once
Pour into the wide hollows of my brain,
And deify me, as if some blithe wine
Or bright elixir peerless I had drunk,
And so become immortal."—Thus the God,   120
While his enkindled eyes, with level glance
Beneath his white soft temples, stedfast kept
Trembling with light upon Mnemosyne.
Soon wild commotions shook him, and made flush
All the immortal fairness of his limbs;
Most like the struggle at the gate of death;
Or liker still to one who should take leave
Of pale immortal death, and with a pang

As hot as death's is chill, with fierce convulse
Die into life: so young Apollo anguish'd: 130
His very hair, his golden tresses famed
Kept undulation round his eager neck.
During the pain Mnemosyne upheld
Her arms as one who prophesied.—At length
Apollo shriek'd;—and lo! from all his limbs
Celestial   .   .   .   .
.   .   .   .

1818–1819 (1820)

FANCY

"Fancy" was written as part of a letter (Dec.-Jan., 1818–19) to Keats's brother and sister-in-law, George and Georgiana, who had emigrated to Kentucky. The theme is the insatiable restlessness of the imagination, desiring to combine all experiences into units of pleasure denser than actual life permits—a theme that recurs in the great odes and *Lamia*.

Ever let the fancy roam,
Pleasure never is at home:
At a touch sweet Pleasure melteth,
Like to bubbles when rain pelteth;
Then let winged Fancy wander
Through the thought still spread beyond her:
Open wide the mind's cage-door,
She'll dart forth, and cloudward soar.
O sweet Fancy! let her loose;
Summer's joys are spoilt by use,   10
And the enjoying of the Spring
Fades as does its blossoming;
Autumn's red-lipp'd fruitage too,
Blushing through the mist and dew,
Cloys with tasting: What do then?
Sit thee by the ingle, when
The sear faggot blazes bright,
Spirit of a winter's night;
When the soundless earth is muffled,
And the caked snow is shuffled   20
From the ploughboy's heavy shoon;
When the Night doth meet the Noon
In a dark conspiracy

86. **dark, dark:** an echo of Milton's *Samson Agonistes*, l. 80.
113. **Knowledge enormous:** sympathetic awareness of the enormously varied experiences of life.

FANCY. 16. **ingle:** fireplace. 21. **shoon:** shoes.

To banish Even from her sky.
Sit thee there, and send abroad,
With a mind self-overaw'd,
Fancy, high-commission'd:—send her!
She has vassals to attend her:
She will bring, in spite of frost,
Beauties that the earth hath lost;   30
She will bring thee, all together,
All delights of summer weather;
All the buds and bells of May,
From dewy sward or thorny spray;
All the heaped Autumn's wealth,
With a still, mysterious stealth:
She will mix these pleasures up
Like three fit wines in a cup,
And thou shalt quaff it:—thou shalt hear
Distant harvest-carols clear;   40
Rustle of the reaped corn;
Sweet birds antheming the morn:
And, in the same moment—hark!
'Tis the early April lark,
Or the rooks, with busy caw,
Foraging for sticks and straw.
Thou shalt, at one glance, behold
The daisy and the marigold;
White-plum'd lillies, and the first
Hedge-grown primrose that hath burst;   50
Shaded hyacinth, alway
Sapphire queen of the mid-May;
And every leaf, and every flower
Pearled with the self-same shower.
Thou shalt see the field-mouse peep
Meagre from its celled sleep;
And the snake all winter-thin
Cast on sunny bank its skin;
Freckled nest-eggs thou shalt see
Hatching in the hawthorn-tree,   60
When the hen-bird's wing doth rest
Quiet on her mossy nest;
Then the hurry and alarm
When the bee-hive casts its swarm;
Acorns ripe down-pattering,
While the autumn breezes sing.

  Oh, sweet Fancy! let her loose;
Every thing is spoilt by use:
Where's the cheek that doth not fade,
Too much gaz'd at? Where's the maid   70
Whose lip mature is ever new?
Where's the eye, however blue,

Doth not weary? Where's the face
One would meet in every place?
Where's the voice, however soft,
One would hear so very oft?
At a touch sweet Pleasure melteth
Like to bubbles when rain pelteth.
Let, then, winged Fancy find
Thee a mistress to thy mind: 80
Dulcet-eyed as Ceres' daughter,
Ere the God of Torment taught her
How to frown and how to chide;
With a waist and with a side
White as Hebe's, when her zone
Slipt its golden clasp, and down
Fell her kirtle to her feet,
While she held the goblet sweet,
And Jove grew languid.—Break the mesh
Of the Fancy's silken leash;   90
Quickly break her prison-string
And such joys as these she'll bring.—
Let the winged Fancy roam,
Pleasure never is at home.

                                    1818 (1820)

ODE
[BARDS OF PASSION]

This ode was sent in the same letter as the lines above.

Bards of Passion and of Mirth,
Ye have left your souls on earth!
Have ye souls in heaven too,
Double-lived in regions new?
Yes, and those of heaven commune
With the spheres of sun and moon;
With the noise of fountains wond'rous,
And the parle of voices thund'rous;
With the whisper of heaven's trees
And one another, in soft ease   10
Seated on Elysian lawns
Brows'd by none but Dian's fawns;
Underneath large blue-bells tented,

81. **Ceres' daughter:** Proserpine. **85. Hebe:** cupbearer
to Jupiter. **zone:** belt. ODE. **8. parle:** speech.

Where the daisies are rose-scented,
And the rose herself has got
Perfume which on earth is not;
Where the nightingale doth sing
Not a senseless, tranced thing,
But divine melodious truth;
Philosophic numbers smooth;   20
Tales and golden histories
Of heaven and its mysteries.

    Thus ye live on high, and then
On the earth ye live again;
And the souls ye left behind you
Teach us, here, the way to find you,
Where your other souls are joying,
Never slumber'd, never cloying.
Here, your earth-born souls still speak
To mortals, of their little week;   30
Of their sorrows and delights;
Of their passions and their spites;
Of their glory and their shame;
What doth strengthen and what maim.
Thus ye teach us, every day,
Wisdom, though fled far away.

    Bards of Passion and of Mirth,
Ye have left your souls on earth!
Ye have souls in heaven too,
Double-lived in regions new!   40

1818 (1820)

THE EVE OF ST. AGNES

In late January and early February, 1819, Keats, unable
to get ahead with *Hyperion*, took off time to write this
romance, the subject of which was suggested by a friend,
Mrs. Isabella Jones. Another romance was not at all what
he wished to write; but he had been unable during the
past month to find an alternative to *Hyperion*; he was
shaken by the death of his brother Tom, tired, and not
completely well himself. Turning to the romance form,
he used the Spenserian stanza and quickly found a very
different style from *Isabella*. Much that he had learned

17–19. nightingale . . . truth: Cf. "Ode to a Nightingale,"
below. **24. live again:** The subject of the poem, said Keats
in a letter, is "the double immortality of Poets."

while writing *Hyperion*—the condensed and richly concrete language, slower movement, and skill in assonance
and vowel-patterning—is put into practice. The story
(which contains echoes of *Romeo and Juliet*, the novels of
Ann Radcliffe, the *Arabian Nights*, and other works) turns
on the legend that, on St. Agnes' Eve (Jan. 20), a maiden
who performed certain rites (those described in stanza
vi) would have a vision of her lover or future husband.
The theme of dream and reality is later treated more
seriously by Keats (especially in *Lamia* and *The Fall of
Hyperion*). Here it is memorable because of the setting,
perhaps the most sumptuously colored and musically
phrased of any narrative poem of the Romantic period.

I

St. Agnes' Eve—Ah, bitter chill it was!
The owl, for all his feathers, was a-cold;
The hare limp'd trembling through the frozen grass,
And silent was the flock in woolly fold:
Numb were the Beadsman's fingers, while he told
His rosary, and while his frosted breath,
Like pious incense from a censer old,
Seem'd taking flight for heaven, without a death,
Past the sweet Virgin's picture, while his prayer he
    saith.

II

His prayer he saith, this patient, holy man;   10
Then takes his lamp, and riseth from his knees,
And back returneth, meagre, barefoot, wan,
Along the chapel aisle by slow degrees:
The sculptur'd dead, on each side, seem to freeze,
Emprison'd in black, purgatorial rails:
Knights, ladies, praying in dumb orat'ries,
He passeth by; and his weak spirit fails
To think how they may ache in icy hoods and mails.

III

Northward he turneth through a little door,
And scarce three steps, ere Music's golden tongue
Flatter'd to tears this aged man and poor;   21
But no—already had his deathbell rung:
The joys of all his life were said and sung:
His was harsh penance on St. Agnes' Eve:
Another way he went, and soon among
Rough ashes sat he for his soul's reprieve,
And all night kept awake, for sinners' sake to grieve.

IV

That ancient Beadsman heard the prelude soft;
And so it chanc'd, for many a door was wide,
From hurry to and fro. Soon, up aloft,   30
The silver, snarling trumpets 'gan to chide:
The level chambers, ready with their pride,
Were glowing to receive a thousand guests:
The carved angels, ever eager-eyed,
Star'd, where upon their heads the cornice rests,
With hair blown back, and wings put cross-wise on
     their breasts.

V

At length burst in the argent revelry,
With plume, tiara, and all rich array,
Numerous as shadows haunting faerily
The brain, new stuff'd, in youth, with triumphs gay
Of old romance. These let us wish away,   41
And turn, sole-thoughted, to one Lady there,
Whose heart had brooded, all that wintry day,
On love, and wing'd St. Agnes' saintly care,
As she had heard old dames full many times declare.

VI

They told her how, upon St. Agnes' Eve,
Young virgins might have visions of delight,
And soft adorings from their loves receive
Upon the honey'd middle of the night,
If ceremonies due they did aright;   50
As, supperless to bed they must retire,
And couch supine their beauties, lilly white;
Nor look behind, nor sideways, but require
Of Heaven with upward eyes for all that they desire.

VII

Full of this whim was thoughtful Madeline:
The music, yearning like a God in pain,
She scarcely heard: her maiden eyes divine,
Fix'd on the floor, saw many a sweeping train
Pass by—she heeded not at all: in vain
Came many a tiptoe, amorous cavalier,   60
And back retir'd; not cool'd by high disdain,
But she saw not: her heart was otherwhere:
She sigh'd for Agnes' dreams, the sweetest of the year.

THE EVE OF ST. AGNES. **58. train:** skirts sweeping the floor.

VIII

She danc'd along with vague, regardless eyes,
Anxious her lips, her breathing quick and short:
The hallow'd hour was near at hand: she sighs
Amid the timbrels, and the throng'd resort
Of whisperers in anger, or in sport;
'Mid looks of love, defiance, hate, and scorn,
Hoodwink'd with faery fancy; all amort,   70
Save to St. Agnes and her lambs unshorn,
And all the bliss to be before to-morrow morn.

IX

So, purposing each moment to retire,
She linger'd still. Meantime, across the moors,
Had come young Porphyro, with heart on fire
For Madeline. Beside the portal doors,
Buttress'd from moonlight, stands he, and implores
All saints to give him sight of Madeline,
But for one moment in the tedious hours,
That he might gaze and worship all unseen;   80
Perchance speak, kneel, touch, kiss—in sooth such
     things have been.

X

He ventures in: let no buzz'd whisper tell:
All eyes be muffled, or a hundred swords
Will storm his heart, Love's fev'rous citadel:
For him, those chambers held barbarian hordes,
Hyena foemen, and hot-blooded lords,
Whose very dogs would execrations howl
Against his lineage: not one breast affords
Him any mercy, in that mansion foul,
Save one old beldame, weak in body and in soul.   90

XI

Ah, happy chance! the aged creature came,
Shuffling along with ivory-headed wand,
To where he stood, hid from the torch's flame,
Behind a broad hall-pillar, far beyond
The sound of merriment and chorus bland:

67. timbrels: small drums. **70. Hoodwink'd:** blinded.
all amort: as if dead. **71. St. Agnes . . . lambs:** On St.
Agnes' day two lambs were offered at the altar, and their
wool was afterwards spun and woven by the nuns.

He startled her; but soon she knew his face,
And grasp'd his fingers in her palsied hand,
Saying, "Mercy, Porphyro! hie thee from this place:
They are all here to-night, the whole blood-thirsty
    race!

XII

"Get hence! get hence! there's dwarfish Hildebrand;
He had a fever late, and in the fit   101
He cursed thee and thine, both house and land:
Then there's that old Lord Maurice, not a whit
More tame for his gray hairs—Alas me! flit!
Flit like a ghost away."—"Ah, Gossip dear,
We're safe enough; here in this arm-chair sit,
And tell me how"—"Good Saints! not here, not
    here;
Follow me, child, or else these stones will be thy bier."

XIII

He follow'd through a lowly arched way,
Brushing the cobwebs with his lofty plume,   110
And as she mutter'd "Well-a—well-a-day!"
He found him in a little moonlight room,
Pale, lattic'd, chill, and silent as a tomb.
"Now tell me where is Madeline," said he,
"O tell me, Angela, by the holy loom
Which none but secret sisterhood may see,
When they St. Agnes' wool are weaving piously."

XIV

"St. Agnes! Ah! it is St. Agnes' Eve—
Yet men will murder upon holy days:
Thou must hold water in a witch's sieve,   120
And be liege-lord of all the Elves and Fays,
To venture so: it fills me with amaze
To see thee, Porphyro!—St. Agnes' Eve!
God's help! my lady fair the conjuror plays
This very night: good angels her deceive!
But let me laugh awhile, I've mickle time to grieve."

XV

Feebly she laugheth in the languid moon,
While Porphyro upon her face doth look,

120. **witch's sieve:** a sieve bewitched so that water cannot
pass through it. 126. **mickle:** much.

Like puzzled urchin on an aged crone
Who keepeth clos'd a wond'rous riddle-book,   130
As spectacled she sits in chimney nook.
But soon his eyes grew brilliant, when she told
His lady's purpose; and he scarce could brook
Tears, at the thought of those enchantments cold,
And Madeline asleep in lap of legends old.

XVI

Sudden a thought came like a full-blown rose,
Flushing his brow, and in his pained heart
Made purple riot: then doth he propose
A stratagem, that makes the beldame start:
"A cruel man and impious thou art: 140
Sweet lady, let her pray, and sleep, and dream
Alone with her good angels, far apart
From wicked men like thee. Go, go!—I deem
Thou canst not surely be the same that thou didst
    seem."

XVII

"I will not harm her, by all saints I swear,"
Quoth Porphyro: "O may I ne'er find grace
When my weak voice shall whisper its last prayer,
If one of her soft ringlets I displace,
Or look with ruffian passion in her face:
Good Angela, believe me by these tears;   150
Or I will, even in a moment's space,
Awake, with horrid shout, my foemen's ears,
And beard them, though they be more fang'd than
    wolves and bears."

XVIII

"Ah! why wilt thou affright a feeble soul?
A poor, weak, palsy-stricken, churchyard thing,
Whose passing-bell may ere the midnight toll;
Whose prayers for thee, each morn and evening,
Were never miss'd."—Thus plaining, doth she bring
A gentler speech from burning Porphyro;
So woful, and of such deep sorrowing,   160
That Angela gives promise she will do
Whatever he shall wish, betide her weal or woe.

XIX

Which was, to lead him, in close secrecy,
Even to Madeline's chamber, and there hide

Him in a closet, of such privacy
That he might see her beauty unespied,
And win perhaps that night a peerless bride,
While legion'd faeries pac'd the coverlet,
And pale enchantment held her sleepy-eyed.
Never on such a night have lovers met,   170
Since Merlin paid his Demon all the monstrous debt.

XX

"It shall be as thou wishest," said the Dame:
"All cates and dainties shall be stored there
Quickly on this feast-night: by the tambour frame
Her own lute thou wilt see: no time to spare,
For I am slow and feeble, and scarce dare
On such a catering trust my dizzy head.
Wait here, my child, with patience; kneel in prayer
The while: Ah! thou must needs the lady wed,
Or may I never leave my grave among the dead." 180

XXI

So saying, she hobbled off with busy fear.
The lover's endless minutes slowly pass'd;
The dame return'd, and whisper'd in his ear
To follow her; with aged eyes aghast
From fright of dim espial. Safe at last,
Through many a dusky gallery, they gain
The maiden's chamber, silken, hush'd, and chaste;
Where Porphyro took covert, pleas'd amain.
His poor guide hurried back with agues in her brain.

XXII

Her falt'ring hand upon the balustrade,   190
Old Angela was feeling for the stair,
When Madeline, St. Agnes' charmed maid,
Rose, like a mission'd spirit, unaware:
With silver taper's light, and pious care,
She turn'd, and down the aged gossip led
To a safe level matting. Now prepare,
Young Porphyro, for gazing on that bed;
She comes, she comes again, like ring-dove fray'd and
    fled.

171. Merlin . . . debt: As the son of a demon, the
Arthurian wizard paid the debt for his existence by performing evil deeds. The line can also be explained by the story
that he was killed, and thus requited for his evil, by the use
of his own spells against him. 173. cates: delicate foods.
174. tambour frame: embroidery frame shaped like a drum.
198. fray'd: alarmed.

XXIII

Out went the taper as she hurried in;
Its little smoke, in pallid moonshine, died: 200
She clos'd the door, she panted, all akin
To spirits of the air, and visions wide:
No uttered syllable, or, woe betide!
But to her heart, her heart was voluble,
Paining with eloquence her balmy side;
As though a tongueless nightingale should swell
Her throat in vain, and die, heart-stifled, in her dell.

XXIV

A casement high and triple-arch'd there was,
All garlanded with carven imag'ries
Of fruits, and flowers, and bunches of knot-grass,
And diamonded with panes of quaint device,   211
Innumerable of stains and splendid dyes,
As are the tiger-moth's deep-damask'd wings;
And in the midst, 'mong thousand heraldries,
And twilight saints, and dim emblazonings,
A shielded scutcheon blush'd with blood of queens
    and kings.

XXV

Full on this casement shone the wintry moon,
And threw warm gules on Madeline's fair breast,
As down she knelt for heaven's grace and boon;
Rose-bloom fell on her hands, together prest,   220
And on her silver cross soft amethyst,
And on her hair a glory, like a saint:
She seem'd a splendid angel, newly drest,
Save wings, for heaven:—Porphyro grew faint:
She knelt, so pure a thing, so free from mortal taint.

XXVI

Anon his heart revives: her vespers done,
Of all its wreathed pearls her hair she frees;
Unclasps her warmed jewels one by one;
Loosens her fragrant boddice; by degrees
Her rich attire creeps rustling to her knees: 230
Half-hidden, like a mermaid in sea-weed,
Pensive awhile she dreams awake, and sees,
In fancy, fair St. Agnes in her bed,
But dares not look behind, or all the charm is fled.

218. gules: the color red in heraldry.

XXVII

Soon, trembling in her soft and chilly nest,
In sort of wakeful swoon, perplex'd she lay,
Until the poppied warmth of sleep oppress'd
Her soothed limbs, and soul fatigued away;
Flown, like a thought, until the morrow-day;
Blissfully haven'd both from joy and pain;   240
Clasp'd like a missal where swart Paynims pray;
Blinded alike from sunshine and from rain,
As though a rose should shut, and be a bud again.

XXVIII

Stol'n to this paradise, and so entranced,
Porphyro gazed upon her empty dress,
And listen'd to her breathing, if it chanced
To wake into a slumberous tenderness;
Which when he heard, that minute did he bless,
And breath'd himself: then from the closet crept,
Noiseless as fear in a wide wilderness,   250
And over the hush'd carpet, silent, stept,
And 'tween the curtains peep'd, where, lo!—how fast
    she slept.

XXIX

Then by the bed-side, where the faded moon
Made a dim, silver twilight, soft he set
A table, and, half anguish'd, threw thereon
A cloth of woven crimson, gold, and jet:—
O for some drowsy Morphean amulet!
The boisterous, midnight, festive clarion,
The kettle-drum, and far-heard clarinet,
Affray his ears, though but in dying tone:—   260
The hall door shuts again, and all the noise is gone.

XXX

And still she slept an azure-lidded sleep,
In blanched linen, smooth, and lavender'd,

241. **Clasp'd . . . pray:** kept shut as a Christian prayer book would be among the paynims, or pagans. 250. **Noiseless as fear:** as a person full of fear. 257. **Morphean:** from Morpheus, god of sleep. 261. **shuts . . . gone:** Keats said that he was thinking here of nights when he would lie awake as a schoolboy at Enfield listening to his friend Cowden Clarke playing the piano downstairs; the door would suddenly be closed and the sound of music cease.

While he from forth the closet brought a heap
Of candied apple, quince, and plum, and gourd;
With jellies soother than the creamy curd,
And lucent syrops, tinct with cinnamon;
Manna and dates, in argosy transferr'd
From Fez; and spiced dainties, every one,
From silken Samarcand to cedar'd Lebanon.   270

XXXI

These delicates he heap'd with glowing hand
On golden dishes and in baskets bright
Of wreathed silver: sumptuous they stand
In the retired quiet of the night,
Filling the chilly room with perfume light.—
"And now, my love, my seraph fair, awake!
Thou art my heaven, and I thine eremite:
Open thine eyes, for meek St. Agnes' sake,
Or I shall drowse beside thee, so my soul doth ache."

XXXII

Thus whispering, his warm, unnerved arm   280
Sank in her pillow. Shaded was her dream
By the dusk curtains:—'twas a midnight charm
Impossible to melt as iced stream:
The lustrous salvers in the moonlight gleam;
Broad golden fringe upon the carpet lies:
It seem'd he never, never could redeem
From such a stedfast spell his lady's eyes;
So mus'd awhile, entoil'd in woofed phantasies.

XXXIII

Awakening up, he took her hollow lute,—
Tumultuous,—and, in chords that tenderest be,   290
He play'd an ancient ditty, long since mute,
In Provence call'd, "La belle dame sans mercy:"
Close to her ear touching the melody;—
Wherewith disturb'd, she utter'd a soft moan:
He ceased—she panted quick—and suddenly
Her blue affrayed eyes wide open shone:
Upon his knees he sank, pale as smooth-sculptured
    stone.

266. **soother:** smoother. 277. **eremite:** hermit. 292. **Provence:** the troubador district in southern France. "**La . . . mercy**": the title of a poem by Alain Chartier ("the fair lady without mercy"). For Keats's poem of the same title, see below.

XXXIV

Her eyes were open, but she still beheld,
Now wide awake, the vision of her sleep:
There was a painful change, that nigh expell'd   300
The blisses of her dreams so pure and deep
At which fair Madeline began to weep,
And moan forth witless words with many a sigh;
While still her gaze on Porphyro would keep;
Who knelt, with joined hands and piteous eye,
Fearing to move or speak, she look'd so dreamingly.

XXXV

"Ah, Porphyro!" said she, "but even now
Thy voice was at sweet tremble in mine ear,
Made tuneable with every sweetest vow;
And those sad eyes were spiritual and clear: 310
How chang'd thou art! how pallid, chill, and drear!
Give me that voice again, my Porphyro,
Those looks immortal, those complainings dear!
Oh leave me not in this eternal woe,
For if thou diest, my Love, I know not where to go."

XXXVI

Beyond a mortal man impassion'd far
At these voluptuous accents, he arose,
Ethereal, flush'd, and like a throbbing star
Seen mid the sapphire heaven's deep repose;
Into her dream he melted, as the rose   320
Blendeth its odour with the violet,—
Solution sweet: meantime the frost-wind blows
Like Love's alarum pattering the sharp sleet
Against the window-panes; St. Agnes' moon hath set.

XXXVII

'Tis dark: quick pattereth the flaw-blown sleet:
"This is no dream, my bride, my Madeline!"
'Tis dark: the iced gusts still rave and beat:
"No dream, alas! alas! and woe is mine!
Porphyro will leave me here to fade and pine.—
Cruel! what traitor could thee hither bring?   330
I curse not, for my heart is lost in thine,
Though thou forsakest a deceived thing;—
A dove forlorn and lost with sick unpruned wing."

325. **flaw-blown:** blown by a brief, strong storm.

XXXVIII

"My Madeline! sweet dreamer! lovely bride!
Say, may I be for aye thy vassal blest?
Thy beauty's shield, heart-shap'd and vermeil dyed?
Ah, silver shrine, here will I take my rest
After so many hours of toil and quest,
A famish'd pilgrim,—sav'd by miracle.
Though I have found, I will not rob thy nest   340
Saving of thy sweet self; if thou think'st well
To trust, fair Madeline, to no rude infidel.

XXXIX

"Hark! 'tis an elfin-storm from faery land,
Of haggard seeming, but a boon indeed:
Arise—arise! the morning is at hand;—
The bloated wassaillers will never heed:—
Let us away, my love, with happy speed;
There are no ears to hear, or eyes to see,—
Drown'd all in Rhenish and the sleepy mead:
Awake! arise! my love, and fearless be,   350
For o'er the southern moors I have a home for thee."

XL

She hurried at his words, beset with fears,
For there were sleeping dragons all around,
At glaring watch, perhaps, with ready spears—
Down the wide stairs a darkling way they found.—
In all the house was heard no human sound.
A chain-droop'd lamp was flickering by each door;
The arras, rich with horseman, hawk, and hound,
Flutter'd in the besieging wind's uproar;
And the long carpets rose along the gusty floor.   360

XLI

They glide, like phantoms, into the wide hall;
Like phantoms, to the iron porch, they glide;
Where lay the Porter, in uneasy sprawl,
With a huge empty flaggon by his side:
The wakeful bloodhound rose, and shook his hide,
But his sagacious eye an inmate owns:
By one, and one, the bolts full easy slide:—
The chains lie silent on the footworn stones;—
The key turns, and the door upon its hinges groans.

336. **vermeil:** vermillion. 349. **Rhenish:** wine from the
Rhine valley. **mead:** fermented honey and malt. 358. **arras:**
tapestry.

XLII

And they are gone: aye, ages long ago   370
These lovers fled away into the storm.
That night the Baron dreamt of many a woe,
And all his warrior-guests, with shade and form
Of witch, and demon, and large coffin-worm,
Were long be-nightmar'd. Angela the old
Died palsy-twitch'd, with meagre face deform;
The Beadsman, after thousand aves told,
For aye unsought for slept among his ashes cold.

1819 (1820)

THE EVE OF SAINT MARK

❧❀❧

The Eve of Saint Mark was written throughout four days
(Feb. 13–17, 1819), just after *The Eve of St. Agnes*, while
Keats was still hoping to get back to *Hyperion* or else
find an alternative that struck him as of equal importance.
In trying his hand at another romance, however briefly,
he created a new style, one that especially attracted the
Pre-Raphaelite poets later in the nineteenth century.
Keats is recalling his visit to the cathedral town of
Chichester, and he later told his brother George that the
lines were written in the "spirit of Town quietude. I think
it will give you the sensation of walking about an old
county Town in a coolish evening." Possibly the story,
if finished, would have turned on the legend that anyone
who stood by the church porch on St. Mark's Eve
(April 24) could see entering the church the apparitions
of those who would become ill throughout the year;
and if the apparitions did not leave the church, this
meant that they would die.

❧❀❧

Upon a Sabbath-day it fell;
Twice holy was the Sabbath-bell,
That call'd the folk to evening prayer;
The city streets were clean and fair
From wholesome drench of April rains;
And, on the western window panes,
The chilly sunset faintly told
Of unmatur'd green vallies cold,
Of the green thorny bloomless hedge,
Of rivers new with spring-tide sedge,   10

Of primroses by shelter'd rills,
And daisies on the aguish hills.
Twice holy was the Sabbath-bell:
The silent streets were crowded well
With staid and pious companies,
Warm from their fire-side orat'ries;
And moving, with demurest air,
To even-song, and vesper prayer.
Each arched porch, and entry low,
Was fill'd with patient folk and slow,   20
With whispers hush, and shuffling feet,
While play'd the organ loud and sweet.

The bells had ceas'd, the prayers begun,
And Bertha had not yet half done
A curious volume, patch'd and torn,
That all day long, from earliest morn,
Had taken captive her two eyes,
Among its golden broideries;
Perplex'd her with a thousand things,—
The stars of Heaven, and angels' wings,   30
Martyrs in a fiery blaze,
Azure saints in silver rays,
Moses' breastplate, and the seven
Candlesticks John saw in Heaven,
The winged Lion of Saint Mark,
And the Covenantal Ark,
With its many mysteries,
Cherubim and golden mice.

Bertha was a maiden fair,
Dwelling in the old Minster-square;   40
From her fire-side she could see,
Sidelong, its rich antiquity,
Far as the Bishop's garden-wall;
Where sycamores and elm-trees tall,
Full-leav'd, the forest had outstript,
By no sharp north-wind ever nipt,
So shelter'd by the mighty pile.
Bertha arose, and read awhile,
With forehead 'gainst the window-pane.
Again she tried, and then again,   50
Until the dusk eve left her dark
Upon the legend of St. Mark.

THE EVE OF SAINT MARK. **12. aguish:** chilly, shivering
(having the qualities of an ague). **33. Moses' breastplate:**
a combined recollection of Moses and Aaron, Exodus 25:7
and 27:4. **34. Candlesticks . . . Heaven:** Revelation 1:12.
36. Covenantal Ark: Exodus 25:10–22. **38. mice:** I Samuel
6:1–11, 18.

From plaited lawn-frill, fine and thin,
She lifted up her soft warm chin,
With aching neck and swimming eyes,
And daz'd with saintly imageries.

All was gloom, and silent all,
Save now and then the still foot-fall
Of one returning homewards late,
Past the echoing minster-gate.   60
The clamorous daws, that all the day
Above tree-tops and towers play,
Pair by pair had gone to rest,
Each in its ancient belfry-nest,
Where asleep they fall betimes,
To music of the drowsy chimes.

All was silent, all was gloom,
Abroad and in the homely room;
Down she sat, poor cheated soul!
And struck a lamp from the dismal coal;   70
Leaned forward, with bright drooping hair
And slant book, full against the glare.
Her shadow, in uneasy guise,
Hover'd about, a giant size,
On ceiling-beam and old oak chair,
The parrot's cage, and panel square;
And the warm angled winter screen,
On which were many monsters seen,
Call'd doves of Siam, Lima mice,
And legless birds of Paradise,   80
Macaw, and tender Av'davat,
And silken-furr'd Angora cat.
Untir'd she read, her shadow still
Glower'd about, as it would fill
The room with wildest forms and shades,
As though some ghostly queen of spades
Had come to mock behind her back,
And dance, and ruffle her garments black.
Untir'd she read the legend page,
Of holy Mark, from youth to age,   90
On land, on sea, in pagan chains,
Rejoicing for his many pains.
Sometimes the learned eremite,
With golden star, or dagger bright,
Referr'd to pious poesies
Written in smallest crow-quill size
Beneath the text; and thus the rhyme
Was parcell'd out from time to time:

81. **Av'davat:** amadavat, an Indian songbird.

——"Als writith he of swevenis,
Men han beforne they wake in bliss,   100
Whanne that hir friendes thinke hem bound
In crimped shroude farre under grounde:
And how a litling child mote be
A saint er its nativitie,
Gif that the modre (God her blesse!)
Kepen in solitarinesse,
And kissen devoute the holy croce.
Of Goddes love, and Sathan's force,—
He writith; and thinges many mo:
Of swiche thinges I may not show.   110
Bot I must tellen verilie
Somdel of Saintè Cicilie,
And chieflie what he auctorethe
Of Saintè Markis life and dethe:"

At length her constant eyelids come
Upon the fervent martyrdom;
Then lastly to his holy shrine,
Exalt amid the tapers' shine
At Venice—

                                   1819 (1848)

WHY DID I LAUGH?

This and the next five poems are enclosed in the long journal-letter (Feb. 14–May 3, 1819) to George and Georgiana Keats in America. For Keats's comment on this sonnet see the letter below under the section for March 19, p. 1224.

Why did I laugh to-night? No voice will tell:
    No God, no Demon of severe response,
Deigns to reply from Heaven or from Hell.
    Then to my human heart I turn at once.
Heart! Thou and I are here sad and alone;
    I say, why did I laugh! O mortal pain!
O Darkness! Darkness! ever must I moan,
    To question Heaven and Hell and Heart in vain.
Why did I laugh? I know this Being's lease,
    My fancy to its utmost blisses spreads;   10

99. **Als . . . swevenis:** also writeth he of dreams. 100. **han:** have. 101. **hir:** their. 102. **crimped:** folded. 103. **mote:** might. 105. **Gif:** if. 112. **Somdel:** something. 113. **auctorethe:** writes.

Yet would I on this very midnight cease,
    And the world's gaudy ensigns see in shreds;
Verse, Fame, and Beauty are intense indeed,
But Death intenser—Death is Life's high meed.

1819 (1848)

ON A DREAM

Keats had been reading the episode of Paolo and Francesca in Dante. He then had the dream that he describes, under the date of April 16, in the journal-letter to George and Georgiana Keats, mentioned above.

As Hermes once took to his feathers light,
    When lulled Argus, baffled, swoon'd and slept,
So on a Delphic reed, my idle spright
    So play'd, so charm'd, so conquer'd, so bereft
The dragon-world of all its hundred eyes;
    And, seeing it asleep, so fled away—
Not to pure Ida with its snow-cold skies,
    Nor unto Tempe where Jove griev'd a day;
But to that second circle of sad hell,
    Where 'mid the gust, the whirlwind, and the flaw
Of rain and hail-stones, lovers need not tell   11
    Their sorrows. Pale were the sweet lips I saw,
Pale were the lips I kiss'd, and fair the form
I floated with, about that melancholy storm.

1819 (1820)

LA BELLE DAME
SANS MERCI

The title is taken from the medieval poem of Alain Chartier. The essence of Keats's ballad (written April 21, 1819) is the mystery of "la belle dame." She is not quite mortal, though she is loved by a mortal. She is not

sinister, though the love of her has led the knight to his present woeful state (a state that might be attributed to the hopelessness of his love for something superhuman rather than to "la belle dame" herself). The theme goes back to Endymion's search for the immortal, and it anticipates the treatment of the mortal's desire to share the existence of the immortal in the "Ode on a Grecian Urn" and, with closer parallels to the present poem, in *Lamia.* A later and inferior version of the ballad incorporates changes suggested by Leigh Hunt.

O, what can ail thee, knight-at-arms,
    Alone and palely loitering?
The sedge has wither'd from the lake,
    And no birds sing.

O, what can ail thee, knight-at-arms,
    So haggard and so woe-begone?
The squirrel's granary is full,
    And the harvest's done.

I see a lilly on thy brow,
    With anguish moist and fever dew,   10
And on thy cheeks a fading rose
    Fast withereth too.

I met a lady in the meads,
    Full beautiful—a faery's child,
Her hair was long, her foot was light,
    And her eyes were wild.

I made a garland for her head,
    And bracelets too, and fragrant zone;
She look'd at me as she did love,
    And made sweet moan.   20

I set her on my pacing steed,
    And nothing else saw all day long,
For sidelong would she bend and sing
    A faery's song.

She found me roots of relish sweet,
    And honey wild, and manna dew,
And sure in language strange she said
    "I love thee true."

She took me to her elfin grot,
    And there she wept and sigh'd full sore,   30
And there I shut her wild wild eyes
    With kisses four.

And there she lulled me asleep,
   And there I dream'd—Ah! woe betide!
The latest dream I ever dream'd
   On the cold hill side.

I saw pale kings and princes too,
   Pale warriors, death-pale were they all;
They cried, "La Belle Dame sans Merci
   Hath thee in thrall!"   40

I saw their starved lips in the gloam,
   With horrid warning gaped wide,
And I awoke, and found me here,
   On the cold hill's side.

And this is why I sojourn here,
   Alone and palely loitering,
Though the sedge is wither'd from the lake,
   And no birds sing.

                                        1819 (1820)

ON FAME

"On Fame" was written extemporaneously, in an experimental rhyme scheme (see headnote to the following poem).

"You cannot eat your cake and have it too."—*Proverb.*

How fever'd is the man, who cannot look
   Upon his mortal days with temperate blood,
Who vexes all the leaves of his life's book,
   And robs his fair name of its maidenhood;
It is as if the rose should pluck herself,
   Or the ripe plum finger its misty bloom,
As if a Naiad, like a meddling elf,
   Should darken her pure grot with muddy gloom,
But the rose leaves herself upon the briar,
   For winds to kiss and grateful bees to feed,   10
And the ripe plum still wears its dim attire,
   The undisturbed lake has crystal space,
   Why then should man, teasing the world for
      grace,
Spoil his salvation for a fierce miscreed?

                                        1819 (1848)

ON THE SONNET

"I have been endeavouring," said Keats (under April 30 in the journal-letter mentioned in the headnote to "Why Did I Laugh") "to discover a better sonnet stanza than we have. The legitimate [Petrarchan form] does not suit the language over-well from the pouncing rhymes —the other kind [Shakespearean form] appears too elegiac—and the couplet at the end of it has seldom a pleasing effect—I do not pretend to have succeeded." The rhyme scheme (*abca bdca bcde de*) tries to avoid both the pouncing couplets of the Petrarchan form and the alternate-rhyming of the Shakespearean form.

If by dull rhymes our English must be chain'd,
   And, like Andromeda, the Sonnet sweet
Fetter'd, in spite of pained loveliness,
   Let us find out, if we must be constrain'd,
Sandals more interwoven and complete
   To fit the naked foot of Poesy:
Let us inspect the Lyre, and weigh the stress
   Of every chord, and see what may be gain'd
By ear industrious, and attention meet;
   Misers of sound and syllable, no less   10
Than Midas of his coinage, let us be
Jealous of dead leaves in the bay wreath crown;
   So, if we may not let the Muse be free,
She will be bound with garlands of her own.

                                        1819 (1848)

TO SLEEP

O soft embalmer of the still midnight,
   Shutting, with careful fingers and benign,
Our gloom-pleas'd eyes, embower'd from the light,
   Enshaded in forgetfulness divine:
O soothest Sleep! if so it please thee, close
   In midst of this thine hymn my willing eyes,
Or wait the amen, ere thy poppy throws
   Around my bed its lulling charities.

ON THE SONNET. **2. Andromeda:** Andromeda was chained to a cliff for a monster to devour, but was rescued by Perseus. **11. Midas:** king of Phrygia. Dionysus granted his prayer that everything he touched might be turned to gold.

Then save me, or the passed day will shine
Upon my pillow, breeding many woes,—   10
  Save me from curious Conscience, that still lords
Its strength for darkness, burrowing like a mole;
  Turn the key deftly in the oiled wards,
And seal the hushed Casket of my Soul.

1819 (1838)

ODE TO PSYCHE

The first of the great odes that Keats wrote during the month between April 20–25 and May 20, the "Ode to Psyche" (finished by April 30) was also, he said, the first poem "with which I have taken even moderate pains." The premise of the poem, as W. J. Bate has shown, is Keats's conviction that the poetry of the future must inevitably turn more to the mind and the "inner life" (see Keats's letter to Reynolds, below, May 3, 1818, p. 1213), and his hope that this can be done without loss. Psyche (the mind) came "too late" to be a part of the Greek hierarchy of gods. Her temple is yet to be built, and this Keats promises to do within an "untrodden region" of the mind. The "pains" Keats took with the poem were partly in order to work out a new form for the lyric that would not have the disadvantages of the two sonnet forms (see "On the Sonnet" above) and permit a longer, though still tightly knit, development (an aim Keats achieves better in the stanzas of the "Ode to a Nightingale" and "Ode on a Grecian Urn" than in the over-elaborate rhyme scheme of the "Ode to Psyche"). He also devoted care to thinking over the symbolic use of the story of Psyche, drawing particularly on the account in Apuleius, where Psyche, arousing the jealousy of Venus, is driven from her home, and Cupid is instructed to make her fall in love with a deformed creature. But Cupid falls in love with Psyche; and, after some difficulties, love (Cupid) becomes united with the mind (Psyche) and proves to be the mind's salvation. For Keats's own comment on the tale, see the journal-letter, below, p. 1226, under April 30, 1819.

O Goddess! hear these tuneless numbers, wrung
  By sweet enforcement and remembrance dear,
And pardon that thy secrets should be sung
  Even into thine own soft-conched ear:
Surely I dreamt to-day, or did I see
  The winged Psyche with awaken'd eyes?
I wander'd in a forest thoughtlessly,
  And, on the sudden, fainting with surprise,

ODE TO PSYCHE. **4. conched:** shell-shaped.

Saw two fair creatures, couched side by side
  In deepest grass, beneath the whisp'ring roof   10
  Of leaves and trembled blossoms, where there ran
    A brooklet, scarce espied:

'Mid hush'd, cool-rooted flowers, fragrant-eyed,
  Blue, silver-white, and budded Tyrian,
They lay calm-breathing on the bedded grass;
  Their arms embraced, and their pinions too;
  Their lips touch'd not, but had not bade adieu,
As if disjoined by soft-handed slumber,
And ready still past kisses to outnumber
  At tender eye-dawn of aurorean love: 20
    The winged boy I knew;
But who wast thou, O happy, happy dove?
    His Psyche true!

O latest born and loveliest vision far
  Of all Olympus' faded hierarchy!
Fairer than Phœbe's sapphire-region'd star,
  Or Vesper, amorous glow-worm of the sky;
Fairer than these, though temple thou hast none,
    Nor altar heap'd with flowers;
Nor virgin-choir to make delicious moan   30
    Upon the midnight hours;
No voice, no lute, no pipe, no incense sweet
  From chain-swung censer teeming;
No shrine, no grove, no oracle, no heat
  Of pale-mouth'd prophet dreaming.

O brightest! though too late for antique vows,
  Too, too late for the fond believing lyre,
When holy were the haunted forest boughs,
  Holy the air, the water, and the fire;
Yet even in these days so far retir'd   40
  From happy pieties, thy lucent fans,
  Fluttering among the faint Olympians,
I see, and sing, by my own eyes inspired.
So let me be thy choir, and make a moan
    Upon the midnight hours;
Thy voice, thy lute, thy pipe, thy incense sweet
  From swinged censer teeming;
Thy shrine, thy grove, thy oracle, thy heat
  Of pale-mouth'd prophet dreaming.

14. Tyrian: a purple dye made in Tyre. **20. aurorean:** dawning. **25. Olympus' . . . hierarchy:** the hierarchy of classical gods. **26. Phœbe's . . . star:** the moon, of which Phoebe was goddess. **27. Vesper:** evening star. **37. fond:** The word may suggest not only affectionate but possibly the older sense of "foolish." **41. fans:** wings.

Yes, I will be thy priest, and build a fane   50
  In some untrodden region of my mind,
Where branched thoughts, new grown with pleasant
          pain,
  Instead of pines shall murmur in the wind:
Far, far around shall those dark-cluster'd trees
  Fledge the wild-ridged mountains steep by steep;
And there by zephyrs, streams, and birds, and bees,
  The moss-lain Dryads shall be lull'd to sleep;
And in the midst of this wide quietness
A rosy sanctuary will I dress
With the wreath'd trellis of a working brain,   60
  With buds, and bells, and stars without a name,
  With all the gardener Fancy e'er could feign,
  Who breeding flowers, will never breed the same:
And there shall be for thee all soft delight
  That shadowy thought can win,
A bright torch, and a casement ope at night,
  To let the warm Love in!

                                        1819 (1820)

ODE TO A NIGHTINGALE

Keats wrote "Ode to a Nightingale" around May 1, 1819, shortly after the "Ode to Psyche." It and the "Grecian Urn" have been discussed as fully as any two lyrics in the English language. Yet, because of their richness and universality, almost every reader continues to find in them new attractions and suggestions. In these two odes Keats develops a prototype for the modern lyric of symbolic debate. The drama is that of a process of changing response to a symbol, a "greeting of the Spirit" and its object (to use Keats's phrase; see the letter to Benjamin Bailey, below, March 13, 1818, p. 1212). The debate lies in the fact that the "greeting" or inter-penetration can never be complete, and there is a continuing tension between what the imagination seeks to make of the object and what the object continues to suggest in its own right. Meanwhile it may be noted that Keats here attains what is to be more or less the norm for his new ode-stanza (an alternate-rhyming quatrain followed, with some variations, by six lines with a rhyme scheme like that of the conventional sestet in the Petrarchan sonnet: *abab cde cde*). The "Ode to a Nightingale" is one of the rare cases where the whole of a major poem was written almost extemporaneously: these eighty lines (apparently with little revision) were composed in a single morning. Charles Brown, with

50. **fane:** temple. 66. **casement ope:** a recurring image in Keats, the most famous use of it being in the "Ode to a Nightingale," l. 69.

whom Keats was living in Hampstead, wrote the following account:

> In the spring of 1819 a nightingale had built her nest near my house. Keats felt a tranquil and continual joy in her song; and one morning he took his chair from the breakfast-table to the grass-plot under a plum-tree, where he sat for two or three hours. When he came into the house, I perceived he had some scraps of paper in his hand, and these he was quietly thrusting behind the books. On inquiry, I found those scraps, four or five in number, contained his poetic feeling on the song of our nightingale.

I

My heart aches, and a drowsy numbness pains
  My sense, as though of hemlock I had drunk,
Or emptied some dull opiate to the drains
  One minute past, and Lethe-wards had sunk:
'Tis not through envy of thy happy lot,
  But being too happy in thine happiness,—
    That thou, light-winged Dryad of the trees,
        In some melodious plot
Of beechen green, and shadows numberless,
  Singest of summer in full-throated ease.   10

II

O, for a draught of vintage! that hath been
  Cool'd a long age in the deep-delved earth,
Tasting of Flora and the country green,
  Dance, and Provençal song, and sunburnt mirth!
O for a beaker full of the warm South,
  Full of the true, the blushful Hippocrene,
    With beaded bubbles winking at the brim,
        And purple-stained mouth;
That I might drink, and leave the world unseen,
  And with thee fade away into the forest dim: 20

III

Fade far away, dissolve, and quite forget
  What thou among the leaves hast never known,
The weariness, the fever, and the fret
  Here, where men sit and hear each other groan;
Where palsy shakes a few, sad, last gray hairs,
  Where youth grows pale, and spectre-thin, and dies;

ODE TO A NIGHTINGALE. 2. **hemlock:** a poisonous herb (not the American hemlock tree). 4. **Lethe-wards:** Lethe is the river of forgetfulness in Hades. 7. **Dryad:** wood nymph. 16. **Hippocrene:** fountain of the muses on Mt. Helicon. 26. **youth . . . dies:** Keats is inevitably recalling the death, the previous December, of his young brother Tom.

Where but to think is to be full of sorrow
    And leaden-eyed despairs,
  Where Beauty cannot keep her lustrous eyes,
    Or new Love pine at them beyond to-morrow.

IV

Away! away! for I will fly to thee,   31
    Not charioted by Bacchus and his pards,
But on the viewless wings of Poesy,
    Though the dull brain perplexes and retards:
Already with thee! tender is the night,
  And haply the Queen-Moon is on her throne,
    Cluster'd around by all her starry Fays;
      But here there is no light,
Save what from heaven is with the breezes blown
  Through verdurous glooms and winding mossy
        ways.   40

V

I cannot see what flowers are at my feet,
  Nor what soft incense hangs upon the boughs,
But, in embalmed darkness, guess each sweet
  Wherewith the seasonable month endows
The grass, the thicket, and the fruit-tree wild;
  White hawthorn, and the pastoral eglantine;
    Fast fading violets cover'd up in leaves;
      And mid-May's eldest child,
The coming musk-rose, full of dewy wine,
  The murmurous haunt of flies on summer eves.

VI

Darkling I listen; and, for many a time   51
  I have been half in love with easeful Death,
Call'd him soft names in many a mused rhyme,
  To take into the air my quiet breath;
Now more than ever seems it rich to die,
  To cease upon the midnight with no pain,
    While thou art pouring forth thy soul abroad
      In such an ecstasy!
Still wouldst thou sing, and I have ears in vain—
  To thy high requiem become a sod.   60

VII

Thou wast not born for death, immortal Bird!
  No hungry generations tread thee down;

---

32. **Bacchus:** god of wine, whose chariot was drawn by leopards. 33. **viewless:** invisible. 37. **Fays:** fairies. 43. **embalmed:** balmy, perfumed. 51. **Darkling:** in the dark.

The voice I hear this passing night was heard
  In ancient days by emperor and clown:
Perhaps the self-same song that found a path
    Through the sad heart of Ruth, when, sick for home,
      She stood in tears amid the alien corn;
      The same that oft-times hath
  Charm'd magic casements, opening on the foam
    Of perilous seas, in faery lands forlorn.   70

VIII

Forlorn! the very word is like a bell
  To toll me back from thee to my sole self!
Adieu! the fancy cannot cheat so well
  As she is fam'd to do, deceiving elf.
Adieu! adieu! thy plaintive anthem fades
  Past the near meadows, over the still stream,
    Up the hill-side; and now 'tis buried deep
      In the next valley-glades:
  Was it a vision, or a waking dream?
    Fled is that music:—Do I wake or sleep?   80
                      1819 (1820)

ODE ON A GRECIAN URN

⤳✺⤳

"Ode on a Grecian Urn" was apparently written within a week or two after the "Ode to a Nightingale," and certainly by the middle of May, 1819. In the two odes the general premise and mode of symbolic debate, mentioned above, are similar; but the progress of this poem differs if only because of the marked dissimilarity of the urn and the song of the nightingale as symbols. Whereas the nightingale's song is by definition fleeting and also spontaneously natural, the urn is stationary, silent, and the product of human art. Hence the confidence with which the urn is approached at first (a "historian" that can reveal a tale) and the quickness with which the poet feels that he can imaginatively share in the life that the urn has captured (sts. ii and iii). Then the limits between the observer's imagination and the work of art emerge. But the poet is not, as at the end of the "Nightingale," left questioning his experience. With the urn still before him, he tries to interpret the experience and put into words the message of this "shape" or "attitude" that, however removed from our personal lives, is still a "friend to man." (See the note below to lines 49–50.)

⤳✺⤳

64–70. ancient . . . forlorn: The associations move from the historical past (of "emperor and clown," or rustic) through the more remote past of biblical legend (Ruth 2) to the realm of pure fancy and magic far removed from human life and thus "forlorn." **79. vision . . . waking dream:** Cf. "Ode to Psyche," above, ll. 5–6.

I

Thou still unravish'd bride of quietness,
   Thou foster-child of silence and slow time,
Sylvan historian, who canst thus express
   A flowery tale more sweetly than our rhyme:
What leaf-fring'd legend haunts about thy shape
   Of deities or mortals, or of both,
      In Tempe or the dales of Arcady?
   What men or gods are these? What maidens loth?
What mad pursuit? What struggle to escape?
      What pipes and timbrels? What wild ecstasy? 10

II

Heard melodies are sweet, but those unheard
   Are sweeter; therefore, ye soft pipes, play on;
Not to the sensual ear, but, more endear'd,
   Pipe to the spirit ditties of no tone:
Fair youth, beneath the trees, thou canst not leave
   Thy song, nor ever can those trees be bare;
      Bold Lover, never, never canst thou kiss,
Though winning near the goal—yet, do not grieve;
   She cannot fade, though thou hast not thy bliss,
      For ever wilt thou love, and she be fair! 20

III

Ah, happy, happy boughs! that cannot shed
   Your leaves, nor ever bid the Spring adieu;
And, happy melodist, unwearied,
   For ever piping songs for ever new;
More happy love! more happy, happy love!
   For ever warm and still to be enjoy'd,
      For ever panting, and for ever young;
All breathing human passion far above,
   That leaves a heart high-sorrowful and cloy'd,
      A burning forehead, and a parching tongue. 30

IV

Who are these coming to the sacrifice?
   To what green altar, O mysterious priest,
Lead'st thou that heifer lowing at the skies,
   And all her silken flanks with garlands drest?

What little town by river or sea shore,
   Or mountain-built with peaceful citadel,
      Is emptied of this folk, this pious morn?
And, little town, thy streets for evermore
   Will silent be; and not a soul to tell
      Why thou art desolate, can e'er return. 40

V

O Attic shape! Fair attitude! with brede
   Of marble men and maidens overwrought,
With forest branches and the trodden weed;
   Thou, silent form, dost tease us out of thought
As doth eternity: Cold Pastoral!
   When old age shall this generation waste,
      Thou shalt remain, in midst of other woe
Than ours, a friend to man, to whom thou say'st,
   "Beauty is truth, truth beauty,—that is all
      Ye know on earth, and all ye need to know." 50

                                            1819 (1820)

ODE ON MELANCHOLY

I

No, no, go not to Lethe, neither twist
   Wolf's-bane, tight-rooted, for its poisonous wine;
Nor suffer thy pale forehead to be kiss'd
   By nightshade, ruby grape of Proserpine;

35. little town: significantly not pictured on the urn; the poet imagines it as the place from which these figures once came, and which, since they can never return to it, is "desolate" (cf. "forlorn" in the "Ode to a Nightingale," above, ll. 70–71). **41. Attic:** from Attica, or Athens. **brede:** embroidery. **44. tease . . . thought:** Cf. "Epistle to J. H. Reynolds," above, l. 77. **49–50. "Beauty . . . know":** The lines have been endlessly disputed, partly because they have usually been punctuated so that only the first five words are enclosed in quotation marks as the comment from the urn. The present punctuation follows that of Douglas Bush in his edition of the poems (see bibliography, above). The message of the urn, then, is one that expresses its *own* special character as a work of art. It is "above" the world of "breathing human passion" (and, to that extent, also limited—a "Cold Pastoral"); and, in offering this message to the observer who lives in a very different world, it serves as a "friend to man." ODE ON MELANCHOLY. **1. Lethe:** See "Ode to a Nightingale," above, l. 4. **2. Wolf's-bane:** like nightshade (l. 4), a poisonous plant. The implication is that such conventional symbols have little to do with true melancholy and sorrow. **4. Proserpine:** wife of Pluto, ruler of the underworld.

ODE ON A GRECIAN URN. **7. Tempe:** a valley in Thessaly. **Arcady:** pastoral district in Greece. **13. sensual:** sensuous— the actual physical capacity to hear. **21–22. happy . . . adieu:** Cf. "In Drear-Nighted December," above, ll. 2–4.

Make not your rosary of yew-berries,
  Nor let the beetle, nor the death-moth be
    Your mournful Psyche, nor the downy owl
A partner in your sorrow's mysteries;
  For shade to shade will come too drowsily,
    And drown the wakeful anguish of the soul.   10

II

But when the melancholy fit shall fall
  Sudden from heaven like a weeping cloud,
That fosters the droop-headed flowers all,
  And hides the green hill in an April shroud;
Then glut thy sorrow on a morning rose,
  Or on the rainbow of the salt sand-wave,
    Or on the wealth of globed peonies;
Or if thy mistress some rich anger shows,
  Emprison her soft hand, and let her rave,
    And feed deep, deep upon her peerless eyes.   20

III

She dwells with Beauty—Beauty that must die;
  And Joy, whose hand is ever at his lips
Bidding adieu; and aching Pleasure nigh,
  Turning to Poison while the bee-mouth sips:
Ay, in the very temple of delight
  Veil'd Melancholy has her sovran shrine,
    Though seen of none save him whose strenuous
      tongue
  Can burst Joy's grape against his palate fine;
His soul shall taste the sadness of her might,
  And be among her cloudy trophies hung.   30

1819 (1820)

---

6. **beetle:** the Egyptian scarab, often placed in coffins. **death-moth:** so called because it has marks similar to a skull. 7. **Psyche:** traditionally symbolized by the butterfly. 21. **She:** not the "mistress" but Melancholy itself. 23-26. **aching . . . shrine:** Keats echoes Hazlitt, ("On Poetry in General"): The sense of beauty seeks in poetry to "*enshrine* itself . . . in the highest forms of fancy, and to relieve the *aching* sense of *pleasure* by expressing it in the boldest manner." 30. **trophies hung:** Cf. Shakespeare's Sonnet XXXI: "Hung with the trophies of my lovers gone."

ODE ON INDOLENCE

"They toil not, neither do they spin."

I

One morn before me were three figures seen,
  With bowed necks, and joined hands, side-faced;
And one behind the other stepp'd serene,
  In placid sandals, and in white robes graced;
They pass'd, like figures on a marble urn,
  When shifted round to see the other side;
    They came again; as when the urn once more
Is shifted round, the first seen shades return;
  And they were strange to me, as may betide
    With vases, to one deep in Phidian lore.   10

II

How is it, Shadows! that I knew ye not?
  How came ye muffled in so hush a mask?
Was it a silent deep-disguised plot
  To steal away, and leave without a task
My idle days? Ripe was the drowsy hour;
  The blissful cloud of summer-indolence
    Benumb'd my eyes; my pulse grew less and less;
Pain had no sting, and pleasure's wreath no flower:
  O, why did ye not melt, and leave my sense
    Unhaunted quite of all but—nothingness?   20

III

A third time came they by;—alas! wherefore?
  My sleep had been embroider'd with dim dreams;
My soul had been a lawn besprinkled o'er
  With flowers, and stirring shades, and baffled beams:
The morn was clouded, but no shower fell,
  Tho' in her lids hung the sweet tears of May;
    The open casement press'd a new-leav'd vine,
Let in the budding warmth and throstle's lay;
  O Shadows! 'twas a time to bid farewell!
    Upon your skirts had fallen no tears of mine.   30

ODE ON INDOLENCE. **1. three figures:** See Keats's remarks (journal-letter, below, March 19, 1819) on his fancying, in a state of indolence, the figures of poetry, ambition, and love passing before him like "figures on a greek vase." **10. Phidian:** referring to Phidias, the great Athenian sculptor.

IV

A third time pass'd they by, and, passing, turn'd
  Each one the face a moment whiles to me;
Then faded, and to follow them I burn'd
  And ached for wings because I knew the three;
The first was a fair Maid, and Love her name;
  The second was Ambition, pale of cheek,
  And ever watchful with fatigued eye;
The last, whom I love more, the more of blame
  Is heap'd upon her, maiden most unmeek,—
    I knew to be my demon Poesy.   40

V

They faded, and, forsooth! I wanted wings:
  O folly! What is Love! and where is it?
And for that poor Ambition! it springs
  From a man's little heart's short fever-fit;
For Poesy!—no,—she has not a joy,—
  At least for me,—so sweet as drowsy noons,
  And evenings steep'd in honied indolence;
O, for an age so shelter'd from annoy,
  That I may never know how change the moons,
    Or hear the voice of busy common-sense!   50

VI

So, ye three Ghosts, adieu! Ye cannot raise
  My head cool-bedded in the flowery grass;
For I would not be dieted with praise,
  A pet-lamb in a sentimental farce!
Fade softly from my eyes, and be once more
  In masque-like figures on the dreamy urn;
    Farewell! I yet have visions for the night,
And for the day faint visions there is store;
  Vanish, ye Phantoms! from my idle spright,
    Into the clouds, and never more return!   60

                                    1819 (1848)

---

44. fever-fit: Cf. "Ode to a Nightingale," above, l. 23.
54. pet-lamb: Cf. Keats's letter to Sarah Jeffrey, below, June 9, 1819: "I hope I am a little more of a Philosopher than I was, consequently a little less of a versifying Pet-lamb."
59. spright: spirit.

LAMIA

Lamia represents a deliberate effort on Keats's part to write a poetic narrative that would move briskly and, as he said, "take hold of people in some way—give them either pleasant or unpleasant sensation. What they want is sensation of some sort." In order to send money to his brother in America, Keats suddenly found (in May, 1819, just as he was finishing the odes) that he either had to earn money through his writing by the end of the summer or else had to turn to other work (see the letters to Sarah Jeffrey, below, for May 31 and June 9, 1819, pp. 1227–29). In late June he began *Lamia*, finished Part I by mid-July, wrote (with Charles Brown) a verse-tragedy, and then turned to his *Fall of Hyperion*, finishing Part II of *Lamia* in early September. Reacting not only against the earlier romances but also against the richly laden style of *Hyperion, The Eve of St. Agnes*, and the odes, he used the crisp neoclassic couplet of Dryden and turned the tale (which he derived from a story of Philostratus, summarized in Robert Burton's *Anatomy of Melancholy*) into a brilliant interplay of contrasts—mortal and immortal, poetry and philosophy, action and retreat. The poem was once regarded as a Romantic attack on science and analytic philosophy (as represented by Apollonius). It has since been often viewed as an attack on Romantic escapism and self-delusion (typified by Lycius). More probably both Lycius and Apollonius represent single-minded attitudes, neither one of which is adequate. Meanwhile the character of Lamia herself is emphasized not only as complex but as changeable, elusive, and to some extent dependent (with all her apparent magic) on the eyes and attitude with which she is viewed. Keats, at the conclusion of his poem, added the following account taken from the *Anatomy of Melancholy*:

Philostratus, in his fourth book *de Vita Apollonii*, hath a memorable instance in this kind, which I may not omit, of one Menippus Lycius, a young man twenty-five years of age, that going betwixt Cenchreas and Corinth, met such a phantasm in the habit of a fair gentlewoman, which, taking him by the hand, carried him home to her house, in the suburbs of Corinth, and told him she was a Phœnician by birth, and if he would tarry with her, he should hear her sing and play, and drink such wine as never any drank, and no man should molest him; but she, being fair and lovely, would live and die with him, that was fair and lovely to behold. The young man, a philosopher, otherwise staid and discreet, able to moderate his passions, though not this of love, tarried with her a while to his great content, and at last married her, to whose wedding, amongst other guests, came Apollonius; who, by some probable conjectures, found her out to be a serpent, a lamia; and that all her furniture was, like Tantalus' gold, described by Homer, no substance but mere illusions. When she saw herself descried, she wept, and desired Apollonius to be silent, but he would not be moved, and thereupon she, plate, house, and all that was in it, vanished in an instant: many thousands took notice of this fact, for it was done in the midst of Greece.

∼∿∽

PART I

Upon a time, before the faery broods
Drove Nymph and Satyr from the prosperous woods,
Before king Oberon's bright diadem,
Sceptre, and mantle, clasp'd with dewy gem,
Frighted away the Dryads and the Fauns
From rushes green, and brakes, and cowslip'd lawns,
The ever-smitten Hermes empty left
His golden throne, bent warm on amorous theft:
From high Olympus had he stolen light,
On this side of Jove's clouds, to escape the sight   10
Of his great summoner, and made retreat
Into a forest on the shores of Crete.
For somewhere in that sacred island dwelt
A nymph, to whom all hoofed Satyrs knelt;
At whose white feet the languid Tritons poured
Pearls, while on land they wither'd and adored.
Fast by the springs where she to bathe was wont,
And in those meads where sometime she might haunt,
Were strewn rich gifts, unknown to any Muse,
Though Fancy's casket were unlock'd to choose.   20
Ah, what a world of love was at her feet!
So Hermes thought, and a celestial heat
Burnt from his winged heels to either ear,
That from a whiteness, as the lilly clear,
Blush'd into roses 'mid his golden hair,
Fallen in jealous curls about his shoulders bare.

From vale to vale, from wood to wood, he flew,
Breathing upon the flowers his passion new,
And wound with many a river to its head,
To find where this sweet nymph prepar'd her secret
      bed: 30

In vain; the sweet nymph might nowhere be found,
And so he rested, on the lonely ground,
Pensive, and full of painful jealousies
Of the Wood-Gods, and even the very trees.
There as he stood, he heard a mournful voice,
Such as once heard, in gentle heart, destroys
All pain but pity: thus the lone voice spake:
"When from this wreathed tomb shall I awake!
When move in a sweet body fit for life,
And love, and pleasure, and the ruddy strife   40
Of hearts and lips! Ah, miserable me!"
The God, dove-footed, glided silently
Round bush and tree, soft-brushing, in his speed,
The taller grasses and full-flowering weed,
Until he found a palpitating snake,
Bright, and cirque-couchant in a dusky brake.

She was a gordian shape of dazzling hue,
Vermilion-spotted, golden, green, and blue;
Striped like a zebra, freckled like a pard,
Eyed like a peacock, and all crimson barr'd;   50
And full of silver moons, that, as she breathed,
Dissolv'd, or brighter shone, or interwreathed
Their lustres with the gloomier tapestries—
So rainbow-sided, touch'd with miseries,
She seem'd, at once, some penanced lady elf,
Some demon's mistress, or the demon's self.
Upon her crest she wore a wannish fire
Sprinkled with stars, like Ariadne's tiar:
Her head was serpent, but ah, bitter-sweet!
She had a woman's mouth with all its pearls complete:
And for her eyes: what could such eyes do there   61
But weep, and weep, that they were born so fair?
As Proserpine still weeps for her Sicilian air.
Her throat was serpent, but the words she spake
Came, as through bubbling honey, for Love's sake,
And thus; while Hermes on his pinions lay,
Like a stoop'd falcon ere he takes his prey.

"Fair Hermes, crown'd with feathers, fluttering light,
I had a splendid dream of thee last night:

46. **cirque-couchant:** lying in the shape of a circle. **47. gordian:** intricately knotted (after Gordius who tied the knot no one could disentangle). **58. Ariadne's tiar:** the crown of Ariadne (daughter of Minos), later transformed into a constellation. **60. pearls:** conventional in Elizabethan love poetry as a synonym for teeth. The intention here is to startle the reader with the incongruity as well as brilliance of Lamia's appearance. **63. Proserpine . . . air:** daughter of Ceres, carried off from the field of Enna, in Sicily, to Hades by Pluto.

LAMIA. PART I. **2–5. Drove . . . Fauns:** the classical deities of the woodlands banished by those of medieval folklore (led by Oberon, king of the fairies). **6. brakes:** thickets. **7. ever-smitten Hermes:** messenger of the gods, who was constantly falling in love. **26. Fallen . . . bare:** the first of several Alexandrines (six-foot lines) with which Keats, following the example of Dryden, varies the flow of the pentameter couplet.

I saw thee sitting, on a throne of gold,   70
Among the Gods, upon Olympus old,
The only sad one; for thou didst not hear
The soft, lute-finger'd Muses chaunting clear,
Nor even Apollo when he sang alone,
Deaf to his throbbing throat's long, long melodious
    moan.
I dreamt I saw thee, robed in purple flakes,
Break amorous through the clouds, as morning breaks,
And, swiftly as a bright Phœbean dart,
Strike for the Cretan isle; and here thou art!
Too gentle Hermes, hast thou found the maid?"   80
Whereat the star of Lethe not delay'd
His rosy eloquence, and thus inquired:
"Thou smooth-lipp'd serpent, surely high inspired!
Thou beauteous wreath, with melancholy eyes,
Possess whatever bliss thou canst devise,
Telling me only where my nymph is fled,—
Where she doth breathe!" "Bright planet, thou hast
    said,"
Return'd the snake, "but seal with oaths, fair God!"
"I swear," said Hermes, "by my serpent rod,
And by thine eyes, and by thy starry crown!"   90
Light flew his earnest words, among the blossoms
    blown.
Then thus again the brilliance feminine:
"Too frail of heart! for this lost nymph of thine,
Free as the air, invisibly, she strays
About these thornless wilds; her pleasant days
She tastes unseen; unseen her nimble feet
Leave traces in the grass and flowers sweet;
From weary tendrils, and bow'd branches green,
She plucks the fruit unseen, she bathes unseen:
And by my power is her beauty veil'd   100
To keep it unaffronted, unassail'd
By the love-glances of unlovely eyes,
Of Satyrs, Fauns, and blear'd Silenus' sighs.
Pale grew her immortality, for woe
Of all these lovers, and she grieved so
I took compassion on her, bade her steep
Her hair in weïrd syrops, that would keep
Her loveliness invisible, yet free
To wander as she loves, in liberty.
Thou shalt behold her, Hermes, thou alone,   110

If thou wilt, as thou swearest, grant my boon!"
Then, once again, the charmed God began
An oath, and through the serpent's ears it ran
Warm, tremulous, devout, psalterian.
Ravish'd, she lifted her Circean head,
Blush'd a live damask, and swift-lisping said,
"I was a woman, let me have once more
A woman's shape, and charming as before.
I love a youth of Corinth—O the bliss!
Give me my woman's form, and place me where he is.
Stoop, Hermes, let me breathe upon thy brow,   121
And thou shalt see thy sweet nymph even now."
The God on half-shut feathers sank serene,
She breath'd upon his eyes, and swift was seen
Of both the guarded nymph near-smiling on the
    green.
It was no dream; or say a dream it was,
Real are the dreams of Gods, and smoothly pass
Their pleasures in a long immortal dream.
One warm, flush'd moment, hovering, it might seem
Dash'd by the wood-nymph's beauty, so he burn'd;
Then, lighting on the printless verdure, turn'd   131
To the swoon'd serpent, and with languid arm,
Delicate, put to proof the lythe Caducean charm.
So done, upon the nymph his eyes he bent
Full of adoring tears and blandishment,
And towards her stept: she, like a moon in wane,
Faded before him, cower'd, nor could restrain
Her fearful sobs, self-folding like a flower
That faints into itself at evening hour:
But the God fostering her chilled hand,   140
She felt the warmth, her eyelids open'd bland,
And, like new flowers at morning song of bees,
Bloom'd, and gave up her honey to the lees.
Into the green-recessed woods they flew;
Nor grew they pale, as mortal lovers do.

    Left to herself, the serpent now began
To change; her elfin blood in madness ran,
Her mouth foam'd, and the grass, therewith besprent,
Wither'd at dew so sweet and virulent;
Her eyes in torture fix'd, and anguish drear,   150
Hot, glaz'd, and wide, with lid-lashes all sear,
Flash'd phosphor and sharp sparks, without one cooling
    tear.

---

78. Phœbean dart: a ray from Phoebus Apollo, god of
the sun. **81. star of Lethe:** Hermes led the souls of the
dead to Hades (where the river of Lethe flowed), and in the
dark he shone like a star. **103. Silenus:** a drunken satyr,
tutor of Bacchus, god of wine. **107. weïrd:** magical.

114. psalterian: musical; from psaltery, a stringed instrument
of the zither type (though possibly Keats is thinking of the
Psalter, or Book of Psalms). **116. damask:** a large pink
rose. **133. put . . . charm:** touched her with the Caduceus
(magic staff).

The colours all inflam'd throughout her train,
She writh'd about, convuls'd with scarlet pain:
A deep volcanian yellow took the place
Of all her milder-mooned body's grace;
And, as the lava ravishes the mead,
Spoilt all her silver mail, and golden brede;
Made gloom of all her frecklings, streaks and bars,
Eclips'd her crescents, and lick'd up her stars: 160
So that, in moments few, she was undrest
Of all her sapphires, greens, and amethyst,
And rubious-argent: of all these bereft,
Nothing but pain and ugliness were left.
Still shone her crown; that vanish'd, also she
Melted and disappear'd as suddenly;
And in the air, her new voice luting soft,
Cried, "Lycius! gentle Lycius!"—Borne aloft
With the bright mists about the mountains hoar
These words dissolv'd: Crete's forests heard no more.

Whither fled Lamia, now a lady bright,   171
A full-born beauty new and exquisite?
She fled into that valley they pass o'er
Who go to Corinth from Cenchreas' shore;
And rested at the foot of those wild hills,
The rugged founts of the Peræan rills,
And of that other ridge whose barren back
Stretches, with all its mist and cloudy rack,
South-westward to Cleone. There she stood
About a young bird's flutter from a wood,   180
Fair, on a sloping green of mossy tread,
By a clear pool, wherein she passioned
To see herself escap'd from so sore ills,
While her robes flaunted with the daffodils.

Ah, happy Lycius!—for she was a maid
More beautiful than ever twisted braid,
Or sigh'd, or blush'd, or on spring-flowered lea
Spread a green kirtle to the minstrelsy:
A virgin purest lipp'd, yet in the lore
Of love deep learned to the red heart's core: 190
Not one hour old, yet of sciential brain
To unperplex bliss from its neighbour pain;
Define their pettish limits, and estrange
Their points of contact, and swift counterchange;

Intrigue with the specious chaos, and dispart
Its most ambiguous atoms with sure art;
As though in Cupid's college she had spent
Sweet days a lovely graduate, still unshent,
And kept his rosy terms in idle languishment.

Why this fair creature chose so faerily   200
By the wayside to linger, we shall see;
But first 'tis fit to tell how she could muse
And dream, when in the serpent prison-house,
Of all she list, strange or magnificent:
How, ever, where she will'd, her spirit went;
Whether to faint Elysium, or where
Down through tress-lifting waves the Nereids fair
Wind into Thetis' bower by many a pearly stair;
Or where God Bacchus drains his cups divine,
Stretch'd out, at ease, beneath a glutinous pine;   210
Or where in Pluto's gardens palatine
Mulciber's columns gleam in far piazzian line.
And sometimes into cities she would send
Her dream, with feast and rioting to blend;
And once, while among mortals dreaming thus,
She saw the young Corinthian Lycius
Charioting foremost in the envious race,
Like a young Jove with calm uneager face,
And fell into a swooning love of him.
Now on the moth-time of that evening dim   220
He would return that way, as well she knew,
To Corinth from the shore; for freshly blew
The eastern soft wind, and his galley now
Grated the quaystones with her brazen prow
In port Cenchreas, from Egina isle
Fresh anchor'd; whither he had been awhile
To sacrifice to Jove, whose temple there
Waits with high marble doors for blood and incense
     rare.
Jove heard his vows, and better'd his desire;
For by some freakful chance he made retire   230
From his companions, and set forth to walk,
Perhaps grown wearied of their Corinth talk:
Over the solitary hills he fared,

---

158. **brede:** embroidery. 163. **rubious-argent:** red and silver. 174. **Cenchreas' shore:** on the eastern harbor of Corinth. 192–94. **unperplex . . . counterchange:** Cf. "On Visiting the Tomb of Burns," above, ll. 8–12, and "Ode on Melancholy," above, ll. 23–30. 193. **pettish:** petulant, because so changeable and uncertain.

198. **unshent:** unspoiled. 199. **terms:** college terms. 203. **serpent prison-house:** Cf. Keats's remark, written in the margin of his copy of *Paradise Lost*, on Milton's empathic description of Satan (IX. 179–91): "Satan having enter'd the Serpent, and inform'd his brutal sense—might seem sufficient—but Milton goes on '*but his sleep disturbed not.*' Whose head is not dizzy at the possible speculations of Satan in the serpent prison? No passage of poetry can ever give a greater pain of suffocation." 207. **Nereids:** sea nymphs 211. **palatine:** palatial. 212. **Mulciber:** Vulcan.

Thoughtless at first, but ere eve's star appeared
His phantasy was lost, where reason fades,
In the calm'd twilight of Platonic shades.
Lamia beheld him coming, near, more near—
Close to her passing, in indifference drear,
His silent sandals swept the mossy green;
So neighbour'd to him, and yet so unseen   240
She stood: he pass'd, shut up in mysteries,
His mind wrapp'd like his mantle, while her eyes
Follow'd his steps, and her neck regal white
Turn'd—syllabling thus, "Ah, Lycius bright,
And will you leave me on the hills alone?
Lycius, look back! and be some pity shown."
He did; not with cold wonder fearingly,
But Orpheus-like at an Eurydice;
For so delicious were the words she sung,   249
It seem'd he had lov'd them a whole summer long:
And soon his eyes had drunk her beauty up,
Leaving no drop in the bewildering cup,
And still the cup was full,—while he, afraid
Lest she should vanish ere his lip had paid
Due adoration, thus began to adore;
Her soft look growing coy, she saw his chain so sure:
"Leave thee alone! Look back! Ah, Goddess, see
Whether my eyes can ever turn from thee!
For pity do not this sad heart belie—
Even as thou vanishest so shall I die.   260
Stay! though a Naiad of the rivers, stay!
To thy far wishes will thy streams obey:
Stay! though the greenest woods be thy domain,
Alone they can drink up the morning rain:
Though a descended Pleiad, will not one
Of thine harmonious sisters keep in tune
Thy spheres, and as thy silver proxy shine?
So sweetly to these ravish'd ears of mine
Came thy sweet greeting, that if thou shouldst fade
Thy memory will waste me to a shade:—   270
For pity do not melt!"—"If I should stay,"
Said Lamia, "here, upon this floor of clay,
And pain my steps upon these flowers too rough,
What canst thou say or do of charm enough
To dull the nice remembrance of my home?
Thou canst not ask me with thee here to roam

Over these hills and vales, where no joy is,—
Empty of immortality and bliss!
Thou art a scholar, Lycius, and must know
That finer spirits cannot breathe below   280
In human climes, and live: Alas! poor youth,
What taste of purer air hast thou to soothe
My essence? What serener palaces,
Where I may all my many senses please,
And by mysterious sleights a hundred thirsts appease?
It cannot be—Adieu!" So said, she rose
Tiptoe with white arms spread. He, sick to lose
The amorous promise of her lone complain,
Swoon'd, murmuring of love, and pale with pain.
The cruel lady, without any show   290
Of sorrow for her tender favourite's woe,
But rather, if her eyes could brighter be,
With brighter eyes and slow amenity,
Put her new lips to his, and gave afresh
The life she had so tangled in her mesh:
And as he from one trance was wakening
Into another, she began to sing,
Happy in beauty, life, and love, and every thing,
A song of love, too sweet for earthly lyres,
While, like held breath, the stars drew in their panting
     fires.   300
And then she whisper'd in such trembling tone,
As those who, safe together met alone
For the first time through many anguish'd days,
Use other speech than looks; bidding him raise
His drooping head, and clear his soul of doubt,
For that she was a woman, and without
Any more subtle fluid in her veins
Than throbbing blood, and that the self-same pains
Inhabited her frail-strung heart as his.
And next she wonder'd how his eyes could miss   310
Her face so long in Corinth, where, she said,
She dwelt but half retir'd, and there had led
Days happy as the gold coin could invent
Without the aid of love; yet in content
Till she saw him, as once she pass'd him by,
Where 'gainst a column he lent thoughtfully
At Venus' temple porch, 'mid baskets heap'd
Of amorous herbs and flowers, newly reap'd
Late on that eve, as 'twas the night before
The Adonian feast; whereof she saw no more,   320
But wept alone those days, for why should she
     adore?

---

248. Orpheus-like . . . Eurydice: Allowed by Pluto to lead
Eurydice back from Hades provided he refrained from
looking at her, Orpheus yielded to the temptation and lost
her. 265. Pleiad: one of the seven Pleiades (daughters of
Atlas) who were changed into a constellation. 275. nice:
exact, detailed.

320. Adonian: feast of Adonis (beloved by Venus).

Lycius from death awoke into amaze,
To see her still, and singing so sweet lays;
Then from amaze into delight he fell
To hear her whisper woman's lore so well;
And every word she spake entic'd him on
To unperplex'd delight and pleasure known.
Let the mad poets say whate'er they please
Of the sweets of Faeries, Peris, Goddesses,
There is not such a treat among them all,   330
Haunters of cavern, lake, and waterfall,
As a real woman, lineal indeed
From Pyrrha's pebbles or old Adam's seed.
Thus gentle Lamia judg'd, and judg'd aright,
That Lycius could not love in half a fright,
So threw the goddess off, and won his heart
More pleasantly by playing woman's part,
With no more awe than what her beauty gave,
That, while it smote, still guaranteed to save.
Lycius to all made eloquent reply,   340
Marrying to every word a twinborn sigh;
And last, pointing to Corinth, ask'd her sweet
If 'twas too far that night for her soft feet.
The way was short, for Lamia's eagerness
Made, by a spell, the triple league decrease
To a few paces; not at all surmised
By blinded Lycius, so in her comprized.
They pass'd the city gates, he knew not how,
So noiseless, and he never thought to know.

As men talk in a dream, so Corinth all,   350
Throughout her palaces imperial,
And all her populous streets and temples lewd,
Mutter'd, like tempest in the distance brew'd,
To the wide-spreaded night above her towers.
Men, women, rich and poor, in the cool hours,
Shuffled their sandals o'er the pavement white
Companion'd or alone; while many a light
Flared, here and there, from wealthy festivals,
And threw their moving shadows on the walls,
Or found them cluster'd in the corniced shade   360
Of some arch'd temple door, or dusky colonade.

Muffling his face, of greeting friends in fear,
Her fingers he press'd hard, as one came near

With curl'd gray beard, sharp eyes, and smooth bald
    crown,
Slow-stepp'd, and robed in philosophic gown:
Lycius shrank closer, as they met and past,
Into his mantle, adding wings to haste,
While hurried Lamia trembled: "Ah," said he,
"Why do you shudder, love, so ruefully?
Why does your tender palm dissolve in dew?"—   370
"I'm wearied," said fair Lamia: "tell me who
Is that old man? I cannot bring to mind
His features:—Lycius! wherefore did you blind
Yourself from his quick eyes?" Lycius replied,
"'Tis Apollonius sage, my trusty guide
And good instructor; but to-night he seems
The ghost of folly haunting my sweet dreams."

While yet he spake they had arrived before
A pillar'd porch, with lofty portal door,
Where hung a silver lamp, whose phosphor glow   380
Reflected in the slabbed steps below,
Mild as a star in water; for so new,
And so unsullied was the marble's hue,
So through the crystal polish, liquid fine,
Ran the dark veins, that none but feet divine
Could e'er have touch'd there. Sounds Æolian
Breath'd from the hinges, as the ample span
Of the wide doors disclos'd a place unknown
Some time to any, but those two alone,
And a few Persian mutes, who that same year   390
Were seen about the markets: none knew where
They could inhabit; the most curious
Were foil'd, who watch'd to trace them to their house:
And but the flitter-winged verse must tell,
For truth's sake, what woe afterwards befel,
'Twould humour many a heart to leave them thus,
Shut from the busy world of more incredulous.

PART II

Love in a hut, with water and a crust,
Is—Love, forgive us!—cinders, ashes, dust;
Love in a palace is perhaps at last
More grievous torment than a hermit's fast:—

---

327. unperplex'd: See above, l. 192. **329. Peris:** fairy-like beings (Persian mythology). **333. Pyrrha's pebbles:** Pyrrha and Deucalion, after the flood, repopulated the earth by casting pebbles that turned into human beings. **347. comprized:** absorbed. **352. temples lewd:** The temples of Venus often provided for ritualistic prostitution and lascivious practices generally.

374. quick eyes: emphasized throughout in the character of Apollonius (cf. below, II. 157, 226, 246, 277, 289, and 299–300). **386. Æolian:** like sounds from an Aeolian harp, a stringed instrument which plays when the wind blows through it (from Aeolus, god of the winds). PART II. **1–9. Love . . . quite:** The light, half-mocking tone of these lines, especially lines 7–9, should be kept in mind in interpreting the character of Lycius.

That is a doubtful tale from faery land,
Hard for the non-elect to understand.
Had Lycius liv'd to hand his story down,
He might have given the moral a fresh frown,
Or clench'd it quite: but too short was their bliss
To breed distrust and hate, that make the soft voice
    hiss.   10
Beside, there, nightly, with terrific glare,
Love, jealous grown of so complete a pair,
Hover'd and buzz'd his wings, with fearful roar,
Above the lintel of their chamber door,
And down the passage cast a glow upon the floor.

    For all this came a ruin: side by side
They were enthroned, in the even tide,
Upon a couch, near to a curtaining
Whose airy texture, from a golden string,
Floated into the room, and let appear   20
Unveil'd the summer heaven, blue and clear,
Betwixt two marble shafts:—there they reposed,
Where use had made it sweet, with eyelids closed,
Saving a tythe which love still open kept,
That they might see each other while they almost slept;
When from the slope side of a suburb hill,
Deafening the swallow's twitter, came a thrill
Of trumpets—Lycius started—the sounds fled,
But left a thought a-buzzing in his head.
For the first time, since first he harbour'd in   30
That purple-lined palace of sweet sin,
His spirit pass'd beyond its golden bourn
Into the noisy world almost forsworn.
The lady, ever watchful, penetrant,
Saw this with pain, so arguing a want
Of something more, more than her empery
Of joys; and she began to moan and sigh
Because he mused beyond her, knowing well
That but a moment's thought is passion's passing bell.
"Why do you sigh, fair creature?" whisper'd he: 40
"Why do you think?" return'd she tenderly:
"You have deserted me;—where am I now?
Not in your heart while care weighs on your brow:
No, no, you have dismiss'd me; and I go
From your breast houseless: aye, it must be so."
He answer'd, bending to her open eyes,
Where he was mirror'd small in paradise,
"My silver planet, both of eve and morn!
Why will you plead yourself so sad forlorn,

While I am striving how to fill my heart   50
With deeper crimson, and a double smart?
How to entangle, trammel up and snare
Your soul in mine, and labyrinth you there
Like the hid scent in an unbudded rose?
Aye, a sweet kiss—you see your mighty woes.
My thoughts! shall I unveil them? Listen then!
What mortal hath a prize, that other men
May be confounded and abash'd withal,
But lets it sometimes pace abroad majestical,
And triumph, as in thee I should rejoice   60
Amid the hoarse alarm of Corinth's voice.
Let my foes choke, and my friends shout afar,
While through the thronged streets your bridal car
Wheels round its dazzling spokes."—The lady's cheek
Trembled; she nothing said, but, pale and meek,
Arose and knelt before him, wept a rain
Of sorrows at his words; at last with pain
Beseeching him, the while his hand she wrung,
To change his purpose. He thereat was stung,
Perverse, with stronger fancy to reclaim   70
Her wild and timid nature to his aim:
Beside, for all his love, in self despite,
Against his better self, he took delight
Luxurious in her sorrows, soft and new.
His passion, cruel grown, took on a hue
Fierce and sanguineous as 'twas possible
In one whose brow had no dark veins to swell.
Fine was the mitigated fury, like
Apollo's presence when in act to strike
The serpent—Ha, the serpent! certes, she   80
Was none. She burnt, she lov'd the tyranny,
And, all subdued, consented to the hour
When to the bridal he should lead his paramour.
Whispering in midnight silence, said the youth,
"Sure some sweet name thou hast, though, by my truth,
I have not ask'd it, ever thinking thee
Not mortal, but of heavenly progeny,
As still I do. Hast any mortal name,
Fit appellation for this dazzling frame?
Or friends or kinsfolk on the citied earth,   90
To share our marriage feast and nuptial mirth?"
"I have no friends," said Lamia, "no, not one;
My presence in wide Corinth hardly known:
My parents' bones are in their dusty urns
Sepulchred, where no kindled incense burns,
Seeing all their luckless race are dead, save me,
And I neglect the holy rite for thee.
Even as you list invite your many guests;
But if, as now it seems, your vision rests

---

36. **empery:** empire. **39. passing bell:** signifying death. **48. eve and morn:** Venus, who is both the morning and evening star.

With any pleasure on me, do not bid　　　　100
Old Apollonius—from him keep me hid."
Lycius, perplex'd at words so blind and blank,
Made close inquiry; from whose touch she shrank,
Feigning a sleep; and he to the dull shade
Of deep sleep in a moment was betray'd.

It was the custom then to bring away
The bride from home at blushing shut of day,
Veil'd, in a chariot, heralded along
By strewn flowers, torches, and a marriage song,
With other pageants: but this fair unknown　　110
Had not a friend. So being left alone,
(Lycius was gone to summon all his kin)
And knowing surely she could never win
His foolish heart from its mad pompousness,
She set herself, high-thoughted, how to dress
The misery in fit magnificence.
She did so, but 'tis doubtful how and whence
Came, and who were her subtle servitors.
About the halls, and to and from the doors,
There was a noise of wings till in short space　　120
The glowing banquet-room shone with wide-arched
　　grace.
A haunting music, sole perhaps and lone
Supportress of the faery-roof, made moan
Throughout, as fearful the whole charm might fade.
Fresh carved cedar, mimicking a glade
Of palm and plantain, met from either side,
High in the midst, in honour of the bride:
Two palms and then two plantains, and so on,
From either side their stems branch'd one to one
All down the aisled place; and beneath all　　130
There ran a stream of lamps straight on from wall to
　　wall.
So canopied, lay an untasted feast
Teeming with odours. Lamia, regal drest,
Silently paced about, and as she went,
In pale contented sort of discontent,
Mission'd her viewless servants to enrich
The fretted splendour of each nook and niche.
Between the tree-stems, marbled plain at first,
Came jasper pannels; then anon, there burst
Forth creeping imagery of slighter trees,　　140
And with the larger wove in small intricacies.
Approving all, she faded at self-will,
And shut the chamber up, close, hush'd and still,
Complete and ready for the revels rude,
When dreadful guests would come to spoil her solitude.

137. **fretted:** with fretwork.

The day appear'd, and all the gossip rout.
O senseless Lycius! Madman! wherefore flout
The silent-blessing fate, warm cloister'd hours,
And show to common eyes these secret bowers?
The herd approach'd; each guest, with busy brain,
Arriving at the portal, gaz'd amain,　　151
And enter'd marveling: for they knew the street,
Remember'd it from childhood all complete
Without a gap, yet ne'er before had seen
That royal porch, that high-built fair demesne;
So in they hurried all, maz'd, curious and keen:
Save one, who look'd thereon with eye severe,
And with calm-planted steps walk'd in austere;
'Twas Apollonius: something too he laugh'd,
As though some knotty problem, that had daft　　160
His patient thought, had now begun to thaw,
And solve and melt:—'twas just as he foresaw.

He met within the murmurous vestibule
His young disciple. "'Tis no common rule,
Lycius," said he, "for uninvited guest
To force himself upon you, and infest
With an unbidden presence the bright throng
Of younger friends; yet must I do this wrong,
And you forgive me." Lycius blush'd, and led
The old man through the inner doors broad-spread;
With reconciling words and courteous mien　　171
Turning into sweet milk the sophist's spleen.

Of wealthy lustre was the banquet-room,
Fill'd with pervading brilliance and perfume:
Before each lucid pannel fuming stood
A censer fed with myrrh and spiced wood,
Each by a sacred tripod held aloft,
Whose slender feet wide-swerv'd upon the soft
Wool-woofed carpets: fifty wreaths of smoke
From fifty censers their light voyage took　　180
To the high roof, still mimick'd as they rose
Along the mirror'd walls by twin-clouds odorous.
Twelve sphered tables, by silk seats insphered,
High as the level of a man's breast rear'd
On libbard's paws, upheld the heavy gold
Of cups and goblets, and the store thrice told
Of Ceres' horn, and, in huge vessels, wine
Come from the gloomy tun with merry shine.
Thus loaded with a feast the tables stood,
Each shrining in the midst the image of a God.　　190
　　When in an antichamber every guest
Had felt the cold full sponge to pleasure press'd,

160. **daft:** puzzled. 185. **libbard:** leopard. 187. **Ceres'
horn:** the horn of plenty.

By minist'ring slaves, upon his hands and feet,
And fragrant oils with ceremony meet
Pour'd on his hair, they all mov'd to the feast
In white robes, and themselves in order placed
Around the silken couches, wondering
Whence all this mighty cost and blaze of wealth could
    spring.

Soft went the music the soft air along,
While fluent Greek a vowel'd undersong   200
Kept up among the guests, discoursing low
At first, for scarcely was the wine at flow;
But when the happy vintage touch'd their brains,
Louder they talk, and louder come the strains
Of powerful instruments:—the gorgeous dyes,
The space, the splendour of the draperies,
The roof of awful richness, nectarous cheer,
Beautiful slaves, and Lamia's self, appear,
Now, when the wine has done its rosy deed,
And every soul from human trammels freed,   210
No more so strange; for merry wine, sweet wine,
Will make Elysian shades not too fair, too divine.
Soon was God Bacchus at meridian height;
Flush'd were their cheeks, and bright eyes double
    bright:
Garlands of every green, and every scent
From vales deflower'd, or forest-trees branch-rent,
In baskets of bright osier'd gold were brought
High as the handles heap'd, to suit the thought
Of every guest; that each, as he did please,
Might fancy-fit his brows, silk-pillow'd at his ease.

What wreath for Lamia? What for Lycius?   221
What for the sage, old Apollonius?
Upon her aching forehead be there hung
The leaves of willow and of adder's tongue;
And for the youth, quick, let us strip for him
The thyrsus, that his watching eyes may swim
Into forgetfulness; and, for the sage,
Let spear-grass and the spiteful thistle wage
War on his temples. Do not all charms fly
At the mere touch of cold philosophy?   230

224. adder's tongue: a fern with tips shaded like adders'
tongues. 226. thyrsus: the vine-leaved staff of Bacchus.
229–37. Do . . . rainbow: See the account of Haydon's
"immortal dinner" (from Haydon's *Autobiography*, above),
in which Keats and Lamb agree that Newton's work on
optics had destroyed the rainbow (by reducing it to prismatic colors) and, while drinking to Newton's health, also
drink confusion to mathematics. The passage above has
something of the same playfulness.

There was an awful rainbow once in heaven:
We know her woof, her texture; she is given
In the dull catalogue of common things.
Philosophy will clip an Angel's wings,
Conquer all mysteries by rule and line,
Empty the haunted air, and gnomed mine—
Unweave a rainbow, as it erewhile made
The tender-person'd Lamia melt into a shade.

By her glad Lycius sitting, in chief place,
Scarce saw in all the room another face,   240
Till, checking his love trance, a cup he took
Full brimm'd, and opposite sent forth a look
'Cross the broad table, to beseech a glance
From his old teacher's wrinkled countenance,
And pledge him. The bald-head philosopher
Had fix'd his eye, without a twinkle or stir
Full on the alarmed beauty of the bride,
Brow-beating her fair form, and troubling her sweet
    pride.
Lycius then press'd her hand, with devout touch,
As pale it lay upon the rosy couch: 250
'Twas icy, and the cold ran through his veins;
Then sudden it grew hot, and all the pains
Of an unnatural heat shot to his heart.
"Lamia, what means this? Wherefore dost thou start?
Know'st thou that man?" Poor Lamia answer'd not.
He gaz'd into her eyes, and not a jot
Own'd they the lovelorn piteous appeal:
More, more he gaz'd: his human senses reel:
Some hungry spell that loveliness absorbs;
There was no recognition in those orbs.   260
"Lamia!" he cried—and no soft-toned reply.
The many heard, and the loud revelry
Grew hush; the stately music no more breathes;
The myrtle sicken'd in a thousand wreaths.
By faint degrees, voice, lute, and pleasure ceased;
A deadly silence step by step increased,
Until it seem'd a horrid presence there,
And not a man but felt the terror in his hair.
"Lamia!" he shriek'd; and nothing but the shriek
With its sad echo did the silence break.   270
"Begone, foul dream!" he cried, gazing again
In the bride's face, where now no azure vein
Wander'd on fair-spaced temples; no soft bloom
Misted the cheek; no passion to illume
The deep-recessed vision:—all was blight;
Lamia, no longer fair, there sat a deadly white.

231. awful: awe-inspiring. 264. myrtle: sacred to Venus
(and hence love).

"Shut, shut those juggling eyes, thou ruthless man!
Turn them aside, wretch! or the righteous ban
Of all the Gods, whose dreadful images
Here represent their shadowy presences,   280
May pierce them on the sudden with the thorn
Of painful blindness; leaving thee forlorn,
In trembling dotage to the feeblest fright
Of conscience, for their long offended might,
For all thine impious proud-heart sophistries,
Unlawful magic, and enticing lies.
Corinthians! look upon that grey-beard wretch!
Mark how, possess'd, his lashless eyelids stretch
Around his demon eyes! Corinthians, see!
My sweet bride withers at their potency."   290
"Fool!" said the sophist, in an under-tone
Gruff with contempt; which a death-nighing moan
From Lycius answer'd, as heart-struck and lost,
He sank supine beside the aching ghost.
"Fool! Fool!" repeated he, while his eyes still
Relented not, nor mov'd; "from every ill
Of life have I preserv'd thee to this day,
And shall I see thee made a serpent's prey?"
Then Lamia breath'd death breath; the sophist's eye,
Like a sharp spear, went through her utterly,   300
Keen, cruel, perceant, stinging: she, as well
As her weak hand could any meaning tell,
Motion'd him to be silent; vainly so,
He look'd and look'd again a level—No!
"A serpent!" echoed he; no sooner said,
Than with a frightful scream she vanished:
And Lycius' arms were empty of delight,
As were his limbs of life, from that same night.
On the high couch he lay!—his friends came round—
Supported him—no pulse, or breath they found,   310
And, in its marriage robe, the heavy body wound.

1819 (1820)

THE FALL OF HYPERION

A DREAM

The Fall of Hyperion was written principally in late July,
August, and early September, 1819. The opening of the

---

301. perceant: piercing.

---

new version (I. 1–293) turns frankly and powerfully on
the nature and uses of the imagination (indeed, as Bate
says, on the nature and uses of the self) from subjective
dream to objective and sympathetic comprehension of
reality and, through this, to a transcendence of self.
Involved in it is the debate on the function and value of
poetry: the poet is forced to justify his existence before
the figure of Moneta (the admonisher). Caught up by
this theme, Keats appears to have been evolving a very
different poem from the first *Hyperion*, and it is doubtful
whether the two could have been combined. True
enough, starting with I. 294, Keats began to take over
passages (with some changes) from the earlier version;
but when he did this (late Oct. and Nov., 1819) his fatal
illness was well under way, and he had little energy for
anything. What the poem would have become, had he
lived, is endlessly debatable. It would probably have
been left as a fragment and been regarded by him as an
exercise while preparing for what he called his "greatest
ambition"—"the writing of a few fine Plays" (see his
letter to Taylor, below, November 17, 1819, p. 1233). It
should be noted that in *The Fall of Hyperion* Keats was
developing still another style. A more relaxed, idiomatic
blank verse replaces the taut, massy splendor of the more
Miltonic first version. A Vergilian mellowness inter-
plays with stark statement, with sharply concrete
imagery, and with a symbolism that occasionally anticipates some twentieth-century poetry. With such a style it
was almost impossible to coalesce that of the first
Hyperion. So Keats himself implied when he wrote to
J. H. Reynolds on September 21, 1819: "I have given up
Hyperion—there were too many Miltonic inversions in
it—Miltonic verse can not be written but in an artful
or rather artist's humour. I wish to give myself up to
other sensations."

CANTO I

Fanatics have their dreams, wherewith they weave
A paradise for a sect; the savage too
From forth the loftiest fashion of his sleep
Guesses at Heaven; pity these have not
Trac'd upon vellum or wild Indian leaf
The shadows of melodious utterance.
But bare of laurel they live, dream, and die;
For Poesy alone can tell her dreams,
With the fine spell of words alone can save
Imagination from the sable charm   10
And dumb enchantment. Who alive can say,
"Thou art no Poet—may'st not tell thy dreams?"
Since every man whose soul is not a clod
Hath visions, and would speak, if he had loved,
And been well nurtured in his mother tongue.

Whether the dream now purpos'd to rehearse
Be poet's or fanatic's will be known
When this warm scribe my hand is in the grave.

Methought I stood where trees of every clime,
Palm, myrtle, oak, and sycamore, and beech,   20
With plantain, and spice-blossoms, made a screen;
In neighbourhood of fountains (by the noise
Soft-showering in my ears), and, (by the touch
Of scent,) not far from roses. Turning round
I saw an arbour with a drooping roof
Of trellis vines, and bells, and larger blooms,
Like floral censers, swinging light in air;
Before its wreathed doorway, on a mound
Of moss, was spread a feast of summer fruits,
Which, nearer seen, seem'd refuse of a meal   30
By angel tasted or our Mother Eve;
For empty shells were scattered on the grass,
And grape-stalks but half bare, and remnants more,
Sweet-smelling, whose pure kinds I could not know.
Still was more plenty than the fabled horn
Thrice emptied could pour forth, at banqueting
For Proserpine return'd to her own fields,
Where the white heifers low. And appetite
More yearning than on Earth I ever felt
Growing within, I ate deliciously;   40
And, after not long, thirsted, for thereby
Stood a cool vessel of transparent juice
Sipp'd by the wander'd bee, the which I took,
And, pledging all the mortals of the world,
And all the dead whose names are in our lips,
Drank. That full draught is parent of my theme.
No Asian poppy nor elixir fine
Of the soon-fading jealous Caliphat;
No poison gender'd in close monkish cell,
To thin the scarlet conclave of old men,   50
Could so have rapt unwilling life away.
Among the fragrant husks and berries crush'd,

Upon the grass I struggled hard against
The domineering potion; but in vain:
The cloudy swoon came on, and down I sank,
Like a Silenus on an antique vase.
How long I slumber'd 'tis a chance to guess.
When sense of life return'd, I started up
As if with wings; but the fair trees were gone,
The mossy mound and arbour were no more: 60
I look'd around upon the carved sides
Of an old sanctuary with roof august,
Builded so high, it seem'd that filmed clouds
Might spread beneath, as o'er the stars of heaven;
So old the place was, I remember'd none
The like upon the Earth: what I had seen
Of grey cathedrals, buttress'd walls, rent towers,
The superannuations of sunk realms,
Or Nature's rocks toil'd hard in waves and winds,
Seem'd but the faulture of decrepit things   70
To that eternal domed Monument.—
Upon the marble at my feet there lay
Store of strange vessels and large draperies,
Which needs had been of dyed asbestos wove,
Or in that place the moth could not corrupt,
So white the linen, so, in some, distinct
Ran imageries from a sombre loom.
All in a mingled heap confus'd there lay
Robes, golden tongs, censer and chafing-dish,
Girdles, and chains, and holy jewelries.   80

Turning from these with awe, once more I rais'd
My eyes to fathom the space every way;
The embossed roof, the silent massy range
Of columns north and south, ending in mist
Of nothing, then to eastward, where black gates
Were shut against the sunrise evermore.—
Then to the west I look'd, and saw far off
An image, huge of feature as a cloud,
At level of whose feet an altar slept,
To be approach'd on either side by steps,   90
And marble balustrade, and patient travail
To count with toil the innumerable degrees.
Towards the altar sober-paced I went,
Repressing haste, as too unholy there;
And, coming nearer, saw beside the shrine
One minist'ring; and there arose a flame.—

THE FALL OF HYPERION. CANTO I. **16. dream:** Though in part an attack on the use of the poetic imagination for dream or escape, the poem is in the form of a dream-vision. **19. every clime:** Cf. "Fancy," above, ll. 31–39. **26. trellis . . . blooms:** Cf. "Ode to Psyche," ll. 60–63. **35. fabled horn:** the horn of plenty; Proserpine's annual return from Hades marks the coming of spring. **48. soon-fading . . . Caliphat:** The caliphs were regarded as both the spiritual and temporal successors of Mohammed; the title was used in various parts of the Mohammedan world. In their rivalry they often plotted against each other, or they were slain by ambitious subordinates. Hence their reigns were often short ("soon-fading"). **50. scarlet conclave:** the College of Cardinals.

56. Silenus: See *Lamia*, above, I. 103, and note. **75. moth . . . corrupt:** Matthew 6 : 19. **96. One minist'ring:** the Roman Moneta, substituted for the Greek goddess Mnemosyne in the first *Hyperion*, probably because (as the name implies) Keats wishes her to serve as an admonisher.

When in mid-May the sickening East wind
Shifts sudden to the south, the small warm rain
Melts out the frozen incense from all flowers,
And fills the air with so much pleasant health   100
That even the dying man forgets his shroud;—
Even so that lofty sacrificial fire,
Sending forth Maian incense, spread around
Forgetfulness of everything but bliss,
And clouded all the altar with soft smoke;
From whose white fragrant curtains thus I heard
Language pronounc'd: "If thou canst not ascend
These steps, die on that marble where thou art.
Thy flesh, near cousin to the common dust,
Will parch for lack of nutriment—thy bones   110
Will wither in few years, and vanish so
That not the quickest eye could find a grain
Of what thou now art on that pavement cold.
The sands of thy short life are spent this hour,
And no hand in the universe can turn
Thy hourglass, if these gummed leaves be burnt
Ere thou canst mount up these immortal steps."
I heard, I look'd: two senses both at once,
So fine, so subtle, felt the tyranny
Of that fierce threat and the hard task proposed.   120
Prodigious seem'd the toil; the leaves were yet
Burning—when suddenly a palsied chill
Struck from the paved level up my limbs,
And was ascending quick to put cold grasp
Upon those streams that pulse beside the throat:
I shriek'd, and the sharp anguish of my shriek
Stung my own ears—I strove hard to escape
The numbness; strove to gain the lowest step.
Slow, heavy, deadly was my pace: the cold
Grew stifling, suffocating, at the heart;   130
And when I clasp'd my hands I felt them not.
One minute before death, my iced foot touch'd
The lowest stair; and as it touch'd, life seem'd
To pour in at the toes: I mounted up,
As once fair angels on a ladder flew
From the green turf to Heaven—"Holy Power,"
Cried I, approaching near the horned shrine,
"What am I that should so be saved from death?

What am I that another death come not
To choke my utterance, sacrilegious, here?"   140
Then said the veiled shadow—"Thou has felt
What 'tis to die and live again before
Thy fated hour, that thou hadst power to do so
Is thy own safety; thou hast dated on
Thy doom."—"High Prophetess," said I, "purge off,
Benign, if so it please thee, my mind's film."—
"None can usurp this height," return'd that shade,
"But those to whom the miseries of the world
Are misery, and will not let them rest.
All else who find a haven in the world,   150
Where they may thoughtless sleep away their days,
If by a chance into this fane they come,
Rot on the pavement where thou rottedst half."—
"Are there not thousands in the world," said I,
Encourag'd by the sooth voice of the shade,
"Who love their fellows even to the death,
Who feel the giant agony of the world,
And more, like slaves to poor humanity,
Labour for mortal good? I sure should see
Other men here; but I am here alone."   160
"Those whom thou spak'st of are no vision'ries,"
Rejoin'd that voice—"They are no dreamers weak,
They seek no wonder but the human face;
No music but a happy-noted voice—
They come not here, they have no thought to come—
And thou art here, for thou art less than they—
What benefit canst thou, or all thy tribe,
To the great world? Thou art a dreaming thing,
A fever of thyself—think of the Earth;
What bliss even in hope is there for thee?   170
What haven? every creature hath its home;
Every sole man hath days of joy and pain,
Whether his labours be sublime or low—
The pain alone; the joy alone; distinct:

142. die . . . again: Cf. *Hyperion*, III. 130 ("Die into life").
147–81. "None . . . knees": a central passage that might be
paraphrased thus: No one even reaches this point except
those endowed with sympathetic awareness of others and of
the miseries of life (the mere "dreamer," who uses the world
as a comfortable "haven" in which to sleep, would rot
entirely on the pavement of trial where the poet half-rotted
before mounting the steps). The poet then asks, "Where
are the thousands who are laboring for humanity?" They
are entirely different: the good they do is direct and concrete;
they accept the actual; they are "no visionaries." The poet,
in short, is in-between the mere "dreamer" and the direct
humanitarian. (His imagination can be used—and is used—
in radically diverse ways. His struggle must be to turn it
from dream or escape, to which it can so easily incline, to the
direct comprehension of life.)

103. Maian: Maia was one of the daughters (the Pleiades) of
Atlas (cf. "Ode to Maia," above). **108. These steps:** suggested
by Dante (*Purgatorio*, IV, IX, XII–XIII), the steps, as Douglas
Bush says, "stand for the climactic stage in the ascent from
the realm of sense to that of 'truth,' tragic understanding, akin
to the changes wrought in Endymion and Apollo." (*John
Keats: Selected Poems and Letters*, ed. Douglas Bush.) **135.
ladder:** the vision of Jacob in Genesis 28:12.

Only the dreamer venoms all his days,
Bearing more woe than all his sins deserve.
Therefore, that happiness be somewhat shar'd,
Such things as thou art are admitted oft
Into like gardens thou didst pass erewhile,
And suffer'd in these temples: for that cause   180
Thou standest safe beneath this statue's knees."
"That I am favour'd for unworthiness,
By such propitious parley medicin'd
In sickness not ignoble, I rejoice,
Aye, and could weep for love of such award."
So answer'd I, continuing, "If it please,
Majestic shadow, tell me: sure not all
Those melodies sung into the World's ear
Are useless: sure a poet is a sage;
A humanist, physician to all men.   190
That I am none I feel, as vultures feel
They are no birds when eagles are abroad.
What am I then: Thou spakest of my tribe:
What tribe?" The tall shade veil'd in drooping white
Then spake, so much more earnest, that the breath
Moved the thin linen folds that drooping hung
About a golden censer from the hand
Pendent—"Art thou not of the dreamer tribe?
The poet and the dreamer are distinct,
Diverse, sheer opposite, antipodes.   200
The one pours out a balm upon the World,
The other vexes it." Then shouted I
Spite of myself, and with a Pythia's spleen,
"Apollo! faded! O far flown Apollo!
Where is thy misty pestilence to creep
Into the dwellings, through the door crannies
Of all mock lyrists, large self worshipers
And careless Hectorers in proud bad verse.
Though I breathe death with them it will be life
To see them sprawl before me into graves.   210
Majestic shadow, tell me where I am,
Whose altar this; for whom this incense curls;

187–210. Majestic . . . graves: Keats's friend, Richard
Woodhouse, left a remark on one of the transcripts saying
that Keats intended to omit these lines. But since the poem
is only a fragment and anything relating to his state of mind
during these crucial last weeks of his writing is of interest,
the lines are customarily printed with the rest of the poem.
203. Pythia: priestess of the oracle of Apollo at Delphi. 205.
pestilence: Apollo was the bringer of pestilence as well as
the god of music and of healing. 207–08. mock . . . verse:
In these bitter lines Keats could have been referring to several
poets of the time, particularly Hunt ("mock lyrists"), Wordsworth ("large self-worshippers"), and Byron ("Hectorers in
proud bad verse").

What image this whose face I cannot see,
For the broad marble knees; and who thou art,
Of accent feminine so courteous?"

Then the tall shade, in drooping linens veil'd,
Spoke out, so much more earnest, that her breath
Stirr'd the thin folds of gauze that drooping hung
About a golden censer from her hand
Pendent; and by her voice I knew she shed   220
Long-treasured tears. "This temple, sad and lone,
Is all spar'd from the thunder of a war
Foughten long since by giant hierarchy
Against rebellion: this old image here,
Whose carved features wrinkled as he fell,
Is Saturn's; I Moneta, left supreme
Sole Priestess of this desolation,"—
I had no words to answer, for my tongue,
Useless, could find about its roofed home
No syllable of a fit majesty   230
To make rejoinder to Moneta's mourn.
There was a silence, while the altar's blaze
Was fainting for sweet food: I look'd thereon,
And on the paved floor, where nigh were piled
Faggots of cinnamon, and many heaps
Of other crisped spice-wood—then again
I look'd upon the altar, and its horns
Whiten'd with ashes, and its lang'rous flame,
And then upon the offerings again;
And so by turns—till sad Moneta cried,   240
"The sacrifice is done, but not the less
Will I be kind to thee for thy good will.
My power, which to me is still a curse,
Shall be to thee a wonder; for the scenes
Still swooning vivid through my globed brain,
With an electral changing misery,
Thou shalt with those dull mortal eyes behold,
Free from all pain, if wonder pain thee not."
As near as an immortal's sphered words
Could to a mother's soften, were these last: 250
And yet I had a terror of her robes,
And chiefly of the veils, that from her brow
Hung pale, and curtain'd her in mysteries,
That made my heart too small to hold its blood.
This saw that Goddess, and with sacred hand
Parted the veils. Then saw I a wan face,
Not pin'd by human sorrows, but bright-blanch'd
By an immortal sickness which kills not;
It works a constant change, which happy death
Can put no end to; deathwards progressing   260
To no death was that visage; it had past

The lilly and the snow; and beyond these
I must not think now, though I saw that face—
But for her eyes I should have fled away.
They held me back, with a benignant light,
Soft mitigated by divinest lids
Half-closed, and visionless entire they seem'd
Of all external things;—they saw me not,
But in blank splendor, beam'd like the mild moon,
Who comforts those she sees not, who knows not 270
What eyes are upward cast. As I had found
A grain of gold upon a mountain side,
And twing'd with avarice strain'd out my eyes
To search its sullen entrails rich with ore,
So at the view of sad Moneta's brow,
I ach'd to see what things the hollow brain
Behind enwombed: what high tragedy
In the dark secret chamber of her skull
Was acting, that could give so dread a stress
To her cold lips, and fill with such a light   280
Her planetary eyes; and touch her voice
With such a sorrow—"Shade of Memory!"—
Cried I, with act adorant at her feet,
"By all the gloom hung round thy fallen house,
By this last temple, by the golden age,
By great Apollo, thy dear Foster Child,
And by thyself, forlorn divinity,
The pale Omega of a withered race,
Let me behold, according as thou saidst,
What in thy brain so ferments to and fro!"   290
No sooner had this conjuration pass'd
My devout lips, than side by side we stood
(Like a stunt bramble by a solemn pine)
Deep in the shady sadness of a vale,
Far sunken from the healthy breath of morn,
Far from the fiery noon and eve's one star.
Onward I look'd beneath the gloomy boughs,
And saw, what first I thought an image huge,
Like to the image pedestal'd so high
In Saturn's temple. Then Moneta's voice   300
Came brief upon mine ear—"So Saturn sat
When he had lost his Realms—" whereon there grew
A power within me of enormous ken
To see as a god sees, and take the depth
Of things as nimbly as the outward eye
Can size and shape pervade. The lofty theme
At those few words hung vast before my mind,

With half-unravel'd web. I set myself
Upon an eagle's watch, that I might see,
And seeing ne'er forget. No stir of life   310
Was in this shrouded vale, not so much air
As in the zoning of a summer's day
Robs not one light seed from the feather'd grass,
But where the dead leaf fell there did it rest:
A stream went voiceless by, still deaden'd more
By reason of the fallen divinity
Spreading more shade; the Naiad 'mid her reeds
Prest her cold finger closer to her lips.

Along the margin-sand large footmarks went
No farther than to where old Saturn's feet   320
Had rested, and there slept, how long a sleep!
Degraded, cold, upon the sodden ground
His old right hand lay nerveless, listless, dead,
Unsceptred; and his realmless eyes were clos'd,
While his bow'd head seem'd listening to the Earth,
His ancient mother, for some comfort yet.

It seem'd no force could wake him from his place;
But there came one who, with a kindred hand
Touch'd his wide shoulders after bending low
With reverence, though to one who knew it not. 330
Then came the griev'd voice of Mnemosyne,
And griev'd I hearken'd. "That divinity
Whom thou saw'st step from yon forlornest wood,
And with slow pace approach our fallen King,
Is Thea, softest-natur'd of our Brood."
I mark'd the Goddess in fair statuary
Surpassing wan Moneta by the head,
And in her sorrow nearer woman's tears.
There was a listening fear in her regard,
As if calamity had but begun;   340
As if the vanward clouds of evil days
Had spent their malice, and the sullen rear
Was with its stored thunder labouring up.
One hand she press'd upon that aching spot
Where beats the human heart, as if just there,
Though an immortal, she felt cruel pain;
The other upon Saturn's bended neck
She laid, and to the level of his hollow ear
Leaning with parted lips, some words she spake
In solemn tenor and deep organ tune;   350
Some mourning words, which in our feeble tongue

---

288. **Omega:** last (after the final letter of the Greek alphabet).
294. **Deep in:** At this point Keats goes back to the first *Hyperion.*

312. **zoning:** course. 331. **Mnemosyne:** Keats means "Moneta" but slips back to the Greek name for her used in the first *Hyperion.*

Would come in this-like accenting; how frail
To that large utterance of the early Gods!

"Saturn! look up—and for what, poor lost King?
I have no comfort for thee; no not one;
I cannot cry, wherefore thus sleepest thou?
For Heaven is parted from thee, and the Earth
Knows thee not, so afflicted, for a God;
And Ocean too, with all its solemn noise,
Has from thy sceptre pass'd, and all the air   360
Is emptied of thine hoary majesty:
Thy thunder, captious at the new command,
Rumbles reluctant o'er our fallen house;
And thy sharp lightning, in unpracticed hands,
Scorches and burns our once serene domain.
With such remorseless speed still come new woes,
That unbelief has not a space to breathe.
Saturn! sleep on:—Me thoughtless, why should I
Thus violate thy slumbrous solitude?
Why should I ope thy melancholy eyes?   370
Saturn, sleep on, while at thy feet I weep."

As when upon a tranced summer-night
Forests, branch-charmed by the earnest stars,
Dream, and so dream all night without a noise,
Save from one gradual solitary gust,
Swelling upon the silence; dying off;
As if the ebbing air had but one wave;
So came these words, and went; the while in tears
She prest her fair large forehead to the earth,
Just where her fallen hair might spread in curls,   380
A soft and silken mat for Saturn's feet.
Long, long those two were postured motionless,
Like sculpture builded-up upon the grave
Of their own power. A long awful time
I look'd upon them: still they were the same;
The frozen God still bending to the earth,
And the sad Goddess weeping at his feet,
Moneta silent. Without stay or prop,
But my own weak mortality, I bore
The load of this eternal quietude,   390
The unchanging gloom, and the three fixed shapes
Ponderous upon my senses, a whole moon.
For by my burning brain I measured sure
Her silver seasons shedded on the night,
And ever day by day methought I grew
More gaunt and ghostly.—Oftentimes I pray'd
Intense, that Death would take me from the Vale
And all its burthens—gasping with despair
Of change, hour after hour I curs'd myself;
Until old Saturn rais'd his faded eyes,   400

And look'd around and saw his kingdom gone,
And all the gloom and sorrow of the place,
And that fair kneeling Goddess at his feet.
As the moist scent of flowers, and grass, and leaves,
Fills forest dells with a pervading air,
Known to the woodland nostril, so the words
Of Saturn fill'd the mossy glooms around,
Even to the hollows of time-eaten oaks,
And to the windings of the foxes' hole,
With sad low tones, while thus he spake, and sent   410
Strange musings to the solitary Pan.
"Moan, brethren, moan; for we are swallow'd up
And buried from all Godlike exercise
Of influence benign on planets pale,
And peaceful sway above man's harvesting,
And all those acts which Deity supreme
Doth ease its heart of love in. Moan and wail,
Moan, brethren, moan; for lo, the rebel spheres
Spin round, the stars their ancient courses keep,
Clouds still with shadowy moisture haunt the earth, 420
Still suck their fill of light from sun and moon;
Still buds the tree, and still the sea-shores murmur;
There is no death in all the Universe,
No smell of death—there shall be death—Moan,
    moan,
Moan, Cybele, moan; for thy pernicious Babes
Have changed a god into a shaking Palsy.
Moan, brethren, moan, for I have no strength left,
Weak as the reed—weak—feeble as my voice—
O, O, the pain, the pain of feebleness.
Moan, moan, for still I thaw—or give me help;   430
Throw down those imps, and give me victory.
Let me hear other groans, and trumpets blown
Of triumph calm, and hymns of festival,
From the gold peaks of Heaven's high-piled clouds;
Voices of soft proclaim, and silver stir
Of strings in hollow shells; and let there be
Beautiful things made new for the surprise
Of the sky-children." So he feebly ceas'd,
With such a poor and sickly sounding pause,
Methought I heard some old man of the earth   440
Bewailing earthly loss; nor could my eyes
And ears act with that pleasant unison of sense
Which marries sweet sound with the grace of form,
And dolorous accent from a tragic harp
With large-limb'd visions.—More I scrutinized:
Still fix'd he sat beneath the sable trees,
Whose arms spread straggling in wild serpent forms,

425. **Cybele:** See *Hyperion*, II. 4, and note.

With leaves all hush'd; his awful presence there
(Now all was silent) gave a deadly lie
To what I erewhile heard—only his lips   450
Trembled amid the white curls of his beard.
They told the truth, though, round, the snowy locks
Hung nobly, as upon the face of heaven
A mid-day fleece of clouds. Thea arose,
And stretched her white arm through the hollow
     dark,
Pointing some whither: whereat he too rose
Like a vast giant, seen by men at sea
To grow pale from the waves at dull midnight.
They melted from my sight into the woods;
Ere I could turn, Moneta cried, "These twain   460
Are speeding to the families of grief,
Where roof'd in by black rocks they waste, in pain
And darkness, for no hope."—And she spake on,
As ye may read who can unwearied pass
Onward from the Antichamber of this dream,
Where even at the open doors awhile
I must delay, and glean my memory
Of her high phrase:—perhaps no further dare.

CANTO II

"Mortal, that thou may'st understand aright,
I humanize my sayings to thine ear,
Making comparisons of earthly things;
Or thou might'st better listen to the wind,
Whose language is to thee a barren noise,
Though it blows legend-laden thro' the trees.—
In melancholy realms big tears are shed,
More sorrow like to this, and such like woe,
Too huge for mortal tongue, or pen of scribe.
The Titans fierce, self hid or prison bound,   10
Groan for the old allegiance once more,
Listening in their doom for Saturn's voice.
But one of our whole eagle-brood still keeps
His sov'reignty, and rule, and majesty;
Blazing Hyperion on his orbed fire
Still sits, still snuffs the incense teeming up
From Man to the Sun's God: yet unsecure.
For as upon the earth dire prodigies
Fright and perplex, so also shudders he:
Nor at dog's howl or gloom-bird's Even screech,   20
Or the familiar visitings of one
Upon the first toll of his passing bell:
But horrors, portioned to a giant nerve,
Make great Hyperion ache. His palace bright,

Bastion'd with pyramids of glowing gold,
And touch'd with shade of bronzed obelisks,
Glares a blood-red thro' all the thousand courts,
Arches, and domes, and fiery galleries:
And all its curtains of Aurorian clouds
Flush angerly; when he would taste the wreaths   30
Of incense breathed aloft from sacred hills,
Instead of sweets, his ample palate takes
Savour of poisonous brass and metals sick.
Wherefore when harbour'd in the sleepy West,
After the full completion of fair day,
For rest divine upon exalted couch
And slumber in the arms of melody,
He paces through the pleasant hours of ease
With strides colossal, on from hall to hall;
While far within each aisle and deep recess   40
His winged minions in close clusters stand
Amaz'd, and full of fear; like anxious men,
Who on a wide plain gather in sad troops,
When earthquakes jar their battlements and towers.
Even now, while Saturn, roused from icy trance,
Goes, step for step, with Thea from yon woods,
Hyperion, leaving twilight in the rear,
Is sloping to the threshold of the West.—
Thither we tend."—Now in clear light I stood,
Reliev'd from the dusk vale. Mnemosyne   50
Was sitting on a square-edg'd polish'd stone,
That in its lucid depth reflected pure
Her priestess-garments.—My quick eyes ran on
From stately nave to nave, from vault to vault,
Through bow'rs of fragrant and enwreathed light
And diamond-paved lustrous long arcades.
Anon rush'd by the bright Hyperion;
His flaming robes stream'd out beyond his heels,
And gave a roar, as if of earthly fire,
That scared away the meek ethereal hours,   60
And made their dove-wings tremble. On he flared.

1819 (1856)

TO AUTUMN

"To Autumn" was written September 19, 1819, after taking the walk near Winchester described in Keats's letter to George Keats (below, p. 1232, under Sept. 21). At the same time he wrote to Reynolds: "How beautiful the season is now—How fine the air. A temperate sharpness about it. . . . I never lik'd stubble fields so much as now—Aye, better than the chilly green of the spring. Somehow a stubble plain looks warm—in the same way that some pictures look warm—this struck me so much in my Sunday's walk that I composed upon it."

I

Season of mists and mellow fruitfulness,
   Close bosom-friend of the maturing sun;
Conspiring with him how to load and bless
   With fruit the vines that round the thatch-eves run;
To bend with apples the moss'd cottage-trees,
   And fill all fruit with ripeness to the core;
     To swell the gourd, and plump the hazel shells
With a sweet kernel; to set budding more,
   And still more, later flowers for the bees,
   Until they think warm days will never cease,   10
     For Summer has o'er-brimm'd their clammy cells.

II

Who hath not seen thee oft amid thy store?
   Sometimes whoever seeks abroad may find
Thee sitting careless on a granary floor,
   Thy hair soft-lifted by the winnowing wind;
Or on a half-reap'd furrow sound asleep,
   Drows'd with the fume of poppies, while thy hook
     Spares the next swath and all its twined flowers:
And sometimes like a gleaner thou dost keep
   Steady thy laden head across a brook;   20
   Or by a cyder-press, with patient look,
     Thou watchest the last oozings hours by hours.

III

Where are the songs of Spring? Ay, where are they?
   Think not of them, thou hast thy music too,—
While barred clouds bloom the soft-dying day,
   And touch the stubble-plains with rosy hue;

Then in a wailful choir the small gnats mourn
   Among the river sallows, borne aloft
     Or sinking as the light wind lives or dies;
And full-grown lambs loud bleat from hilly bourn; 30
   Hedge-crickets sing; and now with treble soft
   The red-breast whistles from a garden-croft;
     And gathering swallows twitter in the skies.

1819 (1820)

THE DAY IS GONE

The day is gone, and all its sweets are gone!
   Sweet voice, sweet lips, soft hand, and softer breast,
Warm breath, light whisper, tender semi-tone,
   Bright eyes, accomplish'd shape, and lang'rous waist!
Faded the flower and all its budded charms,
   Faded the sight of beauty from my eyes,
Faded the shape of beauty from my arms,
   Faded the voice, warmth, whiteness, paradise—
Vanish'd unseasonably at shut of eve,
   When the dusk holiday—or holinight   10
Of fragrant-curtain'd love begins to weave
   The woof of darkness thick, for hid delight;
But, as I've read love's missal through to-day,
He'll let me sleep, seeing I fast and pray.

1819 (1838)

I CRY YOUR MERCY

I cry your mercy—pity—love!—aye, love!
   Merciful love that tantalises not,
One-thoughted, never-wandering, guileless love,
   Unmask'd, and being seen—without a blot!
O! let me have thee whole,—all—all—be mine!
   That shape, that fairness, that sweet minor zest
Of love, your kiss,—those hands, those eyes divine,
   That warm, white, lucent, million-pleasured breast,—

TO AUTUMN. **28. sallows:** willows. **30. bourn:** region. **32. croft:** a fenced plot.

Yourself—your soul—in pity give me all,
    Withhold no atom's atom or I die,   10
Or living on perhaps, your wretched thrall,
    Forget, in the mist of idle misery,
Life's purposes,—the palate of my mind
Losing its gust, and my ambition blind!

                            1819 (1838)

BRIGHT STAR

This sonnet to Fanny Brawne has been dated variously
from the autumn of 1818 to the late autumn of 1819,
generally through speculation about verbal parallels with
the letters, the dates of snowfalls, or similarity of mood
with other poems. It is frequently assigned to April, 1819.
Other probable dates, as far as mood is concerned, are
either the summer of 1819 (though the snow, if taken
literally, presents a problem), or else the late autumn
or early winter of that year, by which time Keats was
mortally ill and beginning to dread that his career as a
poet was finished (in which case the sonnet could be
viewed as something of a companion piece to the two
others printed immediately above).

Bright star, would I were stedfast as thou art—
    Not in lone splendour hung aloft the night
And watching, with eternal lids apart,
    Like nature's patient, sleepless Eremite,
The moving waters at their priestlike task
    Of pure ablution round earth's human shores,
Or gazing on the new soft-fallen mask
    Of snow upon the mountains and the moors—
No—yet still stedfast, still unchangeable,
    Pillow'd upon my fair love's ripening breast,   10
To feel for ever its soft fall and swell,
    Awake for ever in a sweet unrest,
Still, still to hear her tender-taken breath,
And so live ever—or else swoon to death.

                            1819 (1838)

BRIGHT STAR. **3. eternal lids:** Cf. the letter to Tom Keats,
below, June 25–27, 1818: "refine one's sensual vision into a
sort of north star which can never cease to be open lidded
and stedfast over the wonders of the great Power."

THIS LIVING HAND

These are probably the last serious lines that Keats wrote
(around Nov. or Dec., 1819), found on a blank space in
the manuscript of a comic poem, *The Cap and Bells,* that
he was composing in the hope that it would earn money.
They may have been a fragment of a speech for a play
that he was planning or (more probably) have been
written with the thought of Fanny Brawne.

This living hand, now warm and capable
Of earnest grasping, would, if it were cold
And in the icy silence of the tomb,
So haunt thy days and chill thy dreaming nights
That thou wouldst wish thine own heart dry of blood
So in my veins red life might stream again,
And thou be conscience-calm'd—see here it is—
I hold it towards you.

                            1819 (1898)

LETTERS

Few writers and critics in English during the last century,
indeed few readers generally, would disagree with T. S.
Eliot's verdict that Keats's letters are "the most notable
and the most important ever written by any English
poet." Explanations for their almost universal appeal are
ready enough. To begin with, there is the range of
Keats's own nature—its humor, generosity, diversity of
interest, and sympathetic clairvoyance. These and other
qualities constantly interplay in his letters; they are
strewn with observations—on poetry, the imagination,
and indeed life generally—the "brilliance and profundity" of which, as Eliot says, are unexcelled by any
poet, even those who have lived to three times Keats's
age. The letters are altogether spontaneous and not merely
in the sense that they were written without the remotest
thought of publication: there was no self-consciousness
before any of the people to whom they were written.
(Hence the slips of spelling and punctuation in these
letters dashed off so rapidly.) The standard edition is
that of Hyder E. Rollins (2 vols., Harvard University
Press, 1958), from which the texts of the following
selections are taken with a few changes in spelling and
capitalization. The final letter is taken from *The Keats
Circle,* ed. Rollins (2 vols., Harvard, 1948).

THIS LIVING HAND. **1–3. This . . . tomb:** Cf. *Fall of Hyperion,* I. 18.

To J. H. Reynolds
Carisbrooke[1] April 17th [1817]

MY DEAR REYNOLDS,

Ever since I wrote to my Brothers from Southampton I have been in a taking, and at this moment I am about to become settled. for I have unpacked my books, put them into a snug corner—pinned up Haydon— Mary Queen [of] Scotts, and Milton with his daughters in a row. In the passage I found a head of Shakespeare[2] which I had not before seen—It is most likely the same that George spoke so well of; for I like it extremely— Well—this head I have hung over my Books, just above the three in a row . . . From want of regular rest, I have been rather *narvus*—and the passage in Lear[3]— "Do you not hear the Sea?"—has haunted me intensely.

[Here follows the sonnet "On the Sea," printed above.]

. . .

[*April 18th*] Tell George and Tom to write.—I'll tell you what—On the 23rd was Shakespeare born— Now if I should receive a Letter from you and another from my Brothers on that day 'twould be a parlous good thing—Whenever you write say a Word or two on some Passage in Shakespeare that may have come rather new to you; which must be continually happening, notwithstand[g] that we read the same Play forty times—for instance, the following, from the Tempest,[4] never struck me so forcibly as at present,

"Urchins
Shall, for that vast of Night that they may work,
All exercise on thee—"

How can I help bringing to your mind the Line—

In the dark backward and abysm of time—

I find that I cannot exist without poetry—without eternal poetry—half the day will not do—the whole of it—I began with a little, but habit has made me a Leviathan—I had become all in a Tremble from not having written any thing of late—the Sonnet[5] over leaf did me some good. I slept the better last night

for it—this Morning, however, I am nearly as bad again—Just now I opened Spenser, and the first Lines I saw were these.—

"The noble Heart that harbors vertuous thought,
And is with Child of glorious great intent,
Can never rest, until it forth have brought
Th' eternal Brood of Glory excellent—"[6]

. . .

Your sincere friend
JOHN KEATS

To B. R. Haydon
[*May 10–11, 1817*]

MY DEAR HAYDON,

. . . I must think that difficulties nerve the Spirit of a Man—they make our Prime Objects a Refuge as well as a Passion. . . . I read and write about eight hours a day. There is an old saying well begun is half done"— 't is a bad one. I would use instead—Not begun at all 'till half done" so according to that I have not begun my Poem and consequently (a priori) can say nothing about it. Thank God! I do begin arduously where I leave off, notwithstanding occasional depressions: and I hope for the support of a High Power while I clime this little eminence and especially in my Years of more momentous Labor. I remember your saying that you had notions of a good Genius presiding over you—I have of late had the same thought. for things which [I] do half at Random are afterwards confirmed by my judgment in a dozen features of Propriety—Is it too daring to Fancy Shakspeare this Presider? When in the Isle of Wight I met with a Shakspeare in the Passage of the House at which I lodged—it comes nearer to my idea of him than any I have seen—I was but there a Week yet the old Woman made me take it with me though I went off in a hurry—Do you not think this is ominous of good? . . .

[*May 11*] . . . I wrote to Hunt yesterday— scar[c]ely know what I said in it—I could not talk about Poetry in the way I should have liked for I was not in humor with either his or mine. His self delusions are very lamentable they have inticed him into a Situation which I should be less eager after than that of a galley Slave—what you observe thereon is very true must be in time. Perhaps it is a self delusion to say so—but I think I could not be deceived in the Manner that Hunt is—may

LETTERS. **1. Carisbrooke:** in the Isle of Wight, to which Keats had just gone to begin *Endymion.* **2. head of Shakespeare:** an engraving of Shakespeare that Mrs. Cook, the landlady, allowed him to take with him and that he always kept near him thereafter while he wrote. **3. passage in Lear:** *King Lear,* IV. iv. **4. Tempest:** I. ii. 326–28, and I. ii. 50. **5. Sonnet:** "On the Sea."

6. "The . . . excellent": *Faerie Queene,* I. v. i.

I die tomorrow if I am to be. There is no greater Sin after the 7 deadly than to flatter oneself into an idea of being a great Poet—or one of those beings who are privileged to wear out their Lives in the pursuit of Honor—how comfortable a feel it is that such a Crime must bring its heavy Penalty? That if one be a Self-deluder accounts will be balanced? . . . I never quite despair and I read Shakspeare—indeed I shall I think never read any other Book much—Now this might lead me into a long Confab but I desist. I am very near Agreeing with Hazlit that Shakspeare is enough for us . . .

> Your everlasting friend
> JOHN KEATS

To J. H. Reynolds
[*November 22, 1817*]

MY DEAR REYNOLDS,

. . . I like this place[7] very much—There is Hill & Dale and a little River—I went up Box hill this Evening after the Moon—you a' seen the Moon—came down—and wrote some lines. Whenever I am separated from you, and not engaged in a continued Poem—every Letter shall bring you a lyric—but I am too anxious for you to enjoy the whole, to send you a particle. One of the three Books I have with me is Shakespear's Poems: I neer found so many beauties in the sonnets—they seem to be full of fine things said unintentionally—in the intensity of working out conceits—Is this to be borne? Hark ye!

> When lofty trees I see barren of leaves
> Which erst from heat did canopy the herd,
> And Summer's green all girded up in sheaves,
> Borne on the bier with white and bristly beard.[8]

He has left nothing to say about nothing or any thing: for look at Snails, you know what he says about Snails, you know where he talks about "cockled snails"[9]—well, in one of these sonnets, he says—the chap slips into—no! I lie! this is in the Venus and Adonis:[10] the Simile brought it to my Mind.

> Audi—As the snail, whose tender horns being hit,
> Shrinks back into his shelly cave with pain,
> And there all smothered up in shade doth sit,
> Long after fearing to put forth again:

So at his bloody view her eyes are fled,
Into the deep dark Cabins of her head.

He overwhelms a genuine Lover of Poesy with all manner of abuse, talking about—

> "a poets rage
> And stretched metre of an antique song"[11]—

Which by the by will be a capital Motto for my Poem—wont it?—He speaks too of "Time's antique pen"—and "aprils first born flowers"—and "deaths eternal cold"[12] . . .

> Your affectionate friend
> JOHN KEATS

To Benjamin Bailey
November 22, 1817 [*Hampstead*]

MY DEAR BAILEY,

I will get over the first part of this (*unsaid*)[13] Letter as soon as possible for it relates to the affair of poor Crips—To a Man of your nature, such a Letter as Haydon's must have been extremely cutting—What occasions the greater part of the World's Quarrels? simply this, two Minds meet and do not understand each other time enough to p[r]aevent any shock or surprise at the conduct of either party—As soon as I had known Haydon three days I had got enough of his character not to have been surp[r]ised at such a Letter as he has hurt you with. Nor when I knew it was it a principle with me to drop his acquaintance although with you it would have been an imperious feeling. I wish you knew all that I think about Genius and the Heart—and yet I think you are thoroughly acquainted with my innermost breast in that respect or you could not have known me even thus long and still hold me worthy to be your dear friend. In passing however I must say of one thing that has pressed upon me lately and encreased my Humility and capability of sub-mission and that is this truth—Men of Genius are great as certain ethereal Chemicals operating on the Mass of neutral intellect—but they have not any individuality, any determined Character. I would call the top and head of those who have a proper self Men of Power—

But I am running my head into a Subject which I am certain I could not do justice to under five years

---

7. **place:** Keats writes from Burford Bridge, where he had gone to finish *Endymion*. 8. **When . . . beard:** Sonnet XII, ll. 5–8. 9. **"cockled snails":** *Love's Labor's Lost*, IV. iii. 338. 10. **Venus and Adonis:** ll. 1033–38.

11. **"a . . . song":** Sonnet XVII, ll. 11–12. 12. **"Time's . . . cold":** Sonnet XIX, l. 10, XXI, l. 7, and XIII, l. 12. 13. **unsaid:** a play on the legal phrase "said letter" (which would be Haydon's to Bailey, while that of Keats would be "unsaid").

s[t]udy and 3 vols octavo—and moreover long to be talking about the Imagination—so my dear Bailey do not think of this unpleasant affair if possible—do not—I defy any ha[r]m to come of it—I defy—I'll shall write to Crips this Week and reque[s]t him to tell me all his goings on from time to time by Letter wherever I may be—it will all go on well—so dont because you have suddenly discover'd a Coldness in Haydon suffer yourself to be teased. Do not my dear fellow. O I wish I was as certain of the end of all your troubles as that of your momentary start about the authenticity of the Imagination. I am certain of nothing but of the holiness of the Heart's affections and the truth of Imagination—What the imagination seizes as Beauty must be truth[14] —whether it existed before or not—for I have the same Idea of all our Passions as of Love they are all in their sublime, creative of essential Beauty—In a Word, you may know my favorite Speculation by my first Book and the little song I sent in my last—which is a representation from the fancy of the probable mode of operating in these Matters—The Imagination may be compared to Adam's dream[15]—he awoke and found it truth. I am the more zealous in this affair, because I have never yet been able to perceive how any thing can be known for truth by consequitive reasoning—and yet it must be—Can it be that even the greatest Philosopher ever ~~(when)~~ arrived at his goal without putting aside numerous objections—However it may be, O for a Life of Sensations rather than of Thoughts![16] It is 'a Vision in the form of Youth' a Shadow of reality to come—and this consideration had further conv[i]nced me for it has come as auxiliary to another favorite Speculation of mine, that we shall enjoy ourselves here after by having what we called happiness on Earth repeated in a finer tone and so repeated —And yet such a fate can only befall those who delight in sensation rather than hunger as you do after Truth—Adam's dream will do here and seems to be a conviction that Imagination and its empyreal reflection is the same as human Life and its spiritual repetition. But as I was saying—the simple imaginative Mind may have its rewards in the repeti[ti]on of its own silent Working coming continually on the spirit with a fine suddenness—to compare great things with small

—have you never by being surprised with an old Melody—in a delicious place—by a delicious voice, fe[l]t over again your very speculations and surmises at the time it first operated on your soul—do you not remember forming to yourself the singer's face more beautiful than it was possible and yet with the elevation of the Moment you did not think so—even then you were mounted on the Wings of Imagination so high—that the Prototype must be here after—that delicious face you will see—What a time! I am continually running away from the subject—sure this cannot be exactly the case with a complex Mind—one that is imaginative and at the same time careful of its fruits—who would exist partly on sensation partly on thought—to whom it is necessary that years should bring the philosophic Mind[17]—such an one I consider your's and therefore it is necessary to your eternal Happiness that you not only drink this old Wine of Heaven which I shall call the redigestion of our most ethereal Musings on Earth; but also increase in knowledge and know all things. I am glad to hear you are in a fair Way for Easter—you will soon get through your unpleasant reading and then!—but the world is full of troubles and I have not much reason to think myself pesterd with many—I think Jane or Marianne[18] has a better opinion of me than I deserve—for really and truly I do not think my Brothers illness connected with mine—you know more of the real Cause than they do—nor have I any chance of being rack'd as you have been[19]—you perhaps at one time thought there was such a thing as Worldly Happiness to be arrived at, at certain periods of time marked out—you have of necessity from your disposition been thus led away—I scarcely remember counting upon any Happiness—I look not for it if it be not in the present hour—nothing startles me beyond the Moment. The setting sun will always set me to rights —or if a Sparrow come before my Window I take part in its existence and pick about the Gravel.[20] . . .

Your affectionate friend
JOHN KEATS—

14. **Beauty . . . truth:** Cf. "Ode on a Grecian Urn," above, l. 49. 15. **Adam's dream:** See *Paradise Lost*, VIII. 452–90. 16. **Sensations . . . Thoughts:** Sensations is used with the meaning traditional in English empirical philosophy: direct experience—what is learned by means of the senses.

17. **philosophic Mind:** an echo of Wordsworth's ode, "Intimations of Immortality," l. 186. 18. **Jane or Marianne:** sisters of John Hamilton Reynolds. They have a "better opinion" of Keats than he thinks he deserves because they are afraid has he caught tuberculosis (from which his brother Tom is suffering), whereas Keats thinks he is simply being morbid in brooding over troubles and anxieties. 19. **rack'd . . . been:** Bailey was often ill. 20. **Sparrow . . . Gravel:** Cf. "Where's the Poet?" above.

To George and Tom Keats
[*December 21–27, 1817*]
Hampstead Sunday

MY DEAR BROTHERS,

I must crave your pardon for not having written ere this.*** I saw Kean[21] return to the public in Richard III, & finely he did it, & at the request of Reynolds I went to criticise his Luke in Riches—the critique is in todays champion, which I send you with the Examiner in which you will find very proper lamentation on the obsoletion of christmas Gambols & pastimes: but it was mixed up with so much egotism of that drivelling nature that pleasure is entirely lost. Hone the publisher's trial, you must find very amusing; & as Englishmen very ~~amusing~~ encouraging—his *Not Guilty* is a thing, which not to have been, would have dulled still more Liberty's Emblazoning—Lord Ellenborough has been paid in his own coin—Wooler & Hone[22] have done us an essential service—I have had two very pleasant evenings with Dilke yesterday & today; & am at this moment just come from him & feel in the humour to go on with this, began in the morning, & from which he came to fetch me. I spent Friday evening with Wells & went the next morning to see *Death on the Pale horse*. It is a wonderful picture, when West's[23] age is considered; But there is nothing to be intense upon; no women one feels mad to kiss; no face swelling into reality. the excellence of every Art is its intensity, capable of making all disagreeables evaporate, from their being in close relationship with Beauty & Truth[24]—Examine King Lear & you will find this examplified throughout; but in this picture we have unpleasantness without any momentous depth of speculation excited, in which to bury its repulsiveness—The picture is larger than Christ rejected—I dined with Haydon the sunday after you left, & had a very pleasant day, I dined too (for I have been out too much lately) with Horace Smith & met his two Brothers with Hill & Kingston & one Du Bois,[25] they only served to convince me, how superior humour is to wit in respect to enjoyment— These men say things which make one start, without making one feel, they are all alike; their manners are alike; they all know fashionables; they have a mannerism in their very eating & drinking, in their mere handling a Decanter—They talked of Kean & his low company—Would I were with that company instead of yours said I to myself! I know such like acquaintance will never do for me & yet I am going to Reynolds, on wednesday—Brown & Dilke walked with me & back from the Christmas pantomime. I had not a dispute but a disquisition with Dilke, on various subjects; several things dovetailed in my mind, & at once it struck me, what quality went to form a Man of Achievement especially in Literature & which Shakespeare possessed so enormously—I mean *Negative Capability*, that is when man is capable of being in uncertainties, Mysteries, doubts, without any irritable reaching after fact & reason—Coleridge, for instance, would let go by a fine isolated verisimilitude caught from the Penetralium of mystery, from being incapable of remaining content with half knowledge. This pursued through Volumes would perhaps take us no further than this, that with a great poet the sense of Beauty overcomes every other consideration, or rather obliterates all consideration.[26]

21. Kean: Edmund Kean, 1787–1833, a famous Shakespearean actor on whose performance Keats had just written a review in *The Champion*. **22. Wooler & Hone:** liberals just acquitted of libel; Ellenborough, the judge who tried Hone, was a noted conservative. **23. West:** Benjamin West, 1738–1820, president of the Royal Academy. **24. the excellence . . . Truth:** Bate paraphrases this passage: "If the imaginative grasp of an object is sufficiently intense, it takes so strong a hold of the mind that whatever qualities are irrelevant to its central character (the 'disagreeables') evaporate. Its truth (or character) then 'swells into reality' for us so vividly that the dynamic awareness of it is also 'beautiful.' In other words, reality taking form and meaning is 'beauty' if it is vitally enough known and felt" (*Major British Writers*, enlarged ed., 1959, p. 360n.). Cf. the conclusion of the "Ode on a Grecian Urn," above, ll. 49–50.

25. Horace Smith . . . Du Bois: Horace Smith, a poet and novelist, was a friend of Keats, Edward Du Bois a minor writer, and the others were men interested mainly in the more modish uses of literature. **26. Negative Capability . . . consideration:** These difficult remarks are paraphrased by Bate as follows: "Our life is filled with change, uncertainties, mysteries; no one complete system of rigid categories will explain it fully. We can grasp and understand the elusive flux of life only by being imaginatively openminded, sympathetic, receptive—by extending every possible feeler that we may have potentially in us. But we can achieve this active awareness only by *negating* our own *egos*. We must not only rise above our own vanity and prejudices, but resist the temptation to make up our minds on everything, and to have always ready a neat answer. If we discard a momentary insight, for example, because we cannot fit it into a static category or systematic framework, we are selfishly asserting our own 'identity.' A great poet is less concerned with himself, and has his eyes on what is without. With him 'the sense of Beauty'—the capacity to relish concrete reality in its full, if elusive, meaning—'overcomes every other consideration.' In fact, it goes beyond and 'obliterates' the act of 'consideration'—of deliberating, analyzing, and piecing together experience in a logical structure" (*Major British Writers*, p. 361n.). See also Keats's letter to Woodhouse, below, October 27, 1818.

Shelley's poem[27] is out, & there are words about its being objected too, as much as Queen Mab was. Poor Shelley I think he has his Quota of good qualities, in sooth la!! Write soon to your most sincere friend & affectionate Brother.

JOHN

To J. H. Reynolds
[*February 3, 1818*]
Hampstead Tuesday.

MY DEAR REYNOLDS,

I thank you for your dish of Filberts—Would I could get a basket of them by way of desert every day for the sum of two pence—[28]Would we were a sort of ethereal Pigs, & turn'd loose to feed upon spiritual Mast & Acorns—which would be merely being a squirrel & feed upon filberts. for what is a squirrel but an airy pig, or a filbert but a sort of archangelical acorn. About the nuts being worth cracking, all I can say is that where there are a throng of delightful Images ready drawn simplicity is the only thing. the first is the best on account of the first line, and the "arrow—foil'd of its antler'd food"—and moreover (and this is the only word or two I find fault with, the more because I have had so much reason to shun it as a quicksand) the last has "tender and true"—We must cut this, and not be rattlesnaked into any more of the like—It may be said that we ought to read our Contemporaries. that Wordsworth &c should have their due from us. but for the sake of a few fine imaginative or domestic passages, are we to be bullied into a certain Philosophy engendered in the whims of an Egotist[29]—Every man has his speculations, but every man does not brood and peacock over them till he makes a false coinage and deceives himself—Many a man can travel to the very bourne of Heaven,[30] and yet want confidence to put down his halfseeing. Sancho will invent a Journey heavenward as well as any body. We hate poetry that

has a palpable design upon us—and if we do not agree seems to put its hand in its breeches pocket. Poetry should be great & unobtrusive, a thing which enters into one's soul, and does not startle it or amaze it with itself but with its subject.—How beautiful are the retired flowers! how would they lose their beauty were they to throng into the highway crying out, "admire me I am a violet! dote upon me I am a primrose! Modern poets differ from the Elizabethans in this. Each of the moderns like an Elector of Hanover governs his petty state, & knows how many straws are swept daily from the Causeways in all his dominions & has a continual itching that all the Housewives should have their coppers well scoured: the antients were Emperors of vast Provinces, they had only heard of the remote ones and scarcely cared to visit them.—I will cut all this—I will have no more of Wordsworth or Hunt in particular—Why should we be of the tribe of Manasseh, when we can wander with Esau? why should we kick against the Pricks, when we can walk on Roses? Why should we be owls, when we can be Eagles? Why be teased with "nice Eyed wagtails,"[31] when we have in sight "the Cherub Contemplation"?[32] —Why with Wordsworths "Matthew with a bough of wilding in his hand"[33] when we can have Jacques "under an oak &c[34]—The secret of the Bough of Wilding will run through your head faster than I can write it—Old Matthew spoke to him some years ago on some nothing, & because he happens in an Evening Walk to imagine the figure of the old man—he must stamp it down in black & white, and it is henceforth sacred—I don't mean to deny Wordsworth's grandeur & Hunt's merit, but I mean to say we need not be teazed with grandeur & merit—when we can have them uncontaminated & unobtrusive. Let us have the old Poets, & Robin Hood Your letter and its sonnets gave me more pleasure than will the 4th Book of Childe Harold & the whole of any body's life & opinions. . . .

Y^r sincere friend and Coscribbler

John Keats.

---

27. **Shelley's poem:** *Laon and Cythna* (*The Revolt of Islam*). 28. **Filberts . . . pence:** Reynolds had sent Keats two sonnets on "Robin Hood." Hence the rustic allusions. 29. **Wordsworth . . . Egotist:** Here and later, in comparing Wordsworth and the modern poets generally with those of Shakespeare's and Milton's ages, Keats has been partly influenced by Hazlitt, whose *Lectures on the English Poets* (see especially that "On Shakspeare and Milton," above) he was now attending. 30. **bourne of Heaven:** *Hamlet*, III. i. 79ff.

31. **"nice . . . wagtails":** a quotation from Leigh Hunt's *Foliage* (1818), p. xxxiii. **32. "the Cherub Contemplation":** Milton's "Il Penseroso," l. 54. **33. "Matthew . . . hand":** Wordsworth's "Two April Mornings," above, l. 60. **34. Jacques . . . oak:** *As You Like It*, II. i. 31.

To J. H. Reynolds
[*February 19, 1818*]

MY DEAR REYNOLDS,

I have an idea that a Man might pass a very pleasant life in this manner—let him on any certain day read a certain Page of full Poesy or distilled Prose and let him wander with it, and muse upon it, and reflect from it, and bring home to it, and prophesy upon it, and dream upon it—untill it becomes stale—but when will it do so? Never—When Man has arrived at a certain ripeness in intellect any one grand and spiritual passage serves him as a starting post towards all "the two-and-thirty Pallaces" How happy is such a "voyage of conception," what delicious diligent Indolence! A doze upon a Sofa does not hinder it, and a nap upon Clover engenders ethereal finger-pointings—the prattle of a child gives it wings, and the converse of middle age a strength to beat them—a strain of musick conducts to "an odd angle of the Isle" and when the leaves whisper it puts a "girdle round the earth."[35] Nor will this sparing touch of noble Books be any irreverance to their Writers—for perhaps the honors paid by Man to Man are trifles in comparison to the Benefit done by great Works to the "Spirit and pulse of good"[36] by their mere passive existence. Memory should not be called knowledge—Many have original minds who do not think it—they are led away by Custom—Now it appears to me that almost any Man may like the Spider spin from his own inwards his own airy Citadel—the points of leaves and twigs on which the Spider begins her work are few and she fills the Air with a beautiful circuiting: man should be content with as few points to tip with the fine Webb of his Soul and weave a tapestry empyrean—full of Symbols for his spiritual eye, of softness for his spiritual touch, of space for his wandering of distinctness for his Luxury—But the Minds of Mortals are so different and bent on such diverse Journeys that it may at first appear impossible for any common taste and fellowship to exist ~~bettween~~ between two or three under these suppositions—It is however quite the contrary—Minds would leave each other in contrary directions, traverse each other in Numberless points, and all [*for* at] last

greet each other at the Journeys end—An old Man and a child would talk together and the old Man be led on his Path, and the child left thinking—Man should not dispute or assert but whisper results to his neighbour, and thus by every germ of Spirit sucking the Sap from mould ethereal every human might become great, and Humanity instead of being a wide heath of Furse and Briars with here and there a remote Oak or Pine, would become a grand democracy of Forest Trees. . . . Now it is more noble to sit like Jove than to fly like Mercury—let us not therefore go hurrying about and collecting honey-bee like, buzzing here and there impatiently from a knowledge of what is to be arrived at: but let us open our leaves like a flower and be passive and receptive—budding patiently under the eye of Apollo and taking hints from every noble insect that favors us with a visit—sap will be given us for Meat and dew for drink—I was led into these thoughts, my dear Reynolds, by the beauty of the morning operating on a sense of Idleness—I have not read any Books—the Morning said I was right—I had no Idea but of the Morning and the Thrush said I was right—seeming to say—

[Here follows the sonnet "What the Thrush Said," printed above.]

Now I am sensible all this is a mere sophistication, however it may neighbour to any truths, to excuse my own indolence—so I will not deceive myself that Man should be equal with jove—but think himself very well off as a sort of scullion-Mercury or even a humble Bee—It is not matter whether I am right or wrong either one way or another, if there is sufficient to lift a little time from your Shoulders.

Your affectionate friend
JOHN KEATS—

To John Taylor[37]
Hampstead 27 Feby—[*1818*]

MY DEAR TAYLOR,

. . . In *Endymion* I have most likely but moved into the Go-cart from the leading strings. In Poetry I have a few Axioms, and you will see how far I am from their Centre. 1st I think Poetry should surprise by a fine excess and not by Singularity—it should

---

35. "an . . . Isle"; "girdle . . . earth": *The Tempest*, I. ii. 223; *A Midsummer Night's Dream*, II. i. 175. 36. "Spirit . . . good": Wordsworth's "Old Cumberland Beggar," above, l. 77.

37. John Taylor: Keats's publisher (of the firm of Taylor and Hessey).

strike the Reader as a wording of his own highest thoughts, and appear almost a Remembrance—2ⁿᵈ Its touches of Beauty should never be half way therby making the reader breathless instead of content: the rise, the progress, the setting of imagery should like the Sun come ~~natural~~ natural too him—shine over him and set soberly although in magnificence leaving him in the Luxury of twilight—but it is easier to think what Poetry should be than to write it—and this leads me on to another axiom. That if Poetry comes not as naturally as the Leaves to a tree it had better not come at all. However it may be with me I cannot help looking into new countries with 'O for a Muse of fire to ascend !' —If Endymion serves me as a Pioneer perhaps I ought to be content. I have great reason to be content, for thank God I can read and perhaps understand Shakspeare to his depths, and I have I am sure many friends, who, if I fail, will attribute any change in my Life and Temper to Humbleness rather than to Pride—to a cowering under the Wings of great Poets rather than to a Bitterness that I am not appreciated. I am anxious to get Endymion printed that I may forget it and proceed. . . .

Your sincere and obligᵈ friend

JOHN KEATS—

To Benjamin Bailey
[*March 13, 1818*]
Teignmouth Friday

MY DEAR BAILEY,

. . . You know my ideas about Religion—I do not think myself more in the right than other people and that nothing in this world is proveable. I wish I could enter into all your feelings on the subject merely for one short 10 Minutes and give you a Page or two to your liking. I am sometimes so very sceptical as to think Poetry itself a mere Jack a lanthern to amuse whoever may chance to be struck with its brilliance— As Tradesmen say every thing is worth what it will fetch, so probably every mental pursuit takes its reality and worth from the ardour of the pursuer—being in itself a nothing—Ethereal thing[s] may at least be thus real, divided under three heads—Things real—things semireal—and no things—Things real—such as existences of Sun Moon & Stars and passages of Shakspeare —Things semireal such as Love, the Clouds &c which require a greeting of the Spirit to make them wholly exist—and Nothings which are made Great and dignified by an ardent pursuit—Which by the by

stamps the burgundy mark on the bottles of our Minds, insomuch as they are able to "*consec[r]ate whate'er they look upon.*"³⁸ . . .

Aye this may be carried—but what am I talking of— it is an old maxim of mine and of course must be well known that eve[r]y point of thought is the centre of an intellectual world—the two uppermost thoughts in a Man's mind are the two poles of his World he revolves on them and every thing is southward or northward to him through their means—We take but three steps from feathers to iron. Now my dear fellow I must once for all tell you I have not one Idea of the truth of any of my speculations—I shall never be a Reasoner because I care not to be in the right, when retired from bickering and in a proper philosophical temper— . . .

Your affectionate friend

JOHN KEATS—

To B. R. Haydon
[*April 8, 1818*] *Wednesday*—

MY DEAR HAYDON,

. . . I have ever been too sensible of the labyrinthian path to eminence in Art (judging from Poetry) ever to think I understood the emphasis of Painting. The innumerable compositions and decompositions which take place between the intellect and its thousand materials before it arrives at that trembling delicate and snail-horn³⁹ perception of Beauty—I know not your many havens of intenseness—nor ever can know them—but for this I hope not⁴⁰ you atchieve is lost upon me: for when a Schoolboy the abstract Idea I had of an heroic painting—was what I cannot describe I saw it somewhat sideways large prominent round and colour'd with magnificence—somewhat like the feel I have of Anthony and Cleopatra. Or of Alcibiades, leaning on his Crimson Couch in his Galley, his broad shoulders imperceptibly heaving with the Sea—What passage in Shakspeare is finer than this

'See how the surly Warwick mans the Wall' ⁴¹ . . .

Your affectionate friend

JOHN KEATS

---

38. **"consec[r]ate . . . upon"**: Shelley's "Hymn to Intellectual Beauty," above, ll. 13–14. **39. delicate and snail-horn:** Cf. Keats's quotations from Shakespeare in the letter to Reynolds, above, November 22, 1817. **40. not:** a slip for "nought." **41. 'See . . . Wall'**: III *Henry VI*, V. i. 17.

To John Taylor
[*April 24, 1818*]
Teignmouth Friday

MY DEAR TAYLOR,

. . . I was proposing to travel over the north this Summer. There is but one thing to prevent me.—I know nothing—I have read nothing—and I mean to follow Solomon's directions, "get learning—get understanding."[42] I find earlier days are gone by—I find that I can have no enjoyment in the World but continual drinking of Knowledge. I find there is no worthy pursuit but the idea of doing some good to the world. Some do it with their society—some with their wit—some with their benevolence—some with a sort of power of conferring pleasure and good humour on all they meet—and in a thousand ways, all equally dutiful to the command of Great Nature—there is but one way for me. The road lies through application, study, and thought. I will pursue it; and to that end purpose retiring for some years. I have been hovering for some time between an exquisite sense of the luxurious and a love for Philosophy,—were I calculated for the former I should be glad. But as I am not I shall turn my soul to the latter. . . .

Your very sincere friend
JOHN KEATS.

To J. H. Reynolds
Teignmouth, 3 May [*1818*]

MY DEAR REYNOLDS,

What I complain of is that I have been in so uneasy a state of Mind as not to be fit to write to an invalid. I cannot write to any length under a disguised feeling. I should have loaded you with an addition of gloom, which I am sure you do not want. I am now thank God in a humour to give you a good groat's worth—for Tom, after a Night without a Wink of sleep, and overburthened with fever, has got up after a refreshing day-sleep and is better than he has been for a long time; and you I trust have been again round the Common without any effect but refreshment. . . . Were I to study Physic or rather Medicine again, I feel it would not make the least difference in my Poetry; when the Mind is in its infancy a Bias is in reality a Bias, but when we have acquired more strength, a

Bias becomes no Bias. Every department of Knowledge we see excellent and calculated towards a great whole. I am so convinced of this, that I am glad at not having given away my medical Books, which I shall again look over to keep alive the little I know thitherwards; and moreover intend through you and Rice[43] to become a sort of Pip-civilian.[44] An extensive knowlege is needful to thinking people—it takes away the heat and fever; and helps, by widening speculation, to ease the Burden of the Mystery:[45] a thing I begin to understand a little, and which weighed upon you in the most gloomy and true sentence in your Letter. The difference of high Sensations with and without knowledge appears to me this—in the latter case we are falling continually ten thousand fathoms deep and being blown up again without wings and with all [the] horror of a ~~Case~~ bare shoulderd Creature—in the former case, our shoulders are fledged, and we go thro' the same air and space without fear. This is running one's rigs on the score of abstracted benefit—when we come to human Life and the affections it is impossible [to know] how a parallel of breast and head can be drawn—(you will forgive me for thus privately treading out [of] my depth, and take it for treading as schoolboys tread the water)—it is impossible to know how far knowlege will console us for the death of a friend and the ill "that flesh is heir to"—With respect to the affections and Poetry you must know by a sympathy my thoughts that way; and I dare say these few lines will be but a ratification: I wrote them on May-day—and intend to finish the ode all in good time.—

[Here follows "Fragment of an Ode to Maia," printed above.]

You may be anxious to know for fact to what sentence in your Letter I allude. You say "I fear there is little chance of any thing else in this life". You seem by that to have been going through with a more painful and acute zest the same labyrinth that I have—I have come to the same conclusion thus far. My Branchings out therefrom have been numerous: one of them is the consideration of Wordsworth's genius and as a help, in the manner of gold being the meridian Line of worldly wealth,—how he differs from Milton.

---

42. "get . . . understanding": Proverbs 4:5.

43. **Rice:** James Rice, 1792–1832, a lawyer, friend of Keats and Reynolds. See the letter to him, below, February 14, 1820. 44. **Pip-civilian:** an amateur student of civil law (Reynolds, like Rice, was a lawyer). 45. **Burden . . . Mystery:** Wordsworth's "Tintern Abbey," l. 38.

—And here I have nothing but surmises, from an uncertainty whether Miltons apparently less anxiety for Humanity proceeds from his seeing further or no than Wordsworth: And whether Wordsworth has in truth epic passion, and martyrs himself to the human heart, the main region of his song[46]—In regard to his genius alone—we find what he says true as far as we have experienced and we can judge no further but by larger experience—for axioms in philosophy are not axioms until they are proved upon our pulses: We read fine—things but never feel them to the full until we have gone the same steps as the Author.—I know this is not plain; you will know exactly my meaning when I say, that now I shall relish Hamlet more than I ever have done—Or, better—You are sensible no Man can set down Venery as a bestial or joyless thing until he is sick of it and therefore all philosophizing on it would be mere wording. Until we are sick, we understand not;—in fine, as Byron says, "Knowledge is Sorrow";[47] and I go on to say that "Sorrow is Wisdom"—and further for aught we can know for certainty! "Wisdom is folly"—So you see how I have run away from Wordsworth, and Milton, and shall still run away from what was in my head, to observe, that some kind of letters are good squares others handsome ovals, and others some orbicular, others spheroid—and why should there not be another species with two rough edges like a Rat-trap? I hope you will find all my long letters of that species, and all will be well; for by merely touching the spring delicately and etherially, the rough edged will fly immediately into a proper compactness; and thus you may make a good wholesome loaf, with your own le[a]ven in it, of my fragments—If you cannot find this said Rat-trap sufficiently tractable—alas for me, it being an impossibility in grain for my ink to stain otherwise: If I scribble long letters I must play my vagaries. I must be too heavy, or too light, for whole pages—I must be quaint and free of Tropes and figures—I must play my draughts as I please, and for my advantage and your erudition, crown a white with a black, or a black with a white, and move into black or white, far and near as I please—I must go from Hazlitt to Patmore, and make Wordsworth and Coleman play at leap-frog—or keep one of them down a whole half holiday at fly the garter—"From Gray to Gay, from

Little to Shakespeare"[48]—Also, as a long cause requires two or more sittings of the Court, so a long letter will require two or more sittings of the Breech wherefore I shall resume after dinner.—

Have you not seen a Gull, an orc, a sea Mew, or any thing to bring this Line to a proper length, and also fill up this clear part; that like the Gull I may *dip*—I hope, not out of sight—and also, like a Gull, I hope to be lucky in a good sized fish—This crossing a letter is not without its association—for chequer work leads us naturally to a Milkmaid, a Milkmaid to Hogarth Hogarth to Shakespeare Shakespear to Hazlitt—Hazlitt to Shakespeare and thus by merely pulling an apron string we set a pretty peal of Chimes at work—Let them chime on while, with your patience,—I will return to Wordsworth—whether or no he has an extended vision or a circumscribed grandeur—whether he is an eagle in his nest, or on the wing—And to be more explicit and to show you how tall I stand by the giant, I will put down a simile of human life as far as I now perceive it; that is, to the point to which I say we both have arrived at—Well—I compare human life to a large Mansion of Many Apartments, two of which I can only describe, the doors of the rest being as yet shut upon me—The first we step into we call the infant or thoughtless Chamber, in which we remain as long as we do not think—We remain there a long while, and notwithstanding the doors of the second Chamber remain wide open, showing a bright appearance, we care not to hasten to it; but are at length imperceptibly impelled by the awakening of the thinking principle—within us—we no sooner get into the second Chamber, which I shall call the Chamber of Maiden-Thought, than we become intoxicated with the light and the atmosphere, we see nothing but pleasant wonders, and think of delaying there for ever in delight: However among the effects this breathing is father of is that tremendous one of sharpening one's vision into the heart and nature of Man—of convincing ones nerves that the World is full of Misery and Heartbreak, Pain, Sickness and oppression—whereby This Chamber of Maiden Thought becomes gradually darken'd and at the same time on all sides of it many doors are set open—but all dark—all leading to dark passages—We see not the ballance of good and evil.

---

46. **heart . . . song:** paraphrased from Wordsworth's "Prospectus," printed with *The Excursion*, above, l. 41. **47. "Knowledge is Sorrow":** *Manfred*, I. i. 10.

48. **Patmore . . . Shakespeare:** P. G. Patmore, 1786–1855, friend of Hazlitt and Lamb; George Colman the Younger, 1762–1836, playwright; fly the garter was a game; and the quotation is a pun on a line of Pope's, *Essay on Man*, IV. 380: "From grave to gay, from lively to severe."

We are in a Mist—*We* are now in that state—We feel the "burden of the Mystery," To this point was Wordsworth come, as far as I can conceive when he wrote 'Tintern Abbey' and it seems to me that his Genius is explorative of those dark Passages. Now if we live, and go on thinking, we too shall explore them.[49] he is a Genius and superior [to] us, in so far as he can, more than we, make discoveries, and shed a light in them—Here I must think Wordsworth is deeper than Milton—though I think it has depended more upon the general and gregarious advance of intellect, than individual greatness of Mind—From the Paradise Lost and the other Works of Milton, I hope it is not too presuming, even between ourselves to say, his Philosophy, human and divine, may be tolerably understood by one not much advanced in years, In his time Englishmen were just emancipated from a great superstition—and Men had got hold of certain points and resting places in reasoning which were too newly born to be doubted, and too much opposed by the Mass of Europe not to be thought etherial and authentically divine—who could gainsay his ideas on virtue, vice, and Chastity in Comus, just at the time of the dismissal of Cod-pieces and a hundred other disgraces? who would not rest satisfied with his hintings at good and evil in the Paradise Lost, when just free from the inquisition and burning in Smithfield? The Reformation produced such immediate and great benefits, that Protestantism was considered under the immediate eye of heaven, and its own remaining Dogmas and superstitions, then, as it were, regenerated, constituted those resting places and seeming sure points of Reasoning—from that I have mentioned, Milton, whatever he may have thought in the sequel, appears to have been content with these by his writings—He did not think into the human heart, as Wordsworth has done—Yet Milton as a Philosop[h]er, had sure as great powers as Wordsworth—What is then to be inferr'd? O many things—It proves there is really a grand march of intellect—, It proves that a mighty providence subdues the mightiest Minds to the service of the time being, whether it be in human Knowledge or Religion—I have often pitied a Tutor who has to hear "Nom^e: Musa"[50]—so often dinn'd into his ears—I hope you may not have the same pain in this scribbling—I may have read these things before, but I never had even a thus dim perception of them; and moreover I like to say my lesson to one who will endure my tediousness for my own sake—After all there is certainly something real in the World—Moore's present to Hazlitt is real— I like that Moore, and am glad I saw him at the Theatre just before I left Town. Tom has spit a leetle blood this afternoon, and that is rather a damper—but I know— the truth is there is something real in the World Your third Chamber of Life shall be a lucky and a gentle one—stored with the wine of love—and the Bread of Friendship. When you see George if he should not have rec$\bar{e}$d a letter from me tell him he will find one at home most likely—tell Bailey I hope soon to see him—Remember me to all The leaves have been out here, for mony a day—I have written to George for the first stanzas of my Isabel[51]—I shall have them soon and will copy the whole out for you.

> Your affectionate friend
> JOHN KEATS.

To Tom Keats[52]
[*June 25–27, 1818, Ambleside*]

Here beginneth my journal, this Thursday, the 25th day of June, Anno Domini 1818. This morning we arose at 4, and set off in a Scotch mist; put up once under a tree, and in fine, have walked wet and dry to this place, called in the vulgar tongue Endmoor, 17 miles; we have not been incommoded by our knapsacks; they serve capitally, and we shall go on very well.

June 26—I merely put *pro forma*, for there is no such thing as time and space, which by the way came forcibly upon me on seeing for the first hour the Lake and Mountains of Winander—I cannot describe them— they surpass my expectation—beautiful water—shores and islands green to the marge—mountains all round up to the clouds. We set out from Endmoor this morning, breakfasted at Kendal with a soldier who had been in all the wars for the last seventeen years— then we have walked to Bowne's[53] to dinner—said

---

49. **Now . . . them:** another indication of Keats's conviction that the poetry of the future must inevitably turn more to the "inner life," and an anticipation of the promise he makes a year later in the "Ode to Psyche" (see poem and headnote above). 50. **"Nom^e: Musa":** first lesson in Latin grammar: "Nominative: Musa," etc.

51. **Isabel:** Keats's *Isabella*, printed above. 52. **Tom Keats:** Tom (aged eighteen) was remaining alone in the lodgings at Well Walk, Hampstead, where the Keats brothers lived. John, before setting off on the walking trip through northern England and Scotland with Charles Brown, had gone first to Liverpool, whence his brother George was sailing in order to settle in America. 53. **Bowne's:** Bowness, a town on Lake Windermere.

Bowne's situated on the Lake where we have just dined, and I am writing at this present. I took an oar to one of the islands to take up some trout for dinner, which they keep in porous boxes. I enquired of the waiter for Wordsworth—he said he knew him, and that he had been here a few days ago, canvassing for the Lowthers. What think you of that—Wordsworth versus Brougham!![54] Sad—sad—sad—and yet the family has been his friend always. What can we say? We are now about seven miles from Rydale,[55] and expect to see him to-morrow. You shall hear all about our visit.

There are many disfigurements to this Lake—not in the way of land or water. No; the two views we have had of it are of the most noble tenderness—they can never fade away—they make one forget the divisions of life; age, youth, poverty and riches; and refine one's sensual vision into a sort of north star which can never cease to be open lidded and stedfast over the wonders of the great Power.[56] The disfigurement I mean is the miasma of London. I do suppose it contaminated with bucks and soldiers, and women of fashion—and hat-band ignorance. The border inhabitants are quite out of keeping with the romance about them, from a continual intercourse with London rank and fashion. But why should I grumble? They let me have a prime glass of soda water—O they are as good as their neighbors. But Lord Wordsworth, instead of being in retirement, has himself and his house full in the thick of fashionable visitors quite convenient to be pointed at all the summer long. When we had gone about half this morning, we began to get among the hills and to see the mountains grow up before us—the other half brought us to Wynandermere,[57] 14 miles to dinner. The weather is capital for the views, but is now rather misty, and we are in doubt whether to walk to Ambleside to tea—it is five miles along the borders of the Lake. Loughrigg will swell up before us all the way—I have an amazing partiality for mountains in the clouds. There is nothing in Devon like this, and Brown says there is nothing in Wales to be compared to it. I must tell you, that in going through Cheshire and Lancashire, I saw the Welsh mountains at a distance. We have passed the two castles, Lancaster and Kendal.

54. Lowthers . . . Brougham!!: William Lowther was the Tory candidate for Parliament, and Henry Brougham the Whig. 55. Rydale: Wordsworth's home, Rydal Mount, was located there. 56. north . . . Power: Cf. "Bright Star," above, ll. 1–3. 57. Wynandermere: former name of Lake Windermere.

27th—We walked here to Ambleside yesterday along the border of Winandermere all beautiful with wooded shores and Islands—our road was a winding lane, wooded on each side, and green overhead, full of Foxgloves—every now and then a glimpse of the Lake, and all the while Kirkstone and other large hills nestled together in a sort of grey black mist. Ambleside is at the northern extremity of the Lake. We arose this morning at six, because we call it a day of rest, having to call on Wordsworth who lives only two miles hence—before breakfast we went to see the Ambleside water fall. The morning beautiful—the walk easy among the hills. We, I may say, fortunately, missed the direct path, and after wandering a little, found it out by the noise—for, mark you, it is buried in trees, in the bottom of the valley—the stream itself is interesting throughout with "mazy error over pendant shades." Milton meant a smooth river—this is buffetting all the way on a rocky bed ever various—but the waterfall itself, which I came suddenly upon, gave me a pleasant twinge. First we stood a little below the head about half way down the first fall, buried deep in trees, and saw it streaming down two more descents to the depth of near fifty feet—then we went on a jut of rock nearly level with the second fall-head, where the first fall was above us, and the third below our feet still—at the same time we saw that the water was divided by a sort of cataract island on whose other side burst out a glorious stream—then the thunder and the freshness. At the same time the different falls have as different characters; the first darting down the slate-rock like an arrow; the second spreading out like a fan—the third dashed into a mist—and the one on the other side of the rock a sort of mixture of all these. We afterwards moved away a space, and saw nearly the whole more mild, streaming silverly through the trees. What astonishes me more than any thing is the tone, the coloring, the slate, the stone, the moss, the rock-weed; or, if I may so say, the intellect, the countenance of such places. The space, the magnitude of mountains and waterfalls are well imagined before one sees them; but this countenance or intellectual tone must surpass every imagination and defy any remembrance. I shall learn poetry here and shall henceforth write more than ever, for the abstract endeavor of being able to add a mite to that mass of beauty which is harvested from these grand materials, by the finest spirits, and put into etherial existence for the relish of one's fellows. I cannot think with Hazlitt that these scenes make man appear little. I never forgot my stature so completely—I live in the eye; and my imagination, surpassed, is at rest—We shall see another waterfall near Rydal to which we shall proceed after having put these letters in the post office. I long to be at Carlisle, as I expect there a letter from George and one from you. Let any of my friends see my letters—they may not be interested in descriptions—descriptions are bad at all times—I did not intend to give you any; but how can I help it? I am anxious you should taste a little of our pleasure; it may not be an unpleasant thing, as you have not the fatigue. I am well in health. Direct henceforth to Port Patrick till the 12th July. Content that probably three or four pair of eyes whose owners I am rather partial to will run over these lines I remain; and moreover that I am your affectionate brother John.

To Fanny Keats[58]
[*July 2–5, 1818*]
Dumfries July 2nd

MY DEAR FANNY,

I intended to have written to you from Kirkudbright the town I shall be in tomorrow—but I will write now bec[a]use my knapsack has worn my coat in the Seams, my coat has gone to the Taylors and I have but one Coat to my back in these parts. I must tell you how I went to Liverpool with George and our new Sister and the Gentleman my fellow traveller through the Summer and Autumn—We had a tolerable journey to Liverpool—which I left the next morning before George was up for Lancaster—Then we set off from Lancaster on foot with our knapsacks on, and have walked a Little zig zag through the mountains and Lakes of Cumberland and Westmoreland—We came from Carlisle yesterday to this place—We are employed in going up Mountains, looking at Strange towns prying into old ruins and eating very hearty breakfasts. Here we are full in the Midst of broad Scotch 'How is it a' wi yoursel'—the Girls are walking about bare footed and in the worst cottages the Smoke finds its way out of the door—I shall come home full of news for you and for fear I should choak you by too great a dose at once I must make you used to it by a

letter or two—We have been taken for travelling Jewellers, Razor sellers and Spectacle venders because friend Brown wears a pair—The first place we stopped at with our knapsacks contained one Richard Bradshaw a notorious tippler—He stood in the shape of a З and balanced himself as well as he could saying with his nose right in Mr Browns face 'Do—yo u sell Spect—ta—cles?' Mr Abbey says we are Don Quixotes —tell him we are more generally taken for Pedlars—All I hope is that we may not be taken for excisemen in this whiskey country—We are generally up about 5 walking before breakfast and we complete our 20 Miles before dinner—Yesterday we visited Burns's Tomb and this morning the fine Ruins of Lincluden— I had done thus far when my coat came back fortified at all points—so as we lose no time we set forth again through Galloway—all very pleasant and pretty with no fatigue when one is used to it—We are in the midst of Meg Merrilies' country of whom I suppose—you have heard—

[Here follows the ballad of "Meg Merrilies," printed above.]

If you like these sort of Ballads I will now and then scribble one for you—if I send any to Tom I'll tell him to send them to you—I have so many interruptions that I cannot manage to fill a Letter in one day—since I scribbled the Song we have walked through a beautiful Country to Kirkudbright—at which place I will write you a song about myself—

> There was a naughty Boy
>   A naughty boy was he
> He would not stop at home
>   He could not quiet be—
>   He took
>   In his knapsack
>   A Book
>   Full of vowels
>   And a shirt
>   With some towels—
>   A slight cap
>   For night cap—
>   A hair brush
>   Comb ditto
>   New Stockings
>   For old ones
>   Would split O!
>   This knapsack
>   Tight at 's back
>   He revetted close

58. **Fanny Keats:** Keats's young sister (aged fifteen) lived with the Abbeys. Richard Abbey, a tea-broker whom the grandmother of the Keats children had appointed their guardian, and who appropriated much of the money that should have come to them, tried constantly to keep Fanny away from her brothers.

And followe'd his Nose
To the North
To the North
And follow'd his nose
To the North—

There was a naughty boy
　And a naughty boy was he
For nothing would he do
　But scribble poetry—
　　He took
　　An inkstand
　　In his hand
　　And a Pen
　　Big as ten
　　In the other
　　And away
　　In a Pother
　　He ran
　　To the mountains
　　And fountains
　　And ghostes
　　And Postes
　　And witches
　　And ditches
　　And wrote
　　In his coat
　　When the weather
　　Was cool,
　　Fear of gout,
　　And without
　　When the weather
　　Was warm—
　　Och the charm
　　When we choose
　To follow one's nose
　　To the north,
　　To the north,
　To follow one's nose
　　To the north!
There was a naughty boy
　And a naughty boy was he,
He kept little fishes
　In washing tubs three
　　In spite
　　Of the might
　　Of the Maid
　　Nor afraid
　　Of his Granny-good—
　　He often would

Hurly burly
Get up early
And go
By hook or crook
To the brook
And bring home
Miller's thumb,
Tittlebat
Not over fat,
Minnows small
As the stall
Of a glove,
Not above
The size
Of a nice
Little Baby's
Little finger—
O he made
'Twas his trade
Of Fish a pretty Kettle
　A Kettle—
　A Kettle
Of Fish a pretty Kettle
　A Kettle!

There was a naughty Boy,
　And a naughty Boy was he,
He ran away to Scotland
　The people for to see—
　　Then he found
　　That the ground
　　Was as hard,
　　That a yard
　　Was as long,
　　That a song
　　Was as merry,
　　That a cherry
　　Was as red—
　　That lead
　　Was as weighty,
　　That fourscore
　　Was as eighty,
　　That a door
　　Was as wooden
　　As in England—
　So he stood in his shoes
　And he wonder'd,
　　He wonder'd,
　He stood in his
　　Shoes and he wonder'd.

My dear Fanny I am ashamed of writing you such stuff, nor would I if it were not for being tired after my days walking, and ready to tumble int[o bed] so fatigued that when I am asleep you might sew my nose to my great toe and trundle me round the town like a Hoop without waking me—Then I get so hungry—a Ham goes but a very little way and fowls are like Larks to me—A Batch of Bread I make no more ado with than a sheet of parliament; and I can eat a Bull's head as easily as I used to do Bull's eyes—I take a whole string of Pork Sausages down as easily as a Pen'orth of Lady's fingers[59]—Oh dear I must soon be contented with an acre or two of oaten cake a hogshead of Milk and a Cloaths basket of Eggs morning noon and night when I get among the Highlanders—Before we see them we shall pass into Ireland and have a chat with the Paddies, and look at the Giant's Cause-way which you must have heard of—I have not time to tell you particularly for I have to send a Journal to Tom of whom you shall hear all particulars or from me when I return—Since I began this we have walked sixty miles to Newton Stewart at which place I put in this Letter—to night we sleep at Glenluce—tomorrow at Portpatrick and the next day we shall cross in the passage boat to Ireland—I hope Miss Abbey has quite recovered—Present my Respects to her and to Mr And Mrs Abbey—God bless you—

Your affectionate Brother John—
Do write me a Letter directed to *Inverness*. Scotland—

To J. A. Hessey
[*October 8, 1818*]

MY DEAR HESSEY,

You are very good in sending me the letter from the Chronicle—and I am very bad in not acknowledging such a kindness sooner.—pray forgive me.—It has so chanced that I have had that paper every day—I have seen today's. I cannot but feel indebted to those Gentlemen who have taken my part—As for the rest, I begin to get a little acquainted with my own strength and weakness.—Praise or blame has but a momentary effect on the man whose love of beauty in the abstract makes him a severe critic on his own Works. My own

domestic criticism has given me pain without comparison beyond what Blackwood or the Quarterly could possibly inflict. and also when I feel I am right, no external praise can give me such a glow as my own solitary reperception & ratification of what is fine. J. S.[60] is perfectly right in regard to the slip-shod Endymion. That it is so is no fault of mine.—No!—though it may sound a little paradoxical. It is as good as I had power to make it—by myself—Had I been nervous about its being a perfect piece, & with that view asked advice, & trembled over every page, it would not have been written; for it is not in my nature to fumble—I will write independantly.—I have written independently *without Judgment*.—I may write independently & *with judgment* hereafter.—The Genius of Poetry must work out its own salvation in a man: It cannot be matured by law & precept, but by sensation & watchfulness in itself—That which is creative must create itself—In Endymion, I leaped headlong into the Sea, and thereby have become better acquainted with the Soundings, the quicksands, & the rocks, than if I had stayed upon the green shore, and piped a silly pipe, and took tea & comfortable advice.—I was never afraid of failure; for I would sooner fail than not be among the greatest—But I am nigh getting into a rant. So, with remembrances to Taylor and Woodhouse &c I am

Yrs very sincerely
JOHN KEATS.

To Richard Woodhouse[61]
[*October 27, 1818*]

MY DEAR WOODHOUSE,

Your Letter gave me a great satisfaction; more on account of its friendliness, than any relish of that matter in it which is accounted so acceptable in the 'genus irritabile'[62] The best answer I can give you is in a clerklike manner to make some observations on two principle points, which seem to point like indices into the midst of the whole pro and con, about genius, and views and atchievements and ambition and coetera.

---

59. **parliament . . . fingers:** "Parliament" was a gingerbread cake, "bull's eyes" were candy, and "lady's fingers" were white peppermint sticks with pink rings.

60. **J. S.:** possibly John Scott. 61. **Richard Woodhouse:** Woodhouse, 1788–1834, was a scholarly lawyer and reader for Keats's publisher, Taylor and Hessey. An admirer of Keats, he made careful transcripts of many of the poems and recorded much valuable information about them. 62. **'genus irritabile':** the irritable race (of poets), quoted from Horace, *Epistles*, II. ii. 102.

1st As to the poetical Character itself, (I mean that sort of which, if I am any thing, I am a Member; that sort distinguished from the wordsworthian or egotistical sublime; which is a thing per se and stands alone) it is not itself—it has no self—it is every thing and nothing—It has no character—it enjoys light and shade; it lives in gusto, be it foul or fair, high or low, rich or poor, mean or elevated—It has as much delight in conceiving an Iago as an Imogen. What shocks the virtuous philosop[h]er, delights the camelion Poet.[63] It does no harm from its relish of the dark side of things any more than from its taste for the bright one; because they both end in speculation. A Poet is the most unpoetical of any thing in existence; because he has no Identity—he is continually in for—and filling some other Body—The Sun, the Moon, the Sea and Men and Women who are creatures of impulse are poetical and have about them an unchangeable attribute—the poet has none; no identity—he is certainly the most unpoetical of all God's Creatures. If then he has no self, and if I am a Poet, where is the Wonder that I should say I would write no more?[64] Might I not at that very instant [have] been cogitating on the Characters of Saturn and Ops? It is a wretched thing to confess; but is a very fact that not one word I ever utter can be taken for granted as an opinion growing out of my identical nature—how can it, when I have no nature? When I am in a room with People if I ever am free from speculating on creations of my own brain, then not myself goes home to myself: but the identity of every one in the room begins to press upon me that, I am in a very little time anhilated—not only among Men; it would be the same in a Nursery of children: I know not whether I make myself wholly understood: I hope enough so to let you see that no dependence is to be placed on what I said that day.

In the second place I will speak of my views, and of the life I purpose to myself—I am ambitious of doing the world some good: if I should be spared that may be the work of maturer years—in the interval I will assay to reach to as high a summit in Poetry as the nerve bestowed upon me will suffer. The faint conceptions I have of Poems to come brings the blood frequently into my forehead—All I hope is that I may not lose all interest in human affairs—that the solitary indifference I feel for applause even from the finest Spirits, will not blunt any acuteness of vision I may have. I do not think it will—I feel assured I should write from the mere yearning and fondness I have for the Beautiful even if my night's labours should be burnt every morning and no eye ever shine upon them. But even now I am perhaps not speaking from myself; but from some character in whose soul I now live. I am sure however that this next sentence is from myself. I feel your anxiety, good opinion and friendliness in the highest degree, and am

Your's most sincerely

JOHN KEATS

To George and Georgiana Keats
[*December 16, 1818–January 4, 1819*]

MY DEAR BROTHER AND SISTER,

You will have been prepared, before this reaches you for the worst news you could have, nay if Haslam's letter arrives in proper time, I have a consolation in thinking the first shock will be past before you receive this. The last days of poor Tom were of the most distressing nature; but his last moments were not so painful, and his very last was without a pang—I will not enter into any parsonic comments on death—yet the common observations of the commonest people on death are as true as their proverbs. I have scarce a doubt of immortality of some nature or other—neither had Tom. My friends have been exceedingly kind to me every one of them. . . .

Mrs Brawne who took Brown's house for the Summer, still resides in Hampstead—she is a very nice woman—and her daughter senior[65] is I think beautiful and elegant, graceful, silly, fashionable and strange we have a little tiff now and then—and she behaves a little better, or I must have sheered off. . . .

---

63. **camelion Poet:** In these sentences on the ideal of the "characterless" poet, Keats echoes Hazlitt on Shakespeare (cf. "On Shakspeare and Milton," above). For Woodhouse's comments on Keats's remarks, see the account by him printed below. 64. **where . . . more:** During an evening he had spent with Woodhouse, Keats had spoken of the extent to which earlier poets had already appropriated the best subjects and styles, and he added that there was nothing new left to be done. Woodhouse then wrote him a long, encouraging letter. Keats is now trying to reassure Woodhouse: because of his own "chamelion" nature, he is mercurial in his feelings and, having been writing about Saturn and Ops in *Hyperion*, may have been half-identified with their fallen state when he spoke as he did.

65. **her daughter senior:** Keats's first reference to Fanny Brawne (she was the elder of two daughters).

[*December 18*] . . . Shall I give you Miss Brawn[e]? She is about my height—with a fine style of countenance of the lengthen'd sort—she wants sentiment in every feature—she manages to make her hair look well—her nostrills are fine—though a little painful—he[r] mouth is bad and good—he[r] Profil is better than her full-face which indeed is not full but pale and thin without showing any bone—Her shape is very graceful and so are her movements—her Arms are good her hands badish—her feet tolerable—she is not seventeen—but she is ignorant—monstrous in her behaviour flying out in all directions, calling people such names—that I was forced lately to make use of the term *Minx*—this is I think no[t] from any innate vice but from a penchant she has for acting stylishly. I am however tired of such style and shall decline any more of it. . . .

[*December 31*] . . . My thoughts have turned lately this way—The more we know the more inadequacy we discover in the world to satisfy us—this is an old observation; but I have made up my Mind never to take any thing for granted—but even to examine the truth of the commonest proverbs—This however is true—Mrs Tighe and Beattie⁶⁶ once delighted me—now I see through them and can find nothing in them—or weakness—and yet how many they still delight! Perhaps a superior being may look upon Shakspeare in the same light—is it possible? No—This same inadequacy is discovered (forgive me little George you know I don't mean to put you in the mess) in Women with few exceptions—the Dress Maker, the blue Stocking⁶⁷ and the most charming sentimentalist differ but in a Slight degree, and are equally smokeable—But I'll go no further—I may be speaking sacrilegiously—and on my word I have thought so little that I have not one opinion upon any thing except in matters of taste—I never can feel certain of any truth but from a clear perception of its Beauty—and I find myself very young minded even in that perceptive power—which I hope will encrease—A year ago I could not understand in the slightest degree Raphael's cartoons—now I begin to read them a little—and how did I lea[r]n to do so? By seeing something done in quite an opposite spirit—I mean a picture of Guido's in which all the Saints, instead of that heroic simplicity and unaffected grandeur which they inherit from Raphael, had each of them both in countenance and gesture all the canting, solemn melodramatic mawkishness of Mackenzie's father Nicholas⁶⁸—When I was last at Haydon's I looked over a Book of Prints taken from the fresco of the Church at Milan the name of which I forget—in it are comprised Specimens of the first and second age of art in Italy—I do not think I ever had a greater treat out of Shakspeare—Full of Romance and the most tender feeling—magnificence of draperies beyond any I ever saw not excepting Raphael's—But Grotesque to a curious pitch—yet still making up a fine whole—even finer to me than more accomplish'd works—as there was left so much room for Imagination. . . .

> My dearest brother and sister
> Your most affectionate Brother
> JOHN—

To George and Georgiana Keats
[*February 14–May 3, 1819*]
Sunday Morn Feby 14th

MY DEAR BROTHER & SISTER—

. . . I am still at Wentworth Place—indeed I have kept in doors lately, resolved if possible to rid myself of my sore throat—consequently, i have not been to see your Mother since my return from Chichester—but my absence from her has been a great weight upon me—I say since my return from Chichester—I believe I told you I was going thither—I was nearly a fortnight at Mr John Snook's and a few days at old Mr Dilke's—Nothing worth speaking of happened at either place—I took down some of the thin paper and wrote on it a little Poem call'd 'St Agnes Eve'—which you shall have as it is when I have finished the blank part of the rest for you—I went out twice at Chichester to old Dowager card parties—I see very little now, and very few Persons—being almost tired of Men and things—Brown and Dilke are very kind and considerate towards me—The Miss Reynoldses have been stoppi[n]g next door lately—but all very dull—Miss Brawne and I have every now and then a chat and a tiff—Brown and Dilke are walking round their Garden hands in Pockets making observations. The Literary world I know nothing about—There is a Poem

---

66. Mrs Tighe and Beattie: Mary Tighe, 1772–1810, author of *Psyche* (1805), and James Beattie, 1735–1803. **67. blue Stocking:** a woman with literary interests. By "little George" Keats means Georgiana, his sister-in-law.

68. father Nicholas: "The Story of Father Nicholas," by the sentimental writer Henry Mackenzie.

from Rogers dead born—and another Satire is expected from Byron call'd Don Giovanni[69]— . . .

In my next Packet as this is one by the way, I shall send you the Pot of Basil, S[t] Agnes eve, and if I should have finished it a little thing call'd the 'eve of S[t] Mark' you see what fine mother Radcliff[70] names I have—it is not my fault—I did not search for them—I have not gone on with Hyperion—for to tell the truth I have not been in great cue for writing lately—I must wait for the sp[r]ing to rouse me up a little—The only time I went out from Bedhampton was to see a Chapel consecrated—Brown I and John Snook the boy, went in a chaise behind a leaden horse Brown drove, but the horse did not mind him

. . .

[*February 19*] . . . I have not said in any Letter yet a word about my affairs—in a word I am in no despair about them—my poem has not at all succeeded—in the course of a year or so I think I shall try the public again—in a selfish point of view I should suffer my pride and my contempt of public opinion to hold me silent—but for your's and Fanny's sake I will pluck up a spirit, and try again—I have no doubt of success in a course of years if I persevere—but it must be patience—for the Reviews have enervated and made indolent mens minds—few think for themselves—These Reviews too are getting more and more powerful and especially the Quarterly—They are like a superstition which the more it prostrates the Crowd and the longer it continues the more powerful it becomes just in proportion to their increasing weakness—I was in hopes that when people saw, as they must do now, all the trickery and iniquity of these Plagues they would scout them, but no they are like the spectators at the Westminster cock-pit—they like the battle and do not care who wins or who looses. . . . Brown has been walking up and down the room a breeding—now at this moment he is being delivered of a couplet—and I dare say will be as well as can be expected—Gracious—he has twins! I have a long Story to tell you about Bailey[71]—I will say first the circumstances as plainly and as well as I can remember, and then I will make my comment. You know that Bailey was very

much cut up about a little Jilt in the country somewhere; I thought he was in a dying state about it when at Oxford with him: little supposing as I have since heard, that he was at that very time making impatient Love to Marian Reynolds—and guess my astonishment at hearing after this that he had been trying at Miss Martin—So matters have been. So Matters stood—when he got ordained and went to a Curacy near Carlisle where the family of the Gleigs reside— There his susceptible heart was conquered by Miss Gleig—and thereby all his connections in town have been annulled— both male and female. I do not now remember clearly the facts. These however I know— He showed his correspondence with Mariane to Gleig—returned all her Letters and asked for his own— he also wrote very abrupt Letters to Mrs. Reynolds. I do not know any more of the Martin affair than I have written above. No doubt his conduct has been very bad. The great thing to be considered is—whether it is want of delicacy and principle or want of knowledge and polite experience. And again Weakness— yes, that is it; and the want of a Wife—yes, that is it— and then Mariane made great Bones of him although her Mother and sister have teased her very much about it. Her conduct has been very upright throughout the whole affair—She liked Bailey as a Brother—but not as a Husband—especially as he used to woo her with the Bible and Jeremy Taylor under his arm—they walked in no grove but Jeremy Taylor's. Mariane's obstinacy is some excuse—but his so quickly taking to Miss Gleig can have no excuse—except that of a Ploughman who wants a wife. The thing which sways me more against him than any thing else is Rice's conduct on the occasion: Rice would not make an immature resolve: he was ardent in his friendship for Bailey, he examined the whole for and against minutely; and he has abandoned Bailey entirely. All this I am not supposed by the Reynoldses to have any hint of. It will be a good Lesson to the Mother and Daughters—nothing would serve but Bailey. If you mentioned the word Teapot some one of them came out with an à propos about Bailey—noble fellow— fine fellow! was always in their mouths—This may teach them that the man who ridicules romance is the most romantic of Men—that he who abuses women and slights them loves them the most—that he who talks of roasting a Man alive would not do it when it came to the push—and above all, that they are very shallow people who take every thing literally. A Man's life of any worth is a continual allegory, and very few

69. Poem . . . Don Giovanni: *Human Life*, by Samuel Rogers, and the first two cantos of Byron's *Don Juan*. **70. mother Radcliff:** Mrs. Ann Radcliffe, 1764–1823, author of novels with Gothic atmosphere. **71. Bailey:** Keats's friend Benjamin Bailey (see the letters to him above). He was now a clergyman.

eyes can see the Mystery of his life—a life like the scriptures, figurative—which such people can no more make out than they can the Hebrew Bible. Lord Byron cuts a figure—but he is not figurative—Shakespeare led a life of Allegory: his works are the comments on it.

. . .

[*March 19*] Yesterday I got a black eye—the first time I took a Cr[icket] bat—Brown who is always one's friend in a disaster [app]lied a lee[ch to] the eyelid, and there is no infla[mm]ation this morning though the ball hit me dir[ectl]y on the sight—'t was a white ball—I am glad it was not a clout—This is the second black eye I have had since leaving school—during all my [scho]ol days I never had one at all—we must e[a]t a peck before we die—This morning I am in a sort of temper indolent and supremely careless: I long after a stanza or two of Thompson's[72] Castle of indolence—My passions are all asleep from my having slumbered till nearly eleven and weakened the animal fibre all over me to a delightful sensation about three degrees on this side of faintness—if I had teeth of pearl and the breath of lillies I should call it langour—but as I am I must call it Laziness—In this state of effeminacy the fibres of the brain are relaxed in common with the rest of the body, and to such a happy degree that pleasure has no show of enticement and pain no unbearable frown. Neither Poetry, nor Ambition, nor Love have any alertness of countenance as they pass by me: they seem rather like three figures on a greek vase—a Man and two women —whom no one but myself could distinguish in their disguisement.[73] This is the only happiness; and is a rare instance of advantage in the body overpowering the Mind. I have this moment received a note from Haslam in which he expects the death of his Father who has been for some time in a state of insensibility—his mother bears up he says very well—I shall go to town tommorrow to see him. This is the world—thus we cannot expect to give way many hours to pleasure—Circumstances are like Clouds continually gathering and bursting—While we are laughing the seed of some trouble is put into the wide arable land of events—while we are laughing it sprouts it grows and suddenly bears a poison fruit which we must pluck[74]—Even so we have leisure to

reason on the misfortunes of our friends; our own touch us too nearly for words. Very few men have ever arrived at a complete disinterestedness[75] of Mind: very few have been influenced by a pure desire of the benefit of others—in the greater part of the Bene-factors [of]& to Humanity some meretricious motive has sullied their greatness—some melodramatic scenery has facinated them—From the manner in which I feel Haslam's misfortune I perceive how far I am from any humble standard of disinterestedness—Yet this feeling ought to be carried to its highest pitch, as there is no fear of its ever injuring society—which it would do I fear pushed to an extremity—For in wild nature the Hawk would loose his Breakfast of Robins and the Robin his of Worms The Lion must starve as well as the swallow—The greater part of Men make their way with the same instinc-tiveness, the same unwandering eye from their purposes, the same animal eagerness as the Hawk—The Hawk wants a Mate, so does the Man—look at them both they set about it and procure on[e] in the same manner —They want both a nest and they both set about one in the same manner—they get their food in the same manner—The noble animal Man for his amusement smokes his pipe—the Hawk balances about the Clouds —that is the only difference of their leisures. This it is that makes the Amusement of Life—to a speculative Mind. I go among the Fields and catch a glimpse of a stoat or a fieldmouse peeping out of the withered grass —the creature hath a purpose and its eyes are bright with it—I go amongst the buildings of a city and I see a Man hurrying along—to what? The Creature has a pur-pose and his eyes are bright with it. But then as Words-worth says, "we have all one human heart"[76]—there is an ellectric fire in human nature tending to purify—so that among these human creature[s] there is continu-[a]lly some birth of new heroism—The pity is that we must wonder at it: as we should at finding a pearl in rubbish—I have no doubt that thousands of people never heard of have had hearts comp[l]etely disinter-ested: I can remember but two—Socrates and Jesus—

---

72. Thompson: James Thomson, 1700–48, best known for his poems *The Seasons* and *The Castle of Indolence.* **73. Neither . . . disguisement:** Cf. "Ode on Indolence," above. **74. While . . . pluck:** Cf. "Ode on Melancholy," above, ll. 21–24.

75. disinterestedness: freedom from selfish interests of any sort. The word, especially after Keats read Hazlitt's *Principles of Human Action* (see the selection from it above under Hazlitt), became a central ideal of Keats, interrelating with his ideal of Shakespeare as the "characterless" poet. (See the "Negative Capability" letter, above, Dec. 21–27, 1817, and the remarks on the poetic character to Woodhouse, Oct. 27, 1818). **76. "we . . . heart":** Wordsworth's "The Old Cumber-land Beggar," l. 153.

their Histories evince it—What I heard a little time ago, Taylor observe with respect to Socrates, may be said of Jesus—That he was so great as man that though he transmitted no writing of his own to posterity, we have his Mind and his sayings and his greatness handed to us by others. It is to be lamented that the history of the latter was written and revised by Men interested in the pious frauds of Religion. Yet through all this I see his splendour. Even here though I myself am pursueing the same instinctive course as the veriest human animal you can think of—I am however young writing at random—straining at particles of light in the midst of a great darkness—without knowing the bearing of any one assertion of any one opinion. Yet may I not in this be free from sin? May there not be superior beings amused with any graceful, though instinctive attitude my mind m[a]y fall into, as I am entertained with the alertness of a Stoat or the anxiety of a Deer? Though a quarrel in the streets is a thing to be hated, the energies displayed in it are fine; the commonest Man shows a grace in his quarrel—By a superior being our reasoning[s] may take the same tone—though erroneous they may be fine—This is the very thing in which consists poetry; and if so it is not so fine a thing as philosophy—For the same reason that an eagle is not so fine a thing as a truth—Give me this credit—Do you not think I strive—to know myself? Give me this credit—and you will not think that on my own accou[n]t I repeat Milton's lines

"How charming is divine Philosophy
Not harsh and crabbed as dull fools suppose
But musical as is Apollo's lute"—[77]

No—not for myself—feeling grateful as I do to have got into a state of mind to relish them properly—Nothing ever becomes real till it is experienced—Even a Proverb is no proverb to you till your Life has illustrated it—I am ever affraid that your anxiety for me will lead you to fear for the violence of my temperament continually smothered down: for that reason I did not intend to have sent you the following sonnet—but look over the two last pages and ask yourselves whether I have not that in me which will well bear the buffets of the world. It will be the best comment on my sonnet; it will show you that it was written with no Agony but that of ignorance; with no thirst of any thing but knowledge when pushed to the point though the first steps to it were throug[h] my human passions—they went away, and I wrote with my

Mind—and perhaps I must confess a little bit of my heart—

[Here follows the sonnet "Why Did I Laugh?" printed above.]

. . .

[*April 15*] Last Sunday I took a Walk towards Highgate and in the lane that winds by the side of Lord Mansfield's park I met M[r] Green[78] our Demonstrator at Guy's in conversation with Coleridge—I joined them, after enquiring by a look whether it would be agreeable—I walked with him a[t] his alderman-after dinner pace for near two miles I suppose In those two Miles he broached a thousand things—let me see if I can give you a list—Nightingales, Poetry—on Poetical sensation—Metaphysics—Different genera and species of Dreams—Nightmare—a dream accompanied by a sense of touch—single and double touch—A dream related—First and second consciousness—the difference explained between will and Volition—so many metaphysicians from a want of smoking the second consciousness—Monsters—the Kraken[79]—Mermaids— southey believes in them—southeys belief too much diluted—A Ghost story—Good morning—I heard his voice as he came towards me—I heard it as he moved away—I had heard it all the interval—if it may be called so. He was civil enough to ask me to call on him at Highgate Good Night! . . .

[*April 16*] . . . —The fifth canto of Dante pleases me more and more—it is that one in which he meets with Paulo and Francesca—I had passed many days in rather a low state of mind and in the midst of them I dreamt of being in that region of Hell. The dream was one of the most delightful enjoyments I ever had in my life— I floated about the whirling atmosphere as it is described with a beautiful figure to whose lips mine were joined as it seem'd for an age—and in the midst of all this cold and darkness I was warm—even flowery tree tops sprung up and we rested on them sometimes with the lightness of a cloud till the wind blew us away again—I tried a Sonnet upon it—there are fourteen lines but nothing of what I felt in it—o that I could dream it every night—

[Here follows the sonnet "On a Dream," printed above.]

. . .

---

77. "How . . . lute": *Comus*, ll. 475–77.

78. **Mr Green:** Joseph Green of Guy's Hospital. 79. **Kraken:** sea monster in Scandinavian folklore, shaped like a huge octopus.

[*April 21*] I have been reading lately two very different books Robertson's America and Voltaire's Siecle De Louis xiv. It is like walking arm and arm between Pizarro and the great-little Monarch. In How lamentabl[e] a case do we see the great body of the people in both instances: in the first, where Men might seem to inherit quiet of Mind from unsophisticated senses; from uncontamination of civilisation; and especial[l]y from their being as it were estranged from the mutual helps of Society and its mutual injuries— and thereby more immediately under the Protection of Providence—even there they had mortal pains to bear as bad; or even worse than Baliffs, Debts and Poverties of civilised Life—The whole appears to resolve into this—that Man is originally 'a poor forked creature'[80] subject to the same mischances as the beasts of the forest, destined to hardships and disquietude of some kind or other. If he improves by degrees his bodily accomodations and comforts—at each stage, at each accent there are waiting for him a fresh set of annoyances—he is mortal and there is still a heaven with its Stars abov[e] his head. The most interesting question that can come before us is, How far by the persevering endeavours of a seldom appearing Socrates Mankind may be made happy—I can imagine such happiness carried to an extreme—but what must it end in?—Death—and who could in such a case bear with death—the whole troubles of life which are now frittered away in a series of years, would the[n] be accumulated for the last days of a being who instead of hailing its approach, would leave this world as Eve left Paradise—But in truth I do not at all believe in this sort of perfectibility—the nature of the world will not admit of it—the inhabitants of the world will correspond to itself—Let the fish philosophise the ice away from the Rivers in winter time and they shall be at continual play in the tepid delight of summer. Look at the Poles and at the sands of Africa, Whirlpools and volcanoes—Let men exterminate them and I will say that they may arrive at earthly Happiness—The point at which Man may arrive is as far as the paral[l]el state in inanimate nature and no further—For instance suppose a rose to have sensation, it blooms on a beautiful morning it enjoys itself—but there comes a cold wind, a hot sun—it can not escape it, it cannot destroy its annoyances—they are as native to the world as itself: no more can man be happy in spite, the world[l]y elements will prey upon his nature—The

common cognomen of this world among the mis-guided and superstitious is 'a vale of tears' from which we are to be redeemed by a certain arbit[r]ary inter-position of God and taken to Heaven—What a little circumscribe[d] straightened notion! Call the world if you Please "The vale of Soul-making" Then you will find out the use of the world (I am speaking now in the highest terms for human nature admitting it to be immortal which I will here take for granted for the purpose of showing a thought which has struck me concerning it) I say '*Soul making*' Soul as distinguished from an Intelligence—There may be intelligences or sparks of the divinity in millions—but they are not Souls till they acquire identities, till each one is per-sonally itself. I[n]telligences are atoms of perception— they know and they see and they are pure, in short they are God—How then are Souls to be made? How then are these sparks which are God to have identity given them—so as ever to possess a bliss peculiar to each ones individual existence? How, but by the medium of a world like this? This point I sincerely wish to consider because I think it a grander system of salvation than the Christian religion—or rather it is a system of Spirit-creation—This is effected by three grand materials acting the one upon the other for a series of years. These three Materials are the *Intelligence*—the *human heart* (as distinguished from intelligence or Mind) and the *World* or *Elemental space* suited for the proper action of *Mind and Heart* on each other for the purpose of forming the *Soul* or *Intelligence destined to possess the sense of Identity*. I can scarcely express what I but dimly perceive—and yet I think I perceive it— that you may judge the more clearly I will put it in the most homely form possible—I will call the *world* a School instituted for the purpose of teaching little children to read—I will call the *human heart* the *horn Book*[81] used in that School—and I will call the *Child able to read, the Soul* made from that *school* and its *hornbook*. Do you not see how necessary a World of Pains and troubles is to school an Intelligence and make it a soul? A Place where the heart must feel and suffer in a thousand diverse ways! Not merely is the Heart a Hornbook, It is the Minds Bible, it is the Minds ex-perience, it is the teat from which the Mind or intel-ligence sucks its identity—As various as the Lives of Men are—so various become their souls, and thus does

---

80. 'a . . . creature': *King Lear*, III. iv. 111.

81. **horn Book**: child's primer, consisting of a sheet of paper (containing the alphabet and other rudiments) on a wooden board, protected by a covering of transparent horn.

God make individual beings, Souls, Identical Souls of the sparks of his own essence—This appears to me a faint sketch of a system of Salvation which does not affront our reason and humanity—I am convinced that many difficulties which christians labour under would vanish before it—there is one wh[i]ch even now Strikes me—the Salvation of Children—In them the Spark or intelligence returns to God without any identity—it having had no time to learn of, and be altered by, the heart—or seat of the human Passions—It is pretty generally suspected that the chr[i]stian scheme has been coppied from the ancient persian and greek Philosophers. Why may they not have made this simple thing even more simple for common apprehension by introducing Mediators and Personages in the same manner as in the he[a]then mythology abstractions are personified—Seriously I think it probable that this System of Soul-making—may have been the Parent of all the more palpable and personal Schemes of Redemption, among the Zoroastrians the Christians and the Hindoos. For as one part of the human species must have their carved Jupiter; so another part must have the palpable and named Mediatior and saviour, their Christ their Oromanes and their Vishnu[82]—If what I have said should not be plain enough, as I fear it may not be, I will put you in the place where I began in this series of thoughts—I mean, I began by seeing how man was formed by circumstances—and what are circumstances?—but touchstones of his heart—? and what are touchstones?—but proovings of his heart? and what are proovings of his heart but fortifiers or alterers of his nature? and what is his altered nature but his soul?—and what was his soul before it came into the world and had These provings and alterations and perfectionings?—An intelligences—without Identity—and how is this Identity to be made? Through the medium of the Heart? And how is the heart to become this Medium but in a world of Circumstances?

. . .

[April 30] The following Poem—the last I have written is the first and the only one with which I have taken even moderate pains—I have for the most part dash'd off my lines in a hurry—This I have done leisurely—I think it reads the more richly for it and will I hope encourage me to write other thing[s] in even

a more peacable and healthy spirit. You must recollect that Psyche was not embodied as a goddess before the time of Apulieus the Platonist who lived afteir the Augustan age, and consequently the Goddess was never worshipped or sacrificed to with any of the ancient fervour—and perhaps never thought of in the old religion[83]—I am more orthodox than to let a hethen Goddess be so neglected—

[Here follows the "Ode to Psyche," printed above.]

I have been endeavouring to discover a better sonnet stanza than we have. The legitimate does not suit the language over-well from the pouncing rhymes—the other kind appears too elegaic—and the couplet at the end of it has seldom a pleasing effect—I do not pretend to have succeeded—it will explain itself—

[Here follows the sonnet, "If by Dull Rhymes," printed above.]

Here endeth the other Sonnet—

[May 3] this is the 3d of May & every thing is in delightful forwardness; the violets are not withered, before the peeping of the first rose; You must let me know every thing, how parcels go & come, what papers you have, & what Newspapers you want, & other things — God bless you my dear Brother & Sister
Your ever Affectionate Brother
JOHN KEATS—

To Fanny Keats
[May 1, 1819]
Wentworth Place Saturday

MY DEAR FANNY,

. . . I continue increasing my letter to George to send it by one of Birkbeck's[84] sons who is going out soon—so if you will let me have a few more lines, they will be in time—I am glad you got on so well with Monsr le Curè—is he a nice Clergyman—a great deal depends upon a cock'd hat and powder—not gun powder, lord love us, but lady-meal, violet-smooth,

82. **Oromanes . . . Vishnu:** Oromanes (Ahriman) was the evil principle in the ancient Persian religion, Vishnu the "Preserver" in Hinduism.

83. **old religion:** in the Greek religion, as contrasted with the Roman (see the headnote to "Ode to Psyche," above).
84. **Birkbeck:** George Birkbeck, to whose settlement in Illinois George Keats intended to emigrate. Finding most of the better land already appropriated, George went to Kentucky and settled there.

dainty-scented lilly-white, feather-soft, wigsby-dressing, coat-collar-spoiling whisker-reaching, pig-tail loving, swans down-puffing, parson-sweetening powder—I shall call in passing at the Tottenham nursery and see if I can find some seasonable plants for you. That is the nearest place—or by our la' kin or lady kin, that is by the virgin Mary's kindred, is there not a twig-manufacturer in Walthamstow? M*r* & M*rs* Dilke are coming to dine with us to day—they will enjoy the country after Westminster—O there is nothing like fine weather, and health, and Books, and a fine country, and a contented Mind, and Diligent-habit of reading and thinking, and an amulet against the ennui—and, please heaven, a little claret-wine cool out of a cellar a mile deep—with a few or a good many ratafia cakes—a rocky basin to bathe in, a strawberry bed to say your prayers to Flora[85] in, a pad nag to go you ten miles or so; two or three sensible people to chat with; two or th[r]ee spiteful folkes to spar with; two or three odd fishes to laugh at and two or three numskuls to argue with—instead of using dumb bells on a rainy day—

> Two or three Posies
> With two or three simples
> Two or three Noses
> With two or th[r]ee pimples—
> Two or three wise men
> And two or three ninny's
> Two or three purses
> And two or three guineas
> Two or three raps
> At two or three doors
> Two or three naps
> Of two or three hours—
> Two or three Cats
> And two or three mice
> Two or th[r]ee sprats
> At a very great price—
> Two or three sandies
> And two or three tabbies
> Two or th[r]ee dandies—
> And two M*rs*——[86] mum!
> Two or three Smiles
> And two or three frowns
> Two or th[r]ee Miles

> To two or three towns
> Two or three pegs
> For two or three bonnets
> Two or three dove's eggs
> To hatch into sonnets—

Good bye I've an appoantment—can't stop pon word—good bye—now dont get up—open the door myself—go-o-o d bye—see ye Monday

J—K—

To Sarah Jeffrey[87]
[*May 31, 1819*]

*C. Brown Esq*re*'s*
Wentworth Place—Hampstead—

MY DEAR LADY,

   I was making a day or two ago a general conflagra-tion of all old Letters and Memorandums, which had become of no interest to me—I made however, like the Barber-inquisitor in Don Quixote[88] some reser-vations—among the rest your and your Sister's Letters. I assure you you had not entirely vanished from my Mind, or even become shadows in my remembrance: it only needed such a memento as your Letters to bring you back to me—Why have I not written before? Why did I not answer your Honiton Letter? I had no good news for you—every concern of ours, (ours I wish I could say) and still I must say *ours*—though George is in America and I have no Brother left—Though in the midst of my troubles I had no relation except my young sister I have had excellent friends. M*r* B.[89] at whose house I now am, invited me,—I have been with him ever since. I could not make up my mind to let you know these things. Nor should I now—but see what a little interest will do—I want you to do me a Favor; which I will first ask and then tell you the reasons. Enquire in the Villages round Teignmouth if there is any Lodging commodious for

85. **claret-wine . . . Flora:** Cf. in the "Ode to a Nightin-gale," above, ll. 11–13 (the wine "Cool'd . . . in the deep-delved earth, / Tasting of Flora . . ."). **86. M*rs*——:** Abbeys (see above, n. 58).

87. **Sarah Jeffrey:** Keats had met the Jeffrey sisters when he had gone to Teignmouth, Devon (spring, 1818), to nurse Tom. He had just received news that George was in need of money. The alternatives he mentions below were to sign up as a ship's surgeon on an Indiaman or to try, for two or three months more, to write poems that would sell. Since Charles Brown (in whose house Keats now lived) was renting his house for the summer, Keats, if he was to write, had to find a cheap place to stay. **88. Barber-inquisitor . . . Quixote:** Pt. I, chap. VI. **89. M*r* B.:** Charles Brown.

its cheapness; and let me know where it is and what price. I have the choice as it were of two Poisons (yet I ought not to call this a Poison) the one is voyaging to and from India for a few years; the other is leading a fevrous life alone with Poetry—This latter will suit me best—for I cannot resolve to give up my Studies It strikes me it would not be quite so proper for you to make such inquiries—so give my love to your Mother and ask her to do it. Yes, I would rather conquer my indolence and strain my ne[r]ves at some grand Poem—than be in a dunderheaded indiaman—Pray let no one in Teignmouth know any thing of this— Fanny must by this time have altered her name— perhaps you have also[90]—are you all alive? Give my Comp^ts to M^rs—your Sister. I have had good news, (tho' 'tis a queerish world in which such things are call'd good) from George—he and his wife are well—I will tell you more soon—Especially dont let the Newfoundland fisherman know it—and especially no one else—I have been always till now almost as careless of the world as a fly—my troubles were all of the Imagination—My Brother George always stood between me and any dealings with the world—Now I find I must buffet it—I must take my stand upon some vantage ground and begin to fight—I must choose between despair & Energy—I choose the latter—though the world has taken on a quakerish look with me, which I once thought was impossible—

'Nothing can bring back the hour
Of splendour in the grass and glory in the flower' [91]

I once thought this a Melancholist's dream—

But why do I speak to you in this manner? No believe me I do not write for a mere selfish purpose— the manner in which I have written of myself will convince you. I do not do so to Strangers. I have not quite made up my mind—Write me on the receipt of this—and again at your Leisure; between whiles you shall hear from me again—

Your sincere friend
JOHN KEATS

To Sarah Jeffrey
[*June 9, 1819*]
Wentworth Place

MY DEAR YOUNG LADY,

. . . Your advice[92] about the Indiaman is a very wise advice, because it justs suits me, though you are a little in the wrong concerning its destroying the energies of Mind: on the contrary it would be the finest thing in the world to strengthen them—To be thrown among people who care not for you, with whom you have no sympathies forces the Mind upon its own resources, and leaves it free to make its speculations of the differences of human character and to class them with the calmness of a Botanist. An Indiaman is a little world. One of the great reasons that the English have produced the finest writers in the world is, that the English world has ill-treated them during their lives and foster'd them after their deaths. They have in general been trampled aside into the bye paths of life and seen the festerings of Society. They have not been treated like the Raphaels of Italy. And where is the Englishman and Poet who has given a magnificent Entertainment at the christening of one of his Hero's Horses as Boyardo[93] did? He had a Castle in the Appenine. He was a noble Poet of Romance; not a miserable and mighty Poet of the human Heart. The middle age of Shakspeare was all clouded over; his days were not more happy than Hamlet's who is perhaps more like Shakspeare himself in his common every day Life than any other of his Characters—Ben Johnson was a common Soldier and in the Low countries, in the face of two armies, fought a single combat with a French Trooper and slew him—For all this I will not go on board an Indiaman, nor for examples sake run my head into dark alleys: I dare say my discipline is to come, and plenty of it too. I have been very idle lately, very averse to writing; both from the overpowering idea of our dead poets and

---

90. Fanny . . . also: This Fanny was one of the Jeffrey sisters; the implication is that she and perhaps Sarah have married. **91. 'Nothing . . . flower':** Wordsworth's "Ode: Intimations of Immortality," ll. 177–78.

92. advice: not to sign up on an Indiaman. Keats had meanwhile decided to go with his friend James Rice (who was ill and needed companionship) to Shanklin, Isle of Wight, for a month or so. While there Keats plunged into *Lamia* and began *Otho the Great* and *The Fall of Hyperion* (see headnotes to *Lamia* and *The Fall*, above). **93. Boyardo:** Matteo Maria Boiardo, 1434–1494.

from abatement of my love of fame. I hope I am a little more of a Philosopher than I was, consequently a little less of a versifying Pet-lamb.[94] . . .

Ever sincerely yours'
JOHN KEATS.

To J. H. Reynolds
[*July 11, 1819,*]

MY DEAR REYNOLDS,

. . .

You will be glad to hear under my own hand (tho' Rice says we are like sauntering Jack & Idle Joe) how diligent I have been, & am being. I have finish'd the Act,[95] and in the interval of beginning the 2ᵈ have proceeded pretty well with Lamia, finishing the 1ˢᵗ part which consists of about 400 lines. I have great hopes of success, because I make use of my Judgment more deliberately than I yet have done; but in Case of failure with the world, I shall find my content. And here (as I know you have my good at heart as much as a Brother,) I can only repeat to you what I have said to George—that however I shoᵈ like to enjoy what the competences of life procure, I am in no wise dashed at a different prospect. I have spent too many thoughtful days & moralized thro' too many nights for that, and fruitless woᵈ they be indeed, if they did not by degrees make me look upon the affairs of the world with a healthy deliberation. I have of late been moulting: not for fresh feathers & wings: they are gone, and in their stead I hope to have a pair of patient sublunary legs. I have altered, not from a Chrysalis into a butterfly, but the Contrary. having two little loopholes, whence I may look out into the stage of the world: and that world on [my] our coming here I almost forgot. The first time I sat down to write, I coᵈ scarcely believe in the necessity of so doing. It struck me as a great oddity—Yet the very corn which is now so beautiful, as if it had only taken to ripening yesterday, is for the market: So, why shoᵈ I be delicate.—

. . .

To Fanny Brawne
[*July 25, 1819*]
Sunday Night.

MY SWEET GIRL,

I hope you did not blame me much for not obeying your request of a Letter on Saturday: we have had four in our small room playing at cards night and morning leaving me no undisturb'd opportunity to write. Now Rice and Martin[96] are gone I am at liberty. Brown to my sorrow confirms the account you give of your ill health. You cannot conceive how I ache to be with you: how I would die for one hour——for what is in the world? I say you cannot conceive; it is impossible you should look with such eyes upon me as I have upon you: it cannot be. Forgive me if I wander a little this evening, for I have been all day employ'd in a very abstr[a]ct Poem[97] and I am in deep love with you—two things which must excuse me. I have, believe me, not been an age in letting you take possession of me; the very first week I knew you I wrote myself your vassal; but burnt the Letter as the very next time I saw you I thought you manifested some dislike to me. If you should ever feel for Man at the first sight what I did for you, I am lost. Yet I should not quarrel with you, but hate myself if such a thing were to happen—only I should burst if the thing were not as fine as a Man as you are as a Woman. Perhaps I am too vehement, then fancy me on my knees, especially when I mention a part of your Letter which hurt me; you say speaking of Mr. Severn[98] "but you must be satisfied in knowing that I admired you much more than your friend." My dear love, I cannot believe there ever was or ever could be any thing to admire in me especially as far as sight goes—I cannot be admired, I am not a thing to be admired. You are, I love you; all I can bring you is a swooning admiration of your Beauty. I hold that place among Men which snubnos'd brunettes with meeting eyebrows do among women—they are trash to me—unless I should find one among them with a fire in her heart like the one that burns in mine. You absorb me in spite of myself—you alone: for I look not forward

---

94. versifying Pet-lamb: Cf. "Ode on Indolence," above, l. 54 ("A pet-lamb in a sentimental farce"). **95. Act:** the first act of *Otho the Great.*

96. Rice and Martin: James Rice (n. 92, above) and John Martin have now left Shanklin; meanwhile Charles Brown has appeared to work with Keats on *Otho.* **97. Poem:** *The Fall of Hyperion.* **98. Mr. Severn:** Joseph Severn, the friend who accompanied Keats to Rome (see Introduction, above).

with any pleasure to what is call'd being settled in the world; I tremble at domestic cares—yet for you I would meet them, though if it would leave you the happier I would rather die than do so. I have two luxuries to brood over in my walks, your Loveliness and the hour of my death. O that I could have possession of them both in the same minute. I hate the world: it batters too much the wings of my self-will, and would I could take a sweet poison from your lips to send me out of it. From no others would I take it. I am indeed astonish'd to find myself so careless of all cha[r]ms but yours—remembring as I do the time when even a bit of ribband was a matter of interest with me. What softer words can I find for you after this—what it is I will not read. Nor will I say more here, but in a Postscript answer any thing else you may have mentioned in your Letter in so many words—for I am distracted with a thousand thoughts. I will imagine you Venus tonight and pray, pray, pray to your star like a He[a]then.

                              Your's ever, fair Star,
                                      JOHN KEATS.

To Benjamin Bailey
[*August 14, 1819*]

                    . . .

We removed to Winchester for the convenience of a Library and find it an exceeding pleasant Town, enriched with a beautiful Cathedrall and surrounded by a fresh-looking country. We are in tolerably good and cheap Lodgings. Within these two Months I have written 1500 Lines, most of which besides many more of prior composition you will probably see by next Winter. I have written two Tales, one from Boccac[c]io call'd the Pot of Basil; and another call'd St Agnes' Eve on a popular superstition; and a third call'd Lamia—(half finished—I have also been writing parts of my Hyperion[99] and completed 4 Acts of a Tragedy. It was the opinion of most of my friends that I should never be able to write a scene—I will endeavour to wipe away the prejudice—I sincerely hope you will be pleased when my Labours since we last saw each other shall reach you—One of my Ambitions is to make as great a revolution in modern dramatic writing as Kean has done in acting—another to upset the drawling of the blue stocking literary world[100]—if in the course of a few years I do these two things I ought

to die content—and my friends should drink a dozen of Claret on my Tomb—I am convinced more and more every day that (excepting the human friend Philosopher) a fine writer is the most genuine Being in the World. Shakspeare and the paradise Lost every day become greater wonders to me—I look upon fine Phrases like a Lover—I was glad to see, by a Passage in one of Brown's Letters some time ago from the north that you were in such good Spirits—Since that you have been married and in congra[tu]lating you I wish you every continuance of them—Present my Respects to Mrs Bailey. This sounds oddly to me, and I dare say I do it awkwardly enough: but I suppose by this time it is nothing new to you—Brown's remembrances to you—As far as I know we shall remain at Winchester for a goodish while—

                    Ever your sincere friend
                                      JOHN KEATS.

To J. H. Reynolds
[*September 21, 1819*]
Winchester. Tuesday

MY DEAR REYNOLDS,

    . . . How beautiful the season is now—How fine the air. A temperate sharpness about it. Really, without joking, chaste weather—Dian skies—I never lik'd stubble fields so much as now—Aye better than the chilly green of the spring. Somehow a stubble plain looks warm—in the same way that some pictures look warm—this struck me so much in my Sunday's walk that I composed upon it.[101] I hope you are better employed than in gaping after weather. I have been at different times so happy as not to know what weather it was—No I will not copy a parcel of verses. I always somehow associate Chatterton with autumn. He is the purest writer in the English Language. He has no French idiom, or particles like Chaucer—'tis genuine English Idiom in English words. I have given up Hyperion—there were too many Miltonic inversions in it—Miltonic verse cannot be written but in an artful or rather artist's humour. I wish to give myself up to other sensations. English ought to be kept up. It may be interesting to you to pick out some lines from Hyperion and put a mark × to the false beauty proceeding from art, and one ‖ to the true voice of feeling. Upon my soul 'twas imagination I cannot make the distinction—Every now & then there is a

---

99. Hyperion: *The Fall of Hyperion.* **100. Kean . . . world:**
For Kean see n. 21, and for "blue stocking," n. 67, above.

101. Aye . . . it: Cf. the ode "To Autumn," above.

Miltonic intonation—But I cannot make the division properly. The fact is I must take a walk: for I am writing so long a letter to George; and have been employed at it all the morning. You will ask, have I heard from George. I am sorry to say not the best news—I hope for better—This is the reason among others that if I write to you it must be in such a scraplike way. I have no meridian to date Interests from, or measure circumstances—To night I am all in a mist; I scarcely know what's what—But you knowing my unsteady & vagarish disposition, will guess that all this turmoil will be settled by tomorrow morning. It strikes me to night that I have led a very odd sort of life for the two or three last years—Here & there—No anchor—I am glad of it. . . .

Ever your affectionate friend
JOHN KEATS—

To Richard Woodhouse
[September 21–22, 1819] Tuesday—

DEAR WOODHOUSE,

. . . After revolving certain circumstances in my Mind; chiefly connected with a late American letter[102]—I have determined to take up my abode in a cheap Lodging in Town and get employment in some of our elegant Periodical Works—I will no longer live upon hopes—I shall carry my plan into execution speedily—I shall live in Westminster—from which a walk to the British Museum will be noisy and muddy—but otherwise pleasant enough—I shall enquire of Hazlitt how the figures of the market stand.[103] O that I could [write] somthing agrest rural, pleasant, fountain-vo[i]c'd—not plague you with unconnected nonsense—But things won't leave me *alone*. I shall be in Town as soon as either of you—I only wait for an answer from Brown: if he receives mine which is now a very moot point—I will give you a few reasons why I shall persist in not publishing The Pot of Basil—It is too smokeable—I can get it smoak'd at the Carpenters shaving chimney much more cheaply—There is too much inexperience of life, and simplicity of knowlege

in it—which might do very well after one's death—but not while one is alive. There are very few would look to the reality. I intend to use more finesse with the Public. It is possible to write fine things which cannot be laugh'd at in any way. Isabella is what I should call were I a reviewer 'A weaksided Poem' with an amusing sober-sadness about it. Not that I do not think Reynolds and you are quite right about it—it is enough for me. But this will not do to be public—If I may so say, in my dramatic capacity I enter fully into the feeling: but in Propria Persona I should be apt to quiz it myself There is no objection of this kind to Lamia—A good deal to St Agnes Eve—only not so glaring—Would a[s] I say I could write you something sylvestran.[104] But I have no time to think: I am an otiosus-peroccupatus[105] Man—I th[i]nk upon crutches, like the folks in your Pump room[106]—Have you seen old Bramble yet—they say he's on his last legs—The gout did not treat the old Man well so the Physician superseded it, and put the dropsy in office, who gets very fat upon his new employment, and behaves worse than the other to the old Man—But he'll have his house about his ears soon—We shall have another fall of Siege-arms—I suppose Mrs Humphrey persists in a big-belley—poor thing she little thinks how she is spo[i]ling the corners of her mouth—and making her nose quite a piminy.[107] Mr Humphrey I hear was giving a Lecture in the gaming-room—When some one call'd out Spousey! I hear too he has received a challenge from a gentleman who lost that evening—The fact is Mr H. is a mere nothing out of his bed-room.—Old Tabitha died in being bolstered up for a whist-party. They had to cut again—Chowder died long ago—Mrs H. laments that the last time they *put him* (i.e. to breed) he didn't take—They say he was a direct descendent of Cupid and Veney in the Spectator—This may be eisily known by the Parish Books—If you do not write in the course of a day or two: direct to me at Rice's—Let me know how you pass your times and how you are—

Your si[n]cere friend
JOHN KEATS—

102. **letter:** a letter from his brother George stating that he had lost over half of the money he had brought with him by investing in a steamboat that, as it turned out, had already sunk in the Ohio River. Keats, himself lacking money and by now becoming ill, was determined to earn money to send his brother. 103. **figures . . . stand:** writing for periodicals.

104. **sylvestran:** rustic. 105. **otiosus-peroccupatus:** leisurely-terribly-busy man. 106. **Pump room:** the Pump Room at Bath, where Woodhouse was now staying. With the mention of Bath (the scene of part of Smollett's *Humphry Clinker*), Keats suddenly starts, in the sentences that follow, to write a fanciful continuation of the novel. 107. **piminy:** having an affected expression.

To Charles Brown
September 23, 1819

. . .

Do not suffer me to disturb you unpleasantly: I do not mean that you should not suffer me to occupy your thoughts, but to occupy them pleasantly; for, I assure you, I am as far from being unhappy as possible. Imaginary grievances have always been more my torment than real ones. You know this well. Real ones will never have any other effect upon me than to stimulate me to get out of or avoid them. This is easily accounted for. Our imaginary woes are conjured up by our passions, and are fostered by passionate feeling; our real ones come of themselves, and are opposed by an abstract exertion of mind. Real grievances are displacers of passion. The imaginary nail a man down for a sufferer, as on a cross; the real spur him up into an agent. I wish, at one view, you could see my heart towards you. 'Tis only from a high tone of feeling that I can put that word upon paper—out of poetry. I ought to have waited for your answer to my last before I wrote this. I felt, however, compelled to make a rejoinder to your's. I had written to x x x x on the subject of my last,—I scarcely know whether I shall send my letter now. I think he would approve of my plan; it is so evident. Nay, I am convinced, out and out, that by prosing for awhile in periodical works I may maintain myself decently.

. . .

To George and Georgiana Keats
[September 17-27, 1819] Winchester

MY DEAR GEORGE,

. . . Tuesday [*September 21*]—You see I keep adding a sheet daily till I send the packet off—which I shall not do for a few days as I am inclined to write a good deal: for there can be nothing so remembrancing and enchaining as a good long letter be it composed of what it may—From the time you left me, our friends say I have altered completely—am not the same person—perhaps in this letter I am for in a letter one takes up one's existence from the time we last met—I dare say you have altered also—eve[r]y man does—Our bodies every seven years are completely fresh-

materiald—seven years ago it was not this hand that clench'd itself against Hammond[108]—We are like the relict garments of a Saint: the same and not the same: for the careful Monks patch it and patch it: till there's not a thread of the original garment left, and still they show it for St Anthony's shirt. This is the reason why men who had been bosom friends, on being separated for any number of years, afterwards meet coldly, neither of them knowing why—The fact is they are both altered—Men who live together have a silent moulding and influencing power over each other— They interassimulate. 'T is an uneasy thought that in seven years the same hands cannot greet each other again. All this may be obviated by a willful and drama- tic exercise of our Minds towards each other. Some think I have lost that poetic ardour and fire 't is said I once had—the fact is perhaps I have: but instead of that I hope I shall substitute a more thoughtful and quiet power. I am more frequently, now, contented to read and think—but now & then, haunted with ambitious thoughts. Qui[e]ter in my pulse, improved in my digestion; exerting myself against vexing speculations —scarcely content to write the best verses for the fever they leave behind. I want to compose without this fever. I hope I one day shall. You would scarcely imagine I could live alone so comfortably "Kepen in solitarinesse"[109] I told Anne, the servent here, the other day, to say I was not at home if any one should call. I am not certain how I should endu[r]e loneliness and bad weather together. Now the time is beautiful. I take a walk every day for an hour before dinner and this is generally my walk—I go out at the back gate across one street, into the Cathedral yard, which is always interesting; then I pass under the trees along a paved path, pass the beautiful front of the Cathedral, turn to the left under a stone door way—then I am on the other side of the building—which leaving behind me I pass on through two college-like squares seemingly built for the dwelling place of Deans and Prebendaries— garnished with grass and shaded with trees. Then I pass through one of the old city gates and then you are in one College-Street through which I pass and at the end thereof crossing some meadows and at last a country alley of gardens I arrive, that is, my worship

108. **Hammond**: Thomas Hammond, a surgeon to whom Keats was apprenticed after being taken from school. 109. **"Kepen . . . solitarinesse"**: *The Eve of Saint Mark*, above, l. 106.

arrives at the foundation of Saint Cross,[110] which is a very interesting old place, both for its gothic tower and alms-square and for the appropriation of its rich rents to a relation of the Bishop of Winchester—Then I pass across St Cross meadows till you come to the most beautifully clear river.

. . .

[September 24] . . . In the course of a few months I shall be as good an Italian Scholar as I am a French one—I am reading Ariosto[111] at present: not managing more than six or eight stanzas at a time. When I have done this language so as to be able to read it tolerably well—I shall set myself to get complete in latin and there my learning must stop. I do not think of venturing upon Greek. I would not go even so far if I were not persuaded of the power the knowlege of any language gives one. the fact is I like to be acquainted with foreign languages. It is besides a nice way of filling up intervals &c Also the reading of Dante is well worth the while. And in latin there is a fund of curious literature of the middle ages—The Works of many great Men Aretine and Sanazarius and Machievel —I shall never become attach'd to a foreign idiom so as to put it into my writings. The Paradise Lost though so fine in itself is a corruption of our Language—it should be kept as it is unique—a curiosity. a beautiful and grand Curiosity. The most remarkable Production of the world—A northern dialect accommodating itself to greek and latin inversions and intonations. The purest english I think—or what ought to be the purest —is Chatterton's—The Language had existed long enough to be entirely uncorrupted of Chaucer's gallicisms and still the old words are used—Chatterton's language is entirely northern—I prefer the native music of it to Milton's cut by feet I have but lately stood on my guard against Milton. Life to him would be death to me. Miltonic verse cannot be written but in the vein of art. . . .

> Believe me my dear brother and Sister—
> Your affectionate and anxious Brother
>
> JOHN KEATS

To John Taylor
[November 17, 1819]
Wentworth Place
Wednesday

MY DEAR TAYLOR,

I have come to a determination not to publish any thing I have now ready written; but for all that to publish a Poem before long and that I hope to make a fine one. As the marvellous is the most enticing and the surest guarantee of harmonious numbers I have been endeavouring to persuade myself to untether Fancy and let her manage for herself—I and myself cannot agree about this at all. Wonders are no wonders to me. I am more at home amongst Men and women. I would rather read Chaucer than Ariosto—The little dramatic skill I may as yet have however badly it might show in a Drama would I think be sufficient for a Poem—I wish to diffuse the colouring of St Agnes eve throughout a Poem in which Character and Sentiment would be the figures to such drapery—Two or three such Poems, if God should spare me, written in the course of the next six years, would be a famous gradus ad Parnassum altissimum[112]—I mean they would nerve me up to the writing of a few fine Plays—my greatest ambition—when I do feel ambitious. I am sorry to say that is very seldom. The subject we have once or twice talked of appears a promising one, The Earl of Leicester's historry. I am this morning reading Holingshed's Elisabeth,[113] You had some Books awhile ago, you promised to lend me, illustrative of my Subject. If you can lay hold of them or any others which may be serviceable to me I know you will encourage my low-spirited Muse by sending them— or rather by letting me know when our Errand cart Man shall call with my little Box. I will endeavour to set my self selfishly at work on this Poem that is to be—

> Your sincere friend
>
> JOHN KEATS—

110. **I take a walk . . . Cross:** The walk to the monastery of St. Cross is that also mentioned in the letter to Reynolds, above, September 21, and recalled in the ode "To Autumn."
111. **Ariosto:** Ludivico Ariosto, 1474–1533; cf. the remark on him in the next letter.

112. **gradus . . . altissimum:** stair to the top of Parnassus.
113. **Holingshed's Elisabeth:** Raphael Holinshed, *Chronicles of England* (1577).

To James Rice

[February 14–16, 1820]
Wentworth Place
Monday Morn.

MY DEAR RICE,

I have not been well enough to make any tolerable rejoinder to your kind Letter. I will as you advise be very chary of my health and spirits. I am sorry to hear of your relapse and hypochondriac symptoms attending it. Let us hope for the best as you say. I shall follow your example in looking to the future good rather than brooding upon present ill. I have not been so worn with lengthen'd illnesses as you have therefore cannot answer you on your own ground with respect to those haunting and deformed thoughts and feelings you speak of. When I have been or supposed myself in health I have had my share of them, especially within this last year. I may say that for 6 Months before I was taken ill I had not passed a tranquil day—Either that gloom overspre[a]d me or I was suffering under some passionate feeling, or if I turn'd to versify that acerbated the poison of either sensation. The Beauties of Nature had lost their power over me. How astonishingly (here I must premise that illness as far as I can judge in so short a time has relieved my Mind of a load of deceptive thoughts and images and makes me perceive things in a truer light)—How astonishingly does the chance of leaving the world impress a sense of its natural beauties on us. Like poor Falstaff, though I do not babble, I think of green fields.[114] I muse with the greatest affection on every flower I have known from my infancy—their shapes and coulours are as new to me as if I had just created them with a superhuman fancy—It is because they are connected with the most thoughtless and happiest moments of our Lives—I have seen foreign flowers in hothouses of the most beautiful nature, but I do not care a straw for them. The simple flowers of our sp[r]ing are what I want to see again. . . .

I am
my dear Rice
ever most sincer[e]ly yours
JOHN KEATS

To Fanny Brawne[115]

[February (?) 1820]

MY DEAR FANNY,

Do not let your mother suppose that you hurt me by writing at night. For some reason or other your last night's note was not so treasureable as former ones. I would fain that you call me *Love* still. To see you happy and in high spirits is a great consolation to me— still let me believe that you are not half so happy as my restoration would make you. I am nervous, I own, and may think myself worse than I really am; if so you must indulge me, and pamper with that sort of tenderness you have manifested towards me in different Letters. My sweet creature when I look back upon the pains and torments I have suffer'd for you from the day I left you to go to the Isle of Wight; the ecstasies in which I have pass'd some days and the miseries in their turn, I wonder the more at the Beauty which has kept up the spell so fervently. When I send this round I shall be in the front parlour watching to see you show yourself for a minute in the garden. How illness stands as a barrier betwixt me and you! Even if I was well——I must make myself as good a Philosopher as possible. Now I have had opportunities of passing nights anxious and awake I have found other thoughts intrude upon me. "If I should die," said I to myself, "I have left no immortal work behind me—nothing to make my friends proud of my memory—but I have lov'd the principle of beauty in all things, and if I had had time I would have made myself remember'd." Thoughts like these came very feebly whilst I was in health and every pulse beat for you— now you divide with this (may *I* say it?) "last infirmity of noble minds" all my reflection.

God bless you, Love.
J. KEATS.

---

114. **Falstaff . . . fields:** *Henry V*, II. iii. 17.

115. **Fanny Brawne:** Keats, who had a violent hemorrhage from the lungs on February 3, was confined first to bed and then to his room for over a month. During this time he wrote several short letters to Fanny Brawne, who was now living with her family next door, of which this and the following two are typical.

To Fanny Brawne
[February 24(?), 1820]

MY DEAREST GIRL,

Indeed I will not deceive you with respect to my Health. This is the fact as far as I know. I have been confined three weeks and am not yet well—this proves that there is something wrong about me which my constitution will either conquer or give way to—Let us hope for the best. Do you hear the Th[r]ush singing over the field? I think it is a sign of mild weather—so much the better for me. Like all Sinners now I am ill I philosophise aye out of my attachment to every thing, Trees, flowers, Thrushes Sp[r]ing, Summer, Claret &c &c aye [e]very thing but you[116]——my Sister would be glad of my company a little longer. That Thrush is a fine fellow I hope he was fortunate in his choice this year—Do not send any more of my Books home. I have a great pleasure in the thought of you looking on them.

Ever yours
my sweet Fanny
J—K—

To Fanny Brawne
[March 25, 1820]

MY DEAREST FANNY,

Though I shall see you in so short a time I cannot forbear sending you a few lines. You say I did not give you yesterday a minute account of my health. To-day I have left off the Medicine which I took to keep the pulse down[117] and I find I can do very well without it, which is a very favourable sign, as it shows there is no inflammation remaining. You think I may be wearied at night you say: it is my best time; I am at my best about eight o'Clock. I received a Note from Mr. Proctor[118] today. He says he cannot pay me a visit this weather as he is fearful of an inflammation in the Chest. What a horrid climate this is? or what careless inhabitants it has? You are one of them. My dear girl do not make a joke of it: do not expose yourself

to the cold. There's the Thrush again—I can't afford it—he'll run me up a pretty Bill for Music—besides he ought to know I deal at Clementi's.[119] How can you bear so long an imprisonment at Hampstead? I shall always remember it with all the gusto that a monopolizing carle should. I could build an Altar to you for it.

Your affectionate
J. K.

To Percy Bysshe Shelley
Hampstead August 16th [1820]

MY DEAR SHELLEY,

I am very much gratified that you, in a foreign country, and with a mind almost over occupied, should write to me in the strain of the Letter beside me.[120] If I do not take advantage of your invitation it will be prevented by a circumstance I have very much at heart to prophesy—There is no doubt that an english winter would put an end to me, and do so in a lingering hateful manner, therefore I must either voyage or journey to Italy as a soldier marches up to a battery. My nerves at present are the worst part of me, yet they feel soothed when I think that come what extreme may, I shall not be destined to remain in one spot long enough to take a hatred of any four particular bed-posts. I am glad you take any pleasure in my poor Poem;—which I would willingly take the trouble to unwrite, if possible, did I care so much as I have done about Reputation. I received a copy of the Cenci, as from yourself from Hunt. There is only one part of it I am judge of; the Poetry, and dramatic effect, which by many spirits now a days is considered the mammon. A modern work it is said must have a purpose, which may be the God—*an artist* must serve Mammon—he must have "self concentration" selfishness perhaps. You I am sure will forgive me for sincerely remarking that you might curb your magnanimity and be more of an artist, and 'load every rift' of your subject with ore.[121] The thought of such discipline must fall like cold chains upon you, who perhaps never sat with your wings furl'd for six Months together. And is not this extraordina[r]y talk for the writer of Endymion? whose mind was like a pack of scattered cards—I am

---

116. Trees . . . you: Cf. "What the Thrush Said," above, written two years before, and its suggestion of the promise of the year ahead generally. **117. pulse down:** Keats had just had a serious relapse, accompanied by palpitations of the heart (hence the reference to his rapid pulse). **118. Mr. Proctor:** the poet Bryan Waller Procter, who wrote under the name "Barry Cornwall."

119. Clementi's: London music publishers and makers of musical instruments. **120. strain . . . me:** Shelley, now living in Italy, had heard of Keats's illness and invited him to stay with him. **121. load . . . ore:** an echo of the *Faerie Queene*, II. vii. 28 (l. 5).

pick'd up and sorted to a pip. My Imagination is a Monastry and I am its Monk—you must explain my metap^cs to yourself. I am in expectation of Prometheus every day. Could I have my own wish for its interest effected you would have it still in manuscript—or be but now putting an end to the second act. I remember you advising me not to publish my first-blights, on Hampstead heath—I am returning advice upon your hands. Most of the Poems in the volume I send you have been written above two years, and would never have been publish'd but from a hope of gain; so you see I am inclined enough to take your advice now. I must exp[r]ess once more my deep sense of your kindness, adding my sincere thanks and respects for Mrs Shelley. In the hope of soon seeing you [I] remain

most sincerely [yours,]

JOHN KEATS—

To Charles Brown
Rome, 30 November 1820

MY DEAR BROWN,

'Tis the most difficult thing in the world to me to write a letter. My stomach continues so bad, that I feel it worse on opening any book,—yet I am much better than I was in Quarantine. Then I am afraid to encounter the proing and conning of any thing interesting to me in England. I have an habitual feeling of my real life having past, and that I am leading a posthumous existence. God knows how it would have been—but it appears to me—however, I will not speak of that subject. I must have been at Bedhampton nearly at the time you were writing to me from Chichester—how unfortunate—and to pass on the river too! There was my star predominant! I cannot answer any thing in your letter, which followed me from Naples to Rome, because I am afraid to look it over again. I am so weak (in mind) that I cannot bear the sight of any hand writing of a friend I love so much as I do you. Yet I ride the little horse,[122]—and, at my worst, even in Quarantine, summoned up more puns, in a sort of desperation, in one week than in any year of my life. There is one thought enough to kill me—I have been well, healthy, alert&c, walking with her[123]—and now— the knowledge of contrast, feeling for light and shade, all that information (primitive sense) necessary for a

poem are great enemies to the recovery of the stomach. There, you rogue, I put you to the torture,—but you must bring your philosophy to bear—as I do mine, really—or how should I be able to live? Dr Clarke[124] is very attentive to me; he says, there is very little the matter with my lungs, but my stomach, he says, is very bad. I am well disappointed in hearing good news from George,—for it runs in my head we shall all die young. I have not written to x x x x x[125] yet, which he must think very neglectful; being anxious to send him a good account of my health, I have delayed it from week to week. If I recover, I will do all in my power to correct the mistakes made during sickness; and if I should not, all my faults will be forgiven. I shall write to x x x to-morrow, or next day. I will write to x x x x x in the middle of next week. Severn is very well, though he leads so dull a life with me. Remember me to all friends, and tell x x x x I should not have left London without taking leave of him, but from being so low in body and mind. Write to George as soon as you receive this, and tell him how I am, as far as you can guess;—and also a note to my sister—who walks about my imagination like a ghost—she is so like Tom. I can scarcely bid you good bye even in a letter. I always made an awkward bow.

God bless you!

JOHN KEATS.

Joseph Severn to Charles Brown
Rome. 27 February 1821.

MY DEAR BROWN,

He is gone—he died with the most perfect ease—he seemed to go to sleep. On the 23rd, about 4, the approaches of death came on. "Severn—I—lift me up— I am dying—I shall die easy—don't be frightened— be firm, and thank God it has come!" I lifted him up in my arms. The phlegm seemed boiling in his throat, and increased until 11, when he gradually sunk into death—so quiet—that I still thought he slept. I cannot say now—I am broken down from four nights' watching, and no sleep since, and my poor Keats gone. Three days since, the body was opened; the lungs were completely gone. The Doctors could not conceive by what means he had lived these two months. I followed

---

122. little horse: The physicians had prescribed slow horseback-riding as exercise. **123. her:** Fanny Brawne.

124. D^r. Clarke: Dr. James Clark, Keats's physician in Rome. **125. xxxxx:** the names of Keats's friends were deleted by Charles Brown.

his poor body to the grave on Monday, with many English. They take such care of me here—that I must, else, have gone into a fever. I am better now—but still quite disabled.

The Police have been. The furniture, the walls, the floor, every thing must be destroyed by order of the law. But this is well looked to by Dr C.[126]

The letters I put into the coffin with my own hand. I must leave off.

J. S.

CHARLES COWDEN CLARKE

from RECOLLECTIONS OF WRITERS

∽⚬✤⚬∼

Charles Cowden Clarke (1787–1877) was the son of the headmaster of the school at Enfield, which Keats attended from the age of eight to the age of fifteen. After Keats was withdrawn from the school by his guardian and apprenticed to a surgeon, Clarke, who was seven years older than Keats, continued to encourage his interest in literature. His account of Keats is from *Recollections of Writers* (1878).

∽⚬✤⚬∼

In the early part of his school-life John gave no extraordinary indications of intellectual character; but it was remembered of him afterwards, that there was ever present a determined and steady spirit in all his undertakings: I never knew it misdirected in his required pursuit of study. He was a most orderly scholar. The future ramifications of that noble genius were then closely shut in the seed, which was greedily drinking in the moisture which made it afterwards burst forth so kindly into luxuriance and beauty.

My father was in the habit, at each half-year's vacation of bestowing prizes upon those pupils who had performed the greatest quantity of voluntary work; and such was Keats's indefatigable energy for the last two or three successive half-years of his remaining at school, that, upon each occasion, he took the first prize by a considerable distance. He was at work before the first school-hour began, and that was at seven

o'clock; almost all the intervening times of recreation were so devoted; and during the afternoon holidays, when all were at play, he would be in the school—almost the only one—at his Latin or French translation; and so unconscious and regardless was he of the consequences of so close and persevering an application, that he never would have taken the necessary exercise had he not been sometimes driven out for the purpose by one of the masters.

It has just been said that he was a favourite with all. Not the less beloved was he for having a highly pugnacious spirit, which, when roused, was one of the most picturesque exhibitions—off the stage—I ever saw. One of the transports of that marvellous actor, Edmund Kean—whom, by the way, he idolized—was its nearest resmblance; and the two were not very dissimilar in face and figure. Upon one occasion, when an usher, on account of some impertinent behaviour, had boxed his brother Tom's ears, John rushed up, put himself in the received posture of offence, and, it was said, struck the usher—who could, so to say, have put him into his pocket. His passion at times was almost ungovernable; and his brother George, being considerably the taller and stronger, used frequently to hold him down by main force, laughing when John was in "one of his moods," and was endeavouring to beat him. It was all, however, a wisp-of-straw conflagration; for he had an intensely tender affection for his brothers, and proved it upon the most trying occasions. He was not merely the "favourite of all," like a pet prize-fighter, for his terrier courage; but his high-mindedness, his utter unconsciousness of a mean motive, his placability, his generosity, wrought so general a feeling in his behalf, that I never heard a word of disapproval from any one, superior or equal, who had known him.

In the latter part of the time—perhaps eighteen months—that he remained at school, he occupied the hours during meals in reading. Thus, his *whole* time was engrossed. He had a tolerably retentive memory, and the quantity that he read was surprising. He must in those last months have exhausted the school library, which consisted principally of abridgments of all the voyages and travels of any note; Mavor's collection, also his "Universal History;" Robertson's histories of Scotland, America, and Charles the Fifth; all Miss Edgeworth's productions,[1] together with many other

---

126. D^r. C.: Dr. Clark (n. 124, above). By Italian law the furnishings in a room where anyone died of tuberculosis were burned in order to prevent the spread of the infection.

CHARLES COWDEN CLARKE. **1. Mavor's . . . productions:** The references are to William Mavor, 1758–1837, William Robertson, 1721–93, and Maria Edgeworth, 1767–1849.

works equally well calculated for youth. The books, however, that were his constantly recurrent sources of attraction were Tooke's "Pantheon," Lemprière's "Classical Dictionary," which he appeared to *learn*, and Spence's "Polymetis."[2] This was the store whence he acquired his intimacy with the Greek mythology; here was he "suckled in that creed outworn;" for his amount of classical attainment extended no farther than the "Æneid;" with which epic, indeed, he was so fascinated that before leaving school he had *voluntarily* translated in writing a considerable portion. And yet I remember that at that early age—mayhap under fourteen—notwithstanding, and through all its incidental attractiveness, he hazarded the opinion to me (and the expression riveted my surprise), that there was feebleness in the structure of the work. He must have gone through all the better publications in the school library, for he asked me to lend him some of my own books; and, in my "mind's eye," I now see him at supper (we had our meals in the schoolroom), sitting back on the form, from the table, holding the folio volume of Burnet's "History of his Own Time"[3] between himself and the table, eating his meal from beyond it. This work, and Leigh Hunt's *Examiner*—which my father took in, and I used to lend to Keats—no doubt laid the foundation of his love of civil and religious liberty. He once told me, smiling, that one of his guardians, being informed what books I had lent him to read, declared that if he had fifty children he would not send one of them to that school. Bless his patriot head!

When he left Enfield, at fourteen years of age, he was apprenticed to Mr. Thomas Hammond, a medical man, residing in Church Street, Edmonton, and exactly two miles from Enfield. This arrangement evidently gave him satisfaction, and I fear that it was the most placid period of his painful life; for now, with the exception of the duty he had to perform in the surgery—by no means an onerous one—his whole leisure hours were employed in indulging his passion for reading and translating. During his apprenticeship he finished the "Æneid."

The distance between our residences being so short,

I gladly encouraged his inclination to come over when he could claim a leisure hour; and in consequence I saw him about five or six times a month on my own leisure afternoons. He rarely came empty-handed; either he had a book to read, or brought one to be exchanged. When the weather permitted, we always sat in an arbour at the end of a spacious garden, and—in Boswellian dialect—"we had good talk."

It were difficult, at this lapse of time, to note the spark that fired the train of his poetical tendencies; but he must have given unmistakable tokens of his mental bent; otherwise, at that early stage of his career, I never could have read to him the "Epithalamion" of Spenser; and this I remember having done, and in that hallowed old arbour, the scene of many bland and graceful associations—the substances having passed away. At that time he may have been sixteen years old; and at that period of life he certainly appreciated the general beauty of the composition, and felt the more passionate passages; for his features and exclamations were ecstatic. . . .

That night he took away with him the first volume of the "Faerie Queene," and he went through it, as I formerly told his noble biographer, "as a young horse would through a spring meadow—ramping!" Like a true poet, too—a poet "born, not manufactured," a poet in grain, he especially singled out epithets, for that felicity and power in which Spenser is so eminent. He *hoisted* himself up, and looked burly and dominant, as he said, "what an image that is—*'sea-shouldering whales!'*"[4] It was a treat to see as well as hear him read a pathetic passage. Once, when reading the "Cymbeline" aloud, I saw his eyes fill with tears, and his voice faltered when he came to the departure of Posthumus, and Imogen saying she would have watched him—

> 'Till the diminution
> Of space had pointed him sharp as my needle;
> Nay follow'd him till he had *melted from*
> *The smallness of a gnat to air;* and then
> Have turn'd mine eye and wept.[5]

I cannot remember the precise time of our separating at this stage of Keats's career, or which of us first went to London; but it was upon an occasion, when walking thither to see Leigh Hunt, who had just fulfilled his penalty of confinement in Horsemonger Lane Prison for the unwise libel upon the Prince Regent, that

2. **Tooke's . . . Polymetis:** Andrew Tooke's *Pantheon* (1698); John Lemprière's *Classical Dictionary* (1768), and Joseph Spence's *Polymetis* (1747), all standard works on classical mythology. 3. **Burnet's . . . Time:** *History of His Own Time* by Gilbert Burnet, published posthumously in 1724–34, is a notable sourcebook for the history of the late seventeenth century.

4. **'sea-shouldering whales!':** *Faerie Queene*, II. xii. 236. 5. **'Till . . . wept:** *Cymbeline*, I. iii. 18–21.

Keats met me; and, turning, accompanied me back part of the way. At the last field-gate, when taking leave, he gave me the sonnet entitled, "Written on the day that Mr. Leigh Hunt left Prison." This I feel to be the first proof I had received of his having committed himself in verse; and how clearly do I recall the conscious look and hesitation with which he offered it! . . .

When we both had come to London—Keats to enter as a student of St. Thomas's Hospital—he was not long in discovering my abode, which was with a brother-in-law in Clerkenwell; and at that time being housekeeper, and solitary, he would come and renew his loved gossip; till, as the author of the "Urn Burial"[6] says, "we were acting our antipodes—the huntsmen were up in America, and they already were past their first sleep in Persia." At the close of a letter which preceded my appointing him to come and lighten my darkness in Clerkenwell, is his first address upon coming to London. He says,—"Although the Borough is a beastly place in dirt, turnings, and windings, yet No. 8, Dean Street, is not difficult to find; and if you would run the gauntlet over London Bridge, take the first turning to the right, and, moreover, knock at my door, which is nearly opposite a meeting, you would do me a charity, which, as St. Paul saith, is the father of all the virtues. At all events, let me hear from you soon: I say, at all events, not excepting the gout in your fingers." This letter, having no date but the week's day, and no postmark, preceded our first symposium; and a memorable night it was in my life's career.

A beautiful copy of the folio edition of Chapman's translation of Homer had been lent me. It was the property of Mr. Alsager, the gentleman who for years had contributed no small share of celebrity to the great reputation of the *Times* newspaper by the masterly manner in which he conducted the money-market department of that journal. Upon my first introduction to Mr. Alsager he lived opposite to Horsemonger Lane Prison, and upon Mr. Leigh Hunt's being sentenced for the libel, his first day's dinner was sent over by Mr. Alsager.

Well, then, we were put in possession of the Homer of Chapman, and to work we went, turning to some of the "famousest" passages, as we had scrappily known them in Pope's version. There was, for instance, that

perfect scene of the conversation on Troy wall of the old Senators with Helen, who is pointing out to them the several Greek Captains; with the Senator Antenor's vivid portrait of an orator in Ulysses, beginning at the 237th line of the third book:—

But when the prudent Ithacus did to his counsels rise,
He stood a little still, and fix'd upon the earth his eyes,
His sceptre moving neither way, but held it formally,
Like one that vainly doth affect. Of wrathful quality,
And frantic (rashly judging), you would have said he was;
But when out of his ample breast he gave his great voice pass,
And words that flew about our ears like drifts of winter's snow,
None thenceforth might contend with him, though naught admired for show.

The shield and helmet of Diomed, with the accompanying simile, in the opening of the third book; and the prodigious description of Neptune's passage to the Achive ships, in the thirteenth book:—

The woods and all the great hills near trembled beneath the weight
Of his immortal-moving feet. Three steps he only took,
Before he far-off Ægas reach'd, but with the fourth, it shook
With his dread entry.

One scene I could not fail to introduce to him—the shipwreck of Ulysses, in the fifth book of the "Odysseis," and I had the reward of one of his delighted stares, upon reading the following lines:—

Then forth he came, his both knees falt'ring, both
His strong hands hanging down, and all with froth
His cheeks and nostrils flowing, voice and breath
Spent to all use, and down he sank to death.
The sea had soak'd his heart through; all his veins
His toils had rack'd t' a labouring woman's pains.
Dead-weary was he.

On an after-occasion I showed him the couplet, in Pope's translation, upon the same passage:—

From mouth and nose the briny torrent ran,
And *lost in lassitude lay all the man.*

Chapman supplied us with many an after-treat; but it was in the teeming wonderment of this his first introduction, that, when I came down to breakfast the next morning, I found upon my table a letter with no other enclosure than his famous sonnet, "On First

Looking into Chapman's Homer.''[7] We had parted, as I have already said, at day-spring, yet he contrived that I should receive the poem from a distance of, may be, two miles by ten o'clock. In the published copy of this sonnet he made an alteration in the seventh line:—

Yet did I never breathe its pure serene.

The original which he sent me had the phrase—

Yet could I never tell what men could mean;

which he said was bald, and too simply wondering. No one could more earnestly chastise his thoughts than Keats.

. . .

In one of our conversations, about this period, I alluded to his position at St. Thomas's Hospital, coasting and reconnoitring, as it were, for the purpose of discovering what progress he was making in his profession; which I had taken for granted had been his own selection, and not one chosen for him. The total absorption, therefore, of every other mood of his mind than that of imaginative composition, which had now evidently encompassed him, induced me, from a kind motive, to inquire what was his bias of action for the future; and with that transparent candour which formed the mainspring of his rule of conduct, he at once made no secret of his inability to sympathize with the science of anatomy, as a main pursuit in life; for one of the expressions that he used, in describing his unfitness for its mastery, was perfectly characteristic. He said, in illustration of his argument, "The other day, for instance, during the lecture, there came a sunbeam into the room, and with it a whole troop of creatures floating in the ray; and I was off with them to Oberon and fairyland." And yet, with all his self-styled unfitness for the pursuit, I was afterwards informed that at his subsequent examination he displayed an amount of acquirement which surprised his fellow-students, who had scarcely any other association with him than that of a cheerful, crotchety rhymester. He once talked with me, upon my complaining of stomachic derangement, with a remarkable decision of opinion, describing the functions and actions of the organ with the clearness and, as I presume, technical precision of an adult practitioner; casually illustrating the comment, in his characteristic way, with poetical imagery: the stomach, he said, being like

a brood of callow nestlings (opening his capacious mouth) yearning and gaping for sustenance; and, indeed, he merely exemplified what should be, if possible, the "stock in trade" of every poet, viz., to *know* all that is to be known, "in the heaven above, or in the earth beneath, or in the waters under the earth."

It was about this period that, going to call upon Mr. Leigh Hunt, who then occupied a pretty little cottage in the Vale of Health, on Hampstead Heath, I took with me two or three of the poems I had received from Keats. I could not but anticipate that Hunt would speak encouragingly, and indeed approvingly, of the compositions—written, too, by a youth under age; but my partial spirit was not prepared for the unhesitating and prompt admiration which broke forth before he had read twenty lines of the first poem. Horace Smith happened to be there on the occasion, and he was not less demonstrative in his appreciation of their merits. The piece which he read out was the sonnet, "How many Bards gild the Lapses of Time!" marking with particular emphasis and approval the last six lines:—

So the unnumber'd sounds that evening store,
    The songs of birds, the whisp'ring of the leaves,
    The voice of waters, the great bell that heaves
With solemn sound, and thousand others more,
    That distance of recognizance bereaves,
Make pleasing music, and not wild uproar.

Smith repeated with applause the line in italics, saying, "What a well-condensed expression for a youth so young!" After making numerous and eager inquiries about him personally, and with reference to any peculiarities of mind and manner, the visit ended in my being requested to bring him over to the Vale of Health.

That was a "red-letter day" in the young poet's life, and one which will never fade with me while memory lasts.

The character and expression of Keats's features would arrest even the casual passenger in the street; and now they were wrought to a tone of animation that I could not but watch with interest, knowing what was in store for him from the bland encouragement, and Spartan deference in attention, with fascinating conversational eloquence, that he was to encounter and receive. As we approached the Heath, there was the rising and accelerated step, with the gradual subsidence of all talk. The interview, which stretched into three "morning calls," was the prelude to many after-scenes and saunterings about Caen Wood and

7. "On . . . Homer": See above, p. 1126.

its neighbourhood; for Keats was suddenly made a familiar of the household, and was always welcomed.

It was in the library at Hunt's cottage, where an extemporary bed had been made up for him on the sofa, that he composed the frame-work and many lines of the poem on "Sleep and Poetry." . . . The occasion that recurs with the liveliest interest was one evening when—some observations having been made upon the character, habits, and pleasant associations with that reverend denizen of the hearth, the cheerful little grasshopper of the fireside—Hunt proposed to Keats the challenge of writing then, there, and to time, a sonnet "On the Grasshopper and Cricket."[8] No one was present but myself, and they accordingly set to. I, apart, with a book at the end of the sofa, could not avoid furtive glances every now and then at the emulants. I cannot say how long the trial lasted. I was not proposed umpire; and had no stop-watch for the occasion. The time, however, was short for such a performance, and Keats won as to time. But the event of the after-scrutiny was one of many such occurrences which have riveted the memory of Leigh Hunt in my affectionate regard and admiration for unaffected generosity and perfectly unpretentious encouragement. His sincere look of pleasure at the first line—

> The poetry of earth is never dead.

"Such a prosperous opening!" he said; and when he came to the tenth and eleventh lines:—

> On a lone winter evening, *when the frost*
> *Has wrought a silence—*

"Ah! that's perfect! Bravo Keats!" And then he went on in a dilatation upon the dumbness of Nature during the season's suspension and torpidity. With all the kind and gratifying things that were said to him, Keats protested to me, as we were afterwards walking home, that he preferred Hunt's treatment of the subject to his own.

. . .

In my knowledge of fellow-beings, I never knew one who so thoroughly combined the sweetness with the power of gentleness, and the irresistible sway of anger, as Keats. His indignation would have made the boldest grave; and they who had seen him under the influence of injustice and meanness of soul would not forget the expression of his features—"the form of his visage was changed." Upon one occasion, when some local tyranny was being discussed, he amused the party by shouting, "Why is there not a human dust-hole, into which to tumble such fellows?"

Keats had a strong sense of humour, although he was not, in the strict sense of the term, a humorist, still less a farcist. His comic fancy lurked in the outermost and most unlooked-for images of association; which, indeed, may be said to form the components of humour; nevertheless, they did not extend beyond the *quaint* in fulfilment and success. But his perception of humour, with the power of transmitting it by imitation, was both vivid and irresistibly amusing. He once described to me his having gone to see a bear-baiting, the animal the property of a Mr. Tom Oliver. The performance not having begun, Keats was near to, and watched, a young aspirant, who had brought a younger under his wing to witness the solemnity, and whom he oppressively patronized, instructing him in the names and qualities of all the magnates present. Now and then, in his zeal to manifest and impart his knowledge, he would forget himself, and stray beyond the prescribed bounds into the ring, to the lashing resentment of its comptroller, Mr. William Soames, who, after some hints of a practical nature to "keep back," began laying about him with indiscriminate and unmitigable vivacity, the Peripatetic signifying to his pupil, "My eyes! Bill Soames giv' me sich a licker!" evidently grateful, and considering himself complimented upon being included in the general dispensation. Keats's entertainment with and appreciation of this minor scene of low life has often recurred to me. But his concurrent personification of the baiting, with his position—his legs and arms bent and shortened till he looked like Bruin on his hind legs, dabbing his fore paws hither and thither, as the dogs snapped at him, and now and then acting the gasp of one that had been suddenly caught and hugged —his own capacious mouth adding force to the personation, was a remarkable and as memorable a display.

8. **"On . . . Cricket":** See above, p. 1127.

BENJAMIN BAILEY

from REMINISCENCES
OF KEATS

❧

Benjamin Bailey (1791–1853) was a student at Oxford, preparing to enter the clergy, when he first met Keats in London in the spring of 1817. Throughout the summer of that same year, while Keats was writing *Endymion*, Bailey continued to see him frequently and then invited him to stay for a month in Bailey's lodgings at Oxford. There, at Magdalen Hall, Keats wrote Book III of *Endymion* throughout the month of September, 1817, shortly before he became twenty-two. The best account of Bailey and Keats's visit to Oxford is in W. J. Bate, *John Keats*, 1963, pp. 196–218. The following selection, from Bailey's letter to R. M. Milnes (May 7, 1849), the first biographer of Keats, is reprinted from *The Keats Circle*, ed. Hyder E. Rollins, 1948.

❧

It was, I think, about the end of 1816, or the beginning of 1817, that my friend, Mr Reynolds, wrote to me at Oxford respecting Keats, with whom he & his family had just become acquainted. He conveyed to me the same impressions, which the poet made upon the minds of almost all persons who had the happiness of knowing him, & subsequently upon myself. Early in 1817 his first volume of Poems was published by Ollier, which was sent to me. I required no more to satisfy me that he was indeed a Poet of rare and original genius.

On my first visit to London, I believe, after the publication of this Volume—at least not long after—I was introduced to him. I was delighted with the naturalness & simplicity of his character, & was at once drawn to him by his winning & indeed affectionate manner towards those with whom he was himself pleased. Nor was his personal appearance the least charm of a first acquaintance with the young Poet. He bore, along with the strong impress of genius, much beauty of feature & countenance. The Lady's sketch[1] comes very near to my own recollection. The contour of his face was, as she describes it, not square & angular, but circular & oval; & this is the proper shape of a poet's

head. Boccacio's & Spenser's faces & heads are so formed. It is in the character of the countenance what Coleridge would call *femineity* (see his Table Talk) which he thought to be a mental constituent of true genius.[2] His hair was beautiful—a fine brown, rather than auburn, I think; & if you placed your hand upon his head, the silken curls felt like the rich plumage of a bird. I do not particularly remember the thickness of the upper lip, which is so generally discribed, & doubtless correctly;—but the mouth struck me as too wide, both in itself, & as out of harmony with the rest of the face, which, with this single blemish, was eminently beautiful. The eye was full & fine, & softened into tenderness, or beamed with a fiery brightness, according to the current of his thoughts & conversation. Indeed the form of his head was like that of a fine Greek statue:—& he realized to my mind the youthful Apollo, more than any head of a living man whom I have known. Mr Severn's portrait, admirable as it is, does not convey to my mind & memory the peculiar sweetness of expression of John Keats during the,—alas!—short period of my personal intercourse with him. It has the character of more matured thought, with an expression, to me almost painful, of suffering. I know not at what time this likeness was taken: but I should suppose it at least a year or two later than 1817; & a year or two wrought wonders in Keat's mind, & most probably & naturally in the expression of his most expressive features.

At the commencement of the long Vacation I was again in London, on my way to another part of the country: & it was my intention to return to Oxford early in the Vacation for the purpose of reading. I saw much of Keats. And I invited him to return with me to Oxford, & spend as much time as he could afford with me in the silence & solitude of that beautiful place during the absence of the numerous members & students of the University. He accepted my offer, & we returned together—I think in August 1817. It was during this visit, & in my room. that he wrote the third book of Endymion.

BENJAMIN BAILEY. **1. Lady's sketch:** a description of Keats quoted by R. Monckton Milnes in his biography.

2. Coleridge . . . genius: "I cannot hit upon the passage referred to, somewhere in Coleridge's Table Talk. But here is a yet more apposite passage in the same work. 'X——'s face is almost the only exception I know to the observation, that something feminine—not *effeminate*, mind—is discoverable in the countenances of all men of Genius. Look at that face of old Dampier, a rough sailor, but a man of exquisite mind. How soft is the air of his countenance, how delicate the shape of his temples!' " (Bailey's note).

I think he had written the few first introductory lines which he read to me, before he became my guest. I did not then, & I cannot now very much approve that introduction. The "baaing vanities"[3] have something of the character of what was called "the cockney school." Nor do I like many of the forced rhymes, & the apparent effort, by breaking up the lines, to get as far as possible in the opposite direction of the Pope school. But having said this—which was my impression at the time of the composition of this Book, & so remains now—I must repeat at this distance of time, what I always felt & then expressed, that the Poem throughout is full of beauty, both of thought & diction, & rich beyond any poem of the same length in the English language, in the exuberance, even to overflowing, of a fine imagination.

His mode of composition of the third Book, of which I was a witness, is best described by recounting our habits of study for one day during the month he visited me at Oxford. He wrote, & I read, sometimes at the same table, & sometimes at separate desks or tables, from breakfast to the time of our going out for exercise,—generally two or three o'clock. He sat down to his task,—which was about 50 lines a day,— with his paper before him, & wrote with as much regularity, & apparently with as much ease, as he wrote his letters. Indeed he quite acted up to the principle he lays down in the letter of axioms to his publisher, (my old & valued friend Mʳ Taylor) on which you justly set the seal of your approbation—"That if poetry comes not as naturally as the leaves of a tree, it had better not come at all." This axiom he fulfilled to the letter by his own practice, *me teste*,[4] while he composed the third Book of Endymion, in the same room in which I studied daily, until he completed it. Sometimes he fell short of his allotted task,—but not often: & he would make it up another day. But he never forced himself. When he had finished his writing for the day, he usually read it over to me; & he read or wrote letters until we went out for a walk. This was our habit day by day. The rough manuscript was written off daily, & with few erasures.

I remember very distinctly, though at this distance of time, his reading of a few passages; & I almost think I hear his voice, & see his countenance. Most vivid is my recollection of the following passage of the fine & affecting story of the old man, Glaucus, which he read to me immediately after its composition:—

> "The old man raised his hoary head & saw
> The wildered stranger—seeming not to see,
> The features were so lifeless. Suddenly
> He woke as from a trance; his snow white brows
> Went arching up, *& like two magic ploughs*
> *Furrowed deep wrinkles in his forehead large,*
> *Which kept as fixedly as rocky marge,*
> *Till round his withered lips had gone a smile.*"[5]

The lines I have italicised, are those which then forcibly struck me as peculiarly fine, & to my memory have "kept as fixedly as rocky marge." I remember his upward look when he read of the "magic ploughs," which in his hands have turned up so much of the rich soil of Fairyland. When we had finished our studies for the day we took our walk, & sometimes boated on the Isis, as he describes these little excursions very graphically in a letter from Oxford to Mʳ Reynolds. And once we took a longer excursion of a day or two, to Stratford upon Avon, to visit the birthplace of Shakespeare. We went of course to the house visited by so many thousands of all nations of Europe, & inscribed our names in addition to the "numbers numberless" of those which literally blackened the walls: and if those walls have not been washed, or our names wiped out to find place for some others, they will still remain together upon that truly honored wall of a small low attic apartment. We also visited the church, & were pestered with a common place showman of the place. He was struck, I remember, with the simple statue there, which, though rudely executed, we agreed was most probably the best likeness of the many extant, but none very authentic, of the myriad-minded Shakspeare.—His enjoyment was of that genuine, quiet kind which was a part of his gentle nature; deeply feeling what he truly enjoyed, but saying little. On our return to Oxford we renewed our quiet mode of life, until he finished the third Book of Endymion, & the time came that we must part; & I never parted with one whom I had known so short a time, with so much real regret & personal affection, as I did with John Keats when he left Oxford for London at the end of September, or the beginning of October 1817.—

We often projected meeting again with each other. But something or other always intervened to prevent it: &, except now & then I believe in London, this was

3. "baaing vanities": *Endymion*, III. 3. 4. me teste: myself as witness.

5. "The . . . smile": *Endymion*, III. 218–25.

the last, as it was certainly the longest, time I saw Keats. But living as we did for a month or six weeks together (for I do not remember exactly how long) I knew him at that period of his life, perhaps, as well as any one of his friends. There was no reserve of any kind between us.—His health then appeared perfectly good; while mine was quite the reverse. He was soon after seized with the fatal family disease, which terminated in his untimely death. The world may have lost much by the early blight of a blossom of such promise. But for himself, his removal must be accounted a happy one. I could almost say, in submission to the merciful Providence of the Almighty, that my own lot had been far happier, had your sentence of death upon me,[6] along with my friend John Keats, been a true one.—It was a maxim of the ancients—"Quem Dii diligunt, adolescens moritur—" "The favorite of the Gods dies young."[7] And few, like myself, who have lived more than half a century in the world, & have been stript of their nearest & dearest ties to humanity,—do not echo the touching lines of the venerable living Bard, the greatest poet, and one of the best men of his age:—

> "Oh Sir! the good die first
> And they whose hearts are dry as summer dust
> Burn to the socket."
> Wordsworth's Excursion.[8]

I shall not do justice to Keats, if I do not say something further of his temper & manner, & of the style of his conversation, while I enjoyed the happiness & the privilege of his society.—His brother George says of him that to his brothers his temper was uncertain; & he himself confirms this judgment of him in a beautiful passage of a letter to myself (which you have printed, Vol 1 p 146) where he thus speaks of his brothers:—"My love for my brothers, from the early loss of our parents, & even from earlier misfortunes, has grown into an affection, 'passing the love of women.' I have been ill tempered with them, I have vexed them,—but the thought of them has always stifled the impression that any woman might otherwise have made upon me." This might have been so with his brothers. But with his friends, a sweeter tempered man I never knew, than was John Keats. Gentleness

was indeed his proper characteristic, without one particle of dullness, or insipidity, or want of spirit. Quite the contrary. "He was gentle but not fearful," in the chivalric & moral sense of the term "gentle." He was pleased with every thing that occurred in the ordinary mode of life, & a cloud never passed over his face, except of indignation at the wrongs of others.

His conversation was very engaging. He had a sweet-toned voice, "an excellent thing" in *man*, as well as "in woman."[9] A favorite expression of tranquil pleasure & delight at a fine passage of any author, particularly an old poet, was "that it was *nice*," which he pronounced in a gentle undertone. In his letters he talks of *suspecting* everybody. It appeared not in his conversation. On the contrary, he was uniformly the apologist for poor, frail human nature, & allowed for people's faults more than any man I ever knew, (except one, my dear & excellent old friend, Sir William Rough, late Chief Justice of Ceylon) & especially for the faults of his friends. But if any act of wrong or oppression, of fraud or falsehood, was the topic, he rose into sudden & animated indignation. He had a truly poetic feeling for women; & often spoke to me of his sister, who was somehow withholden from him, with great delicacy & tenderness of affection. He had a soul of noble integrity: & his common sense was a conspicuous part of his character. Indeed his character was, in the best sense, manly.

Our conversation rarely or never flagged, during our walks, or boatings, or in the Evening. And I have retained a few of his opinions on Literature & Criticism which I will detail.

The following passage from Wordsworth's ode on Immortality was deeply felt by Keats, who however at this time seemed to me to value this great Poet rather in particular passages than in the full length portrait, as it were, of the great imaginative & philosophic Christian Poet, which he really is, & which Keats obviously, not long afterwards, felt him to be.

> "Not for these I raise
> The song of thanks & praise;
> But for those obstinate questionings
> Of sense & outward things,
> Fallings from us, vanishings;
> Blank misgivings of a creature
> Moving about in worlds not realized,
> *High instincts, before which our mortal nature
> Did tremble like a guilty thing surprized.*"[10]

---

6. sentence . . . me: Milnes, in his biography, had said that Bailey was dead. Bailey, by then an archdeacon in far-off Ceylon, read the remark and then wrote the present account of Keats to Milnes. **7. "The . . . young":** Plautus, *Bacchides*, IV. vii. 16–17. **8.** *The Excursion:* I. 500–02.

9. "an . . . woman": *King Lear*, V. iii. 272–73. **10. "Not . . . surprized":** "Ode: Intimations of Immortality," ll. 139–47.

The last lines he thought were quite awful in their application to a guilty finite creature, like man, in the appalling nature of the feeling which they suggested to a thoughtful mind.

Again, we often talked of that noble passage in the Lines on Tintern Abbey:—

> "That blessed mood,
> In which *the burthen of the mystery*,
> In which the heavy & the weary weight
> Of all this unintelligible world
> Is lightened."[11]

And his references to this passage are frequent in his letters.—But in those exquisite stanzas:—

> "She dwelt among the untrodden ways,
> Beside the springs of Dove—"

ending,—

> "She lived unknown & few could know
> When Lucy ceased to be;
> But she is in her grave, & oh,
> *The difference to me*"—[12]

the simplicity of the last line he declared to be the most perfect pathos.

Among the qualities of high poetic promise in Keats was, even at this time, his correct taste. I remember to have been struck with this by his remarks on that well known & often quoted passage of the Excursion upon the Greek Mythology,—where it is said that

> "Fancy fetched
> Even from the blazing Chariot of the Sun
> A beardless youth who touched a golden lute,
> *And filled the illumined groves with ravishment*."[13]

Keats said this description of Apollo should have ended at the "golden lute," & have left it to the imagination to complete the picture,—*how* he "filled the illumined groves." I think every man of taste will feel the justice of the remark.

Every one now knows what was then known to his friends, that Keats was an ardent admirer of Chatterton. The melody of the verses of "the marvellous Boy who perished in his pride,"[14] enchanted the author

of Endymion. Methinks I now hear him recite, or *chant*, in his peculiar manner, the following stanza of the "Roundelay sung by the minstrels of Ella":

> "*Come with acorn cup & thorn*,
> Drain my hertys blood away;
> Life & all its goods I scorn;
> Dance by night or feast by day."[15]

The first line to his ear possessed the great charm. Indeed his sense of melody was quite exquisite, as is apparent in his own verses; & in none more than in numerous passages of his Endymion.

Another object of his enthusiastic admiration was the Homeric character of Achilles—especially when he is described as "shouting in the trenches."

One of his favorite topics of discourse was the principle of melody in Verse, upon which he had his own notions, particularly in the management of open & close vowels. I think I have seen a somewhat similar theory attributed to Mr Wordsworth. But I do not remember his laying it down in writing. Be this as it may, Keats's theory[16] was worked out by himself. He was himself, as already observed, a master of melody, which may be illustrated by almost numberless passages of his poems. As an instance of this, I may cite a few lines of that most perfect passage of *Hyperion*, which has been quoted by more than one of your Reviewers—the picture of dethroned Satan in his melancholy solitude. Keats's theory was, that the vowels should be so managed as not to clash one with another so as to mar the melody,—& yet that they should be interchanged, like differing notes of music to prevent monotony. The following lines will, I think, illustrate his theory, as I understood him.

> "Dēep ĭn thĕ shădy sădness ŏf ă vāle,
> Fār sūnken from the hēalthy brēath of mōrn—
> Fār frŏm thĕ fĭĕry nōon & ēve's ōne stār—
> Săt grey haired Sāturn, quĭet as a stŏne,
> Stĭll as the sĭlence round about his lāir:
> Fōrest on fōrest hung about his hēad
> Like clōud on clōud."[17]

These lines are exquisitely wrought into melody. They are beautifully varied in their vowel sounds, save when the exception proves the rule, & monotony is a beauty; as in the prolonged breathing, as it were, of the

---

11. **"That . . . lightened"**: "Lines Composed a Few Miles Above Tintern Abbey," ll. 37–41. 12. **"She . . . me"**: "She Dwelt Among the Untrodden Ways," ll. 1–2, 9–12. 13. **"Fancy . . . ravishment"**: *The Excursion*, IV. 857–60. 14. **"the . . . pride"**: Bailey quotes Wordsworth's "Resolution and Independence," ll. 43–44.

15. **"Come . . . day"**: "Mynstrelles Songe," stanza viii. 16. **Keats's theory**: See the detailed discussion in W. J. Bate's *Stylistic Development of Keats*, 1945, pp. 51–56, 65. 17. **"Dēep . . . clōud"**: *Hyperion*, I. 1–7.

similar vowels in "healthy breath of morn," in which we almost inhale the freshness of the morning air; & in the vowel sounds repeated in the words—

> "Sāt grey haired Sāturn"—and
> Fōrest on fōrest—"like cloud on cloud"—

In all which the sameness of the sound increases the melancholy & monotony of the situation of the de-throned Father of the Gods. The rest is beautiful by its skilful variation of the vowel-sounds; as these are touching by their sameness & monotony.

You mention Keats's taste for painting & music. Of the first I remember no more than his general love of the art, & his admiration of Haydon. But I well remember his telling me that, had he studied music, he had some notions of the combinations of sounds, by which he thought he could have done something as original as his poetry.

RICHARD WOODHOUSE

Richard Woodhouse (1788–1834), a lawyer keenly interested in literature, worked as literary and legal adviser for the publishing firm of Taylor and Hessey—the firm that published Keats's *Endymion* (1818) and his third and last volume (1820). He quickly saw Keats's talent and kept a careful record of what Keats wrote. The information he collected has proved invaluable for the study of both Keats's life and poetry.

This letter to Keats's publisher refers to Keats's letter to Woodhouse, October 27, 1818, above, p. 1219. The text is taken from a draft and presumably shows Wood-house in the process of thinking out these ideas, explaining them to himself as well as Taylor. The letter actually sent has not survived.

Richard Woodhouse to John Taylor
October 1818

I believe him to be right with regard to his own Poetical Character—And I perceive clearly the dis-tinction he draws between himself & those of the Wordsworth School.—There are gradations in Poetry & in Poets. One is purely descriptive confining him-self to external nature & visible objects—Another describes in addition the effects of the thoughts of which he is conscious & which others are affected by—Another will soar so far into the regions of imagination as to conceive of beings & substances in situations different from what he has ever seen them but still such as either have actually occurred or may possibly occur—Another will reason in poetry—another be witty.—Another will imagine things that never did nor probably ever will occur, or such as can not in nature occur & yet he will describe them so that you recognize nothing very unnatural in the Descrip-tions when certain principles or powers or conditions are admitted—Another will throw himself into various characters & make them speak as the passions would naturally incite them to do. The highest order of Poet will not only possess all the above powers but will have as high an imagination that he will be able to throw his own soul into any object he sees or imagines, so as to see feel be sensible of, & express, all that the object itself would see feel be sensible of or express—& he will speak out of that object—so that his own self will with the Exception of the Mechanical part be "annihilated."—and it is the excess of this power that I suppose Keats to speaks, when he says he has no identity—As a poet, and when the fit is upon him, this is true—And it is a fact that he does by the power of his Imagination create ideal personages substances & Powers—that he lives for a time in their souls or Essences or ideas—and that occasionally so intensely as to lose consciousness of what is round him. We all do the same in a degree, when we fall into a reverie—(The power of his Imagination is apparent in Every page of his Endymion—& He has affirmed that he can conceive of a billiard Ball that it may have a sense of delight from its own roundness, smoothness volu-bility. & the rapidity of its motion.)

If then his imagination has such power: and he is continually cultivating it, & giving it play. It will acquire strength by the Indulgence & exercise. This in excess is the Case of mad persons. And this may be carried to that extent that he may lose sight of his identity so far as to give him a habit of speaking gener-ally in an assumed character—so that What he says shall be tinged with the Sentiments proper to the Character which at the time has possessed itself of his Imagination.

This being his idea of the Poetical Character—he may well say that a poet has no identity—as a man he must have Identity. But as a poet he need not—And in this Sense a poet is "the most unpoetical of Gods creatures." For his soul has no distinctive character-istic—it can not be itself made the subject of Poetry that is another persons soul can not be thrown into the poet's—for there is no identity or personal impulse to be acted upon—

Shakspear was a poet of the kind above mentioned —and he was perhaps the only one besides Keats who possessed this power in an extraordinary degree, so as to be a feature in his works. He gives a description of his idea of a poet.

The Poets eye &c[1]

Lord Byron does not come up to this Character. He can certainly conceive & describe a dark accomplished vilain in love—& a female tender & kind who loves him. Or a sated & palled Sensualist Misanthrope & Deist[2]—But here his power ends.—The true poet can not only conceive this—but can assume any Character Essence idea or Substance at pleasure. & He has the imaginary faculty not in a limited manner, but in full universality.

Let us pursue Speculation on these Matters: & we shall soon be brought to believe in the truth of every Syllable of Keats's letter, taken as a description of himself & his own Ideas & feelings

CRITICISM
OF A SONNET BY KEATS

∽∾∽

These remarks were jotted down by Woodhouse in a notebook sometime in July, 1820. They refer to Keats's sonnet, "When I Have Fears that I May Cease To Be," ll. 7–8:

> And feel that I may never live to trace
>   Their shadows with the magic hand of chance.

∽∾∽

These lines give some insight into K's mode of writing Poetry. He has repeatedly said in conversation that he never sits down to write, unless he is full of ideas—and then thoughts come about him in troops, as tho' soliciting to be accepted & he selects—one of his Maxims is that if Poetry does not come naturally, it had better not come at all. the moment he feels any dearth he discontinues writing & waits for a happier moment. he is generally more troubled by a redundancy

RICHARD WOODHOUSE. **1. The . . . eye &c.:** Woodhouse alludes to *A Midsummer Night's Dream*, V. i. 12. **2. Deist:** As Woodhouse is using the term, "Deist" is roughly equivalent to "freethinker."

than by a poverty of images, & he culls what appears to him at the time the best.—He never corrects, unless perhaps a word here or there should occur to him as preferable to an expression he has already used—He is impatient of correcting, & says he would rather burn the piece in question & write another or something else—"My judgment, (he says,) is as active while I am actually writing as my imagination In fact all my faculties are strongly excited, & in their full play— And shall I afterwards, when my imagination is idle, & the heat in which I wrote, has gone off, sit down coldly to criticise when in Possession of only one faculty, what I have written, when almost inspired."— This fact explains the reason of the Perfectness, fullness, richness & completion of most that comes from him— He has said, that he has often not been aware of the beauty of some thought or expression until after he has composed & written it down—It has then struck him with astonishment—& seemed rather the production of another person than his own—He has wondered how he came to hit upon it.

J. G. LOCKHART

from ON THE COCKNEY
SCHOOL OF POETRY

NO. IV

∽∾∽

This article, the fourth in a series on Leigh Hunt and his associates, appeared in *Blackwood's Edinburgh Magazine*, August, 1818. It reviews both Keats's *Poems* of 1817 and *Endymion*. It was the anonymous work of John Gibson Lockhart (1794–1854), who was deliberately indulging in personal abuse in order to win readers. He later married the daughter of Sir Walter Scott, became editor of *The Quarterly Review*, and wrote a *Life* of Scott which is one of the famous biographies in English. For further discussion see the Introduction to Keats, above.

∽∾∽

> ————Of Keats,
> The Muses' son of promise, and what feats
> He yet may do, &c.
>
> Cornelius Webb

Of all the manias of this mad age, the most incurable, as well as the most common, seems to be no other than the *Metromanie*. The just celebrity of Robert Burns and Miss Baillie[1] has had the melancholy effect of turning the heads of we know not how many farm-servants and unmarried ladies; our very footmen compose tragedies, and there is scarcely a superannuated governess in the island that does not leave a roll of lyrics behind her in her bandbox. To witness the disease of any human understanding, however feeble, is distressing; but the spectacle of an able mind reduced to a state of insanity is of course ten times more afflicting. It is with such sorrow as this that we have contemplated the case of Mr John Keats. This young man appears to have received from nature talents of an excellent, perhaps even of a superior order—talents which, devoted to the purposes of any useful profession, must have rendered him a respectable, if not an eminent citizen. His friends, we understand, destined him to the career of medicine, and he was bound apprentice some years ago to a worthy apothecary in town. But all has been undone by a sudden attack of the malady to which we have alluded. Whether Mr John had been sent home with a diuretic or composing draught to some patient far gone in the poetical mania, we have not heard. This much is certain, that he has caught the infection, and that thoroughly. For some time we were in hopes, that he might get off with a violent fit or two; but of late the symptoms are terrible. The phrenzy of the "Poems" was bad enough in its way; but it did not alarm us half so seriously as the calm, settled, imperturbable drivelling idiocy of "Endymion." We hope, however, that in so young a person, and with a constitution originally so good, even now the disease is not utterly incurable. Time, firm treatment, and rational restraint, do much for many apparently hopeless invalids, and if Mr Keats should happen, at some interval of reason, to cast his eye upon our pages, he may perhaps be convinced of the existence of his malady, which, in such cases, is often all that is necessary to put the patient in a fair way of being cured.

The readers of the Examiner newspaper were informed, some time ago, by a solemn paragraph, in Mr Hunt's best style, of the appearance of two new stars of glorious magnitude and splendour in the poetical horizon of the land of Cockaigne. One of these turned out, by and by, to be no other than Mr John Keats. This precocious adulation confirmed the wavering apprentice in his desire to quit the gallipots,[2] and at the

same time excited in his too susceptible mind a fatal admiration for the character and talents of the most worthless and affected of all the versifiers of our time. One of his first productions was the following sonnet, "*written on the day when Mr Leigh Hunt left prison.*" It will be recollected, that the cause of Hunt's confinement was a series of libels against his sovereign, and that its fruit was the odious and incestuous "Story of Rimini."[3]

[Lockhart here quotes Keats's sonnet.]

The absurdity of the thought in this sonnet is, however, if possible, surpassed in another, "*addressed to Haydon*" the painter, that clever, but most affected artist, who as little resembles Raphael in genius as he does in person, notwithstanding the foppery of having his hair curled over his shoulders in the old Italian fashion. In this exquisite piece it will be observed, that Mr Keats classes together WORDSWORTH, HUNT, and HAYDON, as the three greatest spirits of the age, and that he alludes to himself, and some others of the rising brood of Cockneys, as likely to attain hereafter an equally honourable elevation. Wordsworth and Hunt! what a juxta-position! The purest, the loftiest, and, we do not fear to say it, the most classical of living English poets, joined together in the same compliment with the meanest, the filthiest, and the most vulgar of Cockney poetasters. No wonder that he who could be guilty of this should class Haydon with Raphael, and himself with Spencer.

> "Great spirits now on earth are sojourning;
>   He of the cloud, the cataract, the lake,
>   Who on Helvellyn's summit, wide awake,
> Catches his freshness from Archangel's wing:
> *He of the rose, the violet, the spring,*
>   *The social smile, the chain for Freedom's sake:*
>   And lo!—whose stedfastness would never take
> A meaner sound than Raphael's whispering.
> And other spirits there are standing apart
>   Upon the forehead of the age to come;
> These, these will give the world another heart,
>   And other pulses. *Hear ye not the hum*
> *Of mighty workings?*————
>   *Listen awhile ye nations, and be dumb.*'

The nations are to listen and be dumb! and why, good Johnny Keats? because Leigh Hunt is editor of the

J. G. LOCKHART. **1. Miss Baillie:** Joanna Baillie, 1762–1851, a minor poetess and playwright of the time. **2. gallipots:** small pots used by apothecaries.

3. Rimini: Lockhart here quotes the sonnet, which was a juvenile effort of Keats and far from typical of the poetry he had been writing the past two years.

Examiner, and Haydon has painted the judgment of Solomon, and you and Cornelius Webb, and a few more city sparks, are pleased to look upon yourselves as so many future Shakspeares and Miltons! The world has really some reason to look to its foundations! . . . His Endymion is not a Greek shepherd, loved by a Grecian goddess; he is merely a young Cockney rhymester, dreaming a phantastic dream at the full of the moon. Costume, were it worth while to notice such a trifle, is violated in every page of this goodly octavo. From his prototype Hunt, John Keats has acquired a sort of vague idea, that the Greeks were a most tasteful people, and that no mythology can be so finely adapted for the purposes of poetry as theirs. It is amusing to see what a hand the two Cockneys make of this mythology; the one confesses that he never read the Greek Tragedians, and the other knows Homer only from Chapman; and both of them write about Apollo, Pan, Nymphs, Muses, and Mysteries, as might be expected from persons of their education. We shall not, however, enlarge at present upon this subject, as we mean to dedicate an entire paper to the classical attainments and attempts of the Cockney poets. As for Mr Keats' "Endymion," it has just as much to do with Greece as it has with "old Tartary the fierce;" no man, whose mind has ever been imbued with the smallest knowledge or feeling of classical poetry or classical history, could have stooped to profane and vulgarise every association in the manner which has been adopted by this "son of promise." Before giving any extracts, we must inform our readers, that this romance is meant to be written in English heroic rhyme. To those who have read any of Hunt's poems, this hint might indeed be needless. Mr Keats has adopted the loose, nerveless versification, and Cockney rhymes of the poet of Rimini; but in fairness to that gentleman, we must add, that the defects of the system are tenfold more conspicuous in his disciple's work than in his own. Mr Hunt is a small poet, but he is a clever man. Mr Keats is a still smaller poet, and he is only a boy of pretty abilities, which he has done every thing in his power to spoil.

THOMAS LOVELL BEDDOES

1803–1849

A GIFTED, original man, Beddoes was the son of a well-known physician and scientific writer. While still a student at Oxford, he published two works that showed a close emotional affinity with Elizabethan and Jacobean tragedy—*The Improvisatore* (1821) and *The Bride's Tragedy* (1822). He published nothing more during his lifetime. Unsure of himself, and particularly hesitant in constructing plot, he began two other plays, *The Second Brother* and *Torrismond*, but they remain only as fragments.

After taking his degree at Oxford (1825), he studied medicine and anatomy in Germany. During these years he wrote his strange drama *Death's Jest Book or the Fool's Tragedy*; but he continued to revise it, and it was not published until a year after his death. His songs and other lyrics, there and elsewhere, are as close to the Elizabethan and Jacobean mode as any that have been written in English since the seventeenth century. He continued to live in Germany and Switzerland, returning to England for only a few brief visits. From Würzberg he moved to Zürich, where he practiced medicine. He was appointed Professor of Anatomy at the university there, but the political authorities refused to confirm his appointment because of his active interest in liberal movements. The thought of death and of his own futility increasingly preyed on him, and at the age of forty-five he committed suicide.

Bibliography

Collected editions include *Poetical Works*, ed. Edmund Gosse, 2 vols., 1890, and *Complete Works*, ed. Gosse, 2 vols., 1928; *Poems*, ed. Ramsay Colles, 1906 (Muses' Library); and *Works*, ed. H. W. Donner, 1935. For biography and criticism, see especially R. H. Snow, *Thomas Lovell Beddoes*, 1928, and H. W. Donner, *Thomas Lovell Beddoes: The Making of a Poet*, 1935.

FROM

THE BRIDE'S TRAGEDY

POOR OLD PILGRIM MISERY

Poor old pilgrim Misery,
  Beneath the silent moon he sate,
A-listening to the screech owl's cry,
  And the cold wind's goblin prate;
Beside him lay his staff of yew
  With withered willow twined,
His scant gray hair all wet with dew,
  His cheeks with grief ybrined;
    And his cry it was ever, alack!
    Alack, and woe is me!   10

Anon a wanton imp astray
  His piteous moaning hears,
And from his bosom steals away
  His rosary of tears:
With his plunder fled that urchin elf,
  And hid it in your eyes,
Then tell me back the stolen pelf,
  Give up the lawless prize;
    Or your cry shall be ever, alack!
    Alack, and woe is me!   20

1822 (1822)

A HO! A HO!

A ho! A ho!
Love's horn doth blow,
  And he will out a-hawking go.
His shafts are light as beauty's sighs,
And bright as midnight's brightest eyes,
  And round his starry way
The swan-winged horses of the skies,
With summer's music in their manes,
Curve their fair necks to zephyr's reins,
  And urge their graceful play.   10

A ho! A ho!
Love's horn doth blow,
  And he will out a-hawking go.
The sparrows flutter round his wrist.

The feathery thieves that Venus kissed
  And taught their morning song,
The linnets seek the airy list,
And swallows too, small pets of Spring,
Beat back the gale with swifter wing,
  And dart and wheel along.   20

A ho! A ho!
Love's horn doth blow,
  And he will out a-hawking go.
Now woe to every gnat that skips
To filch the fruit of ladies' lips,
  His felon blood is shed;
And woe to flies, whose airy ships
On beauty cast their anchoring bite,
And bandit wasp, that naughty wight,
  Whose sting is slaughter-red.   30

1822 (1822)

LINES

WRITTEN IN A BLANK LEAF
OF THE "PROMETHEUS UNBOUND"

Beddoes had just heard of the death of Shelley, whose work he admired more than that of any other contemporary.

Write it in gold—A spirit of the sun,
An intellect ablaze with heavenly thoughts,
A soul with all the dews of pathos shining,
Odorous with love, and sweet to silent woe
With the dark glories of concentrate song,
Was sphered in mortal earth. Angelic sounds
Alive with panting thoughts sunned the dim world.
The bright creations of an human heart
Wrought magic in the bosoms of mankind.
A flooding summer burst on poetry;   10
Of which the crowning sun, the night of beauty,
The dancing showers, the birds, whose anthems wild
Note after note unbind the enchanted leaves
Of breaking buds, eve, and the flow of dawn,
Were centred and condensed in his one name
As in a providence—and that was Shelley.

1822 (1851)

FROM

THE SECOND BROTHER

STREW NOT EARTH
WITH EMPTY STARS

Strew not earth with empty stars,
   Strew it not with roses,
Nor feathers from the crest of Mars,
   Nor summer's idle posies.
'Tis not the primrose-sandalled moon,
   Nor cold and silent morn,
Nor he that climbs the dusty noon,
Nor mower war with scythe that drops,
Stuck with helmed and turbaned tops
   Of enemies new shorn.   10

Ye cups, ye lyres, ye trumpets know,
Pour your music, let it flow,
'Tis Bacchus' son who walks below.

              1825 (1851)

FROM

TORRISMOND

HOW MANY TIMES
DO I LOVE THEE, DEAR?

How many times do I love thee, dear?
   Tell me how many thoughts there be
     In the atmosphere
     Of a new-fall'n year,
Whose white and sable hours appear
   The latest flake of Eternity:
So many times do I love thee, dear.

How many times do I love again?
   Tell me how many beads there are
     In a silver chain   10
     Of evening rain,
Unravelled from the tumbling main,
   And threading the eye of a yellow star:
So many times do I love again.

              1825 (1851)

FROM

DEATH'S JEST BOOK

The next three lyrics are from the play, mentioned in the
Introduction above, that Beddoes continued to alter and
retouch during his last twenty years and that remained
unpublished until 1850, after his death. It was a tragedy
in the "Gothic Elizabethan" vein, with elements that
remind us especially of the plays of John Webster and
Cyril Tourneur. The plot is both rambling and slight,
and the play is remembered only for its lyrics and its
occasionally fine blank verse.

TO SEA, TO SEA!

To sea, to sea! The calm is o'er;
   The wanton water leaps in sport,
And rattles down the pebbly shore;
   The dolphin wheels, the sea-cow snorts,
And unseen mermaids' pearly song
Comes bubbling up, the weeds among.
   Fling broad the sail, dip deep the oar:
   To sea, to sea! the calm is o'er.

To sea, to sea! our wide-winged bark
   Shall billowy cleave its sunny way,   10
And with its shadow, fleet and dark,
   Break the caved Tritons' azure day,
Like mighty eagle soaring light
O'er antelopes on Alpine height.
   The anchor heaves, the ship swings free,
   The sails swell full. To sea, to sea!

            1829–44 (1850)

THE SWALLOW LEAVES HER NEST

The swallow leaves her nest,
The soul my weary breast;
But therefore let the rain
     On my grave
Fall pure; for why complain?
Since both will come again
     O'er the wave.

The wind dead leaves and snow
Doth hurry to and fro;
And, once, a day shall break   10
    O'er the wave,
When a storm of ghosts shall shake
The dead, until they wake
    In the grave.

                    1829–44 (1850)

IF THOU WILT EASE THINE HEART

If thou wilt ease thine heart
Of love and all its smart,
    Then sleep, dear, sleep;
And not a sorrow
    Hang any tear on your eyelashes;
    Lie still and deep,
Sad soul, until the sea-wave washes
    The rim o' the sun tomorrow,
        In eastern sky.

But wilt thou cure thine heart   10
Of love and all its smart,
    Then die, dear, die;
'Tis deeper, sweeter,
    Than on a rose bank to lie dreaming
        With folded eye;
And then alone, amid the beaming
Of love's stars, thou'lt meet her
        In eastern sky.

                    1825–29 (1850)

OLD ADAM, THE CARRION CROW

Old Adam, the carrion crow,
    The old crow of Cairo;
He sat in the shower, and let it flow
    Under his tail and over his crest;
        And through every feather
        Leaked the wet weather;
And the bough swung under his nest;
    For his beak it was heavy with marrow.
    Is that the wind dying? O no;
    It's only two devils, that blow   10
    Through a murderer's bones, to and fro,
        In the ghosts' moonshine.

Ho! Eve, my gray carrion wife,
    When we have supped on kings' marrow,
Where shall we drink and make merry our life?
    Our nest it is queen Cleopatra's skull,
        'Tis cloven and cracked,
        And battered and hacked,
But with tears of blue eyes it is full:
Let us drink then, my raven of Cairo.   20
    Is that the wind dying? O no;
    It's only two devils, that blow
    Through a murderer's bones, to and fro,
        In the ghosts' moonshine.

                    1825–29 (1850)

DREAM-PEDLARY

If there were dreams to sell,
    What would you buy?
Some cost a passing bell;
    Some a light sigh,
That shakes from Life's fresh crown
Only a rose-leaf down.
If there were dreams to sell,
Merry and sad to tell,
And the crier rang the bell,
    What would you buy?   10

A cottage lone and still,
    With bowers nigh,
Shadowy, my woes to still,
    Until I die.
Such pearls from Life's fresh crown
Fain would I shake me down.
Were dreams to have at will,
This would best heal my ill,
    This would I buy.

But there were dreams to sell   20
    Ill didst thou buy;
Life is a dream, they tell,
    Waking, to die.
Dreaming a dream to prize,
Is wishing ghosts to rise;
And if I had the spell
To call the buried well,
    Which one would I?

If there are ghosts to raise,
    What shall I call,   30
Out of hell's murky haze,
    Heaven's blue pall?
Raise my loved long-lost boy,
To lead me to his joy.—
There are no ghosts to raise;
Out of death lead no ways;
    Vain is the call.

Know'st thou not ghosts to sue,
    No love thou hast.
Else lie, as I will do,   40
    And breathe thy last.
So out of Life's fresh crown
Fall like a rose-leaf down.
Thus are the ghosts to woo;
Thus are all dreams made true,
    Ever to last!

             1829–44 (1851)

LET DEW THE FLOWERS FILL

Let dew the flowers fill;
    No need of fell despair,
    Though to the grave you bear
One still of soul—but now too still,
    One fair—but now too fair.
For, beneath your feet, the mound,
    And the waves, that play around,
Have meaning in their grassy, and their watery, smiles;
And, with a thousand sunny wiles,
    Each says, as he reproves,   10
    Death's arrow oft is Love's.

             1830–39 (1851)

INDEX OF AUTHORS, TITLES, AND FIRST LINES

Authors' names are in **boldface** type; titles in *italics*; and first lines in roman.